COMPUTER COMPILATION OF MOLECULAR WEIGHTS AND PERCENTAGE COMPOSITIONS FOR ORGANIC COMPOUNDS

ERRECHNUNG UND ZUSAMMENSTELLUNG VON MOLEKULARGEWICHTEN UND PROZENT-GEHALTEN FUR ORGANISCHE VERBINDUNGEN DURCH COMPUTER

COMPILATION DES POIDS MOLECULAIRES ET DES COMPOSITIONS CENTESIMALES DES PRODUITS ORGANIQUES A L'AIDE D'ORDINATEUR

A

COMPUTER COMPILATION OF MOLECULAR WEIGHTS AND PERCENTAGE COMPOSITIONS FOR ORGANIC COMPOUNDS

MICHAEL J. S. DEWAR
Robert A. Welch Professor of Chemistry, University of Texas

AND

RICHARD JONES
I.C.I. Petrochemical and Polymer Laboratory, Runcorn, Cheshire
formerly of the University of Texas

PERGAMON PRESS

OXFORD · LONDON · EDINBURGH · NEW YORK
TORONTO · SYDNEY · PARIS · BRAUNSCHWEIG

ERRECHNUNG UND ZUSAMMENSTELLUNG VON MOLEKULARGEWICHTEN UND PROZENT-GEHALTEN FUR ORGANISCHE VERBINDUNGEN DURCH COMPUTER

MICHAEL J. S. DEWAR
Robert A. Welch Professor of Chemistry, University of Texas

UND

RICHARD JONES
I.C.I. Petrochemical and Polymer Laboratory, Runcorn, Cheshire
formerly of the University of Texas

PERGAMON PRESS

OXFORD · LONDON · EDINBURGH · NEW YORK
TORONTO · SYDNEY · PARIS · BRAUNSCHWEIG

COMPILATION DES POIDS MOLECULAIRES ET DES COMPOSITIONS CENTESIMALES DES PRODUITS ORGANIQUES A L'AIDE D'ORDINATEUR

MICHAEL J. S. DEWAR

Robert A. Welch Professor of Chemistry, University of Texas

ET

RICHARD JONES

I.C.I. Petrochemical and Polymer Laboratory, Runcorn, Cheshire
formerly of the University of Texas

THE QUEEN'S AWARD TO INDUSTRY 1966

PERGAMON PRESS

OXFORD · LONDON · EDINBURGH · NEW YORK
TORONTO · SYDNEY · PARIS · BRAUNSCHWEIG

Pergamon Press Ltd., Headington Hill Hall, Oxford
4 & 5 Fitzroy Square, London W.1
Pergamon Press (Scotland) Ltd., 2 & 3 Teviot Place, Edinburgh 1
Pergamon Press Inc., Maxwell House, Fairview Park, Elmsford, New York 10523
Pergamon of Canada Ltd., 207 Queen's Quay West, Toronto 1
Pergamon Press (Aust.) Pty. Ltd., 19a Boundary Street, Rushcutters Bay, N.S.W. 2011, Australia
Pergamon Press S.A.R.L., 24 rue des Écoles, Paris 5e
Vieweg & Sohn GmbH, Burgplatz 1, Braunschweig

First edition 1969

Library of Congress Catalog Card No. 69–19974

Printed in Great Britain by A. Wheaton & Co., Exeter,

08 012707 X

CONTENTS

INHALT

TABLE DES MATIERES

PREFACE

THIS volume contains computer output from a program which calculates the molecular weights and percentage compositions of organic compounds allowed by the valency laws. It was decided to limit the program to carbon, hydrogen and two heteroatoms, which could be chosen by the programmer and supplied as data. The number of formulae that can be calculated is, of course, limitless. We have attempted to strike a compromise between quantity and usefulness, the object being to provide the most comprehensive set of tables commensurate with cost. By arranging two pages of computer output on each page of this book, it has been possible to include data for over 100,000 formulae. The elements bromine, chlorine, fluorine, iodine, nitrogen, oxygen and phosphorus are included with a maximum limit of 30 carbon, 2 halogen and phosphorus, 5 nitrogen and 9 oxygen atoms. Compounds containing up to 4 atoms of sulfur can also be handled since the atomic weight of sulfur is double that of oxygen. Atomic weights used for the calculations are the most recent, being those adopted at the 1965 IUPAC conference, based on $^{12}C=12$.

We have included, together with the molecular weight calculated from isotopic natural abundance, the isomolecular weight which is derived from the atomic masses of the most common isotope for each element. This value is for use in conjunction with high resolution mass spectroscopy.

This is the first set of tables to include the halogens or phosphorus. A statistical survey of recent formula indices in *Chemical Abstracts* has revealed that the probability of any organic compound chosen at random from *Chemical Abstracts* appearing in this book is a little greater than 80%. A similar survey on other published tables shows the probability to be much less.

The calculations were performed by a Control Data Corporation 1604 computer, and the results stored on tape and printed. We wish to acknowledge the help afforded to us by the Computation Center Staff at the University of Texas, who arranged for the standard of printing to be as high as possible.

VORWORT

DIESER Band enthält die Computer-Ausgabe von einem Programm, das die Molekulargewichte und Prozent-Gehalte organischer Verbindungen im Rahmen der Valenzgesetze errechnet. Es wurde beschlossen, das Programm auf Kohlenstoff, Wasserstoff und zwei Heteroatome zu begrenzen, die vom Programmierer gewählt und als Daten eingegeben werden konnten. Die Anzahl der Formeln, die berechnet werden können, ist natürlich unbegrenzt. Wir haben versucht, einen Kompromiss zwischen Quantität und Brauchbarkeit zu schliessen, wobei unser Ziel darin bestand, eine möglichst umfassende Tabellensammlung mit vertretbarem Kostenaufwand zu erstellen. Durch die Anordnung von zwei Seiten Computer-Ausgabe auf jeder Buchseite war es möglich, Daten für über 100/000 Formeln aufzunehmen. Die Elemente Brom, Chlor, Fluor, Jod, Stickstoff, Sauerstoff und Phosphor sind darin enthalten; die obere Grenze ist 30 Kohlenstoff-, 2 Halogen- und Phosphor-, 5 Stickstoff- und 9 Sauerstoff-Atome. Verbindungen mit bis zu 4 Schwefelatomen können ebenfalls erfasst werden, da das Atomgewicht von Schwefel doppelt so hoch ist wie das von Sauerstoff. Es wurden die neuesten Atomgewichte für die Berechnung eingesetzt, die bei der IUPAC-Konferenz 1965 angenommen wurden und auf $^{12}C=12$ basieren.

Neben dem Molekulargewicht, das aus der natürlichen Isotopen-Häufigkeit errechnet wurde, haben wir auch das Isomolekulargewicht aufgenommen, das von den Atom-Massen des häufigsten Isotops für jedes Element abgeleitet ist. Dieser Wert ist zur Verwendung im Zusammenhang mit hoch auflösender Massenspektroskopie gedacht.

Die vorliegende Tabellensammlung enthält als erste auch die Halogene und Phosphor. Eine statistische Auswertung neuerer Formel-Indices in *Chemical Abstracts* hat ergeben, dass die Wahrscheinlichkeit, dass eine beliebige in den *Chemical Abstracts* genannte organische Verbindung in diesem Buche erscheint, etwas höher als 80% liegt. Eine ähnliche Auswertung bei anderen veröffentlichten Tabellen zeigt, dass bei ihnen diese Wahrscheinlichkeit sehr viel geringer ist.

Die Berechnungen wurden von einem Computer 1604 der Control Data Corporation ausgeführt und die Ergebnisse auf Band gespeichert und gedruckt. Wir möchten hier den Mitarbeitern des Rechenzentrums an der Universität von Texas für ihre Hilfe danken; sie sorgten dafür, dass der Druck den höchstmöglichen Standard erreichte.

PREFACE

CE VOLUME contient les résultats sortant d'un ordinateur à partir d'un programme de calcul des poids moléculaires et des compositions centésimales des composés organiques permis par les lois de valence. Nous avons décidé de limiter le programme au carbone, à l'hydrogène et à deux hétéroatomes, ceux-ci pouvant être choisis par le programmeur et fournis comme données. Naturellement le nombre de formules que l'on peut calculer est illimité. Nous avons essayé de faire un compromis entre la quantité et l'utilité des résultats à présenter, en ayant pour but de fournir l'ensemble de tables le plus vaste au meilleur prix. En représentant deux pages sortant de la machine par une page de ce livre, nous avons pu inclure des données pour environ 100.000 formules. Les éléments brome, chlore, fluor, iode, azote, oxygène et phosphore sont compris jusqu'à la limite de 30 atomes pour le carbone, 2 pour les halogènes et le phosphore, 5 pour l'azote, et 9 pour l'oxygène au maximum. Ces tables s'appliquent également aux composés contenant jusqu'à 4 atomes de soufre puisque le poids atomique du soufre est le double de celui de l'oxygène. Les poids atomiques utilisés pour les calculs sont les plus récents; ce sont ceux adaptés à la conférence IUPAC de 1965, basés sur $^{12}C = 12$.

Au poids moléculaire calculé à partir de l'abondance des isotopes naturels, nous avons joint le poids isomoléculaire dérivé des masses atomiques de l'isotope le plus courant pour chaque élément. Cette valeur est parallèlement utilisable en spectroscopie de masse à haute résolution.

C'est le premier ensemble de tables qui comporte les halogènes et le phosphore. Une étude statistique des index des formules dans les derniers *Chemical Abstracts* a montré que la probabilité de figurer dans ce livre pour un composé organique quelconque pris au hasard dans les *Chemical Abstracts* est légèrement supérieure à 80%. Une étude semblable faite sur d'autres tables déjà parues montre une probabilité bien moindre.

Les calculs ont été effectués par un ordinateur Control Data Corporation 1604, les résultats emmagasinés sur bande magnétique et imprimés. Nous souhaitons exprimer notre reconnaissance à l'équipe du centre de calcul de l'Université du Texas pour l'aide qu'elle nous a apportée et pour la qualité d'impression qu'elle a rendue la meilleure possible.

EXPLANATORY NOTES FOR THE TABLES

THE compounds are arranged in alphabetical order of heteroatoms in accordance with the formula index of *Chemical Abstracts*. Each page contains four blocks of data since two pages of computer output (each of which contains data for two formulae on every line) have been arranged one on top of the other. Inspection of the layout allows ready recognition of the sequence. Essentially it should be remembered that if the page is bisected with a horizontal line, all the formulae above this line come before any below it.

The tables fall conveniently into three sections:

I. Hydrocarbons.

II. Compounds containing carbon, hydrogen and one heteroatom.

III. Compounds containing carbon, hydrogen and two heteroatoms.

Sections II and III embrace several subsections each one dealing with different heteroatoms. The first page of each subsection contains a title, together with column headings. Subsequent pages do not have column headings but the contents of each column are readily apparent without reference to the first page. It should be noted that columns are in the order:

formula, molecular weight, isomolecular weight (see below for explanation of this term), percent carbon, percent hydrogen, percent element X, percent element Y. In Section III the percentages of the heteroelements X and Y appear in the same order as in the formula, i.e. alphabetically.

All calculations performed by the computer are to eight significant figures. Molecular weights and percentages are rounded to two places of decimals, and isomolecular weights to four places of decimals.

The program rejects any formula which does not obey the valency laws. Therefore the tables include only stable compounds and exclude ions and free radicals.

SECTION I. HYDROCARBONS

Formulae up to $C_{40}H_{82}$ are included, with the restriction that in any compound C_mH_n, $n > \frac{m}{2}$.

SECTION II. COMPOUNDS CONTAINING CARBON, HYDROGEN AND ONE HETEROATOM

The heteroatoms bromine, chlorine, fluorine, iodine, nitrogen, oxygen and phosphorus were chosen since a statistical survey showed them to be the elements which appear most frequently in organic compounds. Formulae up to and including $C_{30}H_nX_x$ are listed, where n is the maximum for saturation, and x is 2 for the halogens and phosphorus, but 5 and 9 for nitrogen and oxygen respectively. The restriction of $n > \frac{2m}{3}$ for any formula $C_mH_nX_x$ is imposed, since the number of realistic compounds thus excluded is few.

SECTION III. COMPOUNDS CONTAINING CARBON, HYDROGEN AND TWO HETEROATOMS

The heteroatoms nitrogen and oxygen appear with each other and separately with each heteroatom included in Section II. Formulae up to $C_{30}H_nX_xY_y$ are calculated, where *n* allows for complete saturation, and *x* and *y* are again 2 for the halogens and phosphorus but 5 and 9 for nitrogen and oxygen respectively. For example, the subsection of compounds with CHNO (the longest) ends with formula $C_{30}H_{68}N_6O_9$. As before, *n* is never less than two-thirds the number of carbon atoms.

MOLECULAR WEIGHTS

The most recent values of atomic weights, which are those adopted at the 23rd IUPAC Conference (1965),* were used in these calculations. At this meeting only one change applicable to these tables was made: the atomic weight 79.904 for bromine was preferred to 79.909. Otherwise those values based on $^{12}C=12$ as adopted at the 21st IUPAC Conference (1961)† remain effective. The values of atomic weights used in the program are given below:

Element	Atomic weight
Carbon	12.01115
Hydrogen	1.00797
Bromine	79.904
Chlorine	35.453
Fluorine	18.9984
Iodine	126.9044
Nitrogen	14.0067
Oxygen	15.9994
Phosphorus	30.9738

ISOMOLECULAR WEIGHT

The term "isomolecular weight" is used for the molecular weight calculated from the atomic mass number of the most common isotope for each element. Thus the isomolecular weight for any compound is the accurate mass (based on $^{12}C=12$) of a molecule of that compound in which every atom is the most common isotope, i.e. the most significant parent peak in the mass spectrum. It is hoped that values for isomolecular weights will be of value in laboratories with access to a high resolution mass spectrometer. Of course, not every fragment peak occurring in the mass spectrum of a compound will be represented in these tables since they include only combinations of atoms which obey the valency laws. However, many will appear and usually the addition or subtraction of 1.0078 for one hydrogen atom will give the required value. The parent ion will always be found in these tables, which is fortunate since, in high resolution mass spectroscopy, parent ions are frequently used to confirm molecular formulae. The atomic mass numbers quoted by Beynon and Williams‡ in their book of tables for mass spectrometry were used in the present calculations, and are listed on page xvii

* See *Comptes Rendus XXIII Conférence IUPAC*, Butterworths Scientific Publications, London, 1965, p. 175.

† See A. E. Cameron and E. Wichers, *J. Am. Chem. Soc.*, **84**, 4175 (1962) for details.

‡ *Mass and Abundance Tables for Use in Mass Spectrometry*, by J. H. Beynon and A. E. Williams, Elsevier Publishing Co., London, 1963.

Element	Atomic mass
Carbon	12
Hydrogen	1.0078246
Bromine	78.91839
Chlorine	34.9688531
Fluorine	18.9984022
Iodine	126.90466
Nitrogen	14.0030738
Oxygen	15.9949141
Phosphorus	30.973764

Each element is represented by the most common isotope. Bromine, which normally contains almost equal amounts of ^{79}Br and ^{81}Br isotopes, has been arbitrarily represented by the ^{79}Br isotope in these tables.

No account has been taken of the fact that in a positive ion, an electron of mass 0.0005843 μ has been removed.*

SULFUR-CONTAINING COMPOUNDS

These tables can be used for compounds containing carbon, hydrogen, oxygen, sulfur and one other heteroatom, since the atomic weight of sulfur is double that of oxygen. The limit for the number of sulfur atoms in a compound is such that

$$2S+O<9,$$

where S and O are the number of atoms of sulfur and oxygen respectively. It can be seen from the relationship above that in the absence of oxygen (O=0), S may be 4.

In order to use the tables for a sulfur compound, the formula should be rewritten with two oxygen atoms in place of each sulfur atom. For example:

$C_4H_9BrS_2$ becomes $C_4H_9BrO_4$

and

$C_6H_7NO_3S$ becomes $C_6H_7NO_5$.

The formula so obtained is found in the tables. For those compounds which do not contain sulfur together with oxygen the molecular weight and percentages are the same, the percent sulfur being that given for oxygen. When sulfur and oxygen appear together the oxygen percent must be allocated in the correct ratio to oxygen and sulfur. A table of factors to be applied to the percentage oxygen for all possible combinations of oxygen and sulfur is given below:

Number of atoms of oxygen and sulfur	Factor to be applied for obtaining %O	Factor to be applied for obtaining %S
OS	0.3333	0.6667
OS_2	0.2000	0.8000
OS_3	0.1429	0.8571
OS_4	0.1111	0.8889
O_2S	0.5000	0.5000
O_2S_2	0.3333	0.6667
O_2S_3	0.2500	0.7500
O_3S	0.6000	0.4000
O_3S_2	0.4286	0.5714
O_3S_3	0.3333	0.6667
O_4S	0.6667	0.3333
O_4S_2	0.5000	0.5000
O_5S	0.7143	0.2857
O_5S_2	0.5556	0.4444
O_6S	0.7500	0.2500
O_7S	0.7778	0.2222

* The errors so incorporated, which are usually negligible, are discussed by J. H. Beynon in *Mass Spectrometry and Its Applications to Organic Chemistry*, Elsevier Publishing Co., London, 1960.

The above explanatory notes for the use of the tables with sulfur-containing compounds are based on the approximation that the atomic weight of sulfur is double that of oxygen. The error due to this can be neglected for most purposes—an example giving observed and calculated values follows:

$C_{16}H_{13}N_3O_2S$	Mol. wt.	%C	%H	%N	%O	%S
Values in tables	311.30	61.73	4.21	13.50	10.28	10.28
Actual values	311.3645	61.7211	4.2085	13.4954	10.2769	10.2978

If required, accurate values can be calculated by using the following formulae:

A. Molecular Weights

$$M_{corr} = M + n\delta,$$

where M_{corr} is the corrected molecular weight, M the molecular weight obtained from the tables, n the number of sulfur atoms in the formula and δ the correction factor 0.0652 (difference between the atomic weight of two oxygens and one sulfur atom). Values of $n\delta$ for $n = 1 \rightarrow 4$ are given below:

n	$n\delta$
1	0.0652
2	0.1304
3	0.1956
4	0.2608

Addition of the appropriate value of $n\delta$ to the molecular weight in the table gives the accurate molecular weight (correct to two places of decimals).

B. Isomolecular Weights

Isomolecular weights are corrected in the same manner as molecular weights, and the corrected value (IM_{corr}) gives the accurate mass of the formula for which one ^{32}S atom has been included in place of two ^{16}O atoms.

$$IM_{corr} = IM + n\delta',$$

where IM is the isomolecular weight found in the tables and δ' is the correction factor -0.0177555 (difference between the weight of two ^{16}O atoms and one ^{32}S atom). Values of $n\delta'$ for $n = 1 \rightarrow 4$ are given below:

n	$n\delta'$
1	−0.0177555
2	−0.0355110
3	−0.0532665
4	−0.0710220

Addition of the appropriate value of $n\delta'$ to the isomolecular weight gives the accurate isomolecular weight (correct to four places of decimals).

C. Percentages of Elements Other than Sulfur

$$P_{corr} = \frac{M}{M+n\delta} P$$

where P_{corr} is the correct percentage, M is the molecular weight and P is the percentage from the tables. Values of $\frac{M}{M+n\delta}$ for various molecular weights and values of n are given below:

Mol. wt.	n			
	1	2	3	4
100	0.9993	0.9987	0.9980	—
200	0.9997	0.9993	0.9990	0.9987
300	0.9998	0.9996	0.9993	0.9991
400	0.9998	0.9997	0.9995	0.9993
500	0.9999	0.9997	0.9996	0.9995

When the appropriate factor is multiplied by the observed percentage the percentage correct to two places of decimals is obtained.

D. Percentages of Sulfur

$$SP_{corr} = \frac{M}{M+n\delta} \times 1.002 \times OP$$

where SP_{corr} is the correct percentage of sulfur and OP is the percentage of oxygen given in the tables for the formula in which each sulfur atom has been replaced by two oxygen atoms. A table of correction factors for various values of n and molecular weight is given below:

Mol. wt.	n			
	1	2	3	4
100	1.0013	1.0007	1.0000	—
200	1.0017	1.0013	1.0010	1.0007
300	1.0018	1.0016	1.0013	1.0011
400	1.0018	1.0017	1.0015	1.0013
500	1.0019	1.0017	1.0016	1.0015

The data given in the tables for the formula $C_9H_{11}NS_4$ ($C_9H_{11}NO_8$) are compared with the accurate data and the data obtained by applying the corrections as an example.

$C_9H_{11}NS_4$	Observed	Actual	Corrected
Molecular weight	261.19	261.4506	261.45
Isomolecular weight	261.0485	260.97744	260.97747
Percent carbon	41.39	41.3462	41.35
Percent hydrogen	4.25	4.2409	4.25
Percent nitrogen	5.36	5.3573	5.36
Percent sulfur	49.00	49.0555	49.05

A list of atomic weights, atomic masses and multiples of atomic weights is supplied in the Appendix at the end of this book.

ERKLARUNGEN ZU DEN TABELLEN

DIE Verbindungen sind in alphabetischer Reihenfolge der Heteroatome angeordnet, entsprechend dem Formel-Index der *Chemical Abstracts*. Jede Seite enthält vier Blocks mit Daten, da jeweils zwei Seiten Computer-Ausgabe (von denen jede wiederum Daten für zwei Formeln auf jeder Zeile enthält) übereinander angeordnet sind. Eine Betrachtung des Layout macht die Reihenfolge deutlich. Im wesentlichen sollte festgehalten werden, dass bei Unterteilung der Seite durch eine horizontale Linie alle Formeln oberhalb der Linie vor denjenigen darunter kommen.

Die Tabellen sind in drei Abschnitte unterteilt:

I. Kohlenwasserstoffe.

II. Verbindungen aus Kohlenstoff, Wasserstoff und einem Heteroatom.

III. Verbindungen aus Kohlenstoff, Wasserstoff und zwei Heteroatomen.

Die Abschnitte II und III enthalten jeweils mehrere Unterabschnitte für die verschiedenen Heteroatome. Die erste Seite jedes Unterabschnitts enthält einen Titel, zusammen mit den Kolonnenüberschriften. Die folgenden Seiten haben keine Kolonnenüberschriften, aber der Inhalt jeder Kolonne ist auch ohne Rückblättern zur ersten Seite klar erkennbar. Die Kolonnen sind wie folgt geordnet:

Formel, Molekulargewicht, Isomolekulargewicht (Erklärung dieses Ausdrucks siehe unten), Prozent Kohlenstoff, Prozent Wasserstoff, Prozent Element X, Prozent Element Y. In Abschnitt III erscheinen die Prozentgehalte der Heteroelemente X und Y in derselben Reihenfolge wie in der Formel, d.h. alphabetisch.

Alle durch den Computer ausgeführten Rechnungen umfassen bis zu acht Ziffern. Die Molekulargewichte und Prozentwerte wurden auf zwei Dezimalen abgerundet, die Isomolekulargewichte auf vier Dezimalen.

Das Programm nimmt keine Formel an, die nicht den Valenzgesetzen entspricht. Die Tabellen enthalten deshalb nur stabile Verbindungen und keine Ionen oder freie Radikale.

ABSCHNITT I. KOHLENWASSERSTOFFE

Es sind Formeln bis $C_{40}H_{82}$ enthalten, mit der Einschränkung, dass in jeder Verbindung mit der Formel C_mH_n $n > \frac{m}{2}$ ist.

ABSCHNITT II. VERBINDUNGEN AUS KOHLENSTOFF, WASSERSTOFF UND EINEM HETEROATOM

Es wurden die Heteroatome Brom, Chlor, Fluor, Jod, Stickstoff, Sauerstoff und Phosphor gewählt, da eine statistische Auswertung zeigte, dass diese Elemente am häufigsten in organischen Verbindungen vorkommen. Es sind Formeln bis einschliesslich $C_{30}H_nX_x$ aufgeführt, wobei der Maximalwert von n Sättigung bedeutet und x für die Halogene und für Phosphor 2 ist, aber 5 für Stickstoff und 9 für Sauerstoff. Es wurde die Einschränkung $n > \frac{2m}{3}$ für jede Formel $C_mH_nX_x$ gemacht, da die Anzahl der möglichen Verbindungen, die damit ausgeschlossen werden, gering ist.

ABSCHNITT III. VERBINDUNGEN AUS KOHLENSTOFF, WASSERSTOFF UND ZWEI HETEROATOMEN

Die Heteroatome Stickstoff und Sauerstoff erscheinen sowohl gemeinsam als auch getrennt mit jedem der in Abschnitt II aufgeführten Heteroatome. Es wurden Formeln bis $C_{30}H_nX_xY_y$ berechnet, wobei n den Maximalwert für Sättigung bedeutet und x und y für die Halogene und Phosphor wiederum jeweils 2 ist, aber 5 bzw. 9 für Stickstoff und Sauerstoff. Der Unterabschnitt für Verbindungen mit CHNO zum Beispiel (der längste) endet mit der Formel $C_{30}H_{68}N_6O_9$. Wie zuvor ist n niemals kleiner als zwei Drittel der Anzahl Kohlenstoff-Atome.

MOLEKULARGEWICHTE

In den vorliegenden Berechnungen wurden die neuesten Werte für die Atomgewichte eingesetzt, die bei der 23. IUPAC-Konferenz (1965)* angenommen wurden. Bei dieser Tagung ergab sich nur eine diese Tabellen betreffende Anderung: das Atomgewicht 79,904 für Brom erhielt den Vorzug über 79,909. Im übrigen blieben die auf $^{12}C=12$ basierenden Werte, wie sie bei der 21. IUPAC-Konferenz (1961)† angenommen wurden, gültig. Die im Programm eingesetzten Werte der Atomgewichte sind nachstehend angeführt:

Element	Atomgewicht
Kohlenstoff	12,01115
Wasserstoff	1,00797
Brom	79,904
Chlor	35,453
Fluor	18,9984
Jod	126,9044
Stickstoff	14,0067
Sauerstoff	15,9994
Phosphor	30,9738

ISOMOLEKULARGEWICHT

Der Ausdruck "Isomolekulargewicht" wird für das Molekulargewicht verwendet, das aus der Atom-Massenzahl des häufigsten Isotops für jedes Element errechnet wurde. Dementsprechend ist das Isomolekulargewicht für eine Verbindung die genaue Masse (auf der Basis von $^{12}C=12$) eines Moleküls dieser Verbindung, in dem jedes Atom das häufigste Isotop ist, d.h. der grösste Molekül-Peak im Massenspektrum. Wir hoffen, dass die Werte für die Isomolekulargewichte in Laboratorien mit Zugang zu einem hoch auflösenden Massenspektrometer von Nutzen sein werden. Natürlich wird nicht jeder Bruchstück-Peak, der im Massenspektrum einer Verbindung erscheint, in den Tabellen enthalten sein, da sie nur Atomkombinationen umfassen, die den Valenzgesetzen gehorchen. Viele werden jedoch vertreten sein, und im allgemeinen wird die Addition oder Subtraktion von 1,0078 für ein Wasserstoff-Atom zu dem gewünschten Wert führen. Das Molekül-Ion findet sich immer in den Tabellen, was von Vorteil ist, weil in der hoch auflösenden Massenspektroskopie die Molekül-Ionen oft zur Bestätigung von Molekülformeln benutzt werden. Die von Beynon und Williams‡ in ihrem Tabellenbuch für Massenspektrometrie angeführten Atom-Massenzahlen wurden in den vorliegenden Berechnungen verwendet und sind nachstehend aufgeführt:

* Siehe *Comptes Rendus XXIII Conférence IUPAC*, Butterworths Scientific Publications, London, 1965, S. 175.

† Siehe A. E. Cameron und E. Wichers, *J. Am. Chem. Soc.*, **84**, 4175 (1962).

‡ *Mass and Abundance Tables for Use in Mass Spectrometry*, von J. H. Beynon und A. E. Williams, Elsevier Publishing Co., London, 1963.

Element	Atom-Masse	Element	Atom-Masse
Kohlenstoff	12	Jod	126,90466
Wasserstoff	1,0078246	Stickstoff	14,0030738
Brom	78,91839	Sauerstoff	15,9949141
Chlor	34,9688531	Phosphor	30,973764
Fluor	18,9984022		

Jedes Element ist durch das häufigste Isotop vertreten. Bei Brom, das normalerweise fast gleiche Mengen ^{79}Br- und ^{81}Br-Isotopen enthält, wurde für diese Tabellen willkürlich das Isotop ^{79}Br eingesetzt.

Die Tatsache, dass bei einem positiven Ion ein Elektron mit der Masse 0,0005843 μ wegfällt, wurde nicht berücksichtigt.*

SCHWEFELHALTIGE VERBINDUNGEN

Die vorliegenden Tabellen können für Verbindungen aus Kohlenstoff, Wasserstoff, Sauerstoff, Schwefel und einem weiteren Heteroatom verwendet werden, da das Atomgewicht von Schwefel doppelt so hoch ist wie das von Sauerstoff. Die obere Grenze für die Anzahl von Schwefelatomen in einer Verbindung ist dadurch gegeben, dass

$$2S+O<9,$$

wobei S und O die Anzahl der Schwefel- bzw. Sauerstoff-Atome bedeuten. Aus der obigen Beziehung ergibt sich, dass bei Abwesenheit von Sauerstoff (O=Null) S=4 betragen kann.

Wenn die Tabellen für eine Schwefelverbindung benutzt werden sollen, dann sollte die Formel so umgeschrieben werden, dass zwei Sauerstoffatome für jedes Schwefelatom stehen. Zum Beispiel:

aus $C_4H_9BrS_2$ wird $C_4H_9BrO_4$

und

aus $C_6H_7NO_3S$ wird $C_6H_7NO_5$.

Die auf diese Weise erhaltene Formel ist in den Tabellen enthalten. In den Verbindungen, die neben dem Schwefel keinen Sauerstoff enthalten, sind Molekulargewichte und Prozent-Gehalte gleich, wobei der für Sauerstoff angegebene Prozent-Gehalt sich auf Schwefel bezieht. Wenn Schwefel und Sauerstoff gemeinsam vorkommen, muss der für Sauerstoff angegebene Prozent-Gehalt im richtigen Verhältnis auf Sauerstoff und Schwefel verteilt werden. Die nachstehende Tabelle enthält die Faktoren zur Berechnung von Sauerstoff und Schwefel aus dem angegebenen Sauerstoff-Gehalt für alle möglichen Kombinationen an Sauerstoff und Schwefel:

Anzahl der Sauerstoff- und Schwefel-Atome	Faktor zur Errechnung von %O	Faktor zur Errechnung von %S
OS	0,3333	0,6667
OS_2	0,2000	0,8000
OS_3	0,1429	0,8571
OS_4	0,1111	0,8889
O_2S	0,5000	0,5000
O_2S_2	0,3333	0,6667
O_2S_3	0,2500	0,7500
O_3S	0,6000	0,4000
O_3S_2	0,4286	0,5714
O_3S_3	0,3333	0,6667
O_4S	0,6667	0,3333
O_4S_2	0,5000	0,5000
O_5S	0,7143	0,2857
O_5S_2	0,5556	0,4444
O_6S	0,7500	0,2500
O_7S	0,7778	0,2222

* Die dadurch auftretenden Fehler, die gewöhnlich vernachlässigbar klein sind, werden von J. H. Beynon in *Mass Spectrometry and Its Applications to Organic Chemistry*, Elsevier Publishing Co., London, 1960, diskutiert.

Die obigen Erklärungen zum Gebrauch der Tabellen für schwefelhaltige Verbindungen stützen sich auf die Annäherung, dass das Atomgewicht von Schwefel doppelt so hoch ist wie das von Sauerstoff. Der dadurch bedingte Fehler kann in den meisten Fällen vernachlässigt werden. Es folgt ein Beispiel, in dem gemessene und errechnete Werte gegenübergestellt sind:

$C_{16}H_{13}N_3O_2S$	Mol. Gew.	%C	%H	%N	%O	%S
Werte in der Tabelle	311,30	61,73	4,21	13,50	10,28	10,28
Tatsächliche Werte	311,3645	61,7211	4,2085	13,4954	10,2769	10,2978

Falls erforderlich, können genaue Werte mit Hilfe der folgenden Formeln errechnet werden.

A. Molekulargewichte

$$M_{\text{corr}} = M + n\delta$$

wobei M_{corr} das korrigierte Molekulargewicht ist, M das aus den Tabellen entnommene Molekulargewicht, n die Anzahl der Schwefelatome in der Formel und δ der Korrekturfaktor 0,0652 (Differenz zwischen dem Atomgewicht von zwei Sauerstoffatomen und einem Schwefelatom). Die Werte von $n\delta$ für $n=1$ bis 4 sind nachstehend angeführt:

n	$n\delta$
1	0,0652
2	0,1304
3	0,1956
4	0,2608

Die Addition des entsprechenden Wertes von $n\delta$ zum in der Tabelle angegebenen Molekulargewicht ergibt das genaue Molekulargewicht (auf zwei Dezimalstellen korrekt).

B. Isomolekulargewichte

Die Isomolekulargewichte werden in gleicher Weise wie die Molekulargewichte korrigiert, und der korrigierte Wert (IM_{corr}) ergibt die genaue Masse der Formel, in die ein ^{32}S-Atom anstelle von zwei ^{16}O-Atomen eingesetzt wurde.

$$IM_{\text{corr}} = IM + n\delta',$$

wobei IM das in den Tabellen gefundene Isomolekulargewicht ist und δ' der Korrekturfaktor $-0{,}0177555$ (Differenz zwischen dem Gewicht von zwei ^{16}O-Atomen und einem ^{32}S-Atom. Nachstehend sind die Werte von $n\delta'$ für $n=1$ bis 4 angeführt:

n	$n\delta'$
1	− 0,0177555
2	− 0,0355110
3	− 0,0532665
4	− 0,0710220

Die Addition des entsprechenden Wertes von $n\delta'$ zum Isomolekulargewicht ergibt das genaue Isomolekulargewicht (bis auf vier Dezimalstellen korrekt).

C. Prozentgehalte von Elementen ausser Schwefel

$$P_{corr} = \frac{M}{M+n\delta} P$$

wobei P_{corr} der richtige Prozent-Gehalt ist, M das Molekulargewicht und P der Prozent-Gehalt aus den Tabellen. Nachstehend sind Werte von $\frac{M}{M+n\delta}$ für verschiedene Molekulargewichte und Werte von n angegeben:

Mol. Gew.	*n* 1	2	3	4
100	0,9993	0,9987	0,9980	—
200	0,9997	0,9993	0,9990	0,9987
300	0,9998	0,9996	0,9993	0,9991
400	0,9998	0,9997	0,9995	0,9993
500	0,9999	0,9997	0,9996	0,9995

Wenn der entsprechende Faktor mit dem abgelesenen Prozentsatz multipliziert wird, dann erhält man den bis auf zwei Dezeimalstellen genauen Prozent-Gehalt.

D. Prozent-Gehalte von Schwefel

$$SP_{corr} = \frac{M}{M+n\delta} \times 1{,}002 \times OP$$

wobei SP_{corr} der richtige Prozent-Gehalt von Schwefel ist und OP der in den Tabellen angegebene Prozent-Gehalt für Sauerstoff in Formeln, in denen jedes Schwefel-Atom durch zwei Sauerstoff-Atome ersetzt wurde. Eine Tabelle von Korrekturfaktoren für verschiedene Werte von n und Molekulargewicht folgt nachstehend:

Mol. Gew.	*n* 1	2	3	4
100	1,0013	1,0007	1,0000	—
200	1,0017	1,0013	1,0010	1,0007
300	1,0018	1,0016	1,0013	1,0011
400	1,0018	1,0017	1,0015	1,0013
500	1,0019	1,0017	1,0016	1,0015

Die in den Tabellen für die Formel $C_9H_{11}NS_4$ ($C_9H_{11}NO_8$) angegebenen Werte werden als Beispiel mit den genauen Werten sowie mit den durch Anwendung der Korrekturfaktoren erhaltenen Werten verglichen.

$C_9H_{11}NS_4$	Abgelesen	Wirklich	Korrigiert
Molekulargewicht	261,19	261,4506	261,45
Isomolekulargewicht	261,0485	260,97744	260,97747
Prozent Kohlenstoff	41,39	41,3462	41,35
Prozent Wasserstoff	4,25	4,2409	4,25
Prozent Stickstoff	5,36	5,3573	5,36
Prozent Schwefel	49,00	49,0555	49,05

Eine Liste von Atomgewichten, Atom-Massen und eine Tabelle für multiple Atomgewichte sind im Anhang am Schluss dieses Buches zu finden.

NOTES EXPLICATIVES DES TABLES

LES composés sont disposés par ordre alphabétique des hétéroatomes conformément à l'index par formules des *Chemical Abstracts*. Les données sont mises en page en quatre groupes, car deux pages de résultats de l'ordinateur sont disposés l'une au-dessus de l'autre et chaque page contient les données à raison de deux formules par ligne. Par l'examen de cet arrangement on reconnaît facilement la séquence. On doit retenir essentiellement que si on divise la page en deux par un trait horizontal, toutes les formules se trouvant au-dessus du trait viennent avant celles qui se trouvent en dessous.

Il est commode de diviser les tables en trois sections:

I. Hydrocarbures.
II. Composés contenant du carbone, de l'hydrogène et un hétéroatome.
III. Composés contenant du carbone, de l'hydrogène et deux hétéroatomes.

Les groupes II et III englobent plusieurs sous-groupes et chaque sous-groupe traite des hétéroatomes différents. La première page de chaque sous-groupe contient un titre, avec des en-têtes pour chaque colonne. Les colonnes des pages suivantes n'ont pas d'en-têtes mais on reconnaît leurs éléments facilement sans avoir besoin de se référer à la première page. Notons que les colonnes sont dans l'ordre:

formule, poids moléculaire, poids isomoléculaire (voir ci-dessous pour une explication de ce terme), pourcentage en carbone, pourcentage en hydrogène, pourcentage en élément X, pourcentage en élément Y. Dans le groupe III les pourcentages en hétéro-éléments X et Y viennent dans le même ordre que dans les formules, ç'est à dire dans l'ordre alphabétique.

Tous les calculs donnés par l'ordinateur sont à huit chiffres significatifs. Les poids moléculaires et les pourcentages sont arrondis à deux décimales près, et les poids isomoléculaires à quatre décimales.

Le programme rejette toute formule n'obéissant pas aux lois de valence. Ainsi, dans ces tables, seuls les composés stables sont inclus; les ions et les radicaux libres sont exclus.

GROUPE I. HYDROCARBURES

Sont insérées les formules allant jusqu'à $C_{40}H_{82}$, avec la restriction que dans un composé donné C_mH_n, $n > \frac{m}{2}$.

GROUPE II. COMPOSÉS CONTENANT DU CARBONE, DE L'HYDROGÈNE ET UN HÉTÉROATOME

On a choisi les hétéroatomes brome, chlore, fluor, iode, azote, oxygène et phosphore, car un examen statistique montre que ce sont les éléments qui apparaissent le plus souvent dans les composés organiques. Figurent les formules jusqu'à et y compris $C_{30}H_nX_x$, où n est la maximum quand le composé est saturé, et x égale 2 pour les halogènes et le phosphore, mais égale 5 et 9 respectivement pour l'azote et l'oxygène. La condition $n > \frac{2m}{3}$ est imposée pour une formule $C_mH_nX_x$ donnée, car ainsi est exclus un nombre faible de composés existant réellement.

GROUPE III. COMPOSÉS CONTENANT DU CARBONE, DE L'HYDROGÈNE ET DEUX HÉTÉROATOMES

Avec chaque hétéroatome compris dans le groupe II figurent les hétéroatomes azote et oxygène, ensemble et séparément. Sont calculées les formules jusqu'à $C_{30}H_nX_xY_y$ où n représente une saturation totale, x et y de nouveau égalent 2 pour les halogènes et le phosphore, mais 5 et 9 respectivement pour l'azote et le phosphore. Par exemple, le sous-groupe des composés avec CHNO (le plus long) se termine par la formule $C_{30}H_{68}N_6O_9$. Comme précédemment, n est jamais inférieur au 2/3 du nombre d'atomes de carbone.

POIDS MOLÉCULAIRES

Dans ces calculs ont été utilisées les valeurs des poids atomiques les plus récentes qui sont celles adoptées à la 23° conférence de l'IUPAC (1965).* Au cours de cette réunion, on a fait un seul changement qui sera appliqué à ces tables; pour le brome on a préféré le poids atomique 79,904 à 79,909. Par contre, les valeurs basées sur $^{12}C=12$ et adoptées à la 21° conférence IUPAC (1961)† restent valables. Les valeurs des poids atomiques utilisées dans le programme sont données ci-dessous:

Elément	Poids atomique
Carbone	12,01115
Hydrogène	1,00797
Brome	79,904
Chlore	35,453
Fluor	18,9984
Iode	126,9044
Azote	14,0067
Oxygène	15,9994
Phosphore	30,9738

POIDS ISOMOLÉCULAIRE

On désigne par "poids isomoléculaire" d'un élément le poids moléculaire calculé à partir du nombre de masse atomique de son isotope le plus courant. Le poids isomoléculaire d'un composé quelconque est ainsi la masse exacte (basée sur $^{12}C=12$) d'une molécule de ce composé dont chaque atome est l'isotope le plus courant, c'est à dire dont le pic de masse la plus élevée est le pic le plus significatif dans le spectre de masse. Nous espérons que les valeurs de poids isomoléculaires seront appréciées dans les laboratoires qui ont accès à un spectromètre de masse à haute résolution. Naturellement nous n'avons pas représenté dans ces tables chaque pic des fragments apparaissant dans le spectre de masse d'un composé puisqu'elles ne comportent que des combinaisons d'atomes qui obéissent aux lois de valence. Cependant, plusieurs de ces pics apparaitront quand-même, et l'addition ou la soustraction de 1,0078 pour un atome d'hydrogène donnera en général la valeur désirée. L'ion de masse la plus élevée figurera toujours dans ces tables, ce qui est une chance, puisque habituellement en spectroscopie de masse de haute résolution on utilise ces ions pour confirmer les formules moléculaires. Dans les calculs présentés ici, on a utilisé les nombres de masses atomiques cités par Benyon et Williams‡ dans leur livre de tables pour la spectrométrie de masse. La liste en est donnée ci-dessous:

* Voir *Comptes Rendus XXIII Conférence IUPAC*, Butterworths Scientific Publications, London, 1965 p. 175.

† Voir A. E. Cameron et E. Wichers, *J. Am. Chem. Soc.*, **84**, 4175 (1962) pour les détails.

‡ *Mass and Abundance Tables for Use in Mass Spectrometry*, par J. H. Benyon et A. E. Williams, Elsevier Publishing Co., London, 1963.

Elément	Masse atomique	Elément	Masse atomique
Carbone	12	Iode	126,90466
Hydrogène	1,0078246	Azote	14,0030738
Brome	78,91839	Oxygène	15,9949141
Chlore	34,9688531	Phosphore	30,973764
Fluor	18,9984022		

Chaque élément est représenté par son isotope le plus courant. Le brome, qui normalement contient les isotopes ^{79}Br et ^{81}Br en quantités presque égales, a été représenté arbitrairement par l'isotope ^{79}Br dans ces tables.

On n'a pas tenu compte du fait que dans chaque ion positif, un électron de masse 0,0005843 μ a été arraché.*

COMPOSÉS CONTENANT DU SOUFRE

Ces tables peuvent être utilisées pour les composés contenant du carbone, de l'hydrogène, de l'oxygène, du soufre, et un autre hétéroatome puisque le poids atomique du soufre est le double de celui de l'oxygène. Dans un composé donné le nombre d'atomes de soufre est limité par la relation

$$2S+O<9$$

où S et O sont respectivement le nombre d'atomes de soufre et d'oxygène. On peut voir d'après cette dernière relation qu'en l'absence d'oxygène (O=0), S peut prendre la valeur 4.

Si on veut utiliser les tables pour un composé soufré, on récrira les formules en remplaçant chaque atome de soufre par deux atomes d'oxygène. Par exemple:

$$C_4H_9BrS_2 \text{ devient } C_4H_9BrO_4$$

et

$$C_6H_7NO_3S \text{ devient } C_6H_7NO_5.$$

La formule ainsi obtenue se trouve dans les tables. Pour les composés qui ne contiennent pas à la fois de soufre et de l'oxygène le poids molèculaire et les pourcentages sont les mêmes, le taux en soufre étant le même que celui donné pour l'oxygène. Lorsque le soufre et l'oxygène apparaissent ensemble, le pourcentage en oxygène doit être pondéré en tenant compte du rapport exact entre l'oxygène et le soufre. On trouvera ci-dessous un tableau de coefficients applicables au pourcentage en oxygène dans toutes les combinaisons possibles de l'oxygène et du soufre:

Nombre d'atomes d'oxygène et de soufre	Facteur à appliquer pour obtenir % en O	Facteur à appliquer pour obtenir % en S
OS	0,3333	0,6667
OS_2	0,2000	0,8000
OS_3	0,1429	0,8571
OS_4	0,1111	0,8889
O_2S	0,5000	0,5000
O_2S_2	0,3333	0,6667
O_2S_3	0,2500	0,7500
O_3S	0,6000	0,4000
O_3S_2	0,4286	0,5714
O_3S_3	0,3333	0,6667
O_4S	0,6667	0,3333
O_4S_2	0,5000	0,5000
O_5S	0,7143	0,2857
O_5S_2	0,5556	0,4444
O_6S	0,7500	0,2500
O_7S	0,7778	0,2222

* Les erreurs généralement négligeables qui sont ainsi incorporés sont discutées par J. H. Benyon dans *Mass Spectrometry and Its Applications to Organic Chemistry*, Elsevier Publishing Co., London, 1960.

Les notes explicatives précédentes sur l'utilisation des tableaux des composés contenant du soufre sont basées sur une approximation, à savoir que le poids atomique du soufre est le double de celui de l'oxygène. L'erreur ainsi commise peut être négligée dans la plupart des cas ; voici un exemple donnant les valeurs observées et calculées :

$C_{16}H_{13}N_3O_2S$	Poids mol.	%C	%H	%N	%O	%S
Valeurs dans les tables	311,30	61,73	4,21	13,50	10,28	10,28
Valeurs réelles	311,3645	61,7211	4,2085	13,4954	10,2769	10,2978

On pourra calculer les valeurs précises, s'il en est besoin, en utilisant les formules suivantes.

A. Poids Moléculaires

$$M_{corr} = M + n\delta$$

où M_{corr} est le poids moléculaire corrigé, M le poids moléculaire qu'on trouve dans les tables, n le nombre d'atomes de soufre dans la formule, et δ le facteur de correction 0,0652 (différence entre le poids atomique de deux oxygènes et un soufre). On trouvera ci-dessous les valeurs de $n\delta$ pour $n = 1 \rightarrow 4$:

n	$n\delta$
1	0,0652
2	0,1304
3	0,1956
4	0,2608

On obtient le poids moléculaire exact corrigé à deux décimales près en ajoutant la valeur appropriée de $n\delta$ au poids moléculaire trouvé dans la table.

B. Poids Isomoléculaire

On corrige les poids isomoléculaires de la même manière que les poids moléculaires, et la valeur corrigée (IM_{corr}) donne la masse exacte des formules pour lesquelles un atome ^{32}S remplace deux atomes ^{16}O.

$$IM_{corr} = IM + n\delta'$$

où IM est le poids isomoléculaire trouvé dans les tables et δ' est le facteur de correction $-0{,}0177555$ (différence entre le poids de deux atomes ^{16}O et un atome ^{32}S). On trouvera ci-dessous les valeurs de $n\delta'$ pour $n = 1 \rightarrow 4$:

n	$n\delta'$
1	−0,0177555
2	−0,0355110
3	−0,0532665
4	−0,0710220

On obtient le poids isomoléculaire exact (corrigé à quatre décimales près) en ajoutant la valeur appropriée de $n\delta'$ au poids isomoléculaire.

C. Pourcentages en Eléments Autres que le Soufre

$$P_{corr} = \frac{M}{M+n\delta} P$$

où P_{corr} est le pourcentage exact, M est le poids moléculaire et P est le pourcentage donné dans les tables. On trouvera ci-dessous les valeurs de $\frac{M}{M+n\delta}$ pour divers poids moléculaires et valeurs de n :

Poids mol.	n			
	1	2	3	4
100	0,9993	0,9987	0,9980	—
200	0,9997	0,9993	0,9990	0,9987
300	0,9998	0,9996	0,9993	0,9991
400	0,9998	0,9997	0,9995	0,9993
500	0,9999	0,9997	0,9996	0,9995

On obtient le pourcentage exact (à deux décimales près) en multipliant le facteur approprié par le pourcentage observé.

D. Pourcentages en Soufre

$$SP_{corr} = \frac{M}{M+n\delta} \times 1{,}002 \times OP$$

où SP_{corr} est le pourcentage exact en soufre et OP est le pourcentage en oxygène relevé dans les tables pour la formule où on a remplacé un atome de soufre par deux atomes d'oxygène. On trouvera ci-dessous un tableau des facteurs de correction pour diverses valeurs de n et de poids moléculaire :

Poids mol.	n			
	1	2	3	4
100	1,0013	1,0007	1,0000	—
200	1,0017	1,0013	1,0010	1,0007
300	1,0018	1,0016	1,0013	1,0011
400	1,0018	1,0017	1,0015	1,0013
500	1,0019	1,0017	1,0016	1,0015

Nous donnons à titre d'exemple les valeurs figurant dans les tables pour la formule $C_9H_{11}NS_4$ ($C_9H_{11}NO_8$) en comparaison avec les valeurs réelles et les valeurs obtenues en effectuant les corrections.

$C_9H_{11}NS_4$	Observé	Réel	Corrigé
Poids moléculaire	261,19	261,4506	261,45
Poids isomoléculaire	261,0485	260,97744	260,97747
Pourcentage carbone	41,39	41,3462	41,35
Pourcentage hydrogène	4,25	4,2409	4,25
Pourcentage azote	5,36	5,3573	5,36
Pourcentage soufre	49,00	49,0555	49,05

En appendice, à la fin du livre, on trouvera une liste des poids atomiques, des masses atomiques et des multiples des poids atomiques.

TABLES OF MOLECULAR WEIGHTS AND PERCENTAGE COMPOSITIONS

TABELLEN DER MOLEKULARGEWICHTE UND PROZENT-GEHALTE

TABLES DES POIDS MOLECULAIRES ET DES COMPOSITIONS CENTESIMALES

I

HYDROCARBONS

KOHLENWASSERSTOFFE

HYDROCARBURES

MOLECULAR WEIGHTS AND PERCENTAGE COMPOSITIONS OF HYDROCARBONS

FORMULA	MOL. WEIGHT	ISOMOL. WEIGHT	PERCENT CARBON	PERCENT HYDROGEN
C 1H 2	14.03	14.0156	85.63	14.37
C 1H 4	16.04	16.0313	74.87	25.13
C 2H 2	26.04	26.0156	92.26	7.74
C 2H 4	28.05	28.0313	85.63	14.37
C 2H 6	30.07	30.0469	79.89	20.11
C 3H 2	38.05	38.0156	94.70	5.30
C 3H 4	40.07	40.0313	89.94	10.06
C 3H 6	42.08	42.0469	85.63	14.37
C 3H 8	44.10	44.0626	81.71	18.29
C 4H 2	50.06	50.0156	95.97	4.03
C 4H 4	52.08	52.0313	92.26	7.74
C 4H 6	54.09	54.0469	88.82	11.18
C 4H 8	56.11	56.0626	85.63	14.37
C 4H10	58.12	58.0782	82.66	17.34
C 5H 2	62.07	62.0156	96.75	3.25
C 5H 4	64.09	64.0313	93.71	6.29
C 5H 6	66.10	66.0469	90.85	9.15
C 5H 8	68.12	68.0626	88.16	11.84
C 5H10	70.14	70.0782	85.63	14.37
C 5H12	72.15	72.0939	83.24	16.76
C 6H 4	76.10	76.0313	94.70	5.30
C 6H 6	78.11	78.0469	92.26	7.74
C 6H 8	80.13	80.0626	89.94	10.06
C 6H10	82.15	82.0782	87.73	12.27
C 6H12	84.16	84.0939	85.63	14.37
C 6H14	86.18	86.1095	83.63	16.37
C 7H 4	88.11	88.0313	95.42	4.58
C 7H 6	90.13	90.0469	93.29	6.71
C 7H 8	92.14	92.0626	91.25	8.75
C 7H10	94.16	94.0782	89.29	10.71
C 7H12	96.17	96.0939	87.42	12.58
C 7H14	98.19	98.1095	85.63	14.37
C 7H16	100.21	100.1252	83.91	16.09
C 8H 4	100.12	100.0313	95.97	4.03
C 8H 6	102.14	102.0469	94.08	5.92
C 8H 8	104.15	104.0626	92.26	7.74
C 8H10	106.17	106.0782	90.51	9.49
C 8H12	108.18	108.0939	88.82	11.18
C 8H14	110.20	110.1095	87.19	12.81
C 8H16	112.22	112.1252	85.63	14.37
C 8H18	114.23	114.1408	84.12	15.88
C 9H 4	112.13	112.0313	96.40	3.60
C 9H 6	114.15	114.0469	94.70	5.30
C 9H 8	116.16	116.0626	93.06	6.94
C 9H10	118.18	118.0782	91.47	8.53
C 9H12	120.20	120.0939	89.94	10.06
C 9H14	122.21	122.1095	88.45	11.55
C 9H16	124.23	124.1252	87.02	12.98
C 9H18	126.24	126.1408	85.63	14.37
C 9H20	128.26	128.1565	84.28	15.72
C10H 6	126.16	126.0469	95.21	4.79
C10H 8	128.18	128.0626	93.71	6.29
C10H10	130.19	130.0782	92.26	7.74
C10H12	132.21	132.0939	90.85	9.15
C10H14	134.22	134.1095	89.49	10.51
C10H16	136.24	136.1252	88.16	11.84
C10H18	138.25	138.1408	86.88	13.12
C10H20	140.27	140.1565	85.63	14.37
C10H22	142.29	142.1721	84.42	15.58
C11H 6	138.17	138.0469	95.62	4.38
C11H 8	140.19	140.0626	94.25	5.75
C11H10	142.20	142.0782	92.91	7.09
C11H12	144.22	144.0939	91.61	8.39
C11H14	146.23	146.1095	90.35	9.65
C11H16	148.25	148.1252	89.12	10.88
C11H18	150.27	150.1408	87.93	12.07
C11H20	152.28	152.1565	86.76	13.24
C11H22	154.30	154.1721	85.63	14.37
C11H24	156.31	156.1878	84.52	15.48
C12H 6	150.18	150.0469	95.97	4.03
C12H 8	152.20	152.0626	94.70	5.30
C12H10	154.21	154.0782	93.46	6.54
C12H12	156.23	156.0939	92.26	7.74
C12H14	158.25	158.1095	91.08	8.92
C12H16	160.26	160.1252	89.94	10.06
C12H18	162.28	162.1408	88.82	11.18
C12H20	164.29	164.1565	87.73	12.27
C12H22	166.31	166.1721	86.67	13.33
C12H24	168.33	168.1878	85.63	14.37
C12H26	170.34	170.2034	84.61	15.39
C13H 6	162.19	162.0469	96.27	3.73
C13H 8	164.21	164.0626	95.09	4.91
C13H10	166.22	166.0782	93.94	6.06
C13H12	168.24	168.0939	92.81	7.19
C13H14	170.26	170.1095	91.71	8.29
C13H16	172.27	172.1252	90.64	9.36
C13H18	174.29	174.1408	89.59	10.41
C13H20	176.30	176.1565	88.57	11.43
C13H22	178.32	178.1721	87.56	12.44
C13H24	180.34	180.1878	86.59	13.41
C13H26	182.35	182.2034	85.63	14.37
C13H28	184.37	184.2191	84.69	15.31
C14H 8	176.22	176.0626	95.42	4.58
C14H10	178.24	178.0782	94.34	5.66
C14H12	180.25	180.0939	93.29	6.71
C14H14	182.27	182.1095	92.26	7.74
C14H16	184.28	184.1252	91.25	8.75
C14H18	186.30	186.1408	90.26	9.74
C14H20	188.32	188.1565	89.29	10.71
C14H22	190.33	190.1721	88.35	11.65

FORMULA	MOL. WEIGHT	ISOMOL. WEIGHT	PERCENT CARBON	PERCENT HYDROGEN
C14H24	192.35	192.1878	87.42	12.58
C14H26	194.36	194.2034	86.52	13.48
C14H28	196.38	196.2191	85.63	14.37
C14H30	198.40	198.2347	84.76	15.24
C15H 8	188.23	188.0626	95.72	4.28
C15H10	190.25	190.0782	94.70	5.30
C15H12	192.26	192.0939	93.71	6.29
C15H14	194.28	194.1095	92.74	7.26
C15H16	196.29	196.1252	91.78	8.22
C15H18	198.31	198.1408	90.85	9.15
C15H20	200.33	200.1565	89.94	10.06
C15H22	202.34	202.1721	89.04	10.96
C15H24	204.36	204.1878	88.16	11.84
C15H26	206.37	206.2034	87.30	12.70
C15H28	208.39	208.2191	86.46	13.54
C15H30	210.41	210.2347	85.63	14.37
C15H32	212.42	212.2504	84.82	15.18
C16H 8	200.24	200.0626	95.97	4.03
C16H10	202.26	202.0782	95.02	4.98
C16H12	204.27	204.0939	94.08	5.92
C16H14	206.29	206.1095	93.16	6.84
C16H16	208.31	208.1252	92.26	7.74
C16H18	210.32	210.1408	91.37	8.63
C16H20	212.34	212.1565	90.51	9.49
C16H22	214.35	214.1721	89.65	10.35
C16H24	216.37	216.1878	88.82	11.18
C16H26	218.39	218.2034	88.00	12.00
C16H28	220.40	220.2191	87.19	12.81
C16H30	222.42	222.2347	86.40	13.60
C16H32	224.43	224.2504	85.63	14.37
C16H34	226.45	226.2660	84.87	15.13
C17H 8	212.25	212.0626	96.20	3.80
C17H10	214.27	214.0782	95.30	4.70
C17H12	216.29	216.0939	94.41	5.59
C17H14	218.30	218.1095	93.54	6.46
C17H16	220.32	220.1252	92.68	7.32
C17H18	222.33	222.1408	91.84	8.16
C17H20	224.35	224.1565	91.01	8.99
C17H22	226.36	226.1721	90.20	9.80
C17H24	228.38	228.1878	89.41	10.59
C17H26	230.40	230.2034	88.63	11.37
C17H28	232.41	232.2191	87.86	12.14
C17H30	234.43	234.2347	87.10	12.90
C17H32	236.44	236.2504	86.36	13.64
C17H34	238.46	238.2660	85.63	14.37
C17H36	240.48	240.2817	84.91	15.09
C18H10	226.28	226.0782	95.55	4.45
C18H12	228.30	228.0939	94.70	5.30
C18H14	230.31	230.1095	93.87	6.13
C18H16	232.33	232.1252	93.06	6.94
C18H18	234.34	234.1408	92.26	7.74
C18H20	236.36	236.1565	91.47	8.53
C18H22	238.38	238.1721	90.70	9.30
C18H24	240.39	240.1878	89.94	10.06
C18H26	242.41	242.2034	89.19	10.81
C18H28	244.42	244.2191	88.45	11.55
C18H30	246.44	246.2347	87.73	12.27
C18H32	248.46	248.2504	87.02	12.98
C18H34	250.47	250.2660	86.32	13.68
C18H36	252.49	252.2817	85.63	14.37
C18H38	254.50	254.2973	84.95	15.05
C19H10	238.29	238.0782	95.77	4.23
C19H12	240.31	240.0939	94.97	5.03
C19H14	242.32	242.1095	94.18	5.82
C19H16	244.34	244.1252	93.40	6.60
C19H18	246.36	246.1408	92.64	7.36
C19H20	248.37	248.1565	91.88	8.12
C19H22	250.39	250.1721	91.14	8.86
C19H24	252.40	252.1878	90.42	9.58
C19H26	254.42	254.2034	89.70	10.30
C19H28	256.44	256.2191	88.99	11.01
C19H30	258.45	258.2347	88.30	11.70
C19H32	260.47	260.2504	87.62	12.38
C19H34	262.48	262.2660	86.94	13.06
C19H36	264.50	264.2817	86.28	13.72
C19H38	266.51	266.2973	85.63	14.37
C19H40	268.53	268.3130	84.99	15.01
C20H10	250.30	250.0782	95.97	4.03
C20H12	252.32	252.0939	95.21	4.79
C20H14	254.33	254.1095	94.45	5.55
C20H16	256.35	256.1252	93.71	6.29
C20H18	258.37	258.1408	92.98	7.02
C20H20	260.38	260.1565	92.26	7.74
C20H22	262.40	262.1721	91.55	8.45
C20H24	264.41	264.1878	90.85	9.15
C20H26	266.43	266.2034	90.16	9.84
C20H28	268.45	268.2191	89.49	10.51
C20H30	270.46	270.2347	88.82	11.18
C20H32	272.48	272.2504	88.16	11.84
C20H34	274.49	274.2660	87.51	12.49
C20H36	276.51	276.2817	86.88	13.12
C20H38	278.53	278.2973	86.25	13.75
C20H40	280.54	280.3130	85.63	14.37
C20H42	282.56	282.3286	85.02	14.98
C21H10	262.31	262.0782	96.16	3.84
C21H12	264.33	264.0939	95.42	4.58
C21H14	266.35	266.1095	94.70	5.30
C21H16	268.36	268.1252	93.99	6.01
C21H18	270.38	270.1408	93.29	6.71
C21H20	272.39	272.1565	92.60	7.40
C21H22	274.41	274.1721	91.92	8.08
C21H24	276.43	276.1878	91.25	8.75
C21H26	278.44	278.2034	90.59	9.41
C21H28	280.46	280.2191	89.94	10.06
C21H30	282.47	282.2347	89.29	10.71
C21H32	284.49	284.2504	88.66	11.34
C21H34	286.51	286.2660	88.04	11.96
C21H36	288.52	288.2817	87.42	12.58
C21H38	290.54	290.2973	86.82	13.18
C21H40	292.55	292.3130	86.22	13.78
C21H42	294.57	294.3286	85.63	14.37
C21H44	296.58	296.3443	85.05	14.95
C22H12	276.34	276.0939	95.62	4.38
C22H14	278.36	278.1095	94.93	5.07
C22H16	280.37	280.1252	94.25	5.75
C22H18	282.39	282.1408	93.58	6.42
C22H20	284.40	284.1565	92.91	7.09
C22H22	286.42	286.1721	92.26	7.74
C22H24	288.44	288.1878	91.61	8.39
C22H26	290.45	290.2034	90.98	9.02

10/21/66 013

C22H28	292.47	292.2191	90.35	9.65
C22H30	294.48	294.2347	89.73	10.27
C22H32	296.50	296.2504	89.12	10.88
C22H34	298.52	298.2660	88.52	11.48
C22H36	300.53	300.2817	87.93	12.07
C22H38	302.55	302.2973	87.34	12.66
C22H40	304.56	304.3130	86.76	13.24
C22H42	306.58	306.3286	86.19	13.81
C22H44	308.60	308.3443	85.63	14.37
C22H46	310.61	310.3599	85.07	14.93
C23H12	288.35	288.0939	95.81	4.19
C23H14	290.37	290.1095	95.14	4.86
C23H16	292.38	292.1252	94.48	5.52
C23H18	294.40	294.1408	93.84	6.16
C23H20	296.42	296.1565	93.20	6.80
C23H22	298.43	298.1721	92.57	7.43
C23H24	300.45	300.1878	91.95	8.05
C23H26	302.46	302.2034	91.34	8.66
C23H28	304.48	304.2191	90.73	9.27
C23H30	306.50	306.2347	90.13	9.87
C23H32	308.51	308.2504	89.54	10.46
C23H34	310.53	310.2660	88.96	11.04
C23H36	312.54	312.2817	88.39	11.61
C23H38	314.56	314.2973	87.82	12.18
C23H40	316.58	316.3130	87.26	12.74
C23H42	318.59	318.3286	86.71	13.29
C23H44	320.61	320.3443	86.17	13.83
C23H46	322.62	322.3599	85.63	14.37
C23H48	324.64	324.3756	85.10	14.90
C24H12	300.36	300.0939	95.97	4.03
C24H14	302.38	302.1095	95.33	4.67
C24H16	304.40	304.1252	94.70	5.30
C24H18	306.41	306.1408	94.08	5.92
C24H20	308.43	308.1565	93.46	6.54
C24H22	310.44	310.1721	92.86	7.14
C24H24	312.46	312.1878	92.26	7.74
C24H26	314.47	314.2034	91.67	8.33
C24H28	316.49	316.2191	91.08	8.92
C24H30	318.51	318.2347	90.51	9.49
C24H32	320.52	320.2504	89.94	10.06
C24H34	322.54	322.2660	89.37	10.63
C24H36	324.55	324.2817	88.82	11.18
C24H38	326.57	326.2973	88.27	11.73
C24H40	328.59	328.3130	87.73	12.27
C24H42	330.60	330.3286	87.19	12.81
C24H44	332.62	332.3443	86.67	13.33
C24H46	334.63	334.3599	86.14	13.86
C24H48	336.65	336.3756	85.63	14.37
C24H50	338.67	338.3912	85.12	14.88
C25H12	312.37	312.0939	96.13	3.87
C25H14	314.39	314.1095	95.51	4.49
C25H16	316.41	316.1252	94.90	5.10
C25H18	318.42	318.1408	94.30	5.70
C25H20	320.44	320.1565	93.71	6.29
C25H22	322.45	322.1721	93.12	6.88
C25H24	324.47	324.1878	92.54	7.46
C25H26	326.49	326.2034	91.97	8.03
C25H28	328.50	328.2191	91.41	8.59
C25H30	330.52	330.2347	90.85	9.15
C25H32	332.53	332.2504	90.30	9.70
C25H34	334.55	334.2660	89.76	10.24
C25H36	336.57	336.2817	89.22	10.78
C25H38	338.58	338.2973	88.69	11.31
C25H40	340.60	340.3130	88.16	11.84
C25H42	342.61	342.3286	87.64	12.36
C25H44	344.63	344.3443	87.13	12.87
C25H46	346.65	346.3599	86.62	13.38
C25H48	348.66	348.3756	86.12	13.88
C25H50	350.68	350.3912	85.63	14.37
C25H52	352.69	352.4069	85.14	14.86
C26H14	326.40	326.1095	95.68	4.32
C26H16	328.42	328.1252	95.09	4.91
C26H18	330.43	330.1408	94.51	5.49
C26H20	332.45	332.1565	93.94	6.06
C26H22	334.47	334.1721	93.37	6.63
C26H24	336.48	336.1878	92.81	7.19
C26H26	338.50	338.2034	92.26	7.74
C26H28	340.51	340.2191	91.71	8.29
C26H30	342.53	342.2347	91.17	8.83
C26H32	344.54	344.2504	90.64	9.36
C26H34	346.56	346.2660	90.11	9.89
C26H36	348.58	348.2817	89.59	10.41
C26H38	350.59	350.2973	89.07	10.93
C26H40	352.61	352.3130	88.57	11.43
C26H42	354.62	354.3286	88.06	11.94
C26H44	356.64	356.3443	87.56	12.44
C26H46	358.66	358.3599	87.07	12.93
C26H48	360.67	360.3756	86.59	13.41
C26H50	362.69	362.3912	86.10	13.90
C26H52	364.70	364.4069	85.63	14.37
C26H54	366.72	366.4225	85.16	14.84
C27H14	338.41	338.1095	95.83	4.17
C27H16	340.43	340.1252	95.26	4.74
C27H18	342.44	342.1408	94.70	5.30
C27H20	344.46	344.1565	94.15	5.85
C27H22	346.48	346.1721	93.60	6.40
C27H24	348.49	348.1878	93.06	6.94
C27H26	350.51	350.2034	92.52	7.48
C27H28	352.52	352.2191	91.99	8.01
C27H30	354.54	354.2347	91.47	8.53
C27H32	356.56	356.2504	90.95	9.05
C27H34	358.57	358.2660	90.44	9.56
C27H36	360.59	360.2817	89.94	10.06
C27H38	362.60	362.2973	89.44	10.56
C27H40	364.62	364.3130	88.94	11.06
C27H42	366.64	366.3286	88.45	11.55
C27H44	368.65	368.3443	87.97	12.03
C27H46	370.67	370.3599	87.49	12.51
C27H48	372.68	372.3756	87.02	12.98
C27H50	374.70	374.3912	86.55	13.45
C27H52	376.72	376.4069	86.09	13.91
C27H54	378.73	378.4225	85.63	14.37
C27H56	380.75	380.4382	85.17	14.83
C28H14	350.42	350.1095	95.97	4.03
C28H16	352.44	352.1252	95.42	4.58
C28H18	354.46	354.1408	94.88	5.12
C28H20	356.47	356.1565	94.34	5.66
C28H22	358.49	358.1721	93.81	6.19
C28H24	360.50	360.1878	93.29	6.71
C28H26	362.52	362.2034	92.77	7.23

10/21/66 014

C28H28	364.54	364.2191	92.26	7.74
C28H30	366.55	366.2347	91.75	8.25
C28H32	368.57	368.2504	91.25	8.75
C28H34	370.58	370.2660	90.75	9.25
C28H36	372.60	372.2817	90.26	9.74
C28H38	374.62	374.2973	89.78	10.22
C28H40	376.63	376.3130	89.29	10.71
C28H42	378.65	378.3286	88.82	11.18
C28H44	380.66	380.3443	88.35	11.65
C28H46	382.68	382.3599	87.88	12.12
C28H48	384.69	384.3756	87.42	12.58
C28H50	386.71	386.3912	86.97	13.03
C28H52	388.73	388.4069	86.52	13.48
C28H54	390.74	390.4225	86.07	13.93
C28H56	392.76	392.4382	85.63	14.37
C28H58	394.77	394.4538	85.19	14.81
C29H14	362.43	362.1095	96.11	3.89
C29H16	364.45	364.1252	95.57	4.43
C29H18	366.47	366.1408	95.05	4.95
C29H20	368.48	368.1565	94.53	5.47
C29H22	370.50	370.1721	94.01	5.99
C29H24	372.51	372.1878	93.51	6.49
C29H26	374.53	374.2034	93.00	7.00
C29H28	376.55	376.2191	92.50	7.50
C29H30	378.56	378.2347	92.01	7.99
C29H32	380.58	380.2504	91.52	8.48
C29H34	382.59	382.2660	91.04	8.96
C29H36	384.61	384.2817	90.57	9.43
C29H38	386.63	386.2973	90.09	9.91
C29H40	388.64	388.3130	89.63	10.37
C29H42	390.66	390.3286	89.16	10.84
C29H44	392.67	392.3443	88.71	11.29
C29H46	394.69	394.3599	88.25	11.75
C29H48	396.71	396.3756	87.80	12.20
C29H50	398.72	398.3912	87.36	12.64
C29H52	400.74	400.4069	86.92	13.08
C29H54	402.75	402.4225	86.49	13.51
C29H56	404.77	404.4382	86.05	13.95
C29H58	406.79	406.4538	85.63	14.37
C29H60	408.80	408.4695	85.21	14.79
C30H16	376.46	376.1252	95.72	4.28
C30H18	378.48	378.1408	95.21	4.79
C30H20	380.49	380.1565	94.70	5.30
C30H22	382.51	382.1721	94.20	5.80
C30H24	384.53	384.1878	93.71	6.29
C30H26	386.54	386.2034	93.22	6.78
C30H28	388.56	388.2191	92.74	7.26
C30H30	390.57	390.2347	92.26	7.74
C30H32	392.59	392.2504	91.78	8.22
C30H34	394.61	394.2660	91.32	8.68
C30H36	396.62	396.2817	90.85	9.15
C30H38	398.64	398.2973	90.39	9.61
C30H40	400.65	400.3130	89.94	10.06
C30H42	402.67	402.3286	89.49	10.51
C30H44	404.69	404.3443	89.04	10.96
C30H46	406.70	406.3599	88.60	11.40
C30H48	408.72	408.3756	88.16	11.84
C30H50	410.73	410.3912	87.73	12.27
C30H52	412.75	412.4069	87.30	12.70
C30H54	414.76	414.4225	86.88	13.12
C30H56	416.78	416.4382	86.46	13.54
C30H58	418.80	418.4538	86.04	13.96
C30H60	420.81	420.4695	85.63	14.37
C30H62	422.83	422.4851	85.22	14.78
C31H16	388.47	388.1252	95.85	4.15
C31H18	390.49	390.1408	95.35	4.65
C31H20	392.51	392.1565	94.86	5.14
C31H22	394.52	394.1721	94.38	5.62
C31H24	396.54	396.1878	93.90	6.10
C31H26	398.55	398.2034	93.42	6.58
C31H28	400.57	400.2191	92.95	7.05
C31H30	402.58	402.2347	92.49	7.51
C31H32	404.60	404.2504	92.03	7.97
C31H34	406.62	406.2660	91.57	8.43
C31H36	408.63	408.2817	91.12	8.88
C31H38	410.65	410.2973	90.67	9.33
C31H40	412.66	412.3130	90.23	9.77
C31H42	414.68	414.3286	89.79	10.21
C31H44	416.70	416.3443	89.36	10.64
C31H46	418.71	418.3599	88.93	11.07
C31H48	420.73	420.3756	88.50	11.50
C31H50	422.74	422.3912	88.08	11.92
C31H52	424.76	424.4069	87.66	12.34
C31H54	426.78	426.4225	87.25	12.75
C31H56	428.79	428.4382	86.84	13.16
C31H58	430.81	430.4538	86.43	13.57
C31H60	432.82	432.4695	86.03	13.97
C31H62	434.84	434.4851	85.63	14.37
C31H64	436.86	436.5008	85.23	14.77
C32H16	400.48	400.1252	95.97	4.03
C32H18	402.50	402.1408	95.49	4.51
C32H20	404.52	404.1565	95.02	4.98
C32H22	406.53	406.1721	94.55	5.45
C32H24	408.55	408.1878	94.08	5.92
C32H26	410.56	410.2034	93.62	6.38
C32H28	412.58	412.2191	93.16	6.84
C32H30	414.60	414.2347	92.71	7.29
C32H32	416.61	416.2504	92.26	7.74
C32H34	418.63	418.2660	91.81	8.19
C32H36	420.64	420.2817	91.37	8.63
C32H38	422.66	422.2973	90.94	9.06
C32H40	424.68	424.3130	90.51	9.49
C32H42	426.69	426.3286	90.08	9.92
C32H44	428.71	428.3443	89.65	10.35
C32H46	430.72	430.3599	89.24	10.76
C32H48	432.74	432.3756	88.82	11.18
C32H50	434.76	434.3912	88.41	11.59
C32H52	436.77	436.4069	88.00	12.00
C32H54	438.79	438.4225	87.60	12.40
C32H56	440.80	440.4382	87.19	12.81
C32H58	442.82	442.4538	86.80	13.20
C32H60	444.84	444.4695	86.40	13.60
C32H62	446.85	446.4851	86.01	13.99
C32H64	448.87	448.5008	85.63	14.37
C32H66	450.88	450.5164	85.25	14.75
C33H16	412.50	412.1252	96.09	3.91
C33H18	414.51	414.1408	95.62	4.38
C33H20	416.53	416.1565	95.16	4.84
C33H22	418.54	418.1721	94.70	5.30
C33H24	420.56	420.1878	94.25	5.75

C33H26	422.58	422.2034	93.80	6.20
C33H28	424.59	424.2191	93.35	6.65
C33H30	426.61	426.2347	92.91	7.09
C33H32	428.62	428.2504	92.47	7.53
C33H34	430.64	430.2660	92.04	7.96
C33H36	432.65	432.2817	91.61	8.39
C33H38	434.67	434.2973	91.19	8.81
C33H40	436.69	436.3130	90.77	9.23
C33H42	438.70	438.3286	90.35	9.65
C33H44	440.72	440.3443	89.94	10.06
C33H46	442.73	442.3599	89.53	10.47
C33H48	444.75	444.3756	89.12	10.88
C33H50	446.77	446.3912	88.72	11.28
C33H52	448.78	448.4069	88.32	11.68
C33H54	450.80	450.4225	87.93	12.07
C33H56	452.81	452.4382	87.53	12.47
C33H58	454.83	454.4538	87.15	12.85
C33H60	456.85	456.4695	86.76	13.24
C33H62	458.86	458.4851	86.38	13.62
C33H64	460.88	460.5008	86.00	14.00
C33H66	462.89	462.5164	85.63	14.37
C33H68	464.91	464.5321	85.26	14.74
C34H18	426.52	426.1408	95.75	4.25
C34H20	428.54	428.1565	95.30	4.70
C34H22	430.55	430.1721	94.85	5.15
C34H24	432.57	432.1878	94.41	5.59
C34H26	434.59	434.2034	93.97	6.03
C34H28	436.60	436.2191	93.54	6.46
C34H30	438.62	438.2347	93.11	6.89
C34H32	440.63	440.2504	92.68	7.32
C34H34	442.65	442.2660	92.26	7.74
C34H36	444.57	444.2817	91.84	8.16
C34H38	446.68	446.2973	91.43	8.57
C34H40	448.70	448.3130	91.01	8.99
C34H42	450.71	450.3286	90.61	9.39
C34H44	452.73	452.3443	90.20	9.80
C34H46	454.75	454.3599	89.80	10.20
C34H48	456.76	456.3756	89.41	10.59
C34H50	458.78	458.3912	89.01	10.99
C34H52	460.79	460.4069	88.63	11.37
C34H54	462.81	462.4225	88.24	11.76
C34H56	464.83	464.4382	87.86	12.14
C34H58	466.84	466.4538	87.48	12.52
C34H60	468.86	468.4695	87.10	12.90
C34H62	470.87	470.4851	86.73	13.27
C34H64	472.89	472.5008	86.36	13.64
C34H66	474.91	474.5164	85.99	14.01
C34H68	476.92	476.5321	85.63	14.37
C34H70	478.94	478.5477	85.27	14.73
C35H18	438.53	438.1408	95.86	4.14
C35H20	440.55	440.1565	95.42	4.58
C35H22	442.57	442.1721	94.99	5.01
C35H24	444.58	444.1878	94.56	5.44
C35H26	446.60	446.2034	94.13	5.87
C35H28	448.61	448.2191	93.71	6.29
C35H30	450.63	450.2347	93.29	6.71
C35H32	452.65	452.2504	92.87	7.13
C35H34	454.66	454.2660	92.46	7.54
C35H36	456.68	456.2817	92.05	7.95
C35H38	458.69	458.2973	91.65	8.35
C35H40	460.71	460.3130	91.25	8.75
C35H42	462.72	462.3286	90.85	9.15
C35H44	464.74	464.3443	90.46	9.54
C35H46	466.75	466.3599	90.07	9.93
C35H48	468.77	468.3756	89.68	10.32
C35H50	470.79	470.3912	89.29	10.71
C35H52	472.80	472.4069	88.91	11.09
C35H54	474.82	474.4225	88.54	11.46
C35H56	476.84	476.4382	88.16	11.84
C35H58	478.85	478.4538	87.79	12.21
C35H60	480.87	480.4695	87.42	12.58
C35H62	482.88	482.4851	87.06	12.94
C35H64	484.90	484.5008	86.70	13.30
C35H66	486.92	486.5164	86.34	13.66
C35H68	488.93	488.5321	85.98	14.02
C35H70	490.95	490.5477	85.63	14.37
C35H72	492.96	492.5634	85.28	14.72
C36H18	450.54	450.1408	95.97	4.03
C36H20	452.56	452.1565	95.55	4.45
C36H22	454.58	454.1721	95.12	4.88
C36H24	456.59	456.1878	94.70	5.30
C36H26	458.61	458.2034	94.29	5.71
C36H28	460.62	460.2191	93.87	6.13
C36H30	462.64	462.2347	93.46	6.54
C36H32	464.65	464.2504	93.06	6.94
C36H34	466.67	466.2660	92.66	7.34
C36H36	468.69	468.2817	92.26	7.74
C36H38	470.70	470.2973	91.86	8.14
C36H40	472.72	472.3130	91.47	8.53
C36H42	474.74	474.3286	91.08	8.92
C36H44	476.75	476.3443	90.70	9.30
C36H46	478.77	478.3599	90.32	9.68
C36H48	480.78	480.3756	89.94	10.06
C36H50	482.80	482.3912	89.56	10.44
C36H52	484.82	484.4069	89.19	10.81
C36H54	486.83	486.4225	88.82	11.18
C36H56	488.85	488.4382	88.45	11.55
C36H58	490.86	490.4538	88.09	11.91
C36H60	492.88	492.4695	87.73	12.27
C36H62	494.90	494.4851	87.37	12.63
C36H64	496.91	496.5008	87.02	12.98
C36H66	498.93	498.5164	86.67	13.33
C36H68	500.94	500.5321	86.32	13.68
C36H70	502.96	502.5477	85.97	14.03
C36H72	504.98	504.5634	85.63	14.37
C36H74	506.99	506.5790	85.29	14.71
C37H18	462.56	462.1408	96.08	3.92
C37H20	464.57	464.1565	95.66	4.34
C37H22	466.59	466.1721	95.25	4.75
C37H24	468.60	468.1878	94.84	5.16
C37H26	470.62	470.2034	94.43	5.57
C37H28	472.64	472.2191	94.03	5.97
C37H30	474.65	474.2347	93.63	6.37
C37H32	476.67	476.2504	93.23	6.77
C37H34	478.68	478.2660	92.84	7.16
C37H36	480.70	480.2817	92.45	7.55
C37H38	482.72	482.2973	92.07	7.93
C37H40	484.73	484.3130	91.68	8.32
C37H42	486.75	486.3286	91.30	8.70
C37H44	488.76	488.3443	90.93	9.07

C37H46	490.78	490.3599	90.55	9.45
C37H48	492.80	492.3756	90.18	9.82
C37H50	494.81	494.3912	89.81	10.19
C37H52	496.83	496.4069	89.45	10.55
C37H54	498.84	498.4225	89.09	10.91
C37H56	500.86	500.4382	88.73	11.27
C37H58	502.87	502.4538	88.37	11.63
C37H60	504.89	504.4695	88.02	11.98
C37H62	506.91	506.4851	87.67	12.33
C37H64	508.92	508.5008	87.32	12.68
C37H66	510.94	510.5164	86.98	13.02
C37H68	512.95	512.5321	86.64	13.36
C37H70	514.97	514.5477	86.30	13.70
C37H72	516.99	516.5634	85.96	14.04
C37H74	519.00	518.5790	85.63	14.37
C37H76	521.02	520.5947	85.30	14.70
C38H20	476.58	476.1565	95.77	4.23
C38H22	478.60	478.1721	95.37	4.63
C38H24	480.61	480.1878	94.97	5.03
C38H26	482.63	482.2034	94.57	5.43
C38H28	484.65	484.2191	94.18	5.82
C38H30	486.66	486.2347	93.79	6.21
C38H32	488.68	488.2504	93.40	6.60
C38H34	490.69	490.2660	93.02	6.98
C38H36	492.71	492.2817	92.64	7.36
C38H38	494.73	494.2973	92.26	7.74
C38H40	496.74	496.3130	91.88	8.12
C38H42	498.76	498.3286	91.51	8.49
C38H44	500.77	500.3443	91.14	8.86
C38H46	502.79	502.3599	90.78	9.22
C38H48	504.81	504.3756	90.42	9.58
C38H50	506.82	506.3912	90.06	9.94
C38H52	508.84	508.4069	89.70	10.30
C38H54	510.85	510.4225	89.35	10.65
C38H56	512.87	512.4382	88.99	11.01
C38H58	514.89	514.4538	88.65	11.35
C38H60	516.90	516.4695	88.30	11.70
C38H62	518.92	518.4851	87.96	12.04
C38H64	520.93	520.5008	87.62	12.38
C38H66	522.95	522.5164	87.28	12.72
C38H68	524.97	524.5321	86.94	13.06
C38H70	526.98	526.5477	86.61	13.39
C38H72	529.00	528.5634	86.28	13.72
C38H74	531.01	530.5790	85.95	14.05
C38H76	533.03	532.5947	85.63	14.37
C38H78	535.05	534.6103	85.31	14.69
C39H20	488.59	488.1565	95.87	4.13
C39H22	490.61	490.1721	95.48	4.52
C39H24	492.63	492.1878	95.09	4.91
C39H26	494.64	494.2034	94.70	5.30
C39H28	496.66	496.2191	94.32	5.68
C39H30	498.67	498.2347	93.94	6.06
C39H32	500.69	500.2504	93.56	6.44
C39H34	502.71	502.2660	93.18	6.82
C39H36	504.72	504.2817	92.81	7.19
C39H38	506.74	506.2973	92.44	7.56
C39H40	508.75	508.3130	92.07	7.93
C39H42	510.77	510.3286	91.71	8.29
C39H44	512.79	512.3443	91.35	8.65
C39H46	514.80	514.3599	90.99	9.01
C39H48	516.82	516.3756	90.64	9.36
C39H50	518.83	518.3912	90.29	9.71
C39H52	520.85	520.4069	89.94	10.06
C39H54	522.87	522.4225	89.59	10.41
C39H56	524.88	524.4382	89.25	10.75
C39H58	526.90	526.4538	88.90	11.10
C39H60	528.91	528.4695	88.57	11.43
C39H62	530.93	530.4851	88.23	11.77
C39H64	532.94	532.5008	87.90	12.10
C39H66	534.96	534.5164	87.56	12.44
C39H68	536.98	536.5321	87.24	12.76
C39H70	538.99	538.5477	86.91	13.09
C39H72	541.01	540.5634	86.59	13.41
C39H74	543.02	542.5790	86.26	13.74
C39H76	545.04	544.5947	85.94	14.06
C39H78	547.06	546.6103	85.63	14.37
C39H80	549.07	548.6260	85.31	14.69
C40H20	500.61	500.1565	95.97	4.03
C40H22	502.62	502.1721	95.59	4.41
C40H24	504.64	504.1878	95.21	4.79
C40H26	506.65	506.2034	94.83	5.17
C40H28	508.67	508.2191	94.45	5.55
C40H30	510.69	510.2347	94.08	5.92
C40H32	512.70	512.2504	93.71	6.29
C40H34	514.72	514.2660	93.34	6.66
C40H36	516.73	516.2817	92.98	7.02
C40H38	518.75	518.2973	92.62	7.38
C40H40	520.76	520.3130	92.26	7.74
C40H42	522.78	522.3286	91.90	8.10
C40H44	524.80	524.3443	91.55	8.45
C40H46	526.81	526.3599	91.20	8.80
C40H48	528.83	528.3756	90.85	9.15
C40H50	530.84	530.3912	90.51	9.49
C40H52	532.86	532.4069	90.16	9.84
C40H54	534.88	534.4225	89.82	10.18
C40H56	536.89	536.4382	89.49	10.51
C40H58	538.91	538.4538	89.15	10.85
C40H60	540.92	540.4695	88.82	11.18
C40H62	542.94	542.4851	88.49	11.51
C40H64	544.96	544.5008	88.16	11.84
C40H66	546.97	546.5164	87.84	12.16
C40H68	548.99	548.5321	87.51	12.49
C40H70	551.00	550.5477	87.19	12.81
C40H72	553.02	552.5634	86.88	13.12
C40H74	555.04	554.5790	86.56	13.44
C40H76	557.05	556.5947	86.25	13.75
C40H78	559.07	558.6103	85.94	14.06
C40H80	561.08	560.6260	85.63	14.37
C40H82	563.10	562.6416	85.32	14.68

II

COMPOUNDS CONTAINING CARBON, HYDROGEN AND ONE HETEROATOM

VERBINDUNGEN AUS KOHLENSTOFF, WASSERSTOFF UND EINEM HETEROATOM

COMPOSES CONTENANT DU CARBONE, DE L'HYDROGENE ET UN HETEROATOME

1. CARBON, HYDROGEN AND BROMINE

1. KOHLENSTOFF, WASSERSTOFF UND BROM

1. CARBONE, HYDROGENE ET BROME

MOLECULAR WEIGHTS AND PERCENTAGE COMPOSITIONS OF COMPOUNDS
CONTAINING CARBON, HYDROGEN AND BROMINE

FORMULA	MOL. WEIGHT	ISOMOL. WEIGHT	PERCENT C	PERCENT H	PERCENT BR
C 1H 0BR2	171.82	169.8368	6.99	0	93.01
C 1H 1BR1	92.92	91.9262	12.93	1.08	85.99
C 1H 2BR2	173.84	171.8524	6.91	1.16	91.93
C 1H 3BR1	94.94	93.9419	12.65	3.19	84.16
C 2H 1BR1	104.93	103.9262	22.89	.96	76.15
C 2H 2BR2	185.85	183.8524	12.93	1.08	85.99
C 2H 3BR1	106.95	105.9419	22.46	2.83	74.71
C 2H 4BR2	187.86	185.8681	12.79	2.15	85.07
C 2H 5BR1	108.97	107.9575	22.05	4.63	73.33
C 3H 1BR1	116.95	115.9262	30.81	.86	68.33
C 3H 2BR2	197.86	195.8524	18.21	1.02	80.77
C 3H 3BR1	118.96	117.9419	30.29	2.54	67.17
C 3H 4BR2	199.87	197.8681	18.03	2.02	79.95
C 3H 5BR1	120.98	119.9575	29.79	4.17	66.05
C 3H 6BR2	201.89	199.8837	17.85	3.00	79.16
C 3H 7BR1	122.99	121.9732	29.30	5.74	64.97
C 4H 2BR2	209.87	207.8524	22.89	.96	76.15
C 4H 3BR1	130.97	129.9419	36.68	2.31	61.01
C 4H 4BR2	211.88	209.8681	22.67	1.90	75.42
C 4H 5BR1	132.99	131.9575	36.13	3.79	60.08
C 4H 6BR2	213.90	211.8837	22.46	2.83	74.71
C 4H 7BR1	135.00	133.9732	35.59	5.23	59.19
C 4H 8BR2	215.92	213.8994	22.25	3.73	74.01
C 4H 9BR1	137.02	135.9888	35.06	6.62	58.32
C 5H 2BR2	221.88	219.8524	27.07	.91	72.02
C 5H 3BR1	142.98	141.9419	42.00	2.11	55.88
C 5H 4BR2	223.90	221.8681	26.82	1.80	71.38
C 5H 5BR1	145.00	143.9575	41.42	3.48	55.11
C 5H 6BR2	225.91	223.8837	26.58	2.68	70.74
C 5H 7BR1	147.02	145.9732	40.85	4.80	54.35
C 5H 8BR2	227.93	225.8994	26.35	3.54	70.11
C 5H 9BR1	149.03	147.9888	40.30	6.09	53.62
C 5H10BR2	229.94	227.9150	26.12	4.38	69.50
C 5H11BR1	151.05	150.0045	39.76	7.34	52.90
C 6H 3BR1	154.99	153.9419	46.50	1.95	51.55
C 6H 4BR2	235.91	233.8681	30.55	1.71	67.74
C 6H 5BR1	157.01	155.9575	45.90	3.21	50.89
C 6H 6BR2	237.92	235.8837	30.29	2.54	67.17
C 6H 7BR1	159.03	157.9732	45.32	4.44	50.25
C 6H 8BR2	239.94	237.8994	30.04	3.36	66.60
C 6H 9BR1	161.04	159.9888	44.75	5.63	49.62
C 6H10BR2	241.95	239.9150	29.79	4.17	66.05
C 6H11BR1	163.06	162.0045	44.20	6.80	49.00
C 6H12BR2	243.97	241.9307	29.54	4.96	65.50
C 6H13BR1	165.07	164.0201	43.66	7.94	48.40
C 7H 3BR1	167.01	165.9419	50.34	1.81	47.84
C 7H 4BR2	247.92	245.8681	33.91	1.63	64.46
C 7H 5BR1	169.02	167.9575	49.74	2.98	47.27
C 7H 6BR2	249.93	247.8837	33.64	2.42	63.94
C 7H 7BR1	171.04	169.9732	49.16	4.13	46.72
C 7H 8BR2	251.95	249.8994	33.37	3.20	63.43
C 7H 9BR1	173.05	171.9888	48.58	5.24	46.17
C 7H10BR2	253.97	251.9150	33.11	3.97	62.93
C 7H11BR1	175.07	174.0045	48.03	6.33	45.64
C 7H12BR2	255.98	253.9307	32.85	4.73	62.43
C 7H13BR1	177.09	176.0201	47.48	7.40	45.12
C 7H14BR2	258.00	255.9463	32.59	5.47	61.94
C 7H15BR1	179.10	178.0358	46.94	8.44	44.61
C 8H 4BR2	259.93	257.8681	36.97	1.55	61.48
C 8H 5BR1	181.03	179.9575	53.08	2.78	44.14
C 8H 6BR2	261.95	259.8837	36.68	2.31	61.01
C 8H 7BR1	183.05	181.9732	52.49	3.85	43.65
C 8H 8BR2	263.96	261.8994	36.40	3.05	60.54
C 8H 9BR1	185.06	183.9888	51.92	4.90	43.18
C 8H10BR2	265.98	263.9150	36.13	3.79	60.08
C 8H11BR1	187.08	186.0045	51.36	5.93	42.71
C 8H12BR2	267.99	265.9307	35.86	4.51	59.63
C 8H13BR1	189.10	188.0201	50.81	6.93	42.26
C 8H14BR2	270.01	267.9463	35.59	5.23	59.19
C 8H15BR1	191.11	190.0358	50.28	7.91	41.81
C 8H16BR2	272.02	269.9620	35.32	5.93	58.75
C 8H17BR1	193.13	192.0514	49.75	8.87	41.37
C 9H 4BR2	271.94	269.8681	39.75	1.48	58.77
C 9H 5BR1	193.04	191.9575	56.00	2.61	41.39
C 9H 6BR2	273.96	271.8837	39.46	2.21	58.33
C 9H 7BR1	195.06	193.9732	55.42	3.62	40.96
C 9H 8BR2	275.97	273.8994	39.17	2.92	57.91
C 9H 9BR1	197.08	195.9888	54.85	4.60	40.54
C 9H10BR2	277.99	275.9150	38.89	3.63	57.49
C 9H11BR1	199.09	198.0045	54.30	5.57	40.13
C 9H12BR2	280.00	277.9307	38.61	4.32	57.07
C 9H13BR1	201.11	200.0201	53.75	6.52	39.73
C 9H14BR2	282.02	279.9463	38.33	5.00	56.67
C 9H15BR1	203.12	202.0358	53.22	7.44	39.34
C 9H16BR2	284.04	281.9620	38.06	5.68	56.26
C 9H17BR1	205.14	204.0514	52.70	8.35	38.95
C 9H18BR2	286.05	283.9776	37.79	6.34	55.87
C 9H19BR1	207.16	206.0671	52.18	9.24	38.57
C10H 5BR1	205.06	203.9575	58.58	2.46	38.97
C10H 6BR2	285.97	283.8837	42.00	2.11	55.88
C10H 7BR1	207.07	205.9732	58.00	3.41	38.59
C10H 8BR2	287.98	285.8994	41.71	2.80	55.49
C10H 9BR1	209.09	207.9888	57.45	4.34	38.22
C10H10BR2	290.00	287.9150	41.42	3.48	55.11
C10H11BR1	211.10	210.0045	56.90	5.25	37.85
C10H12BR2	292.02	289.9307	41.13	4.14	54.73
C10H13BR1	213.12	212.0201	56.36	6.15	37.49
C10H14BR2	294.03	291.9463	40.85	4.80	54.35
C10H15BR1	215.14	214.0358	55.83	7.03	37.14
C10H16BR2	296.05	293.9620	40.57	5.45	53.98

FORMULA	MOL. WEIGHT	ISOMOL. WEIGHT	PERCENT C	PERCENT H	PERCENT BR
C10H17BR1	217.15	216.0514	55.31	7.89	36.80
C10H18BR2	298.06	295.9776	40.30	6.09	53.62
C10H19BR1	219.17	218.0671	54.80	8.74	36.46
C10H20BR2	300.08	297.9933	40.03	6.72	53.26
C10H21BR1	221.18	220.0827	54.30	9.57	36.13
C11H 5BR1	217.07	215.9575	60.87	2.32	36.81
C11H 6BR2	297.98	295.8837	44.34	2.03	53.63
C11H 7BR1	219.08	217.9732	60.31	3.22	36.47
C11H 8BR2	299.99	297.8994	44.04	2.69	53.27
C11H 9BR1	221.10	219.9888	59.76	4.10	36.14
C11H10BR2	302.01	299.9150	43.75	3.34	52.91
C11H11BR1	223.11	222.0045	59.22	4.97	35.81
C11H12BR2	304.03	301.9307	43.46	3.98	52.56
C11H13BR1	225.13	224.0201	58.69	5.82	35.49
C11H14BR2	306.04	303.9463	43.17	4.61	52.22
C11H15BR1	227.15	226.0358	58.17	6.66	35.18
C11H16BR2	308.06	305.9620	42.89	5.24	51.88
C11H17BR1	229.16	228.0514	57.65	7.48	34.87
C11H18BR2	310.07	307.9776	42.61	5.85	51.54
C11H19BR1	231.18	230.0671	57.15	8.28	34.56
C11H20BR2	312.09	309.9933	42.33	6.46	51.21
C11H21BR1	233.19	232.0827	56.66	9.08	34.27
C11H22BR2	314.11	312.0089	42.06	7.06	50.88
C11H23BR1	235.21	234.0984	56.17	9.86	33.97
C12H 6BR2	309.99	307.8837	46.50	1.95	51.55
C12H 7BR1	231.09	229.9732	62.37	3.05	34.58
C12H 8BR2	312.01	309.8994	46.20	2.58	51.22
C12H 9BR1	233.11	231.9888	61.83	3.89	34.28
C12H10BR2	314.02	311.9150	45.90	3.21	50.89
C12H11BR1	235.13	234.0045	61.30	4.72	33.98
C12H12BR2	316.04	313.9307	45.61	3.83	50.57
C12H13BR1	237.14	236.0201	60.78	5.53	33.69
C12H14BR2	318.05	315.9463	45.32	4.44	50.25
C12H15BR1	239.16	238.0358	60.27	6.32	33.41
C12H16BR2	320.07	317.9620	45.03	5.04	49.93
C12H17BR1	241.17	240.0514	59.76	7.11	33.13
C12H18BR2	322.09	319.9776	44.75	5.63	49.62
C12H19BR1	243.19	242.0671	59.27	7.88	32.86
C12H20BR2	324.10	321.9933	44.47	6.22	49.31
C12H21BR1	245.21	244.0827	58.78	8.63	32.59
C12H22BR2	326.12	324.0089	44.20	6.80	49.00
C12H23BR1	247.22	246.0984	58.30	9.38	32.32
C12H24BR2	328.13	326.0246	43.93	7.37	48.70
C12H25BR1	249.24	248.1140	57.83	10.11	32.06
C13H 6BR2	322.00	319.8837	48.49	1.88	49.63
C13H 7BR1	243.10	241.9732	64.23	2.90	32.87
C13H 8BR2	324.02	321.8994	48.19	2.49	49.32
C13H 9BR1	245.12	243.9888	63.70	3.70	32.60
C13H10BR2	326.03	323.9150	47.89	3.09	49.02
C13H11BR1	247.14	246.0045	63.18	4.49	32.33
C13H12BR2	328.05	325.9307	47.60	3.69	48.71
C13H13BR1	249.15	248.0201	62.67	5.26	32.07
C13H14BR2	330.06	327.9463	47.31	4.28	48.42
C13H15BR1	251.17	250.0358	62.17	6.02	31.81
C13H16BR2	332.08	329.9620	47.02	4.86	48.12
C13H17BR1	253.18	252.0514	61.67	6.77	31.56
C13H18BR2	334.10	331.9776	46.74	5.43	47.83
C13H19BR1	255.20	254.0671	61.19	7.50	31.31
C13H20BR2	336.11	333.9933	46.46	6.00	47.55
C13H21BR1	257.22	256.0827	60.71	8.23	31.06
C13H22BR2	338.13	336.0089	46.18	6.56	47.26
C13H23BR1	259.23	258.0984	60.23	8.94	30.82
C13H24BR2	340.14	338.0246	45.91	7.11	46.98
C13H25BR1	261.25	260.1140	59.77	9.65	30.59
C13H26BR2	342.16	340.0402	45.64	7.66	46.71
C13H27BR1	263.26	262.1297	59.31	10.34	30.35
C14H 7BR1	255.12	253.9732	65.91	2.77	31.32
C14H 8BR2	336.03	333.8994	50.04	2.40	47.56
C14H 9BR1	257.13	255.9888	65.40	3.53	31.08
C14H10BR2	338.04	335.9150	49.74	2.98	47.27
C14H11BR1	259.15	258.0045	64.89	4.28	30.83
C14H12BR2	340.06	337.9307	49.45	3.56	46.99
C14H13BR1	261.16	260.0201	64.39	5.02	30.60
C14H14BR2	342.08	339.9463	49.16	4.13	46.72
C14H15BR1	263.18	262.0358	63.89	5.74	30.36
C14H16BR2	344.09	341.9620	48.87	4.69	46.44
C14H17BR1	265.20	264.0514	63.41	6.46	30.13
C14H18BR2	346.11	343.9776	48.58	5.24	46.17
C14H19BR1	267.21	266.0671	62.93	7.17	29.90
C14H20BR2	348.12	345.9933	48.30	5.79	45.91
C14H21BR1	269.23	268.0827	62.46	7.86	29.68
C14H22BR2	350.14	348.0089	48.03	6.33	45.64
C14H23BR1	271.24	270.0984	61.99	8.55	29.46
C14H24BR2	352.16	350.0246	47.75	6.87	45.38
C14H25BR1	273.26	272.1140	61.54	9.22	29.24
C14H26BR2	354.17	352.0402	47.48	7.40	45.12
C14H27BR1	275.28	274.1297	61.09	9.89	29.03
C14H28BR2	356.19	354.0559	47.21	7.92	44.87
C14H29BR1	277.29	276.1453	60.64	10.54	28.82
C15H 7BR1	267.13	265.9732	67.45	2.64	29.91
C15H 8BR2	348.04	345.8994	51.77	2.32	45.92
C15H 9BR1	269.14	267.9888	66.94	3.37	29.69
C15H10BR2	350.05	347.9150	51.47	2.88	45.65
C15H11BR1	271.16	270.0045	66.44	4.09	29.47
C15H12BR2	352.07	349.9307	51.17	3.44	45.39
C15H13BR1	273.17	272.0201	65.95	4.80	29.25
C15H14BR2	354.09	351.9463	50.88	3.99	45.13
C15H15BR1	275.19	274.0358	65.47	5.49	29.04
C15H16BR2	356.10	353.9620	50.59	4.53	44.88
C15H17BR1	277.21	276.0514	64.99	6.18	28.82
C15H18BR2	358.12	355.9776	50.31	5.07	44.62
C15H19BR1	279.22	278.0671	64.52	6.86	28.62
C15H20BR2	360.13	357.9933	50.03	5.60	44.37
C15H21BR1	281.24	280.0827	64.06	7.53	28.41
C15H22BR2	362.15	360.0089	49.75	6.12	44.13
C15H23BR1	283.25	282.0984	63.61	8.18	28.21
C15H24BR2	364.17	362.0246	49.47	6.64	43.88
C15H25BR1	285.27	284.1140	63.16	8.83	28.01
C15H26BR2	366.18	364.0402	49.20	7.16	43.64
C15H27BR1	287.29	286.1297	62.71	9.47	27.81
C15H28BR2	368.20	366.0559	48.93	7.67	43.40
C15H29BR1	289.30	288.1453	62.28	10.10	27.62
C15H30BR2	370.21	368.0715	48.67	8.17	43.17
C15H31BR1	291.32	290.1610	61.85	10.73	27.43
C16H 8BR2	360.05	357.8994	53.38	2.24	44.38
C16H 9BR1	281.15	279.9888	68.35	3.23	28.42
C16H10BR2	362.07	359.9150	53.08	2.78	44.14
C16H11BR1	283.17	282.0045	67.87	3.92	28.22
C16H12BR2	364.08	361.9307	52.78	3.32	43.89
C16H13BR1	285.19	284.0201	67.39	4.59	28.02

C16H14BR2	366.10	363.9463	52.49	3.85	43.65
C16H15BR1	287.20	286.0358	66.91	5.26	27.82
C16H16BR2	368.11	365.9620	52.21	4.38	43.41
C16H17BR1	289.22	288.0514	66.45	5.92	27.63
C16H18BR2	370.13	367.9776	51.92	4.90	43.18
C16H19BR1	291.23	290.0671	65.99	6.58	27.44
C16H20BR2	372.15	369.9933	51.64	5.42	42.94
C16H21BR1	293.25	292.0827	65.53	7.22	27.25
C16H22BR2	374.16	372.0089	51.36	5.93	42.71
C16H23BR1	295.27	294.0984	65.09	7.85	27.06
C16H24BR2	376.18	374.0246	51.09	6.43	42.48
C16H25BR1	297.28	296.1140	64.65	8.49	26.88
C16H26BR2	378.19	376.0402	50.81	6.93	42.26
C16H27BR1	299.30	298.1297	64.21	9.09	26.70
C16H28BR2	380.21	378.0559	50.55	7.42	42.03
C16H29BR1	301.31	300.1453	63.78	9.70	26.52
C16H30BR2	382.23	380.0715	50.28	7.91	41.81
C16H31BR1	303.33	302.1610	63.36	10.30	26.34
C16H32BR2	384.24	382.0872	50.02	8.39	41.59
C16H33BR1	305.35	304.1766	62.94	10.89	26.17
C17H 8BR2	372.06	369.8994	54.88	2.17	42.95
C17H 9BR1	293.17	291.9888	69.65	3.09	27.26
C17H10BR2	374.08	371.9150	54.58	2.69	42.72
C17H11BR1	295.18	294.0045	69.17	3.76	27.07
C17H12BR2	376.09	373.9307	54.29	3.22	42.49
C17H13BR1	297.20	296.0201	68.71	4.41	26.89
C17H14BR2	378.11	375.9463	54.00	3.73	42.27
C17H15BR1	299.21	298.0358	68.24	5.05	26.70
C17H16BR2	380.13	377.9620	53.72	4.24	42.04
C17H17BR1	301.23	300.0514	67.79	5.69	26.53
C17H18BR2	382.14	379.9776	53.43	4.75	41.82
C17H19BR1	303.24	302.0671	67.33	6.32	26.35
C17H20BR2	384.16	381.9933	53.15	5.25	41.60
C17H21BR1	305.26	304.0827	66.89	6.93	26.18
C17H22BR2	386.17	384.0089	52.88	5.74	41.38
C17H23BR1	307.28	306.0984	66.45	7.54	26.00
C17H24BR2	388.19	386.0246	52.60	6.23	41.17
C17H25BR1	309.29	308.1140	66.02	8.15	25.83
C17H26BR2	390.20	388.0402	52.33	6.72	40.95
C17H27BR1	311.31	310.1297	65.59	8.74	25.67
C17H28BR2	392.22	390.0559	52.06	7.20	40.74
C17H29BR1	313.32	312.1453	65.17	9.33	25.50
C17H30BR2	394.24	392.0715	51.79	7.67	40.54
C17H31BR1	315.34	314.1610	64.75	9.91	25.34
C17H32BR2	396.25	394.0872	51.53	8.14	40.33
C17H33BR1	317.36	316.1766	64.34	10.48	25.18
C17H34BR2	398.27	396.1028	51.27	8.60	40.13
C17H35BR1	319.37	318.1923	63.93	11.05	25.02
C18H 9BR1	305.18	303.9888	70.84	2.97	26.18
C18H10BR2	386.09	383.9150	56.00	2.61	41.39
C18H11BR1	307.19	306.0045	70.38	3.61	26.01
C18H12BR2	388.10	385.9307	55.71	3.12	41.18
C18H13BR1	309.21	308.0201	69.92	4.24	25.84
C18H14BR2	390.12	387.9463	55.42	3.62	40.96
C18H15BR1	311.22	310.0358	69.47	4.86	25.67
C18H16BR2	392.14	389.9620	55.13	4.11	40.75
C18H17BR1	313.24	312.0514	69.02	5.47	25.51
C18H18BR2	394.15	391.9776	54.85	4.60	40.54
C18H19BR1	315.26	314.0671	68.58	6.07	25.35
C18H20BR2	396.17	393.9933	54.57	5.09	40.34

011

C18H21BR1	317.27	316.0827	68.14	6.67	25.18
C18H22BR2	398.18	396.0089	54.30	5.57	40.13
C18H23BR1	319.29	318.0984	67.71	7.26	25.03
C18H24BR2	400.20	398.0246	54.02	6.04	39.93
C18H25BR1	321.30	320.1140	67.29	7.84	24.87
C18H26BR2	402.22	400.0402	53.75	6.52	39.73
C18H27BR1	323.32	322.1297	66.87	8.42	24.71
C18H28BR2	404.23	402.0559	53.48	6.98	39.53
C18H29BR1	325.34	324.1453	66.45	8.98	24.56
C18H30BR2	406.25	404.0715	53.22	7.44	39.34
C18H31BR1	327.35	326.1610	66.05	9.55	24.41
C18H32BR2	408.26	406.0872	52.96	7.90	39.14
C18H33BR1	329.37	328.1766	65.64	10.10	24.26
C18H34BR2	410.28	408.1028	52.70	8.35	38.95
C18H35BR1	331.38	330.1923	65.24	10.65	24.11
C18H36BR2	412.30	410.1185	52.44	8.80	38.76
C18H37BR1	333.40	332.2079	64.85	11.19	23.97
C19H 9BR1	317.19	315.9888	71.95	2.86	25.19
C19H10BR2	398.10	395.9150	57.33	2.53	40.14
C19H11BR1	319.20	318.0045	71.49	3.47	25.03
C19H12BR2	400.12	397.9307	57.04	3.02	39.94
C19H13BR1	321.22	320.0201	71.05	4.08	24.88
C19H14BR2	402.13	399.9463	56.75	3.51	39.74
C19H15BR1	323.24	322.0358	70.60	4.68	24.72
C19H16BR2	404.15	401.9620	56.47	3.99	39.54
C19H17BR1	325.25	324.0514	70.16	5.27	24.57
C19H18BR2	406.16	403.9776	56.19	4.47	39.35
C19H19BR1	327.27	326.0671	69.73	5.85	24.42
C19H20BR2	408.18	405.9933	55.91	4.94	39.15
C19H21BR1	329.28	328.0827	69.31	6.43	24.27
C19H22BR2	410.20	408.0089	55.63	5.41	38.96
C19H23BR1	331.30	330.0984	68.88	7.00	24.12
C19H24BR2	412.21	410.0246	55.36	5.87	38.77
C19H25BR1	333.32	332.1140	68.47	7.56	23.97
C19H26BR2	414.23	412.0402	55.09	6.33	38.58
C19H27BR1	335.33	334.1297	68.06	8.12	23.83
C19H28BR2	416.24	414.0559	54.83	6.78	38.39
C19H29BR1	337.35	336.1453	67.65	8.67	23.69
C19H30BR2	418.26	416.0715	54.56	7.23	38.21
C19H31BR1	339.36	338.1610	67.25	9.21	23.55
C19H32BR2	420.27	418.0872	54.30	7.67	38.02
C19H33BR1	341.38	340.1766	66.85	9.74	23.41
C19H34BR2	422.29	420.1028	54.04	8.12	37.84
C19H35BR1	343.39	342.1923	66.46	10.27	23.27
C19H36BR2	424.31	422.1185	53.78	8.55	37.66
C19H37BR1	345.41	344.2079	66.07	10.80	23.13
C19H38BR2	426.32	424.1341	53.53	8.98	37.49
C19H39BR1	347.43	346.2235	65.69	11.31	23.00
C20H10BR2	410.11	407.9150	58.58	2.46	38.97
C20H11BR1	331.21	330.0045	72.53	3.35	24.12
C20H12BR2	412.13	409.9307	58.29	2.93	38.78
C20H13BR1	333.23	332.0201	72.09	3.93	23.98
C20H14BR2	414.14	411.9463	58.00	3.41	38.59
C20H15BR1	335.25	334.0358	71.66	4.51	23.83
C20H16BR2	416.16	413.9620	57.72	3.88	38.40
C20H17BR1	337.26	336.0514	71.23	5.08	23.69
C20H18BR2	418.17	415.9776	57.45	4.34	38.22
C20H19BR1	339.28	338.0671	70.80	5.64	23.55
C20H20BR2	420.19	417.9933	57.17	4.80	38.03
C20H21BR1	341.29	340.0827	70.39	6.20	23.41

C20H22BR2	422.21	420.0089	56.90	5.25	37.85
C20H23BR1	343.31	342.0984	69.97	6.75	23.27
C20H24BR2	424.22	422.0246	56.63	5.70	37.67
C20H25BR1	345.33	344.1140	69.56	7.30	23.14
C20H26BR2	426.24	424.0402	56.36	6.15	37.49
C20H27BR1	347.34	346.1297	69.16	7.84	23.00
C20H28BR2	428.25	426.0559	56.09	6.59	37.32
C20H29BR1	349.36	348.1453	68.76	8.37	22.87
C20H30BR2	430.27	428.0715	55.83	7.03	37.14
C20H31BR1	351.37	350.1610	68.37	8.89	22.74
C20H32BR2	432.29	430.0872	55.57	7.46	36.97
C20H33BR1	353.39	352.1766	67.98	9.41	22.61
C20H34BR2	434.30	432.1028	55.31	7.89	36.80
C20H35BR1	355.41	354.1923	67.59	9.93	22.48
C20H36BR2	436.32	434.1185	55.06	8.32	36.63
C20H37BR1	357.42	356.2079	67.21	10.43	22.36
C20H38BR2	438.33	436.1341	54.80	8.74	36.46
C20H39BR1	359.44	358.2235	66.83	10.94	22.23
C20H40BR2	440.35	438.1498	54.55	9.16	36.29
C20H41BR1	361.45	360.2392	66.46	11.43	22.11
C21H10BR2	422.12	419.9150	59.75	2.39	37.86
C21H11BR1	343.23	342.0045	73.49	3.23	23.28
C21H12BR2	424.14	421.9307	59.47	2.85	37.68
C21H13BR1	345.24	344.0201	73.06	3.80	23.14
C21H14BR2	426.15	423.9463	59.19	3.31	37.50
C21H15BR1	347.26	346.0358	72.64	4.35	23.01
C21H16BR2	428.17	425.9620	58.91	3.77	37.32
C21H17BR1	349.27	348.0514	72.22	4.91	22.88
C21H18BR2	430.19	427.9776	58.63	4.22	37.15
C21H19BR1	351.29	350.0671	71.80	5.45	22.75
C21H20BR2	432.20	429.9933	58.36	4.66	36.98
C21H21BR1	353.31	352.0827	71.39	5.99	22.62
C21H22BR2	434.22	432.0089	58.09	5.11	36.80
C21H23BR1	355.32	354.0984	70.99	6.52	22.49
C21H24BR2	436.23	434.0246	57.82	5.55	36.63
C21H25BR1	357.34	356.1140	70.59	7.05	22.36
C21H26BR2	438.25	436.0402	57.55	5.98	36.47
C21H27BR1	359.35	358.1297	70.19	7.57	22.24
C21H28BR2	440.27	438.0559	57.29	6.41	36.30
C21H29BR1	361.37	360.1453	69.80	8.09	22.11
C21H30BR2	442.28	440.0715	57.03	6.84	36.13
C21H31BR1	363.39	362.1610	69.41	8.60	21.99
C21H32BR2	444.30	442.0872	56.77	7.26	35.97
C21H33BR1	365.40	364.1766	69.03	9.10	21.87
C21H34BR2	446.31	444.1028	56.52	7.68	35.81
C21H35BR1	367.42	366.1923	68.65	9.60	21.75
C21H36BR2	448.33	446.1185	56.26	8.09	35.65
C21H37BR1	369.43	368.2079	68.28	10.10	21.63
C21H38BR2	450.35	448.1341	56.01	8.51	35.49
C21H39BR1	371.45	370.2235	67.91	10.58	21.51
C21H40BR2	452.36	450.1498	55.76	8.91	35.33
C21H41BR1	373.46	372.2392	67.54	11.07	21.40
C21H42BR2	454.38	452.1654	55.51	9.32	35.17
C21H43BR1	375.48	374.2548	67.18	11.54	21.28
C22H11BR1	355.24	354.0045	74.39	3.12	22.49
C22H12BR2	436.15	433.9307	60.59	2.77	36.64
C22H13BR1	357.25	356.0201	73.97	3.67	22.37
C22H14BR2	438.16	435.9463	60.31	3.22	36.47
C22H15BR1	359.27	358.0358	73.55	4.21	22.24
C22H16BR2	440.18	437.9620	60.03	3.66	36.31

012

C22H17BR1	361.28	360.0514	73.14	4.74	22.12
C22H18BR2	442.20	439.9776	59.76	4.10	36.14
C22H19BR1	363.30	362.0671	72.73	5.27	21.99
C22H20BR2	444.21	441.9933	59.49	4.54	35.98
C22H21BR1	365.32	364.0827	72.33	5.79	21.87
C22H22BR2	446.23	444.0089	59.22	4.97	35.81
C22H23BR1	367.33	366.0984	71.94	6.31	21.75
C22H24BR2	448.24	446.0246	58.95	5.40	35.65
C22H25BR1	369.35	368.1140	71.54	6.82	21.63
C22H26BR2	450.26	448.0402	58.69	5.82	35.49
C22H27BR1	371.36	370.1297	71.16	7.33	21.52
C22H28BR2	452.28	450.0559	58.43	6.24	35.33
C22H29BR1	373.38	372.1453	70.77	7.83	21.40
C22H30BR2	454.29	452.0715	58.17	6.66	35.18
C22H31BR1	375.40	374.1610	70.39	8.32	21.29
C22H32BR2	456.31	454.0872	57.91	7.07	35.02
C22H33BR1	377.41	376.1766	70.02	8.81	21.17
C22H34BR2	458.32	456.1028	57.65	7.48	34.87
C22H35BR1	379.43	378.1923	69.64	9.30	21.06
C22H36BR2	460.34	458.1185	57.40	7.88	34.72
C22H37BR1	381.44	380.2079	69.27	9.78	20.95
C22H38BR2	462.36	460.1341	57.15	8.28	34.56
C22H39BR1	383.46	382.2235	68.91	10.25	20.84
C22H40BR2	464.37	462.1498	56.90	8.68	34.41
C22H41BR1	385.48	384.2392	68.55	10.72	20.73
C22H42BR2	466.39	464.1654	56.66	9.08	34.27
C22H43BR1	387.49	386.2548	68.19	11.19	20.62
C22H44BR2	468.40	466.1811	56.41	9.47	34.12
C22H45BR1	389.51	388.2705	67.84	11.65	20.51
C23H11BR1	367.25	366.0045	75.22	3.02	21.76
C23H12BR2	448.16	445.9307	61.64	2.70	35.66
C23H13BR1	369.26	368.0201	74.81	3.55	21.64
C23H14BR2	450.18	447.9463	61.37	3.13	35.50
C23H15BR1	371.28	370.0358	74.41	4.07	21.52
C23H16BR2	452.19	449.9620	61.09	3.57	35.34
C23H17BR1	373.30	372.0514	74.00	4.59	21.41
C23H18BR2	454.21	451.9776	60.82	3.99	35.18
C23H19BR1	375.31	374.0671	73.61	5.10	21.29
C23H20BR2	456.22	453.9933	60.55	4.42	35.03
C23H21BR1	377.33	376.0827	73.21	5.61	21.18
C23H22BR2	458.24	456.0089	60.29	4.84	34.87
C23H23BR1	379.34	378.0984	72.82	6.11	21.06
C23H24BR2	460.26	458.0246	60.02	5.26	34.72
C23H25BR1	381.36	380.1140	72.44	6.61	20.95
C23H26BR2	462.27	460.0402	59.76	5.67	34.57
C23H27BR1	383.38	382.1297	72.06	7.10	20.84
C23H28BR2	464.29	462.0559	59.50	6.08	34.42
C23H29BR1	385.39	384.1453	71.68	7.58	20.73
C23H30BR2	466.30	464.0715	59.24	6.48	34.27
C23H31BR1	387.41	386.1610	71.31	8.07	20.63
C23H32BR2	468.32	466.0872	58.99	6.89	34.12
C23H33BR1	389.42	388.1766	70.94	8.54	20.52
C23H34BR2	470.34	468.1028	58.74	7.29	33.98
C23H35BR1	391.44	390.1923	70.57	9.01	20.41
C23H36BR2	472.35	470.1185	58.49	7.68	33.83
C23H37BR1	393.46	392.2079	70.21	9.48	20.31
C23H38BR2	474.37	472.1341	58.24	8.07	33.69
C23H39BR1	395.47	394.2235	69.85	9.94	20.20
C23H40BR2	476.38	474.1498	57.99	8.46	33.55
C23H41BR1	397.49	396.2392	69.50	10.40	20.10

013

C23H42BR2	478.40	476.1654	57.75	8.85	33.40
C23H43BR1	399.50	398.2548	69.15	10.85	20.00
C23H44BR2	480.42	478.1811	57.50	9.23	33.26
C23H45BR1	401.52	400.2705	68.80	11.30	19.90
C23H46BR2	482.43	480.1967	57.26	9.61	33.13
C23H47BR1	403.54	402.2861	68.46	11.74	19.80
C24H12BR2	460.17	457.9307	62.64	2.63	34.73
C24H13BR1	381.28	380.0201	75.61	3.44	20.96
C24H14BR2	462.19	459.9463	62.37	3.05	34.58
C24H15BR1	383.29	382.0358	75.21	3.94	20.85
C24H16BR2	464.20	461.9620	62.10	3.47	34.43
C24H17BR1	385.31	384.0514	74.82	4.45	20.74
C24H18BR2	466.22	463.9776	61.83	3.89	34.28
C24H19BR1	387.32	386.0671	74.43	4.94	20.63
C24H20BR2	468.23	465.9933	61.56	4.31	34.13
C24H21BR1	389.34	388.0827	74.04	5.44	20.52
C24H22BR2	470.25	468.0089	61.30	4.72	33.98
C24H23BR1	391.35	390.0984	73.66	5.92	20.42
C24H24BR2	472.27	470.0246	61.04	5.12	33.84
C24H25BR1	393.37	392.1140	73.28	6.41	20.31
C24H26BR2	474.28	472.0402	60.78	5.53	33.69
C24H27BR1	395.39	394.1297	72.91	6.88	20.21
C24H28BR2	476.30	474.0559	60.52	5.93	33.55
C24H29BR1	397.40	396.1453	72.54	7.36	20.11
C24H30BR2	478.31	476.0715	60.27	6.32	33.41
C24H31BR1	399.42	398.1610	72.17	7.82	20.01
C24H32BR2	480.33	478.0872	60.01	6.72	33.27
C24H33BR1	401.43	400.1766	71.81	8.29	19.90
C24H34BR2	482.35	480.1028	59.76	7.11	33.13
C24H35BR1	403.45	402.1923	71.45	8.74	19.81
C24H36BR2	484.36	482.1185	59.51	7.49	32.99
C24H37BR1	405.47	404.2079	71.10	9.20	19.71
C24H38BR2	486.38	484.1341	59.27	7.88	32.86
C24H39BR1	407.48	406.2235	70.74	9.65	19.61
C24H40BR2	488.39	486.1498	59.02	8.26	32.72
C24H41BR1	409.50	408.2392	70.40	10.09	19.51
C24H42BR2	490.41	488.1654	58.78	8.63	32.59
C24H43BR1	411.51	410.2548	70.05	10.53	19.42
C24H44BR2	492.43	490.1811	58.54	9.01	32.45
C24H45BR1	413.53	412.2705	69.71	10.97	19.32
C24H46BR2	494.44	492.1967	58.30	9.38	32.32
C24H47BR1	415.55	414.2861	69.37	11.40	19.23
C24H48BR2	496.46	494.2124	58.06	9.75	32.19
C24H49BR1	417.56	416.3018	69.04	11.83	19.14
C25H12BR2	472.18	469.9307	63.59	2.56	33.84
C25H13BR1	393.29	392.0201	76.35	3.33	20.32
C25H14BR2	474.20	471.9463	63.32	2.98	33.70
C25H15BR1	395.30	394.0358	75.96	3.82	20.21
C25H16BR2	476.21	473.9620	63.06	3.39	33.56
C25H17BR1	397.32	396.0514	75.58	4.31	20.11
C25H18BR2	478.23	475.9776	62.79	3.79	33.42
C25H19BR1	399.33	398.0671	75.19	4.80	20.01
C25H20BR2	480.25	477.9933	62.53	4.20	33.28
C25H21BR1	401.35	400.0827	74.82	5.27	19.91
C25H22BR2	482.26	480.0089	62.26	4.60	33.14
C25H23BR1	403.37	402.0984	74.44	5.75	19.81
C25H24BR2	484.28	482.0246	62.01	5.00	33.00
C25H25BR1	405.38	404.1140	74.07	6.22	19.71
C25H26BR2	486.29	484.0402	61.75	5.39	32.86
C25H27BR1	407.40	406.1297	73.71	6.68	19.61

C25H28BR2	488.31	486.0559	61.49	5.78	32.73
C25H29BR1	409.41	408.1453	73.34	7.14	19.52
C25H30BR2	490.33	488.0715	61.24	6.17	32.59
C25H31BR1	411.43	410.1610	72.98	7.59	19.42
C25H32BR2	492.34	490.0872	60.99	6.55	32.46
C25H33BR1	413.45	412.1766	72.63	8.05	19.33
C25H34BR2	494.36	492.1028	60.74	6.93	32.33
C25H35BR1	415.46	414.1923	72.28	8.49	19.23
C25H36BR2	496.37	494.1185	60.49	7.31	32.20
C25H37BR1	417.48	416.2079	71.93	8.93	19.14
C25H38BR2	498.39	496.1341	60.25	7.69	32.06
C25H39BR1	419.49	418.2235	71.58	9.37	19.05
C25H40BR2	500.41	498.1498	60.01	8.06	31.94
C25H41BR1	421.51	420.2392	71.24	9.80	18.96
C25H42BR2	502.42	500.1654	59.77	8.43	31.81
C25H43BR1	423.53	422.2548	70.90	10.23	18.87
C25H44BR2	504.44	502.1811	59.53	8.79	31.68
C25H45BR1	425.54	424.2705	70.56	10.66	18.78
C25H46BR2	506.45	504.1967	59.29	9.16	31.55
C25H47BR1	427.56	426.2861	70.23	11.08	18.69
C25H48BR2	508.47	506.2124	59.06	9.52	31.43
C25H49BR1	429.57	428.3018	69.90	11.50	18.60
C25H50BR2	510.49	508.2280	58.82	9.87	31.31
C25H51BR1	431.59	430.3174	69.58	11.91	18.51
C26H13BR1	405.30	404.0201	77.05	3.23	19.71
C26H14BR2	486.21	483.9463	64.23	2.90	32.87
C26H15BR1	407.31	406.0358	76.67	3.71	19.62
C26H16BR2	488.23	485.9620	63.96	3.30	32.73
C26H17BR1	409.33	408.0514	76.29	4.19	19.52
C26H18BR2	490.24	487.9776	63.70	3.70	32.60
C26H19BR1	411.35	410.0671	75.92	4.66	19.43
C26H20BR2	492.26	489.9933	63.44	4.10	32.46
C26H21BR1	413.36	412.0827	75.55	5.12	19.33
C26H22BR2	494.27	492.0089	63.18	4.49	32.33
C26H23BR1	415.38	414.0984	75.18	5.58	19.24
C26H24BR2	496.29	494.0246	62.92	4.87	32.20
C26H25BR1	417.39	416.1140	74.82	6.04	19.14
C26H26BR2	498.31	496.0402	62.67	5.26	32.07
C26H27BR1	419.41	418.1297	74.46	6.49	19.05
C26H28BR2	500.32	498.0559	62.42	5.64	31.94
C26H29BR1	421.43	420.1453	74.10	6.94	18.96
C26H30BR2	502.34	500.0715	62.17	6.02	31.81
C26H31BR1	423.44	422.1610	73.75	7.38	18.87
C26H32BR2	504.35	502.0872	61.92	6.40	31.69
C26H33BR1	425.46	424.1766	73.40	7.82	18.78
C26H34BR2	506.37	504.1028	61.67	6.77	31.56
C26H35BR1	427.47	426.1923	73.05	8.25	18.69
C26H36BR2	508.38	506.1185	61.43	7.14	31.43
C26H37BR1	429.49	428.2079	72.71	8.68	18.60
C26H38BR2	510.40	508.1341	61.19	7.50	31.31
C26H39BR1	431.50	430.2235	72.37	9.11	18.52
C26H40BR2	512.42	510.1498	60.94	7.87	31.19
C26H41BR1	433.52	432.2392	72.04	9.53	18.43
C26H42BR2	514.43	512.1654	60.71	8.23	31.06
C26H43BR1	435.54	434.2548	71.70	9.95	18.35
C26H44BR2	516.45	514.1811	60.47	8.59	30.94
C26H45BR1	437.55	436.2705	71.37	10.37	18.26
C26H46BR2	518.46	516.1967	60.23	8.94	30.82
C26H47BR1	439.57	438.2861	71.04	10.78	18.18
C26H48BR2	520.48	518.2124	60.00	9.30	30.70

014

C26H49BR1	441.58	440.3018	70.72	11.18	18.09
C26H50BR2	522.50	520.2280	59.77	9.65	30.59
C26H51BR1	443.60	442.3174	70.40	11.59	18.01
C26H52BR2	524.51	522.2437	59.54	9.99	30.47
C26H53BR1	445.62	444.3331	70.08	11.99	17.93
C27H13BR1	417.31	416.0201	77.71	3.14	19.15
C27H14BR2	498.22	495.9463	65.09	2.83	32.08
C27H15BR1	419.32	418.0358	77.34	3.61	19.06
C27H16BR2	500.24	497.9620	64.83	3.22	31.95
C27H17BR1	421.34	420.0514	76.97	4.07	18.96
C27H18BR2	502.25	499.9776	64.57	3.61	31.82
C27H19BR1	423.36	422.0671	76.60	4.52	18.87
C27H20BR2	504.27	501.9933	64.31	4.00	31.69
C27H21BR1	425.37	424.0827	76.24	4.98	18.78
C27H22BR2	506.28	504.0089	64.06	4.38	31.56
C27H23BR1	427.39	426.0984	75.88	5.42	18.70
C27H24BR2	508.30	506.0246	63.80	4.76	31.44
C27H25BR1	429.40	428.1140	75.52	5.87	18.61
C27H26BR2	510.32	508.0402	63.55	5.14	31.32
C27H27BR1	431.42	430.1297	75.17	6.31	18.52
C27H28BR2	512.33	510.0559	63.30	5.51	31.19
C27H29BR1	433.44	432.1453	74.82	6.74	18.44
C27H30BR2	514.35	512.0715	63.05	5.88	31.07
C27H31BR1	435.45	434.1610	74.47	7.18	18.35
C27H32BR2	516.36	514.0872	62.80	6.25	30.95
C27H33BR1	437.47	436.1766	74.13	7.60	18.27
C27H34BR2	518.38	516.1028	62.56	6.61	30.83
C27H35BR1	439.48	438.1923	73.79	8.03	18.18
C27H36BR2	520.40	518.1185	62.32	6.97	30.71
C27H37BR1	441.50	440.2079	73.45	8.45	18.10
C27H38BR2	522.41	520.1341	62.08	7.33	30.59
C27H39BR1	443.52	442.2235	73.12	8.86	18.02
C27H40BR2	524.43	522.1498	61.84	7.69	30.47
C27H41BR1	445.53	444.2392	72.79	9.28	17.93
C27H42BR2	526.44	524.1654	61.60	8.04	30.36
C27H43BR1	447.55	446.2548	72.46	9.68	17.85
C27H44BR2	528.46	526.1811	61.37	8.39	30.24
C27H45BR1	449.56	448.2705	72.14	10.09	17.77
C27H46BR2	530.48	528.1967	61.13	8.74	30.13
C27H47BR1	451.58	450.2861	71.81	10.49	17.69
C27H48BR2	532.49	530.2124	60.90	9.09	30.01
C27H49BR1	453.60	452.3018	71.50	10.89	17.62
C27H50BR2	534.51	532.2280	60.67	9.43	29.90
C27H51BR1	455.61	454.3174	71.18	11.28	17.54
C27H52BR2	536.52	534.2437	60.44	9.77	29.79
C27H53BR1	457.63	456.3331	70.87	11.67	17.46
C27H54BR2	538.54	536.2593	60.22	10.11	29.67
C27H55BR1	459.64	458.3487	70.55	12.06	17.38
C28H14BR2	510.23	507.9463	65.91	2.77	31.32
C28H15BR1	431.34	430.0358	77.97	3.51	18.52
C28H16BR2	512.25	509.9620	65.65	3.15	31.20
C28H17BR1	433.35	432.0514	77.61	3.95	18.44
C28H18BR2	514.26	511.9776	65.40	3.53	31.08
C28H19BR1	435.37	434.0671	77.25	4.40	18.35
C28H20BR2	516.28	513.9933	65.14	3.90	30.95
C28H21BR1	437.38	436.0827	76.89	4.84	18.27
C28H22BR2	518.30	516.0089	64.89	4.28	30.83
C28H23BR1	439.40	438.0984	76.54	5.28	18.18
C28H24BR2	520.31	518.0246	64.64	4.65	30.71
C28H25BR1	441.42	440.1140	76.19	5.71	18.10

C28H26BR2	522.33	520.0402	64.39	5.02	30.60
C28H27BR1	443.43	442.1297	75.84	6.14	18.02
C28H28BR2	524.34	522.0559	64.14	5.38	30.48
C28H29BR1	445.45	444.1453	75.50	6.56	17.94
C28H30BR2	526.36	524.0715	63.89	5.74	30.36
C28H31BR1	447.46	446.1610	75.16	6.98	17.86
C28H32BR2	528.38	526.0872	63.65	6.10	30.25
C28H33BR1	449.48	448.1766	74.82	7.40	17.78
C28H34BR2	530.39	528.1028	63.41	6.46	30.13
C28H35BR1	451.50	450.1923	74.49	7.81	17.70
C28H36BR2	532.41	530.1185	63.17	6.82	30.02
C28H37BR1	453.51	452.2079	74.16	8.22	17.62
C28H38BR2	534.42	532.1341	62.93	7.17	29.90
C28H39BR1	455.53	454.2235	73.83	8.63	17.54
C28H40BR2	536.44	534.1498	62.69	7.52	29.79
C28H41BR1	457.54	456.2392	73.50	9.03	17.46
C28H42BR2	538.45	536.1654	62.46	7.86	29.68
C28H43BR1	459.56	458.2548	73.18	9.43	17.39
C28H44BR2	540.47	538.1811	62.23	8.21	29.57
C28H45BR1	461.57	460.2705	72.86	9.83	17.31
C28H46BR2	542.49	540.1967	61.99	8.55	29.46
C28H47BR1	463.59	462.2861	72.55	10.22	17.24
C28H48BR2	544.50	542.2124	61.77	8.89	29.35
C28H49BR1	465.61	464.3018	72.23	10.61	17.16
C28H50BR2	546.52	544.2280	61.54	9.22	29.24
C28H51BR1	467.62	466.3174	71.92	10.99	17.09
C28H52BR2	548.53	546.2437	61.31	9.56	29.13
C28H53BR1	469.64	468.3331	71.61	11.38	17.01
C28H54BR2	550.55	548.2593	61.09	9.89	29.03
C28H55BR1	471.65	470.3487	71.30	11.75	16.94
C28H56BR2	552.57	550.2750	60.86	10.22	28.92
C28H57BR1	473.67	472.3644	71.00	12.13	16.87
C29H14BR2	522.24	519.9463	66.70	2.70	30.60
C29H15BR1	443.35	442.0358	78.57	3.41	18.02
C29H16BR2	524.26	521.9620	66.44	3.08	30.48
C29H17BR1	445.36	444.0514	78.21	3.85	17.94
C29H18BR2	526.27	523.9776	66.19	3.45	30.37
C29H19BR1	447.38	446.0671	77.86	4.28	17.86
C29H20BR2	528.29	525.9933	65.93	3.82	30.25
C29H21BR1	449.39	448.0827	77.51	4.71	17.78
C29H22BR2	530.31	528.0089	65.68	4.18	30.14
C29H23BR1	451.41	450.0984	77.16	5.14	17.70
C29H24BR2	532.32	530.0246	65.43	4.54	30.02
C29H25BR1	453.43	452.1140	76.82	5.56	17.62
C29H26BR2	534.34	532.0402	65.19	4.90	29.91
C29H27BR1	455.44	454.1297	76.48	5.98	17.54
C29H28BR2	536.35	534.0559	64.94	5.26	29.80
C29H29BR1	457.46	456.1453	76.14	6.39	17.47
C29H30BR2	538.37	536.0715	64.70	5.62	29.68
C29H31BR1	459.47	458.1610	75.81	6.80	17.39
C29H32BR2	540.39	538.0872	64.46	5.97	29.57
C29H33BR1	461.49	460.1766	75.48	7.21	17.31
C29H34BR2	542.40	540.1028	64.22	6.32	29.46
C29H35BR1	463.51	462.1923	75.15	7.61	17.24
C29H36BR2	544.42	542.1185	63.98	6.67	29.35
C29H37BR1	465.52	464.2079	74.82	8.01	17.16
C29H38BR2	546.43	544.1341	63.74	7.01	29.25
C29H39BR1	467.54	466.2235	74.50	8.41	17.09
C29H40BR2	548.45	546.1498	63.51	7.35	29.14
C29H41BR1	469.55	468.2392	74.18	8.80	17.02

C29H42BR2	550.47	548.1654	63.28	7.69	29.03
C29H43BR1	471.57	470.2548	73.86	9.19	16.94
C29H44BR2	552.48	550.1811	63.05	8.03	28.93
C29H45BR1	473.59	472.2705	73.55	9.58	16.87
C29H46BR2	554.50	552.1967	62.82	8.36	28.82
C29H47BR1	475.60	474.2861	73.24	9.96	16.80
C29H48BR2	556.51	554.2124	62.59	8.69	28.72
C29H49BR1	477.62	476.3018	72.93	10.34	16.73
C29H50BR2	558.53	556.2280	62.36	9.02	28.61
C29H51BR1	479.63	478.3174	72.62	10.72	16.66
C29H52BR2	560.55	558.2437	62.14	9.35	28.51
C29H53BR1	481.65	480.3331	72.32	11.09	16.59
C29H54BR2	562.56	560.2593	61.92	9.68	28.41
C29H55BR1	483.67	482.3487	72.02	11.46	16.52
C29H56BR2	564.58	562.2750	61.70	10.00	28.31
C29H57BR1	485.68	484.3644	71.72	11.83	16.45
C29H58BR2	566.59	564.2906	61.48	10.32	28.21
C29H59BR1	487.70	486.3800	71.42	12.19	16.38
C30H15BR1	455.36	454.0358	79.13	3.32	17.55
C30H16BR2	536.27	533.9620	67.19	3.01	29.80
C30H17BR1	457.37	456.0514	78.78	3.75	17.47
C30H18BR2	538.29	535.9776	66.94	3.37	29.69
C30H19BR1	459.39	458.0671	78.44	4.17	17.39
C30H20BR2	540.30	537.9933	66.69	3.73	29.58
C30H21BR1	461.41	460.0827	78.09	4.59	17.32
C30H22BR2	542.32	540.0089	66.44	4.09	29.47
C30H23BR1	463.42	462.0984	77.76	5.00	17.24
C30H24BR2	544.33	542.0246	66.20	4.44	29.36
C30H25BR1	465.44	464.1140	77.42	5.41	17.17
C30H26BR2	546.35	544.0402	65.95	4.80	29.25
C30H27BR1	467.45	466.1297	77.08	5.82	17.09
C30H28BR2	548.37	546.0559	65.71	5.15	29.14
C30H29BR1	469.47	468.1453	76.75	6.23	17.02
C30H30BR2	550.38	548.0715	65.47	5.49	29.04
C30H31BR1	471.49	470.1610	76.43	6.63	16.95
C30H32BR2	552.40	550.0872	65.23	5.84	28.93
C30H33BR1	473.50	472.1766	76.10	7.02	16.88
C30H34BR2	554.41	552.1028	64.99	6.18	28.82
C30H35BR1	475.52	474.1923	75.78	7.42	16.80
C30H36BR2	556.43	554.1185	64.76	6.52	28.72
C30H37BR1	477.53	476.2079	75.46	7.81	16.73
C30H38BR2	558.45	556.1341	64.52	6.86	28.62
C30H39BR1	479.55	478.2235	75.14	8.20	16.66
C30H40BR2	560.46	558.1498	64.29	7.19	28.51
C30H41BR1	481.57	480.2392	74.83	8.58	16.59
C30H42BR2	562.48	560.1654	64.06	7.53	28.41
C30H43BR1	483.58	482.2548	74.51	8.96	16.52
C30H44BR2	564.49	562.1811	63.83	7.86	28.31
C30H45BR1	485.60	484.2705	74.20	9.34	16.45
C30H46BR2	566.51	564.1967	63.61	8.18	28.21
C30H47BR1	487.61	486.2861	73.90	9.72	16.39
C30H48BR2	568.53	566.2124	63.38	8.51	28.11
C30H49BR1	489.63	488.3018	73.59	10.09	16.32
C30H50BR2	570.54	568.2280	63.16	8.83	28.01
C30H51BR1	491.64	490.3174	73.29	10.46	16.25
C30H52BR2	572.56	570.2437	62.93	9.15	27.91
C30H53BR1	493.66	492.3331	72.99	10.82	16.19
C30H54BR2	574.57	572.2593	62.71	9.47	27.81
C30H55BR1	495.68	494.3487	72.70	11.18	16.12
C30H56BR2	576.59	574.2750	62.49	9.79	27.72
C30H57BR1	497.69	496.3644	72.40	11.54	16.05
C30H58BR2	578.60	576.2906	62.28	10.10	27.62
C30H59BR1	499.71	498.3800	72.11	11.90	15.99
C30H60BR2	580.62	578.3063	62.06	10.42	27.52
C30H61BR1	501.72	500.3957	71.82	12.25	15.93

2. CARBON, HYDROGEN AND CHLORINE

2. KOHLENSTOFF, WASSERSTOFF UND CHLOR

2. CARBONE, HYDROGENE ET CHLORE

MOLECULAR WEIGHTS AND PERCENTAGE COMPOSITIONS OF COMPOUNDS
CONTAINING CARBON, HYDROGEN AND CHLORINE

FORMULA	MOL. WEIGHT	ISOMOL. WEIGHT	PERCENT C	PERCENT H	PERCENT Cl	FORMULA	MOL. WEIGHT	ISOMOL. WEIGHT	PERCENT C	PERCENT H	PERCENT Cl
C 1H 0Cl2	82.92	81.9377	14.49	0	85.51	C 7H 8Cl2	163.05	162.0003	51.57	4.95	43.49
C 1H 1Cl1	48.47	47.9767	24.78	2.08	73.14	C 7H 9Cl1	128.60	128.0393	65.38	7.05	27.57
C 1H 2Cl2	84.93	83.9534	14.14	2.37	83.48	C 7H10Cl2	165.06	164.0160	50.94	6.11	42.96
C 1H 3Cl1	50.49	49.9923	23.79	5.99	70.22	C 7H11Cl1	130.62	130.0549	64.37	8.49	27.14
C 2H 1Cl1	60.48	59.9767	39.72	1.67	58.62	C 7H12Cl2	167.08	166.0316	50.32	7.24	42.44
C 2H 2Cl2	96.94	95.9534	24.78	2.08	73.14	C 7H13Cl1	132.63	132.0706	63.39	9.88	26.73
C 2H 3Cl1	62.50	61.9923	38.44	4.84	56.73	C 7H14Cl2	169.10	168.0473	49.72	8.35	41.93
C 2H 4Cl2	98.96	97.9690	24.27	4.07	71.65	C 7H15Cl1	134.65	134.0862	62.44	11.23	26.33
C 2H 5Cl1	64.52	64.0080	37.24	7.81	54.95	C 8H 4Cl2	171.03	169.9690	56.18	2.36	41.46
C 3H 1Cl1	72.49	71.9767	49.71	1.39	48.90	C 8H 5Cl1	136.58	136.0080	70.35	3.69	25.96
C 3H 2Cl2	108.96	107.9534	33.07	1.85	65.08	C 8H 6Cl2	173.04	171.9847	55.53	3.49	40.98
C 3H 3Cl1	74.51	73.9923	48.36	4.06	47.58	C 8H 7Cl1	138.60	138.0236	69.33	5.09	25.58
C 3H 4Cl2	110.97	109.9690	32.47	3.63	63.90	C 8H 8Cl2	175.06	174.0003	54.89	4.61	40.50
C 3H 5Cl1	76.53	76.0080	47.09	6.59	46.33	C 8H 9Cl1	140.61	140.0393	68.34	6.45	25.21
C 3H 6Cl2	112.99	111.9847	31.89	5.35	62.76	C 8H10Cl2	177.07	176.0160	54.26	5.69	40.04
C 3H 7Cl1	78.54	78.0236	45.88	8.98	45.14	C 8H11Cl1	142.63	142.0549	67.37	7.77	24.86
C 4H 2Cl2	120.97	119.9534	39.72	1.67	58.62	C 8H12Cl2	179.09	178.0316	53.65	6.75	39.59
C 4H 3Cl1	86.52	85.9923	55.53	3.49	40.98	C 8H13Cl1	144.65	144.0706	66.43	9.06	24.51
C 4H 4Cl2	122.98	121.9690	39.07	3.28	57.66	C 8H14Cl2	181.11	180.0473	53.06	7.79	39.15
C 4H 5Cl1	88.54	88.0080	54.26	5.69	40.04	C 8H15Cl1	146.66	146.0862	65.52	10.31	24.17
C 4H 6Cl2	125.00	123.9847	38.44	4.84	56.73	C 8H16Cl2	183.12	182.0629	52.47	8.81	38.72
C 4H 7Cl1	90.55	90.0236	53.06	7.79	39.15	C 8H17Cl1	148.68	148.1019	64.63	11.53	23.85
C 4H 8Cl2	127.01	126.0003	37.83	6.35	55.83	C 9H 4Cl2	183.04	181.9690	59.06	2.20	38.74
C 4H 9Cl1	92.57	92.0393	51.90	9.80	38.30	C 9H 5Cl1	148.59	148.0080	72.75	3.39	23.86
C 5H 2Cl2	132.98	131.9534	45.16	1.52	53.32	C 9H 6Cl2	185.05	183.9847	58.42	3.27	38.32
C 5H 3Cl1	98.53	97.9923	60.95	3.07	35.98	C 9H 7Cl1	150.61	150.0236	71.78	4.68	23.54
C 5H 4Cl2	134.99	133.9690	44.49	2.99	52.53	C 9H 8Cl2	187.07	186.0003	57.79	4.31	37.90
C 5H 5Cl1	100.55	100.0080	59.73	5.01	35.26	C 9H 9Cl1	152.63	152.0393	70.83	5.94	23.23
C 5H 6Cl2	137.01	135.9847	43.83	4.41	51.75	C 9H10Cl2	189.09	188.0160	57.17	5.33	37.50
C 5H 7Cl1	102.56	102.0236	58.55	6.88	34.57	C 9H11Cl1	154.64	154.0549	69.90	7.17	22.93
C 5H 8Cl2	139.03	138.0003	43.20	5.80	51.00	C 9H12Cl2	191.10	190.0316	56.57	6.33	37.10
C 5H 9Cl1	104.58	104.0393	57.43	8.67	33.90	C 9H13Cl1	156.66	156.0706	69.00	8.36	22.63
C 5H10Cl2	141.04	140.0160	42.58	7.15	50.27	C 9H14Cl2	193.12	192.0473	55.98	7.31	36.72
C 5H11Cl1	106.60	106.0549	56.34	10.40	33.26	C 9H15Cl1	158.67	158.0862	68.13	9.53	22.34
C 6H 3Cl1	110.54	109.9923	65.19	2.74	32.07	C 9H16Cl2	195.13	194.0629	55.40	8.26	36.34
C 6H 4Cl2	147.00	145.9690	49.02	2.74	48.23	C 9H17Cl1	160.69	160.1019	67.27	10.66	22.06
C 6H 5Cl1	112.56	112.0080	64.03	4.48	31.50	C 9H18Cl2	197.15	196.0785	54.83	9.20	35.97
C 6H 6Cl2	149.02	147.9847	48.36	4.06	47.58	C 9H19Cl1	162.70	162.1175	66.44	11.77	21.79
C 6H 7Cl1	114.58	114.0236	62.90	6.16	30.94	C10H 5Cl1	160.60	160.0080	74.79	3.14	22.07
C 6H 8Cl2	151.04	150.0003	47.71	5.34	46.95	C10H 6Cl2	197.07	195.9847	60.95	3.07	35.98
C 6H 9Cl1	116.59	116.0393	61.81	7.78	30.41	C10H 7Cl1	162.62	162.0236	73.86	4.34	21.80
C 6H10Cl2	153.05	152.0160	47.09	6.59	46.33	C10H 8Cl2	199.08	198.0003	60.33	4.05	35.62
C 6H11Cl1	118.61	118.0549	60.76	9.35	29.89	C10H 9Cl1	164.64	164.0393	72.96	5.51	21.53
C 6H12Cl2	155.07	154.0316	46.47	7.80	45.73	C10H10Cl2	201.10	200.0160	59.73	5.01	35.26
C 6H13Cl1	120.62	120.0706	59.75	10.86	29.39	C10H11Cl1	166.65	166.0549	72.07	6.65	21.27
C 7H 3Cl1	122.55	121.9923	68.60	2.47	28.93	C10H12Cl2	203.11	202.0316	59.14	5.96	34.91
C 7H 4Cl2	159.02	157.9690	52.87	2.54	44.59	C10H13Cl1	168.67	168.0706	71.21	7.77	21.02
C 7H 5Cl1	124.57	124.0080	67.49	4.05	28.46	C10H14Cl2	205.13	204.0473	58.55	6.88	34.57
C 7H 6Cl2	161.03	159.9847	52.21	3.76	44.03	C10H15Cl1	170.68	170.0862	70.37	8.86	20.77
C 7H 7Cl1	126.59	126.0236	66.42	5.57	28.01	C10H16Cl2	207.15	206.0629	57.98	7.79	34.23

C10H17Cl1	172.70	172.1019	69.55	9.92	20.53	C13H22Cl2	249.23	248.1098	62.65	8.90	28.45
C10H18Cl2	209.16	208.0785	57.43	8.67	33.90	C13H23Cl1	214.78	214.1488	72.70	10.79	16.51
C10H19Cl1	174.72	174.1175	68.75	10.96	20.29	C13H24Cl2	251.24	250.1255	62.15	9.63	28.22
C10H20Cl2	211.18	210.0942	56.88	9.55	33.58	C13H25Cl1	216.80	216.1645	72.02	11.62	16.35
C10H21Cl1	176.73	176.1332	67.96	11.98	20.06	C13H26Cl2	253.26	252.1411	61.65	10.35	28.00
C11H 5Cl1	172.62	172.0080	76.54	2.92	20.54	C13H27Cl1	218.81	218.1801	71.36	12.44	16.20
C11H 6Cl2	209.08	207.9847	63.19	2.89	33.91	C14H 7Cl1	210.66	210.0236	79.82	3.35	16.83
C11H 7Cl1	174.63	174.0236	75.66	4.04	20.30	C14H 8Cl2	247.13	246.0003	68.04	3.26	28.69
C11H 8Cl2	211.09	210.0003	62.59	3.82	33.59	C14H 9Cl1	212.68	212.0393	79.07	4.27	16.67
C11H 9Cl1	176.65	176.0393	74.79	5.14	20.07	C14H10Cl2	249.14	248.0160	67.49	4.05	28.46
C11H10Cl2	213.11	212.0160	62.00	4.73	33.27	C14H11Cl1	214.70	214.0549	78.32	5.16	16.51
C11H11Cl1	178.66	178.0549	73.95	6.21	19.84	C14H12Cl2	251.16	250.0316	66.95	4.82	28.23
C11H12Cl2	215.12	214.0316	61.42	5.62	32.96	C14H13Cl1	216.71	216.0706	77.59	6.05	16.36
C11H13Cl1	180.68	180.0706	73.13	7.25	19.62	C14H14Cl2	253.17	252.0473	66.42	5.57	28.01
C11H14Cl2	217.14	216.0473	60.85	6.50	32.65	C14H15Cl1	218.73	218.0862	76.88	6.91	16.21
C11H15Cl1	182.70	182.0862	72.32	8.28	19.41	C14H16Cl2	255.19	254.0629	65.89	6.32	27.79
C11H16Cl2	219.16	218.0629	60.29	7.36	32.35	C14H17Cl1	220.74	220.1019	76.18	7.76	16.06
C11H17Cl1	184.71	184.1019	71.53	9.28	19.19	C14H18Cl2	257.21	256.0785	65.38	7.05	27.57
C11H18Cl2	221.17	220.0785	59.74	8.20	32.06	C14H19Cl1	222.76	222.1175	75.49	8.60	15.92
C11H19Cl1	186.73	186.1175	70.76	10.26	18.99	C14H20Cl2	259.22	258.0942	64.87	7.78	27.35
C11H20Cl2	223.19	222.0942	59.20	9.03	31.77	C14H21Cl1	224.78	224.1332	74.81	9.42	15.77
C11H21Cl1	188.74	188.1332	70.00	11.21	18.78	C14H22Cl2	261.24	260.1098	64.37	8.49	27.14
C11H22Cl2	225.20	224.1098	58.67	9.85	31.49	C14H23Cl1	226.79	226.1488	74.15	10.22	15.63
C11H23Cl1	190.76	190.1488	69.26	12.15	18.59	C14H24Cl2	263.25	262.1255	63.88	9.19	26.93
C12H 6Cl2	221.09	219.9847	65.19	2.74	32.07	C14H25Cl1	228.81	228.1645	73.49	11.01	15.49
C12H 7Cl1	186.64	186.0236	77.22	3.78	19.00	C14H26Cl2	265.27	264.1411	63.39	9.88	26.73
C12H 8Cl2	223.10	222.0003	64.60	3.61	31.78	C14H27Cl1	230.82	230.1801	72.85	11.79	15.36
C12H 9Cl1	188.66	188.0393	76.40	4.81	18.79	C14H28Cl2	267.29	266.1568	62.91	10.56	26.53
C12H10Cl2	225.12	224.0160	64.03	4.48	31.50	C14H29Cl1	232.84	232.1958	72.22	12.55	15.23
C12H11Cl1	190.67	190.0549	75.59	5.81	18.59	C15H 7Cl1	222.68	222.0236	80.91	3.17	15.92
C12H12Cl2	227.14	226.0316	63.46	5.33	31.22	C15H 8Cl2	259.14	258.0003	69.53	3.11	27.36
C12H13Cl1	192.69	192.0706	74.80	6.80	18.40	C15H 9Cl1	224.69	224.0393	80.18	4.04	15.78
C12H14Cl2	229.15	228.0473	62.90	6.16	30.94	C15H10Cl2	261.15	260.0160	68.99	3.86	27.15
C12H15Cl1	194.71	194.0862	74.03	7.77	18.21	C15H11Cl1	226.71	226.0549	79.47	4.89	15.64
C12H16Cl2	231.17	230.0629	62.35	6.98	30.67	C15H12Cl2	263.17	262.0316	68.46	4.60	26.94
C12H17Cl1	196.72	196.1019	73.27	8.71	18.02	C15H13Cl1	228.72	228.0706	78.77	5.73	15.50
C12H18Cl2	233.18	232.0785	61.81	7.78	30.41	C15H14Cl2	265.18	264.0473	67.94	5.32	26.74
C12H19Cl1	198.74	198.1175	72.52	9.64	17.84	C15H15Cl1	230.74	230.0862	78.08	6.55	15.36
C12H20Cl2	235.20	234.0942	61.28	8.57	30.15	C15H16Cl2	267.20	266.0629	67.43	6.04	26.54
C12H21Cl1	200.75	200.1332	71.80	10.54	17.66	C15H17Cl1	232.76	232.1019	77.41	7.36	15.23
C12H22Cl2	237.22	236.1098	60.76	9.35	29.89	C15H18Cl2	269.22	268.0785	66.92	6.74	26.34
C12H23Cl1	202.77	202.1488	71.08	11.43	17.48	C15H19Cl1	234.77	234.1175	76.74	8.16	15.10
C12H24Cl2	239.23	238.1255	60.25	10.11	29.64	C15H20Cl2	271.23	270.0942	66.43	7.43	26.14
C12H25Cl1	204.79	204.1645	70.38	12.31	17.31	C15H21Cl1	236.79	236.1332	76.09	8.94	14.97
C13H 6Cl2	233.10	231.9847	66.99	2.59	30.42	C15H22Cl2	273.25	272.1098	65.94	8.12	25.95
C13H 7Cl1	198.65	198.0236	78.60	3.55	17.85	C15H23Cl1	238.80	238.1488	75.45	9.71	14.85
C13H 8Cl2	235.11	234.0003	66.41	3.43	30.16	C15H24Cl2	275.26	274.1255	65.45	8.79	25.76
C13H 9Cl1	200.67	200.0393	77.81	4.52	17.67	C15H25Cl1	240.82	240.1645	74.81	10.46	14.72
C13H10Cl2	237.13	236.0160	65.85	4.25	29.90	C15H26Cl2	277.28	276.1411	64.98	9.45	25.57
C13H11Cl1	202.69	202.0549	77.04	5.47	17.49	C15H27Cl1	242.84	242.1801	74.19	11.21	14.60
C13H12Cl2	239.15	238.0316	65.29	5.06	29.65	C15H28Cl2	279.30	278.1568	64.51	10.11	25.39
C13H13Cl1	204.70	204.0706	76.28	6.40	17.32	C15H29Cl1	244.85	244.1958	73.58	11.94	14.48
C13H14Cl2	241.16	240.0473	64.75	5.85	29.40	C15H30Cl2	281.31	280.1724	64.05	10.75	25.21
C13H15Cl1	206.72	206.0862	75.54	7.31	17.15	C15H31Cl1	246.87	246.2114	72.98	12.66	14.36
C13H16Cl2	243.18	242.0629	64.21	6.63	29.16	C16H 8Cl2	271.15	270.0003	70.88	2.97	26.15
C13H17Cl1	208.73	208.1019	74.81	8.21	16.98	C16H 9Cl1	236.70	236.0393	81.19	3.83	14.98
C13H18Cl2	245.19	244.0785	63.68	7.40	28.92	C16H10Cl2	273.16	272.0160	70.35	3.69	25.96
C13H19Cl1	210.75	210.1175	74.09	9.09	16.82	C16H11Cl1	238.72	238.0549	80.50	4.64	14.85
C13H20Cl2	247.21	246.0942	63.16	8.15	28.68	C16H12Cl2	275.18	274.0316	69.84	4.40	25.77
C13H21Cl1	212.77	212.1332	73.39	9.95	16.66	C16H13Cl1	240.74	240.0706	79.83	5.44	14.73

C16H14C12	277.20	276.0473	69.33	5.09	25.58
C16H15C11	242.75	242.0862	79.17	6.23	14.60
C16H16C12	279.21	278.0629	68.83	5.78	25.40
C16H17C11	244.77	244.1019	78.51	7.00	14.48
C16H18C12	281.23	280.0785	68.34	6.45	25.21
C16H19C11	246.78	246.1175	77.87	7.76	14.37
C16H20C12	283.24	282.0942	67.85	7.12	25.03
C16H21C11	248.80	248.1332	77.24	8.51	14.25
C16H22C12	285.26	284.1098	67.37	7.77	24.86
C16H23C11	250.81	250.1488	76.62	9.24	14.14
C16H24C12	287.28	286.1255	66.90	8.42	24.68
C16H25C11	252.83	252.1645	76.01	9.97	14.02
C16H26C12	289.29	288.1411	66.43	9.06	24.51
C16H27C11	254.85	254.1801	75.41	10.68	13.91
C16H28C12	291.31	290.1568	65.97	9.69	24.34
C16H29C11	256.86	256.1958	74.82	11.38	13.80
C16H30C12	293.32	292.1724	65.52	10.31	24.17
C16H31C11	258.88	258.2114	74.23	12.07	13.69
C16H32C12	295.34	294.1881	65.07	10.92	24.01
C16H33C11	260.89	260.2271	73.66	12.75	13.59
C17H 8C12	283.16	282.0003	72.11	2.85	25.04
C17H 9C11	248.71	248.0393	82.10	3.65	14.25
C17H10C12	285.18	284.0160	71.60	3.53	24.86
C17H11C11	250.73	250.0549	81.44	4.42	14.14
C17H12C12	287.19	286.0316	71.10	4.21	24.69
C17H13C11	252.75	252.0706	80.79	5.18	14.03
C17H14C12	289.21	288.0473	70.60	4.88	24.52
C17H15C11	254.76	254.0862	80.15	5.93	13.92
C17H16C12	291.22	290.0629	70.11	5.54	24.35
C17H17C11	256.78	256.1019	79.52	6.67	13.81
C17H18C12	293.24	292.0785	69.63	6.19	24.18
C17H19C11	258.79	258.1175	78.90	7.40	13.70
C17H20C12	295.25	294.0942	69.16	6.83	24.02
C17H21C11	260.81	260.1332	78.29	8.12	13.59
C17H22C12	297.27	296.1098	68.69	7.46	23.85
C17H23C11	262.83	262.1488	77.69	8.82	13.49
C17H24C12	299.29	298.1255	68.23	8.08	23.69
C17H25C11	264.84	264.1645	77.10	9.51	13.39
C17H26C12	301.30	300.1411	67.77	8.70	23.53
C17H27C11	266.86	266.1801	76.52	10.20	13.29
C17H28C12	303.32	302.1568	67.32	9.30	23.38
C17H29C11	268.87	268.1958	75.94	10.87	13.19
C17H30C12	305.33	304.1724	66.87	9.90	23.22
C17H31C11	270.89	270.2114	75.38	11.53	13.09
C17H32C12	307.35	306.1881	66.44	10.49	23.07
C17H33C11	272.91	272.2271	74.82	12.19	12.99
C17H34C12	309.37	308.2037	66.00	11.08	22.92
C17H35C11	274.92	274.2427	74.27	12.83	12.90
C18H 9C11	260.73	260.0393	82.92	3.48	13.60
C18H10C12	297.19	296.0160	72.75	3.39	23.86
C18H11C11	262.74	262.0549	82.29	4.22	13.49
C18H12C12	299.20	298.0316	72.26	4.04	23.70
C18H13C11	264.76	264.0706	81.66	4.95	13.39
C18H14C12	301.22	300.0473	71.78	4.68	23.54
C18H15C11	266.77	266.0862	81.04	5.67	13.29
C18H16C12	303.23	302.0629	71.30	5.32	23.38
C18H17C11	268.79	268.1019	80.44	6.38	13.19
C18H18C12	305.25	304.0785	70.83	5.94	23.23
C18H19C11	270.81	270.1175	79.84	7.07	13.09
C18H20C12	307.27	306.0942	70.36	6.56	23.08

C18H21C11	272.82	272.1332	79.25	7.76	12.99
C18H22C12	309.28	308.1098	69.90	7.17	22.93
C18H23C11	274.84	274.1488	78.67	8.44	12.90
C18H24C12	311.30	310.1255	69.45	7.77	22.78
C18H25C11	276.85	276.1645	78.09	9.10	12.81
C18H26C12	313.31	312.1411	69.00	8.36	22.63
C18H27C11	278.87	278.1801	77.53	9.76	12.71
C18H28C12	315.33	314.1568	68.56	8.95	22.49
C18H29C11	280.88	280.1958	76.97	10.41	12.62
C18H30C12	317.35	316.1724	68.13	9.53	22.34
C18H31C11	282.90	282.2114	76.42	11.05	12.53
C18H32C12	319.36	318.1881	67.70	10.10	22.20
C18H33C11	284.92	284.2271	75.88	11.67	12.44
C18H34C12	321.38	320.2037	67.27	10.66	22.06
C18H35C11	286.93	286.2427	75.35	12.30	12.36
C18H36C12	323.39	322.2194	66.85	11.22	21.93
C18H37C11	288.95	288.2584	74.82	12.91	12.27
C19H 9C11	272.74	272.0393	83.67	3.33	13.00
C19H10C12	309.20	308.0160	73.81	3.26	22.93
C19H11C11	274.75	274.0549	83.06	4.04	12.90
C19H12C12	311.21	310.0316	73.33	3.89	22.78
C19H13C11	276.77	276.0706	82.46	4.73	12.81
C19H14C12	313.23	312.0473	72.86	4.51	22.64
C19H15C11	278.78	278.0862	81.86	5.42	12.72
C19H16C12	315.25	314.0629	72.39	5.12	22.49
C19H17C11	280.80	280.1019	81.27	6.10	12.63
C19H18C12	317.26	316.0785	71.93	5.72	22.35
C19H19C11	282.82	282.1175	80.69	6.77	12.54
C19H20C12	319.28	318.0942	71.48	6.31	22.21
C19H21C11	284.83	284.1332	80.12	7.43	12.45
C19H22C12	321.29	320.1098	71.03	6.90	22.07
C19H23C11	286.85	286.1488	79.56	8.08	12.36
C19H24C12	323.31	322.1255	70.59	7.48	21.93
C19H25C11	288.86	288.1645	79.00	8.72	12.27
C19H26C12	325.33	324.1411	70.15	8.06	21.80
C19H27C11	290.88	290.1801	78.46	9.36	12.19
C19H28C12	327.34	326.1568	69.72	8.62	21.66
C19H29C11	292.90	292.1958	77.92	9.98	12.10
C19H30C12	329.36	328.1724	69.29	9.18	21.53
C19H31C11	294.91	294.2114	77.38	10.60	12.02
C19H32C12	331.37	330.1881	68.87	9.73	21.40
C19H33C11	296.93	296.2271	76.86	11.20	11.94
C19H34C12	333.39	332.2037	68.45	10.28	21.27
C19H35C11	298.94	298.2427	76.34	11.80	11.86
C19H36C12	335.40	334.2194	68.04	10.82	21.14
C19H37C11	300.96	300.2584	75.83	12.39	11.78
C19H38C12	337.42	336.2350	67.63	11.35	21.01
C19H39C11	302.98	302.2740	75.32	12.97	11.70
C20H10C12	321.21	320.0160	74.79	3.14	22.07
C20H11C11	286.76	286.0549	83.77	3.87	12.36
C20H12C12	323.22	322.0316	74.32	3.74	21.94
C20H13C11	288.78	288.0706	83.19	4.54	12.28
C20H14C12	325.24	324.0473	73.86	4.34	21.80
C20H15C11	290.80	290.0862	82.61	5.20	12.19
C20H16C12	327.26	326.0629	73.41	4.93	21.67
C20H17C11	292.81	292.1019	82.04	5.85	12.11
C20H18C12	329.27	328.0785	72.96	5.51	21.53
C20H19C11	294.83	294.1175	81.48	6.50	12.03
C20H20C12	331.29	330.0942	72.51	6.09	21.40
C20H21C11	296.84	296.1332	80.93	7.13	11.94

C20H22C12	333.30	332.1098	72.07	6.65	21.27
C20H23C11	298.86	298.1488	80.38	7.76	11.86
C20H24C12	335.32	334.1255	71.64	7.21	21.15
C20H25C11	300.88	300.1645	79.84	8.38	11.78
C20H26C12	337.34	336.1411	71.21	7.77	21.02
C20H27C11	302.89	302.1801	79.31	8.99	11.70
C20H28C12	339.35	338.1568	70.79	8.32	20.89
C20H29C11	304.91	304.1958	78.79	9.59	11.63
C20H30C12	341.37	340.1724	70.37	8.86	20.77
C20H31C11	306.92	306.2114	78.27	10.18	11.55
C20H32C12	343.38	342.1881	69.96	9.39	20.65
C20H33C11	308.94	308.2271	77.76	10.77	11.48
C20H34C12	345.40	344.2037	69.55	9.92	20.53
C20H35C11	310.95	310.2427	77.25	11.35	11.40
C20H36C12	347.42	346.2194	69.15	10.44	20.41
C20H37C11	312.97	312.2584	76.76	11.92	11.33
C20H38C12	349.43	348.2350	68.75	10.96	20.29
C20H39C11	314.99	314.2740	76.26	12.48	11.26
C20H40C12	351.45	350.2507	68.35	11.47	20.18
C20H41C11	317.00	316.2897	75.78	13.04	11.18
C21H10C12	333.22	332.0160	75.70	3.02	21.28
C21H11C11	298.77	298.0549	84.42	3.71	11.87
C21H12C12	335.24	334.0316	75.24	3.61	21.15
C21H13C11	300.79	300.0706	83.86	4.36	11.79
C21H14C12	337.25	336.0473	74.79	4.18	21.02
C21H15C11	302.81	302.0862	83.30	4.99	11.71
C21H16C12	339.27	338.0629	74.35	4.75	20.90
C21H17C11	304.82	304.1019	82.75	5.62	11.63
C21H18C12	341.28	340.0785	73.91	5.32	20.78
C21H19C11	306.84	306.1175	82.20	6.24	11.55
C21H20C12	343.30	342.0942	73.47	5.87	20.65
C21H21C11	308.85	308.1332	81.67	6.85	11.48
C21H22C12	345.32	344.1098	73.04	6.42	20.53
C21H23C11	310.87	310.1488	81.14	7.46	11.40
C21H24C12	347.33	346.1255	72.62	6.96	20.41
C21H25C11	312.89	312.1645	80.62	8.05	11.33
C21H26C12	349.35	348.1411	72.20	7.50	20.30
C21H27C11	314.90	314.1801	80.10	8.64	11.26
C21H28C12	351.36	350.1568	71.79	8.03	20.18
C21H29C11	316.92	316.1958	79.59	9.22	11.19
C21H30C12	353.38	352.1724	71.38	8.56	20.07
C21H31C11	318.93	318.2114	79.09	9.80	11.12
C21H32C12	355.40	354.1881	70.97	9.08	19.95
C21H33C11	320.95	320.2271	78.59	10.36	11.05
C21H34C12	357.41	356.2037	70.57	9.59	19.84
C21H35C11	322.97	322.2427	78.10	10.92	10.98
C21H36C12	359.43	358.2194	70.18	10.10	19.73
C21H37C11	324.98	324.2584	77.61	11.48	10.91
C21H38C12	361.44	360.2350	69.79	10.60	19.62
C21H39C11	327.00	326.2740	77.14	12.02	10.84
C21H40C12	363.46	362.2507	69.40	11.09	19.51
C21H41C11	329.01	328.2897	76.66	12.56	10.78
C21H42C12	365.47	364.2663	69.02	11.58	19.40
C21H43C11	331.03	330.3053	76.20	13.09	10.71
C22H11C11	310.79	310.0549	85.02	3.57	11.41
C22H12C12	347.25	346.0316	76.10	3.48	20.42
C22H13C11	312.80	312.0706	84.48	4.19	11.33
C22H14C12	349.26	348.0473	75.66	4.04	20.30
C22H15C11	314.82	314.0862	83.94	4.80	11.26
C22H16C12	351.28	350.0629	75.22	4.59	20.19

C22H17C11	316.83	316.1019	83.40	5.41	11.19
C22H18C12	353.29	352.0785	74.79	5.14	20.07
C22H19C11	318.85	318.1175	82.87	6.01	11.12
C22H20C12	355.31	354.0942	74.37	5.67	19.96
C22H21C11	320.87	320.1332	82.35	6.60	11.05
C22H22C12	357.33	356.1098	73.95	6.21	19.84
C22H23C11	322.88	322.1488	81.84	7.18	10.98
C22H24C12	359.34	358.1255	73.54	6.73	19.73
C22H25C11	324.90	324.1645	81.33	7.76	10.91
C22H26C12	361.36	360.1411	73.13	7.25	19.62
C22H27C11	326.91	326.1801	80.83	8.32	10.84
C22H28C12	363.37	362.1568	72.72	7.77	19.51
C22H29C11	328.93	328.1958	80.33	8.89	10.78
C22H30C12	365.39	364.1724	72.32	8.28	19.41
C22H31C11	330.95	330.2114	79.85	9.44	10.71
C22H32C12	367.41	366.1881	71.92	8.78	19.30
C22H33C11	332.96	332.2271	79.36	9.99	10.65
C22H34C12	369.42	368.2037	71.53	9.28	19.19
C22H35C11	334.98	334.2427	78.88	10.53	10.58
C22H36C12	371.44	370.2194	71.14	9.77	19.09
C22H37C11	336.99	336.2584	78.41	11.07	10.52
C22H38C12	373.45	372.2350	70.76	10.26	18.99
C22H39C11	339.01	338.2740	77.95	11.60	10.46
C22H40C12	375.47	374.2507	70.38	10.74	18.88
C22H41C11	341.03	340.2897	77.49	12.12	10.40
C22H42C12	377.49	376.2663	70.00	11.21	18.78
C22H43C11	343.04	342.3053	77.03	12.63	10.33
C22H44C12	379.50	378.2820	69.63	11.69	18.68
C22H45C11	345.06	344.3210	76.58	13.15	10.27
C23H11C11	322.80	322.0549	85.58	3.43	10.98
C23H12C12	359.26	358.0316	76.90	3.37	19.74
C23H13C11	324.81	324.0706	85.05	4.03	10.91
C23H14C12	361.27	360.0473	76.47	3.91	19.63
C23H15C11	326.83	326.0862	84.53	4.63	10.85
C23H16C12	363.29	362.0629	76.04	4.44	19.52
C23H17C11	328.84	328.1019	84.01	5.21	10.78
C23H18C12	365.31	364.0785	75.62	4.97	19.41
C23H19C11	330.86	330.1175	83.50	5.79	10.72
C23H20C12	367.32	366.0942	75.21	5.49	19.30
C23H21C11	332.88	332.1332	82.99	6.36	10.65
C23H22C12	369.34	368.1098	74.80	6.00	19.20
C23H23C11	334.89	334.1488	82.49	6.92	10.59
C23H24C12	371.35	370.1255	74.39	6.51	19.09
C23H25C11	336.91	336.1645	82.00	7.48	10.52
C23H26C12	373.37	372.1411	73.99	7.02	18.99
C23H27C11	338.92	338.1801	81.51	8.03	10.46
C23H28C12	375.39	374.1568	73.59	7.52	18.89
C23H29C11	340.94	340.1958	81.03	8.57	10.40
C23H30C12	377.40	376.1724	73.20	8.01	18.79
C23H31C11	342.96	342.2114	80.55	9.11	10.34
C23H32C12	379.42	378.1881	72.81	8.50	18.69
C23H33C11	344.97	344.2271	80.08	9.64	10.28
C23H34C12	381.43	380.2037	72.43	8.98	18.59
C23H35C11	346.99	346.2427	79.62	10.17	10.22
C23H36C12	383.45	382.2194	72.05	9.46	18.49
C23H37C11	349.00	348.2584	79.16	10.69	10.16
C23H38C12	385.47	384.2350	71.67	9.94	18.39
C23H39C11	351.02	350.2740	78.70	11.20	10.10
C23H40C12	387.48	386.2507	71.30	10.41	18.30
C23H41C11	353.04	352.2897	78.25	11.71	10.04

10/21/66 028

C23H42Cl2	389.50	388.2663	70.93	10.87	18.20
C23H43Cl1	355.05	354.3053	77.81	12.21	9.99
C23H44Cl2	391.51	390.2820	70.56	11.33	18.11
C23H45Cl1	357.07	356.3210	77.37	12.70	9.93
C23H46Cl2	393.53	392.2976	70.20	11.78	18.02
C23H47Cl1	359.08	358.3366	76.93	13.19	9.87
C24H12Cl2	371.27	370.0316	77.64	3.26	19.10
C24H13Cl1	336.82	336.0706	85.58	3.89	10.53
C24H14Cl2	373.29	372.0473	77.22	3.78	19.00
C24H15Cl1	338.84	338.0862	85.07	4.46	10.46
C24H16Cl2	375.30	374.0629	76.81	4.30	18.89
C24H17Cl1	340.86	340.1019	84.57	5.03	10.40
C24H18Cl2	377.32	376.0785	76.40	4.81	18.79
C24H19Cl1	342.87	342.1175	84.07	5.59	10.34
C24H20Cl2	379.33	378.0942	75.99	5.31	18.69
C24H21Cl1	344.89	344.1332	83.58	6.14	10.28
C24H22Cl2	381.35	380.1098	75.59	5.81	18.59
C24H23Cl1	346.90	346.1488	83.10	6.68	10.22
C24H24Cl2	383.36	382.1255	75.19	6.31	18.50
C24H25Cl1	348.92	348.1645	82.62	7.22	10.16
C24H26Cl2	385.38	384.1411	74.80	6.80	18.40
C24H27Cl1	350.94	350.1801	82.14	7.76	10.10
C24H28Cl2	387.40	386.1568	74.41	7.29	18.30
C24H29Cl1	352.95	352.1958	81.67	8.28	10.04
C24H30Cl2	389.41	388.1724	74.03	7.77	18.21
C24H31Cl1	354.97	354.2114	81.21	8.80	9.99
C24H32Cl2	391.43	390.1881	73.64	8.24	18.11
C24H33Cl1	356.98	356.2271	80.75	9.32	9.93
C24H34Cl2	393.44	392.2037	73.27	8.71	18.02
C24H35Cl1	359.00	358.2427	80.30	9.83	9.88
C24H36Cl2	395.46	394.2194	72.89	9.18	17.93
C24H37Cl1	361.02	360.2584	79.85	10.33	9.82
C24H38Cl2	397.48	396.2350	72.52	9.64	17.84
C24H39Cl1	363.03	362.2740	79.41	10.83	9.77
C24H40Cl2	399.49	398.2507	72.16	10.09	17.75
C24H41Cl1	365.05	364.2897	78.97	11.32	9.71
C24H42Cl2	401.51	400.2663	71.80	10.54	17.66
C24H43Cl1	367.06	366.3053	78.53	11.81	9.66
C24H44Cl2	403.52	402.2820	71.44	10.99	17.57
C24H45Cl1	369.08	368.3210	78.10	12.29	9.61
C24H46Cl2	405.54	404.2976	71.08	11.43	17.48
C24H47Cl1	371.10	370.3366	77.68	12.77	9.55
C24H48Cl2	407.56	406.3133	70.73	11.87	17.40
C24H49Cl1	373.11	372.3523	77.26	13.24	9.50
C25H12Cl2	383.28	382.0316	78.34	3.16	18.50
C25H13Cl1	348.84	348.0706	86.08	3.76	10.16
C25H14Cl2	385.30	384.0473	77.93	3.66	18.40
C25H15Cl1	350.85	350.0862	85.59	4.31	10.10
C25H16Cl2	387.31	386.0629	77.53	4.16	18.31
C25H17Cl1	352.87	352.1019	85.10	4.86	10.05
C25H18Cl2	389.33	388.0785	77.13	4.66	18.21
C25H19Cl1	354.88	354.1175	84.61	5.40	9.99
C25H20Cl2	391.34	390.0942	76.73	5.15	18.12
C25H21Cl1	356.90	356.1332	84.14	5.93	9.93
C25H22Cl2	393.36	392.1098	76.34	5.64	18.03
C25H23Cl1	358.92	358.1488	83.66	6.46	9.88
C25H24Cl2	395.38	394.1255	75.95	6.12	17.93
C25H25Cl1	360.93	360.1645	83.20	6.98	9.82
C25H26Cl2	397.39	396.1411	75.56	6.59	17.84
C25H27Cl1	362.95	362.1801	82.73	7.50	9.77

C25H28Cl2	399.41	398.1568	75.18	7.07	17.75
C25H29Cl1	364.96	364.1958	82.28	8.01	9.71
C25H30Cl2	401.42	400.1724	74.80	7.53	17.66
C25H31Cl1	366.98	366.2114	81.82	8.51	9.66
C25H32Cl2	403.44	402.1881	74.43	8.00	17.58
C25H33Cl1	368.99	368.2271	81.38	9.01	9.61
C25H34Cl2	405.46	404.2037	74.06	8.45	17.49
C25H35Cl1	371.01	370.2427	80.94	9.51	9.56
C25H36Cl2	407.47	406.2194	73.69	8.91	17.40
C25H37Cl1	373.03	372.2584	80.50	10.00	9.50
C25H38Cl2	409.49	408.2350	73.33	9.35	17.32
C25H39Cl1	375.04	374.2740	80.07	10.48	9.45
C25H40Cl2	411.50	410.2507	72.97	9.80	17.23
C25H41Cl1	377.06	376.2897	79.64	10.96	9.40
C25H42Cl2	413.52	412.2663	72.62	10.24	17.15
C25H43Cl1	379.07	378.3053	79.21	11.43	9.35
C25H44Cl2	415.54	414.2820	72.26	10.67	17.06
C25H45Cl1	381.09	380.3210	78.79	11.90	9.30
C25H46Cl2	417.55	416.2976	71.91	11.10	16.98
C25H47Cl1	383.11	382.3366	78.38	12.37	9.25
C25H48Cl2	419.57	418.3133	71.57	11.53	16.90
C25H49Cl1	385.12	384.3523	77.97	12.82	9.21
C25H50Cl2	421.58	420.3289	71.23	11.95	16.82
C25H51Cl1	387.14	386.3679	77.56	13.28	9.16
C26H13Cl1	360.85	360.0706	86.54	3.63	9.82
C26H14Cl2	397.31	396.0473	78.60	3.55	17.85
C26H15Cl1	362.86	362.0862	86.06	4.17	9.77
C26H16Cl2	399.32	398.0629	78.20	4.04	17.76
C26H17Cl1	364.88	364.1019	85.59	4.70	9.72
C26H18Cl2	401.34	400.0785	77.81	4.52	17.67
C26H19Cl1	366.89	366.1175	85.12	5.22	9.66
C26H20Cl2	403.36	402.0942	77.42	5.00	17.58
C26H21Cl1	368.91	368.1332	84.65	5.74	9.61
C26H22Cl2	405.37	404.1098	77.04	5.47	17.49
C26H23Cl1	370.93	370.1488	84.19	6.25	9.56
C26H24Cl2	407.39	406.1255	76.66	5.94	17.41
C26H25Cl1	372.94	372.1645	83.74	6.76	9.51
C26H26Cl2	409.40	408.1411	76.28	6.40	17.32
C26H27Cl1	374.96	374.1801	83.29	7.26	9.46
C26H28Cl2	411.42	410.1568	75.91	6.86	17.23
C26H29Cl1	376.97	376.1958	82.84	7.75	9.40
C26H30Cl2	413.44	412.1724	75.54	7.31	17.15
C26H31Cl1	378.99	378.2114	82.40	8.24	9.35
C26H32Cl2	415.45	414.1881	75.17	7.76	17.07
C26H33Cl1	381.01	380.2271	81.96	8.73	9.31
C26H34Cl2	417.47	416.2037	74.81	8.21	16.98
C26H35Cl1	383.02	382.2427	81.53	9.21	9.26
C26H36Cl2	419.48	418.2194	74.45	8.65	16.90
C26H37Cl1	385.04	384.2584	81.11	9.69	9.21
C26H38Cl2	421.50	420.2350	74.09	9.09	16.82
C26H39Cl1	387.05	386.2740	80.68	10.16	9.16
C26H40Cl2	423.51	422.2507	73.74	9.52	16.74
C26H41Cl1	389.07	388.2897	80.27	10.62	9.11
C26H42Cl2	425.53	424.2663	73.39	9.95	16.66
C26H43Cl1	391.09	390.3053	79.85	11.08	9.07
C26H44Cl2	427.55	426.2820	73.04	10.37	16.58
C26H45Cl1	393.10	392.3210	79.44	11.54	9.02
C26H46Cl2	429.56	428.2976	72.70	10.79	16.51
C26H47Cl1	395.12	394.3366	79.04	11.99	8.97
C26H48Cl2	431.58	430.3133	72.36	11.21	16.43

10/21/66 029

C26H49Cl1	397.13	396.3523	78.64	12.44	8.93
C26H50Cl2	433.59	432.3289	72.02	11.62	16.35
C26H51Cl1	399.15	398.3679	78.24	12.88	8.88
C26H52Cl2	435.61	434.3446	71.69	12.03	16.28
C26H53Cl1	401.17	400.3836	77.85	13.32	8.84
C27H13Cl1	372.86	372.0706	86.98	3.51	9.51
C27H14Cl2	409.32	408.0473	79.23	3.45	17.32
C27H15Cl1	374.87	374.0862	86.51	4.03	9.46
C27H16Cl2	411.33	410.0629	78.84	3.92	17.24
C27H17Cl1	376.89	376.1019	86.05	4.55	9.41
C27H18Cl2	413.35	412.0785	78.46	4.39	17.15
C27H19Cl1	378.91	378.1175	85.59	5.05	9.36
C27H20Cl2	415.37	414.0942	78.08	4.85	17.07
C27H21Cl1	380.92	380.1332	85.14	5.56	9.31
C27H22Cl2	417.38	416.1098	77.70	5.31	16.99
C27H23Cl1	382.94	382.1488	84.69	6.05	9.26
C27H24Cl2	419.40	418.1255	77.33	5.77	16.91
C27H25Cl1	384.95	384.1645	84.24	6.55	9.21
C27H26Cl2	421.41	420.1411	76.96	6.22	16.83
C27H27Cl1	386.97	386.1801	83.81	7.03	9.16
C27H28Cl2	423.43	422.1568	76.59	6.67	16.75
C27H29Cl1	388.99	388.1958	83.37	7.51	9.11
C27H30Cl2	425.45	424.1724	76.23	7.11	16.67
C27H31Cl1	391.00	390.2114	82.94	7.99	9.07
C27H32Cl2	427.46	426.1881	75.87	7.55	16.59
C27H33Cl1	393.02	392.2271	82.52	8.46	9.02
C27H34Cl2	429.48	428.2037	75.51	7.98	16.51
C27H35Cl1	395.03	394.2427	82.09	8.93	8.97
C27H36Cl2	431.49	430.2194	75.16	8.41	16.43
C27H37Cl1	397.05	396.2584	81.68	9.39	8.93
C27H38Cl2	433.51	432.2350	74.81	8.84	16.36
C27H39Cl1	399.06	398.2740	81.27	9.85	8.88
C27H40Cl2	435.53	434.2507	74.46	9.26	16.28
C27H41Cl1	401.08	400.2897	80.86	10.30	8.84
C27H42Cl2	437.54	436.2663	74.12	9.68	16.21
C27H43Cl1	403.10	402.3053	80.45	10.75	8.80
C27H44Cl2	439.56	438.2820	73.78	10.09	16.13
C27H45Cl1	405.11	404.3210	80.05	11.20	8.75
C27H46Cl2	441.57	440.2976	73.44	10.50	16.06
C27H47Cl1	407.13	406.3366	79.66	11.64	8.71
C27H48Cl2	443.59	442.3133	73.11	10.91	15.98
C27H49Cl1	409.14	408.3523	79.26	12.07	8.67
C27H50Cl2	445.61	444.3289	72.78	11.31	15.91
C27H51Cl1	411.16	410.3679	78.87	12.50	8.62
C27H52Cl2	447.62	446.3446	72.45	11.71	15.84
C27H53Cl1	413.18	412.3836	78.49	12.93	8.58
C27H54Cl2	449.64	448.3602	72.13	12.11	15.77
C27H55Cl1	415.19	414.3992	78.11	13.35	8.54
C28H14Cl2	421.33	420.0473	79.82	3.35	16.83
C28H15Cl1	386.88	386.0862	86.93	3.91	9.16
C28H16Cl2	423.35	422.0629	79.44	3.81	16.75
C28H17Cl1	388.90	388.1019	86.48	4.41	9.12
C28H18Cl2	425.36	424.0785	79.07	4.27	16.67
C28H19Cl1	390.92	390.1175	86.03	4.90	9.07
C28H20Cl2	427.38	426.0942	78.69	4.72	16.59
C28H21Cl1	392.93	392.1332	85.59	5.39	9.02
C28H22Cl2	429.39	428.1098	78.32	5.16	16.51
C28H23Cl1	394.95	394.1488	85.15	5.87	8.98
C28H24Cl2	431.41	430.1255	77.96	5.61	16.44
C28H25Cl1	396.96	396.1645	84.72	6.35	8.93

C28H26Cl2	433.43	432.1411	77.59	6.05	16.36
C28H27Cl1	398.98	398.1801	84.29	6.82	8.89
C28H28Cl2	435.44	434.1568	77.23	6.48	16.28
C28H29Cl1	401.00	400.1958	83.87	7.29	8.84
C28H30Cl2	437.46	436.1724	76.88	6.91	16.21
C28H31Cl1	403.01	402.2114	83.45	7.75	8.80
C28H32Cl2	439.47	438.1881	76.53	7.34	16.13
C28H33Cl1	405.03	404.2271	83.03	8.21	8.75
C28H34Cl2	441.49	440.2037	76.18	7.76	16.06
C28H35Cl1	407.04	406.2427	82.62	8.67	8.71
C28H36Cl2	443.51	442.2194	75.83	8.18	15.99
C28H37Cl1	409.06	408.2584	82.22	9.12	8.67
C28H38Cl2	445.52	444.2350	75.49	8.60	15.92
C28H39Cl1	411.08	410.2740	81.81	9.56	8.62
C28H40Cl2	447.54	446.2507	75.15	9.01	15.84
C28H41Cl1	413.09	412.2897	81.41	10.00	8.58
C28H42Cl2	449.55	448.2663	74.81	9.42	15.77
C28H43Cl1	415.11	414.3053	81.02	10.44	8.54
C28H44Cl2	451.57	450.2820	74.48	9.82	15.70
C28H45Cl1	417.12	416.3210	80.63	10.87	8.50
C28H46Cl2	453.58	452.2976	74.15	10.22	15.63
C28H47Cl1	419.14	418.3366	80.24	11.30	8.46
C28H48Cl2	455.60	454.3133	73.82	10.62	15.56
C28H49Cl1	421.16	420.3523	79.85	11.73	8.42
C28H50Cl2	457.62	456.3289	73.49	11.01	15.49
C28H51Cl1	423.17	422.3679	79.47	12.15	8.38
C28H52Cl2	459.63	458.3446	73.17	11.40	15.43
C28H53Cl1	425.19	424.3836	79.10	12.56	8.34
C28H54Cl2	461.65	460.3602	72.85	11.79	15.36
C28H55Cl1	427.20	426.3992	78.72	12.98	8.30
C28H56Cl2	463.66	462.3759	72.53	12.17	15.29
C28H57Cl1	429.22	428.4149	78.35	13.39	8.26
C29H14Cl2	433.34	432.0473	80.38	3.26	16.36
C29H15Cl1	398.90	398.0862	87.32	3.79	8.89
C29H16Cl2	435.36	434.0629	80.01	3.70	16.29
C29H17Cl1	400.91	400.1019	86.88	4.27	8.84
C29H18Cl2	437.37	436.0785	79.64	4.15	16.21
C29H19Cl1	402.93	402.1175	86.45	4.75	8.80
C29H20Cl2	439.39	438.0942	79.27	4.59	16.14
C29H21Cl1	404.94	404.1332	86.02	5.23	8.76
C29H22Cl2	441.40	440.1098	78.91	5.02	16.06
C29H23Cl1	406.96	406.1488	85.59	5.70	8.71
C29H24Cl2	443.42	442.1255	78.55	5.46	15.99
C29H25Cl1	408.98	408.1645	85.17	6.16	8.67
C29H26Cl2	445.44	444.1411	78.20	5.88	15.92
C29H27Cl1	410.99	410.1801	84.75	6.62	8.63
C29H28Cl2	447.45	446.1568	77.85	6.31	15.85
C29H29Cl1	413.01	412.1958	84.34	7.08	8.58
C29H30Cl2	449.47	448.1724	77.50	6.73	15.78
C29H31Cl1	415.02	414.2114	83.93	7.53	8.54
C29H32Cl2	451.48	450.1881	77.15	7.14	15.71
C29H33Cl1	417.04	416.2271	83.52	7.98	8.50
C29H34Cl2	453.50	452.2037	76.81	7.56	15.64
C29H35Cl1	419.06	418.2427	83.12	8.42	8.46
C29H36Cl2	455.52	454.2194	76.47	7.97	15.57
C29H37Cl1	421.07	420.2584	82.72	8.86	8.42
C29H38Cl2	457.53	456.2350	76.13	8.37	15.50
C29H39Cl1	423.09	422.2740	82.33	9.29	8.38
C29H40Cl2	459.55	458.2507	75.80	8.77	15.43
C29H41Cl1	425.10	424.2897	81.94	9.72	8.34

C29H42Cl2	461.56	460.2663	75.47	9.17	15.36
C29H43Cl1	427.12	426.3053	81.55	10.15	8.30
C29H44Cl2	463.58	462.2820	75.14	9.57	15.30
C29H45Cl1	429.13	428.3210	81.17	10.57	8.26
C29H46Cl2	465.60	464.2976	74.81	9.96	15.23
C29H47Cl1	431.15	430.3366	80.79	10.99	8.22
C29H48Cl2	467.61	466.3133	74.49	10.35	15.16
C29H49Cl1	433.17	432.3523	80.41	11.40	8.18
C29H50Cl2	469.63	468.3289	74.17	10.73	15.10
C29H51Cl1	435.18	434.3679	80.04	11.81	8.15
C29H52Cl2	471.64	470.3446	73.85	11.11	15.03
C29H53Cl1	437.20	436.3836	79.67	12.22	8.11
C29H54Cl2	473.66	472.3602	73.54	11.49	14.97
C29H55Cl1	439.21	438.3992	79.31	12.62	8.07
C29H56Cl2	475.68	474.3759	73.23	11.87	14.91
C29H57Cl1	441.23	440.4149	78.94	13.02	8.04
C29H58Cl2	477.69	476.3915	72.92	12.24	14.84
C29H59Cl1	443.25	442.4305	78.58	13.42	8.00
C30H15Cl1	410.91	410.0862	87.69	3.68	8.63
C30H16Cl2	447.37	446.0629	80.55	3.60	15.85
C30H17Cl1	412.92	412.1019	87.26	4.15	8.59
C30H18Cl2	449.38	448.0785	80.18	4.04	15.78
C30H19Cl1	414.94	414.1175	86.84	4.62	8.54
C30H20Cl2	451.40	450.0942	79.83	4.47	15.71
C30H21Cl1	416.95	416.1332	86.42	5.08	8.50
C30H22Cl2	453.42	452.1098	79.47	4.89	15.64
C30H23Cl1	418.97	418.1488	86.00	5.53	8.46
C30H24Cl2	455.43	454.1255	79.12	5.31	15.57
C30H25Cl1	420.99	420.1645	85.59	5.99	8.42
C30H26Cl2	457.45	456.1411	78.77	5.73	15.50
C30H27Cl1	423.00	422.1801	85.18	6.43	8.38
C30H28Cl2	459.46	458.1568	78.43	6.14	15.43
C30H29Cl1	425.02	424.1958	84.78	6.88	8.34
C30H30Cl2	461.48	460.1724	78.08	6.55	15.36
C30H31Cl1	427.03	426.2114	84.38	7.32	8.30
C30H32Cl2	463.50	462.1881	77.74	6.96	15.30
C30H33Cl1	429.05	428.2271	83.98	7.75	8.26
C30H34Cl2	465.51	464.2037	77.41	7.36	15.23
C30H35Cl1	431.07	430.2427	83.59	8.18	8.22
C30H36Cl2	467.53	466.2194	77.07	7.76	15.17
C30H37Cl1	433.08	432.2584	83.20	8.61	8.19
C30H38Cl2	469.54	468.2350	76.74	8.16	15.10
C30H39Cl1	435.10	434.2740	82.82	9.03	8.15
C30H40Cl2	471.56	470.2507	76.41	8.55	15.04
C30H41Cl1	437.11	436.2897	82.43	9.45	8.11
C30H42Cl2	473.58	472.2663	76.09	8.94	14.97
C30H43Cl1	439.13	438.3053	82.06	9.87	8.07
C30H44Cl2	475.59	474.2820	75.77	9.33	14.91
C30H45Cl1	441.15	440.3210	81.68	10.28	8.04
C30H46Cl2	477.61	476.2976	75.45	9.71	14.85
C30H47Cl1	443.16	442.3366	81.31	10.69	8.00
C30H48Cl2	479.62	478.3133	75.13	10.09	14.78
C30H49Cl1	445.18	444.3523	80.94	11.09	7.96
C30H50Cl2	481.64	480.3289	74.81	10.46	14.72
C30H51Cl1	447.19	446.3679	80.58	11.50	7.93
C30H52Cl2	483.65	482.3446	74.50	10.84	14.66
C30H53Cl1	449.21	448.3836	80.22	11.89	7.89
C30H54Cl2	485.67	484.3602	74.19	11.21	14.60
C30H55Cl1	451.23	450.3992	79.86	12.29	7.86
C30H56Cl2	487.69	486.3759	73.89	11.57	14.54
C30H57Cl1	453.24	452.4149	79.50	12.68	7.82
C30H58Cl2	489.70	488.3915	73.58	11.94	14.48
C30H59Cl1	455.26	454.4305	79.15	13.06	7.79
C30H60Cl2	491.72	490.4072	73.28	12.30	14.42
C30H61Cl1	457.27	456.4462	78.80	13.45	7.75

3. CARBON, HYDROGEN AND FLUORINE

3. KOHLENSTOFF, WASSERSTOFF UND FLUOR

3. CARBONE, HYDROGENE ET FLUOR

MOLECULAR WEIGHTS AND PERCENTAGE COMPOSITIONS OF COMPOUNDS
CONTAINING CARBON, HYDROGEN AND FLUORINE

FORMULA	MOL. WEIGHT	ISOMOL. WEIGHT	PERCENT C	PERCENT H	PERCENT F
C 1H 0F2	50.01	49.9968	24.02	0	75.98
C 1H 1F1	32.02	32.0062	37.51	3.15	59.34
C 1H 2F2	52.02	52.0125	23.09	3.88	73.04
C 1H 3F1	34.03	34.0219	35.29	8.89	55.82
C 2H 1F1	44.03	44.0062	54.56	2.29	43.15
C 2H 2F2	64.04	64.0125	37.51	3.15	59.34
C 2H 3F1	46.04	46.0219	52.17	6.57	41.26
C 2H 4F2	66.05	66.0281	36.37	6.10	57.53
C 2H 5F1	48.06	48.0375	49.98	10.49	39.53
C 3H 1F1	56.04	56.0062	64.30	1.80	33.90
C 3H 2F2	76.05	76.0125	47.38	2.65	49.97
C 3H 3F1	58.06	58.0219	62.07	5.21	32.72
C 3H 4F2	78.06	78.0281	46.16	5.16	48.68
C 3H 5F1	60.07	60.0375	59.98	8.39	31.63
C 3H 6F2	80.08	80.0438	45.00	7.55	47.45
C 3H 7F1	62.09	62.0532	58.04	11.36	30.60
C 4H 2F2	88.06	88.0125	54.56	2.29	43.15
C 4H 3F1	70.07	70.0219	68.57	4.32	27.11
C 4H 4F2	90.07	90.0281	53.34	4.48	42.18
C 4H 5F1	72.08	72.0375	66.65	6.99	26.36
C 4H 6F2	92.09	92.0438	52.17	6.57	41.26
C 4H 7F1	74.10	74.0532	64.84	9.52	25.64
C 4H 8F2	94.11	94.0594	51.05	8.57	40.38
C 4H 9F1	76.11	76.0688	63.12	11.92	24.96
C 5H 2F2	100.07	100.0125	60.01	2.01	37.97
C 5H 3F1	82.08	82.0219	73.17	3.68	23.15
C 5H 4F2	102.08	102.0281	58.83	3.95	37.22
C 5H 5F1	84.09	84.0375	71.42	5.99	22.59
C 5H 6F2	104.10	104.0438	57.69	5.81	36.50
C 5H 7F1	86.11	86.0532	69.74	8.19	22.06
C 5H 8F2	106.12	106.0594	56.59	7.60	35.81
C 5H 9F1	88.13	88.0688	68.15	10.29	21.56
C 5H10F2	108.13	108.0751	55.54	9.32	35.14
C 5H11F1	90.14	90.0845	66.62	12.30	21.08
C 6H 3F1	94.09	94.0219	76.59	3.21	20.19
C 6H 4F2	114.10	114.0281	63.16	3.53	33.30
C 6H 5F1	96.11	96.0375	74.99	5.24	19.77
C 6H 6F2	116.11	116.0438	62.07	5.21	32.72
C 6H 7F1	98.12	98.0532	73.45	7.19	19.36
C 6H 8F2	118.13	118.0594	61.01	6.83	32.17
C 6H 9F1	100.14	100.0688	71.97	9.06	18.97
C 6H10F2	120.14	120.0751	59.98	8.39	31.63
C 6H11F1	102.15	102.0845	70.55	10.85	18.60
C 6H12F2	122.16	122.0907	58.99	9.90	31.10
C 6H13F1	104.17	104.1001	69.18	12.58	18.24
C 7H 3F1	106.10	106.0219	79.24	2.85	17.91
C 7H 4F2	126.11	126.0281	66.67	3.20	30.13
C 7H 5F1	108.12	108.0375	77.77	4.66	17.57
C 7H 6F2	128.12	128.0438	65.62	4.72	29.66
C 7H 7F1	110.13	110.0532	76.34	6.41	17.25

FORMULA	MOL. WEIGHT	ISOMOL. WEIGHT	PERCENT C	PERCENT H	PERCENT F
C 7H 8F2	130.14	130.0594	64.61	6.20	29.20
C 7H 9F1	112.15	112.0688	74.97	8.09	16.94
C 7H10F2	132.15	132.0751	63.62	7.63	28.75
C 7H11F1	114.16	114.0845	73.65	9.71	16.64
C 7H12F2	134.17	134.0907	62.67	9.02	28.32
C 7H13F1	116.18	116.1001	72.37	11.28	16.35
C 7H14F2	136.19	136.1063	61.74	10.36	27.90
C 7H15F1	118.20	118.1158	71.13	12.79	16.07
C 8H 4F2	138.12	138.0281	69.57	2.92	27.51
C 8H 5F1	120.13	120.0375	79.99	4.20	15.82
C 8H 6F2	140.13	140.0438	68.57	4.32	27.11
C 8H 7F1	122.14	122.0532	78.67	5.78	15.55
C 8H 8F2	142.15	142.0594	67.60	5.67	26.73
C 8H 9F1	124.16	124.0688	77.39	7.31	15.30
C 8H10F2	144.17	144.0751	66.65	6.99	26.36
C 8H11F1	126.18	126.0845	76.16	8.79	15.06
C 8H12F2	146.18	146.0907	65.73	8.27	25.99
C 8H13F1	128.19	128.1001	74.96	10.22	14.82
C 8H14F2	148.20	148.1063	64.84	9.52	25.64
C 8H15F1	130.21	130.1158	73.80	11.61	14.59
C 8H16F2	150.21	150.1220	63.97	10.74	25.30
C 8H17F1	132.22	132.1314	72.67	12.96	14.37
C 9H 4F2	150.13	150.0281	72.00	2.69	25.31
C 9H 5F1	132.14	132.0375	81.81	3.81	14.38
C 9H 6F2	152.14	152.0438	71.05	3.98	24.97
C 9H 7F1	134.15	134.0532	80.58	5.26	14.16
C 9H 8F2	154.16	154.0594	70.12	5.23	24.65
C 9H 9F1	136.17	136.0688	79.39	6.66	13.95
C 9H10F2	156.18	156.0751	69.22	6.45	24.33
C 9H11F1	138.19	138.0845	78.23	8.02	13.75
C 9H12F2	158.19	158.0907	68.33	7.65	24.02
C 9H13F1	140.20	140.1001	77.10	9.35	13.55
C 9H14F2	160.21	160.1063	67.47	8.81	23.72
C 9H15F1	142.22	142.1158	76.01	10.63	13.36
C 9H16F2	162.22	162.1220	66.64	9.94	23.42
C 9H17F1	144.23	144.1314	74.95	11.88	13.17
C 9H18F2	164.24	164.1376	65.82	11.05	23.13
C 9H19F1	146.25	146.1471	73.91	13.09	12.99
C10H 5F1	144.15	144.0375	83.32	3.50	13.18
C10H 6F2	164.16	164.0438	73.17	3.68	23.15
C10H 7F1	146.17	146.0532	82.17	4.83	13.00
C10H 8F2	166.17	166.0594	72.28	4.85	22.87
C10H 9F1	148.18	148.0688	81.06	6.12	12.82
C10H10F2	168.19	168.0751	71.42	5.99	22.59
C10H11F1	150.20	150.0845	79.97	7.38	12.65
C10H12F2	170.20	170.0907	70.57	7.11	22.32
C10H13F1	152.21	152.1001	78.91	8.61	12.48
C10H14F2	172.22	172.1063	69.74	8.19	22.06
C10H15F1	154.23	154.1158	77.88	9.80	12.32
C10H16F2	174.24	174.1220	68.94	9.26	21.81

C10H17F1	156.25	156.1314	76.87	10.97	12.16
C10H18F2	176.25	176.1376	68.15	10.29	21.56
C10H19F1	158.26	158.1471	75.89	12.10	12.00
C10H20F2	178.27	178.1533	67.38	11.31	21.31
C10H21F1	160.28	160.1627	74.94	13.21	11.85
C11H 5F1	156.16	156.0375	84.61	3.23	12.17
C11H 6F2	176.17	176.0438	75.00	3.43	21.57
C11H 7F1	158.18	158.0532	83.53	4.46	12.01
C11H 8F2	178.18	178.0594	74.15	4.53	21.32
C11H 9F1	160.19	160.0688	82.48	5.66	11.86
C11H10F2	180.20	180.0751	73.32	5.59	21.09
C11H11F1	162.21	162.0845	81.45	6.84	11.71
C11H12F2	182.22	182.0907	72.51	6.64	20.85
C11H13F1	164.22	164.1001	80.45	7.98	11.57
C11H14F2	184.23	184.1063	71.72	7.66	20.62
C11H15F1	166.24	166.1158	79.48	9.09	11.43
C11H16F2	186.25	186.1220	70.94	8.66	20.40
C11H17F1	168.26	168.1314	78.52	10.18	11.29
C11H18F2	188.26	188.1376	70.18	9.64	20.18
C11H19F1	170.27	170.1471	77.59	11.25	11.16
C11H20F2	190.28	190.1533	69.44	10.59	19.97
C11H21F1	172.29	172.1627	76.69	12.29	11.03
C11H22F2	192.29	192.1689	68.71	11.53	19.76
C11H23F1	174.30	174.1784	75.80	13.30	10.90
C12H 6F2	188.18	188.0438	76.59	3.21	20.19
C12H 7F1	170.19	170.0532	84.69	4.15	11.16
C12H 8F2	190.19	190.0594	75.78	4.24	19.98
C12H 9F1	172.20	172.0688	83.70	5.27	11.03
C12H10F2	192.21	192.0751	74.99	5.24	19.77
C12H11F1	174.22	174.0845	82.73	6.36	10.90
C12H12F2	194.23	194.0907	74.21	6.23	19.56
C12H13F1	176.24	176.1001	81.78	7.44	10.78
C12H14F2	196.24	196.1063	73.45	7.19	19.36
C12H15F1	178.25	178.1158	80.86	8.48	10.66
C12H16F2	198.26	198.1220	72.70	8.13	19.17
C12H17F1	180.27	180.1314	79.96	9.51	10.54
C12H18F2	200.27	200.1376	71.97	9.06	18.97
C12H19F1	182.28	182.1471	79.07	10.51	10.42
C12H20F2	202.29	202.1533	71.25	9.97	18.78
C12H21F1	184.30	184.1627	78.21	11.49	10.31
C12H22F2	204.31	204.1689	70.55	10.85	18.60
C12H23F1	186.32	186.1784	77.36	12.44	10.20
C12H24F2	206.32	206.1846	69.86	11.73	18.42
C12H25F1	188.33	188.1940	76.53	13.38	10.09
C13H 6F2	200.19	200.0438	78.00	3.02	18.98
C13H 7F1	182.20	182.0532	85.70	3.87	10.43
C13H 8F2	202.21	202.0594	77.22	3.99	18.79
C13H 9F1	184.22	184.0688	84.76	4.92	10.31
C13H10F2	204.22	204.0751	76.46	4.94	18.61
C13H11F1	186.23	186.0845	83.84	5.95	10.20
C13H12F2	206.24	206.0907	75.71	5.86	18.42
C13H13F1	188.25	188.1001	82.95	6.96	10.09
C13H14F2	208.25	208.1063	74.98	6.78	18.25
C13H15F1	190.26	190.1158	82.07	7.95	9.99
C13H16F2	210.27	210.1220	74.26	7.67	18.07
C13H17F1	192.28	192.1314	81.21	8.91	9.88
C13H18F2	212.29	212.1376	73.55	8.55	17.90
C13H19F1	194.29	194.1471	80.36	9.86	9.78
C13H20F2	214.30	214.1533	72.86	9.41	17.73
C13H21F1	196.31	196.1627	79.54	10.78	9.68

C13H22F2	216.32	216.1689	72.18	10.25	17.57
C13H23F1	198.33	198.1784	78.73	11.69	9.58
C13H24F2	218.33	218.1846	71.52	11.08	17.40
C13H25F1	200.34	200.1940	77.94	12.58	9.48
C13H26F2	220.35	220.2002	70.86	11.89	17.24
C13H27F1	202.36	202.2097	77.16	13.45	9.39
C14H 7F1	194.21	194.0532	86.58	3.63	9.78
C14H 8F2	214.22	214.0594	78.50	3.76	17.74
C14H 9F1	196.23	196.0688	85.70	4.62	9.68
C14H10F2	216.23	216.0751	77.77	4.66	17.57
C14H11F1	198.24	198.0845	84.82	5.59	9.58
C14H12F2	218.25	218.0907	77.05	5.54	17.41
C14H13F1	200.26	200.1001	83.97	6.54	9.49
C14H14F2	220.26	220.1063	76.34	6.41	17.25
C14H15F1	202.27	202.1158	83.13	7.47	9.39
C14H16F2	222.28	222.1220	75.65	7.26	17.09
C14H17F1	204.29	204.1314	82.31	8.39	9.30
C14H18F2	224.30	224.1376	74.97	8.09	16.94
C14H19F1	206.31	206.1471	81.51	9.28	9.21
C14H20F2	226.31	226.1533	74.30	8.91	16.79
C14H21F1	208.32	208.1627	80.72	10.16	9.12
C14H22F2	228.33	228.1689	73.65	9.71	16.64
C14H23F1	210.34	210.1784	79.95	11.02	9.03
C14H24F2	230.34	230.1846	73.00	10.50	16.50
C14H25F1	212.35	212.1940	79.19	11.87	8.95
C14H26F2	232.36	232.2002	72.37	11.28	16.35
C14H27F1	214.37	214.2097	78.44	12.70	8.86
C14H28F2	234.38	234.2159	71.75	12.04	16.21
C14H29F1	216.39	216.2253	77.71	13.51	8.78
C15H 7F1	206.22	206.0532	87.37	3.42	9.21
C15H 8F2	226.23	226.0594	79.64	3.56	16.80
C15H 9F1	208.24	208.0688	86.52	4.36	9.12
C15H10F2	228.24	228.0751	78.94	4.42	16.65
C15H11F1	210.25	210.0845	85.69	5.27	9.04
C15H12F2	230.26	230.0907	78.25	5.25	16.50
C15H13F1	212.27	212.1001	84.88	6.17	8.95
C15H14F2	232.28	232.1063	77.57	6.08	16.36
C15H15F1	214.29	214.1158	84.08	7.06	8.87
C15H16F2	234.29	234.1220	76.90	6.88	16.22
C15H17F1	216.30	216.1314	83.29	7.92	8.78
C15H18F2	236.31	236.1376	76.24	7.68	16.08
C15H19F1	218.32	218.1471	82.53	8.77	8.70
C15H20F2	238.32	238.1533	75.60	8.46	15.94
C15H21F1	220.33	220.1627	81.77	9.61	8.62
C15H22F2	240.34	240.1689	74.96	9.23	15.81
C15H23F1	222.35	222.1784	81.03	10.43	8.54
C15H24F2	242.36	242.1846	74.34	9.98	15.68
C15H25F1	224.36	224.1940	80.30	11.23	8.47
C15H26F2	244.37	244.2002	73.73	10.72	15.55
C15H27F1	226.38	226.2097	79.59	12.02	8.39
C15H28F2	246.39	246.2159	73.12	11.45	15.42
C15H29F1	228.40	228.2253	78.88	12.80	8.32
C15H30F2	248.40	248.2315	72.53	12.17	15.30
C15H31F1	230.41	230.2410	78.19	13.56	8.25
C16H 8F2	238.24	238.0594	80.67	3.38	15.95
C16H 9F1	220.25	220.0688	87.26	4.12	8.63
C16H10F2	240.25	240.0751	79.99	4.20	15.82
C16H11F1	222.26	222.0845	86.46	4.99	8.55
C16H12F2	242.27	242.0907	79.32	4.99	15.68
C16H13F1	224.28	224.1001	85.69	5.84	8.47

C16H14F2	244.29	244.1063	78.67	5.78	15.55
C16H15F1	226.30	226.1158	84.92	6.68	8.40
C16H16F2	246.30	246.1220	78.03	6.55	15.43
C16H17F1	228.31	228.1314	84.17	7.51	8.32
C16H18F2	248.32	248.1376	77.39	7.31	15.30
C16H19F1	230.33	230.1471	83.44	8.31	8.25
C16H20F2	250.33	250.1533	76.77	8.05	15.18
C16H21F1	232.34	232.1627	82.71	9.11	8.18
C16H22F2	252.35	252.1689	76.16	8.79	15.06
C16H23F1	234.36	234.1784	82.00	9.89	8.11
C16H24F2	254.37	254.1846	75.55	9.51	14.94
C16H25F1	236.38	236.1940	81.30	10.66	8.04
C16H26F2	256.38	256.2002	74.96	10.22	14.82
C16H27F1	238.39	238.2097	80.61	11.42	7.97
C16H28F2	258.40	258.2159	74.37	10.92	14.70
C16H29F1	240.41	240.2253	79.94	12.16	7.90
C16H30F2	260.41	260.2315	73.80	11.61	14.59
C16H31F1	242.42	242.2410	79.27	12.89	7.84
C16H32F2	262.43	262.2472	73.23	12.29	14.48
C16H33F1	244.44	244.2566	78.62	13.61	7.77
C17H 8F2	250.25	250.0594	81.59	3.22	15.18
C17H 9F1	232.26	232.0688	87.91	3.91	8.18
C17H10F2	252.27	252.0751	80.94	4.00	15.06
C17H11F1	234.28	234.0845	87.16	4.73	8.11
C17H12F2	254.28	254.0907	80.30	4.76	14.94
C17H13F1	236.29	236.1001	86.41	5.55	8.04
C17H14F2	256.30	256.1063	79.67	5.51	14.83
C17H15F1	238.31	238.1158	85.68	6.34	7.97
C17H16F2	258.31	258.1220	79.05	6.24	14.71
C17H17F1	240.32	240.1314	84.96	7.13	7.91
C17H18F2	260.33	260.1376	78.43	6.97	14.60
C17H19F1	242.34	242.1471	84.26	7.90	7.84
C17H20F2	262.35	262.1533	77.83	7.68	14.48
C17H21F1	244.36	244.1627	83.56	8.66	7.77
C17H22F2	264.36	264.1689	77.24	8.39	14.37
C17H23F1	246.37	246.1784	82.88	9.41	7.71
C17H24F2	266.38	266.1846	76.65	9.08	14.26
C17H25F1	248.39	248.1940	82.21	10.15	7.65
C17H26F2	268.39	268.2002	76.08	9.76	14.16
C17H27F1	250.40	250.2097	81.54	10.87	7.59
C17H28F2	270.41	270.2159	75.51	10.44	14.05
C17H29F1	252.42	252.2253	80.89	11.58	7.53
C17H30F2	272.43	272.2315	74.95	11.10	13.95
C17H31F1	254.44	254.2410	80.25	12.28	7.47
C17H32F2	274.44	274.2472	74.40	11.75	13.85
C17H33F1	256.45	256.2566	79.62	12.97	7.41
C17H34F2	276.46	276.2628	73.86	12.40	13.74
C17H35F1	258.47	258.2723	79.00	13.65	7.35
C18H 9F1	244.27	244.0688	88.51	3.71	7.78
C18H10F2	264.28	264.0751	81.81	3.81	14.38
C18H11F1	246.29	246.0845	87.78	4.50	7.71
C18H12F2	266.29	266.0907	81.19	4.54	14.27
C18H13F1	248.30	248.1001	87.07	5.28	7.65
C18H14F2	268.31	268.1063	80.58	5.26	14.16
C18H15F1	250.32	250.1158	86.37	6.04	7.59
C18H16F2	270.33	270.1220	79.98	5.97	14.06
C18H17F1	252.33	252.1314	85.68	6.79	7.53
C18H18F2	272.34	272.1376	79.39	6.66	13.95
C18H19F1	254.35	254.1471	85.00	7.53	7.47
C18H20F2	274.36	274.1533	78.80	7.35	13.85

C18H21F1	256.37	256.1627	84.33	8.26	7.41
C18H22F2	276.37	276.1689	78.23	8.02	13.75
C18H23F1	258.38	258.1784	83.67	8.97	7.35
C18H24F2	278.39	278.1846	77.66	8.69	13.65
C18H25F1	260.40	260.1940	83.03	9.68	7.30
C18H26F2	280.40	280.2002	77.10	9.35	13.55
C18H27F1	262.41	262.2097	82.39	10.37	7.24
C18H28F2	282.42	282.2159	76.55	9.99	13.45
C18H29F1	264.43	264.2253	81.76	11.05	7.18
C18H30F2	284.44	284.2315	76.01	10.63	13.36
C18H31F1	266.45	266.2410	81.14	11.73	7.13
C18H32F2	286.45	286.2472	75.48	11.26	13.26
C18H33F1	268.46	268.2566	80.53	12.39	7.08
C18H34F2	288.47	288.2628	74.95	11.88	13.17
C18H35F1	270.48	270.2723	79.93	13.04	7.02
C18H36F2	290.48	290.2785	74.43	12.49	13.08
C18H37F1	272.49	272.2879	79.34	13.69	6.97
C19H 9F1	256.28	256.0688	89.05	3.54	7.41
C19H10F2	276.29	276.0751	82.60	3.65	13.75
C19H11F1	258.30	258.0845	88.35	4.29	7.36
C19H12F2	278.30	278.0907	82.00	4.35	13.65
C19H13F1	260.31	260.1001	87.67	5.03	7.30
C19H14F2	280.32	280.1063	81.41	5.03	13.55
C19H15F1	262.33	262.1158	86.99	5.76	7.24
C19H16F2	282.34	282.1220	80.83	5.71	13.46
C19H17F1	264.35	264.1314	86.33	6.48	7.19
C19H18F2	284.35	284.1376	80.26	6.38	13.36
C19H19F1	266.36	266.1471	85.68	7.19	7.13
C19H20F2	286.37	286.1533	79.69	7.04	13.27
C19H21F1	268.38	268.1627	85.03	7.89	7.08
C19H22F2	288.38	288.1689	79.13	7.69	13.18
C19H23F1	270.39	270.1784	84.40	8.57	7.03
C19H24F2	290.40	290.1846	78.59	8.33	13.08
C19H25F1	272.41	272.1940	83.78	9.25	6.97
C19H26F2	292.42	292.2002	78.04	8.96	12.99
C19H27F1	274.43	274.2097	83.16	9.92	6.92
C19H28F2	294.43	294.2159	77.51	9.59	12.91
C19H29F1	276.44	276.2253	82.55	10.57	6.87
C19H30F2	296.45	296.2315	76.98	10.20	12.82
C19H31F1	278.46	278.2410	81.96	11.22	6.82
C19H32F2	298.46	298.2472	76.46	10.81	12.73
C19H33F1	280.47	280.2566	81.37	11.86	6.77
C19H34F2	300.48	300.2628	75.95	11.41	12.65
C19H35F1	282.49	282.2723	80.79	12.49	6.73
C19H36F2	302.50	302.2785	75.44	12.00	12.56
C19H37F1	284.51	284.2879	80.21	13.11	6.68
C19H38F2	304.51	304.2941	74.94	12.58	12.48
C19H39F1	286.52	286.3036	79.65	13.72	6.63
C20H10F2	288.30	288.0751	83.32	3.50	13.18
C20H11F1	270.31	270.0845	88.87	4.10	7.03
C20H12F2	290.32	290.0907	82.75	4.17	13.09
C20H13F1	272.33	272.1001	88.21	4.81	6.98
C20H14F2	292.33	292.1063	82.17	4.83	13.00
C20H15F1	274.34	274.1158	87.56	5.51	6.93
C20H16F2	294.35	294.1220	81.61	5.48	12.91
C20H17F1	276.36	276.1314	86.92	6.20	6.87
C20H18F2	296.36	296.1376	81.06	6.12	12.82
C20H19F1	278.37	278.1471	86.30	6.88	6.82
C20H20F2	298.38	298.1533	80.51	6.76	12.73
C20H21F1	280.39	280.1627	85.67	7.55	6.78

C20H22F2	300.40	300.1689	79.97	7.38	12.65
C20H23F1	282.40	282.1784	85.06	8.21	6.73
C20H24F2	302.41	302.1846	79.44	8.00	12.56
C20H25F1	284.42	284.1940	84.46	8.86	6.68
C20H26F2	304.43	304.2002	78.91	8.61	12.48
C20H27F1	286.44	286.2097	83.87	9.50	6.63
C20H28F2	306.44	306.2159	78.39	9.21	12.40
C20H29F1	288.45	288.2253	83.28	10.13	6.59
C20H30F2	308.46	308.2315	77.88	9.80	12.32
C20H31F1	290.47	290.2410	82.70	10.76	6.54
C20H32F2	310.47	310.2472	77.37	10.39	12.24
C20H33F1	292.48	292.2566	82.13	11.37	6.50
C20H34F2	312.49	312.2628	76.87	10.97	12.16
C20H35F1	294.50	294.2723	81.57	11.98	6.45
C20H36F2	314.51	314.2785	76.38	11.54	12.08
C20H37F1	296.52	296.2879	81.02	12.58	6.41
C20H38F2	316.52	316.2941	75.89	12.10	12.00
C20H39F1	298.53	298.3036	80.47	13.17	6.36
C20H40F2	318.54	318.3098	75.41	12.66	11.93
C20H41F1	300.55	300.3192	79.93	13.75	6.32
C21H10F2	300.31	300.0751	83.99	3.36	12.65
C21H11F1	282.32	282.0845	89.34	3.93	6.73
C21H12F2	302.33	302.0907	83.43	4.00	12.57
C21H13F1	284.34	284.1001	88.71	4.61	6.68
C21H14F2	304.34	304.1063	82.88	4.64	12.48
C21H15F1	286.35	286.1158	88.09	5.28	6.63
C21H16F2	306.36	306.1220	82.33	5.26	12.40
C21H17F1	288.37	288.1314	87.47	5.94	6.59
C21H18F2	308.37	308.1376	81.79	5.88	12.32
C21H19F1	290.38	290.1471	86.86	6.60	6.54
C21H20F2	310.39	310.1533	81.26	6.49	12.24
C21H21F1	292.40	292.1627	86.26	7.24	6.50
C21H22F2	312.41	312.1689	80.74	7.10	12.16
C21H23F1	294.42	294.1784	85.67	7.87	6.45
C21H24F2	314.42	314.1846	80.22	7.69	12.08
C21H25F1	296.43	296.1940	85.09	8.50	6.41
C21H26F2	316.44	316.2002	79.71	8.28	12.01
C21H27F1	298.45	298.2097	84.52	9.12	6.37
C21H28F2	318.45	318.2159	79.21	8.86	11.93
C21H29F1	300.46	300.2253	83.95	9.73	6.32
C21H30F2	320.47	320.2315	78.71	9.44	11.86
C21H31F1	302.48	302.2410	83.39	10.33	6.28
C21H32F2	322.49	322.2472	78.22	10.00	11.78
C21H33F1	304.50	304.2566	82.84	10.92	6.24
C21H34F2	324.50	324.2628	77.73	10.56	11.71
C21H35F1	306.51	306.2723	82.29	11.51	6.20
C21H36F2	326.52	326.2785	77.25	11.11	11.64
C21H37F1	308.53	308.2879	81.75	12.09	6.16
C21H38F2	328.53	328.2941	76.78	11.66	11.57
C21H39F1	310.54	310.3036	81.22	12.66	6.12
C21H40F2	330.55	330.3098	76.31	12.20	11.50
C21H41F1	312.56	312.3192	80.70	13.22	6.08
C21H42F2	332.57	332.3254	75.84	12.73	11.43
C21H43F1	314.58	314.3349	80.18	13.78	6.04
C22H11F1	294.33	294.0845	89.78	3.77	6.45
C22H12F2	314.34	314.0907	84.06	3.85	12.09
C22H13F1	296.35	296.1001	89.17	4.42	6.41
C22H14F2	316.35	316.1063	83.53	4.46	12.01
C22H15F1	298.36	298.1158	88.56	5.07	6.37
C22H16F2	318.37	318.1220	83.00	5.07	11.93

C22H17F1	300.38	300.1314	87.97	5.70	6.32
C22H18F2	320.39	320.1376	82.48	5.66	11.86
C22H19F1	302.40	302.1471	87.38	6.33	6.28
C22H20F2	322.40	322.1533	81.96	6.25	11.79
C22H21F1	304.41	304.1627	86.81	6.95	6.24
C22H22F2	324.42	324.1689	81.45	6.84	11.71
C22H23F1	306.43	306.1784	86.23	7.57	6.20
C22H24F2	326.43	326.1846	80.95	7.41	11.64
C22H25F1	308.44	308.1940	85.67	8.17	6.16
C22H26F2	328.45	328.2002	80.45	7.98	11.57
C22H27F1	310.46	310.2097	85.11	8.77	6.12
C22H28F2	330.47	330.2159	79.96	8.54	11.50
C22H29F1	312.47	312.2253	84.57	9.35	6.08
C22H30F2	332.48	332.2315	79.48	9.09	11.43
C22H31F1	314.49	314.2410	84.02	9.94	6.04
C22H32F2	334.50	334.2472	79.00	9.64	11.36
C22H33F1	316.51	316.2566	83.49	10.51	6.00
C22H34F2	336.51	336.2628	78.52	10.18	11.29
C22H35F1	318.52	318.2723	82.96	11.08	5.96
C22H36F2	338.53	338.2785	78.06	10.72	11.22
C22H37F1	320.54	320.2879	82.44	11.64	5.93
C22H38F2	340.54	340.2941	77.59	11.25	11.16
C22H39F1	322.55	322.3036	81.92	12.19	5.89
C22H40F2	342.56	342.3098	77.14	11.77	11.09
C22H41F1	324.57	324.3192	81.41	12.73	5.85
C22H42F2	344.58	344.3254	76.69	12.29	11.03
C22H43F1	326.59	326.3349	80.91	13.27	5.82
C22H44F2	346.59	346.3411	76.24	12.80	10.96
C22H45F1	328.60	328.3505	80.41	13.80	5.78
C23H11F1	306.34	306.0845	90.18	3.62	6.20
C23H12F2	326.35	326.0907	84.65	3.71	11.64
C23H13F1	308.36	308.1001	89.59	4.25	6.16
C23H14F2	328.36	328.1063	84.13	4.30	11.57
C23H15F1	310.37	310.1158	89.01	4.87	6.12
C23H16F2	330.38	330.1220	83.62	4.88	11.50
C23H17F1	312.39	312.1314	88.43	5.49	6.08
C23H18F2	332.40	332.1376	83.11	5.46	11.43
C23H19F1	314.41	314.1471	87.87	6.09	6.04
C23H20F2	334.41	334.1533	82.61	6.03	11.36
C23H21F1	316.42	316.1627	87.31	6.69	6.00
C23H22F2	336.43	336.1689	82.11	6.59	11.29
C23H23F1	318.44	318.1784	86.75	7.28	5.97
C23H24F2	338.44	338.1846	81.63	7.15	11.23
C23H25F1	320.45	320.1940	86.21	7.86	5.93
C23H26F2	340.46	340.2002	81.14	7.70	11.16
C23H27F1	322.47	322.2097	85.67	8.44	5.89
C23H28F2	342.48	342.2159	80.66	8.24	11.09
C23H29F1	324.49	324.2253	85.14	9.01	5.85
C23H30F2	344.49	344.2315	80.19	8.78	11.03
C23H31F1	326.50	326.2410	84.61	9.57	5.82
C23H32F2	346.51	346.2472	79.73	9.31	10.97
C23H33F1	328.52	328.2566	84.09	10.13	5.78
C23H34F2	348.52	348.2628	79.26	9.83	10.90
C23H35F1	330.53	330.2723	83.58	10.67	5.75
C23H36F2	350.54	350.2785	78.81	10.35	10.84
C23H37F1	332.55	332.2879	83.07	11.21	5.71
C23H38F2	352.56	352.2941	78.36	10.86	10.78
C23H39F1	334.57	334.3036	82.57	11.75	5.68
C23H40F2	354.57	354.3098	77.91	11.37	10.72
C23H41F1	336.58	336.3192	82.08	12.28	5.64

C23H42F2	356.59	356.3254	77.47	11.87	10.66
C23H43F1	338.60	338.3349	81.59	12.80	5.61
C23H44F2	358.60	358.3411	77.04	12.37	10.60
C23H45F1	340.61	340.3505	81.11	13.32	5.58
C23H46F2	360.62	360.3567	76.61	12.86	10.54
C23H47F1	342.63	342.3662	80.63	13.83	5.54
C24H12F2	338.36	338.0907	85.20	3.57	11.23
C24H13F1	320.37	320.1001	89.98	4.09	5.93
C24H14F2	340.38	340.1063	84.69	4.15	11.16
C24H15F1	322.39	322.1158	89.42	4.69	5.89
C24H16F2	342.39	342.1220	84.19	4.71	11.10
C24H17F1	324.40	324.1314	88.86	5.28	5.86
C24H18F2	344.41	344.1376	83.70	5.27	11.03
C24H19F1	326.42	326.1471	88.31	5.87	5.82
C24H20F2	346.42	346.1533	83.21	5.82	10.97
C24H21F1	328.43	328.1627	87.77	6.44	5.78
C24H22F2	348.44	348.1689	82.73	6.36	10.90
C24H23F1	330.45	330.1784	87.24	7.02	5.75
C24H24F2	350.46	350.1846	82.26	6.90	10.84
C24H25F1	332.47	332.1940	86.71	7.58	5.71
C24H26F2	352.47	352.2002	81.78	7.44	10.78
C24H27F1	334.48	334.2097	86.18	8.14	5.68
C24H28F2	354.49	354.2159	81.32	7.96	10.72
C24H29F1	336.50	336.2253	85.67	8.69	5.65
C24H30F2	356.50	356.2315	80.86	8.48	10.66
C24H31F1	338.51	338.2410	85.16	9.23	5.61
C24H32F2	358.52	358.2472	80.41	9.00	10.60
C24H33F1	340.53	340.2566	84.65	9.77	5.58
C24H34F2	360.54	360.2628	79.96	9.51	10.54
C24H35F1	342.54	342.2723	84.15	10.30	5.55
C24H36F2	362.55	362.2785	79.51	10.01	10.48
C24H37F1	344.56	344.2879	83.66	10.82	5.51
C24H38F2	364.57	364.2941	79.07	10.51	10.42
C24H39F1	346.58	346.3036	83.18	11.34	5.48
C24H40F2	366.58	366.3098	78.64	11.00	10.37
C24H41F1	348.59	348.3192	82.69	11.86	5.45
C24H42F2	368.60	368.3254	78.21	11.49	10.31
C24H43F1	350.61	350.3349	82.22	12.36	5.42
C24H44F2	370.62	370.3411	77.78	11.97	10.25
C24H45F1	352.62	352.3505	81.75	12.86	5.39
C24H46F2	372.63	372.3567	77.36	12.44	10.20
C24H47F1	354.64	354.3662	81.28	13.36	5.36
C24H48F2	374.65	374.3724	76.94	12.91	10.14
C24H49F1	356.66	356.3818	80.82	13.85	5.33
C25H12F2	350.37	350.0907	85.70	3.45	10.84
C25H13F1	332.38	332.1001	90.34	3.94	5.72
C25H14F2	352.39	352.1063	85.21	4.00	10.78
C25H15F1	334.40	334.1158	89.80	4.52	5.68
C25H16F2	354.40	354.1220	84.73	4.55	10.72
C25H17F1	336.41	336.1314	89.26	5.09	5.65
C25H18F2	356.42	356.1376	84.25	5.09	10.66
C25H19F1	338.43	338.1471	88.73	5.66	5.61
C25H20F2	358.43	358.1533	83.77	5.62	10.60
C25H21F1	340.44	340.1627	88.20	6.22	5.58
C25H22F2	360.45	360.1689	83.31	6.15	10.54
C25H23F1	342.46	342.1784	87.68	6.77	5.55
C25H24F2	362.47	362.1846	82.84	6.67	10.48
C25H25F1	344.48	344.1940	87.17	7.32	5.52
C25H26F2	364.48	364.2002	82.38	7.19	10.42
C25H27F1	346.49	346.2097	86.66	7.85	5.48
C25H28F2	366.50	366.2159	81.93	7.70	10.37
C25H29F1	348.51	348.2253	86.16	8.39	5.45
C25H30F2	368.51	368.2315	81.48	8.21	10.31
C25H31F1	350.52	350.2410	85.67	8.91	5.42
C25H32F2	370.53	370.2472	81.04	8.71	10.25
C25H33F1	352.54	352.2566	85.18	9.44	5.39
C25H34F2	372.55	372.2628	80.60	9.20	10.20
C25H35F1	354.56	354.2723	84.69	9.95	5.36
C25H36F2	374.56	374.2785	80.17	9.69	10.14
C25H37F1	356.57	356.2879	84.21	10.46	5.33
C25H38F2	376.58	376.2941	79.74	10.17	10.09
C25H39F1	358.59	358.3036	83.74	10.96	5.30
C25H40F2	378.59	378.3098	79.31	10.65	10.04
C25H41F1	360.60	360.3192	83.27	11.46	5.27
C25H42F2	380.61	380.3254	78.89	11.12	9.98
C25H43F1	362.62	362.3349	82.81	11.95	5.24
C25H44F2	382.63	382.3411	78.48	11.59	9.93
C25H45F1	364.64	364.3505	82.35	12.44	5.21
C25H46F2	384.64	384.3567	78.07	12.05	9.88
C25H47F1	366.65	366.3662	81.90	12.92	5.18
C25H48F2	386.66	386.3724	77.66	12.51	9.83
C25H49F1	368.67	368.3818	81.45	13.40	5.15
C25H50F2	388.67	388.3880	77.26	12.97	9.78
C25H51F1	370.68	370.3975	81.01	13.87	5.13
C26H13F1	344.39	344.1001	90.68	3.80	5.52
C26H14F2	364.40	364.1063	85.70	3.87	10.43
C26H15F1	346.41	346.1158	90.15	4.36	5.48
C26H16F2	366.41	366.1220	85.23	4.40	10.37
C26H17F1	348.42	348.1314	89.63	4.92	5.45
C26H18F2	368.43	368.1376	84.76	4.92	10.31
C26H19F1	350.44	350.1471	89.11	5.46	5.42
C26H20F2	370.45	370.1533	84.30	5.44	10.26
C26H21F1	352.46	352.1627	88.60	6.01	5.39
C26H22F2	372.46	372.1689	83.84	5.95	10.20
C26H23F1	354.47	354.1784	88.10	6.54	5.36
C26H24F2	374.48	374.1846	83.39	6.46	10.15
C26H25F1	356.49	356.1940	87.60	7.07	5.33
C26H26F2	376.49	376.2002	82.95	6.96	10.09
C26H27F1	358.50	358.2097	87.11	7.59	5.30
C26H28F2	378.51	378.2159	82.51	7.46	10.04
C26H29F1	360.52	360.2253	86.62	8.11	5.27
C26H30F2	380.53	380.2315	82.07	7.95	9.99
C26H31F1	362.54	362.2410	86.14	8.62	5.24
C26H32F2	382.54	382.2472	81.64	8.43	9.93
C26H33F1	364.55	364.2566	85.66	9.12	5.21
C26H34F2	384.56	384.2628	81.21	8.91	9.88
C26H35F1	366.57	366.2723	85.19	9.62	5.18
C26H36F2	386.57	386.2785	80.78	9.39	9.83
C26H37F1	368.58	368.2879	84.73	10.12	5.15
C26H38F2	388.59	388.2941	80.36	9.86	9.78
C26H39F1	370.60	370.3036	84.27	10.61	5.13
C26H40F2	390.61	390.3098	79.95	10.32	9.73
C26H41F1	372.62	372.3192	83.81	11.09	5.10
C26H42F2	392.62	392.3254	79.54	10.78	9.68
C26H43F1	374.63	374.3349	83.36	11.57	5.07
C26H44F2	394.64	394.3411	79.13	11.24	9.63
C26H45F1	376.65	376.3505	82.91	12.04	5.04
C26H46F2	396.65	396.3567	78.73	11.69	9.58
C26H47F1	378.66	378.3662	82.47	12.51	5.02
C26H48F2	398.67	398.3724	78.33	12.14	9.53

C26H49F1	380.68	380.3818	82.04	12.97	4.99
C26H50F2	400.69	400.3880	77.94	12.58	9.48
C26H51F1	382.69	382.3975	81.60	13.43	4.96
C26H52F2	402.70	402.4037	77.55	13.02	9.44
C26H53F1	384.71	384.4131	81.18	13.89	4.94
C27H13F1	356.40	356.1001	90.99	3.68	5.33
C27H14F2	376.41	376.1063	86.16	3.75	10.09
C27H15F1	358.42	358.1158	90.48	4.22	5.30
C27H16F2	378.43	378.1220	85.70	4.26	10.04
C27H17F1	360.43	360.1314	89.97	4.75	5.27
C27H18F2	380.44	380.1376	85.24	4.77	9.99
C27H19F1	362.45	362.1471	89.47	5.28	5.24
C27H20F2	382.46	382.1533	84.79	5.27	9.93
C27H21F1	364.47	364.1627	88.98	5.81	5.21
C27H22F2	384.47	384.1689	84.35	5.77	9.88
C27H23F1	366.48	366.1784	88.49	6.33	5.18
C27H24F2	386.49	386.1846	83.91	6.26	9.83
C27H25F1	368.50	368.1940	88.01	6.84	5.16
C27H26F2	388.51	388.2002	83.47	6.75	9.78
C27H27F1	370.51	370.2097	87.53	7.35	5.13
C27H28F2	390.52	390.2159	83.04	7.23	9.73
C27H29F1	372.53	372.2253	87.05	7.85	5.10
C27H30F2	392.54	392.2315	82.62	7.70	9.68
C27H31F1	374.55	374.2410	86.58	8.34	5.07
C27H32F2	394.55	394.2472	82.19	8.18	9.63
C27H33F1	376.56	376.2566	86.12	8.83	5.05
C27H34F2	396.57	396.2628	81.78	8.64	9.58
C27H35F1	378.58	378.2723	85.66	9.32	5.02
C27H36F2	398.58	398.2785	81.36	9.10	9.53
C27H37F1	380.59	380.2879	85.21	9.80	4.99
C27H38F2	400.60	400.2941	80.95	9.56	9.48
C27H39F1	382.61	382.3036	84.76	10.27	4.97
C27H40F2	402.62	402.3098	80.55	10.01	9.44
C27H41F1	384.63	384.3192	84.32	10.74	4.94
C27H42F2	404.63	404.3254	80.15	10.46	9.39
C27H43F1	386.64	386.3349	83.88	11.21	4.91
C27H44F2	406.65	406.3411	79.75	10.91	9.34
C27H45F1	388.66	388.3505	83.44	11.67	4.89
C27H46F2	408.66	408.3567	79.36	11.35	9.30
C27H47F1	390.67	390.3662	83.01	12.13	4.86
C27H48F2	410.68	410.3724	78.97	11.78	9.25
C27H49F1	392.69	392.3818	82.58	12.58	4.84
C27H50F2	412.70	412.3880	78.58	12.21	9.21
C27H51F1	394.71	394.3975	82.16	13.02	4.81
C27H52F2	414.71	414.4037	78.20	12.64	9.16
C27H53F1	396.72	396.4131	81.75	13.47	4.79
C27H54F2	416.73	416.4193	77.82	13.06	9.12
C27H55F1	398.74	398.4288	81.33	13.90	4.76
C28H14F2	388.42	388.1063	86.58	3.63	9.78
C28H15F1	370.43	370.1158	90.79	4.08	5.13
C28H16F2	390.44	390.1220	86.14	4.13	9.73
C28H17F1	372.45	372.1314	90.30	4.60	5.10
C28H18F2	392.45	392.1376	85.70	4.62	9.68
C28H19F1	374.46	374.1471	89.81	5.11	5.07
C28H20F2	394.47	394.1533	85.26	5.11	9.63
C28H21F1	376.48	376.1627	89.33	5.62	5.05
C28H22F2	396.48	396.1689	84.82	5.59	9.58
C28H23F1	378.49	378.1784	88.86	6.13	5.02
C28H24F2	398.50	398.1846	84.39	6.07	9.53
C28H25F1	380.51	380.1940	88.38	6.62	4.99
C28H26F2	400.52	400.2002	83.97	6.54	9.49
C28H27F1	382.53	382.2097	87.92	7.11	4.97
C28H28F2	402.53	402.2159	83.55	7.01	9.44
C28H29F1	384.54	384.2253	87.46	7.60	4.94
C28H30F2	404.55	404.2315	83.13	7.47	9.39
C28H31F1	386.56	386.2410	87.00	8.08	4.91
C28H32F2	406.56	406.2472	82.72	7.93	9.35
C28H33F1	388.57	388.2566	86.55	8.56	4.89
C28H34F2	408.58	408.2628	82.31	8.39	9.30
C28H35F1	390.59	390.2723	86.10	9.03	4.86
C28H36F2	410.60	410.2785	81.91	8.84	9.25
C28H37F1	392.61	392.2879	85.66	9.50	4.84
C28H38F2	412.61	412.2941	81.51	9.28	9.21
C28H39F1	394.62	394.3036	85.22	9.96	4.81
C28H40F2	414.63	414.3098	81.11	9.72	9.16
C28H41F1	396.64	396.3192	84.79	10.42	4.79
C28H42F2	416.64	416.3254	80.72	10.16	9.12
C28H43F1	398.65	398.3349	84.36	10.87	4.77
C28H44F2	418.66	418.3411	80.33	10.59	9.08
C28H45F1	400.67	400.3505	83.94	11.32	4.74
C28H46F2	420.68	420.3567	79.95	11.02	9.03
C28H47F1	402.69	402.3662	83.52	11.76	4.72
C28H48F2	422.69	422.3724	79.56	11.45	8.99
C28H49F1	404.70	404.3818	83.10	12.20	4.69
C28H50F2	424.71	424.3880	79.19	11.87	8.95
C28H51F1	406.72	406.3975	82.69	12.64	4.67
C28H52F2	426.72	426.4037	78.81	12.28	8.90
C28H53F1	408.73	408.4131	82.28	13.07	4.65
C28H54F2	428.74	428.4193	78.44	12.70	8.86
C28H55F1	410.75	410.4288	81.88	13.50	4.63
C28H56F2	430.76	430.4350	78.07	13.10	8.82
C28H57F1	412.76	412.4444	81.48	13.92	4.60
C29H14F2	400.43	400.1063	86.99	3.52	9.49
C29H15F1	382.44	382.1158	91.08	3.95	4.97
C29H16F2	402.45	402.1220	86.55	4.01	9.44
C29H17F1	384.46	384.1314	90.60	4.46	4.94
C29H18F2	404.46	404.1376	86.12	4.49	9.39
C29H19F1	386.47	386.1471	90.13	4.96	4.92
C29H20F2	406.48	406.1533	85.69	4.96	9.35
C29H21F1	388.49	388.1627	89.66	5.45	4.89
C29H22F2	408.50	408.1689	85.27	5.43	9.30
C29H23F1	390.51	390.1784	89.20	5.94	4.87
C29H24F2	410.51	410.1846	84.85	5.89	9.26
C29H25F1	392.52	392.1940	88.74	6.42	4.84
C29H26F2	412.53	412.2002	84.44	6.35	9.21
C29H27F1	394.54	394.2097	88.29	6.90	4.82
C29H28F2	414.54	414.2159	84.03	6.81	9.17
C29H29F1	396.55	396.2253	87.84	7.37	4.79
C29H30F2	416.56	416.2315	83.62	7.26	9.12
C29H31F1	398.57	398.2410	87.39	7.84	4.77
C29H32F2	418.58	418.2472	83.22	7.71	9.08
C29H33F1	400.58	400.2566	86.95	8.30	4.74
C29H34F2	420.59	420.2628	82.82	8.15	9.03
C29H35F1	402.60	402.2723	86.52	8.76	4.72
C29H36F2	422.61	422.2785	82.42	8.59	8.99
C29H37F1	404.62	404.2879	86.09	9.22	4.70
C29H38F2	424.62	424.2941	82.03	9.02	8.95
C29H39F1	406.63	406.3036	85.66	9.67	4.67
C29H40F2	426.64	426.3098	81.64	9.45	8.91
C29H41F1	408.65	408.3192	85.24	10.11	4.65

C29H42F2	428.65	428.3254	81.26	9.88	8.86
C29H43F1	410.66	410.3349	84.82	10.55	4.63
C29H44F2	430.67	430.3411	80.88	10.30	8.82
C29H45F1	412.68	412.3505	84.41	10.99	4.60
C29H46F2	432.69	432.3567	80.50	10.72	8.78
C29H47F1	414.70	414.3662	83.99	11.42	4.58
C29H48F2	434.70	434.3724	80.13	11.13	8.74
C29H49F1	416.71	416.3818	83.59	11.85	4.56
C29H50F2	436.72	436.3880	79.76	11.54	8.70
C29H51F1	418.73	418.3975	83.19	12.28	4.54
C29H52F2	438.73	438.4037	79.39	11.95	8.66
C29H53F1	420.74	420.4131	82.79	12.70	4.52
C29H54F2	440.75	440.4193	79.03	12.35	8.62
C29H55F1	422.76	422.4288	82.39	13.11	4.49
C29H56F2	442.77	442.4350	78.67	12.75	8.58
C29H57F1	424.78	424.4444	82.00	13.53	4.47
C29H58F2	444.78	444.4506	78.31	13.14	8.54
C29H59F1	426.79	426.4601	81.61	13.93	4.45
C30H15F1	394.45	394.1158	91.35	3.83	4.82
C30H16F2	414.46	414.1220	86.94	3.89	9.17
C30H17F1	396.47	396.1314	90.89	4.32	4.79
C30H18F2	416.47	416.1376	86.52	4.36	9.12
C30H19F1	398.48	398.1471	90.43	4.81	4.77
C30H20F2	418.49	418.1533	86.10	4.82	9.08
C30H21F1	400.50	400.1627	89.97	5.29	4.74
C30H22F2	420.51	420.1689	85.69	5.27	9.04
C30H23F1	402.52	402.1784	89.52	5.76	4.72
C30H24F2	422.52	422.1846	85.28	5.73	8.99
C30H25F1	404.53	404.1940	89.07	6.23	4.70
C30H26F2	424.54	424.2002	84.88	6.17	8.95
C30H27F1	406.55	406.2097	88.63	6.69	4.67
C30H28F2	426.55	426.2159	84.48	6.62	8.91
C30H29F1	408.56	408.2253	88.20	7.15	4.65
C30H30F2	428.57	428.2315	84.08	7.06	8.87
C30H31F1	410.58	410.2410	87.76	7.61	4.63
C30H32F2	430.59	430.2472	83.68	7.49	8.82
C30H33F1	412.60	412.2566	87.33	8.06	4.60
C30H34F2	432.60	432.2628	83.29	7.92	8.78
C30H35F1	414.61	414.2723	86.91	8.51	4.58
C30H36F2	434.62	434.2785	82.91	8.35	8.74
C30H37F1	416.63	416.2879	86.49	8.95	4.56
C30H38F2	436.63	436.2941	82.53	8.77	8.70
C30H39F1	418.64	418.3036	86.07	9.39	4.54
C30H40F2	438.65	438.3098	82.15	9.19	8.66
C30H41F1	420.66	420.3192	85.66	9.82	4.52
C30H42F2	440.67	440.3254	81.77	9.61	8.62
C30H43F1	422.68	422.3349	85.25	10.25	4.49
C30H44F2	442.68	442.3411	81.40	10.02	8.58
C30H45F1	424.69	424.3505	84.85	10.68	4.47
C30H46F2	444.70	444.3567	81.03	10.43	8.54
C30H47F1	426.71	426.3662	84.45	11.10	4.45
C30H48F2	446.71	446.3724	80.66	10.83	8.51
C30H49F1	428.72	428.3818	84.05	11.52	4.43
C30H50F2	448.73	448.3880	80.30	11.23	8.47
C30H51F1	430.74	430.3975	83.65	11.93	4.41
C30H52F2	450.75	450.4037	79.94	11.63	8.43
C30H53F1	432.76	432.4131	83.27	12.34	4.39
C30H54F2	452.76	452.4193	79.59	12.02	8.39
C30H55F1	434.77	434.4288	82.88	12.75	4.37
C30H56F2	454.78	454.4350	79.23	12.41	8.36
C30H57F1	436.79	436.4444	82.50	13.15	4.35
C30H58F2	456.79	456.4506	78.88	12.80	8.32
C30H59F1	438.80	438.4601	82.12	13.55	4.33
C30H60F2	458.81	458.4663	78.54	13.18	8.28
C30H61F1	440.82	440.4757	81.74	13.95	4.31

4. CARBON, HYDROGEN AND IODINE

4. KOHLENSTOFF, WASSERSTOFF UND JOD

4. CARBONE, HYDROGENE ET IODE

MOLECULAR WEIGHTS AND PERCENTAGE COMPOSITIONS OF COMPOUNDS
CONTAINING CARBON, HYDROGEN AND IODINE

FORMULA	MOL. WEIGHT	ISOMOL. WEIGHT	PERCENT C	PERCENT H	PERCENT I
C 1H 0I2	265.82	265.8093	4.52	0	95.48
C 1H 1I1	139.92	139.9125	8.58	.72	90.70
C 1H 2I2	267.84	267.8250	4.48	.75	94.76
C 1H 3I1	141.94	141.9281	8.46	2.13	89.41
C 2H 1I1	151.93	151.9125	15.81	.66	83.53
C 2H 2I2	279.85	279.8250	8.58	.72	90.70
C 2H 3I1	153.95	153.9281	15.60	1.96	82.43
C 2H 4I2	281.86	281.8406	8.52	1.43	90.05
C 2H 5I1	155.97	155.9438	15.40	3.23	81.37
C 3H 1I1	163.95	163.9125	21.98	.61	77.41
C 3H 2I2	291.86	291.8250	12.35	.69	86.96
C 3H 3I1	165.96	165.9281	21.71	1.82	76.47
C 3H 4I2	293.87	293.8406	12.26	1.37	86.37
C 3H 5I1	167.98	167.9438	21.45	3.00	75.55
C 3H 6I2	295.89	295.8563	12.18	2.04	85.78
C 3H 7I1	169.99	169.9594	21.20	4.15	74.65
C 4H 2I2	303.87	303.8250	15.81	.66	83.53
C 4H 3I1	177.97	177.9281	27.00	1.70	71.31
C 4H 4I2	305.89	305.8406	15.71	1.32	82.98
C 4H 5I1	179.99	179.9438	26.69	2.80	70.51
C 4H 6I2	307.90	307.8563	15.60	1.96	82.43
C 4H 7I1	182.00	181.9594	26.40	3.88	69.73
C 4H 8I2	309.92	309.8719	15.50	2.60	81.90
C 4H 9I1	184.02	183.9751	26.11	4.93	68.96
C 5H 2I2	315.88	315.8250	19.01	.64	80.35
C 5H 3I1	189.98	189.9281	31.61	1.59	66.80
C 5H 4I2	317.90	317.8406	18.89	1.27	79.84
C 5H 5I1	192.00	191.9438	31.28	2.62	66.10
C 5H 6I2	319.91	319.8563	18.77	1.89	79.34
C 5H 7I1	194.02	193.9594	30.95	3.64	65.41
C 5H 8I2	321.93	321.8719	18.66	2.50	78.84
C 5H 9I1	196.03	195.9751	30.64	4.63	64.74
C 5H10I2	323.94	323.8876	18.54	3.11	78.35
C 5H11I1	198.05	197.9907	30.32	5.60	64.08
C 6H 3I1	202.00	201.9281	35.68	1.50	62.83
C 6H 4I2	329.91	329.8406	21.84	1.22	76.93
C 6H 5I1	204.01	203.9438	35.32	2.47	62.20
C 6H 6I2	331.92	331.8563	21.71	1.82	76.47
C 6H 7I1	206.03	205.9594	34.98	3.42	61.60
C 6H 8I2	333.94	333.8719	21.58	2.41	76.00
C 6H 9I1	208.04	207.9751	34.64	4.36	61.00
C 6H10I2	335.96	335.8876	21.45	3.00	75.55
C 6H11I1	210.06	209.9907	34.31	5.28	60.41
C 6H12I2	337.97	337.9032	21.32	3.58	75.10
C 6H13I1	212.07	212.0064	33.98	6.18	59.84
C 7H 3I1	214.01	213.9281	39.29	1.41	59.30
C 7H 4I2	341.92	341.8406	24.59	1.18	74.23
C 7H 5I1	216.02	215.9438	38.92	2.33	58.75
C 7H 6I2	343.93	343.8563	24.45	1.76	73.80
C 7H 7I1	218.04	217.9594	38.56	3.24	58.20

FORMULA	MOL. WEIGHT	ISOMOL. WEIGHT	PERCENT C	PERCENT H	PERCENT I
C 7H 8I2	345.95	345.8719	24.30	2.33	73.37
C 7H 9I1	220.05	219.9751	38.21	4.12	57.67
C 7H10I2	347.97	347.8876	24.16	2.90	72.94
C 7H11I1	222.07	221.9907	37.86	4.99	57.15
C 7H12I2	349.98	349.9032	24.02	3.46	72.52
C 7H13I1	224.09	224.0064	37.52	5.85	56.63
C 7H14I2	352.00	351.9189	23.89	4.01	72.11
C 7H15I1	226.10	226.0220	37.19	6.69	56.13
C 8H 4I2	353.93	353.8406	27.15	1.14	71.71
C 8H 5I1	228.03	227.9438	42.14	2.21	55.65
C 8H 6I2	355.95	355.8563	27.00	1.70	71.31
C 8H 7I1	230.05	229.9594	41.77	3.07	55.16
C 8H 8I2	357.96	357.8719	26.84	2.25	70.90
C 8H 9I1	232.07	231.9751	41.41	3.91	54.68
C 8H10I2	359.98	359.8876	26.69	2.80	70.51
C 8H11I1	234.08	233.9907	41.05	4.74	54.21
C 8H12I2	361.99	361.9032	26.54	3.34	70.11
C 8H13I1	236.10	236.0064	40.70	5.55	53.75
C 8H14I2	364.01	363.9189	26.40	3.88	69.73
C 8H15I1	238.11	238.0220	40.35	6.35	53.30
C 8H16I2	366.03	365.9345	26.25	4.41	69.34
C 8H17I1	240.13	240.0377	40.02	7.14	52.85
C 9H 4I2	365.94	365.8406	29.54	1.10	69.36
C 9H 5I1	240.04	239.9438	45.03	2.10	52.87
C 9H 6I2	367.96	367.8563	29.38	1.64	68.98
C 9H 7I1	242.06	241.9594	44.66	2.91	52.43
C 9H 8I2	369.97	369.8719	29.22	2.18	68.60
C 9H 9I1	244.08	243.9751	44.29	3.72	51.99
C 9H10I2	371.99	371.8876	29.06	2.71	68.23
C 9H11I1	246.09	245.9907	43.93	4.51	51.57
C 9H12I2	374.00	373.9032	28.90	3.23	67.86
C 9H13I1	248.11	248.0064	43.57	5.28	51.15
C 9H14I2	376.02	375.9189	28.75	3.75	67.50
C 9H15I1	250.12	250.0220	43.22	6.04	50.74
C 9H16I2	378.04	377.9345	28.60	4.27	67.14
C 9H17I1	252.14	252.0377	42.87	6.80	50.33
C 9H18I2	380.05	379.9502	28.44	4.77	66.78
C 9H19I1	254.16	254.0533	42.53	7.54	49.93
C10H 5I1	252.06	251.9438	47.65	2.00	50.35
C10H 6I2	379.97	379.8563	31.61	1.59	66.80
C10H 7I1	254.07	253.9594	47.27	2.78	49.95
C10H 8I2	381.98	381.8719	31.44	2.11	66.44
C10H 9I1	256.09	255.9751	46.90	3.54	49.56
C10H10I2	384.00	383.8876	31.28	2.62	66.10
C10H11I1	258.10	257.9907	46.54	4.30	49.17
C10H12I2	386.02	385.9032	31.12	3.13	65.75
C10H13I1	260.12	260.0064	46.18	5.04	48.79
C10H14I2	388.03	387.9189	30.95	3.64	65.41
C10H15I1	262.14	262.0220	45.82	5.77	48.41
C10H16I2	390.05	389.9345	30.79	4.13	65.07

C10H17I1	264.15	264.0377	45.47	6.49	48.04
C10H18I2	392.06	391.9502	30.64	4.63	64.74
C10H19I1	266.17	266.0533	45.13	7.20	47.68
C10H20I2	394.08	393.9658	30.48	5.12	64.41
C10H21I1	268.18	268.0690	44.79	7.89	47.32
C11H 5I1	264.07	263.9438	50.03	1.91	48.06
C11H 6I2	391.98	391.8563	33.71	1.54	64.75
C11H 7I1	266.08	265.9594	49.65	2.65	47.69
C11H 8I2	394.00	393.8719	33.53	2.05	64.42
C11H 9I1	268.10	267.9751	49.28	3.38	47.33
C11H10I2	396.01	395.8876	33.36	2.55	64.09
C11H11I1	270.11	269.9907	48.91	4.10	46.98
C11H12I2	398.03	397.9032	33.19	3.04	63.77
C11H13I1	272.13	272.0064	48.55	4.82	46.63
C11H14I2	400.04	399.9189	33.03	3.53	63.45
C11H15I1	274.15	274.0220	48.19	5.52	46.29
C11H16I2	402.06	401.9345	32.86	4.01	63.13
C11H17I1	276.16	276.0377	47.84	6.20	45.95
C11H18I2	404.07	403.9502	32.70	4.49	62.81
C11H19I1	278.18	278.0533	47.50	6.88	45.62
C11H20I2	406.09	405.9658	32.54	4.96	62.50
C11H21I1	280.19	280.0690	47.15	7.55	45.29
C11H22I2	408.11	407.9815	32.37	5.43	62.19
C11H23I1	282.21	282.0846	46.82	8.21	44.97
C12H 6I2	403.99	403.8563	35.68	1.50	62.83
C12H 7I1	278.09	277.9594	51.83	2.54	45.63
C12H 8I2	406.01	405.8719	35.50	1.99	62.51
C12H 9I1	280.11	279.9751	51.46	3.24	45.31
C12H10I2	408.02	407.8876	35.32	2.47	62.20
C12H11I1	282.13	281.9907	51.09	3.93	44.98
C12H12I2	410.04	409.9032	35.15	2.95	61.90
C12H13I1	284.14	284.0064	50.73	4.61	44.66
C12H14I2	412.05	411.9189	34.98	3.42	61.60
C12H15I1	286.16	286.0220	50.37	5.28	44.35
C12H16I2	414.07	413.9345	34.81	3.89	61.30
C12H17I1	288.17	288.0377	50.02	5.95	44.04
C12H18I2	416.09	415.9502	34.64	4.36	61.00
C12H19I1	290.19	290.0533	49.67	6.60	43.73
C12H20I2	418.10	417.9658	34.47	4.82	60.70
C12H21I1	292.21	292.0690	49.33	7.24	43.43
C12H22I2	420.12	419.9815	34.31	5.28	60.41
C12H23I1	294.22	294.0846	48.99	7.88	43.13
C12H24I2	422.13	421.9971	34.14	5.73	60.13
C12H25I1	296.24	296.1003	48.65	8.51	42.84
C13H 6I2	416.00	415.8563	37.53	1.45	61.01
C13H 7I1	290.11	289.9594	53.82	2.43	43.74
C13H 8I2	418.02	417.8719	37.35	1.93	60.72
C13H 9I1	292.12	291.9751	53.45	3.11	43.44
C13H10I2	420.03	419.8876	37.17	2.40	60.43
C13H11I1	294.14	293.9907	53.09	3.77	43.14
C13H12I2	422.05	421.9032	37.00	2.87	60.14
C13H13I1	296.15	296.0064	52.72	4.42	42.85
C13H14I2	424.07	423.9189	36.82	3.33	59.85
C13H15I1	298.17	298.0220	52.37	5.07	42.56
C13H16I2	426.08	425.9345	36.65	3.79	59.57
C13H17I1	300.18	300.0377	52.02	5.71	42.28
C13H18I2	428.10	427.9502	36.47	4.24	59.29
C13H19I1	302.20	302.0533	51.67	6.34	41.99
C13H20I2	430.11	429.9658	36.30	4.69	59.01
C13H21I1	304.22	304.0690	51.33	6.96	41.72

C13H22I2	432.13	431.9815	36.13	5.13	58.73
C13H23I1	306.23	306.0846	50.99	7.57	41.44
C13H24I2	434.15	433.9971	35.97	5.57	58.46
C13H25I1	308.25	308.1003	50.66	8.17	41.17
C13H26I2	436.16	436.0128	35.80	6.01	58.19
C13H27I1	310.26	310.1159	50.33	8.77	40.90
C14H 7I1	302.12	301.9594	55.66	2.34	42.01
C14H 8I2	430.03	429.8719	39.10	1.88	59.02
C14H 9I1	304.13	303.9751	55.29	2.98	41.73
C14H10I2	432.04	431.8876	38.92	2.33	58.75
C14H11I1	306.15	305.9907	54.93	3.62	41.45
C14H12I2	434.06	433.9032	38.74	2.79	58.47
C14H13I1	308.16	308.0064	54.57	4.25	41.18
C14H14I2	436.08	435.9189	38.56	3.24	58.20
C14H15I1	310.18	310.0220	54.21	4.87	40.91
C14H16I2	438.09	437.9345	38.38	3.68	57.93
C14H17I1	312.20	312.0377	53.86	5.49	40.65
C14H18I2	440.11	439.9502	38.21	4.12	57.67
C14H19I1	314.21	314.0533	53.52	6.10	40.39
C14H20I2	442.12	441.9658	38.03	4.56	57.41
C14H21I1	316.23	316.0690	53.18	6.69	40.13
C14H22I2	444.14	443.9815	37.86	4.99	57.15
C14H23I1	318.24	318.0846	52.84	7.28	39.88
C14H24I2	446.16	445.9971	37.69	5.42	56.89
C14H25I1	320.26	320.1003	52.51	7.87	39.63
C14H26I2	448.17	448.0128	37.52	5.85	56.63
C14H27I1	322.28	322.1159	52.18	8.44	39.38
C14H28I2	450.19	450.0284	37.35	6.27	56.38
C14H29I1	324.29	324.1316	51.85	9.01	39.13
C15H 7I1	314.13	313.9594	57.35	2.25	40.40
C15H 8I2	442.04	441.8719	40.76	1.82	57.42
C15H 9I1	316.14	315.9751	56.99	2.87	40.14
C15H10I2	444.06	443.8876	40.57	2.27	57.16
C15H11I1	318.16	317.9907	56.63	3.48	39.89
C15H12I2	446.07	445.9032	40.39	2.71	56.90
C15H13I1	320.18	320.0064	56.27	4.09	39.64
C15H14I2	448.09	447.9189	40.21	3.15	56.64
C15H15I1	322.19	322.0220	55.92	4.69	39.39
C15H16I2	450.10	449.9345	40.03	3.58	56.39
C15H17I1	324.21	324.0377	55.57	5.29	39.14
C15H18I2	452.12	451.9502	39.85	4.01	56.14
C15H19I1	326.22	326.0533	55.23	5.87	38.90
C15H20I2	454.14	453.9658	39.67	4.44	55.89
C15H21I1	328.24	328.0690	54.89	6.45	38.66
C15H22I2	456.15	455.9815	39.50	4.86	55.64
C15H23I1	330.25	330.0846	54.55	7.02	38.43
C15H24I2	458.17	457.9971	39.32	5.28	55.40
C15H25I1	332.27	332.1003	54.22	7.58	38.19
C15H26I2	460.18	460.0128	39.15	5.69	55.15
C15H27I1	334.29	334.1159	53.90	8.14	37.96
C15H28I2	462.20	462.0284	38.98	6.11	54.91
C15H29I1	336.30	336.1316	53.57	8.69	37.74
C15H30I2	464.22	464.0441	38.81	6.51	54.67
C15H31I1	338.32	338.1472	53.25	9.24	37.51
C16H 8I2	454.05	453.8719	42.33	1.78	55.90
C16H 9I1	328.15	327.9751	58.56	2.76	38.67
C16H10I2	456.07	455.8876	42.14	2.21	55.65
C16H11I1	330.17	329.9907	58.21	3.36	38.44
C16H12I2	458.08	457.9032	41.95	2.64	55.41
C16H13I1	332.19	332.0064	57.85	3.94	38.20

C16H14I2	460.10	459.9189	41.77	3.07	55.16
C16H15I1	334.20	334.0220	57.50	4.52	37.97
C16H16I2	462.11	461.9345	41.59	3.49	54.92
C16H17I1	336.22	336.0377	57.16	5.10	37.74
C16H18I2	464.13	463.9502	41.41	3.91	54.68
C16H19I1	338.23	338.0533	56.82	5.66	37.52
C16H20I2	466.15	465.9658	41.23	4.32	54.45
C16H21I1	340.25	340.0690	56.48	6.22	37.30
C16H22I2	468.16	467.9815	41.05	4.74	54.21
C16H23I1	342.27	342.0846	56.15	6.77	37.08
C16H24I2	470.18	469.9971	40.87	5.15	53.98
C16H25I1	344.28	344.1003	55.82	7.32	36.86
C16H26I2	472.19	472.0128	40.70	5.55	53.75
C16H27I1	346.30	346.1159	55.50	7.86	36.65
C16H28I2	474.21	474.0284	40.53	5.95	53.52
C16H29I1	348.31	348.1316	55.17	8.39	36.43
C16H30I2	476.23	476.0441	40.35	6.35	53.30
C16H31I1	350.33	350.1472	54.86	8.92	36.22
C16H32I2	478.24	478.0597	40.18	6.74	53.07
C16H33I1	352.35	352.1629	54.54	9.44	36.02
C17H 8I2	466.06	465.8719	43.81	1.73	54.46
C17H 9I1	340.17	339.9751	60.03	2.67	37.31
C17H10I2	468.08	467.8876	43.62	2.15	54.22
C17H11I1	342.18	341.9907	59.67	3.24	37.09
C17H12I2	470.09	469.9032	43.44	2.57	53.99
C17H13I1	344.20	344.0064	59.32	3.81	36.87
C17H14I2	472.11	471.9189	43.25	2.99	53.76
C17H15I1	346.21	346.0220	58.98	4.37	36.65
C17H16I2	474.13	473.9345	43.07	3.40	53.53
C17H17I1	348.23	348.0377	58.64	4.92	36.44
C17H18I2	476.14	475.9502	42.88	3.81	53.31
C17H19I1	350.25	350.0533	58.30	5.47	36.23
C17H20I2	478.16	477.9658	42.70	4.22	53.08
C17H21I1	352.26	352.0690	57.97	6.01	36.03
C17H22I2	480.17	479.9815	42.52	4.62	52.86
C17H23I1	354.28	354.0846	57.64	6.54	35.82
C17H24I2	482.19	481.9971	42.35	5.02	52.64
C17H25I1	356.29	356.1003	57.31	7.07	35.62
C17H26I2	484.21	484.0128	42.17	5.41	52.42
C17H27I1	358.31	358.1159	56.99	7.60	35.42
C17H28I2	486.22	486.0284	42.00	5.80	52.20
C17H29I1	360.33	360.1316	56.67	8.11	35.22
C17H30I2	488.24	488.0441	41.82	6.19	51.98
C17H31I1	362.34	362.1472	56.35	8.62	35.02
C17H32I2	490.25	490.0597	41.65	6.58	51.77
C17H33I1	364.36	364.1629	56.04	9.13	34.83
C17H34I2	492.27	492.0754	41.48	6.96	51.56
C17H35I1	366.37	366.1785	55.73	9.63	34.64
C18H 9I1	352.18	351.9751	61.39	2.58	36.03
C18H10I2	480.09	479.8876	45.03	2.10	52.87
C18H11I1	354.19	353.9907	61.04	3.13	35.83
C18H12I2	482.11	481.9032	44.85	2.51	52.65
C18H13I1	356.21	356.0064	60.69	3.68	35.63
C18H14I2	484.12	483.9189	44.66	2.91	52.43
C18H15I1	358.22	358.0220	60.35	4.22	35.43
C18H16I2	486.14	485.9345	44.47	3.32	52.21
C18H17I1	360.24	360.0377	60.02	4.76	35.23
C18H18I2	488.15	487.9502	44.29	3.72	51.99
C18H19I1	362.26	362.0533	59.68	5.29	35.03
C18H20I2	490.17	489.9658	44.11	4.11	51.78

10/21/66 040

C18H21I1	364.27	364.0690	59.35	5.81	34.84
C18H22I2	492.18	491.9815	43.93	4.51	51.57
C18H23I1	366.29	366.0846	59.02	6.33	34.65
C18H24I2	494.20	493.9971	43.75	4.90	51.36
C18H25I1	368.30	368.1003	58.70	6.84	34.46
C18H26I2	496.22	496.0128	43.57	5.28	51.15
C18H27I1	370.32	370.1159	58.38	7.35	34.27
C18H28I2	498.23	498.0284	43.39	5.66	50.94
C18H29I1	372.34	372.1316	58.07	7.85	34.08
C18H30I2	500.25	500.0441	43.22	6.04	50.74
C18H31I1	374.35	374.1472	57.75	8.35	33.90
C18H32I2	502.26	502.0597	43.05	6.42	50.53
C18H33I1	376.37	376.1629	57.44	8.84	33.72
C18H34I2	504.28	504.0754	42.87	6.80	50.33
C18H35I1	378.38	378.1785	57.14	9.32	33.54
C18H36I2	506.30	506.0910	42.70	7.17	50.13
C18H37I1	380.40	380.1942	56.84	9.80	33.36
C19H 9I1	364.19	363.9751	62.66	2.49	34.85
C19H10I2	492.10	491.8876	46.38	2.05	51.58
C19H11I1	366.20	365.9907	62.32	3.03	34.65
C19H12I2	494.12	493.9032	46.19	2.45	51.37
C19H13I1	368.22	368.0064	61.98	3.56	34.46
C19H14I2	496.13	495.9189	46.00	2.84	51.16
C19H15I1	370.24	370.0220	61.64	4.08	34.28
C19H16I2	498.15	497.9345	45.81	3.24	50.95
C19H17I1	372.25	372.0377	61.31	4.60	34.09
C19H18I2	500.16	499.9502	45.63	3.63	50.75
C19H19I1	374.27	374.0533	60.98	5.12	33.91
C19H20I2	502.18	501.9658	45.44	4.01	50.54
C19H21I1	376.28	376.0690	60.65	5.63	33.73
C19H22I2	504.20	503.9815	45.26	4.40	50.34
C19H23I1	378.30	378.0846	60.33	6.13	33.55
C19H24I2	506.21	505.9971	45.08	4.78	50.14
C19H25I1	380.32	380.1003	60.01	6.63	33.37
C19H26I2	508.23	508.0128	44.90	5.16	49.94
C19H27I1	382.33	382.1159	59.69	7.12	33.19
C19H28I2	510.24	510.0284	44.73	5.53	49.74
C19H29I1	384.35	384.1316	59.38	7.61	33.02
C19H30I2	512.26	512.0441	44.55	5.90	49.55
C19H31I1	386.36	386.1472	59.07	8.09	32.85
C19H32I2	514.28	514.0597	44.38	6.27	49.35
C19H33I1	388.38	388.1629	58.76	8.56	32.68
C19H34I2	516.29	516.0754	44.20	6.64	49.16
C19H35I1	390.40	390.1785	58.46	9.04	32.51
C19H36I2	518.31	518.0910	44.03	7.00	48.97
C19H37I1	392.41	392.1942	58.16	9.50	32.34
C19H38I2	520.32	520.1067	43.86	7.36	48.78
C19H39I1	394.43	394.2098	57.86	9.97	32.17
C20H10I2	504.11	503.8876	47.65	2.00	50.35
C20H11I1	378.22	377.9907	63.51	2.93	33.55
C20H12I2	506.13	505.9032	47.46	2.39	50.15
C20H13I1	380.23	380.0064	63.18	3.45	33.38
C20H14I2	508.14	507.9189	47.27	2.78	49.95
C20H15I1	382.25	382.0220	62.84	3.96	33.20
C20H16I2	510.16	509.9345	47.09	3.16	49.75
C20H17I1	384.26	384.0377	62.52	4.46	33.03
C20H18I2	512.18	511.9502	46.90	3.54	49.56
C20H19I1	386.28	386.0533	62.19	4.96	32.85
C20H20I2	514.19	513.9658	46.72	3.92	49.36
C20H21I1	388.29	388.0690	61.87	5.45	32.68

C20H22I2	516.21	515.9815	46.54	4.30	49.17
C20H23I1	390.31	390.0846	61.55	5.94	32.51
C20H24I2	518.22	517.9971	46.36	4.67	48.98
C20H25I1	392.33	392.1003	61.23	6.42	32.35
C20H26I2	520.24	520.0128	46.18	5.04	48.79
C20H27I1	394.34	394.1159	60.92	6.90	32.18
C20H28I2	522.25	522.0284	46.00	5.40	48.60
C20H29I1	396.36	396.1316	60.61	7.37	32.02
C20H30I2	524.27	524.0441	45.82	5.77	48.41
C20H31I1	398.37	398.1472	60.30	7.84	31.86
C20H32I2	526.29	526.0597	45.64	6.13	48.23
C20H33I1	400.39	400.1629	60.00	8.31	31.70
C20H34I2	528.30	528.0754	45.47	6.49	48.04
C20H35I1	402.41	402.1785	59.70	8.77	31.54
C20H36I2	530.32	530.0910	45.30	6.84	47.86
C20H37I1	404.42	404.1942	59.40	9.22	31.38
C20H38I2	532.33	532.1067	45.13	7.20	47.68
C20H39I1	406.44	406.2098	59.10	9.67	31.22
C20H40I2	534.35	534.1223	44.96	7.55	47.50
C20H41I1	408.45	408.2255	58.81	10.12	31.07
C21H10I2	516.12	515.8876	48.87	1.95	49.18
C21H11I1	390.23	389.9907	64.64	2.84	32.52
C21H12I2	518.14	517.9032	48.68	2.33	48.98
C21H13I1	392.24	392.0064	64.31	3.34	32.35
C21H14I2	520.15	519.9189	48.49	2.71	48.79
C21H15I1	394.26	394.0220	63.98	3.83	32.19
C21H16I2	522.17	521.9345	48.30	3.09	48.61
C21H17I1	396.27	396.0377	63.65	4.32	32.02
C21H18I2	524.19	523.9502	48.12	3.46	48.42
C21H19I1	398.29	398.0533	63.33	4.81	31.86
C21H20I2	526.20	525.9658	47.93	3.83	48.23
C21H21I1	400.31	400.0690	63.01	5.29	31.70
C21H22I2	528.22	527.9815	47.75	4.20	48.05
C21H23I1	402.32	402.0846	62.69	5.76	31.54
C21H24I2	530.23	529.9971	47.57	4.56	47.87
C21H25I1	404.34	404.1003	62.38	6.23	31.39
C21H26I2	532.25	532.0128	47.39	4.92	47.69
C21H27I1	406.35	406.1159	62.07	6.70	31.23
C21H28I2	534.27	534.0284	47.21	5.28	47.51
C21H29I1	408.37	408.1316	61.77	7.16	31.08
C21H30I2	536.28	536.0441	47.03	5.64	47.33
C21H31I1	410.39	410.1472	61.46	7.61	30.92
C21H32I2	538.30	538.0597	46.86	5.99	47.15
C21H33I1	412.40	412.1629	61.16	8.07	30.77
C21H34I2	540.31	540.0754	46.68	6.34	46.97
C21H35I1	414.42	414.1785	60.86	8.51	30.62
C21H36I2	542.33	542.0910	46.51	6.69	46.80
C21H37I1	416.43	416.1942	60.57	8.96	30.47
C21H38I2	544.35	544.1067	46.34	7.04	46.63
C21H39I1	418.45	418.2098	60.28	9.39	30.33
C21H40I2	546.36	546.1223	46.17	7.38	46.45
C21H41I1	420.47	420.2255	59.99	9.83	30.18
C21H42I2	548.38	548.1380	46.00	7.72	46.28
C21H43I1	422.48	422.2411	59.70	10.26	30.04
C22H11I1	402.24	401.9907	65.69	2.76	31.55
C22H12I2	530.15	529.9032	49.84	2.28	47.87
C22H13I1	404.25	404.0064	65.37	3.24	31.39
C22H14I2	532.17	531.9189	49.65	2.65	47.69
C22H15I1	406.27	406.0220	65.04	3.72	31.24
C22H16I2	534.18	533.9345	49.47	3.02	47.51

10/21/66 041

C22H17I1	408.29	408.0377	64.72	4.20	31.08
C22H18I2	536.20	535.9502	49.28	3.38	47.33
C22H19I1	410.30	410.0533	64.40	4.67	30.93
C22H20I2	538.21	537.9658	49.10	3.75	47.16
C22H21I1	412.32	412.0690	64.09	5.13	30.78
C22H22I2	540.23	539.9815	48.91	4.10	46.98
C22H23I1	414.33	414.0846	63.78	5.60	30.63
C22H24I2	542.25	541.9971	48.73	4.46	46.81
C22H25I1	416.35	416.1003	63.47	6.05	30.48
C22H26I2	544.26	544.0128	48.55	4.82	46.63
C22H27I1	418.36	418.1159	63.16	6.51	30.33
C22H28I2	546.28	546.0284	48.37	5.17	46.46
C22H29I1	420.38	420.1316	62.86	6.95	30.19
C22H30I2	548.29	548.0441	48.19	5.52	46.29
C22H31I1	422.40	422.1472	62.56	7.40	30.04
C22H32I2	550.31	550.0597	48.02	5.86	46.12
C22H33I1	424.41	424.1629	62.26	7.84	29.90
C22H34I2	552.33	552.0754	47.84	6.20	45.95
C22H35I1	426.43	426.1785	61.97	8.27	29.76
C22H36I2	554.34	554.0910	47.67	6.55	45.79
C22H37I1	428.44	428.1942	61.68	8.70	29.62
C22H38I2	556.36	556.1067	47.50	6.88	45.62
C22H39I1	430.46	430.2098	61.39	9.13	29.48
C22H40I2	558.37	558.1223	47.32	7.22	45.46
C22H41I1	432.48	432.2255	61.10	9.56	29.34
C22H42I2	560.39	560.1380	47.15	7.55	45.29
C22H43I1	434.49	434.2411	60.82	9.98	29.21
C22H44I2	562.40	562.1536	46.98	7.89	45.13
C22H45I1	436.51	436.2568	60.54	10.39	29.07
C23H11I1	414.25	413.9907	66.69	2.68	30.63
C23H12I2	542.16	541.9032	50.95	2.23	46.81
C23H13I1	416.26	416.0064	66.37	3.15	30.49
C23H14I2	544.18	543.9189	50.77	2.59	46.64
C23H15I1	418.28	418.0220	66.05	3.61	30.34
C23H16I2	546.19	545.9345	50.58	2.95	46.47
C23H17I1	420.30	420.0377	65.73	4.08	30.19
C23H18I2	548.21	547.9502	50.39	3.31	46.30
C23H19I1	422.31	422.0533	65.42	4.53	30.05
C23H20I2	550.22	549.9658	50.21	3.66	46.13
C23H21I1	424.33	424.0690	65.10	4.99	29.91
C23H22I2	552.24	551.9815	50.02	4.02	45.96
C23H23I1	426.34	426.0846	64.80	5.44	29.77
C23H24I2	554.26	553.9971	49.84	4.36	45.79
C23H25I1	428.36	428.1003	64.49	5.88	29.63
C23H26I2	556.27	556.0128	49.66	4.71	45.63
C23H27I1	430.38	430.1159	64.19	6.32	29.49
C23H28I2	558.29	558.0284	49.48	5.06	45.46
C23H29I1	432.39	432.1316	63.89	6.76	29.35
C23H30I2	560.30	560.0441	49.30	5.40	45.30
C23H31I1	434.41	434.1472	63.59	7.19	29.21
C23H32I2	562.32	562.0597	49.13	5.74	45.14
C23H33I1	436.42	436.1629	63.30	7.62	29.08
C23H34I2	564.34	564.0754	48.95	6.07	44.97
C23H35I1	438.44	438.1785	63.01	8.05	28.94
C23H36I2	566.35	566.0910	48.78	6.41	44.81
C23H37I1	440.46	440.1942	62.72	8.47	28.81
C23H38I2	568.37	568.1067	48.61	6.74	44.66
C23H39I1	442.47	442.2098	62.43	8.88	28.68
C23H40I2	570.38	570.1223	48.43	7.07	44.50
C23H41I1	444.49	444.2255	62.15	9.30	28.55

C23H42I2	572.40	572.1380	48.26	7.40	44.34
C23H43I1	446.50	446.2411	61.87	9.71	28.42
C23H44I2	574.42	574.1536	48.09	7.72	44.19
C23H45I1	448.52	448.2568	61.59	10.11	28.29
C23H46I2	576.43	576.1693	47.93	8.04	44.03
C23H47I1	450.54	450.2724	61.32	10.52	28.17
C24H12I2	554.17	553.9032	52.02	2.18	45.80
C24H13I1	428.28	428.0064	67.31	3.06	29.63
C24H14I2	556.19	555.9189	51.83	2.54	45.63
C24H15I1	430.29	430.0220	66.99	3.51	29.49
C24H16I2	558.20	557.9345	51.64	2.89	45.47
C24H17I1	432.31	432.0377	66.68	3.96	29.36
C24H18I2	560.22	559.9502	51.46	3.24	45.31
C24H19I1	434.32	434.0533	66.37	4.41	29.22
C24H20I2	562.24	561.9658	51.27	3.59	45.14
C24H21I1	436.34	436.0690	66.06	4.85	29.08
C24H22I2	564.25	563.9815	51.09	3.93	44.98
C24H23I1	438.36	438.0846	65.76	5.29	28.95
C24H24I2	566.27	565.9971	50.91	4.27	44.82
C24H25I1	440.37	440.1003	65.46	5.72	28.82
C24H26I2	568.28	568.0128	50.73	4.61	44.66
C24H27I1	442.39	442.1159	65.16	6.15	28.69
C24H28I2	570.30	570.0284	50.55	4.95	44.50
C24H29I1	444.40	444.1316	64.87	6.58	28.56
C24H30I2	572.32	572.0441	50.37	5.28	44.35
C24H31I1	446.42	446.1472	64.57	7.00	28.43
C24H32I2	574.33	574.0597	50.19	5.62	44.19
C24H33I1	448.44	448.1629	64.28	7.42	28.30
C24H34I2	576.35	576.0754	50.02	5.95	44.04
C24H35I1	450.45	450.1785	64.00	7.83	28.17
C24H36I2	578.36	578.0910	49.84	6.27	43.88
C24H37I1	452.47	452.1942	63.71	8.24	28.05
C24H38I2	580.38	580.1067	49.67	6.60	43.73
C24H39I1	454.48	454.2098	63.43	8.65	27.92
C24H40I2	582.40	582.1223	49.50	6.92	43.58
C24H41I1	456.50	456.2255	63.15	9.05	27.80
C24H42I2	584.41	584.1380	49.33	7.24	43.43
C24H43I1	458.51	458.2411	62.87	9.45	27.68
C24H44I2	586.43	586.1536	49.16	7.56	43.28
C24H45I1	460.53	460.2568	62.59	9.85	27.56
C24H46I2	588.44	588.1693	48.99	7.88	43.13
C24H47I1	462.55	462.2724	62.32	10.24	27.44
C24H48I2	590.46	590.1849	48.82	8.19	42.99
C24H49I1	464.56	464.2881	62.05	10.63	27.32
C25H12I2	566.18	565.9032	53.04	2.14	44.83
C25H13I1	440.29	440.0064	68.20	2.98	28.82
C25H14I2	568.20	567.9189	52.85	2.48	44.67
C25H15I1	442.30	442.0220	67.89	3.42	28.69
C25H16I2	570.22	569.9345	52.66	2.83	44.51
C25H17I1	444.32	444.0377	67.58	3.86	28.56
C25H18I2	572.23	571.9502	52.48	3.17	44.35
C25H19I1	446.33	446.0533	67.28	4.29	28.43
C25H20I2	574.25	573.9658	52.29	3.51	44.20
C25H21I1	448.35	448.0690	66.97	4.72	28.30
C25H22I2	576.26	575.9815	52.11	3.85	44.04
C25H23I1	450.37	450.0846	66.67	5.15	28.18
C25H24I2	578.28	577.9971	51.93	4.18	43.89
C25H25I1	452.38	452.1003	66.38	5.57	28.05
C25H26I2	580.29	580.0128	51.75	4.52	43.74
C25H27I1	454.40	454.1159	66.08	5.99	27.93

C25H28I2	582.31	582.0284	51.57	4.85	43.59
C25H29I1	456.41	456.1316	65.79	6.40	27.80
C25H30I2	584.33	584.0441	51.39	5.18	43.44
C25H31I1	458.43	458.1472	65.50	6.82	27.68
C25H32I2	586.34	586.0597	51.21	5.50	43.29
C25H33I1	460.45	460.1629	65.21	7.22	27.56
C25H34I2	588.36	588.0754	51.04	5.82	43.14
C25H35I1	462.46	462.1785	64.93	7.63	27.44
C25H36I2	590.37	590.0910	50.86	6.15	42.99
C25H37I1	464.48	464.1942	64.65	8.03	27.32
C25H38I2	592.39	592.1067	50.69	6.47	42.84
C25H39I1	466.49	466.2098	64.37	8.43	27.20
C25H40I2	594.41	594.1223	50.52	6.78	42.70
C25H41I1	468.51	468.2255	64.09	8.82	27.09
C25H42I2	596.42	596.1380	50.35	7.10	42.56
C25H43I1	470.53	470.2411	63.82	9.21	26.97
C25H44I2	598.44	598.1536	50.18	7.41	42.41
C25H45I1	472.54	472.2568	63.55	9.60	26.86
C25H46I2	600.45	600.1693	50.01	7.72	42.27
C25H47I1	474.56	474.2724	63.28	9.98	26.74
C25H48I2	602.47	602.1849	49.84	8.03	42.13
C25H49I1	476.57	476.2881	63.01	10.36	26.63
C25H50I2	604.49	604.2006	49.68	8.34	41.99
C25H51I1	478.59	478.3037	62.74	10.74	26.52
C26H13I1	452.30	452.0064	69.05	2.90	28.06
C26H14I2	580.21	579.9189	53.82	2.43	43.74
C26H15I1	454.31	454.0220	68.74	3.33	27.93
C26H16I2	582.23	581.9345	53.64	2.77	43.59
C26H17I1	456.33	456.0377	68.44	3.76	27.81
C26H18I2	584.24	583.9502	53.45	3.11	43.44
C26H19I1	458.35	458.0533	68.13	4.18	27.69
C26H20I2	586.26	585.9658	53.27	3.44	43.29
C26H21I1	460.36	460.0690	67.84	4.60	27.57
C26H22I2	588.27	587.9815	53.09	3.77	43.14
C26H23I1	462.38	462.0846	67.54	5.01	27.45
C26H24I2	590.29	589.9971	52.90	4.10	43.00
C26H25I1	464.39	464.1003	67.25	5.43	27.33
C26H26I2	592.31	592.0128	52.72	4.42	42.85
C26H27I1	466.41	466.1159	66.96	5.84	27.21
C26H28I2	594.32	594.0284	52.55	4.75	42.71
C26H29I1	468.43	468.1316	66.67	6.24	27.09
C26H30I2	596.34	596.0441	52.37	5.07	42.56
C26H31I1	470.44	470.1472	66.38	6.64	26.98
C26H32I2	598.35	598.0597	52.19	5.39	42.42
C26H33I1	472.46	472.1629	66.10	7.04	26.86
C26H34I2	600.37	600.0754	52.02	5.71	42.28
C26H35I1	474.47	474.1785	65.82	7.44	26.75
C26H36I2	602.39	602.0910	51.84	6.02	42.13
C26H37I1	476.49	476.1942	65.54	7.83	26.63
C26H38I2	604.40	604.1067	51.67	6.34	41.99
C26H39I1	478.51	478.2098	65.26	8.22	26.52
C26H40I2	606.42	606.1223	51.50	6.65	41.85
C26H41I1	480.52	480.2255	64.99	8.60	26.41
C26H42I2	608.43	608.1380	51.33	6.96	41.72
C26H43I1	482.54	482.2411	64.72	8.98	26.30
C26H44I2	610.45	610.1536	51.16	7.27	41.58
C26H45I1	484.55	484.2568	64.45	9.36	26.19
C26H46I2	612.47	612.1693	50.99	7.57	41.44
C26H47I1	486.57	486.2724	64.18	9.74	26.08
C26H48I2	614.48	614.1849	50.82	7.87	41.30

10/21/66 043

C26H49I1	488.58	488.2881	63.92	10.11	25.97
C26H50I2	616.50	616.2006	50.66	8.17	41.17
C26H51I1	490.60	490.3037	63.65	10.48	25.87
C26H52I2	618.51	618.2162	50.49	8.47	41.04
C26H53I1	492.62	492.3194	63.39	10.84	25.76
C27H13I1	464.31	464.0064	69.85	2.82	27.33
C27H14I2	592.22	591.9189	54.76	2.38	42.86
C27H15I1	466.33	466.0220	69.54	3.24	27.21
C27H16I2	594.24	593.9345	54.57	2.71	42.71
C27H17I1	468.34	468.0377	69.24	3.66	27.10
C27H18I2	596.25	595.9502	54.39	3.04	42.57
C27H19I1	470.36	470.0533	68.95	4.07	26.98
C27H20I2	598.27	597.9658	54.21	3.37	42.42
C27H21I1	472.37	472.0690	68.65	4.48	26.87
C27H22I2	600.29	599.9815	54.02	3.69	42.28
C27H23I1	474.39	474.0846	68.36	4.89	26.75
C27H24I2	602.30	601.9971	53.84	4.02	42.14
C27H25I1	476.40	476.1003	68.07	5.29	26.64
C27H26I2	604.32	604.0128	53.66	4.34	42.00
C27H27I1	478.42	478.1159	67.79	5.69	26.53
C27H28I2	606.33	606.0284	53.49	4.65	41.86
C27H29I1	480.44	480.1316	67.50	6.08	26.41
C27H30I2	608.35	608.0441	53.31	4.97	41.72
C27H31I1	482.45	482.1472	67.22	6.48	26.30
C27H32I2	610.36	610.0597	53.13	5.28	41.58
C27H33I1	484.47	484.1629	66.94	6.87	26.19
C27H34I2	612.38	612.0754	52.96	5.60	41.45
C27H35I1	486.48	486.1785	66.66	7.25	26.09
C27H36I2	614.40	614.0910	52.78	5.91	41.31
C27H37I1	488.50	488.1942	66.39	7.63	25.98
C27H38I2	616.41	616.1067	52.61	6.21	41.18
C27H39I1	490.52	490.2098	66.11	8.01	25.87
C27H40I2	618.43	618.1223	52.44	6.52	41.04
C27H41I1	492.53	492.2255	65.84	8.39	25.77
C27H42I2	620.44	620.1380	52.27	6.82	40.91
C27H43I1	494.55	494.2411	65.58	8.76	25.66
C27H44I2	622.46	622.1536	52.10	7.13	40.78
C27H45I1	496.56	496.2568	65.31	9.13	25.56
C27H46I2	624.48	624.1693	51.93	7.42	40.64
C27H47I1	498.58	498.2724	65.04	9.50	25.45
C27H48I2	626.49	626.1849	51.76	7.72	40.51
C27H49I1	500.60	500.2881	64.78	9.87	25.35
C27H50I2	628.51	628.2006	51.60	8.02	40.38
C27H51I1	502.61	502.3037	64.52	10.23	25.25
C27H52I2	630.52	630.2162	51.43	8.31	40.25
C27H53I1	504.63	504.3194	64.27	10.59	25.15
C27H54I2	632.54	632.2318	51.27	8.61	40.13
C27H55I1	506.64	506.3350	64.01	10.94	25.05
C28H14I2	604.23	603.9189	55.66	2.34	42.01
C28H15I1	478.34	478.0220	70.31	3.16	26.53
C28H16I2	606.25	605.9345	55.47	2.66	41.87
C28H17I1	480.35	480.0377	70.01	3.57	26.42
C28H18I2	608.26	607.9502	55.29	2.98	41.73
C28H19I1	482.37	482.0533	69.72	3.97	26.31
C28H20I2	610.28	609.9658	55.11	3.30	41.59
C28H21I1	484.38	484.0690	69.43	4.37	26.20
C28H22I2	612.30	611.9815	54.93	3.62	41.45
C28H23I1	486.40	486.0846	69.14	4.77	26.09
C28H24I2	614.31	613.9971	54.75	3.94	41.32
C28H25I1	488.42	488.1003	68.86	5.16	25.98

C28H26I2	616.33	616.0128	54.57	4.25	41.18
C28H27I1	490.43	490.1159	68.57	5.55	25.88
C28H28I2	618.34	618.0284	54.39	4.56	41.05
C28H29I1	492.45	492.1316	68.29	5.94	25.77
C28H30I2	620.36	620.0441	54.21	4.87	40.91
C28H31I1	494.46	494.1472	68.02	6.32	25.67
C28H32I2	622.38	622.0597	54.04	5.18	40.78
C28H33I1	496.48	496.1629	67.74	6.70	25.56
C28H34I2	624.39	624.0754	53.86	5.49	40.65
C28H35I1	498.50	498.1785	67.47	7.08	25.46
C28H36I2	626.41	626.0910	53.69	5.79	40.52
C28H37I1	500.51	500.1942	67.19	7.45	25.35
C28H38I2	628.42	628.1067	53.52	6.10	40.39
C28H39I1	502.53	502.2098	66.92	7.82	25.25
C28H40I2	630.44	630.1223	53.35	6.40	40.26
C28H41I1	504.54	504.2255	66.66	8.19	25.15
C28H42I2	632.46	632.1380	53.18	6.69	40.13
C28H43I1	506.56	506.2411	66.39	8.56	25.05
C28H44I2	634.47	634.1536	53.01	6.99	40.00
C28H45I1	508.58	508.2568	66.13	8.92	24.95
C28H46I2	636.49	636.1693	52.84	7.28	39.88
C28H47I1	510.59	510.2724	65.87	9.28	24.85
C28H48I2	638.50	638.1849	52.67	7.58	39.75
C28H49I1	512.61	512.2881	65.61	9.64	24.76
C28H50I2	640.52	640.2006	52.51	7.87	39.63
C28H51I1	514.62	514.3037	65.35	9.99	24.66
C28H52I2	642.54	642.2162	52.34	8.16	39.50
C28H53I1	516.64	516.3194	65.10	10.34	24.56
C28H54I2	644.55	644.2318	52.18	8.44	39.38
C28H55I1	518.65	518.3350	64.84	10.69	24.47
C28H56I2	646.57	646.2475	52.02	8.73	39.25
C28H57I1	520.67	520.3507	64.59	11.03	24.37
C29H14I2	616.24	615.9189	56.52	2.29	41.19
C29H15I1	490.35	490.0220	71.04	3.09	25.88
C29H16I2	618.26	617.9345	56.34	2.61	41.05
C29H17I1	492.36	492.0377	70.75	3.48	25.77
C29H18I2	620.28	619.9502	56.16	2.93	40.92
C29H19I1	494.38	494.0533	70.46	3.87	25.67
C29H20I2	622.29	621.9658	55.97	3.24	40.79
C29H21I1	496.40	496.0690	70.17	4.26	25.57
C29H22I2	624.31	623.9815	55.79	3.55	40.65
C29H23I1	498.41	498.0846	69.89	4.65	25.46
C29H24I2	626.32	625.9971	55.61	3.86	40.52
C29H25I1	500.43	500.1003	69.61	5.04	25.36
C29H26I2	628.34	628.0128	55.44	4.17	40.39
C29H27I1	502.44	502.1159	69.33	5.42	25.26
C29H28I2	630.36	630.0284	55.26	4.49	40.26
C29H29I1	504.46	504.1316	69.05	5.79	25.16
C29H30I2	632.37	632.0441	55.08	4.78	40.14
C29H31I1	506.47	506.1472	68.77	6.17	25.06
C29H32I2	634.39	634.0597	54.91	5.08	40.01
C29H33I1	508.49	508.1629	68.50	6.54	24.96
C29H34I2	636.40	636.0754	54.73	5.39	39.88
C29H35I1	510.51	510.1785	68.23	6.91	24.86
C29H36I2	638.42	638.0910	54.56	5.68	39.76
C29H37I1	512.52	512.1942	67.96	7.28	24.76
C29H38I2	640.44	640.1067	54.39	5.98	39.63
C29H39I1	514.54	514.2098	67.70	7.64	24.66
C29H40I2	642.45	642.1223	54.22	6.28	39.51
C29H41I1	516.55	516.2255	67.43	8.00	24.57

C29H42I2	644.47	644.1380	54.05	6.57	39.38
C29H43I1	518.57	518.2411	67.17	8.36	24.47
C29H44I2	646.48	646.1536	53.88	6.86	39.26
C29H45I1	520.59	520.2568	66.91	8.71	24.38
C29H46I2	648.50	648.1693	53.71	7.15	39.14
C29H47I1	522.60	522.2724	66.65	9.07	24.28
C29H48I2	650.51	650.1849	53.55	7.44	39.02
C29H49I1	524.62	524.2881	66.40	9.41	24.19
C29H50I2	652.53	652.2006	53.38	7.72	38.90
C29H51I1	526.63	526.3037	66.14	9.76	24.10
C29H52I2	654.55	654.2162	53.22	8.01	38.78
C29H53I1	528.65	528.3194	65.89	10.11	24.01
C29H54I2	656.56	656.2318	53.05	8.29	38.66
C29H55I1	530.67	530.3350	65.64	10.45	23.91
C29H56I2	658.58	658.2475	52.89	8.57	38.54
C29H57I1	532.68	532.3507	65.39	10.79	23.82
C29H58I2	660.59	660.2631	52.73	8.85	38.42
C29H59I1	534.70	534.3663	65.14	11.12	23.73
C30H15I1	502.36	502.0220	71.73	3.01	25.26
C30H16I2	630.27	629.9345	57.17	2.56	40.27
C30H17I1	504.37	504.0377	71.44	3.40	25.16
C30H18I2	632.29	631.9502	56.99	2.87	40.14
C30H19I1	506.39	506.0533	71.16	3.78	25.06
C30H20I2	634.30	633.9658	56.81	3.18	40.01
C30H21I1	508.41	508.0690	70.88	4.16	24.96
C30H22I2	636.32	635.9815	56.63	3.48	39.89
C30H23I1	510.42	510.0846	70.60	4.54	24.86
C30H24I2	638.33	637.9971	56.45	3.79	39.76
C30H25I1	512.44	512.1003	70.32	4.92	24.76
C30H26I2	640.35	640.0128	56.27	4.09	39.64
C30H27I1	514.45	514.1159	70.04	5.29	24.67
C30H28I2	642.37	642.0284	56.09	4.39	39.51
C30H29I1	516.47	516.1316	69.77	5.66	24.57
C30H30I2	644.38	644.0441	55.92	4.69	39.39
C30H31I1	518.49	518.1472	69.50	6.03	24.48
C30H32I2	646.40	646.0597	55.74	4.99	39.27
C30H33I1	520.50	520.1629	69.23	6.39	24.38
C30H34I2	648.41	648.0754	55.57	5.29	39.14
C30H35I1	522.52	522.1785	68.96	6.75	24.29
C30H36I2	650.43	650.0910	55.40	5.58	39.02
C30H37I1	524.53	524.1942	68.70	7.11	24.19
C30H38I2	652.45	652.1067	55.23	5.87	38.90
C30H39I1	526.55	526.2098	68.43	7.47	24.10
C30H40I2	654.46	654.1223	55.06	6.16	38.78
C30H41I1	528.57	528.2255	68.17	7.82	24.01
C30H42I2	656.48	656.1380	54.89	6.45	38.66
C30H43I1	530.58	530.2411	67.91	8.17	23.92
C30H44I2	658.49	658.1536	54.72	6.74	38.54
C30H45I1	532.60	532.2568	67.66	8.52	23.83
C30H46I2	660.51	660.1693	54.55	7.02	38.43
C30H47I1	534.61	534.2724	67.40	8.86	23.74
C30H48I2	662.53	662.1849	54.39	7.30	38.31
C30H49I1	536.63	536.2881	67.15	9.20	23.65
C30H50I2	664.54	664.2006	54.22	7.58	38.19
C30H51I1	538.65	538.3037	66.90	9.54	23.56
C30H52I2	666.56	666.2162	54.06	7.86	38.08
C30H53I1	540.66	540.3194	66.65	9.88	23.47
C30H54I2	668.57	668.2318	53.90	8.14	37.96
C30H55I1	542.68	542.3350	66.40	10.22	23.38
C30H56I2	670.59	670.2475	53.73	8.42	37.85
C30H57I1	544.69	544.3507	66.15	10.55	23.30
C30H58I2	672.61	672.2631	53.57	8.69	37.74
C30H59I1	546.71	546.3663	65.91	10.88	23.21
C30H60I2	674.62	674.2788	53.41	8.96	37.62
C30H61I1	548.73	548.3820	65.67	11.21	23.13

5. CARBON, HYDROGEN AND NITROGEN

5. KOHLENSTOFF, WASSERSTOFF UND STICKSTOFF

5. CARBONE, HYDROGENE, ET AZOTE

E

MOLECULAR WEIGHTS AND PERCENTAGE COMPOSITIONS OF COMPOUNDS
CONTAINING CARBON, HYDROGEN AND NITROGEN

FORMULA	MOL. WEIGHT	ISOMOL. WEIGHT	PERCENT C	PERCENT H	PERCENT N
C 1H 0N2	40.02	40.0061	30.01	0	69.99
C 1H 0N4	68.04	68.0123	17.65	0	82.35
C 1H 0N6	96.05	96.0184	12.50	0	87.50
C 1H 1N1	27.03	27.0109	44.44	3.73	51.83
C 1H 1N3	55.04	55.0170	21.82	1.83	76.35
C 1H 1N5	83.05	83.0232	14.46	1.21	84.32
C 1H 2N2	42.04	42.0218	28.57	4.80	66.63
C 1H 2N4	70.05	70.0279	17.15	2.88	79.98
C 1H 2N6	98.07	98.0341	12.25	2.06	85.70
C 1H 3N1	29.04	29.0265	41.36	10.41	48.23
C 1H 3N3	57.06	57.0327	21.05	5.30	73.65
C 1H 3N5	85.07	85.0388	14.12	3.55	82.33
C 1H 4N2	44.06	44.0374	27.26	9.15	63.59
C 1H 4N4	72.07	72.0436	16.67	5.59	77.74
C 1H 4N6	100.08	100.0497	12.00	4.03	83.97
C 1H 5N1	31.06	31.0422	38.67	16.23	45.10
C 1H 5N3	59.07	59.0483	20.33	8.53	71.13
C 1H 5N5	87.08	87.0545	13.79	5.79	80.42
C 1H 6N2	46.07	46.0531	26.07	13.13	60.80
C 1H 6N4	74.09	74.0592	16.21	8.16	75.62
C 1H 6N6	102.10	102.0654	11.76	5.92	82.31
C 1H 7N3	61.09	61.0640	19.66	11.55	68.79
C 1H 7N5	89.10	89.0701	13.48	7.92	78.60
C 1H 8N4	76.10	76.0749	15.78	10.60	73.62
C 1H 8N6	104.12	104.0810	11.54	7.75	80.72
C 1H 9N5	91.12	91.0858	13.18	9.96	76.86
C 1H10N6	106.13	106.0967	11.32	9.50	79.19
C 2H 1N1	39.04	39.0109	61.54	2.58	35.88
C 2H 1N3	67.05	67.0170	35.83	1.50	62.67
C 2H 1N5	95.06	95.0232	25.27	1.06	73.67
C 2H 2N2	54.05	54.0218	44.44	3.73	51.83
C 2H 2N4	82.07	82.0279	29.27	2.46	68.27
C 2H 2N6	110.08	110.0341	21.82	1.83	76.35
C 2H 3N1	41.05	41.0265	58.52	7.37	34.12
C 2H 3N3	69.07	69.0327	34.78	4.38	60.84
C 2H 3N5	97.08	97.0388	24.74	3.11	72.14
C 2H 4N2	56.07	56.0374	42.85	7.19	49.96
C 2H 4N4	84.08	84.0436	28.57	4.80	66.63
C 2H 4N6	112.09	112.0497	21.43	3.60	74.97
C 2H 5N1	43.07	43.0422	55.78	11.70	32.52
C 2H 5N3	71.08	71.0483	33.80	7.09	59.11
C 2H 5N5	99.10	99.0545	24.24	5.09	70.67
C 2H 6N2	58.08	58.0531	41.36	10.41	48.23
C 2H 6N4	86.10	86.0592	27.90	7.02	65.07
C 2H 6N6	114.11	114.0654	21.05	5.30	73.65
C 2H 7N1	45.08	45.0578	53.28	15.65	31.07
C 2H 7N3	73.10	73.0640	32.86	9.65	57.48
C 2H 7N5	101.11	101.0701	23.76	6.98	69.26
C 2H 8N2	60.10	60.0687	39.97	13.42	46.61
C 2H 8N4	88.11	88.0749	27.26	9.15	63.59

FORMULA	MOL. WEIGHT	ISOMOL. WEIGHT	PERCENT C	PERCENT H	PERCENT N
C 2H 8N6	116.13	116.0810	20.69	6.94	72.37
C 2H 9N3	75.11	75.0796	31.98	12.08	55.94
C 2H 9N5	103.13	103.0858	23.29	8.80	67.91
C 2H10N4	90.13	90.0905	26.65	11.18	62.16
C 2H10N6	118.14	118.0967	20.33	8.53	71.13
C 2H11N5	105.14	105.1014	22.85	10.55	66.61
C 2H12N6	120.16	120.1123	19.99	10.07	69.94
C 3H 1N1	51.05	51.0109	70.59	1.97	27.44
C 3H 1N3	79.06	79.0170	45.58	1.27	53.15
C 3H 1N5	107.07	107.0232	33.65	.94	65.41
C 3H 2N2	66.06	66.0218	54.54	3.05	42.40
C 3H 2N4	94.08	94.0279	38.30	2.14	59.55
C 3H 2N6	122.09	122.0341	29.51	1.65	68.83
C 3H 3N1	53.06	53.0265	67.91	5.70	26.40
C 3H 3N3	81.08	81.0327	44.44	3.73	51.83
C 3H 3N5	109.09	109.0388	33.03	2.77	64.20
C 3H 4N2	68.08	68.0374	52.93	5.92	41.15
C 3H 4N4	96.09	96.0436	37.50	4.20	58.31
C 3H 4N6	124.11	124.0497	29.03	3.25	67.72
C 3H 5N1	55.08	55.0422	65.42	9.15	25.43
C 3H 5N3	83.09	83.0483	43.36	6.07	50.57
C 3H 5N5	111.11	111.0545	32.43	4.54	63.03
C 3H 6N2	70.09	70.0531	51.41	8.63	39.97
C 3H 6N4	98.11	98.0592	36.73	6.16	57.11
C 3H 6N6	126.12	126.0654	28.57	4.80	66.63
C 3H 7N1	57.10	57.0578	63.11	12.36	24.53
C 3H 7N3	85.11	85.0640	42.34	8.29	49.37
C 3H 7N5	113.12	113.0701	31.85	6.24	61.91
C 3H 8N2	72.11	72.0687	49.97	11.18	38.85
C 3H 8N4	100.12	100.0749	35.99	8.05	55.96
C 3H 8N6	128.14	128.0810	28.12	6.29	65.59
C 3H 9N1	59.11	59.0735	60.96	15.35	23.70
C 3H 9N3	87.13	87.0796	41.36	10.41	48.23
C 3H 9N5	115.14	115.0858	31.30	7.88	60.83
C 3H10N2	74.13	74.0844	48.61	13.60	37.79
C 3H10N4	102.14	102.0905	35.28	9.87	54.85
C 3H10N6	130.15	130.0967	27.69	7.74	64.57
C 3H11N3	89.14	89.0953	40.42	12.44	47.14
C 3H11N5	117.15	117.1014	30.76	9.46	59.78
C 3H12N4	104.16	104.1062	34.60	11.61	53.79
C 3H12N6	132.17	132.1123	27.26	9.15	63.59
C 3H13N5	119.17	119.1171	30.24	11.00	58.77
C 3H14N6	134.19	134.1280	26.85	10.52	62.63
C 4H 2N2	78.07	78.0218	61.54	2.58	35.88
C 4H 2N4	106.09	106.0279	45.29	1.90	52.81
C 4H 2N6	134.10	134.0341	35.83	1.50	62.67
C 4H 3N1	65.08	65.0265	73.83	4.65	21.52
C 4H 3N3	93.09	93.0327	51.61	3.25	45.14
C 4H 3N5	121.10	121.0388	39.67	2.50	57.83
C 4H 4N2	80.09	80.0374	59.99	5.03	34.98

C 4H 4N4	108.10	108.0436	44.44	3.73	51.83
C 4H 4N6	136.12	136.0497	35.30	2.96	61.74
C 4H 5N1	67.09	67.0422	71.61	7.51	20.88
C 4H 5N3	95.10	95.0483	50.52	5.30	44.18
C 4H 5N5	123.12	123.0545	39.02	4.09	56.88
C 4H 6N2	82.11	82.0531	58.52	7.37	34.12
C 4H 6N4	110.12	110.0592	43.63	5.49	50.88
C 4H 6N6	138.13	138.0654	34.78	4.38	60.84
C 4H 7N1	69.11	69.0578	69.52	10.21	20.27
C 4H 7N3	97.12	97.0640	49.47	7.26	43.27
C 4H 7N5	125.13	125.0701	38.39	5.64	55.97
C 4H 8N2	84.12	84.0687	57.11	9.59	33.30
C 4H 8N4	112.14	112.0749	42.85	7.19	49.96
C 4H 8N6	140.15	140.0810	34.28	5.75	59.97
C 4H 9N1	71.12	71.0735	67.55	12.75	19.69
C 4H 9N3	99.14	99.0796	48.46	9.15	42.39
C 4H 9N5	127.15	127.0858	37.79	7.13	55.08
C 4H10N2	86.14	86.0844	55.78	11.70	32.52
C 4H10N4	114.15	114.0905	42.09	8.83	49.08
C 4H10N6	142.16	142.0967	33.80	7.09	59.11
C 4H11N1	73.14	73.0891	65.69	15.16	19.15
C 4H11N3	101.15	101.0953	47.50	10.96	41.54
C 4H11N5	129.17	129.1014	37.20	8.58	54.22
C 4H12N2	88.15	88.1000	54.50	13.72	31.78
C 4H12N4	116.17	116.1062	41.36	10.41	48.23
C 4H12N6	144.18	144.1123	33.32	8.39	58.29
C 4H13N3	103.17	103.1109	46.57	12.70	40.73
C 4H13N5	131.18	131.1171	36.62	9.99	53.39
C 4H14N4	118.18	118.1218	40.65	11.94	47.41
C 4H14N6	146.20	146.1280	32.86	9.65	57.48
C 4H15N5	133.20	133.1327	36.07	11.35	52.58
C 4H16N6	148.21	148.1436	32.42	10.88	56.70
C 5H 2N2	90.09	90.0218	66.67	2.24	31.10
C 5H 2N4	118.10	118.0279	50.85	1.71	47.44
C 5H 2N6	146.11	146.0341	41.10	1.38	57.52
C 5H 3N1	77.09	77.0265	77.91	3.92	18.17
C 5H 3N3	105.10	105.0327	57.14	2.88	39.98
C 5H 3N5	133.11	133.0388	45.12	2.27	52.61
C 5H 4N2	92.10	92.0374	65.21	4.38	30.42
C 5H 4N4	120.11	120.0436	50.00	3.36	46.64
C 5H 4N6	148.13	148.0497	40.54	2.72	56.73
C 5H 5N1	79.10	79.0422	75.92	6.37	17.71
C 5H 5N3	107.12	107.0483	56.07	4.71	39.23
C 5H 5N5	135.13	135.0545	44.44	3.73	51.83
C 5H 6N2	94.12	94.0531	63.81	6.43	29.76
C 5H 6N4	122.13	122.0592	49.17	4.95	45.87
C 5H 6N6	150.14	150.0654	40.00	4.03	55.97
C 5H 7N1	81.12	81.0578	74.03	8.70	17.27
C 5H 7N3	109.13	109.0640	55.03	6.47	38.50
C 5H 7N5	137.15	137.0701	43.79	5.14	51.07
C 5H 8N2	96.13	96.0687	62.47	8.39	29.14
C 5H 8N4	124.15	124.0749	48.37	6.50	45.13
C 5H 8N6	152.16	152.0810	39.47	5.30	55.23
C 5H 9N1	83.13	83.0735	72.24	10.91	16.85
C 5H 9N3	111.15	111.0796	54.03	8.16	37.81
C 5H 9N5	139.16	139.0858	43.16	6.52	50.33
C 5H10N2	98.15	98.0844	61.19	10.27	28.54
C 5H10N4	126.16	126.0905	47.60	7.99	44.41
C 5H10N6	154.18	154.0967	38.95	6.54	54.51
C 5H11N1	85.15	85.0891	70.53	13.02	16.45

C 5H11N3	113.16	113.0953	53.07	9.80	37.13
C 5H11N5	141.18	141.1014	42.54	7.85	49.61
C 5H12N2	100.16	100.1000	59.96	12.08	27.97
C 5H12N4	128.18	128.1062	46.85	9.44	43.71
C 5H12N6	156.19	156.1123	38.45	7.74	53.81
C 5H13N1	87.17	87.1048	68.90	15.03	16.07
C 5H13N3	115.18	115.1109	52.14	11.38	36.48
C 5H13N5	143.19	143.1171	41.94	9.15	48.91
C 5H14N2	102.18	102.1157	58.77	13.81	27.42
C 5H14N4	130.19	130.1218	46.13	10.84	43.03
C 5H14N6	158.21	158.1280	37.96	8.92	53.12
C 5H15N3	117.20	117.1266	51.24	12.90	35.85
C 5H15N5	145.21	145.1327	41.36	10.41	48.23
C 5H16N4	132.21	132.1375	45.42	12.20	42.38
C 5H16N6	160.22	160.1436	37.48	10.07	52.45
C 5H17N5	147.22	147.1484	40.79	11.64	47.57
C 5H18N6	162.24	162.1593	37.02	11.18	51.80
C 6H 3N1	89.10	89.0265	80.89	3.39	15.72
C 6H 3N3	117.11	117.0327	61.54	2.58	35.88
C 6H 3N5	145.12	145.0388	49.66	2.08	48.26
C 6H 4N2	104.11	104.0374	69.22	3.87	26.91
C 6H 4N4	132.13	132.0436	54.54	3.05	42.40
C 6H 4N6	160.14	160.0497	45.00	2.52	52.48
C 6H 5N1	91.11	91.0422	79.10	5.53	15.37
C 6H 5N3	119.13	119.0483	60.50	4.23	35.27
C 6H 5N5	147.14	147.0545	48.98	3.43	47.60
C 6H 6N2	106.13	106.0531	67.91	5.70	26.40
C 6H 6N4	134.14	134.0592	53.72	4.51	41.77
C 6H 6N6	162.15	162.0654	44.44	3.73	51.83
C 6H 7N1	93.13	93.0578	77.38	7.58	15.04
C 6H 7N3	121.14	121.0640	59.49	5.82	34.69
C 6H 7N5	149.16	149.0701	48.32	4.73	46.95
C 6H 8N2	108.14	108.0687	66.64	7.46	25.90
C 6H 8N4	136.16	136.0749	52.93	5.92	41.15
C 6H 8N6	164.17	164.0810	43.90	4.91	51.19
C 6H 9N1	95.15	95.0735	75.74	9.53	14.72
C 6H 9N3	123.16	123.0796	58.52	7.37	34.12
C 6H 9N5	151.17	151.0858	47.67	6.00	46.33
C 6H10N2	110.16	110.0844	65.42	9.15	25.43
C 6H10N4	138.17	138.0905	52.16	7.29	40.55
C 6H10N6	166.19	166.0967	43.36	6.07	50.57
C 6H11N1	97.16	97.0891	74.17	11.41	14.42
C 6H11N3	125.17	125.0953	57.57	8.86	33.57
C 6H11N5	153.19	153.1014	47.04	7.24	45.72
C 6H12N2	112.18	112.1000	64.24	10.78	24.97
C 6H12N4	140.19	140.1062	51.41	8.63	39.97
C 6H12N6	168.20	168.1123	42.85	7.19	49.96
C 6H13N1	99.18	99.1048	72.66	13.21	14.12
C 6H13N3	127.19	127.1109	56.66	10.30	33.04
C 6H13N5	155.20	155.1171	46.43	8.44	45.12
C 6H14N2	114.19	114.1157	63.11	12.36	24.53
C 6H14N4	142.21	142.1218	50.68	9.92	39.40
C 6H14N6	170.22	170.1280	42.34	8.29	49.37
C 6H15N1	101.19	101.1204	71.22	14.94	13.84
C 6H15N3	129.21	129.1266	55.78	11.70	32.52
C 6H15N5	157.22	157.1327	45.84	9.62	44.54
C 6H16N2	116.21	116.1313	62.02	13.88	24.11
C 6H16N4	144.22	144.1375	49.97	11.18	38.85
C 6H16N6	172.23	172.1436	41.84	9.36	48.79
C 6H17N3	131.22	131.1422	54.92	13.06	32.02

C 6H17N5	159.24	159.1484	45.26	10.76	43.98
C 6H18N4	146.24	146.1531	49.28	12.41	38.31
C 6H18N6	174.25	174.1593	41.36	10.41	48.23
C 6H19N5	161.25	161.1640	44.69	11.88	43.43
C 6H20N6	176.27	176.1749	40.89	11.44	47.68
C 7H 3N1	101.11	101.0265	83.16	2.99	13.85
C 7H 3N3	129.12	129.0327	65.12	2.34	32.54
C 7H 3N5	157.14	157.0388	53.51	1.92	44.57
C 7H 4N2	116.12	116.0374	72.40	3.47	24.12
C 7H 4N4	144.14	144.0436	58.33	2.80	38.87
C 7H 4N6	172.15	172.0497	48.84	2.34	48.82
C 7H 5N1	103.12	103.0422	81.53	4.89	13.58
C 7H 5N3	131.14	131.0483	64.11	3.84	32.04
C 7H 5N5	159.15	159.0545	52.83	3.17	44.00
C 7H 6N2	118.14	118.0531	71.17	5.12	23.71
C 7H 6N4	146.15	146.0592	57.53	4.14	38.33
C 7H 6N6	174.17	174.0654	48.27	3.47	48.25
C 7H 7N1	105.14	105.0578	79.97	6.71	13.32
C 7H 7N3	133.15	133.0640	63.14	5.30	31.56
C 7H 7N5	161.17	161.0701	52.17	4.38	43.45
C 7H 8N2	120.16	120.0687	69.97	6.71	23.31
C 7H 8N4	148.17	148.0749	56.74	5.44	37.81
C 7H 8N6	176.18	176.0810	47.72	4.58	47.70
C 7H 9N1	107.16	107.0735	78.46	8.47	13.07
C 7H 9N3	135.17	135.0796	62.20	6.71	31.09
C 7H 9N5	163.18	163.0858	51.52	5.56	42.92
C 7H10N2	122.17	122.0844	68.82	8.25	22.93
C 7H10N4	150.18	150.0905	55.98	6.71	37.31
C 7H10N6	178.20	178.0967	47.18	5.66	47.16
C 7H11N1	109.17	109.0891	77.01	10.16	12.83
C 7H11N3	137.19	137.0953	61.29	8.08	30.63
C 7H11N5	165.20	165.1014	50.89	6.71	42.39
C 7H12N2	124.19	124.1000	67.70	9.74	22.56
C 7H12N4	152.20	152.1062	55.24	7.95	36.81
C 7H12N6	180.21	180.1123	46.65	6.71	46.63
C 7H13N1	111.19	111.1048	75.62	11.79	12.60
C 7H13N3	139.20	139.1109	60.40	9.41	30.19
C 7H13N5	167.22	167.1171	50.28	7.84	41.88
C 7H14N2	126.20	126.1157	66.62	11.18	22.20
C 7H14N4	154.22	154.1218	54.52	9.15	36.33
C 7H14N6	182.23	182.1280	46.14	7.74	46.12
C 7H15N1	113.20	113.1204	74.27	13.36	12.37
C 7H15N3	141.22	141.1266	59.54	10.71	29.76
C 7H15N5	169.23	169.1327	49.68	8.93	41.38
C 7H16N2	128.22	128.1313	65.57	12.58	21.85
C 7H16N4	156.23	156.1375	53.82	10.32	35.86
C 7H16N6	184.25	184.1436	45.63	8.75	45.61
C 7H17N1	115.22	115.1361	72.97	14.87	12.16
C 7H17N3	143.23	143.1422	58.70	11.96	29.34
C 7H17N5	171.25	171.1484	49.10	10.01	40.90
C 7H18N2	130.23	130.1470	64.56	13.93	21.51
C 7H18N4	158.25	158.1531	53.13	11.47	35.40
C 7H18N6	186.26	186.1593	45.14	9.74	45.12
C 7H19N3	145.25	145.1579	57.89	13.19	28.93
C 7H19N5	173.26	173.1640	48.53	11.05	40.42
C 7H20N4	160.26	160.1688	52.46	12.58	34.96
C 7H20N6	188.28	188.1749	44.66	10.71	44.64
C 7H21N5	175.28	175.1797	47.97	12.08	39.96
C 7H22N6	190.29	190.1906	44.18	11.65	44.16
C 8H 4N2	128.13	128.0374	74.99	3.15	21.86

10/21/66 047

C 8H 4N4	156.15	156.0436	61.54	2.58	35.88
C 8H 4N6	184.16	184.0497	52.18	2.19	45.63
C 8H 5N1	115.14	115.0422	83.46	4.38	12.17
C 8H 5N3	143.15	143.0483	67.13	3.52	29.35
C 8H 5N5	171.16	171.0545	56.14	2.94	40.92
C 8H 6N2	130.15	130.0531	73.83	4.65	21.52
C 8H 6N4	158.16	158.0592	60.75	3.82	35.42
C 8H 6N6	186.18	186.0654	51.61	3.25	45.14
C 8H 7N1	117.15	117.0578	82.02	6.02	11.96
C 8H 7N3	145.17	145.0640	66.19	4.86	28.95
C 8H 7N5	173.18	173.0701	55.49	4.07	40.44
C 8H 8N2	132.17	132.0687	72.70	6.10	21.20
C 8H 8N4	160.18	160.0749	59.99	5.03	34.98
C 8H 8N6	188.19	188.0810	51.06	4.28	44.66
C 8H 9N1	119.17	119.0735	80.63	7.61	11.75
C 8H 9N3	147.18	147.0796	65.29	6.16	28.55
C 8H 9N5	175.19	175.0858	54.85	5.18	39.97
C 8H10N2	134.18	134.0844	71.61	7.51	20.88
C 8H10N4	162.20	162.0905	59.24	6.21	34.54
C 8H10N6	190.21	190.0967	50.52	5.30	44.18
C 8H11N1	121.18	121.0891	79.29	9.15	11.56
C 8H11N3	149.20	149.0953	64.40	7.43	28.16
C 8H11N5	177.21	177.1014	54.22	6.26	39.52
C 8H12N2	136.20	136.1000	70.55	8.88	20.57
C 8H12N4	164.21	164.1062	58.52	7.37	34.12
C 8H12N6	192.23	192.1123	49.99	6.29	43.72
C 8H13N1	123.20	123.1048	77.99	10.64	11.37
C 8H13N3	151.21	151.1109	63.55	8.67	27.79
C 8H13N5	179.23	179.1171	53.61	7.31	39.08
C 8H14N2	138.21	138.1157	69.52	10.21	20.27
C 8H14N4	166.23	166.1218	57.81	8.49	33.70
C 8H14N6	194.24	194.1280	49.47	7.26	43.27
C 8H15N1	125.22	125.1204	76.74	12.07	11.19
C 8H15N3	153.23	153.1266	62.71	9.87	27.42
C 8H15N5	181.24	181.1327	53.02	8.34	38.64
C 8H16N2	140.23	140.1313	68.52	11.50	19.98
C 8H16N4	168.24	168.1375	57.11	9.59	33.30
C 8H16N6	196.26	196.1436	48.96	8.22	42.82
C 8H17N1	127.23	127.1361	75.52	13.47	11.01
C 8H17N3	155.24	155.1422	61.90	11.04	27.07
C 8H17N5	183.26	183.1484	52.43	9.35	38.22
C 8H18N2	142.25	142.1470	67.55	12.75	19.69
C 8H18N4	170.26	170.1531	56.44	10.66	32.91
C 8H18N6	198.27	198.1593	48.46	9.15	42.39
C 8H19N1	129.25	129.1517	74.35	14.82	10.84
C 8H19N3	157.26	157.1579	61.10	12.18	26.72
C 8H19N5	185.27	185.1640	51.86	10.34	37.80
C 8H20N2	144.26	144.1626	66.61	13.97	19.42
C 8H20N4	172.28	172.1688	55.78	11.70	32.52
C 8H20N6	200.29	200.1749	47.98	10.07	41.96
C 8H21N3	159.28	159.1735	60.33	13.29	26.38
C 8H21N5	187.29	187.1797	51.31	11.30	37.39
C 8H22N4	174.29	174.1844	55.13	12.72	32.15
C 8H22N6	202.30	202.1906	47.50	10.96	41.54
C 8H23N5	189.31	189.1953	50.76	12.25	36.99
C 8H24N6	204.32	204.2062	47.03	11.84	41.13
C 9H 4N2	140.15	140.0374	77.13	2.88	19.99
C 9H 4N4	168.16	168.0436	64.28	2.40	33.32
C 9H 4N6	196.17	196.0497	55.10	2.06	42.84
C 9H 5N1	127.15	127.0422	85.02	3.96	11.02

C 9H 5N3	155.16	155.0483	69.67	3.25	27.08
C 9H 5N5	183.17	183.0545	59.02	2.75	38.23
C 9H 6N2	142.16	142.0531	76.04	4.25	19.71
C 9H 6N4	170.17	170.0592	63.52	3.55	32.92
C 9H 6N6	198.19	198.0654	54.54	3.05	42.40
C 9H 7N1	129.16	129.0578	83.69	5.46	10.84
C 9H 7N3	157.18	157.0640	68.78	4.49	26.73
C 9H 7N5	185.19	185.0701	58.37	3.81	37.82
C 9H 8N2	144.18	144.0687	74.98	5.59	19.43
C 9H 8N4	172.19	172.0749	62.78	4.68	32.54
C 9H 8N6	200.20	200.0810	54.00	4.03	41.98
C 9H 9N1	131.18	131.0735	82.41	6.92	10.68
C 9H 9N3	159.19	159.0796	67.91	5.70	26.40
C 9H 9N5	187.21	187.0858	57.74	4.85	37.41
C 9H10N2	146.19	146.0844	73.94	6.89	19.16
C 9H10N4	174.21	174.0905	62.05	5.79	32.16
C 9H10N6	202.22	202.0967	53.46	4.98	41.56
C 9H11N1	133.19	133.0891	81.16	8.32	10.52
C 9H11N3	161.21	161.0953	67.06	6.88	26.07
C 9H11N5	189.22	189.1014	57.13	5.86	37.01
C 9H12N2	148.21	148.1000	72.94	8.16	18.90
C 9H12N4	176.22	176.1062	61.34	6.86	31.79
C 9H12N6	204.24	204.1123	52.93	5.92	41.15
C 9H13N1	135.21	135.1048	79.95	9.69	10.36
C 9H13N3	163.22	163.1109	66.23	8.03	25.74
C 9H13N5	191.24	191.1171	56.53	6.85	36.62
C 9H14N2	150.23	150.1157	71.96	9.39	18.65
C 9H14N4	178.24	178.1218	60.65	7.92	31.43
C 9H14N6	206.25	206.1280	52.41	6.84	40.75
C 9H15N1	137.23	137.1204	78.78	11.02	10.21
C 9H15N3	165.24	165.1266	65.42	9.15	25.43
C 9H15N5	193.25	193.1327	55.94	7.82	36.24
C 9H16N2	152.24	152.1313	71.01	10.59	18.40
C 9H16N4	180.25	180.1375	59.97	8.95	31.08
C 9H16N6	208.27	208.1436	51.90	7.74	40.35
C 9H17N1	139.24	139.1361	77.63	12.31	10.06
C 9H17N3	167.26	167.1422	64.63	10.25	25.12
C 9H17N5	195.27	195.1484	55.36	8.78	35.87
C 9H18N2	154.26	154.1470	70.08	11.76	18.16
C 9H18N4	182.27	182.1531	59.31	9.95	30.74
C 9H18N6	210.28	210.1593	51.41	8.63	39.97
C 9H19N1	141.26	141.1517	76.53	13.56	9.92
C 9H19N3	169.27	169.1579	63.86	11.31	24.82
C 9H19N5	197.29	197.1640	54.79	9.71	35.50
C 9H20N2	156.27	156.1626	69.17	12.90	17.93
C 9H20N4	184.29	184.1688	58.66	10.94	30.40
C 9H20N6	212.30	212.1749	50.92	9.50	39.59
C 9H21N1	143.27	143.1674	75.45	14.77	9.78
C 9H21N3	171.29	171.1735	63.11	12.36	24.53
C 9H21N5	199.30	199.1797	54.24	10.62	35.14
C 9H22N2	158.29	158.1783	68.29	14.01	17.70
C 9H22N4	186.30	186.1844	58.02	11.90	30.07
C 9H22N6	214.32	214.1906	50.44	10.35	39.21
C 9H23N3	173.30	173.1892	62.38	13.38	24.25
C 9H23N5	201.32	201.1953	53.70	11.52	34.79
C 9H24N4	188.32	188.2001	57.40	12.85	29.75
C 9H24N6	216.33	216.2062	49.97	11.18	38.85
C 9H25N5	203.33	203.2110	53.16	12.39	34.44
C 9H26N6	218.35	218.2219	49.51	12.00	38.49
C10H 5N1	139.16	139.0422	86.31	3.62	10.07

10/21/66 048

C10H 5N3	167.17	167.0483	71.85	3.01	25.14
C10H 5N5	195.18	195.0545	61.54	2.58	35.88
C10H 6N2	154.17	154.0531	77.91	3.92	18.17
C10H 6N4	182.19	182.0592	65.93	3.32	30.75
C10H 6N6	210.20	210.0654	57.14	2.88	39.98
C10H 7N1	141.17	141.0578	85.08	5.00	9.92
C10H 7N3	169.19	169.0640	70.99	4.17	24.84
C10H 7N5	197.20	197.0701	60.91	3.58	35.51
C10H 8N2	156.19	156.0687	76.90	5.16	17.94
C10H 8N4	184.20	184.0749	65.21	4.38	30.42
C10H 8N6	212.22	212.0810	56.60	3.80	39.60
C10H 9N1	143.19	143.0735	83.88	6.34	9.78
C10H 9N3	171.20	171.0796	70.16	5.30	24.54
C10H 9N5	199.22	199.0858	60.29	4.55	35.15
C10H10N2	158.20	158.0844	75.92	6.37	17.71
C10H10N4	186.22	186.0905	64.50	5.41	30.09
C10H10N6	214.23	214.0967	56.07	4.71	39.23
C10H11N1	145.21	145.0891	82.72	7.64	9.65
C10H11N3	173.22	173.0953	69.34	6.40	24.26
C10H11N5	201.23	201.1014	59.69	5.51	34.80
C10H12N2	160.22	160.1000	74.97	7.55	17.48
C10H12N4	188.23	188.1062	63.81	6.43	29.76
C10H12N6	216.25	216.1123	55.54	5.59	38.86
C10H13N1	147.22	147.1048	81.59	8.90	9.51
C10H13N3	175.24	175.1109	68.54	7.48	23.98
C10H13N5	203.25	203.1171	59.10	6.45	34.46
C10H14N2	162.24	162.1157	74.03	8.70	17.27
C10H14N4	190.25	190.1218	63.13	7.42	29.45
C10H14N6	218.26	218.1280	55.03	6.47	38.50
C10H15N1	149.24	149.1204	80.48	10.13	9.39
C10H15N3	177.25	177.1266	67.76	8.53	23.71
C10H15N5	205.26	205.1327	58.52	7.37	34.12
C10H16N2	164.25	164.1313	73.13	9.82	17.06
C10H16N4	192.27	192.1375	62.47	8.39	29.14
C10H16N6	220.28	220.1436	54.53	7.32	38.15
C10H17N1	151.25	151.1361	79.41	11.33	9.26
C10H17N3	179.27	179.1422	67.00	9.56	23.44
C10H17N5	207.28	207.1484	57.95	8.27	33.79
C10H18N2	166.27	166.1470	72.24	10.91	16.85
C10H18N4	194.28	194.1531	61.82	9.34	28.84
C10H18N6	222.30	222.1593	54.03	8.16	37.81
C10H19N1	153.27	153.1517	78.37	12.50	9.14
C10H19N3	181.28	181.1579	66.26	10.56	23.18
C10H19N5	209.30	209.1640	57.39	9.15	33.46
C10H20N2	168.28	168.1626	71.37	11.98	16.65
C10H20N4	196.30	196.1688	61.19	10.27	28.54
C10H20N6	224.31	224.1749	53.55	8.99	37.47
C10H21N1	155.29	155.1674	77.35	13.63	9.02
C10H21N3	183.30	183.1735	65.53	11.55	22.92
C10H21N5	211.31	211.1797	56.84	10.02	33.14
C10H22N2	170.30	170.1783	70.53	13.02	16.45
C10H22N4	198.31	198.1844	60.57	11.18	28.25
C10H22N6	226.33	226.1906	53.07	9.80	37.13
C10H23N1	157.30	157.1830	76.36	14.74	8.90
C10H23N3	185.31	185.1892	64.81	12.51	22.67
C10H23N5	213.33	213.1953	56.30	10.87	32.83
C10H24N2	172.32	172.1939	69.70	14.04	16.26
C10H24N4	200.33	200.2001	59.96	12.08	27.97
C10H24N6	228.34	228.2062	52.60	10.59	36.80
C10H25N3	187.33	187.2048	64.12	13.45	22.43

10/21/66 049

C10H25N5	215.34	215.2110	55.78	11.70	32.52
C10H26N4	202.35	202.2157	59.36	12.95	27.69
C10H26N6	230.36	230.2219	52.14	11.38	36.48
C10H27N5	217.35	217.2266	55.26	12.52	32.22
C10H28N6	232.37	232.2375	51.69	12.15	36.17
C11H 5N1	151.17	151.0422	87.40	3.33	9.27
C11H 5N3	179.18	179.0483	73.74	2.81	23.45
C11H 5N5	207.20	207.0545	63.77	2.43	33.80
C11H 6N2	166.18	166.0531	79.50	3.64	16.86
C11H 6N4	194.20	194.0592	68.04	3.11	28.85
C11H 6N6	222.21	222.0654	59.46	2.72	37.82
C11H 7N1	153.19	153.0578	86.25	4.61	9.14
C11H 7N3	181.20	181.0640	72.92	3.89	23.19
C11H 7N5	209.21	209.0701	63.15	3.37	33.47
C11H 8N2	168.20	168.0687	78.55	4.79	16.65
C11H 8N4	196.21	196.0749	67.34	4.11	28.55
C11H 8N6	224.23	224.0810	58.92	3.60	37.48
C11H 9N1	155.20	155.0735	85.13	5.85	9.02
C11H 9N3	183.21	183.0796	72.11	4.95	22.93
C11H 9N5	211.23	211.0858	62.55	4.29	33.16
C11H10N2	170.22	170.0844	77.62	5.92	16.46
C11H10N4	198.23	198.0905	66.65	5.08	28.26
C11H10N6	226.24	226.0967	58.40	4.46	37.15
C11H11N1	157.22	157.0891	84.04	7.05	8.91
C11H11N3	185.23	185.0953	71.33	5.99	22.69
C11H11N5	213.24	213.1014	61.96	5.20	32.84
C11H12N2	172.23	172.1000	76.71	7.02	16.26
C11H12N4	200.25	200.1062	65.98	6.04	27.98
C11H12N6	228.26	228.1123	57.88	5.30	36.82
C11H13N1	159.23	159.1048	82.97	8.23	8.80
C11H13N3	187.25	187.1109	70.56	7.00	22.44
C11H13N5	215.26	215.1171	61.38	6.09	32.53
C11H14N2	174.25	174.1157	75.82	8.10	16.08
C11H14N4	202.26	202.1218	65.32	6.98	27.70
C11H14N6	230.27	230.1280	57.38	6.13	36.50
C11H15N1	161.25	161.1204	81.94	9.38	8.69
C11H15N3	189.26	189.1266	69.81	7.99	22.20
C11H15N5	217.28	217.1327	60.81	6.96	32.23
C11H16N2	176.26	176.1313	74.96	9.15	15.89
C11H16N4	204.28	204.1375	64.68	7.89	27.43
C11H16N6	232.29	232.1436	56.88	6.94	36.18
C11H17N1	163.26	163.1361	80.93	10.50	8.58
C11H17N3	191.28	191.1422	69.07	8.96	21.97
C11H17N5	219.29	219.1484	60.25	7.81	31.94
C11H18N2	178.28	178.1470	74.11	10.18	15.71
C11H18N4	206.29	206.1531	64.05	8.79	27.16
C11H18N6	234.31	234.1593	56.39	7.74	35.87
C11H19N1	165.28	165.1517	79.94	11.59	8.47
C11H19N3	193.29	193.1579	68.35	9.91	21.74
C11H19N5	221.31	221.1640	59.70	8.65	31.65
C11H20N2	180.30	180.1626	73.28	11.18	15.54
C11H20N4	208.31	208.1688	63.43	9.68	26.90
C11H20N6	236.32	236.1749	55.91	8.53	35.56
C11H21N1	167.30	167.1674	78.98	12.65	8.37
C11H21N3	195.31	195.1735	67.65	10.84	21.51
C11H21N5	223.32	223.1797	59.16	9.48	31.36
C11H22N2	182.31	182.1783	72.47	12.16	15.37
C11H22N4	210.32	210.1844	62.82	10.54	26.64
C11H22N6	238.34	238.1906	55.43	9.30	35.26
C11H23N1	169.31	169.1830	78.03	13.69	8.27
C11H23N3	197.33	197.1892	66.96	11.75	21.29
C11H23N5	225.34	225.1953	58.63	10.29	31.08
C11H24N2	184.33	184.1939	71.68	13.12	15.20
C11H24N4	212.34	212.2001	62.22	11.39	26.39
C11H24N6	240.35	240.2062	54.97	10.06	34.97
C11H25N1	171.33	171.1987	77.12	14.71	8.18
C11H25N3	199.34	199.2048	66.28	12.64	21.08
C11H25N5	227.36	227.2110	58.11	11.08	30.80
C11H26N2	186.34	186.2096	70.90	14.06	15.03
C11H26N4	214.36	214.2157	61.64	12.23	26.14
C11H26N6	242.37	242.2219	54.51	10.81	34.67
C11H27N3	201.36	201.2205	65.62	13.52	20.87
C11H27N5	229.37	229.2266	57.60	11.87	30.53
C11H28N4	216.37	216.2314	61.06	13.04	25.89
C11H28N6	244.39	244.2375	54.06	11.55	34.39
C11H29N5	231.39	231.2423	57.10	12.63	30.27
C11H30N6	246.40	246.2532	53.62	12.27	34.11
C12H 6N2	178.20	178.0531	80.89	3.39	15.72
C12H 6N4	206.21	206.0592	69.90	2.93	27.17
C12H 6N6	234.22	234.0654	61.54	2.58	35.88
C12H 7N1	165.20	165.0578	87.25	4.27	8.48
C12H 7N3	193.21	193.0640	74.60	3.65	21.75
C12H 7N5	221.22	221.0701	65.15	3.19	31.66
C12H 8N2	180.21	180.0687	79.98	4.47	15.54
C12H 8N4	208.22	208.0749	69.22	3.87	26.91
C12H 8N6	236.24	236.0810	61.01	3.41	35.57
C12H 9N1	167.21	167.0735	86.20	5.43	8.38
C12H 9N3	195.23	195.0796	73.83	4.65	21.52
C12H 9N5	223.24	223.0858	64.56	4.06	31.37
C12H10N2	182.23	182.0844	79.10	5.53	15.37
C12H10N4	210.24	210.0905	68.56	4.79	26.65
C12H10N6	238.25	238.0967	60.50	4.23	35.27
C12H11N1	169.23	169.0891	85.17	6.55	8.28
C12H11N3	197.24	197.0953	73.07	5.62	21.30
C12H11N5	225.25	225.1014	63.99	4.92	31.09
C12H12N2	184.24	184.1000	78.23	6.57	15.20
C12H12N4	212.26	212.1062	67.91	5.70	26.40
C12H12N6	240.27	240.1123	59.99	5.03	34.98
C12H13N1	171.24	171.1048	84.17	7.65	8.18
C12H13N3	199.26	199.1109	72.34	6.58	21.09
C12H13N5	227.27	227.1171	63.42	5.77	30.81
C12H14N2	186.26	186.1157	77.38	7.58	15.04
C12H14N4	214.27	214.1218	67.27	6.59	26.15
C12H14N6	242.29	242.1280	59.49	5.82	34.69
C12H15N1	173.26	173.1204	83.19	8.73	8.08
C12H15N3	201.27	201.1266	71.61	7.51	20.88
C12H15N5	229.29	229.1327	62.86	6.59	30.54
C12H16N2	188.27	188.1313	76.56	8.57	14.88
C12H16N4	216.29	216.1375	66.64	7.46	25.90
C12H16N6	244.30	244.1436	59.00	6.60	34.40
C12H17N1	175.28	175.1361	82.23	9.78	7.99
C12H17N3	203.29	203.1422	70.90	8.43	20.67
C12H17N5	231.30	231.1484	62.31	7.41	30.28
C12H18N2	190.29	190.1470	75.74	9.53	14.72
C12H18N4	218.30	218.1531	66.02	8.31	25.66
C12H18N6	246.32	246.1593	58.52	7.37	34.12
C12H19N1	177.29	177.1517	81.30	10.80	7.90
C12H19N3	205.31	205.1579	70.20	9.33	20.47
C12H19N5	233.32	233.1640	61.78	8.21	30.02
C12H20N2	192.31	192.1626	74.95	10.48	14.57

10/21/66 050

C12H20N4	220.32	220.1688	65.42	9.15	25.43
C12H20N6	248.33	248.1749	58.04	8.12	33.84
C12H21N1	179.31	179.1674	80.38	11.81	7.81
C12H21N3	207.32	207.1735	69.52	10.21	20.27
C12H21N5	235.33	235.1797	61.25	8.99	29.76
C12H22N2	194.32	194.1783	74.17	11.41	14.42
C12H22N4	222.34	222.1844	64.83	9.97	25.20
C12H22N6	250.35	250.1906	57.57	8.86	33.57
C12H23N1	181.32	181.1830	79.49	12.79	7.72
C12H23N3	209.34	209.1892	68.85	11.07	20.07
C12H23N5	237.35	237.1953	60.73	9.77	29.51
C12H24N2	196.34	196.1939	73.41	12.32	14.27
C12H24N4	224.35	224.2001	64.24	10.78	24.97
C12H24N6	252.37	252.2062	57.11	9.59	33.30
C12H25N1	183.34	183.1987	78.62	13.74	7.64
C12H25N3	211.35	211.2048	68.20	11.92	19.88
C12H25N5	239.37	239.2110	60.21	10.53	29.26
C12H26N2	198.35	198.2096	72.66	13.21	14.12
C12H26N4	226.37	226.2157	63.67	11.58	24.75
C12H26N6	254.38	254.2219	56.66	10.30	33.04
C12H27N1	185.36	185.2143	77.76	14.68	7.56
C12H27N3	213.37	213.2205	67.55	12.75	19.69
C12H27N5	241.38	241.2266	59.71	11.27	29.01
C12H28N2	200.37	200.2252	71.93	14.09	13.98
C12H28N4	228.38	228.2314	63.11	12.36	24.53
C12H28N6	256.40	256.2375	56.22	11.01	32.78
C12H29N3	215.39	215.2361	66.92	13.57	19.51
C12H29N5	243.40	243.2423	59.22	12.01	28.77
C12H30N4	230.40	230.2470	62.56	13.12	24.32
C12H30N6	258.41	258.2532	55.78	11.70	32.52
C12H31N5	245.41	245.2579	58.73	12.73	28.54
C12H32N6	260.43	260.2688	55.34	12.39	32.27
C13H 6N2	190.21	190.0531	82.09	3.18	14.73
C13H 6N4	218.22	218.0592	71.55	2.77	25.67
C13H 6N6	246.23	246.0654	63.41	2.46	34.13
C13H 7N1	177.21	177.0578	88.11	3.98	7.90
C13H 7N3	205.22	205.0640	76.09	3.44	20.48
C13H 7N5	233.23	233.0701	66.95	3.03	30.03
C13H 8N2	192.22	192.0687	81.23	4.20	14.57
C13H 8N4	220.24	220.0749	70.90	3.66	25.44
C13H 8N6	248.25	248.0810	62.90	3.25	33.85
C13H 9N1	179.22	179.0735	87.12	5.06	7.82
C13H 9N3	207.24	207.0796	75.35	4.38	20.28
C13H 9N5	235.25	235.0858	66.37	3.86	29.77
C13H10N2	194.24	194.0844	80.39	5.19	14.42
C13H10N4	222.25	222.0905	70.26	4.54	25.21
C13H10N6	250.26	250.0967	62.39	4.03	33.58
C13H11N1	181.24	181.0891	86.15	6.12	7.73
C13H11N3	209.25	209.0953	74.62	5.30	20.08
C13H11N5	237.27	237.1014	65.81	4.67	29.52
C13H12N2	196.25	196.1000	79.56	6.16	14.27
C13H12N4	224.27	224.1062	69.62	5.39	24.98
C13H12N6	252.28	252.1123	61.89	4.79	33.31
C13H13N1	183.26	183.1048	85.21	7.15	7.64
C13H13N3	211.27	211.1109	73.91	6.20	19.89
C13H13N5	239.28	239.1171	65.26	5.48	29.27
C13H14N2	198.27	198.1157	78.75	7.12	14.13
C13H14N4	226.28	226.1218	69.00	6.24	24.76
C13H14N6	254.30	254.1280	61.40	5.55	33.05
C13H15N1	185.27	185.1204	84.28	8.16	7.56
C13H15N3	213.28	213.1266	73.21	7.09	19.70
C13H15N5	241.30	241.1327	64.71	6.27	29.02
C13H16N2	200.29	200.1313	77.96	8.05	13.99
C13H16N4	228.30	228.1375	68.39	7.06	24.54
C13H16N6	256.31	256.1436	60.92	6.29	32.79
C13H17N1	187.29	187.1361	83.37	9.15	7.48
C13H17N3	215.30	215.1422	72.52	7.96	19.52
C13H17N5	243.31	243.1484	64.17	7.04	28.78
C13H18N2	202.30	202.1470	77.18	8.97	13.85
C13H18N4	230.32	230.1531	67.80	7.88	24.33
C13H18N6	258.33	258.1593	60.44	7.02	32.53
C13H19N1	189.30	189.1517	82.48	10.12	7.40
C13H19N3	217.32	217.1579	71.85	8.81	19.34
C13H19N5	245.33	245.1640	63.65	7.81	28.55
C13H20N2	204.32	204.1626	76.42	9.87	13.71
C13H20N4	232.33	232.1688	67.21	8.68	24.12
C13H20N6	260.34	260.1749	59.98	7.74	32.28
C13H21N1	191.32	191.1674	81.61	11.06	7.32
C13H21N3	219.33	219.1735	71.19	9.65	19.16
C13H21N5	247.35	247.1797	63.13	8.56	28.31
C13H22N2	206.33	206.1783	75.68	10.75	13.58
C13H22N4	234.35	234.1844	66.63	9.46	23.91
C13H22N6	262.36	262.1906	59.52	8.45	32.03
C13H23N1	193.33	193.1830	80.76	11.99	7.24
C13H23N3	221.35	221.1892	70.54	10.47	18.98
C13H23N5	249.36	249.1953	62.62	9.30	28.09
C13H24N2	208.35	208.1939	74.94	11.61	13.45
C13H24N4	236.36	236.2001	66.06	10.23	23.70
C13H24N6	264.38	264.2062	59.06	9.15	31.79
C13H25N1	195.35	195.1987	79.93	12.90	7.17
C13H25N3	223.36	223.2048	69.91	11.28	18.81
C13H25N5	251.38	251.2110	62.12	10.02	27.86
C13H26N2	210.37	210.2096	74.23	12.46	13.32
C13H26N4	238.38	238.2157	65.50	10.99	23.50
C13H26N6	266.39	266.2219	58.61	9.84	31.55
C13H27N1	197.37	197.2143	79.11	13.79	7.10
C13H27N3	225.38	225.2205	69.28	12.08	18.64
C13H27N5	253.39	253.2266	61.62	10.74	27.64
C13H28N2	212.38	212.2252	73.52	13.29	13.19
C13H28N4	240.39	240.2314	64.95	11.74	23.31
C13H28N6	268.41	268.2375	58.17	10.52	31.31
C13H29N1	199.38	199.2300	78.31	14.66	7.03
C13H29N3	227.40	227.2361	68.67	12.85	18.48
C13H29N5	255.41	255.2423	61.14	11.44	27.42
C13H30N2	214.40	214.2409	72.83	14.10	13.07
C13H30N4	242.41	242.2470	64.41	12.47	23.11
C13H30N6	270.42	270.2532	57.74	11.18	31.08
C13H31N3	229.41	229.2518	68.06	13.62	18.32
C13H31N5	257.43	257.2579	60.66	12.14	27.21
C13H32N4	244.43	244.2627	63.88	13.20	22.92
C13H32N6	272.44	272.2688	57.31	11.84	30.85
C13H33N5	259.44	259.2736	60.19	12.82	26.99
C13H34N6	274.46	274.2845	56.89	12.49	30.62
C14H 7N1	189.22	189.0578	88.87	3.73	7.40
C14H 7N3	217.23	217.0640	77.41	3.25	19.34
C14H 7N5	245.25	245.0701	68.57	2.88	28.56
C14H 8N2	204.23	204.0687	82.34	3.95	13.72
C14H 8N4	232.25	232.0749	72.40	3.47	24.12
C14H 8N6	260.26	260.0810	64.61	3.10	32.29
C14H 9N1	191.23	191.0735	87.93	4.74	7.32

C14H 9N3	219.25	219.0796	76.70	4.14	19.17
C14H 9N5	247.26	247.0858	68.01	3.67	28.32
C14H10N2	206.25	206.0844	81.53	4.89	13.58
C14H10N4	234.26	234.0905	71.78	4.30	23.92
C14H10N6	262.28	262.0967	64.11	3.84	32.04
C14H11N1	193.25	193.0891	87.01	5.74	7.25
C14H11N3	221.26	221.0953	76.00	5.01	18.99
C14H11N5	249.28	249.1014	67.46	4.45	28.09
C14H12N2	208.27	208.1000	80.74	5.81	13.45
C14H12N4	236.28	236.1062	71.17	5.12	23.71
C14H12N6	264.29	264.1123	63.63	4.58	31.80
C14H13N1	195.27	195.1048	86.12	6.71	7.17
C14H13N3	223.28	223.1109	75.31	5.87	18.82
C14H13N5	251.29	251.1171	66.92	5.21	27.87
C14H14N2	210.28	210.1157	79.97	6.71	13.32
C14H14N4	238.29	238.1218	70.57	5.92	23.51
C14H14N6	266.31	266.1280	63.14	5.30	31.56
C14H15N1	197.28	197.1204	85.24	7.66	7.10
C14H15N3	225.30	225.1266	74.64	6.71	18.65
C14H15N5	253.31	253.1327	66.38	5.97	27.65
C14H16N2	212.30	212.1313	79.21	7.60	13.20
C14H16N4	240.31	240.1375	69.97	6.71	23.31
C14H16N6	268.32	268.1436	62.67	6.01	31.32
C14H17N1	199.30	199.1361	84.37	8.60	7.03
C14H17N3	227.31	227.1422	73.98	7.54	18.49
C14H17N5	255.33	255.1484	65.86	6.71	27.43
C14H18N2	214.31	214.1470	78.46	8.47	13.07
C14H18N4	242.33	242.1531	69.39	7.49	23.12
C14H18N6	270.34	270.1593	62.20	6.71	31.09
C14H19N1	201.31	201.1517	83.53	9.51	6.96
C14H19N3	229.33	229.1579	73.33	8.35	18.32
C14H19N5	257.34	257.1640	65.34	7.44	27.21
C14H20N2	216.33	216.1626	77.73	9.32	12.95
C14H20N4	244.34	244.1688	68.82	8.25	22.93
C14H20N6	272.36	272.1749	61.74	7.40	30.86
C14H21N1	203.33	203.1674	82.70	10.41	6.89
C14H21N3	231.34	231.1735	72.69	9.15	18.16
C14H21N5	259.36	259.1797	64.84	8.16	27.00
C14H22N2	218.34	218.1783	77.01	10.16	12.83
C14H22N4	246.36	246.1844	68.26	9.00	22.74
C14H22N6	274.37	274.1906	61.29	8.08	30.63
C14H23N1	205.35	205.1830	81.89	11.29	6.82
C14H23N3	233.36	233.1892	72.06	9.93	18.01
C14H23N5	261.37	261.1953	64.34	8.87	26.79
C14H24N2	220.36	220.1939	76.31	10.98	12.71
C14H24N4	248.37	248.2001	67.70	9.74	22.56
C14H24N6	276.39	276.2062	60.84	8.75	30.41
C14H25N1	207.36	207.1987	81.09	12.15	6.75
C14H25N3	235.38	235.2048	71.44	10.71	17.85
C14H25N5	263.39	263.2110	63.84	9.57	26.59
C14H26N2	222.38	222.2096	75.62	11.79	12.60
C14H26N4	250.39	250.2157	67.16	10.47	22.38
C14H26N6	278.40	278.2219	60.40	9.41	30.19
C14H27N1	209.38	209.2143	80.31	13.00	6.69
C14H27N3	237.39	237.2205	70.83	11.46	17.70
C14H27N5	265.40	265.2266	63.36	10.25	26.39
C14H28N2	224.39	224.2252	74.94	12.58	12.48
C14H28N4	252.41	252.2314	66.62	11.18	22.20
C14H28N6	280.42	280.2375	59.97	10.06	29.97
C14H29N1	211.39	211.2300	79.55	13.83	6.63

C14H29N3	239.41	239.2361	70.24	12.21	17.55
C14H29N5	267.42	267.2423	62.88	10.93	26.19
C14H30N2	226.41	226.2409	74.27	13.36	12.37
C14H30N4	254.42	254.2470	66.09	11.89	22.02
C14H30N6	282.44	282.2532	59.54	10.71	29.76
C14H31N1	213.41	213.2456	78.79	14.64	6.56
C14H31N3	241.42	241.2518	69.65	12.94	17.41
C14H31N5	269.44	269.2579	62.41	11.60	25.99
C14H32N2	228.42	228.2565	73.62	14.12	12.26
C14H32N4	256.44	256.2627	65.57	12.58	21.85
C14H32N6	284.45	284.2688	59.12	11.34	29.54
C14H33N3	243.44	243.2674	69.08	13.66	17.26
C14H33N5	271.45	271.2736	61.95	12.25	25.80
C14H34N4	258.45	258.2783	65.06	13.26	21.68
C14H34N6	286.47	286.2845	58.70	11.96	29.34
C14H35N5	273.47	273.2892	61.49	12.90	25.61
C14H36N6	288.48	288.3001	58.29	12.58	29.13
C15H 7N1	201.23	201.0578	89.53	3.51	6.96
C15H 7N3	229.24	229.0640	78.59	3.08	18.33
C15H 7N5	257.26	257.0701	70.03	2.74	27.22
C15H 8N2	216.24	216.0687	83.32	3.73	12.95
C15H 8N4	244.26	244.0749	73.76	3.30	22.94
C15H 8N6	272.27	272.0810	66.17	2.96	30.87
C15H 9N1	203.25	203.0735	88.65	4.46	6.89
C15H 9N3	231.26	231.0796	77.91	3.92	18.17
C15H 9N5	259.27	259.0858	69.49	3.50	27.01
C15H10N2	218.26	218.0844	82.55	4.62	12.83
C15H10N4	246.27	246.0905	73.16	4.09	22.75
C15H10N6	274.29	274.0967	65.69	3.67	30.64
C15H11N1	205.26	205.0891	87.77	5.40	6.82
C15H11N3	233.28	233.0953	77.23	4.75	18.01
C15H11N5	261.29	261.1014	68.95	4.24	26.80
C15H12N2	220.28	220.1000	81.79	5.49	12.72
C15H12N4	248.29	248.1062	72.56	4.87	22.57
C15H12N6	276.30	276.1123	65.21	4.38	30.42
C15H13N1	207.28	207.1048	86.92	6.32	6.76
C15H13N3	235.29	235.1109	76.57	5.57	17.86
C15H13N5	263.30	263.1171	68.43	4.98	26.60
C15H14N2	222.29	222.1157	81.05	6.35	12.60
C15H14N4	250.31	250.1218	71.98	5.64	22.38
C15H14N6	278.32	278.1280	64.73	5.07	30.20
C15H15N1	209.29	209.1204	86.08	7.22	6.69
C15H15N3	237.31	237.1266	75.92	6.37	17.71
C15H15N5	265.32	265.1327	67.91	5.70	26.40
C15H16N2	224.31	224.1313	80.32	7.19	12.49
C15H16N4	252.32	252.1375	71.40	6.39	22.20
C15H16N6	280.33	280.1436	64.27	5.75	29.98
C15H17N1	211.31	211.1361	85.26	8.11	6.63
C15H17N3	239.32	239.1422	75.28	7.16	17.56
C15H17N5	267.34	267.1484	67.39	6.41	26.20
C15H18N2	226.32	226.1470	79.61	8.02	12.38
C15H18N4	254.34	254.1531	70.84	7.13	22.03
C15H18N6	282.35	282.1593	63.81	6.43	29.76
C15H19N1	213.33	213.1517	84.46	8.98	6.57
C15H19N3	241.34	241.1579	74.65	7.94	17.41
C15H19N5	269.35	269.1640	66.89	7.11	26.00
C15H20N2	228.34	228.1626	78.90	8.83	12.27
C15H20N4	256.35	256.1688	70.28	7.86	21.86
C15H20N6	284.37	284.1749	63.36	7.09	29.55
C15H21N1	215.34	215.1674	83.67	9.83	6.50

10/21/66 052

C15H21N3	243.35	243.1735	74.03	8.70	17.27
C15H21N5	271.37	271.1797	66.39	7.80	25.81
C15H22N2	230.36	230.1783	78.21	9.63	12.16
C15H22N4	258.37	258.1844	69.73	8.58	21.68
C15H22N6	286.38	286.1906	62.91	7.74	29.35
C15H23N1	217.36	217.1830	82.89	10.67	6.44
C15H23N3	245.37	245.1892	73.43	9.45	17.13
C15H23N5	273.38	273.1953	65.90	8.48	25.62
C15H24N2	232.37	232.1939	77.53	10.41	12.06
C15H24N4	260.39	260.2001	69.19	9.29	21.52
C15H24N6	288.40	288.2062	62.47	8.39	29.14
C15H25N1	219.37	219.1987	82.13	11.49	6.38
C15H25N3	247.39	247.2048	72.83	10.19	16.99
C15H25N5	275.40	275.2110	65.42	9.15	25.43
C15H26N2	234.39	234.2096	76.87	11.18	11.95
C15H26N4	262.40	262.2157	68.66	9.99	21.35
C15H26N6	290.41	290.2219	62.04	9.02	28.94
C15H27N1	221.39	221.2143	81.38	12.29	6.33
C15H27N3	249.40	249.2205	72.24	10.91	16.85
C15H27N5	277.42	277.2266	64.94	9.81	25.24
C15H28N2	236.40	236.2252	76.21	11.94	11.85
C15H28N4	264.42	264.2314	68.14	10.67	21.19
C15H28N6	292.43	292.2375	61.61	9.65	28.74
C15H29N1	223.41	223.2300	80.65	13.08	6.27
C15H29N3	251.42	251.2361	71.66	11.63	16.71
C15H29N5	279.43	279.2423	64.48	10.46	25.06
C15H30N2	238.42	238.2409	75.57	12.68	11.75
C15H30N4	266.43	266.2470	67.62	11.35	21.03
C15H30N6	294.45	294.2532	61.19	10.27	28.54
C15H31N1	225.42	225.2456	79.92	13.86	6.21
C15H31N3	253.43	253.2518	71.09	12.33	16.58
C15H31N5	281.45	281.2579	64.01	11.10	24.88
C15H32N2	240.44	240.2565	74.93	13.42	11.65
C15H32N4	268.45	268.2627	67.11	12.02	20.87
C15H32N6	296.46	296.2688	60.77	10.88	28.35
C15H33N1	227.44	227.2613	79.22	14.63	6.16
C15H33N3	255.45	255.2674	70.53	13.02	16.45
C15H33N5	283.46	283.2736	63.56	11.73	24.71
C15H34N2	242.45	242.2722	74.31	14.14	11.55
C15H34N4	270.47	270.2783	66.61	12.67	20.71
C15H34N6	298.48	298.2845	60.36	11.48	28.16
C15H35N3	257.47	257.2831	69.98	13.70	16.32
C15H35N5	285.48	285.2892	63.11	12.36	24.53
C15H36N4	272.48	272.2940	66.12	13.32	20.56
C15H36N6	300.49	300.3001	59.96	12.08	27.97
C15H37N5	287.50	287.3049	62.67	12.97	24.36
C15H38N6	302.51	302.3158	59.56	12.66	27.78
C16H 8N2	228.26	228.0687	84.19	3.53	12.27
C16H 8N4	256.27	256.0749	74.99	3.15	21.86
C16H 8N6	284.28	284.0810	67.60	2.84	29.56
C16H 9N1	215.26	215.0735	89.28	4.21	6.51
C16H 9N3	243.27	243.0796	79.00	3.73	17.27
C16H 9N5	271.28	271.0858	70.84	3.34	25.82
C16H10N2	230.27	230.0844	83.46	4.38	12.17
C16H10N4	258.28	258.0905	74.41	3.90	21.69
C16H10N6	286.30	286.0967	67.13	3.52	29.35
C16H11N1	217.27	217.0891	88.45	5.10	6.45
C16H11N3	245.29	245.0953	78.35	4.52	17.13
C16H11N5	273.30	273.1014	70.32	4.06	25.63
C16H12N2	232.29	232.1000	82.73	5.21	12.06

C16H12N4	260.30	260.1062	73.83	4.65	21.52
C16H12N6	288.31	288.1123	66.66	4.20	29.15
C16H13N1	219.29	219.1048	87.64	5.98	6.39
C16H13N3	247.30	247.1109	77.71	5.30	16.99
C16H13N5	275.32	275.1171	69.80	4.76	25.44
C16H14N2	234.30	234.1157	82.02	6.02	11.96
C16H14N4	262.32	262.1218	73.26	5.38	21.36
C16H14N6	290.33	290.1280	66.19	4.86	28.95
C16H15N1	221.30	221.1204	86.84	6.83	6.33
C16H15N3	249.32	249.1266	77.08	6.06	16.85
C16H15N5	277.33	277.1327	69.30	5.45	25.25
C16H16N2	236.32	236.1313	81.32	6.82	11.85
C16H16N4	264.33	264.1375	72.70	6.10	21.20
C16H16N6	292.35	292.1436	65.74	5.52	28.75
C16H17N1	223.32	223.1361	86.05	7.67	6.27
C16H17N3	251.33	251.1422	76.46	6.82	16.72
C16H17N5	279.35	279.1484	68.80	6.13	25.07
C16H18N2	238.34	238.1470	80.63	7.61	11.75
C16H18N4	266.35	266.1531	72.15	6.81	21.04
C16H18N6	294.36	294.1593	65.29	6.16	28.55
C16H19N1	225.34	225.1517	85.29	8.50	6.22
C16H19N3	253.35	253.1579	75.85	7.56	16.59
C16H19N5	281.36	281.1640	68.30	6.81	24.89
C16H20N2	240.35	240.1626	79.96	8.39	11.66
C16H20N4	268.36	268.1688	71.61	7.51	20.88
C16H20N6	296.38	296.1749	64.84	6.80	28.36
C16H21N1	227.35	227.1674	84.53	9.31	6.16
C16H21N3	255.37	255.1735	75.26	8.29	16.45
C16H21N5	283.38	283.1797	67.82	7.47	24.71
C16H22N2	242.37	242.1783	79.29	9.15	11.56
C16H22N4	270.38	270.1844	71.08	8.20	20.72
C16H22N6	298.39	298.1906	64.40	7.43	28.16
C16H23N1	229.37	229.1830	83.79	10.11	6.11
C16H23N3	257.38	257.1892	74.67	9.01	16.33
C16H23N5	285.40	285.1953	67.34	8.12	24.54
C16H24N2	244.38	244.1939	78.64	9.90	11.46
C16H24N4	272.40	272.2001	70.55	8.88	20.57
C16H24N6	300.41	300.2062	63.97	8.05	27.98
C16H25N1	231.38	231.1987	83.06	10.89	6.05
C16H25N3	259.40	259.2048	74.09	9.71	16.20
C16H25N5	287.41	287.2110	66.87	8.77	24.37
C16H26N2	246.40	246.2096	77.99	10.64	11.37
C16H26N4	274.41	274.2157	70.03	9.55	20.42
C16H26N6	302.43	302.2219	63.55	8.67	27.79
C16H27N1	233.40	233.2143	82.34	11.66	6.00
C16H27N3	261.41	261.2205	73.52	10.41	16.07
C16H27N5	289.43	289.2266	66.40	9.40	24.20
C16H28N2	248.41	248.2252	77.36	11.36	11.28
C16H28N4	276.43	276.2314	69.52	10.21	20.27
C16H28N6	304.44	304.2375	63.12	9.27	27.60
C16H29N1	235.42	235.2300	81.63	12.42	5.95
C16H29N3	263.43	263.2361	72.95	11.10	15.95
C16H29N5	291.44	291.2423	65.94	10.03	24.03
C16H30N2	250.43	250.2409	76.74	12.07	11.19
C16H30N4	278.44	278.2470	69.02	10.86	20.12
C16H30N6	306.46	306.2532	62.71	9.87	27.42
C16H31N1	237.43	237.2456	80.94	13.16	5.90
C16H31N3	265.45	265.2518	72.40	11.77	15.83
C16H31N5	293.46	293.2579	65.49	10.65	23.86
C16H32N2	252.45	252.2565	76.13	12.78	11.10

C16H32N4	280.46	280.2627	68.52	11.50	19.98
C16H32N6	308.47	308.2688	62.30	10.46	27.24
C16H33N1	239.45	239.2613	80.26	13.89	5.85
C16H33N3	267.46	267.2674	71.85	12.44	15.71
C16H33N5	295.47	295.2736	65.04	11.26	23.70
C16H34N2	254.46	254.2722	75.52	13.47	11.01
C16H34N4	282.48	282.2783	68.03	12.13	19.83
C16H34N6	310.49	310.2845	61.90	11.04	27.07
C16H35N1	241.46	241.2769	79.59	14.61	5.80
C16H35N3	269.48	269.2831	71.32	13.09	15.59
C16H35N5	297.49	297.2892	64.60	11.86	23.54
C16H36N2	256.48	256.2878	74.93	14.15	10.92
C16H36N4	284.49	284.2940	67.55	12.75	19.69
C16H36N6	312.51	312.3001	61.50	11.61	26.89
C16H37N3	271.49	271.2987	70.79	13.74	15.48
C16H37N5	299.51	299.3049	64.16	12.45	23.38
C16H38N4	286.51	286.3096	67.08	13.37	19.56
C16H38N6	314.52	314.3158	61.10	12.18	26.72
C16H39N5	301.52	301.3205	63.74	13.04	23.23
C16H40N6	316.54	316.3314	60.71	12.74	26.55
C17H 8N2	240.27	240.0687	84.98	3.36	11.66
C17H 8N4	268.28	268.0749	76.11	3.01	20.88
C17H 8N6	296.29	296.0810	68.91	2.72	28.36
C17H 9N1	227.27	227.0735	89.85	3.99	6.16
C17H 9N3	255.28	255.0796	79.99	3.55	16.46
C17H 9N5	283.29	283.0858	72.08	3.20	24.72
C17H10N2	242.28	242.0844	84.28	4.16	11.56
C17H10N4	270.30	270.0905	75.54	3.73	20.73
C17H10N6	298.31	298.0967	68.45	3.38	28.17
C17H11N1	229.28	229.0891	89.06	4.84	6.11
C17H11N3	257.30	257.0953	79.36	4.31	16.33
C17H11N5	285.31	285.1014	71.57	3.89	24.55
C17H12N2	244.30	244.1000	83.58	4.95	11.47
C17H12N4	272.31	272.1062	74.98	4.44	20.57
C17H12N6	300.33	300.1123	67.99	4.03	27.98
C17H13N1	231.30	231.1048	88.28	5.67	6.06
C17H13N3	259.31	259.1109	78.74	5.05	16.20
C17H13N5	287.33	287.1171	71.07	4.56	24.37
C17H14N2	246.31	246.1157	82.90	5.73	11.37
C17H14N4	274.33	274.1218	74.43	5.14	20.42
C17H14N6	302.34	302.1280	67.54	4.67	27.80
C17H15N1	233.32	233.1204	87.52	6.48	6.00
C17H15N3	261.33	261.1266	78.13	5.79	16.08
C17H15N5	289.34	289.1327	70.57	5.23	24.20
C17H16N2	248.33	248.1313	82.22	6.49	11.28
C17H16N4	276.34	276.1375	73.89	5.84	20.27
C17H16N6	304.36	304.1436	67.09	5.30	27.61
C17H17N1	235.33	235.1361	86.77	7.28	5.95
C17H17N3	263.35	263.1422	77.54	6.51	15.96
C17H17N5	291.36	291.1484	70.08	5.88	24.04
C17H18N2	250.35	250.1470	81.56	7.25	11.19
C17H18N4	278.36	278.1531	73.35	6.52	20.13
C17H18N6	306.37	306.1593	66.65	5.92	27.43
C17H19N1	237.35	237.1517	86.03	8.07	5.90
C17H19N3	265.36	265.1579	76.95	7.22	15.84
C17H19N5	293.37	293.1640	69.60	6.53	23.87
C17H20N2	252.36	252.1626	80.91	7.99	11.10
C17H20N4	280.38	280.1688	72.83	7.19	19.98
C17H20N6	308.39	308.1749	66.21	6.54	27.25
C17H21N1	239.36	239.1674	85.31	8.84	5.85

C17H21N3	267.38	267.1735	76.37	7.92	15.72
C17H21N5	295.39	295.1797	69.13	7.17	23.71
C17H22N2	254.38	254.1783	80.27	8.72	11.01
C17H22N4	282.39	282.1844	72.31	7.85	19.84
C17H22N6	310.41	310.1906	65.78	7.14	27.07
C17H23N1	241.38	241.1830	84.59	9.60	5.80
C17H23N3	269.39	269.1892	75.80	8.61	15.60
C17H23N5	297.41	297.1953	68.66	7.80	23.55
C17H24N2	256.39	256.1939	79.64	9.44	10.93
C17H24N4	284.41	284.2001	71.79	8.51	19.70
C17H24N6	312.42	312.2062	65.36	7.74	26.90
C17H25N1	243.40	243.1987	83.89	10.35	5.75
C17H25N3	271.41	271.2048	75.23	9.28	15.48
C17H25N5	299.42	299.2110	68.19	8.42	23.39
C17H26N2	258.41	258.2096	79.02	10.14	10.84
C17H26N4	286.42	286.2157	71.29	9.15	19.56
C17H26N6	314.44	314.2219	64.94	8.33	26.73
C17H27N1	245.41	245.2143	83.20	11.09	5.71
C17H27N3	273.42	273.2205	74.68	9.95	15.37
C17H27N5	301.44	301.2266	67.74	9.03	23.23
C17H28N2	260.43	260.2252	78.41	10.84	10.76
C17H28N4	288.44	288.2314	70.79	9.78	19.42
C17H28N6	316.45	316.2375	64.52	8.92	26.56
C17H29N1	247.43	247.2300	82.53	11.81	5.66
C17H29N3	275.44	275.2361	74.13	10.61	15.26
C17H29N5	303.45	303.2423	67.29	9.63	23.08
C17H30N2	262.44	262.2409	77.80	11.52	10.67
C17H30N4	290.46	290.2470	70.30	10.41	19.29
C17H30N6	318.47	318.2532	64.12	9.50	26.39
C17H31N1	249.44	249.2456	81.86	12.53	5.62
C17H31N3	277.46	277.2518	73.59	11.26	15.14
C17H31N5	305.47	305.2579	66.84	10.23	22.93
C17H32N2	264.46	264.2565	77.21	12.20	10.59
C17H32N4	292.47	292.2627	69.82	11.03	19.16
C17H32N6	320.48	320.2688	63.71	10.06	26.22
C17H33N1	251.46	251.2613	81.20	13.23	5.57
C17H33N3	279.47	279.2674	73.06	11.90	15.04
C17H33N5	307.49	307.2736	66.41	10.82	22.78
C17H34N2	266.47	266.2722	76.63	12.86	10.51
C17H34N4	294.49	294.2783	69.34	11.64	19.03
C17H34N6	322.50	322.2845	63.31	10.63	26.06
C17H35N1	253.48	253.2769	80.56	13.92	5.53
C17H35N3	281.49	281.2831	72.54	12.53	14.93
C17H35N5	309.50	309.2892	65.97	11.40	22.63
C17H36N2	268.49	268.2878	76.05	13.52	10.43
C17H36N4	296.50	296.2940	68.87	12.24	18.90
C17H36N6	324.52	324.3001	62.92	11.18	25.90
C17H37N1	255.49	255.2926	79.92	14.60	5.48
C17H37N3	283.50	283.2987	72.02	13.15	14.82
C17H37N5	311.52	311.3049	65.55	11.97	22.48
C17H38N2	270.51	270.3035	75.48	14.16	10.36
C17H38N4	298.52	298.3096	68.40	12.83	18.77
C17H38N6	326.53	326.3158	62.53	11.73	25.74
C17H39N3	285.52	285.3144	71.51	13.77	14.72
C17H39N5	313.53	313.3205	65.13	12.54	22.34
C17H40N4	300.54	300.3253	67.94	13.42	18.64
C17H40N6	328.55	328.3314	62.15	12.27	25.58
C17H41N5	315.55	315.3362	64.71	13.10	22.19
C17H42N6	330.56	330.3471	61.77	12.81	25.42
C18H 9N1	239.28	239.0735	90.36	3.79	5.85

C18H 9N3	267.29	267.0796	80.89	3.39	15.72
C18H 9N5	295.31	295.0858	73.21	3.07	23.72
C18H10N2	254.29	254.0844	85.02	3.96	11.02
C18H10N4	282.31	282.0905	76.58	3.57	19.85
C18H10N6	310.32	310.0967	69.67	3.25	27.08
C18H11N1	241.30	241.0891	89.60	4.60	5.80
C18H11N3	269.31	269.0953	80.28	4.12	15.60
C18H11N5	297.32	297.1014	72.72	3.73	23.55
C18H12N2	256.31	256.1000	84.35	4.72	10.93
C18H12N4	284.32	284.1062	76.04	4.25	19.71
C18H12N6	312.34	312.1123	69.22	3.87	26.91
C18H13N1	243.31	243.1048	88.86	5.39	5.76
C18H13N3	271.32	271.1109	79.68	4.83	15.49
C18H13N5	299.34	299.1171	72.23	4.38	23.40
C18H14N2	258.33	258.1157	83.69	5.46	10.84
C18H14N4	286.34	286.1218	75.51	4.93	19.57
C18H14N6	314.35	314.1280	68.78	4.49	26.73
C18H15N1	245.33	245.1204	88.13	6.16	5.71
C18H15N3	273.34	273.1266	79.10	5.53	15.37
C18H15N5	301.35	301.1327	71.74	5.02	23.24
C18H16N2	260.34	260.1313	83.05	6.19	10.76
C18H16N4	288.36	288.1375	74.98	5.59	19.43
C18H16N6	316.37	316.1436	68.34	5.10	26.56
C18H17N1	247.34	247.1361	87.41	6.93	5.66
C18H17N3	275.36	275.1422	78.52	6.22	15.26
C18H17N5	303.37	303.1484	71.27	5.65	23.09
C18H18N2	262.36	262.1470	82.41	6.92	10.68
C18H18N4	290.37	290.1531	74.46	6.25	19.29
C18H18N6	318.38	318.1593	67.91	5.70	26.40
C18H19N1	249.36	249.1517	86.70	7.68	5.62
C18H19N3	277.37	277.1579	77.95	6.90	15.15
C18H19N5	305.39	305.1640	70.80	6.27	22.93
C18H20N2	264.37	264.1626	81.78	7.63	10.60
C18H20N4	292.39	292.1688	73.94	6.89	19.16
C18H20N6	320.40	320.1749	67.48	6.29	26.23
C18H21N1	251.37	251.1674	86.01	8.42	5.57
C18H21N3	279.39	279.1735	77.38	7.58	15.04
C18H21N5	307.40	307.1797	70.33	6.89	22.78
C18H22N2	266.39	266.1783	81.16	8.32	10.52
C18H22N4	294.40	294.1844	73.44	7.53	19.03
C18H22N6	322.42	322.1906	67.06	6.88	26.07
C18H23N1	253.39	253.1830	85.32	9.15	5.53
C18H23N3	281.40	281.1892	76.83	8.24	14.93
C18H23N5	309.42	309.1953	69.87	7.49	22.63
C18H24N2	268.41	268.1939	80.55	9.01	10.44
C18H24N4	296.42	296.2001	72.94	8.16	18.90
C18H24N6	324.43	324.2062	66.64	7.46	25.90
C18H25N1	255.41	255.1987	84.65	9.87	5.48
C18H25N3	283.42	283.2048	76.28	8.89	14.83
C18H25N5	311.43	311.2110	69.42	8.09	22.49
C18H26N2	270.42	270.2096	79.95	9.69	10.36
C18H26N4	298.43	298.2157	72.44	8.78	18.77
C18H26N6	326.45	326.2219	66.23	8.03	25.74
C18H27N1	257.42	257.2143	83.99	10.57	5.44
C18H27N3	285.44	285.2205	75.74	9.53	14.72
C18H27N5	313.45	313.2266	68.97	8.68	22.34
C18H28N2	272.44	272.2252	79.36	10.36	10.28
C18H28N4	300.45	300.2314	71.96	9.39	18.65
C18H28N6	328.46	328.2375	65.82	8.59	25.59
C18H29N1	259.44	259.2300	83.33	11.27	5.40

C18H29N3	287.45	287.2361	75.21	10.17	14.62
C18H29N5	315.47	315.2423	68.53	9.27	22.20
C18H30N2	274.45	274.2409	78.78	11.02	10.21
C18H30N4	302.47	302.2470	71.48	10.00	18.52
C18H30N6	330.48	330.2532	65.42	9.15	25.43
C18H31N1	261.45	261.2456	82.69	11.95	5.36
C18H31N3	289.47	289.2518	74.69	10.79	14.52
C18H31N5	317.48	317.2579	68.10	9.84	22.06
C18H32N2	276.47	276.2565	78.20	11.67	10.13
C18H32N4	304.48	304.2627	71.01	10.59	18.40
C18H32N6	332.50	332.2688	65.02	9.70	25.28
C18H33N1	263.47	263.2613	82.06	12.62	5.32
C18H33N3	291.48	291.2674	74.17	11.41	14.42
C18H33N5	319.50	319.2736	67.67	10.41	21.92
C18H34N2	278.49	278.2722	77.63	12.31	10.06
C18H34N4	306.50	306.2783	70.54	11.18	18.28
C18H34N6	334.51	334.2845	64.63	10.25	25.12
C18H35N1	265.49	265.2769	81.44	13.29	5.28
C18H35N3	293.50	293.2831	73.66	12.02	14.32
C18H35N5	321.51	321.2892	67.24	10.97	21.78
C18H36N2	280.50	280.2878	77.08	12.94	9.99
C18H36N4	308.51	308.2940	70.08	11.76	18.16
C18H36N6	336.53	336.3001	64.24	10.78	24.97
C18H37N1	267.50	267.2926	80.82	13.94	5.24
C18H37N3	295.52	295.2987	73.16	12.62	14.22
C18H37N5	323.53	323.3049	66.83	11.53	21.65
C18H38N2	282.52	282.3035	76.53	13.56	9.92
C18H38N4	310.53	310.3096	69.62	12.33	18.04
C18H38N6	338.54	338.3158	63.86	11.31	24.82
C18H39N1	269.52	269.3082	80.22	14.59	5.20
C18H39N3	297.53	297.3144	72.66	13.21	14.12
C18H39N5	325.55	325.3205	66.41	12.08	21.51
C18H40N2	284.53	284.3191	75.98	14.17	9.85
C18H40N4	312.55	312.3253	69.17	12.90	17.93
C18H40N6	340.56	340.3314	63.48	11.84	24.68
C18H41N3	299.55	299.3300	72.18	13.80	14.03
C18H41N5	327.56	327.3362	66.00	12.62	21.38
C18H42N4	314.56	314.3409	68.73	13.46	17.81
C18H42N6	342.58	342.3471	63.11	12.36	24.53
C18H43N5	329.58	329.3518	65.60	13.15	21.25
C18H44N6	344.59	344.3627	62.74	12.87	24.39
C19H 9N1	251.29	251.0735	90.82	3.61	5.57
C19H 9N3	279.30	279.0796	81.71	3.25	15.04
C19H 9N5	307.32	307.0858	74.26	2.95	22.79
C19H10N2	266.30	266.0844	85.70	3.79	10.52
C19H10N4	294.32	294.0905	77.54	3.42	19.04
C19H10N6	322.33	322.0967	70.80	3.13	26.07
C19H11N1	253.31	253.0891	90.09	4.38	5.53
C19H11N3	281.32	281.0953	81.12	3.94	14.94
C19H11N5	309.33	309.1014	73.78	3.58	22.64
C19H12N2	268.32	268.1000	85.05	4.51	10.44
C19H12N4	296.33	296.1062	77.01	4.08	18.91
C19H12N6	324.35	324.1123	70.36	3.73	25.91
C19H13N1	255.32	255.1048	89.38	5.13	5.49
C19H13N3	283.34	283.1109	80.54	4.62	14.83
C19H13N5	311.35	311.1171	73.30	4.21	22.49
C19H14N2	270.34	270.1157	84.42	5.22	10.36
C19H14N4	298.35	298.1218	76.49	4.73	18.78
C19H14N6	326.36	326.1280	69.93	4.32	25.75
C19H15N1	257.34	257.1204	88.68	5.88	5.44

C19H15N3	285.35	285.1266	79.98	5.30	14.73
C19H15N5	313.36	313.1327	72.83	4.82	22.35
C19H16N2	272.35	272.1313	83.79	5.92	10.29
C19H16N4	300.37	300.1375	75.98	5.37	18.65
C19H16N6	328.38	328.1436	69.50	4.91	25.59
C19H17N1	259.35	259.1361	87.99	6.61	5.40
C19H17N3	287.37	287.1422	79.41	5.96	14.62
C19H17N5	315.38	315.1484	72.36	5.43	22.21
C19H18N2	274.37	274.1470	83.18	6.61	10.21
C19H18N4	302.38	302.1531	75.47	6.00	18.53
C19H18N6	330.40	330.1593	69.07	5.49	25.44
C19H19N1	261.37	261.1517	87.31	7.33	5.36
C19H19N3	289.38	289.1579	78.86	6.62	14.52
C19H19N5	317.40	317.1640	71.90	6.03	22.06
C19H20N2	276.38	276.1626	82.57	7.29	10.14
C19H20N4	304.40	304.1688	74.97	6.62	18.41
C19H20N6	332.41	332.1749	68.65	6.06	25.28
C19H21N1	263.39	263.1674	86.65	8.04	5.32
C19H21N3	291.40	291.1735	78.32	7.26	14.42
C19H21N5	319.41	319.1797	71.45	6.63	21.93
C19H22N2	278.40	278.1783	81.97	7.97	10.06
C19H22N4	306.41	306.1844	74.48	7.24	18.28
C19H22N6	334.43	334.1906	68.24	6.63	25.13
C19H23N1	265.40	265.1830	85.99	8.74	5.28
C19H23N3	293.42	293.1892	77.78	7.90	14.32
C19H23N5	321.43	321.1953	71.00	7.21	21.79
C19H24N2	280.42	280.1939	81.38	8.63	9.99
C19H24N4	308.43	308.2001	73.99	7.84	18.17
C19H24N6	336.44	336.2062	67.83	7.19	24.98
C19H25N1	267.42	267.1987	85.34	9.42	5.24
C19H25N3	295.43	295.2048	77.25	8.53	14.22
C19H25N5	323.44	323.2110	70.56	7.79	21.65
C19H26N2	282.43	282.2096	80.80	9.28	9.92
C19H26N4	310.45	310.2157	73.51	8.44	18.05
C19H26N6	338.46	338.2219	67.43	7.74	24.83
C19H27N1	269.43	269.2143	84.70	10.10	5.20
C19H27N3	297.45	297.2205	76.72	9.15	14.13
C19H27N5	325.46	325.2266	70.12	8.36	21.52
C19H28N2	284.45	284.2252	80.23	9.92	9.85
C19H28N4	312.46	312.2314	73.04	9.03	17.93
C19H28N6	340.48	340.2375	67.03	8.29	24.68
C19H29N1	271.45	271.2300	84.07	10.77	5.16
C19H29N3	299.46	299.2361	76.21	9.76	14.03
C19H29N5	327.48	327.2423	69.69	8.93	21.39
C19H30N2	286.46	286.2409	79.67	10.56	9.78
C19H30N4	314.48	314.2470	72.57	9.62	17.82
C19H30N6	342.49	342.2532	66.63	8.83	24.54
C19H31N1	273.47	273.2456	83.45	11.43	5.12
C19H31N3	301.48	301.2518	75.70	10.36	13.94
C19H31N5	329.49	329.2579	69.26	9.48	21.25
C19H32N2	288.48	288.2565	79.11	11.18	9.71
C19H32N4	316.49	316.2627	72.11	10.19	17.70
C19H32N6	344.51	344.2688	66.24	9.36	24.39
C19H33N1	275.48	275.2613	82.84	12.07	5.08
C19H33N3	303.49	303.2674	75.19	10.96	13.85
C19H33N5	331.51	331.2736	68.84	10.03	21.13
C19H34N2	290.50	290.2722	78.56	11.80	9.64
C19H34N4	318.51	318.2783	71.65	10.76	17.59
C19H34N6	346.52	346.2845	65.86	9.89	24.25
C19H35N1	277.50	277.2769	82.24	12.71	5.05

C19H35N3	305.51	305.2831	74.70	11.55	13.75
C19H35N5	333.52	333.2892	68.42	10.58	21.00
C19H36N2	292.51	292.2878	78.02	12.41	9.58
C19H36N4	320.53	320.2940	71.20	11.32	17.48
C19H36N6	348.54	348.3001	65.48	10.41	24.11
C19H37N1	279.51	279.2926	81.65	13.34	5.01
C19H37N3	307.53	307.2987	74.21	12.13	13.66
C19H37N5	335.54	335.3049	68.01	11.11	20.87
C19H38N2	294.53	294.3035	77.48	13.00	9.51
C19H38N4	322.54	322.3096	70.75	11.88	17.37
C19H38N6	350.55	350.3158	65.10	10.93	23.97
C19H39N1	281.53	281.3082	81.06	13.96	4.98
C19H39N3	309.54	309.3144	73.73	12.70	13.57
C19H39N5	337.56	337.3205	67.61	11.65	20.75
C19H40N2	296.54	296.3191	76.96	13.60	9.45
C19H40N4	324.56	324.3253	70.31	12.42	17.26
C19H40N6	352.57	352.3314	64.73	11.44	23.84
C19H41N1	283.55	283.3239	80.49	14.58	4.94
C19H41N3	311.56	311.3300	73.25	13.26	13.49
C19H41N5	339.57	339.3362	67.21	12.17	20.62
C19H42N2	298.56	298.3348	76.44	14.18	9.38
C19H42N4	326.57	326.3409	69.88	12.96	17.16
C19H42N6	354.59	354.3471	64.36	11.94	23.70
C19H43N3	313.57	313.3457	72.78	13.82	13.40
C19H43N5	341.59	341.3518	66.81	12.69	20.50
C19H44N4	328.59	328.3566	69.45	13.50	17.05
C19H44N6	356.60	356.3627	64.00	12.44	23.57
C19H45N5	343.60	343.3675	66.42	13.20	20.38
C19H46N6	358.62	358.3784	63.64	12.93	23.43
C20H10N2	278.32	278.0844	86.31	3.62	10.07
C20H10N4	306.33	306.0905	78.42	3.29	18.29
C20H10N6	334.34	334.0967	71.85	3.01	25.14
C20H11N1	265.32	265.0891	90.54	4.18	5.28
C20H11N3	293.33	293.0953	81.89	3.78	14.33
C20H11N5	321.34	321.1014	74.76	3.45	21.79
C20H12N2	280.33	280.1000	85.69	4.31	9.99
C20H12N4	308.35	308.1062	77.91	3.92	18.17
C20H12N6	336.36	336.1123	71.42	3.60	24.99
C20H13N1	267.33	267.1048	89.86	4.90	5.24
C20H13N3	295.35	295.1109	81.34	4.44	14.23
C20H13N5	323.36	323.1171	74.29	4.05	21.66
C20H14N2	282.35	282.1157	85.08	5.00	9.92
C20H14N4	310.36	310.1218	77.40	4.55	18.05
C20H14N6	338.37	338.1280	70.99	4.17	24.84
C20H15N1	269.35	269.1204	89.19	5.61	5.20
C20H15N3	297.36	297.1266	80.78	5.08	14.13
C20H15N5	325.38	325.1327	73.83	4.65	21.52
C20H16N2	284.36	284.1313	84.48	5.67	9.85
C20H16N4	312.38	312.1375	76.90	5.16	17.94
C20H16N6	340.39	340.1436	70.57	4.74	24.69
C20H17N1	271.37	271.1361	88.52	6.31	5.16
C20H17N3	299.38	299.1422	80.24	5.72	14.04
C20H17N5	327.39	327.1484	73.37	5.23	21.39
C20H18N2	286.38	286.1470	83.88	6.34	9.78
C20H18N4	314.39	314.1531	76.41	5.77	17.82
C20H18N6	342.41	342.1593	70.16	5.30	24.54
C20H19N1	273.38	273.1517	87.87	7.01	5.12
C20H19N3	301.39	301.1579	79.70	6.35	13.94
C20H19N5	329.41	329.1640	72.93	5.81	21.26
C20H20N2	288.40	288.1626	83.30	6.99	9.71

C20H20N4	316.41	316.1688	75.92	6.37	17.71
C20H20N6	344.42	344.1749	69.75	5.85	24.40
C20H21N1	275.40	275.1674	87.23	7.69	5.09
C20H21N3	303.41	303.1735	79.17	6.98	13.85
C20H21N5	331.42	331.1797	72.48	6.39	21.13
C20H22N2	290.41	290.1783	82.72	7.64	9.65
C20H22N4	318.43	318.1844	75.44	6.96	17.59
C20H22N6	346.44	346.1906	69.34	6.40	24.26
C20H23N1	277.41	277.1830	86.59	8.36	5.05
C20H23N3	305.43	305.1892	78.65	7.59	13.76
C20H23N5	333.44	333.1953	72.04	6.95	21.00
C20H24N2	292.43	292.1939	82.15	8.27	9.58
C20H24N4	320.44	320.2001	74.97	7.55	17.48
C20H24N6	348.45	348.2062	68.94	6.94	24.12
C20H25N1	279.43	279.1987	85.97	9.02	5.01
C20H25N3	307.44	307.2048	78.14	8.20	13.67
C20H25N5	335.46	335.2110	71.61	7.51	20.88
C20H26N2	294.44	294.2096	81.59	8.90	9.51
C20H26N4	322.46	322.2157	74.50	8.13	17.37
C20H26N6	350.47	350.2219	68.54	7.48	23.98
C20H27N1	281.44	281.2143	85.35	9.67	4.98
C20H27N3	309.46	309.2205	77.63	8.79	13.58
C20H27N5	337.47	337.2266	71.18	8.06	20.75
C20H28N2	296.46	296.2252	81.03	9.52	9.45
C20H28N4	324.47	324.2314	74.03	8.70	17.27
C20H28N6	352.49	352.2375	68.15	8.01	23.84
C20H29N1	283.46	283.2300	84.75	10.31	4.94
C20H29N3	311.47	311.2361	77.12	9.38	13.49
C20H29N5	339.49	339.2423	70.76	8.61	20.63
C20H30N2	298.48	298.2409	80.48	10.13	9.39
C20H30N4	326.49	326.2470	73.58	9.26	17.16
C20H30N6	354.50	354.2532	67.76	8.53	23.71
C20H31N1	285.48	285.2456	84.15	10.95	4.91
C20H31N3	313.49	313.2518	76.63	9.97	13.40
C20H31N5	341.50	341.2579	70.34	9.15	20.51
C20H32N2	300.49	300.2565	79.94	10.73	9.32
C20H32N4	328.50	328.2627	73.13	9.82	17.06
C20H32N6	356.52	356.2688	67.38	9.05	23.57
C20H33N1	287.49	287.2613	83.56	11.57	4.87
C20H33N3	315.51	315.2674	76.14	10.54	13.32
C20H33N5	343.52	343.2736	69.93	9.68	20.39
C20H34N2	302.51	302.2722	79.41	11.33	9.26
C20H34N4	330.52	330.2783	72.68	10.37	16.95
C20H34N6	358.53	358.2845	67.00	9.56	23.44
C20H35N1	289.51	289.2769	82.98	12.19	4.84
C20H35N3	317.52	317.2831	75.66	11.11	13.23
C20H35N5	345.54	345.2892	69.52	10.21	20.27
C20H36N2	304.52	304.2878	78.88	11.92	9.20
C20H36N4	332.54	332.2940	72.24	10.91	16.85
C20H36N6	360.55	360.3001	66.63	10.06	23.31
C20H37N1	291.52	291.2926	82.40	12.79	4.80
C20H37N3	319.54	319.2987	75.18	11.67	13.15
C20H37N5	347.55	347.3049	69.12	10.73	20.15
C20H38N2	306.54	306.3035	78.37	12.50	9.14
C20H38N4	334.55	334.3096	71.80	11.45	16.75
C20H38N6	362.57	362.3158	66.26	10.56	23.18
C20H39N1	293.54	293.3082	81.84	13.39	4.77
C20H39N3	321.55	321.3144	74.71	12.23	13.07
C20H39N5	349.57	349.3205	68.72	11.25	20.03
C20H40N2	308.56	308.3191	77.85	13.07	9.08

C20H40N4	336.57	336.3253	71.37	11.98	16.65
C20H40N6	364.58	364.3314	65.89	11.06	23.05
C20H41N1	295.56	295.3239	81.28	13.98	4.74
C20H41N3	323.57	323.3300	74.24	12.77	12.99
C20H41N5	351.58	351.3362	68.33	11.75	19.92
C20H42N2	310.57	310.3348	77.35	13.63	9.02
C20H42N4	338.58	338.3409	70.95	12.50	16.55
C20H42N6	366.60	366.3471	65.53	11.55	22.92
C20H43N1	297.57	297.3395	80.73	14.57	4.71
C20H43N3	325.59	325.3457	73.78	13.31	12.91
C20H43N5	353.60	353.3518	67.94	12.26	19.81
C20H44N2	312.59	312.3504	76.85	14.19	8.96
C20H44N4	340.60	340.3566	70.53	13.02	16.45
C20H44N6	368.61	368.3627	65.17	12.03	22.80
C20H45N3	327.60	327.3613	73.33	13.85	12.83
C20H45N5	355.62	355.3675	67.55	12.75	19.69
C20H46N4	342.62	342.3722	70.11	13.53	16.35
C20H46N6	370.63	370.3784	64.81	12.51	22.67
C20H47N5	357.63	357.3831	67.17	13.25	19.58
C20H48N6	372.65	372.3940	64.46	12.98	22.55
C21H10N2	290.33	290.0844	86.88	3.47	9.65
C21H10N4	318.34	318.0905	79.23	3.17	17.60
C21H10N6	346.35	346.0967	72.83	2.91	24.26
C21H11N1	277.33	277.0891	90.95	4.00	5.05
C21H11N3	305.34	305.0953	82.61	3.63	13.76
C21H11N5	333.36	333.1014	75.67	3.33	21.01
C21H12N2	292.34	292.1000	86.28	4.14	9.58
C21H12N4	320.36	320.1062	78.74	3.78	17.49
C21H12N6	348.37	348.1123	72.40	3.47	24.12
C21H13N1	279.34	279.1048	90.30	4.69	5.01
C21H13N3	307.36	307.1109	82.07	4.26	13.67
C21H13N5	335.37	335.1171	75.21	3.91	20.88
C21H14N2	294.36	294.1157	85.69	4.79	9.52
C21H14N4	322.37	322.1218	78.24	4.38	17.38
C21H14N6	350.39	350.1280	71.99	4.03	23.99
C21H15N1	281.36	281.1204	89.65	5.37	4.98
C21H15N3	309.37	309.1266	81.53	4.89	13.58
C21H15N5	337.39	337.1327	74.76	4.48	20.76
C21H16N2	296.38	296.1313	85.11	5.44	9.45
C21H16N4	324.39	324.1375	77.76	4.97	17.27
C21H16N6	352.40	352.1436	71.58	4.58	23.85
C21H17N1	283.38	283.1361	89.01	6.05	4.94
C21H17N3	311.39	311.1422	81.00	5.50	13.49
C21H17N5	339.40	339.1484	74.32	5.05	20.63
C21H18N2	298.39	298.1470	84.53	6.08	9.39
C21H18N4	326.40	326.1531	77.28	5.56	17.16
C21H18N6	354.42	354.1593	71.17	5.12	23.71
C21H19N1	285.39	285.1517	88.38	6.71	4.91
C21H19N3	313.41	313.1579	80.48	6.11	13.41
C21H19N5	341.42	341.1640	73.88	5.61	20.51
C21H20N2	300.41	300.1626	83.96	6.71	9.33
C21H20N4	328.42	328.1688	76.80	6.14	17.06
C21H20N6	356.43	356.1749	70.77	5.66	23.58
C21H21N1	287.41	287.1674	87.76	7.36	4.87
C21H21N3	315.42	315.1735	79.97	6.71	13.32
C21H21N5	343.44	343.1797	73.44	6.16	20.39
C21H22N2	302.42	302.1783	83.40	7.33	9.26
C21H22N4	330.44	330.1844	76.33	6.71	16.96
C21H22N6	358.45	358.1906	70.37	6.19	23.45
C21H23N1	289.42	289.1830	87.15	8.01	4.84

C21H23N3	317.44	317.1892	79.46	7.30	13.24
C21H23N5	345.45	345.1953	73.02	6.71	20.27
C21H24N2	304.44	304.1939	82.85	7.95	9.20
C21H24N4	332.45	332.2001	75.87	7.28	16.85
C21H24N6	360.47	360.2062	69.97	6.71	23.31
C21H25N1	291.44	291.1987	86.55	8.65	4.81
C21H25N3	319.45	319.2048	78.96	7.89	13.15
C21H25N5	347.47	347.2110	72.59	7.25	20.16
C21H26N2	306.45	306.2096	82.31	8.55	9.14
C21H26N4	334.47	334.2157	75.41	7.84	16.75
C21H26N6	362.48	362.2219	69.59	7.23	23.18
C21H27N1	293.46	293.2143	85.95	9.27	4.77
C21H27N3	321.47	321.2205	78.46	8.47	13.07
C21H27N5	349.48	349.2266	72.17	7.79	20.04
C21H28N2	308.47	308.2252	81.77	9.15	9.08
C21H28N4	336.48	336.2314	74.96	8.39	16.65
C21H28N6	364.50	364.2375	69.20	7.74	23.06
C21H29N1	295.47	295.2300	85.37	9.89	4.74
C21H29N3	323.49	323.2361	77.97	9.04	12.99
C21H29N5	351.50	351.2423	71.76	8.32	19.92
C21H30N2	310.49	310.2409	81.24	9.74	9.02
C21H30N4	338.50	338.2470	74.52	8.93	16.55
C21H30N6	366.51	366.2532	68.82	8.25	22.93
C21H31N1	297.49	297.2456	84.79	10.50	4.71
C21H31N3	325.50	325.2518	77.49	9.60	12.91
C21H31N5	353.51	353.2579	71.35	8.84	19.81
C21H32N2	312.50	312.2565	80.71	10.32	8.96
C21H32N4	340.52	340.2627	74.07	9.47	16.45
C21H32N6	368.53	368.2688	68.44	8.75	22.80
C21H33N1	299.50	299.2613	84.22	11.11	4.68
C21H33N3	327.52	327.2674	77.01	10.16	12.83
C21H33N5	355.53	355.2736	70.95	9.36	19.70
C21H34N2	314.52	314.2722	80.20	10.90	8.91
C21H34N4	342.53	342.2783	73.64	10.01	16.36
C21H34N6	370.55	370.2845	68.07	9.25	22.68
C21H35N1	301.52	301.2769	83.65	11.70	4.65
C21H35N3	329.53	329.2831	76.54	10.71	12.75
C21H35N5	357.55	357.2892	70.55	9.87	19.59
C21H36N2	316.53	316.2878	79.69	11.46	8.85
C21H36N4	344.55	344.2940	73.21	10.53	16.26
C21H36N6	372.56	372.3001	67.70	9.74	22.56
C21H37N1	303.54	303.2926	83.10	12.29	4.61
C21H37N3	331.55	331.2987	76.08	11.25	12.67
C21H37N5	359.56	359.3049	70.15	10.37	19.48
C21H38N2	318.55	318.3035	79.18	12.02	8.79
C21H38N4	346.56	346.3096	72.78	11.05	16.17
C21H38N6	374.58	374.3158	67.34	10.23	22.44
C21H39N1	305.55	305.3082	82.55	12.87	4.58
C21H39N3	333.57	333.3144	75.62	11.79	12.60
C21H39N5	361.58	361.3205	69.76	10.87	19.37
C21H40N2	320.57	320.3191	78.68	12.58	8.74
C21H40N4	348.58	348.3253	72.36	11.57	16.07
C21H40N6	376.59	376.3314	66.98	10.71	22.32
C21H41N1	307.57	307.3239	82.01	13.44	4.55
C21H41N3	335.58	335.3300	75.16	12.31	12.52
C21H41N5	363.59	363.3362	69.37	11.37	19.26
C21H42N2	322.58	322.3348	78.19	13.12	8.68
C21H42N4	350.60	350.3409	71.94	12.08	15.98
C21H42N6	378.61	378.3471	66.62	11.18	22.20
C21H43N1	309.58	309.3395	81.48	14.00	4.52
C21H43N3	337.60	337.3457	74.71	12.84	12.45
C21H43N5	365.61	365.3518	68.99	11.85	19.16
C21H44N2	324.60	324.3504	77.71	13.66	8.63
C21H44N4	352.61	352.3566	71.53	12.58	15.89
C21H44N6	380.63	380.3627	66.27	11.65	22.08
C21H45N1	311.60	311.3552	80.95	14.56	4.50
C21H45N3	339.61	339.3613	74.27	13.36	12.37
C21H45N5	367.63	367.3675	68.61	12.34	19.05
C21H46N2	326.61	326.3661	77.23	14.20	8.58
C21H46N4	354.63	354.3722	71.13	13.07	15.80
C21H46N6	382.64	382.3784	65.92	12.12	21.96
C21H47N3	341.63	341.3770	73.83	13.87	12.30
C21H47N5	369.64	369.3831	68.24	12.82	18.95
C21H48N4	356.64	356.3879	70.72	13.57	15.71
C21H48N6	384.66	384.3940	65.57	12.58	21.85
C21H49N5	371.66	371.3988	67.87	13.29	18.84
C21H50N6	386.67	386.4097	65.23	13.03	21.73
C22H11N1	289.34	289.0891	91.33	3.83	4.84
C22H11N3	317.35	317.0953	83.27	3.49	13.24
C22H11N5	345.37	345.1014	76.51	3.21	20.28
C22H12N2	304.35	304.1000	86.82	3.97	9.20
C22H12N4	332.37	332.1062	79.50	3.64	16.86
C22H12N6	360.38	360.1123	73.32	3.36	23.32
C22H13N1	291.36	291.1048	90.70	4.50	4.81
C22H13N3	319.37	319.1109	82.74	4.10	13.16
C22H13N5	347.38	347.1171	76.07	3.77	20.16
C22H14N2	306.37	306.1157	86.25	4.61	9.14
C22H14N4	334.38	334.1218	79.02	4.22	16.76
C22H14N6	362.40	362.1280	72.92	3.89	23.19
C22H15N1	293.37	293.1204	90.07	5.15	4.77
C22H15N3	321.38	321.1266	82.22	4.70	13.07
C22H15N5	349.40	349.1327	75.63	4.33	20.04
C22H16N2	308.39	308.1313	85.69	5.23	9.08
C22H16N4	336.40	336.1375	78.55	4.79	16.65
C22H16N6	364.41	364.1436	72.51	4.43	23.06
C22H17N1	295.39	295.1361	89.46	5.80	4.74
C22H17N3	323.40	323.1422	81.71	5.30	12.99
C22H17N5	351.41	351.1484	75.19	4.88	19.93
C22H18N2	310.40	310.1470	85.13	5.85	9.02
C22H18N4	338.42	338.1531	78.08	5.36	16.56
C22H18N6	366.43	366.1593	72.11	4.95	22.93
C22H19N1	297.40	297.1517	88.85	6.44	4.71
C22H19N3	325.42	325.1579	81.20	5.89	12.91
C22H19N5	353.43	353.1640	74.77	5.42	19.82
C22H20N2	312.42	312.1626	84.58	6.45	8.97
C22H20N4	340.43	340.1688	77.62	5.92	16.46
C22H20N6	368.44	368.1749	71.72	5.47	22.81
C22H21N1	299.42	299.1674	88.25	7.07	4.68
C22H21N3	327.43	327.1735	80.70	6.46	12.83
C22H21N5	355.45	355.1797	74.34	5.96	19.70
C22H22N2	314.43	314.1783	84.04	7.05	8.91
C22H22N4	342.45	342.1844	77.16	6.48	16.36
C22H22N6	370.46	370.1906	71.33	5.99	22.69
C22H23N1	301.44	301.1830	87.66	7.69	4.65
C22H23N3	329.45	329.1892	80.21	7.04	12.75
C22H23N5	357.46	357.1953	73.92	6.49	19.59
C22H24N2	316.45	316.1939	83.50	7.64	8.85
C22H24N4	344.46	344.2001	76.71	7.02	16.26
C22H24N6	372.48	372.2062	70.94	6.49	22.56
C22H25N1	303.45	303.1987	87.08	8.30	4.62

C22H25N3	331.46	331.2048	79.72	7.60	12.68
C22H25N5	359.48	359.2110	73.51	7.01	19.48
C22H26N2	318.47	318.2096	82.97	8.23	8.80
C22H26N4	346.48	346.2157	76.27	7.56	16.17
C22H26N6	374.49	374.2219	70.56	7.00	22.44
C22H27N1	305.47	305.2143	86.51	8.91	4.59
C22H27N3	333.48	333.2205	79.24	8.16	12.60
C22H27N5	361.49	361.2266	73.10	7.53	19.37
C22H28N2	320.48	320.2252	82.45	8.81	8.74
C22H28N4	348.50	348.2314	75.82	8.10	16.08
C22H28N6	376.51	376.2375	70.18	7.50	22.32
C22H29N1	307.48	307.2300	85.94	9.51	4.56
C22H29N3	335.50	335.2361	78.76	8.71	12.52
C22H29N5	363.51	363.2423	72.69	8.04	19.27
C22H30N2	322.50	322.2409	81.94	9.38	8.69
C22H30N4	350.51	350.2470	75.39	8.63	15.98
C22H30N6	378.52	378.2532	69.81	7.99	22.20
C22H31N1	309.50	309.2456	85.38	10.10	4.53
C22H31N3	337.51	337.2518	78.29	9.26	12.45
C22H31N5	365.53	365.2579	72.29	8.55	19.16
C22H32N2	324.51	324.2565	81.43	9.94	8.63
C22H32N4	352.53	352.2627	74.96	9.15	15.89
C22H32N6	380.54	380.2688	69.44	8.48	22.08
C22H33N1	311.52	311.2613	84.83	10.68	4.50
C22H33N3	339.53	339.2674	77.83	9.80	12.38
C22H33N5	367.54	367.2736	71.90	9.05	19.05
C22H34N2	326.53	326.2722	80.93	10.50	8.58
C22H34N4	354.54	354.2783	74.53	9.67	15.80
C22H34N6	382.56	382.2845	69.07	8.96	21.97
C22H35N1	313.53	313.2769	84.28	11.25	4.47
C22H35N3	341.54	341.2831	77.37	10.33	12.30
C22H35N5	369.56	369.2892	71.50	9.55	18.95
C22H36N2	328.55	328.2878	80.43	11.04	8.53
C22H36N4	356.56	356.2940	74.11	10.18	15.71
C22H36N6	384.57	384.3001	68.71	9.44	21.85
C22H37N1	315.55	315.2926	83.74	11.82	4.44
C22H37N3	343.56	343.2987	76.91	10.86	12.23
C22H37N5	371.57	371.3049	71.12	10.04	18.85
C22H38N2	330.56	330.3035	79.94	11.59	8.47
C22H38N4	358.57	358.3096	73.69	10.68	15.62
C22H38N6	386.59	386.3158	68.35	9.91	21.74
C22H39N1	317.56	317.3082	83.21	12.38	4.41
C22H39N3	345.58	345.3144	76.47	11.38	12.16
C22H39N5	373.59	373.3205	70.73	10.52	18.75
C22H40N2	332.58	332.3191	79.45	12.12	8.42
C22H40N4	360.59	360.3253	73.28	11.18	15.54
C22H40N6	388.60	388.3314	68.00	10.38	21.63
C22H41N1	319.58	319.3239	82.69	12.93	4.38
C22H41N3	347.59	347.3300	76.02	11.89	12.09
C22H41N5	375.61	375.3362	70.35	11.00	18.65
C22H42N2	334.59	334.3348	78.98	12.65	8.37
C22H42N4	362.61	362.3409	72.87	11.68	15.45
C22H42N6	390.62	390.3471	67.65	10.84	21.51
C22H43N1	321.59	321.3395	82.17	13.48	4.36
C22H43N3	349.61	349.3457	75.58	12.40	12.02
C22H43N5	377.62	377.3518	69.98	11.48	18.55
C22H44N2	336.61	336.3504	78.50	13.18	8.32
C22H44N4	364.62	364.3566	72.47	12.16	15.37
C22H44N6	392.64	392.3627	67.30	11.30	21.40
C22H45N1	323.61	323.3552	81.66	14.02	4.33
C22H45N3	351.62	351.3613	75.15	12.90	11.95
C22H45N5	379.64	379.3675	69.60	11.95	18.45
C22H46N2	338.63	338.3661	78.03	13.69	8.27
C22H46N4	366.64	366.3722	72.07	12.65	15.28
C22H46N6	394.65	394.3784	66.96	11.75	21.29
C22H47N1	325.63	325.3708	81.15	14.55	4.30
C22H47N3	353.64	353.3770	74.72	13.40	11.88
C22H47N5	381.65	381.3831	69.24	12.41	18.35
C22H48N2	340.64	340.3817	77.57	14.20	8.22
C22H48N4	368.65	368.3879	71.68	13.12	15.20
C22H48N6	396.67	396.3940	66.62	12.20	21.19
C22H49N3	355.66	355.3926	74.30	13.89	11.81
C22H49N5	383.67	383.3988	68.87	12.87	18.25
C22H50N4	370.67	370.4035	71.29	13.60	15.11
C22H50N6	398.68	398.4097	66.28	12.64	21.08
C22H51N5	385.69	385.4144	68.51	13.33	18.16
C22H52N6	400.70	400.4253	65.95	13.08	20.97
C23H11N1	301.35	301.0891	91.67	3.68	4.65
C23H11N3	329.36	329.0953	83.88	3.37	12.76
C23H11N5	357.38	357.1014	77.30	3.10	19.60
C23H12N2	316.37	316.1000	87.32	3.82	8.85
C23H12N4	344.38	344.1062	80.22	3.51	16.27
C23H12N6	372.39	372.1123	74.18	3.25	22.57
C23H13N1	303.37	303.1048	91.06	4.32	4.62
C23H13N3	331.38	331.1109	83.37	3.95	12.68
C23H13N5	359.39	359.1171	76.87	3.65	19.49
C23H14N2	318.38	318.1157	86.77	4.43	8.80
C23H14N4	346.39	346.1218	79.75	4.07	16.17
C23H14N6	374.41	374.1280	73.78	3.77	22.45
C23H15N1	305.38	305.1204	90.46	4.95	4.59
C23H15N3	333.40	333.1266	82.86	4.54	12.60
C23H15N5	361.41	361.1327	76.44	4.18	19.38
C23H16N2	320.40	320.1313	86.22	5.03	8.74
C23H16N4	348.41	348.1375	79.29	4.63	16.08
C23H16N6	376.42	376.1436	73.39	4.28	22.33
C23H17N1	307.40	307.1361	89.87	5.57	4.56
C23H17N3	335.41	335.1422	82.36	5.11	12.53
C23H17N5	363.43	363.1484	76.01	4.71	19.27
C23H18N2	322.41	322.1470	85.68	5.63	8.69
C23H18N4	350.43	350.1531	78.83	5.18	15.99
C23H18N6	378.44	378.1593	73.00	4.79	22.21
C23H19N1	309.41	309.1517	89.28	6.19	4.53
C23H19N3	337.43	337.1579	81.87	5.68	12.45
C23H19N5	365.44	365.1640	75.60	5.24	19.16
C23H20N2	324.43	324.1626	85.15	6.21	8.63
C23H20N4	352.44	352.1688	78.38	5.72	15.90
C23H20N6	380.46	380.1749	72.61	5.30	22.09
C23H21N1	311.43	311.1674	88.71	6.80	4.50
C23H21N3	339.44	339.1735	81.39	6.24	12.38
C23H21N5	367.46	367.1797	75.18	5.76	19.06
C23H22N2	326.45	326.1783	84.63	6.79	8.58
C23H22N4	354.46	354.1844	77.94	6.26	15.81
C23H22N6	382.47	382.1906	72.23	5.80	21.97
C23H23N1	313.45	313.1830	88.14	7.40	4.47
C23H23N3	341.46	341.1892	80.90	6.79	12.31
C23H23N5	369.47	369.1953	74.77	6.27	18.95
C23H24N2	328.46	328.1939	84.11	7.37	8.53
C23H24N4	356.47	356.2001	77.50	6.79	15.72
C23H24N6	384.49	384.2062	71.85	6.29	21.86
C23H25N1	315.46	315.1987	87.57	7.99	4.44

C23H25N3	343.48	343.2048	80.43	7.34	12.23
C23H25N5	371.49	371.2110	74.36	6.78	18.85
C23H26N2	330.48	330.2096	83.59	7.93	8.48
C23H26N4	358.49	358.2157	77.06	7.31	15.63
C23H26N6	386.50	386.2219	71.48	6.78	21.74
C23H27N1	317.48	317.2143	87.02	8.57	4.41
C23H27N3	345.49	345.2205	79.96	7.88	12.16
C23H27N5	373.51	373.2266	73.96	7.29	18.75
C23H28N2	332.49	332.2252	83.09	8.49	8.43
C23H28N4	360.51	360.2314	76.63	7.83	15.54
C23H28N6	388.52	388.2375	71.10	7.26	21.63
C23H29N1	319.49	319.2300	86.47	9.15	4.38
C23H29N3	347.51	347.2361	79.50	8.41	12.09
C23H29N5	375.52	375.2423	73.57	7.78	18.65
C23H30N2	334.51	334.2409	82.59	9.04	8.37
C23H30N4	362.52	362.2470	76.20	8.34	15.45
C23H30N6	390.54	390.2532	70.74	7.74	21.52
C23H31N1	321.51	321.2456	85.92	9.72	4.36
C23H31N3	349.52	349.2518	79.04	8.94	12.02
C23H31N5	377.54	377.2579	73.17	8.28	18.55
C23H32N2	336.52	336.2565	82.09	9.58	8.32
C23H32N4	364.54	364.2627	75.78	8.85	15.37
C23H32N6	392.55	392.2688	70.37	8.22	21.41
C23H33N1	323.53	323.2613	85.39	10.28	4.33
C23H33N3	351.54	351.2674	78.58	9.46	11.95
C23H33N5	379.55	379.2736	72.78	8.76	18.45
C23H34N2	338.54	338.2722	81.60	10.12	8.27
C23H34N4	366.55	366.2783	75.37	9.35	15.28
C23H34N6	394.57	394.2845	70.01	8.69	21.30
C23H35N1	325.54	325.2769	84.86	10.84	4.30
C23H35N3	353.56	353.2831	78.14	9.98	11.89
C23H35N5	381.57	381.2892	72.40	9.25	18.35
C23H36N2	340.56	340.2878	81.12	10.66	8.23
C23H36N4	368.57	368.2940	74.95	9.85	15.20
C23H36N6	396.58	396.3001	69.66	9.15	21.19
C23H37N1	327.56	327.2926	84.34	11.39	4.28
C23H37N3	355.57	355.2987	77.69	10.49	11.82
C23H37N5	383.58	383.3049	72.02	9.72	18.26
C23H38N2	342.57	342.3035	80.64	11.18	8.18
C23H38N4	370.59	370.3096	74.55	10.34	15.12
C23H38N6	398.60	398.3158	69.31	9.61	21.08
C23H39N1	329.57	329.3082	83.82	11.93	4.25
C23H39N3	357.59	357.3144	77.26	10.99	11.75
C23H39N5	385.60	385.3205	71.64	10.19	18.16
C23H40N2	344.59	344.3191	80.17	11.70	8.13
C23H40N4	372.60	372.3253	74.14	10.82	15.04
C23H40N6	400.62	400.3314	68.96	10.06	20.98
C23H41N1	331.59	331.3239	83.31	12.46	4.22
C23H41N3	359.60	359.3300	76.82	11.49	11.69
C23H41N5	387.62	387.3362	71.27	10.66	18.07
C23H42N2	346.60	346.3348	79.70	12.21	8.08
C23H42N4	374.62	374.3409	73.74	11.30	14.96
C23H42N6	402.63	402.3471	68.61	10.51	20.87
C23H43N1	333.61	333.3395	82.81	12.99	4.20
C23H43N3	361.62	361.3457	76.39	11.99	11.62
C23H43N5	389.63	389.3518	70.90	11.12	17.97
C23H44N2	348.62	348.3504	79.24	12.72	8.04
C23H44N4	376.63	376.3566	73.35	11.78	14.88
C23H44N6	404.65	404.3627	68.27	10.96	20.77
C23H45N1	335.62	335.3552	82.31	13.51	4.17

10/21/66 059

C23H45N3	363.64	363.3613	75.97	12.47	11.56
C23H45N5	391.65	391.3675	70.54	11.58	17.88
C23H46N2	350.64	350.3661	78.79	13.22	7.99
C23H46N4	378.65	378.3722	72.96	12.25	14.80
C23H46N6	406.66	406.3784	67.93	11.40	20.67
C23H47N1	337.64	337.3708	81.82	14.03	4.15
C23H47N3	365.65	365.3770	75.55	12.96	11.49
C23H47N5	393.66	393.3831	70.18	12.03	17.79
C23H48N2	352.65	352.3817	78.34	13.72	7.94
C23H48N4	380.67	380.3879	72.57	12.71	14.72
C23H48N6	408.68	408.3940	67.60	11.84	20.56
C23H49N1	339.65	339.3865	81.33	14.54	4.12
C23H49N3	367.67	367.3926	75.14	13.43	11.43
C23H49N5	395.68	395.3988	69.82	12.48	17.70
C23H50N2	354.67	354.3974	77.89	14.21	7.90
C23H50N4	382.68	382.4035	72.19	13.17	14.64
C23H50N6	410.70	410.4097	67.27	12.27	20.46
C23H51N3	369.68	369.4083	74.73	13.91	11.37
C23H51N5	397.70	397.4144	69.46	12.93	17.61
C23H52N4	384.70	384.4192	71.81	13.62	14.56
C23H52N6	412.71	412.4253	66.94	12.70	20.36
C23H53N5	399.71	399.4301	69.11	13.37	17.52
C23H54N6	414.73	414.4410	66.61	13.12	20.26
C24H12N2	328.38	328.1000	87.79	3.68	8.53
C24H12N4	356.39	356.1062	80.89	3.39	15.72
C24H12N6	384.40	384.1123	74.99	3.15	21.86
C24H13N1	315.38	315.1048	91.40	4.15	4.44
C24H13N3	343.39	343.1109	83.95	3.82	12.24
C24H13N5	371.40	371.1171	77.62	3.53	18.86
C24H14N2	330.39	330.1157	87.25	4.27	8.48
C24H14N4	358.41	358.1218	80.43	3.94	15.63
C24H14N6	386.42	386.1280	74.60	3.65	21.75
C24H15N1	317.39	317.1204	90.82	4.76	4.41
C24H15N3	345.41	345.1266	83.46	4.38	12.17
C24H15N5	373.42	373.1327	77.20	4.05	18.75
C24H16N2	332.41	332.1313	86.72	4.85	8.43
C24H16N4	360.42	360.1375	79.98	4.47	15.54
C24H16N6	388.44	388.1436	74.21	4.15	21.64
C24H17N1	319.41	319.1361	90.25	5.36	4.39
C24H17N3	347.42	347.1422	82.97	4.93	12.09
C24H17N5	375.44	375.1484	76.78	4.56	18.65
C24H18N2	334.42	334.1470	86.20	5.43	8.38
C24H18N4	362.44	362.1531	79.54	5.01	15.46
C24H18N6	390.45	390.1593	73.83	4.65	21.52
C24H19N1	321.43	321.1517	89.68	5.96	4.36
C24H19N3	349.44	349.1579	82.49	5.48	12.03
C24H19N5	377.45	377.1640	76.37	5.07	18.55
C24H20N2	336.44	336.1626	85.68	5.99	8.33
C24H20N4	364.45	364.1688	79.10	5.53	15.37
C24H20N6	392.47	392.1749	73.45	5.14	21.41
C24H21N1	323.44	323.1674	89.13	6.54	4.33
C24H21N3	351.46	351.1735	82.02	6.02	11.96
C24H21N5	379.47	379.1797	75.97	5.58	18.46
C24H22N2	338.46	338.1783	85.17	6.55	8.28
C24H22N4	366.47	366.1844	78.66	6.05	15.29
C24H22N6	394.48	394.1906	73.07	5.62	21.30
C24H23N1	325.46	325.1830	88.57	7.12	4.30
C24H23N3	353.47	353.1892	81.55	6.56	11.89
C24H23N5	381.48	381.1953	75.56	6.08	18.36
C24H24N2	340.47	340.1939	84.67	7.11	8.23

C24H24N4	368.49	368.2001	78.23	6.57	15.20
C24H24N6	396.50	396.2062	72.70	6.10	21.20
C24H25N1	327.47	327.1987	88.03	7.70	4.28
C24H25N3	355.49	355.2048	81.09	7.09	11.82
C24H25N5	383.50	383.2110	75.17	6.57	18.26
C24H26N2	342.49	342.2096	84.17	7.65	8.18
C24H26N4	370.50	370.2157	77.80	7.07	15.12
C24H26N6	398.52	398.2219	72.34	6.58	21.09
C24H27N1	329.49	329.2143	87.49	8.26	4.25
C24H27N3	357.50	357.2205	80.63	7.61	11.75
C24H27N5	385.52	385.2266	74.77	7.06	18.17
C24H28N2	344.50	344.2252	83.68	8.19	8.13
C24H28N4	372.52	372.2314	77.38	7.58	15.04
C24H28N6	400.53	400.2375	71.97	7.05	20.98
C24H29N1	331.51	331.2300	86.96	8.82	4.23
C24H29N3	359.52	359.2361	80.18	8.13	11.69
C24H29N5	387.53	387.2423	74.39	7.54	18.07
C24H30N2	346.52	346.2409	83.19	8.73	8.08
C24H30N4	374.53	374.2470	76.97	8.07	14.96
C24H30N6	402.55	402.2532	71.61	7.51	20.88
C24H31N1	333.52	333.2456	86.43	9.37	4.20
C24H31N3	361.53	361.2518	79.73	8.64	11.62
C24H31N5	389.55	389.2579	74.00	8.02	17.98
C24H32N2	348.54	348.2565	82.71	9.25	8.04
C24H32N4	376.55	376.2627	76.56	8.57	14.88
C24H32N6	404.56	404.2688	71.25	7.97	20.77
C24H33N1	335.54	335.2613	85.91	9.91	4.17
C24H33N3	363.55	363.2674	79.29	9.15	11.56
C24H33N5	391.56	391.2736	73.62	8.49	17.89
C24H34N2	350.55	350.2722	82.23	9.78	7.99
C24H34N4	378.57	378.2783	76.15	9.05	14.80
C24H34N6	406.58	406.2845	70.90	8.43	20.67
C24H35N1	337.55	337.2769	85.40	10.45	4.15
C24H35N3	365.57	365.2831	78.86	9.65	11.49
C24H35N5	393.58	393.2892	73.24	8.96	17.79
C24H36N2	352.57	352.2878	81.76	10.29	7.95
C24H36N4	380.58	380.2940	75.74	9.53	14.72
C24H36N6	408.59	408.3001	70.55	8.88	20.57
C24H37N1	339.57	339.2926	84.89	10.98	4.12
C24H37N3	367.58	367.2987	78.42	10.15	11.43
C24H37N5	395.60	395.3049	72.87	9.43	17.70
C24H38N2	354.58	354.3035	81.30	10.80	7.90
C24H38N4	382.60	382.3096	75.34	10.01	14.64
C24H38N6	410.61	410.3158	70.20	9.33	20.47
C24H39N1	341.59	341.3082	84.39	11.51	4.10
C24H39N3	369.60	369.3144	77.99	10.64	11.37
C24H39N5	397.61	397.3205	72.50	9.89	17.61
C24H40N2	356.60	356.3191	80.84	11.31	7.86
C24H40N4	384.61	384.3253	74.95	10.48	14.57
C24H40N6	412.63	412.3314	69.86	9.77	20.37
C24H41N1	343.60	343.3239	83.90	12.03	4.08
C24H41N3	371.61	371.3300	77.57	11.12	11.31
C24H41N5	399.63	399.3362	72.13	10.34	17.52
C24H42N2	358.62	358.3348	80.38	11.81	7.81
C24H42N4	386.63	386.3409	74.56	10.95	14.49
C24H42N6	414.64	414.3471	69.52	10.21	20.27
C24H43N1	345.62	345.3395	83.41	12.54	4.05
C24H43N3	373.63	373.3457	77.15	11.60	11.25
C24H43N5	401.64	401.3518	71.77	10.79	17.44
C24H44N2	360.63	360.3504	79.93	12.30	7.77

10/21/66 060

C24H44N4	388.65	388.3566	74.17	11.41	14.42
C24H44N6	416.66	416.3627	69.19	10.64	20.17
C24H45N1	347.63	347.3552	82.92	13.05	4.03
C24H45N3	375.65	375.3613	76.74	12.07	11.19
C24H45N5	403.66	403.3675	71.41	11.24	17.35
C24H46N2	362.65	362.3661	79.49	12.79	7.72
C24H46N4	390.66	390.3722	73.79	11.87	14.34
C24H46N6	418.67	418.3784	68.85	11.07	20.07
C24H47N1	349.65	349.3708	82.44	13.55	4.01
C24H47N3	377.66	377.3770	76.33	12.54	11.13
C24H47N5	405.68	405.3831	71.06	11.68	17.26
C24H48N2	364.66	364.3817	79.05	13.27	7.68
C24H48N4	392.68	392.3879	73.41	12.32	14.27
C24H48N6	420.69	420.3940	68.52	11.50	19.98
C24H49N1	351.66	351.3865	81.97	14.04	3.98
C24H49N3	379.68	379.3926	75.92	13.01	11.07
C24H49N5	407.69	407.3988	70.71	12.11	17.18
C24H50N2	366.68	366.3974	78.62	13.74	7.64
C24H50N4	394.69	394.4035	73.04	12.77	14.20
C24H50N6	422.71	422.4097	68.20	11.92	19.88
C24H51N1	353.68	353.4021	81.51	14.53	3.96
C24H51N3	381.69	381.4083	75.52	13.47	11.01
C24H51N5	409.71	409.4144	70.36	12.55	17.09
C24H52N2	368.70	368.4130	78.19	14.22	7.60
C24H52N4	396.71	396.4192	72.66	13.21	14.12
C24H52N6	424.72	424.4253	67.87	12.34	19.79
C24H53N3	383.71	383.4239	75.13	13.92	10.95
C24H53N5	411.72	411.4301	70.01	12.98	17.01
C24H54N4	398.72	398.4348	72.30	13.65	14.05
C24H54N6	426.74	426.4410	67.55	12.75	19.69
C24H55N5	413.74	413.4457	69.67	13.40	16.93
C24H56N6	428.75	428.4566	67.23	13.17	19.60
C25H12N2	340.39	340.1000	88.22	3.55	8.23
C25H12N4	368.40	368.1062	81.51	3.28	15.21
C25H12N6	396.41	396.1123	75.75	3.05	21.20
C25H13N1	327.39	327.1048	91.72	4.00	4.28
C25H13N3	355.40	355.1109	84.49	3.69	11.82
C25H13N5	383.42	383.1171	78.32	3.42	18.27
C25H14N2	342.40	342.1157	87.70	4.12	8.18
C25H14N4	370.42	370.1218	81.07	3.81	15.13
C25H14N6	398.43	398.1280	75.37	3.54	21.09
C25H15N1	329.40	329.1204	91.16	4.59	4.25
C25H15N3	357.42	357.1266	84.01	4.23	11.76
C25H15N5	385.43	385.1327	77.91	3.92	18.17
C25H16N2	344.42	344.1313	87.18	4.68	8.13
C25H16N4	372.43	372.1375	80.63	4.33	15.04
C25H16N6	400.45	400.1436	74.99	4.03	20.99
C25H17N1	331.42	331.1361	90.60	5.17	4.23
C25H17N3	359.43	359.1422	83.54	4.77	11.69
C25H17N5	387.45	387.1484	77.50	4.42	18.08
C25H18N2	346.44	346.1470	86.68	5.24	8.09
C25H18N4	374.45	374.1531	80.19	4.85	14.96
C25H18N6	402.46	402.1593	74.61	4.51	20.88
C25H19N1	333.44	333.1517	90.06	5.74	4.20
C25H19N3	361.45	361.1579	83.08	5.30	11.63
C25H19N5	389.46	389.1640	77.10	4.92	17.98
C25H20N2	348.45	348.1626	86.18	5.79	8.04
C25H20N4	376.46	376.1688	79.76	5.35	14.88
C25H20N6	404.48	404.1749	74.24	4.98	20.78
C25H21N1	335.45	335.1674	89.51	6.31	4.18

C25H21N3	363.47	363.1735	82.62	5.82	11.56
C25H21N5	391.48	391.1797	76.70	5.41	17.89
C25H22N2	350.47	350.1783	85.68	6.33	7.99
C25H22N4	378.48	378.1844	79.34	5.86	14.80
C25H22N6	406.49	406.1906	73.87	5.46	20.67
C25H23N1	337.47	337.1830	88.98	6.87	4.15
C25H23N3	365.48	365.1892	82.16	6.34	11.50
C25H23N5	393.50	393.1953	76.31	5.89	17.80
C25H24N2	352.48	352.1939	85.19	6.86	7.95
C25H24N4	380.50	380.2001	78.92	6.36	14.72
C25H24N6	408.51	408.2062	73.51	5.92	20.57
C25H25N1	339.48	339.1987	88.45	7.42	4.13
C25H25N3	367.50	367.2048	81.71	6.86	11.43
C25H25N5	395.51	395.2110	75.92	6.37	17.71
C25H26N2	354.50	354.2096	84.71	7.39	7.90
C25H26N4	382.51	382.2157	78.50	6.85	14.65
C25H26N6	410.53	410.2219	73.14	6.38	20.47
C25H27N1	341.50	341.2143	87.93	7.97	4.10
C25H27N3	369.51	369.2205	81.26	7.37	11.37
C25H27N5	397.53	397.2266	75.54	6.85	17.62
C25H28N2	356.52	356.2252	84.23	7.92	7.86
C25H28N4	384.53	384.2314	78.09	7.34	14.57
C25H28N6	412.54	412.2375	72.79	6.84	20.37
C25H29N1	343.52	343.2300	87.41	8.51	4.08
C25H29N3	371.53	371.2361	80.82	7.87	11.31
C25H29N5	399.54	399.2423	75.16	7.32	17.53
C25H30N2	358.53	358.2409	83.75	8.43	7.81
C25H30N4	386.54	386.2470	77.68	7.82	14.49
C25H30N6	414.56	414.2532	72.43	7.29	20.27
C25H31N1	345.53	345.2456	86.90	9.04	4.05
C25H31N3	373.55	373.2518	80.39	8.36	11.25
C25H31N5	401.56	401.2579	74.78	7.78	17.44
C25H32N2	360.55	360.2565	83.28	8.95	7.77
C25H32N4	388.56	388.2627	77.28	8.30	14.42
C25H32N6	416.57	416.2688	72.08	7.74	20.17
C25H33N1	347.55	347.2613	86.40	9.57	4.03
C25H33N3	375.56	375.2674	79.95	8.86	11.19
C25H33N5	403.58	403.2736	74.40	8.24	17.35
C25H34N2	362.56	362.2722	82.82	9.45	7.73
C25H34N4	390.58	390.2783	76.88	8.77	14.34
C25H34N6	418.59	418.2845	71.74	8.19	20.08
C25H35N1	349.56	349.2769	85.90	10.09	4.01
C25H35N3	377.58	377.2831	79.53	9.34	11.13
C25H35N5	405.59	405.2892	74.03	8.70	17.27
C25H36N2	364.58	364.2878	82.36	9.95	7.68
C25H36N4	392.59	392.2940	76.49	9.24	14.27
C25H36N6	420.61	420.3001	71.39	8.63	19.98
C25H37N1	351.58	351.2926	85.41	10.61	3.98
C25H37N3	379.59	379.2987	79.11	9.82	11.07
C25H37N5	407.61	407.3049	73.67	9.15	17.18
C25H38N2	366.60	366.3035	81.91	10.45	7.64
C25H38N4	394.61	394.3096	76.10	9.71	14.20
C25H38N6	422.62	422.3158	71.05	9.06	19.89
C25H39N1	353.60	353.3082	84.92	11.12	3.96
C25H39N3	381.61	381.3144	78.69	10.30	11.01
C25H39N5	409.62	409.3205	73.31	9.60	17.10
C25H40N2	368.61	368.3191	81.46	10.94	7.60
C25H40N4	396.62	396.3253	75.71	10.17	14.13
C25H40N6	424.64	424.3314	70.71	9.49	19.79
C25H41N1	355.61	355.3239	84.44	11.62	3.94

C25H41N3	383.63	383.3300	78.27	10.77	10.95
C25H41N5	411.64	411.3362	72.95	10.04	17.01
C25H42N2	370.63	370.3348	81.02	11.42	7.56
C25H42N4	398.64	398.3409	75.33	10.62	14.05
C25H42N6	426.65	426.3471	70.38	9.92	19.70
C25H43N1	357.63	357.3395	83.96	12.12	3.92
C25H43N3	385.64	385.3457	77.86	11.24	10.90
C25H43N5	413.65	413.3518	72.59	10.48	16.93
C25H44N2	372.64	372.3504	80.58	11.90	7.52
C25H44N4	400.66	400.3566	74.95	11.07	13.98
C25H44N6	428.67	428.3627	70.05	10.35	19.60
C25H45N1	359.64	359.3552	83.49	12.61	3.89
C25H45N3	387.66	387.3613	77.46	11.70	10.84
C25H45N5	415.67	415.3675	72.24	10.91	16.85
C25H46N2	374.66	374.3661	80.15	12.38	7.48
C25H46N4	402.67	402.3722	74.57	11.51	13.91
C25H46N6	430.69	430.3784	69.72	10.77	19.51
C25H47N1	361.66	361.3708	83.03	13.10	3.87
C25H47N3	389.67	389.3770	77.06	12.16	10.78
C25H47N5	417.69	417.3831	71.89	11.34	16.77
C25H48N2	376.67	376.3817	79.72	12.84	7.44
C25H48N4	404.69	404.3879	74.20	11.96	13.84
C25H48N6	432.70	432.3940	69.40	11.18	19.42
C25H49N1	363.68	363.3865	82.57	13.58	3.85
C25H49N3	391.69	391.3926	76.66	12.61	10.73
C25H49N5	419.70	419.3988	71.55	11.77	16.69
C25H50N2	378.69	378.3974	79.29	13.31	7.40
C25H50N4	406.70	406.4035	73.83	12.39	13.78
C25H50N6	434.72	434.4097	69.07	11.59	19.33
C25H51N1	365.69	365.4021	82.11	14.06	3.83
C25H51N3	393.71	393.4083	76.27	13.06	10.67
C25H51N5	421.72	421.4144	71.20	12.19	16.61
C25H52N2	380.71	380.4130	78.87	13.77	7.36
C25H52N4	408.72	408.4192	73.47	12.82	13.71
C25H52N6	436.73	436.4253	68.76	12.00	19.24
C25H53N1	367.71	367.4178	81.66	14.53	3.81
C25H53N3	395.72	395.4239	75.88	13.50	10.62
C25H53N5	423.73	423.4301	70.86	12.61	16.53
C25H54N2	382.72	382.4287	78.46	14.22	7.32
C25H54N4	410.74	410.4348	73.11	13.25	13.64
C25H54N6	438.75	438.4410	68.44	12.41	19.15
C25H55N3	397.74	397.4396	75.50	13.94	10.56
C25H55N5	425.75	425.4457	70.53	13.02	16.45
C25H56N4	412.75	412.4505	72.75	13.68	13.57
C25H56N6	440.77	440.4566	68.13	12.81	19.07
C25H57N5	427.77	427.4614	70.20	13.43	16.37
C25H58N6	442.78	442.4723	67.82	13.20	18.98
C26H13N1	339.40	339.1048	92.01	3.86	4.13
C26H13N3	367.41	367.1109	85.00	3.57	11.44
C26H13N5	395.43	395.1171	78.98	3.31	17.71
C26H14N2	354.41	354.1157	88.11	3.98	7.90
C26H14N4	382.43	382.1218	81.66	3.69	14.65
C26H14N6	410.44	410.1280	76.09	3.44	20.48
C26H15N1	341.42	341.1204	91.47	4.43	4.10
C26H15N3	369.43	369.1266	84.53	4.09	11.37
C26H15N5	397.44	397.1327	78.57	3.80	17.62
C26H16N2	356.43	356.1313	87.62	4.52	7.86
C26H16N4	384.44	384.1375	81.23	4.20	14.57
C26H16N6	412.46	412.1436	75.71	3.91	20.38
C26H17N1	343.43	343.1361	90.93	4.99	4.08

C26H17N3	371.45	371.1422	84.07	4.61	11.31
C26H17N5	399.46	399.1484	78.18	4.29	17.53
C26H18N2	358.45	358.1470	87.12	5.06	7.82
C26H18N4	386.46	386.1531	80.81	4.69	14.50
C26H18N6	414.47	414.1593	75.35	4.38	20.28
C26H19N1	345.45	345.1517	90.40	5.54	4.05
C26H19N3	373.46	373.1579	83.62	5.13	11.25
C26H19N5	401.47	401.1640	77.79	4.77	17.44
C26H20N2	360.46	360.1626	86.64	5.59	7.77
C26H20N4	388.48	388.1688	80.39	5.19	14.42
C26H20N6	416.49	416.1749	74.98	4.84	20.18
C26H21N1	347.46	347.1674	89.88	6.09	4.03
C26H21N3	375.48	375.1735	83.17	5.64	11.19
C26H21N5	403.49	403.1797	77.40	5.25	17.36
C26H22N2	362.48	362.1783	86.15	6.12	7.73
C26H22N4	390.49	390.1844	79.97	5.68	14.35
C26H22N6	418.51	418.1906	74.62	5.30	20.08
C26H23N1	349.48	349.1830	89.36	6.63	4.01
C26H23N3	377.49	377.1892	82.73	6.14	11.13
C26H23N5	405.51	405.1953	77.01	5.72	17.27
C26H24N2	364.49	364.1939	85.68	6.64	7.69
C26H24N4	392.51	392.2001	79.56	6.16	14.27
C26H24N6	420.52	420.2062	74.26	5.75	19.98
C26H25N1	351.50	351.1987	88.85	7.17	3.98
C26H25N3	379.51	379.2048	82.29	6.64	11.07
C26H25N5	407.52	407.2110	76.63	6.18	17.19
C26H26N2	366.51	366.2096	85.21	7.15	7.64
C26H26N4	394.52	394.2157	79.16	6.64	14.20
C26H26N6	422.54	422.2219	73.91	6.20	19.89
C26H27N1	353.51	353.2143	88.34	7.70	3.96
C26H27N3	381.53	381.2205	81.85	7.13	11.01
C26H27N5	409.54	409.2266	76.25	6.65	17.10
C26H28N2	368.53	368.2252	84.74	7.66	7.60
C26H28N4	396.54	396.2314	78.75	7.12	14.13
C26H28N6	424.55	424.2375	73.56	6.65	19.79
C26H29N1	355.53	355.2300	87.84	8.22	3.94
C26H29N3	383.54	383.2361	81.42	7.62	10.96
C26H29N5	411.55	411.2423	75.88	7.10	17.02
C26H30N2	370.54	370.2409	84.28	8.16	7.56
C26H30N4	398.56	398.2470	78.36	7.59	14.06
C26H30N6	426.57	426.2532	73.21	7.09	19.70
C26H31N1	357.54	357.2456	87.34	8.74	3.92
C26H31N3	385.56	385.2518	81.00	8.10	10.90
C26H31N5	413.57	413.2579	75.51	7.56	16.93
C26H32N2	372.56	372.2565	83.82	8.66	7.52
C26H32N4	400.57	400.2627	77.96	8.05	13.99
C26H32N6	428.59	428.2688	72.87	7.53	19.61
C26H33N1	359.56	359.2613	86.85	9.25	3.90
C26H33N3	387.57	387.2674	80.58	8.58	10.84
C26H33N5	415.59	415.2736	75.14	8.00	16.85
C26H34N2	374.57	374.2722	83.37	9.15	7.48
C26H34N4	402.59	402.2783	77.57	8.51	13.92
C26H34N6	430.60	430.2845	72.52	7.96	19.52
C26H35N1	361.58	361.2769	86.37	9.76	3.87
C26H35N3	389.59	389.2831	80.16	9.06	10.79
C26H35N5	417.60	417.2892	74.78	8.45	16.77
C26H36N2	376.59	376.2878	82.93	9.64	7.44
C26H36N4	404.60	404.2940	77.18	8.97	13.85
C26H36N6	432.62	432.3001	72.19	8.39	19.43
C26H37N1	363.59	363.2926	85.89	10.26	3.85

C26H37N3	391.60	391.2987	79.75	9.52	10.73
C26H37N5	419.62	419.3049	74.42	8.89	16.69
C26H38N2	378.61	378.3035	82.48	10.12	7.40
C26H38N4	406.62	406.3096	76.80	9.42	13.78
C26H38N6	434.63	434.3158	71.85	8.81	19.34
C26H39N1	365.61	365.3082	85.42	10.75	3.83
C26H39N3	393.62	393.3144	79.34	9.99	10.68
C26H39N5	421.63	421.3205	74.07	9.32	16.61
C26H40N2	380.62	380.3191	82.05	10.59	7.36
C26H40N4	408.64	408.3253	76.42	9.87	13.71
C26H40N6	436.65	436.3314	71.52	9.23	19.25
C26H41N1	367.62	367.3239	84.95	11.24	3.81
C26H41N3	395.64	395.3300	78.93	10.45	10.62
C26H41N5	423.65	423.3362	73.71	9.75	16.53
C26H42N2	382.64	382.3348	81.61	11.06	7.32
C26H42N4	410.65	410.3409	76.05	10.31	13.64
C26H42N6	438.66	438.3471	71.19	9.65	19.16
C26H43N1	369.64	369.3395	84.49	11.73	3.79
C26H43N3	397.65	397.3457	78.53	10.90	10.57
C26H43N5	425.67	425.3518	73.36	10.18	16.45
C26H44N2	384.65	384.3504	81.19	11.53	7.28
C26H44N4	412.67	412.3566	75.68	10.75	13.58
C26H44N6	440.68	440.3627	70.87	10.06	19.07
C26H45N1	371.66	371.3552	84.03	12.20	3.77
C26H45N3	399.67	399.3613	78.14	11.35	10.51
C26H45N5	427.68	427.3675	73.02	10.61	16.38
C26H46N2	386.67	386.3661	80.76	11.99	7.24
C26H46N4	414.68	414.3722	75.31	11.18	13.51
C26H46N6	442.70	442.3784	70.54	10.47	18.98
C26H47N1	373.67	373.3708	83.57	12.68	3.75
C26H47N3	401.68	401.3770	77.75	11.79	10.46
C26H47N5	429.70	429.3831	72.68	11.03	16.30
C26H48N2	388.69	388.3817	80.35	12.45	7.21
C26H48N4	416.70	416.3879	74.94	11.61	13.45
C26H48N6	444.71	444.3940	70.22	10.88	18.90
C26H49N1	375.69	375.3865	83.12	13.15	3.73
C26H49N3	403.70	403.3926	77.36	12.23	10.41
C26H49N5	431.71	431.3988	72.34	11.44	16.22
C26H50N2	390.70	390.3974	79.93	12.90	7.17
C26H50N4	418.72	418.4035	74.58	12.04	13.38
C26H50N6	446.73	446.4097	69.91	11.28	18.81
C26H51N1	377.70	377.4021	82.68	13.61	3.71
C26H51N3	405.72	405.4083	76.97	12.67	10.36
C26H51N5	433.73	433.4144	72.00	11.85	16.15
C26H52N2	392.72	392.4130	79.52	13.35	7.13
C26H52N4	420.73	420.4192	74.23	12.46	13.32
C26H52N6	448.74	448.4253	69.59	11.68	18.73
C26H53N1	379.72	379.4178	82.24	14.07	3.69
C26H53N3	407.73	407.4239	76.59	13.10	10.31
C26H53N5	435.75	435.4301	71.67	12.26	16.07
C26H54N2	394.73	394.4287	79.11	13.79	7.10
C26H54N4	422.75	422.4348	73.87	12.88	13.25
C26H54N6	450.76	450.4410	69.28	12.08	18.64
C26H55N1	381.73	381.4334	81.81	14.52	3.67
C26H55N3	409.75	409.4396	76.22	13.53	10.26
C26H55N5	437.76	437.4457	71.34	12.66	16.00
C26H56N2	396.75	396.4443	78.71	14.23	7.06
C26H56N4	424.76	424.4505	73.52	13.29	13.19
C26H56N6	452.78	452.4566	68.97	12.47	18.56
C26H57N3	411.76	411.4552	75.84	13.95	10.20

C26H57N5	439.78	439.4614	71.01	13.06	15.92
C26H58N4	426.78	426.4661	73.17	13.70	13.13
C26H58N6	454.79	454.4723	68.67	12.85	18.48
C26H59N5	441.79	441.4770	70.69	13.46	15.85
C26H60N6	456.81	456.4879	68.36	13.24	18.40
C27H13N1	351.41	351.1048	92.29	3.73	3.99
C27H13N3	379.42	379.1109	85.47	3.45	11.07
C27H13N5	407.44	407.1171	79.60	3.22	17.19
C27H14N2	366.43	366.1157	88.50	3.85	7.65
C27H14N4	394.44	394.1218	82.22	3.58	14.20
C27H14N6	422.45	422.1280	76.77	3.34	19.89
C27H15N1	353.43	353.1204	91.76	4.28	3.96
C27H15N3	381.44	381.1266	85.02	3.96	11.02
C27H15N5	409.45	409.1327	79.20	3.69	17.10
C27H16N2	368.44	368.1313	88.02	4.38	7.60
C27H16N4	396.46	396.1375	81.80	4.07	14.13
C27H16N6	424.47	424.1436	76.40	3.80	19.80
C27H17N1	355.44	355.1361	91.24	4.82	3.94
C27H17N3	383.46	383.1422	84.57	4.47	10.96
C27H17N5	411.47	411.1484	78.82	4.16	17.02
C27H18N2	370.46	370.1470	87.54	4.90	7.56
C27H18N4	398.47	398.1531	81.39	4.55	14.06
C27H18N6	426.48	426.1593	76.04	4.25	19.71
C27H19N1	357.46	357.1517	90.72	5.36	3.92
C27H19N3	385.47	385.1579	84.13	4.97	10.90
C27H19N5	413.49	413.1640	78.43	4.63	16.94
C27H20N2	372.47	372.1626	87.07	5.41	7.52
C27H20N4	400.49	400.1688	80.98	5.03	13.99
C27H20N6	428.50	428.1749	75.68	4.70	19.61
C27H21N1	359.48	359.1674	90.22	5.89	3.90
C27H21N3	387.49	387.1735	83.69	5.46	10.84
C27H21N5	415.50	415.1797	78.05	5.09	16.86
C27H22N2	374.49	374.1783	86.60	5.92	7.48
C27H22N4	402.50	402.1844	80.57	5.51	13.92
C27H22N6	430.52	430.1906	75.33	5.15	19.52
C27H23N1	361.49	361.1830	89.71	6.41	3.87
C27H23N3	389.50	389.1892	83.26	5.95	10.79
C27H23N5	417.52	417.1953	77.67	5.55	16.77
C27H24N2	376.51	376.1939	86.13	6.43	7.44
C27H24N4	404.52	404.2001	80.17	5.98	13.85
C27H24N6	432.53	432.2062	74.98	5.59	19.43
C27H25N1	363.51	363.1987	89.21	6.93	3.85
C27H25N3	391.52	391.2048	82.83	6.44	10.73
C27H25N5	419.53	419.2110	77.30	6.01	16.69
C27H26N2	378.52	378.2096	85.68	6.92	7.40
C27H26N4	406.54	406.2157	79.77	6.45	13.78
C27H26N6	434.55	434.2219	74.63	6.03	19.34
C27H27N1	365.52	365.2143	88.72	7.45	3.83
C27H27N3	393.54	393.2205	82.41	6.92	10.68
C27H27N5	421.55	421.2266	76.93	6.46	16.61
C27H28N2	380.54	380.2252	85.22	7.42	7.36
C27H28N4	408.55	408.2314	79.38	6.91	13.71
C27H28N6	436.56	436.2375	74.28	6.46	19.25
C27H29N1	367.54	367.2300	88.24	7.95	3.81
C27H29N3	395.55	395.2361	81.99	7.39	10.62
C27H29N5	423.57	423.2423	76.56	6.90	16.53
C27H30N2	382.55	382.2409	84.77	7.90	7.32
C27H30N4	410.57	410.2470	78.99	7.37	13.65
C27H30N6	438.58	438.2532	73.94	6.89	19.16
C27H31N1	369.55	369.2456	87.75	8.46	3.79
C27H31N3	397.57	397.2518	81.57	7.86	10.57
C27H31N5	425.58	425.2579	76.20	7.34	16.46
C27H32N2	384.57	384.2565	84.33	8.39	7.28
C27H32N4	412.58	412.2627	78.60	7.82	13.58
C27H32N6	440.60	440.2688	73.61	7.32	19.07
C27H33N1	371.57	371.2613	87.28	8.95	3.77
C27H33N3	399.58	399.2674	81.16	8.32	10.52
C27H33N5	427.60	427.2736	75.84	7.78	16.38
C27H34N2	386.59	386.2722	83.89	8.87	7.25
C27H34N4	414.60	414.2783	78.22	8.27	13.51
C27H34N6	442.61	442.2845	73.27	7.74	18.99
C27H35N1	373.59	373.2769	86.81	9.44	3.75
C27H35N3	401.60	401.2831	80.75	8.78	10.46
C27H35N5	429.61	429.2892	75.49	8.21	16.30
C27H36N2	388.60	388.2878	83.45	9.34	7.21
C27H36N4	416.61	416.2940	77.84	8.71	13.45
C27H36N6	444.63	444.3001	72.94	8.16	18.90
C27H37N1	375.60	375.2926	86.34	9.93	3.73
C27H37N3	403.62	403.2987	80.35	9.24	10.41
C27H37N5	431.63	431.3049	75.13	8.64	16.23
C27H38N2	390.62	390.3035	83.02	9.81	7.17
C27H38N4	418.63	418.3096	77.47	9.15	13.38
C27H38N6	446.64	446.3158	72.61	8.58	18.82
C27H39N1	377.62	377.3082	85.88	10.41	3.71
C27H39N3	405.63	405.3144	79.95	9.69	10.36
C27H39N5	433.65	433.3205	74.78	9.07	16.15
C27H40N2	392.63	392.3191	82.60	10.27	7.13
C27H40N4	420.65	420.3253	77.10	9.58	13.32
C27H40N6	448.66	448.3314	72.28	8.99	18.73
C27H41N1	379.63	379.3239	85.42	10.89	3.69
C27H41N3	407.65	407.3300	79.55	10.14	10.31
C27H41N5	435.66	435.3362	74.44	9.49	16.08
C27H42N2	394.65	394.3348	82.17	10.73	7.10
C27H42N4	422.66	422.3409	76.73	10.02	13.26
C27H42N6	450.68	450.3471	71.96	9.39	18.65
C27H43N1	381.65	381.3395	84.97	11.36	3.67
C27H43N3	409.66	409.3457	79.16	10.58	10.26
C27H43N5	437.68	437.3518	74.10	9.90	16.00
C27H44N2	396.67	396.3504	81.76	11.18	7.06
C27H44N4	424.68	424.3566	76.36	10.44	13.19
C27H44N6	452.69	452.3627	71.64	9.80	18.56
C27H45N1	383.67	383.3552	84.53	11.82	3.65
C27H45N3	411.68	411.3613	78.78	11.02	10.21
C27H45N5	439.69	439.3675	73.76	10.32	15.93
C27H46N2	398.68	398.3661	81.34	11.63	7.03
C27H46N4	426.69	426.3722	76.00	10.87	13.13
C27H46N6	454.71	454.3784	71.32	10.20	18.48
C27H47N1	385.68	385.3708	84.09	12.28	3.63
C27H47N3	413.70	413.3770	78.39	11.45	10.16
C27H47N5	441.71	441.3831	73.42	10.73	15.86
C27H48N2	400.70	400.3817	80.93	12.07	6.99
C27H48N4	428.71	428.3879	75.65	11.29	13.07
C27H48N6	456.72	456.3940	71.01	10.59	18.40
C27H49N1	387.70	387.3865	83.65	12.74	3.61
C27H49N3	415.71	415.3926	78.01	11.88	10.11
C27H49N5	443.73	443.3988	73.09	11.13	15.78
C27H50N2	402.71	402.3974	80.53	12.51	6.96
C27H50N4	430.73	430.4035	75.29	11.70	13.01
C27H50N6	458.74	458.4097	70.69	10.99	18.32
C27H51N1	389.71	389.4021	83.22	13.19	3.59

C27H51N3	417.73	417.4083	77.63	12.31	10.06
C27H51N5	445.74	445.4144	72.76	11.53	15.71
C27H52N2	404.73	404.4130	80.13	12.95	6.92
C27H52N4	432.74	432.4192	74.94	12.11	12.95
C27H52N6	460.76	460.4253	70.38	11.38	18.24
C27H53N1	391.73	391.4178	82.79	13.64	3.58
C27H53N3	419.74	419.4239	77.26	12.73	10.01
C27H53N5	447.76	447.4301	72.43	11.93	15.64
C27H54N2	406.74	406.4287	79.73	13.38	6.89
C27H54N4	434.76	434.4348	74.59	12.52	12.89
C27H54N6	462.77	462.4410	70.08	11.76	18.16
C27H55N1	393.75	393.4334	82.36	14.08	3.56
C27H55N3	421.76	421.4396	76.89	13.14	9.96
C27H55N5	449.77	449.4457	72.10	12.33	15.57
C27H56N2	408.76	408.4443	79.34	13.81	6.85
C27H56N4	436.77	436.4505	74.25	12.92	12.83
C27H56N6	464.79	464.4566	69.77	12.14	18.08
C27H57N1	395.76	395.4491	81.94	14.52	3.54
C27H57N3	423.78	423.4552	76.53	13.56	9.92
C27H57N5	451.79	451.4614	71.78	12.72	15.50
C27H58N2	410.78	410.4600	78.95	14.23	6.82
C27H58N4	438.79	438.4661	73.91	13.32	12.77
C27H58N6	466.80	466.4723	69.47	12.52	18.00
C27H59N3	425.79	425.4709	76.16	13.97	9.87
C27H59N5	453.80	453.4770	71.46	13.10	15.43
C27H60N4	440.81	440.4818	73.57	13.72	12.71
C27H60N6	468.82	468.4879	69.17	12.90	17.93
C27H61N5	455.82	455.4927	71.15	13.49	15.36
C27H62N6	470.84	470.5036	68.88	13.27	17.85
C28H14N2	378.44	378.1157	88.87	3.73	7.40
C28H14N4	406.45	406.1218	82.74	3.47	13.78
C28H14N6	434.46	434.1280	77.41	3.25	19.34
C28H15N1	365.44	365.1204	92.03	4.14	3.83
C28H15N3	393.45	393.1266	85.48	3.84	10.68
C28H15N5	421.47	421.1327	79.80	3.59	16.62
C28H16N2	380.45	380.1313	88.40	4.24	7.36
C28H16N4	408.47	408.1375	82.34	3.95	13.72
C28H16N6	436.48	436.1436	77.05	3.69	19.25
C28H17N1	367.45	367.1361	91.52	4.66	3.81
C28H17N3	395.47	395.1422	85.04	4.33	10.63
C28H17N5	423.48	423.1484	79.42	4.05	16.54
C28H18N2	382.47	382.1470	87.93	4.74	7.32
C28H18N4	410.48	410.1531	81.93	4.42	13.65
C28H18N6	438.50	438.1593	76.70	4.14	19.17
C28H19N1	369.47	369.1517	91.03	5.18	3.79
C28H19N3	397.48	397.1579	84.61	4.82	10.57
C28H19N5	425.50	425.1640	79.04	4.50	16.46
C28H20N2	384.48	384.1626	87.47	5.24	7.29
C28H20N4	412.50	412.1688	81.53	4.89	13.58
C28H20N6	440.51	440.1749	76.35	4.58	19.08
C28H21N1	371.49	371.1674	90.53	5.70	3.77
C28H21N3	399.50	399.1735	84.18	5.30	10.52
C28H21N5	427.51	427.1797	78.67	4.95	16.38
C28H22N2	386.50	386.1783	87.01	5.74	7.25
C28H22N4	414.51	414.1844	81.13	5.35	13.52
C28H22N6	442.53	442.1906	76.00	5.01	18.99
C28H23N1	373.50	373.1830	90.04	6.21	3.75
C28H23N3	401.52	401.1892	83.76	5.77	10.47
C28H23N5	429.53	429.1953	78.30	5.40	16.30
C28H24N2	388.52	388.1939	86.56	6.23	7.21
C28H24N4	416.53	416.2001	80.74	5.81	13.45
C28H24N6	444.54	444.2062	75.65	5.44	18.90
C28H25N1	375.52	375.1987	89.56	6.71	3.73
C28H25N3	403.53	403.2048	83.34	6.24	10.41
C28H25N5	431.54	431.2110	77.93	5.84	16.23
C28H26N2	390.53	390.2096	86.12	6.71	7.17
C28H26N4	418.55	418.2157	80.35	6.26	13.39
C28H26N6	446.56	446.2219	75.31	5.87	18.82
C28H27N1	377.53	377.2143	89.08	7.21	3.71
C28H27N3	405.55	405.2205	82.93	6.71	10.36
C28H27N5	433.56	433.2266	77.57	6.28	16.15
C28H28N2	392.55	392.2252	85.67	7.19	7.14
C28H28N4	420.56	420.2314	79.97	6.71	13.32
C28H28N6	448.58	448.2375	74.97	6.29	18.73
C28H29N1	379.55	379.2300	88.61	7.70	3.69
C28H29N3	407.56	407.2361	82.52	7.17	10.31
C28H29N5	435.58	435.2423	77.21	6.71	16.08
C28H30N2	394.56	394.2409	85.24	7.66	7.10
C28H30N4	422.58	422.2470	79.59	7.16	13.26
C28H30N6	450.59	450.2532	74.64	6.71	18.65
C28H31N1	381.57	381.2456	88.14	8.19	3.67
C28H31N3	409.58	409.2518	82.11	7.63	10.26
C28H31N5	437.59	437.2579	76.86	7.14	16.00
C28H32N2	396.58	396.2565	84.80	8.13	7.06
C28H32N4	424.59	424.2627	79.21	7.60	13.20
C28H32N6	452.61	452.2688	74.31	7.13	18.57
C28H33N1	383.58	383.2613	87.68	8.67	3.65
C28H33N3	411.60	411.2674	81.71	8.08	10.21
C28H33N5	439.61	439.2736	76.50	7.57	15.93
C28H34N2	398.60	398.2722	84.37	8.60	7.03
C28H34N4	426.61	426.2783	78.83	8.03	13.13
C28H34N6	454.62	454.2845	73.98	7.54	18.49
C28H35N1	385.60	385.2769	87.22	9.15	3.63
C28H35N3	413.61	413.2831	81.31	8.53	10.16
C28H35N5	441.62	441.2892	76.15	7.99	15.86
C28H36N2	400.61	400.2878	83.95	9.06	6.99
C28H36N4	428.63	428.2940	78.46	8.47	13.07
C28H36N6	456.64	456.3001	73.65	7.95	18.40
C28H37N1	387.61	387.2926	86.76	9.62	3.61
C28H37N3	415.63	415.2987	80.92	8.97	10.11
C28H37N5	443.64	443.3049	75.81	8.41	15.79
C28H38N2	402.63	402.3035	83.53	9.51	6.96
C28H38N4	430.64	430.3096	78.10	8.89	13.01
C28H38N6	458.66	458.3158	73.33	8.35	18.32
C28H39N1	389.63	389.3082	86.32	10.09	3.59
C28H39N3	417.64	417.3144	80.53	9.41	10.06
C28H39N5	445.66	445.3205	75.46	8.82	15.71
C28H40N2	404.64	404.3191	83.11	9.96	6.92
C28H40N4	432.66	432.3253	77.73	9.32	12.95
C28H40N6	460.67	460.3314	73.00	8.75	18.24
C28H41N1	391.65	391.3239	85.87	10.55	3.58
C28H41N3	419.66	419.3300	80.14	9.85	10.01
C28H41N5	447.67	447.3362	75.12	9.23	15.64
C28H42N2	406.66	406.3348	82.70	10.41	6.89
C28H42N4	434.67	434.3409	77.37	9.74	12.89
C28H42N6	462.69	462.3471	72.69	9.15	18.16
C28H43N1	393.66	393.3395	85.43	11.01	3.56
C28H43N3	421.68	421.3457	79.76	10.28	9.97
C28H43N5	449.69	449.3518	74.79	9.64	15.57
C28H44N2	408.68	408.3504	82.29	10.85	6.85

C28H44N4	436.69	436.3566	77.01	10.16	12.83
C28H44N6	464.70	464.3627	72.37	9.54	18.08
C28H45N1	395.68	395.3552	85.00	11.46	3.54
C28H45N3	423.69	423.3613	79.38	10.71	9.92
C28H45N5	451.70	451.3675	74.45	10.04	15.50
C28H46N2	410.69	410.3661	81.89	11.29	6.82
C28H46N4	438.71	438.3722	76.66	10.57	12.77
C28H46N6	466.72	466.3784	72.06	9.93	18.01
C28H47N1	397.69	397.3708	84.57	11.91	3.52
C28H47N3	425.71	425.3770	79.00	11.13	9.87
C28H47N5	453.72	453.3831	74.12	10.44	15.44
C28H48N2	412.71	412.3817	81.49	11.72	6.79
C28H48N4	440.72	440.3879	76.31	10.98	12.71
C28H48N6	468.73	468.3940	71.75	10.32	17.93
C28H49N1	399.71	399.3865	84.14	12.36	3.50
C28H49N3	427.72	427.3926	78.63	11.55	9.82
C28H49N5	455.74	455.3988	73.80	10.84	15.37
C28H50N2	414.72	414.3974	81.09	12.15	6.75
C28H50N4	442.74	442.4035	75.96	11.38	12.65
C28H50N6	470.75	470.4097	71.44	10.71	17.85
C28H51N1	401.73	401.4021	83.72	12.80	3.49
C28H51N3	429.74	429.4083	78.26	11.96	9.78
C28H51N5	457.75	457.4144	73.47	11.23	15.30
C28H52N2	416.74	416.4130	80.70	12.58	6.72
C28H52N4	444.75	444.4192	75.62	11.79	12.60
C28H52N6	472.77	472.4253	71.14	11.09	17.78
C28H53N1	403.74	403.4178	83.30	13.23	3.47
C28H53N3	431.75	431.4239	77.89	12.37	9.73
C28H53N5	459.77	459.4301	73.15	11.62	15.23
C28H54N2	418.76	418.4287	80.31	13.00	6.69
C28H54N4	446.77	446.4348	75.28	12.18	12.54
C28H54N6	474.78	474.4410	70.83	11.46	17.70
C28H55N1	405.76	405.4334	82.89	13.66	3.45
C28H55N3	433.77	433.4396	77.53	12.78	9.69
C28H55N5	461.78	461.4457	72.83	12.01	15.17
C28H56N2	420.77	420.4443	79.93	13.41	6.66
C28H56N4	448.79	448.4505	74.94	12.58	12.48
C28H56N6	476.80	476.4566	70.54	11.84	17.63
C28H57N1	407.77	407.4491	82.48	14.09	3.43
C28H57N3	435.79	435.4552	77.17	13.18	9.64
C28H57N5	463.80	463.4614	72.51	12.39	15.10
C28H58N2	422.79	422.4600	79.55	13.83	6.63
C28H58N4	450.80	450.4661	74.60	12.97	12.43
C28H58N6	478.81	478.4723	70.24	12.21	17.55
C28H59N1	409.79	409.4647	82.07	14.51	3.42
C28H59N3	437.80	437.4709	76.82	13.58	9.60
C28H59N5	465.82	465.4770	72.20	12.77	15.03
C28H60N2	424.80	424.4756	79.17	14.24	6.59
C28H60N4	452.82	452.4818	74.27	13.36	12.37
C28H60N6	480.83	480.4879	69.94	12.58	17.48
C28H61N3	439.82	439.4865	76.47	13.98	9.55
C28H61N5	467.83	467.4927	71.89	13.14	14.97
C28H62N4	454.83	454.4974	73.94	13.74	12.32
C28H62N6	482.85	482.5036	69.65	12.94	17.41
C28H63N5	469.85	469.5083	71.58	13.52	14.91
C28H64N6	484.86	484.5192	69.36	13.30	17.33
C29H14N2	390.45	390.1157	89.21	3.61	7.17
C29H14N4	418.46	418.1218	83.24	3.37	13.39
C29H14N6	446.48	446.1280	78.02	3.16	18.82
C29H15N1	377.45	377.1204	92.28	4.01	3.71
C29H15N3	405.46	405.1266	85.91	3.73	10.36
C29H15N5	433.48	433.1327	80.36	3.49	16.16
C29H16N2	392.46	392.1313	88.75	4.11	7.14
C29H16N4	420.48	420.1375	82.84	3.84	13.32
C29H16N6	448.49	448.1436	77.67	3.60	18.74
C29H17N1	379.47	379.1361	91.79	4.52	3.69
C29H17N3	407.48	407.1422	85.48	4.21	10.31
C29H17N5	435.49	435.1484	79.98	3.93	16.08
C29H18N2	394.48	394.1470	88.30	4.60	7.10
C29H18N4	422.49	422.1531	82.44	4.29	13.26
C29H18N6	450.51	450.1593	77.32	4.03	18.65
C29H19N1	381.48	381.1517	91.31	5.02	3.67
C29H19N3	409.49	409.1579	85.06	4.68	10.26
C29H19N5	437.51	437.1640	79.62	4.38	16.01
C29H20N2	396.50	396.1626	87.85	5.08	7.07
C29H20N4	424.51	424.1688	82.05	4.75	13.20
C29H20N6	452.52	452.1749	76.97	4.45	18.57
C29H21N1	383.50	383.1674	90.83	5.52	3.65
C29H21N3	411.51	411.1735	84.65	5.14	10.21
C29H21N5	439.52	439.1797	79.25	4.82	15.93
C29H22N2	398.51	398.1783	87.41	5.56	7.03
C29H22N4	426.53	426.1844	81.67	5.20	13.14
C29H22N6	454.54	454.1906	76.63	4.88	18.49
C29H23N1	385.51	385.1830	90.35	6.01	3.63
C29H23N3	413.53	413.1892	84.23	5.61	10.16
C29H23N5	441.54	441.1953	78.89	5.25	15.86
C29H24N2	400.53	400.1939	86.97	6.04	6.99
C29H24N4	428.54	428.2001	81.28	5.65	13.07
C29H24N6	456.55	456.2062	76.29	5.30	18.41
C29H25N1	387.53	387.1987	89.88	6.50	3.61
C29H25N3	415.54	415.2048	83.82	6.06	10.11
C29H25N5	443.56	443.2110	78.53	5.68	15.79
C29H26N2	402.54	402.2096	86.53	6.51	6.96
C29H26N4	430.56	430.2157	80.90	6.09	13.01
C29H26N6	458.57	458.2219	75.96	5.71	18.33
C29H27N1	389.55	389.2143	89.42	6.99	3.60
C29H27N3	417.56	417.2205	83.42	6.52	10.06
C29H27N5	445.57	445.2266	78.17	6.11	15.72
C29H28N2	404.56	404.2252	86.10	6.98	6.92
C29H28N4	432.57	432.2314	80.52	6.52	12.95
C29H28N6	460.59	460.2375	75.63	6.13	18.25
C29H29N1	391.56	391.2300	88.96	7.47	3.58
C29H29N3	419.57	419.2361	83.02	6.97	10.01
C29H29N5	447.59	447.2423	77.82	6.53	15.65
C29H30N2	406.58	406.2409	85.67	7.44	6.89
C29H30N4	434.59	434.2470	80.15	6.96	12.89
C29H30N6	462.60	462.2532	75.30	6.54	18.17
C29H31N1	393.58	393.2456	88.50	7.94	3.56
C29H31N3	421.59	421.2518	82.62	7.41	9.97
C29H31N5	449.60	449.2579	77.47	6.95	15.58
C29H32N2	408.59	408.2565	85.25	7.89	6.86
C29H32N4	436.61	436.2627	79.78	7.39	12.83
C29H32N6	464.62	464.2688	74.97	6.94	18.09
C29H33N1	395.59	395.2613	88.05	8.41	3.54
C29H33N3	423.61	423.2674	82.23	7.85	9.92
C29H33N5	451.62	451.2736	77.13	7.37	15.51
C29H34N2	410.61	410.2722	84.83	8.35	6.82
C29H34N4	438.62	438.2783	79.41	7.81	12.77
C29H34N6	466.63	466.2845	74.65	7.34	18.01
C29H35N1	397.61	397.2769	87.60	8.87	3.52

10/21/66 066

C29H35N3	425.62	425.2831	81.84	8.29	9.87
C29H35N5	453.64	453.2892	76.78	7.78	15.44
C29H36N2	412.62	412.2878	84.42	8.79	6.79
C29H36N4	440.64	440.2940	79.05	8.24	12.71
C29H36N6	468.65	468.3001	74.32	7.74	17.93
C29H37N1	399.62	399.2926	87.16	9.33	3.50
C29H37N3	427.64	427.2987	81.45	8.72	9.83
C29H37N5	455.65	455.3049	76.45	8.18	15.37
C29H38N2	414.64	414.3035	84.01	9.24	6.76
C29H38N4	442.65	442.3096	78.69	8.65	12.66
C29H38N6	470.67	470.3158	74.01	8.14	17.86
C29H39N1	401.64	401.3082	86.73	9.79	3.49
C29H39N3	429.65	429.3144	81.07	9.15	9.78
C29H39N5	457.67	457.3205	76.11	8.59	15.30
C29H40N2	416.66	416.3191	83.60	9.68	6.72
C29H40N4	444.67	444.3253	78.33	9.07	12.60
C29H40N6	472.68	472.3314	73.69	8.53	17.78
C29H41N1	403.66	403.3239	86.29	10.24	3.47
C29H41N3	431.67	431.3300	80.69	9.57	9.73
C29H41N5	459.68	459.3362	75.77	8.99	15.24
C29H42N2	418.67	418.3348	83.20	10.11	6.69
C29H42N4	446.68	446.3409	77.98	9.48	12.54
C29H42N6	474.70	474.3471	73.38	8.92	17.70
C29H43N1	405.67	405.3395	85.86	10.68	3.45
C29H43N3	433.69	433.3457	80.32	9.99	9.69
C29H43N5	461.70	461.3518	75.44	9.39	15.17
C29H44N2	420.69	420.3504	82.80	10.54	6.66
C29H44N4	448.70	448.3566	77.63	9.88	12.49
C29H44N6	476.71	476.3627	73.07	9.30	17.63
C29H45N1	407.69	407.3552	85.44	11.13	3.44
C29H45N3	435.70	435.3613	79.95	10.41	9.64
C29H45N5	463.72	463.3675	75.12	9.78	15.10
C29H46N2	422.70	422.3661	82.40	10.97	6.63
C29H46N4	450.72	450.3722	77.28	10.29	12.43
C29H46N6	478.73	478.3784	72.76	9.69	17.55
C29H47N1	409.70	409.3708	85.02	11.56	3.42
C29H47N3	437.72	437.3770	79.58	10.82	9.60
C29H47N5	465.73	465.3831	74.79	10.17	15.04
C29H48N2	424.72	424.3817	82.01	11.39	6.60
C29H48N4	452.73	452.3879	76.94	10.69	12.38
C29H48N6	480.75	480.3940	72.45	10.06	17.48
C29H49N1	411.72	411.3865	84.60	12.00	3.40
C29H49N3	439.73	439.3926	79.21	11.23	9.56
C29H49N5	467.75	467.3988	74.47	10.56	14.97
C29H50N2	426.74	426.3974	81.63	11.81	6.56
C29H50N4	454.75	454.4035	76.60	11.08	12.32
C29H50N6	482.76	482.4097	72.15	10.44	17.41
C29H51N1	413.74	413.4021	84.19	12.42	3.39
C29H51N3	441.75	441.4083	78.85	11.64	9.51
C29H51N5	469.76	469.4144	74.15	10.94	14.91
C29H52N2	428.75	428.4130	81.24	12.22	6.53
C29H52N4	456.76	456.4192	76.26	11.48	12.27
C29H52N6	484.78	484.4253	71.85	10.81	17.34
C29H53N1	415.75	415.4178	83.78	12.85	3.37
C29H53N3	443.77	443.4239	78.49	12.04	9.47
C29H53N5	471.78	471.4301	73.83	11.32	14.84
C29H54N2	430.77	430.4287	80.86	12.64	6.50
C29H54N4	458.78	458.4348	75.92	11.86	12.21
C29H54N6	486.79	486.4410	71.55	11.18	17.26
C29H55N1	417.77	417.4334	83.38	13.27	3.35
C29H55N3	445.78	445.4396	78.14	12.44	9.43
C29H55N5	473.80	473.4457	73.52	11.70	14.78
C29H56N2	432.78	432.4443	80.48	13.04	6.47
C29H56N4	460.80	460.4505	75.59	12.25	12.16
C29H56N6	488.81	488.4566	71.26	11.55	17.19
C29H57N1	419.78	419.4491	82.98	13.69	3.34
C29H57N3	447.80	447.4552	77.79	12.83	9.38
C29H57N5	475.81	475.4614	73.21	12.08	14.72
C29H58N2	434.80	434.4600	80.11	13.45	6.44
C29H58N4	462.81	462.4661	75.26	12.63	12.11
C29H58N6	490.83	490.4723	70.97	11.91	17.12
C29H59N1	421.80	421.4647	82.58	14.10	3.32
C29H59N3	449.81	449.4709	77.44	13.22	9.34
C29H59N5	477.83	477.4770	72.90	12.45	14.66
C29H60N2	436.81	436.4756	79.74	13.85	6.41
C29H60N4	464.83	464.4818	74.94	13.01	12.05
C29H60N6	492.84	492.4879	70.68	12.27	17.05
C29H61N1	423.82	423.4804	82.19	14.51	3.30
C29H61N3	451.83	451.4865	77.09	13.61	9.30
C29H61N5	479.84	479.4927	72.59	12.81	14.60
C29H62N2	438.83	438.4913	79.38	14.24	6.38
C29H62N4	466.84	466.4974	74.61	13.39	12.00
C29H62N6	494.86	494.5036	70.39	12.63	16.98
C29H63N3	453.85	453.5022	76.75	13.99	9.26
C29H63N5	481.86	481.5083	72.29	13.18	14.53
C29H64N4	468.86	468.5131	74.29	13.76	11.95
C29H64N6	496.87	496.5192	70.10	12.98	16.91
C29H65N5	483.87	483.5240	71.99	13.54	14.47
C29H66N6	498.89	498.5349	69.82	13.33	16.85
C30H15N1	389.46	389.1204	92.52	3.88	3.60
C30H15N3	417.47	417.1266	86.31	3.62	10.07
C30H15N5	445.49	445.1327	80.89	3.39	15.72
C30H16N2	404.48	404.1313	89.09	3.99	6.93
C30H16N4	432.49	432.1375	83.32	3.73	12.95
C30H16N6	460.50	460.1436	78.25	3.50	18.25
C30H17N1	391.48	391.1361	92.04	4.38	3.58
C30H17N3	419.49	419.1422	85.90	4.08	10.02
C30H17N5	447.50	447.1484	80.52	3.83	15.65
C30H18N2	406.49	406.1470	88.65	4.46	6.89
C30H18N4	434.50	434.1531	82.93	4.18	12.89
C30H18N6	462.52	462.1593	77.91	3.92	18.17
C30H19N1	393.49	393.1517	91.57	4.87	3.56
C30H19N3	421.51	421.1579	85.49	4.54	9.97
C30H19N5	449.52	449.1640	80.16	4.26	15.58
C30H20N2	408.51	408.1626	88.21	4.93	6.86
C30H20N4	436.52	436.1688	82.55	4.62	12.83
C30H20N6	464.53	464.1749	77.57	4.34	18.09
C30H21N1	395.51	395.1674	91.11	5.35	3.54
C30H21N3	423.52	423.1735	85.08	5.00	9.92
C30H21N5	451.54	451.1797	79.80	4.69	15.51
C30H22N2	410.52	410.1783	87.77	5.40	6.82
C30H22N4	438.54	438.1844	82.17	5.06	12.78
C30H22N6	466.55	466.1906	77.23	4.75	18.01
C30H23N1	397.52	397.1830	90.64	5.83	3.52
C30H23N3	425.54	425.1892	84.68	5.45	9.87
C30H23N5	453.55	453.1953	79.45	5.11	15.44
C30H24N2	412.54	412.1939	87.35	5.86	6.79
C30H24N4	440.55	440.2001	81.79	5.49	12.72
C30H24N6	468.57	468.2062	76.90	5.16	17.94
C30H25N1	399.54	399.1987	90.19	6.31	3.51

C30H25N3	427.55	427.2048	84.28	5.89	9.83
C30H25N5	455.57	455.2110	79.10	5.53	15.37
C30H26N2	414.56	414.2096	86.92	6.32	6.76
C30H26N4	442.57	442.2157	81.42	5.92	12.66
C30H26N6	470.58	470.2219	76.57	5.57	17.86
C30H27N1	401.56	401.2143	89.73	6.78	3.49
C30H27N3	429.57	429.2205	83.88	6.34	9.78
C30H27N5	457.58	457.2266	78.75	5.95	15.31
C30H28N2	416.57	416.2252	86.50	6.78	6.72
C30H28N4	444.58	444.2314	81.05	6.35	12.60
C30H28N6	472.60	472.2375	76.25	5.97	17.78
C30H29N1	403.57	403.2300	89.29	7.24	3.47
C30H29N3	431.59	431.2361	83.49	6.77	9.74
C30H29N5	459.60	459.2423	78.40	6.36	15.24
C30H30N2	418.59	418.2409	86.08	7.22	6.69
C30H30N4	446.60	446.2470	80.68	6.77	12.55
C30H30N6	474.61	474.2532	75.92	6.37	17.71
C30H31N1	405.59	405.2456	88.84	7.70	3.45
C30H31N3	433.60	433.2518	83.10	7.21	9.69
C30H31N5	461.62	461.2579	78.06	6.77	15.17
C30H32N2	420.60	420.2565	85.67	7.67	6.66
C30H32N4	448.62	448.2627	80.32	7.19	12.49
C30H32N6	476.63	476.2688	75.60	6.77	17.63
C30H33N1	407.60	407.2613	88.40	8.16	3.44
C30H33N3	435.62	435.2674	82.72	7.64	9.65
C30H33N5	463.63	463.2736	77.72	7.17	15.11
C30H34N2	422.62	422.2722	85.26	8.11	6.63
C30H34N4	450.63	450.2783	79.96	7.61	12.43
C30H34N6	478.65	478.2845	75.28	7.16	17.56
C30H35N1	409.62	409.2769	87.97	8.61	3.42
C30H35N3	437.63	437.2831	82.34	8.06	9.60
C30H35N5	465.65	465.2892	77.38	7.58	15.04
C30H36N2	424.63	424.2878	84.86	8.55	6.60
C30H36N4	452.65	452.2940	79.61	8.02	12.38
C30H36N6	480.66	480.3001	74.97	7.55	17.48
C30H37N1	411.64	411.2926	87.54	9.06	3.40
C30H37N3	439.65	439.2987	81.96	8.48	9.56
C30H37N5	467.66	467.3049	77.05	7.97	14.98
C30H38N2	426.65	426.3035	84.46	8.98	6.57
C30H38N4	454.66	454.3096	79.25	8.42	12.32
C30H38N6	482.68	482.3158	74.65	7.94	17.41
C30H39N1	413.65	413.3082	87.11	9.50	3.39
C30H39N3	441.67	441.3144	81.59	8.90	9.51
C30H39N5	469.68	469.3205	76.72	8.37	14.91
C30H40N2	428.67	428.3191	84.06	9.41	6.54
C30H40N4	456.68	456.3253	78.90	8.83	12.27
C30H40N6	484.69	484.3314	74.34	8.32	17.34
C30H41N1	415.67	415.3239	86.69	9.94	3.37
C30H41N3	443.68	443.3300	81.21	9.31	9.47
C30H41N5	471.69	471.3362	76.39	8.76	14.85
C30H42N2	430.68	430.3348	83.67	9.83	6.50
C30H42N4	458.70	458.3409	78.56	9.23	12.21
C30H42N6	486.71	486.3471	74.03	8.70	17.27
C30H43N1	417.68	417.3395	86.27	10.38	3.35
C30H43N3	445.70	445.3457	80.85	9.72	9.43
C30H43N5	473.71	473.3518	76.07	9.15	14.78
C30H44N2	432.70	432.3504	83.28	10.25	6.47
C30H44N4	460.71	460.3566	78.21	9.63	12.16
C30H44N6	488.73	488.3627	73.73	9.07	17.20
C30H45N1	419.70	419.3552	85.86	10.81	3.34
C30H45N3	447.71	447.3613	80.48	10.13	9.39
C30H45N5	475.73	475.3675	75.74	9.53	14.72
C30H46N2	434.71	434.3661	82.89	10.67	6.44
C30H46N4	462.73	462.3722	77.87	10.02	12.11
C30H46N6	490.74	490.3784	73.43	9.45	17.13
C30H47N1	421.72	421.3708	85.44	11.23	3.32
C30H47N3	449.73	449.3770	80.12	10.53	9.34
C30H47N5	477.74	477.3831	75.42	9.92	14.65
C30H48N2	436.73	436.3817	82.51	11.08	6.41
C30H48N4	464.74	464.3879	77.53	10.41	12.06
C30H48N6	492.76	492.3940	73.13	9.82	17.06
C30H49N1	423.73	423.3865	85.04	11.66	3.31
C30H49N3	451.75	451.3926	79.76	10.93	9.30
C30H49N5	479.76	479.3988	75.11	10.29	14.60
C30H50N2	438.75	438.3974	82.13	11.49	6.38
C30H50N4	466.76	466.4035	77.20	10.80	12.00
C30H50N6	494.77	494.4097	72.83	10.19	16.99
C30H51N1	425.75	425.4021	84.64	12.07	3.29
C30H51N3	453.76	453.4083	79.41	11.33	9.26
C30H51N5	481.77	481.4144	74.79	10.67	14.54
C30H52N2	440.76	440.4130	81.75	11.89	6.36
C30H52N4	468.78	468.4192	76.87	11.18	11.95
C30H52N6	496.79	496.4253	72.53	10.55	16.92
C30H53N1	427.76	427.4178	84.24	12.49	3.27
C30H53N3	455.78	455.4239	79.06	11.72	9.22
C30H53N5	483.79	483.4301	74.48	11.04	14.48
C30H54N2	442.78	442.4287	81.38	12.29	6.33
C30H54N4	470.79	470.4348	76.54	11.56	11.90
C30H54N6	498.81	498.4410	72.24	10.91	16.85
C30H55N1	429.78	429.4334	83.84	12.90	3.26
C30H55N3	457.79	457.4396	78.71	12.11	9.18
C30H55N5	485.81	485.4457	74.17	11.41	14.42
C30H56N2	444.79	444.4443	81.01	12.69	6.30
C30H56N4	472.81	472.4505	76.21	11.94	11.85
C30H56N6	500.82	500.4566	71.95	11.27	16.78
C30H57N1	431.80	431.4491	83.45	13.31	3.24
C30H57N3	459.81	459.4552	78.37	12.50	9.14
C30H57N5	487.82	487.4614	73.87	11.78	14.36
C30H58N2	446.81	446.4600	80.65	13.08	6.27
C30H58N4	474.82	474.4661	75.89	12.31	11.80
C30H58N6	502.84	502.4723	71.66	11.63	16.71
C30H59N1	433.81	433.4647	83.06	13.71	3.23
C30H59N3	461.82	461.4709	78.02	12.88	9.10
C30H59N5	489.84	489.4770	73.56	12.14	14.30
C30H60N2	448.83	448.4756	80.28	13.47	6.24
C30H60N4	476.84	476.4818	75.57	12.68	11.75
C30H60N6	504.85	504.4879	71.37	11.98	16.65
C30H61N1	435.83	435.4804	82.68	14.11	3.21
C30H61N3	463.84	463.4865	77.68	13.26	9.06
C30H61N5	491.85	491.4927	73.26	12.50	14.24
C30H62N2	450.84	450.4913	79.92	13.86	6.21
C30H62N4	478.86	478.4974	75.25	13.05	11.70
C30H62N6	506.87	506.5036	71.09	12.33	16.58
C30H63N1	437.84	437.4960	82.30	14.50	3.20
C30H63N3	465.86	465.5022	77.35	13.63	9.02
C30H63N5	493.87	493.5083	72.96	12.86	14.18
C30H64N2	452.86	452.5069	79.57	14.25	6.19
C30H64N4	480.87	480.5131	74.93	13.42	11.65
C30H64N6	508.88	508.5192	70.81	12.68	16.51
C30H65N3	467.87	467.5178	77.02	14.00	8.98

C30H65N5	495.89	495.5240	72.66	13.21	14.12
C30H66N4	482.89	482.5287	74.62	13.78	11.60
C30H66N6	510.90	510.5349	70.53	13.02	16.45
C30H67N5	497.90	497.5396	72.37	13.56	14.07
C30H68N6	512.92	512.5505	70.25	13.36	16.38

6. CARBON, HYDROGEN AND OXYGEN

6. KOHLENSTOFF, WASSERSTOFF UND SAUERSTOFF

6. CARBONE, HYDROGENE ET OXYGENE

MOLECULAR WEIGHTS AND PERCENTAGE COMPOSITIONS OF COMPOUNDS
CONTAINING CARBON, HYDROGEN AND OXYGEN

FORMULA	MOL. WEIGHT	ISOMOL. WEIGHT	PERCENT C	PERCENT H	PERCENT O
C 1H 001	28.01	27.9949	42.88	0	57.12
C 1H 002	44.01	43.9898	27.29	0	72.71
C 1H 003	60.01	59.9847	20.02	0	79.98
C 1H 004	76.01	75.9797	15.80	0	84.20
C 1H 005	92.01	91.9746	13.05	0	86.95
C 1H 006	108.01	107.9695	11.12	0	88.88
C 1H 007	124.01	123.9644	9.69	0	90.31
C 1H 008	140.01	139.9593	8.58	0	91.42
C 1H 009	156.01	155.9542	7.70	0	92.30
C 1H 201	30.03	30.0106	40.00	6.71	53.28
C 1H 202	46.03	46.0055	26.10	4.38	69.52
C 1H 203	62.03	62.0004	19.36	3.25	77.38
C 1H 204	78.02	77.9953	15.39	2.58	82.02
C 1H 205	94.02	93.9902	12.77	2.14	85.08
C 1H 206	110.02	109.9851	10.92	1.83	87.25
C 1H 207	126.02	125.9800	9.53	1.60	88.87
C 1H 208	142.02	141.9750	8.46	1.42	90.12
C 1H 209	158.02	157.9699	7.60	1.28	91.12
C 1H 401	32.04	32.0262	37.49	12.58	49.93
C 1H 402	48.04	48.0211	25.00	8.39	66.61
C 1H 403	64.04	64.0160	18.76	6.30	74.95
C 1H 404	80.04	80.0110	15.01	5.04	79.96
C 1H 405	96.04	96.0059	12.51	4.20	83.30
C 1H 406	112.04	112.0008	10.72	3.60	85.68
C 1H 407	128.04	127.9957	9.38	3.15	87.47
C 1H 408	144.04	143.9906	8.34	2.80	88.86
C 1H 409	160.04	159.9855	7.51	2.52	89.98
C 2H 201	42.04	42.0106	57.14	4.80	38.06
C 2H 202	58.04	58.0055	41.39	3.47	55.14
C 2H 203	74.04	74.0004	32.45	2.72	64.83
C 2H 204	90.04	89.9953	26.68	2.24	71.08
C 2H 205	106.04	105.9902	22.66	1.90	75.44
C 2H 206	122.03	121.9851	19.68	1.65	78.66
C 2H 207	138.03	137.9800	17.40	1.46	81.14
C 2H 208	154.03	153.9750	15.60	1.31	83.10
C 2H 209	170.03	169.9699	14.13	1.19	84.69
C 2H 401	44.05	44.0262	54.53	9.15	36.32
C 2H 402	60.05	60.0211	40.00	6.71	53.28
C 2H 403	76.05	76.0160	31.59	5.30	63.11
C 2H 404	92.05	92.0110	26.10	4.38	69.52
C 2H 405	108.05	108.0059	22.23	3.73	74.04
C 2H 406	124.05	124.0008	19.36	3.25	77.38
C 2H 407	140.05	139.9957	17.15	2.88	79.97
C 2H 408	156.05	155.9906	15.39	2.58	82.02
C 2H 409	172.05	171.9855	13.96	2.34	83.69
C 2H 601	46.07	46.0419	52.14	13.13	34.73
C 2H 602	62.07	62.0368	38.70	9.74	51.55
C 2H 603	78.07	78.0317	30.77	7.75	61.48
C 2H 604	94.07	94.0266	25.54	6.43	68.03
C 2H 605	110.07	110.0215	21.83	5.49	72.68

FORMULA	MOL. WEIGHT	ISOMOL. WEIGHT	PERCENT C	PERCENT H	PERCENT O
C 2H 606	126.07	126.0164	19.06	4.80	76.15
C 2H 607	142.07	142.0113	16.91	4.26	78.83
C 2H 608	158.07	158.0063	15.20	3.83	80.98
C 2H 609	174.06	174.0012	13.80	3.47	82.72
C 3H 201	54.05	54.0106	66.67	3.73	29.60
C 3H 202	70.05	70.0055	51.44	2.88	45.68
C 3H 203	86.05	86.0004	41.88	2.34	55.78
C 3H 204	102.05	101.9953	35.31	1.98	62.71
C 3H 205	118.05	117.9902	30.52	1.71	67.77
C 3H 206	134.05	133.9851	26.88	1.50	71.61
C 3H 207	150.05	149.9800	24.02	1.34	74.64
C 3H 208	166.04	165.9750	21.70	1.21	77.08
C 3H 209	182.04	181.9699	19.79	1.11	79.10
C 3H 401	56.06	56.0262	64.27	7.19	28.54
C 3H 402	72.06	72.0211	50.00	5.59	44.40
C 3H 403	88.06	88.0160	40.92	4.58	54.50
C 3H 404	104.06	104.0110	34.63	3.87	61.50
C 3H 405	120.06	120.0059	30.01	3.36	66.63
C 3H 406	136.06	136.0008	26.48	2.96	70.55
C 3H 407	152.06	151.9957	23.70	2.65	73.65
C 3H 408	168.06	167.9906	21.44	2.40	76.16
C 3H 409	184.06	183.9855	19.58	2.19	78.23
C 3H 601	58.08	58.0419	62.04	10.41	27.55
C 3H 602	74.08	74.0368	48.64	8.16	43.19
C 3H 603	90.08	90.0317	40.00	6.71	53.28
C 3H 604	106.08	106.0266	33.97	5.70	60.33
C 3H 605	122.08	122.0215	29.52	4.95	65.53
C 3H 606	138.08	138.0164	26.10	4.38	69.52
C 3H 607	154.08	154.0113	23.39	3.93	72.69
C 3H 608	170.08	170.0063	21.19	3.56	75.26
C 3H 609	186.08	186.0012	19.36	3.25	77.38
C 3H 801	60.10	60.0575	59.96	13.42	26.62
C 3H 802	76.10	76.0524	47.35	10.60	42.05
C 3H 803	92.10	92.0473	39.13	8.76	52.12
C 3H 804	108.09	108.0423	33.34	7.46	59.21
C 3H 805	124.09	124.0372	29.04	6.50	64.46
C 3H 806	140.09	140.0321	25.72	5.76	68.52
C 3H 807	156.09	156.0270	23.08	5.17	71.75
C 3H 808	172.09	172.0219	20.94	4.69	74.38
C 3H 809	188.09	188.0168	19.16	4.29	76.56
C 4H 201	66.06	66.0106	72.73	3.05	24.22
C 4H 202	82.06	82.0055	58.55	2.46	38.99
C 4H 203	98.06	98.0004	49.00	2.06	48.95
C 4H 204	114.06	113.9953	42.12	1.77	56.11
C 4H 205	130.06	129.9902	36.94	1.55	61.51
C 4H 206	146.06	145.9851	32.89	1.38	65.73
C 4H 207	162.06	161.9800	29.65	1.24	69.11
C 4H 208	178.06	177.9750	26.98	1.13	71.88
C 4H 209	194.06	193.9699	24.76	1.04	74.20
C 4H 401	68.08	68.0262	70.58	5.92	23.50

FORMULA	MOL. WEIGHT	ISOMOL. WEIGHT	PERCENT C	PERCENT H	PERCENT O
C 4H 402	84.08	84.0211	57.14	4.80	38.06
C 4H 403	100.07	100.0160	48.01	4.03	47.96
C 4H 404	116.07	116.0110	41.39	3.47	55.14
C 4H 405	132.07	132.0059	36.38	3.05	60.57
C 4H 406	148.07	148.0008	32.45	2.72	64.83
C 4H 407	164.07	163.9957	29.28	2.46	68.26
C 4H 408	180.07	179.9906	26.68	2.24	71.08
C 4H 409	196.07	195.9855	24.50	2.06	73.44
C 4H 601	70.09	70.0419	68.55	8.63	22.83
C 4H 602	86.09	86.0368	55.81	7.02	37.17
C 4H 603	102.09	102.0317	47.06	5.92	47.02
C 4H 604	118.09	118.0266	40.68	5.12	54.19
C 4H 605	134.09	134.0215	35.83	4.51	59.66
C 4H 606	150.09	150.0164	32.01	4.03	63.96
C 4H 607	166.09	166.0113	28.93	3.64	67.43
C 4H 608	182.09	182.0063	26.39	3.32	70.29
C 4H 609	198.09	198.0012	24.25	3.05	72.69
C 4H 801	72.11	72.0575	66.63	11.18	22.19
C 4H 802	88.11	88.0524	54.53	9.15	36.32
C 4H 803	104.11	104.0473	46.15	7.75	46.10
C 4H 804	120.11	120.0423	40.00	6.71	53.28
C 4H 805	136.11	136.0372	35.30	5.92	58.78
C 4H 806	152.10	152.0321	31.59	5.30	63.11
C 4H 807	168.10	168.0270	28.58	4.80	66.62
C 4H 808	184.10	184.0219	26.10	4.38	69.52
C 4H 809	200.10	200.0168	24.01	4.03	71.96
C 4H1001	74.12	74.0732	64.82	13.60	21.58
C 4H1002	90.12	90.0681	53.31	11.18	35.51
C 4H1003	106.12	106.0630	45.27	9.50	45.23
C 4H1004	122.12	122.0579	39.34	8.25	52.40
C 4H1005	138.12	138.0528	34.78	7.30	57.92
C 4H1006	154.12	154.0477	31.17	6.54	62.29
C 4H1007	170.12	170.0426	28.24	5.93	65.83
C 4H1008	186.12	186.0376	25.81	5.42	68.77
C 4H1009	202.12	202.0325	23.77	4.99	71.24
C 5H 201	78.07	78.0106	76.92	2.58	20.49
C 5H 202	94.07	94.0055	63.84	2.14	34.02
C 5H 203	110.07	110.0004	54.56	1.83	43.61
C 5H 204	126.07	125.9953	47.64	1.60	50.76
C 5H 205	142.07	141.9902	42.27	1.42	56.31
C 5H 206	158.07	157.9851	37.99	1.28	60.73
C 5H 207	174.07	173.9800	34.50	1.16	64.34
C 5H 208	190.07	189.9750	31.60	1.06	67.34
C 5H 209	206.07	205.9699	29.14	.98	69.88
C 5H 401	80.09	80.0262	74.99	5.03	19.98
C 5H 402	96.09	96.0211	62.50	4.20	33.30
C 5H 403	112.09	112.0160	53.58	3.60	42.82
C 5H 404	128.09	128.0110	46.89	3.15	49.96
C 5H 405	144.08	144.0059	41.68	2.80	55.52
C 5H 406	160.08	160.0008	37.52	2.52	59.97
C 5H 407	176.08	175.9957	34.11	2.29	63.60
C 5H 408	192.08	191.9906	31.27	2.10	66.64
C 5H 409	208.08	207.9855	28.86	1.94	69.20
C 5H 601	82.10	82.0419	73.15	7.37	19.49
C 5H 602	98.10	98.0368	61.22	6.16	32.62
C 5H 603	114.10	114.0317	52.63	5.30	42.07
C 5H 604	130.10	130.0266	46.16	4.65	49.19
C 5H 605	146.10	146.0215	41.11	4.14	54.75
C 5H 606	162.10	162.0164	37.05	3.73	59.22
C 5H 607	178.10	178.0113	33.72	3.40	62.88

FORMULA	MOL. WEIGHT	ISOMOL. WEIGHT	PERCENT C	PERCENT H	PERCENT O
C 5H 608	194.10	194.0063	30.94	3.12	65.94
C 5H 609	210.10	210.0012	28.58	2.88	68.54
C 5H 801	84.12	84.0575	71.39	9.59	19.02
C 5H 802	100.12	100.0524	59.98	8.05	31.96
C 5H 803	116.12	116.0473	51.72	6.94	41.34
C 5H 804	132.12	132.0423	45.46	6.10	48.44
C 5H 805	148.12	148.0372	40.55	5.44	54.01
C 5H 806	164.12	164.0321	36.59	4.91	58.49
C 5H 807	180.12	180.0270	33.34	4.48	62.18
C 5H 808	196.11	196.0219	30.62	4.11	65.27
C 5H 809	212.11	212.0168	28.31	3.80	67.89
C 5H1001	86.13	86.0732	69.72	11.70	18.57
C 5H1002	102.13	102.0681	58.80	9.87	31.33
C 5H1003	118.13	118.0630	50.84	8.53	40.63
C 5H1004	134.13	134.0579	44.77	7.51	47.71
C 5H1005	150.13	150.0528	40.00	6.71	53.28
C 5H1006	166.13	166.0477	36.15	6.07	57.78
C 5H1007	182.13	182.0426	32.97	5.53	61.49
C 5H1008	198.13	198.0376	30.31	5.09	64.60
C 5H1009	214.13	214.0325	28.05	4.71	67.25
C 5H1201	88.15	88.0888	68.13	13.72	18.15
C 5H1202	104.15	104.0837	57.66	11.61	30.72
C 5H1203	120.15	120.0786	49.98	10.07	39.95
C 5H1204	136.15	136.0736	44.11	8.88	47.01
C 5H1205	152.15	152.0685	39.47	7.95	52.58
C 5H1206	168.15	168.0634	35.72	7.19	57.09
C 5H1207	184.15	184.0583	32.61	6.57	60.82
C 5H1208	200.15	200.0532	30.01	6.04	63.95
C 5H1209	216.15	216.0481	27.78	5.60	66.62
C 6H 401	92.10	92.0262	78.25	4.38	17.37
C 6H 402	108.10	108.0211	66.67	3.73	29.60
C 6H 403	124.10	124.0160	58.07	3.25	38.68
C 6H 404	140.10	140.0110	51.44	2.88	45.68
C 6H 405	156.10	156.0059	46.17	2.58	51.25
C 6H 406	172.10	172.0008	41.88	2.34	55.78
C 6H 407	188.09	187.9957	38.31	2.14	59.54
C 6H 408	204.09	203.9906	35.31	1.98	62.71
C 6H 409	220.09	219.9855	32.74	1.83	65.42
C 6H 601	94.11	94.0419	76.57	6.43	17.00
C 6H 602	110.11	110.0368	65.45	5.49	29.06
C 6H 603	126.11	126.0317	57.14	4.80	38.06
C 6H 604	142.11	142.0266	50.71	4.26	45.03
C 6H 605	158.11	158.0215	45.58	3.83	50.60
C 6H 606	174.11	174.0164	41.39	3.47	55.14
C 6H 607	190.11	190.0113	37.91	3.18	58.91
C 6H 608	206.11	206.0063	34.97	2.93	62.10
C 6H 609	222.11	222.0012	32.45	2.72	64.83
C 6H 801	96.13	96.0575	74.97	8.39	16.64
C 6H 802	112.13	112.0524	64.27	7.19	28.54
C 6H 803	128.13	128.0473	56.25	6.29	37.46
C 6H 804	144.13	144.0423	50.00	5.59	44.40
C 6H 805	160.13	160.0372	45.01	5.04	49.96
C 6H 806	176.13	176.0321	40.92	4.58	54.50
C 6H 807	192.13	192.0270	37.51	4.20	58.29
C 6H 808	208.13	208.0219	34.63	3.87	61.50
C 6H 809	224.13	224.0168	32.15	3.60	64.25
C 6H1001	98.15	98.0732	73.43	10.27	16.30
C 6H1002	114.15	114.0681	63.14	8.83	28.03
C 6H1003	130.14	130.0630	55.37	7.74	36.88
C 6H1004	146.14	146.0579	49.31	6.90	43.79

10/21/66 071

C 6H1005	162.14	162.0528	44.45	6.22	49.34
C 6H1006	178.14	178.0477	40.45	5.66	53.89
C 6H1007	194.14	194.0426	37.12	5.19	57.69
C 6H1008	210.14	210.0376	34.29	4.80	60.91
C 6H1009	226.14	226.0325	31.87	4.46	63.67
C 6H1201	100.16	100.0888	71.95	12.08	15.97
C 6H1202	116.16	116.0837	62.04	10.41	27.55
C 6H1203	132.16	132.0786	54.53	9.15	36.32
C 6H1204	148.16	148.0736	48.64	8.16	43.19
C 6H1205	164.16	164.0685	43.90	7.37	48.73
C 6H1206	180.16	180.0634	40.00	6.71	53.28
C 6H1207	196.16	196.0583	36.74	6.17	57.09
C 6H1208	212.16	212.0532	33.97	5.70	60.33
C 6H1209	228.16	228.0481	31.59	5.30	63.11
C 6H1401	102.18	102.1045	70.53	13.81	15.66
C 6H1402	118.18	118.0994	60.98	11.94	27.08
C 6H1403	134.18	134.0943	53.71	10.52	35.77
C 6H1404	150.18	150.0892	47.99	9.40	42.62
C 6H1405	166.18	166.0841	43.37	8.49	48.14
C 6H1406	182.17	182.0790	39.56	7.75	52.69
C 6H1407	198.17	198.0739	36.37	7.12	56.51
C 6H1408	214.17	214.0689	33.65	6.59	59.76
C 6H1409	230.17	230.0638	31.31	6.13	62.56
C 7H 401	104.11	104.0262	80.76	3.87	15.37
C 7H 402	120.11	120.0211	70.00	3.36	26.64
C 7H 403	136.11	136.0160	61.77	2.96	35.26
C 7H 404	152.11	152.0110	55.28	2.65	42.07
C 7H 405	168.11	168.0059	50.01	2.40	47.59
C 7H 406	184.11	184.0008	45.67	2.19	52.14
C 7H 407	200.11	199.9957	42.02	2.01	55.97
C 7H 408	216.11	215.9906	38.91	1.87	59.23
C 7H 409	232.10	231.9855	36.22	1.74	62.04
C 7H 601	106.13	106.0419	79.23	5.70	15.08
C 7H 602	122.12	122.0368	68.85	4.95	26.20
C 7H 603	138.12	138.0317	60.87	4.38	34.75
C 7H 604	154.12	154.0266	54.55	3.92	41.52
C 7H 605	170.12	170.0215	49.42	3.55	47.02
C 7H 606	186.12	186.0164	45.17	3.25	51.58
C 7H 607	202.12	202.0113	41.60	2.99	55.41
C 7H 608	218.12	218.0063	38.55	2.77	58.68
C 7H 609	234.12	234.0012	35.91	2.58	61.50
C 7H 801	108.14	108.0575	77.75	7.46	14.79
C 7H 802	124.14	124.0524	67.73	6.50	25.78
C 7H 803	140.14	140.0473	60.00	5.75	34.25
C 7H 804	156.14	156.0423	53.85	5.16	40.99
C 7H 805	172.14	172.0372	48.84	4.68	46.47
C 7H 806	188.14	188.0321	44.69	4.29	51.02
C 7H 807	204.14	204.0270	41.19	3.95	54.86
C 7H 808	220.14	220.0219	38.19	3.66	58.14
C 7H 809	236.14	236.0168	35.61	3.41	60.98
C 7H1001	110.16	110.0732	76.33	9.15	14.52
C 7H1002	126.16	126.0681	66.65	7.99	25.36
C 7H1003	142.16	142.0630	59.14	7.09	33.76
C 7H1004	158.16	158.0579	53.16	6.37	40.47
C 7H1005	174.15	174.0528	48.28	5.79	45.93
C 7H1006	190.15	190.0477	44.22	5.30	50.48
C 7H1007	206.15	206.0426	40.78	4.89	54.33
C 7H1008	222.15	222.0376	37.85	4.54	57.62
C 7H1009	238.15	238.0325	35.30	4.23	60.46
C 7H1201	112.17	112.0888	74.95	10.78	14.26

C 7H1202	128.17	128.0837	65.60	9.44	24.97
C 7H1203	144.17	144.0786	58.32	8.39	33.29
C 7H1204	160.17	160.0736	52.49	7.55	39.96
C 7H1205	176.17	176.0685	47.73	6.87	45.41
C 7H1206	192.17	192.0634	43.75	6.29	49.95
C 7H1207	208.17	208.0583	40.39	5.81	53.80
C 7H1208	224.17	224.0532	37.51	5.40	57.10
C 7H1209	240.17	240.0481	35.01	5.04	59.96
C 7H1401	114.19	114.1045	73.63	12.36	14.01
C 7H1402	130.19	130.0994	64.58	10.84	24.58
C 7H1403	146.19	146.0943	57.51	9.65	32.83
C 7H1404	162.19	162.0892	51.84	8.70	39.46
C 7H1405	178.19	178.0841	47.19	7.92	44.90
C 7H1406	194.19	194.0790	43.30	7.27	49.44
C 7H1407	210.19	210.0739	40.00	6.71	53.28
C 7H1408	226.18	226.0689	37.17	6.24	56.59
C 7H1409	242.18	242.0638	34.72	5.83	59.46
C 7H1601	116.20	116.1201	72.35	13.88	13.77
C 7H1602	132.20	132.1150	63.60	12.20	24.20
C 7H1603	148.20	148.1099	56.73	10.88	32.39
C 7H1604	164.20	164.1049	51.20	9.82	38.97
C 7H1605	180.20	180.0998	46.66	8.95	44.39
C 7H1606	196.20	196.0947	42.85	8.22	48.93
C 7H1607	212.20	212.0896	39.62	7.60	52.78
C 7H1608	228.20	228.0845	36.84	7.07	56.09
C 7H1609	244.20	244.0794	34.43	6.60	58.97
C 8H 401	116.12	116.0262	82.75	3.47	13.78
C 8H 402	132.12	132.0211	72.73	3.05	24.22
C 8H 403	148.12	148.0160	64.87	2.72	32.41
C 8H 404	164.12	164.0110	58.55	2.46	38.99
C 8H 405	180.12	180.0059	53.35	2.24	44.41
C 8H 406	196.12	196.0008	49.00	2.06	48.95
C 8H 407	212.12	211.9957	45.30	1.90	52.80
C 8H 408	228.12	227.9906	42.12	1.77	56.11
C 8H 409	244.12	243.9855	39.36	1.65	58.99
C 8H 601	118.14	118.0419	81.34	5.12	13.54
C 8H 602	134.14	134.0368	71.64	4.51	23.86
C 8H 603	150.14	150.0317	64.00	4.03	31.97
C 8H 604	166.13	166.0266	57.84	3.64	38.52
C 8H 605	182.13	182.0215	52.76	3.32	43.92
C 8H 606	198.13	198.0164	48.50	3.05	48.45
C 8H 607	214.13	214.0113	44.87	2.82	52.30
C 8H 608	230.13	230.0063	41.75	2.63	55.62
C 8H 609	246.13	246.0012	39.04	2.46	58.50
C 8H 801	120.15	120.0575	79.97	6.71	13.32
C 8H 802	136.15	136.0524	70.58	5.92	23.50
C 8H 803	152.15	152.0473	63.15	5.30	31.55
C 8H 804	168.15	168.0423	57.14	4.80	38.06
C 8H 805	184.15	184.0372	52.18	4.38	43.44
C 8H 806	200.15	200.0321	48.01	4.03	47.96
C 8H 807	216.15	216.0270	44.46	3.73	51.81
C 8H 808	232.15	232.0219	41.39	3.47	55.14
C 8H 809	248.15	248.0168	38.72	3.25	58.03
C 8H1001	122.17	122.0732	78.65	8.25	13.10
C 8H1002	138.17	138.0681	69.55	7.30	23.16
C 8H1003	154.17	154.0630	62.33	6.54	31.13
C 8H1004	170.17	170.0579	56.47	5.92	37.61
C 8H1005	186.17	186.0528	51.61	5.41	42.97
C 8H1006	202.17	202.0477	47.53	4.99	47.48
C 8H1007	218.16	218.0426	44.04	4.62	51.34

10/21/66 072

C 8H1008	234.16	234.0376	41.03	4.30	54.66
C 8H1009	250.16	250.0325	38.41	4.03	57.56
C 8H1201	124.18	124.0888	77.38	9.74	12.88
C 8H1202	140.18	140.0837	68.55	8.63	22.83
C 8H1203	156.18	156.0786	61.52	7.74	30.73
C 8H1204	172.18	172.0736	55.81	7.02	37.17
C 8H1205	188.18	188.0685	51.06	6.43	42.51
C 8H1206	204.18	204.0634	47.06	5.92	47.02
C 8H1207	220.18	220.0583	43.64	5.49	50.87
C 8H1208	236.18	236.0532	40.68	5.12	54.19
C 8H1209	252.18	252.0481	38.10	4.80	57.10
C 8H1401	126.20	126.1045	76.14	11.18	12.68
C 8H1402	142.20	142.0994	67.57	9.92	22.50
C 8H1403	158.20	158.0943	60.74	8.92	30.34
C 8H1404	174.20	174.0892	55.16	8.10	36.74
C 8H1405	190.20	190.0841	50.52	7.42	42.06
C 8H1406	206.20	206.0790	46.60	6.84	46.56
C 8H1407	222.20	222.0739	43.25	6.35	50.40
C 8H1408	238.20	238.0689	40.34	5.92	53.74
C 8H1409	254.20	254.0638	37.80	5.55	56.65
C 8H1601	128.22	128.1201	74.94	12.58	12.48
C 8H1602	144.22	144.1150	66.63	11.18	22.19
C 8H1603	160.21	160.1099	59.98	10.07	29.96
C 8H1604	176.21	176.1049	54.53	9.15	36.32
C 8H1605	192.21	192.0998	49.99	8.39	41.62
C 8H1606	208.21	208.0947	46.15	7.75	46.10
C 8H1607	224.21	224.0896	42.86	7.19	49.95
C 8H1608	240.21	240.0845	40.00	6.71	53.28
C 8H1609	256.21	256.0794	37.50	6.29	56.20
C 8H1801	130.23	130.1358	73.78	13.93	12.29
C 8H1802	146.23	146.1307	65.71	12.41	21.88
C 8H1803	162.23	162.1256	59.23	11.18	29.59
C 8H1804	178.23	178.1205	53.91	10.18	35.91
C 8H1805	194.23	194.1154	49.47	9.34	41.19
C 8H1806	210.23	210.1103	45.71	8.63	45.66
C 8H1807	226.23	226.1052	42.47	8.02	49.51
C 8H1808	242.23	242.1002	39.67	7.49	52.84
C 8H1809	258.23	258.0951	37.21	7.03	55.76
C 9H 401	128.13	128.0262	84.37	3.15	12.49
C 9H 402	144.13	144.0211	75.00	2.80	22.20
C 9H 403	160.13	160.0160	67.51	2.52	29.97
C 9H 404	176.13	176.0110	61.38	2.29	36.34
C 9H 405	192.13	192.0059	56.26	2.10	41.64
C 9H 406	208.13	208.0008	51.94	1.94	46.12
C 9H 407	224.13	223.9957	48.23	1.80	49.97
C 9H 408	240.13	239.9906	45.02	1.68	53.30
C 9H 409	256.13	255.9855	42.21	1.57	56.22
C 9H 601	130.15	130.0419	83.06	4.65	12.29
C 9H 602	146.15	146.0368	73.97	4.14	21.89
C 9H 603	162.15	162.0317	66.67	3.73	29.60
C 9H 604	178.15	178.0266	60.68	3.39	35.92
C 9H 605	194.15	194.0215	55.68	3.12	41.20
C 9H 606	210.14	210.0164	51.44	2.88	45.68
C 9H 607	226.14	226.0113	47.80	2.67	49.52
C 9H 608	242.14	242.0063	44.64	2.50	52.86
C 9H 609	258.14	258.0012	41.88	2.34	55.78
C 9H 801	132.16	132.0575	81.79	6.10	12.11
C 9H 802	148.16	148.0524	72.96	5.44	21.60
C 9H 803	164.16	164.0473	65.85	4.91	29.24
C 9H 804	180.16	180.0423	60.00	4.48	35.52

C 9H 805	196.16	196.0372	55.11	4.11	40.78
C 9H 806	212.16	212.0321	50.95	3.80	45.25
C 9H 807	228.16	228.0270	47.38	3.53	49.09
C 9H 808	244.16	244.0219	44.27	3.30	52.42
C 9H 809	260.16	260.0168	41.55	3.10	55.35
C 9H1001	134.18	134.0732	80.56	7.51	11.92
C 9H1002	150.18	150.0681	71.98	6.71	21.31
C 9H1003	166.18	166.0630	65.05	6.07	28.88
C 9H1004	182.18	182.0579	59.34	5.53	35.13
C 9H1005	198.18	198.0528	54.55	5.09	40.37
C 9H1006	214.18	214.0477	50.47	4.71	44.82
C 9H1007	230.18	230.0426	46.96	4.38	48.66
C 9H1008	246.18	246.0376	43.91	4.09	51.99
C 9H1009	262.17	262.0325	41.23	3.84	54.92
C 9H1201	136.20	136.0888	79.37	8.88	11.75
C 9H1202	152.19	152.0837	71.03	7.95	21.02
C 9H1203	168.19	168.0786	64.27	7.19	28.54
C 9H1204	184.19	184.0736	58.69	6.57	34.74
C 9H1205	200.19	200.0685	54.00	6.04	39.96
C 9H1206	216.19	216.0634	50.00	5.59	44.40
C 9H1207	232.19	232.0583	46.56	5.21	48.23
C 9H1208	248.19	248.0532	43.56	4.87	51.57
C 9H1209	264.19	264.0481	40.92	4.58	54.50
C 9H1401	138.21	138.1045	78.21	10.21	11.58
C 9H1402	154.21	154.0994	70.10	9.15	20.75
C 9H1403	170.21	170.0943	63.51	8.29	28.20
C 9H1404	186.21	186.0892	58.05	7.58	34.37
C 9H1405	202.21	202.0841	53.46	6.98	39.56
C 9H1406	218.21	218.0790	49.54	6.47	43.99
C 9H1407	234.21	234.0739	46.16	6.03	47.82
C 9H1408	250.21	250.0689	43.20	5.64	51.16
C 9H1409	266.21	266.0638	40.61	5.30	54.09
C 9H1601	140.23	140.1201	77.09	11.50	11.41
C 9H1602	156.23	156.1150	69.19	10.32	20.48
C 9H1603	172.23	172.1099	62.77	9.36	27.87
C 9H1604	188.23	188.1049	57.43	8.57	34.00
C 9H1605	204.22	204.0998	52.93	7.90	39.17
C 9H1606	220.22	220.0947	49.09	7.32	43.59
C 9H1607	236.22	236.0896	45.76	6.83	47.41
C 9H1608	252.22	252.0845	42.86	6.39	50.75
C 9H1609	268.22	268.0794	40.30	6.01	53.68
C 9H1801	142.24	142.1358	76.00	12.76	11.25
C 9H1802	158.24	158.1307	68.31	11.47	20.22
C 9H1803	174.24	174.1256	62.04	10.41	27.55
C 9H1804	190.24	190.1205	56.82	9.54	33.64
C 9H1805	206.24	206.1154	52.41	8.80	38.79
C 9H1806	222.24	222.1103	48.64	8.16	43.19
C 9H1807	238.24	238.1052	45.37	7.62	47.01
C 9H1808	254.24	254.1002	42.52	7.14	50.34
C 9H1809	270.24	270.0951	40.00	6.71	53.28
C 9H2001	144.26	144.1514	74.93	13.97	11.09
C 9H2002	160.26	160.1463	67.45	12.58	19.97
C 9H2003	176.26	176.1412	61.33	11.44	27.23
C 9H2004	192.26	192.1361	56.23	10.49	33.29
C 9H2005	208.26	208.1311	51.91	9.68	38.41
C 9H2006	224.26	224.1260	48.20	8.99	42.81
C 9H2007	240.26	240.1209	44.99	8.39	46.62
C 9H2008	256.25	256.1158	42.18	7.87	49.95
C 9H2009	272.25	272.1107	39.71	7.40	52.89
C10H 601	142.16	142.0419	84.49	4.25	11.25

C10H 602	158.16	158.0368	75.94	3.82	20.23
C10H 603	174.16	174.0317	68.97	3.47	27.56
C10H 604	190.16	190.0266	63.16	3.18	33.66
C10H 605	206.16	206.0215	58.26	2.93	38.80
C10H 606	222.16	222.0164	54.07	2.72	43.21
C10H 607	238.16	238.0113	50.43	2.54	47.03
C10H 608	254.15	254.0063	47.26	2.38	50.36
C10H 609	270.15	270.0012	44.46	2.24	53.30
C10H 801	144.17	144.0575	83.31	5.59	11.10
C10H 802	160.17	160.0524	74.99	5.03	19.98
C10H 803	176.17	176.0473	68.18	4.58	27.24
C10H 804	192.17	192.0423	62.50	4.20	33.30
C10H 805	208.17	208.0372	57.70	3.87	38.43
C10H 806	224.17	224.0321	53.58	3.60	42.82
C10H 807	240.17	240.0270	50.01	3.36	46.63
C10H 808	256.17	256.0219	46.89	3.15	49.96
C10H 809	272.17	272.0168	44.13	2.96	52.91
C10H1001	146.19	146.0732	82.16	6.89	10.94
C10H1002	162.19	162.0681	74.06	6.21	19.73
C10H1003	178.19	178.0630	67.41	5.66	26.94
C10H1004	194.19	194.0579	61.85	5.19	32.96
C10H1005	210.19	210.0528	57.14	4.80	38.06
C10H1006	226.19	226.0477	53.10	4.46	42.44
C10H1007	242.19	242.0426	49.59	4.16	46.24
C10H1008	258.19	258.0376	46.52	3.90	49.57
C10H1009	274.19	274.0325	43.81	3.68	52.52
C10H1201	148.21	148.0888	81.04	8.16	10.80
C10H1202	164.21	164.0837	73.15	7.37	19.49
C10H1203	180.21	180.0786	66.65	6.71	26.64
C10H1204	196.20	196.0736	61.22	6.16	32.62
C10H1205	212.20	212.0685	56.60	5.70	37.70
C10H1206	228.20	228.0634	52.63	5.30	42.07
C10H1207	244.20	244.0583	49.19	4.95	45.86
C10H1208	260.20	260.0532	46.16	4.65	49.19
C10H1209	276.20	276.0481	43.49	4.38	52.13
C10H1401	150.22	150.1045	79.96	9.39	10.65
C10H1402	166.22	166.0994	72.26	8.49	19.25
C10H1403	182.22	182.0943	65.92	7.74	26.34
C10H1404	198.22	198.0892	60.59	7.12	32.29
C10H1405	214.22	214.0841	56.07	6.59	37.34
C10H1406	230.22	230.0790	52.17	6.13	41.70
C10H1407	246.22	246.0739	48.78	5.73	45.49
C10H1408	262.22	262.0689	45.81	5.38	48.81
C10H1409	278.22	278.0638	43.17	5.07	51.76
C10H1601	152.24	152.1201	78.90	10.59	10.51
C10H1602	168.24	168.1150	71.39	9.59	19.02
C10H1603	184.24	184.1099	65.19	8.75	26.05
C10H1604	200.24	200.1049	59.98	8.05	31.96
C10H1605	216.24	216.0998	55.55	7.46	37.00
C10H1606	232.24	232.0947	51.72	6.94	41.34
C10H1607	248.23	248.0896	48.39	6.50	45.12
C10H1608	264.23	264.0845	45.46	6.10	48.44
C10H1609	280.23	280.0794	42.86	5.76	51.38
C10H1801	154.25	154.1358	77.87	11.76	10.37
C10H1802	170.25	170.1307	70.55	10.66	18.79
C10H1803	186.25	186.1256	64.49	9.74	25.77
C10H1804	202.25	202.1205	59.39	8.97	31.64
C10H1805	218.25	218.1154	55.03	8.31	36.65
C10H1806	234.25	234.1103	51.27	7.75	40.98
C10H1807	250.25	250.1052	48.00	7.25	44.75

C10H1808	266.25	266.1002	45.11	6.81	48.07
C10H1809	282.25	282.0951	42.56	6.43	51.02
C10H2001	156.27	156.1514	76.86	12.90	10.24
C10H2002	172.27	172.1463	69.72	11.70	18.57
C10H2003	188.27	188.1412	63.80	10.71	25.49
C10H2004	204.27	204.1361	58.80	9.87	31.33
C10H2005	220.27	220.1311	54.53	9.15	36.32
C10H2006	236.27	236.1260	50.84	8.53	40.63
C10H2007	252.27	252.1209	47.61	7.99	44.40
C10H2008	268.27	268.1158	44.77	7.51	47.71
C10H2009	284.27	284.1107	42.25	7.09	50.65
C10H2201	158.29	158.1671	75.88	14.01	10.11
C10H2202	174.29	174.1620	68.92	12.72	18.36
C10H2203	190.29	190.1569	63.12	11.65	25.22
C10H2204	206.28	206.1518	58.23	10.75	31.02
C10H2205	222.28	222.1467	54.04	9.98	35.99
C10H2206	238.28	238.1416	50.41	9.31	40.29
C10H2207	254.28	254.1365	47.24	8.72	44.04
C10H2208	270.28	270.1315	44.44	8.20	47.36
C10H2209	286.28	286.1264	41.96	7.75	50.30
C11H 601	154.17	154.0419	85.70	3.92	10.38
C11H 602	170.17	170.0368	77.64	3.55	18.80
C11H 603	186.17	186.0317	70.97	3.25	25.78
C11H 604	202.17	202.0266	65.35	2.99	31.66
C11H 605	218.17	218.0215	60.56	2.77	36.67
C11H 606	234.17	234.0164	56.42	2.58	40.99
C11H 607	250.17	250.0113	52.81	2.42	44.77
C11H 608	266.17	266.0063	49.64	2.27	48.09
C11H 609	282.17	282.0012	46.82	2.14	51.03
C11H 801	156.19	156.0575	84.59	5.16	10.24
C11H 802	172.19	172.0524	76.73	4.68	18.58
C11H 803	188.18	188.0473	70.21	4.29	25.51
C11H 804	204.18	204.0423	64.71	3.95	31.34
C11H 805	220.18	220.0372	60.01	3.66	36.33
C11H 806	236.18	236.0321	55.94	3.41	40.64
C11H 807	252.18	252.0270	52.39	3.20	44.41
C11H 808	268.18	268.0219	49.27	3.01	47.73
C11H 809	284.18	284.0168	46.49	2.84	50.67
C11H1001	158.20	158.0732	83.52	6.37	10.11
C11H1002	174.20	174.0681	75.84	5.79	18.37
C11H1003	190.20	190.0630	69.46	5.30	25.24
C11H1004	206.20	206.0579	64.08	4.89	31.04
C11H1005	222.20	222.0528	59.46	4.54	36.00
C11H1006	238.20	238.0477	55.47	4.23	40.30
C11H1007	254.20	254.0426	51.98	3.97	44.06
C11H1008	270.20	270.0376	48.90	3.73	47.37
C11H1009	286.20	286.0325	46.16	3.52	50.31
C11H1201	160.22	160.0888	82.46	7.55	9.99
C11H1202	176.22	176.0837	74.98	6.86	18.16
C11H1203	192.22	192.0786	68.74	6.29	24.97
C11H1204	208.22	208.0736	63.45	5.81	30.74
C11H1205	224.22	224.0685	58.93	5.39	35.68
C11H1206	240.21	240.0634	55.00	5.04	39.96
C11H1207	256.21	256.0583	51.57	4.72	43.71
C11H1208	272.21	272.0532	48.54	4.44	47.02
C11H1209	288.21	288.0481	45.84	4.20	49.96
C11H1401	162.23	162.1045	81.44	8.70	9.86
C11H1402	178.23	178.0994	74.13	7.92	17.95
C11H1403	194.23	194.0943	68.02	7.27	24.71
C11H1404	210.23	210.0892	62.85	6.71	30.44

C11H1405	226.23	226.0841	58.40	6.24	35.36
C11H1406	242.23	242.0790	54.54	5.83	39.63
C11H1407	258.23	258.0739	51.16	5.46	43.37
C11H1408	274.23	274.0689	48.18	5.15	46.67
C11H1409	290.23	290.0638	45.52	4.86	49.61
C11H1601	164.25	164.1201	80.44	9.82	9.74
C11H1602	180.25	180.1150	73.30	8.95	17.75
C11H1603	196.25	196.1099	67.32	8.22	24.46
C11H1604	212.25	212.1049	62.25	7.60	30.15
C11H1605	228.25	228.0998	57.89	7.07	35.05
C11H1606	244.25	244.0947	54.09	6.60	39.30
C11H1607	260.25	260.0896	50.77	6.20	43.03
C11H1608	276.25	276.0845	47.83	5.84	46.33
C11H1609	292.24	292.0794	45.21	5.52	49.27
C11H1801	166.27	166.1358	79.46	10.91	9.62
C11H1802	182.26	182.1307	72.49	9.95	17.56
C11H1803	198.26	198.1256	66.64	9.15	24.21
C11H1804	214.26	214.1205	61.66	8.47	29.87
C11H1805	230.26	230.1154	57.38	7.88	34.74
C11H1806	246.26	246.1103	53.65	7.37	38.98
C11H1807	262.26	262.1052	50.38	6.92	42.70
C11H1808	278.26	278.1002	47.48	6.52	46.00
C11H1809	294.26	294.0951	44.90	6.17	48.93
C11H2001	168.28	168.1514	78.51	11.98	9.51
C11H2002	184.28	184.1463	71.70	10.94	17.36
C11H2003	200.28	200.1412	65.97	10.07	23.97
C11H2004	216.28	216.1361	61.09	9.32	29.59
C11H2005	232.28	232.1311	56.88	8.68	34.44
C11H2006	248.28	248.1260	53.22	8.12	38.66
C11H2007	264.28	264.1209	49.99	7.63	42.38
C11H2008	280.28	280.1158	47.14	7.19	45.67
C11H2009	296.28	296.1107	44.59	6.80	48.60
C11H2201	170.30	170.1671	77.58	13.02	9.39
C11H2202	186.30	186.1620	70.92	11.90	17.18
C11H2203	202.30	202.1569	65.31	10.96	23.73
C11H2204	218.30	218.1518	60.52	10.16	29.32
C11H2205	234.29	234.1467	56.39	9.46	34.14
C11H2206	250.29	250.1416	52.79	8.86	38.35
C11H2207	266.29	266.1365	49.62	8.33	42.06
C11H2208	282.29	282.1315	46.80	7.86	45.34
C11H2209	298.29	298.1264	44.29	7.43	48.27
C11H2401	172.31	172.1827	76.68	14.04	9.29
C11H2402	188.31	188.1776	70.16	12.85	16.99
C11H2403	204.31	204.1725	64.67	11.84	23.49
C11H2404	220.31	220.1674	59.97	10.98	29.05
C11H2405	236.31	236.1624	55.91	10.24	33.85
C11H2406	252.31	252.1573	52.37	9.59	38.05
C11H2407	268.31	268.1522	49.24	9.02	41.74
C11H2408	284.31	284.1471	46.47	8.51	45.02
C11H2409	300.31	300.1420	44.00	8.06	47.95
C12H 601	166.18	166.0419	86.73	3.64	9.63
C12H 602	182.18	182.0368	79.12	3.32	17.56
C12H 603	198.18	198.0317	72.73	3.05	24.22
C12H 604	214.18	214.0266	67.30	2.82	29.88
C12H 605	230.18	230.0215	62.62	2.63	34.75
C12H 606	246.18	246.0164	58.55	2.46	38.99
C12H 607	262.18	262.0113	54.98	2.31	42.72
C12H 608	278.18	278.0063	51.81	2.17	46.01
C12H 609	294.18	294.0012	49.00	2.06	48.95
C12H 801	168.20	168.0575	85.69	4.79	9.51

C12H 802	184.20	184.0524	78.25	4.38	17.37
C12H 803	200.20	200.0473	72.00	4.03	23.98
C12H 804	216.20	216.0423	66.67	3.73	29.60
C12H 805	232.19	232.0372	62.07	3.47	34.45
C12H 806	248.19	248.0321	58.07	3.25	38.68
C12H 807	264.19	264.0270	54.56	3.05	42.39
C12H 808	280.19	280.0219	51.44	2.88	45.68
C12H 809	296.19	296.0168	48.66	2.72	48.62
C12H1001	170.21	170.0732	84.68	5.92	9.40
C12H1002	186.21	186.0681	77.40	5.41	17.18
C12H1003	202.21	202.0630	71.28	4.98	23.74
C12H1004	218.21	218.0579	66.05	4.62	29.33
C12H1005	234.21	234.0528	61.54	4.30	34.16
C12H1006	250.21	250.0477	57.61	4.03	38.37
C12H1007	266.21	266.0426	54.14	3.79	42.07
C12H1008	282.21	282.0376	51.07	3.57	45.35
C12H1009	298.21	298.0325	48.33	3.38	48.29
C12H1201	172.23	172.0888	83.69	7.02	9.29
C12H1202	188.23	188.0837	76.57	6.43	17.00
C12H1203	204.23	204.0786	70.58	5.92	23.50
C12H1204	220.23	220.0736	65.45	5.49	29.06
C12H1205	236.23	236.0685	61.02	5.12	33.86
C12H1206	252.23	252.0634	57.14	4.80	38.06
C12H1207	268.23	268.0583	53.74	4.51	41.75
C12H1208	284.22	284.0532	50.71	4.26	45.03
C12H1209	300.22	300.0481	48.01	4.03	47.96
C12H1401	174.24	174.1045	82.72	8.10	9.18
C12H1402	190.24	190.0994	75.76	7.42	16.82
C12H1403	206.24	206.0943	69.89	6.84	23.27
C12H1404	222.24	222.0892	64.85	6.35	28.80
C12H1405	238.24	238.0841	60.50	5.92	33.58
C12H1406	254.24	254.0790	56.69	5.55	37.76
C12H1407	270.24	270.0739	53.34	5.22	41.44
C12H1408	286.24	286.0689	50.35	4.93	44.72
C12H1409	302.24	302.0638	47.69	4.67	47.64
C12H1601	176.26	176.1201	81.77	9.15	9.08
C12H1602	192.26	192.1150	74.97	8.39	16.64
C12H1603	208.26	208.1099	69.21	7.74	23.05
C12H1604	224.26	224.1049	64.27	7.19	28.54
C12H1605	240.26	240.0998	59.99	6.71	33.30
C12H1606	256.26	256.0947	56.25	6.29	37.46
C12H1607	272.26	272.0896	52.94	5.92	41.14
C12H1608	288.26	288.0845	50.00	5.59	44.40
C12H1609	304.26	304.0794	47.37	5.30	47.33
C12H1801	178.28	178.1358	80.85	10.18	8.97
C12H1802	194.28	194.1307	74.19	9.34	16.47
C12H1803	210.28	210.1256	68.55	8.63	22.83
C12H1804	226.27	226.1205	63.70	8.02	28.28
C12H1805	242.27	242.1154	59.49	7.49	33.02
C12H1806	258.27	258.1103	55.81	7.02	37.17
C12H1807	274.27	274.1052	52.55	6.62	40.83
C12H1808	290.27	290.1002	49.65	6.25	44.09
C12H1809	306.27	306.0951	47.06	5.92	47.02
C12H2001	180.29	180.1514	79.94	11.18	8.87
C12H2002	196.29	196.1463	73.43	10.27	16.30
C12H2003	212.29	212.1412	67.89	9.50	22.61
C12H2004	228.29	228.1361	63.14	8.83	28.03
C12H2005	244.29	244.1311	59.00	8.25	32.75
C12H2006	260.29	260.1260	55.37	7.74	36.88
C12H2007	276.29	276.1209	52.17	7.30	40.54

C12H2008	292.29	292.1158	49.31	6.90	43.79
C12H2009	308.29	308.1107	46.75	6.54	46.71
C12H2201	182.31	182.1671	79.06	12.16	8.78
C12H2202	198.31	198.1620	72.68	11.18	16.14
C12H2203	214.31	214.1569	67.26	10.35	22.40
C12H2204	230.31	230.1518	62.58	9.63	27.79
C12H2205	246.31	246.1467	58.52	9.00	32.48
C12H2206	262.31	262.1416	54.95	8.45	36.60
C12H2207	278.30	278.1365	51.79	7.97	40.24
C12H2208	294.30	294.1315	48.97	7.53	43.49
C12H2209	310.30	310.1264	46.45	7.15	46.40
C12H2401	184.32	184.1827	78.20	13.12	8.68
C12H2402	200.32	200.1776	71.95	12.08	15.97
C12H2403	216.32	216.1725	66.63	11.18	22.19
C12H2404	232.32	232.1674	62.04	10.41	27.55
C12H2405	248.32	248.1624	58.04	9.74	32.22
C12H2406	264.32	264.1573	54.53	9.15	36.32
C12H2407	280.32	280.1522	51.42	8.63	39.95
C12H2408	296.32	296.1471	48.64	8.16	43.19
C12H2409	312.32	312.1420	46.15	7.75	46.10
C12H2601	186.34	186.1984	77.35	14.06	8.59
C12H2602	202.34	202.1933	71.23	12.95	15.81
C12H2603	218.34	218.1882	66.01	12.00	21.98
C12H2604	234.34	234.1831	61.51	11.18	27.31
C12H2605	250.34	250.1780	57.58	10.47	31.96
C12H2606	266.34	266.1729	54.12	9.84	36.04
C12H2607	282.34	282.1678	51.05	9.28	39.67
C12H2608	298.34	298.1628	48.31	8.78	42.90
C12H2609	314.34	314.1577	45.85	8.34	45.81
C13H 601	178.19	178.0419	87.63	3.39	8.98
C13H 602	194.19	194.0368	80.41	3.11	16.48
C13H 603	210.19	210.0317	74.29	2.88	22.84
C13H 604	226.19	226.0266	69.03	2.67	28.29
C13H 605	242.19	242.0215	64.47	2.50	33.03
C13H 606	258.19	258.0164	60.48	2.34	37.18
C13H 607	274.19	274.0113	56.95	2.21	40.85
C13H 608	290.19	290.0063	53.81	2.08	44.11
C13H 609	306.19	306.0012	51.00	1.98	47.03
C13H 801	180.21	180.0575	86.65	4.47	8.88
C13H 802	196.21	196.0524	79.58	4.11	16.31
C13H 803	212.21	212.0473	73.58	3.80	22.62
C13H 804	228.21	228.0423	68.42	3.53	28.04
C13H 805	244.21	244.0372	63.94	3.30	32.76
C13H 806	260.21	260.0321	60.01	3.10	36.89
C13H 807	276.20	276.0270	56.53	2.92	40.55
C13H 808	292.20	292.0219	53.44	2.76	43.80
C13H 809	308.20	308.0168	50.66	2.62	46.72
C13H1001	182.22	182.0732	85.69	5.53	8.78
C13H1002	198.22	198.0681	78.77	5.09	16.14
C13H1003	214.22	214.0630	72.89	4.71	22.41
C13H1004	230.22	230.0579	67.82	4.38	27.80
C13H1005	246.22	246.0528	63.42	4.09	32.49
C13H1006	262.22	262.0477	59.55	3.84	36.61
C13H1007	278.22	278.0426	56.12	3.62	40.25
C13H1008	294.22	294.0376	53.07	3.43	43.50
C13H1009	310.22	310.0325	50.33	3.25	46.42
C13H1201	184.24	184.0888	84.75	6.57	8.68
C13H1202	200.24	200.0837	77.98	6.04	15.98
C13H1203	216.24	216.0786	72.21	5.59	22.20
C13H1204	232.24	232.0736	67.23	5.21	27.56

C13H1205	248.24	248.0685	62.90	4.87	32.23
C13H1206	264.24	264.0634	59.09	4.58	36.33
C13H1207	280.24	280.0583	55.72	4.32	39.96
C13H1208	296.24	296.0532	52.71	4.08	43.21
C13H1209	312.24	312.0481	50.01	3.87	46.12
C13H1401	186.26	186.1045	83.83	7.58	8.59
C13H1402	202.26	202.0994	77.20	6.98	15.82
C13H1403	218.25	218.0943	71.54	6.47	21.99
C13H1404	234.25	234.0892	66.65	6.02	27.32
C13H1405	250.25	250.0841	62.39	5.64	31.97
C13H1406	266.25	266.0790	58.65	5.30	36.05
C13H1407	282.25	282.0739	55.32	5.00	39.68
C13H1408	298.25	298.0689	52.35	4.73	42.92
C13H1409	314.25	314.0638	49.69	4.49	45.82
C13H1601	188.27	188.1201	82.94	8.57	8.50
C13H1602	204.27	204.1150	76.44	7.90	15.66
C13H1603	220.27	220.1099	70.89	7.32	21.79
C13H1604	236.27	236.1049	66.09	6.83	27.09
C13H1605	252.27	252.0998	61.90	6.39	31.71
C13H1606	268.27	268.0947	58.20	6.01	35.78
C13H1607	284.27	284.0896	54.93	5.67	39.40
C13H1608	300.27	300.0845	52.00	5.37	42.63
C13H1609	316.27	316.0794	49.37	5.10	45.53
C13H1801	190.29	190.1358	82.06	9.53	8.41
C13H1802	206.29	206.1307	75.69	8.80	15.51
C13H1803	222.29	222.1256	70.24	8.16	21.59
C13H1804	238.29	238.1205	65.53	7.61	26.86
C13H1805	254.29	254.1154	61.41	7.14	31.46
C13H1806	270.28	270.1103	57.77	6.71	35.52
C13H1807	286.28	286.1052	54.54	6.34	39.12
C13H1808	302.28	302.1002	51.66	6.00	42.34
C13H1809	318.28	318.0951	49.06	5.70	45.24
C13H2001	192.30	192.1514	81.20	10.48	8.32
C13H2002	208.30	208.1463	74.96	9.68	15.36
C13H2003	224.30	224.1412	69.61	8.99	21.40
C13H2004	240.30	240.1361	64.98	8.39	26.63
C13H2005	256.30	256.1311	60.92	7.87	31.21
C13H2006	272.30	272.1260	57.34	7.40	35.25
C13H2007	288.30	288.1209	54.16	6.99	38.85
C13H2008	304.30	304.1158	51.31	6.62	42.06
C13H2009	320.30	320.1107	48.75	6.29	44.96
C13H2201	194.32	194.1671	80.35	11.41	8.23
C13H2202	210.32	210.1620	74.24	10.54	15.21
C13H2203	226.32	226.1569	68.99	9.80	21.21
C13H2204	242.32	242.1518	64.44	9.15	26.41
C13H2205	258.32	258.1467	60.45	8.58	30.97
C13H2206	274.32	274.1416	56.92	8.08	34.99
C13H2207	290.32	290.1365	53.78	7.64	38.58
C13H2208	306.32	306.1315	50.98	7.24	41.79
C13H2209	322.31	322.1264	48.44	6.88	44.68
C13H2401	196.34	196.1827	79.53	12.32	8.15
C13H2402	212.34	212.1776	73.54	11.39	15.07
C13H2403	228.33	228.1725	68.38	10.59	21.02
C13H2404	244.33	244.1674	63.91	9.90	26.19
C13H2405	260.33	260.1624	59.98	9.29	30.73
C13H2406	276.33	276.1573	56.51	8.75	34.74
C13H2407	292.33	292.1522	53.41	8.28	38.31
C13H2408	308.33	308.1471	50.64	7.85	41.51
C13H2409	324.33	324.1420	48.14	7.46	44.40
C13H2601	198.35	198.1984	78.72	13.21	8.07

C13H2602	214.35	214.1933	72.85	12.23	14.93
C13H2603	230.35	230.1882	67.79	11.38	20.84
C13H2604	246.35	246.1831	63.38	10.64	25.98
C13H2605	262.35	262.1780	59.52	9.99	30.49
C13H2606	278.35	278.1729	56.10	9.42	34.49
C13H2607	294.35	294.1678	53.05	8.90	38.05
C13H2608	310.35	310.1628	50.31	8.44	41.24
C13H2609	326.35	326.1577	47.85	8.03	44.12
C13H2801	200.37	200.2140	77.93	14.09	7.99
C13H2802	216.37	216.2089	72.17	13.04	14.79
C13H2803	232.37	232.2038	67.20	12.15	20.66
C13H2804	248.37	248.1987	62.87	11.36	25.77
C13H2805	264.37	264.1937	59.06	10.68	30.26
C13H2806	280.36	280.1886	55.69	10.07	34.24
C13H2807	296.36	296.1835	52.69	9.52	37.79
C13H2808	312.36	312.1784	49.99	9.04	40.98
C13H2809	328.36	328.1733	47.55	8.60	43.85
C14H 801	192.22	192.0575	87.48	4.20	8.32
C14H 802	208.22	208.0524	80.76	3.87	15.37
C14H 803	224.22	224.0473	75.00	3.60	21.41
C14H 804	240.22	240.0423	70.00	3.36	26.64
C14H 805	256.22	256.0372	65.63	3.15	31.22
C14H 806	272.22	272.0321	61.77	2.96	35.26
C14H 807	288.22	288.0270	58.34	2.80	38.86
C14H 808	304.22	304.0219	55.28	2.65	42.07
C14H 809	320.21	320.0168	52.51	2.52	44.97
C14H1001	194.24	194.0732	86.57	5.19	8.24
C14H1002	210.23	210.0681	79.98	4.79	15.22
C14H1003	226.23	226.0630	74.33	4.46	21.22
C14H1004	242.23	242.0579	69.42	4.16	26.42
C14H1005	258.23	258.0528	65.12	3.90	30.98
C14H1006	274.23	274.0477	61.32	3.68	35.01
C14H1007	290.23	290.0426	57.94	3.47	38.59
C14H1008	306.23	306.0376	54.91	3.29	41.80
C14H1009	322.23	322.0325	52.19	3.13	44.69
C14H1201	196.25	196.0888	85.68	6.16	8.15
C14H1202	212.25	212.0837	79.23	5.70	15.08
C14H1203	228.25	228.0786	73.67	5.30	21.03
C14H1204	244.25	244.0736	68.85	4.95	26.20
C14H1205	260.25	260.0685	64.61	4.65	30.74
C14H1206	276.25	276.0634	60.87	4.38	34.75
C14H1207	292.25	292.0583	57.54	4.14	38.32
C14H1208	308.25	308.0532	54.55	3.92	41.52
C14H1209	324.25	324.0481	51.86	3.73	44.41
C14H1401	198.27	198.1045	84.81	7.12	8.07
C14H1402	214.27	214.0994	78.48	6.59	14.93
C14H1403	230.27	230.0943	73.03	6.13	20.84
C14H1404	246.27	246.0892	68.28	5.73	25.99
C14H1405	262.26	262.0841	64.12	5.38	30.50
C14H1406	278.26	278.0790	60.43	5.07	34.50
C14H1407	294.26	294.0739	57.14	4.80	38.06
C14H1408	310.26	310.0689	54.20	4.55	41.25
C14H1409	326.26	326.0638	51.54	4.33	44.13
C14H1601	200.28	200.1201	83.96	8.05	7.99
C14H1602	216.28	216.1150	77.75	7.46	14.79
C14H1603	232.28	232.1099	72.39	6.94	20.66
C14H1604	248.28	248.1049	67.73	6.50	25.78
C14H1605	264.28	264.0998	63.63	6.10	30.27
C14H1606	280.28	280.0947	60.00	5.75	34.25
C14H1607	296.28	296.0896	56.76	5.44	37.80

C14H1608	312.28	312.0845	53.85	5.16	40.99
C14H1609	328.28	328.0794	51.22	4.91	43.86
C14H1801	202.30	202.1358	83.12	8.97	7.91
C14H1802	218.30	218.1307	77.03	8.31	14.66
C14H1803	234.30	234.1256	71.77	7.74	20.49
C14H1804	250.30	250.1205	67.18	7.25	25.57
C14H1805	266.30	266.1154	63.15	6.81	30.04
C14H1806	282.30	282.1103	59.57	6.43	34.01
C14H1807	298.30	298.1052	56.37	6.08	37.55
C14H1808	314.29	314.1002	53.50	5.77	40.72
C14H1809	330.29	330.0951	50.91	5.49	43.60
C14H2001	204.31	204.1514	82.30	9.87	7.83
C14H2002	220.31	220.1463	76.33	9.15	14.52
C14H2003	236.31	236.1412	71.16	8.53	20.31
C14H2004	252.31	252.1361	66.65	7.99	25.36
C14H2005	268.31	268.1311	62.67	7.51	29.81
C14H2006	284.31	284.1260	59.14	7.09	33.76
C14H2007	300.31	300.1209	55.99	6.71	37.29
C14H2008	316.31	316.1158	53.16	6.37	40.47
C14H2009	332.31	332.1107	50.60	6.07	43.33
C14H2201	206.33	206.1671	81.50	10.75	7.75
C14H2202	222.33	222.1620	75.63	9.97	14.39
C14H2203	238.33	238.1569	70.56	9.30	20.14
C14H2204	254.33	254.1518	66.12	8.72	25.16
C14H2205	270.33	270.1467	62.20	8.20	29.59
C14H2206	286.33	286.1416	58.73	7.74	33.53
C14H2207	302.33	302.1365	55.62	7.33	37.04
C14H2208	318.33	318.1315	52.83	6.97	40.21
C14H2209	334.33	334.1264	50.30	6.63	43.07
C14H2401	208.35	208.1827	80.71	11.61	7.68
C14H2402	224.35	224.1776	74.95	10.78	14.26
C14H2403	240.35	240.1725	69.96	10.07	19.97
C14H2404	256.34	256.1674	65.60	9.44	24.97
C14H2405	272.34	272.1624	61.74	8.88	29.37
C14H2406	288.34	288.1573	58.32	8.39	33.29
C14H2407	304.34	304.1522	55.25	7.95	36.80
C14H2408	320.34	320.1471	52.49	7.55	39.96
C14H2409	336.34	336.1420	50.00	7.19	42.81
C14H2601	210.36	210.1984	79.94	12.46	7.61
C14H2602	226.36	226.1933	74.29	11.58	14.14
C14H2603	242.36	242.1882	69.38	10.81	19.80
C14H2604	258.36	258.1831	65.09	10.14	24.77
C14H2605	274.36	274.1780	61.29	9.55	29.16
C14H2606	290.36	290.1729	57.91	9.03	33.06
C14H2607	306.36	306.1678	54.89	8.55	36.56
C14H2608	322.36	322.1628	52.16	8.13	39.71
C14H2609	338.36	338.1577	49.70	7.75	42.56
C14H2801	212.38	212.2140	79.18	13.29	7.53
C14H2802	228.38	228.2089	73.63	12.36	14.01
C14H2803	244.38	244.2038	68.81	11.55	19.64
C14H2804	260.38	260.1987	64.58	10.84	24.58
C14H2805	276.38	276.1937	60.84	10.21	28.94
C14H2806	292.38	292.1886	57.51	9.65	32.83
C14H2807	308.38	308.1835	54.53	9.15	36.32
C14H2808	324.37	324.1784	51.84	8.70	39.46
C14H2809	340.37	340.1733	49.40	8.29	42.30
C14H3001	214.39	214.2297	78.43	14.10	7.46
C14H3002	230.39	230.2246	72.99	13.12	13.89
C14H3003	246.39	246.2195	68.25	12.27	19.48
C14H3004	262.39	262.2144	64.09	11.52	24.39

C14H3005	278.39	278.2093	60.40	10.86	28.74
C14H3006	294.39	294.2042	57.12	10.27	32.61
C14H3007	310.39	310.1991	54.18	9.74	36.08
C14H3008	326.39	326.1941	51.52	9.26	39.22
C14H3009	342.39	342.1890	49.11	8.83	42.06
C15H 801	204.23	204.0575	88.22	3.95	7.83
C15H 802	220.23	220.0524	81.81	3.66	14.53
C15H 803	236.23	236.0473	76.27	3.41	20.32
C15H 804	252.23	252.0423	71.43	3.20	25.37
C15H 805	268.23	268.0372	67.17	3.01	29.82
C15H 806	284.23	284.0321	63.39	2.84	33.77
C15H 807	300.23	300.0270	60.01	2.69	37.30
C15H 808	316.23	316.0219	56.97	2.55	40.48
C15H 809	332.23	332.0168	54.23	2.43	43.34
C15H1001	206.25	206.0732	87.36	4.89	7.76
C15H1002	222.25	222.0681	81.07	4.54	14.40
C15H1003	238.25	238.0630	75.62	4.23	20.15
C15H1004	254.24	254.0579	70.86	3.96	25.17
C15H1005	270.24	270.0528	66.67	3.73	29.60
C15H1006	286.24	286.0477	62.94	3.52	33.54
C15H1007	302.24	302.0426	59.61	3.33	37.05
C15H1008	318.24	318.0376	56.61	3.17	40.22
C15H1009	334.24	334.0325	53.90	3.02	43.08
C15H1201	208.26	208.0888	86.51	5.81	7.68
C15H1202	224.26	224.0837	80.34	5.39	14.27
C15H1203	240.26	240.0786	74.99	5.03	19.98
C15H1204	256.26	256.0736	70.31	4.72	24.97
C15H1205	272.26	272.0685	66.17	4.44	29.38
C15H1206	288.26	288.0634	62.50	4.20	33.30
C15H1207	304.26	304.0583	59.22	3.98	36.81
C15H1208	320.26	320.0532	56.26	3.78	39.97
C15H1209	336.26	336.0481	53.58	3.60	42.82
C15H1401	210.28	210.1045	85.68	6.71	7.61
C15H1402	226.28	226.0994	79.62	6.24	14.14
C15H1403	242.28	242.0943	74.36	5.82	19.81
C15H1404	258.28	258.0892	69.76	5.46	24.78
C15H1405	274.28	274.0841	65.69	5.15	29.17
C15H1406	290.28	290.0790	62.07	4.86	33.07
C15H1407	306.27	306.0739	58.83	4.61	36.57
C15H1408	322.27	322.0689	55.90	4.38	39.72
C15H1409	338.27	338.0638	53.26	4.17	42.57
C15H1601	212.29	212.1201	84.87	7.60	7.54
C15H1602	228.29	228.1150	78.92	7.06	14.02
C15H1603	244.29	244.1099	73.75	6.60	19.65
C15H1604	260.29	260.1049	69.22	6.20	24.59
C15H1605	276.29	276.0998	65.21	5.84	28.95
C15H1606	292.29	292.0947	61.64	5.52	32.84
C15H1607	308.29	308.0896	58.44	5.23	36.33
C15H1608	324.29	324.0845	55.56	4.97	39.47
C15H1609	340.29	340.0794	52.95	4.74	42.32
C15H1801	214.31	214.1358	84.07	8.47	7.47
C15H1802	230.31	230.1307	78.23	7.88	13.89
C15H1803	246.31	246.1256	73.15	7.37	19.49
C15H1804	262.31	262.1205	68.69	6.92	24.40
C15H1805	278.31	278.1154	64.74	6.52	28.74
C15H1806	294.31	294.1103	61.22	6.16	32.62
C15H1807	310.31	310.1052	58.06	5.85	36.09
C15H1808	326.31	326.1002	55.21	5.56	39.23
C15H1809	342.31	342.0951	52.63	5.30	42.07
C15H2001	216.33	216.1514	83.29	9.32	7.40

C15H2002	232.33	232.1463	77.55	8.68	13.77
C15H2003	248.32	248.1412	72.55	8.12	19.33
C15H2004	264.32	264.1361	68.16	7.63	24.21
C15H2005	280.32	280.1311	64.27	7.19	28.54
C15H2006	296.32	296.1260	60.80	6.80	32.40
C15H2007	312.32	312.1209	57.69	6.45	35.86
C15H2008	328.32	328.1158	54.88	6.14	38.98
C15H2009	344.32	344.1107	52.33	5.85	41.82
C15H2201	218.34	218.1671	82.52	10.16	7.33
C15H2202	234.34	234.1620	76.88	9.46	13.65
C15H2203	250.34	250.1569	71.97	8.86	19.17
C15H2204	266.34	266.1518	67.65	8.33	24.03
C15H2205	282.34	282.1467	63.81	7.85	28.33
C15H2206	298.34	298.1416	60.39	7.43	32.18
C15H2207	314.34	314.1365	57.32	7.05	35.63
C15H2208	330.34	330.1315	54.54	6.71	38.75
C15H2209	346.34	346.1264	52.02	6.40	41.58
C15H2401	220.36	220.1827	81.76	10.98	7.26
C15H2402	236.36	236.1776	76.23	10.24	13.54
C15H2403	252.36	252.1725	71.39	9.59	19.02
C15H2404	268.36	268.1674	67.14	9.01	23.85
C15H2405	284.36	284.1624	63.36	8.51	28.13
C15H2406	300.35	300.1573	59.98	8.05	31.96
C15H2407	316.35	316.1522	56.95	7.65	35.40
C15H2408	332.35	332.1471	54.21	7.28	38.51
C15H2409	348.35	348.1420	51.72	6.94	41.34
C15H2601	222.37	222.1984	81.02	11.79	7.19
C15H2602	238.37	238.1933	75.58	10.99	13.42
C15H2603	254.37	254.1882	70.83	10.30	18.87
C15H2604	270.37	270.1831	66.64	9.69	23.67
C15H2605	286.37	286.1780	62.91	9.15	27.93
C15H2606	302.37	302.1729	59.58	8.67	31.75
C15H2607	318.37	318.1678	56.59	8.23	35.18
C15H2608	334.37	334.1628	53.88	7.84	38.28
C15H2609	350.37	350.1577	51.42	7.48	41.10
C15H2801	224.39	224.2140	80.29	12.58	7.13
C15H2802	240.39	240.2089	74.95	11.74	13.31
C15H2803	256.39	256.2038	70.27	11.01	18.72
C15H2804	272.39	272.1987	66.14	10.36	23.50
C15H2805	288.39	288.1937	62.47	9.79	27.74
C15H2806	304.39	304.1886	59.19	9.27	31.54
C15H2807	320.39	320.1835	56.23	8.81	34.96
C15H2808	336.39	336.1784	53.56	8.39	38.05
C15H2809	352.39	352.1733	51.13	8.01	40.86
C15H3001	226.41	226.2297	79.58	13.36	7.07
C15H3002	242.41	242.2246	74.32	12.47	13.20
C15H3003	258.40	258.2195	69.72	11.70	18.57
C15H3004	274.40	274.2144	65.66	11.02	23.32
C15H3005	290.40	290.2093	62.04	10.41	27.55
C15H3006	306.40	306.2042	58.80	9.87	31.33
C15H3007	322.40	322.1991	55.88	9.38	34.74
C15H3008	338.40	338.1941	53.24	8.94	37.82
C15H3009	354.40	354.1890	50.84	8.53	40.63
C15H3201	228.42	228.2453	78.87	14.12	7.00
C15H3202	244.42	244.2402	73.71	13.20	13.09
C15H3203	260.42	260.2351	69.18	12.39	18.43
C15H3204	276.42	276.2300	65.18	11.67	23.15
C15H3205	292.42	292.2250	61.61	11.03	27.36
C15H3206	308.42	308.2199	58.42	10.46	31.13
C15H3207	324.42	324.2148	55.54	9.94	34.52

C15H3208	340.42	340.2097	52.93	9.48	37.60
C15H3209	356.42	356.2046	50.55	9.05	40.40
C16H 801	216.24	216.0575	88.87	3.73	7.40
C16H 802	232.24	232.0524	82.75	3.47	13.78
C16H 803	248.24	248.0473	77.42	3.25	19.34
C16H 804	264.24	264.0423	72.73	3.05	24.22
C16H 805	280.24	280.0372	68.58	2.88	28.55
C16H 806	296.24	296.0321	64.87	2.72	32.41
C16H 807	312.24	312.0270	61.55	2.58	35.87
C16H 808	328.24	328.0219	58.55	2.46	38.99
C16H 809	344.24	344.0168	55.83	2.34	41.83
C16H1001	218.26	218.0732	88.05	4.62	7.33
C16H1002	234.26	234.0681	82.04	4.30	13.66
C16H1003	250.26	250.0630	76.79	4.03	19.18
C16H1004	266.26	266.0579	72.18	3.79	24.04
C16H1005	282.26	282.0528	68.09	3.57	28.34
C16H1006	298.25	298.0477	64.43	3.38	32.19
C16H1007	314.25	314.0426	61.15	3.21	35.64
C16H1008	330.25	330.0376	58.19	3.05	38.76
C16H1009	346.25	346.0325	55.50	2.91	41.59
C16H1201	220.27	220.0888	87.25	5.49	7.26
C16H1202	236.27	236.0837	81.34	5.12	13.54
C16H1203	252.27	252.0786	76.18	4.79	19.03
C16H1204	268.27	268.0736	71.64	4.51	23.86
C16H1205	284.27	284.0685	67.60	4.25	28.14
C16H1206	300.27	300.0634	64.00	4.03	31.97
C16H1207	316.27	316.0583	60.76	3.82	35.41
C16H1208	332.27	332.0532	57.84	3.64	38.52
C16H1209	348.27	348.0481	55.18	3.47	41.35
C16H1401	222.29	222.1045	86.45	6.35	7.20
C16H1402	238.29	238.0994	80.65	5.92	13.43
C16H1403	254.29	254.0943	75.58	5.55	18.88
C16H1404	270.29	270.0892	71.10	5.22	23.68
C16H1405	286.29	286.0841	67.13	4.93	27.94
C16H1406	302.29	302.0790	63.57	4.67	31.76
C16H1407	318.29	318.0739	60.38	4.43	35.19
C16H1408	334.29	334.0689	57.49	4.22	38.29
C16H1409	350.28	350.0638	54.86	4.03	41.11
C16H1601	224.31	224.1201	85.68	7.19	7.13
C16H1602	240.30	240.1150	79.97	6.71	13.32
C16H1603	256.30	256.1099	74.98	6.29	18.73
C16H1604	272.30	272.1049	70.58	5.92	23.50
C16H1605	288.30	288.0998	66.66	5.59	27.75
C16H1606	304.30	304.0947	63.15	5.30	31.55
C16H1607	320.30	320.0896	60.00	5.04	34.97
C16H1608	336.30	336.0845	57.14	4.80	38.06
C16H1609	352.30	352.0794	54.55	4.58	40.87
C16H1801	226.32	226.1358	84.91	8.02	7.07
C16H1802	242.32	242.1307	79.31	7.49	13.21
C16H1803	258.32	258.1256	74.40	7.02	18.58
C16H1804	274.32	274.1205	70.06	6.61	23.33
C16H1805	290.32	290.1154	66.20	6.25	27.55
C16H1806	306.32	306.1103	62.74	5.92	31.34
C16H1807	322.32	322.1052	59.62	5.63	34.75
C16H1808	338.32	338.1002	56.80	5.36	37.83
C16H1809	354.32	354.0951	54.24	5.12	40.64
C16H2001	228.34	228.1514	84.16	8.83	7.01
C16H2002	244.34	244.1463	78.65	8.25	13.10
C16H2003	260.34	260.1412	73.82	7.74	18.44
C16H2004	276.34	276.1361	69.55	7.30	23.16

C16H2005	292.33	292.1311	65.74	6.90	27.36
C16H2006	308.33	308.1260	62.33	6.54	31.13
C16H2007	324.33	324.1209	59.25	6.22	34.53
C16H2008	340.33	340.1158	56.47	5.92	37.61
C16H2009	356.33	356.1107	53.93	5.66	40.41
C16H2201	230.35	230.1671	83.43	9.63	6.95
C16H2202	246.35	246.1620	78.01	9.00	12.99
C16H2203	262.35	262.1569	73.25	8.45	18.30
C16H2204	278.35	278.1518	69.04	7.97	22.99
C16H2205	294.35	294.1467	65.29	7.53	27.18
C16H2206	310.35	310.1416	61.92	7.15	30.93
C16H2207	326.35	326.1365	58.89	6.79	34.32
C16H2208	342.35	342.1315	56.14	6.48	37.39
C16H2209	358.35	358.1264	53.63	6.19	40.18
C16H2401	232.37	232.1827	82.70	10.41	6.89
C16H2402	248.37	248.1776	77.38	9.74	12.88
C16H2403	264.37	264.1725	72.69	9.15	18.16
C16H2404	280.37	280.1674	68.55	8.63	22.83
C16H2405	296.37	296.1624	64.84	8.16	26.99
C16H2406	312.37	312.1573	61.52	7.74	30.73
C16H2407	328.37	328.1522	58.53	7.37	34.11
C16H2408	344.36	344.1471	55.81	7.02	37.17
C16H2409	360.36	360.1420	53.33	6.71	39.96
C16H2601	234.39	234.1984	81.99	11.18	6.83
C16H2602	250.38	250.1933	76.75	10.47	12.78
C16H2603	266.38	266.1882	72.14	9.84	18.02
C16H2604	282.38	282.1831	68.06	9.28	22.66
C16H2605	298.38	298.1780	64.41	8.78	26.81
C16H2606	314.38	314.1729	61.13	8.34	30.53
C16H2607	330.38	330.1678	58.17	7.93	33.90
C16H2608	346.38	346.1628	55.48	7.57	36.95
C16H2609	362.38	362.1577	53.03	7.23	39.74
C16H2801	236.40	236.2140	81.29	11.94	6.77
C16H2802	252.40	252.2089	76.14	11.18	12.68
C16H2803	268.40	268.2038	71.60	10.52	17.88
C16H2804	284.40	284.1987	67.57	9.92	22.50
C16H2805	300.40	300.1937	63.97	9.40	26.63
C16H2806	316.40	316.1886	60.74	8.92	30.34
C16H2807	332.40	332.1835	57.82	8.49	33.69
C16H2808	348.40	348.1784	55.16	8.10	36.74
C16H2809	364.40	364.1733	52.74	7.75	39.52
C16H3001	238.42	238.2297	80.61	12.68	6.71
C16H3002	254.42	254.2246	75.54	11.89	12.58
C16H3003	270.42	270.2195	71.07	11.18	17.75
C16H3004	286.42	286.2144	67.10	10.56	22.34
C16H3005	302.41	302.2093	63.55	10.00	26.45
C16H3006	318.41	318.2042	60.35	9.50	30.15
C16H3007	334.41	334.1991	57.47	9.04	33.49
C16H3008	350.41	350.1941	54.84	8.63	36.53
C16H3009	366.41	366.1890	52.45	8.25	39.30
C16H3201	240.43	240.2453	79.93	13.42	6.65
C16H3202	256.43	256.2402	74.94	12.58	12.48
C16H3203	272.43	272.2351	70.54	11.84	17.62
C16H3204	288.43	288.2300	66.63	11.18	22.19
C16H3205	304.43	304.2250	63.13	10.60	26.28
C16H3206	320.43	320.2199	59.98	10.07	29.96
C16H3207	336.43	336.2148	57.12	9.59	33.29
C16H3208	352.43	352.2097	54.53	9.15	36.32
C16H3209	368.43	368.2046	52.16	8.75	39.08
C16H3401	242.45	242.2610	79.27	14.14	6.60

C16H3402	258.45	258.2559	74.36	13.26	12.38
C16H3403	274.45	274.2508	70.02	12.49	17.49
C16H3404	290.45	290.2457	66.17	11.80	22.03
C16H3405	306.45	306.2406	62.71	11.18	26.10
C16H3406	322.45	322.2355	59.60	10.63	29.77
C16H3407	338.45	338.2304	56.78	10.13	33.09
C16H3408	354.44	354.2253	54.22	9.67	36.11
C16H3409	370.44	370.2203	51.88	9.25	38.87
C17H 801	228.25	228.0575	89.46	3.53	7.01
C17H 802	244.25	244.0524	83.60	3.30	13.10
C17H 803	260.25	260.0473	78.46	3.10	18.44
C17H 804	276.25	276.0423	73.91	2.92	23.17
C17H 805	292.25	292.0372	69.87	2.76	27.37
C17H 806	308.25	308.0321	66.24	2.62	31.14
C17H 807	324.25	324.0270	62.97	2.49	34.54
C17H 808	340.25	340.0219	60.01	2.37	37.62
C17H 809	356.25	356.0168	57.32	2.26	40.42
C17H1001	230.27	230.0732	88.67	4.38	6.95
C17H1002	246.27	246.0681	82.91	4.09	12.99
C17H1003	262.27	262.0630	77.86	3.84	18.30
C17H1004	278.27	278.0579	73.38	3.62	23.00
C17H1005	294.27	294.0528	69.39	3.43	27.19
C17H1006	310.27	310.0477	65.81	3.25	30.94
C17H1007	326.27	326.0426	62.58	3.09	34.33
C17H1008	342.26	342.0376	59.66	2.95	37.40
C17H1009	358.26	358.0325	56.99	2.81	40.19
C17H1201	232.28	232.0888	87.90	5.21	6.89
C17H1202	248.28	248.0837	82.24	4.87	12.89
C17H1203	264.28	264.0786	77.26	4.58	18.16
C17H1204	280.28	280.0736	72.85	4.32	22.83
C17H1205	296.28	296.0685	68.92	4.08	27.00
C17H1206	312.28	312.0634	65.39	3.87	30.74
C17H1207	328.28	328.0583	62.20	3.68	34.12
C17H1208	344.28	344.0532	59.31	3.51	37.18
C17H1209	360.28	360.0481	56.68	3.36	39.97
C17H1401	234.30	234.1045	87.15	6.02	6.83
C17H1402	250.30	250.0994	81.58	5.64	12.78
C17H1403	266.30	266.0943	76.68	5.30	18.02
C17H1404	282.30	282.0892	72.33	5.00	22.67
C17H1405	298.30	298.0841	68.45	4.73	26.82
C17H1406	314.30	314.0790	64.97	4.49	30.54
C17H1407	330.30	330.0739	61.82	4.27	33.91
C17H1408	346.30	346.0689	58.96	4.08	36.96
C17H1409	362.30	362.0638	56.36	3.90	39.75
C17H1601	236.32	236.1201	86.41	6.82	6.77
C17H1602	252.32	252.1150	80.93	6.39	12.68
C17H1603	268.32	268.1099	76.10	6.01	17.89
C17H1604	284.31	284.1049	71.82	5.67	22.51
C17H1605	300.31	300.0998	67.99	5.37	26.64
C17H1606	316.31	316.0947	64.55	5.10	30.35
C17H1607	332.31	332.0896	61.44	4.85	33.70
C17H1608	348.31	348.0845	58.62	4.63	36.75
C17H1609	364.31	364.0794	56.05	4.43	39.53
C17H1801	238.33	238.1358	85.67	7.61	6.71
C17H1802	254.33	254.1307	80.28	7.13	12.58
C17H1803	270.33	270.1256	75.53	6.71	17.76
C17H1804	286.33	286.1205	71.31	6.34	22.35
C17H1805	302.33	302.1154	67.54	6.00	26.46
C17H1806	318.33	318.1103	64.14	5.70	30.16
C17H1807	334.33	334.1052	61.07	5.43	33.50

C17H1808	350.33	350.1002	58.29	5.18	36.54
C17H1809	366.33	366.0951	55.74	4.95	39.31
C17H2001	240.35	240.1514	84.96	8.39	6.66
C17H2002	256.35	256.1463	79.65	7.86	12.48
C17H2003	272.35	272.1412	74.97	7.40	17.62
C17H2004	288.35	288.1361	70.81	6.99	22.19
C17H2005	304.35	304.1311	67.09	6.62	26.28
C17H2006	320.35	320.1260	63.74	6.29	29.97
C17H2007	336.34	336.1209	60.71	5.99	33.30
C17H2008	352.34	352.1158	57.95	5.72	36.33
C17H2009	368.34	368.1107	55.43	5.47	39.09
C17H2201	242.36	242.1671	84.25	9.15	6.60
C17H2202	258.36	258.1620	79.03	8.58	12.39
C17H2203	274.36	274.1569	74.42	8.08	17.49
C17H2204	290.36	290.1518	70.32	7.64	22.04
C17H2205	306.36	306.1467	66.65	7.24	26.11
C17H2206	322.36	322.1416	63.34	6.88	29.78
C17H2207	338.36	338.1365	60.35	6.55	33.10
C17H2208	354.36	354.1315	57.62	6.26	36.12
C17H2209	370.36	370.1264	55.13	5.99	38.88
C17H2401	244.38	244.1827	83.55	9.90	6.55
C17H2402	260.38	260.1776	78.42	9.29	12.29
C17H2403	276.38	276.1725	73.88	8.75	17.37
C17H2404	292.38	292.1674	69.84	8.27	21.89
C17H2405	308.38	308.1624	66.21	7.84	25.94
C17H2406	324.38	324.1573	62.95	7.46	29.59
C17H2407	340.38	340.1522	59.99	7.11	32.90
C17H2408	356.38	356.1471	57.30	6.79	35.92
C17H2409	372.38	372.1420	54.83	6.50	38.67
C17H2601	246.40	246.1984	82.87	10.64	6.49
C17H2602	262.40	262.1933	77.82	9.99	12.19
C17H2603	278.39	278.1882	73.35	9.41	17.24
C17H2604	294.39	294.1831	69.36	8.90	21.74
C17H2605	310.39	310.1780	65.78	8.44	25.77
C17H2606	326.39	326.1729	62.56	8.03	29.41
C17H2607	342.39	342.1678	59.64	7.65	32.71
C17H2608	358.39	358.1628	56.97	7.31	35.71
C17H2609	374.39	374.1577	54.54	7.00	38.46
C17H2801	248.41	248.2140	82.20	11.36	6.44
C17H2802	264.41	264.2089	77.22	10.67	12.10
C17H2803	280.41	280.2038	72.82	10.06	17.12
C17H2804	296.41	296.1987	68.89	9.52	21.59
C17H2805	312.41	312.1937	65.36	9.03	25.61
C17H2806	328.41	328.1886	62.18	8.59	29.23
C17H2807	344.41	344.1835	59.29	8.19	32.52
C17H2808	360.41	360.1784	56.66	7.83	35.51
C17H2809	376.41	376.1733	54.25	7.50	38.25
C17H3001	250.43	250.2297	81.54	12.07	6.39
C17H3002	266.43	266.2246	76.64	11.35	12.01
C17H3003	282.43	282.2195	72.30	10.71	16.99
C17H3004	298.43	298.2144	68.42	10.13	21.45
C17H3005	314.43	314.2093	64.94	9.62	25.44
C17H3006	330.43	330.2042	61.80	9.15	29.05
C17H3007	346.42	346.1991	58.94	8.73	32.33
C17H3008	362.42	362.1941	56.34	8.34	35.32
C17H3009	378.42	378.1890	53.96	7.99	38.05
C17H3201	252.44	252.2453	80.89	12.78	6.34
C17H3202	268.44	268.2402	76.06	12.02	11.92
C17H3203	284.44	284.2351	71.79	11.34	16.87
C17H3204	300.44	300.2300	67.96	10.74	21.30

C17H3205	316.44	316.2250	64.53	10.19	25.28
C17H3206	332.44	332.2199	61.42	9.70	28.88
C17H3207	348.44	348.2148	58.60	9.26	32.14
C17H3208	364.44	364.2097	56.03	8.85	35.12
C17H3209	380.44	380.2046	53.67	8.48	37.85
C17H3401	254.46	254.2610	80.24	13.47	6.29
C17H3402	270.46	270.2559	75.50	12.67	11.83
C17H3403	286.46	286.2508	71.28	11.96	16.76
C17H3404	302.46	302.2457	67.51	11.33	21.16
C17H3405	318.46	318.2406	64.12	10.76	25.12
C17H3406	334.46	334.2355	61.05	10.25	28.70
C17H3407	350.46	350.2304	58.26	9.78	31.96
C17H3408	366.46	366.2253	55.72	9.35	34.93
C17H3409	382.46	382.2203	53.39	8.96	37.65
C17H3601	256.48	256.2766	79.61	14.15	6.24
C17H3602	272.48	272.2715	74.94	13.32	11.74
C17H3603	288.47	288.2664	70.78	12.58	16.64
C17H3604	304.47	304.2613	67.06	11.92	21.02
C17H3605	320.47	320.2563	63.71	11.32	24.96
C17H3606	336.47	336.2512	60.69	10.78	28.53
C17H3607	352.47	352.2461	57.93	10.29	31.77
C17H3608	368.47	368.2410	55.42	9.85	34.74
C17H3609	384.47	384.2359	53.11	9.44	37.45
C18H1001	242.28	242.0732	89.24	4.16	6.60
C18H1002	258.28	258.0681	83.71	3.90	12.39
C18H1003	274.28	274.0630	78.83	3.67	17.50
C18H1004	290.28	290.0579	74.48	3.47	22.05
C18H1005	306.28	306.0528	70.59	3.29	26.12
C18H1006	322.28	322.0477	67.09	3.13	29.79
C18H1007	338.28	338.0426	63.91	2.98	33.11
C18H1008	354.28	354.0376	61.03	2.85	36.13
C18H1009	370.27	370.0325	58.39	2.72	38.89
C18H1201	244.30	244.0888	88.50	4.95	6.55
C18H1202	260.30	260.0837	83.06	4.65	12.29
C18H1203	276.29	276.0786	78.25	4.38	17.37
C18H1204	292.29	292.0736	73.97	4.14	21.89
C18H1205	308.29	308.0685	70.13	3.92	25.95
C18H1206	324.29	324.0634	66.67	3.73	29.60
C18H1207	340.29	340.0583	63.53	3.55	32.91
C18H1208	356.29	356.0532	60.68	3.39	35.92
C18H1209	372.29	372.0481	58.07	3.25	38.68
C18H1401	246.31	246.1045	87.78	5.73	6.50
C18H1402	262.31	262.0994	82.42	5.38	12.20
C18H1403	278.31	278.0943	77.68	5.07	17.25
C18H1404	294.31	294.0892	73.46	4.79	21.74
C18H1405	310.31	310.0841	69.67	4.55	25.78
C18H1406	326.31	326.0790	66.26	4.32	29.42
C18H1407	342.31	342.0739	63.16	4.12	32.72
C18H1408	358.31	358.0689	60.34	3.94	35.72
C18H1409	374.31	374.0638	57.76	3.77	38.47
C18H1601	248.33	248.1201	87.06	6.49	6.44
C18H1602	264.33	264.1150	81.79	6.10	12.11
C18H1603	280.33	280.1099	77.12	5.75	17.12
C18H1604	296.33	296.1049	72.96	5.44	21.60
C18H1605	312.33	312.0998	69.22	5.16	25.61
C18H1606	328.32	328.0947	65.85	4.91	29.24
C18H1607	344.32	344.0896	62.79	4.68	32.53
C18H1608	360.32	360.0845	60.00	4.48	35.52
C18H1609	376.32	376.0794	57.45	4.29	38.26
C18H1801	250.34	250.1358	86.36	7.25	6.39

C18H1802	266.34	266.1307	81.17	6.81	12.01
C18H1803	282.34	282.1256	76.57	6.43	17.00
C18H1804	298.34	298.1205	72.47	6.08	21.45
C18H1805	314.34	314.1154	68.78	5.77	25.45
C18H1806	330.34	330.1103	65.45	5.49	29.06
C18H1807	346.34	346.1052	62.42	5.24	32.34
C18H1808	362.34	362.1002	59.67	5.01	35.32
C18H1809	378.34	378.0951	57.14	4.80	38.06
C18H2001	252.36	252.1514	85.67	7.99	6.34
C18H2002	268.36	268.1463	80.56	7.51	11.92
C18H2003	284.36	284.1412	76.03	7.09	16.88
C18H2004	300.36	300.1361	71.98	6.71	21.31
C18H2005	316.36	316.1311	68.34	6.37	25.29
C18H2006	332.36	332.1260	65.05	6.07	28.88
C18H2007	348.36	348.1209	62.06	5.79	32.15
C18H2008	364.36	364.1158	59.34	5.53	35.13
C18H2009	380.35	380.1107	56.84	5.30	37.86
C18H2201	254.38	254.1671	84.99	8.72	6.29
C18H2202	270.37	270.1620	79.96	8.20	11.83
C18H2203	286.37	286.1569	75.50	7.74	16.76
C18H2204	302.37	302.1518	71.50	7.33	21.17
C18H2205	318.37	318.1467	67.91	6.97	25.13
C18H2206	334.37	334.1416	64.66	6.63	28.71
C18H2207	350.37	350.1365	61.71	6.33	31.96
C18H2208	366.37	366.1315	59.01	6.05	34.94
C18H2209	382.37	382.1264	56.54	5.80	37.66
C18H2401	256.39	256.1827	84.32	9.44	6.24
C18H2402	272.39	272.1776	79.37	8.88	11.75
C18H2403	288.39	288.1725	74.97	8.39	16.64
C18H2404	304.39	304.1674	71.03	7.95	21.02
C18H2405	320.39	320.1624	67.48	7.55	24.97
C18H2406	336.39	336.1573	64.27	7.19	28.54
C18H2407	352.39	352.1522	61.35	6.86	31.78
C18H2408	368.39	368.1471	58.69	6.57	34.74
C18H2409	384.39	384.1420	56.25	6.29	37.46
C18H2601	258.41	258.1984	83.67	10.14	6.19
C18H2602	274.41	274.1933	78.79	9.55	11.66
C18H2603	290.41	290.1882	74.45	9.02	16.53
C18H2604	306.41	306.1831	70.56	8.55	20.89
C18H2605	322.40	322.1780	67.06	8.13	24.81
C18H2606	338.40	338.1729	63.89	7.74	28.37
C18H2607	354.40	354.1678	61.00	7.39	31.60
C18H2608	370.40	370.1628	58.37	7.08	34.56
C18H2609	386.40	386.1577	55.95	6.78	37.27
C18H2801	260.42	260.2140	83.02	10.84	6.14
C18H2802	276.42	276.2089	78.21	10.21	11.58
C18H2803	292.42	292.2038	73.93	9.65	16.41
C18H2804	308.42	308.1987	70.10	9.15	20.75
C18H2805	324.42	324.1937	66.64	8.70	24.66
C18H2806	340.42	340.1886	63.51	8.29	28.20
C18H2807	356.42	356.1835	60.66	7.92	31.42
C18H2808	372.42	372.1784	58.05	7.58	34.37
C18H2809	388.42	388.1733	55.66	7.27	37.07
C18H3001	262.44	262.2297	82.38	11.52	6.10
C18H3002	278.44	278.2246	77.65	10.86	11.49
C18H3003	294.44	294.2195	73.43	10.27	16.30
C18H3004	310.44	310.2144	69.64	9.74	20.62
C18H3005	326.44	326.2093	66.23	9.26	24.51
C18H3006	342.44	342.2042	63.14	8.83	28.03
C18H3007	358.44	358.1991	60.32	8.44	31.25

C18H3008	374.44	374.1941	57.74	8.08	34.18
C18H3009	390.43	390.1890	55.37	7.74	36.88
C18H3201	264.46	264.2453	81.75	12.20	6.05
C18H3202	280.45	280.2402	77.09	11.50	11.41
C18H3203	296.45	296.2351	72.93	10.88	16.19
C18H3204	312.45	312.2300	69.19	10.32	20.48
C18H3205	328.45	328.2250	65.82	9.82	24.36
C18H3206	344.45	344.2199	62.77	9.36	27.87
C18H3207	360.45	360.2148	59.98	8.95	31.07
C18H3208	376.45	376.2097	57.43	8.57	34.00
C18H3209	392.45	392.2046	55.09	8.22	36.69
C18H3401	266.47	266.2610	81.13	12.86	6.00
C18H3402	282.47	282.2559	76.54	12.13	11.33
C18H3403	298.47	298.2508	72.44	11.48	16.08
C18H3404	314.47	314.2457	68.75	10.90	20.35
C18H3405	330.47	330.2406	65.42	10.37	24.21
C18H3406	346.47	346.2355	62.40	9.89	27.71
C18H3407	362.47	362.2304	59.65	9.45	30.90
C18H3408	378.47	378.2253	57.13	9.06	33.82
C18H3409	394.47	394.2203	54.81	8.69	36.50
C18H3601	268.49	268.2766	80.53	13.52	5.96
C18H3602	284.49	284.2715	76.00	12.76	11.25
C18H3603	300.49	300.2664	71.95	12.08	15.97
C18H3604	316.49	316.2613	68.31	11.47	20.22
C18H3605	332.48	332.2563	65.03	10.91	24.06
C18H3606	348.48	348.2512	62.04	10.41	27.55
C18H3607	364.48	364.2461	59.32	9.96	30.73
C18H3608	380.48	380.2410	56.82	9.54	33.64
C18H3609	396.48	396.2359	54.53	9.15	36.32
C18H3801	270.50	270.2922	79.93	14.16	5.91
C18H3802	286.50	286.2872	75.46	13.37	11.17
C18H3803	302.50	302.2821	71.47	12.66	15.87
C18H3804	318.50	318.2770	67.88	12.03	20.09
C18H3805	334.50	334.2719	64.63	11.45	23.92
C18H3806	350.50	350.2668	61.68	10.93	27.39
C18H3807	366.50	366.2617	58.99	10.45	30.56
C18H3808	382.50	382.2566	56.52	10.01	33.46
C18H3809	398.50	398.2516	54.25	9.61	36.13
C19H1001	254.29	254.0732	89.74	3.96	6.29
C19H1002	270.29	270.0681	84.43	3.73	11.84
C19H1003	286.29	286.0630	79.71	3.52	16.77
C19H1004	302.29	302.0579	75.49	3.33	21.17
C19H1005	318.29	318.0528	71.70	3.17	25.13
C19H1006	334.29	334.0477	68.27	3.02	28.72
C19H1007	350.29	350.0426	65.15	2.88	31.97
C19H1008	366.29	366.0376	62.30	2.75	34.94
C19H1009	382.29	382.0325	59.70	2.64	37.67
C19H1201	256.31	256.0888	89.04	4.72	6.24
C19H1202	272.31	272.0837	83.81	4.44	11.75
C19H1203	288.31	288.0786	79.16	4.20	16.65
C19H1204	304.31	304.0736	74.99	3.97	21.03
C19H1205	320.30	320.0685	71.25	3.78	24.98
C19H1206	336.30	336.0634	67.86	3.60	28.54
C19H1207	352.30	352.0583	64.78	3.43	31.79
C19H1208	368.30	368.0532	61.96	3.28	34.75
C19H1209	384.30	384.0481	59.38	3.15	37.47
C19H1401	258.32	258.1045	88.34	5.46	6.19
C19H1402	274.32	274.0994	83.19	5.14	11.66
C19H1403	290.32	290.0943	78.61	4.86	16.53
C19H1404	306.32	306.0892	74.50	4.61	20.89
C19H1405	322.32	322.0841	70.80	4.38	24.82
C19H1406	338.32	338.0790	67.45	4.17	28.37
C19H1407	354.32	354.0739	64.41	3.98	31.61
C19H1408	370.32	370.0689	61.63	3.81	34.56
C19H1409	386.32	386.0638	59.07	3.65	37.27
C19H1601	260.34	260.1201	87.66	6.19	6.15
C19H1602	276.34	276.1150	82.58	5.84	11.58
C19H1603	292.34	292.1099	78.06	5.52	16.42
C19H1604	308.34	308.1049	74.01	5.23	20.76
C19H1605	324.34	324.0998	70.36	4.97	24.66
C19H1606	340.34	340.0947	67.05	4.74	28.21
C19H1607	356.34	356.0896	64.04	4.53	31.43
C19H1608	372.33	372.0845	61.29	4.33	34.38
C19H1609	388.33	388.0794	58.77	4.15	37.08
C19H1801	262.35	262.1358	86.99	6.92	6.10
C19H1802	278.35	278.1307	81.99	6.52	11.50
C19H1803	294.35	294.1256	77.53	6.16	16.31
C19H1804	310.35	310.1205	73.53	5.85	20.62
C19H1805	326.35	326.1154	69.93	5.56	24.51
C19H1806	342.35	342.1103	66.66	5.30	28.04
C19H1807	358.35	358.1052	63.68	5.06	31.25
C19H1808	374.35	374.1002	60.96	4.85	34.19
C19H1809	390.35	390.0951	58.46	4.65	36.89
C19H2001	264.37	264.1514	86.32	7.63	6.05
C19H2002	280.37	280.1463	81.40	7.19	11.41
C19H2003	296.37	296.1412	77.00	6.80	16.20
C19H2004	312.37	312.1361	73.06	6.45	20.49
C19H2005	328.37	328.1311	69.50	6.14	24.36
C19H2006	344.37	344.1260	66.27	5.85	27.88
C19H2007	360.37	360.1209	63.33	5.59	31.08
C19H2008	376.37	376.1158	60.64	5.36	34.01
C19H2009	392.37	392.1107	58.16	5.14	36.70
C19H2201	266.39	266.1671	85.67	8.32	6.01
C19H2202	282.39	282.1620	80.82	7.85	11.33
C19H2203	298.39	298.1569	76.48	7.43	16.09
C19H2204	314.38	314.1518	72.59	7.05	20.36
C19H2205	330.38	330.1467	69.07	6.71	24.21
C19H2206	346.38	346.1416	65.88	6.40	27.71
C19H2207	362.38	362.1365	62.98	6.12	30.91
C19H2208	378.38	378.1315	60.31	5.86	33.83
C19H2209	394.38	394.1264	57.87	5.62	36.51
C19H2401	268.40	268.1827	85.03	9.01	5.96
C19H2402	284.40	284.1776	80.24	8.51	11.25
C19H2403	300.40	300.1725	75.97	8.05	15.98
C19H2404	316.40	316.1674	72.13	7.65	20.23
C19H2405	332.40	332.1624	68.66	7.28	24.07
C19H2406	348.40	348.1573	65.50	6.94	27.55
C19H2407	364.40	364.1522	62.63	6.64	30.73
C19H2408	380.40	380.1471	59.99	6.36	33.65
C19H2409	396.40	396.1420	57.57	6.10	36.33
C19H2601	270.42	270.1984	84.39	9.69	5.92
C19H2602	286.42	286.1933	79.68	9.15	11.17
C19H2603	302.42	302.1882	75.46	8.67	15.87
C19H2604	318.42	318.1831	71.67	8.23	20.10
C19H2605	334.42	334.1780	68.24	7.84	23.92
C19H2606	350.42	350.1729	65.13	7.48	27.40
C19H2607	366.41	366.1678	62.28	7.15	30.57
C19H2608	382.41	382.1628	59.68	6.85	33.47
C19H2609	398.41	398.1577	57.28	6.58	36.14
C19H2801	272.43	272.2140	83.77	10.36	5.87

C19H2802	288.43	288.2089	79.12	9.78	11.09
C19H2803	304.43	304.2038	74.96	9.27	15.77
C19H2804	320.43	320.1987	71.22	8.81	19.97
C19H2805	336.43	336.1937	67.83	8.39	23.78
C19H2806	352.43	352.1886	64.75	8.01	27.24
C19H2807	368.43	368.1835	61.94	7.66	30.40
C19H2808	384.43	384.1784	59.36	7.34	33.29
C19H2809	400.43	400.1733	56.99	7.05	35.96
C19H3001	274.45	274.2297	83.15	11.02	5.83
C19H3002	290.45	290.2246	78.57	10.41	11.02
C19H3003	306.45	306.2195	74.47	9.87	15.66
C19H3004	322.45	322.2144	70.77	9.38	19.85
C19H3005	338.45	338.2093	67.43	8.93	23.64
C19H3006	354.45	354.2042	64.39	8.53	27.08
C19H3007	370.45	370.1991	61.60	8.16	30.23
C19H3008	386.45	386.1941	59.05	7.82	33.12
C19H3009	402.45	402.1890	56.71	7.51	35.78
C19H3201	276.47	276.2453	82.55	11.67	5.79
C19H3202	292.47	292.2402	78.03	11.03	10.94
C19H3203	308.47	308.2351	73.98	10.46	15.56
C19H3204	324.46	324.2300	70.33	9.94	19.72
C19H3205	340.46	340.2250	67.03	9.47	23.50
C19H3206	356.46	356.2199	64.02	9.05	26.93
C19H3207	372.46	372.2148	61.27	8.66	30.07
C19H3208	388.46	388.2097	58.75	8.30	32.95
C19H3209	404.46	404.2046	56.42	7.97	35.60
C19H3401	278.48	278.2610	81.95	12.31	5.75
C19H3402	294.48	294.2559	77.50	11.64	10.87
C19H3403	310.48	310.2508	73.50	11.04	15.46
C19H3404	326.48	326.2457	69.90	10.50	19.60
C19H3405	342.48	342.2406	66.64	10.01	23.36
C19H3406	358.48	358.2355	63.66	9.56	26.78
C19H3407	374.48	374.2304	60.94	9.15	29.91
C19H3408	390.48	390.2253	58.44	8.78	32.78
C19H3409	406.48	406.2203	56.14	8.43	35.42
C19H3601	280.50	280.2766	81.36	12.94	5.70
C19H3602	296.50	296.2715	76.97	12.24	10.79
C19H3603	312.50	312.2664	73.03	11.61	15.36
C19H3604	328.50	328.2613	69.47	11.05	19.48
C19H3605	344.50	344.2563	66.25	10.53	23.22
C19H3606	360.50	360.2512	63.31	10.07	26.63
C19H3607	376.49	376.2461	60.61	9.64	29.75
C19H3608	392.49	392.2410	58.14	9.25	32.61
C19H3609	408.49	408.2359	55.87	8.88	35.25
C19H3801	282.51	282.2922	80.78	13.56	5.66
C19H3802	298.51	298.2872	76.45	12.83	10.72
C19H3803	314.51	314.2821	72.56	12.18	15.26
C19H3804	330.51	330.2770	69.05	11.59	19.36
C19H3805	346.51	346.2719	65.86	11.05	23.09
C19H3806	362.51	362.2668	62.95	10.57	26.48
C19H3807	378.51	378.2617	60.29	10.12	29.59
C19H3808	394.51	394.2566	57.85	9.71	32.44
C19H3809	410.51	410.2516	55.59	9.33	35.08
C19H4001	284.53	284.3079	80.21	14.17	5.62
C19H4002	300.53	300.3028	75.94	13.42	10.65
C19H4003	316.53	316.2977	72.10	12.74	15.16
C19H4004	332.53	332.2926	68.63	12.12	19.25
C19H4005	348.53	348.2876	65.48	11.57	22.95
C19H4006	364.53	364.2825	62.60	11.06	26.33
C19H4007	380.53	380.2774	59.97	10.60	29.43
C19H4008	396.53	396.2723	57.55	10.17	32.28
C19H4009	412.53	412.2672	55.32	9.77	34.91
C20H1001	266.30	266.0732	90.21	3.79	6.01
C20H1002	282.30	282.0681	85.09	3.57	11.33
C20H1003	298.30	298.0630	80.53	3.38	16.09
C20H1004	314.30	314.0579	76.43	3.21	20.36
C20H1005	330.30	330.0528	72.73	3.05	24.22
C20H1006	346.30	346.0477	69.37	2.91	27.72
C20H1007	362.30	362.0426	66.31	2.78	30.91
C20H1008	378.30	378.0376	63.50	2.66	33.83
C20H1009	394.30	394.0325	60.92	2.56	36.52
C20H1201	268.32	268.0888	89.53	4.51	5.96
C20H1202	284.32	284.0837	84.49	4.25	11.25
C20H1203	300.32	300.0786	79.99	4.03	15.98
C20H1204	316.32	316.0736	75.94	3.82	20.23
C20H1205	332.32	332.0685	72.29	3.64	24.07
C20H1206	348.32	348.0634	68.97	3.47	27.56
C20H1207	364.31	364.0583	65.94	3.32	30.74
C20H1208	380.31	380.0532	63.16	3.18	33.66
C20H1209	396.31	396.0481	60.61	3.05	36.33
C20H1401	270.33	270.1045	88.86	5.22	5.92
C20H1402	286.33	286.0994	83.90	4.93	11.18
C20H1403	302.33	302.0943	79.46	4.67	15.88
C20H1404	318.33	318.0892	75.46	4.43	20.10
C20H1405	334.33	334.0841	71.85	4.22	23.93
C20H1406	350.33	350.0790	68.57	4.03	27.40
C20H1407	366.33	366.0739	65.58	3.85	30.57
C20H1408	382.33	382.0689	62.83	3.69	33.48
C20H1409	398.33	398.0638	60.31	3.54	36.15
C20H1601	272.35	272.1201	88.20	5.92	5.87
C20H1602	288.35	288.1150	83.31	5.59	11.10
C20H1603	304.35	304.1099	78.93	5.30	15.77
C20H1604	320.35	320.1048	74.99	5.03	19.98
C20H1605	336.35	336.0998	71.42	4.79	23.78
C20H1606	352.35	352.0947	68.18	4.58	27.24
C20H1607	368.35	368.0896	65.22	4.38	30.41
C20H1608	384.35	384.0845	62.50	4.20	33.30
C20H1609	400.35	400.0794	60.00	4.03	35.97
C20H1801	274.37	274.1358	87.56	6.61	5.83
C20H1802	290.37	290.1307	82.73	6.25	11.02
C20H1803	306.36	306.1256	78.41	5.92	15.67
C20H1804	322.36	322.1205	74.52	5.63	19.85
C20H1805	338.36	338.1154	71.00	5.36	23.64
C20H1806	354.36	354.1103	67.79	5.12	27.09
C20H1807	370.36	370.1052	64.86	4.90	30.24
C20H1808	386.36	386.1002	62.18	4.70	33.13
C20H1809	402.36	402.0951	59.70	4.51	35.79
C20H2001	276.38	276.1514	86.92	7.29	5.79
C20H2002	292.38	292.1463	82.16	6.89	10.94
C20H2003	308.38	308.1412	77.90	6.54	15.56
C20H2004	324.38	324.1361	74.06	6.21	19.73
C20H2005	340.38	340.1311	70.58	5.92	23.50
C20H2006	356.38	356.1260	67.41	5.66	26.94
C20H2007	372.38	372.1209	64.51	5.41	30.08
C20H2008	388.38	388.1158	61.85	5.19	32.96
C20H2009	404.38	404.1107	59.41	4.99	35.61
C20H2201	278.40	278.1671	86.29	7.97	5.75
C20H2202	294.40	294.1620	81.60	7.53	10.87
C20H2203	310.40	310.1569	77.39	7.14	15.46
C20H2204	326.40	326.1518	73.60	6.79	19.61

10/21/66 083

C20H2205	342.40	342.1467	70.16	6.48	23.36
C20H2206	358.39	358.1416	67.03	6.19	26.79
C20H2207	374.39	374.1365	64.16	5.92	29.91
C20H2208	390.39	390.1315	61.53	5.68	32.79
C20H2209	406.39	406.1264	59.11	5.46	35.43
C20H2401	280.41	280.1827	85.67	8.63	5.71
C20H2402	296.41	296.1776	81.04	8.16	10.80
C20H2403	312.41	312.1725	76.89	7.74	15.36
C20H2404	328.41	328.1674	73.15	7.37	19.49
C20H2405	344.41	344.1624	69.75	7.02	23.23
C20H2406	360.41	360.1573	66.65	6.71	26.64
C20H2407	376.41	376.1522	63.82	6.43	29.75
C20H2408	392.41	392.1471	61.22	6.16	32.62
C20H2409	408.41	408.1420	58.82	5.92	35.26
C20H2601	282.43	282.1984	85.06	9.28	5.66
C20H2602	298.43	298.1933	80.50	8.78	10.72
C20H2603	314.43	314.1882	76.40	8.33	15.27
C20H2604	330.43	330.1831	72.70	7.93	19.37
C20H2605	346.43	346.1780	69.34	7.57	23.09
C20H2606	362.43	362.1729	66.28	7.23	26.49
C20H2607	378.43	378.1678	63.48	6.93	29.60
C20H2608	394.43	394.1628	60.90	6.64	32.45
C20H2609	410.42	410.1577	58.53	6.39	35.08
C20H2801	284.45	284.2140	84.45	9.92	5.62
C20H2802	300.44	300.2089	79.96	9.39	10.65
C20H2803	316.44	316.2038	75.91	8.92	15.17
C20H2804	332.44	332.1987	72.26	8.49	19.25
C20H2805	348.44	348.1937	68.94	8.10	22.96
C20H2806	364.44	364.1886	65.92	7.74	26.34
C20H2807	380.44	380.1835	63.14	7.42	29.44
C20H2808	396.44	396.1784	60.59	7.12	32.29
C20H2809	412.44	412.1733	58.24	6.84	34.91
C20H3001	286.46	286.2297	83.86	10.56	5.59
C20H3002	302.46	302.2246	79.42	10.00	10.58
C20H3003	318.46	318.2195	75.43	9.50	15.07
C20H3004	334.46	334.2144	71.82	9.04	19.13
C20H3005	350.46	350.2093	68.55	8.63	22.83
C20H3006	366.46	366.2042	65.55	8.25	26.20
C20H3007	382.46	382.1991	62.81	7.91	29.28
C20H3008	398.46	398.1941	60.29	7.59	32.12
C20H3009	414.46	414.1890	57.96	7.30	34.74
C20H3201	288.48	288.2453	83.27	11.18	5.55
C20H3202	304.48	304.2402	78.90	10.59	10.51
C20H3203	320.48	320.2351	74.96	10.06	14.98
C20H3204	336.48	336.2300	71.39	9.59	19.02
C20H3205	352.48	352.2250	68.15	9.15	22.70
C20H3206	368.47	368.2199	65.19	8.75	26.05
C20H3207	384.47	384.2148	62.48	8.39	29.13
C20H3208	400.47	400.2097	59.98	8.05	31.96
C20H3209	416.47	416.2046	57.68	7.74	34.57
C20H3401	290.49	290.2610	82.69	11.80	5.51
C20H3402	306.49	306.2559	78.38	11.18	10.44
C20H3403	322.49	322.2508	74.49	10.63	14.88
C20H3404	338.49	338.2457	70.97	10.12	18.91
C20H3405	354.49	354.2406	67.77	9.67	22.57
C20H3406	370.49	370.2355	64.84	9.25	25.91
C20H3407	386.49	386.2304	62.16	8.87	28.98
C20H3408	402.49	402.2253	59.68	8.51	31.80
C20H3409	418.49	418.2203	57.40	8.19	34.41
C20H3601	292.51	292.2766	82.12	12.41	5.47
C20H3602	308.51	308.2715	77.87	11.76	10.37
C20H3603	324.51	324.2664	74.03	11.18	14.79
C20H3604	340.51	340.2613	70.55	10.66	18.79
C20H3605	356.51	356.2563	67.38	10.18	22.44
C20H3606	372.51	372.2512	64.49	9.74	25.77
C20H3607	388.51	388.2461	61.83	9.34	28.83
C20H3608	404.51	404.2410	59.39	8.97	31.64
C20H3609	420.50	420.2359	57.13	8.63	34.24
C20H3801	294.53	294.2922	81.56	13.00	5.43
C20H3802	310.52	310.2872	77.36	12.33	10.30
C20H3803	326.52	326.2821	73.57	11.73	14.70
C20H3804	342.52	342.2770	70.13	11.18	18.68
C20H3805	358.52	358.2719	67.00	10.68	22.31
C20H3806	374.52	374.2668	64.14	10.23	25.63
C20H3807	390.52	390.2617	61.51	9.81	28.68
C20H3808	406.52	406.2566	59.09	9.42	31.49
C20H3809	422.52	422.2516	56.85	9.07	34.08
C20H4001	296.54	296.3079	81.01	13.60	5.40
C20H4002	312.54	312.3028	76.86	12.90	10.24
C20H4003	328.54	328.2977	73.12	12.27	14.61
C20H4004	344.54	344.2926	69.72	11.70	18.57
C20H4005	360.54	360.2876	66.63	11.18	22.19
C20H4006	376.54	376.2825	63.80	10.71	25.49
C20H4007	392.54	392.2774	61.20	10.27	28.53
C20H4008	408.54	408.2723	58.80	9.87	31.33
C20H4009	424.54	424.2672	56.58	9.50	33.92
C20H4201	298.56	298.3235	80.46	14.18	5.36
C20H4202	314.56	314.3185	76.37	13.46	10.17
C20H4203	330.56	330.3134	72.67	12.81	14.52
C20H4204	346.56	346.3083	69.32	12.22	18.47
C20H4205	362.55	362.3032	66.26	11.68	22.06
C20H4206	378.55	378.2981	63.46	11.18	25.36
C20H4207	394.55	394.2930	60.88	10.73	28.39
C20H4208	410.55	410.2879	58.51	10.31	31.18
C20H4209	426.55	426.2829	56.32	9.92	33.76
C21H1001	278.31	278.0732	90.63	3.62	5.75
C21H1002	294.31	294.0681	85.70	3.42	10.87
C21H1003	310.31	310.0630	81.28	3.25	15.47
C21H1004	326.31	326.0579	77.30	3.09	19.61
C21H1005	342.31	342.0528	73.69	2.94	23.37
C21H1006	358.31	358.0477	70.40	2.81	26.79
C21H1007	374.31	374.0426	67.39	2.69	29.92
C21H1008	390.31	390.0376	64.62	2.58	32.79
C21H1009	406.31	406.0325	62.08	2.48	35.44
C21H1201	280.33	280.0888	89.98	4.31	5.71
C21H1202	296.33	296.0837	85.12	4.08	10.80
C21H1203	312.33	312.0786	80.76	3.87	15.37
C21H1204	328.33	328.0736	76.82	3.68	19.49
C21H1205	344.33	344.0685	73.25	3.51	23.23
C21H1206	360.33	360.0634	70.00	3.36	26.64
C21H1207	376.33	376.0583	67.03	3.21	29.76
C21H1208	392.32	392.0532	64.29	3.08	32.62
C21H1209	408.32	408.0481	61.77	2.96	35.26
C21H1401	282.35	282.1045	89.34	5.00	5.67
C21H1402	298.34	298.0994	84.54	4.73	10.73
C21H1403	314.34	314.0943	80.24	4.49	15.27
C21H1404	330.34	330.0892	76.36	4.27	19.37
C21H1405	346.34	346.0841	72.83	4.07	23.10
C21H1406	362.34	362.0790	69.61	3.89	26.49
C21H1407	378.34	378.0739	66.67	3.73	29.60

10/21/66 084

C21H1408	394.34	394.0689	63.96	3.58	32.46
C21H1409	410.34	410.0638	61.47	3.44	35.09
C21H1601	284.36	284.1201	88.70	5.67	5.63
C21H1602	300.36	300.1150	83.98	5.37	10.65
C21H1603	316.36	316.1099	79.73	5.10	15.17
C21H1604	332.36	332.1048	75.89	4.85	19.26
C21H1605	348.36	348.0998	72.41	4.63	22.96
C21H1606	364.36	364.0947	69.23	4.43	26.35
C21H1607	380.36	380.0896	66.32	4.24	29.44
C21H1608	396.36	396.0845	63.64	4.07	32.29
C21H1609	412.36	412.0794	61.17	3.91	34.92
C21H1801	286.38	286.1358	88.08	6.34	5.59
C21H1802	302.38	302.1307	83.42	6.00	10.58
C21H1803	318.38	318.1256	79.23	5.70	15.08
C21H1804	334.38	334.1205	75.43	5.43	19.14
C21H1805	350.37	350.1154	71.99	5.18	22.83
C21H1806	366.37	366.1103	68.85	4.95	26.20
C21H1807	382.37	382.1052	65.97	4.74	29.29
C21H1808	398.37	398.1002	63.32	4.55	32.13
C21H1809	414.37	414.0951	60.87	4.38	34.75
C21H2001	288.39	288.1514	87.46	6.99	5.55
C21H2002	304.39	304.1463	82.86	6.62	10.51
C21H2003	320.39	320.1412	78.73	6.29	14.98
C21H2004	336.39	336.1361	74.98	5.99	19.02
C21H2005	352.39	352.1311	71.58	5.72	22.70
C21H2006	368.39	368.1260	68.47	5.47	26.06
C21H2007	384.39	384.1209	65.62	5.24	29.14
C21H2008	400.39	400.1158	63.00	5.03	31.97
C21H2009	416.39	416.1107	60.58	4.84	34.58
C21H2201	290.41	290.1671	86.85	7.64	5.51
C21H2202	306.41	306.1620	82.32	7.24	10.44
C21H2203	322.41	322.1569	78.23	6.88	14.89
C21H2204	338.41	338.1518	74.54	6.55	18.91
C21H2205	354.41	354.1467	71.17	6.26	22.57
C21H2206	370.41	370.1416	68.10	5.99	25.92
C21H2207	386.41	386.1365	65.28	5.74	28.98
C21H2208	402.40	402.1315	62.68	5.51	31.81
C21H2209	418.40	418.1264	60.28	5.30	34.42
C21H2401	292.42	292.1827	86.26	8.27	5.47
C21H2402	308.42	308.1776	81.78	7.84	10.37
C21H2403	324.42	324.1725	77.75	7.46	14.79
C21H2404	340.42	340.1674	74.09	7.11	18.80
C21H2405	356.42	356.1624	70.77	6.79	22.44
C21H2406	372.42	372.1573	67.73	6.50	25.78
C21H2407	388.42	388.1522	64.94	6.23	28.83
C21H2408	404.42	404.1471	62.37	5.98	31.65
C21H2409	420.42	420.1420	60.00	5.75	34.25
C21H2601	294.44	294.1984	85.67	8.90	5.43
C21H2602	310.44	310.1933	81.25	8.44	10.31
C21H2603	326.44	326.1882	77.27	8.03	14.70
C21H2604	342.44	342.1831	73.66	7.65	18.69
C21H2605	358.44	358.1780	70.37	7.31	22.32
C21H2606	374.44	374.1729	67.36	7.00	25.64
C21H2607	390.44	390.1678	64.60	6.71	28.68
C21H2608	406.44	406.1628	62.06	6.45	31.49
C21H2609	422.44	422.1577	59.71	6.20	34.09
C21H2801	296.46	296.2140	85.08	9.52	5.40
C21H2802	312.46	312.2089	80.73	9.03	10.24
C21H2803	328.46	328.2038	76.79	8.59	14.61
C21H2804	344.45	344.1987	73.23	8.19	18.58
C21H2805	360.45	360.1937	69.98	7.83	22.19
C21H2806	376.45	376.1886	67.00	7.50	25.50
C21H2807	392.45	392.1835	64.27	7.19	28.54
C21H2808	408.45	408.1784	61.75	6.91	31.34
C21H2809	424.45	424.1733	59.43	6.65	33.92
C21H3001	298.47	298.2297	84.51	10.13	5.36
C21H3002	314.47	314.2246	80.21	9.62	10.18
C21H3003	330.47	330.2195	76.33	9.15	14.52
C21H3004	346.47	346.2144	72.80	8.73	18.47
C21H3005	362.47	362.2093	69.59	8.34	22.07
C21H3006	378.47	378.2042	66.65	7.99	25.36
C21H3007	394.47	394.1991	63.94	7.67	28.39
C21H3008	410.47	410.1941	61.45	7.37	31.18
C21H3009	426.47	426.1890	59.14	7.09	33.76
C21H3201	300.49	300.2453	83.94	10.73	5.32
C21H3202	316.49	316.2402	79.70	10.19	10.11
C21H3203	332.49	332.2351	75.86	9.70	14.44
C21H3204	348.49	348.2300	72.38	9.26	18.36
C21H3205	364.49	364.2250	69.20	8.85	21.95
C21H3206	380.49	380.2199	66.29	8.48	25.23
C21H3207	396.48	396.2148	63.62	8.14	28.25
C21H3208	412.48	412.2097	61.15	7.82	31.03
C21H3209	428.48	428.2046	58.87	7.53	33.61
C21H3401	302.50	302.2610	83.38	11.33	5.29
C21H3402	318.50	318.2559	79.19	10.76	10.05
C21H3403	334.50	334.2508	75.41	10.25	14.35
C21H3404	350.50	350.2457	71.96	9.78	18.26
C21H3405	366.50	366.2406	68.82	9.35	21.83
C21H3406	382.50	382.2355	65.94	8.96	25.10
C21H3407	398.50	398.2304	63.30	8.60	28.10
C21H3408	414.50	414.2253	60.85	8.27	30.88
C21H3409	430.50	430.2203	58.59	7.96	33.45
C21H3601	304.52	304.2766	82.83	11.92	5.25
C21H3602	320.52	320.2715	78.70	11.32	9.98
C21H3603	336.52	336.2664	74.95	10.78	14.26
C21H3604	352.52	352.2613	71.55	10.29	18.15
C21H3605	368.52	368.2563	68.45	9.85	21.71
C21H3606	384.52	384.2512	65.60	9.44	24.97
C21H3607	400.52	400.2461	62.98	9.06	27.96
C21H3608	416.52	416.2410	60.56	8.71	30.73
C21H3609	432.52	432.2359	58.32	8.39	33.29
C21H3801	306.54	306.2922	82.29	12.50	5.22
C21H3802	322.54	322.2872	78.20	11.88	9.92
C21H3803	338.54	338.2821	74.51	11.31	14.18
C21H3804	354.53	354.2770	71.15	10.80	18.05
C21H3805	370.53	370.2719	68.07	10.34	21.59
C21H3806	386.53	386.2668	65.26	9.91	24.84
C21H3807	402.53	402.2617	62.66	9.52	27.82
C21H3808	418.53	418.2566	60.27	9.15	30.58
C21H3809	434.53	434.2516	58.05	8.81	33.14
C21H4001	308.55	308.3079	81.75	13.07	5.19
C21H4002	324.55	324.3028	77.72	12.42	9.86
C21H4003	340.55	340.2977	74.07	11.84	14.09
C21H4004	356.55	356.2926	70.74	11.31	17.95
C21H4005	372.55	372.2876	67.70	10.82	21.47
C21H4006	388.55	388.2825	64.92	10.38	24.71
C21H4007	404.55	404.2774	62.35	9.97	27.68
C21H4008	420.55	420.2723	59.98	9.59	30.44
C21H4009	436.55	436.2672	57.78	9.24	32.98
C21H4201	310.57	310.3235	81.22	13.63	5.15

C21H4202	326.57	326.3185	77.24	12.96	9.80
C21H4203	342.57	342.3134	73.63	12.36	14.01
C21H4204	358.57	358.3083	70.35	11.81	17.85
C21H4205	374.57	374.3032	67.34	11.30	21.36
C21H4206	390.57	390.2981	64.58	10.84	24.58
C21H4207	406.56	406.2930	62.04	10.41	27.55
C21H4208	422.56	422.2879	59.69	10.02	30.29
C21H4209	438.56	438.2829	57.51	9.65	32.83
C21H4401	312.58	312.3392	80.69	14.19	5.12
C21H4402	328.58	328.3341	76.76	13.50	9.74
C21H4403	344.58	344.3290	73.20	12.87	13.93
C21H4404	360.58	360.3239	69.95	12.30	17.75
C21H4405	376.58	376.3189	66.98	11.78	21.24
C21H4406	392.58	392.3138	64.25	11.30	24.45
C21H4407	408.58	408.3087	61.73	10.85	27.41
C21H4408	424.58	424.3036	59.41	10.45	30.15
C21H4409	440.58	440.2985	57.25	10.07	32.68
C22H1201	292.34	292.0888	90.39	4.14	5.47
C22H1202	308.34	308.0837	85.70	3.92	10.38
C22H1203	324.34	324.0786	81.47	3.73	14.80
C22H1204	340.34	340.0736	77.64	3.55	18.80
C22H1205	356.34	356.0685	74.16	3.39	22.45
C22H1206	372.34	372.0634	70.97	3.25	25.78
C22H1207	388.34	388.0583	68.05	3.11	28.84
C22H1208	404.34	404.0532	65.35	2.99	31.66
C22H1209	420.34	420.0481	62.87	2.88	34.26
C22H1401	294.36	294.1045	89.77	4.79	5.44
C22H1402	310.36	310.0994	85.14	4.55	10.31
C22H1403	326.36	326.0943	80.97	4.32	14.71
C22H1404	342.35	342.0892	77.18	4.12	18.69
C22H1405	358.35	358.0841	73.74	3.94	22.32
C22H1406	374.35	374.0790	70.59	3.77	25.64
C22H1407	390.35	390.0739	67.69	3.62	28.69
C22H1408	406.35	406.0689	65.03	3.47	31.50
C22H1409	422.35	422.0638	62.57	3.34	34.09
C22H1601	296.37	296.1201	89.16	5.44	5.40
C22H1602	312.37	312.1150	84.59	5.16	10.24
C22H1603	328.37	328.1099	80.47	4.91	14.62
C22H1604	344.37	344.1049	76.73	4.68	18.58
C22H1605	360.37	360.0998	73.33	4.48	22.20
C22H1606	376.37	376.0947	70.21	4.29	25.51
C22H1607	392.37	392.0896	67.35	4.11	28.54
C22H1608	408.37	408.0845	64.71	3.95	31.34
C22H1609	424.37	424.0794	62.27	3.80	33.93
C22H1801	298.39	298.1358	88.56	6.08	5.36
C22H1802	314.39	314.1307	84.05	5.77	10.18
C22H1803	330.39	330.1256	79.98	5.49	14.53
C22H1804	346.39	346.1205	76.29	5.24	18.48
C22H1805	362.39	362.1154	72.92	5.01	22.08
C22H1806	378.39	378.1103	69.84	4.79	25.37
C22H1807	394.38	394.1052	67.00	4.60	28.40
C22H1808	410.38	410.1002	64.39	4.42	31.19
C22H1809	426.38	426.0951	61.97	4.26	33.77
C22H2001	300.40	300.1514	87.96	6.71	5.33
C22H2002	316.40	316.1463	83.52	6.37	10.11
C22H2003	332.40	332.1412	79.50	6.06	14.44
C22H2004	348.40	348.1361	75.84	5.79	18.37
C22H2005	364.40	364.1311	72.51	5.53	21.95
C22H2006	380.40	380.1260	69.46	5.30	25.24
C22H2007	396.40	396.1209	66.66	5.09	28.25
C22H2008	412.40	412.1158	64.08	4.89	31.04
C22H2009	428.40	428.1107	61.68	4.71	33.61
C22H2201	302.42	302.1671	87.38	7.33	5.29
C22H2202	318.42	318.1620	82.99	6.96	10.05
C22H2203	334.42	334.1569	79.02	6.63	14.35
C22H2204	350.42	350.1518	75.41	6.33	18.26
C22H2205	366.42	366.1467	72.12	6.05	21.83
C22H2206	382.42	382.1416	69.10	5.80	25.10
C22H2207	398.42	398.1365	66.32	5.57	28.11
C22H2208	414.42	414.1315	63.76	5.35	30.89
C22H2209	430.42	430.1264	61.39	5.15	33.45
C22H2401	304.44	304.1827	86.80	7.95	5.26
C22H2402	320.44	320.1776	82.46	7.55	9.99
C22H2403	336.43	336.1725	78.54	7.19	14.27
C22H2404	352.43	352.1674	74.98	6.86	18.16
C22H2405	368.43	368.1624	71.72	6.57	21.71
C22H2406	384.43	384.1573	68.74	6.29	24.97
C22H2407	400.43	400.1522	65.99	6.04	27.97
C22H2408	416.43	416.1471	63.45	5.81	30.74
C22H2409	432.43	432.1420	61.11	5.59	33.30
C22H2601	306.45	306.1984	86.23	8.55	5.22
C22H2602	322.45	322.1933	81.95	8.13	9.92
C22H2603	338.45	338.1882	78.07	7.74	14.18
C22H2604	354.45	354.1831	74.55	7.39	18.06
C22H2605	370.45	370.1780	71.33	7.07	21.59
C22H2606	386.45	386.1729	68.38	6.78	24.84
C22H2607	402.45	402.1678	65.66	6.51	27.83
C22H2608	418.45	418.1628	63.15	6.26	30.59
C22H2609	434.45	434.1577	60.82	6.03	33.14
C22H2801	308.47	308.2140	85.66	9.15	5.19
C22H2802	324.47	324.2089	81.44	8.70	9.86
C22H2803	340.47	340.2038	77.61	8.29	14.10
C22H2804	356.47	356.1987	74.13	7.92	17.95
C22H2805	372.47	372.1937	70.94	7.58	21.48
C22H2806	388.46	388.1886	68.02	7.27	24.71
C22H2807	404.46	404.1835	65.33	6.98	27.69
C22H2808	420.46	420.1784	62.85	6.71	30.44
C22H2809	436.46	436.1733	60.54	6.47	32.99
C22H3001	310.48	310.2297	85.11	9.74	5.15
C22H3002	326.48	326.2246	80.94	9.26	9.80
C22H3003	342.48	342.2195	77.16	8.83	14.01
C22H3004	358.48	358.2144	73.71	8.44	17.85
C22H3005	374.48	374.2093	70.56	8.07	21.36
C22H3006	390.48	390.2042	67.67	7.74	24.58
C22H3007	406.48	406.1991	65.01	7.44	27.55
C22H3008	422.48	422.1941	62.55	7.16	30.30
C22H3009	438.48	438.1890	60.26	6.90	32.84
C22H3201	312.50	312.2453	84.56	10.32	5.12
C22H3202	328.50	328.2402	80.44	9.82	9.74
C22H3203	344.50	344.2351	76.70	9.36	13.93
C22H3204	360.50	360.2300	73.30	8.95	17.75
C22H3205	376.50	376.2250	70.19	8.57	21.25
C22H3206	392.50	392.2199	67.32	8.22	24.46
C22H3207	408.50	408.2148	64.69	7.90	27.42
C22H3208	424.50	424.2097	62.25	7.60	30.15
C22H3209	440.49	440.2046	59.99	7.32	32.69
C22H3401	314.52	314.2610	84.02	10.90	5.09
C22H3402	330.52	330.2559	79.95	10.37	9.68
C22H3403	346.51	346.2508	76.26	9.89	13.85
C22H3404	362.51	362.2457	72.89	9.45	17.65

C22H3405	378.51	378.2406	69.81	9.05	21.13
C22H3406	394.51	394.2355	66.98	8.69	24.33
C22H3407	410.51	410.2304	64.37	8.35	27.28
C22H3408	426.51	426.2253	61.96	8.04	30.01
C22H3409	442.51	442.2203	59.71	7.74	32.54
C22H3601	316.53	316.2766	83.48	11.46	5.05
C22H3602	332.53	332.2715	79.46	10.91	9.62
C22H3603	348.53	348.2664	75.82	10.41	13.77
C22H3604	364.53	364.2613	72.49	9.95	17.56
C22H3605	380.53	380.2563	69.44	9.54	21.02
C22H3606	396.53	396.2512	66.64	9.15	24.21
C22H3607	412.53	412.2461	64.06	8.80	27.15
C22H3608	428.53	428.2410	61.66	8.47	29.87
C22H3609	444.53	444.2359	59.44	8.16	32.39
C22H3801	318.55	318.2922	82.95	12.02	5.02
C22H3802	334.55	334.2872	78.99	11.45	9.56
C22H3803	350.55	350.2821	75.38	10.93	13.69
C22H3804	366.55	366.2770	72.09	10.45	17.46
C22H3805	382.55	382.2719	69.08	10.01	20.91
C22H3806	398.54	398.2668	66.30	9.61	24.09
C22H3807	414.54	414.2617	63.74	9.24	27.02
C22H3808	430.54	430.2566	61.37	8.90	29.73
C22H3809	446.54	446.2516	59.18	8.58	32.25
C22H4001	320.56	320.3079	82.43	12.58	4.99
C22H4002	336.56	336.3028	78.51	11.98	9.51
C22H4003	352.56	352.2977	74.95	11.44	13.61
C22H4004	368.56	368.2926	71.70	10.94	17.36
C22H4005	384.56	384.2876	68.71	10.48	20.80
C22H4006	400.56	400.2825	65.97	10.07	23.97
C22H4007	416.56	416.2774	63.44	9.68	26.89
C22H4008	432.56	432.2723	61.09	9.32	29.59
C22H4009	448.56	448.2672	58.91	8.99	32.10
C22H4201	322.58	322.3235	81.92	13.12	4.96
C22H4202	338.58	338.3185	78.05	12.50	9.45
C22H4203	354.58	354.3134	74.52	11.94	13.54
C22H4204	370.58	370.3083	71.31	11.42	17.27
C22H4205	386.58	386.3032	68.36	10.95	20.69
C22H4206	402.58	402.2981	65.64	10.52	23.85
C22H4207	418.58	418.2930	63.13	10.11	26.76
C22H4208	434.58	434.2879	60.81	9.74	29.45
C22H4209	450.57	450.2829	58.65	9.40	31.96
C22H4401	324.60	324.3392	81.41	13.66	4.93
C22H4402	340.59	340.3341	77.58	13.02	9.39
C22H4403	356.59	356.3290	74.10	12.44	13.46
C22H4404	372.59	372.3239	70.92	11.90	17.18
C22H4405	388.59	388.3189	68.00	11.41	20.59
C22H4406	404.59	404.3138	65.31	10.96	23.73
C22H4407	420.59	420.3087	62.83	10.54	26.63
C22H4408	436.59	436.3036	60.52	10.16	29.32
C22H4409	452.59	452.2985	58.39	9.80	31.82
C22H4601	326.61	326.3548	80.91	14.20	4.90
C22H4602	342.61	342.3498	77.13	13.53	9.34
C22H4603	358.61	358.3447	73.69	12.93	13.38
C22H4604	374.61	374.3396	70.54	12.38	17.08
C22H4605	390.61	390.3345	67.65	11.87	20.48
C22H4606	406.61	406.3294	64.99	11.40	23.61
C22H4607	422.61	422.3243	62.53	10.97	26.50
C22H4608	438.61	438.3192	60.25	10.57	29.18
C22H4609	454.61	454.3142	58.13	10.20	31.67
C23H1201	304.35	304.0888	90.77	3.97	5.26
C23H1202	320.35	320.0837	86.24	3.78	9.99
C23H1203	336.35	336.0786	82.13	3.60	14.27
C23H1204	352.35	352.0736	78.40	3.43	18.16
C23H1205	368.35	368.0685	75.00	3.28	21.72
C23H1206	384.35	384.0634	71.88	3.15	24.98
C23H1207	400.35	400.0583	69.00	3.02	27.97
C23H1208	416.35	416.0532	66.35	2.91	30.74
C23H1209	432.35	432.0481	63.90	2.80	33.31
C23H1401	306.37	306.1045	90.17	4.61	5.22
C23H1402	322.37	322.0994	85.70	4.38	9.93
C23H1403	338.37	338.0943	81.64	4.17	14.19
C23H1404	354.37	354.0892	77.96	3.98	18.06
C23H1405	370.37	370.0841	74.59	3.81	21.60
C23H1406	386.36	386.0790	71.50	3.65	24.85
C23H1407	402.36	402.0739	68.66	3.51	27.83
C23H1408	418.36	418.0689	66.03	3.37	30.59
C23H1409	434.36	434.0638	63.60	3.25	33.15
C23H1601	308.38	308.1201	89.58	5.23	5.19
C23H1602	324.38	324.1150	85.16	4.97	9.86
C23H1603	340.38	340.1099	81.16	4.74	14.10
C23H1604	356.38	356.1049	77.52	4.53	17.96
C23H1605	372.38	372.0998	74.19	4.33	21.48
C23H1606	388.38	388.0947	71.13	4.15	24.72
C23H1607	404.38	404.0896	68.32	3.99	27.70
C23H1608	420.38	420.0845	65.72	3.84	30.45
C23H1609	436.38	436.0794	63.31	3.70	33.00
C23H1801	310.40	310.1358	89.00	5.85	5.15
C23H1802	326.40	326.1307	84.64	5.56	9.80
C23H1803	342.40	342.1256	80.68	5.30	14.02
C23H1804	358.40	358.1205	77.08	5.06	17.86
C23H1805	374.40	374.1154	73.79	4.85	21.37
C23H1806	390.40	390.1103	70.76	4.65	24.59
C23H1807	406.40	406.1052	67.98	4.46	27.56
C23H1808	422.40	422.1002	65.40	4.30	30.30
C23H1809	438.39	438.0951	63.02	4.14	32.85
C23H2001	312.42	312.1514	88.43	6.45	5.12
C23H2002	328.41	328.1463	84.12	6.14	9.74
C23H2003	344.41	344.1412	80.21	5.85	13.94
C23H2004	360.41	360.1361	76.65	5.59	17.76
C23H2005	376.41	376.1311	73.39	5.36	21.25
C23H2006	392.41	392.1260	70.40	5.14	24.46
C23H2007	408.41	408.1209	67.64	4.94	27.42
C23H2008	424.41	424.1158	65.09	4.75	30.16
C23H2009	440.41	440.1107	62.73	4.58	32.70
C23H2201	314.43	314.1671	87.86	7.05	5.09
C23H2202	330.43	330.1620	83.60	6.71	9.68
C23H2203	346.43	346.1569	79.74	6.40	13.86
C23H2204	362.43	362.1518	76.22	6.12	17.66
C23H2205	378.43	378.1467	73.00	5.86	21.14
C23H2206	394.43	394.1416	70.04	5.62	24.34
C23H2207	410.43	410.1365	67.31	5.40	27.29
C23H2208	426.43	426.1315	64.78	5.20	30.02
C23H2209	442.43	442.1264	62.44	5.01	32.55
C23H2401	316.45	316.1827	87.30	7.64	5.06
C23H2402	332.45	332.1776	83.10	7.28	9.63
C23H2403	348.45	348.1725	79.28	6.94	13.77
C23H2404	364.45	364.1674	75.80	6.64	17.56
C23H2405	380.44	380.1624	72.61	6.36	21.03
C23H2406	396.44	396.1573	69.68	6.10	24.21
C23H2407	412.44	412.1522	66.98	5.87	27.15

C23H2408	428.44	428.1471	64.48	5.65	29.87
C23H2409	444.44	444.1420	62.16	5.44	32.40
C23H2601	318.46	318.1984	86.75	8.23	5.02
C23H2602	334.46	334.1933	82.60	7.84	9.57
C23H2603	350.46	350.1882	78.83	7.48	13.70
C23H2604	366.46	366.1831	75.38	7.15	17.46
C23H2605	382.46	382.1780	72.23	6.85	20.92
C23H2606	398.46	398.1729	69.33	6.58	24.09
C23H2607	414.46	414.1678	66.65	6.32	27.02
C23H2608	430.46	430.1628	64.18	6.09	29.73
C23H2609	446.46	446.1577	61.88	5.87	32.25
C23H2801	320.48	320.2140	86.20	8.81	4.99
C23H2802	336.48	336.2089	82.10	8.39	9.51
C23H2803	352.48	352.2038	78.38	8.01	13.62
C23H2804	368.48	368.1987	74.97	7.66	17.37
C23H2805	384.48	384.1937	71.85	7.34	20.81
C23H2806	400.48	400.1886	68.98	7.05	23.97
C23H2807	416.48	416.1835	66.33	6.78	26.89
C23H2808	432.47	432.1784	63.88	6.53	29.60
C23H2809	448.47	448.1733	61.60	6.29	32.11
C23H3001	322.49	322.2297	85.66	9.38	4.96
C23H3002	338.49	338.2246	81.61	8.93	9.45
C23H3003	354.49	354.2195	77.93	8.53	13.54
C23H3004	370.49	370.2144	74.56	8.16	17.27
C23H3005	386.49	386.2093	71.48	7.82	20.70
C23H3006	402.49	402.2042	68.64	7.51	23.85
C23H3007	418.49	418.1991	66.01	7.23	26.76
C23H3008	434.49	434.1941	63.58	6.96	29.46
C23H3009	450.49	450.1890	61.32	6.71	31.96
C23H3201	324.51	324.2453	85.13	9.94	4.93
C23H3202	340.51	340.2402	81.13	9.47	9.40
C23H3203	356.51	356.2351	77.49	9.05	13.46
C23H3204	372.51	372.2300	74.16	8.66	17.18
C23H3205	388.51	388.2250	71.11	8.30	20.59
C23H3206	404.51	404.2199	68.29	7.97	23.73
C23H3207	420.51	420.2148	65.70	7.67	26.63
C23H3208	436.51	436.2097	63.29	7.39	29.32
C23H3209	452.51	452.2046	61.05	7.13	31.82
C23H3401	326.53	326.2610	84.60	10.50	4.90
C23H3402	342.53	342.2559	80.65	10.01	9.34
C23H3403	358.53	358.2508	77.05	9.56	13.39
C23H3404	374.53	374.2457	73.76	9.15	17.09
C23H3405	390.52	390.2406	70.74	8.78	20.48
C23H3406	406.52	406.2355	67.96	8.43	23.61
C23H3407	422.52	422.2304	65.38	8.11	26.51
C23H3408	438.52	438.2253	63.00	7.82	29.19
C23H3409	454.52	454.2203	60.78	7.54	31.68
C23H3601	328.54	328.2766	84.09	11.04	4.87
C23H3602	344.54	344.2715	80.18	10.53	9.29
C23H3603	360.54	360.2664	76.62	10.06	13.31
C23H3604	376.54	376.2613	73.37	9.64	17.00
C23H3605	392.54	392.2563	70.38	9.24	20.38
C23H3606	408.54	408.2512	67.62	8.88	23.50
C23H3607	424.54	424.2461	65.07	8.55	26.38
C23H3608	440.54	440.2410	62.71	8.24	29.05
C23H3609	456.54	456.2359	60.51	7.95	31.54
C23H3801	330.56	330.2922	83.57	11.59	4.84
C23H3802	346.56	346.2872	79.71	11.05	9.23
C23H3803	362.56	362.2821	76.20	10.56	13.24
C23H3804	378.56	378.2770	72.98	10.12	16.91

C23H3805	394.56	394.2719	70.02	9.71	20.28
C23H3806	410.56	410.2668	67.29	9.33	23.38
C23H3807	426.56	426.2617	64.76	8.98	26.26
C23H3808	442.55	442.2566	62.42	8.65	28.92
C23H3809	458.55	458.2516	60.25	8.35	31.40
C23H4001	332.57	332.3079	83.07	12.12	4.81
C23H4002	348.57	348.3028	79.25	11.57	9.18
C23H4003	364.57	364.2977	75.78	11.06	13.17
C23H4004	380.57	380.2926	72.59	10.59	16.82
C23H4005	396.57	396.2876	69.66	10.17	20.17
C23H4006	412.57	412.2825	66.96	9.77	23.27
C23H4007	428.57	428.2774	64.46	9.41	26.13
C23H4008	444.57	444.2723	62.14	9.07	28.79
C23H4009	460.57	460.2672	59.98	8.75	31.26
C23H4201	334.59	334.3235	82.57	12.65	4.78
C23H4202	350.59	350.3185	78.80	12.08	9.13
C23H4203	366.59	366.3134	75.36	11.55	13.09
C23H4204	382.59	382.3083	72.21	11.07	16.73
C23H4205	398.59	398.3032	69.31	10.62	20.07
C23H4206	414.59	414.2981	66.63	10.21	23.15
C23H4207	430.59	430.2930	64.16	9.83	26.01
C23H4208	446.59	446.2879	61.86	9.48	28.66
C23H4209	462.59	462.2829	59.72	9.15	31.13
C23H4401	336.61	336.3392	82.07	13.18	4.75
C23H4402	352.61	352.3341	78.35	12.58	9.07
C23H4403	368.61	368.3290	74.95	12.03	13.02
C23H4404	384.60	384.3239	71.83	11.53	16.64
C23H4405	400.60	400.3189	68.96	11.07	19.97
C23H4406	416.60	416.3138	66.31	10.65	23.04
C23H4407	432.60	432.3087	63.86	10.25	25.89
C23H4408	448.60	448.3036	61.58	9.89	28.53
C23H4409	464.60	464.2985	59.46	9.55	30.99
C23H4601	338.62	338.3548	81.58	13.69	4.72
C23H4602	354.62	354.3498	77.90	13.07	9.02
C23H4603	370.62	370.3447	74.54	12.51	12.95
C23H4604	386.62	386.3396	71.45	11.99	16.55
C23H4605	402.62	402.3345	68.61	11.52	19.87
C23H4606	418.62	418.3294	65.99	11.08	22.93
C23H4607	434.62	434.3243	63.56	10.67	25.77
C23H4608	450.62	450.3192	61.31	10.29	28.40
C23H4609	466.62	466.3142	59.20	9.94	30.86
C23H4801	340.64	340.3705	81.10	14.20	4.70
C23H4802	356.64	356.3654	77.46	13.57	8.97
C23H4803	372.64	372.3603	74.14	12.98	12.88
C23H4804	388.64	388.3552	71.08	12.45	16.47
C23H4805	404.64	404.3502	68.27	11.96	19.77
C23H4806	420.64	420.3451	65.68	11.50	22.82
C23H4807	436.63	436.3400	63.27	11.08	25.65
C23H4808	452.63	452.3349	61.03	10.69	28.28
C23H4809	468.63	468.3298	58.95	10.32	30.73
C24H1201	316.36	316.0888	91.12	3.82	5.06
C24H1202	332.36	332.0837	86.73	3.64	9.63
C24H1203	348.36	348.0786	82.75	3.47	13.78
C24H1204	364.36	364.0736	79.12	3.32	17.56
C24H1205	380.36	380.0685	75.79	3.18	21.03
C24H1206	396.36	396.0634	72.73	3.05	24.22
C24H1207	412.36	412.0583	69.91	2.93	27.16
C24H1208	428.36	428.0532	67.30	2.82	29.88
C24H1209	444.36	444.0481	64.87	2.72	32.41
C24H1401	318.38	318.1045	90.54	4.43	5.03

C24H1402	334.38	334.0994	86.21	4.22	9.57
C24H1403	350.38	350.0943	82.27	4.03	13.70
C24H1404	366.38	366.0892	78.68	3.85	17.47
C24H1405	382.38	382.0841	75.39	3.69	20.92
C24H1406	398.38	398.0790	72.36	3.54	24.10
C24H1407	414.37	414.0739	69.57	3.41	27.03
C24H1408	430.37	430.0689	66.98	3.28	29.74
C24H1409	446.37	446.0638	64.58	3.16	32.26
C24H1601	320.39	320.1201	89.97	5.03	4.99
C24H1602	336.39	336.1150	85.69	4.79	9.51
C24H1603	352.39	352.1099	81.80	4.58	13.62
C24H1604	368.39	368.1049	78.25	4.38	17.37
C24H1605	384.39	384.0998	74.99	4.20	20.81
C24H1606	400.39	400.0947	72.00	4.03	23.98
C24H1607	416.39	416.0896	69.23	3.87	26.90
C24H1608	432.39	432.0845	66.67	3.73	29.60
C24H1609	448.39	448.0794	64.29	3.60	32.11
C24H1801	322.41	322.1358	89.41	5.63	4.96
C24H1802	338.41	338.1307	85.18	5.36	9.46
C24H1803	354.41	354.1256	81.34	5.12	13.54
C24H1804	370.41	370.1205	77.82	4.90	17.28
C24H1805	386.41	386.1154	74.60	4.70	20.70
C24H1806	402.41	402.1103	71.64	4.51	23.86
C24H1807	418.41	418.1052	68.90	4.34	26.77
C24H1808	434.41	434.1002	66.36	4.18	29.46
C24H1809	450.41	450.0951	64.00	4.03	31.97
C24H2001	324.43	324.1514	88.85	6.21	4.93
C24H2002	340.43	340.1463	84.68	5.92	9.40
C24H2003	356.43	356.1412	80.88	5.66	13.47
C24H2004	372.42	372.1361	77.40	5.41	17.18
C24H2005	388.42	388.1311	74.21	5.19	20.60
C24H2006	404.42	404.1260	71.28	4.98	23.74
C24H2007	420.42	420.1209	68.57	4.80	26.64
C24H2008	436.42	436.1158	66.05	4.62	29.33
C24H2009	452.42	452.1107	63.72	4.46	31.83
C24H2201	326.44	326.1671	88.31	6.79	4.90
C24H2202	342.44	342.1620	84.18	6.48	9.34
C24H2203	358.44	358.1569	80.42	6.19	13.39
C24H2204	374.44	374.1518	76.99	5.92	17.09
C24H2205	390.44	390.1467	73.83	5.68	20.49
C24H2206	406.44	406.1416	70.93	5.46	23.62
C24H2207	422.44	422.1365	68.24	5.25	26.51
C24H2208	438.44	438.1315	65.75	5.06	29.19
C24H2209	454.44	454.1264	63.43	4.88	31.69
C24H2401	328.46	328.1827	87.76	7.37	4.87
C24H2402	344.46	344.1776	83.69	7.02	9.29
C24H2403	360.46	360.1725	79.97	6.71	13.32
C24H2404	376.46	376.1674	76.57	6.43	17.00
C24H2405	392.46	392.1624	73.45	6.16	20.38
C24H2406	408.46	408.1573	70.58	5.92	23.50
C24H2407	424.45	424.1522	67.91	5.70	26.39
C24H2408	440.45	440.1471	65.45	5.49	29.06
C24H2409	456.45	456.1420	63.15	5.30	31.55
C24H2601	330.47	330.1984	87.23	7.93	4.84
C24H2602	346.47	346.1933	83.20	7.56	9.24
C24H2603	362.47	362.1882	79.53	7.23	13.24
C24H2604	378.47	378.1831	76.17	6.92	16.91
C24H2605	394.47	394.1780	73.08	6.64	20.28
C24H2606	410.47	410.1729	70.23	6.38	23.39
C24H2607	426.47	426.1678	67.59	6.15	26.26

C24H2608	442.47	442.1628	65.15	5.92	28.93
C24H2609	458.47	458.1577	62.88	5.72	31.41
C24H2801	332.49	332.2140	86.70	8.49	4.81
C24H2802	348.49	348.2089	82.72	8.10	9.18
C24H2803	364.49	364.2038	79.09	7.74	13.17
C24H2804	380.49	380.1987	75.76	7.42	16.82
C24H2805	396.49	396.1937	72.71	7.12	20.18
C24H2806	412.49	412.1886	69.89	6.84	23.27
C24H2807	428.49	428.1835	67.28	6.59	26.14
C24H2808	444.49	444.1784	64.85	6.35	28.80
C24H2809	460.49	460.1733	62.60	6.13	31.27
C24H3001	334.51	334.2297	86.18	9.04	4.78
C24H3002	350.51	350.2246	82.24	8.63	9.13
C24H3003	366.50	366.2195	78.65	8.25	13.10
C24H3004	382.50	382.2144	75.36	7.91	16.73
C24H3005	398.50	398.2093	72.34	7.59	20.07
C24H3006	414.50	414.2042	69.55	7.30	23.16
C24H3007	430.50	430.1991	66.96	7.02	26.02
C24H3008	446.50	446.1941	64.56	6.77	28.67
C24H3009	462.50	462.1890	62.33	6.54	31.13
C24H3201	336.52	336.2453	85.66	9.58	4.75
C24H3202	352.52	352.2402	81.77	9.15	9.08
C24H3203	368.52	368.2351	78.22	8.75	13.02
C24H3204	384.52	384.2300	74.97	8.39	16.64
C24H3205	400.52	400.2250	71.97	8.05	19.97
C24H3206	416.52	416.2199	69.21	7.74	23.05
C24H3207	432.52	432.2148	66.65	7.46	25.89
C24H3208	448.52	448.2097	64.27	7.19	28.54
C24H3209	464.52	464.2046	62.06	6.94	31.00
C24H3401	338.54	338.2610	85.15	10.12	4.73
C24H3402	354.54	354.2559	81.31	9.67	9.03
C24H3403	370.54	370.2508	77.80	9.25	12.95
C24H3404	386.54	386.2457	74.58	8.87	16.56
C24H3405	402.54	402.2406	71.61	8.51	19.87
C24H3406	418.53	418.2355	68.88	8.19	22.94
C24H3407	434.53	434.2304	66.34	7.89	25.77
C24H3408	450.53	450.2253	63.98	7.61	28.41
C24H3409	466.53	466.2203	61.79	7.35	30.86
C24H3601	340.55	340.2766	84.65	10.66	4.70
C24H3602	356.55	356.2715	80.85	10.18	8.97
C24H3603	372.55	372.2664	77.38	9.74	12.88
C24H3604	388.55	388.2613	74.19	9.34	16.47
C24H3605	404.55	404.2563	71.26	8.97	19.77
C24H3606	420.55	420.2512	68.55	8.63	22.83
C24H3607	436.55	436.2461	66.03	8.31	25.65
C24H3608	452.55	452.2410	63.70	8.02	28.28
C24H3609	468.55	468.2359	61.52	7.74	30.73
C24H3801	342.57	342.2922	84.15	11.18	4.67
C24H3802	358.57	358.2872	80.39	10.68	8.92
C24H3803	374.57	374.2821	76.96	10.23	12.81
C24H3804	390.57	390.2770	73.81	9.81	16.39
C24H3805	406.57	406.2719	70.90	9.42	19.68
C24H3806	422.57	422.2668	68.22	9.06	22.72
C24H3807	438.57	438.2617	65.73	8.73	25.54
C24H3808	454.57	454.2566	63.42	8.43	28.16
C24H3809	470.57	470.2516	61.26	8.14	30.60
C24H4001	344.59	344.3079	83.66	11.70	4.64
C24H4002	360.59	360.3028	79.94	11.18	8.87
C24H4003	376.58	376.2977	76.55	10.71	12.75
C24H4004	392.58	392.2926	73.43	10.27	16.30

C24H4005	408.58	408.2876	70.55	9.87	19.58
C24H4006	424.58	424.2825	67.89	9.50	22.61
C24H4007	440.58	440.2774	65.43	9.15	25.42
C24H4008	456.58	456.2723	63.14	8.83	28.03
C24H4009	472.58	472.2672	61.00	8.53	30.47
C24H4201	346.60	346.3235	83.17	12.21	4.62
C24H4202	362.60	362.3185	79.50	11.68	8.82
C24H4203	378.60	378.3134	76.14	11.18	12.68
C24H4204	394.60	394.3083	73.05	10.73	16.22
C24H4205	410.60	410.3032	70.21	10.31	19.48
C24H4206	426.60	426.2981	67.57	9.92	22.50
C24H4207	442.60	442.2930	65.13	9.57	25.30
C24H4208	458.60	458.2879	62.86	9.23	27.91
C24H4209	474.60	474.2829	60.74	8.92	30.34
C24H4401	348.62	348.3392	82.69	12.72	4.59
C24H4402	364.62	364.3341	79.06	12.16	8.78
C24H4403	380.62	380.3290	75.74	11.65	12.61
C24H4404	396.62	396.3239	72.68	11.18	16.14
C24H4405	412.62	412.3189	69.86	10.75	19.39
C24H4406	428.61	428.3138	67.26	10.35	22.40
C24H4407	444.61	444.3087	64.84	9.98	25.19
C24H4408	460.61	460.3036	62.58	9.63	27.79
C24H4409	476.61	476.2985	60.48	9.31	30.21
C24H4601	350.63	350.3548	82.21	13.22	4.56
C24H4602	366.63	366.3498	78.63	12.65	8.73
C24H4603	382.63	382.3447	75.34	12.12	12.54
C24H4604	398.63	398.3396	72.31	11.63	16.05
C24H4605	414.63	414.3345	69.52	11.18	19.29
C24H4606	430.63	430.3294	66.94	10.77	22.29
C24H4607	446.63	446.3243	64.54	10.38	25.08
C24H4608	462.63	462.3192	62.31	10.02	27.67
C24H4609	478.63	478.3142	60.23	9.69	30.08
C24H4801	352.65	352.3705	81.74	13.72	4.54
C24H4802	368.65	368.3654	78.20	13.12	8.68
C24H4803	384.65	384.3603	74.94	12.58	12.48
C24H4804	400.65	400.3552	71.95	12.08	15.97
C24H4805	416.65	416.3502	69.19	11.61	19.20
C24H4806	432.65	432.3451	66.63	11.18	22.19
C24H4807	448.65	448.3400	64.25	10.78	24.96
C24H4808	464.65	464.3349	62.04	10.41	27.55
C24H4809	480.64	480.3298	59.98	10.07	29.96
C24H5001	354.67	354.3861	81.28	14.21	4.51
C24H5002	370.66	370.3811	77.77	13.60	8.63
C24H5003	386.66	386.3760	74.55	13.03	12.41
C24H5004	402.66	402.3709	71.59	12.52	15.89
C24H5005	418.66	418.3658	68.85	12.04	19.11
C24H5006	434.66	434.3607	66.32	11.59	22.09
C24H5007	450.66	450.3556	63.97	11.18	24.85
C24H5008	466.66	466.3505	61.77	10.80	27.43
C24H5009	482.66	482.3455	59.72	10.44	29.83
C25H1201	328.37	328.0888	91.44	3.68	4.87
C25H1202	344.37	344.0837	87.20	3.51	9.29
C25H1203	360.37	360.0786	83.32	3.36	13.32
C25H1204	376.37	376.0736	79.78	3.21	17.00
C25H1205	392.37	392.0685	76.53	3.08	20.39
C25H1206	408.37	408.0634	73.53	2.96	23.51
C25H1207	424.37	424.0583	70.76	2.85	26.39
C25H1208	440.37	440.0532	68.19	2.75	29.07
C25H1209	456.37	456.0481	65.80	2.65	31.55
C25H1401	330.39	330.1045	90.89	4.27	4.84

C25H1402	346.39	346.0994	86.69	4.07	9.24
C25H1403	362.39	362.0943	82.86	3.89	13.24
C25H1404	378.39	378.0892	79.36	3.73	16.91
C25H1405	394.39	394.0841	76.14	3.58	20.28
C25H1406	410.39	410.0790	73.17	3.44	23.39
C25H1407	426.39	426.0739	70.42	3.31	26.27
C25H1408	442.39	442.0689	67.88	3.19	28.93
C25H1409	458.38	458.0638	65.51	3.08	31.41
C25H1601	332.41	332.1201	90.34	4.85	4.81
C25H1602	348.41	348.1150	86.19	4.63	9.18
C25H1603	364.40	364.1099	82.40	4.43	13.17
C25H1604	380.40	380.1049	78.94	4.24	16.82
C25H1605	396.40	396.0998	75.75	4.07	20.18
C25H1606	412.40	412.0947	72.81	3.91	23.28
C25H1607	428.40	428.0896	70.09	3.76	26.14
C25H1608	444.40	444.0845	67.57	3.63	28.80
C25H1609	460.40	460.0794	65.22	3.50	31.28
C25H1801	334.42	334.1358	89.79	5.43	4.78
C25H1802	350.42	350.1307	85.69	5.18	9.13
C25H1803	366.42	366.1256	81.95	4.95	13.10
C25H1804	382.42	382.1205	78.52	4.74	16.73
C25H1805	398.42	398.1154	75.37	4.55	20.08
C25H1806	414.42	414.1103	72.46	4.38	23.16
C25H1807	430.42	430.1052	69.76	4.22	26.02
C25H1808	446.42	446.1002	67.26	4.06	28.67
C25H1809	462.42	462.0951	64.94	3.92	31.14
C25H2001	336.44	336.1514	89.25	5.99	4.76
C25H2002	352.44	352.1463	85.20	5.72	9.08
C25H2003	368.44	368.1412	81.50	5.47	13.03
C25H2004	384.44	384.1361	78.11	5.24	16.65
C25H2005	400.44	400.1311	74.99	5.03	19.98
C25H2006	416.43	416.1260	72.11	4.84	23.05
C25H2007	432.43	432.1209	69.44	4.66	25.90
C25H2008	448.43	448.1158	66.96	4.50	28.54
C25H2009	464.43	464.1107	64.65	4.34	31.00
C25H2201	338.45	338.1671	88.72	6.55	4.73
C25H2202	354.45	354.1620	84.72	6.26	9.03
C25H2203	370.45	370.1569	81.06	5.99	12.96
C25H2204	386.45	386.1518	77.70	5.74	16.56
C25H2205	402.45	402.1467	74.61	5.51	19.88
C25H2206	418.45	418.1416	71.76	5.30	22.94
C25H2207	434.45	434.1365	69.12	5.10	25.78
C25H2208	450.45	450.1315	66.66	4.92	28.42
C25H2209	466.45	466.1264	64.38	4.75	30.87
C25H2401	340.47	340.1827	88.20	7.11	4.70
C25H2402	356.47	356.1776	84.24	6.79	8.98
C25H2403	372.47	372.1725	80.62	6.49	12.89
C25H2404	388.47	388.1674	77.30	6.23	16.47
C25H2405	404.47	404.1624	74.24	5.98	19.78
C25H2406	420.47	420.1573	71.42	5.75	22.83
C25H2407	436.47	436.1522	68.80	5.54	25.66
C25H2408	452.47	452.1471	66.37	5.35	28.29
C25H2409	468.46	468.1420	64.10	5.16	30.74
C25H2601	342.49	342.1984	87.68	7.65	4.67
C25H2602	358.48	358.1933	83.76	7.31	8.93
C25H2603	374.48	374.1882	80.18	7.00	12.82
C25H2604	390.48	390.1831	76.90	6.71	16.39
C25H2605	406.48	406.1780	73.87	6.45	19.68
C25H2606	422.48	422.1729	71.07	6.20	22.72
C25H2607	438.48	438.1678	68.48	5.98	25.54

C25H2608	454.48	454.1628	66.07	5.77	28.16
C25H2609	470.48	470.1577	63.82	5.57	30.61
C25H2801	344.50	344.2140	87.16	8.19	4.64
C25H2802	360.50	360.2089	83.29	7.83	8.88
C25H2803	376.50	376.2038	79.76	7.50	12.75
C25H2804	392.50	392.1987	76.50	7.19	16.31
C25H2805	408.50	408.1937	73.51	6.91	19.58
C25H2806	424.50	424.1886	70.74	6.65	22.61
C25H2807	440.50	440.1835	68.17	6.41	25.42
C25H2808	456.50	456.1784	65.78	6.18	28.04
C25H2809	472.50	472.1733	63.55	5.97	30.48
C25H3001	346.52	346.2297	86.66	8.73	4.62
C25H3002	362.52	362.2246	82.83	8.34	8.83
C25H3003	378.52	378.2195	79.33	7.99	12.68
C25H3004	394.52	394.2144	76.11	7.66	16.22
C25H3005	410.51	410.2093	73.15	7.37	19.49
C25H3006	426.51	426.2042	70.40	7.09	22.51
C25H3007	442.51	442.1991	67.86	6.83	25.31
C25H3008	458.51	458.1941	65.49	6.60	27.92
C25H3009	474.51	474.1890	63.28	6.37	30.35
C25H3201	348.53	348.2453	86.15	9.25	4.59
C25H3202	364.53	364.2402	82.37	8.85	8.78
C25H3203	380.53	380.2351	78.91	8.48	12.61
C25H3204	396.53	396.2300	75.73	8.13	16.14
C25H3205	412.53	412.2250	72.79	7.82	19.39
C25H3206	428.53	428.2199	70.07	7.53	22.40
C25H3207	444.53	444.2148	67.55	7.26	25.19
C25H3208	460.53	460.2097	65.20	7.00	27.79
C25H3209	476.53	476.2046	63.01	6.77	30.22
C25H3401	350.55	350.2610	85.66	9.78	4.56
C25H3402	366.55	366.2559	81.92	9.35	8.73
C25H3403	382.55	382.2508	78.49	8.96	12.55
C25H3404	398.55	398.2457	75.34	8.60	16.06
C25H3405	414.55	414.2406	72.44	8.27	19.30
C25H3406	430.55	430.2355	69.74	7.96	22.30
C25H3407	446.55	446.2304	67.24	7.67	25.08
C25H3408	462.54	462.2253	64.92	7.41	27.67
C25H3409	478.54	478.2203	62.75	7.16	30.09
C25H3601	352.57	352.2766	85.17	10.29	4.54
C25H3602	368.56	368.2715	81.47	9.85	8.68
C25H3603	384.56	384.2664	78.08	9.44	12.48
C25H3604	400.56	400.2613	74.96	9.06	15.98
C25H3605	416.56	416.2563	72.08	8.71	19.20
C25H3606	432.56	432.2512	69.42	8.39	22.19
C25H3607	448.56	448.2461	66.94	8.09	24.97
C25H3608	464.56	464.2410	64.64	7.81	27.55
C25H3609	480.56	480.2359	62.49	7.55	29.96
C25H3801	354.58	354.2922	84.69	10.80	4.51
C25H3802	370.58	370.2872	81.03	10.34	8.63
C25H3803	386.58	386.2821	77.68	9.91	12.42
C25H3804	402.58	402.2770	74.59	9.51	15.90
C25H3805	418.58	418.2719	71.74	9.15	19.11
C25H3806	434.58	434.2668	69.10	8.81	22.09
C25H3807	450.58	450.2617	66.64	8.50	24.86
C25H3808	466.58	466.2566	64.36	8.21	27.43
C25H3809	482.58	482.2516	62.22	7.94	29.84
C25H4001	356.60	356.3079	84.21	11.31	4.49
C25H4002	372.60	372.3028	80.59	10.82	8.59
C25H4003	388.60	388.2977	77.27	10.38	12.35
C25H4004	404.60	404.2926	74.22	9.97	15.82

C25H4005	420.59	420.2876	71.39	9.59	19.02
C25H4006	436.59	436.2825	68.78	9.23	21.99
C25H4007	452.59	452.2774	66.35	8.91	24.75
C25H4008	468.59	468.2723	64.08	8.60	27.31
C25H4009	484.59	484.2672	61.97	8.32	29.71
C25H4201	358.61	358.3235	83.73	11.81	4.46
C25H4202	374.61	374.3185	80.16	11.30	8.54
C25H4203	390.61	390.3134	76.87	10.84	12.29
C25H4204	406.61	406.3083	73.85	10.41	15.74
C25H4205	422.61	422.3032	71.05	10.02	18.93
C25H4206	438.61	438.2981	68.46	9.65	21.89
C25H4207	454.61	454.2930	66.05	9.31	24.64
C25H4208	470.61	470.2879	63.81	9.00	27.20
C25H4209	486.61	486.2829	61.71	8.70	29.59
C25H4401	360.63	360.3392	83.27	12.30	4.44
C25H4402	376.63	376.3341	79.73	11.78	8.50
C25H4403	392.63	392.3290	76.48	11.30	12.22
C25H4404	408.63	408.3239	73.48	10.85	15.66
C25H4405	424.63	424.3189	70.72	10.44	18.84
C25H4406	440.63	440.3138	68.15	10.07	21.79
C25H4407	456.63	456.3087	65.76	9.71	24.53
C25H4408	472.62	472.3036	63.53	9.38	27.08
C25H4409	488.62	488.2985	61.45	9.08	29.47
C25H4601	362.64	362.3548	82.80	12.79	4.41
C25H4602	378.64	378.3498	79.30	12.25	8.45
C25H4603	394.64	394.3447	76.09	11.75	12.16
C25H4604	410.64	410.3396	73.12	11.29	15.58
C25H4605	426.64	426.3345	70.38	10.87	18.75
C25H4606	442.64	442.3294	67.84	10.47	21.69
C25H4607	458.64	458.3243	65.47	10.11	24.42
C25H4608	474.64	474.3192	63.26	9.77	26.97
C25H4609	490.64	490.3142	61.20	9.45	29.35
C25H4801	364.66	364.3705	82.34	13.27	4.39
C25H4802	380.66	380.3654	78.88	12.71	8.41
C25H4803	396.66	396.3603	75.70	12.20	12.10
C25H4804	412.66	412.3552	72.77	11.72	15.51
C25H4805	428.66	428.3502	70.05	11.29	18.66
C25H4806	444.66	444.3451	67.53	10.88	21.59
C25H4807	460.66	460.3400	65.18	10.50	24.31
C25H4808	476.66	476.3349	63.00	10.15	26.85
C25H4809	492.66	492.3298	60.95	9.82	29.23
C25H5001	366.68	366.3861	81.89	13.74	4.36
C25H5002	382.68	382.3811	78.47	13.17	8.36
C25H5003	398.68	398.3760	75.32	12.64	12.04
C25H5004	414.67	414.3709	72.41	12.15	15.43
C25H5005	430.67	430.3658	69.72	11.70	18.57
C25H5006	446.67	446.3607	67.23	11.28	21.49
C25H5007	462.67	462.3556	64.90	10.89	24.21
C25H5008	478.67	478.3505	62.73	10.53	26.74
C25H5009	494.67	494.3455	60.70	10.19	29.11
C25H5201	368.69	368.4018	81.44	14.22	4.34
C25H5202	384.69	384.3967	78.06	13.63	8.32
C25H5203	400.69	400.3916	74.94	13.08	11.98
C25H5204	416.69	416.3865	72.06	12.58	15.36
C25H5205	432.69	432.3814	69.40	12.11	18.49
C25H5206	448.69	448.3764	66.92	11.68	21.39
C25H5207	464.69	464.3713	64.62	11.28	24.10
C25H5208	480.69	480.3662	62.47	10.90	26.63
C25H5209	496.69	496.3611	60.46	10.55	28.99
C26H1401	342.40	342.1045	91.21	4.12	4.67

C26H1402	358.40	358.0994	87.13	3.94	8.93
C26H1403	374.40	374.0943	83.41	3.77	12.82
C26H1404	390.40	390.0892	79.99	3.61	16.39
C26H1405	406.40	406.0841	76.84	3.47	19.68
C26H1406	422.40	422.0790	73.93	3.34	22.73
C26H1407	438.40	438.0739	71.23	3.22	25.55
C26H1408	454.40	454.0689	68.73	3.11	28.17
C26H1409	470.40	470.0638	66.39	3.00	30.61
C26H1601	344.42	344.1201	90.67	4.68	4.65
C26H1602	360.42	360.1150	86.65	4.47	8.88
C26H1603	376.42	376.1099	82.96	4.28	12.75
C26H1604	392.42	392.1049	79.58	4.11	16.31
C26H1605	408.41	408.0998	76.46	3.95	19.59
C26H1606	424.41	424.0947	73.58	3.80	22.62
C26H1607	440.41	440.0896	70.91	3.66	25.43
C26H1608	456.41	456.0845	68.42	3.53	28.04
C26H1609	472.41	472.0794	66.11	3.41	30.48
C26H1801	346.43	346.1358	90.14	5.24	4.62
C26H1802	362.43	362.1307	86.17	5.01	8.83
C26H1803	378.43	378.1256	82.52	4.79	12.68
C26H1804	394.43	394.1205	79.17	4.60	16.23
C26H1805	410.43	410.1154	76.09	4.42	19.49
C26H1806	426.43	426.1103	73.23	4.25	22.51
C26H1807	442.43	442.1052	70.59	4.10	25.31
C26H1808	458.43	458.1002	68.12	3.96	27.92
C26H1809	474.43	474.0951	65.82	3.82	30.35
C26H2001	348.45	348.1514	89.62	5.79	4.59
C26H2002	364.45	364.1463	85.69	5.53	8.78
C26H2003	380.45	380.1412	82.08	5.30	12.62
C26H2004	396.45	396.1361	78.77	5.09	16.14
C26H2005	412.45	412.1311	75.72	4.89	19.40
C26H2006	428.45	428.1260	72.89	4.71	22.41
C26H2007	444.45	444.1209	70.27	4.54	25.20
C26H2008	460.44	460.1158	67.82	4.38	27.80
C26H2009	476.44	476.1107	65.55	4.23	30.22
C26H2201	350.46	350.1671	89.11	6.33	4.57
C26H2202	366.46	366.1620	85.22	6.05	8.73
C26H2203	382.46	382.1569	81.65	5.80	12.55
C26H2204	398.46	398.1518	78.37	5.57	16.06
C26H2205	414.46	414.1467	75.35	5.35	19.30
C26H2206	430.46	430.1416	72.55	5.15	22.30
C26H2207	446.46	446.1365	69.95	4.97	25.09
C26H2208	462.46	462.1315	67.53	4.80	27.68
C26H2209	478.46	478.1264	65.27	4.63	30.10
C26H2401	352.48	352.1827	88.60	6.86	4.54
C26H2402	368.48	368.1776	84.75	6.57	8.68
C26H2403	384.48	384.1725	81.22	6.29	12.48
C26H2404	400.48	400.1674	77.98	6.04	15.98
C26H2405	416.48	416.1624	74.98	5.81	19.21
C26H2406	432.48	432.1573	72.21	5.59	22.20
C26H2407	448.48	448.1522	69.63	5.39	24.97
C26H2408	464.48	464.1471	67.23	5.21	27.56
C26H2409	480.48	480.1420	65.00	5.03	29.97
C26H2601	354.50	354.1984	88.09	7.39	4.51
C26H2602	370.50	370.1933	84.29	7.07	8.64
C26H2603	386.50	386.1882	80.80	6.78	12.42
C26H2604	402.49	402.1831	77.59	6.51	15.90
C26H2605	418.49	418.1780	74.62	6.26	19.12
C26H2606	434.49	434.1729	71.87	6.03	22.09
C26H2607	450.49	450.1678	69.32	5.82	24.86

C26H2608	466.49	466.1628	66.94	5.62	27.44
C26H2609	482.49	482.1577	64.72	5.43	29.84
C26H2801	356.51	356.2140	87.60	7.92	4.49
C26H2802	372.51	372.2089	83.83	7.58	8.59
C26H2803	388.51	388.2038	80.38	7.26	12.35
C26H2804	404.51	404.1987	77.20	6.98	15.82
C26H2805	420.51	420.1937	74.26	6.71	19.02
C26H2806	436.51	436.1886	71.54	6.47	21.99
C26H2807	452.51	452.1835	69.01	6.24	24.75
C26H2808	468.51	468.1784	66.66	6.02	27.32
C26H2809	484.51	484.1733	64.46	5.83	29.72
C26H3001	358.53	358.2297	87.10	8.43	4.46
C26H3002	374.53	374.2246	83.38	8.07	8.54
C26H3003	390.53	390.2195	79.97	7.74	12.29
C26H3004	406.53	406.2144	76.82	7.44	15.74
C26H3005	422.53	422.2093	73.91	7.16	18.93
C26H3006	438.53	438.2042	71.21	6.90	21.89
C26H3007	454.52	454.1991	68.71	6.65	24.64
C26H3008	470.52	470.1941	66.37	6.43	27.20
C26H3009	486.52	486.1890	64.19	6.22	29.60
C26H3201	360.54	360.2453	86.62	8.95	4.44
C26H3202	376.54	376.2402	82.94	8.57	8.50
C26H3203	392.54	392.2351	79.56	8.22	12.23
C26H3204	408.54	408.2300	76.44	7.90	15.66
C26H3205	424.54	424.2250	73.56	7.60	18.84
C26H3206	440.54	440.2199	70.89	7.32	21.79
C26H3207	456.54	456.2148	68.40	7.07	24.53
C26H3208	472.54	472.2097	66.09	6.83	27.09
C26H3209	488.54	488.2046	63.92	6.60	29.47
C26H3401	362.56	362.2610	86.13	9.45	4.41
C26H3402	378.56	378.2559	82.49	9.05	8.45
C26H3403	394.56	394.2508	79.15	8.69	12.17
C26H3404	410.56	410.2457	76.06	8.35	15.59
C26H3405	426.56	426.2406	73.21	8.03	18.75
C26H3406	442.56	442.2355	70.56	7.74	21.69
C26H3407	458.56	458.2304	68.10	7.47	24.42
C26H3408	474.56	474.2253	65.81	7.22	26.97
C26H3409	490.56	490.2203	63.66	6.99	29.35
C26H3601	364.58	364.2766	85.66	9.95	4.39
C26H3602	380.58	380.2715	82.06	9.53	8.41
C26H3603	396.58	396.2664	78.75	9.15	12.10
C26H3604	412.57	412.2613	75.69	8.80	15.51
C26H3605	428.57	428.2563	72.87	8.47	18.67
C26H3606	444.57	444.2512	70.24	8.16	21.59
C26H3607	460.57	460.2461	67.80	7.88	24.32
C26H3608	476.57	476.2410	65.53	7.61	26.86
C26H3609	492.57	492.2359	63.40	7.37	29.23
C26H3801	366.59	366.2922	85.19	10.45	4.36
C26H3802	382.59	382.2872	81.62	10.01	8.36
C26H3803	398.59	398.2821	78.35	9.61	12.04
C26H3804	414.59	414.2770	75.32	9.24	15.44
C26H3805	430.59	430.2719	72.53	8.90	18.58
C26H3806	446.59	446.2668	69.93	8.58	21.50
C26H3807	462.59	462.2617	67.51	8.28	24.21
C26H3808	478.59	478.2566	65.25	8.00	26.74
C26H3809	494.59	494.2516	63.14	7.74	29.11
C26H4001	368.61	368.3079	84.72	10.94	4.34
C26H4002	384.61	384.3028	81.20	10.48	8.32
C26H4003	400.61	400.2977	77.95	10.06	11.98
C26H4004	416.61	416.2926	74.96	9.68	15.36

C26H4005	432.61	432.2876	72.19	9.32	18.49
C26H4006	448.61	448.2825	69.61	8.99	21.40
C26H4007	464.60	464.2774	67.22	8.68	24.11
C26H4008	480.60	480.2723	64.98	8.39	26.63
C26H4009	496.60	496.2672	62.89	8.12	29.00
C26H4201	370.62	370.3235	84.26	11.42	4.32
C26H4202	386.62	386.3185	80.77	10.95	8.28
C26H4203	402.62	402.3134	77.56	10.51	11.92
C26H4204	418.62	418.3083	74.60	10.11	15.29
C26H4205	434.62	434.3032	71.85	9.74	18.41
C26H4206	450.62	450.2981	69.30	9.39	21.30
C26H4207	466.62	466.2930	66.93	9.07	24.00
C26H4208	482.62	482.2879	64.71	8.77	26.52
C26H4209	498.62	498.2829	62.63	8.49	28.88
C26H4401	372.64	372.3392	83.80	11.90	4.29
C26H4402	388.64	388.3341	80.35	11.41	8.23
C26H4403	404.64	404.3290	77.18	10.96	11.86
C26H4404	420.64	420.3239	74.24	10.54	15.21
C26H4405	436.64	436.3189	71.52	10.16	18.32
C26H4406	452.64	452.3138	68.99	9.80	21.21
C26H4407	468.64	468.3087	66.64	9.46	23.90
C26H4408	484.64	484.3036	64.44	9.15	26.41
C26H4409	500.64	500.2985	62.38	8.86	28.76
C26H4601	374.66	374.3548	83.35	12.38	4.27
C26H4602	390.66	390.3498	79.94	11.87	8.19
C26H4603	406.65	406.3447	76.79	11.40	11.80
C26H4604	422.65	422.3396	73.89	10.97	15.14
C26H4605	438.65	438.3345	71.19	10.57	18.24
C26H4606	454.65	454.3294	68.69	10.20	21.11
C26H4607	470.65	470.3243	66.35	9.85	23.80
C26H4608	486.65	486.3192	64.17	9.53	26.30
C26H4609	502.65	502.3142	62.13	9.22	28.65
C26H4801	376.67	376.3705	82.91	12.84	4.25
C26H4802	392.67	392.3654	79.53	12.32	8.15
C26H4803	408.67	408.3603	76.42	11.84	11.74
C26H4804	424.67	424.3552	73.54	11.39	15.07
C26H4805	440.67	440.3502	70.87	10.98	18.15
C26H4806	456.67	456.3451	68.38	10.59	21.02
C26H4807	472.67	472.3400	66.07	10.24	23.69
C26H4808	488.67	488.3349	63.91	9.90	26.19
C26H4809	504.67	504.3298	61.88	9.59	28.53
C26H5001	378.69	378.3861	82.47	13.31	4.22
C26H5002	394.69	394.3811	79.12	12.77	8.11
C26H5003	410.69	410.3760	76.04	12.27	11.69
C26H5004	426.69	426.3709	73.19	11.81	15.00
C26H5005	442.69	442.3658	70.54	11.38	18.07
C26H5006	458.68	458.3607	68.08	10.99	20.93
C26H5007	474.68	474.3556	65.79	10.62	23.59
C26H5008	490.68	490.3505	63.64	10.27	26.09
C26H5009	506.68	506.3455	61.63	9.95	28.42
C26H5201	380.70	380.4018	82.03	13.77	4.20
C26H5202	396.70	396.3967	78.72	13.21	8.07
C26H5203	412.70	412.3916	75.67	12.70	11.63
C26H5204	428.70	428.3865	72.85	12.23	14.93
C26H5205	444.70	444.3814	70.22	11.79	17.99
C26H5206	460.70	460.3764	67.79	11.38	20.84
C26H5207	476.70	476.3713	65.51	11.00	23.49
C26H5208	492.70	492.3662	63.38	10.64	25.98
C26H5209	508.70	508.3611	61.39	10.30	28.31
C26H5401	382.72	382.4174	81.60	14.22	4.18

C26H5402	398.72	398.4124	78.32	13.65	8.03
C26H5403	414.72	414.4073	75.30	13.12	11.57
C26H5404	430.72	430.4022	72.50	12.64	14.86
C26H5405	446.72	446.3971	69.91	12.18	17.91
C26H5406	462.72	462.3920	67.49	11.76	20.75
C26H5407	478.72	478.3869	65.23	11.37	23.40
C26H5408	494.72	494.3818	63.13	11.00	25.87
C26H5409	510.71	510.3768	61.15	10.66	28.19
C27H1401	354.41	354.1045	91.50	3.98	4.51
C27H1402	370.41	370.0994	87.55	3.81	8.64
C27H1403	386.41	386.0943	83.93	3.65	12.42
C27H1404	402.41	402.0892	80.59	3.51	15.90
C27H1405	418.41	418.0841	77.51	3.37	19.12
C27H1406	434.41	434.0790	74.65	3.25	22.10
C27H1407	450.41	450.0739	72.00	3.13	24.87
C27H1408	466.41	466.0689	69.53	3.03	27.44
C27H1409	482.41	482.0638	67.23	2.93	29.85
C27H1601	356.43	356.1201	90.99	4.52	4.49
C27H1602	372.43	372.1150	87.08	4.33	8.59
C27H1603	388.43	388.1099	83.49	4.15	12.36
C27H1604	404.43	404.1049	80.19	3.99	15.82
C27H1605	420.43	420.0998	77.14	3.84	19.03
C27H1606	436.42	436.0947	74.31	3.70	22.00
C27H1607	452.42	452.0896	71.68	3.56	24.75
C27H1608	468.42	468.0845	69.23	3.44	27.32
C27H1609	484.42	484.0794	66.95	3.33	29.72
C27H1801	358.44	358.1358	90.47	5.06	4.46
C27H1802	374.44	374.1307	86.61	4.85	8.55
C27H1803	390.44	390.1256	83.06	4.65	12.29
C27H1804	406.44	406.1205	79.79	4.46	15.75
C27H1805	422.44	422.1154	76.77	4.29	18.94
C27H1806	438.44	438.1103	73.97	4.14	21.89
C27H1807	454.44	454.1052	71.36	3.99	24.64
C27H1808	470.44	470.1002	68.94	3.86	27.21
C27H1809	486.44	486.0951	66.67	3.73	29.60
C27H2001	360.46	360.1514	89.97	5.59	4.44
C27H2002	376.46	376.1463	86.15	5.36	8.50
C27H2003	392.46	392.1412	82.63	5.14	12.23
C27H2004	408.46	408.1361	79.40	4.94	15.67
C27H2005	424.46	424.1311	76.40	4.75	18.85
C27H2006	440.46	440.1260	73.63	4.58	21.79
C27H2007	456.46	456.1209	71.05	4.42	24.54
C27H2008	472.46	472.1158	68.64	4.27	27.09
C27H2009	488.46	488.1107	66.39	4.13	29.48
C27H2201	362.48	362.1671	89.47	6.12	4.41
C27H2202	378.48	378.1620	85.69	5.86	8.45
C27H2203	394.47	394.1569	82.21	5.62	12.17
C27H2204	410.47	410.1518	79.01	5.40	15.59
C27H2205	426.47	426.1467	76.04	5.20	18.76
C27H2206	442.47	442.1416	73.29	5.01	21.70
C27H2207	458.47	458.1365	70.74	4.84	24.43
C27H2208	474.47	474.1315	68.35	4.67	26.98
C27H2209	490.47	490.1264	66.12	4.52	29.36
C27H2401	364.49	364.1827	88.97	6.64	4.39
C27H2402	380.49	380.1776	85.23	6.36	8.41
C27H2403	396.49	396.1725	81.79	6.10	12.11
C27H2404	412.49	412.1674	78.62	5.86	15.51
C27H2405	428.49	428.1624	75.68	5.65	18.67
C27H2406	444.49	444.1573	72.96	5.44	21.60
C27H2407	460.49	460.1522	70.43	5.25	24.32

C27H2408	476.49	476.1471	68.06	5.08	26.86
C27H2409	492.49	492.1420	65.85	4.91	29.24
C27H2601	366.51	366.1984	88.48	7.15	4.37
C27H2602	382.51	382.1933	84.78	6.85	8.37
C27H2603	398.51	398.1882	81.38	6.58	12.04
C27H2604	414.51	414.1831	78.24	6.32	15.44
C27H2605	430.51	430.1780	75.33	6.09	18.58
C27H2606	446.50	446.1729	72.63	5.87	21.50
C27H2607	462.50	462.1678	70.12	5.67	24.22
C27H2608	478.50	478.1628	67.77	5.48	26.75
C27H2609	494.50	494.1577	65.58	5.30	29.12
C27H2801	368.52	368.2140	88.00	7.66	4.34
C27H2802	384.52	384.2089	84.34	7.34	8.32
C27H2803	400.52	400.2038	80.97	7.05	11.98
C27H2804	416.52	416.1987	77.86	6.78	15.36
C27H2805	432.52	432.1937	74.98	6.53	18.50
C27H2806	448.52	448.1886	72.30	6.29	21.40
C27H2807	464.52	464.1835	69.81	6.08	24.11
C27H2808	480.52	480.1784	67.49	5.87	26.64
C27H2809	496.52	496.1733	65.31	5.68	29.00
C27H3001	370.54	370.2297	87.52	8.16	4.32
C27H3002	386.54	386.2246	83.90	7.82	8.28
C27H3003	402.54	402.2195	80.56	7.51	11.92
C27H3004	418.54	418.2144	77.48	7.22	15.29
C27H3005	434.54	434.2093	74.63	6.96	18.41
C27H3006	450.54	450.2042	71.98	6.71	21.31
C27H3007	466.54	466.1991	69.51	6.48	24.01
C27H3008	482.54	482.1941	67.21	6.27	26.53
C27H3009	498.53	498.1890	65.05	6.07	28.88
C27H3201	372.56	372.2453	87.05	8.66	4.29
C27H3202	388.55	388.2402	83.46	8.30	8.24
C27H3203	404.55	404.2351	80.16	7.97	11.86
C27H3204	420.55	420.2300	77.11	7.67	15.22
C27H3205	436.55	436.2250	74.29	7.39	18.32
C27H3206	452.55	452.2199	71.66	7.13	21.21
C27H3207	468.55	468.2148	69.21	6.88	23.90
C27H3208	484.55	484.2097	66.93	6.66	26.42
C27H3209	500.55	500.2046	64.79	6.44	28.77
C27H3401	374.57	374.2610	86.58	9.15	4.27
C27H3402	390.57	390.2559	83.03	8.77	8.19
C27H3403	406.57	406.2508	79.77	8.43	11.81
C27H3404	422.57	422.2457	76.74	8.11	15.14
C27H3405	438.57	438.2406	73.95	7.81	18.24
C27H3406	454.57	454.2355	71.34	7.54	21.12
C27H3407	470.57	470.2304	68.92	7.28	23.80
C27H3408	486.57	486.2253	66.65	7.04	26.31
C27H3409	502.57	502.2203	64.53	6.82	28.65
C27H3601	376.59	376.2766	86.12	9.64	4.25
C27H3602	392.59	392.2715	82.61	9.24	8.15
C27H3603	408.59	408.2664	79.37	8.88	11.75
C27H3604	424.59	424.2613	76.38	8.55	15.07
C27H3605	440.58	440.2563	73.61	8.24	18.16
C27H3606	456.58	456.2512	71.03	7.95	21.02
C27H3607	472.58	472.2461	68.62	7.68	23.70
C27H3608	488.58	488.2410	66.38	7.43	26.20
C27H3609	504.58	504.2359	64.27	7.19	28.54
C27H3801	378.60	378.2922	85.66	10.12	4.23
C27H3802	394.60	394.2872	82.18	9.71	8.11
C27H3803	410.60	410.2821	78.98	9.33	11.69
C27H3804	426.60	426.2770	76.02	8.98	15.00

C27H3805	442.60	442.2719	73.27	8.65	18.07
C27H3806	458.60	458.2668	70.72	8.35	20.93
C27H3807	474.60	474.2617	68.33	8.07	23.60
C27H3808	490.60	490.2566	66.10	7.81	26.09
C27H3809	506.60	506.2516	64.02	7.56	28.42
C27H4001	380.62	380.3079	85.20	10.59	4.20
C27H4002	396.62	396.3028	81.77	10.17	8.07
C27H4003	412.62	412.2977	78.60	9.77	11.63
C27H4004	428.62	428.2926	75.66	9.41	14.93
C27H4005	444.62	444.2876	72.94	9.07	17.99
C27H4006	460.62	460.2825	70.41	8.75	20.84
C27H4007	476.62	476.2774	68.04	8.46	23.50
C27H4008	492.62	492.2723	65.83	8.18	25.98
C27H4009	508.61	508.2672	63.76	7.93	28.31
C27H4201	382.64	382.3235	84.75	11.06	4.18
C27H4202	398.63	398.3185	81.35	10.62	8.03
C27H4203	414.63	414.3134	78.21	10.21	11.58
C27H4204	430.63	430.3083	75.31	9.83	14.86
C27H4205	446.63	446.3032	72.61	9.48	17.91
C27H4206	462.63	462.2981	70.10	9.15	20.75
C27H4207	478.63	478.2930	67.76	8.84	23.40
C27H4208	494.63	494.2879	65.56	8.56	25.88
C27H4209	510.63	510.2829	63.51	8.29	28.20
C27H4401	384.65	384.3392	84.31	11.53	4.16
C27H4402	400.65	400.3341	80.94	11.07	7.99
C27H4403	416.65	416.3290	77.84	10.64	11.52
C27H4404	432.65	432.3239	74.96	10.25	14.79
C27H4405	448.65	448.3189	72.28	9.89	17.83
C27H4406	464.65	464.3138	69.79	9.55	20.66
C27H4407	480.65	480.3087	67.47	9.23	23.30
C27H4408	496.65	496.3036	65.30	8.93	25.77
C27H4409	512.65	512.2985	63.26	8.65	28.09
C27H4601	386.67	386.3548	83.87	11.99	4.14
C27H4602	402.67	402.3498	80.54	11.51	7.95
C27H4603	418.67	418.3447	77.46	11.07	11.46
C27H4604	434.67	434.3396	74.61	10.67	14.72
C27H4605	450.66	450.3345	71.96	10.29	17.75
C27H4606	466.66	466.3294	69.49	9.94	20.57
C27H4607	482.66	482.3243	67.19	9.61	23.20
C27H4608	498.66	498.3192	65.03	9.30	25.67
C27H4609	514.66	514.3142	63.01	9.01	27.98
C27H4801	388.68	388.3705	83.44	12.45	4.12
C27H4802	404.68	404.3654	80.14	11.96	7.91
C27H4803	420.68	420.3603	77.09	11.50	11.41
C27H4804	436.68	436.3552	74.26	11.08	14.66
C27H4805	452.68	452.3502	71.64	10.69	17.67
C27H4806	468.68	468.3451	69.19	10.32	20.48
C27H4807	484.68	484.3400	66.91	9.98	23.11
C27H4808	500.68	500.3349	64.77	9.66	25.56
C27H4809	516.68	516.3298	62.77	9.36	27.87
C27H5001	390.70	390.3861	83.01	12.90	4.10
C27H5002	406.70	406.3811	79.74	12.39	7.87
C27H5003	422.70	422.3760	76.72	11.92	11.36
C27H5004	438.70	438.3709	73.92	11.49	14.59
C27H5005	454.70	454.3658	71.32	11.08	17.59
C27H5006	470.70	470.3607	68.90	10.71	20.39
C27H5007	486.70	486.3556	66.63	10.36	23.01
C27H5008	502.69	502.3505	64.51	10.03	25.46
C27H5009	518.69	518.3455	62.52	9.72	27.76
C27H5201	392.71	392.4018	82.58	13.35	4.07

C27H5202	408.71	408.3967	79.35	12.82	7.83
C27H5203	424.71	424.3916	76.36	12.34	11.30
C27H5204	440.71	440.3865	73.59	11.89	14.52
C27H5205	456.71	456.3814	71.01	11.48	17.52
C27H5206	472.71	472.3764	68.60	11.09	20.31
C27H5207	488.71	488.3713	66.36	10.73	22.92
C27H5208	504.71	504.3662	64.25	10.39	25.36
C27H5209	520.71	520.3611	62.28	10.07	27.65
C27H5401	394.73	394.4174	82.16	13.79	4.05
C27H5402	410.73	410.4124	78.96	13.25	7.79
C27H5403	426.73	426.4073	76.00	12.76	11.25
C27H5404	442.73	442.4022	73.25	12.29	14.46
C27H5405	458.73	458.3971	70.70	11.87	17.44
C27H5406	474.73	474.3920	68.31	11.47	20.22
C27H5407	490.73	490.3869	66.09	11.09	22.82
C27H5408	506.73	506.3818	64.00	10.74	25.26
C27H5409	522.73	522.3768	62.04	10.41	27.55
C27H5601	396.75	396.4331	81.74	14.23	4.03
C27H5602	412.75	412.4280	78.57	13.68	7.75
C27H5603	428.75	428.4229	75.64	13.17	11.20
C27H5604	444.74	444.4178	72.92	12.69	14.39
C27H5605	460.74	460.4127	70.39	12.25	17.36
C27H5606	476.74	476.4077	68.02	11.84	20.14
C27H5607	492.74	492.4026	65.82	11.46	22.73
C27H5608	508.74	508.3975	63.75	11.10	25.16
C27H5609	524.74	524.3924	61.80	10.76	27.44
C28H1401	366.42	366.1045	91.78	3.85	4.37
C28H1402	382.42	382.0994	87.94	3.69	8.37
C28H1403	398.42	398.0943	84.41	3.54	12.05
C28H1404	414.42	414.0892	81.15	3.41	15.44
C28H1405	430.42	430.0841	78.14	3.28	18.59
C28H1406	446.42	446.0790	75.34	3.16	21.50
C28H1407	462.42	462.0739	72.73	3.05	24.22
C28H1408	478.42	478.0689	70.30	2.95	26.75
C28H1409	494.42	494.0638	68.02	2.85	29.12
C28H1601	368.44	368.1201	91.28	4.38	4.34
C28H1602	384.44	384.1150	87.48	4.20	8.32
C28H1603	400.44	400.1099	83.99	4.03	11.99
C28H1604	416.44	416.1049	80.76	3.87	15.37
C28H1605	432.44	432.0998	77.77	3.73	18.50
C28H1606	448.44	448.0947	75.00	3.60	21.41
C28H1607	464.44	464.0896	72.41	3.47	24.11
C28H1608	480.43	480.0845	70.00	3.36	26.64
C28H1609	496.43	496.0794	67.75	3.25	29.01
C28H1801	370.46	370.1358	90.78	4.90	4.32
C28H1802	386.45	386.1307	87.03	4.69	8.28
C28H1803	402.45	402.1256	83.57	4.51	11.93
C28H1804	418.45	418.1205	80.37	4.34	15.29
C28H1805	434.45	434.1154	77.41	4.18	18.41
C28H1806	450.45	450.1103	74.66	4.03	21.31
C28H1807	466.45	466.1052	72.10	3.89	24.01
C28H1808	482.45	482.1002	69.71	3.76	26.53
C28H1809	498.45	498.0951	67.47	3.64	28.89
C28H2001	372.47	372.1514	90.29	5.41	4.30
C28H2002	388.47	388.1463	86.57	5.19	8.24
C28H2003	404.47	404.1412	83.15	4.98	11.87
C28H2004	420.47	420.1361	79.98	4.79	15.22
C28H2005	436.47	436.1311	77.05	4.62	18.33
C28H2006	452.47	452.1260	74.33	4.46	21.22
C28H2007	468.47	468.1209	71.79	4.30	23.91

10/21/66 094

C28H2008	484.47	484.1158	69.42	4.16	26.42
C28H2009	500.47	500.1107	67.20	4.03	28.77
C28H2201	374.49	374.1671	89.81	5.92	4.27
C28H2202	390.49	390.1620	86.13	5.68	8.19
C28H2203	406.49	406.1569	82.74	5.46	11.81
C28H2204	422.49	422.1518	79.60	5.25	15.15
C28H2205	438.48	438.1467	76.70	5.06	18.24
C28H2206	454.48	454.1416	74.00	4.88	21.12
C28H2207	470.48	470.1365	71.48	4.71	23.80
C28H2208	486.48	486.1315	69.13	4.56	26.31
C28H2209	502.48	502.1264	66.93	4.41	28.66
C28H2401	376.50	376.1827	89.33	6.43	4.25
C28H2402	392.50	392.1776	85.68	6.16	8.15
C28H2403	408.50	408.1725	82.33	5.92	11.75
C28H2404	424.50	424.1674	79.23	5.70	15.08
C28H2405	440.50	440.1624	76.35	5.49	18.16
C28H2406	456.50	456.1573	73.67	5.30	21.03
C28H2407	472.50	472.1522	71.18	5.12	23.70
C28H2408	488.50	488.1471	68.85	4.95	26.20
C28H2409	504.50	504.1420	66.66	4.80	28.54
C28H2601	378.52	378.1984	88.85	6.92	4.23
C28H2602	394.52	394.1933	85.25	6.64	8.11
C28H2603	410.52	410.1882	81.92	6.38	11.69
C28H2604	426.52	426.1831	78.85	6.14	15.00
C28H2605	442.52	442.1780	76.00	5.92	18.08
C28H2606	458.52	458.1729	73.35	5.72	20.94
C28H2607	474.52	474.1678	70.87	5.52	23.60
C28H2608	490.51	490.1628	68.56	5.34	26.09
C28H2609	506.51	506.1577	66.40	5.17	28.43
C28H2801	380.53	380.2140	88.38	7.42	4.20
C28H2802	396.53	396.2089	84.81	7.12	8.07
C28H2803	412.53	412.2038	81.52	6.84	11.63
C28H2804	428.53	428.1987	78.48	6.59	14.93
C28H2805	444.53	444.1937	75.66	6.35	18.00
C28H2806	460.53	460.1886	73.03	6.13	20.84
C28H2807	476.53	476.1835	70.58	5.92	23.50
C28H2808	492.53	492.1784	68.28	5.73	25.99
C28H2809	508.53	508.1733	66.13	5.55	28.32
C28H3001	382.55	382.2297	87.91	7.90	4.18
C28H3002	398.55	398.2246	84.38	7.59	8.03
C28H3003	414.55	414.2195	81.13	7.29	11.58
C28H3004	430.55	430.2144	78.11	7.02	14.86
C28H3005	446.55	446.2093	75.31	6.77	17.91
C28H3006	462.55	462.2042	72.71	6.54	20.75
C28H3007	478.55	478.1991	70.28	6.32	23.40
C28H3008	494.55	494.1941	68.00	6.11	25.88
C28H3009	510.55	510.1890	65.87	5.92	28.20
C28H3201	384.57	384.2453	87.45	8.39	4.16
C28H3202	400.57	400.2402	83.96	8.05	7.99
C28H3203	416.57	416.2351	80.73	7.74	11.52
C28H3204	432.56	432.2300	77.75	7.46	14.79
C28H3205	448.56	448.2250	74.98	7.19	17.83
C28H3206	464.56	464.2199	72.39	6.94	20.66
C28H3207	480.56	480.2148	69.98	6.71	23.31
C28H3208	496.56	496.2097	67.73	6.50	25.78
C28H3209	512.56	512.2046	65.61	6.29	28.09
C28H3401	386.58	386.2610	87.00	8.87	4.14
C28H3402	402.58	402.2559	83.54	8.51	7.95
C28H3403	418.58	418.2508	80.35	8.19	11.47
C28H3404	434.58	434.2457	77.39	7.89	14.73

C28H3405	450.58	450.2406	74.64	7.61	17.75
C28H3406	466.58	466.2355	72.08	7.35	20.57
C28H3407	482.58	482.2304	69.69	7.10	23.21
C28H3408	498.58	498.2253	67.45	6.87	25.67
C28H3409	514.58	514.2203	65.36	6.66	27.98
C28H3601	388.60	388.2766	86.54	9.34	4.12
C28H3602	404.60	404.2715	83.12	8.97	7.91
C28H3603	420.60	420.2664	79.96	8.63	11.41
C28H3604	436.60	436.2613	77.03	8.31	14.66
C28H3605	452.60	452.2563	74.31	8.02	17.68
C28H3606	468.60	468.2512	71.77	7.74	20.49
C28H3607	484.59	484.2461	69.40	7.49	23.11
C28H3608	500.59	500.2410	67.18	7.25	25.57
C28H3609	516.59	516.2359	65.10	7.02	27.87
C28H3801	390.61	390.2922	86.10	9.81	4.10
C28H3802	406.61	406.2872	82.71	9.42	7.87
C28H3803	422.61	422.2821	79.58	9.06	11.36
C28H3804	438.61	438.2770	76.68	8.73	14.59
C28H3805	454.61	454.2719	73.98	8.43	17.60
C28H3806	470.61	470.2668	71.46	8.14	20.40
C28H3807	486.61	486.2617	69.11	7.87	23.02
C28H3808	502.61	502.2566	66.91	7.62	25.47
C28H3809	518.61	518.2516	64.85	7.39	27.77
C28H4001	392.63	392.3079	85.66	10.27	4.07
C28H4002	408.63	408.3028	82.30	9.87	7.83
C28H4003	424.63	424.2977	79.20	9.50	11.30
C28H4004	440.63	440.2926	76.33	9.15	14.52
C28H4005	456.63	456.2876	73.65	8.83	17.52
C28H4006	472.63	472.2825	71.16	8.53	20.31
C28H4007	488.63	488.2774	68.83	8.25	22.92
C28H4008	504.63	504.2723	66.65	7.99	25.36
C28H4009	520.63	520.2672	64.60	7.74	27.66
C28H4201	394.65	394.3235	85.22	10.73	4.05
C28H4202	410.65	410.3185	81.90	10.31	7.79
C28H4203	426.65	426.3134	78.83	9.92	11.25
C28H4204	442.64	442.3083	75.98	9.56	14.46
C28H4205	458.64	458.3032	73.33	9.23	17.44
C28H4206	474.64	474.2981	70.86	8.92	20.22
C28H4207	490.64	490.2930	68.55	8.63	22.83
C28H4208	506.64	506.2879	66.38	8.36	25.26
C28H4209	522.64	522.2829	64.35	8.10	27.55
C28H4401	396.66	396.3392	84.79	11.18	4.03
C28H4402	412.66	412.3341	81.50	10.75	7.75
C28H4403	428.66	428.3290	78.46	10.35	11.20
C28H4404	444.66	444.3239	75.63	9.97	14.39
C28H4405	460.66	460.3189	73.01	9.63	17.37
C28H4406	476.66	476.3138	70.56	9.30	20.14
C28H4407	492.66	492.3087	68.26	9.00	22.73
C28H4408	508.66	508.3036	66.12	8.72	25.16
C28H4409	524.66	524.2985	64.10	8.45	27.45
C28H4601	398.68	398.3548	84.36	11.63	4.01
C28H4602	414.68	414.3498	81.10	11.18	7.72
C28H4603	430.68	430.3447	78.09	10.77	11.14
C28H4604	446.68	446.3396	75.29	10.38	14.33
C28H4605	462.68	462.3345	72.69	10.02	17.29
C28H4606	478.68	478.3294	70.26	9.69	20.05
C28H4607	494.67	494.3243	67.99	9.37	22.64
C28H4608	510.67	510.3192	65.86	9.08	25.06
C28H4609	526.67	526.3142	63.86	8.80	27.34
C28H4801	400.69	400.3705	83.93	12.07	3.99

10/21/66 095

C28H4802	416.69	416.3654	80.71	11.61	7.68
C28H4803	432.69	432.3603	77.73	11.18	11.09
C28H4804	448.69	448.3552	74.95	10.78	14.26
C28H4805	464.69	464.3502	72.37	10.41	17.22
C28H4806	480.69	480.3451	69.96	10.07	19.97
C28H4807	496.69	496.3400	67.71	9.74	22.55
C28H4808	512.69	512.3349	65.60	9.44	24.97
C28H4809	528.69	528.3298	63.61	9.15	27.24
C28H5001	402.71	402.3861	83.51	12.51	3.97
C28H5002	418.71	418.3811	80.32	12.04	7.64
C28H5003	434.71	434.3760	77.36	11.59	11.04
C28H5004	450.71	450.3709	74.62	11.18	14.20
C28H5005	466.71	466.3658	72.06	10.80	17.14
C28H5006	482.71	482.3607	69.67	10.44	19.89
C28H5007	498.71	498.3556	67.44	10.11	22.46
C28H5008	514.71	514.3505	65.34	9.79	24.87
C28H5009	530.71	530.3455	63.37	9.50	27.13
C28H5201	404.73	404.4018	83.10	12.95	3.95
C28H5202	420.73	420.3967	79.94	12.46	7.61
C28H5203	436.72	436.3916	77.01	12.00	10.99
C28H5204	452.72	452.3865	74.29	11.58	14.14
C28H5205	468.72	468.3814	71.75	11.18	17.07
C28H5206	484.72	484.3764	69.38	10.81	19.80
C28H5207	500.72	500.3713	67.17	10.47	22.37
C28H5208	516.72	516.3662	65.09	10.14	24.77
C28H5209	532.72	532.3611	63.13	9.84	27.03
C28H5401	406.74	406.4174	82.68	13.38	3.93
C28H5402	422.74	422.4124	79.56	12.88	7.57
C28H5403	438.74	438.4073	76.65	12.41	10.94
C28H5404	454.74	454.4022	73.96	11.97	14.07
C28H5405	470.74	470.3971	71.44	11.56	16.99
C28H5406	486.74	486.3920	69.09	11.18	19.72
C28H5407	502.74	502.3869	66.90	10.83	22.28
C28H5408	518.74	518.3818	64.83	10.49	24.67
C28H5409	534.74	534.3768	62.89	10.18	26.93
C28H5601	408.76	408.4331	82.28	13.81	3.91
C28H5602	424.76	424.4280	79.18	13.29	7.53
C28H5603	440.76	440.4229	76.30	12.81	10.89
C28H5604	456.76	456.4178	73.63	12.36	14.01
C28H5605	472.76	472.4127	71.14	11.94	16.92
C28H5606	488.75	488.4077	68.81	11.55	19.64
C28H5607	504.75	504.4026	66.63	11.18	22.19
C28H5608	520.75	520.3975	64.58	10.84	24.58
C28H5609	536.75	536.3924	62.66	10.52	26.83
C28H5801	410.77	410.4487	81.87	14.23	3.89
C28H5802	426.77	426.4437	78.80	13.70	7.50
C28H5803	442.77	442.4386	75.96	13.20	10.84
C28H5804	458.77	458.4335	73.31	12.74	13.95
C28H5805	474.77	474.4284	70.84	12.31	16.85
C28H5806	490.77	490.4233	68.53	11.91	19.56
C28H5807	506.77	506.4182	66.36	11.54	22.10
C28H5808	522.77	522.4131	64.33	11.18	24.48
C28H5809	538.77	538.4081	62.42	10.85	26.73
C29H1401	378.43	378.1045	92.04	3.73	4.23
C29H1402	394.43	394.0994	88.31	3.58	8.11
C29H1403	410.43	410.0943	84.87	3.44	11.69
C29H1404	426.43	426.0892	81.68	3.31	15.01
C29H1405	442.43	442.0841	78.73	3.19	18.08
C29H1406	458.43	458.0790	75.98	3.08	20.94
C29H1407	474.43	474.0739	73.42	2.97	23.61

C29H1408	490.43	490.0689	71.02	2.88	26.10
C29H1409	506.43	506.0638	68.78	2.79	28.43
C29H1601	380.45	380.1201	91.56	4.24	4.21
C29H1602	396.45	396.1150	87.86	4.07	8.07
C29H1603	412.45	412.1099	84.45	3.91	11.64
C29H1604	428.45	428.1049	81.30	3.76	14.94
C29H1605	444.45	444.0998	78.37	3.63	18.00
C29H1606	460.45	460.0947	75.65	3.50	20.85
C29H1607	476.45	476.0896	73.11	3.38	23.51
C29H1608	492.45	492.0845	70.73	3.27	25.99
C29H1609	508.45	508.0794	68.51	3.17	28.32
C29H1801	382.47	382.1358	91.07	4.74	4.18
C29H1802	398.47	398.1307	87.42	4.55	8.03
C29H1803	414.47	414.1256	84.04	4.38	11.58
C29H1804	430.46	430.1205	80.92	4.21	14.87
C29H1805	446.46	446.1154	78.02	4.06	17.92
C29H1806	462.46	462.1103	75.32	3.92	20.76
C29H1807	478.46	478.1052	72.80	3.79	23.41
C29H1808	494.46	494.1002	70.44	3.67	25.89
C29H1809	510.46	510.0951	68.24	3.55	28.21
C29H2001	384.48	384.1514	90.60	5.24	4.16
C29H2002	400.48	400.1463	86.98	5.03	7.99
C29H2003	416.48	416.1412	83.63	4.84	11.52
C29H2004	432.48	432.1361	80.54	4.66	14.80
C29H2005	448.48	448.1311	77.67	4.50	17.84
C29H2006	464.48	464.1260	74.99	4.34	20.67
C29H2007	480.48	480.1209	72.50	4.20	23.31
C29H2008	496.48	496.1158	70.16	4.06	25.78
C29H2009	512.48	512.1107	67.97	3.93	28.10
C29H2201	386.50	386.1671	90.12	5.74	4.14
C29H2202	402.50	402.1620	86.54	5.51	7.95
C29H2203	418.50	418.1569	83.23	5.30	11.47
C29H2204	434.50	434.1518	80.17	5.10	14.73
C29H2205	450.50	450.1467	77.32	4.92	17.76
C29H2206	466.50	466.1416	74.67	4.75	20.58
C29H2207	482.49	482.1365	72.19	4.60	23.21
C29H2208	498.49	498.1315	69.88	4.45	25.68
C29H2209	514.49	514.1264	67.70	4.31	27.99
C29H2401	388.51	388.1827	89.66	6.23	4.12
C29H2402	404.51	404.1776	86.11	5.98	7.91
C29H2403	420.51	420.1725	82.83	5.75	11.41
C29H2404	436.51	436.1674	79.80	5.54	14.66
C29H2405	452.51	452.1624	76.98	5.35	17.68
C29H2406	468.51	468.1573	74.35	5.16	20.49
C29H2407	484.51	484.1522	71.89	4.99	23.12
C29H2408	500.51	500.1471	69.59	4.83	25.57
C29H2409	516.51	516.1420	67.44	4.68	27.88
C29H2601	390.53	390.1984	89.19	6.71	4.10
C29H2602	406.53	406.1933	85.68	6.45	7.87
C29H2603	422.53	422.1882	82.44	6.20	11.36
C29H2604	438.53	438.1831	79.43	5.98	14.59
C29H2605	454.53	454.1780	76.63	5.77	17.60
C29H2606	470.53	470.1729	74.03	5.57	20.40
C29H2607	486.53	486.1678	71.59	5.39	23.02
C29H2608	502.53	502.1628	69.31	5.22	25.47
C29H2609	518.53	518.1577	67.18	5.05	27.77
C29H2801	392.55	392.2140	88.73	7.19	4.08
C29H2802	408.55	408.2089	85.26	6.91	7.83
C29H2803	424.54	424.2038	82.05	6.65	11.31
C29H2804	440.54	440.1987	79.07	6.41	14.53

10/21/66 096

C29H2805	456.54	456.1937	76.30	6.18	17.52
C29H2806	472.54	472.1886	73.71	5.97	20.31
C29H2807	488.54	488.1835	71.30	5.78	22.92
C29H2808	504.54	504.1784	69.04	5.59	25.37
C29H2809	520.54	520.1733	66.92	5.42	27.66
C29H3001	394.56	394.2297	88.28	7.66	4.05
C29H3002	410.56	410.2246	84.84	7.37	7.79
C29H3003	426.56	426.2195	81.66	7.09	11.25
C29H3004	442.56	442.2144	78.71	6.83	14.46
C29H3005	458.56	458.2093	75.96	6.59	17.45
C29H3006	474.56	474.2042	73.40	6.37	20.23
C29H3007	490.56	490.1991	71.01	6.16	22.83
C29H3008	506.56	506.1941	68.76	5.97	25.27
C29H3009	522.56	522.1890	66.66	5.79	27.56
C29H3201	396.58	396.2453	87.83	8.13	4.03
C29H3202	412.58	412.2402	84.43	7.82	7.76
C29H3203	428.58	428.2351	81.27	7.53	11.20
C29H3204	444.58	444.2300	78.35	7.26	14.40
C29H3205	460.58	460.2250	75.63	7.00	17.37
C29H3206	476.57	476.2199	73.09	6.77	20.14
C29H3207	492.57	492.2148	70.71	6.55	22.74
C29H3208	508.57	508.2097	68.49	6.34	25.17
C29H3209	524.57	524.2046	66.40	6.15	27.45
C29H3401	398.59	398.2610	87.39	8.60	4.01
C29H3402	414.59	414.2559	84.02	8.27	7.72
C29H3403	430.59	430.2508	80.89	7.96	11.15
C29H3404	446.59	446.2457	78.00	7.67	14.33
C29H3405	462.59	462.2406	75.30	7.41	17.29
C29H3406	478.59	478.2355	72.78	7.16	20.06
C29H3407	494.59	494.2304	70.43	6.93	22.64
C29H3408	510.59	510.2253	68.22	6.71	25.07
C29H3409	526.59	526.2203	66.15	6.51	27.34
C29H3601	400.61	400.2766	86.95	9.06	3.99
C29H3602	416.61	416.2715	83.61	8.71	7.68
C29H3603	432.61	432.2664	80.52	8.39	11.10
C29H3604	448.61	448.2613	77.65	8.09	14.27
C29H3605	464.61	464.2563	74.97	7.81	17.22
C29H3606	480.61	480.2512	72.48	7.55	19.97
C29H3607	496.61	496.2461	70.14	7.31	22.55
C29H3608	512.61	512.2410	67.95	7.08	24.97
C29H3609	528.60	528.2359	65.89	6.86	27.24
C29H3801	402.63	402.2922	86.51	9.51	3.97
C29H3802	418.63	418.2872	83.21	9.15	7.64
C29H3803	434.62	434.2821	80.14	8.81	11.04
C29H3804	450.62	450.2770	77.30	8.50	14.20
C29H3805	466.62	466.2719	74.65	8.21	17.14
C29H3806	482.62	482.2668	72.17	7.94	19.89
C29H3807	498.62	498.2617	69.86	7.68	22.46
C29H3808	514.62	514.2566	67.69	7.44	24.87
C29H3809	530.62	530.2516	65.64	7.22	27.14
C29H4001	404.64	404.3079	86.08	9.96	3.95
C29H4002	420.64	420.3028	82.81	9.59	7.61
C29H4003	436.64	436.2977	79.77	9.23	10.99
C29H4004	452.64	452.2926	76.95	8.91	14.14
C29H4005	468.64	468.2876	74.33	8.60	17.07
C29H4006	484.64	484.2825	71.87	8.32	19.81
C29H4007	500.64	500.2774	69.58	8.05	22.37
C29H4008	516.64	516.2723	67.42	7.80	24.77
C29H4009	532.64	532.2672	65.40	7.57	27.03
C29H4201	406.66	406.3235	85.66	10.41	3.93

C29H4202	422.66	422.3185	82.41	10.02	7.57
C29H4203	438.66	438.3134	79.41	9.65	10.94
C29H4204	454.66	454.3083	76.61	9.31	14.08
C29H4205	470.66	470.3032	74.01	8.99	17.00
C29H4206	486.65	486.2981	71.58	8.70	19.73
C29H4207	502.65	502.2930	69.30	8.42	22.28
C29H4208	518.65	518.2879	67.16	8.16	24.68
C29H4209	534.65	534.2829	65.15	7.92	26.93
C29H4401	408.67	408.3392	85.23	10.85	3.91
C29H4402	424.67	424.3341	82.02	10.44	7.53
C29H4403	440.67	440.3290	79.04	10.06	10.89
C29H4404	456.67	456.3239	76.27	9.71	14.01
C29H4405	472.67	472.3189	73.69	9.38	16.92
C29H4406	488.67	488.3138	71.28	9.08	19.64
C29H4407	504.67	504.3087	69.02	8.79	22.19
C29H4408	520.67	520.3036	66.90	8.52	24.58
C29H4409	536.67	536.2985	64.90	8.26	26.83
C29H4601	410.69	410.3548	84.81	11.29	3.90
C29H4602	426.69	426.3498	81.63	10.87	7.50
C29H4603	442.69	442.3447	78.68	10.47	10.84
C29H4604	458.69	458.3396	75.94	10.11	13.95
C29H4605	474.69	474.3345	73.38	9.77	16.85
C29H4606	490.69	490.3294	70.99	9.45	19.56
C29H4607	506.69	506.3243	68.75	9.15	22.10
C29H4608	522.69	522.3192	66.64	8.87	24.49
C29H4609	538.68	538.3142	64.66	8.61	26.73
C29H4801	412.71	412.3705	84.40	11.72	3.88
C29H4802	428.70	428.3654	81.25	11.29	7.46
C29H4803	444.70	444.3603	78.33	10.88	10.79
C29H4804	460.70	460.3552	75.61	10.50	13.89
C29H4805	476.70	476.3502	73.07	10.15	16.78
C29H4806	492.70	492.3451	70.70	9.82	19.48
C29H4807	508.70	508.3400	68.47	9.51	22.02
C29H4808	524.70	524.3349	66.39	9.22	24.39
C29H4809	540.70	540.3298	64.42	8.95	26.63
C29H5001	414.72	414.3861	83.99	12.15	3.86
C29H5002	430.72	430.3811	80.87	11.70	7.43
C29H5003	446.72	446.3760	77.97	11.28	10.74
C29H5004	462.72	462.3709	75.28	10.89	13.83
C29H5005	478.72	478.3658	72.76	10.53	16.71
C29H5006	494.72	494.3607	70.41	10.19	19.40
C29H5007	510.72	510.3556	68.20	9.87	21.93
C29H5008	526.72	526.3505	66.13	9.57	24.30
C29H5009	542.72	542.3455	64.18	9.29	26.53
C29H5201	416.74	416.4018	83.58	12.58	3.84
C29H5202	432.74	432.3967	80.49	12.11	7.39
C29H5203	448.74	448.3916	77.62	11.68	10.70
C29H5204	464.74	464.3865	74.95	11.28	13.77
C29H5205	480.73	480.3814	72.46	10.90	16.64
C29H5206	496.73	496.3764	70.12	10.55	19.33
C29H5207	512.73	512.3713	67.93	10.22	21.84
C29H5208	528.73	528.3662	65.88	9.91	24.21
C29H5209	544.73	544.3611	63.94	9.62	26.43
C29H5401	418.75	418.4174	83.18	13.00	3.82
C29H5402	434.75	434.4124	80.12	12.52	7.36
C29H5403	450.75	450.4073	77.28	12.08	10.65
C29H5404	466.75	466.4022	74.63	11.66	13.71
C29H5405	482.75	482.3971	72.15	11.28	16.57
C29H5406	498.75	498.3920	69.84	10.91	19.25
C29H5407	514.75	514.3869	67.67	10.57	21.76
C29H5408	530.75	530.3818	65.63	10.26	24.12
C29H5409	546.75	546.3768	63.71	9.96	26.34
C29H5601	420.77	420.4331	82.78	13.42	3.80
C29H5602	436.77	436.4280	79.75	12.92	7.33
C29H5603	452.77	452.4229	76.93	12.47	10.60
C29H5604	468.77	468.4178	74.31	12.04	13.65
C29H5605	484.77	484.4127	71.85	11.64	16.50
C29H5606	500.77	500.4077	69.56	11.27	19.17
C29H5607	516.77	516.4026	67.40	10.92	21.67
C29H5608	532.76	532.3975	65.38	10.59	24.02
C29H5609	548.76	548.3924	63.47	10.29	26.24
C29H5801	422.79	422.4487	82.39	13.83	3.78
C29H5802	438.78	438.4437	79.38	13.32	7.29
C29H5803	454.78	454.4386	76.59	12.85	10.55
C29H5804	470.78	470.4335	73.99	12.42	13.59
C29H5805	486.78	486.4284	71.56	12.01	16.43
C29H5806	502.78	502.4233	69.28	11.63	19.09
C29H5807	518.78	518.4182	67.14	11.27	21.59
C29H5808	534.78	534.4131	65.13	10.93	23.93
C29H5809	550.78	550.4081	63.24	10.61	26.14
C29H6001	424.80	424.4644	82.00	14.24	3.77
C29H6002	440.80	440.4593	79.02	13.72	7.26
C29H6003	456.80	456.4542	76.25	13.24	10.51
C29H6004	472.80	472.4491	73.67	12.79	13.54
C29H6005	488.80	488.4440	71.26	12.37	16.37
C29H6006	504.80	504.4390	69.00	11.98	19.02
C29H6007	520.80	520.4339	66.88	11.61	21.50
C29H6008	536.80	536.4288	64.89	11.27	23.84
C29H6009	552.80	552.4237	63.01	10.94	26.05
C30H1601	392.46	392.1201	91.81	4.11	4.08
C30H1602	408.46	408.1150	88.22	3.95	7.83
C30H1603	424.46	424.1099	84.89	3.80	11.31
C30H1604	440.46	440.1049	81.81	3.66	14.53
C30H1605	456.46	456.0998	78.94	3.53	17.53
C30H1606	472.46	472.0947	76.27	3.41	20.32
C30H1607	488.46	488.0896	73.77	3.30	22.93
C30H1608	504.46	504.0845	71.43	3.20	25.37
C30H1609	520.46	520.0794	69.23	3.10	27.67
C30H1801	394.48	394.1358	91.34	4.60	4.06
C30H1802	410.48	410.1307	87.78	4.42	7.80
C30H1803	426.48	426.1256	84.49	4.25	11.25
C30H1804	442.48	442.1205	81.44	4.10	14.46
C30H1805	458.47	458.1154	78.59	3.96	17.45
C30H1806	474.47	474.1103	75.94	3.82	20.23
C30H1807	490.47	490.1052	73.47	3.70	22.83
C30H1808	506.47	506.1002	71.15	3.58	25.27
C30H1809	522.47	522.0951	68.97	3.47	27.56
C30H2001	396.49	396.1514	90.88	5.08	4.04
C30H2002	412.49	412.1463	87.36	4.89	7.76
C30H2003	428.49	428.1412	84.09	4.70	11.20
C30H2004	444.49	444.1361	81.07	4.54	14.40
C30H2005	460.49	460.1311	78.25	4.38	17.37
C30H2006	476.49	476.1260	75.62	4.23	20.15
C30H2007	492.49	492.1209	73.17	4.09	22.74
C30H2008	508.49	508.1158	70.86	3.96	25.17
C30H2009	524.49	524.1107	68.70	3.84	27.45
C30H2201	398.51	398.1671	90.42	5.56	4.01
C30H2202	414.51	414.1620	86.93	5.35	7.72
C30H2203	430.51	430.1569	83.70	5.15	11.15
C30H2204	446.51	446.1518	80.70	4.97	14.33

C30H2205	462.51	462.1467	77.91	4.79	17.30
C30H2206	478.51	478.1416	75.30	4.63	20.06
C30H2207	494.51	494.1365	72.87	4.48	22.65
C30H2208	510.51	510.1315	70.58	4.34	25.07
C30H2209	526.50	526.1264	68.44	4.21	27.35
C30H2401	400.53	400.1827	89.97	6.04	3.99
C30H2402	416.52	416.1776	86.51	5.81	7.68
C30H2403	432.52	432.1725	83.31	5.59	11.10
C30H2404	448.52	448.1674	80.34	5.39	14.27
C30H2405	464.52	464.1624	77.57	5.21	17.22
C30H2406	480.52	480.1573	74.99	5.03	19.98
C30H2407	496.52	496.1522	72.57	4.87	22.56
C30H2408	512.52	512.1471	70.31	4.72	24.97
C30H2409	528.52	528.1420	68.18	4.58	27.24
C30H2601	402.54	402.1984	89.51	6.51	3.97
C30H2602	418.54	418.1933	86.09	6.26	7.65
C30H2603	434.54	434.1882	82.92	6.03	11.05
C30H2604	450.54	450.1831	79.98	5.82	14.20
C30H2605	466.54	466.1780	77.24	5.62	17.15
C30H2606	482.54	482.1729	74.67	5.43	19.89
C30H2607	498.54	498.1678	72.28	5.26	22.46
C30H2608	514.54	514.1628	70.03	5.09	24.88
C30H2609	530.54	530.1577	67.92	4.94	27.14
C30H2801	404.56	404.2140	89.07	6.98	3.95
C30H2802	420.56	420.2089	85.68	6.71	7.61
C30H2803	436.56	436.2038	82.54	6.46	10.99
C30H2804	452.56	452.1987	79.62	6.24	14.14
C30H2805	468.55	468.1937	76.90	6.02	17.07
C30H2806	484.55	484.1886	74.36	5.82	19.81
C30H2807	500.55	500.1835	71.99	5.64	22.37
C30H2808	516.55	516.1784	69.76	5.46	24.78
C30H2809	532.55	532.1733	67.66	5.30	27.04
C30H3001	406.57	406.2297	88.63	7.44	3.94
C30H3002	422.57	422.2246	85.27	7.16	7.57
C30H3003	438.57	438.2195	82.16	6.89	10.94
C30H3004	454.57	454.2144	79.27	6.65	14.08
C30H3005	470.57	470.2093	76.57	6.43	17.00
C30H3006	486.57	486.2042	74.06	6.21	19.73
C30H3007	502.57	502.1991	71.70	6.02	22.28
C30H3008	518.57	518.1941	69.49	5.83	24.68
C30H3009	534.57	534.1890	67.41	5.66	26.94
C30H3201	408.59	408.2453	88.19	7.89	3.92
C30H3202	424.59	424.2402	84.87	7.60	7.54
C30H3203	440.59	440.2351	81.78	7.32	10.89
C30H3204	456.59	456.2300	78.92	7.06	14.02
C30H3205	472.59	472.2250	76.25	6.83	16.93
C30H3206	488.59	488.2199	73.75	6.60	19.65
C30H3207	504.59	504.2148	71.41	6.39	22.20
C30H3208	520.58	520.2097	69.22	6.20	24.59
C30H3209	536.58	536.2046	67.15	6.01	26.84
C30H3401	410.60	410.2610	87.76	8.35	3.90
C30H3402	426.60	426.2559	84.47	8.03	7.50
C30H3403	442.60	442.2508	81.41	7.74	10.84
C30H3404	458.60	458.2457	78.57	7.47	13.95
C30H3405	474.60	474.2406	75.92	7.22	16.86
C30H3406	490.60	490.2355	73.45	6.99	19.57
C30H3407	506.60	506.2304	71.13	6.76	22.11
C30H3408	522.60	522.2253	68.95	6.56	24.49
C30H3409	538.60	538.2203	66.90	6.36	26.73
C30H3601	412.62	412.2766	87.33	8.79	3.88
C30H3602	428.62	428.2715	84.07	8.47	7.47
C30H3603	444.62	444.2664	81.04	8.16	10.80
C30H3604	460.62	460.2613	78.23	7.88	13.89
C30H3605	476.62	476.2563	75.60	7.61	16.78
C30H3606	492.62	492.2512	73.15	7.37	19.49
C30H3607	508.62	508.2461	70.85	7.13	22.02
C30H3608	524.62	524.2410	68.69	6.92	24.40
C30H3609	540.62	540.2359	66.65	6.71	26.64
C30H3801	414.64	414.2922	86.90	9.24	3.86
C30H3802	430.64	430.2872	83.67	8.89	7.43
C30H3803	446.64	446.2821	80.68	8.58	10.75
C30H3804	462.63	462.2770	77.89	8.28	13.83
C30H3805	478.63	478.2719	75.28	8.00	16.71
C30H3806	494.63	494.2668	72.85	7.74	19.41
C30H3807	510.63	510.2617	70.57	7.50	21.93
C30H3808	526.63	526.2566	68.42	7.27	24.30
C30H3809	542.63	542.2516	66.40	7.06	26.54
C30H4001	416.65	416.3079	86.48	9.68	3.84
C30H4002	432.65	432.3028	83.29	9.32	7.40
C30H4003	448.65	448.2977	80.32	8.99	10.70
C30H4004	464.65	464.2926	77.55	8.68	13.77
C30H4005	480.65	480.2876	74.97	8.39	16.64
C30H4006	496.65	496.2825	72.55	8.12	19.33
C30H4007	512.65	512.2774	70.29	7.86	21.85
C30H4008	528.65	528.2723	68.16	7.63	24.21
C30H4009	544.65	544.2672	66.16	7.40	26.44
C30H4201	418.67	418.3235	86.07	10.11	3.82
C30H4202	434.67	434.3185	82.90	9.74	7.36
C30H4203	450.67	450.3134	79.96	9.39	10.65
C30H4204	466.67	466.3083	77.21	9.07	13.71
C30H4205	482.67	482.3032	74.66	8.77	16.57
C30H4206	498.67	498.2981	72.26	8.49	19.25
C30H4207	514.67	514.2930	70.01	8.23	21.76
C30H4208	530.66	530.2879	67.90	7.98	24.12
C30H4209	546.66	546.2829	65.92	7.74	26.34
C30H4401	420.68	420.3392	85.65	10.54	3.80
C30H4402	436.68	436.3341	82.52	10.16	7.33
C30H4403	452.68	452.3290	79.60	9.80	10.60
C30H4404	468.68	468.3239	76.88	9.46	13.65
C30H4405	484.68	484.3189	74.34	9.15	16.51
C30H4406	500.68	500.3138	71.97	8.86	19.17
C30H4407	516.68	516.3087	69.74	8.58	21.68
C30H4408	532.68	532.3036	67.65	8.33	24.03
C30H4409	548.68	548.2985	65.67	8.08	26.24
C30H4601	422.70	422.3548	85.25	10.97	3.79
C30H4602	438.70	438.3498	82.14	10.57	7.29
C30H4603	454.70	454.3447	79.25	10.20	10.56
C30H4604	470.70	470.3396	76.55	9.85	13.60
C30H4605	486.70	486.3345	74.04	9.53	16.44
C30H4606	502.70	502.3294	71.68	9.22	19.10
C30H4607	518.70	518.3243	69.47	8.94	21.59
C30H4608	534.70	534.3192	67.39	8.67	23.94
C30H4609	550.70	550.3142	65.43	8.42	26.15
C30H4801	424.72	424.3705	84.84	11.39	3.77
C30H4802	440.72	440.3654	81.76	10.98	7.26
C30H4803	456.72	456.3603	78.90	10.59	10.51
C30H4804	472.71	472.3552	76.23	10.24	13.54
C30H4805	488.71	488.3502	73.73	9.90	16.37
C30H4806	504.71	504.3451	71.39	9.59	19.02
C30H4807	520.71	520.3400	69.20	9.29	21.51

C30H4808	536.71	536.3349	67.14	9.01	23.85
C30H4809	552.71	552.3298	65.19	8.75	26.05
C30H5001	426.73	426.3861	84.44	11.81	3.75
C30H5002	442.73	442.3811	81.39	11.38	7.23
C30H5003	458.73	458.3760	78.55	10.99	10.46
C30H5004	474.73	474.3709	75.90	10.62	13.48
C30H5005	490.73	490.3658	73.43	10.27	16.30
C30H5006	506.73	506.3607	71.11	9.95	18.94
C30H5007	522.73	522.3556	68.93	9.64	21.43
C30H5008	538.73	538.3505	66.89	9.36	23.76
C30H5009	554.73	554.3455	64.96	9.09	25.96
C30H5201	428.75	428.4018	84.04	12.22	3.73
C30H5202	444.75	444.3967	81.02	11.79	7.19
C30H5203	460.75	460.3916	78.21	11.38	10.42
C30H5204	476.75	476.3865	75.58	10.99	13.42
C30H5205	492.75	492.3814	73.13	10.64	16.23
C30H5206	508.75	508.3764	70.83	10.30	18.87
C30H5207	524.74	524.3713	68.67	9.99	21.34
C30H5208	540.74	540.3662	66.64	9.69	23.67
C30H5209	556.74	556.3611	64.72	9.41	25.86
C30H5401	430.76	430.4174	83.65	12.64	3.71
C30H5402	446.76	446.4124	80.65	12.18	7.16
C30H5403	462.76	462.4073	77.87	11.76	10.37
C30H5404	478.76	478.4022	75.26	11.37	13.37
C30H5405	494.76	494.3971	72.83	11.00	16.17
C30H5406	510.76	510.3920	70.55	10.66	18.79
C30H5407	526.76	526.3869	68.41	10.33	21.26
C30H5408	542.76	542.3818	66.39	10.03	23.58
C30H5409	558.76	558.3768	64.49	9.74	25.77
C30H5601	432.78	432.4331	83.26	13.04	3.70
C30H5602	448.78	448.4280	80.29	12.58	7.13
C30H5603	464.78	464.4229	77.53	12.14	10.33
C30H5604	480.78	480.4178	74.95	11.74	13.31
C30H5605	496.78	496.4127	72.53	11.36	16.10
C30H5606	512.78	512.4077	70.27	11.01	18.72
C30H5607	528.78	528.4026	68.14	10.67	21.18
C30H5608	544.78	544.3975	66.14	10.36	23.50
C30H5609	560.78	560.3924	64.26	10.07	25.68
C30H5801	434.80	434.4487	82.87	13.45	3.68
C30H5802	450.80	450.4437	79.93	12.97	7.10
C30H5803	466.79	466.4386	77.19	12.52	10.28
C30H5804	482.79	482.4335	74.64	12.11	13.26
C30H5805	498.79	498.4284	72.24	11.72	16.04
C30H5806	514.79	514.4233	70.00	11.36	18.65
C30H5807	530.79	530.4182	67.89	11.01	21.10
C30H5808	546.79	546.4131	65.90	10.69	23.41
C30H5809	562.79	562.4081	64.03	10.39	25.59
C30H6001	436.81	436.4644	82.49	13.85	3.66
C30H6002	452.81	452.4593	79.58	13.36	7.07
C30H6003	468.81	468.4542	76.86	12.90	10.24
C30H6004	484.81	484.4491	74.32	12.47	13.20
C30H6005	500.81	500.4440	71.95	12.08	15.97
C30H6006	516.81	516.4390	69.72	11.70	18.57
C30H6007	532.81	532.4339	67.63	11.35	21.02
C30H6008	548.81	548.4288	65.66	11.02	23.32
C30H6009	564.81	564.4237	63.80	10.71	25.49
C30H6201	438.83	438.4800	82.11	14.24	3.65
C30H6202	454.83	454.4750	79.22	13.74	7.04
C30H6203	470.83	470.4699	76.53	13.27	10.19
C30H6204	486.83	486.4648	74.02	12.84	13.15
C30H6205	502.83	502.4597	71.66	12.43	15.91
C30H6206	518.83	518.4546	69.45	12.05	18.50
C30H6207	534.82	534.4495	67.37	11.68	20.94
C30H6208	550.82	550.4444	65.42	11.35	23.24
C30H6209	566.82	566.4394	63.57	11.03	25.40

7. CARBON, HYDROGEN AND PHOSPHORUS

7. KOHLENSTOFF, WASSERSTOFF UND PHOSPHOR

7. CARBONE, HYDROGENE ET PHOSPHORE

MOLECULAR WEIGHTS AND PERCENTAGE COMPOSITIONS OF COMPOUNDS
CONTAINING CARBON, HYDROGEN AND PHOSPHORUS

FORMULA	MOL. WEIGHT	ISOMOL. WEIGHT	PERCENT C	PERCENT H	PERCENT P
C 1H 0P2	73.96	73.9475	16.24	0	83.76
C 1H 1P1	43.99	43.9816	27.30	2.29	70.41
C 1H 2P2	75.97	75.9632	15.81	2.65	81.54
C 1H 3P1	46.01	45.9972	26.11	6.57	67.32
C 1H 4P2	77.99	77.9788	15.40	5.17	79.43
C 1H 5P1	48.02	48.0129	25.01	10.49	64.50
C 1H 6P2	80.01	79.9945	15.01	7.56	77.43
C 2H 1P1	56.00	55.9816	42.89	1.80	55.31
C 2H 2P2	87.99	87.9632	27.30	2.29	70.41
C 2H 3P1	58.02	57.9972	41.40	5.21	53.38
C 2H 4P2	90.00	89.9788	26.69	4.48	68.83
C 2H 5P1	60.04	60.0129	40.01	8.39	51.59
C 2H 6P2	92.02	91.9945	26.11	6.57	67.32
C 2H 7P1	62.05	62.0285	38.71	11.37	49.92
C 2H 8P2	94.03	94.0101	25.55	8.58	65.88
C 3H 1P1	68.02	67.9816	52.98	1.48	45.54
C 3H 2P2	100.00	99.9632	36.03	2.02	61.95
C 3H 3P1	70.03	69.9972	51.45	4.32	44.23
C 3H 4P2	102.01	101.9788	35.32	3.95	60.73
C 3H 5P1	72.05	72.0129	50.01	7.00	42.99
C 3H 6P2	104.03	103.9945	34.64	5.81	59.55
C 3H 7P1	74.06	74.0285	48.65	9.53	41.82
C 3H 8P2	106.04	106.0101	33.98	7.60	58.42
C 3H 9P1	76.08	76.0442	47.36	11.92	40.71
C 3H10P2	108.06	108.0258	33.35	9.33	57.33
C 4H 2P2	112.01	111.9632	42.89	1.80	55.31
C 4H 3P1	82.04	81.9972	58.56	3.69	37.75
C 4H 4P2	114.02	113.9788	42.14	3.54	54.33
C 4H 5P1	84.06	84.0129	57.16	6.00	36.85
C 4H 6P2	116.04	115.9945	41.40	5.21	53.38
C 4H 7P1	86.07	86.0285	55.82	8.20	35.99
C 4H 8P2	118.06	118.0101	40.70	6.83	52.47
C 4H 9P1	88.09	88.0442	54.54	10.30	35.16
C 4H10P2	120.07	120.0258	40.01	8.39	51.59
C 4H11P1	90.11	90.0598	53.32	12.31	34.37
C 4H12P2	122.09	122.0414	39.35	9.91	50.74
C 5H 2P2	124.02	123.9632	48.42	1.63	49.95
C 5H 3P1	94.05	93.9972	63.85	3.22	32.93
C 5H 4P2	126.04	125.9788	47.65	3.20	49.15
C 5H 5P1	96.07	96.0129	62.51	5.25	32.24
C 5H 6P2	128.05	127.9945	46.90	4.72	48.38
C 5H 7P1	98.09	98.0285	61.23	7.19	31.58
C 5H 8P2	130.07	130.0101	46.17	6.20	47.63
C 5H 9P1	100.10	100.0442	59.99	9.06	30.94
C 5H10P2	132.08	132.0258	45.47	7.63	46.90
C 5H11P1	102.12	102.0598	58.81	10.86	30.33
C 5H12P2	134.10	134.0414	44.78	9.02	46.20
C 5H13P1	104.13	104.0755	57.67	12.58	29.74
C 5H14P2	136.11	136.0571	44.12	10.37	45.51
C 6H 3P1	106.06	105.9972	67.95	2.85	29.20

FORMULA	MOL. WEIGHT	ISOMOL. WEIGHT	PERCENT C	PERCENT H	PERCENT P
C 6H 4P2	138.05	137.9788	52.20	2.92	44.87
C 6H 5P1	108.08	108.0129	66.68	4.66	28.66
C 6H 6P2	140.06	139.9945	51.45	4.32	44.23
C 6H 7P1	110.10	110.0285	65.46	6.41	28.13
C 6H 8P2	142.08	142.0101	50.72	5.68	43.60
C 6H 9P1	112.11	112.0442	64.28	8.09	27.63
C 6H10P2	144.09	144.0258	50.01	7.00	42.99
C 6H11P1	114.13	114.0598	63.15	9.72	27.14
C 6H12P2	146.11	146.0414	49.32	8.28	42.40
C 6H13P1	116.14	116.0755	62.05	11.28	26.67
C 6H14P2	148.13	148.0571	48.65	9.53	41.82
C 6H15P1	118.16	118.0911	60.99	12.80	26.21
C 6H16P2	150.14	150.0727	48.00	10.74	41.26
C 7H 3P1	118.08	117.9972	71.21	2.56	26.23
C 7H 4P2	150.06	149.9788	56.03	2.69	41.28
C 7H 5P1	120.09	120.0129	70.01	4.20	25.79
C 7H 6P2	152.07	151.9945	55.29	3.98	40.74
C 7H 7P1	122.11	122.0285	68.86	5.78	25.37
C 7H 8P2	154.09	154.0101	54.56	5.23	40.20
C 7H 9P1	124.12	124.0442	67.74	7.31	24.95
C 7H10P2	156.11	156.0258	53.86	6.46	39.68
C 7H11P1	126.14	126.0598	66.65	8.79	24.56
C 7H12P2	158.12	158.0414	53.17	7.65	39.18
C 7H13P1	128.16	128.0755	65.61	10.22	24.17
C 7H14P2	160.14	160.0571	52.50	8.81	38.68
C 7H15P1	130.17	130.0911	64.59	11.62	23.79
C 7H16P2	162.15	162.0727	51.85	9.95	38.20
C 7H17P1	132.19	132.1068	63.61	12.96	23.43
C 7H18P2	164.17	164.0884	51.21	11.05	37.73
C 8H 4P2	162.07	161.9788	59.29	2.49	38.22
C 8H 5P1	132.10	132.0129	72.74	3.82	23.45
C 8H 6P2	164.08	163.9945	58.56	3.69	37.75
C 8H 7P1	134.12	134.0285	71.64	5.26	23.09
C 8H 8P2	166.10	166.0101	57.85	4.85	37.30
C 8H 9P1	136.13	136.0442	70.58	6.66	22.75
C 8H10P2	168.12	168.0258	57.16	6.00	36.85
C 8H11P1	138.15	138.0598	69.55	8.03	22.42
C 8H12P2	170.13	170.0414	56.48	7.11	36.41
C 8H13P1	140.17	140.0755	68.55	9.35	22.10
C 8H14P2	172.15	172.0571	55.82	8.20	35.99
C 8H15P1	142.18	142.0911	67.58	10.63	21.78
C 8H16P2	174.16	174.0727	55.17	9.26	35.57
C 8H17P1	144.20	144.1068	66.64	11.88	21.48
C 8H18P2	176.18	176.0884	54.54	10.30	35.16
C 8H19P1	146.21	146.1224	65.72	13.10	21.18
C 8H20P2	178.20	178.1040	53.92	11.31	34.76
C 9H 4P2	174.08	173.9788	62.10	2.32	35.59
C 9H 5P1	144.11	144.0129	75.01	3.50	21.49
C 9H 6P2	176.10	175.9945	61.39	3.43	35.18
C 9H 7P1	146.13	146.0285	73.98	4.83	21.20

FORMULA	MOL. WEIGHT	ISOMOL. WEIGHT	PERCENT C	PERCENT H	PERCENT P
C 9H 8P2	178.11	178.0101	60.69	4.53	34.78
C 9H 9P1	148.15	148.0442	72.97	6.12	20.91
C 9H10P2	180.13	180.0258	60.01	5.60	34.39
C 9H11P1	150.16	150.0598	71.99	7.38	20.63
C 9H12P2	182.14	182.0414	59.35	6.64	34.01
C 9H13P1	152.18	152.0755	71.04	8.61	20.35
C 9H14P2	184.16	184.0571	58.70	7.66	33.64
C 9H15P1	154.19	154.0911	70.11	9.81	20.09
C 9H16P2	186.18	186.0727	58.06	8.66	33.27
C 9H17P1	156.21	156.1068	69.20	10.97	19.83
C 9H18P2	188.19	188.0884	57.44	9.64	32.92
C 9H19P1	158.23	158.1224	68.32	12.10	19.58
C 9H20P2	190.21	190.1040	56.83	10.60	32.57
C 9H21P1	160.24	160.1381	67.46	13.21	19.33
C 9H22P2	192.22	192.1197	56.24	11.54	32.23
C10H 5P1	156.13	156.0129	76.93	3.23	19.84
C10H 6P2	188.11	187.9945	63.85	3.22	32.93
C10H 7P1	158.14	158.0285	75.95	4.46	19.59
C10H 8P2	190.12	190.0101	63.18	4.24	32.58
C10H 9P1	160.16	160.0442	75.00	5.66	19.34
C10H10P2	192.14	192.0258	62.51	5.25	32.24
C10H11P1	162.17	162.0598	74.06	6.84	19.10
C10H12P2	194.15	194.0414	61.86	6.23	31.91
C10H13P1	164.19	164.0755	73.15	7.98	18.86
C10H14P2	196.17	196.0571	61.23	7.19	31.58
C10H15P1	166.20	166.0911	72.27	9.10	18.64
C10H16P2	198.19	198.0727	60.61	8.14	31.26
C10H17P1	168.22	168.1068	71.40	10.19	18.41
C10H18P2	200.20	200.0884	59.99	9.06	30.94
C10H19P1	170.24	170.1224	70.56	11.25	18.19
C10H20P2	202.22	202.1040	59.40	9.97	30.63
C10H21P1	172.25	172.1381	69.73	12.29	17.98
C10H22P2	204.23	204.1197	58.81	10.86	30.33
C10H23P1	174.27	174.1537	68.92	13.30	17.77
C10H24P2	206.25	206.1353	58.24	11.73	30.04
C11H 5P1	168.14	168.0129	78.58	3.00	18.42
C11H 6P2	200.12	199.9945	66.02	3.02	30.96
C11H 7P1	170.15	170.0285	77.65	4.15	18.20
C11H 8P2	202.13	202.0101	65.36	3.99	30.65
C11H 9P1	172.17	172.0442	76.74	5.27	17.99
C11H10P2	204.15	204.0258	64.72	4.94	30.34
C11H11P1	174.18	174.0598	75.85	6.37	17.78
C11H12P2	206.17	206.0414	64.09	5.87	30.05
C11H13P1	176.20	176.0755	74.98	7.44	17.58
C11H14P2	208.18	208.0571	63.47	6.78	29.76
C11H15P1	178.22	178.0911	74.14	8.48	17.38
C11H16P2	210.20	210.0727	62.86	7.67	29.47
C11H17P1	180.23	180.1068	73.31	9.51	17.19
C11H18P2	212.21	212.0884	62.26	8.55	29.19
C11H19P1	182.25	182.1224	72.50	10.51	17.00
C11H20P2	214.23	214.1040	61.67	9.41	28.92
C11H21P1	184.26	184.1381	71.70	11.49	16.81
C11H22P2	216.25	216.1197	61.10	10.25	28.65
C11H23P1	186.28	186.1537	70.93	12.45	16.63
C11H24P2	218.26	218.1353	60.53	11.08	28.38
C11H25P1	188.30	188.1694	70.17	13.38	16.45
C11H26P2	220.28	220.1510	59.98	11.90	28.12
C12H 6P2	212.13	211.9945	67.95	2.85	29.20
C12H 7P1	182.16	182.0285	79.12	3.87	17.00
C12H 8P2	214.15	214.0101	67.31	3.77	28.93

FORMULA	MOL. WEIGHT	ISOMOL. WEIGHT	PERCENT C	PERCENT H	PERCENT P
C12H 9P1	184.18	184.0442	78.26	4.93	16.82
C12H10P2	216.16	216.0258	66.68	4.66	28.66
C12H11P1	186.20	186.0598	77.41	5.95	16.64
C12H12P2	218.18	218.0414	66.06	5.54	28.39
C12H13P1	188.21	188.0755	76.58	6.96	16.46
C12H14P2	220.19	220.0571	65.46	6.41	28.13
C12H15P1	190.23	190.0911	75.77	7.95	16.28
C12H16P2	222.21	222.0727	64.86	7.26	27.88
C12H17P1	192.24	192.1068	74.97	8.91	16.11
C12H18P2	224.22	224.0884	64.28	8.09	27.63
C12H19P1	194.26	194.1224	74.20	9.86	15.94
C12H20P2	226.24	226.1040	63.71	8.91	27.38
C12H21P1	196.27	196.1381	73.43	10.78	15.78
C12H22P2	228.26	228.1197	63.15	9.72	27.14
C12H23P1	198.29	198.1537	72.69	11.69	15.62
C12H24P2	230.27	230.1353	62.59	10.51	26.90
C12H25P1	200.31	200.1694	71.96	12.58	15.46
C12H26P2	232.29	232.1510	62.05	11.28	26.67
C12H27P1	202.32	202.1850	71.24	13.45	15.31
C12H28P2	234.30	234.1666	61.52	12.05	26.44
C13H 6P2	224.14	223.9945	69.66	2.70	27.64
C13H 7P1	194.17	194.0285	80.41	3.63	15.95
C13H 8P2	226.16	226.0101	69.04	3.57	27.39
C13H 9P1	196.19	196.0442	79.59	4.62	15.79
C13H10P2	228.17	228.0258	68.43	4.42	27.15
C13H11P1	198.21	198.0598	78.78	5.59	15.63
C13H12P2	230.19	230.0414	67.83	5.25	26.91
C13H13P1	200.22	200.0755	77.99	6.54	15.47
C13H14P2	232.20	232.0571	67.24	6.08	26.68
C13H15P1	202.24	202.0911	77.21	7.48	15.32
C13H16P2	234.22	234.0727	66.67	6.89	26.45
C13H17P1	204.25	204.1068	76.45	8.39	15.16
C13H18P2	236.24	236.0884	66.10	7.68	26.22
C13H19P1	206.27	206.1224	75.70	9.28	15.02
C13H20P2	238.25	238.1040	65.54	8.46	26.00
C13H21P1	208.29	208.1381	74.97	10.16	14.87
C13H22P2	240.27	240.1197	64.99	9.23	25.78
C13H23P1	210.30	210.1537	74.25	11.02	14.73
C13H24P2	242.28	242.1353	64.45	9.98	25.57
C13H25P1	212.32	212.1694	73.54	11.87	14.59
C13H26P2	244.30	244.1510	63.92	10.73	25.36
C13H27P1	214.33	214.1850	72.85	12.70	14.45
C13H28P2	246.32	246.1666	63.39	11.46	25.15
C13H29P1	216.35	216.2007	72.17	13.51	14.32
C13H30P2	248.33	248.1823	62.88	12.18	24.95
C14H 7P1	206.19	206.0285	81.56	3.42	15.02
C14H 8P2	238.17	238.0101	70.60	3.39	26.01
C14H 9P1	208.20	208.0442	80.77	4.36	14.88
C14H10P2	240.18	240.0258	70.01	4.20	25.79
C14H11P1	210.22	210.0598	79.99	5.27	14.73
C14H12P2	242.20	242.0414	69.43	4.99	25.58
C14H13P1	212.23	212.0755	79.23	6.17	14.59
C14H14P2	244.22	244.0571	68.86	5.78	25.37
C14H15P1	214.25	214.0911	78.49	7.06	14.46
C14H16P2	246.23	246.0727	68.29	6.55	25.16
C14H17P1	216.27	216.1068	77.75	7.92	14.32
C14H18P2	248.25	248.0884	67.74	7.31	24.95
C14H19P1	218.28	218.1224	77.04	8.77	14.19
C14H20P2	250.26	250.1040	67.19	8.06	24.75
C14H21P1	220.30	220.1381	76.33	9.61	14.06

C14H22P2	252.28	252.1197	66.65	8.79	24.56
C14H23P1	222.31	222.1537	75.64	10.43	13.93
C14H24P2	254.29	254.1353	66.13	9.51	24.36
C14H25P1	224.33	224.1694	74.96	11.23	13.81
C14H26P2	256.31	256.1510	65.61	10.22	24.17
C14H27P1	226.35	226.1850	74.29	12.02	13.68
C14H28P2	258.33	258.1666	65.09	10.93	23.98
C14H29P1	228.36	228.2007	73.64	12.80	13.56
C14H30P2	260.34	260.1823	64.59	11.62	23.79
C14H31P1	230.38	230.2163	72.99	13.56	13.44
C14H32P2	262.36	262.1979	64.09	12.29	23.61
C15H 7P1	218.20	218.0285	82.57	3.23	14.20
C15H 8P2	250.18	250.0101	72.02	3.22	24.76
C15H 9P1	220.21	220.0442	81.82	4.12	14.07
C15H10P2	252.19	252.0258	71.44	4.00	24.56
C15H11P1	222.23	222.0598	81.07	4.99	13.94
C15H12P2	254.21	254.0414	70.87	4.76	24.37
C15H13P1	224.24	224.0755	80.34	5.84	13.81
C15H14P2	256.23	256.0571	70.32	5.51	24.18
C15H15P1	226.26	226.0911	79.63	6.68	13.69
C15H16P2	258.24	258.0727	69.77	6.25	23.99
C15H17P1	228.28	228.1068	78.92	7.51	13.57
C15H18P2	260.26	260.0884	69.23	6.97	23.80
C15H19P1	230.29	230.1224	78.23	8.32	13.45
C15H20P2	262.27	262.1040	68.69	7.69	23.62
C15H21P1	232.31	232.1381	77.56	9.11	13.33
C15H22P2	264.29	264.1197	68.17	8.39	23.44
C15H23P1	234.32	234.1537	76.89	9.89	13.22
C15H24P2	266.31	266.1353	67.65	9.08	23.26
C15H25P1	236.34	236.1694	76.23	10.66	13.11
C15H26P2	268.32	268.1510	67.15	9.77	23.09
C15H27P1	238.36	238.1850	75.59	11.42	12.99
C15H28P2	270.34	270.1666	66.65	10.44	22.91
C15H29P1	240.37	240.2007	74.95	12.16	12.89
C15H30P2	272.35	272.1823	66.15	11.10	22.75
C15H31P1	242.39	242.2163	74.33	12.89	12.78
C15H32P2	274.37	274.1979	65.67	11.76	22.58
C15H33P1	244.40	244.2320	73.72	13.61	12.67
C15H34P2	276.39	276.2136	65.19	12.40	22.41
C16H 8P2	262.19	262.0101	73.30	3.08	23.63
C16H 9P1	232.22	232.0442	82.76	3.91	13.34
C16H10P2	264.21	264.0258	72.74	3.82	23.45
C16H11P1	234.24	234.0598	82.04	4.73	13.22
C16H12P2	266.22	266.0414	72.19	4.54	23.27
C16H13P1	236.26	236.0755	81.34	5.55	13.11
C16H14P2	268.24	268.0571	71.64	5.26	23.09
C16H15P1	238.27	238.0911	80.66	6.35	13.00
C16H16P2	270.25	270.0727	71.11	5.97	22.92
C16H17P1	240.29	240.1068	79.98	7.13	12.89
C16H18P2	272.27	272.0884	70.58	6.66	22.75
C16H19P1	242.30	242.1224	79.31	7.90	12.78
C16H20P2	274.29	274.1040	70.07	7.35	22.59
C16H21P1	244.32	244.1381	78.66	8.66	12.68
C16H22P2	276.30	276.1197	69.55	8.03	22.42
C16H23P1	246.34	246.1537	78.01	9.41	12.57
C16H24P2	278.32	278.1353	69.05	8.69	22.26
C16H25P1	248.35	248.1694	77.38	10.15	12.47
C16H26P2	280.33	280.1510	68.55	9.35	22.10
C16H27P1	250.37	250.1850	76.76	10.87	12.37
C16H28P2	282.35	282.1666	68.06	10.00	21.94

C16H29P1	252.38	252.2007	76.15	11.58	12.27
C16H30P2	284.37	284.1823	67.58	10.63	21.78
C16H31P1	254.40	254.2163	75.54	12.28	12.18
C16H32P2	286.38	286.1979	67.11	11.26	21.63
C16H33P1	256.42	256.2320	74.95	12.97	12.08
C16H34P2	288.40	288.2136	66.64	11.88	21.48
C16H35P1	258.43	258.2476	74.36	13.65	11.99
C16H36P2	290.41	290.2292	66.17	12.49	21.33
C17H 8P2	274.20	274.0101	74.47	2.94	22.59
C17H 9P1	244.24	244.0442	83.60	3.71	12.68
C17H10P2	276.22	276.0258	73.92	3.65	22.43
C17H11P1	246.25	246.0598	82.92	4.50	12.58
C17H12P2	278.23	278.0414	73.39	4.35	22.26
C17H13P1	248.27	248.0755	82.25	5.28	12.48
C17H14P2	280.25	280.0571	72.86	5.04	22.10
C17H15P1	250.28	250.0911	81.58	6.04	12.38
C17H16P2	282.26	282.0727	72.34	5.71	21.95
C17H17P1	252.30	252.1068	80.93	6.79	12.28
C17H18P2	284.28	284.0884	71.83	6.38	21.79
C17H19P1	254.31	254.1224	80.29	7.53	12.18
C17H20P2	286.30	286.1040	71.32	7.04	21.64
C17H21P1	256.33	256.1381	79.66	8.26	12.08
C17H22P2	288.31	288.1197	70.82	7.69	21.49
C17H23P1	258.35	258.1537	79.04	8.97	11.99
C17H24P2	290.33	290.1353	70.33	8.33	21.34
C17H25P1	260.36	260.1694	78.43	9.68	11.90
C17H26P2	292.34	292.1510	69.85	8.96	21.19
C17H27P1	262.38	262.1850	77.82	10.37	11.81
C17H28P2	294.36	294.1666	69.37	9.59	21.04
C17H29P1	264.39	264.2007	77.23	11.06	11.71
C17H30P2	296.38	296.1823	68.90	10.20	20.90
C17H31P1	266.41	266.2163	76.64	11.73	11.63
C17H32P2	298.39	298.1979	68.43	10.81	20.76
C17H33P1	268.43	268.2320	76.07	12.39	11.54
C17H34P2	300.41	300.2136	67.97	11.41	20.62
C17H35P1	270.44	270.2476	75.50	13.04	11.45
C17H36P2	302.42	302.2292	67.52	12.00	20.48
C17H37P1	272.46	272.2633	74.94	13.69	11.37
C17H38P2	304.44	304.2449	67.07	12.58	20.35
C18H 9P1	256.25	256.0442	84.37	3.54	12.09
C18H10P2	288.23	288.0258	75.01	3.50	21.49
C18H11P1	258.26	258.0598	83.71	4.29	11.99
C18H12P2	290.24	290.0414	74.49	4.17	21.34
C18H13P1	260.28	260.0755	83.07	5.03	11.90
C18H14P2	292.26	292.0571	73.98	4.83	21.20
C18H15P1	262.29	262.0911	82.43	5.76	11.81
C18H16P2	294.28	294.0727	73.47	5.48	21.05
C18H17P1	264.31	264.1068	81.80	6.48	11.72
C18H18P2	296.29	296.0884	72.97	6.12	20.91
C18H19P1	266.33	266.1224	81.18	7.19	11.63
C18H20P2	298.31	298.1040	72.48	6.76	20.77
C18H21P1	268.34	268.1381	80.57	7.89	11.54
C18H22P2	300.32	300.1197	71.99	7.38	20.63
C18H23P1	270.36	270.1537	79.97	8.58	11.46
C18H24P2	302.34	302.1353	71.51	8.00	20.49
C18H25P1	272.37	272.1694	79.38	9.25	11.37
C18H26P2	304.36	304.1510	71.04	8.61	20.35
C18H27P1	274.39	274.1850	78.79	9.92	11.29
C18H28P2	306.37	306.1666	70.57	9.21	20.22
C18H29P1	276.41	276.2007	78.22	10.58	11.21

C18H30P2	308.39	308.1823	70.11	9.81	20.09
C18H31P1	278.42	278.2163	77.65	11.22	11.12
C18H32P2	310.40	310.1979	69.65	10.39	19.96
C18H33P1	280.44	280.2320	77.09	11.86	11.04
C18H34P2	312.42	312.2136	69.20	10.97	19.83
C18H35P1	282.45	282.2476	76.54	12.49	10.97
C18H36P2	314.44	314.2292	68.76	11.54	19.70
C18H37P1	284.47	284.2633	76.00	13.11	10.89
C18H38P2	316.45	316.2449	68.32	12.10	19.58
C18H39P1	286.49	286.2789	75.47	13.72	10.81
C18H40P2	318.47	318.2605	67.89	12.66	19.45
C19H 9P1	268.26	268.0442	85.07	3.38	11.55
C19H10P2	300.24	300.0258	76.01	3.36	20.63
C19H11P1	270.27	270.0598	84.44	4.10	11.46
C19H12P2	302.26	302.0414	75.50	4.00	20.50
C19H13P1	272.29	272.0755	83.81	4.81	11.38
C19H14P2	304.27	304.0571	75.00	4.64	20.36
C19H15P1	274.31	274.0911	83.20	5.51	11.29
C19H16P2	306.29	306.0727	74.51	5.27	20.23
C19H17P1	276.32	276.1068	82.59	6.20	11.21
C19H18P2	308.30	308.0884	74.02	5.88	20.09
C19H19P1	278.34	278.1224	81.99	6.88	11.13
C19H20P2	310.32	310.1040	73.54	6.50	19.96
C19H21P1	280.35	280.1381	81.40	7.55	11.05
C19H22P2	312.33	312.1197	73.07	7.10	19.83
C19H23P1	282.37	282.1537	80.82	8.21	10.97
C19H24P2	314.35	314.1353	72.60	7.70	19.71
C19H25P1	284.38	284.1694	80.25	8.86	10.89
C19H26P2	316.37	316.1510	72.14	8.28	19.58
C19H27P1	286.40	286.1850	79.68	9.50	10.81
C19H28P2	318.38	318.1666	71.68	8.86	19.46
C19H29P1	288.42	288.2007	79.13	10.14	10.74
C19H30P2	320.40	320.1823	71.23	9.44	19.33
C19H31P1	290.43	290.2163	78.58	10.76	10.66
C19H32P2	322.41	322.1979	70.78	10.00	19.21
C19H33P1	292.45	292.2320	78.03	11.37	10.59
C19H34P2	324.43	324.2136	70.34	10.56	19.09
C19H35P1	294.46	294.2476	77.50	11.98	10.52
C19H36P2	326.45	326.2292	69.91	11.12	18.98
C19H37P1	296.48	296.2633	76.97	12.58	10.45
C19H38P2	328.46	328.2449	69.48	11.66	18.86
C19H39P1	298.50	298.2789	76.45	13.17	10.38
C19H40P2	330.48	330.2605	69.06	12.20	18.74
C19H41P1	300.51	300.2946	75.94	13.75	10.31
C19H42P2	332.49	332.2762	68.64	12.73	18.63
C20H10P2	312.25	312.0258	76.93	3.23	19.84
C20H11P1	282.28	282.0598	85.10	3.93	10.97
C20H12P2	314.27	314.0414	76.44	3.85	19.71
C20H13P1	284.30	284.0755	84.50	4.61	10.89
C20H14P2	316.28	316.0571	75.95	4.46	19.59
C20H15P1	286.32	286.0911	83.90	5.28	10.82
C20H16P2	318.30	318.0727	75.47	5.07	19.46
C20H17P1	288.33	288.1068	83.31	5.94	10.74
C20H18P2	320.31	320.0884	75.00	5.66	19.34
C20H19P1	290.35	290.1224	82.74	6.60	10.67
C20H20P2	322.33	322.1040	74.53	6.25	19.22
C20H21P1	292.36	292.1381	82.17	7.24	10.59
C20H22P2	324.35	324.1197	74.06	6.84	19.10
C20H23P1	294.38	294.1537	81.60	7.88	10.52
C20H24P2	326.36	326.1353	73.61	7.41	18.98

C20H25P1	296.40	296.1694	81.05	8.50	10.45
C20H26P2	328.38	328.1510	73.15	7.98	18.86
C20H27P1	298.41	298.1850	80.50	9.12	10.38
C20H28P2	330.39	330.1666	72.71	8.54	18.75
C20H29P1	300.43	300.2007	79.96	9.73	10.31
C20H30P2	332.41	332.1823	72.27	9.10	18.64
C20H31P1	302.44	302.2163	79.43	10.33	10.24
C20H32P2	334.43	334.1979	71.83	9.64	18.52
C20H33P1	304.46	304.2320	78.90	10.93	10.17
C20H34P2	336.44	336.2136	71.40	10.19	18.41
C20H35P1	306.48	306.2476	78.38	11.51	10.11
C20H36P2	338.46	338.2292	70.98	10.72	18.30
C20H37P1	308.49	308.2633	77.87	12.09	10.04
C20H38P2	340.47	340.2449	70.56	11.25	18.19
C20H39P1	310.51	310.2789	77.36	12.66	9.98
C20H40P2	342.49	342.2605	70.14	11.77	18.09
C20H41P1	312.52	312.2946	76.87	13.22	9.91
C20H42P2	344.51	344.2762	69.73	12.29	17.98
C20H43P1	314.54	314.3102	76.37	13.78	9.85
C20H44P2	346.52	346.2918	69.32	12.80	17.88
C21H10P2	324.26	324.0258	77.79	3.11	19.10
C21H11P1	294.30	294.0598	85.71	3.77	10.52
C21H12P2	326.28	326.0414	77.31	3.71	18.99
C21H13P1	296.31	296.0755	85.12	4.42	10.45
C21H14P2	328.29	328.0571	76.83	4.30	18.87
C21H15P1	298.33	298.0911	84.55	5.07	10.38
C21H16P2	330.31	330.0727	76.36	4.88	18.75
C21H17P1	300.34	300.1068	83.98	5.71	10.31
C21H18P2	332.33	332.0884	75.90	5.46	18.64
C21H19P1	302.36	302.1224	83.42	6.33	10.24
C21H20P2	334.34	334.1040	75.44	6.03	18.53
C21H21P1	304.38	304.1381	82.87	6.95	10.18
C21H22P2	336.36	336.1197	74.99	6.59	18.42
C21H23P1	306.39	306.1537	82.32	7.57	10.11
C21H24P2	338.37	338.1353	74.54	7.15	18.31
C21H25P1	308.41	308.1694	81.79	8.17	10.04
C21H26P2	340.39	340.1510	74.10	7.70	18.20
C21H27P1	310.42	310.1850	81.25	8.77	9.98
C21H28P2	342.40	342.1666	73.67	8.24	18.09
C21H29P1	312.44	312.2007	80.73	9.36	9.91
C21H30P2	344.42	344.1823	73.23	8.78	17.99
C21H31P1	314.46	314.2163	80.21	9.94	9.85
C21H32P2	346.44	346.1979	72.81	9.31	17.88
C21H33P1	316.47	316.2320	79.70	10.51	9.79
C21H34P2	348.45	348.2136	72.39	9.84	17.78
C21H35P1	318.49	318.2476	79.20	11.08	9.73
C21H36P2	350.47	350.2292	71.97	10.35	17.68
C21H37P1	320.50	320.2633	78.70	11.64	9.66
C21H38P2	352.48	352.2449	71.56	10.87	17.57
C21H39P1	322.52	322.2789	78.21	12.19	9.60
C21H40P2	354.50	354.2605	71.15	11.37	17.47
C21H41P1	324.53	324.2946	77.72	12.73	9.54
C21H42P2	356.52	356.2762	70.75	11.87	17.38
C21H43P1	326.55	326.3102	77.24	13.27	9.49
C21H44P2	358.53	358.2918	70.35	12.37	17.28
C21H45P1	328.57	328.3259	76.77	13.81	9.43
C21H46P2	360.55	360.3075	69.96	12.86	17.18
C22H11P1	306.31	306.0598	86.27	3.62	10.11
C22H12P2	338.29	338.0414	78.11	3.58	18.31
C22H13P1	308.32	308.0755	85.70	4.25	10.05

C22H14P2	340.30	340.0571	77.65	4.15	18.20
C22H15P1	310.34	310.0911	85.15	4.87	9.98
C22H16P2	342.32	342.0727	77.19	4.71	18.10
C22H17P1	312.35	312.1068	84.60	5.49	9.92
C22H18P2	344.34	344.0884	76.74	5.27	17.99
C22H19P1	314.37	314.1224	84.06	6.09	9.85
C22H20P2	346.35	346.1040	76.29	5.82	17.89
C22H21P1	316.39	316.1381	83.52	6.69	9.79
C22H22P2	348.37	348.1197	75.85	6.37	17.78
C22H23P1	318.40	318.1537	82.99	7.28	9.73
C22H24P2	350.38	350.1353	75.42	6.90	17.68
C22H25P1	320.42	320.1694	82.47	7.86	9.67
C22H26P2	352.40	352.1510	74.98	7.44	17.58
C22H27P1	322.43	322.1850	81.95	8.44	9.61
C22H28P2	354.42	354.1666	74.56	7.96	17.48
C22H29P1	324.45	324.2007	81.44	9.01	9.55
C22H30P2	356.43	356.1823	74.14	8.48	17.38
C22H31P1	326.47	326.2163	80.94	9.57	9.49
C22H32P2	358.45	358.1979	73.72	9.00	17.28
C22H33P1	328.48	328.2320	80.44	10.13	9.43
C22H34P2	360.46	360.2136	73.31	9.51	17.19
C22H35P1	330.50	330.2476	79.95	10.67	9.37
C22H36P2	362.48	362.2292	72.90	10.01	17.09
C22H37P1	332.51	332.2633	79.47	11.22	9.32
C22H38P2	364.50	364.2449	72.50	10.51	17.00
C22H39P1	334.53	334.2789	78.99	11.75	9.26
C22H40P2	366.51	366.2605	72.10	11.00	16.90
C22H41P1	336.55	336.2946	78.52	12.28	9.20
C22H42P2	368.53	368.2762	71.70	11.49	16.81
C22H43P1	338.56	338.3102	78.05	12.80	9.15
C22H44P2	370.54	370.2918	71.31	11.97	16.72
C22H45P1	340.58	340.3259	77.59	13.32	9.09
C22H46P2	372.56	372.3075	70.93	12.45	16.63
C22H47P1	342.59	342.3415	77.13	13.83	9.04
C22H48P2	374.58	374.3231	70.55	12.92	16.54
C23H11P1	318.32	318.0598	86.79	3.48	9.73
C23H12P2	350.30	350.0414	78.86	3.45	17.68
C23H13P1	320.33	320.0755	86.24	4.09	9.67
C23H14P2	352.32	352.0571	78.41	4.01	17.58
C23H15P1	322.35	322.0911	85.70	4.69	9.61
C23H16P2	354.33	354.0727	77.97	4.55	17.48
C23H17P1	324.37	324.1068	85.17	5.28	9.55
C23H18P2	356.35	356.0884	77.52	5.09	17.38
C23H19P1	326.38	326.1224	84.64	5.87	9.49
C23H20P2	358.36	358.1040	77.09	5.63	17.29
C23H21P1	328.40	328.1381	84.12	6.45	9.43
C23H22P2	360.38	360.1197	76.66	6.15	17.19
C23H23P1	330.41	330.1537	83.61	7.02	9.37
C23H24P2	362.40	362.1353	76.23	6.68	17.09
C23H25P1	332.43	332.1694	83.10	7.58	9.32
C23H26P2	364.41	364.1510	75.81	7.19	17.00
C23H27P1	334.45	334.1850	82.60	8.14	9.26
C23H28P2	366.43	366.1666	75.39	7.70	16.91
C23H29P1	336.46	336.2007	82.11	8.69	9.21
C23H30P2	368.44	368.1823	74.98	8.21	16.81
C23H31P1	338.48	338.2163	81.62	9.23	9.15
C23H32P2	370.46	370.1979	74.57	8.71	16.72
C23H33P1	340.49	340.2320	81.13	9.77	9.10
C23H34P2	372.48	372.2136	74.17	9.20	16.63
C23H35P1	342.51	342.2476	80.66	10.30	9.04

C23H36P2	374.49	374.2292	73.77	9.69	16.54
C23H37P1	344.53	344.2633	80.18	10.83	8.99
C23H38P2	376.51	376.2449	73.37	10.17	16.45
C23H39P1	346.54	346.2789	79.72	11.34	8.94
C23H40P2	378.52	378.2605	72.98	10.65	16.37
C23H41P1	348.56	348.2946	79.26	11.86	8.89
C23H42P2	380.54	380.2762	72.60	11.12	16.28
C23H43P1	350.57	350.3102	78.80	12.36	8.84
C23H44P2	382.55	382.2918	72.21	11.59	16.19
C23H45P1	352.59	352.3259	78.35	12.86	8.78
C23H46P2	384.57	384.3075	71.84	12.06	16.11
C23H47P1	354.60	354.3415	77.91	13.36	8.73
C23H48P2	386.59	386.3231	71.46	12.52	16.02
C23H49P1	356.62	356.3572	77.47	13.85	8.69
C23H50P2	388.60	388.3388	71.09	12.97	15.94
C24H12P2	362.31	362.0414	79.56	3.34	17.10
C24H13P1	332.35	332.0755	86.74	3.94	9.32
C24H14P2	364.33	364.0571	79.12	3.87	17.00
C24H15P1	334.36	334.0911	86.21	4.52	9.26
C24H16P2	366.34	366.0727	78.69	4.40	16.91
C24H17P1	336.38	336.1068	85.70	5.09	9.21
C24H18P2	368.36	368.0884	78.26	4.93	16.82
C24H19P1	338.39	338.1224	85.19	5.66	9.15
C24H20P2	370.37	370.1040	77.83	5.44	16.73
C24H21P1	340.41	340.1381	84.68	6.22	9.10
C24H22P2	372.39	372.1197	77.41	5.95	16.64
C24H23P1	342.42	342.1537	84.18	6.77	9.05
C24H24P2	374.41	374.1353	76.99	6.46	16.55
C24H25P1	344.44	344.1694	83.69	7.32	8.99
C24H26P2	376.42	376.1510	76.58	6.96	16.46
C24H27P1	346.46	346.1850	83.20	7.86	8.94
C24H28P2	378.44	378.1666	76.17	7.46	16.37
C24H29P1	348.47	348.2007	82.72	8.39	8.89
C24H30P2	380.45	380.1823	75.77	7.95	16.28
C24H31P1	350.49	350.2163	82.25	8.92	8.84
C24H32P2	382.47	382.1979	75.37	8.43	16.20
C24H33P1	352.50	352.2320	81.78	9.44	8.79
C24H34P2	384.49	384.2136	74.97	8.91	16.11
C24H35P1	354.52	354.2476	81.31	9.95	8.74
C24H36P2	386.50	386.2292	74.58	9.39	16.03
C24H37P1	356.54	356.2633	80.85	10.46	8.69
C24H38P2	388.52	388.2449	74.20	9.86	15.94
C24H39P1	358.55	358.2789	80.40	10.96	8.64
C24H40P2	390.53	390.2605	73.81	10.32	15.86
C24H41P1	360.57	360.2946	79.95	11.46	8.59
C24H42P2	392.55	392.2762	73.43	10.78	15.78
C24H43P1	362.58	362.3102	79.50	11.95	8.54
C24H44P2	394.57	394.2918	73.06	11.24	15.70
C24H45P1	364.60	364.3259	79.06	12.44	8.50
C24H46P2	396.58	396.3075	72.69	11.69	15.62
C24H47P1	366.62	366.3415	78.63	12.92	8.45
C24H48P2	398.60	398.3231	72.32	12.14	15.54
C24H49P1	368.63	368.3572	78.20	13.40	8.40
C24H50P2	400.61	400.3388	71.96	12.58	15.46
C24H51P1	370.65	370.3728	77.77	13.87	8.36
C24H52P2	402.63	402.3544	71.60	13.02	15.39
C25H12P2	374.32	374.0414	80.22	3.23	16.55
C25H13P1	344.36	344.0755	87.20	3.81	8.99
C25H14P2	376.34	376.0571	79.79	3.75	16.46
C25H15P1	346.37	346.0911	86.69	4.37	8.94

C25H16P2	378.35	378.0727	79.36	4.26	16.37
C25H17P1	348.39	348.1068	86.19	4.92	8.89
C25H18P2	380.37	380.0884	78.94	4.77	16.29
C25H19P1	350.40	350.1224	85.70	5.47	8.84
C25H20P2	382.39	382.1040	78.53	5.27	16.20
C25H21P1	352.42	352.1381	85.20	6.01	8.79
C25H22P2	384.40	384.1197	78.12	5.77	16.12
C25H23P1	354.44	354.1537	84.72	6.54	8.74
C25H24P2	386.42	386.1353	77.71	6.26	16.03
C25H25P1	356.45	356.1694	84.24	7.07	8.69
C25H26P2	388.43	388.1510	77.31	6.75	15.95
C25H27P1	358.47	358.1850	83.77	7.59	8.64
C25H28P2	390.45	390.1666	76.91	7.23	15.87
C25H29P1	360.48	360.2007	83.30	8.11	8.59
C25H30P2	392.47	392.1823	76.51	7.70	15.78
C25H31P1	362.50	362.2163	82.84	8.62	8.54
C25H32P2	394.48	394.1979	76.12	8.18	15.70
C25H33P1	364.52	364.2320	82.38	9.13	8.50
C25H34P2	396.50	396.2136	75.73	8.64	15.62
C25H35P1	366.53	366.2476	81.92	9.63	8.45
C25H36P2	398.51	398.2292	75.35	9.11	15.54
C25H37P1	368.55	368.2633	81.48	10.12	8.40
C25H38P2	400.53	400.2449	74.97	9.56	15.47
C25H39P1	370.56	370.2789	81.03	10.61	8.36
C25H40P2	402.55	402.2605	74.60	10.02	15.39
C25H41P1	372.58	372.2946	80.59	11.09	8.31
C25H42P2	404.56	404.2762	74.22	10.46	15.31
C25H43P1	374.60	374.3102	80.16	11.57	8.27
C25H44P2	406.58	406.2918	73.86	10.91	15.24
C25H45P1	376.61	376.3259	79.73	12.04	8.22
C25H46P2	408.59	408.3075	73.49	11.35	15.16
C25H47P1	378.63	378.3415	79.31	12.51	8.18
C25H48P2	410.61	410.3231	73.13	11.78	15.09
C25H49P1	380.64	380.3572	78.89	12.98	8.14
C25H50P2	412.62	412.3388	72.77	12.21	15.01
C25H51P1	382.66	382.3728	78.47	13.43	8.09
C25H52P2	414.64	414.3544	72.42	12.64	14.94
C25H53P1	384.67	384.3885	78.06	13.89	8.05
C25H54P2	416.66	416.3701	72.07	13.06	14.87
C26H13P1	356.37	356.0755	87.63	3.68	8.69
C26H14P2	388.35	388.0571	80.41	3.63	15.95
C26H15P1	358.38	358.0911	87.14	4.22	8.64
C26H16P2	390.37	390.0727	80.00	4.13	15.87
C26H17P1	360.40	360.1068	86.65	4.75	8.59
C26H18P2	392.38	392.0884	79.59	4.62	15.79
C26H19P1	362.42	362.1224	86.17	5.28	8.55
C26H20P2	394.40	394.1040	79.18	5.11	15.71
C26H21P1	364.43	364.1381	85.69	5.81	8.50
C26H22P2	396.41	396.1197	78.78	5.59	15.63
C26H23P1	366.45	366.1537	85.22	6.33	8.45
C26H24P2	398.43	398.1353	78.38	6.07	15.55
C26H25P1	368.46	368.1694	84.75	6.84	8.41
C26H26P2	400.44	400.1510	77.99	6.54	15.47
C26H27P1	370.48	370.1850	84.29	7.35	8.36
C26H28P2	402.46	402.1666	77.60	7.01	15.39
C26H29P1	372.49	372.2007	83.84	7.85	8.32
C26H30P2	404.48	404.1823	77.21	7.48	15.32
C26H31P1	374.51	374.2163	83.39	8.34	8.27
C26H32P2	406.49	406.1979	76.83	7.93	15.24
C26H33P1	376.53	376.2320	82.94	8.83	8.23

C26H34P2	408.51	408.2136	76.45	8.39	15.16
C26H35P1	378.54	378.2476	82.50	9.32	8.18
C26H36P2	410.52	410.2292	76.07	8.84	15.09
C26H37P1	380.56	380.2633	82.06	9.80	8.14
C26H38P2	412.54	412.2449	75.70	9.28	15.02
C26H39P1	382.57	382.2789	81.63	10.28	8.10
C26H40P2	414.56	414.2605	75.33	9.73	14.94
C26H41P1	384.59	384.2946	81.20	10.75	8.05
C26H42P2	416.57	416.2762	74.97	10.16	14.87
C26H43P1	386.61	386.3102	80.78	11.21	8.01
C26H44P2	418.59	418.2918	74.61	10.60	14.80
C26H45P1	388.62	388.3259	80.36	11.67	7.97
C26H46P2	420.60	420.3075	74.25	11.02	14.73
C26H47P1	390.64	390.3415	79.94	12.13	7.93
C26H48P2	422.62	422.3231	73.89	11.45	14.66
C26H49P1	392.65	392.3572	79.53	12.58	7.89
C26H50P2	424.64	424.3388	73.54	11.87	14.59
C26H51P1	394.67	394.3728	79.13	13.03	7.85
C26H52P2	426.65	426.3544	73.20	12.29	14.52
C26H53P1	396.69	396.3885	78.72	13.47	7.81
C26H54P2	428.67	428.3701	72.85	12.70	14.45
C26H55P1	398.70	398.4041	78.33	13.90	7.77
C26H56P2	430.68	430.3857	72.51	13.11	14.38
C27H13P1	368.38	368.0755	88.03	3.56	8.41
C27H14P2	400.36	400.0571	81.00	3.52	15.47
C27H15P1	370.39	370.0911	87.56	4.08	8.36
C27H16P2	402.38	402.0727	80.60	4.01	15.40
C27H17P1	372.41	372.1068	87.08	4.60	8.32
C27H18P2	404.39	404.0884	80.19	4.49	15.32
C27H19P1	374.43	374.1224	86.61	5.11	8.27
C27H20P2	406.41	406.1040	79.80	4.96	15.24
C27H21P1	376.44	376.1381	86.15	5.62	8.23
C27H22P2	408.42	408.1197	79.40	5.43	15.17
C27H23P1	378.46	378.1537	85.69	6.13	8.18
C27H24P2	410.44	410.1353	79.01	5.89	15.09
C27H25P1	380.47	380.1694	85.24	6.62	8.14
C27H26P2	412.46	412.1510	78.63	6.35	15.02
C27H27P1	382.49	382.1850	84.79	7.12	8.10
C27H28P2	414.47	414.1666	78.24	6.81	14.95
C27H29P1	384.51	384.2007	84.34	7.60	8.06
C27H30P2	416.49	416.1823	77.87	7.26	14.87
C27H31P1	386.52	386.2163	83.90	8.08	8.01
C27H32P2	418.50	418.1979	77.49	7.71	14.80
C27H33P1	388.54	388.2320	83.47	8.56	7.97
C27H34P2	420.52	420.2136	77.12	8.15	14.73
C27H35P1	390.55	390.2476	83.04	9.03	7.93
C27H36P2	422.54	422.2292	76.75	8.59	14.66
C27H37P1	392.57	392.2633	82.61	9.50	7.89
C27H38P2	424.55	424.2449	76.39	9.02	14.59
C27H39P1	394.59	394.2789	82.19	9.96	7.85
C27H40P2	426.57	426.2605	76.03	9.45	14.52
C27H41P1	396.60	396.2946	81.77	10.42	7.81
C27H42P2	428.58	428.2762	75.67	9.88	14.45
C27H43P1	398.62	398.3102	81.36	10.87	7.77
C27H44P2	430.60	430.2918	75.31	10.30	14.39
C27H45P1	400.63	400.3259	80.95	11.32	7.73
C27H46P2	432.62	432.3075	74.96	10.72	14.32
C27H47P1	402.65	402.3415	80.54	11.77	7.69
C27H48P2	434.63	434.3231	74.62	11.13	14.25
C27H49P1	404.67	404.3572	80.14	12.21	7.65

C27H50P2	436.65	436.3388	74.27	11.54	14.19
C27H51P1	406.68	406.3728	79.74	12.64	7.62
C27H52P2	438.66	438.3544	73.93	11.95	14.12
C27H53P1	408.70	408.3885	79.35	13.07	7.58
C27H54P2	440.68	440.3701	73.59	12.35	14.06
C27H55P1	410.71	410.4041	78.96	13.50	7.54
C27H56P2	442.69	442.3857	73.26	12.75	13.99
C27H57P1	412.73	412.4198	78.57	13.92	7.50
C27H58P2	444.71	444.4014	72.92	13.15	13.93
C28H14P2	412.37	412.0571	81.56	3.42	15.02
C28H15P1	382.41	382.0911	87.95	3.95	8.10
C28H16P2	414.39	414.0727	81.16	3.89	14.95
C28H17P1	384.42	384.1068	87.49	4.46	8.06
C28H18P2	416.40	416.0884	80.77	4.36	14.88
C28H19P1	386.44	386.1224	87.03	4.96	8.02
C28H20P2	418.42	418.1040	80.38	4.82	14.81
C28H21P1	388.45	388.1381	86.58	5.45	7.97
C28H22P2	420.44	420.1197	79.99	5.27	14.73
C28H23P1	390.47	390.1537	86.13	5.94	7.93
C28H24P2	422.45	422.1353	79.61	5.73	14.66
C28H25P1	392.49	392.1694	85.69	6.42	7.89
C28H26P2	424.47	424.1510	79.23	6.17	14.59
C28H27P1	394.50	394.1850	85.25	6.90	7.85
C28H28P2	426.48	426.1666	78.86	6.62	14.53
C28H29P1	396.52	396.2007	84.82	7.37	7.81
C28H30P2	428.50	428.1823	78.49	7.06	14.46
C28H31P1	398.53	398.2163	84.39	7.84	7.77
C28H32P2	430.51	430.1979	78.12	7.49	14.39
C28H33P1	400.55	400.2320	83.96	8.30	7.73
C28H34P2	432.53	432.2136	77.75	7.92	14.32
C28H35P1	402.56	402.2476	83.54	8.76	7.69
C28H36P2	434.55	434.2292	77.39	8.35	14.26
C28H37P1	404.58	404.2633	83.13	9.22	7.66
C28H38P2	436.56	436.2449	77.04	8.77	14.19
C28H39P1	406.60	406.2789	82.71	9.67	7.62
C28H40P2	438.58	438.2605	76.68	9.19	14.12
C28H41P1	408.61	408.2946	82.31	10.11	7.58
C28H42P2	440.59	440.2762	76.33	9.61	14.06
C28H43P1	410.63	410.3102	81.90	10.56	7.54
C28H44P2	442.61	442.2918	75.98	10.02	14.00
C28H45P1	412.64	412.3259	81.50	10.99	7.51
C28H46P2	444.63	444.3075	75.64	10.43	13.93
C28H47P1	414.66	414.3415	81.11	11.42	7.47
C28H48P2	446.64	446.3231	75.30	10.83	13.87
C28H49P1	416.68	416.3572	80.71	11.85	7.43
C28H50P2	448.66	448.3388	74.96	11.23	13.81
C28H51P1	418.69	418.3728	80.32	12.28	7.40
C28H52P2	450.67	450.3544	74.62	11.63	13.75
C28H53P1	420.71	420.3885	79.94	12.70	7.36
C28H54P2	452.69	452.3701	74.29	12.02	13.68
C28H55P1	422.72	422.4041	79.56	13.11	7.33
C28H56P2	454.71	454.3857	73.96	12.41	13.62
C28H57P1	424.74	424.4198	79.18	13.53	7.29
C28H58P2	456.72	456.4014	73.64	12.80	13.56
C28H59P1	426.76	426.4354	78.81	13.94	7.26
C28H60P2	458.74	458.4170	73.31	13.18	13.50
C29H14P2	424.38	424.0571	82.08	3.33	14.60
C29H15P1	394.42	394.0911	88.31	3.83	7.85
C29H16P2	426.40	426.0727	81.69	3.78	14.53
C29H17P1	396.43	396.1068	87.86	4.32	7.81
C29H18P2	428.41	428.0884	81.31	4.24	14.46
C29H19P1	398.45	398.1224	87.42	4.81	7.77
C29H20P2	430.43	430.1040	80.92	4.68	14.39
C29H21P1	400.46	400.1381	86.98	5.29	7.73
C29H22P2	432.45	432.1197	80.55	5.13	14.32
C29H23P1	402.48	402.1537	86.54	5.76	7.70
C29H24P2	434.46	434.1353	80.17	5.57	14.26
C29H25P1	404.50	404.1694	86.11	6.23	7.66
C29H26P2	436.48	436.1510	79.80	6.00	14.19
C29H27P1	406.51	406.1850	85.69	6.69	7.62
C29H28P2	438.49	438.1666	79.44	6.44	14.13
C29H29P1	408.53	408.2007	85.26	7.16	7.58
C29H30P2	440.51	440.1823	79.07	6.86	14.06
C29H31P1	410.54	410.2163	84.84	7.61	7.54
C29H32P2	442.53	442.1979	78.71	7.29	14.00
C29H33P1	412.56	412.2320	84.43	8.06	7.51
C29H34P2	444.54	444.2136	78.36	7.71	13.94
C29H35P1	414.58	414.2476	84.02	8.51	7.47
C29H36P2	446.56	446.2292	78.00	8.13	13.87
C29H37P1	416.59	416.2633	83.61	8.95	7.44
C29H38P2	448.57	448.2449	77.65	8.54	13.81
C29H39P1	418.61	418.2789	83.21	9.39	7.40
C29H40P2	450.59	450.2605	77.30	8.95	13.75
C29H41P1	420.62	420.2946	82.81	9.83	7.36
C29H42P2	452.61	452.2762	76.96	9.35	13.69
C29H43P1	422.64	422.3102	82.42	10.26	7.33
C29H44P2	454.62	454.2918	76.62	9.76	13.63
C29H45P1	424.66	424.3259	82.02	10.68	7.29
C29H46P2	456.64	456.3075	76.28	10.15	13.57
C29H47P1	426.67	426.3415	81.64	11.10	7.26
C29H48P2	458.65	458.3231	75.94	10.55	13.51
C29H49P1	428.69	428.3572	81.25	11.52	7.23
C29H50P2	460.67	460.3388	75.61	10.94	13.45
C29H51P1	430.70	430.3728	80.87	11.94	7.19
C29H52P2	462.69	462.3544	75.28	11.33	13.39
C29H53P1	432.72	432.3885	80.50	12.35	7.16
C29H54P2	464.70	464.3701	74.96	11.71	13.33
C29H55P1	434.74	434.4041	80.12	12.75	7.12
C29H56P2	466.72	466.3857	74.63	12.09	13.27
C29H57P1	436.75	436.4198	79.75	13.15	7.09
C29H58P2	468.73	468.4014	74.31	12.47	13.22
C29H59P1	438.77	438.4354	79.39	13.55	7.06
C29H60P2	470.75	470.4170	73.99	12.85	13.16
C29H61P1	440.78	440.4511	79.02	13.95	7.03
C29H62P2	472.77	472.4327	73.68	13.22	13.10
C30H15P1	406.43	406.0911	88.66	3.72	7.62
C30H16P2	438.41	438.0727	82.19	3.68	14.13
C30H17P1	408.44	408.1068	88.22	4.20	7.58
C30H18P2	440.43	440.0884	81.82	4.12	14.07
C30H19P1	410.46	410.1224	87.79	4.67	7.55
C30H20P2	442.44	442.1040	81.44	4.56	14.00
C30H21P1	412.48	412.1381	87.36	5.13	7.51
C30H22P2	444.46	444.1197	81.07	4.99	13.94
C30H23P1	414.49	414.1537	86.93	5.59	7.47
C30H24P2	446.47	446.1353	80.71	5.42	13.87
C30H25P1	416.51	416.1694	86.51	6.05	7.44
C30H26P2	448.49	448.1510	80.34	5.84	13.81
C30H27P1	418.52	418.1850	86.10	6.50	7.40
C30H28P2	450.51	450.1666	79.98	6.26	13.75
C30H29P1	420.54	420.2007	85.68	6.95	7.37

10/21/66 107

C30H30P2	452.52	452.1823	79.63	6.68	13.69
C30H31P1	422.56	422.2163	85.28	7.39	7.33
C30H32P2	454.54	454.1979	79.28	7.10	13.63
C30H33P1	424.57	424.2320	84.87	7.83	7.30
C30H34P2	456.55	456.2136	78.92	7.51	13.57
C30H35P1	426.59	426.2476	84.47	8.27	7.26
C30H36P2	458.57	458.2292	78.58	7.91	13.51
C30H37P1	428.60	428.2633	84.07	8.70	7.23
C30H38P2	460.58	460.2449	78.23	8.32	13.45
C30H39P1	430.62	430.2789	83.68	9.13	7.19
C30H40P2	462.60	462.2605	77.89	8.72	13.39
C30H41P1	432.64	432.2946	83.29	9.55	7.16
C30H42P2	464.62	464.2762	77.56	9.11	13.33
C30H43P1	434.65	434.3102	82.90	9.97	7.13
C30H44P2	466.63	466.2918	77.22	9.50	13.28
C30H45P1	436.67	436.3259	82.52	10.39	7.09
C30H46P2	468.65	468.3075	76.89	9.89	13.22
C30H47P1	438.68	438.3415	82.14	10.80	7.06
C30H48P2	470.66	470.3231	76.56	10.28	13.16
C30H49P1	440.70	440.3572	81.76	11.21	7.03
C30H50P2	472.68	472.3388	76.23	10.66	13.11
C30H51P1	442.71	442.3728	81.39	11.61	7.00
C30H52P2	474.70	474.3544	75.91	11.04	13.05
C30H53P1	444.73	444.3885	81.02	12.01	6.96
C30H54P2	476.71	476.3701	75.59	11.42	12.99
C30H55P1	446.75	446.4041	80.66	12.41	6.93
C30H56P2	478.73	478.3857	75.27	11.79	12.94
C30H57P1	448.76	448.4198	80.30	12.80	6.90
C30H58P2	480.74	480.4014	74.95	12.16	12.89
C30H59P1	450.78	450.4354	79.94	13.19	6.87
C30H60P2	482.76	482.4170	74.64	12.53	12.83
C30H61P1	452.79	452.4511	79.58	13.58	6.84
C30H62P2	484.78	484.4327	74.33	12.89	12.78
C30H63P1	454.81	454.4667	79.23	13.96	6.81
C30H64P2	486.79	486.4483	74.02	13.25	12.73

III

COMPOUNDS CONTAINING CARBON, HYDROGEN AND TWO HETEROATOMS

VERBINDUNGEN AUS KOHLENSTOFF, WASSERSTOFF UND ZWEI HETEROATOMEN

COMPOSES CONTENANT DU CARBONE, DE L'HYDROGENE ET DEUX HETEROATOMES

1. CARBON, HYDROGEN, BROMINE AND NITROGEN

1. KOHLENSTOFF, WASSERSTOFF, BROM UND STICKSTOFF

1. CARBONE, HYDROGENE, BROME ET AZOTE

MOLECULAR WEIGHTS AND PERCENTAGE COMPOSITIONS OF COMPOUNDS
CONTAINING CARBON, HYDROGEN, BROMINE
AND NITROGEN

FORMULA	MOL. WEIGHT	ISOMOL. WEIGHT	PERC. C	PERC. H	PERC. BR	PERC. N	FORMULA	MOL. WEIGHT	ISOMOL. WEIGHT	PERC. C	PERC. H	PERC. BR	PERC. N
C 1H 0BR1N1	105.92	104.9215	11.34	0	75.44	13.22	C 2H 1BR2N5	254.87	252.8600	9.43	.40	62.70	27.48
C 1H 0BR1N3	133.94	132.9276	8.97	0	59.66	31.37	C 2H 2BR1N1	119.95	118.9371	20.03	1.68	66.62	11.68
C 1H 0BR1N5	161.95	160.9338	7.42	0	49.34	43.24	C 2H 2BR1N3	147.96	146.9433	16.24	1.36	54.00	28.40
C 1H 0BR2N2	199.83	197.8429	6.01	0	79.97	14.02	C 2H 2BR1N5	175.98	174.9494	13.65	1.15	45.41	39.80
C 1H 0BR2N4	227.85	225.8491	5.27	0	70.14	24.59	C 2H 2BR2N2	213.86	211.8586	11.23	.94	74.73	13.10
C 1H 0BR2N6	255.86	253.8552	4.69	0	62.46	32.85	C 2H 2BR2N4	241.87	239.8647	9.93	.83	66.07	23.16
C 1H 1BR1N2	120.94	119.9324	9.93	.83	66.07	23.16	C 2H 2BR2N6	269.89	267.8709	8.90	.75	59.21	31.14
C 1H 1BR1N4	148.95	147.9385	8.06	.68	53.64	37.61	C 2H 3BR1N2	134.96	133.9480	17.80	2.24	59.20	20.76
C 1H 1BR1N6	176.96	175.9447	6.79	.57	45.15	47.49	C 2H 3BR1N4	162.98	161.9542	14.74	1.86	49.03	34.38
C 1H 1BR2N1	186.83	184.8477	6.43	.54	85.53	7.50	C 2H 3BR1N6	190.99	189.9603	12.58	1.58	41.84	44.00
C 1H 1BR2N3	214.85	212.8538	5.59	.47	74.38	19.56	C 2H 3BR2N1	200.86	198.8633	11.96	1.51	79.56	6.97
C 1H 1BR2N5	242.86	240.8600	4.95	.42	65.80	28.84	C 2H 3BR2N3	228.87	226.8695	10.50	1.32	69.82	18.36
C 1H 2BR1N1	107.94	106.9371	11.13	1.87	74.03	12.98	C 2H 3BR2N5	256.89	254.8756	9.35	1.18	62.21	27.26
C 1H 2BR1N3	135.95	134.9433	8.83	1.48	58.77	30.91	C 2H 4BR1N1	121.96	120.9528	19.70	3.31	65.51	11.48
C 1H 2BR1N5	163.96	162.9494	7.33	1.23	48.73	42.71	C 2H 4BR1N3	149.98	148.9589	16.02	2.69	53.28	28.02
C 1H 2BR2N2	201.85	199.8586	5.95	1.00	79.17	13.88	C 2H 4BR1N5	177.99	176.9651	13.50	2.27	44.89	39.35
C 1H 2BR2N4	229.86	227.8647	5.23	.88	69.52	24.37	C 2H 4BR2N2	215.88	213.8742	11.13	1.87	74.03	12.98
C 1H 2BR2N6	257.88	255.8709	4.66	.78	61.97	32.59	C 2H 4BR2N4	243.89	241.8804	9.85	1.65	65.52	22.97
C 1H 3BR1N2	122.95	121.9480	9.77	2.46	64.99	22.78	C 2H 4BR2N6	271.90	269.8865	8.83	1.48	58.77	30.91
C 1H 3BR1N4	150.97	149.9542	7.96	2.00	52.93	37.11	C 2H 5BR1N2	136.98	135.9637	17.54	3.68	58.33	20.45
C 1H 3BR1N6	178.98	177.9603	6.71	1.69	44.64	46.96	C 2H 5BR1N4	164.99	163.9698	14.56	3.05	48.43	33.96
C 1H 3BR2N1	188.85	186.8633	6.36	1.60	84.62	7.42	C 2H 5BR1N6	193.01	191.9760	12.45	2.61	41.40	43.54
C 1H 3BR2N3	216.86	214.8695	5.54	1.39	73.69	19.38	C 2H 5BR2N1	202.88	200.8790	11.84	2.48	78.77	6.90
C 1H 3BR2N5	244.88	242.8756	4.90	1.23	65.26	28.60	C 2H 5BR2N3	230.89	228.8851	10.40	2.18	69.21	18.20
C 1H 4BR1N1	109.95	108.9528	10.92	3.67	72.67	12.74	C 2H 5BR2N5	258.90	256.8913	9.28	1.95	61.72	27.05
C 1H 4BR1N3	137.97	136.9589	8.71	2.92	57.92	30.46	C 2H 6BR1N1	123.98	122.9684	19.38	4.88	64.45	11.30
C 1H 4BR1N5	165.98	164.9651	7.24	2.43	48.14	42.19	C 2H 6BR1N3	151.99	150.9746	15.80	3.98	52.57	27.65
C 1H 4BR2N2	203.86	201.8742	5.89	1.98	78.39	13.74	C 2H 6BR1N5	180.01	178.9807	13.35	3.36	44.39	38.91
C 1H 4BR2N4	231.88	229.8804	5.18	1.74	68.92	24.16	C 2H 6BR2N2	217.89	215.8899	11.02	2.78	73.34	12.86
C 1H 4BR2N6	259.89	257.8865	4.62	1.55	61.49	32.34	C 2H 6BR2N4	245.90	243.8960	9.77	2.46	64.99	22.78
C 1H 5BR1N2	124.97	123.9637	9.61	4.03	63.94	22.42	C 2H 6BR2N6	273.92	271.9022	8.77	2.21	58.34	30.68
C 1H 5BR1N4	152.98	151.9698	7.85	3.29	52.23	36.62	C 2H 7BR1N2	139.00	137.9793	17.28	5.08	57.49	20.15
C 1H 5BR1N6	181.00	179.9760	6.64	2.78	44.15	46.43	C 2H 7BR1N4	167.01	165.9855	14.38	4.22	47.84	33.55
C 1H 5BR2N3	218.88	216.8851	5.49	2.30	73.01	19.20	C 2H 7BR1N6	195.02	193.9916	12.32	3.62	40.97	43.09
C 1H 5BR2N5	246.89	244.8913	4.86	2.04	64.73	28.37	C 2H 7BR2N3	232.91	230.9008	10.31	3.03	68.61	18.04
C 1H 6BR1N3	139.98	138.9746	8.58	4.32	57.08	30.02	C 2H 7BR2N5	260.92	258.9069	9.21	2.70	61.25	26.84
C 1H 6BR1N5	168.00	166.9807	7.15	3.60	47.56	41.69	C 2H 8BR1N3	154.01	152.9902	15.60	5.24	51.88	27.28
C 1H 6BR2N4	233.89	231.8960	5.14	2.59	68.33	23.95	C 2H 8BR1N5	182.02	180.9964	13.20	4.43	43.90	38.47
C 1H 6BR2N6	261.91	259.9022	4.59	2.31	61.02	32.09	C 2H 8BR2N4	247.92	245.9117	9.69	3.25	64.46	22.60
C 1H 7BR1N4	155.00	153.9855	7.75	4.55	51.55	36.15	C 2H 8BR2N6	275.93	273.9178	8.71	2.92	57.92	30.46
C 1H 7BR1N6	183.01	181.9916	6.56	3.86	43.66	45.92	C 2H 9BR1N4	169.02	168.0011	14.21	5.37	47.27	33.15
C 1H 7BR2N5	248.91	246.9069	4.83	2.83	64.20	28.14	C 2H 9BR1N6	197.04	196.0073	12.19	4.60	40.55	42.65
C 1H 8BR1N5	170.01	168.9964	7.06	4.74	47.00	41.19	C 2H 9BR2N5	262.94	260.9226	9.14	3.45	60.78	26.64
C 1H 8BR2N6	263.92	261.9178	4.55	3.06	60.55	31.84	C 2H10BR1N5	184.04	183.0120	13.05	5.48	43.42	38.05
C 1H 9BR1N6	185.03	184.0073	6.49	4.90	43.19	45.42	C 2H10BR2N6	277.95	275.9335	8.64	3.63	57.50	30.24
C 2H 1BR1N2	132.95	131.9324	18.07	.76	60.10	21.07	C 2H11BR1N6	199.05	198.0229	12.07	5.57	40.14	42.22
C 2H 1BR1N4	160.96	159.9385	14.92	.63	49.64	34.81	C 3H 1BR1N2	144.96	143.9324	24.86	.70	55.12	19.33
C 2H 1BR1N6	188.97	187.9447	12.71	.53	42.28	44.47	C 3H 1BR1N4	172.97	171.9385	20.83	.58	46.19	32.39
C 2H 1BR2N1	198.84	196.8477	12.08	.51	80.37	7.04	C 3H 1BR1N6	200.99	199.9447	17.93	.50	39.76	41.81
C 2H 1BR2N3	226.86	224.8538	10.59	.44	70.44	18.52	C 3H 1BR2N1	210.86	208.8477	17.09	.48	75.79	6.64

C 3H 1BR2N3	238.87	236.8538	15.08	.42	66.90	17.59	C 4H 2BR1N3	171.98	170.9433	27.94	1.17	46.46	24.43
C 3H 1BR2N5	266.88	264.8600	13.50	.38	59.88	26.24	C 4H 2BR1N5	200.00	198.9494	24.02	1.01	39.95	35.02
C 3H 2BR1N1	131.96	130.9371	27.31	1.53	60.55	10.61	C 4H 2BR2N2	237.88	235.8586	20.20	.85	67.18	11.78
C 3H 2BR1N3	159.97	158.9433	22.52	1.26	49.95	26.27	C 4H 2BR2N4	265.90	263.8647	18.07	.76	60.10	21.07
C 3H 2BR1N5	187.99	186.9494	19.17	1.07	42.51	37.25	C 4H 2BR2N6	293.91	291.8709	16.35	.69	54.37	28.59
C 3H 2BR2N2	225.87	223.8586	15.95	.89	70.75	12.40	C 4H 3BR1N2	158.99	157.9480	30.22	1.90	50.26	17.62
C 3H 2BR2N4	253.88	251.8647	14.19	.79	62.95	22.07	C 4H 3BR1N4	187.00	185.9542	25.69	1.62	42.73	29.96
C 3H 2BR2N6	281.90	279.8709	12.78	.72	56.69	29.81	C 4H 3BR1N6	215.01	213.9603	22.35	1.41	37.16	39.09
C 3H 3BR1N2	146.97	145.9480	24.52	2.06	54.37	19.06	C 4H 3BR2N1	224.88	222.8633	21.36	1.34	71.06	6.23
C 3H 3BR1N4	174.99	173.9542	20.59	1.73	45.66	32.02	C 4H 3BR2N3	252.90	250.8695	19.00	1.20	63.19	16.62
C 3H 3BR1N6	203.00	201.9603	17.75	1.49	39.36	41.40	C 4H 3BR2N5	280.91	278.8756	17.10	1.08	56.89	24.93
C 3H 3BR2N1	212.87	210.8633	16.93	1.42	75.07	6.58	C 4H 4BR1N1	145.99	144.9528	32.91	2.76	54.73	9.59
C 3H 3BR2N3	240.89	238.8695	14.96	1.26	66.34	17.44	C 4H 4BR1N3	174.00	172.9589	27.61	2.32	45.92	24.15
C 3H 3BR2N5	268.90	266.8756	13.40	1.12	59.43	26.04	C 4H 4BR1N5	202.01	200.9651	23.78	2.00	39.55	34.67
C 3H 4BR1N1	133.98	132.9528	26.90	3.01	59.64	10.45	C 4H 4BR2N2	239.90	237.8742	20.03	1.68	66.62	11.68
C 3H 4BR1N3	161.99	160.9589	22.24	2.49	49.33	25.94	C 4H 4BR2N4	267.91	265.8804	17.93	1.50	59.65	20.91
C 3H 4BR1N5	190.00	188.9651	18.96	2.12	42.05	36.86	C 4H 4BR2N6	295.92	293.8865	16.24	1.36	54.00	28.40
C 3H 4BR2N2	227.89	225.8742	15.81	1.77	70.13	12.29	C 4H 5BR1N2	161.00	159.9637	29.84	3.13	49.63	17.40
C 3H 4BR2N4	255.90	253.8804	14.08	1.58	62.45	21.89	C 4H 5BR1N4	189.02	187.9698	25.42	2.67	42.27	29.64
C 3H 4BR2N6	283.91	281.8865	12.69	1.42	56.29	29.60	C 4H 5BR1N6	217.03	215.9760	22.14	2.32	36.82	38.72
C 3H 5BR1N2	148.99	147.9637	24.19	3.38	53.63	18.80	C 4H 5BR2N1	226.90	224.8790	21.17	2.22	70.43	6.17
C 3H 5BR1N4	177.00	175.9698	20.36	2.85	45.14	31.65	C 4H 5BR2N3	254.91	252.8851	18.85	1.98	62.69	16.48
C 3H 5BR1N6	205.02	203.9760	17.58	2.46	38.97	40.99	C 4H 5BR2N5	282.93	280.8913	16.98	1.78	56.48	24.75
C 3H 5BR2N1	214.89	212.8790	16.77	2.35	74.37	6.52	C 4H 6BR1N1	148.00	146.9684	32.46	4.09	53.99	9.46
C 3H 5BR2N3	242.90	240.8851	14.83	2.07	65.79	17.30	C 4H 6BR1N3	176.02	174.9746	27.30	3.44	45.40	23.87
C 3H 5BR2N5	270.91	268.8913	13.30	1.86	58.99	25.85	C 4H 6BR1N5	204.03	202.9807	23.55	2.96	39.16	34.33
C 3H 6BR1N1	135.99	134.9684	26.50	4.45	58.76	10.30	C 4H 6BR2N2	241.91	239.8899	19.86	2.50	66.06	11.58
C 3H 6BR1N3	164.01	162.9746	21.97	3.69	48.72	25.62	C 4H 6BR2N4	269.93	267.8960	17.80	2.24	59.20	20.76
C 3H 6BR1N5	192.02	190.9807	18.77	3.15	41.61	36.47	C 4H 6BR2N6	297.94	295.9022	16.13	2.03	53.64	28.21
C 3H 6BR2N2	229.90	227.8899	15.67	2.63	69.51	12.18	C 4H 7BR1N2	163.02	161.9793	29.47	4.33	49.02	17.18
C 3H 6BR2N4	257.92	255.8960	13.97	2.34	61.96	21.72	C 4H 7BR1N4	191.03	189.9855	25.15	3.69	41.83	29.33
C 3H 6BR2N6	285.93	283.9022	12.60	2.12	55.89	29.39	C 4H 7BR1N6	219.04	217.9916	21.93	3.22	36.48	38.37
C 3H 7BR1N2	151.01	149.9793	23.86	4.67	52.91	18.55	C 4H 7BR2N1	228.92	226.8946	20.99	3.08	69.81	6.12
C 3H 7BR1N4	179.02	177.9855	20.13	3.94	44.63	31.30	C 4H 7BR2N3	256.93	254.9008	18.70	2.75	62.20	16.35
C 3H 7BR1N6	207.03	205.9916	17.40	3.41	38.59	40.59	C 4H 7BR2N5	284.94	282.9069	16.86	2.48	56.08	24.58
C 3H 7BR2N1	216.90	214.8946	16.61	3.25	73.68	6.46	C 4H 8BR1N1	150.02	148.9841	32.03	5.38	53.26	9.34
C 3H 7BR2N3	244.92	242.9008	14.71	2.88	65.25	17.16	C 4H 8BR1N3	178.03	176.9902	26.99	4.53	44.88	23.60
C 3H 7BR2N5	272.93	270.9069	13.20	2.59	58.55	25.66	C 4H 8BR1N5	206.05	204.9964	23.32	3.91	38.78	33.99
C 3H 8BR1N1	138.01	136.9841	26.11	5.84	57.90	10.15	C 4H 8BR2N2	243.93	241.9055	19.70	3.31	65.51	11.48
C 3H 8BR1N3	166.02	164.9902	21.70	4.86	48.13	25.31	C 4H 8BR2N4	271.94	269.9117	17.67	2.97	58.77	20.60
C 3H 8BR1N5	194.03	192.9964	18.57	4.16	41.18	36.09	C 4H 8BR2N6	299.96	297.9178	16.02	2.69	53.28	28.02
C 3H 8BR2N2	231.92	229.9055	15.54	3.48	68.91	12.08	C 4H 9BR1N2	165.03	163.9950	29.11	5.50	48.42	16.97
C 3H 8BR2N4	259.93	257.9117	13.86	3.10	61.48	21.55	C 4H 9BR1N4	193.05	192.0011	24.89	4.70	41.39	29.02
C 3H 8BR2N6	287.95	285.9178	12.51	2.80	55.50	29.19	C 4H 9BR1N6	221.06	220.0073	21.73	4.10	36.15	38.02
C 3H 9BR1N2	153.02	151.9950	23.55	5.93	52.22	18.31	C 4H 9BR2N1	230.93	228.9103	20.80	3.93	69.20	6.07
C 3H 9BR1N4	181.04	180.0011	19.90	5.01	44.14	30.95	C 4H 9BR2N3	258.94	256.9164	18.55	3.50	61.72	16.23
C 3H 9BR1N6	209.05	208.0073	17.24	4.34	38.22	40.20	C 4H 9BR2N5	286.96	284.9226	16.74	3.16	55.69	24.41
C 3H 9BR2N3	246.93	244.9164	14.59	3.67	64.72	17.02	C 4H10BR1N1	152.04	150.9997	31.60	6.63	52.56	9.21
C 3H 9BR2N5	274.95	272.9226	13.11	3.30	58.12	25.47	C 4H10BR1N3	180.05	179.0059	26.68	5.60	44.38	23.34
C 3H10BR1N3	168.04	167.0059	21.44	6.00	47.55	25.01	C 4H10BR1N5	208.06	207.0120	23.09	4.84	38.40	33.66
C 3H10BR1N5	196.05	195.0120	18.38	5.14	40.76	35.72	C 4H10BR2N2	245.95	243.9212	19.53	4.10	64.98	11.39
C 3H10BR2N4	261.95	259.9273	13.76	3.85	61.01	21.39	C 4H10BR2N4	273.96	271.9273	17.54	3.68	58.33	20.45
C 3H10BR2N6	289.96	287.9335	12.43	3.48	55.11	28.98	C 4H10BR2N6	301.97	299.9335	15.91	3.34	52.92	27.83
C 3H11BR1N4	183.05	182.0168	19.68	6.06	43.65	30.61	C 4H11BR1N2	167.05	166.0106	28.76	6.64	47.83	16.77
C 3H11BR1N6	211.07	210.0229	17.07	5.25	37.86	39.82	C 4H11BR1N4	195.06	194.0168	24.63	5.68	40.96	28.72
C 3H11BR2N5	276.96	274.9382	13.01	4.00	57.70	25.29	C 4H11BR1N6	223.08	222.0229	21.54	4.97	35.82	37.67
C 3H12BR1N5	198.07	197.0277	18.19	6.11	40.34	35.36	C 4H11BR2N3	260.96	258.9321	18.41	4.25	61.24	16.10
C 3H12BR2N6	291.98	289.9491	12.34	4.14	54.73	28.78	C 4H11BR2N5	288.97	286.9382	16.63	3.84	55.30	24.24
C 3H13BR1N6	213.08	212.0386	16.91	6.15	37.50	39.44	C 4H12BR1N3	182.06	181.0215	26.39	6.64	43.89	23.08
C 4H 2BR1N1	143.97	142.9371	33.37	1.40	55.50	9.73	C 4H12BR1N5	210.08	209.0277	22.87	5.76	38.04	33.34

243

C 4H12BR2N4	275.98	273.9430	17.41	4.38	57.91	20.30
C 4H12BR2N6	303.99	301.9491	15.80	3.98	52.57	27.65
C 4H13BR1N4	197.08	196.0324	24.38	6.65	40.54	28.43
C 4H13BR1N6	225.09	224.0386	21.34	5.82	35.50	37.34
C 4H13BR2N5	290.99	288.9539	16.51	4.50	54.92	24.07
C 4H14BR1N5	212.09	211.0433	22.65	6.65	37.67	33.02
C 4H14BR2N6	306.00	303.9648	15.70	4.61	52.22	27.46
C 4H15BR1N6	227.11	226.0542	21.15	6.66	35.18	37.00
C 5H 2BR1N1	155.98	154.9371	38.50	1.29	51.23	8.98
C 5H 2BR1N3	184.00	182.9433	32.64	1.10	43.43	22.84
C 5H 2BR1N5	212.01	210.9494	28.33	.95	37.69	33.03
C 5H 2BR2N2	249.89	247.8586	24.03	.81	63.95	11.21
C 5H 2BR2N4	277.91	275.8647	21.61	.73	57.50	20.16
C 5H 2BR2N6	305.92	303.8709	19.63	.66	52.24	27.47
C 5H 3BR1N2	171.00	169.9480	35.12	1.77	46.73	16.38
C 5H 3BR1N4	199.01	197.9542	30.18	1.52	40.15	28.15
C 5H 3BR1N6	227.02	225.9603	26.45	1.33	35.20	37.02
C 5H 3BR2N1	236.89	234.8633	25.35	1.28	67.46	5.91
C 5H 3BR2N3	264.91	262.8695	22.67	1.14	60.33	15.86
C 5H 3BR2N5	292.92	290.8756	20.50	1.03	54.56	23.91
C 5H 4BR1N1	158.00	156.9528	38.01	2.55	50.57	8.87
C 5H 4BR1N3	186.01	184.9589	32.29	2.17	42.96	22.59
C 5H 4BR1N5	214.03	212.9651	28.06	1.88	37.33	32.72
C 5H 4BR2N2	251.91	249.8742	23.84	1.60	63.44	11.12
C 5H 4BR2N4	279.92	277.8804	21.45	1.44	57.09	20.02
C 5H 4BR2N6	307.94	305.8865	19.50	1.31	51.90	27.29
C 5H 5BR1N2	173.01	171.9637	34.71	2.91	46.18	16.19
C 5H 5BR1N4	201.03	199.9698	29.87	2.51	39.75	27.87
C 5H 5BR1N6	229.04	227.9760	26.22	2.20	34.89	36.69
C 5H 5BR2N1	238.91	236.8790	25.14	2.11	66.89	5.86
C 5H 5BR2N3	266.92	264.8851	22.50	1.89	59.87	15.74
C 5H 5BR2N5	294.94	292.8913	20.36	1.71	54.18	23.75
C 5H 6BR1N1	160.01	158.9684	37.53	3.78	49.94	8.75
C 5H 6BR1N3	188.03	186.9746	31.94	3.22	42.50	22.35
C 5H 6BR1N5	216.04	214.9807	27.80	2.80	36.99	32.42
C 5H 6BR2N2	253.92	251.8899	23.65	2.38	62.94	11.03
C 5H 6BR2N4	281.94	279.8960	21.30	2.15	56.68	19.87
C 5H 6BR2N6	309.95	307.9022	19.38	1.95	51.56	27.11
C 5H 7BR1N2	175.03	173.9793	34.31	4.03	45.65	16.01
C 5H 7BR1N4	203.04	201.9855	29.58	3.48	39.35	27.59
C 5H 7BR1N6	231.06	229.9916	25.99	3.05	34.58	36.37
C 5H 7BR2N1	240.93	238.8946	24.93	2.93	66.33	5.81
C 5H 7BR2N3	268.94	266.9008	22.33	2.62	59.42	15.62
C 5H 7BR2N5	296.95	294.9069	20.22	2.38	53.82	23.58
C 5H 8BR1N1	162.03	160.9841	37.06	4.98	49.31	8.64
C 5H 8BR1N3	190.04	188.9902	31.60	4.24	42.05	22.11
C 5H 8BR1N5	218.06	216.9964	27.54	3.70	36.64	32.12
C 5H 8BR2N2	255.94	253.9055	23.46	3.15	62.44	10.95
C 5H 8BR2N4	283.95	281.9117	21.15	2.84	56.28	19.73
C 5H 8BR2N6	311.97	309.9178	19.25	2.58	51.23	26.94
C 5H 9BR1N2	177.04	175.9950	33.92	5.12	45.13	15.82
C 5H 9BR1N4	205.06	204.0011	29.29	4.42	38.97	27.32
C 5H 9BR1N6	233.07	232.0073	25.77	3.89	34.28	36.06
C 5H 9BR2N1	242.94	240.9103	24.72	3.73	65.78	5.77
C 5H 9BR2N3	270.96	268.9164	22.16	3.35	58.98	15.51
C 5H 9BR2N5	298.97	296.9226	20.09	3.03	53.45	23.43
C 5H10BR1N1	164.05	162.9997	36.61	6.14	48.71	8.54
C 5H10BR1N3	192.06	191.0059	31.27	5.25	41.60	21.88
C 5H10BR1N5	220.07	219.0120	27.29	4.58	36.31	31.82
C 5H10BR2N2	257.96	255.9212	23.28	3.91	61.95	10.86
C 5H10BR2N4	285.97	283.9273	21.00	3.52	55.88	19.59
C 5H10BR2N6	313.98	311.9335	19.13	3.21	50.90	26.77
C 5H11BR1N2	179.06	178.0106	33.54	6.19	44.62	15.64
C 5H11BR1N4	207.07	206.0168	29.00	5.35	38.59	27.06
C 5H11BR1N6	235.09	234.0229	25.55	4.72	33.99	35.75
C 5H11BR2N1	244.96	242.9259	24.52	4.53	65.24	5.72
C 5H11BR2N3	272.97	270.9321	22.00	4.06	58.54	15.39
C 5H11BR2N5	300.98	298.9382	19.95	3.68	53.10	23.27
C 5H12BR1N1	166.06	165.0154	36.16	7.28	48.12	8.43
C 5H12BR1N3	194.08	193.0215	30.94	6.23	41.17	21.65
C 5H12BR1N5	222.09	221.0277	27.04	5.45	35.98	31.53
C 5H12BR2N2	259.97	257.9368	23.10	4.65	61.47	10.78
C 5H12BR2N4	287.99	285.9430	20.85	4.20	55.49	19.45
C 5H12BR2N6	316.00	313.9491	19.01	3.83	50.57	26.60
C 5H13BR1N2	181.08	180.0263	33.17	7.24	44.13	15.47
C 5H13BR1N4	209.09	208.0324	28.72	6.27	38.22	26.80
C 5H13BR1N6	237.10	236.0386	25.33	5.53	33.70	35.44
C 5H13BR2N3	274.99	272.9477	21.84	4.77	58.11	15.28
C 5H13BR2N5	303.00	300.9539	19.82	4.32	52.74	23.11
C 5H14BR1N3	196.09	195.0372	30.63	7.20	40.75	21.43
C 5H14BR1N5	224.10	223.0433	26.80	6.30	35.65	31.25
C 5H14BR2N4	290.00	287.9586	20.71	4.87	55.11	19.32
C 5H14BR2N6	318.02	315.9648	18.88	4.44	50.25	26.43
C 5H15BR1N4	211.11	210.0481	28.45	7.16	37.85	26.54
C 5H15BR1N6	239.12	238.0542	25.12	6.32	33.42	35.15
C 5H15BR2N5	305.02	302.9695	19.69	4.96	52.39	22.96
C 5H16BR1N5	226.12	225.0590	26.56	7.13	35.34	30.97
C 5H16BR2N6	320.03	317.9804	18.77	5.04	49.94	26.26
C 5H17BR1N6	241.14	240.0699	24.91	7.11	33.14	34.85
C 6H 3BR1N2	183.01	181.9480	39.38	1.65	43.66	15.31
C 6H 3BR1N4	211.02	209.9542	34.15	1.43	37.87	26.55
C 6H 3BR1N6	239.04	237.9603	30.15	1.27	33.43	35.16
C 6H 3BR2N1	248.91	246.8633	28.95	1.21	64.20	5.63
C 6H 3BR2N3	276.92	274.8695	26.02	1.09	57.71	15.17
C 6H 3BR2N5	304.93	302.8756	23.63	.99	52.41	22.97
C 6H 4BR1N1	170.01	168.9528	42.39	2.37	47.00	8.24
C 6H 4BR1N3	198.02	196.9589	36.39	2.04	40.35	21.22
C 6H 4BR1N5	226.04	224.9651	31.88	1.78	35.35	30.98
C 6H 4BR2N2	263.92	261.8742	27.31	1.53	60.55	10.61
C 6H 4BR2N4	291.93	289.8804	24.69	1.38	54.74	19.19
C 6H 4BR2N6	319.95	317.8865	22.52	1.26	49.95	26.27
C 6H 5BR1N2	185.02	183.9637	38.95	2.72	43.19	15.14
C 6H 5BR1N4	213.04	211.9698	33.83	2.37	37.51	26.30
C 6H 5BR1N6	241.05	239.9760	29.90	2.09	33.15	34.86
C 6H 5BR2N1	250.92	248.8790	28.72	2.01	63.69	5.58
C 6H 5BR2N3	278.93	276.8851	25.84	1.81	57.29	15.06
C 6H 5BR2N5	306.95	304.8913	23.48	1.64	52.06	22.82
C 6H 6BR1N1	172.03	170.9684	41.89	3.52	46.45	8.14
C 6H 6BR1N3	200.04	198.9746	36.03	3.02	39.94	21.01
C 6H 6BR1N5	228.05	226.9807	31.60	2.65	35.04	30.71
C 6H 6BR2N2	265.94	263.8899	27.10	2.27	60.09	10.53
C 6H 6BR2N4	293.95	291.8960	24.52	2.06	54.37	19.06
C 6H 6BR2N6	321.96	319.9022	22.38	1.88	49.64	26.10
C 6H 7BR1N2	187.04	185.9793	38.53	3.77	42.72	14.98
C 6H 7BR1N4	215.05	213.9855	33.51	3.28	37.16	26.05
C 6H 7BR1N6	243.07	241.9916	29.65	2.90	32.87	34.57
C 6H 7BR2N1	252.94	250.8946	28.49	2.79	63.18	5.54
C 6H 7BR2N3	280.95	278.9008	25.65	2.51	56.88	14.96
C 6H 7BR2N5	308.96	306.9069	23.33	2.28	51.72	22.67
C 6H 8BR1N1	174.04	172.9841	41.41	4.63	45.91	8.05

244

C 6H 8BR1N3	202.05	200.9902	35.67	3.99	39.55	20.80
C 6H 8BR1N5	230.07	228.9964	31.32	3.50	34.73	30.44
C 6H 8BR2N2	267.95	265.9055	26.90	3.01	59.64	10.45
C 6H 8BR2N4	295.97	293.9117	24.35	2.72	54.00	18.93
C 6H 8BR2N6	323.98	321.9178	22.24	2.49	49.33	25.94
C 6H 9BR1N2	189.06	187.9950	38.12	4.80	42.26	14.82
C 6H 9BR1N4	217.07	216.0011	33.20	4.18	36.81	25.81
C 6H 9BR1N6	245.08	244.0073	29.41	3.70	32.60	34.29
C 6H 9BR2N1	254.95	252.9103	28.27	3.56	62.68	5.49
C 6H 9BR2N3	282.97	280.9164	25.47	3.21	56.48	14.85
C 6H 9BR2N5	310.98	308.9226	23.17	2.92	51.39	22.52
C 6H10BR1N1	176.06	174.9997	40.93	5.73	45.39	7.96
C 6H10BR1N3	204.07	203.0059	35.31	4.94	39.16	20.59
C 6H10BR1N5	232.08	231.0120	31.05	4.34	34.43	30.18
C 6H10BR2N2	269.97	267.9212	26.69	3.73	59.20	10.38
C 6H10BR2N4	297.98	295.9273	24.19	3.38	53.63	18.80
C 6H10BR2N6	325.99	323.9335	22.11	3.09	49.02	25.78
C 6H11BR1N2	191.07	190.0106	37.72	5.80	41.82	14.66
C 6H11BR1N4	219.09	218.0168	32.89	5.06	36.47	25.57
C 6H11BR1N6	247.10	246.0229	29.17	4.49	32.34	34.01
C 6H11BR2N1	256.97	254.9259	28.04	4.31	62.19	5.45
C 6H11BR2N3	284.98	282.9321	25.29	3.89	56.08	14.74
C 6H11BR2N5	313.00	310.9382	23.02	3.54	51.06	22.38
C 6H12BR1N1	178.07	177.0154	40.47	6.79	44.87	7.87
C 6H12BR1N3	206.09	205.0215	34.97	5.87	38.77	20.39
C 6H12BR1N5	234.10	233.0277	30.78	5.17	34.13	29.92
C 6H12BR2N2	271.98	269.9368	26.50	4.45	58.76	10.30
C 6H12BR2N4	300.00	297.9430	24.02	4.03	53.27	18.68
C 6H12BR2N6	328.01	325.9491	21.97	3.69	48.72	25.62
C 6H13BR1N2	193.09	192.0263	37.32	6.79	41.38	14.51
C 6H13BR1N4	221.10	220.0324	32.59	5.93	36.14	25.34
C 6H13BR1N6	249.11	248.0386	28.93	5.26	32.08	33.74
C 6H13BR2N1	258.99	256.9416	27.83	5.06	61.71	5.41
C 6H13BR2N3	287.00	284.9477	25.11	4.57	55.68	14.64
C 6H13BR2N5	315.01	312.9539	22.88	4.16	50.73	22.23
C 6H14BR1N1	180.09	179.0310	40.02	7.84	44.37	7.78
C 6H14BR1N3	208.10	207.0372	34.63	6.78	38.40	20.19
C 6H14BR1N5	236.12	235.0433	30.52	5.98	33.84	29.66
C 6H14BR2N2	274.00	271.9525	26.30	5.15	58.32	10.22
C 6H14BR2N4	302.01	299.9586	23.86	4.67	52.91	18.55
C 6H14BR2N6	330.03	327.9648	21.84	4.28	48.42	25.46
C 6H15BR1N2	195.10	194.0419	36.94	7.75	40.95	14.36
C 6H15BR1N4	223.12	222.0481	32.30	6.78	35.81	25.11
C 6H15BR1N6	251.13	250.0542	28.70	6.02	31.82	33.46
C 6H15BR2N3	289.01	286.9634	24.94	5.23	55.29	14.54
C 6H15BR2N5	317.03	314.9695	22.73	4.77	50.41	22.09
C 6H16BR1N3	210.12	209.0528	34.30	7.68	38.03	20.00
C 6H16BR1N5	238.13	237.0590	30.26	6.77	33.55	29.41
C 6H16BR2N4	304.03	301.9743	23.70	5.30	52.56	18.43
C 6H16BR2N6	332.04	329.9804	21.70	4.86	48.13	25.31
C 6H17BR1N4	225.13	224.0637	32.01	7.61	35.49	24.89
C 6H17BR1N6	253.15	252.0699	28.47	6.77	31.56	33.20
C 6H17BR2N5	319.04	316.9852	22.59	5.37	50.09	21.95
C 6H18BR1N5	240.15	239.0746	30.01	7.56	33.27	29.16
C 6H18BR2N6	334.06	331.9961	21.57	5.43	47.84	25.16
C 6H19BR1N6	255.16	254.0855	28.24	7.51	31.31	32.94
C 7H 3BR1N2	195.02	193.9480	43.11	1.55	40.97	14.36
C 7H 3BR1N4	223.03	221.9542	37.70	1.36	35.83	25.12
C 7H 3BR1N6	251.05	249.9603	33.49	1.20	31.83	33.48
C 7H 3BR2N1	260.92	258.8633	32.22	1.16	61.25	5.37
C 7H 3BR2N3	288.93	286.8695	29.10	1.05	55.31	14.54
C 7H 3BR2N5	316.94	314.8756	26.53	.95	50.42	22.10
C 7H 4BR1N1	182.02	180.9528	46.19	2.22	43.90	7.70
C 7H 4BR1N3	210.03	208.9589	40.03	1.92	38.04	20.01
C 7H 4BR1N5	238.05	236.9651	35.32	1.69	33.57	29.42
C 7H 4BR2N2	275.93	273.8742	30.47	1.46	57.92	10.15
C 7H 4BR2N4	303.94	301.8804	27.66	1.33	52.58	18.43
C 7H 4BR2N6	331.96	329.8865	25.33	1.21	48.14	25.32
C 7H 5BR1N2	197.04	195.9637	42.67	2.56	40.55	14.22
C 7H 5BR1N4	225.05	223.9698	37.36	2.24	35.51	24.90
C 7H 5BR1N6	253.06	251.9760	33.22	1.99	31.57	33.21
C 7H 5BR2N1	262.93	260.8790	31.98	1.92	60.78	5.33
C 7H 5BR2N3	290.95	288.8851	28.90	1.73	54.93	14.44
C 7H 5BR2N5	318.96	316.8913	26.36	1.58	50.10	21.96
C 7H 6BR1N1	184.04	182.9684	45.69	3.29	43.42	7.61
C 7H 6BR1N3	212.05	210.9746	39.65	2.85	37.68	19.82
C 7H 6BR1N5	240.06	238.9807	35.02	2.52	33.28	29.17
C 7H 6BR2N2	277.95	275.8899	30.25	2.18	57.50	10.08
C 7H 6BR2N4	305.96	303.8960	27.48	1.98	52.23	18.31
C 7H 6BR2N6	333.97	331.9022	25.18	1.81	47.85	25.16
C 7H 7BR1N2	199.05	197.9793	42.24	3.54	40.14	14.07
C 7H 7BR1N4	227.06	225.9855	37.03	3.11	35.19	24.67
C 7H 7BR1N6	255.08	253.9916	32.96	2.77	31.33	32.95
C 7H 7BR2N1	264.95	262.8946	31.73	2.66	60.32	5.29
C 7H 7BR2N3	292.96	290.9008	28.70	2.41	54.55	14.34
C 7H 7BR2N5	320.98	318.9069	26.19	2.20	49.79	21.82
C 7H 8BR1N1	186.05	184.9841	45.19	4.33	42.95	7.53
C 7H 8BR1N3	214.07	212.9902	39.28	3.77	37.33	19.63
C 7H 8BR1N5	242.08	240.9964	34.73	3.33	33.01	28.93
C 7H 8BR2N2	279.96	277.9055	30.03	2.88	57.08	10.01
C 7H 8BR2N4	307.98	305.9117	27.30	2.62	51.89	18.19
C 7H 8BR2N6	335.99	333.9178	25.02	2.40	47.56	25.01
C 7H 9BR1N2	201.07	199.9950	41.82	4.51	39.74	13.93
C 7H 9BR1N4	229.08	228.0011	36.70	3.96	34.88	24.46
C 7H 9BR1N6	257.09	256.0073	32.70	3.53	31.08	32.69
C 7H 9BR2N1	266.96	264.9103	31.49	3.40	59.86	5.25
C 7H 9BR2N3	294.98	292.9164	28.50	3.08	54.18	14.25
C 7H 9BR2N5	322.99	320.9226	26.03	2.81	49.48	21.68
C 7H10BR1N1	188.07	186.9997	44.71	5.36	42.49	7.45
C 7H10BR1N3	216.08	215.0059	38.91	4.66	36.98	19.45
C 7H10BR1N5	244.10	243.0120	34.44	4.13	32.73	28.69
C 7H10BR2N2	281.98	279.9212	29.82	3.57	56.67	9.93
C 7H10BR2N4	309.99	307.9273	27.12	3.25	51.55	18.07
C 7H10BR2N6	338.01	335.9335	24.87	2.98	47.28	24.86
C 7H11BR1N2	203.08	202.0106	41.40	5.46	39.35	13.79
C 7H11BR1N4	231.10	230.0168	36.38	4.80	34.58	24.24
C 7H11BR1N6	259.11	258.0229	32.45	4.28	30.84	32.43
C 7H11BR2N1	268.98	266.9259	31.26	4.12	59.41	5.21
C 7H11BR2N3	296.99	294.9321	28.31	3.73	53.81	14.15
C 7H11BR2N5	325.01	322.9382	25.87	3.41	49.17	21.55
C 7H12BR1N1	190.08	189.0154	44.23	6.36	42.04	7.37
C 7H12BR1N3	218.10	217.0215	38.55	5.55	36.64	19.27
C 7H12BR1N5	246.11	245.0277	34.16	4.91	32.47	28.46
C 7H12BR2N2	284.00	281.9368	29.61	4.26	56.27	9.86
C 7H12BR2N4	312.01	309.9430	26.95	3.88	51.22	17.96
C 7H12BR2N6	340.02	337.9491	24.73	3.56	47.00	24.72
C 7H13BR1N2	205.10	204.0263	40.99	6.39	38.96	13.66
C 7H13BR1N4	233.11	232.0324	36.07	5.62	34.28	24.03
C 7H13BR1N6	261.13	260.0386	32.20	5.02	30.60	32.18
C 7H13BR2N1	271.00	268.9416	31.03	4.84	58.97	5.17

C 7H13BR2N3	299.01	296.9477	28.12	4.38	53.45	14.05
C 7H13BR2N5	327.02	324.9539	25.71	4.01	48.87	21.42
C 7H14BR1N1	192.10	191.0310	43.77	7.35	41.59	7.29
C 7H14BR1N3	220.11	219.0372	38.20	6.41	36.30	19.09
C 7H14BR1N5	248.13	247.0433	33.89	5.69	32.20	28.22
C 7H14BR2N2	286.01	283.9525	29.40	4.93	55.87	9.79
C 7H14BR2N4	314.02	311.9586	26.77	4.49	50.89	17.84
C 7H14BR2N6	342.04	339.9648	24.58	4.13	46.72	24.57
C 7H15BR1N2	207.11	206.0419	40.59	7.30	38.58	13.53
C 7H15BR1N4	235.13	234.0481	35.76	6.43	33.98	23.83
C 7H15BR1N6	263.14	262.0542	31.95	5.75	30.37	31.94
C 7H15BR2N1	273.01	270.9572	30.80	5.54	58.54	5.13
C 7H15BR2N3	301.03	298.9634	27.93	5.02	53.09	13.96
C 7H15BR2N5	329.04	326.9695	25.55	4.60	48.57	21.28
C 7H16BR1N1	194.12	193.0467	43.31	8.31	41.16	7.22
C 7H16BR1N3	222.13	221.0528	37.85	7.26	35.97	18.92
C 7H16BR1N5	250.14	249.0590	33.61	6.45	31.94	28.00
C 7H16BR2N2	288.03	285.9681	29.19	5.60	55.48	9.73
C 7H16BR2N4	316.04	313.9743	26.60	5.10	50.57	17.73
C 7H16BR2N6	344.05	341.9804	24.44	4.69	46.45	24.43
C 7H17BR1N2	209.13	208.0576	40.20	8.19	38.21	13.40
C 7H17BR1N4	237.14	236.0637	35.45	7.23	33.69	23.63
C 7H17BR1N6	265.16	264.0699	31.71	6.46	30.13	31.69
C 7H17BR2N3	303.04	300.9790	27.74	5.65	52.73	13.87
C 7H17BR2N5	331.06	328.9852	25.40	5.18	48.27	21.15
C 7H18BR1N3	224.15	223.0685	37.51	8.09	35.65	18.75
C 7H18BR1N5	252.16	251.0746	33.34	7.20	31.69	27.77
C 7H18BR2N4	318.06	315.9899	26.43	5.70	50.25	17.62
C 7H18BR2N6	346.07	343.9961	24.30	5.24	46.18	24.28
C 7H19BR1N4	239.16	238.0794	35.16	8.01	33.41	23.43
C 7H19BR1N6	267.17	266.0855	31.47	7.17	29.91	31.46
C 7H19BR2N5	333.07	331.0008	25.24	5.75	47.98	21.03
C 7H20BR1N5	254.17	253.0903	33.08	7.93	31.44	27.55
C 7H20BR2N6	348.09	346.0117	24.15	5.79	45.91	24.14
C 7H21BR1N6	269.19	268.1011	31.23	7.86	29.68	31.22
C 8H 4BR1N1	194.03	192.9528	49.52	2.08	41.18	7.22
C 8H 4BR1N3	222.05	220.9589	43.27	1.82	35.99	18.92
C 8H 4BR1N5	250.06	248.9651	38.43	1.61	31.95	28.01
C 8H 4BR2N2	287.94	285.8742	33.37	1.40	55.50	9.73
C 8H 4BR2N4	315.96	313.8804	30.41	1.28	50.58	17.73
C 8H 4BR2N6	343.97	341.8865	27.94	1.17	46.46	24.43
C 8H 5BR1N2	209.05	207.9637	45.97	2.41	38.22	13.40
C 8H 5BR1N4	237.06	235.9698	40.53	2.13	33.71	23.63
C 8H 5BR1N6	265.07	263.9760	36.25	1.90	30.14	31.70
C 8H 5BR2N1	274.94	272.8790	34.95	1.83	58.12	5.09
C 8H 5BR2N3	302.96	300.8851	31.72	1.66	52.75	13.87
C 8H 5BR2N5	330.97	328.8913	29.03	1.52	48.28	21.16
C 8H 6BR1N1	196.05	194.9684	49.01	3.08	40.76	7.14
C 8H 6BR1N3	224.06	222.9746	42.89	2.70	35.66	18.75
C 8H 6BR1N5	252.07	250.9807	38.12	2.40	31.70	27.78
C 8H 6BR2N2	289.96	287.8899	33.14	2.09	55.11	9.66
C 8H 6BR2N4	317.97	315.8960	30.22	1.90	50.26	17.62
C 8H 6BR2N6	345.99	343.9022	27.77	1.75	46.19	24.29
C 8H 7BR1N2	211.06	209.9793	45.53	3.34	37.86	13.27
C 8H 7BR1N4	239.08	237.9855	40.19	2.95	33.42	23.43
C 8H 7BR1N6	267.09	265.9916	35.98	2.64	29.92	31.47
C 8H 7BR2N1	276.96	274.8946	34.69	2.55	57.70	5.06
C 8H 7BR2N3	304.97	302.9008	31.51	2.31	52.40	13.78
C 8H 7BR2N5	332.99	330.9069	28.86	2.12	47.99	21.03
C 8H 8BR1N1	198.06	196.9841	48.51	4.07	40.34	7.07
C 8H 8BR1N3	226.08	224.9902	42.50	3.57	35.34	18.59
C 8H 8BR1N5	254.09	252.9964	37.82	3.17	31.45	27.56
C 8H 8BR2N2	291.97	289.9055	32.91	2.76	54.73	9.59
C 8H 8BR2N4	319.99	317.9117	30.03	2.52	49.94	17.51
C 8H 8BR2N6	348.00	345.9178	27.61	2.32	45.92	24.15
C 8H 9BR1N2	213.08	211.9950	45.10	4.26	37.50	13.15
C 8H 9BR1N4	241.09	240.0011	39.86	3.76	33.14	23.24
C 8H 9BR1N6	269.11	268.0073	35.71	3.37	29.69	31.23
C 8H 9BR2N1	278.98	276.9103	34.44	3.25	57.28	5.02
C 8H 9BR2N3	306.99	304.9164	31.30	2.96	52.06	13.69
C 8H 9BR2N5	335.00	332.9226	28.68	2.71	47.70	20.91
C 8H10BR1N1	200.08	198.9997	48.03	5.04	39.94	7.00
C 8H10BR1N3	228.09	227.0059	42.13	4.42	35.03	18.42
C 8H10BR1N5	256.11	255.0120	37.52	3.94	31.20	27.35
C 8H10BR2N2	293.99	291.9212	32.68	3.43	54.36	9.53
C 8H10BR2N4	322.00	319.9273	29.84	3.13	49.63	17.40
C 8H10BR2N6	350.02	347.9335	27.45	2.88	45.66	24.01
C 8H11BR1N2	215.09	214.0106	44.67	5.15	37.15	13.02
C 8H11BR1N4	243.11	242.0168	39.53	4.56	32.87	23.05
C 8H11BR1N6	271.12	270.0229	35.44	4.09	29.47	31.00
C 8H11BR2N1	280.99	278.9259	34.20	3.95	56.87	4.98
C 8H11BR2N3	309.00	306.9321	31.10	3.59	51.72	13.60
C 8H11BR2N5	337.02	334.9382	28.51	3.29	47.42	20.78
C 8H12BR1N1	202.10	201.0154	47.55	5.99	39.54	6.93
C 8H12BR1N3	230.11	229.0215	41.76	5.26	34.72	18.26
C 8H12BR1N5	258.12	257.0277	37.23	4.69	30.96	27.13
C 8H12BR2N2	296.01	293.9368	32.46	4.09	53.99	9.46
C 8H12BR2N4	324.02	321.9430	29.66	3.73	49.32	17.29
C 8H12BR2N6	352.03	349.9491	27.30	3.44	45.40	23.87
C 8H13BR1N2	217.11	216.0263	44.26	6.04	36.80	12.90
C 8H13BR1N4	245.12	244.0324	39.20	5.35	32.60	22.86
C 8H13BR1N6	273.14	272.0386	35.18	4.80	29.25	30.77
C 8H13BR2N1	283.01	280.9416	33.95	4.63	56.47	4.95
C 8H13BR2N3	311.02	308.9477	30.89	4.21	51.38	13.51
C 8H13BR2N5	339.03	336.9539	28.34	3.86	47.14	20.66
C 8H14BR1N1	204.11	203.0310	47.08	6.91	39.15	6.86
C 8H14BR1N3	232.12	231.0372	41.40	6.08	34.42	18.10
C 8H14BR1N5	260.14	259.0433	36.94	5.42	30.72	26.92
C 8H14BR2N2	298.02	295.9525	32.24	4.74	53.62	9.40
C 8H14BR2N4	326.04	323.9586	29.47	4.33	49.02	17.18
C 8H14BR2N6	354.05	351.9648	27.14	3.99	45.14	23.74
C 8H15BR1N2	219.13	218.0419	43.85	6.90	36.46	12.78
C 8H15BR1N4	247.14	246.0481	38.88	6.12	32.33	22.67
C 8H15BR1N6	275.15	274.0542	34.92	5.49	29.04	30.54
C 8H15BR2N1	285.02	282.9572	33.71	5.30	56.07	4.91
C 8H15BR2N3	313.04	310.9634	30.70	4.83	51.05	13.42
C 8H15BR2N5	341.05	338.9695	28.17	4.43	46.86	20.53
C 8H16BR1N1	206.13	205.0467	46.62	7.82	38.76	6.80
C 8H16BR1N3	234.14	233.0528	41.04	6.89	34.13	17.95
C 8H16BR1N5	262.15	261.0590	36.65	6.15	30.48	26.71
C 8H16BR2N2	300.04	297.9681	32.03	5.38	53.26	9.34
C 8H16BR2N4	328.05	325.9743	29.29	4.92	48.71	17.08
C 8H16BR2N6	356.06	353.9804	26.99	4.53	44.88	23.60
C 8H17BR1N2	221.14	220.0576	43.45	7.75	36.13	12.67
C 8H17BR1N4	249.16	248.0637	38.57	6.88	32.07	22.49
C 8H17BR1N6	277.17	276.0699	34.67	6.18	28.83	30.32
C 8H17BR2N1	287.04	284.9729	33.48	5.97	55.67	4.88
C 8H17BR2N3	315.05	312.9790	30.50	5.44	50.72	13.34
C 8H17BR2N5	343.07	340.9852	28.01	4.99	46.58	20.41
C 8H18BR1N1	208.14	207.0623	46.16	8.72	38.39	6.73

C 8H18BR1N3	236.16	235.0685	40.69	7.68	33.84	17.79
C 8H18BR1N5	264.17	263.0746	36.37	6.87	30.25	26.51
C 8H18BR2N2	302.05	299.9838	31.81	6.01	52.91	9.27
C 8H18BR2N4	330.07	327.9899	29.11	5.50	48.42	16.97
C 8H18BR2N6	358.08	355.9961	26.83	5.07	44.63	23.47
C 8H19BR1N2	223.16	222.0732	43.06	8.58	35.81	12.55
C 8H19BR1N4	251.17	250.0794	38.26	7.62	31.81	22.31
C 8H19BR1N6	279.18	278.0855	34.42	6.86	28.62	30.10
C 8H19BR2N3	317.07	314.9947	30.31	6.04	50.40	13.25
C 8H19BR2N5	345.08	343.0008	27.85	5.55	46.31	20.29
C 8H20BR1N3	238.17	237.0841	40.34	8.46	33.55	17.64
C 8H20BR1N5	266.19	265.0903	36.10	7.57	30.02	26.31
C 8H20BR2N4	332.08	330.0056	28.94	6.07	48.12	16.87
C 8H20BR2N6	360.10	358.0117	26.68	5.60	44.38	23.34
C 8H21BR1N4	253.19	252.0950	37.95	8.36	31.56	22.13
C 8H21BR1N6	281.20	280.1011	34.17	7.53	28.42	29.89
C 8H21BR2N5	347.10	345.0165	27.68	6.10	46.04	20.18
C 8H22BR1N5	268.20	267.1059	35.83	8.27	29.79	26.11
C 8H22BR2N6	362.11	360.0274	26.54	6.12	44.13	23.21
C 8H23BR1N6	283.22	282.1168	33.93	8.19	28.21	29.67
C 9H 4BR1N1	206.04	204.9528	52.46	1.96	38.78	6.80
C 9H 4BR1N3	234.06	232.9589	46.19	1.72	34.14	17.95
C 9H 4BR1N5	262.07	260.9651	41.25	1.54	30.49	26.72
C 9H 4BR2N2	299.95	297.8742	36.04	1.34	53.28	9.34
C 9H 4BR2N4	327.97	325.8804	32.96	1.23	48.73	17.08
C 9H 4BR2N6	355.98	353.8865	30.37	1.13	44.89	23.61
C 9H 5BR1N2	221.06	219.9637	48.90	2.28	36.15	12.67
C 9H 5BR1N4	249.07	247.9698	43.40	2.02	32.08	22.49
C 9H 5BR1N6	277.08	275.9760	39.01	1.82	28.84	30.33
C 9H 5BR2N1	286.95	284.8790	37.67	1.76	55.69	4.88
C 9H 5BR2N3	314.97	312.8851	34.32	1.60	50.74	13.34
C 9H 5BR2N5	342.98	340.8913	31.52	1.47	46.59	20.42
C 9H 6BR1N1	208.06	206.9684	51.96	2.91	38.40	6.73
C 9H 6BR1N3	236.07	234.9746	45.79	2.56	33.85	17.80
C 9H 6BR1N5	264.09	262.9807	40.93	2.29	30.26	26.52
C 9H 6BR2N2	301.97	299.8899	35.80	2.00	52.92	9.28
C 9H 6BR2N4	329.98	327.8960	32.76	1.83	48.43	16.98
C 9H 6BR2N6	358.00	355.9022	30.20	1.69	44.64	23.48
C 9H 7BR1N2	223.07	221.9793	48.46	3.16	35.82	12.56
C 9H 7BR1N4	251.09	249.9855	43.05	2.81	31.82	22.31
C 9H 7BR1N6	279.10	277.9916	38.73	2.53	28.63	30.11
C 9H 7BR2N1	288.97	286.8946	37.41	2.44	55.30	4.85
C 9H 7BR2N3	316.98	314.9008	34.10	2.23	50.42	13.26
C 9H 7BR2N5	345.00	342.9069	31.33	2.05	46.32	20.30
C 9H 8BR1N1	210.07	208.9841	51.46	3.84	38.04	6.67
C 9H 8BR1N3	238.09	236.9902	45.40	3.39	33.56	17.65
C 9H 8BR1N5	266.10	264.9964	40.62	3.03	30.03	26.32
C 9H 8BR2N2	303.99	301.9055	35.56	2.65	52.57	9.22
C 9H 8BR2N4	332.00	329.9117	32.56	2.43	48.14	16.88
C 9H 8BR2N6	360.01	357.9178	30.03	2.24	44.39	23.34
C 9H 9BR1N2	225.09	223.9950	48.03	4.03	35.50	12.45
C 9H 9BR1N4	253.10	252.0011	42.71	3.58	31.57	22.14
C 9H 9BR1N6	281.12	280.0073	38.45	3.23	28.42	29.90
C 9H 9BR2N1	290.99	288.9103	37.15	3.12	54.92	4.81
C 9H 9BR2N3	319.00	316.9164	33.89	2.84	50.10	13.17
C 9H 9BR2N5	347.01	344.9226	31.15	2.61	46.05	20.18
C 9H10BR1N1	212.09	210.9997	50.97	4.75	37.67	6.60
C 9H10BR1N3	240.10	239.0059	45.02	4.20	33.28	17.50
C 9H10BR1N5	268.12	267.0120	40.32	3.76	29.80	26.12
C 9H10BR2N2	306.00	303.9212	35.33	3.29	52.22	9.15
C 9H10BR2N4	334.01	331.9273	32.36	3.02	47.84	16.77
C 9H10BR2N6	362.03	359.9335	29.86	2.78	44.14	23.21
C 9H11BR1N2	227.11	226.0106	47.60	4.88	35.18	12.33
C 9H11BR1N4	255.12	254.0168	42.37	4.35	31.32	21.96
C 9H11BR1N6	283.13	282.0229	38.18	3.92	28.22	29.68
C 9H11BR2N1	293.00	290.9259	36.89	3.78	54.54	4.78
C 9H11BR2N3	321.02	318.9321	33.67	3.45	49.78	13.09
C 9H11BR2N5	349.03	346.9382	30.97	3.18	45.79	20.07
C 9H12BR1N1	214.11	213.0154	50.49	5.65	37.32	6.54
C 9H12BR1N3	242.12	241.0215	44.65	5.00	33.00	17.36
C 9H12BR1N5	270.13	269.0277	40.02	4.48	29.58	25.93
C 9H12BR2N2	308.02	305.9368	35.10	3.93	51.88	9.09
C 9H12BR2N4	336.03	333.9430	32.17	3.60	47.56	16.67
C 9H12BR2N6	364.04	361.9491	29.69	3.32	43.90	23.09
C 9H13BR1N2	229.12	228.0263	47.18	5.72	34.87	12.23
C 9H13BR1N4	257.13	256.0324	42.04	5.10	31.07	21.79
C 9H13BR1N6	285.15	284.0386	37.91	4.60	28.02	29.47
C 9H13BR2N1	295.02	292.9416	36.64	4.44	54.17	4.75
C 9H13BR2N3	323.03	320.9477	33.46	4.06	49.47	13.01
C 9H13BR2N5	351.05	348.9539	30.79	3.73	45.52	19.95
C 9H14BR1N1	216.12	215.0310	50.02	6.53	36.97	6.48
C 9H14BR1N3	244.14	243.0372	44.28	5.78	32.73	17.21
C 9H14BR1N5	272.15	271.0433	39.72	5.19	29.36	25.73
C 9H14BR2N2	310.03	307.9525	34.87	4.55	51.55	9.04
C 9H14BR2N4	338.05	335.9586	31.98	4.17	47.27	16.57
C 9H14BR2N6	366.06	363.9648	29.53	3.85	43.66	22.96
C 9H15BR1N2	231.14	230.0419	46.77	6.54	34.57	12.12
C 9H15BR1N4	259.15	258.0481	41.71	5.83	30.83	21.62
C 9H15BR1N6	287.16	286.0542	37.64	5.27	27.83	29.27
C 9H15BR2N1	297.03	294.9572	36.39	5.09	53.80	4.72
C 9H15BR2N3	325.05	322.9634	33.26	4.65	49.16	12.93
C 9H15BR2N5	353.06	350.9695	30.62	4.28	45.26	19.84
C 9H16BR1N1	218.14	217.0467	49.56	7.39	36.63	6.42
C 9H16BR1N3	246.15	245.0528	43.92	6.55	32.46	17.07
C 9H16BR1N5	274.17	273.0590	39.43	5.88	29.14	25.54
C 9H16BR2N2	312.05	309.9681	34.64	5.17	51.21	8.98
C 9H16BR2N4	340.06	337.9743	31.79	4.74	46.99	16.48
C 9H16BR2N6	368.08	365.9804	29.37	4.38	43.42	22.83
C 9H17BR1N2	233.15	232.0576	46.36	7.35	34.27	12.02
C 9H17BR1N4	261.17	260.0637	41.39	6.56	30.60	21.45
C 9H17BR1N6	289.18	288.0699	37.38	5.93	27.63	29.06
C 9H17BR2N1	299.05	296.9729	36.15	5.73	53.44	4.68
C 9H17BR2N3	327.06	324.9790	33.05	5.24	48.86	12.85
C 9H17BR2N5	355.08	352.9852	30.44	4.83	45.01	19.72
C 9H18BR1N1	220.15	219.0623	49.10	8.24	36.29	6.36
C 9H18BR1N3	248.17	247.0685	43.56	7.31	32.20	16.93
C 9H18BR1N5	276.18	275.0746	39.14	6.57	28.93	25.36
C 9H18BR2N2	314.07	311.9838	34.42	5.78	50.88	8.92
C 9H18BR2N4	342.08	339.9899	31.60	5.30	46.72	16.38
C 9H18BR2N6	370.09	367.9961	29.21	4.90	43.18	22.71
C 9H19BR1N2	235.17	234.0732	45.97	8.14	33.98	11.91
C 9H19BR1N4	263.18	262.0794	41.07	7.28	30.36	21.29
C 9H19BR1N6	291.20	290.0855	37.12	6.58	27.44	28.86
C 9H19BR2N1	301.07	298.9885	35.91	6.36	53.08	4.65
C 9H19BR2N3	329.08	326.9947	32.85	5.82	48.56	12.77
C 9H19BR2N5	357.09	355.0008	30.27	5.36	44.75	19.61
C 9H20BR1N1	222.17	221.0780	48.66	9.07	35.97	6.30
C 9H20BR1N3	250.18	249.0841	43.21	8.06	31.94	16.80
C 9H20BR1N5	278.20	277.0903	38.86	7.25	28.72	25.17
C 9H20BR2N2	316.08	313.9994	34.20	6.38	50.56	8.86

C 9H20BR2N4	344.09	342.0056	31.42	5.86	46.44	16.28
C 9H20BR2N6	372.11	370.0117	29.05	5.42	42.95	22.58
C 9H21BR1N2	237.19	236.0889	45.58	8.92	33.69	11.81
C 9H21BR1N4	265.20	264.0950	40.76	7.98	30.13	21.13
C 9H21BR1N6	293.21	292.1011	36.87	7.22	27.25	28.66
C 9H21BR2N3	331.10	329.0103	32.65	6.39	48.27	12.69
C 9H21BR2N5	359.11	357.0165	30.10	5.89	44.50	19.50
C 9H22BR1N3	252.20	251.0998	42.86	8.79	31.68	16.66
C 9H22BR1N5	280.21	279.1059	38.58	7.91	28.52	24.99
C 9H22BR2N4	346.11	344.0212	31.23	6.41	46.17	16.19
C 9H22BR2N6	374.12	372.0274	28.89	5.93	42.72	22.46
C 9H23BR1N4	267.21	266.1107	40.45	8.68	29.90	20.97
C 9H23BR1N6	295.23	294.1168	36.62	7.85	27.07	28.47
C 9H23BR2N5	361.13	359.0321	29.93	6.42	44.25	19.39
C 9H24BR1N5	282.23	281.1215	38.30	8.57	28.31	24.81
C 9H24BR2N6	376.14	374.0430	28.74	6.43	42.49	22.34
C 9H25BR1N6	297.24	296.1324	36.37	8.48	26.88	28.27
C10H 5BR1N2	233.07	231.9637	51.53	2.16	34.28	12.02
C10H 5BR1N4	261.08	259.9698	46.01	1.93	30.60	21.46
C10H 5BR1N6	289.10	287.9760	41.55	1.74	27.64	29.07
C10H 5BR2N1	298.97	296.8790	40.18	1.69	53.45	4.69
C10H 5BR2N3	326.98	324.8851	36.73	1.54	48.87	12.85
C10H 5BR2N5	354.99	352.8913	33.83	1.42	45.02	19.73
C10H 6BR1N1	220.07	218.9684	54.58	2.75	36.31	6.36
C10H 6BR1N3	248.08	246.9746	48.42	2.44	32.21	16.94
C10H 6BR1N5	276.10	274.9807	43.50	2.19	28.94	25.37
C10H 6BR2N2	313.98	311.8899	38.25	1.93	50.90	8.92
C10H 6BR2N4	341.99	339.8960	35.12	1.77	46.73	16.38
C10H 6BR2N6	370.01	367.9022	32.46	1.63	43.19	22.71
C10H 7BR1N2	235.08	233.9793	51.09	3.00	33.99	11.92
C10H 7BR1N4	263.10	261.9855	45.65	2.68	30.37	21.30
C10H 7BR1N6	291.11	289.9916	41.26	2.42	27.45	28.87
C10H 7BR2N1	300.98	298.8946	39.91	2.34	53.10	4.65
C10H 7BR2N3	329.00	326.9008	36.51	2.14	48.57	12.77
C10H 7BR2N5	357.01	354.9069	33.64	1.98	44.76	19.62
C10H 8BR1N1	222.09	220.9841	54.08	3.63	35.98	6.31
C10H 8BR1N3	250.10	248.9902	48.03	3.22	31.95	16.80
C10H 8BR1N5	278.11	276.9964	43.19	2.90	28.73	25.18
C10H 8BR2N2	316.00	313.9055	38.01	2.55	50.57	8.87
C10H 8BR2N4	344.01	341.9117	34.92	2.34	46.45	16.29
C10H 8BR2N6	372.02	369.9178	32.29	2.17	42.96	22.59
C10H 9BR1N2	237.10	235.9950	50.66	3.83	33.70	11.81
C10H 9BR1N4	265.11	264.0011	45.31	3.42	30.14	21.13
C10H 9BR1N6	293.13	292.0073	40.98	3.09	27.26	28.67
C10H 9BR2N1	303.00	300.9103	39.64	2.99	52.74	4.62
C10H 9BR2N3	331.01	328.9164	36.29	2.74	48.28	12.69
C10H 9BR2N5	359.02	356.9226	33.45	2.53	44.51	19.51
C10H10BR1N1	224.10	222.9997	53.60	4.50	35.66	6.25
C10H10BR1N3	252.12	251.0059	47.64	4.00	31.69	16.67
C10H10BR1N5	280.13	279.0120	42.88	3.60	28.52	25.00
C10H10BR2N2	318.01	315.9212	37.77	3.17	50.25	8.81
C10H10BR2N4	346.03	343.9273	34.71	2.91	46.18	16.19
C10H10BR2N6	374.04	371.9335	32.11	2.69	42.72	22.47
C10H11BR1N2	239.12	238.0106	50.23	4.64	33.42	11.72
C10H11BR1N4	267.13	266.0168	44.96	4.15	29.91	20.97
C10H11BR1N6	295.14	294.0229	40.70	3.76	27.07	28.47
C10H11BR2N1	305.01	302.9259	39.38	3.64	52.39	4.59
C10H11BR2N3	333.03	330.9321	36.07	3.33	47.99	12.62
C10H11BR2N5	361.04	358.9382	33.27	3.07	44.26	19.40
C10H12BR1N1	226.12	225.0154	53.12	5.35	35.34	6.19
C10H12BR1N3	254.13	253.0215	47.26	4.76	31.44	16.53
C10H12BR1N5	282.14	281.0277	42.57	4.29	28.32	24.82
C10H12BR2N2	320.03	317.9368	37.53	3.78	49.94	8.75
C10H12BR2N4	348.04	345.9430	34.51	3.48	45.92	16.10
C10H12BR2N6	376.06	373.9491	31.94	3.22	42.50	22.35
C10H13BR1N2	241.13	240.0263	49.81	5.43	33.14	11.62
C10H13BR1N4	269.15	268.0324	44.63	4.87	29.69	20.82
C10H13BR1N6	297.16	296.0386	40.42	4.41	26.89	28.28
C10H13BR2N1	307.03	304.9416	39.12	4.27	52.05	4.56
C10H13BR2N3	335.04	332.9477	35.85	3.91	47.70	12.54
C10H13BR2N5	363.06	360.9539	33.08	3.61	44.02	19.29
C10H14BR1N1	228.13	227.0310	52.65	6.19	35.03	6.14
C10H14BR1N3	256.15	255.0372	46.89	5.51	31.19	16.40
C10H14BR1N5	284.16	283.0433	42.27	4.97	28.12	24.65
C10H14BR2N2	322.04	319.9525	37.30	4.38	49.62	8.70
C10H14BR2N4	350.06	347.9586	34.31	4.03	45.65	16.01
C10H14BR2N6	378.07	375.9648	31.77	3.73	42.27	22.23
C10H15BR1N2	243.15	242.0419	49.40	6.22	32.86	11.52
C10H15BR1N4	271.16	270.0481	44.30	5.58	29.47	20.66
C10H15BR1N6	299.18	298.0542	40.15	5.05	26.71	28.09
C10H15BR2N1	309.05	306.9572	38.87	4.89	51.71	4.53
C10H15BR2N3	337.06	334.9634	35.64	4.49	47.41	12.47
C10H15BR2N5	365.07	362.9695	32.90	4.14	43.77	19.18
C10H16BR1N1	230.15	229.0467	52.19	7.01	34.72	6.09
C10H16BR1N3	258.16	257.0528	46.53	6.25	30.95	16.28
C10H16BR1N5	286.18	285.0590	41.97	5.64	27.92	24.47
C10H16BR2N2	324.06	321.9681	37.06	4.98	49.31	8.64
C10H16BR2N4	352.07	349.9743	34.12	4.58	45.39	15.91
C10H16BR2N6	380.09	377.9804	31.60	4.24	42.05	22.11
C10H17BR1N2	245.16	244.0576	48.99	6.99	32.59	11.43
C10H17BR1N4	273.18	272.0637	43.97	6.27	29.25	20.51
C10H17BR1N6	301.19	300.0699	39.88	5.69	26.53	27.90
C10H17BR2N1	311.06	308.9729	38.61	5.51	51.38	4.50
C10H17BR2N3	339.08	336.9790	35.42	5.05	47.13	12.39
C10H17BR2N5	367.09	364.9852	32.72	4.67	43.53	19.08
C10H18BR1N1	232.17	231.0623	51.74	7.81	34.42	6.03
C10H18BR1N3	260.18	259.0685	46.16	6.97	30.71	16.15
C10H18BR1N5	288.19	287.0746	41.68	6.30	27.73	24.30
C10H18BR2N2	326.08	323.9838	36.84	5.56	49.01	8.59
C10H18BR2N4	354.09	351.9899	33.92	5.12	45.13	15.82
C10H18BR2N6	382.10	379.9961	31.43	4.75	41.82	21.99
C10H19BR1N2	247.18	246.0732	48.59	7.75	32.33	11.33
C10H19BR1N4	275.19	274.0794	43.65	6.96	29.04	20.36
C10H19BR1N6	303.21	302.0855	39.61	6.32	26.35	27.72
C10H19BR2N1	313.08	310.9885	38.36	6.12	51.04	4.47
C10H19BR2N3	341.09	338.9947	35.21	5.61	46.85	12.32
C10H19BR2N5	369.10	367.0008	32.54	5.19	43.30	18.97
C10H20BR1N1	234.18	233.0780	51.29	8.61	34.12	5.98
C10H20BR1N3	262.20	261.0841	45.81	7.69	30.48	16.03
C10H20BR1N5	290.21	289.0903	41.39	6.95	27.53	24.13
C10H20BR2N2	328.09	325.9994	36.61	6.14	48.71	8.54
C10H20BR2N4	356.11	354.0056	33.73	5.66	44.88	15.73
C10H20BR2N6	384.12	382.0117	31.27	5.25	41.60	21.88
C10H21BR1N2	249.20	248.0889	48.20	8.49	32.06	11.24
C10H21BR1N4	277.21	276.0950	43.33	7.64	28.82	20.21
C10H21BR1N6	305.22	304.1011	39.35	6.94	26.18	27.53
C10H21BR2N1	315.09	313.0042	38.12	6.72	50.72	4.45
C10H21BR2N3	343.11	341.0103	35.01	6.17	46.58	12.25
C10H21BR2N5	371.12	369.0165	32.36	5.70	43.06	18.87
C10H22BR1N1	236.20	235.0936	50.85	9.39	33.83	5.93

C10H22BR1N3	264.21	263.0998	45.46	8.39	30.24	15.90
C10H22BR1N5	292.22	291.1059	41.10	7.59	27.34	23.97
C10H22BR2N2	330.11	328.0151	36.39	6.72	48.41	8.49
C10H22BR2N4	358.12	356.0212	33.54	6.19	44.62	15.64
C10H22BR2N6	386.14	384.0274	31.11	5.74	41.39	21.76
C10H23BR1N2	251.21	250.1045	47.81	9.23	31.81	11.15
C10H23BR1N4	279.23	278.1107	43.02	8.30	28.62	20.07
C10H23BR1N6	307.24	306.1168	39.09	7.55	26.01	27.35
C10H23BR2N3	345.12	343.0260	34.80	6.72	46.30	12.18
C10H23BR2N5	373.14	371.0321	32.19	6.21	42.83	18.77
C10H24BR1N3	266.23	265.1154	45.12	9.09	30.01	15.78
C10H24BR1N5	294.24	293.1215	40.82	8.22	27.16	23.80
C10H24BR2N4	360.14	358.0369	33.35	6.72	44.37	15.56
C10H24BR2N6	388.15	386.0430	30.94	6.23	41.17	21.65
C10H25BR1N4	281.24	280.1263	42.71	8.96	28.41	19.92
C10H25BR1N6	309.25	308.1324	38.84	8.15	25.84	27.18
C10H25BR2N5	375.15	373.0478	32.02	6.72	42.60	18.67
C10H26BR1N5	296.26	295.1372	40.54	8.85	26.97	23.64
C10H26BR2N6	390.17	388.0587	30.78	6.72	40.96	21.54
C10H27BR1N6	311.27	310.1481	38.59	8.74	25.67	27.00
C11H 5BR1N2	245.08	243.9637	53.91	2.06	32.60	11.43
C11H 5BR1N4	273.09	271.9698	48.38	1.85	29.26	20.52
C11H 5BR1N6	301.11	299.9760	43.88	1.67	26.54	27.91
C11H 5BR2N1	310.98	308.8790	42.49	1.62	51.39	4.50
C11H 5BR2N3	338.99	336.8851	38.98	1.49	47.14	12.40
C11H 5BR2N5	367.00	364.8913	36.00	1.37	43.54	19.08
C11H 6BR1N1	232.08	230.9684	56.93	2.61	34.43	6.04
C11H 6BR1N3	260.09	258.9746	50.80	2.33	30.72	16.16
C11H 6BR1N5	288.11	286.9807	45.86	2.10	27.73	24.31
C11H 6BR2N2	325.99	323.8899	40.53	1.86	49.02	8.59
C11H 6BR2N4	354.01	351.8960	37.32	1.71	45.14	15.83
C11H 6BR2N6	382.02	379.9022	34.59	1.58	41.83	22.00
C11H 7BR1N2	247.10	245.9793	53.47	2.86	32.34	11.34
C11H 7BR1N4	275.11	273.9855	48.03	2.56	29.04	20.37
C11H 7BR1N6	303.12	301.9916	43.59	2.33	26.36	27.72
C11H 7BR2N1	312.99	310.8946	42.21	2.25	51.06	4.48
C11H 7BR2N3	341.01	338.9008	38.74	2.07	46.86	12.32
C11H 7BR2N5	369.02	366.9069	35.80	1.91	43.31	18.98
C11H 8BR1N1	234.10	232.9841	56.44	3.44	34.13	5.98
C11H 8BR1N3	262.11	260.9902	50.41	3.08	30.48	16.03
C11H 8BR1N5	290.12	288.9964	45.54	2.78	27.54	24.14
C11H 8BR2N2	328.01	325.9055	40.28	2.46	48.72	8.54
C11H 8BR2N4	356.02	353.9117	37.11	2.26	44.89	15.74
C11H 8BR2N6	384.03	381.9178	34.40	2.10	41.61	21.88
C11H 9BR1N2	249.11	247.9950	53.04	3.64	32.08	11.25
C11H 9BR1N4	277.13	276.0011	47.68	3.27	28.83	20.22
C11H 9BR1N6	305.14	304.0073	43.30	2.97	26.19	27.54
C11H 9BR2N1	315.01	312.9103	41.94	2.88	50.73	4.45
C11H 9BR2N3	343.02	340.9164	38.52	2.64	46.59	12.25
C11H 9BR2N5	371.04	368.9226	35.61	2.44	43.07	18.88
C11H10BR1N1	236.11	234.9997	55.96	4.27	33.84	5.93
C11H10BR1N3	264.13	263.0059	50.02	3.82	30.25	15.91
C11H10BR1N5	292.14	291.0120	45.23	3.45	27.35	23.97
C11H10BR2N2	330.02	327.9212	40.03	3.05	48.42	8.49
C11H10BR2N4	358.04	355.9273	36.90	2.82	44.63	15.65
C11H10BR2N6	386.05	383.9335	34.22	2.61	41.40	21.77
C11H11BR1N2	251.13	250.0106	52.61	4.42	31.82	11.16
C11H11BR1N4	279.14	278.0168	47.33	3.97	28.62	20.07
C11H11BR1N6	307.15	306.0229	43.02	3.61	26.01	27.36
C11H11BR2N1	317.03	314.9259	41.68	3.50	50.41	4.42
C11H11BR2N3	345.04	342.9321	38.29	3.21	46.32	12.18
C11H11BR2N5	373.05	370.9382	35.42	2.97	42.84	18.77
C11H12BR1N1	238.13	237.0154	55.48	5.08	33.55	5.88
C11H12BR1N3	266.14	265.0215	49.64	4.54	30.02	15.79
C11H12BR1N5	294.16	293.0277	44.92	4.11	27.16	23.81
C11H12BR2N2	332.04	329.9368	39.79	3.64	48.13	8.44
C11H12BR2N4	360.05	357.9430	36.70	3.36	44.38	15.56
C11H12BR2N6	388.07	385.9491	34.05	3.12	41.18	21.66
C11H13BR1N2	253.14	252.0263	52.19	5.18	31.56	11.07
C11H13BR1N4	281.16	280.0324	46.99	4.66	28.42	19.93
C11H13BR1N6	309.17	308.0386	42.73	4.24	25.84	27.18
C11H13BR2N1	319.04	316.9416	41.41	4.11	50.09	4.39
C11H13BR2N3	347.05	344.9477	38.07	3.78	46.05	12.11
C11H13BR2N5	375.07	372.9539	35.23	3.49	42.61	18.67
C11H14BR1N1	240.14	239.0310	55.02	5.88	33.27	5.83
C11H14BR1N3	268.16	267.0372	49.27	5.26	29.80	15.67
C11H14BR1N5	296.17	295.0433	44.61	4.76	26.98	23.65
C11H14BR2N2	334.06	331.9525	39.55	4.22	47.84	8.39
C11H14BR2N4	362.07	359.9586	36.49	3.90	44.14	15.47
C11H14BR2N6	390.08	387.9648	33.87	3.62	40.97	21.54
C11H15BR1N2	255.16	254.0419	51.78	5.93	31.32	10.98
C11H15BR1N4	283.17	282.0481	46.66	5.34	28.22	19.79
C11H15BR1N6	311.19	310.0542	42.46	4.86	25.68	27.01
C11H15BR2N1	321.06	318.9572	41.15	4.71	49.78	4.36
C11H15BR2N3	349.07	346.9634	37.85	4.33	45.78	12.04
C11H15BR2N5	377.08	374.9695	35.04	4.01	42.38	18.57
C11H16BR1N1	242.16	241.0467	54.56	6.66	33.00	5.78
C11H16BR1N3	270.17	269.0528	48.90	5.97	29.57	15.55
C11H16BR1N5	298.19	297.0590	44.31	5.41	26.80	23.49
C11H16BR2N2	336.07	333.9681	39.31	4.80	47.55	8.34
C11H16BR2N4	364.08	361.9743	36.29	4.43	43.89	15.39
C11H16BR2N6	392.10	389.9804	33.70	4.11	40.76	21.43
C11H17BR1N2	257.18	256.0576	51.37	6.66	31.07	10.89
C11H17BR1N4	285.19	284.0637	46.33	6.01	28.02	19.65
C11H17BR1N6	313.20	312.0699	42.18	5.47	25.51	26.83
C11H17BR2N1	323.07	320.9729	40.90	5.30	49.47	4.34
C11H17BR2N3	351.09	348.9790	37.63	4.88	45.52	11.97
C11H17BR2N5	379.10	376.9852	34.85	4.52	42.15	18.47
C11H18BR1N1	244.18	243.0623	54.11	7.43	32.72	5.74
C11H18BR1N3	272.19	271.0685	48.54	6.67	29.36	15.44
C11H18BR1N5	300.20	299.0746	44.01	6.04	26.62	23.33
C11H18BR2N2	338.09	335.9838	39.08	5.37	47.27	8.29
C11H18BR2N4	366.10	363.9899	36.09	4.96	43.65	15.30
C11H18BR2N6	394.11	391.9961	33.52	4.60	40.55	21.32
C11H19BR1N2	259.19	258.0732	50.97	7.39	30.83	10.81
C11H19BR1N4	287.20	286.0794	46.00	6.67	27.82	19.51
C11H19BR1N6	315.22	314.0855	41.91	6.08	25.35	26.66
C11H19BR2N1	325.09	322.9885	40.64	5.89	49.16	4.31
C11H19BR2N3	353.10	350.9947	37.42	5.42	45.26	11.90
C11H19BR2N5	381.12	379.0008	34.67	5.03	41.93	18.38
C11H20BR1N1	246.19	245.0780	53.67	8.19	32.46	5.69
C11H20BR1N3	274.21	273.0841	48.18	7.35	29.14	15.32
C11H20BR1N5	302.22	301.0903	43.72	6.67	26.44	23.17
C11H20BR2N2	340.10	337.9994	38.85	5.93	46.99	8.24
C11H20BR2N4	368.12	366.0056	35.89	5.48	43.41	15.22
C11H20BR2N6	396.13	394.0117	33.35	5.09	40.34	21.22
C11H21BR1N2	261.21	260.0889	50.58	8.10	30.59	10.72
C11H21BR1N4	289.22	288.0950	45.68	7.32	27.63	19.37
C11H21BR1N6	317.23	316.1011	41.65	6.67	25.19	26.49
C11H21BR2N1	327.10	325.0042	40.39	6.47	48.86	4.28

249

C11H21BR2N3	355.12	353.0103	37.21	5.96	45.00	11.83
C11H21BR2N5	383.13	381.0165	34.48	5.52	41.71	18.28
C11H22BR1N1	248.21	247.0936	53.23	8.93	32.19	5.64
C11H22BR1N3	276.22	275.0998	47.83	8.03	28.93	15.21
C11H22BR1N5	304.24	303.1059	43.43	7.29	26.26	23.02
C11H22BR2N2	342.12	340.0151	38.62	6.48	46.71	8.19
C11H22BR2N4	370.13	368.0212	35.70	5.99	43.18	15.14
C11H22BR2N6	398.15	396.0274	33.18	5.57	40.14	21.11
C11H23BR1N2	263.22	262.1045	50.19	8.81	30.36	10.64
C11H23BR1N4	291.24	290.1107	45.37	7.96	27.44	19.24
C11H23BR1N6	319.25	318.1168	41.39	7.26	25.03	26.32
C11H23BR2N1	329.12	327.0198	40.14	7.04	48.56	4.26
C11H23BR2N3	357.13	355.0260	37.00	6.49	44.75	11.77
C11H23BR2N5	385.15	383.0321	34.30	6.02	41.49	18.18
C11H24BR1N1	250.22	249.1093	52.80	9.67	31.93	5.60
C11H24BR1N3	278.24	277.1154	47.49	8.69	28.72	15.10
C11H24BR1N5	306.25	305.1215	43.14	7.90	26.09	22.87
C11H24BR2N2	344.14	342.0307	38.39	7.03	46.44	8.14
C11H24BR2N4	372.15	370.0369	35.50	6.50	42.94	15.05
C11H24BR2N6	400.16	398.0430	33.02	6.05	39.94	21.00
C11H25BR1N2	265.24	264.1202	49.81	9.50	30.13	10.56
C11H25BR1N4	293.25	292.1263	45.05	8.59	27.25	19.11
C11H25BR1N6	321.27	320.1324	41.13	7.84	24.87	26.16
C11H25BR2N3	359.15	357.0416	36.79	7.02	44.50	11.70
C11H25BR2N5	387.16	385.0478	34.13	6.51	41.28	18.09
C11H26BR1N3	280.25	279.1311	47.14	9.35	28.51	14.99
C11H26BR1N5	308.27	307.1372	42.86	8.50	25.92	22.72
C11H26BR2N4	374.16	372.0525	35.31	7.00	42.71	14.97
C11H26BR2N6	402.18	400.0587	32.85	6.52	39.74	20.90
C11H27BR1N4	295.27	294.1419	44.75	9.22	27.06	18.97
C11H27BR1N6	323.28	322.1481	40.87	8.42	24.72	26.00
C11H27BR2N5	389.18	387.0634	33.95	6.99	41.06	18.00
C11H28BR1N5	310.28	309.1528	42.58	9.10	25.75	22.57
C11H28BR2N6	404.19	402.0743	32.69	6.98	39.54	20.79
C11H29BR1N6	325.30	324.1637	40.62	8.99	24.56	25.83
C12H 6BR1N1	244.09	242.9684	59.05	2.48	32.74	5.74
C12H 6BR1N3	272.11	270.9746	52.97	2.22	29.37	15.44
C12H 6BR1N5	300.12	298.9807	48.03	2.02	26.62	23.34
C12H 6BR2N2	338.00	335.8899	42.64	1.79	47.28	8.29
C12H 6BR2N4	366.02	363.8960	39.38	1.65	43.66	15.31
C12H 6BR2N6	394.03	391.9022	36.58	1.53	40.56	21.33
C12H 7BR1N2	259.11	257.9793	55.63	2.72	30.84	10.81
C12H 7BR1N4	287.12	285.9855	50.20	2.46	27.83	19.51
C12H 7BR1N6	315.13	313.9916	45.74	2.24	25.36	26.67
C12H 7BR2N1	325.00	322.8946	44.35	2.17	49.17	4.31
C12H 7BR2N3	353.02	350.9008	40.83	2.00	45.27	11.90
C12H 7BR2N5	381.03	378.9069	37.83	1.85	41.94	18.38
C12H 8BR1N1	246.11	244.9841	58.57	3.28	32.47	5.69
C12H 8BR1N3	274.12	272.9902	52.58	2.94	29.15	15.33
C12H 8BR1N5	302.14	300.9964	47.71	2.67	26.45	23.18
C12H 8BR2N2	340.02	337.9055	42.39	2.37	47.00	8.24
C12H 8BR2N4	368.03	365.9117	39.16	2.19	43.42	15.22
C12H 8BR2N6	396.05	393.9178	36.39	2.04	40.35	21.22
C12H 9BR1N2	261.12	259.9950	55.20	3.47	30.60	10.73
C12H 9BR1N4	289.14	288.0011	49.85	3.14	27.64	19.38
C12H 9BR1N6	317.15	316.0073	45.45	2.86	25.19	26.50
C12H 9BR2N1	327.02	324.9103	44.07	2.77	48.87	4.28
C12H 9BR2N3	355.03	352.9164	40.60	2.56	45.01	11.84
C12H 9BR2N5	383.05	380.9226	37.63	2.37	41.72	18.28
C12H10BR1N1	248.12	246.9997	58.09	4.06	32.20	5.65

C12H10BR1N3	276.14	275.0059	52.20	3.65	28.94	15.22
C12H10BR1N5	304.15	303.0120	47.39	3.31	26.27	23.03
C12H10BR2N2	342.03	339.9212	42.14	2.95	46.72	8.19
C12H10BR2N4	370.05	367.9273	38.95	2.72	43.19	15.14
C12H10BR2N6	398.06	395.9335	36.21	2.53	40.15	21.11
C12H11BR1N2	263.14	262.0106	54.77	4.21	30.37	10.65
C12H11BR1N4	291.15	290.0168	49.50	3.81	27.44	19.24
C12H11BR1N6	319.17	318.0229	45.16	3.47	25.04	26.33
C12H11BR2N1	329.04	326.9259	43.80	3.37	48.57	4.26
C12H11BR2N3	357.05	354.9321	40.37	3.11	44.76	11.77
C12H11BR2N5	385.06	382.9382	37.43	2.88	41.50	18.19
C12H12BR1N1	250.14	249.0154	57.62	4.84	31.94	5.60
C12H12BR1N3	278.15	277.0215	51.82	4.35	28.73	15.11
C12H12BR1N5	306.17	305.0277	47.08	3.95	26.10	22.87
C12H12BR2N2	344.05	341.9368	41.89	3.52	46.45	8.14
C12H12BR2N4	372.06	369.9430	38.74	3.25	42.95	15.06
C12H12BR2N6	400.08	397.9491	36.03	3.02	39.94	21.01
C12H13BR1N2	265.15	264.0263	54.36	4.94	30.13	10.56
C12H13BR1N4	293.17	292.0324	49.16	4.47	27.26	19.11
C12H13BR1N6	321.18	320.0386	44.88	4.08	24.88	26.17
C12H13BR2N1	331.05	328.9416	43.54	3.96	48.27	4.23
C12H13BR2N3	359.07	356.9477	40.14	3.65	44.51	11.70
C12H13BR2N5	387.08	384.9539	37.24	3.39	41.29	18.09
C12H14BR1N1	252.16	251.0310	57.16	5.60	31.69	5.55
C12H14BR1N3	280.17	279.0372	51.45	5.04	28.52	15.00
C12H14BR1N5	308.18	307.0433	46.77	4.58	25.93	22.72
C12H14BR2N2	346.07	343.9525	41.65	4.08	46.18	8.09
C12H14BR2N4	374.08	371.9586	38.53	3.77	42.72	14.98
C12H14BR2N6	402.09	399.9648	35.85	3.51	39.74	20.90
C12H15BR1N2	267.17	266.0419	53.95	5.66	29.91	10.49
C12H15BR1N4	295.18	294.0481	48.83	5.12	27.07	18.98
C12H15BR1N6	323.20	322.0542	44.60	4.68	24.72	26.00
C12H15BR2N1	333.07	330.9572	43.27	4.54	47.98	4.21
C12H15BR2N3	361.08	358.9634	39.92	4.19	44.26	11.64
C12H15BR2N5	389.09	386.9695	37.04	3.89	41.07	18.00
C12H16BR1N1	254.17	253.0467	56.71	6.35	31.44	5.51
C12H16BR1N3	282.19	281.0528	51.08	5.72	28.32	14.89
C12H16BR1N5	310.20	309.0590	46.46	5.20	25.76	22.58
C12H16BR2N2	348.08	345.9681	41.41	4.63	45.91	8.05
C12H16BR2N4	376.10	373.9743	38.32	4.29	42.49	14.90
C12H16BR2N6	404.11	401.9804	35.67	3.99	39.55	20.80
C12H17BR1N2	269.19	268.0576	53.54	6.37	29.68	10.41
C12H17BR1N4	297.20	296.0637	48.50	5.77	26.89	18.85
C12H17BR1N6	325.21	324.0699	44.32	5.27	24.57	25.84
C12H17BR2N1	335.08	332.9729	43.01	5.11	47.69	4.18
C12H17BR2N3	363.10	360.9790	39.70	4.72	44.01	11.57
C12H17BR2N5	391.11	388.9852	36.85	4.38	40.86	17.91
C12H18BR1N1	256.19	255.0623	56.26	7.08	31.19	5.47
C12H18BR1N3	284.20	283.0685	50.72	6.38	28.12	14.79
C12H18BR1N5	312.21	311.0746	46.16	5.81	25.59	22.43
C12H18BR2N2	350.10	347.9838	41.17	5.18	45.65	8.00
C12H18BR2N4	378.11	375.9899	38.12	4.80	42.26	14.82
C12H18BR2N6	406.13	403.9961	35.49	4.47	39.35	20.69
C12H19BR1N2	271.20	270.0732	53.15	7.06	29.46	10.33
C12H19BR1N4	299.22	298.0794	48.17	6.40	26.70	18.72
C12H19BR1N6	327.23	326.0855	44.05	5.85	24.42	25.68
C12H19BR2N1	337.10	334.9885	42.76	5.68	47.41	4.16
C12H19BR2N3	365.11	362.9947	39.48	5.25	43.77	11.51
C12H19BR2N5	393.13	391.0008	36.66	4.87	40.65	17.81
C12H20BR1N1	258.20	257.0780	55.82	7.81	30.95	5.42

250

C12H20BR1N3	286.22	285.0841	50.36	7.04	27.92	14.68
C12H20BR1N5	314.23	313.0903	45.87	6.42	25.43	22.29
C12H20BR2N2	352.11	349.9994	40.93	5.73	45.39	7.96
C12H20BR2N4	380.13	378.0056	37.92	5.30	42.04	14.74
C12H20BR2N6	408.14	406.0117	35.31	4.94	39.16	20.59
C12H21BR1N2	273.22	272.0889	52.75	7.75	29.25	10.25
C12H21BR1N4	301.23	300.0950	47.85	7.03	26.53	18.60
C12H21BR1N6	329.25	328.1011	43.78	6.43	24.27	25.53
C12H21BR2N1	339.12	337.0042	42.50	6.24	47.12	4.13
C12H21BR2N3	367.13	365.0103	39.26	5.77	43.53	11.45
C12H21BR2N5	395.14	393.0165	36.48	5.36	40.44	17.72
C12H22BR1N1	260.22	259.0936	55.39	8.52	30.71	5.38
C12H22BR1N3	288.23	287.0998	50.01	7.69	27.72	14.58
C12H22BR1N5	316.25	315.1059	45.58	7.01	25.27	22.15
C12H22BR2N2	354.13	352.0151	40.70	6.26	45.13	7.91
C12H22BR2N4	382.14	380.0212	37.72	5.80	41.82	14.66
C12H22BR2N6	410.16	408.0274	35.14	5.41	38.96	20.49
C12H23BR1N2	275.23	274.1045	52.37	8.42	29.03	10.18
C12H23BR1N4	303.25	302.1107	47.53	7.65	26.35	18.48
C12H23BR1N6	331.26	330.1168	43.51	7.00	24.12	25.37
C12H23BR2N1	341.13	339.0198	42.25	6.80	46.85	4.11
C12H23BR2N3	369.15	367.0260	39.05	6.28	43.29	11.38
C12H23BR2N5	397.16	395.0321	36.29	5.84	40.24	17.63
C12H24BR1N1	262.24	261.1093	54.96	9.23	30.47	5.34
C12H24BR1N3	290.25	289.1154	49.66	8.33	27.53	14.48
C12H24BR1N5	318.26	317.1215	45.29	7.60	25.11	22.00
C12H24BR2N2	356.15	354.0307	40.47	6.79	44.87	7.87
C12H24BR2N4	384.16	382.0369	37.52	6.30	41.60	14.58
C12H24BR2N6	412.17	410.0430	34.97	5.87	38.77	20.39
C12H25BR1N2	277.25	276.1202	51.99	9.09	28.82	10.10
C12H25BR1N4	305.26	304.1263	47.22	8.25	26.18	18.35
C12H25BR1N6	333.28	332.1324	43.25	7.56	23.98	25.22
C12H25BR2N1	343.15	341.0355	42.00	7.34	46.57	4.08
C12H25BR2N3	371.16	369.0416	38.83	6.79	43.06	11.32
C12H25BR2N5	399.17	397.0478	36.11	6.31	40.03	17.54
C12H26BR1N1	264.25	263.1249	54.54	9.92	30.24	5.30
C12H26BR1N3	292.27	291.1311	49.32	8.97	27.34	14.38
C12H26BR1N5	320.28	319.1372	45.00	8.18	24.95	21.87
C12H26BR2N2	358.16	356.0464	40.24	7.32	44.62	7.82
C12H26BR2N4	386.18	384.0525	37.32	6.79	41.38	14.51
C12H26BR2N6	414.19	412.0587	34.80	6.33	38.58	20.29
C12H27BR1N2	279.27	278.1358	51.61	9.75	28.61	10.03
C12H27BR1N4	307.28	306.1419	46.91	8.86	26.00	18.23
C12H27BR1N6	335.29	334.1481	42.99	8.12	23.83	25.06
C12H27BR2N3	373.18	371.0573	38.62	7.29	42.82	11.26
C12H27BR2N5	401.19	399.0634	35.93	6.78	39.83	17.46
C12H28BR1N3	294.28	293.1467	48.98	9.59	27.15	14.28
C12H28BR1N5	322.29	321.1528	44.72	8.76	24.79	21.73
C12H28BR2N4	388.19	386.0682	37.13	7.27	41.17	14.43
C12H28BR2N6	416.21	414.0743	34.63	6.78	38.40	20.19
C12H29BR1N4	309.30	308.1576	46.60	9.45	25.83	18.11
C12H29BR1N6	337.31	336.1637	42.73	8.67	23.69	24.91
C12H29BR2N5	403.21	401.0791	35.75	7.25	39.63	17.37
C12H30BR1N5	324.31	323.1685	44.44	9.32	24.64	21.59
C12H30BR2N6	418.22	416.0900	34.46	7.23	38.21	20.09
C12H31BR1N6	339.33	338.1794	42.48	9.21	23.55	24.77
C13H 6BR1N1	256.10	254.9684	60.97	2.36	31.20	5.47
C13H 6BR1N3	284.12	282.9746	54.96	2.13	28.12	14.79
C13H 6BR1N5	312.13	310.9807	50.03	1.94	25.60	22.44
C13H 6BR2N2	350.01	347.8899	44.61	1.73	45.66	8.00

C13H 6BR2N4	378.03	375.8960	41.31	1.60	42.27	14.82
C13H 6BR2N6	406.04	403.9022	38.46	1.49	39.36	20.70
C13H 7BR1N2	271.12	269.9793	57.59	2.60	29.47	10.33
C13H 7BR1N4	299.13	297.9855	52.20	2.36	26.71	18.73
C13H 7BR1N6	327.14	325.9916	47.73	2.16	24.42	25.69
C13H 7BR2N1	337.02	334.8946	46.33	2.09	47.42	4.16
C13H 7BR2N3	365.03	362.9008	42.78	1.93	43.78	11.51
C13H 7BR2N5	393.04	390.9069	39.73	1.80	40.66	17.82
C13H 8BR1N1	258.12	256.9841	60.49	3.12	30.96	5.43
C13H 8BR1N3	286.13	284.9902	54.57	2.82	27.93	14.69
C13H 8BR1N5	314.15	312.9964	49.70	2.57	25.44	22.29
C13H 8BR2N2	352.03	349.9055	44.36	2.29	45.40	7.96
C13H 8BR2N4	380.04	377.9117	41.09	2.12	42.05	14.74
C13H 8BR2N6	408.06	405.9178	38.27	1.98	39.16	20.60
C13H 9BR1N2	273.13	271.9950	57.17	3.32	29.25	10.26
C13H 9BR1N4	301.15	300.0011	51.85	3.01	26.53	18.60
C13H 9BR1N6	329.16	328.0073	47.44	2.76	24.28	25.53
C13H 9BR2N1	339.03	336.9103	46.06	2.68	47.14	4.13
C13H 9BR2N3	367.04	364.9164	42.54	2.47	43.54	11.45
C13H 9BR2N5	395.06	392.9226	39.52	2.30	40.45	17.73
C13H10BR1N1	260.14	258.9997	60.02	3.87	30.72	5.38
C13H10BR1N3	288.15	287.0059	54.19	3.50	27.73	14.58
C13H10BR1N5	316.16	315.0120	49.39	3.19	25.27	22.15
C13H10BR2N2	354.05	351.9212	44.10	2.85	45.14	7.91
C13H10BR2N4	382.06	379.9273	40.87	2.64	41.83	14.66
C13H10BR2N6	410.07	407.9335	38.08	2.46	38.97	20.49
C13H11BR1N2	275.15	274.0106	56.75	4.03	29.04	10.18
C13H11BR1N4	303.16	302.0168	51.51	3.66	26.36	18.48
C13H11BR1N6	331.18	330.0229	47.15	3.35	24.13	25.38
C13H11BR2N1	341.05	338.9259	45.78	3.25	46.86	4.11
C13H11BR2N3	369.06	366.9321	42.31	3.00	43.30	11.39
C13H11BR2N5	397.07	394.9382	39.32	2.79	40.25	17.64
C13H12BR1N1	262.15	261.0154	59.56	4.61	30.48	5.34
C13H12BR1N3	290.16	289.0215	53.81	4.17	27.54	14.48
C13H12BR1N5	318.18	317.0277	49.07	3.80	25.11	22.01
C13H12BR2N2	356.06	353.9368	43.85	3.40	44.88	7.87
C13H12BR2N4	384.08	381.9430	40.65	3.15	41.61	14.59
C13H12BR2N6	412.09	409.9491	37.89	2.94	38.78	20.39
C13H13BR1N2	277.17	276.0263	56.34	4.73	28.83	10.11
C13H13BR1N4	305.18	304.0324	51.16	4.29	26.18	18.36
C13H13BR1N6	333.19	332.0386	46.86	3.93	23.98	25.22
C13H13BR2N1	343.06	340.9416	45.51	3.82	46.58	4.08
C13H13BR2N3	371.08	368.9477	42.08	3.53	43.07	11.32
C13H13BR2N5	399.09	396.9539	39.13	3.28	40.04	17.55
C13H14BR1N1	264.17	263.0310	59.11	5.34	30.25	5.30
C13H14BR1N3	292.18	291.0372	53.44	4.83	27.35	14.38
C13H14BR1N5	320.19	319.0433	48.77	4.41	24.95	21.87
C13H14BR2N2	358.08	355.9525	43.61	3.94	44.63	7.82
C13H14BR2N4	386.09	383.9586	40.44	3.65	41.39	14.51
C13H14BR2N6	414.10	411.9648	37.71	3.41	38.59	20.29
C13H15BR1N2	279.18	278.0419	55.93	5.42	28.62	10.03
C13H15BR1N4	307.20	306.0481	50.83	4.92	26.01	18.24
C13H15BR1N6	335.21	334.0542	46.58	4.51	23.84	25.07
C13H15BR2N1	345.08	342.9572	45.25	4.38	46.31	4.06
C13H15BR2N3	373.09	370.9634	41.85	4.05	42.83	11.26
C13H15BR2N5	401.11	398.9695	38.93	3.77	39.84	17.46
C13H16BR1N1	266.18	265.0467	58.66	6.06	30.02	5.26
C13H16BR1N3	294.20	293.0528	53.08	5.48	27.16	14.28
C13H16BR1N5	322.21	321.0590	48.46	5.01	24.80	21.74
C13H16BR2N2	360.09	357.9681	43.36	4.48	44.38	7.78

C13H16BR2N4	388.11	385.9743	40.23	4.16	41.18	14.44
C13H16BR2N6	416.12	413.9804	37.52	3.88	38.40	20.20
C13H17BR1N2	281.20	280.0576	55.53	6.09	28.42	9.96
C13H17BR1N4	309.21	308.0637	50.50	5.54	25.84	18.12
C13H17BR1N6	337.22	336.0699	46.30	5.08	23.69	24.92
C13H17BR2N1	347.10	344.9729	44.99	4.94	46.04	4.04
C13H17BR2N3	375.11	372.9790	41.63	4.57	42.60	11.20
C13H17BR2N5	403.12	400.9852	38.73	4.25	39.64	17.37
C13H18BR1N1	268.20	267.0623	58.22	6.76	29.79	5.22
C13H18BR1N3	296.21	295.0685	52.71	6.13	26.98	14.19
C13H18BR1N5	324.23	323.0746	48.16	5.60	24.64	21.60
C13H18BR2N2	362.11	359.9838	43.12	5.01	44.13	7.74
C13H18BR2N4	390.12	387.9899	40.02	4.65	40.96	14.36
C13H18BR2N6	418.14	415.9961	37.34	4.34	38.22	20.10
C13H19BR1N2	283.21	282.0732	55.13	6.76	28.21	9.89
C13H19BR1N4	311.23	310.0794	50.17	6.15	25.67	18.00
C13H19BR1N6	339.24	338.0855	46.03	5.65	23.55	24.77
C13H19BR2N1	349.11	346.9885	44.73	5.49	45.78	4.01
C13H19BR2N3	377.12	374.9947	41.40	5.08	42.38	11.14
C13H19BR2N5	405.14	403.0008	38.54	4.73	39.45	17.29
C13H20BR1N1	270.22	269.0780	57.79	7.46	29.57	5.18
C13H20BR1N3	298.23	297.0841	52.36	6.76	26.79	14.09
C13H20BR1N5	326.24	325.0903	47.86	6.18	24.49	21.47
C13H20BR2N2	364.13	361.9994	42.88	5.54	43.89	7.69
C13H20BR2N4	392.14	390.0056	39.82	5.14	40.75	14.29
C13H20BR2N6	420.15	418.0117	37.16	4.80	38.04	20.00
C13H21BR1N2	285.23	284.0889	54.74	7.42	28.01	9.82
C13H21BR1N4	313.24	312.0950	49.85	6.76	25.51	17.89
C13H21BR1N6	341.26	340.1011	45.76	6.20	23.41	24.63
C13H21BR2N1	351.13	349.0042	44.47	6.03	45.51	3.99
C13H21BR2N3	379.14	377.0103	41.18	5.58	42.15	11.08
C13H21BR2N5	407.15	405.0165	38.35	5.20	39.25	17.20
C13H22BR1N1	272.23	271.0936	57.36	8.15	29.35	5.15
C13H22BR1N3	300.24	299.0998	52.01	7.39	26.61	14.00
C13H22BR1N5	328.26	327.1059	47.57	6.76	24.34	21.33
C13H22BR2N2	366.14	364.0151	42.65	6.06	43.65	7.65
C13H22BR2N4	394.16	392.0212	39.62	5.63	40.54	14.21
C13H22BR2N6	422.17	420.0274	36.99	5.25	37.85	19.91
C13H23BR1N2	287.25	286.1045	54.36	8.07	27.82	9.75
C13H23BR1N4	315.26	314.1107	49.53	7.35	25.35	17.77
C13H23BR1N6	343.27	342.1168	45.49	6.75	23.28	24.48
C13H23BR2N1	353.14	351.0198	44.22	6.56	45.25	3.97
C13H23BR2N3	381.16	379.0260	40.97	6.08	41.93	11.02
C13H23BR2N5	409.17	407.0321	38.16	5.67	39.06	17.12
C13H24BR1N1	274.25	273.1093	56.94	8.82	29.14	5.11
C13H24BR1N3	302.26	301.1154	51.66	8.00	26.44	13.90
C13H24BR1N5	330.27	329.1215	47.28	7.32	24.19	21.20
C13H24BR2N2	368.16	366.0307	42.41	6.57	43.41	7.61
C13H24BR2N4	396.17	394.0369	39.41	6.11	40.34	14.14
C13H24BR2N6	424.18	422.0430	36.81	5.70	37.67	19.81
C13H25BR1N2	289.26	288.1202	53.98	8.71	27.62	9.68
C13H25BR1N4	317.27	316.1263	49.21	7.94	25.18	17.66
C13H25BR1N6	345.29	344.1324	45.22	7.30	23.14	24.34
C13H25BR2N1	355.16	353.0355	43.96	7.10	45.00	3.94
C13H25BR2N3	383.17	381.0416	40.75	6.58	41.71	10.97
C13H25BR2N5	411.19	409.0478	37.97	6.13	38.87	17.03
C13H26BR1N1	276.26	275.1249	56.52	9.49	28.92	5.07
C13H26BR1N3	304.28	303.1311	51.32	8.61	26.26	13.81
C13H26BR1N5	332.29	331.1372	46.99	7.89	24.05	21.08
C13H26BR2N2	370.17	368.0464	42.18	7.08	43.17	7.57

C13H26BR2N4	398.19	396.0525	39.21	6.58	40.13	14.07
C13H26BR2N6	426.20	424.0587	36.64	6.15	37.50	19.72
C13H27BR1N2	291.28	290.1358	53.61	9.34	27.43	9.62
C13H27BR1N4	319.29	318.1419	48.90	8.52	25.03	17.55
C13H27BR1N6	347.30	346.1481	44.96	7.84	23.01	24.20
C13H27BR2N1	357.17	355.0511	43.72	7.62	44.74	3.92
C13H27BR2N3	385.19	383.0573	40.54	7.07	41.49	10.91
C13H27BR2N5	413.20	411.0634	37.79	6.59	38.68	16.95
C13H28BR1N1	278.28	277.1406	56.11	10.14	28.71	5.03
C13H28BR1N3	306.29	305.1467	50.98	9.21	26.09	13.72
C13H28BR1N5	334.31	333.1528	46.71	8.44	23.90	20.95
C13H28BR2N2	372.19	370.0620	41.95	7.58	42.94	7.53
C13H28BR2N4	400.20	398.0682	39.02	7.05	39.93	14.00
C13H28BR2N6	428.22	426.0743	36.46	6.59	37.32	19.63
C13H29BR1N2	293.29	292.1515	53.24	9.97	27.24	9.55
C13H29BR1N4	321.31	320.1576	48.60	9.10	24.87	17.44
C13H29BR1N6	349.32	348.1637	44.70	8.37	22.87	24.06
C13H29BR2N3	387.20	385.0729	40.33	7.55	41.27	10.85
C13H29BR2N5	415.22	413.0791	37.61	7.04	38.49	16.87
C13H30BR1N3	308.31	307.1623	50.65	9.81	25.92	13.63
C13H30BR1N5	336.32	335.1685	46.43	8.99	23.76	20.82
C13H30BR2N4	402.22	400.0838	38.82	7.52	39.73	13.93
C13H30BR2N6	430.23	428.0900	36.29	7.03	37.14	19.53
C13H31BR1N4	323.32	322.1732	48.29	9.66	24.71	17.33
C13H31BR1N6	351.34	350.1794	44.44	8.89	22.74	23.92
C13H31BR2N5	417.23	415.0947	37.42	7.49	38.30	16.79
C13H32BR1N5	338.34	337.1841	46.15	9.53	23.62	20.70
C13H32BR2N6	432.25	430.1056	36.12	7.46	36.97	19.44
C13H33BR1N6	353.35	352.1950	44.19	9.41	22.61	23.78
C14H 7BR1N2	283.13	281.9793	59.39	2.49	28.22	9.89
C14H 7BR1N4	311.14	309.9855	54.04	2.27	25.68	18.01
C14H 7BR1N6	339.16	337.9916	49.58	2.08	23.56	24.78
C14H 7BR2N1	349.03	346.8946	48.18	2.02	45.79	4.01
C14H 7BR2N3	377.04	374.9008	44.60	1.87	42.38	11.14
C14H 7BR2N5	405.05	402.9069	41.51	1.74	39.45	17.29
C14H 8BR1N1	270.13	268.9841	62.25	2.99	29.58	5.19
C14H 8BR1N3	298.14	296.9902	56.40	2.70	26.80	14.09
C14H 8BR1N5	326.16	324.9964	51.56	2.47	24.50	21.47
C14H 8BR2N2	364.04	361.9055	46.19	2.22	43.90	7.70
C14H 8BR2N4	392.05	389.9117	42.89	2.06	40.76	14.29
C14H 8BR2N6	420.07	417.9178	40.03	1.92	38.04	20.01
C14H 9BR1N2	285.15	283.9950	58.97	3.18	28.02	9.82
C14H 9BR1N4	313.16	312.0011	53.70	2.90	25.52	17.89
C14H 9BR1N6	341.17	340.0073	49.29	2.66	23.42	24.63
C14H 9BR2N1	351.04	348.9103	47.90	2.58	45.52	3.99
C14H 9BR2N3	379.06	376.9164	44.36	2.39	42.16	11.09
C14H 9BR2N5	407.07	404.9226	41.31	2.23	39.26	17.20
C14H10BR1N1	272.15	270.9997	61.79	3.70	29.36	5.15
C14H10BR1N3	300.16	299.0059	56.02	3.36	26.62	14.00
C14H10BR1N5	328.17	327.0120	51.24	3.07	24.35	21.34
C14H10BR2N2	366.06	363.9212	45.94	2.75	43.66	7.65
C14H10BR2N4	394.07	391.9273	42.67	2.56	40.55	14.22
C14H10BR2N6	422.08	419.9335	39.84	2.39	37.86	19.91
C14H11BR1N2	287.16	286.0106	58.56	3.86	27.83	9.76
C14H11BR1N4	315.17	314.0168	53.35	3.52	25.35	17.78
C14H11BR1N6	343.19	342.0229	49.00	3.23	23.28	24.49
C14H11BR2N1	353.06	350.9259	47.63	3.14	45.26	3.97
C14H11BR2N3	381.07	378.9321	44.13	2.91	41.94	11.03
C14H11BR2N5	409.09	406.9382	41.11	2.71	39.06	17.12
C14H12BR1N1	274.16	273.0154	61.33	4.41	29.14	5.11

C14H12BR1N3	302.18	301.0215	55.65	4.00	26.44	13.91
C14H12BR1N5	330.19	329.0277	50.93	3.66	24.20	21.21
C14H12BR2N2	368.07	365.9368	45.65	3.29	43.42	7.61
C14H12BR2N4	396.09	393.9430	42.45	3.05	40.35	14.15
C14H12BR2N6	424.10	421.9491	39.65	2.85	37.68	19.82
C14H13BR1N2	289.18	288.0263	58.15	4.53	27.63	9.69
C14H13BR1N4	317.19	316.0324	53.01	4.13	25.19	17.66
C14H13BR1N6	345.20	344.0386	48.71	3.80	23.15	24.35
C14H13BR2N1	355.07	352.9416	47.36	3.69	45.01	3.94
C14H13BR2N3	383.09	380.9477	43.89	3.42	41.72	10.97
C14H13BR2N5	411.10	408.9539	40.90	3.19	38.87	17.04
C14H14BR1N1	276.18	275.0310	60.89	5.11	28.93	5.07
C14H14BR1N3	304.19	303.0372	55.28	4.64	26.27	13.81
C14H14BR1N5	332.21	331.0433	50.62	4.25	24.05	21.08
C14H14BR2N2	370.09	367.9525	45.44	3.81	43.18	7.57
C14H14BR2N4	398.10	395.9586	42.24	3.54	40.14	14.07
C14H14BR2N6	426.12	423.9648	39.46	3.31	37.50	19.72
C14H15BR1N2	291.19	290.0419	57.75	5.19	27.44	9.62
C14H15BR1N4	319.21	318.0481	52.68	4.74	25.03	17.55
C14H15BR1N6	347.22	346.0542	48.43	4.35	23.01	24.20
C14H15BR2N1	357.09	354.9572	47.09	4.23	44.75	3.92
C14H15BR2N3	385.10	382.9634	43.67	3.93	41.50	10.91
C14H15BR2N5	413.12	410.9695	40.70	3.66	38.68	16.95
C14H16BR1N1	278.19	277.0467	60.45	5.80	28.72	5.03
C14H16BR1N3	306.21	305.0528	54.92	5.27	26.09	13.72
C14H16BR1N5	334.22	333.0590	50.31	4.83	23.91	20.95
C14H16BR2N2	372.11	369.9681	45.19	4.33	42.95	7.53
C14H16BR2N4	400.12	397.9743	42.03	4.03	39.94	14.00
C14H16BR2N6	428.13	425.9804	39.28	3.77	37.33	19.63
C14H17BR1N2	293.21	292.0576	57.35	5.84	27.25	9.55
C14H17BR1N4	321.22	320.0637	52.35	5.33	24.87	17.44
C14H17BR1N6	349.24	348.0699	48.15	4.91	22.88	24.06
C14H17BR2N1	359.11	356.9729	46.83	4.77	44.50	3.90
C14H17BR2N3	387.12	384.9790	43.44	4.43	41.28	10.85
C14H17BR2N5	415.13	412.9852	40.51	4.13	38.50	16.87
C14H18BR1N1	280.21	279.0623	60.01	6.47	28.52	5.00
C14H18BR1N3	308.22	307.0685	54.56	5.89	25.92	13.63
C14H18BR1N5	336.24	335.0746	50.01	5.40	23.76	20.83
C14H18BR2N2	374.12	371.9838	44.95	4.85	42.72	7.49
C14H18BR2N4	402.13	399.9899	41.82	4.51	39.74	13.93
C14H18BR2N6	430.15	427.9961	39.09	4.22	37.15	19.54
C14H19BR1N2	295.22	294.0732	56.96	6.49	27.07	9.49
C14H19BR1N4	323.24	322.0794	52.02	5.92	24.72	17.33
C14H19BR1N6	351.25	350.0855	47.87	5.45	22.75	23.93
C14H19BR2N1	361.12	358.9885	46.56	5.30	44.25	3.88
C14H19BR2N3	389.14	386.9947	43.21	4.92	41.07	10.80
C14H19BR2N5	417.15	415.0008	40.31	4.59	38.31	16.79
C14H20BR1N1	282.23	281.0780	59.58	7.14	28.31	4.96
C14H20BR1N3	310.24	309.0841	54.20	6.50	25.76	13.54
C14H20BR1N5	338.25	337.0903	49.71	5.96	23.62	20.70
C14H20BR2N2	376.14	373.9994	44.71	5.36	42.49	7.45
C14H20BR2N4	404.15	402.0056	41.61	4.99	39.54	13.86
C14H20BR2N6	432.16	430.0117	38.91	4.66	36.98	19.45
C14H21BR1N2	297.24	296.0889	56.57	7.12	26.88	9.42
C14H21BR1N4	325.25	324.0950	51.70	6.51	24.57	17.23
C14H21BR1N6	353.27	352.1011	47.60	5.99	22.62	23.79
C14H21BR2N1	363.14	361.0042	46.31	5.83	44.01	3.86
C14H21BR2N3	391.15	389.0103	42.99	5.41	40.86	10.74
C14H21BR2N5	419.16	417.0165	40.12	5.05	38.13	16.71
C14H22BR1N1	284.24	283.0936	59.16	7.80	28.11	4.93

C14H22BR1N3	312.26	311.0998	53.85	7.10	25.59	13.46
C14H22BR1N5	340.27	339.1059	49.42	6.52	23.48	20.58
C14H22BR2N2	378.15	376.0151	44.47	5.86	42.26	7.41
C14H22BR2N4	406.17	404.0212	41.40	5.46	39.35	13.79
C14H22BR2N6	434.18	432.0274	38.73	5.11	36.81	19.36
C14H23BR1N2	299.26	298.1045	56.19	7.75	26.70	9.36
C14H23BR1N4	327.27	326.1107	51.38	7.08	24.42	17.12
C14H23BR1N6	355.28	354.1168	47.33	6.53	22.49	23.65
C14H23BR2N1	365.15	363.0198	46.05	6.35	43.76	3.84
C14H23BR2N3	393.17	391.0260	42.77	5.90	40.65	10.69
C14H23BR2N5	421.18	419.0321	39.92	5.50	37.94	16.63
C14H24BR1N1	286.26	285.1093	58.74	8.45	27.91	4.89
C14H24BR1N3	314.27	313.1154	53.51	7.70	25.43	13.37
C14H24BR1N5	342.28	341.1215	49.13	7.07	23.34	20.46
C14H24BR2N2	380.17	378.0307	44.23	6.36	42.04	7.37
C14H24BR2N4	408.18	406.0369	41.20	5.93	39.15	13.73
C14H24BR2N6	436.20	434.0430	38.55	5.55	36.64	19.27
C14H25BR1N2	301.27	300.1202	55.82	8.36	26.52	9.30
C14H25BR1N4	329.29	328.1263	51.07	7.65	24.27	17.01
C14H25BR1N6	357.30	356.1324	47.06	7.05	22.36	23.52
C14H25BR2N1	367.17	365.0355	45.80	6.86	43.52	3.81
C14H25BR2N3	395.18	393.0416	42.55	6.38	40.44	10.63
C14H25BR2N5	423.20	421.0478	39.73	5.95	37.76	16.55
C14H26BR1N1	288.27	287.1249	58.33	9.09	27.72	4.86
C14H26BR1N3	316.29	315.1311	53.17	8.29	25.26	13.29
C14H26BR1N5	344.30	343.1372	48.84	7.61	23.21	20.34
C14H26BR2N2	382.18	380.0464	44.00	6.86	41.81	7.33
C14H26BR2N4	410.20	408.0525	40.99	6.39	38.96	13.66
C14H26BR2N6	438.21	436.0587	38.37	5.98	36.47	19.18
C14H27BR1N2	303.29	302.1358	55.44	8.97	26.35	9.24
C14H27BR1N4	331.30	330.1419	50.76	8.21	24.12	16.91
C14H27BR1N6	359.32	358.1481	46.80	7.57	22.24	23.39
C14H27BR2N1	369.19	367.0511	45.55	7.37	43.29	3.79
C14H27BR2N3	397.20	395.0573	42.34	6.85	40.23	10.58
C14H27BR2N5	425.21	423.0634	39.55	6.40	37.58	16.47
C14H28BR1N1	290.29	289.1406	57.93	9.72	27.53	4.83
C14H28BR1N3	318.30	317.1467	52.83	8.87	25.10	13.20
C14H28BR1N5	346.32	345.1528	48.56	8.15	23.07	20.22
C14H28BR2N2	384.20	382.0620	43.77	7.35	41.59	7.29
C14H28BR2N4	412.21	410.0682	40.79	6.85	38.77	13.59
C14H28BR2N6	440.23	438.0743	38.20	6.41	36.30	19.09
C14H29BR1N2	305.30	304.1515	55.08	9.57	26.17	9.18
C14H29BR1N4	333.32	332.1576	50.45	8.77	23.97	16.81
C14H29BR1N6	361.33	360.1637	46.54	8.09	22.11	23.26
C14H29BR2N1	371.20	369.0668	45.30	7.87	43.05	3.77
C14H29BR2N3	399.22	397.0729	42.12	7.32	40.03	10.53
C14H29BR2N5	427.23	425.0791	39.36	6.84	37.41	16.39
C14H30BR1N1	292.31	291.1562	57.53	10.35	27.34	4.79
C14H30BR1N3	320.32	319.1623	52.50	9.44	24.95	13.12
C14H30BR1N5	348.33	347.1685	48.27	8.68	22.94	20.11
C14H30BR2N2	386.22	384.0777	43.54	7.83	41.38	7.25
C14H30BR2N4	414.23	412.0838	40.59	7.30	38.58	13.53
C14H30BR2N6	442.24	440.0900	38.02	6.84	36.14	19.00
C14H31BR1N2	307.32	306.1671	54.72	10.17	26.00	9.12
C14H31BR1N4	335.33	334.1732	50.15	9.32	23.83	16.71
C14H31BR1N6	363.35	362.1794	46.28	8.60	21.99	23.13
C14H31BR2N3	401.23	399.0886	41.91	7.79	39.83	10.47
C14H31BR2N5	429.24	427.0947	39.17	7.28	37.23	16.32
C14H32BR1N3	322.34	321.1780	52.17	10.01	24.79	13.04
C14H32BR1N5	350.35	349.1841	48.00	9.21	22.81	19.99

253

C14H32BR2N4	416.25	414.0595	40.40	7.75	38.39	13.46
C14H32BR2N6	444.26	442.1056	37.85	7.26	35.97	18.92
C14H33BR1N4	337.35	336.1889	49.85	9.86	23.69	16.61
C14H33BR1N6	365.36	364.1950	46.02	9.10	21.87	23.00
C14H33BR2N5	431.26	429.1104	38.99	7.71	37.06	16.24
C14H34BR1N5	352.36	351.1598	47.72	9.73	22.68	19.88
C14H34BR2N6	446.28	444.1213	37.68	7.68	35.81	18.83
C14H35BR1N6	367.38	366.2107	45.77	9.60	21.75	22.88
C15H 7BR1N2	295.14	293.9793	61.04	2.39	27.07	9.49
C15H 7BR1N4	323.15	321.9855	55.75	2.18	24.73	17.34
C15H 7BR1N6	351.17	349.9916	51.31	2.01	22.75	23.93
C15H 7BR2N1	361.04	358.8946	49.90	1.95	44.26	3.88
C15H 7BR2N3	389.05	386.9008	46.31	1.81	41.08	10.80
C15H 7BR2N5	417.06	414.9069	43.20	1.69	38.32	16.79
C15H 8BR1N1	282.14	280.9841	63.86	2.86	28.32	4.96
C15H 8BR1N3	310.16	308.9902	58.09	2.60	25.76	13.55
C15H 8BR1N5	338.17	336.9964	53.28	2.38	23.63	20.71
C15H 8BR2N2	376.05	373.9055	47.91	2.14	42.50	7.45
C15H 8BR2N4	404.07	401.9117	44.59	2.00	39.55	13.87
C15H 8BR2N6	432.08	429.9178	41.70	1.87	36.99	19.45
C15H 9BR1N2	297.16	295.9950	60.63	3.05	26.89	9.43
C15H 9BR1N4	325.17	324.0011	55.41	2.79	24.57	17.23
C15H 9BR1N6	353.18	352.0073	51.01	2.57	22.62	23.80
C15H 9BR2N1	363.05	360.9103	49.63	2.50	44.02	3.86
C15H 9BR2N3	391.07	388.9164	46.07	2.32	40.86	10.74
C15H 9BR2N5	419.08	416.9226	42.99	2.16	38.13	16.71
C15H10BR1N1	284.16	282.9997	63.40	3.55	28.12	4.93
C15H10BR1N3	312.17	311.0059	57.71	3.23	25.60	13.46
C15H10BR1N5	340.18	339.0120	52.96	2.96	23.49	20.59
C15H10BR2N2	378.07	375.9212	47.65	2.67	42.27	7.41
C15H10BR2N4	406.08	403.9273	44.37	2.48	39.35	13.80
C15H10BR2N6	434.10	431.9335	41.50	2.32	36.81	19.36
C15H11BR1N2	299.17	298.0106	60.22	3.71	26.71	9.36
C15H11BR1N4	327.19	326.0168	55.07	3.39	24.42	17.12
C15H11BR1N6	355.20	354.0229	50.72	3.12	22.50	23.66
C15H11BR2N1	365.07	362.9259	49.35	3.04	43.77	3.84
C15H11BR2N3	393.08	390.9321	45.83	2.82	40.66	10.69
C15H11BR2N5	421.10	418.9382	42.79	2.63	37.95	16.63
C15H12BR1N1	286.17	285.0154	62.96	4.23	27.92	4.89
C15H12BR1N3	314.19	313.0215	57.34	3.85	25.43	13.37
C15H12BR1N5	342.20	341.0277	52.65	3.53	23.35	20.47
C15H12BR2N2	380.08	377.9368	47.40	3.18	42.05	7.37
C15H12BR2N4	408.10	405.9430	44.15	2.96	39.16	13.73
C15H12BR2N6	436.11	433.9491	41.31	2.77	36.64	19.27
C15H13BR1N2	301.19	300.0263	59.82	4.35	26.53	9.30
C15H13BR1N4	329.20	328.0324	54.73	3.98	24.27	17.02
C15H13BR1N6	357.22	356.0386	50.44	3.67	22.37	23.53
C15H13BR2N1	367.09	364.9416	49.08	3.57	43.53	3.82
C15H13BR2N3	395.10	392.9477	45.60	3.32	40.45	10.64
C15H13BR2N5	423.11	420.9539	42.58	3.10	37.77	16.55
C15H14BR1N1	288.19	287.0310	62.52	4.90	27.73	4.86
C15H14BR1N3	316.20	315.0372	56.98	4.46	25.27	13.29
C15H14BR1N5	344.22	343.0433	52.34	4.10	23.21	20.35
C15H14BR2N2	382.10	379.9525	47.15	3.69	41.82	7.33
C15H14BR2N4	410.11	407.9586	43.93	3.44	38.97	13.66
C15H14BR2N6	438.13	435.9648	41.12	3.22	36.48	19.18
C15H15BR1N2	303.20	302.0419	59.42	4.99	26.35	9.24
C15H15BR1N4	331.22	330.0481	54.40	4.56	24.12	16.92
C15H15BR1N6	359.23	358.0542	50.15	4.21	22.24	23.39
C15H15BR2N1	369.10	366.9572	48.81	4.10	43.30	3.79
C15H15BR2N3	397.11	394.9634	45.37	3.81	40.24	10.58
C15H15BR2N5	425.13	422.9695	42.38	3.56	37.59	16.47
C15H16BR1N1	290.21	289.0467	62.08	5.56	27.53	4.83
C15H16BR1N3	318.22	317.0528	56.62	5.07	25.11	13.20
C15H16BR1N5	346.23	345.0590	52.04	4.66	23.08	20.23
C15H16BR2N2	384.12	381.9681	46.90	4.20	41.60	7.29
C15H16BR2N4	412.13	409.9743	43.72	3.91	38.78	13.59
C15H16BR2N6	440.14	437.9804	40.93	3.66	36.31	19.09
C15H17BR1N2	305.22	304.0576	59.03	5.61	26.18	9.18
C15H17BR1N4	333.23	332.0637	54.07	5.14	23.98	16.81
C15H17BR1N6	361.25	360.0699	49.87	4.74	22.12	23.26
C15H17BR2N1	371.12	368.9729	48.55	4.62	43.06	3.77
C15H17BR2N3	399.13	396.9790	45.14	4.29	40.04	10.53
C15H17BR2N5	427.14	424.9852	42.18	4.01	37.41	16.40
C15H18BR1N1	292.22	291.0623	61.65	6.21	27.34	4.79
C15H18BR1N3	320.23	319.0685	56.26	5.67	24.95	13.12
C15H18BR1N5	348.25	347.0746	51.74	5.21	22.94	20.11
C15H18BR2N2	386.13	383.9838	46.66	4.70	41.39	7.25
C15H18BR2N4	414.15	411.9899	43.50	4.38	38.59	13.53
C15H18BR2N6	442.16	439.9961	40.75	4.10	36.14	19.01
C15H19BR1N2	307.24	306.0732	58.64	6.23	26.01	9.12
C15H19BR1N4	335.25	334.0794	53.74	5.71	23.83	16.71
C15H19BR1N6	363.26	362.0855	49.60	5.27	22.00	23.13
C15H19BR2N1	373.13	370.9885	48.28	5.13	42.83	3.75
C15H19BR2N3	401.15	398.9947	44.91	4.77	39.84	10.47
C15H19BR2N5	429.16	427.0008	41.98	4.46	37.24	16.32
C15H20BR1N1	294.24	293.0780	61.23	6.85	27.16	4.76
C15H20BR1N3	322.25	321.0841	55.91	6.26	24.80	13.04
C15H20BR1N5	350.26	349.0903	51.44	5.76	22.81	19.99
C15H20BR2N2	388.15	385.9994	46.42	5.19	41.17	7.22
C15H20BR2N4	416.16	414.0056	43.29	4.84	38.40	13.46
C15H20BR2N6	444.17	442.0117	40.56	4.54	35.98	18.92
C15H21BR1N2	309.25	308.0889	58.26	6.84	25.84	9.06
C15H21BR1N4	337.27	336.0950	53.42	6.28	23.69	16.61
C15H21BR1N6	365.28	364.1011	49.32	5.79	21.87	23.01
C15H21BR2N1	375.15	373.0042	48.03	5.64	42.60	3.73
C15H21BR2N3	403.16	401.0103	44.69	5.25	39.64	10.42
C15H21BR2N5	431.18	429.0165	41.79	4.91	37.06	16.24
C15H22BR1N1	296.25	295.0936	60.82	7.49	26.97	4.73
C15H22BR1N3	324.27	323.0998	55.56	6.84	24.64	12.96
C15H22BR1N5	352.28	351.1059	51.14	6.29	22.68	19.88
C15H22BR2N2	390.16	388.0151	46.18	5.68	40.96	7.18
C15H22BR2N4	418.18	416.0212	43.08	5.30	38.22	13.40
C15H22BR2N6	446.19	444.0274	40.38	4.97	35.82	18.84
C15H23BR1N2	311.27	310.1045	57.88	7.45	25.67	9.00
C15H23BR1N4	339.28	338.1107	53.10	6.83	23.55	16.51
C15H23BR1N6	367.29	366.1168	49.05	6.31	21.75	22.88
C15H23BR2N1	377.17	375.0198	47.77	6.15	42.37	3.71
C15H23BR2N3	405.18	403.0260	44.47	5.72	39.44	10.37
C15H23BR2N5	433.19	431.0321	41.59	5.35	36.89	16.17
C15H24BR1N1	298.27	297.1093	60.40	8.11	26.79	4.70
C15H24BR1N3	326.28	325.1154	55.22	7.41	24.49	12.88
C15H24BR1N5	354.30	353.1215	50.85	6.83	22.55	19.77
C15H24BR2N2	392.18	390.0307	45.94	6.17	40.75	7.14
C15H24BR2N4	420.19	418.0369	42.88	5.76	38.03	13.33
C15H24BR2N6	448.21	446.0430	40.20	5.40	35.65	18.75
C15H25BR1N2	313.28	312.1202	57.51	8.04	25.51	8.94
C15H25BR1N4	341.30	340.1263	52.79	7.38	23.41	16.42
C15H25BR1N6	369.31	368.1324	48.78	6.82	21.64	22.76
C15H25BR2N1	379.18	377.0355	47.51	6.65	42.15	3.69

254

C15H25BR2N3	407.19	405.0416	44.25	6.19	39.25	10.32
C15H25BR2N5	435.21	433.0478	41.40	5.79	36.72	16.09
C15H26BR1N1	300.29	299.1249	60.00	8.73	26.61	4.66
C15H26BR1N3	328.30	327.1311	54.88	7.98	24.34	12.80
C15H26BR1N5	356.31	355.1372	50.56	7.36	22.43	19.66
C15H26BR2N2	394.20	392.0464	45.71	6.65	40.54	7.11
C15H26BR2N4	422.21	420.0525	42.67	6.21	37.85	13.27
C15H26BR2N6	450.22	448.0587	40.02	5.82	35.50	18.67
C15H27BR1N2	315.30	314.1358	57.14	8.63	25.34	8.88
C15H27BR1N4	343.31	342.1419	52.48	7.93	23.27	16.32
C15H27BR1N6	371.33	370.1481	48.52	7.33	21.52	22.63
C15H27BR2N1	381.20	379.0511	47.26	7.14	41.92	3.67
C15H27BR2N3	409.21	407.0573	44.03	6.65	39.05	10.27
C15H27BR2N5	437.22	435.0634	41.21	6.22	36.55	16.02
C15H28BR1N1	302.30	301.1406	59.60	9.34	26.43	4.63
C15H28BR1N3	330.31	329.1467	54.54	8.54	24.19	12.72
C15H28BR1N5	358.33	357.1528	50.28	7.88	22.30	19.54
C15H28BR2N2	396.21	394.0620	45.47	7.12	40.33	7.07
C15H28BR2N4	424.23	422.0682	42.47	6.65	37.67	13.21
C15H28BR2N6	452.24	450.0743	39.84	6.24	35.34	18.58
C15H29BR1N2	317.32	316.1515	56.78	9.21	25.18	8.83
C15H29BR1N4	345.33	344.1576	52.17	8.46	23.14	16.22
C15H29BR1N6	373.34	372.1637	48.26	7.83	21.40	22.51
C15H29BR2N1	383.21	381.0668	47.01	7.63	41.70	3.66
C15H29BR2N3	411.23	409.0729	43.81	7.11	38.86	10.22
C15H29BR2N5	439.24	437.0791	41.02	6.65	36.38	15.94
C15H30BR1N1	304.32	303.1562	59.20	9.94	26.26	4.60
C15H30BR1N3	332.33	331.1623	54.21	9.10	24.04	12.64
C15H30BR1N5	360.34	359.1685	50.00	8.39	22.17	19.44
C15H30BR2N2	398.23	396.0777	45.24	7.59	40.13	7.03
C15H30BR2N4	426.24	424.0838	42.27	7.09	37.49	13.14
C15H30BR2N6	454.25	452.0900	39.66	6.66	35.18	18.50
C15H31BR1N2	319.33	318.1671	56.42	9.79	25.02	8.77
C15H31BR1N4	347.35	346.1732	51.87	9.00	23.00	16.13
C15H31BR1N6	375.36	374.1794	48.00	8.32	21.29	22.39
C15H31BR2N1	385.23	383.0824	46.77	8.11	41.48	3.64
C15H31BR2N3	413.24	411.0886	43.60	7.56	38.67	10.17
C15H31BR2N5	441.26	439.0947	40.83	7.08	36.22	15.87
C15H32BR1N1	306.33	305.1719	58.81	10.53	26.08	4.57
C15H32BR1N3	334.35	333.1780	53.89	9.65	23.90	12.57
C15H32BR1N5	362.36	361.1841	49.72	8.90	22.05	19.33
C15H32BR2N2	400.24	398.0933	45.01	8.06	39.93	7.00
C15H32BR2N4	428.26	426.0995	42.07	7.53	37.32	13.08
C15H32BR2N6	456.27	454.1056	39.49	7.07	35.02	18.42
C15H33BR1N2	321.35	320.1827	56.07	10.35	24.87	8.72
C15H33BR1N4	349.36	348.1889	51.57	9.52	22.87	16.04
C15H33BR1N6	377.37	376.1950	47.74	8.81	21.17	22.27
C15H33BR2N3	415.26	413.1042	43.39	8.01	38.48	10.12
C15H33BR2N5	443.27	441.1104	40.64	7.50	36.05	15.80
C15H34BR1N3	336.36	335.1936	53.56	10.19	23.76	12.49
C15H34BR1N5	364.38	363.1998	49.45	9.41	21.93	19.22
C15H34BR2N4	430.27	428.1151	41.87	7.96	37.14	13.02
C15H34BR2N6	458.29	456.1213	39.31	7.48	34.87	18.34
C15H35BR1N4	351.38	350.2045	51.27	10.04	22.74	15.94
C15H35BR1N6	379.39	378.2107	47.49	9.30	21.06	22.15
C15H35BR2N5	445.29	443.1260	40.46	7.92	35.89	15.73
C15H36BR1N5	366.39	365.2154	49.17	9.90	21.81	19.11
C15H36BR2N6	460.30	458.1369	39.14	7.88	34.72	18.26
C15H37BR1N6	381.41	380.2263	47.24	9.78	20.95	22.03
C16H 8BR1N1	294.15	292.9841	65.33	2.74	27.16	4.76
C16H 8BR1N3	322.17	320.9902	59.65	2.50	24.80	13.04
C16H 8BR1N5	350.18	348.9964	54.88	2.30	22.82	20.00
C16H 8BR2N2	388.06	385.9055	49.52	2.08	41.18	7.22
C16H 8BR2N4	416.08	413.9117	46.19	1.94	38.41	13.47
C16H 8BR2N6	444.09	441.9178	43.27	1.82	35.99	18.92
C16H 9BR1N2	309.17	307.9950	62.16	2.93	25.84	9.06
C16H 9BR1N4	337.18	336.0011	57.00	2.69	23.70	16.62
C16H 9BR1N6	365.19	364.0073	52.62	2.48	21.88	23.01
C16H 9BR2N1	375.06	372.9103	51.24	2.42	42.61	3.73
C16H 9BR2N3	403.08	400.9164	47.68	2.25	39.65	10.42
C16H 9BR2N5	431.09	428.9226	44.58	2.10	37.07	16.25
C16H10BR1N1	296.17	294.9997	64.89	3.40	26.98	4.73
C16H10BR1N3	324.18	323.0059	59.28	3.11	24.65	12.96
C16H10BR1N5	352.20	351.0120	54.57	2.86	22.69	19.88
C16H10BR2N2	390.08	387.9212	49.27	2.58	40.97	7.18
C16H10BR2N4	418.09	415.9273	45.97	2.41	38.22	13.40
C16H10BR2N6	446.11	443.9335	43.08	2.26	35.82	18.84
C16H11BR1N2	311.18	310.0106	61.76	3.56	25.68	9.00
C16H11BR1N4	339.20	338.0168	56.66	3.27	23.56	16.52
C16H11BR1N6	367.21	366.0229	52.33	3.02	21.76	22.89
C16H11BR2N1	377.08	374.9259	50.96	2.94	42.38	3.71
C16H11BR2N3	405.09	402.9321	47.44	2.74	39.45	10.37
C16H11BR2N5	433.11	430.9382	44.37	2.56	36.90	16.17
C16H12BR1N1	298.18	297.0154	64.45	4.06	26.80	4.70
C16H12BR1N3	326.20	325.0215	58.91	3.71	24.50	12.88
C16H12BR1N5	354.21	353.0277	54.26	3.41	22.56	19.77
C16H12BR2N2	392.10	389.9368	49.01	3.08	40.76	7.14
C16H12BR2N4	420.11	417.9430	45.74	2.88	38.04	13.34
C16H12BR2N6	448.12	445.9491	42.89	2.70	35.66	18.75
C16H13BR1N2	313.20	312.0263	61.36	4.18	25.51	8.94
C16H13BR1N4	341.21	340.0324	56.32	3.84	23.42	16.42
C16H13BR1N6	369.23	368.0386	52.05	3.55	21.64	22.76
C16H13BR2N1	379.10	376.9416	50.69	3.46	42.15	3.69
C16H13BR2N3	407.11	404.9477	47.21	3.22	39.25	10.32
C16H13BR2N5	435.12	432.9539	44.17	3.01	36.73	16.10
C16H14BR1N1	300.20	299.0310	64.02	4.70	26.62	4.67
C16H14BR1N3	328.21	327.0372	58.55	4.30	24.35	12.80
C16H14BR1N5	356.23	355.0433	53.95	3.96	22.43	19.66
C16H14BR2N2	394.11	391.9525	48.76	3.58	40.55	7.11
C16H14BR2N4	422.12	419.9586	45.53	3.34	37.86	13.27
C16H14BR2N6	450.14	447.9648	42.69	3.13	35.50	18.67
C16H15BR1N2	315.22	314.0419	60.97	4.80	25.35	8.89
C16H15BR1N4	343.23	342.0481	55.99	4.41	23.28	16.32
C16H15BR1N6	371.24	370.0542	51.77	4.07	21.52	22.64
C16H15BR2N1	381.11	378.9572	50.43	3.97	41.93	3.68
C16H15BR2N3	409.13	406.9634	46.97	3.70	39.06	10.27
C16H15BR2N5	437.14	434.9695	43.96	3.46	36.56	16.02
C16H16BR1N1	302.22	301.0467	63.59	5.34	26.44	4.63
C16H16BR1N3	330.23	329.0528	58.20	4.88	24.20	12.72
C16H16BR1N5	358.24	357.0590	53.64	4.50	22.30	19.55
C16H16BR2N2	396.13	393.9681	48.51	4.07	40.34	7.07
C16H16BR2N4	424.14	421.9743	45.31	3.80	37.68	13.21
C16H16BR2N6	452.15	449.9804	42.50	3.57	35.34	18.59
C16H17BR1N2	317.23	316.0576	60.58	5.40	25.19	8.83
C16H17BR1N4	345.24	344.0637	55.66	4.96	23.14	16.23
C16H17BR1N6	373.26	372.0699	51.49	4.59	21.41	22.52
C16H17BR2N1	383.13	380.9729	50.16	4.47	41.71	3.66
C16H17BR2N3	411.14	408.9790	46.74	4.17	38.87	10.22
C16H17BR2N5	439.16	436.9852	43.76	3.90	36.39	15.95
C16H18BR1N1	304.23	303.0623	63.17	5.96	26.26	4.60

255

C16H18BR1N3	332.25	331.0685	57.84	5.46	24.05	12.65
C16H18BR1N5	360.26	359.0746	53.34	5.04	22.18	19.44
C16H18BR2N2	398.14	395.9838	48.27	4.56	40.14	7.04
C16H18BR2N4	426.16	423.9899	45.10	4.26	37.50	13.15
C16H18BR2N6	454.17	451.9961	42.31	3.99	35.19	18.50
C16H19BR1N2	319.25	318.0732	60.20	6.00	25.03	8.77
C16H19BR1N4	347.26	346.0794	55.34	5.52	23.01	16.13
C16H19BR1N6	375.27	374.0855	51.21	5.10	21.29	22.39
C16H19BR2N1	385.14	382.9885	49.90	4.97	41.49	3.64
C16H19BR2N3	413.16	410.9947	46.51	4.64	38.68	10.17
C16H19BR2N5	441.17	439.0008	43.56	4.34	36.22	15.87
C16H20BR1N1	306.25	305.0780	62.75	6.58	26.09	4.57
C16H20BR1N3	334.26	333.0841	57.49	6.03	23.90	12.57
C16H20BR1N5	362.28	361.0903	53.05	5.56	22.06	19.33
C16H20BR2N2	400.16	397.9994	48.03	5.04	39.94	7.00
C16H20BR2N4	428.17	426.0056	44.88	4.71	37.32	13.09
C16H20BR2N6	456.19	454.0117	42.13	4.42	35.03	18.42
C16H21BR1N2	321.26	320.0889	59.82	6.59	24.87	8.72
C16H21BR1N4	349.28	348.0950	55.02	6.06	22.88	16.04
C16H21BR1N6	377.29	376.1011	50.94	5.61	21.18	22.27
C16H21BR2N1	387.16	385.0042	49.64	5.47	41.28	3.62
C16H21BR2N3	415.17	413.0103	46.29	5.10	38.49	10.12
C16H21BR2N5	443.19	441.0165	43.36	4.78	36.06	15.80
C16H22BR1N1	308.26	307.0936	62.34	7.19	25.92	4.54
C16H22BR1N3	336.28	335.0998	57.15	6.59	23.76	12.50
C16H22BR1N5	364.29	363.1059	52.75	6.09	21.93	19.22
C16H22BR2N2	402.18	400.0151	47.78	5.51	39.74	6.97
C16H22BR2N4	430.19	428.0212	44.67	5.15	37.15	13.02
C16H22BR2N6	458.20	456.0274	41.94	4.84	34.88	18.34
C16H23BR1N2	323.28	322.1045	59.45	7.17	24.72	8.67
C16H23BR1N4	351.29	350.1107	54.71	6.60	22.75	15.95
C16H23BR1N6	379.31	378.1168	50.67	6.11	21.07	22.16
C16H23BR2N1	389.18	387.0198	49.38	5.96	41.06	3.60
C16H23BR2N3	417.19	415.0260	46.06	5.56	38.31	10.07
C16H23BR2N5	445.20	443.0321	43.17	5.21	35.90	15.73
C16H24BR1N1	310.28	309.1093	61.94	7.80	25.75	4.51
C16H24BR1N3	338.29	337.1154	56.81	7.15	23.62	12.42
C16H24BR1N5	366.31	365.1215	52.46	6.60	21.81	19.12
C16H24BR2N2	404.19	402.0307	47.55	5.99	39.54	6.93
C16H24BR2N4	432.20	430.0369	44.46	5.60	36.98	12.96
C16H24BR2N6	460.22	458.0430	41.76	5.26	34.72	18.26
C16H25BR1N2	325.30	324.1202	59.08	7.75	24.56	8.61
C16H25BR1N4	353.31	352.1263	54.39	7.13	22.62	15.86
C16H25BR1N6	381.32	380.1324	50.40	6.61	20.95	22.04
C16H25BR2N1	391.19	389.0355	49.13	6.44	40.85	3.58
C16H25BR2N3	419.21	417.0416	45.84	6.01	38.12	10.02
C16H25BR2N5	447.22	445.0478	42.97	5.63	35.73	15.66
C16H26BR1N1	312.30	311.1249	61.54	8.39	25.59	4.49
C16H26BR1N3	340.31	339.1311	56.47	7.70	23.48	12.35
C16H26BR1N5	368.32	367.1372	52.18	7.12	21.69	19.01
C16H26BR2N2	406.21	404.0464	47.31	6.45	39.34	6.90
C16H26BR2N4	434.22	432.0525	44.26	6.04	36.80	12.90
C16H26BR2N6	462.23	460.0587	41.58	5.67	34.57	18.18
C16H27BR1N2	327.31	326.1358	58.71	8.31	24.41	8.56
C16H27BR1N4	355.32	354.1419	54.09	7.66	22.49	15.77
C16H27BR1N6	383.34	382.1481	50.13	7.10	20.84	21.92
C16H27BR2N1	393.21	391.0511	48.87	6.92	40.64	3.56
C16H27BR2N3	421.22	419.0573	45.62	6.46	37.94	9.98
C16H27BR2N5	449.24	447.0634	42.78	6.06	35.57	15.59
C16H28BR1N1	314.31	313.1406	61.14	8.98	25.42	4.46
C16H28BR1N3	342.33	341.1467	56.14	8.24	23.34	12.27
C16H28BR1N5	370.34	369.1528	51.89	7.62	21.58	18.91
C16H28BR2N2	408.22	406.0620	47.08	6.91	39.15	6.86
C16H28BR2N4	436.24	434.0682	44.05	6.47	36.63	12.84
C16H28BR2N6	464.25	462.0743	41.40	6.08	34.42	18.10
C16H29BR1N2	329.33	328.1515	58.35	8.88	24.26	8.51
C16H29BR1N4	357.34	356.1576	53.78	8.18	22.36	15.68
C16H29BR1N6	385.35	384.1637	49.87	7.59	20.74	21.81
C16H29BR2N1	395.22	393.0668	48.63	7.40	40.43	3.54
C16H29BR2N3	423.24	421.0729	45.41	6.91	37.76	9.93
C16H29BR2N5	451.25	449.0791	42.59	6.48	35.41	15.52
C16H30BR1N1	316.33	315.1562	60.75	9.56	25.26	4.43
C16H30BR1N3	344.34	343.1623	55.81	8.78	23.20	12.20
C16H30BR1N5	372.36	371.1685	51.61	8.12	21.46	18.81
C16H30BR2N2	410.24	408.0777	46.85	7.37	38.95	6.83
C16H30BR2N4	438.25	436.0838	43.85	6.90	36.46	12.78
C16H30BR2N6	466.27	464.0900	41.22	6.49	34.27	18.02
C16H31BR1N2	331.34	330.1671	58.00	9.43	24.12	8.45
C16H31BR1N4	359.36	358.1732	53.48	8.70	22.24	15.59
C16H31BR1N6	387.37	386.1794	49.61	8.07	20.63	21.70
C16H31BR2N1	397.24	395.0824	48.38	7.87	40.23	3.53
C16H31BR2N3	425.25	423.0886	45.19	7.35	37.58	9.88
C16H31BR2N5	453.27	451.0947	42.40	6.89	35.26	15.45
C16H32BR1N1	318.34	317.1719	60.37	10.13	25.10	4.40
C16H32BR1N3	346.36	345.1780	55.49	9.31	23.07	12.13
C16H32BR1N5	374.37	373.1841	51.33	8.62	21.34	18.71
C16H32BR2N2	412.25	410.0933	46.62	7.82	38.76	6.80
C16H32BR2N4	440.27	438.0995	43.65	7.33	36.30	12.73
C16H32BR2N6	468.28	466.1056	41.04	6.89	34.13	17.95
C16H33BR1N2	333.36	332.1827	57.65	9.98	23.97	8.40
C16H33BR1N4	361.37	360.1889	53.18	9.20	22.11	15.50
C16H33BR1N6	389.39	388.1950	49.35	8.54	20.52	21.58
C16H33BR2N1	399.26	397.0981	48.13	8.33	40.03	3.51
C16H33BR2N3	427.27	425.1042	44.98	7.79	37.40	9.83
C16H33BR2N5	455.28	453.1104	42.21	7.31	35.10	15.38
C16H34BR1N1	320.36	319.1875	59.99	10.70	24.94	4.37
C16H34BR1N3	348.37	347.1936	55.16	9.84	22.94	12.06
C16H34BR1N5	376.39	375.1998	51.06	9.11	21.23	18.61
C16H34BR2N2	414.27	412.1090	46.39	8.27	38.58	6.76
C16H34BR2N4	442.28	440.1151	43.45	7.75	36.13	12.67
C16H34BR2N6	470.30	468.1213	40.86	7.29	33.98	17.87
C16H35BR1N2	335.37	334.1984	57.30	10.52	23.83	8.35
C16H35BR1N4	363.39	362.2045	52.89	9.71	21.99	15.42
C16H35BR1N6	391.40	390.2107	49.10	9.01	20.41	21.47
C16H35BR2N3	429.29	427.1199	44.77	8.22	37.23	9.79
C16H35BR2N5	457.30	455.1260	42.02	7.71	34.95	15.31
C16H36BR1N3	350.39	349.2093	54.85	10.36	22.80	11.99
C16H36BR1N5	378.40	377.2154	50.79	9.59	21.12	18.51
C16H36BR2N4	444.30	442.1308	43.25	8.17	35.97	12.61
C16H36BR2N6	472.31	470.1369	40.69	7.68	33.84	17.79
C16H37BR1N4	365.40	364.2202	52.59	10.21	21.87	15.33
C16H37BR1N6	393.42	392.2263	48.85	9.48	20.31	21.36
C16H37BR2N5	459.31	457.1417	41.84	8.12	34.79	15.25
C16H38BR1N5	380.42	379.2311	50.52	10.07	21.00	18.41
C16H38BR2N6	474.33	472.1526	40.52	8.08	33.69	17.72
C16H39BR1N6	395.43	394.2420	48.60	9.94	20.21	21.25
C17H 8BR1N1	306.16	304.9841	66.69	2.63	26.10	4.57
C17H 8BR1N3	334.18	332.9902	61.10	2.41	23.91	12.57
C17H 8BR1N5	362.19	360.9964	56.38	2.23	22.06	19.34
C17H 8BR2N2	400.07	397.9055	51.04	2.02	39.94	7.00

256

C17H 8BR2N4	428.09	425.9117	47.70	1.88	37.33	13.09
C17H 8BR2N6	456.10	453.9178	44.77	1.77	35.04	18.43
C17H 9BR1N2	321.18	319.9950	63.58	2.82	24.88	8.72
C17H 9BR1N4	349.19	348.0011	58.47	2.60	22.88	16.04
C17H 9BR1N6	377.21	376.0073	54.13	2.40	21.18	22.28
C17H 9BR2N1	387.08	384.9103	52.75	2.34	41.29	3.62
C17H 9BR2N3	415.09	412.9164	49.19	2.19	38.50	10.12
C17H 9BR2N5	443.10	440.9226	46.08	2.05	36.07	15.81
C17H10BR1N1	308.18	306.9997	66.26	3.27	25.93	4.54
C17H10BR1N3	336.19	335.0059	60.74	3.00	23.77	12.50
C17H10BR1N5	364.21	363.0120	56.06	2.77	21.94	19.23
C17H10BR2N2	402.09	399.9212	50.78	2.51	39.74	6.97
C17H10BR2N4	430.10	427.9273	47.47	2.34	37.16	13.03
C17H10BR2N6	458.12	455.9335	44.57	2.20	34.88	18.34
C17H11BR1N2	323.19	322.0106	63.18	3.43	24.72	8.67
C17H11BR1N4	351.21	350.0168	58.14	3.16	22.75	15.95
C17H11BR1N6	379.22	378.0229	53.84	2.92	21.07	22.16
C17H11BR2N1	389.09	386.9259	52.48	2.85	41.07	3.60
C17H11BR2N3	417.11	414.9321	48.95	2.66	38.31	10.07
C17H11BR2N5	445.12	442.9382	45.87	2.49	35.90	15.73
C17H12BR1N1	310.20	309.0154	65.83	3.90	25.76	4.52
C17H12BR1N3	338.21	337.0215	60.37	3.58	23.63	12.42
C17H12BR1N5	366.22	365.0277	55.76	3.30	21.82	19.12
C17H12BR2N2	404.11	401.9368	50.53	2.99	39.55	6.93
C17H12BR2N4	432.12	429.9430	47.25	2.80	36.98	12.97
C17H12BR2N6	460.13	457.9491	44.38	2.63	34.73	18.26
C17H13BR1N2	325.21	324.0263	62.79	4.03	24.57	8.61
C17H13BR1N4	353.22	352.0324	57.81	3.71	22.62	15.86
C17H13BR1N6	381.24	380.0386	53.56	3.44	20.96	22.04
C17H13BR2N1	391.11	388.9416	52.21	3.35	40.86	3.58
C17H13BR2N3	419.12	416.9477	48.72	3.13	38.13	10.03
C17H13BR2N5	447.13	444.9539	45.67	2.93	35.74	15.66
C17H14BR1N1	312.21	311.0310	65.40	4.52	25.59	4.49
C17H14BR1N3	340.23	339.0372	60.02	4.15	23.49	12.35
C17H14BR1N5	368.24	367.0433	55.45	3.83	21.70	19.02
C17H14BR2N2	406.12	403.9525	50.28	3.47	39.35	6.90
C17H14BR2N4	434.14	431.9586	47.03	3.25	36.81	12.91
C17H14BR2N6	462.15	459.9648	44.18	3.05	34.58	18.18
C17H15BR1N2	327.23	326.0419	62.40	4.62	24.42	8.56
C17H15BR1N4	355.24	354.0481	57.48	4.26	22.49	15.77
C17H15BR1N6	383.25	382.0542	53.28	3.95	20.85	21.93
C17H15BR2N1	393.12	390.9572	51.94	3.85	40.65	3.56
C17H15BR2N3	421.14	418.9634	48.49	3.59	37.95	9.98
C17H15BR2N5	449.15	446.9695	45.46	3.37	35.58	15.59
C17H16BR1N1	314.23	313.0467	64.98	5.13	25.43	4.46
C17H16BR1N3	342.24	341.0528	59.66	4.71	23.35	12.28
C17H16BR1N5	370.25	369.0590	55.15	4.36	21.58	18.91
C17H16BR2N2	408.14	405.9681	50.03	3.95	39.16	6.86
C17H16BR2N4	436.15	433.9743	46.82	3.70	36.64	12.85
C17H16BR2N6	464.17	461.9804	43.99	3.47	34.43	18.11
C17H17BR1N2	329.24	328.0576	62.02	5.20	24.27	8.51
C17H17BR1N4	357.26	356.0637	57.15	4.80	22.37	15.68
C17H17BR1N6	385.27	384.0699	53.00	4.45	20.74	21.81
C17H17BR2N1	395.14	392.9729	51.68	4.34	40.44	3.54
C17H17BR2N3	423.15	420.9790	48.25	4.05	37.77	9.93
C17H17BR2N5	451.17	448.9852	45.26	3.80	35.42	15.52
C17H18BR1N1	316.24	315.0623	64.57	5.74	25.27	4.43
C17H18BR1N3	344.26	343.0685	59.31	5.27	23.21	12.21
C17H18BR1N5	372.27	371.0746	54.85	4.87	21.46	18.81
C17H18BR2N2	410.15	407.9838	49.78	4.42	38.96	6.83
C17H18BR2N4	438.17	435.9899	46.60	4.14	36.47	12.79
C17H18BR2N6	466.18	463.9961	43.80	3.89	34.28	18.03
C17H19BR1N2	331.26	330.0732	61.64	5.78	24.12	8.46
C17H19BR1N4	359.27	358.0794	56.83	5.33	22.24	15.59
C17H19BR1N6	387.29	386.0855	52.72	4.95	20.63	21.70
C17H19BR2N1	397.16	394.9885	51.41	4.82	40.24	3.53
C17H19BR2N3	425.17	422.9947	48.03	4.50	37.59	9.88
C17H19BR2N5	453.18	451.0008	45.06	4.23	35.26	15.45
C17H20BR1N1	318.26	317.0780	64.16	6.33	25.11	4.40
C17H20BR1N3	346.27	345.0841	58.97	5.82	23.08	12.13
C17H20BR1N5	374.29	373.0903	54.55	5.39	21.35	18.71
C17H20BR2N2	412.17	409.9994	49.54	4.89	38.77	6.80
C17H20BR2N4	440.18	438.0056	46.39	4.58	36.30	12.73
C17H20BR2N6	468.20	466.0117	43.61	4.31	34.13	17.95
C17H21BR1N2	333.27	332.0889	61.27	6.35	23.98	8.41
C17H21BR1N4	361.29	360.0950	56.52	5.86	22.12	15.51
C17H21BR1N6	389.30	388.1011	52.45	5.44	20.52	21.59
C17H21BR2N1	399.17	397.0042	51.15	5.30	40.03	3.51
C17H21BR2N3	427.19	425.0103	47.80	4.96	37.41	9.84
C17H21BR2N5	455.20	453.0165	44.86	4.65	35.11	15.39
C17H22BR1N1	320.28	319.0936	63.75	6.92	24.95	4.37
C17H22BR1N3	348.29	347.0998	58.63	6.37	22.94	12.06
C17H22BR1N5	376.30	375.1059	54.26	5.89	21.23	18.61
C17H22BR2N2	414.19	412.0151	49.30	5.35	38.58	6.76
C17H22BR2N4	442.20	440.0212	46.18	5.01	36.14	12.67
C17H22BR2N6	470.21	468.0274	43.42	4.72	33.99	17.87
C17H23BR1N2	335.29	334.1045	60.90	6.91	23.83	8.35
C17H23BR1N4	363.30	362.1107	56.20	6.38	21.99	15.42
C17H23BR1N6	391.32	390.1168	52.18	5.92	20.42	21.48
C17H23BR2N1	401.19	399.0198	50.90	5.78	39.83	3.49
C17H23BR2N3	429.20	427.0260	47.57	5.40	37.23	9.79
C17H23BR2N5	457.21	455.0321	44.66	5.07	34.95	15.32
C17H24BR1N1	322.29	321.1093	63.36	7.51	24.79	4.35
C17H24BR1N3	350.30	349.1154	58.29	6.91	22.81	12.00
C17H24BR1N5	378.32	377.1215	53.97	6.39	21.12	18.51
C17H24BR2N2	416.20	414.0307	49.06	5.81	38.40	6.73
C17H24BR2N4	444.22	442.0369	45.97	5.45	35.98	12.61
C17H24BR2N6	472.23	470.0430	43.24	5.12	33.84	17.80
C17H25BR1N2	337.31	336.1202	60.54	7.47	23.69	8.31
C17H25BR1N4	365.32	364.1263	55.89	6.90	21.87	15.34
C17H25BR1N6	393.33	392.1324	51.91	6.41	20.31	21.37
C17H25BR2N1	403.20	401.0355	50.64	6.25	39.63	3.47
C17H25BR2N3	431.22	429.0416	47.35	5.84	37.06	9.74
C17H25BR2N5	459.23	457.0478	44.46	5.49	34.80	15.25
C17H26BR1N1	324.31	323.1249	62.96	8.08	24.64	4.32
C17H26BR1N3	352.32	351.1311	57.96	7.44	22.68	11.93
C17H26BR1N5	380.33	379.1372	53.69	6.89	21.01	18.41
C17H26BR2N2	418.22	416.0464	48.82	6.27	38.21	6.70
C17H26BR2N4	446.23	444.0525	45.76	5.87	35.81	12.56
C17H26BR2N6	474.24	472.0587	43.06	5.53	33.70	17.72
C17H27BR1N2	339.32	338.1358	60.18	8.02	23.55	8.26
C17H27BR1N4	367.34	366.1419	55.59	7.41	21.75	15.25
C17H27BR1N6	395.35	394.1481	51.65	6.88	20.21	21.26
C17H27BR2N1	405.22	403.0511	50.39	6.72	39.44	3.46
C17H27BR2N3	433.23	431.0573	47.13	6.28	36.89	9.70
C17H27BR2N5	461.25	459.0634	44.27	5.90	34.65	15.18
C17H28BR1N1	326.32	325.1406	62.57	8.65	24.49	4.29
C17H28BR1N3	354.34	353.1467	57.63	7.97	22.55	11.86
C17H28BR1N5	382.35	381.1528	53.40	7.38	20.90	18.32
C17H28BR2N2	420.23	418.0620	48.59	6.72	38.03	6.67

C17H28BR2N4	448.25	446.0682	45.55	6.30	35.65	12.50
C17H28BR2N6	476.26	474.0743	42.87	5.93	33.55	17.65
C17H29BR1N2	341.34	340.1515	59.82	8.56	23.41	8.21
C17H29BR1N4	369.35	368.1576	55.28	7.91	21.63	15.17
C17H29BR1N6	397.36	396.1637	51.39	7.36	20.11	21.15
C17H29BR2N1	407.24	405.0668	50.14	7.18	39.24	3.44
C17H29BR2N3	435.25	433.0729	46.91	6.72	36.72	9.65
C17H29BR2N5	463.26	461.0791	44.08	6.31	34.50	15.12
C17H30BR1N1	328.34	327.1562	62.19	9.21	24.34	4.27
C17H30BR1N3	356.35	355.1623	57.30	8.49	22.42	11.79
C17H30BR1N5	384.37	383.1685	53.12	7.87	20.79	18.22
C17H30BR2N2	422.25	420.0777	48.36	7.16	37.85	6.63
C17H30BR2N4	450.26	448.0838	45.35	6.72	35.49	12.44
C17H30BR2N6	478.28	476.0900	42.69	6.32	33.41	17.57
C17H31BR1N2	343.35	342.1671	59.47	9.10	23.27	8.16
C17H31BR1N4	371.37	370.1732	54.98	8.41	21.52	15.09
C17H31BR1N6	399.38	398.1794	51.13	7.82	20.01	21.04
C17H31BR2N1	409.25	407.0824	49.89	7.64	39.05	3.42
C17H31BR2N3	437.26	435.0886	46.70	7.15	36.55	9.61
C17H31BR2N5	465.28	463.0947	43.89	6.72	34.35	15.05
C17H32BR1N1	330.36	329.1719	61.81	9.76	24.19	4.24
C17H32BR1N3	358.37	357.1780	56.98	9.00	22.30	11.73
C17H32BR1N5	386.38	385.1841	52.85	8.35	20.68	18.13
C17H32BR2N2	424.27	422.0933	48.13	7.60	37.67	6.60
C17H32BR2N4	452.28	450.0995	45.15	7.13	35.33	12.39
C17H32BR2N6	480.29	478.1056	42.51	6.72	33.27	17.50
C17H33BR1N2	345.37	344.1827	59.12	9.63	23.14	8.11
C17H33BR1N4	373.38	372.1889	54.69	8.91	21.40	15.01
C17H33BR1N6	401.40	400.1950	50.87	8.29	19.91	20.94
C17H33BR2N1	411.27	409.0981	49.65	8.09	38.86	3.41
C17H33BR2N3	439.28	437.1042	46.48	7.57	36.38	9.57
C17H33BR2N5	467.29	465.1104	43.70	7.12	34.20	14.99
C17H34BR1N1	332.37	331.1875	61.43	10.31	24.04	4.21
C17H34BR1N3	360.38	359.1936	56.66	9.51	22.17	11.66
C17H34BR1N5	388.40	387.1998	52.57	8.82	20.57	18.03
C17H34BR2N2	426.28	424.1090	47.90	8.04	37.49	6.57
C17H34BR2N4	454.30	452.1151	44.95	7.54	35.18	12.33
C17H34BR2N6	482.31	480.1213	42.34	7.11	33.13	17.42
C17H35BR1N2	347.39	346.1984	58.78	10.16	23.00	8.06
C17H35BR1N4	375.40	374.2045	54.39	9.40	21.29	14.92
C17H35BR1N6	403.41	402.2107	50.62	8.75	19.81	20.83
C17H35BR2N1	413.28	411.1137	49.41	8.54	38.67	3.39
C17H35BR2N3	441.30	439.1199	46.27	7.99	36.21	9.52
C17H35BR2N5	469.31	467.1260	43.51	7.52	34.05	14.92
C17H36BR1N1	334.39	333.2031	61.06	10.85	23.90	4.19
C17H36BR1N3	362.40	361.2093	56.34	10.01	22.05	11.59
C17H36BR1N5	390.41	389.2154	52.30	9.29	20.47	17.94
C17H36BR2N2	428.30	426.1246	47.67	8.47	37.31	6.54
C17H36BR2N4	456.31	454.1308	44.75	7.95	35.02	12.28
C17H36BR2N6	484.32	482.1369	42.16	7.49	33.00	17.35
C17H37BR1N2	349.40	348.2140	58.44	10.67	22.87	8.02
C17H37BR1N4	377.42	376.2202	54.10	9.88	21.17	14.84
C17H37BR1N6	405.43	404.2263	50.36	9.20	19.71	20.73
C17H37BR2N3	443.31	441.1355	46.06	8.41	36.05	9.48
C17H37BR2N5	471.33	469.1417	43.32	7.91	33.91	14.86
C17H38BR1N3	364.42	363.2249	56.03	10.51	21.93	11.53
C17H38BR1N5	392.43	391.2311	52.03	9.76	20.36	17.85
C17H38BR2N4	458.33	456.1464	44.55	8.36	34.87	12.22
C17H38BR2N6	486.34	484.1526	41.98	7.88	32.86	17.28
C17H39BR1N4	379.43	378.2358	53.81	10.36	21.06	14.77

C17H39BR1N6	407.44	406.2420	50.11	9.65	19.61	20.63
C17H39BR2N5	473.34	471.1573	43.14	8.30	33.76	14.80
C17H40BR1N5	394.45	393.2467	51.77	10.22	20.26	17.75
C17H40BR2N6	488.36	486.1682	41.81	8.26	32.72	17.21
C17H41BR1N6	409.46	408.2576	49.87	10.09	19.51	20.52
C18H 9BR1N2	333.19	331.9950	64.89	2.72	23.98	8.41
C18H 9BR1N4	361.20	360.0011	59.86	2.51	22.12	15.51
C18H 9BR1N6	389.22	388.0073	55.55	2.33	20.53	21.59
C18H 9BR2N1	399.09	396.9103	54.17	2.27	40.04	3.51
C18H 9BR2N3	427.10	424.9164	50.62	2.12	37.42	9.84
C18H 9BR2N5	455.11	452.9226	47.50	1.99	35.11	15.39
C18H10BR1N1	320.19	318.9997	67.52	3.15	24.96	4.37
C18H10BR1N3	348.20	347.0059	62.09	2.89	22.95	12.07
C18H10BR1N5	376.22	375.0120	57.47	2.68	21.24	18.62
C18H10BR2N2	414.10	411.9212	52.21	2.43	38.59	6.76
C18H10BR2N4	442.12	439.9273	48.90	2.28	36.15	12.67
C18H10BR2N6	470.13	467.9335	45.99	2.14	33.99	17.88
C18H11BR1N2	335.21	334.0106	64.50	3.31	23.84	8.36
C18H11BR1N4	363.22	362.0168	59.52	3.05	22.00	15.43
C18H11BR1N6	391.23	390.0229	55.26	2.83	20.42	21.48
C18H11BR2N1	401.10	398.9259	53.90	2.76	39.84	3.49
C18H11BR2N3	429.12	426.9321	50.38	2.58	37.24	9.79
C18H11BR2N5	457.13	454.9382	47.30	2.43	34.96	15.32
C18H12BR1N1	322.21	321.0154	67.10	3.75	24.80	4.35
C18H12BR1N3	350.22	349.0215	61.73	3.45	22.82	12.00
C18H12BR1N5	378.23	377.0277	57.16	3.20	21.13	18.52
C18H12BR2N2	416.12	413.9368	51.96	2.91	38.40	6.73
C18H12BR2N4	444.13	441.9430	48.68	2.72	35.98	12.61
C18H12BR2N6	472.14	469.9491	45.79	2.56	33.85	17.80
C18H13BR1N2	337.22	336.0263	64.11	3.89	23.69	8.31
C18H13BR1N4	365.24	364.0324	59.19	3.59	21.88	15.34
C18H13BR1N6	393.25	392.0386	54.98	3.33	20.32	21.37
C18H13BR2N1	403.12	400.9416	53.63	3.25	39.64	3.47
C18H13BR2N3	431.13	428.9477	50.15	3.04	37.07	9.75
C18H13BR2N5	459.15	456.9539	47.09	2.85	34.81	15.25
C18H14BR1N1	324.22	323.0310	66.68	4.35	24.64	4.32
C18H14BR1N3	352.24	351.0372	61.38	4.01	22.68	11.93
C18H14BR1N5	380.25	379.0433	56.86	3.71	21.01	18.42
C18H14BR2N2	418.13	415.9525	51.71	3.37	38.22	6.70
C18H14BR2N4	446.15	443.9586	48.46	3.16	35.82	12.56
C18H14BR2N6	474.16	471.9648	45.60	2.98	33.70	17.72
C18H15BR1N2	339.24	338.0419	63.73	4.46	23.55	8.26
C18H15BR1N4	367.25	366.0481	58.87	4.12	21.76	15.26
C18H15BR1N6	395.26	394.0542	54.70	3.83	20.22	21.26
C18H15BR2N1	405.13	402.9572	53.37	3.73	39.45	3.46
C18H15BR2N3	433.15	430.9634	49.91	3.49	36.89	9.70
C18H15BR2N5	461.16	458.9695	46.88	3.28	34.65	15.19
C18H16BR1N1	326.24	325.0467	66.27	4.94	24.49	4.29
C18H16BR1N3	354.25	353.0528	61.03	4.55	22.56	11.86
C18H16BR1N5	382.27	381.0590	56.56	4.22	20.90	18.32
C18H16BR2N2	420.15	417.9681	51.46	3.84	38.04	6.67
C18H16BR2N4	448.16	445.9743	48.24	3.60	35.66	12.50
C18H16BR2N6	476.18	473.9804	45.40	3.39	33.56	17.65
C18H17BR1N2	341.25	340.0576	63.35	5.02	23.41	8.21
C18H17BR1N4	369.27	368.0637	58.55	4.64	21.64	15.17
C18H17BR1N6	397.28	396.0699	54.42	4.31	20.11	21.15
C18H17BR2N1	407.15	404.9729	53.10	4.21	39.25	3.44
C18H17BR2N3	435.16	432.9790	49.68	3.94	36.72	9.66
C18H17BR2N5	463.18	460.9852	46.68	3.70	34.50	15.12
C18H18BR1N1	328.25	327.0623	65.86	5.53	24.34	4.27

C18H18BR1N3	356.27	355.0685	60.68	5.09	22.43	11.79
C18H18BR1N5	384.28	383.0746	56.26	4.72	20.79	18.22
C18H18BR2N2	422.17	419.9838	51.21	4.30	37.85	6.64
C18H18BR2N4	450.18	447.9899	48.03	4.03	35.50	12.45
C18H18BR2N6	478.19	475.9961	45.21	3.79	33.42	17.57
C18H19BR1N2	343.27	342.0732	62.98	5.58	23.28	8.16
C18H19BR1N4	371.28	370.0794	58.23	5.16	21.52	15.09
C18H19BR1N6	399.30	398.0855	54.15	4.80	20.01	21.05
C18H19BR2N1	409.17	406.9885	52.84	4.68	39.06	3.42
C18H19BR2N3	437.18	434.9947	49.45	4.38	36.55	9.61
C18H19BR2N5	465.19	463.0008	46.48	4.12	34.35	15.05
C18H20BR1N1	330.27	329.0780	65.46	6.10	24.19	4.24
C18H20BR1N3	358.28	357.0841	60.34	5.63	22.30	11.73
C18H20BR1N5	386.30	385.0903	55.97	5.22	20.68	18.13
C18H20BR2N2	424.18	421.9994	50.97	4.75	37.67	6.60
C18H20BR2N4	452.19	450.0056	47.81	4.46	35.34	12.39
C18H20BR2N6	480.21	478.0117	45.02	4.20	33.28	17.50
C18H21BR1N2	345.29	344.0889	62.62	6.13	23.14	8.11
C18H21BR1N4	373.30	372.0950	57.92	5.67	21.40	15.01
C18H21BR1N6	401.31	400.1011	53.87	5.27	19.91	20.94
C18H21BR2N1	411.18	409.0042	52.58	5.15	38.87	3.41
C18H21BR2N3	439.20	437.0103	49.23	4.82	36.39	9.57
C18H21BR2N5	467.21	465.0165	46.27	4.53	34.20	14.99
C18H22BR1N1	332.29	331.0936	65.06	6.67	24.05	4.22
C18H22BR1N3	360.30	359.0998	60.01	6.15	22.18	11.66
C18H22BR1N5	388.31	387.1059	55.68	5.71	20.58	18.04
C18H22BR2N2	426.20	424.0151	50.73	5.20	37.50	6.57
C18H22BR2N4	454.21	452.0212	47.60	4.88	35.18	12.33
C18H22BR2N6	482.22	480.0274	44.83	4.60	33.14	17.43
C18H23BR1N2	347.30	346.1045	62.25	6.68	23.01	8.07
C18H23BR1N4	375.31	374.1107	57.61	6.18	21.29	14.93
C18H23BR1N6	403.33	402.1168	53.60	5.75	19.81	20.84
C18H23BR2N1	413.20	411.0198	52.32	5.61	38.68	3.39
C18H23BR2N3	441.21	439.0260	49.00	5.25	36.22	9.52
C18H23BR2N5	469.23	467.0321	46.08	4.94	34.06	14.93
C18H24BR1N1	334.30	333.1093	64.67	7.24	23.90	4.19
C18H24BR1N3	362.32	361.1154	59.67	6.68	22.05	11.60
C18H24BR1N5	390.33	389.1215	55.39	6.20	20.47	17.94
C18H24BR2N2	428.21	426.0307	50.49	5.65	37.32	6.54
C18H24BR2N4	456.23	454.0369	47.39	5.30	35.03	12.28
C18H24BR2N6	484.24	482.0430	44.65	5.00	33.00	17.36
C18H25BR1N2	349.32	348.1202	61.89	7.21	22.87	8.02
C18H25BR1N4	377.33	376.1263	57.30	6.68	21.18	14.85
C18H25BR1N6	405.34	404.1324	53.34	6.22	19.71	20.73
C18H25BR2N1	415.21	413.0355	52.07	6.07	38.49	3.37
C18H25BR2N3	443.23	441.0416	48.78	5.69	36.06	9.48
C18H25BR2N5	471.24	469.0478	45.88	5.35	33.91	14.86
C18H26BR1N1	336.32	335.1249	64.28	7.79	23.76	4.16
C18H26BR1N3	364.33	363.1311	59.34	7.19	21.93	11.53
C18H26BR1N5	392.35	391.1372	55.10	6.68	20.37	17.85
C18H26BR2N2	430.23	428.0464	50.25	6.09	37.14	6.51
C18H26BR2N4	458.24	456.0525	47.18	5.72	34.87	12.23
C18H26BR2N6	486.26	484.0587	44.46	5.39	32.86	17.28
C18H27BR1N2	351.33	350.1358	61.54	7.75	22.74	7.97
C18H27BR1N4	379.35	378.1419	56.99	7.17	21.06	14.77
C18H27BR1N6	407.36	406.1481	53.07	6.68	19.62	20.63
C18H27BR2N1	417.23	415.0511	51.82	6.52	38.30	3.36
C18H27BR2N3	445.24	443.0573	48.56	6.11	35.89	9.44
C18H27BR2N5	473.26	471.0634	45.68	5.75	33.77	14.80
C18H28BR1N1	338.33	337.1406	63.90	8.34	23.62	4.14

C18H28BR1N3	366.35	365.1467	59.02	7.70	21.81	11.47
C18H28BR1N5	394.36	393.1528	54.82	7.16	20.26	17.76
C18H28BR2N2	432.25	430.0620	50.02	6.53	36.97	6.48
C18H28BR2N4	460.26	458.0682	46.97	6.13	34.72	12.17
C18H28BR2N6	488.27	486.0743	44.28	5.78	32.73	17.21
C18H29BR1N2	353.35	352.1515	61.19	8.27	22.61	7.93
C18H29BR1N4	381.36	380.1576	56.69	7.66	20.95	14.69
C18H29BR1N6	409.38	408.1637	52.81	7.14	19.52	20.53
C18H29BR2N1	419.25	417.0668	51.57	6.97	38.12	3.34
C18H29BR2N3	447.26	445.0729	48.34	6.54	35.73	9.40
C18H29BR2N5	475.27	473.0791	45.49	6.15	33.62	14.74
C18H30BR1N1	340.35	339.1562	63.52	8.88	23.48	4.12
C18H30BR1N3	368.36	367.1623	58.69	8.21	21.69	11.41
C18H30BR1N5	396.38	395.1685	54.54	7.63	20.16	17.67
C18H30BR2N2	434.26	432.0777	49.79	6.96	36.80	6.45
C18H30BR2N4	462.27	460.0838	46.77	6.54	34.57	12.12
C18H30BR2N6	490.29	488.0900	44.10	6.17	32.59	17.14
C18H31BR1N2	355.37	354.1671	60.84	8.79	22.49	7.88
C18H31BR1N4	383.38	382.1732	56.39	8.15	20.84	14.61
C18H31BR1N6	411.39	410.1794	52.55	7.60	19.42	20.43
C18H31BR2N1	421.26	419.0824	51.32	7.42	37.94	3.32
C18H31BR2N3	449.28	447.0886	48.12	6.95	35.57	9.35
C18H31BR2N5	477.29	475.0947	45.30	6.55	33.48	14.67
C18H32BR1N1	342.37	341.1719	63.15	9.42	23.34	4.09
C18H32BR1N3	370.38	369.1780	58.37	8.71	21.57	11.35
C18H32BR1N5	398.39	397.1841	54.27	8.10	20.06	17.58
C18H32BR2N2	436.28	434.0933	49.56	7.39	36.63	6.42
C18H32BR2N4	464.29	462.0995	46.57	6.95	34.42	12.07
C18H32BR2N6	492.30	490.1056	43.92	6.55	32.46	17.07
C18H33BR1N2	357.38	356.1827	60.50	9.31	22.36	7.84
C18H33BR1N4	385.39	384.1889	56.10	8.63	20.73	14.54
C18H33BR1N6	413.41	412.1950	52.30	8.05	19.33	20.33
C18H33BR2N1	423.28	421.0981	51.08	7.86	37.75	3.31
C18H33BR2N3	451.29	449.1042	47.91	7.37	35.41	9.31
C18H33BR2N5	479.31	477.1104	45.11	6.94	33.34	14.61
C18H34BR1N1	344.38	343.1875	62.78	9.95	23.20	4.07
C18H34BR1N3	372.40	371.1936	58.06	9.20	21.46	11.28
C18H34BR1N5	400.41	399.1998	53.99	8.56	19.96	17.49
C18H34BR2N2	438.29	436.1090	49.33	7.82	36.46	6.39
C18H34BR2N4	466.31	464.1151	46.36	7.35	34.27	12.02
C18H34BR2N6	494.32	492.1213	43.74	6.93	32.33	17.00
C18H35BR1N2	359.40	358.1984	60.16	9.82	22.23	7.79
C18H35BR1N4	387.41	386.2045	55.81	9.11	20.63	14.46
C18H35BR1N6	415.42	414.2107	52.04	8.49	19.23	20.23
C18H35BR2N1	425.29	423.1137	50.84	8.30	37.58	3.29
C18H35BR2N3	453.31	451.1199	47.69	7.78	35.25	9.27
C18H35BR2N5	481.32	479.1260	44.92	7.33	33.20	14.55
C18H36BR1N1	346.40	345.2031	62.41	10.48	23.07	4.04
C18H36BR1N3	374.41	373.2093	57.74	9.69	21.34	11.22
C18H36BR1N5	402.43	401.2154	53.72	9.02	19.86	17.40
C18H36BR2N2	440.31	438.1246	49.10	8.24	36.29	6.36
C18H36BR2N4	468.32	466.1308	46.16	7.75	34.12	11.96
C18H36BR2N6	496.34	494.1369	43.56	7.31	32.20	16.93
C18H37BR1N2	361.41	360.2140	59.82	10.32	22.11	7.75
C18H37BR1N4	389.43	388.2202	55.52	9.58	20.52	14.39
C18H37BR1N6	417.44	416.2263	51.79	8.93	19.14	20.13
C18H37BR2N1	427.31	425.1294	50.60	8.73	37.40	3.28
C18H37BR2N3	455.32	453.1355	47.48	8.19	35.10	9.23
C18H37BR2N5	483.34	481.1417	44.73	7.72	33.06	14.49
C18H38BR1N1	348.41	347.2188	62.05	10.99	22.93	4.02

259

C18H38BR1N3	376.43	375.2249	57.43	10.18	21.23	11.16
C18H38BR1N5	404.44	403.2311	53.46	9.47	19.76	17.32
C18H38BR2N2	442.32	440.1403	48.88	8.66	36.13	6.33
C18H38BR2N4	470.34	468.1464	45.97	8.14	33.98	11.91
C18H38BR2N6	498.35	496.1526	43.38	7.69	32.07	16.86
C18H39BR1N2	363.43	362.2297	59.49	10.82	21.99	7.71
C18H39BR1N4	391.44	390.2358	55.23	10.04	20.41	14.31
C18H39BR1N6	419.46	418.2420	51.54	9.37	19.05	20.04
C18H39BR2N3	457.34	455.1512	47.27	8.60	34.94	9.19
C18H39BR2N5	485.35	483.1573	44.55	8.10	32.93	14.43
C18H40BR1N3	378.44	377.2406	57.13	10.65	21.11	11.10
C18H40BR1N5	406.46	405.2467	53.19	9.92	19.66	17.23
C18H40BR2N4	472.35	470.1621	45.77	8.54	33.83	11.86
C18H40BR2N6	500.37	498.1682	43.21	8.06	31.94	16.80
C18H41BR1N4	393.46	392.2515	54.95	10.50	20.31	14.24
C18H41BR1N6	421.47	420.2576	51.30	9.81	18.96	19.94
C18H41BR2N5	487.37	485.1730	44.36	8.48	32.79	14.37
C18H42BR1N5	408.47	407.2624	52.93	10.36	19.56	17.15
C18H42BR2N6	502.38	500.1839	43.03	8.43	31.81	16.73
C18H43BR1N6	423.49	422.2733	51.05	10.23	18.87	19.84
C19H 9BR1N2	345.20	343.9950	66.11	2.63	23.15	8.12
C19H 9BR1N4	373.21	372.0011	61.15	2.43	21.41	15.01
C19H 9BR1N6	401.23	400.0073	56.88	2.26	19.91	20.95
C19H 9BR2N1	411.10	408.9103	55.51	2.21	38.87	3.41
C19H 9BR2N3	439.11	436.9164	51.97	2.07	36.39	9.57
C19H 9BR2N5	467.13	464.9226	48.85	1.94	34.21	14.99
C19H10BR1N1	332.20	330.9997	68.70	3.03	24.05	4.22
C19H10BR1N3	360.22	359.0059	63.35	2.80	22.18	11.67
C19H10BR1N5	388.23	387.0120	58.78	2.60	20.58	18.04
C19H10BR2N2	426.11	423.9212	53.56	2.37	37.50	6.57
C19H10BR2N4	454.13	451.9273	50.25	2.22	35.19	12.34
C19H10BR2N6	482.14	479.9335	47.33	2.09	33.15	17.43
C19H11BR1N2	347.22	346.0106	65.73	3.19	23.01	8.07
C19H11BR1N4	375.23	374.0168	60.82	2.95	21.29	14.93
C19H11BR1N6	403.24	402.0229	56.59	2.75	19.82	20.84
C19H11BR2N1	413.11	410.9259	55.24	2.68	38.68	3.39
C19H11BR2N3	441.13	438.9321	51.73	2.51	36.23	9.53
C19H11BR2N5	469.14	466.9382	48.64	2.36	34.06	14.93
C19H12BR1N1	334.22	333.0154	68.28	3.62	23.91	4.19
C19H12BR1N3	362.23	361.0215	63.00	3.34	22.06	11.60
C19H12BR1N5	390.24	389.0277	58.48	3.10	20.48	17.95
C19H12BR2N2	428.13	425.9368	53.30	2.83	37.33	6.54
C19H12BR2N4	456.14	453.9430	50.03	2.65	35.03	12.28
C19H12BR2N6	484.16	481.9491	47.14	2.50	33.01	17.36
C19H13BR1N2	349.23	348.0263	65.35	3.75	22.88	8.02
C19H13BR1N4	377.25	376.0324	60.49	3.47	21.18	14.85
C19H13BR1N6	405.26	404.0386	56.31	3.23	19.72	20.74
C19H13BR2N1	415.13	412.9416	54.97	3.16	38.50	3.37
C19H13BR2N3	443.14	440.9477	51.50	2.96	36.06	9.48
C19H13BR2N5	471.16	468.9539	48.44	2.78	33.92	14.86
C19H14BR1N1	336.23	335.0310	67.87	4.20	23.76	4.17
C19H14BR1N3	364.25	363.0372	62.65	3.87	21.94	11.54
C19H14BR1N5	392.26	391.0433	58.18	3.60	20.37	17.85
C19H14BR2N2	430.14	427.9525	53.05	3.28	37.15	6.51
C19H14BR2N4	458.16	455.9586	49.81	3.08	34.88	12.23
C19H14BR2N6	486.17	483.9648	46.94	2.90	32.87	17.29
C19H15BR1N2	351.25	350.0419	64.97	4.30	22.75	7.98
C19H15BR1N4	379.26	378.0481	60.17	3.99	21.07	14.77
C19H15BR1N6	407.28	406.0542	56.03	3.71	19.62	20.63
C19H15BR2N1	417.15	414.9572	54.71	3.62	38.31	3.36
C19H15BR2N3	445.16	442.9634	51.27	3.40	35.90	9.44
C19H15BR2N5	473.17	470.9695	48.23	3.20	33.77	14.80
C19H16BR1N1	338.25	337.0467	67.47	4.77	23.62	4.14
C19H16BR1N3	366.26	365.0528	62.31	4.40	21.82	11.47
C19H16BR1N5	394.28	393.0590	57.88	4.09	20.27	17.76
C19H16BR2N2	432.16	429.9681	52.81	3.73	36.98	6.48
C19H16BR2N4	460.17	457.9743	49.59	3.50	34.73	12.18
C19H16BR2N6	488.19	485.9804	46.75	3.30	32.73	17.21
C19H17BR1N2	353.26	352.0576	64.60	4.85	22.62	7.93
C19H17BR1N4	381.28	380.0637	59.85	4.49	20.96	14.69
C19H17BR1N6	409.29	408.0699	55.76	4.19	19.52	20.53
C19H17BR2N1	419.16	416.9729	54.44	4.09	38.13	3.34
C19H17BR2N3	447.18	444.9790	51.03	3.83	35.74	9.40
C19H17BR2N5	475.19	472.9852	48.03	3.61	33.63	14.74
C19H18BR1N1	340.27	339.0623	67.07	5.33	23.48	4.12
C19H18BR1N3	368.28	367.0685	61.97	4.93	21.70	11.41
C19H18BR1N5	396.29	395.0746	57.59	4.58	20.16	17.67
C19H18BR2N2	434.18	431.9838	52.56	4.18	36.81	6.45
C19H18BR2N4	462.19	459.9899	49.38	3.93	34.58	12.12
C19H18BR2N6	490.20	487.9961	46.55	3.70	32.60	17.14
C19H19BR1N2	355.28	354.0732	64.23	5.39	22.49	7.88
C19H19BR1N4	383.29	382.0794	59.54	5.00	20.85	14.62
C19H19BR1N6	411.31	410.0855	55.48	4.66	19.43	20.43
C19H19BR2N1	421.18	418.9885	54.18	4.55	37.94	3.33
C19H19BR2N3	449.19	446.9947	50.81	4.26	35.58	9.35
C19H19BR2N5	477.20	475.0008	47.82	4.01	33.49	14.68
C19H20BR1N1	342.28	341.0780	66.67	5.89	23.34	4.09
C19H20BR1N3	370.30	369.0841	61.63	5.44	21.58	11.35
C19H20BR1N5	398.31	397.0903	57.30	5.06	20.06	17.58
C19H20BR2N2	436.19	433.9994	52.32	4.62	36.64	6.42
C19H20BR2N4	464.21	462.0056	49.16	4.34	34.43	12.07
C19H20BR2N6	492.22	490.0117	46.36	4.10	32.47	17.07
C19H21BR1N2	357.30	356.0889	63.87	5.92	22.36	7.84
C19H21BR1N4	385.31	384.0950	59.23	5.49	20.74	14.54
C19H21BR1N6	413.32	412.1011	55.21	5.12	19.33	20.33
C19H21BR2N1	423.19	421.0042	53.93	5.00	37.76	3.31
C19H21BR2N3	451.21	449.0103	50.58	4.69	35.42	9.31
C19H21BR2N5	479.22	477.0165	47.62	4.42	33.35	14.61
C19H22BR1N1	344.30	343.0936	66.28	6.44	23.21	4.07
C19H22BR1N3	372.31	371.0998	61.30	5.96	21.46	11.29
C19H22BR1N5	400.32	399.1059	57.01	5.54	19.96	17.49
C19H22BR2N2	438.21	436.0151	52.08	5.06	36.47	6.39
C19H22BR2N4	466.22	464.0212	48.95	4.76	34.28	12.02
C19H22BR2N6	494.24	492.0274	46.17	4.49	32.33	17.00
C19H23BR1N2	359.31	358.1045	63.51	6.45	22.24	7.80
C19H23BR1N4	387.33	386.1107	58.92	5.99	20.63	14.47
C19H23BR1N6	415.34	414.1168	54.95	5.58	19.24	20.23
C19H23BR2N1	425.21	423.0198	53.67	5.45	37.58	3.29
C19H23BR2N3	453.22	451.0260	50.35	5.12	35.26	9.27
C19H23BR2N5	481.24	479.0321	47.42	4.82	33.21	14.55
C19H24BR1N1	346.31	345.1093	65.90	6.99	23.07	4.04
C19H24BR1N3	374.33	373.1154	60.97	6.46	21.35	11.23
C19H24BR1N5	402.34	401.1215	56.72	6.01	19.86	17.41
C19H24BR2N2	440.22	438.0307	51.84	5.50	36.30	6.36
C19H24BR2N4	468.24	466.0369	48.74	5.17	34.13	11.97
C19H24BR2N6	496.25	494.0430	45.99	4.87	32.20	16.94
C19H25BR1N2	361.33	360.1202	63.16	6.97	22.11	7.75
C19H25BR1N4	389.34	388.1263	58.61	6.47	20.52	14.39
C19H25BR1N6	417.36	416.1324	54.68	6.04	19.15	20.14
C19H25BR2N1	427.23	425.0355	53.42	5.90	37.41	3.28

260

C19H25BR2N3	455.24	453.0416	50.13	5.54	35.10	9.23
C19H25BR2N5	483.25	481.0478	47.22	5.21	33.07	14.49
C19H26BR1N1	348.33	347.1249	65.52	7.52	22.94	4.02
C19H26BR1N3	376.34	375.1311	60.64	6.96	21.23	11.17
C19H26BR1N5	404.36	403.1372	56.44	6.48	19.76	17.32
C19H26BR2N2	442.24	440.0464	51.60	5.93	36.14	6.33
C19H26BR2N4	470.25	468.0525	48.53	5.57	33.98	11.91
C19H26BR2N6	498.27	496.0587	45.80	5.26	32.07	16.87
C19H27BR1N2	363.34	362.1358	62.81	7.49	21.99	7.71
C19H27BR1N4	391.36	390.1419	58.31	6.95	20.42	14.32
C19H27BR1N6	419.37	418.1481	54.42	6.49	19.05	20.04
C19H27BR2N1	429.24	427.0511	53.17	6.34	37.23	3.26
C19H27BR2N3	457.26	455.0573	49.91	5.95	34.95	9.19
C19H27BR2N5	485.27	483.0634	47.03	5.61	32.93	14.43
C19H28BR1N1	350.35	349.1406	65.14	8.06	22.81	4.00
C19H28BR1N3	378.36	377.1467	60.32	7.46	21.12	11.11
C19H28BR1N5	406.37	405.1528	56.16	6.95	19.66	17.23
C19H28BR2N2	444.26	442.0620	51.37	6.35	35.97	6.31
C19H28BR2N4	472.27	470.0682	48.32	5.98	33.84	11.86
C19H28BR2N6	500.28	498.0743	45.62	5.64	31.94	16.80
C19H29BR1N2	365.36	364.1515	62.46	8.00	21.87	7.67
C19H29BR1N4	393.37	392.1576	58.01	7.43	20.31	14.24
C19H29BR1N6	421.39	420.1637	54.16	6.94	18.96	19.94
C19H29BR2N1	431.26	429.0668	52.92	6.78	37.06	3.25
C19H29BR2N3	459.27	457.0729	49.69	6.36	34.80	9.15
C19H29BR2N5	487.28	485.0791	46.83	6.00	32.80	14.37
C19H30BR1N1	352.36	351.1562	64.77	8.58	22.68	3.98
C19H30BR1N3	380.38	379.1623	60.00	7.95	21.01	11.05
C19H30BR1N5	408.39	407.1685	55.88	7.40	19.57	17.15
C19H30BR2N2	446.27	444.0777	51.14	6.78	35.81	6.28
C19H30BR2N4	474.29	472.0838	48.12	6.38	33.69	11.81
C19H30BR2N6	502.30	500.0900	45.43	6.02	31.82	16.73
C19H31BR1N2	367.38	366.1671	62.12	8.51	21.75	7.63
C19H31BR1N4	395.39	394.1732	57.72	7.90	20.21	14.17
C19H31BR1N6	423.40	422.1794	53.90	7.38	18.87	19.85
C19H31BR2N1	433.27	431.0824	52.67	7.21	36.88	3.23
C19H31BR2N3	461.29	459.0886	49.47	6.77	34.64	9.11
C19H31BR2N5	489.30	487.0947	46.64	6.39	32.66	14.31
C19H32BR1N1	354.38	353.1719	64.40	9.10	22.55	3.95
C19H32BR1N3	382.39	381.1780	59.68	8.44	20.90	10.99
C19H32BR1N5	410.40	409.1841	55.61	7.86	19.47	17.06
C19H32BR2N2	448.29	446.0933	50.91	7.20	35.65	6.25
C19H32BR2N4	476.30	474.0995	47.91	6.77	33.55	11.76
C19H32BR2N6	504.32	502.1056	45.25	6.40	31.69	16.66
C19H33BR1N2	369.39	368.1827	61.78	9.00	21.63	7.58
C19H33BR1N4	397.41	396.1889	57.43	8.37	20.11	14.10
C19H33BR1N6	425.42	424.1950	53.64	7.82	18.78	19.75
C19H33BR2N1	435.29	433.0981	52.43	7.64	36.71	3.22
C19H33BR2N3	463.30	461.1042	49.26	7.18	34.49	9.07
C19H33BR2N5	491.32	489.1104	46.45	6.77	32.53	14.25
C19H34BR1N1	356.39	355.1875	64.03	9.62	22.42	3.93
C19H34BR1N3	384.41	383.1936	59.37	8.92	20.79	10.93
C19H34BR1N5	412.42	411.1998	55.33	8.31	19.37	16.98
C19H34BR2N2	450.30	448.1090	50.68	7.61	35.49	6.22
C19H34BR2N4	478.32	476.1151	47.71	7.16	33.41	11.71
C19H34BR2N6	506.33	504.1213	45.07	6.77	31.56	16.60
C19H35BR1N2	371.41	370.1984	61.45	9.50	21.51	7.54
C19H35BR1N4	399.42	398.2045	57.14	8.83	20.00	14.03
C19H35BR1N6	427.43	426.2107	53.39	8.25	18.69	19.66
C19H35BR2N1	437.31	435.1137	52.19	8.07	36.54	3.20
C19H35BR2N3	465.32	463.1199	49.04	7.58	34.34	9.03
C19H35BR2N5	493.33	491.1260	46.26	7.15	32.39	14.20
C19H36BR1N1	358.41	357.2031	63.67	10.12	22.29	3.91
C19H36BR1N3	386.42	385.2093	59.06	9.39	20.68	10.87
C19H36BR1N5	414.44	413.2154	55.07	8.76	19.28	16.90
C19H36BR2N2	452.32	450.1246	50.45	8.02	35.33	6.19
C19H36BR2N4	480.33	478.1308	47.51	7.55	33.27	11.66
C19H36BR2N6	508.35	506.1369	44.89	7.14	31.44	16.53
C19H37BR1N2	373.42	372.2140	61.11	9.99	21.40	7.50
C19H37BR1N4	401.44	400.2202	56.85	9.29	19.90	13.96
C19H37BR1N6	429.45	428.2263	53.14	8.68	18.61	19.57
C19H37BR2N1	439.32	437.1294	51.95	8.49	36.38	3.19
C19H37BR2N3	467.33	465.1355	48.83	7.98	34.20	8.99
C19H37BR2N5	495.35	493.1417	46.07	7.53	32.26	14.14
C19H38BR1N1	360.43	359.2188	63.32	10.63	22.17	3.89
C19H38BR1N3	388.44	387.2249	58.75	9.86	20.57	10.82
C19H38BR1N5	416.45	415.2311	54.80	9.20	19.19	16.82
C19H38BR2N2	454.34	452.1403	50.23	8.43	35.17	6.17
C19H38BR2N4	482.35	480.1464	47.31	7.94	33.13	11.62
C19H38BR2N6	510.36	508.1526	44.72	7.51	31.31	16.47
C19H39BR1N2	375.44	374.2297	60.79	10.47	21.28	7.46
C19H39BR1N4	403.45	402.2358	56.56	9.74	19.81	13.89
C19H39BR1N6	431.47	430.2420	52.89	9.11	18.52	19.48
C19H39BR2N1	441.34	439.1450	51.71	8.91	36.21	3.17
C19H39BR2N3	469.35	467.1512	48.62	8.38	34.05	8.95
C19H39BR2N5	497.36	495.1573	45.88	7.90	32.13	14.08
C19H40BR1N1	362.44	361.2344	62.97	11.12	22.05	3.86
C19H40BR1N3	390.45	389.2406	58.45	10.33	20.46	10.76
C19H40BR1N5	418.47	417.2467	54.54	9.63	19.09	16.74
C19H40BR2N2	456.35	454.1559	50.01	8.84	35.02	6.14
C19H40BR2N4	484.37	482.1621	47.12	8.32	32.99	11.57
C19H40BR2N6	512.38	510.1682	44.54	7.87	31.19	16.40
C19H41BR1N2	377.46	376.2453	60.46	10.95	21.17	7.42
C19H41BR1N4	405.47	404.2515	56.28	10.19	19.71	13.82
C19H41BR1N6	433.48	432.2576	52.65	9.53	18.43	19.39
C19H41BR2N3	471.37	469.1668	48.41	8.77	33.90	8.91
C19H41BR2N5	499.38	497.1730	45.70	8.28	32.00	14.02
C19H42BR1N3	392.47	391.2562	58.15	10.79	20.36	10.71
C19H42BR1N5	420.48	419.2624	54.27	10.07	19.00	16.66
C19H42BR2N4	486.38	484.1777	46.92	8.70	32.86	11.52
C19H42BR2N6	514.39	512.1839	44.37	8.23	31.07	16.34
C19H43BR1N4	407.49	406.2671	56.00	10.64	19.61	13.75
C19H43BR1N6	435.50	434.2733	52.40	9.95	18.35	19.30
C19H43BR2N5	501.40	499.1886	45.52	8.64	31.87	13.97
C19H44BR1N5	422.50	421.2780	54.01	10.50	18.91	16.58
C19H44BR2N6	516.41	514.1995	44.19	8.59	30.95	16.27
C19H45BR1N6	437.51	436.2889	52.16	10.37	18.26	19.21
C20H10BR1N1	344.21	342.9997	69.79	2.93	23.21	4.07
C20H10BR1N3	372.23	371.0059	64.54	2.71	21.47	11.29
C20H10BR1N5	400.24	399.0120	60.02	2.52	19.96	17.50
C20H10BR2N2	438.12	435.9212	54.83	2.30	36.48	6.39
C20H10BR2N4	466.14	463.9273	51.53	2.16	34.28	12.02
C20H10BR2N6	494.15	491.9335	48.61	2.04	32.34	17.01
C20H11BR1N2	359.23	358.0106	66.87	3.09	22.24	7.80
C20H11BR1N4	387.24	386.0168	62.03	2.86	20.63	14.47
C20H11BR1N6	415.25	414.0229	57.85	2.67	19.24	20.24
C20H11BR2N1	425.13	422.9259	56.51	2.61	37.59	3.29
C20H11BR2N3	453.14	450.9321	53.01	2.45	35.27	9.27
C20H11BR2N5	481.15	478.9382	49.93	2.30	33.21	14.56
C20H12BR1N1	346.23	345.0154	69.38	3.49	23.08	4.05

C20H12BR1N3	374.24	373.0215	64.19	3.23	21.35	11.23
C20H12BR1N5	402.26	401.0277	59.72	3.01	19.86	17.41
C20H12BR2N2	440.14	437.9368	54.58	2.75	36.31	6.36
C20H12BR2N4	468.15	465.9430	51.31	2.58	34.14	11.97
C20H12BR2N6	496.17	493.9491	48.42	2.44	32.21	16.94
C20H13BR1N2	361.24	360.0263	66.50	3.63	22.12	7.75
C20H13BR1N4	389.26	388.0324	61.71	3.37	20.53	14.39
C20H13BR1N6	417.27	416.0386	57.57	3.14	19.15	20.14
C20H13BR2N1	427.14	424.9416	56.24	3.07	37.41	3.28
C20H13BR2N3	455.15	452.9477	52.78	2.88	35.11	9.23
C20H13BR2N5	483.17	480.9539	49.72	2.71	33.08	14.49
C20H14BR1N1	348.25	347.0310	68.98	4.05	22.94	4.02
C20H14BR1N3	376.26	375.0372	63.85	3.75	21.24	11.17
C20H14BR1N5	404.27	403.0433	59.42	3.49	19.76	17.32
C20H14BR2N2	442.16	439.9525	54.33	3.19	36.14	6.34
C20H14BR2N4	470.17	467.9586	51.09	3.00	33.99	11.92
C20H14BR2N6	498.18	495.9648	48.22	2.83	32.08	16.87
C20H15BR1N2	363.26	362.0419	66.13	4.16	22.00	7.71
C20H15BR1N4	391.27	390.0481	61.40	3.86	20.42	14.32
C20H15BR1N6	419.29	418.0542	57.29	3.61	19.06	20.04
C20H15BR2N1	429.16	426.9572	55.98	3.52	37.24	3.26
C20H15BR2N3	457.17	454.9634	52.55	3.31	34.96	9.19
C20H15BR2N5	485.18	482.9695	49.51	3.12	32.94	14.43
C20H16BR1N1	350.26	349.0467	68.58	4.60	22.81	4.00
C20H16BR1N3	378.27	377.0528	63.50	4.26	21.12	11.11
C20H16BR1N5	406.29	405.0590	59.13	3.97	19.67	17.24
C20H16BR2N2	444.17	441.9681	54.08	3.63	35.98	6.31
C20H16BR2N4	472.19	469.9743	50.87	3.42	33.84	11.87
C20H16BR2N6	500.20	497.9804	48.03	3.22	31.95	16.80
C20H17BR1N2	365.28	364.0576	65.76	4.69	21.87	7.67
C20H17BR1N4	393.29	392.0637	61.08	4.36	20.32	14.25
C20H17BR1N6	421.30	420.0699	57.02	4.07	18.97	19.95
C20H17BR2N1	431.17	428.9729	55.71	3.97	37.06	3.25
C20H17BR2N3	459.19	456.9790	52.31	3.73	34.80	9.15
C20H17BR2N5	487.20	484.9852	49.31	3.52	32.80	14.37
C20H18BR1N1	352.28	351.0623	68.19	5.15	22.68	3.98
C20H18BR1N3	380.29	379.0685	63.17	4.77	21.01	11.05
C20H18BR1N5	408.30	407.0746	58.83	4.44	19.57	17.15
C20H18BR2N2	446.19	443.9838	53.84	4.07	35.82	6.28
C20H18BR2N4	474.20	471.9899	50.66	3.83	33.70	11.81
C20H18BR2N6	502.21	499.9961	47.83	3.61	31.82	16.73
C20H19BR1N2	367.29	366.0732	65.40	5.21	21.75	7.63
C20H19BR1N4	395.31	394.0794	60.77	4.84	20.21	14.17
C20H19BR1N6	423.32	422.0855	56.75	4.52	18.88	19.85
C20H19BR2N1	433.19	430.9885	55.45	4.42	36.89	3.23
C20H19BR2N3	461.20	458.9947	52.09	4.15	34.65	9.11
C20H19BR2N5	489.22	487.0008	49.10	3.91	32.67	14.32
C20H20BR1N1	354.29	353.0780	67.80	5.69	22.55	3.95
C20H20BR1N3	382.31	381.0841	62.84	5.27	20.90	10.99
C20H20BR1N5	410.32	409.0903	58.55	4.91	19.47	17.07
C20H20BR2N2	448.20	445.9994	53.60	4.50	35.66	6.25
C20H20BR2N4	476.22	474.0056	50.44	4.23	33.56	11.76
C20H20BR2N6	504.23	502.0117	47.64	4.00	31.69	16.67
C20H21BR1N2	369.31	368.0889	65.05	5.73	21.64	7.59
C20H21BR1N4	397.32	396.0950	60.46	5.33	20.11	14.10
C20H21BR1N6	425.33	424.1011	56.48	4.98	18.79	19.76
C20H21BR2N1	435.21	433.0042	55.20	4.86	36.72	3.22
C20H21BR2N3	463.22	461.0103	51.86	4.57	34.50	9.07
C20H21BR2N5	491.23	489.0165	48.90	4.31	32.53	14.26
C20H22BR1N1	356.31	355.0936	67.42	6.22	22.43	3.93
C20H22BR1N3	384.32	383.0998	62.51	5.77	20.79	10.93
C20H22BR1N5	412.34	411.1059	58.26	5.38	19.38	16.98
C20H22BR2N2	450.22	448.0151	53.36	4.93	35.50	6.22
C20H22BR2N4	478.23	476.0212	50.23	4.64	33.42	11.72
C20H22BR2N6	506.25	504.0274	47.45	4.38	31.57	16.60
C20H23BR1N2	371.32	370.1045	64.69	6.24	21.52	7.54
C20H23BR1N4	399.34	398.1107	60.16	5.81	20.01	14.03
C20H23BR1N6	427.35	426.1168	56.21	5.42	18.70	19.67
C20H23BR2N1	437.22	435.0198	54.94	5.30	36.55	3.20
C20H23BR2N3	465.23	463.0260	51.63	4.98	34.35	9.03
C20H23BR2N5	493.25	491.0321	48.70	4.70	32.40	14.20
C20H24BR1N1	358.32	357.1093	67.04	6.75	22.30	3.91
C20H24BR1N3	386.34	385.1154	62.18	6.26	20.68	10.88
C20H24BR1N5	414.35	413.1215	57.98	5.84	19.28	16.90
C20H24BR2N2	452.24	450.0307	53.12	5.35	35.34	6.19
C20H24BR2N4	480.25	478.0369	50.02	5.04	33.28	11.67
C20H24BR2N6	508.26	506.0430	47.26	4.76	31.44	16.53
C20H25BR1N2	373.34	372.1202	64.34	6.75	21.40	7.50
C20H25BR1N4	401.35	400.1263	59.85	6.28	19.91	13.96
C20H25BR1N6	429.37	428.1324	55.95	5.87	18.61	19.57
C20H25BR2N1	439.24	437.0355	54.69	5.74	36.38	3.19
C20H25BR2N3	467.25	465.0416	51.41	5.39	34.20	8.99
C20H25BR2N5	495.26	493.0478	48.50	5.09	32.27	14.14
C20H26BR1N1	360.34	359.1249	66.67	7.27	22.17	3.89
C20H26BR1N3	388.35	387.1311	61.86	6.75	20.58	10.82
C20H26BR1N5	416.37	415.1372	57.69	6.29	19.19	16.82
C20H26BR2N2	454.25	452.0464	52.88	5.77	35.18	6.17
C20H26BR2N4	482.27	480.0525	49.81	5.43	33.14	11.62
C20H26BR2N6	510.28	508.0587	47.08	5.14	31.32	16.47
C20H27BR1N2	375.36	374.1358	64.00	7.25	21.29	7.46
C20H27BR1N4	403.37	402.1419	59.55	6.75	19.81	13.89
C20H27BR1N6	431.38	430.1481	55.69	6.31	18.52	19.48
C20H27BR2N1	441.25	439.0511	54.44	6.17	36.22	3.17
C20H27BR2N3	469.27	467.0573	51.19	5.80	34.05	8.95
C20H27BR2N5	497.28	495.0634	48.31	5.47	32.14	14.08
C20H28BR1N1	362.36	361.1406	66.29	7.79	22.05	3.87
C20H28BR1N3	390.37	389.1467	61.54	7.23	20.47	10.76
C20H28BR1N5	418.38	417.1528	57.42	6.75	19.10	16.74
C20H28BR2N2	456.27	454.0620	52.65	6.19	35.03	6.14
C20H28BR2N4	484.28	482.0682	49.60	5.83	33.00	11.57
C20H28BR2N6	512.29	510.0743	46.89	5.51	31.19	16.40
C20H29BR1N2	377.37	376.1515	63.66	7.75	21.17	7.42
C20H29BR1N4	405.38	404.1576	59.26	7.21	19.71	13.82
C20H29BR1N6	433.40	432.1637	55.43	6.74	18.44	19.39
C20H29BR2N1	443.27	441.0668	54.19	6.59	36.05	3.16
C20H29BR2N3	471.28	469.0729	50.97	6.20	33.91	8.92
C20H29BR2N5	499.30	497.0791	48.11	5.85	32.01	14.03
C20H30BR1N1	364.37	363.1562	65.93	8.30	21.93	3.84
C20H30BR1N3	392.39	391.1623	61.22	7.71	20.36	10.71
C20H30BR1N5	420.40	419.1685	57.14	7.19	19.01	16.66
C20H30BR2N2	458.28	456.0777	52.42	6.60	34.87	6.11
C20H30BR2N4	486.30	484.0838	49.40	6.22	32.86	11.52
C20H30BR2N6	514.31	512.0900	46.71	5.88	31.07	16.34
C20H31BR1N2	379.39	378.1671	63.32	8.24	21.06	7.38
C20H31BR1N4	407.40	406.1732	58.96	7.67	19.61	13.75
C20H31BR1N6	435.41	434.1794	55.17	7.18	18.35	19.30
C20H31BR2N1	445.28	443.0824	53.95	7.02	35.89	3.15
C20H31BR2N3	473.30	471.0886	50.76	6.60	33.76	8.88
C20H31BR2N5	501.31	499.0947	47.92	6.23	31.88	13.97
C20H32BR1N1	366.39	365.1719	65.57	8.80	21.81	3.82

C20H32BR1N3	394.40	393.1780	60.91	8.18	20.26	10.65
C20H32BR1N5	422.42	421.1841	56.87	7.64	18.92	16.58
C20H32BR2N2	460.30	458.0933	52.19	7.01	34.72	6.09
C20H32BR2N4	488.31	486.0995	49.19	6.61	32.73	11.47
C20H32BR2N6	516.33	514.1056	46.53	6.25	30.95	16.28
C20H33BR1N2	381.40	380.1827	62.98	8.72	20.95	7.34
C20H33BR1N4	409.42	408.1889	58.67	8.12	19.52	13.68
C20H33BR1N6	437.43	436.1950	54.92	7.60	18.27	19.21
C20H33BR2N1	447.30	445.0981	53.71	7.44	35.73	3.13
C20H33BR2N3	475.31	473.1042	50.54	7.00	33.62	8.84
C20H33BR2N5	503.33	501.1104	47.73	6.61	31.75	13.91
C20H34BR1N1	368.40	367.1875	65.21	9.30	21.69	3.80
C20H34BR1N3	396.42	395.1936	60.60	8.65	20.16	10.60
C20H34BR1N5	424.43	423.1998	56.60	8.07	18.83	16.50
C20H34BR2N2	462.32	460.1090	51.96	7.41	34.57	6.06
C20H34BR2N4	490.33	488.1151	48.99	6.99	32.59	11.43
C20H34BR2N6	518.34	516.1213	46.34	6.61	30.83	16.21
C20H35BR1N2	383.42	382.1984	62.65	9.20	20.84	7.31
C20H35BR1N4	411.43	410.2045	58.39	8.57	19.42	13.62
C20H35BR1N6	439.45	438.2107	54.66	8.03	18.18	19.12
C20H35BR2N1	449.32	447.1137	53.46	7.85	35.57	3.12
C20H35BR2N3	477.33	475.1199	50.33	7.39	33.48	8.80
C20H35BR2N5	505.34	503.1260	47.54	6.98	31.62	13.86
C20H36BR1N1	370.42	369.2031	64.85	9.80	21.57	3.78
C20H36BR1N3	398.43	397.2093	60.29	9.11	20.05	10.55
C20H36BR1N5	426.45	425.2154	56.33	8.51	18.74	16.42
C20H36BR2N2	464.33	462.1246	51.74	7.81	34.42	6.03
C20H36BR2N4	492.34	490.1308	48.79	7.37	32.46	11.38
C20H36BR2N6	520.36	518.1369	46.16	6.97	30.71	16.15
C20H37BR1N2	385.44	384.2140	62.33	9.68	20.73	7.27
C20H37BR1N4	413.45	412.2202	58.10	9.02	19.33	13.55
C20H37BR1N6	441.46	440.2263	54.42	8.45	18.10	19.04
C20H37BR2N1	451.33	449.1294	53.23	8.26	35.41	3.10
C20H37BR2N3	479.35	477.1355	50.11	7.78	33.34	8.77
C20H37BR2N5	507.36	505.1417	47.35	7.35	31.50	13.80
C20H38BR1N1	372.44	371.2188	64.50	10.28	21.45	3.76
C20H38BR1N3	400.45	399.2249	59.99	9.56	19.95	10.49
C20H38BR1N5	428.46	427.2311	56.07	8.94	18.65	16.35
C20H38BR2N2	466.35	464.1403	51.51	8.21	34.27	6.01
C20H38BR2N4	494.36	492.1464	48.59	7.75	32.33	11.33
C20H38BR2N6	522.37	520.1526	45.99	7.33	30.59	16.09
C20H39BR1N2	387.45	386.2297	62.00	10.15	20.62	7.23
C20H39BR1N4	415.46	414.2358	57.82	9.46	19.23	13.49
C20H39BR1N6	443.48	442.2420	54.17	8.86	18.02	18.95
C20H39BR2N1	453.35	451.1450	52.99	8.67	35.25	3.09
C20H39BR2N3	481.36	479.1512	49.90	8.17	33.20	8.73
C20H39BR2N5	509.38	507.1573	47.16	7.72	31.37	13.75
C20H40BR1N1	374.45	373.2344	64.15	10.77	21.34	3.74
C20H40BR1N3	402.47	401.2406	59.69	10.02	19.85	10.44
C20H40BR1N5	430.48	429.2467	55.80	9.37	18.56	16.27
C20H40BR2N2	468.36	466.1559	51.29	8.61	34.12	5.98
C20H40BR2N4	496.38	494.1621	48.40	8.12	32.19	11.29
C20H40BR2N6	524.39	522.1682	45.81	7.69	30.48	16.03
C20H41BR1N2	389.47	388.2453	61.68	10.61	20.52	7.19
C20H41BR1N4	417.48	416.2515	57.54	9.90	19.14	13.42
C20H41BR1N6	445.49	444.2576	53.92	9.28	17.94	18.86
C20H41BR2N1	455.36	453.1607	52.75	9.08	35.09	3.08
C20H41BR2N3	483.38	481.1668	49.70	8.55	33.06	8.69
C20H41BR2N5	511.39	509.1730	46.97	8.08	31.25	13.69
C20H42BR1N1	376.47	375.2501	63.81	11.25	21.22	3.72
C20H42BR1N3	404.48	403.2562	59.39	10.47	19.75	10.39
C20H42BR1N5	432.50	431.2624	55.54	9.79	18.48	16.19
C20H42BR2N2	470.38	468.1716	51.07	9.00	33.97	5.96
C20H42BR2N4	498.39	496.1777	48.20	8.49	32.06	11.24
C20H42BR2N6	526.41	524.1839	45.63	8.04	30.36	15.96
C20H43BR1N2	391.48	390.2610	61.36	11.07	20.41	7.16
C20H43BR1N4	419.50	418.2671	57.26	10.33	19.05	13.36
C20H43BR1N6	447.51	446.2733	53.68	9.69	17.86	18.78
C20H43BR2N3	485.39	483.1825	49.49	8.93	32.92	8.66
C20H43BR2N5	513.41	511.1886	46.79	8.44	31.13	13.64
C20H44BR1N3	406.50	405.2719	59.10	10.91	19.66	10.34
C20H44BR1N5	434.51	433.2780	55.29	10.21	18.39	16.12
C20H44BR2N4	500.41	498.1934	48.01	8.86	31.94	11.20
C20H44BR2N6	528.42	526.1995	45.46	8.39	30.24	15.90
C20H45BR1N4	421.51	420.2828	56.99	10.76	18.96	13.29
C20H45BR1N6	449.53	448.2889	53.44	10.09	17.78	18.70
C20H45BR2N5	515.42	513.2043	46.61	8.80	31.01	13.59
C20H46BR1N5	436.53	435.2937	55.03	10.62	18.30	16.04
C20H46BR2N6	530.44	528.2152	45.29	8.74	30.13	15.84
C20H47BR1N6	451.54	450.3046	53.20	10.49	17.70	18.61
C21H10BR1N1	356.22	354.9997	70.81	2.83	22.43	3.93
C21H10BR1N3	384.24	383.0059	65.65	2.62	20.80	10.94
C21H10BR1N5	412.25	411.0120	61.18	2.45	19.38	16.99
C21H10BR2N2	450.14	447.9212	56.04	2.24	35.50	6.22
C21H10BR2N4	478.15	475.9273	52.75	2.11	33.42	11.72
C21H10BR2N6	506.16	503.9335	49.83	1.99	31.57	16.60
C21H11BR1N2	371.24	370.0106	67.94	2.99	21.52	7.55
C21H11BR1N4	399.25	398.0168	63.18	2.78	20.01	14.03
C21H11BR1N6	427.27	426.0229	59.03	2.60	18.70	19.67
C21H11BR2N1	437.14	434.9259	57.70	2.54	36.56	3.20
C21H11BR2N3	465.15	462.9321	54.23	2.38	34.36	9.03
C21H11BR2N5	493.16	490.9382	51.15	2.25	32.40	14.20
C21H12BR1N1	358.24	357.0154	70.41	3.38	22.30	3.91
C21H12BR1N3	386.25	385.0215	65.30	3.13	20.69	10.88
C21H12BR1N5	414.27	413.0277	60.89	2.92	19.29	16.91
C21H12BR2N2	452.15	449.9368	55.79	2.68	35.34	6.20
C21H12BR2N4	480.16	477.9430	52.53	2.52	33.28	11.67
C21H12BR2N6	508.18	505.9491	49.64	2.38	31.45	16.54
C21H13BR1N2	373.26	372.0263	67.58	3.51	21.41	7.51
C21H13BR1N4	401.27	400.0324	62.86	3.27	19.91	13.96
C21H13BR1N6	429.28	428.0386	58.76	3.05	18.61	19.58
C21H13BR2N1	439.15	436.9416	57.44	2.98	36.39	3.19
C21H13BR2N3	467.17	464.9477	53.99	2.80	34.21	8.99
C21H13BR2N5	495.18	492.9539	50.94	2.65	32.27	14.14
C21H14BR1N1	360.26	359.0310	70.02	3.92	22.18	3.89
C21H14BR1N3	388.27	387.0372	64.96	3.63	20.58	10.82
C21H14BR1N5	416.28	415.0433	60.59	3.39	19.19	16.82
C21H14BR2N2	454.17	451.9525	55.54	3.11	35.19	6.17
C21H14BR2N4	482.18	479.9586	52.31	2.93	33.14	11.62
C21H14BR2N6	510.19	507.9648	49.44	2.77	31.32	16.47
C21H15BR1N2	375.27	374.0419	67.21	4.03	21.29	7.46
C21H15BR1N4	403.29	402.0481	62.54	3.75	19.81	13.89
C21H15BR1N6	431.30	430.0542	58.48	3.51	18.53	19.49
C21H15BR2N1	441.17	438.9572	57.17	3.43	36.22	3.17
C21H15BR2N3	469.18	466.9634	53.76	3.22	34.06	8.96
C21H15BR2N5	497.20	494.9695	50.73	3.04	32.14	14.09
C21H16BR1N1	362.27	361.0467	69.63	4.45	22.06	3.87
C21H16BR1N3	390.29	389.0528	64.63	4.13	20.47	10.77
C21H16BR1N5	418.30	417.0590	60.30	3.86	19.10	16.74
C21H16BR2N2	456.18	453.9681	55.29	3.54	35.03	6.14

C21H16BR2N4	484.20	481.9743	52.09	3.33	33.00	11.57
C21H16BR2N6	512.21	509.9804	49.24	3.15	31.20	16.41
C21H17BR1N2	377.29	376.0576	66.85	4.54	21.18	7.42
C21H17BR1N4	405.30	404.0637	62.23	4.23	19.71	13.82
C21H17BR1N6	433.31	432.0699	58.21	3.95	18.44	19.39
C21H17BR2N1	443.18	440.9729	56.91	3.87	36.06	3.16
C21H17BR2N3	471.20	468.9790	53.53	3.64	33.92	8.92
C21H17BR2N5	499.21	496.9852	50.53	3.43	32.01	14.03
C21H18BR1N1	364.29	363.0623	69.24	4.98	21.93	3.84
C21H18BR1N3	392.30	391.0685	64.30	4.62	20.37	10.71
C21H18BR1N5	420.32	419.0746	60.01	4.32	19.01	16.66
C21H18BR2N2	458.20	455.9838	55.05	3.96	34.88	6.11
C21H18BR2N4	486.21	483.9899	51.88	3.73	32.87	11.52
C21H18BR2N6	514.23	511.9961	49.05	3.53	31.08	16.34
C21H19BR1N2	379.30	378.0732	66.50	5.05	21.07	7.39
C21H19BR1N4	407.32	406.0794	61.93	4.70	19.62	13.76
C21H19BR1N6	435.33	434.0855	57.94	4.40	18.35	19.30
C21H19BR2N1	445.20	442.9885	56.66	4.30	35.90	3.15
C21H19BR2N3	473.21	470.9947	53.30	4.05	33.77	8.88
C21H19BR2N5	501.23	499.0008	50.32	3.82	31.88	13.97
C21H20BR1N1	366.30	365.0780	68.86	5.50	21.81	3.82
C21H20BR1N3	394.32	393.0841	63.97	5.11	20.26	10.66
C21H20BR1N5	422.33	421.0903	59.72	4.77	18.92	16.58
C21H20BR2N2	460.21	457.9994	54.81	4.38	34.72	6.09
C21H20BR2N4	488.23	486.0056	51.66	4.13	32.73	11.48
C21H20BR2N6	516.24	514.0117	48.86	3.91	30.96	16.28
C21H21BR1N2	381.32	380.0889	66.15	5.55	20.95	7.35
C21H21BR1N4	409.33	408.0950	61.62	5.17	19.52	13.69
C21H21BR1N6	437.35	436.1011	57.67	4.84	18.27	19.22
C21H21BR2N1	447.22	445.0042	56.40	4.73	35.73	3.13
C21H21BR2N3	475.23	473.0103	53.08	4.45	33.63	8.84
C21H21BR2N5	503.24	501.0165	50.12	4.21	31.76	13.92
C21H22BR1N1	368.32	367.0936	68.48	6.02	21.69	3.80
C21H22BR1N3	396.33	395.0998	63.64	5.60	20.16	10.60
C21H22BR1N5	424.35	423.1059	59.44	5.23	18.83	16.50
C21H22BR2N2	462.23	460.0151	54.57	4.80	34.57	6.06
C21H22BR2N4	490.24	488.0212	51.45	4.52	32.60	11.43
C21H22BR2N6	518.26	516.0274	48.67	4.28	30.84	16.22
C21H23BR1N2	383.33	382.1045	65.80	6.05	20.84	7.31
C21H23BR1N4	411.35	410.1107	61.32	5.64	19.42	13.62
C21H23BR1N6	439.36	438.1168	57.41	5.28	18.19	19.13
C21H23BR2N1	449.23	447.0198	56.15	5.16	35.57	3.12
C21H23BR2N3	477.25	475.0260	52.85	4.86	33.49	8.80
C21H23BR2N5	505.26	503.0321	49.92	4.59	31.63	13.86
C21H24BR1N1	370.34	369.1093	68.11	6.53	21.58	3.78
C21H24BR1N3	398.35	397.1154	63.32	6.07	20.06	10.55
C21H24BR1N5	426.36	425.1215	59.16	5.67	18.74	16.43
C21H24BR2N2	464.25	462.0307	54.33	5.21	34.42	6.03
C21H24BR2N4	492.26	490.0369	51.24	4.91	32.46	11.38
C21H24BR2N6	520.27	518.0430	48.48	4.65	30.72	16.15
C21H25BR1N2	385.35	384.1202	65.46	6.54	20.74	7.27
C21H25BR1N4	413.36	412.1263	61.02	6.10	19.33	13.55
C21H25BR1N6	441.38	440.1324	57.15	5.71	18.10	19.04
C21H25BR2N1	451.25	449.0355	55.90	5.58	35.41	3.10
C21H25BR2N3	479.26	477.0416	52.63	5.26	33.34	8.77
C21H25BR2N5	507.27	505.0478	49.72	4.97	31.50	13.81
C21H26BR1N1	372.35	371.1249	67.74	7.04	21.46	3.76
C21H26BR1N3	400.37	399.1311	63.00	6.55	19.96	10.50
C21H26BR1N5	428.38	427.1372	58.88	6.12	18.65	16.35
C21H26BR2N2	466.26	464.0464	54.10	5.62	34.27	6.01

C21H26BR2N4	494.28	492.0525	51.03	5.30	32.33	11.34
C21H26BR2N6	522.29	520.0587	48.29	5.02	30.60	16.09
C21H27BR1N2	387.37	386.1358	65.12	7.03	20.63	7.23
C21H27BR1N4	415.38	414.1419	60.72	6.55	19.24	13.49
C21H27BR1N6	443.39	442.1481	56.89	6.14	18.02	18.95
C21H27BR2N1	453.26	451.0511	55.65	6.00	35.26	3.09
C21H27BR2N3	481.28	479.0573	52.41	5.65	33.20	8.73
C21H27BR2N5	509.29	507.0634	49.53	5.34	31.38	13.75
C21H28BR1N1	374.37	373.1406	67.38	7.54	21.34	3.74
C21H28BR1N3	402.38	401.1467	62.69	7.01	19.86	10.44
C21H28BR1N5	430.39	429.1528	58.61	6.56	18.57	16.27
C21H28BR2N2	468.28	466.0620	53.86	6.03	34.13	5.98
C21H28BR2N4	496.29	494.0682	50.82	5.69	32.20	11.29
C21H28BR2N6	524.31	522.0743	48.11	5.38	30.48	16.03
C21H29BR1N2	389.38	388.1515	64.78	7.51	20.52	7.19
C21H29BR1N4	417.40	416.1576	60.43	7.00	19.14	13.42
C21H29BR1N6	445.41	444.1637	56.63	6.56	17.94	18.87
C21H29BR2N1	455.28	453.0668	55.40	6.42	35.10	3.08
C21H29BR2N3	483.29	481.0729	52.19	6.05	33.07	8.69
C21H29BR2N5	511.31	509.0791	49.33	5.72	31.25	13.70
C21H30BR1N1	376.38	375.1562	67.02	8.03	21.23	3.72
C21H30BR1N3	404.40	403.1623	62.37	7.48	19.76	10.39
C21H30BR1N5	432.41	431.1685	58.33	6.99	18.48	16.20
C21H30BR2N2	470.29	468.0777	53.63	6.43	33.98	5.96
C21H30BR2N4	498.31	496.0838	50.62	6.07	32.07	11.24
C21H30BR2N6	526.32	524.0900	47.92	5.75	30.36	15.97
C21H31BR1N2	391.40	390.1671	64.44	7.98	20.41	7.16
C21H31BR1N4	419.41	418.1732	60.14	7.45	19.05	13.36
C21H31BR1N6	447.43	446.1794	56.37	6.98	17.86	18.78
C21H31BR2N1	457.30	455.0824	55.16	6.83	34.95	3.06
C21H31BR2N3	485.31	483.0886	51.97	6.44	32.93	8.66
C21H31BR2N5	513.32	511.0947	49.14	6.09	31.13	13.64
C21H32BR1N1	378.40	377.1719	66.66	8.52	21.12	3.70
C21H32BR1N3	406.41	405.1780	62.06	7.94	19.66	10.34
C21H32BR1N5	434.43	433.1841	58.06	7.42	18.39	16.12
C21H32BR2N2	472.31	470.0933	53.40	6.83	33.84	5.93
C21H32BR2N4	500.32	498.0995	50.41	6.45	31.94	11.20
C21H32BR2N6	528.34	526.1056	47.74	6.11	30.25	15.91
C21H33BR1N2	393.41	392.1827	64.11	8.45	20.31	7.12
C21H33BR1N4	421.43	420.1889	59.85	7.89	18.96	13.29
C21H33BR1N6	449.44	448.1950	56.12	7.40	17.78	18.70
C21H33BR2N1	459.31	457.0981	54.92	7.24	34.79	3.05
C21H33BR2N3	487.33	485.1042	51.76	6.83	32.79	8.62
C21H33BR2N5	515.34	513.1104	48.95	6.45	31.01	13.59
C21H34BR1N1	380.42	379.1875	66.30	9.01	21.00	3.68
C21H34BR1N3	408.43	407.1936	61.76	8.39	19.56	10.29
C21H34BR1N5	436.44	435.1998	57.79	7.85	18.31	16.05
C21H34BR2N2	474.33	472.1090	53.18	7.23	33.69	5.91
C21H34BR2N4	502.34	500.1151	50.21	6.82	31.81	11.15
C21H34BR2N6	530.35	528.1213	47.56	6.46	30.13	15.85
C21H35BR1N2	395.43	394.1984	63.79	8.92	20.21	7.08
C21H35BR1N4	423.44	422.2045	59.57	8.33	18.87	13.23
C21H35BR1N6	451.46	450.2107	55.87	7.81	17.70	18.62
C21H35BR2N1	461.33	459.1137	54.68	7.65	34.64	3.04
C21H35BR2N3	489.34	487.1199	51.55	7.21	32.66	8.59
C21H35BR2N5	517.35	515.1260	48.75	6.82	30.89	13.54
C21H36BR1N1	382.43	381.2031	65.96	9.49	20.89	3.66
C21H36BR1N3	410.45	409.2093	61.45	8.84	19.47	10.24
C21H36BR1N5	438.46	437.2154	57.53	8.28	18.22	15.97
C21H36BR2N2	476.34	474.1246	52.95	7.62	33.55	5.88

C21H36BR2N4	504.36	502.1308	50.01	7.19	31.69	11.11
C21H36BR2N6	532.37	530.1369	47.38	6.82	30.02	15.79
C21H37BR1N2	397.45	396.2140	63.46	9.38	20.10	7.05
C21H37BR1N4	425.46	424.2202	59.29	8.77	18.78	13.17
C21H37BR1N6	453.47	452.2263	55.62	8.22	17.62	18.53
C21H37BR2N1	463.34	461.1294	54.44	8.05	34.49	3.02
C21H37BR2N3	491.36	489.1355	51.33	7.59	32.52	8.55
C21H37BR2N5	519.37	517.1417	48.57	7.18	30.77	13.48
C21H38BR1N1	384.45	383.2188	65.61	9.96	20.78	3.64
C21H38BR1N3	412.46	411.2249	61.15	9.29	19.37	10.19
C21H38BR1N5	440.47	439.2311	57.26	8.70	18.14	15.90
C21H38BR2N2	478.36	476.1403	52.73	8.01	33.41	5.86
C21H38BR2N4	506.37	504.1464	49.81	7.56	31.56	11.06
C21H38BR2N6	534.39	532.1526	47.20	7.17	29.91	15.73
C21H39BR1N2	399.46	398.2297	63.14	9.84	20.00	7.01
C21H39BR1N4	427.48	426.2358	59.01	9.20	18.69	13.11
C21H39BR1N6	455.49	454.2420	55.38	8.63	17.54	18.45
C21H39BR2N1	465.36	463.1450	54.20	8.45	34.34	3.01
C21H39BR2N3	493.37	491.1512	51.12	7.97	32.39	8.52
C21H39BR2N5	521.39	519.1573	48.38	7.54	30.65	13.43
C21H40BR1N1	386.46	385.2344	65.27	10.43	20.68	3.62
C21H40BR1N3	414.48	413.2406	60.86	9.73	19.28	10.14
C21H40BR1N5	442.49	441.2467	57.00	9.11	18.06	15.83
C21H40BR2N2	480.37	478.1559	52.51	8.39	33.27	5.83
C21H40BR2N4	508.39	506.1621	49.61	7.93	31.43	11.02
C21H40BR2N6	536.40	534.1682	47.02	7.52	29.79	15.67
C21H41BR1N2	401.48	400.2453	62.83	10.29	19.90	6.98
C21H41BR1N4	429.49	428.2515	58.73	9.62	18.60	13.04
C21H41BR1N6	457.51	456.2576	55.13	9.03	17.47	18.37
C21H41BR2N1	467.38	465.1607	53.97	8.84	34.19	3.00
C21H41BR2N3	495.39	493.1668	50.92	8.34	32.26	8.48
C21H41BR2N5	523.40	521.1730	48.19	7.90	30.53	13.38
C21H42BR1N1	388.48	387.2501	64.93	10.90	20.57	3.61
C21H42BR1N3	416.49	415.2562	60.56	10.16	19.18	10.09
C21H42BR1N5	444.51	443.2624	56.74	9.52	17.98	15.76
C21H42BR2N2	482.39	480.1716	52.29	8.78	33.13	5.81
C21H42BR2N4	510.40	508.1777	49.42	8.29	31.31	10.98
C21H42BR2N6	538.42	536.1839	46.85	7.86	29.68	15.61
C21H43BR1N2	403.49	402.2610	62.51	10.74	19.80	6.94
C21H43BR1N4	431.51	430.2671	58.45	10.04	18.52	12.98
C21H43BR1N6	459.52	458.2733	54.89	9.43	17.39	18.29
C21H43BR2N1	469.39	467.1763	53.74	9.23	34.05	2.98
C21H43BR2N3	497.40	495.1825	50.71	8.71	32.13	8.45
C21H43BR2N5	525.42	523.1886	48.01	8.25	30.42	13.33
C21H44BR1N1	390.50	389.2657	64.59	11.36	20.46	3.59
C21H44BR1N3	418.51	417.2719	60.27	10.60	19.09	10.04
C21H44BR1N5	446.52	445.2780	56.49	9.93	17.89	15.68
C21H44BR2N2	484.41	482.1872	52.07	9.16	32.99	5.78
C21H44BR2N4	512.42	510.1934	49.22	8.66	31.19	10.93
C21H44BR2N6	540.43	538.1995	46.67	8.21	29.57	15.55
C21H45BR1N2	405.51	404.2766	62.20	11.19	19.70	6.91
C21H45BR1N4	433.52	432.2828	58.18	10.46	18.43	12.92
C21H45BR1N6	461.54	460.2889	54.65	9.83	17.31	18.21
C21H45BR2N3	499.42	497.1981	50.51	9.08	32.00	8.41
C21H45BR2N5	527.43	525.2043	47.82	8.60	30.30	13.28
C21H46BR1N3	420.52	419.2875	59.98	11.03	19.00	9.99
C21H46BR1N5	448.54	447.2937	56.23	10.34	17.81	15.61
C21H46BR2N4	514.44	512.2090	49.03	9.01	31.06	10.89
C21H46BR2N6	542.45	540.2152	46.50	8.55	29.46	15.49
C21H47BR1N4	435.54	434.2984	57.91	10.88	18.35	12.86

C21H47BR1N6	463.55	462.3046	54.41	10.22	17.24	18.13
C21H47BR2N5	529.45	527.2199	47.64	8.95	30.18	13.23
C21H48BR1N5	450.55	449.3093	55.98	10.74	17.73	15.54
C21H48BR2N6	544.46	542.2308	46.33	8.89	29.35	15.44
C21H49BR1N6	465.57	464.3202	54.18	10.61	17.16	18.05
C22H11BR1N2	383.25	382.0106	68.95	2.89	20.85	7.31
C22H11BR1N4	411.26	410.0168	64.25	2.70	19.43	13.62
C22H11BR1N6	439.28	438.0229	60.15	2.52	18.19	19.13
C22H11BR2N1	449.15	446.9259	58.83	2.47	35.58	3.12
C22H11BR2N3	477.16	474.9321	55.38	2.32	33.49	8.81
C22H11BR2N5	505.17	502.9382	52.31	2.19	31.63	13.86
C22H12BR1N1	370.25	369.0154	71.37	3.27	21.58	3.78
C22H12BR1N3	398.27	397.0215	66.35	3.04	20.06	10.55
C22H12BR1N5	426.28	425.0277	61.99	2.84	18.74	16.43
C22H12BR2N2	464.16	461.9368	56.93	2.61	34.43	6.04
C22H12BR2N4	492.18	489.9430	53.69	2.46	32.47	11.38
C22H12BR2N6	520.19	517.9491	50.80	2.33	30.72	16.16
C22H13BR1N2	385.27	384.0263	68.59	3.40	20.74	7.27
C22H13BR1N4	413.28	412.0324	63.94	3.17	19.33	13.56
C22H13BR1N6	441.29	440.0386	59.88	2.97	18.11	19.04
C22H13BR2N1	451.16	448.9416	58.57	2.90	35.42	3.10
C22H13BR2N3	479.18	476.9477	55.15	2.73	33.35	8.77
C22H13BR2N5	507.19	504.9539	52.10	2.58	31.51	13.81
C22H14BR1N1	372.27	371.0310	70.98	3.79	21.46	3.76
C22H14BR1N3	400.28	399.0372	66.01	3.53	19.96	10.50
C22H14BR1N5	428.29	427.0433	61.70	3.29	18.66	16.35
C22H14BR2N2	466.18	463.9525	56.68	3.03	34.28	6.01
C22H14BR2N4	494.19	491.9586	53.47	2.86	32.34	11.34
C22H14BR2N6	522.21	519.9648	50.60	2.70	30.60	16.09
C22H15BR1N2	387.28	386.0419	68.23	3.90	20.63	7.23
C22H15BR1N4	415.30	414.0481	63.63	3.64	19.24	13.49
C22H15BR1N6	443.31	442.0542	59.61	3.41	18.02	18.96
C22H15BR2N1	453.18	450.9572	58.31	3.34	35.26	3.09
C22H15BR2N3	481.19	478.9634	54.91	3.14	33.21	8.73
C22H15BR2N5	509.21	506.9695	51.89	2.97	31.38	13.75
C22H16BR1N1	374.28	373.0467	70.60	4.31	21.35	3.74
C22H16BR1N3	402.30	401.0528	65.68	4.01	19.86	10.45
C22H16BR1N5	430.31	429.0590	61.41	3.75	18.57	16.28
C22H16BR2N2	468.19	465.9681	56.44	3.44	34.13	5.98
C22H16BR2N4	496.21	493.9743	53.25	3.25	32.21	11.29
C22H16BR2N6	524.22	521.9804	50.41	3.08	30.48	16.03
C22H17BR1N2	389.30	388.0576	67.88	4.40	20.53	7.20
C22H17BR1N4	417.31	416.0637	63.32	4.11	19.15	13.43
C22H17BR1N6	445.32	444.0699	59.34	3.85	17.94	18.87
C22H17BR2N1	455.20	452.9729	58.05	3.76	35.11	3.08
C22H17BR2N3	483.21	480.9790	54.69	3.55	33.07	8.70
C22H17BR2N5	511.22	508.9852	51.69	3.35	31.26	13.70
C22H18BR1N1	376.30	375.0623	70.22	4.82	21.23	3.72
C22H18BR1N3	404.31	403.0685	65.36	4.49	19.76	10.39
C22H18BR1N5	432.33	431.0746	61.12	4.20	18.48	16.20
C22H18BR2N2	470.21	467.9838	56.20	3.86	33.99	5.96
C22H18BR2N4	498.22	495.9899	53.04	3.64	32.08	11.25
C22H18BR2N6	526.24	523.9961	50.21	3.45	30.37	15.97
C22H19BR1N2	391.31	390.0732	67.53	4.89	20.42	7.16
C22H19BR1N4	419.33	418.0794	63.02	4.57	19.06	13.36
C22H19BR1N6	447.34	446.0855	59.07	4.28	17.86	18.79
C22H19BR2N1	457.21	454.9885	57.79	4.19	34.95	3.06
C22H19BR2N3	485.22	482.9947	54.46	3.95	32.93	8.66
C22H19BR2N5	513.24	511.0008	51.49	3.73	31.14	13.65
C22H20BR1N1	378.32	377.0780	69.85	5.33	21.12	3.70

C22H20BR1N3	406.33	405.0841	65.03	4.96	19.66	10.34
C22H20BR1N5	434.34	433.0903	60.84	4.64	18.40	16.12
C22H20BR2N2	472.23	469.9994	55.96	4.27	33.84	5.93
C22H20BR2N4	500.24	498.0056	52.82	4.03	31.95	11.20
C22H20BR2N6	528.25	526.0117	50.02	3.82	30.25	15.91
C22H21BR1N2	393.33	392.0889	67.18	5.38	20.31	7.12
C22H21BR1N4	421.34	420.0950	62.71	5.02	18.96	13.30
C22H21BR1N6	449.36	448.1011	58.81	4.71	17.78	18.70
C22H21BR2N1	459.23	457.0042	57.54	4.61	34.80	3.05
C22H21BR2N3	487.24	485.0103	54.23	4.34	32.80	8.62
C22H21BR2N5	515.25	513.0165	51.28	4.11	31.02	13.59
C22H22BR1N1	380.33	379.0936	69.48	5.83	21.01	3.68
C22H22BR1N3	408.34	407.0998	64.71	5.43	19.57	10.29
C22H22BR1N5	436.36	435.1059	60.56	5.08	18.31	16.05
C22H22BR2N2	474.24	472.0151	55.72	4.68	33.70	5.91
C22H22BR2N4	502.26	500.0212	52.61	4.42	31.82	11.16
C22H22BR2N6	530.27	528.0274	49.83	4.18	30.14	15.85
C22H23BR1N2	395.35	394.1045	66.84	5.86	20.21	7.09
C22H23BR1N4	423.36	422.1107	62.42	5.48	18.87	13.23
C22H23BR1N6	451.37	450.1168	58.54	5.14	17.70	18.62
C22H23BR2N1	461.24	459.0198	57.29	5.03	34.65	3.04
C22H23BR2N3	489.26	487.0260	54.01	4.74	32.66	8.59
C22H23BR2N5	517.27	515.0321	51.08	4.48	30.89	13.54
C22H24BR1N1	382.35	381.1093	69.11	6.33	20.90	3.66
C22H24BR1N3	410.36	409.1154	64.39	5.90	19.47	10.24
C22H24BR1N5	438.37	437.1215	60.28	5.52	18.23	15.98
C22H24BR2N2	476.26	474.0307	55.48	5.08	33.55	5.88
C22H24BR2N4	504.27	502.0369	52.40	4.80	31.69	11.11
C22H24BR2N6	532.28	530.0430	49.64	4.54	30.02	15.79
C22H25BR1N2	397.36	396.1202	66.50	6.34	20.11	7.05
C22H25BR1N4	425.38	424.1263	62.12	5.92	18.78	13.17
C22H25BR1N6	453.39	452.1324	58.28	5.56	17.62	18.54
C22H25BR2N1	463.26	461.0355	57.04	5.44	34.50	3.02
C22H25BR2N3	491.27	489.0416	53.79	5.13	32.53	8.55
C22H25BR2N5	519.29	517.0478	50.89	4.85	30.77	13.49
C22H26BR1N1	384.36	383.1249	68.75	6.82	20.79	3.64
C22H26BR1N3	412.38	411.1311	64.08	6.36	19.38	10.19
C22H26BR1N5	440.39	439.1372	60.00	5.95	18.14	15.90
C22H26BR2N2	478.27	476.0464	55.25	5.48	33.41	5.86
C22H26BR2N4	506.29	504.0525	52.19	5.18	31.56	11.07
C22H26BR2N6	534.30	532.0587	49.46	4.90	29.91	15.73
C22H27BR1N2	399.38	398.1358	66.16	6.81	20.01	7.01
C22H27BR1N4	427.39	426.1419	61.83	6.37	18.70	13.11
C22H27BR1N6	455.40	454.1481	58.02	5.98	17.55	18.45
C22H27BR2N1	465.28	463.0511	56.79	5.85	34.35	3.01
C22H27BR2N3	493.29	491.0573	53.57	5.52	32.40	8.52
C22H27BR2N5	521.30	519.0634	50.69	5.22	30.66	13.43
C22H28BR1N1	386.38	385.1406	68.39	7.30	20.68	3.63
C22H28BR1N3	414.39	413.1467	63.77	6.81	19.28	10.14
C22H28BR1N5	442.41	441.1528	59.73	6.38	18.06	15.83
C22H28BR2N2	480.29	478.0620	55.02	5.88	33.27	5.83
C22H28BR2N4	508.30	506.0682	51.99	5.55	31.44	11.02
C22H28BR2N6	536.32	534.0743	49.27	5.26	29.80	15.67
C22H29BR1N2	401.39	400.1515	65.83	7.28	19.91	6.98
C22H29BR1N4	429.41	428.1576	61.54	6.81	18.61	13.05
C22H29BR1N6	457.42	456.1637	57.77	6.39	17.47	18.37
C22H29BR2N1	467.29	465.0668	56.55	6.26	34.20	3.00
C22H29BR2N3	495.30	493.0729	53.35	5.90	32.26	8.48
C22H29BR2N5	523.32	521.0791	50.49	5.59	30.54	13.38
C22H30BR1N1	388.40	387.1562	68.04	7.79	20.57	3.61

C22H30BR1N3	416.41	415.1623	63.46	7.26	19.19	10.09
C22H30BR1N5	444.42	443.1685	59.46	6.80	17.98	15.76
C22H30BR2N2	482.31	480.0777	54.79	6.27	33.13	5.81
C22H30BR2N4	510.32	508.0838	51.78	5.93	31.32	10.98
C22H30BR2N6	538.33	536.0900	49.09	5.62	29.69	15.61
C22H31BR1N2	403.41	402.1671	65.50	7.75	19.81	6.94
C22H31BR1N4	431.42	430.1732	61.25	7.24	18.52	12.99
C22H31BR1N6	459.44	458.1794	57.52	6.80	17.39	18.29
C22H31BR2N1	469.31	467.0824	56.31	6.66	34.05	2.98
C22H31BR2N3	497.32	495.0886	53.13	6.28	32.13	8.45
C22H31BR2N5	525.33	523.0947	50.30	5.95	30.42	13.33
C22H32BR1N1	390.41	389.1719	67.68	8.26	20.47	3.59
C22H32BR1N3	418.42	417.1780	63.15	7.71	19.10	10.04
C22H32BR1N5	446.44	445.1841	59.19	7.22	17.90	15.69
C22H32BR2N2	484.32	482.0933	54.56	6.66	33.00	5.78
C22H32BR2N4	512.34	510.0995	51.58	6.30	31.19	10.94
C22H32BR2N6	540.35	538.1056	48.90	5.97	29.57	15.55
C22H33BR1N2	405.43	404.1827	65.18	8.20	19.71	6.91
C22H33BR1N4	433.44	432.1889	60.96	7.67	18.43	12.93
C22H33BR1N6	461.45	460.1950	57.26	7.21	17.32	18.21
C22H33BR2N1	471.32	469.0981	56.06	7.06	33.91	2.97
C22H33BR2N3	499.34	497.1042	52.92	6.66	32.00	8.42
C22H33BR2N5	527.35	525.1104	50.11	6.31	30.30	13.28
C22H34BR1N1	392.43	391.1875	67.34	8.73	20.36	3.57
C22H34BR1N3	420.44	419.1936	62.85	8.15	19.00	9.99
C22H34BR1N5	448.45	447.1998	58.92	7.64	17.82	15.62
C22H34BR2N2	486.34	484.1090	54.33	7.05	32.86	5.76
C22H34BR2N4	514.35	512.1151	51.37	6.66	31.07	10.89
C22H34BR2N6	542.36	540.1213	48.72	6.32	29.47	15.50
C22H35BR1N2	407.44	406.1984	64.85	8.66	19.61	6.88
C22H35BR1N4	435.46	434.2045	60.68	8.10	18.35	12.87
C22H35BR1N6	463.47	462.2107	57.01	7.61	17.24	18.13
C22H35BR2N1	473.34	471.1137	55.83	7.45	33.76	2.96
C22H35BR2N3	501.35	499.1199	52.71	7.04	31.88	8.38
C22H35BR2N5	529.37	527.1260	49.92	6.66	30.19	13.23
C22H36BR1N1	394.44	393.2031	66.99	9.20	20.26	3.55
C22H36BR1N3	422.46	421.2093	62.55	8.59	18.91	9.95
C22H36BR1N5	450.47	449.2154	58.66	8.06	17.74	15.55
C22H36BR2N2	488.35	486.1246	54.11	7.43	32.72	5.74
C22H36BR2N4	516.37	514.1308	51.17	7.03	30.95	10.85
C22H36BR2N6	544.38	542.1369	48.54	6.67	29.36	15.44
C22H37BR1N2	409.46	408.2140	64.54	9.11	19.51	6.84
C22H37BR1N4	437.47	436.2202	60.40	8.53	18.26	12.81
C22H37BR1N6	465.48	464.2263	56.77	8.01	17.17	18.05
C22H37BR2N1	475.35	473.1294	55.59	7.85	33.62	2.95
C22H37BR2N3	503.37	501.1355	52.50	7.41	31.75	8.35
C22H37BR2N5	531.38	529.1417	49.73	7.02	30.07	13.18
C22H38BR1N1	396.46	395.2188	66.65	9.66	20.15	3.53
C22H38BR1N3	424.47	423.2249	62.25	9.02	18.82	9.90
C22H38BR1N5	452.49	451.2311	58.40	8.46	17.66	15.48
C22H38BR2N2	490.37	488.1403	53.89	7.81	32.59	5.71
C22H38BR2N4	518.38	516.1464	50.97	7.39	30.83	10.81
C22H38BR2N6	546.40	544.1526	48.36	7.01	29.25	15.38
C22H39BR1N2	411.47	410.2297	64.22	9.55	19.42	6.81
C22H39BR1N4	439.49	438.2358	60.13	8.94	18.18	12.75
C22H39BR1N6	467.50	466.2420	56.52	8.41	17.09	17.98
C22H39BR2N1	477.37	475.1450	55.35	8.23	33.48	2.93
C22H39BR2N3	505.38	503.1512	52.29	7.78	31.62	8.31
C22H39BR2N5	533.40	531.1573	49.54	7.37	29.96	13.13
C22H40BR1N1	398.47	397.2344	66.31	10.12	20.05	3.52

C22H40BR1N3	426.49	425.2406	61.96	9.45	18.74	9.85
C22H40BR1N5	454.50	453.2467	58.14	8.87	17.58	15.41
C22H40BR2N2	492.39	490.1559	53.67	8.19	32.46	5.69
C22H40BR2N4	520.40	518.1621	50.78	7.75	30.71	10.77
C22H40BR2N6	548.41	546.1682	48.18	7.35	29.14	15.32
C22H41BR1N2	413.49	412.2453	63.91	9.99	19.32	6.77
C22H41BR1N4	441.50	440.2515	59.85	9.36	18.10	12.69
C22H41BR1N6	469.52	468.2576	56.28	8.80	17.02	17.90
C22H41BR2N1	479.39	477.1607	55.12	8.62	33.34	2.92
C22H41BR2N3	507.40	505.1668	52.08	8.14	31.50	8.28
C22H41BR2N5	535.41	533.1730	49.35	7.72	29.85	13.08
C22H42BR1N1	400.49	399.2501	65.98	10.57	19.95	3.50
C22H42BR1N3	428.50	427.2562	61.67	9.88	18.65	9.81
C22H42BR1N5	456.52	455.2624	57.88	9.27	17.50	15.34
C22H42BR2N2	494.40	492.1716	53.45	8.56	32.32	5.67
C22H42BR2N4	522.41	520.1777	50.58	8.10	30.59	10.72
C22H42BR2N6	550.43	548.1839	48.01	7.69	29.03	15.27
C22H43BR1N2	415.51	414.2610	63.60	10.43	19.23	6.74
C22H43BR1N4	443.52	442.2671	59.58	9.77	18.02	12.63
C22H43BR1N6	471.53	470.2733	56.04	9.19	16.95	17.82
C22H43BR2N1	481.40	479.1763	54.89	9.00	33.20	2.91
C22H43BR2N3	509.42	507.1825	51.87	8.51	31.37	8.25
C22H43BR2N5	537.43	535.1886	49.17	8.06	29.74	13.03
C22H44BR1N1	402.51	401.2657	65.65	11.02	19.85	3.48
C22H44BR1N3	430.52	429.2719	61.38	10.30	18.56	9.76
C22H44BR1N5	458.53	457.2780	57.63	9.67	17.43	15.27
C22H44BR2N2	496.42	494.1872	53.23	8.93	32.19	5.64
C22H44BR2N4	524.43	522.1934	50.39	8.46	30.47	10.68
C22H44BR2N6	552.44	550.1995	47.83	8.03	28.93	15.21
C22H45BR1N2	417.52	416.2766	63.29	10.86	19.14	6.71
C22H45BR1N4	445.53	444.2828	59.31	10.18	17.93	12.58
C22H45BR1N6	473.55	472.2889	55.80	9.58	16.87	17.75
C22H45BR2N1	483.42	481.1920	54.66	9.38	33.06	2.90
C22H45BR2N3	511.43	509.1981	51.67	8.87	31.25	8.22
C22H45BR2N5	539.45	537.2043	48.98	8.41	29.62	12.98
C22H46BR1N1	404.52	403.2814	65.32	11.46	19.75	3.46
C22H46BR1N3	432.54	431.2875	61.09	10.72	18.47	9.71
C22H46BR1N5	460.55	459.2937	57.38	10.07	17.35	15.21
C22H46BR2N2	498.43	496.2029	53.02	9.30	32.06	5.62
C22H46BR2N4	526.45	524.2090	50.19	8.81	30.36	10.64
C22H46BR2N6	554.46	552.2152	47.66	8.36	28.82	15.16
C22H47BR1N2	419.54	418.2923	62.98	11.29	19.05	6.68
C22H47BR1N4	447.55	446.2984	59.04	10.59	17.85	12.52
C22H47BR1N6	475.56	474.3046	55.56	9.96	16.80	17.67
C22H47BR2N3	513.45	511.2138	51.46	9.23	31.12	8.18
C22H47BR2N5	541.46	539.2199	48.80	8.75	29.51	12.93
C22H48BR1N3	434.55	433.3032	60.81	11.13	18.39	9.67
C22H48BR1N5	462.57	461.3093	57.13	10.46	17.27	15.14
C22H48BR2N4	528.46	526.2247	50.00	9.16	30.24	10.60
C22H48BR2N6	556.48	554.2308	47.49	8.69	28.72	15.10
C22H49BR1N4	449.57	448.3141	58.78	10.99	17.77	12.46
C22H49BR1N6	477.58	476.3202	55.33	10.34	16.73	17.60
C22H49BR2N5	543.48	541.2356	48.62	9.09	29.40	12.89
C22H50BR1N5	464.58	463.3250	56.88	10.85	17.20	15.07
C22H50BR2N6	558.49	556.2465	47.31	9.02	28.61	15.05
C22H51BR1N6	479.60	478.3359	55.10	10.72	16.66	17.52
C23H11BR1N2	395.26	394.0106	69.89	2.81	20.22	7.09
C23H11BR1N4	423.27	422.0168	65.27	2.62	18.88	13.24
C23H11BR1N6	451.29	450.0229	61.22	2.46	17.71	18.62
C23H11BR2N1	461.16	458.9259	59.90	2.40	34.65	3.04

C23H11BR2N3	489.17	486.9321	56.47	2.27	32.67	8.59
C23H11BR2N5	517.19	514.9382	53.42	2.14	30.90	13.54
C23H12BR1N1	382.26	381.0154	72.27	3.16	20.90	3.66
C23H12BR1N3	410.28	409.0215	67.33	2.95	19.48	10.24
C23H12BR1N5	438.29	437.0277	63.03	2.76	18.23	15.98
C23H12BR2N2	476.17	473.9368	58.02	2.54	33.56	5.88
C23H12BR2N4	504.19	501.9430	54.79	2.40	31.70	11.11
C23H12BR2N6	532.20	529.9491	51.91	2.27	30.03	15.79
C23H13BR1N2	397.28	396.0263	69.54	3.30	20.11	7.05
C23H13BR1N4	425.29	424.0324	64.96	3.08	18.79	13.17
C23H13BR1N6	453.30	452.0386	60.94	2.89	17.63	18.54
C23H13BR2N1	463.17	460.9416	59.64	2.83	34.50	3.02
C23H13BR2N3	491.19	488.9477	56.24	2.67	32.53	8.55
C23H13BR2N5	519.20	516.9539	53.21	2.52	30.78	13.49
C23H14BR1N1	384.28	383.0310	71.89	3.67	20.79	3.64
C23H14BR1N3	412.29	411.0372	67.01	3.42	19.38	10.19
C23H14BR1N5	440.31	439.0433	62.74	3.20	18.15	15.91
C23H14BR2N2	478.19	475.9525	57.77	2.95	33.42	5.86
C23H14BR2N4	506.20	503.9586	54.57	2.79	31.57	11.07
C23H14BR2N6	534.22	531.9648	51.71	2.64	29.91	15.73
C23H15BR1N2	399.29	398.0419	69.19	3.79	20.01	7.02
C23H15BR1N4	427.31	426.0481	64.65	3.54	18.70	13.11
C23H15BR1N6	455.32	454.0542	60.67	3.32	17.55	18.46
C23H15BR2N1	465.19	462.9572	59.39	3.25	34.35	3.01
C23H15BR2N3	493.20	490.9634	56.01	3.07	32.40	8.52
C23H15BR2N5	521.22	518.9695	53.00	2.90	30.66	13.44
C23H16BR1N1	386.29	385.0467	71.51	4.17	20.68	3.63
C23H16BR1N3	414.31	413.0528	66.68	3.89	19.29	10.14
C23H16BR1N5	442.32	441.0590	62.46	3.65	18.06	15.83
C23H16BR2N2	480.21	477.9681	57.53	3.36	33.28	5.83
C23H16BR2N4	508.22	505.9743	54.36	3.17	31.44	11.02
C23H16BR2N6	536.23	533.9804	51.52	3.01	29.80	15.67
C23H17BR1N2	401.31	400.0576	68.84	4.27	19.91	6.98
C23H17BR1N4	429.32	428.0637	64.35	3.99	18.61	13.05
C23H17BR1N6	457.34	456.0699	60.41	3.75	17.47	18.38
C23H17BR2N1	467.21	464.9729	59.13	3.67	34.20	3.00
C23H17BR2N3	495.22	492.9790	55.78	3.46	32.27	8.49
C23H17BR2N5	523.23	520.9852	52.80	3.27	30.54	13.38
C23H18BR1N1	388.31	387.0623	71.14	4.67	20.58	3.61
C23H18BR1N3	416.32	415.0685	66.36	4.36	19.19	10.09
C23H18BR1N5	444.34	443.0746	62.17	4.08	17.98	15.76
C23H18BR2N2	482.22	479.9838	57.29	3.76	33.14	5.81
C23H18BR2N4	510.23	507.9899	54.14	3.56	31.32	10.98
C23H18BR2N6	538.25	535.9961	51.33	3.37	29.69	15.61
C23H19BR1N2	403.33	402.0732	68.49	4.75	19.81	6.95
C23H19BR1N4	431.34	430.0794	64.05	4.44	18.52	12.99
C23H19BR1N6	459.35	458.0855	60.14	4.17	17.39	18.30
C23H19BR2N1	469.22	466.9885	58.88	4.08	34.06	2.99
C23H19BR2N3	497.24	494.9947	55.56	3.85	32.14	8.45
C23H19BR2N5	525.25	523.0008	52.60	3.65	30.43	13.33
C23H20BR1N1	390.33	389.0780	70.78	5.16	20.47	3.59
C23H20BR1N3	418.34	417.0841	66.04	4.82	19.10	10.04
C23H20BR1N5	446.35	445.0903	61.89	4.52	17.90	15.69
C23H20BR2N2	484.24	481.9994	57.05	4.16	33.00	5.79
C23H20BR2N4	512.25	510.0056	53.93	3.94	31.20	10.94
C23H20BR2N6	540.26	538.0117	51.13	3.73	29.58	15.56
C23H21BR1N2	405.34	404.0889	68.15	5.22	19.71	6.91
C23H21BR1N4	433.35	432.0950	63.75	4.88	18.44	12.93
C23H21BR1N6	461.37	460.1011	59.88	4.59	17.32	18.22
C23H21BR2N1	471.24	469.0042	58.62	4.49	33.91	2.97

C23H21BR2N3	499.25	497.0103	55.33	4.24	32.01	8.42
C23H21BR2N5	527.27	525.0165	52.39	4.01	30.31	13.28
C23H22BR1N1	392.34	391.0936	70.41	5.65	20.37	3.57
C23H22BR1N3	420.36	419.0998	65.72	5.28	19.01	10.00
C23H22BR1N5	448.37	447.1059	61.61	4.95	17.82	15.62
C23H22BR2N2	486.25	484.0151	56.81	4.56	32.87	5.76
C23H22BR2N4	514.27	512.0212	53.72	4.31	31.07	10.89
C23H22BR2N6	542.28	540.0274	50.94	4.09	29.47	15.50
C23H23BR1N2	407.36	406.1045	67.82	5.69	19.62	6.88
C23H23BR1N4	435.37	434.1107	63.45	5.32	18.35	12.87
C23H23BR1N6	463.38	462.1168	59.62	5.00	17.24	18.14
C23H23BR2N1	473.25	471.0198	58.37	4.90	33.77	2.96
C23H23BR2N3	501.27	499.0260	55.11	4.62	31.88	8.38
C23H23BR2N5	529.28	527.0321	52.19	4.38	30.19	13.23
C23H24BR1N1	394.36	393.1093	70.05	6.13	20.26	3.55
C23H24BR1N3	422.37	421.1154	65.41	5.73	18.92	9.95
C23H24BR1N5	450.39	449.1215	61.34	5.37	17.74	15.55
C23H24BR2N2	488.27	486.0307	56.58	4.95	32.73	5.74
C23H24BR2N4	516.28	514.0369	53.51	4.69	30.95	10.85
C23H24BR2N6	544.30	542.0430	50.75	4.44	29.36	15.44
C23H25BR1N2	409.37	408.1202	67.48	6.16	19.52	6.84
C23H25BR1N4	437.39	436.1263	63.16	5.76	18.27	12.81
C23H25BR1N6	465.40	464.1324	59.36	5.41	17.17	18.06
C23H25BR2N1	475.27	473.0355	58.13	5.30	33.62	2.95
C23H25BR2N3	503.28	501.0416	54.89	5.01	31.75	8.35
C23H25BR2N5	531.30	529.0478	52.00	4.74	30.08	13.18
C23H26BR1N1	396.37	395.1249	69.70	6.61	20.16	3.53
C23H26BR1N3	424.39	423.1311	65.10	6.18	18.83	9.90
C23H26BR1N5	452.40	451.1372	61.06	5.79	17.66	15.48
C23H26BR2N2	490.29	488.0464	56.35	5.35	32.59	5.71
C23H26BR2N4	518.30	516.0525	53.30	5.06	30.83	10.81
C23H26BR2N6	546.31	544.0587	50.57	4.80	29.25	15.38
C23H27BR1N2	411.39	410.1358	67.15	6.62	19.42	6.81
C23H27BR1N4	439.40	438.1419	62.87	6.19	18.18	12.75
C23H27BR1N6	467.42	466.1481	59.10	5.82	17.09	17.98
C23H27BR2N1	477.29	475.0511	57.88	5.70	33.48	2.93
C23H27BR2N3	505.30	503.0573	54.67	5.39	31.63	8.32
C23H27BR2N5	533.31	531.0634	51.80	5.10	29.97	13.13
C23H28BR1N1	398.39	397.1406	69.34	7.08	20.06	3.52
C23H28BR1N3	426.40	425.1467	64.79	6.62	18.74	9.85
C23H28BR1N5	454.42	453.1528	60.79	6.21	17.58	15.41
C23H28BR2N2	492.30	490.0620	56.12	5.73	32.46	5.69
C23H28BR2N4	520.31	518.0682	53.09	5.42	30.71	10.77
C23H28BR2N6	548.33	546.0743	50.38	5.15	29.14	15.33
C23H29BR1N2	413.40	412.1515	66.82	7.07	19.33	6.78
C23H29BR1N4	441.42	440.1576	62.58	6.62	18.10	12.69
C23H29BR1N6	469.43	468.1637	58.85	6.23	17.02	17.90
C23H29BR2N1	479.30	477.0668	57.64	6.10	33.34	2.92
C23H29BR2N3	507.32	505.0729	54.45	5.76	31.50	8.28
C23H29BR2N5	535.33	533.0791	51.60	5.46	29.85	13.08
C23H30BR1N1	400.41	399.1562	68.99	7.55	19.96	3.50
C23H30BR1N3	428.42	427.1623	64.48	7.06	18.65	9.81
C23H30BR1N5	456.43	455.1685	60.53	6.63	17.51	15.34
C23H30BR2N2	494.32	492.0777	55.89	6.12	32.33	5.67
C23H30BR2N4	522.33	520.0838	52.89	5.79	30.60	10.73
C23H30BR2N6	550.34	548.0900	50.20	5.49	29.04	15.27
C23H31BR1N2	415.42	414.1671	66.50	7.52	19.23	6.74
C23H31BR1N4	443.43	442.1732	62.30	7.05	18.02	12.63
C23H31BR1N6	471.45	470.1794	58.60	6.63	16.95	17.83
C23H31BR2N1	481.32	479.0824	57.40	6.49	33.20	2.91

C23H31BR2N3	509.33	507.0886	54.24	6.13	31.38	8.25
C23H31BR2N5	537.35	535.0947	51.41	5.82	29.74	13.03
C23H32BR1N1	402.42	401.1719	68.65	8.02	19.86	3.48
C23H32BR1N3	430.44	429.1780	64.18	7.49	18.56	9.76
C23H32BR1N5	458.45	457.1841	60.26	7.04	17.43	15.28
C23H32BR2N2	496.33	494.0933	55.66	6.50	32.20	5.64
C23H32BR2N4	524.35	522.0995	52.69	6.15	30.48	10.69
C23H32BR2N6	552.36	550.1056	50.01	5.84	28.93	15.21
C23H33BR1N2	417.44	416.1827	66.18	7.97	19.14	6.71
C23H33BR1N4	445.45	444.1889	62.02	7.47	17.94	12.58
C23H33BR1N6	473.46	472.1950	58.35	7.03	16.88	17.75
C23H33BR2N1	483.33	481.0981	57.16	6.88	33.06	2.90
C23H33BR2N3	511.35	509.1042	54.03	6.50	31.25	8.22
C23H33BR2N5	539.36	537.1104	51.22	6.17	29.63	12.98
C23H34BR1N1	404.44	403.1875	68.31	8.47	19.76	3.46
C23H34BR1N3	432.45	431.1936	63.88	7.92	18.48	9.72
C23H34BR1N5	460.46	459.1998	60.00	7.44	17.35	15.21
C23H34BR2N2	498.35	496.1090	55.43	6.88	32.07	5.62
C23H34BR2N4	526.36	524.1151	52.48	6.51	30.36	10.64
C23H34BR2N6	554.38	552.1213	49.83	6.18	28.83	15.16
C23H35BR1N2	419.45	418.1984	65.86	8.41	19.05	6.68
C23H35BR1N4	447.47	446.2045	61.74	7.88	17.86	12.52
C23H35BR1N6	475.48	474.2107	58.10	7.42	16.80	17.67
C23H35BR2N1	485.35	483.1137	56.92	7.27	32.93	2.89
C23H35BR2N3	513.36	511.1199	53.81	6.87	31.13	8.19
C23H35BR2N5	541.38	539.1260	51.03	6.52	29.52	12.94
C23H36BR1N1	406.45	405.2031	67.97	8.93	19.66	3.45
C23H36BR1N3	434.47	433.2093	63.59	8.35	18.39	9.67
C23H36BR1N5	462.48	461.2154	59.73	7.85	17.28	15.14
C23H36BR2N2	500.36	498.1246	55.21	7.25	31.94	5.60
C23H36BR2N4	528.38	526.1308	52.28	6.87	30.25	10.60
C23H36BR2N6	556.39	554.1369	49.65	6.52	28.72	15.10
C23H37BR1N2	421.47	420.2140	65.55	8.85	18.96	6.65
C23H37BR1N4	449.48	448.2202	61.46	8.30	17.78	12.46
C23H37BR1N6	477.50	476.2263	57.86	7.81	16.73	17.60
C23H37BR2N1	487.37	485.1294	56.68	7.65	32.79	2.87
C23H37BR2N3	515.38	513.1355	53.60	7.24	31.01	8.15
C23H37BR2N5	543.39	541.1417	50.84	6.86	29.41	12.89
C23H38BR1N1	408.47	407.2188	67.63	9.38	19.56	3.43
C23H38BR1N3	436.48	435.2249	63.29	8.78	18.31	9.63
C23H38BR1N5	464.50	463.2311	59.47	8.25	17.20	15.08
C23H38BR2N2	502.38	500.1403	54.99	7.62	31.81	5.58
C23H38BR2N4	530.39	528.1464	52.09	7.22	30.13	10.56
C23H38BR2N6	558.41	556.1526	49.47	6.86	28.62	15.05
C23H39BR1N2	423.48	422.2297	65.23	9.28	18.87	6.61
C23H39BR1N4	451.50	450.2358	61.19	8.71	17.70	12.41
C23H39BR1N6	479.51	478.2420	57.61	8.20	16.66	17.53
C23H39BR2N1	489.38	487.1450	56.45	8.03	32.66	2.86
C23H39BR2N3	517.40	515.1512	53.39	7.60	30.89	8.12
C23H39BR2N5	545.41	543.1573	50.65	7.21	29.30	12.84
C23H40BR1N1	410.49	409.2344	67.30	9.82	19.47	3.41
C23H40BR1N3	438.50	437.2406	63.00	9.19	18.22	9.58
C23H40BR1N5	466.51	465.2467	59.22	8.64	17.13	15.01
C23H40BR2N2	504.40	502.1559	54.77	7.99	31.68	5.55
C23H40BR2N4	532.41	530.1621	51.89	7.57	30.02	10.52
C23H40BR2N6	560.42	558.1682	49.29	7.19	28.52	15.00
C23H41BR1N2	425.50	424.2453	64.93	9.71	18.78	6.58
C23H41BR1N4	453.51	452.2515	60.91	9.11	17.62	12.35
C23H41BR1N6	481.53	480.2576	57.37	8.58	16.59	17.45
C23H41BR2N1	491.40	489.1607	56.22	8.41	32.52	2.85

C23H41BR2N3	519.41	517.1668	53.19	7.96	30.77	8.09
C23H41BR2N5	547.42	545.1730	50.46	7.55	29.19	12.79
C23H42BR1N1	412.50	411.2501	66.97	10.26	19.37	3.40
C23H42BR1N3	440.52	439.2562	62.71	9.61	18.14	9.54
C23H42BR1N5	468.53	467.2624	58.96	9.04	17.05	14.95
C23H42BR2N2	506.41	504.1716	54.55	8.36	31.56	5.53
C23H42BR2N4	534.43	532.1777	51.69	7.92	29.90	10.48
C23H42BR2N6	562.44	560.1839	49.12	7.53	28.41	14.94
C23H43BR1N2	427.52	426.2610	64.62	10.14	18.69	6.55
C23H43BR1N4	455.53	454.2671	60.65	9.51	17.54	12.30
C23H43BR1N6	483.54	482.2733	57.13	8.96	16.52	17.38
C23H43BR2N1	493.41	491.1763	55.99	8.78	32.39	2.84
C23H43BR2N3	521.43	519.1825	52.98	8.31	30.65	8.06
C23H43BR2N5	549.44	547.1886	50.28	7.89	29.09	12.75
C23H44BR1N1	414.52	413.2657	66.65	10.70	19.28	3.38
C23H44BR1N3	442.53	441.2719	62.43	10.02	18.06	9.50
C23H44BR1N5	470.54	469.2780	58.71	9.43	16.98	14.88
C23H44BR2N2	508.43	506.1872	54.34	8.72	31.43	5.51
C23H44BR2N4	536.44	534.1934	51.50	8.27	29.79	10.44
C23H44BR2N6	564.46	562.1995	48.94	7.86	28.31	14.89
C23H45BR1N2	429.53	428.2766	64.32	10.56	18.60	6.52
C23H45BR1N4	457.55	456.2828	60.38	9.91	17.46	12.25
C23H45BR1N6	485.56	484.2889	56.89	9.34	16.46	17.31
C23H45BR2N1	495.43	493.1920	55.76	9.16	32.26	2.83
C23H45BR2N3	523.44	521.1981	52.78	8.67	30.53	8.03
C23H45BR2N5	551.46	549.2043	50.10	8.23	28.98	12.70
C23H46BR1N1	416.53	415.2814	66.32	11.13	19.18	3.36
C23H46BR1N3	444.55	443.2875	62.14	10.43	17.97	9.45
C23H46BR1N5	472.56	471.2937	58.46	9.81	16.91	14.82
C23H46BR2N2	510.44	508.2029	54.12	9.08	31.31	5.49
C23H46BR2N4	538.46	536.2090	51.31	8.61	29.68	10.41
C23H46BR2N6	566.47	564.2152	48.77	8.19	28.21	14.84
C23H47BR1N2	431.55	430.2923	64.02	10.98	18.52	6.49
C23H47BR1N4	459.56	458.2984	60.11	10.31	17.39	12.19
C23H47BR1N6	487.58	486.3046	56.66	9.72	16.39	17.24
C23H47BR2N1	497.45	495.2076	55.53	9.52	32.13	2.82
C23H47BR2N3	525.46	523.2138	52.57	9.02	30.41	8.00
C23H47BR2N5	553.47	551.2199	49.91	8.56	28.87	12.65
C23H48BR1N1	418.55	417.2970	66.00	11.56	19.09	3.35
C23H48BR1N3	446.56	445.3032	61.86	10.83	17.89	9.41
C23H48BR1N5	474.58	473.3093	58.21	10.19	16.84	14.76
C23H48BR2N2	512.46	510.2185	53.91	9.44	31.18	5.47
C23H48BR2N4	540.47	538.2247	51.11	8.95	29.57	10.37
C23H48BR2N6	568.49	566.2308	48.60	8.51	28.11	14.78
C23H49BR1N2	433.56	432.3079	63.72	11.39	18.43	6.46
C23H49BR1N4	461.58	460.3141	59.85	10.70	17.31	12.14
C23H49BR1N6	489.59	488.3202	56.43	10.09	16.32	17.17
C23H49BR2N3	527.48	525.2294	52.37	9.36	30.30	7.97
C23H49BR2N5	555.49	553.2356	49.73	8.89	28.77	12.61
C23H50BR1N3	448.58	447.3188	61.58	11.24	17.81	9.37
C23H50BR1N5	476.59	475.3250	57.96	10.57	16.77	14.69
C23H50BR2N4	542.49	540.2403	50.92	9.29	29.46	10.33
C23H50BR2N6	570.50	568.2465	48.42	8.83	28.01	14.73
C23H51BR1N4	463.59	462.3297	59.59	11.09	17.24	12.09
C23H51BR1N6	491.61	490.3359	56.19	10.46	16.25	17.09
C23H51BR2N5	557.50	555.2512	49.55	9.22	28.66	12.56
C23H52BR1N5	478.61	477.3406	57.72	10.95	16.70	14.63
C23H52BR2N6	572.52	570.2621	48.25	9.16	27.91	14.68
C23H53BR1N6	493.62	492.3515	55.97	10.82	16.19	17.03
C24H12BR1N1	394.27	393.0154	73.11	3.07	20.27	3.55

C24H12BR1N3	422.29	421.0215	68.26	2.86	18.92	9.95
C24H12BR1N5	450.30	449.0277	64.02	2.69	17.74	15.55
C24H12BR2N2	488.18	485.9368	59.05	2.48	32.74	5.74
C24H12BR2N4	516.20	513.9430	55.84	2.34	30.96	10.85
C24H12BR2N6	544.21	541.9491	52.97	2.22	29.37	15.44
C24H13BR1N2	409.29	408.0263	70.43	3.20	19.52	6.84
C24H13BR1N4	437.30	436.0324	65.92	3.00	18.27	12.81
C24H13BR1N6	465.32	464.0386	61.95	2.82	17.17	18.06
C24H13BR2N1	475.19	472.9416	60.66	2.76	33.63	2.95
C24H13BR2N3	503.20	500.9477	57.29	2.60	31.76	8.35
C24H13BR2N5	531.21	528.9539	54.27	2.47	30.08	13.18
C24H14BR1N1	396.29	395.0310	72.74	3.56	20.16	3.53
C24H14BR1N3	424.30	423.0372	67.94	3.33	18.83	9.90
C24H14BR1N5	452.32	451.0433	63.73	3.12	17.67	15.48
C24H14BR2N2	490.20	487.9525	58.81	2.88	32.60	5.71
C24H14BR2N4	518.21	515.9586	55.63	2.72	30.84	10.81
C24H14BR2N6	546.23	543.9648	52.77	2.58	29.26	15.39
C24H15BR1N2	411.30	410.0419	70.09	3.68	19.43	6.81
C24H15BR1N4	439.32	438.0481	65.62	3.44	18.19	12.75
C24H15BR1N6	467.33	466.0542	61.68	3.24	17.10	17.98
C24H15BR2N1	477.20	474.9572	60.41	3.17	33.49	2.94
C24H15BR2N3	505.22	502.9634	57.06	2.99	31.63	8.32
C24H15BR2N5	533.23	530.9695	54.06	2.84	29.97	13.13
C24H16BR1N1	398.31	397.0467	72.37	4.05	20.06	3.52
C24H16BR1N3	426.32	425.0528	67.62	3.78	18.74	9.86
C24H16BR1N5	454.33	453.0590	63.45	3.55	17.59	15.41
C24H16BR2N2	492.22	489.9681	58.57	3.28	32.47	5.69
C24H16BR2N4	520.23	517.9743	55.41	3.10	30.72	10.77
C24H16BR2N6	548.24	545.9804	52.58	2.94	29.15	15.33
C24H17BR1N2	413.32	412.0576	69.74	4.15	19.33	6.78
C24H17BR1N4	441.33	440.0637	65.32	3.88	18.11	12.69
C24H17BR1N6	469.35	468.0699	61.42	3.65	17.02	17.91
C24H17BR2N1	479.22	476.9729	60.15	3.58	33.35	2.92
C24H17BR2N3	507.23	504.9790	56.83	3.38	31.51	8.28
C24H17BR2N5	535.24	532.9852	53.86	3.20	29.86	13.08
C24H18BR1N1	400.32	399.0623	72.01	4.53	19.96	3.50
C24H18BR1N3	428.34	427.0685	67.30	4.24	18.65	9.81
C24H18BR1N5	456.35	455.0746	63.17	3.98	17.51	15.35
C24H18BR2N2	494.23	491.9838	58.33	3.67	32.33	5.67
C24H18BR2N4	522.25	519.9899	55.20	3.47	30.60	10.73
C24H18BR2N6	550.26	547.9961	52.39	3.30	29.04	15.27
C24H19BR1N2	415.34	414.0732	69.41	4.61	19.24	6.74
C24H19BR1N4	443.35	442.0794	65.02	4.32	18.02	12.64
C24H19BR1N6	471.36	470.0855	61.16	4.06	16.95	17.83
C24H19BR2N1	481.23	478.9885	59.90	3.98	33.21	2.91
C24H19BR2N3	509.25	506.9947	56.61	3.76	31.38	8.25
C24H19BR2N5	537.26	535.0008	53.66	3.56	29.74	13.04
C24H20BR1N1	402.34	401.0780	71.65	5.01	19.86	3.48
C24H20BR1N3	430.35	429.0841	66.98	4.68	18.57	9.76
C24H20BR1N5	458.36	457.0903	62.89	4.40	17.43	15.28
C24H20BR2N2	496.25	493.9994	58.09	4.06	32.20	5.65
C24H20BR2N4	524.26	522.0056	54.99	3.85	30.48	10.69
C24H20BR2N6	552.28	550.0117	52.20	3.65	28.94	15.22
C24H21BR1N2	417.35	416.0889	69.07	5.07	19.15	6.71
C24H21BR1N4	445.37	444.0950	64.73	4.75	17.94	12.58
C24H21BR1N6	473.38	472.1011	60.90	4.47	16.88	17.75
C24H21BR2N1	483.25	481.0042	59.65	4.38	33.07	2.90
C24H21BR2N3	511.26	509.0103	56.38	4.14	31.26	8.22
C24H21BR2N5	539.28	537.0165	53.45	3.93	29.63	12.99
C24H22BR1N1	404.35	403.0936	71.29	5.48	19.76	3.46

C24H22BR1N3	432.37	431.0998	66.67	5.13	18.48	9.72
C24H22BR1N5	460.38	459.1059	62.62	4.82	17.36	15.21
C24H22BR2N2	498.26	496.0151	57.85	4.45	32.07	5.62
C24H22BR2N4	526.28	524.0212	54.77	4.21	30.37	10.65
C24H22BR2N6	554.29	552.0274	52.01	4.00	28.83	15.16
C24H23BR1N2	419.37	418.1045	68.74	5.53	19.05	6.68
C24H23BR1N4	447.38	446.1107	64.43	5.18	17.86	12.52
C24H23BR1N6	475.40	474.1168	60.64	4.88	16.81	17.68
C24H23BR2N1	485.27	483.0198	59.40	4.78	32.93	2.89
C24H23BR2N3	513.28	511.0260	56.16	4.52	31.13	8.19
C24H23BR2N5	541.29	539.0321	53.26	4.28	29.52	12.94
C24H24BR1N1	406.37	405.1093	70.94	5.95	19.66	3.45
C24H24BR1N3	434.38	433.1154	66.36	5.57	18.39	9.67
C24H24BR1N5	462.40	461.1215	62.34	5.23	17.28	15.15
C24H24BR2N2	500.28	498.0307	57.62	4.84	31.94	5.60
C24H24BR2N4	528.29	526.0369	54.57	4.58	30.25	10.61
C24H24BR2N6	556.31	554.0430	51.82	4.35	28.73	15.11
C24H25BR1N2	421.38	420.1202	68.41	5.98	18.96	6.65
C24H25BR1N4	449.40	448.1263	64.15	5.61	17.78	12.47
C24H25BR1N6	477.41	476.1324	60.38	5.28	16.74	17.60
C24H25BR2N1	487.28	485.0355	59.16	5.17	32.80	2.87
C24H25BR2N3	515.29	513.0416	55.94	4.89	31.01	8.15
C24H25BR2N5	543.31	541.0478	53.06	4.64	29.41	12.89
C24H26BR1N1	408.39	407.1249	70.59	6.42	19.57	3.43
C24H26BR1N3	436.40	435.1311	66.06	6.01	18.31	9.63
C24H26BR1N5	464.41	463.1372	62.07	5.64	17.21	15.08
C24H26BR2N2	502.30	500.0464	57.39	5.22	31.82	5.58
C24H26BR2N4	530.31	528.0525	54.36	4.94	30.13	10.56
C24H26BR2N6	558.32	556.0587	51.63	4.69	28.62	15.05
C24H27BR1N2	423.40	422.1358	68.08	6.43	18.87	6.62
C24H27BR1N4	451.41	450.1419	63.86	6.03	17.70	12.41
C24H27BR1N6	479.43	478.1481	60.13	5.68	16.67	17.53
C24H27BR2N1	489.30	487.0511	58.91	5.56	32.66	2.86
C24H27BR2N3	517.31	515.0573	55.72	5.26	30.89	8.12
C24H27BR2N5	545.32	543.0634	52.86	4.99	29.31	12.84
C24H28BR1N1	410.40	409.1406	70.24	6.88	19.47	3.41
C24H28BR1N3	438.41	437.1467	65.75	6.44	18.23	9.58
C24H28BR1N5	466.43	465.1528	61.80	6.05	17.13	15.01
C24H28BR2N2	504.31	502.0620	57.16	5.60	31.69	5.55
C24H28BR2N4	532.33	530.0682	54.15	5.30	30.02	10.52
C24H28BR2N6	560.34	558.0743	51.45	5.04	28.52	15.00
C24H29BR1N2	425.42	424.1515	67.76	6.87	18.78	6.58
C24H29BR1N4	453.43	452.1576	63.57	6.45	17.62	12.36
C24H29BR1N6	481.44	480.1637	59.88	6.07	16.60	17.46
C24H29BR2N1	491.31	489.0668	58.67	5.95	32.53	2.85
C24H29BR2N3	519.33	517.0729	55.51	5.63	30.77	8.09
C24H29BR2N5	547.34	545.0791	52.67	5.34	29.20	12.80
C24H30BR1N1	412.42	411.1562	69.90	7.33	19.37	3.40
C24H30BR1N3	440.43	439.1623	65.45	6.87	18.14	9.54
C24H30BR1N5	468.44	467.1685	61.54	6.46	17.06	14.95
C24H30BR2N2	506.33	504.0777	56.93	5.97	31.56	5.53
C24H30BR2N4	534.34	532.0838	53.95	5.66	29.91	10.49
C24H30BR2N6	562.35	560.0900	51.26	5.38	28.42	14.94
C24H31BR1N2	427.43	426.1671	67.44	7.31	18.69	6.55
C24H31BR1N4	455.45	454.1732	63.29	6.86	17.54	12.30
C24H31BR1N6	483.46	482.1794	59.63	6.46	16.53	17.38
C24H31BR2N1	493.33	491.0824	58.43	6.33	32.39	2.84
C24H31BR2N3	521.34	519.0886	55.29	5.99	30.65	8.06
C24H31BR2N5	549.36	547.0947	52.47	5.69	29.09	12.75
C24H32BR1N1	414.43	413.1719	69.56	7.78	19.28	3.38
C24H32BR1N3	442.45	441.1780	65.15	7.29	18.06	9.50
C24H32BR1N5	470.46	469.1841	61.27	6.86	16.98	14.89
C24H32BR2N2	508.34	506.0933	56.71	6.35	31.44	5.51
C24H32BR2N4	536.36	534.0995	53.75	6.01	29.80	10.45
C24H32BR2N6	564.37	562.1056	51.08	5.72	28.32	14.89
C24H33BR1N2	429.45	428.1827	67.13	7.75	18.61	6.52
C24H33BR1N4	457.46	456.1889	63.01	7.27	17.47	12.25
C24H33BR1N6	485.47	484.1950	59.38	6.85	16.46	17.31
C24H33BR2N1	495.35	493.0981	58.20	6.72	32.26	2.83
C24H33BR2N3	523.36	521.1042	55.08	6.36	30.54	8.03
C24H33BR2N5	551.37	549.1104	52.28	6.03	28.98	12.70
C24H34BR1N1	416.45	415.1875	69.22	8.23	19.19	3.36
C24H34BR1N3	444.46	443.1936	64.86	7.71	17.98	9.45
C24H34BR1N5	472.48	471.1998	61.01	7.25	16.91	14.82
C24H34BR2N2	510.36	508.1090	56.48	6.72	31.31	5.49
C24H34BR2N4	538.37	536.1151	53.54	6.37	29.68	10.41
C24H34BR2N6	566.39	564.1213	50.90	6.05	28.22	14.84
C24H35BR1N2	431.46	430.1984	66.81	8.18	18.52	6.49
C24H35BR1N4	459.48	458.2045	62.74	7.68	17.39	12.19
C24H35BR1N6	487.49	486.2107	59.13	7.24	16.39	17.24
C24H35BR2N1	497.36	495.1137	57.96	7.09	32.13	2.82
C24H35BR2N3	525.37	523.1199	54.87	6.72	30.42	8.00
C24H35BR2N5	553.39	551.1260	52.09	6.38	28.88	12.66
C24H36BR1N1	418.47	417.2031	68.89	8.67	19.09	3.35
C24H36BR1N3	446.48	445.2093	64.56	8.13	17.90	9.41
C24H36BR1N5	474.49	473.2154	60.75	7.65	16.84	14.76
C24H36BR2N2	512.38	510.1246	56.26	7.08	31.19	5.47
C24H36BR2N4	540.39	538.1308	53.34	6.71	29.57	10.37
C24H36BR2N6	568.40	566.1369	50.72	6.38	28.12	14.79
C24H37BR1N2	433.48	432.2140	66.50	8.60	18.43	6.46
C24H37BR1N4	461.49	460.2202	62.46	8.08	17.31	12.14
C24H37BR1N6	489.51	488.2263	58.89	7.62	16.32	17.17
C24H37BR2N1	499.38	497.1294	57.73	7.47	32.00	2.80
C24H37BR2N3	527.39	525.1355	54.66	7.07	30.30	7.97
C24H37BR2N5	555.40	553.1417	51.90	6.71	28.77	12.61
C24H38BR1N1	420.48	419.2188	68.56	9.11	19.00	3.33
C24H38BR1N3	448.49	447.2249	64.27	8.54	17.82	9.37
C24H38BR1N5	476.51	475.2311	60.50	8.04	16.77	14.70
C24H38BR2N2	514.39	512.1403	56.04	7.45	31.07	5.45
C24H38BR2N4	542.41	540.1464	53.15	7.06	29.46	10.33
C24H38BR2N6	570.42	568.1526	50.54	6.71	28.02	14.73
C24H39BR1N2	435.50	434.2297	66.19	9.03	18.35	6.43
C24H39BR1N4	463.51	462.2358	62.19	8.48	17.24	12.09
C24H39BR1N6	491.52	490.2420	58.65	8.00	16.26	17.10
C24H39BR2N1	501.39	499.1450	57.49	7.84	31.87	2.79
C24H39BR2N3	529.41	527.1512	54.45	7.43	30.19	7.94
C24H39BR2N5	557.42	555.1573	51.71	7.05	28.67	12.56
C24H40BR1N1	422.50	421.2344	68.23	9.54	18.91	3.32
C24H40BR1N3	450.51	449.2406	63.99	8.95	17.74	9.33
C24H40BR1N5	478.52	477.2467	60.24	8.43	16.70	14.64
C24H40BR2N2	516.41	514.1559	55.82	7.81	30.95	5.42
C24H40BR2N4	544.42	542.1621	52.95	7.41	29.35	10.29
C24H40BR2N6	572.43	570.1682	50.36	7.04	27.92	14.68
C24H41BR1N2	437.51	436.2453	65.89	9.45	18.26	6.40
C24H41BR1N4	465.53	464.2515	61.92	8.88	17.16	12.04
C24H41BR1N6	493.54	492.2576	58.41	8.37	16.19	17.03
C24H41BR2N1	503.41	501.1607	57.26	8.21	31.75	2.78
C24H41BR2N3	531.42	529.1668	54.24	7.78	30.07	7.91
C24H41BR2N5	559.44	557.1730	51.53	7.39	28.57	12.52
C24H42BR1N1	424.51	423.2501	67.91	9.97	18.82	3.30

C24H42BR1N3	452.53	451.2562	63.70	9.36	17.66	9.29
C24H42BR1N5	480.54	479.2624	59.99	8.81	16.63	14.57
C24H42BR2N2	518.42	516.1716	55.60	8.17	30.83	5.40
C24H42BR2N4	546.44	544.1777	52.75	7.75	29.25	10.25
C24H42BR2N6	574.45	572.1839	50.18	7.37	27.82	14.63
C24H43BR1N2	439.53	438.2610	65.59	9.86	18.18	6.37
C24H43BR1N4	467.54	466.2671	61.66	9.27	17.09	11.98
C24H43BR1N6	495.55	494.2733	58.17	8.75	16.12	16.96
C24H43BR2N1	505.43	503.1763	57.03	8.58	31.62	2.77
C24H43BR2N3	533.44	531.1825	54.04	8.13	29.96	7.88
C24H43BR2N5	561.45	559.1886	51.34	7.72	28.46	12.47
C24H44BR1N1	426.53	425.2657	67.58	10.40	18.73	3.28
C24H44BR1N3	454.54	453.2719	63.42	9.76	17.58	9.24
C24H44BR1N5	482.56	481.2780	59.74	9.19	16.56	14.51
C24H44BR2N2	520.44	518.1872	55.39	8.52	30.71	5.38
C24H44BR2N4	548.45	546.1934	52.56	8.09	29.14	10.22
C24H44BR2N6	576.47	574.1995	50.01	7.69	27.72	14.58
C24H45BR1N2	441.54	440.2766	65.29	10.27	18.10	6.34
C24H45BR1N4	469.56	468.2828	61.39	9.66	17.02	11.93
C24H45BR1N6	497.57	496.2889	57.94	9.12	16.06	16.89
C24H45BR2N1	507.44	505.1920	56.81	8.94	31.49	2.76
C24H45BR2N3	535.45	533.1981	53.84	8.47	29.85	7.85
C24H45BR2N5	563.47	561.2043	51.16	8.05	28.36	12.43
C24H46BR1N1	428.54	427.2814	67.27	10.82	18.65	3.27
C24H46BR1N3	456.56	455.2875	63.14	10.16	17.50	9.20
C24H46BR1N5	484.57	483.2937	59.49	9.57	16.49	14.45
C24H46BR2N2	522.46	520.2029	55.18	8.87	30.59	5.36
C24H46BR2N4	550.47	548.2090	52.37	8.42	29.03	10.18
C24H46BR2N6	578.48	576.2152	49.83	8.02	27.63	14.53
C24H47BR1N2	443.56	442.2923	64.99	10.68	18.01	6.32
C24H47BR1N4	471.57	470.2984	61.13	10.05	16.94	11.88
C24H47BR1N6	499.59	498.3046	57.70	9.48	15.99	16.82
C24H47BR2N1	509.46	507.2076	56.58	9.30	31.37	2.75
C24H47BR2N3	537.47	535.2138	53.63	8.81	29.73	7.82
C24H47BR2N5	565.48	563.2199	50.98	8.38	28.26	12.38
C24H48BR1N1	430.56	429.2970	66.95	11.24	18.56	3.25
C24H48BR1N3	458.57	457.3032	62.86	10.55	17.42	9.16
C24H48BR1N5	486.59	485.3093	59.24	9.94	16.42	14.39
C24H48BR2N2	524.47	522.2185	54.96	9.23	30.47	5.34
C24H48BR2N4	552.48	550.2247	52.18	8.76	28.93	10.14
C24H48BR2N6	580.50	578.2308	49.66	8.33	27.53	14.48
C24H49BR1N2	445.58	444.3079	64.70	11.08	17.93	6.29
C24H49BR1N4	473.59	472.3141	60.87	10.43	16.87	11.83
C24H49BR1N6	501.60	500.3202	57.47	9.85	15.93	16.75
C24H49BR2N1	511.47	509.2233	56.36	9.66	31.24	2.74
C24H49BR2N3	539.49	537.2294	53.43	9.16	29.62	7.79
C24H49BR2N5	567.50	565.2356	50.80	8.70	28.16	12.34
C24H50BR1N1	432.58	431.3127	66.64	11.65	18.47	3.24
C24H50BR1N3	460.59	459.3188	62.59	10.94	17.35	9.12
C24H50BR1N5	488.60	487.3250	59.00	10.31	16.35	14.33
C24H50BR2N2	526.49	524.2342	54.75	9.57	30.35	5.32
C24H50BR2N4	554.50	552.2403	51.99	9.09	28.82	10.10
C24H50BR2N6	582.51	580.2465	49.49	8.65	27.43	14.43
C24H51BR1N2	447.59	446.3236	64.40	11.49	17.85	6.26
C24H51BR1N4	475.60	474.3297	60.61	10.81	16.80	11.78
C24H51BR1N6	503.62	502.3359	57.24	10.21	15.87	16.69
C24H51BR2N3	541.50	539.2451	53.23	9.49	29.51	7.76
C24H51BR2N5	569.52	567.2512	50.62	9.03	28.06	12.30
C24H52BR1N3	462.61	461.3345	62.31	11.33	17.27	9.08
C24H52BR1N5	490.62	489.3406	58.76	10.68	16.29	14.27
C24H52BR2N4	556.52	554.2560	51.80	9.42	28.72	10.07
C24H52BR2N6	584.53	582.2621	49.32	8.97	27.34	14.38
C24H53BR1N4	477.62	476.3454	60.35	11.19	16.73	11.73
C24H53BR1N6	505.63	504.3515	57.01	10.57	15.80	16.62
C24H53BR2N5	571.53	569.2669	50.44	9.35	27.96	12.25
C24H54BR1N5	492.64	491.3563	58.52	11.05	16.22	14.22
C24H54BR2N6	586.55	584.2778	49.15	9.28	27.25	14.33
C24H55BR1N6	507.65	506.3672	56.78	10.92	15.74	16.55
C25H12BR1N1	406.29	405.0154	73.91	2.98	19.67	3.45
C25H12BR1N3	434.30	433.0215	69.14	2.79	18.40	9.68
C25H12BR1N5	462.31	461.0277	64.95	2.62	17.28	15.15
C25H12BR2N2	500.20	497.9368	60.03	2.42	31.95	5.60
C25H12BR2N4	528.21	525.9430	56.85	2.29	30.25	10.61
C25H12BR2N6	556.22	553.9491	53.99	2.17	28.73	15.11
C25H13BR1N2	421.30	420.0263	71.27	3.11	18.97	6.65
C25H13BR1N4	449.31	448.0324	66.83	2.92	17.78	12.47
C25H13BR1N6	477.33	476.0386	62.91	2.75	16.74	17.61
C25H13BR2N1	487.20	484.9416	61.63	2.69	32.80	2.87
C25H13BR2N3	515.21	512.9477	58.28	2.54	31.02	8.16
C25H13BR2N5	543.22	540.9539	55.28	2.41	29.42	12.89
C25H14BR1N1	408.30	407.0310	73.54	3.46	19.57	3.43
C25H14BR1N3	436.31	435.0372	68.82	3.23	18.31	9.63
C25H14BR1N5	464.33	463.0433	64.67	3.04	17.21	15.08
C25H14BR2N2	502.21	499.9525	59.79	2.81	31.82	5.58
C25H14BR2N4	530.23	527.9586	56.63	2.66	30.14	10.57
C25H14BR2N6	558.24	555.9648	53.79	2.53	28.63	15.05
C25H15BR1N2	423.32	422.0419	70.93	3.57	18.88	6.62
C25H15BR1N4	451.33	450.0481	66.53	3.35	17.70	12.41
C25H15BR1N6	479.34	478.0542	62.64	3.15	16.67	17.53
C25H15BR2N1	489.21	486.9572	61.38	3.09	32.67	2.86
C25H15BR2N3	517.23	514.9634	58.06	2.92	30.90	8.12
C25H15BR2N5	545.24	542.9695	55.07	2.77	29.31	12.84
C25H16BR1N1	410.32	409.0467	73.18	3.93	19.47	3.41
C25H16BR1N3	438.33	437.0528	68.51	3.68	18.23	9.59
C25H16BR1N5	466.34	465.0590	64.39	3.46	17.13	15.02
C25H16BR2N2	504.23	501.9681	59.55	3.20	31.69	5.56
C25H16BR2N4	532.24	529.9743	56.42	3.03	30.03	10.53
C25H16BR2N6	560.25	557.9804	53.60	2.88	28.52	15.00
C25H17BR1N2	425.33	424.0576	70.60	4.03	18.79	6.59
C25H17BR1N4	453.35	452.0637	66.24	3.78	17.63	12.36
C25H17BR1N6	481.36	480.0699	62.38	3.56	16.60	17.46
C25H17BR2N1	491.23	488.9729	61.13	3.49	32.53	2.85
C25H17BR2N3	519.24	516.9790	57.83	3.30	30.78	8.09
C25H17BR2N5	547.26	544.9852	54.87	3.13	29.20	12.80
C25H18BR1N1	412.33	411.0623	72.82	4.40	19.38	3.40
C25H18BR1N3	440.35	439.0685	68.19	4.12	18.15	9.54
C25H18BR1N5	468.36	467.0746	64.11	3.87	17.06	14.95
C25H18BR2N2	506.24	503.9838	59.32	3.58	31.57	5.53
C25H18BR2N4	534.26	531.9899	56.20	3.40	29.91	10.49
C25H18BR2N6	562.27	559.9961	53.40	3.23	28.42	14.95
C25H19BR1N2	427.35	426.0732	70.27	4.48	18.70	6.56
C25H19BR1N4	455.36	454.0794	65.94	4.21	17.55	12.30
C25H19BR1N6	483.37	482.0855	62.12	3.96	16.53	17.39
C25H19BR2N1	493.24	490.9885	60.88	3.88	32.40	2.84
C25H19BR2N3	521.26	518.9947	57.61	3.67	30.66	8.06
C25H19BR2N5	549.27	547.0008	54.67	3.49	29.09	12.75
C25H20BR1N1	414.35	413.0780	72.47	4.87	19.28	3.38
C25H20BR1N3	442.36	441.0841	67.88	4.56	18.06	9.50
C25H20BR1N5	470.38	469.0903	63.84	4.29	16.99	14.89
C25H20BR2N2	508.26	505.9994	59.08	3.97	31.44	5.51

271

C25H20BR2N4	536.27	534.0056	55.99	3.76	29.80	10.45
C25H20BR2N6	564.29	562.0117	53.21	3.57	28.32	14.89
C25H21BR1N2	429.36	428.0889	69.94	4.93	18.61	6.52
C25H21BR1N4	457.38	456.0950	65.65	4.63	17.47	12.25
C25H21BR1N6	485.39	484.1011	61.86	4.36	16.46	17.31
C25H21BR2N1	495.26	493.0042	60.63	4.27	32.27	2.83
C25H21BR2N3	523.27	521.0103	57.38	4.05	30.54	8.03
C25H21BR2N5	551.29	549.0165	54.47	3.84	28.99	12.70
C25H22BR1N1	416.36	415.0936	72.12	5.33	19.19	3.36
C25H22BR1N3	444.38	443.0998	67.57	4.99	17.98	9.46
C25H22BR1N5	472.39	471.1059	63.57	4.69	16.91	14.83
C25H22BR2N2	510.28	508.0151	58.85	4.35	31.32	5.49
C25H22BR2N4	538.29	536.0212	55.78	4.12	29.69	10.41
C25H22BR2N6	566.30	564.0274	53.02	3.92	28.22	14.84
C25H23BR1N2	431.38	430.1045	69.61	5.37	18.52	6.49
C25H23BR1N4	459.39	458.1107	65.36	5.05	17.39	12.20
C25H23BR1N6	487.41	486.1168	61.61	4.76	16.39	17.24
C25H23BR2N1	497.28	495.0198	60.38	4.66	32.14	2.82
C25H23BR2N3	525.29	523.0260	57.16	4.41	30.42	8.00
C25H23BR2N5	553.30	551.0321	54.27	4.19	28.88	12.66
C25H24BR1N1	418.38	417.1093	71.77	5.78	19.10	3.35
C25H24BR1N3	446.39	445.1154	67.27	5.42	17.90	9.41
C25H24BR1N5	474.41	473.1215	63.30	5.10	16.84	14.76
C25H24BR2N2	512.29	510.0307	58.61	4.72	31.19	5.47
C25H24BR2N4	540.30	538.0369	55.58	4.48	29.58	10.37
C25H24BR2N6	568.32	566.0430	52.84	4.26	28.12	14.79
C25H25BR1N2	433.40	432.1202	69.29	5.81	18.44	6.46
C25H25BR1N4	461.41	460.1263	65.08	5.46	17.32	12.14
C25H25BR1N6	489.42	488.1324	61.35	5.15	16.33	17.17
C25H25BR2N1	499.29	497.0355	60.14	5.05	32.01	2.81
C25H25BR2N3	527.31	525.0416	56.95	4.78	30.31	7.97
C25H25BR2N5	555.32	553.0478	54.07	4.54	28.78	12.61
C25H26BR1N1	420.40	419.1249	71.43	6.23	19.01	3.33
C25H26BR1N3	448.41	447.1311	66.97	5.84	17.82	9.37
C25H26BR1N5	476.42	475.1372	63.03	5.50	16.77	14.70
C25H26BR2N2	514.31	512.0464	58.39	5.10	31.07	5.45
C25H26BR2N4	542.32	540.0525	55.37	4.83	29.47	10.33
C25H26BR2N6	570.33	568.0587	52.65	4.60	28.02	14.74
C25H27BR1N2	435.41	434.1358	68.96	6.25	18.35	6.43
C25H27BR1N4	463.42	462.1419	64.80	5.87	17.24	12.09
C25H27BR1N6	491.44	490.1481	61.10	5.54	16.26	17.10
C25H27BR2N1	501.31	499.0511	59.90	5.43	31.88	2.79
C25H27BR2N3	529.32	527.0573	56.73	5.14	30.19	7.94
C25H27BR2N5	557.34	555.0634	53.88	4.88	28.67	12.57
C25H28BR1N1	422.41	421.1406	71.09	6.68	18.92	3.32
C25H28BR1N3	450.43	449.1467	66.67	6.27	17.74	9.33
C25H28BR1N5	478.44	477.1528	62.76	5.90	16.70	14.64
C25H28BR2N2	516.32	514.0620	58.16	5.47	30.95	5.43
C25H28BR2N4	544.34	542.0682	55.16	5.18	29.36	10.29
C25H28BR2N6	572.35	570.0743	52.46	4.93	27.92	14.68
C25H29BR1N2	437.43	436.1515	68.65	6.68	18.27	6.40
C25H29BR1N4	465.44	464.1576	64.51	6.28	17.17	12.04
C25H29BR1N6	493.45	492.1637	60.85	5.92	16.19	17.03
C25H29BR2N1	503.32	501.0668	59.66	5.81	31.75	2.78
C25H29BR2N3	531.34	529.0729	56.51	5.50	30.08	7.91
C25H29BR2N5	559.35	557.0791	53.68	5.23	28.57	12.52
C25H30BR1N1	424.43	423.1562	70.75	7.12	18.83	3.30
C25H30BR1N3	452.44	451.1623	66.37	6.68	17.66	9.29
C25H30BR1N5	480.46	479.1685	62.50	6.29	16.63	14.58
C25H30BR2N2	518.34	516.0777	57.93	5.83	30.83	5.40
C25H30BR2N4	546.35	544.0838	54.96	5.53	29.25	10.25
C25H30BR2N6	574.37	572.0900	52.28	5.26	27.82	14.63
C25H31BR1N2	439.44	438.1671	68.33	7.11	18.18	6.37
C25H31BR1N4	467.46	466.1732	64.24	6.68	17.09	11.99
C25H31BR1N6	495.47	494.1794	60.60	6.31	16.13	16.96
C25H31BR2N1	505.34	503.0824	59.42	6.18	31.62	2.77
C25H31BR2N3	533.35	531.0886	56.30	5.86	29.96	7.88
C25H31BR2N5	561.37	559.0947	53.49	5.57	28.47	12.48
C25H32BR1N1	426.44	425.1719	70.41	7.56	18.74	3.28
C25H32BR1N3	454.46	453.1780	66.07	7.10	17.58	9.25
C25H32BR1N5	482.47	481.1841	62.24	6.69	16.56	14.52
C25H32BR2N2	520.36	518.0933	57.71	6.20	30.71	5.38
C25H32BR2N4	548.37	546.0995	54.76	5.88	29.14	10.22
C25H32BR2N6	576.38	574.1056	52.10	5.60	27.73	14.58
C25H33BR1N2	441.46	440.1827	68.02	7.53	18.10	6.35
C25H33BR1N4	469.47	468.1889	63.96	7.09	17.02	11.93
C25H33BR1N6	497.49	496.1950	60.36	6.69	16.06	16.89
C25H33BR2N1	507.36	505.0981	59.18	6.56	31.50	2.76
C25H33BR2N3	535.37	533.1042	56.09	6.21	29.85	7.85
C25H33BR2N5	563.38	561.1104	53.30	5.90	28.37	12.43
C25H34BR1N1	428.46	427.1875	70.08	8.00	18.65	3.27
C25H34BR1N3	456.47	455.1936	65.78	7.51	17.50	9.21
C25H34BR1N5	484.49	483.1998	61.98	7.07	16.49	14.46
C25H34BR2N2	522.37	520.1090	57.48	6.56	30.59	5.36
C25H34BR2N4	550.38	548.1151	54.56	6.23	29.04	10.18
C25H34BR2N6	578.40	576.1213	51.92	5.93	27.63	14.53
C25H35BR1N2	443.48	442.1984	67.71	7.96	18.02	6.32
C25H35BR1N4	471.49	470.2045	63.69	7.48	16.95	11.88
C25H35BR1N6	499.50	498.2107	60.12	7.06	16.00	16.82
C25H35BR2N1	509.37	507.1137	58.95	6.93	31.37	2.75
C25H35BR2N3	537.39	535.1199	55.88	6.56	29.74	7.82
C25H35BR2N5	565.40	563.1260	53.11	6.24	28.26	12.39
C25H36BR1N1	430.48	429.2031	69.75	8.43	18.56	3.25
C25H36BR1N3	458.49	457.2093	65.49	7.91	17.43	9.16
C25H36BR1N5	486.50	485.2154	61.72	7.46	16.42	14.40
C25H36BR2N2	524.39	522.1246	57.26	6.92	30.48	5.34
C25H36BR2N4	552.40	550.1308	54.36	6.57	28.93	10.14
C25H36BR2N6	580.41	578.1369	51.74	6.25	27.53	14.48
C25H37BR1N2	445.49	444.2140	67.40	8.37	17.94	6.29
C25H37BR1N4	473.50	472.2202	63.42	7.88	16.88	11.83
C25H37BR1N6	501.52	500.2263	59.87	7.44	15.93	16.76
C25H37BR2N1	511.39	509.1294	58.72	7.29	31.25	2.74
C25H37BR2N3	539.40	537.1355	55.67	6.91	29.63	7.79
C25H37BR2N5	567.42	565.1417	52.92	6.57	28.16	12.34
C25H38BR1N1	432.49	431.2188	69.43	8.86	18.48	3.24
C25H38BR1N3	460.51	459.2249	65.21	8.32	17.35	9.12
C25H38BR1N5	488.52	487.2311	61.47	7.84	16.36	14.34
C25H38BR2N2	526.40	524.1403	57.04	7.28	30.36	5.32
C25H38BR2N4	554.42	552.1464	54.16	6.91	28.82	10.11
C25H38BR2N6	582.43	580.1526	51.56	6.58	27.44	14.43
C25H39BR1N2	447.51	446.2297	67.10	8.78	17.86	6.26
C25H39BR1N4	475.52	474.2358	63.15	8.27	16.80	11.78
C25H39BR1N6	503.53	502.2420	59.63	7.81	15.87	16.69
C25H39BR2N1	513.40	511.1450	58.49	7.66	31.13	2.73
C25H39BR2N3	541.42	539.1512	55.46	7.26	29.52	7.76
C25H39BR2N5	569.43	567.1573	52.73	6.90	28.06	12.30
C25H40BR1N1	434.51	433.2344	69.11	9.28	18.39	3.22
C25H40BR1N3	462.52	461.2406	64.92	8.72	17.28	9.09
C25H40BR1N5	490.54	489.2467	61.21	8.22	16.29	14.28
C25H40BR2N2	528.42	526.1559	56.83	7.63	30.24	5.30

272

C25H40BR2N4	556.43	554.1621	53.97	7.25	28.72	10.07
C25H40BR2N6	584.45	582.1682	51.38	6.90	27.34	14.38
C25H41BR1N2	449.52	448.2453	66.80	9.19	17.78	6.23
C25H41BR1N4	477.54	476.2515	62.88	8.65	16.73	11.73
C25H41BR1N6	505.55	504.2576	59.40	8.17	15.81	16.62
C25H41BR2N1	515.42	513.1607	58.26	8.02	31.01	2.72
C25H41BR2N3	543.43	541.1668	55.26	7.60	29.41	7.73
C25H41BR2N5	571.45	569.1730	52.55	7.23	27.97	12.26
C25H42BR1N1	436.52	435.2501	68.79	9.70	18.30	3.21
C25H42BR1N3	464.54	463.2562	64.64	9.11	17.20	9.05
C25H42BR1N5	492.55	491.2624	60.96	8.59	16.22	14.22
C25H42BR2N2	530.43	528.1716	56.61	7.98	30.13	5.28
C25H42BR2N4	558.45	556.1777	53.77	7.58	28.62	10.03
C25H42BR2N6	586.46	584.1839	51.20	7.22	27.25	14.33
C25H43BR1N2	451.54	450.2610	66.50	9.60	17.70	6.20
C25H43BR1N4	479.55	478.2671	62.62	9.04	16.66	11.68
C25H43BR1N6	507.57	506.2733	59.16	8.54	15.74	16.56
C25H43BR2N1	517.44	515.1763	58.03	8.38	30.88	2.71
C25H43BR2N3	545.45	543.1825	55.05	7.95	29.30	7.70
C25H43BR2N5	573.46	571.1886	52.36	7.56	27.87	12.21
C25H44BR1N1	438.54	437.2657	68.47	10.11	18.22	3.19
C25H44BR1N3	466.55	465.2719	64.36	9.51	17.13	9.01
C25H44BR1N5	494.57	493.2780	60.72	8.97	16.16	14.16
C25H44BR2N2	532.45	530.1872	56.40	8.33	30.01	5.26
C25H44BR2N4	560.46	558.1934	53.58	7.91	28.51	10.00
C25H44BR2N6	588.48	586.1995	51.03	7.54	27.16	14.28
C25H45BR1N2	453.55	452.2766	66.21	10.00	17.62	6.18
C25H45BR1N4	481.57	480.2828	62.35	9.42	16.59	11.63
C25H45BR1N6	509.58	508.2889	58.93	8.90	15.68	16.49
C25H45BR2N1	519.45	517.1920	57.81	8.73	30.76	2.70
C25H45BR2N3	547.47	545.1981	54.85	8.29	29.19	7.68
C25H45BR2N5	575.48	573.2043	52.18	7.88	27.77	12.17
C25H46BR1N1	440.56	439.2814	68.16	10.52	18.14	3.18
C25H46BR1N3	468.57	467.2875	64.08	9.90	17.05	8.97
C25H46BR1N5	496.58	495.2937	60.47	9.34	16.09	14.10
C25H46BR2N2	534.47	532.2029	56.18	8.68	29.90	5.24
C25H46BR2N4	562.48	560.2090	53.38	8.24	28.41	9.96
C25H46BR2N6	590.49	588.2152	50.85	7.85	27.06	14.23
C25H47BR1N2	455.57	454.2923	65.91	10.40	17.54	6.15
C25H47BR1N4	483.58	482.2984	62.09	9.80	16.52	11.59
C25H47BR1N6	511.60	510.3046	58.69	9.26	15.62	16.43
C25H47BR2N1	521.47	519.2076	57.58	9.08	30.65	2.69
C25H47BR2N3	549.48	547.2138	54.65	8.62	29.08	7.65
C25H47BR2N5	577.49	575.2199	52.00	8.20	27.67	12.13
C25H48BR1N1	442.57	441.2970	67.85	10.93	18.05	3.16
C25H48BR1N3	470.59	469.3032	63.81	10.28	16.98	8.93
C25H48BR1N5	498.60	497.3093	60.22	9.70	16.03	14.05
C25H48BR2N2	536.48	534.2185	55.97	9.02	29.79	5.22
C25H48BR2N4	564.50	562.2247	53.19	8.57	28.31	9.93
C25H48BR2N6	592.51	590.2308	50.68	8.17	26.97	14.18
C25H49BR1N2	457.59	456.3079	65.62	10.79	17.46	6.12
C25H49BR1N4	485.60	484.3141	61.84	10.17	16.45	11.54
C25H49BR1N6	513.61	512.3202	58.46	9.62	15.56	16.36
C25H49BR2N1	523.48	521.2233	57.36	9.43	30.53	2.68
C25H49BR2N3	551.50	549.2294	54.45	8.96	28.98	7.62
C25H49BR2N5	579.51	577.2356	51.82	8.52	27.58	12.08
C25H50BR1N1	444.59	443.3127	67.54	11.34	17.97	3.15
C25H50BR1N3	472.60	471.3188	63.54	10.66	16.91	8.89
C25H50BR1N5	500.61	499.3250	59.98	10.07	15.96	13.99
C25H50BR2N2	538.50	536.2342	55.76	9.36	29.68	5.20
C25H50BR2N4	566.51	564.2403	53.00	8.90	28.21	9.89
C25H50BR2N6	594.53	592.2465	50.51	8.48	26.88	14.14
C25H51BR1N2	459.60	458.3236	65.33	11.18	17.39	6.10
C25H51BR1N4	487.62	486.3297	61.58	10.54	16.39	11.49
C25H51BR1N6	515.63	514.3359	58.24	9.97	15.50	16.30
C25H51BR2N1	525.50	523.2389	57.14	9.78	30.41	2.67
C25H51BR2N3	553.51	551.2451	54.25	9.29	28.87	7.59
C25H51BR2N5	581.53	579.2512	51.64	8.84	27.48	12.04
C25H52BR1N1	446.60	445.3283	67.24	11.74	17.89	3.14
C25H52BR1N3	474.62	473.3345	63.27	11.04	16.84	8.85
C25H52BR1N5	502.63	501.3406	59.74	10.43	15.90	13.93
C25H52BR2N2	540.51	538.2498	55.55	9.70	29.57	5.18
C25H52BR2N4	568.53	566.2560	52.82	9.22	28.11	9.85
C25H52BR2N6	596.54	594.2621	50.34	8.79	26.79	14.09
C25H53BR1N2	461.62	460.3392	65.05	11.57	17.31	6.07
C25H53BR1N4	489.63	488.3454	61.33	10.91	16.32	11.44
C25H53BR1N6	517.65	516.3515	58.01	10.32	15.44	16.24
C25H53BR2N3	555.53	553.2607	54.05	9.62	28.77	7.56
C25H53BR2N5	583.54	581.2669	51.46	9.15	27.39	12.00
C25H54BR1N3	476.63	475.3501	63.00	11.42	16.76	8.82
C25H54BR1N5	504.65	503.3563	59.50	10.79	15.83	13.88
C25H54BR2N4	570.54	568.2716	52.63	9.54	28.01	9.82
C25H54BR2N6	598.56	596.2778	50.17	9.09	26.70	14.04
C25H55BR1N4	491.65	490.3610	61.08	11.28	16.25	11.40
C25H55BR1N6	519.66	518.3672	57.78	10.67	15.38	16.17
C25H55BR2N5	585.56	583.2825	51.28	9.47	27.29	11.96
C25H56BR1N5	506.66	505.3719	59.27	11.14	15.77	13.82
C25H56BR2N6	600.57	598.2934	50.00	9.40	26.61	13.99
C25H57BR1N6	521.68	520.3828	57.56	11.01	15.32	16.11
C26H13BR1N2	433.31	432.0263	72.07	3.02	18.44	6.46
C26H13BR1N4	461.32	460.0324	67.69	2.84	17.32	12.14
C26H13BR1N6	489.34	488.0386	63.82	2.68	16.33	17.17
C26H13BR2N1	499.21	496.9416	62.56	2.62	32.01	2.81
C26H13BR2N3	527.22	524.9477	59.23	2.49	30.31	7.97
C26H13BR2N5	555.24	552.9539	56.24	2.36	28.78	12.61
C26H14BR1N1	420.31	419.0310	74.30	3.36	19.01	3.33
C26H14BR1N3	448.33	447.0372	69.66	3.15	17.82	9.37
C26H14BR1N5	476.34	475.0433	65.56	2.96	16.77	14.70
C26H14BR2N2	514.22	511.9525	60.73	2.74	31.08	5.45
C26H14BR2N4	542.24	539.9586	57.59	2.60	29.47	10.33
C26H14BR2N6	570.25	567.9648	54.76	2.47	28.02	14.74
C26H15BR1N2	435.33	434.0419	71.74	3.47	18.35	6.44
C26H15BR1N4	463.34	462.0481	67.40	3.26	17.25	12.09
C26H15BR1N6	491.35	490.0542	63.56	3.08	16.26	17.10
C26H15BR2N1	501.22	498.9572	62.31	3.02	31.88	2.79
C26H15BR2N3	529.24	526.9634	59.01	2.86	30.20	7.94
C26H15BR2N5	557.25	554.9695	56.04	2.71	28.68	12.57
C26H16BR1N1	422.33	421.0467	73.94	3.82	18.92	3.32
C26H16BR1N3	450.34	449.0528	69.35	3.58	17.74	9.33
C26H16BR1N5	478.35	477.0590	65.28	3.37	16.70	14.64
C26H16BR2N2	516.24	513.9681	60.49	3.12	30.96	5.43
C26H16BR2N4	544.25	541.9743	57.38	2.96	29.36	10.29
C26H16BR2N6	572.27	569.9804	54.57	2.82	27.93	14.69
C26H17BR1N2	437.34	436.0576	71.41	3.92	18.27	6.41
C26H17BR1N4	465.36	464.0637	67.11	3.68	17.17	12.04
C26H17BR1N6	493.37	492.0699	63.30	3.47	16.20	17.03
C26H17BR2N1	503.24	500.9729	62.06	3.41	31.76	2.78
C26H17BR2N3	531.25	528.9790	58.78	3.23	30.08	7.91
C26H17BR2N5	559.27	556.9852	55.84	3.06	28.57	12.52
C26H18BR1N1	424.34	423.0623	73.59	4.28	18.83	3.30

C26H18BR1N3	452.36	451.0685	69.04	4.01	17.66	9.29
C26H18BR1N5	480.37	479.0746	65.01	3.78	16.63	14.58
C26H18BR2N2	518.25	515.9838	60.26	3.50	30.84	5.41
C26H18BR2N4	546.27	543.9899	57.17	3.32	29.25	10.26
C26H18BR2N6	574.28	571.9961	54.38	3.16	27.83	14.63
C26H19BR1N2	439.36	438.0732	71.08	4.36	18.19	6.38
C26H19BR1N4	467.37	466.0794	66.82	4.10	17.10	11.99
C26H19BR1N6	495.39	494.0855	63.04	3.87	16.13	16.96
C26H19BR2N1	505.26	502.9885	61.81	3.79	31.63	2.77
C26H19BR2N3	533.27	530.9947	58.56	3.59	29.97	7.88
C26H19BR2N5	561.28	559.0008	55.64	3.41	28.47	12.48
C26H20BR1N1	426.36	425.0780	73.25	4.73	18.74	3.29
C26H20BR1N3	454.37	453.0841	68.73	4.44	17.59	9.25
C26H20BR1N5	482.39	481.0903	64.74	4.18	16.56	14.52
C26H20BR2N2	520.27	517.9994	60.02	3.87	30.72	5.38
C26H20BR2N4	548.28	546.0056	56.96	3.68	29.15	10.22
C26H20BR2N6	576.30	574.0117	54.19	3.50	27.73	14.58
C26H21BR1N2	441.37	440.0889	70.75	4.80	18.10	6.35
C26H21BR1N4	469.39	468.0950	66.53	4.51	17.02	11.94
C26H21BR1N6	497.40	496.1011	62.78	4.26	16.06	16.90
C26H21BR2N1	507.27	505.0042	61.56	4.17	31.50	2.76
C26H21BR2N3	535.29	533.0103	58.34	3.95	29.85	7.85
C26H21BR2N5	563.30	561.0165	55.44	3.76	28.37	12.43
C26H22BR1N1	428.38	427.0936	72.90	5.18	18.65	3.27
C26H22BR1N3	456.39	455.0998	68.43	4.86	17.51	9.21
C26H22BR1N5	484.40	483.1059	64.47	4.58	16.50	14.46
C26H22BR2N2	522.29	520.0151	59.79	4.25	30.60	5.36
C26H22BR2N4	550.30	548.0212	56.75	4.03	29.04	10.18
C26H22BR2N6	578.31	576.0274	54.00	3.83	27.63	14.53
C26H23BR1N2	443.39	442.1045	70.43	5.23	18.02	6.32
C26H23BR1N4	471.40	470.1107	66.25	4.92	16.95	11.89
C26H23BR1N6	499.42	498.1168	62.53	4.64	16.00	16.83
C26H23BR2N1	509.29	507.0198	61.32	4.55	31.38	2.75
C26H23BR2N3	537.30	535.0260	58.12	4.31	29.74	7.82
C26H23BR2N5	565.31	563.0321	55.24	4.10	28.27	12.39
C26H24BR1N1	430.39	429.1093	72.56	5.62	18.57	3.25
C26H24BR1N3	458.41	457.1154	68.13	5.28	17.43	9.17
C26H24BR1N5	486.42	485.1215	64.20	4.97	16.43	14.40
C26H24BR2N2	524.30	522.0307	59.56	4.61	30.48	5.34
C26H24BR2N4	552.32	550.0369	56.54	4.38	28.93	10.14
C26H24BR2N6	580.33	578.0430	53.81	4.17	27.54	14.48
C26H25BR1N2	445.41	444.1202	70.11	5.66	17.94	6.29
C26H25BR1N4	473.42	472.1263	65.96	5.32	16.88	11.83
C26H25BR1N6	501.43	500.1324	62.28	5.03	15.94	16.76
C26H25BR2N1	511.30	509.0355	61.08	4.93	31.25	2.74
C26H25BR2N3	539.32	537.0416	57.90	4.67	29.63	7.79
C26H25BR2N5	567.33	565.0478	55.05	4.44	28.17	12.34
C26H26BR1N1	432.41	431.1249	72.22	6.06	18.48	3.24
C26H26BR1N3	460.42	459.1311	67.83	5.69	17.35	9.13
C26H26BR1N5	488.43	487.1372	63.94	5.37	16.36	14.34
C26H26BR2N2	526.32	524.0464	59.33	4.98	30.36	5.32
C26H26BR2N4	554.33	552.0525	56.34	4.73	28.83	10.11
C26H26BR2N6	582.35	580.0587	53.63	4.50	27.44	14.43
C26H27BR1N2	447.42	446.1358	69.80	6.08	17.86	6.26
C26H27BR1N4	475.44	474.1419	65.68	5.72	16.81	11.78
C26H27BR1N6	503.45	502.1481	62.03	5.41	15.87	16.69
C26H27BR2N1	513.32	511.0511	60.84	5.30	31.13	2.73
C26H27BR2N3	541.33	539.0573	57.69	5.03	29.52	7.76
C26H27BR2N5	569.35	567.0634	54.85	4.78	28.07	12.30
C26H28BR1N1	434.42	433.1406	71.89	6.50	18.39	3.22

C26H28BR1N3	462.44	461.1467	67.53	6.10	17.28	9.09
C26H28BR1N5	490.45	489.1528	63.67	5.75	16.29	14.28
C26H28BR2N2	528.33	526.0620	59.11	5.34	30.25	5.30
C26H28BR2N4	556.35	554.0682	56.13	5.07	28.72	10.07
C26H28BR2N6	584.36	582.0743	53.44	4.83	27.35	14.38
C26H29BR1N2	449.44	448.1515	69.48	6.50	17.78	6.23
C26H29BR1N4	477.45	476.1576	65.41	6.12	16.74	11.73
C26H29BR1N6	505.47	504.1637	61.78	5.78	15.81	16.63
C26H29BR2N1	515.34	513.0668	60.60	5.67	31.01	2.72
C26H29BR2N3	543.35	541.0729	57.47	5.38	29.41	7.73
C26H29BR2N5	571.36	569.0791	54.66	5.12	27.97	12.26
C26H30BR1N1	436.44	435.1562	71.55	6.93	18.31	3.21
C26H30BR1N3	464.45	463.1623	67.24	6.51	17.20	9.05
C26H30BR1N5	492.47	491.1685	63.41	6.14	16.23	14.22
C26H30BR2N2	530.35	528.0777	58.88	5.70	30.13	5.28
C26H30BR2N4	558.36	556.0838	55.93	5.42	28.62	10.03
C26H30BR2N6	586.38	584.0900	53.26	5.16	27.25	14.33
C26H31BR1N2	451.45	450.1671	69.17	6.92	17.70	6.21
C26H31BR1N4	479.47	478.1732	65.13	6.52	16.67	11.69
C26H31BR1N6	507.48	506.1794	61.54	6.16	15.75	16.56
C26H31BR2N1	517.35	515.0824	60.36	6.04	30.89	2.71
C26H31BR2N3	545.37	543.0886	57.26	5.73	29.30	7.70
C26H31BR2N5	573.38	571.0947	54.46	5.45	27.87	12.21
C26H32BR1N1	438.46	437.1719	71.22	7.36	18.22	3.19
C26H32BR1N3	466.47	465.1780	66.95	6.91	17.13	9.01
C26H32BR1N5	494.48	493.1841	63.15	6.52	16.16	14.16
C26H32BR2N2	532.37	530.0933	58.66	6.06	30.02	5.26
C26H32BR2N4	560.38	558.0995	55.73	5.76	28.52	10.00
C26H32BR2N6	588.39	586.1056	53.08	5.48	27.16	14.28
C26H33BR1N2	453.47	452.1827	68.87	7.34	17.62	6.18
C26H33BR1N4	481.48	480.1889	64.86	6.91	16.60	11.64
C26H33BR1N6	509.50	508.1950	61.29	6.53	15.68	16.49
C26H33BR2N1	519.37	517.0981	60.13	6.40	30.77	2.70
C26H33BR2N3	547.38	545.1042	57.05	6.08	29.20	7.68
C26H33BR2N5	575.39	573.1104	54.27	5.78	27.77	12.17
C26H34BR1N1	440.47	439.1875	70.90	7.78	18.14	3.18
C26H34BR1N3	468.48	467.1936	66.66	7.32	17.06	8.97
C26H34BR1N5	496.50	495.1998	62.90	6.90	16.09	14.11
C26H34BR2N2	534.38	532.1090	58.44	6.41	29.91	5.24
C26H34BR2N4	562.40	560.1151	55.53	6.09	28.42	9.96
C26H34BR2N6	590.41	588.1213	52.89	5.80	27.07	14.23
C26H35BR1N2	455.49	454.1984	68.56	7.75	17.54	6.15
C26H35BR1N4	483.50	482.2045	64.59	7.30	16.53	11.59
C26H35BR1N6	511.51	510.2107	61.05	6.90	15.62	16.43
C26H35BR2N1	521.38	519.1137	59.90	6.77	30.65	2.69
C26H35BR2N3	549.40	547.1199	56.84	6.42	29.09	7.65
C26H35BR2N5	577.41	575.1260	54.08	6.11	27.68	12.13
C26H36BR1N1	442.49	441.2031	70.58	8.20	18.06	3.17
C26H36BR1N3	470.50	469.2093	66.37	7.71	16.98	8.93
C26H36BR1N5	498.51	497.2154	62.64	7.28	16.03	14.05
C26H36BR2N2	536.40	534.1246	58.22	6.76	29.79	5.22
C26H36BR2N4	564.41	562.1308	55.33	6.43	28.31	9.93
C26H36BR2N6	592.43	590.1369	52.71	6.13	26.98	14.19
C26H37BR1N2	457.50	456.2140	68.26	8.15	17.47	6.12
C26H37BR1N4	485.52	484.2202	64.32	7.68	16.46	11.54
C26H37BR1N6	513.53	512.2263	60.81	7.26	15.56	16.37
C26H37BR2N1	523.40	521.1294	59.67	7.13	30.53	2.68
C26H37BR2N3	551.41	549.1355	56.63	6.76	28.98	7.62
C26H37BR2N5	579.43	577.1417	53.90	6.44	27.58	12.09
C26H38BR1N1	444.50	443.2188	70.26	8.62	17.98	3.15

C26H38BR1N3	472.52	471.2249	66.09	8.11	16.91	8.89
C26H38BR1N5	500.53	499.2311	62.39	7.65	15.96	13.99
C26H38BR2N2	538.41	536.1403	58.00	7.11	29.68	5.20
C26H38BR2N4	566.43	564.1464	55.13	6.76	28.21	9.89
C26H38BR2N6	594.44	592.1526	52.54	6.44	26.88	14.14
C26H39BR1N2	459.52	458.2297	67.96	8.55	17.39	6.10
C26H39BR1N4	487.53	486.2358	64.06	8.06	16.39	11.49
C26H39BR1N6	515.54	514.2420	60.57	7.63	15.50	16.30
C26H39BR2N1	525.42	523.1450	59.44	7.48	30.42	2.67
C26H39BR2N3	553.43	551.1512	56.43	7.10	28.88	7.59
C26H39BR2N5	581.44	579.1573	53.71	6.76	27.48	12.04
C26H40BR1N1	446.52	445.2344	69.94	9.03	17.89	3.14
C26H40BR1N3	474.53	473.2406	65.81	8.50	16.84	8.86
C26H40BR1N5	502.55	501.2467	62.14	8.02	15.90	13.94
C26H40BR2N2	540.43	538.1559	57.79	7.46	29.57	5.18
C26H40BR2N4	568.44	566.1621	54.94	7.09	28.11	9.86
C26H40BR2N6	596.46	594.1682	52.36	6.76	26.79	14.09
C26H41BR1N2	461.53	460.2453	67.66	8.95	17.31	6.07
C26H41BR1N4	489.55	488.2515	63.79	8.44	16.32	11.44
C26H41BR1N6	517.56	516.2576	60.34	7.98	15.44	16.24
C26H41BR2N1	527.43	525.1607	59.21	7.84	30.30	2.66
C26H41BR2N3	555.44	553.1668	56.22	7.44	28.77	7.57
C26H41BR2N5	583.46	581.1730	53.52	7.08	27.39	12.00
C26H42BR1N1	448.54	447.2501	69.62	9.44	17.81	3.12
C26H42BR1N3	476.55	475.2562	65.53	8.88	16.77	8.82
C26H42BR1N5	504.56	503.2624	61.89	8.39	15.84	13.88
C26H42BR2N2	542.45	540.1716	57.57	7.80	29.46	5.16
C26H42BR2N4	570.46	568.1777	54.74	7.42	28.01	9.82
C26H42BR2N6	598.47	596.1839	52.18	7.07	26.70	14.04
C26H43BR1N2	463.55	462.2610	67.37	9.35	17.24	6.04
C26H43BR1N4	491.56	490.2671	63.53	8.82	16.26	11.40
C26H43BR1N6	519.58	518.2733	60.10	8.34	15.38	16.17
C26H43BR2N1	529.45	527.1763	58.98	8.19	30.18	2.65
C26H43BR2N3	557.46	555.1825	56.02	7.78	28.67	7.54
C26H43BR2N5	585.47	583.1886	53.34	7.40	27.30	11.96
C26H44BR1N1	450.55	449.2657	69.31	9.84	17.73	3.11
C26H44BR1N3	478.56	477.2719	65.26	9.27	16.70	8.78
C26H44BR1N5	506.58	505.2780	61.65	8.75	15.77	13.82
C26H44BR2N2	544.46	542.1872	57.36	8.15	29.35	5.15
C26H44BR2N4	572.48	570.1934	54.55	7.75	27.92	9.79
C26H44BR2N6	600.49	598.1995	52.01	7.39	26.61	14.00
C26H45BR1N2	465.57	464.2766	67.08	9.74	17.16	6.02
C26H45BR1N4	493.58	492.2828	63.27	9.19	16.19	11.35
C26H45BR1N6	521.59	520.2889	59.87	8.70	15.32	16.11
C26H45BR2N1	531.46	529.1920	58.76	8.53	30.07	2.64
C26H45BR2N3	559.48	557.1981	55.82	8.11	28.56	7.51
C26H45BR2N5	587.49	585.2043	53.16	7.72	27.20	11.92
C26H46BR1N1	452.57	451.2814	69.00	10.25	17.66	3.09
C26H46BR1N3	480.58	479.2875	64.98	9.65	16.63	8.74
C26H46BR1N5	508.59	507.2937	61.40	9.12	15.71	13.77
C26H46BR2N2	546.48	544.2029	57.15	8.48	29.24	5.13
C26H46BR2N4	574.49	572.2090	54.36	8.07	27.82	9.75
C26H46BR2N6	602.50	600.2152	51.83	7.70	26.52	13.95
C26H47BR1N2	467.58	466.2923	66.79	10.13	17.09	5.99
C26H47BR1N4	495.60	494.2984	63.01	9.56	16.12	11.30
C26H47BR1N6	523.61	522.3046	59.64	9.05	15.26	16.05
C26H47BR2N1	533.48	531.2076	58.54	8.88	29.96	2.63
C26H47BR2N3	561.49	559.2138	55.62	8.44	28.46	7.48
C26H47BR2N5	589.51	587.2199	52.97	8.04	27.11	11.88
C26H48BR1N1	454.58	453.2970	68.70	10.64	17.58	3.08

C26H48BR1N3	482.60	481.3032	64.71	10.03	16.56	8.71
C26H48BR1N5	510.61	509.3093	61.16	9.48	15.65	13.72
C26H48BR2N2	548.49	546.2185	56.94	8.82	29.14	5.11
C26H48BR2N4	576.51	574.2247	54.17	8.39	27.72	9.72
C26H48BR2N6	604.52	602.2308	51.66	8.00	26.44	13.90
C26H49BR1N2	469.60	468.3079	66.50	10.52	17.02	5.97
C26H49BR1N4	497.61	496.3141	62.76	9.93	16.06	11.26
C26H49BR1N6	525.62	524.3202	59.41	9.40	15.20	15.99
C26H49BR2N1	535.50	533.2233	58.32	9.22	29.84	2.62
C26H49BR2N3	563.51	561.2294	55.42	8.76	28.36	7.46
C26H49BR2N5	591.52	589.2356	52.79	8.35	27.02	11.84
C26H50BR1N1	456.60	455.3127	68.39	11.04	17.50	3.07
C26H50BR1N3	484.61	483.3188	64.44	10.40	16.49	8.67
C26H50BR1N5	512.63	511.3250	60.92	9.83	15.59	13.66
C26H50BR2N2	550.51	548.2342	56.73	9.15	29.03	5.09
C26H50BR2N4	578.52	576.2403	53.98	8.71	27.62	9.68
C26H50BR2N6	606.54	604.2465	51.49	8.31	26.35	13.86
C26H51BR1N2	471.61	470.3236	66.22	10.90	16.94	5.94
C26H51BR1N4	499.63	498.3297	62.50	10.29	15.99	11.21
C26H51BR1N6	527.64	526.3359	59.19	9.74	15.14	15.93
C26H51BR2N1	537.51	535.2389	58.10	9.56	29.73	2.61
C26H51BR2N3	565.52	563.2451	55.22	9.09	28.26	7.43
C26H51BR2N5	593.54	591.2512	52.61	8.66	26.92	11.80
C26H52BR1N1	458.62	457.3283	68.09	11.43	17.42	3.05
C26H52BR1N3	486.63	485.3345	64.17	10.77	16.42	8.63
C26H52BR1N5	514.64	513.3406	60.68	10.18	15.53	13.61
C26H52BR2N2	552.53	550.2498	56.52	9.49	28.92	5.07
C26H52BR2N4	580.54	578.2560	53.79	9.03	27.53	9.65
C26H52BR2N6	608.55	606.2621	51.32	8.61	26.26	13.81
C26H53BR1N2	473.63	472.3392	65.94	11.28	16.87	5.91
C26H53BR1N4	501.64	500.3454	62.25	10.65	15.93	11.17
C26H53BR1N6	529.66	528.3515	58.96	10.09	15.09	15.87
C26H53BR2N1	539.53	537.2546	57.88	9.90	29.62	2.60
C26H53BR2N3	567.54	565.2607	55.03	9.41	28.16	7.40
C26H53BR2N5	595.55	593.2669	52.44	8.97	26.83	11.76
C26H54BR1N1	460.63	459.3440	67.80	11.82	17.35	3.04
C26H54BR1N3	488.64	487.3501	63.91	11.14	16.35	8.60
C26H54BR1N5	516.66	515.3563	60.44	10.54	15.47	13.56
C26H54BR2N2	554.54	552.2655	56.31	9.82	28.82	5.05
C26H54BR2N4	582.56	580.2716	53.61	9.34	27.43	9.62
C26H54BR2N6	610.57	608.2778	51.15	8.91	26.17	13.76
C26H55BR1N2	475.65	474.3549	65.66	11.66	16.80	5.89
C26H55BR1N4	503.66	502.3610	62.00	11.01	15.86	11.12
C26H55BR1N6	531.67	530.3672	58.74	10.43	15.03	15.81
C26H55BR2N3	569.56	567.2764	54.83	9.73	28.06	7.38
C26H55BR2N5	597.57	595.2825	52.26	9.28	26.74	11.72
C26H56BR1N3	490.66	489.3658	63.65	11.50	16.28	8.56
C26H56BR1N5	518.67	517.3719	60.21	10.88	15.41	13.50
C26H56BR2N4	584.57	582.2873	53.42	9.66	27.34	9.58
C26H56BR2N6	612.58	610.2934	50.98	9.21	26.09	13.72
C26H57BR1N4	505.67	504.3767	61.76	11.36	15.80	11.08
C26H57BR1N6	533.69	532.3828	58.52	10.77	14.97	15.75
C26H57BR2N5	599.59	597.2982	52.08	9.58	26.65	11.68
C26H58BR1N5	520.69	519.3876	59.98	11.23	15.35	13.45
C26H58BR2N6	614.60	612.3090	50.81	9.51	26.00	13.67
C26H59BR1N6	535.70	534.3985	58.30	11.10	14.92	15.69
C27H13BR1N2	445.32	444.0263	72.82	2.94	17.94	6.29
C27H13BR1N4	473.34	472.0324	68.51	2.77	16.88	11.84
C27H13BR1N6	501.35	500.0386	64.69	2.61	15.94	16.76
C27H13BR2N1	511.22	508.9416	63.44	2.56	31.26	2.74

275

C27H13BR2N3	539.23	536.9477	60.14	2.43	29.64	7.79
C27H13BR2N5	567.25	564.9539	57.17	2.31	28.17	12.35
C27H14BR1N1	432.32	431.0310	75.01	3.26	18.48	3.24
C27H14BR1N3	460.34	459.0372	70.45	3.07	17.36	9.13
C27H14BR1N5	488.35	487.0433	66.41	2.89	16.36	14.34
C27H14BR2N2	526.23	523.9525	61.63	2.68	30.37	5.32
C27H14BR2N4	554.25	551.9586	58.51	2.55	28.83	10.11
C27H14BR2N6	582.26	579.9648	55.70	2.42	27.45	14.43
C27H15BR1N2	447.34	446.0419	72.50	3.38	17.86	6.26
C27H15BR1N4	475.35	474.0481	68.22	3.18	16.81	11.79
C27H15BR1N6	503.36	502.0542	64.43	3.00	15.87	16.70
C27H15BR2N1	513.24	510.9572	63.19	2.95	31.14	2.73
C27H15BR2N3	541.25	538.9634	59.92	2.79	29.53	7.76
C27H15BR2N5	569.26	566.9695	56.97	2.66	28.07	12.30
C27H16BR1N1	434.34	433.0467	74.67	3.71	18.40	3.22
C27H16BR1N3	462.35	461.0528	70.14	3.49	17.28	9.09
C27H16BR1N5	490.37	489.0590	66.13	3.29	16.29	14.28
C27H16BR2N2	528.25	525.9681	61.39	3.05	30.25	5.30
C27H16BR2N4	556.26	553.9743	58.30	2.90	28.73	10.07
C27H16BR2N6	584.28	581.9804	55.50	2.76	27.35	14.38
C27H17BR1N2	449.35	448.0576	72.17	3.81	17.78	6.23
C27H17BR1N4	477.37	476.0637	67.94	3.59	16.74	11.74
C27H17BR1N6	505.38	504.0699	64.17	3.39	15.81	16.63
C27H17BR2N1	515.25	512.9729	62.94	3.33	31.02	2.72
C27H17BR2N3	543.26	540.9790	59.69	3.15	29.42	7.73
C27H17BR2N5	571.28	568.9852	56.77	3.00	27.97	12.26
C27H18BR1N1	436.36	435.0623	74.32	4.16	18.31	3.21
C27H18BR1N3	464.37	463.0685	69.84	3.91	17.21	9.05
C27H18BR1N5	492.38	491.0746	65.86	3.68	16.23	14.22
C27H18BR2N2	530.27	527.9838	61.16	3.42	30.14	5.28
C27H18BR2N4	558.28	555.9899	58.09	3.25	28.63	10.04
C27H18BR2N6	586.29	583.9961	55.31	3.09	27.26	14.33
C27H19BR1N2	451.37	450.0732	71.85	4.24	17.70	6.21
C27H19BR1N4	479.38	478.0794	67.65	4.00	16.67	11.69
C27H19BR1N6	507.40	506.0855	63.91	3.77	15.75	16.56
C27H19BR2N1	517.27	514.9885	62.70	3.70	30.89	2.71
C27H19BR2N3	545.28	542.9947	59.47	3.51	29.31	7.71
C27H19BR2N5	573.29	571.0008	56.57	3.34	27.88	12.22
C27H20BR1N1	438.37	437.0780	73.98	4.60	18.23	3.20
C27H20BR1N3	466.38	465.0841	69.54	4.32	17.13	9.01
C27H20BR1N5	494.40	493.0903	65.60	4.08	16.16	14.17
C27H20BR2N2	532.28	529.9994	60.93	3.79	30.02	5.26
C27H20BR2N4	560.30	558.0056	57.88	3.60	28.52	10.00
C27H20BR2N6	588.31	586.0117	55.12	3.43	27.16	14.29
C27H21BR1N2	453.39	452.0889	71.53	4.67	17.62	6.18
C27H21BR1N4	481.40	480.0950	67.37	4.40	16.60	11.64
C27H21BR1N6	509.41	508.1011	63.66	4.16	15.69	16.50
C27H21BR2N1	519.28	517.0042	62.45	4.08	30.77	2.70
C27H21BR2N3	547.30	545.0103	59.26	3.87	29.20	7.68
C27H21BR2N5	575.31	573.0165	56.37	3.68	27.78	12.17
C27H22BR1N1	440.39	439.0936	73.64	5.04	18.14	3.18
C27H22BR1N3	468.40	467.0998	69.24	4.73	17.06	8.97
C27H22BR1N5	496.41	495.1059	65.33	4.47	16.10	14.11
C27H22BR2N2	534.30	532.0151	60.70	4.15	29.91	5.24
C27H22BR2N4	562.31	560.0212	57.67	3.94	28.42	9.96
C27H22BR2N6	590.32	588.0274	54.94	3.76	27.07	14.24
C27H23BR1N2	455.40	454.1045	71.21	5.09	17.55	6.15
C27H23BR1N4	483.42	482.1107	67.09	4.80	16.53	11.59
C27H23BR1N6	511.43	510.1168	63.41	4.53	15.62	16.43
C27H23BR2N1	521.30	519.0198	62.21	4.45	30.66	2.69

C27H23BR2N3	549.31	547.0260	59.04	4.22	29.09	7.65
C27H23BR2N5	577.33	575.0321	56.17	4.02	27.68	12.13
C27H24BR1N1	442.40	441.1093	73.30	5.47	18.06	3.17
C27H24BR1N3	470.42	469.1154	68.94	5.14	16.99	8.93
C27H24BR1N5	498.43	497.1215	65.06	4.85	16.03	14.05
C27H24BR2N2	536.31	534.0307	60.47	4.51	29.80	5.22
C27H24BR2N4	564.33	562.0369	57.47	4.29	28.32	9.93
C27H24BR2N6	592.34	590.0430	54.75	4.08	26.98	14.19
C27H25BR1N2	457.42	456.1202	70.90	5.51	17.47	6.12
C27H25BR1N4	485.43	484.1263	66.81	5.19	16.46	11.54
C27H25BR1N6	513.44	512.1324	63.16	4.91	15.56	16.37
C27H25BR2N1	523.31	521.0355	61.97	4.82	30.54	2.68
C27H25BR2N3	551.33	549.0416	58.82	4.57	28.99	7.62
C27H25BR2N5	579.34	577.0478	55.98	4.35	27.58	12.09
C27H26BR1N1	444.42	443.1249	72.97	5.90	17.98	3.15
C27H26BR1N3	472.43	471.1311	68.64	5.55	16.91	8.89
C27H26BR1N5	500.45	499.1372	64.80	5.24	15.97	13.99
C27H26BR2N2	538.33	536.0464	60.24	4.87	29.69	5.20
C27H26BR2N4	566.34	564.0525	57.26	4.63	28.22	9.89
C27H26BR2N6	594.36	592.0587	54.56	4.41	26.89	14.14
C27H27BR1N2	459.43	458.1358	70.59	5.92	17.39	6.10
C27H27BR1N4	487.45	486.1419	66.53	5.58	16.39	11.49
C27H27BR1N6	515.46	514.1481	62.91	5.28	15.50	16.30
C27H27BR2N1	525.33	523.0511	61.73	5.18	30.42	2.67
C27H27BR2N3	553.34	551.0573	58.61	4.92	28.88	7.59
C27H27BR2N5	581.36	579.0634	55.78	4.68	27.49	12.05
C27H28BR1N1	446.43	445.1406	72.64	6.32	17.90	3.14
C27H28BR1N3	474.45	473.1467	68.35	5.95	16.84	8.86
C27H28BR1N5	502.46	501.1528	64.54	5.62	15.90	13.94
C27H28BR2N2	540.35	538.0620	60.02	5.22	29.58	5.18
C27H28BR2N4	568.36	566.0682	57.06	4.97	28.12	9.86
C27H28BR2N6	596.37	594.0743	54.38	4.73	26.80	14.09
C27H29BR1N2	461.45	460.1515	70.28	6.33	17.32	6.07
C27H29BR1N4	489.46	488.1576	66.26	5.97	16.32	11.45
C27H29BR1N6	517.48	516.1637	62.67	5.65	15.44	16.24
C27H29BR2N1	527.35	525.0668	61.50	5.54	30.30	2.66
C27H29BR2N3	555.36	553.0729	58.39	5.26	28.78	7.57
C27H29BR2N5	583.37	581.0791	55.59	5.01	27.39	12.00
C27H30BR1N1	448.45	447.1562	72.32	6.74	17.82	3.12
C27H30BR1N3	476.46	475.1623	68.06	6.35	16.77	8.82
C27H30BR1N5	504.48	503.1685	64.28	5.99	15.84	13.88
C27H30BR2N2	542.36	540.0777	59.79	5.58	29.47	5.17
C27H30BR2N4	570.37	568.0838	56.86	5.30	28.02	9.82
C27H30BR2N6	598.39	596.0900	54.20	5.05	26.71	14.04
C27H31BR1N2	463.47	462.1671	69.97	6.74	17.24	6.04
C27H31BR1N4	491.48	490.1732	65.98	6.36	16.26	11.40
C27H31BR1N6	519.49	518.1794	62.43	6.01	15.38	16.18
C27H31BR2N1	529.36	527.0824	61.26	5.90	30.19	2.65
C27H31BR2N3	557.38	555.0886	58.18	5.61	28.67	7.54
C27H31BR2N5	585.39	583.0947	55.40	5.34	27.30	11.96
C27H32BR1N1	450.47	449.1719	71.99	7.16	17.74	3.11
C27H32BR1N3	478.48	477.1780	67.78	6.74	16.70	8.78
C27H32BR1N5	506.49	505.1841	64.03	6.37	15.78	13.83
C27H32BR2N2	544.38	542.0933	59.57	5.93	29.36	5.15
C27H32BR2N4	572.39	570.0995	56.66	5.64	27.92	9.79
C27H32BR2N6	600.40	598.1056	54.01	5.37	26.62	14.00
C27H33BR1N2	465.48	464.1827	69.67	7.15	17.17	6.02
C27H33BR1N4	493.49	492.1889	65.72	6.74	16.19	11.35
C27H33BR1N6	521.51	520.1950	62.19	6.38	15.32	16.11
C27H33BR2N1	531.38	529.0981	61.03	6.26	30.07	2.64

276

C27H33BR2N3	559.39	557.1042	57.97	5.95	28.57	7.51
C27H33BR2N5	587.41	585.1104	55.21	5.66	27.21	11.92
C27H34BR1N1	452.48	451.1875	71.67	7.57	17.66	3.10
C27H34BR1N3	480.50	479.1936	67.49	7.13	16.63	8.75
C27H34BR1N5	508.51	507.1998	63.77	6.74	15.71	13.77
C27H34BR2N2	546.39	544.1090	59.35	6.27	29.25	5.13
C27H34BR2N4	574.41	572.1151	56.46	5.97	27.82	9.75
C27H34BR2N6	602.42	600.1213	53.83	5.69	26.53	13.95
C27H35BR1N2	467.50	466.1984	69.37	7.55	17.09	5.99
C27H35BR1N4	495.51	494.2045	65.45	7.12	16.13	11.31
C27H35BR1N6	523.52	522.2107	61.95	6.74	15.26	16.05
C27H35BR2N1	533.39	531.1137	60.80	6.61	29.96	2.63
C27H35BR2N3	561.41	559.1199	57.77	6.28	28.47	7.48
C27H35BR2N5	589.42	587.1260	55.02	5.99	27.11	11.88
C27H36BR1N1	454.50	453.2031	71.35	7.98	17.58	3.08
C27H36BR1N3	482.51	481.2093	67.21	7.52	16.56	8.71
C27H36BR1N5	510.53	509.2154	63.52	7.11	15.65	13.72
C27H36BR2N2	548.41	546.1246	59.13	6.62	29.14	5.11
C27H36BR2N4	576.42	574.1308	56.26	6.30	27.72	9.72
C27H36BR2N6	604.44	602.1369	53.65	6.00	26.44	13.90
C27H37BR1N2	469.51	468.2140	69.07	7.94	17.02	5.97
C27H37BR1N4	497.53	496.2202	65.18	7.50	16.06	11.26
C27H37BR1N6	525.54	524.2263	61.71	7.10	15.20	15.99
C27H37BR2N1	535.41	533.1294	60.57	6.97	29.85	2.62
C27H37BR2N3	563.42	561.1355	57.56	6.62	28.36	7.46
C27H37BR2N5	591.44	589.1417	54.83	6.31	27.02	11.84
C27H38BR1N1	456.51	455.2188	71.04	8.39	17.50	3.07
C27H38BR1N3	484.53	483.2249	66.93	7.91	16.49	8.67
C27H38BR1N5	512.54	511.2311	63.27	7.47	15.59	13.66
C27H38BR2N2	550.43	548.1403	58.92	6.96	29.03	5.09
C27H38BR2N4	578.44	576.1464	56.06	6.62	27.63	9.69
C27H38BR2N6	606.45	604.1526	53.48	6.32	26.35	13.86
C27H39BR1N2	471.53	470.2297	68.78	8.34	16.95	5.94
C27H39BR1N4	499.54	498.2358	64.92	7.87	16.00	11.22
C27H39BR1N6	527.56	526.2420	61.47	7.45	15.15	15.93
C27H39BR2N1	537.43	535.1450	60.34	7.31	29.74	2.61
C27H39BR2N3	565.44	563.1512	57.35	6.95	28.26	7.43
C27H39BR2N5	593.45	591.1573	54.65	6.62	26.93	11.80
C27H40BR1N1	458.53	457.2344	70.73	8.79	17.43	3.05
C27H40BR1N3	486.54	485.2406	66.65	8.29	16.42	8.64
C27H40BR1N5	514.56	513.2467	63.03	7.84	15.53	13.61
C27H40BR2N2	552.44	550.1559	58.70	7.30	28.93	5.07
C27H40BR2N4	580.45	578.1621	55.87	6.95	27.53	9.65
C27H40BR2N6	608.47	606.1682	53.30	6.63	26.26	13.81
C27H41BR1N2	473.55	472.2453	68.48	8.73	16.87	5.92
C27H41BR1N4	501.56	500.2515	64.66	8.24	15.93	11.17
C27H41BR1N6	529.57	528.2576	61.24	7.80	15.09	15.87
C27H41BR2N1	539.44	537.1607	60.12	7.66	29.62	2.60
C27H41BR2N3	567.46	565.1668	57.15	7.28	28.16	7.40
C27H41BR2N5	595.47	593.1730	54.46	6.94	26.84	11.76
C27H42BR1N1	460.55	459.2501	70.42	9.19	17.35	3.04
C27H42BR1N3	488.56	487.2562	66.38	8.67	16.36	8.60
C27H42BR1N5	516.57	515.2624	62.78	8.20	15.47	13.56
C27H42BR2N2	554.46	552.1716	58.49	7.64	28.82	5.05
C27H42BR2N4	582.47	580.1777	55.68	7.27	27.44	9.62
C27H42BR2N6	610.48	608.1839	53.12	6.93	26.18	13.77
C27H43BR1N2	475.56	474.2610	68.19	9.11	16.80	5.89
C27H43BR1N4	503.57	502.2671	64.40	8.61	15.87	11.13
C27H43BR1N6	531.59	530.2733	61.01	8.15	15.03	15.81
C27H43BR2N1	541.46	539.1763	59.89	8.00	29.51	2.59

C27H43BR2N3	569.47	567.1825	56.95	7.61	28.06	7.38
C27H43BR2N5	597.49	595.1886	54.28	7.25	26.75	11.72
C27H44BR1N1	462.56	461.2657	70.11	9.59	17.27	3.03
C27H44BR1N3	490.58	489.2719	66.11	9.04	16.29	8.57
C27H44BR1N5	518.59	517.2780	62.54	8.55	15.41	13.50
C27H44BR2N2	556.47	554.1872	58.28	7.97	28.72	5.03
C27H44BR2N4	584.49	582.1934	55.48	7.59	27.34	9.59
C27H44BR2N6	612.50	610.1995	52.95	7.24	26.09	13.72
C27H45BR1N2	477.58	476.2766	67.91	9.50	16.73	5.87
C27H45BR1N4	505.59	504.2828	64.14	8.97	15.80	11.08
C27H45BR1N6	533.60	532.2889	60.78	8.50	14.97	15.75
C27H45BR2N1	543.47	541.1920	59.67	8.35	29.40	2.58
C27H45BR2N3	571.49	569.1981	56.75	7.94	27.96	7.35
C27H45BR2N5	599.50	597.2043	54.10	7.57	26.66	11.68
C27H46BR1N1	464.58	463.2814	69.81	9.98	17.20	3.01
C27H46BR1N3	492.59	491.2875	65.84	9.41	16.22	8.53
C27H46BR1N5	520.61	519.2937	62.29	8.91	15.35	13.45
C27H46BR2N2	558.49	556.2029	58.07	8.30	28.61	5.02
C27H46BR2N4	586.50	584.2090	55.29	7.91	27.25	9.55
C27H46BR2N6	614.52	612.2152	52.77	7.55	26.01	13.68
C27H47BR1N2	479.59	478.2923	67.62	9.88	16.66	5.84
C27H47BR1N4	507.61	506.2984	63.89	9.33	15.74	11.04
C27H47BR1N6	535.62	534.3046	60.55	8.84	14.92	15.69
C27H47BR2N1	545.49	543.2076	59.45	8.68	29.30	2.57
C27H47BR2N3	573.50	571.2138	56.55	8.26	27.87	7.33
C27H47BR2N5	601.52	599.2199	53.91	7.88	26.57	11.64
C27H48BR1N1	466.59	465.2970	69.50	10.37	17.12	3.00
C27H48BR1N3	494.61	493.3032	65.57	9.78	16.16	8.50
C27H48BR1N5	522.62	521.3093	62.05	9.26	15.29	13.40
C27H48BR2N2	560.51	558.2185	57.86	8.63	28.51	5.00
C27H48BR2N4	588.52	586.2247	55.10	8.22	27.15	9.52
C27H48BR2N6	616.53	614.2308	52.60	7.85	25.92	13.63
C27H49BR1N2	481.61	480.3079	67.34	10.26	16.59	5.82
C27H49BR1N4	509.62	508.3141	63.64	9.69	15.68	10.99
C27H49BR1N6	537.64	536.3202	60.32	9.19	14.86	15.63
C27H49BR2N1	547.51	545.2233	59.23	9.02	29.19	2.56
C27H49BR2N3	575.52	573.2294	56.35	8.58	27.77	7.30
C27H49BR2N5	603.53	601.2356	53.73	8.18	26.48	11.60
C27H50BR1N1	468.61	467.3127	69.20	10.75	17.05	2.99
C27H50BR1N3	496.62	495.3188	65.30	10.15	16.09	8.46
C27H50BR1N5	524.64	523.3250	61.81	9.61	15.23	13.35
C27H50BR2N2	562.52	560.2342	57.65	8.96	28.41	4.98
C27H50BR2N4	590.53	588.2403	54.92	8.53	27.06	9.49
C27H50BR2N6	618.55	616.2465	52.43	8.15	25.84	13.59
C27H51BR1N2	483.62	482.3236	67.06	10.63	16.52	5.79
C27H51BR1N4	511.64	510.3297	63.38	10.05	15.62	10.95
C27H51BR1N6	539.65	538.3359	60.09	9.53	14.81	15.57
C27H51BR2N1	549.52	547.2389	59.02	9.35	29.08	2.55
C27H51BR2N3	577.54	575.2451	56.15	8.90	27.67	7.28
C27H51BR2N5	605.55	603.2512	53.55	8.49	26.39	11.57
C27H52BR1N1	470.63	469.3283	68.91	11.14	16.98	2.98
C27H52BR1N3	498.64	497.3345	65.04	10.51	16.02	8.43
C27H52BR1N5	526.65	525.3406	61.58	9.95	15.17	13.30
C27H52BR2N2	564.54	562.2498	57.45	9.28	28.31	4.96
C27H52BR2N4	592.55	590.2560	54.73	8.85	26.97	9.46
C27H52BR2N6	620.56	618.2621	52.26	8.45	25.75	13.54
C27H53BR1N2	485.64	484.3392	66.78	11.00	16.45	5.77
C27H53BR1N4	513.65	512.3454	63.14	10.40	15.56	10.91
C27H53BR1N6	541.67	540.3515	59.87	9.86	14.75	15.52
C27H53BR2N1	551.54	549.2546	58.80	9.69	28.97	2.54

C27H53BR2N3	579.55	577.2607	55.96	9.22	27.57	7.25
C27H53BR2N5	607.56	605.2669	53.38	8.79	26.30	11.53
C27H54BR1N1	472.64	471.3440	68.61	11.52	16.91	2.96
C27H54BR1N3	500.66	499.3501	64.78	10.87	15.96	8.39
C27H54BR1N5	528.67	527.3563	61.34	10.30	15.11	13.25
C27H54BR2N2	566.55	564.2655	57.24	9.61	28.21	4.94
C27H54BR2N4	594.57	592.2716	54.54	9.15	26.88	9.42
C27H54BR2N6	622.58	620.2778	52.09	8.74	25.67	13.50
C27H55BR1N2	487.66	486.3549	66.50	11.37	16.39	5.74
C27H55BR1N4	515.67	514.3610	62.89	10.75	15.50	10.86
C27H55BR1N6	543.68	542.3672	59.65	10.20	14.70	15.46
C27H55BR2N1	553.55	551.2702	58.59	10.01	28.87	2.53
C27H55BR2N3	581.57	579.2764	55.76	9.53	27.48	7.23
C27H55BR2N5	609.58	607.2825	53.20	9.09	26.22	11.49
C27H56BR1N1	474.66	473.3596	68.32	11.89	16.83	2.95
C27H56BR1N3	502.67	501.3658	64.52	11.23	15.90	8.36
C27H56BR1N5	530.68	529.3719	61.11	10.64	15.06	13.20
C27H56BR2N2	568.57	566.2811	57.04	9.93	28.11	4.93
C27H56BR2N4	596.58	594.2873	54.36	9.46	26.79	9.39
C27H56BR2N6	624.60	622.2934	51.92	9.04	25.59	13.46
C27H57BR1N2	489.67	488.3705	66.23	11.73	16.32	5.72
C27H57BR1N4	517.69	516.3767	62.64	11.10	15.43	10.82
C27H57BR1N6	545.70	544.3828	59.43	10.53	14.64	15.40
C27H57BR2N3	583.58	581.2920	55.57	9.85	27.38	7.20
C27H57BR2N5	611.60	609.2982	53.03	9.39	26.13	11.45
C27H58BR1N3	504.69	503.3814	64.26	11.58	15.83	8.33
C27H58BR1N5	532.70	531.3876	60.88	10.97	15.00	13.15
C27H58BR2N4	598.60	596.3029	54.18	9.77	26.70	9.36
C27H58BR2N6	626.61	624.3090	51.75	9.33	25.50	13.41
C27H59BR1N4	519.70	518.3923	62.40	11.44	15.37	10.78
C27H59BR1N6	547.72	546.3985	59.21	10.86	14.59	15.34
C27H59BR2N5	613.61	611.3138	52.85	9.69	26.04	11.41
C27H60BR1N5	534.72	533.4032	60.65	11.31	14.94	13.10
C27H60BR2N6	628.63	626.3247	51.59	9.62	25.42	13.37
C27H61BR1N6	549.73	548.4141	58.99	11.18	14.54	15.29
C28H14BR1N1	444.33	443.0310	75.69	3.18	17.98	3.15
C28H14BR1N3	472.35	471.0372	71.20	2.99	16.92	8.90
C28H14BR1N5	500.36	499.0433	67.21	2.82	15.97	14.00
C28H14BR2N2	538.25	535.9525	62.48	2.62	29.69	5.20
C28H14BR2N4	566.26	563.9586	59.39	2.49	28.22	9.89
C28H14BR2N6	594.27	591.9648	56.59	2.37	26.89	14.14
C28H15BR1N2	459.35	458.0419	73.21	3.29	17.40	6.10
C28H15BR1N4	487.36	486.0481	69.01	3.10	16.40	11.50
C28H15BR1N6	515.38	514.0542	65.26	2.93	15.50	16.31
C28H15BR2N1	525.25	522.9572	64.03	2.88	30.43	2.67
C28H15BR2N3	553.26	550.9634	60.79	2.73	28.88	7.60
C28H15BR2N5	581.27	578.9695	57.86	2.60	27.49	12.05
C28H16BR1N1	446.35	445.0467	75.35	3.61	17.90	3.14
C28H16BR1N3	474.36	473.0528	70.90	3.40	16.84	8.86
C28H16BR1N5	502.38	501.0590	66.94	3.21	15.91	13.94
C28H16BR2N2	540.26	537.9681	62.25	2.99	29.58	5.19
C28H16BR2N4	568.27	565.9743	59.18	2.84	28.12	9.86
C28H16BR2N6	596.29	593.9804	56.40	2.70	26.80	14.09
C28H17BR1N2	461.37	460.0576	72.90	3.71	17.32	6.07
C28H17BR1N4	489.38	488.0637	68.72	3.50	16.33	11.45
C28H17BR1N6	517.39	516.0699	65.00	3.31	15.44	16.24
C28H17BR2N1	527.26	524.9729	63.78	3.25	30.31	2.66
C28H17BR2N3	555.28	552.9790	60.57	3.09	28.78	7.57
C28H17BR2N5	583.29	580.9852	57.66	2.94	27.40	12.01
C28H18BR1N1	448.37	447.0623	75.01	4.05	17.82	3.12

C28H18BR1N3	476.38	475.0685	70.60	3.81	16.77	8.82
C28H18BR1N5	504.39	503.0746	66.68	3.60	15.84	13.88
C28H18BR2N2	542.28	539.9838	62.02	3.35	29.47	5.17
C28H18BR2N4	570.29	567.9899	58.97	3.18	28.02	9.82
C28H18BR2N6	598.30	595.9961	56.21	3.03	26.71	14.05
C28H19BR1N2	463.38	462.0732	72.58	4.13	17.24	6.05
C28H19BR1N4	491.39	490.0794	68.44	3.90	16.26	11.40
C28H19BR1N6	519.41	518.0855	64.75	3.69	15.38	16.18
C28H19BR2N1	529.28	526.9885	63.54	3.62	30.19	2.65
C28H19BR2N3	557.29	554.9947	60.35	3.44	28.68	7.54
C28H19BR2N5	585.31	583.0008	57.46	3.27	27.30	11.97
C28H20BR1N1	450.38	449.0780	74.67	4.48	17.74	3.11
C28H20BR1N3	478.40	477.0841	70.30	4.21	16.70	8.78
C28H20BR1N5	506.41	505.0903	66.41	3.98	15.78	13.83
C28H20BR2N2	544.29	541.9994	61.79	3.70	29.36	5.15
C28H20BR2N4	572.31	570.0056	58.76	3.52	27.92	9.79
C28H20BR2N6	600.32	598.0117	56.02	3.36	26.62	14.00
C28H21BR1N2	465.40	464.0889	72.26	4.55	17.17	6.02
C28H21BR1N4	493.41	492.0950	68.16	4.29	16.19	11.36
C28H21BR1N6	521.42	520.1011	64.50	4.06	15.32	16.12
C28H21BR2N1	531.29	529.0042	63.30	3.98	30.08	2.64
C28H21BR2N3	559.31	557.0103	60.13	3.78	28.57	7.51
C28H21BR2N5	587.32	585.0165	57.26	3.60	27.21	11.92
C28H22BR1N1	452.40	451.0936	74.34	4.90	17.66	3.10
C28H22BR1N3	480.41	479.0998	70.01	4.62	16.63	8.75
C28H22BR1N5	508.43	507.1059	66.15	4.36	15.72	13.77
C28H22BR2N2	546.31	544.0151	61.56	4.06	29.25	5.13
C28H22BR2N4	574.32	572.0212	58.56	3.86	27.83	9.76
C28H22BR2N6	602.34	600.0274	55.83	3.68	26.53	13.95
C28H23BR1N2	467.41	466.1045	71.95	4.96	17.09	5.99
C28H23BR1N4	495.43	494.1107	67.88	4.68	16.13	11.31
C28H23BR1N6	523.44	522.1168	64.25	4.43	15.27	16.06
C28H23BR2N1	533.31	531.0198	63.06	4.35	29.97	2.63
C28H23BR2N3	561.32	559.0260	59.91	4.13	28.47	7.49
C28H23BR2N5	589.34	587.0321	57.07	3.93	27.12	11.88
C28H24BR1N1	454.41	453.1093	74.01	5.32	17.58	3.08
C28H24BR1N3	482.43	481.1154	69.71	5.01	16.56	8.71
C28H24BR1N5	510.44	509.1215	65.89	4.74	15.65	13.72
C28H24BR2N2	548.32	546.0307	61.33	4.41	29.14	5.11
C28H24BR2N4	576.34	574.0369	58.35	4.20	27.73	9.72
C28H24BR2N6	604.35	602.0430	55.65	4.00	26.44	13.91
C28H25BR1N2	469.43	468.1202	71.64	5.37	17.02	5.97
C28H25BR1N4	497.44	496.1263	67.61	5.07	16.06	11.26
C28H25BR1N6	525.46	524.1324	64.00	4.80	15.21	15.99
C28H25BR2N1	535.33	533.0355	62.82	4.71	29.85	2.62
C28H25BR2N3	563.34	561.0416	59.70	4.47	28.37	7.46
C28H25BR2N5	591.35	589.0478	56.87	4.26	27.02	11.84
C28H26BR1N1	456.43	455.1249	73.68	5.74	17.51	3.07
C28H26BR1N3	484.44	483.1311	69.42	5.41	16.49	8.67
C28H26BR1N5	512.46	511.1372	65.63	5.11	15.59	13.67
C28H26BR2N2	550.34	548.0464	61.11	4.76	29.04	5.09
C28H26BR2N4	578.35	576.0525	58.15	4.53	27.63	9.69
C28H26BR2N6	606.37	604.0587	55.46	4.32	26.35	13.86
C28H27BR1N2	471.44	470.1358	71.34	5.77	16.95	5.94
C28H27BR1N4	499.46	498.1419	67.34	5.45	16.00	11.22
C28H27BR1N6	527.47	526.1481	63.76	5.16	15.15	15.93
C28H27BR2N1	537.34	535.0511	62.59	5.06	29.74	2.61
C28H27BR2N3	565.36	563.0573	59.49	4.81	28.27	7.43
C28H27BR2N5	593.37	591.0634	56.68	4.59	26.93	11.80
C28H28BR1N1	458.45	457.1406	73.36	6.16	17.43	3.06

C28H28BR1N3	486.46	485.1467	69.13	5.80	16.43	8.64
C28H28BR1N5	514.47	513.1528	65.37	5.49	15.53	13.61
C28H28BR2N2	552.36	550.0620	60.89	5.11	28.93	5.07
C28H28BR2N4	580.37	578.0682	57.95	4.86	27.54	9.65
C28H28BR2N6	608.38	606.0743	55.28	4.64	26.27	13.81
C28H29BR1N2	473.46	472.1515	71.03	6.17	16.88	5.92
C28H29BR1N4	501.47	500.1576	67.06	5.83	15.93	11.17
C28H29BR1N6	529.49	528.1637	63.52	5.52	15.09	15.87
C28H29BR2N1	539.36	537.0668	62.35	5.42	29.63	2.60
C28H29BR2N3	567.37	565.0729	59.28	5.15	28.17	7.41
C28H29BR2N5	595.38	593.0791	56.49	4.91	26.84	11.76
C28H30BR1N1	460.46	459.1562	73.04	6.57	17.35	3.04
C28H30BR1N3	488.48	487.1623	68.85	6.19	16.36	8.60
C28H30BR1N5	516.49	515.1685	65.12	5.85	15.47	13.56
C28H30BR2N2	554.37	552.0777	60.67	5.45	28.83	5.05
C28H30BR2N4	582.39	580.0838	57.75	5.19	27.44	9.62
C28H30BR2N6	610.40	608.0900	55.10	4.95	26.18	13.77
C28H31BR1N2	475.48	474.1671	70.73	6.57	16.81	5.89
C28H31BR1N4	503.49	502.1732	66.80	6.21	15.87	11.13
C28H31BR1N6	531.50	530.1794	63.28	5.88	15.03	15.81
C28H31BR2N1	541.37	539.0824	62.12	5.77	29.52	2.59
C28H31BR2N3	569.39	567.0886	59.07	5.49	28.07	7.38
C28H31BR2N5	597.40	595.0947	56.30	5.23	26.75	11.72
C28H32BR1N1	462.48	461.1719	72.72	6.97	17.28	3.03
C28H32BR1N3	490.49	489.1780	68.57	6.58	16.29	8.57
C28H32BR1N5	518.50	517.1841	64.86	6.22	15.41	13.51
C28H32BR2N2	556.39	554.0933	60.45	5.80	28.72	5.03
C28H32BR2N4	584.40	582.0995	57.55	5.52	27.35	9.59
C28H32BR2N6	612.42	610.1056	54.92	5.27	26.09	13.72
C28H33BR1N2	477.49	476.1827	70.43	6.97	16.73	5.87
C28H33BR1N4	505.51	504.1889	66.53	6.58	15.81	11.08
C28H33BR1N6	533.52	532.1950	63.04	6.23	14.98	15.75
C28H33BR2N1	543.39	541.0981	61.89	6.12	29.41	2.58
C28H33BR2N3	571.40	569.1042	58.86	5.82	27.97	7.35
C28H33BR2N5	599.42	597.1104	56.11	5.55	26.66	11.68
C28H34BR1N1	464.49	463.1875	72.40	7.38	17.20	3.02
C28H34BR1N3	492.51	491.1936	68.29	6.96	16.22	8.53
C28H34BR1N5	520.52	519.1998	64.61	6.58	15.35	13.45
C28H34BR2N2	558.40	556.1090	60.23	6.14	28.62	5.02
C28H34BR2N4	586.42	584.1151	57.35	5.84	27.25	9.55
C28H34BR2N6	614.43	612.1213	54.74	5.58	26.01	13.68
C28H35BR1N2	479.51	478.1984	70.14	7.36	16.66	5.84
C28H35BR1N4	507.52	506.2045	66.27	6.95	15.74	11.04
C28H35BR1N6	535.54	534.2107	62.80	6.59	14.92	15.69
C28H35BR2N1	545.41	543.1137	61.66	6.47	29.30	2.57
C28H35BR2N3	573.42	571.1199	58.65	6.15	27.87	7.33
C28H35BR2N5	601.43	599.1260	55.92	5.87	26.57	11.64
C28H36BR1N1	466.51	465.2031	72.09	7.78	17.13	3.00
C28H36BR1N3	494.52	493.2093	68.01	7.34	16.16	8.50
C28H36BR1N5	522.54	521.2154	64.36	6.94	15.29	13.40
C28H36BR2N2	560.42	558.1246	60.01	6.47	28.52	5.00
C28H36BR2N4	588.43	586.1308	57.15	6.17	27.16	9.52
C28H36BR2N6	616.45	614.1369	54.56	5.89	25.92	13.63
C28H37BR1N2	481.52	480.2140	69.84	7.75	16.59	5.82
C28H37BR1N4	509.54	508.2202	66.00	7.32	15.68	11.00
C28H37BR1N6	537.55	536.2263	62.56	6.94	14.86	15.63
C28H37BR2N1	547.42	545.1294	61.44	6.81	29.19	2.56
C28H37BR2N3	575.44	573.1355	58.44	6.48	27.77	7.30
C28H37BR2N5	603.45	601.1417	55.73	6.18	26.48	11.61
C28H38BR1N1	468.53	467.2188	71.78	8.18	17.05	2.99

C28H38BR1N3	496.54	495.2249	67.73	7.71	16.09	8.46
C28H38BR1N5	524.55	523.2311	64.11	7.30	15.23	13.35
C28H38BR2N2	562.44	560.1403	59.80	6.81	28.41	4.98
C28H38BR2N4	590.45	588.1464	56.96	6.49	27.07	9.49
C28H38BR2N6	618.46	616.1526	54.38	6.19	25.84	13.59
C28H39BR1N2	483.54	482.2297	69.55	8.13	16.52	5.79
C28H39BR1N4	511.55	510.2358	65.74	7.68	15.62	10.95
C28H39BR1N6	539.57	538.2420	62.33	7.29	14.81	15.58
C28H39BR2N1	549.44	547.1450	61.21	7.15	29.09	2.55
C28H39BR2N3	577.45	575.1512	58.24	6.81	27.67	7.28
C28H39BR2N5	605.46	603.1573	55.55	6.49	26.39	11.57
C28H40BR1N1	470.54	469.2344	71.47	8.57	16.98	2.98
C28H40BR1N3	498.56	497.2406	67.46	8.09	16.03	8.43
C28H40BR1N5	526.57	525.2467	63.87	7.66	15.17	13.30
C28H40BR2N2	564.45	562.1559	59.58	7.14	28.31	4.96
C28H40BR2N4	592.47	590.1621	56.76	6.81	26.97	9.46
C28H40BR2N6	620.48	618.1682	54.20	6.50	25.76	13.54
C28H41BR1N2	485.56	484.2453	69.26	8.51	16.46	5.77
C28H41BR1N4	513.57	512.2515	65.49	8.05	15.56	10.91
C28H41BR1N6	541.58	540.2576	62.10	7.63	14.75	15.52
C28H41BR2N1	551.45	549.1607	60.99	7.49	28.98	2.54
C28H41BR2N3	579.47	577.1668	58.04	7.13	27.58	7.25
C28H41BR2N5	607.48	605.1730	55.36	6.80	26.31	11.53
C28H42BR1N1	472.56	471.2501	71.17	8.96	16.91	2.96
C28H42BR1N3	500.57	499.2562	67.19	8.46	15.96	8.39
C28H42BR1N5	528.58	527.2624	63.63	8.01	15.12	13.25
C28H42BR2N2	566.47	564.1716	59.37	7.47	28.21	4.95
C28H42BR2N4	594.48	592.1777	56.57	7.12	26.88	9.42
C28H42BR2N6	622.50	620.1839	54.03	6.80	25.67	13.50
C28H43BR1N2	487.57	486.2610	68.98	8.89	16.39	5.75
C28H43BR1N4	515.59	514.2671	65.23	8.41	15.50	10.87
C28H43BR1N6	543.60	542.2733	61.87	7.97	14.70	15.46
C28H43BR2N1	553.47	551.1763	60.76	7.83	28.87	2.53
C28H43BR2N3	581.48	579.1825	57.84	7.45	27.48	7.23
C28H43BR2N5	609.50	607.1886	55.18	7.11	26.22	11.49
C28H44BR1N1	474.57	473.2657	70.87	9.35	16.84	2.95
C28H44BR1N3	502.59	501.2719	66.92	8.82	15.90	8.36
C28H44BR1N5	530.60	529.2780	63.38	8.36	15.06	13.20
C28H44BR2N2	568.48	566.1872	59.16	7.80	28.11	4.93
C28H44BR2N4	596.50	594.1934	56.38	7.44	26.79	9.39
C28H44BR2N6	624.51	622.1995	53.85	7.10	25.59	13.46
C28H45BR1N2	489.59	488.2766	68.69	9.26	16.32	5.72
C28H45BR1N4	517.60	516.2828	64.98	8.76	15.44	10.82
C28H45BR1N6	545.62	544.2889	61.64	8.31	14.64	15.40
C28H45BR2N1	555.49	553.1920	60.54	8.17	28.77	2.52
C28H45BR2N3	583.50	581.1981	57.64	7.77	27.39	7.20
C28H45BR2N5	611.51	609.2043	55.00	7.42	26.13	11.45
C28H46BR1N1	476.59	475.2814	70.57	9.73	16.77	2.94
C28H46BR1N3	504.60	503.2875	66.65	9.19	15.84	8.33
C28H46BR1N5	532.62	531.2937	63.14	8.71	15.00	13.15
C28H46BR2N2	570.50	568.2029	58.95	8.13	28.01	4.91
C28H46BR2N4	598.51	596.2090	56.19	7.75	26.70	9.36
C28H46BR2N6	626.53	624.2152	53.68	7.40	25.51	13.41
C28H47BR1N2	491.60	490.2923	68.41	9.64	16.25	5.70
C28H47BR1N4	519.62	518.2984	64.72	9.12	15.38	10.78
C28H47BR1N6	547.63	546.3046	61.41	8.65	14.59	15.35
C28H47BR2N1	557.50	555.2076	60.32	8.50	28.67	2.51
C28H47BR2N3	585.51	583.2138	57.44	8.09	27.29	7.18
C28H47BR2N5	613.53	611.2199	54.82	7.72	26.05	11.41
C28H48BR1N1	478.61	477.2970	70.27	10.11	16.70	2.93

C28H48BR1N3	506.62	505.3032	66.38	9.55	15.77	8.29
C28H48BR1N5	534.63	533.3093	62.91	9.05	14.95	13.10
C28H48BR2N2	572.52	570.2185	58.74	8.45	27.91	4.89
C28H48BR2N4	600.53	598.2247	56.00	8.06	26.61	9.33
C28H48BR2N6	628.54	626.2308	53.51	7.70	25.43	13.37
C28H49BR1N2	493.62	492.3079	68.13	10.01	16.19	5.68
C28H49BR1N4	521.63	520.3141	64.47	9.47	15.32	10.74
C28H49BR1N6	549.65	548.3202	61.19	8.99	14.54	15.29
C28H49BR2N1	559.52	557.2233	60.11	8.83	28.56	2.50
C28H49BR2N3	587.53	585.2294	57.24	8.41	27.20	7.15
C28H49BR2N5	615.54	613.2356	54.64	8.02	25.96	11.38
C28H50BR1N1	480.62	479.3127	69.97	10.49	16.63	2.91
C28H50BR1N3	508.63	507.3188	66.12	9.91	15.71	8.26
C28H50BR1N5	536.65	535.3250	62.67	9.39	14.89	13.05
C28H50BR2N2	574.53	572.2342	58.54	8.77	27.82	4.88
C28H50BR2N4	602.55	600.2403	55.82	8.36	26.52	9.30
C28H50BR2N6	630.56	628.2465	53.34	7.99	25.34	13.33
C28H51BR1N2	495.64	494.3236	67.85	10.37	16.12	5.65
C28H51BR1N4	523.65	522.3297	64.22	9.82	15.26	10.70
C28H51BR1N6	551.66	550.3359	60.96	9.32	14.48	15.23
C28H51BR2N1	561.53	559.2389	59.89	9.15	28.46	2.49
C28H51BR2N3	589.55	587.2451	57.05	8.72	27.11	7.13
C28H51BR2N5	617.56	615.2512	54.46	8.32	25.88	11.34
C28H52BR1N1	482.64	481.3283	69.68	10.86	16.56	2.90
C28H52BR1N3	510.65	509.3345	65.86	10.26	15.65	8.23
C28H52BR1N5	538.66	537.3406	62.43	9.73	14.83	13.00
C28H52BR2N2	576.55	574.2498	58.33	9.09	27.72	4.86
C28H52BR2N4	604.56	602.2560	55.63	8.67	26.43	9.27
C28H52BR2N6	632.57	630.2621	53.17	8.29	25.26	13.29
C28H53BR1N2	497.65	496.3392	67.58	10.73	16.06	5.63
C28H53BR1N4	525.67	524.3454	63.98	10.16	15.20	10.66
C28H53BR1N6	553.68	552.3515	60.74	9.65	14.43	15.18
C28H53BR2N1	563.55	561.2546	59.68	9.48	28.36	2.49
C28H53BR2N3	591.56	589.2607	56.85	9.03	27.01	7.10
C28H53BR2N5	619.58	617.2669	54.28	8.62	25.79	11.30
C28H54BR1N1	484.65	483.3440	69.39	11.23	16.49	2.89
C28H54BR1N3	512.67	511.3501	65.60	10.62	15.59	8.20
C28H54BR1N5	540.68	539.3563	62.20	10.07	14.78	12.95
C28H54BR2N2	578.56	576.2655	58.13	9.41	27.62	4.84
C28H54BR2N4	606.58	604.2716	55.44	8.97	26.35	9.24
C28H54BR2N6	634.59	632.2778	53.00	8.58	25.18	13.24
C28H55BR1N2	499.67	498.3549	67.31	11.10	15.99	5.61
C28H55BR1N4	527.68	526.3610	63.73	10.51	15.14	10.62
C28H55BR1N6	555.69	554.3672	60.52	9.98	14.38	15.12
C28H55BR2N1	565.57	563.2702	59.46	9.80	28.26	2.48
C28H55BR2N3	593.58	591.2764	56.66	9.34	26.92	7.08
C28H55BR2N5	621.59	619.2825	54.10	8.92	25.71	11.27
C28H56BR1N1	486.67	485.3596	69.10	11.60	16.42	2.88
C28H56BR1N3	514.68	513.3658	65.34	10.97	15.52	8.16
C28H56BR1N5	542.70	541.3719	61.97	10.40	14.72	12.90
C28H56BR2N2	580.58	578.2811	57.93	9.72	27.53	4.83
C28H56BR2N4	608.59	606.2873	55.26	9.27	26.26	9.21
C28H56BR2N6	636.61	634.2934	52.83	8.87	25.10	13.20
C28H57BR1N2	501.68	500.3705	67.04	11.45	15.93	5.58
C28H57BR1N4	529.70	528.3767	63.49	10.85	15.08	10.58
C28H57BR1N6	557.71	556.3828	60.30	10.30	14.33	15.07
C28H57BR2N1	567.58	565.2859	59.25	10.12	28.16	2.47
C28H57BR2N3	595.59	593.2920	56.47	9.65	26.83	7.06
C28H57BR2N5	623.61	621.2982	53.93	9.21	25.63	11.23
C28H58BR1N1	488.69	487.3753	68.82	11.96	16.35	2.87

C28H58BR1N3	516.70	515.3814	65.09	11.31	15.46	8.13
C28H58BR1N5	544.71	543.3876	61.74	10.73	14.67	12.86
C28H58BR2N2	582.60	580.2968	57.73	10.03	27.43	4.81
C28H58BR2N4	610.61	608.3029	55.08	9.57	26.17	9.18
C28H58BR2N6	638.62	636.3090	52.66	9.15	25.02	13.16
C28H59BR1N2	503.70	502.3862	66.77	11.81	15.86	5.56
C28H59BR1N4	531.71	530.3923	63.25	11.18	15.03	10.54
C28H59BR1N6	559.73	558.3985	60.09	10.62	14.28	15.01
C28H59BR2N3	597.61	595.3077	56.28	9.95	26.74	7.03
C28H59BR2N5	625.62	623.3138	53.76	9.51	25.54	11.19
C28H60BR1N3	518.71	517.3971	64.84	11.66	15.40	8.10
C28H60BR1N5	546.73	545.4032	61.51	11.06	14.61	12.81
C28H60BR2N4	612.63	610.3186	54.90	9.87	26.09	9.15
C28H60BR2N6	640.64	638.3247	52.50	9.44	24.95	13.12
C28H61BR1N4	533.73	532.4080	63.01	11.52	14.97	10.50
C28H61BR1N6	561.74	560.4141	59.87	10.95	14.22	14.96
C28H61BR2N5	627.64	625.3294	53.58	9.80	25.46	11.16
C28H62BR1N5	548.74	547.4189	61.29	11.39	14.56	12.76
C28H62BR2N6	642.65	640.3403	52.33	9.72	24.87	13.08
C28H63BR1N6	563.76	562.4298	59.66	11.26	14.17	14.91
C29H14BR1N1	456.35	455.0310	76.33	3.09	17.51	3.07
C29H14BR1N3	484.36	483.0372	71.91	2.91	16.50	8.68
C29H14BR1N5	512.37	511.0433	67.98	2.75	15.59	13.67
C29H14BR2N2	550.26	547.9525	63.30	2.56	29.04	5.09
C29H14BR2N4	578.27	575.9586	60.24	2.44	27.64	9.69
C29H14BR2N6	606.28	603.9648	57.45	2.33	26.36	13.86
C29H15BR1N2	471.36	470.0419	73.90	3.21	16.95	5.94
C29H15BR1N4	499.37	498.0481	69.75	3.03	16.00	11.22
C29H15BR1N6	527.39	526.0542	66.05	2.87	15.15	15.94
C29H15BR2N1	537.26	534.9572	64.83	2.81	29.75	2.61
C29H15BR2N3	565.27	562.9634	61.62	2.67	28.27	7.43
C29H15BR2N5	593.28	590.9695	58.71	2.55	26.94	11.80
C29H16BR1N1	458.36	457.0467	75.99	3.52	17.43	3.06
C29H16BR1N3	486.37	485.0528	71.62	3.32	16.43	8.64
C29H16BR1N5	514.39	513.0590	67.72	3.14	15.53	13.61
C29H16BR2N2	552.27	549.9681	63.07	2.92	28.94	5.07
C29H16BR2N4	580.29	577.9743	60.03	2.78	27.54	9.66
C29H16BR2N6	608.30	605.9804	57.26	2.65	26.27	13.82
C29H17BR1N2	473.38	472.0576	73.58	3.62	16.88	5.92
C29H17BR1N4	501.39	500.0637	69.47	3.42	15.94	11.17
C29H17BR1N6	529.40	528.0699	65.80	3.24	15.09	15.87
C29H17BR2N1	539.27	536.9729	64.59	3.18	29.63	2.60
C29H17BR2N3	567.29	564.9790	61.40	3.02	28.17	7.41
C29H17BR2N5	595.30	592.9852	58.51	2.88	26.84	11.76
C29H18BR1N1	460.38	459.0623	75.66	3.94	17.36	3.04
C29H18BR1N3	488.39	487.0685	71.32	3.71	16.36	8.60
C29H18BR1N5	516.40	515.0746	67.45	3.51	15.47	13.56
C29H18BR2N2	554.29	551.9838	62.84	3.27	28.83	5.05
C29H18BR2N4	582.30	579.9899	59.82	3.12	27.44	9.62
C29H18BR2N6	610.32	607.9961	57.07	2.97	26.18	13.77
C29H19BR1N2	475.39	474.0732	73.27	4.03	16.81	5.89
C29H19BR1N4	503.41	502.0794	69.19	3.80	15.87	11.13
C29H19BR1N6	531.42	530.0855	65.55	3.60	15.04	15.81
C29H19BR2N1	541.29	538.9885	64.35	3.54	29.52	2.59
C29H19BR2N3	569.30	566.9947	61.18	3.36	28.07	7.38
C29H19BR2N5	597.32	595.0008	58.31	3.21	26.75	11.72
C29H20BR1N1	462.39	461.0780	75.33	4.36	17.28	3.03
C29H20BR1N3	490.41	489.0841	71.03	4.11	16.29	8.57
C29H20BR1N5	518.42	517.0903	67.19	3.89	15.41	13.51
C29H20BR2N2	556.30	553.9994	62.61	3.62	28.73	5.04

C29H20BR2N4	584.32	582.0056	59.61	3.45	27.35	9.59
C29H20BR2N6	612.33	610.0117	56.88	3.29	26.10	13.72
C29H21BR1N2	477.41	476.0889	72.96	4.43	16.74	5.87
C29H21BR1N4	505.42	504.0950	68.92	4.19	15.81	11.09
C29H21BR1N6	533.43	532.1011	65.30	3.97	14.98	15.75
C29H21BR2N1	543.31	541.0042	64.11	3.90	29.41	2.58
C29H21BR2N3	571.32	569.0103	60.97	3.71	27.97	7.35
C29H21BR2N5	599.33	597.0165	58.12	3.53	26.66	11.69
C29H22BR1N1	464.41	463.0936	75.00	4.77	17.21	3.02
C29H22BR1N3	492.42	491.0998	70.74	4.50	16.23	8.53
C29H22BR1N5	520.44	519.1059	66.93	4.26	15.35	13.46
C29H22BR2N2	558.32	556.0151	62.39	3.97	28.62	5.02
C29H22BR2N4	586.33	584.0212	59.41	3.78	27.26	9.56
C29H22BR2N6	614.35	612.0274	56.70	3.61	26.01	13.68
C29H23BR1N2	479.42	478.1045	72.65	4.84	16.67	5.84
C29H23BR1N4	507.44	506.1107	68.64	4.57	15.75	11.04
C29H23BR1N6	535.45	534.1168	65.05	4.33	14.92	15.70
C29H23BR2N1	545.32	543.0198	63.87	4.25	29.31	2.57
C29H23BR2N3	573.33	571.0260	60.75	4.04	27.87	7.33
C29H23BR2N5	601.35	599.0321	57.92	3.86	26.57	11.65
C29H24BR1N1	466.43	465.1093	74.68	5.19	17.13	3.00
C29H24BR1N3	494.44	493.1154	70.45	4.89	16.16	8.50
C29H24BR1N5	522.45	521.1215	66.67	4.63	15.29	13.40
C29H24BR2N2	560.34	558.0307	62.16	4.32	28.52	5.00
C29H24BR2N4	588.35	586.0369	59.20	4.11	27.16	9.52
C29H24BR2N6	616.36	614.0430	56.51	3.92	25.93	13.63
C29H25BR1N2	481.44	480.1202	72.35	5.23	16.60	5.82
C29H25BR1N4	509.45	508.1263	68.37	4.95	15.68	11.00
C29H25BR1N6	537.47	536.1324	64.81	4.69	14.87	15.64
C29H25BR2N1	547.34	545.0355	63.64	4.60	29.20	2.56
C29H25BR2N3	575.35	573.0416	60.54	4.38	27.78	7.30
C29H25BR2N5	603.36	601.0478	57.73	4.18	26.49	11.61
C29H26BR1N1	468.44	467.1249	74.36	5.59	17.06	2.99
C29H26BR1N3	496.45	495.1311	70.16	5.28	16.09	8.46
C29H26BR1N5	524.47	523.1372	66.41	5.00	15.24	13.35
C29H26BR2N2	562.35	560.0464	61.94	4.66	28.42	4.98
C29H26BR2N4	590.37	588.0525	59.00	4.44	27.07	9.49
C29H26BR2N6	618.38	616.0587	56.33	4.24	25.84	13.59
C29H27BR1N2	483.46	482.1358	72.05	5.63	16.53	5.79
C29H27BR1N4	511.47	510.1419	68.10	5.32	15.62	10.95
C29H27BR1N6	539.48	538.1481	64.57	5.04	14.81	15.58
C29H27BR2N1	549.35	547.0511	63.41	4.95	29.09	2.55
C29H27BR2N3	577.37	575.0573	60.33	4.71	27.68	7.28
C29H27BR2N5	605.38	603.0634	57.54	4.50	26.40	11.57
C29H28BR1N1	470.46	469.1406	74.04	6.00	16.98	2.98
C29H28BR1N3	498.47	497.1467	69.88	5.66	16.03	8.43
C29H28BR1N5	526.48	525.1528	66.16	5.36	15.18	13.30
C29H28BR2N2	564.37	562.0620	61.72	5.00	28.32	4.96
C29H28BR2N4	592.38	590.0682	58.80	4.76	26.98	9.46
C29H28BR2N6	620.39	618.0743	56.15	4.55	25.76	13.55
C29H29BR1N2	485.47	484.1515	71.75	6.02	16.46	5.77
C29H29BR1N4	513.49	512.1576	67.84	5.69	15.56	10.91
C29H29BR1N6	541.50	540.1637	64.33	5.40	14.76	15.52
C29H29BR2N1	551.37	549.0668	63.17	5.30	28.98	2.54
C29H29BR2N3	579.38	577.0729	60.12	5.05	27.58	7.25
C29H29BR2N5	607.40	605.0791	57.35	4.81	26.31	11.53
C29H30BR1N1	472.47	471.1562	73.72	6.40	16.91	2.96
C29H30BR1N3	500.49	499.1623	69.60	6.04	15.97	8.40
C29H30BR1N5	528.50	527.1685	65.91	5.72	15.12	13.25
C29H30BR2N2	566.38	564.0777	61.50	5.34	28.22	4.95

C29H30BR2N4	594.40	592.0838	58.60	5.09	26.89	9.43
C29H30BR2N6	622.41	620.0900	55.96	4.86	25.68	13.50
C29H31BR1N2	487.49	486.1671	71.45	6.41	16.39	5.75
C29H31BR1N4	515.50	514.1732	67.57	6.06	15.50	10.87
C29H31BR1N6	543.51	542.1794	64.09	5.75	14.70	15.46
C29H31BR2N1	553.39	551.0824	62.94	5.65	28.88	2.53
C29H31BR2N3	581.40	579.0886	59.91	5.37	27.49	7.23
C29H31BR2N5	609.41	607.0947	57.16	5.13	26.22	11.49
C29H32BR1N1	474.49	473.1719	73.41	6.80	16.84	2.95
C29H32BR1N3	502.50	501.1780	69.32	6.42	15.90	8.36
C29H32BR1N5	530.52	529.1841	65.66	6.08	15.06	13.20
C29H32BR2N2	568.40	566.0933	61.28	5.67	28.12	4.93
C29H32BR2N4	596.41	594.0995	58.40	5.41	26.79	9.39
C29H32BR2N6	624.43	622.1056	55.78	5.17	25.59	13.46
C29H33BR1N2	489.50	488.1827	71.16	6.80	16.32	5.72
C29H33BR1N4	517.52	516.1889	67.31	6.43	15.44	10.83
C29H33BR1N6	545.53	544.1950	63.85	6.10	14.65	15.41
C29H33BR2N1	555.40	553.0981	62.72	5.99	28.77	2.52
C29H33BR2N3	583.41	581.1042	59.70	5.70	27.39	7.20
C29H33BR2N5	611.43	609.1104	56.97	5.44	26.14	11.45
C29H34BR1N1	476.51	475.1875	73.10	7.19	16.77	2.94
C29H34BR1N3	504.52	503.1936	69.04	6.79	15.84	8.33
C29H34BR1N5	532.53	531.1998	65.41	6.44	15.00	13.15
C29H34BR2N2	570.42	568.1090	61.06	6.01	28.02	4.91
C29H34BR2N4	598.43	596.1151	58.21	5.73	26.70	9.36
C29H34BR2N6	626.44	624.1213	55.60	5.47	25.51	13.42
C29H35BR1N2	491.52	490.1984	70.87	7.18	16.26	5.70
C29H35BR1N4	519.53	518.2045	67.05	6.79	15.38	10.78
C29H35BR1N6	547.55	546.2107	63.62	6.44	14.59	15.35
C29H35BR2N1	557.42	555.1137	62.49	6.33	28.67	2.51
C29H35BR2N3	585.43	583.1199	59.50	6.03	27.30	7.18
C29H35BR2N5	613.44	611.1260	56.78	5.75	26.05	11.42
C29H36BR1N1	478.52	477.2031	72.79	7.58	16.70	2.93
C29H36BR1N3	506.53	505.2093	68.77	7.16	15.77	8.30
C29H36BR1N5	534.55	533.2154	65.16	6.79	14.95	13.10
C29H36BR2N2	572.43	570.1246	60.85	6.34	27.92	4.89
C29H36BR2N4	600.45	598.1308	58.01	6.04	26.61	9.33
C29H36BR2N6	628.46	626.1369	55.43	5.77	25.43	13.37
C29H37BR1N2	493.54	492.2140	70.58	7.56	16.19	5.68
C29H37BR1N4	521.55	520.2202	66.79	7.15	15.32	10.74
C29H37BR1N6	549.56	548.2263	63.38	6.79	14.54	15.29
C29H37BR2N1	559.43	557.1294	62.26	6.67	28.57	2.50
C29H37BR2N3	587.45	585.1355	59.29	6.35	27.20	7.15
C29H37BR2N5	615.46	613.1417	56.60	6.06	25.97	11.38
C29H38BR1N1	480.54	479.2188	72.49	7.97	16.63	2.91
C29H38BR1N3	508.55	507.2249	68.49	7.53	15.71	8.26
C29H38BR1N5	536.56	535.2311	64.92	7.14	14.89	13.05
C29H38BR2N2	574.45	572.1403	60.64	6.67	27.82	4.88
C29H38BR2N4	602.46	600.1464	57.82	6.36	26.53	9.30
C29H38BR2N6	630.47	628.1526	55.25	6.08	25.35	13.33
C29H39BR1N2	495.55	494.2297	70.29	7.93	16.12	5.65
C29H39BR1N4	523.56	522.2358	66.53	7.51	15.26	10.70
C29H39BR1N6	551.58	550.2420	63.15	7.13	14.49	15.24
C29H39BR2N1	561.45	559.1450	62.04	7.00	28.46	2.49
C29H39BR2N3	589.46	587.1512	59.09	6.67	27.11	7.13
C29H39BR2N5	617.48	615.1573	56.41	6.37	25.88	11.34
C29H40BR1N1	482.55	481.2344	72.18	8.36	16.56	2.90
C29H40BR1N3	510.57	509.2406	68.22	7.90	15.65	8.23
C29H40BR1N5	538.58	537.2467	64.67	7.49	14.84	13.00
C29H40BR2N2	576.46	574.1559	60.42	6.99	27.72	4.86

C29H40BR2N4	604.48	602.1621	57.62	6.67	26.44	9.27
C29H40BR2N6	632.49	630.1682	55.07	6.37	25.27	13.29
C29H41BR1N2	497.57	496.2453	70.01	8.31	16.06	5.63
C29H41BR1N4	525.58	524.2515	66.27	7.86	15.20	10.66
C29H41BR1N6	553.59	552.2576	62.92	7.47	14.43	15.18
C29H41BR2N1	563.46	561.1607	61.82	7.33	28.36	2.49
C29H41BR2N3	591.48	589.1668	58.89	6.99	27.02	7.10
C29H41BR2N5	619.49	617.1730	56.23	6.67	25.80	11.30
C29H42BR1N1	484.57	483.2501	71.88	8.74	16.49	2.89
C29H42BR1N3	512.58	511.2562	67.95	8.26	15.59	8.20
C29H42BR1N5	540.60	539.2624	64.43	7.83	14.78	12.95
C29H42BR2N2	578.48	576.1716	60.21	7.32	27.63	4.84
C29H42BR2N4	606.49	604.1777	57.43	6.98	26.35	9.24
C29H42BR2N6	634.51	632.1839	54.90	6.67	25.19	13.24
C29H43BR1N2	499.58	498.2610	69.72	8.68	15.99	5.61
C29H43BR1N4	527.60	526.2671	66.02	8.22	15.14	10.62
C29H43BR1N6	555.61	554.2733	62.69	7.80	14.38	15.13
C29H43BR2N1	565.48	563.1763	61.60	7.66	28.26	2.48
C29H43BR2N3	593.49	591.1825	58.69	7.30	26.93	7.08
C29H43BR2N5	621.51	619.1886	56.04	6.97	25.71	11.27
C29H44BR1N1	486.58	485.2657	71.59	9.11	16.42	2.88
C29H44BR1N3	514.60	513.2719	67.69	8.62	15.53	8.17
C29H44BR1N5	542.61	541.2780	64.19	8.17	14.73	12.91
C29H44BR2N2	580.50	578.1872	60.00	7.64	27.53	4.83
C29H44BR2N4	608.51	606.1934	57.24	7.29	26.26	9.21
C29H44BR2N6	636.52	634.1995	54.72	6.97	25.11	13.20
C29H45BR1N2	501.60	500.2766	69.44	9.04	15.93	5.58
C29H45BR1N4	529.61	528.2828	65.77	8.56	15.09	10.58
C29H45BR1N6	557.63	556.2889	62.47	8.13	14.33	15.07
C29H45BR2N1	567.50	565.1920	61.38	7.99	28.16	2.47
C29H45BR2N3	595.51	593.1981	58.49	7.62	26.84	7.06
C29H45BR2N5	623.52	621.2043	55.86	7.27	25.63	11.23
C29H46BR1N1	488.60	487.2814	71.29	9.49	16.35	2.87
C29H46BR1N3	516.61	515.2875	67.42	8.98	15.47	8.13
C29H46BR1N5	544.63	543.2937	63.96	8.51	14.67	12.86
C29H46BR2N2	582.51	580.2029	59.80	7.96	27.43	4.81
C29H46BR2N4	610.52	608.2090	57.05	7.59	26.18	9.18
C29H46BR2N6	638.54	636.2152	54.55	7.26	25.03	13.16
C29H47BR1N2	503.62	502.2923	69.16	9.41	15.87	5.56
C29H47BR1N4	531.63	530.2984	65.52	8.91	15.03	10.54
C29H47BR1N6	559.64	558.3046	62.24	8.47	14.28	15.02
C29H47BR2N1	569.51	567.2076	61.16	8.32	28.06	2.46
C29H47BR2N3	597.53	595.2138	58.29	7.93	26.74	7.03
C29H47BR2N5	625.54	623.2199	55.68	7.57	25.55	11.20
C29H48BR1N1	490.62	489.2970	71.00	9.86	16.29	2.85
C29H48BR1N3	518.63	517.3032	67.16	9.33	15.41	8.10
C29H48BR1N5	546.64	545.3093	63.72	8.85	14.62	12.81
C29H48BR2N2	584.53	582.2185	59.59	8.28	27.34	4.79
C29H48BR2N4	612.54	610.2247	56.87	7.90	26.09	9.15
C29H48BR2N6	640.55	638.2308	54.38	7.55	24.95	13.12
C29H49BR1N2	505.63	504.3079	68.89	9.77	15.80	5.54
C29H49BR1N4	533.64	532.3141	65.27	9.26	14.97	10.50
C29H49BR1N6	561.66	560.3202	62.02	8.79	14.23	14.96
C29H49BR2N1	571.53	569.2233	60.95	8.64	27.96	2.45
C29H49BR2N3	599.54	597.2294	58.10	8.24	26.66	7.01
C29H49BR2N5	627.56	625.2356	55.50	7.87	25.47	11.16
C29H50BR1N1	492.63	491.3127	70.71	10.23	16.22	2.84
C29H50BR1N3	520.65	519.3188	66.90	9.68	15.35	8.07
C29H50BR1N5	548.66	547.3250	63.49	9.19	14.56	12.76
C29H50BR2N2	586.54	584.2342	59.39	8.59	27.25	4.78
C29H50BR2N4	614.56	612.2403	56.68	8.20	26.00	9.12
C29H50BR2N6	642.57	640.2465	54.21	7.84	24.87	13.08
C29H51BR1N2	507.65	506.3236	68.62	10.13	15.74	5.52
C29H51BR1N4	535.66	534.3297	65.03	9.60	14.92	10.46
C29H51BR1N6	563.67	562.3359	61.80	9.12	14.18	14.91
C29H51BR2N1	573.54	571.2389	60.73	8.96	27.86	2.44
C29H51BR2N3	601.56	599.2451	57.90	8.55	26.57	6.99
C29H51BR2N5	629.57	627.2512	55.33	8.17	25.38	11.12
C29H52BR1N1	494.65	493.3283	70.42	10.60	16.15	2.83
C29H52BR1N3	522.66	521.3345	66.64	10.03	15.29	8.04
C29H52BR1N5	550.68	549.3406	63.25	9.52	14.51	12.72
C29H52BR2N2	588.56	586.2498	59.18	8.91	27.15	4.76
C29H52BR2N4	616.57	614.2560	56.49	8.50	25.92	9.09
C29H52BR2N6	644.59	642.2621	54.04	8.13	24.79	13.04
C29H53BR1N2	509.66	508.3392	68.34	10.48	15.68	5.50
C29H53BR1N4	537.68	536.3454	64.78	9.94	14.86	10.42
C29H53BR1N6	565.69	564.3515	61.57	9.44	14.13	14.86
C29H53BR2N1	575.56	573.2546	60.52	9.28	27.77	2.43
C29H53BR2N3	603.57	601.2607	57.71	8.85	26.48	6.96
C29H53BR2N5	631.59	629.2669	55.15	8.46	25.30	11.09
C29H54BR1N1	496.66	495.3440	70.13	10.96	16.09	2.82
C29H54BR1N3	524.68	523.3501	66.39	10.37	15.23	8.01
C29H54BR1N5	552.69	551.3563	63.02	9.85	14.46	12.67
C29H54BR2N2	590.58	588.2655	58.98	9.22	27.06	4.74
C29H54BR2N4	618.59	616.2716	56.31	8.80	25.83	9.06
C29H54BR2N6	646.60	644.2778	53.87	8.42	24.72	13.00
C29H55BR1N2	511.68	510.3549	68.07	10.83	15.62	5.47
C29H55BR1N4	539.69	538.3610	64.54	10.27	14.81	10.38
C29H55BR1N6	567.71	566.3672	61.36	9.77	14.07	14.80
C29H55BR2N1	577.58	575.2702	60.31	9.60	27.67	2.43
C29H55BR2N3	605.59	603.2764	57.52	9.15	26.39	6.94
C29H55BR2N5	633.60	631.2825	54.97	8.75	25.22	11.05
C29H56BR1N1	498.68	497.3596	69.85	11.32	16.02	2.81
C29H56BR1N3	526.69	525.3658	66.13	10.72	15.17	7.98
C29H56BR1N5	554.71	553.3719	62.79	10.18	14.40	12.63
C29H56BR2N2	592.59	590.2811	58.78	9.53	26.97	4.73
C29H56BR2N4	620.60	618.2873	56.13	9.10	25.75	9.03
C29H56BR2N6	648.62	646.2934	53.70	8.70	24.64	12.96
C29H57BR1N2	513.70	512.3705	67.81	11.18	15.55	5.45
C29H57BR1N4	541.71	540.3767	64.30	10.61	14.75	10.34
C29H57BR1N6	569.72	568.3828	61.14	10.08	14.03	14.75
C29H57BR2N1	579.59	577.2859	60.10	9.91	27.57	2.42
C29H57BR2N3	607.61	605.2920	57.33	9.46	26.30	6.92
C29H57BR2N5	635.62	633.2982	54.80	9.04	25.14	11.02
C29H58BR1N1	500.70	499.3753	69.57	11.68	15.96	2.80
C29H58BR1N3	528.71	527.3814	65.88	11.06	15.11	7.95
C29H58BR1N5	556.72	555.3876	62.57	10.50	14.35	12.58
C29H58BR2N2	594.61	592.2968	58.58	9.83	26.88	4.71
C29H58BR2N4	622.62	620.3029	55.94	9.39	25.67	9.00
C29H58BR2N6	650.63	648.3090	53.54	8.99	24.56	12.92
C29H59BR1N2	515.71	514.3862	67.54	11.53	15.49	5.43
C29H59BR1N4	543.72	542.3923	64.06	10.94	14.70	10.30
C29H59BR1N6	571.74	570.3985	60.92	10.40	13.98	14.70
C29H59BR2N1	581.61	579.3015	59.89	10.23	27.48	2.41
C29H59BR2N3	609.62	607.3077	57.14	9.76	26.21	6.89
C29H59BR2N5	637.64	635.3138	54.63	9.33	25.06	10.98
C29H60BR1N1	502.71	501.3909	69.29	12.03	15.89	2.79
C29H60BR1N3	530.73	529.3971	65.63	11.40	15.06	7.92
C29H60BR1N5	558.74	557.4032	62.34	10.82	14.30	12.53
C29H60BR2N2	596.62	594.3124	58.38	10.14	26.79	4.70

C29H60BR2N4	624.64	622.3186	55.76	9.68	25.58	8.97
C29H60BR2N6	652.65	650.3247	53.37	9.27	24.49	12.88
C29H61BR1N2	517.73	516.4018	67.28	11.88	15.43	5.41
C29H61BR1N4	545.74	544.4080	63.83	11.27	14.64	10.27
C29H61BR1N6	573.75	572.4141	60.71	10.72	13.93	14.65
C29H61BR2N3	611.64	609.3233	56.95	10.05	26.13	6.87
C29H61BR2N5	639.65	637.3294	54.46	9.61	24.98	10.95
C29H62BR1N3	532.74	531.4127	65.38	11.73	15.00	7.89
C29H62BR1N5	560.75	559.4189	62.12	11.14	14.25	12.49
C29H62BR2N4	626.65	624.3342	55.58	9.97	25.50	8.94
C29H62BR2N6	654.67	652.3403	53.21	9.55	24.41	12.84
C29H63BR1N4	547.76	546.4236	63.59	11.59	14.59	10.23
C29H63BR1N6	575.77	574.4298	60.50	11.03	13.88	14.60
C29H63BR2N5	641.67	639.3451	54.28	9.90	24.91	10.91
C29H64BR1N5	562.77	561.4345	61.89	11.46	14.20	12.44
C29H64BR2N6	656.68	654.3560	53.04	9.82	24.34	12.80
C29H65BR1N6	577.79	576.4454	60.29	11.34	13.83	14.55
C30H15BR1N2	483.37	482.0419	74.55	3.13	16.53	5.80
C30H15BR1N4	511.38	510.0481	70.46	2.96	15.63	10.96
C30H15BR1N6	539.40	538.0542	66.80	2.80	14.81	15.58
C30H15BR2N1	549.27	546.9572	65.60	2.75	29.09	2.55
C30H15BR2N3	577.28	574.9634	62.42	2.62	27.68	7.28
C30H15BR2N5	605.30	602.9695	59.53	2.50	26.40	11.57
C30H16BR1N1	470.37	469.0467	76.61	3.43	16.99	2.98
C30H16BR1N3	498.39	497.0528	72.30	3.24	16.03	8.43
C30H16BR1N5	526.40	525.0590	68.45	3.06	15.18	13.30
C30H16BR2N2	564.28	561.9681	63.86	2.86	28.32	4.96
C30H16BR2N4	592.30	589.9743	60.84	2.72	26.98	9.46
C30H16BR2N6	620.31	617.9804	58.09	2.60	25.76	13.55
C30H17BR1N2	485.39	484.0576	74.24	3.53	16.46	5.77
C30H17BR1N4	513.40	512.0637	70.19	3.34	15.56	10.91
C30H17BR1N6	541.41	540.0699	66.55	3.16	14.76	15.52
C30H17BR2N1	551.28	548.9729	65.36	3.11	28.99	2.54
C30H17BR2N3	579.30	576.9790	62.20	2.96	27.59	7.25
C30H17BR2N5	607.31	604.9852	59.33	2.82	26.31	11.53
C30H18BR1N1	472.39	471.0623	76.28	3.84	16.91	2.97
C30H18BR1N3	500.40	499.0685	72.01	3.63	15.97	8.40
C30H18BR1N5	528.42	527.0746	68.19	3.43	15.12	13.25
C30H18BR2N2	566.30	563.9838	63.63	3.20	28.22	4.95
C30H18BR2N4	594.31	591.9899	60.63	3.05	26.89	9.43
C30H18BR2N6	622.33	619.9961	57.90	2.92	25.68	13.50
C30H19BR1N2	487.40	486.0732	73.93	3.93	16.39	5.75
C30H19BR1N4	515.42	514.0794	69.91	3.72	15.50	10.87
C30H19BR1N6	543.43	542.0855	66.31	3.52	14.70	15.46
C30H19BR2N1	553.30	550.9885	65.12	3.46	28.88	2.53
C30H19BR2N3	581.31	578.9947	61.99	3.29	27.49	7.23
C30H19BR2N5	609.33	607.0008	59.14	3.14	26.23	11.49
C30H20BR1N1	474.40	473.0780	75.96	4.25	16.84	2.95
C30H20BR1N3	502.42	501.0841	71.72	4.01	15.90	8.36
C30H20BR1N5	530.43	529.0903	67.93	3.80	15.06	13.20
C30H20BR2N2	568.32	565.9994	63.40	3.55	28.12	4.93
C30H20BR2N4	596.33	594.0056	60.43	3.38	26.80	9.40
C30H20BR2N6	624.34	622.0117	57.71	3.23	25.60	13.46
C30H21BR1N2	489.42	488.0889	73.62	4.32	16.33	5.72
C30H21BR1N4	517.43	516.0950	69.64	4.09	15.44	10.83
C30H21BR1N6	545.45	544.1011	66.06	3.88	14.65	15.41
C30H21BR2N1	555.32	553.0042	64.89	3.81	28.78	2.52
C30H21BR2N3	583.33	581.0103	61.77	3.63	27.40	7.20
C30H21BR2N5	611.34	609.0165	58.94	3.46	26.14	11.46
C30H22BR1N1	476.42	475.0936	75.63	4.65	16.77	2.94
C30H22BR1N3	504.43	503.0998	71.43	4.40	15.84	8.33
C30H22BR1N5	532.45	531.1059	67.68	4.16	15.01	13.15
C30H22BR2N2	570.33	568.0151	63.18	3.89	28.02	4.91
C30H22BR2N4	598.34	596.0212	60.22	3.71	26.71	9.36
C30H22BR2N6	626.36	624.0274	57.53	3.54	25.51	13.42
C30H23BR1N2	491.44	490.1045	73.32	4.72	16.26	5.70
C30H23BR1N4	519.45	518.1107	69.37	4.46	15.38	10.79
C30H23BR1N6	547.46	546.1168	65.82	4.23	14.60	15.35
C30H23BR2N1	557.33	555.0198	64.65	4.16	28.67	2.51
C30H23BR2N3	585.35	583.0260	61.56	3.96	27.30	7.18
C30H23BR2N5	613.36	611.0321	58.75	3.78	26.05	11.42
C30H24BR1N1	478.44	477.1093	75.32	5.06	16.70	2.93
C30H24BR1N3	506.45	505.1154	71.15	4.78	15.78	8.30
C30H24BR1N5	534.46	533.1215	67.42	4.53	14.95	13.10
C30H24BR2N2	572.35	570.0307	62.96	4.23	27.92	4.89
C30H24BR2N4	600.36	598.0369	60.02	4.03	26.62	9.33
C30H24BR2N6	628.37	626.0430	57.34	3.85	25.43	13.37
C30H25BR1N2	493.45	492.1202	73.02	5.11	16.19	5.68
C30H25BR1N4	521.46	520.1263	69.10	4.83	15.32	10.74
C30H25BR1N6	549.48	548.1324	65.58	4.59	14.54	15.29
C30H25BR2N1	559.35	557.0355	64.42	4.51	28.57	2.50
C30H25BR2N3	587.36	585.0416	61.35	4.29	27.21	7.15
C30H25BR2N5	615.38	613.0478	58.56	4.09	25.97	11.38
C30H26BR1N1	480.45	479.1249	75.00	5.45	16.63	2.92
C30H26BR1N3	508.47	507.1311	70.87	5.15	15.71	8.26
C30H26BR1N5	536.48	535.1372	67.17	4.89	14.89	13.05
C30H26BR2N2	574.36	572.0464	62.74	4.56	27.82	4.88
C30H26BR2N4	602.38	600.0525	59.82	4.35	26.53	9.30
C30H26BR2N6	630.39	628.0587	57.16	4.16	25.35	13.33
C30H27BR1N2	495.47	494.1358	72.73	5.49	16.13	5.65
C30H27BR1N4	523.48	522.1419	68.83	5.20	15.26	10.70
C30H27BR1N6	551.49	550.1481	65.34	4.93	14.49	15.24
C30H27BR2N1	561.36	559.0511	64.19	4.85	28.47	2.50
C30H27BR2N3	589.38	587.0573	61.14	4.62	27.11	7.13
C30H27BR2N5	617.39	615.0634	58.36	4.41	25.88	11.34
C30H28BR1N1	482.47	481.1406	74.69	5.85	16.56	2.90
C30H28BR1N3	510.48	509.1467	70.59	5.53	15.65	8.23
C30H28BR1N5	538.50	537.1528	66.92	5.24	14.84	13.01
C30H28BR2N2	576.38	574.0620	62.52	4.90	27.73	4.86
C30H28BR2N4	604.39	602.0682	59.62	4.67	26.44	9.27
C30H28BR2N6	632.41	630.0743	56.98	4.46	25.27	13.29
C30H29BR1N2	497.48	496.1515	72.43	5.88	16.06	5.63
C30H29BR1N4	525.50	524.1576	68.57	5.56	15.21	10.66
C30H29BR1N6	553.51	552.1637	65.10	5.28	14.44	15.18
C30H29BR2N1	563.38	561.0668	63.96	5.19	28.37	2.49
C30H29BR2N3	591.39	589.0729	60.93	4.94	27.02	7.11
C30H29BR2N5	619.41	617.0791	58.17	4.72	25.80	11.31
C30H30BR1N1	484.48	483.1562	74.37	6.24	16.49	2.89
C30H30BR1N3	512.50	511.1623	70.31	5.90	15.59	8.20
C30H30BR1N5	540.51	539.1685	66.67	5.59	14.78	12.96
C30H30BR2N2	578.39	576.0777	62.30	5.23	27.63	4.84
C30H30BR2N4	606.41	604.0838	59.42	4.99	26.35	9.24
C30H30BR2N6	634.42	632.0900	56.80	4.77	25.19	13.25
C30H31BR1N2	499.50	498.1671	72.14	6.26	16.00	5.61
C30H31BR1N4	527.51	526.1732	68.31	5.92	15.15	10.62
C30H31BR1N6	555.53	554.1794	64.86	5.62	14.38	15.13
C30H31BR2N1	565.40	563.0824	63.73	5.53	28.26	2.48
C30H31BR2N3	593.41	591.0886	60.72	5.27	26.93	7.08
C30H31BR2N5	621.42	619.0947	57.99	5.03	25.72	11.27
C30H32BR1N1	486.50	485.1719	74.07	6.63	16.42	2.88

283

C30H32BR1N3	514.51	513.1780	70.03	6.27	15.53	8.17
C30H32BR1N5	542.53	541.1841	66.42	5.95	14.73	12.91
C30H32BR2N2	580.41	578.0933	62.08	5.56	27.53	4.83
C30H32BR2N4	608.42	606.0995	59.22	5.30	26.27	9.21
C30H32BR2N6	636.44	634.1056	56.62	5.07	25.11	13.20
C30H33BR1N2	501.51	500.1827	71.85	6.63	15.93	5.59
C30H33BR1N4	529.53	528.1889	68.05	6.28	15.09	10.58
C30H33BR1N6	557.54	556.1950	64.63	5.97	14.33	15.07
C30H33BR2N1	567.41	565.0981	63.50	5.86	28.16	2.47
C30H33BR2N3	595.43	593.1042	60.52	5.59	26.84	7.06
C30H33BR2N5	623.44	621.1104	57.80	5.34	25.63	11.23
C30H34BR1N1	488.52	487.1875	73.76	7.02	16.36	2.87
C30H34BR1N3	516.53	515.1936	69.76	6.63	15.47	8.14
C30H34BR1N5	544.54	543.1998	66.17	6.29	14.67	12.86
C30H34BR2N2	582.43	580.1090	61.87	5.88	27.44	4.81
C30H34BR2N4	610.44	608.1151	59.03	5.61	26.18	9.18
C30H34BR2N6	638.45	636.1213	56.44	5.37	25.03	13.16
C30H35BR1N2	503.53	502.1984	71.56	7.01	15.87	5.56
C30H35BR1N4	531.54	530.2045	67.79	6.64	15.03	10.54
C30H35BR1N6	559.56	558.2107	64.40	6.30	14.28	15.02
C30H35BR2N1	569.43	567.1137	63.28	6.20	28.06	2.46
C30H35BR2N3	597.44	595.1199	60.31	5.91	26.75	7.03
C30H35BR2N5	625.45	623.1260	57.61	5.64	25.55	11.20
C30H36BR1N1	490.53	489.2031	73.46	7.40	16.29	2.86
C30H36BR1N3	518.55	517.2093	69.49	7.00	15.41	8.10
C30H36BR1N5	546.56	545.2154	65.93	6.64	14.62	12.81
C30H36BR2N2	584.44	582.1246	61.65	6.21	27.34	4.79
C30H36BR2N4	612.46	610.1308	58.83	5.92	26.09	9.15
C30H36BR2N6	640.47	638.1369	56.26	5.67	24.95	13.12
C30H37BR1N2	505.55	504.2140	71.28	7.38	15.81	5.54
C30H37BR1N4	533.56	532.2202	67.53	6.99	14.98	10.50
C30H37BR1N6	561.57	560.2263	64.17	6.64	14.23	14.97
C30H37BR2N1	571.44	569.1294	63.06	6.53	27.97	2.45
C30H37BR2N3	599.46	597.1355	60.11	6.22	26.66	7.01
C30H37BR2N5	627.47	625.1417	57.43	5.94	25.47	11.16
C30H38BR1N1	492.55	491.2188	73.16	7.78	16.22	2.84
C30H38BR1N3	520.56	519.2249	69.22	7.36	15.35	8.07
C30H38BR1N5	548.57	547.2311	65.69	6.98	14.57	12.77
C30H38BR2N2	586.46	584.1403	61.44	6.53	27.25	4.78
C30H38BR2N4	614.47	612.1464	58.64	6.23	26.01	9.12
C30H38BR2N6	642.49	640.1526	56.08	5.96	24.87	13.08
C30H39BR1N2	507.56	506.2297	70.99	7.75	15.74	5.52
C30H39BR1N4	535.58	534.2358	67.28	7.34	14.92	10.46
C30H39BR1N6	563.59	562.2420	63.94	6.98	14.18	14.91
C30H39BR2N1	573.46	571.1450	62.84	6.86	27.87	2.44
C30H39BR2N3	601.47	599.1512	59.91	6.54	26.57	6.99
C30H39BR2N5	629.49	627.1573	57.24	6.24	25.39	11.13
C30H40BR1N1	494.56	493.2344	72.86	8.15	16.16	2.83
C30H40BR1N3	522.58	521.2406	68.95	7.72	15.29	8.04
C30H40BR1N5	550.59	549.2467	65.45	7.32	14.51	12.72
C30H40BR2N2	588.47	586.1559	61.23	6.85	27.16	4.76
C30H40BR2N4	616.49	614.1621	58.45	6.54	25.92	9.09
C30H40BR2N6	644.50	642.1682	55.91	6.26	24.80	13.04
C30H41BR1N2	509.58	508.2453	70.71	8.11	15.68	5.50
C30H41BR1N4	537.59	536.2515	67.03	7.69	14.86	10.42
C30H41BR1N6	565.61	564.2576	63.71	7.31	14.13	14.86
C30H41BR2N1	575.48	573.1607	62.62	7.18	27.77	2.43
C30H41BR2N3	603.49	601.1668	59.71	6.85	26.48	6.96
C30H41BR2N5	631.50	629.1730	57.06	6.54	25.31	11.09
C30H42BR1N1	496.58	495.2501	72.56	8.53	16.09	2.82

C30H42BR1N3	524.59	523.2562	68.69	8.07	15.23	8.01
C30H42BR1N5	552.61	551.2624	65.21	7.66	14.46	12.67
C30H42BR2N2	590.49	588.1716	61.02	7.17	27.06	4.74
C30H42BR2N4	618.50	616.1777	58.26	6.84	25.84	9.06
C30H42BR2N6	646.52	644.1839	55.73	6.55	24.72	13.00
C30H43BR1N2	511.59	510.2610	70.43	8.47	15.62	5.48
C30H43BR1N4	539.61	538.2671	66.78	8.03	14.81	10.38
C30H43BR1N6	567.62	566.2733	63.48	7.64	14.08	14.81
C30H43BR2N1	577.49	575.1763	62.40	7.51	27.67	2.43
C30H43BR2N3	605.51	603.1825	59.51	7.16	26.39	6.94
C30H43BR2N5	633.52	631.1886	56.88	6.84	25.23	11.05
C30H44BR1N1	498.60	497.2657	72.27	8.90	16.03	2.81
C30H44BR1N3	526.61	525.2719	68.43	8.42	15.17	7.98
C30H44BR1N5	554.62	553.2780	64.97	8.00	14.41	12.63
C30H44BR2N2	592.51	590.1872	60.82	7.49	26.97	4.73
C30H44BR2N4	620.52	618.1934	58.07	7.15	25.75	9.03
C30H44BR2N6	648.53	646.1995	55.56	6.84	24.64	12.96
C30H45BR1N2	513.61	512.2766	70.16	8.83	15.56	5.45
C30H45BR1N4	541.62	540.2828	66.53	8.37	14.75	10.34
C30H45BR1N6	569.64	568.2889	63.26	7.96	14.03	14.75
C30H45BR2N1	579.51	577.1920	62.18	7.83	27.58	2.42
C30H45BR2N3	607.52	605.1981	59.31	7.47	26.30	6.92
C30H45BR2N5	635.53	633.2043	56.70	7.14	25.15	11.02
C30H46BR1N1	500.61	499.2814	71.98	9.26	15.96	2.80
C30H46BR1N3	528.63	527.2875	68.16	8.77	15.12	7.95
C30H46BR1N5	556.64	555.2937	64.73	8.33	14.35	12.58
C30H46BR2N2	594.52	592.2029	60.61	7.80	26.88	4.71
C30H46BR2N4	622.54	620.2090	57.88	7.45	25.67	9.00
C30H46BR2N6	650.55	648.2152	55.39	7.13	24.57	12.92
C30H47BR1N2	515.63	514.2923	69.88	9.19	15.50	5.43
C30H47BR1N4	543.64	542.2984	66.28	8.71	14.70	10.31
C30H47BR1N6	571.65	570.3046	63.03	8.29	13.98	14.70
C30H47BR2N1	581.52	579.2076	61.96	8.15	27.48	2.41
C30H47BR2N3	609.54	607.2138	59.12	7.77	26.22	6.89
C30H47BR2N5	637.55	635.2199	56.52	7.43	25.07	10.98
C30H48BR1N1	502.63	501.2970	71.69	9.63	15.90	2.79
C30H48BR1N3	530.64	529.3032	67.91	9.12	15.06	7.92
C30H48BR1N5	558.65	557.3093	64.50	8.66	14.30	12.54
C30H48BR2N2	596.54	594.2185	60.40	8.11	26.79	4.70
C30H48BR2N4	624.55	622.2247	57.69	7.75	25.59	8.97
C30H48BR2N6	652.57	650.2308	55.22	7.41	24.49	12.88
C30H49BR1N2	517.64	516.3079	69.61	9.54	15.44	5.41
C30H49BR1N4	545.66	544.3141	66.04	9.05	14.64	10.27
C30H49BR1N6	573.67	572.3202	62.81	8.61	13.93	14.65
C30H49BR2N1	583.54	581.2233	61.75	8.46	27.39	2.40
C30H49BR2N3	611.55	609.2294	58.92	8.08	26.13	6.87
C30H49BR2N5	639.57	637.2356	56.34	7.72	24.99	10.95
C30H50BR1N1	504.64	503.3127	71.40	9.99	15.83	2.78
C30H50BR1N3	532.66	531.3188	67.65	9.46	15.00	7.89
C30H50BR1N5	560.67	559.3250	64.27	8.99	14.25	12.49
C30H50BR2N2	598.55	596.2342	60.20	8.42	26.70	4.68
C30H50BR2N4	626.57	624.2403	57.51	8.04	25.51	8.94
C30H50BR2N6	654.58	652.2465	55.05	7.70	24.41	12.84
C30H51BR1N2	519.66	518.3236	69.34	9.89	15.38	5.39
C30H51BR1N4	547.67	546.3297	65.79	9.39	14.59	10.23
C30H51BR1N6	575.69	574.3359	62.59	8.93	13.88	14.60
C30H51BR2N1	585.56	583.2389	61.54	8.78	27.29	2.39
C30H51BR2N3	613.57	611.2451	58.73	8.38	26.05	6.85
C30H51BR2N5	641.58	639.2512	56.16	8.01	24.91	10.92
C30H52BR1N1	506.66	505.3283	71.12	10.35	15.77	2.76

284

C30H52BR1N3	534.67	533.3345	67.39	9.80	14.94	7.86
C30H52BR1N5	562.69	561.3406	64.04	9.32	14.20	12.45
C30H52BR2N2	600.57	598.2498	60.00	8.73	26.61	4.66
C30H52BR2N4	628.58	626.2560	57.32	8.34	25.42	8.91
C30H52BR2N6	656.60	654.2621	54.88	7.98	24.34	12.80
C30H53BR1N2	521.67	520.3392	69.07	10.24	15.32	5.37
C30H53BR1N4	549.69	548.3454	65.55	9.72	14.54	10.19
C30H53BR1N6	577.70	576.3515	62.37	9.25	13.83	14.55
C30H53BR2N1	587.57	585.2546	61.33	9.09	27.20	2.38
C30H53BR2N3	615.59	613.2607	58.54	8.68	25.96	6.83
C30H53BR2N5	643.60	641.2669	55.99	8.30	24.83	10.88
C30H54BR1N1	508.68	507.3440	70.84	10.70	15.71	2.75
C30H54BR1N3	536.69	535.3501	67.14	10.14	14.89	7.83
C30H54BR1N5	564.70	563.3563	63.81	9.64	14.15	12.40
C30H54BR2N2	602.59	600.2655	59.80	9.03	26.52	4.65
C30H54BR2N4	630.60	628.2716	57.14	8.63	25.34	8.88
C30H54BR2N6	658.61	656.2778	54.71	8.26	24.26	12.76
C30H55BR1N2	523.69	522.3549	68.81	10.59	15.26	5.35
C30H55BR1N4	551.70	550.3610	65.31	10.05	14.48	10.16
C30H55BR1N6	579.72	578.3672	62.16	9.56	13.78	14.50
C30H55BR2N1	589.59	587.2702	61.12	9.40	27.11	2.38
C30H55BR2N3	617.60	615.2764	58.34	8.98	25.88	6.80
C30H55BR2N5	645.61	643.2825	55.81	8.59	24.75	10.85
C30H56BR1N1	510.69	509.3596	70.56	11.05	15.65	2.74
C30H56BR1N3	538.70	537.3658	66.89	10.48	14.83	7.80
C30H56BR1N5	566.72	565.3719	63.58	9.96	14.10	12.36
C30H56BR2N2	604.60	602.2811	59.60	9.34	26.43	4.63
C30H56BR2N4	632.62	630.2873	56.96	8.92	25.26	8.86
C30H56BR2N6	660.63	658.2934	54.54	8.54	24.19	12.72
C30H57BR1N2	525.71	524.3705	68.54	10.93	15.20	5.33
C30H57BR1N4	553.72	552.3767	65.08	10.38	14.43	10.12
C30H57BR1N6	581.73	580.3828	61.94	9.88	13.74	14.45
C30H57BR2N1	591.60	589.2859	60.91	9.71	27.01	2.37
C30H57BR2N3	619.62	617.2920	58.15	9.27	25.79	6.78
C30H57BR2N5	647.63	645.2982	55.64	8.87	24.68	10.81
C30H58BR1N1	512.71	511.3753	70.28	11.40	15.58	2.73
C30H58BR1N3	540.72	539.3814	66.64	10.81	14.78	7.77
C30H58BR1N5	568.73	567.3876	63.36	10.28	14.05	12.31
C30H58BR2N2	606.62	604.2968	59.40	9.64	26.34	4.62
C30H58BR2N4	634.63	632.3029	56.78	9.21	25.18	8.83
C30H58BR2N6	662.64	660.3090	54.38	8.82	24.12	12.68
C30H59BR1N2	527.72	526.3862	68.28	11.27	15.14	5.31
C30H59BR1N4	555.74	554.3923	64.84	10.70	14.38	10.08
C30H59BR1N6	583.75	582.3985	61.73	10.19	13.69	14.40
C30H59BR2N1	593.62	591.3015	60.70	10.02	26.92	2.36
C30H59BR2N3	621.63	619.3077	57.97	9.57	25.71	6.76
C30H59BR2N5	649.65	647.3138	55.47	9.15	24.60	10.78
C30H60BR1N1	514.72	513.3909	70.01	11.75	15.52	2.72
C30H60BR1N3	542.74	541.3971	66.39	11.14	14.72	7.74
C30H60BR1N5	570.75	569.4032	63.13	10.60	14.00	12.27
C30H60BR2N2	608.63	606.3124	59.20	9.94	26.26	4.60
C30H60BR2N4	636.65	634.3186	56.60	9.50	25.10	8.80
C30H60BR2N6	664.66	662.3247	54.21	9.10	24.04	12.64
C30H61BR1N2	529.74	528.4018	68.02	11.61	15.08	5.29
C30H61BR1N4	557.75	556.4080	64.60	11.02	14.33	10.05
C30H61BR1N6	585.76	584.4141	61.52	10.50	13.64	14.35
C30H61BR2N1	595.64	593.3172	60.50	10.32	26.83	2.35
C30H61BR2N3	623.65	621.3233	57.78	9.86	25.62	6.74
C30H61BR2N5	651.66	649.3294	55.29	9.44	24.52	10.75
C30H62BR1N1	516.74	515.4066	69.73	12.09	15.46	2.71

C30H62BR1N3	544.75	543.4127	66.15	11.47	14.67	7.71
C30H62BR1N5	572.77	571.4189	62.91	10.91	13.95	12.23
C30H62BR2N2	610.65	608.3281	59.01	10.23	26.17	4.59
C30H62BR2N4	638.66	636.3342	56.42	9.79	25.02	8.77
C30H62BR2N6	666.68	664.3403	54.05	9.37	23.97	12.61
C30H63BR1N2	531.75	530.4175	67.76	11.94	15.03	5.27
C30H63BR1N4	559.77	558.4236	64.37	11.34	14.27	10.01
C30H63BR1N6	587.78	586.4298	61.30	10.80	13.59	14.30
C30H63BR2N3	625.66	623.3390	57.59	10.15	25.54	6.72
C30H63BR2N5	653.68	651.3451	55.12	9.71	24.45	10.71
C30H64BR1N3	546.77	545.4284	65.90	11.80	14.61	7.69
C30H64BR1N5	574.78	573.4345	62.69	11.22	13.90	12.18
C30H64BR2N4	640.68	638.3498	56.24	10.07	24.94	8.74
C30H64BR2N6	668.69	666.3560	53.89	9.65	23.90	12.57
C30H65BR1N4	561.78	560.4393	64.14	11.66	14.22	9.97
C30H65BR1N6	589.80	588.4454	61.09	11.11	13.55	14.25
C30H65BR2N5	655.69	653.3607	54.95	9.99	24.37	10.68
C30H66BR1N5	576.80	575.4502	62.47	11.53	13.85	12.14
C30H66BR2N6	670.71	668.3716	53.72	9.92	23.83	12.53
C30H67BR1N6	591.81	590.4611	60.89	11.41	13.50	14.20

2. CARBON, HYDROGEN, BROMINE AND OXYGEN

2. KOHLENSTOFF, WASSERSTOFF, BROM UND SAUERSTOFF

2. CARBONE, HYDROGENE, BROME ET OXYGENE

MOLECULAR WEIGHTS AND PERCENTAGE COMPOSITIONS OF COMPOUNDS
CONTAINING CARBON, HYDROGEN, BROMINE
AND OXYGEN

FORMULA	MOL. WEIGHT	ISOMOL. WEIGHT	PERC. C	PERC. H	PERC. BR	PERC. O
C 1H 0BR201	187.82	185.8317	6.40	0	85.09	8.52
C 1H 0BR202	203.82	201.8266	5.89	0	78.41	15.70
C 1H 0BR203	219.82	217.8215	5.46	0	72.70	21.84
C 1H 0BR204	235.82	233.8164	5.09	0	67.77	27.14
C 1H 0BR205	251.82	249.8114	4.77	0	63.46	31.77
C 1H 0BR206	267.82	265.8063	4.48	0	59.67	35.84
C 1H 0BR207	283.81	281.8012	4.23	0	56.31	39.46
C 1H 0BR208	299.81	297.7961	4.01	0	53.30	42.69
C 1H 0BR209	315.81	313.7910	3.80	0	50.60	45.59
C 1H 1BR101	108.92	107.9211	11.03	.93	73.36	14.69
C 1H 1BR102	124.92	123.9160	9.61	.81	63.96	25.62
C 1H 1BR103	140.92	139.9110	8.52	.72	56.70	34.06
C 1H 1BR104	156.92	155.9059	7.65	.64	50.92	40.78
C 1H 1BR105	172.92	171.9008	6.95	.58	46.21	46.26
C 1H 1BR106	188.92	187.8957	6.36	.53	42.30	50.81
C 1H 1BR107	204.92	203.8906	5.86	.49	38.99	54.65
C 1H 1BR108	220.92	219.8855	5.44	.46	36.17	57.94
C 1H 1BR109	236.92	235.8804	5.07	.43	33.73	60.78
C 1H 2BR201	189.83	187.8473	6.33	1.06	84.18	8.43
C 1H 2BR202	205.83	203.8423	5.84	.98	77.64	15.55
C 1H 2BR203	221.83	219.8372	5.41	.91	72.04	21.64
C 1H 2BR204	237.83	235.8321	5.05	.85	67.19	26.91
C 1H 2BR205	253.83	251.8270	4.73	.79	62.96	31.52
C 1H 2BR206	269.83	267.8219	4.45	.75	59.23	35.58
C 1H 2BR207	285.83	283.8168	4.20	.71	55.91	39.18
C 1H 2BR208	301.83	299.8117	3.98	.67	52.95	42.41
C 1H 2BR209	317.83	315.8067	3.78	.63	50.28	45.31
C 1H 3BR101	110.94	109.9368	10.83	2.73	72.03	14.42
C 1H 3BR102	126.94	125.9317	9.46	2.38	62.95	25.21
C 1H 3BR103	142.94	141.9266	8.40	2.12	55.90	33.58
C 1H 3BR104	158.94	157.9215	7.56	1.90	50.27	40.27
C 1H 3BR105	174.94	173.9164	6.87	1.73	45.68	45.73
C 1H 3BR106	190.94	189.9113	6.29	1.58	41.85	50.28
C 1H 3BR107	206.93	205.9063	5.80	1.46	38.61	54.12
C 1H 3BR108	222.93	221.9012	5.39	1.36	35.84	57.41
C 1H 3BR109	238.93	237.8961	5.03	1.27	33.44	60.27
C 2H 1BR101	120.93	119.9211	19.86	.83	66.07	13.23
C 2H 1BR102	136.93	135.9160	17.54	.74	58.35	23.37
C 2H 1BR103	152.93	151.9110	15.71	.66	52.25	31.39
C 2H 1BR104	168.93	167.9059	14.22	.60	47.30	37.88
C 2H 1BR105	184.93	183.9008	12.99	.55	43.21	43.26
C 2H 1BR106	200.93	199.8957	11.96	.50	39.77	47.78
C 2H 1BR107	216.93	215.8906	11.07	.46	36.83	51.63
C 2H 1BR108	232.93	231.8855	10.31	.43	34.30	54.95
C 2H 1BR109	248.93	247.8804	9.65	.40	32.10	57.85
C 2H 2BR201	201.85	199.8473	11.90	1.00	79.17	7.93
C 2H 2BR202	217.85	215.8423	11.03	.93	73.36	14.69
C 2H 2BR203	233.84	231.8372	10.27	.86	68.34	20.53
C 2H 2BR204	249.84	247.8321	9.61	.81	63.96	25.62
C 2H 2BR205	265.84	263.8270	9.04	.76	60.11	30.09

FORMULA	MOL. WEIGHT	ISOMOL. WEIGHT	PERC. C	PERC. H	PERC. BR	PERC. O
C 2H 2BR206	281.84	279.8219	8.52	.72	56.70	34.06
C 2H 2BR207	297.84	295.8168	8.07	.68	53.66	37.60
C 2H 2BR208	313.84	311.8117	7.65	.64	50.92	40.78
C 2H 2BR209	329.84	327.8067	7.28	.61	48.45	43.66
C 2H 3BR101	122.95	121.9368	19.54	2.46	64.99	13.01
C 2H 3BR102	138.95	137.9317	17.29	2.18	57.51	23.03
C 2H 3BR103	154.95	153.9266	15.50	1.95	51.57	30.98
C 2H 3BR104	170.95	169.9215	14.05	1.77	46.74	37.44
C 2H 3BR105	186.95	185.9164	12.85	1.62	42.74	42.79
C 2H 3BR106	202.95	201.9113	11.84	1.49	39.37	47.30
C 2H 3BR107	218.95	217.9063	10.97	1.38	36.49	51.15
C 2H 3BR108	234.95	233.9012	10.22	1.29	34.01	54.48
C 2H 3BR109	250.94	249.8961	9.57	1.21	31.84	57.38
C 2H 4BR201	203.86	201.8630	11.78	1.98	78.39	7.85
C 2H 4BR202	219.86	217.8579	10.93	1.83	72.69	14.55
C 2H 4BR203	235.86	233.8528	10.18	1.71	67.76	20.35
C 2H 4BR204	251.86	249.8477	9.54	1.60	63.45	25.41
C 2H 4BR205	267.86	265.8426	8.97	1.51	59.66	29.87
C 2H 4BR206	283.86	281.8376	8.46	1.42	56.30	33.82
C 2H 4BR207	299.86	297.8325	8.01	1.34	53.29	37.35
C 2H 4BR208	315.86	313.8274	7.61	1.28	50.59	40.52
C 2H 4BR209	331.86	329.8223	7.24	1.21	48.16	43.39
C 2H 5BR101	124.97	123.9524	19.22	4.03	63.94	12.80
C 2H 5BR102	140.96	139.9473	17.04	3.58	56.68	22.70
C 2H 5BR103	156.96	155.9423	15.30	3.21	50.91	30.58
C 2H 5BR104	172.96	171.9372	13.89	2.91	46.20	37.00
C 2H 5BR105	188.96	187.9321	12.71	2.67	42.29	42.33
C 2H 5BR106	204.96	203.9270	11.72	2.46	38.98	46.84
C 2H 5BR107	220.96	219.9219	10.87	2.28	36.16	50.69
C 2H 5BR108	236.96	235.9168	10.14	2.13	33.72	54.02
C 2H 5BR109	252.96	251.9117	9.50	1.99	31.59	56.92
C 3H 1BR101	132.94	131.9211	27.10	.76	60.10	12.03
C 3H 1BR102	148.94	147.9160	24.19	.68	53.65	21.48
C 3H 1BR103	164.94	163.9110	21.85	.61	48.44	29.10
C 3H 1BR104	180.94	179.9059	19.91	.56	44.16	35.37
C 3H 1BR105	196.94	195.9008	18.30	.51	40.57	40.62
C 3H 1BR106	212.94	211.8957	16.92	.47	37.52	45.08
C 3H 1BR107	228.94	227.8906	15.74	.44	34.90	48.92
C 3H 1BR108	244.94	243.8855	14.71	.41	32.62	52.26
C 3H 1BR109	260.94	259.8804	13.81	.39	30.62	55.18
C 3H 2BR201	213.86	211.8473	16.85	.94	74.73	7.48
C 3H 2BR202	229.86	227.8423	15.68	.88	69.53	13.92
C 3H 2BR203	245.86	243.8372	14.66	.82	65.00	19.52
C 3H 2BR204	261.85	259.8321	13.76	.77	61.03	24.44
C 3H 2BR205	277.85	275.8270	12.97	.73	57.52	28.79
C 3H 2BR206	293.85	291.8219	12.26	.69	54.38	32.67
C 3H 2BR207	309.85	307.8168	11.63	.65	51.58	36.14
C 3H 2BR208	325.85	323.8117	11.06	.62	49.04	39.28
C 3H 2BR209	341.85	339.8067	10.54	.59	46.75	42.12
C 3H 3BR101	134.96	133.9368	26.70	2.24	59.21	11.85

C 3H 3BR102	150.96	149.9317	23.87	2.00	52.93	21.20
C 3H 3BR103	166.96	165.9266	21.58	1.81	47.86	28.75
C 3H 3BR104	182.96	181.9215	19.69	1.65	43.67	34.98
C 3H 3BR105	198.96	197.9164	18.11	1.52	40.16	40.21
C 3H 3BR106	214.96	213.9113	16.76	1.41	37.17	44.66
C 3H 3BR107	230.96	229.9063	15.60	1.31	34.60	48.49
C 3H 3BR108	246.96	245.9012	14.59	1.22	32.36	51.83
C 3H 3BR109	262.96	261.8961	13.70	1.15	30.39	54.76
C 3H 4BR201	215.87	213.8630	16.69	1.87	74.03	7.41
C 3H 4BR202	231.87	229.8579	15.54	1.74	68.92	13.80
C 3H 4BR203	247.87	245.8528	14.54	1.63	64.47	19.36
C 3H 4BR204	263.87	261.8477	13.66	1.53	60.56	24.25
C 3H 4BR205	279.87	277.8426	12.88	1.44	57.10	28.58
C 3H 4BR206	295.87	293.8376	12.18	1.36	54.01	32.45
C 3H 4BR207	311.87	309.8325	11.55	1.29	51.24	35.91
C 3H 4BR208	327.87	325.8274	10.99	1.23	48.74	39.04
C 3H 4BR209	343.87	341.8223	10.48	1.17	46.47	41.87
C 3H 5BR101	136.98	135.9524	26.31	3.68	58.33	11.68
C 3H 5BR102	152.98	151.9473	23.55	3.29	52.23	20.92
C 3H 5BR103	168.98	167.9423	21.32	2.98	47.29	28.41
C 3H 5BR104	184.97	183.9372	19.48	2.72	43.20	34.60
C 3H 5BR105	200.97	199.9321	17.93	2.51	39.76	39.80
C 3H 5BR106	216.97	215.9270	16.61	2.32	36.83	44.24
C 3H 5BR107	232.97	231.9219	15.47	2.16	34.30	48.07
C 3H 5BR108	248.97	247.9168	14.47	2.02	32.09	51.41
C 3H 5BR109	264.97	263.9117	13.60	1.90	30.16	54.34
C 3H 6BR201	217.89	215.8786	16.54	2.78	73.34	7.34
C 3H 6BR202	233.89	231.8736	15.41	2.59	68.33	13.68
C 3H 6BR203	249.89	247.8685	14.42	2.42	63.95	19.21
C 3H 6BR204	265.89	263.8634	13.55	2.27	60.10	24.07
C 3H 6BR205	281.89	279.8583	12.78	2.15	56.69	28.38
C 3H 6BR206	297.89	295.8532	12.10	2.03	53.65	32.23
C 3H 6BR207	313.89	311.8481	11.48	1.93	50.91	35.68
C 3H 6BR208	329.88	327.8430	10.92	1.83	48.44	38.80
C 3H 6BR209	345.88	343.8380	10.42	1.75	46.20	41.63
C 3H 7BR101	138.99	137.9681	25.92	5.08	57.49	11.51
C 3H 7BR102	154.99	153.9630	23.25	4.55	51.55	20.65
C 3H 7BR103	170.99	169.9579	21.07	4.13	46.73	28.07
C 3H 7BR104	186.99	185.9528	19.27	3.77	42.73	34.22
C 3H 7BR105	202.99	201.9477	17.75	3.48	39.36	39.41
C 3H 7BR106	218.99	217.9426	16.45	3.22	36.49	43.84
C 3H 7BR107	234.99	233.9376	15.33	3.00	34.00	47.66
C 3H 7BR108	250.99	249.9325	14.36	2.81	31.84	51.00
C 3H 7BR109	266.99	265.9274	13.50	2.64	29.93	53.93
C 4H 2BR201	225.87	223.8473	21.27	.89	70.75	7.08
C 4H 2BR202	241.87	239.8423	19.86	.83	66.07	13.23
C 4H 2BR203	257.87	255.8372	18.63	.78	61.97	18.61
C 4H 2BR204	273.87	271.8321	17.54	.74	58.35	23.37
C 4H 2BR205	289.87	287.8270	16.57	.70	55.13	27.60
C 4H 2BR206	305.86	303.8219	15.71	.66	52.25	31.39
C 4H 2BR207	321.86	319.8168	14.93	.63	49.65	34.80
C 4H 2BR208	337.86	335.8117	14.22	.60	47.30	37.88
C 4H 2BR209	353.86	351.8067	13.58	.57	45.16	40.69
C 4H 3BR101	146.97	145.9368	32.69	2.06	54.37	10.89
C 4H 3BR102	162.97	161.9317	29.48	1.86	49.03	19.63
C 4H 3BR103	178.97	177.9266	26.84	1.69	44.65	26.82
C 4H 3BR104	194.97	193.9215	24.64	1.55	40.98	32.82
C 4H 3BR105	210.97	209.9164	22.77	1.43	37.87	37.92
C 4H 3BR106	226.97	225.9113	21.17	1.33	35.20	42.29
C 4H 3BR107	242.97	241.9063	19.77	1.24	32.89	46.09

C 4H 3BR108	258.97	257.9012	18.55	1.17	30.85	49.43
C 4H 3BR109	274.97	273.8961	17.47	1.10	29.06	52.37
C 4H 4BR201	227.88	225.8630	21.08	1.77	70.13	7.02
C 4H 4BR202	243.88	241.8579	19.70	1.65	65.53	13.12
C 4H 4BR203	259.88	257.8528	18.49	1.55	61.49	18.47
C 4H 4BR204	275.88	273.8477	17.41	1.46	57.93	23.20
C 4H 4BR205	291.88	289.8426	16.46	1.38	54.75	27.41
C 4H 4BR206	307.88	305.8376	15.60	1.31	51.91	31.18
C 4H 4BR207	323.88	321.8325	14.83	1.24	49.34	34.58
C 4H 4BR208	339.88	337.8274	14.14	1.19	47.02	37.66
C 4H 4BR209	355.88	353.8223	13.50	1.13	44.91	40.46
C 4H 5BR101	148.99	147.9524	32.25	3.38	53.63	10.74
C 4H 5BR102	164.99	163.9473	29.12	3.05	48.43	19.39
C 4H 5BR103	180.99	179.9423	26.55	2.78	44.15	26.52
C 4H 5BR104	196.99	195.9372	24.39	2.56	40.56	32.49
C 4H 5BR105	212.99	211.9321	22.56	2.37	37.52	37.56
C 4H 5BR106	228.98	227.9270	20.98	2.20	34.89	41.92
C 4H 5BR107	244.98	243.9219	19.61	2.06	32.62	45.72
C 4H 5BR108	260.98	259.9168	18.41	1.93	30.62	49.04
C 4H 5BR109	276.98	275.9117	17.35	1.82	28.85	51.99
C 4H 6BR201	229.90	227.8786	20.90	2.63	69.51	6.96
C 4H 6BR202	245.90	243.8736	19.54	2.46	64.99	13.01
C 4H 6BR203	261.90	259.8685	18.34	2.31	61.02	18.33
C 4H 6BR204	277.90	275.8634	17.29	2.18	57.51	23.03
C 4H 6BR205	293.90	291.8583	16.35	2.06	54.38	27.22
C 4H 6BR206	309.90	307.8532	15.50	1.95	51.57	30.98
C 4H 6BR207	325.90	323.8481	14.74	1.86	49.04	34.37
C 4H 6BR208	341.90	339.8430	14.05	1.77	46.74	37.44
C 4H 6BR209	357.90	355.8380	13.42	1.69	44.65	40.23
C 4H 7BR101	151.00	149.9681	31.82	4.67	52.92	10.60
C 4H 7BR102	167.00	165.9630	28.77	4.22	47.85	19.16
C 4H 7BR103	183.00	181.9579	26.25	3.86	43.66	26.23
C 4H 7BR104	199.00	197.9528	24.14	3.55	40.15	32.16
C 4H 7BR105	215.00	213.9477	22.35	3.28	37.16	37.21
C 4H 7BR106	231.00	229.9426	20.80	3.05	34.59	41.56
C 4H 7BR107	247.00	245.9376	19.45	2.86	32.35	45.34
C 4H 7BR108	263.00	261.9325	18.27	2.68	30.38	48.67
C 4H 7BR109	279.00	277.9274	17.22	2.53	28.64	51.61
C 4H 8BR201	231.92	229.8943	20.72	3.48	68.91	6.90
C 4H 8BR202	247.92	245.8892	19.38	3.25	64.46	12.91
C 4H 8BR203	263.91	261.8841	18.20	3.06	60.55	18.19
C 4H 8BR204	279.91	277.8790	17.16	2.88	57.09	22.86
C 4H 8BR205	295.91	293.8739	16.24	2.73	54.00	27.03
C 4H 8BR206	311.91	309.8689	15.40	2.59	51.23	30.78
C 4H 8BR207	327.91	325.8638	14.65	2.46	48.74	34.15
C 4H 8BR208	343.91	341.8587	13.97	2.34	46.47	37.22
C 4H 8BR209	359.91	357.8536	13.35	2.24	44.40	40.01
C 4H 9BR101	153.02	151.9837	31.40	5.93	52.22	10.46
C 4H 9BR102	169.02	167.9786	28.43	5.37	47.28	18.93
C 4H 9BR103	185.02	183.9736	25.97	4.90	43.19	25.94
C 4H 9BR104	201.02	199.9685	23.90	4.51	39.75	31.84
C 4H 9BR105	217.02	215.9634	22.14	4.18	36.82	36.86
C 4H 9BR106	233.02	231.9583	20.62	3.89	34.29	41.20
C 4H 9BR107	249.02	247.9532	19.29	3.64	32.09	44.98
C 4H 9BR108	265.02	263.9481	18.13	3.42	30.15	48.30
C 4H 9BR109	281.01	279.9430	17.10	3.23	28.43	51.24
C 5H 2BR201	237.88	235.8473	25.25	.85	67.18	6.73
C 5H 2BR202	253.88	251.8423	23.66	.79	62.95	12.60
C 5H 2BR203	269.88	267.8372	22.25	.75	59.21	17.79
C 5H 2BR204	285.88	283.8321	21.01	.71	55.90	22.39

011

C 5H 2BR205	301.88	299.8270	19.89	.67	52.94	26.50
C 5H 2BR206	317.88	315.8219	18.89	.63	50.27	30.20
C 5H 2BR207	333.88	331.8168	17.99	.60	47.86	33.54
C 5H 2BR208	349.87	347.8117	17.16	.58	45.68	36.58
C 5H 2BR209	365.87	363.8067	16.41	.55	43.68	39.36
C 5H 3BR101	158.98	157.9368	37.77	1.90	50.26	10.06
C 5H 3BR102	174.98	173.9317	34.32	1.73	45.66	18.29
C 5H 3BR103	190.98	189.9266	31.45	1.58	41.84	25.13
C 5H 3BR104	206.98	205.9215	29.02	1.46	38.60	30.92
C 5H 3BR105	222.98	221.9164	26.93	1.36	35.83	35.88
C 5H 3BR106	238.98	237.9113	25.13	1.27	33.44	40.17
C 5H 3BR107	254.98	253.9063	23.55	1.19	31.34	43.92
C 5H 3BR108	270.98	269.9012	22.16	1.12	29.49	47.23
C 5H 3BR109	286.98	285.8961	20.93	1.05	27.84	50.18
C 5H 4BR201	239.90	237.8630	25.03	1.68	66.62	6.67
C 5H 4BR202	255.89	253.8579	23.47	1.58	62.45	12.50
C 5H 4BR203	271.89	269.8528	22.09	1.48	58.78	17.65
C 5H 4BR204	287.89	285.8477	20.86	1.40	55.51	22.23
C 5H 4BR205	303.89	301.8426	19.76	1.33	52.59	26.32
C 5H 4BR206	319.89	317.8376	18.77	1.26	49.96	30.01
C 5H 4BR207	335.89	333.8325	17.88	1.20	47.58	33.34
C 5H 4BR208	351.89	349.8274	17.07	1.15	45.41	36.37
C 5H 4BR209	367.89	365.8223	16.32	1.10	43.44	39.14
C 5H 5BR101	161.00	159.9524	37.30	3.13	49.63	9.94
C 5H 5BR102	177.00	175.9473	33.93	2.85	45.14	18.08
C 5H 5BR103	193.00	191.9423	31.12	2.61	41.40	24.87
C 5H 5BR104	209.00	207.9372	28.74	2.41	38.23	30.62
C 5H 5BR105	225.00	223.9321	26.69	2.24	35.51	35.55
C 5H 5BR106	241.00	239.9270	24.92	2.09	33.16	39.83
C 5H 5BR107	257.00	255.9219	23.37	1.96	31.09	43.58
C 5H 5BR108	272.99	271.9168	22.00	1.85	29.27	46.89
C 5H 5BR109	288.99	287.9117	20.78	1.74	27.65	49.83
C 5H 6BR201	241.91	239.8786	24.83	2.50	66.06	6.61
C 5H 6BR202	257.91	255.8736	23.29	2.34	61.96	12.41
C 5H 6BR203	273.91	271.8685	21.93	2.21	58.34	17.52
C 5H 6BR204	289.91	287.8634	20.72	2.09	55.12	22.08
C 5H 6BR205	305.91	303.8583	19.63	1.98	52.24	26.15
C 5H 6BR206	321.91	319.8532	18.66	1.88	49.64	29.82
C 5H 6BR207	337.91	335.8481	17.77	1.79	47.29	33.14
C 5H 6BR208	353.91	351.8430	16.97	1.71	45.16	36.17
C 5H 6BR209	369.91	367.8380	16.24	1.63	43.20	38.93
C 5H 7BR101	163.01	161.9681	36.84	4.33	49.02	9.81
C 5H 7BR102	179.01	177.9630	33.55	3.94	44.64	17.87
C 5H 7BR103	195.01	193.9579	30.80	3.62	40.97	24.61
C 5H 7BR104	211.01	209.9528	28.46	3.34	37.87	30.33
C 5H 7BR105	227.01	225.9477	26.45	3.11	35.20	35.24
C 5H 7BR106	243.01	241.9426	24.71	2.90	32.88	39.50
C 5H 7BR107	259.01	257.9376	23.19	2.72	30.85	43.24
C 5H 7BR108	275.01	273.9325	21.84	2.57	29.05	46.54
C 5H 7BR109	291.01	289.9274	20.64	2.42	27.46	49.48
C 5H 8BR201	243.93	241.8943	24.62	3.31	65.51	6.56
C 5H 8BR202	259.93	257.8892	23.10	3.10	61.48	12.31
C 5H 8BR203	275.93	273.8841	21.77	2.92	57.92	17.40
C 5H 8BR204	291.93	289.8790	20.57	2.76	54.74	21.92
C 5H 8BR205	307.92	305.8739	19.50	2.62	51.90	25.98
C 5H 8BR206	323.92	321.8689	18.54	2.49	49.34	29.64
C 5H 8BR207	339.92	337.8638	17.67	2.37	47.01	32.95
C 5H 8BR208	355.92	353.8587	16.87	2.27	44.90	35.96
C 5H 8BR209	371.92	369.8536	16.15	2.17	42.97	38.72
C 5H 9BR101	165.03	163.9837	36.39	5.50	48.42	9.69
C 5H 9BR102	181.03	179.9786	33.17	5.01	44.14	17.68
C 5H 9BR103	197.03	195.9736	30.48	4.60	40.55	24.36
C 5H 9BR104	213.03	211.9685	28.19	4.26	37.51	30.04
C 5H 9BR105	229.03	227.9634	26.22	3.96	34.89	34.93
C 5H 9BR106	245.03	243.9583	24.51	3.70	32.61	39.18
C 5H 9BR107	261.03	259.9532	23.01	3.48	30.61	42.91
C 5H 9BR108	277.03	275.9481	21.68	3.27	28.84	46.20
C 5H 9BR109	293.03	291.9430	20.50	3.10	27.27	49.14
C 5H10BR201	245.94	243.9099	24.42	4.10	64.98	6.51
C 5H10BR202	261.94	259.9049	22.93	3.85	61.01	12.22
C 5H10BR203	277.94	275.8998	21.61	3.63	57.50	17.27
C 5H10BR204	293.94	291.8947	20.43	3.43	54.37	21.77
C 5H10BR205	309.94	307.8896	19.38	3.25	51.56	25.81
C 5H10BR206	325.94	323.8845	18.43	3.09	49.03	29.45
C 5H10BR207	341.94	339.8794	17.56	2.95	46.74	32.75
C 5H10BR208	357.94	355.8743	16.78	2.82	44.65	35.76
C 5H10BR209	373.94	371.8693	16.06	2.70	42.74	38.51
C 5H11BR101	167.05	165.9994	35.95	6.64	47.83	9.58
C 5H11BR102	183.05	181.9943	32.81	6.06	43.65	17.48
C 5H11BR103	199.05	197.9892	30.17	5.57	40.14	24.11
C 5H11BR104	215.05	213.9841	27.93	5.16	37.16	29.76
C 5H11BR105	231.04	229.9790	25.99	4.80	34.58	34.62
C 5H11BR106	247.04	245.9739	24.31	4.49	32.34	38.86
C 5H11BR107	263.04	261.9689	22.83	4.22	30.38	42.58
C 5H11BR108	279.04	277.9638	21.52	3.97	28.64	45.87
C 5H11BR109	295.04	293.9587	20.35	3.76	27.08	48.80
C 6H 3BR101	170.99	169.9368	42.15	1.77	46.73	9.36
C 6H 3BR102	186.99	185.9317	38.54	1.62	42.73	17.11
C 6H 3BR103	202.99	201.9266	35.50	1.49	39.36	23.65
C 6H 3BR104	218.99	217.9215	32.91	1.38	36.49	29.22
C 6H 3BR105	234.99	233.9164	30.67	1.29	34.00	34.04
C 6H 3BR106	250.99	249.9113	28.71	1.20	31.84	38.25
C 6H 3BR107	266.99	265.9063	26.99	1.13	29.93	41.95
C 6H 3BR108	282.99	281.9012	25.47	1.07	28.24	45.23
C 6H 3BR109	298.99	297.8961	24.10	1.01	26.72	48.16
C 6H 4BR201	251.91	249.8630	28.61	1.60	63.44	6.35
C 6H 4BR202	267.91	265.8579	26.90	1.50	59.65	11.94
C 6H 4BR203	283.90	281.8528	25.38	1.42	56.29	16.91
C 6H 4BR204	299.90	297.8477	24.03	1.34	53.29	21.34
C 6H 4BR205	315.90	313.8426	22.81	1.28	50.59	25.32
C 6H 4BR206	331.90	329.8376	21.71	1.21	48.15	28.92
C 6H 4BR207	347.90	345.8325	20.71	1.16	45.93	32.19
C 6H 4BR208	363.90	361.8274	19.80	1.11	43.92	35.17
C 6H 4BR209	379.90	377.8223	18.97	1.06	42.07	37.90
C 6H 5BR101	173.01	171.9524	41.65	2.91	46.18	9.25
C 6H 5BR102	189.01	187.9473	38.13	2.67	42.28	16.93
C 6H 5BR103	205.01	203.9423	35.15	2.46	38.98	23.41
C 6H 5BR104	221.01	219.9372	32.61	2.28	36.15	28.96
C 6H 5BR105	237.01	235.9321	30.41	2.13	33.71	33.75
C 6H 5BR106	253.01	251.9270	28.48	1.99	31.58	37.94
C 6H 5BR107	269.01	267.9219	26.79	1.87	29.70	41.63
C 6H 5BR108	285.01	283.9168	25.29	1.77	28.04	44.91
C 6H 5BR109	301.01	299.9117	23.94	1.67	26.55	47.84
C 6H 6BR201	253.92	251.8786	28.38	2.38	62.94	6.30
C 6H 6BR202	269.92	267.8736	26.70	2.24	59.21	11.85
C 6H 6BR203	285.92	283.8685	25.21	2.12	55.89	16.79
C 6H 6BR204	301.92	299.8634	23.87	2.00	52.93	21.20
C 6H 6BR205	317.92	315.8583	22.67	1.90	50.27	25.16
C 6H 6BR206	333.92	331.8532	21.58	1.81	47.86	28.75
C 6H 6BR207	349.92	347.8481	20.60	1.73	45.67	32.01

012

C 6H 6BR208	365.92	363.8430	19.69	1.65	43.67	34.98
C 6H 6BR209	381.92	379.8380	18.87	1.58	41.84	37.70
C 6H 7BR101	175.03	173.9681	41.17	4.03	45.65	9.14
C 6H 7BR102	191.03	189.9630	37.73	3.69	41.83	16.75
C 6H 7BR103	207.02	205.9579	34.81	3.41	38.60	23.18
C 6H 7BR104	223.02	221.9528	32.31	3.16	35.83	28.70
C 6H 7BR105	239.02	237.9477	30.15	2.95	33.43	33.47
C 6H 7BR106	255.02	253.9426	28.26	2.77	31.33	37.64
C 6H 7BR107	271.02	269.9376	26.59	2.60	29.48	41.32
C 6H 7BR108	287.02	285.9325	25.11	2.46	27.84	44.59
C 6H 7BR109	303.02	301.9274	23.78	2.33	26.37	47.52
C 6H 8BR201	255.94	253.8943	28.16	3.15	62.44	6.25
C 6H 8BR202	271.94	269.8892	26.50	2.97	58.77	11.77
C 6H 8BR203	287.94	285.8841	25.03	2.80	55.50	16.67
C 6H 8BR204	303.94	301.8790	23.71	2.65	52.58	21.06
C 6H 8BR205	319.94	317.8739	22.53	2.52	49.95	25.00
C 6H 8BR206	335.94	333.8689	21.45	2.40	47.57	28.58
C 6H 8BR207	351.93	349.8638	20.48	2.29	45.41	31.82
C 6H 8BR208	367.93	365.8587	19.59	2.19	43.43	34.79
C 6H 8BR209	383.93	381.8536	18.77	2.10	41.62	37.51
C 6H 9BR101	177.04	175.9837	40.71	5.12	45.13	9.04
C 6H 9BR102	193.04	191.9786	37.33	4.70	41.39	16.58
C 6H 9BR103	209.04	207.9736	34.48	4.34	38.22	22.96
C 6H 9BR104	225.04	223.9685	32.02	4.03	35.51	28.44
C 6H 9BR105	241.04	239.9634	29.90	3.76	33.15	33.19
C 6H 9BR106	257.04	255.9583	28.04	3.53	31.09	37.35
C 6H 9BR107	273.04	271.9532	26.39	3.32	29.26	41.02
C 6H 9BR108	289.04	287.9481	24.93	3.14	27.64	44.28
C 6H 9BR109	305.04	303.9430	23.63	2.97	26.19	47.21
C 6H10BR201	257.95	255.9099	27.94	3.91	61.95	6.20
C 6H10BR202	273.95	271.9049	26.31	3.68	58.33	11.68
C 6H10BR203	289.95	287.8998	24.85	3.48	55.12	16.55
C 6H10BR204	305.95	303.8947	23.55	3.29	52.23	20.92
C 6H10BR205	321.95	319.8896	22.38	3.13	49.64	24.85
C 6H10BR206	337.95	335.8845	21.32	2.98	47.29	28.41
C 6H10BR207	353.95	351.8794	20.36	2.85	45.15	31.64
C 6H10BR208	369.95	367.8743	19.48	2.72	43.20	34.60
C 6H10BR209	385.95	383.8693	18.67	2.61	41.41	37.31
C 6H11BR101	179.06	177.9994	40.25	6.19	44.62	8.94
C 6H11BR102	195.06	193.9943	36.95	5.68	40.96	16.40
C 6H11BR103	211.06	209.9892	34.15	5.25	37.86	22.74
C 6H11BR104	227.06	225.9841	31.74	4.88	35.19	28.19
C 6H11BR105	243.06	241.9790	29.65	4.56	32.87	32.91
C 6H11BR106	259.05	257.9739	27.82	4.28	30.84	37.06
C 6H11BR107	275.05	273.9689	26.20	4.03	29.05	40.72
C 6H11BR108	291.05	289.9638	24.76	3.81	27.45	43.98
C 6H11BR109	307.05	305.9587	23.47	3.61	26.02	46.90
C 6H12BR201	259.97	257.9256	27.72	4.65	61.47	6.15
C 6H12BR202	275.97	273.9205	26.11	4.38	57.91	11.60
C 6H12BR203	291.97	289.9154	24.68	4.14	54.73	16.44
C 6H12BR204	307.97	305.9103	23.40	3.93	51.89	20.78
C 6H12BR205	323.97	321.9052	22.25	3.73	49.33	24.69
C 6H12BR206	339.97	337.9002	21.20	3.56	47.01	28.24
C 6H12BR207	355.97	353.8951	20.25	3.40	44.89	31.46
C 6H12BR208	371.97	369.8900	19.37	3.25	42.96	34.41
C 6H12BR209	387.97	385.8849	18.58	3.12	41.16	37.12
C 6H13BR101	181.07	180.0150	39.80	7.24	44.13	8.84
C 6H13BR102	197.07	196.0099	36.57	6.65	40.55	16.24
C 6H13BR103	213.07	212.0049	33.82	6.15	37.50	22.53
C 6H13BR104	229.07	227.9998	31.46	5.72	34.88	27.94
C 6H13BR105	245.07	243.9947	29.41	5.35	32.60	32.64
C 6H13BR106	261.07	259.9896	27.60	5.02	30.61	36.77
C 6H13BR107	277.07	275.9845	26.01	4.73	28.84	40.42
C 6H13BR108	293.07	291.9794	24.59	4.47	27.26	43.67
C 6H13BR109	309.07	307.9743	23.32	4.24	25.85	46.59
C 7H 3BR101	183.01	181.9368	45.94	1.65	43.66	8.74
C 7H 3BR102	199.00	197.9317	42.25	1.52	40.15	16.08
C 7H 3BR103	215.00	213.9266	39.11	1.41	37.16	22.32
C 7H 3BR104	231.00	229.9215	36.40	1.31	34.59	27.70
C 7H 3BR105	247.00	245.9164	34.04	1.22	32.35	32.39
C 7H 3BR106	263.00	261.9113	31.97	1.15	30.38	36.50
C 7H 3BR107	279.00	277.9063	30.14	1.08	28.64	40.14
C 7H 3BR108	295.00	293.9012	28.50	1.03	27.09	43.39
C 7H 3BR109	311.00	309.8961	27.03	.97	25.69	46.30
C 7H 4BR201	263.92	261.8630	31.86	1.53	60.55	6.06
C 7H 4BR202	279.92	277.8579	30.04	1.44	57.09	11.43
C 7H 4BR203	295.92	293.8528	28.41	1.36	54.00	16.22
C 7H 4BR204	311.92	309.8477	26.96	1.29	51.23	20.52
C 7H 4BR205	327.91	325.8426	25.64	1.23	48.73	24.40
C 7H 4BR206	343.91	341.8376	24.45	1.17	46.47	27.91
C 7H 4BR207	359.91	357.8325	23.36	1.12	44.40	31.12
C 7H 4BR208	375.91	373.8274	22.37	1.07	42.51	34.05
C 7H 4BR209	391.91	389.8223	21.45	1.03	40.78	36.74
C 7H 5BR101	185.02	183.9524	45.44	2.72	43.19	8.65
C 7H 5BR102	201.02	199.9473	41.83	2.51	39.75	15.92
C 7H 5BR103	217.02	215.9423	38.74	2.32	36.82	22.12
C 7H 5BR104	233.02	231.9372	36.08	2.16	34.29	27.46
C 7H 5BR105	249.02	247.9321	33.76	2.02	32.09	32.12
C 7H 5BR106	265.02	263.9270	31.73	1.90	30.15	36.22
C 7H 5BR107	281.02	279.9219	29.92	1.79	28.43	39.85
C 7H 5BR108	297.02	295.9168	28.31	1.70	26.90	43.09
C 7H 5BR109	313.02	311.9117	26.86	1.61	25.53	46.00
C 7H 6BR201	265.93	263.8786	31.62	2.27	60.09	6.02
C 7H 6BR202	281.93	279.8736	29.82	2.15	56.68	11.35
C 7H 6BR203	297.93	295.8685	28.22	2.03	53.64	16.11
C 7H 6BR204	313.93	311.8634	26.78	1.93	50.91	20.39
C 7H 6BR205	329.93	327.8583	25.48	1.83	48.44	24.25
C 7H 6BR206	345.93	343.8532	24.30	1.75	46.20	27.75
C 7H 6BR207	361.93	359.8481	23.23	1.67	44.15	30.94
C 7H 6BR208	377.93	375.8430	22.25	1.60	42.29	33.87
C 7H 6BR209	393.93	391.8380	21.34	1.54	40.57	36.55
C 7H 7BR101	187.04	185.9681	44.95	3.77	42.72	8.55
C 7H 7BR102	203.04	201.9630	41.41	3.48	39.35	15.76
C 7H 7BR103	219.04	217.9579	38.39	3.22	36.48	21.91
C 7H 7BR104	235.04	233.9528	35.77	3.00	34.00	27.23
C 7H 7BR105	251.03	249.9477	33.49	2.81	31.83	31.87
C 7H 7BR106	267.03	265.9426	31.49	2.64	29.92	35.95
C 7H 7BR107	283.03	281.9376	29.71	2.49	28.23	39.57
C 7H 7BR108	299.03	297.9325	28.12	2.36	26.72	42.80
C 7H 7BR109	315.03	313.9274	26.69	2.24	25.36	45.71
C 7H 8BR201	267.95	265.8943	31.38	3.01	59.64	5.97
C 7H 8BR202	283.95	281.8892	29.61	2.84	56.28	11.27
C 7H 8BR203	299.95	297.8841	28.03	2.69	53.28	16.00
C 7H 8BR204	315.95	313.8790	26.61	2.55	50.58	20.26
C 7H 8BR205	331.95	329.8739	25.33	2.43	48.14	24.10
C 7H 8BR206	347.95	345.8689	24.16	2.32	45.93	27.59
C 7H 8BR207	363.95	361.8638	23.10	2.22	43.91	30.77
C 7H 8BR208	379.95	377.8587	22.13	2.12	42.06	33.69
C 7H 8BR209	395.94	393.8536	21.23	2.04	40.36	36.37
C 7H 9BR101	189.05	187.9837	44.47	4.80	42.27	8.46

C 7H 9BR102	205.05	203.9786	41.00	4.42	38.97	15.61
C 7H 9BR103	221.05	219.9736	38.04	4.10	36.15	21.71
C 7H 9BR104	237.05	235.9685	35.47	3.83	33.71	27.00
C 7H 9BR105	253.05	251.9634	33.23	3.58	31.58	31.61
C 7H 9BR106	269.05	267.9583	31.25	3.37	29.70	35.68
C 7H 9BR107	285.05	283.9532	29.50	3.18	28.03	39.29
C 7H 9BR108	301.05	299.9481	27.93	3.01	26.54	42.52
C 7H 9BR109	317.05	315.9430	26.52	2.86	25.20	45.42
C 7H10BR201	269.97	267.9099	31.14	3.73	59.20	5.93
C 7H10BR202	285.96	283.9049	29.40	3.52	55.88	11.19
C 7H10BR203	301.96	299.8598	27.84	3.34	52.92	15.90
C 7H10BR204	317.96	315.8947	26.44	3.17	50.26	20.13
C 7H10BR205	333.96	331.8896	25.18	3.02	47.85	23.95
C 7H10BR206	349.96	347.8845	24.02	2.88	45.66	27.43
C 7H10BR207	365.96	363.8794	22.97	2.75	43.67	30.60
C 7H10BR208	381.96	379.8743	22.01	2.64	41.84	33.51
C 7H10BR209	397.96	395.8693	21.13	2.53	40.16	36.18
C 7H11BR101	191.07	189.9994	44.00	5.80	41.82	8.37
C 7H11BR102	207.07	205.9943	40.60	5.35	38.59	15.45
C 7H11BR103	223.07	221.9892	37.69	4.97	35.82	21.52
C 7H11BR104	239.07	237.9841	35.17	4.64	33.42	26.77
C 7H11BR105	255.07	253.9790	32.96	4.35	31.33	31.36
C 7H11BR106	271.07	269.9739	31.02	4.09	29.48	35.41
C 7H11BR107	287.07	285.9689	29.29	3.86	27.83	39.01
C 7H11BR108	303.06	301.9638	27.74	3.66	26.37	42.23
C 7H11BR109	319.06	317.9587	26.35	3.48	25.04	45.13
C 7H12BR201	271.98	269.9256	30.91	4.45	58.76	5.88
C 7H12BR202	287.98	285.9205	29.20	4.20	55.49	11.11
C 7H12BR203	303.98	301.9154	27.66	3.98	52.57	15.79
C 7H12BR204	319.98	317.9103	26.28	3.78	49.94	20.00
C 7H12BR205	335.98	333.9052	25.02	3.60	47.56	23.81
C 7H12BR206	351.98	349.9002	23.89	3.44	45.40	27.27
C 7H12BR207	367.98	365.8951	22.85	3.29	43.43	30.44
C 7H12BR208	383.98	381.8900	21.90	3.15	41.62	33.33
C 7H12BR209	399.98	397.8849	21.02	3.02	39.95	36.00
C 7H13BR101	193.09	192.0150	43.54	6.79	41.38	8.29
C 7H13BR102	209.08	208.0099	40.21	6.27	38.22	15.30
C 7H13BR103	225.08	224.0049	37.35	5.82	35.50	21.32
C 7H13BR104	241.08	239.9998	34.88	5.44	33.14	26.55
C 7H13BR105	257.08	255.9947	32.70	5.10	31.08	31.12
C 7H13BR106	273.08	271.9896	30.79	4.80	29.26	35.15
C 7H13BR107	289.08	287.9845	29.08	4.53	27.64	38.74
C 7H13BR108	305.08	303.9794	27.56	4.30	26.19	41.95
C 7H13BR109	321.08	319.9743	26.19	4.08	24.89	44.85
C 7H14BR201	274.00	271.9412	30.69	5.15	58.32	5.84
C 7H14BR202	290.00	287.9362	28.99	4.87	55.11	11.03
C 7H14BR203	306.00	303.9311	27.48	4.61	52.23	15.69
C 7H14BR204	322.00	319.9260	26.11	4.38	49.63	19.88
C 7H14BR205	337.99	335.9209	24.88	4.18	47.28	23.67
C 7H14BR206	353.99	351.9158	23.75	3.99	45.14	27.12
C 7H14BR207	369.99	367.9107	22.72	3.81	43.19	30.27
C 7H14BR208	385.99	383.9056	21.78	3.66	41.40	33.16
C 7H14BR209	401.99	399.9006	20.92	3.51	39.75	35.82
C 7H15BR101	195.10	194.0307	43.09	7.75	40.96	8.20
C 7H15BR102	211.10	210.0256	39.83	7.16	37.85	15.16
C 7H15BR103	227.10	226.0205	37.02	6.66	35.18	21.14
C 7H15BR104	243.10	242.0154	34.59	6.22	32.87	26.33
C 7H15BR105	259.10	258.0103	32.45	5.84	30.84	30.88
C 7H15BR106	275.10	274.0052	30.56	5.50	29.05	34.90
C 7H15BR107	291.10	290.0002	28.88	5.19	27.45	38.47

C 7H15BR108	307.10	305.9951	27.38	4.92	26.02	41.68
C 7H15BR109	323.10	321.9900	26.02	4.68	24.73	44.57
C 8H 4BR201	275.93	273.8630	34.82	1.46	57.92	5.80
C 8H 4BR202	291.93	289.8579	32.92	1.38	54.74	10.96
C 8H 4BR203	307.93	305.8528	31.21	1.31	51.90	15.59
C 8H 4BR204	323.93	321.8477	29.66	1.24	49.33	19.76
C 8H 4BR205	339.93	337.8426	28.27	1.19	47.01	23.53
C 8H 4BR206	355.93	353.8376	27.00	1.13	44.90	26.97
C 8H 4BR207	371.92	369.8325	25.84	1.08	42.97	30.11
C 8H 4BR208	387.92	385.8274	24.77	1.04	41.20	32.99
C 8H 4BR209	403.92	401.8223	23.79	1.00	39.56	35.65
C 8H 5BR101	197.03	195.9524	48.77	2.56	40.55	8.12
C 8H 5BR102	213.03	211.9473	45.11	2.37	37.51	15.02
C 8H 5BR103	229.03	227.9423	41.95	2.20	34.89	20.96
C 8H 5BR104	245.03	243.9372	39.22	2.06	32.61	26.12
C 8H 5BR105	261.03	259.9321	36.81	1.93	30.61	30.65
C 8H 5BR106	277.03	275.9270	34.69	1.82	28.84	34.65
C 8H 5BR107	293.03	291.9219	32.79	1.72	27.27	38.22
C 8H 5BR108	309.03	307.9168	31.09	1.63	25.86	41.42
C 8H 5BR109	325.03	323.9117	29.56	1.55	24.58	44.30
C 8H 6BR201	277.94	275.8786	34.57	2.18	57.50	5.76
C 8H 6BR202	293.94	291.8736	32.69	2.06	54.37	10.89
C 8H 6BR203	309.94	307.8685	31.00	1.95	51.56	15.49
C 8H 6BR204	325.94	323.8634	29.48	1.86	49.03	19.63
C 8H 6BR205	341.94	339.8583	28.10	1.77	46.74	23.39
C 8H 6BR206	357.94	355.8532	26.84	1.69	44.65	26.82
C 8H 6BR207	373.94	371.8481	25.70	1.62	42.74	29.95
C 8H 6BR208	389.94	387.8430	24.64	1.55	40.98	32.82
C 8H 6BR209	405.94	403.8380	23.67	1.49	39.37	35.47
C 8H 7BR101	199.05	197.9681	48.27	3.54	40.14	8.04
C 8H 7BR102	215.05	213.9630	44.68	3.28	37.16	14.88
C 8H 7BR103	231.05	229.9579	41.59	3.05	34.58	20.77
C 8H 7BR104	247.05	245.9528	38.90	2.86	32.34	25.91
C 8H 7BR105	263.05	261.9477	36.53	2.68	30.38	30.41
C 8H 7BR106	279.05	277.9426	34.43	2.53	28.63	34.40
C 8H 7BR107	295.04	293.9376	32.57	2.39	27.08	37.96
C 8H 7BR108	311.04	309.9325	30.89	2.27	25.69	41.15
C 8H 7BR109	327.04	325.9274	29.38	2.16	24.43	44.03
C 8H 8BR201	279.96	277.8943	34.32	2.88	57.08	5.71
C 8H 8BR202	295.96	293.8892	32.47	2.72	54.00	10.81
C 8H 8BR203	311.96	309.8841	30.80	2.58	51.23	15.39
C 8H 8BR204	327.96	325.8790	29.30	2.46	48.73	19.51
C 8H 8BR205	343.96	341.8739	27.94	2.34	46.46	23.26
C 8H 8BR206	359.96	357.8689	26.69	2.24	44.40	26.67
C 8H 8BR207	375.96	373.8638	25.56	2.14	42.51	29.79
C 8H 8BR208	391.96	389.8587	24.52	2.06	40.77	32.66
C 8H 8BR209	407.96	405.8536	23.55	1.98	39.17	35.30
C 8H 9BR101	201.06	199.9837	47.79	4.51	39.74	7.96
C 8H 9BR102	217.06	215.9786	44.27	4.18	36.81	14.74
C 8H 9BR103	233.06	231.9736	41.23	3.89	34.28	20.59
C 8H 9BR104	249.06	247.9685	38.58	3.64	32.08	25.70
C 8H 9BR105	265.06	263.9634	36.25	3.42	30.15	30.18
C 8H 9BR106	281.06	279.9583	34.19	3.23	28.43	34.15
C 8H 9BR107	297.06	295.9532	32.35	3.05	26.90	37.70
C 8H 9BR108	313.06	311.9481	30.69	2.90	25.52	40.89
C 8H 9BR109	329.06	327.9430	29.20	2.76	24.28	43.76
C 8H10BR201	281.98	279.9099	34.08	3.57	56.67	5.67
C 8H10BR202	297.98	295.9049	32.25	3.38	53.63	10.74
C 8H10BR203	313.98	311.8998	30.60	3.21	50.90	15.29
C 8H10BR204	329.97	327.8947	29.12	3.05	48.43	19.39

014

C 8H10BR205	345.97	343.8896	27.77	2.91	46.19	23.12
C 8H10BR206	361.97	359.8845	26.55	2.78	44.15	26.52
C 8H10BR207	377.97	375.8794	25.42	2.67	42.28	29.63
C 8H10BR208	393.97	391.8743	24.39	2.56	40.56	32.49
C 8H10BR209	409.97	407.8693	23.44	2.46	38.98	35.12
C 8H11BR101	203.08	201.9994	47.32	5.46	39.35	7.88
C 8H11BR102	219.08	217.9943	43.86	5.06	36.47	14.61
C 8H11BR103	235.08	233.9892	40.88	4.72	33.99	20.42
C 8H11BR104	251.08	249.9841	38.27	4.42	31.82	25.49
C 8H11BR105	267.08	265.9790	35.98	4.15	29.92	29.95
C 8H11BR106	283.08	281.9739	33.94	3.92	28.23	33.91
C 8H11BR107	299.08	297.9689	32.13	3.71	26.72	37.45
C 8H11BR108	315.08	313.9638	30.50	3.52	25.36	40.62
C 8H11BR109	331.08	329.9587	29.02	3.35	24.13	43.49
C 8H12BR201	283.99	281.9256	33.84	4.26	56.27	5.63
C 8H12BR202	299.99	297.9205	32.03	4.03	53.27	10.67
C 8H12BR203	315.99	313.9154	30.41	3.83	50.57	15.19
C 8H12BR204	331.99	329.9103	28.94	3.64	48.14	19.28
C 8H12BR205	347.99	345.9052	27.61	3.48	45.92	22.99
C 8H12BR206	363.99	361.9002	26.40	3.32	43.90	26.37
C 8H12BR207	379.99	377.8951	25.29	3.18	42.06	29.47
C 8H12BR208	395.99	393.8900	24.27	3.05	40.36	32.32
C 8H12BR209	411.99	409.8849	23.32	2.94	38.79	34.95
C 8H13BR101	205.10	204.0150	46.85	6.39	38.96	7.80
C 8H13BR102	221.10	220.0099	43.46	5.93	36.14	14.47
C 8H13BR103	237.10	236.0049	40.53	5.53	33.70	20.24
C 8H13BR104	253.09	251.9998	37.97	5.18	31.57	25.29
C 8H13BR105	269.09	267.9947	35.71	4.87	29.69	29.73
C 8H13BR106	285.09	283.9896	33.70	4.60	28.03	33.67
C 8H13BR107	301.09	299.9845	31.91	4.35	26.54	37.20
C 8H13BR108	317.09	315.9794	30.30	4.13	25.20	40.37
C 8H13BR109	333.09	331.9743	28.85	3.93	23.99	43.23
C 8H14BR201	286.01	283.9412	33.60	4.93	55.88	5.59
C 8H14BR202	302.01	299.9362	31.82	4.67	52.92	10.60
C 8H14BR203	318.01	315.9311	30.22	4.44	50.25	15.09
C 8H14BR204	334.01	331.9260	28.77	4.22	47.85	19.16
C 8H14BR205	350.01	347.9209	27.45	4.03	45.66	22.86
C 8H14BR206	366.01	363.9158	26.25	3.86	43.66	26.23
C 8H14BR207	382.00	379.9107	25.15	3.69	41.83	29.32
C 8H14BR208	398.00	395.9056	24.14	3.55	40.15	32.16
C 8H14BR209	414.00	411.9006	23.21	3.41	38.60	34.78
C 8H15BR101	207.11	206.0307	46.39	7.30	38.58	7.72
C 8H15BR102	223.11	222.0256	43.07	6.78	35.81	14.34
C 8H15BR103	239.11	238.0205	40.19	6.32	33.42	20.07
C 8H15BR104	255.11	254.0154	37.67	5.93	31.32	25.09
C 8H15BR105	271.11	270.0103	35.44	5.58	29.47	29.51
C 8H15BR106	287.11	286.0052	33.47	5.27	27.83	33.44
C 8H15BR107	303.11	302.0002	31.70	4.99	26.36	36.95
C 8H15BR108	319.11	317.9951	30.11	4.74	25.04	40.11
C 8H15BR109	335.11	333.9900	28.67	4.51	23.84	42.97
C 8H16BR201	288.02	285.9569	33.36	5.60	55.48	5.55
C 8H16BR202	304.02	301.9518	31.61	5.30	52.56	10.53
C 8H16BR203	320.02	317.9467	30.03	5.04	49.94	15.00
C 8H16BR204	336.02	333.9416	28.60	4.80	47.56	19.05
C 8H16BR205	352.02	349.9365	27.30	4.58	45.40	22.73
C 8H16BR206	368.02	365.9315	26.11	4.38	43.42	26.08
C 8H16BR207	384.02	381.9264	25.02	4.20	41.61	29.16
C 8H16BR208	400.02	397.9213	24.02	4.03	39.95	32.00
C 8H16BR209	416.02	413.9162	23.10	3.88	38.41	34.61
C 8H17BR101	209.13	208.0463	45.95	8.19	38.21	7.65

C 8H17BR102	225.13	224.0412	42.68	7.61	35.49	14.21
C 8H17BR103	241.13	240.0362	39.85	7.11	33.14	19.91
C 8H17BR104	257.13	256.0311	37.37	6.66	31.08	24.89
C 8H17BR105	273.13	272.0260	35.18	6.27	29.26	29.29
C 8H17BR106	289.13	288.0209	33.23	5.93	27.64	33.20
C 8H17BR107	305.12	304.0158	31.49	5.62	26.19	36.70
C 8H17BR108	321.12	320.0107	29.92	5.34	24.88	39.86
C 8H17BR109	337.12	336.0056	28.50	5.08	23.70	42.71
C 9H 4BR201	287.94	285.8630	37.54	1.40	55.50	5.56
C 9H 4BR202	303.94	301.8579	35.57	1.33	52.58	10.53
C 9H 4BR203	319.94	317.8528	33.79	1.26	49.95	15.00
C 9H 4BR204	335.94	333.8477	32.18	1.20	47.57	19.05
C 9H 4BR205	351.94	349.8426	30.72	1.15	45.41	22.73
C 9H 4BR206	367.94	365.8376	29.38	1.10	43.43	26.09
C 9H 4BR207	383.94	381.8325	28.16	1.05	41.62	29.17
C 9H 4BR208	399.94	397.8274	27.03	1.01	39.96	32.00
C 9H 4BR209	415.93	413.8223	25.99	.97	38.42	34.62
C 9H 5BR101	209.04	207.9524	51.71	2.41	38.22	7.65
C 9H 5BR102	225.04	223.9473	48.04	2.24	35.51	14.22
C 9H 5BR103	241.04	239.9423	44.85	2.09	33.15	19.91
C 9H 5BR104	257.04	255.9372	42.06	1.96	31.09	24.90
C 9H 5BR105	273.04	271.9321	39.59	1.85	29.26	29.30
C 9H 5BR106	289.04	287.9270	37.40	1.74	27.64	33.21
C 9H 5BR107	305.04	303.9219	35.44	1.65	26.19	36.72
C 9H 5BR108	321.04	319.9168	33.67	1.57	24.89	39.87
C 9H 5BR109	337.04	335.9117	32.07	1.50	23.71	42.72
C 9H 6BR201	289.96	287.8786	37.28	2.09	55.11	5.52
C 9H 6BR202	305.95	303.8736	35.33	1.98	52.23	10.46
C 9H 6BR203	321.95	319.8685	33.58	1.88	49.64	14.91
C 9H 6BR204	337.95	335.8634	31.99	1.79	47.29	18.94
C 9H 6BR205	353.95	351.8583	30.54	1.71	45.15	22.60
C 9H 6BR206	369.95	367.8532	29.22	1.63	43.20	25.95
C 9H 6BR207	385.95	383.8481	28.01	1.57	41.41	29.02
C 9H 6BR208	401.95	399.8430	26.89	1.50	39.76	31.84
C 9H 6BR209	417.95	415.8380	25.86	1.45	38.24	34.45
C 9H 7BR101	211.06	209.9681	51.22	3.34	37.86	7.58
C 9H 7BR102	227.06	225.9630	47.61	3.11	35.19	14.09
C 9H 7BR103	243.06	241.9579	44.48	2.90	32.87	19.75
C 9H 7BR104	259.06	257.9528	41.73	2.72	30.84	24.70
C 9H 7BR105	275.06	273.9477	39.30	2.57	29.05	29.08
C 9H 7BR106	291.06	289.9426	37.14	2.42	27.45	32.98
C 9H 7BR107	307.06	305.9376	35.21	2.30	26.02	36.47
C 9H 7BR108	323.06	321.9325	33.46	2.18	24.73	39.62
C 9H 7BR109	339.05	337.9274	31.88	2.08	23.57	42.47
C 9H 8BR201	291.97	289.8943	37.02	2.76	54.73	5.48
C 9H 8BR202	307.97	305.8892	35.10	2.62	51.89	10.39
C 9H 8BR203	323.97	321.8841	33.37	2.49	49.33	14.82
C 9H 8BR204	339.97	337.8790	31.80	2.37	47.01	18.82
C 9H 8BR205	355.97	353.8739	30.37	2.27	44.89	22.47
C 9H 8BR206	371.97	369.8689	29.06	2.17	42.96	25.81
C 9H 8BR207	387.97	385.8638	27.86	2.08	41.19	28.87
C 9H 8BR208	403.97	401.8587	26.76	2.00	39.56	31.68
C 9H 8BR209	419.97	417.8536	25.74	1.92	38.05	34.29
C 9H 9BR101	213.08	211.9837	50.73	4.26	37.50	7.51
C 9H 9BR102	229.07	227.9786	47.19	3.96	34.88	13.97
C 9H 9BR103	245.07	243.9736	44.11	3.70	32.60	19.59
C 9H 9BR104	261.07	259.9685	41.41	3.47	30.61	24.51
C 9H 9BR105	277.07	275.9634	39.02	3.27	28.84	28.87
C 9H 9BR106	293.07	291.9583	36.89	3.10	27.26	32.76
C 9H 9BR107	309.07	307.9532	34.98	2.94	25.85	36.24

015

C 9H 9BR108	325.07	323.9481	33.25	2.79	24.58	39.37
C 9H 9BR109	341.07	339.9430	31.69	2.66	23.43	42.22
C 9H10BR201	293.99	291.9099	36.77	3.43	54.36	5.44
C 9H10BR202	309.99	307.9049	34.87	3.25	51.55	10.32
C 9H10BR203	325.99	323.8998	33.16	3.09	49.02	14.72
C 9H10BR204	341.99	339.8947	31.61	2.95	46.73	18.71
C 9H10BR205	357.99	355.8896	30.20	2.82	44.64	22.35
C 9H10BR206	373.98	371.8845	28.91	2.70	42.73	25.67
C 9H10BR207	389.98	387.8794	27.72	2.58	40.98	28.72
C 9H10BR208	405.98	403.8743	26.63	2.48	39.36	31.53
C 9H10BR209	421.98	419.8693	25.62	2.39	37.87	34.12
C 9H11BR101	215.09	213.9994	50.26	5.15	37.15	7.44
C 9H11BR102	231.09	229.9943	46.78	4.80	34.58	13.85
C 9H11BR103	247.09	245.9892	43.75	4.49	32.34	19.43
C 9H11BR104	263.09	261.9841	41.09	4.21	30.37	24.33
C 9H11BR105	279.09	277.9790	38.73	3.97	28.63	28.66
C 9H11BR106	295.09	293.9739	36.63	3.76	27.08	32.53
C 9H11BR107	311.09	309.9689	34.75	3.56	25.69	36.00
C 9H11BR108	327.09	325.9638	33.05	3.39	24.43	39.13
C 9H11BR109	343.09	341.9587	31.51	3.23	23.29	41.97
C 9H12BR201	296.00	293.9256	36.52	4.09	53.99	5.41
C 9H12BR202	312.00	309.9205	34.65	3.88	51.22	10.26
C 9H12BR203	328.00	325.9154	32.96	3.69	48.72	14.63
C 9H12BR204	344.00	341.9103	31.42	3.52	46.46	18.60
C 9H12BR205	360.00	357.9052	30.03	3.36	44.39	22.22
C 9H12BR206	376.00	373.9002	28.75	3.22	42.50	25.53
C 9H12BR207	392.00	389.8951	27.58	3.09	40.77	28.57
C 9H12BR208	408.00	405.8900	26.50	2.96	39.17	31.37
C 9H12BR209	424.00	421.8849	25.50	2.85	37.69	33.96
C 9H13BR101	217.11	216.0150	49.79	6.04	36.80	7.37
C 9H13BR102	233.11	232.0099	46.37	5.62	34.28	13.73
C 9H13BR103	249.11	248.0049	43.40	5.26	32.08	19.27
C 9H13BR104	265.11	263.9998	40.78	4.94	30.14	24.14
C 9H13BR105	281.10	279.9947	38.46	4.66	28.42	28.46
C 9H13BR106	297.10	295.9896	36.38	4.41	26.89	32.31
C 9H13BR107	313.10	311.9845	34.53	4.19	25.52	35.77
C 9H13BR108	329.10	327.9794	32.85	3.98	24.28	38.89
C 9H13BR109	345.10	343.9743	31.32	3.80	23.15	41.73
C 9H14BR201	298.02	295.9412	36.27	4.74	53.62	5.37
C 9H14BR202	314.02	311.9362	34.42	4.49	50.89	10.19
C 9H14BR203	330.02	327.9311	32.76	4.28	48.42	14.54
C 9H14BR204	346.02	343.9260	31.24	4.08	46.18	18.50
C 9H14BR205	362.02	359.9209	29.86	3.90	44.14	22.10
C 9H14BR206	378.02	375.9158	28.60	3.73	42.28	25.39
C 9H14BR207	394.02	391.9107	27.44	3.58	40.56	28.42
C 9H14BR208	410.02	407.9056	26.36	3.44	38.98	31.22
C 9H14BR209	426.01	423.9006	25.37	3.31	37.51	33.80
C 9H15BR101	219.12	218.0307	49.33	6.90	36.47	7.30
C 9H15BR102	235.12	234.0256	45.98	6.43	33.98	13.61
C 9H15BR103	251.12	250.0205	43.05	6.02	31.82	19.11
C 9H15BR104	267.12	266.0154	40.47	5.66	29.91	23.96
C 9H15BR105	283.12	282.0103	38.18	5.34	28.22	28.26
C 9H15BR106	299.12	298.0052	36.14	5.05	26.71	32.09
C 9H15BR107	315.12	314.0002	34.30	4.80	25.36	35.54
C 9H15BR108	331.12	329.9951	32.65	4.57	24.13	38.66
C 9H15BR109	347.12	345.9900	31.14	4.36	23.02	41.48
C 9H16BR201	300.04	297.9569	36.03	5.38	53.26	5.33
C 9H16BR202	316.03	313.9518	34.21	5.10	50.57	10.13
C 9H16BR203	332.03	329.9467	32.56	4.86	48.13	14.46
C 9H16BR204	348.03	345.9416	31.06	4.63	45.92	18.39

C 9H16BR205	364.03	361.9365	29.70	4.43	43.90	21.98
C 9H16BR206	380.03	377.9315	28.45	4.24	42.05	25.26
C 9H16BR207	396.03	393.9264	27.30	4.07	40.35	28.28
C 9H16BR208	412.03	409.9213	26.24	3.91	38.79	31.06
C 9H16BR209	428.03	425.9162	25.26	3.77	37.34	33.64
C 9H17BR101	221.14	220.0463	48.88	7.75	36.13	7.23
C 9H17BR102	237.14	236.0412	45.59	7.23	33.70	13.49
C 9H17BR103	253.14	252.0362	42.70	6.77	31.57	18.96
C 9H17BR104	269.14	268.0311	40.17	6.37	29.69	23.78
C 9H17BR105	285.14	284.0260	37.91	6.01	28.02	28.06
C 9H17BR106	301.14	300.0209	35.90	5.69	26.53	31.88
C 9H17BR107	317.14	316.0158	34.09	5.40	25.20	35.31
C 9H17BR108	333.14	332.0107	32.45	5.14	23.99	38.42
C 9H17BR109	349.13	348.0056	30.96	4.91	22.89	41.24
C 9H18BR201	302.05	299.9725	35.79	6.01	52.91	5.30
C 9H18BR202	318.05	315.9675	33.99	5.70	50.25	10.06
C 9H18BR203	334.05	331.9624	32.36	5.43	47.84	14.37
C 9H18BR204	350.05	347.9573	30.88	5.18	45.65	18.28
C 9H18BR205	366.05	363.9522	29.53	4.96	43.66	21.85
C 9H18BR206	382.05	379.9471	28.29	4.75	41.83	25.13
C 9H18BR207	398.05	395.9420	27.16	4.56	40.15	28.14
C 9H18BR208	414.05	411.9369	26.11	4.38	38.60	30.91
C 9H18BR209	430.05	427.9318	25.14	4.22	37.16	33.48
C 9H19BR101	223.16	222.0620	48.44	8.58	35.81	7.17
C 9H19BR102	239.15	238.0569	45.20	8.01	33.41	13.38
C 9H19BR103	255.15	254.0518	42.37	7.51	31.32	18.81
C 9H19BR104	271.15	270.0467	39.87	7.06	29.47	23.60
C 9H19BR105	287.15	286.0416	37.65	6.67	27.83	27.86
C 9H19BR106	303.15	302.0365	35.66	6.32	26.36	31.67
C 9H19BR107	319.15	318.0315	33.87	6.00	25.04	35.09
C 9H19BR108	335.15	334.0264	32.25	5.71	23.84	38.19
C 9H19BR109	351.15	350.0213	30.78	5.45	22.75	41.01
C10H 5BR101	221.05	219.9524	54.34	2.28	36.15	7.24
C10H 5BR102	237.05	235.9473	50.67	2.13	33.71	13.50
C10H 5BR103	253.05	251.9423	47.46	1.99	31.58	18.97
C10H 5BR104	269.05	267.9372	44.64	1.87	29.70	23.79
C10H 5BR105	285.05	283.9321	42.14	1.77	28.03	28.06
C10H 5BR106	301.05	299.9270	39.90	1.67	26.54	31.89
C10H 5BR107	317.05	315.9219	37.88	1.59	25.20	35.32
C10H 5BR108	333.05	331.9168	36.06	1.51	23.99	38.43
C10H 5BR109	349.05	347.9117	34.41	1.44	22.89	41.25
C10H 6BR201	301.97	299.8786	39.78	2.00	52.92	5.30
C10H 6BR202	317.97	315.8736	37.77	1.90	50.26	10.06
C10H 6BR203	333.97	331.8685	35.97	1.81	47.85	14.37
C10H 6BR204	349.96	347.8634	34.32	1.73	45.66	18.29
C10H 6BR205	365.96	363.8583	32.82	1.65	43.67	21.86
C10H 6BR206	381.96	379.8532	31.45	1.58	41.84	25.13
C10H 6BR207	397.96	395.8481	30.18	1.52	40.16	28.14
C10H 6BR208	413.96	411.8430	29.02	1.46	38.60	30.92
C10H 6BR209	429.96	427.8380	27.94	1.41	37.17	33.49
C10H 7BR101	223.07	221.9681	53.84	3.16	35.82	7.17
C10H 7BR102	239.07	237.9630	50.24	2.95	33.42	13.38
C10H 7BR103	255.07	253.9579	47.09	2.77	31.33	18.82
C10H 7BR104	271.07	269.9528	44.31	2.60	29.48	23.61
C10H 7BR105	287.07	285.9477	41.84	2.46	27.83	27.87
C10H 7BR106	303.07	301.9426	39.63	2.33	26.37	31.67
C10H 7BR107	319.07	317.9376	37.64	2.21	25.04	35.10
C10H 7BR108	335.07	333.9325	35.85	2.11	23.85	38.20
C10H 7BR109	351.07	349.9274	34.21	2.01	22.76	41.02
C10H 8BR201	303.98	301.8943	39.51	2.65	52.57	5.26

016

C10H 8BR202	319.98	317.8892	37.54	2.52	49.94	10.00
C10H 8BR203	335.98	333.8841	35.75	2.40	47.56	14.29
C10H 8BR204	351.98	349.8790	34.12	2.29	45.40	18.18
C10H 8BR205	367.98	365.8739	32.64	2.19	43.43	21.74
C10H 8BR206	383.98	381.8689	31.28	2.10	41.62	25.00
C10H 8BR207	399.98	397.8638	30.03	2.02	39.95	28.00
C10H 8BR208	415.98	413.8587	28.87	1.94	38.42	30.77
C10H 8BR209	431.98	429.8536	27.81	1.87	36.99	33.33
C10H 9BR101	225.09	223.9837	53.36	4.03	35.50	7.11
C10H 9BR102	241.09	239.9786	49.82	3.76	33.14	13.27
C10H 9BR103	257.09	255.9736	46.72	3.53	31.08	18.67
C10H 9BR104	273.08	271.9685	43.98	3.32	29.26	23.44
C10H 9BR105	289.08	287.9634	41.55	3.14	27.64	27.67
C10H 9BR106	305.08	303.9583	39.37	2.97	26.19	31.47
C10H 9BR107	321.08	319.9532	37.41	2.83	24.89	34.88
C10H 9BR108	337.08	335.9481	35.63	2.69	23.70	37.97
C10H 9BR109	353.08	351.9430	34.02	2.57	22.63	40.78
C10H10BR201	306.00	303.9099	39.25	3.29	52.23	5.23
C10H10BR202	322.00	319.9049	37.30	3.13	49.63	9.94
C10H10BR203	338.00	335.8998	35.54	2.98	47.28	14.20
C10H10BR204	354.00	351.8947	33.93	2.85	45.14	18.08
C10H10BR205	370.00	367.8896	32.46	2.72	43.19	21.62
C10H10BR206	386.00	383.8845	31.12	2.61	41.40	24.87
C10H10BR207	401.99	399.8794	29.88	2.51	39.75	27.86
C10H10BR208	417.99	415.8743	28.74	2.41	38.23	30.62
C10H10BR209	433.99	431.8693	27.68	2.32	36.82	33.18
C10H11BR101	227.10	225.9994	52.89	4.88	35.18	7.05
C10H11BR102	243.10	241.9943	49.41	4.56	32.87	13.16
C10H11BR103	259.10	257.9892	46.36	4.28	30.84	18.52
C10H11BR104	275.10	273.9841	43.66	4.03	29.05	23.26
C10H11BR105	291.10	289.9790	41.26	3.81	27.45	27.48
C10H11BR106	307.10	305.9739	39.11	3.61	26.02	31.26
C10H11BR107	323.10	321.9689	37.17	3.43	24.73	34.66
C10H11BR108	339.10	337.9638	35.42	3.27	23.56	37.75
C10H11BR109	355.10	353.9587	33.82	3.12	22.50	40.55
C10H12BR201	308.01	305.9256	39.00	3.93	51.88	5.19
C10H12BR202	324.01	321.9205	37.07	3.73	49.32	9.88
C10H12BR203	340.01	337.9154	35.33	3.56	47.00	14.12
C10H12BR204	356.01	353.9103	33.74	3.40	44.89	17.98
C10H12BR205	372.01	369.9052	32.29	3.25	42.96	21.50
C10H12BR206	388.01	385.9002	30.96	3.12	41.19	24.74
C10H12BR207	404.01	401.8951	29.73	2.99	39.56	27.72
C10H12BR208	420.01	417.8900	28.60	2.88	38.05	30.47
C10H12BR209	436.01	433.8849	27.55	2.77	36.65	33.03
C10H13BR101	229.12	228.0150	52.42	5.72	34.87	6.98
C10H13BR102	245.12	244.0099	49.00	5.35	32.60	13.05
C10H13BR103	261.12	260.0049	46.00	5.02	30.60	18.38
C10H13BR104	277.12	275.9998	43.34	4.73	28.83	23.09
C10H13BR105	293.12	291.9947	40.98	4.47	27.26	27.29
C10H13BR106	309.12	307.9896	38.86	4.24	25.85	31.06
C10H13BR107	325.11	323.9845	36.94	4.03	24.58	34.45
C10H13BR108	341.11	339.9794	35.21	3.84	23.42	37.52
C10H13BR109	357.11	355.9743	33.63	3.67	22.37	40.32
C10H14BR201	310.03	307.9412	38.74	4.55	51.55	5.16
C10H14BR202	326.03	323.9362	36.84	4.33	49.02	9.81
C10H14BR203	342.03	339.9311	35.12	4.13	46.72	14.03
C10H14BR204	358.03	355.9260	33.55	3.94	44.64	17.87
C10H14BR205	374.03	371.9209	32.11	3.77	42.73	21.39
C10H14BR206	390.03	387.9158	30.80	3.62	40.97	24.61
C10H14BR207	406.03	403.9107	29.58	3.48	39.36	27.58

C10H14BR208	422.03	419.9056	28.46	3.34	37.87	30.33
C10H14BR209	438.03	435.9006	27.42	3.22	36.48	32.87
C10H15BR101	231.13	230.0307	51.97	6.54	34.57	6.92
C10H15BR102	247.13	246.0256	48.60	6.12	32.33	12.95
C10H15BR103	263.13	262.0205	45.65	5.75	30.37	18.24
C10H15BR104	279.13	278.0154	43.03	5.42	28.63	22.93
C10H15BR105	295.13	294.0103	40.70	5.12	27.07	27.11
C10H15BR106	311.13	310.0052	38.60	4.86	25.68	30.85
C10H15BR107	327.13	326.0002	36.72	4.62	24.43	34.24
C10H15BR108	343.13	341.9951	35.00	4.41	23.29	37.30
C10H15BR109	359.13	357.9900	33.45	4.21	22.25	40.10
C10H16BR201	312.05	309.9569	38.49	5.17	51.21	5.13
C10H16BR202	328.05	325.9518	36.61	4.92	48.72	9.75
C10H16BR203	344.05	341.9467	34.91	4.69	46.45	13.95
C10H16BR204	360.04	357.9416	33.36	4.48	44.39	17.77
C10H16BR205	376.04	373.9365	31.94	4.29	42.50	21.27
C10H16BR206	392.04	389.9315	30.64	4.11	40.76	24.49
C10H16BR207	408.04	405.9264	29.44	3.95	39.16	27.45
C10H16BR208	424.04	421.9213	28.33	3.80	37.69	30.18
C10H16BR209	440.04	437.9162	27.30	3.66	36.32	32.72
C10H17BR101	233.15	232.0463	51.52	7.35	34.27	6.86
C10H17BR102	249.15	248.0412	48.21	6.88	32.07	12.84
C10H17BR103	265.15	264.0362	45.30	6.46	30.14	18.10
C10H17BR104	281.15	280.0311	42.72	6.09	28.42	22.76
C10H17BR105	297.15	296.0260	40.42	5.77	26.89	26.92
C10H17BR106	313.15	312.0209	38.36	5.47	25.52	30.66
C10H17BR107	329.15	328.0158	36.49	5.21	24.28	34.03
C10H17BR108	345.15	344.0107	34.80	4.96	23.15	37.08
C10H17BR109	361.15	360.0056	33.26	4.74	22.13	39.87
C10H18BR201	314.06	311.9725	38.24	5.78	50.88	5.09
C10H18BR202	330.06	327.9675	36.39	5.50	48.42	9.69
C10H18BR203	346.06	343.9624	34.71	5.24	46.18	13.87
C10H18BR204	362.06	359.9573	33.17	5.01	44.14	17.68
C10H18BR205	378.06	375.9522	31.77	4.80	42.27	21.16
C10H18BR206	394.06	391.9471	30.48	4.60	40.55	24.36
C10H18BR207	410.06	407.9420	29.29	4.42	38.97	27.31
C10H18BR208	426.06	423.9369	28.19	4.26	37.51	30.04
C10H18BR209	442.06	439.9318	27.17	4.10	36.15	32.57
C10H19BR101	235.17	234.0620	51.08	8.14	33.98	6.80
C10H19BR102	251.17	250.0569	47.82	7.63	31.81	12.74
C10H19BR103	267.17	266.0518	44.96	7.17	29.91	17.97
C10H19BR104	283.16	282.0467	42.42	6.76	28.22	22.60
C10H19BR105	299.16	298.0416	40.15	6.40	26.71	26.74
C10H19BR106	315.16	314.0365	38.11	6.08	25.35	30.46
C10H19BR107	331.16	330.0315	36.27	5.78	24.13	33.82
C10H19BR108	347.16	346.0264	34.60	5.52	23.02	36.87
C10H19BR109	363.16	362.0213	33.07	5.27	22.00	39.65
C10H20BR201	316.08	313.9882	38.00	6.38	50.56	5.06
C10H20BR202	332.08	329.9831	36.17	6.07	48.12	9.64
C10H20BR203	348.08	345.9780	34.51	5.79	45.91	13.79
C10H20BR204	364.08	361.9729	32.99	5.54	43.89	17.58
C10H20BR205	380.08	377.9678	31.60	5.30	42.05	21.05
C10H20BR206	396.08	393.9628	30.33	5.09	40.35	24.24
C10H20BR207	412.07	409.9577	29.15	4.89	38.78	27.18
C10H20BR208	428.07	425.9526	28.06	4.71	37.33	29.90
C10H20BR209	444.07	441.9475	27.05	4.54	35.99	32.43
C10H21BR101	237.18	236.0776	50.64	8.92	33.69	6.75
C10H21BR102	253.18	252.0725	47.44	8.36	31.56	12.64
C10H21BR103	269.18	268.0674	44.62	7.86	29.68	17.83
C10H21BR104	285.18	284.0624	42.12	7.42	28.02	22.44

C10H21BR105	301.18	300.0573	39.88	7.03	26.53	26.56
C10H21BR106	317.18	316.0522	37.87	6.67	25.19	30.27
C10H21BR107	333.18	332.0471	36.05	6.35	23.98	33.61
C10H21BR108	349.18	348.0420	34.40	6.06	22.88	36.66
C10H21BR109	365.18	364.0369	32.89	5.80	21.88	39.43
C11H 5BR101	233.07	231.9524	56.69	2.16	34.28	6.86
C11H 5BR102	249.07	247.9473	53.05	2.02	32.08	12.85
C11H 5BR103	265.06	263.9423	49.85	1.90	30.15	18.11
C11H 5BR104	281.06	279.9372	47.01	1.79	28.43	22.77
C11H 5BR105	297.06	295.9321	44.48	1.70	26.90	26.93
C11H 5BR106	313.06	311.9270	42.20	1.61	25.52	30.66
C11H 5BR107	329.06	327.9219	40.15	1.53	24.28	34.03
C11H 5BR108	345.06	343.9168	38.29	1.46	23.16	37.09
C11H 5BR109	361.06	359.9117	36.59	1.40	22.13	39.88
C11H 6BR201	313.98	311.8786	42.08	1.93	50.90	5.10
C11H 6BR202	329.98	327.8736	40.04	1.83	48.43	9.70
C11H 6BR203	345.98	343.8685	38.19	1.75	46.19	13.87
C11H 6BR204	361.98	359.8634	36.50	1.67	44.15	17.68
C11H 6BR205	377.98	375.8583	34.96	1.60	42.28	21.16
C11H 6BR206	393.97	391.8532	33.54	1.54	40.56	24.37
C11H 6BR207	409.97	407.8481	32.23	1.48	38.98	27.32
C11H 6BR208	425.97	423.8430	31.02	1.42	37.52	30.05
C11H 6BR209	441.97	439.8380	29.89	1.37	36.16	32.58
C11H 7BR101	235.08	233.9681	56.20	3.00	33.99	6.81
C11H 7BR102	251.08	249.9630	52.62	2.81	31.82	12.74
C11H 7BR103	267.08	265.9579	49.47	2.64	29.92	17.97
C11H 7BR104	283.08	281.9528	46.67	2.49	28.23	22.61
C11H 7BR105	299.08	297.9477	44.18	2.36	26.72	26.75
C11H 7BR106	315.08	313.9426	41.93	2.24	25.36	30.47
C11H 7BR107	331.08	329.9376	39.91	2.13	24.13	33.83
C11H 7BR108	347.08	345.9325	38.07	2.03	23.02	36.88
C11H 7BR109	363.08	361.9274	36.39	1.94	22.01	39.66
C11H 8BR201	315.99	313.8943	41.81	2.55	50.57	5.06
C11H 8BR202	331.99	329.8892	39.80	2.43	48.14	9.64
C11H 8BR203	347.99	345.8841	37.97	2.32	45.92	13.79
C11H 8BR204	363.99	361.8790	36.30	2.22	43.90	17.58
C11H 8BR205	379.99	377.8739	34.77	2.12	42.06	21.05
C11H 8BR206	395.99	393.8689	33.37	2.04	40.36	24.24
C11H 8BR207	411.99	409.8638	32.07	1.96	38.79	27.18
C11H 8BR208	427.99	425.8587	30.87	1.88	37.34	29.91
C11H 8BR209	443.99	441.8536	29.76	1.82	35.99	32.43
C11H 9BR101	237.10	235.9837	55.72	3.83	33.70	6.75
C11H 9BR102	253.10	251.9786	52.20	3.58	31.57	12.64
C11H 9BR103	269.10	267.9736	49.10	3.37	29.69	17.84
C11H 9BR104	285.10	283.9685	46.34	3.18	28.03	22.45
C11H 9BR105	301.10	299.9634	43.88	3.01	26.54	26.57
C11H 9BR106	317.09	315.9583	41.67	2.86	25.20	30.27
C11H 9BR107	333.09	331.9532	39.67	2.72	23.99	33.62
C11H 9BR108	349.09	347.9481	37.85	2.60	22.89	36.67
C11H 9BR109	365.09	363.9430	36.19	2.48	21.89	39.44
C11H10BR201	318.01	315.9099	41.45	3.17	50.25	5.03
C11H10BR202	334.01	331.9049	39.56	3.02	47.85	9.58
C11H10BR203	350.01	347.8998	37.75	2.88	45.66	13.71
C11H10BR204	366.01	363.8947	36.10	2.75	43.66	17.49
C11H10BR205	382.01	379.8896	34.59	2.64	41.83	20.94
C11H10BR206	398.01	395.8845	33.20	2.53	40.15	24.12
C11H10BR207	414.01	411.8794	31.91	2.43	38.60	27.05
C11H10BR208	430.01	427.8743	30.73	2.34	37.16	29.77
C11H10BR209	446.00	443.8693	29.62	2.26	35.83	32.29
C11H11BR101	239.11	237.9994	55.26	4.64	33.42	6.69
C11H11BR102	255.11	253.9943	51.79	4.35	31.32	12.54
C11H11BR103	271.11	269.9892	48.73	4.09	29.47	17.70
C11H11BR104	287.11	285.9841	46.02	3.86	27.83	22.29
C11H11BR105	303.11	301.9790	43.59	3.66	26.36	26.39
C11H11BR106	319.11	317.9739	41.40	3.47	25.04	30.08
C11H11BR107	335.11	333.9689	39.43	3.31	23.84	33.42
C11H11BR108	351.11	349.9638	37.63	3.16	22.76	36.45
C11H11BR109	367.11	365.9587	35.99	3.02	21.77	39.22
C11H12BR201	320.03	317.9256	41.29	3.78	49.94	5.00
C11H12BR202	336.03	333.9205	39.32	3.60	47.56	9.52
C11H12BR203	352.02	349.9154	37.53	3.44	45.40	13.63
C11H12BR204	368.02	365.9103	35.90	3.29	43.42	17.39
C11H12BR205	384.02	381.9052	34.40	3.15	41.61	20.83
C11H12BR206	400.02	397.9002	33.03	3.02	39.95	24.00
C11H12BR207	416.02	413.8951	31.76	2.91	38.41	26.92
C11H12BR208	432.02	429.8900	30.58	2.80	36.99	29.63
C11H12BR209	448.02	445.8849	29.49	2.70	35.67	32.14
C11H13BR101	241.13	240.0150	54.79	5.43	33.14	6.64
C11H13BR102	257.13	256.0099	51.38	5.10	31.08	12.44
C11H13BR103	273.13	272.0049	48.37	4.80	29.26	17.57
C11H13BR104	289.13	287.9998	45.70	4.53	27.64	22.13
C11H13BR105	305.13	303.9947	43.30	4.29	26.19	26.22
C11H13BR106	321.13	319.9896	41.14	4.08	24.88	29.89
C11H13BR107	337.13	335.9845	39.19	3.89	23.70	33.22
C11H13BR108	353.13	351.9794	37.42	3.71	22.63	36.25
C11H13BR109	369.12	367.9743	35.79	3.55	21.65	39.01
C11H14BR201	322.04	319.9412	41.03	4.38	49.62	4.97
C11H14BR202	338.04	335.9362	39.08	4.17	47.27	9.47
C11H14BR203	354.04	351.9311	37.32	3.99	45.14	13.56
C11H14BR204	370.04	367.9260	35.70	3.81	43.19	17.29
C11H14BR205	386.04	383.9209	34.23	3.66	41.40	20.72
C11H14BR206	402.04	399.9158	32.86	3.51	39.75	23.88
C11H14BR207	418.04	415.9107	31.61	3.38	38.23	26.79
C11H14BR208	434.04	431.9056	30.44	3.25	36.82	29.49
C11H14BR209	450.04	447.9006	29.36	3.14	35.51	32.00
C11H15BR101	243.15	242.0307	54.34	6.22	32.86	6.58
C11H15BR102	259.14	258.0256	50.98	5.83	30.83	12.35
C11H15BR103	275.14	274.0205	48.02	5.50	29.04	17.44
C11H15BR104	291.14	290.0154	45.38	5.19	27.44	21.98
C11H15BR105	307.14	306.0103	43.02	4.92	26.02	26.05
C11H15BR106	323.14	322.0052	40.89	4.68	24.73	29.71
C11H15BR107	339.14	338.0002	38.96	4.46	23.56	33.02
C11H15BR108	355.14	353.9951	37.20	4.26	22.50	36.04
C11H15BR109	371.14	369.9900	35.60	4.07	21.53	38.80
C11H16BR201	324.06	321.9569	40.77	4.98	49.31	4.94
C11H16BR202	340.06	337.9518	38.85	4.74	46.99	9.41
C11H16BR203	356.06	353.9467	37.11	4.53	44.88	13.48
C11H16BR204	372.06	369.9416	35.51	4.33	42.95	17.20
C11H16BR205	388.06	385.9365	34.05	4.16	41.18	20.61
C11H16BR206	404.05	401.9315	32.70	3.99	39.55	23.76
C11H16BR207	420.05	417.9264	31.45	3.84	38.04	26.66
C11H16BR208	436.05	433.9213	30.30	3.70	36.65	29.35
C11H16BR209	452.05	449.9162	29.23	3.57	35.35	31.85
C11H17BR101	245.16	244.0463	53.89	6.99	32.59	6.53
C11H17BR102	261.16	260.0412	50.59	6.56	30.60	12.25
C11H17BR103	277.16	276.0362	47.67	6.18	28.83	17.32
C11H17BR104	293.16	292.0311	45.07	5.85	27.26	21.83
C11H17BR105	309.16	308.0260	42.74	5.54	25.85	25.88
C11H17BR106	325.16	324.0209	40.63	5.27	24.57	29.52
C11H17BR107	341.16	340.0158	38.73	5.02	23.42	32.83

018

C11H17BR108	357.16	356.0107	36.99	4.80	22.37	35.84
C11H17BR109	373.16	372.0056	35.41	4.59	21.41	38.59
C11H18BR201	326.07	323.9725	40.52	5.56	49.01	4.91
C11H18BR202	342.07	339.9675	38.62	5.30	46.72	9.35
C11H18BR203	358.07	355.9624	36.90	5.07	44.63	13.40
C11H18BR204	374.07	371.9573	35.32	4.85	42.72	17.11
C11H18BR205	390.07	387.9522	33.87	4.65	40.97	20.51
C11H18BR206	406.07	403.9471	32.54	4.47	39.35	23.64
C11H18BR207	422.07	419.9420	31.30	4.30	37.86	26.53
C11H18BR208	438.07	435.9369	30.16	4.14	36.48	29.22
C11H18BR209	454.07	451.9318	29.10	4.00	35.19	31.71
C11H19BR101	247.18	246.0620	53.45	7.75	32.33	6.47
C11H19BR102	263.18	262.0569	50.20	7.28	30.36	12.16
C11H19BR103	279.18	278.0518	47.33	6.86	28.62	17.19
C11H19BR104	295.18	294.0467	44.76	6.49	27.07	21.68
C11H19BR105	311.18	310.0416	42.46	6.15	25.68	25.71
C11H19BR106	327.17	326.0365	40.38	5.85	24.42	29.34
C11H19BR107	343.17	342.0315	38.50	5.58	23.28	32.64
C11H19BR108	359.17	358.0264	36.79	5.33	22.25	35.64
C11H19BR109	375.17	374.0213	35.22	5.10	21.30	38.38
C11H20BR201	328.09	325.9882	40.27	6.14	48.71	4.88
C11H20BR202	344.09	341.9831	38.40	5.86	46.44	9.30
C11H20BR203	360.09	357.9780	36.69	5.60	44.38	13.33
C11H20BR204	376.09	373.9729	35.13	5.36	42.49	17.02
C11H20BR205	392.09	389.9678	33.70	5.14	40.76	20.40
C11H20BR206	408.09	405.9628	32.38	4.94	39.16	23.52
C11H20BR207	424.09	421.9577	31.15	4.75	37.68	26.41
C11H20BR208	440.09	437.9526	30.02	4.58	36.31	29.08
C11H20BR209	456.08	453.9475	28.97	4.42	35.04	31.57
C11H21BR101	249.19	248.0776	53.02	8.49	32.07	6.42
C11H21BR102	265.19	264.0725	49.82	7.98	30.13	12.07
C11H21BR103	281.19	280.0674	46.99	7.53	28.42	17.07
C11H21BR104	297.19	296.0624	44.46	7.12	26.89	21.53
C11H21BR105	313.19	312.0573	42.19	6.76	25.51	25.54
C11H21BR106	329.19	328.0522	40.14	6.43	24.27	29.16
C11H21BR107	345.19	344.0471	38.28	6.13	23.15	32.44
C11H21BR108	361.19	360.0420	36.58	5.86	22.12	35.44
C11H21BR109	377.19	376.0369	35.03	5.61	21.18	38.18
C11H22BR201	330.11	328.0038	40.02	6.72	48.41	4.85
C11H22BR202	346.10	343.9987	38.17	6.41	46.17	9.25
C11H22BR203	362.10	359.9937	36.49	6.12	44.13	13.26
C11H22BR204	378.10	375.9886	34.94	5.86	42.27	16.93
C11H22BR205	394.10	391.9835	33.52	5.63	40.55	20.30
C11H22BR206	410.10	407.9784	32.22	5.41	38.97	23.41
C11H22BR207	426.10	423.9733	31.01	5.20	37.50	26.28
C11H22BR208	442.10	439.9682	29.89	5.02	36.15	28.95
C11H22BR209	458.10	455.9631	28.84	4.84	34.88	31.43
C11H23BR101	251.21	250.0933	52.59	9.23	31.81	6.37
C11H23BR102	267.21	266.0882	49.45	8.68	29.90	11.98
C11H23BR103	283.21	282.0831	46.65	8.19	28.21	16.95
C11H23BR104	299.21	298.0780	44.16	7.75	26.71	21.39
C11H23BR105	315.21	314.0729	41.92	7.35	25.35	25.38
C11H23BR106	331.21	330.0678	39.89	7.00	24.13	28.98
C11H23BR107	347.21	346.0628	38.05	6.68	23.01	32.26
C11H23BR108	363.21	362.0577	36.38	6.38	22.00	35.24
C11H23BR109	379.20	378.0526	34.84	6.11	21.07	37.97
C12H 6BR201	325.99	323.8786	44.21	1.86	49.02	4.91
C12H 6BR202	341.99	339.8736	42.15	1.77	46.73	9.36
C12H 6BR203	357.99	355.8685	40.26	1.69	44.64	13.41
C12H 6BR204	373.99	371.8634	38.54	1.62	42.73	17.11
C12H 6BR205	389.99	387.8583	36.96	1.55	40.98	20.51
C12H 6BR206	405.99	403.8532	35.50	1.49	39.36	23.65
C12H 6BR207	421.99	419.8481	34.16	1.43	37.87	26.54
C12H 6BR208	437.98	435.8430	32.91	1.38	36.49	29.22
C12H 6BR209	453.98	451.8380	31.75	1.33	35.20	31.72
C12H 7BR101	247.09	245.9681	58.33	2.86	32.34	6.48
C12H 7BR102	263.09	261.9630	54.78	2.68	30.37	12.16
C12H 7BR103	279.09	277.9579	51.64	2.53	28.63	17.20
C12H 7BR104	295.09	293.9528	48.84	2.39	27.08	21.69
C12H 7BR105	311.09	309.9477	46.33	2.27	25.69	25.72
C12H 7BR106	327.09	325.9426	44.07	2.16	24.43	29.35
C12H 7BR107	343.09	341.9376	42.01	2.06	23.29	32.64
C12H 7BR108	359.09	357.9325	40.14	1.96	22.25	35.64
C12H 7BR109	375.09	373.9274	38.43	1.88	21.30	38.39
C12H 8BR201	328.00	325.8943	43.94	2.46	48.72	4.88
C12H 8BR202	344.00	341.8892	41.90	2.34	46.46	9.30
C12H 8BR203	360.00	357.8841	40.04	2.24	44.39	13.33
C12H 8BR204	376.00	373.8790	38.33	2.14	42.50	17.02
C12H 8BR205	392.00	389.8739	36.77	2.06	40.77	20.41
C12H 8BR206	408.00	405.8689	35.33	1.98	39.17	23.53
C12H 8BR207	424.00	421.8638	33.99	1.90	37.69	26.41
C12H 8BR208	440.00	437.8587	32.76	1.83	36.32	29.09
C12H 8BR209	456.00	453.8536	31.61	1.77	35.05	31.58
C12H 9BR101	249.11	247.9837	57.86	3.64	32.08	6.42
C12H 9BR102	265.11	263.9786	54.37	3.42	30.14	12.07
C12H 9BR103	281.11	279.9736	51.27	3.23	28.42	17.07
C12H 9BR104	297.11	295.9685	48.51	3.05	26.89	21.54
C12H 9BR105	313.11	311.9634	46.03	2.90	25.52	25.55
C12H 9BR106	329.11	327.9583	43.80	2.76	24.28	29.17
C12H 9BR107	345.11	343.9532	41.77	2.63	23.15	32.45
C12H 9BR108	361.10	359.9481	39.91	2.51	22.13	35.45
C12H 9BR109	377.10	375.9430	38.22	2.41	21.19	38.18
C12H10BR201	330.02	327.9099	43.67	3.05	48.42	4.85
C12H10BR202	346.02	343.9049	41.65	2.91	46.18	9.25
C12H10BR203	362.02	359.8998	39.81	2.78	44.14	13.26
C12H10BR204	378.02	375.8947	38.13	2.67	42.28	16.93
C12H10BR205	394.02	391.8896	36.58	2.56	40.56	20.30
C12H10BR206	410.02	407.8845	35.15	2.46	38.98	23.41
C12H10BR207	426.02	423.8794	33.83	2.37	37.51	26.29
C12H10BR208	442.02	439.8743	32.61	2.28	36.15	28.96
C12H10BR209	458.02	455.8693	31.47	2.20	34.89	31.44
C12H11BR101	251.12	249.9994	57.40	4.42	31.82	6.37
C12H11BR102	267.12	265.9943	53.96	4.15	29.91	11.98
C12H11BR103	283.12	281.9892	50.91	3.92	28.22	16.95
C12H11BR104	299.12	297.9841	48.19	3.71	26.71	21.40
C12H11BR105	315.12	313.9790	45.74	3.52	25.36	25.39
C12H11BR106	331.12	329.9739	43.53	3.35	24.13	28.99
C12H11BR107	347.12	345.9689	41.52	3.19	23.02	32.26
C12H11BR108	363.12	361.9638	39.69	3.05	22.00	35.25
C12H11BR109	379.12	377.9587	38.02	2.92	21.08	37.98
C12H12BR201	332.04	329.9256	43.41	3.64	48.13	4.82
C12H12BR202	348.04	345.9205	41.41	3.48	45.92	9.19
C12H12BR203	364.04	361.9154	39.59	3.32	43.90	13.19
C12H12BR204	380.04	377.9103	37.93	3.18	42.05	16.84
C12H12BR205	396.03	393.9052	36.39	3.05	40.35	20.20
C12H12BR206	412.03	409.9002	34.98	2.94	38.79	23.30
C12H12BR207	428.03	425.8951	33.67	2.83	37.34	26.17
C12H12BR208	444.03	441.8900	32.46	2.72	35.99	28.83
C12H12BR209	460.03	457.8849	31.33	2.63	34.74	31.30
C12H13BR101	253.14	252.0150	56.94	5.18	31.57	6.32

019

C12H13BR102	269.14	268.0099	53.55	4.87	29.69	11.89
C12H13BR103	285.14	284.0049	50.55	4.60	28.02	16.83
C12H13BR104	301.14	299.9998	47.86	4.35	26.53	21.25
C12H13BR105	317.14	315.9947	45.45	4.13	25.20	25.22
C12H13BR106	333.14	331.9896	43.27	3.93	23.99	28.82
C12H13BR107	349.14	347.9845	41.28	3.75	22.89	32.08
C12H13BR108	365.14	363.9794	39.47	3.59	21.88	35.05
C12H13BR109	381.14	379.9743	37.82	3.44	20.96	37.78
C12H14BR201	334.05	331.9412	43.15	4.22	47.84	4.79
C12H14BR202	350.05	347.9362	41.17	4.03	45.65	9.14
C12H14BR203	366.05	363.9311	39.38	3.86	43.66	13.11
C12H14BR204	382.05	379.9260	37.73	3.69	41.83	16.75
C12H14BR205	398.05	395.9209	36.21	3.55	40.15	20.10
C12H14BR206	414.05	411.9158	34.81	3.41	38.60	23.18
C12H14BR207	430.05	427.9107	33.52	3.28	37.16	26.04
C12H14BR208	446.05	443.9056	32.31	3.16	35.83	28.70
C12H14BR209	462.05	459.9006	31.19	3.05	34.59	31.16
C12H15BR101	255.16	254.0307	56.45	5.93	31.32	6.27
C12H15BR102	271.16	270.0256	53.16	5.58	29.47	11.80
C12H15BR103	287.16	286.0205	50.19	5.27	27.83	16.72
C12H15BR104	303.15	302.0154	47.54	4.99	26.36	21.11
C12H15BR105	319.15	318.0103	45.16	4.74	25.04	25.07
C12H15BR106	335.15	334.0052	43.01	4.51	23.84	28.64
C12H15BR107	351.15	350.0002	41.05	4.31	22.75	31.89
C12H15BR108	367.15	365.9951	39.26	4.12	21.76	34.86
C12H15BR109	383.15	381.9900	37.62	3.95	20.85	37.58
C12H16BR201	336.07	333.9569	42.89	4.80	47.55	4.76
C12H16BR202	352.07	349.9518	40.94	4.58	45.39	9.09
C12H16BR203	368.07	365.9467	39.16	4.38	43.42	13.04
C12H16BR204	384.07	381.9416	37.53	4.20	41.61	16.66
C12H16BR205	400.07	397.9365	36.03	4.03	39.95	20.00
C12H16BR206	416.07	413.9315	34.64	3.88	38.41	23.07
C12H16BR207	432.07	429.9264	33.36	3.73	36.99	25.92
C12H16BR208	448.06	445.9213	32.17	3.60	35.67	28.57
C12H16BR209	464.06	461.9162	31.06	3.48	34.44	31.03
C12H17BR101	257.17	256.0463	56.05	6.66	31.07	6.22
C12H17BR102	273.17	272.0412	52.76	6.27	29.25	11.71
C12H17BR103	289.17	288.0362	49.84	5.93	27.63	16.60
C12H17BR104	305.17	304.0311	47.23	5.62	26.18	20.97
C12H17BR105	321.17	320.0260	44.88	5.34	24.88	24.91
C12H17BR106	337.17	336.0209	42.75	5.08	23.70	28.47
C12H17BR107	353.17	352.0158	40.81	4.85	22.62	31.71
C12H17BR108	369.17	368.0107	39.04	4.64	21.64	34.67
C12H17BR109	385.17	384.0056	37.42	4.45	20.75	37.38
C12H18BR201	338.08	335.9725	42.63	5.37	47.27	4.73
C12H18BR202	354.08	351.9675	40.71	5.12	45.13	9.04
C12H18BR203	370.08	367.9624	38.95	4.90	43.18	12.97
C12H18BR204	386.08	383.9573	37.33	4.70	41.39	16.58
C12H18BR205	402.08	399.9522	35.85	4.51	39.75	19.90
C12H18BR206	418.08	415.9471	34.48	4.34	38.22	22.96
C12H18BR207	434.08	431.9420	33.20	4.18	36.82	25.80
C12H18BR208	450.08	447.9369	32.02	4.03	35.51	28.44
C12H18BR209	466.08	463.9318	30.92	3.89	34.29	30.89
C12H19BR101	259.19	258.0620	55.61	7.39	30.83	6.17
C12H19BR102	275.19	274.0569	52.38	6.96	29.04	11.63
C12H19BR103	291.19	290.0518	49.50	6.58	27.44	16.48
C12H19BR104	307.19	306.0467	46.92	6.23	26.01	20.83
C12H19BR105	323.19	322.0416	44.60	5.93	24.72	24.75
C12H19BR106	339.19	338.0365	42.49	5.65	23.56	28.30
C12H19BR107	355.19	354.0315	40.58	5.39	22.50	31.53
C12H19BR108	371.18	370.0264	38.83	5.16	21.53	34.48
C12H19BR109	387.18	386.0213	37.23	4.95	20.64	37.19
C12H20BR201	340.10	337.9882	42.38	5.93	46.99	4.70
C12H20BR202	356.10	353.9831	40.48	5.66	44.88	8.99
C12H20BR203	372.10	369.9780	38.74	5.42	42.95	12.90
C12H20BR204	388.10	385.9729	37.14	5.19	41.18	16.49
C12H20BR205	404.10	401.9678	35.67	4.99	39.55	19.80
C12H20BR206	420.10	417.9628	34.31	4.80	38.04	22.85
C12H20BR207	436.10	433.9577	33.05	4.62	36.65	25.68
C12H20BR208	452.10	449.9526	31.88	4.46	35.35	28.31
C12H20BR209	468.10	465.9475	30.79	4.31	34.14	30.76
C12H21BR101	261.20	260.0776	55.18	8.10	30.59	6.13
C12H21BR102	277.20	276.0725	52.00	7.64	28.82	11.54
C12H21BR103	293.20	292.0674	49.16	7.22	27.25	16.37
C12H21BR104	309.20	308.0624	46.61	6.85	25.84	20.70
C12H21BR105	325.20	324.0573	44.32	6.51	24.57	24.60
C12H21BR106	341.20	340.0522	42.24	6.20	23.42	28.13
C12H21BR107	357.20	356.0471	40.35	5.93	22.37	31.35
C12H21BR108	373.20	372.0420	38.62	5.67	21.41	34.30
C12H21BR109	389.20	388.0369	37.03	5.44	20.53	37.00
C12H22BR201	342.12	340.0038	42.13	6.48	46.71	4.68
C12H22BR202	358.12	355.9987	40.25	6.19	44.62	8.94
C12H22BR203	374.12	371.9937	38.53	5.93	42.72	12.83
C12H22BR204	390.11	387.9886	36.95	5.68	40.96	16.40
C12H22BR205	406.11	403.9835	35.49	5.46	39.35	19.70
C12H22BR206	422.11	419.9784	34.15	5.25	37.86	22.74
C12H22BR207	438.11	435.9733	32.90	5.06	36.48	25.56
C12H22BR208	454.11	451.9682	31.74	4.88	35.19	28.19
C12H22BR209	470.11	467.9631	30.66	4.72	33.99	30.63
C12H23BR101	263.22	262.0933	54.76	8.81	30.36	6.08
C12H23BR102	279.22	278.0882	51.62	8.30	28.62	11.46
C12H23BR103	295.22	294.0831	48.82	7.85	27.07	16.26
C12H23BR104	311.22	310.0780	46.31	7.45	25.67	20.56
C12H23BR105	327.22	326.0729	44.05	7.08	24.42	24.45
C12H23BR106	343.22	342.0678	41.99	6.75	23.28	27.97
C12H23BR107	359.22	358.0628	40.12	6.45	22.24	31.18
C12H23BR108	375.22	374.0577	38.41	6.18	21.30	34.11
C12H23BR109	391.22	390.0526	36.84	5.93	20.42	36.81
C12H24BR201	344.13	342.0195	41.88	7.03	46.44	4.65
C12H24BR202	360.13	358.0144	40.02	6.72	44.37	8.89
C12H24BR203	376.13	374.0093	38.32	6.43	42.49	12.76
C12H24BR204	392.13	390.0042	36.76	6.17	40.75	16.32
C12H24BR205	408.13	405.9991	35.32	5.93	39.16	19.60
C12H24BR206	424.13	421.9941	33.98	5.70	37.68	22.63
C12H24BR207	440.13	437.9890	32.75	5.50	36.31	25.45
C12H24BR208	456.13	453.9839	31.60	5.30	35.04	28.06
C12H24BR209	472.13	469.9788	30.53	5.12	33.85	30.50
C12H25BR101	265.24	264.1089	54.34	9.50	30.13	6.03
C12H25BR102	281.24	280.1038	51.25	8.96	28.41	11.38
C12H25BR103	297.24	296.0987	48.49	8.48	26.88	16.15
C12H25BR104	313.23	312.0937	46.01	8.04	25.51	20.43
C12H25BR105	329.23	328.0886	43.78	7.65	24.27	24.30
C12H25BR106	345.23	344.0835	41.75	7.30	23.14	27.81
C12H25BR107	361.23	360.0784	39.90	6.98	22.12	31.00
C12H25BR108	377.23	376.0733	38.21	6.68	21.18	33.93
C12H25BR109	393.23	392.0682	36.65	6.41	20.32	36.62
C13H 6BR201	338.00	335.8786	46.20	1.79	47.28	4.73
C13H 6BR202	354.00	351.8736	44.11	1.71	45.14	9.04
C13H 6BR203	370.00	367.8685	42.20	1.63	43.19	12.97
C13H 6BR204	386.00	383.8634	40.45	1.57	41.40	16.58

020

C13H 6BR205	402.00	399.8583	38.84	1.50	39.75	19.90
C13H 6BR206	418.00	415.8532	37.36	1.45	38.23	22.97
C13H 6BR207	434.00	431.8481	35.98	1.39	36.82	25.81
C13H 6BR208	450.00	447.8430	34.70	1.34	35.51	28.44
C13H 6BR209	466.00	463.8380	33.51	1.30	34.29	30.90
C13H 7BR101	259.10	257.9681	60.26	2.72	30.84	6.17
C13H 7BR102	275.10	273.9630	56.76	2.56	29.05	11.63
C13H 7BR103	291.10	289.9579	53.64	2.42	27.45	16.49
C13H 7BR104	307.10	305.9528	50.84	2.30	26.02	20.84
C13H 7BR105	323.10	321.9477	48.33	2.18	24.73	24.76
C13H 7BR106	339.10	337.9426	46.05	2.08	23.56	28.31
C13H 7BR107	355.10	353.9376	43.97	1.99	22.50	31.54
C13H 7BR108	371.10	369.9325	42.08	1.90	21.53	34.49
C13H 7BR109	387.10	385.9274	40.34	1.82	20.64	37.20
C13H 8BR201	340.02	337.8943	45.92	2.37	47.00	4.71
C13H 8BR202	356.02	353.8892	43.86	2.27	44.89	8.99
C13H 8BR203	372.01	369.8841	41.97	2.17	42.96	12.90
C13H 8BR204	388.01	385.8790	40.24	2.08	41.19	16.49
C13H 8BR205	404.01	401.8739	38.65	2.00	39.56	19.80
C13H 8BR206	420.01	417.8689	37.18	1.92	38.05	22.86
C13H 8BR207	436.01	433.8638	35.81	1.85	36.65	25.69
C13H 8BR208	452.01	449.8587	34.54	1.78	35.35	28.32
C13H 8BR209	468.01	465.8536	33.36	1.72	34.15	30.77
C13H 9BR101	261.12	259.9837	59.80	3.47	30.60	6.13
C13H 9BR102	277.12	275.9786	56.35	3.27	28.83	11.55
C13H 9BR103	293.12	291.9736	53.27	3.09	27.26	16.37
C13H 9BR104	309.12	307.9685	50.51	2.93	25.85	20.70
C13H 9BR105	325.12	323.9634	48.03	2.79	24.58	24.61
C13H 9BR106	341.12	339.9583	45.77	2.66	23.42	28.14
C13H 9BR107	357.12	355.9532	43.72	2.54	22.37	31.36
C13H 9BR108	373.12	371.9481	41.85	2.43	21.42	34.30
C13H 9BR109	389.12	387.9430	40.13	2.33	20.53	37.01
C13H10BR201	342.03	339.9099	45.65	2.95	46.72	4.68
C13H10BR202	358.03	355.9049	43.61	2.82	44.64	8.94
C13H10BR203	374.03	371.8998	41.75	2.69	42.73	12.83
C13H10BR204	390.03	387.8947	40.03	2.58	40.97	16.41
C13H10BR205	406.03	403.8896	38.46	2.48	39.36	19.70
C13H10BR206	422.03	419.8845	37.00	2.39	37.87	22.75
C13H10BR207	438.03	435.8794	35.65	2.30	36.48	25.57
C13H10BR208	454.03	451.8743	34.39	2.22	35.20	28.19
C13H10BR209	470.03	467.8693	33.22	2.14	34.00	30.64
C13H11BR101	263.14	261.9994	59.34	4.21	30.37	6.08
C13H11BR102	279.14	277.9943	55.94	3.97	28.63	11.46
C13H11BR103	295.13	293.9892	52.91	3.76	27.07	16.26
C13H11BR104	311.13	309.9841	50.19	3.56	25.68	20.57
C13H11BR105	327.13	325.9790	47.73	3.39	24.43	24.45
C13H11BR106	343.13	341.9739	45.51	3.23	23.29	27.98
C13H11BR107	359.13	357.9689	43.48	3.09	22.25	31.19
C13H11BR108	375.13	373.9638	41.62	2.96	21.30	34.12
C13H11BR109	391.13	389.9587	39.92	2.83	20.43	36.81
C13H12BR201	344.05	341.9256	45.38	3.52	46.45	4.65
C13H12BR202	360.05	357.9205	43.37	3.36	44.39	8.89
C13H12BR203	376.05	373.9154	41.52	3.22	42.50	12.76
C13H12BR204	392.05	389.9103	39.83	3.09	40.76	16.32
C13H12BR205	408.05	405.9052	38.27	2.96	39.16	19.60
C13H12BR206	424.04	421.9002	36.82	2.85	37.69	22.64
C13H12BR207	440.04	437.8951	35.48	2.75	36.32	25.45
C13H12BR208	456.04	453.8900	34.24	2.65	35.04	28.07
C13H12BR209	472.04	469.8849	33.08	2.56	33.85	30.50
C13H13BR101	265.15	264.0150	58.89	4.94	30.14	6.03
C13H13BR102	281.15	280.0099	55.54	4.66	28.42	11.38
C13H13BR103	297.15	296.0049	52.55	4.41	26.89	16.15
C13H13BR104	313.15	311.9998	49.86	4.18	25.52	20.44
C13H13BR105	329.15	327.9947	47.44	3.98	24.28	24.30
C13H13BR106	345.15	343.9896	45.24	3.80	23.15	27.81
C13H13BR107	361.15	359.9845	43.24	3.63	22.12	31.01
C13H13BR108	377.15	375.9794	41.40	3.47	21.19	33.94
C13H13BR109	393.15	391.9743	39.72	3.33	20.32	36.63
C13H14BR201	346.06	343.9412	45.12	4.08	46.18	4.62
C13H14BR202	362.06	359.9362	43.13	3.90	44.14	8.84
C13H14BR203	378.06	375.9311	41.30	3.73	42.27	12.70
C13H14BR204	394.06	391.9260	39.62	3.58	40.55	16.24
C13H14BR205	410.06	407.9209	38.08	3.44	38.97	19.51
C13H14BR206	426.06	423.9158	36.65	3.31	37.51	22.53
C13H14BR207	442.06	439.9107	35.32	3.19	36.15	25.33
C13H14BR208	458.06	455.9056	34.09	3.08	34.89	27.94
C13H14BR209	474.06	471.9006	32.94	2.98	33.71	30.37
C13H15BR101	267.17	266.0307	58.44	5.66	29.91	5.99
C13H15BR102	283.17	282.0256	55.14	5.34	28.22	11.30
C13H15BR103	299.17	298.0205	52.19	5.05	26.71	16.04
C13H15BR104	315.17	314.0154	49.54	4.80	25.35	20.31
C13H15BR105	331.17	330.0103	47.15	4.57	24.13	24.16
C13H15BR106	347.16	346.0052	44.98	4.36	23.02	27.65
C13H15BR107	363.16	362.0002	43.00	4.16	22.00	30.84
C13H15BR108	379.16	377.9951	41.18	3.99	21.07	33.76
C13H15BR109	395.16	393.9900	39.51	3.83	20.22	36.44
C13H16BR201	348.08	345.9569	44.86	4.63	45.91	4.60
C13H16BR202	364.08	361.9518	42.89	4.43	43.89	8.79
C13H16BR203	380.08	377.9467	41.08	4.24	42.05	12.63
C13H16BR204	396.08	393.9416	39.42	4.07	40.35	16.16
C13H16BR205	412.08	409.9365	37.89	3.91	38.78	19.41
C13H16BR206	428.08	425.9315	36.48	3.77	37.33	22.43
C13H16BR207	444.08	441.9264	35.16	3.63	35.99	25.22
C13H16BR208	460.08	457.9213	33.94	3.51	34.74	27.82
C13H16BR209	476.08	473.9162	32.80	3.39	33.57	30.25
C13H17BR101	269.18	268.0463	58.01	6.37	29.68	5.94
C13H17BR102	285.18	284.0412	54.75	6.01	28.02	11.22
C13H17BR103	301.18	300.0362	51.84	5.69	26.53	15.94
C13H17BR104	317.18	316.0311	49.23	5.40	25.19	20.18
C13H17BR105	333.18	332.0260	46.86	5.14	23.98	24.01
C13H17BR106	349.18	348.0209	44.72	4.91	22.88	27.49
C13H17BR107	365.18	364.0158	42.76	4.69	21.88	30.67
C13H17BR108	381.18	380.0107	40.96	4.50	20.96	33.58
C13H17BR109	397.18	396.0056	39.31	4.31	20.12	36.25
C13H18BR201	350.10	347.9725	44.60	5.18	45.65	4.57
C13H18BR202	366.10	363.9675	42.65	4.96	43.65	8.74
C13H18BR203	382.09	379.9624	40.87	4.75	41.82	12.56
C13H18BR204	398.09	395.9573	39.22	4.56	40.14	16.08
C13H18BR205	414.09	411.9522	37.71	4.38	38.59	19.32
C13H18BR206	430.09	427.9471	36.30	4.22	37.16	22.32
C13H18BR207	446.09	443.9420	35.00	4.07	35.82	25.11
C13H18BR208	462.09	459.9369	33.79	3.93	34.58	27.70
C13H18BR209	478.09	475.9318	32.66	3.79	33.43	30.12
C13H19BR101	271.20	270.0620	57.58	7.06	29.46	5.90
C13H19BR102	287.20	286.0569	54.37	6.67	27.82	11.14
C13H19BR103	303.20	302.0518	51.50	6.32	26.35	15.83
C13H19BR104	319.20	318.0467	48.92	6.00	25.03	20.05
C13H19BR105	335.20	334.0416	46.58	5.71	23.84	23.87
C13H19BR106	351.20	350.0365	44.46	5.45	22.75	27.33
C13H19BR107	367.20	366.0315	42.52	5.22	21.76	30.50

C13H19BR108	383.20	382.0264	40.75	5.00	20.85	33.40
C13H19BR109	399.19	398.0213	39.11	4.80	20.02	36.07
C13H20BR201	352.11	349.9882	44.35	5.73	45.39	4.54
C13H20BR202	368.11	365.9831	42.42	5.48	43.41	8.69
C13H20BR203	384.11	381.9780	40.65	5.25	41.60	12.50
C13H20BR204	400.11	397.9729	39.03	5.04	39.94	16.00
C13H20BR205	416.11	413.9678	37.52	4.84	38.41	19.22
C13H20BR206	432.11	429.9628	36.14	4.67	36.98	22.22
C13H20BR207	448.11	445.9577	34.85	4.50	35.66	24.99
C13H20BR208	464.11	461.9526	33.64	4.34	34.43	27.58
C13H20BR209	480.11	477.9475	32.52	4.20	33.29	29.99
C13H21BR101	273.22	272.0776	57.15	7.75	29.25	5.86
C13H21BR102	289.22	288.0725	53.99	7.32	27.63	11.06
C13H21BR103	305.21	304.0674	51.16	6.94	26.18	15.73
C13H21BR104	321.21	320.0624	48.61	6.59	24.88	19.92
C13H21BR105	337.21	336.0573	46.30	6.28	23.70	23.72
C13H21BR106	353.21	352.0522	44.21	5.99	22.62	27.18
C13H21BR107	369.21	368.0471	42.29	5.73	21.64	30.33
C13H21BR108	385.21	384.0420	40.53	5.49	20.74	33.23
C13H21BR109	401.21	400.0369	38.92	5.28	19.92	35.89
C13H22BR201	354.13	352.0038	44.09	6.26	45.13	4.52
C13H22BR202	370.13	367.9987	42.19	5.99	43.18	8.65
C13H22BR203	386.13	383.9937	40.44	5.74	41.39	12.43
C13H22BR204	402.13	399.9886	38.83	5.51	39.74	15.91
C13H22BR205	418.13	415.9835	37.34	5.30	38.22	19.13
C13H22BR206	434.12	431.9784	35.97	5.11	36.81	22.11
C13H22BR207	450.12	447.9733	34.69	4.93	35.50	24.88
C13H22BR208	466.12	463.9682	33.50	4.76	34.28	27.46
C13H22BR209	482.12	479.9631	32.39	4.60	33.15	29.87
C13H23BR101	275.23	274.0933	56.73	8.42	29.03	5.81
C13H23BR102	291.23	290.0882	53.62	7.96	27.44	10.99
C13H23BR103	307.23	306.0831	50.82	7.55	26.01	15.62
C13H23BR104	323.23	322.0780	48.31	7.17	24.72	19.80
C13H23BR105	339.23	338.0729	46.03	6.83	23.55	23.58
C13H23BR106	355.23	354.0678	43.96	6.53	22.49	27.02
C13H23BR107	371.23	370.0628	42.06	6.25	21.52	30.17
C13H23BR108	387.23	386.0577	40.32	5.99	20.63	33.05
C13H23BR109	403.23	402.0526	38.72	5.75	19.82	35.71
C13H24BR201	356.14	354.0195	43.84	6.79	44.87	4.49
C13H24BR202	372.14	370.0144	41.96	6.50	42.94	8.60
C13H24BR203	388.14	386.0093	40.23	6.23	41.17	12.37
C13H24BR204	404.14	402.0042	38.64	5.99	39.54	15.84
C13H24BR205	420.14	417.9991	37.16	5.76	38.04	19.04
C13H24BR206	436.14	433.9941	35.80	5.55	36.64	22.01
C13H24BR207	452.14	449.9890	34.53	5.35	35.34	24.77
C13H24BR208	468.14	465.9839	33.35	5.17	34.14	27.34
C13H24BR209	484.14	481.9788	32.25	5.00	33.01	29.74
C13H25BR101	277.25	276.1089	56.32	9.09	28.82	5.77
C13H25BR102	293.25	292.1038	53.25	8.59	27.25	10.91
C13H25BR103	309.25	308.0987	50.49	8.15	25.84	15.52
C13H25BR104	325.25	324.0937	48.01	7.75	24.57	19.68
C13H25BR105	341.25	340.0886	45.76	7.38	23.42	23.44
C13H25BR106	357.24	356.0835	43.71	7.05	22.37	26.87
C13H25BR107	373.24	372.0784	41.83	6.75	21.41	30.01
C13H25BR108	389.24	388.0733	40.11	6.47	20.53	32.88
C13H25BR109	405.24	404.0682	38.53	6.22	19.72	35.53
C13H26BR201	358.16	356.0351	43.60	7.32	44.62	4.47
C13H26BR202	374.16	372.0300	41.73	7.00	42.71	8.55
C13H26BR203	390.16	388.0250	40.02	6.72	40.96	12.30
C13H26BR204	406.16	404.0199	38.44	6.45	39.35	15.76

C13H26BR205	422.16	420.0148	36.99	6.21	37.86	18.95
C13H26BR206	438.16	436.0097	35.64	5.98	36.47	21.91
C13H26BR207	454.16	452.0046	34.38	5.77	35.19	24.66
C13H26BR208	470.16	467.9995	33.21	5.57	33.99	27.22
C13H26BR209	486.15	483.9944	32.12	5.39	32.87	29.62
C13H27BR101	279.26	278.1246	55.91	9.75	28.61	5.73
C13H27BR102	295.26	294.1195	52.88	9.22	27.06	10.84
C13H27BR103	311.26	310.1144	50.17	8.74	25.67	15.42
C13H27BR104	327.26	326.1093	47.71	8.32	24.42	19.56
C13H27BR105	343.26	342.1042	45.49	7.93	23.28	23.30
C13H27BR106	359.26	358.0991	43.46	7.58	22.24	26.72
C13H27BR107	375.26	374.0941	41.61	7.25	21.29	29.84
C13H27BR108	391.26	390.0890	39.91	6.96	20.42	32.71
C13H27BR109	407.26	406.0839	38.34	6.68	19.62	35.36
C14H 7BR101	271.12	269.9681	62.02	2.60	29.47	5.90
C14H 7BR102	287.11	285.9630	58.57	2.46	27.83	11.14
C14H 7BR103	303.11	301.9579	55.48	2.33	26.36	15.84
C14H 7BR104	319.11	317.9528	52.69	2.21	25.04	20.05
C14H 7BR105	335.11	333.9477	50.18	2.11	23.84	23.87
C14H 7BR106	351.11	349.9426	47.89	2.01	22.76	27.34
C14H 7BR107	367.11	365.9376	45.81	1.92	21.77	30.51
C14H 7BR108	383.11	381.9325	43.89	1.84	20.86	33.41
C14H 7BR109	399.11	397.9274	42.13	1.77	20.02	36.08
C14H 8BR201	352.03	349.8943	47.77	2.29	45.40	4.54
C14H 8BR202	368.03	365.8892	45.69	2.19	43.42	8.69
C14H 8BR203	384.03	381.8841	43.79	2.10	41.61	12.50
C14H 8BR204	400.03	397.8790	42.04	2.02	39.95	16.00
C14H 8BR205	416.02	413.8739	40.42	1.94	38.41	19.23
C14H 8BR206	432.02	429.8689	38.92	1.87	36.99	22.22
C14H 8BR207	448.02	445.8638	37.53	1.80	35.67	25.00
C14H 8BR208	464.02	461.8587	36.24	1.74	34.44	27.58
C14H 8BR209	480.02	477.8536	35.03	1.68	33.29	30.00
C14H 9BR101	273.13	271.9837	61.57	3.32	29.25	5.86
C14H 9BR102	289.13	287.9786	58.16	3.14	27.64	11.07
C14H 9BR103	305.13	303.9736	55.11	2.97	26.19	15.73
C14H 9BR104	321.13	319.9685	52.36	2.82	24.88	19.93
C14H 9BR105	337.13	335.9634	49.88	2.69	23.70	23.73
C14H 9BR106	353.13	351.9583	47.62	2.57	22.63	27.18
C14H 9BR107	369.13	367.9532	45.56	2.46	21.65	30.34
C14H 9BR108	385.13	383.9481	43.66	2.36	20.75	33.23
C14H 9BR109	401.13	399.9430	41.92	2.26	19.92	35.90
C14H10BR201	354.04	351.9099	47.50	2.85	45.14	4.52
C14H10BR202	370.04	367.9049	45.44	2.72	43.19	8.65
C14H10BR203	386.04	383.8998	43.56	2.61	41.40	12.43
C14H10BR204	402.04	399.8947	41.83	2.51	39.75	15.92
C14H10BR205	418.04	415.8896	40.22	2.41	38.23	19.14
C14H10BR206	434.04	431.8845	38.74	2.32	36.82	22.12
C14H10BR207	450.04	447.8794	37.36	2.24	35.51	24.89
C14H10BR208	466.04	463.8743	36.08	2.16	34.29	27.46
C14H10BR209	482.04	479.8693	34.88	2.09	33.15	29.87
C14H11BR101	275.15	273.9994	61.11	4.03	29.04	5.81
C14H11BR102	291.15	289.9943	57.76	3.81	27.44	10.99
C14H11BR103	307.15	305.9892	54.75	3.61	26.01	15.63
C14H11BR104	323.15	321.9841	52.04	3.43	24.73	19.80
C14H11BR105	339.14	337.9790	49.58	3.27	23.56	23.59
C14H11BR106	355.14	353.9739	47.35	3.12	22.50	27.03
C14H11BR107	371.14	369.9689	45.31	2.99	21.53	30.18
C14H11BR108	387.14	385.9638	43.44	2.86	20.64	33.06
C14H11BR109	403.14	401.9587	41.71	2.75	19.82	35.72
C14H12BR201	356.06	353.9256	47.23	3.40	44.88	4.49

C14H12BR202	372.06	369.9205	45.20	3.25	42.95	8.60
C14H12BR203	388.06	385.9154	43.33	3.12	41.18	12.37
C14H12BR204	404.06	401.9103	41.62	2.99	39.55	15.84
C14H12BR205	420.06	417.9052	40.03	2.88	38.04	19.04
C14H12BR206	436.06	433.9002	38.56	2.77	36.65	22.01
C14H12BR207	452.06	449.8951	37.20	2.68	35.35	24.77
C14H12BR208	468.05	465.8900	35.93	2.58	34.14	27.35
C14H12BR209	484.05	481.8849	34.74	2.50	33.01	29.75
C14H13BR101	277.16	276.0150	60.67	4.73	28.83	5.77
C14H13BR102	293.16	292.0099	57.36	4.47	27.26	10.92
C14H13BR103	309.16	308.0049	54.39	4.24	25.85	15.53
C14H13BR104	325.16	323.9998	51.71	4.03	24.57	19.68
C14H13BR105	341.16	339.9947	49.29	3.84	23.42	23.45
C14H13BR106	357.16	355.9896	47.08	3.67	22.37	26.88
C14H13BR107	373.16	371.9845	45.06	3.51	21.41	30.01
C14H13BR108	389.16	387.9794	43.21	3.37	20.53	32.89
C14H13BR109	405.16	403.9743	41.50	3.23	19.72	35.54
C14H14BR201	358.08	355.9412	46.96	3.94	44.63	4.47
C14H14BR202	374.07	371.9362	44.95	3.77	42.72	8.55
C14H14BR203	390.07	387.9311	43.11	3.62	40.97	12.30
C14H14BR204	406.07	403.9260	41.41	3.48	39.35	15.76
C14H14BR205	422.07	419.9209	39.84	3.34	37.86	18.95
C14H14BR206	438.07	435.9158	38.39	3.22	36.48	21.91
C14H14BR207	454.07	451.9107	37.03	3.11	35.19	24.66
C14H14BR208	470.07	467.9056	35.77	3.00	34.00	27.23
C14H14BR209	486.07	483.9006	34.60	2.90	32.88	29.62
C14H15BR101	279.18	278.0307	60.23	5.42	28.62	5.73
C14H15BR102	295.18	294.0256	56.97	5.12	27.07	10.84
C14H15BR103	311.18	310.0205	54.04	4.86	25.68	15.42
C14H15BR104	327.18	326.0154	51.40	4.62	24.42	19.56
C14H15BR105	343.18	342.0103	49.00	4.41	23.28	23.31
C14H15BR106	359.18	358.0052	46.82	4.21	22.25	26.73
C14H15BR107	375.18	374.0002	44.82	4.03	21.30	29.85
C14H15BR108	391.17	389.9951	42.99	3.87	20.43	32.72
C14H15BR109	407.17	405.9900	41.30	3.71	19.62	35.36
C14H16BR201	360.09	357.9569	46.70	4.48	44.38	4.44
C14H16BR202	376.09	373.9518	44.71	4.29	42.49	8.51
C14H16BR203	392.09	389.9467	42.89	4.11	40.76	12.24
C14H16BR204	408.09	405.9416	41.21	3.95	39.16	15.68
C14H16BR205	424.09	421.9365	39.65	3.80	37.68	18.86
C14H16BR206	440.09	437.9315	38.21	3.66	36.31	21.81
C14H16BR207	456.09	453.9264	36.87	3.54	35.04	24.56
C14H16BR208	472.09	469.9213	35.62	3.42	33.85	27.11
C14H16BR209	488.09	485.9162	34.45	3.30	32.74	29.50
C14H17BR101	281.19	280.0463	59.80	6.09	28.42	5.69
C14H17BR102	297.19	296.0412	56.58	5.77	26.89	10.77
C14H17BR103	313.19	312.0362	53.69	5.47	25.51	15.33
C14H17BR104	329.19	328.0311	51.08	5.21	24.27	19.44
C14H17BR105	345.19	344.0260	48.71	4.96	23.15	23.17
C14H17BR106	361.19	360.0209	46.56	4.74	22.12	26.58
C14H17BR107	377.19	376.0158	44.58	4.54	21.18	29.69
C14H17BR108	393.19	392.0107	42.77	4.36	20.32	32.55
C14H17BR109	409.19	408.0056	41.09	4.19	19.53	35.19
C14H18BR201	362.11	359.9725	46.44	5.01	44.13	4.42
C14H18BR202	378.11	375.9675	44.47	4.80	42.27	8.46
C14H18BR203	394.11	391.9624	42.67	4.60	40.55	12.18
C14H18BR204	410.11	407.9573	41.00	4.42	38.97	15.61
C14H18BR205	426.10	423.9522	39.46	4.26	37.50	18.77
C14H18BR206	442.10	439.9471	38.04	4.10	36.15	21.71
C14H18BR207	458.10	455.9420	36.71	3.96	34.88	24.45

C14H18BR208	474.10	471.9369	35.47	3.83	33.71	27.00
C14H18BR209	490.10	487.9318	34.31	3.70	32.61	29.38
C14H19BR101	283.21	282.0620	59.37	6.76	28.21	5.65
C14H19BR102	299.21	298.0569	56.20	6.40	26.70	10.69
C14H19BR103	315.21	314.0518	53.35	6.08	25.35	15.23
C14H19BR104	331.21	330.0467	50.77	5.78	24.12	19.32
C14H19BR105	347.21	346.0416	48.43	5.52	23.01	23.04
C14H19BR106	363.21	362.0365	46.30	5.27	22.00	26.43
C14H19BR107	379.21	378.0315	44.34	5.05	21.07	29.53
C14H19BR108	395.21	394.0264	42.55	4.85	20.22	32.39
C14H19BR109	411.21	410.0213	40.89	4.66	19.43	35.02
C14H20BR201	364.12	361.9882	46.18	5.54	43.89	4.39
C14H20BR202	380.12	377.9831	44.24	5.30	42.04	8.42
C14H20BR203	396.12	393.9780	42.45	5.09	40.34	12.12
C14H20BR204	412.12	409.9729	40.80	4.89	38.78	15.53
C14H20BR205	428.12	425.9678	39.28	4.71	37.33	18.69
C14H20BR206	444.12	441.9628	37.86	4.54	35.98	21.61
C14H20BR207	460.12	457.9577	36.55	4.38	34.73	24.34
C14H20BR208	476.12	473.9526	35.32	4.23	33.56	26.88
C14H20BR209	492.12	489.9475	34.17	4.10	32.47	29.26
C14H21BR101	285.23	284.0776	58.96	7.42	28.01	5.61
C14H21BR102	301.23	300.0725	55.82	7.03	26.53	10.62
C14H21BR103	317.23	316.0674	53.01	6.67	25.19	15.13
C14H21BR104	333.23	332.0624	50.46	6.35	23.98	19.21
C14H21BR105	349.22	348.0573	48.15	6.06	22.88	22.91
C14H21BR106	365.22	364.0522	46.04	5.80	21.88	26.28
C14H21BR107	381.22	380.0471	44.11	5.55	20.96	29.38
C14H21BR108	397.22	396.0420	42.33	5.33	20.12	32.22
C14H21BR109	413.22	412.0369	40.69	5.12	19.34	34.85
C14H22BR201	366.14	364.0038	45.93	6.06	43.65	4.37
C14H22BR202	382.14	379.9987	44.00	5.80	41.82	8.37
C14H22BR203	398.14	395.9937	42.24	5.57	40.14	12.06
C14H22BR204	414.14	411.9886	40.60	5.35	38.59	15.45
C14H22BR205	430.14	427.9835	39.09	5.16	37.15	18.60
C14H22BR206	446.14	443.9784	37.69	4.97	35.82	21.52
C14H22BR207	462.14	459.9733	36.39	4.80	34.58	24.23
C14H22BR208	478.13	475.9682	35.17	4.64	33.42	26.77
C14H22BR209	494.13	491.9631	34.03	4.49	32.34	29.14
C14H23BR101	287.24	286.0933	58.54	8.07	27.82	5.57
C14H23BR102	303.24	302.0882	55.45	7.65	26.35	10.55
C14H23BR103	319.24	318.0831	52.67	7.26	25.03	15.04
C14H23BR104	335.24	334.0780	50.16	6.92	23.83	19.09
C14H23BR105	351.24	350.0729	47.87	6.60	22.75	22.78
C14H23BR106	367.24	366.0678	45.79	6.31	21.76	26.14
C14H23BR107	383.24	382.0628	43.88	6.05	20.85	29.22
C14H23BR108	399.24	398.0577	42.12	5.81	20.01	32.06
C14H23BR109	415.24	414.0526	40.50	5.58	19.24	34.68
C14H24BR201	368.15	366.0195	45.68	6.57	43.41	4.35
C14H24BR202	384.15	382.0144	43.77	6.30	41.60	8.33
C14H24BR203	400.15	398.0093	42.02	6.05	39.94	11.99
C14H24BR204	416.15	414.0042	40.41	5.81	38.40	15.38
C14H24BR205	432.15	429.9991	38.91	5.60	36.98	18.51
C14H24BR206	448.15	445.9941	37.52	5.40	35.66	21.42
C14H24BR207	464.15	461.9890	36.23	5.21	34.43	24.13
C14H24BR208	480.15	477.9839	35.02	5.04	33.28	26.66
C14H24BR209	496.15	493.9788	33.89	4.88	32.21	29.02
C14H25BR101	289.26	288.1089	58.13	8.71	27.62	5.53
C14H25BR102	305.26	304.1038	55.09	8.26	26.18	10.48
C14H25BR103	321.26	320.0987	52.34	7.84	24.87	14.94
C14H25BR104	337.26	336.0937	49.86	7.47	23.69	18.98

023

C14H25BR105	353.26	352.0886	47.60	7.13	22.62	22.65
C14H25BR106	369.26	368.0835	45.54	6.82	21.64	26.00
C14H25BR107	385.26	384.0784	43.65	6.54	20.74	29.07
C14H25BR108	401.25	400.0733	41.91	6.28	19.91	31.90
C14H25BR109	417.25	416.0682	40.30	6.04	19.15	34.51
C14H26BR201	370.17	368.0351	45.43	7.08	43.17	4.32
C14H26BR202	386.17	384.0300	43.54	6.79	41.38	8.29
C14H26BR203	402.17	400.0250	41.81	6.52	39.74	11.93
C14H26BR204	418.17	416.0199	40.21	6.27	38.22	15.30
C14H26BR205	434.17	432.0148	38.73	6.04	36.81	18.43
C14H26BR206	450.17	448.0097	37.35	5.82	35.50	21.32
C14H26BR207	466.17	464.0046	36.07	5.62	34.28	24.02
C14H26BR208	482.17	479.9995	34.88	5.44	33.14	26.55
C14H26BR209	498.17	495.9944	33.76	5.26	32.08	28.90
C14H27BR101	291.27	290.1246	57.73	9.34	27.43	5.49
C14H27BR102	307.27	306.1195	54.73	8.86	26.00	10.41
C14H27BR103	323.27	322.1144	52.02	8.42	24.72	14.85
C14H27BR104	339.27	338.1093	49.56	8.02	23.55	18.86
C14H27BR105	355.27	354.1042	47.33	7.66	22.49	22.52
C14H27BR106	371.27	370.0991	45.29	7.33	21.52	25.86
C14H27BR107	387.27	386.0941	43.42	7.03	20.63	28.92
C14H27BR108	403.27	402.0890	41.70	6.75	19.81	31.74
C14H27BR109	419.27	418.0839	40.11	6.49	19.06	34.34
C14H28BR201	372.19	370.0508	45.18	7.58	42.94	4.30
C14H28BR202	388.19	386.0457	43.32	7.27	41.17	8.24
C14H28BR203	404.19	402.0406	41.60	6.98	39.54	11.88
C14H28BR204	420.18	418.0355	40.02	6.72	38.03	15.23
C14H28BR205	436.18	434.0304	38.55	6.47	36.64	18.34
C14H28BR206	452.18	450.0254	37.19	6.24	35.34	21.23
C14H28BR207	468.18	466.0203	35.92	6.03	34.13	23.92
C14H28BR208	484.18	482.0152	34.73	5.83	33.01	26.44
C14H28BR209	500.18	498.0101	33.62	5.64	31.95	28.79
C14H29BR101	293.29	292.1402	57.33	9.97	27.24	5.46
C14H29BR102	309.29	308.1351	54.37	9.45	25.83	10.35
C14H29BR103	325.29	324.1300	51.69	8.99	24.56	14.76
C14H29BR104	341.29	340.1250	49.27	8.56	23.41	18.75
C14H29BR105	357.29	356.1199	47.06	8.18	22.36	22.39
C14H29BR106	373.29	372.1148	45.05	7.83	21.41	25.72
C14H29BR107	389.29	388.1097	43.20	7.51	20.53	28.77
C14H29BR108	405.29	404.1046	41.49	7.21	19.72	31.58
C14H29BR109	421.29	420.0995	39.91	6.94	18.97	34.18
C15H 7BR101	283.13	281.9681	63.63	2.49	28.22	5.65
C15H 7BR102	299.13	297.9630	60.23	2.36	26.71	10.70
C15H 7BR103	315.13	313.9579	57.17	2.24	25.36	15.23
C15H 7BR104	331.12	329.9528	54.41	2.13	24.13	19.33
C15H 7BR105	347.12	345.9477	51.90	2.03	23.02	23.05
C15H 7BR106	363.12	361.9426	49.62	1.94	22.00	26.44
C15H 7BR107	379.12	377.9376	47.52	1.86	21.08	29.54
C15H 7BR108	395.12	393.9325	45.60	1.79	20.22	32.39
C15H 7BR109	411.12	409.9274	43.82	1.72	19.44	35.02
C15H 8BR201	364.04	361.8943	49.49	2.22	43.90	4.39
C15H 8BR202	380.04	377.8892	47.41	2.12	42.05	8.42
C15H 8BR203	396.04	393.8841	45.49	2.04	40.35	12.12
C15H 8BR204	412.04	409.8790	43.73	1.96	38.78	15.53
C15H 8BR205	428.04	425.8739	42.09	1.88	37.34	18.69
C15H 8BR206	444.04	441.8689	40.57	1.82	35.99	21.62
C15H 8BR207	460.03	457.8638	39.16	1.75	34.74	24.35
C15H 8BR208	476.03	473.8587	37.85	1.69	33.57	26.89
C15H 8BR209	492.03	489.8536	36.62	1.64	32.48	29.27
C15H 9BR101	285.14	283.9837	63.15	3.18	28.02	5.61

C15H 9BR102	301.14	299.9786	59.83	3.01	26.53	10.63
C15H 9BR103	317.14	315.9736	56.81	2.86	25.20	15.13
C15H 9BR104	333.14	331.9685	54.08	2.72	23.99	19.21
C15H 9BR105	349.14	347.9634	51.60	2.60	22.89	22.91
C15H 9BR106	365.14	363.9583	49.34	2.48	21.88	26.29
C15H 9BR107	381.14	379.9532	47.27	2.38	20.96	29.38
C15H 9BR108	397.14	395.9481	45.37	2.28	20.12	32.23
C15H 9BR109	413.14	411.9430	43.61	2.20	19.34	34.85
C15H10BR201	366.05	363.9099	49.22	2.75	43.66	4.37
C15H10BR202	382.05	379.9049	47.16	2.64	41.83	8.38
C15H10BR203	398.05	395.8998	45.26	2.53	40.15	12.06
C15H10BR204	414.05	411.8947	43.51	2.43	38.60	15.46
C15H10BR205	430.05	427.8896	41.89	2.34	37.16	18.60
C15H10BR206	446.05	443.8845	40.39	2.26	35.83	21.52
C15H10BR207	462.05	459.8794	38.99	2.18	34.59	24.24
C15H10BR208	478.05	475.8743	37.69	2.11	33.43	26.77
C15H10BR209	494.05	491.8693	36.47	2.04	32.35	29.15
C15H11BR101	287.16	285.9994	62.74	3.86	27.83	5.57
C15H11BR102	303.16	301.9943	59.43	3.66	26.36	10.56
C15H11BR103	319.16	317.9892	56.45	3.47	25.04	15.04
C15H11BR104	335.16	333.9841	53.76	3.31	23.84	19.09
C15H11BR105	351.16	349.9790	51.31	3.16	22.75	22.78
C15H11BR106	367.16	365.9739	49.07	3.02	21.76	26.15
C15H11BR107	383.15	381.9689	47.02	2.89	20.85	29.23
C15H11BR108	399.15	397.9638	45.14	2.78	20.02	32.07
C15H11BR109	415.15	413.9587	43.40	2.67	19.25	34.68
C15H12BR201	368.07	365.9256	48.95	3.29	43.42	4.35
C15H12BR202	384.07	381.9205	46.91	3.15	41.61	8.33
C15H12BR203	400.07	397.9154	45.03	3.02	39.95	12.00
C15H12BR204	416.07	413.9103	43.30	2.91	38.41	15.38
C15H12BR205	432.07	429.9052	41.70	2.80	36.99	18.51
C15H12BR206	448.07	445.9002	40.21	2.70	35.67	21.42
C15H12BR207	464.07	461.8951	38.82	2.61	34.44	24.13
C15H12BR208	480.07	477.8900	37.53	2.52	33.29	26.66
C15H12BR209	496.07	493.8849	36.32	2.44	32.22	29.03
C15H13BR101	289.17	288.0150	62.30	4.53	27.63	5.53
C15H13BR102	305.17	304.0099	59.04	4.29	26.18	10.49
C15H13BR103	321.17	320.0049	56.10	4.08	24.88	14.94
C15H13BR104	337.17	335.9998	53.43	3.89	23.70	18.98
C15H13BR105	353.17	351.9947	51.01	3.71	22.62	22.65
C15H13BR106	369.17	367.9896	48.80	3.55	21.64	26.00
C15H13BR107	385.17	383.9845	46.78	3.40	20.75	29.08
C15H13BR108	401.17	399.9794	44.91	3.27	19.92	31.91
C15H13BR109	417.17	415.9743	43.19	3.14	19.15	34.52
C15H14BR201	370.09	367.9412	48.68	3.81	43.18	4.32
C15H14BR202	386.09	383.9362	46.67	3.66	41.39	8.29
C15H14BR203	402.09	399.9311	44.81	3.51	39.74	11.94
C15H14BR204	418.08	415.9260	43.09	3.38	38.22	15.31
C15H14BR205	434.08	431.9209	41.51	3.25	36.82	18.43
C15H14BR206	450.08	447.9158	40.03	3.14	35.51	21.33
C15H14BR207	466.08	463.9107	38.66	3.03	34.29	24.03
C15H14BR208	482.08	479.9056	37.37	2.93	33.15	26.55
C15H14BR209	498.08	495.9006	36.17	2.83	32.08	28.91
C15H15BR101	291.19	290.0307	61.87	5.19	27.44	5.49
C15H15BR102	307.19	306.0256	58.65	4.92	26.01	10.42
C15H15BR103	323.19	322.0205	55.75	4.68	24.72	14.85
C15H15BR104	339.19	338.0154	53.12	4.46	23.56	18.87
C15H15BR105	355.19	354.0103	50.72	4.26	22.50	22.52
C15H15BR106	371.19	370.0052	48.54	4.07	21.53	25.86
C15H15BR107	387.19	386.0002	46.53	3.90	20.64	28.93

024

C15H15BR108	403.19	401.9951	44.69	3.75	19.82	31.75
C15H15BR109	419.19	417.9900	42.98	3.61	19.06	34.35
C15H16BR201	372.10	369.9569	48.42	4.33	42.95	4.30
C15H16BR202	388.10	385.9518	46.42	4.16	41.18	8.24
C15H16BR203	404.10	401.9467	44.58	3.99	39.55	11.88
C15H16BR204	420.10	417.9416	42.89	3.84	38.04	15.23
C15H16BR205	436.10	433.9365	41.31	3.70	36.64	18.34
C15H16BR206	452.10	449.9315	39.85	3.57	35.35	21.23
C15H16BR207	468.10	465.9264	38.49	3.45	34.14	23.93
C15H16BR208	484.10	481.9213	37.22	3.33	33.01	26.44
C15H16BR209	500.10	497.9162	36.03	3.22	31.96	28.79
C15H17BR101	293.21	292.0463	61.45	5.84	27.25	5.46
C15H17BR102	309.21	308.0412	58.27	5.54	25.84	10.35
C15H17BR103	325.20	324.0362	55.40	5.27	24.57	14.76
C15H17BR104	341.20	340.0311	52.80	5.02	23.42	18.76
C15H17BR105	357.20	356.0260	50.44	4.80	22.37	22.40
C15H17BR106	373.20	372.0209	48.28	4.59	21.41	25.72
C15H17BR107	389.20	388.0158	46.29	4.40	20.53	28.78
C15H17BR108	405.20	404.0107	44.46	4.23	19.72	31.59
C15H17BR109	421.20	420.0056	42.77	4.07	18.97	34.19
C15H18BR201	374.12	371.9725	48.16	4.85	42.72	4.28
C15H18BR202	390.12	387.9675	46.18	4.65	40.96	8.20
C15H18BR203	406.12	403.9624	44.36	4.47	39.35	11.82
C15H18BR204	422.12	419.9573	42.68	4.30	37.86	15.16
C15H18BR205	438.12	435.9522	41.12	4.14	36.48	18.26
C15H18BR206	454.12	451.9471	39.67	4.00	35.19	21.14
C15H18BR207	470.11	467.9420	38.32	3.86	33.99	23.82
C15H18BR208	486.11	483.9369	37.06	3.73	32.87	26.33
C15H18BR209	502.11	499.9318	35.88	3.61	31.83	28.68
C15H19BR101	295.22	294.0620	61.03	6.49	27.07	5.42
C15H19BR102	311.22	310.0569	57.89	6.15	25.67	10.28
C15H19BR103	327.22	326.0518	55.06	5.85	24.42	14.67
C15H19BR104	343.22	342.0467	52.49	5.58	23.28	18.65
C15H19BR105	359.22	358.0416	50.16	5.33	22.24	22.27
C15H19BR106	375.22	374.0365	48.02	5.10	21.30	25.58
C15H19BR107	391.22	390.0315	46.05	4.90	20.42	28.63
C15H19BR108	407.22	406.0264	44.24	4.70	19.62	31.43
C15H19BR109	423.22	422.0213	42.57	4.53	18.88	34.02
C15H20BR201	376.13	373.9882	47.90	5.36	42.49	4.25
C15H20BR202	392.13	389.9831	45.95	5.14	40.75	8.16
C15H20BR203	408.13	405.9780	44.14	4.94	39.16	11.76
C15H20BR204	424.13	421.9729	42.48	4.75	37.68	15.09
C15H20BR205	440.13	437.9678	40.93	4.58	36.31	18.18
C15H20BR206	456.13	453.9628	39.50	4.42	35.04	21.05
C15H20BR207	472.13	469.9577	38.16	4.27	33.85	23.72
C15H20BR208	488.13	485.9526	36.91	4.13	32.74	26.22
C15H20BR209	504.13	501.9475	35.74	4.00	31.70	28.56
C15H21BR101	297.24	296.0776	60.61	7.12	26.88	5.38
C15H21BR102	313.24	312.0725	57.52	6.76	25.51	10.22
C15H21BR103	329.24	328.0674	54.72	6.43	24.27	14.58
C15H21BR104	345.24	344.0624	52.19	6.13	23.14	18.54
C15H21BR105	361.24	360.0573	49.88	5.86	22.12	22.15
C15H21BR106	377.24	376.0522	47.76	5.61	21.18	25.45
C15H21BR107	393.23	392.0471	45.82	5.38	20.32	28.48
C15H21BR108	409.23	408.0420	44.03	5.17	19.53	31.28
C15H21BR109	425.23	424.0369	42.37	4.98	18.79	33.86
C15H22BR201	378.15	376.0038	47.64	5.86	42.26	4.23
C15H22BR202	394.15	391.9987	45.71	5.63	40.55	8.12
C15H22BR203	410.15	407.9937	43.93	5.41	38.96	11.70
C15H22BR204	426.15	423.9886	42.28	5.20	37.50	15.02

C15H22BR205	442.15	439.9835	40.75	5.02	36.14	18.09
C15H22BR206	458.15	455.9784	39.33	4.84	34.88	20.95
C15H22BR207	474.15	471.9733	38.00	4.68	33.70	23.62
C15H22BR208	490.15	487.9682	36.76	4.52	32.60	26.11
C15H22BR209	506.15	503.9631	35.60	4.38	31.57	28.45
C15H23BR101	299.25	298.0933	60.21	7.75	26.70	5.35
C15H23BR102	315.25	314.0882	57.15	7.35	25.35	10.15
C15H23BR103	331.25	330.0831	54.39	7.00	24.12	14.49
C15H23BR104	347.25	346.0780	51.88	6.68	23.01	18.43
C15H23BR105	363.25	362.0729	49.60	6.38	22.00	22.02
C15H23BR106	379.25	378.0678	47.51	6.11	21.07	25.31
C15H23BR107	395.25	394.0628	45.58	5.87	20.22	28.34
C15H23BR108	411.25	410.0577	43.81	5.64	19.43	31.12
C15H23BR109	427.25	426.0526	42.17	5.43	18.70	33.70
C15H24BR201	380.17	378.0195	47.39	6.36	42.04	4.21
C15H24BR202	396.17	394.0144	45.48	6.11	40.34	8.08
C15H24BR203	412.16	410.0093	43.71	5.87	38.77	11.65
C15H24BR204	428.16	426.0042	42.08	5.65	37.32	14.95
C15H24BR205	444.16	441.9991	40.56	5.45	35.98	18.01
C15H24BR206	460.16	457.9941	39.15	5.26	34.73	20.86
C15H24BR207	476.16	473.9890	37.84	5.08	33.56	23.52
C15H24BR208	492.16	489.9839	36.61	4.92	32.47	26.01
C15H24BR209	508.16	505.9788	35.45	4.76	31.45	28.34
C15H25BR101	301.27	300.1089	59.80	8.36	26.52	5.31
C15H25BR102	317.27	316.1038	56.79	7.94	25.18	10.09
C15H25BR103	333.27	332.0987	54.06	7.56	23.98	14.40
C15H25BR104	349.27	348.0937	51.58	7.21	22.88	18.32
C15H25BR105	365.27	364.0886	49.32	6.90	21.88	21.90
C15H25BR106	381.27	380.0835	47.25	6.61	20.96	25.18
C15H25BR107	397.27	396.0784	45.35	6.34	20.11	28.19
C15H25BR108	413.27	412.0733	43.60	6.10	19.33	30.97
C15H25BR109	429.27	428.0682	41.97	5.87	18.61	33.54
C15H26BR201	382.18	380.0351	47.14	6.86	41.81	4.19
C15H26BR202	398.18	396.0300	45.25	6.58	40.13	8.04
C15H26BR203	414.18	412.0250	43.50	6.33	38.58	11.59
C15H26BR204	430.18	428.0199	41.88	6.09	37.15	14.88
C15H26BR205	446.18	444.0148	40.38	5.87	35.82	17.93
C15H26BR206	462.18	460.0097	38.98	5.67	34.58	20.77
C15H26BR207	478.18	476.0046	37.68	5.48	33.42	23.42
C15H26BR208	494.18	491.9995	36.46	5.30	32.34	25.90
C15H26BR209	510.18	507.9944	35.31	5.14	31.32	28.22
C15H27BR101	303.29	302.1246	59.41	8.97	26.35	5.28
C15H27BR102	319.29	318.1195	56.43	8.52	25.03	10.02
C15H27BR103	335.28	334.1144	53.74	8.12	23.83	14.32
C15H27BR104	351.28	350.1093	51.29	7.75	22.75	18.22
C15H27BR105	367.28	366.1042	49.05	7.41	21.76	21.78
C15H27BR106	383.28	382.0991	47.01	7.10	20.85	25.05
C15H27BR107	399.28	398.0941	45.12	6.82	20.01	28.05
C15H27BR108	415.28	414.0890	43.38	6.55	19.24	30.82
C15H27BR109	431.28	430.0839	41.77	6.31	18.53	33.39
C15H28BR201	384.20	382.0508	46.89	7.35	41.60	4.16
C15H28BR202	400.20	398.0457	45.02	7.05	39.93	8.00
C15H28BR203	416.20	414.0406	43.29	6.78	38.40	11.53
C15H28BR204	432.20	430.0355	41.69	6.53	36.98	14.81
C15H28BR205	448.20	446.0304	40.20	6.30	35.66	17.85
C15H28BR206	464.19	462.0254	38.81	6.08	34.43	20.68
C15H28BR207	480.19	478.0203	37.52	5.88	33.28	23.32
C15H28BR208	496.19	494.0152	36.31	5.69	32.21	25.80
C15H28BR209	512.19	510.0101	35.18	5.51	31.20	28.11
C15H29BR101	305.30	304.1402	59.01	9.57	26.17	5.24

C15H29BR102	321.30	320.1351	56.07	9.10	24.87	9.96
C15H29BR103	337.30	336.1300	53.41	8.67	23.69	14.23
C15H29BR104	353.30	352.1250	51.00	8.27	22.62	18.11
C15H29BR105	369.30	368.1199	48.79	7.92	21.64	21.66
C15H29BR106	385.30	384.1148	46.76	7.59	20.74	24.91
C15H29BR107	401.30	400.1097	44.90	7.28	19.91	27.91
C15H29BR108	417.30	416.1046	43.17	7.00	19.15	30.67
C15H29BR109	433.30	432.0995	41.58	6.75	18.44	33.23
C15H30BR201	386.21	384.0664	46.65	7.83	41.38	4.14
C15H30BR202	402.21	400.0613	44.79	7.52	39.73	7.96
C15H30BR203	418.21	416.0563	43.08	7.23	38.21	11.48
C15H30BR204	434.21	432.0512	41.49	6.96	36.80	14.74
C15H30BR205	450.21	448.0461	40.02	6.72	35.50	17.77
C15H30BR206	466.21	464.0410	38.65	6.49	34.28	20.59
C15H30BR207	482.21	480.0359	37.36	6.27	33.14	23.23
C15H30BR208	498.21	496.0308	36.16	6.07	32.08	25.69
C15H30BR209	514.21	512.0257	35.04	5.88	31.08	28.00
C15H31BR101	307.32	306.1559	58.63	10.17	26.00	5.21
C15H31BR102	323.32	322.1508	55.72	9.66	24.71	9.90
C15H31BR103	339.32	338.1457	53.10	9.21	23.55	14.15
C15H31BR104	355.32	354.1406	50.71	8.79	22.49	18.01
C15H31BR105	371.32	370.1355	48.52	8.42	21.52	21.54
C15H31BR106	387.31	386.1304	46.52	8.07	20.63	24.79
C15H31BR107	403.31	402.1254	44.67	7.75	19.81	27.77
C15H31BR108	419.31	418.1203	42.97	7.45	19.06	30.52
C15H31BR109	435.31	434.1152	41.39	7.18	18.36	33.08
C16H 8BR201	376.05	373.8943	51.10	2.14	42.50	4.25
C16H 8BR202	392.05	389.8892	49.02	2.06	40.76	8.16
C16H 8BR203	408.05	405.8841	47.10	1.98	39.16	11.76
C16H 8BR204	424.05	421.8790	45.32	1.90	37.69	15.09
C16H 8BR205	440.05	437.8739	43.67	1.83	36.32	18.18
C16H 8BR206	456.05	453.8689	42.14	1.77	35.04	21.05
C16H 8BR207	472.05	469.8638	40.71	1.71	33.85	23.73
C16H 8BR208	488.05	485.8587	39.38	1.65	32.74	26.23
C16H 8BR209	504.04	501.8536	38.13	1.60	31.71	28.57
C16H 9BR101	297.15	295.9837	64.67	3.05	26.89	5.38
C16H 9BR102	313.15	311.9786	61.37	2.90	25.52	10.22
C16H 9BR103	329.15	327.9736	58.39	2.76	24.28	14.58
C16H 9BR104	345.15	343.9685	55.68	2.63	23.15	18.54
C16H 9BR105	361.15	359.9634	53.21	2.51	22.12	22.15
C16H 9BR106	377.15	375.9583	50.96	2.41	21.19	25.45
C16H 9BR107	393.15	391.9532	48.88	2.31	20.32	28.49
C16H 9BR108	409.15	407.9481	46.97	2.22	19.53	31.28
C16H 9BR109	425.15	423.9430	45.20	2.13	18.79	33.87
C16H10BR201	378.07	375.9099	50.83	2.67	42.27	4.23
C16H10BR202	394.06	391.9049	48.77	2.56	40.55	8.12
C16H10BR203	410.06	407.8998	46.87	2.46	38.97	11.71
C16H10BR204	426.06	423.8947	45.11	2.37	37.51	15.02
C16H10BR205	442.06	439.8896	43.47	2.28	36.15	18.10
C16H10BR206	458.06	455.8845	41.95	2.20	34.89	20.96
C16H10BR207	474.06	471.8794	40.54	2.13	33.71	23.62
C16H10BR208	490.06	487.8743	39.22	2.06	32.61	26.12
C16H10BR209	506.06	503.8693	37.98	1.99	31.58	28.45
C16H11BR101	299.17	297.9994	64.24	3.71	26.71	5.35
C16H11BR102	315.17	313.9943	60.98	3.52	25.35	10.15
C16H11BR103	331.17	329.9892	58.03	3.35	24.13	14.49
C16H11BR104	347.17	345.9841	55.36	3.19	23.02	18.43
C16H11BR105	363.17	361.9790	52.92	3.05	22.00	22.03
C16H11BR106	379.17	377.9739	50.68	2.92	21.07	25.32
C16H11BR107	395.17	393.9689	48.63	2.81	20.22	28.34

C16H11BR108	411.17	409.9638	46.74	2.70	19.43	31.13
C16H11BR109	427.16	425.9587	44.99	2.60	18.71	33.71
C16H12BR201	380.08	377.9256	50.56	3.18	42.05	4.21
C16H12BR202	396.08	393.9205	48.52	3.05	40.35	8.08
C16H12BR203	412.08	409.9154	46.64	2.94	38.78	11.65
C16H12BR204	428.08	425.9103	44.89	2.83	37.33	14.95
C16H12BR205	444.08	441.9052	43.28	2.72	35.99	18.01
C16H12BR206	460.08	457.9002	41.77	2.63	34.73	20.87
C16H12BR207	476.08	473.8951	40.37	2.54	33.57	23.52
C16H12BR208	492.08	489.8900	39.05	2.46	32.48	26.01
C16H12BR209	508.08	505.8849	37.82	2.38	31.45	28.34
C16H13BR101	301.19	300.0150	63.81	4.35	26.53	5.31
C16H13BR102	317.18	316.0099	60.59	4.13	25.19	10.09
C16H13BR103	333.18	332.0049	57.68	3.93	23.98	14.41
C16H13BR104	349.18	347.9998	55.04	3.75	22.88	18.33
C16H13BR105	365.18	363.9947	52.63	3.59	21.88	21.91
C16H13BR106	381.18	379.9896	50.42	3.44	20.96	25.18
C16H13BR107	397.18	395.9845	48.39	3.30	20.12	28.20
C16H13BR108	413.18	411.9794	46.51	3.17	19.34	30.98
C16H13BR109	429.18	427.9743	44.78	3.05	18.62	33.55
C16H14BR201	382.10	379.9412	50.30	3.69	41.82	4.19
C16H14BR202	398.10	395.9362	48.27	3.54	40.14	8.04
C16H14BR203	414.10	411.9311	46.41	3.41	38.59	11.59
C16H14BR204	430.10	427.9260	44.68	3.28	37.16	14.88
C16H14BR205	446.09	443.9209	43.08	3.16	35.82	17.93
C16H14BR206	462.09	459.9158	41.59	3.05	34.58	20.77
C16H14BR207	478.09	475.9107	40.20	2.95	33.43	23.43
C16H14BR208	494.09	491.9056	38.90	2.86	32.34	25.91
C16H14BR209	510.09	507.9006	37.68	2.77	31.33	28.23
C16H15BR101	303.20	302.0307	63.38	4.99	26.35	5.28
C16H15BR102	319.20	318.0256	60.21	4.74	25.03	10.02
C16H15BR103	335.20	334.0205	57.33	4.51	23.84	14.32
C16H15BR104	351.20	350.0154	54.72	4.31	22.75	18.22
C16H15BR105	367.20	366.0103	52.34	4.12	21.76	21.79
C16H15BR106	383.20	382.0052	50.15	3.95	20.85	25.05
C16H15BR107	399.20	398.0002	48.14	3.79	20.02	28.06
C16H15BR108	415.20	413.9951	46.29	3.64	19.24	30.83
C16H15BR109	431.20	429.9900	44.57	3.51	18.53	33.39
C16H16BR201	384.11	381.9569	50.03	4.20	41.60	4.17
C16H16BR202	400.11	397.9518	48.03	4.03	39.94	8.00
C16H16BR203	416.11	413.9467	46.18	3.88	38.41	11.53
C16H16BR204	432.11	429.9416	44.47	3.73	36.98	14.81
C16H16BR205	448.11	445.9365	42.89	3.60	35.66	17.85
C16H16BR206	464.11	461.9315	41.41	3.47	34.43	20.68
C16H16BR207	480.11	477.9264	40.03	3.36	33.29	23.33
C16H16BR208	496.11	493.9213	38.74	3.25	32.21	25.80
C16H16BR209	512.11	509.9162	37.53	3.15	31.21	28.12
C16H17BR101	305.22	304.0463	62.96	5.61	26.18	5.24
C16H17BR102	321.22	320.0412	59.83	5.33	24.88	9.96
C16H17BR103	337.22	336.0362	56.99	5.08	23.70	14.23
C16H17BR104	353.22	352.0311	54.41	4.85	22.62	18.12
C16H17BR105	369.21	368.0260	52.05	4.64	21.64	21.67
C16H17BR106	385.21	384.0209	49.89	4.45	20.74	24.92
C16H17BR107	401.21	400.0158	47.90	4.27	19.92	27.91
C16H17BR108	417.21	416.0107	46.06	4.11	19.15	30.68
C16H17BR109	433.21	432.0056	44.36	3.96	18.44	33.24
C16H18BR201	386.13	383.9725	49.77	4.70	41.39	4.14
C16H18BR202	402.13	399.9675	47.79	4.51	39.74	7.96
C16H18BR203	418.13	415.9624	45.96	4.34	38.22	11.48
C16H18BR204	434.13	431.9573	44.27	4.18	36.81	14.74

026

C16H18BR205	450.13	447.9522	42.69	4.03	35.50	17.77
C16H18BR206	466.13	463.9471	41.23	3.89	34.28	20.59
C16H18BR207	482.13	479.9420	39.86	3.76	33.15	23.23
C16H18BR208	498.13	495.9369	38.58	3.64	32.08	25.70
C16H18BR209	514.12	511.9318	37.38	3.53	31.08	28.01
C16H19BR101	307.23	306.0620	62.55	6.23	26.01	5.21
C16H19BR102	323.23	322.0569	59.46	5.92	24.72	9.90
C16H19BR103	339.23	338.0518	56.65	5.65	23.55	14.15
C16H19BR104	355.23	354.0467	54.10	5.39	22.49	18.02
C16H19BR105	371.23	370.0416	51.77	5.16	21.52	21.55
C16H19BR106	387.23	386.0365	49.63	4.95	20.63	24.79
C16H19BR107	403.23	402.0315	47.66	4.75	19.82	27.77
C16H19BR108	419.23	418.0264	45.84	4.57	19.06	30.53
C16H19BR109	435.23	434.0213	44.16	4.40	18.36	33.08
C16H20BR201	388.15	385.9882	49.51	5.19	41.17	4.12
C16H20BR202	404.14	401.9831	47.55	4.99	39.54	7.92
C16H20BR203	420.14	417.9780	45.74	4.80	38.04	11.42
C16H20BR204	436.14	433.9729	44.06	4.62	36.64	14.67
C16H20BR205	452.14	449.9678	42.50	4.46	35.34	17.69
C16H20BR206	468.14	465.9628	41.05	4.31	34.14	20.51
C16H20BR207	484.14	481.9577	39.69	4.16	33.01	23.13
C16H20BR208	500.14	497.9526	38.42	4.03	31.95	25.59
C16H20BR209	516.14	513.9475	37.23	3.91	30.96	27.90
C16H21BR101	309.25	308.0776	62.14	6.84	25.84	5.17
C16H21BR102	325.25	324.0725	59.09	6.51	24.57	9.84
C16H21BR103	341.25	340.0674	56.32	6.20	23.42	14.07
C16H21BR104	357.25	356.0624	53.79	5.93	22.37	17.91
C16H21BR105	373.25	372.0573	51.49	5.67	21.41	21.43
C16H21BR106	389.25	388.0522	49.37	5.44	20.53	24.66
C16H21BR107	405.25	404.0471	47.42	5.22	19.72	27.64
C16H21BR108	421.24	420.0420	45.62	5.02	18.97	30.38
C16H21BR109	437.24	436.0369	43.95	4.84	18.27	32.93
C16H22BR201	390.16	388.0038	49.26	5.68	40.96	4.10
C16H22BR202	406.16	403.9987	47.32	5.46	39.35	7.88
C16H22BR203	422.16	419.9937	45.52	5.25	37.85	11.37
C16H22BR204	438.16	435.9886	43.86	5.06	36.47	14.61
C16H22BR205	454.16	451.9835	42.32	4.88	35.19	17.61
C16H22BR206	470.16	467.9784	40.88	4.72	33.99	20.42
C16H22BR207	486.16	483.9733	39.53	4.56	32.87	23.04
C16H22BR208	502.16	499.9682	38.27	4.42	31.82	25.49
C16H22BR209	518.16	515.9631	37.09	4.28	30.84	27.79
C16H23BR101	311.27	310.0933	61.74	7.45	25.67	5.14
C16H23BR102	327.26	326.0882	58.72	7.08	24.42	9.78
C16H23BR103	343.26	342.0831	55.99	6.75	23.28	13.98
C16H23BR104	359.26	358.0780	53.49	6.45	22.24	17.81
C16H23BR105	375.26	374.0729	51.21	6.18	21.29	21.32
C16H23BR106	391.26	390.0678	49.12	5.93	20.42	24.54
C16H23BR107	407.26	406.0628	47.19	5.69	19.62	27.50
C16H23BR108	423.26	422.0577	45.40	5.48	18.88	30.24
C16H23BR109	439.26	438.0526	43.75	5.28	18.19	32.78
C16H24BR201	392.18	390.0195	49.00	6.17	40.75	4.08
C16H24BR202	408.18	406.0144	47.08	5.93	39.15	7.84
C16H24BR203	424.18	422.0093	45.31	5.70	37.67	11.32
C16H24BR204	440.18	438.0042	43.66	5.50	36.31	14.54
C16H24BR205	456.17	453.9991	42.13	5.30	35.03	17.54
C16H24BR206	472.17	469.9941	40.70	5.12	33.85	20.33
C16H24BR207	488.17	485.9890	39.37	4.96	32.74	22.94
C16H24BR208	504.17	501.9839	38.12	4.80	31.70	25.39
C16H24BR209	520.17	517.9788	36.95	4.65	30.72	27.68
C16H25BR101	313.28	312.1089	61.34	8.04	25.51	5.11

C16H25BR102	329.28	328.1038	58.36	7.65	24.27	9.72
C16H25BR103	345.28	344.0987	55.66	7.30	23.14	13.90
C16H25BR104	361.28	360.0937	53.19	6.98	22.12	17.71
C16H25BR105	377.28	376.0886	50.94	6.68	21.18	21.20
C16H25BR106	393.28	392.0835	48.87	6.41	20.32	24.41
C16H25BR107	409.28	408.0784	46.96	6.16	19.52	27.36
C16H25BR108	425.28	424.0733	45.19	5.93	18.79	30.10
C16H25BR109	441.28	440.0682	43.55	5.71	18.11	32.63
C16H26BR201	394.19	392.0351	48.75	6.65	40.54	4.06
C16H26BR202	410.19	408.0300	46.85	6.39	38.96	7.80
C16H26BR203	426.19	424.0250	45.09	6.15	37.50	11.26
C16H26BR204	442.19	440.0199	43.46	5.93	36.14	14.47
C16H26BR205	458.19	456.0148	41.94	5.72	34.88	17.46
C16H26BR206	474.19	472.0097	40.53	5.53	33.70	20.24
C16H26BR207	490.19	488.0046	39.20	5.35	32.60	22.85
C16H26BR208	506.19	503.9995	37.97	5.18	31.57	25.29
C16H26BR209	522.19	519.9944	36.80	5.02	30.60	27.58
C16H27BR101	315.30	314.1246	60.95	8.63	25.34	5.07
C16H27BR102	331.30	330.1195	58.01	8.21	24.12	9.66
C16H27BR103	347.30	346.1144	55.34	7.84	23.01	13.82
C16H27BR104	363.30	362.1093	52.90	7.49	21.99	17.62
C16H27BR105	379.29	378.1042	50.67	7.18	21.07	21.09
C16H27BR106	395.29	394.0991	48.62	6.88	20.21	24.28
C16H27BR107	411.29	410.0941	46.73	6.62	19.43	27.23
C16H27BR108	427.29	426.0890	44.98	6.37	18.70	29.95
C16H27BR109	443.29	442.0839	43.35	6.14	18.03	32.48
C16H28BR201	396.21	394.0508	48.50	7.12	40.33	4.04
C16H28BR202	412.21	410.0457	46.62	6.85	38.77	7.76
C16H28BR203	428.21	426.0406	44.88	6.59	37.32	11.21
C16H28BR204	444.21	442.0355	43.26	6.35	35.98	14.41
C16H28BR205	460.21	458.0304	41.76	6.13	34.73	17.38
C16H28BR206	476.21	474.0254	40.36	5.93	33.56	20.16
C16H28BR207	492.21	490.0203	39.04	5.73	32.47	22.75
C16H28BR208	508.20	506.0152	37.82	5.55	31.45	25.19
C16H28BR209	524.20	522.0101	36.66	5.38	30.49	27.47
C16H29BR101	317.31	316.1402	60.56	9.21	25.18	5.04
C16H29BR102	333.31	332.1351	57.66	8.77	23.97	9.60
C16H29BR103	349.31	348.1300	55.02	8.37	22.87	13.74
C16H29BR104	365.31	364.1250	52.61	8.00	21.87	17.52
C16H29BR105	381.31	380.1199	50.40	7.67	20.96	20.98
C16H29BR106	397.31	396.1148	48.37	7.36	20.11	24.16
C16H29BR107	413.31	412.1097	46.50	7.07	19.33	27.10
C16H29BR108	429.31	428.1046	44.76	6.81	18.61	29.81
C16H29BR109	445.31	444.0995	43.16	6.56	17.94	32.34
C16H30BR201	398.22	396.0664	48.26	7.59	40.13	4.02
C16H30BR202	414.22	412.0613	46.39	7.30	38.58	7.72
C16H30BR203	430.22	428.0563	44.67	7.03	37.15	11.16
C16H30BR204	446.22	444.0512	43.07	6.78	35.81	14.34
C16H30BR205	462.22	460.0461	41.58	6.54	34.57	17.31
C16H30BR206	478.22	476.0410	40.19	6.32	33.42	20.07
C16H30BR207	494.22	492.0359	38.89	6.12	32.34	22.66
C16H30BR208	510.22	508.0308	37.67	5.93	31.32	25.09
C16H30BR209	526.22	524.0257	36.52	5.75	30.37	27.36
C16H31BR101	319.33	318.1559	60.18	9.79	25.02	5.01
C16H31BR102	335.33	334.1508	57.31	9.32	23.83	9.54
C16H31BR103	351.33	350.1457	54.70	8.89	22.74	13.66
C16H31BR104	367.33	366.1406	52.32	8.51	21.75	17.42
C16H31BR105	383.33	382.1355	50.13	8.15	20.84	20.87
C16H31BR106	399.33	398.1304	48.13	7.82	20.01	24.04
C16H31BR107	415.33	414.1254	46.27	7.52	19.24	26.97

027

C16H31BR108	431.32	430.1203	44.56	7.24	18.53	29.67
C16H31BR109	447.32	446.1152	42.96	6.99	17.86	32.19
C16H32BR201	400.24	398.0821	48.02	8.06	39.93	4.00
C16H32BR202	416.24	414.0770	46.17	7.75	38.39	7.69
C16H32BR203	432.24	430.0719	44.46	7.46	36.97	11.10
C16H32BR204	448.24	446.0668	42.87	7.20	35.65	14.28
C16H32BR205	464.24	462.0617	41.40	6.95	34.42	17.23
C16H32BR206	480.24	478.0567	40.02	6.72	33.28	19.99
C16H32BR207	496.24	494.0516	38.73	6.50	32.20	22.57
C16H32BR208	512.24	510.0465	37.52	6.30	31.20	24.99
C16H32BR209	528.24	526.0414	36.38	6.11	30.25	27.26
C16H33BR101	321.34	320.1715	59.80	10.35	24.87	4.98
C16H33BR102	337.34	336.1664	56.97	9.86	23.69	9.49
C16H33BR103	353.34	352.1613	54.39	9.41	22.61	13.58
C16H33BR104	369.34	368.1563	52.03	9.01	21.63	17.33
C16H33BR105	385.34	384.1512	49.87	8.63	20.74	20.76
C16H33BR106	401.34	400.1461	47.88	8.29	19.91	23.92
C16H33BR107	417.34	416.1410	46.05	7.97	19.15	26.84
C16H33BR108	433.34	432.1359	44.35	7.68	18.44	29.54
C16H33BR109	449.34	448.1308	42.77	7.40	17.78	32.05
C17H 8BR201	388.06	385.8943	52.62	2.08	41.18	4.12
C17H 8BR202	404.06	401.8892	50.53	2.00	39.55	7.92
C17H 8BR203	420.06	417.8841	48.61	1.92	38.04	11.43
C17H 8BR204	436.06	433.8790	46.83	1.85	36.65	14.68
C17H 8BR205	452.06	449.8739	45.17	1.78	35.35	17.70
C17H 8BR206	468.06	465.8689	43.62	1.72	34.14	20.51
C17H 8BR207	484.06	481.8638	42.18	1.67	33.01	23.14
C17H 8BR208	500.06	497.8587	40.83	1.61	31.96	25.60
C17H 8BR209	516.06	513.8536	39.57	1.56	30.97	27.90
C17H 9BR101	309.16	307.9837	66.05	2.93	25.85	5.18
C17H 9BR102	325.16	323.9786	62.80	2.79	24.57	9.84
C17H 9BR103	341.16	339.9736	59.85	2.66	23.42	14.07
C17H 9BR104	357.16	355.9685	57.17	2.54	22.37	17.92
C17H 9BR105	373.16	371.9634	54.72	2.43	21.41	21.44
C17H 9BR106	389.16	387.9583	52.47	2.33	20.53	24.67
C17H 9BR107	405.16	403.9532	50.40	2.24	19.72	27.64
C17H 9BR108	421.16	419.9481	48.48	2.15	18.97	30.39
C17H 9BR109	437.16	435.9430	46.71	2.08	18.28	32.94
C17H10BR201	390.08	387.9099	52.35	2.58	40.97	4.10
C17H10BR202	406.08	403.9049	50.28	2.48	39.35	7.88
C17H10BR203	422.08	419.8998	48.38	2.39	37.86	11.37
C17H10BR204	438.07	435.8947	46.61	2.30	36.48	14.61
C17H10BR205	454.07	451.8896	44.97	2.22	35.19	17.62
C17H10BR206	470.07	467.8845	43.44	2.14	34.00	20.42
C17H10BR207	486.07	483.8794	42.01	2.07	32.88	23.04
C17H10BR208	502.07	499.8743	40.67	2.01	31.83	25.49
C17H10BR209	518.07	515.8693	39.41	1.95	30.85	27.79
C17H11BR101	311.18	309.9994	65.62	3.56	25.68	5.14
C17H11BR102	327.18	325.9943	62.41	3.39	24.42	9.78
C17H11BR103	343.18	341.9892	59.50	3.23	23.28	13.99
C17H11BR104	359.18	357.9841	56.85	3.09	22.25	17.82
C17H11BR105	375.18	373.9790	54.42	2.96	21.30	21.32
C17H11BR106	391.18	389.9739	52.20	2.83	20.43	24.54
C17H11BR107	407.18	405.9689	50.15	2.72	19.62	27.51
C17H11BR108	423.18	421.9638	48.25	2.62	18.88	30.25
C17H11BR109	439.18	437.9587	46.49	2.52	18.19	32.79
C17H12BR201	392.09	389.9256	52.08	3.08	40.76	4.08
C17H12BR202	408.09	405.9205	50.04	2.96	39.16	7.84
C17H12BR203	424.09	421.9154	48.15	2.85	37.68	11.32
C17H12BR204	440.09	437.9103	46.40	2.75	36.31	14.54
C17H12BR205	456.09	453.9052	44.77	2.65	35.04	17.54
C17H12BR206	472.09	469.9002	43.25	2.56	33.85	20.33
C17H12BR207	488.09	485.8951	41.83	2.48	32.74	22.95
C17H12BR208	504.09	501.8900	40.51	2.40	31.70	25.39
C17H12BR209	520.09	517.8849	39.26	2.33	30.73	27.69
C17H13BR101	313.20	312.0150	65.20	4.18	25.51	5.11
C17H13BR102	329.20	328.0099	62.03	3.98	24.27	9.72
C17H13BR103	345.20	344.0049	59.15	3.80	23.15	13.90
C17H13BR104	361.19	359.9998	56.53	3.63	22.12	17.72
C17H13BR105	377.19	375.9947	54.13	3.47	21.18	21.21
C17H13BR106	393.19	391.9896	51.93	3.33	20.32	24.41
C17H13BR107	409.19	407.9845	49.90	3.20	19.53	27.37
C17H13BR108	425.19	423.9794	48.02	3.08	18.79	30.10
C17H13BR109	441.19	439.9743	46.28	2.97	18.11	32.64
C17H14BR201	394.11	391.9412	51.81	3.58	40.55	4.06
C17H14BR202	410.11	407.9362	49.79	3.44	38.97	7.80
C17H14BR203	426.11	423.9311	47.92	3.31	37.50	11.26
C17H14BR204	442.11	439.9260	46.19	3.19	36.15	14.48
C17H14BR205	458.11	455.9209	44.57	3.08	34.88	17.46
C17H14BR206	474.11	471.9158	43.07	2.98	33.71	20.25
C17H14BR207	490.10	487.9107	41.66	2.88	32.61	22.85
C17H14BR208	506.10	503.9056	40.35	2.79	31.58	25.29
C17H14BR209	522.10	519.9006	39.11	2.70	30.61	27.58
C17H15BR101	315.21	314.0307	64.78	4.80	25.35	5.08
C17H15BR102	331.21	330.0256	61.65	4.56	24.12	9.66
C17H15BR103	347.21	346.0205	58.81	4.35	23.01	13.82
C17H15BR104	363.21	362.0154	56.22	4.16	22.00	17.62
C17H15BR105	379.21	378.0103	53.85	3.99	21.07	21.10
C17H15BR106	395.21	394.0052	51.67	3.83	20.22	24.29
C17H15BR107	411.21	410.0002	49.66	3.68	19.43	27.24
C17H15BR108	427.21	425.9951	47.80	3.54	18.70	29.96
C17H15BR109	443.21	441.9900	46.07	3.41	18.03	32.49
C17H16BR201	396.12	393.9569	51.55	4.07	40.34	4.04
C17H16BR202	412.12	409.9518	49.55	3.91	38.78	7.76
C17H16BR203	428.12	425.9467	47.69	3.77	37.33	11.21
C17H16BR204	444.12	441.9416	45.98	3.63	35.98	14.41
C17H16BR205	460.12	457.9365	44.38	3.51	34.73	17.39
C17H16BR206	476.12	473.9315	42.89	3.39	33.56	20.16
C17H16BR207	492.12	489.9264	41.49	3.28	32.47	22.76
C17H16BR208	508.12	505.9213	40.19	3.17	31.45	25.19
C17H16BR209	524.12	521.9162	38.96	3.08	30.49	27.47
C17H17BR101	317.23	316.0463	64.37	5.40	25.19	5.04
C17H17BR102	333.23	332.0412	61.28	5.14	23.98	9.60
C17H17BR103	349.23	348.0362	58.47	4.91	22.88	13.74
C17H17BR104	365.23	364.0311	55.91	4.69	21.88	17.52
C17H17BR105	381.23	380.0260	53.56	4.49	20.96	20.98
C17H17BR106	397.23	396.0209	51.40	4.31	20.12	24.17
C17H17BR107	413.22	412.0158	49.41	4.15	19.34	27.10
C17H17BR108	429.22	428.0107	47.57	3.99	18.62	29.82
C17H17BR109	445.22	444.0056	45.86	3.85	17.95	32.34
C17H18BR201	398.14	395.9725	51.29	4.56	40.14	4.02
C17H18BR202	414.14	411.9675	49.30	4.38	38.59	7.73
C17H18BR203	430.14	427.9624	47.47	4.22	37.15	11.16
C17H18BR204	446.14	443.9573	45.77	4.07	35.82	14.34
C17H18BR205	462.14	459.9522	44.18	3.93	34.58	17.31
C17H18BR206	478.14	475.9471	42.71	3.79	33.42	20.08
C17H18BR207	494.14	491.9420	41.32	3.67	32.34	22.66
C17H18BR208	510.14	507.9369	40.03	3.56	31.33	25.09
C17H18BR209	526.14	523.9318	38.81	3.45	30.37	27.37
C17H19BR101	319.24	318.0620	63.96	6.00	25.03	5.01

028

C17H19BR102	335.24	334.0569	60.91	5.71	23.83	9.54
C17H19BR103	351.24	350.0518	58.13	5.45	22.75	13.67
C17H19BR104	367.24	366.0467	55.60	5.21	21.76	17.43
C17H19BR105	383.24	382.0416	53.28	5.00	20.85	20.87
C17H19BR106	399.24	398.0365	51.14	4.80	20.01	24.04
C17H19BR107	415.24	414.0315	49.17	4.61	19.24	26.97
C17H19BR108	431.24	430.0264	47.35	4.44	18.53	29.68
C17H19BR109	447.24	446.0213	45.66	4.28	17.87	32.20
C17H20BR201	400.16	397.9882	51.03	5.04	39.94	4.00
C17H20BR202	416.16	413.9831	49.07	4.84	38.40	7.69
C17H20BR203	432.16	429.9780	47.25	4.66	36.98	11.11
C17H20BR204	448.15	445.9729	45.56	4.50	35.66	14.28
C17H20BR205	464.15	461.9678	43.99	4.34	34.43	17.24
C17H20BR206	480.15	477.9628	42.53	4.20	33.28	19.99
C17H20BR207	496.15	493.9577	41.15	4.06	32.21	22.57
C17H20BR208	512.15	509.9526	39.87	3.94	31.20	24.99
C17H20BR209	528.15	525.9475	38.66	3.82	30.26	27.26
C17H21BR101	321.26	320.0776	63.56	6.59	24.87	4.98
C17H21BR102	337.26	336.0725	60.54	6.28	23.69	9.49
C17H21BR103	353.26	352.0674	57.80	5.99	22.62	13.59
C17H21BR104	369.26	368.0624	55.30	5.73	21.64	17.33
C17H21BR105	385.26	384.0573	53.00	5.49	20.74	20.76
C17H21BR106	401.26	400.0522	50.89	5.28	19.91	23.92
C17H21BR107	417.26	416.0471	48.94	5.07	19.15	26.84
C17H21BR108	433.26	432.0420	47.13	4.89	18.44	29.54
C17H21BR109	449.26	448.0369	45.45	4.71	17.79	32.05
C17H22BR201	402.17	400.0038	50.77	5.51	39.74	3.98
C17H22BR202	418.17	415.9987	48.83	5.30	38.22	7.65
C17H22BR203	434.17	431.9937	47.03	5.11	36.81	11.06
C17H22BR204	450.17	447.9886	45.36	4.93	35.50	14.22
C17H22BR205	466.17	463.9835	43.80	4.76	34.28	17.16
C17H22BR206	482.17	479.9784	42.35	4.60	33.14	19.91
C17H22BR207	498.17	495.9733	40.99	4.45	32.08	22.48
C17H22BR208	514.17	511.9682	39.71	4.31	31.08	24.89
C17H22BR209	530.17	527.9631	38.51	4.18	30.14	27.16
C17H23BR101	323.28	322.0933	63.16	7.17	24.72	4.95
C17H23BR102	339.28	338.0882	60.18	6.83	23.55	9.43
C17H23BR103	355.28	354.0831	57.47	6.53	22.49	13.51
C17H23BR104	371.27	370.0780	55.00	6.24	21.52	17.24
C17H23BR105	387.27	386.0729	52.72	5.99	20.63	20.66
C17H23BR106	403.27	402.0678	50.63	5.75	19.81	23.80
C17H23BR107	419.27	418.0628	48.70	5.53	19.06	26.71
C17H23BR108	435.27	434.0577	46.91	5.33	18.36	29.41
C17H23BR109	451.27	450.0526	45.25	5.14	17.71	31.91
C17H24BR201	404.19	402.0195	50.52	5.99	39.54	3.96
C17H24BR202	420.19	418.0144	48.59	5.76	38.03	7.62
C17H24BR203	436.19	434.0093	46.81	5.55	36.64	11.00
C17H24BR204	452.19	450.0042	45.16	5.35	35.34	14.15
C17H24BR205	468.19	465.9991	43.61	5.17	34.13	17.09
C17H24BR206	484.19	481.9941	42.17	5.00	33.01	19.83
C17H24BR207	500.18	497.9890	40.82	4.84	31.95	22.39
C17H24BR208	516.18	513.9839	39.56	4.69	30.96	24.80
C17H24BR209	532.18	529.9788	38.37	4.55	30.03	27.06
C17H25BR101	325.29	324.1089	62.77	7.75	24.56	4.92
C17H25BR102	341.29	340.1038	59.83	7.38	23.41	9.38
C17H25BR103	357.29	356.0987	57.15	7.05	22.36	13.43
C17H25BR104	373.29	372.0937	54.70	6.75	21.41	17.14
C17H25BR105	389.29	388.0886	52.45	6.47	20.53	20.55
C17H25BR106	405.29	404.0835	50.38	6.22	19.72	23.69
C17H25BR107	421.29	420.0784	48.47	5.98	18.97	26.58
C17H25BR108	437.29	436.0733	46.69	5.76	18.27	29.27
C17H25BR109	453.29	452.0682	45.05	5.56	17.63	31.77
C17H26BR201	406.20	404.0351	50.27	6.45	39.34	3.94
C17H26BR202	422.20	420.0300	48.36	6.21	37.85	7.58
C17H26BR203	438.20	436.0250	46.60	5.98	36.47	10.95
C17H26BR204	454.20	452.0199	44.96	5.77	35.18	14.09
C17H26BR205	470.20	468.0148	43.43	5.57	33.99	17.01
C17H26BR206	486.20	484.0097	42.00	5.39	32.87	19.74
C17H26BR207	502.20	500.0046	40.66	5.22	31.82	22.30
C17H26BR208	518.20	515.9995	39.40	5.06	30.84	24.70
C17H26BR209	534.20	531.9944	38.22	4.91	29.92	26.96
C17H27BR101	327.31	326.1246	62.38	8.31	24.41	4.89
C17H27BR102	343.31	342.1195	59.48	7.93	23.27	9.32
C17H27BR103	359.31	358.1144	56.83	7.57	22.24	13.36
C17H27BR104	375.31	374.1093	54.41	7.25	21.29	17.05
C17H27BR105	391.31	390.1042	52.18	6.95	20.42	20.44
C17H27BR106	407.31	406.0991	50.13	6.68	19.62	23.57
C17H27BR107	423.30	422.0941	48.24	6.43	18.88	26.46
C17H27BR108	439.30	438.0890	46.48	6.20	18.19	29.14
C17H27BR109	455.30	454.0839	44.85	5.98	17.55	31.63
C17H28BR201	408.22	406.0508	50.02	6.91	39.15	3.92
C17H28BR202	424.22	422.0457	48.13	6.65	37.67	7.54
C17H28BR203	440.22	438.0406	46.38	6.41	36.30	10.90
C17H28BR204	456.22	454.0355	44.76	6.19	35.03	14.03
C17H28BR205	472.22	470.0304	43.24	5.98	33.84	16.94
C17H28BR206	488.22	486.0254	41.82	5.78	32.73	19.66
C17H28BR207	504.22	502.0203	40.50	5.60	31.69	22.21
C17H28BR208	520.22	518.0152	39.25	5.43	30.72	24.60
C17H28BR209	536.22	534.0101	38.08	5.26	29.80	26.85
C17H29BR101	329.32	328.1402	62.00	8.88	24.26	4.86
C17H29BR102	345.32	344.1351	59.13	8.46	23.14	9.27
C17H29BR103	361.32	360.1300	56.51	8.09	22.11	13.28
C17H29BR104	377.32	376.1250	54.12	7.75	21.18	16.96
C17H29BR105	393.32	392.1199	51.91	7.43	20.32	20.34
C17H29BR106	409.32	408.1148	49.88	7.14	19.52	23.45
C17H29BR107	425.32	424.1097	48.01	6.87	18.79	26.33
C17H29BR108	441.32	440.1046	46.27	6.62	18.11	29.00
C17H29BR109	457.32	456.0995	44.65	6.39	17.47	31.49
C17H30BR201	410.24	408.0664	49.77	7.37	38.96	3.90
C17H30BR202	426.24	424.0613	47.91	7.09	37.49	7.51
C17H30BR203	442.23	440.0563	46.17	6.84	36.14	10.85
C17H30BR204	458.23	456.0512	44.56	6.60	34.87	13.97
C17H30BR205	474.23	472.0461	43.06	6.38	33.70	16.87
C17H30BR206	490.23	488.0410	41.65	6.17	32.60	19.58
C17H30BR207	506.23	504.0359	40.34	5.97	31.57	22.12
C17H30BR208	522.23	520.0308	39.10	5.79	30.60	24.51
C17H30BR209	538.23	536.0257	37.94	5.62	29.69	26.75
C17H31BR101	331.34	330.1559	61.63	9.43	24.12	4.83
C17H31BR102	347.34	346.1508	58.79	9.00	23.00	9.21
C17H31BR103	363.34	362.1457	56.20	8.60	21.99	13.21
C17H31BR104	379.34	378.1406	53.83	8.24	21.06	16.87
C17H31BR105	395.34	394.1355	51.65	7.90	20.21	20.24
C17H31BR106	411.34	410.1304	49.64	7.60	19.43	23.34
C17H31BR107	427.34	426.1254	47.78	7.31	18.70	26.21
C17H31BR108	443.34	442.1203	46.06	7.05	18.02	28.87
C17H31BR109	459.34	458.1152	44.45	6.80	17.40	31.35
C17H32BR201	412.25	410.0821	49.53	7.82	38.76	3.88
C17H32BR202	428.25	426.0770	47.68	7.53	37.32	7.47
C17H32BR203	444.25	442.0719	45.96	7.26	35.97	10.80
C17H32BR204	460.25	458.0668	44.36	7.01	34.72	13.90

C17H32BR205	476.25	474.0617	42.87	6.77	33.56	16.80
C17H32BR206	492.25	490.0567	41.48	6.55	32.46	19.50
C17H32BR207	508.25	506.0516	40.18	6.35	31.44	22.04
C17H32BR208	524.25	522.0465	38.95	6.15	30.48	24.42
C17H32BR209	540.25	538.0414	37.80	5.97	29.58	26.65
C17H33BR101	333.36	332.1715	61.25	9.98	23.97	4.80
C17H33BR102	349.36	348.1664	58.45	9.52	22.87	9.16
C17H33BR103	365.35	364.1613	55.89	9.10	21.87	13.14
C17H33BR104	381.35	380.1563	53.54	8.72	20.95	16.78
C17H33BR105	397.35	396.1512	51.39	8.37	20.11	20.13
C17H33BR106	413.35	412.1461	49.40	8.05	19.33	23.22
C17H33BR107	429.35	428.1410	47.56	7.75	18.61	26.08
C17H33BR108	445.35	444.1359	45.85	7.47	17.94	28.74
C17H33BR109	461.35	460.1308	44.26	7.21	17.32	31.21
C17H34BR201	414.27	412.0977	49.29	8.27	38.58	3.86
C17H34BR202	430.27	428.0926	47.46	7.97	37.14	7.44
C17H34BR203	446.27	444.0876	45.76	7.68	35.81	10.76
C17H34BR204	462.27	460.0825	44.17	7.41	34.57	13.84
C17H34BR205	478.27	476.0774	42.69	7.17	33.41	16.73
C17H34BR206	494.26	492.0723	41.31	6.93	32.33	19.42
C17H34BR207	510.26	508.0672	40.02	6.72	31.32	21.95
C17H34BR208	526.26	524.0621	38.80	6.51	30.37	24.32
C17H34BR209	542.26	540.0570	37.66	6.32	29.47	26.55
C17H35BR101	335.37	334.1872	60.88	10.52	23.83	4.77
C17H35BR102	351.37	350.1821	58.11	10.04	22.74	9.11
C17H35BR103	367.37	366.1770	55.58	9.60	21.75	13.07
C17H35BR104	383.37	382.1719	53.26	9.20	20.84	16.69
C17H35BR105	399.37	398.1668	51.13	8.83	20.01	20.03
C17H35BR106	415.37	414.1617	49.16	8.49	19.24	23.11
C17H35BR107	431.37	430.1566	47.34	8.18	18.52	25.96
C17H35BR108	447.37	446.1516	45.64	7.89	17.86	28.61
C17H35BR109	463.37	462.1465	44.07	7.61	17.24	31.08
C18H 9BR101	321.18	319.9837	67.32	2.82	24.88	4.98
C18H 9BR102	337.18	335.9786	64.12	2.69	23.70	9.49
C18H 9BR103	353.17	351.9736	61.22	2.57	22.62	13.59
C18H 9BR104	369.17	367.9685	58.56	2.46	21.64	17.34
C18H 9BR105	385.17	383.9634	56.13	2.36	20.74	20.77
C18H 9BR106	401.17	399.9583	53.89	2.26	19.92	23.93
C18H 9BR107	417.17	415.9532	51.83	2.17	19.15	26.85
C18H 9BR108	433.17	431.9481	49.91	2.09	18.45	29.55
C18H 9BR109	449.17	447.9430	48.13	2.02	17.79	32.06
C18H10BR201	402.09	399.9099	53.77	2.51	39.74	3.98
C18H10BR202	418.09	415.9049	51.71	2.41	38.22	7.65
C18H10BR203	434.09	431.8998	49.81	2.32	36.81	11.06
C18H10BR204	450.09	447.8947	48.04	2.24	35.51	14.22
C18H10BR205	466.09	463.8896	46.39	2.16	34.29	17.16
C18H10BR206	482.08	479.8845	44.85	2.09	33.15	19.91
C18H10BR207	498.08	495.8794	43.41	2.02	32.08	22.49
C18H10BR208	514.08	511.8743	42.06	1.96	31.09	24.90
C18H10BR209	530.08	527.8693	40.79	1.90	30.15	27.16
C18H11BR101	323.19	321.9994	66.90	3.43	24.72	4.95
C18H11BR102	339.19	337.9943	63.74	3.27	23.56	9.43
C18H11BR103	355.19	353.9892	60.87	3.12	22.50	13.51
C18H11BR104	371.19	369.9841	58.25	2.99	21.53	17.24
C18H11BR105	387.19	385.9790	55.84	2.86	20.64	20.66
C18H11BR106	403.19	401.9739	53.62	2.75	19.82	23.81
C18H11BR107	419.19	417.9689	51.58	2.65	19.06	26.72
C18H11BR108	435.19	433.9638	49.68	2.55	18.36	29.41
C18H11BR109	451.19	449.9587	47.92	2.46	17.71	31.91
C18H12BR201	404.10	401.9256	53.50	2.99	39.55	3.96
C18H12BR202	420.10	417.9205	51.46	2.88	38.04	7.62
C18H12BR203	436.10	433.9154	49.58	2.77	36.64	11.01
C18H12BR204	452.10	449.9103	47.82	2.68	35.35	14.16
C18H12BR205	468.10	465.9052	46.19	2.58	34.14	17.09
C18H12BR206	484.10	481.9002	44.66	2.50	33.01	19.83
C18H12BR207	500.10	497.8951	43.23	2.42	31.96	22.39
C18H12BR208	516.10	513.8900	41.89	2.34	30.96	24.80
C18H12BR209	532.10	529.8849	40.63	2.27	30.03	27.06
C18H13BR101	325.21	324.0150	66.48	4.03	24.57	4.92
C18H13BR102	341.21	340.0099	63.36	3.84	23.42	9.38
C18H13BR103	357.21	356.0049	60.53	3.67	22.37	13.44
C18H13BR104	373.21	371.9998	57.93	3.51	21.41	17.15
C18H13BR105	389.21	387.9947	55.55	3.37	20.53	20.55
C18H13BR106	405.20	403.9896	53.36	3.23	19.72	23.69
C18H13BR107	421.20	419.9845	51.33	3.11	18.97	26.59
C18H13BR108	437.20	435.9794	49.45	3.00	18.28	29.28
C18H13BR109	453.20	451.9743	47.71	2.89	17.63	31.77
C18H14BR201	406.12	403.9412	53.24	3.47	39.35	3.94
C18H14BR202	422.12	419.9362	51.22	3.34	37.86	7.58
C18H14BR203	438.12	435.9311	49.35	3.22	36.48	10.96
C18H14BR204	454.12	451.9260	47.61	3.11	35.19	14.09
C18H14BR205	470.12	467.9209	45.99	3.00	33.99	17.02
C18H14BR206	486.12	483.9158	44.48	2.90	32.87	19.75
C18H14BR207	502.12	499.9107	43.06	2.81	31.83	22.30
C18H14BR208	518.12	515.9056	41.73	2.72	30.84	24.70
C18H14BR209	534.11	531.9006	40.48	2.64	29.92	26.96
C18H15BR101	327.22	326.0307	66.07	4.62	24.42	4.89
C18H15BR102	343.22	342.0256	62.99	4.41	23.28	9.32
C18H15BR103	359.22	358.0205	60.19	4.21	22.24	13.36
C18H15BR104	375.22	374.0154	57.62	4.03	21.30	17.06
C18H15BR105	391.22	390.0103	55.26	3.86	20.42	20.45
C18H15BR106	407.22	406.0052	53.09	3.71	19.62	23.57
C18H15BR107	423.22	422.0002	51.08	3.57	18.88	26.46
C18H15BR108	439.22	437.9951	49.22	3.44	18.19	29.14
C18H15BR109	455.22	453.9900	47.49	3.32	17.55	31.63
C18H16BR201	408.14	405.9569	52.97	3.95	39.16	3.92
C18H16BR202	424.14	421.9518	50.97	3.80	37.68	7.54
C18H16BR203	440.13	437.9467	49.12	3.66	36.31	10.91
C18H16BR204	456.13	453.9416	47.40	3.54	35.04	14.03
C18H16BR205	472.13	469.9365	45.79	3.42	33.85	16.94
C18H16BR206	488.13	485.9315	44.29	3.30	32.74	19.67
C18H16BR207	504.13	501.9264	42.89	3.20	31.70	22.22
C18H16BR208	520.13	517.9213	41.57	3.10	30.72	24.61
C18H16BR209	536.13	533.9162	40.33	3.01	29.81	26.86
C18H17BR101	329.24	328.0463	65.67	5.20	24.27	4.86
C18H17BR102	345.24	344.0412	62.62	4.96	23.14	9.27
C18H17BR103	361.24	360.0362	59.85	4.74	22.12	13.29
C18H17BR104	377.24	376.0311	57.31	4.54	21.18	16.96
C18H17BR105	393.24	392.0260	54.98	4.36	20.32	20.34
C18H17BR106	409.24	408.0209	52.83	4.19	19.53	23.46
C18H17BR107	425.24	424.0158	50.84	4.03	18.79	26.34
C18H17BR108	441.24	440.0107	49.00	3.88	18.11	29.01
C18H17BR109	457.23	456.0056	47.28	3.75	17.48	31.49
C18H18BR201	410.15	407.9725	52.71	4.42	38.96	3.90
C18H18BR202	426.15	423.9675	50.73	4.26	37.50	7.51
C18H18BR203	442.15	439.9624	48.90	4.10	36.14	10.86
C18H18BR204	458.15	455.9573	47.19	3.96	34.88	13.97
C18H18BR205	474.15	471.9522	45.60	3.83	33.70	16.87
C18H18BR206	490.15	487.9471	44.11	3.70	32.60	19.59
C18H18BR207	506.15	503.9420	42.71	3.58	31.57	22.13

030

C18H18BR208	522.15	519.9369	41.41	3.47	30.61	24.51
C18H18BR209	538.15	535.9318	40.18	3.37	29.70	26.76
C18H19BR101	331.26	330.0620	65.27	5.78	24.12	4.83
C18H19BR102	347.25	346.0569	62.26	5.52	23.01	9.21
C18H19BR103	363.25	362.0518	59.52	5.27	22.00	13.21
C18H19BR104	379.25	378.0467	57.01	5.05	21.07	16.87
C18H19BR105	395.25	394.0416	54.70	4.85	20.22	20.24
C18H19BR106	411.25	410.0365	52.57	4.66	19.43	23.34
C18H19BR107	427.25	426.0315	50.60	4.48	18.70	26.21
C18H19BR108	443.25	442.0264	48.78	4.32	18.03	28.88
C18H19BR109	459.25	458.0213	47.08	4.17	17.40	31.35
C18H20BR201	412.17	409.9882	52.45	4.89	38.77	3.88
C18H20BR202	428.17	425.9831	50.49	4.71	37.32	7.47
C18H20BR203	444.17	441.9780	48.68	4.54	35.98	10.81
C18H20BR204	460.17	457.9729	46.98	4.38	34.73	13.91
C18H20BR205	476.17	473.9678	45.40	4.23	33.56	16.80
C18H20BR206	492.16	489.9628	43.93	4.10	32.47	19.50
C18H20BR207	508.16	505.9577	42.55	3.97	31.45	22.04
C18H20BR208	524.16	521.9526	41.25	3.85	30.49	24.42
C18H20BR209	540.16	537.9475	40.03	3.73	29.59	26.66
C18H21BR101	333.27	332.0776	64.87	6.35	23.98	4.80
C18H21BR102	349.27	348.0725	61.90	6.06	22.88	9.16
C18H21BR103	365.27	364.0674	59.19	5.79	21.88	13.14
C18H21BR104	381.27	380.0624	56.71	5.55	20.96	16.79
C18H21BR105	397.27	396.0573	54.42	5.33	20.11	20.14
C18H21BR106	413.27	412.0522	52.31	5.12	19.33	23.23
C18H21BR107	429.27	428.0471	50.36	4.93	18.61	26.09
C18H21BR108	445.27	444.0420	48.56	4.75	17.95	28.75
C18H21BR109	461.27	460.0369	46.87	4.59	17.32	31.22
C18H22BR201	414.18	412.0038	52.20	5.35	38.58	3.86
C18H22BR202	430.18	427.9987	50.26	5.15	37.15	7.44
C18H22BR203	446.18	443.9937	48.46	4.97	35.82	10.76
C18H22BR204	462.18	459.9886	46.78	4.80	34.58	13.85
C18H22BR205	478.18	475.9835	45.21	4.64	33.42	16.73
C18H22BR206	494.18	491.9784	43.75	4.49	32.34	19.43
C18H22BR207	510.18	507.9733	42.38	4.35	31.32	21.95
C18H22BR208	526.18	523.9682	41.09	4.21	30.37	24.33
C18H22BR209	542.18	539.9631	39.88	4.09	29.48	26.56
C18H23BR101	335.29	334.0933	64.48	6.91	23.83	4.77
C18H23BR102	351.29	350.0882	61.55	6.60	22.75	9.11
C18H23BR103	367.29	366.0831	58.86	6.31	21.76	13.07
C18H23BR104	383.29	382.0780	56.41	6.05	20.85	16.70
C18H23BR105	399.29	398.0729	54.15	5.81	20.01	20.04
C18H23BR106	415.28	414.0678	52.06	5.58	19.24	23.12
C18H23BR107	431.28	430.0628	50.13	5.38	18.53	25.97
C18H23BR108	447.28	446.0577	48.34	5.18	17.86	28.62
C18H23BR109	463.28	462.0526	46.67	5.00	17.25	31.08
C18H24BR201	416.20	414.0195	51.95	5.81	38.40	3.84
C18H24BR202	432.20	430.0144	50.02	5.60	36.98	7.40
C18H24BR203	448.20	446.0093	48.24	5.40	35.66	10.71
C18H24BR204	464.20	462.0042	46.58	5.21	34.43	13.79
C18H24BR205	480.20	477.9991	45.02	5.04	33.28	16.66
C18H24BR206	496.20	493.9941	43.57	4.88	32.21	19.35
C18H24BR207	512.20	509.9890	42.21	4.72	31.20	21.87
C18H24BR208	528.20	525.9839	40.93	4.58	30.26	24.23
C18H24BR209	544.19	541.9788	39.73	4.45	29.37	26.46
C18H25BR101	337.30	336.1089	64.10	7.47	23.69	4.74
C18H25BR102	353.30	352.1038	61.19	7.13	22.62	9.06
C18H25BR103	369.30	368.0987	58.54	6.82	21.64	13.00
C18H25BR104	385.30	384.0937	56.11	6.54	20.74	16.61
C18H25BR105	401.30	400.0886	53.87	6.28	19.91	19.93
C18H25BR106	417.30	416.0835	51.81	6.04	19.15	23.00
C18H25BR107	433.30	432.0784	49.90	5.82	18.44	25.85
C18H25BR108	449.30	448.0733	48.12	5.61	17.78	28.49
C18H25BR109	465.30	464.0682	46.46	5.42	17.17	30.95
C18H26BR201	418.22	416.0351	51.70	6.27	38.21	3.83
C18H26BR202	434.21	432.0300	49.79	6.04	36.80	7.37
C18H26BR203	450.21	448.0250	48.02	5.82	35.50	10.66
C18H26BR204	466.21	464.0199	46.37	5.62	34.28	13.73
C18H26BR205	482.21	480.0148	44.84	5.43	33.14	16.59
C18H26BR206	498.21	496.0097	43.40	5.26	32.08	19.27
C18H26BR207	514.21	512.0046	42.05	5.10	31.08	21.78
C18H26BR208	530.21	527.9995	40.78	4.94	30.14	24.14
C18H26BR209	546.21	543.9944	39.58	4.80	29.26	26.36
C18H27BR101	339.32	338.1246	63.72	8.02	23.55	4.72
C18H27BR102	355.32	354.1195	60.85	7.66	22.49	9.01
C18H27BR103	371.32	370.1144	58.23	7.33	21.52	12.93
C18H27BR104	387.32	386.1093	55.82	7.03	20.63	16.52
C18H27BR105	403.32	402.1042	53.61	6.75	19.81	19.83
C18H27BR106	419.32	418.0991	51.56	6.49	19.06	22.89
C18H27BR107	435.32	434.0941	49.67	6.25	18.36	25.73
C18H27BR108	451.32	450.0890	47.90	6.03	17.70	28.36
C18H27BR109	467.31	466.0839	46.26	5.82	17.10	30.81
C18H28BR201	420.23	418.0508	51.45	6.72	38.03	3.81
C18H28BR202	436.23	434.0457	49.56	6.47	36.63	7.34
C18H28BR203	452.23	450.0406	47.81	6.24	35.34	10.61
C18H28BR204	468.23	466.0355	46.17	6.03	34.13	13.67
C18H28BR205	484.23	482.0304	44.65	5.83	33.00	16.52
C18H28BR206	500.23	498.0254	43.22	5.64	31.95	19.19
C18H28BR207	516.23	514.0203	41.88	5.47	30.96	21.70
C18H28BR208	532.23	530.0152	40.62	5.30	30.03	24.05
C18H28BR209	548.23	546.0101	39.44	5.15	29.15	26.27
C18H29BR101	341.34	340.1402	63.34	8.56	23.41	4.69
C18H29BR102	357.33	356.1351	60.50	8.18	22.36	8.95
C18H29BR103	373.33	372.1300	57.91	7.83	21.40	12.86
C18H29BR104	389.33	388.1250	55.53	7.51	20.52	16.44
C18H29BR105	405.33	404.1199	53.34	7.21	19.71	19.74
C18H29BR106	421.33	420.1148	51.31	6.94	18.96	22.78
C18H29BR107	437.33	436.1097	49.44	6.68	18.27	25.61
C18H29BR108	453.33	452.1046	47.69	6.45	17.63	28.23
C18H29BR109	469.33	468.0995	46.07	6.23	17.03	30.68
C18H30BR201	422.25	420.0664	51.20	7.16	37.85	3.79
C18H30BR202	438.25	436.0613	49.33	6.90	36.47	7.30
C18H30BR203	454.25	452.0563	47.60	6.66	35.18	10.57
C18H30BR204	470.25	468.0512	45.98	6.43	33.98	13.61
C18H30BR205	486.24	484.0461	44.46	6.22	32.87	16.45
C18H30BR206	502.24	500.0410	43.05	6.02	31.82	19.11
C18H30BR207	518.24	516.0359	41.72	5.83	30.84	21.61
C18H30BR208	534.24	532.0308	40.47	5.66	29.91	23.96
C18H30BR209	550.24	548.0257	39.29	5.50	29.04	26.17
C18H31BR101	343.35	342.1559	62.97	9.10	23.27	4.66
C18H31BR102	359.35	358.1508	60.16	8.70	22.24	8.90
C18H31BR103	375.35	374.1457	57.60	8.32	21.29	12.79
C18H31BR104	391.35	390.1406	55.24	7.98	20.42	16.35
C18H31BR105	407.35	406.1355	53.08	7.67	19.62	19.64
C18H31BR106	423.35	422.1304	51.07	7.38	18.87	22.68
C18H31BR107	439.35	438.1254	49.21	7.11	18.19	25.49
C18H31BR108	455.35	454.1203	47.48	6.86	17.55	28.11
C18H31BR109	471.35	470.1152	45.87	6.63	16.95	30.55
C18H32BR201	424.26	422.0821	50.96	7.60	37.67	3.77

031

C18H32BR202	440.26	438.0770	49.11	7.33	36.30	7.27
C18H32BR203	456.26	454.0719	47.39	7.07	35.03	10.52
C18H32BR204	472.26	470.0668	45.78	6.83	33.84	13.55
C18H32BR205	488.26	486.0617	44.28	6.61	32.73	16.38
C18H32BR206	504.26	502.0567	42.87	6.40	31.69	19.04
C18H32BR207	520.26	518.0516	41.56	6.20	30.72	21.53
C18H32BR208	536.26	534.0465	40.32	6.01	29.80	23.87
C18H32BR209	552.26	550.0414	39.15	5.84	28.94	26.07
C18H33BR101	345.37	344.1715	62.60	9.63	23.14	4.63
C18H33BR102	361.37	360.1664	59.83	9.20	22.11	8.85
C18H33BR103	377.37	376.1613	57.29	8.81	21.17	12.72
C18H33BR104	393.37	392.1563	54.96	8.46	20.31	16.27
C18H33BR105	409.36	408.1512	52.81	8.13	19.52	19.54
C18H33BR106	425.36	424.1461	50.83	7.82	18.78	22.57
C18H33BR107	441.36	440.1410	48.98	7.54	18.10	25.37
C18H33BR108	457.36	456.1359	47.27	7.27	17.47	27.99
C18H33BR109	473.36	472.1308	45.67	7.03	16.88	30.42
C18H34BR201	426.28	424.0977	50.72	8.04	37.49	3.75
C18H34BR202	442.28	440.0926	48.88	7.75	36.13	7.23
C18H34BR203	458.28	456.0876	47.18	7.48	34.87	10.47
C18H34BR204	474.28	472.0825	45.59	7.23	33.70	13.49
C18H34BR205	490.28	488.0774	44.10	6.99	32.60	16.32
C18H34BR206	506.28	504.0723	42.70	6.77	31.57	18.96
C18H34BR207	522.28	520.0672	41.40	6.56	30.60	21.44
C18H34BR208	538.27	536.0621	40.17	6.37	29.69	23.78
C18H34BR209	554.27	552.0570	39.01	6.18	28.83	25.98
C18H35BR101	347.38	346.1872	62.24	10.16	23.00	4.61
C18H35BR102	363.38	362.1821	59.50	9.71	21.99	8.81
C18H35BR103	379.38	378.1770	56.99	9.30	21.06	12.65
C18H35BR104	395.38	394.1719	54.68	8.92	20.21	16.19
C18H35BR105	411.38	410.1668	52.55	8.58	19.42	19.45
C18H35BR106	427.38	426.1617	50.59	8.25	18.70	22.46
C18H35BR107	443.38	442.1566	48.76	7.96	18.02	25.26
C18H35BR108	459.38	458.1516	47.06	7.68	17.39	27.86
C18H35BR109	475.38	474.1465	45.48	7.42	16.81	30.29
C18H36BR201	428.30	426.1134	50.48	8.47	37.31	3.74
C18H36BR202	444.29	442.1083	48.66	8.17	35.97	7.20
C18H36BR203	460.29	458.1032	46.97	7.88	34.72	10.43
C18H36BR204	476.29	474.0981	45.39	7.62	33.55	13.44
C18H36BR205	492.29	490.0930	43.92	7.37	32.46	16.25
C18H36BR206	508.29	506.0880	42.53	7.14	31.44	18.89
C18H36BR207	524.29	522.0829	41.24	6.92	30.48	21.36
C18H36BR208	540.29	538.0778	40.02	6.72	29.58	23.69
C18H36BR209	556.29	554.0727	38.86	6.52	28.73	25.88
C18H37BR101	349.40	348.2028	61.88	10.67	22.87	4.58
C18H37BR102	365.40	364.1977	59.17	10.21	21.87	8.76
C18H37BR103	381.40	380.1926	56.69	9.78	20.95	12.58
C18H37BR104	397.40	396.1876	54.40	9.38	20.11	16.10
C18H37BR105	413.40	412.1825	52.30	9.02	19.33	19.35
C18H37BR106	429.40	428.1774	50.35	8.69	18.61	22.36
C18H37BR107	445.40	444.1723	48.54	8.37	17.94	25.15
C18H37BR108	461.39	460.1672	46.86	8.08	17.32	27.74
C18H37BR109	477.39	476.1621	45.29	7.81	16.74	30.16
C19H 9BR101	333.19	331.9837	68.49	2.72	23.98	4.80
C19H 9BR102	349.19	347.9786	65.36	2.60	22.88	9.16
C19H 9BR103	365.19	363.9736	62.49	2.48	21.88	13.14
C19H 9BR104	381.19	379.9685	59.87	2.38	20.96	16.79
C19H 9BR105	397.18	395.9634	57.46	2.28	20.12	20.14
C19H 9BR106	413.18	411.9583	55.23	2.20	19.34	23.23
C19H 9BR107	429.18	427.9532	53.17	2.11	18.62	26.10
C19H 9BR108	445.18	443.9481	51.26	2.04	17.95	28.75
C19H 9BR109	461.18	459.9430	49.48	1.97	17.33	31.22
C19H10BR201	414.10	411.9099	55.11	2.43	38.59	3.86
C19H10BR202	430.10	427.9049	53.06	2.34	37.16	7.44
C19H10BR203	446.10	443.8998	51.16	2.26	35.82	10.76
C19H10BR204	462.10	459.8947	49.39	2.18	34.58	13.85
C19H10BR205	478.10	475.8896	47.73	2.11	33.43	16.73
C19H10BR206	494.10	491.8845	46.19	2.04	32.34	19.43
C19H10BR207	510.10	507.8794	44.74	1.98	31.33	21.96
C19H10BR208	526.09	523.8743	43.38	1.92	30.38	24.33
C19H10BR209	542.09	539.8693	42.10	1.86	29.48	26.56
C19H11BR101	335.20	333.9994	68.08	3.31	23.84	4.77
C19H11BR102	351.20	349.9943	64.98	3.16	22.75	9.11
C19H11BR103	367.20	365.9892	62.15	3.02	21.76	13.07
C19H11BR104	383.20	381.9841	59.55	2.89	20.85	16.70
C19H11BR105	399.20	397.9790	57.17	2.78	20.02	20.04
C19H11BR106	415.20	413.9739	54.96	2.67	19.24	23.12
C19H11BR107	431.20	429.9689	52.92	2.57	18.53	25.97
C19H11BR108	447.20	445.9638	51.03	2.48	17.87	28.62
C19H11BR109	463.20	461.9587	49.27	2.39	17.25	31.09
C19H12BR201	416.11	413.9256	54.84	2.91	38.40	3.84
C19H12BR202	432.11	429.9205	52.81	2.80	36.98	7.41
C19H12BR203	448.11	445.9154	50.93	2.70	35.66	10.71
C19H12BR204	464.11	461.9103	49.17	2.61	34.43	13.79
C19H12BR205	480.11	477.9052	47.53	2.52	33.29	16.66
C19H12BR206	496.11	493.9002	46.00	2.44	32.21	19.35
C19H12BR207	512.11	509.8951	44.56	2.36	31.21	21.87
C19H12BR208	528.11	525.8900	43.21	2.29	30.26	24.24
C19H12BR209	544.11	541.8849	41.94	2.22	29.37	26.46
C19H13BR101	337.22	336.0150	67.67	3.89	23.69	4.74
C19H13BR102	353.22	352.0099	64.61	3.71	22.62	9.06
C19H13BR103	369.22	368.0049	61.81	3.55	21.64	13.00
C19H13BR104	385.22	383.9998	59.24	3.40	20.74	16.61
C19H13BR105	401.22	399.9947	56.88	3.27	19.92	19.94
C19H13BR106	417.22	415.9896	54.70	3.14	19.15	23.01
C19H13BR107	433.22	431.9845	52.68	3.02	18.44	25.85
C19H13BR108	449.21	447.9794	50.80	2.92	17.79	28.49
C19H13BR109	465.21	463.9743	49.06	2.82	17.18	30.95
C19H14BR201	418.13	415.9412	54.58	3.37	38.22	3.83
C19H14BR202	434.13	431.9362	52.57	3.25	36.81	7.37
C19H14BR203	450.13	447.9311	50.70	3.14	35.50	10.66
C19H14BR204	466.13	463.9260	48.96	3.03	34.28	13.73
C19H14BR205	482.13	479.9209	47.33	2.93	33.15	16.59
C19H14BR206	498.13	495.9158	45.81	2.83	32.08	19.27
C19H14BR207	514.13	511.9107	44.39	2.74	31.08	21.78
C19H14BR208	530.13	527.9056	43.05	2.66	30.15	24.14
C19H14BR209	546.13	543.9006	41.79	2.58	29.26	26.37
C19H15BR101	339.23	338.0307	67.27	4.46	23.55	4.72
C19H15BR102	355.23	354.0256	64.24	4.26	22.49	9.01
C19H15BR103	371.23	370.0205	61.47	4.07	21.52	12.93
C19H15BR104	387.23	386.0154	58.93	3.90	20.63	16.53
C19H15BR105	403.23	402.0103	56.60	3.75	19.82	19.84
C19H15BR106	419.23	418.0052	54.44	3.61	19.06	22.90
C19H15BR107	435.23	434.0002	52.43	3.47	18.36	25.73
C19H15BR108	451.23	449.9951	50.58	3.35	17.71	28.37
C19H15BR109	467.23	465.9900	48.84	3.24	17.10	30.82
C19H16BR201	420.15	417.9569	54.32	3.84	38.04	3.81
C19H16BR202	436.15	433.9518	52.32	3.70	36.64	7.34
C19H16BR203	452.15	449.9467	50.47	3.57	35.34	10.62
C19H16BR204	468.14	465.9416	48.75	3.44	34.14	13.67

032

C19H16BR205	484.14	481.9365	47.14	3.33	33.01	16.52
C19H16BR206	500.14	497.9315	45.63	3.22	31.95	19.19
C19H16BR207	516.14	513.9264	44.21	3.12	30.96	21.70
C19H16BR208	532.14	529.9213	42.89	3.03	30.03	24.05
C19H16BR209	548.14	545.9162	41.63	2.94	29.15	26.27
C19H17BR101	341.25	340.0463	66.88	5.02	23.42	4.69
C19H17BR102	357.25	356.0412	63.88	4.80	22.37	8.96
C19H17BR103	373.25	372.0362	61.14	4.59	21.41	12.86
C19H17BR104	389.25	388.0311	58.63	4.40	20.53	16.44
C19H17BR105	405.25	404.0260	56.31	4.23	19.72	19.74
C19H17BR106	421.25	420.0209	54.18	4.07	18.97	22.79
C19H17BR107	437.25	436.0158	52.19	3.92	18.27	25.61
C19H17BR108	453.25	452.0107	50.35	3.78	17.63	28.24
C19H17BR109	469.25	468.0056	48.63	3.65	17.03	30.69
C19H18BR201	422.16	419.9725	54.06	4.30	37.85	3.79
C19H18BR202	438.16	435.9675	52.08	4.14	36.47	7.30
C19H18BR203	454.16	451.9624	50.25	3.99	35.19	10.57
C19H18BR204	470.16	467.9573	48.54	3.86	33.99	13.61
C19H18BR205	486.16	483.9522	46.94	3.73	32.87	16.45
C19H18BR206	502.16	499.9471	45.45	3.61	31.82	19.12
C19H18BR207	518.16	515.9420	44.04	3.50	30.84	21.61
C19H18BR208	534.16	531.9369	42.72	3.40	29.92	23.96
C19H18BR209	550.16	547.9318	41.48	3.30	29.05	26.17
C19H19BR101	343.27	342.0620	66.48	5.58	23.28	4.66
C19H19BR102	359.27	358.0569	63.52	5.33	22.24	8.91
C19H19BR103	375.27	374.0518	60.81	5.10	21.29	12.79
C19H19BR104	391.26	390.0467	58.33	4.89	20.42	16.36
C19H19BR105	407.26	406.0416	56.04	4.70	19.62	19.64
C19H19BR106	423.26	422.0365	53.92	4.52	18.88	22.68
C19H19BR107	439.26	438.0315	51.95	4.36	18.19	25.50
C19H19BR108	455.26	454.0264	50.13	4.21	17.55	28.11
C19H19BR109	471.26	470.0213	48.43	4.06	16.96	30.56
C19H20BR201	424.18	421.9882	53.80	4.75	37.67	3.77
C19H20BR202	440.18	437.9831	51.85	4.58	36.31	7.27
C19H20BR203	456.18	453.9780	50.03	4.42	35.03	10.52
C19H20BR204	472.18	469.9729	48.33	4.27	33.84	13.55
C19H20BR205	488.18	485.9678	46.75	4.13	32.74	16.39
C19H20BR206	504.18	501.9628	45.26	4.00	31.70	19.04
C19H20BR207	520.18	517.9577	43.87	3.88	30.72	21.53
C19H20BR208	536.17	533.9526	42.56	3.76	29.81	23.87
C19H20BR209	552.17	549.9475	41.33	3.65	28.94	26.08
C19H21BR101	345.28	344.0776	66.09	6.13	23.14	4.63
C19H21BR102	361.28	360.0725	63.17	5.86	22.12	8.86
C19H21BR103	377.28	376.0674	60.49	5.61	21.18	12.72
C19H21BR104	393.28	392.0624	58.03	5.38	20.32	16.27
C19H21BR105	409.28	408.0573	55.76	5.17	19.52	19.55
C19H21BR106	425.28	424.0522	53.66	4.98	18.79	22.57
C19H21BR107	441.28	440.0471	51.72	4.80	18.11	25.38
C19H21BR108	457.28	456.0420	49.91	4.63	17.47	27.99
C19H21BR109	473.28	472.0369	48.22	4.47	16.88	30.42
C19H22BR201	426.19	424.0038	53.55	5.20	37.50	3.75
C19H22BR202	442.19	439.9987	51.61	5.01	36.14	7.24
C19H22BR203	458.19	455.9937	49.81	4.84	34.88	10.48
C19H22BR204	474.19	471.9886	48.13	4.68	33.70	13.50
C19H22BR205	490.19	487.9835	46.56	4.52	32.60	16.32
C19H22BR206	506.19	503.9784	45.08	4.38	31.57	18.96
C19H22BR207	522.19	519.9733	43.70	4.25	30.60	21.45
C19H22BR208	538.19	535.9682	42.40	4.12	29.69	23.78
C19H22BR209	554.19	551.9631	41.18	4.00	28.84	25.98
C19H23BR101	347.30	346.0933	65.71	6.68	23.01	4.61
C19H23BR102	363.30	362.0882	62.82	6.38	21.99	8.81
C19H23BR103	379.30	378.0831	60.17	6.11	21.07	12.65
C19H23BR104	395.30	394.0780	57.73	5.86	20.21	16.19
C19H23BR105	411.30	410.0729	55.49	5.64	19.43	19.45
C19H23BR106	427.30	426.0678	53.41	5.43	18.70	22.47
C19H23BR107	443.29	442.0628	51.48	5.23	18.03	25.26
C19H23BR108	459.29	458.0577	49.69	5.05	17.40	27.87
C19H23BR109	475.29	474.0526	48.01	4.88	16.81	30.30
C19H24BR201	428.21	426.0195	53.29	5.65	37.32	3.74
C19H24BR202	444.21	442.0144	51.37	5.45	35.98	7.20
C19H24BR203	460.21	458.0093	49.59	5.26	34.73	10.43
C19H24BR204	476.21	474.0042	47.92	5.08	33.56	13.44
C19H24BR205	492.21	489.9991	46.36	4.91	32.47	16.25
C19H24BR206	508.21	505.9941	44.91	4.76	31.45	18.89
C19H24BR207	524.21	521.9890	43.53	4.61	30.49	21.36
C19H24BR208	540.21	537.9839	42.25	4.48	29.58	23.69
C19H24BR209	556.21	553.9788	41.03	4.35	28.73	25.89
C19H25BR101	349.31	348.1089	65.33	7.21	22.87	4.58
C19H25BR102	365.31	364.1038	62.47	6.90	21.87	8.76
C19H25BR103	381.31	380.0987	59.85	6.61	20.95	12.59
C19H25BR104	397.31	396.0937	57.44	6.34	20.11	16.11
C19H25BR105	413.31	412.0886	55.22	6.10	19.33	19.36
C19H25BR106	429.31	428.0835	53.16	5.87	18.61	22.36
C19H25BR107	445.31	444.0784	51.25	5.66	17.94	25.15
C19H25BR108	461.31	460.0733	49.47	5.46	17.32	27.75
C19H25BR109	477.31	476.0682	47.81	5.28	16.74	30.17
C19H26BR201	430.23	428.0351	53.04	6.09	37.15	3.72
C19H26BR202	446.23	444.0300	51.14	5.87	35.81	7.17
C19H26BR203	462.23	460.0250	49.37	5.67	34.57	10.38
C19H26BR204	478.22	476.0199	47.72	5.48	33.42	13.38
C19H26BR205	494.22	492.0148	46.18	5.30	32.34	16.19
C19H26BR206	510.22	508.0097	44.73	5.14	31.32	18.81
C19H26BR207	526.22	524.0046	43.37	4.98	30.37	21.28
C19H26BR208	542.22	539.9995	42.09	4.83	29.47	23.61
C19H26BR209	558.22	555.9944	40.88	4.69	28.63	25.80
C19H27BR101	351.33	350.1246	64.96	7.75	22.74	4.55
C19H27BR102	367.33	366.1195	62.13	7.41	21.75	8.71
C19H27BR103	383.33	382.1144	59.53	7.10	20.84	12.52
C19H27BR104	399.33	398.1093	57.15	6.82	20.01	16.03
C19H27BR105	415.33	414.1042	54.95	6.55	19.24	19.26
C19H27BR106	431.33	430.0991	52.91	6.31	18.53	22.26
C19H27BR107	447.33	446.0941	51.02	6.08	17.86	25.04
C19H27BR108	463.33	462.0890	49.26	5.87	17.25	27.63
C19H27BR109	479.33	478.0839	47.61	5.68	16.67	30.04
C19H28BR201	432.24	430.0508	52.80	6.53	36.97	3.70
C19H28BR202	448.24	446.0457	50.91	6.30	35.65	7.14
C19H28BR203	464.24	462.0406	49.16	6.08	34.42	10.34
C19H28BR204	480.24	478.0355	47.52	5.88	33.28	13.33
C19H28BR205	496.24	494.0304	45.99	5.69	32.20	16.12
C19H28BR206	512.24	510.0254	44.55	5.51	31.20	18.74
C19H28BR207	528.24	526.0203	43.20	5.34	30.25	21.20
C19H28BR208	544.24	542.0152	41.93	5.19	29.36	23.52
C19H28BR209	560.24	558.0101	40.73	5.04	28.53	25.70
C19H29BR101	353.35	352.1402	64.59	8.27	22.61	4.53
C19H29BR102	369.35	368.1351	61.79	7.91	21.63	8.66
C19H29BR103	385.35	384.1300	59.22	7.59	20.74	12.46
C19H29BR104	401.34	400.1250	56.86	7.28	19.91	15.95
C19H29BR105	417.34	416.1199	54.68	7.00	19.15	19.17
C19H29BR106	433.34	432.1148	52.66	6.75	18.44	22.15
C19H29BR107	449.34	448.1097	50.79	6.51	17.78	24.92

033

C19H29BR108	465.34	464.1046	49.04	6.28	17.17	27.51
C19H29BR109	481.34	480.0595	47.41	6.07	16.60	29.92
C19H30BR201	434.26	432.0664	52.55	6.96	36.80	3.68
C19H30BR202	450.26	448.0613	50.68	6.72	35.49	7.11
C19H30BR203	466.26	464.0563	48.95	6.49	34.27	10.29
C19H30BR204	482.26	480.0512	47.32	6.27	33.14	13.27
C19H30BR205	498.26	496.0461	45.80	6.07	32.07	16.06
C19H30BR206	514.26	512.0410	44.38	5.88	31.08	18.67
C19H30BR207	530.25	528.0359	43.04	5.70	30.14	21.12
C19H30BR208	546.25	544.0308	41.78	5.54	29.26	23.43
C19H30BR209	562.25	560.0257	40.59	5.38	28.42	25.61
C19H31BR101	355.36	354.1559	64.22	8.79	22.49	4.50
C19H31BR102	371.36	370.1508	61.45	8.41	21.52	8.62
C19H31BR103	387.36	386.1457	58.91	8.07	20.63	12.39
C19H31BR104	403.36	402.1406	56.58	7.75	19.81	15.87
C19H31BR105	419.36	418.1355	54.42	7.45	19.05	19.08
C19H31BR106	435.36	434.1304	52.42	7.18	18.35	22.05
C19H31BR107	451.36	450.1254	50.56	6.92	17.70	24.81
C19H31BR108	467.36	466.1203	48.83	6.69	17.10	27.39
C19H31BR109	483.36	482.1152	47.21	6.46	16.53	29.79
C19H32BR201	436.27	434.0821	52.31	7.39	36.63	3.67
C19H32BR202	452.27	450.0770	50.46	7.13	35.33	7.08
C19H32BR203	468.27	466.0719	48.73	6.89	34.13	10.25
C19H32BR204	484.27	482.0668	47.12	6.66	33.00	13.22
C19H32BR205	500.27	498.0617	45.62	6.45	31.94	15.99
C19H32BR206	516.27	514.0567	44.20	6.25	30.95	18.59
C19H32BR207	532.27	530.0516	42.88	6.06	30.02	21.04
C19H32BR208	548.27	546.0465	41.62	5.88	29.15	23.35
C19H32BR209	564.27	562.0414	40.44	5.72	28.32	25.52
C19H33BR101	357.38	356.1715	63.86	9.31	22.36	4.48
C19H33BR102	373.38	372.1664	61.12	8.91	21.40	8.57
C19H33BR103	389.38	388.1613	58.61	8.54	20.52	12.33
C19H33BR104	405.38	404.1563	56.30	8.21	19.71	15.79
C19H33BR105	421.38	420.1512	54.16	7.89	18.96	18.98
C19H33BR106	437.38	436.1461	52.18	7.61	18.27	21.95
C19H33BR107	453.37	452.1410	50.34	7.34	17.62	24.70
C19H33BR108	469.37	468.1359	48.62	7.09	17.02	27.27
C19H33BR109	485.37	484.1308	47.02	6.85	16.46	29.67
C19H34BR201	438.29	436.0977	52.07	7.82	36.46	3.65
C19H34BR202	454.29	452.0926	50.23	7.54	35.18	7.04
C19H34BR203	470.29	468.0876	48.53	7.29	33.98	10.21
C19H34BR204	486.29	484.0825	46.93	7.05	32.86	13.16
C19H34BR205	502.29	500.0774	45.43	6.82	31.82	15.93
C19H34BR206	518.29	516.0723	44.03	6.61	30.83	18.52
C19H34BR207	534.29	532.0672	42.71	6.41	29.91	20.96
C19H34BR208	550.29	548.0621	41.47	6.23	29.04	23.26
C19H34BR209	566.29	564.0570	40.30	6.05	28.22	25.43
C19H35BR101	359.39	358.1872	63.50	9.82	22.23	4.45
C19H35BR102	375.39	374.1821	60.79	9.40	21.29	8.52
C19H35BR103	391.39	390.1770	58.31	9.01	20.42	12.26
C19H35BR104	407.39	406.1719	56.02	8.66	19.61	15.71
C19H35BR105	423.39	422.1668	53.90	8.33	18.87	18.89
C19H35BR106	439.39	438.1617	51.94	8.03	18.19	21.85
C19H35BR107	455.39	454.1566	50.11	7.75	17.55	24.59
C19H35BR108	471.39	470.1516	48.41	7.48	16.95	27.15
C19H35BR109	487.39	486.1465	46.82	7.24	16.39	29.54
C19H36BR201	440.31	438.1134	51.83	8.24	36.29	3.63
C19H36BR202	456.31	454.1083	50.01	7.95	35.02	7.01
C19H36BR203	472.30	470.1032	48.32	7.68	33.84	10.16
C19H36BR204	488.30	486.0981	46.74	7.43	32.73	13.11

C19H36BR205	504.30	502.0930	45.25	7.20	31.69	15.86
C19H36BR206	520.30	518.0880	43.86	6.97	30.71	18.45
C19H36BR207	536.30	534.0829	42.55	6.77	29.80	20.88
C19H36BR208	552.30	550.0778	41.32	6.57	28.93	23.17
C19H36BR209	568.30	566.0727	40.16	6.39	28.12	25.34
C19H37BR101	361.41	360.2028	63.14	10.32	22.11	4.43
C19H37BR102	377.41	376.1977	60.47	9.88	21.17	8.48
C19H37BR103	393.41	392.1926	58.01	9.48	20.31	12.20
C19H37BR104	409.41	408.1876	55.74	9.11	19.52	15.63
C19H37BR105	425.41	424.1825	53.65	8.77	18.78	18.80
C19H37BR106	441.41	440.1774	51.70	8.45	18.10	21.75
C19H37BR107	457.41	456.1723	49.89	8.15	17.47	24.48
C19H37BR108	473.41	472.1672	48.21	7.88	16.88	27.04
C19H37BR109	489.41	488.1621	46.63	7.62	16.33	29.42
C19H38BR201	442.32	440.1290	51.59	8.66	36.13	3.62
C19H38BR202	458.32	456.1239	49.79	8.36	34.87	6.98
C19H38BR203	474.32	472.1189	48.11	8.08	33.69	10.12
C19H38BR204	490.32	488.1138	46.54	7.81	32.59	13.05
C19H38BR205	506.32	504.1087	45.07	7.56	31.56	15.80
C19H38BR206	522.32	520.1036	43.69	7.33	30.60	18.38
C19H38BR207	538.32	536.0985	42.39	7.12	29.69	20.80
C19H38BR208	554.32	552.0934	41.17	6.91	28.83	23.09
C19H38BR209	570.32	568.0883	40.01	6.72	28.02	25.25
C19H39BR101	363.43	362.2185	62.79	10.82	21.99	4.40
C19H39BR102	379.43	378.2134	60.15	10.36	21.06	8.43
C19H39BR103	395.42	394.2083	57.71	9.94	20.21	12.14
C19H39BR104	411.42	410.2032	55.47	9.55	19.42	15.56
C19H39BR105	427.42	426.1981	53.39	9.20	18.69	18.72
C19H39BR106	443.42	442.1930	51.47	8.87	18.02	21.65
C19H39BR107	459.42	458.1879	49.67	8.56	17.39	24.38
C19H39BR108	475.42	474.1829	48.00	8.27	16.81	26.92
C19H39BR109	491.42	490.1778	46.44	8.00	16.26	29.30
C20H10BR201	426.11	423.9099	56.38	2.37	37.50	3.75
C20H10BR202	442.11	439.9049	54.34	2.28	36.15	7.24
C20H10BR203	458.11	455.8998	52.44	2.20	34.88	10.48
C20H10BR204	474.11	471.8947	50.67	2.13	33.71	13.50
C20H10BR205	490.11	487.8896	49.01	2.06	32.61	16.32
C20H10BR206	506.11	503.8845	47.46	1.99	31.58	18.97
C20H10BR207	522.11	519.8794	46.01	1.93	30.61	21.45
C20H10BR208	538.11	535.8743	44.64	1.87	29.70	23.79
C20H10BR209	554.11	551.8693	43.35	1.82	28.84	25.99
C20H11BR101	347.21	345.9994	69.19	3.19	23.01	4.61
C20H11BR102	363.21	361.9943	66.14	3.05	22.00	8.81
C20H11BR103	379.21	377.9892	63.35	2.92	21.07	12.66
C20H11BR104	395.21	393.9841	60.78	2.81	20.22	16.19
C20H11BR105	411.21	409.9790	58.42	2.70	19.43	19.45
C20H11BR106	427.21	425.9739	56.23	2.60	18.70	22.47
C20H11BR107	443.21	441.9689	54.20	2.50	18.03	25.27
C20H11BR108	459.21	457.9638	52.31	2.41	17.40	27.87
C20H11BR109	475.21	473.9587	50.55	2.33	16.81	30.30
C20H12BR201	428.13	425.9256	56.11	2.83	37.33	3.74
C20H12BR202	444.13	441.9205	54.09	2.72	35.98	7.20
C20H12BR203	460.12	457.9154	52.21	2.63	34.73	10.43
C20H12BR204	476.12	473.9103	50.45	2.54	33.56	13.44
C20H12BR205	492.12	489.9052	48.81	2.46	32.47	16.26
C20H12BR206	508.12	505.9002	47.28	2.38	31.45	18.89
C20H12BR207	524.12	521.8951	45.83	2.31	30.49	21.37
C20H12BR208	540.12	537.8900	44.48	2.24	29.59	23.70
C20H12BR209	556.12	553.8849	43.20	2.18	28.74	25.89
C20H13BR101	349.23	348.0150	68.79	3.75	22.88	4.58

034

C20H13BR102	365.23	364.0099	65.77	3.59	21.88	8.76
C20H13BR103	381.23	380.0049	63.01	3.44	20.96	12.59
C20H13BR104	397.23	395.9998	60.47	3.30	20.12	16.11
C20H13BR105	413.23	411.9947	58.13	3.17	19.34	19.36
C20H13BR106	429.23	427.9896	55.97	3.05	18.62	22.36
C20H13BR107	445.23	443.9845	53.96	2.94	17.95	25.15
C20H13BR108	461.23	459.9794	52.08	2.84	17.32	27.75
C20H13BR109	477.23	475.9743	50.34	2.75	16.74	30.17
C20H14BR201	430.14	427.9412	55.85	3.28	37.15	3.72
C20H14BR202	446.14	443.9362	53.84	3.16	35.82	7.17
C20H14BR203	462.14	459.9311	51.98	3.05	34.58	10.39
C20H14BR204	478.14	475.9260	50.24	2.95	33.42	13.38
C20H14BR205	494.14	491.9209	48.61	2.86	32.34	16.19
C20H14BR206	510.14	507.9158	47.09	2.77	31.33	18.82
C20H14BR207	526.14	523.9107	45.66	2.68	30.37	21.29
C20H14BR208	542.14	539.9056	44.31	2.60	29.48	23.61
C20H14BR209	558.14	555.9006	43.04	2.53	28.63	25.80
C20H15BR101	351.25	350.0307	68.39	4.30	22.75	4.56
C20H15BR102	367.25	366.0256	65.41	4.12	21.76	8.71
C20H15BR103	383.24	382.0205	62.68	3.95	20.85	12.52
C20H15BR104	399.24	398.0154	60.17	3.79	20.01	16.03
C20H15BR105	415.24	414.0103	57.85	3.64	19.24	19.27
C20H15BR106	431.24	430.0052	55.70	3.51	18.53	22.26
C20H15BR107	447.24	446.0002	53.71	3.38	17.87	25.04
C20H15BR108	463.24	461.9951	51.86	3.26	17.25	27.63
C20H15BR109	479.24	477.9900	50.13	3.15	16.67	30.05
C20H16BR201	432.16	429.9569	55.59	3.73	36.98	3.70
C20H16BR202	448.16	445.9518	53.60	3.60	35.66	7.14
C20H16BR203	464.16	461.9467	51.75	3.47	34.43	10.34
C20H16BR204	480.16	477.9416	50.03	3.36	33.28	13.33
C20H16BR205	496.16	493.9365	48.42	3.25	32.21	16.12
C20H16BR206	512.15	509.9315	46.90	3.15	31.20	18.74
C20H16BR207	528.15	525.9264	45.48	3.05	30.26	21.21
C20H16BR208	544.15	541.9213	44.15	2.96	29.37	23.52
C20H16BR209	560.15	557.9162	42.89	2.88	28.53	25.71
C20H17BR101	353.26	352.0463	68.00	4.85	22.62	4.53
C20H17BR102	369.26	368.0412	65.06	4.64	21.64	8.67
C20H17BR103	385.26	384.0362	62.35	4.45	20.74	12.46
C20H17BR104	401.26	400.0311	59.87	4.27	19.91	15.95
C20H17BR105	417.26	416.0260	57.57	4.11	19.15	19.17
C20H17BR106	433.26	432.0209	55.45	3.96	18.44	22.16
C20H17BR107	449.26	448.0158	53.47	3.81	17.79	24.93
C20H17BR108	465.26	464.0107	51.63	3.68	17.17	27.51
C20H17BR109	481.26	480.0056	49.92	3.56	16.60	29.92
C20H18BR201	434.17	431.9725	55.33	4.18	36.81	3.69
C20H18BR202	450.17	447.9675	53.36	4.03	35.50	7.11
C20H18BR203	466.17	463.9624	51.53	3.89	34.28	10.30
C20H18BR204	482.17	479.9573	49.82	3.76	33.14	13.27
C20H18BR205	498.17	495.9522	48.22	3.64	32.08	16.06
C20H18BR206	514.17	511.9471	46.72	3.53	31.08	18.67
C20H18BR207	530.17	527.9420	45.31	3.42	30.14	21.12
C20H18BR208	546.17	543.9369	43.98	3.32	29.26	23.44
C20H18BR209	562.17	559.9318	42.73	3.23	28.43	25.61
C20H19BR101	355.28	354.0620	67.62	5.39	22.49	4.50
C20H19BR102	371.28	370.0569	64.70	5.16	21.52	8.62
C20H19BR103	387.28	386.0518	62.03	4.95	20.63	12.39
C20H19BR104	403.28	402.0467	59.57	4.75	19.81	15.87
C20H19BR105	419.28	418.0416	57.29	4.57	19.06	19.08
C20H19BR106	435.27	434.0365	55.19	4.40	18.36	22.05
C20H19BR107	451.27	450.0315	53.23	4.24	17.71	24.82

C20H19BR108	467.27	466.0264	51.41	4.10	17.10	27.39
C20H19BR109	483.27	482.0213	49.71	3.96	16.53	29.80
C20H20BR201	436.19	433.9882	55.07	4.62	36.64	3.67
C20H20BR202	452.19	449.9831	53.12	4.46	35.34	7.08
C20H20BR203	468.19	465.9780	51.31	4.31	34.13	10.25
C20H20BR204	484.19	481.9729	49.61	4.16	33.01	13.22
C20H20BR205	500.19	497.9678	48.03	4.03	31.95	15.99
C20H20BR206	516.19	513.9628	46.54	3.91	30.96	18.60
C20H20BR207	532.19	529.9577	45.14	3.79	30.03	21.04
C20H20BR208	548.19	545.9526	43.82	3.68	29.15	23.35
C20H20BR209	564.19	561.9475	42.58	3.57	28.33	25.52
C20H21BR101	357.29	356.0776	67.23	5.92	22.36	4.48
C20H21BR102	373.29	372.0725	64.35	5.67	21.41	8.57
C20H21BR103	389.29	388.0674	61.71	5.44	20.53	12.33
C20H21BR104	405.29	404.0624	59.27	5.22	19.72	15.79
C20H21BR105	421.29	420.0573	57.02	5.02	18.97	18.99
C20H21BR106	437.29	436.0522	54.93	4.84	18.27	21.95
C20H21BR107	453.29	452.0471	53.00	4.67	17.63	24.71
C20H21BR108	469.29	468.0420	51.19	4.51	17.03	27.27
C20H21BR109	485.29	484.0369	49.50	4.36	16.47	29.67
C20H22BR201	438.21	436.0038	54.82	5.06	36.47	3.65
C20H22BR202	454.21	451.9987	52.89	4.88	35.18	7.05
C20H22BR203	470.20	467.9937	51.09	4.72	33.99	10.21
C20H22BR204	486.20	483.9886	49.41	4.56	32.87	13.16
C20H22BR205	502.20	499.9835	47.83	4.42	31.82	15.93
C20H22BR206	518.20	515.9784	46.36	4.28	30.84	18.52
C20H22BR207	534.20	531.9733	44.97	4.15	29.92	20.97
C20H22BR208	550.20	547.9682	43.66	4.03	29.05	23.26
C20H22BR209	566.20	563.9631	42.43	3.92	28.22	25.43
C20H23BR101	359.31	358.0933	66.86	6.45	22.24	4.45
C20H23BR102	375.31	374.0882	64.01	6.18	21.29	8.53
C20H23BR103	391.31	390.0831	61.39	5.92	20.42	12.27
C20H23BR104	407.31	406.0780	58.98	5.69	19.62	15.71
C20H23BR105	423.31	422.0729	56.75	5.48	18.88	18.90
C20H23BR106	439.31	438.0678	54.68	5.28	18.19	21.85
C20H23BR107	455.31	454.0628	52.76	5.09	17.55	24.60
C20H23BR108	471.31	470.0577	50.97	4.92	16.95	27.16
C20H23BR109	487.30	486.0526	49.30	4.76	16.40	29.55
C20H24BR201	440.22	438.0195	54.57	5.50	36.30	3.63
C20H24BR202	456.22	454.0144	52.65	5.30	35.03	7.01
C20H24BR203	472.22	470.0093	50.87	5.12	33.84	10.16
C20H24BR204	488.22	486.0042	49.20	4.95	32.73	13.11
C20H24BR205	504.22	501.9991	47.64	4.80	31.69	15.87
C20H24BR206	520.22	517.9941	46.18	4.65	30.72	18.45
C20H24BR207	536.22	533.9890	44.80	4.51	29.80	20.89
C20H24BR208	552.22	549.9839	43.50	4.38	28.94	23.18
C20H24BR209	568.22	565.9788	42.28	4.26	28.12	25.34
C20H25BR101	361.33	360.1089	66.48	6.97	22.11	4.43
C20H25BR102	377.33	376.1038	63.36	6.68	21.18	8.48
C20H25BR103	393.32	392.0987	61.08	6.41	20.32	12.20
C20H25BR104	409.32	408.0937	58.69	6.16	19.52	15.63
C20H25BR105	425.32	424.0886	56.48	5.92	18.79	18.81
C20H25BR106	441.32	440.0835	54.43	5.71	18.11	21.75
C20H25BR107	457.32	456.0784	52.53	5.51	17.47	24.49
C20H25BR108	473.32	472.0733	50.75	5.32	16.88	27.04
C20H25BR109	489.32	488.0682	49.09	5.15	16.33	29.43
C20H26BR201	442.24	440.0351	54.32	5.93	36.14	3.62
C20H26BR202	458.24	456.0300	52.42	5.72	34.87	6.98
C20H26BR203	474.24	472.0250	50.65	5.53	33.70	10.12
C20H26BR204	490.24	488.0199	49.00	5.35	32.60	13.05

035

C20H26BR205	506.24	504.0148	47.45	5.18	31.57	15.80
C20H26BR206	522.23	520.0097	46.00	5.02	30.60	18.38
C20H26BR207	538.23	536.0046	44.63	4.87	29.69	20.81
C20H26BR208	554.23	551.9995	43.34	4.73	28.83	23.09
C20H26BR209	570.23	567.9944	42.13	4.60	28.03	25.25
C20H27BR101	363.34	362.1246	66.11	7.49	21.99	4.40
C20H27BR102	379.34	378.1195	63.33	7.17	21.06	8.44
C20H27BR103	395.34	394.1144	60.76	6.88	20.21	12.14
C20H27BR104	411.34	410.1093	58.40	6.62	19.43	15.56
C20H27BR105	427.34	426.1042	56.21	6.37	18.70	18.72
C20H27BR106	443.34	442.0991	54.18	6.14	18.02	21.65
C20H27BR107	459.34	458.0941	52.30	5.92	17.40	24.38
C20H27BR108	475.34	474.0890	50.54	5.73	16.81	26.93
C20H27BR109	491.34	490.0839	48.89	5.54	16.26	29.31
C20H28BR201	444.25	442.0508	54.07	6.35	35.97	3.60
C20H28BR202	460.25	458.0457	52.19	6.13	34.72	6.95
C20H28BR203	476.25	474.0406	50.44	5.93	33.56	10.08
C20H28BR204	492.25	490.0355	48.80	5.73	32.46	13.00
C20H28BR205	508.25	506.0304	47.26	5.55	31.44	15.74
C20H28BR206	524.25	522.0254	45.82	5.38	30.48	18.31
C20H28BR207	540.25	538.0203	44.47	5.22	29.58	20.73
C20H28BR208	556.25	554.0152	43.19	5.07	28.73	23.01
C20H28BR209	572.25	570.0101	41.98	4.93	27.93	25.16
C20H29BR101	365.36	364.1402	65.75	8.00	21.87	4.38
C20H29BR102	381.36	380.1351	62.99	7.67	20.95	8.39
C20H29BR103	397.36	396.1300	60.46	7.36	20.11	12.08
C20H29BR104	413.36	412.1250	58.12	7.07	19.33	15.48
C20H29BR105	429.36	428.1199	55.95	6.81	18.61	18.63
C20H29BR106	445.35	444.1148	53.94	6.56	17.94	21.56
C20H29BR107	461.35	460.1097	52.07	6.34	17.32	24.28
C20H29BR108	477.35	476.1046	50.32	6.12	16.74	26.81
C20H29BR109	493.35	492.0995	48.69	5.92	16.20	29.19
C20H30BR201	446.27	444.0664	53.83	6.78	35.81	3.59
C20H30BR202	462.27	460.0613	51.97	6.54	34.57	6.92
C20H30BR203	478.27	476.0563	50.23	6.32	33.41	10.04
C20H30BR204	494.27	492.0512	48.60	6.12	32.33	12.95
C20H30BR205	510.27	508.0461	47.08	5.93	31.32	15.68
C20H30BR206	526.27	524.0410	45.65	5.75	30.37	18.24
C20H30BR207	542.27	540.0359	44.30	5.58	29.47	20.65
C20H30BR208	558.27	556.0308	43.03	5.42	28.63	22.93
C20H30BR209	574.26	572.0257	41.83	5.27	27.83	25.07
C20H31BR101	367.37	366.1559	65.39	8.51	21.75	4.36
C20H31BR102	383.37	382.1508	62.66	8.15	20.84	8.35
C20H31BR103	399.37	398.1457	60.15	7.82	20.01	12.02
C20H31BR104	415.37	414.1406	57.83	7.52	19.24	15.41
C20H31BR105	431.37	430.1355	55.69	7.24	18.52	18.54
C20H31BR106	447.37	446.1304	53.70	6.98	17.86	21.46
C20H31BR107	463.37	462.1254	51.84	6.74	17.24	24.17
C20H31BR108	479.37	478.1203	50.11	6.52	16.67	26.70
C20H31BR109	495.37	494.1152	48.49	6.31	16.13	29.07
C20H32BR201	448.29	446.0821	53.59	7.20	35.65	3.57
C20H32BR202	464.28	462.0770	51.74	6.95	34.42	6.89
C20H32BR203	480.28	478.0719	50.02	6.72	33.27	9.99
C20H32BR204	496.28	494.0668	48.40	6.50	32.20	12.90
C20H32BR205	512.28	510.0617	46.89	6.30	31.20	15.62
C20H32BR206	528.28	526.0567	45.47	6.11	30.25	18.17
C20H32BR207	544.28	542.0516	44.14	5.93	29.36	20.58
C20H32BR208	560.28	558.0465	42.88	5.76	28.52	22.84
C20H32BR209	576.28	574.0414	41.69	5.60	27.73	24.99
C20H33BR101	369.39	368.1715	65.03	9.00	21.63	4.33

C20H33BR102	385.39	384.1664	62.33	8.63	20.73	8.30
C20H33BR103	401.39	400.1613	59.85	8.29	19.91	11.96
C20H33BR104	417.39	416.1563	57.55	7.97	19.14	15.33
C20H33BR105	433.39	432.1512	55.43	7.68	18.44	18.46
C20H33BR106	449.39	448.1461	53.46	7.40	17.78	21.36
C20H33BR107	465.39	464.1410	51.62	7.15	17.17	24.07
C20H33BR108	481.39	480.1359	49.90	6.91	16.60	26.59
C20H33BR109	497.38	496.1308	48.30	6.69	16.06	28.95
C20H34BR201	450.30	448.0977	53.35	7.61	35.49	3.55
C20H34BR202	466.30	464.0926	51.52	7.35	34.27	6.86
C20H34BR203	482.30	480.0876	49.81	7.11	33.13	9.95
C20H34BR204	498.30	496.0825	48.21	6.88	32.07	12.84
C20H34BR205	514.30	512.0774	46.71	6.66	31.07	15.55
C20H34BR206	530.30	528.0723	45.30	6.46	30.14	18.10
C20H34BR207	546.30	544.0672	43.97	6.27	29.25	20.50
C20H34BR208	562.30	560.0621	42.72	6.09	28.42	22.76
C20H34BR209	578.30	576.0570	41.54	5.93	27.63	24.90
C20H35BR101	371.41	370.1872	64.68	9.50	21.51	4.31
C20H35BR102	387.40	386.1821	62.01	9.11	20.63	8.26
C20H35BR103	403.40	402.1770	59.55	8.75	19.81	11.90
C20H35BR104	419.40	418.1719	57.28	8.41	19.05	15.26
C20H35BR105	435.40	434.1668	55.17	8.10	18.35	18.37
C20H35BR106	451.40	450.1617	53.22	7.82	17.70	21.27
C20H35BR107	467.40	466.1566	51.40	7.55	17.10	23.96
C20H35BR108	483.40	482.1516	49.69	7.30	16.53	26.48
C20H35BR109	499.40	498.1465	48.10	7.06	16.00	28.83
C20H36BR201	452.32	450.1134	53.11	8.02	35.33	3.54
C20H36BR202	468.32	466.1083	51.29	7.75	34.12	6.83
C20H36BR203	484.32	482.1032	49.60	7.49	33.00	9.91
C20H36BR204	500.32	498.0981	48.01	7.25	31.94	12.79
C20H36BR205	516.31	514.0930	46.53	7.03	30.95	15.49
C20H36BR206	532.31	530.0880	45.13	6.82	30.02	18.03
C20H36BR207	548.31	546.0829	43.81	6.62	29.15	20.43
C20H36BR208	564.31	562.0778	42.57	6.43	28.32	22.68
C20H36BR209	580.31	578.0727	41.40	6.25	27.54	24.81
C20H37BR101	373.42	372.2028	64.33	9.99	21.40	4.28
C20H37BR102	389.42	388.1977	61.69	9.58	20.52	8.22
C20H37BR103	405.42	404.1926	59.25	9.20	19.71	11.84
C20H37BR104	421.42	420.1876	57.00	8.85	18.96	15.19
C20H37BR105	437.42	436.1825	54.92	8.53	18.27	18.29
C20H37BR106	453.42	452.1774	52.98	8.23	17.62	21.17
C20H37BR107	469.42	468.1723	51.17	7.94	17.02	23.86
C20H37BR108	485.42	484.1672	49.49	7.68	16.46	26.37
C20H37BR109	501.42	500.1621	47.91	7.44	15.94	28.72
C20H38BR201	454.33	452.1290	52.87	8.43	35.17	3.52
C20H38BR202	470.33	468.1239	51.08	8.14	33.98	6.80
C20H38BR203	486.33	484.1189	49.39	7.88	32.86	9.87
C20H38BR204	502.33	500.1138	47.82	7.63	31.81	12.74
C20H38BR205	518.33	516.1087	46.35	7.39	30.83	15.43
C20H38BR206	534.33	532.1036	44.96	7.17	29.91	17.97
C20H38BR207	550.33	548.0985	43.65	6.96	29.04	20.35
C20H38BR208	566.33	564.0934	42.42	6.76	28.22	22.60
C20H38BR209	582.33	580.0883	41.25	6.58	27.44	24.73
C20H39BR101	375.44	374.2185	63.98	10.47	21.28	4.26
C20H39BR102	391.44	390.2134	61.37	10.04	20.41	8.17
C20H39BR103	407.44	406.2083	58.96	9.65	19.61	11.78
C20H39BR104	423.44	422.2032	56.73	9.28	18.87	15.11
C20H39BR105	439.43	438.1981	54.67	8.95	18.18	18.20
C20H39BR106	455.43	454.1930	52.75	8.63	17.54	21.08
C20H39BR107	471.43	470.1879	50.96	8.34	16.95	23.76

036

C20H39BR108	487.43	486.1829	49.28	8.06	16.39	26.26
C20H39BR109	503.43	502.1778	47.72	7.81	15.87	28.60
C20H40BR201	456.35	454.1447	52.64	8.84	35.02	3.51
C20H40BR202	472.35	470.1396	50.86	8.54	33.83	6.77
C20H40BR203	488.35	486.1345	49.19	8.26	32.72	9.83
C20H40BR204	504.35	502.1294	47.63	7.99	31.69	12.69
C20H40BR205	520.35	518.1243	46.17	7.75	30.71	15.37
C20H40BR206	536.35	534.1192	44.79	7.52	29.80	17.90
C20H40BR207	552.35	550.1142	43.49	7.30	28.93	20.28
C20H40BR208	568.34	566.1091	42.27	7.09	28.12	22.52
C20H40BR209	584.34	582.1040	41.11	6.90	27.35	24.64
C20H41BR101	377.45	376.2341	63.64	10.95	21.17	4.24
C20H41BR102	393.45	392.2290	61.06	10.50	20.31	8.13
C20H41BR103	409.45	408.2239	58.67	10.09	19.51	11.72
C20H41BR104	425.45	424.2189	56.46	9.71	18.78	15.04
C20H41BR105	441.45	440.2138	54.42	9.36	18.10	18.12
C20H41BR106	457.45	456.2087	52.51	9.03	17.47	20.99
C20H41BR107	473.45	472.2036	50.74	8.73	16.88	23.66
C20H41BR108	489.45	488.1985	49.08	8.44	16.33	26.15
C20H41BR109	505.45	504.1934	47.53	8.18	15.81	28.49
C21H10BR201	438.12	435.9099	57.57	2.30	36.48	3.65
C21H10BR202	454.12	451.9049	55.54	2.22	35.19	7.05
C21H10BR203	470.12	467.8998	53.65	2.14	33.99	10.21
C21H10BR204	486.12	483.8947	51.89	2.07	32.87	13.16
C21H10BR205	502.12	499.8896	50.23	2.01	31.83	15.93
C21H10BR206	518.12	515.8845	48.68	1.95	30.84	18.53
C21H10BR207	534.12	531.8794	47.22	1.89	29.92	20.97
C21H10BR208	550.12	547.8743	45.85	1.83	29.05	23.27
C21H10BR209	566.12	563.8693	44.56	1.78	28.23	25.44
C21H11BR101	359.23	357.9994	70.22	3.09	22.24	4.45
C21H11BR102	375.22	373.9943	67.22	2.95	21.29	8.53
C21H11BR103	391.22	389.9892	64.47	2.83	20.42	12.27
C21H11BR104	407.22	405.9841	61.94	2.72	19.62	15.72
C21H11BR105	423.22	421.9790	59.60	2.62	18.88	18.90
C21H11BR106	439.22	437.9739	57.43	2.52	18.19	21.86
C21H11BR107	455.22	453.9689	55.41	2.44	17.55	24.60
C21H11BR108	471.22	469.9638	53.53	2.35	16.96	27.16
C21H11BR109	487.22	485.9587	51.77	2.28	16.40	29.55
C21H12BR201	440.14	437.9256	57.31	2.75	36.31	3.64
C21H12BR202	456.14	453.9205	55.30	2.65	35.04	7.02
C21H12BR203	472.14	469.9154	53.42	2.56	33.85	10.17
C21H12BR204	488.14	485.9103	51.67	2.48	32.74	13.11
C21H12BR205	504.13	501.9052	50.03	2.40	31.70	15.87
C21H12BR206	520.13	517.9002	48.49	2.33	30.72	18.46
C21H12BR207	536.13	533.8951	47.05	2.26	29.81	20.89
C21H12BR208	552.13	549.8900	45.68	2.19	28.94	23.18
C21H12BR209	568.13	565.8849	44.40	2.13	28.13	25.35
C21H13BR101	361.24	360.0150	69.82	3.63	22.12	4.43
C21H13BR102	377.24	376.0099	66.86	3.47	21.18	8.48
C21H13BR103	393.24	392.0049	64.14	3.33	20.32	12.21
C21H13BR104	409.24	407.9998	61.63	3.20	19.53	15.64
C21H13BR105	425.24	423.9947	59.32	3.08	18.79	18.81
C21H13BR106	441.24	439.9896	57.17	2.97	18.11	21.76
C21H13BR107	457.24	455.9845	55.16	2.87	17.48	24.49
C21H13BR108	473.24	471.9794	53.30	2.77	16.88	27.05
C21H13BR109	489.24	487.9743	51.56	2.68	16.33	29.43
C21H14BR201	442.15	439.9412	57.05	3.19	36.14	3.62
C21H14BR202	458.15	455.9362	55.05	3.08	34.88	6.98
C21H14BR203	474.15	471.9311	53.20	2.98	33.70	10.12
C21H14BR204	490.15	487.9260	51.46	2.88	32.60	13.06

C21H14BR205	506.15	503.9209	49.83	2.79	31.57	15.80
C21H14BR206	522.15	519.9158	48.31	2.70	30.61	18.38
C21H14BR207	538.15	535.9107	46.87	2.62	29.70	20.81
C21H14BR208	554.15	551.9056	45.52	2.55	28.84	23.10
C21H14BR209	570.15	567.9006	44.24	2.48	28.03	25.26
C21H15BR101	363.26	362.0307	69.44	4.16	22.00	4.40
C21H15BR102	379.26	378.0256	66.51	3.99	21.07	8.44
C21H15BR103	395.26	394.0205	63.82	3.83	20.22	12.14
C21H15BR104	411.26	410.0154	61.33	3.68	19.43	15.56
C21H15BR105	427.25	426.0103	59.04	3.54	18.70	18.72
C21H15BR106	443.25	442.0052	56.91	3.41	18.03	21.66
C21H15BR107	459.25	458.0002	54.92	3.29	17.40	24.39
C21H15BR108	475.25	473.9951	53.07	3.18	16.81	26.93
C21H15BR109	491.25	489.9900	51.35	3.08	16.27	29.31
C21H16BR201	444.17	441.9569	56.79	3.63	35.98	3.60
C21H16BR202	460.17	457.9518	54.81	3.50	34.73	6.95
C21H16BR203	476.17	473.9467	52.97	3.39	33.56	10.08
C21H16BR204	492.17	489.9416	51.25	3.28	32.47	13.00
C21H16BR205	508.17	505.9365	49.64	3.17	31.45	15.74
C21H16BR206	524.17	521.9315	48.12	3.08	30.49	18.31
C21H16BR207	540.17	537.9264	46.70	2.99	29.59	20.73
C21H16BR208	556.16	553.9213	45.35	2.90	28.73	23.01
C21H16BR209	572.16	569.9162	44.08	2.82	27.93	25.17
C21H17BR101	365.27	364.0463	69.05	4.69	21.88	4.38
C21H17BR102	381.27	380.0412	66.16	4.49	20.96	8.39
C21H17BR103	397.27	396.0362	63.49	4.31	20.11	12.08
C21H17BR104	413.27	412.0311	61.03	4.15	19.33	15.49
C21H17BR105	429.27	428.0260	58.76	3.99	18.61	18.64
C21H17BR106	445.27	444.0209	56.65	3.85	17.95	21.56
C21H17BR107	461.27	460.0158	54.68	3.71	17.32	24.28
C21H17BR108	477.27	476.0107	52.85	3.59	16.74	26.82
C21H17BR109	493.27	492.0056	51.14	3.47	16.20	29.19
C21H18BR201	446.19	443.9725	56.53	4.07	35.82	3.59
C21H18BR202	462.18	459.9675	54.57	3.93	34.58	6.92
C21H18BR203	478.18	475.9624	52.75	3.79	33.42	10.04
C21H18BR204	494.18	491.9573	51.04	3.67	32.34	12.95
C21H18BR205	510.18	507.9522	49.44	3.56	31.32	15.68
C21H18BR206	526.18	523.9471	47.94	3.45	30.37	18.24
C21H18BR207	542.18	539.9420	46.52	3.35	29.48	20.66
C21H18BR208	558.18	555.9369	45.19	3.25	28.63	22.93
C21H18BR209	574.18	571.9318	43.93	3.16	27.83	25.08
C21H19BR101	367.29	366.0620	68.67	5.21	21.76	4.36
C21H19BR102	383.29	382.0569	65.81	5.00	20.85	8.35
C21H19BR103	399.29	398.0518	63.17	4.80	20.01	12.02
C21H19BR104	415.29	414.0467	60.74	4.61	19.24	15.41
C21H19BR105	431.29	430.0416	58.48	4.44	18.53	18.55
C21H19BR106	447.29	446.0365	56.39	4.28	17.86	21.46
C21H19BR107	463.29	462.0315	54.44	4.13	17.25	24.17
C21H19BR108	479.28	478.0264	52.63	4.00	16.67	26.71
C21H19BR109	495.28	494.0213	50.93	3.87	16.13	29.07
C21H20BR201	448.20	445.9882	56.28	4.50	35.66	3.57
C21H20BR202	464.20	461.9831	54.34	4.34	34.43	6.89
C21H20BR203	480.20	477.9780	52.53	4.20	33.28	10.00
C21H20BR204	496.20	493.9729	50.83	4.06	32.21	12.90
C21H20BR205	512.20	509.9678	49.25	3.94	31.20	15.62
C21H20BR206	528.20	525.9628	47.75	3.82	30.26	18.17
C21H20BR207	544.20	541.9577	46.35	3.70	29.37	20.58
C21H20BR208	560.20	557.9526	45.03	3.60	28.53	22.85
C21H20BR209	576.20	573.9475	43.78	3.50	27.73	24.99
C21H21BR101	369.30	368.0776	68.30	5.73	21.64	4.33

C21H21BR102	385.30	384.0725	65.46	5.49	20.74	8.30
C21H21BR103	401.30	400.0674	62.85	5.27	19.91	11.96
C21H21BR104	417.30	416.0624	60.44	5.07	19.15	15.34
C21H21BR105	433.30	432.0573	58.21	4.89	18.44	18.46
C21H21BR106	449.30	448.0522	56.14	4.71	17.78	21.37
C21H21BR107	465.30	464.0471	54.21	4.55	17.17	24.07
C21H21BR108	481.30	480.0420	52.41	4.40	16.60	26.59
C21H21BR109	497.30	496.0369	50.72	4.26	16.07	28.96
C21H22BR201	450.22	448.0038	56.03	4.93	35.50	3.55
C21H22BR202	466.22	463.9987	54.10	4.76	34.28	6.86
C21H22BR203	482.22	479.9937	52.31	4.60	33.14	9.95
C21H22BR204	498.22	495.9886	50.63	4.45	32.08	12.85
C21H22BR205	514.21	511.9835	49.05	4.31	31.08	15.56
C21H22BR206	530.21	527.9784	47.57	4.18	30.14	18.11
C21H22BR207	546.21	543.9733	46.18	4.06	29.26	20.50
C21H22BR208	562.21	559.9682	44.86	3.94	28.42	22.77
C21H22BR209	578.21	575.9631	43.62	3.84	27.64	24.90
C21H23BR101	371.32	370.0933	67.93	6.24	21.52	4.31
C21H23BR102	387.32	386.0882	65.12	5.99	20.63	8.26
C21H23BR103	403.32	402.0831	62.54	5.75	19.81	11.90
C21H23BR104	419.32	418.0780	60.15	5.53	19.06	15.26
C21H23BR105	435.32	434.0729	57.94	5.33	18.36	18.38
C21H23BR106	451.32	450.0678	55.89	5.14	17.70	21.27
C21H23BR107	467.32	466.0628	53.97	4.96	17.10	23.97
C21H23BR108	483.32	482.0577	52.19	4.80	16.53	26.48
C21H23BR109	499.32	498.0526	50.52	4.64	16.00	28.84
C21H24BR201	452.23	450.0195	55.78	5.35	35.34	3.54
C21H24BR202	468.23	466.0144	53.87	5.17	34.13	6.83
C21H24BR203	484.23	482.0093	52.09	5.00	33.00	9.91
C21H24BR204	500.23	498.0042	50.42	4.84	31.95	12.79
C21H24BR205	516.23	513.9991	48.86	4.69	30.96	15.50
C21H24BR206	532.23	529.9941	47.39	4.55	30.03	18.04
C21H24BR207	548.23	545.9890	46.01	4.41	29.15	20.43
C21H24BR208	564.23	561.9839	44.70	4.29	28.32	22.68
C21H24BR209	580.23	577.9788	43.47	4.17	27.54	24.82
C21H25BR101	373.34	372.1089	67.56	6.75	21.40	4.29
C21H25BR102	389.34	388.1038	64.79	6.47	20.52	8.22
C21H25BR103	405.34	404.0987	62.23	6.22	19.71	11.84
C21H25BR104	421.34	420.0937	59.87	5.98	18.96	15.19
C21H25BR105	437.33	436.0886	57.68	5.76	18.27	18.29
C21H25BR106	453.33	452.0835	55.64	5.56	17.63	21.18
C21H25BR107	469.33	468.0784	53.74	5.37	17.03	23.86
C21H25BR108	485.33	484.0733	51.97	5.19	16.46	26.37
C21H25BR109	501.33	500.0682	50.31	5.03	15.94	28.72
C21H26BR201	454.25	452.0351	55.53	5.77	35.18	3.52
C21H26BR202	470.25	468.0300	53.64	5.57	33.98	6.80
C21H26BR203	486.25	484.0250	51.87	5.39	32.87	9.87
C21H26BR204	502.25	500.0199	50.22	5.22	31.82	12.74
C21H26BR205	518.25	516.0148	48.67	5.06	30.84	15.44
C21H26BR206	534.25	532.0097	47.21	4.91	29.91	17.97
C21H26BR207	550.25	548.0046	45.84	4.76	29.04	20.35
C21H26BR208	566.24	563.9995	44.55	4.63	28.22	22.60
C21H26BR209	582.24	579.9944	43.32	4.50	27.45	24.73
C21H27BR101	375.35	374.1246	67.20	7.25	21.29	4.26
C21H27BR102	391.35	390.1195	64.45	6.95	20.42	8.18
C21H27BR103	407.35	406.1144	61.92	6.68	19.62	11.78
C21H27BR104	423.35	422.1093	59.58	6.43	18.87	15.12
C21H27BR105	439.35	438.1042	57.41	6.19	18.19	18.21
C21H27BR106	455.35	454.0991	55.39	5.98	17.55	21.08
C21H27BR107	471.35	470.0941	53.51	5.77	16.95	23.76

C21H27BR108	487.35	486.0890	51.76	5.58	16.40	26.26
C21H27BR109	503.35	502.0839	50.11	5.41	15.87	28.61
C21H28BR201	456.26	454.0508	55.28	6.19	35.03	3.51
C21H28BR202	472.26	470.0457	53.41	5.98	33.84	6.78
C21H28BR203	488.26	486.0406	51.66	5.78	32.73	9.83
C21H28BR204	504.26	502.0355	50.02	5.60	31.69	12.69
C21H28BR205	520.26	518.0304	48.48	5.42	30.72	15.38
C21H28BR206	536.26	534.0254	47.04	5.26	29.80	17.90
C21H28BR207	552.26	550.0203	45.67	5.11	28.94	20.28
C21H28BR208	568.26	566.0152	44.39	4.97	28.12	22.52
C21H28BR209	584.26	582.0101	43.17	4.83	27.35	24.65
C21H29BR101	377.37	376.1402	66.84	7.75	21.17	4.24
C21H29BR102	393.37	392.1351	64.12	7.43	20.31	8.13
C21H29BR103	409.37	408.1300	61.62	7.14	19.52	11.72
C21H29BR104	425.37	424.1250	59.30	6.87	18.78	15.05
C21H29BR105	441.37	440.1199	57.15	6.62	18.10	18.12
C21H29BR106	457.37	456.1148	55.15	6.39	17.47	20.99
C21H29BR107	473.37	472.1097	53.29	6.18	16.88	23.66
C21H29BR108	489.36	488.1046	51.54	5.97	16.33	26.16
C21H29BR109	505.36	504.0995	49.91	5.78	15.81	28.49
C21H30BR201	458.28	456.0664	55.04	6.60	34.87	3.49
C21H30BR202	474.28	472.0613	53.18	6.38	33.69	6.75
C21H30BR203	490.28	488.0563	51.45	6.17	32.60	9.79
C21H30BR204	506.28	504.0512	49.82	5.97	31.57	12.64
C21H30BR205	522.28	520.0461	48.29	5.79	30.60	15.32
C21H30BR206	538.28	536.0410	46.86	5.62	29.69	17.83
C21H30BR207	554.28	552.0359	45.51	5.46	28.83	20.21
C21H30BR208	570.28	568.0308	44.23	5.30	28.02	22.44
C21H30BR209	586.28	584.0257	43.02	5.16	27.26	24.56
C21H31BR101	379.38	378.1559	66.49	8.24	21.06	4.22
C21H31BR102	395.38	394.1508	63.79	7.90	20.21	8.09
C21H31BR103	411.38	410.1457	61.31	7.60	19.42	11.67
C21H31BR104	427.38	426.1406	59.02	7.31	18.70	14.97
C21H31BR105	443.38	442.1355	56.89	7.05	18.02	18.04
C21H31BR106	459.38	458.1304	54.91	6.80	17.39	20.90
C21H31BR107	475.38	474.1254	53.06	6.57	16.81	23.56
C21H31BR108	491.38	490.1203	51.33	6.36	16.26	26.05
C21H31BR109	507.38	506.1152	49.71	6.16	15.75	28.38
C21H32BR201	460.30	458.0821	54.80	7.01	34.72	3.48
C21H32BR202	476.30	474.0770	52.96	6.77	33.55	6.72
C21H32BR203	492.30	490.0719	51.24	6.55	32.46	9.75
C21H32BR204	508.29	506.0668	49.62	6.35	31.44	12.59
C21H32BR205	524.29	522.0617	48.11	6.15	30.48	15.26
C21H32BR206	540.29	538.0567	46.68	5.97	29.58	17.77
C21H32BR207	556.29	554.0516	45.34	5.80	28.73	20.13
C21H32BR208	572.29	570.0465	44.07	5.64	27.92	22.37
C21H32BR209	588.29	586.0414	42.88	5.48	27.16	24.48
C21H33BR101	381.40	380.1715	66.13	8.72	20.95	4.19
C21H33BR102	397.40	396.1664	63.47	8.37	20.11	8.05
C21H33BR103	413.40	412.1613	61.01	8.05	19.33	11.61
C21H33BR104	429.40	428.1563	58.74	7.75	18.61	14.90
C21H33BR105	445.40	444.1512	56.63	7.47	17.94	17.96
C21H33BR106	461.40	460.1461	54.67	7.21	17.32	20.81
C21H33BR107	477.40	476.1410	52.84	6.97	16.74	23.46
C21H33BR108	493.40	492.1359	51.12	6.74	16.19	25.94
C21H33BR109	509.40	508.1308	49.52	6.53	15.69	28.27
C21H34BR201	462.31	460.0977	54.56	7.41	34.57	3.46
C21H34BR202	478.31	476.0926	52.73	7.16	33.41	6.69
C21H34BR203	494.31	492.0876	51.03	6.93	32.33	9.71
C21H34BR204	510.31	508.0825	49.43	6.72	31.32	12.54

038

C21H34BR205	526.31	524.0774	47.93	6.51	30.36	15.20
C21H34BR206	542.31	540.0723	46.51	6.32	29.47	17.70
C21H34BR207	558.31	556.0672	45.18	6.14	28.62	20.06
C21H34BR208	574.31	572.0621	43.92	5.97	27.83	22.29
C21H34BR209	590.31	588.0570	42.73	5.81	27.07	24.39
C21H35BR101	383.42	382.1872	65.79	9.20	20.84	4.17
C21H35BR102	399.42	398.1821	63.15	8.83	20.01	8.01
C21H35BR103	415.42	414.1770	60.72	8.49	19.23	11.55
C21H35BR104	431.41	430.1719	58.47	8.18	18.52	14.83
C21H35BR105	447.41	446.1668	56.38	7.89	17.86	17.88
C21H35BR106	463.41	462.1617	54.43	7.61	17.24	20.72
C21H35BR107	479.41	478.1566	52.61	7.36	16.67	23.36
C21H35BR108	495.41	494.1516	50.91	7.12	16.13	25.84
C21H35BR109	511.41	510.1465	49.32	6.90	15.62	28.16
C21H36BR201	464.33	462.1134	54.32	7.81	34.42	3.45
C21H36BR202	480.33	478.1083	52.51	7.55	33.27	6.66
C21H36BR203	496.33	494.1032	50.82	7.31	32.20	9.67
C21H36BR204	512.33	510.0981	49.23	7.08	31.19	12.49
C21H36BR205	528.33	526.0930	47.74	6.87	30.25	15.14
C21H36BR206	544.33	542.0880	46.34	6.67	29.36	17.64
C21H36BR207	560.32	558.0829	45.02	6.48	28.52	19.99
C21H36BR208	576.32	574.0778	43.77	6.30	27.73	22.21
C21H36BR209	592.32	590.0727	42.58	6.13	26.98	24.31
C21H37BR101	385.43	384.2028	65.44	9.68	20.73	4.15
C21H37BR102	401.43	400.1977	62.83	9.29	19.90	7.97
C21H37BR103	417.43	416.1926	60.43	8.93	19.14	11.50
C21H37BR104	433.43	432.1876	58.19	8.60	18.44	14.77
C21H37BR105	449.43	448.1825	56.12	8.30	17.78	17.80
C21H37BR106	465.43	464.1774	54.19	8.01	17.17	20.63
C21H37BR107	481.43	480.1723	52.39	7.75	16.60	23.26
C21H37BR108	497.43	496.1672	50.71	7.50	16.06	25.73
C21H37BR109	513.43	512.1621	49.13	7.26	15.56	28.05
C21H38BR201	466.34	464.1290	54.09	8.21	34.27	3.43
C21H38BR202	482.34	480.1239	52.29	7.94	33.13	6.63
C21H38BR203	498.34	496.1189	50.61	7.69	32.07	9.63
C21H38BR204	514.34	512.1138	49.04	7.45	31.07	12.44
C21H38BR205	530.34	528.1087	47.56	7.22	30.13	15.08
C21H38BR206	546.34	544.1036	46.17	7.01	29.25	17.57
C21H38BR207	562.34	560.0985	44.85	6.81	28.42	19.92
C21H38BR208	578.34	576.0934	43.61	6.62	27.63	22.13
C21H38BR209	594.34	592.0883	42.44	6.44	26.89	24.23
C21H39BR101	387.45	386.2185	65.10	10.15	20.62	4.13
C21H39BR102	403.45	402.2134	62.52	9.74	19.81	7.93
C21H39BR103	419.45	418.2083	60.13	9.37	19.05	11.44
C21H39BR104	435.45	434.2032	57.93	9.03	18.35	14.70
C21H39BR105	451.45	450.1981	55.87	8.71	17.70	17.72
C21H39BR106	467.45	466.1930	53.96	8.41	17.09	20.54
C21H39BR107	483.44	482.1879	52.17	8.13	16.53	23.17
C21H39BR108	499.44	498.1829	50.50	7.87	16.00	25.63
C21H39BR109	515.44	514.1778	48.94	7.63	15.50	27.94
C21H40BR201	468.36	466.1447	53.85	8.61	34.12	3.42
C21H40BR202	484.36	482.1396	52.08	8.32	32.99	6.61
C21H40BR203	500.36	498.1345	50.41	8.06	31.94	9.59
C21H40BR204	516.36	514.1294	48.85	7.81	30.95	12.39
C21H40BR205	532.36	530.1243	47.38	7.57	30.02	15.03
C21H40BR206	548.36	546.1192	46.00	7.35	29.14	17.51
C21H40BR207	564.36	562.1142	44.69	7.14	28.32	19.84
C21H40BR208	580.36	578.1091	43.46	6.95	27.54	22.05
C21H40BR209	596.36	594.1040	42.30	6.76	26.80	24.15
C21H41BR101	389.46	388.2341	64.76	10.61	20.52	4.11

C21H41BR102	405.46	404.2290	62.21	10.19	19.71	7.89
C21H41BR103	421.46	420.2239	59.85	9.81	18.96	11.39
C21H41BR104	437.46	436.2189	57.66	9.45	18.27	14.63
C21H41BR105	453.46	452.2138	55.62	9.11	17.62	17.64
C21H41BR106	469.46	468.2087	53.73	8.80	17.02	20.45
C21H41BR107	485.46	484.2036	51.96	8.51	16.46	23.07
C21H41BR108	501.46	500.1985	50.30	8.24	15.93	25.52
C21H41BR109	517.46	516.1934	48.74	7.99	15.44	27.83
C21H42BR201	470.38	468.1603	53.62	9.00	33.97	3.40
C21H42BR202	486.38	484.1552	51.86	8.70	32.86	6.58
C21H42BR203	502.38	500.1502	50.21	8.43	31.81	9.55
C21H42BR204	518.37	516.1451	48.66	8.17	30.83	12.35
C21H42BR205	534.37	532.1400	47.20	7.92	29.91	14.97
C21H42BR206	550.37	548.1349	45.83	7.69	29.04	17.44
C21H42BR207	566.37	564.1298	44.54	7.47	28.22	19.77
C21H42BR208	582.37	580.1247	43.31	7.27	27.44	21.98
C21H42BR209	598.37	596.1196	42.15	7.07	26.71	24.06
C21H43BR101	391.48	390.2498	64.43	11.07	20.41	4.09
C21H43BR102	407.48	406.2447	61.90	10.64	19.61	7.85
C21H43BR103	423.48	422.2396	59.56	10.23	18.87	11.33
C21H43BR104	439.48	438.2345	57.39	9.86	18.18	14.56
C21H43BR105	455.48	454.2294	55.38	9.52	17.54	17.56
C21H43BR106	471.48	470.2243	53.50	9.19	16.95	20.36
C21H43BR107	487.48	486.2192	51.74	8.89	16.39	22.97
C21H43BR108	503.48	502.2142	50.10	8.61	15.87	25.42
C21H43BR109	519.48	518.2091	48.56	8.34	15.38	27.72
C22H11BR101	371.24	369.9994	71.18	2.99	21.52	4.31
C22H11BR102	387.24	385.9943	68.24	2.86	20.63	8.26
C22H11BR103	403.24	401.9892	65.53	2.75	19.82	11.90
C22H11BR104	419.23	417.9841	63.03	2.64	19.06	15.27
C22H11BR105	435.23	433.9790	60.71	2.55	18.36	18.38
C22H11BR106	451.23	449.9739	58.56	2.46	17.71	21.27
C22H11BR107	467.23	465.9689	56.56	2.37	17.10	23.97
C22H11BR108	483.23	481.9638	54.68	2.29	16.54	26.49
C22H11BR109	499.23	497.9587	52.93	2.22	16.01	28.84
C22H12BR201	452.15	449.9256	58.44	2.68	35.34	3.54
C22H12BR202	468.15	465.9205	56.44	2.58	34.14	6.84
C22H12BR203	484.15	481.9154	54.58	2.50	33.01	9.91
C22H12BR204	500.15	497.9103	52.83	2.42	31.95	12.80
C22H12BR205	516.15	513.9052	51.20	2.34	30.96	15.50
C22H12BR206	532.15	529.9002	49.66	2.27	30.03	18.04
C22H12BR207	548.14	545.8951	48.21	2.21	29.15	20.43
C22H12BR208	564.14	561.8900	46.84	2.14	28.33	22.69
C22H12BR209	580.14	577.8849	45.55	2.08	27.55	24.82
C22H13BR101	373.25	372.0150	70.80	3.51	21.41	4.29
C22H13BR102	389.25	388.0099	67.89	3.37	20.53	8.22
C22H13BR103	405.25	404.0049	65.21	3.23	19.72	11.84
C22H13BR104	421.25	419.9998	62.73	3.11	18.97	15.19
C22H13BR105	437.25	435.9947	60.43	3.00	18.27	18.30
C22H13BR106	453.25	451.9896	58.30	2.89	17.63	21.18
C22H13BR107	469.25	467.9845	56.31	2.79	17.03	23.87
C22H13BR108	485.25	483.9794	54.46	2.70	16.47	26.38
C22H13BR109	501.25	499.9743	52.72	2.61	15.94	28.73
C22H14BR201	454.16	451.9412	58.18	3.11	35.19	3.52
C22H14BR202	470.16	467.9362	56.20	3.00	33.99	6.81
C22H14BR203	486.16	483.9311	54.35	2.90	32.87	9.87
C22H14BR204	502.16	499.9260	52.62	2.81	31.82	12.74
C22H14BR205	518.16	515.9209	51.00	2.72	30.84	15.44
C22H14BR206	534.16	531.9158	49.47	2.64	29.92	17.97
C22H14BR207	550.16	547.9107	48.03	2.56	29.05	20.36

039

C22H14BR208	566.16	563.9056	46.67	2.49	28.23	22.61
C22H14BR209	582.16	579.9006	45.39	2.42	27.45	24.73
C22H15BR101	375.27	374.0307	70.42	4.03	21.29	4.26
C22H15BR102	391.27	390.0256	67.54	3.86	20.42	8.18
C22H15BR103	407.27	406.0205	64.88	3.71	19.62	11.79
C22H15BR104	423.27	422.0154	62.43	3.57	18.88	15.12
C22H15BR105	439.27	438.0103	60.16	3.44	18.19	18.21
C22H15BR106	455.27	454.0052	58.04	3.32	17.55	21.09
C22H15BR107	471.26	470.0002	56.07	3.21	16.96	23.76
C22H15BR108	487.26	485.9951	54.23	3.10	16.40	26.27
C22H15BR109	503.26	501.9900	52.51	3.00	15.88	28.61
C22H16BR201	456.18	453.9569	57.93	3.54	35.03	3.51
C22H16BR202	472.18	469.9518	55.96	3.42	33.84	6.78
C22H16BR203	488.18	485.9467	54.13	3.30	32.74	9.83
C22H16BR204	504.18	501.9416	52.41	3.20	31.70	12.69
C22H16BR205	520.18	517.9365	50.80	3.10	30.72	15.38
C22H16BR206	536.18	533.9315	49.28	3.01	29.81	17.90
C22H16BR207	552.18	549.9264	47.86	2.92	28.94	20.28
C22H16BR208	568.18	565.9213	46.51	2.84	28.13	22.53
C22H16BR209	584.18	581.9162	45.23	2.76	27.36	24.65
C22H17BR101	377.28	376.0463	70.04	4.54	21.18	4.24
C22H17BR102	393.28	392.0412	67.19	4.36	20.32	8.14
C22H17BR103	409.28	408.0362	64.56	4.19	19.52	11.73
C22H17BR104	425.28	424.0311	62.13	4.03	18.79	15.05
C22H17BR105	441.28	440.0260	59.88	3.88	18.11	18.13
C22H17BR106	457.28	456.0209	57.79	3.75	17.47	20.99
C22H17BR107	473.28	472.0158	55.83	3.62	16.88	23.66
C22H17BR108	489.28	488.0107	54.01	3.50	16.33	26.16
C22H17BR109	505.28	504.0056	52.30	3.39	15.81	28.50
C22H18BR201	458.20	455.9725	57.67	3.96	34.88	3.49
C22H18BR202	474.20	471.9675	55.72	3.83	33.70	6.75
C22H18BR203	490.19	487.9624	53.91	3.70	32.60	9.79
C22H18BR204	506.19	503.9573	52.20	3.58	31.57	12.64
C22H18BR205	522.19	519.9522	50.60	3.47	30.60	15.32
C22H18BR206	538.19	535.9471	49.10	3.37	29.69	17.84
C22H18BR207	554.19	551.9420	47.68	3.27	28.84	20.21
C22H18BR208	570.19	567.9369	46.34	3.18	28.03	22.45
C22H18BR209	586.19	583.9318	45.08	3.10	27.26	24.56
C22H19BR101	379.30	378.0620	69.67	5.05	21.07	4.22
C22H19BR102	395.30	394.0569	66.85	4.84	20.21	8.09
C22H19BR103	411.30	410.0518	64.25	4.66	19.43	11.67
C22H19BR104	427.30	426.0467	61.84	4.48	18.70	14.98
C22H19BR105	443.30	442.0416	59.61	4.32	18.02	18.05
C22H19BR106	459.30	458.0365	57.53	4.17	17.40	20.90
C22H19BR107	475.30	474.0315	55.60	4.03	16.81	23.56
C22H19BR108	491.30	490.0264	53.79	3.90	16.26	26.05
C22H19BR109	507.30	506.0213	52.09	3.78	15.75	28.38
C22H20BR201	460.21	457.9882	57.42	4.38	34.72	3.48
C22H20BR202	476.21	473.9831	55.49	4.23	33.56	6.72
C22H20BR203	492.21	489.9780	53.69	4.10	32.47	9.75
C22H20BR204	508.21	505.9729	52.00	3.97	31.45	12.59
C22H20BR205	524.21	521.9678	50.41	3.85	30.49	15.26
C22H20BR206	540.21	537.9628	48.92	3.73	29.58	17.77
C22H20BR207	556.21	553.9577	47.51	3.62	28.73	20.14
C22H20BR208	572.21	569.9526	46.18	3.52	27.93	22.37
C22H20BR209	588.21	585.9475	44.92	3.43	27.17	24.48
C22H21BR101	381.32	380.0776	69.30	5.55	20.95	4.20
C22H21BR102	397.32	396.0725	66.51	5.33	20.11	8.05
C22H21BR103	413.31	412.0674	63.93	5.12	19.33	11.61
C22H21BR104	429.31	428.0624	61.55	4.93	18.61	14.91
C22H21BR105	445.31	444.0573	59.34	4.75	17.94	17.96
C22H21BR106	461.31	460.0522	57.28	4.59	17.32	20.81
C22H21BR107	477.31	476.0471	55.36	4.43	16.74	23.46
C22H21BR108	493.31	492.0420	53.57	4.29	16.20	25.95
C22H21BR109	509.31	508.0369	51.88	4.16	15.69	28.27
C22H22BR201	462.23	460.0038	57.17	4.80	34.57	3.46
C22H22BR202	478.23	475.9987	55.26	4.64	33.42	6.69
C22H22BR203	494.23	491.9937	53.47	4.49	32.33	9.71
C22H22BR204	510.23	507.9886	51.79	4.35	31.32	12.54
C22H22BR205	526.23	523.9835	50.22	4.21	30.37	15.20
C22H22BR206	542.23	539.9784	48.73	4.09	29.47	17.70
C22H22BR207	558.22	555.9733	47.34	3.97	28.63	20.06
C22H22BR208	574.22	571.9682	46.02	3.86	27.83	22.29
C22H22BR209	590.22	587.9631	44.77	3.76	27.08	24.40
C22H23BR101	383.33	382.0933	68.93	6.05	20.84	4.17
C22H23BR102	399.33	398.0882	66.17	5.81	20.01	8.01
C22H23BR103	415.33	414.0831	63.62	5.58	19.24	11.56
C22H23BR104	431.33	430.0780	61.26	5.37	18.53	14.84
C22H23BR105	447.33	446.0729	59.07	5.18	17.86	17.88
C22H23BR106	463.33	462.0678	57.03	5.00	17.25	20.72
C22H23BR107	479.33	478.0628	55.13	4.84	16.67	23.37
C22H23BR108	495.33	494.0577	53.35	4.68	16.13	25.84
C22H23BR109	511.33	510.0526	51.68	4.53	15.63	28.16
C22H24BR201	464.24	462.0195	56.92	5.21	34.42	3.45
C22H24BR202	480.24	478.0144	55.02	5.04	33.28	6.66
C22H24BR203	496.24	494.0093	53.25	4.87	32.20	9.67
C22H24BR204	512.24	510.0042	51.59	4.72	31.20	12.49
C22H24BR205	528.24	525.9991	50.02	4.58	30.25	15.14
C22H24BR206	544.24	541.9941	48.55	4.44	29.36	17.64
C22H24BR207	560.24	557.9890	47.17	4.32	28.52	19.99
C22H24BR208	576.24	573.9839	45.86	4.20	27.73	22.21
C22H24BR209	592.24	589.9788	44.62	4.08	26.98	24.31
C22H25BR101	385.35	384.1089	68.57	6.54	20.74	4.15
C22H25BR102	401.35	400.1038	65.84	6.28	19.91	7.97
C22H25BR103	417.35	416.0987	63.32	6.04	19.15	11.50
C22H25BR104	433.35	432.0937	60.98	5.82	18.44	14.77
C22H25BR105	449.35	448.0886	58.81	5.61	17.78	17.80
C22H25BR106	465.34	464.0835	56.78	5.42	17.17	20.63
C22H25BR107	481.34	480.0784	54.90	5.24	16.60	23.27
C22H25BR108	497.34	496.0733	53.13	5.07	16.07	25.74
C22H25BR109	513.34	512.0682	51.48	4.91	15.57	28.05
C22H26BR201	466.26	464.0351	56.67	5.62	34.27	3.43
C22H26BR202	482.26	480.0300	54.79	5.43	33.14	6.64
C22H26BR203	498.26	496.0250	53.03	5.26	32.07	9.63
C22H26BR204	514.26	512.0199	51.38	5.10	31.08	12.44
C22H26BR205	530.26	528.0148	49.83	4.94	30.14	15.09
C22H26BR206	546.26	544.0097	48.37	4.80	29.26	17.57
C22H26BR207	562.26	560.0046	47.00	4.66	28.42	19.92
C22H26BR208	578.26	575.9995	45.70	4.53	27.64	22.13
C22H26BR209	594.26	591.9944	44.47	4.41	26.89	24.23
C22H27BR101	387.36	386.1246	68.22	7.03	20.63	4.13
C22H27BR102	403.36	402.1195	65.51	6.75	19.81	7.93
C22H27BR103	419.36	418.1144	63.01	6.49	19.05	11.45
C22H27BR104	435.36	434.1093	60.70	6.25	18.35	14.70
C22H27BR105	451.36	450.1042	58.54	6.03	17.70	17.72
C22H27BR106	467.36	466.0991	56.54	5.82	17.10	20.54
C22H27BR107	483.36	482.0941	54.67	5.63	16.53	23.17
C22H27BR108	499.36	498.0890	52.92	5.45	16.00	25.63
C22H27BR109	515.36	514.0839	51.27	5.28	15.50	27.94
C22H28BR201	468.28	466.0508	56.43	6.03	34.13	3.42

040

C22H28BR202	484.28	482.0457	54.57	5.83	33.00	6.61
C22H28BR203	500.27	498.0406	52.82	5.64	31.94	9.59
C22H28BR204	516.27	514.0355	51.18	5.47	30.95	12.40
C22H28BR205	532.27	530.0304	49.64	5.30	30.02	15.03
C22H28BR206	548.27	546.0254	48.20	5.15	29.15	17.51
C22H28BR207	564.27	562.0203	46.83	5.00	28.32	19.85
C22H28BR208	580.27	578.0152	45.54	4.86	27.54	22.06
C22H28BR209	596.27	594.0101	44.32	4.73	26.80	24.15
C22H29BR101	389.38	388.1402	67.86	7.51	20.52	4.11
C22H29BR102	405.38	404.1351	65.18	7.21	19.71	7.89
C22H29BR103	421.38	420.1300	62.71	6.94	18.96	11.39
C22H29BR104	437.38	436.1250	60.42	6.68	18.27	14.63
C22H29BR105	453.38	452.1199	58.28	6.45	17.62	17.64
C22H29BR106	469.38	468.1148	56.30	6.23	17.02	20.45
C22H29BR107	485.38	484.1097	54.44	6.02	16.46	23.07
C22H29BR108	501.38	500.1046	52.70	5.83	15.94	25.53
C22H29BR109	517.38	516.0995	51.07	5.65	15.44	27.83
C22H30BR201	470.29	468.0664	56.19	6.43	33.98	3.40
C22H30BR202	486.29	484.0613	54.34	6.22	32.86	6.58
C22H30BR203	502.29	500.0563	52.61	6.02	31.82	9.56
C22H30BR204	518.29	516.0512	50.98	5.83	30.83	12.35
C22H30BR205	534.29	532.0461	49.46	5.66	29.91	14.97
C22H30BR206	550.29	548.0410	48.02	5.50	29.04	17.44
C22H30BR207	566.29	564.0359	46.66	5.34	28.22	19.78
C22H30BR208	582.29	580.0308	45.38	5.19	27.44	21.98
C22H30BR209	598.29	596.0257	44.17	5.05	26.71	24.07
C22H31BR101	391.40	390.1559	67.51	7.98	20.42	4.09
C22H31BR102	407.40	406.1508	64.86	7.67	19.61	7.85
C22H31BR103	423.39	422.1457	62.41	7.38	18.87	11.34
C22H31BR104	439.39	438.1406	60.14	7.11	18.19	14.56
C22H31BR105	455.39	454.1355	58.03	6.86	17.55	17.57
C22H31BR106	471.39	470.1304	56.06	6.63	16.95	20.36
C22H31BR107	487.39	486.1254	54.22	6.41	16.39	22.98
C22H31BR108	503.39	502.1203	52.49	6.21	15.87	25.43
C22H31BR109	519.39	518.1152	50.88	6.02	15.38	27.72
C22H32BR201	472.31	470.0821	55.95	6.83	33.84	3.39
C22H32BR202	488.31	486.0770	54.11	6.61	32.73	6.55
C22H32BR203	504.31	502.0719	52.40	6.40	31.69	9.52
C22H32BR204	520.31	518.0668	50.79	6.20	30.71	12.30
C22H32BR205	536.31	534.0617	49.27	6.01	29.80	14.92
C22H32BR206	552.30	550.0567	47.84	5.84	28.93	17.38
C22H32BR207	568.30	566.0516	46.50	5.68	28.12	19.71
C22H32BR208	584.30	582.0465	45.22	5.52	27.35	21.91
C22H32BR209	600.30	598.0414	44.02	5.37	26.62	23.99
C22H33BR101	393.41	392.1715	67.17	8.46	20.31	4.07
C22H33BR102	409.41	408.1664	64.54	8.12	19.52	7.82
C22H33BR103	425.41	424.1613	62.12	7.82	18.78	11.28
C22H33BR104	441.41	440.1563	59.86	7.54	18.10	14.50
C22H33BR105	457.41	456.1512	57.77	7.27	17.47	17.49
C22H33BR106	473.41	472.1461	55.82	7.03	16.88	20.28
C22H33BR107	489.41	488.1410	53.99	6.80	16.33	22.88
C22H33BR108	505.41	504.1359	52.28	6.58	15.81	25.33
C22H33BR109	521.41	520.1308	50.68	6.38	15.32	27.62
C22H34BR201	474.32	472.0977	55.71	7.23	33.69	3.37
C22H34BR202	490.32	488.0926	53.89	6.99	32.59	6.53
C22H34BR203	506.32	504.0876	52.19	6.77	31.56	9.48
C22H34BR204	522.32	520.0825	50.59	6.56	30.60	12.25
C22H34BR205	538.32	536.0774	49.09	6.37	29.69	14.86
C22H34BR206	554.32	552.0723	47.67	6.18	28.83	17.32
C22H34BR207	570.32	568.0672	46.33	6.01	28.02	19.64
C22H34BR208	586.32	584.0621	45.07	5.85	27.26	21.83
C22H34BR209	602.32	600.0570	43.87	5.69	26.53	23.91
C22H35BR101	395.43	394.1872	66.83	8.92	20.21	4.05
C22H35BR102	411.43	410.1821	64.23	8.57	19.42	7.78
C22H35BR103	427.43	426.1770	61.82	8.25	18.69	11.23
C22H35BR104	443.43	442.1719	59.59	7.96	18.02	14.43
C22H35BR105	459.43	458.1668	57.52	7.68	17.39	17.41
C22H35BR106	475.42	474.1617	55.58	7.42	16.81	20.19
C22H35BR107	491.42	490.1566	53.77	7.18	16.26	22.79
C22H35BR108	507.42	506.1516	52.08	6.95	15.75	25.22
C22H35BR109	523.42	522.1465	50.48	6.74	15.27	27.51
C22H36BR201	476.34	474.1134	55.47	7.62	33.55	3.36
C22H36BR202	492.34	490.1083	53.67	7.37	32.46	6.50
C22H36BR203	508.34	506.1032	51.98	7.14	31.44	9.44
C22H36BR204	524.34	522.0981	50.40	6.92	30.48	12.21
C22H36BR205	540.34	538.0930	48.90	6.72	29.58	14.81
C22H36BR206	556.34	554.0880	47.50	6.52	28.73	17.26
C22H36BR207	572.34	570.0829	46.17	6.34	27.92	19.57
C22H36BR208	588.34	586.0778	44.91	6.17	27.16	21.76
C22H36BR209	604.33	602.0727	43.72	6.00	26.44	23.83
C22H37BR101	397.44	396.2028	66.49	9.38	20.10	4.03
C22H37BR102	413.44	412.1977	63.91	9.02	19.33	7.74
C22H37BR103	429.44	428.1926	61.53	8.68	18.61	11.18
C22H37BR104	445.44	444.1876	59.32	8.37	17.94	14.37
C22H37BR105	461.44	460.1825	57.27	8.08	17.32	17.34
C22H37BR106	477.44	476.1774	55.35	7.81	16.74	20.11
C22H37BR107	493.44	492.1723	53.55	7.56	16.19	22.70
C22H37BR108	509.44	508.1672	51.87	7.32	15.68	25.12
C22H37BR109	525.44	524.1621	50.29	7.10	15.21	27.40
C22H38BR201	478.36	476.1290	55.24	8.01	33.41	3.34
C22H38BR202	494.35	492.1239	53.45	7.75	32.33	6.47
C22H38BR203	510.35	508.1189	51.78	7.51	31.31	9.40
C22H38BR204	526.35	524.1138	50.20	7.28	30.36	12.16
C22H38BR205	542.35	540.1087	48.72	7.06	29.47	14.75
C22H38BR206	558.35	556.1036	47.33	6.86	28.62	17.19
C22H38BR207	574.35	572.0985	46.01	6.67	27.82	19.50
C22H38BR208	590.35	588.0934	44.76	6.49	27.07	21.68
C22H38BR209	606.35	604.0883	43.58	6.32	26.36	23.75
C22H39BR101	399.46	398.2185	66.15	9.84	20.00	4.01
C22H39BR102	415.46	414.2134	63.60	9.46	19.23	7.70
C22H39BR103	431.46	430.2083	61.24	9.11	18.52	11.12
C22H39BR104	447.46	446.2032	59.05	8.79	17.86	14.30
C22H39BR105	463.46	462.1981	57.02	8.48	17.24	17.26
C22H39BR106	479.46	478.1930	55.11	8.20	16.67	20.02
C22H39BR107	495.46	494.1879	53.33	7.93	16.13	22.60
C22H39BR108	511.46	510.1829	51.67	7.69	15.62	25.03
C22H39BR109	527.45	526.1778	50.10	7.45	15.15	27.30
C22H40BR201	480.37	478.1447	55.01	8.39	33.27	3.33
C22H40BR202	496.37	494.1396	53.24	8.12	32.20	6.45
C22H40BR203	512.37	510.1345	51.57	7.87	31.19	9.37
C22H40BR204	528.37	526.1294	50.01	7.63	30.25	12.11
C22H40BR205	544.37	542.1243	48.54	7.41	29.36	14.70
C22H40BR206	560.37	558.1192	47.16	7.20	28.52	17.13
C22H40BR207	576.37	574.1142	45.85	7.00	27.73	19.43
C22H40BR208	592.37	590.1091	44.61	6.81	26.98	21.61
C22H40BR209	608.37	606.1040	43.44	6.63	26.27	23.67
C22H41BR101	401.48	400.2341	65.82	10.29	19.90	3.99
C22H41BR102	417.47	416.2290	63.30	9.90	19.14	7.66
C22H41BR103	433.47	432.2239	60.96	9.53	18.43	11.07
C22H41BR104	449.47	448.2189	58.79	9.19	17.78	14.24

C22H41BR105	465.47	464.2138	56.77	8.88	17.17	17.19
C22H41BR106	481.47	480.2087	54.88	8.58	16.60	19.94
C22H41BR107	497.47	496.2036	53.12	8.31	16.06	22.51
C22H41BR108	513.47	512.1985	51.46	8.05	15.56	24.93
C22H41BR109	529.47	528.1934	49.91	7.81	15.09	27.20
C22H42BR201	482.39	480.1603	54.78	8.78	33.13	3.32
C22H42BR202	498.39	496.1552	53.02	8.49	32.07	6.42
C22H42BR203	514.39	512.1502	51.37	8.23	31.07	9.33
C22H42BR204	530.39	528.1451	49.82	7.98	30.13	12.07
C22H42BR205	546.39	544.1400	48.36	7.75	29.25	14.64
C22H42BR206	562.38	560.1349	46.99	7.53	28.42	17.07
C22H42BR207	578.38	576.1298	45.69	7.32	27.63	19.36
C22H42BR208	594.38	592.1247	44.46	7.12	26.89	21.53
C22H42BR209	610.38	608.1196	43.29	6.94	26.18	23.59
C22H43BR101	403.49	402.2498	65.49	10.74	19.80	3.97
C22H43BR102	419.49	418.2447	62.99	10.33	19.05	7.63
C22H43BR103	435.49	434.2396	60.68	9.95	18.35	11.02
C22H43BR104	451.49	450.2345	58.53	9.60	17.70	14.17
C22H43BR105	467.49	466.2294	56.52	9.27	17.09	17.11
C22H43BR106	483.49	482.2243	54.65	8.96	16.53	19.85
C22H43BR107	499.49	498.2192	52.90	8.68	16.00	22.42
C22H43BR108	515.49	514.2142	51.26	8.41	15.50	24.83
C22H43BR109	531.49	530.2091	49.72	8.15	15.03	27.09
C22H44BR201	484.40	482.1760	54.55	9.16	32.99	3.30
C22H44BR202	500.40	498.1709	52.81	8.86	31.94	6.39
C22H44BR203	516.40	514.1658	51.17	8.59	30.95	9.29
C22H44BR204	532.40	530.1607	49.63	8.33	30.02	12.02
C22H44BR205	548.40	546.1556	48.18	8.09	29.14	14.59
C22H44BR206	564.40	562.1505	46.82	7.86	28.31	17.01
C22H44BR207	580.40	578.1455	45.53	7.64	27.53	19.30
C22H44BR208	596.40	594.1404	44.31	7.44	26.80	21.46
C22H44BR209	612.40	610.1353	43.15	7.24	26.10	23.51
C22H45BR101	405.51	404.2654	65.16	11.19	19.70	3.95
C22H45BR102	421.51	420.2603	62.69	10.76	18.96	7.59
C22H45BR103	437.51	436.2552	60.40	10.37	18.26	10.97
C22H45BR104	453.51	452.2502	58.27	10.00	17.62	14.11
C22H45BR105	469.50	468.2451	56.28	9.66	17.02	17.04
C22H45BR106	485.50	484.2400	54.43	9.34	16.46	19.77
C22H45BR107	501.50	500.2349	52.69	9.04	15.93	22.33
C22H45BR108	517.50	516.2298	51.06	8.76	15.44	24.73
C22H45BR109	533.50	532.2247	49.53	8.50	14.98	26.99
C23H11BR101	383.25	381.9594	72.08	2.89	20.85	4.17
C23H11BR102	399.25	397.9543	69.19	2.78	20.01	8.01
C23H11BR103	415.25	413.9892	66.53	2.67	19.24	11.56
C23H11BR104	431.25	429.9841	64.06	2.57	18.53	14.84
C23H11BR105	447.25	445.9790	61.77	2.48	17.87	17.89
C23H11BR106	463.24	461.9739	59.64	2.39	17.25	20.72
C23H11BR107	479.24	477.9689	57.64	2.31	16.67	23.37
C23H11BR108	495.24	493.9638	55.78	2.24	16.13	25.84
C23H11BR109	511.24	509.9587	54.04	2.17	15.63	28.17
C23H12BR201	464.16	461.9256	59.52	2.61	34.43	3.45
C23H12BR202	480.16	477.9205	57.53	2.52	33.28	6.66
C23H12BR203	496.16	493.9154	55.68	2.44	32.21	9.67
C23H12BR204	512.16	509.9103	53.94	2.36	31.20	12.50
C23H12BR205	528.16	525.9052	52.31	2.29	30.26	15.15
C23H12BR206	544.16	541.9002	50.77	2.22	29.37	17.64
C23H12BR207	560.16	557.8951	49.32	2.16	28.53	19.99
C23H12BR208	576.16	573.8900	47.95	2.10	27.74	22.22
C23H12BR209	592.15	589.8849	46.65	2.04	26.99	24.32
C23H13BR101	385.26	384.0150	71.71	3.40	20.74	4.15
C23H13BR102	401.26	400.0099	68.85	3.27	19.91	7.97
C23H13BR103	417.26	416.0049	66.21	3.14	19.15	11.50
C23H13BR104	433.26	431.9998	63.76	3.02	18.44	14.77
C23H13BR105	449.26	447.9947	61.49	2.92	17.79	17.81
C23H13BR106	465.26	463.9896	59.38	2.82	17.17	20.63
C23H13BR107	481.26	479.9845	57.40	2.72	16.60	23.27
C23H13BR108	497.26	495.9794	55.56	2.64	16.07	25.74
C23H13BR109	513.26	511.9743	53.82	2.55	15.57	28.05
C23H14BR201	466.18	463.9412	59.26	3.03	34.28	3.43
C23H14BR202	482.17	479.9362	57.29	2.93	33.14	6.64
C23H14BR203	498.17	495.9311	55.45	2.83	32.08	9.63
C23H14BR204	514.17	511.9260	53.73	2.74	31.08	12.45
C23H14BR205	530.17	527.9209	52.11	2.66	30.14	15.09
C23H14BR206	546.17	543.9158	50.58	2.58	29.26	17.58
C23H14BR207	562.17	559.9107	49.14	2.51	28.43	19.92
C23H14BR208	578.17	575.9056	47.78	2.44	27.64	22.14
C23H14BR209	594.17	591.9006	46.49	2.38	26.90	24.23
C23H15BR101	387.28	386.0307	71.33	3.90	20.63	4.13
C23H15BR102	403.28	402.0256	68.50	3.75	19.81	7.93
C23H15BR103	419.28	418.0205	65.89	3.61	19.06	11.45
C23H15BR104	435.28	434.0154	63.47	3.47	18.36	14.70
C23H15BR105	451.28	450.0103	61.22	3.35	17.71	17.73
C23H15BR106	467.28	466.0052	59.12	3.24	17.10	20.54
C23H15BR107	483.28	482.0002	57.16	3.13	16.53	23.17
C23H15BR108	499.28	497.9951	55.33	3.03	16.00	25.64
C23H15BR109	515.27	513.9900	53.61	2.93	15.51	27.95
C23H16BR201	468.19	465.9569	59.01	3.44	34.13	3.42
C23H16BR202	484.19	481.9518	57.06	3.33	33.01	6.61
C23H16BR203	500.19	497.9467	55.23	3.22	31.95	9.60
C23H16BR204	516.19	513.9416	53.52	3.12	30.96	12.40
C23H16BR205	532.19	529.9365	51.91	3.03	30.03	15.03
C23H16BR206	548.19	545.9315	50.39	2.94	29.15	17.51
C23H16BR207	564.19	561.9264	48.97	2.86	28.33	19.85
C23H16BR208	580.19	577.9213	47.62	2.78	27.54	22.06
C23H16BR209	596.19	593.9162	46.34	2.71	26.81	24.15
C23H17BR101	389.30	388.0463	70.96	4.40	20.53	4.11
C23H17BR102	405.29	404.0412	68.16	4.23	19.72	7.90
C23H17BR103	421.29	420.0362	65.57	4.07	18.97	11.39
C23H17BR104	437.29	436.0311	63.17	3.92	18.27	14.63
C23H17BR105	453.29	452.0260	60.94	3.78	17.63	17.65
C23H17BR106	469.29	468.0209	58.87	3.65	17.03	20.46
C23H17BR107	485.29	484.0158	56.93	3.53	16.47	23.08
C23H17BR108	501.29	500.0107	55.11	3.42	15.94	25.53
C23H17BR109	517.29	516.0056	53.40	3.31	15.45	27.84
C23H18BR201	470.21	467.9725	58.75	3.86	33.99	3.40
C23H18BR202	486.21	483.9675	56.82	3.73	32.87	6.58
C23H18BR203	502.21	499.9624	55.01	3.61	31.82	9.56
C23H18BR204	518.21	515.9573	53.31	3.50	30.84	12.35
C23H18BR205	534.20	531.9522	51.71	3.40	29.92	14.97
C23H18BR206	550.20	547.9471	50.21	3.30	29.05	17.45
C23H18BR207	566.20	563.9420	48.79	3.20	28.22	19.78
C23H18BR208	582.20	579.9369	47.45	3.12	27.45	21.98
C23H18BR209	598.20	595.9318	46.18	3.03	26.71	24.07
C23H19BR101	391.31	390.0620	70.60	4.89	20.42	4.09
C23H19BR102	407.31	406.0569	67.82	4.70	19.62	7.86
C23H19BR103	423.31	422.0518	65.26	4.52	18.88	11.34
C23H19BR104	439.31	438.0467	62.88	4.36	18.19	14.57
C23H19BR105	455.31	454.0416	60.67	4.21	17.55	17.57
C23H19BR106	471.31	470.0365	58.61	4.06	16.95	20.37
C23H19BR107	487.31	486.0315	56.69	3.93	16.40	22.98

C23H19BR108	503.31	502.0264	54.89	3.81	15.88	25.43
C23H19BR109	519.31	518.0213	53.20	3.69	15.39	27.73
C23H20BR201	472.22	469.9882	58.50	4.27	33.84	3.39
C23H20BR202	488.22	485.9831	56.58	4.13	32.73	6.55
C23H20BR203	504.22	501.9780	54.79	4.00	31.69	9.52
C23H20BR204	520.22	517.9729	53.10	3.88	30.72	12.30
C23H20BR205	536.22	533.9678	51.52	3.76	29.80	14.92
C23H20BR206	552.22	549.9628	50.03	3.65	28.94	17.38
C23H20BR207	568.22	565.9577	48.62	3.55	28.12	19.71
C23H20BR208	584.22	581.9526	47.29	3.45	27.35	21.91
C23H20BR209	600.22	597.9475	46.03	3.36	26.62	23.99
C23H21BR101	393.33	392.0776	70.24	5.38	20.31	4.07
C23H21BR102	409.33	408.0725	67.49	5.17	19.52	7.82
C23H21BR103	425.33	424.0674	64.95	4.98	18.79	11.29
C23H21BR104	441.33	440.0624	62.60	4.80	18.11	14.50
C23H21BR105	457.32	456.0573	60.41	4.63	17.47	17.49
C23H21BR106	473.32	472.0522	58.37	4.47	16.88	20.28
C23H21BR107	489.32	488.0471	56.46	4.33	16.33	22.89
C23H21BR108	505.32	504.0420	54.67	4.19	15.81	25.33
C23H21BR109	521.32	520.0369	52.99	4.06	15.33	27.62
C23H22BR201	474.24	472.0038	58.25	4.68	33.70	3.37
C23H22BR202	490.24	487.9987	56.35	4.52	32.60	6.53
C23H22BR203	506.24	503.9937	54.57	4.38	31.57	9.48
C23H22BR204	522.24	519.9886	52.90	4.25	30.60	12.25
C23H22BR205	538.24	535.9835	51.33	4.12	29.69	14.86
C23H22BR206	554.24	551.9784	49.84	4.00	28.83	17.32
C23H22BR207	570.24	567.9733	48.45	3.89	28.02	19.64
C23H22BR208	586.23	583.9682	47.12	3.78	27.26	21.83
C23H22BR209	602.23	599.9631	45.87	3.68	26.54	23.91
C23H23BR101	395.34	394.0933	69.88	5.86	20.21	4.05
C23H23BR102	411.34	410.0882	67.16	5.64	19.43	7.78
C23H23BR103	427.34	426.0831	64.65	5.43	18.70	11.23
C23H23BR104	443.34	442.0780	62.31	5.23	18.02	14.44
C23H23BR105	459.34	458.0729	60.14	5.05	17.40	17.42
C23H23BR106	475.34	474.0678	58.12	4.88	16.81	20.20
C23H23BR107	491.34	490.0628	56.23	4.72	16.26	22.79
C23H23BR108	507.34	506.0577	54.45	4.57	15.75	25.23
C23H23BR109	523.34	522.0526	52.79	4.43	15.27	27.51
C23H24BR201	476.26	474.0195	58.01	5.08	33.56	3.36
C23H24BR202	492.25	490.0144	56.12	4.91	32.46	6.50
C23H24BR203	508.25	506.0093	54.35	4.76	31.44	9.44
C23H24BR204	524.25	522.0042	52.70	4.61	30.48	12.21
C23H24BR205	540.25	537.9991	51.13	4.48	29.58	14.81
C23H24BR206	556.25	553.9941	49.66	4.35	28.73	17.26
C23H24BR207	572.25	569.9890	48.28	4.23	27.93	19.57
C23H24BR208	588.25	585.9839	46.96	4.11	27.17	21.76
C23H24BR209	604.25	601.9788	45.72	4.00	26.45	23.83
C23H25BR101	397.36	396.1089	69.52	6.34	20.11	4.03
C23H25BR102	413.36	412.1038	66.83	6.10	19.33	7.74
C23H25BR103	429.36	428.0987	64.34	5.87	18.61	11.18
C23H25BR104	445.36	444.0937	62.03	5.66	17.94	14.37
C23H25BR105	461.36	460.0886	59.88	5.46	17.32	17.34
C23H25BR106	477.36	476.0835	57.87	5.28	16.74	20.11
C23H25BR107	493.36	492.0784	56.00	5.11	16.20	22.70
C23H25BR108	509.35	508.0733	54.24	4.95	15.69	25.13
C23H25BR109	525.35	524.0682	52.58	4.80	15.21	27.41
C23H26BR201	478.27	476.0351	57.76	5.48	33.41	3.35
C23H26BR202	494.27	492.0300	55.89	5.30	32.33	6.47
C23H26BR203	510.27	508.0250	54.14	5.14	31.32	9.41
C23H26BR204	526.27	524.0199	52.49	4.98	30.37	12.16
C23H26BR205	542.27	540.0148	50.94	4.83	29.47	14.75
C23H26BR206	558.27	556.0097	49.48	4.69	28.63	17.20
C23H26BR207	574.27	572.0046	48.11	4.56	27.83	19.50
C23H26BR208	590.27	587.9995	46.80	4.44	27.07	21.68
C23H26BR209	606.27	603.9944	45.57	4.32	26.36	23.75
C23H27BR101	399.38	398.1246	69.17	6.81	20.01	4.01
C23H27BR102	415.37	414.1195	66.51	6.55	19.24	7.70
C23H27BR103	431.37	430.1144	64.04	6.31	18.52	11.13
C23H27BR104	447.37	446.1093	61.75	6.08	17.86	14.31
C23H27BR105	463.37	462.1042	59.62	5.87	17.24	17.26
C23H27BR106	479.37	478.0991	57.63	5.68	16.67	20.03
C23H27BR107	495.37	494.0941	55.77	5.49	16.13	22.61
C23H27BR108	511.37	510.0890	54.02	5.32	15.63	25.03
C23H27BR109	527.37	526.0839	52.38	5.16	15.15	27.30
C23H28BR201	480.29	478.0508	57.52	5.88	33.27	3.33
C23H28BR202	496.29	494.0457	55.66	5.69	32.20	6.45
C23H28BR203	512.29	510.0406	53.93	5.51	31.20	9.37
C23H28BR204	528.29	526.0355	52.29	5.34	30.25	12.11
C23H28BR205	544.28	542.0304	50.76	5.19	29.36	14.70
C23H28BR206	560.28	558.0254	49.31	5.04	28.52	17.13
C23H28BR207	576.28	574.0203	47.94	4.90	27.73	19.43
C23H28BR208	592.28	590.0152	46.64	4.77	26.98	21.61
C23H28BR209	608.28	606.0101	45.42	4.64	26.27	23.67
C23H29BR101	401.39	400.1402	68.82	7.28	19.91	3.99
C23H29BR102	417.39	416.1351	66.19	7.00	19.14	7.67
C23H29BR103	433.39	432.1300	63.74	6.74	18.44	11.08
C23H29BR104	449.39	448.1250	61.47	6.50	17.78	14.24
C23H29BR105	465.39	464.1199	59.36	6.28	17.17	17.19
C23H29BR106	481.39	480.1148	57.39	6.07	16.60	19.94
C23H29BR107	497.39	496.1097	55.54	5.88	16.06	22.52
C23H29BR108	513.39	512.1046	53.81	5.69	15.56	24.93
C23H29BR109	529.39	528.0995	52.18	5.52	15.09	27.20
C23H30BR201	482.30	480.0664	57.28	6.27	33.13	3.32
C23H30BR202	498.30	496.0613	55.44	6.07	32.07	6.42
C23H30BR203	514.30	512.0563	53.71	5.88	31.07	9.33
C23H30BR204	530.30	528.0512	52.09	5.70	30.14	12.07
C23H30BR205	546.30	544.0461	50.57	5.54	29.25	14.64
C23H30BR206	562.30	560.0410	49.13	5.38	28.42	17.07
C23H30BR207	578.30	576.0359	47.77	5.23	27.63	19.37
C23H30BR208	594.30	592.0308	46.48	5.09	26.89	21.54
C23H30BR209	610.30	608.0257	45.27	4.95	26.19	23.59
C23H31BR101	403.41	402.1559	68.48	7.75	19.81	3.97
C23H31BR102	419.41	418.1508	65.87	7.45	19.05	7.63
C23H31BR103	435.41	434.1457	63.45	7.18	18.35	11.02
C23H31BR104	451.41	450.1406	61.20	6.92	17.70	14.18
C23H31BR105	467.40	466.1355	59.10	6.69	17.10	17.12
C23H31BR106	483.40	482.1304	57.15	6.46	16.53	19.86
C23H31BR107	499.40	498.1254	55.32	6.26	16.00	22.43
C23H31BR108	515.40	514.1203	53.60	6.06	15.50	24.83
C23H31BR109	531.40	530.1152	51.99	5.88	15.04	27.10
C23H32BR201	484.32	482.0821	57.04	6.66	33.00	3.30
C23H32BR202	500.32	498.0770	55.22	6.45	31.94	6.40
C23H32BR203	516.32	514.0719	53.51	6.25	30.95	9.30
C23H32BR204	532.32	530.0668	51.90	6.06	30.02	12.02
C23H32BR205	548.32	546.0617	50.38	5.88	29.15	14.59
C23H32BR206	564.32	562.0567	48.95	5.72	28.32	17.01
C23H32BR207	580.32	578.0516	47.60	5.56	27.54	19.30
C23H32BR208	596.31	594.0465	46.33	5.41	26.80	21.46
C23H32BR209	612.31	610.0414	45.12	5.27	26.10	23.52
C23H33BR101	405.42	404.1715	68.14	8.20	19.71	3.95

043

C23H33BR102	421.42	420.1664	65.55	7.89	18.96	7.59
C23H33BR103	437.42	436.1613	63.16	7.60	18.27	10.97
C23H33BR104	453.42	452.1563	60.93	7.34	17.62	14.11
C23H33BR105	469.42	468.1512	58.85	7.09	17.02	17.04
C23H33BR106	485.42	484.1461	56.91	6.85	16.46	19.78
C23H33BR107	501.42	500.1410	55.09	6.63	15.94	22.34
C23H33BR108	517.42	516.1359	53.39	6.43	15.44	24.74
C23H33BR109	533.42	532.1308	51.79	6.24	14.98	26.99
C23H34BR201	486.33	484.0977	56.80	7.05	32.86	3.29
C23H34BR202	502.33	500.0926	54.95	6.82	31.81	6.37
C23H34BR203	518.33	516.0876	53.30	6.61	30.83	9.26
C23H34BR204	534.33	532.0825	51.70	6.41	29.91	11.98
C23H34BR205	550.33	548.0774	50.20	6.23	29.04	14.54
C23H34BR206	566.33	564.0723	48.78	6.05	28.22	16.95
C23H34BR207	582.33	580.0672	47.44	5.89	27.44	19.23
C23H34BR208	598.33	596.0621	46.17	5.73	26.71	21.39
C23H34BR209	614.33	612.0570	44.97	5.58	26.01	23.44
C23H35BR101	407.44	406.1872	67.80	8.66	19.61	3.93
C23H35BR102	423.44	422.1821	65.24	8.33	18.87	7.56
C23H35BR103	439.44	438.1770	62.87	8.03	18.18	10.92
C23H35BR104	455.44	454.1719	60.66	7.75	17.54	14.05
C23H35BR105	471.44	470.1668	58.60	7.48	16.95	16.97
C23H35BR106	487.44	486.1617	56.68	7.24	16.39	19.69
C23H35BR107	503.44	502.1566	54.87	7.01	15.87	22.25
C23H35BR108	519.43	518.1516	53.18	6.79	15.38	24.64
C23H35BR109	535.43	534.1465	51.59	6.59	14.92	26.89
C23H36BR201	488.35	486.1134	56.57	7.43	32.72	3.28
C23H36BR202	504.35	502.1083	54.77	7.19	31.69	6.34
C23H36BR203	520.35	518.1032	53.09	6.97	30.71	9.22
C23H36BR204	536.35	534.0981	51.51	6.77	29.80	11.93
C23H36BR205	552.35	550.0930	50.01	6.57	28.93	14.48
C23H36BR206	568.35	566.0880	48.61	6.38	28.12	16.89
C23H36BR207	584.35	582.0829	47.28	6.21	27.35	19.17
C23H36BR208	600.35	598.0778	46.02	6.04	26.62	21.32
C23H36BR209	616.35	614.0727	44.82	5.89	25.93	23.36
C23H37BR101	409.45	408.2028	67.47	9.11	19.51	3.91
C23H37BR102	425.45	424.1977	64.93	8.77	18.78	7.52
C23H37BR103	441.45	440.1926	62.58	8.45	18.10	10.87
C23H37BR104	457.45	456.1876	60.39	8.15	17.47	13.99
C23H37BR105	473.45	472.1825	58.35	7.88	16.88	16.90
C23H37BR106	489.45	488.1774	56.44	7.62	16.33	19.61
C23H37BR107	505.45	504.1723	54.66	7.38	15.81	22.16
C23H37BR108	521.45	520.1672	52.98	7.15	15.32	24.55
C23H37BR109	537.45	536.1621	51.40	6.94	14.87	26.79
C23H38BR201	490.37	488.1290	56.34	7.81	32.59	3.26
C23H38BR202	506.37	504.1239	54.56	7.56	31.56	6.32
C23H38BR203	522.37	520.1189	52.89	7.33	30.59	9.19
C23H38BR204	538.36	536.1138	51.31	7.11	29.68	11.89
C23H38BR205	554.36	552.1087	49.83	6.91	28.83	14.43
C23H38BR206	570.36	568.1036	48.44	6.72	28.02	16.83
C23H38BR207	586.36	584.0985	47.11	6.53	27.25	19.10
C23H38BR208	602.36	600.0934	45.86	6.36	26.53	21.25
C23H38BR209	618.36	616.0883	44.68	6.19	25.84	23.29
C23H39BR101	411.47	410.2185	67.14	9.55	19.42	3.89
C23H39BR102	427.47	426.2134	64.63	9.20	18.69	7.49
C23H39BR103	443.47	442.2083	62.29	8.86	18.02	10.82
C23H39BR104	459.47	458.2032	60.13	8.56	17.39	13.93
C23H39BR105	475.47	474.1981	58.10	8.27	16.81	16.82
C23H39BR106	491.47	490.1930	56.21	8.00	16.26	19.53
C23H39BR107	507.47	506.1879	54.44	7.75	15.75	22.07
C23H39BR108	523.47	522.1829	52.77	7.51	15.26	24.45
C23H39BR109	539.47	538.1778	51.21	7.29	14.81	26.69
C23H40BR201	492.38	490.1447	56.11	8.19	32.46	3.25
C23H40BR202	508.38	506.1396	54.34	7.93	31.43	6.29
C23H40BR203	524.38	522.1345	52.68	7.69	30.48	9.15
C23H40BR204	540.38	538.1294	51.12	7.46	29.57	11.84
C23H40BR205	556.38	554.1243	49.65	7.25	28.72	14.38
C23H40BR206	572.38	570.1192	48.26	7.04	27.92	16.77
C23H40BR207	588.38	586.1142	46.95	6.85	27.16	19.03
C23H40BR208	604.38	602.1091	45.71	6.67	26.44	21.18
C23H40BR209	620.38	618.1040	44.53	6.50	25.76	23.21
C23H41BR101	413.49	412.2341	66.81	9.99	19.32	3.87
C23H41BR102	429.49	428.2290	64.32	9.62	18.60	7.45
C23H41BR103	445.49	444.2239	62.01	9.28	17.94	10.77
C23H41BR104	461.48	460.2189	59.86	8.96	17.31	13.87
C23H41BR105	477.48	476.2138	57.86	8.66	16.73	16.75
C23H41BR106	493.48	492.2087	55.98	8.37	16.19	19.45
C23H41BR107	509.48	508.2036	54.22	8.11	15.68	21.98
C23H41BR108	525.48	524.1985	52.57	7.86	15.21	24.36
C23H41BR109	541.48	540.1934	51.02	7.63	14.76	26.59
C23H42BR201	494.40	492.1603	55.88	8.56	32.32	3.24
C23H42BR202	510.40	508.1552	54.13	8.29	31.31	6.27
C23H42BR203	526.40	524.1502	52.48	8.04	30.36	9.12
C23H42BR204	542.40	540.1451	50.93	7.81	29.46	11.80
C23H42BR205	558.40	556.1400	49.47	7.58	28.62	14.33
C23H42BR206	574.40	572.1349	48.10	7.37	27.82	16.71
C23H42BR207	590.39	588.1298	46.79	7.17	27.07	18.97
C23H42BR208	606.39	604.1247	45.56	6.98	26.35	21.11
C23H42BR209	622.39	620.1196	44.39	6.80	25.68	23.14
C23H43BR101	415.50	414.2498	66.49	10.43	19.23	3.85
C23H43BR102	431.50	430.2447	64.02	10.04	18.52	7.42
C23H43BR103	447.50	446.2396	61.73	9.69	17.86	10.73
C23H43BR104	463.50	462.2345	59.60	9.35	17.24	13.81
C23H43BR105	479.50	478.2294	57.61	9.04	16.66	16.68
C23H43BR106	495.50	494.2243	55.75	8.75	16.13	19.37
C23H43BR107	511.50	510.2192	54.01	8.47	15.62	21.90
C23H43BR108	527.50	526.2142	52.37	8.22	15.15	24.26
C23H43BR109	543.50	542.2091	50.83	7.97	14.70	26.49
C23H44BR201	496.41	494.1760	55.65	8.93	32.19	3.22
C23H44BR202	512.41	510.1709	53.91	8.66	31.19	6.24
C23H44BR203	528.41	526.1658	52.28	8.39	30.24	9.08
C23H44BR204	544.41	542.1607	50.74	8.15	29.35	11.76
C23H44BR205	560.41	558.1556	49.30	7.91	28.52	14.27
C23H44BR206	576.41	574.1505	47.93	7.69	27.72	16.65
C23H44BR207	592.41	590.1455	46.63	7.49	26.98	18.91
C23H44BR208	608.41	606.1404	45.41	7.29	26.27	21.04
C23H44BR209	624.41	622.1353	44.24	7.10	25.59	23.06
C23H45BR101	417.52	416.2654	66.17	10.86	19.14	3.83
C23H45BR102	433.52	432.2603	63.72	10.46	18.43	7.38
C23H45BR103	449.52	448.2552	61.46	10.09	17.78	10.68
C23H45BR104	465.52	464.2502	59.34	9.74	17.16	13.75
C23H45BR105	481.52	480.2451	57.37	9.42	16.59	16.61
C23H45BR106	497.52	496.2400	55.53	9.12	16.06	19.30
C23H45BR107	513.51	512.2349	53.80	8.83	15.56	21.81
C23H45BR108	529.51	528.2298	52.17	8.57	15.09	24.17
C23H45BR109	545.51	544.2247	50.64	8.31	14.65	26.40
C23H46BR201	498.43	496.1916	55.43	9.30	32.06	3.21
C23H46BR202	514.43	512.1865	53.70	9.01	31.07	6.22
C23H46BR203	530.43	528.1815	52.08	8.74	30.13	9.05
C23H46BR204	546.43	544.1764	50.56	8.49	29.25	11.71

044

C23H46BR205	562.43	560.1713	49.12	8.24	28.41	14.22
C23H46BR206	578.43	576.1662	47.76	8.02	27.63	16.60
C23H46BR207	594.43	592.1611	46.47	7.80	26.88	18.84
C23H46BR208	610.43	608.1560	45.26	7.60	26.18	20.97
C23H46BR209	626.43	624.1509	44.10	7.40	25.51	22.99
C23H47BR101	419.53	418.2811	65.85	11.29	19.05	3.81
C23H47BR102	435.53	434.2760	63.43	10.88	18.35	7.35
C23H47BR103	451.53	450.2709	61.18	10.49	17.70	10.63
C23H47BR104	467.53	466.2658	59.09	10.13	17.09	13.69
C23H47BR105	483.53	482.2607	57.13	9.80	16.53	16.54
C23H47BR106	499.53	498.2556	55.30	9.48	16.00	19.22
C23H47BR107	515.53	514.2505	53.59	9.19	15.50	21.72
C23H47BR108	531.53	530.2455	51.97	8.91	15.03	24.08
C23H47BR109	547.53	546.2404	50.46	8.65	14.59	26.30
C24H12BR201	476.17	473.9256	60.54	2.54	33.56	3.36
C24H12BR202	492.17	489.9205	58.57	2.46	32.47	6.50
C24H12BR203	508.17	505.9154	56.73	2.38	31.45	9.45
C24H12BR204	524.17	521.9103	55.00	2.31	30.49	12.21
C24H12BR205	540.17	537.9052	53.37	2.24	29.58	14.81
C24H12BR206	556.17	553.9002	51.83	2.17	28.73	17.26
C24H12BR207	572.17	569.8951	50.38	2.11	27.93	19.57
C24H12BR208	588.17	585.8900	49.01	2.06	27.17	21.76
C24H12BR209	604.17	601.8849	47.71	2.00	26.45	23.83
C24H13BR101	397.27	396.0150	72.56	3.30	20.11	4.03
C24H13BR102	413.27	412.0099	69.75	3.17	19.33	7.74
C24H13BR103	429.27	428.0049	67.15	3.05	18.61	11.18
C24H13BR104	445.27	443.9998	64.74	2.94	17.94	14.37
C24H13BR105	461.27	459.9947	62.49	2.84	17.32	17.34
C24H13BR106	477.27	475.9896	60.40	2.75	16.74	20.11
C24H13BR107	493.27	491.9845	58.44	2.66	16.20	22.70
C24H13BR108	509.27	507.9794	56.60	2.57	15.69	25.13
C24H13BR109	525.27	523.9743	54.88	2.49	15.21	27.41
C24H14BR201	478.19	475.9412	60.28	2.95	33.42	3.35
C24H14BR202	494.19	491.9362	58.33	2.86	32.34	6.48
C24H14BR203	510.19	507.9311	56.50	2.77	31.32	9.41
C24H14BR204	526.18	523.9260	54.78	2.68	30.37	12.16
C24H14BR205	542.18	539.9209	53.17	2.60	29.47	14.75
C24H14BR206	558.18	555.9158	51.64	2.53	28.63	17.20
C24H14BR207	574.18	571.9107	50.20	2.46	27.83	19.51
C24H14BR208	590.18	587.9056	48.84	2.39	27.08	21.69
C24H14BR209	606.18	603.9006	47.55	2.33	26.36	23.75
C24H15BR101	399.29	398.0307	72.19	3.79	20.01	4.01
C24H15BR102	415.29	414.0256	69.41	3.64	19.24	7.71
C24H15BR103	431.29	430.0205	66.84	3.51	18.53	11.13
C24H15BR104	447.29	446.0154	64.45	3.38	17.86	14.31
C24H15BR105	463.29	462.0103	62.22	3.26	17.25	17.27
C24H15BR106	479.29	478.0052	60.15	3.15	16.67	20.03
C24H15BR107	495.29	494.0002	58.20	3.05	16.13	22.61
C24H15BR108	511.29	509.9951	56.38	2.96	15.63	25.03
C24H15BR109	527.29	525.9900	54.67	2.87	15.15	27.31
C24H16BR201	480.20	477.9569	60.03	3.36	33.28	3.33
C24H16BR202	496.20	493.9518	58.09	3.25	32.21	6.45
C24H16BR203	512.20	509.9467	56.28	3.15	31.20	9.37
C24H16BR204	528.20	525.9416	54.58	3.05	30.26	12.12
C24H16BR205	544.20	541.9365	52.97	2.96	29.37	14.70
C24H16BR206	560.20	557.9315	51.46	2.88	28.53	17.14
C24H16BR207	576.20	573.9264	50.03	2.80	27.73	19.44
C24H16BR208	592.20	589.9213	48.68	2.72	26.99	21.61
C24H16BR209	608.20	605.9162	47.40	2.65	26.28	23.68
C24H17BR101	401.31	400.0463	71.83	4.27	19.91	3.99
C24H17BR102	417.31	416.0412	69.08	4.11	19.15	7.67
C24H17BR103	433.31	432.0362	66.53	3.95	18.44	11.08
C24H17BR104	449.30	448.0311	64.16	3.81	17.78	14.24
C24H17BR105	465.30	464.0260	61.95	3.68	17.17	17.19
C24H17BR106	481.30	480.0209	59.89	3.56	16.60	19.95
C24H17BR107	497.30	496.0158	57.97	3.45	16.07	22.52
C24H17BR108	513.30	512.0107	56.16	3.34	15.57	24.94
C24H17BR109	529.30	528.0056	54.46	3.24	15.10	27.20
C24H18BR201	482.22	479.9725	59.78	3.76	33.14	3.32
C24H18BR202	498.22	495.9675	57.86	3.64	32.08	6.42
C24H18BR203	514.22	511.9624	56.06	3.53	31.08	9.33
C24H18BR204	530.22	527.9573	54.37	3.42	30.14	12.07
C24H18BR205	546.22	543.9522	52.78	3.32	29.26	14.65
C24H18BR206	562.22	559.9471	51.27	3.23	28.42	17.07
C24H18BR207	578.21	575.9420	49.85	3.14	27.64	19.37
C24H18BR208	594.21	591.9369	48.51	3.05	26.89	21.54
C24H18BR209	610.21	607.9318	47.24	2.97	26.19	23.60
C24H19BR101	403.32	402.0620	71.47	4.75	19.81	3.97
C24H19BR102	419.32	418.0569	68.75	4.57	19.06	7.63
C24H19BR103	435.32	434.0518	66.22	4.40	18.36	11.03
C24H19BR104	451.32	450.0467	63.87	4.24	17.70	14.18
C24H19BR105	467.32	466.0416	61.69	4.10	17.10	17.12
C24H19BR106	483.32	482.0365	59.64	3.96	16.53	19.86
C24H19BR107	499.32	498.0315	57.73	3.84	16.00	22.43
C24H19BR108	515.32	514.0264	55.94	3.72	15.51	24.84
C24H19BR109	531.32	530.0213	54.26	3.60	15.04	27.10
C24H20BR201	484.23	481.9882	59.53	4.16	33.00	3.30
C24H20BR202	500.23	497.9831	57.63	4.03	31.95	6.40
C24H20BR203	516.23	513.9780	55.84	3.91	30.96	9.30
C24H20BR204	532.23	529.9729	54.16	3.79	30.03	12.02
C24H20BR205	548.23	545.9678	52.58	3.68	29.15	14.59
C24H20BR206	564.23	561.9628	51.09	3.57	28.32	17.01
C24H20BR207	580.23	577.9577	49.68	3.47	27.54	19.30
C24H20BR208	596.23	593.9526	48.35	3.38	26.80	21.47
C24H20BR209	612.23	609.9475	47.08	3.29	26.10	23.52
C24H21BR101	405.34	404.0776	71.12	5.22	19.71	3.95
C24H21BR102	421.34	420.0725	68.42	5.02	18.96	7.59
C24H21BR103	437.34	436.0674	65.91	4.84	18.27	10.98
C24H21BR104	453.34	452.0624	63.59	4.67	17.63	14.12
C24H21BR105	469.34	468.0573	61.42	4.51	17.02	17.04
C24H21BR106	485.34	484.0522	59.40	4.36	16.46	19.78
C24H21BR107	501.33	500.0471	57.50	4.22	15.94	22.34
C24H21BR108	517.33	516.0420	55.72	4.09	15.45	24.74
C24H21BR109	533.33	532.0369	54.05	3.97	14.98	27.00
C24H22BR201	486.25	484.0038	59.28	4.56	32.87	3.29
C24H22BR202	502.25	499.9987	57.40	4.42	31.82	6.37
C24H22BR203	518.25	515.9937	55.62	4.28	30.84	9.26
C24H22BR204	534.25	531.9886	53.96	4.15	29.91	11.98
C24H22BR205	550.25	547.9835	52.39	4.03	29.04	14.54
C24H22BR206	566.25	563.9784	50.91	3.92	28.22	16.95
C24H22BR207	582.25	579.9733	49.51	3.81	27.45	19.24
C24H22BR208	598.25	595.9682	48.19	3.71	26.71	21.40
C24H22BR209	614.25	611.9631	46.93	3.61	26.02	23.44
C24H23BR101	407.35	406.0933	70.77	5.69	19.62	3.93
C24H23BR102	423.35	422.0882	68.09	5.48	18.87	7.56
C24H23BR103	439.35	438.0831	65.61	5.28	18.19	10.92
C24H23BR104	455.35	454.0780	63.31	5.09	17.55	14.05
C24H23BR105	471.35	470.0729	61.16	4.92	16.95	16.97
C24H23BR106	487.35	486.0678	59.15	4.76	16.40	19.70
C24H23BR107	503.35	502.0628	57.27	4.61	15.87	22.25

C24H23BR108	519.35	518.0577	55.51	4.46	15.39	24.65
C24H23BR109	535.35	534.0526	53.85	4.33	14.93	26.90
C24H24BR201	488.27	486.0195	59.04	4.95	32.73	3.28
C24H24BR202	5C4.27	502.0144	57.17	4.80	31.69	6.35
C24H24BR203	520.27	518.0093	55.41	4.65	30.72	9.23
C24H24BR204	536.26	534.0042	53.75	4.51	29.80	11.93
C24H24BR205	552.26	549.9991	52.20	4.38	28.94	14.49
C24H24BR206	568.26	565.9941	50.73	4.26	28.12	16.89
C24H24BR207	584.26	581.9890	49.34	4.14	27.35	19.17
C24H24BR208	600.26	597.9839	48.02	4.03	26.62	21.32
C24H24BR209	616.26	613.9788	46.78	3.93	25.93	23.37
C24H25BR101	409.37	408.1089	70.42	6.16	19.52	3.91
C24H25BR102	425.37	424.1038	67.77	5.92	18.78	7.52
C24H25BR103	441.37	440.0987	65.31	5.71	18.10	10.87
C24H25BR104	457.37	456.0937	63.03	5.51	17.47	13.99
C24H25BR105	473.37	472.0886	60.90	5.32	16.88	16.90
C24H25BR106	489.37	488.0835	58.91	5.15	16.33	19.62
C24H25BR107	505.37	504.0784	57.04	4.99	15.81	22.16
C24H25BR108	521.37	520.0733	55.29	4.83	15.33	24.55
C24H25BR109	537.37	536.0682	53.64	4.69	14.87	26.80
C24H26BR201	490.28	488.0351	58.80	5.35	32.60	3.26
C24H26BR202	506.28	504.0300	56.94	5.18	31.57	6.32
C24H26BR203	522.28	520.0250	55.19	5.02	30.60	9.19
C24H26BR204	538.28	536.0199	53.55	4.87	29.69	11.89
C24H26BR205	554.28	552.0148	52.01	4.73	28.83	14.43
C24H26BR206	570.28	568.0097	50.55	4.60	28.02	16.83
C24H26BR207	586.28	584.0046	49.17	4.47	27.26	19.10
C24H26BR208	602.28	599.9995	47.86	4.35	26.53	21.25
C24H26BR209	618.28	615.9944	46.62	4.24	25.85	23.29
C24H27BR101	411.39	410.1246	70.07	6.62	19.42	3.89
C24H27BR102	427.39	426.1195	67.45	6.37	18.70	7.49
C24H27BR103	443.38	442.1144	65.02	6.14	18.02	10.83
C24H27BR104	459.38	458.1093	62.75	5.92	17.39	13.93
C24H27BR105	475.38	474.1042	60.64	5.72	16.81	16.83
C24H27BR106	491.38	490.0991	58.66	5.54	16.26	19.54
C24H27BR107	507.38	506.0941	56.81	5.36	15.75	22.07
C24H27BR108	523.38	522.0890	55.08	5.20	15.27	24.46
C24H27BR109	539.38	538.0839	53.44	5.05	14.81	26.70
C24H28BR201	492.30	490.0508	58.56	5.73	32.46	3.25
C24H28BR202	508.30	506.0457	56.71	5.55	31.44	6.30
C24H28BR203	524.30	522.0406	54.98	5.38	30.48	9.15
C24H28BR204	540.30	538.0355	53.35	5.22	29.58	11.84
C24H28BR205	556.30	554.0304	51.82	5.07	28.73	14.38
C24H28BR206	572.30	570.0254	50.37	4.93	27.92	16.77
C24H28BR207	588.29	586.0203	49.00	4.80	27.16	19.04
C24H28BR208	604.29	602.0152	47.70	4.67	26.45	21.18
C24H28BR209	620.29	618.0101	46.47	4.55	25.76	23.21
C24H29BR101	413.40	412.1402	69.73	7.07	19.33	3.87
C24H29BR102	429.40	428.1351	67.13	6.81	18.61	7.45
C24H29BR103	445.40	444.1300	64.72	6.56	17.94	10.78
C24H29BR104	461.40	460.1250	62.48	6.34	17.32	13.87
C24H29BR105	477.40	476.1199	60.38	6.12	16.74	16.76
C24H29BR106	493.40	492.1148	58.42	5.92	16.19	19.46
C24H29BR107	509.40	508.1097	56.59	5.74	15.69	21.99
C24H29BR108	525.40	524.1046	54.87	5.56	15.21	24.36
C24H29BR109	541.40	540.0995	53.25	5.40	14.76	26.60
C24H30BR201	494.31	492.0664	58.32	6.12	32.33	3.24
C24H30BR202	510.31	508.0613	56.49	5.93	31.32	6.27
C24H30BR203	526.31	524.0563	54.77	5.75	30.36	9.12
C24H30BR204	542.31	540.0512	53.16	5.58	29.47	11.80
C24H30BR205	558.31	556.0461	51.63	5.42	28.62	14.33
C24H30BR206	574.31	572.0410	50.19	5.27	27.83	16.72
C24H30BR207	590.31	588.0359	48.83	5.12	27.07	18.97
C24H30BR208	606.31	604.0308	47.54	4.99	26.36	21.11
C24H30BR209	622.31	620.0257	46.32	4.86	25.68	23.14
C24H31BR101	415.42	414.1559	69.39	7.52	19.23	3.85
C24H31BR102	431.42	430.1508	66.82	7.24	18.52	7.42
C24H31BR103	447.42	446.1457	64.43	6.98	17.86	10.73
C24H31BR104	463.42	462.1406	62.20	6.74	17.24	13.81
C24H31BR105	479.42	478.1355	60.13	6.52	16.67	16.69
C24H31BR106	495.42	494.1304	58.19	6.31	16.13	19.38
C24H31BR107	511.41	510.1254	56.37	6.11	15.62	21.90
C24H31BR108	527.41	526.1203	54.66	5.92	15.15	24.27
C24H31BR109	543.41	542.1152	53.05	5.75	14.70	26.50
C24H32BR201	496.33	494.0821	58.08	6.50	32.20	3.22
C24H32BR202	512.33	510.0770	56.27	6.30	31.19	6.25
C24H32BR203	528.33	526.0719	54.56	6.11	30.25	9.08
C24H32BR204	544.33	542.0668	52.96	5.93	29.36	11.76
C24H32BR205	560.33	558.0617	51.45	5.76	28.52	14.28
C24H32BR206	576.33	574.0567	50.02	5.60	27.73	16.66
C24H32BR207	592.33	590.0516	48.67	5.45	26.98	18.91
C24H32BR208	608.33	606.0465	47.39	5.30	26.27	21.04
C24H32BR209	624.33	622.0414	46.17	5.17	25.60	23.06
C24H33BR101	417.43	416.1715	69.06	7.97	19.14	3.83
C24H33BR102	433.43	432.1664	66.51	7.67	18.44	7.38
C24H33BR103	449.43	448.1613	64.14	7.40	17.78	10.68
C24H33BR104	465.43	464.1563	61.94	7.15	17.17	13.75
C24H33BR105	481.43	480.1512	59.88	6.91	16.60	16.62
C24H33BR106	497.43	496.1461	57.95	6.69	16.06	19.30
C24H33BR107	513.43	512.1410	56.15	6.48	15.56	21.81
C24H33BR108	529.43	528.1359	54.45	6.28	15.09	24.18
C24H33BR109	545.43	544.1308	52.85	6.10	14.65	26.40
C24H34BR201	498.35	496.0977	57.84	6.88	32.07	3.21
C24H34BR202	514.35	512.0926	56.05	6.66	31.07	6.22
C24H34BR203	530.34	528.0876	54.35	6.46	30.13	9.05
C24H34BR204	546.34	544.0825	52.76	6.27	29.25	11.71
C24H34BR205	562.34	560.0774	51.26	6.09	28.42	14.23
C24H34BR206	578.34	576.0723	49.84	5.93	27.63	16.60
C24H34BR207	594.34	592.0672	48.50	5.77	26.89	18.84
C24H34BR208	610.34	608.0621	47.23	5.62	26.18	20.97
C24H34BR209	626.34	624.0570	46.02	5.47	25.51	22.99
C24H35BR101	419.45	418.1872	68.73	8.41	19.05	3.81
C24H35BR102	435.45	434.1821	66.20	8.10	18.35	7.35
C24H35BR103	451.45	450.1770	63.85	7.81	17.70	10.63
C24H35BR104	467.45	466.1719	61.67	7.55	17.09	13.69
C24H35BR105	483.45	482.1668	59.63	7.30	16.53	16.55
C24H35BR106	499.45	498.1617	57.72	7.06	16.00	19.22
C24H35BR107	515.45	514.1566	55.93	6.84	15.50	21.73
C24H35BR108	531.45	530.1516	54.24	6.64	15.04	24.08
C24H35BR109	547.45	546.1465	52.66	6.44	14.60	26.30
C24H36BR201	500.36	498.1134	57.61	7.25	31.94	3.20
C24H36BR202	516.36	514.1083	55.83	7.03	30.95	6.20
C24H36BR203	532.36	530.1032	54.15	6.82	30.02	9.02
C24H36BR204	548.36	546.0981	52.57	6.62	29.14	11.67
C24H36BR205	564.36	562.0930	51.08	6.43	28.32	14.17
C24H36BR206	580.36	578.0880	49.67	6.25	27.54	16.54
C24H36BR207	596.36	594.0829	48.34	6.08	26.80	18.78
C24H36BR208	612.36	610.0778	47.08	5.93	26.10	20.90
C24H36BR209	628.36	626.0727	45.88	5.77	25.43	22.92
C24H37BR101	421.47	420.2028	68.40	8.85	18.96	3.80

046

C24H37BR102	437.47	436.1977	65.89	8.53	18.27	7.31
C24H37BR103	453.46	452.1926	63.57	8.22	17.62	10.58
C24H37BR104	469.46	468.1876	61.40	7.94	17.02	13.63
C24H37BR105	485.46	484.1825	59.38	7.68	16.46	16.48
C24H37BR106	501.46	500.1774	57.49	7.44	15.93	19.14
C24H37BR107	517.46	516.1723	55.71	7.21	15.44	21.64
C24H37BR108	533.46	532.1672	54.04	6.99	14.98	23.99
C24H37BR109	549.46	548.1621	52.46	6.79	14.54	26.21
C24H38BR201	502.38	500.1290	57.38	7.62	31.81	3.18
C24H38BR202	518.38	516.1239	55.61	7.39	30.83	6.17
C24H38BR203	534.38	532.1189	53.94	7.17	29.91	8.98
C24H38BR204	550.38	548.1138	52.38	6.96	29.04	11.63
C24H38BR205	566.38	564.1087	50.90	6.76	28.22	14.12
C24H38BR206	582.37	580.1036	49.50	6.58	27.44	16.48
C24H38BR207	598.37	596.0985	48.18	6.40	26.71	18.72
C24H38BR208	614.37	612.0934	46.92	6.23	26.01	20.83
C24H38BR209	630.37	628.0883	45.73	6.08	25.35	22.84
C24H39BR101	423.48	422.2185	68.07	9.28	18.87	3.78
C24H39BR102	439.48	438.2134	65.59	8.94	18.18	7.28
C24H39BR103	455.48	454.2083	63.29	8.63	17.54	10.54
C24H39BR104	471.48	470.2032	61.14	8.34	16.95	13.57
C24H39BR105	487.48	486.1981	59.13	8.06	16.39	16.41
C24H39BR106	503.48	502.1930	57.26	7.81	15.87	19.07
C24H39BR107	519.48	518.1879	55.49	7.57	15.38	21.56
C24H39BR108	535.48	534.1829	53.83	7.34	14.92	23.90
C24H39BR109	551.48	550.1778	52.27	7.13	14.49	26.11
C24H40BR201	504.39	502.1447	57.15	7.99	31.68	3.17
C24H40BR202	520.39	518.1396	55.39	7.75	30.71	6.15
C24H40BR203	536.39	534.1345	53.74	7.52	29.79	8.95
C24H40BR204	552.39	550.1294	52.19	7.30	28.93	11.59
C24H40BR205	568.39	566.1243	50.72	7.09	28.12	14.07
C24H40BR206	584.39	582.1192	49.33	6.90	27.35	16.43
C24H40BR207	600.39	598.1142	48.01	6.72	26.62	18.65
C24H40BR208	616.39	614.1091	46.77	6.54	25.93	20.77
C24H40BR209	632.39	630.1040	45.58	6.38	25.27	22.77
C24H41BR101	425.50	424.2341	67.75	9.71	18.78	3.76
C24H41BR102	441.50	440.2290	65.29	9.36	18.10	7.25
C24H41BR103	457.50	456.2239	63.01	9.03	17.47	10.49
C24H41BR104	473.50	472.2189	60.88	8.73	16.88	13.52
C24H41BR105	489.50	488.2138	58.89	8.44	16.32	16.34
C24H41BR106	505.49	504.2087	57.03	8.18	15.81	18.99
C24H41BR107	521.49	520.2036	55.28	7.92	15.32	21.48
C24H41BR108	537.49	536.1985	53.63	7.69	14.87	23.81
C24H41BR109	553.49	552.1934	52.08	7.47	14.44	26.02
C24H42BR201	506.41	504.1603	56.92	8.36	31.56	3.16
C24H42BR202	522.41	520.1552	55.18	8.10	30.59	6.13
C24H42BR203	538.41	536.1502	53.54	7.86	29.68	8.91
C24H42BR204	554.41	552.1451	52.00	7.64	28.82	11.54
C24H42BR205	570.41	568.1400	50.54	7.42	28.02	14.02
C24H42BR206	586.41	584.1349	49.16	7.22	27.25	16.37
C24H42BR207	602.41	600.1298	47.85	7.03	26.53	18.59
C24H42BR208	618.41	616.1247	46.61	6.85	25.84	20.70
C24H42BR209	634.40	632.1196	45.44	6.67	25.19	22.70
C24H43BR101	427.51	426.2498	67.43	10.14	18.69	3.74
C24H43BR102	443.51	442.2447	65.00	9.77	18.02	7.21
C24H43BR103	459.51	458.2396	62.73	9.43	17.39	10.45
C24H43BR104	475.51	474.2345	60.62	9.11	16.80	13.46
C24H43BR105	491.51	490.2294	58.65	8.82	16.26	16.28
C24H43BR106	507.51	506.2243	56.80	8.54	15.74	18.92
C24H43BR107	523.51	522.2192	55.06	8.28	15.26	21.39
C24H43BR108	539.51	538.2142	53.43	8.03	14.81	23.72
C24H43BR109	555.51	554.2091	51.89	7.80	14.38	25.92
C24H44BR201	508.43	506.1760	56.70	8.72	31.43	3.15
C24H44BR202	524.43	522.1709	54.97	8.46	30.47	6.10
C24H44BR203	540.42	538.1658	53.34	8.21	29.57	8.88
C24H44BR204	556.42	554.1607	51.81	7.97	28.72	11.50
C24H44BR205	572.42	570.1556	50.36	7.75	27.92	13.98
C24H44BR206	588.42	586.1505	48.99	7.54	27.16	16.31
C24H44BR207	604.42	602.1455	47.69	7.34	26.44	18.53
C24H44BR208	620.42	618.1404	46.46	7.15	25.76	20.63
C24H44BR209	636.42	634.1353	45.30	6.97	25.11	22.63
C24H45BR101	429.53	428.2654	67.11	10.56	18.60	3.72
C24H45BR102	445.53	444.2603	64.70	10.18	17.93	7.18
C24H45BR103	461.53	460.2552	62.46	9.83	17.31	10.40
C24H45BR104	477.53	476.2502	60.37	9.50	16.73	13.40
C24H45BR105	493.53	492.2451	58.41	9.19	16.19	16.21
C24H45BR106	509.53	508.2400	56.58	8.90	15.68	18.84
C24H45BR107	525.53	524.2349	54.85	8.63	15.20	21.31
C24H45BR108	541.53	540.2298	53.23	8.38	14.76	23.64
C24H45BR109	557.52	556.2247	51.70	8.14	14.33	25.83
C24H46BR201	510.44	508.1916	56.47	9.08	31.31	3.13
C24H46BR202	526.44	524.1865	54.76	8.81	30.36	6.08
C24H46BR203	542.44	540.1815	53.14	8.55	29.46	8.85
C24H46BR204	558.44	556.1764	51.62	8.30	28.62	11.46
C24H46BR205	574.44	572.1713	50.18	8.07	27.82	13.93
C24H46BR206	590.44	588.1662	48.82	7.85	27.07	16.26
C24H46BR207	606.44	604.1611	47.53	7.65	26.35	18.47
C24H46BR208	622.44	620.1560	46.31	7.45	25.67	20.56
C24H46BR209	638.44	636.1509	45.15	7.26	25.03	22.55
C24H47BR101	431.55	430.2811	66.80	10.98	18.52	3.71
C24H47BR102	447.54	446.2760	64.41	10.59	17.85	7.15
C24H47BR103	463.54	462.2709	62.19	10.22	17.24	10.35
C24H47BR104	479.54	478.2658	60.11	9.88	16.66	13.35
C24H47BR105	495.54	494.2607	58.17	9.56	16.12	16.14
C24H47BR106	511.54	510.2556	56.35	9.26	15.62	18.77
C24H47BR107	527.54	526.2505	54.64	8.98	15.15	21.23
C24H47BR108	543.54	542.2455	53.04	8.72	14.70	23.55
C24H47BR109	559.54	558.2404	51.52	8.47	14.28	25.73
C24H48BR201	512.46	510.2073	56.25	9.44	31.18	3.12
C24H48BR202	528.46	526.2022	54.55	9.16	30.24	6.06
C24H48BR203	544.46	542.1971	52.95	8.89	29.35	8.82
C24H48BR204	560.46	558.1920	51.43	8.63	28.51	11.42
C24H48BR205	576.46	574.1869	50.01	8.39	27.72	13.88
C24H48BR206	592.45	590.1818	48.66	8.17	26.97	16.20
C24H48BR207	608.45	606.1768	47.38	7.95	26.26	18.41
C24H48BR208	624.45	622.1717	46.16	7.75	25.59	20.50
C24H48BR209	640.45	638.1666	45.01	7.55	24.95	22.48
C24H49BR101	433.56	432.2967	66.49	11.39	18.43	3.69
C24H49BR102	449.56	448.2916	64.12	10.99	17.77	7.12
C24H49BR103	465.56	464.2865	61.92	10.61	17.16	10.31
C24H49BR104	481.56	480.2815	59.86	10.26	16.59	13.29
C24H49BR105	497.56	496.2764	57.94	9.93	16.06	16.08
C24H49BR106	513.56	512.2713	56.13	9.62	15.56	18.69
C24H49BR107	529.56	528.2662	54.44	9.33	15.09	21.15
C24H49BR108	545.56	544.2611	52.84	9.05	14.65	23.46
C24H49BR109	561.56	560.2560	51.33	8.80	14.23	25.64
C25H12BR201	488.18	485.9256	61.51	2.48	32.74	3.28
C25H12BR202	504.18	501.9205	59.56	2.40	31.70	6.35
C25H12BR203	520.18	517.9154	57.73	2.33	30.72	9.23
C25H12BR204	536.18	533.9103	56.00	2.26	29.80	11.94

047

C25H12BR205	552.18	549.9052	54.38	2.19	28.94	14.49
C25H12BR206	568.18	565.9002	52.85	2.13	28.13	16.90
C25H12BR207	584.18	581.8951	51.40	2.07	27.36	19.17
C25H12BR208	600.18	597.8900	50.03	2.02	26.63	21.33
C25H12BR209	616.18	613.8849	48.73	1.96	25.94	23.37
C25H13BR101	409.29	408.0150	73.37	3.20	19.52	3.91
C25H13BR102	425.29	424.0099	70.61	3.08	18.79	7.52
C25H13BR103	441.28	440.0049	68.05	2.97	18.11	10.88
C25H13BR104	457.28	455.9998	65.67	2.87	17.47	14.00
C25H13BR105	473.28	471.9947	63.45	2.77	16.88	16.90
C25H13BR106	489.28	487.9896	61.37	2.68	16.33	19.62
C25H13BR107	505.28	503.9845	59.43	2.59	15.81	22.17
C25H13BR108	521.28	519.9794	57.60	2.51	15.33	24.55
C25H13BR109	537.28	535.9743	55.89	2.44	14.87	26.80
C25H14BR201	490.20	487.9412	61.26	2.88	32.60	3.26
C25H14BR202	506.20	503.9362	59.32	2.79	31.57	6.32
C25H14BR203	522.20	519.9311	57.50	2.70	30.60	9.19
C25H14BR204	538.20	535.9260	55.79	2.62	29.69	11.89
C25H14BR205	554.20	551.9209	54.18	2.55	28.84	14.43
C25H14BR206	570.19	567.9158	52.66	2.47	28.03	16.84
C25H14BR207	586.19	583.9107	51.23	2.41	27.26	19.11
C25H14BR208	602.19	599.9056	49.86	2.34	26.54	21.25
C25H14BR209	618.19	615.9006	48.57	2.28	25.85	23.29
C25H15BR101	411.30	410.0307	73.01	3.68	19.43	3.89
C25H15BR102	427.30	426.0256	70.27	3.54	18.70	7.49
C25H15BR103	443.30	442.0205	67.74	3.41	18.02	10.83
C25H15BR104	459.30	458.0154	65.38	3.29	17.40	13.93
C25H15BR105	475.30	474.0103	63.18	3.18	16.81	16.83
C25H15BR106	491.30	490.0052	61.12	3.08	16.26	19.54
C25H15BR107	507.30	506.0002	59.19	2.98	15.75	22.08
C25H15BR108	523.30	521.9951	57.38	2.89	15.27	24.46
C25H15BR109	539.30	537.9900	55.68	2.80	14.82	26.70
C25H16BR201	492.21	489.9569	61.01	3.28	32.47	3.25
C25H16BR202	508.21	505.9518	59.09	3.17	31.45	6.30
C25H16BR203	524.21	521.9467	57.28	3.08	30.49	9.16
C25H16BR204	540.21	537.9416	55.59	2.99	29.58	11.85
C25H16BR205	556.21	553.9365	53.99	2.90	28.73	14.38
C25H16BR206	572.21	569.9315	52.48	2.82	27.93	16.78
C25H16BR207	588.21	585.9264	51.05	2.74	27.17	19.04
C25H16BR208	604.21	601.9213	49.70	2.67	26.45	21.18
C25H16BR209	620.21	617.9162	48.42	2.60	25.77	23.22
C25H17BR101	413.32	412.0463	72.65	4.15	19.33	3.87
C25H17BR102	429.32	428.0412	69.94	3.99	18.61	7.45
C25H17BR103	445.32	444.0362	67.43	3.85	17.94	10.78
C25H17BR104	461.32	460.0311	65.09	3.71	17.32	13.87
C25H17BR105	477.32	476.0260	62.91	3.59	16.74	16.76
C25H17BR106	493.31	492.0209	60.87	3.47	16.20	19.46
C25H17BR107	509.31	508.0158	58.96	3.36	15.69	21.99
C25H17BR108	525.31	524.0107	57.16	3.26	15.21	24.37
C25H17BR109	541.31	540.0056	55.47	3.17	14.76	26.60
C25H18BR201	494.23	491.9725	60.76	3.67	32.33	3.24
C25H18BR202	510.23	507.9675	58.85	3.56	31.32	6.27
C25H18BR203	526.23	523.9624	57.06	3.45	30.37	9.12
C25H18BR204	542.23	539.9573	55.38	3.35	29.47	11.80
C25H18BR205	558.23	555.9522	53.79	3.25	28.63	14.33
C25H18BR206	574.23	571.9471	52.29	3.16	27.83	16.72
C25H18BR207	590.23	587.9420	50.88	3.07	27.08	18.98
C25H18BR208	606.23	603.9369	49.53	2.99	26.36	21.11
C25H18BR209	622.22	619.9318	48.26	2.92	25.68	23.14
C25H19BR101	415.33	414.0620	72.30	4.61	19.24	3.85
C25H19BR102	431.33	430.0569	69.62	4.44	18.52	7.42
C25H19BR103	447.33	446.0518	67.13	4.28	17.86	10.73
C25H19BR104	463.33	462.0467	64.81	4.13	17.25	13.81
C25H19BR105	479.33	478.0416	62.65	4.00	16.67	16.69
C25H19BR106	495.33	494.0365	60.62	3.87	16.13	19.38
C25H19BR107	511.33	510.0315	58.73	3.75	15.63	21.90
C25H19BR108	527.33	526.0264	56.94	3.63	15.15	24.27
C25H19BR109	543.33	542.0213	55.27	3.52	14.71	26.50
C25H20BR201	496.25	493.9882	60.51	4.06	32.20	3.22
C25H20BR202	512.24	509.9831	58.62	3.94	31.20	6.25
C25H20BR203	528.24	525.9780	56.84	3.82	30.25	9.09
C25H20BR204	544.24	541.9729	55.17	3.70	29.36	11.76
C25H20BR205	560.24	557.9678	53.60	3.60	28.52	14.28
C25H20BR206	576.24	573.9628	52.11	3.50	27.73	16.66
C25H20BR207	592.24	589.9577	50.70	3.40	26.98	18.91
C25H20BR208	608.24	605.9526	49.37	3.31	26.27	21.04
C25H20BR209	624.24	621.9475	48.10	3.23	25.60	23.07
C25H21BR101	417.35	416.0776	71.95	5.07	19.15	3.83
C25H21BR102	433.35	432.0725	69.29	4.88	18.44	7.38
C25H21BR103	449.35	448.0674	66.83	4.71	17.78	10.68
C25H21BR104	465.35	464.0624	64.53	4.55	17.17	13.75
C25H21BR105	481.35	480.0573	62.38	4.40	16.60	16.62
C25H21BR106	497.35	496.0522	60.38	4.26	16.07	19.30
C25H21BR107	513.35	512.0471	58.49	4.12	15.57	21.82
C25H21BR108	529.35	528.0420	56.73	4.00	15.09	24.18
C25H21BR109	545.34	544.0369	55.06	3.88	14.65	26.40
C25H22BR201	498.26	496.0038	60.27	4.45	32.07	3.21
C25H22BR202	514.26	511.9987	58.39	4.31	31.08	6.22
C25H22BR203	530.26	527.9937	56.63	4.18	30.14	9.05
C25H22BR204	546.26	543.9886	54.97	4.06	29.25	11.72
C25H22BR205	562.26	559.9835	53.41	3.94	28.42	14.23
C25H22BR206	578.26	575.9784	51.93	3.83	27.64	16.60
C25H22BR207	594.26	591.9733	50.53	3.73	26.89	18.85
C25H22BR208	610.26	607.9682	49.21	3.63	26.19	20.97
C25H22BR209	626.26	623.9631	47.95	3.54	25.52	22.99
C25H23BR101	419.37	418.0933	71.60	5.53	19.05	3.82
C25H23BR102	435.36	434.0882	68.97	5.33	18.35	7.35
C25H23BR103	451.36	450.0831	66.53	5.14	17.70	10.63
C25H23BR104	467.36	466.0780	64.25	4.96	17.10	13.69
C25H23BR105	483.36	482.0729	62.12	4.80	16.53	16.55
C25H23BR106	499.36	498.0678	60.13	4.64	16.00	19.22
C25H23BR107	515.36	514.0628	58.27	4.50	15.50	21.73
C25H23BR108	531.36	530.0577	56.51	4.36	15.04	24.09
C25H23BR109	547.36	546.0526	54.86	4.24	14.60	26.31
C25H24BR201	500.28	498.0195	60.02	4.84	31.94	3.20
C25H24BR202	516.28	514.0144	58.16	4.69	30.95	6.20
C25H24BR203	532.28	530.0093	56.41	4.54	30.02	9.02
C25H24BR204	548.28	546.0042	54.77	4.41	29.15	11.67
C25H24BR205	564.28	561.9991	53.21	4.29	28.32	14.18
C25H24BR206	580.27	577.9941	51.75	4.17	27.54	16.54
C25H24BR207	596.27	593.9890	50.36	4.06	26.80	18.78
C25H24BR208	612.27	609.9839	49.04	3.95	26.10	20.90
C25H24BR209	628.27	625.9788	47.79	3.85	25.44	22.92
C25H25BR101	421.38	420.1089	71.26	5.98	18.96	3.80
C25H25BR102	437.38	436.1038	68.65	5.76	18.27	7.32
C25H25BR103	453.38	452.0987	66.23	5.56	17.62	10.59
C25H25BR104	469.38	468.0937	63.97	5.37	17.02	13.63
C25H25BR105	485.38	484.0886	61.86	5.19	16.46	16.48
C25H25BR106	501.38	500.0835	59.89	5.03	15.94	19.15
C25H25BR107	517.38	516.0784	58.04	4.87	15.44	21.65

048

C25H25BR108	533.38	532.0733	56.30	4.72	14.98	24.00
C25H25BR109	549.38	548.0682	54.66	4.59	14.54	26.21
C25H26BR201	502.29	500.0351	59.78	5.22	31.82	3.19
C25H26BR202	518.29	516.0300	57.94	5.06	30.83	6.17
C25H26BR203	534.29	532.0250	56.20	4.91	29.91	8.98
C25H26BR204	550.29	548.0199	54.57	4.76	29.04	11.63
C25H26BR205	566.29	564.0148	53.03	4.63	28.22	14.13
C25H26BR206	582.29	580.0097	51.57	4.50	27.44	16.49
C25H26BR207	598.29	596.0046	50.19	4.38	26.71	18.72
C25H26BR208	614.29	611.9995	48.88	4.27	26.02	20.84
C25H26BR209	630.29	627.9944	47.64	4.16	25.35	22.85
C25H27BR101	423.40	422.1246	70.92	6.43	18.87	3.78
C25H27BR102	439.40	438.1195	68.34	6.19	18.18	7.28
C25H27BR103	455.40	454.1144	65.94	5.98	17.55	10.54
C25H27BR104	471.40	470.1093	63.70	5.77	16.95	13.58
C25H27BR105	487.39	486.1042	61.61	5.58	16.39	16.41
C25H27BR106	503.39	502.0991	59.65	5.41	15.87	19.07
C25H27BR107	519.39	518.0941	57.81	5.24	15.38	21.56
C25H27BR108	535.39	534.0890	56.09	5.08	14.92	23.91
C25H27BR109	551.39	550.0839	54.46	4.94	14.49	26.11
C25H28BR201	504.31	502.0508	59.54	5.60	31.69	3.17
C25H28BR202	520.31	518.0457	57.71	5.42	30.71	6.15
C25H28BR203	536.31	534.0406	55.99	5.26	29.80	8.95
C25H28BR204	552.31	550.0355	54.37	5.11	28.93	11.59
C25H28BR205	568.31	566.0304	52.84	4.97	28.12	14.08
C25H28BR206	584.31	582.0254	51.39	4.83	27.35	16.43
C25H28BR207	600.31	598.0203	50.02	4.70	26.62	18.66
C25H28BR208	616.31	614.0152	48.72	4.58	25.93	20.77
C25H28BR209	632.30	630.0101	47.49	4.46	25.27	22.77
C25H29BR101	425.41	424.1402	70.59	6.87	18.78	3.76
C25H29BR102	441.41	440.1351	68.03	6.62	18.10	7.25
C25H29BR103	457.41	456.1300	65.65	6.39	17.47	10.49
C25H29BR104	473.41	472.1250	63.43	6.17	16.88	13.52
C25H29BR105	489.41	488.1199	61.36	5.97	16.33	16.35
C25H29BR106	505.41	504.1148	59.41	5.78	15.81	18.99
C25H29BR107	521.41	520.1097	57.59	5.61	15.32	21.48
C25H29BR108	537.41	536.1046	55.88	5.44	14.87	23.82
C25H29BR109	553.41	552.0995	54.26	5.28	14.44	26.02
C25H30BR201	506.33	504.0664	59.31	5.97	31.56	3.16
C25H30BR202	522.32	520.0613	57.49	5.79	30.60	6.13
C25H30BR203	538.32	536.0563	55.78	5.62	29.69	8.92
C25H30BR204	554.32	552.0512	54.17	5.46	28.83	11.55
C25H30BR205	570.32	568.0461	52.65	5.30	28.02	14.03
C25H30BR206	586.32	584.0410	51.21	5.16	27.26	16.37
C25H30BR207	602.32	600.0359	49.85	5.02	26.53	18.59
C25H30BR208	618.32	616.0308	48.56	4.89	25.85	20.70
C25H30BR209	634.32	632.0257	47.34	4.77	25.19	22.70
C25H31BR101	427.43	426.1559	70.25	7.31	18.69	3.74
C25H31BR102	443.43	442.1508	67.72	7.05	18.02	7.22
C25H31BR103	459.43	458.1457	65.36	6.80	17.39	10.45
C25H31BR104	475.43	474.1406	63.16	6.57	16.81	13.46
C25H31BR105	491.43	490.1355	61.10	6.36	16.26	16.28
C25H31BR106	507.43	506.1304	59.18	6.16	15.75	18.92
C25H31BR107	523.43	522.1254	57.37	5.97	15.27	21.40
C25H31BR108	539.43	538.1203	55.67	5.79	14.81	23.73
C25H31BR109	555.42	554.1152	54.06	5.63	14.39	25.93
C25H32BR201	508.34	506.0821	59.07	6.35	31.44	3.15
C25H32BR202	524.34	522.0770	57.27	6.15	30.48	6.10
C25H32BR203	540.34	538.0719	55.57	5.97	29.58	8.88
C25H32BR204	556.34	554.0668	53.97	5.80	28.72	11.50
C25H32BR205	572.34	570.0617	52.47	5.64	27.92	13.98
C25H32BR206	588.34	586.0567	51.04	5.48	27.16	16.32
C25H32BR207	604.34	602.0516	49.69	5.34	26.44	18.53
C25H32BR208	620.34	618.0465	48.41	5.20	25.76	20.63
C25H32BR209	636.34	634.0414	47.19	5.07	25.11	22.63
C25H33BR101	429.45	428.1715	69.92	7.75	18.61	3.73
C25H33BR102	445.44	444.1664	67.41	7.47	17.94	7.18
C25H33BR103	461.44	460.1613	65.07	7.21	17.32	10.40
C25H33BR104	477.44	476.1563	62.89	6.97	16.74	13.40
C25H33BR105	493.44	492.1512	60.85	6.74	16.19	16.21
C25H33BR106	509.44	508.1461	58.94	6.53	15.68	18.84
C25H33BR107	525.44	524.1410	57.15	6.33	15.21	21.31
C25H33BR108	541.44	540.1359	55.46	6.14	14.76	23.64
C25H33BR109	557.44	556.1308	53.87	5.97	14.33	25.83
C25H34BR201	510.36	508.0977	58.84	6.72	31.31	3.13
C25H34BR202	526.36	524.0926	57.05	6.51	30.36	6.08
C25H34BR203	542.36	540.0876	55.37	6.32	29.47	8.85
C25H34BR204	558.36	556.0825	53.78	6.14	28.62	11.46
C25H34BR205	574.35	572.0774	52.28	5.97	27.82	13.93
C25H34BR206	590.35	588.0723	50.86	5.81	27.07	16.26
C25H34BR207	606.35	604.0672	49.52	5.65	26.36	18.47
C25H34BR208	622.35	620.0621	48.25	5.51	25.68	20.57
C25H34BR209	638.35	636.0570	47.04	5.37	25.03	22.56
C25H35BR101	431.46	430.1872	69.60	8.18	18.52	3.71
C25H35BR102	447.46	446.1821	67.11	7.88	17.86	7.15
C25H35BR103	463.46	462.1770	64.79	7.61	17.24	10.36
C25H35BR104	479.46	478.1719	62.63	7.36	16.67	13.35
C25H35BR105	495.46	494.1668	60.61	7.12	16.13	16.15
C25H35BR106	511.46	510.1617	58.71	6.90	15.62	18.77
C25H35BR107	527.46	526.1566	56.93	6.69	15.15	21.23
C25H35BR108	543.46	542.1516	55.25	6.49	14.70	23.55
C25H35BR109	559.46	558.1465	53.67	6.31	14.28	25.74
C25H36BR201	512.37	510.1134	58.61	7.08	31.19	3.12
C25H36BR202	528.37	526.1083	56.83	6.87	30.25	6.06
C25H36BR203	544.37	542.1032	55.16	6.67	29.36	8.82
C25H36BR204	560.37	558.0981	53.59	6.48	28.52	11.42
C25H36BR205	576.37	574.0930	52.10	6.30	27.73	13.88
C25H36BR206	592.37	590.0880	50.69	6.13	26.98	16.21
C25H36BR207	608.37	606.0829	49.36	5.96	26.27	18.41
C25H36BR208	624.37	622.0778	48.09	5.81	25.60	20.50
C25H36BR209	640.37	638.0727	46.89	5.67	24.96	22.49
C25H37BR101	433.48	432.2028	69.27	8.60	18.43	3.69
C25H37BR102	449.48	448.1977	66.81	8.30	17.78	7.12
C25H37BR103	465.48	464.1926	64.51	8.01	17.17	10.31
C25H37BR104	481.48	480.1876	62.37	7.75	16.60	13.29
C25H37BR105	497.47	496.1825	60.36	7.50	16.06	16.08
C25H37BR106	513.47	512.1774	58.48	7.26	15.56	18.70
C25H37BR107	529.47	528.1723	56.71	7.04	15.09	21.15
C25H37BR108	545.47	544.1672	55.05	6.84	14.65	23.46
C25H37BR109	561.47	560.1621	53.48	6.64	14.23	25.65
C25H38BR201	514.39	512.1290	58.38	7.45	31.07	3.11
C25H38BR202	530.39	528.1239	56.61	7.22	30.13	6.03
C25H38BR203	546.39	544.1189	54.96	7.01	29.25	8.78
C25H38BR204	562.39	560.1138	53.39	6.81	28.42	11.38
C25H38BR205	578.39	576.1087	51.92	6.62	27.63	13.83
C25H38BR206	594.39	592.1036	50.52	6.44	26.89	16.15
C25H38BR207	610.39	608.0985	49.19	6.28	26.18	18.35
C25H38BR208	626.38	624.0934	47.94	6.11	25.51	20.43
C25H38BR209	642.38	640.0883	46.74	5.96	24.88	22.42
C25H39BR101	435.49	434.2185	68.95	9.03	18.35	3.67

C25H39BR102	451.49	450.2134	66.51	8.71	17.70	7.09
C25H39BR103	467.49	466.2083	64.23	8.41	17.09	10.27
C25H39BR104	483.49	482.2032	62.11	8.13	16.53	13.24
C25H39BR105	499.49	498.1981	60.12	7.87	16.00	16.02
C25H39BR106	515.49	514.1930	58.25	7.63	15.50	18.62
C25H39BR107	531.49	530.1879	56.50	7.40	15.03	21.07
C25H39BR108	547.49	546.1829	54.85	7.18	14.59	23.38
C25H39BR109	563.49	562.1778	53.29	6.98	14.18	25.55
C25H40BR201	516.40	514.1447	58.15	7.81	30.95	3.10
C25H40BR202	532.40	530.1396	56.40	7.57	30.02	6.01
C25H40BR203	548.40	546.1345	54.76	7.35	29.14	8.75
C25H40BR204	564.40	562.1294	53.20	7.14	28.31	11.34
C25H40BR205	580.40	578.1243	51.74	6.95	27.53	13.78
C25H40BR206	596.40	594.1192	50.35	6.76	26.80	16.10
C25H40BR207	612.40	610.1142	49.03	6.58	26.10	18.29
C25H40BR208	628.40	626.1091	47.78	6.42	25.43	20.37
C25H40BR209	644.40	642.1040	46.60	6.26	24.80	22.35
C25H41BR101	437.51	436.2341	68.63	9.45	18.26	3.66
C25H41BR102	453.51	452.2290	66.21	9.11	17.62	7.06
C25H41BR103	469.51	468.2239	63.96	8.80	17.02	10.22
C25H41BR104	485.51	484.2189	61.85	8.51	16.46	13.18
C25H41BR105	501.51	500.2138	59.88	8.24	15.93	15.95
C25H41BR106	517.51	516.2087	58.02	7.99	15.44	18.55
C25H41BR107	533.51	532.2036	56.28	7.75	14.98	20.99
C25H41BR108	549.50	548.1985	54.65	7.52	14.54	23.29
C25H41BR109	565.50	564.1934	53.10	7.31	14.13	25.46
C25H42BR201	518.42	516.1603	57.92	8.17	30.83	3.09
C25H42BR202	534.42	532.1552	56.19	7.92	29.90	5.99
C25H42BR203	550.42	548.1502	54.55	7.69	29.03	8.72
C25H42BR204	566.42	564.1451	53.01	7.47	28.21	11.30
C25H42BR205	582.42	580.1400	51.56	7.27	27.44	13.74
C25H42BR206	598.42	596.1349	50.18	7.07	26.71	16.04
C25H42BR207	614.42	612.1298	48.87	6.89	26.01	18.23
C25H42BR208	630.42	628.1247	47.63	6.72	25.35	20.30
C25H42BR209	646.42	644.1196	46.45	6.55	24.72	22.28
C25H43BR101	439.52	438.2498	68.32	9.86	18.18	3.64
C25H43BR102	455.52	454.2447	65.92	9.51	17.54	7.02
C25H43BR103	471.52	470.2396	63.68	9.19	16.95	10.18
C25H43BR104	487.52	486.2345	61.59	8.89	16.39	13.13
C25H43BR105	503.52	502.2294	59.64	8.61	15.87	15.89
C25H43BR106	519.52	518.2243	57.80	8.34	15.38	18.48
C25H43BR107	535.52	534.2192	56.07	8.09	14.92	20.91
C25H43BR108	551.52	550.2142	54.45	7.86	14.49	23.21
C25H43BR109	567.52	566.2091	52.91	7.64	14.08	25.37
C25H44BR201	520.44	518.1760	57.70	8.52	30.71	3.07
C25H44BR202	536.44	534.1709	55.98	8.27	29.79	5.97
C25H44BR203	552.44	550.1658	54.36	8.03	28.93	8.69
C25H44BR204	568.44	566.1607	52.83	7.80	28.11	11.26
C25H44BR205	584.43	582.1556	51.38	7.59	27.34	13.69
C25H44BR206	600.43	598.1505	50.01	7.39	26.62	15.99
C25H44BR207	616.43	614.1455	48.71	7.19	25.92	18.17
C25H44BR208	632.43	630.1404	47.48	7.01	25.27	20.24
C25H44BR209	648.43	646.1353	46.31	6.84	24.65	22.21
C25H45BR101	441.54	440.2654	68.01	10.27	18.10	3.62
C25H45BR102	457.54	456.2603	65.63	9.91	17.46	6.99
C25H45BR103	473.54	472.2552	63.41	9.58	16.87	10.14
C25H45BR104	489.54	488.2502	61.34	9.27	16.32	13.07
C25H45BR105	505.54	504.2451	59.40	8.97	15.81	15.82
C25H45BR106	521.54	520.2400	57.58	8.70	15.32	18.41
C25H45BR107	537.54	536.2349	55.86	8.44	14.86	20.83
C25H45BR108	553.54	552.2298	54.25	8.19	14.44	23.12
C25H45BR109	569.54	568.2247	52.72	7.96	14.03	25.28
C25H46BR201	522.45	520.1916	57.47	8.87	30.59	3.06
C25H46BR202	538.45	536.1865	55.77	8.61	29.68	5.94
C25H46BR203	554.45	552.1815	54.16	8.36	28.82	8.66
C25H46BR204	570.45	568.1764	52.64	8.13	28.01	11.22
C25H46BR205	586.45	584.1713	51.20	7.91	27.25	13.64
C25H46BR206	602.45	600.1662	49.84	7.70	26.53	15.93
C25H46BR207	618.45	616.1611	48.55	7.50	25.84	18.11
C25H46BR208	634.45	632.1560	47.33	7.31	25.19	20.17
C25H46BR209	650.45	648.1509	46.16	7.13	24.57	22.14
C25H47BR101	443.56	442.2811	67.70	10.68	18.01	3.61
C25H47BR102	459.56	458.2760	65.34	10.31	17.39	6.96
C25H47BR103	475.56	474.2709	63.14	9.96	16.80	10.09
C25H47BR104	491.55	490.2658	61.09	9.64	16.26	13.02
C25H47BR105	507.55	506.2607	59.16	9.33	15.74	15.76
C25H47BR106	523.55	522.2556	57.35	9.05	15.26	18.34
C25H47BR107	539.55	538.2505	55.65	8.78	14.81	20.76
C25H47BR108	555.55	554.2455	54.05	8.53	14.38	23.04
C25H47BR109	571.55	570.2404	52.54	8.29	13.98	25.19
C25H48BR201	524.47	522.2073	57.25	9.23	30.47	3.05
C25H48BR202	540.47	538.2022	55.56	8.95	29.57	5.92
C25H48BR203	556.47	554.1971	53.96	8.69	28.72	8.63
C25H48BR204	572.47	570.1920	52.45	8.45	27.92	11.18
C25H48BR205	588.47	586.1869	51.03	8.22	27.16	13.59
C25H48BR206	604.47	602.1818	49.68	8.00	26.44	15.88
C25H48BR207	620.47	618.1768	48.40	7.80	25.76	18.05
C25H48BR208	636.46	634.1717	47.18	7.60	25.11	20.11
C25H48BR209	652.46	650.1666	46.02	7.42	24.49	22.07
C25H49BR101	445.57	444.2967	67.39	11.08	17.93	3.59
C25H49BR102	461.57	460.2916	65.06	10.70	17.31	6.93
C25H49BR103	477.57	476.2865	62.88	10.34	16.73	10.05
C25H49BR104	493.57	492.2815	60.84	10.01	16.19	12.97
C25H49BR105	509.57	508.2764	58.93	9.69	15.68	15.70
C25H49BR106	525.57	524.2713	57.13	9.40	15.20	18.27
C25H49BR107	541.57	540.2662	55.45	9.12	14.75	20.68
C25H49BR108	557.57	556.2611	53.86	8.86	14.33	22.96
C25H49BR109	573.57	572.2560	52.35	8.61	13.93	25.11
C25H50BR201	526.48	524.2229	57.03	9.57	30.35	3.04
C25H50BR202	542.48	540.2178	55.35	9.29	29.46	5.90
C25H50BR203	558.48	556.2128	53.77	9.02	28.61	8.59
C25H50BR204	574.48	572.2077	52.27	8.77	27.82	11.14
C25H50BR205	590.48	588.2026	50.85	8.54	27.06	13.55
C25H50BR206	606.48	604.1975	49.51	8.31	26.35	15.83
C25H50BR207	622.48	620.1924	48.24	8.10	25.67	17.99
C25H50BR208	638.48	636.1873	47.03	7.89	25.03	20.05
C25H50BR209	654.48	652.1822	45.88	7.70	24.42	22.00
C25H51BR101	447.59	446.3124	67.09	11.49	17.85	3.57
C25H51BR102	463.59	462.3073	64.77	11.09	17.24	6.90
C25H51BR103	479.59	478.3022	62.61	10.72	16.66	10.01
C25H51BR104	495.59	494.2971	60.59	10.37	16.12	12.91
C25H51BR105	511.59	510.2920	58.70	10.05	15.62	15.64
C25H51BR106	527.59	526.2869	56.92	9.74	15.15	18.20
C25H51BR107	543.59	542.2818	55.24	9.46	14.70	20.60
C25H51BR108	559.58	558.2768	53.66	9.19	14.28	22.87
C25H51BR109	575.58	574.2717	52.17	8.93	13.88	25.02
C26H13BR101	421.30	420.0150	74.13	3.11	18.97	3.80
C26H13BR102	437.30	436.0099	71.41	3.00	18.27	7.32
C26H13BR103	453.30	452.0049	68.89	2.89	17.63	10.59
C26H13BR104	469.30	467.9998	66.54	2.79	17.03	13.64

050

C26H13BR105	485.29	483.9947	64.35	2.70	16.47	16.48
C26H13BR106	501.29	499.9896	62.30	2.61	15.94	19.15
C26H13BR107	517.29	515.9845	60.37	2.53	15.45	21.65
C26H13BR108	533.29	531.9794	58.56	2.46	14.98	24.00
C26H13BR109	549.29	547.9743	56.85	2.39	14.55	26.21
C26H14BR201	502.21	499.9412	62.18	2.81	31.82	3.19
C26H14BR202	518.21	515.9362	60.26	2.72	30.84	6.17
C26H14BR203	534.21	531.9311	58.46	2.64	29.91	8.98
C26H14BR204	550.21	547.9260	56.76	2.56	29.05	11.63
C26H14BR205	566.21	563.9209	55.15	2.49	28.22	14.13
C26H14BR206	582.21	579.9158	53.64	2.42	27.45	16.49
C26H14BR207	598.21	595.9107	52.20	2.36	26.71	18.72
C26H14BR208	614.20	611.9056	50.84	2.30	26.02	20.84
C26H14BR209	630.20	627.9006	49.55	2.24	25.36	22.85
C26H15BR101	423.31	422.0307	73.77	3.57	18.88	3.78
C26H15BR102	439.31	438.0256	71.09	3.44	18.19	7.28
C26H15BR103	455.31	454.0205	68.59	3.32	17.55	10.54
C26H15BR104	471.31	470.0154	66.26	3.21	16.95	13.58
C26H15BR105	487.31	486.0103	64.08	3.10	16.40	16.42
C26H15BR106	503.31	502.0052	62.05	3.00	15.88	19.07
C26H15BR107	519.31	518.0002	60.14	2.91	15.39	21.57
C26H15BR108	535.31	533.9951	58.34	2.82	14.93	23.91
C26H15BR109	551.31	549.9900	56.65	2.74	14.49	26.12
C26H16BR201	504.22	501.9569	61.93	3.20	31.69	3.17
C26H16BR202	520.22	517.9518	60.03	3.10	30.72	6.15
C26H16BR203	536.22	533.9467	58.24	3.01	29.80	8.95
C26H16BR204	552.22	549.9416	56.55	2.92	28.94	11.59
C26H16BR205	568.22	565.9365	54.96	2.84	28.12	14.08
C26H16BR206	584.22	581.9315	53.45	2.76	27.35	16.43
C26H16BR207	600.22	597.9264	52.03	2.69	26.62	18.66
C26H16BR208	616.22	613.9213	50.68	2.62	25.93	20.77
C26H16BR209	632.22	629.9162	49.40	2.55	25.28	22.78
C26H17BR101	425.33	424.0463	73.42	4.03	18.79	3.76
C26H17BR102	441.33	440.0412	70.76	3.88	18.11	7.25
C26H17BR103	457.33	456.0362	68.29	3.75	17.47	10.50
C26H17BR104	473.33	472.0311	65.98	3.62	16.88	13.52
C26H17BR105	489.33	488.0260	63.82	3.50	16.33	16.35
C26H17BR106	505.33	504.0209	61.80	3.39	15.81	19.00
C26H17BR107	521.33	520.0158	59.90	3.29	15.33	21.48
C26H17BR108	537.32	536.0107	58.12	3.19	14.87	23.82
C26H17BR109	553.32	552.0056	56.44	3.10	14.44	26.02
C26H18BR201	506.24	503.9725	61.69	3.58	31.57	3.16
C26H18BR202	522.24	519.9675	59.80	3.47	30.60	6.13
C26H18BR203	538.24	535.9624	58.02	3.37	29.69	8.92
C26H18BR204	554.24	551.9573	56.35	3.27	28.83	11.55
C26H18BR205	570.24	567.9522	54.76	3.18	28.02	14.03
C26H18BR206	586.24	583.9471	53.27	3.09	27.26	16.37
C26H18BR207	602.24	599.9420	51.85	3.01	26.54	18.60
C26H18BR208	618.24	615.9369	50.51	2.93	25.85	20.70
C26H18BR209	634.24	631.9318	49.24	2.86	25.20	22.70
C26H19BR101	427.34	426.0620	73.08	4.48	18.70	3.74
C26H19BR102	443.34	442.0569	70.44	4.32	18.02	7.22
C26H19BR103	459.34	458.0518	67.99	4.17	17.40	10.45
C26H19BR104	475.34	474.0467	65.70	4.03	16.81	13.46
C26H19BR105	491.34	490.0416	63.56	3.90	16.26	16.28
C26H19BR106	507.34	506.0365	61.55	3.77	15.75	18.92
C26H19BR107	523.34	522.0315	59.67	3.66	15.27	21.40
C26H19BR108	539.34	538.0264	57.90	3.55	14.82	23.73
C26H19BR109	555.34	554.0213	56.23	3.45	14.39	25.93
C26H20BR201	508.26	505.9882	61.44	3.97	31.44	3.15
C26H20BR202	524.26	521.9831	59.57	3.85	30.48	6.10
C26H20BR203	540.26	537.9780	57.80	3.73	29.58	8.88
C26H20BR204	556.25	553.9729	56.14	3.62	28.73	11.51
C26H20BR205	572.25	569.9678	54.57	3.52	27.93	13.98
C26H20BR206	588.25	585.9628	53.09	3.43	27.17	16.32
C26H20BR207	604.25	601.9577	51.68	3.34	26.45	18.53
C26H20BR208	620.25	617.9526	50.35	3.25	25.76	20.64
C26H20BR209	636.25	633.9475	49.08	3.17	25.12	22.63
C26H21BR101	429.36	428.0776	72.73	4.93	18.61	3.73
C26H21BR102	445.36	444.0725	70.12	4.75	17.94	7.18
C26H21BR103	461.36	460.0674	67.69	4.59	17.32	10.40
C26H21BR104	477.36	476.0624	65.42	4.43	16.74	13.41
C26H21BR105	493.36	492.0573	63.30	4.29	16.20	16.21
C26H21BR106	509.36	508.0522	61.31	4.16	15.69	18.85
C26H21BR107	525.36	524.0471	59.44	4.03	15.21	21.32
C26H21BR108	541.36	540.0420	57.69	3.91	14.76	23.64
C26H21BR109	557.36	556.0369	56.03	3.80	14.34	25.84
C26H22BR201	510.27	508.0038	61.20	4.35	31.32	3.14
C26H22BR202	526.27	523.9987	59.34	4.21	30.37	6.08
C26H22BR203	542.27	539.9937	57.59	4.09	29.47	8.85
C26H22BR204	558.27	555.9886	55.94	3.97	28.63	11.46
C26H22BR205	574.27	571.9835	54.38	3.86	27.83	13.93
C26H22BR206	590.27	587.9784	52.91	3.76	27.07	16.26
C26H22BR207	606.27	603.9733	51.51	3.66	26.36	18.47
C26H22BR208	622.27	619.9682	50.19	3.56	25.68	20.57
C26H22BR209	638.27	635.9631	48.93	3.47	25.04	22.56
C26H23BR101	431.38	430.0933	72.39	5.37	18.52	3.71
C26H23BR102	447.38	446.0882	69.80	5.18	17.86	7.15
C26H23BR103	463.38	462.0831	67.39	5.00	17.24	10.36
C26H23BR104	479.37	478.0780	65.15	4.84	16.67	13.35
C26H23BR105	495.37	494.0729	63.04	4.68	16.13	16.15
C26H23BR106	511.37	510.0678	61.07	4.53	15.63	18.77
C26H23BR107	527.37	526.0628	59.22	4.40	15.15	21.24
C26H23BR108	543.37	542.0577	57.47	4.27	14.71	23.56
C26H23BR109	559.37	558.0526	55.83	4.14	14.28	25.74
C26H24BR201	512.29	510.0195	60.96	4.72	31.19	3.12
C26H24BR202	528.29	526.0144	59.11	4.58	30.25	6.06
C26H24BR203	544.29	542.0093	57.38	4.44	29.36	8.82
C26H24BR204	560.29	558.0042	55.74	4.32	28.52	11.42
C26H24BR205	576.29	573.9991	54.19	4.20	27.73	13.88
C26H24BR206	592.29	589.9941	52.73	4.08	26.98	16.21
C26H24BR207	608.28	605.9890	51.34	3.98	26.27	18.41
C26H24BR208	624.28	621.9839	50.02	3.88	25.60	20.50
C26H24BR209	640.28	637.9788	48.77	3.78	24.96	22.49
C26H25BR101	433.39	432.1089	72.06	5.81	18.44	3.69
C26H25BR102	449.39	448.1038	69.49	5.61	17.78	7.12
C26H25BR103	465.39	464.0987	67.10	5.41	17.17	10.31
C26H25BR104	481.39	480.0937	64.87	5.23	16.60	13.29
C26H25BR105	497.39	496.0886	62.79	5.07	16.06	16.08
C26H25BR106	513.39	512.0835	60.83	4.91	15.56	18.70
C26H25BR107	529.39	528.0784	58.99	4.76	15.09	21.16
C26H25BR108	545.39	544.0733	57.26	4.62	14.65	23.47
C26H25BR109	561.39	560.0682	55.63	4.49	14.23	25.65
C26H26BR201	514.30	512.0351	60.72	5.10	31.07	3.11
C26H26BR202	530.30	528.0300	58.89	4.94	30.14	6.03
C26H26BR203	546.30	544.0250	57.16	4.80	29.25	8.79
C26H26BR204	562.30	560.0199	55.54	4.66	28.42	11.38
C26H26BR205	578.30	576.0148	54.00	4.53	27.63	13.83
C26H26BR206	594.30	592.0097	52.55	4.41	26.89	16.15
C26H26BR207	610.30	608.0046	51.17	4.29	26.19	18.35

051

C26H26BR208	626.30	623.9995	49.86	4.18	25.52	20.44
C26H26BR209	642.30	639.9944	48.62	4.08	24.88	22.42
C26H27BR101	435.41	434.1246	71.72	6.25	18.35	3.67
C26H27BR102	451.41	450.1195	69.18	6.03	17.70	7.09
C26H27BR103	467.41	466.1144	66.81	5.82	17.10	10.27
C26H27BR104	483.41	482.1093	64.60	5.63	16.53	13.24
C26H27BR105	499.41	498.1042	62.53	5.45	16.00	16.02
C26H27BR106	515.41	514.0991	60.59	5.28	15.50	18.63
C26H27BR107	531.40	530.0941	58.77	5.12	15.04	21.08
C26H27BR108	547.40	546.0890	57.05	4.97	14.60	23.38
C26H27BR109	563.40	562.0839	55.43	4.83	14.18	25.56
C26H28BR201	516.32	514.0508	60.48	5.47	30.95	3.10
C26H28BR202	532.32	530.0457	58.67	5.30	30.02	6.01
C26H28BR203	548.32	546.0406	56.95	5.15	29.15	8.75
C26H28BR204	564.32	562.0355	55.34	5.00	28.32	11.34
C26H28BR205	580.32	578.0304	53.81	4.86	27.54	13.79
C26H28BR206	596.32	594.0254	52.37	4.73	26.80	16.10
C26H28BR207	612.32	610.0203	51.00	4.61	26.10	18.29
C26H28BR208	628.32	626.0152	49.70	4.49	25.43	20.37
C26H28BR209	644.32	642.0101	48.47	4.38	24.80	22.35
C26H29BR101	437.42	436.1402	71.39	6.68	18.27	3.66
C26H29BR102	453.42	452.1351	68.87	6.45	17.62	7.06
C26H29BR103	469.42	468.1300	66.53	6.23	17.02	10.22
C26H29BR104	485.42	484.1250	64.33	6.02	16.46	13.18
C26H29BR105	501.42	500.1199	62.28	5.83	15.94	15.95
C26H29BR106	517.42	516.1148	60.36	5.65	15.44	18.55
C26H29BR107	533.42	532.1097	58.54	5.48	14.98	21.00
C26H29BR108	549.42	548.1046	56.84	5.32	14.54	23.30
C26H29BR109	565.42	564.0995	55.23	5.17	14.13	25.47
C26H30BR201	518.34	516.0664	60.25	5.83	30.83	3.09
C26H30BR202	534.34	532.0613	58.44	5.66	29.91	5.99
C26H30BR203	550.34	548.0563	56.75	5.49	29.04	8.72
C26H30BR204	566.33	564.0512	55.14	5.34	28.22	11.30
C26H30BR205	582.33	580.0461	53.63	5.19	27.44	13.74
C26H30BR206	598.33	596.0410	52.19	5.05	26.71	16.04
C26H30BR207	614.33	612.0359	50.83	4.92	26.01	18.23
C26H30BR208	630.33	628.0308	49.54	4.80	25.35	20.31
C26H30BR209	646.33	644.0257	48.32	4.68	24.73	22.28
C26H31BR101	439.44	438.1559	71.07	7.11	18.18	3.64
C26H31BR102	455.44	454.1508	68.57	6.86	17.54	7.03
C26H31BR103	471.44	470.1457	66.24	6.63	16.95	10.18
C26H31BR104	487.44	486.1406	64.07	6.41	16.39	13.13
C26H31BR105	503.44	502.1355	62.03	6.21	15.87	15.89
C26H31BR106	519.44	518.1304	60.12	6.02	15.38	18.48
C26H31BR107	535.44	534.1254	58.32	5.84	14.92	20.92
C26H31BR108	551.44	550.1203	56.63	5.67	14.49	23.21
C26H31BR109	567.44	566.1152	55.04	5.51	14.08	25.38
C26H32BR201	520.35	518.0821	60.02	6.20	30.71	3.07
C26H32BR202	536.35	534.0770	58.22	6.01	29.80	5.97
C26H32BR203	552.35	550.0719	56.54	5.84	28.93	8.69
C26H32BR204	568.35	566.0668	54.95	5.68	28.12	11.26
C26H32BR205	584.35	582.0617	53.44	5.52	27.35	13.69
C26H32BR206	600.35	598.0567	52.02	5.37	26.62	15.99
C26H32BR207	616.35	614.0516	50.67	5.23	25.93	18.17
C26H32BR208	632.35	630.0465	49.39	5.10	25.27	20.24
C26H32BR209	648.35	646.0414	48.17	4.97	24.65	22.21
C26H33BR101	441.46	440.1715	70.74	7.53	18.10	3.62
C26H33BR102	457.46	456.1664	68.27	7.27	17.47	6.99
C26H33BR103	473.46	472.1613	65.96	7.03	16.88	10.14
C26H33BR104	489.45	488.1563	63.80	6.80	16.33	13.08
C26H33BR105	505.45	504.1512	61.78	6.58	15.81	15.83
C26H33BR106	521.45	520.1461	59.89	6.38	15.32	18.41
C26H33BR107	537.45	536.1410	58.11	6.19	14.87	20.84
C26H33BR108	553.45	552.1359	56.43	6.01	14.44	23.13
C26H33BR109	569.45	568.1308	54.84	5.84	14.03	25.29
C26H34BR201	522.37	520.0977	59.78	6.56	30.59	3.06
C26H34BR202	538.37	536.0926	58.01	6.37	29.68	5.94
C26H34BR203	554.37	552.0876	56.33	6.18	28.83	8.66
C26H34BR204	570.37	568.0825	54.75	6.01	28.02	11.22
C26H34BR205	586.37	584.0774	53.26	5.84	27.25	13.64
C26H34BR206	602.37	600.0723	51.84	5.69	26.53	15.94
C26H34BR207	618.36	616.0672	50.50	5.54	25.84	18.11
C26H34BR208	634.36	632.0621	49.23	5.40	25.19	20.18
C26H34BR209	650.36	648.0570	48.02	5.27	24.57	22.14
C26H35BR101	443.47	442.1872	70.42	7.96	18.02	3.61
C26H35BR102	459.47	458.1821	67.97	7.68	17.39	6.96
C26H35BR103	475.47	474.1770	65.68	7.42	16.81	10.09
C26H35BR104	491.47	490.1719	63.54	7.18	16.26	13.02
C26H35BR105	507.47	506.1668	61.54	6.95	15.75	15.76
C26H35BR106	523.47	522.1617	59.66	6.74	15.26	18.34
C26H35BR107	539.47	538.1566	57.89	6.54	14.81	20.76
C26H35BR108	555.47	554.1516	56.22	6.35	14.38	23.04
C26H35BR109	571.47	570.1465	54.65	6.17	13.98	25.20
C26H36BR201	524.38	522.1134	59.55	6.92	30.48	3.05
C26H36BR202	540.38	538.1083	57.79	6.72	29.57	5.92
C26H36BR203	556.38	554.1032	56.13	6.52	28.72	8.63
C26H36BR204	572.38	570.0981	54.56	6.34	27.92	11.18
C26H36BR205	588.38	586.0930	53.08	6.17	27.16	13.60
C26H36BR206	604.38	602.0880	51.67	6.00	26.44	15.88
C26H36BR207	620.38	618.0829	50.34	5.85	25.76	18.05
C26H36BR208	636.38	634.0778	49.07	5.70	25.11	20.11
C26H36BR209	652.38	650.0727	47.87	5.56	24.50	22.07
C26H37BR101	445.49	444.2028	70.10	8.37	17.94	3.59
C26H37BR102	461.49	460.1977	67.67	8.08	17.31	6.93
C26H37BR103	477.49	476.1926	65.40	7.81	16.73	10.05
C26H37BR104	493.49	492.1876	63.28	7.56	16.19	12.97
C26H37BR105	509.49	508.1825	61.30	7.32	15.68	15.70
C26H37BR106	525.49	524.1774	59.43	7.10	15.21	18.27
C26H37BR107	541.48	540.1723	57.67	6.89	14.76	20.68
C26H37BR108	557.48	556.1672	56.02	6.69	14.33	22.96
C26H37BR109	573.48	572.1621	54.45	6.50	13.93	25.11
C26H38BR201	526.40	524.1290	59.33	7.28	30.36	3.04
C26H38BR202	542.40	540.1239	57.58	7.06	29.46	5.90
C26H38BR203	558.40	556.1189	55.93	6.86	28.62	8.60
C26H38BR204	574.40	572.1138	54.37	6.67	27.82	11.14
C26H38BR205	590.40	588.1087	52.89	6.49	27.07	13.55
C26H38BR206	606.40	604.1036	51.50	6.32	26.35	15.83
C26H38BR207	622.40	620.0985	50.18	6.15	25.68	17.99
C26H38BR208	638.40	636.0934	48.92	6.00	25.03	20.05
C26H38BR209	654.40	652.0883	47.72	5.85	24.42	22.00
C26H39BR101	447.50	446.2185	69.78	8.78	17.86	3.58
C26H39BR102	463.50	462.2134	67.38	8.48	17.24	6.90
C26H39BR103	479.50	478.2083	65.13	8.20	16.66	10.01
C26H39BR104	495.50	494.2032	63.02	7.93	16.13	12.92
C26H39BR105	511.50	510.1981	61.05	7.69	15.62	15.64
C26H39BR106	527.50	526.1930	59.20	7.45	15.15	18.20
C26H39BR107	543.50	542.1879	57.46	7.23	14.70	20.61
C26H39BR108	559.50	558.1829	55.82	7.03	14.28	22.88
C26H39BR109	575.50	574.1778	54.26	6.83	13.88	25.02
C26H40BR201	528.42	526.1447	59.10	7.63	30.24	3.03

052

C26H40BR202	544.42	542.1396	57.36	7.41	29.35	5.88
C26H40BR203	560.41	558.1345	55.72	7.19	28.52	8.56
C26H40BR204	576.41	574.1294	54.18	6.99	27.72	11.10
C26H40BR205	592.41	590.1243	52.71	6.81	26.98	13.50
C26H40BR206	608.41	606.1192	51.33	6.63	26.27	15.78
C26H40BR207	624.41	622.1142	50.01	6.46	25.59	17.94
C26H40BR208	640.41	638.1091	48.76	6.30	24.95	19.99
C26H40BR209	656.41	654.1040	47.58	6.14	24.35	21.94
C26H41BR101	449.52	448.2341	69.47	9.19	17.78	3.56
C26H41BR102	465.52	464.2290	67.08	8.88	17.16	6.87
C26H41BR103	481.52	480.2239	64.86	8.58	16.59	9.97
C26H41BR104	497.52	496.2189	62.77	8.31	16.06	12.86
C26H41BR105	513.52	512.2138	60.81	8.05	15.56	15.58
C26H41BR106	529.52	528.2087	58.98	7.80	15.09	18.13
C26H41BR107	545.52	544.2036	57.25	7.58	14.65	20.53
C26H41BR108	561.52	560.1985	55.62	7.36	14.23	22.79
C26H41BR109	577.52	576.1934	54.07	7.16	13.84	24.93
C26H42BR201	530.43	528.1603	58.87	7.98	30.13	3.02
C26H42BR202	546.43	544.1552	57.15	7.75	29.25	5.86
C26H42BR203	562.43	560.1502	55.53	7.53	28.41	8.53
C26H42BR204	578.43	576.1451	53.99	7.32	27.63	11.06
C26H42BR205	594.43	592.1400	52.54	7.12	26.88	13.46
C26H42BR206	610.43	608.1349	51.16	6.94	26.18	15.73
C26H42BR207	626.43	624.1298	49.85	6.76	25.51	17.88
C26H42BR208	642.43	640.1247	48.61	6.59	24.88	19.92
C26H42BR209	658.43	656.1196	47.43	6.43	24.27	21.87
C26H43BR101	451.54	450.2498	69.16	9.60	17.70	3.54
C26H43BR102	467.54	466.2447	66.79	9.27	17.09	6.84
C26H43BR103	483.53	482.2396	64.58	8.96	16.52	9.93
C26H43BR104	499.53	498.2345	62.52	8.68	16.00	12.81
C26H43BR105	515.53	514.2294	60.58	8.41	15.50	15.52
C26H43BR106	531.53	530.2243	58.75	8.15	15.03	18.06
C26H43BR107	547.53	546.2192	57.04	7.92	14.59	20.45
C26H43BR108	563.53	562.2142	55.42	7.69	14.18	22.71
C26H43BR109	579.53	578.2091	53.89	7.48	13.79	24.85
C26H44BR201	532.45	530.1760	58.65	8.33	30.01	3.00
C26H44BR202	548.45	546.1709	56.94	8.09	29.14	5.83
C26H44BR203	564.45	562.1658	55.33	7.86	28.31	8.50
C26H44BR204	580.45	578.1607	53.80	7.64	27.53	11.03
C26H44BR205	596.45	594.1556	52.36	7.44	26.79	13.41
C26H44BR206	612.44	610.1505	50.99	7.24	26.09	15.67
C26H44BR207	628.44	626.1455	49.69	7.06	25.43	17.82
C26H44BR208	644.44	642.1404	48.46	6.88	24.80	19.86
C26H44BR209	660.44	658.1353	47.28	6.72	24.20	21.80
C26H45BR101	453.55	452.2654	68.85	10.00	17.62	3.53
C26H45BR102	469.55	468.2603	66.51	9.66	17.02	6.81
C26H45BR103	485.55	484.2552	64.32	9.34	16.46	9.89
C26H45BR104	501.55	500.2502	62.26	9.04	15.93	12.76
C26H45BR105	517.55	516.2451	60.34	8.76	15.44	15.46
C26H45BR106	533.55	532.2400	58.53	8.50	14.98	17.99
C26H45BR107	549.55	548.2349	56.83	8.25	14.54	20.38
C26H45BR108	565.55	564.2298	55.22	8.02	14.13	22.63
C26H45BR109	581.55	580.2247	53.70	7.80	13.74	24.76
C26H46BR201	534.46	532.1916	58.43	8.68	29.90	2.99
C26H46BR202	550.46	548.1865	56.73	8.42	29.03	5.81
C26H46BR203	566.46	564.1815	55.13	8.19	28.21	8.47
C26H46BR204	582.46	580.1764	53.62	7.96	27.44	10.99
C26H46BR205	598.46	596.1713	52.18	7.75	26.70	13.37
C26H46BR206	614.46	612.1662	50.82	7.55	26.01	15.62
C26H46BR207	630.46	628.1611	49.53	7.35	25.35	17.76
C26H46BR208	646.46	644.1560	48.31	7.17	24.72	19.80
C26H46BR209	662.46	660.1509	47.14	7.00	24.12	21.74
C26H47BR101	455.57	454.2811	68.55	10.40	17.54	3.51
C26H47BR102	471.57	470.2760	66.22	10.05	16.94	6.79
C26H47BR103	487.57	486.2709	64.05	9.72	16.39	9.84
C26H47BR104	503.57	502.2658	62.02	9.41	15.87	12.71
C26H47BR105	519.57	518.2607	60.11	9.12	15.38	15.40
C26H47BR106	535.56	534.2556	58.31	8.85	14.92	17.92
C26H47BR107	551.56	550.2505	56.62	8.59	14.49	20.31
C26H47BR108	567.56	566.2455	55.02	8.35	14.08	22.55
C26H47BR109	583.56	582.2404	53.51	8.12	13.69	24.68
C26H48BR201	536.48	534.2073	58.21	9.02	29.79	2.98
C26H48BR202	552.48	550.2022	56.53	8.76	28.93	5.79
C26H48BR203	568.48	566.1971	54.93	8.51	28.11	8.44
C26H48BR204	584.48	582.1920	53.43	8.28	27.34	10.95
C26H48BR205	600.48	598.1869	52.01	8.06	26.61	13.32
C26H48BR206	616.48	614.1818	50.66	7.85	25.92	15.57
C26H48BR207	632.48	630.1768	49.38	7.65	25.27	17.71
C26H48BR208	648.48	646.1717	48.16	7.46	24.64	19.74
C26H48BR209	664.48	662.1666	47.00	7.28	24.05	21.67
C26H49BR101	457.58	456.2967	68.25	10.79	17.46	3.50
C26H49BR102	473.58	472.2916	65.94	10.43	16.87	6.76
C26H49BR103	489.58	488.2865	63.79	10.09	16.32	9.80
C26H49BR104	505.58	504.2815	61.77	9.77	15.80	12.66
C26H49BR105	521.58	520.2764	59.87	9.47	15.32	15.34
C26H49BR106	537.58	536.2713	58.09	9.19	14.86	17.86
C26H49BR107	553.58	552.2662	56.41	8.92	14.43	20.23
C26H49BR108	569.58	568.2611	54.83	8.67	14.03	22.47
C26H49BR109	585.58	584.2560	53.33	8.43	13.65	24.59
C26H50BR201	538.50	536.2229	57.99	9.36	29.68	2.97
C26H50BR202	554.50	552.2178	56.32	9.09	28.82	5.77
C26H50BR203	570.49	568.2128	54.74	8.83	28.01	8.41
C26H50BR204	586.49	584.2077	53.25	8.59	27.25	10.91
C26H50BR205	602.49	600.2026	51.83	8.36	26.52	13.28
C26H50BR206	618.49	616.1975	50.49	8.15	25.84	15.52
C26H50BR207	634.49	632.1924	49.22	7.94	25.19	17.65
C26H50BR208	650.49	648.1873	48.01	7.75	24.57	19.68
C26H50BR209	666.49	664.1822	46.86	7.56	23.98	21.60
C26H51BR101	459.60	458.3124	67.95	11.19	17.39	3.48
C26H51BR102	475.60	474.3073	65.66	10.81	16.80	6.73
C26H51BR103	491.60	490.3022	63.53	10.46	16.25	9.76
C26H51BR104	507.60	506.2971	61.52	10.13	15.74	12.61
C26H51BR105	523.60	522.2920	59.64	9.82	15.26	15.28
C26H51BR106	539.60	538.2869	57.87	9.53	14.81	17.79
C26H51BR107	555.60	554.2818	56.21	9.25	14.38	20.16
C26H51BR108	571.60	570.2768	54.63	8.99	13.98	22.39
C26H51BR109	587.59	586.2717	53.15	8.75	13.60	24.51
C26H52BR201	540.51	538.2386	57.78	9.70	29.57	2.96
C26H52BR202	556.51	554.2335	56.12	9.42	28.72	5.75
C26H52BR203	572.51	570.2284	54.55	9.16	27.91	8.38
C26H52BR204	588.51	586.2233	53.06	8.91	27.15	10.87
C26H52BR205	604.51	602.2182	51.66	8.67	26.44	13.23
C26H52BR206	620.51	618.2131	50.33	8.45	25.75	15.47
C26H52BR207	636.51	634.2081	49.06	8.23	25.11	17.60
C26H52BR208	652.51	650.2030	47.86	8.03	24.49	19.62
C26H52BR209	668.51	666.1979	46.71	7.84	23.91	21.54
C26H53BR101	461.62	460.3280	67.65	11.57	17.31	3.47
C26H53BR102	477.62	476.3229	65.39	11.19	16.73	6.70
C26H53BR103	493.61	492.3178	63.27	10.82	16.19	9.72
C26H53BR104	509.61	508.3128	61.28	10.48	15.68	12.56

C26H53BR105	525.61	524.3077	59.41	10.16	15.20	15.22
C26H53BR106	541.61	540.3026	57.66	9.86	14.75	17.72
C26H53BR107	557.61	556.2975	56.00	9.58	14.33	20.08
C26H53BR108	573.61	572.2924	54.44	9.31	13.93	22.31
C26H53BR109	589.61	588.2873	52.97	9.06	13.55	24.42
C27H13BR101	433.31	432.0150	74.84	3.02	18.44	3.69
C27H13BR102	449.31	448.0099	72.18	2.92	17.78	7.12
C27H13BR103	465.31	464.0049	69.70	2.82	17.17	10.32
C27H13BR104	481.31	479.9998	67.38	2.72	16.60	13.30
C27H13BR105	497.31	495.9947	65.21	2.63	16.07	16.09
C27H13BR106	513.31	511.9896	63.18	2.55	15.57	18.70
C27H13BR107	529.30	527.9845	61.27	2.48	15.10	21.16
C27H13BR108	545.30	543.9794	59.47	2.40	14.65	23.47
C27H13BR109	561.30	559.9743	57.78	2.33	14.24	25.65
C27H14BR201	514.22	511.9412	63.07	2.74	31.08	3.11
C27H14BR202	530.22	527.9362	61.16	2.66	30.14	6.04
C27H14BR203	546.22	543.9311	59.37	2.58	29.26	8.79
C27H14BR204	562.22	559.9260	57.68	2.51	28.42	11.38
C27H14BR205	578.22	575.9209	56.09	2.44	27.64	13.84
C27H14BR206	594.22	591.9158	54.58	2.37	26.89	16.16
C27H14BR207	610.22	607.9107	53.15	2.31	26.19	18.35
C27H14BR208	626.22	623.9056	51.79	2.25	25.52	20.44
C27H14BR209	642.22	639.9006	50.50	2.20	24.88	22.42
C27H15BR101	435.32	434.0307	74.50	3.47	18.36	3.68
C27H15BR102	451.32	450.0256	71.86	3.35	17.70	7.09
C27H15BR103	467.32	466.0205	69.40	3.24	17.10	10.27
C27H15BR104	483.32	482.0154	67.10	3.13	16.53	13.24
C27H15BR105	499.32	498.0103	64.95	3.03	16.00	16.02
C27H15BR106	515.32	514.0052	62.93	2.93	15.51	18.63
C27H15BR107	531.32	530.0002	61.04	2.85	15.04	21.08
C27H15BR108	547.32	545.9951	59.25	2.76	14.60	23.39
C27H15BR109	563.32	561.9900	57.57	2.68	14.18	25.56
C27H16BR201	516.24	513.9569	62.82	3.12	30.96	3.10
C27H16BR202	532.24	529.9518	60.93	3.03	30.03	6.01
C27H16BR203	548.23	545.9467	59.15	2.94	29.15	8.76
C27H16BR204	564.23	561.9416	57.48	2.86	28.32	11.34
C27H16BR205	580.23	577.9365	55.89	2.78	27.54	13.79
C27H16BR206	596.23	593.9315	54.39	2.70	26.80	16.10
C27H16BR207	612.23	609.9264	52.97	2.63	26.10	18.29
C27H16BR208	628.23	625.9213	51.62	2.57	25.44	20.37
C27H16BR209	644.23	641.9162	50.34	2.50	24.81	22.35
C27H17BR101	437.34	436.0463	74.15	3.92	18.27	3.66
C27H17BR102	453.34	452.0412	71.54	3.78	17.63	7.06
C27H17BR103	469.34	468.0362	69.10	3.65	17.02	10.23
C27H17BR104	485.34	484.0311	66.82	3.53	16.46	13.19
C27H17BR105	501.34	500.0260	64.69	3.42	15.94	15.96
C27H17BR106	517.34	516.0209	62.69	3.31	15.45	18.56
C27H17BR107	533.34	532.0158	60.81	3.21	14.98	21.00
C27H17BR108	549.34	548.0107	59.04	3.12	14.55	23.30
C27H17BR109	565.34	564.0056	57.36	3.03	14.13	25.47
C27H18BR201	518.25	515.9725	62.58	3.50	30.84	3.09
C27H18BR202	534.25	531.9675	60.70	3.40	29.91	5.99
C27H18BR203	550.25	547.9624	58.94	3.30	29.04	8.72
C27H18BR204	566.25	563.9573	57.27	3.20	28.22	11.30
C27H18BR205	582.25	579.9522	55.70	3.12	27.45	13.74
C27H18BR206	598.25	595.9471	54.21	3.03	26.71	16.05
C27H18BR207	614.25	611.9420	52.80	2.95	26.02	18.23
C27H18BR208	630.25	627.9369	51.46	2.88	25.36	20.31
C27H18BR209	646.25	643.9318	50.18	2.81	24.73	22.28
C27H19BR101	439.36	438.0620	73.81	4.36	18.19	3.64

C27H19BR102	455.36	454.0569	71.22	4.21	17.55	7.03
C27H19BR103	471.35	470.0518	68.80	4.06	16.95	10.18
C27H19BR104	487.35	486.0467	66.54	3.93	16.40	13.13
C27H19BR105	503.35	502.0416	64.43	3.80	15.87	15.89
C27H19BR106	519.35	518.0365	62.44	3.69	15.39	18.48
C27H19BR107	535.35	534.0315	60.58	3.58	14.93	20.92
C27H19BR108	551.35	550.0264	58.82	3.47	14.49	23.21
C27H19BR109	567.35	566.0213	57.16	3.38	14.08	25.38
C27H20BR201	520.27	517.9882	62.33	3.87	30.72	3.08
C27H20BR202	536.27	533.9831	60.47	3.76	29.80	5.97
C27H20BR203	552.27	549.9780	58.72	3.65	28.94	8.69
C27H20BR204	568.27	565.9729	57.07	3.55	28.12	11.26
C27H20BR205	584.27	581.9678	55.51	3.45	27.35	13.69
C27H20BR206	600.26	597.9628	54.03	3.36	26.62	15.99
C27H20BR207	616.26	613.9577	52.62	3.27	25.93	18.17
C27H20BR208	632.26	629.9526	51.29	3.19	25.28	20.24
C27H20BR209	648.26	645.9475	50.03	3.11	24.65	22.21
C27H21BR101	441.37	440.0776	73.48	4.80	18.10	3.62
C27H21BR102	457.37	456.0725	70.91	4.63	17.47	7.00
C27H21BR103	473.37	472.0674	68.51	4.47	16.88	10.14
C27H21BR104	489.37	488.0624	66.27	4.33	16.33	13.08
C27H21BR105	505.37	504.0573	64.17	4.19	15.81	15.83
C27H21BR106	521.37	520.0522	62.20	4.06	15.33	18.41
C27H21BR107	537.37	536.0471	60.35	3.94	14.87	20.84
C27H21BR108	553.37	552.0420	58.60	3.83	14.44	23.13
C27H21BR109	569.37	568.0369	56.96	3.72	14.03	25.29
C27H22BR201	522.28	520.0038	62.09	4.25	30.60	3.06
C27H22BR202	538.28	535.9987	60.25	4.12	29.69	5.94
C27H22BR203	554.28	551.9937	58.51	4.00	28.83	8.66
C27H22BR204	570.28	567.9886	56.87	3.89	28.02	11.22
C27H22BR205	586.28	583.9835	55.31	3.78	27.26	13.64
C27H22BR206	602.28	599.9784	53.85	3.68	26.53	15.94
C27H22BR207	618.28	615.9733	52.45	3.59	25.85	18.11
C27H22BR208	634.28	631.9682	51.13	3.50	25.20	20.18
C27H22BR209	650.28	647.9631	49.87	3.41	24.58	22.14
C27H23BR101	443.39	442.0933	73.14	5.23	18.02	3.61
C27H23BR102	459.39	458.0882	70.59	5.05	17.39	6.97
C27H23BR103	475.39	474.0831	68.22	4.88	16.81	10.10
C27H23BR104	491.39	490.0780	66.00	4.72	16.26	13.02
C27H23BR105	507.39	506.0729	63.92	4.57	15.75	15.77
C27H23BR106	523.38	522.0678	61.96	4.43	15.27	18.34
C27H23BR107	539.38	538.0628	60.12	4.30	14.81	20.76
C27H23BR108	555.38	554.0577	58.39	4.17	14.39	23.05
C27H23BR109	571.38	570.0526	56.76	4.06	13.98	25.20
C27H24BR201	524.30	522.0195	61.85	4.61	30.48	3.05
C27H24BR202	540.30	538.0144	60.02	4.48	29.58	5.92
C27H24BR203	556.30	554.0093	58.30	4.35	28.73	8.63
C27H24BR204	572.30	570.0042	56.67	4.23	27.92	11.18
C27H24BR205	588.30	585.9991	55.13	4.11	27.16	13.60
C27H24BR206	604.30	601.9941	53.67	4.00	26.45	15.89
C27H24BR207	620.30	617.9890	52.28	3.90	25.76	18.06
C27H24BR208	636.30	633.9839	50.97	3.80	25.12	20.12
C27H24BR209	652.29	649.9788	49.72	3.71	24.50	22.08
C27H25BR101	445.40	444.1089	72.81	5.66	17.94	3.59
C27H25BR102	461.40	460.1038	70.29	5.46	17.32	6.94
C27H25BR103	477.40	476.0987	67.93	5.28	16.74	10.05
C27H25BR104	493.40	492.0937	65.73	5.11	16.19	12.97
C27H25BR105	509.40	508.0886	63.66	4.95	15.69	15.70
C27H25BR106	525.40	524.0835	61.72	4.80	15.21	18.27
C27H25BR107	541.40	540.0784	59.90	4.65	14.76	20.69

054

C27H25BR108	557.40	556.0733	58.18	4.52	14.34	22.96
C27H25BR109	573.40	572.0682	56.56	4.39	13.94	25.11
C27H26BR201	526.32	524.0351	61.62	4.98	30.36	3.04
C27H26BR202	542.32	540.0300	59.80	4.83	29.47	5.90
C27H26BR203	558.31	556.0250	58.09	4.69	28.62	8.60
C27H26BR204	574.31	572.0199	56.47	4.56	27.83	11.14
C27H26BR205	590.31	588.0148	54.94	4.44	27.07	13.55
C27H26BR206	606.31	604.0097	53.49	4.32	26.36	15.83
C27H26BR207	622.31	620.0046	52.11	4.21	25.68	18.00
C27H26BR208	638.31	635.9995	50.81	4.11	25.04	20.05
C27H26BR209	654.31	651.9944	49.56	4.01	24.42	22.01
C27H27BR101	447.42	446.1246	72.48	6.08	17.86	3.58
C27H27BR102	463.42	462.1195	69.98	5.87	17.24	6.90
C27H27BR103	479.42	478.1144	67.64	5.68	16.67	10.01
C27H27BR104	495.42	494.1093	65.46	5.49	16.13	12.92
C27H27BR105	511.42	510.1042	63.41	5.32	15.62	15.64
C27H27BR106	527.42	526.0991	61.49	5.16	15.15	18.20
C27H27BR107	543.42	542.0941	59.68	5.01	14.70	20.61
C27H27BR108	559.42	558.0890	57.97	4.86	14.28	22.88
C27H27BR109	575.41	574.0839	56.36	4.73	13.89	25.02
C27H28BR201	528.33	526.0508	61.38	5.34	30.25	3.03
C27H28BR202	544.33	542.0457	59.58	5.18	29.36	5.88
C27H28BR203	560.33	558.0406	57.88	5.04	28.52	8.57
C27H28BR204	576.33	574.0355	56.27	4.90	27.73	11.10
C27H28BR205	592.33	590.0304	54.75	4.76	26.98	13.51
C27H28BR206	608.33	606.0254	53.31	4.64	26.27	15.78
C27H28BR207	624.33	622.0203	51.94	4.52	25.60	17.94
C27H28BR208	640.33	638.0152	50.65	4.41	24.96	19.99
C27H28BR209	656.33	654.0101	49.41	4.30	24.35	21.94
C27H29BR101	449.44	448.1402	72.16	6.50	17.78	3.56
C27H29BR102	465.43	464.1351	69.68	6.28	17.17	6.88
C27H29BR103	481.43	480.1300	67.36	6.07	16.60	9.97
C27H29BR104	497.43	496.1250	65.19	5.88	16.06	12.87
C27H29BR105	513.43	512.1199	63.16	5.69	15.56	15.58
C27H29BR106	529.43	528.1148	61.25	5.52	15.09	18.13
C27H29BR107	545.43	544.1097	59.46	5.36	14.65	20.53
C27H29BR108	561.43	560.1046	57.76	5.21	14.23	22.80
C27H29BR109	577.43	576.0995	56.16	5.06	13.84	24.94
C27H30BR201	530.35	528.0664	61.15	5.70	30.13	3.02
C27H30BR202	546.35	544.0613	59.36	5.53	29.25	5.86
C27H30BR203	562.35	560.0563	57.67	5.38	28.42	8.54
C27H30BR204	578.35	576.0512	56.07	5.23	27.63	11.07
C27H30BR205	594.35	592.0461	54.56	5.09	26.89	13.46
C27H30BR206	610.34	608.0410	53.13	4.95	26.18	15.73
C27H30BR207	626.34	624.0359	51.78	4.83	25.51	17.88
C27H30BR208	642.34	640.0308	50.49	4.71	24.88	19.93
C27H30BR209	658.34	656.0257	49.26	4.59	24.27	21.87
C27H31BR101	451.45	450.1559	71.84	6.92	17.70	3.54
C27H31BR102	467.45	466.1508	69.38	6.68	17.09	6.85
C27H31BR103	483.45	482.1457	67.08	6.46	16.53	9.93
C27H31BR104	499.45	498.1406	64.93	6.26	16.00	12.81
C27H31BR105	515.45	514.1355	62.92	6.06	15.50	15.52
C27H31BR106	531.45	530.1304	61.02	5.88	15.04	18.06
C27H31BR107	547.45	546.1254	59.24	5.71	14.60	20.46
C27H31BR108	563.45	562.1203	57.56	5.55	14.18	22.72
C27H31BR109	579.45	578.1152	55.97	5.39	13.79	24.85
C27H32BR201	532.36	530.0821	60.92	6.06	30.02	3.01
C27H32BR202	548.36	546.0770	59.14	5.88	29.14	5.84
C27H32BR203	564.36	562.0719	57.46	5.72	28.32	8.50
C27H32BR204	580.36	578.0668	55.88	5.56	27.54	11.03

C27H32BR205	596.36	594.0617	54.38	5.41	26.80	13.41
C27H32BR206	612.36	610.0567	52.96	5.27	26.10	15.68
C27H32BR207	628.36	626.0516	51.61	5.13	25.43	17.82
C27H32BR208	644.36	642.0465	50.33	5.01	24.80	19.86
C27H32BR209	660.36	658.0414	49.11	4.88	24.20	21.81
C27H33BR101	453.47	452.1715	71.52	7.34	17.62	3.53
C27H33BR102	469.47	468.1664	69.08	7.09	17.02	6.82
C27H33BR103	485.47	484.1613	66.80	6.85	16.46	9.89
C27H33BR104	501.47	500.1563	64.67	6.63	15.93	12.76
C27H33BR105	517.47	516.1512	62.67	6.43	15.44	15.46
C27H33BR106	533.46	532.1461	60.79	6.24	14.98	17.99
C27H33BR107	549.46	548.1410	59.02	6.05	14.54	20.38
C27H33BR108	565.46	564.1359	57.35	5.88	14.13	22.64
C27H33BR109	581.46	580.1308	55.77	5.72	13.74	24.76
C27H34BR201	534.38	532.0977	60.69	6.41	29.91	2.99
C27H34BR202	550.38	548.0926	58.92	6.23	29.04	5.81
C27H34BR203	566.38	564.0876	57.26	6.05	28.22	8.47
C27H34BR204	582.38	580.0825	55.69	5.88	27.44	10.99
C27H34BR205	598.38	596.0774	54.20	5.73	26.71	13.37
C27H34BR206	614.38	612.0723	52.79	5.58	26.01	15.63
C27H34BR207	630.38	628.0672	51.45	5.44	25.35	17.77
C27H34BR208	646.38	644.0621	50.17	5.30	24.72	19.80
C27H34BR209	662.37	660.0570	48.96	5.17	24.13	21.74
C27H35BR101	455.48	454.1872	71.20	7.75	17.54	3.51
C27H35BR102	471.48	470.1821	68.78	7.48	16.95	6.79
C27H35BR103	487.48	486.1770	66.53	7.24	16.39	9.85
C27H35BR104	503.48	502.1719	64.41	7.01	15.87	12.71
C27H35BR105	519.48	518.1668	62.43	6.79	15.38	15.40
C27H35BR106	535.48	534.1617	60.56	6.59	14.92	17.93
C27H35BR107	551.48	550.1566	58.81	6.40	14.49	20.31
C27H35BR108	567.48	566.1516	57.15	6.22	14.08	22.56
C27H35BR109	583.48	582.1465	55.58	6.05	13.69	24.68
C27H36BR201	536.40	534.1134	60.46	6.76	29.79	2.98
C27H36BR202	552.39	550.1083	58.71	6.57	28.93	5.79
C27H36BR203	568.39	566.1032	57.06	6.38	28.12	8.44
C27H36BR204	584.39	582.0981	55.49	6.21	27.35	10.95
C27H36BR205	600.39	598.0930	54.01	6.04	26.62	13.32
C27H36BR206	616.39	614.0880	52.61	5.89	25.93	15.57
C27H36BR207	632.39	630.0829	51.28	5.74	25.27	17.71
C27H36BR208	648.39	646.0778	50.02	5.60	24.65	19.74
C27H36BR209	664.39	662.0727	48.81	5.46	24.05	21.67
C27H37BR101	457.50	456.2028	70.89	8.15	17.47	3.50
C27H37BR102	473.50	472.1977	68.49	7.88	16.88	6.76
C27H37BR103	489.50	488.1926	66.25	7.62	16.32	9.81
C27H37BR104	505.50	504.1876	64.15	7.38	15.81	12.66
C27H37BR105	521.50	520.1825	62.19	7.15	15.32	15.34
C27H37BR106	537.50	536.1774	60.34	6.94	14.87	17.86
C27H37BR107	553.50	552.1723	58.59	6.74	14.44	20.23
C27H37BR108	569.50	568.1672	56.95	6.55	14.03	22.48
C27H37BR109	585.49	584.1621	55.39	6.37	13.65	24.59
C27H38BR201	538.41	536.1290	60.23	7.11	29.68	2.97
C27H38BR202	554.41	552.1239	58.49	6.91	28.82	5.77
C27H38BR203	570.41	568.1189	56.85	6.71	28.02	8.41
C27H38BR204	586.41	584.1138	55.30	6.53	27.25	10.91
C27H38BR205	602.41	600.1087	53.83	6.36	26.53	13.28
C27H38BR206	618.41	616.1036	52.44	6.19	25.84	15.52
C27H38BR207	634.41	632.0985	51.12	6.04	25.19	17.65
C27H38BR208	650.41	648.0934	49.86	5.89	24.57	19.68
C27H38BR209	666.41	664.0883	48.66	5.75	23.98	21.61
C27H39BR101	459.52	458.2185	70.57	8.55	17.39	3.48

055

C27H39BR102	475.51	474.2134	68.20	8.27	16.80	6.73
C27H39BR103	491.51	490.2083	65.98	8.00	16.26	9.77
C27H39BR104	507.51	506.2032	63.90	7.75	15.74	12.61
C27H39BR105	523.51	522.1981	61.95	7.51	15.26	15.28
C27H39BR106	539.51	538.1930	60.11	7.29	14.81	17.79
C27H39BR107	555.51	554.1879	58.38	7.08	14.38	20.16
C27H39BR108	571.51	570.1829	56.74	6.88	13.98	22.40
C27H39BR109	587.51	586.1778	55.20	6.69	13.60	24.51
C27H40BR201	540.43	538.1447	60.01	7.46	29.57	2.96
C27H40BR202	556.43	554.1396	58.28	7.25	28.72	5.75
C27H40BR203	572.43	570.1345	56.65	7.04	27.92	8.39
C27H40BR204	588.43	586.1294	55.11	6.85	27.16	10.88
C27H40BR205	604.42	602.1243	53.65	6.67	26.44	13.24
C27H40BR206	620.42	618.1192	52.27	6.50	25.76	15.47
C27H40BR207	636.42	634.1142	50.96	6.34	25.11	17.60
C27H40BR208	652.42	650.1091	49.71	6.18	24.49	19.62
C27H40BR209	668.42	666.1040	48.52	6.03	23.91	21.54
C27H41BR101	461.53	460.2341	70.27	8.95	17.31	3.47
C27H41BR102	477.53	476.2290	67.91	8.65	16.73	6.70
C27H41BR103	493.53	492.2239	65.71	8.37	16.19	9.73
C27H41BR104	509.53	508.2189	63.65	8.11	15.68	12.56
C27H41BR105	525.53	524.2138	61.71	7.86	15.20	15.22
C27H41BR106	541.53	540.2087	59.89	7.63	14.76	17.73
C27H41BR107	557.53	556.2036	58.17	7.41	14.33	20.09
C27H41BR108	573.53	572.1985	56.55	7.21	13.93	22.32
C27H41BR109	589.53	588.1934	55.01	7.01	13.55	24.43
C27H42BR201	542.44	540.1603	59.79	7.80	29.46	2.95
C27H42BR202	558.44	556.1552	58.07	7.58	28.62	5.73
C27H42BR203	574.44	572.1502	56.45	7.37	27.82	8.36
C27H42BR204	590.44	588.1451	54.93	7.17	27.07	10.84
C27H42BR205	606.44	604.1400	53.48	6.98	26.35	13.19
C27H42BR206	622.44	620.1349	52.10	6.80	25.67	15.42
C27H42BR207	638.44	636.1298	50.80	6.63	25.03	17.54
C27H42BR208	654.44	652.1247	49.55	6.47	24.42	19.56
C27H42BR209	670.44	668.1196	48.37	6.31	23.84	21.48
C27H43BR101	463.55	462.2498	69.96	9.35	17.24	3.45
C27H43BR102	479.55	478.2447	67.63	9.04	16.66	6.67
C27H43BR103	495.55	494.2396	65.44	8.75	16.12	9.69
C27H43BR104	511.55	510.2345	63.40	8.47	15.62	12.51
C27H43BR105	527.54	526.2294	61.47	8.22	15.15	15.16
C27H43BR106	543.54	542.2243	59.66	7.97	14.70	17.66
C27H43BR107	559.54	558.2192	57.96	7.75	14.28	20.02
C27H43BR108	575.54	574.2142	56.35	7.53	13.88	22.24
C27H43BR109	591.54	590.2091	54.82	7.33	13.51	24.34
C27H44BR201	544.46	542.1760	59.56	8.15	29.35	2.94
C27H44BR202	560.46	558.1709	57.86	7.91	28.51	5.71
C27H44BR203	576.46	574.1658	56.26	7.69	27.72	8.33
C27H44BR204	592.46	590.1607	54.74	7.49	26.97	10.80
C27H44BR205	608.46	606.1556	53.30	7.29	26.26	13.15
C27H44BR206	624.46	622.1505	51.93	7.10	25.59	15.37
C27H44BR207	640.46	638.1455	50.64	6.92	24.95	17.49
C27H44BR208	656.45	654.1404	49.40	6.76	24.34	19.50
C27H44BR209	672.45	670.1353	48.23	6.60	23.76	21.41
C27H45BR101	465.56	464.2654	69.66	9.74	17.16	3.44
C27H45BR102	481.56	480.2603	67.34	9.42	16.59	6.64
C27H45BR103	497.56	496.2552	65.18	9.12	16.06	9.65
C27H45BR104	513.56	512.2502	63.15	8.83	15.56	12.46
C27H45BR105	529.56	528.2451	61.24	8.57	15.09	15.11
C27H45BR106	545.56	544.2400	59.44	8.31	14.65	17.60
C27H45BR107	561.56	560.2349	57.75	8.08	14.23	19.94
C27H45BR108	577.56	576.2298	56.15	7.85	13.83	22.16
C27H45BR109	593.56	592.2247	54.64	7.64	13.46	24.26
C27H46BR201	546.48	544.1916	59.34	8.48	29.24	2.93
C27H46BR202	562.47	560.1865	57.66	8.24	28.41	5.69
C27H46BR203	578.47	576.1815	56.06	8.02	27.63	8.30
C27H46BR204	594.47	592.1764	54.55	7.80	26.88	10.77
C27H46BR205	610.47	608.1713	53.12	7.60	26.18	13.10
C27H46BR206	626.47	624.1662	51.77	7.40	25.51	15.32
C27H46BR207	642.47	640.1611	50.48	7.22	24.87	17.43
C27H46BR208	658.47	656.1560	49.25	7.04	24.27	19.44
C27H46BR209	674.47	672.1509	48.08	6.87	23.69	21.35
C27H47BR101	467.58	466.2811	69.36	10.13	17.09	3.42
C27H47BR102	483.58	482.2760	67.06	9.80	16.52	6.62
C27H47BR103	499.58	498.2709	64.92	9.48	15.99	9.61
C27H47BR104	515.58	514.2658	62.90	9.19	15.50	12.41
C27H47BR105	531.58	530.2607	61.01	8.91	15.03	15.05
C27H47BR106	547.58	546.2556	59.22	8.65	14.59	17.53
C27H47BR107	563.58	562.2505	57.54	8.41	14.18	19.87
C27H47BR108	579.57	578.2455	55.95	8.17	13.79	22.08
C27H47BR109	595.57	594.2404	54.45	7.95	13.42	24.18
C27H48BR201	548.49	546.2073	59.13	8.82	29.14	2.92
C27H48BR202	564.49	562.2022	57.45	8.57	28.31	5.67
C27H48BR203	580.49	578.1971	55.87	8.33	27.53	8.27
C27H48BR204	596.49	594.1920	54.37	8.11	26.79	10.73
C27H48BR205	612.49	610.1869	52.95	7.90	26.09	13.06
C27H48BR206	628.49	626.1818	51.60	7.70	25.43	15.27
C27H48BR207	644.49	642.1768	50.32	7.51	24.80	17.38
C27H48BR208	660.49	658.1717	49.10	7.33	24.20	19.38
C27H48BR209	676.49	674.1666	47.94	7.15	23.62	21.29
C27H49BR101	469.59	468.2967	69.06	10.52	17.02	3.41
C27H49BR102	485.59	484.2916	66.78	10.17	16.45	6.59
C27H49BR103	501.59	500.2865	64.65	9.85	15.93	9.57
C27H49BR104	517.59	516.2815	62.66	9.54	15.44	12.36
C27H49BR105	533.59	532.2764	60.78	9.26	14.97	14.99
C27H49BR106	549.59	548.2713	59.01	8.99	14.54	17.47
C27H49BR107	565.59	564.2662	57.34	8.73	14.13	19.80
C27H49BR108	581.59	580.2611	55.76	8.49	13.74	22.01
C27H49BR109	597.59	596.2560	54.27	8.26	13.37	24.10
C27H50BR201	550.51	548.2229	58.91	9.15	29.03	2.91
C27H50BR202	566.51	564.2178	57.25	8.90	28.21	5.65
C27H50BR203	582.51	580.2128	55.67	8.65	27.43	8.24
C27H50BR204	598.51	596.2077	54.19	8.42	26.70	10.69
C27H50BR205	614.50	612.2026	52.77	8.20	26.01	13.02
C27H50BR206	630.50	628.1975	51.44	7.99	25.35	15.23
C27H50BR207	646.50	644.1924	50.16	7.80	24.72	17.32
C27H50BR208	662.50	660.1873	48.95	7.61	24.12	19.32
C27H50BR209	678.50	676.1822	47.80	7.43	23.55	21.22
C27H51BR101	471.61	470.3124	68.76	10.90	16.94	3.39
C27H51BR102	487.61	486.3073	66.51	10.54	16.39	6.56
C27H51BR103	503.61	502.3022	64.40	10.21	15.87	9.53
C27H51BR104	519.61	518.2971	62.41	9.89	15.38	12.32
C27H51BR105	535.61	534.2920	60.55	9.60	14.92	14.94
C27H51BR106	551.61	550.2869	58.79	9.32	14.49	17.40
C27H51BR107	567.61	566.2818	57.13	9.06	14.08	19.73
C27H51BR108	583.61	582.2768	55.57	8.81	13.69	21.93
C27H51BR109	599.61	598.2717	54.09	8.57	13.33	24.01
C27H52BR201	552.52	550.2386	58.69	9.49	28.92	2.90
C27H52BR202	568.52	566.2335	57.04	9.22	28.11	5.63
C27H52BR203	584.52	582.2284	55.48	8.97	27.34	8.21
C27H52BR204	600.52	598.2233	54.00	8.73	26.61	10.66

056

C27H52BR205	616.52	614.2182	52.60	8.50	25.92	12.98
C27H52BR206	632.52	630.2131	51.27	8.29	25.27	15.18
C27H52BR207	648.52	646.2081	50.01	8.08	24.64	17.27
C27H52BR208	664.52	662.2030	48.80	7.89	24.05	19.26
C27H52BR209	680.52	678.1979	47.66	7.70	23.48	21.16
C27H53BR101	473.63	472.3280	68.47	11.28	16.87	3.38
C27H53BR102	489.63	488.3229	66.23	10.91	16.32	6.54
C27H53BR103	505.63	504.3178	64.14	10.57	15.80	9.49
C27H53BR104	521.63	520.3128	62.17	10.24	15.32	12.27
C27H53BR105	537.62	536.3077	60.32	9.94	14.86	14.88
C27H53BR106	553.62	552.3026	58.58	9.65	14.43	17.34
C27H53BR107	569.62	568.2975	56.93	9.38	14.03	19.66
C27H53BR108	585.62	584.2924	55.38	9.12	13.64	21.86
C27H53BR109	601.62	600.2873	53.90	8.88	13.28	23.93
C27H54BR201	554.54	552.2542	58.48	9.82	28.82	2.89
C27H54BR202	570.54	568.2491	56.84	9.54	28.01	5.61
C27H54BR203	586.54	584.2441	55.29	9.28	27.25	8.18
C27H54BR204	602.54	600.2390	53.82	9.03	26.52	10.62
C27H54BR205	618.54	616.2339	52.43	8.80	25.84	12.93
C27H54BR206	634.54	632.2288	51.11	8.58	25.19	15.13
C27H54BR207	650.54	648.2237	49.85	8.37	24.57	17.22
C27H54BR208	666.53	664.2186	48.65	8.17	23.98	19.20
C27H54BR209	682.53	680.2135	47.51	7.97	23.41	21.10
C27H55BR101	475.64	474.3437	68.18	11.66	16.80	3.36
C27H55BR102	491.64	490.3386	65.96	11.28	16.25	6.51
C27H55BR103	507.64	506.3335	63.88	10.92	15.74	9.46
C27H55BR104	523.64	522.3284	61.93	10.59	15.26	12.22
C27H55BR105	539.64	538.3233	60.10	10.27	14.81	14.82
C27H55BR106	555.64	554.3182	58.37	9.98	14.38	17.28
C27H55BR107	571.64	570.3131	56.73	9.70	13.98	19.59
C27H55BR108	587.64	586.3081	55.19	9.43	13.60	21.78
C27H55BR109	603.64	602.3030	53.72	9.18	13.24	23.85
C28H14BR201	526.23	523.9412	63.91	2.68	30.37	3.04
C28H14BR202	542.23	539.9362	62.02	2.60	29.47	5.90
C28H14BR203	558.23	555.9311	60.25	2.53	28.63	8.60
C28H14BR204	574.23	571.9260	58.57	2.46	27.83	11.14
C28H14BR205	590.23	587.9209	56.98	2.39	27.08	13.55
C28H14BR206	606.23	603.9158	55.48	2.33	26.36	15.84
C28H14BR207	622.23	619.9107	54.05	2.27	25.68	18.00
C28H14BR208	638.23	635.9056	52.69	2.21	25.04	20.05
C28H14BR209	654.23	651.9006	51.41	2.16	24.43	22.01
C28H15BR101	447.34	446.0307	75.18	3.38	17.86	3.58
C28H15BR102	463.33	462.0256	72.59	3.26	17.25	6.91
C28H15BR103	479.33	478.0205	70.16	3.15	16.67	10.01
C28H15BR104	495.33	494.0154	67.90	3.05	16.13	12.92
C28H15BR105	511.33	510.0103	65.77	2.96	15.63	15.64
C28H15BR106	527.33	526.0052	63.78	2.87	15.15	18.20
C28H15BR107	543.33	542.0002	61.90	2.78	14.71	20.61
C28H15BR108	559.33	557.9951	60.13	2.70	14.29	22.88
C28H15BR109	575.33	573.9900	58.46	2.63	13.89	25.03
C28H16BR201	528.25	525.9569	63.67	3.05	30.25	3.03
C28H16BR202	544.25	541.9518	61.79	2.96	29.36	5.88
C28H16BR203	560.25	557.9467	60.03	2.88	28.52	8.57
C28H16BR204	576.25	573.9416	58.36	2.80	27.73	11.11
C28H16BR205	592.24	589.9365	56.79	2.72	26.98	13.51
C28H16BR206	608.24	605.9315	55.29	2.65	26.27	15.78
C28H16BR207	624.24	621.9264	53.88	2.58	25.60	17.94
C28H16BR208	640.24	637.9213	52.53	2.52	24.96	19.99
C28H16BR209	656.24	653.9162	51.25	2.46	24.35	21.94
C28H17BR101	449.35	448.0463	74.84	3.81	17.78	3.56
C28H17BR102	465.35	464.0412	72.27	3.68	17.17	6.88
C28H17BR103	481.35	480.0362	69.87	3.56	16.60	9.97
C28H17BR104	497.35	496.0311	67.62	3.45	16.07	12.87
C28H17BR105	513.35	512.0260	65.51	3.34	15.57	15.58
C28H17BR106	529.35	528.0209	63.53	3.24	15.09	18.13
C28H17BR107	545.35	544.0158	61.67	3.14	14.65	20.54
C28H17BR108	561.35	560.0107	59.91	3.05	14.23	22.80
C28H17BR109	577.35	576.0056	58.25	2.97	13.84	24.94
C28H18BR201	530.26	527.9725	63.42	3.42	30.14	3.02
C28H18BR202	546.26	543.9675	61.57	3.32	29.25	5.86
C28H18BR203	562.26	559.9624	59.81	3.23	28.42	8.54
C28H18BR204	578.26	575.9573	58.16	3.14	27.64	11.07
C28H18BR205	594.26	591.9522	56.59	3.05	26.89	13.46
C28H18BR206	610.26	607.9471	55.11	2.97	26.19	15.73
C28H18BR207	626.26	623.9420	53.70	2.90	25.52	17.88
C28H18BR208	642.26	639.9369	52.36	2.82	24.88	19.93
C28H18BR209	658.26	655.9318	51.09	2.76	24.28	21.88
C28H19BR101	451.37	450.0620	74.51	4.24	17.70	3.54
C28H19BR102	467.37	466.0569	71.96	4.10	17.10	6.85
C28H19BR103	483.37	482.0518	69.58	3.96	16.53	9.93
C28H19BR104	499.37	498.0467	67.35	3.84	16.00	12.82
C28H19BR105	515.36	514.0416	65.26	3.72	15.50	15.52
C28H19BR106	531.36	530.0365	63.29	3.60	15.04	18.07
C28H19BR107	547.36	546.0315	61.44	3.50	14.60	20.46
C28H19BR108	563.36	562.0264	59.70	3.40	14.18	22.72
C28H19BR109	579.36	578.0213	58.05	3.31	13.79	24.85
C28H20BR201	532.28	529.9882	63.18	3.79	30.02	3.01
C28H20BR202	548.28	545.9831	61.34	3.68	29.15	5.84
C28H20BR203	564.28	561.9780	59.60	3.57	28.32	8.51
C28H20BR204	580.28	577.9729	57.96	3.47	27.54	11.03
C28H20BR205	596.28	593.9678	56.40	3.38	26.80	13.42
C28H20BR206	612.28	609.9628	54.93	3.29	26.10	15.68
C28H20BR207	628.28	625.9577	53.53	3.21	25.44	17.83
C28H20BR208	644.27	641.9526	52.20	3.13	24.80	19.87
C28H20BR209	660.27	657.9475	50.94	3.05	24.20	21.81
C28H21BR101	453.38	452.0776	74.18	4.67	17.62	3.53
C28H21BR102	469.38	468.0725	71.65	4.51	17.02	6.82
C28H21BR103	485.38	484.0674	69.29	4.36	16.46	9.89
C28H21BR104	501.38	500.0624	67.08	4.22	15.94	12.76
C28H21BR105	517.38	516.0573	65.00	4.09	15.44	15.46
C28H21BR106	533.38	532.0522	63.05	3.97	14.98	18.00
C28H21BR107	549.38	548.0471	61.22	3.85	14.54	20.39
C28H21BR108	565.38	564.0420	59.48	3.74	14.13	22.64
C28H21BR109	581.38	580.0369	57.85	3.64	13.74	24.77
C28H22BR201	534.29	532.0038	62.95	4.15	29.91	2.99
C28H22BR202	550.29	547.9987	61.11	4.03	29.04	5.81
C28H22BR203	566.29	563.9937	59.39	3.92	28.22	8.48
C28H22BR204	582.29	579.9886	57.76	3.81	27.44	10.99
C28H22BR205	598.29	595.9835	56.21	3.71	26.71	13.37
C28H22BR206	614.29	611.9784	54.75	3.61	26.01	15.63
C28H22BR207	630.29	627.9733	53.36	3.52	25.35	17.77
C28H22BR208	646.29	643.9682	52.04	3.43	24.73	19.80
C28H22BR209	662.29	659.9631	50.78	3.35	24.13	21.74
C28H23BR101	455.40	454.0933	73.85	5.09	17.55	3.51
C28H23BR102	471.40	470.0882	71.34	4.92	16.95	6.79
C28H23BR103	487.40	486.0831	69.00	4.76	16.39	9.85
C28H23BR104	503.40	502.0780	66.81	4.61	15.87	12.71
C28H23BR105	519.40	518.0729	64.75	4.46	15.38	15.40
C28H23BR106	535.40	534.0678	62.82	4.33	14.92	17.93
C28H23BR107	551.40	550.0628	60.99	4.20	14.49	20.31

C28H23BR108	567.39	566.0577	59.27	4.09	14.08	22.56
C28H23BR109	583.39	582.0526	57.65	3.97	13.70	24.68
C28H24BR201	536.31	534.0195	62.71	4.51	29.80	2.98
C28H24BR202	552.31	550.0144	60.89	4.38	28.93	5.79
C28H24BR203	568.31	566.0093	59.18	4.26	28.12	8.45
C28H24BR204	584.31	582.0042	57.56	4.14	27.35	10.95
C28H24BR205	600.31	597.9991	56.02	4.03	26.62	13.33
C28H24BR206	616.31	613.9941	54.57	3.93	25.93	15.58
C28H24BR207	632.31	629.9890	53.19	3.83	25.27	17.71
C28H24BR208	648.31	645.9839	51.88	3.73	24.65	19.74
C28H24BR209	664.31	661.9788	50.63	3.64	24.06	21.68
C28H25BR101	457.41	456.1089	73.52	5.51	17.47	3.50
C28H25BR102	473.41	472.1038	71.04	5.32	16.88	6.76
C28H25BR103	489.41	488.0987	68.72	5.15	16.33	9.81
C28H25BR104	505.41	504.0937	66.54	4.99	15.81	12.66
C28H25BR105	521.41	520.0886	64.50	4.83	15.32	15.34
C28H25BR106	537.41	536.0835	62.58	4.69	14.87	17.86
C28H25BR107	553.41	552.0784	60.77	4.55	14.44	20.24
C28H25BR108	569.41	568.0733	59.06	4.43	14.03	22.48
C28H25BR109	585.41	584.0682	57.45	4.30	13.65	24.60
C28H26BR201	538.33	536.0351	62.47	4.87	29.69	2.97
C28H26BR202	554.33	552.0300	60.67	4.73	28.83	5.77
C28H26BR203	570.33	568.0250	58.97	4.60	28.02	8.42
C28H26BR204	586.33	584.0199	57.36	4.47	27.26	10.92
C28H26BR205	602.32	600.0148	55.84	4.35	26.53	13.28
C28H26BR206	618.32	616.0097	54.39	4.24	25.85	15.53
C28H26BR207	634.32	632.0046	53.02	4.13	25.19	17.66
C28H26BR208	650.32	647.9995	51.71	4.03	24.57	19.68
C28H26BR209	666.32	663.9944	50.47	3.93	23.98	21.61
C28H27BR101	459.43	458.1246	73.20	5.92	17.39	3.48
C28H27BR102	475.43	474.1195	70.74	5.72	16.81	6.73
C28H27BR103	491.43	490.1144	68.44	5.54	16.26	9.77
C28H27BR104	507.43	506.1093	66.28	5.36	15.75	12.61
C28H27BR105	523.43	522.1042	64.25	5.20	15.27	15.28
C28H27BR106	539.43	538.0991	62.35	5.05	14.81	17.80
C28H27BR107	555.43	554.0941	60.55	4.90	14.39	20.16
C28H27BR108	571.43	570.0890	58.85	4.76	13.98	22.40
C28H27BR109	587.43	586.0839	57.25	4.63	13.60	24.51
C28H28BR201	540.34	538.0508	62.24	5.22	29.58	2.96
C28H28BR202	556.34	554.0457	60.45	5.07	28.72	5.75
C28H28BR203	572.34	570.0406	58.76	4.93	27.92	8.39
C28H28BR204	588.34	586.0355	57.16	4.80	27.16	10.88
C28H28BR205	604.34	602.0304	55.65	4.67	26.44	13.24
C28H28BR206	620.34	618.0254	54.21	4.55	25.76	15.47
C28H28BR207	636.34	634.0203	52.85	4.44	25.11	17.60
C28H28BR208	652.34	650.0152	51.55	4.33	24.50	19.62
C28H28BR209	668.34	666.0101	50.32	4.22	23.91	21.55
C28H29BR101	461.45	460.1402	72.88	6.33	17.32	3.47
C28H29BR102	477.45	476.1351	70.44	6.12	16.74	6.70
C28H29BR103	493.45	492.1300	68.16	5.92	16.19	9.73
C28H29BR104	509.44	508.1250	66.02	5.74	15.68	12.56
C28H29BR105	525.44	524.1199	64.01	5.56	15.21	15.22
C28H29BR106	541.44	540.1148	62.11	5.40	14.76	17.73
C28H29BR107	557.44	556.1097	60.33	5.24	14.33	20.09
C28H29BR108	573.44	572.1046	58.65	5.10	13.93	22.32
C28H29BR109	589.44	588.0995	57.06	4.96	13.56	24.43
C28H30BR201	542.36	540.0664	62.01	5.58	29.47	2.95
C28H30BR202	558.36	556.0613	60.23	5.42	28.62	5.73
C28H30BR203	574.36	572.0563	58.55	5.26	27.82	8.36
C28H30BR204	590.36	588.0512	56.97	5.12	27.07	10.84
C28H30BR205	606.36	604.0461	55.46	4.99	26.36	13.19
C28H30BR206	622.36	620.0410	54.04	4.86	25.68	15.42
C28H30BR207	638.36	636.0359	52.68	4.74	25.03	17.54
C28H30BR208	654.35	652.0308	51.40	4.62	24.42	19.56
C28H30BR209	670.35	668.0257	50.17	4.51	23.84	21.48
C28H31BR101	463.46	462.1559	72.57	6.74	17.24	3.45
C28H31BR102	479.46	478.1508	70.14	6.52	16.67	6.67
C28H31BR103	495.46	494.1457	67.88	6.31	16.13	9.69
C28H31BR104	511.46	510.1406	65.76	6.11	15.62	12.51
C28H31BR105	527.46	526.1355	63.76	5.92	15.15	15.17
C28H31BR106	543.46	542.1304	61.88	5.75	14.70	17.66
C28H31BR107	559.46	558.1254	60.11	5.59	14.28	20.02
C28H31BR108	575.46	574.1203	58.44	5.43	13.89	22.24
C28H31BR109	591.46	590.1152	56.86	5.28	13.51	24.35
C28H32BR201	544.37	542.0821	61.78	5.93	29.36	2.94
C28H32BR202	560.37	558.0770	60.02	5.76	28.52	5.71
C28H32BR203	576.37	574.0719	58.35	5.60	27.73	8.33
C28H32BR204	592.37	590.0668	56.77	5.45	26.98	10.80
C28H32BR205	608.37	606.0617	55.28	5.30	26.27	13.15
C28H32BR206	624.37	622.0567	53.86	5.17	25.60	15.37
C28H32BR207	640.37	638.0516	52.52	5.04	24.96	17.49
C28H32BR208	656.37	654.0465	51.24	4.91	24.35	19.50
C28H32BR209	672.37	670.0414	50.02	4.80	23.77	21.42
C28H33BR101	465.48	464.1715	72.25	7.15	17.17	3.44
C28H33BR102	481.48	480.1664	69.85	6.91	16.60	6.65
C28H33BR103	497.48	496.1613	67.60	6.69	16.06	9.65
C28H33BR104	513.48	512.1563	65.50	6.48	15.56	12.46
C28H33BR105	529.48	528.1512	63.52	6.28	15.09	15.11
C28H33BR106	545.48	544.1461	61.65	6.10	14.65	17.60
C28H33BR107	561.48	560.1410	59.90	5.92	14.23	19.95
C28H33BR108	577.47	576.1359	58.24	5.76	13.84	22.16
C28H33BR109	593.47	592.1308	56.67	5.60	13.46	24.26
C28H34BR201	546.39	544.0977	61.55	6.27	29.25	2.93
C28H34BR202	562.39	560.0926	59.80	6.09	28.42	5.69
C28H34BR203	578.39	576.0876	58.15	5.93	27.63	8.30
C28H34BR204	594.39	592.0825	56.58	5.77	26.89	10.77
C28H34BR205	610.39	608.0774	55.10	5.61	26.18	13.11
C28H34BR206	626.39	624.0723	53.69	5.47	25.51	15.33
C28H34BR207	642.39	640.0672	52.35	5.33	24.88	17.43
C28H34BR208	658.39	656.0621	51.08	5.21	24.27	19.44
C28H34BR209	674.39	672.0570	49.87	5.08	23.70	21.35
C28H35BR101	467.49	466.1872	71.94	7.55	17.09	3.42
C28H35BR102	483.49	482.1821	69.56	7.30	16.53	6.62
C28H35BR103	499.49	498.1770	67.33	7.06	16.00	9.61
C28H35BR104	515.49	514.1719	65.24	6.84	15.50	12.41
C28H35BR105	531.49	530.1668	63.28	6.64	15.03	15.05
C28H35BR106	547.49	546.1617	61.43	6.44	14.59	17.53
C28H35BR107	563.49	562.1566	59.68	6.26	14.18	19.88
C28H35BR108	579.49	578.1516	58.04	6.09	13.79	22.09
C28H35BR109	595.49	594.1465	56.48	5.92	13.42	24.18
C28H36BR201	548.41	546.1134	61.33	6.62	29.14	2.92
C28H36BR202	564.41	562.1083	59.59	6.43	28.31	5.67
C28H36BR203	580.41	578.1032	57.94	6.25	27.53	8.27
C28H36BR204	596.40	594.0981	56.39	6.08	26.80	10.73
C28H36BR205	612.40	610.0930	54.92	5.93	26.10	13.06
C28H36BR206	628.40	626.0880	53.52	5.77	25.43	15.28
C28H36BR207	644.40	642.0829	52.19	5.63	24.80	17.38
C28H36BR208	660.40	658.0778	50.93	5.49	24.20	19.38
C28H36BR209	676.40	674.0727	49.72	5.36	23.63	21.29
C28H37BR101	469.51	468.2028	71.63	7.94	17.02	3.41

058

C28H37BR102	485.51	484.1977	69.27	7.68	16.46	6.59
C28H37BR103	501.51	500.1926	67.06	7.44	15.93	9.57
C28H37BR104	517.51	516.1876	64.95	7.21	15.44	12.37
C28H37BR105	533.51	532.1825	63.04	6.99	14.98	14.99
C28H37BR106	549.51	548.1774	61.20	6.79	14.54	17.47
C28H37BR107	565.51	564.1723	59.47	6.59	14.13	19.80
C28H37BR108	581.51	580.1672	57.83	6.41	13.74	22.01
C28H37BR109	597.51	596.1621	56.29	6.24	13.37	24.10
C28H38BR201	550.42	548.1290	61.10	6.96	29.03	2.91
C28H38BR202	566.42	564.1239	59.37	6.76	28.21	5.65
C28H38BR203	582.42	580.1189	57.74	6.58	27.44	8.24
C28H38BR204	598.42	596.1138	56.20	6.40	26.70	10.69
C28H38BR205	614.42	612.1087	54.74	6.23	26.01	13.02
C28H38BR206	630.42	628.1036	53.35	6.08	25.35	15.23
C28H38BR207	646.42	644.0985	52.03	5.93	24.72	17.33
C28H38BR208	662.42	660.0934	50.77	5.78	24.12	19.32
C28H38BR209	678.42	676.0883	49.57	5.65	23.56	21.23
C28H39BR101	471.53	470.2185	71.32	8.34	16.95	3.39
C28H39BR102	487.53	486.2134	68.98	8.06	16.39	6.56
C28H39BR103	503.53	502.2083	66.79	7.81	15.87	9.53
C28H39BR104	519.52	518.2032	64.73	7.57	15.38	12.32
C28H39BR105	535.52	534.1981	62.80	7.34	14.92	14.94
C28H39BR106	551.52	550.1930	60.98	7.13	14.49	17.41
C28H39BR107	567.52	566.1879	59.26	6.93	14.08	19.73
C28H39BR108	583.52	582.1829	57.63	6.74	13.69	21.93
C28H39BR109	599.52	598.1778	56.10	6.56	13.33	24.02
C28H40BR201	552.44	550.1447	60.88	7.30	28.93	2.90
C28H40BR202	568.44	566.1396	59.16	7.09	28.11	5.63
C28H40BR203	584.44	582.1345	57.54	6.90	27.34	8.21
C28H40BR204	600.44	598.1294	56.01	6.71	26.62	10.66
C28H40BR205	616.44	614.1243	54.56	6.54	25.92	12.98
C28H40BR206	632.44	630.1192	53.18	6.38	25.27	15.18
C28H40BR207	648.43	646.1142	51.87	6.22	24.65	17.27
C28H40BR208	664.43	662.1091	50.62	6.07	24.05	19.26
C28H40BR209	680.43	678.1040	49.43	5.93	23.49	21.16
C28H41BR101	473.54	472.2341	71.02	8.73	16.87	3.38
C28H41BR102	489.54	488.2290	68.70	8.44	16.32	6.54
C28H41BR103	505.54	504.2239	66.53	8.17	15.81	9.49
C28H41BR104	521.54	520.2189	64.48	7.92	15.32	12.27
C28H41BR105	537.54	536.2138	62.57	7.69	14.86	14.88
C28H41BR106	553.54	552.2087	60.76	7.47	14.44	17.34
C28H41BR107	569.54	568.2036	59.05	7.26	14.03	19.66
C28H41BR108	585.54	584.1985	57.44	7.06	13.65	21.86
C28H41BR109	601.54	600.1934	55.91	6.87	13.28	23.94
C28H42BR201	554.45	552.1603	60.66	7.64	28.82	2.89
C28H42BR202	570.45	568.1552	58.96	7.42	28.01	5.61
C28H42BR203	586.45	584.1502	57.35	7.22	27.25	8.18
C28H42BR204	602.45	600.1451	55.82	7.03	26.53	10.62
C28H42BR205	618.45	616.1400	54.38	6.85	25.84	12.94
C28H42BR206	634.45	632.1349	53.01	6.67	25.19	15.13
C28H42BR207	650.45	648.1298	51.70	6.51	24.57	17.22
C28H42BR208	666.45	664.1247	50.46	6.35	23.98	19.21
C28H42BR209	682.45	680.1196	49.28	6.20	23.42	21.10
C28H43BR101	475.56	474.2498	70.72	9.11	16.80	3.36
C28H43BR102	491.56	490.2447	68.42	8.82	16.26	6.51
C28H43BR103	507.56	506.2396	66.26	8.54	15.74	9.46
C28H43BR104	523.56	522.2345	64.24	8.28	15.26	12.22
C28H43BR105	539.56	538.2294	62.33	8.03	14.81	14.83
C28H43BR106	555.56	554.2243	60.54	7.80	14.38	17.28
C28H43BR107	571.55	570.2192	58.84	7.58	13.98	19.59
C28H43BR108	587.55	586.2142	57.24	7.38	13.60	21.78
C28H43BR109	603.55	602.2091	55.72	7.18	13.24	23.86
C28H44BR201	556.47	554.1760	60.44	7.97	28.72	2.88
C28H44BR202	572.47	570.1709	58.75	7.75	27.92	5.59
C28H44BR203	588.47	586.1658	57.15	7.54	27.16	8.16
C28H44BR204	604.47	602.1607	55.64	7.34	26.44	10.59
C28H44BR205	620.47	618.1556	54.20	7.15	25.76	12.89
C28H44BR206	636.47	634.1505	52.84	6.97	25.11	15.08
C28H44BR207	652.47	650.1455	51.54	6.80	24.49	17.16
C28H44BR208	668.47	666.1404	50.31	6.63	23.91	19.15
C28H44BR209	684.47	682.1353	49.14	6.48	23.35	21.04
C28H45BR101	477.57	476.2654	70.42	9.50	16.73	3.35
C28H45BR102	493.57	492.2603	68.14	9.19	16.19	6.48
C28H45BR103	509.57	508.2552	66.00	8.90	15.68	9.42
C28H45BR104	525.57	524.2502	63.99	8.63	15.20	12.18
C28H45BR105	541.57	540.2451	62.10	8.38	14.75	14.77
C28H45BR106	557.57	556.2400	60.32	8.14	14.33	17.22
C28H45BR107	573.57	572.2349	58.63	7.91	13.93	19.53
C28H45BR108	589.57	588.2298	57.04	7.69	13.55	21.71
C28H45BR109	605.57	604.2247	55.54	7.49	13.19	23.78
C28H46BR201	558.49	556.1916	60.22	8.30	28.61	2.86
C28H46BR202	574.49	572.1865	58.54	8.07	27.82	5.57
C28H46BR203	590.49	588.1815	56.96	7.85	27.06	8.13
C28H46BR204	606.48	604.1764	55.45	7.65	26.35	10.55
C28H46BR205	622.48	620.1713	54.03	7.45	25.67	12.85
C28H46BR206	638.48	636.1662	52.67	7.26	25.03	15.04
C28H46BR207	654.48	652.1611	51.39	7.08	24.42	17.11
C28H46BR208	670.48	668.1560	50.16	6.92	23.83	19.09
C28H46BR209	686.48	684.1509	48.99	6.75	23.28	20.98
C28H47BR101	479.59	478.2811	70.12	9.88	16.66	3.34
C28H47BR102	495.59	494.2760	67.86	9.56	16.12	6.46
C28H47BR103	511.59	510.2709	65.74	9.26	15.62	9.38
C28H47BR104	527.59	526.2658	63.75	8.98	15.15	12.13
C28H47BR105	543.59	542.2607	61.87	8.72	14.70	14.72
C28H47BR106	559.59	558.2556	60.10	8.47	14.28	17.15
C28H47BR107	575.59	574.2505	58.43	8.23	13.88	19.46
C28H47BR108	591.59	590.2455	56.85	8.01	13.51	21.64
C28H47BR109	607.59	606.2404	55.35	7.80	13.15	23.70
C28H48BR201	560.50	558.2073	60.00	8.63	28.51	2.85
C28H48BR202	576.50	574.2022	58.34	8.39	27.72	5.55
C28H48BR203	592.50	590.1971	56.76	8.17	26.97	8.10
C28H48BR204	608.50	606.1920	55.27	7.95	26.26	10.52
C28H48BR205	624.50	622.1869	53.85	7.75	25.59	12.81
C28H48BR206	640.50	638.1818	52.51	7.55	24.95	14.99
C28H48BR207	656.50	654.1768	51.23	7.37	24.34	17.06
C28H48BR208	672.50	670.1717	50.01	7.19	23.76	19.03
C28H48BR209	688.50	686.1666	48.85	7.03	23.21	20.91
C28H49BR101	481.61	480.2967	69.83	10.26	16.59	3.32
C28H49BR102	497.61	496.2916	67.59	9.93	16.06	6.43
C28H49BR103	513.60	512.2865	65.48	9.62	15.56	9.35
C28H49BR104	529.60	528.2815	63.50	9.33	15.09	12.08
C28H49BR105	545.60	544.2764	61.64	9.05	14.65	14.66
C28H49BR106	561.60	560.2713	59.88	8.79	14.23	17.09
C28H49BR107	577.60	576.2662	58.23	8.55	13.83	19.39
C28H49BR108	593.60	592.2611	56.66	8.32	13.46	21.56
C28H49BR109	609.60	608.2560	55.17	8.10	13.11	23.62
C28H50BR201	562.52	560.2229	59.79	8.96	28.41	2.84
C28H50BR202	578.52	576.2178	58.13	8.71	27.62	5.53
C28H50BR203	594.52	592.2128	56.57	8.48	26.88	8.07
C28H50BR204	610.52	608.2077	55.09	8.26	26.18	10.48

059

C28H50BR205	626.52	624.2026	53.68	8.04	25.51	12.77
C28H50BR206	642.52	640.1975	52.34	7.84	24.87	14.94
C28H50BR207	658.51	656.1924	51.07	7.65	24.27	17.01
C28H50BR208	674.51	672.1873	49.86	7.47	23.69	18.98
C28H50BR209	690.51	688.1822	48.70	7.30	23.14	20.85
C28H51BR101	483.62	482.3124	69.54	10.63	16.52	3.31
C28H51BR102	499.62	498.3073	67.31	10.29	15.99	6.40
C28H51BR103	515.62	514.3022	65.22	9.97	15.50	9.31
C28H51BR104	531.62	530.2971	63.26	9.67	15.03	12.04
C28H51BR105	547.62	546.2920	61.41	9.39	14.59	14.61
C28H51BR106	563.62	562.2869	59.67	9.12	14.18	17.03
C28H51BR107	579.62	578.2818	58.02	8.87	13.79	19.32
C28H51BR108	595.62	594.2768	56.46	8.63	13.42	21.49
C28H51BR109	611.62	610.2717	54.99	8.41	13.06	23.54
C28H52BR201	564.53	562.2386	59.57	9.28	28.31	2.83
C28H52BR202	580.53	578.2335	57.93	9.03	27.53	5.51
C28H52BR203	596.53	594.2284	56.38	8.79	26.79	8.05
C28H52BR204	612.53	610.2233	54.91	8.56	26.09	10.45
C28H52BR205	628.53	626.2182	53.51	8.34	25.43	12.73
C28H52BR206	644.53	642.2131	52.18	8.13	24.79	14.89
C28H52BR207	660.53	658.2081	50.92	7.94	24.19	16.96
C28H52BR208	676.53	674.2030	49.71	7.75	23.62	18.92
C28H52BR209	692.53	690.1979	48.56	7.57	23.08	20.79
C28H53BR101	485.64	484.3280	69.25	11.00	16.45	3.29
C28H53BR102	501.64	500.3229	67.04	10.65	15.93	6.38
C28H53BR103	517.64	516.3178	64.97	10.32	15.44	9.27
C28H53BR104	533.64	532.3128	63.02	10.01	14.97	11.99
C28H53BR105	549.64	548.3077	61.19	9.72	14.54	14.55
C28H53BR106	565.64	564.3026	59.46	9.44	14.13	16.97
C28H53BR107	581.63	580.2975	57.82	9.18	13.74	19.26
C28H53BR108	597.63	596.2924	56.27	8.94	13.37	21.42
C28H53BR109	613.63	612.2873	54.81	8.71	13.02	23.47
C28H54BR201	566.55	564.2542	59.36	9.61	28.21	2.82
C28H54BR202	582.55	580.2491	57.73	9.34	27.43	5.49
C28H54BR203	598.55	596.2441	56.19	9.09	26.70	8.02
C28H54BR204	614.55	612.2390	54.73	8.86	26.00	10.41
C28H54BR205	630.55	628.2339	53.34	8.63	25.34	12.69
C28H54BR206	646.55	644.2288	52.02	8.42	24.72	14.85
C28H54BR207	662.55	660.2237	50.76	8.22	24.12	16.90
C28H54BR208	678.55	676.2186	49.56	8.02	23.55	18.86
C28H54BR209	694.55	692.2135	48.42	7.84	23.01	20.73
C28H55BR101	487.65	486.3437	68.97	11.37	16.39	3.28
C28H55BR102	503.65	502.3386	66.77	11.01	15.86	6.35
C28H55BR103	519.65	518.3335	64.72	10.67	15.38	9.24
C28H55BR104	535.65	534.3284	62.79	10.35	14.92	11.95
C28H55BR105	551.65	550.3233	60.96	10.05	14.48	14.50
C28H55BR106	567.65	566.3182	59.25	9.77	14.08	16.91
C28H55BR107	583.65	582.3131	57.62	9.50	13.69	19.19
C28H55BR108	599.65	598.3081	56.08	9.25	13.33	21.34
C28H55BR109	615.65	614.3030	54.63	9.00	12.98	23.39
C28H56BR201	568.57	566.2699	59.15	9.93	28.11	2.81
C28H56BR202	584.57	582.2648	57.53	9.66	27.34	5.47
C28H56BR203	600.56	598.2597	56.00	9.40	26.61	7.99
C28H56BR204	616.56	614.2546	54.55	9.15	25.92	10.38
C28H56BR205	632.56	630.2495	53.17	8.92	25.26	12.65
C28H56BR206	648.56	646.2444	51.85	8.70	24.64	14.80
C28H56BR207	664.56	662.2394	50.61	8.49	24.05	16.85
C28H56BR208	680.56	678.2343	49.42	8.29	23.48	18.81
C28H56BR209	696.56	694.2292	48.28	8.10	22.94	20.67
C28H57BR101	489.67	488.3593	68.68	11.73	16.32	3.27
C28H57BR102	505.67	504.3542	66.51	11.36	15.80	6.33
C28H57BR103	521.67	520.3491	64.47	11.01	15.32	9.20
C28H57BR104	537.67	536.3440	62.55	10.69	14.86	11.90
C28H57BR105	553.67	552.3390	60.74	10.38	14.43	14.45
C28H57BR106	569.67	568.3339	59.04	10.09	14.03	16.85
C28H57BR107	585.67	584.3288	57.42	9.81	13.64	19.12
C28H57BR108	601.67	600.3237	55.90	9.55	13.28	21.27
C28H57BR109	617.67	616.3186	54.45	9.30	12.94	23.31
C29H14BR201	538.24	535.9412	64.71	2.62	29.69	2.97
C29H14BR202	554.24	551.9362	62.85	2.55	28.83	5.77
C29H14BR203	570.24	567.9311	61.08	2.47	28.02	8.42
C29H14BR204	586.24	583.9260	59.42	2.41	27.26	10.92
C29H14BR205	602.24	599.9209	57.84	2.34	26.54	13.28
C29H14BR206	618.24	615.9158	56.34	2.28	25.85	15.53
C29H14BR207	634.24	631.9107	54.92	2.22	25.20	17.66
C29H14BR208	650.24	647.9056	53.57	2.17	24.58	19.68
C29H14BR209	666.24	663.9006	52.28	2.12	23.99	21.61
C29H15BR101	459.35	458.0307	75.83	3.29	17.40	3.48
C29H15BR102	475.35	474.0256	73.28	3.18	16.81	6.73
C29H15BR103	491.35	490.0205	70.89	3.08	16.26	9.77
C29H15BR104	507.34	506.0154	68.66	2.98	15.75	12.61
C29H15BR105	523.34	522.0103	66.56	2.89	15.27	15.29
C29H15BR106	539.34	538.0052	64.58	2.80	14.82	17.80
C29H15BR107	555.34	554.0002	62.72	2.72	14.39	20.17
C29H15BR108	571.34	569.9951	60.97	2.65	13.99	22.40
C29H15BR109	587.34	585.9900	59.31	2.57	13.60	24.52
C29H16BR201	540.26	537.9569	64.47	2.99	29.58	2.96
C29H16BR202	556.26	553.9518	62.62	2.90	28.73	5.75
C29H16BR203	572.26	569.9467	60.87	2.82	27.93	8.39
C29H16BR204	588.26	585.9416	59.21	2.74	27.17	10.88
C29H16BR205	604.26	601.9365	57.65	2.67	26.45	13.24
C29H16BR206	620.26	617.9315	56.16	2.60	25.76	15.48
C29H16BR207	636.25	633.9264	54.75	2.53	25.12	17.60
C29H16BR208	652.25	649.9213	53.40	2.47	24.50	19.62
C29H16BR209	668.25	665.9162	52.12	2.41	23.91	21.55
C29H17BR101	461.36	460.0463	75.50	3.71	17.32	3.47
C29H17BR102	477.36	476.0412	72.97	3.59	16.74	6.70
C29H17BR103	493.36	492.0362	70.60	3.47	16.20	9.73
C29H17BR104	509.36	508.0311	68.38	3.36	15.69	12.56
C29H17BR105	525.36	524.0260	66.30	3.26	15.21	15.23
C29H17BR106	541.36	540.0209	64.34	3.17	14.76	17.73
C29H17BR107	557.36	556.0158	62.50	3.07	14.34	20.09
C29H17BR108	573.36	572.0107	60.75	2.99	13.94	22.32
C29H17BR109	589.36	588.0056	59.10	2.91	13.56	24.43
C29H18BR201	542.27	539.9725	64.23	3.35	29.47	2.95
C29H18BR202	558.27	555.9675	62.39	3.25	28.63	5.73
C29H18BR203	574.27	571.9624	60.65	3.16	27.83	8.36
C29H18BR204	590.27	587.9573	59.01	3.07	27.07	10.84
C29H18BR205	606.27	603.9522	57.45	2.99	26.36	13.19
C29H18BR206	622.27	619.9471	55.98	2.92	25.68	15.43
C29H18BR207	638.27	635.9420	54.57	2.84	25.04	17.55
C29H18BR208	654.27	651.9369	53.24	2.77	24.43	19.56
C29H18BR209	670.27	667.9318	51.97	2.71	23.84	21.48
C29H19BR101	463.38	462.0620	75.17	4.13	17.24	3.45
C29H19BR102	479.38	478.0569	72.66	4.00	16.67	6.68
C29H19BR103	495.38	494.0518	70.31	3.87	16.13	9.69
C29H19BR104	511.38	510.0467	68.11	3.75	15.63	12.51
C29H19BR105	527.38	526.0416	66.05	3.63	15.15	15.17
C29H19BR106	543.38	542.0365	64.10	3.52	14.71	17.67
C29H19BR107	559.37	558.0315	62.27	3.42	14.28	20.02

060

C29H19BR108	575.37	574.0264	60.54	3.33	13.89	22.25
C29H19BR109	591.37	590.0213	58.90	3.24	13.51	24.35
C29H20BR201	544.29	541.9882	64.00	3.70	29.36	2.94
C29H20BR202	560.29	557.9831	62.17	3.60	28.52	5.71
C29H20BR203	576.29	573.9780	60.44	3.50	27.73	8.33
C29H20BR204	592.29	589.9729	58.81	3.40	26.98	10.81
C29H20BR205	608.29	605.9678	57.26	3.31	26.27	13.15
C29H20BR206	624.29	621.9628	55.80	3.23	25.60	15.38
C29H20BR207	640.29	637.9577	54.40	3.15	24.96	17.49
C29H20BR208	656.29	653.9526	53.07	3.07	24.35	19.50
C29H20BR209	672.29	669.9475	51.81	3.00	23.77	21.42
C29H21BR101	465.39	464.0776	74.84	4.55	17.17	3.44
C29H21BR102	481.39	480.0725	72.36	4.40	16.60	6.65
C29H21BR103	497.39	496.0674	70.03	4.26	16.06	9.65
C29H21BR104	513.39	512.0624	67.85	4.12	15.56	12.47
C29H21BR105	529.39	528.0573	65.80	4.00	15.09	15.11
C29H21BR106	545.39	544.0522	63.87	3.88	14.65	17.60
C29H21BR107	561.39	560.0471	62.05	3.77	14.23	19.95
C29H21BR108	577.39	576.0420	60.33	3.67	13.84	22.17
C29H21BR109	593.39	592.0369	58.70	3.57	13.47	24.27
C29H22BR201	546.31	544.0038	63.76	4.06	29.25	2.93
C29H22BR202	562.31	559.9987	61.95	3.94	28.42	5.69
C29H22BR203	578.30	575.9937	60.23	3.83	27.63	8.30
C29H22BR204	594.30	591.9886	58.61	3.73	26.89	10.77
C29H22BR205	610.30	607.9835	57.07	3.63	26.18	13.11
C29H22BR206	626.30	623.9784	55.62	3.54	25.52	15.33
C29H22BR207	642.30	639.9733	54.23	3.45	24.88	17.44
C29H22BR208	658.30	655.9682	52.91	3.37	24.28	19.44
C29H22BR209	674.30	671.9631	51.66	3.29	23.70	21.35
C29H23BR101	467.41	466.0933	74.52	4.96	17.10	3.42
C29H23BR102	483.41	482.0882	72.06	4.80	16.53	6.62
C29H23BR103	499.41	498.0831	69.75	4.64	16.00	9.61
C29H23BR104	515.41	514.0780	67.58	4.50	15.50	12.42
C29H23BR105	531.41	530.0729	65.55	4.36	15.04	15.05
C29H23BR106	547.41	546.0678	63.63	4.24	14.60	17.54
C29H23BR107	563.41	562.0628	61.82	4.11	14.18	19.88
C29H23BR108	579.41	578.0577	60.12	4.00	13.79	22.09
C29H23BR109	595.41	594.0526	58.50	3.89	13.42	24.18
C29H24BR201	548.32	546.0195	63.53	4.41	29.14	2.92
C29H24BR202	564.32	562.0144	61.72	4.29	28.32	5.67
C29H24BR203	580.32	578.0093	60.02	4.17	27.54	8.27
C29H24BR204	596.32	594.0042	58.41	4.06	26.80	10.73
C29H24BR205	612.32	609.9991	56.89	3.95	26.10	13.06
C29H24BR206	628.32	625.9941	55.44	3.85	25.43	15.28
C29H24BR207	644.32	641.9890	54.06	3.75	24.80	17.38
C29H24BR208	660.32	657.9839	52.75	3.66	24.20	19.38
C29H24BR209	676.32	673.9788	51.50	3.58	23.63	21.29
C29H25BR101	469.43	468.1089	74.20	5.37	17.02	3.41
C29H25BR102	485.43	484.1038	71.76	5.19	16.46	6.59
C29H25BR103	501.42	500.0987	69.47	5.03	15.94	9.57
C29H25BR104	517.42	516.0937	67.32	4.87	15.44	12.37
C29H25BR105	533.42	532.0886	65.30	4.72	14.98	15.00
C29H25BR106	549.42	548.0835	63.40	4.59	14.54	17.47
C29H25BR107	565.42	564.0784	61.60	4.46	14.13	19.81
C29H25BR108	581.42	580.0733	59.91	4.33	13.74	22.01
C29H25BR109	597.42	596.0682	58.30	4.22	13.37	24.10
C29H26BR201	550.34	548.0351	63.29	4.76	29.04	2.91
C29H26BR202	566.34	564.0300	61.50	4.63	28.22	5.65
C29H26BR203	582.34	580.0250	59.81	4.50	27.44	8.24
C29H26BR204	598.34	596.0199	58.22	4.38	26.71	10.70
C29H26BR205	614.34	612.0148	56.70	4.27	26.01	13.02
C29H26BR206	630.33	628.0097	55.26	4.16	25.35	15.23
C29H26BR207	646.33	644.0046	53.89	4.05	24.73	17.33
C29H26BR208	662.33	659.9995	52.59	3.96	24.13	19.32
C29H26BR209	678.33	675.9944	51.35	3.86	23.56	21.23
C29H27BR101	471.44	470.1246	73.88	5.77	16.95	3.39
C29H27BR102	487.44	486.1195	71.46	5.58	16.39	6.56
C29H27BR103	503.44	502.1144	69.19	5.41	15.87	9.53
C29H27BR104	519.44	518.1093	67.06	5.24	15.38	12.32
C29H27BR105	535.44	534.1042	65.05	5.08	14.92	14.94
C29H27BR106	551.44	550.0991	63.17	4.94	14.49	17.41
C29H27BR107	567.44	566.0941	61.39	4.80	14.08	19.74
C29H27BR108	583.44	582.0890	59.70	4.66	13.70	21.94
C29H27BR109	599.44	598.0839	58.11	4.54	13.33	24.02
C29H28BR201	552.35	550.0508	63.06	5.11	28.93	2.90
C29H28BR202	568.35	566.0457	61.29	4.97	28.12	5.63
C29H28BR203	584.35	582.0406	59.61	4.83	27.35	8.21
C29H28BR204	600.35	598.0355	58.02	4.70	26.62	10.66
C29H28BR205	616.35	614.0304	56.51	4.58	25.93	12.98
C29H28BR206	632.35	630.0254	55.08	4.46	25.27	15.18
C29H28BR207	648.35	646.0203	53.72	4.35	24.65	17.27
C29H28BR208	664.35	662.0152	52.43	4.25	24.05	19.27
C29H28BR209	680.35	678.0101	51.20	4.15	23.49	21.16
C29H29BR101	473.46	472.1402	73.57	6.17	16.88	3.38
C29H29BR102	489.46	488.1351	71.17	5.97	16.33	6.54
C29H29BR103	505.46	504.1300	68.91	5.78	15.81	9.50
C29H29BR104	521.46	520.1250	66.80	5.61	15.32	12.27
C29H29BR105	537.46	536.1199	64.81	5.44	14.87	14.88
C29H29BR106	553.45	552.1148	62.94	5.28	14.44	17.34
C29H29BR107	569.45	568.1097	61.17	5.13	14.03	19.67
C29H29BR108	585.45	584.1046	59.50	4.99	13.65	21.86
C29H29BR109	601.45	600.0995	57.91	4.86	13.29	23.94
C29H30BR201	554.37	552.0664	62.83	5.45	28.83	2.89
C29H30BR202	570.37	568.0613	61.07	5.30	28.02	5.61
C29H30BR203	586.37	584.0563	59.40	5.16	27.25	8.19
C29H30BR204	602.37	600.0512	57.83	5.02	26.53	10.62
C29H30BR205	618.37	616.0461	56.33	4.89	25.84	12.94
C29H30BR206	634.37	632.0410	54.91	4.77	25.19	15.13
C29H30BR207	650.37	648.0359	53.56	4.65	24.57	17.22
C29H30BR208	666.37	664.0308	52.27	4.54	23.98	19.21
C29H30BR209	682.37	680.0257	51.05	4.43	23.42	21.10
C29H31BR101	475.47	474.1559	73.26	6.57	16.81	3.36
C29H31BR102	491.47	490.1508	70.87	6.36	16.26	6.51
C29H31BR103	507.47	506.1457	68.64	6.16	15.75	9.46
C29H31BR104	523.47	522.1406	66.54	5.97	15.26	12.23
C29H31BR105	539.47	538.1355	64.57	5.79	14.81	14.83
C29H31BR106	555.47	554.1304	62.71	5.63	14.38	17.28
C29H31BR107	571.47	570.1254	60.95	5.47	13.98	19.60
C29H31BR108	587.47	586.1203	59.29	5.32	13.60	21.79
C29H31BR109	603.47	602.1152	57.72	5.18	13.24	23.86
C29H32BR201	556.39	554.0821	62.60	5.80	28.72	2.88
C29H32BR202	572.39	570.0770	60.85	5.64	27.92	5.59
C29H32BR203	588.38	586.0719	59.20	5.48	27.16	8.16
C29H32BR204	604.38	602.0668	57.63	5.34	26.44	10.59
C29H32BR205	620.38	618.0617	56.15	5.20	25.76	12.89
C29H32BR206	636.38	634.0567	54.73	5.07	25.11	15.08
C29H32BR207	652.38	650.0516	53.39	4.94	24.50	17.17
C29H32BR208	668.38	666.0465	52.11	4.83	23.91	19.15
C29H32BR209	684.38	682.0414	50.90	4.71	23.35	21.04
C29H33BR101	477.49	476.1715	72.95	6.97	16.73	3.35

C29H33BR102	493.49	492.1664	70.58	6.74	16.19	6.48
C29H33BR103	509.49	508.1613	68.37	6.53	15.68	9.42
C29H33BR104	525.49	524.1563	66.29	6.33	15.21	12.18
C29H33BR105	541.49	540.1512	64.33	6.14	14.76	14.77
C29H33BR106	557.49	556.1461	62.48	5.97	14.33	17.22
C29H33BR107	573.49	572.1410	60.74	5.80	13.93	19.53
C29H33BR108	589.49	588.1359	59.09	5.64	13.55	21.71
C29H33BR109	605.48	604.1308	57.53	5.49	13.20	23.78
C29H34BR201	558.40	556.0977	62.38	6.14	28.62	2.87
C29H34BR202	574.40	572.0926	60.64	5.97	27.82	5.57
C29H34BR203	590.40	588.0876	59.00	5.80	27.07	8.13
C29H34BR204	606.40	604.0825	57.44	5.65	26.35	10.55
C29H34BR205	622.40	620.0774	55.96	5.51	25.68	12.85
C29H34BR206	638.40	636.0723	54.56	5.37	25.03	15.04
C29H34BR207	654.40	652.0672	53.23	5.24	24.42	17.11
C29H34BR208	670.40	668.0621	51.96	5.11	23.84	19.09
C29H34BR209	686.40	684.0570	50.75	4.99	23.28	20.98
C29H35BR101	479.51	478.1872	72.64	7.36	16.66	3.34
C29H35BR102	495.51	494.1821	70.30	7.12	16.13	6.46
C29H35BR103	511.50	510.1770	68.10	6.90	15.62	9.38
C29H35BR104	527.50	526.1719	66.03	6.69	15.15	12.13
C29H35BR105	543.50	542.1668	64.09	6.49	14.70	14.72
C29H35BR106	559.50	558.1617	62.26	6.31	14.28	17.16
C29H35BR107	575.50	574.1566	60.53	6.13	13.88	19.46
C29H35BR108	591.50	590.1516	58.89	5.96	13.51	21.64
C29H35BR109	607.50	606.1465	57.34	5.81	13.15	23.70
C29H36BR201	560.42	558.1134	62.15	6.47	28.52	2.85
C29H36BR202	576.42	574.1083	60.43	6.30	27.72	5.55
C29H36BR203	592.42	590.1032	58.80	6.13	26.98	8.10
C29H36BR204	608.42	606.0981	57.25	5.96	26.27	10.52
C29H36BR205	624.42	622.0930	55.78	5.81	25.59	12.81
C29H36BR206	640.41	638.0880	54.39	5.67	24.95	14.99
C29H36BR207	656.41	654.0829	53.06	5.53	24.35	17.06
C29H36BR208	672.41	670.0778	51.80	5.40	23.77	19.04
C29H36BR209	688.41	686.0727	50.60	5.27	23.21	20.92
C29H37BR101	481.52	480.2028	72.34	7.75	16.59	3.32
C29H37BR102	497.52	496.1977	70.01	7.50	16.06	6.43
C29H37BR103	513.52	512.1926	67.83	7.26	15.56	9.35
C29H37BR104	529.52	528.1876	65.78	7.04	15.09	12.09
C29H37BR105	545.52	544.1825	63.85	6.84	14.65	14.66
C29H37BR106	561.52	560.1774	62.03	6.64	14.23	17.10
C29H37BR107	577.52	576.1723	60.31	6.46	13.84	19.39
C29H37BR108	593.52	592.1672	58.69	6.28	13.46	21.57
C29H37BR109	609.52	608.1621	57.15	6.12	13.11	23.62
C29H38BR201	562.43	560.1290	61.93	6.81	28.41	2.84
C29H38BR202	578.43	576.1239	60.22	6.62	27.63	5.53
C29H38BR203	594.43	592.1189	58.60	6.44	26.88	8.07
C29H38BR204	610.43	608.1138	57.06	6.27	26.18	10.48
C29H38BR205	626.43	624.1087	55.60	6.11	25.51	12.77
C29H38BR206	642.43	640.1036	54.22	5.96	24.88	14.94
C29H38BR207	658.43	656.0985	52.90	5.82	24.27	17.01
C29H38BR208	674.43	672.0934	51.65	5.68	23.70	18.98
C29H38BR209	690.43	688.0883	50.45	5.55	23.15	20.86
C29H39BR101	483.54	482.2185	72.04	8.13	16.52	3.31
C29H39BR102	499.54	498.2134	69.73	7.87	16.00	6.41
C29H39BR103	515.54	514.2083	67.57	7.63	15.50	9.31
C29H39BR104	531.54	530.2032	65.53	7.40	15.03	12.04
C29H39BR105	547.54	546.1981	63.62	7.18	14.59	14.61
C29H39BR106	563.53	562.1930	61.81	6.98	14.18	17.03
C29H39BR107	579.53	578.1879	60.10	6.78	13.79	19.33
C29H39BR108	595.53	594.1829	58.49	6.60	13.42	21.49
C29H39BR109	611.53	610.1778	56.96	6.43	13.07	23.55
C29H40BR201	564.45	562.1447	61.71	7.14	28.31	2.83
C29H40BR202	580.45	578.1396	60.01	6.95	27.53	5.51
C29H40BR203	596.45	594.1345	58.40	6.76	26.79	8.05
C29H40BR204	612.45	610.1294	56.87	6.58	26.09	10.45
C29H40BR205	628.45	626.1243	55.43	6.42	25.43	12.73
C29H40BR206	644.45	642.1192	54.05	6.26	24.80	14.90
C29H40BR207	660.45	658.1142	52.74	6.10	24.20	16.96
C29H40BR208	676.45	674.1091	51.49	5.96	23.62	18.92
C29H40BR209	692.44	690.1040	50.30	5.82	23.08	20.80
C29H41BR101	485.55	484.2341	71.74	8.51	16.46	3.30
C29H41BR102	501.55	500.2290	69.45	8.24	15.93	6.38
C29H41BR103	517.55	516.2239	67.30	7.99	15.44	9.27
C29H41BR104	533.55	532.2189	65.28	7.75	14.98	11.99
C29H41BR105	549.55	548.2138	63.38	7.52	14.54	14.56
C29H41BR106	565.55	564.2087	61.59	7.31	14.13	16.97
C29H41BR107	581.55	580.2036	59.90	7.11	13.74	19.26
C29H41BR108	597.55	596.1985	58.29	6.92	13.37	21.42
C29H41BR109	613.55	612.1934	56.77	6.74	13.02	23.47
C29H42BR201	566.47	564.1603	61.49	7.47	28.21	2.82
C29H42BR202	582.46	580.1552	59.80	7.27	27.44	5.49
C29H42BR203	598.46	596.1502	58.20	7.07	26.70	8.02
C29H42BR204	614.46	612.1451	56.69	6.89	26.01	10.42
C29H42BR205	630.46	628.1400	55.25	6.71	25.35	12.69
C29H42BR206	646.46	644.1349	53.88	6.55	24.72	14.85
C29H42BR207	662.46	660.1298	52.58	6.39	24.12	16.91
C29H42BR208	678.46	676.1247	51.34	6.24	23.55	18.87
C29H42BR209	694.46	692.1196	50.16	6.10	23.01	20.73
C29H43BR101	487.57	486.2498	71.44	8.89	16.39	3.28
C29H43BR102	503.57	502.2447	69.17	8.61	15.87	6.35
C29H43BR103	519.57	518.2396	67.04	8.34	15.38	9.24
C29H43BR104	535.57	534.2345	65.04	8.09	14.92	11.95
C29H43BR105	551.57	550.2294	63.15	7.86	14.49	14.50
C29H43BR106	567.57	566.2243	61.37	7.64	14.08	16.91
C29H43BR107	583.57	582.2192	59.69	7.43	13.69	19.19
C29H43BR108	599.57	598.2142	58.10	7.23	13.33	21.35
C29H43BR109	615.56	614.2091	56.59	7.04	12.98	23.39
C29H44BR201	568.48	566.1760	61.27	7.80	28.11	2.81
C29H44BR202	584.48	582.1709	59.60	7.59	27.34	5.47
C29H44BR203	600.48	598.1658	58.01	7.39	26.61	7.99
C29H44BR204	616.48	614.1607	56.50	7.19	25.92	10.38
C29H44BR205	632.48	630.1556	55.07	7.01	25.27	12.65
C29H44BR206	648.48	646.1505	53.71	6.84	24.64	14.80
C29H44BR207	664.48	662.1455	52.42	6.67	24.05	16.85
C29H44BR208	680.48	678.1404	51.19	6.52	23.48	18.81
C29H44BR209	696.48	694.1353	50.01	6.37	22.95	20.67
C29H45BR101	489.59	488.2654	71.15	9.26	16.32	3.27
C29H45BR102	505.58	504.2603	68.90	8.97	15.80	6.33
C29H45BR103	521.58	520.2552	66.78	8.70	15.32	9.20
C29H45BR104	537.58	536.2502	64.79	8.44	14.86	11.90
C29H45BR105	553.58	552.2451	62.92	8.19	14.43	14.45
C29H45BR106	569.58	568.2400	61.15	7.96	14.03	16.85
C29H45BR107	585.58	584.2349	59.48	7.75	13.65	19.13
C29H45BR108	601.58	600.2298	57.90	7.54	13.28	21.28
C29H45BR109	617.58	616.2247	56.40	7.34	12.94	23.32
C29H46BR201	570.50	568.1916	61.06	8.13	28.01	2.80
C29H46BR202	586.50	584.1865	59.39	7.91	27.25	5.46
C29H46BR203	602.50	600.1815	57.81	7.70	26.52	7.97
C29H46BR204	618.50	616.1764	56.32	7.50	25.84	10.35

062

C29H46BR205	634.49	632.1713	54.90	7.31	25.19	12.61
C29H46BR206	650.49	648.1662	53.55	7.13	24.57	14.76
C29H46BR207	666.49	664.1611	52.26	6.96	23.98	16.80
C29H46BR208	682.49	680.1560	51.04	6.79	23.42	18.75
C29H46BR209	698.49	696.1509	49.87	6.64	22.88	20.62
C29H47BR101	491.60	490.2811	70.85	9.64	16.25	3.25
C29H47BR102	507.60	506.2760	68.62	9.33	15.74	6.30
C29H47BR103	523.60	522.2709	66.52	9.05	15.26	9.17
C29H47BR104	539.60	538.2658	64.55	8.78	14.81	11.86
C29H47BR105	555.60	554.2607	62.69	8.53	14.38	14.40
C29H47BR106	571.60	570.2556	60.94	8.29	13.98	16.79
C29H47BR107	587.60	586.2505	59.28	8.06	13.60	19.06
C29H47BR108	603.60	602.2455	57.71	7.85	13.24	21.21
C29H47BR109	619.60	618.2404	56.22	7.65	12.90	23.24
C29H48BR201	572.51	570.2073	60.84	8.45	27.91	2.79
C29H48BR202	588.51	586.2022	59.19	8.22	27.15	5.44
C29H48BR203	604.51	602.1971	57.62	8.00	26.44	7.94
C29H48BR204	620.51	618.1920	56.13	7.80	25.75	10.31
C29H48BR205	636.51	634.1869	54.72	7.60	25.11	12.57
C29H48BR206	652.51	650.1818	53.38	7.41	24.49	14.71
C29H48BR207	668.51	666.1768	52.10	7.24	23.91	16.75
C29H48BR208	684.51	682.1717	50.89	7.07	23.35	18.70
C29H48BR209	700.51	698.1666	49.72	6.91	22.81	20.56
C29H49BR101	493.62	492.2967	70.57	10.01	16.19	3.24
C29H49BR102	509.62	508.2916	68.35	9.69	15.68	6.28
C29H49BR103	525.62	524.2865	66.27	9.40	15.20	9.13
C29H49BR104	541.62	540.2815	64.31	9.12	14.75	11.82
C29H49BR105	557.61	556.2764	62.47	8.86	14.33	14.35
C29H49BR106	573.61	572.2713	60.72	8.61	13.93	16.74
C29H49BR107	589.61	588.2662	59.08	8.38	13.55	18.99
C29H49BR108	605.61	604.2611	57.52	8.16	13.19	21.13
C29H49BR109	621.61	620.2560	56.04	7.95	12.85	23.16
C29H50BR201	574.53	572.2229	60.63	8.77	27.82	2.78
C29H50BR202	590.53	588.2178	58.99	8.53	27.06	5.42
C29H50BR203	606.53	604.2128	57.43	8.31	26.35	7.91
C29H50BR204	622.53	620.2077	55.95	8.10	25.67	10.28
C29H50BR205	638.53	636.2026	54.55	7.89	25.03	12.53
C29H50BR206	654.53	652.1975	53.22	7.70	24.42	14.67
C29H50BR207	670.53	668.1924	51.95	7.52	23.83	16.70
C29H50BR208	686.53	684.1873	50.74	7.34	23.28	18.64
C29H50BR209	702.52	700.1822	49.58	7.17	22.75	20.50
C29H51BR101	495.63	494.3124	70.28	10.37	16.12	3.23
C29H51BR102	511.63	510.3073	68.08	10.05	15.62	6.25
C29H51BR103	527.63	526.3022	66.02	9.74	15.14	9.10
C29H51BR104	543.63	542.2971	64.07	9.46	14.70	11.77
C29H51BR105	559.63	558.2920	62.24	9.19	14.28	14.29
C29H51BR106	575.63	574.2869	60.51	8.93	13.88	16.68
C29H51BR107	591.63	590.2818	58.88	8.69	13.51	18.93
C29H51BR108	607.63	606.2768	57.33	8.46	13.15	21.06
C29H51BR109	623.63	622.2717	55.85	8.24	12.81	23.09
C29H52BR201	576.55	574.2386	60.42	9.09	27.72	2.78
C29H52BR202	592.54	590.2335	58.78	8.85	26.97	5.40
C29H52BR203	608.54	606.2284	57.24	8.61	26.26	7.89
C29H52BR204	624.54	622.2233	55.77	8.39	25.59	10.25
C29H52BR205	640.54	638.2182	54.38	8.18	24.95	12.49
C29H52BR206	656.54	654.2131	53.05	7.98	24.34	14.62
C29H52BR207	672.54	670.2081	51.79	7.79	23.76	16.65
C29H52BR208	688.54	686.2030	50.59	7.61	23.21	18.59
C29H52BR209	704.54	702.1979	49.44	7.44	22.68	20.44
C29H53BR101	497.65	496.3280	69.99	10.73	16.06	3.21
C29H53BR102	513.65	512.3229	67.81	10.40	15.56	6.23
C29H53BR103	529.65	528.3178	65.77	10.09	15.09	9.06
C29H53BR104	545.65	544.3128	63.84	9.79	14.64	11.73
C29H53BR105	561.65	560.3077	62.02	9.51	14.23	14.24
C29H53BR106	577.65	576.3026	60.30	9.25	13.83	16.62
C29H53BR107	593.65	592.2975	58.68	9.00	13.46	18.87
C29H53BR108	609.64	608.2924	57.14	8.76	13.11	21.00
C29H53BR109	625.64	624.2873	55.67	8.54	12.77	23.02
C29H54BR201	578.56	576.2542	60.21	9.41	27.62	2.77
C29H54BR202	594.56	592.2491	58.59	9.15	26.88	5.38
C29H54BR203	610.56	608.2441	57.05	8.91	26.17	7.86
C29H54BR204	626.56	624.2390	55.59	8.69	25.51	10.21
C29H54BR205	642.56	640.2339	54.21	8.47	24.87	12.45
C29H54BR206	658.56	656.2288	52.89	8.27	24.27	14.58
C29H54BR207	674.56	672.2237	51.64	8.07	23.69	16.60
C29H54BR208	690.56	688.2186	50.44	7.88	23.14	18.54
C29H54BR209	706.56	704.2135	49.30	7.70	22.62	20.38
C29H55BR101	499.67	498.3437	69.71	11.10	15.99	3.20
C29H55BR102	515.66	514.3386	67.55	10.75	15.50	6.21
C29H55BR103	531.66	530.3335	65.52	10.43	15.03	9.03
C29H55BR104	547.66	546.3284	63.60	10.12	14.59	11.69
C29H55BR105	563.66	562.3233	61.80	9.84	14.18	14.19
C29H55BR106	579.66	578.3182	60.09	9.56	13.78	16.56
C29H55BR107	595.66	594.3131	58.48	9.31	13.41	18.80
C29H55BR108	611.66	610.3081	56.95	9.06	13.06	20.93
C29H55BR109	627.66	626.3030	55.50	8.83	12.73	22.94
C29H56BR201	580.58	578.2699	60.00	9.72	27.53	2.76
C29H56BR202	596.58	594.2648	58.39	9.46	26.79	5.36
C29H56BR203	612.58	610.2597	56.86	9.21	26.09	7.84
C29H56BR204	628.58	626.2546	55.41	8.98	25.42	10.18
C29H56BR205	644.57	642.2495	54.04	8.76	24.79	12.41
C29H56BR206	660.57	658.2444	52.73	8.55	24.19	14.53
C29H56BR207	676.57	674.2394	51.48	8.34	23.62	16.55
C29H56BR208	692.57	690.2343	50.29	8.15	23.07	18.48
C29H56BR209	708.57	706.2292	49.16	7.97	22.55	20.32
C29H57BR101	501.68	500.3593	69.43	11.45	15.93	3.19
C29H57BR102	517.68	516.3542	67.29	11.10	15.44	6.18
C29H57BR103	533.68	532.3491	65.27	10.77	14.97	8.99
C29H57BR104	549.68	548.3440	63.37	10.45	14.54	11.64
C29H57BR105	565.68	564.3390	61.58	10.16	14.13	14.14
C29H57BR106	581.68	580.3339	59.88	9.88	13.74	16.50
C29H57BR107	597.68	596.3288	58.28	9.61	13.37	18.74
C29H57BR108	613.68	612.3237	56.76	9.36	13.02	20.86
C29H57BR109	629.68	628.3186	55.32	9.12	12.69	22.87
C29H58BR201	582.59	580.2855	59.79	10.03	27.43	2.75
C29H58BR202	598.59	596.2804	58.19	9.77	26.70	5.35
C29H58BR203	614.59	612.2753	56.68	9.51	26.00	7.81
C29H58BR204	630.59	628.2703	55.24	9.27	25.34	10.15
C29H58BR205	646.59	644.2652	53.87	9.04	24.72	12.37
C29H58BR206	662.59	660.2601	52.57	8.82	24.12	14.49
C29H58BR207	678.59	676.2550	51.33	8.62	23.55	16.50
C29H58BR208	694.59	692.2499	50.15	8.42	23.01	18.43
C29H58BR209	710.59	708.2448	49.02	8.23	22.49	20.26
C29H59BR101	503.70	502.3750	69.15	11.81	15.86	3.18
C29H59BR102	519.70	518.3699	67.02	11.44	15.38	6.16
C29H59BR103	535.70	534.3648	65.02	11.10	14.92	8.96
C29H59BR104	551.70	550.3597	63.14	10.78	14.48	11.60
C29H59BR105	567.69	566.3546	61.36	10.48	14.08	14.09
C29H59BR106	583.69	582.3495	59.68	10.19	13.69	16.45
C29H59BR107	599.69	598.3444	58.08	9.92	13.32	18.68

063

C29H59BR108	615.69	614.3394	56.57	9.66	12.98	20.79
C29H59BR109	631.69	630.3343	55.14	9.41	12.65	22.80
C30H15BR101	471.36	470.0307	76.45	3.21	16.95	3.39
C30H15BR102	487.36	486.0256	73.94	3.10	16.40	6.57
C30H15BR103	503.36	502.0205	71.59	3.00	15.87	9.54
C30H15BR104	519.36	518.0154	69.38	2.91	15.39	12.32
C30H15BR105	535.36	534.0103	67.31	2.82	14.93	14.94
C30H15BR106	551.35	550.0052	65.35	2.74	14.49	17.41
C30H15BR107	567.35	566.0002	63.51	2.66	14.08	19.74
C30H15BR108	583.35	581.9951	61.77	2.59	13.70	21.94
C30H15BR109	599.35	597.9900	60.12	2.52	13.33	24.03
C30H16BR201	552.27	549.9569	65.25	2.92	28.94	2.90
C30H16BR202	568.27	565.9518	63.41	2.84	28.12	5.63
C30H16BR203	584.27	581.9467	61.67	2.76	27.35	8.22
C30H16BR204	600.27	597.9416	60.03	2.69	26.62	10.66
C30H16BR205	616.27	613.9365	58.47	2.62	25.93	12.98
C30H16BR206	632.27	629.9315	56.99	2.55	25.28	15.18
C30H16BR207	648.27	645.9264	55.58	2.49	24.65	17.28
C30H16BR208	664.27	661.9213	54.25	2.43	24.06	19.27
C30H16BR209	680.26	677.9162	52.97	2.37	23.49	21.17
C30H17BR101	473.37	472.0463	76.12	3.62	16.88	3.38
C30H17BR102	489.37	488.0412	73.63	3.50	16.33	6.54
C30H17BR103	505.37	504.0362	71.30	3.39	15.81	9.50
C30H17BR104	521.37	520.0311	69.11	3.29	15.33	12.27
C30H17BR105	537.37	536.0260	67.06	3.19	14.87	14.89
C30H17BR106	553.37	552.0209	65.12	3.10	14.44	17.35
C30H17BR107	569.37	568.0158	63.29	3.01	14.03	19.67
C30H17BR108	585.37	584.0107	61.56	2.93	13.65	21.87
C30H17BR109	601.37	600.0056	59.92	2.85	13.29	23.94
C30H18BR201	554.29	551.9725	65.01	3.27	28.83	2.89
C30H18BR202	570.28	567.9675	63.19	3.18	28.02	5.61
C30H18BR203	586.28	583.9624	61.46	3.09	27.26	8.19
C30H18BR204	602.28	599.9573	59.83	3.01	26.53	10.63
C30H18BR205	618.28	615.9522	58.28	2.93	25.85	12.94
C30H18BR206	634.28	631.9471	56.81	2.86	25.20	15.13
C30H18BR207	650.28	647.9420	55.41	2.79	24.58	17.22
C30H18BR208	666.28	663.9369	54.08	2.72	23.99	19.21
C30H18BR209	682.28	679.9318	52.81	2.66	23.42	21.10
C30H19BR101	475.39	474.0620	75.80	4.03	16.81	3.37
C30H19BR102	491.39	490.0569	73.33	3.90	16.26	6.51
C30H19BR103	507.39	506.0518	71.02	3.77	15.75	9.46
C30H19BR104	523.39	522.0467	68.85	3.66	15.27	12.23
C30H19BR105	539.39	538.0416	66.80	3.55	14.81	14.83
C30H19BR106	555.39	554.0365	64.88	3.45	14.39	17.28
C30H19BR107	571.39	570.0315	63.06	3.35	13.98	19.60
C30H19BR108	587.39	586.0264	61.35	3.26	13.60	21.79
C30H19BR109	603.38	602.0213	59.72	3.17	13.24	23.86
C30H20BR201	556.30	553.9882	64.77	3.62	28.73	2.88
C30H20BR202	572.30	569.9831	62.96	3.52	27.92	5.59
C30H20BR203	588.30	585.9780	61.25	3.43	27.16	8.16
C30H20BR204	604.30	601.9729	59.63	3.34	26.45	10.59
C30H20BR205	620.30	617.9678	58.09	3.25	25.76	12.90
C30H20BR206	636.30	633.9628	56.63	3.17	25.12	15.09
C30H20BR207	652.30	649.9577	55.24	3.09	24.50	17.17
C30H20BR208	668.30	665.9526	53.92	3.02	23.91	19.15
C30H20BR209	684.30	681.9475	52.66	2.95	23.35	21.04
C30H21BR101	477.41	476.0776	75.48	4.43	16.74	3.35
C30H21BR102	493.40	492.0725	73.03	4.29	16.19	6.49
C30H21BR103	509.40	508.0674	70.74	4.16	15.69	9.42
C30H21BR104	525.40	524.0624	68.58	4.03	15.21	12.18
C30H21BR105	541.40	540.0573	66.56	3.91	14.76	14.78
C30H21BR106	557.40	556.0522	64.65	3.80	14.34	17.22
C30H21BR107	573.40	572.0471	62.84	3.69	13.94	19.53
C30H21BR108	589.40	588.0420	61.14	3.59	13.56	21.72
C30H21BR109	605.40	604.0369	59.52	3.50	13.20	23.79
C30H22BR201	558.32	556.0038	64.54	3.97	28.62	2.87
C30H22BR202	574.32	571.9987	62.74	3.86	27.83	5.57
C30H22BR203	590.32	587.9937	61.04	3.76	27.07	8.13
C30H22BR204	606.32	603.9886	59.43	3.66	26.36	10.56
C30H22BR205	622.31	619.9835	57.90	3.56	25.68	12.85
C30H22BR206	638.31	635.9784	56.45	3.47	25.04	15.04
C30H22BR207	654.31	651.9733	55.07	3.39	24.42	17.12
C30H22BR208	670.31	667.9682	53.76	3.31	23.84	19.09
C30H22BR209	686.31	683.9631	52.50	3.23	23.29	20.98
C30H23BR101	479.42	478.0933	75.16	4.84	16.67	3.34
C30H23BR102	495.42	494.0882	72.73	4.68	16.13	6.46
C30H23BR103	511.42	510.0831	70.46	4.53	15.62	9.39
C30H23BR104	527.42	526.0780	68.32	4.40	15.15	12.13
C30H23BR105	543.42	542.0729	66.31	4.27	14.70	14.72
C30H23BR106	559.42	558.0678	64.41	4.14	14.28	17.16
C30H23BR107	575.42	574.0628	62.62	4.03	13.89	19.46
C30H23BR108	591.42	590.0577	60.93	3.92	13.51	21.64
C30H23BR109	607.42	606.0526	59.32	3.82	13.15	23.71
C30H24BR201	560.33	558.0195	64.31	4.32	28.52	2.86
C30H24BR202	576.33	574.0144	62.52	4.20	27.73	5.55
C30H24BR203	592.33	590.0093	60.83	4.08	26.98	8.10
C30H24BR204	608.33	606.0042	59.23	3.98	26.27	10.52
C30H24BR205	624.33	621.9991	57.72	3.87	25.60	12.81
C30H24BR206	640.33	637.9941	56.27	3.78	24.96	14.99
C30H24BR207	656.33	653.9890	54.90	3.69	24.35	17.06
C30H24BR208	672.33	669.9839	53.59	3.60	23.77	19.04
C30H24BR209	688.33	685.9788	52.35	3.51	23.22	20.92
C30H25BR101	481.44	480.1089	74.85	5.23	16.60	3.32
C30H25BR102	497.44	496.1038	72.44	5.07	16.06	6.43
C30H25BR103	513.44	512.0987	70.18	4.91	15.56	9.35
C30H25BR104	529.44	528.0937	68.06	4.76	15.09	12.09
C30H25BR105	545.43	544.0886	66.06	4.62	14.65	14.67
C30H25BR106	561.43	560.0835	64.18	4.49	14.23	17.10
C30H25BR107	577.43	576.0784	62.40	4.36	13.84	19.40
C30H25BR108	593.43	592.0733	60.72	4.25	13.46	21.57
C30H25BR109	609.43	608.0682	59.13	4.13	13.11	23.63
C30H26BR201	562.35	560.0351	64.08	4.66	28.42	2.85
C30H26BR202	578.35	576.0300	62.30	4.53	27.63	5.53
C30H26BR203	594.35	592.0250	60.63	4.41	26.89	8.08
C30H26BR204	610.35	608.0199	59.04	4.29	26.18	10.49
C30H26BR205	626.35	624.0148	57.53	4.18	25.51	12.77
C30H26BR206	642.35	640.0097	56.10	4.08	24.88	14.94
C30H26BR207	658.35	656.0046	54.73	3.98	24.27	17.01
C30H26BR208	674.34	671.9995	53.43	3.89	23.70	18.98
C30H26BR209	690.34	687.9944	52.20	3.80	23.15	20.86
C30H27BR101	483.45	482.1246	74.53	5.63	16.53	3.31
C30H27BR102	499.45	498.1195	72.15	5.45	16.00	6.41
C30H27BR103	515.45	514.1144	69.91	5.28	15.50	9.31
C30H27BR104	531.45	530.1093	67.80	5.12	15.04	12.04
C30H27BR105	547.45	546.1042	65.82	4.97	14.60	14.61
C30H27BR106	563.45	562.0991	63.95	4.83	14.18	17.04
C30H27BR107	579.45	578.0941	62.19	4.70	13.79	19.33
C30H27BR108	595.45	594.0890	60.51	4.57	13.42	21.50
C30H27BR109	611.45	610.0839	58.93	4.45	13.07	23.55
C30H28BR201	564.37	562.0508	63.85	5.00	28.32	2.83

064

C30H28BR202	580.36	578.0457	62.09	4.86	27.54	5.51
C30H28BR203	596.36	594.0406	60.42	4.73	26.80	8.05
C30H28BR204	612.36	610.0355	58.84	4.61	26.10	10.45
C30H28BR205	628.36	626.0304	57.34	4.49	25.43	12.73
C30H28BR206	644.36	642.0254	55.92	4.38	24.80	14.90
C30H28BR207	660.36	658.0203	54.57	4.27	24.20	16.96
C30H28BR208	676.36	674.0152	53.28	4.17	23.63	18.92
C30H28BR209	692.36	690.0101	52.04	4.08	23.08	20.80
C30H29BR101	485.47	484.1402	74.22	6.02	16.46	3.30
C30H29BR102	501.47	500.1351	71.86	5.83	15.93	6.38
C30H29BR103	517.47	516.1300	69.63	5.65	15.44	9.28
C30H29BR104	533.47	532.1250	67.55	5.48	14.98	12.00
C30H29BR105	549.47	548.1199	65.58	5.32	14.54	14.56
C30H29BR106	565.47	564.1148	63.72	5.17	14.13	16.98
C30H29BR107	581.47	580.1097	61.97	5.03	13.74	19.26
C30H29BR108	597.46	596.1046	60.31	4.89	13.37	21.42
C30H29BR109	613.46	612.0995	58.74	4.76	13.03	23.47
C30H30BR201	566.38	564.0664	63.62	5.34	28.22	2.82
C30H30BR202	582.38	580.0613	61.87	5.19	27.44	5.49
C30H30BR203	598.38	596.0563	60.22	5.05	26.71	8.02
C30H30BR204	614.38	612.0512	58.65	4.92	26.01	10.42
C30H30BR205	630.38	628.0461	57.16	4.80	25.35	12.69
C30H30BR206	646.38	644.0410	55.75	4.68	24.72	14.85
C30H30BR207	662.38	660.0359	54.40	4.57	24.13	16.91
C30H30BR208	678.38	676.0308	53.12	4.46	23.56	18.87
C30H30BR209	694.38	692.0257	51.89	4.35	23.01	20.74
C30H31BR101	487.48	486.1559	73.92	6.41	16.39	3.28
C30H31BR102	503.48	502.1508	71.57	6.21	15.87	6.36
C30H31BR103	519.48	518.1457	69.36	6.02	15.38	9.24
C30H31BR104	535.48	534.1406	67.29	5.84	14.92	11.95
C30H31BR105	551.48	550.1355	65.34	5.67	14.49	14.51
C30H31BR106	567.48	566.1304	63.50	5.51	14.08	16.92
C30H31BR107	583.48	582.1254	61.76	5.36	13.69	19.19
C30H31BR108	599.48	598.1203	60.11	5.21	13.33	21.35
C30H31BR109	615.48	614.1152	58.55	5.08	12.98	23.40
C30H32BR201	568.40	566.0821	63.39	5.67	28.12	2.81
C30H32BR202	584.40	582.0770	61.66	5.52	27.35	5.48
C30H32BR203	600.40	598.0719	60.02	5.37	26.62	7.99
C30H32BR204	616.40	614.0668	58.46	5.23	25.93	10.38
C30H32BR205	632.39	630.0617	56.98	5.10	25.27	12.65
C30H32BR206	648.39	646.0567	55.57	4.97	24.65	14.81
C30H32BR207	664.39	662.0516	54.24	4.85	24.05	16.86
C30H32BR208	680.39	678.0465	52.96	4.74	23.49	18.81
C30H32BR209	696.39	694.0414	51.74	4.63	22.95	20.68
C30H33BR101	489.50	488.1715	73.61	6.80	16.32	3.27
C30H33BR102	505.50	504.1664	71.28	6.58	15.81	6.33
C30H33BR103	521.50	520.1613	69.10	6.38	15.32	9.20
C30H33BR104	537.50	536.1563	67.04	6.19	14.87	11.91
C30H33BR105	553.50	552.1512	65.10	6.01	14.44	14.45
C30H33BR106	569.50	568.1461	63.27	5.84	14.03	16.86
C30H33BR107	585.50	584.1410	61.54	5.68	13.65	19.13
C30H33BR108	601.50	600.1359	59.91	5.53	13.28	21.28
C30H33BR109	617.50	616.1308	58.35	5.39	12.94	23.32
C30H34BR201	570.41	568.0977	63.17	6.01	28.02	2.80
C30H34BR202	586.41	584.0926	61.45	5.84	27.25	5.46
C30H34BR203	602.41	600.0876	59.82	5.69	26.53	7.97
C30H34BR204	618.41	616.0825	58.27	5.54	25.84	10.35
C30H34BR205	634.41	632.0774	56.80	5.40	25.19	12.61
C30H34BR206	650.41	648.0723	55.40	5.27	24.57	14.76
C30H34BR207	666.41	664.0672	54.07	5.14	23.98	16.81
C30H34BR208	682.41	680.0621	52.80	5.02	23.42	18.76
C30H34BR209	698.41	696.0570	51.59	4.91	22.88	20.62
C30H35BR101	491.52	490.1872	73.31	7.18	16.26	3.26
C30H35BR102	507.52	506.1821	71.00	6.95	15.74	6.30
C30H35BR103	523.52	522.1770	68.83	6.74	15.26	9.17
C30H35BR104	539.52	538.1719	66.79	6.54	14.81	11.86
C30H35BR105	555.51	554.1668	64.87	6.35	14.38	14.40
C30H35BR106	571.51	570.1617	63.05	6.17	13.98	16.80
C30H35BR107	587.51	586.1566	61.33	6.00	13.60	19.06
C30H35BR108	603.51	602.1516	59.71	5.85	13.24	21.21
C30H35BR109	619.51	618.1465	58.16	5.69	12.90	23.24
C30H36BR201	572.43	570.1134	62.95	6.34	27.92	2.80
C30H36BR202	588.43	586.1083	61.24	6.17	27.16	5.44
C30H36BR203	604.43	602.1032	59.62	6.00	26.44	7.94
C30H36BR204	620.43	618.0981	58.08	5.85	25.76	10.32
C30H36BR205	636.43	634.0930	56.62	5.70	25.11	12.57
C30H36BR206	652.43	650.0880	55.23	5.56	24.49	14.71
C30H36BR207	668.43	666.0829	53.91	5.43	23.91	16.76
C30H36BR208	684.42	682.0778	52.65	5.30	23.35	18.70
C30H36BR209	700.42	698.0727	51.45	5.18	22.82	20.56
C30H37BR101	493.53	492.2028	73.01	7.56	16.19	3.24
C30H37BR102	509.53	508.1977	70.72	7.32	15.68	6.28
C30H37BR103	525.53	524.1926	68.57	7.10	15.20	9.13
C30H37BR104	541.53	540.1876	66.54	6.89	14.76	11.82
C30H37BR105	557.53	556.1825	64.63	6.69	14.33	14.35
C30H37BR106	573.53	572.1774	62.83	6.50	13.93	16.74
C30H37BR107	589.53	588.1723	61.12	6.33	13.55	19.00
C30H37BR108	605.53	604.1672	59.51	6.16	13.20	21.14
C30H37BR109	621.53	620.1621	57.98	6.00	12.86	23.17
C30H38BR201	574.44	572.1290	62.73	6.67	27.82	2.79
C30H38BR202	590.44	588.1239	61.03	6.49	27.07	5.42
C30H38BR203	606.44	604.1189	59.42	6.32	26.35	7.91
C30H38BR204	622.44	620.1138	57.89	6.15	25.67	10.28
C30H38BR205	638.44	636.1087	56.44	6.00	25.03	12.53
C30H38BR206	654.44	652.1036	55.06	5.85	24.42	14.67
C30H38BR207	670.44	668.0985	53.75	5.71	23.84	16.70
C30H38BR208	686.44	684.0934	52.49	5.58	23.28	18.65
C30H38BR209	702.44	700.0883	51.30	5.45	22.75	20.50
C30H39BR101	495.55	494.2185	72.71	7.93	16.12	3.23
C30H39BR102	511.55	510.2134	70.44	7.68	15.62	6.26
C30H39BR103	527.55	526.2083	68.30	7.45	15.15	9.10
C30H39BR104	543.55	542.2032	66.29	7.23	14.70	11.77
C30H39BR105	559.55	558.1981	64.40	7.03	14.28	14.30
C30H39BR106	575.55	574.1930	62.61	6.83	13.88	16.68
C30H39BR107	591.55	590.1879	60.91	6.65	13.51	18.93
C30H39BR108	607.54	606.1829	59.31	6.47	13.15	21.07
C30H39BR109	623.54	622.1778	57.79	6.30	12.81	23.09
C30H40BR201	576.46	574.1447	62.51	6.99	27.72	2.78
C30H40BR202	592.46	590.1396	60.82	6.81	26.97	5.40
C30H40BR203	608.46	606.1345	59.22	6.63	26.26	7.89
C30H40BR204	624.46	622.1294	57.70	6.46	25.59	10.25
C30H40BR205	640.46	638.1243	56.26	6.30	24.95	12.49
C30H40BR206	656.46	654.1192	54.89	6.14	24.34	14.62
C30H40BR207	672.46	670.1142	53.58	6.00	23.76	16.65
C30H40BR208	688.46	686.1091	52.34	5.86	23.21	18.59
C30H40BR209	704.46	702.1040	51.15	5.72	22.69	20.44
C30H41BR101	497.56	496.2341	72.42	8.31	16.06	3.22
C30H41BR102	513.56	512.2290	70.16	8.05	15.56	6.23
C30H41BR103	529.56	528.2239	68.04	7.80	15.09	9.06
C30H41BR104	545.56	544.2189	66.05	7.58	14.65	11.73

C30H41BR105	561.56	560.2138	64.17	7.36	14.23	14.25
C30H41BR106	577.56	576.2087	62.39	7.16	13.83	16.62
C30H41BR107	593.56	592.2036	60.71	6.96	13.46	18.87
C30H41BR108	609.56	608.1985	59.11	6.78	13.11	21.00
C30H41BR109	625.56	624.1934	57.60	6.61	12.77	23.02
C30H42BR201	578.48	576.1603	62.29	7.32	27.63	2.77
C30H42BR202	594.48	592.1552	60.61	7.12	26.88	5.38
C30H42BR203	610.48	608.1502	59.03	6.93	26.18	7.86
C30H42BR204	626.47	624.1451	57.52	6.76	25.51	10.22
C30H42BR205	642.47	640.1400	56.09	6.59	24.87	12.45
C30H42BR206	658.47	656.1349	54.72	6.43	24.27	14.58
C30H42BR207	674.47	672.1298	53.42	6.28	23.69	16.60
C30H42BR208	690.47	688.1247	52.19	6.13	23.14	18.54
C30H42BR209	706.47	704.1196	51.00	5.99	22.62	20.38
C30H43BR101	499.58	498.2498	72.13	8.68	15.99	3.20
C30H43BR102	515.58	514.2447	69.89	8.41	15.50	6.21
C30H43BR103	531.58	530.2396	67.79	8.15	15.03	9.03
C30H43BR104	547.58	546.2345	65.81	7.92	14.59	11.69
C30H43BR105	563.58	562.2294	63.94	7.69	14.18	14.19
C30H43BR106	579.58	578.2243	62.17	7.48	13.79	16.56
C30H43BR107	595.58	594.2192	60.50	7.28	13.42	18.80
C30H43BR108	611.58	610.2142	58.92	7.09	13.07	20.93
C30H43BR109	627.58	626.2091	57.42	6.91	12.73	22.94
C30H44BR201	580.49	578.1760	62.07	7.64	27.53	2.76
C30H44BR202	596.49	594.1709	60.41	7.44	26.79	5.36
C30H44BR203	612.49	610.1658	58.83	7.24	26.09	7.84
C30H44BR204	628.49	626.1607	57.33	7.06	25.43	10.18
C30H44BR205	644.49	642.1556	55.91	6.88	24.80	12.41
C30H44BR206	660.49	658.1505	54.56	6.71	24.20	14.53
C30H44BR207	676.49	674.1455	53.27	6.56	23.62	16.56
C30H44BR208	692.49	690.1404	52.03	6.40	23.08	18.48
C30H44BR209	708.49	706.1353	50.86	6.26	22.56	20.32
C30H45BR101	501.60	500.2654	71.84	9.04	15.93	3.19
C30H45BR102	517.60	516.2603	69.62	8.76	15.44	6.18
C30H45BR103	533.60	532.2552	67.53	8.50	14.97	9.00
C30H45BR104	549.59	548.2502	65.56	8.25	14.54	11.64
C30H45BR105	565.59	564.2451	63.71	8.02	14.13	14.14
C30H45BR106	581.59	580.2400	61.96	7.80	13.74	16.51
C30H45BR107	597.59	596.2349	60.30	7.59	13.37	18.74
C30H45BR108	613.59	612.2298	58.73	7.39	13.02	20.86
C30H45BR109	629.59	628.2247	57.23	7.20	12.69	22.87
C30H46BR201	582.51	580.1916	61.86	7.96	27.43	2.75
C30H46BR202	598.51	596.1865	60.21	7.75	26.70	5.35
C30H46BR203	614.51	612.1815	58.64	7.55	26.01	7.81
C30H46BR204	630.51	628.1764	57.15	7.35	25.35	10.15
C30H46BR205	646.51	644.1713	55.74	7.17	24.72	12.37
C30H46BR206	662.51	660.1662	54.39	7.00	24.12	14.49
C30H46BR207	678.50	676.1611	53.11	6.83	23.55	16.51
C30H46BR208	694.50	692.1560	51.88	6.68	23.01	18.43
C30H46BR209	710.50	708.1509	50.72	6.53	22.49	20.27
C30H47BR101	503.61	502.2811	71.55	9.41	15.87	3.18
C30H47BR102	519.61	518.2760	69.35	9.12	15.38	6.16
C30H47BR103	535.61	534.2709	67.28	8.84	14.92	8.96
C30H47BR104	551.61	550.2658	65.32	8.59	14.49	11.60
C30H47BR105	567.61	566.2607	63.48	8.35	14.08	14.09
C30H47BR106	583.61	582.2556	61.74	8.12	13.69	16.45
C30H47BR107	599.61	598.2505	60.09	7.90	13.33	18.68
C30H47BR108	615.61	614.2455	58.53	7.70	12.98	20.79
C30H47BR109	631.61	630.2404	57.05	7.50	12.65	22.80
C30H48BR201	584.52	582.2073	61.65	8.28	27.34	2.74
C30H48BR202	600.52	598.2022	60.00	8.06	26.61	5.33
C30H48BR203	616.52	614.1971	58.45	7.85	25.92	7.79
C30H48BR204	632.52	630.1920	56.97	7.65	25.27	10.12
C30H48BR205	648.52	646.1869	55.56	7.46	24.64	12.34
C30H48BR206	664.52	662.1818	54.22	7.28	24.05	14.45
C30H48BR207	680.52	678.1768	52.95	7.11	23.48	16.46
C30H48BR208	696.52	694.1717	51.73	6.95	22.94	18.38
C30H48BR209	712.52	710.1666	50.57	6.79	22.43	20.21
C30H49BR101	505.63	504.2967	71.26	9.77	15.80	3.16
C30H49BR102	521.63	520.2916	69.08	9.47	15.32	6.13
C30H49BR103	537.63	536.2865	67.02	9.19	14.86	8.93
C30H49BR104	553.63	552.2815	65.09	8.92	14.43	11.56
C30H49BR105	569.63	568.2764	63.26	8.67	14.03	14.04
C30H49BR106	585.63	584.2713	61.53	8.43	13.64	16.39
C30H49BR107	601.62	600.2662	59.89	8.21	13.28	18.62
C30H49BR108	617.62	616.2611	58.34	8.00	12.94	20.72
C30H49BR109	633.62	632.2560	56.87	7.79	12.61	22.73
C30H50BR201	586.54	584.2229	61.43	8.59	27.25	2.73
C30H50BR202	602.54	600.2178	59.80	8.36	26.52	5.31
C30H50BR203	618.54	616.2128	58.26	8.15	25.84	7.76
C30H50BR204	634.54	632.2077	56.79	7.94	25.18	10.09
C30H50BR205	650.54	648.2026	55.39	7.75	24.57	12.30
C30H50BR206	666.54	664.1975	54.06	7.56	23.98	14.40
C30H50BR207	682.54	680.1924	52.79	7.38	23.41	16.41
C30H50BR208	698.54	696.1873	51.58	7.21	22.88	18.32
C30H50BR209	714.54	712.1822	50.43	7.05	22.37	20.15
C30H51BR101	507.64	506.3124	70.98	10.13	15.74	3.15
C30H51BR102	523.64	522.3073	68.81	9.82	15.26	6.11
C30H51BR103	539.64	538.3022	66.77	9.53	14.81	8.89
C30H51BR104	555.64	554.2971	64.85	9.25	14.38	11.52
C30H51BR105	571.64	570.2920	63.03	8.99	13.98	13.99
C30H51BR106	587.64	586.2869	61.32	8.75	13.60	16.34
C30H51BR107	603.64	602.2818	59.69	8.52	13.24	18.55
C30H51BR108	619.64	618.2768	58.15	8.30	12.90	20.66
C30H51BR109	635.64	634.2717	56.69	8.09	12.57	22.65
C30H52BR201	588.56	586.2386	61.22	8.91	27.15	2.72
C30H52BR202	604.56	602.2335	59.60	8.67	26.43	5.29
C30H52BR203	620.56	618.2284	58.07	8.45	25.75	7.73
C30H52BR204	636.55	634.2233	56.61	8.23	25.11	10.05
C30H52BR205	652.55	650.2182	55.22	8.03	24.49	12.26
C30H52BR206	668.55	666.2131	53.90	7.84	23.90	14.36
C30H52BR207	684.55	682.2081	52.64	7.66	23.34	16.36
C30H52BR208	700.55	698.2030	51.44	7.48	22.81	18.27
C30H52BR209	716.55	714.1979	50.29	7.31	22.30	20.10
C30H53BR101	509.66	508.3280	70.70	10.48	15.68	3.14
C30H53BR102	525.66	524.3229	68.55	10.16	15.20	6.09
C30H53BR103	541.66	540.3178	66.52	9.86	14.75	8.86
C30H53BR104	557.66	556.3128	64.62	9.58	14.33	11.48
C30H53BR105	573.66	572.3077	62.81	9.31	13.93	13.95
C30H53BR106	589.66	588.3026	61.11	9.06	13.55	16.28
C30H53BR107	605.66	604.2975	59.49	8.82	13.19	18.49
C30H53BR108	621.66	620.2924	57.96	8.59	12.85	20.59
C30H53BR109	637.66	636.2873	56.51	8.38	12.53	22.58
C30H54BR201	590.57	588.2542	61.01	9.22	27.06	2.71
C30H54BR202	606.57	604.2491	59.41	8.97	26.35	5.28
C30H54BR203	622.57	620.2441	57.88	8.74	25.67	7.71
C30H54BR204	638.57	636.2390	56.43	8.52	25.03	10.02
C30H54BR205	654.57	652.2339	55.05	8.32	24.41	12.22
C30H54BR206	670.57	668.2288	53.74	8.12	23.83	14.32
C30H54BR207	686.57	684.2237	52.48	7.93	23.28	16.31

066

C30H54BR208	702.57	700.2186	51.29	7.75	22.75	18.22
C30H54BR209	718.57	716.2135	50.15	7.57	22.24	20.04
C30H55BR101	511.68	510.3437	70.42	10.83	15.62	3.13
C30H55BR102	527.68	526.3386	68.29	10.51	15.14	6.06
C30H55BR103	543.68	542.3335	66.28	10.20	14.70	8.83
C30H55BR104	559.67	558.3284	64.38	9.91	14.28	11.43
C30H55BR105	575.67	574.3233	62.59	9.63	13.88	13.90
C30H55BR106	591.67	590.3182	60.90	9.37	13.50	16.22
C30H55BR107	607.67	606.3131	59.30	9.12	13.15	18.43
C30H55BR108	623.67	622.3081	57.78	8.89	12.81	20.52
C30H55BR109	639.67	638.3030	56.33	8.67	12.49	22.51
C30H56BR201	592.59	590.2699	60.81	9.53	26.97	2.70
C30H56BR202	608.59	606.2648	59.21	9.27	26.26	5.26
C30H56BR203	624.59	622.2597	57.69	9.04	25.59	7.68
C30H56BR204	640.59	638.2546	56.25	8.81	24.95	9.99
C30H56BR205	656.59	654.2495	54.88	8.60	24.34	12.18
C30H56BR206	672.59	670.2444	53.57	8.39	23.76	14.27
C30H56BR207	688.58	686.2394	52.33	8.20	23.21	16.26
C30H56BR208	704.58	702.2343	51.14	8.01	22.68	18.17
C30H56BR209	720.58	718.2292	50.01	7.83	22.18	19.98
C30H57BR101	513.69	512.3593	70.15	11.18	15.55	3.11
C30H57BR102	529.69	528.3542	68.03	10.85	15.09	6.04
C30H57BR103	545.69	544.3491	66.03	10.53	14.64	8.80
C30H57BR104	561.69	560.3440	64.15	10.23	14.23	11.39
C30H57BR105	577.69	576.3390	62.38	9.95	13.83	13.85
C30H57BR106	593.69	592.3339	60.69	9.68	13.46	16.17
C30H57BR107	609.69	608.3288	59.10	9.42	13.11	18.37
C30H57BR108	625.69	624.3237	57.59	9.18	12.77	20.46
C30H57BR109	641.69	640.3186	56.15	8.95	12.45	22.44
C30H58BR201	594.60	592.2855	60.60	9.83	26.88	2.69
C30H58BR202	610.60	608.2804	59.01	9.57	26.17	5.24
C30H58BR203	626.60	624.2753	57.51	9.33	25.50	7.66
C30H58BR204	642.60	640.2703	56.07	9.10	24.87	9.96
C30H58BR205	658.60	656.2652	54.71	8.88	24.26	12.15
C30H58BR206	674.60	672.2601	53.41	8.67	23.69	14.23
C30H58BR207	690.60	688.2550	52.18	8.47	23.14	16.22
C30H58BR208	706.60	704.2499	51.00	8.27	22.62	18.11
C30H58BR209	722.60	720.2448	49.87	8.09	22.12	19.93
C30H59BR101	515.71	514.3750	69.87	11.53	15.49	3.10
C30H59BR102	531.71	530.3699	67.77	11.18	15.03	6.02
C30H59BR103	547.71	546.3648	65.79	10.86	14.59	8.76
C30H59BR104	563.71	562.3597	63.92	10.55	14.17	11.35
C30H59BR105	579.71	578.3546	62.16	10.26	13.78	13.80
C30H59BR106	595.71	594.3495	60.49	9.98	13.41	16.11
C30H59BR107	611.70	610.3444	58.91	9.72	13.06	18.31
C30H59BR108	627.70	626.3394	57.41	9.47	12.73	20.39
C30H59BR109	643.70	642.3343	55.98	9.24	12.41	22.37
C30H60BR201	596.62	594.3012	60.40	10.14	26.79	2.68
C30H60BR202	612.62	610.2961	58.82	9.87	26.09	5.22
C30H60BR203	628.62	626.2910	57.32	9.62	25.42	7.64
C30H60BR204	644.62	642.2859	55.90	9.38	24.79	9.93
C30H60BR205	660.62	658.2808	54.55	9.15	24.19	12.11
C30H60BR206	676.62	674.2757	53.26	8.94	23.62	14.19
C30H60BR207	692.62	690.2707	52.03	8.73	23.07	16.17
C30H60BR208	708.62	706.2656	50.85	8.53	22.55	18.06
C30H60BR209	724.62	722.2605	49.73	8.35	22.05	19.87
C30H61BR101	517.72	516.3906	69.60	11.88	15.43	3.09
C30H61BR102	533.72	532.3855	67.51	11.52	14.97	6.00
C30H61BR103	549.72	548.3804	65.55	11.18	14.54	8.73
C30H61BR104	565.72	564.3753	63.69	10.87	14.12	11.31
C30H61BR105	581.72	580.3703	61.94	10.57	13.74	13.75
C30H61BR106	597.72	596.3652	60.28	10.29	13.37	16.06
C30H61BR107	613.72	612.3601	58.71	10.02	13.02	18.25
C30H61BR108	629.72	628.3550	57.22	9.76	12.69	20.33
C30H61BR109	645.72	644.3499	55.80	9.52	12.37	22.30

3. CARBON, HYDROGEN, CHLORINE AND NITROGEN

3. KOHLENSTOFF, WASSERSTOFF, CHLOR UND STICKSTOFF

3. CARBONE, HYDROGENE, CHLORE ET AZOTE

MOLECULAR WEIGHTS AND PERCENTAGE COMPOSITIONS OF COMPOUNDS
CONTAINING CARBON, HYDROGEN, CHLORINE
AND NITROGEN

FORMULA	MOL. WEIGHT	ISOMOL. WEIGHT	PERC. C	PERC. H	PERC. Cl	PERC. N
C 1H 0Cl1N1	61.47	60.9719	19.54	0	57.67	22.79
C 1H 0Cl1N3	89.48	88.9781	13.42	0	39.62	46.96
C 1H 0Cl1N5	117.50	116.9842	10.22	0	30.17	59.60
C 1H 0Cl2N2	110.93	109.9439	10.83	0	63.92	25.25
C 1H 0Cl2N4	138.94	137.9500	8.64	0	51.03	40.32
C 1H 0Cl2N6	166.96	165.9561	7.19	0	42.47	50.34
C 1H 1Cl1N2	76.49	75.9828	15.70	1.32	46.35	36.63
C 1H 1Cl1N4	104.50	103.9890	11.49	.96	33.93	53.61
C 1H 1Cl1N6	132.51	131.9951	9.06	.76	26.75	63.42
C 1H 1Cl2N1	97.93	96.9486	12.26	1.03	72.40	14.30
C 1H 1Cl2N3	125.95	124.9548	9.54	.80	56.30	33.36
C 1H 1Cl2N5	153.96	152.9609	7.80	.65	46.06	45.49
C 1H 2Cl1N1	63.49	62.9876	18.92	3.18	55.84	22.06
C 1H 2Cl1N3	91.50	90.9937	13.13	2.20	38.75	45.92
C 1H 2Cl1N5	119.51	118.9999	10.05	1.69	29.66	58.60
C 1H 2Cl2N2	112.95	111.9595	10.63	1.78	62.78	24.80
C 1H 2Cl2N4	140.96	139.9657	8.52	1.43	50.30	39.75
C 1H 2Cl2N6	168.97	167.9718	7.11	1.19	41.96	49.74
C 1H 3Cl1N2	78.50	77.9985	15.30	3.85	45.16	35.69
C 1H 3Cl1N4	106.51	106.0046	11.28	2.84	33.28	52.60
C 1H 3Cl1N6	134.53	134.0108	8.93	2.25	26.35	62.47
C 1H 3Cl2N1	99.95	98.9643	12.02	3.03	70.94	14.01
C 1H 3Cl2N3	127.96	126.9704	9.39	2.36	55.41	32.84
C 1H 3Cl2N5	155.97	154.9765	7.70	1.94	45.46	44.90
C 1H 4Cl1N1	65.50	65.0032	18.34	6.16	54.12	21.38
C 1H 4Cl1N3	93.52	93.0094	12.84	4.31	37.91	44.93
C 1H 4Cl1N5	121.53	121.0155	9.88	3.32	29.17	57.63
C 1H 4Cl2N2	114.96	113.9752	10.45	3.51	61.68	24.37
C 1H 4Cl2N4	142.98	141.9813	8.40	2.82	49.59	39.19
C 1H 4Cl2N6	170.99	169.9874	7.02	2.36	41.47	49.15
C 1H 5Cl1N2	80.52	80.0141	14.92	6.26	44.03	34.79
C 1H 5Cl1N4	108.53	108.0203	11.07	4.64	32.67	51.62
C 1H 5Cl1N6	136.54	136.0264	8.80	3.69	25.96	61.55
C 1H 5Cl2N3	129.98	128.9861	9.24	3.88	54.55	32.33
C 1H 5Cl2N5	157.99	156.9922	7.60	3.19	44.88	44.33
C 1H 6Cl1N3	95.53	95.0250	12.57	6.33	37.11	43.99
C 1H 6Cl1N5	123.55	123.0312	9.72	4.90	28.70	56.69
C 1H 6Cl2N4	144.99	143.9969	8.28	4.17	48.90	38.64
C 1H 6Cl2N6	173.01	172.0031	6.94	3.50	40.98	48.58
C 1H 7Cl1N4	110.55	110.0359	10.87	6.38	32.07	50.68
C 1H 7Cl1N6	138.56	138.0421	8.67	5.09	25.59	60.65
C 1H 7Cl2N5	160.01	159.0078	7.51	4.41	44.31	43.77
C 1H 8Cl1N5	125.56	125.0468	9.57	6.42	28.24	55.78
C 1H 8Cl2N6	175.02	174.0187	6.86	4.61	40.51	48.02
C 1H 9Cl1N6	140.58	140.0577	8.54	6.45	25.22	59.78
C 2H 1Cl1N2	88.50	87.9828	27.14	1.14	40.06	31.65
C 2H 1Cl1N4	116.51	115.9890	20.62	.87	30.43	48.09
C 2H 1Cl1N6	144.52	143.9951	16.62	.70	24.53	58.15
C 2H 1Cl2N1	109.94	108.9486	21.85	.92	64.49	12.74
C 2H 1Cl2N3	137.96	136.9548	17.41	.73	51.40	30.46

FORMULA	MOL. WEIGHT	ISOMOL. WEIGHT	PERC. C	PERC. H	PERC. Cl	PERC. N
C 2H 1Cl2N5	165.97	164.9609	14.47	.61	42.72	42.20
C 2H 2Cl1N1	75.50	74.9876	31.82	2.67	46.96	18.55
C 2H 2Cl1N3	103.51	102.9937	23.21	1.95	34.25	40.59
C 2H 2Cl1N5	131.52	130.9999	18.26	1.53	26.96	53.25
C 2H 2Cl2N2	124.96	123.9595	19.22	1.61	56.74	22.42
C 2H 2Cl2N4	152.97	151.9657	15.70	1.32	46.35	36.63
C 2H 2Cl2N6	180.98	179.9718	13.27	1.11	39.18	46.44
C 2H 3Cl1N2	90.51	89.9985	26.54	3.34	39.17	30.95
C 2H 3Cl1N4	118.53	118.0046	20.27	2.55	29.91	47.27
C 2H 3Cl1N6	146.54	146.0108	16.39	2.06	24.19	57.35
C 2H 3Cl2N1	111.96	110.9643	21.46	2.70	63.33	12.51
C 2H 3Cl2N3	139.97	138.9704	17.16	2.16	50.66	30.02
C 2H 3Cl2N5	167.99	166.9765	14.30	1.80	42.21	41.69
C 2H 4Cl1N1	77.51	77.0032	30.99	5.20	45.74	18.07
C 2H 4Cl1N3	105.53	105.0094	22.76	3.82	33.60	39.82
C 2H 4Cl1N5	133.54	133.0155	17.99	3.02	26.55	52.44
C 2H 4Cl2N2	126.97	125.9752	18.92	3.18	55.84	22.06
C 2H 4Cl2N4	154.99	153.9813	15.50	2.60	45.75	36.15
C 2H 4Cl2N6	183.00	181.9874	13.13	2.20	38.75	45.92
C 2H 5Cl1N2	92.53	92.0141	25.96	5.45	38.32	30.28
C 2H 5Cl1N4	120.54	120.0203	19.93	4.18	29.41	46.48
C 2H 5Cl1N6	148.56	148.0264	16.17	3.39	23.87	56.57
C 2H 5Cl2N1	113.97	112.9799	21.08	4.42	62.21	12.29
C 2H 5Cl2N3	141.99	140.9861	16.92	3.55	49.94	29.59
C 2H 5Cl2N5	170.00	168.9922	14.13	2.96	41.71	41.20
C 2H 6Cl1N1	79.53	79.0189	30.21	7.60	44.58	17.61
C 2H 6Cl1N3	107.54	107.0250	22.34	5.62	32.97	39.07
C 2H 6Cl1N5	135.56	135.0312	17.72	4.46	26.15	51.66
C 2H 6Cl2N2	128.99	127.9908	18.62	4.69	54.97	21.72
C 2H 6Cl2N4	157.00	155.9969	15.30	3.85	45.16	35.69
C 2H 6Cl2N6	185.02	184.0031	12.98	3.27	38.32	45.42
C 2H 7Cl1N2	94.54	94.0298	25.41	7.46	37.50	29.63
C 2H 7Cl1N4	122.56	122.0359	19.60	5.76	28.93	45.71
C 2H 7Cl1N6	150.57	150.0421	15.95	4.69	23.55	55.81
C 2H 7Cl2N3	144.00	143.0017	16.68	4.90	49.24	29.18
C 2H 7Cl2N5	172.02	171.0078	13.97	4.10	41.22	40.71
C 2H 8Cl1N3	109.56	109.0407	21.93	7.36	32.36	38.35
C 2H 8Cl1N5	137.57	137.0468	17.46	5.86	25.77	50.91
C 2H 8Cl2N4	159.02	158.0126	15.11	5.07	44.59	35.23
C 2H 8Cl2N6	187.03	186.0187	12.84	4.31	37.91	44.93
C 2H 9Cl1N4	124.57	124.0516	19.28	7.28	28.46	44.97
C 2H 9Cl1N6	152.59	152.0577	15.74	5.95	23.23	55.08
C 2H 9Cl2N5	174.03	173.0235	13.80	5.21	40.74	40.24
C 2H10Cl1N5	139.59	139.0625	17.21	7.22	25.40	50.17
C 2H10Cl2N6	189.05	188.0344	12.71	5.33	37.51	44.45
C 2H11Cl1N6	154.60	154.0734	15.54	7.17	22.93	54.36
C 3H 1Cl1N2	100.51	99.9828	35.85	1.00	35.27	27.87
C 3H 1Cl1N4	128.52	127.9890	28.04	.78	27.59	43.59
C 3H 1Cl1N6	156.53	155.9951	23.02	.64	22.65	53.69
C 3H 1Cl2N1	121.95	120.9486	29.55	.83	58.14	11.49

FORMULA	MOL. WEIGHT	ISOMOL. WEIGHT	PERC. C	PERC. H	PERC. Cl	PERC. N
C 3H 1Cl2N3	149.97	148.9548	24.03	.67	47.28	28.02
C 3H 1Cl2N5	177.98	176.9609	20.25	.57	39.84	39.35
C 3H 2Cl1N1	87.51	86.9876	41.18	2.30	40.51	16.01
C 3H 2Cl1N3	115.52	114.9937	31.19	1.75	30.69	36.37
C 3H 2Cl1N5	143.54	142.9999	25.10	1.40	24.70	48.79
C 3H 2Cl2N2	136.97	135.9595	26.31	1.47	51.77	20.45
C 3H 2Cl2N4	164.98	163.9657	21.84	1.22	42.98	33.96
C 3H 2Cl2N6	193.00	191.9718	18.67	1.04	36.74	43.55
C 3H 3Cl1N2	102.52	101.9985	35.15	2.95	34.58	27.32
C 3H 3Cl1N4	130.54	130.0046	27.60	2.32	27.16	42.92
C 3H 3Cl1N6	158.55	158.0108	22.73	1.91	22.36	53.01
C 3H 3Cl2N1	123.97	122.9643	29.07	2.44	57.20	11.30
C 3H 3Cl2N3	151.98	150.9704	23.71	1.99	46.65	27.65
C 3H 3Cl2N5	180.00	178.9765	20.02	1.68	39.39	38.91
C 3H 4Cl1N1	89.53	89.0032	40.25	4.50	39.60	15.65
C 3H 4Cl1N3	117.54	117.0094	30.66	3.43	30.16	35.75
C 3H 4Cl1N5	145.55	145.0155	24.76	2.77	24.36	48.12
C 3H 4Cl2N2	138.98	137.9752	25.93	2.90	51.02	20.16
C 3H 4Cl2N4	167.00	165.9813	21.58	2.41	42.46	33.55
C 3H 4Cl2N6	195.01	193.9874	18.48	2.07	36.36	43.09
C 3H 5Cl1N2	104.54	104.0141	34.47	4.82	33.91	26.80
C 3H 5Cl1N4	132.55	132.0203	27.18	3.80	26.75	42.27
C 3H 5Cl1N6	160.57	160.0264	22.44	3.14	22.08	52.34
C 3H 5Cl2N1	125.99	124.9799	28.60	4.00	56.28	11.12
C 3H 5Cl2N3	154.00	152.9861	23.40	3.27	46.04	27.29
C 3H 5Cl2N5	182.01	180.9922	19.80	2.77	38.96	38.48
C 3H 6Cl1N1	91.54	91.0189	39.36	6.61	38.73	15.30
C 3H 6Cl1N3	119.55	119.0250	30.14	5.06	29.65	35.15
C 3H 6Cl1N5	147.57	147.0312	24.42	4.10	24.02	47.46
C 3H 6Cl2N2	141.00	139.9908	25.56	4.29	50.29	19.87
C 3H 6Cl2N4	169.01	167.9969	21.32	3.58	41.95	33.15
C 3H 6Cl2N6	197.03	196.0031	18.29	3.07	35.99	42.65
C 3H 7Cl1N2	106.56	106.0298	33.82	6.62	33.27	26.29
C 3H 7Cl1N4	134.57	134.0359	26.78	5.24	26.35	41.63
C 3H 7Cl1N6	162.58	162.0421	22.16	4.34	21.81	51.69
C 3H 7Cl2N1	128.00	126.9956	28.15	5.51	55.39	10.94
C 3H 7Cl2N3	156.02	155.0017	23.10	4.52	45.45	26.93
C 3H 7Cl2N5	184.03	183.0078	19.58	3.83	38.53	38.06
C 3H 8Cl1N1	93.56	93.0345	38.52	8.62	37.89	14.97
C 3H 8Cl1N3	121.57	121.0407	29.64	6.63	29.16	34.56
C 3H 8Cl1N5	149.58	149.0468	24.09	5.39	23.70	46.82
C 3H 8Cl2N2	143.02	142.0065	25.20	5.64	49.58	19.59
C 3H 8Cl2N4	171.03	170.0126	21.07	4.71	41.46	32.76
C 3H 8Cl2N6	199.04	198.0187	18.10	4.05	35.62	42.22
C 3H 9Cl1N2	108.57	108.0454	33.19	8.36	32.65	25.80
C 3H 9Cl1N4	136.58	136.0516	26.38	6.64	25.96	41.02
C 3H 9Cl1N6	164.60	164.0577	21.89	5.51	21.54	51.06
C 3H 9Cl2N3	158.03	157.0173	22.80	5.74	44.87	26.59
C 3H 9Cl2N5	186.04	185.0235	19.37	4.88	38.11	37.64
C 3H10Cl1N3	123.59	123.0563	29.16	8.16	28.69	34.00
C 3H10Cl1N5	151.60	151.0625	23.77	6.65	23.39	46.20
C 3H10Cl2N4	173.05	172.0282	20.82	5.82	40.98	32.38
C 3H10Cl2N6	201.06	200.0344	17.92	5.01	35.27	41.80
C 3H11Cl1N4	138.60	138.0672	26.00	8.00	25.58	40.42
C 3H11Cl1N6	166.61	166.0734	21.63	6.65	21.28	50.44
C 3H11Cl2N5	188.06	187.0391	19.16	5.90	37.70	37.24
C 3H12Cl1N5	153.62	153.0781	23.46	7.87	23.08	45.59
C 3H12Cl2N6	203.08	202.0500	17.74	5.96	34.92	41.38
C 3H13Cl1N6	168.63	168.0890	21.37	7.77	21.02	49.84
C 4H 2Cl1N1	99.52	98.9876	48.28	2.03	35.62	14.07

FORMULA	MOL. WEIGHT	ISOMOL. WEIGHT	PERC. C	PERC. H	PERC. Cl	PERC. N
C 4H 2Cl1N3	127.53	126.9937	37.67	1.58	27.80	32.95
C 4H 2Cl1N5	155.55	154.9999	30.89	1.30	22.79	45.02
C 4H 2Cl2N2	148.98	147.9595	32.25	1.35	47.59	18.80
C 4H 2Cl2N4	176.99	175.9657	27.14	1.14	40.06	31.65
C 4H 2Cl2N6	205.01	203.9718	23.44	.98	34.59	40.99
C 4H 3Cl1N2	114.53	113.9985	41.95	2.64	30.95	24.46
C 4H 3Cl1N4	142.55	142.0046	33.70	2.12	24.87	39.30
C 4H 3Cl1N6	170.56	170.0108	28.17	1.77	20.79	49.27
C 4H 3Cl2N1	135.98	134.9643	35.33	2.22	52.14	10.30
C 4H 3Cl2N3	163.99	162.9704	29.30	1.84	43.24	25.62
C 4H 3Cl2N5	192.01	190.9765	25.02	1.57	36.93	36.47
C 4H 4Cl1N1	101.54	101.0032	47.32	3.97	34.92	13.79
C 4H 4Cl1N3	129.55	129.0094	37.09	3.11	27.37	32.44
C 4H 4Cl1N5	157.56	157.0155	30.49	2.56	22.50	44.45
C 4H 4Cl2N2	151.00	149.9752	31.82	2.67	46.96	18.55
C 4H 4Cl2N4	179.01	177.9813	26.84	2.25	39.61	31.30
C 4H 4Cl2N6	207.02	205.9874	23.21	1.95	34.25	40.59
C 4H 5Cl1N2	116.55	116.0141	41.22	4.32	30.42	24.04
C 4H 5Cl1N4	144.56	144.0203	33.23	3.49	24.52	38.76
C 4H 5Cl1N6	172.58	172.0264	27.84	2.92	20.54	48.70
C 4H 5Cl2N1	138.00	136.9799	34.82	3.65	51.38	10.15
C 4H 5Cl2N3	166.01	164.9861	28.94	3.04	42.71	25.31
C 4H 5Cl2N5	194.02	192.9922	24.76	2.60	36.54	36.10
C 4H 6Cl1N1	103.55	103.0189	46.40	5.84	34.24	13.53
C 4H 6Cl1N3	131.57	131.0250	36.52	4.60	26.95	31.94
C 4H 6Cl1N5	159.58	159.0312	30.11	3.79	22.22	43.89
C 4H 6Cl2N2	153.01	151.9908	31.40	3.95	46.34	18.31
C 4H 6Cl2N4	181.03	179.9969	26.54	3.34	39.17	30.95
C 4H 6Cl2N6	209.04	208.0031	22.98	2.89	33.92	40.20
C 4H 7Cl1N2	118.57	118.0298	40.52	5.95	29.90	23.63
C 4H 7Cl1N4	146.58	146.0359	32.78	4.81	24.19	38.22
C 4H 7Cl1N6	174.59	174.0421	27.52	4.04	20.31	48.13
C 4H 7Cl2N1	140.01	138.9956	34.31	5.04	50.64	10.00
C 4H 7Cl2N3	168.03	167.0017	28.59	4.20	42.20	25.01
C 4H 7Cl2N5	196.04	195.0078	24.51	3.60	36.17	35.72
C 4H 8Cl1N1	105.57	105.0345	45.51	7.64	33.58	13.27
C 4H 8Cl1N3	133.58	133.0407	35.97	6.04	26.54	31.46
C 4H 8Cl1N5	161.59	161.0468	29.73	4.99	21.94	43.34
C 4H 8Cl2N2	155.03	154.0065	30.99	5.20	45.74	18.07
C 4H 8Cl2N4	183.04	182.0126	26.25	4.41	38.74	30.61
C 4H 8Cl2N6	211.05	210.0187	22.76	3.82	33.60	39.82
C 4H 9Cl1N2	120.58	120.0454	39.84	7.52	29.40	23.23
C 4H 9Cl1N4	148.60	148.0516	32.33	6.10	23.86	37.70
C 4H 9Cl1N6	176.61	176.0577	27.20	5.14	20.07	47.59
C 4H 9Cl2N1	142.03	141.0112	33.83	6.39	49.92	9.86
C 4H 9Cl2N3	170.04	169.0173	28.25	5.33	41.70	24.71
C 4H 9Cl2N5	198.06	197.0235	24.26	4.58	35.80	35.36
C 4H10Cl1N1	107.58	107.0502	44.66	9.37	32.95	13.02
C 4H10Cl1N3	135.60	135.0563	35.43	7.43	26.15	30.99
C 4H10Cl1N5	163.61	163.0625	29.37	6.16	21.67	42.80
C 4H10Cl2N2	157.04	156.0221	30.59	6.42	45.15	17.84
C 4H10Cl2N4	185.06	184.0282	25.96	5.45	38.32	30.28
C 4H10Cl2N6	213.07	212.0344	22.55	4.73	33.28	39.44
C 4H11Cl1N2	122.60	122.0611	39.19	9.04	28.92	22.85
C 4H11Cl1N4	150.61	150.0672	31.90	7.36	23.54	37.20
C 4H11Cl1N6	178.63	178.0734	26.90	6.21	19.85	47.05
C 4H11Cl2N3	172.06	171.0330	27.92	6.44	41.21	24.42
C 4H11Cl2N5	200.07	199.0391	24.01	5.54	35.44	35.00
C 4H12Cl1N3	137.61	137.0720	34.91	8.79	25.76	30.53
C 4H12Cl1N5	165.63	165.0781	29.01	7.30	21.41	42.28

C 4H12C12N4	187.07	186.0439	25.68	6.47	37.90	29.95
C 4H12C12N6	215.09	214.0500	22.34	5.62	32.97	39.07
C 4H13C11N4	152.63	152.0829	31.48	8.59	23.23	36.71
C 4H13C11N6	180.64	180.0890	26.60	7.25	19.63	46.52
C 4H13C12N5	202.09	201.0548	23.77	6.48	35.09	34.66
C 4H14C11N5	167.64	167.0938	28.66	8.42	21.15	41.78
C 4H14C12N6	217.10	216.0657	22.13	6.50	32.66	38.71
C 4H15C11N6	182.66	182.1047	26.30	8.28	19.41	46.01
C 5H 2C11N1	111.53	110.9876	53.85	1.81	31.79	12.56
C 5H 2C11N3	139.54	138.9937	43.04	1.44	25.41	30.11
C 5H 2C11N5	167.56	166.9999	35.84	1.20	21.16	41.80
C 5H 2C12N2	160.99	159.9595	37.30	1.25	44.04	17.40
C 5H 2C12N4	189.00	187.9657	31.77	1.07	37.52	29.64
C 5H 2C12N6	217.02	215.9718	27.67	.93	32.67	38.73
C 5H 3C11N2	126.55	125.9985	47.46	2.39	28.02	22.14
C 5H 3C11N4	154.56	154.0046	38.86	1.96	22.94	36.25
C 5H 3C11N6	182.57	182.0108	32.89	1.66	19.42	46.03
C 5H 3C12N1	147.99	146.9643	40.58	2.04	47.91	9.46
C 5H 3C12N3	176.01	174.9704	34.12	1.72	40.29	23.87
C 5H 3C12N5	204.02	202.9765	29.44	1.48	34.75	34.33
C 5H 4C11N1	113.55	113.0032	52.89	3.55	31.22	12.34
C 5H 4C11N3	141.56	141.0094	42.42	2.85	25.04	29.68
C 5H 4C11N5	169.57	169.0155	35.42	2.38	20.91	41.30
C 5H 4C12N2	163.01	161.9752	36.84	2.47	43.50	17.19
C 5H 4C12N4	191.02	189.9813	31.44	2.11	37.12	29.33
C 5H 4C12N6	219.03	217.9874	27.42	1.84	32.37	38.37
C 5H 5C11N2	128.56	128.0141	46.71	3.92	27.58	21.79
C 5H 5C11N4	156.58	156.0203	38.36	3.22	22.64	35.78
C 5H 5C11N6	184.59	184.0264	32.53	2.73	19.21	45.53
C 5H 5C12N1	150.01	148.9799	40.03	3.36	47.27	9.34
C 5H 5C12N3	178.02	176.9861	33.74	2.83	39.83	23.60
C 5H 5C12N5	206.04	204.9922	29.15	2.45	34.41	33.99
C 5H 6C11N1	115.56	115.0189	51.97	5.23	30.68	12.12
C 5H 6C11N3	143.58	143.0250	41.83	4.21	24.69	29.27
C 5H 6C11N5	171.59	171.0312	35.00	3.52	20.66	40.81
C 5H 6C12N2	165.02	163.9908	36.39	3.66	42.97	16.98
C 5H 6C12N4	193.04	191.9969	31.11	3.13	36.73	29.02
C 5H 6C12N6	221.05	220.0031	27.17	2.74	32.08	38.02
C 5H 7C11N2	130.58	130.0298	45.99	5.40	27.15	21.45
C 5H 7C11N4	158.59	158.0359	37.87	4.45	22.35	35.33
C 5H 7C11N6	186.60	186.0421	32.18	3.78	19.00	45.04
C 5H 7C12N1	152.02	150.9956	39.50	4.64	46.64	9.21
C 5H 7C12N3	180.04	179.0017	33.36	3.92	39.38	23.34
C 5H 7C12N5	208.05	207.0078	28.87	3.39	34.08	33.66
C 5H 8C11N1	117.58	117.0345	51.08	6.86	30.15	11.91
C 5H 8C11N3	145.59	145.0407	41.25	5.54	24.35	28.86
C 5H 8C11N5	173.61	173.0468	34.59	4.64	20.42	40.34
C 5H 8C12N2	167.04	166.0065	35.95	4.83	42.45	16.77
C 5H 8C12N4	195.05	194.0126	30.79	4.13	36.35	28.72
C 5H 8C12N6	223.07	222.0187	26.92	3.61	31.79	37.68
C 5H 9C11N2	132.59	132.0454	45.29	6.84	26.74	21.13
C 5H 9C11N4	160.61	160.0516	37.39	5.65	22.07	34.88
C 5H 9C11N6	188.62	188.0577	31.84	4.81	18.80	44.56
C 5H 9C12N1	154.04	153.0112	38.99	5.89	46.03	9.09
C 5H 9C12N3	182.05	181.0173	32.99	4.98	38.95	23.08
C 5H 9C12N5	210.07	209.0235	28.59	4.32	33.75	33.34
C 5H10C11N1	119.60	119.0502	50.22	8.43	29.64	11.71
C 5H10C11N3	147.61	147.0563	40.69	6.83	24.02	28.47
C 5H10C11N5	175.62	175.0625	34.20	5.74	20.19	39.88
C 5H10C12N2	169.05	168.0221	35.52	5.96	41.94	16.57

C 5H10C12N4	197.07	196.0282	30.47	5.11	35.98	28.43
C 5H10C12N6	225.08	224.0344	26.68	4.48	31.50	37.34
C 5H11C11N2	134.61	134.0611	44.61	8.24	26.34	20.81
C 5H11C11N4	162.62	162.0672	36.93	6.82	21.80	34.45
C 5H11C11N6	190.64	190.0734	31.50	5.82	18.60	44.08
C 5H11C12N1	156.06	155.0269	38.48	7.10	45.44	8.98
C 5H11C12N3	184.07	183.0330	32.63	6.02	38.52	22.83
C 5H11C12N5	212.08	211.0391	28.32	5.23	33.43	33.02
C 5H12C11N1	121.61	121.0658	49.38	9.95	29.15	11.52
C 5H12C11N3	149.62	149.0720	40.14	8.08	23.69	28.08
C 5H12C11N5	177.64	177.0781	33.81	6.81	19.96	39.42
C 5H12C12N2	171.07	170.0377	35.11	7.07	41.45	16.38
C 5H12C12N4	199.08	198.0439	30.17	6.08	35.62	28.14
C 5H12C12N6	227.10	226.0500	26.44	5.33	31.22	37.01
C 5H13C11N2	136.63	136.0767	43.96	9.59	25.95	20.50
C 5H13C11N4	164.64	164.0829	36.48	7.96	21.53	34.03
C 5H13C11N6	192.65	192.0890	31.17	6.80	18.40	43.62
C 5H13C12N3	186.09	185.0486	32.27	7.04	38.10	22.58
C 5H13C12N5	214.10	213.0548	28.05	6.12	33.12	32.71
C 5H14C11N3	151.64	151.0876	39.60	9.31	23.38	27.71
C 5H14C11N5	179.65	179.0938	33.43	7.85	19.73	38.98
C 5H14C12N4	201.10	200.0595	29.86	7.02	35.26	27.86
C 5H14C12N6	229.11	228.0657	26.21	6.16	30.95	36.68
C 5H15C11N4	166.66	166.0985	36.04	9.07	21.27	33.62
C 5H15C11N6	194.67	194.1047	30.85	7.77	18.21	43.17
C 5H15C12N5	216.11	215.0704	27.79	7.00	32.81	32.41
C 5H16C11N5	181.67	181.1094	33.06	8.88	19.52	38.55
C 5H16C12N6	231.13	230.0813	25.98	6.98	30.68	36.36
C 5H17C11N6	196.68	196.1203	30.53	8.71	18.03	42.73
C 6H 3C11N2	138.56	137.9985	52.01	2.18	25.59	20.22
C 6H 3C11N4	166.57	166.0046	43.27	1.82	21.28	33.64
C 6H 3C11N6	194.58	194.0108	37.04	1.55	18.22	43.19
C 6H 3C12N1	160.00	158.9643	45.04	1.89	44.32	8.75
C 6H 3C12N3	188.02	186.9704	38.33	1.61	37.71	22.35
C 6H 3C12N5	216.03	214.9765	33.36	1.40	32.82	32.42
C 6H 4C11N1	125.56	125.0032	57.40	3.21	28.24	11.16
C 6H 4C11N3	153.57	153.0094	46.93	2.63	23.09	27.36
C 6H 4C11N5	181.59	181.0155	39.69	2.22	19.52	38.57
C 6H 4C12N2	175.02	173.9752	41.18	2.30	40.51	16.01
C 6H 4C12N4	203.03	201.9813	35.50	1.99	34.92	27.60
C 6H 4C12N6	231.04	229.9874	31.19	1.75	30.69	36.37
C 6H 5C11N2	140.57	140.0141	51.27	3.59	25.22	19.93
C 6H 5C11N4	168.59	168.0203	42.75	2.99	21.03	33.23
C 6H 5C11N6	196.60	196.0264	36.66	2.56	18.03	42.75
C 6H 5C12N1	162.02	160.9799	44.48	3.11	43.76	8.65
C 6H 5C12N3	190.03	188.9861	37.92	2.65	37.31	22.11
C 6H 5C12N5	218.05	216.9922	33.05	2.31	32.52	32.12
C 6H 6C11N1	127.57	127.0189	56.49	4.74	27.79	10.98
C 6H 6C11N3	155.59	155.0250	46.32	3.89	22.79	27.01
C 6H 6C11N5	183.60	183.0312	39.25	3.29	19.31	38.14
C 6H 6C12N2	177.03	175.9908	40.71	3.42	40.05	15.82
C 6H 6C12N4	205.05	203.9969	35.15	2.95	34.58	27.32
C 6H 6C12N6	233.06	232.0031	30.92	2.59	30.42	36.06
C 6H 7C11N2	142.59	142.0298	50.54	4.95	24.86	19.65
C 6H 7C11N4	170.60	170.0359	42.24	4.14	20.78	32.84
C 6H 7C11N6	198.62	198.0421	36.28	3.55	17.85	42.31
C 6H 7C12N1	164.04	162.9956	43.93	4.30	43.23	8.54
C 6H 7C12N3	192.05	191.0017	37.53	3.67	36.92	21.88
C 6H 7C12N5	220.06	219.0078	32.75	3.21	32.22	31.82
C 6H 8C11N1	129.59	129.0345	55.61	6.22	27.36	10.81

C 6H 8C11N3	157.60	157.0407	45.73	5.12	22.50	26.66
C 6H 8C11N5	185.62	185.0468	38.83	4.34	19.10	37.73
C 6H 8C12N2	179.05	178.0065	40.25	4.50	39.60	15.65
C 6H 8C12N4	207.06	206.0126	34.80	3.89	34.24	27.06
C 6H 8C12N6	235.08	234.0187	30.66	3.43	30.16	35.75
C 6H 9C11N2	144.61	144.0454	49.84	6.27	24.52	19.37
C 6H 9C11N4	172.62	172.0516	41.75	5.26	20.54	32.46
C 6H 9C11N6	200.63	200.0577	35.92	4.52	17.67	41.89
C 6H 9C12N1	166.05	165.0112	43.40	5.46	42.70	8.44
C 6H 9C12N3	194.06	193.0173	37.14	4.67	36.54	21.65
C 6H 9C12N5	222.08	221.0235	32.45	4.08	31.93	31.54
C 6H10C11N1	131.61	131.0502	54.76	7.66	26.94	10.64
C 6H10C11N3	159.62	159.0563	45.15	6.31	22.21	26.33
C 6H10C11N5	187.63	187.0625	38.41	5.37	18.89	37.32
C 6H10C12N2	181.07	180.0221	39.80	5.57	39.16	15.47
C 6H10C12N4	209.08	208.0282	34.47	4.82	33.91	26.80
C 6H10C12N6	237.09	236.0344	30.40	4.25	29.91	35.45
C 6H11C11N2	146.62	146.0611	49.15	7.56	24.18	19.11
C 6H11C11N4	174.63	174.0672	41.27	6.35	20.30	32.08
C 6H11C11N6	202.65	202.0734	35.56	5.47	17.49	41.47
C 6H11C12N1	168.07	167.0269	42.88	6.60	42.19	8.33
C 6H11C12N3	196.08	195.0330	36.75	5.65	36.16	21.43
C 6H11C12N5	224.09	223.0391	32.16	4.95	31.64	31.25
C 6H12C11N1	133.62	133.0658	53.93	9.05	26.53	10.48
C 6H12C11N3	161.64	161.0720	44.59	7.48	21.93	26.00
C 6H12C11N5	189.65	189.0781	38.00	6.38	18.69	36.93
C 6H12C12N2	183.08	182.0377	39.36	6.61	38.73	15.30
C 6H12C12N4	211.10	210.0439	34.14	5.73	33.59	26.54
C 6H12C12N6	239.11	238.0500	30.14	5.06	29.65	35.15
C 6H13C11N2	148.64	148.0767	48.49	8.82	23.85	18.85
C 6H13C11N4	176.65	176.0829	40.80	7.42	20.07	31.72
C 6H13C11N6	204.66	204.0890	35.21	6.40	17.32	41.06
C 6H13C12N1	170.08	169.0425	42.37	7.70	41.69	8.24
C 6H13C12N3	198.10	197.0486	36.38	6.61	35.79	21.21
C 6H13C12N5	226.11	225.0548	31.87	5.80	31.36	30.97
C 6H14C11N1	135.64	135.0815	53.13	10.40	26.14	10.33
C 6H14C11N3	163.65	163.0876	44.04	8.62	21.66	25.68
C 6H14C11N5	191.66	191.0938	37.60	7.36	18.50	36.54
C 6H14C12N2	185.10	184.0534	38.93	7.62	38.31	15.13
C 6H14C12N4	213.11	212.0595	33.82	6.62	33.27	26.29
C 6H14C12N6	241.12	240.0657	29.89	5.85	29.41	34.85
C 6H15C11N2	150.65	150.0924	47.84	10.04	23.53	18.59
C 6H15C11N4	178.67	178.0985	40.34	8.46	19.84	31.36
C 6H15C11N6	206.68	206.1047	34.87	7.32	17.15	40.66
C 6H15C12N3	200.11	199.0643	36.01	7.56	35.43	21.00
C 6H15C12N5	228.13	227.0704	31.59	6.63	31.08	30.70
C 6H16C11N3	165.67	165.1033	43.50	9.73	21.40	25.36
C 6H16C11N5	193.68	193.1094	37.21	8.33	18.30	36.16
C 6H16C12N4	215.13	214.0752	33.50	7.50	32.96	26.04
C 6H16C12N6	243.14	242.0813	29.64	6.63	29.16	34.56
C 6H17C11N4	180.68	180.1142	39.89	9.48	19.62	31.01
C 6H17C11N6	208.70	208.1203	34.53	8.21	16.99	40.27
C 6H17C12N5	230.14	229.0861	31.31	7.45	30.81	30.43
C 6H18C11N5	195.70	195.1251	36.83	9.27	18.12	35.79
C 6H18C12N6	245.16	244.0970	29.40	7.40	28.92	34.28
C 6H19C11N6	210.71	210.1360	34.20	9.09	16.83	39.88
C 7H 3C11N2	150.57	149.9985	55.84	2.01	23.55	18.61
C 7H 3C11N4	178.58	178.0046	47.08	1.69	19.85	31.37
C 7H 3C11N6	206.60	206.0108	40.70	1.46	17.16	40.68
C 7H 3C12N1	172.01	170.9643	48.88	1.76	41.22	8.14

C 7H 3C12N3	200.03	198.9704	42.03	1.51	35.45	21.01
C 7H 3C12N5	228.04	226.9765	36.87	1.33	31.09	30.71
C 7H 4C11N1	137.57	137.0032	61.12	2.93	25.77	10.18
C 7H 4C11N3	165.58	165.0094	50.78	2.43	21.41	25.38
C 7H 4C11N5	193.60	193.0155	43.43	2.08	18.31	36.17
C 7H 4C12N2	187.03	185.9752	44.95	2.16	37.91	14.98
C 7H 4C12N4	215.04	213.9813	39.10	1.87	32.97	26.05
C 7H 4C12N6	243.06	241.9874	34.59	1.66	29.17	34.58
C 7H 5C11N2	152.58	152.0141	55.10	3.30	23.24	18.36
C 7H 5C11N4	180.60	180.0203	46.56	2.79	19.63	31.02
C 7H 5C11N6	208.61	208.0264	40.30	2.42	16.99	40.29
C 7H 5C12N1	174.03	172.9799	48.31	2.90	40.74	8.05
C 7H 5C12N3	202.04	200.9861	41.61	2.49	35.09	20.80
C 7H 5C12N5	230.06	228.9922	36.55	2.19	30.82	30.44
C 7H 6C11N1	139.59	139.0189	60.23	4.33	25.40	10.03
C 7H 6C11N3	167.60	167.0250	50.17	3.61	21.15	25.07
C 7H 6C11N5	195.61	195.0312	42.98	3.09	18.12	35.80
C 7H 6C12N2	189.05	187.9908	44.48	3.20	37.51	14.82
C 7H 6C12N4	217.06	215.9969	38.74	2.79	32.67	25.81
C 7H 6C12N6	245.07	244.0031	34.31	2.47	28.93	34.29
C 7H 7C11N2	154.60	154.0298	54.38	4.56	22.93	18.12
C 7H 7C11N4	182.61	182.0359	46.04	3.86	19.41	30.68
C 7H 7C11N6	210.63	210.0421	39.92	3.35	16.83	39.90
C 7H 7C12N1	176.05	174.9956	47.76	4.01	40.28	7.96
C 7H 7C12N3	204.06	203.0017	41.20	3.46	34.75	20.59
C 7H 7C12N5	232.07	231.0078	36.23	3.04	30.55	30.18
C 7H 8C11N1	141.60	141.0345	59.38	5.69	25.04	9.89
C 7H 8C11N3	169.61	169.0407	49.57	4.75	20.90	24.77
C 7H 8C11N5	197.63	197.0468	42.54	4.08	17.94	35.44
C 7H 8C12N2	191.06	190.0065	44.01	4.22	37.11	14.66
C 7H 8C12N4	219.07	218.0126	38.38	3.68	32.37	25.57
C 7H 8C12N6	247.09	246.0187	34.03	3.26	28.70	34.01
C 7H 9C11N2	156.62	156.0454	53.68	5.79	22.64	17.89
C 7H 9C11N4	184.63	184.0516	45.54	4.91	19.20	30.35
C 7H 9C11N6	212.64	212.0577	39.54	4.27	16.67	39.52
C 7H 9C12N1	178.06	177.0112	47.22	5.09	39.82	7.87
C 7H 9C12N3	206.08	205.0173	40.80	4.40	34.41	20.39
C 7H 9C12N5	234.09	233.0235	35.92	3.88	30.29	29.92
C 7H10C11N1	143.62	143.0502	58.54	7.02	24.69	9.75
C 7H10C11N3	171.63	171.0563	48.99	5.87	20.66	24.48
C 7H10C11N5	199.64	199.0625	42.11	5.05	17.76	35.08
C 7H10C12N2	193.08	192.0221	43.55	5.22	36.72	14.51
C 7H10C12N4	221.09	220.0282	38.03	4.56	32.07	25.34
C 7H10C12N6	249.10	248.0344	33.75	4.05	28.46	33.74
C 7H11C11N2	158.63	158.0611	53.00	6.99	22.35	17.66
C 7H11C11N4	186.65	186.0672	45.05	5.94	18.99	30.02
C 7H11C11N6	214.66	214.0734	39.17	5.17	16.52	39.15
C 7H11C12N1	180.08	179.0269	46.69	6.16	39.38	7.78
C 7H11C12N3	208.09	207.0330	40.40	5.33	34.07	20.19
C 7H11C12N5	236.11	235.0391	35.61	4.70	30.03	29.66
C 7H12C11N1	145.63	145.0658	57.73	8.31	24.34	9.62
C 7H12C11N3	173.65	173.0720	48.42	6.97	20.42	24.20
C 7H12C11N5	201.66	201.0781	41.69	6.00	17.58	34.73
C 7H12C12N2	195.09	194.0377	43.10	6.20	36.34	14.36
C 7H12C12N4	223.11	222.0439	37.69	5.42	31.78	25.11
C 7H12C12N6	251.12	250.0500	33.48	4.82	28.24	33.47
C 7H13C11N2	160.65	160.0767	52.34	8.16	22.07	17.44
C 7H13C11N4	188.66	188.0829	44.57	6.95	18.79	29.70
C 7H13C11N6	216.67	216.0890	38.80	6.05	16.36	38.79
C 7H13C12N1	182.09	181.0425	46.17	7.20	38.94	7.69

C 7H13C12N3	210.11	209.0486	40.02	6.24	33.75	20.00
C 7H13C12N5	238.12	237.0548	35.31	5.50	29.78	29.41
C 7H14C11N1	147.65	147.0815	56.94	9.56	24.01	9.49
C 7H14C11N3	175.66	175.0876	47.86	8.03	20.18	23.92
C 7H14C11N5	203.68	203.0938	41.28	6.93	17.41	34.38
C 7H14C12N2	197.11	196.0534	42.66	7.16	35.97	14.21
C 7H14C12N4	225.12	224.0595	37.35	6.27	31.50	24.89
C 7H14C12N6	253.14	252.0657	33.21	5.57	28.01	33.20
C 7H15C11N2	162.66	162.0924	51.69	9.29	21.80	17.22
C 7H15C11N4	190.68	190.0985	44.09	7.93	18.59	29.38
C 7H15C11N6	218.69	218.1047	38.45	6.91	16.21	38.43
C 7H15C12N1	184.11	183.0581	45.67	8.21	38.51	7.61
C 7H15C12N3	212.12	211.0643	39.64	7.13	33.43	19.81
C 7H15C12N5	240.14	239.0704	35.01	6.30	29.53	29.16
C 7H16C11N1	149.67	149.0971	56.18	10.78	23.69	9.36
C 7H16C11N3	177.68	177.1033	47.32	9.08	19.95	23.65
C 7H16C11N5	205.69	205.1094	40.88	7.84	17.24	34.05
C 7H16C12N2	199.12	198.0690	42.22	8.10	35.61	14.07
C 7H16C12N4	227.14	226.0752	37.02	7.10	31.22	24.67
C 7H16C12N6	255.15	254.0813	32.95	6.32	27.79	32.94
C 7H17C11N2	164.68	164.1080	51.06	10.41	21.53	17.01
C 7H17C11N4	192.69	192.1142	43.63	8.89	18.40	29.08
C 7H17C11N6	220.71	220.1203	38.09	7.76	16.06	38.08
C 7H17C12N3	214.14	213.0799	39.26	8.00	33.11	19.62
C 7H17C12N5	242.15	241.0861	34.72	7.08	29.28	28.92
C 7H18C11N3	179.69	179.1189	46.79	10.10	19.73	23.38
C 7H18C11N5	207.71	207.1251	40.48	8.74	17.07	33.72
C 7H18C12N4	229.15	228.0908	36.69	7.92	30.94	24.45
C 7H18C12N6	257.17	256.0970	32.69	7.06	27.57	32.68
C 7H19C11N4	194.71	194.1298	43.18	9.84	18.21	28.77
C 7H19C11N6	222.72	222.1360	37.75	8.60	15.92	37.73
C 7H19C12N5	244.17	243.1017	34.43	7.84	29.04	28.68
C 7H20C11N5	209.72	209.1407	40.09	9.61	16.90	33.39
C 7H20C12N6	259.18	258.1126	32.44	7.78	27.36	32.42
C 7H21C11N6	224.74	224.1516	37.41	9.42	15.78	37.39
C 8H 4C11N1	149.58	149.0032	64.24	2.70	23.70	9.36
C 8H 4C11N3	177.59	177.0094	54.11	2.27	19.96	23.66
C 8H 4C11N5	205.61	205.0155	46.73	1.96	17.24	34.06
C 8H 4C12N2	199.04	197.9752	48.28	2.03	35.62	14.07
C 8H 4C12N4	227.05	225.9813	42.32	1.78	31.23	24.68
C 8H 4C12N6	255.07	253.9874	37.67	1.58	27.80	32.95
C 8H 5C11N2	164.60	164.0141	58.38	3.06	21.54	17.02
C 8H 5C11N4	192.61	192.0203	49.89	2.62	18.41	29.09
C 8H 5C11N6	220.62	220.0264	43.55	2.28	16.07	38.09
C 8H 5C12N1	186.04	184.9799	51.65	2.71	38.11	7.53
C 8H 5C12N3	214.06	212.9861	44.89	2.35	33.13	19.63
C 8H 5C12N5	242.07	240.9922	39.70	2.08	29.29	28.93
C 8H 6C11N1	151.60	151.0189	63.38	3.99	23.39	9.24
C 8H 6C11N3	179.61	179.0250	53.50	3.37	19.74	23.40
C 8H 6C11N5	207.62	207.0312	46.28	2.91	17.08	33.73
C 8H 6C12N2	201.06	199.9908	47.79	3.01	35.27	13.93
C 8H 6C12N4	229.07	227.9969	41.95	2.64	30.95	24.46
C 8H 6C12N6	257.08	256.0031	37.38	2.35	27.58	32.69
C 8H 7C11N2	166.61	166.0298	57.67	4.23	21.28	16.81
C 8H 7C11N4	194.62	194.0359	49.37	3.63	18.22	28.79
C 8H 7C11N6	222.64	222.0421	43.16	3.17	15.92	37.75
C 8H 7C12N1	188.06	186.9956	51.10	3.75	37.70	7.45
C 8H 7C12N3	216.07	215.0017	44.47	3.27	32.82	19.45
C 8H 7C12N5	244.08	243.0078	39.37	2.89	29.05	28.69
C 8H 8C11N1	153.61	153.0345	62.55	5.25	23.08	9.12

C 8H 8C11N3	181.63	181.0407	52.90	4.44	19.52	23.14
C 8H 8C11N5	209.64	209.0468	45.84	3.85	16.91	33.41
C 8H 8C12N2	203.07	202.0065	47.32	3.97	34.92	13.79
C 8H 8C12N4	231.09	230.0126	41.58	3.49	30.68	24.25
C 8H 8C12N6	259.10	258.0187	37.09	3.11	27.37	32.44
C 8H 9C11N2	168.63	168.0454	56.98	5.38	21.02	16.61
C 8H 9C11N4	196.64	196.0516	48.87	4.61	18.03	28.49
C 8H 9C11N6	224.65	224.0577	42.77	4.04	15.78	37.41
C 8H 9C12N1	190.07	189.0112	50.55	4.77	37.30	7.37
C 8H 9C12N3	218.09	217.0173	44.06	4.16	32.51	19.27
C 8H 9C12N5	246.10	245.0235	39.04	3.69	28.81	28.46
C 8H10C11N1	155.63	155.0502	61.74	6.48	22.78	9.00
C 8H10C11N3	183.64	183.0563	52.32	5.49	19.31	22.88
C 8H10C11N5	211.66	211.0625	45.40	4.76	16.75	33.09
C 8H10C12N2	205.09	204.0221	46.85	4.91	34.57	13.66
C 8H10C12N4	233.10	232.0282	41.22	4.32	30.42	24.04
C 8H10C12N6	261.12	260.0344	36.80	3.86	27.16	32.19
C 8H11C11N2	170.64	170.0611	56.31	6.50	20.78	16.42
C 8H11C11N4	198.66	198.0672	48.37	5.58	17.85	28.20
C 8H11C11N6	226.67	226.0734	42.39	4.89	15.64	37.08
C 8H11C12N1	192.09	191.0269	50.02	5.77	36.91	7.29
C 8H11C12N3	220.10	219.0330	43.66	5.04	32.21	19.09
C 8H11C12N5	248.12	247.0391	38.73	4.47	28.58	28.23
C 8H12C11N1	157.64	157.0658	60.95	7.67	22.49	8.88
C 8H12C11N3	185.66	185.0720	51.76	6.52	19.10	22.63
C 8H12C11N5	213.67	213.0781	44.97	5.66	16.59	32.78
C 8H12C12N2	207.10	206.0377	46.40	5.84	34.24	13.53
C 8H12C12N4	235.12	234.0439	40.87	5.14	30.16	23.83
C 8H12C12N6	263.13	262.0500	36.52	4.60	26.95	31.94
C 8H13C11N2	172.66	172.0767	55.65	7.59	20.53	16.22
C 8H13C11N4	200.67	200.0829	47.88	6.53	17.67	27.92
C 8H13C11N6	228.69	228.0890	42.02	5.73	15.50	36.75
C 8H13C12N1	194.11	193.0425	49.50	6.75	36.53	7.22
C 8H13C12N3	222.12	221.0486	43.26	5.90	31.92	18.92
C 8H13C12N5	250.13	249.0548	38.42	5.24	28.35	28.00
C 8H14C11N1	159.66	159.0815	60.18	8.84	22.21	8.77
C 8H14C11N3	187.67	187.0876	51.20	7.52	18.89	22.39
C 8H14C11N5	215.69	215.0938	44.55	6.54	16.44	32.47
C 8H14C12N2	209.12	208.0534	45.95	6.75	33.91	13.40
C 8H14C12N4	237.13	236.0595	40.52	5.95	29.90	23.63
C 8H14C12N6	265.15	264.0657	36.24	5.32	26.74	31.70
C 8H15C11N2	174.68	174.0924	55.01	8.66	20.30	16.04
C 8H15C11N4	202.69	202.0985	47.41	7.46	17.49	27.64
C 8H15C11N6	230.70	230.1047	41.65	6.55	15.37	36.43
C 8H15C12N1	196.12	195.0581	48.99	7.71	36.15	7.14
C 8H15C12N3	224.13	223.0643	42.87	6.75	31.64	18.75
C 8H15C12N5	252.15	251.0704	38.11	6.00	28.12	27.77
C 8H16C11N1	161.68	161.0971	59.43	9.98	21.93	8.66
C 8H16C11N3	189.69	189.1033	50.66	8.50	18.69	22.15
C 8H16C11N5	217.70	217.1094	44.14	7.41	16.29	32.17
C 8H16C12N2	211.14	210.0690	45.51	7.64	33.58	13.27
C 8H16C12N4	239.15	238.0752	40.18	6.74	29.65	23.43
C 8H16C12N6	267.16	266.0813	35.97	6.04	26.54	31.46
C 8H17C11N2	176.69	176.1080	54.38	9.70	20.06	15.85
C 8H17C11N4	204.70	204.1142	46.94	8.37	17.32	27.37
C 8H17C11N6	232.72	232.1203	41.29	7.36	15.23	36.11
C 8H17C12N1	198.14	197.0738	48.50	8.65	35.79	7.07
C 8H17C12N3	226.15	225.0799	42.49	7.58	31.35	18.58
C 8H17C12N5	254.16	253.0861	37.81	6.74	27.90	27.55
C 8H18C11N1	163.69	163.1128	58.70	11.08	21.66	8.56

C 8H18C11N3	191.71	191.1189	50.12	9.46	18.49	21.92
C 8H18C11N5	219.72	219.1251	43.73	8.26	16.14	31.87
C 8H18C12N2	213.15	212.0847	45.08	8.51	33.27	13.14
C 8H18C12N4	241.17	240.0908	39.84	7.52	29.40	23.23
C 8H18C12N6	269.18	268.0970	35.70	6.74	26.34	31.22
C 8H19C11N2	178.71	178.1237	53.77	10.72	19.84	15.68
C 8H19C11N4	206.72	206.1298	46.48	9.26	17.15	27.10
C 8H19C11N6	234.73	234.1360	40.94	8.16	15.10	35.80
C 8H19C12N3	228.17	227.0956	42.11	8.39	31.08	18.42
C 8H19C12N5	256.18	255.1017	37.51	7.48	27.68	27.34
C 8H20C11N3	193.72	193.1346	49.60	10.41	18.30	21.69
C 8H20C11N5	221.74	221.1407	43.34	9.09	15.99	31.58
C 8H20C12N4	243.18	242.1065	39.51	8.29	29.16	23.04
C 8H20C12N6	271.19	270.1126	35.43	7.43	26.15	30.99
C 8H21C11N4	208.74	208.1455	46.03	10.14	16.98	26.84
C 8H21C11N6	236.75	236.1516	40.59	8.94	14.97	35.50
C 8H21C12N5	258.20	257.1174	37.22	8.20	27.46	27.12
C 8H22C11N5	223.75	223.1564	42.94	9.91	15.84	31.30
C 8H22C12N6	273.21	272.1283	35.17	8.12	25.95	30.76
C 8H23C11N6	238.77	238.1673	40.24	9.71	14.85	35.20
C 9H 4C11N1	161.59	161.0032	66.90	2.50	21.94	8.67
C 9H 4C11N3	189.61	189.0094	57.01	2.13	18.70	22.16
C 9H 4C11N5	217.62	217.0155	49.67	1.85	16.29	32.18
C 9H 4C12N2	211.05	209.9752	51.22	1.91	33.60	13.27
C 9H 4C12N4	239.07	237.9813	45.22	1.69	29.66	23.44
C 9H 4C12N6	267.08	265.9874	40.48	1.51	26.55	31.47
C 9H 5C11N2	176.61	176.0141	61.21	2.85	20.07	15.86
C 9H 5C11N4	204.62	204.0203	52.83	2.46	17.33	27.38
C 9H 5C11N6	232.63	232.0264	46.47	2.17	15.24	36.13
C 9H 5C12N1	198.05	196.9799	54.58	2.54	35.80	7.07
C 9H 5C12N3	226.07	224.9861	47.82	2.23	31.37	18.59
C 9H 5C12N5	254.08	252.9922	42.55	1.98	27.91	27.56
C 9H 6C11N1	163.61	163.0189	66.07	3.70	21.67	8.56
C 9H 6C11N3	191.62	191.0250	56.41	3.16	18.50	21.93
C 9H 6C11N5	219.63	219.0312	49.22	2.75	16.14	31.89
C 9H 6C12N2	213.07	211.9908	50.74	2.84	33.28	13.15
C 9H 6C12N4	241.08	239.9969	44.84	2.51	29.41	23.24
C 9H 6C12N6	269.09	268.0031	40.17	2.25	26.35	31.23
C 9H 7C11N2	178.62	178.0298	60.52	3.95	19.85	15.68
C 9H 7C11N4	206.64	206.0359	52.31	3.41	17.16	27.11
C 9H 7C11N6	234.65	234.0421	46.07	3.01	15.11	35.82
C 9H 7C12N1	200.07	198.9956	54.03	3.53	35.44	7.00
C 9H 7C12N3	228.08	227.0017	47.40	3.09	31.09	18.42
C 9H 7C12N5	256.10	255.0078	42.21	2.76	27.69	27.35
C 9H 8C11N1	165.62	165.0345	65.27	4.87	21.41	8.46
C 9H 8C11N3	193.64	193.0407	55.83	4.16	18.31	21.70
C 9H 8C11N5	221.65	221.0468	48.77	3.64	15.99	31.60
C 9H 8C12N2	215.08	214.0065	50.26	3.75	32.97	13.02
C 9H 8C12N4	243.10	242.0126	44.47	3.32	29.17	23.05
C 9H 8C12N6	271.11	270.0187	39.87	2.97	26.15	31.00
C 9H 9C11N2	180.64	180.0454	59.84	5.02	19.63	15.51
C 9H 9C11N4	208.65	208.0516	51.81	4.35	16.99	26.85
C 9H 9C11N6	236.67	236.0577	45.68	3.83	14.98	35.51
C 9H 9C12N1	202.08	201.0112	53.49	4.49	35.09	6.93
C 9H 9C12N3	230.10	229.0173	46.98	3.94	30.82	18.26
C 9H 9C12N5	258.11	257.0235	41.88	3.51	27.47	27.13
C 9H10C11N1	167.64	167.0502	64.48	6.01	21.15	8.36
C 9H10C11N3	195.65	195.0563	55.25	5.15	18.12	21.48
C 9H10C11N5	223.67	223.0625	48.33	4.51	15.85	31.31
C 9H10C12N2	217.10	216.0221	49.79	4.64	32.66	12.90

C 9H10C12N4	245.11	244.0282	44.10	4.11	28.93	22.86
C 9H10C12N6	273.13	272.0344	39.58	3.69	25.96	30.77
C 9H11C11N2	182.65	182.0611	59.18	6.07	19.41	15.34
C 9H11C11N4	210.67	210.0672	51.31	5.26	16.83	26.59
C 9H11C11N6	238.68	238.0734	45.29	4.65	14.85	35.21
C 9H11C12N1	204.10	203.0269	52.96	5.43	34.74	6.86
C 9H11C12N3	232.11	231.0330	46.57	4.78	30.55	18.10
C 9H11C12N5	260.13	259.0391	41.56	4.26	27.26	26.92
C 9H12C11N1	169.66	169.0658	63.72	7.13	20.90	8.26
C 9H12C11N3	197.67	197.0720	54.69	6.12	17.94	21.26
C 9H12C11N5	225.68	225.0781	47.90	5.36	15.71	31.03
C 9H12C12N2	219.12	218.0377	49.33	5.52	32.36	12.78
C 9H12C12N4	247.13	246.0439	43.74	4.89	28.69	22.67
C 9H12C12N6	275.14	274.0500	39.29	4.40	25.77	30.54
C 9H13C11N2	184.67	184.0767	58.54	7.10	19.20	15.17
C 9H13C11N4	212.68	212.0829	50.83	6.16	16.67	26.34
C 9H13C11N6	240.70	240.0890	44.91	5.44	14.73	34.92
C 9H13C12N1	206.12	205.0425	52.45	6.36	34.40	6.80
C 9H13C12N3	234.13	233.0486	46.17	5.60	30.28	17.95
C 9H13C12N5	262.14	261.0548	41.24	5.00	27.05	26.72
C 9H14C11N1	171.67	171.0815	62.97	8.22	20.65	8.16
C 9H14C11N3	199.69	199.0876	54.14	7.07	17.75	21.04
C 9H14C11N5	227.70	227.0938	47.48	6.20	15.57	30.76
C 9H14C12N2	221.13	220.0534	48.89	6.38	32.07	12.67
C 9H14C12N4	249.14	248.0595	43.39	5.66	28.46	22.49
C 9H14C12N6	277.16	276.0657	39.00	5.09	25.58	30.32
C 9H15C11N2	186.69	186.0924	57.90	8.10	18.99	15.01
C 9H15C11N4	214.70	214.0985	50.35	7.04	16.51	26.10
C 9H15C11N6	242.71	242.1047	44.54	6.23	14.61	34.63
C 9H15C12N1	208.13	207.0581	51.94	7.26	34.07	6.73
C 9H15C12N3	236.15	235.0643	45.78	6.40	30.03	17.79
C 9H15C12N5	264.16	263.0704	40.92	5.72	26.84	26.51
C 9H16C11N1	173.69	173.0971	62.24	9.29	20.41	8.06
C 9H16C11N3	201.70	201.1033	53.59	8.00	17.58	20.83
C 9H16C11N5	229.71	229.1094	47.06	7.02	15.43	30.49
C 9H16C12N2	223.15	222.0690	48.44	7.23	31.78	12.55
C 9H16C12N4	251.16	250.0752	43.04	6.42	28.23	22.31
C 9H16C12N6	279.17	278.0813	38.72	5.78	25.40	30.10
C 9H17C11N2	188.70	188.1080	57.29	9.08	18.79	14.85
C 9H17C11N4	216.72	216.1142	49.88	7.91	16.36	25.85
C 9H17C11N6	244.73	244.1203	44.17	7.00	14.49	34.34
C 9H17C12N1	210.15	209.0738	51.44	8.15	33.74	6.67
C 9H17C12N3	238.16	237.0799	45.39	7.19	29.77	17.64
C 9H17C12N5	266.18	265.0861	40.61	6.44	26.64	26.31
C 9H18C11N1	175.70	175.1128	61.52	10.33	20.18	7.97
C 9H18C11N3	203.72	203.1189	53.06	8.91	17.40	20.63
C 9H18C11N5	231.73	231.1251	46.65	7.83	15.30	30.22
C 9H18C12N2	225.16	224.0847	48.01	8.06	31.49	12.44
C 9H18C12N4	253.18	252.0908	42.70	7.17	28.01	22.13
C 9H18C12N6	281.19	280.0970	38.44	6.45	25.22	29.89
C 9H19C11N2	190.72	190.1237	56.68	10.04	18.59	14.69
C 9H19C11N4	218.73	218.1298	49.42	8.76	16.21	25.61
C 9H19C11N6	246.74	246.1360	43.81	7.76	14.37	34.06
C 9H19C12N1	212.16	211.0894	50.95	9.03	33.42	6.60
C 9H19C12N3	240.18	239.0956	45.01	7.97	29.52	17.50
C 9H19C12N5	268.19	267.1017	40.31	7.14	26.44	26.11
C 9H20C11N1	177.72	177.1284	60.83	11.34	19.95	7.88
C 9H20C11N3	205.73	205.1346	52.54	9.80	17.23	20.42
C 9H20C11N5	233.75	233.1407	46.25	8.62	15.17	29.96
C 9H20C12N2	227.18	226.1003	47.58	8.87	31.21	12.33

C 9H20C12N4	255.19	254.1065	42.36	7.90	27.79	21.95
C 9H20C12N6	283.21	282.1126	38.17	7.12	25.04	29.67
C 9H21C11N2	192.73	192.1393	56.09	10.98	18.39	14.53
C 9H21C11N4	220.75	220.1455	48.97	9.59	16.06	25.38
C 9H21C11N6	248.76	248.1516	43.46	8.51	14.25	33.78
C 9H21C12N3	242.19	241.1112	44.63	8.74	29.28	17.35
C 9H21C12N5	270.21	269.1174	40.01	7.83	26.24	25.92
C 9H22C11N3	207.75	207.1502	52.03	10.67	17.07	20.23
C 9H22C11N5	235.76	235.1564	45.85	9.41	15.04	29.71
C 9H22C12N4	257.21	256.1221	42.03	8.62	27.57	21.78
C 9H22C12N6	285.22	284.1283	37.90	7.77	24.86	29.46
C 9H23C11N4	222.76	222.1611	48.53	10.41	15.92	25.15
C 9H23C11N6	250.78	250.1673	43.11	9.24	14.14	33.51
C 9H23C12N5	272.22	271.1330	39.71	8.52	26.05	25.73
C 9H24C11N5	237.78	237.1720	45.46	10.17	14.91	29.45
C 9H24C12N6	287.24	286.1439	37.63	8.42	24.69	29.26
C 9H25C11N6	252.79	252.1829	42.76	9.97	14.02	33.24
C10H 5C11N2	188.62	188.0141	63.68	2.67	18.80	14.85
C10H 5C11N4	216.63	216.0203	55.45	2.33	16.37	25.86
C10H 5C11N6	244.64	244.0264	49.10	2.06	14.49	34.35
C10H 5C12N1	210.06	208.9799	57.18	2.40	33.75	6.67
C10H 5C12N3	238.08	236.9861	50.45	2.12	29.78	17.65
C10H 5C12N5	266.09	264.9922	45.14	1.89	26.65	26.32
C10H 6C11N1	175.62	175.0189	68.39	3.44	20.19	7.98
C10H 6C11N3	203.63	203.0250	58.98	2.97	17.41	20.64
C10H 6C11N5	231.65	231.0312	51.85	2.61	15.30	30.23
C10H 6C12N2	225.08	223.9908	53.36	2.69	31.50	12.45
C10H 6C12N4	253.09	251.9969	47.46	2.39	28.02	22.14
C10H 6C12N6	281.11	280.0031	42.73	2.15	25.22	29.90
C10H 7C11N2	190.63	190.0298	63.01	3.70	18.60	14.69
C10H 7C11N4	218.65	218.0359	54.93	3.23	16.21	25.62
C10H 7C11N6	246.66	246.0421	48.70	2.86	14.37	34.07
C10H 7C12N1	212.08	210.9956	56.63	3.33	33.43	6.60
C10H 7C12N3	240.09	239.0017	50.03	2.94	29.53	17.50
C10H 7C12N5	268.11	267.0078	44.80	2.63	26.45	26.12
C10H 8C11N1	177.63	177.0345	67.62	4.54	19.96	7.89
C10H 8C11N3	205.65	205.0407	58.41	3.92	17.24	20.43
C10H 8C11N5	233.66	233.0468	51.40	3.45	15.17	29.97
C10H 8C12N2	227.09	226.0065	52.89	3.55	31.22	12.34
C10H 8C12N4	255.11	254.0126	47.08	3.16	27.79	21.96
C10H 8C12N6	283.12	282.0187	42.42	2.85	25.04	29.68
C10H 9C11N2	192.65	192.0454	62.35	4.71	18.40	14.54
C10H 9C11N4	220.66	220.0516	54.43	4.11	16.07	25.39
C10H 9C11N6	248.68	248.0577	48.30	3.65	14.26	33.80
C10H 9C12N1	214.10	213.0112	56.10	4.24	33.12	6.54
C10H 9C12N3	242.11	241.0173	49.61	3.75	29.29	17.36
C10H 9C12N5	270.12	269.0235	44.47	3.36	26.25	25.93
C10H10C11N1	179.65	179.0502	66.86	5.61	19.73	7.80
C10H10C11N3	207.66	207.0563	57.84	4.85	17.07	20.23
C10H10C11N5	235.68	235.0625	50.96	4.28	15.04	29.72
C10H10C12N2	229.11	228.0221	52.43	4.40	30.95	12.23
C10H10C12N4	257.12	256.0282	46.71	3.92	27.58	21.79
C10H10C12N6	285.14	284.0344	42.12	3.54	24.87	29.47
C10H11C11N2	194.67	194.0611	61.70	5.70	18.21	14.39
C10H11C11N4	222.68	222.0672	53.94	4.98	15.92	25.16
C10H11C11N6	250.69	250.0734	47.91	4.42	14.14	33.52
C10H11C12N1	216.11	215.0269	55.58	5.13	32.81	6.48
C10H11C12N3	244.13	243.0330	49.20	4.54	29.04	17.21
C10H11C12N5	272.14	271.0391	44.14	4.07	26.06	25.73
C10H12C11N1	181.67	181.0658	66.12	6.66	19.52	7.71
C10H12C11N3	209.68	209.0720	57.28	5.77	16.91	20.04
C10H12C11N5	237.69	237.0781	50.53	5.09	14.92	29.46
C10H12C12N2	231.13	230.0377	51.97	5.23	30.68	12.12
C10H12C12N4	259.14	258.0439	46.35	4.67	27.36	21.62
C10H12C12N6	287.15	286.0500	41.83	4.21	24.69	29.27
C10H13C11N2	196.68	196.0767	61.07	6.66	18.03	14.24
C10H13C11N4	224.69	224.0829	53.46	5.83	15.78	24.93
C10H13C11N6	252.71	252.0890	47.53	5.19	14.03	33.26
C10H13C12N1	218.13	217.0425	55.06	6.01	32.51	6.42
C10H13C12N3	246.14	245.0486	48.80	5.32	28.81	17.07
C10H13C12N5	274.15	273.0548	43.81	4.78	25.86	25.55
C10H14C11N1	183.68	183.0815	65.39	7.68	19.30	7.63
C10H14C11N3	211.70	211.0876	56.74	6.67	16.75	19.85
C10H14C11N5	239.71	239.0938	50.11	5.89	14.79	29.22
C10H14C12N2	233.14	232.0534	51.52	6.05	30.41	12.02
C10H14C12N4	261.16	260.0595	45.99	5.40	27.15	21.45
C10H14C12N6	289.17	288.0657	41.54	4.88	24.52	29.06
C10H15C11N2	198.70	198.0924	60.45	7.61	17.84	14.10
C10H15C11N4	226.71	226.0985	52.98	6.67	15.64	24.71
C10H15C11N6	254.72	254.1047	47.15	5.94	13.92	32.99
C10H15C12N1	220.14	219.0581	54.56	6.87	32.21	6.36
C10H15C12N3	248.16	247.0643	48.40	6.09	28.57	16.93
C10H15C12N5	276.17	275.0704	43.49	5.47	25.67	25.36
C10H16C11N1	185.70	185.0971	64.68	8.68	19.09	7.54
C10H16C11N3	213.71	213.1033	56.20	7.55	16.59	19.66
C10H16C11N5	241.73	241.1094	49.69	6.67	14.67	28.97
C10H16C12N2	235.16	234.0690	51.08	6.86	30.15	11.91
C10H16C12N4	263.17	262.0752	45.64	6.13	26.94	21.29
C10H16C12N6	291.19	290.0813	41.25	5.54	24.35	28.86
C10H17C11N2	200.71	200.1080	59.84	8.54	17.66	13.96
C10H17C11N4	228.73	228.1142	52.51	7.49	15.50	24.50
C10H17C11N6	256.74	256.1203	46.78	6.67	13.81	32.73
C10H17C12N1	222.16	221.0738	54.07	7.71	31.92	6.30
C10H17C12N3	250.17	249.0799	48.01	6.85	28.34	16.80
C10H17C12N5	278.19	277.0861	43.18	6.16	25.49	25.18
C10H18C11N1	187.71	187.1128	63.99	9.67	18.89	7.46
C10H18C11N3	215.73	215.1189	55.68	8.41	16.43	19.48
C10H18C11N5	243.74	243.1251	49.28	7.44	14.55	28.73
C10H18C12N2	237.17	236.0847	50.64	7.65	29.90	11.81
C10H18C12N4	265.19	264.0908	45.29	6.84	26.74	21.13
C10H18C12N6	293.20	292.0970	40.97	6.19	24.18	28.66
C10H19C11N2	202.73	202.1237	59.25	9.45	17.49	13.82
C10H19C11N4	230.74	230.1298	52.05	8.30	15.36	24.28
C10H19C11N6	258.76	258.1360	46.42	7.40	13.70	32.48
C10H19C12N1	224.18	223.0894	53.58	8.54	31.63	6.25
C10H19C12N3	252.19	251.0956	47.63	7.59	28.12	16.66
C10H19C12N5	280.20	279.1017	42.87	6.83	25.31	24.99
C10H20C11N1	189.73	189.1284	63.31	10.63	18.69	7.38
C10H20C11N3	217.74	217.1346	55.16	9.26	16.28	19.30
C10H20C11N5	245.76	245.1407	48.87	8.20	14.43	28.50
C10H20C12N2	239.19	238.1003	50.22	8.43	29.64	11.71
C10H20C12N4	267.20	266.1065	44.95	7.54	26.54	20.97
C10H20C12N6	295.22	294.1126	40.69	6.83	24.02	28.47
C10H21C11N2	204.75	204.1393	58.66	10.34	17.32	13.68
C10H21C11N4	232.76	232.1455	51.60	9.09	15.23	24.07
C10H21C11N6	260.77	260.1516	46.06	8.12	13.60	32.23
C10H21C12N1	226.19	225.1051	53.10	9.36	31.35	6.19
C10H21C12N3	254.20	253.1112	47.25	8.33	27.89	16.53
C10H21C12N5	282.22	281.1174	42.56	7.50	25.12	24.82
C10H22C11N1	191.75	191.1441	62.64	11.56	18.49	7.30

C10H22C11N3	219.76	219.1502	54.66	10.09	16.13	19.12
C10H22C11N5	247.77	247.1564	48.48	8.95	14.31	28.27
C10H22C12N2	241.21	240.1160	49.80	9.19	29.40	11.61
C10H22C12N4	269.22	268.1221	44.61	8.24	26.34	20.81
C10H22C12N6	297.23	296.1283	40.41	7.46	23.86	28.27
C10H23C11N2	206.76	206.1550	58.09	11.21	17.15	13.55
C10H23C11N4	234.77	234.1611	51.16	9.87	15.10	23.86
C10H23C11N6	262.79	262.1673	45.71	8.82	13.49	31.98
C10H23C12N3	256.22	255.1269	46.88	9.05	27.67	16.40
C10H23C12N5	284.23	283.1330	42.26	8.16	24.95	24.64
C10H24C11N3	221.78	221.1659	54.16	10.91	15.99	18.95
C10H24C11N5	249.79	249.1720	48.09	9.68	14.19	28.04
C10H24C12N4	271.24	270.1378	44.28	8.92	26.14	20.66
C10H24C12N6	299.25	298.1439	40.14	8.08	23.69	28.08
C10H25C11N4	236.79	236.1768	50.72	10.64	14.97	23.66
C10H25C11N6	264.80	264.1829	45.36	9.52	13.39	31.74
C10H25C12N5	286.25	285.1487	41.96	8.80	24.77	24.47
C10H26C11N5	251.81	251.1877	47.70	10.41	14.08	27.81
C10H26C12N6	301.26	300.1596	39.87	8.70	23.54	27.90
C10H27C11N6	266.82	266.1986	45.02	10.20	13.29	31.50
C11H 5C11N2	200.63	200.0141	65.85	2.51	17.67	13.96
C11H 5C11N4	228.64	228.0203	57.79	2.20	15.51	24.50
C11H 5C11N6	256.66	256.0264	51.48	1.96	13.81	32.74
C11H 5C12N1	222.08	220.9799	59.49	2.27	31.93	6.31
C11H 5C12N3	250.09	248.9861	52.83	2.02	28.35	16.80
C11H 5C12N5	278.10	276.9922	47.51	1.81	25.50	25.18
C11H 6C11N1	187.63	187.0189	70.42	3.22	18.90	7.47
C11H 6C11N3	215.64	215.0250	61.27	2.80	16.44	19.49
C11H 6C11N5	243.66	243.0312	54.22	2.48	14.55	28.74
C11H 6C12N2	237.09	235.9908	55.73	2.55	29.91	11.82
C11H 6C12N4	265.10	263.9969	49.84	2.28	26.75	21.13
C11H 6C12N6	293.12	292.0031	45.08	2.06	24.19	28.67
C11H 7C11N2	202.64	202.0298	65.20	3.48	17.50	13.82
C11H 7C11N4	230.66	230.0359	57.28	3.06	15.37	24.29
C11H 7C11N6	258.67	258.0421	51.08	2.73	13.71	32.49
C11H 7C12N1	224.09	222.9956	58.96	3.15	31.64	6.25
C11H 7C12N3	252.10	251.0017	52.41	2.80	28.13	16.67
C11H 7C12N5	280.12	279.0078	47.17	2.52	25.31	25.00
C11H 8C11N1	189.65	189.0345	69.67	4.25	18.69	7.39
C11H 8C11N3	217.66	217.0407	60.70	3.70	16.29	19.31
C11H 8C11N5	245.67	245.0468	53.78	3.28	14.43	28.51
C11H 8C12N2	239.11	238.0065	55.26	3.37	29.65	11.72
C11H 8C12N4	267.12	266.0126	49.46	3.02	26.54	20.97
C11H 8C12N6	295.13	294.0187	44.77	2.73	24.03	28.48
C11H 9C11N2	204.66	204.0454	64.56	4.43	17.32	13.69
C11H 9C11N4	232.67	232.0516	56.78	3.90	15.24	24.08
C11H 9C11N6	260.69	260.0577	50.68	3.48	13.60	32.24
C11H 9C12N1	226.11	225.0112	58.43	4.01	31.36	6.19
C11H 9C12N3	254.12	253.0173	51.99	3.57	27.90	16.54
C11H 9C12N5	282.13	281.0235	46.83	3.22	25.13	24.82
C11H10C11N1	191.66	191.0502	68.94	5.26	18.50	7.31
C11H10C11N3	219.68	219.0563	60.14	4.59	16.14	19.13
C11H10C11N5	247.69	247.0625	53.34	4.07	14.31	28.27
C11H10C12N2	241.12	240.0221	54.79	4.18	29.41	11.62
C11H10C12N4	269.14	268.0282	49.09	3.75	26.35	20.82
C11H10C12N6	297.15	296.0344	44.46	3.39	23.86	28.28
C11H11C11N2	206.68	206.0611	63.93	5.36	17.15	13.55
C11H11C11N4	234.69	234.0672	56.30	4.72	15.11	23.87
C11H11C11N6	262.70	262.0734	50.29	4.22	13.50	31.99
C11H11C12N1	228.12	227.0269	57.92	4.86	31.08	6.14
C11H11C12N3	256.14	255.0330	51.58	4.33	27.68	16.41
C11H11C12N5	284.15	283.0391	46.50	3.90	24.95	24.65
C11H12C11N1	193.68	193.0658	68.22	6.25	18.31	7.23
C11H12C11N3	221.69	221.0720	59.60	5.46	15.99	18.95
C11H12C11N5	249.70	249.0781	52.91	4.84	14.20	28.05
C11H12C12N2	243.14	242.0377	54.34	4.97	29.16	11.52
C11H12C12N4	271.15	270.0439	48.73	4.46	26.15	20.66
C11H12C12N6	299.16	298.0500	44.16	4.04	23.70	28.09
C11H13C11N2	208.69	208.0767	63.31	6.28	16.99	13.42
C11H13C11N4	236.71	236.0829	55.82	5.54	14.98	23.67
C11H13C11N6	264.72	264.0890	49.91	4.95	13.39	31.75
C11H13C12N1	230.14	229.0425	57.41	5.69	30.81	6.09
C11H13C12N3	258.15	257.0486	51.18	5.08	27.47	16.28
C11H13C12N5	286.17	285.0548	46.17	4.58	24.78	24.47
C11H14C11N1	195.69	195.0815	67.51	7.21	18.12	7.16
C11H14C11N3	223.71	223.0876	59.06	6.31	15.85	18.78
C11H14C11N5	251.72	251.0938	52.49	5.61	14.08	27.82
C11H14C12N2	245.15	244.0534	53.89	5.76	28.92	11.43
C11H14C12N4	273.17	272.0595	48.37	5.17	25.96	20.51
C11H14C12N6	301.18	300.0657	43.87	4.69	23.54	27.90
C11H15C11N2	210.71	210.0924	62.70	7.18	16.83	13.29
C11H15C11N4	238.72	238.0985	55.35	6.33	14.85	23.47
C11H15C11N6	266.74	266.1047	49.53	5.67	13.29	31.51
C11H15C12N1	232.15	231.0581	56.91	6.51	30.54	6.03
C11H15C12N3	260.17	259.0643	50.78	5.81	27.25	16.15
C11H15C12N5	288.18	287.0704	45.85	5.25	24.60	24.30
C11H16C11N1	197.71	197.0971	66.83	8.16	17.93	7.08
C11H16C11N3	225.72	225.1033	58.53	7.14	15.71	18.62
C11H16C11N5	253.74	253.1094	52.07	6.36	13.97	27.60
C11H16C12N2	247.17	246.0690	53.45	6.52	28.69	11.33
C11H16C12N4	275.18	274.0752	48.01	5.86	25.77	20.36
C11H16C12N6	303.20	302.0813	43.58	5.32	23.39	27.72
C11H17C11N2	212.72	212.1080	62.11	8.06	16.67	13.17
C11H17C11N4	240.74	240.1142	54.88	7.12	14.73	23.27
C11H17C11N6	268.75	268.1203	49.16	6.38	13.19	31.27
C11H17C12N1	234.17	233.0738	56.42	7.32	30.28	5.98
C11H17C12N3	262.18	261.0799	50.39	6.54	27.04	16.03
C11H17C12N5	290.20	289.0861	45.53	5.90	24.43	24.13
C11H18C11N1	199.73	199.1128	66.15	9.08	17.75	7.01
C11H18C11N3	227.74	227.1189	58.01	7.97	15.57	18.45
C11H18C11N5	255.75	255.1251	51.66	7.09	13.86	27.38
C11H18C12N2	249.19	248.0847	53.02	7.28	28.46	11.24
C11H18C12N4	277.20	276.0908	47.66	6.55	25.58	20.21
C11H18C12N6	305.21	304.0970	43.29	5.94	23.23	27.53
C11H19C11N2	214.74	214.1237	61.53	8.92	16.51	13.05
C11H19C11N4	242.75	242.1298	54.43	7.89	14.60	23.08
C11H19C11N6	270.77	270.1360	48.80	7.07	13.09	31.04
C11H19C12N1	236.19	235.0894	55.94	8.11	30.02	5.93
C11H19C12N3	264.20	263.0956	50.01	7.25	26.84	15.90
C11H19C12N5	292.21	291.1017	45.21	6.55	24.27	23.97
C11H20C11N1	201.74	201.1284	65.49	9.99	17.57	6.94
C11H20C11N3	229.76	229.1346	57.51	8.77	15.43	18.29
C11H20C11N5	257.77	257.1407	51.26	7.82	13.75	27.17
C11H20C12N2	251.20	250.1003	52.60	8.03	28.23	11.15
C11H20C12N4	279.21	278.1065	47.32	7.22	25.39	20.07
C11H20C12N6	307.23	306.1126	43.00	6.56	23.08	27.35
C11H21C11N2	216.76	216.1393	60.95	9.77	16.36	12.92
C11H21C11N4	244.77	244.1455	53.98	8.65	14.48	22.89
C11H21C11N6	272.78	272.1516	48.44	7.76	13.00	30.81
C11H21C12N1	238.20	237.1051	55.47	8.89	29.77	5.88

C11H21Cl2N3	266.22	265.1112	49.63	7.95	26.63	15.78
C11H21Cl2N5	294.23	293.1174	44.90	7.19	24.10	23.80
C11H22Cl1N1	203.76	203.1441	64.84	10.88	17.40	6.87
C11H22Cl1N3	231.77	231.1502	57.01	9.57	15.30	18.13
C11H22Cl1N5	259.78	259.1564	50.86	8.54	13.65	26.96
C11H22Cl2N2	253.22	252.1160	52.18	8.76	28.00	11.06
C11H22Cl2N4	281.23	280.1221	46.98	7.89	25.21	19.92
C11H22Cl2N6	309.24	308.1283	42.72	7.17	22.93	27.18
C11H23Cl1N2	218.77	218.1550	60.39	10.60	16.21	12.80
C11H23Cl1N4	246.79	246.1611	53.54	9.39	14.37	22.70
C11H23Cl1N6	274.80	274.1673	48.08	8.44	12.90	30.58
C11H23Cl2N1	240.22	239.1207	55.00	9.65	29.52	5.83
C11H23Cl2N3	268.23	267.1269	49.26	8.64	26.43	15.67
C11H23Cl2N5	296.25	295.1330	44.60	7.83	23.93	23.64
C11H24Cl1N1	205.77	205.1597	64.21	11.76	17.23	6.81
C11H24Cl1N3	233.79	233.1659	56.51	10.35	15.16	17.97
C11H24Cl1N5	261.80	261.1720	50.47	9.24	13.54	26.75
C11H24Cl2N2	255.23	254.1316	51.77	9.48	27.78	10.98
C11H24Cl2N4	283.25	282.1378	46.65	8.54	25.03	19.78
C11H24Cl2N6	311.26	310.1439	42.45	7.77	22.78	27.00
C11H25Cl1N2	220.79	220.1706	59.84	11.41	16.06	12.69
C11H25Cl1N4	248.80	248.1768	53.10	10.13	14.25	22.52
C11H25Cl1N6	276.82	276.1829	47.73	9.10	12.81	30.36
C11H25Cl2N3	270.25	269.1425	48.89	9.32	26.24	15.55
C11H25Cl2N5	298.26	297.1487	44.30	8.45	23.77	23.48
C11H26Cl1N3	235.80	235.1815	56.03	11.11	15.04	17.82
C11H26Cl1N5	263.82	263.1877	50.08	9.93	13.44	26.55
C11H26Cl2N4	285.26	284.1534	46.32	9.19	24.86	19.64
C11H26Cl2N6	313.28	312.1596	42.17	8.37	22.63	26.83
C11H27Cl1N4	250.82	250.1924	52.68	10.85	14.13	22.34
C11H27Cl1N6	278.83	278.1986	47.38	9.76	12.71	30.14
C11H27Cl2N5	300.28	299.1643	44.00	9.06	23.61	23.32
C11H28Cl1N5	265.83	265.2033	49.70	10.62	13.34	26.34
C11H28Cl2N6	315.29	314.1752	41.90	8.95	22.49	26.65
C11H29Cl1N6	280.85	280.2142	47.04	10.41	12.62	29.92
C12H 6Cl1N1	199.64	199.0189	72.20	3.03	17.76	7.02
C12H 6Cl1N3	227.65	227.0250	63.31	2.66	15.57	18.46
C12H 6Cl1N5	255.67	255.0312	56.38	2.37	13.87	27.39
C12H 6Cl2N2	249.10	247.9908	57.86	2.43	28.46	11.25
C12H 6Cl2N4	277.11	275.9969	52.01	2.18	25.59	20.22
C12H 6Cl2N6	305.13	304.0031	47.24	1.98	23.24	27.54
C12H 7Cl1N2	214.66	214.0298	67.15	3.29	16.52	13.05
C12H 7Cl1N4	242.67	242.0359	59.40	2.91	14.61	23.09
C12H 7Cl1N6	270.68	270.0421	53.25	2.61	13.10	31.05
C12H 7Cl2N1	236.10	234.9956	61.05	2.99	30.03	5.93
C12H 7Cl2N3	264.12	263.0017	54.57	2.67	26.85	15.91
C12H 7Cl2N5	292.13	291.0078	49.34	2.42	24.27	23.97
C12H 8Cl1N1	201.66	201.0345	71.47	4.00	17.58	6.95
C12H 8Cl1N3	229.67	229.0407	62.76	3.51	15.44	18.30
C12H 8Cl1N5	257.68	257.0468	55.93	3.13	13.76	27.18
C12H 8Cl2N2	251.12	250.0065	57.40	3.21	28.24	11.16
C12H 8Cl2N4	279.13	278.0126	51.64	2.89	25.40	20.07
C12H 8Cl2N6	307.14	306.0187	46.93	2.63	23.09	27.36
C12H 9Cl1N2	216.67	216.0454	66.52	4.19	16.36	12.93
C12H 9Cl1N4	244.69	244.0516	58.91	3.71	14.49	22.90
C12H 9Cl1N6	272.70	272.0577	52.85	3.33	13.00	30.82
C12H 9Cl2N1	238.12	237.0112	60.53	3.81	29.78	5.88
C12H 9Cl2N3	266.13	265.0173	54.16	3.41	26.64	15.79
C12H 9Cl2N5	294.15	293.0235	49.00	3.08	24.11	23.81
C12H10Cl1N1	203.67	203.0502	70.77	4.95	17.41	6.88
C12H10Cl1N3	231.69	231.0563	62.21	4.35	15.30	18.14
C12H10Cl1N5	259.70	259.0625	55.50	3.88	13.65	26.97
C12H10Cl2N2	253.13	252.0221	56.94	3.98	28.01	11.07
C12H10Cl2N4	281.15	280.0282	51.27	3.59	25.22	19.93
C12H10Cl2N6	309.16	308.0344	46.62	3.26	22.94	27.18
C12H11Cl1N2	218.69	218.0611	65.91	5.07	16.21	12.81
C12H11Cl1N4	246.70	246.0672	58.42	4.49	14.37	22.71
C12H11Cl1N6	274.71	274.0734	52.47	4.04	12.91	30.59
C12H11Cl2N1	240.13	239.0269	60.02	4.62	29.53	5.83
C12H11Cl2N3	268.15	267.0330	53.75	4.13	26.44	15.67
C12H11Cl2N5	296.16	295.0391	48.67	3.74	23.94	23.65
C12H12Cl1N1	205.69	205.0658	70.07	5.88	17.24	6.81
C12H12Cl1N3	233.70	233.0720	61.67	5.18	15.17	17.98
C12H12Cl1N5	261.72	261.0781	55.07	4.62	13.55	26.76
C12H12Cl2N2	255.15	254.0377	56.49	4.74	27.79	10.98
C12H12Cl2N4	283.16	282.0439	50.90	4.27	25.04	19.79
C12H12Cl2N6	311.18	310.0500	46.32	3.89	22.79	27.01
C12H13Cl1N2	220.70	220.0767	65.31	5.94	16.06	12.69
C12H13Cl1N4	248.72	248.0829	57.95	5.27	14.25	22.53
C12H13Cl1N6	276.73	276.0890	52.08	4.74	12.81	30.37
C12H13Cl2N1	242.15	241.0425	59.52	5.41	29.28	5.78
C12H13Cl2N3	270.16	269.0486	53.35	4.85	26.25	15.55
C12H13Cl2N5	298.18	297.0548	48.34	4.39	23.78	23.49
C12H14Cl1N1	207.71	207.0815	69.39	6.79	17.07	6.74
C12H14Cl1N3	235.72	235.0876	61.15	5.99	15.04	17.83
C12H14Cl1N5	263.73	263.0938	54.65	5.35	13.44	26.55
C12H14Cl2N2	257.16	256.0534	56.05	5.49	27.57	10.89
C12H14Cl2N4	285.18	284.0595	50.54	4.95	24.86	19.65
C12H14Cl2N6	313.19	312.0657	46.02	4.51	22.64	26.83
C12H15Cl1N2	222.72	222.0924	64.72	6.79	15.92	12.58
C12H15Cl1N4	250.73	250.0985	57.48	6.03	14.14	22.35
C12H15Cl1N6	278.75	278.1047	51.71	5.42	12.72	30.15
C12H15Cl2N1	244.17	243.0581	59.03	6.19	29.04	5.74
C12H15Cl2N3	272.18	271.0643	52.96	5.55	26.05	15.44
C12H15Cl2N5	300.19	299.0704	48.01	5.04	23.62	23.33
C12H16Cl1N1	209.72	209.0971	68.73	7.69	16.90	6.68
C12H16Cl1N3	237.73	237.1033	60.63	6.78	14.91	17.68
C12H16Cl1N5	265.75	265.1094	54.24	6.07	13.34	26.35
C12H16Cl2N2	259.18	258.0690	55.61	6.22	27.36	10.81
C12H16Cl2N4	287.19	286.0752	50.19	5.62	24.69	19.51
C12H16Cl2N6	315.21	314.0813	45.73	5.12	22.50	26.66
C12H17Cl1N2	224.74	224.1080	64.13	7.62	15.78	12.47
C12H17Cl1N4	252.75	252.1142	57.03	6.78	14.03	22.17
C12H17Cl1N6	280.76	280.1203	51.34	6.10	12.63	29.93
C12H17Cl2N1	246.18	245.0738	58.55	6.96	28.80	5.69
C12H17Cl2N3	274.20	273.0799	52.57	6.25	25.86	15.32
C12H17Cl2N5	302.21	301.0861	47.69	5.67	23.46	23.17
C12H18Cl1N1	211.74	211.1128	68.07	8.57	16.74	6.62
C12H18Cl1N3	239.75	239.1189	60.12	7.57	14.79	17.53
C12H18Cl1N5	267.76	267.1251	53.83	6.78	13.24	26.15
C12H18Cl2N2	261.20	260.0847	55.18	6.95	27.15	10.73
C12H18Cl2N4	289.21	288.0908	49.84	6.27	24.52	19.37
C12H18Cl2N6	317.22	316.0970	45.44	5.72	22.35	26.49
C12H19Cl1N2	226.75	226.1237	63.56	8.45	15.64	12.35
C12H19Cl1N4	254.77	254.1298	56.58	7.52	13.92	21.99
C12H19Cl1N6	282.78	282.1360	50.97	6.77	12.54	29.72
C12H19Cl2N1	248.20	247.0894	58.07	7.72	28.57	5.64
C12H19Cl2N3	276.21	275.0956	52.18	6.93	25.67	15.21
C12H19Cl2N5	304.22	303.1017	47.38	6.30	23.31	23.02
C12H20Cl1N1	213.75	213.1284	67.43	9.43	16.59	6.55

C12H20Cl1N3	241.77	241.1346	59.62	8.34	14.66	17.38
C12H20Cl1N5	269.78	269.1407	53.43	7.47	13.14	25.96
C12H20Cl2N2	263.21	262.1003	54.76	7.66	26.94	10.64
C12H20Cl2N4	291.23	290.1065	49.49	6.92	24.35	19.24
C12H20Cl2N6	319.24	318.1126	45.15	6.31	22.21	26.33
C12H21Cl1N2	228.77	228.1393	63.00	9.25	15.50	12.25
C12H21Cl1N4	256.78	256.1455	56.13	8.24	13.81	21.82
C12H21Cl1N6	284.79	284.1516	50.61	7.43	12.45	29.51
C12H21Cl2N1	250.21	249.1051	57.60	8.46	28.34	5.60
C12H21Cl2N3	278.23	277.1112	51.80	7.61	25.48	15.10
C12H21Cl2N5	306.24	305.1174	47.07	6.91	23.15	22.87
C12H22Cl1N1	215.77	215.1441	66.80	10.28	16.43	6.49
C12H22Cl1N3	243.78	243.1502	59.12	9.10	14.54	17.24
C12H22Cl1N5	271.80	271.1564	53.03	8.16	13.04	25.77
C12H22Cl2N2	265.23	264.1160	54.34	8.36	26.73	10.56
C12H22Cl2N4	293.24	292.1221	49.15	7.56	24.18	19.11
C12H22Cl2N6	321.26	320.1283	44.87	6.90	22.07	26.16
C12H23Cl1N2	230.78	230.1550	62.45	10.05	15.36	12.14
C12H23Cl1N4	258.80	258.1611	55.69	8.96	13.70	21.65
C12H23Cl1N6	286.81	286.1673	50.25	8.08	12.36	29.30
C12H23Cl2N1	252.23	251.1207	57.14	9.19	28.11	5.55
C12H23Cl2N3	280.24	279.1269	51.43	8.27	25.30	14.99
C12H23Cl2N5	308.26	307.1330	46.76	7.52	23.00	22.72
C12H24Cl1N1	217.78	217.1597	66.18	11.11	16.28	6.43
C12H24Cl1N3	245.80	245.1659	58.64	9.84	14.42	17.10
C12H24Cl1N5	273.81	273.1720	52.64	8.84	12.95	25.58
C12H24Cl2N2	267.24	266.1316	53.93	9.05	26.53	10.48
C12H24Cl2N4	295.26	294.1378	48.82	8.19	24.01	18.98
C12H24Cl2N6	323.27	322.1439	44.59	7.48	21.93	26.00
C12H25Cl1N2	232.80	232.1706	61.91	10.82	15.23	12.03
C12H25Cl1N4	260.81	260.1768	55.26	9.66	13.59	21.48
C12H25Cl1N6	288.83	288.1829	49.90	8.72	12.27	29.10
C12H25Cl2N1	254.25	253.1364	56.69	9.91	27.89	5.51
C12H25Cl2N3	282.26	281.1425	51.06	8.93	25.12	14.89
C12H25Cl2N5	310.27	309.1487	46.45	8.12	22.85	22.57
C12H26Cl1N1	219.80	219.1754	65.57	11.92	16.13	6.37
C12H26Cl1N3	247.81	247.1815	58.16	10.58	14.31	16.96
C12H26Cl1N5	275.83	275.1877	52.26	9.50	12.85	25.39
C12H26Cl2N2	269.26	268.1473	53.53	9.73	26.33	10.40
C12H26Cl2N4	297.27	296.1534	48.49	8.82	23.85	18.85
C12H26Cl2N6	325.29	324.1596	44.31	8.06	21.80	25.84
C12H27Cl1N2	234.82	234.1863	61.38	11.59	15.10	11.93
C12H27Cl1N4	262.83	262.1924	54.84	10.35	13.49	21.32
C12H27Cl1N6	290.84	290.1986	49.56	9.36	12.19	28.90
C12H27Cl2N3	284.28	283.1582	50.70	9.57	24.94	14.78
C12H27Cl2N5	312.29	311.1643	46.15	8.71	22.71	22.43
C12H28Cl1N3	249.83	249.1972	57.69	11.30	14.19	16.82
C12H28Cl1N5	277.84	277.2033	51.88	10.16	12.76	25.21
C12H28Cl2N4	299.29	298.1691	48.16	9.43	23.69	18.72
C12H28Cl2N6	327.30	326.1752	44.04	8.62	21.66	25.68
C12H29Cl1N4	264.84	264.2081	54.42	11.04	13.39	21.15
C12H29Cl1N6	292.86	292.2142	49.22	9.98	12.11	28.70
C12H29Cl2N5	314.30	313.1800	45.86	9.30	22.56	22.28
C12H30Cl1N5	279.86	279.2190	51.50	10.81	12.67	25.02
C12H30Cl2N6	329.32	328.1909	43.77	9.18	21.53	25.52
C12H31Cl1N6	294.87	294.2299	48.88	10.60	12.02	28.50
C13H 6Cl1N1	211.65	211.0189	73.77	2.86	16.75	6.62
C13H 6Cl1N3	239.67	239.0250	65.15	2.52	14.79	17.53
C13H 6Cl1N5	267.68	267.0312	58.33	2.26	13.24	26.16
C13H 6Cl2N2	261.11	259.9908	59.80	2.32	27.16	10.73
C13H 6Cl2N4	289.13	287.9969	54.01	2.09	24.52	19.38
C13H 6Cl2N6	317.14	316.0031	49.24	1.91	22.36	26.50
C13H 7Cl1N2	226.67	226.0298	68.89	3.11	15.64	12.36
C13H 7Cl1N4	254.68	254.0359	61.31	2.77	13.92	22.00
C13H 7Cl1N6	282.69	282.0421	55.23	2.50	12.54	29.73
C13H 7Cl2N1	248.11	246.9956	62.93	2.84	28.58	5.65
C13H 7Cl2N3	276.13	275.0017	56.55	2.56	25.68	15.22
C13H 7Cl2N5	304.14	303.0078	51.34	2.32	23.31	23.03
C13H 8Cl1N1	213.67	213.0345	73.08	3.77	16.59	6.56
C13H 8Cl1N3	241.68	241.0407	64.61	3.34	14.67	17.39
C13H 8Cl1N5	269.70	269.0468	57.90	2.99	13.15	25.97
C13H 8Cl2N2	263.13	262.0065	59.34	3.06	26.95	10.65
C13H 8Cl2N4	291.14	290.0126	53.63	2.77	24.35	19.24
C13H 8Cl2N6	319.15	318.0187	48.92	2.53	22.22	26.33
C13H 9Cl1N2	228.68	228.0454	68.28	3.97	15.50	12.25
C13H 9Cl1N4	256.70	256.0516	60.83	3.53	13.81	21.83
C13H 9Cl1N6	284.71	284.0577	54.84	3.19	12.45	29.52
C13H 9Cl2N1	250.13	249.0112	62.43	3.63	28.35	5.60
C13H 9Cl2N3	278.14	277.0173	56.14	3.26	25.49	15.11
C13H 9Cl2N5	306.16	305.0235	51.00	2.96	23.16	22.88
C13H10Cl1N1	215.68	215.0502	72.40	4.67	16.44	6.49
C13H10Cl1N3	243.70	243.0563	64.07	4.14	14.55	17.24
C13H10Cl1N5	271.71	271.0625	57.47	3.71	13.05	25.77
C13H10Cl2N2	265.14	264.0221	58.89	3.80	26.74	10.57
C13H10Cl2N4	293.16	292.0282	53.26	3.44	24.19	19.11
C13H10Cl2N6	321.17	320.0344	48.62	3.14	22.08	26.17
C13H11Cl1N2	230.70	230.0611	67.68	4.81	15.37	12.14
C13H11Cl1N4	258.71	258.0672	60.35	4.29	13.70	21.66
C13H11Cl1N6	286.73	286.0734	54.46	3.87	12.36	29.31
C13H11Cl2N1	252.15	251.0269	61.93	4.40	28.12	5.56
C13H11Cl2N3	280.16	279.0330	55.73	3.96	25.31	15.00
C13H11Cl2N5	308.17	307.0391	50.67	3.60	23.01	22.73
C13H12Cl1N1	217.70	217.0658	71.72	5.56	16.29	6.43
C13H12Cl1N3	245.71	245.0720	63.55	4.92	14.43	17.10
C13H12Cl1N5	273.73	273.0781	57.04	4.42	12.95	25.59
C13H12Cl2N2	267.16	266.0377	58.45	4.53	26.54	10.49
C13H12Cl2N4	295.17	294.0439	52.90	4.10	24.02	18.98
C13H12Cl2N6	323.19	322.0500	48.31	3.74	21.94	26.00
C13H13Cl1N2	232.71	232.0767	67.10	5.63	15.23	12.04
C13H13Cl1N4	260.73	260.0829	59.89	5.03	13.60	21.49
C13H13Cl1N6	288.74	288.0890	54.08	4.54	12.28	29.11
C13H13Cl2N1	254.16	253.0425	61.44	5.16	27.90	5.51
C13H13Cl2N3	282.17	281.0486	55.34	4.64	25.13	14.89
C13H13Cl2N5	310.19	309.0548	50.34	4.22	22.86	22.58
C13H14Cl1N1	219.72	219.0815	71.07	6.42	16.14	6.37
C13H14Cl1N3	247.73	247.0876	63.03	5.70	14.31	16.96
C13H14Cl1N5	275.74	275.0938	56.63	5.12	12.86	25.40
C13H14Cl2N2	269.18	268.0534	58.01	5.24	26.34	10.41
C13H14Cl2N4	297.19	296.0595	52.54	4.75	23.86	18.85
C13H14Cl2N6	325.20	324.0657	48.01	4.34	21.80	25.84
C13H15Cl1N2	234.73	234.0924	66.52	6.44	15.10	11.93
C13H15Cl1N4	262.74	262.0985	59.43	5.75	13.49	21.32
C13H15Cl1N6	290.76	290.1047	53.70	5.20	12.19	28.90
C13H15Cl2N1	256.18	255.0581	60.95	5.90	27.68	5.47
C13H15Cl2N3	284.19	283.0643	54.94	5.32	24.95	14.79
C13H15Cl2N5	312.20	311.0704	50.01	4.84	22.71	22.43
C13H16Cl1N1	221.73	221.0971	70.42	7.27	15.99	6.32
C13H16Cl1N3	249.75	249.1033	62.52	6.46	14.20	16.83
C13H16Cl1N5	277.76	277.1094	56.22	5.81	12.76	25.21
C13H16Cl2N2	271.19	270.0690	57.58	5.95	26.15	10.33

C13H16C12N4	299.21	298.0752	52.19	5.39	23.70	18.73
C13H16C12N6	327.22	326.0813	47.72	4.93	21.67	25.68
C13H17C11N2	236.75	236.1080	65.95	7.24	14.98	11.83
C13H17C11N4	264.76	264.1142	58.98	6.47	13.39	21.16
C13H17C11N6	292.77	292.1203	53.33	5.85	12.11	28.70
C13H17C12N1	258.19	257.0738	60.48	6.64	27.46	5.42
C13H17C12N3	286.21	285.0799	54.56	5.99	24.77	14.68
C13H17C12N5	314.22	313.0861	49.69	5.45	22.57	22.29
C13H18C11N1	223.75	223.1128	69.79	8.11	15.85	6.26
C13H18C11N3	251.76	251.1189	62.02	7.21	14.08	16.69
C13H18C11N5	279.77	279.1251	55.81	6.49	12.67	25.03
C13H18C12N2	273.21	272.0847	57.15	6.64	25.95	10.25
C13H18C12N4	301.22	300.0908	51.84	6.02	23.54	18.60
C13H18C12N6	329.23	328.0970	47.43	5.51	21.54	25.53
C13H19C11N2	238.76	238.1237	65.40	8.02	14.85	11.73
C13H19C11N4	266.78	266.1298	58.53	7.18	13.29	21.00
C13H19C11N6	294.79	294.1360	52.97	6.50	12.03	28.51
C13H19C12N1	260.21	259.0894	60.01	7.36	27.25	5.38
C13H19C12N3	288.22	287.0956	54.18	6.64	24.60	14.58
C13H19C12N5	316.24	315.1017	49.38	6.06	22.42	22.15
C13H20C11N1	225.76	225.1284	69.16	8.93	15.70	6.20
C13H20C11N3	253.78	253.1346	61.53	7.94	13.97	16.56
C13H20C11N5	281.79	281.1407	55.41	7.15	12.58	24.85
C13H20C12N2	275.22	274.1003	56.73	7.32	25.76	10.18
C13H20C12N4	303.24	302.1065	51.49	6.65	23.38	18.48
C13H20C12N6	331.25	330.1126	47.14	6.09	21.41	25.37
C13H21C11N2	240.78	240.1393	64.85	8.79	14.72	11.63
C13H21C11N4	268.79	268.1455	58.09	7.87	13.19	20.84
C13H21C11N6	296.81	296.1516	52.61	7.13	11.94	28.31
C13H21C12N1	262.23	261.1051	59.55	8.07	27.04	5.34
C13H21C12N3	290.24	289.1112	53.80	7.29	24.43	14.48
C13H21C12N5	318.25	317.1174	49.06	6.65	22.28	22.01
C13H22C11N1	227.78	227.1441	68.55	9.74	15.56	6.15
C13H22C11N3	255.79	255.1502	61.04	8.67	13.86	16.43
C13H22C11N5	283.81	283.1564	55.02	7.81	12.49	24.68
C13H22C12N2	277.24	276.1160	56.32	8.00	25.58	10.10
C13H22C12N4	305.25	304.1221	51.15	7.26	23.23	18.35
C13H22C12N6	333.27	332.1283	46.85	6.65	21.28	25.22
C13H23C11N2	242.79	242.1550	64.31	9.55	14.60	11.54
C13H23C11N4	270.81	270.1611	57.66	8.56	13.09	20.69
C13H23C11N6	298.82	298.1673	52.25	7.76	11.86	28.12
C13H23C12N1	264.24	263.1207	59.09	8.77	26.83	5.30
C13H23C12N3	292.25	291.1269	53.43	7.93	24.26	14.38
C13H23C12N5	320.27	319.1330	48.75	7.24	22.14	21.87
C13H24C11N1	229.80	229.1597	67.95	10.53	15.43	6.10
C13H24C11N3	257.81	257.1659	60.57	9.38	13.75	16.30
C13H24C11N5	285.82	285.1720	54.63	8.46	12.40	24.50
C13H24C12N2	279.26	278.1316	55.91	8.66	25.39	10.03
C13H24C12N4	307.27	306.1378	50.82	7.87	23.08	18.23
C13H24C12N6	335.28	334.1439	46.57	7.22	21.15	25.07
C13H25C11N2	244.81	244.1706	63.78	10.29	14.48	11.44
C13H25C11N4	272.82	272.1768	57.23	9.24	12.99	20.54
C13H25C11N6	300.84	300.1829	51.90	8.38	11.78	27.94
C13H25C12N1	266.26	265.1364	58.64	9.46	26.63	5.26
C13H25C12N3	294.27	293.1425	53.06	8.56	24.10	14.28
C13H25C12N5	322.28	321.1487	48.45	7.82	22.00	21.73
C13H26C11N1	231.81	231.1754	67.36	11.31	15.29	6.04
C13H26C11N3	259.83	259.1815	60.10	10.09	13.64	16.17
C13H26C11N5	287.84	287.1877	54.25	9.10	12.32	24.33
C13H26C12N2	281.27	280.1473	55.51	9.32	25.21	9.96
C13H26C12N4	309.28	308.1534	50.49	8.47	22.93	18.11
C13H26C12N6	337.30	336.1596	46.29	7.77	21.02	24.92
C13H27C11N2	246.83	246.1863	63.26	11.03	14.36	11.35
C13H27C11N4	274.84	274.1924	56.81	9.90	12.90	20.39
C13H27C11N6	302.85	302.1986	51.56	8.99	11.71	27.75
C13H27C12N1	268.27	267.1520	58.20	10.14	26.43	5.22
C13H27C12N3	296.29	295.1582	52.70	9.19	23.93	14.18
C13H27C12N5	324.30	323.1643	48.15	8.39	21.86	21.60
C13H28C11N1	233.83	233.1910	66.78	12.07	15.16	5.99
C13H28C11N3	261.84	261.1972	59.63	10.78	13.54	16.05
C13H28C11N5	289.85	289.2033	53.87	9.74	12.23	24.16
C13H28C12N2	283.29	282.1629	55.12	9.96	25.03	9.89
C13H28C12N4	311.30	310.1691	50.16	9.07	22.78	18.00
C13H28C12N6	339.31	338.1752	46.02	8.32	20.90	24.77
C13H29C11N2	248.84	248.2019	62.75	11.75	14.25	11.26
C13H29C11N4	276.86	276.2081	56.40	10.56	12.81	20.24
C13H29C11N6	304.87	304.2142	51.22	9.59	11.63	27.57
C13H29C12N3	298.30	297.1738	52.34	9.80	23.77	14.09
C13H29C12N5	326.32	325.1800	47.85	8.96	21.73	21.46
C13H30C11N3	263.86	263.2128	59.18	11.46	13.44	15.93
C13H30C11N5	291.87	291.2190	53.50	10.36	12.15	23.99
C13H30C12N4	313.32	312.1847	49.84	9.65	22.63	17.88
C13H30C12N6	341.33	340.1909	45.75	8.86	20.77	24.62
C13H31C11N4	278.87	278.2237	55.99	11.20	12.71	20.09
C13H31C11N6	306.89	306.2299	50.88	10.18	11.55	27.38
C13H31C12N5	328.33	327.1956	47.56	9.52	21.60	21.33
C13H32C11N5	293.89	293.2346	53.13	10.98	12.06	23.83
C13H32C12N6	343.35	342.2065	45.48	9.39	20.65	24.48
C13H33C11N6	308.90	308.2455	50.55	10.77	11.48	27.21
C14H 7C11N2	238.68	238.0298	70.45	2.96	14.85	11.74
C14H 7C11N4	266.69	266.0359	63.05	2.65	13.29	21.01
C14H 7C11N6	294.71	294.0421	57.06	2.39	12.03	28.52
C14H 7C12N1	260.12	258.9956	64.64	2.71	27.26	5.38
C14H 7C12N3	288.14	287.0017	58.36	2.45	24.61	14.58
C14H 7C12N5	316.15	315.0078	53.19	2.23	22.43	22.15
C14H 8C11N1	225.68	225.0345	74.51	3.57	15.71	6.21
C14H 8C11N3	253.69	253.0407	66.28	3.18	13.97	16.56
C14H 8C11N5	281.71	281.0468	59.69	2.86	12.59	24.86
C14H 8C12N2	275.14	274.0065	61.12	2.93	25.77	10.18
C14H 8C12N4	303.15	302.0126	55.47	2.66	23.39	18.48
C14H 8C12N6	331.17	330.0187	50.78	2.43	21.41	25.38
C14H 9C11N2	240.69	240.0454	69.86	3.77	14.73	11.64
C14H 9C11N4	268.71	268.0516	62.58	3.38	13.19	20.85
C14H 9C11N6	296.72	296.0577	56.67	3.06	11.95	28.32
C14H 9C12N1	262.14	261.0112	64.15	3.46	27.05	5.34
C14H 9C12N3	290.15	289.0173	57.95	3.13	24.44	14.48
C14H 9C12N5	318.17	317.0235	52.85	2.85	22.29	22.01
C14H10C11N1	227.70	227.0502	73.85	4.43	15.57	6.15
C14H10C11N3	255.71	255.0563	65.76	3.94	13.86	16.43
C14H10C11N5	283.72	283.0625	59.27	3.55	12.50	24.68
C14H10C12N2	277.16	276.0221	60.67	3.64	25.58	10.11
C14H10C12N4	305.17	304.0282	55.10	3.30	23.24	18.36
C14H10C12N6	333.18	332.0344	50.47	3.03	21.28	25.22
C14H11C11N2	242.71	242.0611	69.28	4.57	14.61	11.54
C14H11C11N4	270.72	270.0672	62.11	4.10	13.10	20.70
C14H11C11N6	298.74	298.0734	56.29	3.71	11.87	28.13
C14H11C12N1	264.16	263.0269	63.66	4.20	26.84	5.30
C14H11C12N3	292.17	291.0330	57.55	3.79	24.27	14.38
C14H11C12N5	320.18	319.0391	52.52	3.46	22.15	21.87
C14H12C11N1	229.71	229.0658	73.20	5.27	15.43	6.10

C14H12C11N3	257.72	257.0720	65.25	4.69	13.76	16.30
C14H12C11N5	285.74	285.0781	58.85	4.23	12.41	24.51
C14H12C12N2	279.17	278.0377	60.23	4.33	25.40	10.03
C14H12C12N4	307.18	306.0439	54.74	3.94	23.08	18.24
C14H12C12N6	335.20	334.0500	50.17	3.61	21.15	25.07
C14H13C11N2	244.73	244.0767	68.71	5.35	14.49	11.45
C14H13C11N4	272.74	272.0829	61.65	4.80	13.00	20.54
C14H13C11N6	300.75	300.0890	55.91	4.36	11.79	27.94
C14H13C12N1	266.17	265.0425	63.18	4.92	26.64	5.26
C14H13C12N3	294.19	293.0486	57.16	4.45	24.10	14.28
C14H13C12N5	322.20	321.0548	52.19	4.07	22.01	21.74
C14H14C11N1	231.73	231.0815	72.57	6.09	15.30	6.04
C14H14C11N3	259.74	259.0876	64.74	5.43	13.65	16.18
C14H14C11N5	287.75	287.0938	58.44	4.90	12.32	24.34
C14H14C12N2	281.19	280.0534	59.80	5.02	25.22	9.96
C14H14C12N4	309.20	308.0595	54.38	4.56	22.93	18.12
C14H14C12N6	337.21	336.0657	49.87	4.18	21.03	24.92
C14H15C11N2	246.74	246.0924	68.15	6.13	14.37	11.35
C14H15C11N4	274.76	274.0985	61.20	5.50	12.90	20.39
C14H15C11N6	302.77	302.1047	55.54	4.99	11.71	27.76
C14H15C12N1	268.19	267.0581	62.70	5.64	26.44	5.22
C14H15C12N3	296.20	295.0643	56.77	5.10	23.94	14.19
C14H15C12N5	324.22	323.0704	51.87	4.66	21.87	21.60
C14H16C11N1	233.74	233.0971	71.94	6.90	15.17	5.99
C14H16C11N3	261.76	261.1033	64.24	6.16	13.54	16.05
C14H16C11N5	289.77	289.1094	58.03	5.57	12.23	24.17
C14H16C12N2	283.20	282.0690	59.38	5.69	25.04	9.89
C14H16C12N4	311.22	310.0752	54.03	5.18	22.78	18.00
C14H16C12N6	339.23	338.0813	49.57	4.75	20.90	24.77
C14H17C11N2	248.76	248.1080	67.60	6.89	14.25	11.26
C14H17C11N4	276.77	276.1142	60.76	6.19	12.81	20.24
C14H17C11N6	304.78	304.1203	55.17	5.62	11.63	27.57
C14H17C12N1	270.20	269.0738	62.23	6.34	26.24	5.18
C14H17C12N3	298.22	297.0799	56.39	5.75	23.78	14.09
C14H17C12N5	326.23	325.0861	51.55	5.25	21.73	21.47
C14H18C11N1	235.76	235.1128	71.33	7.70	15.04	5.94
C14H18C11N3	263.77	263.1189	63.75	6.88	13.44	15.93
C14H18C11N5	291.79	291.1251	57.63	6.22	12.15	24.00
C14H18C12N2	285.22	284.0847	58.96	6.36	24.86	9.82
C14H18C12N4	313.23	312.0908	53.68	5.79	22.64	17.89
C14H18C12N6	341.25	340.0970	49.28	5.32	20.78	24.63
C14H19C11N2	250.77	250.1237	67.05	7.64	14.14	11.17
C14H19C11N4	278.79	278.1298	60.32	6.87	12.72	20.10
C14H19C11N6	306.80	306.1360	54.81	6.24	11.56	27.39
C14H19C12N1	272.22	271.0894	61.77	7.04	26.05	5.15
C14H19C12N3	300.23	299.0956	56.01	6.38	23.62	14.00
C14H19C12N5	328.25	327.1017	51.23	5.83	21.60	21.34
C14H20C11N1	237.78	237.1284	70.72	8.48	14.91	5.89
C14H20C11N3	265.79	265.1346	63.27	7.58	13.34	15.81
C14H20C11N5	293.80	293.1407	57.23	6.86	12.07	23.84
C14H20C12N2	287.23	286.1003	58.54	7.02	24.69	9.75
C14H20C12N4	315.25	314.1065	53.34	6.39	22.49	17.77
C14H20C12N6	343.26	342.1126	48.99	5.87	20.66	24.48
C14H21C11N2	252.79	252.1393	66.52	8.37	14.02	11.08
C14H21C11N4	280.80	280.1455	59.88	7.54	12.63	19.95
C14H21C11N6	308.82	308.1516	54.45	6.85	11.48	27.21
C14H21C12N1	274.24	273.1051	61.32	7.72	25.86	5.11
C14H21C12N3	302.25	301.1112	55.63	7.00	23.46	13.90
C14H21C12N5	330.26	329.1174	50.92	6.41	21.47	21.21
C14H22C11N1	239.79	239.1441	70.13	9.25	14.78	5.84
C14H22C11N3	267.80	267.1502	62.79	8.28	13.24	15.69
C14H22C11N5	295.82	295.1564	56.84	7.50	11.98	23.67
C14H22C12N2	289.25	288.1160	58.14	7.67	24.51	9.68
C14H22C12N4	317.26	316.1221	53.00	6.99	22.35	17.66
C14H22C12N6	345.28	344.1283	48.70	6.42	20.54	24.34
C14H23C11N2	254.81	254.1550	65.99	9.10	13.91	10.99
C14H23C11N4	282.82	282.1611	59.46	8.20	12.54	19.81
C14H23C11N6	310.83	310.1673	54.10	7.46	11.41	27.04
C14H23C12N1	276.25	275.1207	60.87	8.39	25.67	5.07
C14H23C12N3	304.27	303.1269	55.27	7.62	23.30	13.81
C14H23C12N5	332.28	331.1330	50.61	6.98	21.34	21.08
C14H24C11N1	241.81	241.1597	69.54	10.00	14.66	5.79
C14H24C11N3	269.82	269.1659	62.32	8.97	13.14	15.57
C14H24C11N5	297.83	297.1720	56.46	8.12	11.90	23.51
C14H24C12N2	291.27	290.1316	57.73	8.31	24.34	9.62
C14H24C12N4	319.28	318.1378	52.67	7.58	22.21	17.55
C14H24C12N6	347.29	346.1439	48.42	6.97	20.42	24.20
C14H25C11N2	256.82	256.1706	65.48	9.81	13.80	10.91
C14H25C11N4	284.84	284.1768	59.04	8.85	12.45	19.67
C14H25C11N6	312.85	312.1829	53.75	8.05	11.33	26.86
C14H25C12N1	278.27	277.1364	60.43	9.06	25.48	5.03
C14H25C12N3	306.28	305.1425	54.90	8.23	23.15	13.72
C14H25C12N5	334.29	333.1487	50.30	7.54	21.21	20.95
C14H26C11N1	243.82	243.1754	68.97	10.75	14.54	5.74
C14H26C11N3	271.84	271.1815	61.86	9.64	13.04	15.46
C14H26C11N5	299.85	299.1877	56.08	8.74	11.82	23.36
C14H26C12N2	293.28	292.1473	57.34	8.94	24.18	9.55
C14H26C12N4	321.30	320.1534	52.34	8.16	22.07	17.44
C14H26C12N6	349.31	348.1596	48.14	7.50	20.30	24.06
C14H27C11N2	258.84	258.1863	64.97	10.51	13.70	10.82
C14H27C11N4	286.85	286.1924	58.62	9.49	12.36	19.53
C14H27C11N6	314.86	314.1986	53.41	8.64	11.26	26.69
C14H27C12N1	280.28	279.1520	59.99	9.71	25.30	5.00
C14H27C12N3	308.30	307.1582	54.54	8.83	23.00	13.63
C14H27C12N5	336.31	335.1643	50.00	8.09	21.08	20.82
C14H28C11N1	245.84	245.1910	68.40	11.48	14.42	5.70
C14H28C11N3	273.85	273.1972	61.40	10.31	12.95	15.34
C14H28C11N5	301.87	301.2033	55.71	9.35	11.74	23.20
C14H28C12N2	295.30	294.1629	56.94	9.56	24.01	9.49
C14H28C12N4	323.31	322.1691	52.01	8.73	21.93	17.33
C14H28C12N6	351.33	350.1752	47.86	8.03	20.18	23.92
C14H29C11N2	260.85	260.2019	64.46	11.21	13.59	10.74
C14H29C11N4	288.87	288.2081	58.21	10.12	12.27	19.40
C14H29C11N6	316.88	316.2142	53.07	9.22	11.19	26.52
C14H29C12N1	282.30	281.1677	59.57	10.35	25.12	4.96
C14H29C12N3	310.31	309.1738	54.19	9.42	22.85	13.54
C14H29C12N5	338.33	337.1800	49.70	8.64	20.96	20.70
C14H30C11N1	247.85	247.2067	67.84	12.20	14.30	5.65
C14H30C11N3	275.87	275.2128	60.96	10.96	12.85	15.23
C14H30C11N5	303.88	303.2190	55.34	9.95	11.67	23.05
C14H30C12N2	297.31	296.1786	56.56	10.17	23.85	9.42
C14H30C12N4	325.33	324.1847	51.69	9.29	21.80	17.22
C14H30C12N6	353.34	352.1909	47.59	8.56	20.07	23.78
C14H31C11N2	262.87	262.2176	63.97	11.89	13.49	10.66
C14H31C11N4	290.88	290.2237	57.81	10.74	12.19	19.26
C14H31C11N6	318.90	318.2299	52.73	9.80	11.12	26.35
C14H31C12N3	312.33	311.1895	53.84	10.00	22.70	13.45
C14H31C12N5	340.34	339.1956	49.41	9.18	20.83	20.58
C14H32C11N3	277.88	277.2285	60.51	11.61	12.76	15.12
C14H32C11N5	305.90	305.2346	54.97	10.54	11.59	22.89

C14H32C12N4	327.34	326.2004	51.37	9.85	21.66	17.12
C14H32C12N6	355.36	354.2065	47.32	9.08	19.95	23.65
C14H33C11N4	292.90	292.2394	57.41	11.36	12.10	19.13
C14H33C11N6	320.91	320.2455	52.40	10.37	11.05	26.19
C14H33C12N5	342.36	341.2113	49.12	9.72	20.71	20.46
C14H34C11N5	307.91	307.2503	54.61	11.13	11.51	22.74
C14H34C12N6	357.37	356.2222	47.05	9.59	19.84	23.52
C14H35C11N6	322.93	322.2612	52.07	10.92	10.98	26.02
C15H 7C11N2	250.69	250.0298	71.87	2.81	14.14	11.17
C15H 7C11N4	278.70	278.0359	64.64	2.53	12.72	20.10
C15H 7C11N6	306.72	306.0421	58.74	2.30	11.56	27.40
C15H 7C12N1	272.14	270.9956	66.20	2.59	26.06	5.15
C15H 7C12N3	300.15	299.0017	60.03	2.35	23.62	14.00
C15H 7C12N5	328.16	327.0078	54.90	2.15	21.61	21.34
C15H 8C11N1	237.69	237.0345	75.80	3.39	14.92	5.89
C15H 8C11N3	265.70	265.0407	67.81	3.03	13.34	15.81
C15H 8C11N5	293.72	293.0468	61.34	2.75	12.07	23.84
C15H 8C12N2	287.15	286.0065	62.74	2.81	24.69	9.76
C15H 8C12N4	315.16	314.0126	57.17	2.56	22.50	17.78
C15H 8C12N6	343.18	342.0187	52.50	2.35	20.66	24.49
C15H 9C11N2	252.71	252.0454	71.30	3.59	14.03	11.09
C15H 9C11N4	280.72	280.0516	64.18	3.23	12.63	19.96
C15H 9C11N6	308.73	308.0577	58.36	2.94	11.48	27.22
C15H 9C12N1	274.15	273.0112	65.72	3.31	25.86	5.11
C15H 9C12N3	302.17	301.0173	59.63	3.00	23.47	13.91
C15H 9C12N5	330.18	329.0235	54.57	2.75	21.48	21.21
C15H10C11N1	239.71	239.0502	75.16	4.21	14.79	5.84
C15H10C11N3	267.72	267.0563	67.30	3.77	13.24	15.70
C15H10C11N5	295.73	295.0625	60.92	3.41	11.99	23.68
C15H10C12N2	289.17	288.0221	62.31	3.49	24.52	9.69
C15H10C12N4	317.18	316.0282	56.80	3.18	22.36	17.66
C15H10C12N6	345.19	344.0344	52.19	2.92	20.54	24.35
C15H11C11N2	254.72	254.0611	70.73	4.35	13.92	11.00
C15H11C11N4	282.73	282.0672	63.72	3.92	12.54	19.82
C15H11C11N6	310.75	310.0734	57.98	3.57	11.41	27.04
C15H11C12N1	276.17	275.0269	65.24	4.01	25.67	5.07
C15H11C12N3	304.18	303.0330	59.23	3.65	23.31	13.81
C15H11C12N5	332.19	331.0391	54.24	3.34	21.34	21.08
C15H12C11N1	241.72	241.0658	74.53	5.00	14.67	5.79
C15H12C11N3	269.74	269.0720	66.79	4.48	13.14	15.58
C15H12C11N5	297.75	297.0781	60.51	4.06	11.91	23.52
C15H12C12N2	291.18	290.0377	61.87	4.15	24.35	9.62
C15H12C12N4	319.20	318.0439	56.44	3.79	22.21	17.55
C15H12C12N6	347.21	346.0500	51.89	3.48	20.42	24.20
C15H13C11N2	256.74	256.0767	70.18	5.10	13.81	10.91
C15H13C11N4	284.75	284.0829	63.27	4.60	12.45	19.68
C15H13C11N6	312.76	312.0890	57.60	4.19	11.34	26.87
C15H13C12N1	278.18	277.0425	64.77	4.71	25.49	5.04
C15H13C12N3	306.20	305.0486	58.84	4.28	23.16	13.72
C15H13C12N5	334.21	333.0548	53.91	3.92	21.22	20.95
C15H14C11N1	243.74	243.0815	73.92	5.79	14.55	5.75
C15H14C11N3	271.75	271.0876	66.30	5.19	13.05	15.46
C15H14C11N5	299.77	299.0938	60.10	4.71	11.83	23.36
C15H14C12N2	293.20	292.0534	61.45	4.81	24.18	9.55
C15H14C12N4	321.21	320.0595	56.09	4.39	22.07	17.44
C15H14C12N6	349.23	348.0657	51.59	4.04	20.30	24.06
C15H15C11N2	258.75	258.0924	69.63	5.84	13.70	10.83
C15H15C11N4	286.77	286.0985	62.83	5.27	12.36	19.54
C15H15C11N6	314.78	314.1047	57.24	4.80	11.26	26.70
C15H15C12N1	280.20	279.0581	64.30	5.40	25.31	5.00
C15H15C12N3	308.21	307.0643	58.46	4.91	23.01	13.63
C15H15C12N5	336.23	335.0704	53.59	4.50	21.09	20.83
C15H16C11N1	245.75	245.0971	73.31	6.56	14.43	5.70
C15H16C11N3	273.77	273.1033	65.81	5.89	12.95	15.35
C15H16C11N5	301.78	301.1094	59.70	5.34	11.75	23.21
C15H16C12N2	295.21	294.0690	61.03	5.46	24.02	9.49
C15H16C12N4	323.23	322.0752	55.74	4.99	21.94	17.33
C15H16C12N6	351.24	350.0813	51.29	4.59	20.19	23.93
C15H17C11N2	260.77	260.1080	69.09	6.57	13.60	10.74
C15H17C11N4	288.78	288.1142	62.39	5.93	12.28	19.40
C15H17C11N6	316.80	316.1203	56.87	5.41	11.19	26.53
C15H17C12N1	282.22	281.0738	63.84	6.07	25.12	4.96
C15H17C12N3	310.23	309.0799	58.08	5.52	22.86	13.54
C15H17C12N5	338.24	337.0861	53.27	5.07	20.96	20.71
C15H18C11N1	247.77	247.1128	72.72	7.32	14.31	5.65
C15H18C11N3	275.78	275.1189	65.33	6.58	12.86	15.24
C15H18C11N5	303.80	303.1251	59.31	5.97	11.67	23.05
C15H18C12N2	297.23	296.0847	60.62	6.10	23.86	9.42
C15H18C12N4	325.24	324.0908	55.39	5.58	21.80	17.23
C15H18C12N6	353.26	352.0970	51.00	5.14	20.07	23.79
C15H19C11N2	262.79	262.1237	68.56	7.29	13.49	10.66
C15H19C11N4	290.80	290.1298	61.96	6.59	12.19	19.27
C15H19C11N6	318.81	318.1360	56.51	6.01	11.12	26.36
C15H19C12N1	284.23	283.0894	63.39	6.74	24.95	4.93
C15H19C12N3	312.24	311.0956	57.70	6.13	22.71	13.46
C15H19C12N5	340.26	339.1017	52.95	5.63	20.84	20.58
C15H20C11N1	249.79	249.1284	72.13	8.07	14.19	5.61
C15H20C11N3	277.80	277.1346	64.86	7.26	12.76	15.13
C15H20C11N5	305.81	305.1407	58.91	6.59	11.59	22.90
C15H20C12N2	299.25	298.1003	60.21	6.74	23.69	9.36
C15H20C12N4	327.26	326.1065	55.05	6.16	21.67	17.12
C15H20C12N6	355.27	354.1126	50.71	5.67	19.96	23.66
C15H21C11N2	264.80	264.1393	68.04	7.99	13.39	10.58
C15H21C11N4	292.81	292.1455	61.53	7.23	12.11	19.13
C15H21C11N6	320.83	320.1516	56.16	6.60	11.05	26.19
C15H21C12N1	286.25	285.1051	62.94	7.39	24.77	4.89
C15H21C12N3	314.26	313.1112	57.33	6.74	22.56	13.37
C15H21C12N5	342.27	341.1174	52.64	6.18	20.72	20.46
C15H22C11N1	251.80	251.1441	71.55	8.81	14.08	5.56
C15H22C11N3	279.82	279.1502	64.39	7.92	12.67	15.02
C15H22C11N5	307.83	307.1564	58.53	7.20	11.52	22.75
C15H22C12N2	301.26	300.1160	59.80	7.36	23.54	9.30
C15H22C12N4	329.28	328.1221	54.72	6.73	21.53	17.02
C15H22C12N6	357.29	356.1283	50.43	6.21	19.85	23.52
C15H23C11N2	266.82	266.1550	67.52	8.69	13.29	10.50
C15H23C11N4	294.83	294.1611	61.11	7.86	12.02	19.00
C15H23C11N6	322.84	322.1673	55.81	7.18	10.98	26.03
C15H23C12N1	288.26	287.1207	62.50	8.04	24.60	4.86
C15H23C12N3	316.28	315.1269	56.97	7.33	22.42	13.29
C15H23C12N5	344.29	343.1330	52.33	6.73	20.59	20.34
C15H24C11N1	253.82	253.1597	70.98	9.53	13.97	5.52
C15H24C11N3	281.83	281.1659	63.93	8.58	12.58	14.91
C15H24C11N5	309.85	309.1720	58.15	7.81	11.44	22.60
C15H24C12N2	303.28	302.1316	59.41	7.98	23.38	9.24
C15H24C12N4	331.29	330.1378	54.38	7.30	21.40	16.91
C15H24C12N6	359.30	358.1439	50.14	6.73	19.73	23.39
C15H25C11N2	268.83	268.1706	67.02	9.37	13.19	10.42
C15H25C11N4	296.85	296.1768	60.69	8.49	11.94	18.87
C15H25C11N6	324.86	324.1829	55.46	7.76	10.91	25.87
C15H25C12N1	290.28	289.1364	62.07	8.68	24.43	4.83

C15H25C12N3	318.29	317.1425	56.60	7.92	22.28	13.20
C15H25C12N5	346.31	345.1487	52.03	7.28	20.47	20.22
C15H26C11N1	255.83	255.1754	70.42	10.24	13.86	5.47
C15H26C11N3	283.85	283.1815	63.47	9.23	12.49	14.80
C15H26C11N5	311.86	311.1877	57.77	8.40	11.37	22.46
C15H26C12N2	305.29	304.1473	59.01	8.58	23.23	9.18
C15H26C12N4	333.31	332.1534	54.05	7.86	21.27	16.81
C15H26C12N6	361.32	360.1596	49.86	7.25	19.62	23.26
C15H27C11N2	270.85	270.1863	66.52	10.05	13.09	10.34
C15H27C11N4	298.86	298.1924	60.28	9.11	11.86	18.75
C15H27C11N6	326.88	326.1986	55.12	8.33	10.85	25.71
C15H27C12N1	292.30	291.1520	61.64	9.31	24.26	4.79
C15H27C12N3	320.31	319.1582	56.25	8.50	22.14	13.12
C15H27C12N5	348.32	347.1643	51.72	7.81	20.36	20.11
C15H28C11N1	257.85	257.1910	69.87	10.95	13.75	5.43
C15H28C11N3	285.86	285.1972	63.03	9.87	12.40	14.70
C15H28C11N5	313.88	313.2033	57.40	8.99	11.30	22.31
C15H28C12N2	307.31	306.1629	58.63	9.18	23.07	9.12
C15H28C12N4	335.32	334.1691	53.73	8.42	21.15	16.71
C15H28C12N6	363.34	362.1752	49.59	7.77	19.52	23.13
C15H29C11N2	272.86	272.2019	66.03	10.71	12.99	10.27
C15H29C11N4	300.88	300.2081	59.88	9.72	11.78	18.62
C15H29C11N6	328.89	328.2142	54.78	8.89	10.78	25.55
C15H29C12N1	294.31	293.1677	61.22	9.93	24.09	4.76
C15H29C12N3	322.32	321.1738	55.90	9.07	22.00	13.04
C15H29C12N5	350.34	349.1800	51.43	8.34	20.24	19.99
C15H30C11N1	259.87	259.2067	69.33	11.64	13.64	5.39
C15H30C11N3	287.88	287.2128	62.58	10.50	12.32	14.60
C15H30C11N5	315.89	315.2190	57.03	9.57	11.22	22.17
C15H30C12N2	309.33	308.1786	58.25	9.78	22.92	9.06
C15H30C12N4	337.34	336.1847	53.41	8.96	21.02	16.61
C15H30C12N6	365.35	364.1909	49.31	8.28	19.41	23.00
C15H31C11N2	274.88	274.2176	65.54	11.37	12.90	10.19
C15H31C11N4	302.89	302.2237	59.48	10.32	11.70	18.50
C15H31C11N6	330.91	330.2299	54.45	9.44	10.71	25.40
C15H31C12N1	296.33	295.1833	60.80	10.54	23.93	4.73
C15H31C12N3	324.34	323.1895	55.55	9.63	21.86	12.96
C15H31C12N5	352.35	351.1956	51.13	8.87	20.12	19.88
C15H32C11N1	261.88	261.2223	68.80	12.32	13.54	5.35
C15H32C11N3	289.90	289.2285	62.15	11.13	12.23	14.49
C15H32C11N5	317.91	317.2346	56.67	10.15	11.15	22.03
C15H32C12N2	311.34	310.1942	57.87	10.36	22.77	9.00
C15H32C12N4	339.36	338.2004	53.09	9.50	20.89	16.51
C15H32C12N6	367.37	366.2065	49.04	8.78	19.30	22.88
C15H33C11N2	276.90	276.2332	65.07	12.01	12.80	10.12
C15H33C11N4	304.91	304.2394	59.09	10.91	11.63	18.37
C15H33C11N6	332.92	332.2455	54.12	9.99	10.65	25.24
C15H33C12N3	326.36	325.2051	55.21	10.19	21.73	12.88
C15H33C12N5	354.37	353.2113	50.84	9.39	20.01	19.76
C15H34C11N3	291.91	291.2441	61.72	11.74	12.15	14.39
C15H34C11N5	319.92	319.2503	56.32	10.71	11.08	21.89
C15H34C12N4	341.37	340.2160	52.78	10.04	20.77	16.41
C15H34C12N6	369.38	368.2222	48.77	9.28	19.20	22.75
C15H35C11N4	306.93	306.2550	58.70	11.49	11.55	18.25
C15H35C11N6	334.94	334.2612	53.79	10.53	10.58	25.09
C15H35C12N5	356.39	355.2269	50.55	9.90	19.90	19.65
C15H36C11N5	321.94	321.2659	55.96	11.27	11.01	21.75
C15H36C12N6	371.40	370.2378	48.51	9.77	19.09	22.63
C15H37C11N6	336.96	336.2768	53.47	11.07	10.52	24.94
C16H 8C11N1	249.70	249.0345	76.96	3.23	14.20	5.61
C16H 8C11N3	277.72	277.0407	69.20	2.90	12.77	15.13
C16H 8C11N5	305.73	305.0468	62.86	2.64	11.60	22.91
C16H 8C12N2	299.16	298.0065	64.24	2.70	23.70	9.36
C16H 8C12N4	327.17	326.0126	58.74	2.46	21.67	17.12
C16H 8C12N6	355.19	354.0187	54.11	2.27	19.96	23.66
C16H 9C11N2	264.72	264.0454	72.60	3.43	13.39	10.58
C16H 9C11N4	292.73	292.0516	65.65	3.10	12.11	19.14
C16H 9C11N6	320.74	320.0577	59.92	2.83	11.05	26.20
C16H 9C12N1	286.16	285.0112	67.16	3.17	24.78	4.89
C16H 9C12N3	314.18	313.0173	61.17	2.89	22.57	13.37
C16H 9C12N5	342.19	341.0235	56.16	2.65	20.72	20.47
C16H10C11N1	251.72	251.0502	76.35	4.00	14.08	5.56
C16H10C11N3	279.73	279.0563	68.70	3.60	12.67	15.02
C16H10C11N5	307.74	307.0625	62.45	3.28	11.52	22.76
C16H10C12N2	301.18	300.0221	63.81	3.35	23.54	9.30
C16H10C12N4	329.19	328.0282	58.38	3.06	21.54	17.02
C16H10C12N6	357.20	356.0344	53.80	2.82	19.85	23.53
C16H11C11N2	266.73	266.0611	72.05	4.16	13.29	10.50
C16H11C11N4	294.75	294.0672	65.20	3.76	12.03	19.01
C16H11C11N6	322.76	322.0734	59.54	3.44	10.98	26.04
C16H11C12N1	288.18	287.0269	66.69	3.85	24.60	4.86
C16H11C12N3	316.19	315.0330	60.78	3.51	22.42	13.29
C16H11C12N5	344.21	343.0391	55.83	3.22	20.60	20.35
C16H12C11N1	253.73	253.0658	75.74	4.77	13.97	5.52
C16H12C11N3	281.75	281.0720	68.21	4.29	12.58	14.91
C16H12C11N5	309.76	309.0781	62.04	3.90	11.45	22.61
C16H12C12N2	303.19	302.0377	63.38	3.99	23.39	9.24
C16H12C12N4	331.21	330.0439	58.02	3.65	21.41	16.92
C16H12C12N6	359.22	358.0500	53.50	3.37	19.74	23.40
C16H13C11N2	268.75	268.0767	71.51	4.88	13.19	10.42
C16H13C11N4	296.76	296.0829	64.76	4.42	11.95	18.88
C16H13C11N6	324.78	324.0890	59.17	4.03	10.92	25.88
C16H13C12N1	290.19	289.0425	66.22	4.52	24.43	4.83
C16H13C12N3	318.21	317.0486	60.39	4.12	22.28	13.21
C16H13C12N5	346.22	345.0548	55.51	3.78	20.48	20.23
C16H14C11N1	255.75	255.0815	75.14	5.52	13.86	5.48
C16H14C11N3	283.76	283.0876	67.72	4.97	12.49	14.81
C16H14C11N5	311.78	311.0938	61.64	4.53	11.37	22.46
C16H14C12N2	305.21	304.0534	62.97	4.62	23.23	9.18
C16H14C12N4	333.22	332.0595	57.67	4.23	21.28	16.81
C16H14C12N6	361.24	360.0657	53.20	3.91	19.63	23.26
C16H15C11N2	270.76	270.0924	70.98	5.58	13.09	10.35
C16H15C11N4	298.78	298.0985	64.32	5.06	11.87	18.75
C16H15C11N6	326.79	326.1047	58.81	4.63	10.85	25.72
C16H15C12N1	292.21	291.0581	65.77	5.17	24.27	4.79
C16H15C12N3	320.22	319.0643	60.01	4.72	22.14	13.12
C16H15C12N5	348.24	347.0704	55.19	4.34	20.36	20.11
C16H16C11N1	257.77	257.0971	74.56	6.26	13.75	5.43
C16H16C11N3	285.78	285.1033	67.25	5.64	12.41	14.70
C16H16C11N5	313.79	313.1094	61.24	5.14	11.30	22.32
C16H16C12N2	307.23	306.0690	62.55	5.25	23.08	9.12
C16H16C12N4	335.24	334.0752	57.33	4.81	21.15	16.71
C16H16C12N6	363.25	362.0813	52.90	4.44	19.52	23.14
C16H17C11N2	272.78	272.1080	70.45	6.28	13.00	10.27
C16H17C11N4	300.79	300.1142	63.89	5.70	11.79	18.63
C16H17C11N6	328.81	328.1203	58.45	5.21	10.78	25.56
C16H17C12N1	294.23	293.0738	65.32	5.82	24.10	4.76
C16H17C12N3	322.24	321.0799	59.64	5.32	22.00	13.04
C16H17C12N5	350.25	349.0861	54.87	4.89	20.24	20.00
C16H18C11N1	259.78	259.1128	73.98	6.98	13.65	5.39

C16H18C11N3	287.79	287.1189	66.78	6.30	12.32	14.60
C16H18C11N5	315.81	315.1251	60.85	5.75	11.23	22.18
C16H18C12N2	309.24	308.0847	62.15	5.87	22.93	9.06
C16H18C12N4	337.25	336.0908	56.98	5.38	21.02	16.61
C16H18C12N6	365.27	364.0970	52.61	4.97	19.41	23.01
C16H19C11N2	274.80	274.1237	69.93	6.97	12.90	10.19
C16H19C11N4	302.81	302.1298	63.47	6.32	11.71	18.50
C16H19C11N6	330.82	330.1360	58.09	5.79	10.72	25.40
C16H19C12N1	296.24	295.0894	64.87	6.46	23.94	4.73
C16H19C12N3	324.26	323.0956	59.27	5.91	21.87	12.96
C16H19C12N5	352.27	351.1017	54.55	5.44	20.13	19.88
C16H20C11N1	261.80	261.1284	73.41	7.70	13.54	5.35
C16H20C11N3	289.81	289.1346	66.31	6.96	12.23	14.50
C16H20C11N5	317.82	317.1407	60.47	6.34	11.15	22.04
C16H20C12N2	311.26	310.1003	61.74	6.48	22.78	9.00
C16H20C12N4	339.27	338.1065	56.64	5.94	20.90	16.51
C16H20C12N6	367.28	366.1126	52.32	5.49	19.31	22.88
C16H21C11N2	276.81	276.1393	69.43	7.65	12.81	10.12
C16H21C11N4	304.83	304.1455	63.05	6.94	11.63	18.38
C16H21C11N6	332.84	332.1516	57.74	6.36	10.65	25.25
C16H21C12N1	298.26	297.1051	64.43	7.10	23.77	4.70
C16H21C12N3	326.27	325.1112	58.90	6.49	21.73	12.88
C16H21C12N5	354.29	353.1174	54.24	5.97	20.01	19.77
C16H22C11N1	263.81	263.1441	72.85	8.41	13.44	5.31
C16H22C11N3	291.83	291.1502	65.85	7.60	12.15	14.40
C16H22C11N5	319.84	319.1564	60.09	6.93	11.08	21.90
C16H22C12N2	313.27	312.1160	61.35	7.08	22.63	8.94
C16H22C12N4	341.29	340.1221	56.31	6.50	20.78	16.42
C16H22C12N6	369.30	368.1283	52.04	6.00	19.20	22.76
C16H23C11N2	278.83	278.1550	68.92	8.31	12.72	10.05
C16H23C11N4	306.84	306.1611	62.63	7.56	11.55	18.26
C16H23C11N6	334.85	334.1673	57.39	6.92	10.59	25.10
C16H23C12N1	300.27	299.1207	64.00	7.72	23.61	4.66
C16H23C12N3	328.29	327.1269	58.54	7.06	21.60	12.80
C16H23C12N5	356.30	355.1330	53.94	6.51	19.90	19.66
C16H24C11N1	265.83	265.1597	72.29	9.10	13.34	5.27
C16H24C11N3	293.84	293.1659	65.40	8.23	12.07	14.30
C16H24C11N5	321.86	321.1720	59.71	7.52	11.02	21.76
C16H24C12N2	315.29	314.1316	60.95	7.67	22.49	8.88
C16H24C12N4	343.30	342.1378	55.98	7.05	20.65	16.32
C16H24C12N6	371.32	370.1439	51.76	6.52	19.10	22.63
C16H25C11N2	280.84	280.1706	68.43	8.97	12.62	9.97
C16H25C11N4	308.86	308.1768	62.22	8.16	11.48	18.14
C16H25C11N6	336.87	336.1829	57.05	7.48	10.52	24.95
C16H25C12N1	302.29	301.1364	63.57	8.34	23.46	4.63
C16H25C12N3	330.30	329.1425	58.18	7.63	21.47	12.72
C16H25C12N5	358.32	357.1487	53.63	7.03	19.79	19.55
C16H26C11N1	267.85	267.1754	71.75	9.78	13.24	5.23
C16H26C11N3	295.86	295.1815	64.96	8.86	11.98	14.20
C16H26C11N5	323.87	323.1877	59.34	8.09	10.95	21.62
C16H26C12N2	317.31	316.1473	60.57	8.26	22.35	8.83
C16H26C12N4	345.32	344.1534	55.65	7.59	20.53	16.22
C16H26C12N6	373.33	372.1596	51.48	7.02	18.99	22.51
C16H27C11N2	282.86	282.1863	67.94	9.62	12.53	9.90
C16H27C11N4	310.87	310.1924	61.82	8.75	11.40	18.02
C16H27C11N6	338.89	338.1986	56.71	8.03	10.46	24.80
C16H27C12N1	304.31	303.1520	63.15	8.94	23.30	4.60
C16H27C12N3	332.32	331.1582	57.83	8.19	21.34	12.64
C16H27C12N5	360.33	359.1643	53.33	7.55	19.68	19.44
C16H28C11N1	269.86	269.1910	71.21	10.46	13.14	5.19
C16H28C11N3	297.87	297.1972	64.52	9.47	11.90	14.11
C16H28C11N5	325.89	325.2033	58.97	8.66	10.88	21.49
C16H28C12N2	319.32	318.1629	60.18	8.84	22.21	8.77
C16H28C12N4	347.33	346.1691	55.33	8.13	20.41	16.13
C16H28C12N6	375.35	374.1752	51.20	7.52	18.89	22.39
C16H29C11N2	284.88	284.2019	67.46	10.26	12.45	9.83
C16H29C11N4	312.89	312.2081	61.42	9.34	11.33	17.91
C16H29C11N6	340.90	340.2142	56.37	8.57	10.40	24.65
C16H29C12N1	306.32	305.1677	62.74	9.54	23.15	4.57
C16H29C12N3	334.34	333.1738	57.48	8.74	21.21	12.57
C16H29C12N5	362.35	361.1800	53.04	8.07	19.57	19.33
C16H30C11N1	271.88	271.2067	70.69	11.12	13.04	5.15
C16H30C11N3	299.89	299.2128	64.08	10.08	11.82	14.01
C16H30C11N5	327.90	327.2190	58.61	9.22	10.81	21.36
C16H30C12N2	321.34	320.1786	59.81	9.41	22.07	8.72
C16H30C12N4	349.35	348.1847	55.01	8.66	20.30	16.04
C16H30C12N6	377.36	376.1909	50.93	8.01	18.79	22.27
C16H31C11N2	286.89	286.2176	66.99	10.89	12.36	9.76
C16H31C11N4	314.91	314.2237	61.03	9.92	11.26	17.79
C16H31C11N6	342.92	342.2299	56.04	9.11	10.34	24.51
C16H31C12N1	308.34	307.1833	62.33	10.13	23.00	4.54
C16H31C12N3	336.35	335.1895	57.14	9.29	21.08	12.49
C16H31C12N5	364.36	363.1956	52.74	8.58	19.46	19.22
C16H32C11N1	273.89	273.2223	70.17	11.78	12.94	5.11
C16H32C11N3	301.91	301.2285	63.65	10.68	11.74	13.92
C16H32C11N5	329.92	329.2346	58.25	9.78	10.75	21.23
C16H32C12N2	323.35	322.1942	59.43	9.98	21.93	8.66
C16H32C12N4	351.37	350.2004	54.69	9.18	20.18	15.95
C16H32C12N6	379.38	378.2065	50.66	8.50	18.69	22.15
C16H33C11N2	288.91	288.2332	66.52	11.51	12.27	9.70
C16H33C11N4	316.92	316.2394	60.64	10.50	11.19	17.68
C16H33C11N6	344.93	344.2455	55.71	9.64	10.28	24.36
C16H33C12N1	310.35	309.1990	61.92	10.72	22.85	4.51
C16H33C12N3	338.37	337.2051	56.80	9.83	20.96	12.42
C16H33C12N5	366.38	365.2113	52.45	9.08	19.35	19.11
C16H34C11N1	275.91	275.2380	69.65	12.42	12.85	5.08
C16H34C11N3	303.92	303.2441	63.23	11.28	11.67	13.83
C16H34C11N5	331.94	331.2503	57.90	10.32	10.68	21.10
C16H34C12N2	325.37	324.2099	59.06	10.53	21.79	8.61
C16H34C12N4	353.38	352.2160	54.38	9.70	20.06	15.85
C16H34C12N6	381.40	380.2222	50.39	8.99	18.59	22.03
C16H35C11N2	290.92	290.2489	66.06	12.13	12.19	9.63
C16H35C11N4	318.94	318.2550	60.26	11.06	11.12	17.57
C16H35C11N6	346.95	346.2612	55.39	10.17	10.22	24.22
C16H35C12N3	340.38	339.2208	56.46	10.36	20.83	12.34
C16H35C12N5	368.40	367.2269	52.17	9.58	19.25	19.01
C16H36C11N3	305.94	305.2598	62.82	11.86	11.59	13.73
C16H36C11N5	333.95	333.2659	57.55	10.87	10.62	20.97
C16H36C12N4	355.40	354.2317	54.07	10.21	19.95	15.76
C16H36C12N6	383.41	382.2378	50.12	9.46	18.49	21.92
C16H37C11N4	320.95	320.2707	59.88	11.62	11.05	17.46
C16H37C11N6	348.97	348.2768	55.07	10.69	10.16	24.08
C16H37C12N5	370.41	369.2426	51.88	10.07	19.14	18.91
C16H38C11N5	335.97	335.2816	57.20	11.40	10.55	20.85
C16H38C12N6	385.43	384.2535	49.86	9.94	18.40	21.80
C16H39C11N6	350.98	350.2925	54.75	11.20	10.10	23.94
C17H 8C11N1	261.71	261.0345	78.02	3.08	13.55	5.35
C17H 8C11N3	289.73	289.0407	70.48	2.78	12.24	14.50
C17H 8C11N5	317.74	317.0468	64.26	2.54	11.16	22.04
C17H 8C12N2	311.17	310.0065	65.62	2.59	22.79	9.00

C17H 8C12N4	339.19	338.0126	60.20	2.38	20.90	16.52
C17H 8C12N6	367.20	366.0187	55.61	2.20	19.31	22.89
C17H 9C11N2	276.73	276.0454	73.79	3.28	12.81	10.12
C17H 9C11N4	304.74	304.0516	67.00	2.98	11.63	18.39
C17H 9C11N6	332.75	332.0577	61.36	2.73	10.65	25.26
C17H 9C12N1	298.17	297.0112	68.48	3.04	23.78	4.70
C17H 9C12N3	326.19	325.0173	62.60	2.78	21.74	12.88
C17H 9C12N5	354.20	353.0235	57.65	2.56	20.02	19.77
C17H10C11N1	263.73	263.0502	77.42	3.82	13.44	5.31
C17H10C11N3	291.74	291.0563	69.99	3.46	12.15	14.40
C17H10C11N5	319.76	319.0625	63.86	3.15	11.09	21.90
C17H10C12N2	313.19	312.0221	65.20	3.22	22.64	8.94
C17H10C12N4	341.20	340.0282	59.84	2.95	20.78	16.42
C17H10C12N6	369.22	368.0344	55.30	2.73	19.20	22.76
C17H11C11N2	278.74	278.0611	73.25	3.98	12.72	10.05
C17H11C11N4	306.76	306.0672	66.56	3.61	11.56	18.26
C17H11C11N6	334.77	334.0734	60.99	3.31	10.59	25.10
C17H11C12N1	300.19	299.0269	68.02	3.69	23.62	4.67
C17H11C12N3	328.20	327.0330	62.21	3.38	21.60	12.80
C17H11C12N5	356.22	355.0391	57.32	3.11	19.91	19.66
C17H12C11N1	265.74	265.0658	76.84	4.55	13.34	5.27
C17H12C11N3	293.76	293.0720	69.51	4.12	12.07	14.30
C17H12C11N5	321.77	321.0781	63.46	3.76	11.02	21.76
C17H12C12N2	315.20	314.0377	64.78	3.84	22.50	8.89
C17H12C12N4	343.22	342.0439	59.49	3.52	20.66	16.32
C17H12C12N6	371.23	370.0500	55.00	3.26	19.10	22.64
C17H13C11N2	280.76	280.0767	72.73	4.67	12.63	9.98
C17H13C11N4	308.77	308.0829	66.13	4.24	11.48	18.14
C17H13C11N6	336.79	336.0890	60.63	3.89	10.53	24.95
C17H13C12N1	302.21	301.0425	67.57	4.34	23.46	4.63
C17H13C12N3	330.22	329.0486	61.83	3.97	21.47	12.72
C17H13C12N5	358.23	357.0548	57.00	3.66	19.79	19.55
C17H14C11N1	267.76	267.0815	76.26	5.27	13.24	5.23
C17H14C11N3	295.77	295.0876	69.04	4.77	11.99	14.21
C17H14C11N5	323.79	323.0938	63.06	4.36	10.95	21.63
C17H14C12N2	317.22	316.0534	64.37	4.45	22.35	8.83
C17H14C12N4	345.23	344.0595	59.15	4.09	20.54	16.23
C17H14C12N6	373.25	372.0657	54.71	3.78	19.00	22.52
C17H15C11N2	282.78	282.0924	72.21	5.35	12.54	9.91
C17H15C11N4	310.79	310.0985	65.70	4.86	11.41	18.03
C17H15C11N6	338.80	338.1047	60.27	4.46	10.46	24.81
C17H15C12N1	304.22	303.0581	67.12	4.97	23.31	4.60
C17H15C12N3	332.24	331.0643	61.46	4.55	21.34	12.65
C17H15C12N5	360.25	359.0704	56.68	4.20	19.68	19.44
C17H16C11N1	269.78	269.0971	75.69	5.98	13.14	5.19
C17H16C11N3	297.79	297.1033	68.57	5.42	11.91	14.11
C17H16C11N5	325.80	325.1094	62.67	4.95	10.88	21.50
C17H16C12N2	319.24	318.0690	63.96	5.05	22.21	8.78
C17H16C12N4	347.25	346.0752	58.80	4.64	20.42	16.13
C17H16C12N6	375.26	374.0813	54.41	4.30	18.90	22.39
C17H17C11N2	284.79	284.1080	71.70	6.02	12.45	9.84
C17H17C11N4	312.80	312.1142	65.28	5.48	11.33	17.91
C17H17C11N6	340.82	340.1203	59.91	5.03	10.40	24.66
C17H17C12N1	306.24	305.0738	66.68	5.60	23.15	4.57
C17H17C12N3	334.25	333.0799	61.09	5.13	21.21	12.57
C17H17C12N5	362.26	361.0861	56.36	4.73	19.57	19.33
C17H18C11N1	271.79	271.1128	75.13	6.68	13.04	5.15
C17H18C11N3	299.81	299.1189	68.11	6.05	11.83	14.02
C17H18C11N5	327.82	327.1251	62.29	5.53	10.81	21.36
C17H18C12N2	321.25	320.0847	63.56	5.65	22.07	8.72
C17H18C12N4	349.27	348.0908	58.46	5.19	20.30	16.04
C17H18C12N6	377.28	376.0970	54.12	4.81	18.79	22.28
C17H19C11N2	286.81	286.1237	71.19	6.68	12.36	9.77
C17H19C11N4	314.82	314.1298	64.86	6.08	11.26	17.80
C17H19C11N6	342.83	342.1360	59.56	5.59	10.34	24.51
C17H19C12N1	308.25	307.0894	66.24	6.21	23.00	4.54
C17H19C12N3	336.27	335.0956	60.72	5.70	21.09	12.50
C17H19C12N5	364.28	363.1017	56.05	5.26	19.46	19.23
C17H20C11N1	273.81	273.1284	74.57	7.36	12.95	5.12
C17H20C11N3	301.82	301.1346	67.65	6.68	11.75	13.92
C17H20C11N5	329.84	329.1407	61.91	6.11	10.75	21.23
C17H20C12N2	323.27	322.1003	63.16	6.24	21.93	8.67
C17H20C12N4	351.28	350.1065	58.13	5.74	20.18	15.95
C17H20C12N6	379.30	378.1126	53.83	5.31	18.69	22.16
C17H21C11N2	288.82	288.1393	70.70	7.33	12.27	9.70
C17H21C11N4	316.84	316.1455	64.45	6.68	11.19	17.68
C17H21C11N6	344.85	344.1516	59.21	6.14	10.28	24.37
C17H21C12N1	310.27	309.1051	65.81	6.82	22.85	4.51
C17H21C12N3	338.28	337.1112	60.36	6.26	20.96	12.42
C17H21C12N5	366.30	365.1174	55.74	5.78	19.36	19.12
C17H22C11N1	275.82	275.1441	74.03	8.04	12.85	5.08
C17H22C11N3	303.84	303.1502	67.20	7.30	11.67	13.83
C17H22C11N5	331.85	331.1564	61.53	6.68	10.68	21.10
C17H22C12N2	325.28	324.1160	62.77	6.82	21.80	8.61
C17H22C12N4	353.30	352.1221	57.80	6.28	20.07	15.86
C17H22C12N6	381.31	380.1283	53.55	5.82	18.60	22.04
C17H23C11N2	290.84	290.1550	70.21	7.97	12.19	9.63
C17H23C11N4	318.85	318.1611	64.04	7.27	11.12	17.57
C17H23C11N6	346.87	346.1673	58.87	6.68	10.22	24.23
C17H23C12N1	312.29	311.1207	65.39	7.42	22.71	4.49
C17H23C12N3	340.30	339.1269	60.00	6.81	20.84	12.35
C17H23C12N5	368.31	367.1330	55.44	6.29	19.25	19.01
C17H24C11N1	277.84	277.1597	73.49	8.71	12.76	5.04
C17H24C11N3	305.85	305.1659	66.76	7.91	11.59	13.74
C17H24C11N5	333.87	333.1720	61.16	7.25	10.62	20.98
C17H24C12N2	327.30	326.1316	62.39	7.39	21.66	8.56
C17H24C12N4	355.31	354.1378	57.47	6.81	19.96	15.77
C17H24C12N6	383.33	382.1439	53.27	6.31	18.50	21.92
C17H25C11N2	292.86	292.1706	69.72	8.60	12.11	9.57
C17H25C11N4	320.87	320.1768	63.64	7.85	11.05	17.46
C17H25C11N6	348.88	348.1829	58.53	7.22	10.16	24.09
C17H25C12N1	314.30	313.1364	64.97	8.02	22.56	4.46
C17H25C12N3	342.31	341.1425	59.65	7.36	20.71	12.28
C17H25C12N5	370.33	369.1487	55.14	6.80	19.15	18.91
C17H26C11N1	279.86	279.1754	72.96	9.36	12.67	5.00
C17H26C11N3	307.87	307.1815	66.32	8.51	11.52	13.65
C17H26C11N5	335.88	335.1877	60.79	7.80	10.56	20.85
C17H26C12N2	329.32	328.1473	62.00	7.96	21.53	8.51
C17H26C12N4	357.33	356.1534	57.14	7.33	19.84	15.68
C17H26C12N6	385.34	384.1596	52.99	6.80	18.40	21.81
C17H27C11N2	294.87	294.1863	69.25	9.23	12.02	9.50
C17H27C11N4	322.88	322.1924	63.24	8.43	10.98	17.35
C17H27C11N6	350.90	350.1986	58.19	7.76	10.10	23.95
C17H27C12N1	316.32	315.1520	64.55	8.60	22.42	4.43
C17H27C12N3	344.33	343.1582	59.30	7.90	20.59	12.20
C17H27C12N5	372.34	371.1643	54.84	7.31	19.04	18.81
C17H28C11N1	281.87	281.1910	72.44	10.01	12.58	4.97
C17H28C11N3	309.89	309.1972	65.89	9.11	11.44	13.56
C17H28C11N5	337.90	337.2033	60.43	8.35	10.49	20.73
C17H28C12N2	331.33	330.1629	61.63	8.52	21.40	8.45

C17H28C12N4	359.35	358.1691	56.82	7.85	19.73	15.59
C17H28C12N6	387.36	386.1752	52.71	7.29	18.30	21.70
C17H29C11N2	296.89	296.2019	68.78	9.85	11.94	9.44
C17H29C11N4	324.90	324.2081	62.85	9.00	10.91	17.24
C17H29C11N6	352.91	352.2142	57.86	8.28	10.05	23.81
C17H29C12N1	318.33	317.1677	64.14	9.18	22.27	4.40
C17H29C12N3	346.35	345.1738	58.96	8.44	20.47	12.13
C17H29C12N5	374.36	373.1800	54.54	7.81	18.94	18.71
C17H30C11N1	283.89	283.2067	71.93	10.65	12.49	4.93
C17H30C11N3	311.90	311.2128	65.47	9.70	11.37	13.47
C17H30C11N5	339.92	339.2190	60.07	8.90	10.43	20.60
C17H30C12N2	333.35	332.1786	61.25	9.07	21.27	8.40
C17H30C12N4	361.36	360.1847	56.51	8.37	19.62	15.50
C17H30C12N6	389.37	388.1909	52.44	7.77	18.21	21.58
C17H31C11N2	298.90	298.2176	68.31	10.45	11.86	9.37
C17H31C11N4	326.92	326.2237	62.46	9.56	10.84	17.14
C17H31C11N6	354.93	354.2299	57.53	8.80	9.99	23.68
C17H31C12N1	320.35	319.1833	63.74	9.75	22.13	4.37
C17H31C12N3	348.36	347.1895	58.61	8.97	20.35	12.06
C17H31C12N5	376.38	375.1956	54.25	8.30	18.84	18.61
C17H32C11N1	285.90	285.2223	71.42	11.28	12.40	4.90
C17H32C11N3	313.92	313.2285	65.05	10.27	11.29	13.39
C17H32C11N5	341.93	341.2346	59.72	9.43	10.37	20.48
C17H32C12N2	335.36	334.1942	60.89	9.62	21.14	8.35
C17H32C12N4	363.38	362.2004	56.19	8.88	19.51	15.42
C17H32C12N6	391.39	390.2065	52.17	8.24	18.12	21.47
C17H33C11N2	300.92	300.2332	67.86	11.05	11.78	9.31
C17H33C11N4	328.93	328.2394	62.08	10.11	10.78	17.03
C17H33C11N6	356.95	356.2455	57.20	9.32	9.93	23.54
C17H33C12N1	322.37	321.1990	63.34	10.32	22.00	4.34
C17H33C12N3	350.38	349.2051	58.28	9.49	20.24	11.99
C17H33C12N5	378.39	377.2113	53.96	8.79	18.74	18.51
C17H34C11N1	287.92	287.2380	70.92	11.90	12.31	4.86
C17H34C11N3	315.93	315.2441	64.63	10.85	11.22	13.30
C17H34C11N5	343.95	343.2503	59.37	9.96	10.31	20.36
C17H34C12N2	337.38	336.2099	60.52	10.16	21.02	8.30
C17H34C12N4	365.39	364.2160	55.88	9.38	19.41	15.33
C17H34C12N6	393.41	392.2222	51.90	8.71	18.02	21.36
C17H35C11N2	302.93	302.2489	67.40	11.65	11.70	9.25
C17H35C11N4	330.95	330.2550	61.70	10.66	10.71	16.93
C17H35C11N6	358.96	358.2612	56.88	9.83	9.88	23.41
C17H35C12N1	324.38	323.2146	62.95	10.88	21.86	4.32
C17H35C12N3	352.39	351.2208	57.94	10.01	20.12	11.92
C17H35C12N5	380.41	379.2269	53.68	9.27	18.64	18.41
C17H36C11N1	289.94	289.2536	70.43	12.52	12.23	4.83
C17H36C11N3	317.95	317.2598	64.22	11.41	11.15	13.22
C17H36C11N5	345.96	345.2659	59.02	10.49	10.25	20.24
C17H36C12N2	339.40	338.2255	60.16	10.69	20.89	8.25
C17H36C12N4	367.41	366.2317	55.58	9.88	19.30	15.25
C17H36C12N6	395.42	394.2378	51.64	9.18	17.93	21.25
C17H37C11N2	304.95	304.2645	66.96	12.23	11.63	9.19
C17H37C11N4	332.96	332.2707	61.32	11.20	10.65	16.83
C17H37C11N6	360.98	360.2768	56.57	10.33	9.82	23.28
C17H37C12N3	354.41	353.2364	57.61	10.52	20.01	11.86
C17H37C12N5	382.42	381.2426	53.39	9.75	18.54	18.31
C17H38C11N3	319.97	319.2754	63.82	11.97	11.08	13.13
C17H38C11N5	347.98	347.2816	58.68	11.01	10.19	20.13
C17H38C12N4	369.43	368.2473	55.27	10.37	19.19	15.17
C17H38C12N6	397.44	396.2535	51.38	9.64	17.84	21.15
C17H39C11N4	334.98	334.2363	60.96	11.74	10.58	16.73
C17H39C11N6	362.99	362.2925	56.25	10.83	9.77	23.15
C17H39C12N5	384.44	383.2582	53.11	10.23	18.44	18.22
C17H40C11N5	349.99	349.2972	58.34	11.52	10.13	20.01
C17H40C12N6	399.45	398.2691	51.12	10.09	17.75	21.04
C17H41C11N6	365.01	364.3081	55.94	11.32	9.71	23.02
C18H 9C11N2	288.74	288.0454	74.88	3.14	12.28	9.70
C18H 9C11N4	316.75	316.0516	68.26	2.86	11.19	17.69
C18H 9C11N6	344.77	344.0577	62.71	2.63	10.28	24.38
C18H 9C12N1	310.19	309.0112	69.70	2.92	22.86	4.52
C18H 9C12N3	338.20	337.0173	63.93	2.68	20.97	12.42
C18H 9C12N5	366.21	365.0235	59.04	2.48	19.36	19.12
C18H10C11N1	275.74	275.0502	78.41	3.66	12.86	5.08
C18H10C11N3	303.75	303.0563	71.18	3.32	11.67	13.83
C18H10C11N5	331.77	331.0625	65.17	3.04	10.69	21.11
C18H10C12N2	325.20	324.0221	66.48	3.10	21.80	8.61
C18H10C12N4	353.21	352.0282	61.21	2.85	20.07	15.86
C18H10C12N6	381.23	380.0344	56.71	2.64	18.60	22.04
C18H11C11N2	290.75	290.0611	74.36	3.81	12.19	9.63
C18H11C11N4	318.77	318.0672	67.82	3.48	11.12	17.58
C18H11C11N6	346.78	346.0734	62.34	3.20	10.22	24.23
C18H11C12N1	312.20	311.0269	69.25	3.55	22.71	4.49
C18H11C12N3	340.21	339.0330	63.55	3.26	20.84	12.35
C18H11C12N5	368.23	367.0391	58.71	3.01	19.26	19.02
C18H12C11N1	277.76	277.0658	77.84	4.35	12.76	5.04
C18H12C11N3	305.77	305.0720	70.71	3.96	11.59	13.74
C18H12C11N5	333.78	333.0781	64.77	3.62	10.62	20.98
C18H12C12N2	327.22	326.0377	66.07	3.70	21.67	8.56
C18H12C12N4	355.23	354.0439	60.86	3.41	19.96	15.77
C18H12C12N6	383.24	382.0500	56.41	3.16	18.50	21.93
C18H13C11N2	292.77	292.0767	73.85	4.48	12.11	9.57
C18H13C11N4	320.78	320.0829	67.40	4.08	11.05	17.47
C18H13C11N6	348.80	348.0890	61.98	3.76	10.16	24.09
C18H13C12N1	314.22	313.0425	68.81	4.17	22.57	4.46
C18H13C12N3	342.23	341.0486	63.17	3.83	20.72	12.28
C18H13C12N5	370.24	369.0548	58.39	3.54	19.15	18.92
C18H14C11N1	279.77	279.0815	77.28	5.04	12.67	5.01
C18H14C11N3	307.79	307.0876	70.24	4.58	11.52	13.65
C18H14C11N5	335.80	335.0938	64.38	4.20	10.56	20.86
C18H14C12N2	329.23	328.0534	65.67	4.29	21.54	8.51
C18H14C12N4	357.25	356.0595	60.52	3.95	19.85	15.68
C18H14C12N6	385.26	384.0657	56.12	3.66	18.40	21.81
C18H15C11N2	294.79	294.0924	73.34	5.13	12.03	9.50
C18H15C11N4	322.80	322.0985	66.98	4.68	10.98	17.36
C18H15C11N6	350.81	350.1047	61.63	4.31	10.11	23.96
C18H15C12N1	316.23	315.0581	68.37	4.78	22.42	4.43
C18H15C12N3	344.25	343.0643	62.80	4.39	20.60	12.21
C18H15C12N5	372.26	371.0704	58.08	4.06	19.05	18.81
C18H16C11N1	281.79	281.0971	76.72	5.72	12.58	4.97
C18H16C11N3	309.80	309.1033	69.79	5.21	11.44	13.56
C18H16C11N5	337.81	337.1094	64.00	4.77	10.49	20.73
C18H16C12N2	331.25	330.0690	65.27	4.87	21.41	8.46
C18H16C12N4	359.26	358.0752	60.18	4.49	19.74	15.60
C18H16C12N6	387.27	386.0813	55.83	4.16	18.31	21.70
C18H17C11N2	296.80	296.1080	72.84	5.77	11.94	9.44
C18H17C11N4	324.82	324.1142	66.56	5.28	10.91	17.25
C18H17C11N6	352.83	352.1203	61.28	4.86	10.05	23.82
C18H17C12N1	318.25	317.0738	67.93	5.38	22.28	4.40
C18H17C12N3	346.26	345.0799	62.44	4.95	20.48	12.14
C18H17C12N5	374.28	373.0861	57.77	4.58	18.94	18.71
C18H18C11N1	283.80	283.1128	76.18	6.39	12.49	4.94

C18H18C11N3	311.82	311.1189	69.34	5.82	11.37	13.48
C18H18C11N5	339.83	339.1251	63.62	5.34	10.43	20.61
C18H18C12N2	333.26	332.0847	64.87	5.44	21.28	8.41
C18H18C12N4	361.28	360.0908	59.84	5.02	19.63	15.51
C18H18C12N6	389.29	388.0970	55.54	4.66	18.21	21.59
C18H19C11N2	298.82	298.1237	72.35	6.41	11.86	9.37
C18H19C11N4	326.83	326.1298	66.15	5.86	10.85	17.14
C18H19C11N6	354.85	354.1360	60.93	5.40	9.99	23.68
C18H19C12N1	320.26	319.0894	67.51	5.98	22.14	4.37
C18H19C12N3	348.28	347.0956	62.08	5.50	20.36	12.07
C18H19C12N5	376.29	375.1017	57.46	5.09	18.84	18.61
C18H20C11N1	285.82	285.1284	75.64	7.05	12.40	4.90
C18H20C11N3	313.83	313.1346	68.89	6.42	11.30	13.39
C18H20C11N5	341.85	341.1407	63.24	5.90	10.37	20.49
C18H20C12N2	335.28	334.1003	64.48	6.01	21.15	8.36
C18H20C12N4	363.29	362.1065	59.51	5.55	19.52	15.42
C18H20C12N6	391.31	390.1126	55.25	5.15	18.12	21.48
C18H21C11N2	300.83	300.1393	71.87	7.04	11.78	9.31
C18H21C11N4	328.85	328.1455	65.74	6.44	10.78	17.04
C18H21C11N6	356.86	356.1516	60.58	5.93	9.93	23.55
C18H21C12N1	322.28	321.1051	67.08	6.57	22.00	4.35
C18H21C12N3	350.29	349.1112	61.72	6.04	20.24	12.00
C18H21C12N5	378.31	377.1174	57.15	5.60	18.74	18.51
C18H22C11N1	287.84	287.1441	75.11	7.70	12.32	4.87
C18H22C11N3	315.85	315.1502	68.45	7.02	11.22	13.30
C18H22C11N5	343.86	343.1564	62.87	6.45	10.31	20.37
C18H22C12N2	337.30	336.1160	64.10	6.57	21.02	8.31
C18H22C12N4	365.31	364.1221	59.18	6.07	19.41	15.34
C18H22C12N6	393.32	392.1283	54.97	5.64	18.03	21.37
C18H23C11N2	302.85	302.1550	71.39	7.66	11.71	9.25
C18H23C11N4	330.86	330.1611	65.34	7.01	10.72	16.93
C18H23C11N6	358.88	358.1673	60.24	6.46	9.88	23.42
C18H23C12N1	324.30	323.1207	66.67	7.15	21.86	4.32
C18H23C12N3	352.31	351.1269	61.37	6.58	20.13	11.93
C18H23C12N5	380.32	379.1330	56.85	6.10	18.64	18.41
C18H24C11N1	289.85	289.1597	74.59	8.35	12.23	4.83
C18H24C11N3	317.87	317.1659	68.02	7.61	11.15	13.22
C18H24C11N5	345.88	345.1720	62.51	6.99	10.25	20.25
C18H24C12N2	339.31	338.1316	63.72	7.13	20.90	8.26
C18H24C12N4	367.32	366.1378	58.86	6.59	19.30	15.25
C18H24C12N6	395.34	394.1439	54.69	6.12	17.94	21.26
C18H25C11N2	304.87	304.1706	70.92	8.27	11.63	9.19
C18H25C11N4	332.88	332.1768	64.95	7.57	10.65	16.83
C18H25C11N6	360.89	360.1829	59.91	6.98	9.82	23.29
C18H25C12N1	326.31	325.1364	66.26	7.72	21.73	4.29
C18H25C12N3	354.33	353.1425	61.02	7.11	20.01	11.86
C18H25C12N5	382.34	381.1487	56.55	6.59	18.55	18.32
C18H26C11N1	291.87	291.1754	74.07	8.98	12.15	4.80
C18H26C11N3	319.88	319.1815	67.59	8.19	11.08	13.14
C18H26C11N5	347.89	347.1877	62.15	7.53	10.19	20.13
C18H26C12N2	341.33	340.1473	63.34	7.68	20.77	8.21
C18H26C12N4	369.34	368.1534	58.54	7.10	19.20	15.17
C18H26C12N6	397.35	396.1596	54.41	6.60	17.84	21.15
C18H27C11N2	306.88	306.1863	70.45	8.87	11.55	9.13
C18H27C11N4	334.90	334.1924	64.56	8.13	10.59	16.73
C18H27C11N6	362.91	362.1986	59.57	7.50	9.77	23.16
C18H27C12N1	328.33	327.1520	65.85	8.29	21.60	4.27
C18H27C12N3	356.34	355.1582	60.67	7.64	19.90	11.79
C18H27C12N5	384.36	383.1643	56.25	7.08	18.45	18.22
C18H28C11N1	293.88	293.1910	73.57	9.60	12.06	4.77
C18H28C11N3	321.90	321.1972	67.16	8.77	11.01	13.05
C18H28C11N5	349.91	349.2033	61.79	8.07	10.13	20.01
C18H28C12N2	343.34	342.1629	62.97	8.22	20.65	8.16
C18H28C12N4	371.36	370.1691	58.22	7.60	19.09	15.09
C18H28C12N6	399.37	398.1752	54.14	7.07	17.75	21.04
C18H29C11N2	308.90	308.2019	69.99	9.46	11.48	9.07
C18H29C11N4	336.91	336.2081	64.17	8.68	10.52	16.63
C18H29C11N6	364.93	364.2142	59.25	8.01	9.72	23.03
C18H29C12N1	330.34	329.1677	65.45	8.85	21.46	4.24
C18H29C12N3	358.36	357.1738	60.33	8.16	19.79	11.73
C18H29C12N5	386.37	385.1800	55.96	7.57	18.35	18.13
C18H30C11N1	295.90	295.2067	73.07	10.22	11.98	4.73
C18H30C11N3	323.91	323.2128	66.75	9.34	10.95	12.97
C18H30C11N5	351.93	351.2190	61.43	8.59	10.07	19.90
C18H30C12N2	345.36	344.1786	62.60	8.76	20.53	8.11
C18H30C12N4	373.37	372.1847	57.90	8.10	18.99	15.01
C18H30C12N6	401.39	400.1909	53.86	7.53	17.67	20.94
C18H31C11N2	310.91	310.2176	69.54	10.05	11.40	9.01
C18H31C11N4	338.93	338.2237	63.79	9.22	10.46	16.53
C18H31C11N6	366.94	366.2299	58.92	8.52	9.66	22.90
C18H31C12N1	332.36	331.1833	65.05	9.40	21.33	4.21
C18H31C12N3	360.37	359.1895	59.99	8.67	19.68	11.66
C18H31C12N5	388.39	387.1956	55.67	8.05	18.26	18.03
C18H32C11N1	297.92	297.2223	72.57	10.83	11.90	4.70
C18H32C11N3	325.93	325.2285	66.33	9.90	10.88	12.89
C18H32C11N5	353.94	353.2346	61.08	9.11	10.02	19.79
C18H32C12N2	347.38	346.1942	62.24	9.29	20.41	8.06
C18H32C12N4	375.39	374.2004	57.59	8.59	18.89	14.93
C18H32C12N6	403.40	402.2065	53.59	8.00	17.58	20.83
C18H33C11N2	312.93	312.2332	69.09	10.63	11.33	8.95
C18H33C11N4	340.94	340.2394	63.41	9.76	10.40	16.43
C18H33C11N6	368.96	368.2455	58.60	9.02	9.61	22.78
C18H33C12N1	334.38	333.1990	64.66	9.95	21.21	4.19
C18H33C12N3	362.39	361.2051	59.66	9.18	19.57	11.60
C18H33C12N5	390.40	389.2113	55.38	8.52	18.16	17.94
C18H34C11N1	299.93	299.2380	72.08	11.43	11.82	4.67
C18H34C11N3	327.94	327.2441	65.93	10.45	10.81	12.81
C18H34C11N5	355.96	355.2503	60.74	9.63	9.96	19.67
C18H34C12N2	349.39	348.2099	61.88	9.81	20.29	8.02
C18H34C12N4	377.40	376.2160	57.29	9.08	18.79	14.85
C18H34C12N6	405.42	404.2222	53.33	8.45	17.49	20.73
C18H35C11N2	314.95	314.2489	68.65	11.20	11.26	8.89
C18H35C11N4	342.96	342.2550	63.04	10.29	10.34	16.34
C18H35C11N6	370.97	370.2612	58.28	9.51	9.56	22.65
C18H35C12N1	336.39	335.2146	64.27	10.49	21.08	4.16
C18H35C12N3	364.41	363.2208	59.33	9.68	19.46	11.53
C18H35C12N5	392.42	391.2269	55.09	8.99	18.07	17.85
C18H36C11N1	301.95	301.2536	71.60	12.02	11.74	4.64
C18H36C11N3	329.96	329.2598	65.52	11.00	10.74	12.73
C18H36C11N5	357.97	357.2659	60.40	10.14	9.90	19.56
C18H36C12N2	351.41	350.2255	61.52	10.33	20.18	7.97
C18H36C12N4	379.42	378.2317	56.98	9.56	18.69	14.77
C18H36C12N6	407.43	406.2378	53.06	8.91	17.40	20.63
C18H37C11N2	316.96	316.2645	68.21	11.77	11.19	8.84
C18H37C11N4	344.98	344.2707	62.67	10.81	10.28	16.24
C18H37C11N6	372.99	372.2768	57.96	10.00	9.51	22.53
C18H37C12N1	338.41	337.2303	63.89	11.02	20.95	4.14
C18H37C12N3	366.42	365.2364	59.00	10.18	19.35	11.47
C18H37C12N5	394.44	393.2426	54.81	9.46	17.98	17.76
C18H38C11N1	303.96	303.2693	71.13	12.60	11.66	4.61

C18H38C11N3	331.98	331.2754	65.13	11.54	10.68	12.66
C18H38C11N5	359.99	359.2816	60.06	10.64	9.85	19.45
C18H38C12N2	353.42	352.2412	61.17	10.84	20.06	7.93
C18H38C12N4	381.44	380.2473	56.68	10.04	18.59	14.69
C18H38C12N6	409.45	408.2535	52.80	9.35	17.32	20.53
C18H39C11N2	318.98	318.2802	67.78	12.32	11.11	8.78
C18H39C11N4	346.99	346.2863	62.31	11.33	10.22	16.15
C18H39C11N6	375.00	374.2925	57.65	10.48	9.45	22.41
C18H39C12N3	368.44	367.2521	58.68	10.67	19.25	11.40
C18H39C12N5	396.45	395.2582	54.53	9.92	17.89	17.67
C18H40C11N3	333.99	333.2911	64.73	12.07	10.61	12.58
C18H40C11N5	362.01	361.2972	59.72	11.14	9.79	19.35
C18H40C12N4	383.45	382.2630	56.38	10.51	18.49	14.61
C18H40C12N6	411.47	410.2691	52.54	9.80	17.23	20.42
C18H41C11N4	349.01	348.3020	61.95	11.84	10.16	16.05
C18H41C11N6	377.02	376.3081	57.34	10.96	9.40	22.29
C18H41C12N5	398.47	397.2739	54.26	10.37	17.79	17.58
C18H42C11N5	364.02	363.3129	59.39	11.63	9.74	19.24
C18H42C12N6	413.48	412.2848	52.29	10.24	17.15	20.33
C18H43C11N6	379.04	378.3238	57.04	11.43	9.35	22.17
C19H 9C11N2	300.75	300.0454	75.88	3.02	11.79	9.31
C19H 9C11N4	328.76	328.0516	69.42	2.76	10.78	17.04
C19H 9C11N6	356.78	356.0577	63.96	2.54	9.94	23.56
C19H 9C12N1	322.20	321.0112	70.83	2.82	22.01	4.35
C19H 9C12N3	350.21	349.0173	65.16	2.59	20.25	12.00
C19H 9C12N5	378.22	377.0235	60.34	2.40	18.75	18.52
C19H10C11N1	287.75	287.0502	79.31	3.50	12.32	4.87
C19H10C11N3	315.76	315.0563	72.27	3.19	11.23	13.31
C19H10C11N5	343.78	343.0625	66.38	2.93	10.31	20.37
C19H10C12N2	337.21	336.0221	67.68	2.99	21.03	8.31
C19H10C12N4	365.22	364.0282	62.49	2.76	19.41	15.34
C19H10C12N6	393.24	392.0344	58.03	2.56	18.03	21.37
C19H11C11N2	302.77	302.0611	75.38	3.66	11.71	9.25
C19H11C11N4	330.78	330.0672	68.99	3.35	10.72	16.94
C19H11C11N6	358.79	358.0734	63.61	3.09	9.88	23.42
C19H11C12N1	324.21	323.0269	70.39	3.42	21.87	4.32
C19H11C12N3	352.23	351.0330	64.79	3.15	20.13	11.93
C19H11C12N5	380.24	379.0391	60.02	2.92	18.65	18.42
C19H12C11N1	289.77	289.0658	78.76	4.17	12.23	4.83
C19H12C11N3	317.78	317.0720	71.81	3.81	11.16	13.22
C19H12C11N5	345.79	345.0781	66.00	3.50	10.25	20.25
C19H12C12N2	339.23	338.0377	67.27	3.57	20.90	8.26
C19H12C12N4	367.24	366.0439	62.14	3.29	19.31	15.26
C19H12C12N6	395.25	394.0500	57.74	3.06	17.94	21.26
C19H13C11N2	304.78	304.0767	74.88	4.30	11.63	9.19
C19H13C11N4	332.80	332.0829	68.57	3.94	10.65	16.84
C19H13C11N6	360.81	360.0890	63.25	3.63	9.83	23.29
C19H13C12N1	326.23	325.0425	69.95	4.02	21.74	4.29
C19H13C12N3	354.24	353.0486	64.42	3.70	20.02	11.86
C19H13C12N5	382.25	381.0548	59.70	3.43	18.55	18.32
C19H14C11N1	291.78	291.0815	78.21	4.84	12.15	4.80
C19H14C11N3	319.80	319.0876	71.36	4.41	11.09	13.14
C19H14C11N5	347.81	347.0938	65.61	4.06	10.19	20.14
C19H14C12N2	341.24	340.0534	66.88	4.14	20.78	8.21
C19H14C12N4	369.26	368.0595	61.80	3.82	19.20	15.17
C19H14C12N6	397.27	396.0657	57.45	3.55	17.85	21.15
C19H15C11N2	306.80	306.0924	74.39	4.93	11.56	9.13
C19H15C11N4	334.81	334.0985	68.16	4.52	10.59	16.73
C19H15C11N6	362.82	362.1047	62.90	4.17	9.77	23.16
C19H15C12N1	328.24	327.0581	69.53	4.61	21.60	4.27
C19H15C12N3	356.26	355.0643	64.06	4.24	19.90	11.79
C19H15C12N5	384.27	383.0704	59.39	3.93	18.45	18.23
C19H16C11N1	293.80	293.0971	77.68	5.49	12.07	4.77
C19H16C11N3	321.81	321.1033	70.91	5.01	11.02	13.06
C19H16C11N5	349.83	349.1094	65.24	4.61	10.13	20.02
C19H16C12N2	343.26	342.0690	66.48	4.70	20.66	8.16
C19H16C12N4	371.27	370.0752	61.47	4.34	19.10	15.09
C19H16C12N6	399.29	398.0813	57.16	4.04	17.76	21.05
C19H17C11N2	308.81	308.1080	73.90	5.55	11.48	9.07
C19H17C11N4	336.83	336.1142	67.75	5.09	10.53	16.63
C19H17C11N6	364.84	364.1203	62.55	4.70	9.72	23.03
C19H17C12N1	330.26	329.0738	69.10	5.19	21.47	4.24
C19H17C12N3	358.27	357.0799	63.70	4.78	19.79	11.73
C19H17C12N5	386.29	385.0861	59.08	4.44	18.36	18.13
C19H18C11N1	295.82	295.1128	77.15	6.13	11.98	4.73
C19H18C11N3	323.83	323.1189	70.47	5.60	10.95	12.98
C19H18C11N5	351.84	351.1251	64.86	5.16	10.08	19.90
C19H18C12N2	345.27	344.0847	66.10	5.25	20.54	8.11
C19H18C12N4	373.29	372.0908	61.14	4.86	18.99	15.01
C19H18C12N6	401.30	400.0970	56.87	4.52	17.67	20.94
C19H19C11N2	310.83	310.1237	73.42	6.16	11.41	9.01
C19H19C11N4	338.84	338.1298	67.35	5.65	10.46	16.53
C19H19C11N6	366.86	366.1360	62.21	5.22	9.66	22.91
C19H19C12N1	332.28	331.0894	68.68	5.76	21.34	4.22
C19H19C12N3	360.29	359.0956	63.34	5.32	19.68	11.66
C19H19C12N5	388.30	387.1017	58.77	4.93	18.26	18.04
C19H20C11N1	297.83	297.1284	76.62	6.77	11.90	4.70
C19H20C11N3	325.84	325.1346	70.04	6.19	10.88	12.90
C19H20C11N5	353.86	353.1407	64.49	5.70	10.02	19.79
C19H20C12N2	347.29	346.1003	65.71	5.80	20.42	8.07
C19H20C12N4	375.30	374.1065	60.81	5.37	18.89	14.93
C19H20C12N6	403.32	402.1126	56.58	5.00	17.58	20.84
C19H21C11N2	312.85	312.1393	72.95	6.77	11.33	8.95
C19H21C11N4	340.86	340.1455	66.95	6.21	10.40	16.44
C19H21C11N6	368.87	368.1516	61.87	5.74	9.61	22.78
C19H21C12N1	334.29	333.1051	68.27	6.33	21.21	4.19
C19H21C12N3	362.31	361.1112	62.99	5.84	19.57	11.60
C19H21C12N5	390.32	389.1174	58.47	5.42	18.17	17.94
C19H22C11N1	299.85	299.1441	76.11	7.40	11.82	4.67
C19H22C11N3	327.86	327.1502	69.61	6.76	10.81	12.82
C19H22C11N5	355.87	355.1564	64.13	6.23	9.96	19.68
C19H22C12N2	349.31	348.1160	65.33	6.35	20.30	8.02
C19H22C12N4	377.32	376.1221	60.48	5.88	18.79	14.85
C19H22C12N6	405.33	404.1283	56.30	5.47	17.49	20.73
C19H23C11N2	314.86	314.1550	72.48	7.36	11.26	8.90
C19H23C11N4	342.87	342.1611	66.56	6.76	10.34	16.34
C19H23C11N6	370.89	370.1673	61.53	6.25	9.56	22.66
C19H23C12N1	336.31	335.1207	67.86	6.89	21.08	4.16
C19H23C12N3	364.32	363.1269	62.64	6.36	19.46	11.53
C19H23C12N5	392.33	391.1330	58.17	5.91	18.07	17.85
C19H24C11N1	301.86	301.1597	75.60	8.01	11.74	4.64
C19H24C11N3	329.88	329.1659	69.18	7.33	10.75	12.74
C19H24C11N5	357.89	357.1720	63.77	6.76	9.91	19.57
C19H24C12N2	351.32	350.1316	64.96	6.89	20.18	7.97
C19H24C12N4	379.34	378.1378	60.16	6.38	18.69	14.77
C19H24C12N6	407.35	406.1439	56.02	5.94	17.41	20.63
C19H25C11N2	316.88	316.1706	72.02	7.95	11.19	8.84
C19H25C11N4	344.89	344.1768	66.17	7.31	10.28	16.24
C19H25C11N6	372.90	372.1829	61.20	6.76	9.51	22.54
C19H25C12N1	338.32	337.1364	67.45	7.45	20.96	4.14

C19H25C12N3	366.34	365.1425	62.30	6.88	19.36	11.47
C19H25C12N5	394.35	393.1487	57.87	6.39	17.98	17.76
C19H26C11N1	303.88	303.1754	75.10	8.62	11.67	4.61
C19H26C11N3	331.89	331.1815	68.76	7.90	10.68	12.66
C19H26C11N5	359.91	359.1877	63.41	7.28	9.85	19.46
C19H26C12N2	353.34	352.1473	64.59	7.42	20.07	7.93
C19H26C12N4	381.35	380.1534	59.84	6.87	18.59	14.69
C19H26C12N6	409.37	408.1596	55.75	6.40	17.32	20.53
C19H27C11N2	318.89	318.1863	71.56	8.53	11.12	8.78
C19H27C11N4	346.91	346.1924	65.78	7.85	10.22	16.15
C19H27C11N6	374.92	374.1986	60.87	7.26	9.46	22.42
C19H27C12N1	340.34	339.1520	67.05	8.00	20.83	4.12
C19H27C12N3	368.35	367.1582	61.95	7.39	19.25	11.41
C19H27C12N5	396.37	395.1643	57.58	6.87	17.89	17.67
C19H28C11N1	305.89	305.1910	74.60	9.23	11.59	4.58
C19H28C11N3	333.91	333.1972	68.35	8.45	10.62	12.58
C19H28C11N5	361.92	361.2033	63.06	7.80	9.80	19.35
C19H28C12N2	355.35	354.1629	64.22	7.94	19.95	7.88
C19H28C12N4	383.37	382.1691	59.53	7.36	18.50	14.61
C19H28C12N6	411.38	410.1752	55.47	6.86	17.24	20.43
C19H29C11N2	320.91	320.2019	71.11	9.11	11.05	8.73
C19H29C11N4	348.92	348.2081	65.40	8.38	10.16	16.06
C19H29C11N6	376.94	376.2142	60.54	7.75	9.41	22.30
C19H29C12N1	342.36	341.1677	66.66	8.54	20.71	4.09
C19H29C12N3	370.37	369.1738	61.62	7.89	19.14	11.35
C19H29C12N5	398.38	397.1800	57.28	7.34	17.80	17.58
C19H30C11N1	307.91	307.2067	74.12	9.82	11.51	4.55
C19H30C11N3	335.92	335.2128	67.94	9.00	10.55	12.51
C19H30C11N5	363.94	363.2190	62.71	8.31	9.74	19.24
C19H30C12N2	357.37	356.1786	63.86	8.46	19.84	7.84
C19H30C12N4	385.38	384.1847	59.22	7.85	18.40	14.54
C19H30C12N6	413.40	412.1909	55.20	7.31	17.15	20.33
C19H31C11N2	322.93	322.2176	70.67	9.68	10.98	8.67
C19H31C11N4	350.94	350.2237	65.03	8.90	10.10	15.96
C19H31C11N6	378.95	378.2299	60.22	8.25	9.36	22.18
C19H31C12N1	344.37	343.1833	66.27	9.07	20.59	4.07
C19H31C12N3	372.39	371.1895	61.28	8.39	19.04	11.28
C19H31C12N5	400.40	399.1956	57.00	7.80	17.71	17.49
C19H32C11N1	309.93	309.2223	73.63	10.41	11.44	4.52
C19H32C11N3	337.94	337.2285	67.53	9.54	10.49	12.43
C19H32C11N5	365.95	365.2346	62.36	8.81	9.69	19.14
C19H32C12N2	359.39	358.1942	63.50	8.98	19.73	7.79
C19H32C12N4	387.40	386.2004	58.91	8.33	18.30	14.46
C19H32C12N6	415.41	414.2065	54.94	7.76	17.07	20.23
C19H33C11N2	324.94	324.2332	70.23	10.24	10.91	8.62
C19H33C11N4	352.95	352.2394	64.66	9.42	10.04	15.87
C19H33C11N6	380.97	380.2455	59.90	8.73	9.31	22.06
C19H33C12N1	346.39	345.1990	65.88	9.60	20.47	4.04
C19H33C12N3	374.40	373.2051	60.95	8.88	18.94	11.22
C19H33C12N5	402.41	401.2113	56.71	8.27	17.62	17.40
C19H34C11N1	311.94	311.2380	73.16	10.99	11.37	4.49
C19H34C11N3	339.96	339.2441	67.13	10.08	10.43	12.36
C19H34C11N5	367.97	367.2503	62.02	9.31	9.63	19.03
C19H34C12N2	361.40	360.2099	63.15	9.48	19.62	7.75
C19H34C12N4	389.42	388.2160	58.60	8.80	18.21	14.39
C19H34C12N6	417.43	416.2222	54.67	8.21	16.99	20.13
C19H35C11N2	326.96	326.2489	69.80	10.79	10.84	8.57
C19H35C11N4	354.97	354.2550	64.29	9.94	9.99	15.78
C19H35C11N6	382.98	382.2612	59.59	9.21	9.26	21.94
C19H35C12N1	348.40	347.2146	65.50	10.13	20.35	4.02
C19H35C12N3	376.42	375.2208	60.63	9.37	18.84	11.16
C19H35C12N5	404.43	403.2269	56.43	8.72	17.53	17.32
C19H36C11N1	313.96	313.2536	72.69	11.56	11.29	4.46
C19H36C11N3	341.97	341.2598	66.73	10.61	10.37	12.29
C19H36C11N5	369.99	369.2659	61.68	9.81	9.58	18.93
C19H36C12N2	363.42	362.2255	62.80	9.98	19.51	7.71
C19H36C12N4	391.43	390.2317	58.30	9.27	18.11	14.31
C19H36C12N6	419.44	418.2378	54.41	8.65	16.90	20.04
C19H37C11N2	328.97	328.2645	69.37	11.34	10.78	8.52
C19H37C11N4	356.99	356.2707	63.93	10.45	9.93	15.69
C19H37C11N6	385.00	384.2768	59.28	9.69	9.21	21.83
C19H37C12N1	350.42	349.2303	65.13	10.64	20.23	4.00
C19H37C12N3	378.43	377.2364	60.30	9.86	18.74	11.10
C19H37C12N5	406.45	405.2426	56.15	9.18	17.45	17.23
C19H38C11N1	315.97	315.2693	72.22	12.12	11.22	4.43
C19H38C11N3	343.99	343.2754	66.34	11.13	10.31	12.22
C19H38C11N5	372.00	371.2816	61.35	10.30	9.53	18.83
C19H38C12N2	365.43	364.2412	62.45	10.48	19.40	7.67
C19H38C12N4	393.45	392.2473	58.00	9.74	18.02	14.24
C19H38C12N6	421.46	420.2535	54.15	9.09	16.82	19.94
C19H39C11N2	330.99	330.2802	68.95	11.88	10.71	8.46
C19H39C11N4	359.00	358.2863	63.57	10.95	9.88	15.61
C19H39C11N6	387.02	386.2925	58.97	10.16	9.16	21.71
C19H39C12N1	352.44	351.2459	64.75	11.15	20.12	3.97
C19H39C12N3	380.45	379.2521	59.98	10.33	18.64	11.04
C19H39C12N5	408.46	407.2582	55.87	9.62	17.36	17.15
C19H40C11N1	317.99	317.2849	71.77	12.68	11.15	4.40
C19H40C11N3	346.00	345.2911	65.96	11.65	10.25	12.14
C19H40C11N5	374.02	373.2972	61.02	10.78	9.48	18.72
C19H40C12N2	367.45	366.2568	62.11	10.97	19.30	7.62
C19H40C12N4	395.46	394.2630	57.71	10.20	17.93	14.17
C19H40C12N6	423.48	422.2691	53.89	9.52	16.74	19.85
C19H41C11N2	333.01	332.2958	68.53	12.41	10.65	8.41
C19H41C11N4	361.02	360.3020	63.21	11.45	9.82	15.52
C19H41C11N6	389.03	388.3081	58.66	10.62	9.11	21.60
C19H41C12N3	382.46	381.2677	59.67	10.81	18.54	10.99
C19H41C12N5	410.48	409.2739	55.60	10.07	17.27	17.06
C19H42C11N3	348.02	347.3067	65.57	12.16	10.19	12.07
C19H42C11N5	376.03	375.3129	60.69	11.26	9.43	18.62
C19H42C12N4	397.48	396.2786	57.41	10.65	17.84	14.10
C19H42C12N6	425.49	424.2848	53.63	9.95	16.66	19.75
C19H43C11N4	363.03	362.3176	62.86	11.94	9.77	15.43
C19H43C11N6	391.05	390.3238	58.36	11.08	9.07	21.49
C19H43C12N5	412.49	411.2895	55.32	10.51	17.19	16.98
C19H44C11N5	378.05	377.3285	60.37	11.73	9.38	18.52
C19H44C12N6	427.51	426.3004	53.38	10.37	16.59	19.66
C19H45C11N6	393.06	392.3394	58.06	11.54	9.02	21.38
C20H10C11N1	299.76	299.0502	80.14	3.36	11.83	4.67
C20H10C11N3	327.78	327.0563	73.29	3.08	10.82	12.82
C20H10C11N5	355.79	355.0625	67.52	2.83	9.96	19.68
C20H10C12N2	349.22	348.0221	68.79	2.89	20.30	8.02
C20H10C12N4	377.24	376.0282	63.68	2.67	18.80	14.85
C20H10C12N6	405.25	404.0344	59.28	2.49	17.50	20.74
C20H11C11N2	314.78	314.0611	76.32	3.52	11.26	8.90
C20H11C11N4	342.79	342.0672	70.08	3.23	10.34	16.34
C20H11C11N6	370.80	370.0734	64.78	2.99	9.56	22.66
C20H11C12N1	336.22	335.0269	71.45	3.30	21.09	4.17
C20H11C12N3	364.24	363.0330	65.95	3.04	19.47	11.54
C20H11C12N5	392.25	391.0391	61.24	2.83	18.08	17.85
C20H12C11N1	301.78	301.0658	79.60	4.01	11.75	4.64

C20H12Cl1N3	329.79	329.0720	72.84	3.67	10.75	12.74
C20H12Cl1N5	357.81	357.0781	67.14	3.38	9.91	19.57
C20H12Cl2N2	351.24	350.0377	68.39	3.44	20.19	7.98
C20H12Cl2N4	379.25	378.0439	63.34	3.19	18.70	14.77
C20H12Cl2N6	407.26	406.0500	58.98	2.97	17.41	20.64
C20H13Cl1N2	316.79	316.0767	75.83	4.14	11.19	8.84
C20H13Cl1N4	344.81	344.0829	69.67	3.80	10.28	16.25
C20H13Cl1N6	372.82	372.0890	64.43	3.51	9.51	22.54
C20H13Cl2N1	338.24	337.0425	71.02	3.87	20.96	4.14
C20H13Cl2N3	366.25	365.0486	65.59	3.58	19.36	11.47
C20H13Cl2N5	394.27	393.0548	60.93	3.32	17.98	17.76
C20H14Cl1N1	3C3.79	303.0815	79.07	4.65	11.67	4.61
C20H14Cl1N3	331.81	331.0876	72.40	4.25	10.68	12.66
C20H14Cl1N5	359.82	359.0938	66.76	3.92	9.85	19.46
C20H14Cl2N2	353.25	352.0534	68.00	3.99	20.07	7.93
C20H14Cl2N4	381.27	380.0595	63.01	3.70	18.60	14.69
C20H14Cl2N6	409.28	408.0657	58.69	3.45	17.32	20.53
C20H15Cl1N2	318.81	318.0924	75.35	4.74	11.12	8.79
C20H15Cl1N4	346.82	346.0985	69.26	4.36	10.22	16.15
C20H15Cl1N6	374.84	374.1047	64.09	4.03	9.46	22.42
C20H15Cl2N1	340.26	339.0581	70.60	4.44	20.84	4.12
C20H15Cl2N3	368.27	367.0643	65.23	4.11	19.25	11.41
C20H15Cl2N5	396.28	395.0704	60.62	3.82	17.89	17.67
C20H16Cl1N1	305.81	305.0971	78.55	5.27	11.59	4.58
C20H16Cl1N3	333.82	333.1033	71.96	4.83	10.62	12.59
C20H16Cl1N5	361.84	361.1094	66.39	4.46	9.80	19.35
C20H16Cl2N2	355.27	354.0690	67.62	4.54	19.96	7.89
C20H16Cl2N4	383.28	382.0752	62.68	4.21	18.50	14.62
C20H16Cl2N6	411.30	410.0813	58.41	3.92	17.24	20.43
C20H17Cl1N2	320.82	320.1080	74.88	5.34	11.05	8.73
C20H17Cl1N4	348.84	348.1142	68.86	4.91	10.16	16.06
C20H17Cl1N6	376.85	376.1203	63.74	4.55	9.41	22.30
C20H17Cl2N1	342.27	341.0738	70.18	5.01	20.72	4.09
C20H17Cl2N3	370.28	369.0799	64.88	4.63	19.15	11.35
C20H17Cl2N5	398.30	397.0861	60.31	4.30	17.80	17.58
C20H18Cl1N1	307.83	307.1128	78.04	5.89	11.52	4.55
C20H18Cl1N3	335.84	335.1189	71.53	5.40	10.56	12.51
C20H18Cl1N5	363.85	363.1251	66.02	4.99	9.74	19.25
C20H18Cl2N2	357.29	356.0847	67.24	5.08	19.85	7.84
C20H18Cl2N4	385.30	384.0908	62.35	4.71	18.40	14.54
C20H18Cl2N6	413.31	412.0970	58.12	4.39	17.16	20.33
C20H19Cl1N2	322.84	322.1237	74.41	5.93	10.98	8.68
C20H19Cl1N4	350.85	350.1298	68.47	5.46	10.10	15.97
C20H19Cl1N6	378.87	378.1360	63.41	5.05	9.36	22.18
C20H19Cl2N1	344.29	343.0894	69.77	5.56	20.60	4.07
C20H19Cl2N3	372.30	371.0956	64.52	5.14	19.05	11.29
C20H19Cl2N5	400.31	399.1017	60.01	4.78	17.71	17.49
C20H20Cl1N1	309.84	309.1284	77.53	6.51	11.44	4.52
C20H20Cl1N3	337.86	337.1346	71.10	5.97	10.49	12.44
C20H20Cl1N5	365.87	365.1407	65.66	5.51	9.69	19.14
C20H20Cl2N2	359.30	358.1003	66.86	5.61	19.73	7.80
C20H20Cl2N4	387.32	386.1065	62.02	5.20	18.31	14.47
C20H20Cl2N6	415.33	414.1126	57.84	4.85	17.07	20.23
C20H21Cl1N2	324.86	324.1393	73.95	6.52	10.91	8.62
C20H21Cl1N4	352.87	352.1455	68.08	6.00	10.05	15.88
C20H21Cl1N6	380.88	380.1516	63.07	5.56	9.31	22.06
C20H21Cl2N1	346.30	345.1051	69.37	6.11	20.48	4.04
C20H21Cl2N3	374.32	373.1112	64.18	5.65	18.94	11.23
C20H21Cl2N5	402.33	401.1174	59.71	5.26	17.62	17.41
C20H22Cl1N1	311.86	311.1441	77.03	7.11	11.37	4.49
C20H22Cl1N3	339.87	339.1502	70.68	6.52	10.43	12.36
C20H22Cl1N5	367.88	367.1564	65.30	6.03	9.64	19.04
C20H22Cl2N2	361.32	360.1160	66.49	6.14	19.62	7.75
C20H22Cl2N4	389.33	388.1221	61.70	5.70	18.21	14.39
C20H22Cl2N6	417.34	416.1283	57.56	5.31	16.99	20.14
C20H23Cl1N2	326.87	326.1550	73.49	7.09	10.85	8.57
C20H23Cl1N4	354.89	354.1611	67.69	6.53	9.99	15.79
C20H23Cl1N6	382.90	382.1673	62.74	6.05	9.26	21.95
C20H23Cl2N1	348.32	347.1207	68.97	6.66	20.36	4.02
C20H23Cl2N3	376.33	375.1269	63.83	6.16	18.84	11.17
C20H23Cl2N5	404.35	403.1330	59.41	5.73	17.54	17.32
C20H24Cl1N1	313.87	313.1597	76.53	7.71	11.30	4.46
C20H24Cl1N3	341.89	341.1659	70.26	7.08	10.37	12.29
C20H24Cl1N5	369.90	369.1720	64.94	6.54	9.58	18.93
C20H24Cl2N2	363.33	362.1316	66.12	6.66	19.52	7.71
C20H24Cl2N4	391.35	390.1378	61.38	6.18	18.12	14.32
C20H24Cl2N6	419.36	418.1439	57.28	5.77	16.91	20.04
C20H25Cl1N2	328.89	328.1706	73.04	7.66	10.78	8.52
C20H25Cl1N4	356.90	356.1768	67.31	7.06	9.93	15.70
C20H25Cl1N6	384.92	384.1829	62.41	6.55	9.21	21.83
C20H25Cl2N1	350.33	349.1364	68.57	7.19	20.24	4.00
C20H25Cl2N3	378.35	377.1425	63.49	6.66	18.74	11.11
C20H25Cl2N5	406.36	405.1487	59.12	6.20	17.45	17.23
C20H26Cl1N1	315.89	315.1754	76.05	8.30	11.22	4.43
C20H26Cl1N3	343.90	343.1815	69.85	7.62	10.31	12.22
C20H26Cl1N5	371.92	371.1877	64.59	7.05	9.53	18.83
C20H26Cl2N2	365.35	364.1473	65.75	7.17	19.41	7.67
C20H26Cl2N4	393.36	392.1534	61.07	6.66	18.03	14.24
C20H26Cl2N6	421.38	420.1596	57.01	6.22	16.83	19.94
C20H27Cl1N2	330.90	330.1863	72.60	8.22	10.71	8.47
C20H27Cl1N4	358.92	358.1924	66.93	7.58	9.88	15.61
C20H27Cl1N6	386.93	386.1986	62.08	7.03	9.16	21.72
C20H27Cl2N1	352.35	351.1520	68.18	7.72	20.12	3.98
C20H27Cl2N3	380.36	379.1582	63.16	7.16	18.64	11.05
C20H27Cl2N5	408.38	407.1643	58.82	6.66	17.36	17.15
C20H28Cl1N1	317.91	317.1910	75.56	8.88	11.15	4.41
C20H28Cl1N3	345.92	345.1972	69.44	8.16	10.25	12.15
C20H28Cl1N5	373.93	373.2033	64.24	7.55	9.48	18.73
C20H28Cl2N2	367.37	366.1629	65.39	7.68	19.30	7.63
C20H28Cl2N4	395.38	394.1691	60.76	7.14	17.93	14.17
C20H28Cl2N6	423.39	422.1752	56.74	6.67	16.75	19.85
C20H29Cl1N2	332.92	332.2019	72.16	8.78	10.65	8.41
C20H29Cl1N4	360.93	360.2081	66.56	8.10	9.82	15.52
C20H29Cl1N6	388.95	388.2142	61.76	7.52	9.12	21.61
C20H29Cl2N1	354.37	353.1677	67.79	8.25	20.01	3.95
C20H29Cl2N3	382.38	381.1738	62.82	7.64	18.54	10.99
C20H29Cl2N5	410.39	409.1800	58.53	7.12	17.28	17.06
C20H30Cl1N1	319.92	319.2067	75.09	9.45	11.08	4.38
C20H30Cl1N3	347.94	347.2128	69.04	8.69	10.19	12.08
C20H30Cl1N5	375.95	375.2190	63.90	8.04	9.43	18.63
C20H30Cl2N2	369.38	368.1786	65.03	8.19	19.20	7.58
C20H30Cl2N4	397.39	396.1847	60.45	7.61	17.84	14.10
C20H30Cl2N6	425.41	424.1909	56.47	7.11	16.67	19.76
C20H31Cl1N2	334.94	334.2176	71.72	9.33	10.58	8.36
C20H31Cl1N4	362.95	362.2237	66.19	8.61	9.77	15.44
C20H31Cl1N6	390.96	390.2299	61.44	7.99	9.07	21.50
C20H31Cl2N1	356.38	355.1833	67.41	8.77	19.90	3.93
C20H31Cl2N3	384.40	383.1895	62.49	8.13	18.45	10.93
C20H31Cl2N5	412.41	411.1956	58.25	7.58	17.19	16.98
C20H32Cl1N1	321.94	321.2223	74.62	10.02	11.01	4.35

C20H32Cl1N3	349.95	349.2285	68.64	9.22	10.13	12.01
C20H32Cl1N5	377.96	377.2346	63.56	8.53	9.38	18.53
C20H32Cl2N2	371.40	370.1942	64.68	8.68	19.09	7.54
C20H32Cl2N4	399.41	398.2004	60.14	8.08	17.75	14.03
C20H32Cl2N6	427.42	426.2065	56.20	7.55	16.59	19.66
C20H33Cl1N2	336.95	336.2332	71.29	9.87	10.52	8.31
C20H33Cl1N4	364.97	364.2394	65.82	9.11	9.71	15.35
C20H33Cl1N6	392.98	392.2455	61.13	8.46	9.02	21.39
C20H33Cl2N1	358.40	357.1990	67.03	9.28	19.78	3.91
C20H33Cl2N3	386.41	385.2051	62.17	8.61	18.35	10.87
C20H33Cl2N5	414.43	413.2113	57.97	8.03	17.11	16.90
C20H34Cl1N1	323.95	323.2380	74.15	10.58	10.94	4.32
C20H34Cl1N3	351.97	351.2441	68.25	9.74	10.07	11.94
C20H34Cl1N5	379.98	379.2503	63.22	9.02	9.33	18.43
C20H34Cl2N2	373.41	372.2099	64.33	9.18	18.99	7.50
C20H34Cl2N4	401.43	400.2160	59.84	8.54	17.66	13.96
C20H34Cl2N6	429.44	428.2222	55.94	7.98	16.51	19.57
C20H35Cl1N2	338.97	338.2489	70.87	10.41	10.46	8.26
C20H35Cl1N4	366.98	366.2550	65.46	9.61	9.66	15.27
C20H35Cl1N6	395.00	394.2612	60.82	8.93	8.98	21.28
C20H35Cl2N1	360.41	359.2146	66.65	9.79	19.67	3.89
C20H35Cl2N3	388.43	387.2208	61.84	9.08	18.25	10.82
C20H35Cl2N5	416.44	415.2269	57.68	8.47	17.03	16.82
C20H36Cl1N1	325.97	325.2536	73.69	11.13	10.88	4.30
C20H36Cl1N3	353.98	353.2598	67.86	10.25	10.02	11.87
C20H36Cl1N5	382.00	381.2659	62.89	9.50	9.28	18.33
C20H36Cl2N2	375.43	374.2255	63.99	9.67	18.89	7.46
C20H36Cl2N4	403.44	402.2317	59.54	8.99	17.58	13.89
C20H36Cl2N6	431.46	430.2378	55.68	8.41	16.43	19.48
C20H37Cl1N2	340.98	340.2645	70.45	10.94	10.40	8.22
C20H37Cl1N4	369.00	368.2707	65.10	10.11	9.61	15.18
C20H37Cl1N6	397.01	396.2768	60.51	9.39	8.93	21.17
C20H37Cl2N1	362.43	361.2303	66.28	10.29	19.56	3.86
C20H37Cl2N3	390.44	389.2364	61.53	9.55	18.16	10.76
C20H37Cl2N5	418.46	417.2426	57.41	8.91	16.94	16.74
C20H38Cl1N1	327.99	327.2693	73.24	11.68	10.81	4.27
C20H38Cl1N3	356.00	355.2754	67.48	10.76	9.96	11.80
C20H38Cl1N5	384.01	383.2816	62.56	9.97	9.23	18.24
C20H38Cl2N2	377.45	376.2412	63.64	10.15	18.79	7.42
C20H38Cl2N4	405.46	404.2473	59.25	9.45	17.49	13.82
C20H38Cl2N6	433.47	432.2535	55.42	8.84	16.36	19.39
C20H39Cl1N2	343.00	342.2802	70.04	11.46	10.34	8.17
C20H39Cl1N4	371.01	370.2863	64.75	10.60	9.56	15.10
C20H39Cl1N6	399.03	398.2925	60.20	9.85	8.88	21.06
C20H39Cl2N1	364.45	363.2459	65.91	10.79	19.46	3.84
C20H39Cl2N3	392.46	391.2521	61.21	10.02	18.07	10.71
C20H39Cl2N5	420.47	419.2582	57.13	9.35	16.86	16.66
C20H40Cl1N1	330.00	329.2849	72.79	12.22	10.74	4.24
C20H40Cl1N3	358.01	357.2911	67.10	11.26	9.90	11.74
C20H40Cl1N5	386.03	385.2972	62.23	10.44	9.18	18.14
C20H40Cl2N2	379.46	378.2568	63.31	10.63	18.69	7.38
C20H40Cl2N4	407.47	406.2630	58.95	9.89	17.40	13.75
C20H40Cl2N6	435.49	434.2691	55.16	9.26	16.28	19.30
C20H41Cl1N2	345.02	344.2958	69.63	11.98	10.28	8.12
C20H41Cl1N4	373.03	372.3020	64.40	11.08	9.50	15.02
C20H41Cl1N6	401.04	400.3081	59.90	10.30	8.84	20.96
C20H41Cl2N1	366.46	365.2616	65.55	11.28	19.35	3.82
C20H41Cl2N3	394.48	393.2677	60.90	10.48	17.97	10.65
C20H41Cl2N5	422.49	421.2739	56.86	9.78	16.78	16.58
C20H42Cl1N1	332.02	331.3006	72.35	12.75	10.68	4.22
C20H42Cl1N3	360.03	359.3067	66.72	11.76	9.85	11.67
C20H42Cl1N5	388.04	387.3129	61.91	10.91	9.14	18.05
C20H42Cl2N2	381.48	380.2725	62.97	11.10	18.59	7.34
C20H42Cl2N4	409.49	408.2786	58.66	10.34	17.32	13.68
C20H42Cl2N6	437.50	436.2848	54.91	9.68	16.21	19.21
C20H43Cl1N2	347.03	346.3115	69.22	12.49	10.22	8.07
C20H43Cl1N4	375.05	374.3176	64.05	11.56	9.45	14.94
C20H43Cl1N6	403.06	402.3238	59.60	10.75	8.80	20.85
C20H43Cl2N3	396.49	395.2834	60.59	10.93	17.88	10.60
C20H43Cl2N5	424.51	423.2895	56.59	10.21	16.70	16.50
C20H44Cl1N3	362.05	361.3224	66.35	12.25	9.79	11.61
C20H44Cl1N5	390.06	389.3285	61.59	11.37	9.09	17.95
C20H44Cl2N4	411.51	410.2943	58.38	10.78	17.23	13.62
C20H44Cl2N6	439.52	438.3004	54.66	10.09	16.13	19.12
C20H45Cl1N4	377.06	376.3333	63.71	12.03	9.40	14.86
C20H45Cl1N6	405.07	404.3394	59.30	11.20	8.75	20.75
C20H45Cl2N5	426.52	425.3052	56.32	10.63	16.62	16.42
C20H46Cl1N5	392.08	391.3442	61.27	11.83	9.04	17.86
C20H46Cl2N6	441.54	440.3161	54.41	10.50	16.06	19.03
C20H47Cl1N6	407.09	406.3551	59.01	11.64	8.71	20.64
C21H10Cl1N1	311.77	311.0502	80.90	3.23	11.37	4.49
C21H10Cl1N3	339.79	339.0563	74.23	2.97	10.43	12.37
C21H10Cl1N5	367.80	367.0625	68.58	2.74	9.64	19.04
C21H10Cl2N2	361.23	360.0221	69.83	2.79	19.63	7.75
C21H10Cl2N4	389.25	388.0282	64.80	2.59	18.22	14.39
C21H10Cl2N6	417.26	416.0344	60.45	2.42	16.99	20.14
C21H11Cl1N2	326.79	326.0611	77.19	3.39	10.85	8.57
C21H11Cl1N4	354.80	354.0672	71.09	3.13	9.99	15.79
C21H11Cl1N6	382.82	382.0734	65.89	2.90	9.26	21.95
C21H11Cl2N1	348.23	347.0269	72.43	3.18	20.36	4.02
C21H11Cl2N3	376.25	375.0330	67.04	2.95	18.85	11.17
C21H11Cl2N5	404.26	403.0391	62.39	2.74	17.54	17.32
C21H12Cl1N1	313.79	313.0658	80.38	3.85	11.30	4.46
C21H12Cl1N3	341.80	341.0720	73.80	3.54	10.37	12.29
C21H12Cl1N5	369.82	369.0781	68.21	3.27	9.59	18.94
C21H12Cl2N2	363.25	362.0377	69.44	3.33	19.52	7.71
C21H12Cl2N4	391.26	390.0439	64.47	3.09	18.12	14.32
C21H12Cl2N6	419.28	418.0500	60.16	2.88	16.91	20.04
C21H13Cl1N2	328.80	328.0767	76.71	3.99	10.78	8.52
C21H13Cl1N4	356.82	356.0829	70.69	3.67	9.94	15.70
C21H13Cl1N6	384.83	384.0890	65.54	3.41	9.21	21.84
C21H13Cl2N1	350.25	349.0425	72.02	3.74	20.24	4.00
C21H13Cl2N3	378.26	377.0486	66.68	3.46	18.75	11.11
C21H13Cl2N5	406.28	405.0548	62.08	3.23	17.45	17.24
C21H14Cl1N1	315.81	315.0815	79.87	4.47	11.23	4.44
C21H14Cl1N3	343.82	343.0876	73.36	4.10	10.31	12.22
C21H14Cl1N5	371.83	371.0938	67.84	3.80	9.53	18.83
C21H14Cl2N2	365.27	364.0534	69.06	3.86	19.41	7.67
C21H14Cl2N4	393.28	392.0595	64.14	3.59	18.03	14.25
C21H14Cl2N6	421.29	420.0657	59.87	3.35	16.83	19.95
C21H15Cl1N2	330.82	330.0924	76.25	4.57	10.72	8.47
C21H15Cl1N4	358.83	358.0985	70.29	4.21	9.88	15.61
C21H15Cl1N6	386.85	386.1047	65.20	3.91	9.16	21.72
C21H15Cl2N1	352.27	351.0581	71.60	4.29	20.13	3.98
C21H15Cl2N3	380.28	379.0643	66.33	3.98	18.65	11.05
C21H15Cl2N5	408.29	407.0704	61.78	3.70	17.37	17.15
C21H16Cl1N1	317.82	317.0971	79.36	5.07	11.16	4.41
C21H16Cl1N3	345.83	345.1033	72.93	4.66	10.25	12.15
C21H16Cl1N5	373.85	373.1033	67.47	4.31	9.48	18.73
C21H16Cl2N2	367.28	366.0690	68.68	4.39	19.31	7.63

C21H16Cl2N4	395.29	394.0752	63.81	4.08	17.94	14.17
C21H16Cl2N6	423.31	422.0813	59.59	3.81	16.75	19.85
C21H17ClN2	332.84	332.1080	75.78	5.15	10.65	8.42
C21H17ClN4	360.85	360.1142	69.90	4.75	9.82	15.53
C21H17ClN6	388.86	388.1203	64.86	4.41	9.12	21.61
C21H17Cl2N1	354.28	353.0738	71.20	4.84	20.01	3.95
C21H17Cl2N3	382.30	381.0799	65.98	4.48	18.55	10.99
C21H17Cl2N5	410.31	409.0861	61.47	4.18	17.28	17.07
C21H18ClN1	319.84	319.1128	78.86	5.67	11.08	4.38
C21H18ClN3	347.85	347.1189	72.51	5.22	10.19	12.08
C21H18ClN5	375.86	375.1251	67.11	4.83	9.43	18.63
C21H18Cl2N2	369.30	368.0847	68.30	4.91	19.20	7.59
C21H18Cl2N4	397.31	396.0908	63.49	4.57	17.85	14.10
C21H18Cl2N6	425.32	424.0970	59.30	4.27	16.67	19.76
C21H19ClN2	334.85	334.1237	75.33	5.72	10.59	8.37
C21H19ClN4	362.87	362.1298	69.51	5.28	9.77	15.44
C21H19ClN6	390.88	390.1360	64.53	4.90	9.07	21.50
C21H19Cl2N1	356.30	355.0894	70.79	5.38	19.90	3.93
C21H19Cl2N3	384.31	383.0956	65.63	4.98	18.45	10.93
C21H19Cl2N5	412.33	411.1017	61.17	4.64	17.20	16.99
C21H20ClN1	321.85	321.1284	78.37	6.26	11.02	4.35
C21H20ClN3	349.87	349.1346	72.09	5.76	10.13	12.01
C21H20ClN5	377.88	377.1407	66.75	5.33	9.38	18.53
C21H20Cl2N2	371.31	370.1003	67.93	5.43	19.10	7.54
C21H20Cl2N4	399.33	398.1065	63.16	5.05	17.76	14.03
C21H20Cl2N6	427.34	426.1126	59.02	4.72	16.59	19.67
C21H21ClN2	336.87	336.1393	74.88	6.28	10.52	8.32
C21H21ClN4	364.88	364.1455	69.13	5.80	9.72	15.35
C21H21ClN6	392.89	392.1516	64.20	5.39	9.02	21.39
C21H21Cl2N1	358.31	357.1051	70.39	5.91	19.79	3.91
C21H21Cl2N3	386.33	385.1112	65.29	5.48	18.35	10.88
C21H21Cl2N5	414.34	413.1174	60.88	5.11	17.11	16.90
C21H22ClN1	323.87	323.1441	77.88	6.85	10.95	4.32
C21H22ClN3	351.88	351.1502	71.68	6.30	10.08	11.94
C21H22ClN5	379.90	379.1564	66.40	5.84	9.33	18.43
C21H22Cl2N2	373.33	372.1160	67.56	5.94	18.99	7.50
C21H22Cl2N4	401.34	400.1221	62.85	5.53	17.67	13.96
C21H22Cl2N6	429.36	428.1283	58.75	5.16	16.51	19.57
C21H23ClN2	338.88	338.1550	74.43	6.84	10.46	8.27
C21H23ClN4	366.90	366.1611	68.75	6.32	9.66	15.27
C21H23ClN6	394.91	394.1673	63.87	5.87	8.98	21.28
C21H23Cl2N1	360.33	359.1207	70.00	6.43	19.68	3.89
C21H23Cl2N3	388.34	387.1269	64.95	5.97	18.26	10.82
C21H23Cl2N5	416.36	415.1330	60.58	5.57	17.03	16.82
C21H24ClN1	325.89	325.1597	77.40	7.42	10.88	4.30
C21H24ClN3	353.90	353.1659	71.27	6.84	10.02	11.87
C21H24ClN5	381.91	381.1720	66.05	6.33	9.28	18.34
C21H24Cl2N2	375.34	374.1316	67.20	6.45	18.89	7.46
C21H24Cl2N4	403.36	402.1378	62.53	6.00	17.58	13.89
C21H24Cl2N6	431.37	430.1439	58.47	5.61	16.44	19.48
C21H25ClN2	340.90	340.1706	73.99	7.39	10.40	8.22
C21H25ClN4	368.91	368.1768	68.37	6.83	9.61	15.19
C21H25ClN6	396.93	396.1829	63.55	6.35	8.93	21.17
C21H25Cl2N1	362.35	361.1364	69.61	6.95	19.57	3.87
C21H25Cl2N3	390.36	389.1425	64.62	6.46	18.16	10.76
C21H25Cl2N5	418.37	417.1487	60.29	6.02	16.95	16.74
C21H26ClN1	327.90	327.1754	76.92	7.99	10.81	4.27
C21H26ClN3	355.91	355.1815	70.87	7.36	9.96	11.81
C21H26ClN5	383.93	383.1877	65.70	6.83	9.23	18.24
C21H26Cl2N2	377.36	376.1473	66.84	6.94	18.79	7.42

C21H26Cl2N4	405.37	404.1534	62.22	6.46	17.49	13.82
C21H26Cl2N6	433.39	432.1596	58.20	6.05	16.36	19.39
C21H27ClN2	342.92	342.1863	73.56	7.94	10.34	8.17
C21H27ClN4	370.93	370.1924	68.00	7.34	9.56	15.10
C21H27ClN6	398.94	398.1986	63.23	6.82	8.89	21.07
C21H27Cl2N1	364.36	363.1520	69.23	7.47	19.46	3.84
C21H27Cl2N3	392.38	391.1582	64.28	6.94	18.07	10.71
C21H27Cl2N5	420.39	419.1643	60.00	6.47	16.87	16.66
C21H28ClN1	329.92	329.1910	76.45	8.55	10.75	4.25
C21H28ClN3	357.93	357.1972	70.47	7.89	9.90	11.74
C21H28ClN5	385.94	385.2033	65.36	7.31	9.19	18.15
C21H28Cl2N2	379.38	378.1629	66.49	7.44	18.69	7.38
C21H28Cl2N4	407.39	406.1691	61.91	6.93	17.40	13.75
C21H28Cl2N6	435.40	434.1752	57.93	6.48	16.29	19.30
C21H29ClN2	344.93	344.2019	73.13	8.47	10.28	8.12
C21H29ClN4	372.95	372.2081	67.63	7.84	9.51	15.02
C21H29ClN6	400.96	400.2142	62.91	7.29	8.84	20.96
C21H29Cl2N1	366.38	365.1677	68.85	7.98	19.35	3.82
C21H29Cl2N3	394.39	393.1738	63.96	7.41	17.98	10.65
C21H29Cl2N5	422.40	421.1800	59.71	6.92	16.79	16.58
C21H30ClN1	331.93	331.2067	75.99	9.11	10.68	4.22
C21H30ClN3	359.95	359.2128	70.08	8.40	9.85	11.67
C21H30ClN5	387.96	387.2190	65.02	7.79	9.14	18.05
C21H30Cl2N2	381.39	380.1786	66.14	7.93	18.59	7.35
C21H30Cl2N4	409.41	408.1847	61.61	7.39	17.32	13.68
C21H30Cl2N6	437.42	436.1909	57.66	6.91	16.21	19.21
C21H31ClN2	346.95	346.2176	72.70	9.01	10.22	8.07
C21H31ClN4	374.96	374.2237	67.27	8.33	9.46	14.94
C21H31ClN6	402.97	402.2299	62.59	7.75	8.80	20.85
C21H31Cl2N1	368.39	367.1833	68.47	8.48	19.25	3.80
C21H31Cl2N3	396.41	395.1895	63.63	7.88	17.89	10.60
C21H31Cl2N5	424.42	423.1956	59.43	7.36	16.71	16.50
C21H32ClN1	333.95	333.2223	75.53	9.66	10.62	4.19
C21H32ClN3	361.96	361.2285	69.69	8.91	9.79	11.61
C21H32ClN5	389.98	389.2346	64.68	8.27	9.09	17.96
C21H32Cl2N2	383.41	382.1942	65.79	8.41	18.49	7.31
C21H32Cl2N4	411.42	410.2004	61.31	7.84	17.23	13.62
C21H32Cl2N6	439.44	438.2065	57.40	7.34	16.14	19.12
C21H33ClN2	348.96	348.2332	72.28	9.53	10.16	8.03
C21H33ClN4	376.98	376.2394	66.91	8.82	9.40	14.86
C21H33ClN6	404.99	404.2455	62.28	8.21	8.75	20.75
C21H33Cl2N1	370.41	369.1990	68.10	8.98	19.14	3.78
C21H33Cl2N3	398.42	397.2051	63.31	8.35	17.80	10.55
C21H33Cl2N5	426.44	425.2113	59.15	7.80	16.63	16.42
C21H34ClN1	335.96	335.2380	75.08	10.20	10.55	4.17
C21H34ClN3	363.98	363.2441	69.30	9.42	9.74	11.54
C21H34ClN5	391.99	391.2503	64.35	8.74	9.04	17.87
C21H34Cl2N2	385.42	384.2099	65.44	8.89	18.40	7.27
C21H34Cl2N4	413.44	412.2160	61.01	8.29	17.15	13.55
C21H34Cl2N6	441.45	440.2222	57.14	7.76	16.06	19.04
C21H35ClN2	350.98	350.2489	71.87	10.05	10.10	7.98
C21H35ClN4	378.99	378.2550	66.55	9.31	9.35	14.78
C21H35ClN6	407.01	406.2612	61.97	8.67	8.71	20.65
C21H35Cl2N1	372.43	371.2146	67.73	9.47	19.04	3.76
C21H35Cl2N3	400.44	399.2208	62.99	8.81	17.71	10.49
C21H35Cl2N5	428.45	427.2269	58.87	8.23	16.55	16.35
C21H36ClN1	337.98	337.2536	74.63	10.74	10.49	4.14
C21H36ClN3	365.99	365.2598	68.92	9.91	9.69	11.48
C21H36ClN5	394.01	393.2659	64.02	9.21	9.00	17.77
C21H36Cl2N2	387.44	386.2255	65.10	9.37	18.30	7.23

C21H36Cl2N4	415.45	414.2317	60.71	8.73	17.07	13.49
C21H36Cl2N6	443.47	442.2378	56.88	8.18	15.99	18.95
C21H37ClN2	353.00	352.2645	71.46	10.57	10.04	7.94
C21H37ClN4	381.01	380.2707	66.20	9.79	9.31	14.70
C21H37ClN6	409.02	408.2768	61.67	9.12	8.67	20.55
C21H37Cl2N1	374.44	373.2303	67.36	9.96	18.94	3.74
C21H37Cl2N3	402.46	401.2364	62.67	9.27	17.62	10.44
C21H37Cl2N5	430.47	429.2426	58.60	8.66	16.47	16.27
C21H38ClN1	340.00	339.2693	74.19	11.27	10.43	4.12
C21H38ClN3	368.01	367.2754	68.54	10.41	9.63	11.42
C21H38ClN5	396.02	395.2816	63.69	9.67	8.95	17.68
C21H38Cl2N2	389.46	388.2412	64.77	9.83	18.21	7.19
C21H38Cl2N4	417.47	416.2473	60.42	9.18	16.98	13.42
C21H38Cl2N6	445.48	444.2535	56.62	8.60	15.92	18.86
C21H39ClN2	355.01	354.2802	71.05	11.07	9.99	7.89
C21H39ClN4	383.02	382.2863	65.85	10.26	9.26	14.63
C21H39ClN6	411.04	410.2925	61.37	9.56	8.63	20.45
C21H39Cl2N1	376.46	375.2459	67.00	10.44	18.84	3.72
C21H39Cl2N3	404.47	403.2521	62.36	9.72	17.53	10.39
C21H39Cl2N5	432.48	431.2582	58.32	9.09	16.40	16.19
C21H40ClN1	342.01	341.2849	73.75	11.79	10.37	4.10
C21H40ClN3	370.03	369.2911	68.17	10.90	9.58	11.36
C21H40ClN5	398.04	397.2972	63.37	10.13	8.91	17.59
C21H40Cl2N2	391.47	390.2568	64.43	10.30	18.11	7.16
C21H40Cl2N4	419.49	418.2630	60.13	9.61	16.90	13.36
C21H40Cl2N6	447.50	446.2691	56.37	9.01	15.84	18.78
C21H41ClN2	357.03	356.2958	70.65	11.58	9.93	7.85
C21H41ClN4	385.04	384.3020	65.51	10.73	9.21	14.55
C21H41ClN6	413.05	412.3081	61.07	10.01	8.58	20.35
C21H41Cl2N1	378.47	377.2616	66.65	10.92	18.73	3.70
C21H41Cl2N3	406.49	405.2677	62.05	10.17	17.44	10.34
C21H41Cl2N5	434.50	433.2739	58.05	9.51	16.32	16.12
C21H42ClN1	344.03	343.3006	73.32	12.31	10.31	4.07
C21H42ClN3	372.04	371.3067	67.80	11.38	9.53	11.29
C21H42ClN5	400.06	399.3129	63.05	10.58	8.86	17.51
C21H42Cl2N2	393.49	392.2725	64.10	10.76	18.02	7.12
C21H42Cl2N4	421.50	420.2786	59.84	10.04	16.82	13.29
C21H42Cl2N6	449.52	448.2848	56.11	9.42	15.77	18.70
C21H43ClN2	359.04	358.3115	70.25	12.07	9.87	7.80
C21H43ClN4	387.06	386.3176	65.17	11.20	9.16	14.48
C21H43ClN6	415.07	414.3238	60.77	10.44	8.54	20.25
C21H43Cl2N1	380.49	379.2772	66.29	11.39	18.64	3.68
C21H43Cl2N3	408.50	407.2834	61.75	10.61	17.36	10.29
C21H43Cl2N5	436.52	435.2895	57.78	9.93	16.24	16.04
C21H44ClN1	346.04	345.3162	72.89	12.82	10.25	4.05
C21H44ClN3	374.06	373.3224	67.43	11.86	9.48	11.23
C21H44ClN5	402.07	401.3285	62.73	11.03	8.82	17.42
C21H44Cl2N2	395.50	394.2881	63.78	11.21	17.93	7.08
C21H44Cl2N4	423.52	422.2943	59.56	10.47	16.74	13.23
C21H44Cl2N6	451.53	450.3004	55.86	9.82	15.70	18.61
C21H45ClN2	361.06	360.3271	69.86	12.56	9.82	7.76
C21H45ClN4	389.07	388.3333	64.83	11.66	9.11	14.40
C21H45ClN6	417.09	416.3394	60.48	10.88	8.50	20.15
C21H45Cl2N3	410.52	409.2990	61.44	11.05	17.27	10.24
C21H45Cl2N5	438.53	437.3052	57.52	10.34	16.17	15.97
C21H46ClN3	376.07	375.3380	67.07	12.33	9.43	11.17
C21H46ClN5	404.09	403.3442	62.42	11.47	8.77	17.33
C21H46Cl2N4	425.53	424.3099	59.27	10.90	16.66	13.17
C21H46Cl2N6	453.55	452.3161	55.61	10.22	15.63	18.53
C21H47ClN4	391.09	390.3489	64.50	12.11	9.07	14.33

C21H47ClN6	419.10	418.3551	60.18	11.30	8.46	20.05
C21H47Cl2N5	440.55	439.3208	57.25	10.75	16.09	15.90
C21H48ClN5	406.10	405.3598	62.11	11.91	8.73	17.25
C21H48Cl2N6	455.56	454.3317	55.37	10.62	15.56	18.45
C21H49ClN6	421.12	420.3707	59.90	11.73	8.42	19.96
C22H11ClN2	338.80	338.0611	77.99	3.27	10.46	8.27
C22H11ClN4	366.81	366.0672	72.04	3.02	9.67	15.27
C22H11ClN6	394.83	394.0734	66.93	2.81	8.98	21.29
C22H11Cl2N1	360.25	359.0269	73.35	3.08	19.68	3.89
C22H11Cl2N3	388.26	387.0330	68.06	2.86	18.26	10.82
C22H11Cl2N5	416.27	415.0391	63.48	2.66	17.03	16.82
C22H12ClN1	325.80	325.0658	81.11	3.71	10.88	4.30
C22H12ClN3	353.81	353.0720	74.68	3.42	10.02	11.88
C22H12ClN5	381.83	381.0781	69.21	3.17	9.29	18.34
C22H12Cl2N2	375.26	374.0377	70.42	3.22	18.90	7.47
C22H12Cl2N4	403.27	402.0439	65.53	3.00	17.58	13.89
C22H12Cl2N6	431.29	430.0500	61.27	2.80	16.44	19.49
C22H13ClN2	340.82	340.0767	77.53	3.84	10.40	8.22
C22H13ClN4	368.83	368.0829	71.64	3.55	9.61	15.19
C22H13ClN6	396.84	396.0890	66.59	3.30	8.93	21.18
C22H13Cl2N1	362.26	361.0425	72.94	3.62	19.57	3.87
C22H13Cl2N3	390.28	389.0486	67.71	3.36	18.17	10.77
C22H13Cl2N5	418.29	417.0548	63.17	3.13	16.95	16.74
C22H14ClN1	327.82	327.0815	80.61	4.30	10.81	4.27
C22H14ClN3	355.83	355.0876	74.26	3.97	9.96	11.81
C22H14ClN5	383.84	383.0938	68.84	3.68	9.24	18.25
C22H14Cl2N2	377.28	376.0534	70.04	3.74	18.79	7.43
C22H14Cl2N4	405.29	404.0595	65.20	3.48	17.50	13.82
C22H14Cl2N6	433.30	432.0657	60.98	3.26	16.36	19.40
C22H15ClN2	342.83	342.0924	77.08	4.41	10.34	8.17
C22H15ClN4	370.84	370.0985	71.25	4.08	9.56	15.11
C22H15ClN6	398.86	398.1047	66.25	3.79	8.89	21.07
C22H15Cl2N1	364.28	363.0581	72.54	4.15	19.46	3.85
C22H15Cl2N3	392.29	391.0643	67.36	3.85	18.07	10.71
C22H15Cl2N5	420.30	419.0704	62.87	3.60	16.87	16.66
C22H16ClN1	329.83	329.0971	80.11	4.89	10.75	4.25
C22H16ClN3	357.85	357.1033	73.84	4.51	9.91	11.74
C22H16ClN5	385.86	385.1094	68.48	4.18	9.19	18.15
C22H16Cl2N2	379.29	378.0690	69.67	4.25	18.69	7.39
C22H16Cl2N4	407.31	406.0752	64.88	3.96	17.41	13.76
C22H16Cl2N6	435.32	434.0813	60.70	3.70	16.29	19.31
C22H17ClN2	344.85	344.1080	76.63	4.97	10.28	8.12
C22H17ClN4	372.86	372.1142	70.87	4.60	9.51	15.03
C22H17ClN6	400.87	400.1203	65.92	4.27	8.84	20.96
C22H17Cl2N1	366.29	365.0738	72.14	4.68	19.36	3.82
C22H17Cl2N3	394.31	393.0799	67.02	4.35	17.98	10.66
C22H17Cl2N5	422.32	421.0861	62.57	4.06	16.79	16.58
C22H18ClN1	331.85	331.1128	79.63	5.47	10.68	4.22
C22H18ClN3	359.86	359.1189	73.43	5.04	9.85	11.68
C22H18ClN5	387.88	387.1251	68.13	4.68	9.14	18.06
C22H18Cl2N2	381.31	380.0847	69.30	4.76	18.60	7.35
C22H18Cl2N4	409.32	408.0908	64.56	4.43	17.32	13.69
C22H18Cl2N6	437.33	436.0970	60.42	4.15	16.21	19.22
C22H19ClN2	346.86	346.1237	76.18	5.52	10.22	8.08
C22H19ClN4	374.88	374.1298	70.49	5.11	9.46	14.95
C22H19ClN6	402.89	402.1360	65.59	4.75	8.80	20.86
C22H19Cl2N1	368.31	367.0894	71.75	5.20	19.25	3.80
C22H19Cl2N3	396.32	395.0956	66.67	4.83	17.89	10.60
C22H19Cl2N5	424.34	423.1017	62.27	4.51	16.71	16.50
C22H20ClN1	333.86	333.1284	79.15	6.04	10.62	4.20

C22H20C11N3	361.88	361.1346	73.02	5.57	9.80	11.61
C22H20C11N5	389.89	389.1407	67.77	5.17	9.09	17.96
C22H20C12N2	383.32	382.1003	68.94	5.26	18.50	7.31
C22H20C12N4	411.34	410.1065	64.24	4.90	17.24	13.62
C22H20C12N6	439.35	438.1126	60.14	4.59	16.14	19.13
C22H21C11N2	348.88	348.1393	75.74	6.07	10.16	8.03
C22H21C11N4	376.89	376.1455	70.11	5.62	9.41	14.87
C22H21C11N6	404.91	404.1516	65.26	5.23	8.76	20.76
C22H21C12N1	370.33	369.1051	71.35	5.72	19.15	3.78
C22H21C12N3	398.34	397.1112	66.34	5.31	17.80	10.55
C22H21C12N5	426.35	425.1174	61.98	4.96	16.63	16.43
C22H22C11N1	335.88	335.1441	78.67	6.60	10.56	4.17
C22H22C11N3	363.89	363.1502	72.62	6.09	9.74	11.55
C22H22C11N5	391.91	391.1564	67.43	5.66	9.05	17.87
C22H22C12N2	385.34	384.1160	68.57	5.75	18.40	7.27
C22H22C12N4	413.35	412.1221	63.93	5.36	17.15	13.55
C22H22C12N6	441.37	440.1283	59.87	5.02	16.07	19.04
C22H23C11N2	350.90	350.1550	75.31	6.61	10.10	7.98
C22H23C11N4	378.91	378.1611	69.74	6.12	9.36	14.79
C22H23C11N6	406.92	406.1673	64.94	5.70	8.71	20.65
C22H23C12N1	372.34	371.1207	70.97	6.23	19.04	3.76
C22H23C12N3	400.35	399.1269	66.00	5.79	17.71	10.50
C22H23C12N5	428.37	427.1330	61.69	5.41	16.55	16.35
C22H24C11N1	337.90	337.1597	78.20	7.16	10.49	4.15
C22H24C11N3	365.91	365.1659	72.22	6.61	9.69	11.48
C22H24C11N5	393.92	393.1720	67.08	6.14	9.00	17.78
C22H24C12N2	387.36	386.1316	68.22	6.25	18.31	7.23
C22H24C12N4	415.37	414.1378	63.62	5.82	17.07	13.49
C22H24C12N6	443.38	442.1439	59.60	5.46	15.99	18.95
C22H25C11N2	352.91	352.1706	74.88	7.14	10.05	7.94
C22H25C11N4	380.92	380.1768	69.37	6.62	9.31	14.71
C22H25C11N6	408.94	408.1829	64.62	6.16	8.67	20.55
C22H25C12N1	374.36	373.1364	70.59	6.73	18.94	3.74
C22H25C12N3	402.37	401.1425	65.67	6.26	17.62	10.44
C22H25C12N5	430.38	429.1487	61.40	5.86	16.48	16.27
C22H26C11N1	339.91	339.1754	77.74	7.71	10.43	4.12
C22H26C11N3	367.93	367.1815	71.82	7.12	9.64	11.42
C22H26C11N5	395.94	395.1877	66.74	6.62	8.95	17.69
C22H26C12N2	389.37	388.1473	67.86	6.73	18.21	7.19
C22H26C12N4	417.39	416.1534	63.31	6.28	16.99	13.42
C22H26C12N6	445.40	444.1596	59.33	5.88	15.92	18.87
C22H27C11N2	354.93	354.1863	74.45	7.67	9.99	7.89
C22H27C11N4	382.94	382.1924	69.00	7.11	9.26	14.63
C22H27C11N6	410.95	410.1986	64.30	6.62	8.63	20.45
C22H27C12N1	376.37	375.1520	70.21	7.23	18.84	3.72
C22H27C12N3	404.39	403.1582	65.34	6.73	17.53	10.39
C22H27C12N5	432.40	431.1643	61.11	6.29	16.40	16.20
C22H28C11N1	341.93	341.1910	77.28	8.25	10.37	4.10
C22H28C11N3	369.94	369.1972	71.43	7.63	9.58	11.36
C22H28C11N5	397.95	397.2033	66.40	7.09	8.91	17.60
C22H28C12N2	391.39	390.1629	67.51	7.21	18.12	7.16
C22H28C12N4	419.40	418.1691	63.01	6.73	16.91	13.36
C22H28C12N6	447.41	446.1752	59.06	6.31	15.85	18.78
C22H29C11N2	356.94	356.2019	74.03	8.19	9.93	7.85
C22H29C11N4	384.96	384.2081	68.64	7.59	9.21	14.55
C22H29C11N6	412.97	412.2142	63.99	7.08	8.58	20.35
C22H29C12N1	378.39	377.1677	69.83	7.73	18.74	3.70
C22H29C12N3	406.40	405.1738	65.02	7.19	17.45	10.34
C22H29C12N5	434.42	433.1800	60.83	6.73	16.32	16.12
C22H30C11N1	343.94	343.2067	76.83	8.79	10.31	4.07

C22H30C11N3	371.96	371.2128	71.04	8.13	9.53	11.30
C22H30C11N5	399.97	399.2190	66.07	7.56	8.86	17.51
C22H30C12N2	393.40	392.1786	67.17	7.69	18.02	7.12
C22H30C12N4	421.42	420.1847	62.70	7.18	16.83	13.29
C22H30C12N6	449.43	448.1909	58.80	6.73	15.78	18.70
C22H31C11N2	358.96	358.2176	73.61	8.70	9.88	7.80
C22H31C11N4	386.97	386.2237	68.29	8.07	9.16	14.48
C22H31C11N6	414.99	414.2299	63.68	7.53	8.54	20.25
C22H31C12N1	380.41	379.1833	69.46	8.21	18.64	3.68
C22H31C12N3	408.42	407.1895	64.70	7.65	17.36	10.29
C22H31C12N5	436.43	435.1956	60.55	7.16	16.25	16.05
C22H32C11N1	345.96	345.2223	76.38	9.32	10.25	4.05
C22H32C11N3	373.97	373.2285	70.66	8.62	9.48	11.24
C22H32C11N5	401.99	401.2346	65.73	8.02	8.82	17.42
C22H32C12N2	395.42	394.1942	66.83	8.16	17.93	7.08
C22H32C12N4	423.43	422.2004	62.41	7.62	16.75	13.23
C22H32C12N6	451.45	450.2065	58.53	7.14	15.71	18.62
C22H33C11N2	360.97	360.2332	73.20	9.21	9.82	7.76
C22H33C11N4	388.99	388.2394	67.93	8.55	9.11	14.40
C22H33C11N6	417.00	416.2455	63.37	7.98	8.50	20.15
C22H33C12N1	382.42	381.1990	69.10	8.70	18.54	3.66
C22H33C12N3	410.43	409.2051	64.38	8.10	17.28	10.24
C22H33C12N5	438.45	437.2113	60.27	7.59	16.17	15.97
C22H34C11N1	347.98	347.2380	75.94	9.85	10.19	4.03
C22H34C11N3	375.99	375.2441	70.28	9.11	9.43	11.18
C22H34C11N5	404.00	403.2503	65.41	8.48	8.78	17.33
C22H34C12N2	397.44	396.2099	66.49	8.62	17.84	7.05
C22H34C12N4	425.45	424.2160	62.11	8.06	16.67	13.17
C22H34C12N6	453.46	452.2222	58.27	7.56	15.64	18.53
C22H35C11N2	362.99	362.2489	72.80	9.72	9.77	7.72
C22H35C11N4	391.00	390.2550	67.58	9.02	9.07	14.33
C22H35C11N6	419.02	418.2612	63.06	8.42	8.46	20.06
C22H35C12N1	384.44	383.2146	68.74	9.18	18.44	3.64
C22H35C12N3	412.45	411.2208	64.07	8.55	17.19	10.19
C22H35C12N5	440.46	439.2269	59.99	8.01	16.10	15.90
C22H36C11N1	349.99	349.2536	75.50	10.37	10.13	4.00
C22H36C11N3	378.01	377.2598	69.91	9.60	9.38	11.12
C22H36C11N5	406.02	405.2659	65.08	8.94	8.73	17.25
C22H36C12N2	399.45	398.2255	66.15	9.08	17.75	7.01
C22H36C12N4	427.47	426.2317	61.82	8.49	16.59	13.11
C22H36C12N6	455.48	454.2378	58.01	7.97	15.57	18.45
C22H37C11N2	365.01	364.2645	72.39	10.22	9.71	7.67
C22H37C11N4	393.02	392.2707	67.23	9.49	9.02	14.26
C22H37C11N6	421.03	420.2768	62.76	8.86	8.42	19.96
C22H37C12N1	386.45	385.2303	68.38	9.65	18.35	3.62
C22H37C12N3	414.47	413.2364	63.76	9.00	17.11	10.14
C22H37C12N5	442.48	441.2426	59.72	8.43	16.02	15.83
C22H38C11N1	352.01	351.2693	75.07	10.88	10.07	3.98
C22H38C11N3	380.02	379.2754	69.53	10.08	9.33	11.06
C22H38C11N5	408.03	407.2816	64.76	9.39	8.69	17.16
C22H38C12N2	401.47	400.2412	65.82	9.54	17.66	6.98
C22H38C12N4	429.48	428.2473	61.53	8.92	16.51	13.05
C22H38C12N6	457.49	456.2535	57.76	8.37	15.50	18.37
C22H39C11N2	367.02	366.2802	72.00	10.71	9.66	7.63
C22H39C11N4	395.04	394.2863	66.89	9.95	8.97	14.18
C22H39C11N6	423.05	422.2925	62.46	9.29	8.38	19.87
C22H39C12N1	388.47	387.2459	68.02	10.12	18.25	3.61
C22H39C12N3	416.48	415.2521	63.45	9.44	17.02	10.09
C22H39C12N5	444.50	443.2582	59.45	8.84	15.95	15.76
C22H40C11N1	354.02	353.2849	74.64	11.39	10.01	3.96

10/21/66 177

C22H40C11N3	382.04	381.2911	69.17	10.55	9.28	11.00
C22H40C11N5	410.05	409.2972	64.44	9.83	8.65	17.08
C22H40C12N2	403.48	402.2568	65.49	9.99	17.57	6.94
C22H40C12N4	431.50	430.2630	61.24	9.34	16.43	12.98
C22H40C12N6	459.51	458.2691	57.51	8.77	15.43	18.29
C22H41C11N2	369.04	368.2958	71.60	11.20	9.61	7.59
C22H41C11N4	397.05	396.3020	66.55	10.41	8.93	14.11
C22H41C11N6	425.07	424.3081	62.17	9.72	8.34	19.77
C22H41C12N1	390.48	389.2616	67.67	10.58	18.16	3.59
C22H41C12N3	418.50	417.2677	63.14	9.88	16.94	10.04
C22H41C12N5	446.51	445.2739	59.18	9.26	15.88	15.68
C22H42C11N1	356.04	355.3006	74.22	11.89	9.96	3.93
C22H42C11N3	384.05	383.3067	68.80	11.02	9.23	10.94
C22H42C11N5	412.07	411.3129	64.13	10.27	8.60	17.00
C22H42C12N2	405.50	404.2725	65.17	10.44	17.49	6.91
C22H42C12N4	433.51	432.2786	60.95	9.77	16.36	12.92
C22H42C12N6	461.53	460.2848	57.25	9.17	15.36	18.21
C22H43C11N2	371.05	370.3115	71.21	11.68	9.55	7.55
C22H43C11N4	399.07	398.3176	66.22	10.86	8.88	14.04
C22H43C11N6	427.08	426.3238	61.87	10.15	8.30	19.68
C22H43C12N1	392.50	391.2772	67.32	11.04	18.07	3.57
C22H43C12N3	420.51	419.2834	62.84	10.31	16.86	9.99
C22H43C12N5	448.53	447.2895	58.91	9.66	15.81	15.61
C22H44C11N1	358.06	357.3162	73.80	12.39	9.90	3.91
C22H44C11N3	386.07	385.3224	68.45	11.49	9.18	10.88
C22H44C11N5	414.08	413.3285	63.81	10.71	8.56	16.91
C22H44C12N2	407.52	406.2881	64.84	10.88	17.40	6.87
C22H44C12N4	435.53	434.2943	60.67	10.18	16.28	12.86
C22H44C12N6	463.54	462.3004	57.01	9.57	15.30	18.13
C22H45C11N2	373.07	372.3271	70.83	12.16	9.50	7.51
C22H45C11N4	401.08	400.3333	65.88	11.31	8.84	13.97
C22H45C11N6	429.10	428.3394	61.58	10.57	8.26	19.59
C22H45C12N1	394.52	393.2929	66.98	11.50	17.97	3.55
C22H45C12N3	422.53	421.2990	62.54	10.74	16.78	9.94
C22H45C12N5	450.54	449.3052	58.65	10.07	15.74	15.54
C22H46C11N1	360.07	359.3319	73.39	12.88	9.85	3.89
C22H46C11N3	388.09	387.3380	68.09	11.95	9.14	10.83
C22H46C11N5	416.10	415.3442	63.51	11.14	8.52	16.83
C22H46C12N2	409.53	408.3038	64.52	11.32	17.31	6.84
C22H46C12N4	437.54	436.3099	60.39	10.60	16.21	12.80
C22H46C12N6	465.56	464.3161	56.76	9.96	15.23	18.05
C22H47C11N2	375.09	374.3428	70.45	12.63	9.45	7.47
C22H47C11N4	403.10	402.3489	65.55	11.75	8.80	13.90
C22H47C11N6	431.11	430.3551	61.29	10.99	8.22	19.49
C22H47C12N3	424.55	423.3147	62.24	11.16	16.70	9.90
C22H47C12N5	452.56	451.3208	58.39	10.47	15.67	15.47
C22H48C11N3	390.10	389.3537	67.74	12.40	9.09	10.77
C22H48C11N5	418.11	417.3598	63.20	11.57	8.48	16.75
C22H48C12N4	439.56	438.3256	60.12	11.01	16.13	12.75
C22H48C12N6	467.57	466.3317	56.51	10.35	15.16	17.97
C22H49C11N4	405.12	404.3646	65.23	12.19	8.75	13.83
C22H49C11N6	433.13	432.3707	61.01	11.40	8.19	19.40
C22H49C12N5	454.58	453.3365	58.13	10.87	15.60	15.41
C22H50C11N5	420.13	419.3755	62.90	12.00	8.44	16.67
C22H50C12N6	469.59	468.3474	56.27	10.73	15.10	17.90
C22H51C11N6	435.14	434.3864	60.73	11.81	8.15	19.31
C23H11C11N2	350.81	350.0611	78.75	3.16	10.11	7.99
C23H11C11N4	378.82	378.0672	72.92	2.93	9.36	14.79
C23H11C11N6	406.84	406.0734	67.90	2.73	8.71	20.66
C23H11C12N1	372.26	371.0269	74.21	2.98	19.05	3.76

C23H11C12N3	400.27	399.0330	69.02	2.77	17.71	10.50
C23H11C12N5	428.28	427.0391	64.50	2.59	16.56	16.35
C23H12C11N1	337.81	337.0658	81.78	3.58	10.49	4.15
C23H12C11N3	365.83	365.0720	75.52	3.31	9.69	11.49
C23H12C11N5	393.84	393.0781	70.14	3.07	9.00	17.78
C23H12C12N2	387.27	386.0377	71.33	3.12	18.31	7.23
C23H12C12N4	415.28	414.0439	66.52	2.91	17.07	13.49
C23H12C12N6	443.30	442.0500	62.32	2.73	16.00	18.96
C23H13C11N2	352.83	352.0767	78.30	3.71	10.05	7.94
C23H13C11N4	380.84	380.0829	72.54	3.44	9.31	14.71
C23H13C11N6	408.85	408.0890	67.57	3.20	8.67	20.56
C23H13C12N1	374.27	373.0425	73.81	3.50	18.95	3.74
C23H13C12N3	402.29	401.0486	68.67	3.26	17.63	10.45
C23H13C12N5	430.30	429.0548	64.20	3.05	16.48	16.28
C23H14C11N1	339.83	339.0815	81.29	4.15	10.43	4.12
C23H14C11N3	367.84	367.0876	75.10	3.84	9.64	11.42
C23H14C11N5	395.85	395.0938	69.79	3.56	8.96	17.69
C23H14C12N2	389.29	388.0534	70.96	3.62	18.21	7.20
C23H14C12N4	417.30	416.0595	66.20	3.38	16.99	13.43
C23H14C12N6	445.31	444.0657	62.04	3.17	15.92	18.87
C23H15C11N2	354.84	354.0924	77.85	4.26	9.99	7.89
C23H15C11N4	382.86	382.0985	72.16	3.95	9.26	14.63
C23H15C11N6	410.87	410.1047	67.24	3.68	8.63	20.45
C23H15C12N1	376.29	375.0581	73.42	4.02	18.84	3.72
C23H15C12N3	404.30	403.0643	68.33	3.74	17.54	10.39
C23H15C12N5	432.32	431.0704	63.90	3.50	16.40	16.20
C23H16C11N1	341.84	341.0971	80.81	4.72	10.37	4.10
C23H16C11N3	369.86	369.1033	74.69	4.36	9.59	11.36
C23H16C11N5	397.87	397.1094	69.43	4.05	8.91	17.60
C23H16C12N2	391.30	390.0690	70.60	4.12	18.12	7.16
C23H16C12N4	419.32	418.0752	65.88	3.85	16.91	13.36
C23H16C12N6	447.33	446.0813	61.76	3.61	15.85	18.79
C23H17C11N2	356.86	356.1080	77.41	4.80	9.93	7.85
C23H17C11N4	384.87	384.1142	71.78	4.45	9.21	14.56
C23H17C11N6	412.89	412.1203	66.91	4.15	8.59	20.35
C23H17C12N1	378.30	377.0738	73.02	4.53	18.74	3.70
C23H17C12N3	406.32	405.0799	67.99	4.22	17.45	10.34
C23H17C12N5	434.33	433.0861	63.60	3.95	16.33	16.12
C23H18C11N1	343.86	343.1128	80.34	5.28	10.31	4.07
C23H18C11N3	371.87	371.1189	74.29	4.88	9.53	11.30
C23H18C11N5	399.89	399.1251	69.08	4.54	8.87	17.51
C23H18C12N2	393.32	392.0847	70.24	4.61	18.03	7.12
C23H18C12N4	421.33	420.0908	65.57	4.31	16.83	13.30
C23H18C12N6	449.35	448.0970	61.48	4.04	15.78	18.70
C23H19C11N2	358.87	358.1237	76.98	5.34	9.88	7.81
C23H19C11N4	386.89	386.1298	71.40	4.95	9.16	14.48
C23H19C11N6	414.90	414.1360	66.58	4.62	8.54	20.26
C23H19C12N1	380.32	379.0894	72.64	5.04	18.64	3.68
C23H19C12N3	408.33	407.0956	67.65	4.69	17.36	10.29
C23H19C12N5	436.35	435.1017	63.31	4.39	16.25	16.05
C23H20C11N1	345.88	345.1284	79.87	5.83	10.25	4.05
C23H20C11N3	373.89	373.1346	73.89	5.39	9.48	11.24
C23H20C11N5	401.90	401.1407	68.74	5.02	8.82	17.43
C23H20C12N2	395.34	394.1003	69.88	5.10	17.94	7.09
C23H20C12N4	423.35	422.1065	65.26	4.76	16.75	13.23
C23H20C12N6	451.36	450.1126	61.21	4.47	15.71	18.62
C23H21C11N2	360.89	360.1393	76.55	5.87	9.82	7.76
C23H21C11N4	388.90	388.1455	71.03	5.44	9.12	14.41
C23H21C11N6	416.92	416.1516	66.26	5.08	8.50	20.16
C23H21C12N1	382.34	381.1051	72.25	5.54	18.55	3.66

C23H21Cl2N3	410.35	409.1112	67.32	5.16	17.28	10.24
C23H21Cl2N5	438.36	437.1174	63.02	4.83	16.18	15.98
C23H22Cl1N1	347.89	347.1441	79.41	6.37	10.19	4.03
C23H22Cl1N3	375.90	375.1502	73.49	5.90	9.43	11.18
C23H22Cl1N5	403.92	403.1564	68.39	5.49	8.78	17.34
C23H22Cl2N2	397.35	396.1160	69.52	5.58	17.84	7.05
C23H22Cl2N4	425.36	424.1221	64.95	5.21	16.67	13.17
C23H22Cl2N6	453.38	452.1283	60.93	4.89	15.64	18.54
C23H23Cl1N2	362.91	362.1550	76.12	6.39	9.77	7.72
C23H23Cl1N4	390.92	390.1611	70.67	5.93	9.07	14.33
C23H23Cl1N6	418.93	418.1673	65.94	5.53	8.46	20.06
C23H23Cl2N1	384.35	383.1207	71.88	6.03	18.45	3.64
C23H23Cl2N3	412.37	411.1269	66.99	5.62	17.19	10.19
C23H23Cl2N5	440.38	439.1330	62.73	5.26	16.10	15.90
C23H24Cl1N1	349.91	349.1597	78.95	6.91	10.13	4.00
C23H24Cl1N3	377.92	377.1659	73.10	6.40	9.38	11.12
C23H24Cl1N5	405.93	405.1720	68.05	5.96	8.73	17.25
C23H24Cl2N2	399.37	398.1316	69.17	6.06	17.75	7.01
C23H24Cl2N4	427.38	426.1378	64.64	5.66	16.59	13.11
C23H24Cl2N6	455.39	454.1439	60.66	5.31	15.57	18.45
C23H25Cl1N2	364.92	364.1706	75.70	6.91	9.72	7.68
C23H25Cl1N4	392.94	392.1768	70.31	6.41	9.02	14.26
C23H25Cl1N6	420.95	420.1829	65.63	5.99	8.42	19.96
C23H25Cl2N1	386.37	385.1364	71.50	6.52	18.35	3.63
C23H25Cl2N3	414.38	413.1425	66.67	6.08	17.11	10.14
C23H25Cl2N5	442.40	441.1487	62.45	5.70	16.03	15.83
C23H26Cl1N1	351.92	351.1754	78.50	7.45	10.07	3.98
C23H26Cl1N3	379.94	379.1815	72.71	6.90	9.33	11.06
C23H26Cl1N5	407.95	407.1877	67.72	6.42	8.69	17.17
C23H26Cl2N2	401.38	400.1473	68.83	6.53	17.67	6.98
C23H26Cl2N4	429.40	428.1534	64.34	6.10	16.51	13.05
C23H26Cl2N6	457.41	456.1596	60.40	5.73	15.50	18.37
C23H27Cl1N2	366.94	366.1863	75.29	7.42	9.66	7.63
C23H27Cl1N4	394.95	394.1924	69.95	6.89	8.98	14.19
C23H27Cl1N6	422.96	422.1986	65.31	6.43	8.38	19.87
C23H27Cl2N1	388.38	387.1520	71.13	7.01	18.26	3.61
C23H27Cl2N3	416.40	415.1582	66.34	6.54	17.03	10.09
C23H27Cl2N5	444.41	443.1643	62.16	6.12	15.96	15.76
C23H28Cl1N1	353.94	353.1910	78.05	7.97	10.02	3.96
C23H28Cl1N3	381.95	381.1972	72.33	7.39	9.28	11.00
C23H28Cl1N5	409.97	409.2033	67.39	6.88	8.65	17.08
C23H28Cl2N2	403.40	402.1629	68.48	7.00	17.58	6.94
C23H28Cl2N4	431.41	430.1691	64.04	6.54	16.44	12.99
C23H28Cl2N6	459.43	458.1752	60.13	6.14	15.43	18.29
C23H29Cl1N2	368.95	368.2019	74.88	7.92	9.61	7.59
C23H29Cl1N4	396.97	396.2081	69.59	7.36	8.93	14.11
C23H29Cl1N6	424.98	424.2142	65.00	6.88	8.34	19.78
C23H29Cl2N1	390.40	389.1677	70.76	7.49	18.16	3.59
C23H29Cl2N3	418.41	417.1738	66.02	6.99	16.95	10.04
C23H29Cl2N5	446.43	445.1800	61.88	6.55	15.88	15.69
C23H30Cl1N1	355.96	355.2067	77.61	8.50	9.96	3.93
C23H30Cl1N3	383.97	383.2128	71.95	7.88	9.23	10.94
C23H30Cl1N5	411.98	411.2190	67.06	7.34	8.61	17.00
C23H30Cl2N2	405.41	404.1786	68.14	7.46	17.49	6.91
C23H30Cl2N4	433.43	432.1847	63.74	6.98	16.36	12.93
C23H30Cl2N6	461.44	460.1909	59.87	6.55	15.37	18.21
C23H31Cl1N2	370.97	370.2176	74.47	8.42	9.56	7.55
C23H31Cl1N4	398.98	398.2237	69.24	7.83	8.89	14.04
C23H31Cl1N6	427.00	426.2299	64.70	7.32	8.30	19.68
C23H31Cl2N1	392.42	391.1833	70.40	7.96	18.07	3.57
C23H31Cl2N3	420.43	419.1895	65.71	7.43	16.87	9.99
C23H31Cl2N5	448.44	447.1956	61.60	6.97	15.81	15.62
C23H32Cl1N1	357.97	357.2223	77.17	9.01	9.90	3.91
C23H32Cl1N3	385.98	385.2285	71.57	8.36	9.19	10.89
C23H32Cl1N5	414.00	413.2346	66.73	7.79	8.56	16.92
C23H32Cl2N2	407.43	406.1942	67.80	7.92	17.40	6.88
C23H32Cl2N4	435.44	434.2004	63.44	7.41	16.28	12.87
C23H32Cl2N6	463.46	462.2065	59.61	6.96	15.30	18.13
C23H33Cl1N2	372.99	372.2332	74.07	8.92	9.51	7.51
C23H33Cl1N4	401.00	400.2394	68.89	8.30	8.84	13.97
C23H33Cl1N6	429.01	428.2455	64.39	7.75	8.26	19.59
C23H33Cl2N1	394.43	393.1990	70.04	8.43	17.98	3.55
C23H33Cl2N3	422.45	421.2051	65.39	7.87	16.78	9.95
C23H33Cl2N5	450.46	449.2113	61.33	7.38	15.74	15.55
C23H34Cl1N1	359.99	359.2380	76.74	9.52	9.85	3.89
C23H34Cl1N3	388.00	387.2441	71.20	8.83	9.14	10.83
C23H34Cl1N5	416.01	415.2503	66.41	8.24	8.52	16.83
C23H34Cl2N2	409.45	408.2099	67.47	8.37	17.32	6.84
C23H34Cl2N4	437.46	436.2160	63.15	7.83	16.21	12.81
C23H34Cl2N6	465.47	464.2222	59.35	7.36	15.23	18.05
C23H35Cl1N2	375.00	374.2489	73.67	9.41	9.45	7.47
C23H35Cl1N4	403.02	402.2550	68.55	8.75	8.80	13.90
C23H35Cl1N6	431.03	430.2612	64.09	8.18	8.23	19.50
C23H35Cl2N1	396.45	395.2146	69.68	8.90	17.89	3.53
C23H35Cl2N3	424.46	423.2208	65.08	8.31	16.70	9.90
C23H35Cl2N5	452.47	451.2269	61.05	7.80	15.67	15.48
C23H36Cl1N1	362.00	361.2536	76.31	10.02	9.79	3.87
C23H36Cl1N3	390.02	389.2598	70.83	9.30	9.09	10.77
C23H36Cl1N5	418.03	417.2659	66.09	8.68	8.48	16.75
C23H36Cl2N2	411.46	410.2255	67.14	8.82	17.23	6.81
C23H36Cl2N4	439.48	438.2317	62.86	8.26	16.13	12.75
C23H36Cl2N6	467.49	466.2378	59.09	7.76	15.17	17.98
C23H37Cl1N2	377.02	376.2645	73.27	9.89	9.40	7.43
C23H37Cl1N4	405.03	404.2707	68.21	9.21	8.75	13.83
C23H37Cl1N6	433.04	432.2768	63.79	8.61	8.19	19.41
C23H37Cl2N1	398.46	397.2303	69.33	9.36	17.79	3.52
C23H37Cl2N3	426.48	425.2364	64.78	8.74	16.63	9.85
C23H37Cl2N5	454.49	453.2426	60.78	8.21	15.60	15.41
C23H38Cl1N1	364.02	363.2693	75.89	10.52	9.74	3.85
C23H38Cl1N3	392.03	391.2754	70.47	9.77	9.04	10.72
C23H38Cl1N5	420.05	419.2816	65.77	9.12	8.44	16.67
C23H38Cl2N2	413.48	412.2412	66.81	9.26	17.15	6.78
C23H38Cl2N4	441.49	440.2473	62.57	8.68	16.06	12.69
C23H38Cl2N6	469.51	468.2535	58.84	8.16	15.10	17.90
C23H39Cl1N2	379.03	378.2802	72.88	10.37	9.35	7.39
C23H39Cl1N4	407.05	406.2863	67.87	9.66	8.71	13.76
C23H39Cl1N6	435.06	434.2925	63.50	9.04	8.15	19.32
C23H39Cl2N1	400.48	399.2459	68.98	9.82	17.71	3.50
C23H39Cl2N3	428.49	427.2521	64.47	9.17	16.55	9.81
C23H39Cl2N5	456.51	455.2582	60.52	8.61	15.53	15.34
C23H40Cl1N1	366.03	365.2849	75.47	11.02	9.69	3.83
C23H40Cl1N3	394.05	393.2911	70.11	10.23	9.00	10.66
C23H40Cl1N5	422.06	421.2972	65.45	9.55	8.40	16.59
C23H40Cl2N2	415.49	414.2568	66.49	9.70	17.07	6.74
C23H40Cl2N4	443.51	442.2630	62.29	9.09	15.99	12.63
C23H40Cl2N6	471.52	470.2691	58.59	8.55	15.04	17.82
C23H41Cl1N2	381.05	380.2958	72.50	10.85	9.30	7.35
C23H41Cl1N4	409.06	408.3020	67.53	10.10	8.67	13.70
C23H41Cl1N6	437.08	436.3081	63.21	9.46	8.11	19.23
C23H41Cl2N1	402.50	401.2616	68.64	10.27	17.62	3.48

C23H41Cl2N3	430.51	429.2677	64.17	9.60	16.47	9.76
C23H41Cl2N5	458.52	457.2739	60.25	9.01	15.46	15.27
C23H42Cl1N1	368.05	367.3006	75.06	11.50	9.63	3.81
C23H42Cl1N3	396.06	395.3067	69.75	10.69	8.95	10.61
C23H42Cl1N5	424.08	423.3129	65.14	9.98	8.36	16.51
C23H42Cl2N2	417.51	416.2725	66.17	10.14	16.98	6.71
C23H42Cl2N4	445.52	444.2786	62.01	9.50	15.92	12.58
C23H42Cl2N6	473.54	472.2848	58.34	8.94	14.97	17.75
C23H43Cl1N2	383.07	382.3115	72.12	11.31	9.26	7.31
C23H43Cl1N4	411.08	410.3176	67.20	10.54	8.62	13.63
C23H43Cl1N6	439.09	438.3238	62.92	9.87	8.07	19.14
C23H43Cl2N1	404.51	403.2772	68.29	10.71	17.53	3.46
C23H43Cl2N3	432.53	431.2834	63.87	10.02	16.39	9.72
C23H43Cl2N5	460.54	459.2895	59.99	9.41	15.40	15.21
C23H44Cl1N1	370.07	369.3162	74.65	11.98	9.58	3.78
C23H44Cl1N3	398.08	397.3224	69.40	11.14	8.91	10.56
C23H44Cl1N5	426.09	425.3285	64.83	10.41	8.32	16.44
C23H44Cl2N2	419.53	418.2881	65.85	10.57	16.90	6.68
C23H44Cl2N4	447.54	446.2943	61.73	9.91	15.84	12.52
C23H44Cl2N6	475.55	474.3004	58.09	9.33	14.91	17.67
C23H45Cl1N2	385.08	384.3271	71.74	11.78	9.21	7.27
C23H45Cl1N4	413.09	412.3333	66.87	10.98	8.58	13.56
C23H45Cl1N6	441.11	440.3394	62.63	10.28	8.04	19.05
C23H45Cl2N1	406.53	405.2929	67.96	11.16	17.44	3.45
C23H45Cl2N3	434.54	433.2990	63.57	10.44	16.32	9.67
C23H45Cl2N5	462.55	461.3052	59.72	9.81	15.33	15.14
C23H46Cl1N1	372.08	371.3319	74.25	12.46	9.53	3.76
C23H46Cl1N3	400.10	399.3380	69.05	11.59	8.86	10.50
C23H46Cl1N5	428.11	427.3442	64.53	10.83	8.28	16.36
C23H46Cl2N2	421.54	420.3038	65.53	11.00	16.82	6.65
C23H46Cl2N4	449.56	448.3099	61.45	10.31	15.77	12.46
C23H46Cl2N6	477.57	476.3161	57.85	9.71	14.85	17.60
C23H47Cl1N2	387.10	386.3428	71.37	12.24	9.16	7.24
C23H47Cl1N4	415.11	414.3489	66.55	11.41	8.54	13.50
C23H47Cl1N6	443.12	442.3551	62.34	10.69	8.00	18.97
C23H47Cl2N1	408.54	407.3085	67.62	11.60	17.36	3.43
C23H47Cl2N3	436.56	435.3147	63.28	10.85	16.24	9.63
C23H47Cl2N5	464.57	463.3208	59.46	10.20	15.26	15.07
C23H48Cl1N1	374.10	373.3475	73.85	12.93	9.48	3.74
C23H48Cl1N3	402.11	401.3537	68.70	12.03	8.82	10.45
C23H48Cl1N5	430.13	429.3598	64.23	11.25	8.24	16.28
C23H48Cl2N2	423.56	422.3194	65.22	11.42	16.74	6.61
C23H48Cl2N4	451.57	450.3256	61.18	10.71	15.70	12.41
C23H48Cl2N6	479.59	478.3317	57.60	10.09	14.78	17.52
C23H49Cl1N2	389.11	388.3584	71.00	12.69	9.11	7.20
C23H49Cl1N4	417.13	416.3646	66.23	11.84	8.50	13.43
C23H49Cl1N6	445.14	444.3707	62.06	11.10	7.96	18.88
C23H49Cl2N3	438.57	437.3303	62.99	11.26	16.17	9.58
C23H49Cl2N5	466.59	465.3365	59.21	10.59	15.20	15.01
C23H50Cl1N3	404.13	403.3693	68.36	12.47	8.77	10.40
C23H50Cl1N5	432.14	431.3755	63.93	11.66	8.20	16.21
C23H50Cl2N4	453.59	452.3412	60.90	11.11	15.63	12.35
C23H50Cl2N6	481.60	480.3474	57.36	10.46	14.72	17.45
C23H51Cl1N4	419.14	418.3802	65.91	12.26	8.46	13.37
C23H51Cl1N6	447.16	446.3864	61.78	11.50	7.93	18.79
C23H51Cl2N5	468.60	467.3521	58.95	10.97	15.13	14.95
C23H52Cl1N5	434.16	433.3911	63.63	12.07	8.17	16.13
C23H52Cl2N6	483.62	482.3630	57.12	10.84	14.66	17.38
C23H53Cl1N6	449.17	448.4020	61.50	11.89	7.89	18.71
C24H12Cl1N1	349.82	349.0658	82.40	3.46	10.13	4.00
C24H12Cl1N3	377.84	377.0720	76.29	3.20	9.38	11.12
C24H12Cl1N5	405.85	405.0781	71.03	2.98	8.74	17.26
C24H12Cl2N2	399.28	398.0377	72.20	3.03	17.76	7.02
C24H12Cl2N4	427.30	426.0439	67.46	2.83	16.59	13.11
C24H12Cl2N6	455.31	454.0500	63.31	2.66	15.57	18.46
C24H13Cl1N2	364.84	364.0767	79.01	3.59	9.72	7.68
C24H13Cl1N4	392.85	392.0829	73.38	3.34	9.02	14.26
C24H13Cl1N6	420.86	420.0890	68.49	3.11	8.42	19.97
C24H13Cl2N1	386.28	385.0425	74.63	3.39	18.36	3.63
C24H13Cl2N3	414.30	413.0486	69.58	3.16	17.11	10.14
C24H13Cl2N5	442.31	441.0548	65.17	2.96	16.03	15.83
C24H14Cl1N1	351.84	351.0815	81.93	4.01	10.08	3.98
C24H14Cl1N3	379.85	379.0876	75.89	3.72	9.33	11.06
C24H14Cl1N5	407.87	407.0938	70.68	3.46	8.69	17.17
C24H14Cl2N2	401.30	400.0534	71.83	3.52	17.67	6.98
C24H14Cl2N4	429.31	428.0595	67.15	3.29	16.52	13.05
C24H14Cl2N6	457.33	456.0657	63.03	3.09	15.50	18.38
C24H15Cl1N2	366.85	366.0924	78.58	4.12	9.66	7.64
C24H15Cl1N4	394.87	394.0985	73.00	3.83	8.98	14.19
C24H15Cl1N6	422.88	422.1047	68.17	3.58	8.38	19.87
C24H15Cl2N1	388.30	387.0581	74.24	3.89	18.26	3.61
C24H15Cl2N3	416.31	415.0643	69.24	3.63	17.03	10.09
C24H15Cl2N5	444.33	443.0704	64.88	3.40	15.96	15.76
C24H16Cl1N1	353.85	353.0971	81.46	4.56	10.02	3.96
C24H16Cl1N3	381.87	381.1033	75.49	4.22	9.28	11.00
C24H16Cl1N5	409.88	409.1094	70.33	3.93	8.65	17.09
C24H16Cl2N2	403.31	402.0690	71.47	4.00	17.58	6.95
C24H16Cl2N4	431.33	430.0752	66.83	3.74	16.44	12.99
C24H16Cl2N6	459.34	458.0813	62.76	3.51	15.44	18.30
C24H17Cl1N2	368.87	368.1080	78.15	4.65	9.61	7.59
C24H17Cl1N4	396.88	396.1142	72.63	4.32	8.93	14.12
C24H17Cl1N6	424.90	424.1203	67.84	4.03	8.34	19.78
C24H17Cl2N1	390.32	389.0738	73.85	4.39	18.17	3.59
C24H17Cl2N3	418.33	417.0799	68.91	4.10	16.95	10.04
C24H17Cl2N5	446.34	445.0861	64.58	3.84	15.89	15.69
C24H18Cl1N1	355.87	355.1128	81.00	5.10	9.96	3.94
C24H18Cl1N3	383.88	383.1189	75.09	4.73	9.24	10.95
C24H18Cl1N5	411.90	411.1251	69.99	4.40	8.61	17.00
C24H18Cl2N2	405.33	404.0847	71.12	4.48	17.49	6.91
C24H18Cl2N4	433.34	432.0908	66.52	4.19	16.36	12.93
C24H18Cl2N6	461.36	460.0970	62.48	3.93	15.37	18.22
C24H19Cl1N2	370.89	370.1237	77.72	5.16	9.56	7.55
C24H19Cl1N4	398.90	398.1298	72.27	4.80	8.89	14.05
C24H19Cl1N6	426.91	426.1360	67.52	4.49	8.30	19.69
C24H19Cl2N1	392.33	391.0894	73.48	4.88	18.07	3.57
C24H19Cl2N3	420.35	419.0956	68.58	4.56	16.87	10.00
C24H19Cl2N5	448.36	447.1017	64.29	4.27	15.81	15.62
C24H20Cl1N1	357.89	357.1284	80.55	5.63	9.91	3.91
C24H20Cl1N3	385.90	385.1346	74.70	5.22	9.19	10.89
C24H20Cl1N5	413.91	413.1407	69.64	4.87	8.57	16.92
C24H20Cl2N2	407.35	406.1003	70.77	4.95	17.41	6.88
C24H20Cl2N4	435.36	434.1065	66.21	4.63	16.29	12.87
C24H20Cl2N6	463.37	462.1126	62.21	4.35	15.30	18.14
C24H21Cl1N2	372.90	372.1393	77.30	5.68	9.51	7.51
C24H21Cl1N4	400.91	400.1455	71.90	5.28	8.84	13.97
C24H21Cl1N6	428.93	428.1516	67.21	4.93	8.27	19.59
C24H21Cl2N1	394.35	393.1051	73.10	5.37	17.98	3.55
C24H21Cl2N3	422.36	421.1112	68.25	5.01	16.79	9.95
C24H21Cl2N5	450.37	449.1174	64.01	4.70	15.74	15.55
C24H22Cl1N1	359.90	359.1441	80.10	6.16	9.85	3.89

C24H22C11N3	387.92	387.1502	74.31	5.72	9.14	10.83
C24H22C11N5	415.93	415.1564	69.31	5.33	8.52	16.84
C24H22C12N2	409.36	408.1160	70.42	5.42	17.32	6.84
C24H22C12N4	437.38	436.1221	65.91	5.07	16.21	12.81
C24H22C12N6	465.39	464.1283	61.94	4.76	15.24	18.06
C24H23C11N2	374.92	374.1550	76.89	6.18	9.46	7.47
C24H23C11N4	402.93	402.1611	71.54	5.75	8.80	13.90
C24H23C11N6	430.94	430.1673	66.89	5.38	8.23	19.50
C24H23C12N1	396.36	395.1207	72.73	5.85	17.89	3.53
C24H23C12N3	424.38	423.1269	67.93	5.46	16.71	9.90
C24H23C12N5	452.39	451.1330	63.72	5.12	15.67	15.48
C24H24C11N1	361.92	361.1597	79.65	6.68	9.80	3.87
C24H24C11N3	389.93	389.1659	73.93	6.20	9.09	10.78
C24H24C11N5	417.95	417.1720	68.97	5.79	8.48	16.76
C24H24C12N2	411.38	410.1316	70.07	5.88	17.24	6.81
C24H24C12N4	439.39	438.1378	65.61	5.51	16.14	12.75
C24H24C12N6	467.41	466.1439	61.67	5.18	15.17	17.98
C24H25C11N2	376.93	376.1706	76.48	6.69	9.41	7.43
C24H25C11N4	404.95	404.1768	71.19	6.22	8.75	13.84
C24H25C11N6	432.96	432.1829	66.58	5.82	8.19	19.41
C24H25C12N1	398.38	397.1364	72.36	6.33	17.80	3.52
C24H25C12N3	426.39	425.1425	67.61	5.91	16.63	9.85
C24H25C12N5	454.41	453.1487	63.44	5.55	15.60	15.41
C24H26C11N1	363.93	363.1754	79.21	7.20	9.74	3.85
C24H26C11N3	391.95	391.1815	73.55	6.69	9.05	10.72
C24H26C11N5	419.96	419.1877	68.64	6.24	8.44	16.68
C24H26C12N2	413.39	412.1473	69.73	6.34	17.15	6.78
C24H26C12N4	441.41	440.1534	65.31	5.94	16.06	12.69
C24H26C12N6	469.42	468.1596	61.41	5.58	15.10	17.90
C24H27C11N2	378.95	378.1863	76.07	7.18	9.36	7.39
C24H27C11N4	406.96	406.1924	70.83	6.69	8.71	13.77
C24H27C11N6	434.98	434.1986	66.27	6.26	8.15	19.32
C24H27C12N1	400.40	399.1520	72.00	6.80	17.71	3.50
C24H27C12N3	428.41	427.1582	67.29	6.35	16.55	9.81
C24H27C12N5	456.42	455.1643	63.16	5.96	15.54	15.34
C24H28C11N1	365.95	365.1910	78.77	7.71	9.69	3.83
C24H28C11N3	393.96	393.1972	73.17	7.16	9.00	10.67
C24H28C11N5	421.98	421.2033	68.31	6.69	8.40	16.60
C24H28C12N2	415.41	414.1629	69.39	6.79	17.07	6.74
C24H28C12N4	443.42	442.1691	65.01	6.36	15.99	12.64
C24H28C12N6	471.44	470.1752	61.15	5.99	15.04	17.83
C24H29C11N2	380.97	380.2019	75.67	7.67	9.31	7.35
C24H29C11N4	408.98	408.2081	70.48	7.15	8.67	13.70
C24H29C11N6	436.99	436.2142	65.97	6.69	8.11	19.23
C24H29C12N1	402.41	401.1677	71.64	7.26	17.62	3.48
C24H29C12N3	430.42	429.1738	66.97	6.79	16.47	9.76
C24H29C12N5	458.44	457.1800	62.88	6.38	15.47	15.28
C24H30C11N1	367.97	367.2067	78.34	8.22	9.63	3.81
C24H30C11N3	395.98	395.2128	72.80	7.64	8.95	10.61
C24H30C11N5	423.99	423.2190	67.99	7.13	8.36	16.52
C24H30C12N2	417.43	416.1786	69.06	7.24	16.99	6.71
C24H30C12N4	445.44	444.1847	64.72	6.79	15.92	12.58
C24H30C12N6	473.45	472.1909	60.89	6.39	14.98	17.75
C24H31C11N2	382.98	382.2176	75.27	8.16	9.26	7.31
C24H31C11N4	410.99	410.2237	70.14	7.60	8.63	13.63
C24H31C11N6	439.01	438.2299	65.66	7.12	8.08	19.14
C24H31C12N1	404.43	403.1833	71.28	7.73	17.53	3.46
C24H31C12N3	432.44	431.1895	66.66	7.23	16.40	9.72
C24H31C12N5	460.45	459.1956	62.61	6.79	15.40	15.21
C24H32C11N1	369.98	369.2223	77.91	8.72	9.58	3.79

C24H32C11N3	398.00	397.2285	72.43	8.10	8.91	10.56
C24H32C11N5	426.01	425.2346	67.67	7.57	8.32	16.44
C24H32C12N2	419.44	418.1942	68.73	7.69	16.90	6.68
C24H32C12N4	447.46	446.2004	64.42	7.21	15.85	12.52
C24H32C12N6	475.47	474.2065	60.63	6.78	14.91	17.68
C24H33C11N2	385.00	384.2332	74.88	8.64	9.21	7.28
C24H33C11N4	413.01	412.2394	69.80	8.05	8.58	13.57
C24H33C11N6	441.02	440.2455	65.36	7.54	8.04	19.06
C24H33C12N1	406.44	405.1990	70.92	8.18	17.45	3.45
C24H33C12N3	434.46	433.2051	66.35	7.66	16.32	9.67
C24H33C12N5	462.47	461.2113	62.33	7.19	15.33	15.14
C24H34C11N1	372.00	371.2380	77.49	9.21	9.53	3.77
C24H34C11N3	400.01	399.2441	72.06	8.57	8.86	10.50
C24H34C11N5	428.03	427.2503	67.35	8.01	8.28	16.36
C24H34C12N2	421.46	420.2099	68.40	8.13	16.82	6.65
C24H34C12N4	449.47	448.2160	64.13	7.62	15.78	12.47
C24H34C12N6	477.48	476.2222	60.37	7.18	14.85	17.60
C24H35C11N2	387.01	386.2489	74.49	9.12	9.16	7.24
C24H35C11N4	415.03	414.2550	69.46	8.50	8.54	13.50
C24H35C11N6	443.04	442.2612	65.07	7.96	8.00	18.97
C24H35C12N1	408.46	407.2146	70.57	8.64	17.36	3.43
C24H35C12N3	436.47	435.2208	66.04	8.08	16.25	9.63
C24H35C12N5	464.49	463.2269	62.06	7.60	15.27	15.08
C24H36C11N1	374.01	373.2536	77.07	9.70	9.48	3.74
C24H36C11N3	402.03	401.2598	71.70	9.03	8.82	10.45
C24H36C11N5	430.04	429.2659	67.03	8.44	8.24	16.29
C24H36C12N2	423.47	422.2255	68.07	8.57	16.74	6.62
C24H36C12N4	451.49	450.2317	63.85	8.04	15.70	12.41
C24H36C12N6	479.50	478.2378	60.12	7.57	14.79	17.53
C24H37C11N2	389.03	388.2645	74.10	9.59	9.11	7.20
C24H37C11N4	417.04	416.2707	69.12	8.94	8.50	13.43
C24H37C11N6	445.06	444.2768	64.77	8.38	7.97	18.88
C24H37C12N1	410.48	409.2303	70.23	9.09	17.27	3.41
C24H37C12N3	438.49	437.2364	65.74	8.51	16.17	9.58
C24H37C12N5	466.50	465.2426	61.79	7.99	15.20	15.01
C24H38C11N1	376.03	375.2693	76.66	10.19	9.43	3.72
C24H38C11N3	404.04	403.2754	71.35	9.48	8.77	10.40
C24H38C11N5	432.06	431.2816	66.72	8.87	8.21	16.21
C24H38C12N2	425.49	424.2412	67.75	9.00	16.66	6.58
C24H38C12N4	453.50	452.2473	63.56	8.45	15.64	12.35
C24H38C12N6	481.52	480.2535	59.87	7.95	14.73	17.45
C24H39C11N2	391.04	390.2802	73.72	10.05	9.07	7.16
C24H39C11N4	419.06	418.2863	68.79	9.38	8.46	13.37
C24H39C11N6	447.07	446.2925	64.48	8.79	7.93	18.80
C24H39C12N1	412.49	411.2459	69.88	9.53	17.19	3.40
C24H39C12N3	440.50	439.2521	65.44	8.92	16.10	9.54
C24H39C12N5	468.52	467.2582	61.53	8.39	15.13	14.95
C24H40C11N1	378.05	377.2849	76.25	10.67	9.38	3.71
C24H40C11N3	406.06	405.2911	70.99	9.93	8.73	10.35
C24H40C11N5	434.07	433.2972	66.41	9.29	8.17	16.13
C24H40C12N2	427.51	426.2568	67.43	9.43	16.59	6.55
C24H40C12N4	455.52	454.2630	63.28	8.85	15.57	12.30
C24H40C12N6	483.53	482.2691	59.62	8.34	14.66	17.38
C24H41C11N2	393.06	392.2958	73.34	10.51	9.02	7.13
C24H41C11N4	421.07	420.3020	68.46	9.81	8.42	13.31
C24H41C11N6	449.09	448.3081	64.19	9.20	7.89	18.71
C24H41C12N1	414.51	413.2616	69.54	9.97	17.11	3.38
C24H41C12N3	442.52	441.2677	65.14	9.34	16.02	9.50
C24H41C12N5	470.53	469.2739	61.26	8.78	15.07	14.88
C24H42C11N1	380.06	379.3006	75.85	11.14	9.33	3.69

C24H42C11N3	408.08	407.3067	70.64	10.37	8.69	10.30
C24H42C11N5	436.09	435.3129	66.10	9.71	8.13	16.06
C24H42C12N2	429.52	428.2725	67.11	9.86	16.51	6.52
C24H42C12N4	457.54	456.2786	63.00	9.25	15.50	12.25
C24H42C12N6	485.55	484.2848	59.37	8.72	14.60	17.31
C24H43C11N2	395.08	394.3115	72.96	10.97	8.97	7.09
C24H43C11N4	423.09	422.3176	68.13	10.24	8.38	13.24
C24H43C11N6	451.10	450.3238	63.90	9.61	7.86	18.63
C24H43C12N1	416.52	415.2772	69.21	10.41	17.02	3.36
C24H43C12N3	444.54	443.2834	64.85	9.75	15.95	9.45
C24H43C12N5	472.55	471.2895	61.00	9.17	15.00	14.82
C24H44C11N1	382.08	381.3162	75.45	11.61	9.28	3.67
C24H44C11N3	410.09	409.3224	70.29	10.81	8.65	10.25
C24H44C11N5	438.10	437.3285	65.80	10.12	8.09	15.99
C24H44C12N2	431.54	430.2881	66.80	10.28	16.43	6.49
C24H44C12N4	459.55	458.2943	62.73	9.65	15.43	12.19
C24H44C12N6	487.56	486.3004	59.12	9.10	14.54	17.24
C24H45C11N2	397.09	396.3271	72.59	11.42	8.93	7.05
C24H45C11N4	425.11	424.3333	67.81	10.67	8.34	13.18
C24H45C11N6	453.12	452.3394	63.62	10.01	7.82	18.55
C24H45C12N1	418.54	417.2929	68.87	10.84	16.94	3.35
C24H45C12N3	446.55	445.2990	64.55	10.16	15.88	9.41
C24H45C12N5	474.57	473.3052	60.74	9.56	14.94	14.76
C24H46C11N1	384.09	383.3319	75.05	12.07	9.23	3.65
C24H46C11N3	412.11	411.3380	69.95	11.25	8.60	10.20
C24H46C11N5	440.12	439.3442	65.50	10.53	8.06	15.91
C24H46C12N2	433.55	432.3038	66.49	10.69	16.35	6.46
C24H46C12N4	461.57	460.3099	62.45	10.05	15.36	12.14
C24H46C12N6	489.58	488.3161	58.88	9.47	14.48	17.17
C24H47C11N2	399.11	398.3428	72.23	11.87	8.88	7.02
C24H47C11N4	427.12	426.3489	67.49	11.09	8.30	13.12
C24H47C11N6	455.14	454.3551	63.34	10.41	7.79	18.46
C24H47C12N1	420.55	419.3085	68.54	11.26	16.86	3.33
C24H47C12N3	448.57	447.3147	64.26	10.56	15.81	9.37
C24H47C12N5	476.58	475.3208	60.49	9.94	14.88	14.69
C24H48C11N1	386.11	385.3475	74.66	12.53	9.18	3.63
C24H48C11N3	414.12	413.3537	69.61	11.68	8.56	10.15
C24H48C11N5	442.14	441.3598	65.20	10.94	8.02	15.84
C24H48C12N2	435.57	434.3194	66.18	11.11	16.28	6.43
C24H48C12N4	463.58	462.3256	62.18	10.44	15.30	12.09
C24H48C12N6	491.60	490.3317	58.64	9.84	14.42	17.10
C24H49C11N2	401.12	400.3584	71.86	12.31	8.84	6.98
C24H49C11N4	429.14	428.3646	67.17	11.51	8.26	13.06
C24H49C11N6	457.15	456.3707	63.06	10.80	7.76	18.38
C24H49C12N1	422.57	421.3242	68.22	11.69	16.78	3.31
C24H49C12N3	450.58	449.3303	63.98	10.96	15.74	9.33
C24H49C12N5	478.60	477.3365	60.23	10.32	14.82	14.63
C24H50C11N1	388.13	387.3632	74.27	12.99	9.13	3.61
C24H50C11N3	416.14	415.3693	69.27	12.11	8.52	10.10
C24H50C11N5	444.15	443.3755	64.90	11.35	7.98	15.77
C24H50C12N2	437.59	436.3351	65.88	11.52	16.20	6.40
C24H50C12N4	465.60	464.3412	61.91	10.82	15.23	12.03
C24H50C12N6	493.61	492.3474	58.40	10.21	14.36	17.03
C24H51C11N2	403.14	402.3741	71.51	12.75	8.79	6.95
C24H51C11N4	431.15	430.3802	66.86	11.92	8.22	12.99
C24H51C11N6	459.17	458.3864	62.78	11.20	7.72	18.30
C24H51C12N3	452.60	451.3460	63.69	11.36	15.67	9.28
C24H51C12N5	480.61	479.3521	59.98	10.70	14.75	14.57
C24H52C11N3	418.16	417.3850	68.94	12.53	8.48	10.05
C24H52C11N5	446.17	445.3911	64.61	11.75	7.95	15.70

C24H52C12N4	467.61	466.3569	61.65	11.21	15.16	11.98
C24H52C12N6	495.63	494.3630	58.16	10.58	14.31	16.96
C24H53C11N4	433.17	432.3959	66.55	12.33	8.18	12.93
C24H53C11N6	461.18	460.4020	62.51	11.58	7.69	18.22
C24H53C12N5	482.63	481.3678	59.73	11.07	14.69	14.51
C24H54C11N5	448.18	447.4068	64.32	12.14	7.91	15.63
C24H54C12N6	497.64	496.3787	57.93	10.94	14.25	16.89
C24H55C11N6	463.20	462.4176	62.23	11.97	7.65	18.14
C25H12C11N1	361.83	361.0658	82.99	3.34	9.80	3.87
C25H12C11N3	389.85	389.0720	77.02	3.10	9.09	10.78
C25H12C11N5	417.86	417.0781	71.86	2.89	8.48	16.76
C25H12C12N2	411.29	410.0377	73.01	2.94	17.24	6.81
C25H12C12N4	439.31	438.0439	68.35	2.75	16.14	12.75
C25H12C12N6	467.32	466.0500	64.26	2.59	15.17	17.98
C25H13C11N2	376.85	376.0767	79.68	3.48	9.41	7.43
C25H13C11N4	404.86	404.0829	74.17	3.24	8.76	13.84
C25H13C11N6	432.88	432.0890	69.37	3.03	8.19	19.41
C25H13C12N1	398.30	397.0425	75.39	3.29	17.80	3.52
C25H13C12N3	426.31	425.0486	70.44	3.07	16.63	9.86
C25H13C12N5	454.32	453.0548	66.09	2.88	15.61	15.41
C25H14C11N1	363.85	363.0815	82.53	3.88	9.74	3.85
C25H14C11N3	391.86	391.0876	76.63	3.60	9.05	10.72
C25H14C11N5	419.88	419.0938	71.52	3.36	8.44	16.68
C25H14C12N2	413.31	412.0534	72.65	3.41	17.16	6.78
C25H14C12N4	441.32	440.0595	68.04	3.20	16.07	12.70
C25H14C12N6	469.34	468.0657	63.98	3.01	15.11	17.91
C25H15C11N2	378.86	378.0924	79.26	3.99	9.36	7.39
C25H15C11N4	406.88	406.0985	73.80	3.72	8.71	13.77
C25H15C11N6	434.89	434.1047	69.05	3.48	8.15	19.32
C25H15C12N1	400.31	399.0581	75.01	3.78	17.71	3.50
C25H15C12N3	428.32	427.0643	70.11	3.53	16.55	9.81
C25H15C12N5	456.34	455.0704	65.80	3.31	15.54	15.35
C25H16C11N1	365.87	365.0971	82.07	4.41	9.69	3.83
C25H16C11N3	393.88	393.1033	76.24	4.09	9.00	10.67
C25H16C11N5	421.89	421.1094	71.17	3.82	8.40	16.60
C25H16C12N2	415.33	414.0690	72.30	3.88	17.07	6.74
C25H16C12N4	443.34	442.0752	67.73	3.64	15.99	12.64
C25H16C12N6	471.35	470.0813	63.71	3.42	15.04	17.83
C25H17C11N2	380.88	380.1080	78.84	4.50	9.31	7.35
C25H17C11N4	408.89	408.1142	73.44	4.19	8.67	13.70
C25H17C11N6	436.91	436.1203	68.73	3.92	8.11	19.24
C25H17C12N1	402.33	401.0738	74.64	4.26	17.62	3.48
C25H17C12N3	430.34	429.0799	69.78	3.98	16.48	9.76
C25H17C12N5	458.35	457.0861	65.51	3.74	15.47	15.28
C25H18C11N1	367.88	367.1128	81.62	4.93	9.64	3.81
C25H18C11N3	395.90	395.1189	75.85	4.58	8.96	10.61
C25H18C11N5	423.91	423.1251	70.84	4.28	8.36	16.52
C25H18C12N2	417.34	416.0847	71.95	4.35	16.99	6.71
C25H18C12N4	445.36	444.0908	67.42	4.07	15.92	12.58
C25H18C12N6	473.37	472.0970	63.43	3.83	14.98	17.75
C25H19C11N2	382.90	382.1237	78.42	5.00	9.26	7.32
C25H19C11N4	410.91	410.1298	73.08	4.66	8.63	13.63
C25H19C11N6	438.92	438.1360	68.41	4.36	8.08	19.15
C25H19C12N1	404.34	403.0894	74.26	4.74	17.54	3.46
C25H19C12N3	432.36	431.0956	69.45	4.43	16.40	9.72
C25H19C12N5	460.37	459.1017	65.23	4.16	15.40	15.21
C25H20C11N1	369.90	369.1284	81.18	5.45	9.58	3.79
C25H20C11N3	397.91	397.1346	75.46	5.07	8.91	10.56
C25H20C11N5	425.92	425.1407	70.50	4.73	8.32	16.44
C25H20C12N2	419.36	418.1003	71.60	4.81	16.91	6.68

L

C25H20C12N4	447.37	446.1065	67.12	4.51	15.85	12.52
C25H20C12N6	475.38	474.1126	63.17	4.24	14.92	17.68
C25H21C11N2	384.91	384.1393	78.01	5.50	9.21	7.28
C25H21C11N4	412.93	412.1455	72.72	5.13	8.59	13.57
C25H21C11N6	440.94	440.1516	68.10	4.80	8.04	19.06
C25H21C12N1	406.36	405.1051	73.89	5.21	17.45	3.45
C25H21C12N3	434.37	433.1112	69.13	4.87	16.32	9.67
C25H21C12N5	462.39	461.1174	64.94	4.58	15.33	15.15
C25H22C11N1	371.91	371.1441	80.74	5.96	9.53	3.77
C25H22C11N3	399.93	399.1502	75.08	5.54	8.86	10.51
C25H22C11N5	427.94	427.1564	70.17	5.18	8.28	16.37
C25H22C12N2	421.37	420.1160	71.26	5.26	16.83	6.65
C25H22C12N4	449.39	448.1221	66.82	4.93	15.78	12.47
C25H22C12N6	477.40	476.1283	62.90	4.65	14.85	17.60
C25H23C11N2	386.93	386.1550	77.61	5.99	9.16	7.24
C25H23C11N4	414.94	414.1611	72.37	5.59	8.54	13.50
C25H23C11N6	442.96	442.1673	67.79	5.23	8.00	18.97
C25H23C12N1	408.37	407.1207	73.53	5.68	17.36	3.43
C25H23C12N3	436.39	435.1269	68.81	5.31	16.25	9.63
C25H23C12N5	464.40	463.1330	64.66	4.99	15.27	15.08
C25H24C11N1	373.93	373.1597	80.30	6.47	9.48	3.75
C25H24C11N3	401.94	401.1659	74.71	6.02	8.82	10.45
C25H24C11N5	429.96	429.1720	69.84	5.63	8.25	16.29
C25H24C12N2	423.39	422.1316	70.92	5.71	16.75	6.62
C25H24C12N4	451.40	450.1378	66.52	5.36	15.71	12.41
C25H24C12N6	479.42	478.1439	62.63	5.05	14.79	17.53
C25H25C11N2	388.94	388.1706	77.20	6.48	9.12	7.20
C25H25C11N4	416.96	416.1768	72.02	6.04	8.50	13.44
C25H25C11N6	444.97	444.1829	67.48	5.66	7.97	18.89
C25H25C12N1	410.39	409.1364	73.17	6.14	17.28	3.41
C25H25C12N3	438.40	437.1425	68.49	5.75	16.17	9.58
C25H25C12N5	466.42	465.1487	64.38	5.40	15.20	15.02
C25H26C11N1	375.95	375.1754	79.87	6.97	9.43	3.73
C25H26C11N3	403.96	403.1815	74.33	6.49	8.78	10.40
C25H26C11N5	431.97	431.1877	69.51	6.07	8.21	16.21
C25H26C12N2	425.41	424.1473	70.59	6.16	16.67	6.59
C25H26C12N4	453.42	452.1534	66.23	5.78	15.64	12.36
C25H26C12N6	481.43	480.1596	62.37	5.44	14.73	17.46
C25H27C11N2	390.96	390.1863	76.81	6.96	9.07	7.17
C25H27C11N4	418.97	418.1924	71.67	6.50	8.46	13.37
C25H27C11N6	446.99	446.1986	67.18	6.09	7.93	18.80
C25H27C12N1	412.41	411.1520	72.81	6.60	17.19	3.40
C25H27C12N3	440.42	439.1582	68.18	6.18	16.10	9.54
C25H27C12N5	468.43	467.1643	64.10	5.81	15.14	14.95
C25H28C11N1	377.96	377.1910	79.45	7.47	9.38	3.71
C25H28C11N3	405.98	405.1972	73.96	6.95	8.73	10.35
C25H28C11N5	433.99	433.2033	69.19	6.50	8.17	16.14
C25H28C12N2	427.42	426.1629	70.25	6.60	16.59	6.55
C25H28C12N4	455.43	454.1691	65.93	6.20	15.57	12.30
C25H28C12N6	483.45	482.1752	62.11	5.84	14.67	17.38
C25H29C11N2	392.98	392.2019	76.41	7.44	9.02	7.13
C25H29C11N4	420.99	420.2081	71.33	6.94	8.42	13.31
C25H29C11N6	449.00	448.2142	66.88	6.51	7.90	18.72
C25H29C12N1	414.42	413.1677	72.46	7.05	17.11	3.38
C25H29C12N3	442.44	441.1738	67.87	6.61	16.03	9.50
C25H29C12N5	470.45	469.1800	63.83	6.21	15.07	14.89
C25H30C11N1	379.98	379.2067	79.03	7.96	9.33	3.69
C25H30C11N3	407.99	407.2128	73.60	7.41	8.69	10.30
C25H30C11N5	436.00	435.2190	68.87	6.94	8.13	16.06
C25H30C12N2	429.44	428.1786	69.92	7.04	16.51	6.52

C25H30C12N4	457.45	456.1847	65.64	6.61	15.50	12.25
C25H30C12N6	485.46	484.1909	61.85	6.23	14.61	17.31
C25H31C11N2	394.99	394.2176	76.02	7.91	8.98	7.09
C25H31C11N4	423.01	422.2237	70.99	7.39	8.38	13.24
C25H31C11N6	451.02	450.2299	66.58	6.93	7.86	18.63
C25H31C12N1	416.44	415.1833	72.11	7.50	17.03	3.36
C25H31C12N3	444.45	443.1895	67.56	7.03	15.95	9.45
C25H31C12N5	472.47	471.1956	63.56	6.61	15.01	14.82
C25H32C11N1	381.99	381.2223	78.61	8.44	9.28	3.67
C25H32C11N3	410.01	409.2285	73.24	7.87	8.65	10.25
C25H32C11N5	438.02	437.2346	68.55	7.36	8.09	15.99
C25H32C12N2	431.45	430.1942	69.60	7.48	16.43	6.49
C25H32C12N4	459.47	458.2004	65.35	7.02	15.43	12.19
C25H32C12N6	487.48	486.2065	61.60	6.62	14.55	17.24
C25H33C11N2	397.01	396.2332	75.64	8.38	8.93	7.06
C25H33C11N4	425.02	424.2394	70.65	7.83	8.34	13.18
C25H33C11N6	453.03	452.2455	66.28	7.34	7.83	18.55
C25H33C12N1	418.45	417.1990	71.76	7.95	16.94	3.35
C25H33C12N3	446.47	445.2051	67.26	7.45	15.88	9.41
C25H33C12N5	474.48	473.2113	63.29	7.01	14.94	14.76
C25H34C11N1	384.01	383.2380	78.20	8.92	9.23	3.65
C25H34C11N3	412.02	411.2441	72.88	8.32	8.60	10.20
C25H34C11N5	440.04	439.2503	68.24	7.79	8.06	15.92
C25H34C12N2	433.47	432.2099	69.27	7.91	16.36	6.46
C25H34C12N4	461.48	460.2160	65.07	7.43	15.36	12.14
C25H34C12N6	489.50	488.2222	61.34	7.00	14.49	17.17
C25H35C11N2	399.02	398.2489	75.25	8.84	8.88	7.02
C25H35C11N4	427.04	426.2550	70.32	8.26	8.30	13.12
C25H35C11N6	455.05	454.2612	65.99	7.75	7.79	18.47
C25H35C12N1	420.47	419.2146	71.41	8.39	16.86	3.33
C25H35C12N3	448.48	447.2208	66.95	7.87	15.81	9.37
C25H35C12N5	476.50	475.2269	63.02	7.40	14.88	14.70
C25H36C11N1	386.03	385.2536	77.79	9.40	9.18	3.63
C25H36C11N3	414.04	413.2598	72.52	8.76	8.56	10.15
C25H36C11N5	442.05	441.2659	67.93	8.21	8.02	15.84
C25H36C12N2	435.49	434.2255	68.95	8.33	16.28	6.43
C25H36C12N4	463.50	462.2317	64.79	7.83	15.30	12.09
C25H36C12N6	491.51	490.2378	61.09	7.38	14.43	17.10
C25H37C11N2	401.04	400.2645	74.88	9.30	8.84	6.99
C25H37C11N4	429.05	428.2707	69.99	8.69	8.26	13.06
C25H37C11N6	457.07	456.2768	65.70	8.16	7.76	18.39
C25H37C12N1	422.49	421.2303	71.07	8.83	16.78	3.32
C25H37C12N3	450.50	449.2364	66.65	8.28	15.74	9.33
C25H37C12N5	478.51	477.2426	62.75	7.79	14.82	14.64
C25H38C11N1	388.04	387.2693	77.38	9.87	9.14	3.61
C25H38C11N3	416.05	415.2754	72.17	9.21	8.52	10.10
C25H38C11N5	444.07	443.2816	67.62	8.63	7.98	15.77
C25H38C12N2	437.50	436.2412	68.63	8.75	16.21	6.40
C25H38C12N4	465.51	464.2473	64.50	8.23	15.23	12.04
C25H38C12N6	493.53	492.2535	60.84	7.76	14.37	17.03
C25H39C11N2	403.06	402.2802	74.50	9.75	8.80	6.95
C25H39C11N4	431.07	430.2863	69.66	9.12	8.22	13.00
C25H39C11N6	459.08	458.2925	65.41	8.56	7.72	18.31
C25H39C12N1	424.50	423.2459	70.74	9.26	16.70	3.30
C25H39C12N3	452.52	451.2521	66.36	8.69	15.67	9.29
C25H39C12N5	480.53	479.2582	62.49	8.18	14.76	14.57
C25H40C11N1	390.06	389.2849	76.98	10.34	9.09	3.59
C25H40C11N3	418.07	417.2911	71.82	9.64	8.48	10.05
C25H40C11N5	446.08	445.2972	67.31	9.04	7.95	15.70
C25H40C12N2	439.52	438.2568	68.32	9.17	16.13	6.37

C25H40C12N4	467.53	466.2630	64.23	8.62	15.17	11.98
C25H40C12N6	495.54	494.2691	60.60	8.14	14.31	16.96
C25H41C11N2	405.07	404.2958	74.13	10.20	8.75	6.92
C25H41C11N4	433.09	432.3020	69.33	9.54	8.19	12.94
C25H41C11N6	461.10	460.3081	65.12	8.96	7.69	18.23
C25H41C12N1	426.52	425.2616	70.40	9.69	16.62	3.28
C25H41C12N3	454.53	453.2677	66.06	9.09	15.60	9.24
C25H41C12N5	482.55	481.2739	62.23	8.56	14.69	14.51
C25H42C11N1	392.07	391.3006	76.59	10.80	9.04	3.57
C25H42C11N3	420.09	419.3067	71.48	10.08	8.44	10.00
C25H42C11N5	448.10	447.3129	67.01	9.45	7.91	15.63
C25H42C12N2	441.53	440.2725	68.01	9.59	16.06	6.34
C25H42C12N4	469.55	468.2786	63.95	9.02	15.10	11.93
C25H42C12N6	497.56	496.2848	60.35	8.51	14.25	16.89
C25H43C11N2	407.09	406.3115	73.76	10.65	8.71	6.88
C25H43C11N4	435.10	434.3176	69.01	9.96	8.15	12.88
C25H43C11N6	463.11	462.3238	64.84	9.36	7.66	18.15
C25H43C12N1	428.53	427.2772	70.07	10.11	16.55	3.27
C25H43C12N3	456.55	455.2834	65.77	9.49	15.53	9.20
C25H43C12N5	484.56	483.2895	61.97	8.94	14.63	14.45
C25H44C11N1	394.09	393.3162	76.20	11.25	9.00	3.55
C25H44C11N3	422.10	421.3224	71.14	10.51	8.40	9.95
C25H44C11N5	450.12	449.3285	66.71	9.85	7.88	15.56
C25H44C12N2	443.55	442.2881	67.70	10.00	15.99	6.32
C25H44C12N4	471.56	470.2943	63.68	9.41	15.04	11.88
C25H44C12N6	499.58	498.3004	60.11	8.88	14.19	16.82
C25H45C11N2	409.10	408.3271	73.40	11.09	8.67	6.85
C25H45C11N4	437.12	436.3333	68.70	10.38	8.11	12.82
C25H45C11N6	465.13	464.3394	64.56	9.75	7.62	18.07
C25H45C12N1	430.55	429.2929	69.74	10.54	16.47	3.25
C25H45C12N3	458.56	457.2990	65.48	9.89	15.46	9.16
C25H45C12N5	486.58	485.3052	61.71	9.32	14.57	14.39
C25H46C11N1	396.11	395.3319	75.81	11.71	8.95	3.54
C25H46C11N3	424.12	423.3380	70.80	10.93	8.36	9.91
C25H46C11N5	452.13	451.3442	66.41	10.26	7.84	15.49
C25H46C12N2	445.56	444.3038	67.39	10.41	15.91	6.29
C25H46C12N4	473.58	472.3099	63.41	9.79	14.97	11.83
C25H46C12N6	501.59	500.3161	59.87	9.24	14.14	16.75
C25H47C11N2	411.12	410.3428	73.04	11.52	8.62	6.81
C25H47C11N4	439.13	438.3489	68.38	10.79	8.07	12.76
C25H47C11N6	467.15	466.3551	64.28	10.14	7.59	17.99
C25H47C12N1	432.57	431.3085	69.42	10.95	16.39	3.24
C25H47C12N3	460.58	459.3147	65.20	10.29	15.39	9.12
C25H47C12N5	488.59	487.3208	61.46	9.70	14.51	14.33
C25H48C11N1	398.12	397.3475	75.42	12.15	8.91	3.52
C25H48C11N3	426.13	425.3537	70.47	11.35	8.32	9.86
C25H48C11N5	454.15	453.3598	66.12	10.65	7.81	15.42
C25H48C12N2	447.58	446.3194	67.09	10.81	15.84	6.26
C25H48C12N4	475.59	474.3256	63.14	10.17	14.91	11.78
C25H48C12N6	503.61	502.3317	59.63	9.61	14.08	16.69
C25H49C11N2	413.14	412.3584	72.68	11.96	8.58	6.78
C25H49C11N4	441.15	440.3646	68.07	11.20	8.04	12.70
C25H49C11N6	469.16	468.3707	64.00	10.53	7.56	17.91
C25H49C12N1	434.58	433.3242	69.10	11.37	16.32	3.22
C25H49C12N3	462.60	461.3303	64.91	10.68	15.33	9.08
C25H49C12N5	490.61	489.3365	61.21	10.07	14.45	14.27
C25H50C11N1	400.14	399.3632	75.04	12.60	8.86	3.50
C25H50C11N3	428.15	427.3693	70.13	11.77	8.28	9.81
C25H50C11N5	456.16	455.3755	65.83	11.05	7.77	15.35
C25H50C12N2	449.60	448.3351	66.79	11.21	15.77	6.23

C25H50C12N4	477.61	476.3412	62.87	10.55	14.85	11.73
C25H50C12N6	505.62	504.3474	59.39	9.97	14.02	16.62
C25H51C11N2	415.15	414.3741	72.33	12.38	8.54	6.75
C25H51C11N4	443.17	442.3802	67.76	11.60	8.00	12.64
C25H51C11N6	471.18	470.3864	63.73	10.91	7.52	17.84
C25H51C12N1	436.60	435.3398	68.78	11.77	16.24	3.21
C25H51C12N3	464.61	463.3460	64.63	11.06	15.26	9.04
C25H51C12N5	492.62	491.3521	60.95	10.44	14.39	14.22
C25H52C11N1	402.15	401.3788	74.67	13.03	8.82	3.48
C25H52C11N3	430.17	429.3850	69.81	12.18	8.24	9.77
C25H52C11N5	458.18	457.3911	65.54	11.44	7.74	15.29
C25H52C12N2	451.61	450.3507	66.49	11.61	15.70	6.20
C25H52C12N4	479.63	478.3569	62.61	10.93	14.78	11.68
C25H52C12N6	507.64	506.3630	59.15	10.33	13.97	16.56
C25H53C11N2	417.17	416.3897	71.98	12.81	8.50	6.72
C25H53C11N4	445.18	444.3959	67.45	12.00	7.96	12.59
C25H53C11N6	473.19	472.4020	63.46	11.29	7.49	17.76
C25H53C12N3	466.63	465.3616	64.35	11.45	15.20	9.01
C25H53C12N5	494.64	493.3678	60.71	10.80	14.33	14.16
C25H54C11N3	432.18	431.4006	69.48	12.59	8.20	9.72
C25H54C11N5	460.20	459.4068	65.25	11.83	7.70	15.22
C25H54C12N4	481.64	480.3725	62.34	11.30	14.72	11.63
C25H54C12N6	509.66	508.3787	58.92	10.68	13.91	16.49
C25H55C11N4	447.20	446.4115	67.15	12.40	7.93	12.53
C25H55C11N6	475.21	474.4176	63.19	11.67	7.46	17.68
C25H55C12N5	496.66	495.3834	60.46	11.16	14.28	14.10
C25H56C11N5	462.21	461.4224	64.97	12.21	7.67	15.15
C25H56C12N6	511.67	510.3943	58.69	11.03	13.86	16.42
C25H57C11N6	477.23	476.4333	62.92	12.04	7.43	17.61
C26H13C11N2	388.86	388.0767	80.31	3.37	9.12	7.20
C26H13C11N4	416.87	416.0829	74.91	3.14	8.50	13.44
C26H13C11N6	444.89	444.0890	70.20	2.95	7.97	18.89
C26H13C12N1	410.31	409.0425	76.11	3.19	17.28	3.41
C26H13C12N3	438.32	437.0486	71.25	2.99	16.18	9.59
C26H13C12N5	466.33	465.0548	66.97	2.81	15.21	15.02
C26H14C11N1	375.86	375.0815	83.09	3.75	9.43	3.73
C26H14C11N3	403.87	403.0876	77.32	3.49	8.78	10.40
C26H14C11N5	431.89	431.0938	72.31	3.27	8.21	16.22
C26H14C12N2	425.32	424.0534	73.42	3.32	16.67	6.59
C26H14C12N4	453.33	452.0595	68.89	3.11	15.64	12.36
C26H14C12N6	481.35	480.0657	64.88	2.93	14.73	17.46
C26H15C11N2	390.88	390.0924	79.89	3.87	9.07	7.17
C26H15C11N4	418.89	418.0985	74.55	3.61	8.46	13.38
C26H15C11N6	446.90	446.1047	69.88	3.38	7.93	18.81
C26H15C12N1	412.32	411.0581	75.74	3.67	17.20	3.40
C26H15C12N3	440.34	439.0643	70.92	3.43	16.10	9.54
C26H15C12N5	468.35	467.0704	66.68	3.23	15.14	14.95
C26H16C11N1	377.88	377.0971	82.64	4.27	9.38	3.71
C26H16C11N3	405.89	405.1033	76.94	3.97	8.73	10.35
C26H16C11N5	433.90	433.1094	71.97	3.72	8.17	16.14
C26H16C12N2	427.34	426.0690	73.08	3.77	16.59	6.56
C26H16C12N4	455.35	454.0752	68.58	3.54	15.57	12.30
C26H16C12N6	483.36	482.0813	64.61	3.34	14.67	17.39
C26H17C11N2	392.89	392.1080	79.48	4.36	9.02	7.13
C26H17C11N4	420.91	420.1142	74.19	4.07	8.42	13.31
C26H17C11N6	448.92	448.1203	69.56	3.82	7.90	18.72
C26H17C12N1	414.34	413.0738	75.37	4.14	17.11	3.38
C26H17C12N3	442.35	441.0799	70.60	3.87	16.03	9.50
C26H17C12N5	470.36	469.0861	66.39	3.64	15.07	14.89
C26H18C11N1	379.89	379.1128	82.20	4.78	9.33	3.69

C26H18C11N3	407.91	407.1189	76.56	4.45	8.69	10.30	C26H28C11N3	417.99	417.1972	74.71	6.75	8.48	10.05
C26H18C11N5	435.92	435.1251	71.64	4.16	8.13	16.07	C26H28C11N5	446.00	445.2033	70.02	6.33	7.95	15.70
C26H18C12N2	429.35	428.0847	72.74	4.23	16.51	6.52	C26H28C12N2	439.43	438.1629	71.07	6.42	16.14	6.37
C26H18C12N4	457.37	456.0908	68.28	3.97	15.50	12.25	C26H28C12N4	467.45	466.1691	66.81	6.04	15.17	11.99
C26H18C12N6	485.38	484.0970	64.34	3.74	14.61	17.31	C26H28C12N6	495.46	494.1752	63.03	5.70	14.31	16.96
C26H19C11N2	394.91	394.1237	79.08	4.85	8.98	7.09	C26H29C11N2	404.99	404.2019	77.11	7.22	8.75	6.92
C26H19C11N4	422.92	422.1298	73.84	4.53	8.38	13.25	C26H29C11N4	433.00	432.2081	72.12	6.75	8.19	12.94
C26H19C11N6	450.93	450.1360	69.25	4.25	7.86	18.64	C26H29C11N6	461.01	460.2142	67.74	6.34	7.69	18.23
C26H19C12N1	416.35	415.0894	75.01	4.60	17.03	3.36	C26H29C12N1	426.43	425.1677	73.23	6.85	16.63	3.28
C26H19C12N3	444.37	443.0956	70.28	4.31	15.96	9.46	C26H29C12N3	454.45	453.1738	68.72	6.43	15.60	9.25
C26H19C12N5	472.38	471.1017	66.11	4.05	15.01	14.83	C26H29C12N5	482.46	481.1800	64.73	6.06	14.70	14.52
C26H20C11N1	381.91	381.1284	81.77	5.28	9.28	3.67	C26H30C11N1	391.99	391.2067	79.67	7.71	9.04	3.57
C26H20C11N3	409.92	409.1346	76.18	4.92	8.65	10.25	C26H30C11N3	420.00	419.2128	74.35	7.20	8.44	10.00
C26H20C11N5	437.94	437.1407	71.31	4.60	8.10	15.99	C26H30C11N5	448.02	447.2190	69.71	6.75	7.91	15.63
C26H20C12N2	431.37	430.1003	72.40	4.67	16.44	6.49	C26H30C12N2	441.45	440.1786	70.74	6.85	16.06	6.35
C26H20C12N4	459.38	458.1065	67.98	4.39	15.44	12.20	C26H30C12N4	469.46	468.1847	66.52	6.44	15.10	11.93
C26H20C12N6	487.40	486.1126	64.07	4.14	14.55	17.24	C26H30C12N6	497.48	496.1909	62.77	6.08	14.25	16.89
C26H21C11N2	396.92	396.1393	78.68	5.33	8.93	7.06	C26H31C11N2	407.00	406.2176	76.73	7.68	8.71	6.88
C26H21C11N4	424.94	424.1455	73.49	4.98	8.34	13.18	C26H31C11N4	435.02	434.2237	71.79	7.18	8.15	12.88
C26H21C11N6	452.95	452.1516	68.95	4.67	7.83	18.55	C26H31C11N6	463.03	462.2299	67.44	6.75	7.66	18.15
C26H21C12N1	418.37	417.1051	74.64	5.06	16.95	3.35	C26H31C12N1	428.45	427.1833	72.89	7.29	16.55	3.27
C26H21C12N3	446.38	445.1112	69.96	4.74	15.88	9.41	C26H31C12N3	456.46	455.1895	68.42	6.85	15.53	9.21
C26H21C12N5	474.40	473.1174	65.83	4.46	14.95	14.76	C26H31C12N5	484.48	483.1956	64.46	6.45	14.64	14.46
C26H22C11N1	383.92	383.1441	81.34	5.78	9.23	3.65	C26H32C11N1	394.00	393.2223	79.26	8.19	9.00	3.55
C26H22C11N3	411.94	411.1502	75.81	5.38	8.61	10.20	C26H32C11N3	422.02	421.2285	74.00	7.64	8.40	9.96
C26H22C11N5	439.95	439.1564	70.98	5.04	8.06	15.92	C26H32C11N5	450.03	449.2346	69.39	7.17	7.88	15.56
C26H22C12N2	433.38	432.1160	72.06	5.12	16.36	6.46	C26H32C12N2	443.46	442.1942	70.42	7.27	15.99	6.32
C26H22C12N4	461.40	460.1221	67.68	4.81	15.37	12.14	C26H32C12N4	471.48	470.2004	66.24	6.84	15.04	11.88
C26H22C12N6	489.41	488.1283	63.81	4.53	14.49	17.17	C26H32C12N6	499.49	498.2065	62.52	6.46	14.20	16.83
C26H23C11N2	398.94	398.1550	78.28	5.81	8.89	7.02	C26H33C11N2	409.02	408.2332	76.35	8.13	8.67	6.85
C26H23C11N4	426.95	426.1611	73.14	5.43	8.30	13.12	C26H33C11N4	437.03	436.2394	71.46	7.61	8.11	12.82
C26H23C11N6	454.97	454.1673	68.64	5.10	7.79	18.47	C26H33C11N6	465.05	464.2455	67.15	7.15	7.62	18.07
C26H23C12N1	420.39	419.1207	74.29	5.51	16.87	3.33	C26H33C12N1	430.47	429.1990	72.55	7.73	16.47	3.25
C26H23C12N3	448.40	447.1269	69.65	5.17	15.81	9.37	C26H33C12N3	458.48	457.2051	68.11	7.26	15.47	9.17
C26H23C12N5	476.41	475.1330	65.55	4.87	14.88	14.70	C26H33C12N5	486.49	485.2113	64.19	6.84	14.57	14.40
C26H24C11N1	385.94	385.1597	80.92	6.27	9.19	3.63	C26H34C11N1	396.02	395.2380	78.86	8.65	8.95	3.54
C26H24C11N3	413.95	413.1659	75.44	5.84	8.56	10.15	C26H34C11N3	424.03	423.2441	73.65	8.08	8.36	9.91
C26H24C11N5	441.97	441.1720	70.66	5.47	8.02	15.85	C26H34C11N5	452.05	451.2503	69.08	7.58	7.84	15.49
C26H24C12N2	435.40	434.1316	71.72	5.56	16.29	6.43	C26H34C12N2	445.48	444.2099	70.10	7.69	15.92	6.29
C26H24C12N4	463.41	462.1378	67.39	5.22	15.30	12.09	C26H34C12N4	473.49	472.2160	65.95	7.24	14.98	11.83
C26H24C12N6	491.43	490.1439	63.55	4.92	14.43	17.10	C26H34C12N6	501.51	500.2222	62.27	6.83	14.14	16.76
C26H25C11N2	400.96	400.1706	77.89	6.28	8.84	6.99	C26H35C11N2	411.04	410.2489	75.98	8.58	8.63	6.82
C26H25C11N4	428.97	428.1768	72.80	5.87	8.26	13.06	C26H35C11N4	439.05	438.2550	71.13	8.04	8.07	12.76
C26H25C11N6	456.98	456.1829	68.34	5.51	7.76	18.39	C26H35C11N6	467.06	466.2612	66.86	7.55	7.59	17.99
C26H25C12N1	422.40	421.1364	73.93	5.97	16.79	3.32	C26H35C12N1	432.48	431.2146	72.21	8.16	16.40	3.24
C26H25C12N3	450.42	449.1425	69.33	5.59	15.74	9.33	C26H35C12N3	460.49	459.2208	67.82	7.66	15.40	9.12
C26H25C12N5	478.43	477.1487	65.27	5.27	14.82	14.64	C26H35C12N5	488.51	487.2269	63.93	7.22	14.51	14.34
C26H26C11N1	387.96	387.1754	80.50	6.76	9.14	3.61	C26H36C11N1	398.04	397.2536	78.46	9.12	8.91	3.52
C26H26C11N3	415.97	415.1815	75.08	6.30	8.52	10.10	C26H36C11N3	426.05	425.2598	73.30	8.52	8.32	9.86
C26H26C11N5	443.98	443.1877	70.34	5.90	7.99	15.77	C26H36C11N5	454.06	453.2659	68.78	7.99	7.81	15.42
C26H26C12N2	437.42	436.1473	71.39	5.99	16.21	6.40	C26H36C12N2	447.50	446.2255	69.79	8.11	15.85	6.26
C26H26C12N4	465.43	464.1534	67.10	5.63	15.23	12.04	C26H36C12N4	475.51	474.2317	65.67	7.63	14.91	11.78
C26H26C12N6	493.44	492.1596	63.29	5.31	14.37	17.03	C26H36C12N6	503.52	502.2378	62.02	7.21	14.08	16.69
C26H27C11N2	402.97	402.1863	77.50	6.75	8.80	6.95	C26H37C11N2	413.05	412.2645	75.61	9.03	8.58	6.78
C26H27C11N4	430.98	430.1924	72.46	6.31	8.23	13.00	C26H37C11N4	441.06	440.2707	70.80	8.46	8.04	12.70
C26H27C11N6	459.00	458.1986	68.04	5.93	7.72	18.31	C26H37C11N6	469.08	468.2768	66.58	7.95	7.56	17.92
C26H27C12N1	424.42	423.1520	73.58	6.41	16.71	3.30	C26H37C12N1	434.50	433.2303	71.87	8.58	16.32	3.22
C26H27C12N3	452.43	451.1582	69.02	6.02	15.67	9.29	C26H37C12N3	462.51	461.2364	67.52	8.06	15.33	9.09
C26H27C12N5	480.44	479.1643	65.00	5.66	14.76	14.58	C26H37C12N5	490.52	489.2426	63.66	7.60	14.46	14.28
C26H28C11N1	389.97	389.1910	80.08	7.24	9.09	3.59	C26H38C11N1	400.05	399.2693	78.06	9.57	8.86	3.50

C26H38C11N3	428.07	427.2754	72.95	8.95	8.28	9.82	C26H48C11N3	438.15	437.3537	71.28	11.04	8.09	9.59
C26H38C11N5	456.08	455.2816	68.47	8.40	7.77	15.36	C26H48C11N5	466.16	465.3598	66.99	10.38	7.61	15.02
C26H38C12N2	449.51	448.2412	69.47	8.52	15.77	6.23	C26H48C12N2	459.59	458.3194	67.95	10.53	15.43	6.10
C26H38C12N4	477.53	476.2473	65.40	8.02	14.85	11.73	C26H48C12N4	487.61	486.3256	64.05	9.92	14.54	11.49
C26H38C12N6	505.54	504.2535	61.77	7.58	14.03	16.62	C26H48C12N6	515.62	514.3317	60.57	9.38	13.75	16.30
C26H39C11N2	415.07	414.2802	75.24	9.47	8.54	6.75	C26H49C11N2	425.15	424.3584	73.45	11.62	8.34	6.59
C26H39C11N4	443.08	442.2863	70.48	8.87	8.00	12.64	C26H49C11N4	453.16	452.3646	68.91	10.90	7.82	12.36
C26H39C11N6	471.09	470.2925	66.29	8.34	7.53	17.84	C26H49C11N6	481.17	480.3707	64.90	10.26	7.37	17.47
C26H39C12N1	436.51	435.2459	71.54	9.01	16.24	3.21	C26H49C12N1	446.59	445.3242	69.93	11.06	15.88	3.14
C26H39C12N3	464.53	463.2521	67.23	8.46	15.26	9.05	C26H49C12N3	474.61	473.3303	65.80	10.41	14.94	8.85
C26H39C12N5	492.54	491.2582	63.40	7.98	14.40	14.22	C26H49C12N5	502.62	501.3365	62.13	9.83	14.11	13.93
C26H40C11N1	402.07	401.2849	77.67	10.03	8.82	3.48	C26H50C11N1	412.15	411.3632	75.77	12.23	8.60	3.40
C26H40C11N3	430.08	429.2911	72.61	9.37	8.24	9.77	C26H50C11N3	440.16	439.3693	70.95	11.45	8.05	9.55
C26H40C11N5	458.10	457.2972	68.17	8.80	7.74	15.29	C26H50C11N5	468.17	467.3755	66.70	10.76	7.57	14.96
C26H40C12N2	451.53	450.2568	69.16	8.93	15.70	6.20	C26H50C12N2	461.61	460.3351	67.65	10.92	15.36	6.07
C26H40C12N4	479.54	478.2630	65.12	8.41	14.79	11.68	C26H50C12N4	489.62	488.3412	63.78	10.29	14.48	11.44
C26H40C12N6	507.55	506.2691	61.53	7.94	13.97	16.56	C26H50C12N6	517.63	516.3474	60.33	9.74	13.70	16.24
C26H41C11N2	417.08	416.2958	74.87	9.91	8.50	6.72	C26H51C11N2	427.16	426.3741	73.11	12.03	8.30	6.56
C26H41C11N4	445.10	444.3020	70.16	9.28	7.97	12.59	C26H51C11N4	455.18	454.3802	68.61	11.29	7.79	12.31
C26H41C11N6	473.11	472.3081	66.01	8.74	7.49	17.76	C26H51C11N6	483.19	482.3864	64.63	10.64	7.34	17.39
C26H41C12N1	438.53	437.2616	71.21	9.42	16.17	3.19	C26H51C12N1	448.61	447.3398	69.61	11.46	15.81	3.12
C26H41C12N3	466.54	465.2677	66.94	8.86	15.20	9.01	C26H51C12N3	476.62	475.3460	65.52	10.79	14.88	8.82
C26H41C12N5	494.56	493.2739	63.15	8.36	14.34	14.16	C26H51C12N5	504.64	503.3521	61.88	10.19	14.05	13.88
C26H42C11N1	404.08	403.3006	77.28	10.48	8.77	3.47	C26H52C11N1	414.16	413.3788	75.40	12.66	8.56	3.38
C26H42C11N3	432.10	431.3067	72.27	9.80	8.20	9.72	C26H52C11N3	442.18	441.3850	70.63	11.85	8.02	9.50
C26H42C11N5	460.11	459.3129	67.87	9.20	7.71	15.22	C26H52C11N5	470.19	469.3911	66.42	11.15	7.54	14.89
C26H42C12N2	453.54	452.2725	68.86	9.33	15.63	6.18	C26H52C12N2	463.62	462.3507	67.36	11.31	15.29	6.04
C26H42C12N4	481.56	480.2786	64.85	8.79	14.72	11.63	C26H52C12N4	491.64	490.3569	63.52	10.66	14.42	11.40
C26H42C12N6	509.57	508.2848	61.28	8.31	13.91	16.49	C26H52C12N6	519.65	518.3630	60.10	10.09	13.64	16.17
C26H43C11N2	419.10	418.3115	74.51	10.34	8.46	6.68	C26H53C11N2	429.18	428.3897	72.76	12.45	8.26	6.53
C26H43C11N4	447.11	446.3176	69.85	9.69	7.93	12.53	C26H53C11N4	457.19	456.3959	68.31	11.68	7.75	12.25
C26H43C11N6	475.13	474.3238	65.73	9.12	7.46	17.69	C26H53C11N6	485.21	484.4020	64.36	11.01	7.31	17.32
C26H43C12N1	440.55	439.2772	70.89	9.84	16.10	3.18	C26H53C12N1	450.63	449.3555	69.30	11.86	15.74	3.11
C26H43C12N3	468.56	467.2834	66.65	9.25	15.13	8.97	C26H53C12N3	478.64	477.3616	65.25	11.16	14.81	8.78
C26H43C12N5	496.57	495.2895	62.89	8.73	14.28	14.10	C26H53C12N5	506.65	505.3678	61.64	10.54	14.00	13.82
C26H44C11N1	406.10	405.3162	76.90	10.92	8.73	3.45	C26H54C11N1	416.18	415.3945	75.04	13.08	8.52	3.37
C26H44C11N3	434.11	433.3224	71.94	10.22	8.17	9.68	C26H54C11N3	444.19	443.4006	70.30	12.25	7.98	9.46
C26H44C11N5	462.13	461.3285	67.58	9.60	7.67	15.15	C26H54C11N5	472.21	471.4068	66.13	11.53	7.51	14.83
C26H44C12N2	455.56	454.2881	68.55	9.74	15.56	6.15	C26H54C12N2	465.64	464.3664	67.07	11.69	15.23	6.02
C26H44C12N4	483.57	482.2943	64.58	9.17	14.66	11.59	C26H54C12N4	493.65	492.3725	63.26	11.03	14.36	11.35
C26H44C12N6	511.59	510.3004	61.04	8.67	13.86	16.43	C26H54C12N6	521.67	520.3787	59.86	10.43	13.59	16.11
C26H45C11N2	421.11	420.3271	74.16	10.77	8.42	6.65	C26H55C11N2	431.19	430.4054	72.42	12.86	8.22	6.50
C26H45C11N4	449.13	448.3333	69.53	10.10	7.89	12.47	C26H55C11N4	459.21	458.4115	68.01	12.07	7.72	12.20
C26H45C11N6	477.14	476.3394	65.45	9.51	7.43	17.61	C26H55C11N6	487.22	486.4176	64.10	11.38	7.28	17.25
C26H45C12N1	442.56	441.2929	70.56	10.25	16.02	3.16	C26H55C12N3	480.65	479.3773	64.97	11.53	14.75	8.74
C26H45C12N3	470.57	469.2990	66.36	9.64	15.07	8.93	C26H55C12N5	508.67	507.3834	61.39	10.90	13.94	13.77
C26H45C12N5	498.59	497.3052	62.63	9.10	14.22	14.05	C26H56C11N3	446.21	445.4163	69.99	12.65	7.95	9.42
C26H46C11N1	408.12	407.3319	76.52	11.36	8.69	3.43	C26H56C11N5	474.22	473.4224	65.85	11.90	7.48	14.77
C26H46C11N3	436.13	435.3380	71.60	10.63	8.13	9.63	C26H56C12N4	495.67	494.3882	63.00	11.39	14.31	11.30
C26H46C11N5	464.14	463.3442	67.28	9.99	7.64	15.09	C26H56C12N6	523.68	522.3943	59.63	10.78	13.54	16.05
C26H46C12N2	457.58	456.3038	68.25	10.13	15.50	6.12	C26H57C11N4	461.22	460.4272	67.71	12.46	7.69	12.15
C26H46C12N4	485.59	484.3099	64.31	9.55	14.60	11.54	C26H57C11N6	489.24	488.4333	63.83	11.74	7.25	17.18
C26H46C12N6	513.60	512.3161	60.80	9.03	13.81	16.36	C26H57C12N5	510.68	509.3991	61.15	11.25	13.88	13.71
C26H47C11N2	423.13	422.3428	73.80	11.20	8.38	6.62	C26H58C11N5	476.24	475.4380	65.57	12.28	7.44	14.71
C26H47C11N4	451.14	450.3489	69.22	10.50	7.86	12.42	C26H58C12N6	525.70	524.4100	59.40	11.12	13.49	15.99
C26H47C11N6	479.16	478.3551	65.17	9.89	7.40	17.54	C26H59C11N6	491.25	490.4489	63.57	12.11	7.22	17.11
C26H47C12N1	444.58	443.3085	70.24	10.66	15.95	3.15	C27H13C11N2	400.87	400.0767	80.90	3.27	8.84	6.99
C26H47C12N3	472.59	471.3147	66.08	10.02	15.00	8.89	C27H13C11N4	428.88	428.0829	75.62	3.06	8.27	13.06
C26H47C12N5	500.60	499.3208	62.38	9.46	14.16	13.99	C27H13C11N6	456.90	456.0890	70.98	2.87	7.76	18.39
C26H48C11N1	410.13	409.3475	76.14	11.80	8.64	3.42	C27H13C12N1	422.32	421.0425	76.79	3.10	16.79	3.32

C27H13C12N3	450.33	449.0486	72.01	2.91	15.75	9.33
C27H13C12N5	478.34	477.0548	67.80	2.74	14.82	14.64
C27H14C11N1	387.87	387.0815	83.61	3.64	9.14	3.61
C27H14C11N3	415.89	415.0876	77.98	3.39	8.52	10.10
C27H14C11N5	443.90	443.0938	73.06	3.18	7.99	15.78
C27H14C12N2	437.33	436.0534	74.15	3.23	16.21	6.41
C27H14C12N4	465.35	464.0595	69.69	3.03	15.24	12.04
C27H14C12N6	493.36	492.0657	65.73	2.86	14.37	17.03
C27H15C11N2	402.89	402.0924	80.49	3.75	8.80	6.95
C27H15C11N4	430.90	430.0985	75.26	3.51	8.23	13.00
C27H15C11N6	458.91	458.1047	70.67	3.29	7.73	18.31
C27H15C12N1	424.33	423.0581	76.43	3.56	16.71	3.30
C27H15C12N3	452.35	451.0643	71.69	3.34	15.68	9.29
C27H15C12N5	480.36	479.0704	67.51	3.15	14.76	14.58
C27H16C11N1	389.89	389.0971	83.18	4.14	9.09	3.59
C27H16C11N3	417.90	417.1033	77.60	3.86	8.48	10.06
C27H16C11N5	445.92	445.1094	72.73	3.62	7.95	15.71
C27H16C12N2	439.35	438.0690	73.81	3.67	16.14	6.38
C27H16C12N4	467.36	466.0752	69.39	3.45	15.17	11.99
C27H16C12N6	495.37	494.0813	65.47	3.26	14.31	16.96
C27H17C11N2	404.90	404.1080	80.09	4.23	8.76	6.92
C27H17C11N4	432.92	432.1142	74.91	3.96	8.19	12.94
C27H17C11N6	460.93	460.1203	70.36	3.72	7.69	18.23
C27H17C12N1	426.35	425.0738	76.06	4.02	16.63	3.29
C27H17C12N3	454.36	453.0799	71.37	3.77	15.61	9.25
C27H17C12N5	482.38	481.0861	67.23	3.55	14.70	14.52
C27H18C11N1	391.90	391.1128	82.75	4.63	9.05	3.57
C27H18C11N3	419.92	419.1189	77.23	4.32	8.44	10.01
C27H18C11N5	447.93	447.1251	72.40	4.05	7.91	15.63
C27H18C12N2	441.36	440.0847	73.48	4.11	16.07	6.35
C27H18C12N4	469.38	468.0908	69.09	3.87	15.11	11.94
C27H18C12N6	497.39	496.0970	65.20	3.65	14.26	16.90
C27H19C11N2	406.92	406.1237	79.70	4.71	8.71	6.88
C27H19C11N4	434.93	434.1298	74.56	4.40	8.15	12.88
C27H19C11N6	462.95	462.1360	70.05	4.14	7.66	18.15
C27H19C12N1	428.37	427.0894	75.71	4.47	16.55	3.27
C27H19C12N3	456.38	455.0956	71.06	4.20	15.54	9.21
C27H19C12N5	484.39	483.1017	66.95	3.95	14.64	14.46
C27H20C11N1	393.92	393.1284	82.33	5.12	9.00	3.56
C27H20C11N3	421.93	421.1346	76.86	4.78	8.40	9.96
C27H20C11N5	449.95	449.1407	72.08	4.48	7.88	15.56
C27H20C12N2	443.38	442.1003	73.14	4.55	15.99	6.32
C27H20C12N4	471.39	470.1065	68.80	4.28	15.04	11.89
C27H20C12N6	499.41	498.1126	64.94	4.04	14.20	16.83
C27H21C11N2	408.93	408.1393	79.30	5.18	8.67	6.85
C27H21C11N4	436.95	436.1455	74.22	4.84	8.11	12.82
C27H21C11N6	464.96	464.1516	69.75	4.55	7.62	18.07
C27H21C12N1	430.38	429.1051	75.35	4.92	16.48	3.25
C27H21C12N3	458.39	457.1112	70.75	4.62	15.47	9.17
C27H21C12N5	486.41	485.1174	66.67	4.35	14.58	14.40
C27H22C11N1	395.94	395.1441	81.91	5.60	8.95	3.54
C27H22C11N3	423.95	423.1502	76.50	5.23	8.36	9.91
C27H22C11N5	451.96	451.1564	71.75	4.91	7.84	15.50
C27H22C12N2	445.40	444.1160	72.81	4.98	15.92	6.29
C27H22C12N4	473.41	472.1221	68.50	4.68	14.98	11.83
C27H22C12N6	501.42	500.1283	64.68	4.42	14.14	16.76
C27H23C11N2	410.95	410.1550	78.91	5.64	8.63	6.82
C27H23C11N4	438.96	438.1611	73.88	5.28	8.08	12.76
C27H23C11N6	466.98	466.1673	69.45	4.96	7.59	18.00
C27H23C12N1	432.40	431.1207	75.00	5.36	16.40	3.24

C27H23C12N3	460.41	459.1269	70.44	5.04	15.40	9.13
C27H23C12N5	488.42	487.1330	66.40	4.75	14.52	14.34
C27H24C11N1	397.95	397.1597	81.49	6.08	8.91	3.52
C27H24C11N3	425.97	425.1659	76.13	5.68	8.32	9.86
C27H24C11N5	453.98	453.1720	71.44	5.33	7.81	15.43
C27H24C12N2	447.41	446.1316	72.48	5.41	15.85	6.26
C27H24C12N4	475.43	474.1378	68.21	5.09	14.91	11.78
C27H24C12N6	503.44	502.1439	64.42	4.81	14.08	16.69
C27H25C11N2	412.97	412.1706	78.53	6.10	8.58	6.78
C27H25C11N4	440.98	440.1768	73.54	5.71	8.04	12.71
C27H25C11N6	468.99	468.1829	69.15	5.37	7.56	17.92
C27H25C12N1	434.41	433.1364	74.65	5.80	16.32	3.22
C27H25C12N3	462.43	461.1425	70.13	5.45	15.33	9.09
C27H25C12N5	490.44	489.1487	66.12	5.14	14.46	14.28
C27H26C11N1	399.97	399.1754	81.08	6.55	8.86	3.50
C27H26C11N3	427.98	427.1815	75.77	6.12	8.28	9.82
C27H26C11N5	455.99	455.1877	71.12	5.75	7.77	15.36
C27H26C12N2	449.43	448.1473	72.16	5.83	15.78	6.23
C27H26C12N4	477.44	476.1534	67.92	5.49	14.85	11.73
C27H26C12N6	505.45	504.1596	64.16	5.18	14.03	16.63
C27H27C11N2	414.98	414.1863	78.15	6.56	8.54	6.75
C27H27C11N4	443.00	442.1924	73.21	6.14	8.00	12.65
C27H27C11N6	471.01	470.1986	68.85	5.78	7.53	17.84
C27H27C12N1	436.43	435.1520	74.31	6.24	16.25	3.21
C27H27C12N3	464.44	463.1582	69.83	5.86	15.27	9.05
C27H27C12N5	492.46	491.1643	65.85	5.53	14.40	14.22
C27H28C11N1	401.98	401.1910	80.68	7.02	8.82	3.48
C27H28C11N3	430.00	429.1972	75.42	6.56	8.24	9.77
C27H28C11N5	458.01	457.2033	70.81	6.16	7.74	15.29
C27H28C12N2	451.44	450.1629	71.84	6.25	15.71	6.21
C27H28C12N4	479.46	478.1691	67.64	5.89	14.79	11.69
C27H28C12N6	507.47	506.1752	63.91	5.56	13.97	16.56
C27H29C11N2	417.00	416.2019	77.77	7.01	8.50	6.72
C27H29C11N4	445.01	444.2081	72.87	6.57	7.97	12.59
C27H29C11N6	473.03	472.2142	68.56	6.18	7.49	17.77
C27H29C12N1	438.44	437.1677	73.97	6.67	16.17	3.19
C27H29C12N3	466.46	465.1738	69.52	6.27	15.20	9.01
C27H29C12N5	494.47	493.1800	65.59	5.91	14.34	14.16
C27H30C11N1	404.00	403.2067	80.27	7.48	8.78	3.47
C27H30C11N3	432.01	431.2128	75.07	7.00	8.21	9.73
C27H30C11N5	460.03	459.2190	70.50	6.57	7.71	15.22
C27H30C12N2	453.46	452.1786	71.52	6.67	15.64	6.18
C27H30C12N4	481.47	480.1847	67.36	6.28	14.73	11.64
C27H30C12N6	509.49	508.1909	63.65	5.94	13.92	16.50
C27H31C11N2	419.01	418.2176	77.40	7.46	8.46	6.69
C27H31C11N4	447.03	446.2237	72.55	6.99	7.93	12.53
C27H31C11N6	475.04	474.2299	68.27	6.58	7.46	17.69
C27H31C12N1	440.46	439.1833	73.63	7.09	16.10	3.18
C27H31C12N3	468.47	467.1895	69.22	6.67	15.14	8.97
C27H31C12N5	496.49	495.1956	65.32	6.29	14.28	14.11
C27H32C11N1	406.02	405.2223	79.87	7.94	8.73	3.45
C27H32C11N3	434.03	433.2285	74.72	7.43	8.17	9.68
C27H32C11N5	462.04	461.2346	70.19	6.98	7.67	15.16
C27H32C12N2	455.48	454.1942	71.20	7.08	15.57	6.15
C27H32C12N4	483.49	482.2004	67.08	6.67	14.67	11.59
C27H32C12N6	511.50	510.2065	63.40	6.31	13.86	16.43
C27H33C11N2	421.03	420.2332	77.03	7.90	8.42	6.65
C27H33C11N4	449.04	448.2394	72.22	7.41	7.90	12.48
C27H33C11N6	477.06	476.2455	67.98	6.97	7.43	17.62
C27H33C12N1	442.48	441.1990	73.29	7.52	16.02	3.17

C27H33C12N3	470.49	469.2051	68.93	7.07	15.07	8.93
C27H33C12N5	498.50	497.2113	65.05	6.67	14.22	14.05
C27H34C11N1	408.03	407.2380	79.48	8.40	8.69	3.43
C27H34C11N3	436.05	435.2441	74.37	7.86	8.13	9.64
C27H34C11N5	464.06	463.2503	69.88	7.39	7.64	15.09
C27H34C12N2	457.49	456.2099	70.89	7.49	15.50	6.12
C27H34C12N4	485.50	484.2160	66.80	7.06	14.60	11.54
C27H34C12N6	513.52	512.2222	63.15	6.67	13.81	16.37
C27H35C11N2	423.05	422.2489	76.66	8.34	8.38	6.62
C27H35C11N4	451.06	450.2550	71.90	7.82	7.86	12.42
C27H35C11N6	479.07	478.2612	67.69	7.36	7.40	17.54
C27H35C12N1	444.49	443.2146	72.96	7.94	15.95	3.15
C27H35C12N3	472.51	471.2208	68.63	7.47	15.01	8.89
C27H35C12N5	500.52	499.2269	64.79	7.05	14.17	13.99
C27H36C11N1	410.05	409.2536	79.09	8.85	8.65	3.42
C27H36C11N3	438.06	437.2598	74.03	8.28	8.09	9.59
C27H36C11N5	466.07	465.2659	69.58	7.79	7.61	15.03
C27H36C12N2	459.51	458.2255	70.58	7.90	15.43	6.10
C27H36C12N4	487.52	486.2317	66.52	7.44	14.54	11.49
C27H36C12N6	515.53	514.2378	62.91	7.04	13.75	16.30
C27H37C11N2	425.06	424.2645	76.29	8.77	8.34	6.59
C27H37C11N4	453.08	452.2707	71.58	8.23	7.82	12.37
C27H37C11N6	481.09	480.2768	67.41	7.75	7.37	17.47
C27H37C12N1	446.51	445.2303	72.63	8.35	15.88	3.14
C27H37C12N3	474.52	473.2364	68.34	7.86	14.94	8.86
C27H37C12N5	502.54	501.2426	64.53	7.42	14.11	13.94
C27H38C11N1	412.06	411.2693	78.70	9.30	8.60	3.40
C27H38C11N3	440.08	439.2754	73.69	8.70	8.06	9.55
C27H38C11N5	468.09	467.2816	69.28	8.18	7.57	14.96
C27H38C12N2	461.52	460.2412	70.27	8.30	15.36	6.07
C27H38C12N4	489.54	488.2473	66.25	7.82	14.48	11.44
C27H38C12N6	517.55	516.2535	62.66	7.40	13.70	16.24
C27H39C11N2	427.08	426.2802	75.93	9.20	8.30	6.56
C27H39C11N4	455.09	454.2863	71.26	8.64	7.79	12.31
C27H39C11N6	483.11	482.2925	67.13	8.14	7.34	17.40
C27H39C12N1	448.52	447.2459	72.30	8.76	15.81	3.12
C27H39C12N3	476.54	475.2521	68.05	8.25	14.88	8.82
C27H39C12N5	504.55	503.2582	64.28	7.79	14.05	13.88
C27H40C11N1	414.08	413.2849	78.32	9.74	8.56	3.38
C27H40C11N3	442.09	441.2911	73.36	9.12	8.02	9.50
C27H40C11N5	470.11	469.2972	68.98	8.58	7.54	14.90
C27H40C12N2	463.54	462.2568	69.96	8.70	15.30	6.04
C27H40C12N4	491.55	490.2630	65.97	8.20	14.42	11.40
C27H40C12N6	519.57	518.2691	62.42	7.76	13.65	16.18
C27H41C11N2	429.09	428.2958	75.58	9.63	8.26	6.53
C27H41C11N4	457.11	456.3020	70.95	9.04	7.76	12.26
C27H41C11N6	485.12	484.3081	66.85	8.52	7.31	17.32
C27H41C12N1	450.54	449.2616	71.98	9.17	15.74	3.11
C27H41C12N3	478.55	477.2677	67.77	8.64	14.82	8.78
C27H41C12N5	506.57	505.2739	64.02	8.16	14.00	13.83
C27H42C11N1	416.10	415.3006	77.94	10.17	8.52	3.37
C27H42C11N3	444.11	443.3067	73.02	9.53	7.98	9.46
C27H42C11N5	472.12	471.3129	68.69	8.97	7.51	14.83
C27H42C12N2	465.56	464.2725	69.66	9.09	15.23	6.02
C27H42C12N4	493.57	492.2786	65.71	8.58	14.37	11.35
C27H42C12N6	521.58	520.2848	62.18	8.12	13.59	16.11
C27H43C11N2	431.11	430.3115	75.22	10.05	8.22	6.50
C27H43C11N4	459.12	458.3176	70.63	9.44	7.72	12.20
C27H43C11N6	487.14	486.3238	66.57	8.90	7.28	17.25
C27H43C12N1	452.56	451.2772	71.66	9.58	15.67	3.10

C27H43C12N3	480.57	479.2834	67.48	9.02	14.75	8.74
C27H43C12N5	508.58	507.2895	63.77	8.52	13.94	13.77
C27H44C11N1	418.11	417.3162	77.56	10.61	8.48	3.35
C27H44C11N3	446.12	445.3224	72.69	9.94	7.95	9.42
C27H44C11N5	474.14	473.3285	68.40	9.35	7.48	14.77
C27H44C12N2	467.57	466.2881	69.36	9.49	15.16	5.99
C27H44C12N4	495.58	494.2943	65.44	8.95	14.31	11.31
C27H44C12N6	523.60	522.3004	61.94	8.47	13.54	16.05
C27H45C11N2	433.13	432.3271	74.87	10.47	8.19	6.47
C27H45C11N4	461.14	460.3333	70.33	9.84	7.69	12.15
C27H45C11N6	489.15	488.3394	66.30	9.27	7.25	17.18
C27H45C12N1	454.57	453.2929	71.34	9.98	15.60	3.08
C27H45C12N3	482.59	481.2990	67.20	9.40	14.69	8.71
C27H45C12N5	510.60	509.3052	63.51	8.88	13.89	13.72
C27H46C11N1	420.13	419.3319	77.19	11.04	8.44	3.33
C27H46C11N3	448.14	447.3380	72.37	10.35	7.91	9.38
C27H46C11N5	476.15	475.3442	68.11	9.74	7.45	14.71
C27H46C12N2	469.59	468.3038	69.06	9.87	15.10	5.97
C27H46C12N4	497.60	496.3099	65.17	9.32	14.25	11.26
C27H46C12N6	525.61	524.3161	61.70	8.82	13.49	15.99
C27H47C11N2	435.14	434.3428	74.53	10.89	8.15	6.44
C27H47C11N4	463.16	462.3489	70.02	10.23	7.65	12.10
C27H47C11N6	491.17	490.3551	66.03	9.65	7.22	17.11
C27H47C12N1	456.59	455.3085	71.03	10.38	15.53	3.07
C27H47C12N3	484.60	483.3147	66.92	9.78	14.63	8.67
C27H47C12N5	512.62	511.3208	63.26	9.24	13.83	13.66
C27H48C11N1	422.14	421.3475	76.82	11.46	8.40	3.32
C27H48C11N3	450.16	449.3537	72.04	10.75	7.88	9.33
C27H48C11N5	478.17	477.3598	67.82	10.12	7.41	14.65
C27H48C12N2	471.60	470.3194	68.77	10.26	15.04	5.94
C27H48C12N4	499.62	498.3256	64.91	9.68	14.19	11.21
C27H48C12N6	527.63	526.3317	61.46	9.17	13.44	15.93
C27H49C11N2	437.16	436.3584	74.18	11.30	8.11	6.41
C27H49C11N4	465.17	464.3646	69.72	10.62	7.62	12.04
C27H49C11N6	493.18	492.3707	65.76	10.01	7.19	17.04
C27H49C12N1	458.60	457.3242	70.71	10.77	15.46	3.05
C27H49C12N3	486.62	485.3303	66.64	10.15	14.57	8.64
C27H49C12N5	514.63	513.3365	63.02	9.60	13.78	13.61
C27H50C11N1	424.16	423.3632	76.46	11.88	8.36	3.30
C27H50C11N3	452.17	451.3693	71.72	11.15	7.84	9.29
C27H50C11N5	480.19	479.3755	67.54	10.50	7.38	14.58
C27H50C12N2	473.62	472.3351	68.47	10.64	14.97	5.91
C27H50C12N4	501.63	500.3412	64.65	10.05	14.14	11.17
C27H50C12N6	529.65	528.3474	61.23	9.52	13.39	15.87
C27H51C11N2	439.17	438.3741	73.84	11.71	8.07	6.38
C27H51C11N4	467.19	466.3802	69.42	11.00	7.59	11.99
C27H51C11N6	495.20	494.3864	65.49	10.38	7.16	16.97
C27H51C12N1	460.62	459.3398	70.41	11.16	15.39	3.04
C27H51C12N3	488.63	487.3460	66.37	10.52	14.51	8.60
C27H51C12N5	516.65	515.3521	62.77	9.95	13.72	13.56
C27H52C11N1	426.18	425.3788	76.10	12.30	8.32	3.29
C27H52C11N3	454.19	453.3850	71.40	11.54	7.81	9.25
C27H52C11N5	482.20	481.3911	67.25	10.87	7.35	14.52
C27H52C12N2	475.63	474.3507	68.18	11.02	14.91	5.89
C27H52C12N4	503.65	502.3569	64.39	10.41	14.08	11.12
C27H52C12N6	531.66	530.3630	61.00	9.86	13.34	15.81
C27H53C11N2	441.19	440.3897	73.51	12.11	8.04	6.35
C27H53C11N4	469.20	468.3959	69.12	11.39	7.56	11.94
C27H53C11N6	497.22	496.4020	65.22	10.74	7.13	16.90
C27H53C12N1	462.64	461.3555	70.10	11.55	15.33	3.03

C27H53C12N3	490.65	489.3616	66.10	10.89	14.45	8.56
C27H53C12N5	518.66	517.3678	62.53	10.30	13.67	13.50
C27H54C11N1	428.19	427.3945	75.74	12.71	8.28	3.27
C27H54C11N3	456.20	455.4006	71.09	11.93	7.77	9.21
C27H54C11N5	484.22	483.4068	66.97	11.24	7.32	14.46
C27H54C12N2	477.65	476.3664	67.90	11.40	14.84	5.86
C27H54C12N4	505.66	504.3725	64.13	10.76	14.02	11.08
C27H54C12N6	533.68	532.3787	60.77	10.20	13.29	15.75
C27H55C11N2	443.21	442.4054	73.17	12.51	8.00	6.32
C27H55C11N4	471.22	470.4115	68.82	11.76	7.52	11.89
C27H55C11N6	499.23	498.4176	64.96	11.10	7.10	16.83
C27H55C12N1	464.65	463.3711	69.79	11.93	15.26	3.01
C27H55C12N3	492.67	491.3773	65.83	11.25	14.39	8.53
C27H55C12N5	520.68	519.3834	62.28	10.65	13.62	13.45
C27H56C11N1	430.21	429.4101	75.38	13.12	8.24	3.26
C27H56C11N3	458.22	457.4163	70.77	12.32	7.74	9.17
C27H56C11N5	486.23	485.4224	66.70	11.61	7.29	14.40
C27H56C12N2	479.67	478.3820	67.61	11.77	14.78	5.84
C27H56C12N4	507.68	506.3882	63.88	11.12	13.97	11.04
C27H56C12N6	535.69	534.3943	60.54	10.54	13.24	15.69
C27H57C11N2	445.22	444.4210	72.84	12.90	7.96	6.29
C27H57C11N4	473.24	472.4272	68.53	12.14	7.49	11.84
C27H57C11N6	501.25	500.4333	64.70	11.46	7.07	16.77
C27H57C12N3	494.68	493.3929	65.56	11.61	14.33	8.49
C27H57C12N5	522.69	521.3991	62.04	10.99	13.57	13.40
C27H58C11N3	460.24	459.4319	70.46	12.70	7.70	9.13
C27H58C11N5	488.25	487.4380	66.42	11.97	7.26	14.34
C27H58C12N4	509.70	508.4038	63.63	11.47	13.91	10.99
C27H58C12N6	537.71	536.4100	60.31	10.87	13.19	15.63
C27H59C11N4	475.25	474.4428	68.24	12.51	7.46	11.79
C27H59C11N6	503.26	502.4489	64.44	11.82	7.04	16.70
C27H59C12N5	524.71	523.4147	61.81	11.33	13.51	13.35
C27H60C11N5	490.27	489.4537	66.15	12.34	7.23	14.28
C27H60C12N6	539.73	538.4256	60.09	11.21	13.14	15.57
C27H61C11N6	505.28	504.4646	64.18	12.17	7.02	16.63
C28H14C11N1	399.88	399.0815	84.10	3.53	8.87	3.50
C28H14C11N3	427.90	427.0876	78.60	3.30	8.29	9.82
C28H14C11N5	455.91	455.0938	73.77	3.10	7.78	15.36
C28H14C12N2	449.34	448.0534	74.85	3.14	15.78	6.23
C28H14C12N4	477.36	476.0595	70.45	2.96	14.85	11.74
C28H14C12N6	505.37	504.0657	66.55	2.79	14.03	16.63
C28H15C11N2	414.90	414.0924	81.06	3.64	8.54	6.75
C28H15C11N4	442.91	442.0985	75.93	3.41	8.00	12.65
C28H15C11N6	470.92	470.1047	71.42	3.21	7.53	17.85
C28H15C12N1	436.34	435.0581	77.07	3.47	16.25	3.21
C28H15C12N3	464.36	463.0643	72.43	3.26	15.27	9.05
C28H15C12N5	492.37	491.0704	68.30	3.07	14.40	14.22
C28H16C11N1	401.90	401.0971	83.68	4.01	8.82	3.49
C28H16C11N3	429.91	429.1033	78.23	3.75	8.25	9.77
C28H16C11N5	457.93	457.1094	73.44	3.52	7.74	15.29
C28H16C12N2	451.36	450.0690	74.51	3.57	15.71	6.21
C28H16C12N4	479.37	478.0752	70.16	3.36	14.79	11.69
C28H16C12N6	507.39	506.0813	66.28	3.18	13.97	16.56
C28H17C11N2	416.91	416.1080	80.67	4.11	8.50	6.72
C28H17C11N4	444.93	444.1142	75.59	3.85	7.97	12.59
C28H17C11N6	472.94	472.1203	71.11	3.62	7.50	17.77
C28H17C12N1	438.36	437.0738	76.72	3.91	16.18	3.20
C28H17C12N3	466.37	465.0799	72.11	3.67	15.20	9.01
C28H17C12N5	494.39	493.0861	68.03	3.47	14.34	14.17
C28H18C11N1	403.92	403.1128	83.26	4.49	8.78	3.47

C28H18C11N3	431.93	431.1189	77.86	4.20	8.21	9.73
C28H18C11N5	459.94	459.1251	73.12	3.94	7.71	15.23
C28H18C12N2	453.38	452.0847	74.18	4.00	15.64	6.18
C28H18C12N4	481.39	480.0908	69.86	3.77	14.73	11.64
C28H18C12N6	509.40	508.0970	66.02	3.56	13.92	16.50
C28H19C11N2	418.93	418.1237	80.28	4.57	8.46	6.69
C28H19C11N4	446.94	446.1298	75.25	4.28	7.93	12.54
C28H19C11N6	474.96	474.1360	70.81	4.03	7.46	17.69
C28H19C12N1	440.38	439.0894	76.37	4.35	16.10	3.18
C28H19C12N3	468.39	467.0956	71.80	4.09	15.14	8.97
C28H19C12N5	496.40	495.1017	67.75	3.86	14.28	14.11
C28H20C11N1	405.93	405.1284	82.85	4.97	8.73	3.45
C28H20C11N3	433.94	433.1346	77.50	4.65	8.17	9.68
C28H20C11N5	461.96	461.1407	72.80	4.36	7.67	15.16
C28H20C12N2	455.39	454.1003	73.85	4.43	15.57	6.15
C28H20C12N4	483.40	482.1065	69.57	4.17	14.67	11.59
C28H20C12N6	511.42	510.1126	65.76	3.94	13.86	16.43
C28H21C11N2	420.95	420.1393	79.89	5.03	8.42	6.65
C28H21C11N4	448.96	448.1455	74.91	4.71	7.90	12.48
C28H21C11N6	476.97	476.1516	70.51	4.44	7.43	17.62
C28H21C12N1	442.39	441.1051	76.02	4.78	16.03	3.17
C28H21C12N3	470.41	469.1112	71.49	4.50	15.07	8.93
C28H21C12N5	498.42	497.1174	67.48	4.25	14.23	14.05
C28H22C11N1	407.95	407.1441	82.44	5.44	8.69	3.43
C28H22C11N3	435.96	435.1502	77.14	5.09	8.13	9.64
C28H22C11N5	463.97	463.1564	72.49	4.78	7.64	15.09
C28H22C12N2	457.41	456.1160	73.53	4.85	15.50	6.12
C28H22C12N4	485.42	484.1221	69.28	4.57	14.61	11.54
C28H22C12N6	513.43	512.1283	65.50	4.32	13.81	16.37
C28H23C11N2	422.96	422.1550	79.51	5.48	8.38	6.62
C28H23C11N4	450.98	450.1611	74.57	5.14	7.86	12.42
C28H23C11N6	478.99	478.1673	70.21	4.84	7.40	17.55
C28H23C12N1	444.41	443.1207	75.68	5.22	15.96	3.15
C28H23C12N3	472.42	471.1269	71.19	4.91	15.01	8.89
C28H23C12N5	500.44	499.1330	67.20	4.63	14.17	13.99
C28H24C11N1	409.96	409.1597	82.03	5.90	8.65	3.42
C28H24C11N3	437.98	437.1659	76.79	5.52	8.09	9.59
C28H24C11N5	465.99	465.1720	72.17	5.19	7.61	15.03
C28H24C12N2	459.42	458.1316	73.20	5.27	15.43	6.10
C28H24C12N4	487.44	486.1378	69.00	4.96	14.55	11.49
C28H24C12N6	515.45	514.1439	65.25	4.69	13.76	16.30
C28H25C11N2	424.98	424.1706	79.14	5.93	8.34	6.59
C28H25C11N4	452.99	452.1768	74.24	5.56	7.83	12.37
C28H25C11N6	481.00	480.1829	69.92	5.24	7.37	17.47
C28H25C12N1	446.42	445.1364	75.33	5.64	15.88	3.14
C28H25C12N3	474.44	473.1425	70.89	5.31	14.95	8.86
C28H25C12N5	502.45	501.1487	66.93	5.02	14.11	13.94
C28H26C11N1	411.98	411.1754	81.63	6.36	8.61	3.40
C28H26C11N3	439.99	439.1815	76.44	5.96	8.06	9.55
C28H26C11N5	468.01	467.1877	71.86	5.60	7.58	14.96
C28H26C12N2	461.44	460.1473	72.88	5.68	15.37	6.07
C28H26C12N4	489.45	488.1534	68.71	5.35	14.49	11.45
C28H26C12N6	517.47	516.1596	64.99	5.06	13.70	16.24
C28H27C11N2	426.99	426.1863	78.76	6.37	8.30	6.56
C28H27C11N4	455.01	454.1924	73.91	5.98	7.79	12.31
C28H27C11N6	483.02	482.1986	69.63	5.63	7.34	17.40
C28H27C12N1	448.44	447.1520	75.00	6.07	15.81	3.12
C28H27C12N3	476.45	475.1582	70.59	5.71	14.88	8.82
C28H27C12N5	504.47	503.1643	66.67	5.39	14.06	13.88
C28H28C11N1	414.00	413.1910	81.24	6.82	8.56	3.38

C28H28C11N3	442.01	441.1972	76.09	6.39	8.02	9.51
C28H28C11N5	470.02	469.2033	71.55	6.00	7.54	14.90
C28H28C12N2	463.45	462.1629	72.57	6.09	15.30	6.04
C28H28C12N4	491.47	490.1691	68.43	5.74	14.43	11.40
C28H28C12N6	519.48	518.1752	64.74	5.43	13.65	16.18
C28H29C11N2	429.01	428.2019	78.39	6.81	8.26	6.53
C28H29C11N4	457.02	456.2081	73.59	6.40	7.76	12.26
C28H29C11N6	485.04	484.2142	69.34	6.03	7.31	17.33
C28H29C12N1	450.46	449.1677	74.66	6.49	15.74	3.11
C28H29C12N3	478.47	477.1738	70.29	6.11	14.82	8.78
C28H29C12N5	506.48	505.1800	66.40	5.77	14.00	13.83
C28H30C11N1	416.01	415.2067	80.84	7.27	8.52	3.37
C28H30C11N3	444.02	443.2128	75.74	6.81	7.98	9.46
C28H30C11N5	472.04	471.2190	71.25	6.41	7.51	14.84
C28H30C12N2	465.47	464.1786	72.25	6.50	15.23	6.02
C28H30C12N4	493.48	492.1847	68.15	6.13	14.37	11.35
C28H30C12N6	521.50	520.1909	64.49	5.80	13.60	16.12
C28H31C11N2	431.03	430.2176	78.03	7.25	8.23	6.50
C28H31C11N4	459.04	458.2237	73.26	6.81	7.72	12.21
C28H31C11N6	487.05	486.2299	69.05	6.42	7.28	17.25
C28H31C12N1	452.47	451.1833	74.33	6.91	15.67	3.10
C28H31C12N3	480.49	479.1895	69.99	6.50	14.76	8.75
C28H31C12N5	508.50	507.1956	66.14	6.14	13.94	13.77
C28H32C11N1	418.03	417.2223	80.45	7.72	8.48	3.35
C28H32C11N3	446.04	445.2285	75.40	7.23	7.95	9.42
C28H32C11N5	474.05	473.2346	70.94	6.80	7.48	14.77
C28H32C12N2	467.49	466.1942	71.94	6.90	15.17	5.99
C28H32C12N4	495.50	494.2004	67.87	6.51	14.31	11.31
C28H32C12N6	523.51	522.2065	64.24	6.16	13.54	16.05
C28H33C11N2	433.04	432.2332	77.66	7.68	8.19	6.47
C28H33C11N4	461.06	460.2394	72.94	7.21	7.69	12.15
C28H33C11N6	489.07	488.2455	68.77	6.80	7.25	17.18
C28H33C12N1	454.49	453.1990	74.00	7.32	15.60	3.08
C28H33C12N3	482.50	481.2051	69.70	6.89	14.70	8.71
C28H33C12N5	510.51	509.2113	65.88	6.52	13.89	13.72
C28H34C11N1	420.04	419.2380	80.07	8.16	8.44	3.33
C28H34C11N3	448.06	447.2441	75.06	7.65	7.91	9.38
C28H34C11N5	476.07	475.2503	70.64	7.20	7.45	14.71
C28H34C12N2	469.50	468.2099	71.63	7.30	15.10	5.97
C28H34C12N4	497.52	496.2160	67.60	6.89	14.25	11.26
C28H34C12N6	525.53	524.2222	63.99	6.52	13.49	15.99
C28H35C11N2	435.06	434.2489	77.30	8.11	8.15	6.44
C28H35C11N4	463.07	462.2550	72.63	7.62	7.66	12.10
C28H35C11N6	491.08	490.2612	68.48	7.18	7.22	17.11
C28H35C12N1	456.50	455.2146	73.67	7.73	15.53	3.07
C28H35C12N3	484.52	483.2208	69.41	7.28	14.63	8.67
C28H35C12N5	512.53	511.2269	65.62	6.88	13.83	13.66
C28H36C11N1	422.06	421.2536	79.68	8.60	8.40	3.32
C28H36C11N3	450.07	449.2598	74.72	8.06	7.88	9.34
C28H36C11N5	478.09	477.2659	70.35	7.59	7.42	14.65
C28H36C12N2	471.52	470.2255	71.33	7.70	15.04	5.94
C28H36C12N4	499.53	498.2317	67.33	7.26	14.19	11.22
C28H36C12N6	527.55	526.2378	63.75	6.88	13.44	15.93
C28H37C11N2	437.07	436.2645	76.95	8.53	8.11	6.41
C28H37C11N4	465.09	464.2707	72.31	8.02	7.62	12.05
C28H37C11N6	493.10	492.2768	68.20	7.56	7.19	17.04
C28H37C12N1	458.52	457.2303	73.35	8.13	15.46	3.05
C28H37C12N3	486.53	485.2364	69.12	7.67	14.57	8.64
C28H37C12N5	514.55	513.2426	65.36	7.25	13.78	13.61
C28H38C11N1	424.07	423.2693	79.30	9.03	8.36	3.30

C28H38C11N3	452.09	451.2754	74.39	8.47	7.84	9.29
C28H38C11N5	480.10	479.2816	70.05	7.98	7.38	14.59
C28H38C12N2	473.53	472.2412	71.02	8.09	14.97	5.92
C28H38C12N4	501.55	500.2473	67.05	7.64	14.14	11.17
C28H38C12N6	529.56	528.2535	63.51	7.23	13.39	15.87
C28H39C11N2	439.09	438.2802	76.59	8.95	8.07	6.38
C28H39C11N4	467.10	466.2863	72.00	8.42	7.59	11.99
C28H39C11N6	495.12	494.2925	67.93	7.94	7.16	16.97
C28H39C12N1	460.54	459.2459	73.03	8.54	15.40	3.04
C28H39C12N3	488.55	487.2521	68.84	8.05	14.51	8.60
C28H39C12N5	516.56	515.2582	65.11	7.61	13.73	13.56
C28H40C11N1	426.09	425.2849	78.93	9.46	8.32	3.29
C28H40C11N3	454.10	453.2911	74.06	8.88	7.81	9.25
C28H40C11N5	482.12	481.2972	69.76	8.36	7.35	14.53
C28H40C12N2	475.55	474.2568	70.72	8.48	14.91	5.89
C28H40C12N4	503.56	502.2630	66.79	8.01	14.08	11.13
C28H40C12N6	531.58	530.2691	63.27	7.58	13.34	15.81
C28H41C11N2	441.11	440.2958	76.24	9.37	8.04	6.35
C28H41C11N4	469.12	468.3020	71.69	8.81	7.56	11.94
C28H41C11N6	497.13	496.3081	67.65	8.31	7.13	16.91
C28H41C12N1	462.55	461.2616	72.71	8.93	15.33	3.03
C28H41C12N3	490.57	489.2677	68.56	8.42	14.45	8.57
C28H41C12N5	518.58	517.2739	64.85	7.97	13.67	13.50
C28H42C11N1	428.11	427.3006	78.56	9.89	8.28	3.27
C28H42C11N3	456.12	455.3067	73.73	9.28	7.77	9.21
C28H42C11N5	484.13	483.3129	69.47	8.74	7.32	14.47
C28H42C12N2	477.57	476.2725	70.42	8.86	14.85	5.87
C28H42C12N4	505.58	504.2786	66.52	8.37	14.02	11.08
C28H42C12N6	533.59	532.2848	63.03	7.93	13.29	15.75
C28H43C11N2	443.12	442.3115	75.90	9.78	8.00	6.32
C28H43C11N4	471.13	470.3176	71.38	9.20	7.53	11.89
C28H43C11N6	499.15	498.3238	67.38	8.68	7.10	16.84
C28H43C12N1	464.57	463.2772	72.39	9.33	15.26	3.01
C28H43C12N3	492.58	491.2834	68.28	8.80	14.39	8.53
C28H43C12N5	520.59	519.2895	64.60	8.33	13.62	13.45
C28H44C11N1	430.12	429.3162	78.19	10.31	8.24	3.26
C28H44C11N3	458.14	457.3224	73.41	9.68	7.74	9.17
C28H44C11N5	486.15	485.3285	69.18	9.12	7.29	14.41
C28H44C12N2	479.58	478.2881	70.13	9.25	14.78	5.84
C28H44C12N4	507.60	506.2943	66.26	8.74	13.97	11.04
C28H44C12N6	535.61	534.3004	62.79	8.28	13.24	15.69
C28H45C11N2	445.14	444.3271	75.55	10.19	7.96	6.29
C28H45C11N4	473.15	472.3333	71.08	9.59	7.49	11.84
C28H45C11N6	501.16	500.3394	67.11	9.05	7.07	16.77
C28H45C12N1	466.58	465.2929	72.08	9.72	15.20	3.00
C28H45C12N3	494.60	493.2990	68.00	9.17	14.34	8.50
C28H45C12N5	522.61	521.3052	64.35	8.68	13.57	13.40
C28H46C11N1	432.14	431.3319	77.83	10.73	8.20	3.24
C28H46C11N3	460.15	459.3380	73.09	10.08	7.70	9.13
C28H46C11N5	488.17	487.3442	68.89	9.50	7.26	14.35
C28H46C12N2	481.60	480.3038	69.83	9.63	14.72	5.82
C28H46C12N4	509.61	508.3099	65.99	9.10	13.91	10.99
C28H46C12N6	537.63	536.3161	62.56	8.62	13.19	15.63
C28H47C11N2	447.15	446.3428	75.21	10.59	7.93	6.26
C28H47C11N4	475.17	474.3489	70.78	9.97	7.46	11.79
C28H47C11N6	503.18	502.3551	66.84	9.42	7.05	16.70
C28H47C12N1	468.60	467.3085	71.77	10.11	15.13	2.99
C28H47C12N3	496.61	495.3147	67.72	9.54	14.28	8.46
C28H47C12N5	524.63	523.3208	64.11	9.03	13.52	13.35
C28H48C11N1	434.15	433.3475	77.46	11.14	8.17	3.23

C28H48C11N3	462.17	461.3537	72.77	10.47	7.67	9.09
C28H48C11N5	490.18	489.3598	68.61	9.87	7.23	14.29
C28H48C12N2	483.61	482.3194	69.54	10.00	14.66	5.79
C28H48C12N4	511.63	510.3256	65.73	9.46	13.86	10.95
C28H48C12N6	539.64	538.3317	62.32	8.97	13.14	15.57
C28H49C11N2	449.17	448.3584	74.87	11.00	7.89	6.24
C28H49C11N4	477.18	476.3646	70.48	10.35	7.43	11.74
C28H49C11N6	505.20	504.3707	66.57	9.78	7.02	16.64
C28H49C12N1	470.62	469.3242	71.46	10.49	15.07	2.98
C28H49C12N3	498.63	497.3303	67.45	9.91	14.22	8.43
C28H49C12N5	526.64	525.3365	63.86	9.38	13.46	13.30
C28H50C11N1	436.17	435.3632	77.11	11.55	8.13	3.21
C28H50C11N3	464.18	463.3693	72.45	10.86	7.64	9.05
C28H50C11N5	492.20	491.3755	68.33	10.24	7.20	14.23
C28H50C12N2	485.63	484.3351	69.25	10.38	14.60	5.77
C28H50C12N4	513.64	512.3412	65.48	9.81	13.80	10.91
C28H50C12N6	541.66	540.3474	62.09	9.30	13.09	15.52
C28H51C11N2	451.19	450.3741	74.54	11.39	7.86	6.21
C28H51C11N4	479.20	478.3802	70.18	10.73	7.40	11.69
C28H51C11N6	507.21	506.3864	66.31	10.14	6.99	16.57
C28H51C12N1	472.63	471.3398	71.16	10.88	15.00	2.96
C28H51C12N3	500.64	499.3460	67.18	10.27	14.16	8.39
C28H51C12N5	528.66	527.3521	63.62	9.72	13.41	13.25
C28H52C11N1	438.19	437.3788	76.75	11.96	8.09	3.20
C28H52C11N3	466.20	465.3850	72.14	11.24	7.60	9.01
C28H52C11N5	494.21	493.3911	68.05	10.61	7.17	14.17
C28H52C12N2	487.65	486.3507	68.97	10.75	14.54	5.74
C28H52C12N4	515.66	514.3569	65.22	10.16	13.75	10.87
C28H52C12N6	543.67	542.3630	61.86	9.64	13.04	15.46
C28H53C11N2	453.20	452.3897	74.21	11.79	7.82	6.18
C28H53C11N4	481.21	480.3959	69.89	11.10	7.37	11.64
C28H53C11N6	509.23	508.4020	66.04	10.49	6.96	16.50
C28H53C12N1	474.65	473.3555	70.86	11.26	14.94	2.95
C28H53C12N3	502.66	501.3616	66.91	10.63	14.11	8.36
C28H53C12N5	530.67	529.3678	63.37	10.07	13.36	13.20
C28H54C11N1	440.20	439.3945	76.40	12.36	8.05	3.18
C28H54C11N3	468.22	467.4006	71.83	11.63	7.57	8.97
C28H54C11N5	496.23	495.4068	67.77	10.97	7.14	14.11
C28H54C12N2	489.66	488.3664	68.68	11.12	14.48	5.72
C28H54C12N4	517.68	516.3725	64.97	10.51	13.70	10.82
C28H54C12N6	545.69	544.3787	61.63	9.97	12.99	15.40
C28H55C11N2	455.22	454.4054	73.88	12.18	7.79	6.15
C28H55C11N4	483.23	482.4115	69.60	11.47	7.34	11.59
C28H55C11N6	511.24	510.4176	65.78	10.84	6.93	16.44
C28H55C12N1	476.66	475.3711	70.56	11.63	14.88	2.94
C28H55C12N3	504.68	503.3773	66.64	10.98	14.05	8.33
C28H55C12N5	532.69	531.3834	63.13	10.41	13.31	13.15
C28H56C11N1	442.22	441.4101	76.05	12.76	8.02	3.17
C28H56C11N3	470.23	469.4163	71.52	12.00	7.54	8.94
C28H56C11N5	498.25	497.4224	67.50	11.33	7.12	14.06
C28H56C12N2	491.68	490.3820	68.40	11.48	14.42	5.70
C28H56C12N4	519.69	518.3882	64.71	10.86	13.64	10.78
C28H56C12N6	547.70	546.3943	61.40	10.31	12.95	15.34
C28H57C11N2	457.23	456.4210	73.55	12.57	7.75	6.13
C28H57C11N4	485.25	484.4272	69.31	11.84	7.31	11.55
C28H57C11N6	513.26	512.4333	65.52	11.19	6.91	16.37
C28H57C12N1	478.68	477.3868	70.26	12.00	14.81	2.93
C28H57C12N3	506.69	505.3929	66.37	11.34	13.99	8.29
C28H57C12N5	534.71	533.3991	62.90	10.75	13.26	13.10
C28H58C11N1	444.23	443.4258	75.71	13.16	7.98	3.15

C28H58C11N3	472.25	471.4319	71.22	12.38	7.51	8.90
C28H58C11N5	500.26	499.4380	67.23	11.69	7.09	14.00
C28H58C12N2	493.69	492.3977	68.12	11.84	14.36	5.67
C28H58C12N4	521.71	520.4038	64.46	11.21	13.59	10.74
C28H58C12N6	549.72	548.4100	61.18	10.63	12.90	15.29
C28H59C11N2	459.25	458.4367	73.23	12.95	7.72	6.10
C28H59C11N4	487.26	486.4428	69.02	12.20	7.28	11.50
C28H59C11N6	515.28	514.4489	65.27	11.54	6.88	16.31
C28H59C12N3	508.71	507.4086	66.11	11.69	13.94	8.26
C28H59C12N5	536.72	535.4147	62.66	11.08	13.21	13.05
C28H60C11N3	474.26	473.4476	70.91	12.75	7.48	8.86
C28H60C11N5	502.28	501.4537	66.96	12.04	7.06	13.94
C28H60C12N4	523.72	522.4195	64.22	11.55	13.54	10.70
C28H60C12N6	551.74	550.4256	60.96	10.96	12.85	15.23
C28H61C11N4	489.28	488.4584	68.74	12.57	7.25	11.45
C28H61C11N6	517.29	516.4646	65.01	11.89	6.85	16.25
C28H61C12N5	538.74	537.4304	62.43	11.41	13.16	13.00
C28H62C11N5	504.29	503.4693	66.69	12.39	7.03	13.89
C28H62C12N6	553.75	552.4413	60.73	11.29	12.80	15.18
C28H63C11N6	519.31	518.4802	64.76	12.23	6.83	16.18
C29H14C11N1	411.89	411.0815	84.57	3.43	8.61	3.40
C29H14C11N3	439.91	439.0876	79.18	3.21	8.06	9.55
C29H14C11N5	467.92	467.0938	74.44	3.02	7.58	14.97
C29H14C12N2	461.35	460.0534	75.50	3.06	15.37	6.07
C29H14C12N4	489.37	488.0595	71.18	2.88	14.49	11.45
C29H14C12N6	517.38	516.0657	67.32	2.73	13.70	16.24
C29H15C11N2	426.91	426.0924	81.59	3.54	8.30	6.56
C29H15C11N4	454.92	454.0985	76.57	3.32	7.79	12.32
C29H15C11N6	482.94	482.1047	72.13	3.13	7.34	17.40
C29H15C12N1	448.36	447.0581	77.69	3.37	15.81	3.12
C29H15C12N3	476.37	475.0643	73.12	3.17	14.88	8.82
C29H15C12N5	504.38	503.0704	69.06	3.00	14.06	13.89
C29H16C11N1	413.91	413.0971	84.15	3.90	8.57	3.38
C29H16C11N3	441.92	441.1033	78.82	3.65	8.02	9.51
C29H16C11N5	469.94	469.1094	74.12	3.43	7.54	14.90
C29H16C12N2	463.37	462.0690	75.17	3.48	15.30	6.05
C29H16C12N4	491.38	490.0752	70.89	3.28	14.43	11.40
C29H16C12N6	519.40	518.0813	67.06	3.11	13.65	16.18
C29H17C11N2	428.93	428.1080	81.21	3.99	8.27	6.53
C29H17C11N4	456.94	456.1142	76.23	3.75	7.76	12.26
C29H17C11N6	484.95	484.1203	71.83	3.53	7.31	17.33
C29H17C12N1	450.37	449.0738	77.34	3.80	15.74	3.11
C29H17C12N3	478.38	477.0799	72.81	3.58	14.82	8.78
C29H17C12N5	506.40	505.0861	68.78	3.38	14.00	13.83
C29H18C11N1	415.93	415.1128	83.75	4.36	8.52	3.37
C29H18C11N3	443.94	443.1189	78.46	4.09	7.99	9.47
C29H18C11N5	471.95	471.1251	73.80	3.84	7.51	14.84
C29H18C12N2	465.39	464.0847	74.85	3.90	15.24	6.02
C29H18C12N4	493.40	492.0908	70.60	3.68	14.37	11.36
C29H18C12N6	521.41	520.0970	66.80	3.48	13.60	16.12
C29H19C11N2	430.94	430.1237	80.83	4.44	8.23	6.50
C29H19C11N4	458.95	458.1298	75.89	4.17	7.72	12.21
C29H19C11N6	486.97	486.1360	71.53	3.93	7.28	17.26
C29H19C12N1	452.39	451.0894	77.00	4.23	15.67	3.10
C29H19C12N3	480.40	479.0956	72.51	3.99	14.76	8.75
C29H19C12N5	508.41	507.1017	68.51	3.77	13.95	13.77
C29H20C11N1	417.94	417.1284	83.34	4.82	8.48	3.35
C29H20C11N3	445.96	445.1346	78.11	4.52	7.95	9.42
C29H20C11N5	473.97	473.1407	73.49	4.25	7.48	14.78
C29H20C12N2	467.40	466.1003	74.52	4.31	15.17	5.99

C29H20C12N4	495.42	494.1065	70.31	4.07	14.31	11.31
C29H20C12N6	523.43	522.1126	66.55	3.85	13.55	16.06
C29H21C11N2	432.96	432.1393	80.45	4.89	8.19	6.47
C29H21C11N4	460.97	460.1455	75.56	4.59	7.69	12.15
C29H21C11N6	488.98	488.1516	71.23	4.33	7.25	17.19
C29H21C12N1	454.40	453.1051	76.66	4.66	15.60	3.08
C29H21C12N3	482.42	481.1112	72.20	4.39	14.70	8.71
C29H21C12N5	510.43	509.1174	68.24	4.15	13.89	13.72
C29H22C11N1	419.96	419.1441	82.94	5.28	8.44	3.34
C29H22C11N3	447.97	447.1502	77.76	4.95	7.91	9.38
C29H22C11N5	475.99	475.1564	73.18	4.66	7.45	14.71
C29H22C12N2	469.42	468.1160	74.20	4.72	15.11	5.97
C29H22C12N4	497.43	496.1221	70.02	4.46	14.25	11.26
C29H22C12N6	525.44	524.1283	66.29	4.22	13.49	15.99
C29H23C11N2	434.97	434.1550	80.08	5.33	8.15	6.44
C29H23C11N4	462.99	462.1611	75.23	5.01	7.66	12.10
C29H23C11N6	491.00	490.1673	70.94	4.72	7.22	17.12
C29H23C12N1	456.42	455.1207	76.32	5.08	15.54	3.07
C29H23C12N3	484.43	483.1269	71.90	4.79	14.64	8.67
C29H23C12N5	512.45	511.1330	67.97	4.52	13.84	13.67
C29H24C11N1	421.97	421.1597	82.55	5.73	8.40	3.32
C29H24C11N3	449.99	449.1659	77.41	5.38	7.88	9.34
C29H24C11N5	478.00	477.1720	72.87	5.06	7.42	14.65
C29H24C12N2	471.43	470.1316	73.89	5.13	15.04	5.94
C29H24C12N4	499.45	498.1378	69.74	4.84	14.20	11.22
C29H24C12N6	527.46	526.1439	66.04	4.59	13.44	15.93
C29H25C11N2	436.99	436.1706	79.71	5.77	8.11	6.41
C29H25C11N4	465.00	464.1768	74.91	5.42	7.62	12.05
C29H25C11N6	493.02	492.1829	70.65	5.11	7.19	17.05
C29H25C12N1	458.44	457.1364	75.98	5.50	15.47	3.06
C29H25C12N3	486.45	485.1425	71.61	5.18	14.58	8.64
C29H25C12N5	514.46	513.1487	67.71	4.90	13.78	13.61
C29H26C11N1	423.99	423.1754	82.15	6.18	8.36	3.30
C29H26C11N3	452.00	451.1815	77.06	5.80	7.84	9.30
C29H26C11N5	480.02	479.1877	72.56	5.46	7.39	14.59
C29H26C12N2	473.45	472.1473	73.57	5.54	14.98	5.92
C29H26C12N4	501.46	500.1534	69.46	5.23	14.14	11.17
C29H26C12N6	529.48	528.1596	65.79	4.95	13.39	15.87
C29H27C11N2	439.00	438.1863	79.34	6.20	8.08	6.38
C29H27C11N4	467.02	466.1924	74.58	5.83	7.59	12.00
C29H27C11N6	495.03	494.1986	70.36	5.50	7.16	16.98
C29H27C12N1	460.45	459.1520	75.65	5.91	15.40	3.04
C29H27C12N3	488.46	487.1582	71.31	5.57	14.52	8.60
C29H27C12N5	516.48	515.1643	67.44	5.27	13.73	13.56
C29H28C11N1	426.01	425.1910	81.76	6.63	8.32	3.29
C29H28C11N3	454.02	453.1972	76.72	6.22	7.81	9.26
C29H28C11N5	482.03	481.2033	72.26	5.86	7.35	14.53
C29H28C12N2	475.47	474.1629	73.26	5.94	14.91	5.89
C29H28C12N4	503.48	502.1691	69.18	5.61	14.08	11.13
C29H28C12N6	531.49	530.1752	65.54	5.31	13.34	15.81
C29H29C11N2	441.02	440.2019	78.98	6.63	8.04	6.35
C29H29C11N4	469.03	468.2081	74.26	6.23	7.56	11.95
C29H29C11N6	497.05	496.2142	70.08	5.88	7.13	16.91
C29H29C12N1	462.47	461.1677	75.32	6.32	15.33	3.03
C29H29C12N3	490.48	489.1738	71.02	5.96	14.46	8.57
C29H29C12N5	518.49	517.1800	67.18	5.64	13.68	13.51
C29H30C11N1	428.02	427.2067	81.38	7.06	8.28	3.27
C29H30C11N3	456.04	455.2128	76.38	6.63	7.77	9.21
C29H30C11N5	484.05	483.2190	71.96	6.25	7.32	14.47
C29H30C12N2	477.48	476.1786	72.95	6.33	14.85	5.87

C29H30C12N4	505.50	504.1847	68.91	5.98	14.03	11.08
C29H30C12N6	533.51	532.1909	65.29	5.67	13.29	15.75
C29H31C11N2	443.04	442.2176	78.62	7.05	8.00	6.32
C29H31C11N4	471.05	470.2237	73.95	6.63	7.53	11.89
C29H31C11N6	499.06	498.2299	69.80	6.26	7.10	16.84
C29H31C12N1	464.48	463.1833	74.99	6.73	15.27	3.02
C29H31C12N3	492.50	491.1895	70.73	6.34	14.40	8.53
C29H31C12N5	520.51	519.1956	66.92	6.00	13.62	13.45
C29H32C11N1	430.04	429.2223	81.00	7.50	8.24	3.26
C29H32C11N3	458.05	457.2285	76.04	7.04	7.74	9.17
C29H32C11N5	486.06	485.2346	71.66	6.64	7.29	14.41
C29H32C12N2	479.50	478.1942	72.64	6.73	14.79	5.84
C29H32C12N4	507.51	506.2004	68.63	6.36	13.97	11.04
C29H32C12N6	535.52	534.2065	65.04	6.02	13.24	15.69
C29H33C11N2	445.05	444.2332	78.27	7.47	7.97	6.29
C29H33C11N4	473.07	472.2394	73.63	7.03	7.49	11.84
C29H33C11N6	501.08	500.2455	69.51	6.64	7.08	16.77
C29H33C12N1	466.50	465.1990	74.67	7.13	15.20	3.00
C29H33C12N3	494.51	493.2051	70.44	6.73	14.34	8.50
C29H33C12N5	522.53	521.2113	66.66	6.37	13.57	13.40
C29H34C11N1	432.05	431.2380	80.62	7.93	8.21	3.24
C29H34C11N3	460.07	459.2441	75.71	7.45	7.71	9.13
C29H34C11N5	488.08	487.2503	71.37	7.02	7.26	14.35
C29H34C12N2	481.51	480.2099	72.34	7.12	14.73	5.82
C29H34C12N4	509.53	508.2160	68.36	6.73	13.92	11.00
C29H34C12N6	537.54	536.2222	64.80	6.38	13.19	15.63
C29H35C11N2	447.07	446.2489	77.91	7.89	7.93	6.27
C29H35C11N4	475.08	474.2550	73.32	7.43	7.46	11.79
C29H35C11N6	503.10	502.2612	69.24	7.01	7.05	16.70
C29H35C12N1	468.52	467.2146	74.35	7.53	15.13	2.99
C29H35C12N3	496.53	495.2208	70.15	7.11	14.28	8.46
C29H35C12N5	524.54	523.2269	66.41	6.73	13.52	13.35
C29H36C11N1	434.07	433.2536	80.25	8.36	8.17	3.23
C29H36C11N3	462.08	461.2598	75.38	7.85	7.67	9.09
C29H36C11N5	490.10	489.2659	71.07	7.40	7.23	14.29
C29H36C12N2	483.53	482.2255	72.04	7.50	14.66	5.79
C29H36C12N4	511.54	510.2317	68.09	7.09	13.86	10.95
C29H36C12N6	539.56	538.2378	64.56	6.73	13.14	15.58
C29H37C11N2	449.08	448.2645	77.56	8.30	7.89	6.24
C29H37C11N4	477.10	476.2707	73.01	7.82	7.43	11.74
C29H37C11N6	505.11	504.2768	68.96	7.38	7.02	16.64
C29H37C12N1	470.53	469.2303	74.03	7.93	15.07	2.98
C29H37C12N3	498.54	497.2364	69.87	7.48	14.22	8.43
C29H37C12N5	526.56	525.2426	66.15	7.08	13.47	13.30
C29H38C11N1	436.09	435.2693	79.87	8.78	8.13	3.21
C29H38C11N3	464.10	463.2754	75.05	8.25	7.64	9.05
C29H38C11N5	492.11	491.2816	70.78	7.78	7.20	14.23
C29H38C12N2	485.55	484.2412	71.74	7.89	14.60	5.77
C29H38C12N4	513.56	512.2473	67.83	7.46	13.81	10.91
C29H38C12N6	541.57	540.2535	64.32	7.07	13.09	15.52
C29H39C11N2	451.10	450.2802	77.22	8.71	7.86	6.21
C29H39C11N4	479.11	478.2863	72.70	8.20	7.40	11.69
C29H39C11N6	507.13	506.2925	68.69	7.75	6.99	16.57
C29H39C12N1	472.55	471.2459	73.71	8.32	15.01	2.96
C29H39C12N3	500.56	499.2521	69.59	7.85	14.17	8.39
C29H39C12N5	528.57	527.2582	65.90	7.44	13.41	13.25
C29H40C11N1	438.10	437.2849	79.51	9.20	8.09	3.20
C29H40C11N3	466.12	465.2911	74.73	8.65	7.61	9.01
C29H40C11N5	494.13	493.2972	70.49	8.16	7.17	14.17
C29H40C12N2	487.56	486.2568	71.44	8.27	14.54	5.75

C29H40C12N4	515.57	514.2630	67.56	7.82	13.75	10.87	C29H50C12N4	525.65	524.3412	66.26	9.59	13.49	10.66
C29H40C12N6	543.59	542.2691	64.08	7.42	13.04	15.46	C29H50C12N6	553.67	552.3474	62.91	9.10	12.81	15.18
C29H41C11N2	453.12	452.2958	76.87	9.12	7.82	6.18	C29H51C11N2	463.20	462.3741	75.20	11.10	7.65	6.05
C29H41C11N4	481.13	480.3020	72.40	8.59	7.37	11.64	C29H51C11N4	491.21	490.3802	70.91	10.47	7.22	11.41
C29H41C11N6	509.14	508.3081	68.41	8.12	6.96	16.51	C29H51C11N6	519.22	518.3864	67.09	9.90	6.83	16.19
C29H41C12N1	474.56	473.2616	73.40	8.71	14.94	2.95	C29H51C12N1	484.64	483.3398	71.87	10.61	14.63	2.89
C29H41C12N3	502.58	501.2677	69.31	8.22	14.11	8.36	C29H51C12N3	512.66	511.3460	67.94	10.03	13.83	8.20
C29H41C12N5	530.59	529.2739	65.65	7.79	13.36	13.20	C29H51C12N5	540.67	539.3521	64.42	9.51	13.11	12.95
C29H42C11N1	440.12	439.3006	79.14	9.62	8.06	3.18	C29H52C11N1	450.20	449.3788	77.37	11.64	7.87	3.11
C29H42C11N3	468.13	467.3067	74.41	9.04	7.57	8.98	C29H52C11N3	478.21	477.3850	72.84	10.96	7.41	8.79
C29H42C11N5	496.14	495.3129	70.21	8.53	7.15	14.12	C29H52C11N5	506.22	505.3911	68.81	10.35	7.00	13.83
C29H42C12N2	489.58	488.2725	71.15	8.65	14.48	5.72	C29H52C12N2	499.66	498.3507	69.71	10.49	14.19	5.61
C29H42C12N4	517.59	516.2786	67.30	8.18	13.70	10.82	C29H52C12N4	527.67	526.3569	66.01	9.93	13.44	10.62
C29H42C12N6	545.60	544.2848	63.84	7.76	13.00	15.40	C29H52C12N6	555.68	554.3630	62.68	9.43	12.76	15.12
C29H43C11N2	455.13	454.3115	76.53	9.52	7.79	6.15	C29H53C11N2	465.21	464.3897	74.87	11.48	7.62	6.02
C29H43C11N4	483.15	482.3176	72.09	8.97	7.34	11.60	C29H53C11N4	493.23	492.3959	70.62	10.83	7.19	11.36
C29H43C11N6	511.15	510.3238	68.14	8.48	6.94	16.44	C29H53C11N6	521.24	520.4020	66.83	10.25	6.80	16.12
C29H43C12N1	476.58	475.2772	73.09	9.09	14.88	2.94	C29H53C12N1	486.66	485.3555	71.57	10.98	14.57	2.88
C29H43C12N3	504.59	503.2834	69.03	8.59	14.05	8.33	C29H53C12N3	514.67	513.3616	67.68	10.38	13.78	8.16
C29H43C12N5	532.61	531.2895	65.40	8.14	13.31	13.15	C29H53C12N5	542.69	541.3678	64.19	9.84	13.07	12.90
C29H44C11N1	442.13	441.3162	78.78	10.03	8.02	3.17	C29H54C11N1	452.21	451.3945	77.03	12.04	7.84	3.10
C29H44C11N3	470.15	469.3224	74.09	9.43	7.54	8.94	C29H54C11N3	480.23	479.4006	72.53	11.33	7.38	8.75
C29H44C11N5	498.16	497.3285	69.92	8.90	7.12	14.06	C29H54C11N5	508.24	507.4068	68.54	10.71	6.98	13.78
C29H44C12N2	491.59	490.2881	70.86	9.02	14.42	5.70	C29H54C12N2	501.67	500.3664	69.43	10.85	14.13	5.58
C29H44C12N4	519.61	518.2943	67.04	8.54	13.65	10.78	C29H54C12N4	529.69	528.3725	65.76	10.28	13.39	10.58
C29H44C12N6	547.62	546.3004	63.61	8.10	12.95	15.35	C29H54C12N6	557.70	556.3787	62.46	9.76	12.71	15.07
C29H45C11N2	457.15	456.3271	76.19	9.92	7.76	6.13	C29H55C11N2	467.23	466.4054	74.55	11.87	7.59	6.00
C29H45C11N4	485.16	484.3333	71.80	9.35	7.31	11.55	C29H55C11N4	495.24	494.4115	70.33	11.19	7.16	11.31
C29H45C11N6	513.18	512.3394	67.88	8.84	6.91	16.38	C29H55C11N6	523.25	522.4176	66.57	10.59	6.78	16.06
C29H45C12N1	478.59	477.2929	72.78	9.48	14.82	2.93	C29H55C12N1	488.67	487.3711	71.28	11.34	14.51	2.87
C29H45C12N3	506.61	505.2990	68.76	8.95	14.00	8.29	C29H55C12N3	516.69	515.3773	67.41	10.73	13.72	8.13
C29H45C12N5	534.62	533.3052	65.15	8.48	13.26	13.10	C29H55C12N5	544.70	543.3834	63.95	10.18	13.02	12.86
C29H46C11N1	444.15	443.3319	78.42	10.44	7.98	3.15	C29H56C11N1	454.23	453.4101	76.68	12.43	7.81	3.08
C29H46C11N3	472.16	471.3380	73.77	9.82	7.51	8.90	C29H56C11N3	482.24	481.4163	72.23	11.70	7.35	8.71
C29H46C11N5	500.18	499.3442	69.64	9.27	7.09	14.00	C29H56C11N5	510.26	509.4224	68.26	11.06	6.95	13.73
C29H46C12N2	493.61	492.3038	70.57	9.39	14.36	5.68	C29H56C12N2	503.69	502.3820	69.15	11.21	14.08	5.56
C29H46C12N4	521.62	520.3099	66.78	8.89	13.59	10.74	C29H56C12N4	531.70	530.3882	65.51	10.62	13.34	10.54
C29H46C12N6	549.64	548.3161	63.37	8.44	12.90	15.29	C29H56C12N6	559.72	558.3943	62.23	10.08	12.67	15.01
C29H47C11N2	459.16	458.3428	75.86	10.32	7.72	6.10	C29H57C11N2	469.24	468.4210	74.23	12.24	7.56	5.97
C29H47C11N4	487.18	486.3489	71.50	9.72	7.28	11.50	C29H57C11N4	497.26	496.4272	70.05	11.55	7.13	11.27
C29H47C11N6	515.19	514.3551	67.61	9.20	6.88	16.31	C29H57C11N6	525.27	524.4333	66.31	10.94	6.75	16.00
C29H47C12N1	480.61	479.3085	72.48	9.86	14.75	2.91	C29H57C12N1	490.69	489.3868	70.99	11.71	14.45	2.85
C29H47C12N3	508.62	507.3147	68.48	9.31	13.94	8.26	C29H57C12N3	518.70	517.3929	67.15	11.08	13.67	8.10
C29H47C12N5	536.64	535.3208	64.91	8.83	13.21	13.05	C29H57C12N5	546.72	545.3991	63.71	10.51	12.97	12.81
C29H48C11N1	446.17	445.3475	78.07	10.84	7.95	3.14	C29H58C11N1	456.25	455.4258	76.35	12.81	7.77	3.07
C29H48C11N3	474.18	473.3537	73.46	10.20	7.48	8.86	C29H58C11N3	484.26	483.4319	71.93	12.07	7.32	8.68
C29H48C11N5	502.19	501.3598	69.36	9.63	7.06	13.95	C29H58C11N5	512.27	511.4380	68.00	11.41	6.92	13.67
C29H48C12N2	495.63	494.3194	70.28	9.76	14.31	5.65	C29H58C12N2	505.71	504.3977	68.88	11.56	14.02	5.54
C29H48C12N4	523.64	522.3256	66.52	9.24	13.54	10.70	C29H58C12N4	533.72	532.4038	65.26	10.95	13.29	10.50
C29H48C12N6	551.65	550.3317	63.14	8.77	12.85	15.23	C29H58C12N6	561.73	560.4100	62.01	10.41	12.62	14.96
C29H49C11N2	461.18	460.3584	75.53	10.71	7.69	6.07	C29H59C11N2	471.26	470.4367	73.91	12.62	7.52	5.94
C29H49C11N4	489.19	488.3646	71.20	10.10	7.25	11.45	C29H59C11N4	499.27	498.4428	69.77	11.91	7.10	11.22
C29H49C11N6	517.21	516.3707	67.35	9.55	6.85	16.25	C29H59C11N6	527.29	526.4489	66.06	11.28	6.72	15.94
C29H49C12N1	482.63	481.3242	72.17	10.23	14.69	2.90	C29H59C12N1	492.71	491.4024	70.70	12.07	14.39	2.84
C29H49C12N3	510.64	509.3303	68.21	9.67	13.89	8.23	C29H59C12N3	520.72	519.4086	66.89	11.42	13.62	8.07
C29H49C12N5	538.65	537.3365	64.67	9.17	13.16	13.00	C29H59C12N5	548.73	547.4147	63.48	10.84	12.92	12.76
C29H50C11N1	448.18	447.3632	77.72	11.25	7.91	3.13	C29H60C11N1	458.26	457.4414	76.01	13.20	7.74	3.06
C29H50C11N3	476.19	475.3693	73.15	10.58	7.45	8.82	C29H60C11N3	486.27	485.4476	71.63	12.44	7.29	8.64
C29H50C11N5	504.21	503.3755	69.08	10.00	7.03	13.89	C29H60C11N5	514.29	513.4537	67.73	11.76	6.89	13.62
C29H50C12N2	497.64	496.3351	69.99	10.13	14.25	5.63	C29H60C12N2	507.72	506.4133	68.61	11.91	13.97	5.52

C29H60C12N4	535.73	534.4195	65.02	11.29	13.24	10.46	C30H22C11N3	459.98	459.1502	78.34	4.82	7.71	9.14
C29H60C12N6	563.75	562.4256	61.79	10.73	12.58	14.91	C30H22C11N5	488.00	487.1564	73.84	4.54	7.27	14.35
C29H61C11N2	473.28	472.4523	73.60	12.99	7.49	5.92	C30H22C12N2	481.43	480.1160	74.85	4.61	14.73	5.82
C29H61C11N4	501.29	500.4584	69.49	12.27	7.07	11.18	C30H22C12N4	509.44	508.1221	70.73	4.35	13.92	11.00
C29H61C11N6	529.30	528.4646	65.81	11.62	6.70	15.88	C30H22C12N6	537.46	536.1283	67.04	4.13	13.19	15.64
C29H61C12N3	522.74	521.4242	66.63	11.76	13.56	8.04	C30H23C11N2	446.98	446.1550	80.61	5.19	7.93	6.27
C29H61C12N5	550.75	549.4304	63.25	11.16	12.87	12.72	C30H23C11N4	475.00	474.1611	75.86	4.88	7.46	11.80
C29H62C11N3	488.29	487.4632	71.34	12.80	7.26	8.61	C30H23C11N6	503.01	502.1673	71.64	4.61	7.05	16.71
C29H62C11N5	516.30	515.4693	67.46	12.10	6.87	13.56	C30H23C12N1	468.43	467.1207	76.92	4.95	15.14	2.99
C29H62C12N4	537.75	536.4351	64.77	11.62	13.19	10.42	C30H23C12N3	496.44	495.1269	72.58	4.67	14.28	8.46
C29H62C12N6	565.76	564.4413	61.57	11.05	12.53	14.85	C30H23C12N5	524.46	523.1330	68.71	4.42	13.52	13.35
C29H63C11N4	503.31	502.4741	69.21	12.62	7.04	11.13	C30H24C11N1	433.99	433.1597	83.03	5.57	8.17	3.23
C29H63C11N6	531.32	530.4802	65.56	11.95	6.67	15.82	C30H24C11N3	462.00	461.1659	77.99	5.24	7.67	9.10
C29H63C12N5	552.76	551.4460	63.01	11.49	12.83	12.67	C30H24C11N5	490.01	489.1720	73.54	4.94	7.24	14.29
C29H64C11N5	518.32	517.4850	67.20	12.45	6.84	13.51	C30H24C12N2	483.45	482.1316	74.53	5.00	14.67	5.79
C29H64C12N6	567.78	566.4569	61.35	11.36	12.49	14.80	C30H24C12N4	511.46	510.1378	70.45	4.73	13.86	10.95
C29H65C11N6	533.33	532.4959	65.31	12.28	6.65	15.76	C30H24C12N6	539.47	538.1439	66.79	4.48	13.14	15.58
C30H15C11N2	438.92	438.0924	82.10	3.44	8.08	6.38	C30H25C11N2	449.00	448.1706	80.25	5.61	7.90	6.24
C30H15C11N4	466.93	466.0985	77.17	3.24	7.59	12.00	C30H25C11N4	477.01	476.1768	75.54	5.28	7.43	11.75
C30H15C11N6	494.95	494.1047	72.80	3.05	7.16	16.98	C30H25C11N6	505.03	504.1829	71.35	4.99	7.02	16.64
C30H15C12N1	460.37	459.0581	78.27	3.28	15.40	3.04	C30H25C12N1	470.45	469.1364	76.59	5.36	15.07	2.98
C30H15C12N3	488.38	487.0643	73.78	3.10	14.52	8.60	C30H25C12N3	498.46	497.1425	72.29	5.06	14.23	8.43
C30H15C12N5	516.39	515.0704	69.78	2.93	13.73	13.56	C30H25C12N5	526.47	525.1487	68.44	4.79	13.47	13.30
C30H16C11N1	425.92	425.0971	84.60	3.79	8.32	3.29	C30H26C11N1	436.00	435.1754	82.65	6.01	8.13	3.21
C30H16C11N3	453.94	453.1033	79.38	3.55	7.81	9.26	C30H26C11N3	464.01	463.1815	77.66	5.65	7.64	9.06
C30H16C11N5	481.95	481.1094	74.77	3.35	7.36	14.53	C30H26C11N5	492.03	491.1877	73.23	5.33	7.21	14.23
C30H16C12N2	475.38	474.0690	75.80	3.39	14.92	5.89	C30H26C12N2	485.46	484.1473	74.23	5.40	14.61	5.77
C30H16C12N4	503.39	502.0752	71.58	3.20	14.09	11.13	C30H26C12N4	513.47	512.1534	70.18	5.10	13.81	10.91
C30H16C12N6	531.41	530.0813	67.81	3.03	13.34	15.81	C30H26C12N6	541.49	540.1596	66.55	4.84	13.09	15.52
C30H17C11N2	440.94	440.1080	81.72	3.89	8.04	6.35	C30H27C11N2	451.02	450.1863	79.89	6.03	7.86	6.21
C30H17C11N4	468.95	468.1142	76.84	3.65	7.56	11.95	C30H27C11N4	479.03	478.1924	75.22	5.68	7.40	11.70
C30H17C11N6	496.96	496.1203	72.51	3.45	7.13	16.91	C30H27C11N6	507.04	506.1986	71.07	5.37	6.99	16.57
C30H17C12N1	462.38	461.0738	77.93	3.71	15.33	3.03	C30H27C12N1	472.46	471.1520	76.27	5.76	15.01	2.96
C30H17C12N3	490.40	489.0799	73.48	3.49	14.46	8.57	C30H27C12N3	500.48	499.1582	72.00	5.44	14.17	8.40
C30H17C12N5	518.41	517.0861	69.51	3.31	13.68	13.51	C30H27C12N5	528.49	527.1643	68.18	5.15	13.42	13.25
C30H18C11N1	427.94	427.1128	84.20	4.24	8.28	3.27	C30H28C11N1	438.02	437.1910	82.26	6.44	8.09	3.20
C30H18C11N3	455.95	455.1189	79.03	3.98	7.78	9.22	C30H28C11N3	466.03	465.1972	77.32	6.06	7.61	9.02
C30H18C11N5	483.96	483.1251	74.45	3.75	7.33	14.47	C30H28C11N5	494.04	493.2033	72.94	5.71	7.18	14.18
C30H18C12N2	477.40	476.0847	75.48	3.80	14.85	5.87	C30H28C12N2	487.48	486.1629	73.92	5.79	14.55	5.75
C30H18C12N4	505.41	504.0908	71.30	3.59	14.03	11.09	C30H28C12N4	515.49	514.1691	69.90	5.48	13.76	10.87
C30H18C12N6	533.42	532.0970	67.55	3.40	13.29	15.75	C30H28C12N6	543.50	542.1752	66.30	5.19	13.05	15.46
C30H19C11N2	442.95	442.1237	81.35	4.32	8.00	6.32	C30H29C11N2	453.03	452.2019	79.54	6.45	7.83	6.18
C30H19C11N4	470.97	470.1298	76.51	4.07	7.53	11.90	C30H29C11N4	481.05	480.2081	74.91	6.08	7.37	11.65
C30H19C11N6	498.98	498.1360	72.21	3.84	7.11	16.84	C30H29C11N6	509.06	508.2142	70.78	5.74	6.96	16.51
C30H19C12N1	464.40	463.0894	77.59	4.12	15.27	3.02	C30H29C12N1	474.48	473.1677	75.94	6.16	14.94	2.95
C30H19C12N3	492.41	491.0956	73.18	3.89	14.40	8.53	C30H29C12N3	502.49	501.1738	71.71	5.82	14.11	8.36
C30H19C12N5	520.43	519.1017	69.24	3.68	13.62	13.46	C30H29C12N5	530.51	529.1800	67.92	5.51	13.37	13.20
C30H20C11N1	429.95	429.1284	83.81	4.69	8.25	3.26	C30H30C11N1	440.03	439.2067	81.89	6.87	8.06	3.18
C30H20C11N3	457.97	457.1346	78.68	4.40	7.74	9.18	C30H30C11N3	468.05	467.2128	76.99	6.46	7.57	8.98
C30H20C11N5	485.98	485.1407	74.15	4.15	7.30	14.41	C30H30C11N5	496.06	495.2190	72.64	6.10	7.15	14.12
C30H20C12N2	479.41	478.1003	75.16	4.21	14.79	5.84	C30H30C12N2	489.49	488.1786	73.61	6.18	14.49	5.72
C30H20C12N4	507.43	506.1065	71.01	3.97	13.97	11.04	C30H30C12N4	517.51	516.1847	69.63	5.84	13.70	10.83
C30H20C12N6	535.44	534.1126	67.30	3.77	13.24	15.70	C30H30C12N6	545.52	544.1909	66.05	5.54	13.00	15.41
C30H21C11N2	444.97	444.1393	80.98	4.76	7.97	6.30	C30H31C11N2	455.05	454.2176	79.19	6.87	7.79	6.16
C30H21C11N4	472.98	472.1455	76.18	4.48	7.50	11.85	C30H31C11N4	483.06	482.2237	74.59	6.47	7.34	11.60
C30H21C11N6	501.00	500.1516	71.92	4.23	7.08	16.77	C30H31C11N6	511.07	510.2299	70.51	6.11	6.94	16.44
C30H21C12N1	466.41	465.1051	77.26	4.54	15.20	3.00	C30H31C12N1	476.49	475.1833	75.62	6.56	14.88	2.94
C30H21C12N3	494.43	493.1112	72.88	4.28	14.34	8.50	C30H31C12N3	504.51	503.1895	71.42	6.19	14.05	8.33
C30H21C12N5	522.44	521.1174	68.97	4.05	13.57	13.41	C30H31C12N5	532.52	531.1956	67.67	5.87	13.32	13.15
C30H22C11N1	431.97	431.1441	83.42	5.13	8.21	3.24	C30H32C11N1	442.05	441.2223	81.51	7.30	8.02	3.17

C30H32Cl1N3	470.06	469.2285	76.66	6.86	7.54	8.94
C30H32Cl1N5	498.08	497.2346	72.35	6.48	7.12	14.06
C30H32Cl2N2	491.51	490.1942	73.31	6.56	14.43	5.70
C30H32Cl2N4	519.52	518.2004	69.36	6.21	13.65	10.78
C30H32Cl2N6	547.54	546.2065	65.81	5.89	12.95	15.35
C30H33Cl1N2	457.06	456.2332	78.84	7.28	7.76	6.13
C30H33Cl1N4	485.08	484.2394	74.28	6.86	7.31	11.55
C30H33Cl1N6	513.09	512.2455	70.23	6.48	6.91	16.38
C30H33Cl2N1	478.51	477.1990	75.30	6.95	14.82	2.93
C30H33Cl2N3	506.52	505.2051	71.14	6.57	14.00	8.30
C30H33Cl2N5	534.54	533.2113	67.41	6.22	13.26	13.10
C30H34Cl1N1	444.07	443.2380	81.14	7.72	7.98	3.15
C30H34Cl1N3	472.08	471.2441	76.33	7.26	7.51	8.90
C30H34Cl1N5	500.09	499.2503	72.05	6.85	7.09	14.00
C30H34Cl2N2	493.52	492.2099	73.01	6.94	14.37	5.68
C30H34Cl2N4	521.54	520.2160	69.09	6.57	13.60	10.74
C30H34Cl2N6	549.55	548.2222	65.57	6.24	12.90	15.29
C30H35Cl1N2	459.08	458.2489	78.49	7.68	7.72	6.10
C30H35Cl1N4	487.09	486.2550	73.98	7.24	7.28	11.50
C30H35Cl1N6	515.11	514.2612	69.95	6.85	6.88	16.32
C30H35Cl2N1	480.53	479.2146	74.99	7.34	14.76	2.91
C30H35Cl2N3	508.54	507.2208	70.86	6.94	13.94	8.26
C30H35Cl2N5	536.55	535.2269	67.16	6.58	13.22	13.05
C30H36Cl1N1	446.08	445.2536	80.78	8.13	7.95	3.14
C30H36Cl1N3	474.09	473.2598	76.00	7.65	7.48	8.86
C30H36Cl1N5	502.11	501.2659	71.76	7.23	7.06	13.95
C30H36Cl2N2	495.54	494.2255	72.72	7.32	14.31	5.65
C30H36Cl2N4	523.55	522.2317	68.82	6.93	13.54	10.70
C30H36Cl2N6	551.57	550.2378	65.33	6.58	12.86	15.24
C30H37Cl1N2	461.10	460.2645	78.15	8.09	7.69	6.08
C30H37Cl1N4	489.11	488.2707	73.67	7.63	7.25	11.45
C30H37Cl1N6	517.12	516.2768	69.68	7.21	6.86	16.25
C30H37Cl2N1	482.54	481.2303	74.67	7.73	14.69	2.90
C30H37Cl2N3	510.56	509.2364	70.58	7.30	13.89	8.23
C30H37Cl2N5	538.57	537.2426	66.91	6.92	13.17	13.00
C30H38Cl1N1	448.10	447.2693	80.41	8.55	7.91	3.13
C30H38Cl1N3	476.11	475.2754	75.68	8.04	7.45	8.83
C30H38Cl1N5	504.12	503.2816	71.48	7.60	7.03	13.89
C30H38Cl2N2	497.56	496.2412	72.42	7.70	14.25	5.63
C30H38Cl2N4	525.57	524.2473	68.56	7.29	13.49	10.66
C30H38Cl2N6	553.58	552.2535	65.09	6.92	12.81	15.18
C30H39Cl1N2	463.11	462.2802	77.81	8.49	7.66	6.05
C30H39Cl1N4	491.13	490.2863	73.37	8.00	7.22	11.41
C30H39Cl1N6	519.14	518.2925	69.41	7.57	6.83	16.19
C30H39Cl2N1	484.56	483.2459	74.36	8.11	14.63	2.89
C30H39Cl2N3	512.57	511.2521	70.30	7.67	13.83	8.20
C30H39Cl2N5	540.58	539.2582	66.66	7.27	13.12	12.96
C30H40Cl1N1	450.11	449.2849	80.05	8.96	7.88	3.11
C30H40Cl1N3	478.13	477.2911	75.36	8.43	7.41	8.79
C30H40Cl1N5	506.14	505.2972	71.19	7.97	7.00	13.84
C30H40Cl2N2	499.57	498.2568	72.13	8.07	14.19	5.61
C30H40Cl2N4	527.59	526.2630	68.30	7.64	13.44	10.62
C30H40Cl2N6	555.60	554.2691	64.86	7.26	12.76	15.13
C30H41Cl1N2	465.13	464.2958	77.47	8.89	7.62	6.02
C30H41Cl1N4	493.14	492.3020	73.07	8.38	7.19	11.36
C30H41Cl1N6	521.15	520.3081	69.14	7.93	6.80	16.13
C30H41Cl2N1	486.57	485.2616	74.06	8.49	14.57	2.88
C30H41Cl2N3	514.59	513.2677	70.02	8.03	13.78	8.17
C30H41Cl2N5	542.60	541.2739	66.41	7.62	13.07	12.91
C30H42Cl1N1	452.13	451.3006	79.70	9.36	7.84	3.10
C30H42Cl1N3	480.14	479.3067	75.05	8.82	7.38	8.75
C30H42Cl1N5	508.16	507.3129	70.91	8.33	6.98	13.78
C30H42Cl2N2	501.59	500.2725	71.84	8.44	14.14	5.58
C30H42Cl2N4	529.60	528.2786	68.04	7.99	13.39	10.58
C30H42Cl2N6	557.62	556.2848	64.62	7.59	12.72	15.07
C30H43Cl1N2	467.14	466.3115	77.14	9.28	7.59	6.00
C30H43Cl1N4	495.16	494.3176	72.77	8.75	7.16	11.31
C30H43Cl1N6	523.17	522.3238	68.88	8.28	6.78	16.06
C30H43Cl2N1	488.59	487.2772	73.75	8.87	14.51	2.87
C30H43Cl2N3	516.60	515.2834	69.75	8.39	13.73	8.13
C30H43Cl2N5	544.62	543.2895	66.16	7.96	13.02	12.86
C30H44Cl1N1	454.14	453.3162	79.34	9.77	7.81	3.08
C30H44Cl1N3	482.16	481.3224	74.73	9.20	7.35	8.72
C30H44Cl1N5	510.17	509.3285	70.63	8.69	6.95	13.73
C30H44Cl2N2	503.60	502.2881	71.55	8.81	14.08	5.56
C30H44Cl2N4	531.62	530.2943	67.78	8.34	13.34	10.54
C30H44Cl2N6	559.63	558.3004	64.39	7.92	12.67	15.02
C30H45Cl1N2	469.16	468.3271	76.80	9.67	7.56	5.97
C30H45Cl1N4	497.17	496.3333	72.48	9.12	7.13	11.27
C30H45Cl1N6	525.19	524.3394	68.61	8.64	6.75	16.00
C30H45Cl2N1	490.61	489.2929	73.45	9.25	14.45	2.85
C30H45Cl2N3	518.62	517.2990	69.48	8.75	13.67	8.10
C30H45Cl2N5	546.63	545.3052	65.92	8.30	12.97	12.81
C30H46Cl1N1	456.16	455.3319	78.99	10.16	7.77	3.07
C30H46Cl1N3	484.17	483.3380	74.42	9.58	7.32	8.68
C30H46Cl1N5	512.19	511.3442	70.35	9.05	6.92	13.67
C30H46Cl2N2	505.62	504.3038	71.27	9.17	14.02	5.54
C30H46Cl2N4	533.63	532.3099	67.52	8.69	13.29	10.50
C30H46Cl2N6	561.65	560.3161	64.16	8.26	12.62	14.96
C30H47Cl1N2	471.18	470.3428	76.48	10.05	7.52	5.95
C30H47Cl1N4	499.19	498.3489	72.18	9.49	7.10	11.22
C30H47Cl1N6	527.20	526.3551	68.35	8.99	6.72	15.94
C30H47Cl2N1	492.62	491.3085	73.15	9.62	14.39	2.84
C30H47Cl2N3	520.64	519.3147	69.21	9.10	13.62	8.07
C30H47Cl2N5	548.65	547.3208	65.68	8.63	12.92	12.76
C30H48Cl1N1	458.18	457.3475	78.65	10.56	7.74	3.06
C30H48Cl1N3	486.19	485.3537	74.11	9.95	7.29	8.64
C30H48Cl1N5	514.20	513.3598	70.08	9.41	6.89	13.62
C30H48Cl2N2	507.64	506.3194	70.98	9.53	13.97	5.52
C30H48Cl2N4	535.65	534.3256	67.27	9.03	13.24	10.46
C30H48Cl2N6	563.66	562.3317	63.93	8.58	12.58	14.91
C30H49Cl1N2	473.19	472.3584	76.15	10.44	7.49	5.92
C30H49Cl1N4	501.20	500.3646	71.89	9.85	7.07	11.18
C30H49Cl1N6	529.22	528.3707	68.09	9.33	6.70	15.88
C30H49Cl2N1	494.64	493.3242	72.85	9.99	14.33	2.83
C30H49Cl2N3	522.65	521.3303	68.94	9.45	13.57	8.04
C30H49Cl2N5	550.66	549.3365	65.44	8.97	12.88	12.72
C30H50Cl1N1	460.19	459.3632	78.30	10.95	7.70	3.04
C30H50Cl1N3	488.21	487.3693	73.81	10.32	7.26	8.61
C30H50Cl1N5	516.22	515.3755	69.80	9.76	6.87	13.57
C30H50Cl2N2	509.65	508.3351	70.70	9.89	13.91	5.50
C30H50Cl2N4	537.67	536.3412	67.02	9.37	13.19	10.42
C30H50Cl2N6	565.68	564.3474	63.70	8.91	12.53	14.86
C30H51Cl1N2	475.21	474.3741	75.83	10.82	7.46	5.89
C30H51Cl1N4	503.22	502.3802	71.61	10.22	7.05	11.13
C30H51Cl1N6	531.23	530.3864	67.83	9.68	6.67	15.82
C30H51Cl2N1	496.65	495.3398	72.55	10.35	14.28	2.82
C30H51Cl2N3	524.67	523.3460	68.68	9.80	13.51	8.01
C30H51Cl2N5	552.68	551.3521	65.20	9.30	12.83	12.67
C30H52Cl1N1	462.21	461.3788	77.96	11.34	7.67	3.03

C30H52Cl1N3	490.22	489.3850	73.50	10.69	7.23	8.57
C30H52Cl1N5	518.24	517.3911	69.53	10.11	6.84	13.51
C30H52Cl2N2	511.67	510.3507	70.42	10.24	13.86	5.47
C30H52Cl2N4	539.68	538.3569	66.77	9.71	13.14	10.38
C30H52Cl2N6	567.70	566.3630	63.47	9.23	12.49	14.80
C30H53Cl1N2	477.22	476.3897	75.51	11.19	7.43	5.87
C30H53Cl1N4	505.24	504.3959	71.32	10.57	7.02	11.09
C30H53Cl1N6	533.25	532.4020	67.57	10.02	6.65	15.76
C30H53Cl2N1	498.67	497.3555	72.26	10.71	14.22	2.81
C30H53Cl2N3	526.68	525.3616	68.42	10.14	13.46	7.98
C30H53Cl2N5	554.70	553.3678	64.96	9.63	12.78	12.63
C30H54Cl1N1	464.22	463.3945	77.62	11.73	7.64	3.02
C30H54Cl1N3	492.24	491.4006	73.20	11.06	7.20	8.54
C30H54Cl1N5	520.25	519.4068	69.26	10.46	6.81	13.46
C30H54Cl2N2	513.68	512.3664	70.15	10.60	13.80	5.45
C30H54Cl2N4	541.70	540.3725	66.52	10.05	13.09	10.34
C30H54Cl2N6	569.71	568.3787	63.25	9.55	12.45	14.75
C30H55Cl1N2	479.24	478.4054	75.19	11.57	7.40	5.85
C30H55Cl1N4	507.25	506.4115	71.04	10.93	6.99	11.05
C30H55Cl1N6	535.27	534.4176	67.32	10.36	6.62	15.70
C30H55Cl2N1	500.69	499.3711	71.97	11.07	14.16	2.80
C30H55Cl2N3	528.70	527.3773	68.15	10.49	13.41	7.95
C30H55Cl2N5	556.71	555.3834	64.73	9.96	12.74	12.58
C30H56Cl1N1	466.24	465.4101	77.29	12.11	7.60	3.00
C30H56Cl1N3	494.25	493.4163	72.90	11.42	7.17	8.50
C30H56Cl1N5	522.27	521.4224	68.99	10.81	6.79	13.41
C30H56Cl2N2	515.70	514.3820	69.87	10.95	13.75	5.43
C30H56Cl2N4	543.71	542.3882	66.27	10.38	13.04	10.30
C30H56Cl2N6	571.73	570.3943	63.03	9.87	12.40	14.70
C30H57Cl1N2	481.26	480.4210	74.87	11.94	7.37	5.82
C30H57Cl1N4	509.27	508.4272	70.76	11.28	6.96	11.00
C30H57Cl1N6	537.28	536.4333	67.07	10.69	6.60	15.64
C30H57Cl2N1	502.70	501.3868	71.68	11.43	14.10	2.79
C30H57Cl2N3	530.71	529.3929	67.90	10.83	13.36	7.92
C30H57Cl2N5	558.73	557.3991	64.49	10.28	12.69	12.53
C30H58Cl1N1	468.26	467.4258	76.95	12.49	7.57	2.99
C30H58Cl1N3	496.27	495.4319	72.61	11.78	7.14	8.47
C30H58Cl1N5	524.28	523.4380	68.73	11.15	6.76	13.36
C30H58Cl2N2	517.72	516.3977	69.60	11.29	13.70	5.41
C30H58Cl2N4	545.73	544.4038	66.03	10.71	12.99	10.27
C30H58Cl2N6	573.74	572.4100	62.80	10.19	12.36	14.65
C30H59Cl1N2	483.27	482.4367	74.56	12.31	7.34	5.80
C30H59Cl1N4	511.28	510.4428	70.48	11.63	6.93	10.96
C30H59Cl1N6	539.30	538.4489	66.82	11.03	6.57	15.58
C30H59Cl2N1	504.72	503.4024	71.39	11.78	14.05	2.78
C30H59Cl2N3	532.73	531.4086	67.64	11.16	13.31	7.89
C30H59Cl2N5	560.74	559.4147	64.26	10.61	12.64	12.49
C30H60Cl1N1	470.27	469.4414	76.62	12.86	7.54	2.98
C30H60Cl1N3	498.29	497.4476	72.31	12.14	7.11	8.43
C30H60Cl1N5	526.30	525.4537	68.47	11.49	6.74	13.31
C30H60Cl2N2	519.73	518.4133	69.33	11.64	13.64	5.39
C30H60Cl2N4	547.75	546.4195	65.79	11.04	12.95	10.23
C30H60Cl2N6	575.76	574.4256	62.58	10.50	12.32	14.60
C30H61Cl1N2	485.29	484.4523	74.25	12.67	7.31	5.77
C30H61Cl1N4	513.30	512.4584	70.20	11.98	6.91	10.92
C30H61Cl1N6	541.31	540.4646	66.57	11.36	6.55	15.53
C30H61Cl2N1	506.73	505.4181	71.11	12.13	13.99	2.76
C30H61Cl2N3	534.75	533.4242	67.38	11.50	13.26	7.86
C30H61Cl2N5	562.76	561.4304	64.03	10.93	12.60	12.44
C30H62Cl1N1	472.29	471.4571	76.30	13.23	7.51	2.97
C30H62Cl1N3	500.30	499.4632	72.02	12.49	7.09	8.40
C30H62Cl1N5	528.32	527.4693	68.20	11.83	6.71	13.26
C30H62Cl2N2	521.75	520.4290	69.06	11.98	13.59	5.37
C30H62Cl2N4	549.76	548.4351	65.54	11.37	12.90	10.19
C30H62Cl2N6	577.77	576.4413	62.37	10.82	12.27	14.55
C30H63Cl1N2	487.30	486.4680	73.94	13.03	7.28	5.75
C30H63Cl1N4	515.32	514.4741	69.92	12.32	6.88	10.87
C30H63Cl1N6	543.33	542.4802	66.32	11.69	6.53	15.47
C30H63Cl2N3	536.76	535.4399	67.13	11.83	13.21	7.83
C30H63Cl2N5	564.78	563.4460	63.80	11.24	12.55	12.40
C30H64Cl1N3	502.32	501.4788	71.73	12.84	7.06	8.37
C30H64Cl1N5	530.33	529.4850	67.95	12.16	6.69	13.21
C30H64Cl2N4	551.78	550.4508	65.30	11.69	12.85	10.15
C30H64Cl2N6	579.79	578.4569	62.15	11.13	12.23	14.49
C30H65Cl1N4	517.33	516.4897	69.65	12.66	6.85	10.83
C30H65Cl1N6	545.35	544.4959	66.07	12.01	6.50	15.41
C30H65Cl2N5	566.79	565.4617	63.57	11.56	12.51	12.36
C30H66Cl1N5	532.35	531.5006	67.69	12.50	6.66	13.16
C30H66Cl2N6	581.81	580.4726	61.93	11.43	12.19	14.44
C30H67Cl1N6	547.36	546.5115	65.83	12.34	6.48	15.35

4. CARBON, HYDROGEN, CHLORINE AND OXYGEN

4. KOHLENSTOFF, WASSERSTOFF, CHLOR UND SAUERSTOFF

4. CARBONE, HYDROGENE, CHLORE ET OXYGENE

MOLECULAR WEIGHTS AND PERCENTAGE COMPOSITIONS OF COMPOUNDS CONTAINING CARBON, HYDROGEN, CHLORINE AND OXYGEN

FORMULA	MOL. WEIGHT	ISOMOL. WEIGHT	PERC. C	PERC. H	PERC. Cl	PERC. O	FORMULA	MOL. WEIGHT	ISOMOL. WEIGHT	PERC. C	PERC. H	PERC. Cl	PERC. O
C 1H 0C1201	98.92	97.9326	12.14	0	71.68	16.17	C 2H 2C1206	192.94	191.9228	12.45	1.04	36.75	49.75
C 1H 0C1202	114.92	113.9275	10.45	0	61.70	27.85	C 2H 2C1207	208.94	207.9178	11.50	.96	33.94	53.60
C 1H 0C1203	130.92	129.9224	9.17	0	54.16	36.66	C 2H 2C1208	224.94	223.9127	10.68	.90	31.52	56.90
C 1H 0C1204	146.91	145.9174	8.18	0	48.26	43.56	C 2H 2C1209	240.94	239.9076	9.97	.84	29.43	59.76
C 1H 0C1205	162.91	161.9123	7.37	0	43.52	49.10	C 2H 3C1101	78.50	77.9872	30.60	3.85	45.16	20.38
C 1H 0C1206	178.91	177.9072	6.71	0	39.63	53.66	C 2H 3C1102	94.50	93.9822	25.42	3.20	37.52	33.86
C 1H 0C1207	194.91	193.9021	6.16	0	36.38	57.46	C 2H 3C1103	110.50	109.9771	21.74	2.74	32.08	43.44
C 1H 0C1208	210.91	209.8970	5.69	0	33.62	60.69	C 2H 3C1104	126.50	125.9720	18.99	2.39	28.03	50.59
C 1H 0C1209	226.91	225.8919	5.29	0	31.25	63.46	C 2H 3C1105	142.50	141.9669	16.86	2.12	24.88	56.14
C 1H 1C1101	64.47	63.9716	18.63	1.56	54.99	24.82	C 2H 3C1106	158.50	157.9618	15.16	1.91	22.37	60.57
C 1H 1C1102	80.47	79.9665	14.93	1.25	44.06	39.76	C 2H 3C1107	174.50	173.9567	13.77	1.73	20.32	64.18
C 1H 1C1103	96.47	95.9614	12.45	1.04	36.75	49.75	C 2H 3C1108	190.49	189.9516	12.61	1.59	18.61	67.19
C 1H 1C1104	112.47	111.9563	10.68	.90	31.52	56.90	C 2H 3C1109	206.49	205.9466	11.63	1.46	17.17	69.73
C 1H 1C1105	128.47	127.9512	9.35	.78	27.60	62.27	C 2H 4C1201	114.96	113.9639	20.90	3.51	61.68	13.92
C 1H 1C1106	144.47	143.9462	8.31	.70	24.54	66.45	C 2H 4C1202	130.96	129.9588	18.34	3.08	54.14	24.43
C 1H 1C1107	160.47	159.9411	7.49	.63	22.09	69.79	C 2H 4C1203	146.96	145.9537	16.35	2.74	48.25	32.66
C 1H 1C1108	176.47	175.9360	6.81	.57	20.09	72.53	C 2H 4C1204	162.96	161.9487	14.74	2.47	43.51	39.27
C 1H 1C1109	192.47	191.9309	6.24	.52	18.42	74.82	C 2H 4C1205	178.96	177.9436	13.42	2.25	39.62	44.70
C 1H 2C1201	100.93	99.9483	11.90	2.00	70.25	15.85	C 2H 4C1206	194.96	193.9385	12.32	2.07	36.37	49.24
C 1H 2C1202	116.93	115.9432	10.27	1.72	60.64	27.37	C 2H 4C1207	210.96	209.9334	11.39	1.91	33.61	53.09
C 1H 2C1203	132.93	131.9381	9.04	1.52	53.34	36.11	C 2H 4C1208	226.96	225.9283	10.58	1.78	31.24	56.40
C 1H 2C1204	148.93	147.9330	8.06	1.35	47.61	42.97	C 2H 4C1209	242.95	241.9232	9.89	1.66	29.18	59.27
C 1H 2C1205	164.93	163.9279	7.28	1.22	42.99	48.50	C 2H 5C1101	80.51	80.0029	29.84	6.26	44.03	19.87
C 1H 2C1206	180.93	179.9228	6.64	1.11	39.19	53.06	C 2H 5C1102	96.51	95.9978	24.89	5.22	36.73	33.15
C 1H 2C1207	196.93	195.9178	6.10	1.02	36.01	56.87	C 2H 5C1103	112.51	111.9927	21.35	4.48	31.51	42.66
C 1H 2C1208	212.93	211.9127	5.64	.95	33.30	60.11	C 2H 5C1104	128.51	127.9876	18.69	3.92	27.59	49.80
C 1H 2C1209	228.93	227.9076	5.25	.88	30.97	62.90	C 2H 5C1105	144.51	143.9825	16.62	3.49	24.53	55.36
C 1H 3C1101	66.49	65.9872	18.07	4.55	53.32	24.06	C 2H 5C1106	160.51	159.9775	14.97	3.14	22.09	59.81
C 1H 3C1102	82.49	81.9822	14.56	3.67	42.98	38.79	C 2H 5C1107	176.51	175.9724	13.61	2.86	20.09	63.45
C 1H 3C1103	98.49	97.9771	12.20	3.07	36.00	48.74	C 2H 5C1108	192.51	191.9673	12.48	2.62	18.42	66.49
C 1H 3C1104	114.49	113.9720	10.49	2.64	30.97	55.90	C 2H 5C1109	208.51	207.9622	11.52	2.42	17.00	69.06
C 1H 3C1105	130.49	129.9669	9.21	2.32	27.17	61.31	C 3H 1C1101	88.49	87.9716	40.72	1.14	40.06	18.08
C 1H 3C1106	146.48	145.9618	8.20	2.06	24.20	65.53	C 3H 1C1102	104.49	103.9665	34.48	.96	33.93	30.62
C 1H 3C1107	162.48	161.9567	7.39	1.86	21.82	68.93	C 3H 1C1103	120.49	119.9614	29.91	.84	29.42	39.83
C 1H 3C1108	178.48	177.9516	6.73	1.69	19.86	71.71	C 3H 1C1104	136.49	135.9563	26.40	.74	25.97	46.89
C 1H 3C1109	194.48	193.9466	6.18	1.55	18.23	74.04	C 3H 1C1105	152.49	151.9512	23.63	.66	23.25	52.46
C 2H 1C1101	76.48	75.9716	31.41	1.32	46.35	20.92	C 3H 1C1106	168.49	167.9462	21.39	.60	21.04	56.97
C 2H 1C1102	92.48	91.9665	25.98	1.09	38.33	34.60	C 3H 1C1107	184.49	183.9411	19.53	.55	19.22	60.71
C 2H 1C1103	108.48	107.9614	22.14	.93	32.68	44.25	C 3H 1C1108	200.49	199.9360	17.97	.50	17.68	63.84
C 2H 1C1104	124.48	123.9563	19.30	.81	28.48	51.41	C 3H 1C1109	216.49	215.9309	16.64	.47	16.38	66.51
C 2H 1C1105	140.48	139.9512	17.10	.72	25.24	56.95	C 3H 2C1201	124.95	123.9483	28.84	1.61	56.75	12.80
C 2H 1C1106	156.48	155.9462	15.35	.64	22.66	61.35	C 3H 2C1202	140.95	139.9432	25.56	1.43	50.30	22.70
C 2H 1C1107	172.48	171.9411	13.93	.58	20.55	64.93	C 3H 2C1203	156.95	155.9381	22.96	1.28	45.18	30.58
C 2H 1C1108	188.48	187.9360	12.75	.53	18.81	67.91	C 3H 2C1204	172.95	171.9330	20.83	1.17	41.00	37.00
C 2H 1C1109	204.48	203.9309	11.75	.49	17.34	70.42	C 3H 2C1205	188.95	187.9279	19.07	1.07	37.53	42.34
C 2H 2C1201	112.94	111.9483	21.27	1.78	62.78	14.17	C 3H 2C1206	204.95	203.9228	17.58	.98	34.60	46.84
C 2H 2C1202	128.94	127.9432	18.63	1.56	54.99	24.82	C 3H 2C1207	220.95	219.9178	16.31	.91	32.09	50.69
C 2H 2C1203	144.94	143.9381	16.57	1.39	48.92	33.12	C 3H 2C1208	236.95	235.9127	15.21	.85	29.92	54.02
C 2H 2C1204	160.94	159.9330	14.93	1.25	44.06	39.76	C 3H 2C1209	252.95	251.9076	14.25	.80	28.03	56.93
C 2H 2C1205	176.94	175.9279	13.58	1.14	40.07	45.21	C 3H 3C1101	90.51	89.9872	39.81	3.34	39.17	17.68

068

C 3H 3C1102	106.51	105.9822	33.83	2.84	33.29	30.04	C 4H 3C1108	214.52	213.9516	22.40	1.41	16.53	59.67
C 3H 3C1103	122.51	121.9771	29.41	2.47	28.94	39.18	C 4H 3C1109	230.52	229.9466	20.84	1.31	15.38	62.47
C 3H 3C1104	138.51	137.9720	26.02	2.18	25.60	46.20	C 4H 4C1201	138.98	137.9639	34.57	2.90	51.02	11.51
C 3H 3C1105	154.51	153.9669	23.32	1.96	22.95	51.78	C 4H 4C1202	154.98	153.9588	31.00	2.60	45.75	20.65
C 3H 3C1106	170.51	169.9618	21.13	1.77	20.79	56.30	C 4H 4C1203	170.98	169.9537	28.10	2.36	41.47	28.07
C 3H 3C1107	186.51	185.9567	19.32	1.62	19.01	60.05	C 4H 4C1204	186.98	185.9487	25.70	2.16	37.92	34.23
C 3H 3C1108	202.51	201.9516	17.79	1.49	17.51	63.21	C 4H 4C1205	202.98	201.9436	23.67	1.99	34.93	39.41
C 3H 3C1109	218.50	217.9466	16.49	1.38	16.23	65.90	C 4H 4C1206	218.98	217.9385	21.94	1.84	32.38	43.84
C 3H 4C1201	126.97	125.9639	28.38	3.18	55.84	12.60	C 4H 4C1207	234.98	233.9334	20.45	1.72	30.18	47.66
C 3H 4C1202	142.97	141.9588	25.20	2.82	49.59	22.38	C 4H 4C1208	250.98	249.9283	19.14	1.61	28.25	51.00
C 3H 4C1203	158.97	157.9537	22.67	2.54	44.60	30.19	C 4H 4C1209	266.98	265.9232	18.00	1.51	26.56	53.94
C 3H 4C1204	174.97	173.9487	20.59	2.30	40.52	36.58	C 4H 5C1101	104.54	104.0029	45.96	4.82	33.91	15.31
C 3H 4C1205	190.97	189.9436	18.87	2.11	37.13	41.89	C 4H 5C1102	120.54	119.9978	39.86	4.18	29.41	26.55
C 3H 4C1206	206.97	205.9385	17.41	1.95	34.26	46.38	C 4H 5C1103	136.54	135.9927	35.19	3.69	25.97	35.15
C 3H 4C1207	222.97	221.9334	16.16	1.81	31.80	50.23	C 4H 5C1104	152.54	151.9876	31.50	3.30	23.24	41.96
C 3H 4C1208	238.97	237.9283	15.08	1.69	29.67	53.56	C 4H 5C1105	168.53	167.9825	28.51	2.99	21.04	47.47
C 3H 4C1209	254.97	253.9232	14.13	1.58	27.81	56.48	C 4H 5C1106	184.53	183.9775	26.04	2.73	19.21	52.02
C 3H 5C1101	92.53	92.0029	38.94	5.45	38.32	17.29	C 4H 5C1107	200.53	199.9724	23.96	2.51	17.68	55.85
C 3H 5C1102	108.53	107.9978	33.20	4.64	32.67	29.49	C 4H 5C1108	216.53	215.9673	22.19	2.33	16.37	59.11
C 3H 5C1103	124.52	123.9927	28.94	4.05	28.47	38.55	C 4H 5C1109	232.53	231.9622	20.66	2.17	15.25	61.92
C 3H 5C1104	140.52	139.9876	25.64	3.59	25.23	45.54	C 4H 6C1201	141.00	139.9796	34.07	4.29	50.29	11.35
C 3H 5C1105	156.52	155.9825	23.02	3.22	22.65	51.11	C 4H 6C1202	157.00	155.9745	30.60	3.85	45.16	20.38
C 3H 5C1106	172.52	171.9775	20.89	2.92	20.55	55.64	C 4H 6C1203	173.00	171.9694	27.77	3.50	40.99	27.75
C 3H 5C1107	188.52	187.9724	19.11	2.67	18.81	59.41	C 4H 6C1204	189.00	187.9643	25.42	3.20	37.52	33.86
C 3H 5C1108	204.52	203.9673	17.62	2.46	17.33	62.58	C 4H 6C1205	205.00	203.9592	23.44	2.95	34.59	39.02
C 3H 5C1109	220.52	219.9622	16.34	2.29	16.08	65.30	C 4H 6C1206	220.99	219.9541	21.74	2.74	32.08	43.44
C 3H 6C1201	128.99	127.9796	27.94	4.69	54.97	12.40	C 4H 6C1207	236.99	235.9491	20.27	2.55	29.92	47.26
C 3H 6C1202	144.99	143.9745	24.85	4.17	48.91	22.07	C 4H 6C1208	252.99	251.9440	18.99	2.39	28.03	50.59
C 3H 6C1203	160.99	159.9694	22.38	3.76	44.04	29.82	C 4H 6C1209	268.99	267.9389	17.86	2.25	26.36	53.53
C 3H 6C1204	176.98	175.9643	20.36	3.42	40.06	36.16	C 4H 7C1101	106.55	106.0185	45.09	6.62	33.27	15.02
C 3H 6C1205	192.98	191.9592	18.67	3.13	36.74	41.45	C 4H 7C1102	122.55	122.0135	39.20	5.76	28.93	26.11
C 3H 6C1206	208.98	207.9541	17.24	2.89	33.93	45.93	C 4H 7C1103	138.55	138.0084	34.68	5.09	25.59	34.64
C 3H 6C1207	224.98	223.9491	16.02	2.69	31.52	49.78	C 4H 7C1104	154.55	154.0033	31.09	4.57	22.94	41.41
C 3H 6C1208	240.98	239.9440	14.95	2.51	29.42	53.11	C 4H 7C1105	170.55	169.9982	28.17	4.14	20.79	46.91
C 3H 6C1209	256.98	255.9389	14.02	2.35	27.59	56.03	C 4H 7C1106	186.55	185.9931	25.75	3.78	19.00	51.46
C 3H 7C1101	94.54	94.0185	38.11	7.46	37.50	16.92	C 4H 7C1107	202.55	201.9880	23.72	3.48	17.50	55.29
C 3H 7C1102	110.54	110.0135	32.60	6.38	32.07	28.95	C 4H 7C1108	218.55	217.9829	21.98	3.23	16.22	58.57
C 3H 7C1103	126.54	126.0084	28.48	5.58	28.02	37.93	C 4H 7C1109	234.55	233.9779	20.48	3.01	15.12	61.39
C 3H 7C1104	142.54	142.0033	25.28	4.95	24.87	44.90	C 4H 8C1201	143.01	141.9952	33.59	5.64	49.58	11.19
C 3H 7C1105	158.54	157.9982	22.73	4.45	22.36	50.46	C 4H 8C1202	159.01	157.9901	30.21	5.07	44.59	20.12
C 3H 7C1106	174.54	173.9931	20.64	4.04	20.31	55.00	C 4H 8C1203	175.01	173.9850	27.45	4.61	40.51	27.43
C 3H 7C1107	190.54	189.9880	18.91	3.70	18.61	58.78	C 4H 8C1204	191.01	189.9800	25.15	4.22	37.12	33.50
C 3H 7C1108	206.54	205.9829	17.45	3.42	17.17	61.97	C 4H 8C1205	207.01	205.9749	23.21	3.90	34.25	38.64
C 3H 7C1109	222.54	221.9779	16.19	3.17	15.93	64.71	C 4H 8C1206	223.01	221.9698	21.54	3.62	31.79	43.05
C 4H 2C1201	136.97	135.9483	35.08	1.47	51.77	11.68	C 4H 8C1207	239.01	237.9647	20.10	3.37	29.67	46.86
C 4H 2C1202	152.97	151.9432	31.41	1.32	46.35	20.92	C 4H 8C1208	255.01	253.9596	18.84	3.16	27.81	50.19
C 4H 2C1203	168.96	167.9381	28.43	1.19	41.96	28.41	C 4H 8C1209	271.01	269.9545	17.73	2.98	26.16	53.13
C 4H 2C1204	184.96	183.9330	25.98	1.09	38.33	34.60	C 4H 9C1101	108.57	108.0342	44.25	8.36	32.65	14.74
C 4H 2C1205	200.96	199.9279	23.91	1.00	35.28	39.81	C 4H 9C1102	124.57	124.0291	38.57	7.28	28.46	25.69
C 4H 2C1206	216.96	215.9228	22.14	.93	32.68	44.25	C 4H 9C1103	140.57	140.0240	34.18	6.45	25.22	34.15
C 4H 2C1207	232.96	231.9178	20.62	.87	30.44	48.07	C 4H 9C1104	156.57	156.0189	30.69	5.79	22.64	40.88
C 4H 2C1208	248.96	247.9127	19.30	.81	28.48	51.41	C 4H 9C1105	172.57	172.0138	27.84	5.26	20.54	46.36
C 4H 2C1209	264.96	263.9076	18.13	.76	26.76	54.35	C 4H 9C1106	188.57	188.0088	25.48	4.81	18.80	50.91
C 4H 3C1101	102.52	101.9872	46.86	2.95	34.58	15.61	C 4H 9C1107	204.57	204.0037	23.49	4.43	17.33	54.75
C 4H 3C1102	118.52	117.9822	40.54	2.55	29.91	27.00	C 4H 9C1108	220.56	219.9986	21.78	4.11	16.07	58.03
C 4H 3C1103	134.52	133.9771	35.72	2.25	26.36	35.68	C 4H 9C1109	236.56	235.9935	20.31	3.83	14.99	60.87
C 4H 3C1104	150.52	149.9720	31.92	2.01	23.55	42.52	C 5H 2C1201	148.98	147.9483	40.31	1.35	47.60	10.74
C 4H 3C1105	166.52	165.9669	28.85	1.82	21.29	48.04	C 5H 2C1202	164.98	163.9432	36.40	1.22	42.98	19.40
C 4H 3C1106	182.52	181.9618	26.32	1.66	19.42	52.60	C 5H 2C1203	180.98	179.9381	33.18	1.11	39.18	26.52
C 4H 3C1107	198.52	197.9567	24.20	1.52	17.86	56.42	C 5H 2C1204	196.98	195.9330	30.49	1.02	36.00	32.49

069

C 5H 2C1205	212.97	211.9279	28.20	.95	33.29	37.56
C 5H 2C1206	228.97	227.9228	26.23	.88	30.97	41.92
C 5H 2C1207	244.97	243.9178	24.52	.82	28.94	45.72
C 5H 2C1208	260.97	259.9127	23.01	.77	27.17	49.05
C 5H 2C1209	276.97	275.9076	21.68	.73	25.60	51.99
C 5H 3C1101	114.53	113.9872	52.44	2.64	30.95	13.97
C 5H 3C1102	130.53	129.9822	46.01	2.32	27.16	24.51
C 5H 3C1103	146.53	145.9771	40.99	2.06	24.19	32.76
C 5H 3C1104	162.53	161.9720	36.95	1.86	21.81	39.38
C 5H 3C1105	178.53	177.9669	33.64	1.69	19.86	44.81
C 5H 3C1106	194.53	193.9618	30.87	1.55	18.23	49.35
C 5H 3C1107	210.53	209.9567	28.53	1.44	16.84	53.20
C 5H 3C1108	226.53	225.9516	26.51	1.33	15.65	56.50
C 5H 3C1109	242.53	241.9466	24.76	1.25	14.62	59.37
C 5H 4C1201	150.99	149.9639	39.77	2.67	46.96	10.60
C 5H 4C1202	166.99	165.9588	35.96	2.41	42.46	19.16
C 5H 4C1203	182.99	181.9537	32.82	2.20	38.75	26.23
C 5H 4C1204	198.99	197.9487	30.18	2.03	35.63	32.16
C 5H 4C1205	214.99	213.9436	27.93	1.88	32.98	37.21
C 5H 4C1206	230.99	229.9385	26.00	1.75	30.70	41.56
C 5H 4C1207	246.99	245.9334	24.32	1.63	28.71	45.34
C 5H 4C1208	262.99	261.9283	22.84	1.53	26.96	48.67
C 5H 4C1209	278.99	277.9232	21.53	1.45	25.42	51.61
C 5H 5C1101	116.55	116.0029	51.53	4.32	30.42	13.73
C 5H 5C1102	132.55	131.9978	45.31	3.80	26.75	24.14
C 5H 5C1103	148.55	147.9927	40.43	3.39	23.87	32.31
C 5H 5C1104	164.55	163.9876	36.50	3.06	21.55	38.89
C 5H 5C1105	180.55	179.9825	33.26	2.79	19.64	44.31
C 5H 5C1106	196.54	195.9775	30.56	2.56	18.04	48.84
C 5H 5C1107	212.54	211.9724	28.26	2.37	16.68	52.69
C 5H 5C1108	228.54	227.9673	26.28	2.21	15.51	56.00
C 5H 5C1109	244.54	243.9622	24.56	2.06	14.50	58.88
C 5H 6C1201	153.01	151.9796	39.25	3.95	46.34	10.46
C 5H 6C1202	169.01	167.9745	35.53	3.58	41.95	18.93
C 5H 6C1203	185.01	183.9694	32.46	3.27	38.33	25.94
C 5H 6C1204	201.01	199.9643	29.88	3.01	35.28	31.84
C 5H 6C1205	217.01	215.9592	27.67	2.79	32.67	36.86
C 5H 6C1206	233.01	231.9541	25.77	2.60	30.43	41.20
C 5H 6C1207	249.01	247.9491	24.12	2.43	28.48	44.98
C 5H 6C1208	265.00	263.9440	22.66	2.28	26.76	48.30
C 5H 6C1209	281.00	279.9389	21.37	2.15	25.23	51.24
C 5H 7C1101	118.56	118.0185	50.65	5.95	29.90	13.49
C 5H 7C1102	134.56	134.0135	44.63	5.24	26.35	23.78
C 5H 7C1103	150.56	150.0084	39.89	4.69	23.55	31.88
C 5H 7C1104	166.56	166.0033	36.06	4.24	21.29	38.42
C 5H 7C1105	182.56	181.9982	32.90	3.86	19.42	43.82
C 5H 7C1106	198.56	197.9931	30.25	3.55	17.85	48.35
C 5H 7C1107	214.56	213.9880	27.99	3.29	16.52	52.20
C 5H 7C1108	230.56	229.9829	26.05	3.06	15.38	55.51
C 5H 7C1109	246.56	245.9779	24.36	2.86	14.38	58.40
C 5H 8C1201	155.02	153.9952	38.74	5.20	45.74	10.32
C 5H 8C1202	171.02	169.9901	35.12	4.71	41.46	18.71
C 5H 8C1203	187.02	185.9850	32.11	4.31	37.91	25.66
C 5H 8C1204	203.02	201.9800	29.58	3.97	34.93	31.52
C 5H 8C1205	219.02	217.9749	27.42	3.68	32.37	36.52
C 5H 8C1206	235.02	233.9698	25.55	3.43	30.17	40.85
C 5H 8C1207	251.02	249.9647	23.92	3.21	28.25	44.62
C 5H 8C1208	267.02	265.9596	22.49	3.02	26.55	47.93
C 5H 8C1209	283.02	281.9545	21.22	2.85	25.05	50.88
C 5H 9C1101	120.58	120.0342	49.81	7.52	29.40	13.27
C 5H 9C1102	136.58	136.0291	43.97	6.64	25.96	23.43
C 5H 9C1103	152.58	152.0240	39.36	5.95	23.24	31.46
C 5H 9C1104	168.58	168.0189	35.62	5.38	21.03	37.96
C 5H 9C1105	184.58	184.0138	32.54	4.91	19.21	43.34
C 5H 9C1106	200.58	200.0088	29.94	4.52	17.68	47.86
C 5H 9C1107	216.58	216.0037	27.73	4.19	16.37	51.71
C 5H 9C1108	232.58	231.9986	25.82	3.90	15.24	55.03
C 5H 9C1109	248.58	247.9935	24.16	3.65	14.26	57.93
C 5H10C1201	157.04	156.0109	38.24	6.42	45.15	10.19
C 5H10C1202	173.04	172.0058	34.71	5.83	40.98	18.49
C 5H10C1203	189.04	188.0007	31.77	5.33	37.51	25.39
C 5H10C1204	205.04	203.9956	29.29	4.92	34.58	31.21
C 5H10C1205	221.04	219.9905	27.17	4.56	32.08	36.19
C 5H10C1206	237.04	235.9854	25.34	4.25	29.91	40.50
C 5H10C1207	253.04	251.9804	23.73	3.98	28.02	44.26
C 5H10C1208	269.04	267.9753	22.32	3.75	26.36	47.58
C 5H10C1209	285.04	283.9702	21.07	3.54	24.88	50.52
C 5H11C1101	122.60	122.0498	48.99	9.04	28.92	13.05
C 5H11C1102	138.60	138.0448	43.33	8.00	25.58	23.09
C 5H11C1103	154.59	154.0397	38.85	7.17	22.93	31.05
C 5H11C1104	170.59	170.0346	35.20	6.50	20.78	37.51
C 5H11C1105	186.59	186.0295	32.19	5.94	19.00	42.87
C 5H11C1106	202.59	202.0244	29.64	5.47	17.50	47.38
C 5H11C1107	218.59	218.0193	27.47	5.07	16.22	51.24
C 5H11C1108	234.59	234.0142	25.60	4.73	15.11	54.56
C 5H11C1109	250.59	250.0092	23.97	4.42	14.15	57.46
C 6H 3C1101	126.54	125.9872	56.95	2.39	28.02	12.64
C 6H 3C1102	142.54	141.9822	50.56	2.12	24.87	22.45
C 6H 3C1103	158.54	157.9771	45.46	1.91	22.36	30.27
C 6H 3C1104	174.54	173.9720	41.29	1.73	20.31	36.67
C 6H 3C1105	190.54	189.9669	37.82	1.59	18.61	41.98
C 6H 3C1106	206.54	205.9618	34.89	1.46	17.17	46.48
C 6H 3C1107	222.54	221.9567	32.38	1.36	15.93	50.33
C 6H 3C1108	238.54	237.9516	30.21	1.27	14.86	53.66
C 6H 3C1109	254.54	253.9466	28.31	1.19	13.93	56.57
C 6H 4C1201	163.00	161.9639	44.21	2.47	43.50	9.82
C 6H 4C1202	179.00	177.9588	40.26	2.25	39.61	17.88
C 6H 4C1203	195.00	193.9537	36.96	2.07	36.36	24.61
C 6H 4C1204	211.00	209.9487	34.15	1.91	33.60	30.33
C 6H 4C1205	227.00	225.9436	31.75	1.78	31.24	35.24
C 6H 4C1206	243.00	241.9385	29.66	1.66	29.18	39.50
C 6H 4C1207	259.00	257.9334	27.82	1.56	27.38	43.24
C 6H 4C1208	275.00	273.9283	26.21	1.47	25.78	46.54
C 6H 4C1209	291.00	289.9232	24.77	1.39	24.37	49.48
C 6H 5C1101	128.56	128.0029	56.06	3.92	27.58	12.45
C 6H 5C1102	144.56	143.9978	49.85	3.49	24.53	22.14
C 6H 5C1103	160.56	159.9927	44.89	3.14	22.08	29.89
C 6H 5C1104	176.56	175.9876	40.82	2.85	20.08	36.25
C 6H 5C1105	192.56	191.9825	37.43	2.62	18.41	41.54
C 6H 5C1106	208.56	207.9775	34.56	2.42	17.00	46.03
C 6H 5C1107	224.56	223.9724	32.09	2.24	15.79	49.87
C 6H 5C1108	240.55	239.9673	29.96	2.10	14.74	53.21
C 6H 5C1109	256.55	255.9622	28.09	1.96	13.82	56.13
C 6H 6C1201	165.02	163.9796	43.67	3.66	42.97	9.70
C 6H 6C1202	181.02	179.9745	39.81	3.34	39.17	17.68
C 6H 6C1203	197.02	195.9694	36.58	3.07	35.99	24.36
C 6H 6C1204	213.02	211.9643	33.83	2.84	33.29	30.04
C 6H 6C1205	229.02	227.9592	31.47	2.64	30.96	34.93
C 6H 6C1206	245.02	243.9541	29.41	2.47	28.94	39.18
C 6H 6C1207	261.02	259.9491	27.61	2.32	27.17	42.91

070

C 6H 6C1208	277.02	275.9440	26.02	2.18	25.60	46.20
C 6H 6C1209	293.02	291.9389	24.59	2.06	24.20	49.14
C 6H 7C1101	130.58	130.0185	55.19	5.40	27.15	12.25
C 6H 7C1102	146.57	146.0135	49.17	4.81	24.19	21.83
C 6H 7C1103	162.57	162.0084	44.33	4.34	21.81	29.52
C 6H 7C1104	178.57	178.0033	40.36	3.95	19.85	35.84
C 6H 7C1105	194.57	193.9982	37.04	3.63	18.22	41.11
C 6H 7C1106	210.57	209.9931	34.22	3.35	16.84	45.59
C 6H 7C1107	226.57	225.9880	31.81	3.11	15.65	49.43
C 6H 7C1108	242.57	241.9829	29.71	2.91	14.62	52.77
C 6H 7C1109	258.57	257.9779	27.87	2.73	13.71	55.69
C 6H 8C1201	167.04	165.9952	43.14	4.83	42.45	9.58
C 6H 8C1202	183.04	181.9901	39.37	4.41	38.74	17.48
C 6H 8C1203	199.03	197.9850	36.21	4.05	35.62	24.12
C 6H 8C1204	215.03	213.9800	33.51	3.75	32.97	29.76
C 6H 8C1205	231.03	229.9749	31.19	3.49	30.69	34.63
C 6H 8C1206	247.03	245.9698	29.17	3.26	28.70	38.86
C 6H 8C1207	263.03	261.9647	27.40	3.07	26.96	42.58
C 6H 8C1208	279.03	277.9596	25.83	2.89	25.41	45.87
C 6H 8C1209	295.03	293.9545	24.43	2.73	24.03	48.81
C 6H 9C1101	132.59	132.0342	54.35	6.84	26.74	12.07
C 6H 9C1102	148.59	148.0291	48.50	6.11	23.86	21.53
C 6H 9C1103	164.59	164.0240	43.79	5.51	21.54	29.16
C 6H 9C1104	180.59	180.0189	39.91	5.02	19.63	35.44
C 6H 9C1105	196.59	196.0138	36.66	4.61	18.03	40.69
C 6H 9C1106	212.59	212.0088	33.90	4.27	16.68	45.16
C 6H 9C1107	228.59	228.0037	31.53	3.97	15.51	48.99
C 6H 9C1108	244.59	243.9986	29.46	3.71	14.50	52.33
C 6H 9C1109	260.59	259.9935	27.66	3.48	13.61	55.26
C 6H10C1201	169.05	168.0109	42.63	5.96	41.94	9.46
C 6H10C1202	185.05	184.0058	38.94	5.45	38.32	17.29
C 6H10C1203	201.05	200.0007	35.85	5.01	35.27	23.87
C 6H10C1204	217.05	215.9956	33.20	4.64	32.67	29.49
C 6H10C1205	233.05	231.9905	30.92	4.33	30.43	34.33
C 6H10C1206	249.05	247.9854	28.94	4.05	28.47	38.55
C 6H10C1207	265.05	263.9804	27.19	3.80	26.75	42.25
C 6H10C1208	281.05	279.9753	25.64	3.59	25.23	45.54
C 6H10C1209	297.05	295.9702	24.26	3.39	23.87	48.48
C 6H11C1101	134.61	134.0498	53.54	8.24	26.34	11.89
C 6H11C1102	150.61	150.0448	47.85	7.36	23.54	21.25
C 6H11C1103	166.61	166.0397	43.26	6.66	21.28	28.81
C 6H11C1104	182.61	182.0346	39.47	6.07	19.42	35.05
C 6H11C1105	198.60	198.0295	36.29	5.58	17.85	40.28
C 6H11C1106	214.60	214.0244	33.58	5.17	16.52	44.73
C 6H11C1107	230.60	230.0193	31.25	4.81	15.37	48.57
C 6H11C1108	246.60	246.0142	29.22	4.50	14.38	51.90
C 6H11C1109	262.60	262.0092	27.44	4.22	13.50	54.83
C 6H12C1201	171.07	170.0265	42.13	7.07	41.45	9.35
C 6H12C1202	187.07	186.0214	38.52	6.47	37.90	17.11
C 6H12C1203	203.07	202.0163	35.49	5.96	34.92	23.64
C 6H12C1204	219.07	218.0113	32.90	5.52	32.37	29.21
C 6H12C1205	235.07	234.0062	30.66	5.15	30.16	34.03
C 6H12C1206	251.06	250.0011	28.70	4.82	28.24	38.24
C 6H12C1207	267.06	265.9960	26.98	4.53	26.55	41.94
C 6H12C1208	283.06	281.9909	25.46	4.27	25.05	45.22
C 6H12C1209	299.06	297.9858	24.10	4.04	23.71	48.15
C 6H13C1101	136.62	136.0655	52.75	9.59	25.95	11.71
C 6H13C1102	152.62	152.0604	47.22	8.59	23.23	20.97
C 6H13C1103	168.62	168.0553	42.74	7.77	21.03	28.47
C 6H13C1104	184.62	184.0502	39.04	7.10	19.20	34.66
C 6H13C1105	200.62	200.0451	35.92	6.53	17.67	39.87
C 6H13C1106	216.62	216.0401	33.27	6.05	16.37	44.32
C 6H13C1107	232.62	232.0350	30.98	5.63	15.24	48.15
C 6H13C1108	248.62	248.0299	28.99	5.27	14.26	51.48
C 6H13C1109	264.62	264.0248	27.23	4.95	13.40	54.42
C 7H 3C1101	138.55	137.9872	60.68	2.18	25.59	11.55
C 7H 3C1102	154.55	153.9822	54.40	1.96	22.94	20.70
C 7H 3C1103	170.55	169.9771	49.30	1.77	20.79	28.14
C 7H 3C1104	186.55	185.9720	45.07	1.62	19.00	34.31
C 7H 3C1105	202.55	201.9669	41.51	1.49	17.50	39.49
C 7H 3C1106	218.55	217.9618	38.47	1.38	16.22	43.92
C 7H 3C1107	234.55	233.9567	35.85	1.29	15.12	47.75
C 7H 3C1108	250.55	249.9516	33.56	1.21	14.15	51.09
C 7H 3C1109	266.55	265.9466	31.54	1.13	13.30	54.02
C 7H 4C1201	175.02	173.9639	48.04	2.30	40.51	9.14
C 7H 4C1202	191.01	189.9588	44.02	2.11	37.12	16.75
C 7H 4C1203	207.01	205.9537	40.61	1.95	34.25	23.19
C 7H 4C1204	223.01	221.9487	37.70	1.81	31.79	28.70
C 7H 4C1205	239.01	237.9436	35.18	1.69	29.67	33.47
C 7H 4C1206	255.01	253.9385	32.97	1.58	27.80	37.64
C 7H 4C1207	271.01	269.9334	31.02	1.49	26.16	41.33
C 7H 4C1208	287.01	285.9283	29.29	1.40	24.70	44.60
C 7H 4C1209	303.01	301.9232	27.75	1.33	23.40	47.52
C 7H 5C1101	140.57	140.0029	59.81	3.59	25.22	11.38
C 7H 5C1102	156.57	155.9978	53.70	3.22	22.64	20.44
C 7H 5C1103	172.57	171.9927	48.72	2.92	20.54	27.81
C 7H 5C1104	188.57	187.9876	44.59	2.67	18.80	33.94
C 7H 5C1105	204.57	203.9825	41.10	2.46	17.33	39.11
C 7H 5C1106	220.57	219.9775	38.12	2.28	16.07	43.52
C 7H 5C1107	236.57	235.9724	35.54	2.13	14.99	47.34
C 7H 5C1108	252.57	251.9673	33.29	2.00	14.04	50.68
C 7H 5C1109	268.57	267.9622	31.31	1.88	13.20	53.62
C 7H 6C1201	177.03	175.9796	47.49	3.42	40.05	9.04
C 7H 6C1202	193.03	191.9745	43.56	3.13	36.73	16.58
C 7H 6C1203	209.03	207.9694	40.22	2.89	33.92	22.96
C 7H 6C1204	225.03	223.9643	37.36	2.69	31.51	28.44
C 7H 6C1205	241.03	239.9592	34.88	2.51	29.42	33.19
C 7H 6C1206	257.03	255.9541	32.71	2.35	27.59	37.35
C 7H 6C1207	273.03	271.9491	30.79	2.22	25.97	41.02
C 7H 6C1208	289.03	287.9440	29.09	2.09	24.53	44.28
C 7H 6C1209	305.03	303.9389	27.56	1.98	23.25	47.21
C 7H 7C1101	142.59	142.0185	58.97	4.95	24.86	11.22
C 7H 7C1102	158.59	158.0135	53.02	4.45	22.36	20.18
C 7H 7C1103	174.59	174.0084	48.16	4.04	20.31	27.49
C 7H 7C1104	190.58	190.0033	44.12	3.70	18.60	33.58
C 7H 7C1105	206.58	205.9982	40.70	3.42	17.16	38.72
C 7H 7C1106	222.58	221.9931	37.77	3.17	15.93	43.13
C 7H 7C1107	238.58	237.9880	35.24	2.96	14.86	46.94
C 7H 7C1108	254.58	253.9829	33.03	2.77	13.93	50.28
C 7H 7C1109	270.58	269.9779	31.07	2.61	13.10	53.22
C 7H 8C1201	179.05	177.9952	46.96	4.50	39.60	8.94
C 7H 8C1202	195.05	193.9901	43.11	4.13	36.35	16.41
C 7H 8C1203	211.05	209.9850	39.84	3.82	33.60	22.74
C 7H 8C1204	227.05	225.9800	37.03	3.55	31.23	28.19
C 7H 8C1205	243.04	241.9749	34.59	3.32	29.17	32.91
C 7H 8C1206	259.04	257.9698	32.46	3.11	27.37	37.06
C 7H 8C1207	275.04	273.9647	30.57	2.93	25.78	40.72
C 7H 8C1208	291.04	289.9596	28.89	2.77	24.36	43.98
C 7H 8C1209	307.04	305.9545	27.38	2.63	23.09	46.90
C 7H 9C1101	144.60	144.0342	58.14	6.27	24.52	11.06

C 7H 9C1102	160.60	160.0291	52.35	5.65	22.08	19.92
C 7H 9C1103	176.60	176.0240	47.61	5.14	20.08	27.18
C 7H 9C1104	192.60	192.0189	43.65	4.71	18.41	33.23
C 7H 9C1105	208.60	208.0138	40.31	4.35	17.00	38.35
C 7H 9C1106	224.60	224.0088	37.43	4.04	15.79	42.74
C 7H 9C1107	240.60	240.0037	34.95	3.77	14.74	46.55
C 7H 9C1108	256.60	255.9986	32.77	3.54	13.82	49.88
C 7H 9C1109	272.60	271.9935	30.84	3.33	13.01	52.82
C 7H10C1201	181.06	180.0109	46.44	5.57	39.16	8.84
C 7H10C1202	197.06	196.0058	42.67	5.11	35.98	16.24
C 7H10C1203	213.06	212.0007	39.46	4.73	33.28	22.53
C 7H10C1204	229.06	227.9956	36.71	4.40	30.96	27.94
C 7H10C1205	245.06	243.9905	34.31	4.11	28.93	32.64
C 7H10C1206	261.06	259.9854	32.21	3.86	27.16	36.77
C 7H10C1207	277.06	275.9804	30.35	3.64	25.59	40.42
C 7H10C1208	293.06	291.9753	28.69	3.44	24.20	43.68
C 7H10C1209	309.06	307.9702	27.20	3.26	22.94	46.59
C 7H11C1101	146.62	146.0498	57.34	7.56	24.18	10.91
C 7H11C1102	162.62	162.0448	51.70	6.82	21.80	19.68
C 7H11C1103	178.62	178.0397	47.07	6.21	19.85	26.87
C 7H11C1104	194.62	194.0346	43.20	5.70	18.22	32.88
C 7H11C1105	210.62	210.0295	39.92	5.26	16.83	37.98
C 7H11C1106	226.62	226.0244	37.10	4.89	15.64	42.36
C 7H11C1107	242.61	242.0193	34.65	4.57	14.61	46.16
C 7H11C1108	258.61	258.0142	32.51	4.29	13.71	49.49
C 7H11C1109	274.61	274.0092	30.62	4.04	12.91	52.44
C 7H12C1201	183.08	182.0265	45.92	6.61	38.73	8.74
C 7H12C1202	199.08	198.0214	42.23	6.08	35.62	16.07
C 7H12C1203	215.08	214.0163	39.09	5.62	32.97	22.32
C 7H12C1204	231.08	230.0113	36.39	5.23	30.68	27.70
C 7H12C1205	247.08	246.0062	34.03	4.90	28.70	32.38
C 7H12C1206	263.08	262.0011	31.96	4.60	26.95	36.49
C 7H12C1207	279.08	277.9960	30.13	4.33	25.41	40.13
C 7H12C1208	295.07	293.9909	28.49	4.10	24.03	43.38
C 7H12C1209	311.07	309.9858	27.03	3.89	22.79	46.29
C 7H13C1101	148.63	148.0655	56.57	8.82	23.85	10.76
C 7H13C1102	164.63	164.0604	51.07	7.96	21.53	19.44
C 7H13C1103	180.63	180.0553	46.55	7.25	19.63	26.57
C 7H13C1104	196.63	196.0502	42.76	6.66	18.03	32.55
C 7H13C1105	212.63	212.0451	39.54	6.16	16.67	37.62
C 7H13C1106	228.63	228.0401	36.77	5.73	15.51	41.99
C 7H13C1107	244.63	244.0350	34.37	5.36	14.49	45.78
C 7H13C1108	260.63	260.0299	32.26	5.03	13.60	49.11
C 7H13C1109	276.63	276.0248	30.39	4.74	12.82	52.05
C 7H14C1201	185.10	184.0422	45.42	7.62	38.31	8.64
C 7H14C1202	201.09	200.0371	41.81	7.02	35.26	15.91
C 7H14C1203	217.09	216.0320	38.73	6.50	32.66	22.11
C 7H14C1204	233.09	232.0269	36.07	6.05	30.42	27.46
C 7H14C1205	249.09	248.0218	33.75	5.67	28.47	32.12
C 7H14C1206	265.09	264.0167	31.72	5.32	26.75	36.21
C 7H14C1207	281.09	280.0116	29.91	5.02	25.23	39.84
C 7H14C1208	297.09	296.0066	28.30	4.75	23.87	43.08
C 7H14C1209	313.09	312.0015	26.85	4.51	22.65	45.99
C 7H15C1101	150.65	150.0811	55.81	10.04	23.53	10.62
C 7H15C1102	166.65	166.0761	50.45	9.07	21.27	19.20
C 7H15C1103	182.65	182.0710	46.03	8.28	19.41	26.28
C 7H15C1104	198.65	198.0659	42.33	7.61	17.85	32.22
C 7H15C1105	214.65	214.0608	39.17	7.04	16.52	37.27
C 7H15C1106	230.65	230.0557	36.45	6.56	15.37	41.62
C 7H15C1107	246.65	246.0506	34.09	6.13	14.37	45.41
C 7H15C1108	262.65	262.0455	32.01	5.76	13.50	48.73
C 7H15C1109	278.65	278.0404	30.17	5.43	12.72	51.68
C 8H 4C1201	187.03	185.9639	51.38	2.16	37.91	8.55
C 8H 4C1202	203.03	201.9588	47.33	1.99	34.92	15.76
C 8H 4C1203	219.03	217.9537	43.87	1.84	32.37	21.91
C 8H 4C1204	235.02	233.9487	40.88	1.72	30.17	27.23
C 8H 4C1205	251.02	249.9436	38.28	1.61	28.25	31.87
C 8H 4C1206	267.02	265.9385	35.99	1.51	26.55	35.95
C 8H 4C1207	283.02	281.9334	33.95	1.42	25.05	39.57
C 8H 4C1208	299.02	297.9283	32.13	1.35	23.71	42.80
C 8H 4C1209	315.02	313.9232	30.50	1.28	22.51	45.71
C 8H 5C1101	152.58	152.0029	62.98	3.30	23.24	10.49
C 8H 5C1102	168.58	167.9978	57.00	2.99	21.03	18.98
C 8H 5C1103	184.58	183.9927	52.06	2.73	19.21	26.00
C 8H 5C1104	200.58	199.9876	47.91	2.51	17.68	31.91
C 8H 5C1105	216.58	215.9825	44.37	2.33	16.37	36.94
C 8H 5C1106	232.58	231.9775	41.31	2.17	15.24	41.27
C 8H 5C1107	248.58	247.9724	38.66	2.03	14.26	45.05
C 8H 5C1108	264.58	263.9673	36.32	1.90	13.40	48.38
C 8H 5C1109	280.58	279.9622	34.25	1.80	12.64	51.32
C 8H 6C1201	189.04	187.9796	50.83	3.20	37.51	8.46
C 8H 6C1202	205.04	203.9745	46.86	2.95	34.58	15.61
C 8H 6C1203	221.04	219.9694	43.47	2.74	32.08	21.71
C 8H 6C1204	237.04	235.9643	40.54	2.55	29.91	27.00
C 8H 6C1205	253.04	251.9592	37.97	2.39	28.02	31.61
C 8H 6C1206	269.04	267.9541	35.72	2.25	26.36	35.68
C 8H 6C1207	285.04	283.9491	33.71	2.12	24.88	39.29
C 8H 6C1208	301.04	299.9440	31.92	2.01	23.55	42.52
C 8H 6C1209	317.04	315.9389	30.31	1.91	22.37	45.42
C 8H 7C1101	154.60	154.0185	62.15	4.56	22.93	10.35
C 8H 7C1102	170.60	170.0135	56.33	4.14	20.78	18.76
C 8H 7C1103	186.60	186.0084	51.50	3.78	19.00	25.72
C 8H 7C1104	202.60	202.0033	47.43	3.48	17.50	31.59
C 8H 7C1105	218.59	217.9982	43.96	3.23	16.22	36.60
C 8H 7C1106	234.59	233.9931	40.96	3.01	15.11	40.92
C 8H 7C1107	250.59	249.9880	38.34	2.82	14.15	44.69
C 8H 7C1108	266.59	265.9829	36.04	2.65	13.30	48.01
C 8H 7C1109	282.59	281.9779	34.00	2.50	12.55	50.95
C 8H 8C1201	191.06	189.9952	50.29	4.22	37.11	8.37
C 8H 8C1202	207.06	205.9901	46.41	3.89	34.24	15.45
C 8H 8C1203	223.06	221.9850	43.08	3.62	31.79	21.52
C 8H 8C1204	239.06	237.9800	40.20	3.37	29.66	26.77
C 8H 8C1205	255.06	253.9749	37.67	3.16	27.80	31.36
C 8H 8C1206	271.06	269.9698	35.45	2.97	26.16	35.42
C 8H 8C1207	287.05	285.9647	33.47	2.81	24.70	39.02
C 8H 8C1208	303.05	301.9596	31.71	2.66	23.40	42.24
C 8H 8C1209	319.05	317.9545	30.12	2.53	22.22	45.13
C 8H 9C1101	156.61	156.0342	61.35	5.79	22.64	10.22
C 8H 9C1102	172.61	172.0291	55.67	5.26	20.54	18.54
C 8H 9C1103	188.61	188.0240	50.95	4.81	18.80	25.45
C 8H 9C1104	204.61	204.0189	46.96	4.43	17.33	31.28
C 8H 9C1105	220.61	220.0138	43.56	4.11	16.07	36.26
C 8H 9C1106	236.61	236.0088	40.61	3.83	14.98	40.57
C 8H 9C1107	252.61	252.0037	38.04	3.59	14.03	44.34
C 8H 9C1108	268.61	267.9986	35.77	3.38	13.20	47.65
C 8H 9C1109	284.61	283.9935	33.76	3.19	12.46	50.59
C 8H10C1201	193.07	192.0109	49.77	5.22	36.72	8.29
C 8H10C1202	209.07	208.0058	45.96	4.82	33.91	15.31
C 8H10C1203	225.07	224.0007	42.69	4.48	31.50	21.33
C 8H10C1204	241.07	239.9956	39.86	4.18	29.41	26.55

072

C 8H10C1205	257.07	255.9905	37.38	3.92	27.58	31.12
C 8H10C1206	273.07	271.9854	35.19	3.69	25.97	35.15
C 8H10C1207	289.07	287.9804	33.24	3.49	24.53	38.74
C 8H10C1208	305.07	303.9753	31.50	3.30	23.24	41.96
C 8H10C1209	321.07	319.9702	29.93	3.14	22.08	44.85
C 8H11C1101	158.63	158.0498	60.57	6.99	22.35	10.09
C 8H11C1102	174.63	174.0448	55.02	6.35	20.30	18.32
C 8H11C1103	190.63	190.0397	50.41	5.82	18.60	25.18
C 8H11C1104	206.63	206.0346	46.50	5.37	17.16	30.97
C 8H11C1105	222.63	222.0295	43.16	4.98	15.92	35.93
C 8H11C1106	238.63	238.0244	40.27	4.65	14.86	40.23
C 8H11C1107	254.63	254.0193	37.74	4.35	13.92	43.98
C 8H11C1108	270.63	270.0142	35.51	4.10	13.10	47.30
C 8H11C1109	286.62	286.0092	33.52	3.87	12.37	50.24
C 8H12C1201	195.09	194.0265	49.25	6.20	36.35	8.20
C 8H12C1202	211.09	210.0214	45.52	5.73	33.59	15.16
C 8H12C1203	227.09	226.0163	42.31	5.33	31.22	21.14
C 8H12C1204	243.09	242.0113	39.53	4.98	29.17	26.33
C 8H12C1205	259.09	258.0062	37.09	4.67	27.37	30.88
C 8H12C1206	275.09	274.0011	34.93	4.40	25.78	34.90
C 8H12C1207	291.09	289.9960	33.01	4.16	24.36	38.48
C 8H12C1208	307.09	305.9909	31.29	3.94	23.09	41.68
C 8H12C1209	323.09	321.9858	29.74	3.74	21.95	44.57
C 8H13C1101	160.65	160.0655	59.81	8.16	22.07	9.96
C 8H13C1102	176.64	176.0604	54.40	7.42	20.07	18.11
C 8H13C1103	192.64	192.0553	49.88	6.80	18.40	24.92
C 8H13C1104	208.64	208.0502	46.05	6.28	16.99	30.67
C 8H13C1105	224.64	224.0451	42.77	5.83	15.78	35.61
C 8H13C1106	240.64	240.0401	39.93	5.45	14.73	39.89
C 8H13C1107	256.64	256.0350	37.44	5.11	13.81	43.64
C 8H13C1108	272.64	272.0299	35.24	4.81	13.00	46.95
C 8H13C1109	288.64	288.0248	33.29	4.54	12.28	49.89
C 8H14C1201	197.11	196.0422	48.75	7.16	35.97	8.12
C 8H14C1202	213.11	212.0371	45.09	6.62	33.27	15.02
C 8H14C1203	229.10	228.0320	41.94	6.16	30.95	20.95
C 8H14C1204	245.10	244.0269	39.20	5.76	28.93	26.11
C 8H14C1205	261.10	260.0218	36.80	5.40	27.16	30.64
C 8H14C1206	277.10	276.0167	34.68	5.09	25.59	34.64
C 8H14C1207	293.10	292.0116	32.78	4.81	24.19	38.21
C 8H14C1208	309.10	308.0066	31.09	4.57	22.94	41.41
C 8H14C1209	325.10	324.0015	29.56	4.34	21.81	44.29
C 8H15C1101	162.66	162.0811	59.07	9.30	21.80	9.84
C 8H15C1102	178.66	178.0761	53.78	8.46	19.84	17.91
C 8H15C1103	194.66	194.0710	49.36	7.77	18.21	24.66
C 8H15C1104	210.66	210.0659	45.61	7.18	16.83	30.38
C 8H15C1105	226.66	226.0608	42.39	6.67	15.64	35.29
C 8H15C1106	242.66	242.0557	39.60	6.23	14.61	39.56
C 8H15C1107	258.66	258.0506	37.15	5.85	13.71	43.30
C 8H15C1108	274.66	274.0455	34.99	5.50	12.91	46.60
C 8H15C1109	290.66	290.0404	33.06	5.20	12.20	49.54
C 8H16C1201	199.12	198.0578	48.26	8.10	35.61	8.03
C 8H16C1202	215.12	214.0527	44.67	7.50	32.96	14.87
C 8H16C1203	231.12	230.0476	41.58	6.98	30.68	20.77
C 8H16C1204	247.12	246.0426	38.88	6.53	28.69	25.90
C 8H16C1205	263.12	262.0375	36.52	6.13	26.95	30.40
C 8H16C1206	279.12	278.0324	34.43	5.78	25.40	34.39
C 8H16C1207	295.12	294.0273	32.56	5.46	24.03	37.95
C 8H16C1208	311.12	310.0222	30.89	5.18	22.79	41.14
C 8H16C1209	327.12	326.0171	29.37	4.93	21.68	44.02
C 8H17C1101	164.68	164.0968	58.35	10.41	21.53	9.72
C 8H17C1102	180.68	180.0917	53.18	9.48	19.62	17.71
C 8H17C1103	196.68	196.0866	48.86	8.71	18.03	24.40
C 8H17C1104	212.68	212.0815	45.18	8.06	16.67	30.09
C 8H17C1105	228.67	228.0764	42.02	7.49	15.50	34.98
C 8H17C1106	244.67	244.0714	39.27	7.00	14.49	39.23
C 8H17C1107	260.67	260.0663	36.86	6.57	13.60	42.96
C 8H17C1108	276.67	276.0612	34.73	6.19	12.81	46.26
C 8H17C1109	292.67	292.0561	32.83	5.85	12.11	49.20
C 9H 4C1201	199.04	197.9639	54.31	2.03	35.62	8.04
C 9H 4C1202	215.04	213.9588	50.27	1.87	32.97	14.88
C 9H 4C1203	231.04	229.9537	46.79	1.75	30.69	20.78
C 9H 4C1204	247.04	245.9487	43.76	1.63	28.70	25.91
C 9H 4C1205	263.04	261.9436	41.10	1.53	26.96	30.41
C 9H 4C1206	279.03	277.9385	38.74	1.44	25.41	34.40
C 9H 4C1207	295.03	293.9334	36.64	1.37	24.03	37.96
C 9H 4C1208	311.03	309.9283	34.76	1.30	22.80	41.15
C 9H 4C1209	327.03	325.9232	33.05	1.23	21.68	44.03
C 9H 5C1101	164.59	164.0029	65.68	3.06	21.54	9.72
C 9H 5C1102	180.59	179.9978	59.86	2.79	19.63	17.72
C 9H 5C1103	196.59	195.9927	54.99	2.56	18.03	24.42
C 9H 5C1104	212.59	211.9876	50.85	2.37	16.68	30.10
C 9H 5C1105	228.59	227.9825	47.29	2.20	15.51	35.00
C 9H 5C1106	244.59	243.9775	44.20	2.06	14.49	39.25
C 9H 5C1107	260.59	259.9724	41.48	1.93	13.60	42.98
C 9H 5C1108	276.59	275.9673	39.08	1.82	12.82	46.28
C 9H 5C1109	292.59	291.9622	36.95	1.72	12.12	49.21
C 9H 6C1201	201.05	199.9796	53.77	3.01	35.27	7.96
C 9H 6C1202	217.05	215.9745	49.80	2.79	32.67	14.74
C 9H 6C1203	233.05	231.9694	46.38	2.60	30.42	20.60
C 9H 6C1204	249.05	247.9643	43.40	2.43	28.47	25.70
C 9H 6C1205	265.05	263.9592	40.78	2.28	26.75	30.18
C 9H 6C1206	281.05	279.9541	38.46	2.15	25.23	34.16
C 9H 6C1207	297.05	295.9491	36.39	2.04	23.87	37.70
C 9H 6C1208	313.05	311.9440	34.53	1.93	22.65	40.89
C 9H 6C1209	329.05	327.9389	32.85	1.84	21.55	43.76
C 9H 7C1101	166.61	166.0185	64.88	4.23	21.28	9.60
C 9H 7C1102	182.61	182.0135	59.20	3.86	19.41	17.52
C 9H 7C1103	198.61	198.0084	54.43	3.55	17.85	24.17
C 9H 7C1104	214.61	214.0033	50.37	3.29	16.52	29.82
C 9H 7C1105	230.61	229.9982	46.88	3.06	15.37	34.69
C 9H 7C1106	246.61	245.9931	43.84	2.86	14.38	38.93
C 9H 7C1107	262.60	261.9880	41.16	2.69	13.50	42.65
C 9H 7C1108	278.60	277.9829	38.80	2.53	12.73	45.94
C 9H 7C1109	294.60	293.9779	36.69	2.40	12.03	48.88
C 9H 8C1201	203.07	201.9952	53.23	3.97	34.92	7.88
C 9H 8C1202	219.07	217.9901	49.35	3.68	32.37	14.61
C 9H 8C1203	235.07	233.9850	45.99	3.43	30.16	20.42
C 9H 8C1204	251.07	249.9800	43.06	3.21	28.24	25.49
C 9H 8C1205	267.07	265.9749	40.48	3.02	26.55	29.95
C 9H 8C1206	283.07	281.9698	38.19	2.85	25.05	33.91
C 9H 8C1207	299.07	297.9647	36.15	2.70	23.71	37.45
C 9H 8C1208	315.07	313.9596	34.31	2.56	22.51	40.62
C 9H 8C1209	331.06	329.9545	32.65	2.44	21.42	43.49
C 9H 9C1101	168.62	168.0342	64.11	5.38	21.02	9.49
C 9H 9C1102	184.62	184.0291	58.55	4.91	19.20	17.33
C 9H 9C1103	200.62	200.0240	53.88	4.52	17.67	23.92
C 9H 9C1104	216.62	216.0189	49.90	4.19	16.37	29.54
C 9H 9C1105	232.62	232.0138	46.47	3.90	15.24	34.39
C 9H 9C1106	248.62	248.0088	43.48	3.65	14.26	38.61
C 9H 9C1107	264.62	264.0037	40.85	3.43	13.40	42.32

073

C 9H 9Cl1O8	280.62	279.9986	38.52	3.23	12.63	45.61
C 9H 9Cl1O9	296.62	295.9935	36.44	3.06	11.95	48.55
C 9H10Cl2O1	205.09	204.0109	52.71	4.91	34.57	7.80
C 9H10Cl2O2	221.08	220.0058	48.90	4.56	32.07	14.47
C 9H10Cl2O3	237.08	236.0007	45.60	4.25	29.91	20.25
C 9H10Cl2O4	253.08	251.9956	42.71	3.98	28.02	25.29
C 9H10Cl2O5	269.08	267.9905	40.17	3.75	26.35	29.73
C 9H10Cl2O6	285.08	283.9854	37.92	3.54	24.87	33.67
C 9H10Cl2O7	301.08	299.9804	35.90	3.35	23.55	37.20
C 9H10Cl2O8	317.08	315.9753	34.09	3.18	22.36	40.37
C 9H10Cl2O9	333.08	331.9702	32.45	3.03	21.29	43.23
C 9H11Cl1O1	170.64	170.0498	63.35	6.50	20.78	9.38
C 9H11Cl1O2	186.64	186.0448	57.92	5.94	19.00	17.14
C 9H11Cl1O3	202.64	202.0397	53.35	5.47	17.50	23.69
C 9H11Cl1O4	218.64	218.0346	49.44	5.07	16.22	29.27
C 9H11Cl1O5	234.64	234.0295	46.07	4.73	15.11	34.09
C 9H11Cl1O6	250.64	250.0244	43.13	4.42	14.15	38.30
C 9H11Cl1O7	266.64	266.0193	40.54	4.16	13.30	42.00
C 9H11Cl1O8	282.64	282.0142	38.25	3.92	12.54	45.29
C 9H11Cl1O9	298.64	298.0092	36.20	3.71	11.87	48.22
C 9H12Cl2O1	207.10	206.0265	52.20	5.84	34.24	7.73
C 9H12Cl2O2	223.10	222.0214	48.45	5.42	31.78	14.34
C 9H12Cl2O3	239.10	238.0163	45.21	5.06	29.66	20.07
C 9H12Cl2O4	255.10	254.0113	42.38	4.74	27.80	25.09
C 9H12Cl2O5	271.10	270.0062	39.87	4.46	26.16	29.51
C 9H12Cl2O6	287.10	286.0011	37.65	4.21	24.70	33.44
C 9H12Cl2O7	303.10	301.9960	35.67	3.99	23.39	36.95
C 9H12Cl2O8	319.10	317.9909	33.88	3.79	22.22	40.11
C 9H12Cl2O9	335.10	333.9858	32.26	3.61	21.16	42.97
C 9H13Cl1O1	172.66	172.0655	62.61	7.59	20.53	9.27
C 9H13Cl1O2	188.66	188.0604	57.30	6.95	18.79	16.96
C 9H13Cl1O3	204.66	204.0553	52.82	6.40	17.32	23.45
C 9H13Cl1O4	220.65	220.0502	48.99	5.94	16.07	29.00
C 9H13Cl1O5	236.65	236.0451	45.68	5.54	14.98	33.80
C 9H13Cl1O6	252.65	252.0401	42.79	5.19	14.03	38.00
C 9H13Cl1O7	268.65	268.0350	40.24	4.88	13.20	41.69
C 9H13Cl1O8	284.65	284.0299	37.98	4.60	12.45	44.97
C 9H13Cl1O9	300.65	300.0248	35.96	4.36	11.79	47.89
C 9H14Cl2O1	209.12	208.0422	51.69	6.75	33.91	7.65
C 9H14Cl2O2	225.12	224.0371	48.02	6.27	31.50	14.21
C 9H14Cl2O3	241.12	240.0320	44.83	5.85	29.41	19.91
C 9H14Cl2O4	257.12	256.0269	42.04	5.49	27.58	24.89
C 9H14Cl2O5	273.11	272.0218	39.58	5.17	25.96	29.29
C 9H14Cl2O6	289.11	288.0167	37.39	4.88	24.53	33.20
C 9H14Cl2O7	305.11	304.0116	35.43	4.63	23.24	36.71
C 9H14Cl2O8	321.11	320.0066	33.66	4.39	22.08	39.86
C 9H14Cl2O9	337.11	336.0015	32.07	4.19	21.03	42.71
C 9H15Cl1O1	174.67	174.0811	61.89	8.66	20.30	9.16
C 9H15Cl1O2	190.67	190.0761	56.69	7.93	18.59	16.78
C 9H15Cl1O3	206.67	206.0710	52.31	7.32	17.15	23.22
C 9H15Cl1O4	222.67	222.0659	48.55	6.79	15.92	28.74
C 9H15Cl1O5	238.67	238.0608	45.29	6.33	14.85	33.52
C 9H15Cl1O6	254.67	254.0557	42.45	5.94	13.92	37.69
C 9H15Cl1O7	270.67	270.0506	39.94	5.59	13.10	41.38
C 9H15Cl1O8	286.67	286.0455	37.71	5.27	12.37	44.65
C 9H15Cl1O9	302.67	302.0404	35.72	5.00	11.71	47.58
C 9H16Cl2O1	211.13	210.0578	51.20	7.64	33.58	7.58
C 9H16Cl2O2	227.13	226.0527	47.59	7.10	31.22	14.09
C 9H16Cl2O3	243.13	242.0476	44.46	6.63	29.16	19.74
C 9H16Cl2O4	259.13	258.0426	41.72	6.22	27.36	24.70
C 9H16Cl2O5	275.13	274.0375	39.29	5.86	25.77	29.08
C 9H16Cl2O6	291.13	290.0324	37.13	5.54	24.36	32.97
C 9H16Cl2O7	307.13	306.0273	35.20	5.25	23.09	36.47
C 9H16Cl2O8	323.13	322.0222	33.45	4.99	21.94	39.61
C 9H16Cl2O9	339.13	338.0171	31.88	4.76	20.91	42.46
C 9H17Cl1O1	176.69	176.0968	61.18	9.70	20.07	9.06
C 9H17Cl1O2	192.69	192.0917	56.10	8.89	18.40	16.61
C 9H17Cl1O3	208.69	208.0866	51.80	8.21	16.99	23.00
C 9H17Cl1O4	224.69	224.0815	48.11	7.63	15.78	28.48
C 9H17Cl1O5	240.69	240.0764	44.91	7.12	14.73	33.24
C 9H17Cl1O6	256.69	256.0714	42.11	6.68	13.81	37.40
C 9H17Cl1O7	272.68	272.0663	39.64	6.28	13.00	41.07
C 9H17Cl1O8	288.68	288.0612	37.45	5.94	12.28	44.34
C 9H17Cl1O9	304.68	304.0561	35.48	5.62	11.64	47.26
C 9H18Cl2O1	213.15	212.0735	50.72	8.51	33.27	7.51
C 9H18Cl2O2	229.15	228.0684	47.17	7.92	30.94	13.96
C 9H18Cl2O3	245.15	244.0633	44.10	7.40	28.92	19.58
C 9H18Cl2O4	261.15	260.0582	41.39	6.95	27.15	24.51
C 9H18Cl2O5	277.15	276.0531	39.00	6.55	25.58	28.86
C 9H18Cl2O6	293.15	292.0480	36.88	6.19	24.19	32.75
C 9H18Cl2O7	309.15	308.0429	34.97	5.87	22.94	36.23
C 9H18Cl2O8	325.15	324.0379	33.25	5.58	21.81	39.37
C 9H18Cl2O9	341.14	340.0328	31.69	5.32	20.78	42.21
C 9H19Cl1O1	178.70	178.1124	60.49	10.72	19.84	8.95
C 9H19Cl1O2	194.70	194.1073	55.52	9.84	18.21	16.43
C 9H19Cl1O3	210.70	210.1023	51.30	9.09	16.83	22.78
C 9H19Cl1O4	226.70	226.0972	47.68	8.45	15.64	28.23
C 9H19Cl1O5	242.70	242.0921	44.54	7.89	14.61	32.96
C 9H19Cl1O6	258.70	258.0870	41.79	7.40	13.70	37.11
C 9H19Cl1O7	274.70	274.0819	39.35	6.97	12.91	40.77
C 9H19Cl1O8	290.70	290.0768	37.19	6.59	12.20	44.03
C 9H19Cl1O9	306.70	306.0717	35.25	6.24	11.56	46.95
C10H 5Cl1O1	176.60	176.0029	68.01	2.85	20.07	9.06
C10H 5Cl1O2	192.60	191.9978	62.36	2.62	18.41	16.61
C10H 5Cl1O3	208.60	207.9927	57.58	2.42	17.00	23.01
C10H 5Cl1O4	224.60	223.9876	53.48	2.24	15.78	28.49
C10H 5Cl1O5	240.60	239.9825	49.92	2.09	14.74	33.25
C10H 5Cl1O6	256.60	255.9775	46.81	1.96	13.82	37.41
C10H 5Cl1O7	272.60	271.9724	44.06	1.85	13.01	41.08
C10H 5Cl1O8	288.60	287.9673	41.62	1.75	12.28	44.35
C10H 5Cl1O9	304.60	303.9622	39.43	1.65	11.64	47.27
C10H 6Cl2O1	213.06	211.9796	56.37	2.84	33.28	7.51
C10H 6Cl2O2	229.06	227.9745	52.44	2.64	30.95	13.97
C10H 6Cl2O3	245.06	243.9694	49.01	2.47	28.93	19.59
C10H 6Cl2O4	261.06	259.9643	46.01	2.32	27.16	24.51
C10H 6Cl2O5	277.06	275.9592	43.35	2.18	25.59	28.87
C10H 6Cl2O6	293.06	291.9541	40.99	2.06	24.19	32.76
C10H 6Cl2O7	309.06	307.9491	38.86	1.96	22.94	36.24
C10H 6Cl2O8	325.06	323.9440	36.95	1.86	21.81	39.38
C10H 6Cl2O9	341.06	339.9389	35.22	1.77	20.79	42.22
C10H 7Cl1O1	178.62	178.0185	67.24	3.95	19.85	8.96
C10H 7Cl1O2	194.62	194.0135	61.72	3.63	18.22	16.44
C10H 7Cl1O3	210.62	210.0084	57.03	3.35	16.83	22.79
C10H 7Cl1O4	226.62	226.0033	53.00	3.11	15.64	28.24
C10H 7Cl1O5	242.62	241.9982	49.51	2.91	14.61	32.97
C10H 7Cl1O6	258.62	257.9931	46.44	2.73	13.71	37.12
C10H 7Cl1O7	274.62	273.9880	43.74	2.57	12.91	40.78
C10H 7Cl1O8	290.62	289.9829	41.33	2.43	12.20	44.04
C10H 7Cl1O9	306.61	305.9779	39.17	2.30	11.56	46.96
C10H 8Cl2O1	215.08	213.9952	55.84	3.75	32.97	7.44

074

C10H 8Cl2O2	231.08	229.9901	51.98	3.49	30.68	13.85
C10H 8Cl2O3	247.08	245.9850	48.61	3.26	28.70	19.43
C10H 8Cl2O4	263.08	261.9800	45.66	3.07	26.95	24.33
C10H 8Cl2O5	279.08	277.9749	43.04	2.89	25.41	28.66
C10H 8Cl2O6	295.08	293.9698	40.71	2.73	24.03	32.53
C10H 8Cl2O7	311.08	309.9647	38.61	2.59	22.79	36.00
C10H 8Cl2O8	327.08	325.9596	36.72	2.47	21.68	39.13
C10H 8Cl2O9	343.08	341.9545	35.01	2.35	20.67	41.97
C10H 9Cl1O1	180.64	180.0342	66.49	5.02	19.63	8.86
C10H 9Cl1O2	196.64	196.0291	61.08	4.61	18.03	16.27
C10H 9Cl1O3	212.63	212.0240	56.49	4.27	16.67	22.57
C10H 9Cl1O4	228.63	228.0189	52.53	3.97	15.51	27.99
C10H 9Cl1O5	244.63	244.0138	49.10	3.71	14.49	32.70
C10H 9Cl1O6	260.63	260.0088	46.08	3.48	13.60	36.83
C10H 9Cl1O7	276.63	276.0037	43.42	3.28	12.82	40.49
C10H 9Cl1O8	292.63	291.9986	41.05	3.10	12.12	43.74
C10H 9Cl1O9	308.63	307.9935	38.92	2.94	11.49	46.66
C10H10Cl2O1	217.10	216.0109	55.33	4.64	32.66	7.37
C10H10Cl2O2	233.10	232.0058	51.53	4.32	30.42	13.73
C10H10Cl2O3	249.10	248.0007	48.22	4.05	28.47	19.27
C10H10Cl2O4	265.09	263.9956	45.31	3.80	26.75	24.14
C10H10Cl2O5	281.09	279.9905	42.73	3.59	25.22	28.46
C10H10Cl2O6	297.09	295.9854	40.43	3.39	23.87	32.31
C10H10Cl2O7	313.09	311.9804	38.36	3.22	22.65	35.77
C10H10Cl2O8	329.09	327.9753	36.50	3.06	21.55	38.89
C10H10Cl2O9	345.09	343.9702	34.81	2.92	20.55	41.73
C10H11Cl1O1	182.65	182.0498	65.76	6.07	19.41	8.76
C10H11Cl1O2	198.65	198.0448	60.46	5.58	17.85	16.11
C10H11Cl1O3	214.65	214.0397	55.96	5.17	16.52	22.36
C10H11Cl1O4	230.65	230.0346	52.08	4.81	15.37	27.75
C10H11Cl1O5	246.65	246.0295	48.70	4.50	14.37	32.43
C10H11Cl1O6	262.65	262.0244	45.73	4.22	13.50	36.55
C10H11Cl1O7	278.65	278.0193	43.11	3.98	12.72	40.19
C10H11Cl1O8	294.65	294.0142	40.76	3.76	12.03	43.44
C10H11Cl1O9	310.65	310.0092	38.66	3.57	11.41	46.35
C10H12Cl2O1	219.11	218.0265	54.82	5.52	32.36	7.30
C10H12Cl2O2	235.11	234.0214	51.09	5.14	30.16	13.61
C10H12Cl2O3	251.11	250.0163	47.83	4.82	28.24	19.11
C10H12Cl2O4	267.11	266.0113	44.97	4.53	26.55	23.96
C10H12Cl2O5	283.11	282.0062	42.43	4.27	25.05	28.26
C10H12Cl2O6	299.11	298.0011	40.16	4.04	23.71	32.09
C10H12Cl2O7	315.11	313.9960	38.12	3.84	22.50	35.54
C10H12Cl2O8	331.11	329.9909	36.28	3.65	21.41	38.66
C10H12Cl2O9	347.11	345.9858	34.60	3.48	20.43	41.48
C10H13Cl1O1	184.67	184.0655	65.04	7.10	19.20	8.66
C10H13Cl1O2	200.67	200.0604	59.86	6.53	17.67	15.95
C10H13Cl1O3	216.67	216.0553	55.44	6.05	16.36	22.15
C10H13Cl1O4	232.67	232.0502	51.62	5.63	15.24	27.51
C10H13Cl1O5	248.67	248.0451	48.30	5.27	14.26	32.17
C10H13Cl1O6	264.66	264.0401	45.38	4.95	13.40	36.27
C10H13Cl1O7	280.66	280.0350	42.80	4.67	12.63	39.90
C10H13Cl1O8	296.66	296.0299	40.49	4.42	11.95	43.14
C10H13Cl1O9	312.66	312.0248	38.42	4.19	11.34	46.05
C10H14Cl2O1	221.13	220.0422	54.32	6.38	32.07	7.24
C10H14Cl2O2	237.13	236.0371	50.65	5.95	29.90	13.49
C10H14Cl2O3	253.13	252.0320	47.45	5.57	28.01	18.96
C10H14Cl2O4	269.13	268.0269	44.63	5.24	26.35	23.78
C10H14Cl2O5	285.13	284.0218	42.13	4.95	24.87	28.06
C10H14Cl2O6	301.13	300.0167	39.89	4.69	23.55	31.88
C10H14Cl2O7	317.12	316.0116	37.88	4.45	22.36	35.32
C10H14Cl2O8	333.12	332.0066	36.06	4.24	21.29	38.42
C10H14Cl2O9	349.12	348.0015	34.40	4.04	20.31	41.24
C10H15Cl1O1	186.68	186.0811	64.34	8.10	18.99	8.57
C10H15Cl1O2	202.68	202.0761	59.26	7.46	17.49	15.79
C10H15Cl1O3	218.68	218.0710	54.93	6.91	16.21	21.95
C10H15Cl1O4	234.68	234.0659	51.18	6.44	15.11	27.27
C10H15Cl1O5	250.68	250.0608	47.91	6.03	14.14	31.91
C10H15Cl1O6	266.68	266.0557	45.04	5.67	13.29	36.00
C10H15Cl1O7	282.68	282.0506	42.49	5.35	12.54	39.62
C10H15Cl1O8	298.68	298.0455	40.21	5.06	11.87	42.85
C10H15Cl1O9	314.68	314.0404	38.17	4.80	11.27	45.76
C10H16Cl2O1	223.14	222.0578	53.83	7.23	31.78	7.17
C10H16Cl2O2	239.14	238.0527	50.23	6.74	29.65	13.38
C10H16Cl2O3	255.14	254.0476	47.08	6.32	27.79	18.81
C10H16Cl2O4	271.14	270.0426	44.30	5.95	26.15	23.60
C10H16Cl2O5	287.14	286.0375	41.83	5.62	24.69	27.86
C10H16Cl2O6	303.14	302.0324	39.62	5.32	23.39	31.67
C10H16Cl2O7	319.14	318.0273	37.64	5.05	22.22	35.09
C10H16Cl2O8	335.14	334.0222	35.84	4.81	21.16	38.19
C10H16Cl2O9	351.14	350.0171	34.21	4.59	20.19	41.01
C10H17Cl1O1	188.70	188.0968	63.65	9.08	18.79	8.48
C10H17Cl1O2	204.70	204.0917	58.68	8.37	17.32	15.63
C10H17Cl1O3	220.70	220.0866	54.42	7.76	16.06	21.75
C10H17Cl1O4	236.70	236.0815	50.74	7.24	14.98	27.04
C10H17Cl1O5	252.70	252.0764	47.53	6.78	14.03	31.66
C10H17Cl1O6	268.70	268.0714	44.70	6.38	13.19	35.73
C10H17Cl1O7	284.70	284.0663	42.19	6.02	12.45	39.34
C10H17Cl1O8	300.70	300.0612	39.94	5.70	11.79	42.57
C10H17Cl1O9	316.69	316.0561	37.93	5.41	11.19	45.47
C10H18Cl2O1	225.16	224.0735	53.34	8.06	31.49	7.11
C10H18Cl2O2	241.16	240.0684	49.81	7.52	29.40	13.27
C10H18Cl2O3	257.16	256.0633	46.71	7.06	27.57	18.66
C10H18Cl2O4	273.16	272.0582	43.97	6.64	25.96	23.43
C10H18Cl2O5	289.16	288.0531	41.54	6.27	24.52	27.67
C10H18Cl2O6	305.16	304.0480	39.36	5.95	23.24	31.46
C10H18Cl2O7	321.16	320.0429	37.40	5.65	22.08	34.87
C10H18Cl2O8	337.16	336.0379	35.62	5.38	21.03	37.96
C10H18Cl2O9	353.16	352.0328	34.01	5.14	20.08	40.77
C10H19Cl1O1	190.72	190.1124	62.98	10.04	18.59	8.39
C10H19Cl1O2	206.71	206.1073	58.10	9.26	17.15	15.48
C10H19Cl1O3	222.71	222.1023	53.93	8.60	15.92	21.55
C10H19Cl1O4	238.71	238.0972	50.32	8.02	14.85	26.81
C10H19Cl1O5	254.71	254.0921	47.16	7.52	13.92	31.41
C10H19Cl1O6	270.71	270.0870	44.37	7.07	13.10	35.46
C10H19Cl1O7	286.71	286.0819	41.89	6.68	12.37	39.06
C10H19Cl1O8	302.71	302.0768	39.68	6.33	11.71	42.28
C10H19Cl1O9	318.71	318.0717	37.69	6.01	11.12	45.18
C10H20Cl2O1	227.18	226.0891	52.87	8.87	31.21	7.04
C10H20Cl2O2	243.18	242.0840	49.39	8.29	29.16	13.16
C10H20Cl2O3	259.18	258.0789	46.34	7.78	27.36	18.52
C10H20Cl2O4	275.17	274.0739	43.65	7.33	25.77	23.26
C10H20Cl2O5	291.17	290.0688	41.25	6.92	24.35	27.47
C10H20Cl2O6	307.17	306.0637	39.10	6.56	23.08	31.25
C10H20Cl2O7	323.17	322.0586	37.17	6.24	21.94	34.66
C10H20Cl2O8	339.17	338.0535	35.41	5.94	20.91	37.74
C10H20Cl2O9	355.17	354.0484	33.82	5.68	19.96	40.54
C10H21Cl1O1	192.73	192.1281	62.32	10.98	18.40	8.30
C10H21Cl1O2	208.73	208.1230	57.54	10.14	16.99	15.33
C10H21Cl1O3	224.73	224.1179	53.45	9.42	15.78	21.36
C10H21Cl1O4	240.73	240.1128	49.89	8.79	14.73	26.58

C10H21C1105	256.73	256.1077	46.79	8.25	13.81	31.16
C10H21C1106	272.73	272.1027	44.04	7.76	13.00	35.20
C10H21C1107	288.73	288.0976	41.60	7.33	12.28	38.79
C10H21C1108	304.73	304.0925	39.42	6.95	11.63	42.00
C10H21C1109	320.73	320.0874	37.45	6.60	11.05	44.90
C11H 5C1101	188.61	188.0029	70.05	2.67	18.80	8.48
C11H 5C1102	204.61	203.9978	64.57	2.46	17.33	15.64
C11H 5C1103	220.61	219.9927	59.89	2.28	16.07	21.76
C11H 5C1104	236.61	235.9876	55.84	2.13	14.98	27.05
C11H 5C1105	252.61	251.9825	52.30	2.00	14.03	31.67
C11H 5C1106	268.61	267.9775	49.19	1.88	13.20	35.74
C11H 5C1107	284.61	283.9724	46.42	1.77	12.46	39.35
C11H 5C1108	300.61	299.9673	43.95	1.68	11.79	42.58
C11H 5C1109	316.61	315.9622	41.73	1.59	11.20	45.48
C11H 6C1201	225.08	223.9796	58.70	2.69	31.50	7.11
C11H 6C1202	241.08	239.9745	54.81	2.51	29.41	13.27
C11H 6C1203	257.07	255.9694	51.39	2.35	27.58	18.67
C11H 6C1204	273.07	271.9643	48.38	2.21	25.97	23.44
C11H 6C1205	289.07	287.9592	45.71	2.09	24.53	27.67
C11H 6C1206	305.07	303.9541	43.31	1.98	23.24	31.47
C11H 6C1207	321.07	319.9491	41.15	1.88	22.08	34.88
C11H 6C1208	337.07	335.9440	39.20	1.79	21.04	37.97
C11H 6C1209	353.07	351.9389	37.42	1.71	20.08	40.78
C11H 7C1101	190.63	190.0185	69.31	3.70	18.60	8.39
C11H 7C1102	206.63	206.0135	63.94	3.41	17.16	15.49
C11H 7C1103	222.63	222.0084	59.35	3.17	15.92	21.56
C11H 7C1104	238.63	238.0033	55.37	2.96	14.86	26.82
C11H 7C1105	254.63	253.9982	51.89	2.77	13.92	31.42
C11H 7C1106	270.63	269.9931	48.82	2.61	13.10	35.47
C11H 7C1107	286.63	285.9880	46.10	2.46	12.37	39.07
C11H 7C1108	302.63	301.9829	43.66	2.33	11.72	42.29
C11H 7C1109	318.63	317.9779	41.47	2.21	11.13	45.19
C11H 8C1201	227.09	225.9952	58.18	3.55	31.22	7.05
C11H 8C1202	243.09	241.9901	54.35	3.32	29.17	13.16
C11H 8C1203	259.09	257.9850	50.99	3.11	27.37	18.53
C11H 8C1204	275.09	273.9800	48.03	2.93	25.78	23.26
C11H 8C1205	291.09	289.9749	45.39	2.77	24.36	27.48
C11H 8C1206	307.09	305.9698	43.02	2.63	23.09	31.26
C11H 8C1207	323.09	321.9647	40.89	2.50	21.95	34.66
C11H 8C1208	339.09	337.9596	38.96	2.38	20.91	37.75
C11H 8C1209	355.09	353.9545	37.21	2.27	19.97	40.55
C11H 9C1101	192.65	192.0342	68.58	4.71	18.40	8.31
C11H 9C1102	208.65	208.0291	63.32	4.35	16.99	15.34
C11H 9C1103	224.65	224.0240	58.81	4.04	15.78	21.37
C11H 9C1104	240.64	240.0189	54.90	3.77	14.73	26.59
C11H 9C1105	256.64	256.0138	51.48	3.53	13.81	31.17
C11H 9C1106	272.64	272.0088	48.46	3.33	13.00	35.21
C11H 9C1107	288.64	288.0037	45.77	3.14	12.28	38.80
C11H 9C1108	304.64	303.9986	43.37	2.98	11.63	42.01
C11H 9C1109	320.64	319.9935	41.21	2.83	11.06	44.91
C11H10C1201	229.11	228.0109	57.67	4.40	30.95	6.98
C11H10C1202	245.11	244.0058	53.90	4.11	28.93	13.06
C11H10C1203	261.11	260.0007	50.60	3.86	27.16	18.38
C11H10C1204	277.11	275.9956	47.68	3.64	25.59	23.09
C11H10C1205	293.11	291.9905	45.08	3.44	24.19	27.29
C11H10C1206	309.10	307.9854	42.74	3.26	22.94	31.06
C11H10C1207	325.10	323.9804	40.64	3.10	21.81	34.45
C11H10C1208	341.10	339.9753	38.73	2.96	20.79	37.52
C11H10C1209	357.10	355.9702	37.00	2.82	19.86	40.32
C11H11C1101	194.66	194.0498	67.87	5.70	18.21	8.22

C11H11C1102	210.66	210.0448	62.72	5.26	16.83	15.19
C11H11C1103	226.66	226.0397	58.29	4.89	15.64	21.18
C11H11C1104	242.66	242.0346	54.45	4.57	14.61	26.37
C11H11C1105	258.66	258.0295	51.08	4.29	13.71	30.93
C11H11C1106	274.66	274.0244	48.10	4.04	12.91	34.95
C11H11C1107	290.66	290.0193	45.46	3.81	12.20	38.53
C11H11C1108	306.66	306.0142	43.08	3.62	11.56	41.74
C11H11C1109	322.66	322.0092	40.95	3.44	10.99	44.63
C11H12C1201	231.12	230.0265	57.17	5.23	30.68	6.92
C11H12C1202	247.12	246.0214	53.46	4.89	28.69	12.95
C11H12C1203	263.12	262.0163	50.21	4.60	26.95	18.24
C11H12C1204	279.12	278.0113	47.34	4.33	25.40	22.93
C11H12C1205	295.12	294.0062	44.77	4.10	24.03	27.11
C11H12C1206	311.12	310.0011	42.47	3.89	22.79	30.86
C11H12C1207	327.12	325.9960	40.39	3.70	21.68	34.24
C11H12C1208	343.12	341.9909	38.51	3.53	20.67	37.30
C11H12C1209	359.12	357.9858	36.79	3.37	19.74	40.10
C11H13C1101	196.68	196.0655	67.18	6.66	18.03	8.13
C11H13C1102	212.68	212.0604	62.12	6.16	16.67	15.05
C11H13C1103	228.68	228.0553	57.78	5.73	15.50	20.99
C11H13C1104	244.68	244.0502	54.00	5.36	14.49	26.16
C11H13C1105	260.68	260.0451	50.68	5.03	13.60	30.69
C11H13C1106	276.68	276.0401	47.75	4.74	12.81	34.70
C11H13C1107	292.68	292.0350	45.14	4.48	12.11	38.27
C11H13C1108	308.67	308.0299	42.80	4.25	11.49	41.47
C11H13C1109	324.67	324.0248	40.69	4.04	10.92	44.35
C11H14C1201	233.14	232.0422	56.67	6.05	30.41	6.86
C11H14C1202	249.14	248.0371	53.03	5.66	28.46	12.84
C11H14C1203	265.14	264.0320	49.83	5.32	26.74	18.10
C11H14C1204	281.14	280.0269	47.00	5.02	25.22	22.76
C11H14C1205	297.14	296.0218	44.47	4.75	23.86	26.92
C11H14C1206	313.14	312.0167	42.19	4.51	22.64	30.66
C11H14C1207	329.14	328.0116	40.14	4.29	21.54	34.03
C11H14C1208	345.14	344.0066	38.28	4.09	20.54	37.09
C11H14C1209	361.13	360.0015	36.59	3.91	19.63	39.87
C11H15C1101	198.69	198.0811	66.50	7.61	17.84	8.05
C11H15C1102	214.69	214.0761	61.54	7.04	16.51	14.90
C11H15C1103	230.69	230.0710	57.27	6.55	15.37	20.81
C11H15C1104	246.69	246.0659	53.56	6.13	14.37	25.94
C11H15C1105	262.69	262.0608	50.30	5.76	13.50	30.45
C11H15C1106	278.69	278.0557	47.41	5.43	12.72	34.45
C11H15C1107	294.69	294.0506	44.83	5.13	12.03	38.00
C11H15C1108	310.69	310.0455	42.53	4.87	11.41	41.20
C11H15C1109	326.69	326.0404	40.44	4.63	10.85	44.08
C11H16C1201	235.16	234.0578	56.19	6.86	30.15	6.80
C11H16C1202	251.15	250.0527	52.61	6.42	28.23	12.74
C11H16C1203	267.15	266.0476	49.46	6.04	26.54	17.97
C11H16C1204	283.15	282.0426	46.66	5.70	25.04	22.60
C11H16C1205	299.15	298.0375	44.17	5.39	23.70	26.74
C11H16C1206	315.15	314.0324	41.92	5.12	22.50	30.46
C11H16C1207	331.15	330.0273	39.90	4.87	21.41	33.82
C11H16C1208	347.15	346.0222	38.06	4.65	20.43	36.87
C11H16C1209	363.15	362.0171	36.38	4.44	19.53	39.65
C11H17C1101	200.71	200.0968	65.83	8.54	17.66	7.97
C11H17C1102	216.71	216.0917	60.97	7.91	16.36	14.77
C11H17C1103	232.71	232.0866	56.78	7.36	15.23	20.63
C11H17C1104	248.71	248.0815	53.12	6.89	14.25	25.73
C11H17C1105	264.71	264.0764	49.91	6.47	13.39	30.22
C11H17C1106	280.71	280.0714	47.07	6.10	12.63	34.20
C11H17C1107	296.71	296.0663	44.53	5.78	11.95	37.75

C11H17C1108	312.71	312.0612	42.25	5.48	11.34	40.93
C11H17C1109	328.71	328.0561	40.19	5.21	10.79	43.81
C11H18C1201	237.17	236.0735	55.71	7.65	29.90	6.75
C11H18C1202	253.17	252.0684	52.19	7.17	28.01	12.64
C11H18C1203	269.17	268.0633	49.09	6.74	26.34	17.83
C11H18C1204	285.17	284.0582	46.33	6.36	24.86	22.44
C11H18C1205	301.17	300.0531	43.87	6.02	23.54	26.56
C11H18C1206	317.17	316.0480	41.66	5.72	22.36	30.27
C11H18C1207	333.17	332.0429	39.66	5.45	21.28	33.62
C11H18C1208	349.17	348.0379	37.84	5.20	20.31	36.66
C11H18C1209	365.17	364.0328	36.18	4.97	19.42	39.43
C11H19C1101	202.73	202.1124	65.17	9.45	17.49	7.89
C11H19C1102	218.73	218.1073	60.41	8.76	16.21	14.63
C11H19C1103	234.73	234.1023	56.29	8.16	15.10	20.45
C11H19C1104	250.72	250.0972	52.70	7.64	14.14	25.53
C11H19C1105	266.72	266.0921	49.54	7.18	13.29	29.99
C11H19C1106	282.72	282.0870	46.73	6.77	12.54	33.95
C11H19C1107	298.72	298.0819	44.23	6.41	11.87	37.49
C11H19C1108	314.72	314.0768	41.98	6.09	11.26	40.67
C11H19C1109	330.72	330.0717	39.95	5.79	10.72	43.54
C11H20C1201	239.19	238.0891	55.24	8.43	29.64	6.69
C11H20C1202	255.19	254.0840	51.77	7.90	27.79	12.54
C11H20C1203	271.19	270.0789	48.72	7.43	26.15	17.70
C11H20C1204	287.19	286.0739	46.01	7.02	24.69	22.28
C11H20C1205	303.19	302.0688	43.58	6.65	23.39	26.39
C11H20C1206	319.18	318.0637	41.39	6.32	22.21	30.08
C11H20C1207	335.18	334.0586	39.42	6.01	21.15	33.41
C11H20C1208	351.18	350.0535	37.62	5.74	20.19	36.45
C11H20C1209	367.18	366.0484	35.98	5.49	19.31	39.22
C11H21C1101	204.74	204.1281	64.53	10.34	17.32	7.81
C11H21C1102	220.74	220.1230	59.85	9.59	16.06	14.50
C11H21C1103	236.74	236.1179	55.81	8.94	14.98	20.27
C11H21C1104	252.74	252.1128	52.28	8.38	14.03	25.32
C11H21C1105	268.74	268.1077	49.16	7.88	13.19	29.77
C11H21C1106	284.74	284.1027	46.40	7.43	12.45	33.71
C11H21C1107	300.74	300.0976	43.93	7.04	11.79	37.24
C11H21C1108	316.74	316.0925	41.71	6.68	11.19	40.41
C11H21C1109	332.74	332.0874	39.71	6.36	10.65	43.28
C11H22C1201	241.20	240.1048	54.78	9.19	29.40	6.63
C11H22C1202	257.20	256.0997	51.37	8.62	27.57	12.44
C11H22C1203	273.20	272.0946	48.36	8.12	25.95	17.57
C11H22C1204	289.20	288.0895	45.69	7.67	24.52	22.13
C11H22C1205	305.20	304.0844	43.29	7.27	23.23	26.21
C11H22C1206	321.20	320.0793	41.13	6.90	22.08	29.89
C11H22C1207	337.20	336.0742	39.18	6.58	21.03	33.21
C11H22C1208	353.20	352.0692	37.41	6.28	20.08	36.24
C11H22C1209	369.20	368.0641	35.79	6.01	19.21	39.00
C11H23C1101	206.76	206.1437	63.90	11.21	17.15	7.74
C11H23C1102	222.76	222.1386	59.31	10.41	15.92	14.36
C11H23C1103	238.76	238.1336	55.34	9.71	14.85	20.10
C11H23C1104	254.76	254.1285	51.86	9.10	13.92	25.12
C11H23C1105	270.76	270.1234	48.80	8.56	13.09	29.55
C11H23C1106	286.76	286.1183	46.08	8.08	12.36	33.48
C11H23C1107	302.75	302.1132	43.64	7.66	11.71	36.99
C11H23C1108	318.75	318.1081	41.45	7.27	11.12	40.15
C11H23C1109	334.75	334.1030	39.47	6.93	10.59	43.02
C12H 6C1201	237.09	235.9796	60.79	2.55	29.91	6.75
C12H 6C1202	253.09	251.9745	56.95	2.39	28.02	12.64
C12H 6C1203	269.09	267.9694	53.56	2.25	26.35	17.84
C12H 6C1204	285.09	283.9643	50.56	2.12	24.87	22.45

C12H 6C1205	301.08	299.9592	47.87	2.01	23.55	26.57
C12H 6C1206	317.08	315.9541	45.46	1.91	22.36	30.27
C12H 6C1207	333.08	331.9491	43.27	1.82	21.29	33.62
C12H 6C1208	349.08	347.9440	41.29	1.73	20.31	36.67
C12H 6C1209	365.08	363.9389	39.48	1.66	19.42	39.44
C12H 7C1101	202.64	202.0185	71.13	3.48	17.50	7.90
C12H 7C1102	218.64	218.0135	65.92	3.23	16.22	14.64
C12H 7C1103	234.64	234.0084	61.43	3.01	15.11	20.46
C12H 7C1104	250.64	250.0033	57.51	2.82	14.14	25.53
C12H 7C1105	266.64	265.9982	54.06	2.65	13.30	30.00
C12H 7C1106	282.64	281.9931	51.00	2.50	12.54	33.96
C12H 7C1107	298.64	297.9880	48.26	2.36	11.87	37.50
C12H 7C1108	314.64	313.9829	45.81	2.24	11.27	40.68
C12H 7C1109	330.64	329.9779	43.59	2.13	10.72	43.55
C12H 8C1201	239.10	237.9952	60.28	3.37	29.66	6.69
C12H 8C1202	255.10	253.9901	56.50	3.16	27.80	12.54
C12H 8C1203	271.10	269.9850	53.17	2.97	26.15	17.70
C12H 8C1204	287.10	285.9800	50.20	2.81	24.70	22.29
C12H 8C1205	303.10	301.9749	47.55	2.66	23.39	26.39
C12H 8C1206	319.10	317.9698	45.17	2.53	22.22	30.08
C12H 8C1207	335.10	333.9647	43.01	2.41	21.16	33.42
C12H 8C1208	351.10	349.9596	41.05	2.30	20.20	36.46
C12H 8C1209	367.10	365.9545	39.26	2.20	19.32	39.23
C12H 9C1101	204.66	204.0342	70.43	4.43	17.32	7.82
C12H 9C1102	220.66	220.0291	65.32	4.11	16.07	14.50
C12H 9C1103	236.66	236.0240	60.90	3.83	14.98	20.28
C12H 9C1104	252.66	252.0189	57.05	3.59	14.03	25.33
C12H 9C1105	268.66	268.0138	53.65	3.38	13.20	29.78
C12H 9C1106	284.65	284.0088	50.63	3.19	12.45	33.72
C12H 9C1107	300.65	300.0037	47.94	3.02	11.79	37.25
C12H 9C1108	316.65	315.9986	45.52	2.86	11.20	40.42
C12H 9C1109	332.65	331.9935	43.33	2.73	10.66	43.29
C12H10C1201	241.12	240.0109	59.78	4.18	29.41	6.64
C12H10C1202	257.12	256.0058	56.06	3.92	27.58	12.45
C12H10C1203	273.12	272.0007	52.77	3.69	25.96	17.57
C12H10C1204	289.12	287.9956	49.85	3.49	24.53	22.14
C12H10C1205	305.12	303.9905	47.24	3.30	23.24	26.22
C12H10C1206	321.12	319.9854	44.89	3.14	22.08	29.89
C12H10C1207	337.12	335.9804	42.76	2.99	21.03	33.22
C12H10C1208	353.11	351.9753	40.82	2.85	20.08	36.25
C12H10C1209	369.11	367.9702	39.05	2.73	19.21	39.01
C12H11C1101	206.67	206.0498	69.74	5.36	17.15	7.74
C12H11C1102	222.67	222.0448	64.73	4.98	15.92	14.37
C12H11C1103	238.67	238.0397	60.39	4.65	14.85	20.11
C12H11C1104	254.67	254.0346	56.60	4.35	13.92	25.13
C12H11C1105	270.67	270.0295	53.25	4.10	13.10	29.56
C12H11C1106	286.67	286.0244	50.28	3.87	12.37	33.49
C12H11C1107	302.67	302.0193	47.62	3.66	11.71	37.00
C12H11C1108	318.67	318.0142	45.23	3.48	11.13	40.17
C12H11C1109	334.67	334.0092	43.07	3.31	10.59	43.03
C12H12C1201	243.13	242.0265	59.28	4.97	29.16	6.58
C12H12C1202	259.13	258.0214	55.62	4.67	27.36	12.35
C12H12C1203	275.13	274.0163	52.39	4.40	25.77	17.45
C12H12C1204	291.13	290.0113	49.51	4.15	24.36	21.98
C12H12C1205	307.13	306.0062	46.93	3.94	23.09	26.05
C12H12C1206	323.13	322.0011	44.61	3.74	21.94	29.71
C12H12C1207	339.13	337.9960	42.50	3.57	20.91	33.02
C12H12C1208	355.13	353.9909	40.59	3.41	19.97	36.04
C12H12C1209	371.13	369.9858	38.84	3.26	19.11	38.80
C12H13C1101	208.69	208.0655	69.07	6.28	16.99	7.67

077

C12H13C1102	224.69	224.0604	64.15	5.83	15.78	14.24
C12H13C1103	240.69	240.0553	59.88	5.44	14.73	19.94
C12H13C1104	256.69	256.0502	56.15	5.10	13.81	24.93
C12H13C1105	272.69	272.0451	52.86	4.81	13.00	29.34
C12H13C1106	288.69	288.0401	49.93	4.54	12.28	33.25
C12H13C1107	304.69	304.0350	47.31	4.30	11.64	36.76
C12H13C1108	320.69	320.0299	44.95	4.09	11.06	39.91
C12H13C1109	336.69	336.0248	42.81	3.89	10.53	42.77
C12H14C1201	245.15	244.0422	58.79	5.76	28.92	6.53
C12H14C1202	261.15	260.0371	55.19	5.40	27.15	12.25
C12H14C1203	277.15	276.0320	52.01	5.09	25.58	17.32
C12H14C1204	293.15	292.0269	49.17	4.81	24.19	21.83
C12H14C1205	309.15	308.0218	46.62	4.56	22.94	25.88
C12H14C1206	325.15	324.0167	44.33	4.34	21.81	29.52
C12H14C1207	341.15	340.0116	42.25	4.14	20.78	32.83
C12H14C1208	357.15	356.0066	40.36	3.95	19.85	35.84
C12H14C1209	373.15	372.0015	38.63	3.78	19.00	38.59
C12H15C1101	210.71	210.0811	68.41	7.18	16.83	7.59
C12H15C1102	226.71	226.0761	63.58	6.67	15.64	14.11
C12H15C1103	242.70	242.0710	59.39	6.23	14.61	19.78
C12H15C1104	258.70	258.0659	55.71	5.84	13.70	24.74
C12H15C1105	274.70	274.0608	52.47	5.50	12.91	29.12
C12H15C1106	290.70	290.0557	49.58	5.20	12.20	33.02
C12H15C1107	306.70	306.0506	46.99	4.93	11.56	36.52
C12H15C1108	322.70	322.0455	44.66	4.69	10.99	39.66
C12H15C1109	338.70	338.0404	42.55	4.46	10.47	42.51
C12H16C1201	247.17	246.0578	58.31	6.52	28.69	6.47
C12H16C1202	263.17	262.0527	54.77	6.13	26.94	12.16
C12H16C1203	279.17	278.0476	51.63	5.78	25.40	17.19
C12H16C1204	295.16	294.0426	48.83	5.46	24.02	21.68
C12H16C1205	311.16	310.0375	46.32	5.18	22.79	25.71
C12H16C1206	327.16	326.0324	44.06	4.93	21.67	29.34
C12H16C1207	343.16	342.0273	42.00	4.70	20.66	32.64
C12H16C1208	359.16	358.0222	40.13	4.49	19.74	35.64
C12H16C1209	375.16	374.0171	38.42	4.30	18.90	38.38
C12H17C1101	212.72	212.0968	67.76	8.06	16.67	7.52
C12H17C1102	228.72	228.0917	63.02	7.49	15.50	13.99
C12H17C1103	244.72	244.0866	58.90	7.00	14.49	19.61
C12H17C1104	260.72	260.0815	55.28	6.57	13.60	24.55
C12H17C1105	276.72	276.0764	52.09	6.19	12.81	28.91
C12H17C1106	292.72	292.0714	49.24	5.85	12.11	32.79
C12H17C1107	308.72	308.0663	46.69	5.55	11.48	36.28
C12H17C1108	324.72	324.0612	44.39	5.28	10.92	39.42
C12H17C1109	340.72	340.0561	42.30	5.03	10.41	42.26
C12H18C1201	249.18	248.0735	57.84	7.28	28.46	6.42
C12H18C1202	265.18	264.0684	54.35	6.84	26.74	12.07
C12H18C1203	281.18	280.0633	51.26	6.45	25.22	17.07
C12H18C1204	297.18	296.0582	48.50	6.11	23.86	21.53
C12H18C1205	313.18	312.0531	46.02	5.79	22.64	25.54
C12H18C1206	329.18	328.0480	43.79	5.51	21.54	29.16
C12H18C1207	345.18	344.0429	41.76	5.26	20.54	32.45
C12H18C1208	361.18	360.0379	39.91	5.02	19.63	35.44
C12H18C1209	377.18	376.0328	38.21	4.81	18.80	38.18
C12H19C1101	214.74	214.1124	67.12	8.92	16.51	7.45
C12H19C1102	230.74	230.1073	62.47	8.30	15.37	13.87
C12H19C1103	246.74	246.1023	58.42	7.76	14.37	19.45
C12H19C1104	262.74	262.0972	54.86	7.29	13.49	24.36
C12H19C1105	278.74	278.0921	51.71	6.87	12.72	28.70
C12H19C1106	294.73	294.0870	48.90	6.50	12.03	32.57
C12H19C1107	310.73	310.0819	46.38	6.16	11.41	36.04

C12H19C1108	326.73	326.0768	44.11	5.86	10.85	39.17
C12H19C1109	342.73	342.0717	42.05	5.59	10.34	42.01
C12H20C1201	251.20	250.0891	57.38	8.03	28.23	6.37
C12H20C1202	267.20	266.0840	53.94	7.54	26.54	11.98
C12H20C1203	283.20	282.0789	50.90	7.12	25.04	16.95
C12H20C1204	299.20	298.0739	48.17	6.74	23.70	21.39
C12H20C1205	315.20	314.0688	45.73	6.40	22.50	25.38
C12H20C1206	331.20	330.0637	43.52	6.09	21.41	28.98
C12H20C1207	347.20	346.0586	41.51	5.81	20.42	32.26
C12H20C1208	363.19	362.0535	39.69	5.55	19.52	35.24
C12H20C1209	379.19	378.0484	38.01	5.32	18.70	37.97
C12H21C1101	216.75	216.1281	66.50	9.77	16.36	7.38
C12H21C1102	232.75	232.1230	61.93	9.09	15.23	13.75
C12H21C1103	248.75	248.1179	57.94	8.51	14.25	19.30
C12H21C1104	264.75	264.1128	54.44	8.00	13.39	24.17
C12H21C1105	280.75	280.1077	51.34	7.54	12.63	28.49
C12H21C1106	296.75	296.1027	48.57	7.13	11.95	32.35
C12H21C1107	312.75	312.0976	46.09	6.77	11.34	35.81
C12H21C1108	328.75	328.0925	43.84	6.44	10.78	38.93
C12H21C1109	344.75	344.0874	41.81	6.14	10.28	41.77
C12H22C1201	253.21	252.1048	56.92	8.76	28.00	6.32
C12H22C1202	269.21	268.0997	53.54	8.24	26.34	11.89
C12H22C1203	285.21	284.0946	50.54	7.78	24.86	16.83
C12H22C1204	301.21	300.0895	47.85	7.36	23.54	21.25
C12H22C1205	317.21	316.0844	45.44	6.99	22.35	25.22
C12H22C1206	333.21	332.0793	43.26	6.66	21.28	28.81
C12H22C1207	349.21	348.0742	41.27	6.35	20.30	32.07
C12H22C1208	365.21	364.0692	39.47	6.07	19.42	35.05
C12H22C1209	381.21	380.0641	37.81	5.82	18.60	37.77
C12H23C1101	218.77	218.1437	65.88	10.60	16.21	7.31
C12H23C1102	234.77	234.1386	61.39	9.87	15.10	13.63
C12H23C1103	250.77	250.1336	57.48	9.24	14.14	19.14
C12H23C1104	266.77	266.1285	54.03	8.69	13.29	23.99
C12H23C1105	282.77	282.1234	50.97	8.20	12.54	28.29
C12H23C1106	298.77	298.1183	48.24	7.76	11.87	32.13
C12H23C1107	314.77	314.1132	45.79	7.37	11.26	35.58
C12H23C1108	330.77	330.1081	43.58	7.01	10.72	38.70
C12H23C1109	346.76	346.1030	41.57	6.69	10.22	41.53
C12H24C1201	255.23	254.1204	56.47	9.48	27.78	6.27
C12H24C1202	271.23	270.1153	53.14	8.92	26.14	11.80
C12H24C1203	287.23	286.1102	50.18	8.42	24.69	16.71
C12H24C1204	303.23	302.1052	47.53	7.98	23.38	21.11
C12H24C1205	319.23	318.1001	45.15	7.58	22.21	25.06
C12H24C1206	335.23	334.0950	43.00	7.22	21.15	28.64
C12H24C1207	351.23	350.0899	41.04	6.89	20.19	31.89
C12H24C1208	367.23	366.0848	39.25	6.59	19.31	34.85
C12H24C1209	383.23	382.0797	37.61	6.31	18.50	37.57
C12H25C1101	220.79	220.1594	65.28	11.41	16.06	7.25
C12H25C1102	236.78	236.1543	60.87	10.64	14.97	13.51
C12H25C1103	252.78	252.1492	57.02	9.97	14.03	18.99
C12H25C1104	268.78	268.1441	53.62	9.38	13.19	23.81
C12H25C1105	284.78	284.1390	50.61	8.85	12.45	28.09
C12H25C1106	300.78	300.1340	47.92	8.38	11.79	31.92
C12H25C1107	316.78	316.1289	45.50	7.95	11.19	35.35
C12H25C1108	332.78	332.1238	43.31	7.57	10.65	38.46
C12H25C1109	348.78	348.1187	41.33	7.22	10.16	41.29
C13H 6C1201	249.10	247.9796	62.68	2.43	28.47	6.42
C13H 6C1202	265.10	263.9745	58.90	2.28	26.75	12.07
C13H 6C1203	281.10	279.9694	55.55	2.15	25.22	17.08
C13H 6C1204	297.10	295.9643	52.56	2.04	23.87	21.54

078

C13H 6C1205	313.10	311.9592	49.87	1.93	22.65	25.55
C13H 6C1206	329.10	327.9541	47.45	1.84	21.55	29.17
C13H 6C1207	345.09	343.9491	45.25	1.75	20.55	32.45
C13H 6C1208	361.09	359.9440	43.24	1.67	19.64	35.45
C13H 6C1209	377.09	375.9389	41.41	1.60	18.80	38.19
C13H 7C1101	214.65	214.0185	72.74	3.29	16.52	7.45
C13H 7C1102	230.65	230.0135	67.70	3.06	15.37	13.87
C13H 7C1103	246.65	246.0084	63.31	2.86	14.37	19.46
C13H 7C1104	262.65	262.0033	59.45	2.69	13.50	24.37
C13H 7C1105	278.65	277.9982	56.04	2.53	12.72	28.71
C13H 7C1106	294.65	293.9931	52.99	2.39	12.03	32.58
C13H 7C1107	310.65	309.9880	50.26	2.27	11.41	36.05
C13H 7C1108	326.65	325.9829	47.80	2.16	10.85	39.18
C13H 7C1109	342.65	341.9779	45.57	2.06	10.35	42.02
C13H 8C1201	251.11	249.9952	62.18	3.21	28.24	6.37
C13H 8C1202	267.11	265.9901	58.46	3.02	26.55	11.98
C13H 8C1203	283.11	281.9850	55.15	2.85	25.05	16.95
C13H 8C1204	299.11	297.9800	52.20	2.70	23.71	21.40
C13H 8C1205	315.11	313.9749	49.55	2.56	22.50	25.39
C13H 8C1206	331.11	329.9698	47.16	2.44	21.41	28.99
C13H 8C1207	347.11	345.9647	44.98	2.32	20.43	32.27
C13H 8C1208	363.11	361.9596	43.00	2.22	19.53	35.25
C13H 8C1209	379.11	377.9545	41.19	2.13	18.70	37.98
C13H 9C1101	216.67	216.0342	72.07	4.19	16.36	7.38
C13H 9C1102	232.67	232.0291	67.11	3.90	15.24	13.75
C13H 9C1103	248.67	248.0240	62.79	3.65	14.26	19.30
C13H 9C1104	264.67	264.0189	59.00	3.43	13.40	24.18
C13H 9C1105	280.67	280.0138	55.63	3.23	12.63	28.50
C13H 9C1106	296.67	296.0088	52.63	3.06	11.95	32.36
C13H 9C1107	312.67	312.0037	49.94	2.90	11.34	35.82
C13H 9C1108	328.66	327.9986	47.51	2.76	10.79	38.94
C13H 9C1109	344.66	343.9935	45.30	2.63	10.29	41.78
C13H10C1201	253.13	252.0109	61.69	3.98	28.01	6.32
C13H10C1202	269.13	268.0058	58.02	3.75	26.35	11.89
C13H10C1203	285.13	284.0007	54.76	3.54	24.87	16.83
C13H10C1204	301.13	299.9956	51.85	3.35	23.55	21.25
C13H10C1205	317.13	315.9905	49.24	3.18	22.36	25.23
C13H10C1206	333.13	331.9854	46.87	3.03	21.28	28.82
C13H10C1207	349.13	347.9804	44.72	2.89	20.31	32.08
C13H10C1208	365.13	363.9753	42.76	2.76	19.42	35.06
C13H10C1209	381.13	379.9702	40.97	2.64	18.60	37.78
C13H11C1101	218.69	218.0498	71.40	5.07	16.21	7.32
C13H11C1102	234.68	234.0448	66.53	4.72	15.11	13.63
C13H11C1103	250.68	250.0397	62.29	4.42	14.14	19.15
C13H11C1104	266.68	266.0346	58.55	4.16	13.29	24.00
C13H11C1105	282.68	282.0295	55.24	3.92	12.54	28.30
C13H11C1106	298.68	298.0244	52.28	3.71	11.87	32.14
C13H11C1107	314.68	314.0193	49.62	3.52	11.27	35.59
C13H11C1108	330.68	330.0142	47.22	3.35	10.72	38.71
C13H11C1109	346.68	346.0092	45.04	3.20	10.23	41.54
C13H12C1201	255.15	254.0265	61.20	4.74	27.79	6.27
C13H12C1202	271.15	270.0214	57.59	4.46	26.15	11.80
C13H12C1203	287.14	286.0163	54.38	4.21	24.69	16.72
C13H12C1204	303.14	302.0113	51.51	3.99	23.39	21.11
C13H12C1205	319.14	318.0062	48.93	3.79	22.22	25.07
C13H12C1206	335.14	334.0011	46.59	3.61	21.16	28.64
C13H12C1207	351.14	349.9960	44.47	3.44	20.19	31.89
C13H12C1208	367.14	365.9909	42.53	3.29	19.31	34.86
C13H12C1209	383.14	381.9858	40.75	3.16	18.51	37.58
C13H13C1101	220.70	220.0655	70.75	5.94	16.06	7.25

C13H13C1102	236.70	236.0604	65.97	5.54	14.98	13.52
C13H13C1103	252.70	252.0553	61.79	5.19	14.03	18.99
C13H13C1104	268.70	268.0502	58.11	4.88	13.19	23.82
C13H13C1105	284.70	284.0451	54.85	4.60	12.45	28.10
C13H13C1106	300.70	300.0401	51.93	4.36	11.79	31.92
C13H13C1107	316.70	316.0350	49.30	4.14	11.19	35.36
C13H13C1108	332.70	332.0299	46.93	3.94	10.66	38.47
C13H13C1109	348.70	348.0248	44.78	3.76	10.17	41.30
C13H14C1201	257.16	256.0422	60.72	5.49	27.57	6.22
C13H14C1202	273.16	272.0371	57.16	5.17	25.96	11.71
C13H14C1203	289.16	288.0320	54.00	4.88	24.52	16.60
C13H14C1204	305.16	304.0269	51.17	4.62	23.24	20.97
C13H14C1205	321.16	320.0218	48.62	4.39	22.08	24.91
C13H14C1206	337.16	336.0167	46.31	4.19	21.03	28.47
C13H14C1207	353.16	352.0116	44.21	4.00	20.08	31.71
C13H14C1208	369.16	368.0066	42.30	3.82	19.21	34.67
C13H14C1209	385.16	384.0015	40.54	3.66	18.41	37.39
C13H15C1101	222.72	222.0811	70.11	6.79	15.92	7.18
C13H15C1102	238.72	238.0761	65.41	6.33	14.85	13.40
C13H15C1103	254.72	254.0710	61.30	5.94	13.92	18.84
C13H15C1104	270.72	270.0659	57.68	5.59	13.10	23.64
C13H15C1105	286.71	286.0608	54.46	5.27	12.37	27.90
C13H15C1106	302.71	302.0557	51.58	4.99	11.71	31.71
C13H15C1107	318.71	318.0506	48.99	4.74	11.12	35.14
C13H15C1108	334.71	334.0455	46.65	4.52	10.59	38.24
C13H15C1109	350.71	350.0404	44.52	4.31	10.11	41.06
C13H16C1201	259.18	258.0578	60.25	6.22	27.36	6.17
C13H16C1202	275.18	274.0527	56.74	5.86	25.77	11.63
C13H16C1203	291.18	290.0476	53.63	5.54	24.35	16.48
C13H16C1204	307.18	306.0426	50.83	5.25	23.08	20.83
C13H16C1205	323.18	322.0375	48.32	4.99	21.94	24.75
C13H16C1206	339.17	338.0324	46.04	4.75	20.91	28.30
C13H16C1207	355.17	354.0273	43.96	4.54	19.96	31.53
C13H16C1208	371.17	370.0222	42.07	4.35	19.10	34.48
C13H16C1209	387.17	386.0171	40.33	4.17	18.31	37.19
C13H17C1101	224.73	224.0968	69.48	7.62	15.78	7.12
C13H17C1102	240.73	240.0917	64.86	7.12	14.73	13.29
C13H17C1103	256.73	256.0866	60.82	6.67	13.81	18.70
C13H17C1104	272.73	272.0815	57.25	6.28	13.00	23.47
C13H17C1105	288.73	288.0764	54.08	5.93	12.28	27.71
C13H17C1106	304.73	304.0714	51.24	5.62	11.63	31.50
C13H17C1107	320.73	320.0663	48.68	5.34	11.05	34.92
C13H17C1108	336.73	336.0612	46.37	5.09	10.53	38.01
C13H17C1109	352.73	352.0561	44.27	4.86	10.05	40.82
C13H18C1201	261.19	260.0735	59.78	6.95	27.15	6.13
C13H18C1202	277.19	276.0684	56.33	6.55	25.58	11.54
C13H18C1203	293.19	292.0633	53.26	6.19	24.18	16.37
C13H18C1204	309.19	308.0582	50.50	5.87	22.93	20.70
C13H18C1205	325.19	324.0531	48.02	5.58	21.80	24.60
C13H18C1206	341.19	340.0480	45.76	5.32	20.78	28.14
C13H18C1207	357.19	356.0429	43.71	5.08	19.85	31.35
C13H18C1208	373.19	372.0379	41.84	4.86	19.00	34.30
C13H18C1209	389.19	388.0328	40.12	4.66	18.22	37.00
C13H19C1101	226.75	226.1124	68.86	8.45	15.64	7.06
C13H19C1102	242.75	242.1073	64.32	7.89	14.60	13.18
C13H19C1103	258.75	258.1023	60.35	7.40	13.70	18.55
C13H19C1104	274.75	274.0972	56.83	6.97	12.90	23.29
C13H19C1105	290.75	290.0921	53.70	6.59	12.19	27.51
C13H19C1106	306.75	306.0870	50.90	6.24	11.56	31.30
C13H19C1107	322.75	322.0819	48.38	5.93	10.98	34.70

C13H19C1108	338.74	338.0768	46.10	5.65	10.47	37.79
C13H19C1109	354.74	354.0717	44.02	5.40	9.99	40.59
C13H20C1201	263.21	262.0891	59.32	7.66	26.94	6.08
C13H20C1202	279.21	278.0840	55.92	7.22	25.40	11.46
C13H20C1203	295.21	294.0789	52.85	6.83	24.02	16.26
C13H20C1204	311.21	310.0739	50.17	6.48	22.78	20.56
C13H20C1205	327.21	326.0688	47.72	6.16	21.67	24.45
C13H20C1206	343.21	342.0637	45.50	5.87	20.66	27.97
C13H20C1207	359.21	358.0586	43.47	5.61	19.74	31.18
C13H20C1208	375.21	374.0535	41.62	5.37	18.90	34.11
C13H20C1209	391.20	390.0484	39.91	5.15	18.13	36.81
C13H21C1101	228.76	228.1281	68.26	9.25	15.50	6.99
C13H21C1102	244.76	244.1230	63.79	8.65	14.48	13.07
C13H21C1103	260.76	260.1179	59.88	8.12	13.60	18.41
C13H21C1104	276.76	276.1128	56.42	7.65	12.81	23.12
C13H21C1105	292.76	292.1077	53.34	7.23	12.11	27.32
C13H21C1106	308.76	308.1027	50.57	6.86	11.48	31.09
C13H21C1107	324.76	324.0976	48.08	6.52	10.92	34.49
C13H21C1108	340.76	340.0925	45.82	6.21	10.40	37.56
C13H21C1109	356.76	356.0874	43.77	5.93	9.94	40.36
C13H22C1201	265.23	264.1048	58.87	8.36	26.73	6.03
C13H22C1202	281.23	280.0997	55.52	7.89	25.21	11.38
C13H22C1203	297.22	296.0946	52.53	7.46	23.86	16.15
C13H22C1204	313.22	312.0895	49.85	7.08	22.64	20.43
C13H22C1205	329.22	328.0844	47.43	6.74	21.54	24.30
C13H22C1206	345.22	344.0793	45.23	6.42	20.54	27.81
C13H22C1207	361.22	360.0742	43.23	6.14	19.63	31.00
C13H22C1208	377.22	376.0692	41.39	5.88	18.80	33.93
C13H22C1209	393.22	392.0641	39.71	5.64	18.03	36.62
C13H23C1101	230.78	230.1437	67.66	10.05	15.36	6.93
C13H23C1102	246.78	246.1386	63.27	9.39	14.37	12.97
C13H23C1103	262.78	262.1336	59.42	8.82	13.49	18.27
C13H23C1104	278.78	278.1285	56.01	8.32	12.72	22.96
C13H23C1105	294.78	294.1234	52.97	7.86	12.03	27.14
C13H23C1106	310.78	310.1183	50.24	7.46	11.41	30.89
C13H23C1107	326.78	326.1132	47.78	7.09	10.85	34.27
C13H23C1108	342.78	342.1081	45.55	6.76	10.34	37.34
C13H23C1109	358.78	358.1030	43.52	6.46	9.88	40.13
C13H24C1201	267.24	266.1204	58.43	9.05	26.53	5.99
C13H24C1202	283.24	282.1153	55.13	8.54	25.03	11.30
C13H24C1203	299.24	298.1102	52.18	8.08	23.70	16.04
C13H24C1204	315.24	314.1052	49.53	7.67	22.49	20.30
C13H24C1205	331.24	330.1001	47.14	7.30	21.41	24.15
C13H24C1206	347.24	346.0950	44.97	6.97	20.42	27.65
C13H24C1207	363.24	362.0899	42.99	6.66	19.52	30.83
C13H24C1208	379.24	378.0848	41.17	6.38	18.70	33.75
C13H24C1209	395.24	394.0797	39.51	6.12	17.94	36.43
C13H25C1101	232.80	232.1594	67.07	10.82	15.23	6.87
C13H25C1102	248.80	248.1543	62.76	10.13	14.25	12.86
C13H25C1103	264.80	264.1492	58.97	9.52	13.39	18.13
C13H25C1104	280.79	280.1441	55.61	8.97	12.63	22.79
C13H25C1105	296.79	296.1390	52.61	8.49	11.95	26.95
C13H25C1106	312.79	312.1340	49.92	8.06	11.33	30.69
C13H25C1107	328.79	328.1289	47.49	7.66	10.78	34.06
C13H25C1108	344.79	344.1238	45.29	7.31	10.28	37.12
C13H25C1109	360.79	360.1187	43.28	6.98	9.83	39.91
C13H26C1201	269.26	268.1361	57.99	9.73	26.33	5.94
C13H26C1202	285.26	284.1310	54.74	9.19	24.86	11.22
C13H26C1203	301.26	300.1259	51.83	8.70	23.54	15.93
C13H26C1204	317.26	316.1208	49.22	8.26	22.35	20.17
C13H26C1205	333.26	332.1157	46.85	7.86	21.28	24.00
C13H26C1206	349.25	348.1106	44.71	7.50	20.30	27.49
C13H26C1207	365.25	364.1055	42.75	7.18	19.41	30.66
C13H26C1208	381.25	380.1005	40.96	6.87	18.60	33.57
C13H26C1209	397.25	396.0954	39.31	6.60	17.85	36.25
C13H27C1101	234.81	234.1750	66.50	11.59	15.10	6.81
C13H27C1102	250.81	250.1699	62.26	10.85	14.14	12.76
C13H27C1103	266.81	266.1649	58.52	10.20	13.29	17.99
C13H27C1104	282.81	282.1598	55.21	9.62	12.54	22.63
C13H27C1105	298.81	298.1547	52.26	9.11	11.86	26.77
C13H27C1106	314.81	314.1496	49.60	8.64	11.26	30.49
C13H27C1107	330.81	330.1445	47.20	8.23	10.72	33.86
C13H27C1108	346.81	346.1394	45.02	7.85	10.22	36.91
C13H27C1109	362.81	362.1343	43.04	7.50	9.77	39.69
C14H 7C1101	226.66	226.0185	74.19	3.11	15.64	7.06
C14H 7C1102	242.66	242.0135	69.30	2.91	14.61	13.19
C14H 7C1103	258.66	258.0084	65.01	2.73	13.71	18.56
C14H 7C1104	274.66	274.0033	61.22	2.57	12.91	23.30
C14H 7C1105	290.66	289.9982	57.85	2.43	12.20	27.52
C14H 7C1106	306.66	305.9931	54.83	2.30	11.56	31.30
C14H 7C1107	322.66	321.9880	52.12	2.19	10.99	34.71
C14H 7C1108	338.66	337.9829	49.65	2.08	10.47	37.79
C14H 7C1109	354.66	353.9779	47.41	1.99	10.00	40.60
C14H 8C1201	263.13	261.9952	63.91	3.06	26.95	6.08
C14H 8C1202	279.12	277.9901	60.24	2.89	25.40	11.46
C14H 8C1203	295.12	293.9850	56.98	2.73	24.03	16.26
C14H 8C1204	311.12	309.9800	54.05	2.59	22.79	20.57
C14H 8C1205	327.12	325.9749	51.40	2.47	21.68	24.45
C14H 8C1206	343.12	341.9698	49.01	2.35	20.66	27.98
C14H 8C1207	359.12	357.9647	46.82	2.25	19.74	31.19
C14H 8C1208	375.12	373.9596	44.83	2.15	18.90	34.12
C14H 8C1209	391.12	389.9545	42.99	2.06	18.13	36.82
C14H 9C1101	228.68	228.0342	73.53	3.97	15.50	7.00
C14H 9C1102	244.68	244.0291	68.73	3.71	14.49	13.08
C14H 9C1103	260.68	260.0240	64.51	3.48	13.60	18.41
C14H 9C1104	276.68	276.0189	60.78	3.28	12.81	23.13
C14H 9C1105	292.68	292.0138	57.45	3.10	12.11	27.33
C14H 9C1106	308.68	308.0088	54.48	2.94	11.49	31.10
C14H 9C1107	324.68	324.0037	51.79	2.79	10.92	34.49
C14H 9C1108	340.68	339.9986	49.36	2.66	10.41	37.57
C14H 9C1109	356.68	355.9935	47.15	2.54	9.94	40.37
C14H10C1201	265.14	264.0109	63.42	3.80	26.74	6.03
C14H10C1202	281.14	280.0058	59.81	3.59	25.22	11.38
C14H10C1203	297.14	296.0007	56.59	3.39	23.86	16.15
C14H10C1204	313.14	311.9956	53.70	3.22	22.64	20.44
C14H10C1205	329.14	327.9905	51.09	3.06	21.54	24.30
C14H10C1206	345.14	343.9854	48.72	2.92	20.54	27.81
C14H10C1207	361.14	359.9804	46.56	2.79	19.63	31.01
C14H10C1208	377.14	375.9753	44.59	2.67	18.80	33.94
C14H10C1209	393.14	391.9702	42.77	2.56	18.04	36.63
C14H11C1101	230.70	230.0498	72.89	4.81	15.37	6.94
C14H11C1102	246.70	246.0448	68.16	4.49	14.37	12.97
C14H11C1103	262.69	262.0397	64.01	4.22	13.50	18.27
C14H11C1104	278.69	278.0346	60.34	3.98	12.72	22.96
C14H11C1105	294.69	294.0295	57.06	3.76	12.03	27.15
C14H11C1106	310.69	310.0244	54.12	3.57	11.41	30.90
C14H11C1107	326.69	326.0193	51.47	3.39	10.85	34.28
C14H11C1108	342.69	342.0142	49.07	3.24	10.35	37.35
C14H11C1109	358.69	358.0092	46.88	3.09	9.88	40.14
C14H12C1201	267.16	266.0265	62.94	4.53	26.54	5.99

080

C14H12C1202	283.16	282.0214	59.39	4.27	25.04	11.30
C14H12C1203	299.16	298.0163	56.21	4.04	23.70	16.04
C14H12C1204	315.16	314.0113	53.36	3.84	22.50	20.31
C14H12C1205	331.15	330.0062	50.78	3.65	21.41	24.16
C14H12C1206	347.15	346.0011	48.44	3.48	20.42	27.65
C14H12C1207	363.15	361.9960	46.30	3.33	19.53	30.84
C14H12C1208	379.15	377.9909	44.35	3.19	18.70	33.76
C14H12C1209	395.15	393.9858	42.55	3.06	17.94	36.44
C14H13C1101	232.71	232.0655	72.26	5.63	15.23	6.88
C14H13C1102	248.71	248.0604	67.61	5.27	14.25	12.87
C14H13C1103	264.71	264.0553	63.52	4.95	13.39	18.13
C14H13C1104	280.71	280.0502	59.90	4.67	12.63	22.80
C14H13C1105	296.71	296.0451	56.67	4.42	11.95	26.96
C14H13C1106	312.71	312.0401	53.77	4.19	11.34	30.70
C14H13C1107	328.71	328.0350	51.16	3.99	10.79	34.07
C14H13C1108	344.71	344.0299	48.78	3.80	10.28	37.13
C14H13C1109	360.71	360.0248	46.62	3.63	9.83	39.92
C14H14C1201	269.17	268.0422	62.47	5.24	26.34	5.94
C14H14C1202	285.17	284.0371	58.97	4.95	24.86	11.22
C14H14C1203	301.17	300.0320	55.83	4.69	23.54	15.94
C14H14C1204	317.17	316.0269	53.02	4.45	22.36	20.18
C14H14C1205	333.17	332.0218	50.47	4.24	21.28	24.01
C14H14C1206	349.17	348.0167	48.16	4.04	20.31	27.49
C14H14C1207	365.17	364.0116	46.05	3.86	19.42	30.67
C14H14C1208	381.17	380.0066	44.12	3.70	18.60	33.58
C14H14C1209	397.17	396.0015	42.34	3.55	17.85	36.26
C14H15C1101	234.73	234.0811	71.64	6.44	15.10	6.82
C14H15C1102	250.73	250.0761	67.07	6.03	14.14	12.76
C14H15C1103	266.73	266.0710	63.04	5.67	13.29	18.00
C14H15C1104	282.73	282.0659	59.48	5.35	12.54	22.64
C14H15C1105	298.73	298.0608	56.29	5.06	11.87	26.78
C14H15C1106	314.73	314.0557	53.43	4.80	11.26	30.50
C14H15C1107	330.72	330.0506	50.84	4.57	10.72	33.86
C14H15C1108	346.72	346.0455	48.50	4.36	10.23	36.92
C14H15C1109	362.72	362.0404	46.36	4.17	9.77	39.70
C14H16C1201	271.19	270.0578	62.01	5.95	26.15	5.90
C14H16C1202	287.19	286.0527	58.55	5.62	24.69	11.14
C14H16C1203	303.19	302.0476	55.46	5.32	23.39	15.83
C14H16C1204	319.19	318.0426	52.68	5.05	22.21	20.05
C14H16C1205	335.19	334.0375	50.17	4.81	21.15	23.87
C14H16C1206	351.19	350.0324	47.88	4.59	20.19	27.33
C14H16C1207	367.19	366.0273	45.80	4.39	19.31	30.50
C14H16C1208	383.18	382.0222	43.88	4.21	18.50	33.40
C14H16C1209	399.18	398.0171	42.12	4.04	17.76	36.07
C14H17C1101	236.74	236.0968	71.03	7.24	14.98	6.76
C14H17C1102	252.74	252.0917	66.53	6.78	14.03	12.66
C14H17C1103	268.74	268.0866	62.57	6.38	13.19	17.86
C14H17C1104	284.74	284.0815	59.06	6.02	12.45	22.48
C14H17C1105	300.74	300.0764	55.91	5.70	11.79	26.60
C14H17C1106	316.74	316.0714	53.09	5.41	11.19	30.31
C14H17C1107	332.74	332.0663	50.54	5.15	10.65	33.66
C14H17C1108	348.74	348.0612	48.22	4.91	10.17	36.70
C14H17C1109	364.74	364.0561	46.10	4.70	9.72	39.48
C14H18C1201	273.20	272.0735	61.55	6.64	25.95	5.86
C14H18C1202	289.20	288.0684	58.14	6.27	24.52	11.06
C14H18C1203	305.20	304.0633	55.10	5.94	23.23	15.73
C14H18C1204	321.20	320.0582	52.35	5.65	22.08	19.92
C14H18C1205	337.20	336.0531	49.87	5.38	21.03	23.72
C14H18C1206	353.20	352.0480	47.61	5.14	20.08	27.18
C14H18C1207	369.20	368.0429	45.55	4.91	19.21	30.33
C14H18C1208	385.20	384.0379	43.65	4.71	18.41	33.23
C14H18C1209	401.20	400.0328	41.91	4.52	17.67	35.89
C14H19C1101	238.76	238.1124	70.43	8.02	14.85	6.70
C14H19C1102	254.76	254.1073	66.01	7.52	13.92	12.56
C14H19C1103	270.76	270.1023	62.11	7.07	13.09	17.73
C14H19C1104	286.76	286.0972	58.64	6.68	12.36	22.32
C14H19C1105	302.76	302.0921	55.54	6.33	11.71	26.42
C14H19C1106	318.76	318.0870	52.75	6.01	11.12	30.12
C14H19C1107	334.76	334.0819	50.23	5.72	10.59	33.46
C14H19C1108	350.76	350.0768	47.94	5.46	10.11	36.49
C14H19C1109	366.76	366.0717	45.85	5.22	9.67	39.26
C14H20C1201	275.22	274.0891	61.10	7.32	25.76	5.81
C14H20C1202	291.22	290.0840	57.74	6.92	24.35	10.99
C14H20C1203	307.22	306.0789	54.73	6.56	23.08	15.62
C14H20C1204	323.22	322.0739	52.03	6.24	21.94	19.80
C14H20C1205	339.22	338.0688	49.57	5.94	20.90	23.58
C14H20C1206	355.22	354.0637	47.34	5.68	19.96	27.02
C14H20C1207	371.22	370.0586	45.30	5.43	19.10	30.17
C14H20C1208	387.22	386.0535	43.43	5.21	18.31	33.06
C14H20C1209	403.22	402.0484	41.70	5.00	17.59	35.71
C14H21C1101	240.78	240.1281	69.84	8.79	14.72	6.64
C14H21C1102	256.78	256.1230	65.49	8.24	13.81	12.46
C14H21C1103	272.77	272.1179	61.65	7.76	13.00	17.60
C14H21C1104	288.77	288.1128	58.23	7.33	12.28	22.16
C14H21C1105	304.77	304.1077	55.17	6.95	11.63	26.25
C14H21C1106	320.77	320.1027	52.42	6.60	11.05	29.93
C14H21C1107	336.77	336.0976	49.93	6.29	10.53	33.26
C14H21C1108	352.77	352.0925	47.67	6.00	10.05	36.28
C14H21C1109	368.77	368.0874	45.60	5.74	9.61	39.05
C14H22C1201	277.24	276.1048	60.65	8.00	25.58	5.77
C14H22C1202	293.24	292.0997	57.34	7.56	24.18	10.91
C14H22C1203	309.24	308.0946	54.38	7.17	22.93	15.52
C14H22C1204	325.24	324.0895	51.70	6.82	21.80	19.68
C14H22C1205	341.23	340.0844	49.28	6.50	20.78	23.44
C14H22C1206	357.23	356.0793	47.07	6.21	19.85	26.87
C14H22C1207	373.23	372.0742	45.05	5.94	19.00	30.01
C14H22C1208	389.23	388.0692	43.20	5.70	18.22	32.88
C14H22C1209	405.23	404.0641	41.50	5.47	17.50	35.53
C14H23C1101	242.79	242.1437	69.26	9.55	14.60	6.59
C14H23C1102	258.79	258.1386	64.98	8.96	13.70	12.36
C14H23C1103	274.79	274.1336	61.19	8.44	12.90	17.47
C14H23C1104	290.79	290.1285	57.83	7.97	12.19	22.01
C14H23C1105	306.79	306.1234	54.81	7.56	11.56	26.08
C14H23C1106	322.79	322.1183	52.09	7.18	10.98	29.74
C14H23C1107	338.79	338.1132	49.63	6.84	10.46	33.06
C14H23C1108	354.79	354.1081	47.40	6.53	9.99	36.08
C14H23C1109	370.79	370.1030	45.35	6.25	9.56	38.83
C14H24C1201	279.25	278.1204	60.22	8.66	25.39	5.73
C14H24C1202	295.25	294.1153	56.95	8.19	24.02	10.84
C14H24C1203	311.25	310.1102	54.03	7.77	22.78	15.42
C14H24C1204	327.25	326.1052	51.38	7.39	21.67	19.56
C14H24C1205	343.25	342.1001	48.99	7.05	20.66	23.31
C14H24C1206	359.25	358.0950	46.81	6.73	19.74	26.72
C14H24C1207	375.25	374.0899	44.81	6.45	18.90	29.85
C14H24C1208	391.25	390.0848	42.98	6.18	18.12	32.71
C14H24C1209	407.25	406.0797	41.29	5.94	17.41	35.36
C14H25C1101	244.81	244.1594	68.69	10.29	14.48	6.54
C14H25C1102	260.81	260.1543	64.48	9.66	13.59	12.27
C14H25C1103	276.81	276.1492	60.75	9.10	12.81	17.34
C14H25C1104	292.81	292.1441	57.43	8.61	12.11	21.86

081

C14H25C1105	308.81	308.1390	54.45	8.16	11.48	25.91
C14H25C1106	324.80	324.1340	51.77	7.76	10.92	29.56
C14H25C1107	340.80	340.1289	49.34	7.39	10.40	32.86
C14H25C1108	356.80	356.1238	47.13	7.06	9.94	35.87
C14H25C1109	372.80	372.1187	45.11	6.76	9.51	38.62
C14H26C1201	281.27	280.1361	59.78	9.32	25.21	5.69
C14H26C1202	297.27	296.1310	56.57	8.82	23.85	10.76
C14H26C1203	313.27	312.1259	53.68	8.37	22.63	15.32
C14H26C1204	329.27	328.1208	51.07	7.96	21.53	19.44
C14H26C1205	345.27	344.1157	48.70	7.59	20.54	23.17
C14H26C1206	361.27	360.1106	46.55	7.25	19.63	26.57
C14H26C1207	377.27	376.1055	44.57	6.95	18.79	29.69
C14H26C1208	393.26	392.1005	42.76	6.66	18.03	32.55
C14H26C1209	409.26	408.0954	41.09	6.40	17.33	35.18
C14H27C1101	246.82	246.1750	68.13	11.03	14.36	6.48
C14H27C1102	262.82	262.1699	63.98	10.35	13.49	12.18
C14H27C1103	278.82	278.1649	60.31	9.76	12.72	17.21
C14H27C1104	294.82	294.1598	57.04	9.23	12.03	21.71
C14H27C1105	310.82	310.1547	54.10	8.76	11.41	25.74
C14H27C1106	326.82	326.1496	51.45	8.33	10.85	29.37
C14H27C1107	342.82	342.1445	49.05	7.94	10.34	32.67
C14H27C1108	358.82	358.1394	46.86	7.58	9.88	35.67
C14H27C1109	374.82	374.1343	44.86	7.26	9.46	38.42
C14H28C1201	283.28	282.1517	59.36	9.96	25.03	5.65
C14H28C1202	299.28	298.1466	56.19	9.43	23.69	10.69
C14H28C1203	315.28	314.1415	53.33	8.95	22.49	15.22
C14H28C1204	331.28	330.1365	50.76	8.52	21.40	19.32
C14H28C1205	347.28	346.1314	48.42	8.13	20.42	23.04
C14H28C1206	363.28	362.1263	46.29	7.77	19.52	26.42
C14H28C1207	379.28	378.1212	44.34	7.44	18.69	29.53
C14H28C1208	395.28	394.1161	42.54	7.14	17.94	32.38
C14H28C1209	411.28	410.1110	40.89	6.86	17.24	35.01
C14H29C1101	248.84	248.1907	67.58	11.75	14.25	6.43
C14H29C1102	264.84	264.1856	63.49	11.04	13.39	12.08
C14H29C1103	280.84	280.1805	59.88	10.41	12.62	17.09
C14H29C1104	296.84	296.1754	56.65	9.85	11.94	21.56
C14H29C1105	312.84	312.1703	53.75	9.34	11.33	25.57
C14H29C1106	328.84	328.1653	51.14	8.89	10.78	29.19
C14H29C1107	344.84	344.1602	48.76	8.48	10.28	32.48
C14H29C1108	360.84	360.1551	46.60	8.10	9.83	35.47
C14H29C1109	376.83	376.1500	44.62	7.76	9.41	38.21
C15H 7C1101	238.68	238.0185	75.49	2.96	14.85	6.70
C15H 7C1102	254.67	254.0135	70.74	2.77	13.92	12.56
C15H 7C1103	270.67	270.0084	66.56	2.61	13.10	17.73
C15H 7C1104	286.67	286.0033	62.85	2.46	12.37	22.32
C15H 7C1105	302.67	301.9982	59.53	2.33	11.71	26.43
C15H 7C1106	318.67	317.9931	56.54	2.21	11.13	30.12
C15H 7C1107	334.67	333.9880	53.83	2.11	10.59	33.46
C15H 7C1108	350.67	349.9829	51.38	2.01	10.11	36.50
C15H 7C1109	366.67	365.9779	49.14	1.92	9.67	39.27
C15H 8C1201	275.14	273.9952	65.48	2.93	25.77	5.82
C15H 8C1202	291.14	289.9901	61.88	2.77	24.35	10.99
C15H 8C1203	307.14	305.9850	58.66	2.63	23.09	15.63
C15H 8C1204	323.13	321.9800	55.76	2.50	21.94	19.81
C15H 8C1205	339.13	337.9749	53.13	2.38	20.91	23.59
C15H 8C1206	355.13	353.9698	50.73	2.27	19.97	27.03
C15H 8C1207	371.13	369.9647	48.55	2.17	19.11	30.18
C15H 8C1208	387.13	385.9596	46.54	2.08	18.32	33.06
C15H 8C1209	403.13	401.9545	44.69	2.00	17.59	35.72
C15H 9C1101	240.69	240.0342	74.85	3.77	14.73	6.65

C15H 9C1102	256.69	256.0291	70.19	3.53	13.81	12.47
C15H 9C1103	272.69	272.0240	66.07	3.33	13.00	17.60
C15H 9C1104	288.69	288.0189	62.41	3.14	12.28	22.17
C15H 9C1105	304.69	304.0138	59.13	2.98	11.64	26.26
C15H 9C1106	320.69	320.0088	56.18	2.83	11.06	29.93
C15H 9C1107	336.69	336.0037	53.51	2.69	10.53	33.26
C15H 9C1108	352.69	351.9986	51.08	2.57	10.05	36.29
C15H 9C1109	368.69	367.9935	48.87	2.46	9.62	39.06
C15H10C1201	277.15	276.0109	65.01	3.64	25.58	5.77
C15H10C1202	293.15	292.0058	61.46	3.44	24.19	10.92
C15H10C1203	309.15	308.0007	58.28	3.26	22.94	15.53
C15H10C1204	325.15	323.9956	55.41	3.10	21.81	19.68
C15H10C1205	341.15	339.9905	52.81	2.95	20.78	23.45
C15H10C1206	357.15	355.9854	50.45	2.82	19.85	26.88
C15H10C1207	373.15	371.9804	48.28	2.70	19.00	30.01
C15H10C1208	389.15	387.9753	46.30	2.59	18.22	32.89
C15H10C1209	405.15	403.9702	44.47	2.49	17.50	35.54
C15H11C1101	242.71	242.0498	74.23	4.57	14.61	6.59
C15H11C1102	258.71	258.0448	69.64	4.29	13.70	12.37
C15H11C1103	274.71	274.0397	65.59	4.04	12.91	17.47
C15H11C1104	290.71	290.0346	61.98	3.81	12.20	22.01
C15H11C1105	306.70	306.0295	58.74	3.62	11.56	26.08
C15H11C1106	322.70	322.0244	55.83	3.44	10.99	29.75
C15H11C1107	338.70	338.0193	53.19	3.27	10.47	33.07
C15H11C1108	354.70	354.0142	50.79	3.13	10.00	36.09
C15H11C1109	370.70	370.0092	48.60	2.99	9.56	38.84
C15H12C1201	279.17	278.0265	64.54	4.33	25.40	5.73
C15H12C1202	295.17	294.0214	61.04	4.10	24.02	10.84
C15H12C1203	311.17	310.0163	57.90	3.89	22.79	15.43
C15H12C1204	327.17	326.0113	55.07	3.70	21.67	19.56
C15H12C1205	343.17	342.0062	52.50	3.52	20.66	23.31
C15H12C1206	359.17	358.0011	50.16	3.37	19.74	26.73
C15H12C1207	375.16	373.9960	48.02	3.22	18.90	29.85
C15H12C1208	391.16	389.9909	46.06	3.09	18.13	32.72
C15H12C1209	407.16	405.9858	44.25	2.97	17.41	35.37
C15H13C1101	244.72	244.0655	73.62	5.35	14.49	6.54
C15H13C1102	260.72	260.0604	69.10	5.03	13.60	12.27
C15H13C1103	276.72	276.0553	65.11	4.74	12.81	17.35
C15H13C1104	292.72	292.0502	61.55	4.48	12.11	21.86
C15H13C1105	308.72	308.0451	58.36	4.24	11.48	25.91
C15H13C1106	324.72	324.0401	55.48	4.04	10.92	29.56
C15H13C1107	340.72	340.0350	52.88	3.85	10.41	32.87
C15H13C1108	356.72	356.0299	50.51	3.67	9.94	35.88
C15H13C1109	372.72	372.0248	48.34	3.52	9.51	38.63
C15H14C1201	281.18	280.0422	64.07	5.02	25.22	5.69
C15H14C1202	297.18	296.0371	60.62	4.75	23.86	10.77
C15H14C1203	313.18	312.0320	57.53	4.51	22.64	15.33
C15H14C1204	329.18	328.0269	54.73	4.29	21.54	19.44
C15H14C1205	345.18	344.0218	52.19	4.09	20.54	23.18
C15H14C1206	361.18	360.0167	49.88	3.91	19.63	26.58
C15H14C1207	377.18	376.0116	47.77	3.74	18.80	29.69
C15H14C1208	393.18	392.0066	45.82	3.59	18.03	32.55
C15H14C1209	409.18	408.0015	44.03	3.45	17.33	35.19
C15H15C1101	246.74	246.0811	73.02	6.13	14.37	6.48
C15H15C1102	262.74	262.0761	68.57	5.75	13.49	12.18
C15H15C1103	278.74	278.0710	64.64	5.42	12.72	17.22
C15H15C1104	294.74	294.0659	61.13	5.13	12.03	21.71
C15H15C1105	310.74	310.0608	57.98	4.87	11.41	25.74
C15H15C1106	326.74	326.0557	55.14	4.63	10.85	29.38
C15H15C1107	342.74	342.0506	52.57	4.41	10.34	32.68

082

C15H15C1108	358.73	358.0455	50.22	4.21	9.88	35.68
C15H15C1109	374.73	374.0404	48.08	4.03	9.46	38.43
C15H16C1201	283.20	282.0578	63.62	5.69	25.04	5.65
C15H16C1202	299.20	298.0527	60.22	5.39	23.70	10.69
C15H16C1203	315.20	314.0476	57.16	5.12	22.50	15.23
C15H16C1204	331.20	330.0426	54.40	4.87	21.41	19.32
C15H16C1205	347.20	346.0375	51.89	4.65	20.42	23.04
C15H16C1206	363.20	362.0324	49.61	4.44	19.52	26.43
C15H16C1207	379.20	378.0273	47.51	4.25	18.70	29.54
C15H16C1208	395.20	394.0222	45.59	4.08	17.94	32.39
C15H16C1209	411.20	410.0171	43.82	3.92	17.24	35.02
C15H17C1101	248.76	248.0968	72.43	6.89	14.25	6.43
C15H17C1102	264.75	264.0917	68.05	6.47	13.39	12.09
C15H17C1103	280.75	280.0866	64.17	6.10	12.63	17.10
C15H17C1104	296.75	296.0815	60.71	5.77	11.95	21.57
C15H17C1105	312.75	312.0764	57.61	5.48	11.34	25.58
C15H17C1106	328.75	328.0714	54.80	5.21	10.78	29.20
C15H17C1107	344.75	344.0663	52.26	4.97	10.28	32.49
C15H17C1108	360.75	360.0612	49.94	4.75	9.83	35.48
C15H17C1109	376.75	376.0561	47.82	4.55	9.41	38.22
C15H18C1201	285.22	284.0735	63.17	6.36	24.86	5.61
C15H18C1202	301.22	300.0684	59.81	6.02	23.54	10.62
C15H18C1203	317.21	316.0633	56.80	5.72	22.35	15.13
C15H18C1204	333.21	332.0582	54.07	5.44	21.28	19.21
C15H18C1205	349.21	348.0531	51.59	5.20	20.30	22.91
C15H18C1206	365.21	364.0480	49.33	4.97	19.41	26.29
C15H18C1207	381.21	380.0429	47.26	4.76	18.60	29.38
C15H18C1208	397.21	396.0379	45.36	4.57	17.85	32.22
C15H18C1209	413.21	412.0328	43.60	4.39	17.16	34.85
C15H19C1101	250.77	250.1124	71.85	7.64	14.14	6.38
C15H19C1102	266.77	266.1073	67.54	7.18	13.29	11.99
C15H19C1103	282.77	282.1023	63.72	6.77	12.54	16.97
C15H19C1104	298.77	298.0972	60.30	6.41	11.87	21.42
C15H19C1105	314.77	314.0921	57.24	6.08	11.26	25.41
C15H19C1106	330.77	330.0870	54.47	5.79	10.72	29.02
C15H19C1107	346.77	346.0819	51.96	5.52	10.22	32.30
C15H19C1108	362.77	362.0768	49.66	5.28	9.77	35.28
C15H19C1109	378.77	378.0717	47.57	5.06	9.36	38.02
C15H20C1201	287.23	286.0891	62.73	7.02	24.69	5.57
C15H20C1202	303.23	302.0840	59.42	6.65	23.38	10.55
C15H20C1203	319.23	318.0789	56.44	6.31	22.21	15.04
C15H20C1204	335.23	334.0739	53.74	6.01	21.15	19.09
C15H20C1205	351.23	350.0688	51.30	5.74	20.19	22.78
C15H20C1206	367.23	366.0637	49.06	5.49	19.31	26.14
C15H20C1207	383.23	382.0586	47.01	5.26	18.50	29.22
C15H20C1208	399.23	398.0535	45.13	5.05	17.76	32.06
C15H20C1209	415.23	414.0484	43.39	4.86	17.08	34.68
C15H21C1101	252.79	252.1281	71.27	8.37	14.02	6.33
C15H21C1102	268.79	268.1230	67.03	7.88	13.19	11.90
C15H21C1103	284.79	284.1179	63.26	7.43	12.45	16.85
C15H21C1104	300.79	300.1128	59.90	7.04	11.79	21.28
C15H21C1105	316.78	316.1077	56.87	6.68	11.19	25.25
C15H21C1106	332.78	332.1027	54.14	6.36	10.65	28.85
C15H21C1107	348.78	348.0976	51.66	6.07	10.16	32.11
C15H21C1108	364.78	364.0925	49.39	5.80	9.72	35.09
C15H21C1109	380.78	380.0874	47.32	5.56	9.31	37.82
C15H22C1201	289.25	288.1048	62.29	7.67	24.51	5.53
C15H22C1202	305.25	304.0997	59.02	7.26	23.23	10.48
C15H22C1203	321.25	320.0946	56.08	6.90	22.07	14.94
C15H22C1204	337.25	336.0895	53.42	6.58	21.02	18.98

C15H22C1205	353.25	352.0844	51.00	6.28	20.07	22.65
C15H22C1206	369.24	368.0793	48.79	6.01	19.20	26.00
C15H22C1207	385.24	384.0742	46.77	5.76	18.41	29.07
C15H22C1208	401.24	400.0692	44.90	5.53	17.67	31.90
C15H22C1209	417.24	416.0641	43.18	5.31	16.99	34.51
C15H23C1101	254.80	254.1437	70.71	9.10	13.91	6.28
C15H23C1102	270.80	270.1386	66.53	8.56	13.09	11.82
C15H23C1103	286.80	286.1336	62.82	8.08	12.36	16.74
C15H23C1104	302.80	302.1285	59.50	7.66	11.71	21.14
C15H23C1105	318.80	318.1234	56.51	7.27	11.12	25.09
C15H23C1106	334.80	334.1183	53.81	6.92	10.59	28.67
C15H23C1107	350.80	350.1132	51.36	6.61	10.11	31.93
C15H23C1108	366.80	366.1081	49.12	6.32	9.67	34.90
C15H23C1109	382.80	382.1030	47.07	6.06	9.26	37.62
C15H24C1201	291.26	290.1204	61.86	8.31	24.34	5.49
C15H24C1202	307.26	306.1153	58.64	7.87	23.08	10.41
C15H24C1203	323.26	322.1102	55.73	7.48	21.93	14.85
C15H24C1204	339.26	338.1052	53.11	7.13	20.90	18.86
C15H24C1205	355.26	354.1001	50.71	6.81	19.96	22.52
C15H24C1206	371.26	370.0950	48.53	6.52	19.10	25.86
C15H24C1207	387.26	386.0899	46.52	6.25	18.31	28.92
C15H24C1208	403.26	402.0848	44.68	6.00	17.58	31.74
C15H24C1209	419.26	418.0797	42.97	5.77	16.91	34.35
C15H25C1101	256.82	256.1594	70.15	9.81	13.80	6.23
C15H25C1102	272.82	272.1543	66.04	9.24	13.00	11.73
C15H25C1103	288.82	288.1492	62.38	8.72	12.28	16.62
C15H25C1104	304.82	304.1441	59.11	8.27	11.63	21.00
C15H25C1105	320.82	320.1390	56.16	7.85	11.05	24.94
C15H25C1106	336.82	336.1340	53.49	7.48	10.53	28.50
C15H25C1107	352.82	352.1289	51.07	7.14	10.05	31.74
C15H25C1108	368.81	368.1238	48.85	6.83	9.61	34.70
C15H25C1109	384.81	384.1187	46.82	6.55	9.21	37.42
C15H26C1201	293.28	292.1361	61.43	8.94	24.18	5.46
C15H26C1202	309.28	308.1310	58.25	8.47	22.93	10.35
C15H26C1203	325.28	324.1259	55.39	8.06	21.80	14.76
C15H26C1204	341.28	340.1208	52.79	7.68	20.78	18.75
C15H26C1205	357.28	356.1157	50.43	7.34	19.85	22.39
C15H26C1206	373.28	372.1106	48.27	7.02	19.00	25.72
C15H26C1207	389.28	388.1055	46.28	6.73	18.21	28.77
C15H26C1208	405.28	404.1005	44.46	6.47	17.50	31.58
C15H26C1209	421.28	420.0954	42.77	6.22	16.83	34.18
C15H27C1101	258.83	258.1750	69.61	10.51	13.70	6.18
C15H27C1102	274.83	274.1699	65.55	9.90	12.90	11.64
C15H27C1103	290.83	290.1649	61.95	9.36	12.19	16.50
C15H27C1104	306.83	306.1598	58.72	8.87	11.55	20.86
C15H27C1105	322.83	322.1547	55.81	8.43	10.98	24.78
C15H27C1106	338.83	338.1496	53.17	8.03	10.46	28.33
C15H27C1107	354.83	354.1445	50.78	7.67	9.99	31.56
C15H27C1108	370.83	370.1394	48.58	7.34	9.56	34.52
C15H27C1109	386.83	386.1343	46.58	7.04	9.17	37.22
C15H28C1201	295.30	294.1517	61.01	9.56	24.01	5.42
C15H28C1202	311.30	310.1466	57.88	9.07	22.78	10.28
C15H28C1203	327.29	326.1415	55.05	8.62	21.66	14.67
C15H28C1204	343.29	342.1365	52.48	8.22	20.65	18.64
C15H28C1205	359.29	358.1314	50.14	7.86	19.73	22.27
C15H28C1206	375.29	374.1263	48.01	7.52	18.89	25.58
C15H28C1207	391.29	390.1212	46.04	7.21	18.12	28.62
C15H28C1208	407.29	406.1161	44.24	6.93	17.41	31.43
C15H28C1209	423.29	422.1110	42.56	6.67	16.75	34.02
C15H29C1101	260.85	260.1907	69.07	11.21	13.59	6.13

C15H29C1102	276.85	276.1856	65.08	10.56	12.81	11.56
C15H29C1103	292.85	292.1805	61.52	9.98	12.11	16.39
C15H29C1104	308.85	308.1754	58.34	9.46	11.48	20.72
C15H29C1105	324.85	324.1703	55.46	9.00	10.91	24.63
C15H29C1106	340.85	340.1653	52.86	8.58	10.40	28.16
C15H29C1107	356.85	356.1602	50.49	8.19	9.94	31.38
C15H29C1108	372.85	372.1551	48.32	7.84	9.51	34.33
C15H29C1109	388.85	388.1500	46.33	7.52	9.12	37.03
C15H30C1201	297.31	296.1674	60.60	10.17	23.85	5.38
C15H30C1202	313.31	312.1623	57.50	9.65	22.63	10.21
C15H30C1203	329.31	328.1572	54.71	9.18	21.53	14.58
C15H30C1204	345.31	344.1521	52.18	8.76	20.53	18.53
C15H30C1205	361.31	360.1470	49.87	8.37	19.62	22.14
C15H30C1206	377.31	376.1419	47.75	8.01	18.79	25.44
C15H30C1207	393.31	392.1368	45.81	7.69	18.03	28.48
C15H30C1208	409.31	408.1318	44.02	7.39	17.32	31.27
C15H30C1209	425.31	424.1267	42.36	7.11	16.67	33.86
C15H31C1101	262.87	262.2063	68.54	11.89	13.49	6.09
C15H31C1102	278.87	278.2012	64.61	11.21	12.71	11.47
C15H31C1103	294.87	294.1962	61.10	10.60	12.02	16.28
C15H31C1104	310.86	310.1911	57.96	10.05	11.40	20.59
C15H31C1105	326.86	326.1860	55.12	9.56	10.85	24.47
C15H31C1106	342.86	342.1809	52.55	9.11	10.34	28.00
C15H31C1107	358.86	358.1758	50.21	8.71	9.88	31.21
C15H31C1108	374.86	374.1707	48.06	8.34	9.46	34.14
C15H31C1109	390.86	390.1656	46.09	7.99	9.07	36.84
C16H 8C1201	287.15	285.9952	66.93	2.81	24.69	5.57
C16H 8C1202	303.15	301.9901	63.39	2.66	23.39	10.56
C16H 8C1203	319.15	317.9850	60.22	2.53	22.22	15.04
C16H 8C1204	335.15	333.9800	57.34	2.41	21.16	19.10
C16H 8C1205	351.15	349.9749	54.73	2.30	20.19	22.78
C16H 8C1206	367.14	365.9698	52.34	2.20	19.31	26.15
C16H 8C1207	383.14	381.9647	50.16	2.10	18.51	29.23
C16H 8C1208	399.14	397.9596	48.15	2.02	17.76	32.07
C16H 8C1209	415.14	413.9545	46.29	1.94	17.08	34.69
C16H 9C1101	252.70	252.0342	76.05	3.59	14.03	6.33
C16H 9C1102	268.70	268.0291	71.52	3.38	13.19	11.91
C16H 9C1103	284.70	284.0240	67.50	3.19	12.45	16.86
C16H 9C1104	300.70	300.0189	63.91	3.02	11.79	21.28
C16H 9C1105	316.70	316.0138	60.68	2.86	11.19	25.26
C16H 9C1106	332.70	332.0088	57.76	2.73	10.66	28.85
C16H 9C1107	348.70	348.0037	55.11	2.60	10.17	32.12
C16H 9C1108	364.70	363.9986	52.70	2.49	9.72	35.10
C16H 9C1109	380.70	379.9935	50.48	2.38	9.31	37.82
C16H10C1201	289.16	288.0109	66.46	3.49	24.52	5.53
C16H10C1202	305.16	304.0058	62.98	3.30	23.24	10.49
C16H10C1203	321.16	320.0007	59.84	3.14	22.08	14.95
C16H10C1204	337.16	335.9956	57.00	2.99	21.03	18.98
C16H10C1205	353.16	351.9905	54.42	2.85	20.08	22.65
C16H10C1206	369.16	367.9854	52.06	2.73	19.21	26.00
C16H10C1207	385.16	383.9804	49.90	2.62	18.41	29.08
C16H10C1208	401.16	399.9753	47.91	2.51	17.68	31.91
C16H10C1209	417.16	415.9702	46.07	2.42	17.00	34.52
C16H11C1101	254.72	254.0498	75.45	4.35	13.92	6.28
C16H11C1102	270.72	270.0448	70.99	4.10	13.10	11.82
C16H11C1103	286.72	286.0397	67.03	3.87	12.37	16.74
C16H11C1104	302.72	302.0346	63.48	3.66	11.71	21.14
C16H11C1105	318.72	318.0295	60.30	3.48	11.12	25.10
C16H11C1106	334.72	334.0244	57.42	3.31	10.59	28.68
C16H11C1107	350.71	350.0193	54.80	3.16	10.11	31.93
C16H11C1108	366.71	366.0142	52.41	3.02	9.67	34.90
C16H11C1109	382.71	382.0092	50.21	2.90	9.26	37.62
C16H12C1201	291.18	290.0265	66.00	4.15	24.35	5.49
C16H12C1202	307.18	306.0214	62.56	3.94	23.08	10.42
C16H12C1203	323.18	322.0163	59.47	3.74	21.94	14.85
C16H12C1204	339.18	338.0113	56.66	3.57	20.91	18.87
C16H12C1205	355.18	354.0062	54.11	3.41	19.96	22.52
C16H12C1206	371.18	370.0011	51.78	3.26	19.10	25.86
C16H12C1207	387.18	385.9960	49.64	3.12	18.31	28.93
C16H12C1208	403.18	401.9909	47.67	3.00	17.59	31.75
C16H12C1209	419.17	417.9858	45.85	2.89	16.92	34.35
C16H13C1101	256.73	256.0655	74.85	5.10	13.81	6.23
C16H13C1102	272.73	272.0604	70.46	4.80	13.00	11.73
C16H13C1103	288.73	288.0553	66.56	4.54	12.28	16.62
C16H13C1104	304.73	304.0502	63.06	4.30	11.63	21.00
C16H13C1105	320.73	320.0451	59.92	4.09	11.05	24.94
C16H13C1106	336.73	336.0401	57.07	3.89	10.53	28.51
C16H13C1107	352.73	352.0350	54.48	3.71	10.05	31.75
C16H13C1108	368.73	368.0299	52.12	3.55	9.61	34.71
C16H13C1109	384.73	384.0248	49.95	3.41	9.22	37.43
C16H14C1201	293.20	292.0422	65.55	4.81	24.18	5.46
C16H14C1202	309.19	308.0371	62.15	4.56	22.93	10.35
C16H14C1203	325.19	324.0320	59.10	4.34	21.80	14.76
C16H14C1204	341.19	340.0269	56.33	4.14	20.78	18.76
C16H14C1205	357.19	356.0218	53.80	3.95	19.85	22.40
C16H14C1206	373.19	372.0167	51.50	3.78	19.00	25.72
C16H14C1207	389.19	388.0116	49.38	3.63	18.22	28.78
C16H14C1208	405.19	404.0066	47.43	3.48	17.50	31.59
C16H14C1209	421.19	420.0015	45.63	3.35	16.83	34.19
C16H15C1101	258.75	258.0811	74.27	5.84	13.70	6.18
C16H15C1102	274.75	274.0761	69.95	5.50	12.90	11.65
C16H15C1103	290.75	290.0710	66.10	5.20	12.19	16.51
C16H15C1104	306.75	306.0659	62.65	4.93	11.56	20.86
C16H15C1105	322.75	322.0608	59.54	4.68	10.98	24.79
C16H15C1106	338.75	338.0557	56.73	4.46	10.47	28.34
C16H15C1107	354.75	354.0506	54.17	4.26	9.99	31.57
C16H15C1108	370.75	370.0455	51.84	4.08	9.56	34.52
C16H15C1109	386.75	386.0404	49.69	3.91	9.17	37.23
C16H16C1201	295.21	294.0578	65.10	5.46	24.02	5.42
C16H16C1202	311.21	310.0527	61.75	5.18	22.78	10.28
C16H16C1203	327.21	326.0476	58.73	4.93	21.67	14.67
C16H16C1204	343.21	342.0426	55.99	4.70	20.66	18.65
C16H16C1205	359.21	358.0375	53.50	4.49	19.74	22.27
C16H16C1206	375.21	374.0324	51.22	4.30	18.90	25.58
C16H16C1207	391.21	390.0273	49.12	4.12	18.12	28.63
C16H16C1208	407.21	406.0222	47.19	3.96	17.41	31.43
C16H16C1209	423.21	422.0171	45.41	3.81	16.75	34.02
C16H17C1101	260.77	260.0968	73.70	6.57	13.60	6.14
C16H17C1102	276.77	276.0917	69.44	6.19	12.81	11.56
C16H17C1103	292.77	292.0866	65.64	5.85	12.11	16.39
C16H17C1104	308.76	308.0815	62.24	5.55	11.48	20.73
C16H17C1105	324.76	324.0764	59.17	5.28	10.92	24.63
C16H17C1106	340.76	340.0714	56.40	5.03	10.40	28.17
C16H17C1107	356.76	356.0663	53.87	4.80	9.94	31.39
C16H17C1108	372.76	372.0612	51.56	4.60	9.51	34.34
C16H17C1109	388.76	388.0561	49.43	4.41	9.12	37.04
C16H18C1201	297.23	296.0735	64.66	6.10	23.86	5.38
C16H18C1202	313.23	312.0684	61.35	5.79	22.64	10.22
C16H18C1203	329.23	328.0633	58.37	5.51	21.54	14.58
C16H18C1204	345.23	344.0582	55.67	5.26	20.54	18.54

084

C16H18C1205	361.22	360.0531	53.20	5.02	19.63	22.15
C16H18C1206	377.22	376.0480	50.95	4.81	18.80	25.45
C16H18C1207	393.22	392.0429	48.87	4.61	18.03	28.48
C16H18C1208	409.22	408.0379	46.96	4.43	17.33	31.28
C16H18C1209	425.22	424.0328	45.19	4.27	16.68	33.86
C16H19C1101	262.78	262.1124	73.13	7.29	13.49	6.09
C16H19C1102	278.78	278.1073	68.94	6.87	12.72	11.48
C16H19C1103	294.78	294.1023	65.19	6.50	12.03	16.28
C16H19C1104	310.78	310.0972	61.84	6.16	11.41	20.59
C16H19C1105	326.78	326.0921	58.81	5.86	10.85	24.48
C16H19C1106	342.78	342.0870	56.06	5.59	10.34	28.01
C16H19C1107	358.78	358.0819	53.56	5.34	9.88	31.22
C16H19C1108	374.78	374.0768	51.28	5.11	9.46	34.15
C16H19C1109	390.78	390.0717	49.18	4.90	9.07	36.85
C16H20C1201	299.24	298.0891	64.22	6.74	23.70	5.35
C16H20C1202	315.24	314.0840	60.96	6.39	22.49	10.15
C16H20C1203	331.24	330.0789	58.02	6.09	21.41	14.49
C16H20C1204	347.24	346.0739	55.34	5.81	20.42	18.43
C16H20C1205	363.24	362.0688	52.91	5.55	19.52	22.02
C16H20C1206	379.24	378.0637	50.67	5.32	18.70	25.31
C16H20C1207	395.24	394.0586	48.62	5.10	17.94	28.34
C16H20C1208	411.24	410.0535	46.73	4.90	17.24	31.12
C16H20C1209	427.24	426.0484	44.98	4.72	16.60	33.70
C16H21C1101	264.80	264.1281	72.58	7.99	13.39	6.04
C16H21C1102	280.80	280.1230	68.44	7.54	12.63	11.40
C16H21C1103	296.80	296.1179	64.75	7.13	11.95	16.17
C16H21C1104	312.80	312.1128	61.44	6.77	11.33	20.46
C16H21C1105	328.80	328.1077	58.45	6.44	10.78	24.33
C16H21C1106	344.80	344.1027	55.74	6.14	10.28	27.84
C16H21C1107	360.79	360.0976	53.27	5.87	9.83	31.04
C16H21C1108	376.79	376.0925	51.00	5.62	9.41	33.97
C16H21C1109	392.79	392.0874	48.93	5.39	9.03	36.66
C16H22C1201	301.26	300.1048	63.79	7.36	23.54	5.31
C16H22C1202	317.26	316.0997	60.57	6.99	22.35	10.09
C16H22C1203	333.26	332.0946	57.67	6.65	21.28	14.40
C16H22C1204	349.26	348.0895	55.02	6.35	20.30	18.32
C16H22C1205	365.26	364.0844	52.61	6.07	19.41	21.90
C16H22C1206	381.26	380.0793	50.41	5.82	18.60	25.18
C16H22C1207	397.26	396.0742	48.38	5.58	17.85	28.19
C16H22C1208	413.25	412.0692	46.50	5.37	17.16	30.97
C16H22C1209	429.25	428.0641	44.77	5.17	16.52	33.55
C16H23C1101	266.81	266.1437	72.03	8.69	13.29	6.00
C16H23C1102	282.81	282.1386	67.95	8.20	12.54	11.31
C16H23C1103	298.81	298.1336	64.31	7.76	11.86	16.06
C16H23C1104	314.81	314.1285	61.05	7.36	11.26	20.33
C16H23C1105	330.81	330.1234	58.09	7.01	10.72	24.18
C16H23C1106	346.81	346.1183	55.41	6.68	10.22	27.68
C16H23C1107	362.81	362.1132	52.97	6.39	9.77	30.87
C16H23C1108	378.81	378.1081	50.73	6.12	9.36	33.79
C16H23C1109	394.81	394.1030	48.68	5.87	8.98	36.47
C16H24C1201	303.28	302.1204	63.37	7.98	23.38	5.28
C16H24C1202	319.27	318.1153	60.19	7.58	22.21	10.02
C16H24C1203	335.27	334.1102	57.32	7.22	21.15	14.32
C16H24C1204	351.27	350.1052	54.71	6.89	20.19	18.22
C16H24C1205	367.27	366.1001	52.33	6.59	19.31	21.78
C16H24C1206	383.27	382.0950	50.14	6.31	18.50	25.05
C16H24C1207	399.27	398.0899	48.13	6.06	17.76	28.05
C16H24C1208	415.27	414.0848	46.28	5.83	17.07	30.82
C16H24C1209	431.27	430.0797	44.56	5.61	16.44	33.39
C16H25C1101	268.83	268.1594	71.49	9.37	13.19	5.95
C16H25C1102	284.83	284.1543	67.47	8.85	12.45	11.23
C16H25C1103	300.83	300.1492	63.88	8.38	11.79	15.96
C16H25C1104	316.83	316.1441	60.66	7.95	11.19	20.20
C16H25C1105	332.83	332.1390	57.74	7.57	10.65	24.04
C16H25C1106	348.83	348.1340	55.09	7.22	10.16	27.52
C16H25C1107	364.83	364.1289	52.68	6.91	9.72	30.70
C16H25C1108	380.83	380.1238	50.46	6.62	9.31	33.61
C16H25C1109	396.83	396.1187	48.43	6.35	8.93	36.29
C16H26C1201	305.29	304.1361	62.95	8.58	23.23	5.24
C16H26C1202	321.29	320.1310	59.81	8.16	22.07	9.96
C16H26C1203	337.29	336.1259	56.98	7.77	21.02	14.23
C16H26C1204	353.29	352.1208	54.40	7.42	20.07	18.11
C16H26C1205	369.29	368.1157	52.04	7.10	19.20	21.66
C16H26C1206	385.29	384.1106	49.88	6.80	18.40	24.92
C16H26C1207	401.29	400.1055	47.89	6.53	17.67	27.91
C16H26C1208	417.29	416.1005	46.05	6.28	16.99	30.67
C16H26C1209	433.29	432.0954	44.35	6.05	16.36	33.23
C16H27C1101	270.85	270.1750	70.95	10.05	13.09	5.91
C16H27C1102	286.85	286.1699	67.00	9.49	12.36	11.16
C16H27C1103	302.84	302.1649	63.46	8.99	11.71	15.85
C16H27C1104	318.84	318.1598	60.27	8.54	11.12	20.07
C16H27C1105	334.84	334.1547	57.39	8.13	10.59	23.89
C16H27C1106	350.84	350.1496	54.78	7.76	10.11	27.36
C16H27C1107	366.84	366.1445	52.39	7.42	9.66	30.53
C16H27C1108	382.84	382.1394	50.20	7.11	9.26	33.43
C16H27C1109	398.84	398.1343	48.18	6.82	8.89	36.10
C16H28C1201	307.31	306.1517	62.54	9.18	23.07	5.21
C16H28C1202	323.31	322.1466	59.44	8.73	21.93	9.90
C16H28C1203	339.31	338.1415	56.64	8.32	20.90	14.15
C16H28C1204	355.31	354.1365	54.09	7.94	19.96	18.01
C16H28C1205	371.30	370.1314	51.76	7.60	19.10	21.54
C16H28C1206	387.30	386.1263	49.62	7.29	18.31	24.79
C16H28C1207	403.30	402.1212	47.65	7.00	17.58	27.77
C16H28C1208	419.30	418.1161	45.83	6.73	16.91	30.53
C16H28C1209	435.30	434.1110	44.15	6.48	16.29	33.08
C16H29C1101	272.86	272.1907	70.43	10.71	12.99	5.86
C16H29C1102	288.86	288.1856	66.53	10.12	12.27	11.08
C16H29C1103	304.86	304.1805	63.04	9.59	11.63	15.74
C16H29C1104	320.86	320.1754	59.89	9.11	11.05	19.95
C16H29C1105	336.86	336.1703	57.05	8.68	10.52	23.75
C16H29C1106	352.86	352.1653	54.46	8.28	10.05	27.21
C16H29C1107	368.86	368.1602	52.10	7.92	9.61	30.36
C16H29C1108	384.86	384.1551	49.93	7.60	9.21	33.26
C16H29C1109	400.86	400.1500	47.94	7.29	8.84	35.92
C16H30C1201	309.32	308.1674	62.13	9.78	22.92	5.17
C16H30C1202	325.32	324.1623	59.07	9.30	21.80	9.84
C16H30C1203	341.32	340.1572	56.30	8.86	20.77	14.06
C16H30C1204	357.32	356.1521	53.78	8.46	19.84	17.91
C16H30C1205	373.32	372.1470	51.48	8.10	18.99	21.43
C16H30C1206	389.32	388.1419	49.36	7.77	18.21	24.66
C16H30C1207	405.32	404.1368	47.41	7.46	17.49	27.63
C16H30C1208	421.32	420.1318	45.61	7.18	16.83	30.38
C16H30C1209	437.32	436.1267	43.94	6.91	16.21	32.93
C16H31C1101	274.88	274.2063	69.91	11.37	12.90	5.82
C16H31C1102	290.88	290.2012	66.07	10.74	12.19	11.00
C16H31C1103	306.88	306.1962	62.62	10.18	11.55	15.64
C16H31C1104	322.88	322.1911	59.52	9.68	10.98	19.82
C16H31C1105	338.88	338.1860	56.71	9.22	10.46	23.61
C16H31C1106	354.87	354.1809	54.15	8.81	9.99	27.05
C16H31C1107	370.87	370.1758	51.82	8.43	9.56	30.20

085

C16H31C1108	386.87	386.1707	49.67	8.08	9.16	33.08
C16H31C1109	402.87	402.1656	47.70	7.76	8.80	35.74
C16H32C1201	311.34	310.1830	61.73	10.36	22.77	5.14
C16H32C1202	327.34	326.1779	58.71	9.85	21.66	9.78
C16H32C1203	343.34	342.1728	55.97	9.39	20.65	13.98
C16H32C1204	359.34	358.1677	53.48	8.98	19.73	17.81
C16H32C1205	375.34	374.1627	51.20	8.59	18.89	21.31
C16H32C1206	391.34	390.1576	49.11	8.24	18.12	24.53
C16H32C1207	407.34	406.1525	47.18	7.92	17.41	27.49
C16H32C1208	423.33	422.1474	45.40	7.62	16.75	30.23
C16H32C1209	439.33	438.1423	43.74	7.34	16.14	32.78
C16H33C1101	276.89	276.2220	69.41	12.01	12.80	5.78
C16H33C1102	292.89	292.2169	65.61	11.36	12.10	10.93
C16H33C1103	308.89	308.2118	62.22	10.77	11.48	15.54
C16H33C1104	324.89	324.2067	59.15	10.24	10.91	19.70
C16H33C1105	340.89	340.2016	56.38	9.76	10.40	23.47
C16H33C1106	356.89	356.1965	53.85	9.32	9.93	26.90
C16H33C1107	372.89	372.1915	51.54	8.92	9.51	30.03
C16H33C1108	388.89	388.1864	49.42	8.55	9.12	32.91
C16H33C1109	404.89	404.1813	47.46	8.22	8.76	35.56
C17H 8C1201	299.16	297.9952	68.25	2.70	23.70	5.35
C17H 8C1202	315.16	313.9901	64.79	2.56	22.50	10.15
C17H 8C1203	331.16	329.9850	61.66	2.44	21.41	14.49
C17H 8C1204	347.16	345.9800	58.82	2.32	20.42	18.43
C17H 8C1205	363.16	361.9749	56.23	2.22	19.52	22.03
C17H 8C1206	379.16	377.9698	53.85	2.13	18.70	25.32
C17H 8C1207	395.16	393.9647	51.67	2.04	17.94	28.34
C17H 8C1208	411.15	409.9596	49.66	1.96	17.25	31.13
C17H 8C1209	427.15	425.9545	47.80	1.89	16.60	33.71
C17H 9C1101	264.71	264.0342	77.14	3.43	13.39	6.04
C17H 9C1102	280.71	280.0291	72.74	3.23	12.63	11.40
C17H 9C1103	296.71	296.0240	68.82	3.06	11.95	16.18
C17H 9C1104	312.71	312.0189	65.30	2.90	11.34	20.47
C17H 9C1105	328.71	328.0138	62.12	2.76	10.79	24.34
C17H 9C1106	344.71	344.0088	59.24	2.63	10.28	27.85
C17H 9C1107	360.71	360.0037	56.61	2.51	9.83	31.05
C17H 9C1108	376.71	375.9986	54.20	2.41	9.41	33.98
C17H 9C1109	392.71	391.9935	52.00	2.31	9.03	36.67
C17H10C1201	301.17	300.0109	67.80	3.35	23.54	5.31
C17H10C1202	317.17	316.0058	64.38	3.18	22.36	10.09
C17H10C1203	333.17	332.0007	61.29	3.03	21.28	14.41
C17H10C1204	349.17	347.9956	58.48	2.89	20.31	18.33
C17H10C1205	365.17	363.9905	55.92	2.76	19.42	21.91
C17H10C1206	381.17	379.9854	53.57	2.64	18.60	25.18
C17H10C1207	397.17	395.9804	51.41	2.54	17.85	28.20
C17H10C1208	413.17	411.9753	49.42	2.44	17.16	30.98
C17H10C1209	429.17	427.9702	47.58	2.35	16.52	33.55
C17H11C1101	266.73	266.0498	76.55	4.16	13.29	6.00
C17H11C1102	282.73	282.0448	72.22	3.92	12.54	11.32
C17H11C1103	298.73	298.0397	68.35	3.71	11.87	16.07
C17H11C1104	314.73	314.0346	64.88	3.52	11.26	20.33
C17H11C1105	330.73	330.0295	61.74	3.35	10.72	24.19
C17H11C1106	346.73	346.0244	58.89	3.20	10.23	27.69
C17H11C1107	362.73	362.0193	56.29	3.06	9.77	30.88
C17H11C1108	378.73	378.0142	53.91	2.93	9.36	33.80
C17H11C1109	394.72	394.0092	51.73	2.81	8.98	36.48
C17H12C1201	303.19	302.0265	67.35	3.99	23.39	5.28
C17H12C1202	319.19	318.0214	63.97	3.79	22.21	10.03
C17H12C1203	335.19	334.0163	60.92	3.61	21.15	14.32
C17H12C1204	351.19	350.0113	58.14	3.44	20.19	18.22
C17H12C1205	367.19	366.0062	55.61	3.29	19.31	21.79
C17H12C1206	383.19	382.0011	53.29	3.16	18.50	25.05
C17H12C1207	399.19	397.9960	51.15	3.03	17.76	28.06
C17H12C1208	415.19	413.9909	49.18	2.91	17.08	30.83
C17H12C1209	431.19	429.9858	47.36	2.81	16.44	33.40
C17H13C1101	268.75	268.0655	75.98	4.88	13.19	5.95
C17H13C1102	284.74	284.0604	71.71	4.60	12.45	11.24
C17H13C1103	300.74	300.0553	67.89	4.36	11.79	15.96
C17H13C1104	316.74	316.0502	64.47	4.14	11.19	20.20
C17H13C1105	332.74	332.0451	61.37	3.94	10.65	24.04
C17H13C1106	348.74	348.0401	58.55	3.76	10.17	27.53
C17H13C1107	364.74	364.0350	55.98	3.59	9.72	30.71
C17H13C1108	380.74	380.0299	53.63	3.44	9.31	33.62
C17H13C1109	396.74	396.0248	51.47	3.30	8.94	36.29
C17H14C1201	305.21	304.0422	66.90	4.62	23.23	5.24
C17H14C1202	321.21	320.0371	63.57	4.39	22.07	9.96
C17H14C1203	337.21	336.0320	60.55	4.18	21.03	14.23
C17H14C1204	353.20	352.0269	57.81	4.00	20.08	18.12
C17H14C1205	369.20	368.0218	55.31	3.82	19.21	21.67
C17H14C1206	385.20	384.0167	53.01	3.66	18.41	24.92
C17H14C1207	401.20	400.0116	50.89	3.52	17.67	27.92
C17H14C1208	417.20	416.0066	48.94	3.38	17.00	30.68
C17H14C1209	433.20	432.0015	47.13	3.26	16.37	33.24
C17H15C1101	270.76	270.0811	75.41	5.58	13.09	5.91
C17H15C1102	286.76	286.0761	71.21	5.27	12.36	11.16
C17H15C1103	302.76	302.0710	67.44	4.99	11.71	15.85
C17H15C1104	318.76	318.0659	64.06	4.74	11.12	20.08
C17H15C1105	334.76	334.0608	61.00	4.52	10.59	23.90
C17H15C1106	350.76	350.0557	58.21	4.31	10.11	27.37
C17H15C1107	366.76	366.0506	55.67	4.12	9.67	30.54
C17H15C1108	382.76	382.0455	53.35	3.95	9.26	33.44
C17H15C1109	398.76	398.0404	51.21	3.79	8.89	36.11
C17H16C1201	307.22	306.0578	66.46	5.25	23.08	5.21
C17H16C1202	323.22	322.0527	63.17	4.99	21.94	9.90
C17H16C1203	339.22	338.0476	60.19	4.75	20.90	14.15
C17H16C1204	355.22	354.0426	57.48	4.54	19.96	18.02
C17H16C1205	371.22	370.0375	55.00	4.34	19.10	21.55
C17H16C1206	387.22	386.0324	52.73	4.16	18.31	24.79
C17H16C1207	403.22	402.0273	50.64	4.00	17.58	27.78
C17H16C1208	419.22	418.0222	48.71	3.85	16.91	30.53
C17H16C1209	435.22	434.0171	46.92	3.71	16.29	33.09
C17H17C1101	272.78	272.0968	74.86	6.28	13.00	5.87
C17H17C1102	288.78	288.0917	70.71	5.93	12.28	11.08
C17H17C1103	304.78	304.0866	67.00	5.62	11.63	15.75
C17H17C1104	320.78	320.0815	63.65	5.34	11.05	19.95
C17H17C1105	336.78	336.0764	60.63	5.09	10.53	23.75
C17H17C1106	352.77	352.0714	57.88	4.86	10.05	27.21
C17H17C1107	368.77	368.0663	55.37	4.65	9.61	30.37
C17H17C1108	384.77	384.0612	53.07	4.45	9.21	33.27
C17H17C1109	400.77	400.0561	50.95	4.28	8.85	35.93
C17H18C1201	309.24	308.0735	66.03	5.87	22.93	5.17
C17H18C1202	325.24	324.0684	62.78	5.58	21.80	9.84
C17H18C1203	341.24	340.0633	59.84	5.32	20.78	14.07
C17H18C1204	357.24	356.0582	57.16	5.08	19.85	17.91
C17H18C1205	373.24	372.0531	54.71	4.86	19.00	21.43
C17H18C1206	389.24	388.0480	52.46	4.66	18.22	24.66
C17H18C1207	405.23	404.0429	50.39	4.48	17.50	27.64
C17H18C1208	421.23	420.0379	48.47	4.31	16.83	30.39
C17H18C1209	437.23	436.0328	46.70	4.15	16.22	32.93
C17H19C1101	274.79	274.1124	74.31	6.97	12.90	5.82

086

C17H19C1102	290.79	290.1073	70.22	6.59	12.19	11.00
C17H19C1103	306.79	306.1023	66.56	6.24	11.56	15.65
C17H19C1104	322.79	322.0972	63.26	5.93	10.98	19.83
C17H19C1105	338.79	338.0921	60.27	5.65	10.46	23.61
C17H19C1106	354.79	354.0870	57.55	5.40	9.99	27.06
C17H19C1107	370.79	370.0819	55.07	5.17	9.56	30.20
C17H19C1108	386.79	386.0768	52.79	4.95	9.17	33.09
C17H19C1109	402.79	402.0717	50.69	4.75	8.80	35.75
C17H20C1201	311.25	310.0891	65.60	6.48	22.78	5.14
C17H20C1202	327.25	326.0840	62.39	6.16	21.67	9.78
C17H20C1203	343.25	342.0789	59.49	5.87	20.66	13.98
C17H20C1204	359.25	358.0739	56.84	5.61	19.74	17.81
C17H20C1205	375.25	374.0688	54.41	5.37	18.90	21.32
C17H20C1206	391.25	390.0637	52.19	5.15	18.12	24.54
C17H20C1207	407.25	406.0586	50.14	4.95	17.41	27.50
C17H20C1208	423.25	422.0535	48.24	4.76	16.75	30.24
C17H20C1209	439.25	438.0484	46.49	4.59	16.14	32.78
C17H21C1101	276.81	276.1281	73.77	7.65	12.81	5.78
C17H21C1102	292.81	292.1230	69.73	7.23	12.11	10.93
C17H21C1103	308.81	308.1179	66.12	6.85	11.48	15.54
C17H21C1104	324.81	324.1128	62.86	6.52	10.92	19.70
C17H21C1105	340.81	340.1077	59.91	6.21	10.40	23.47
C17H21C1106	356.81	356.1027	57.23	5.93	9.94	26.90
C17H21C1107	372.81	372.0976	54.77	5.68	9.51	30.04
C17H21C1108	388.81	388.0925	52.52	5.44	9.12	32.92
C17H21C1109	404.80	404.0874	50.44	5.23	8.76	35.57
C17H22C1201	313.27	312.1048	65.18	7.08	22.63	5.11
C17H22C1202	329.27	328.0997	62.01	6.73	21.53	9.72
C17H22C1203	345.27	344.0946	59.14	6.42	20.54	13.90
C17H22C1204	361.27	360.0895	56.52	6.14	19.63	17.71
C17H22C1205	377.27	376.0844	54.12	5.88	18.79	21.20
C17H22C1206	393.27	392.0793	51.92	5.64	18.03	24.41
C17H22C1207	409.27	408.0742	49.89	5.42	17.33	27.36
C17H22C1208	425.27	424.0692	48.01	5.21	16.67	30.10
C17H22C1209	441.27	440.0641	46.27	5.03	16.07	32.63
C17H23C1101	278.83	278.1437	73.23	8.31	12.72	5.74
C17H23C1102	294.82	294.1386	69.26	7.86	12.03	10.85
C17H23C1103	310.82	310.1336	65.69	7.46	11.41	15.44
C17H23C1104	326.82	326.1285	62.48	7.09	10.85	19.58
C17H23C1105	342.82	342.1234	59.56	6.76	10.34	23.33
C17H23C1106	358.82	358.1183	56.91	6.46	9.88	26.75
C17H23C1107	374.82	374.1132	54.48	6.19	9.46	29.88
C17H23C1108	390.82	390.1081	52.25	5.93	9.07	32.75
C17H23C1109	406.82	406.1030	50.19	5.70	8.71	35.40
C17H24C1201	315.29	314.1204	64.76	7.67	22.49	5.07
C17H24C1202	331.29	330.1153	61.64	7.30	21.40	9.66
C17H24C1203	347.29	346.1102	58.80	6.97	20.42	13.82
C17H24C1204	363.28	362.1052	56.21	6.66	19.52	17.62
C17H24C1205	379.28	378.1001	53.84	6.38	18.69	21.09
C17H24C1206	395.28	394.0950	51.66	6.12	17.94	24.29
C17H24C1207	411.28	410.0899	49.65	5.88	17.24	27.23
C17H24C1208	427.28	426.0848	47.79	5.66	16.59	29.96
C17H24C1209	443.28	442.0797	46.06	5.46	16.00	32.48
C17H25C1101	280.84	280.1594	72.71	8.97	12.62	5.70
C17H25C1102	296.84	296.1543	68.79	8.49	11.94	10.78
C17H25C1103	312.84	312.1492	65.27	8.05	11.33	15.34
C17H25C1104	328.84	328.1441	62.09	7.66	10.78	19.46
C17H25C1105	344.84	344.1390	59.21	7.31	10.28	23.20
C17H25C1106	360.84	360.1340	56.59	6.98	9.83	26.60
C17H25C1107	376.84	376.1289	54.19	6.69	9.41	29.72
C17H25C1108	392.84	392.1238	51.98	6.41	9.02	32.58
C17H25C1109	408.84	408.1187	49.94	6.16	8.67	35.22
C17H26C1201	317.30	316.1361	64.35	8.26	22.35	5.04
C17H26C1202	333.30	332.1310	61.26	7.86	21.27	9.60
C17H26C1203	349.30	348.1259	58.46	7.50	20.30	13.74
C17H26C1204	365.30	364.1208	55.90	7.17	19.41	17.52
C17H26C1205	381.30	380.1157	53.55	6.87	18.60	20.98
C17H26C1206	397.30	396.1106	51.39	6.60	17.85	24.16
C17H26C1207	413.30	412.1055	49.40	6.34	17.16	27.10
C17H26C1208	429.30	428.1005	47.56	6.10	16.52	29.82
C17H26C1209	445.30	444.0954	45.85	5.89	15.92	32.34
C17H27C1101	282.86	282.1750	72.19	9.62	12.53	5.66
C17H27C1102	298.86	298.1699	68.32	9.11	11.86	10.71
C17H27C1103	314.86	314.1649	64.85	8.64	11.26	15.24
C17H27C1104	330.86	330.1598	61.72	8.23	10.72	19.34
C17H27C1105	346.85	346.1547	58.87	7.85	10.22	23.06
C17H27C1106	362.85	362.1496	56.27	7.50	9.77	26.46
C17H27C1107	378.85	378.1445	53.90	7.18	9.36	29.56
C17H27C1108	394.85	394.1394	51.71	6.89	8.98	32.42
C17H27C1109	410.85	410.1343	49.70	6.62	8.63	35.05
C17H28C1201	319.32	318.1517	63.95	8.84	22.21	5.01
C17H28C1202	335.32	334.1466	60.89	8.42	21.15	9.54
C17H28C1203	351.32	350.1415	58.12	8.03	20.18	13.66
C17H28C1204	367.32	366.1365	55.59	7.68	19.30	17.42
C17H28C1205	383.32	382.1314	53.27	7.36	18.50	20.87
C17H28C1206	399.32	398.1263	51.13	7.07	17.76	24.04
C17H28C1207	415.31	414.1212	49.17	6.80	17.07	26.97
C17H28C1208	431.31	430.1161	47.34	6.54	16.44	29.68
C17H28C1209	447.31	446.1110	45.65	6.31	15.85	32.19
C17H29C1101	284.87	284.1907	71.68	10.26	12.45	5.62
C17H29C1102	300.87	300.1856	67.87	9.72	11.78	10.64
C17H29C1103	316.87	316.1805	64.44	9.22	11.19	15.15
C17H29C1104	332.87	332.1754	61.34	8.78	10.65	19.23
C17H29C1105	348.87	348.1703	58.53	8.38	10.16	22.93
C17H29C1106	364.87	364.1653	55.96	8.01	9.72	26.31
C17H29C1107	380.87	380.1602	53.61	7.67	9.31	29.41
C17H29C1108	396.87	396.1551	51.45	7.37	8.93	32.25
C17H29C1109	412.87	412.1500	49.46	7.08	8.59	34.88
C17H30C1201	321.33	320.1674	63.54	9.41	22.07	4.98
C17H30C1202	337.33	336.1623	60.53	8.96	21.02	9.49
C17H30C1203	353.33	352.1572	57.79	8.56	20.07	13.58
C17H30C1204	369.33	368.1521	55.29	8.19	19.20	17.33
C17H30C1205	385.33	384.1470	52.99	7.85	18.40	20.76
C17H30C1206	401.33	400.1419	50.88	7.53	17.67	23.92
C17H30C1207	417.33	416.1368	48.93	7.25	16.99	26.84
C17H30C1208	433.33	432.1318	47.12	6.98	16.36	29.54
C17H30C1209	449.33	448.1267	45.44	6.73	15.78	32.05
C17H31C1101	286.89	286.2063	71.17	10.89	12.36	5.58
C17H31C1102	302.89	302.2012	67.41	10.32	11.70	10.56
C17H31C1103	318.89	318.1962	64.03	9.80	11.12	15.05
C17H31C1104	334.89	334.1911	60.97	9.33	10.59	19.11
C17H31C1105	350.89	350.1860	58.19	8.91	10.10	22.80
C17H31C1106	366.89	366.1809	55.65	8.52	9.66	26.17
C17H31C1107	382.89	382.1758	53.33	8.16	9.26	29.25
C17H31C1108	398.88	398.1707	51.19	7.83	8.89	32.09
C17H31C1109	414.88	414.1656	49.22	7.53	8.55	34.71
C17H32C1201	323.35	322.1830	63.15	9.98	21.93	4.95
C17H32C1202	339.35	338.1779	60.17	9.50	20.89	9.43
C17H32C1203	355.35	354.1728	57.46	9.08	19.95	13.51
C17H32C1204	371.35	370.1677	54.99	8.69	19.09	17.23

C17H32C1205	387.35	386.1627	52.71	8.33	18.31	20.65
C17H32C1206	403.35	402.1576	50.62	8.00	17.58	23.80
C17H32C1207	419.35	418.1525	48.69	7.69	16.91	26.71
C17H32C1208	435.35	434.1474	46.90	7.41	16.29	29.40
C17H32C1209	451.35	450.1423	45.24	7.15	15.71	31.90
C17H33C1101	288.90	288.2220	70.68	11.51	12.27	5.54
C17H33C1102	304.90	304.2169	66.97	10.91	11.63	10.49
C17H33C1103	320.90	320.2118	63.63	10.37	11.05	14.96
C17H33C1104	336.90	336.2067	60.61	9.87	10.52	19.00
C17H33C1105	352.90	352.2016	57.86	9.43	10.05	22.67
C17H33C1106	368.90	368.1965	55.35	9.02	9.61	26.02
C17H33C1107	384.90	384.1915	53.05	8.64	9.21	29.10
C17H33C1108	400.90	400.1864	50.93	8.30	8.84	31.93
C17H33C1109	416.90	416.1813	48.98	7.98	8.50	34.54
C17H34C1201	325.37	324.1987	62.76	10.53	21.79	4.92
C17H34C1202	341.37	340.1936	59.82	10.04	20.77	9.37
C17H34C1203	357.36	356.1885	57.14	9.59	19.84	13.43
C17H34C1204	373.36	372.1834	54.69	9.18	18.99	17.14
C17H34C1205	389.36	388.1783	52.44	8.80	18.21	20.55
C17H34C1206	405.36	404.1732	50.37	8.45	17.49	23.68
C17H34C1207	421.36	420.1681	48.46	8.13	16.83	26.58
C17H34C1208	437.36	436.1631	46.69	7.84	16.21	29.27
C17H34C1209	453.36	452.1580	45.04	7.56	15.64	31.76
C17H35C1101	290.92	290.2376	70.19	12.13	12.19	5.50
C17H35C1102	306.92	306.2325	66.53	11.49	11.55	10.43
C17H35C1103	322.92	322.2275	63.23	10.92	10.98	14.86
C17H35C1104	338.92	338.2224	60.25	10.41	10.46	18.88
C17H35C1105	354.92	354.2173	57.53	9.94	9.99	22.54
C17H35C1106	370.92	370.2122	55.05	9.51	9.56	25.88
C17H35C1107	386.92	386.2071	52.77	9.12	9.16	28.95
C17H35C1108	402.92	402.2020	50.68	8.76	8.80	31.77
C17H35C1109	418.92	418.1969	48.74	8.42	8.46	34.37
C18H 9C1101	276.72	276.0342	78.13	3.28	12.81	5.78
C18H 9C1102	292.72	292.0291	73.86	3.10	12.11	10.93
C18H 9C1103	308.72	308.0240	70.03	2.94	11.48	15.55
C18H 9C1104	324.72	324.0189	66.58	2.79	10.92	19.71
C18H 9C1105	340.72	340.0138	63.45	2.66	10.41	23.48
C18H 9C1106	356.72	356.0088	60.61	2.54	9.94	26.91
C18H 9C1107	372.72	372.0037	58.01	2.43	9.51	30.05
C18H 9C1108	388.72	387.9986	55.62	2.33	9.12	32.93
C18H 9C1109	404.72	403.9935	53.42	2.24	8.76	35.58
C18H10C1201	313.19	312.0109	69.03	3.22	22.64	5.11
C18H10C1202	329.19	328.0058	65.68	3.06	21.54	9.72
C18H10C1203	345.18	344.0007	62.63	2.92	20.54	13.91
C18H10C1204	361.18	359.9956	59.86	2.79	19.63	17.72
C18H10C1205	377.18	375.9905	57.32	2.67	18.80	21.21
C18H10C1206	393.18	391.9854	54.99	2.56	18.03	24.42
C18H10C1207	409.18	407.9804	52.84	2.46	17.33	27.37
C18H10C1208	425.18	423.9753	50.85	2.37	16.68	30.10
C18H10C1209	441.18	439.9702	49.00	2.28	16.07	32.64
C18H11C1101	278.74	278.0498	77.56	3.98	12.72	5.74
C18H11C1102	294.74	294.0448	73.35	3.76	12.03	10.86
C18H11C1103	310.74	310.0397	69.58	3.57	11.41	15.45
C18H11C1104	326.74	326.0346	66.17	3.39	10.85	19.59
C18H11C1105	342.74	342.0295	63.08	3.24	10.34	23.34
C18H11C1106	358.74	358.0244	60.27	3.09	9.88	26.76
C18H11C1107	374.74	374.0193	57.69	2.96	9.46	29.89
C18H11C1108	390.74	390.0142	55.33	2.84	9.07	32.76
C18H11C1109	406.74	406.0092	53.16	2.73	8.72	35.40
C18H12C1201	315.20	314.0265	68.59	3.84	22.50	5.08

C18H12C1202	331.20	330.0214	65.28	3.65	21.41	9.66
C18H12C1203	347.20	346.0163	62.27	3.48	20.42	13.82
C18H12C1204	363.20	362.0113	59.53	3.33	19.52	17.62
C18H12C1205	379.20	378.0062	57.02	3.19	18.70	21.10
C18H12C1206	395.20	394.0011	54.71	3.06	17.94	24.29
C18H12C1207	411.20	409.9960	52.58	2.94	17.24	27.24
C18H12C1208	427.20	425.9909	50.61	2.83	16.60	29.96
C18H12C1209	443.20	441.9858	48.78	2.73	16.00	32.49
C18H13C1101	280.76	280.0655	77.01	4.67	12.63	5.70
C18H13C1102	296.76	296.0604	72.85	4.42	11.95	10.78
C18H13C1103	312.76	312.0553	69.13	4.19	11.34	15.35
C18H13C1104	328.75	328.0502	65.76	3.99	10.78	19.47
C18H13C1105	344.75	344.0451	62.71	3.80	10.28	23.20
C18H13C1106	360.75	360.0401	59.93	3.63	9.83	26.61
C18H13C1107	376.75	376.0350	57.39	3.48	9.41	29.73
C18H13C1108	392.75	392.0299	55.05	3.34	9.03	32.59
C18H13C1109	408.75	408.0248	52.89	3.21	8.67	35.23
C18H14C1201	317.22	316.0422	68.16	4.45	22.35	5.04
C18H14C1202	333.22	332.0371	64.88	4.23	21.28	9.60
C18H14C1203	349.22	348.0320	61.91	4.04	20.30	13.74
C18H14C1204	365.22	364.0269	59.20	3.86	19.41	17.52
C18H14C1205	381.22	380.0218	56.71	3.70	18.60	20.98
C18H14C1206	397.21	396.0167	54.43	3.55	17.85	24.17
C18H14C1207	413.21	412.0116	52.32	3.42	17.16	27.10
C18H14C1208	429.21	428.0066	50.37	3.29	16.52	29.82
C18H14C1209	445.21	444.0015	48.56	3.17	15.93	32.34
C18H15C1101	282.77	282.0811	76.46	5.35	12.54	5.66
C18H15C1102	298.77	298.0761	72.36	5.06	11.87	10.71
C18H15C1103	314.77	314.0710	68.68	4.80	11.26	15.25
C18H15C1104	330.77	330.0659	65.36	4.57	10.72	19.35
C18H15C1105	346.77	346.0608	62.35	4.36	10.22	23.07
C18H15C1106	362.77	362.0557	59.60	4.17	9.77	26.46
C18H15C1107	378.77	378.0506	57.08	3.99	9.36	29.57
C18H15C1108	394.77	394.0455	54.77	3.83	8.98	32.42
C18H15C1109	410.77	410.0404	52.63	3.68	8.63	35.05
C18H16C1201	319.23	318.0578	67.72	5.05	22.21	5.01
C18H16C1202	335.23	334.0527	64.49	4.81	21.15	9.55
C18H16C1203	351.23	350.0476	61.55	4.59	20.19	13.67
C18H16C1204	367.23	366.0426	58.87	4.39	19.31	17.43
C18H16C1205	383.23	382.0375	56.42	4.21	18.50	20.87
C18H16C1206	399.23	398.0324	54.15	4.04	17.76	24.05
C18H16C1207	415.23	414.0273	52.07	3.88	17.08	26.97
C18H16C1208	431.23	430.0222	50.14	3.74	16.44	29.68
C18H16C1209	447.23	446.0171	48.34	3.61	15.85	32.20
C18H17C1101	284.79	284.0968	75.92	6.02	12.45	5.62
C18H17C1102	300.79	300.0917	71.88	5.70	11.79	10.64
C18H17C1103	316.79	316.0866	68.25	5.41	11.19	15.15
C18H17C1104	332.79	332.0815	64.97	5.15	10.65	19.23
C18H17C1105	348.79	348.0764	61.99	4.91	10.16	22.94
C18H17C1106	364.79	364.0714	59.27	4.70	9.72	26.32
C18H17C1107	380.78	380.0663	56.78	4.50	9.31	29.41
C18H17C1108	396.78	396.0612	54.49	4.32	8.94	32.26
C18H17C1109	412.78	412.0561	52.38	4.15	8.59	34.88
C18H18C1201	321.25	320.0735	67.30	5.65	22.07	4.98
C18H18C1202	337.25	336.0684	64.11	5.38	21.02	9.49
C18H18C1203	353.25	352.0633	61.20	5.14	20.07	13.59
C18H18C1204	369.25	368.0582	58.55	4.91	19.20	17.33
C18H18C1205	385.25	384.0531	56.12	4.71	18.41	20.77
C18H18C1206	401.25	400.0480	53.88	4.52	17.67	23.92
C18H18C1207	417.25	416.0429	51.82	4.35	16.99	26.84

C18H18C1208	433.25	432.0379	49.90	4.19	16.37	29.54
C18H18C1209	449.24	448.0328	48.13	4.04	15.78	32.05
C18H19C1101	286.80	286.1124	75.38	6.68	12.36	5.58
C18H19C1102	302.80	302.1073	71.40	6.32	11.71	10.57
C18H19C1103	318.80	318.1023	67.82	6.01	11.12	15.06
C18H19C1104	334.80	334.0972	64.58	5.72	10.59	19.12
C18H19C1105	350.80	350.0921	61.63	5.46	10.11	22.80
C18H19C1106	366.80	366.0870	58.94	5.22	9.67	26.17
C18H19C1107	382.80	382.0819	56.48	5.00	9.26	29.26
C18H19C1108	398.80	398.0768	54.21	4.80	8.89	32.10
C18H19C1109	414.80	414.0717	52.12	4.62	8.55	34.71
C18H20C1201	323.27	322.0891	66.88	6.24	21.93	4.95
C18H20C1202	339.26	338.0840	63.73	5.94	20.90	9.43
C18H20C1203	355.26	354.0789	60.86	5.67	19.96	13.51
C18H20C1204	371.26	370.0739	58.23	5.43	19.10	17.24
C18H20C1205	387.26	386.0688	55.83	5.21	18.31	20.66
C18H20C1206	403.26	402.0637	53.61	5.00	17.58	23.80
C18H20C1207	419.26	418.0586	51.57	4.81	16.91	26.71
C18H20C1208	435.26	434.0535	49.67	4.63	16.29	29.41
C18H20C1209	451.26	450.0484	47.91	4.47	15.71	31.91
C18H21C1101	288.82	288.1281	74.86	7.33	12.28	5.54
C18H21C1102	304.82	304.1230	70.93	6.94	11.63	10.50
C18H21C1103	320.82	320.1179	67.39	6.60	11.05	14.96
C18H21C1104	336.82	336.1128	64.19	6.28	10.53	19.00
C18H21C1105	352.82	352.1077	61.28	6.00	10.05	22.67
C18H21C1106	368.82	368.1027	58.62	5.74	9.61	26.03
C18H21C1107	384.82	384.0976	56.18	5.50	9.21	29.10
C18H21C1108	400.82	400.0925	53.94	5.28	8.85	31.93
C18H21C1109	416.82	416.0874	51.87	5.08	8.51	34.55
C18H22C1201	325.28	324.1048	66.47	6.82	21.80	4.92
C18H22C1202	341.28	340.0997	63.35	6.50	20.78	9.38
C18H22C1203	357.28	356.0946	60.51	6.21	19.85	13.43
C18H22C1204	373.28	372.0895	57.92	5.94	19.00	17.14
C18H22C1205	389.28	388.0844	55.54	5.70	18.21	20.55
C18H22C1206	405.28	404.0793	53.35	5.47	17.50	23.69
C18H22C1207	421.28	420.0742	51.32	5.26	16.83	26.58
C18H22C1208	437.28	436.0692	49.44	5.07	16.22	29.27
C18H22C1209	453.28	452.0641	47.70	4.89	15.64	31.77
C18H23C1101	290.84	290.1437	74.34	7.97	12.19	5.50
C18H23C1102	306.84	306.1386	70.46	7.56	11.55	10.43
C18H23C1103	322.84	322.1336	66.97	7.18	10.98	14.87
C18H23C1104	338.83	338.1285	63.81	6.84	10.46	18.89
C18H23C1105	354.83	354.1234	60.93	6.53	9.99	22.54
C18H23C1106	370.83	370.1183	58.30	6.25	9.56	25.89
C18H23C1107	386.83	386.1132	55.89	5.99	9.16	28.95
C18H23C1108	402.83	402.1081	53.67	5.76	8.80	31.77
C18H23C1109	418.83	418.1030	51.62	5.54	8.46	34.38
C18H24C1201	327.30	326.1204	66.06	7.39	21.66	4.89
C18H24C1202	343.30	342.1153	62.98	7.05	20.65	9.32
C18H24C1203	359.30	358.1102	60.17	6.73	19.73	13.36
C18H24C1204	375.30	374.1052	57.61	6.45	18.89	17.05
C18H24C1205	391.29	390.1001	55.25	6.18	18.12	20.44
C18H24C1206	407.29	406.0950	53.08	5.94	17.41	23.57
C18H24C1207	423.29	422.0899	51.08	5.72	16.75	26.46
C18H24C1208	439.29	438.0848	49.22	5.51	16.14	29.14
C18H24C1209	455.29	454.0797	47.49	5.31	15.57	31.63
C18H25C1101	292.85	292.1594	73.83	8.60	12.11	5.46
C18H25C1102	308.85	308.1543	70.00	8.16	11.48	10.36
C18H25C1103	324.85	324.1492	66.55	7.76	10.91	14.78
C18H25C1104	340.85	340.1441	63.43	7.39	10.40	18.78

C18H25C1105	356.85	356.1390	60.59	7.06	9.93	22.42
C18H25C1106	372.85	372.1340	57.99	6.76	9.51	25.75
C18H25C1107	388.85	388.1289	55.60	6.48	9.12	28.80
C18H25C1108	404.85	404.1238	53.40	6.22	8.76	31.62
C18H25C1109	420.85	420.1187	51.37	5.99	8.42	34.22
C18H26C1201	329.31	328.1361	65.65	7.96	21.53	4.86
C18H26C1202	345.31	344.1310	62.61	7.59	20.53	9.27
C18H26C1203	361.31	360.1259	59.84	7.25	19.62	13.28
C18H26C1204	377.31	376.1208	57.30	6.95	18.79	16.96
C18H26C1205	393.31	392.1157	54.97	6.66	18.03	20.34
C18H26C1206	409.31	408.1106	52.82	6.40	17.32	23.45
C18H26C1207	425.31	424.1055	50.83	6.16	16.67	26.33
C18H26C1208	441.31	440.1005	48.99	5.94	16.07	29.00
C18H26C1209	457.31	456.0954	47.28	5.73	15.51	31.49
C18H27C1101	294.87	294.1750	73.32	9.23	12.02	5.43
C18H27C1102	310.87	310.1699	69.55	8.75	11.40	10.29
C18H27C1103	326.87	326.1649	66.14	8.33	10.85	14.68
C18H27C1104	342.87	342.1598	63.06	7.94	10.34	18.67
C18H27C1105	358.87	358.1547	60.25	7.58	9.88	22.29
C18H27C1106	374.87	374.1496	57.67	7.26	9.46	25.61
C18H27C1107	390.86	390.1445	55.31	6.96	9.07	28.65
C18H27C1108	406.86	406.1394	53.14	6.69	8.71	31.46
C18H27C1109	422.86	422.1343	51.13	6.44	8.38	34.05
C18H28C1201	331.33	330.1517	65.25	8.52	21.40	4.83
C18H28C1202	347.33	346.1466	62.25	8.13	20.41	9.21
C18H28C1203	363.33	362.1415	59.51	7.77	19.52	13.21
C18H28C1204	379.33	378.1365	57.00	7.44	18.69	16.87
C18H28C1205	395.33	394.1314	54.69	7.14	17.94	20.24
C18H28C1206	411.33	410.1263	52.56	6.86	17.24	23.34
C18H28C1207	427.33	426.1212	50.59	6.60	16.59	26.21
C18H28C1208	443.33	442.1161	48.77	6.37	15.99	28.87
C18H28C1209	459.32	458.1110	47.07	6.14	15.44	31.35
C18H29C1101	296.88	296.1907	72.82	9.85	11.94	5.39
C18H29C1102	312.88	312.1856	69.10	9.34	11.33	10.23
C18H29C1103	328.88	328.1805	65.74	8.89	10.78	14.59
C18H29C1104	344.88	344.1754	62.69	8.48	10.28	18.56
C18H29C1105	360.88	360.1703	59.91	8.10	9.82	22.17
C18H29C1106	376.88	376.1653	57.37	7.76	9.41	25.47
C18H29C1107	392.88	392.1602	55.03	7.44	9.02	28.51
C18H29C1108	408.88	408.1551	52.88	7.15	8.67	31.30
C18H29C1109	424.88	424.1500	50.89	6.88	8.34	33.89
C18H30C1201	333.35	332.1674	64.86	9.07	21.27	4.80
C18H30C1202	349.34	348.1623	61.89	8.66	20.30	9.16
C18H30C1203	365.34	364.1572	59.18	8.28	19.41	13.14
C18H30C1204	381.34	380.1521	56.69	7.93	18.59	16.78
C18H30C1205	397.34	396.1470	54.41	7.61	17.85	20.13
C18H30C1206	413.34	412.1419	52.31	7.32	17.15	23.22
C18H30C1207	429.34	428.1368	50.36	7.04	16.52	26.09
C18H30C1208	445.34	444.1318	48.55	6.79	15.92	28.74
C18H30C1209	461.34	460.1267	46.86	6.55	15.37	31.21
C18H31C1101	298.90	298.2063	72.33	10.45	11.86	5.35
C18H31C1102	314.90	314.2012	68.66	9.92	11.26	10.16
C18H31C1103	330.90	330.1962	65.34	9.44	10.71	14.51
C18H31C1104	346.90	346.1911	62.32	9.01	10.22	18.45
C18H31C1105	362.90	362.1860	59.58	8.61	9.77	22.04
C18H31C1106	378.90	378.1809	57.06	8.25	9.36	25.34
C18H31C1107	394.90	394.1758	54.75	7.91	8.98	28.36
C18H31C1108	410.90	410.1707	52.62	7.60	8.63	31.15
C18H31C1109	426.90	426.1656	50.64	7.32	8.30	33.73
C18H32C1201	335.36	334.1830	64.47	9.62	21.14	4.77

089

C18H32C1202	351.36	350.1779	61.53	9.18	20.18	9.11
C18H32C1203	367.36	366.1728	58.85	8.78	19.30	13.07
C18H32C1204	383.36	382.1677	56.40	8.41	18.50	16.69
C18H32C1205	399.36	398.1627	54.14	8.08	17.75	20.03
C18H32C1206	415.36	414.1576	52.05	7.77	17.07	23.11
C18H32C1207	431.36	430.1525	50.12	7.48	16.44	25.96
C18H32C1208	447.36	446.1474	48.33	7.21	15.85	28.61
C18H32C1209	463.36	462.1423	46.66	6.96	15.30	31.08
C18H33C1101	300.92	300.2220	71.85	11.05	11.78	5.32
C18H33C1102	316.92	316.2169	68.22	10.50	11.19	10.10
C18H33C1103	332.91	332.2118	64.94	9.99	10.65	14.42
C18H33C1104	348.91	348.2067	61.96	9.53	10.16	18.34
C18H33C1105	364.91	364.2016	59.25	9.12	9.72	21.92
C18H33C1106	380.91	380.1965	56.76	8.73	9.31	25.20
C18H33C1107	396.91	396.1915	54.47	8.38	8.93	28.22
C18H33C1108	412.91	412.1864	52.36	8.06	8.59	31.00
C18H33C1109	428.91	428.1813	50.41	7.76	8.27	33.57
C18H34C1201	337.38	336.1987	64.08	10.16	21.02	4.74
C18H34C1202	353.38	352.1936	61.18	9.70	20.07	9.06
C18H34C1203	369.38	368.1885	58.53	9.28	19.20	12.99
C18H34C1204	385.38	384.1834	56.10	8.89	18.40	16.61
C18H34C1205	401.37	400.1783	53.87	8.54	17.67	19.93
C18H34C1206	417.37	416.1732	51.80	8.21	16.99	23.00
C18H34C1207	433.37	432.1681	49.89	7.91	16.36	25.84
C18H34C1208	449.37	448.1631	48.11	7.63	15.78	28.48
C18H34C1209	465.37	464.1580	46.46	7.36	15.24	30.94
C18H35C1101	302.93	302.2376	71.37	11.65	11.70	5.28
C18H35C1102	318.93	318.2325	67.79	11.06	11.12	10.03
C18H35C1103	334.93	334.2275	64.55	10.53	10.59	14.33
C18H35C1104	350.93	350.2224	61.61	10.05	10.10	18.24
C18H35C1105	366.93	366.2173	58.92	9.61	9.66	21.80
C18H35C1106	382.93	382.2122	56.46	9.21	9.26	25.07
C18H35C1107	398.93	398.2071	54.20	8.84	8.89	28.07
C18H35C1108	414.93	414.2020	52.11	8.50	8.54	30.85
C18H35C1109	430.93	430.1969	50.17	8.19	8.23	33.42
C18H36C1201	339.39	338.2143	63.70	10.69	20.89	4.71
C18H36C1202	355.39	354.2092	60.83	10.21	19.95	9.00
C18H36C1203	371.39	370.2041	58.21	9.77	19.09	12.92
C18H36C1204	387.39	386.1990	55.81	9.37	18.30	16.52
C18H36C1205	403.39	402.1940	53.60	9.00	17.58	19.83
C18H36C1206	419.39	418.1889	51.55	8.65	16.91	22.89
C18H36C1207	435.39	434.1838	49.66	8.33	16.29	25.72
C18H36C1208	451.39	450.1787	47.90	8.04	15.71	28.36
C18H36C1209	467.39	466.1736	46.26	7.76	15.17	30.81
C18H37C1101	304.95	304.2533	70.90	12.23	11.63	5.25
C18H37C1102	320.95	320.2482	67.36	11.62	11.05	9.97
C18H37C1103	336.95	336.2431	64.16	11.07	10.52	14.25
C18H37C1104	352.95	352.2380	61.26	10.57	10.04	18.13
C18H37C1105	368.95	368.2329	58.60	10.11	9.61	21.68
C18H37C1106	384.94	384.2278	56.16	9.69	9.21	24.94
C18H37C1107	400.94	400.2228	53.92	9.30	8.84	27.93
C18H37C1108	416.94	416.2177	51.85	8.94	8.50	30.70
C18H37C1109	432.94	432.2126	49.94	8.61	8.19	33.26
C19H 9C1101	288.74	288.0342	79.04	3.14	12.28	5.54
C19H 9C1102	304.74	304.0291	74.89	2.98	11.63	10.50
C19H 9C1103	320.73	320.0240	71.15	2.83	11.05	14.97
C19H 9C1104	336.73	336.0189	67.77	2.69	10.53	19.01
C19H 9C1105	352.73	352.0138	64.70	2.57	10.05	22.68
C19H 9C1106	368.73	368.0088	61.89	2.46	9.61	26.03
C19H 9C1107	384.73	384.0037	59.32	2.36	9.21	29.11
C19H 9C1108	400.73	399.9986	56.95	2.26	8.85	31.94
C19H 9C1109	416.73	415.9935	54.76	2.18	8.51	34.55
C19H10C1201	325.20	324.0109	70.18	3.10	21.80	4.92
C19H10C1202	341.20	340.0058	66.89	2.95	20.78	9.38
C19H10C1203	357.20	356.0007	63.89	2.82	19.85	13.44
C19H10C1204	373.20	371.9956	61.15	2.70	19.00	17.15
C19H10C1205	389.19	387.9905	58.64	2.59	18.22	20.55
C19H10C1206	405.19	403.9854	56.32	2.49	17.50	23.69
C19H10C1207	421.19	419.9804	54.18	2.39	16.83	26.59
C19H10C1208	437.19	435.9753	52.20	2.31	16.22	29.28
C19H10C1209	453.19	451.9702	50.36	2.22	15.65	31.77
C19H11C1101	290.75	290.0498	78.49	3.81	12.19	5.50
C19H11C1102	306.75	306.0448	74.40	3.61	11.56	10.43
C19H11C1103	322.75	322.0397	70.71	3.44	10.98	14.87
C19H11C1104	338.75	338.0346	67.37	3.27	10.47	18.89
C19H11C1105	354.75	354.0295	64.33	3.13	9.99	22.55
C19H11C1106	370.75	370.0244	61.55	2.99	9.56	25.89
C19H11C1107	386.75	386.0193	59.01	2.87	9.17	28.96
C19H11C1108	402.75	402.0142	56.66	2.75	8.80	31.78
C19H11C1109	418.75	418.0092	54.50	2.65	8.47	34.39
C19H12C1201	327.21	326.0265	69.74	3.70	21.67	4.89
C19H12C1202	343.21	342.0214	66.49	3.52	20.66	9.32
C19H12C1203	359.21	358.0163	63.53	3.37	19.74	13.36
C19H12C1204	375.21	374.0113	60.82	3.22	18.90	17.06
C19H12C1205	391.21	390.0062	58.33	3.09	18.12	20.45
C19H12C1206	407.21	406.0011	56.04	2.97	17.41	23.57
C19H12C1207	423.21	421.9960	53.92	2.86	16.75	26.46
C19H12C1208	439.21	437.9909	51.96	2.75	16.14	29.14
C19H12C1209	455.21	453.9858	50.13	2.66	15.58	31.63
C19H13C1101	292.77	292.0655	77.95	4.48	12.11	5.46
C19H13C1102	308.77	308.0604	73.91	4.24	11.48	10.36
C19H13C1103	324.77	324.0553	70.27	4.03	10.92	14.78
C19H13C1104	340.77	340.0502	66.97	3.85	10.40	18.78
C19H13C1105	356.77	356.0451	63.97	3.67	9.94	22.42
C19H13C1106	372.76	372.0401	61.22	3.52	9.51	25.75
C19H13C1107	388.76	388.0350	58.70	3.37	9.12	28.81
C19H13C1108	404.76	404.0299	56.38	3.24	8.76	31.62
C19H13C1109	420.76	420.0248	54.24	3.11	8.43	34.22
C19H14C1201	329.23	328.0422	69.32	4.29	21.54	4.86
C19H14C1202	345.23	344.0371	66.10	4.09	20.54	9.27
C19H14C1203	361.23	360.0320	63.18	3.91	19.63	13.29
C19H14C1204	377.23	376.0269	60.50	3.74	18.80	16.97
C19H14C1205	393.23	392.0218	58.04	3.59	18.03	20.34
C19H14C1206	409.23	408.0167	55.77	3.45	17.33	23.46
C19H14C1207	425.23	424.0116	53.67	3.32	16.67	26.34
C19H14C1208	441.22	440.0066	51.72	3.20	16.07	29.01
C19H14C1209	457.22	456.0015	49.91	3.09	15.51	31.49
C19H15C1101	294.78	294.0811	77.42	5.13	12.03	5.43
C19H15C1102	310.78	310.0761	73.43	4.86	11.41	10.30
C19H15C1103	326.78	326.0710	69.84	4.63	10.85	14.69
C19H15C1104	342.78	342.0659	66.58	4.41	10.34	18.67
C19H15C1105	358.78	358.0608	63.61	4.21	9.88	22.30
C19H15C1106	374.78	374.0557	60.89	4.03	9.46	25.61
C19H15C1107	390.78	390.0506	58.40	3.87	9.07	28.66
C19H15C1108	406.78	406.0455	56.10	3.72	8.72	31.47
C19H15C1109	422.78	422.0404	53.98	3.58	8.39	34.06
C19H16C1201	331.24	330.0578	68.90	4.87	21.41	4.83
C19H16C1202	347.24	346.0527	65.72	4.64	20.42	9.22
C19H16C1203	363.24	362.0476	62.83	4.44	19.52	13.21
C19H16C1204	379.24	378.0426	60.18	4.25	18.70	16.88

090

C19H16C1205	395.24	394.0375	57.74	4.08	17.94	20.24
C19H16C1206	411.24	410.0324	55.49	3.92	17.24	23.34
C19H16C1207	427.24	426.0273	53.42	3.77	16.60	26.21
C19H16C1208	443.24	442.0222	51.49	3.64	16.00	28.88
C19H16C1209	459.24	458.0171	49.69	3.51	15.44	31.35
C19H17C1101	296.80	296.0968	76.89	5.77	11.95	5.39
C19H17C1102	312.80	312.0917	72.96	5.48	11.33	10.23
C19H17C1103	328.80	328.0866	69.41	5.21	10.78	14.60
C19H17C1104	344.80	344.0815	66.19	4.97	10.28	18.56
C19H17C1105	360.80	360.0764	63.25	4.75	9.83	22.17
C19H17C1106	376.80	376.0714	60.57	4.55	9.41	25.48
C19H17C1107	392.80	392.0663	58.10	4.36	9.03	28.51
C19H17C1108	408.80	408.0612	55.83	4.19	8.67	31.31
C19H17C1109	424.79	424.0561	53.72	4.03	8.35	33.90
C19H18C1201	333.26	332.0735	68.48	5.44	21.28	4.80
C19H18C1202	349.26	348.0684	65.34	5.19	20.30	9.16
C19H18C1203	365.26	364.0633	62.48	4.97	19.41	13.14
C19H18C1204	381.26	380.0582	59.86	4.76	18.60	16.79
C19H18C1205	397.26	396.0531	57.45	4.57	17.85	20.14
C19H18C1206	413.26	412.0480	55.22	4.39	17.16	23.23
C19H18C1207	429.26	428.0429	53.16	4.23	16.52	26.09
C19H18C1208	445.26	444.0379	51.25	4.07	15.92	28.75
C19H18C1209	461.26	460.0328	49.48	3.93	15.37	31.22
C19H19C1101	298.82	298.1124	76.37	6.41	11.86	5.35
C19H19C1102	314.82	314.1073	72.49	6.08	11.26	10.16
C19H19C1103	330.81	330.1023	68.98	5.79	10.72	14.51
C19H19C1104	346.81	346.0972	65.80	5.52	10.22	18.45
C19H19C1105	362.81	362.0921	62.90	5.28	9.77	22.05
C19H19C1106	378.81	378.0870	60.24	5.06	9.36	25.34
C19H19C1107	394.81	394.0819	57.80	4.85	8.98	28.37
C19H19C1108	410.81	410.0768	55.55	4.66	8.63	31.16
C19H19C1109	426.81	426.0717	53.47	4.49	8.31	33.74
C19H20C1201	335.28	334.0891	68.07	6.01	21.15	4.77
C19H20C1202	351.28	350.0840	64.97	5.74	20.19	9.11
C19H20C1203	367.28	366.0789	62.14	5.49	19.31	13.07
C19H20C1204	383.27	382.0739	59.54	5.26	18.50	16.70
C19H20C1205	399.27	398.0688	57.16	5.05	17.76	20.04
C19H20C1206	415.27	414.0637	54.95	4.85	17.07	23.12
C19H20C1207	431.27	430.0586	52.92	4.67	16.44	25.97
C19H20C1208	447.27	446.0535	51.02	4.51	15.85	28.62
C19H20C1209	463.27	462.0484	49.26	4.35	15.31	31.08
C19H21C1101	300.83	300.1281	75.86	7.04	11.78	5.32
C19H21C1102	316.83	316.1230	72.03	6.68	11.19	10.10
C19H21C1103	332.83	332.1179	68.57	6.36	10.65	14.42
C19H21C1104	348.83	348.1128	65.42	6.07	10.16	18.35
C19H21C1105	364.83	364.1077	62.55	5.80	9.72	21.93
C19H21C1106	380.83	380.1027	59.93	5.56	9.31	25.21
C19H21C1107	396.83	396.0976	57.51	5.33	8.93	28.22
C19H21C1108	412.83	412.0925	55.28	5.13	8.59	31.00
C19H21C1109	428.83	428.0874	53.22	4.94	8.27	33.58
C19H22C1201	337.29	336.1048	67.66	6.57	21.02	4.74
C19H22C1202	353.29	352.0997	64.60	6.28	20.07	9.06
C19H22C1203	369.29	368.0946	61.80	6.00	19.20	13.00
C19H22C1204	385.29	384.0895	59.23	5.76	18.40	16.61
C19H22C1205	401.29	400.0844	56.87	5.53	17.67	19.93
C19H22C1206	417.29	416.0793	54.69	5.31	16.99	23.00
C19H22C1207	433.29	432.0742	52.67	5.12	16.36	25.85
C19H22C1208	449.29	448.0692	50.79	4.94	15.78	28.49
C19H22C1209	465.29	464.0641	49.05	4.77	15.24	30.95
C19H23C1101	302.85	302.1437	75.36	7.66	11.71	5.28
C19H23C1102	318.85	318.1386	71.57	7.27	11.12	10.04
C19H23C1103	334.85	334.1336	68.15	6.92	10.59	14.33
C19H23C1104	350.85	350.1285	65.05	6.61	10.11	18.24
C19H23C1105	366.85	366.1234	62.21	6.32	9.66	21.81
C19H23C1106	382.84	382.1183	59.61	6.06	9.26	25.07
C19H23C1107	398.84	398.1132	57.22	5.81	8.89	28.08
C19H23C1108	414.84	414.1081	55.01	5.59	8.55	30.85
C19H23C1109	430.84	430.1030	52.97	5.38	8.23	33.42
C19H24C1201	339.31	338.1204	67.26	7.13	20.90	4.72
C19H24C1202	355.31	354.1153	64.23	6.81	19.96	9.01
C19H24C1203	371.31	370.1102	61.46	6.52	19.10	12.93
C19H24C1204	387.31	386.1052	58.92	6.25	18.31	16.52
C19H24C1205	403.31	402.1001	56.59	6.00	17.58	19.84
C19H24C1206	419.31	418.0950	54.43	5.77	16.91	22.89
C19H24C1207	435.30	434.0899	52.43	5.56	16.29	25.73
C19H24C1208	451.30	450.0848	50.57	5.36	15.71	28.36
C19H24C1209	467.30	466.0797	48.84	5.18	15.17	30.81
C19H25C1101	304.86	304.1594	74.86	8.27	11.63	5.25
C19H25C1102	320.86	320.1543	71.12	7.85	11.05	9.97
C19H25C1103	336.86	336.1492	67.75	7.48	10.52	14.25
C19H25C1104	352.86	352.1441	64.67	7.14	10.05	18.14
C19H25C1105	368.86	368.1390	61.87	6.83	9.61	21.69
C19H25C1106	384.86	384.1340	59.30	6.55	9.21	24.94
C19H25C1107	400.86	400.1289	56.93	6.29	8.84	27.94
C19H25C1108	416.86	416.1238	54.75	6.05	8.50	30.70
C19H25C1109	432.86	432.1187	52.72	5.82	8.19	33.27
C19H26C1201	341.32	340.1361	66.86	7.68	20.77	4.69
C19H26C1202	357.32	356.1310	63.87	7.33	19.84	8.96
C19H26C1203	373.32	372.1259	61.13	7.02	18.99	12.86
C19H26C1204	389.32	388.1208	58.62	6.73	18.21	16.44
C19H26C1205	405.32	404.1157	56.30	6.47	17.49	19.74
C19H26C1206	421.32	420.1106	54.17	6.22	16.83	22.78
C19H26C1207	437.32	436.1055	52.18	5.99	16.21	25.61
C19H26C1208	453.32	452.1005	50.34	5.78	15.64	28.24
C19H26C1209	469.32	468.0954	48.63	5.58	15.11	30.68
C19H27C1101	306.88	306.1750	74.37	8.87	11.55	5.21
C19H27C1102	322.88	322.1699	70.68	8.43	10.98	9.91
C19H27C1103	338.88	338.1649	67.34	8.03	10.46	14.16
C19H27C1104	354.88	354.1598	64.31	7.67	9.99	18.03
C19H27C1105	370.88	370.1547	61.53	7.34	9.56	21.57
C19H27C1106	386.88	386.1496	58.99	7.03	9.16	24.81
C19H27C1107	402.88	402.1445	56.65	6.76	8.80	27.80
C19H27C1108	418.88	418.1394	54.48	6.50	8.46	30.56
C19H27C1109	434.87	434.1343	52.48	6.26	8.15	33.11
C19H28C1201	343.34	342.1517	66.47	8.22	20.65	4.66
C19H28C1202	359.34	358.1466	63.51	7.85	19.73	8.90
C19H28C1203	375.34	374.1415	60.80	7.52	18.89	12.79
C19H28C1204	391.34	390.1365	58.32	7.21	18.12	16.35
C19H28C1205	407.34	406.1314	56.03	6.93	17.41	19.64
C19H28C1206	423.34	422.1263	53.91	6.67	16.75	22.68
C19H28C1207	439.34	438.1212	51.94	6.42	16.14	25.49
C19H28C1208	455.34	454.1161	50.12	6.20	15.57	28.11
C19H28C1209	471.34	470.1110	48.42	5.99	15.04	30.55
C19H29C1101	308.90	308.1907	73.88	9.46	11.48	5.18
C19H29C1102	324.89	324.1856	70.24	9.00	10.91	9.85
C19H29C1103	340.89	340.1805	66.95	8.57	10.40	14.08
C19H29C1104	356.89	356.1754	63.94	8.19	9.93	17.93
C19H29C1105	372.89	372.1703	61.20	7.84	9.51	21.45
C19H29C1106	388.89	388.1653	58.68	7.52	9.12	24.68
C19H29C1107	404.89	404.1602	56.36	7.22	8.76	27.66

C19H29C1108	420.89	420.1551	54.22	6.95	8.42	30.41
C19H29C1109	436.89	436.1500	52.24	6.69	8.11	32.96
C19H30C1201	345.36	344.1674	66.08	8.76	20.53	4.63
C19H30C1202	361.36	360.1623	63.15	8.37	19.62	8.86
C19H30C1203	377.36	376.1572	60.48	8.01	18.79	12.72
C19H30C1204	393.35	392.1521	58.02	7.69	18.03	16.27
C19H30C1205	409.35	408.1470	55.75	7.39	17.32	19.54
C19H30C1206	425.35	424.1419	53.65	7.11	16.67	22.57
C19H30C1207	441.35	440.1368	51.71	6.85	16.07	25.38
C19H30C1208	457.35	456.1318	49.90	6.61	15.50	27.99
C19H30C1209	473.35	472.1267	48.21	6.39	14.98	30.42
C19H31C1101	310.91	310.2063	73.40	10.05	11.40	5.15
C19H31C1102	326.91	326.2012	69.81	9.56	10.84	9.79
C19H31C1103	342.91	342.1962	66.55	9.11	10.34	14.00
C19H31C1104	358.91	358.1911	63.58	8.71	9.88	17.83
C19H31C1105	374.91	374.1860	60.87	8.33	9.46	21.34
C19H31C1106	390.91	390.1809	58.38	7.99	9.07	24.56
C19H31C1107	406.91	406.1758	56.08	7.68	8.71	27.52
C19H31C1108	422.91	422.1707	53.96	7.39	8.38	30.27
C19H31C1109	438.91	438.1656	52.00	7.12	8.08	32.81
C19H32C1201	347.37	346.1830	65.70	9.29	20.41	4.61
C19H32C1202	363.37	362.1779	62.80	8.88	19.51	8.81
C19H32C1203	379.37	378.1728	60.16	8.50	18.69	12.65
C19H32C1204	395.37	394.1677	57.72	8.16	17.93	16.19
C19H32C1205	411.37	410.1627	55.48	7.84	17.24	19.45
C19H32C1206	427.37	426.1576	53.40	7.55	16.59	22.46
C19H32C1207	443.37	442.1525	51.47	7.27	15.99	25.26
C19H32C1208	459.37	458.1474	49.68	7.02	15.44	27.86
C19H32C1209	475.37	474.1423	48.01	6.79	14.92	30.29
C19H33C1101	312.93	312.2220	72.93	10.63	11.33	5.11
C19H33C1102	328.93	328.2169	69.38	10.11	10.78	9.73
C19H33C1103	344.93	344.2118	66.16	9.64	10.28	13.92
C19H33C1104	360.93	360.2067	63.23	9.22	9.82	17.73
C19H33C1105	376.92	376.2016	60.55	8.82	9.41	21.22
C19H33C1106	392.92	392.1965	58.08	8.47	9.02	24.43
C19H33C1107	408.92	408.1915	55.81	8.13	8.67	27.39
C19H33C1108	424.92	424.1864	53.71	7.83	8.34	30.12
C19H33C1109	440.92	440.1813	51.76	7.54	8.04	32.66
C19H34C1201	349.39	348.1987	65.32	9.81	20.29	4.58
C19H34C1202	365.39	364.1936	62.46	9.38	19.41	8.76
C19H34C1203	381.39	380.1885	59.84	8.99	18.59	12.59
C19H34C1204	397.39	396.1834	57.43	8.62	17.84	16.10
C19H34C1205	413.39	412.1783	55.21	8.29	17.15	19.35
C19H34C1206	429.39	428.1732	53.15	7.98	16.51	22.36
C19H34C1207	445.38	444.1681	51.24	7.69	15.92	25.15
C19H34C1208	461.38	460.1631	49.46	7.43	15.37	27.74
C19H34C1209	477.38	476.1580	47.80	7.18	14.85	30.16
C19H35C1101	314.94	314.2376	72.46	11.20	11.26	5.08
C19H35C1102	330.94	330.2325	68.96	10.66	10.71	9.67
C19H35C1103	346.94	346.2275	65.78	10.17	10.22	13.83
C19H35C1104	362.94	362.2224	62.88	9.72	9.77	17.63
C19H35C1105	378.94	378.2173	60.22	9.31	9.36	21.11
C19H35C1106	394.94	394.2122	57.78	8.93	8.98	24.31
C19H35C1107	410.94	410.2071	55.53	8.58	8.63	27.25
C19H35C1108	426.94	426.2020	53.45	8.26	8.30	29.98
C19H35C1109	442.94	442.1969	51.52	7.96	8.00	32.51
C19H36C1201	351.40	350.2143	64.94	10.33	20.18	4.55
C19H36C1202	367.40	366.2092	62.11	9.88	19.30	8.71
C19H36C1203	383.40	382.2041	59.52	9.46	18.49	12.52
C19H36C1204	399.40	398.1990	57.14	9.09	17.75	16.02
C19H36C1205	415.40	414.1940	54.94	8.74	17.07	19.26
C19H36C1206	431.40	430.1889	52.90	8.41	16.44	22.25
C19H36C1207	447.40	446.1838	51.01	8.11	15.83	25.03
C19H36C1208	463.40	462.1787	49.25	7.83	15.30	27.62
C19H36C1209	479.40	478.1736	47.60	7.57	14.79	30.04
C19H37C1101	316.96	316.2533	72.00	11.77	11.19	5.05
C19H37C1102	332.96	332.2482	68.54	11.20	10.65	9.61
C19H37C1103	348.96	348.2431	65.40	10.69	10.16	13.75
C19H37C1104	364.96	364.2380	62.53	10.22	9.71	17.54
C19H37C1105	380.96	380.2329	59.90	9.79	9.31	21.00
C19H37C1106	396.96	396.2278	57.49	9.40	8.93	24.18
C19H37C1107	412.96	412.2228	55.26	9.03	8.59	27.12
C19H37C1108	428.95	428.2177	53.20	8.69	8.26	29.84
C19H37C1109	444.95	444.2126	51.29	8.38	7.97	32.36
C19H38C1201	353.42	352.2300	64.57	10.84	20.06	4.53
C19H38C1202	369.42	368.2249	61.78	10.37	19.19	8.66
C19H38C1203	385.42	384.2198	59.21	9.94	18.40	12.45
C19H38C1204	401.42	400.2147	56.85	9.54	17.66	15.94
C19H38C1205	417.42	416.2096	54.67	9.18	16.99	19.16
C19H38C1206	433.42	432.2045	52.65	8.84	16.36	22.15
C19H38C1207	449.42	448.1994	50.78	8.52	15.78	24.92
C19H38C1208	465.42	464.1944	49.03	8.23	15.23	27.50
C19H38C1209	481.42	480.1893	47.40	7.96	14.73	29.91
C19H39C1101	318.98	318.2689	71.55	12.32	11.11	5.02
C19H39C1102	334.97	334.2638	68.13	11.74	10.58	9.55
C19H39C1103	350.97	350.2588	65.02	11.20	10.10	13.68
C19H39C1104	366.97	366.2537	62.19	10.71	9.66	17.44
C19H39C1105	382.97	382.2486	59.59	10.26	9.26	20.89
C19H39C1106	398.97	398.2435	57.20	9.85	8.89	24.06
C19H39C1107	414.97	414.2384	54.99	9.47	8.54	26.99
C19H39C1108	430.97	430.2333	52.95	9.12	8.23	29.70
C19H39C1109	446.97	446.2282	51.06	8.79	7.93	32.22
C20H10C1201	337.21	336.0109	71.24	2.99	21.03	4.74
C20H10C1202	353.21	352.0058	68.01	2.85	20.07	9.06
C20H10C1203	369.21	368.0007	65.06	2.73	19.20	13.00
C20H10C1204	385.21	383.9956	62.36	2.62	18.41	16.61
C20H10C1205	401.21	399.9905	59.88	2.51	17.67	19.94
C20H10C1206	417.21	415.9854	57.58	2.42	17.00	23.01
C20H10C1207	433.20	431.9804	55.45	2.33	16.37	25.85
C20H10C1208	449.20	447.9753	53.48	2.24	15.78	28.49
C20H10C1209	465.20	463.9702	51.64	2.17	15.24	30.95
C20H11C1101	302.76	302.0498	79.34	3.66	11.71	5.28
C20H11C1102	318.76	318.0448	75.36	3.48	11.12	10.04
C20H11C1103	334.76	334.0397	71.76	3.31	10.59	14.34
C20H11C1104	350.76	350.0346	68.49	3.16	10.11	18.25
C20H11C1105	366.76	366.0295	65.50	3.02	9.67	21.81
C20H11C1106	382.76	382.0244	62.76	2.90	9.26	25.08
C20H11C1107	398.76	398.0193	60.24	2.78	8.89	28.09
C20H11C1108	414.76	414.0142	57.92	2.67	8.55	30.86
C20H11C1109	430.76	430.0092	55.77	2.57	8.23	33.43
C20H12C1201	339.22	338.0265	70.82	3.57	20.90	4.72
C20H12C1202	355.22	354.0214	67.63	3.41	19.96	9.01
C20H12C1203	371.22	370.0163	64.71	3.26	19.10	12.93
C20H12C1204	387.22	386.0113	62.04	3.12	18.31	16.53
C20H12C1205	403.22	402.0062	59.58	3.00	17.58	19.84
C20H12C1206	419.22	418.0011	57.30	2.89	16.91	22.90
C20H12C1207	435.22	433.9960	55.20	2.78	16.29	25.73
C20H12C1208	451.22	449.9909	53.24	2.68	15.71	28.37
C20H12C1209	467.22	465.9858	51.42	2.59	15.18	30.82
C20H13C1101	304.78	304.0655	78.82	4.30	11.63	5.25

092

C20H13C1102	320.78	320.0604	74.89	4.08	11.05	9.98
C20H13C1103	336.78	336.0553	71.33	3.89	10.53	14.25
C20H13C1104	352.78	352.0502	68.09	3.71	10.05	18.14
C20H13C1105	368.78	368.0451	65.14	3.55	9.61	21.69
C20H13C1106	384.78	384.0401	62.43	3.41	9.21	24.95
C20H13C1107	400.78	400.0350	59.94	3.27	8.85	27.94
C20H13C1108	416.77	416.0299	57.64	3.14	8.51	30.71
C20H13C1109	432.77	432.0248	55.51	3.03	8.19	33.27
C20H14C1201	341.24	340.0422	70.40	4.14	20.78	4.69
C20H14C1202	357.24	356.0371	67.24	3.95	19.85	8.96
C20H14C1203	373.24	372.0320	64.36	3.78	19.00	12.86
C20H14C1204	389.24	388.0269	61.72	3.63	18.22	16.44
C20H14C1205	405.24	404.0218	59.28	3.48	17.50	19.74
C20H14C1206	421.24	420.0167	57.03	3.35	16.83	22.79
C20H14C1207	437.24	436.0116	54.94	3.23	16.22	25.61
C20H14C1208	453.24	452.0066	53.00	3.11	15.64	28.24
C20H14C1209	469.24	468.0015	51.19	3.01	15.11	30.69
C20H15C1101	306.79	306.0811	78.30	4.93	11.56	5.22
C20H15C1102	322.79	322.0761	74.42	4.68	10.98	9.91
C20H15C1103	338.79	338.0710	70.91	4.46	10.46	14.17
C20H15C1104	354.79	354.0659	67.71	4.26	9.99	18.04
C20H15C1105	370.79	370.0608	64.79	4.08	9.56	21.57
C20H15C1106	386.79	386.0557	62.11	3.91	9.17	24.82
C20H15C1107	402.79	402.0506	59.64	3.75	8.80	27.80
C20H15C1108	418.79	418.0455	57.36	3.61	8.47	30.56
C20H15C1109	434.79	434.0404	55.25	3.48	8.15	33.12
C20H16C1201	343.26	342.0578	69.98	4.70	20.66	4.66
C20H16C1202	359.26	358.0527	66.87	4.49	19.74	8.91
C20H16C1203	375.25	374.0476	64.02	4.30	18.90	12.79
C20H16C1204	391.25	390.0426	61.40	4.12	18.12	16.36
C20H16C1205	407.25	406.0375	58.99	3.96	17.41	19.64
C20H16C1206	423.25	422.0324	56.76	3.81	16.75	22.68
C20H16C1207	439.25	438.0273	54.69	3.67	16.14	25.50
C20H16C1208	455.25	454.0222	52.77	3.54	15.58	28.12
C20H16C1209	471.25	470.0171	50.98	3.42	15.05	30.56
C20H17C1101	308.81	308.0968	77.79	5.55	11.48	5.18
C20H17C1102	324.81	324.0917	73.96	5.28	10.91	9.85
C20H17C1103	340.81	340.0866	70.49	5.03	10.40	14.08
C20H17C1104	356.81	356.0815	67.33	4.80	9.94	17.94
C20H17C1105	372.81	372.0764	64.44	4.60	9.51	21.46
C20H17C1106	388.81	388.0714	61.78	4.41	9.12	24.69
C20H17C1107	404.81	404.0663	59.34	4.23	8.76	27.67
C20H17C1108	420.81	420.0612	57.09	4.07	8.43	30.42
C20H17C1109	436.81	436.0561	55.00	3.92	8.12	32.97
C20H18C1201	345.27	344.0735	69.58	5.25	20.54	4.63
C20H18C1202	361.27	360.0684	66.49	5.02	19.63	8.86
C20H18C1203	377.27	376.0633	63.67	4.81	18.79	12.72
C20H18C1204	393.27	392.0582	61.08	4.61	18.03	16.27
C20H18C1205	409.27	408.0531	58.70	4.43	17.33	19.55
C20H18C1206	425.27	424.0480	56.49	4.27	16.67	22.57
C20H18C1207	441.27	440.0429	54.44	4.11	16.07	25.38
C20H18C1208	457.27	456.0379	52.53	3.97	15.51	27.99
C20H18C1209	473.27	472.0328	50.76	3.83	14.98	30.43
C20H19C1101	310.83	310.1124	77.29	6.16	11.41	5.15
C20H19C1102	326.83	326.1073	73.50	5.86	10.85	9.79
C20H19C1103	342.83	342.1023	70.07	5.59	10.34	14.00
C20H19C1104	358.83	358.0972	66.95	5.34	9.88	17.84
C20H19C1105	374.82	374.0921	64.09	5.11	9.46	21.34
C20H19C1106	390.82	390.0870	61.47	4.90	9.07	24.56
C20H19C1107	406.82	406.0819	59.05	4.71	8.71	27.53
C20H19C1108	422.82	422.0768	56.81	4.53	8.38	30.27
C20H19C1109	438.82	438.0717	54.74	4.36	8.08	32.81
C20H20C1201	347.29	346.0891	69.17	5.80	20.42	4.61
C20H20C1202	363.29	362.0840	66.12	5.55	19.52	8.81
C20H20C1203	379.29	378.0789	63.34	5.32	18.69	12.65
C20H20C1204	395.29	394.0739	60.77	5.10	17.94	16.19
C20H20C1205	411.29	410.0688	58.41	4.90	17.24	19.45
C20H20C1206	427.28	426.0637	56.22	4.72	16.59	22.47
C20H20C1207	443.28	442.0586	54.19	4.55	16.00	25.27
C20H20C1208	459.28	458.0535	52.30	4.39	15.44	27.87
C20H20C1209	475.28	474.0484	50.54	4.24	14.92	30.30
C20H21C1101	312.84	312.1281	76.79	6.77	11.33	5.11
C20H21C1102	328.84	328.1230	73.05	6.44	10.78	9.73
C20H21C1103	344.84	344.1179	69.66	6.14	10.28	13.92
C20H21C1104	360.84	360.1128	66.57	5.87	9.83	17.74
C20H21C1105	376.84	376.1077	63.75	5.62	9.41	21.23
C20H21C1106	392.84	392.1027	61.15	5.39	9.02	24.44
C20H21C1107	408.84	408.0976	58.76	5.18	8.67	27.39
C20H21C1108	424.84	424.0925	56.54	4.98	8.35	30.13
C20H21C1109	440.84	440.0874	54.49	4.80	8.04	32.66
C20H22C1201	349.30	348.1048	68.77	6.35	20.30	4.58
C20H22C1202	365.30	364.0997	65.76	6.07	19.41	8.76
C20H22C1203	381.30	380.0946	63.00	5.82	18.60	12.59
C20H22C1204	397.30	396.0895	60.46	5.58	17.85	16.11
C20H22C1205	413.30	412.0844	58.12	5.37	17.16	19.36
C20H22C1206	429.30	428.0793	55.96	5.17	16.52	22.36
C20H22C1207	445.30	444.0742	53.95	4.98	15.92	25.15
C20H22C1208	461.30	460.0692	52.08	4.81	15.37	27.75
C20H22C1209	477.30	476.0641	50.33	4.65	14.86	30.17
C20H23C1101	314.86	314.1437	76.30	7.36	11.26	5.08
C20H23C1102	330.86	330.1386	72.61	7.01	10.72	9.67
C20H23C1103	346.86	346.1336	69.26	6.68	10.22	13.84
C20H23C1104	362.86	362.1285	66.20	6.39	9.77	17.64
C20H23C1105	378.86	378.1234	63.41	6.12	9.36	21.12
C20H23C1106	394.86	394.1183	60.84	5.87	8.98	24.31
C20H23C1107	410.86	410.1132	58.47	5.64	8.63	27.26
C20H23C1108	426.85	426.1081	56.28	5.43	8.31	29.99
C20H23C1109	442.85	442.1030	54.24	5.23	8.01	32.52
C20H24C1201	351.32	350.1204	68.38	6.89	20.18	4.55
C20H24C1202	367.32	366.1153	65.40	6.59	19.30	8.71
C20H24C1203	383.32	382.1102	62.67	6.31	18.50	12.52
C20H24C1204	399.32	398.1052	60.16	6.06	17.76	16.03
C20H24C1205	415.32	414.1001	57.84	5.82	17.07	19.26
C20H24C1206	431.32	430.0950	55.70	5.61	16.44	22.26
C20H24C1207	447.32	446.0899	53.70	5.41	15.85	25.04
C20H24C1208	463.32	462.0848	51.85	5.22	15.30	27.63
C20H24C1209	479.31	478.0797	50.12	5.05	14.79	30.04
C20H25C1101	316.87	316.1594	75.81	7.95	11.19	5.05
C20H25C1102	332.87	332.1543	72.17	7.57	10.65	9.61
C20H25C1103	348.87	348.1492	68.86	7.22	10.16	13.76
C20H25C1104	364.87	364.1441	65.84	6.91	9.72	17.54
C20H25C1105	380.87	380.1390	63.07	6.62	9.31	21.00
C20H25C1106	396.87	396.1340	60.53	6.35	8.93	24.19
C20H25C1107	412.87	412.1289	58.18	6.10	8.59	27.13
C20H25C1108	428.87	428.1238	56.01	5.88	8.27	29.84
C20H25C1109	444.87	444.1187	54.00	5.66	7.97	32.37
C20H26C1201	353.34	352.1361	67.99	7.42	20.07	4.53
C20H26C1202	369.34	368.1310	65.04	7.10	19.20	8.66
C20H26C1203	385.33	384.1259	62.34	6.80	18.40	12.46
C20H26C1204	401.33	400.1208	59.86	6.53	17.67	15.95

093

C20H26C1205	417.33	416.1157	57.56	6.28	16.99	19.17
C20H26C1206	433.33	432.1106	55.44	6.05	16.36	22.15
C20H26C1207	449.33	448.1055	53.46	5.83	15.78	24.92
C20H26C1208	465.33	464.1005	51.62	5.63	15.24	27.51
C20H26C1209	481.33	480.0954	49.91	5.44	14.73	29.92
C20H27C1101	318.89	318.1750	75.33	8.53	11.12	5.02
C20H27C1102	334.89	334.1699	71.73	8.13	10.59	9.56
C20H27C1103	350.89	350.1649	68.46	7.76	10.10	13.68
C20H27C1104	366.89	366.1598	65.48	7.42	9.66	17.44
C20H27C1105	382.89	382.1547	62.74	7.11	9.26	20.89
C20H27C1106	398.89	398.1496	60.22	6.82	8.89	24.07
C20H27C1107	414.89	414.1445	57.90	6.56	8.55	26.99
C20H27C1108	430.89	430.1394	55.75	6.32	8.23	29.71
C20H27C1109	446.89	446.1343	53.75	6.09	7.93	32.22
C20H28C1201	355.35	354.1517	67.60	7.94	19.95	4.50
C20H28C1202	371.35	370.1466	64.69	7.60	19.09	8.62
C20H28C1203	387.35	386.1415	62.02	7.29	18.31	12.39
C20H28C1204	403.35	402.1365	59.56	7.00	17.58	15.87
C20H28C1205	419.35	418.1314	57.28	6.73	16.91	19.08
C20H28C1206	435.35	434.1263	55.18	6.48	16.29	22.05
C20H28C1207	451.35	450.1212	53.22	6.25	15.71	24.81
C20H28C1208	467.35	466.1161	51.40	6.04	15.17	27.39
C20H28C1209	483.35	482.1110	49.70	5.84	14.67	29.79
C20H29C1101	320.91	320.1907	74.86	9.11	11.05	4.99
C20H29C1102	336.91	336.1856	71.30	8.68	10.52	9.50
C20H29C1103	352.91	352.1805	68.07	8.28	10.05	13.60
C20H29C1104	368.90	368.1754	65.12	7.92	9.61	17.35
C20H29C1105	384.90	384.1703	62.41	7.59	9.21	20.78
C20H29C1106	400.90	400.1653	59.92	7.29	8.84	23.95
C20H29C1107	416.90	416.1602	57.62	7.01	8.50	26.86
C20H29C1108	432.90	432.1551	55.49	6.75	8.19	29.57
C20H29C1109	448.90	448.1500	53.51	6.51	7.90	32.08
C20H30C1201	357.37	356.1674	67.22	8.46	19.84	4.48
C20H30C1202	373.37	372.1623	64.34	8.10	18.99	8.57
C20H30C1203	389.37	388.1572	61.70	7.77	18.21	12.33
C20H30C1204	405.37	404.1521	59.26	7.46	17.49	15.79
C20H30C1205	421.37	420.1470	57.01	7.18	16.83	18.99
C20H30C1206	437.36	436.1419	54.93	6.91	16.21	21.95
C20H30C1207	453.36	452.1368	52.99	6.67	15.64	24.70
C20H30C1208	469.36	468.1318	51.18	6.44	15.11	27.27
C20H30C1209	485.36	484.1267	49.49	6.23	14.61	29.67
C20H31C1101	322.92	322.2063	74.39	9.68	10.98	4.95
C20H31C1102	338.92	338.2012	70.88	9.22	10.46	9.44
C20H31C1103	354.92	354.1962	67.68	8.80	9.99	13.52
C20H31C1104	370.92	370.1911	64.76	8.42	9.56	17.25
C20H31C1105	386.92	386.1860	62.09	8.08	9.16	20.68
C20H31C1106	402.92	402.1809	59.62	7.76	8.80	23.83
C20H31C1107	418.92	418.1758	57.34	7.46	8.46	26.73
C20H31C1108	434.92	434.1707	55.23	7.18	8.15	29.43
C20H31C1109	450.92	450.1656	53.27	6.93	7.86	31.93
C20H32C1201	359.38	358.1830	66.84	8.98	19.73	4.45
C20H32C1202	375.38	374.1779	63.99	8.59	18.89	8.52
C20H32C1203	391.38	390.1728	61.38	8.24	18.12	12.26
C20H32C1204	407.38	406.1677	58.97	7.92	17.41	15.71
C20H32C1205	423.38	422.1627	56.74	7.62	16.75	18.89
C20H32C1206	439.38	438.1576	54.67	7.34	16.14	21.85
C20H32C1207	455.38	454.1525	52.75	7.08	15.57	24.59
C20H32C1208	471.38	470.1474	50.96	6.84	15.04	27.15
C20H32C1209	487.38	486.1423	49.29	6.62	14.55	29.54
C20H33C1101	324.94	324.2220	73.93	10.24	10.91	4.92

C20H33C1102	340.94	340.2169	70.46	9.76	10.40	9.39
C20H33C1103	356.94	356.2118	67.30	9.32	9.93	13.45
C20H33C1104	372.94	372.2067	64.41	8.92	9.51	17.16
C20H33C1105	388.94	388.2016	61.76	8.55	9.12	20.57
C20H33C1106	404.94	404.1965	59.32	8.21	8.76	23.71
C20H33C1107	420.93	420.1915	57.07	7.90	8.42	26.61
C20H33C1108	436.93	436.1864	54.98	7.61	8.11	29.29
C20H33C1109	452.93	452.1813	53.04	7.34	7.83	31.79
C20H34C1201	361.40	360.1987	66.47	9.48	19.62	4.43
C20H34C1202	377.40	376.1936	63.65	9.08	18.79	8.48
C20H34C1203	393.40	392.1885	61.06	8.71	18.02	12.20
C20H34C1204	409.40	408.1834	58.68	8.37	17.32	15.63
C20H34C1205	425.40	424.1783	56.47	8.06	16.67	18.81
C20H34C1206	441.40	440.1732	54.42	7.76	16.06	21.75
C20H34C1207	457.40	456.1681	52.52	7.49	15.50	24.49
C20H34C1208	473.40	472.1631	50.74	7.24	14.98	27.04
C20H34C1209	489.39	488.1580	49.09	7.00	14.49	29.42
C20H35C1101	326.95	326.2376	73.47	10.79	10.84	4.89
C20H35C1102	342.95	342.2325	70.05	10.29	10.34	9.33
C20H35C1103	358.95	358.2275	66.92	9.83	9.88	13.37
C20H35C1104	374.95	374.2224	64.07	9.41	9.46	17.07
C20H35C1105	390.95	390.2173	61.45	9.02	9.07	20.46
C20H35C1106	406.95	406.2122	59.03	8.67	8.71	23.59
C20H35C1107	422.95	422.2071	56.80	8.34	8.38	26.48
C20H35C1108	438.95	438.2020	54.73	8.04	8.08	29.16
C20H35C1109	454.95	454.1969	52.80	7.75	7.79	31.65
C20H36C1201	363.42	362.2143	66.10	9.98	19.51	4.40
C20H36C1202	379.41	378.2092	63.31	9.56	18.69	8.43
C20H36C1203	395.41	394.2041	60.75	9.18	17.93	12.14
C20H36C1204	411.41	410.1990	58.39	8.82	17.23	15.56
C20H36C1205	427.41	426.1940	56.20	8.49	16.59	18.72
C20H36C1206	443.41	442.1889	54.18	8.18	15.99	21.65
C20H36C1207	459.41	458.1838	52.29	7.90	15.43	24.38
C20H36C1208	475.41	474.1787	50.53	7.63	14.91	26.92
C20H36C1209	491.41	490.1736	48.88	7.38	14.43	29.30
C20H37C1101	328.97	328.2533	73.02	11.34	10.78	4.86
C20H37C1102	344.97	344.2482	69.64	10.81	10.28	9.28
C20H37C1103	360.97	360.2431	66.55	10.33	9.82	13.30
C20H37C1104	376.97	376.2380	63.72	9.89	9.40	16.98
C20H37C1105	392.97	392.2329	61.13	9.49	9.02	20.36
C20H37C1106	408.97	408.2278	58.74	9.12	8.67	23.47
C20H37C1107	424.97	424.2228	56.53	8.78	8.34	26.35
C20H37C1108	440.97	440.2177	54.48	8.46	8.04	29.03
C20H37C1109	456.97	456.2126	52.57	8.16	7.76	31.51
C20H38C1201	365.43	364.2300	65.74	10.48	19.40	4.38
C20H38C1202	381.43	380.2249	62.98	10.04	18.59	8.39
C20H38C1203	397.43	396.2198	60.44	9.64	17.84	12.08
C20H38C1204	413.43	412.2147	58.10	9.26	17.15	15.48
C20H38C1205	429.43	428.2096	55.94	8.92	16.51	18.63
C20H38C1206	445.43	444.2045	53.93	8.60	15.92	21.55
C20H38C1207	461.43	460.1994	52.06	8.30	15.37	24.27
C20H38C1208	477.43	476.1944	50.32	8.02	14.85	26.81
C20H38C1209	493.43	492.1893	48.68	7.76	14.37	29.18
C20H39C1101	330.99	330.2689	72.58	11.88	10.71	4.83
C20H39C1102	346.99	346.2638	69.23	11.33	10.22	9.22
C20H39C1103	362.99	362.2588	66.18	10.83	9.77	13.22
C20H39C1104	378.98	378.2537	63.39	10.37	9.35	16.89
C20H39C1105	394.98	394.2486	60.82	9.95	8.98	20.25
C20H39C1106	410.98	410.2435	58.45	9.57	8.63	23.36
C20H39C1107	426.98	426.2384	56.26	9.21	8.30	26.23

094

C20H39C1108	442.98	442.2333	54.23	8.87	8.00	28.89
C20H39C1109	458.98	458.2282	52.34	8.56	7.72	31.37
C20H40C1201	367.45	366.2456	65.38	10.97	19.30	4.35
C20H40C1202	383.45	382.2405	62.65	10.51	18.49	8.35
C20H40C1203	399.45	398.2354	60.14	10.09	17.75	12.02
C20H40C1204	415.45	414.2303	57.82	9.70	17.07	15.40
C20H40C1205	431.44	430.2253	55.68	9.35	16.43	18.54
C20H40C1206	447.44	446.2202	53.69	9.01	15.85	21.45
C20H40C1207	463.44	462.2151	51.83	8.70	15.30	24.17
C20H40C1208	479.44	478.2100	50.10	8.41	14.79	26.70
C20H40C1209	495.44	494.2049	48.49	8.14	14.31	29.06
C20H41C1101	333.00	332.2846	72.14	12.41	10.65	4.80
C20H41C1102	349.00	348.2795	68.83	11.84	10.16	9.17
C20H41C1103	365.00	364.2744	65.81	11.32	9.71	13.15
C20H41C1104	381.00	380.2693	63.05	10.85	9.31	16.80
C20H41C1105	397.00	396.2642	60.51	10.41	8.93	20.15
C20H41C1106	413.00	412.2591	58.17	10.01	8.58	23.24
C20H41C1107	429.00	428.2541	56.00	9.63	8.26	26.11
C20H41C1108	445.00	444.2490	53.98	9.29	7.97	28.76
C20H41C1109	461.00	460.2439	52.11	8.96	7.69	31.24
C21H10C1201	349.22	348.0109	72.23	2.89	20.30	4.58
C21H10C1202	365.22	364.0058	69.06	2.76	19.41	8.76
C21H10C1203	381.22	380.0007	66.17	2.64	18.60	12.59
C21H10C1204	397.22	395.9956	63.50	2.54	17.85	16.11
C21H10C1205	413.22	411.9905	61.04	2.44	17.16	19.36
C21H10C1206	429.22	427.9854	58.77	2.35	16.52	22.37
C21H10C1207	445.22	443.9804	56.65	2.26	15.93	25.16
C21H10C1208	461.22	459.9753	54.69	2.19	15.37	27.75
C21H10C1209	477.21	475.9702	52.86	2.11	14.86	30.17
C21H11C1101	314.77	314.0498	80.13	3.52	11.26	5.08
C21H11C1102	330.77	330.0448	76.26	3.35	10.72	9.67
C21H11C1103	346.77	346.0397	72.74	3.20	10.22	13.84
C21H11C1104	362.77	362.0346	69.53	3.06	9.77	17.64
C21H11C1105	378.77	378.0295	66.59	2.93	9.36	21.12
C21H11C1106	394.77	394.0244	63.89	2.81	8.98	24.32
C21H11C1107	410.77	410.0193	61.41	2.70	8.63	27.26
C21H11C1108	426.77	426.0142	59.10	2.60	8.31	29.99
C21H11C1109	442.77	442.0092	56.97	2.50	8.01	32.52
C21H12C1201	351.24	350.0265	71.81	3.44	20.19	4.56
C21H12C1202	367.23	366.0214	68.68	3.29	19.31	8.71
C21H12C1203	383.23	382.0163	65.82	3.16	18.50	12.52
C21H12C1204	399.23	398.0113	63.18	3.03	17.76	16.03
C21H12C1205	415.23	414.0062	60.75	2.91	17.08	19.27
C21H12C1206	431.23	430.0011	58.49	2.80	16.44	22.26
C21H12C1207	447.23	445.9960	56.40	2.70	15.85	25.04
C21H12C1208	463.23	461.9909	54.45	2.61	15.31	27.63
C21H12C1209	479.23	477.9858	52.63	2.52	14.80	30.05
C21H13C1101	316.79	316.0655	79.62	4.14	11.19	5.05
C21H13C1102	332.79	332.0604	75.79	3.94	10.65	9.62
C21H13C1103	348.79	348.0553	72.32	3.76	10.16	13.76
C21H13C1104	364.79	364.0502	69.15	3.59	9.72	17.54
C21H13C1105	380.79	380.0451	66.24	3.44	9.31	21.01
C21H13C1106	396.79	396.0401	63.57	3.30	8.94	24.19
C21H13C1107	412.79	412.0350	61.11	3.17	8.59	27.13
C21H13C1108	428.79	428.0299	58.83	3.06	8.27	29.85
C21H13C1109	444.79	444.0248	56.71	2.95	7.97	32.37
C21H14C1201	353.25	352.0422	71.40	3.99	20.07	4.53
C21H14C1202	369.25	368.0371	68.31	3.82	19.20	8.67
C21H14C1203	385.25	384.0320	65.47	3.66	18.41	12.46
C21H14C1204	401.25	400.0269	62.86	3.52	17.67	15.95

C21H14C1205	417.25	416.0218	60.45	3.38	16.99	19.17
C21H14C1206	433.25	432.0167	58.22	3.26	16.37	22.16
C21H14C1207	449.25	448.0116	56.15	3.14	15.78	24.93
C21H14C1208	465.25	464.0066	54.22	3.03	15.24	27.51
C21H14C1209	481.25	480.0015	52.41	2.93	14.73	29.92
C21H15C1101	318.81	318.0811	79.12	4.74	11.12	5.02
C21H15C1102	334.81	334.0761	75.34	4.52	10.59	9.56
C21H15C1103	350.80	350.0710	71.90	4.31	10.11	13.68
C21H15C1104	366.80	366.0659	68.77	4.12	9.67	17.45
C21H15C1105	382.80	382.0608	65.89	3.95	9.26	20.90
C21H15C1106	398.80	398.0557	63.25	3.79	8.89	24.07
C21H15C1107	414.80	414.0506	60.81	3.64	8.55	27.00
C21H15C1108	430.80	430.0455	58.55	3.51	8.23	29.71
C21H15C1109	446.80	446.0404	56.45	3.38	7.93	32.23
C21H16C1201	355.27	354.0578	71.00	4.54	19.96	4.50
C21H16C1202	371.27	370.0527	67.94	4.34	19.10	8.62
C21H16C1203	387.27	386.0476	65.13	4.16	18.31	12.39
C21H16C1204	403.27	402.0426	62.55	4.00	17.58	15.87
C21H16C1205	419.26	418.0375	60.16	3.85	16.91	19.08
C21H16C1206	435.26	434.0324	57.95	3.71	16.29	22.05
C21H16C1207	451.26	450.0273	55.90	3.57	15.71	24.82
C21H16C1208	467.26	466.0222	53.98	3.45	15.17	27.39
C21H16C1209	483.26	482.0171	52.19	3.34	14.67	29.80
C21H17C1101	320.82	320.0968	78.62	5.34	11.05	4.99
C21H17C1102	336.82	336.0917	74.89	5.09	10.53	9.50
C21H17C1103	352.82	352.0866	71.49	4.86	10.05	13.60
C21H17C1104	368.82	368.0815	68.39	4.65	9.61	17.35
C21H17C1105	384.82	384.0764	65.55	4.45	9.21	20.79
C21H17C1106	400.82	400.0714	62.93	4.28	8.85	23.95
C21H17C1107	416.82	416.0663	60.51	4.11	8.51	26.87
C21H17C1108	432.82	432.0612	58.28	3.96	8.19	29.57
C21H17C1109	448.82	448.0561	56.20	3.82	7.90	32.08
C21H18C1201	357.28	356.0735	70.60	5.08	19.85	4.48
C21H18C1202	373.28	372.0684	67.57	4.86	19.00	8.57
C21H18C1203	389.28	388.0633	64.79	4.66	18.21	12.33
C21H18C1204	405.28	404.0582	62.24	4.48	17.50	15.79
C21H18C1205	421.28	420.0531	59.87	4.31	16.83	18.99
C21H18C1206	437.28	436.0480	57.68	4.15	16.22	21.95
C21H18C1207	453.28	452.0429	55.65	4.00	15.64	24.71
C21H18C1208	469.28	468.0379	53.75	3.87	15.11	27.27
C21H18C1209	485.28	484.0328	51.98	3.74	14.61	29.67
C21H19C1101	322.84	322.1124	78.13	5.93	10.98	4.96
C21H19C1102	338.84	338.1073	74.44	5.65	10.46	9.44
C21H19C1103	354.84	354.1023	71.08	5.40	9.99	13.53
C21H19C1104	370.84	370.0972	68.02	5.16	9.56	17.26
C21H19C1105	386.84	386.0921	65.20	4.95	9.16	20.68
C21H19C1106	402.83	402.0870	62.61	4.75	8.80	23.83
C21H19C1107	418.83	418.0819	60.22	4.57	8.46	26.74
C21H19C1108	434.83	434.0768	58.01	4.40	8.15	29.44
C21H19C1109	450.83	450.0717	55.95	4.25	7.86	31.94
C21H20C1201	359.30	358.0891	70.20	5.61	19.73	4.45
C21H20C1202	375.30	374.0840	67.21	5.37	18.89	8.53
C21H20C1203	391.30	390.0789	64.46	5.15	18.12	12.27
C21H20C1204	407.30	406.0739	61.93	4.95	17.41	15.71
C21H20C1205	423.30	422.0688	59.59	4.76	16.75	18.90
C21H20C1206	439.30	438.0637	57.42	4.59	16.14	21.85
C21H20C1207	455.30	454.0586	55.40	4.43	15.57	24.60
C21H20C1208	471.29	470.0535	53.52	4.28	15.04	27.16
C21H20C1209	487.29	486.0484	51.76	4.14	14.55	29.55
C21H21C1101	324.85	324.1281	77.65	6.52	10.91	4.93

C21H21C1102	340.85	340.1230	74.00	6.21	10.40	9.39
C21H21C1103	356.85	356.1179	70.68	5.93	9.93	13.45
C21H21C1104	372.85	372.1128	67.65	5.68	9.51	17.16
C21H21C1105	388.85	388.1077	64.87	5.44	9.12	20.57
C21H21C1106	404.85	404.1027	62.30	5.23	8.76	23.71
C21H21C1107	420.85	420.0976	59.93	5.03	8.42	26.61
C21H21C1108	436.85	436.0925	57.74	4.85	8.12	29.30
C21H21C1109	452.85	452.0874	55.70	4.67	7.83	31.80
C21H22C1201	361.31	360.1048	69.81	6.14	19.62	4.43
C21H22C1202	377.31	376.0997	66.85	5.88	18.79	8.48
C21H22C1203	393.31	392.0946	64.13	5.64	18.03	12.20
C21H22C1204	409.31	408.0895	61.62	5.42	17.32	15.64
C21H22C1205	425.31	424.0844	59.31	5.21	16.67	18.81
C21H22C1206	441.31	440.0793	57.16	5.02	16.07	21.75
C21H22C1207	457.31	456.0742	55.16	4.85	15.50	24.49
C21H22C1208	473.31	472.0692	53.29	4.69	14.98	27.04
C21H22C1209	489.31	488.0641	51.55	4.53	14.49	29.43
C21H23C1101	326.87	326.1437	77.17	7.09	10.85	4.89
C21H23C1102	342.87	342.1386	73.57	6.76	10.34	9.33
C21H23C1103	358.87	358.1336	70.29	6.46	9.88	13.37
C21H23C1104	374.87	374.1285	67.29	6.18	9.46	17.07
C21H23C1105	390.87	390.1234	64.53	5.93	9.07	20.47
C21H23C1106	406.87	406.1183	61.99	5.70	8.71	23.59
C21H23C1107	422.87	422.1132	59.65	5.48	8.38	26.48
C21H23C1108	438.87	438.1081	57.47	5.28	8.08	29.17
C21H23C1109	454.87	454.1030	55.45	5.10	7.79	31.66
C21H24C1201	363.33	362.1204	69.42	6.66	19.52	4.40
C21H24C1202	379.33	378.1153	66.49	6.38	18.69	8.44
C21H24C1203	395.33	394.1102	63.80	6.12	17.94	12.14
C21H24C1204	411.33	410.1052	61.32	5.88	17.24	15.56
C21H24C1205	427.33	426.1001	59.03	5.66	16.59	18.72
C21H24C1206	443.33	442.0950	56.90	5.46	15.99	21.65
C21H24C1207	459.33	458.0899	54.91	5.27	15.44	24.38
C21H24C1208	475.33	474.0848	53.07	5.09	14.92	26.93
C21H24C1209	491.33	490.0797	51.34	4.92	14.43	29.31
C21H25C1101	328.89	328.1594	76.69	7.66	10.78	4.86
C21H25C1102	344.89	344.1543	73.14	7.31	10.28	9.28
C21H25C1103	360.88	360.1492	69.89	6.98	9.82	13.30
C21H25C1104	376.88	376.1441	66.93	6.69	9.41	16.98
C21H25C1105	392.88	392.1390	64.20	6.41	9.02	20.36
C21H25C1106	408.88	408.1340	61.69	6.16	8.67	23.48
C21H25C1107	424.88	424.1289	59.37	5.93	8.34	26.36
C21H25C1108	440.88	440.1238	57.21	5.72	8.04	29.03
C21H25C1109	456.88	456.1187	55.21	5.52	7.76	31.52
C21H26C1201	365.35	364.1361	69.04	7.17	19.41	4.38
C21H26C1202	381.35	380.1310	66.14	6.87	18.59	8.39
C21H26C1203	397.35	396.1259	63.48	6.60	17.84	12.08
C21H26C1204	413.34	412.1208	61.02	6.34	17.15	15.48
C21H26C1205	429.34	428.1157	58.75	6.10	16.51	18.63
C21H26C1206	445.34	444.1106	56.64	5.88	15.92	21.56
C21H26C1207	461.34	460.1055	54.67	5.68	15.37	24.28
C21H26C1208	477.34	476.1005	52.84	5.49	14.85	26.81
C21H26C1209	493.34	492.0954	51.13	5.31	14.37	29.19
C21H27C1101	330.90	330.1750	76.23	8.22	10.71	4.84
C21H27C1102	346.90	346.1699	72.71	7.85	10.22	9.22
C21H27C1103	362.90	362.1649	69.51	7.50	9.77	13.23
C21H27C1104	378.90	378.1598	66.57	7.18	9.36	16.89
C21H27C1105	394.90	394.1547	63.87	6.89	8.98	20.26
C21H27C1106	410.90	410.1496	61.39	6.62	8.63	23.36
C21H27C1107	426.90	426.1445	59.09	6.38	8.30	26.23
C21H27C1108	442.90	442.1394	56.95	6.14	8.00	28.90
C21H27C1109	458.90	458.1343	54.97	5.93	7.73	31.38
C21H28C1201	367.36	366.1517	68.66	7.68	19.30	4.36
C21H28C1202	383.36	382.1466	65.80	7.36	18.50	8.35
C21H28C1203	399.36	398.1415	63.16	7.07	17.75	12.02
C21H28C1204	415.36	414.1365	60.73	6.79	17.07	15.41
C21H28C1205	431.36	430.1314	58.47	6.54	16.44	18.55
C21H28C1206	447.36	446.1263	56.38	6.31	15.85	21.46
C21H28C1207	463.36	462.1212	54.44	6.09	15.30	24.17
C21H28C1208	479.36	478.1161	52.62	5.89	14.79	26.70
C21H28C1209	495.36	494.1110	50.92	5.70	14.31	29.07
C21H29C1101	332.92	332.1907	75.76	8.78	10.65	4.81
C21H29C1102	348.92	348.1856	72.29	8.38	10.16	9.17
C21H29C1103	364.92	364.1805	69.12	8.01	9.72	13.15
C21H29C1104	380.92	380.1754	66.22	7.67	9.31	16.80
C21H29C1105	396.92	396.1703	63.55	7.36	8.93	20.15
C21H29C1106	412.91	412.1653	61.09	7.08	8.59	23.25
C21H29C1107	428.91	428.1602	58.81	6.82	8.27	26.11
C21H29C1108	444.91	444.1551	56.69	6.57	7.97	28.77
C21H29C1109	460.91	460.1500	54.72	6.34	7.69	31.24
C21H30C1201	369.38	368.1674	68.29	8.19	19.20	4.33
C21H30C1202	385.38	384.1623	65.45	7.85	18.40	8.30
C21H30C1203	401.38	400.1572	62.84	7.53	17.67	11.96
C21H30C1204	417.38	416.1521	60.43	7.25	16.99	15.33
C21H30C1205	433.38	432.1470	58.20	6.98	16.36	18.46
C21H30C1206	449.38	448.1419	56.13	6.73	15.78	21.36
C21H30C1207	465.38	464.1368	54.20	6.50	15.24	24.07
C21H30C1208	481.37	480.1318	52.40	6.28	14.73	26.59
C21H30C1209	497.37	496.1267	50.71	6.08	14.26	28.95
C21H31C1101	334.93	334.2063	75.31	9.33	10.59	4.78
C21H31C1102	350.93	350.2012	71.88	8.90	10.10	9.12
C21H31C1103	366.93	366.1962	68.74	8.52	9.66	13.08
C21H31C1104	382.93	382.1911	65.87	8.16	9.26	16.71
C21H31C1105	398.93	398.1860	63.23	7.83	8.89	20.05
C21H31C1106	414.93	414.1809	60.79	7.53	8.54	23.14
C21H31C1107	430.93	430.1758	58.53	7.25	8.23	25.99
C21H31C1108	446.93	446.1707	56.44	6.99	7.93	28.64
C21H31C1109	462.93	462.1656	54.49	6.75	7.66	31.11
C21H32C1201	371.39	370.1830	67.92	8.68	19.09	4.31
C21H32C1202	387.39	386.1779	65.11	8.33	18.30	8.26
C21H32C1203	403.39	402.1728	62.53	8.00	17.58	11.90
C21H32C1204	419.39	418.1677	60.14	7.69	16.91	15.26
C21H32C1205	435.39	434.1627	57.93	7.41	16.29	18.37
C21H32C1206	451.39	450.1576	55.88	7.15	15.71	21.27
C21H32C1207	467.39	466.1525	53.97	6.90	15.17	23.96
C21H32C1208	483.39	482.1474	52.18	6.67	14.67	26.48
C21H32C1209	499.39	498.1423	50.51	6.46	14.20	28.83
C21H33C1101	336.95	336.2220	74.86	9.87	10.52	4.75
C21H33C1102	352.95	352.2169	71.46	9.42	10.04	9.07
C21H33C1103	368.95	368.2118	68.37	9.02	9.61	13.01
C21H33C1104	384.95	384.2067	65.52	8.64	9.21	16.63
C21H33C1105	400.95	400.2016	62.91	8.30	8.84	19.95
C21H33C1106	416.95	416.1965	60.50	7.98	8.50	23.02
C21H33C1107	432.95	432.1915	58.26	7.68	8.19	25.87
C21H33C1108	448.95	448.1864	56.18	7.41	7.90	28.51
C21H33C1109	464.94	464.1813	54.25	7.15	7.63	30.97
C21H34C1201	373.41	372.1987	67.55	9.18	18.99	4.28
C21H34C1202	389.41	388.1936	64.77	8.80	18.21	8.22
C21H34C1203	405.41	404.1885	62.22	8.45	17.49	11.84
C21H34C1204	421.41	420.1834	59.85	8.13	16.83	15.19

C21H34C1205	437.41	436.1783	57.67	7.84	16.21	18.29
C21H34C1206	453.41	452.1732	55.63	7.56	15.64	21.17
C21H34C1207	469.41	468.1681	53.73	7.30	15.11	23.86
C21H34C1208	485.41	484.1631	51.96	7.06	14.61	26.37
C21H34C1209	501.41	500.1580	50.31	6.83	14.14	28.72
C21H35C1101	338.97	338.2376	74.41	10.41	10.46	4.72
C21H35C1102	354.96	354.2325	71.06	9.94	9.99	9.01
C21H35C1103	370.96	370.2275	67.99	9.51	9.56	12.94
C21H35C1104	386.96	386.2224	65.18	9.12	9.16	16.54
C21H35C1105	402.96	402.2173	62.59	8.75	8.80	19.85
C21H35C1106	418.96	418.2122	60.20	8.42	8.46	22.91
C21H35C1107	434.96	434.2071	57.99	8.11	8.15	25.75
C21H35C1108	450.96	450.2020	55.93	7.82	7.86	28.38
C21H35C1109	466.96	466.1969	54.02	7.56	7.59	30.84
C21H36C1201	375.43	374.2143	67.19	9.67	18.89	4.26
C21H36C1202	391.43	390.2092	64.44	9.27	18.11	8.17
C21H36C1203	407.43	406.2041	61.91	8.91	17.40	11.78
C21H36C1204	423.42	422.1990	59.57	8.57	16.75	15.11
C21H36C1205	439.42	438.1940	57.40	8.26	16.14	18.20
C21H36C1206	455.42	454.1889	55.38	7.97	15.57	21.08
C21H36C1207	471.42	470.1838	53.50	7.70	15.04	23.76
C21H36C1208	487.42	486.1787	51.75	7.44	14.55	26.26
C21H36C1209	503.42	502.1736	50.10	7.21	14.08	28.60
C21H37C1101	340.98	340.2533	73.97	10.94	10.40	4.69
C21H37C1102	356.98	356.2482	70.66	10.45	9.93	8.96
C21H37C1103	372.98	372.2431	67.63	10.00	9.51	12.87
C21H37C1104	388.98	388.2380	64.85	9.59	9.11	16.45
C21H37C1105	404.98	404.2329	62.28	9.21	8.75	19.75
C21H37C1106	420.98	420.2278	59.92	8.86	8.42	22.80
C21H37C1107	436.98	436.2228	57.72	8.53	8.11	25.63
C21H37C1108	452.98	452.2177	55.68	8.23	7.83	28.26
C21H37C1109	468.98	468.2126	53.78	7.95	7.56	30.70
C21H38C1201	377.44	376.2300	66.83	10.15	18.79	4.24
C21H38C1202	393.44	392.2249	64.11	9.74	18.02	8.13
C21H38C1203	409.44	408.2198	61.60	9.35	17.32	11.72
C21H38C1204	425.44	424.2147	59.29	9.00	16.67	15.04
C21H38C1205	441.44	440.2096	57.14	8.68	16.06	18.12
C21H38C1206	457.44	456.2045	55.14	8.37	15.50	20.99
C21H38C1207	473.44	472.1994	53.28	8.09	14.98	23.66
C21H38C1208	489.44	488.1944	51.54	7.83	14.49	26.15
C21H38C1209	505.44	504.1893	49.90	7.58	14.03	28.49
C21H39C1101	343.00	342.2689	73.54	11.46	10.34	4.66
C21H39C1102	359.00	358.2638	70.26	10.95	9.88	8.91
C21H39C1103	375.00	374.2588	67.26	10.48	9.45	12.80
C21H39C1104	391.00	390.2537	64.51	10.05	9.07	16.37
C21H39C1105	406.99	406.2486	61.97	9.66	8.71	19.66
C21H39C1106	422.99	422.2435	59.63	9.29	8.38	22.69
C21H39C1107	438.99	438.2384	57.46	8.95	8.08	25.51
C21H39C1108	454.99	454.2333	55.44	8.64	7.79	28.13
C21H39C1109	470.99	470.2282	53.55	8.35	7.53	30.57
C21H40C1201	379.46	378.2456	66.47	10.63	18.69	4.22
C21H40C1202	395.46	394.2405	63.78	10.20	17.93	8.09
C21H40C1203	411.46	410.2354	61.30	9.80	17.23	11.67
C21H40C1204	427.46	426.2303	59.01	9.43	16.59	14.97
C21H40C1205	443.46	442.2253	56.88	9.09	15.99	18.04
C21H40C1206	459.46	458.2202	54.90	8.78	15.43	20.89
C21H40C1207	475.45	474.2151	53.05	8.48	14.91	23.56
C21H40C1208	491.45	490.2100	51.32	8.20	14.43	26.04
C21H40C1209	507.45	506.2049	49.71	7.95	13.97	28.38
C21H41C1101	345.01	344.2846	73.11	11.98	10.28	4.64
C21H41C1102	361.01	360.2795	69.87	11.45	9.82	8.86
C21H41C1103	377.01	376.2744	66.90	10.96	9.40	12.73
C21H41C1104	393.01	392.2693	64.18	10.52	9.02	16.28
C21H41C1105	409.01	408.2642	61.67	10.10	8.67	19.56
C21H41C1106	425.01	424.2591	59.35	9.72	8.34	22.59
C21H41C1107	441.01	440.2541	57.19	9.37	8.04	25.40
C21H41C1108	457.01	456.2490	55.19	9.04	7.76	28.01
C21H41C1109	473.01	472.2439	53.33	8.74	7.50	30.44
C21H42C1201	381.47	380.2613	66.12	11.10	18.59	4.19
C21H42C1202	397.47	396.2562	63.46	10.65	17.84	8.05
C21H42C1203	413.47	412.2511	61.00	10.24	17.15	11.61
C21H42C1204	429.47	428.2460	58.73	9.86	16.51	14.90
C21H42C1205	445.47	444.2409	56.62	9.50	15.92	17.96
C21H42C1206	461.47	460.2358	54.66	9.17	15.37	20.80
C21H42C1207	477.47	476.2307	52.83	8.87	14.85	23.46
C21H42C1208	493.47	492.2257	51.11	8.58	14.37	25.94
C21H42C1209	509.47	508.2206	49.51	8.31	13.92	28.26
C21H43C1101	347.03	346.3002	72.68	12.49	10.22	4.61
C21H43C1102	363.03	362.2951	69.48	11.94	9.77	8.81
C21H43C1103	379.03	378.2901	66.55	11.44	9.35	12.66
C21H43C1104	395.03	394.2850	63.85	10.97	8.97	16.20
C21H43C1105	411.03	410.2799	61.37	10.54	8.63	19.46
C21H43C1106	427.03	426.2748	59.07	10.15	8.30	22.48
C21H43C1107	443.03	442.2697	56.93	9.78	8.00	25.28
C21H43C1108	459.03	458.2646	54.95	9.44	7.72	27.88
C21H43C1109	475.02	474.2595	53.10	9.12	7.46	30.31
C22H11C1101	326.79	326.0498	80.86	3.39	10.85	4.90
C22H11C1102	342.78	342.0448	77.09	3.23	10.34	9.33
C22H11C1103	358.78	358.0397	73.65	3.09	9.88	13.38
C22H11C1104	374.78	374.0346	70.51	2.96	9.46	17.08
C22H11C1105	390.78	390.0295	67.62	2.84	9.07	20.47
C22H11C1106	406.78	406.0244	64.96	2.73	8.72	23.60
C22H11C1107	422.78	422.0193	62.50	2.62	8.39	26.49
C22H11C1108	438.78	438.0142	60.22	2.53	8.08	29.17
C22H11C1109	454.78	454.0092	58.10	2.44	7.80	31.66
C22H12C1201	363.25	362.0265	72.75	3.33	19.52	4.40
C22H12C1202	379.25	378.0214	69.68	3.19	18.70	8.44
C22H12C1203	395.25	394.0163	66.86	3.06	17.94	12.14
C22H12C1204	411.24	410.0113	64.26	2.94	17.24	15.56
C22H12C1205	427.24	426.0062	61.85	2.83	16.60	18.72
C22H12C1206	443.24	442.0011	59.62	2.73	16.00	21.66
C22H12C1207	459.24	457.9960	57.54	2.63	15.44	24.39
C22H12C1208	475.24	473.9909	55.60	2.55	14.92	26.93
C22H12C1209	491.24	489.9858	53.79	2.46	14.43	29.31
C22H13C1101	328.80	328.0655	80.37	3.99	10.78	4.87
C22H13C1102	344.80	344.0604	76.64	3.80	10.28	9.28
C22H13C1103	360.80	360.0553	73.24	3.63	9.83	13.30
C22H13C1104	376.80	376.0502	70.13	3.48	9.41	16.98
C22H13C1105	392.80	392.0451	67.27	3.34	9.03	20.37
C22H13C1106	408.80	408.0401	64.64	3.21	8.67	23.48
C22H13C1107	424.80	424.0350	62.20	3.08	8.35	26.36
C22H13C1108	440.80	440.0299	59.95	2.97	8.04	29.04
C22H13C1109	456.80	456.0248	57.85	2.87	7.76	31.52
C22H14C1201	365.26	364.0422	72.34	3.86	19.41	4.38
C22H14C1202	381.26	380.0371	69.31	3.70	18.60	8.39
C22H14C1203	397.26	396.0320	66.52	3.55	17.85	12.08
C22H14C1204	413.26	412.0269	63.94	3.41	17.16	15.49
C22H14C1205	429.26	428.0218	61.56	3.29	16.52	18.64
C22H14C1206	445.26	444.0167	59.35	3.17	15.92	21.56
C22H14C1207	461.26	460.0116	57.29	3.06	15.37	24.28

097

C22H14C1208	477.26	476.0066	55.37	2.96	14.86	26.82
C22H14C1209	493.26	492.0015	53.57	2.86	14.38	29.19
C22H15C1101	330.82	330.0811	79.88	4.57	10.72	4.84
C22H15C1102	346.82	346.0761	76.19	4.36	10.22	9.23
C22H15C1103	362.82	362.0710	72.83	4.17	9.77	13.23
C22H15C1104	378.82	378.0659	69.76	3.99	9.36	16.89
C22H15C1105	394.81	394.0608	66.93	3.83	8.98	20.26
C22H15C1106	410.81	410.0557	64.32	3.68	8.63	23.37
C22H15C1107	426.81	426.0506	61.91	3.54	8.31	26.24
C22H15C1108	442.81	442.0455	59.67	3.41	8.01	28.91
C22H15C1109	458.81	458.0404	57.59	3.30	7.73	31.38
C22H16C1201	367.28	366.0578	71.95	4.39	19.31	4.36
C22H16C1202	383.28	382.0527	68.94	4.21	18.50	8.35
C22H16C1203	399.28	398.0476	66.18	4.04	17.76	12.02
C22H16C1204	415.28	414.0426	63.63	3.88	17.07	15.41
C22H16C1205	431.28	430.0375	61.27	3.74	16.44	18.55
C22H16C1206	447.28	446.0324	59.08	3.61	15.85	21.46
C22H16C1207	463.27	462.0273	57.04	3.48	15.31	24.17
C22H16C1208	479.27	478.0222	55.13	3.36	14.79	26.71
C22H16C1209	495.27	494.0171	53.35	3.26	14.32	29.07
C22H17C1101	332.83	332.0968	79.39	5.15	10.65	4.81
C22H17C1102	348.83	348.0917	75.75	4.91	10.16	9.17
C22H17C1103	364.83	364.0866	72.43	4.70	9.72	13.16
C22H17C1104	380.83	380.0815	69.39	4.50	9.31	16.80
C22H17C1105	396.83	396.0764	66.59	4.32	8.93	20.16
C22H17C1106	412.83	412.0714	64.01	4.15	8.59	23.25
C22H17C1107	428.83	428.0663	61.62	4.00	8.27	26.12
C22H17C1108	444.83	444.0612	59.40	3.85	7.97	28.77
C22H17C1109	460.83	460.0561	57.34	3.72	7.69	31.25
C22H18C1201	369.29	368.0735	71.55	4.91	19.20	4.33
C22H18C1202	385.29	384.0684	68.58	4.71	18.40	8.31
C22H18C1203	401.29	400.0633	65.85	4.52	17.67	11.96
C22H18C1204	417.29	416.0582	63.32	4.35	16.99	15.34
C22H18C1205	433.29	432.0531	60.99	4.19	16.36	18.46
C22H18C1206	449.29	448.0480	58.81	4.04	15.78	21.37
C22H18C1207	465.29	464.0429	56.79	3.90	15.24	24.07
C22H18C1208	481.29	480.0379	54.90	3.77	14.73	26.59
C22H18C1209	497.29	496.0328	53.14	3.65	14.26	28.96
C22H19C1101	334.85	334.1124	78.91	5.72	10.59	4.78
C22H19C1102	350.85	350.1073	75.32	5.46	10.10	9.12
C22H19C1103	366.85	366.1023	72.03	5.22	9.66	13.08
C22H19C1104	382.85	382.0972	69.02	5.00	9.26	16.72
C22H19C1105	398.85	398.0921	66.25	4.80	8.89	20.06
C22H19C1106	414.85	414.0870	63.70	4.62	8.55	23.14
C22H19C1107	430.85	430.0819	61.33	4.45	8.23	25.99
C22H19C1108	446.84	446.0768	59.14	4.29	7.93	28.64
C22H19C1109	462.84	462.0717	57.09	4.14	7.66	31.11
C22H20C1201	371.31	370.0891	71.17	5.43	19.10	4.31
C22H20C1202	387.31	386.0840	68.23	5.20	18.31	8.26
C22H20C1203	403.31	402.0789	65.52	5.00	17.58	11.90
C22H20C1204	419.31	418.0739	63.02	4.81	16.91	15.26
C22H20C1205	435.31	434.0688	60.70	4.63	16.29	18.38
C22H20C1206	451.31	450.0637	58.55	4.47	15.71	21.27
C22H20C1207	467.31	466.0586	56.55	4.31	15.17	23.97
C22H20C1208	483.31	482.0535	54.67	4.17	14.67	26.48
C22H20C1209	499.31	498.0484	52.92	4.04	14.20	28.84
C22H21C1101	336.87	336.1281	78.44	6.28	10.52	4.75
C22H21C1102	352.86	352.1230	74.89	6.00	10.05	9.07
C22H21C1103	368.86	368.1179	71.64	5.74	9.61	13.01
C22H21C1104	384.86	384.1128	68.66	5.50	9.21	16.63

C22H21C1105	400.86	400.1077	65.92	5.28	8.84	19.96
C22H21C1106	416.86	416.1027	63.39	5.08	8.50	23.03
C22H21C1107	432.86	432.0976	61.05	4.89	8.19	25.87
C22H21C1108	448.86	448.0925	58.87	4.72	7.90	28.52
C22H21C1109	464.86	464.0874	56.84	4.55	7.63	30.98
C22H22C1201	373.33	372.1048	70.78	5.94	18.99	4.29
C22H22C1202	389.33	388.0997	67.87	5.70	18.21	8.22
C22H22C1203	405.32	404.0946	65.19	5.47	17.49	11.84
C22H22C1204	421.32	420.0895	62.72	5.26	16.83	15.19
C22H22C1205	437.32	436.0844	60.42	5.07	16.21	18.29
C22H22C1206	453.32	452.0793	58.29	4.89	15.64	21.18
C22H22C1207	469.32	468.0742	56.30	4.72	15.11	23.86
C22H22C1208	485.32	484.0692	54.45	4.57	14.61	26.37
C22H22C1209	501.32	500.0641	52.71	4.42	14.14	28.72
C22H23C1101	338.88	338.1437	77.98	6.84	10.46	4.72
C22H23C1102	354.88	354.1386	74.46	6.53	9.99	9.02
C22H23C1103	370.88	370.1336	71.25	6.25	9.56	12.94
C22H23C1104	386.88	386.1285	68.30	5.99	9.16	16.54
C22H23C1105	402.88	402.1234	65.59	5.75	8.80	19.86
C22H23C1106	418.88	418.1183	63.08	5.53	8.46	22.92
C22H23C1107	434.88	434.1132	60.76	5.33	8.15	25.75
C22H23C1108	450.88	450.1081	58.61	5.14	7.86	28.39
C22H23C1109	466.88	466.1030	56.60	4.97	7.59	30.84
C22H24C1201	375.34	374.1204	70.40	6.45	18.89	4.26
C22H24C1202	391.34	390.1153	67.52	6.18	18.12	8.18
C22H24C1203	407.34	406.1102	64.87	5.94	17.41	11.78
C22H24C1204	423.34	422.1052	62.42	5.71	16.75	15.12
C22H24C1205	439.34	438.1001	60.15	5.51	16.14	18.21
C22H24C1206	455.34	454.0950	58.03	5.31	15.57	21.08
C22H24C1207	471.34	470.0899	56.06	5.13	15.04	23.76
C22H24C1208	487.34	486.0848	54.22	4.96	14.55	26.26
C22H24C1209	503.34	502.0797	52.50	4.81	14.09	28.61
C22H25C1101	340.90	340.1594	77.51	7.39	10.40	4.69
C22H25C1102	356.90	356.1543	74.04	7.06	9.93	8.97
C22H25C1103	372.90	372.1492	70.86	6.76	9.51	12.87
C22H25C1104	388.90	388.1441	67.95	6.48	9.12	16.46
C22H25C1105	404.89	404.1390	65.26	6.22	8.76	19.76
C22H25C1106	420.89	420.1340	62.78	5.99	8.42	22.81
C22H25C1107	436.89	436.1289	60.48	5.77	8.11	25.63
C22H25C1108	452.89	452.1238	58.35	5.56	7.83	28.26
C22H25C1109	468.89	468.1187	56.36	5.37	7.56	30.71
C22H26C1201	377.36	376.1361	70.03	6.94	18.79	4.24
C22H26C1202	393.36	392.1310	67.18	6.66	18.03	8.13
C22H26C1203	409.36	408.1259	64.55	6.40	17.32	11.73
C22H26C1204	425.36	424.1208	62.12	6.16	16.67	15.05
C22H26C1205	441.36	440.1157	59.87	5.94	16.07	18.13
C22H26C1206	457.35	456.1106	57.78	5.73	15.50	20.99
C22H26C1207	473.35	472.1055	55.82	5.54	14.98	23.66
C22H26C1208	489.35	488.1005	54.00	5.36	14.49	26.16
C22H26C1209	505.35	504.0954	52.29	5.19	14.03	28.49
C22H27C1101	342.91	342.1750	77.06	7.94	10.34	4.67
C22H27C1102	358.91	358.1699	73.62	7.58	9.88	8.92
C22H27C1103	374.91	374.1649	70.48	7.26	9.46	12.80
C22H27C1104	390.91	390.1598	67.60	6.96	9.07	16.37
C22H27C1105	406.91	406.1547	64.94	6.69	8.71	19.66
C22H27C1106	422.91	422.1496	62.48	6.44	8.38	22.70
C22H27C1107	438.91	438.1445	60.20	6.20	8.08	25.52
C22H27C1108	454.91	454.1394	58.09	5.98	7.79	28.14
C22H27C1109	470.91	470.1343	56.11	5.78	7.53	30.58
C22H28C1201	379.37	378.1517	69.65	7.44	18.69	4.22

098

C22H28C1202	395.37	394.1466	66.83	7.14	17.93	8.09
C22H28C1203	411.37	410.1415	64.24	6.86	17.24	11.67
C22H28C1204	427.37	426.1365	61.83	6.60	16.59	14.97
C22H28C1205	443.37	442.1314	59.60	6.37	15.99	18.04
C22H28C1206	459.37	458.1263	57.52	6.14	15.44	20.90
C22H28C1207	475.37	474.1212	55.59	5.94	14.92	23.56
C22H28C1208	491.37	490.1161	53.78	5.74	14.43	26.05
C22H28C1209	507.37	506.1110	52.08	5.56	13.98	28.38
C22H29C1101	344.93	344.1907	76.61	8.47	10.28	4.64
C22H29C1102	360.93	360.1856	73.21	8.10	9.82	8.87
C22H29C1103	376.93	376.1805	70.11	7.76	9.41	12.73
C22H29C1104	392.93	392.1754	67.25	7.44	9.02	16.29
C22H29C1105	408.93	408.1703	64.62	7.15	8.67	19.56
C22H29C1106	424.93	424.1653	62.19	6.88	8.34	22.59
C22H29C1107	440.93	440.1602	59.93	6.63	8.04	25.40
C22H29C1108	456.92	456.1551	57.83	6.40	7.76	28.01
C22H29C1109	472.92	472.1500	55.87	6.18	7.50	30.45
C22H30C1201	381.39	380.1674	69.28	7.93	18.59	4.20
C22H30C1202	397.39	396.1623	66.50	7.61	17.84	8.05
C22H30C1203	413.39	412.1572	63.92	7.31	17.15	11.61
C22H30C1204	429.39	428.1521	61.54	7.04	16.51	14.90
C22H30C1205	445.39	444.1470	59.33	6.79	15.92	17.96
C22H30C1206	461.39	460.1419	57.27	6.55	15.37	20.81
C22H30C1207	477.39	476.1368	55.35	6.33	14.85	23.46
C22H30C1208	493.39	492.1318	53.56	6.13	14.37	25.94
C22H30C1209	509.39	508.1267	51.88	5.94	13.92	28.27
C22H31C1101	346.94	346.2063	76.16	9.01	10.22	4.61
C22H31C1102	362.94	362.2012	72.81	8.61	9.77	8.82
C22H31C1103	378.94	378.1962	69.73	8.25	9.36	12.67
C22H31C1104	394.94	394.1911	66.91	7.91	8.98	16.20
C22H31C1105	410.94	410.1860	64.30	7.60	8.63	19.47
C22H31C1106	426.94	426.1809	61.89	7.32	8.30	22.48
C22H31C1107	442.94	442.1758	59.66	7.05	8.00	25.28
C22H31C1108	458.94	458.1707	57.58	6.81	7.72	27.89
C22H31C1109	474.94	474.1656	55.64	6.58	7.46	30.32
C22H32C1201	383.41	382.1830	68.92	8.41	18.49	4.17
C22H32C1202	399.41	398.1779	66.16	8.08	17.75	8.01
C22H32C1203	415.40	414.1728	63.61	7.76	17.07	11.55
C22H32C1204	431.40	430.1677	61.25	7.48	16.44	14.83
C22H32C1205	447.40	446.1627	59.06	7.21	15.85	17.88
C22H32C1206	463.40	462.1576	57.02	6.96	15.30	20.72
C22H32C1207	479.40	478.1525	55.12	6.73	14.79	23.36
C22H32C1208	495.40	494.1474	53.34	6.51	14.31	25.84
C22H32C1209	511.40	510.1423	51.67	6.31	13.87	28.16
C22H33C1101	348.96	348.2220	75.72	9.53	10.16	4.58
C22H33C1102	364.96	364.2169	72.40	9.11	9.71	8.77
C22H33C1103	380.96	380.2118	69.36	8.73	9.31	12.60
C22H33C1104	396.96	396.2067	66.57	8.38	8.93	16.12
C22H33C1105	412.96	412.2016	63.99	8.05	8.59	19.37
C22H33C1106	428.96	428.1965	61.60	7.75	8.26	22.38
C22H33C1107	444.96	444.1915	59.39	7.48	7.97	25.17
C22H33C1108	460.96	460.1864	57.33	7.22	7.69	27.77
C22H33C1109	476.96	476.1813	55.40	6.97	7.43	30.19
C22H34C1201	385.42	384.1987	68.56	8.89	18.40	4.15
C22H34C1202	401.42	400.1936	65.83	8.54	17.66	7.97
C22H34C1203	417.42	416.1885	63.30	8.21	16.99	11.50
C22H34C1204	433.42	432.1834	60.97	7.91	16.36	14.77
C22H34C1205	449.42	448.1783	58.80	7.63	15.78	17.80
C22H34C1206	465.42	464.1732	56.78	7.36	15.23	20.63
C22H34C1207	481.42	480.1681	54.85	7.12	14.73	23.26

C22H34C1208	497.42	496.1631	53.12	6.89	14.25	25.73
C22H34C1209	513.42	512.1580	51.47	6.68	13.81	28.05
C22H35C1101	350.98	350.2376	75.29	10.05	10.10	4.56
C22H35C1102	366.98	366.2325	72.01	9.61	9.66	8.72
C22H35C1103	382.98	382.2275	69.00	9.21	9.26	12.53
C22H35C1104	398.97	398.2224	66.23	8.84	8.89	16.04
C22H35C1105	414.97	414.2173	63.68	8.50	8.54	19.28
C22H35C1106	430.97	430.2122	61.31	8.19	8.23	22.27
C22H35C1107	446.97	446.2071	59.12	7.89	7.93	25.06
C22H35C1108	462.97	462.2020	57.08	7.62	7.66	27.65
C22H35C1109	478.97	478.1969	55.17	7.37	7.40	30.06
C22H36C1201	387.44	386.2143	68.20	9.37	18.30	4.13
C22H36C1202	403.44	402.2092	65.50	8.99	17.58	7.93
C22H36C1203	419.44	418.2041	63.00	8.65	16.91	11.44
C22H36C1204	435.44	434.1990	60.69	8.33	16.28	14.70
C22H36C1205	451.44	450.1940	58.53	8.04	15.71	17.72
C22H36C1206	467.43	466.1889	56.53	7.76	15.17	20.54
C22H36C1207	483.43	482.1838	54.66	7.51	14.67	23.17
C22H36C1208	499.43	498.1787	52.91	7.27	14.20	25.63
C22H36C1209	515.43	514.1736	51.27	7.04	13.76	27.94
C22H37C1101	352.99	352.2533	74.86	10.57	10.04	4.53
C22H37C1102	368.99	368.2482	71.61	10.11	9.61	8.67
C22H37C1103	384.99	384.2431	68.64	9.69	9.21	12.47
C22H37C1104	400.99	400.2380	65.90	9.30	8.84	15.96
C22H37C1105	416.99	416.2329	63.37	8.94	8.50	19.18
C22H37C1106	432.99	432.2278	61.03	8.61	8.19	22.17
C22H37C1107	448.99	448.2228	58.85	8.31	7.90	24.94
C22H37C1108	464.99	464.2177	56.83	8.02	7.62	27.53
C22H37C1109	480.99	480.2126	54.94	7.75	7.37	29.94
C22H38C1201	389.45	388.2300	67.85	9.84	18.21	4.11
C22H38C1202	405.45	404.2249	65.17	9.45	17.49	7.89
C22H38C1203	421.45	420.2198	62.70	9.09	16.82	11.39
C22H38C1204	437.45	436.2147	60.41	8.76	16.21	14.63
C22H38C1205	453.45	452.2096	58.27	8.45	15.64	17.64
C22H38C1206	469.45	468.2045	56.29	8.16	15.10	20.45
C22H38C1207	485.45	484.1994	54.43	7.89	14.61	23.07
C22H38C1208	501.45	500.1944	52.70	7.64	14.14	25.53
C22H38C1209	517.45	516.1893	51.07	7.40	13.70	27.83
C22H39C1101	355.01	354.2689	74.43	11.07	9.99	4.51
C22H39C1102	371.01	370.2638	71.22	10.60	9.56	8.62
C22H39C1103	387.01	386.2588	68.28	10.16	9.16	12.40
C22H39C1104	403.01	402.2537	65.57	9.75	8.80	15.88
C22H39C1105	419.01	418.2486	63.06	9.38	8.46	19.09
C22H39C1106	435.01	434.2435	60.75	9.04	8.15	22.07
C22H39C1107	451.00	450.2384	58.59	8.72	7.86	24.83
C22H39C1108	467.00	466.2333	56.58	8.42	7.59	27.41
C22H39C1109	483.00	482.2282	54.71	8.14	7.34	29.81
C22H40C1201	391.47	390.2456	67.50	10.30	18.11	4.09
C22H40C1202	407.47	406.2405	64.85	9.89	17.40	7.85
C22H40C1203	423.47	422.2354	62.40	9.52	16.74	11.33
C22H40C1204	439.47	438.2303	60.13	9.17	16.13	14.56
C22H40C1205	455.47	454.2253	58.02	8.85	15.57	17.56
C22H40C1206	471.47	470.2202	56.05	8.55	15.04	20.36
C22H40C1207	487.47	486.2151	54.21	8.27	14.55	22.98
C22H40C1208	503.47	502.2100	52.49	8.01	14.08	25.42
C22H40C1209	519.46	518.2049	50.87	7.76	13.65	27.72
C22H41C1101	357.02	356.2846	74.01	11.58	9.93	4.48
C22H41C1102	373.02	372.2795	70.84	11.08	9.50	8.58
C22H41C1103	389.02	388.2744	67.93	10.62	9.11	12.34
C22H41C1104	405.02	404.2693	65.24	10.20	8.75	15.80

C22H41C1105	421.02	420.2642	62.76	9.82	8.42	19.00
C22H41C1106	437.02	436.2591	60.47	9.46	8.11	21.97
C22H41C1107	453.02	452.2541	58.33	9.12	7.83	24.72
C22H41C1108	469.02	468.2490	56.34	8.81	7.56	27.29
C22H41C1109	485.02	484.2439	54.48	8.52	7.31	29.69
C22H42C1201	393.49	392.2613	67.16	10.76	18.02	4.07
C22H42C1202	409.48	408.2562	64.53	10.34	17.32	7.81
C22H42C1203	425.48	424.2511	62.10	9.95	16.66	11.28
C22H42C1204	441.48	440.2460	59.85	9.59	16.06	14.50
C22H42C1205	457.48	456.2409	57.76	9.25	15.50	17.49
C22H42C1206	473.48	472.2358	55.81	8.94	14.98	20.27
C22H42C1207	489.48	488.2307	53.98	8.65	14.49	22.88
C22H42C1208	505.48	504.2257	52.28	8.38	14.03	25.32
C22H42C1209	521.48	520.2206	50.67	8.12	13.60	27.61
C22H43C1101	359.04	358.3002	73.60	12.07	9.87	4.46
C22H43C1102	375.04	374.2951	70.46	11.56	9.45	8.53
C22H43C1103	391.04	390.2901	67.58	11.08	9.07	12.27
C22H43C1104	407.04	406.2850	64.92	10.65	8.71	15.72
C22H43C1105	423.04	422.2799	62.46	10.25	8.38	18.91
C22H43C1106	439.04	438.2748	60.19	9.87	8.08	21.87
C22H43C1107	455.04	454.2697	58.07	9.53	7.79	24.61
C22H43C1108	471.04	470.2646	56.10	9.20	7.53	27.17
C22H43C1109	487.04	486.2595	54.26	8.90	7.28	29.57
C22H44C1201	395.50	394.2769	66.81	11.21	17.93	4.05
C22H44C1202	411.50	410.2718	64.22	10.78	17.23	7.78
C22H44C1203	427.50	426.2667	61.81	10.37	16.59	11.23
C22H44C1204	443.50	442.2616	59.58	10.00	15.99	14.43
C22H44C1205	459.50	458.2566	57.51	9.65	15.43	17.41
C22H44C1206	475.50	474.2515	55.57	9.33	14.91	20.19
C22H44C1207	491.50	490.2464	53.76	9.02	14.43	22.79
C22H44C1208	507.50	506.2413	52.07	8.74	13.97	25.22
C22H44C1209	523.50	522.2362	50.48	8.47	13.54	27.51
C22H45C1101	361.06	360.3159	73.19	12.56	9.82	4.43
C22H45C1102	377.06	376.3108	70.08	12.03	9.40	8.49
C22H45C1103	393.06	392.3057	67.23	11.54	9.02	12.21
C22H45C1104	409.05	408.3006	64.60	11.09	8.67	15.65
C22H45C1105	425.05	424.2955	62.17	10.67	8.34	18.82
C22H45C1106	441.05	440.2904	59.91	10.28	8.04	21.77
C22H45C1107	457.05	456.2854	57.82	9.92	7.76	24.50
C22H45C1108	473.05	472.2803	55.86	9.59	7.49	27.06
C22H45C1109	489.05	488.2752	54.03	9.27	7.25	29.44
C23H11C1101	338.80	338.0498	81.54	3.27	10.46	4.72
C23H11C1102	354.80	354.0448	77.86	3.13	9.99	9.02
C23H11C1103	370.80	370.0397	74.50	2.99	9.56	12.94
C23H11C1104	386.79	386.0346	71.42	2.87	9.17	16.55
C23H11C1105	402.79	402.0295	68.59	2.75	8.80	19.86
C23H11C1106	418.79	418.0244	65.96	2.65	8.47	22.92
C23H11C1107	434.79	434.0193	63.54	2.55	8.15	25.76
C23H11C1108	450.79	450.0142	61.28	2.46	7.86	28.39
C23H11C1109	466.79	466.0092	59.18	2.38	7.60	30.85
C23H12C1201	375.26	374.0265	73.62	3.22	18.90	4.26
C23H12C1202	391.26	390.0214	70.61	3.09	18.12	8.18
C23H12C1203	407.26	406.0163	67.83	2.97	17.41	11.79
C23H12C1204	423.26	422.0113	65.27	2.86	16.75	15.12
C23H12C1205	439.26	438.0062	62.89	2.75	16.14	18.21
C23H12C1206	455.25	454.0011	60.68	2.66	15.58	21.09
C23H12C1207	471.25	469.9960	58.62	2.57	15.05	23.77
C23H12C1208	487.25	485.9909	56.70	2.48	14.55	26.27
C23H12C1209	503.25	501.9858	54.89	2.40	14.09	28.61
C23H13C1101	340.81	340.0655	81.06	3.84	10.40	4.69

C23H13C1102	356.81	356.0604	77.42	3.67	9.94	8.97
C23H13C1103	372.81	372.0553	74.10	3.51	9.51	12.87
C23H13C1104	388.81	388.0502	71.05	3.37	9.12	16.46
C23H13C1105	404.81	404.0451	68.24	3.24	8.76	19.76
C23H13C1106	420.81	420.0401	65.65	3.11	8.42	22.81
C23H13C1107	436.81	436.0350	63.24	3.00	8.12	25.64
C23H13C1108	452.81	452.0299	61.01	2.89	7.83	28.27
C23H13C1109	468.81	468.0248	58.93	2.80	7.56	30.72
C23H14C1201	377.27	376.0422	73.22	3.74	18.79	4.24
C23H14C1202	393.27	392.0371	70.25	3.59	18.03	8.14
C23H14C1203	409.27	408.0320	67.50	3.45	17.32	11.73
C23H14C1204	425.27	424.0269	64.96	3.32	16.67	15.05
C23H14C1205	441.27	440.0218	62.60	3.20	16.07	18.13
C23H14C1206	457.27	456.0167	60.41	3.09	15.51	20.99
C23H14C1207	473.27	472.0116	58.37	2.98	14.98	23.66
C23H14C1208	489.27	488.0066	56.46	2.88	14.49	26.16
C23H14C1209	505.27	504.0015	54.68	2.79	14.03	28.50
C23H15C1101	342.83	342.0811	80.58	4.41	10.34	4.67
C23H15C1102	358.83	358.0761	76.99	4.21	9.88	8.92
C23H15C1103	374.83	374.0710	73.70	4.03	9.46	12.81
C23H15C1104	390.83	390.0659	70.69	3.87	9.07	16.37
C23H15C1105	406.83	406.0608	67.91	3.72	8.71	19.66
C23H15C1106	422.83	422.0557	65.34	3.58	8.38	22.70
C23H15C1107	438.82	438.0506	62.95	3.45	8.08	25.52
C23H15C1108	454.82	454.0455	60.74	3.32	7.79	28.14
C23H15C1109	470.82	470.0404	58.68	3.21	7.53	30.58
C23H16C1201	379.29	378.0578	72.84	4.25	18.69	4.22
C23H16C1202	395.29	394.0527	69.89	4.08	17.94	8.10
C23H16C1203	411.29	410.0476	67.17	3.92	17.24	11.67
C23H16C1204	427.29	426.0426	64.65	3.77	16.59	14.98
C23H16C1205	443.29	442.0375	62.32	3.64	16.00	18.05
C23H16C1206	459.29	458.0324	60.15	3.51	15.44	20.90
C23H16C1207	475.29	474.0273	58.12	3.39	14.92	23.56
C23H16C1208	491.29	490.0222	56.23	3.28	14.43	26.05
C23H16C1209	507.28	506.0171	54.46	3.18	13.98	28.39
C23H17C1101	344.84	344.0968	80.11	4.97	10.28	4.64
C23H17C1102	360.84	360.0917	76.56	4.75	9.83	8.87
C23H17C1103	376.84	376.0866	73.31	4.55	9.41	12.74
C23H17C1104	392.84	392.0815	70.32	4.36	9.02	16.29
C23H17C1105	408.84	408.0764	67.57	4.19	8.67	19.57
C23H17C1106	424.84	424.0714	65.03	4.03	8.34	22.60
C23H17C1107	440.84	440.0663	62.67	3.89	8.04	25.41
C23H17C1108	456.84	456.0612	60.47	3.75	7.76	28.02
C23H17C1109	472.84	472.0561	58.42	3.62	7.50	30.45
C23H18C1201	381.31	380.0735	72.45	4.76	18.60	4.20
C23H18C1202	397.30	396.0684	69.53	4.57	17.85	8.05
C23H18C1203	413.30	412.0633	66.84	4.39	17.16	11.61
C23H18C1204	429.30	428.0582	64.35	4.23	16.52	14.91
C23H18C1205	445.30	444.0531	62.04	4.07	15.92	17.96
C23H18C1206	461.30	460.0480	59.89	3.93	15.37	20.81
C23H18C1207	477.30	476.0429	57.88	3.80	14.86	23.46
C23H18C1208	493.30	492.0379	56.00	3.68	14.37	25.95
C23H18C1209	509.30	508.0328	54.24	3.56	13.92	28.27
C23H19C1101	346.86	346.1124	79.64	5.52	10.22	4.61
C23H19C1102	362.86	362.1073	76.13	5.28	9.77	8.82
C23H19C1103	378.86	378.1023	72.92	5.06	9.36	12.67
C23H19C1104	394.86	394.0972	69.96	4.85	8.98	16.21
C23H19C1105	410.86	410.0921	67.24	4.66	8.63	19.47
C23H19C1106	426.86	426.0870	64.72	4.49	8.31	22.49
C23H19C1107	442.86	442.0819	62.38	4.32	8.01	25.29

100

C23H19C1108	458.86	458.0768	60.21	4.17	7.73	27.89
C23H19C1109	474.86	474.0717	58.18	4.03	7.47	30.32
C23H20C1201	383.32	382.0891	72.07	5.26	18.50	4.17
C23H20C1202	399.32	398.0840	69.18	5.05	17.76	8.01
C23H20C1203	415.32	414.0789	66.52	4.85	17.07	11.56
C23H20C1204	431.32	430.0739	64.05	4.67	16.44	14.84
C23H20C1205	447.32	446.0688	61.76	4.51	15.85	17.88
C23H20C1206	463.32	462.0637	59.63	4.35	15.30	20.72
C23H20C1207	479.32	478.0586	57.64	4.21	14.79	23.37
C23H20C1208	495.32	494.0535	55.77	4.07	14.32	25.84
C23H20C1209	511.32	510.0484	54.03	3.94	13.87	28.16
C23H21C1101	348.88	348.1281	79.18	6.07	10.16	4.59
C23H21C1102	364.88	364.1230	75.71	5.80	9.72	8.77
C23H21C1103	380.88	380.1179	72.53	5.56	9.31	12.60
C23H21C1104	396.87	396.1128	69.61	5.33	8.93	16.13
C23H21C1105	412.87	412.1077	66.91	5.13	8.59	19.38
C23H21C1106	428.87	428.1027	64.41	4.94	8.27	22.38
C23H21C1107	444.87	444.0976	62.10	4.76	7.97	25.17
C23H21C1108	460.87	460.0925	59.94	4.59	7.69	27.77
C23H21C1109	476.87	476.0874	57.93	4.44	7.43	30.20
C23H22C1201	385.34	384.1048	71.69	5.75	18.40	4.15
C23H22C1202	401.34	400.0997	68.83	5.53	17.67	7.97
C23H22C1203	417.34	416.0946	66.20	5.31	16.99	11.50
C23H22C1204	433.34	432.0895	63.75	5.12	16.36	14.77
C23H22C1205	449.33	448.0844	61.48	4.94	15.78	17.80
C23H22C1206	465.33	464.0793	59.37	4.77	15.24	20.63
C23H22C1207	481.33	480.0742	57.39	4.61	14.73	23.27
C23H22C1208	497.33	496.0692	55.55	4.46	14.26	25.74
C23H22C1209	513.33	512.0641	53.82	4.32	13.81	28.05
C23H23C1101	350.89	350.1437	78.73	6.61	10.10	4.56
C23H23C1102	366.89	366.1386	75.30	6.32	9.66	8.72
C23H23C1103	382.89	382.1336	72.15	6.05	9.26	12.54
C23H23C1104	398.89	398.1285	69.26	5.81	8.89	16.04
C23H23C1105	414.89	414.1234	66.59	5.59	8.55	19.28
C23H23C1106	430.89	430.1183	64.11	5.38	8.23	22.28
C23H23C1107	446.89	446.1132	61.82	5.19	7.93	25.06
C23H23C1108	462.89	462.1081	59.68	5.01	7.66	27.65
C23H23C1109	478.89	478.1030	57.69	4.84	7.40	30.07
C23H24C1201	387.35	386.1204	71.32	6.25	18.31	4.13
C23H24C1202	403.35	402.1153	68.49	6.00	17.58	7.93
C23H24C1203	419.35	418.1102	65.88	5.77	16.91	11.45
C23H24C1204	435.35	434.1052	63.46	5.56	16.29	14.70
C23H24C1205	451.35	450.1001	61.21	5.36	15.71	17.72
C23H24C1206	467.35	466.0950	59.11	5.18	15.17	20.54
C23H24C1207	483.35	482.0899	57.15	5.00	14.67	23.17
C23H24C1208	499.35	498.0848	55.32	4.84	14.20	25.63
C23H24C1209	515.35	514.0797	53.61	4.69	13.76	27.94
C23H25C1101	352.91	352.1594	78.28	7.14	10.05	4.53
C23H25C1102	368.91	368.1543	74.89	6.83	9.61	8.67
C23H25C1103	384.91	384.1492	71.77	6.55	9.21	12.47
C23H25C1104	400.91	400.1441	68.91	6.29	8.84	15.96
C23H25C1105	416.91	416.1390	66.26	6.04	8.50	19.19
C23H25C1106	432.91	432.1340	63.81	5.82	8.19	22.17
C23H25C1107	448.90	448.1289	61.54	5.61	7.90	24.95
C23H25C1108	464.90	464.1238	59.42	5.42	7.63	27.53
C23H25C1109	480.90	480.1187	57.45	5.24	7.37	29.94
C23H26C1201	389.37	388.1361	70.95	6.73	18.21	4.11
C23H26C1202	405.37	404.1310	68.15	6.47	17.49	7.89
C23H26C1203	421.37	420.1259	65.56	6.22	16.83	11.39
C23H26C1204	437.37	436.1208	63.16	5.99	16.21	14.63

C23H26C1205	453.37	452.1157	60.93	5.78	15.64	17.65
C23H26C1206	469.37	468.1106	58.86	5.58	15.11	20.45
C23H26C1207	485.37	484.1055	56.92	5.40	14.61	23.07
C23H26C1208	501.36	500.1005	55.10	5.23	14.14	25.53
C23H26C1209	517.36	516.0954	53.40	5.07	13.71	27.83
C23H27C1101	354.92	354.1750	77.84	7.67	9.99	4.51
C23H27C1102	370.92	370.1699	74.48	7.34	9.56	8.63
C23H27C1103	386.92	386.1649	71.40	7.03	9.16	12.41
C23H27C1104	402.92	402.1598	68.56	6.75	8.80	15.88
C23H27C1105	418.92	418.1547	65.94	6.50	8.46	19.10
C23H27C1106	434.92	434.1496	63.52	6.26	8.15	22.07
C23H27C1107	450.92	450.1445	61.27	6.04	7.86	24.84
C23H27C1108	466.92	466.1394	59.17	5.83	7.59	27.41
C23H27C1109	482.92	482.1343	57.21	5.64	7.34	29.82
C23H28C1201	391.39	390.1517	70.58	7.21	18.12	4.09
C23H28C1202	407.38	406.1466	67.81	6.93	17.41	7.85
C23H28C1203	423.38	422.1415	65.25	6.67	16.75	11.34
C23H28C1204	439.38	438.1365	62.87	6.42	16.14	14.57
C23H28C1205	455.38	454.1314	60.66	6.20	15.57	17.57
C23H28C1206	471.38	470.1263	58.61	5.99	15.04	20.36
C23H28C1207	487.38	486.1212	56.68	5.79	14.55	22.98
C23H28C1208	503.38	502.1161	54.88	5.61	14.09	25.43
C23H28C1209	519.38	518.1110	53.19	5.43	13.65	27.72
C23H29C1101	356.94	356.1907	77.40	8.19	9.93	4.48
C23H29C1102	372.94	372.1856	74.08	7.84	9.51	8.58
C23H29C1103	388.94	388.1805	71.03	7.52	9.12	12.34
C23H29C1104	404.94	404.1754	68.22	7.22	8.76	15.80
C23H29C1105	420.94	420.1703	65.63	6.94	8.42	19.00
C23H29C1106	436.94	436.1653	63.23	6.69	8.11	21.97
C23H29C1107	452.94	452.1602	60.99	6.45	7.83	24.73
C23H29C1108	468.94	468.1551	58.91	6.23	7.56	27.29
C23H29C1109	484.94	484.1500	56.97	6.03	7.31	29.69
C23H30C1201	393.40	392.1674	70.22	7.69	18.02	4.07
C23H30C1202	409.40	408.1623	67.48	7.39	17.32	7.82
C23H30C1203	425.40	424.1572	64.94	7.11	16.67	11.28
C23H30C1204	441.40	440.1521	62.59	6.85	16.06	14.50
C23H30C1205	457.40	456.1470	60.40	6.61	15.50	17.49
C23H30C1206	473.40	472.1419	58.36	6.39	14.98	20.28
C23H30C1207	489.40	488.1368	56.45	6.18	14.49	22.88
C23H30C1208	505.40	504.1318	54.66	5.98	14.03	25.33
C23H30C1209	521.40	520.1267	52.98	5.80	13.60	27.62
C23H31C1101	358.96	358.2063	76.96	8.70	9.88	4.46
C23H31C1102	374.96	374.2012	73.68	8.33	9.46	8.53
C23H31C1103	390.95	390.1962	70.66	7.99	9.07	12.28
C23H31C1104	406.95	406.1911	67.88	7.68	8.71	15.73
C23H31C1105	422.95	422.1860	65.32	7.39	8.38	18.91
C23H31C1106	438.95	438.1809	62.94	7.12	8.08	21.87
C23H31C1107	454.95	454.1758	60.72	6.87	7.79	24.62
C23H31C1108	470.95	470.1707	58.66	6.63	7.53	27.18
C23H31C1109	486.95	486.1656	56.73	6.42	7.28	29.57
C23H32C1201	395.42	394.1830	69.86	8.16	17.93	4.05
C23H32C1202	411.42	410.1779	67.15	7.84	17.23	7.78
C23H32C1203	427.42	426.1728	64.63	7.55	16.59	11.23
C23H32C1204	443.42	442.1677	62.30	7.27	15.99	14.43
C23H32C1205	459.41	458.1627	60.13	7.02	15.43	17.41
C23H32C1206	475.41	474.1576	58.11	6.78	14.91	20.19
C23H32C1207	491.41	490.1525	56.22	6.56	14.43	22.79
C23H32C1208	507.41	506.1474	54.44	6.36	13.97	25.23
C23H32C1209	523.41	522.1423	52.78	6.16	13.55	27.51
C23H33C1101	360.97	360.2220	76.53	9.21	9.82	4.43

101

C23H33C1102	376.97	376.2169	73.28	8.82	9.40	8.49
C23H33C1103	392.97	392.2118	70.30	8.46	9.02	12.21
C23H33C1104	408.97	408.2067	67.55	8.13	8.67	15.65
C23H33C1105	424.97	424.2016	65.01	7.83	8.34	18.82
C23H33C1106	440.97	440.1965	62.65	7.54	8.04	21.77
C23H33C1107	456.97	456.1915	60.45	7.28	7.76	24.51
C23H33C1108	472.97	472.1864	58.41	7.03	7.50	27.06
C23H33C1109	488.97	488.1813	56.50	6.80	7.25	29.45
C23H34C1201	397.43	396.1987	69.51	8.62	17.84	4.03
C23H34C1202	413.43	412.1936	66.82	8.29	17.15	7.74
C23H34C1203	429.43	428.1885	64.33	7.98	16.51	11.18
C23H34C1204	445.43	444.1834	62.02	7.69	15.92	14.37
C23H34C1205	461.43	460.1783	59.87	7.43	15.37	17.34
C23H34C1206	477.43	476.1732	57.86	7.18	14.85	20.11
C23H34C1207	493.43	492.1681	55.99	6.95	14.37	22.70
C23H34C1208	509.43	508.1631	54.23	6.73	13.92	25.13
C23H34C1209	525.43	524.1580	52.58	6.52	13.49	27.41
C23H35C1101	362.99	362.2376	76.11	9.72	9.77	4.41
C23H35C1102	378.99	378.2325	72.89	9.31	9.35	8.44
C23H35C1103	394.99	394.2275	69.94	8.93	8.98	12.15
C23H35C1104	410.99	410.2224	67.22	8.58	8.63	15.57
C23H35C1105	426.99	426.2173	64.70	8.26	8.30	18.74
C23H35C1106	442.98	442.2122	62.36	7.96	8.00	21.67
C23H35C1107	458.98	458.2071	60.19	7.69	7.72	24.40
C23H35C1108	474.98	474.2020	58.16	7.43	7.46	26.95
C23H35C1109	490.98	490.1969	56.27	7.19	7.22	29.33
C23H36C1201	399.45	398.2143	69.16	9.08	17.75	4.01
C23H36C1202	415.45	414.2092	66.50	8.73	17.07	7.70
C23H36C1203	431.45	430.2041	64.03	8.41	16.43	11.12
C23H36C1204	447.45	446.1990	61.74	8.11	15.85	14.30
C23H36C1205	463.45	462.1940	59.61	7.83	15.30	17.26
C23H36C1206	479.45	478.1889	57.62	7.57	14.79	20.02
C23H36C1207	495.45	494.1838	55.76	7.32	14.31	22.61
C23H36C1208	511.44	510.1787	54.01	7.09	13.86	25.03
C23H36C1209	527.44	526.1736	52.38	6.88	13.44	27.30
C23H37C1101	365.00	364.2533	75.65	10.22	9.71	4.38
C23H37C1102	381.00	380.2482	72.51	9.79	9.31	8.40
C23H37C1103	397.00	396.2431	69.59	9.39	8.93	12.09
C23H37C1104	413.00	412.2380	66.89	9.03	8.58	15.50
C23H37C1105	429.00	428.2329	64.40	8.69	8.26	18.65
C23H37C1106	445.00	444.2278	62.08	8.38	7.97	21.57
C23H37C1107	461.00	460.2228	59.93	8.09	7.69	24.29
C23H37C1108	477.00	476.2177	57.92	7.82	7.43	26.83
C23H37C1109	493.00	492.2126	56.04	7.56	7.19	29.21
C23H38C1201	401.46	400.2300	68.81	9.54	17.66	3.99
C23H38C1202	417.46	416.2249	66.17	9.18	16.98	7.67
C23H38C1203	433.46	432.2198	63.73	8.84	16.36	11.07
C23H38C1204	449.46	448.2147	61.46	8.52	15.78	14.24
C23H38C1205	465.46	464.2096	59.35	8.23	15.23	17.19
C23H38C1206	481.46	480.2045	57.38	7.96	14.73	19.94
C23H38C1207	497.46	496.1994	55.53	7.70	14.25	22.51
C23H38C1208	513.46	512.1944	53.80	7.46	13.81	24.93
C23H38C1209	529.46	528.1893	52.18	7.23	13.39	27.20
C23H39C1101	367.02	366.2689	75.27	10.71	9.66	4.36
C23H39C1102	383.02	382.2638	72.13	10.26	9.26	8.35
C23H39C1103	399.02	398.2588	69.23	9.85	8.89	12.03
C23H39C1104	415.02	414.2537	66.56	9.47	8.54	15.42
C23H39C1105	431.02	430.2486	64.09	9.12	8.23	18.56
C23H39C1106	447.02	446.2435	61.80	8.79	7.93	21.47
C23H39C1107	463.02	462.2384	59.66	8.49	7.66	24.19

C23H39C1108	479.02	478.2333	57.67	8.21	7.40	26.72
C23H39C1109	495.01	494.2282	55.81	7.94	7.16	29.09
C23H40C1201	403.48	402.2456	68.47	9.99	17.57	3.97
C23H40C1202	419.48	418.2405	65.86	9.61	16.90	7.63
C23H40C1203	435.48	434.2354	63.44	9.26	16.28	11.02
C23H40C1204	451.48	450.2303	61.19	8.93	15.71	14.18
C23H40C1205	467.48	466.2253	59.10	8.62	15.17	17.11
C23H40C1206	483.48	482.2202	57.14	8.34	14.67	19.86
C23H40C1207	499.48	498.2151	55.31	8.07	14.20	22.42
C23H40C1208	515.48	514.2100	53.59	7.82	13.76	24.83
C23H40C1209	531.48	530.2049	51.98	7.59	13.34	27.09
C23H41C1101	369.04	368.2846	74.86	11.20	9.61	4.34
C23H41C1102	385.04	384.2795	71.75	10.73	9.21	8.31
C23H41C1103	401.03	400.2744	68.89	10.31	8.84	11.97
C23H41C1104	417.03	416.2693	66.24	9.91	8.50	15.35
C23H41C1105	433.03	432.2642	63.80	9.54	8.19	18.47
C23H41C1106	449.03	448.2591	61.52	9.20	7.90	21.38
C23H41C1107	465.03	464.2541	59.41	8.89	7.62	24.08
C23H41C1108	481.03	480.2490	57.43	8.59	7.37	26.61
C23H41C1109	497.03	496.2439	55.58	8.31	7.13	28.97
C23H42C1201	405.50	404.2613	68.13	10.44	17.49	3.95
C23H42C1202	421.50	420.2562	65.54	10.04	16.82	7.59
C23H42C1203	437.50	436.2511	63.14	9.68	16.21	10.97
C23H42C1204	453.49	452.2460	60.92	9.34	15.64	14.11
C23H42C1205	469.49	468.2409	58.84	9.02	15.10	17.04
C23H42C1206	485.49	484.2358	56.90	8.72	14.60	19.77
C23H42C1207	501.49	500.2307	55.09	8.44	14.14	22.33
C23H42C1208	517.49	516.2257	53.38	8.18	13.70	24.73
C23H42C1209	533.49	532.2206	51.78	7.94	13.29	26.99
C23H43C1101	371.05	370.3002	74.45	11.68	9.55	4.31
C23H43C1102	387.05	386.2951	71.37	11.20	9.16	8.27
C23H43C1103	403.05	402.2901	68.54	10.75	8.80	11.91
C23H43C1104	419.05	418.2850	65.92	10.34	8.46	15.27
C23H43C1105	435.05	434.2799	63.50	9.96	8.15	18.39
C23H43C1106	451.05	450.2748	61.25	9.61	7.86	21.28
C23H43C1107	467.05	466.2697	59.15	9.28	7.59	23.98
C23H43C1108	483.05	482.2646	57.19	8.97	7.34	26.50
C23H43C1109	499.05	498.2595	55.36	8.69	7.10	28.85
C23H44C1201	407.51	406.2769	67.79	10.88	17.40	3.93
C23H44C1202	423.51	422.2718	65.23	10.47	16.74	7.56
C23H44C1203	439.51	438.2667	62.86	10.09	16.13	10.92
C23H44C1204	455.51	454.2616	60.65	9.74	15.57	14.05
C23H44C1205	471.51	470.2566	58.59	9.41	15.04	16.97
C23H44C1206	487.51	486.2515	56.67	9.10	14.54	19.69
C23H44C1207	503.51	502.2464	54.87	8.81	14.08	22.24
C23H44C1208	519.51	518.2413	53.18	8.54	13.65	24.64
C23H44C1209	535.51	534.2362	51.59	8.28	13.24	26.89
C23H45C1101	373.07	372.3159	74.05	12.16	9.50	4.29
C23H45C1102	389.07	388.3108	71.00	11.66	9.11	8.22
C23H45C1103	405.07	404.3057	68.20	11.20	8.75	11.85
C23H45C1104	421.07	420.3006	65.61	10.77	8.42	15.20
C23H45C1105	437.07	436.2955	63.21	10.38	8.11	18.30
C23H45C1106	453.06	452.2904	60.98	10.01	7.83	21.19
C23H45C1107	469.06	468.2854	58.90	9.67	7.56	23.88
C23H45C1108	485.06	484.2803	56.95	9.35	7.31	26.39
C23H45C1109	501.06	500.2752	55.13	9.05	7.08	28.74
C23H46C1201	409.53	408.2926	67.46	11.32	17.31	3.91
C23H46C1202	425.53	424.2875	64.92	10.90	16.66	7.52
C23H46C1203	441.53	440.2824	62.57	10.50	16.06	10.87
C23H46C1204	457.53	456.2773	60.38	10.13	15.50	13.99

102

C23H46C1205	473.53	472.2722	58.34	9.79	14.97	16.89
C23H46C1206	489.53	488.2671	56.43	9.47	14.48	19.61
C23H46C1207	505.52	504.2620	54.65	9.17	14.03	22.15
C23H46C1208	521.52	520.2570	52.97	8.89	13.60	24.54
C23H46C1209	537.52	536.2519	51.39	8.63	13.19	26.79
C23H47C1101	375.08	374.3315	73.65	12.63	9.45	4.27
C23H47C1102	391.08	390.3264	70.64	12.11	9.07	8.18
C23H47C1103	407.08	406.3214	67.86	11.64	8.71	11.79
C23H47C1104	423.08	422.3163	65.30	11.20	8.38	15.13
C23H47C1105	439.08	438.3112	62.92	10.79	8.07	18.22
C23H47C1106	455.08	454.3061	60.70	10.41	7.79	21.09
C23H47C1107	471.08	470.3010	58.64	10.06	7.53	23.77
C23H47C1108	487.08	486.2959	56.72	9.73	7.28	26.28
C23H47C1109	503.08	502.2908	54.91	9.42	7.05	28.62
C24H12C1201	387.27	386.0265	74.44	3.12	18.31	4.13
C24H12C1202	403.27	402.0214	71.48	3.00	17.58	7.93
C24H12C1203	419.27	418.0163	68.76	2.88	16.91	11.45
C24H12C1204	435.27	434.0113	66.23	2.78	16.29	14.70
C24H12C1205	451.27	450.0062	63.88	2.68	15.71	17.73
C24H12C1206	467.27	466.0011	61.69	2.59	15.17	20.54
C24H12C1207	483.27	481.9960	59.65	2.50	14.67	23.17
C24H12C1208	499.26	497.9909	57.74	2.42	14.20	25.64
C24H12C1209	515.26	513.9858	55.95	2.35	13.76	27.95
C24H13C1101	352.82	352.0655	81.70	3.71	10.05	4.53
C24H13C1102	368.82	368.0604	78.16	3.55	9.61	8.68
C24H13C1103	384.82	384.0553	74.91	3.41	9.21	12.47
C24H13C1104	400.82	400.0502	71.92	3.27	8.85	15.97
C24H13C1105	416.82	416.0451	69.16	3.14	8.51	19.19
C24H13C1106	432.82	432.0401	66.60	3.03	8.19	22.18
C24H13C1107	448.82	448.0350	64.23	2.92	7.90	24.95
C24H13C1108	464.82	464.0299	62.02	2.82	7.63	27.54
C24H13C1109	480.82	480.0248	59.95	2.73	7.37	29.95
C24H14C1201	389.28	388.0422	74.05	3.63	18.21	4.11
C24H14C1202	405.28	404.0371	71.13	3.48	17.50	7.90
C24H14C1203	421.28	420.0320	68.43	3.35	16.83	11.39
C24H14C1204	437.28	436.0269	65.92	3.23	16.22	14.64
C24H14C1205	453.28	452.0218	63.60	3.11	15.64	17.65
C24H14C1206	469.28	468.0167	61.43	3.01	15.11	20.46
C24H14C1207	485.28	484.0116	59.40	2.91	14.61	23.08
C24H14C1208	501.28	500.0066	57.51	2.82	14.14	25.53
C24H14C1209	517.28	516.0015	55.73	2.73	13.71	27.84
C24H15C1101	354.84	354.0811	81.24	4.26	9.99	4.51
C24H15C1102	370.84	370.0761	77.73	4.08	9.56	8.63
C24H15C1103	386.84	386.0710	74.52	3.91	9.16	12.41
C24H15C1104	402.84	402.0659	71.56	3.75	8.80	15.89
C24H15C1105	418.84	418.0608	68.83	3.61	8.46	19.10
C24H15C1106	434.84	434.0557	66.29	3.48	8.15	22.08
C24H15C1107	450.84	450.0506	63.94	3.35	7.86	24.84
C24H15C1108	466.84	466.0455	61.75	3.24	7.59	27.42
C24H15C1109	482.83	482.0404	59.70	3.13	7.34	29.82
C24H16C1201	391.30	390.0578	73.67	4.12	18.12	4.09
C24H16C1202	407.30	406.0527	70.78	3.96	17.41	7.86
C24H16C1203	423.30	422.0476	68.10	3.81	16.75	11.34
C24H16C1204	439.30	438.0426	65.62	3.67	16.14	14.57
C24H16C1205	455.30	454.0375	63.31	3.54	15.57	17.57
C24H16C1206	471.30	470.0324	61.16	3.42	15.04	20.37
C24H16C1207	487.30	486.0273	59.16	3.31	14.55	22.98
C24H16C1208	503.30	502.0222	57.28	3.20	14.09	25.43
C24H16C1209	519.30	518.0171	55.51	3.11	13.65	27.73
C24H17C1101	356.86	356.0968	80.78	4.80	9.93	4.48

C24H17C1102	372.85	372.0917	77.31	4.60	9.51	8.58
C24H17C1103	388.85	388.0866	74.13	4.41	9.12	12.34
C24H17C1104	404.85	404.0815	71.20	4.23	8.76	15.81
C24H17C1105	420.85	420.0764	68.50	4.07	8.42	19.01
C24H17C1106	436.85	436.0714	65.99	3.92	8.12	21.97
C24H17C1107	452.85	452.0663	63.66	3.78	7.83	24.73
C24H17C1108	468.85	468.0612	61.48	3.65	7.56	27.30
C24H17C1109	484.85	484.0561	59.45	3.53	7.31	29.70
C24H18C1201	393.32	392.0735	73.29	4.61	18.03	4.07
C24H18C1202	409.32	408.0684	70.43	4.43	17.32	7.82
C24H18C1203	425.32	424.0633	67.78	4.27	16.67	11.29
C24H18C1204	441.31	440.0582	65.32	4.11	16.07	14.50
C24H18C1205	457.31	456.0531	63.03	3.97	15.50	17.49
C24H18C1206	473.31	472.0480	60.90	3.83	14.98	20.28
C24H18C1207	489.31	488.0429	58.91	3.71	14.49	22.89
C24H18C1208	505.31	504.0379	57.05	3.59	14.03	25.33
C24H18C1209	521.31	520.0328	55.30	3.48	13.60	27.62
C24H19C1101	358.87	358.1124	80.33	5.34	9.88	4.46
C24H19C1102	374.87	374.1073	76.90	5.11	9.46	8.54
C24H19C1103	390.87	390.1023	73.75	4.90	9.07	12.28
C24H19C1104	406.87	406.0972	70.85	4.71	8.71	15.73
C24H19C1105	422.87	422.0921	68.17	4.53	8.38	18.92
C24H19C1106	438.87	438.0870	65.68	4.36	8.08	21.87
C24H19C1107	454.87	454.0819	63.37	4.21	7.79	24.62
C24H19C1108	470.87	470.0768	61.22	4.07	7.53	27.18
C24H19C1109	486.87	486.0717	59.21	3.93	7.28	29.58
C24H20C1201	395.33	394.0891	72.92	5.10	17.94	4.05
C24H20C1202	411.33	410.0840	70.08	4.90	17.24	7.78
C24H20C1203	427.33	426.0789	67.46	4.72	16.59	11.23
C24H20C1204	443.33	442.0739	65.02	4.55	15.99	14.44
C24H20C1205	459.33	458.0688	62.76	4.39	15.44	17.42
C24H20C1206	475.33	474.0637	60.65	4.24	14.92	20.20
C24H20C1207	491.33	490.0586	58.67	4.10	14.43	22.79
C24H20C1208	507.33	506.0535	56.82	3.97	13.98	25.23
C24H20C1209	523.33	522.0484	55.08	3.85	13.55	27.52
C24H21C1101	360.89	360.1281	79.88	5.87	9.82	4.43
C24H21C1102	376.89	376.1230	76.49	5.62	9.41	8.49
C24H21C1103	392.89	392.1179	73.37	5.39	9.02	12.22
C24H21C1104	408.89	408.1128	70.50	5.18	8.67	15.65
C24H21C1105	424.88	424.1077	67.85	4.98	8.34	18.83
C24H21C1106	440.88	440.1027	65.38	4.80	8.04	21.77
C24H21C1107	456.88	456.0976	63.09	4.63	7.76	24.51
C24H21C1108	472.88	472.0925	60.96	4.48	7.50	27.07
C24H21C1109	488.88	488.0874	58.96	4.33	7.25	29.45
C24H22C1201	397.35	396.1048	72.55	5.58	17.84	4.03
C24H22C1202	413.35	412.0997	69.74	5.36	17.15	7.74
C24H22C1203	429.35	428.0946	67.14	5.16	16.51	11.18
C24H22C1204	445.35	444.0895	64.73	4.98	15.92	14.37
C24H22C1205	461.35	460.0844	62.48	4.81	15.37	17.34
C24H22C1206	477.35	476.0793	60.39	4.65	14.85	20.11
C24H22C1207	493.34	492.0742	58.43	4.49	14.37	22.70
C24H22C1208	509.34	508.0692	56.60	4.35	13.92	25.13
C24H22C1209	525.34	524.0641	54.87	4.22	13.50	27.41
C24H23C1101	362.90	362.1437	79.43	6.39	9.77	4.41
C24H23C1102	378.90	378.1386	76.08	6.12	9.36	8.45
C24H23C1103	394.90	394.1336	73.00	5.87	8.98	12.15
C24H23C1104	410.90	410.1285	70.15	5.64	8.63	15.57
C24H23C1105	426.90	426.1234	67.53	5.43	8.30	18.74
C24H23C1106	442.90	442.1183	65.09	5.23	8.00	21.67
C24H23C1107	458.90	458.1132	62.82	5.05	7.73	24.41

103

C24H23C1108	474.90	474.1081	60.70	4.88	7.47	26.95
C24H23C1109	490.90	490.1030	58.72	4.72	7.22	29.33
C24H24C1201	399.36	398.1204	72.18	6.06	17.75	4.01
C24H24C1202	415.36	414.1153	69.40	5.82	17.07	7.70
C24H24C1203	431.36	430.1102	66.83	5.61	16.44	11.13
C24H24C1204	447.36	446.1052	64.44	5.41	15.85	14.31
C24H24C1205	463.36	462.1001	62.21	5.22	15.30	17.26
C24H24C1206	479.36	478.0950	60.14	5.05	14.79	20.03
C24H24C1207	495.36	494.0899	58.19	4.88	14.31	22.61
C24H24C1208	511.36	510.0848	56.37	4.73	13.87	25.03
C24H24C1209	527.36	526.0797	54.66	4.59	13.45	27.30
C24H25C1101	364.92	364.1594	78.99	6.91	9.72	4.38
C24H25C1102	380.92	380.1543	75.68	6.62	9.31	8.40
C24H25C1103	396.92	396.1492	72.63	6.35	8.93	12.09
C24H25C1104	412.92	412.1441	69.81	6.10	8.59	15.50
C24H25C1105	428.92	428.1390	67.21	5.88	8.27	18.65
C24H25C1106	444.92	444.1340	64.79	5.66	7.97	21.58
C24H25C1107	460.92	460.1289	62.54	5.47	7.69	24.30
C24H25C1108	476.92	476.1238	60.44	5.28	7.43	26.84
C24H25C1109	492.91	492.1187	58.48	5.11	7.19	29.21
C24H26C1201	401.38	400.1361	71.82	6.53	17.67	3.99
C24H26C1202	417.38	416.1310	69.07	6.28	16.99	7.67
C24H26C1203	433.38	432.1259	66.52	6.05	16.36	11.08
C24H26C1204	449.38	448.1208	64.15	5.83	15.78	14.24
C24H26C1205	465.38	464.1157	61.94	5.63	15.24	17.19
C24H26C1206	481.38	480.1106	59.88	5.44	14.73	19.94
C24H26C1207	497.38	496.1055	57.96	5.27	14.26	22.52
C24H26C1208	513.38	512.1005	56.15	5.10	13.81	24.93
C24H26C1209	529.38	528.0954	54.45	4.95	13.39	27.20
C24H27C1101	366.94	366.1750	78.56	7.42	9.66	4.36
C24H27C1102	382.93	382.1699	75.28	7.11	9.26	8.36
C24H27C1103	398.93	398.1649	72.26	6.82	8.89	12.03
C24H27C1104	414.93	414.1598	69.47	6.56	8.54	15.42
C24H27C1105	430.93	430.1547	66.89	6.32	8.23	18.56
C24H27C1106	446.93	446.1496	64.50	6.09	7.93	21.48
C24H27C1107	462.93	462.1445	62.27	5.88	7.66	24.19
C24H27C1108	478.93	478.1394	60.19	5.68	7.40	26.73
C24H27C1109	494.93	494.1343	58.24	5.50	7.16	29.09
C24H28C1201	403.40	402.1517	71.46	7.00	17.58	3.97
C24H28C1202	419.40	418.1466	68.73	6.73	16.91	7.63
C24H28C1203	435.39	434.1415	66.21	6.48	16.29	11.02
C24H28C1204	451.39	450.1365	63.86	6.25	15.71	14.18
C24H28C1205	467.39	466.1314	61.68	6.04	15.17	17.12
C24H28C1206	483.39	482.1263	59.63	5.84	14.67	19.86
C24H28C1207	499.39	498.1212	57.72	5.65	14.20	22.43
C24H28C1208	515.39	514.1161	55.93	5.48	13.76	24.83
C24H28C1209	531.39	530.1110	54.25	5.31	13.34	27.10
C24H29C1101	368.95	368.1907	78.13	7.92	9.61	4.34
C24H29C1102	384.95	384.1856	74.88	7.59	9.21	8.31
C24H29C1103	400.95	400.1805	71.90	7.29	8.84	11.97
C24H29C1104	416.95	416.1754	69.14	7.01	8.50	15.35
C24H29C1105	432.95	432.1703	66.58	6.75	8.19	18.48
C24H29C1106	448.95	448.1653	64.21	6.51	7.90	21.38
C24H29C1107	464.95	464.1602	62.00	6.29	7.63	24.09
C24H29C1108	480.95	480.1551	59.94	6.08	7.37	26.61
C24H29C1109	496.95	496.1500	58.01	5.88	7.13	28.98
C24H30C1201	405.41	404.1674	71.10	7.46	17.49	3.95
C24H30C1202	421.41	420.1623	68.41	7.18	16.83	7.59
C24H30C1203	437.41	436.1572	65.90	6.91	16.21	10.97
C24H30C1204	453.41	452.1521	63.58	6.67	15.64	14.11
C24H30C1205	469.41	468.1470	61.41	6.44	15.11	17.04
C24H30C1206	485.41	484.1419	59.39	6.23	14.61	19.78
C24H30C1207	501.41	500.1368	57.49	6.03	14.14	22.34
C24H30C1208	517.41	516.1318	55.71	5.84	13.70	24.74
C24H30C1209	533.41	532.1267	54.04	5.67	13.29	27.00
C24H31C1101	370.97	370.2063	77.71	8.42	9.56	4.31
C24H31C1102	386.97	386.2012	74.49	8.07	9.16	8.27
C24H31C1103	402.97	402.1962	71.54	7.75	8.80	11.91
C24H31C1104	418.97	418.1911	68.80	7.46	8.46	15.28
C24H31C1105	434.96	434.1860	66.27	7.18	8.15	18.39
C24H31C1106	450.96	450.1809	63.92	6.93	7.86	21.29
C24H31C1107	466.96	466.1758	61.73	6.69	7.59	23.98
C24H31C1108	482.96	482.1707	59.69	6.47	7.34	26.50
C24H31C1109	498.96	498.1656	57.77	6.26	7.11	28.86
C24H32C1201	407.43	406.1830	70.75	7.92	17.40	3.93
C24H32C1202	423.43	422.1779	68.08	7.62	16.75	7.56
C24H32C1203	439.43	438.1728	65.60	7.34	16.14	10.92
C24H32C1204	455.43	454.1677	63.30	7.08	15.57	14.05
C24H32C1205	471.43	470.1627	61.15	6.84	15.04	16.97
C24H32C1206	487.43	486.1576	59.14	6.62	14.55	19.69
C24H32C1207	503.42	502.1525	57.26	6.41	14.08	22.25
C24H32C1208	519.42	518.1474	55.50	6.21	13.65	24.64
C24H32C1209	535.42	534.1423	53.84	6.02	13.24	26.89
C24H33C1101	372.98	372.2220	77.29	8.92	9.51	4.29
C24H33C1102	388.98	388.2169	74.11	8.55	9.11	8.23
C24H33C1103	404.98	404.2118	71.18	8.21	8.75	11.85
C24H33C1104	420.98	420.2067	68.48	7.90	8.42	15.20
C24H33C1105	436.98	436.2016	65.97	7.61	8.11	18.31
C24H33C1106	452.98	452.1965	63.64	7.34	7.83	21.19
C24H33C1107	468.98	468.1915	61.47	7.09	7.56	23.88
C24H33C1108	484.98	484.1864	59.44	6.86	7.31	26.39
C24H33C1109	500.98	500.1813	57.54	6.64	7.08	28.74
C24H34C1201	409.44	408.1987	70.40	8.37	17.32	3.91
C24H34C1202	425.44	424.1936	67.76	8.06	16.67	7.52
C24H34C1203	441.44	440.1885	65.30	7.76	16.06	10.87
C24H34C1204	457.44	456.1834	63.02	7.49	15.50	13.99
C24H34C1205	473.44	472.1783	60.89	7.24	14.98	16.90
C24H34C1206	489.44	488.1732	58.90	7.00	14.49	19.61
C24H34C1207	505.44	504.1681	57.03	6.78	14.03	22.16
C24H34C1208	521.44	520.1631	55.28	6.57	13.60	24.55
C24H34C1209	537.44	536.1580	53.64	6.38	13.19	26.79
C24H35C1101	375.00	374.2376	76.87	9.41	9.45	4.27
C24H35C1102	391.00	390.2325	73.73	9.02	9.07	8.18
C24H35C1103	407.00	406.2275	70.83	8.67	8.71	11.79
C24H35C1104	423.00	422.2224	68.15	8.34	8.38	15.13
C24H35C1105	439.00	438.2173	65.67	8.04	8.08	18.22
C24H35C1106	455.00	454.2122	63.36	7.75	7.79	21.10
C24H35C1107	471.00	470.2071	61.20	7.49	7.53	23.78
C24H35C1108	486.99	486.2020	59.19	7.24	7.28	26.28
C24H35C1109	502.99	502.1969	57.31	7.01	7.05	28.63
C24H36C1201	411.46	410.2143	70.06	8.82	17.23	3.89
C24H36C1202	427.46	426.2092	67.44	8.49	16.59	7.49
C24H36C1203	443.46	442.2041	65.00	8.18	15.99	10.82
C24H36C1204	459.46	458.1990	62.74	7.90	15.43	13.93
C24H36C1205	475.46	474.1940	60.63	7.63	14.91	16.83
C24H36C1206	491.46	490.1889	58.66	7.38	14.43	19.53
C24H36C1207	507.46	506.1838	56.81	7.15	13.97	22.07
C24H36C1208	523.46	522.1787	55.07	6.93	13.55	24.45
C24H36C1209	539.46	538.1736	53.44	6.73	13.14	26.69
C24H37C1101	377.01	376.2533	76.46	9.89	9.40	4.24

104

C24H37C1102	393.01	392.2482	73.35	9.49	9.02	8.14
C24H37C1103	409.01	408.2431	70.48	9.12	8.67	11.74
C24H37C1104	425.01	424.2380	67.83	8.77	8.34	15.06
C24H37C1105	441.01	440.2329	65.36	8.46	8.04	18.14
C24H37C1106	457.01	456.2278	63.08	8.16	7.76	21.01
C24H37C1107	473.01	472.2228	60.94	7.88	7.50	23.68
C24H37C1108	489.01	488.2177	58.95	7.63	7.25	26.17
C24H37C1109	505.01	504.2126	57.08	7.38	7.02	28.51
C24H38C1201	413.48	412.2300	69.72	9.26	17.15	3.87
C24H38C1202	429.48	428.2249	67.12	8.92	16.51	7.45
C24H38C1203	445.47	444.2198	64.71	8.60	15.92	10.77
C24H38C1204	461.47	460.2147	62.47	8.30	15.37	13.87
C24H38C1205	477.47	476.2096	60.37	8.02	14.85	16.75
C24H38C1206	493.47	492.2045	58.42	7.76	14.37	19.45
C24H38C1207	509.47	508.1994	56.58	7.52	13.92	21.98
C24H38C1208	525.47	524.1944	54.86	7.29	13.49	24.36
C24H38C1209	541.47	540.1893	53.24	7.07	13.10	26.59
C24H39C1101	379.03	378.2689	76.05	10.37	9.35	4.22
C24H39C1102	395.03	394.2638	72.97	9.95	8.97	8.10
C24H39C1103	411.03	410.2588	70.13	9.56	8.63	11.68
C24H39C1104	427.03	426.2537	67.51	9.21	8.30	14.99
C24H39C1105	443.03	442.2486	65.07	8.87	8.00	18.06
C24H39C1106	459.03	458.2435	62.80	8.56	7.72	20.91
C24H39C1107	475.03	474.2384	60.68	8.28	7.46	23.58
C24H39C1108	491.03	490.2333	58.71	8.01	7.22	26.07
C24H39C1109	507.03	506.2282	56.85	7.75	6.99	28.40
C24H40C1201	415.49	414.2456	69.38	9.70	17.07	3.85
C24H40C1202	431.49	430.2405	66.81	9.34	16.43	7.42
C24H40C1203	447.49	446.2354	64.42	9.01	15.85	10.73
C24H40C1204	463.49	462.2303	62.19	8.70	15.30	13.81
C24H40C1205	479.49	478.2253	60.12	8.41	14.79	16.68
C24H40C1206	495.49	494.2202	58.18	8.14	14.31	19.37
C24H40C1207	511.49	510.2151	56.36	7.88	13.86	21.90
C24H40C1208	527.49	526.2100	54.65	7.64	13.44	24.27
C24H40C1209	543.49	542.2049	53.04	7.42	13.05	26.49
C24H41C1101	381.05	380.2846	75.65	10.85	9.30	4.20
C24H41C1102	397.05	396.2795	72.60	10.41	8.93	8.06
C24H41C1103	413.05	412.2744	69.79	10.01	8.58	11.62
C24H41C1104	429.04	428.2693	67.19	9.63	8.26	14.92
C24H41C1105	445.04	444.2642	64.77	9.29	7.97	17.98
C24H41C1106	461.04	460.2591	62.52	8.96	7.69	20.82
C24H41C1107	477.04	476.2541	60.43	8.66	7.43	23.48
C24H41C1108	493.04	492.2490	58.47	8.38	7.19	25.96
C24H41C1109	509.04	508.2439	56.63	8.12	6.96	28.29
C24H42C1201	417.51	416.2613	69.04	10.14	16.98	3.83
C24H42C1202	433.51	432.2562	66.50	9.77	16.36	7.38
C24H42C1203	449.51	448.2511	64.13	9.42	15.77	10.68
C24H42C1204	465.51	464.2460	61.93	9.09	15.23	13.75
C24H42C1205	481.51	480.2409	59.87	8.79	14.73	16.61
C24H42C1206	497.50	496.2358	57.94	8.51	14.25	19.30
C24H42C1207	513.50	512.2307	56.14	8.24	13.81	21.81
C24H42C1208	529.50	528.2257	54.44	8.00	13.39	24.17
C24H42C1209	545.50	544.2206	52.84	7.76	13.00	26.40
C24H43C1101	383.06	382.3002	75.25	11.31	9.26	4.18
C24H43C1102	399.06	398.2951	72.24	10.86	8.88	8.02
C24H43C1103	415.06	414.2901	69.45	10.44	8.54	11.56
C24H43C1104	431.06	430.2850	66.87	10.05	8.22	14.85
C24H43C1105	447.06	446.2799	64.48	9.70	7.93	17.89
C24H43C1106	463.06	462.2748	62.25	9.36	7.66	20.73
C24H43C1107	479.06	478.2697	60.17	9.05	7.40	23.38
C24H43C1108	495.06	494.2646	58.23	8.76	7.16	25.85
C24H43C1109	511.06	510.2595	56.41	8.48	6.94	28.18
C24H44C1201	419.52	418.2769	68.71	10.57	16.90	3.81
C24H44C1202	435.52	434.2718	66.19	10.18	16.28	7.35
C24H44C1203	451.52	450.2667	63.84	9.82	15.70	10.63
C24H44C1204	467.52	466.2616	61.66	9.49	15.17	13.69
C24H44C1205	483.52	482.2566	59.62	9.17	14.66	16.54
C24H44C1206	499.52	498.2515	57.71	8.88	14.19	19.22
C24H44C1207	515.52	514.2464	55.92	8.60	13.75	21.72
C24H44C1208	531.52	530.2413	54.23	8.34	13.34	24.08
C24H44C1209	547.52	546.2362	52.65	8.10	12.95	26.30
C24H45C1101	385.08	384.3159	74.86	11.78	9.21	4.15
C24H45C1102	401.08	400.3108	71.87	11.31	8.84	7.98
C24H45C1103	417.08	416.3057	69.12	10.88	8.50	11.51
C24H45C1104	433.08	432.3006	66.56	10.47	8.19	14.78
C24H45C1105	449.08	448.2955	64.19	10.10	7.89	17.81
C24H45C1106	465.08	464.2904	61 98	9.75	7.62	20.64
C24H45C1107	481.08	480.2854	59.92	9.43	7.37	23.28
C24H45C1108	497.07	496.2803	57.99	9.13	7.13	25.75
C24H45C1109	513.07	512.2752	56.18	8.84	6.91	28.07
C24H46C1201	421.54	420.2926	68.38	11.00	16.82	3.80
C24H46C1202	437.54	436.2875	65.88	10.60	16.21	7.31
C24H46C1203	453.54	452.2824	63.56	10.22	15.63	10.58
C24H46C1204	469.54	468.2773	61.39	9.87	15.10	13.63
C24H46C1205	485.54	484.2722	59.37	9.55	14.60	16.48
C24H46C1206	501.54	500.2671	57.48	9.24	14.14	19.14
C24H46C1207	517.54	516.2620	55.70	8.96	13.70	21.64
C24H46C1208	533.54	532.2570	54.03	8.69	13.29	23.99
C24H46C1209	549.53	548.2519	52.46	8.44	12.90	26.20
C24H47C1101	387.09	386.3315	74.47	12.24	9.16	4.13
C24H47C1102	403.09	402.3264	71.51	11.75	8.80	7.94
C24H47C1103	419.09	418.3214	68.78	11.30	8.46	11.45
C24H47C1104	435.09	434.3163	66.25	10.89	8.15	14.71
C24H47C1105	451.09	450.3112	63.90	10.50	7.86	17.73
C24H47C1106	467.09	466.3061	61.72	10.14	7.59	20.55
C24H47C1107	483.09	482.3010	59.67	9.81	7.34	23.18
C24H47C1108	499.09	498.2959	57.76	9.49	7.10	25.65
C24H47C1109	515.09	514.2908	55.96	9.20	6.88	27.96
C24H48C1201	423.56	422.3082	68.06	11.42	16.74	3.78
C24H48C1202	439.55	438.3031	65.58	11.01	16.13	7.28
C24H48C1203	455.55	454.2980	63.28	10.62	15.56	10.54
C24H48C1204	471.55	470.2929	61.13	10.26	15.04	13.57
C24H48C1205	487.55	486.2879	59.13	9.92	14.54	16.41
C24H48C1206	503.55	502.2828	57.25	9.61	14.08	19.06
C24H48C1207	519.55	518.2777	55.48	9.31	13.65	21.56
C24H48C1208	535.55	534.2726	53.83	9.03	13.24	23.90
C24H48C1209	551.55	550.2675	52.26	8.77	12.86	26.11
C24H49C1101	389.11	388.3472	74.08	12.69	9.11	4.11
C24H49C1102	405.11	404.3421	71.16	12.19	8.75	7.90
C24H49C1103	421.11	420.3370	68.45	11.73	8.42	11.40
C24H49C1104	437.11	436.3319	65.95	11.30	8.11	14.64
C24H49C1105	453.11	452.3268	63.62	10.90	7.82	17.66
C24H49C1106	469.11	468.3217	61.45	10.53	7.56	20.46
C24H49C1107	485.11	484.3167	59.42	10.18	7.31	23.09
C24H49C1108	501.11	500.3116	57.53	9.86	7.07	25.54
C24H49C1109	517.11	516.3065	55.75	9.55	6.86	27.85
C25H12C1201	399.28	398.0265	75.21	3.03	17.76	4.01
C25H12C1202	415.28	414.0214	72.31	2.91	17.07	7.71
C25H12C1203	431.28	430.0163	69.63	2.80	16.44	11.13
C25H12C1204	447.28	446.0113	67.13	2.70	15.85	14.31

105

C25H12C1205	463.28	462.0062	64.82	2.61	15.31	17.27
C25H12C1206	479.28	478.0011	62.65	2.52	14.79	20.03
C25H12C1207	495.28	493.9960	60.63	2.44	14.32	22.61
C25H12C1208	511.28	509.9909	58.73	2.37	13.87	25.03
C25H12C1209	527.27	525.9858	56.95	2.29	13.45	27.31
C25H13C1101	364.83	364.0655	82.31	3.59	9.72	4.39
C25H13C1102	380.83	380.0604	78.85	3.44	9.31	8.40
C25H13C1103	396.83	396.0553	75.67	3.30	8.93	12.10
C25H13C1104	412.83	412.0502	72.74	3.17	8.59	15.50
C25H13C1105	428.83	428.0451	70.02	3.06	8.27	18.65
C25H13C1106	444.83	444.0401	67.50	2.95	7.97	21.58
C25H13C1107	460.83	460.0350	65.16	2.84	7.69	24.30
C25H13C1108	476.83	476.0299	62.97	2.75	7.44	26.84
C25H13C1109	492.83	492.0248	60.93	2.66	7.19	29.22
C25H14C1201	401.30	400.0422	74.83	3.52	17.67	3.99
C25H14C1202	417.30	416.0371	71.96	3.38	16.99	7.67
C25H14C1203	433.29	432.0320	69.30	3.26	16.36	11.08
C25H14C1204	449.29	448.0269	66.83	3.14	15.78	14.24
C25H14C1205	465.29	464.0218	64.54	3.03	15.24	17.19
C25H14C1206	481.29	480.0167	62.39	2.93	14.73	19.95
C25H14C1207	497.29	496.0116	60.38	2.84	14.26	22.52
C25H14C1208	513.29	512.0066	58.50	2.75	13.81	24.94
C25H14C1209	529.29	528.0015	56.73	2.67	13.40	27.21
C25H15C1101	366.85	366.0811	81.85	4.12	9.66	4.36
C25H15C1102	382.85	382.0761	78.43	3.95	9.26	8.36
C25H15C1103	398.85	398.0710	75.29	3.79	8.89	12.03
C25H15C1104	414.85	414.0659	72.38	3.64	8.55	15.43
C25H15C1105	430.85	430.0608	69.69	3.51	8.23	18.57
C25H15C1106	446.85	446.0557	67.20	3.38	7.93	21.48
C25H15C1107	462.85	462.0506	64.88	3.27	7.66	24.20
C25H15C1108	478.85	478.0455	62.71	3.16	7.40	26.73
C25H15C1109	494.85	494.0404	60.68	3.06	7.16	29.10
C25H16C1201	403.31	402.0578	74.45	4.00	17.58	3.97
C25H16C1202	419.31	418.0527	71.61	3.85	16.91	7.63
C25H16C1203	435.31	434.0476	68.98	3.70	16.29	11.03
C25H16C1204	451.31	450.0426	66.53	3.57	15.71	14.18
C25H16C1205	467.31	466.0375	64.26	3.45	15.17	17.12
C25H16C1206	483.31	482.0324	62.13	3.34	14.67	19.86
C25H16C1207	499.31	498.0273	60.14	3.23	14.20	22.43
C25H16C1208	515.31	514.0222	58.27	3.13	13.76	24.84
C25H16C1209	531.31	530.0171	56.52	3.04	13.35	27.10
C25H17C1101	368.87	368.0968	81.41	4.65	9.61	4.34
C25H17C1102	384.87	384.0917	78.02	4.45	9.21	8.31
C25H17C1103	400.87	400.0866	74.91	4.27	8.84	11.97
C25H17C1104	416.86	416.0815	72.03	4.11	8.50	15.35
C25H17C1105	432.86	432.0764	69.37	3.96	8.19	18.48
C25H17C1106	448.86	448.0714	66.90	3.82	7.90	21.39
C25H17C1107	464.86	464.0663	64.60	3.69	7.63	24.09
C25H17C1108	480.86	480.0612	62.45	3.56	7.37	26.62
C25H17C1109	496.86	496.0561	60.44	3.45	7.14	28.98
C25H18C1201	405.33	404.0735	74.08	4.48	17.49	3.95
C25H18C1202	421.33	420.0684	71.27	4.31	16.83	7.59
C25H18C1203	437.33	436.0633	68.66	4.15	16.21	10.98
C25H18C1204	453.33	452.0582	66.24	4.00	15.64	14.12
C25H18C1205	469.33	468.0531	63.98	3.87	15.11	17.05
C25H18C1206	485.32	484.0480	61.87	3.74	14.61	19.78
C25H18C1207	501.32	500.0429	59.90	3.62	14.14	22.34
C25H18C1208	517.32	516.0379	58.04	3.51	13.71	24.74
C25H18C1209	533.32	532.0328	56.30	3.40	13.30	27.00
C25H19C1101	370.88	370.1124	80.96	5.16	9.56	4.31

C25H19C1102	386.88	386.1073	77.62	4.95	9.16	8.27
C25H19C1103	402.88	402.1023	74.53	4.75	8.80	11.91
C25H19C1104	418.88	418.0972	71.69	4.57	8.46	15.28
C25H19C1105	434.88	434.0921	69.05	4.40	8.15	18.40
C25H19C1106	450.88	450.0870	66.60	4.25	7.86	21.29
C25H19C1107	466.88	466.0819	64.32	4.10	7.59	23.99
C25H19C1108	482.88	482.0768	62.19	3.97	7.34	26.51
C25H19C1109	498.88	498.0717	60.19	3.84	7.11	28.86
C25H20C1201	407.34	406.0891	73.72	4.95	17.41	3.93
C25H20C1202	423.34	422.0840	70.93	4.76	16.75	7.56
C25H20C1203	439.34	438.0789	68.35	4.59	16.14	10.93
C25H20C1204	455.34	454.0739	65.95	4.43	15.57	14.05
C25H20C1205	471.34	470.0688	63.71	4.28	15.04	16.97
C25H20C1206	487.34	486.0637	61.62	4.14	14.55	19.70
C25H20C1207	503.34	502.0586	59.66	4.01	14.09	22.25
C25H20C1208	519.34	518.0535	57.82	3.88	13.65	24.65
C25H20C1209	535.34	534.0484	56.09	3.77	13.25	26.90
C25H21C1101	372.90	372.1281	80.53	5.68	9.51	4.29
C25H21C1102	388.90	388.1230	77.21	5.44	9.12	8.23
C25H21C1103	404.90	404.1179	74.16	5.23	8.76	11.85
C25H21C1104	420.90	420.1128	71.34	5.03	8.42	15.21
C25H21C1105	436.90	436.1077	68.73	4.84	8.11	18.31
C25H21C1106	452.90	452.1027	66.30	4.67	7.83	21.20
C25H21C1107	468.89	468.0976	64.04	4.51	7.56	23.89
C25H21C1108	484.89	484.0925	61.93	4.37	7.31	26.40
C25H21C1109	500.89	500.0874	59.95	4.23	7.08	28.75
C25H22C1201	409.36	408.1048	73.35	5.42	17.32	3.91
C25H22C1202	425.36	424.0997	70.59	5.21	16.67	7.52
C25H22C1203	441.36	440.0946	68.04	5.02	16.07	10.88
C25H22C1204	457.36	456.0895	65.66	4.85	15.50	13.99
C25H22C1205	473.36	472.0844	63.44	4.68	14.98	16.90
C25H22C1206	489.36	488.0793	61.36	4.53	14.49	19.62
C25H22C1207	505.36	504.0742	59.42	4.39	14.03	22.16
C25H22C1208	521.36	520.0692	57.60	4.25	13.60	24.55
C25H22C1209	537.35	536.0641	55.88	4.13	13.20	26.80
C25H23C1101	374.91	374.1437	80.09	6.18	9.46	4.27
C25H23C1102	390.91	390.1386	76.81	5.93	9.07	8.19
C25H23C1103	406.91	406.1336	73.79	5.70	8.71	11.80
C25H23C1104	422.91	422.1285	71.00	5.48	8.38	15.13
C25H23C1105	438.91	438.1234	68.41	5.28	8.08	18.23
C25H23C1106	454.91	454.1183	66.01	5.10	7.79	21.10
C25H23C1107	470.91	470.1132	63.77	4.92	7.53	23.78
C25H23C1108	486.91	486.1081	61.67	4.76	7.28	26.29
C25H23C1109	502.91	502.1030	59.71	4.61	7.05	28.63
C25H24C1201	411.38	410.1204	72.99	5.88	17.24	3.89
C25H24C1202	427.37	426.1153	70.26	5.66	16.59	7.49
C25H24C1203	443.37	442.1102	67.73	5.46	15.99	10.83
C25H24C1204	459.37	458.1052	65.37	5.27	15.44	13.93
C25H24C1205	475.37	474.1001	63.17	5.09	14.92	16.83
C25H24C1206	491.37	490.0950	61.11	4.92	14.43	19.54
C25H24C1207	507.37	506.0899	59.18	4.77	13.98	22.07
C25H24C1208	523.37	522.0848	57.37	4.62	13.55	24.46
C25H24C1209	539.37	538.0797	55.67	4.49	13.15	26.70
C25H25C1101	376.93	376.1594	79.66	6.69	9.41	4.24
C25H25C1102	392.93	392.1543	76.42	6.41	9.02	8.14
C25H25C1103	408.93	408.1492	73.43	6.16	8.67	11.74
C25H25C1104	424.93	424.1441	70.67	5.93	8.34	15.06
C25H25C1105	440.93	440.1390	68.10	5.72	8.04	18.14
C25H25C1106	456.93	456.1340	65.72	5.51	7.76	21.01
C25H25C1107	472.93	472.1289	63.49	5.33	7.50	23.68

106

C25H25C1108	488.93	488.1238	61.42	5.15	7.25	26.18
C25H25C1109	504.93	504.1187	59.47	4.99	7.02	28.52
C25H26C1201	413.39	412.1361	72.64	6.34	17.15	3.87
C25H26C1202	429.39	428.1310	69.93	6.10	16.51	7.45
C25H26C1203	445.39	444.1259	67.42	5.88	15.92	10.78
C25H26C1204	461.39	460.1208	65.08	5.68	15.37	13.87
C25H26C1205	477.39	476.1157	62.90	5.49	14.85	16.76
C25H26C1206	493.39	492.1106	60.86	5.31	14.37	19.46
C25H26C1207	509.39	508.1055	58.95	5.14	13.92	21.99
C25H26C1208	525.39	524.1005	57.15	4.99	13.50	24.36
C25H26C1209	541.39	540.0954	55.46	4.84	13.10	26.60
C25H27C1101	378.95	378.1750	79.24	7.18	9.36	4.22
C25H27C1102	394.95	394.1699	76.03	6.89	8.98	8.10
C25H27C1103	410.95	410.1649	73.07	6.62	8.63	11.68
C25H27C1104	426.94	426.1598	70.33	6.37	8.30	14.99
C25H27C1105	442.94	442.1547	67.79	6.14	8.00	18.06
C25H27C1106	458.94	458.1496	65.43	5.93	7.72	20.92
C25H27C1107	474.94	474.1445	63.22	5.73	7.46	23.58
C25H27C1108	490.94	490.1394	61.16	5.54	7.22	26.07
C25H27C1109	506.94	506.1343	59.23	5.37	6.99	28.40
C25H28C1201	415.41	414.1517	72.29	6.79	17.07	3.85
C25H28C1202	431.41	430.1466	69.60	6.54	16.44	7.42
C25H28C1203	447.41	446.1415	67.12	6.31	15.85	10.73
C25H28C1204	463.41	462.1365	64.80	6.09	15.30	13.81
C25H28C1205	479.40	478.1314	62.64	5.89	14.79	16.69
C25H28C1206	495.40	494.1263	60.61	5.70	14.31	19.38
C25H28C1207	511.40	510.1212	58.72	5.52	13.86	21.90
C25H28C1208	527.40	526.1161	56.94	5.35	13.44	24.27
C25H28C1209	543.40	542.1110	55.26	5.19	13.05	26.50
C25H29C1101	380.96	380.1907	78.82	7.67	9.31	4.20
C25H29C1102	396.96	396.1856	75.64	7.36	8.93	8.06
C25H29C1103	412.96	412.1805	72.71	7.08	8.59	11.62
C25H29C1104	428.96	428.1754	70.00	6.81	8.26	14.92
C25H29C1105	444.96	444.1703	67.48	6.57	7.97	17.98
C25H29C1106	460.96	460.1653	65.14	6.34	7.69	20.83
C25H29C1107	476.96	476.1602	62.96	6.13	7.43	23.48
C25H29C1108	492.96	492.1551	60.91	5.93	7.19	25.96
C25H29C1109	508.96	508.1500	59.00	5.74	6.97	28.29
C25H30C1201	417.42	416.1674	71.94	7.24	16.99	3.83
C25H30C1202	433.42	432.1623	69.28	6.98	16.36	7.38
C25H30C1203	449.42	448.1572	66.81	6.73	15.78	10.68
C25H30C1204	465.42	464.1521	64.52	6.50	15.23	13.75
C25H30C1205	481.42	480.1470	62.37	6.28	14.73	16.62
C25H30C1206	497.42	496.1419	60.37	6.08	14.25	19.30
C25H30C1207	513.42	512.1368	58.49	5.89	13.81	21.81
C25H30C1208	529.42	528.1318	56.72	5.71	13.39	24.18
C25H30C1209	545.42	544.1267	55.05	5.54	13.00	26.40
C25H31C1101	382.98	382.2063	78.41	8.16	9.26	4.18
C25H31C1102	398.98	398.2012	75.26	7.83	8.89	8.02
C25H31C1103	414.98	414.1962	72.36	7.53	8.54	11.57
C25H31C1104	430.98	430.1911	69.67	7.25	8.23	14.85
C25H31C1105	446.98	446.1860	67.18	6.99	7.93	17.90
C25H31C1106	462.98	462.1809	64.86	6.75	7.66	20.73
C25H31C1107	478.97	478.1758	62.69	6.52	7.40	23.38
C25H31C1108	494.97	494.1707	60.67	6.31	7.16	25.86
C25H31C1109	510.97	510.1656	58.77	6.12	6.94	28.18
C25H32C1201	419.44	418.1830	71.59	7.69	16.90	3.81
C25H32C1202	435.44	434.1779	68.96	7.41	16.28	7.35
C25H32C1203	451.44	450.1728	66.52	7.14	15.71	10.63
C25H32C1204	467.44	466.1677	64.24	6.90	15.17	13.69

C25H32C1205	483.44	482.1627	62.11	6.67	14.67	16.55
C25H32C1206	499.44	498.1576	60.12	6.46	14.20	19.22
C25H32C1207	515.44	514.1525	58.26	6.26	13.76	21.73
C25H32C1208	531.43	530.1474	56.50	6.07	13.34	24.08
C25H32C1209	547.43	546.1423	54.85	5.89	12.95	26.30
C25H33C1101	384.99	384.2220	78.00	8.64	9.21	4.16
C25H33C1102	400.99	400.2169	74.88	8.30	8.84	7.98
C25H33C1103	416.99	416.2118	72.01	7.98	8.50	11.51
C25H33C1104	432.99	432.2067	69.35	7.68	8.19	14.78
C25H33C1105	448.99	448.2016	66.88	7.41	7.90	17.82
C25H33C1106	464.99	464.1965	64.58	7.15	7.62	20.64
C25H33C1107	480.99	480.1915	62.43	6.92	7.37	23.28
C25H33C1108	496.99	496.1864	60.42	6.69	7.13	25.75
C25H33C1109	512.99	512.1813	58.54	6.48	6.91	28.07
C25H34C1201	421.46	420.1987	71.25	8.13	16.82	3.80
C25H34C1202	437.45	436.1936	68.64	7.83	16.21	7.31
C25H34C1203	453.45	452.1885	66.22	7.56	15.64	10.59
C25H34C1204	469.45	468.1834	63.96	7.30	15.10	13.63
C25H34C1205	485.45	484.1783	61.86	7.06	14.61	16.48
C25H34C1206	501.45	500.1732	59.88	6.83	14.14	19.14
C25H34C1207	517.45	516.1681	58.03	6.62	13.70	21.64
C25H34C1208	533.45	532.1631	56.29	6.42	13.29	23.99
C25H34C1209	549.45	548.1580	54.65	6.24	12.90	26.21
C25H35C1101	387.01	386.2376	77.59	9.12	9.16	4.13
C25H35C1102	403.01	402.2325	74.51	8.75	8.80	7.94
C25H35C1103	419.01	418.2275	71.66	8.42	8.46	11.46
C25H35C1104	435.01	434.2224	69.03	8.11	8.15	14.71
C25H35C1105	451.01	450.2173	66.58	7.82	7.86	17.74
C25H35C1106	467.01	466.2122	64.30	7.55	7.59	20.56
C25H35C1107	483.01	482.2071	62.17	7.30	7.34	23.19
C25H35C1108	499.01	498.2020	60.18	7.07	7.10	25.65
C25H35C1109	515.01	514.1969	58.31	6.85	6.88	27.96
C25H36C1201	423.47	422.2143	70.91	8.57	16.74	3.78
C25H36C1202	439.47	438.2092	68.33	8.26	16.13	7.28
C25H36C1203	455.47	454.2041	65.93	7.97	15.57	10.54
C25H36C1204	471.47	470.1990	63.69	7.70	15.04	13.57
C25H36C1205	487.47	486.1940	61.60	7.44	14.55	16.41
C25H36C1206	503.47	502.1889	59.64	7.21	14.08	19.07
C25H36C1207	519.47	518.1838	57.81	6.99	13.65	21.56
C25H36C1208	535.47	534.1787	56.08	6.78	13.24	23.90
C25H36C1209	551.47	550.1736	54.45	6.58	12.86	26.11
C25H37C1101	389.03	388.2533	77.19	9.59	9.11	4.11
C25H37C1102	405.03	404.2482	74.14	9.21	8.75	7.90
C25H37C1103	421.02	420.2431	71.32	8.86	8.42	11.40
C25H37C1104	437.02	436.2380	68.71	8.53	8.11	14.64
C25H37C1105	453.02	452.2329	66.28	8.23	7.83	17.66
C25H37C1106	469.02	468.2278	64.02	7.95	7.56	20.47
C25H37C1107	485.02	484.2228	61.91	7.69	7.31	23.09
C25H37C1108	501.02	500.2177	59.93	7.44	7.08	25.55
C25H37C1109	517.02	516.2126	58.08	7.21	6.86	27.85
C25H38C1201	425.49	424.2300	70.57	9.00	16.66	3.76
C25H38C1202	441.49	440.2249	68.02	8.68	16.06	7.25
C25H38C1203	457.49	456.2198	65.64	8.37	15.50	10.49
C25H38C1204	473.49	472.2147	63.42	8.09	14.98	13.52
C25H38C1205	489.49	488.2096	61.35	7.83	14.49	16.34
C25H38C1206	505.48	504.2045	59.40	7.58	14.03	18.99
C25H38C1207	521.48	520.1994	57.58	7.34	13.60	21.48
C25H38C1208	537.48	536.1944	55.87	7.13	13.19	23.81
C25H38C1209	553.48	552.1893	54.25	6.92	12.81	26.02
C25H39C1101	391.04	390.2689	76.79	10.05	9.07	4.09

C25H39C11O2	407.04	406.2638	73.77	9.66	8.71	7.86
C25H39C11O3	423.04	422.2588	70.98	9.29	8.38	11.35
C25H39C11O4	439.04	438.2537	68.39	8.95	8.08	14.58
C25H39C11O5	455.04	454.2486	65.99	8.64	7.79	17.58
C25H39C11O6	471.04	470.2435	63.75	8.35	7.53	20.38
C25H39C11O7	487.04	486.2384	61.65	8.07	7.28	23.00
C25H39C11O8	503.04	502.2333	59.69	7.81	7.05	25.44
C25H39C11O9	519.04	518.2282	57.85	7.57	6.83	27.74
C25H40C12O1	427.50	426.2456	70.24	9.43	16.59	3.74
C25H40C12O2	443.50	442.2405	67.71	9.09	15.99	7.22
C25H40C12O3	459.50	458.2354	65.35	8.77	15.43	10.45
C25H40C12O4	475.50	474.2303	63.15	8.48	14.91	13.46
C25H40C12O5	491.50	490.2253	61.09	8.20	14.43	16.28
C25H40C12O6	507.50	506.2202	59.17	7.94	13.97	18.92
C25H40C12O7	523.50	522.2151	57.36	7.70	13.54	21.39
C25H40C12O8	539.50	538.2100	55.66	7.47	13.14	23.72
C25H40C12O9	555.50	554.2049	54.06	7.26	12.76	25.92
C25H41C11O1	393.06	392.2846	76.40	10.51	9.02	4.07
C25H41C11O2	409.06	408.2795	73.41	10.10	8.67	7.82
C25H41C11O3	425.06	424.2744	70.64	9.72	8.34	11.29
C25H41C11O4	441.06	440.2693	68.08	9.37	8.04	14.51
C25H41C11O5	457.06	456.2642	65.70	9.04	7.76	17.50
C25H41C11O6	473.05	472.2591	63.48	8.74	7.49	20.29
C25H41C11O7	489.05	488.2541	61.40	8.45	7.25	22.90
C25H41C11O8	505.05	504.2490	59.45	8.18	7.02	25.34
C25H41C11O9	521.05	520.2439	57.63	7.93	6.80	27.64
C25H42C12O1	429.52	428.2613	69.91	9.86	16.51	3.72
C25H42C12O2	445.52	444.2562	67.40	9.50	15.92	7.18
C25H42C12O3	461.52	460.2511	65.06	9.17	15.36	10.40
C25H42C12O4	477.52	476.2460	62.88	8.87	14.85	13.40
C25H42C12O5	493.52	492.2409	60.84	8.58	14.37	16.21
C25H42C12O6	509.52	508.2358	58.93	8.31	13.92	18.84
C25H42C12O7	525.52	524.2307	57.14	8.06	13.49	21.31
C25H42C12O8	541.51	540.2257	55.45	7.82	13.09	23.64
C25H42C12O9	557.51	556.2206	53.86	7.59	12.72	25.83
C25H43C11O1	395.07	394.3002	76.01	10.97	8.97	4.05
C25H43C11O2	411.07	410.2951	73.05	10.54	8.62	7.78
C25H43C11O3	427.07	426.2901	70.31	10.15	8.30	11.24
C25H43C11O4	443.07	442.2850	67.77	9.78	8.00	14.44
C25H43C11O5	459.07	458.2799	65.41	9.44	7.72	17.43
C25H43C11O6	475.07	474.2748	63.21	9.12	7.46	20.21
C25H43C11O7	491.07	490.2697	61.15	8.83	7.22	22.81
C25H43C11O8	507.07	506.2646	59.22	8.55	6.99	25.24
C25H43C11O9	523.07	522.2595	57.41	8.29	6.78	27.53
C25H44C12O1	431.53	430.2769	69.58	10.28	16.43	3.71
C25H44C12O2	447.53	446.2718	67.10	9.91	15.84	7.15
C25H44C12O3	463.53	462.2667	64.78	9.57	15.30	10.35
C25H44C12O4	479.53	478.2616	62.62	9.25	14.79	13.35
C25H44C12O5	495.53	494.2566	60.60	8.95	14.31	16.14
C25H44C12O6	511.53	510.2515	58.70	8.67	13.86	18.77
C25H44C12O7	527.53	526.2464	56.92	8.41	13.44	21.23
C25H44C12O8	543.53	542.2413	55.25	8.16	13.05	23.55
C25H44C12O9	559.53	558.2362	53.67	7.93	12.67	25.73
C25H45C11O1	397.09	396.3159	75.62	11.42	8.93	4.03
C25H45C11O2	413.09	412.3108	72.69	10.98	8.58	7.75
C25H45C11O3	429.09	428.3057	69.98	10.57	8.26	11.19
C25H45C11O4	445.09	444.3006	67.47	10.19	7.97	14.38
C25H45C11O5	461.09	460.2955	65.12	9.84	7.69	17.35
C25H45C11O6	477.09	476.2904	62.94	9.51	7.43	20.12
C25H45C11O7	493.09	492.2854	60.90	9.20	7.19	22.71
C25H45C11O8	509.09	508.2803	58.98	8.91	6.96	25.14
C25H45C11O9	525.08	524.2752	57.19	8.64	6.75	27.42
C25H46C12O1	433.55	432.2926	69.26	10.69	16.35	3.69
C25H46C12O2	449.55	448.2875	66.80	10.31	15.77	7.12
C25H46C12O3	465.55	464.2824	64.50	9.96	15.23	10.31
C25H46C12O4	481.55	480.2773	62.36	9.63	14.72	13.29
C25H46C12O5	497.55	496.2722	60.35	9.32	14.25	16.08
C25H46C12O6	513.55	512.2671	58.47	9.03	13.81	18.69
C25H46C12O7	529.55	528.2620	56.70	8.76	13.39	21.15
C25H46C12O8	545.55	544.2570	55.04	8.50	13.00	23.46
C25H46C12O9	561.55	560.2519	53.47	8.26	12.63	25.64
C25H47C11O1	399.11	398.3315	75.24	11.87	8.88	4.01
C25H47C11O2	415.11	414.3264	72.34	11.41	8.54	7.71
C25H47C11O3	431.10	430.3214	69.65	10.99	8.22	11.13
C25H47C11O4	447.10	446.3163	67.16	10.60	7.93	14.31
C25H47C11O5	463.10	462.3112	64.84	10.23	7.66	17.27
C25H47C11O6	479.10	478.3061	62.68	9.89	7.40	20.04
C25H47C11O7	495.10	494.3010	60.65	9.57	7.16	22.62
C25H47C11O8	511.10	510.2959	58.75	9.27	6.94	25.04
C25H47C11O9	527.10	526.2908	56.97	8.99	6.73	27.32
C25H48C12O1	435.57	434.3082	68.94	11.11	16.28	3.67
C25H48C12O2	451.57	450.3031	66.50	10.71	15.70	7.09
C25H48C12O3	467.57	466.2980	64.22	10.35	15.16	10.27
C25H48C12O4	483.56	482.2929	62.10	10.01	14.66	13.23
C25H48C12O5	499.56	498.2879	60.11	9.68	14.19	16.01
C25H48C12O6	515.56	514.2828	58.24	9.38	13.75	18.62
C25H48C12O7	531.56	530.2777	56.49	9.10	13.34	21.07
C25H48C12O8	547.56	546.2726	54.84	8.84	12.95	23.38
C25H48C12O9	563.56	562.2675	53.28	8.59	12.58	25.55
C25H49C11O1	401.12	400.3472	74.86	12.31	8.84	3.99
C25H49C11O2	417.12	416.3421	71.99	11.84	8.50	7.67
C25H49C11O3	433.12	432.3370	69.33	11.40	8.19	11.08
C25H49C11O4	449.12	448.3319	66.86	11.00	7.89	14.25
C25H49C11O5	465.12	464.3268	64.56	10.62	7.62	17.20
C25H49C11O6	481.12	480.3217	62.41	10.27	7.37	19.95
C25H49C11O7	497.12	496.3167	60.40	9.94	7.13	22.53
C25H49C11O8	513.12	512.3116	58.52	9.63	6.91	24.94
C25H49C11O9	529.12	528.3065	56.75	9.33	6.70	27.21
C25H50C12O1	437.58	436.3239	68.62	11.52	16.20	3.66
C25H50C12O2	453.58	452.3188	66.20	11.11	15.63	7.05
C25H50C12O3	469.58	468.3137	63.95	10.73	15.10	10.22
C25H50C12O4	485.58	484.3086	61.84	10.38	14.60	13.18
C25H50C12O5	501.58	500.3035	59.87	10.05	14.14	15.95
C25H50C12O6	517.58	516.2984	58.02	9.74	13.70	18.55
C25H50C12O7	533.58	532.2933	56.28	9.45	13.29	20.99
C25H50C12O8	549.58	548.2882	54.64	9.17	12.90	23.29
C25H50C12O9	565.58	564.2832	53.09	8.91	12.54	25.46
C25H51C11O1	403.14	402.3628	74.49	12.75	8.79	3.97
C25H51C11O2	419.14	418.3577	71.64	12.26	8.46	7.63
C25H51C11O3	435.14	434.3527	69.01	11.81	8.15	11.03
C25H51C11O4	451.14	450.3476	66.56	11.39	7.86	14.19
C25H51C11O5	467.14	466.3425	64.28	11.00	7.59	17.13
C25H51C11O6	483.13	482.3374	62.15	10.64	7.34	19.87
C25H51C11O7	499.13	498.3323	60.16	10.30	7.10	22.44
C25H51C11O8	515.13	514.3272	58.29	9.98	6.88	24.85
C25H51C11O9	531.13	530.3221	56.54	9.68	6.67	27.11
C26H13C11O1	376.85	376.0655	82.87	3.48	9.41	4.25
C26H13C11O2	392.85	392.0604	79.49	3.34	9.02	8.15
C26H13C11O3	408.84	408.0553	76.38	3.21	8.67	11.74
C26H13C11O4	424.84	424.0502	73.51	3.08	8.34	15.06

108

C26H13C11O5	440.84	440.0451	70.84	2.97	8.04	18.15
C26H13C11O6	456.84	456.0401	68.36	2.87	7.76	21.01
C26H13C11O7	472.84	472.0350	66.05	2.77	7.50	23.69
C26H13C11O8	488.84	488.0299	63.88	2.68	7.25	26.18
C26H13C11O9	504.84	504.0248	61.86	2.60	7.02	28.52
C26H14C12O1	413.31	412.0422	75.56	3.41	17.16	3.87
C26H14C12O2	429.31	428.0371	72.74	3.29	16.52	7.45
C26H14C12O3	445.31	444.0320	70.13	3.17	15.92	10.78
C26H14C12O4	461.31	460.0269	67.70	3.06	15.37	13.87
C26H14C12O5	477.30	476.0218	65.43	2.96	14.86	16.76
C26H14C12O6	493.30	492.0167	63.31	2.86	14.37	19.46
C26H14C12O7	509.30	508.0116	61.32	2.77	13.92	21.99
C26H14C12O8	525.30	524.0066	59.45	2.69	13.50	24.37
C26H14C12O9	541.30	540.0015	57.69	2.61	13.10	26.60
C26H15C11O1	378.86	378.0811	82.43	3.99	9.36	4.22
C26H15C11O2	394.86	394.0761	79.09	3.83	8.98	8.10
C26H15C11O3	410.86	410.0710	76.01	3.68	8.63	11.68
C26H15C11O4	426.86	426.0659	73.16	3.54	8.31	14.99
C26H15C11O5	442.86	442.0608	70.52	3.41	8.01	18.06
C26H15C11O6	458.86	458.0557	68.06	3.30	7.73	20.92
C26H15C11O7	474.86	474.0506	65.76	3.18	7.47	23.59
C26H15C11O8	490.86	490.0455	63.62	3.08	7.22	26.08
C26H15C11O9	506.86	506.0404	61.61	2.98	6.99	28.41
C26H16C12O1	415.32	414.0578	75.19	3.88	17.07	3.85
C26H16C12O2	431.32	430.0527	72.40	3.74	16.44	7.42
C26H16C12O3	447.32	446.0476	69.81	3.61	15.85	10.73
C26H16C12O4	463.32	462.0426	67.40	3.48	15.30	13.81
C26H16C12O5	479.32	478.0375	65.15	3.36	14.79	16.69
C26H16C12O6	495.32	494.0324	63.05	3.26	14.32	19.38
C26H16C12O7	511.32	510.0273	61.08	3.15	13.87	21.90
C26H16C12O8	527.32	526.0222	59.22	3.06	13.45	24.27
C26H16C12O9	543.32	542.0171	57.48	2.97	13.05	26.50
C26H17C11O1	380.88	380.0968	81.99	4.50	9.31	4.20
C26H17C11O2	396.88	396.0917	78.69	4.32	8.93	8.06
C26H17C11O3	412.88	412.0866	75.64	4.15	8.59	11.63
C26H17C11O4	428.88	428.0815	72.82	4.00	8.27	14.92
C26H17C11O5	444.88	444.0764	70.20	3.85	7.97	17.98
C26H17C11O6	460.87	460.0714	67.76	3.72	7.69	20.83
C26H17C11O7	476.87	476.0663	65.49	3.59	7.43	23.49
C26H17C11O8	492.87	492.0612	63.36	3.48	7.19	25.97
C26H17C11O9	508.87	508.0561	61.37	3.37	6.97	28.30
C26H18C12O1	417.34	416.0735	74.83	4.35	16.99	3.83
C26H18C12O2	433.34	432.0684	72.07	4.19	16.36	7.38
C26H18C12O3	449.34	448.0633	69.50	4.04	15.78	10.68
C26H18C12O4	465.34	464.0582	67.11	3.90	15.24	13.75
C26H18C12O5	481.34	480.0531	64.88	3.77	14.73	16.62
C26H18C12O6	497.34	496.0480	62.79	3.65	14.26	19.30
C26H18C12O7	513.34	512.0429	60.84	3.53	13.81	21.82
C26H18C12O8	529.33	528.0379	59.00	3.43	13.40	24.18
C26H18C12O9	545.33	544.0328	57.27	3.33	13.00	26.40
C26H19C11O1	382.89	382.1124	81.56	5.00	9.26	4.18
C26H19C11O2	398.89	398.1073	78.29	4.80	8.89	8.02
C26H19C11O3	414.89	414.1023	75.27	4.62	8.55	11.57
C26H19C11O4	430.89	430.0972	72.48	4.44	8.23	14.85
C26H19C11O5	446.89	446.0921	69.88	4.29	7.93	17.90
C26H19C11O6	462.89	462.0870	67.47	4.14	7.66	20.74
C26H19C11O7	478.89	478.0819	65.21	4.00	7.40	23.39
C26H19C11O8	494.89	494.0768	63.10	3.87	7.16	25.86
C26H19C11O9	510.89	510.0717	61.13	3.75	6.94	28.19
C26H20C12O1	419.35	418.0891	74.47	4.81	16.91	3.82
C26H20C12O2	435.35	434.0840	71.73	4.63	16.29	7.35
C26H20C12O3	451.35	450.0789	69.19	4.47	15.71	10.63
C26H20C12O4	467.35	466.0739	66.82	4.31	15.17	13.69
C26H20C12O5	483.35	482.0688	64.61	4.17	14.67	16.55
C26H20C12O6	499.35	498.0637	62.54	4.04	14.20	19.22
C26H20C12O7	515.35	514.0586	60.60	3.91	13.76	21.73
C26H20C12O8	531.35	530.0535	58.77	3.79	13.34	24.09
C26H20C12O9	547.35	546.0484	57.05	3.68	12.95	26.31
C26H21C11O1	384.91	384.1281	81.13	5.50	9.21	4.16
C26H21C11O2	400.91	400.1230	77.90	5.28	8.84	7.98
C26H21C11O3	416.91	416.1179	74.91	5.08	8.50	11.51
C26H21C11O4	432.91	432.1128	72.14	4.89	8.19	14.78
C26H21C11O5	448.91	448.1077	69.57	4.72	7.90	17.82
C26H21C11O6	464.91	464.1027	67.17	4.55	7.63	20.65
C26H21C11O7	480.91	480.0976	64.94	4.40	7.37	23.29
C26H21C11O8	496.91	496.0925	62.85	4.26	7.13	25.76
C26H21C11O9	512.90	512.0874	60.89	4.13	6.91	28.07
C26H22C12O1	421.37	420.1048	74.11	5.26	16.83	3.80
C26H22C12O2	437.37	436.0997	71.40	5.07	16.21	7.32
C26H22C12O3	453.37	452.0946	68.88	4.89	15.64	10.59
C26H22C12O4	469.37	468.0895	66.53	4.72	15.11	13.63
C26H22C12O5	485.37	484.0844	64.34	4.57	14.61	16.48
C26H22C12O6	501.37	500.0793	62.29	4.42	14.14	19.15
C26H22C12O7	517.37	516.0742	60.36	4.29	13.71	21.65
C26H22C12O8	533.37	532.0692	58.55	4.16	13.29	24.00
C26H22C12O9	549.37	548.0641	56.85	4.04	12.91	26.21
C26H23C11O1	386.93	386.1437	80.71	5.99	9.16	4.14
C26H23C11O2	402.93	402.1386	77.51	5.75	8.80	7.94
C26H23C11O3	418.92	418.1336	74.55	5.53	8.46	11.46
C26H23C11O4	434.92	434.1285	71.80	5.33	8.15	14.71
C26H23C11O5	450.92	450.1234	69.26	5.14	7.86	17.74
C26H23C11O6	466.92	466.1183	66.88	4.97	7.59	20.56
C26H23C11O7	482.92	482.1132	64.67	4.80	7.34	23.19
C26H23C11O8	498.92	498.1081	62.59	4.65	7.11	25.65
C26H23C11O9	514.92	514.1030	60.65	4.50	6.89	27.96
C26H24C12O1	423.39	422.1204	73.76	5.71	16.75	3.78
C26H24C12O2	439.39	438.1153	71.07	5.51	16.14	7.28
C26H24C12O3	455.39	454.1102	68.58	5.31	15.57	10.54
C26H24C12O4	471.38	470.1052	66.25	5.13	15.04	13.58
C26H24C12O5	487.38	486.1001	64.07	4.96	14.55	16.41
C26H24C12O6	503.38	502.0950	62.04	4.81	14.09	19.07
C26H24C12O7	519.38	518.0899	60.13	4.66	13.65	21.56
C26H24C12O8	535.38	534.0848	58.33	4.52	13.24	23.91
C26H24C12O9	551.38	550.0797	56.64	4.39	12.86	26.12
C26H25C11O1	388.94	388.1594	80.29	6.48	9.12	4.11
C26H25C11O2	404.94	404.1543	77.12	6.22	8.76	7.90
C26H25C11O3	420.94	420.1492	74.19	5.99	8.42	11.40
C26H25C11O4	436.94	436.1441	71.47	5.77	8.11	14.65
C26H25C11O5	452.94	452.1390	68.95	5.56	7.83	17.66
C26H25C11O6	468.94	468.1340	66.60	5.37	7.56	20.47
C26H25C11O7	484.94	484.1289	64.40	5.20	7.31	23.09
C26H25C11O8	500.94	500.1238	62.34	5.03	7.08	25.55
C26H25C11O9	516.94	516.1187	60.41	4.87	6.86	27.86
C26H26C12O1	425.40	424.1361	73.41	6.16	16.67	3.76
C26H26C12O2	441.40	440.1310	70.75	5.94	16.06	7.25
C26H26C12O3	457.40	456.1259	68.27	5.73	15.50	10.49
C26H26C12O4	473.40	472.1208	65.97	5.54	14.98	13.52
C26H26C12O5	489.40	488.1157	63.81	5.35	14.49	16.35
C26H26C12O6	505.40	504.1106	61.79	5.19	14.03	18.99
C26H26C12O7	521.40	520.1055	59.89	5.03	13.60	21.48

109

C26H26C1208	537.40	536.1005	58.11	4.88	13.19	23.82
C26H26C1209	553.40	552.0954	56.43	4.74	12.81	26.02
C26H27C1101	390.96	390.1750	79.88	6.96	9.07	4.09
C26H27C1102	406.96	406.1699	76.74	6.69	8.71	7.86
C26H27C1103	422.96	422.1649	73.84	6.43	8.38	11.35
C26H27C1104	438.96	438.1598	71.14	6.20	8.08	14.58
C26H27C1105	454.96	454.1547	68.64	5.98	7.79	17.58
C26H27C1106	470.95	470.1496	66.31	5.78	7.53	20.38
C26H27C1107	486.95	486.1445	64.13	5.59	7.28	23.00
C26H27C1108	502.95	502.1394	62.09	5.41	7.05	25.45
C26H27C1109	518.95	518.1343	60.18	5.24	6.83	27.75
C26H28C1201	427.42	426.1517	73.06	6.60	16.59	3.74
C26H28C1202	443.42	442.1466	70.43	6.36	15.99	7.22
C26H28C1203	459.42	458.1415	67.98	6.14	15.43	10.45
C26H28C1204	475.42	474.1365	65.69	5.94	14.91	13.46
C26H28C1205	491.42	490.1314	63.55	5.74	14.43	16.28
C26H28C1206	507.42	506.1263	61.55	5.56	13.97	18.92
C26H28C1207	523.41	522.1212	59.66	5.39	13.55	21.40
C26H28C1208	539.41	538.1161	57.89	5.23	13.14	23.73
C26H28C1209	555.41	554.1110	56.23	5.08	12.77	25.93
C26H29C1101	392.97	392.1907	79.47	7.44	9.02	4.07
C26H29C1102	408.97	408.1856	76.36	7.15	8.67	7.82
C26H29C1103	424.97	424.1805	73.48	6.88	8.34	11.29
C26H29C1104	440.97	440.1754	70.82	6.63	8.04	14.51
C26H29C1105	456.97	456.1703	68.34	6.40	7.76	17.51
C26H29C1106	472.97	472.1653	66.03	6.18	7.50	20.30
C26H29C1107	488.97	488.1602	63.87	5.98	7.25	22.90
C26H29C1108	504.97	504.1551	61.84	5.79	7.02	25.35
C26H29C1109	520.97	520.1500	59.94	5.61	6.81	27.64
C26H30C1201	429.43	428.1674	72.72	7.04	16.51	3.73
C26H30C1202	445.43	444.1623	70.11	6.79	15.92	7.18
C26H30C1203	461.43	460.1572	67.68	6.55	15.37	10.40
C26H30C1204	477.43	476.1521	65.41	6.33	14.85	13.40
C26H30C1205	493.43	492.1470	63.29	6.13	14.37	16.21
C26H30C1206	509.43	508.1419	61.30	5.94	13.92	18.84
C26H30C1207	525.43	524.1368	59.44	5.76	13.49	21.32
C26H30C1208	541.43	540.1318	57.68	5.59	13.10	23.64
C26H30C1209	557.43	556.1267	56.02	5.42	12.72	25.83
C26H31C1101	394.99	394.2063	79.06	7.91	8.98	4.05
C26H31C1102	410.99	410.2012	75.99	7.60	8.63	7.79
C26H31C1103	426.99	426.1962	73.14	7.32	8.30	11.24
C26H31C1104	442.99	442.1911	70.50	7.05	8.00	14.45
C26H31C1105	458.99	458.1860	68.04	6.81	7.72	17.43
C26H31C1106	474.99	474.1809	65.75	6.58	7.46	20.21
C26H31C1107	490.99	490.1758	63.60	6.36	7.22	22.81
C26H31C1108	506.99	506.1707	61.60	6.16	6.99	25.25
C26H31C1109	522.98	522.1656	59.71	5.97	6.78	27.53
C26H32C1201	431.45	430.1830	72.38	7.48	16.43	3.71
C26H32C1202	447.45	446.1779	69.79	7.21	15.85	7.15
C26H32C1203	463.45	462.1728	67.38	6.96	15.30	10.36
C26H32C1204	479.45	478.1677	65.14	6.73	14.79	13.35
C26H32C1205	495.45	494.1627	63.03	6.51	14.31	16.15
C26H32C1206	511.45	510.1576	61.06	6.31	13.86	18.77
C26H32C1207	527.45	526.1525	59.21	6.12	13.44	21.23
C26H32C1208	543.45	542.1474	57.46	5.94	13.05	23.55
C26H32C1209	559.45	558.1423	55.82	5.77	12.67	25.74
C26H33C1101	397.01	396.2220	78.66	8.38	8.93	4.03
C26H33C1102	413.00	412.2169	75.61	8.05	8.58	7.75
C26H33C1103	429.00	428.2118	72.79	7.75	8.26	11.19
C26H33C1104	445.00	444.2067	70.18	7.47	7.97	14.38

C26H33C1105	461.00	460.2016	67.74	7.22	7.69	17.35
C26H33C1106	477.00	476.1965	65.47	6.97	7.43	20.12
C26H33C1107	493.00	492.1915	63.34	6.75	7.19	22.72
C26H33C1108	509.00	508.1864	61.35	6.53	6.97	25.15
C26H33C1109	525.00	524.1813	59.48	6.34	6.75	27.43
C26H34C1201	433.47	432.1987	72.04	7.91	16.36	3.69
C26H34C1202	449.47	448.1936	69.48	7.62	15.78	7.12
C26H34C1203	465.47	464.1885	67.09	7.36	15.23	10.31
C26H34C1204	481.46	480.1834	64.86	7.12	14.73	13.29
C26H34C1205	497.46	496.1783	62.78	6.89	14.25	16.08
C26H34C1206	513.46	512.1732	60.82	6.67	13.81	18.70
C26H34C1207	529.46	528.1681	58.98	6.47	13.39	21.15
C26H34C1208	545.46	544.1631	57.25	6.28	13.00	23.47
C26H34C1209	561.46	560.1580	55.62	6.10	12.63	25.65
C26H35C1101	399.02	398.2376	78.26	8.84	8.88	4.01
C26H35C1102	415.02	414.2325	75.25	8.50	8.54	7.71
C26H35C1103	431.02	430.2275	72.45	8.18	8.23	11.14
C26H35C1104	447.02	446.2224	69.86	7.89	7.93	14.32
C26H35C1105	463.02	462.2173	67.45	7.62	7.66	17.28
C26H35C1106	479.02	478.2122	65.19	7.36	7.40	20.04
C26H35C1107	495.02	494.2071	63.09	7.13	7.16	22.62
C26H35C1108	511.02	510.2020	61.11	6.90	6.94	25.05
C26H35C1109	527.02	526.1969	59.26	6.69	6.73	27.32
C26H36C1201	435.48	434.2143	71.71	8.33	16.28	3.67
C26H36C1202	451.48	450.2092	69.17	8.04	15.71	7.09
C26H36C1203	467.48	466.2041	66.80	7.76	15.17	10.27
C26H36C1204	483.48	482.1990	64.59	7.51	14.67	13.24
C26H36C1205	499.48	498.1940	62.52	7.26	14.20	16.02
C26H36C1206	515.48	514.1889	60.58	7.04	13.76	18.62
C26H36C1207	531.48	530.1838	58.76	6.83	13.34	21.07
C26H36C1208	547.48	546.1787	57.04	6.63	12.95	23.38
C26H36C1209	563.48	562.1736	55.42	6.44	12.58	25.55
C26H37C1101	401.04	400.2533	77.87	9.30	8.84	3.99
C26H37C1102	417.04	416.2482	74.88	8.94	8.50	7.67
C26H37C1103	433.04	432.2431	72.12	8.61	8.19	11.08
C26H37C1104	449.04	448.2380	69.55	8.31	7.90	14.25
C26H37C1105	465.03	464.2329	67.15	8.02	7.62	17.20
C26H37C1106	481.03	480.2278	64.92	7.75	7.37	19.96
C26H37C1107	497.03	496.2228	62.83	7.50	7.13	22.53
C26H37C1108	513.03	512.2177	60.87	7.27	6.91	24.95
C26H37C1109	529.03	528.2126	59.03	7.05	6.70	27.22
C26H38C1201	437.50	436.2300	71.38	8.75	16.21	3.66
C26H38C1202	453.50	452.2249	68.86	8.45	15.64	7.06
C26H38C1203	469.50	468.2198	66.52	8.16	15.10	10.22
C26H38C1204	485.50	484.2147	64.32	7.89	14.60	13.18
C26H38C1205	501.50	500.2096	62.27	7.64	14.14	15.95
C26H38C1206	517.50	516.2045	60.35	7.40	13.70	18.55
C26H38C1207	533.49	532.1994	58.54	7.18	13.29	20.99
C26H38C1208	549.49	548.1944	56.83	6.97	12.90	23.29
C26H38C1209	565.49	564.1893	55.22	6.77	12.54	25.46
C26H39C1101	403.05	402.2689	77.48	9.75	8.80	3.97
C26H39C1102	419.05	418.2638	74.52	9.38	8.46	7.64
C26H39C1103	435.05	434.2588	71.78	9.04	8.15	11.03
C26H39C1104	451.05	450.2537	69.24	8.72	7.86	14.19
C26H39C1105	467.05	466.2486	66.86	8.42	7.59	17.13
C26H39C1106	483.05	482.2435	64.65	8.14	7.34	19.87
C26H39C1107	499.05	498.2384	62.58	7.88	7.10	22.44
C26H39C1108	515.05	514.2333	60.63	7.63	6.88	24.85
C26H39C1109	531.05	530.2282	58.81	7.40	6.68	27.12
C26H40C1201	439.51	438.2456	71.05	9.17	16.13	3.64

110

C26H40C1202	455.51	454.2405	68.56	8.85	15.57	7.02
C26H40C1203	471.51	470.2354	66.23	8.55	15.04	10.18
C26H40C1204	487.51	486.2303	64.06	8.27	14.54	13.13
C26H40C1205	503.51	502.2253	62.02	8.01	14.08	15.89
C26H40C1206	519.51	518.2202	60.11	7.76	13.65	18.48
C26H40C1207	535.51	534.2151	58.32	7.53	13.24	20.91
C26H40C1208	551.51	550.2100	56.62	7.31	12.86	23.21
C26H40C1209	567.51	566.2049	55.03	7.10	12.49	25.37
C26H41C1101	405.07	404.2846	77.10	10.20	8.75	3.95
C26H41C1102	421.07	420.2795	74.17	9.81	8.42	7.60
C26H41C1103	437.07	436.2744	71.45	9.46	8.11	10.98
C26H41C1104	453.07	452.2693	68.93	9.12	7.83	14.13
C26H41C1105	469.07	468.2642	66.58	8.81	7.56	17.05
C26H41C1106	485.07	484.2591	64.38	8.52	7.31	19.79
C26H41C1107	501.07	500.2541	62.33	8.25	7.08	22.35
C26H41C1108	517.06	516.2490	60.40	7.99	6.86	24.75
C26H41C1109	533.06	532.2439	58.58	7.75	6.65	27.01
C26H42C1201	441.53	440.2613	70.73	9.59	16.06	3.62
C26H42C1202	457.53	456.2562	68.26	9.25	15.50	6.99
C26H42C1203	473.53	472.2511	65.95	8.94	14.97	10.14
C26H42C1204	489.53	488.2460	63.79	8.65	14.48	13.07
C26H42C1205	505.53	504.2409	61.78	8.37	14.03	15.82
C26H42C1206	521.53	520.2358	59.88	8.12	13.60	18.41
C26H42C1207	537.53	536.2307	58.10	7.88	13.19	20.84
C26H42C1208	553.53	552.2257	56.42	7.65	12.81	23.12
C26H42C1209	569.53	568.2206	54.83	7.43	12.45	25.28
C26H43C1101	407.09	406.3002	76.71	10.65	8.71	3.93
C26H43C1102	423.08	422.2951	73.81	10.24	8.38	7.56
C26H43C1103	439.08	438.2901	71.12	9.87	8.07	10.93
C26H43C1104	455.08	454.2850	68.62	9.52	7.79	14.06
C26H43C1105	471.08	470.2799	66.29	9.20	7.53	16.98
C26H43C1106	487.08	486.2748	64.11	8.90	7.28	19.71
C26H43C1107	503.08	502.2697	62.08	8.62	7.05	22.26
C26H43C1108	519.08	518.2646	60.16	8.35	6.83	24.66
C26H43C1109	535.08	534.2595	58.36	8.10	6.63	26.91
C26H44C1201	443.55	442.2769	70.41	10.00	15.99	3.61
C26H44C1202	459.55	458.2718	67.96	9.65	15.43	6.96
C26H44C1203	475.54	474.2667	65.67	9.33	14.91	10.09
C26H44C1204	491.54	490.2616	63.53	9.02	14.43	13.02
C26H44C1205	507.54	506.2566	61.53	8.74	13.97	15.76
C26H44C1206	523.54	522.2515	59.65	8.47	13.54	18.34
C26H44C1207	539.54	538.2464	57.88	8.22	13.14	20.76
C26H44C1208	555.54	554.2413	56.21	7.98	12.76	23.04
C26H44C1209	571.54	570.2362	54.64	7.76	12.41	25.19
C26H45C1101	409.10	408.3159	76.34	11.09	8.67	3.91
C26H45C1102	425.10	424.3108	73.46	10.67	8.34	7.53
C26H45C1103	441.10	440.3057	70.80	10.28	8.04	10.88
C26H45C1104	457.10	456.3006	68.32	9.92	7.76	14.00
C26H45C1105	473.10	472.2955	66.01	9.59	7.49	16.91
C26H45C1106	489.10	488.2904	63.85	9.27	7.25	19.63
C26H45C1107	505.10	504.2854	61.83	8.98	7.02	22.17
C26H45C1108	521.10	520.2803	59.93	8.70	6.80	24.56
C26H45C1109	537.10	536.2752	58.14	8.45	6.60	26.81
C26H46C1201	445.56	444.2926	70.09	10.41	15.91	3.59
C26H46C1202	461.56	460.2875	67.66	10.05	15.36	6.93
C26H46C1203	477.56	476.2824	65.39	9.71	14.85	10.05
C26H46C1204	493.56	492.2773	63.27	9.39	14.37	12.97
C26H46C1205	509.56	508.2722	61.29	9.10	13.92	15.70
C26H46C1206	525.56	524.2671	59.42	8.82	13.49	18.27
C26H46C1207	541.56	540.2620	57.67	8.56	13.09	20.68

C26H46C1208	557.56	556.2570	56.01	8.32	12.72	22.96
C26H46C1209	573.56	572.2519	54.45	8.08	12.36	25.11
C26H47C1101	411.12	410.3315	75.96	11.52	8.62	3.89
C26H47C1102	427.12	426.3264	73.12	11.09	8.30	7.49
C26H47C1103	443.12	442.3214	70.48	10.69	8.00	10.83
C26H47C1104	459.12	458.3163	68.02	10.32	7.72	13.94
C26H47C1105	475.11	474.3112	65.73	9.97	7.46	16.84
C26H47C1106	491.11	490.3061	63.59	9.65	7.22	19.55
C26H47C1107	507.11	506.3010	61.58	9.34	6.99	22.08
C26H47C1108	523.11	522.2959	59.70	9.06	6.78	24.47
C26H47C1109	539.11	538.2908	57.93	8.79	6.58	26.71
C26H48C1201	447.58	446.3082	69.77	10.81	15.84	3.57
C26H48C1202	463.58	462.3031	67.37	10.44	15.30	6.90
C26H48C1203	479.58	478.2980	65.12	10.09	14.79	10.01
C26H48C1204	495.58	494.2929	63.02	9.76	14.31	12.91
C26H48C1205	511.58	510.2879	61.04	9.46	13.86	15.64
C26H48C1206	527.57	526.2828	59.19	9.17	13.44	18.20
C26H48C1207	543.57	542.2777	57.45	8.90	13.04	20.60
C26H48C1208	559.57	558.2726	55.81	8.65	12.67	22.87
C26H48C1209	575.57	574.2675	54.26	8.41	12.32	25.02
C26H49C1101	413.13	412.3472	75.59	11.96	8.58	3.87
C26H49C1102	429.13	428.3421	72.77	11.51	8.26	7.46
C26H49C1103	445.13	444.3370	70.16	11.10	7.96	10.78
C26H49C1104	461.13	460.3319	67.72	10.71	7.69	13.88
C26H49C1105	477.13	476.3268	65.45	10.35	7.43	16.77
C26H49C1106	493.13	492.3217	63.33	10.02	7.19	19.47
C26H49C1107	509.13	508.3167	61.34	9.70	6.96	22.00
C26H49C1108	525.13	524.3116	59.47	9.41	6.75	24.37
C26H49C1109	541.13	540.3065	57.71	9.13	6.55	26.61
C26H50C1201	449.59	448.3239	69.46	11.21	15.77	3.56
C26H50C1202	465.59	464.3188	67.07	10.82	15.23	6.87
C26H50C1203	481.59	480.3137	64.85	10.46	14.72	9.97
C26H50C1204	497.59	496.3086	62.76	10.13	14.25	12.86
C26H50C1205	513.59	512.3035	60.81	9.81	13.81	15.58
C26H50C1206	529.59	528.2984	58.97	9.52	13.39	18.13
C26H50C1207	545.59	544.2933	57.24	9.24	13.00	20.53
C26H50C1208	561.59	560.2882	55.61	8.97	12.63	22.79
C26H50C1209	577.59	576.2832	54.07	8.73	12.28	24.93
C26H51C1101	415.15	414.3628	75.22	12.38	8.54	3.85
C26H51C1102	431.15	430.3577	72.43	11.92	8.22	7.42
C26H51C1103	447.15	446.3527	69.84	11.50	7.93	10.73
C26H51C1104	463.15	462.3476	67.43	11.10	7.65	13.82
C26H51C1105	479.15	478.3425	65.18	10.73	7.40	16.70
C26H51C1106	495.15	494.3374	63.07	10.38	7.16	19.39
C26H51C1107	511.15	510.3323	61.10	10.06	6.94	21.91
C26H51C1108	527.14	526.3272	59.24	9.75	6.73	24.28
C26H51C1109	543.14	542.3221	57.50	9.46	6.53	26.51
C26H52C1201	451.61	450.3395	69.15	11.61	15.70	3.54
C26H52C1202	467.61	466.3344	66.78	11.21	15.16	6.84
C26H52C1203	483.61	482.3293	64.57	10.84	14.66	9.93
C26H52C1204	499.61	498.3242	62.51	10.49	14.19	12.81
C26H52C1205	515.61	514.3192	60.57	10.17	13.75	15.52
C26H52C1206	531.61	530.3141	58.74	9.86	13.34	18.06
C26H52C1207	547.61	546.3090	57.03	9.57	12.95	20.45
C26H52C1208	563.61	562.3039	55.41	9.30	12.58	22.71
C26H52C1209	579.60	578.2988	53.88	9.04	12.23	24.84
C26H53C1101	417.16	416.3785	74.86	12.81	8.50	3.84
C26H53C1102	433.16	432.3734	72.10	12.33	8.18	7.39
C26H53C1103	449.16	448.3683	69.53	11.89	7.89	10.69
C26H53C1104	465.16	464.3632	67.14	11.48	7.62	13.76

C26H53C1105	481.16	480.3581	64.90	11.10	7.37	16.63
C26H53C1106	497.16	496.3530	62.81	10.75	7.13	19.31
C26H53C1107	513.16	512.3480	60.86	10.41	6.91	21.82
C26H53C1108	529.16	528.3429	59.02	10.10	6.70	24.19
C26H53C1109	545.16	544.3378	57.28	9.80	6.50	26.41
C27H13C1101	388.86	388.0655	83.40	3.37	9.12	4.11
C27H13C1102	404.86	404.0604	80.10	3.24	8.76	7.90
C27H13C1103	420.86	420.0553	77.06	3.11	8.42	11.40
C27H13C1104	436.86	436.0502	74.24	3.00	8.12	14.65
C27H13C1105	452.85	452.0451	71.61	2.89	7.83	17.67
C27H13C1106	468.85	468.0401	69.17	2.79	7.56	20.47
C27H13C1107	484.85	484.0350	66.89	2.70	7.31	23.10
C27H13C1108	500.85	500.0299	64.75	2.62	7.08	25.56
C27H13C1109	516.85	516.0248	62.75	2.54	6.86	27.86
C27H14C1201	425.32	424.0422	76.25	3.32	16.67	3.76
C27H14C1202	441.32	440.0371	73.48	3.20	16.07	7.25
C27H14C1203	457.32	456.0320	70.91	3.09	15.50	10.50
C27H14C1204	473.32	472.0269	68.52	2.98	14.98	13.52
C27H14C1205	489.32	488.0218	66.28	2.88	14.49	16.35
C27H14C1206	505.32	504.0167	64.18	2.79	14.03	19.00
C27H14C1207	521.31	520.0116	62.21	2.71	13.60	21.48
C27H14C1208	537.31	536.0066	60.36	2.63	13.20	23.82
C27H14C1209	553.31	552.0015	58.61	2.55	12.81	26.02
C27H15C1101	390.87	390.0811	82.97	3.87	9.07	4.09
C27H15C1102	406.87	406.0761	79.71	3.72	8.71	7.86
C27H15C1103	422.87	422.0710	76.69	3.58	8.38	11.35
C27H15C1104	438.87	438.0659	73.89	3.45	8.08	14.58
C27H15C1105	454.87	454.0608	71.30	3.32	7.79	17.59
C27H15C1106	470.87	470.0557	68.87	3.21	7.53	20.39
C27H15C1107	486.87	486.0506	66.61	3.11	7.28	23.00
C27H15C1108	502.87	502.0455	64.49	3.01	7.05	25.45
C27H15C1109	518.87	518.0404	62.50	2.91	6.83	27.75
C27H16C1201	427.33	426.0578	75.89	3.77	16.59	3.74
C27H16C1202	443.33	442.0527	73.15	3.64	15.99	7.22
C27H16C1203	459.33	458.0476	70.60	3.51	15.44	10.45
C27H16C1204	475.33	474.0426	68.23	3.39	14.92	13.46
C27H16C1205	491.33	490.0375	66.00	3.28	14.43	16.28
C27H16C1206	507.33	506.0324	63.92	3.18	13.98	18.92
C27H16C1207	523.33	522.0273	61.97	3.08	13.55	21.40
C27H16C1208	539.33	538.0222	60.13	2.99	13.15	23.73
C27H16C1209	555.33	554.0171	58.40	2.90	12.77	25.93
C27H17C1101	392.89	392.0968	82.54	4.36	9.02	4.07
C27H17C1102	408.89	408.0917	79.31	4.19	8.67	7.83
C27H17C1103	424.89	424.0866	76.33	4.03	8.34	11.30
C27H17C1104	440.89	440.0815	73.56	3.89	8.04	14.52
C27H17C1105	456.89	456.0764	70.98	3.75	7.76	17.51
C27H17C1106	472.89	472.0714	68.58	3.62	7.50	20.30
C27H17C1107	488.89	488.0663	66.33	3.51	7.25	22.91
C27H17C1108	504.88	504.0612	64.23	3.39	7.02	25.35
C27H17C1109	520.88	520.0561	62.26	3.29	6.81	27.64
C27H18C1201	429.35	428.0735	75.53	4.23	16.51	3.73
C27H18C1202	445.35	444.0684	72.82	4.07	15.92	7.19
C27H18C1203	461.35	460.0633	70.29	3.93	15.37	10.40
C27H18C1204	477.35	476.0582	67.94	3.80	14.85	13.41
C27H18C1205	493.35	492.0531	65.73	3.68	14.37	16.22
C27H18C1206	509.35	508.0480	63.67	3.56	13.92	18.85
C27H18C1207	525.35	524.0429	61.73	3.45	13.50	21.32
C27H18C1208	541.35	540.0379	59.91	3.35	13.10	23.64
C27H18C1209	557.35	556.0328	58.19	3.26	12.72	25.84
C27H19C1101	394.90	394.1124	82.12	4.85	8.98	4.05

C27H19C1102	410.90	410.1073	78.92	4.66	8.63	7.79
C27H19C1103	426.90	426.1023	75.97	4.49	8.30	11.24
C27H19C1104	442.90	442.0972	73.22	4.32	8.00	14.45
C27H19C1105	458.90	458.0921	70.67	4.17	7.73	17.43
C27H19C1106	474.90	474.0870	68.29	4.03	7.47	20.21
C27H19C1107	490.90	490.0819	66.06	3.90	7.22	22.81
C27H19C1108	506.90	506.0768	63.98	3.78	6.99	25.25
C27H19C1109	522.90	522.0717	62.02	3.66	6.78	27.54
C27H20C1201	431.37	430.0891	75.18	4.67	16.44	3.71
C27H20C1202	447.37	446.0840	72.49	4.51	15.85	7.15
C27H20C1203	463.36	462.0789	69.99	4.35	15.30	10.36
C27H20C1204	479.36	478.0739	67.65	4.21	14.79	13.35
C27H20C1205	495.36	494.0688	65.47	4.07	14.31	16.15
C27H20C1206	511.36	510.0637	63.42	3.94	13.87	18.77
C27H20C1207	527.36	526.0586	61.49	3.82	13.45	21.24
C27H20C1208	543.36	542.0535	59.68	3.71	13.05	23.56
C27H20C1209	559.36	558.0484	57.98	3.60	12.68	25.74
C27H21C1101	396.92	396.1281	81.70	5.33	8.93	4.03
C27H21C1102	412.92	412.1230	78.54	5.13	8.59	7.75
C27H21C1103	428.92	428.1179	75.61	4.94	8.27	11.19
C27H21C1104	444.92	444.1128	72.89	4.76	7.97	14.38
C27H21C1105	460.92	460.1077	70.36	4.59	7.69	17.36
C27H21C1106	476.92	476.1027	68.00	4.44	7.43	20.13
C27H21C1107	492.92	492.0976	65.79	4.29	7.19	22.72
C27H21C1108	508.92	508.0925	63.72	4.16	6.97	25.15
C27H21C1109	524.92	524.0874	61.78	4.03	6.75	27.43
C27H22C1201	433.38	432.1048	74.83	5.12	16.36	3.69
C27H22C1202	449.38	448.0997	72.17	4.93	15.78	7.12
C27H22C1203	465.38	464.0946	69.69	4.76	15.24	10.31
C27H22C1204	481.38	480.0895	67.37	4.61	14.73	13.29
C27H22C1205	497.38	496.0844	65.20	4.46	14.26	16.08
C27H22C1206	513.38	512.0793	63.17	4.32	13.81	18.70
C27H22C1207	529.38	528.0742	61.26	4.19	13.39	21.16
C27H22C1208	545.38	544.0692	59.46	4.07	13.00	23.47
C27H22C1209	561.38	560.0641	57.77	3.95	12.63	25.65
C27H23C1101	398.94	398.1437	81.29	5.81	8.89	4.01
C27H23C1102	414.94	414.1386	78.16	5.59	8.54	7.71
C27H23C1103	430.94	430.1336	75.26	5.38	8.23	11.14
C27H23C1104	446.93	446.1285	72.56	5.19	7.93	14.32
C27H23C1105	462.93	462.1234	70.05	5.01	7.66	17.28
C27H23C1106	478.93	478.1183	67.71	4.84	7.40	20.04
C27H23C1107	494.93	494.1132	65.52	4.68	7.16	22.63
C27H23C1108	510.93	510.1081	63.47	4.54	6.94	25.05
C27H23C1109	526.93	526.1030	61.55	4.40	6.73	27.33
C27H24C1201	435.40	434.1204	74.48	5.56	16.29	3.67
C27H24C1202	451.40	450.1153	71.84	5.36	15.71	7.09
C27H24C1203	467.40	466.1102	69.38	5.18	15.17	10.27
C27H24C1204	483.40	482.1052	67.09	5.00	14.67	13.24
C27H24C1205	499.40	498.1001	64.94	4.84	14.20	16.02
C27H24C1206	515.39	514.0950	62.92	4.69	13.76	18.63
C27H24C1207	531.39	530.0899	61.03	4.55	13.34	21.08
C27H24C1208	547.39	546.0848	59.24	4.42	12.95	23.38
C27H24C1209	563.39	562.0797	57.56	4.29	12.59	25.56
C27H25C1101	400.95	400.1594	80.88	6.28	8.84	3.99
C27H25C1102	416.95	416.1543	77.78	6.04	8.50	7.67
C27H25C1103	432.95	432.1492	74.90	5.82	8.19	11.09
C27H25C1104	448.95	448.1441	72.24	5.61	7.90	14.25
C27H25C1105	464.95	464.1390	69.75	5.42	7.63	17.21
C27H25C1106	480.95	480.1340	67.43	5.24	7.37	19.96
C27H25C1107	496.95	496.1289	65.26	5.07	7.13	22.54

112

C27H25C1108	512.95	512.1238	63.22	4.91	6.91	24.95
C27H25C1109	528.95	528.1187	61.31	4.76	6.70	27.22
C27H26C1201	437.41	436.1361	74.14	5.99	16.21	3.66
C27H26C1202	453.41	452.1310	71.52	5.78	15.64	7.06
C27H26C1203	469.41	468.1259	69.09	5.58	15.11	10.23
C27H26C1204	485.41	484.1208	66.81	5.40	14.61	13.18
C27H26C1205	501.41	500.1157	64.68	5.23	14.14	15.95
C27H26C1206	517.41	516.1106	62.68	5.07	13.70	18.55
C27H26C1207	533.41	532.1055	60.80	4.91	13.29	21.00
C27H26C1208	549.41	548.1005	59.03	4.77	12.91	23.30
C27H26C1209	565.41	564.0954	57.36	4.64	12.54	25.47
C27H27C1101	402.97	402.1750	80.48	6.75	8.80	3.97
C27H27C1102	418.97	418.1699	77.40	6.50	8.46	7.64
C27H27C1103	434.97	434.1649	74.56	6.26	8.15	11.03
C27H27C1104	450.97	450.1598	71.91	6.03	7.86	14.19
C27H27C1105	466.97	466.1547	69.45	5.83	7.59	17.13
C27H27C1106	482.97	482.1496	67.15	5.64	7.34	19.88
C27H27C1107	498.97	498.1445	64.99	5.45	7.11	22.45
C27H27C1108	514.96	514.1394	62.98	5.28	6.88	24.86
C27H27C1109	530.96	530.1343	61.08	5.13	6.68	27.12
C27H28C1201	439.43	438.1517	73.80	6.42	16.14	3.64
C27H28C1202	455.43	454.1466	71.21	6.20	15.57	7.03
C27H28C1203	471.43	470.1415	68.79	5.99	15.04	10.18
C27H28C1204	487.43	486.1365	66.53	5.79	14.55	13.13
C27H28C1205	503.43	502.1314	64.42	5.61	14.08	15.89
C27H28C1206	519.43	518.1263	62.43	5.43	13.65	18.48
C27H28C1207	535.43	534.1212	60.57	5.27	13.24	20.92
C27H28C1208	551.43	550.1161	58.81	5.12	12.86	23.21
C27H28C1209	567.42	566.1110	57.15	4.97	12.50	25.38
C27H29C1101	404.98	404.1907	80.08	7.22	8.75	3.95
C27H29C1102	420.98	420.1856	77.03	6.94	8.42	7.60
C27H29C1103	436.98	436.1805	74.21	6.69	8.11	10.98
C27H29C1104	452.98	452.1754	71.59	6.45	7.83	14.13
C27H29C1105	468.98	468.1703	69.15	6.23	7.56	17.06
C27H29C1106	484.98	484.1653	66.87	6.03	7.31	19.79
C27H29C1107	500.98	500.1602	64.73	5.83	7.08	22.36
C27H29C1108	516.98	516.1551	62.73	5.65	6.86	24.76
C27H29C1109	532.98	532.1500	60.85	5.48	6.65	27.02
C27H30C1201	441.45	440.1674	73.46	6.85	16.06	3.62
C27H30C1202	457.44	456.1623	70.89	6.61	15.50	7.00
C27H30C1203	473.44	472.1572	68.50	6.39	14.98	10.14
C27H30C1204	489.44	488.1521	66.26	6.18	14.49	13.08
C27H30C1205	505.44	504.1470	64.16	5.98	14.03	15.83
C27H30C1206	521.44	520.1419	62.19	5.80	13.60	18.41
C27H30C1207	537.44	536.1368	60.34	5.63	13.19	20.84
C27H30C1208	553.44	552.1318	58.60	5.46	12.81	23.13
C27H30C1209	569.44	568.1267	56.95	5.31	12.45	25.29
C27H31C1101	407.00	406.2063	79.68	7.68	8.71	3.93
C27H31C1102	423.00	422.2012	76.67	7.39	8.38	7.56
C27H31C1103	439.00	438.1962	73.87	7.12	8.08	10.93
C27H31C1104	455.00	454.1911	71.28	6.87	7.79	14.07
C27H31C1105	471.00	470.1860	68.85	6.63	7.53	16.98
C27H31C1106	487.00	486.1809	66.59	6.42	7.28	19.71
C27H31C1107	503.00	502.1758	64.47	6.21	7.05	22.27
C27H31C1108	519.00	518.1707	62.49	6.02	6.83	24.66
C27H31C1109	535.00	534.1656	60.62	5.84	6.63	26.92
C27H32C1201	443.46	442.1830	73.13	7.27	15.99	3.61
C27H32C1202	459.46	458.1779	70.58	7.02	15.43	6.96
C27H32C1203	475.46	474.1728	68.21	6.78	14.91	10.10
C27H32C1204	491.46	490.1677	65.99	6.56	14.43	13.02

C27H32C1205	507.46	506.1627	63.91	6.36	13.97	15.76
C27H32C1206	523.46	522.1576	61.95	6.16	13.55	18.34
C27H32C1207	539.46	538.1525	60.12	5.98	13.14	20.76
C27H32C1208	555.46	554.1474	58.38	5.81	12.77	23.04
C27H32C1209	571.46	570.1423	56.75	5.64	12.41	25.20
C27H33C1101	409.02	408.2220	79.29	8.13	8.67	3.91
C27H33C1102	425.02	424.2169	76.30	7.83	8.34	7.53
C27H33C1103	441.02	440.2118	73.54	7.54	8.04	10.88
C27H33C1104	457.01	456.2067	70.96	7.28	7.76	14.00
C27H33C1105	473.01	472.2016	68.56	7.03	7.50	16.91
C27H33C1106	489.01	488.1965	66.32	6.80	7.25	19.63
C27H33C1107	505.01	504.1915	64.22	6.59	7.02	22.18
C27H33C1108	521.01	520.1864	62.24	6.38	6.80	24.57
C27H33C1109	537.01	536.1813	60.39	6.19	6.60	26.81
C27H34C1201	445.48	444.1987	72.80	7.69	15.92	3.59
C27H34C1202	461.48	460.1936	70.27	7.43	15.37	6.93
C27H34C1203	477.48	476.1885	67.92	7.18	14.85	10.05
C27H34C1204	493.48	492.1834	65.72	6.94	14.37	12.97
C27H34C1205	509.48	508.1783	63.65	6.73	13.92	15.70
C27H34C1206	525.47	524.1732	61.72	6.52	13.49	18.27
C27H34C1207	541.47	540.1681	59.89	6.33	13.10	20.68
C27H34C1208	557.47	556.1631	58.17	6.15	12.72	22.96
C27H34C1209	573.47	572.1580	56.55	5.98	12.36	25.11
C27H35C1101	411.03	410.2376	78.90	8.58	8.63	3.89
C27H35C1102	427.03	426.2325	75.94	8.26	8.30	7.49
C27H35C1103	443.03	442.2275	73.20	7.96	8.00	10.83
C27H35C1104	459.03	458.2224	70.65	7.69	7.72	13.94
C27H35C1105	475.03	474.2173	68.27	7.43	7.46	16.84
C27H35C1106	491.03	490.2122	66.05	7.18	7.22	19.55
C27H35C1107	507.03	506.2071	63.96	6.96	6.99	22.09
C27H35C1108	523.03	522.2020	62.00	6.75	6.78	24.47
C27H35C1109	539.03	538.1969	60.16	6.54	6.58	26.71
C27H36C1201	447.49	446.2143	72.47	8.11	15.85	3.58
C27H36C1202	463.49	462.2092	69.97	7.83	15.30	6.90
C27H36C1203	479.49	478.2041	67.63	7.57	14.79	10.01
C27H36C1204	495.49	494.1990	65.45	7.32	14.31	12.92
C27H36C1205	511.49	510.1940	63.40	7.09	13.86	15.64
C27H36C1206	527.49	526.1889	61.48	6.88	13.44	18.20
C27H36C1207	543.49	542.1838	59.67	6.68	13.05	20.61
C27H36C1208	559.49	558.1787	57.96	6.49	12.67	22.88
C27H36C1209	575.49	574.1736	56.35	6.31	12.32	25.02
C27H37C1101	413.05	412.2533	78.51	9.03	8.58	3.87
C27H37C1102	429.05	428.2482	75.59	8.69	8.26	7.46
C27H37C1103	445.05	444.2431	72.87	8.38	7.97	10.78
C27H37C1104	461.05	460.2380	70.34	8.09	7.69	13.88
C27H37C1105	477.05	476.2329	67.98	7.82	7.43	16.77
C27H37C1106	493.05	492.2278	65.78	7.56	7.19	19.47
C27H37C1107	509.04	508.2228	63.71	7.33	6.96	22.00
C27H37C1108	525.04	524.2177	61.77	7.10	6.75	24.38
C27H37C1109	541.04	540.2126	59.94	6.89	6.55	26.61
C27H38C1201	449.51	448.2300	72.15	8.52	15.77	3.56
C27H38C1202	465.51	464.2249	69.67	8.23	15.23	6.87
C27H38C1203	481.51	480.2198	67.35	7.95	14.73	9.97
C27H38C1204	497.51	496.2147	65.19	7.70	14.25	12.86
C27H38C1205	513.51	512.2096	63.15	7.46	13.81	15.58
C27H38C1206	529.51	528.2045	61.25	7.23	13.39	18.13
C27H38C1207	545.51	544.1994	59.45	7.02	13.00	20.53
C27H38C1208	561.51	560.1944	57.76	6.82	12.63	22.80
C27H38C1209	577.50	576.1893	56.16	6.63	12.28	24.93
C27H39C1101	415.06	414.2689	78.13	9.47	8.54	3.85

C27H39C1102	431.06	430.2638	75.23	9.12	8.22	7.42
C27H39C1103	447.06	446.2588	72.54	8.79	7.93	10.74
C27H39C1104	463.06	462.2537	70.03	8.49	7.66	13.82
C27H39C1105	479.06	478.2486	67.70	8.21	7.40	16.70
C27H39C1106	495.06	494.2435	65.51	7.94	7.16	19.39
C27H39C1107	511.06	510.2384	63.46	7.69	6.94	21.91
C27H39C1108	527.06	526.2333	61.53	7.46	6.73	24.28
C27H39C1109	543.06	542.2282	59.72	7.24	6.53	26.52
C27H40C1201	451.53	450.2456	71.82	8.93	15.70	3.54
C27H40C1202	467.52	466.2405	69.37	8.62	15.17	6.84
C27H40C1203	483.52	482.2354	67.07	8.34	14.66	9.93
C27H40C1204	499.52	498.2303	64.92	8.07	14.19	12.81
C27H40C1205	515.52	514.2253	62.91	7.82	13.75	15.52
C27H40C1206	531.52	530.2202	61.01	7.59	13.34	18.06
C27H40C1207	547.52	546.2151	59.23	7.36	12.95	20.46
C27H40C1208	563.52	562.2100	57.55	7.15	12.58	22.71
C27H40C1209	579.52	578.2049	55.96	6.96	12.24	24.85
C27H41C1101	417.08	416.2846	77.76	9.91	8.50	3.84
C27H41C1102	433.08	432.2795	74.88	9.54	8.19	7.39
C27H41C1103	449.08	448.2744	72.21	9.20	7.89	10.69
C27H41C1104	465.08	464.2693	69.73	8.89	7.62	13.76
C27H41C1105	481.08	480.2642	67.41	8.59	7.37	16.63
C27H41C1106	497.08	496.2591	65.24	8.31	7.13	19.31
C27H41C1107	513.08	512.2541	63.21	8.05	6.91	21.83
C27H41C1108	529.08	528.2490	61.30	7.81	6.70	24.19
C27H41C1109	545.08	544.2439	59.50	7.58	6.50	26.42
C27H42C1201	453.54	452.2613	71.50	9.33	15.63	3.53
C27H42C1202	469.54	468.2562	69.07	9.02	15.10	6.81
C27H42C1203	485.54	484.2511	66.79	8.72	14.60	9.89
C27H42C1204	501.54	500.2460	64.66	8.44	14.14	12.76
C27H42C1205	517.54	516.2409	62.66	8.18	13.70	15.46
C27H42C1206	533.54	532.2358	60.78	7.93	13.29	17.99
C27H42C1207	549.54	548.2307	59.01	7.70	12.90	20.38
C27H42C1208	565.54	564.2257	57.34	7.49	12.54	22.63
C27H42C1209	581.54	580.2206	55.77	7.28	12.19	24.76
C27H43C1101	419.10	418.3002	77.38	10.34	8.46	3.82
C27H43C1102	435.10	434.2951	74.54	9.96	8.15	7.35
C27H43C1103	451.09	450.2901	71.89	9.61	7.86	10.64
C27H43C1104	467.09	466.2850	69.43	9.28	7.59	13.70
C27H43C1105	483.09	482.2799	67.13	8.97	7.34	16.56
C27H43C1106	499.09	498.2748	64.98	8.68	7.10	19.23
C27H43C1107	515.09	514.2697	62.96	8.41	6.88	21.74
C27H43C1108	531.09	530.2646	61.06	8.16	6.68	24.10
C27H43C1109	547.09	546.2595	59.28	7.92	6.48	26.32
C27H44C1201	455.56	454.2769	71.19	9.74	15.56	3.51
C27H44C1202	471.56	470.2718	68.77	9.41	15.04	6.79
C27H44C1203	487.56	486.2667	66.52	9.10	14.54	9.84
C27H44C1204	503.56	502.2616	64.40	8.81	14.08	12.71
C27H44C1205	519.55	518.2566	62.42	8.54	13.65	15.40
C27H44C1206	535.55	534.2515	60.55	8.28	13.24	17.92
C27H44C1207	551.55	550.2464	58.80	8.04	12.86	20.31
C27H44C1208	567.55	566.2413	57.14	7.81	12.49	22.55
C27H44C1209	583.55	582.2362	55.57	7.60	12.15	24.68
C27H45C1101	421.11	420.3159	77.01	10.77	8.42	3.80
C27H45C1102	437.11	436.3108	74.19	10.38	8.11	7.32
C27H45C1103	453.11	452.3057	71.57	10.01	7.82	10.59
C27H45C1104	469.11	468.3006	69.13	9.67	7.56	13.64
C27H45C1105	485.11	484.2955	66.85	9.35	7.31	16.49
C27H45C1106	501.11	500.2904	64.72	9.05	7.07	19.16
C27H45C1107	517.11	516.2854	62.71	8.77	6.86	21.66

C27H45C1108	533.11	532.2803	60.83	8.51	6.65	24.01
C27H45C1109	549.11	548.2752	59.06	8.26	6.46	26.22
C27H46C1201	457.57	456.2926	70.87	10.13	15.50	3.50
C27H46C1202	473.57	472.2875	68.48	9.79	14.97	6.76
C27H46C1203	489.57	488.2824	66.24	9.47	14.48	9.80
C27H46C1204	505.57	504.2773	64.15	9.17	14.02	12.66
C27H46C1205	521.57	520.2722	62.18	8.89	13.59	15.34
C27H46C1206	537.57	536.2671	60.33	8.63	13.19	17.86
C27H46C1207	553.57	552.2620	58.58	8.38	12.81	20.23
C27H46C1208	569.57	568.2570	56.94	8.14	12.45	22.47
C27H46C1209	585.57	584.2519	55.38	7.92	12.11	24.59
C27H47C1101	423.13	422.3315	76.64	11.20	8.38	3.78
C27H47C1102	439.13	438.3264	73.85	10.79	8.07	7.29
C27H47C1103	455.13	454.3214	71.26	10.41	7.79	10.55
C27H47C1104	471.13	470.3163	68.84	10.06	7.53	13.58
C27H47C1105	487.13	486.3112	66.57	9.73	7.28	16.42
C27H47C1106	503.13	502.3061	64.46	9.42	7.05	19.08
C27H47C1107	519.12	518.3010	62.47	9.13	6.83	21.57
C27H47C1108	535.12	534.2959	60.60	8.85	6.63	23.92
C27H47C1109	551.12	550.2908	58.84	8.60	6.43	26.13
C27H48C1201	459.59	458.3082	70.56	10.53	15.43	3.48
C27H48C1202	475.59	474.3031	68.19	10.17	14.91	6.73
C27H48C1203	491.59	490.2980	65.97	9.84	14.42	9.76
C27H48C1204	507.59	506.2929	63.89	9.53	13.97	12.61
C27H48C1205	523.59	522.2879	61.94	9.24	13.54	15.28
C27H48C1206	539.59	538.2828	60.10	8.97	13.14	17.79
C27H48C1207	555.59	554.2777	58.37	8.71	12.76	20.16
C27H48C1208	571.58	570.2726	56.74	8.46	12.41	22.39
C27H48C1209	587.58	586.2675	55.19	8.23	12.07	24.51
C27H49C1101	425.14	424.3472	76.28	11.62	8.34	3.76
C27H49C1102	441.14	440.3421	73.51	11.20	8.04	7.25
C27H49C1103	457.14	456.3370	70.94	10.80	7.76	10.50
C27H49C1104	473.14	472.3319	68.54	10.44	7.49	13.53
C27H49C1105	489.14	488.3268	66.30	10.10	7.25	16.35
C27H49C1106	505.14	504.3217	64.20	9.78	7.02	19.00
C27H49C1107	521.14	520.3167	62.23	9.48	6.80	21.49
C27H49C1108	537.14	536.3116	60.38	9.20	6.60	23.83
C27H49C1109	553.14	552.3065	58.63	8.93	6.41	26.03
C27H50C1201	461.60	460.3239	70.26	10.92	15.36	3.47
C27H50C1202	477.60	476.3188	67.90	10.55	14.85	6.70
C27H50C1203	493.60	492.3137	65.70	10.21	14.36	9.72
C27H50C1204	509.60	508.3086	63.64	9.89	13.91	12.56
C27H50C1205	525.60	524.3035	61.70	9.59	13.49	15.22
C27H50C1206	541.60	540.2984	59.88	9.31	13.09	17.72
C27H50C1207	557.60	556.2933	58.16	9.04	12.72	20.09
C27H50C1208	573.60	572.2882	56.54	8.79	12.36	22.31
C27H50C1209	589.60	588.2832	55.00	8.55	12.03	24.42
C27H51C1101	427.16	426.3628	75.92	12.03	8.30	3.75
C27H51C1102	443.16	442.3577	73.18	11.60	8.00	7.22
C27H51C1103	459.16	458.3527	70.63	11.20	7.72	10.45
C27H51C1104	475.16	474.3476	68.25	10.82	7.46	13.47
C27H51C1105	491.16	490.3425	66.03	10.47	7.22	16.29
C27H51C1106	507.16	506.3374	63.94	10.14	6.99	18.93
C27H51C1107	523.16	522.3323	61.99	9.83	6.78	21.41
C27H51C1108	539.16	538.3272	60.15	9.53	6.58	23.74
C27H51C1109	555.16	554.3221	58.42	9.26	6.39	25.94
C27H52C1201	463.62	462.3395	69.95	11.31	15.29	3.45
C27H52C1202	479.62	478.3344	67.62	10.93	14.78	6.67
C27H52C1203	.95.62	494.3293	65.43	10.58	14.31	9.68
C27H52C1204	511.62	510.3242	63.39	10.24	13.86	12.51

C27H52C1205	527.62	526.3192	61.47	9.93	13.44	15.16
C27H52C1206	543.62	542.3141	59.66	9.64	13.04	17.66
C27H52C1207	559.62	558.3090	57.95	9.37	12.67	20.01
C27H52C1208	575.62	574.3039	56.34	9.11	12.32	22.24
C27H52C1209	591.62	590.2988	54.82	8.86	11.99	24.34
C27H53C1101	429.18	428.3785	75.56	12.45	8.26	3.73
C27H53C1102	445.18	444.3734	72.85	12.00	7.96	7.19
C27H53C1103	461.17	460.3683	70.32	11.58	7.69	10.41
C27H53C1104	477.17	476.3632	67.96	11.20	7.43	13.41
C27H53C1105	493.17	492.3581	65.76	10.83	7.19	16.22
C27H53C1106	509.17	508.3530	63.69	10.49	6.96	18.85
C27H53C1107	525.17	524.3480	61.75	10.17	6.75	21.33
C27H53C1108	541.17	540.3429	59.93	9.87	6.55	23.65
C27H53C1109	557.17	556.3378	58.20	9.59	6.36	25.84
C27H54C1201	465.64	464.3551	69.65	11.69	15.23	3.44
C27H54C1202	481.64	480.3501	67.33	11.30	14.72	6.64
C27H54C1203	497.64	496.3450	65.17	10.94	14.25	9.65
C27H54C1204	513.64	512.3399	63.14	10.60	13.80	12.46
C27H54C1205	529.63	528.3348	61.23	10.28	13.39	15.10
C27H54C1206	545.63	544.3297	59.44	9.98	13.00	17.59
C27H54C1207	561.63	560.3246	57.74	9.69	12.62	19.94
C27H54C1208	577.63	576.3195	56.14	9.42	12.28	22.16
C27H54C1209	593.63	592.3145	54.63	9.17	11.94	24.26
C27H55C1101	431.19	430.3941	75.21	12.86	8.22	3.71
C27H55C1102	447.19	446.3890	72.52	12.40	7.93	7.16
C27H55C1103	463.19	462.3839	70.01	11.97	7.65	10.36
C27H55C1104	479.19	478.3789	67.68	11.57	7.40	13.36
C27H55C1105	495.19	494.3738	65.49	11.20	7.16	16.15
C27H55C1106	511.19	510.3687	63.44	10.84	6.94	18.78
C27H55C1107	527.19	526.3636	61.52	10.52	6.72	21.24
C27H55C1108	543.19	542.3585	59.70	10.21	6.53	23.56
C27H55C1109	559.19	558.3534	58.00	9.91	6.34	25.75
C28H14C1201	437.33	436.0422	76.90	3.23	16.21	3.66
C28H14C1202	453.33	452.0371	74.19	3.11	15.64	7.06
C28H14C1203	469.33	468.0320	71.66	3.01	15.11	10.23
C28H14C1204	485.33	484.0269	69.30	2.91	14.61	13.19
C28H14C1205	501.33	500.0218	67.08	2.81	14.14	15.96
C28H14C1206	517.33	516.0167	65.01	2.73	13.71	18.56
C28H14C1207	533.33	532.0116	63.06	2.65	13.30	21.00
C28H14C1208	549.32	548.0066	61.22	2.57	12.91	23.30
C28H14C1209	565.32	564.0015	59.49	2.50	12.54	25.47
C28H15C1101	402.88	402.0811	83.48	3.75	8.80	3.97
C28H15C1102	418.88	418.0761	80.29	3.61	8.46	7.64
C28H15C1103	434.88	434.0710	77.33	3.48	8.15	11.04
C28H15C1104	450.88	450.0659	74.59	3.35	7.86	14.19
C28H15C1105	466.88	466.0608	72.03	3.24	7.59	17.13
C28H15C1106	482.88	482.0557	69.65	3.13	7.34	19.88
C28H15C1107	498.88	498.0506	67.41	3.03	7.11	22.45
C28H15C1108	514.88	514.0455	65.32	2.94	6.89	24.86
C28H15C1109	530.88	530.0404	63.35	2.85	6.68	27.12
C28H16C1201	439.35	438.0578	76.55	3.67	16.14	3.64
C28H16C1202	455.34	454.0527	73.86	3.54	15.57	7.03
C28H16C1203	471.34	470.0476	71.35	3.42	15.04	10.18
C28H16C1204	487.34	486.0426	69.01	3.31	14.55	13.13
C28H16C1205	503.34	502.0375	66.82	3.20	14.09	15.89
C28H16C1206	519.34	518.0324	64.76	3.11	13.65	18.48
C28H16C1207	535.34	534.0273	62.82	3.01	13.25	20.92
C28H16C1208	551.34	550.0222	61.00	2.93	12.86	23.22
C28H16C1209	567.34	566.0171	59.28	2.84	12.50	25.38
C28H17C1101	404.90	404.0968	83.06	4.23	8.76	3.95

C28H17C1102	420.90	420.0917	79.90	4.07	8.42	7.60
C28H17C1103	436.90	436.0866	76.98	3.92	8.11	10.99
C28H17C1104	452.90	452.0815	74.26	3.78	7.83	14.13
C28H17C1105	468.90	468.0764	71.72	3.65	7.56	17.06
C28H17C1106	484.90	484.0714	69.36	3.53	7.31	19.80
C28H17C1107	500.90	500.0663	67.14	3.42	7.08	22.36
C28H17C1108	516.90	516.0612	65.06	3.32	6.86	24.76
C28H17C1109	532.90	532.0561	63.11	3.22	6.65	27.02
C28H18C1201	441.36	440.0735	76.20	4.11	16.07	3.63
C28H18C1202	457.36	456.0684	73.53	3.97	15.50	7.00
C28H18C1203	473.36	472.0633	71.05	3.83	14.98	10.14
C28H18C1204	489.36	488.0582	68.73	3.71	14.49	13.08
C28H18C1205	505.36	504.0531	66.55	3.59	14.03	15.83
C28H18C1206	521.36	520.0480	64.51	3.48	13.60	18.41
C28H18C1207	537.36	536.0429	62.59	3.38	13.20	20.84
C28H18C1208	553.36	552.0379	60.78	3.28	12.81	23.13
C28H18C1209	569.36	568.0328	59.07	3.19	12.45	25.29
C28H19C1101	406.92	406.1124	82.65	4.71	8.71	3.93
C28H19C1102	422.92	422.1073	79.52	4.53	8.38	7.57
C28H19C1103	438.91	438.1023	76.62	4.36	8.08	10.94
C28H19C1104	454.91	454.0972	73.93	4.21	7.79	14.07
C28H19C1105	470.91	470.0921	71.42	4.07	7.53	16.99
C28H19C1106	486.91	486.0870	69.07	3.93	7.28	19.72
C28H19C1107	502.91	502.0819	66.87	3.81	7.05	22.27
C28H19C1108	518.91	518.0768	64.81	3.69	6.83	24.67
C28H19C1109	534.91	534.0717	62.87	3.58	6.63	26.92
C28H20C1201	443.38	442.0891	75.85	4.55	15.99	3.61
C28H20C1202	459.38	458.0840	73.21	4.39	15.44	6.97
C28H20C1203	475.38	474.0789	70.75	4.24	14.92	10.10
C28H20C1204	491.38	490.0739	68.44	4.10	14.43	13.02
C28H20C1205	507.37	506.0688	66.28	3.97	13.98	15.77
C28H20C1206	523.37	522.0637	64.26	3.85	13.55	18.34
C28H20C1207	539.37	538.0586	62.35	3.74	13.15	20.76
C28H20C1208	555.37	554.0535	60.56	3.63	12.77	23.05
C28H20C1209	571.37	570.0484	58.86	3.53	12.41	25.20
C28H21C1101	408.93	408.1281	82.24	5.18	8.67	3.91
C28H21C1102	424.93	424.1230	79.15	4.98	8.34	7.53
C28H21C1103	440.93	440.1179	76.27	4.80	8.04	10.89
C28H21C1104	456.93	456.1128	73.60	4.63	7.76	14.01
C28H21C1105	472.93	472.1077	71.11	4.48	7.50	16.92
C28H21C1106	488.93	488.1027	68.79	4.33	7.25	19.63
C28H21C1107	504.93	504.0976	66.61	4.19	7.02	22.18
C28H21C1108	520.93	520.0925	64.56	4.06	6.81	24.57
C28H21C1109	536.93	536.0874	62.64	3.94	6.60	26.82
C28H22C1201	445.39	444.1048	75.51	4.98	15.92	3.59
C28H22C1202	461.39	460.0997	72.89	4.81	15.37	6.94
C28H22C1203	477.39	476.0946	70.45	4.65	14.85	10.05
C28H22C1204	493.39	492.0895	68.16	4.49	14.37	12.97
C28H22C1205	509.39	508.0844	66.02	4.35	13.92	15.70
C28H22C1206	525.39	524.0793	64.01	4.22	13.50	18.27
C28H22C1207	541.39	540.0742	62.12	4.10	13.10	20.69
C28H22C1208	557.39	556.0692	60.34	3.98	12.72	22.96
C28H22C1209	573.39	572.0641	58.65	3.87	12.37	25.11
C28H23C1101	410.95	410.1437	81.84	5.64	8.63	3.89
C28H23C1102	426.95	426.1386	78.77	5.43	8.30	7.49
C28H23C1103	442.95	442.1336	75.93	5.23	8.00	10.84
C28H23C1104	458.95	458.1285	73.28	5.05	7.72	13.94
C28H23C1105	474.95	474.1234	70.81	4.88	7.46	16.84
C28H23C1106	490.94	490.1183	68.50	4.72	7.22	19.55
C28H23C1107	506.94	506.1132	66.34	4.57	6.99	22.09

C28H23C1108	522.94	522.1081	64.31	4.43	6.78	24.48
C28H23C1109	538.94	538.1030	62.40	4.30	6.58	26.72
C28H24C1201	447.41	446.1204	75.17	5.41	15.85	3.58
C28H24C1202	463.41	462.1153	72.57	5.22	15.30	6.91
C28H24C1203	479.41	478.1102	70.15	5.05	14.79	10.01
C28H24C1204	495.41	494.1052	67.85	4.88	14.31	12.92
C28H24C1205	511.41	510.1001	65.76	4.73	13.86	15.64
C28H24C1206	527.41	526.0950	63.77	4.59	13.44	18.20
C28H24C1207	543.41	542.0899	61.89	4.45	13.05	20.61
C28H24C1208	559.40	558.0848	60.12	4.32	12.68	22.88
C28H24C1209	575.40	574.0797	58.45	4.20	12.32	25.02
C28H25C1101	412.96	412.1594	81.44	6.10	8.59	3.87
C28H25C1102	428.96	428.1543	78.40	5.87	8.26	7.46
C28H25C1103	444.96	444.1492	75.58	5.66	7.97	10.79
C28H25C1104	460.96	460.1441	72.96	5.47	7.69	13.88
C28H25C1105	476.96	476.1390	70.51	5.28	7.43	16.77
C28H25C1106	492.96	492.1340	68.22	5.11	7.19	19.47
C28H25C1107	508.96	508.1289	66.08	4.95	6.97	22.00
C28H25C1108	524.96	524.1238	64.06	4.80	6.75	24.38
C28H25C1109	540.96	540.1187	62.17	4.66	6.55	26.62
C28H26C1201	449.42	448.1361	74.83	5.83	15.78	3.56
C28H26C1202	465.42	464.1310	72.26	5.63	15.23	6.88
C28H26C1203	481.42	480.1259	69.86	5.44	14.73	9.97
C28H26C1204	497.42	496.1208	67.61	5.27	14.25	12.87
C28H26C1205	513.42	512.1157	65.50	5.10	13.81	15.58
C28H26C1206	529.42	528.1106	63.52	4.95	13.39	18.13
C28H26C1207	545.42	544.1055	61.66	4.80	13.00	20.53
C28H26C1208	561.42	560.1005	59.90	4.67	12.63	22.80
C28H26C1209	577.42	576.0954	58.24	4.54	12.28	24.94
C28H27C1101	414.98	414.1750	81.04	6.56	8.54	3.86
C28H27C1102	430.98	430.1699	78.03	6.31	8.23	7.42
C28H27C1103	446.98	446.1649	75.24	6.09	7.93	10.74
C28H27C1104	462.98	462.1598	72.64	5.88	7.66	13.82
C28H27C1105	478.98	478.1547	70.21	5.68	7.40	16.70
C28H27C1106	494.98	494.1496	67.95	5.50	7.16	19.39
C28H27C1107	510.98	510.1445	65.82	5.33	6.94	21.92
C28H27C1108	526.98	526.1394	63.82	5.16	6.73	24.29
C28H27C1109	542.97	542.1343	61.94	5.01	6.53	26.52
C28H28C1201	451.44	450.1517	74.50	6.25	15.71	3.54
C28H28C1202	467.44	466.1466	71.95	6.04	15.17	6.85
C28H28C1203	483.44	482.1415	69.57	5.84	14.67	9.93
C28H28C1204	499.44	498.1365	67.34	5.65	14.20	12.81
C28H28C1205	515.44	514.1314	65.25	5.48	13.76	15.52
C28H28C1206	531.44	530.1263	63.28	5.31	13.34	18.06
C28H28C1207	547.44	546.1212	61.43	5.16	12.95	20.46
C28H28C1208	563.44	562.1161	59.69	5.01	12.58	22.72
C28H28C1209	579.44	578.1110	58.04	4.87	12.24	24.85
C28H29C1101	417.00	416.1907	80.65	7.01	8.50	3.84
C28H29C1102	433.00	432.1856	77.67	6.75	8.19	7.39
C28H29C1103	448.99	448.1805	74.90	6.51	7.90	10.69
C28H29C1104	464.99	464.1754	72.33	6.29	7.62	13.76
C28H29C1105	480.99	480.1703	69.92	6.08	7.37	16.63
C28H29C1106	496.99	496.1653	67.67	5.88	7.13	19.32
C28H29C1107	512.99	512.1602	65.56	5.70	6.91	21.83
C28H29C1108	528.99	528.1551	63.58	5.53	6.70	24.20
C28H29C1109	544.99	544.1500	61.71	5.36	6.51	26.42
C28H30C1201	453.46	452.1674	74.17	6.67	15.64	3.53
C28H30C1202	469.46	468.1623	71.64	6.44	15.10	6.82
C28H30C1203	485.46	484.1572	69.28	6.23	14.61	9.89
C28H30C1204	501.45	500.1521	67.07	6.03	14.14	12.76
C28H30C1205	517.45	516.1470	64.99	5.84	13.70	15.46
C28H30C1206	533.45	532.1419	63.04	5.67	13.29	18.00
C28H30C1207	549.45	548.1368	61.21	5.50	12.90	20.38
C28H30C1208	565.45	564.1318	59.48	5.35	12.54	22.64
C28H30C1209	581.45	580.1267	57.84	5.20	12.19	24.76
C28H31C1101	419.01	418.2063	80.26	7.46	8.46	3.82
C28H31C1102	435.01	434.2012	77.31	7.18	8.15	7.36
C28H31C1103	451.01	450.1962	74.57	6.93	7.86	10.64
C28H31C1104	467.01	466.1911	72.01	6.69	7.59	13.70
C28H31C1105	483.01	482.1860	69.63	6.47	7.34	16.56
C28H31C1106	499.01	498.1809	67.40	6.26	7.10	19.24
C28H31C1107	515.01	514.1758	65.30	6.07	6.88	21.75
C28H31C1108	531.01	530.1707	63.33	5.88	6.68	24.10
C28H31C1109	547.01	546.1656	61.48	5.71	6.48	26.32
C28H32C1201	455.47	454.1830	73.84	7.08	15.57	3.51
C28H32C1202	471.47	470.1779	71.33	6.84	15.04	6.79
C28H32C1203	487.47	486.1728	68.99	6.62	14.55	9.85
C28H32C1204	503.47	502.1677	66.80	6.41	14.08	12.71
C28H32C1205	519.47	518.1627	64.74	6.21	13.65	15.40
C28H32C1206	535.47	534.1576	62.81	6.02	13.24	17.93
C28H32C1207	551.47	550.1525	60.98	5.85	12.86	20.31
C28H32C1208	567.47	566.1474	59.27	5.68	12.50	22.56
C28H32C1209	583.47	582.1423	57.64	5.53	12.15	24.68
C28H33C1101	421.03	420.2220	79.88	7.90	8.42	3.80
C28H33C1102	437.03	436.2169	76.95	7.61	8.11	7.32
C28H33C1103	453.03	452.2118	74.24	7.34	7.83	10.60
C28H33C1104	469.03	468.2067	71.70	7.09	7.56	13.64
C28H33C1105	485.03	484.2016	69.34	6.86	7.31	16.49
C28H33C1106	501.02	500.1965	67.12	6.64	7.08	19.16
C28H33C1107	517.02	516.1915	65.05	6.43	6.86	21.66
C28H33C1108	533.02	532.1864	63.10	6.24	6.65	24.01
C28H33C1109	549.02	548.1813	61.26	6.06	6.46	26.23
C28H34C1201	457.49	456.1987	73.51	7.49	15.50	3.50
C28H34C1202	473.49	472.1936	71.03	7.24	14.98	6.76
C28H34C1203	489.49	488.1885	68.71	7.00	14.49	9.81
C28H34C1204	505.49	504.1834	66.53	6.78	14.03	12.66
C28H34C1205	521.49	520.1783	64.49	6.57	13.60	15.34
C28H34C1206	537.49	536.1732	62.57	6.38	13.19	17.86
C28H34C1207	553.48	552.1681	60.76	6.19	12.81	20.23
C28H34C1208	569.48	568.1631	59.06	6.02	12.45	22.48
C28H34C1209	585.48	584.1580	57.44	5.85	12.11	24.59
C28H35C1101	423.04	422.2376	79.50	8.34	8.38	3.78
C28H35C1102	439.04	438.2325	76.60	8.04	8.08	7.29
C28H35C1103	455.04	454.2275	73.91	7.75	7.79	10.55
C28H35C1104	471.04	470.2224	71.40	7.49	7.53	13.59
C28H35C1105	487.04	486.2173	69.05	7.24	7.28	16.43
C28H35C1106	503.04	502.2122	66.86	7.01	7.05	19.08
C28H35C1107	519.04	518.2071	64.80	6.80	6.83	21.58
C28H35C1108	535.04	534.2020	62.86	6.59	6.63	23.92
C28H35C1109	551.04	550.1969	61.03	6.40	6.43	26.13
C28H36C1201	459.50	458.2143	73.19	7.90	15.43	3.48
C28H36C1202	475.50	474.2092	70.73	7.63	14.91	6.73
C28H36C1203	491.50	490.2041	68.43	7.38	14.43	9.77
C28H36C1204	507.50	506.1990	66.27	7.15	13.97	12.61
C28H36C1205	523.50	522.1940	64.24	6.93	13.54	15.28
C28H36C1206	539.50	538.1889	62.34	6.73	13.14	17.79
C28H36C1207	555.50	554.1838	60.54	6.53	12.76	20.16
C28H36C1208	571.50	570.1787	58.85	6.35	12.41	22.40
C28H36C1209	587.50	586.1736	57.24	6.18	12.07	24.51
C28H37C1101	425.06	424.2533	79.12	8.77	8.34	3.76

C28H37C1102	441.06	440.2482	76.25	8.46	8.04	7.25
C28H37C1103	457.06	456.2431	73.58	8.16	7.76	10.50
C28H37C1104	473.06	472.2380	71.09	7.88	7.49	13.53
C28H37C1105	489.06	488.2329	68.77	7.63	7.25	16.36
C28H37C1106	505.06	504.2278	66.59	7.38	7.02	19.01
C28H37C1107	521.06	520.2228	64.54	7.16	6.80	21.49
C28H37C1108	537.06	536.2177	62.62	6.94	6.60	23.83
C28H37C1109	553.05	552.2126	60.81	6.74	6.41	26.04
C28H38C1201	461.52	460.2300	72.87	8.30	15.36	3.47
C28H38C1202	477.52	476.2249	70.43	8.02	14.85	6.70
C28H38C1203	493.52	492.2198	68.15	7.76	14.37	9.73
C28H38C1204	509.52	508.2147	66.01	7.52	13.92	12.56
C28H38C1205	525.52	524.2096	64.00	7.29	13.49	15.22
C28H38C1206	541.52	540.2045	62.11	7.07	13.09	17.73
C28H38C1207	557.52	556.1994	60.32	6.87	12.72	20.09
C28H38C1208	573.52	572.1944	58.64	6.68	12.36	22.32
C28H38C1209	589.52	588.1893	57.05	6.50	12.03	24.43
C28H39C1101	427.08	426.2689	78.75	9.20	8.30	3.75
C28H39C1102	443.07	442.2638	75.90	8.87	8.00	7.22
C28H39C1103	459.07	458.2588	73.26	8.56	7.72	10.46
C28H39C1104	475.07	474.2537	70.79	8.27	7.46	13.47
C28H39C1105	491.07	490.2486	68.49	8.01	7.22	16.29
C28H39C1106	507.07	506.2435	66.32	7.75	6.99	18.93
C28H39C1107	523.07	522.2384	64.30	7.52	6.78	21.41
C28H39C1108	539.07	538.2333	62.39	7.29	6.58	23.74
C28H39C1109	555.07	554.2282	60.59	7.08	6.39	25.94
C28H40C1201	463.54	462.2456	72.55	8.70	15.30	3.45
C28H40C1202	479.54	478.2405	70.13	8.41	14.79	6.67
C28H40C1203	495.54	494.2354	67.87	8.14	14.31	9.69
C28H40C1204	511.53	510.2303	65.75	7.88	13.86	12.51
C28H40C1205	527.53	526.2253	63.75	7.64	13.44	15.16
C28H40C1206	543.53	542.2202	61.88	7.42	13.05	17.66
C28H40C1207	559.53	558.2151	60.11	7.21	12.67	20.02
C28H40C1208	575.53	574.2100	58.43	7.01	12.32	22.24
C28H40C1209	591.53	590.2049	56.85	6.82	11.99	24.34
C28H41C1101	429.09	428.2846	78.38	9.63	8.26	3.73
C28H41C1102	445.09	444.2795	75.56	9.29	7.97	7.19
C28H41C1103	461.09	460.2744	72.94	8.96	7.69	10.41
C28H41C1104	477.09	476.2693	70.49	8.66	7.43	13.41
C28H41C1105	493.09	492.2642	68.21	8.38	7.19	16.22
C28H41C1106	509.09	508.2591	66.06	8.12	6.96	18.86
C28H41C1107	525.09	524.2541	64.05	7.87	6.75	21.33
C28H41C1108	541.09	540.2490	62.15	7.64	6.55	23.66
C28H41C1109	557.09	556.2439	60.37	7.42	6.36	25.85
C28H42C1201	465.55	464.2613	72.24	9.09	15.23	3.44
C28H42C1202	481.55	480.2562	69.84	8.79	14.72	6.64
C28H42C1203	497.55	496.2511	67.59	8.51	14.25	9.65
C28H42C1204	513.55	512.2460	65.49	8.24	13.81	12.46
C28H42C1205	529.55	528.2409	63.51	7.99	13.39	15.11
C28H42C1206	545.55	544.2358	61.65	7.76	13.00	17.60
C28H42C1207	561.55	560.2307	59.89	7.54	12.63	19.94
C28H42C1208	577.55	576.2257	58.23	7.33	12.28	22.16
C28H42C1209	593.55	592.2206	56.66	7.13	11.95	24.26
C28H43C1101	431.11	430.3002	78.01	10.05	8.22	3.71
C28H43C1102	447.11	446.2951	75.22	9.69	7.93	7.16
C28H43C1103	463.11	462.2901	72.62	9.36	7.66	10.36
C28H43C1104	479.11	478.2850	70.20	9.05	7.40	13.36
C28H43C1105	495.10	494.2799	67.93	8.75	7.16	16.16
C28H43C1106	511.10	510.2748	65.80	8.48	6.94	18.78
C28H43C1107	527.10	526.2697	63.80	8.22	6.73	21.25
C28H43C1108	543.10	542.2646	61.92	7.98	6.53	23.57
C28H43C1109	559.10	558.2595	60.15	7.75	6.34	25.75
C28H44C1201	467.57	466.2769	71.93	9.49	15.16	3.42
C28H44C1202	483.57	482.2718	69.55	9.17	14.66	6.62
C28H44C1203	499.57	498.2667	67.32	8.88	14.19	9.61
C28H44C1204	515.57	514.2616	65.23	8.60	13.75	12.41
C28H44C1205	531.57	530.2566	63.27	8.34	13.34	15.05
C28H44C1206	547.57	546.2515	61.42	8.10	12.95	17.53
C28H44C1207	563.56	562.2464	59.68	7.87	12.58	19.87
C28H44C1208	579.56	578.2413	58.03	7.65	12.23	22.08
C28H44C1209	595.56	594.2362	56.47	7.45	11.91	24.18
C28H45C1101	433.12	432.3159	77.65	10.47	8.19	3.69
C28H45C1102	449.12	448.3108	74.88	10.10	7.89	7.12
C28H45C1103	465.12	464.3057	72.31	9.75	7.62	10.32
C28H45C1104	481.12	480.3006	69.90	9.43	7.37	13.30
C28H45C1105	497.12	496.2955	67.65	9.12	7.13	16.09
C28H45C1106	513.12	512.2904	65.54	8.84	6.91	18.71
C28H45C1107	529.12	528.2854	63.56	8.57	6.70	21.17
C28H45C1108	545.12	544.2803	61.70	8.32	6.50	23.48
C28H45C1109	561.12	560.2752	59.94	8.08	6.32	25.66
C28H46C1201	469.58	468.2926	71.62	9.87	15.10	3.41
C28H46C1202	485.58	484.2875	69.26	9.55	14.60	6.59
C28H46C1203	501.58	500.2824	67.05	9.24	14.14	9.57
C28H46C1204	517.58	516.2773	64.98	8.96	13.70	12.36
C28H46C1205	533.58	532.2722	63.03	8.69	13.29	14.99
C28H46C1206	549.58	548.2671	61.19	8.44	12.90	17.47
C28H46C1207	565.58	564.2620	59.46	8.20	12.54	19.80
C28H46C1208	581.58	580.2570	57.83	7.97	12.19	22.01
C28H46C1209	597.58	596.2519	56.28	7.76	11.87	24.10
C28H47C1101	435.14	434.3315	77.29	10.89	8.15	3.68
C28H47C1102	451.14	450.3264	74.55	10.50	7.86	7.09
C28H47C1103	467.14	466.3214	71.99	10.14	7.59	10.27
C28H47C1104	483.14	482.3163	69.61	9.81	7.34	13.25
C28H47C1105	499.14	498.3112	67.38	9.49	7.10	16.03
C28H47C1106	515.14	514.3061	65.29	9.20	6.88	18.64
C28H47C1107	531.14	530.3010	63.32	8.92	6.67	21.09
C28H47C1108	547.13	546.2959	61.47	8.66	6.48	23.39
C28H47C1109	563.13	562.2908	59.72	8.41	6.30	25.57
C28H48C1201	471.60	470.3082	71.31	10.26	15.04	3.39
C28H48C1202	487.60	486.3031	68.97	9.92	14.54	6.56
C28H48C1203	503.60	502.2980	66.78	9.61	14.08	9.53
C28H48C1204	519.60	518.2929	64.73	9.31	13.65	12.32
C28H48C1205	535.60	534.2879	62.79	9.03	13.24	14.94
C28H48C1206	551.60	550.2828	60.97	8.77	12.85	17.40
C28H48C1207	567.60	566.2777	59.25	8.52	12.49	19.73
C28H48C1208	583.60	582.2726	57.63	8.29	12.15	21.93
C28H48C1209	599.60	598.2675	56.09	8.07	11.83	24.02
C28H49C1101	437.16	436.3472	76.93	11.30	8.11	3.66
C28H49C1102	453.15	452.3421	74.22	10.90	7.82	7.06
C28H49C1103	469.15	468.3370	71.68	10.53	7.56	10.23
C28H49C1104	485.15	484.3319	69.32	10.18	7.31	13.19
C28H49C1105	501.15	500.3268	67.11	9.86	7.07	15.96
C28H49C1106	517.15	516.3217	65.03	9.55	6.86	18.56
C28H49C1107	533.15	532.3167	63.08	9.26	6.65	21.01
C28H49C1108	549.15	548.3116	61.24	8.99	6.46	23.31
C28H49C1109	565.15	564.3065	59.51	8.74	6.27	25.48
C28H50C1201	473.62	472.3239	71.01	10.64	14.97	3.38
C28H50C1202	489.62	488.3188	68.69	10.29	14.48	6.54
C28H50C1203	505.61	504.3137	66.52	9.97	14.02	9.49
C28H50C1204	521.61	520.3086	64.48	9.66	13.59	12.27

117

C28H50C1205	537.61	536.3035	62.56	9.37	13.19	14.88
C28H50C1206	553.61	552.2984	60.75	9.10	12.81	17.34
C28H50C1207	569.61	568.2933	59.04	8.85	12.45	19.66
C28H50C1208	585.61	584.2882	57.43	8.61	12.11	21.86
C28H50C1209	601.61	600.2832	55.90	8.38	11.79	23.93
C28H51C1101	439.17	438.3628	76.58	11.71	8.07	3.64
C28H51C1102	455.17	454.3577	73.89	11.29	7.79	7.03
C28H51C1103	471.17	470.3527	71.38	10.91	7.52	10.19
C28H51C1104	487.17	486.3476	69.03	10.55	7.28	13.14
C28H51C1105	503.17	502.3425	66.84	10.22	7.05	15.90
C28H51C1106	519.17	518.3374	64.78	9.90	6.83	18.49
C28H51C1107	535.17	534.3323	62.84	9.61	6.62	20.93
C28H51C1108	551.17	550.3272	61.02	9.33	6.43	23.22
C28H51C1109	567.17	566.3221	59.30	9.06	6.25	25.39
C28H52C1201	475.63	474.3395	70.71	11.02	14.91	3.36
C28H52C1202	491.63	490.3344	68.41	10.66	14.42	6.51
C28H52C1203	507.63	506.3293	66.25	10.33	13.97	9.46
C28H52C1204	523.63	522.3242	64.23	10.01	13.54	12.22
C28H52C1205	539.63	538.3192	62.32	9.71	13.14	14.82
C28H52C1206	555.63	554.3141	60.53	9.43	12.76	17.28
C28H52C1207	571.63	570.3090	58.83	9.17	12.40	19.59
C28H52C1208	587.63	586.3039	57.23	8.92	12.07	21.78
C28H52C1209	603.63	602.2988	55.72	8.68	11.75	23.85
C28H53C1101	441.19	440.3785	76.23	12.11	8.04	3.63
C28H53C1102	457.19	456.3734	73.56	11.69	7.75	7.00
C28H53C1103	473.19	472.3683	71.07	11.29	7.49	10.14
C28H53C1104	489.19	488.3632	68.75	10.92	7.25	13.08
C28H53C1105	505.18	504.3581	66.57	10.57	7.02	15.84
C28H53C1106	521.18	520.3530	64.53	10.25	6.80	18.42
C28H53C1107	537.18	536.3480	62.61	9.94	6.60	20.85
C28H53C1108	553.18	552.3429	60.80	9.66	6.41	23.14
C28H53C1109	569.18	568.3378	59.09	9.39	6.23	25.30
C28H54C1201	477.65	476.3551	70.41	11.40	14.84	3.35
C28H54C1202	493.65	492.3501	68.13	11.03	14.36	6.48
C28H54C1203	509.65	508.3450	65.99	10.68	13.91	9.42
C28H54C1204	525.65	524.3399	63.98	10.35	13.49	12.18
C28H54C1205	541.65	540.3348	62.09	10.05	13.09	14.77
C28H54C1206	557.64	556.3297	60.31	9.76	12.72	17.21
C28H54C1207	573.64	572.3246	58.63	9.49	12.36	19.52
C28H54C1208	589.64	588.3195	57.04	9.23	12.03	21.71
C28H54C1209	605.64	604.3145	55.53	8.99	11.71	23.78
C28H55C1101	443.20	442.3941	75.88	12.51	8.00	3.61
C28H55C1102	459.20	458.3890	73.24	12.07	7.72	6.97
C28H55C1103	475.20	474.3839	70.77	11.67	7.46	10.10
C28H55C1104	491.20	490.3789	68.47	11.29	7.22	13.03
C28H55C1105	507.20	506.3738	66.31	10.93	6.99	15.77
C28H55C1106	523.20	522.3687	64.28	10.60	6.78	18.35
C28H55C1107	539.20	538.3636	62.37	10.28	6.58	20.77
C28H55C1108	555.20	554.3585	60.58	9.99	6.39	23.05
C28H55C1109	571.20	570.3534	58.88	9.71	6.21	25.21
C28H56C1201	479.66	478.3708	70.11	11.77	14.78	3.34
C28H56C1202	495.66	494.3657	67.85	11.39	14.31	6.46
C28H56C1203	511.66	510.3606	65.73	11.03	13.86	9.38
C28H56C1204	527.66	526.3555	63.74	10.70	13.44	12.13
C28H56C1205	543.66	542.3505	61.86	10.38	13.04	14.71
C28H56C1206	559.66	558.3454	60.09	10.09	12.67	17.15
C28H56C1207	575.66	574.3403	58.42	9.81	12.32	19.46
C28H56C1208	591.66	590.3352	56.84	9.54	11.98	21.63
C28H56C1209	607.66	606.3301	55.35	9.29	11.67	23.70
C28H57C1101	445.22	444.4098	75.54	12.90	7.96	3.59
C28H57C1102	461.22	460.4047	72.92	12.46	7.69	6.94
C28H57C1103	477.22	476.3996	70.47	12.04	7.43	10.06
C28H57C1104	493.22	492.3945	68.19	11.65	7.19	12.98
C28H57C1105	509.22	508.3894	66.05	11.28	6.96	15.71
C28H57C1106	525.22	524.3843	64.03	10.94	6.75	18.28
C28H57C1107	541.22	540.3793	62.14	10.62	6.55	20.69
C28H57C1108	557.21	556.3742	60.36	10.31	6.36	22.97
C28H57C1109	573.21	572.3691	58.67	10.02	6.18	25.12
C29H14C1201	449.34	448.0422	77.52	3.14	15.78	3.56
C29H14C1202	465.34	464.0371	74.85	3.03	15.24	6.88
C29H14C1203	481.34	480.0320	72.37	2.93	14.73	9.97
C29H14C1204	497.34	496.0269	70.04	2.84	14.26	12.87
C29H14C1205	513.34	512.0218	67.85	2.75	13.81	15.58
C29H14C1206	529.34	528.0167	65.80	2.67	13.40	18.14
C29H14C1207	545.34	544.0116	63.87	2.59	13.00	20.54
C29H14C1208	561.34	560.0066	62.05	2.51	12.63	22.80
C29H14C1209	577.34	576.0015	60.33	2.44	12.28	24.94
C29H15C1101	414.90	414.0811	83.95	3.64	8.55	3.86
C29H15C1102	430.89	430.0761	80.84	3.51	8.23	7.43
C29H15C1103	446.89	446.0710	77.94	3.38	7.93	10.74
C29H15C1104	462.89	462.0659	75.25	3.27	7.66	13.83
C29H15C1105	478.89	478.0608	72.74	3.16	7.40	16.70
C29H15C1106	494.89	494.0557	70.38	3.06	7.16	19.40
C29H15C1107	510.89	510.0506	68.18	2.96	6.94	21.92
C29H15C1108	526.89	526.0455	66.11	2.87	6.73	24.29
C29H15C1109	542.89	542.0404	64.16	2.79	6.53	26.52
C29H16C1201	451.36	450.0578	77.17	3.57	15.71	3.54
C29H16C1202	467.36	466.0527	74.53	3.45	15.17	6.85
C29H16C1203	483.36	482.0476	72.06	3.34	14.67	9.93
C29H16C1204	499.35	498.0426	69.75	3.23	14.20	12.82
C29H16C1205	515.35	514.0375	67.59	3.13	13.76	15.52
C29H16C1206	531.35	530.0324	65.55	3.04	13.34	18.07
C29H16C1207	547.35	546.0273	63.64	2.95	12.95	20.46
C29H16C1208	563.35	562.0222	61.83	2.86	12.59	22.72
C29H16C1209	579.35	578.0171	60.12	2.78	12.24	24.85
C29H17C1101	416.91	416.0968	83.55	4.11	8.50	3.84
C29H17C1102	432.91	432.0917	80.46	3.96	8.19	7.39
C29H17C1103	448.91	448.0866	77.59	3.82	7.90	10.69
C29H17C1104	464.91	464.0815	74.92	3.69	7.63	13.77
C29H17C1105	480.91	480.0764	72.43	3.56	7.37	16.63
C29H17C1106	496.91	496.0714	70.10	3.45	7.13	19.32
C29H17C1107	512.91	512.0663	67.91	3.34	6.91	21.84
C29H17C1108	528.91	528.0612	65.86	3.24	6.70	24.20
C29H17C1109	544.91	544.0561	63.92	3.14	6.51	26.43
C29H18C1201	453.37	452.0735	76.83	4.00	15.64	3.53
C29H18C1202	469.37	468.0684	74.21	3.87	15.11	6.82
C29H18C1203	485.37	484.0633	71.76	3.74	14.61	9.89
C29H18C1204	501.37	500.0582	69.47	3.62	14.14	12.76
C29H18C1205	517.37	516.0531	67.33	3.51	13.71	15.46
C29H18C1206	533.37	532.0480	65.31	3.40	13.29	18.00
C29H18C1207	549.37	548.0429	63.40	3.30	12.91	20.39
C29H18C1208	565.37	564.0379	61.61	3.21	12.54	22.64
C29H18C1209	581.37	580.0328	59.91	3.12	12.20	24.77
C29H19C1101	418.93	418.1124	83.15	4.57	8.46	3.82
C29H19C1102	434.93	434.1073	80.09	4.40	8.15	7.36
C29H19C1103	450.93	450.1023	77.25	4.25	7.86	10.64
C29H19C1104	466.93	466.0972	74.60	4.10	7.59	13.71
C29H19C1105	482.92	482.0921	72.13	3.97	7.34	16.57
C29H19C1106	498.92	498.0870	69.81	3.84	7.11	19.24
C29H19C1107	514.92	514.0819	67.65	3.72	6.89	21.75

118

C29H19C1108	530.92	530.0768	65.61	3.61	6.68	24.11
C29H19C1109	546.92	546.0717	63.69	3.50	6.48	26.33
C29H20C1201	455.39	454.0891	76.49	4.43	15.57	3.51
C29H20C1202	471.39	470.0840	73.89	4.28	15.04	6.79
C29H20C1203	487.39	486.0789	71.47	4.14	14.55	9.85
C29H20C1204	503.39	502.0739	69.20	4.00	14.09	12.71
C29H20C1205	519.39	518.0688	67.06	3.88	13.65	15.40
C29H20C1206	535.39	534.0637	65.06	3.77	13.24	17.93
C29H20C1207	551.38	550.0586	63.17	3.66	12.86	20.31
C29H20C1208	567.38	566.0535	61.39	3.55	12.50	22.56
C29H20C1209	583.38	582.0484	59.71	3.46	12.15	24.68
C29H21C1101	420.94	420.1281	82.75	5.03	8.42	3.80
C29H21C1102	436.94	436.1230	79.72	4.84	8.11	7.32
C29H21C1103	452.94	452.1179	76.90	4.67	7.83	10.60
C29H21C1104	468.94	468.1128	74.28	4.51	7.56	13.65
C29H21C1105	484.94	484.1077	71.83	4.36	7.31	16.50
C29H21C1106	500.94	500.1027	69.53	4.23	7.08	19.16
C29H21C1107	516.94	516.0976	67.38	4.09	6.86	21.67
C29H21C1108	532.94	532.0925	65.36	3.97	6.65	24.02
C29H21C1109	548.94	548.0874	63.45	3.86	6.46	26.23
C29H22C1201	457.40	456.1048	76.15	4.85	15.50	3.50
C29H22C1202	473.40	472.0997	73.58	4.68	14.98	6.76
C29H22C1203	489.40	488.0946	71.17	4.53	14.49	9.81
C29H22C1204	505.40	504.0895	68.92	4.39	14.03	12.66
C29H22C1205	521.40	520.0844	66.81	4.25	13.60	15.34
C29H22C1206	537.40	536.0793	64.82	4.13	13.19	17.86
C29H22C1207	553.40	552.0742	62.94	4.01	12.81	20.24
C29H22C1208	569.40	568.0692	61.17	3.89	12.45	22.48
C29H22C1209	585.40	584.0641	59.50	3.79	12.11	24.60
C29H23C1101	422.96	422.1437	82.35	5.48	8.38	3.78
C29H23C1102	438.96	438.1386	79.35	5.28	8.08	7.29
C29H23C1103	454.96	454.1336	76.56	5.10	7.79	10.55
C29H23C1104	470.96	470.1285	73.96	4.92	7.53	13.59
C29H23C1105	486.96	486.1234	71.53	4.76	7.28	16.43
C29H23C1106	502.96	502.1183	69.26	4.61	7.05	19.09
C29H23C1107	518.96	518.1132	67.12	4.47	6.83	21.58
C29H23C1108	534.95	534.1081	65.11	4.33	6.63	23.93
C29H23C1109	550.95	550.1030	63.22	4.21	6.43	26.14
C29H24C1201	459.42	458.1204	75.82	5.27	15.43	3.48
C29H24C1202	475.42	474.1153	73.27	5.09	14.91	6.73
C29H24C1203	491.42	490.1102	70.88	4.92	14.43	9.77
C29H24C1204	507.42	506.1052	68.65	4.77	13.97	12.61
C29H24C1205	523.42	522.1001	66.55	4.62	13.55	15.28
C29H24C1206	539.42	538.0950	64.57	4.48	13.14	17.80
C29H24C1207	555.42	554.0899	62.71	4.36	12.77	20.16
C29H24C1208	571.42	570.0848	60.96	4.23	12.41	22.40
C29H24C1209	587.42	586.0797	59.30	4.12	12.07	24.51
C29H25C1101	424.97	424.1594	81.96	5.93	8.34	3.76
C29H25C1102	440.97	440.1543	78.99	5.71	8.04	7.26
C29H25C1103	456.97	456.1492	76.22	5.51	7.76	10.50
C29H25C1104	472.97	472.1441	73.65	5.33	7.50	13.53
C29H25C1105	488.97	488.1390	71.24	5.15	7.25	16.36
C29H25C1106	504.97	504.1340	68.98	4.99	7.02	19.01
C29H25C1107	520.97	520.1289	66.86	4.84	6.81	21.50
C29H25C1108	536.97	536.1238	64.87	4.69	6.60	23.84
C29H25C1109	552.97	552.1187	62.99	4.56	6.41	26.04
C29H26C1201	461.44	460.1361	75.49	5.68	15.37	3.47
C29H26C1202	477.44	476.1310	72.96	5.49	14.85	6.70
C29H26C1203	493.43	492.1259	70.59	5.31	14.37	9.73
C29H26C1204	509.43	508.1208	68.37	5.14	13.92	12.56
C29H26C1205	525.43	524.1157	66.29	4.99	13.49	15.22
C29H26C1206	541.43	540.1106	64.33	4.84	13.10	17.73
C29H26C1207	557.43	556.1055	62.49	4.70	12.72	20.09
C29H26C1208	573.43	572.1005	60.74	4.57	12.37	22.32
C29H26C1209	589.43	588.0954	59.09	4.45	12.03	24.43
C29H27C1101	426.99	426.1750	81.58	6.37	8.30	3.75
C29H27C1102	442.99	442.1699	78.63	6.14	8.00	7.22
C29H27C1103	458.99	458.1649	75.89	5.93	7.72	10.46
C29H27C1104	474.99	474.1598	73.33	5.73	7.46	13.47
C29H27C1105	490.99	490.1547	70.94	5.54	7.22	16.29
C29H27C1106	506.99	506.1496	68.70	5.37	6.99	18.93
C29H27C1107	522.99	522.1445	66.60	5.20	6.78	21.41
C29H27C1108	538.99	538.1394	64.63	5.05	6.58	23.75
C29H27C1109	554.99	554.1343	62.76	4.90	6.39	25.95
C29H28C1201	463.45	462.1517	75.16	6.09	15.30	3.45
C29H28C1202	479.45	478.1466	72.65	5.89	14.79	6.67
C29H28C1203	495.45	494.1415	70.30	5.70	14.31	9.69
C29H28C1204	511.45	510.1365	68.11	5.52	13.86	12.51
C29H28C1205	527.45	526.1314	66.04	5.35	13.44	15.17
C29H28C1206	543.45	542.1263	64.09	5.19	13.05	17.66
C29H28C1207	559.45	558.1212	62.26	5.04	12.67	20.02
C29H28C1208	575.45	574.1161	60.53	4.90	12.32	22.24
C29H28C1209	591.45	590.1110	58.89	4.77	11.99	24.35
C29H29C1101	429.01	428.1907	81.19	6.81	8.26	3.73
C29H29C1102	445.01	444.1856	78.27	6.57	7.97	7.19
C29H29C1103	461.01	460.1805	75.56	6.34	7.69	10.41
C29H29C1104	477.01	476.1754	73.02	6.13	7.43	13.42
C29H29C1105	493.00	492.1703	70.65	5.93	7.19	16.23
C29H29C1106	509.00	508.1653	68.43	5.74	6.97	18.86
C29H29C1107	525.00	524.1602	66.35	5.57	6.75	21.33
C29H29C1108	541.00	540.1551	64.38	5.40	6.55	23.66
C29H29C1109	557.00	556.1500	62.54	5.25	6.36	25.85
C29H30C1201	465.47	464.1674	74.83	6.50	15.23	3.44
C29H30C1202	481.47	480.1623	72.35	6.28	14.73	6.65
C29H30C1203	497.47	496.1572	70.02	6.08	14.25	9.65
C29H30C1204	513.47	512.1521	67.84	5.89	13.81	12.46
C29H30C1205	529.47	528.1470	65.79	5.71	13.39	15.11
C29H30C1206	545.46	544.1419	63.86	5.54	13.00	17.60
C29H30C1207	561.46	560.1368	62.04	5.39	12.63	19.95
C29H30C1208	577.46	576.1318	60.32	5.24	12.28	22.17
C29H30C1209	593.46	592.1267	58.69	5.10	11.95	24.26
C29H31C1101	431.02	430.2063	80.81	7.25	8.23	3.71
C29H31C1102	447.02	446.2012	77.92	6.99	7.93	7.16
C29H31C1103	463.02	462.1962	75.23	6.75	7.66	10.37
C29H31C1104	479.02	478.1911	72.72	6.52	7.40	13.36
C29H31C1105	495.02	494.1860	70.37	6.31	7.16	16.16
C29H31C1106	511.02	510.1809	68.16	6.11	6.94	18.79
C29H31C1107	527.02	526.1758	66.09	5.93	6.73	21.25
C29H31C1108	543.02	542.1707	64.15	5.75	6.53	23.57
C29H31C1109	559.02	558.1656	62.31	5.59	6.34	25.76
C29H32C1201	467.48	466.1830	74.51	6.90	15.17	3.42
C29H32C1202	483.48	482.1779	72.04	6.67	14.67	6.62
C29H32C1203	499.48	498.1728	69.74	6.46	14.20	9.61
C29H32C1204	515.48	514.1677	67.57	6.26	13.76	12.42
C29H32C1205	531.48	530.1627	65.54	6.07	13.34	15.05
C29H32C1206	547.48	546.1576	63.62	5.89	12.95	17.53
C29H32C1207	563.48	562.1525	61.82	5.72	12.58	19.88
C29H32C1208	579.48	578.1474	60.11	5.57	12.24	22.09
C29H32C1209	595.48	594.1423	58.49	5.42	11.91	24.18
C29H33C1101	433.04	432.2220	80.44	7.68	8.19	3.69

119

C29H33C1102	449.04	448.2169	77.57	7.41	7.90	7.13
C29H33C1103	465.04	464.2118	74.90	7.15	7.62	10.32
C29H33C1104	481.04	480.2067	72.41	6.91	7.37	13.30
C29H33C1105	497.04	496.2016	70.08	6.69	7.13	16.09
C29H33C1106	513.04	512.1965	67.89	6.48	6.91	18.71
C29H33C1107	529.04	528.1915	65.84	6.29	6.70	21.17
C29H33C1108	545.03	544.1864	63.91	6.10	6.50	23.48
C29H33C1109	561.03	560.1813	62.09	5.93	6.32	25.67
C29H34C1201	469.50	468.1987	74.19	7.30	15.10	3.41
C29H34C1202	485.50	484.1936	71.75	7.06	14.60	6.59
C29H34C1203	501.50	500.1885	69.46	6.83	14.14	9.57
C29H34C1204	517.50	516.1834	67.31	6.62	13.70	12.37
C29H34C1205	533.50	532.1783	65.29	6.42	13.29	14.99
C29H34C1206	549.50	548.1732	63.39	6.24	12.90	17.47
C29H34C1207	565.50	564.1681	61.60	6.06	12.54	19.80
C29H34C1208	581.50	580.1631	59.90	5.89	12.19	22.01
C29H34C1209	597.49	596.1580	58.30	5.74	11.87	24.10
C29H35C1101	435.05	434.2376	80.06	8.11	8.15	3.68
C29H35C1102	451.05	450.2325	77.22	7.82	7.86	7.09
C29H35C1103	467.05	466.2275	74.58	7.55	7.59	10.28
C29H35C1104	483.05	482.2224	72.11	7.30	7.34	13.25
C29H35C1105	499.05	498.2173	69.80	7.07	7.10	16.03
C29H35C1106	515.05	514.2122	67.63	6.85	6.88	18.64
C29H35C1107	531.05	530.2071	65.59	6.64	6.68	21.09
C29H35C1108	547.05	546.2020	63.67	6.45	6.48	23.40
C29H35C1109	563.05	562.1969	61.86	6.27	6.30	25.57
C29H36C1201	471.52	470.2143	73.87	7.70	15.04	3.39
C29H36C1202	487.52	486.2092	71.45	7.44	14.54	6.56
C29H36C1203	503.51	502.2041	69.18	7.21	14.08	9.53
C29H36C1204	519.51	518.1990	67.05	6.98	13.65	12.32
C29H36C1205	535.51	534.1940	65.04	6.78	13.24	14.94
C29H36C1206	551.51	550.1889	63.16	6.58	12.86	17.41
C29H36C1207	567.51	566.1838	61.38	6.39	12.49	19.73
C29H36C1208	583.51	582.1787	59.69	6.22	12.15	21.94
C29H36C1209	599.51	598.1736	58.10	6.05	11.83	24.02
C29H37C1101	437.07	436.2533	79.69	8.53	8.11	3.66
C29H37C1102	453.07	452.2482	76.88	8.23	7.83	7.06
C29H37C1103	469.07	468.2431	74.26	7.95	7.56	10.23
C29H37C1104	485.07	484.2380	71.81	7.69	7.31	13.19
C29H37C1105	501.07	500.2329	69.52	7.44	7.08	15.97
C29H37C1106	517.07	516.2278	67.37	7.21	6.86	18.57
C29H37C1107	533.07	532.2228	65.34	7.00	6.65	21.01
C29H37C1108	549.07	548.2177	63.44	6.79	6.46	23.31
C29H37C1109	565.07	564.2126	61.64	6.60	6.27	25.48
C29H38C1201	473.53	472.2300	73.56	8.09	14.97	3.38
C29H38C1202	489.53	488.2249	71.15	7.82	14.48	6.54
C29H38C1203	505.53	504.2198	68.90	7.58	14.03	9.49
C29H38C1204	521.53	520.2147	66.79	7.34	13.60	12.27
C29H38C1205	537.53	536.2096	64.80	7.13	13.19	14.88
C29H38C1206	553.53	552.2045	62.93	6.92	12.81	17.34
C29H38C1207	569.53	568.1994	61.16	6.73	12.45	19.66
C29H38C1208	585.53	584.1944	59.49	6.54	12.11	21.86
C29H38C1209	601.53	600.1893	57.91	6.37	11.79	23.94
C29H39C1101	439.09	438.2689	79.33	8.95	8.07	3.64
C29H39C1102	455.09	454.2638	76.54	8.64	7.79	7.03
C29H39C1103	471.09	470.2588	73.94	8.34	7.53	10.19
C29H39C1104	487.08	486.2537	71.51	8.07	7.28	13.14
C29H39C1105	503.08	502.2486	69.24	7.81	7.05	15.90
C29H39C1106	519.08	518.2435	67.10	7.57	6.83	18.49
C29H39C1107	535.08	534.2384	65.10	7.35	6.63	20.93
C29H39C1108	551.08	550.2333	63.21	7.13	6.43	23.23
C29H39C1109	567.08	566.2282	61.42	6.93	6.25	25.39
C29H40C1201	475.55	474.2456	73.25	8.48	14.91	3.36
C29H40C1202	491.55	490.2405	70.86	8.20	14.43	6.51
C29H40C1203	507.55	506.2354	68.63	7.94	13.97	9.46
C29H40C1204	523.55	522.2303	66.53	7.70	13.54	12.22
C29H40C1205	539.55	538.2253	64.56	7.47	13.14	14.83
C29H40C1206	555.54	554.2202	62.70	7.26	12.76	17.28
C29H40C1207	571.54	570.2151	60.94	7.05	12.41	19.60
C29H40C1208	587.54	586.2100	59.28	6.86	12.07	21.78
C29H40C1209	603.54	602.2049	57.71	6.68	11.75	23.86
C29H41C1101	441.10	440.2846	78.97	9.37	8.04	3.63
C29H41C1102	457.10	456.2795	76.20	9.04	7.76	7.00
C29H41C1103	473.10	472.2744	73.63	8.74	7.49	10.15
C29H41C1104	489.10	488.2693	71.22	8.45	7.25	13.08
C29H41C1105	505.10	504.2642	68.96	8.18	7.02	15.84
C29H41C1106	521.10	520.2591	66.84	7.93	6.80	18.42
C29H41C1107	537.10	536.2541	64.85	7.69	6.60	20.85
C29H41C1108	553.10	552.2490	62.98	7.47	6.41	23.14
C29H41C1109	569.10	568.2439	61.21	7.26	6.23	25.30
C29H42C1201	477.56	476.2613	72.94	8.86	14.85	3.35
C29H42C1202	493.56	492.2562	70.57	8.58	14.37	6.48
C29H42C1203	509.56	508.2511	68.36	8.31	13.92	9.42
C29H42C1204	525.56	524.2460	66.28	8.06	13.49	12.18
C29H42C1205	541.56	540.2409	64.32	7.82	13.09	14.77
C29H42C1206	557.56	556.2358	62.47	7.59	12.72	17.22
C29H42C1207	573.56	572.2307	60.73	7.38	12.36	19.53
C29H42C1208	589.56	588.2257	59.08	7.18	12.03	21.71
C29H42C1209	605.56	604.2206	57.52	6.99	11.71	23.78
C29H43C1101	443.12	442.3002	78.61	9.78	8.00	3.61
C29H43C1102	459.12	458.2951	75.87	9.44	7.72	6.97
C29H43C1103	475.12	474.2901	73.31	9.12	7.46	10.10
C29H43C1104	491.12	490.2850	70.92	8.83	7.22	13.03
C29H43C1105	507.12	506.2799	68.69	8.55	6.99	15.77
C29H43C1106	523.12	522.2748	66.59	8.29	6.78	18.35
C29H43C1107	539.11	538.2697	64.61	8.04	6.58	20.77
C29H43C1108	555.11	554.2646	62.75	7.81	6.39	23.06
C29H43C1109	571.11	570.2595	60.99	7.59	6.21	25.21
C29H44C1201	479.58	478.2769	72.63	9.25	14.79	3.34
C29H44C1202	495.58	494.2718	70.29	8.95	14.31	6.46
C29H44C1203	511.58	510.2667	68.09	8.67	13.86	9.38
C29H44C1204	527.58	526.2616	66.02	8.41	13.44	12.13
C29H44C1205	543.58	542.2566	64.08	8.16	13.04	14.72
C29H44C1206	559.58	558.2515	62.25	7.93	12.67	17.16
C29H44C1207	575.58	574.2464	60.52	7.71	12.32	19.46
C29H44C1208	591.58	590.2413	58.88	7.50	11.99	21.64
C29H44C1209	607.57	606.2362	57.33	7.30	11.67	23.70
C29H45C1101	445.13	444.3159	78.25	10.19	7.96	3.59
C29H45C1102	461.13	460.3108	75.54	9.84	7.69	6.94
C29H45C1103	477.13	476.3057	73.00	9.51	7.43	10.06
C29H45C1104	493.13	492.3006	70.63	9.20	7.19	12.98
C29H45C1105	509.13	508.2955	68.42	8.91	6.96	15.71
C29H45C1106	525.13	524.2904	66.33	8.64	6.75	18.28
C29H45C1107	541.13	540.2854	64.37	8.38	6.55	20.70
C29H45C1108	557.13	556.2803	62.52	8.14	6.36	22.97
C29H45C1109	573.13	572.2752	60.78	7.91	6.19	25.12
C29H46C1201	481.60	480.2926	72.33	9.63	14.72	3.32
C29H46C1202	497.59	496.2875	70.00	9.32	14.25	6.43
C29H46C1203	513.59	512.2824	67.82	9.03	13.81	9.35
C29H46C1204	529.59	528.2773	65.77	8.76	13.39	12.08

120

C29H46C1205	545.59	544.2722	63.84	8.50	13.00	14.66
C29H46C1206	561.59	560.2671	62.02	8.26	12.63	17.09
C29H46C1207	577.59	576.2620	60.31	8.03	12.28	19.39
C29H46C1208	593.59	592.2570	58.68	7.81	11.95	21.56
C29H46C1209	609.59	608.2519	57.14	7.61	11.63	23.62
C29H47C1101	447.15	446.3315	77.90	10.59	7.93	3.58
C29H47C1102	463.15	462.3264	75.21	10.23	7.65	6.91
C29H47C1103	479.15	478.3214	72.70	9.89	7.40	10.02
C29H47C1104	495.15	494.3163	70.35	9.57	7.16	12.92
C29H47C1105	511.15	510.3112	68.15	9.27	6.94	15.65
C29H47C1106	527.15	526.3061	66.08	8.99	6.73	18.21
C29H47C1107	543.15	542.3010	64.13	8.72	6.53	20.62
C29H47C1108	559.15	558.2959	62.30	8.47	6.34	22.89
C29H47C1109	575.15	574.2908	60.56	8.24	6.16	25.04
C29H48C1201	483.61	482.3082	72.03	10.00	14.66	3.31
C29H48C1202	499.61	498.3031	69.72	9.68	14.19	6.40
C29H48C1203	515.61	514.2980	67.56	9.38	13.75	9.31
C29H48C1204	531.61	530.2929	65.52	9.10	13.34	12.04
C29H48C1205	547.61	546.2879	63.61	8.84	12.95	14.61
C29H48C1206	563.61	562.2828	61.80	8.58	12.58	17.03
C29H48C1207	579.61	578.2777	60.10	8.35	12.23	19.32
C29H48C1208	595.61	594.2726	58.48	8.12	11.90	21.49
C29H48C1209	611.61	610.2675	56.95	7.91	11.59	23.54
C29H49C1101	449.17	448.3472	77.55	11.00	7.89	3.56
C29H49C1102	465.17	464.3421	74.88	10.62	7.62	6.88
C29H49C1103	481.17	480.3370	72.39	10.26	7.37	9.98
C29H49C1104	497.16	496.3319	70.06	9.93	7.13	12.87
C29H49C1105	513.16	512.3268	67.88	9.62	6.91	15.59
C29H49C1106	529.16	528.3217	65.83	9.33	6.70	18.14
C29H49C1107	545.16	544.3167	63.89	9.06	6.50	20.54
C29H49C1108	561.16	560.3116	62.07	8.80	6.32	22.81
C29H49C1109	577.16	576.3065	60.35	8.56	6.14	24.95
C29H50C1201	485.63	484.3239	71.73	10.38	14.60	3.29
C29H50C1202	501.63	500.3188	69.44	10.05	14.14	6.38
C29H50C1203	517.63	516.3137	67.29	9.74	13.70	9.27
C29H50C1204	533.63	532.3086	65.27	9.44	13.29	11.99
C29H50C1205	549.62	548.3035	63.37	9.17	12.90	14.55
C29H50C1206	565.62	564.2984	61.58	8.91	12.54	16.97
C29H50C1207	581.62	580.2933	59.89	8.67	12.19	19.26
C29H50C1208	597.62	596.2882	58.28	8.43	11.86	21.42
C29H50C1209	613.62	612.2832	56.77	8.21	11.56	23.47
C29H51C1101	451.18	450.3628	77.20	11.39	7.86	3.55
C29H51C1102	467.18	466.3577	74.56	11.00	7.59	6.85
C29H51C1103	483.18	482.3527	72.09	10.64	7.34	9.93
C29H51C1104	499.18	498.3476	69.78	10.30	7.10	12.82
C29H51C1105	515.18	514.3425	67.61	9.98	6.88	15.53
C29H51C1106	531.18	530.3374	65.58	9.68	6.67	18.07
C29H51C1107	547.18	546.3323	63.66	9.39	6.48	20.47
C29H51C1108	563.18	562.3272	61.85	9.13	6.30	22.73
C29H51C1109	579.18	578.3221	60.14	8.88	6.12	24.86
C29H52C1201	487.64	486.3395	71.43	10.75	14.54	3.28
C29H52C1202	503.64	502.3344	69.16	10.41	14.08	6.35
C29H52C1203	519.64	518.3293	67.03	10.09	13.65	9.24
C29H52C1204	535.64	534.3242	65.03	9.79	13.24	11.95
C29H52C1205	551.64	550.3192	63.14	9.50	12.85	14.50
C29H52C1206	567.64	566.3141	61.36	9.23	12.49	16.91
C29H52C1207	583.64	582.3090	59.68	8.98	12.15	19.19
C29H52C1208	599.64	598.3039	58.09	8.74	11.82	21.35
C29H52C1209	615.64	614.2988	56.58	8.51	11.52	23.39
C29H53C1101	453.20	452.3785	76.86	11.79	7.82	3.53
C29H53C1102	469.20	468.3734	74.24	11.39	7.56	6.82
C29H53C1103	485.20	484.3683	71.79	11.01	7.31	9.89
C29H53C1104	501.20	500.3632	69.50	10.66	7.07	12.77
C29H53C1105	517.20	516.3581	67.35	10.33	6.85	15.47
C29H53C1106	533.20	532.3530	65.33	10.02	6.65	18.00
C29H53C1107	549.19	548.3480	63.42	9.73	6.46	20.39
C29H53C1108	565.19	564.3429	61.63	9.45	6.27	22.65
C29H53C1109	581.19	580.3378	59.93	9.19	6.10	24.78
C29H54C1201	489.66	488.3551	71.14	11.12	14.48	3.27
C29H54C1202	505.66	504.3501	68.89	10.76	14.02	6.33
C29H54C1203	521.66	520.3450	66.77	10.43	13.59	9.20
C29H54C1204	537.66	536.3399	64.79	10.12	13.19	11.90
C29H54C1205	553.66	552.3348	62.91	9.83	12.81	14.45
C29H54C1206	569.66	568.3297	61.15	9.55	12.45	16.85
C29H54C1207	585.66	584.3246	59.48	9.29	12.11	19.12
C29H54C1208	601.65	600.3195	57.89	9.05	11.79	21.27
C29H54C1209	617.65	616.3145	56.39	8.81	11.48	23.31
C29H55C1101	455.21	454.3941	76.52	12.18	7.79	3.51
C29H55C1102	471.21	470.3890	73.92	11.77	7.52	6.79
C29H55C1103	487.21	486.3839	71.49	11.38	7.28	9.85
C29H55C1104	503.21	502.3789	69.22	11.02	7.05	12.72
C29H55C1105	519.21	518.3738	67.09	10.68	6.83	15.41
C29H55C1106	535.21	534.3687	65.08	10.36	6.62	17.94
C29H55C1107	551.21	550.3636	63.19	10.06	6.43	20.32
C29H55C1108	567.21	566.3585	61.41	9.77	6.25	22.57
C29H55C1109	583.21	582.3534	59.73	9.51	6.08	24.69
C29H56C1201	491.68	490.3708	70.84	11.48	14.42	3.25
C29H56C1202	507.67	506.3657	68.61	11.12	13.97	6.30
C29H56C1203	523.67	522.3606	66.52	10.78	13.54	9.17
C29H56C1204	539.67	538.3555	64.54	10.46	13.14	11.86
C29H56C1205	555.67	554.3505	62.68	10.16	12.76	14.40
C29H56C1206	571.67	570.3454	60.93	9.87	12.40	16.79
C29H56C1207	587.67	586.3403	59.27	9.61	12.07	19.06
C29H56C1208	603.67	602.3352	57.70	9.35	11.75	21.20
C29H56C1209	619.67	618.3301	56.21	9.11	11.44	23.24
C29H57C1101	457.23	456.4098	76.18	12.57	7.75	3.50
C29H57C1102	473.23	472.4047	73.61	12.14	7.49	6.76
C29H57C1103	489.23	488.3996	71.20	11.74	7.25	9.81
C29H57C1104	505.23	504.3945	68.94	11.37	7.02	12.67
C29H57C1105	521.23	520.3894	66.83	11.02	6.80	15.35
C29H57C1106	537.23	536.3843	64.84	10.69	6.60	17.87
C29H57C1107	553.23	552.3793	62.96	10.39	6.41	20.24
C29H57C1108	569.23	568.3742	61.19	10.09	6.23	22.49
C29H57C1109	585.23	584.3691	59.52	9.82	6.06	24.60
C29H58C1201	493.69	492.3864	70.55	11.84	14.36	3.24
C29H58C1202	509.69	508.3814	68.34	11.47	13.91	6.28
C29H58C1203	525.69	524.3763	66.26	11.12	13.49	9.13
C29H58C1204	541.69	540.3712	64.30	10.79	13.09	11.81
C29H58C1205	557.69	556.3661	62.46	10.48	12.71	14.34
C29H58C1206	573.69	572.3610	60.72	10.19	12.36	16.73
C29H58C1207	589.69	588.3559	59.07	9.91	12.02	18.99
C29H58C1208	605.69	604.3508	57.51	9.65	11.71	21.13
C29H58C1209	621.69	620.3458	56.03	9.40	11.41	23.16
C29H59C1101	459.25	458.4254	75.85	12.95	7.72	3.48
C29H59C1102	475.25	474.4203	73.29	12.51	7.46	6.73
C29H59C1103	491.24	490.4152	70.91	12.11	7.22	9.77
C29H59C1104	507.24	506.4102	68.67	11.72	6.99	12.62
C29H59C1105	523.24	522.4051	66.57	11.37	6.78	15.29
C29H59C1106	539.24	538.4000	64.59	11.03	6.57	17.80
C29H59C1107	555.24	554.3949	62.73	10.71	6.39	20.17

121

C29H59C1108	571.24	570.3898	60.98	10.41	6.21	22.41
C29H59C1109	587.24	586.3847	59.32	10.13	6.04	24.52
C30H15C1101	426.91	426.0811	84.41	3.54	8.30	3.75
C30H15C1102	442.91	442.0761	81.36	3.41	8.00	7.22
C30H15C1103	458.91	458.0710	78.52	3.29	7.73	10.46
C30H15C1104	474.90	474.0659	75.88	3.18	7.47	13.48
C30H15C1105	490.90	490.0608	73.40	3.08	7.22	16.30
C30H15C1106	506.90	506.0557	71.09	2.98	6.99	18.94
C30H15C1107	522.90	522.0506	68.91	2.89	6.78	21.42
C30H15C1108	538.90	538.0455	66.86	2.81	6.58	23.75
C30H15C1109	554.90	554.0404	64.94	2.72	6.39	25.95
C30H16C1201	463.37	462.0578	77.76	3.48	15.30	3.45
C30H16C1202	479.37	478.0527	75.17	3.36	14.79	6.68
C30H16C1203	495.37	494.0476	72.74	3.26	14.31	9.69
C30H16C1204	511.37	510.0426	70.47	3.15	13.87	12.52
C30H16C1205	527.37	526.0375	68.33	3.06	13.45	15.17
C30H16C1206	543.36	542.0324	66.32	2.97	13.05	17.67
C30H16C1207	559.36	558.0273	64.42	2.88	12.68	20.02
C30H16C1208	575.36	574.0222	62.63	2.80	12.32	22.25
C30H16C1209	591.36	590.0171	60.93	2.73	11.99	24.35
C30H17C1101	428.92	428.0968	84.01	4.00	8.27	3.73
C30H17C1102	444.92	444.0917	80.99	3.85	7.97	7.19
C30H17C1103	460.92	460.0866	78.18	3.72	7.69	10.41
C30H17C1104	476.92	476.0815	75.55	3.59	7.43	13.42
C30H17C1105	492.92	492.0764	73.10	3.48	7.19	16.23
C30H17C1106	508.92	508.0714	70.80	3.37	6.97	18.86
C30H17C1107	524.92	524.0663	68.65	3.26	6.75	21.34
C30H17C1108	540.92	540.0612	66.62	3.17	6.55	23.66
C30H17C1109	556.92	556.0561	64.70	3.08	6.37	25.86
C30H18C1201	465.38	464.0735	77.43	3.90	15.24	3.44
C30H18C1202	481.38	480.0684	74.85	3.77	14.73	6.65
C30H18C1203	497.38	496.0633	72.45	3.65	14.26	9.65
C30H18C1204	513.38	512.0582	70.19	3.53	13.81	12.47
C30H18C1205	529.38	528.0531	68.07	3.43	13.39	15.11
C30H18C1206	545.38	544.0480	66.07	3.33	13.00	17.60
C30H18C1207	561.38	560.0429	64.19	3.23	12.63	19.95
C30H18C1208	577.38	576.0379	62.41	3.14	12.28	22.17
C30H18C1209	593.38	592.0328	60.73	3.06	11.95	24.27
C30H19C1101	430.94	430.1124	83.62	4.44	8.23	3.71
C30H19C1102	446.94	446.1073	80.62	4.29	7.93	7.16
C30H19C1103	462.94	462.1023	77.84	4.14	7.66	10.37
C30H19C1104	478.94	478.0972	75.24	4.00	7.40	13.36
C30H19C1105	494.94	494.0921	72.80	3.87	7.16	16.16
C30H19C1106	510.94	510.0870	70.52	3.75	6.94	18.79
C30H19C1107	526.93	526.0819	68.38	3.63	6.73	21.25
C30H19C1108	542.93	542.0768	66.37	3.53	6.53	23.57
C30H19C1109	558.93	558.0717	64.47	3.43	6.34	25.76
C30H20C1201	467.40	466.0891	77.09	4.31	15.17	3.42
C30H20C1202	483.40	482.0840	74.54	4.17	14.67	6.62
C30H20C1203	499.40	498.0789	72.15	4.04	14.20	9.61
C30H20C1204	515.40	514.0739	69.91	3.91	13.76	12.42
C30H20C1205	531.40	530.0688	67.81	3.79	13.34	15.05
C30H20C1206	547.40	546.0637	65.83	3.68	12.95	17.54
C30H20C1207	563.40	562.0586	63.96	3.58	12.59	19.88
C30H20C1208	579.40	578.0535	62.19	3.48	12.24	22.09
C30H20C1209	595.39	594.0484	60.52	3.39	11.91	24.18
C30H21C1101	432.95	432.1281	83.23	4.89	8.19	3.70
C30H21C1102	448.95	448.1230	80.26	4.71	7.90	7.13
C30H21C1103	464.95	464.1179	77.50	4.55	7.63	10.32
C30H21C1104	480.95	480.1128	74.92	4.40	7.37	13.31
C30H21C1105	496.95	496.1077	72.51	4.26	7.13	16.10
C30H21C1106	512.95	512.1027	70.25	4.13	6.91	18.71
C30H21C1107	528.95	528.0976	68.12	4.00	6.70	21.17
C30H21C1108	544.95	544.0925	66.12	3.88	6.51	23.49
C30H21C1109	560.95	560.0874	64.24	3.77	6.32	25.67
C30H22C1201	469.42	468.1048	76.76	4.72	15.11	3.41
C30H22C1202	485.41	484.0997	74.23	4.57	14.61	6.59
C30H22C1203	501.41	500.0946	71.86	4.42	14.14	9.57
C30H22C1204	517.41	516.0895	69.64	4.29	13.70	12.37
C30H22C1205	533.41	532.0844	67.55	4.16	13.29	15.00
C30H22C1206	549.41	548.0793	65.59	4.04	12.91	17.47
C30H22C1207	565.41	564.0742	63.73	3.92	12.54	19.81
C30H22C1208	581.41	580.0692	61.98	3.81	12.20	22.01
C30H22C1209	597.41	596.0641	60.32	3.71	11.87	24.10
C30H23C1101	434.97	434.1437	82.84	5.33	8.15	3.68
C30H23C1102	450.97	450.1386	79.90	5.14	7.86	7.10
C30H23C1103	466.97	466.1336	77.16	4.96	7.59	10.28
C30H23C1104	482.97	482.1285	74.61	4.80	7.34	13.25
C30H23C1105	498.97	498.1234	72.22	4.65	7.11	16.03
C30H23C1106	514.97	514.1183	69.97	4.50	6.88	18.64
C30H23C1107	530.97	530.1132	67.86	4.37	6.68	21.09
C30H23C1108	546.97	546.1081	65.88	4.24	6.48	23.40
C30H23C1109	562.97	562.1030	64.01	4.12	6.30	25.58
C30H24C1201	471.43	470.1204	76.43	5.13	15.04	3.39
C30H24C1202	487.43	486.1153	73.93	4.96	14.55	6.56
C30H24C1203	503.43	502.1102	71.58	4.81	14.08	9.53
C30H24C1204	519.43	518.1052	69.37	4.66	13.65	12.32
C30H24C1205	535.43	534.1001	67.30	4.52	13.24	14.94
C30H24C1206	551.43	550.0950	65.35	4.39	12.86	17.41
C30H24C1207	567.43	566.0899	63.50	4.26	12.50	19.74
C30H24C1208	583.43	582.0848	61.76	4.15	12.15	21.94
C30H24C1209	599.43	598.0797	60.11	4.04	11.83	24.02
C30H25C1101	436.99	436.1594	82.46	5.77	8.11	3.66
C30H25C1102	452.99	452.1543	79.55	5.56	7.83	7.06
C30H25C1103	468.98	468.1492	76.83	5.37	7.56	10.23
C30H25C1104	484.98	484.1441	74.30	5.20	7.31	13.20
C30H25C1105	500.98	500.1390	71.93	5.03	7.08	15.97
C30H25C1106	516.98	516.1340	69.70	4.87	6.86	18.57
C30H25C1107	532.98	532.1289	67.61	4.73	6.65	21.01
C30H25C1108	548.98	548.1238	65.64	4.59	6.46	23.32
C30H25C1109	564.98	564.1187	63.78	4.46	6.28	25.49
C30H26C1201	473.45	472.1361	76.11	5.54	14.98	3.38
C30H26C1202	489.45	488.1310	73.62	5.35	14.49	6.54
C30H26C1203	505.45	504.1259	71.29	5.18	14.03	9.50
C30H26C1204	521.45	520.1208	69.10	5.03	13.60	12.27
C30H26C1205	537.44	536.1157	67.05	4.88	13.19	14.88
C30H26C1206	553.44	552.1106	65.11	4.74	12.81	17.35
C30H26C1207	569.44	568.1055	63.28	4.60	12.45	19.67
C30H26C1208	585.44	584.1005	61.55	4.48	12.11	21.86
C30H26C1209	601.44	600.0954	59.91	4.36	11.79	23.94
C30H27C1101	439.00	438.1750	82.08	6.20	8.08	3.64
C30H27C1102	455.00	454.1699	79.19	5.98	7.79	7.03
C30H27C1103	471.00	470.1649	76.50	5.78	7.53	10.19
C30H27C1104	487.00	486.1598	73.99	5.59	7.28	13.14
C30H27C1105	503.00	502.1547	71.64	5.41	7.05	15.90
C30H27C1106	519.00	518.1496	69.43	5.24	6.83	18.50
C30H27C1107	535.00	534.1445	67.35	5.09	6.63	20.93
C30H27C1108	551.00	550.1394	65.40	4.94	6.43	23.23
C30H27C1109	567.00	566.1343	63.55	4.80	6.25	25.40
C30H28C1201	475.46	474.1517	75.79	5.94	14.91	3.37

122

C30H28C1202	491.46	490.1466	73.32	5.74	14.43	6.51
C30H28C1203	507.46	506.1415	71.01	5.56	13.97	9.46
C30H28C1204	523.46	522.1365	68.84	5.39	13.55	12.23
C30H28C1205	539.46	538.1314	66.80	5.23	13.14	14.83
C30H28C1206	555.46	554.1263	64.87	5.08	12.77	17.28
C30H28C1207	571.46	570.1212	63.06	4.94	12.41	19.60
C30H28C1208	587.46	586.1161	61.34	4.80	12.07	21.79
C30H28C1209	603.46	602.1110	59.71	4.68	11.75	23.86
C30H29C1101	441.02	440.1907	81.71	6.63	8.04	3.63
C30H29C1102	457.02	456.1856	78.84	6.40	7.76	7.00
C30H29C1103	473.02	472.1805	76.18	6.18	7.50	10.15
C30H29C1104	489.02	488.1754	73.69	5.98	7.25	13.09
C30H29C1105	505.02	504.1703	71.35	5.79	7.02	15.84
C30H29C1106	521.02	520.1653	69.16	5.61	6.80	18.42
C30H29C1107	537.01	536.1602	67.10	5.44	6.60	20.86
C30H29C1108	553.01	552.1551	65.16	5.29	6.41	23.15
C30H29C1109	569.01	568.1500	63.33	5.14	6.23	25.31
C30H30C1201	477.48	476.1674	75.47	6.33	14.85	3.35
C30H30C1202	493.48	492.1623	73.02	6.13	14.37	6.48
C30H30C1203	509.48	508.1572	70.73	5.94	13.92	9.42
C30H30C1204	525.48	524.1521	68.57	5.75	13.49	12.18
C30H30C1205	541.48	540.1470	66.55	5.58	13.09	14.77
C30H30C1206	557.48	556.1419	64.64	5.42	12.72	17.22
C30H30C1207	573.48	572.1368	62.83	5.27	12.36	19.53
C30H30C1208	589.47	588.1318	61.13	5.13	12.03	21.71
C30H30C1209	605.47	604.1267	59.51	4.99	11.71	23.78
C30H31C1101	443.03	442.2063	81.33	7.05	8.00	3.61
C30H31C1102	459.03	458.2012	78.50	6.81	7.72	6.97
C30H31C1103	475.03	474.1962	75.85	6.58	7.46	10.10
C30H31C1104	491.03	490.1911	73.38	6.36	7.22	13.03
C30H31C1105	507.03	506.1860	71.07	6.16	6.99	15.78
C30H31C1106	523.03	522.1809	68.89	5.97	6.78	18.35
C30H31C1107	539.03	538.1758	66.85	5.80	6.58	20.78
C30H31C1108	555.03	554.1707	64.92	5.63	6.39	23.06
C30H31C1109	571.03	570.1656	63.10	5.47	6.21	25.22
C30H32C1201	479.49	478.1830	75.15	6.73	14.79	3.34
C30H32C1202	495.49	494.1779	72.72	6.51	14.31	6.46
C30H32C1203	511.49	510.1728	70.45	6.31	13.86	9.38
C30H32C1204	527.49	526.1677	68.31	6.11	13.44	12.13
C30H32C1205	543.49	542.1627	66.30	5.93	13.05	14.72
C30H32C1206	559.49	558.1576	64.40	5.77	12.67	17.16
C30H32C1207	575.49	574.1525	62.61	5.60	12.32	19.46
C30H32C1208	591.49	590.1474	60.92	5.45	11.99	21.64
C30H32C1209	607.49	606.1423	59.32	5.31	11.67	23.70
C30H33C1101	445.05	444.2220	80.96	7.47	7.97	3.59
C30H33C1102	461.05	460.2169	78.16	7.21	7.69	6.94
C30H33C1103	477.05	476.2118	75.53	6.97	7.43	10.06
C30H33C1104	493.05	492.2067	73.08	6.75	7.19	12.98
C30H33C1105	509.05	508.2016	70.79	6.53	6.96	15.72
C30H33C1106	525.05	524.1965	68.63	6.34	6.75	18.28
C30H33C1107	541.05	540.1915	66.60	6.15	6.55	20.70
C30H33C1108	557.05	556.1864	64.69	5.97	6.36	22.98
C30H33C1109	573.05	572.1813	62.88	5.80	6.19	25.13
C30H34C1201	481.51	480.1987	74.83	7.12	14.73	3.32
C30H34C1202	497.51	496.1936	72.43	6.89	14.25	6.43
C30H34C1203	513.51	512.1885	70.17	6.67	13.81	9.35
C30H34C1204	529.51	528.1834	68.05	6.47	13.39	12.09
C30H34C1205	545.51	544.1783	66.05	6.28	13.00	14.66
C30H34C1206	561.51	560.1732	64.17	6.10	12.63	17.10
C30H34C1207	577.51	576.1681	62.39	5.93	12.28	19.39
C30H34C1208	593.51	592.1631	60.71	5.77	11.95	21.57
C30H34C1209	609.51	608.1580	59.12	5.62	11.63	23.62
C30H35C1101	447.07	446.2376	80.60	7.89	7.93	3.58
C30H35C1102	463.07	462.2325	77.82	7.62	7.66	6.91
C30H35C1103	479.06	478.2275	75.22	7.36	7.40	10.02
C30H35C1104	495.06	494.2224	72.79	7.13	7.16	12.93
C30H35C1105	511.06	510.2173	70.51	6.90	6.94	15.65
C30H35C1106	527.06	526.2122	68.37	6.69	6.73	18.21
C30H35C1107	543.06	542.2071	66.35	6.50	6.53	20.62
C30H35C1108	559.06	558.2020	64.45	6.31	6.34	22.89
C30H35C1109	575.06	574.1969	62.66	6.13	6.17	25.04
C30H36C1201	483.53	482.2143	74.52	7.50	14.66	3.31
C30H36C1202	499.53	498.2092	72.14	7.26	14.19	6.41
C30H36C1203	515.53	514.2041	69.90	7.04	13.75	9.31
C30H36C1204	531.53	530.1990	67.79	6.83	13.34	12.04
C30H36C1205	547.52	546.1940	65.81	6.63	12.95	14.61
C30H36C1206	563.52	562.1889	63.94	6.44	12.58	17.04
C30H36C1207	579.52	578.1838	62.18	6.26	12.24	19.33
C30H36C1208	595.52	594.1787	60.51	6.09	11.91	21.49
C30H36C1209	611.52	610.1736	58.92	5.93	11.60	23.55
C30H37C1101	449.08	448.2533	80.24	8.30	7.89	3.56
C30H37C1102	465.08	464.2482	77.48	8.02	7.62	6.88
C30H37C1103	481.08	480.2431	74.90	7.75	7.37	9.98
C30H37C1104	497.08	496.2380	72.49	7.50	7.13	12.87
C30H37C1105	513.08	512.2329	70.23	7.27	6.91	15.59
C30H37C1106	529.08	528.2278	68.11	7.05	6.70	18.14
C30H37C1107	545.08	544.2228	66.11	6.84	6.50	20.55
C30H37C1108	561.08	560.2177	64.22	6.65	6.32	22.81
C30H37C1109	577.08	576.2126	62.44	6.46	6.14	24.95
C30H38C1201	485.54	484.2300	74.21	7.89	14.60	3.30
C30H38C1202	501.54	500.2249	71.85	7.64	14.14	6.38
C30H38C1203	517.54	516.2198	69.62	7.40	13.70	9.27
C30H38C1204	533.54	532.2147	67.54	7.18	13.29	11.99
C30H38C1205	549.54	548.2096	65.57	6.97	12.90	14.56
C30H38C1206	565.54	564.2045	63.72	6.77	12.54	16.97
C30H38C1207	581.54	580.1994	61.96	6.59	12.19	19.26
C30H38C1208	597.54	596.1944	60.30	6.41	11.87	21.42
C30H38C1209	613.54	612.1893	58.73	6.24	11.56	23.47
C30H39C1101	451.10	450.2689	79.88	8.71	7.86	3.55
C30H39C1102	467.10	466.2638	77.14	8.42	7.59	6.85
C30H39C1103	483.10	482.2588	74.59	8.14	7.34	9.94
C30H39C1104	499.10	498.2537	72.20	7.88	7.10	12.82
C30H39C1105	515.10	514.2486	69.95	7.63	6.88	15.53
C30H39C1106	531.09	530.2435	67.85	7.40	6.68	18.08
C30H39C1107	547.09	546.2384	65.86	7.19	6.48	20.47
C30H39C1108	563.09	562.2333	63.99	6.98	6.30	22.73
C30H39C1109	579.09	578.2282	62.22	6.79	6.12	24.87
C30H40C1201	487.56	486.2456	73.91	8.27	14.54	3.28
C30H40C1202	503.56	502.2405	71.56	8.01	14.08	6.35
C30H40C1203	519.56	518.2354	69.35	7.76	13.65	9.24
C30H40C1204	535.56	534.2303	67.28	7.53	13.24	11.95
C30H40C1205	551.56	550.2253	65.33	7.31	12.86	14.50
C30H40C1206	567.56	566.2202	63.49	7.10	12.49	16.91
C30H40C1207	583.56	582.2151	61.75	6.91	12.15	19.19
C30H40C1208	599.55	598.2100	60.10	6.72	11.83	21.35
C30H40C1209	615.55	614.2049	58.54	6.55	11.52	23.39
C30H41C1101	453.11	452.2846	79.52	9.12	7.82	3.53
C30H41C1102	469.11	468.2795	76.81	8.81	7.56	6.82
C30H41C1103	485.11	484.2744	74.28	8.52	7.31	9.89
C30H41C1104	501.11	500.2693	71.91	8.25	7.07	12.77

C30H41C1105	517.11	516.2642	69.68	7.99	6.86	15.47
C30H41C1106	533.11	532.2591	67.59	7.75	6.65	18.01
C30H41C1107	549.11	548.2541	65.62	7.53	6.46	20.40
C30H41C1108	565.11	564.2490	63.76	7.31	6.27	22.65
C30H41C1109	581.11	580.2439	62.01	7.11	6.10	24.78
C30H42C1201	489.57	488.2613	73.60	8.65	14.48	3.27
C30H42C1202	505.57	504.2562	71.27	8.37	14.02	6.33
C30H42C1203	521.57	520.2511	69.09	8.12	13.59	9.20
C30H42C1204	537.57	536.2460	67.03	7.88	13.19	11.90
C30H42C1205	553.57	552.2409	65.09	7.65	12.81	14.45
C30H42C1206	569.57	568.2358	63.26	7.43	12.45	16.85
C30H42C1207	585.57	584.2307	61.54	7.23	12.11	19.13
C30H42C1208	601.57	600.2257	59.90	7.04	11.79	21.28
C30H42C1209	617.57	616.2206	58.35	6.86	11.48	23.32
C30H43C1101	455.13	454.3002	79.17	9.52	7.79	3.52
C30H43C1102	471.13	470.2951	76.48	9.20	7.53	6.79
C30H43C1103	487.13	486.2901	73.97	8.90	7.28	9.85
C30H43C1104	503.13	502.2850	71.62	8.61	7.05	12.72
C30H43C1105	519.13	518.2799	69.41	8.35	6.83	15.41
C30H43C1106	535.13	534.2748	67.34	8.10	6.63	17.94
C30H43C1107	551.13	550.2697	65.38	7.86	6.43	20.32
C30H43C1108	567.13	566.2646	63.54	7.64	6.25	22.57
C30H43C1109	583.12	582.2595	61.79	7.43	6.08	24.69
C30H44C1201	491.59	490.2769	73.30	9.02	14.42	3.25
C30H44C1202	507.59	506.2718	70.99	8.74	13.97	6.30
C30H44C1203	523.59	522.2667	68.82	8.47	13.54	9.17
C30H44C1204	539.59	538.2616	66.78	8.22	13.14	11.86
C30H44C1205	555.59	554.2566	64.86	7.98	12.76	14.40
C30H44C1206	571.59	570.2515	63.04	7.76	12.41	16.79
C30H44C1207	587.59	586.2464	61.32	7.55	12.07	19.06
C30H44C1208	603.59	602.2413	59.70	7.35	11.75	21.21
C30H44C1209	619.59	618.2362	58.16	7.16	11.44	23.24
C30H45C1101	457.15	456.3159	78.82	9.92	7.76	3.50
C30H45C1102	473.14	472.3108	76.16	9.59	7.49	6.76
C30H45C1103	489.14	488.3057	73.67	9.27	7.25	9.81
C30H45C1104	505.14	504.3006	71.33	8.98	7.02	12.67
C30H45C1105	521.14	520.2955	69.14	8.70	6.80	15.35
C30H45C1106	537.14	536.2904	67.08	8.44	6.60	17.87
C30H45C1107	553.14	552.2854	65.14	8.20	6.41	20.25
C30H45C1108	569.14	568.2803	63.31	7.97	6.23	22.49
C30H45C1109	585.14	584.2752	61.58	7.75	6.06	24.61
C30H46C1201	493.61	492.2926	73.00	9.39	14.36	3.24
C30H46C1202	509.61	508.2875	70.71	9.10	13.91	6.28
C30H46C1203	525.61	524.2824	68.56	8.82	13.49	9.13
C30H46C1204	541.60	540.2773	66.53	8.56	13.09	11.82
C30H46C1205	557.60	556.2722	64.62	8.32	12.72	14.35
C30H46C1206	573.60	572.2671	62.82	8.08	12.36	16.74
C30H46C1207	589.60	588.2620	61.11	7.86	12.03	19.00
C30H46C1208	605.60	604.2570	59.50	7.66	11.71	21.14
C30H46C1209	621.60	620.2519	57.97	7.46	11.41	23.17
C30H47C1101	459.16	458.3315	78.48	10.32	7.72	3.48
C30H47C1102	475.16	474.3264	75.83	9.97	7.46	6.73
C30H47C1103	491.16	490.3214	73.36	9.65	7.22	9.77
C30H47C1104	507.16	506.3163	71.05	9.34	6.99	12.62
C30H47C1105	523.16	522.3112	68.88	9.06	6.78	15.29
C30H47C1106	539.16	538.3061	66.83	8.79	6.58	17.80
C30H47C1107	555.16	554.3010	64.91	8.53	6.39	20.17
C30H47C1108	571.16	570.2959	63.09	8.29	6.21	22.41
C30H47C1109	587.16	586.2908	61.37	8.07	6.04	24.52
C30H48C1201	495.62	494.3082	72.70	9.76	14.31	3.23
C30H48C1202	511.62	510.3031	70.43	9.46	13.86	6.25
C30H48C1203	527.62	526.2980	68.29	9.17	13.44	9.10
C30H48C1204	543.62	542.2929	66.28	8.90	13.04	11.77
C30H48C1205	559.62	558.2879	64.39	8.65	12.67	14.29
C30H48C1206	575.62	574.2828	62.60	8.41	12.32	16.68
C30H48C1207	591.62	590.2777	60.91	8.18	11.99	18.93
C30H48C1208	607.62	606.2726	59.30	7.96	11.67	21.07
C30H48C1209	623.62	622.2675	57.78	7.76	11.37	23.09
C30H49C1101	461.18	460.3472	78.13	10.71	7.69	3.47
C30H49C1102	477.18	476.3421	75.51	10.35	7.43	6.71
C30H49C1103	493.18	492.3370	73.06	10.01	7.19	9.73
C30H49C1104	509.18	508.3319	70.77	9.70	6.96	12.57
C30H49C1105	525.18	524.3268	68.61	9.40	6.75	15.23
C30H49C1106	541.17	540.3217	66.58	9.13	6.55	17.74
C30H49C1107	557.17	556.3167	64.67	8.86	6.36	20.10
C30H49C1108	573.17	572.3116	62.87	8.62	6.19	22.33
C30H49C1109	589.17	588.3065	61.16	8.38	6.02	24.44
C30H50C1201	497.64	496.3239	72.41	10.13	14.25	3.22
C30H50C1202	513.64	512.3188	70.15	9.81	13.80	6.23
C30H50C1203	529.64	528.3137	68.03	9.52	13.39	9.06
C30H50C1204	545.64	544.3086	66.04	9.24	13.00	11.73
C30H50C1205	561.64	560.3035	64.16	8.97	12.62	14.24
C30H50C1206	577.64	576.2984	62.38	8.72	12.28	16.62
C30H50C1207	593.63	592.2933	60.70	8.49	11.94	18.87
C30H50C1208	609.63	608.2882	59.11	8.27	11.63	21.00
C30H50C1209	625.63	624.2832	57.60	8.06	11.33	23.02
C30H51C1101	463.19	462.3628	77.79	11.10	7.65	3.45
C30H51C1102	479.19	478.3577	75.20	10.73	7.40	6.68
C30H51C1103	495.19	494.3527	72.77	10.38	7.16	9.69
C30H51C1104	511.19	510.3476	70.49	10.06	6.94	12.52
C30H51C1105	527.19	526.3425	68.35	9.75	6.72	15.17
C30H51C1106	543.19	542.3374	66.34	9.46	6.53	17.67
C30H51C1107	559.19	558.3323	64.44	9.19	6.34	20.03
C30H51C1108	575.19	574.3272	62.65	8.94	6.16	22.25
C30H51C1109	591.19	590.3221	60.95	8.70	6.00	24.36
C30H52C1201	499.65	498.3395	72.12	10.49	14.19	3.20
C30H52C1202	515.65	514.3344	69.88	10.16	13.75	6.21
C30H52C1203	531.65	530.3293	67.78	9.86	13.34	9.03
C30H52C1204	547.65	546.3242	65.80	9.57	12.95	11.69
C30H52C1205	563.65	562.3192	63.93	9.30	12.58	14.19
C30H52C1206	579.65	578.3141	62.16	9.04	12.23	16.56
C30H52C1207	595.65	594.3090	60.49	8.80	11.90	18.80
C30H52C1208	611.65	610.3039	58.91	8.57	11.59	20.93
C30H52C1209	627.65	626.2988	57.41	8.35	11.30	22.94
C30H53C1101	465.21	464.3785	77.46	11.48	7.62	3.44
C30H53C1102	481.21	480.3734	74.88	11.10	7.37	6.65
C30H53C1103	497.21	496.3683	72.47	10.74	7.13	9.65
C30H53C1104	513.21	512.3632	70.21	10.41	6.91	12.47
C30H53C1105	529.21	528.3581	68.09	10.09	6.70	15.12
C30H53C1106	545.21	544.3530	66.09	9.80	6.50	17.61
C30H53C1107	561.21	560.3480	64.21	9.52	6.32	19.96
C30H53C1108	577.21	576.3429	62.43	9.26	6.14	22.17
C30H53C1109	593.20	592.3378	60.74	9.01	5.98	24.27
C30H54C1201	501.67	500.3551	71.83	10.85	14.13	3.19
C30H54C1202	517.67	516.3501	69.61	10.51	13.70	6.18
C30H54C1203	533.67	532.3450	67.52	10.20	13.29	8.99
C30H54C1204	549.67	548.3399	65.55	9.90	12.90	11.64
C30H54C1205	565.67	564.3348	63.70	9.62	12.53	14.14
C30H54C1206	581.67	580.3297	61.95	9.36	12.19	16.50
C30H54C1207	597.67	596.3246	60.29	9.11	11.86	18.74

124

C30H54C1208	613.67	612.3195	58.72	8.87	11.55	20.86
C30H54C1209	629.67	628.3145	57.23	8.64	11.26	22.87
C30H55C1101	467.23	466.3941	77.12	11.87	7.59	3.42
C30H55C1102	483.22	482.3890	74.57	11.47	7.34	6.62
C30H55C1103	499.22	498.3839	72.18	11.10	7.10	9.61
C30H55C1104	515.22	514.3789	69.94	10.76	6.88	12.42
C30H55C1105	531.22	530.3738	67.83	10.44	6.67	15.06
C30H55C1106	547.22	546.3687	65.85	10.13	6.48	17.54
C30H55C1107	563.22	562.3636	63.98	9.84	6.29	19.88
C30H55C1108	579.22	578.3585	62.21	9.57	6.12	22.10
C30H55C1109	595.22	594.3534	60.54	9.31	5.96	24.19
C30H56C1201	503.69	502.3708	71.54	11.21	14.08	3.18
C30H56C1202	519.69	518.3657	69.34	10.86	13.64	6.16
C30H56C1203	535.69	534.3606	67.27	10.54	13.24	8.96
C30H56C1204	551.68	550.3555	65.32	10.23	12.85	11.60
C30H56C1205	567.68	566.3505	63.47	9.94	12.49	14.09
C30H56C1206	583.68	582.3454	61.73	9.67	12.15	16.45
C30H56C1207	599.68	598.3403	60.09	9.41	11.82	18.68
C30H56C1208	615.68	614.3352	58.53	9.17	11.52	20.79
C30H56C1209	631.68	630.3301	57.04	8.94	11.22	22.80
C30H57C1101	469.24	468.4098	76.79	12.24	7.56	3.41
C30H57C1102	485.24	484.4047	74.26	11.84	7.31	6.59
C30H57C1103	501.24	500.3996	71.89	11.46	7.07	9.58
C30H57C1104	517.24	516.3945	69.66	11.11	6.85	12.37
C30H57C1105	533.24	532.3894	67.57	10.77	6.65	15.00
C30H57C1106	549.24	548.3843	65.61	10.46	6.45	17.48
C30H57C1107	565.24	564.3793	63.75	10.16	6.27	19.81
C30H57C1108	581.24	580.3742	61.99	9.88	6.10	22.02
C30H57C1109	597.24	596.3691	60.33	9.62	5.94	24.11
C30H58C1201	505.70	504.3864	71.25	11.56	14.02	3.16
C30H58C1202	521.70	520.3814	69.07	11.21	13.59	6.13
C30H58C1203	537.70	536.3763	67.01	10.87	13.19	8.93
C30H58C1204	553.70	552.3712	65.08	10.56	12.81	11.56
C30H58C1205	569.70	568.3661	63.25	10.26	12.45	14.04
C30H58C1206	585.70	584.3610	61.52	9.98	12.11	16.39
C30H58C1207	601.70	600.3559	59.89	9.72	11.78	18.61
C30H58C1208	617.70	616.3508	58.34	9.46	11.48	20.72
C30H58C1209	633.70	632.3458	56.86	9.23	11.19	22.72
C30H59C1101	471.26	470.4254	76.46	12.62	7.52	3.40
C30H59C1102	487.26	486.4203	73.95	12.21	7.28	6.57
C30H59C1103	503.26	502.4152	71.60	11.82	7.04	9.54
C30H59C1104	519.26	518.4102	69.39	11.45	6.83	12.32
C30H59C1105	535.25	534.4051	67.32	11.11	6.62	14.95
C30H59C1106	551.25	550.4000	65.37	10.79	6.43	17.41
C30H59C1107	567.25	566.3949	63.52	10.48	6.25	19.74
C30H59C1108	583.25	582.3898	61.78	10.20	6.08	21.95
C30H59C1109	599.25	598.3847	60.13	9.92	5.92	24.03
C30H60C1201	507.72	506.4021	70.97	11.91	13.97	3.15
C30H60C1202	523.72	522.3970	68.80	11.55	13.54	6.11
C30H60C1203	539.72	538.3919	66.76	11.21	13.14	8.89
C30H60C1204	555.72	554.3868	64.84	10.88	12.76	11.52
C30H60C1205	571.72	570.3818	63.03	10.58	12.40	13.99
C30H60C1206	587.72	586.3767	61.31	10.29	12.06	16.33
C30H60C1207	603.71	602.3716	59.69	10.02	11.74	18.55
C30H60C1208	619.71	618.3665	58.15	9.76	11.44	20.65
C30H60C1209	635.71	634.3614	56.68	9.51	11.15	22.65
C30H61C1101	473.27	472.4411	76.14	12.99	7.49	3.38
C30H61C1102	489.27	488.4360	73.65	12.57	7.25	6.54
C30H61C1103	505.27	504.4309	71.31	12.17	7.02	9.50
C30H61C1104	521.27	520.4258	69.13	11.80	6.80	12.28
C30H61C1105	537.27	536.4207	67.07	11.44	6.60	14.89
C30H61C1106	553.27	552.4156	65.13	11.11	6.41	17.35
C30H61C1107	569.27	568.4106	63.30	10.80	6.23	19.67
C30H61C1108	585.27	584.4055	61.57	10.51	6.06	21.87
C30H61C1109	601.27	600.4004	59.93	10.23	5.90	23.95

5. CARBON, HYDROGEN, FLUORINE AND NITROGEN

5. KOHLENSTOFF, WASSERSTOFF, FLUOR UND STICKSTOFF

5. CARBONE, HYDROGENE, FLUOR ET AZOTE

MOLECULAR WEIGHTS AND PERCENTAGE COMPOSITIONS OF COMPOUNDS CONTAINING CARBON, HYDROGEN, FLUORINE AND NITROGEN

FORMULA	MOL. WEIGHT	ISOMOL. WEIGHT	PERC. C	PERC. H	PERC. F	PERC. N
C 1H 0F1N1	45.02	45.0015	26.68	0	42.20	31.11
C 1H 0F1N3	73.03	73.0076	16.45	0	26.01	57.54
C 1H 0F1N5	101.04	101.0138	11.89	0	18.80	69.31
C 1H 0F2N2	78.02	78.0030	15.39	0	48.70	35.90
C 1H 0F2N4	106.03	106.0091	11.33	0	35.83	52.84
C 1H 0F2N6	134.05	134.0152	8.96	0	28.35	62.69
C 1H 1F1N2	60.03	60.0124	20.01	1.68	31.65	46.66
C 1H 1F1N4	88.04	88.0185	13.64	1.14	21.58	63.63
C 1H 1F1N6	116.06	116.0247	10.35	.87	16.37	72.41
C 1H 1F2N1	65.02	65.0077	18.47	1.55	58.44	21.54
C 1H 1F2N3	93.04	93.0139	12.91	1.08	40.84	45.17
C 1H 1F2N5	121.05	121.0200	9.92	.83	31.39	57.86
C 1H 2F1N1	47.03	47.0171	25.54	4.29	40.39	29.78
C 1H 2F1N3	75.05	75.0233	16.01	2.69	25.32	55.99
C 1H 2F1N5	103.06	103.0294	11.65	1.96	18.43	67.95
C 1H 2F2N2	80.04	80.0186	15.01	2.52	47.47	35.00
C 1H 2F2N4	108.05	108.0247	11.12	1.87	35.17	51.85
C 1H 2F2N6	136.06	136.0309	8.83	1.48	27.93	61.77
C 1H 3F1N2	62.05	62.0280	19.36	4.87	30.62	45.15
C 1H 3F1N4	90.06	90.0342	13.34	3.36	21.10	62.21
C 1H 3F1N6	118.07	118.0403	10.17	2.56	16.09	71.18
C 1H 3F2N1	67.04	67.0234	17.92	4.51	56.68	20.89
C 1H 3F2N3	95.05	95.0295	12.64	3.18	39.97	44.21
C 1H 3F2N5	123.07	123.0356	9.76	2.46	30.88	56.91
C 1H 4F1N1	49.05	49.0328	24.49	8.22	38.73	28.56
C 1H 4F1N3	77.06	77.0389	15.59	5.23	24.65	54.53
C 1H 4F1N5	105.07	105.0451	11.43	3.84	18.08	66.65
C 1H 4F2N2	82.05	82.0343	14.64	4.91	46.31	34.14
C 1H 4F2N4	110.07	110.0404	10.91	3.66	34.52	50.90
C 1H 4F2N6	138.08	138.0465	8.70	2.92	27.52	60.86
C 1H 5F1N2	64.06	64.0437	18.75	7.87	29.66	43.73
C 1H 5F1N4	92.08	92.0498	13.04	5.47	20.63	60.85
C 1H 5F1N6	120.09	120.0560	10.00	4.20	15.82	69.98
C 1H 5F2N3	97.07	97.0451	12.37	5.19	39.14	43.29
C 1H 5F2N5	125.08	125.0513	9.60	4.03	30.38	55.99
C 1H 6F1N3	79.08	79.0546	15.19	7.65	24.03	53.14
C 1H 6F1N5	107.09	107.0607	11.22	5.65	17.74	65.40
C 1H 6F2N4	112.08	112.0560	10.72	5.40	33.90	49.99
C 1H 6F2N6	140.10	140.0622	8.57	4.32	27.12	59.99
C 1H 7F1N4	94.09	94.0655	12.77	7.50	20.19	59.54
C 1H 7F1N6	122.11	122.0716	9.84	5.78	15.56	68.83
C 1H 7F2N5	127.10	127.0669	9.45	5.55	29.90	55.10
C 1H 8F1N5	109.11	109.0764	11.01	7.39	17.41	64.19
C 1H 8F2N6	142.11	142.0778	8.45	5.67	26.74	59.14
C 1H 9F1N6	124.12	124.0873	9.68	7.31	15.31	67.71
C 2H 1F1N2	72.04	72.0124	33.34	1.40	26.37	38.88
C 2H 1F1N4	100.06	100.0185	24.01	1.01	18.99	56.00
C 2H 1F1N6	128.07	128.0247	18.76	.79	14.83	65.62
C 2H 1F2N1	77.03	77.0077	31.18	1.31	49.32	18.18
C 2H 1F2N3	105.05	105.0139	22.87	.96	36.17	40.00

FORMULA	MOL. WEIGHT	ISOMOL. WEIGHT	PERC. C	PERC. H	PERC. F	PERC. N
C 2H 1F2N5	133.06	133.0200	18.05	.76	28.56	52.63
C 2H 2F1N1	59.04	59.0171	40.69	3.41	32.18	23.72
C 2H 2F1N3	87.06	87.0233	27.59	2.32	21.82	48.27
C 2H 2F1N5	115.07	115.0294	20.88	1.75	16.51	60.86
C 2H 2F2N2	92.05	92.0186	26.10	2.19	41.28	30.43
C 2H 2F2N4	120.06	120.0247	20.01	1.68	31.65	46.66
C 2H 2F2N6	148.08	148.0309	16.22	1.36	25.66	56.76
C 2H 3F1N2	74.06	74.0280	32.44	4.08	25.65	37.83
C 2H 3F1N4	102.07	102.0342	23.53	2.96	18.61	54.89
C 2H 3F1N6	130.08	130.0403	18.47	2.32	14.60	64.60
C 2H 3F2N1	79.05	79.0234	30.39	3.83	48.07	17.72
C 2H 3F2N3	107.06	107.0295	22.44	2.82	35.49	39.25
C 2H 3F2N5	135.08	135.0356	17.78	2.24	28.13	51.85
C 2H 4F1N1	61.06	61.0328	39.34	6.60	31.11	22.94
C 2H 4F1N3	89.07	89.0389	26.97	4.53	21.33	47.18
C 2H 4F1N5	117.09	117.0451	20.52	3.44	16.23	59.81
C 2H 4F2N2	94.06	94.0343	25.54	4.29	40.39	29.78
C 2H 4F2N4	122.08	122.0404	19.68	3.30	31.13	45.89
C 2H 4F2N6	150.09	150.0465	16.01	2.69	25.32	55.99
C 2H 5F1N2	76.07	76.0437	31.58	6.62	24.97	36.82
C 2H 5F1N4	104.09	104.0498	23.08	4.84	18.25	53.83
C 2H 5F1N6	132.10	132.0560	18.18	3.82	14.38	63.62
C 2H 5F2N1	81.07	81.0390	29.63	6.22	46.87	17.28
C 2H 5F2N3	109.08	109.0451	22.02	4.62	34.83	38.52
C 2H 5F2N5	137.09	137.0513	17.52	3.68	27.72	51.08
C 2H 6F1N1	63.08	63.0484	38.09	9.59	30.12	22.21
C 2H 6F1N3	91.09	91.0546	26.37	6.64	20.86	46.13
C 2H 6F1N5	119.10	119.0607	20.17	5.08	15.95	58.80
C 2H 6F2N2	96.08	96.0499	25.00	6.29	39.55	29.16
C 2H 6F2N4	124.09	124.0560	19.36	4.87	30.62	45.15
C 2H 6F2N6	152.11	152.0622	15.79	3.98	24.98	55.25
C 2H 7F1N2	78.09	78.0593	30.76	9.04	24.33	35.87
C 2H 7F1N4	106.10	106.0655	22.64	6.65	17.91	52.80
C 2H 7F1N6	134.12	134.0716	17.91	5.26	14.17	62.66
C 2H 7F2N3	111.09	111.0608	21.62	6.35	34.20	37.82
C 2H 7F2N5	139.11	139.0669	17.27	5.07	27.31	50.34
C 2H 8F1N3	93.10	93.0702	25.80	8.66	20.41	45.13
C 2H 8F1N5	121.12	121.0764	19.83	6.66	15.69	57.82
C 2H 8F2N4	126.11	126.0717	19.05	6.39	30.13	44.43
C 2H 8F2N6	154.12	154.0778	15.59	5.23	24.65	54.53
C 2H 9F1N4	108.12	108.0811	22.22	8.39	17.57	51.82
C 2H 9F1N6	136.13	136.0873	17.65	6.66	13.96	61.73
C 2H 9F2N5	141.12	141.0826	17.02	6.43	26.92	49.63
C 2H10F1N5	123.13	123.0920	19.51	8.19	15.43	56.88
C 2H10F2N6	156.14	156.0935	15.39	6.46	24.34	53.82
C 2H11F1N6	138.15	138.1029	17.39	8.03	13.75	60.83
C 3H 1F1N2	84.05	84.0124	42.87	1.20	22.60	33.33
C 3H 1F1N4	112.07	112.0185	32.15	.90	16.95	49.99
C 3H 1F1N6	140.08	140.0247	25.72	.72	13.56	59.99
C 3H 1F2N1	89.04	89.0077	40.47	1.13	42.67	15.73

C 3H 1F2N3	117.06	117.0139	30.78	.86	32.46	35.90
C 3H 1F2N5	145.07	145.0200	24.84	.69	26.19	48.28
C 3H 2F1N1	71.05	71.0171	50.71	2.84	26.74	19.71
C 3H 2F1N3	99.07	99.0233	36.37	2.03	19.18	42.42
C 3H 2F1N5	127.08	127.0294	28.35	1.59	14.95	55.11
C 3H 2F2N2	104.06	104.0186	34.63	1.94	36.51	26.92
C 3H 2F2N4	132.07	132.0247	27.28	1.53	28.77	42.42
C 3H 2F2N6	160.09	160.0309	22.51	1.26	23.74	52.50
C 3H 3F1N2	86.07	86.0280	41.87	3.51	22.07	32.55
C 3H 3F1N4	114.08	114.0342	31.59	2.65	16.65	49.11
C 3H 3F1N6	142.10	142.0403	25.36	2.13	13.37	59.14
C 3H 3F2N1	91.06	91.0234	39.57	3.32	41.73	15.38
C 3H 3F2N3	119.07	119.0295	30.26	2.54	31.91	35.29
C 3H 3F2N5	147.09	147.0356	24.50	2.06	25.83	47.61
C 3H 4F1N1	73.07	73.0328	49.31	5.52	26.00	19.17
C 3H 4F1N3	101.08	101.0389	35.65	3.99	18.79	41.57
C 3H 4F1N5	129.10	129.0451	27.91	3.12	14.72	54.25
C 3H 4F2N2	106.08	106.0343	33.97	3.80	35.82	26.41
C 3H 4F2N4	134.09	134.0404	26.87	3.01	28.34	41.78
C 3H 4F2N6	162.10	162.0465	22.23	2.49	23.44	51.84
C 3H 5F1N2	88.09	88.0437	40.91	5.72	21.57	31.80
C 3H 5F1N4	116.10	116.0498	31.04	4.34	16.36	48.26
C 3H 5F1N6	144.11	144.0560	25.00	3.50	13.18	58.32
C 3H 5F2N1	93.08	93.0390	38.71	5.41	40.82	15.05
C 3H 5F2N3	121.09	121.0451	29.76	4.16	31.38	34.70
C 3H 5F2N5	149.10	149.0513	24.17	3.38	25.48	46.97
C 3H 6F1N1	75.09	75.0484	47.99	8.05	25.30	18.65
C 3H 6F1N3	103.10	103.0546	34.95	5.87	18.43	40.76
C 3H 6F1N5	131.11	131.0607	27.48	4.61	14.49	53.41
C 3H 6F2N2	108.09	108.0499	33.34	5.60	35.15	25.92
C 3H 6F2N4	136.10	136.0560	26.47	4.44	27.92	41.16
C 3H 6F2N6	164.12	164.0622	21.96	3.69	23.15	51.21
C 3H 7F1N2	90.10	90.0593	39.99	7.83	21.09	31.09
C 3H 7F1N4	118.11	118.0655	30.51	5.97	16.08	47.43
C 3H 7F1N6	146.13	146.0716	24.66	4.83	13.00	57.51
C 3H 7F2N1	95.09	95.0547	37.89	7.42	39.96	14.73
C 3H 7F2N3	123.11	123.0608	29.27	5.73	30.87	34.13
C 3H 7F2N5	151.12	151.0669	23.84	4.67	25.14	46.34
C 3H 8F1N1	77.10	77.0641	46.73	10.46	24.64	18.17
C 3H 8F1N3	105.12	105.0702	34.28	7.67	18.07	39.98
C 3H 8F1N5	133.13	133.0764	27.07	6.06	14.27	52.61
C 3H 8F2N2	110.11	110.0655	32.73	7.32	34.51	25.44
C 3H 8F2N4	138.12	138.0717	26.09	5.84	27.51	40.56
C 3H 8F2N6	166.13	166.0778	21.69	4.85	22.87	50.59
C 3H 9F1N2	92.12	92.0750	39.12	9.85	20.62	30.41
C 3H 9F1N4	120.13	120.0811	30.00	7.55	15.81	46.64
C 3H 9F1N6	148.14	148.0873	24.32	6.12	12.82	56.73
C 3H 9F2N3	125.12	125.0764	28.80	7.25	30.37	33.58
C 3H 9F2N5	153.14	153.0826	23.53	5.92	24.81	45.73
C 3H10F1N3	107.13	107.0859	33.63	9.41	17.73	39.22
C 3H10F1N5	135.15	135.0920	26.66	7.46	14.06	51.82
C 3H10F2N4	140.14	140.0873	25.71	7.19	27.11	39.98
C 3H10F2N6	168.15	168.0935	21.43	5.99	22.60	49.98
C 3H11F1N4	122.15	122.0968	29.50	9.08	15.55	45.87
C 3H11F1N6	150.16	150.1029	24.00	7.38	12.65	55.97
C 3H11F2N5	155.15	155.0982	23.22	7.15	24.49	45.14
C 3H12F1N5	137.16	137.1077	26.27	8.82	13.85	51.06
C 3H12F2N6	170.17	170.1091	21.18	7.11	22.33	49.39
C 3H13F1N6	152.18	152.1186	23.68	8.61	12.48	55.23
C 4H 2F1N1	83.07	83.0171	57.84	2.43	22.87	16.86

C 4H 2F1N3	111.08	111.0233	43.25	1.81	17.10	37.83
C 4H 2F1N5	139.09	139.0294	34.54	1.45	13.66	50.35
C 4H 2F2N2	116.07	116.0186	41.39	1.74	32.74	24.13
C 4H 2F2N4	144.08	144.0247	33.34	1.40	26.37	38.88
C 4H 2F2N6	172.10	172.0309	27.92	1.17	22.08	48.83
C 4H 3F1N2	98.08	98.0280	48.98	3.08	19.37	28.56
C 4H 3F1N4	126.09	126.0342	38.10	2.40	15.07	44.43
C 4H 3F1N6	154.11	154.0403	31.18	1.96	12.33	54.53
C 4H 3F2N1	103.07	103.0234	46.61	2.93	36.86	13.59
C 4H 3F2N3	131.09	131.0295	36.65	2.31	28.99	32.06
C 4H 3F2N5	159.10	159.0356	30.20	1.90	23.88	44.02
C 4H 4F1N1	85.08	85.0328	56.47	4.74	22.33	16.46
C 4H 4F1N3	113.09	113.0389	42.48	3.57	16.80	37.15
C 4H 4F1N5	141.11	141.0451	34.05	2.86	13.46	49.63
C 4H 4F2N2	118.09	118.0343	40.69	3.41	32.18	23.72
C 4H 4F2N4	146.10	146.0404	32.88	2.76	26.01	38.35
C 4H 4F2N6	174.11	174.0465	27.59	2.32	21.82	48.27
C 4H 5F1N2	100.10	100.0437	48.00	5.04	18.98	27.99
C 4H 5F1N4	128.11	128.0498	37.50	3.93	14.83	43.73
C 4H 5F1N6	156.12	156.0560	30.77	3.23	12.17	53.83
C 4H 5F2N1	105.09	105.0390	45.72	4.80	36.16	13.33
C 4H 5F2N3	133.10	133.0451	36.10	3.79	28.55	31.57
C 4H 5F2N5	161.11	161.0513	29.82	3.13	23.58	43.47
C 4H 6F1N1	87.10	87.0484	55.16	6.94	21.81	16.08
C 4H 6F1N3	115.11	115.0546	41.74	5.25	16.50	36.50
C 4H 6F1N5	143.12	143.0607	33.57	4.23	13.27	48.93
C 4H 6F2N2	120.10	120.0499	40.00	5.04	31.64	23.32
C 4H 6F2N4	148.12	148.0560	32.44	4.08	25.65	37.83
C 4H 6F2N6	176.13	176.0622	27.28	3.43	21.57	47.72
C 4H 7F1N2	102.11	102.0593	47.05	6.91	18.61	27.43
C 4H 7F1N4	130.13	130.0655	36.92	5.42	14.60	43.06
C 4H 7F1N6	158.14	158.0716	30.38	4.46	12.01	53.14
C 4H 7F2N1	107.10	107.0547	44.86	6.59	35.48	13.08
C 4H 7F2N3	135.12	135.0608	35.56	5.22	28.12	31.10
C 4H 7F2N5	163.13	163.0669	29.45	4.33	23.29	42.93
C 4H 8F1N1	89.11	89.0641	53.91	9.05	21.32	15.72
C 4H 8F1N3	117.13	117.0702	41.02	6.88	16.22	35.88
C 4H 8F1N5	145.14	145.0764	33.10	5.56	13.09	48.25
C 4H 8F2N2	122.12	122.0655	39.34	6.60	31.11	22.94
C 4H 8F2N4	150.13	150.0717	32.00	5.37	25.31	37.32
C 4H 8F2N6	178.15	178.0778	26.97	4.53	21.33	47.18
C 4H 9F1N2	104.13	104.0750	46.14	8.71	18.25	26.90
C 4H 9F1N4	132.14	132.0811	36.36	6.87	14.38	42.40
C 4H 9F1N6	160.15	160.0873	30.00	5.66	11.86	52.47
C 4H 9F2N1	109.12	109.0703	44.03	8.31	34.82	12.84
C 4H 9F2N3	137.13	137.0764	35.03	6.62	27.71	30.64
C 4H 9F2N5	165.15	165.0826	29.09	5.49	23.01	42.41
C 4H10F1N1	91.13	91.0797	52.72	11.06	20.85	15.37
C 4H10F1N3	119.14	119.0859	40.33	8.46	15.95	35.27
C 4H10F1N5	147.16	147.0920	32.65	6.85	12.91	47.59
C 4H10F2N2	124.13	124.0812	38.70	8.12	30.61	22.57
C 4H10F2N4	152.15	152.0873	31.58	6.62	24.97	36.82
C 4H10F2N6	180.16	180.0935	26.67	5.59	21.09	46.65
C 4H11F1N2	106.14	106.0906	45.26	10.45	17.90	26.39
C 4H11F1N4	134.16	134.0968	35.81	8.26	14.16	41.76
C 4H11F1N6	162.17	162.1029	29.63	6.84	11.72	51.82
C 4H11F2N3	139.15	139.0921	34.53	7.97	27.31	30.20
C 4H11F2N5	167.16	167.0982	28.74	6.63	22.73	41.90
C 4H12F1N3	121.16	121.1015	39.65	9.98	15.68	34.68
C 4H12F1N5	149.17	149.1077	32.21	8.11	12.74	46.95

C 4H12F2N4	154.16	154.1030	31.16	7.85	24.65	36.34
C 4H12F2N6	182.18	182.1091	26.37	6.64	20.86	46.13
C 4H13F1N4	136.17	136.1124	35.28	9.62	13.95	41.14
C 4H13F1N6	164.19	164.1186	29.26	7.98	11.57	51.19
C 4H13F2N5	169.18	169.1139	28.40	7.75	22.46	41.40
C 4H14F1N5	151.19	151.1233	31.78	9.33	12.57	46.32
C 4H14F2N6	184.19	184.1248	26.08	7.66	20.63	45.63
C 4H15F1N6	166.20	166.1342	28.91	9.10	11.43	50.56
C 5H 2F1N1	95.08	95.0171	63.17	2.12	19.98	14.73
C 5H 2F1N3	123.09	123.0233	48.79	1.64	15.43	34.14
C 5H 2F1N5	151.10	151.0294	39.74	1.33	12.57	46.35
C 5H 2F2N2	128.08	128.0186	46.89	1.57	29.67	21.87
C 5H 2F2N4	156.10	156.0247	38.47	1.29	24.34	35.89
C 5H 2F2N6	184.11	184.0309	32.62	1.09	20.64	45.65
C 5H 3F1N2	110.09	110.0280	54.55	2.75	17.26	25.45
C 5H 3F1N4	138.10	138.0342	43.49	2.19	13.76	40.57
C 5H 3F1N6	166.12	166.0403	36.15	1.82	11.44	50.59
C 5H 3F2N1	115.08	115.0234	52.18	2.63	33.02	12.17
C 5H 3F2N3	143.10	143.0295	41.97	2.11	26.55	29.36
C 5H 3F2N5	171.11	171.0356	35.10	1.77	22.21	40.93
C 5H 4F1N1	97.09	97.0328	61.85	4.15	19.57	14.43
C 5H 4F1N3	125.11	125.0389	48.00	3.22	15.19	33.59
C 5H 4F1N5	153.12	153.0451	39.22	2.63	12.41	45.74
C 5H 4F2N2	130.10	130.0343	46.16	3.10	29.21	21.53
C 5H 4F2N4	158.11	158.0404	37.98	2.55	24.03	35.44
C 5H 4F2N6	186.12	186.0465	32.27	2.17	20.41	45.15
C 5H 5F1N2	112.11	112.0437	53.57	4.50	16.95	24.99
C 5H 5F1N4	140.12	140.0498	42.86	3.60	13.56	39.98
C 5H 5F1N6	168.13	168.0560	35.72	3.00	11.30	49.98
C 5H 5F2N1	117.10	117.0390	51.29	4.30	32.45	11.96
C 5H 5F2N3	145.11	145.0451	41.39	3.47	26.18	28.96
C 5H 5F2N5	173.13	173.0513	34.69	2.91	21.95	40.45
C 5H 6F1N1	99.11	99.0484	60.60	6.10	19.17	14.13
C 5H 6F1N3	127.12	127.0546	47.24	4.76	14.95	33.05
C 5H 6F1N5	155.14	155.0607	38.71	3.90	12.25	45.14
C 5H 6F2N2	132.11	132.0499	45.46	4.58	28.76	21.20
C 5H 6F2N4	160.13	160.0560	37.51	3.78	23.73	34.99
C 5H 6F2N6	188.14	188.0622	31.92	3.21	20.20	44.67
C 5H 7F1N2	114.12	114.0593	52.62	6.18	16.65	24.55
C 5H 7F1N4	142.14	142.0655	42.25	4.96	13.37	39.42
C 5H 7F1N6	170.15	170.0716	35.30	4.15	11.17	49.39
C 5H 7F2N1	119.12	119.0547	50.42	5.92	31.90	11.76
C 5H 7F2N3	147.13	147.0608	40.82	4.80	25.83	28.56
C 5H 7F2N5	175.14	175.0669	34.29	4.03	21.69	39.99
C 5H 8F1N1	101.12	101.0641	59.39	7.97	18.79	13.85
C 5H 8F1N3	129.14	129.0702	46.51	6.24	14.71	32.54
C 5H 8F1N5	157.15	157.0764	38.22	5.13	12.09	44.56
C 5H 8F2N2	134.13	134.0655	44.77	6.01	28.33	20.89
C 5H 8F2N4	162.14	162.0717	37.04	4.97	23.43	34.55
C 5H 8F2N6	190.16	190.0778	31.58	4.24	19.98	44.20
C 5H 9F1N2	116.14	116.0750	51.71	7.81	16.36	24.12
C 5H 9F1N4	144.15	144.0811	41.66	6.29	13.18	38.87
C 5H 9F1N6	172.17	172.0873	34.88	5.27	11.03	48.81
C 5H 9F2N1	121.13	121.0703	49.58	7.49	31.37	11.56
C 5H 9F2N3	149.14	149.0764	40.27	6.08	25.48	28.17
C 5H 9F2N5	177.16	177.0826	33.90	5.12	21.45	39.53
C 5H10F1N1	103.14	103.0797	58.23	9.77	18.42	13.58
C 5H10F1N3	131.15	131.0859	45.79	7.69	14.49	32.04
C 5H10F1N5	159.17	159.0920	37.73	6.33	11.94	44.00
C 5H10F2N2	136.15	136.0812	44.11	7.40	27.91	20.58

C 5H10F2N4	164.16	164.0873	36.58	6.14	23.15	34.13
C 5H10F2N6	192.17	192.0935	31.25	5.25	19.77	43.73
C 5H11F1N2	118.16	118.0906	50.83	9.38	16.08	23.71
C 5H11F1N4	146.17	146.0968	41.09	7.59	13.00	38.33
C 5H11F1N6	174.18	174.1029	34.48	6.37	10.91	48.25
C 5H11F2N1	123.15	123.0859	48.77	9.00	30.85	11.37
C 5H11F2N3	151.16	151.0921	39.73	7.34	25.14	27.80
C 5H11F2N5	179.17	179.0982	33.52	6.19	21.21	39.09
C 5H12F1N1	105.16	105.0954	57.11	11.50	18.07	13.32
C 5H12F1N3	133.17	133.1015	45.10	9.08	14.27	31.55
C 5H12F1N5	161.18	161.1077	37.26	7.50	11.79	43.45
C 5H12F2N2	138.16	138.0968	43.47	8.75	27.50	20.28
C 5H12F2N4	166.17	166.1030	36.14	7.28	22.87	33.72
C 5H12F2N6	194.19	194.1091	30.93	6.23	19.57	43.28
C 5H13F1N2	120.17	120.1063	49.98	10.90	15.81	23.31
C 5H13F1N4	148.18	148.1124	40.53	8.84	12.82	37.81
C 5H13F1N6	176.20	176.1186	34.08	7.44	10.78	47.70
C 5H13F2N3	153.18	153.1077	39.21	8.55	24.81	27.43
C 5H13F2N5	181.19	181.1139	33.15	7.23	20.97	38.65
C 5H14F1N3	135.19	135.1172	44.42	10.44	14.05	31.08
C 5H14F1N5	163.20	163.1233	36.80	8.65	11.64	42.91
C 5H14F2N4	168.19	168.1186	35.71	8.39	22.59	33.31
C 5H14F2N6	196.20	196.1248	30.61	7.19	19.37	42.83
C 5H15F1N4	150.20	150.1281	39.98	10.07	12.65	37.30
C 5H15F1N6	178.21	178.1342	33.70	8.48	10.66	47.16
C 5H15F2N5	183.21	183.1295	32.78	8.25	20.74	38.23
C 5H16F1N5	165.22	165.1390	36.35	9.76	11.50	42.39
C 5H16F2N6	198.22	198.1404	30.30	8.14	19.17	42.40
C 5H17F1N6	180.23	180.1499	33.32	9.51	10.54	46.63
C 6H 3F1N2	122.10	122.0280	59.02	2.48	15.56	22.94
C 6H 3F1N4	150.12	150.0342	48.01	2.01	12.66	37.32
C 6H 3F1N6	178.13	178.0403	40.46	1.70	10.67	47.18
C 6H 3F2N1	127.09	127.0234	56.70	2.38	29.90	11.02
C 6H 3F2N3	155.11	155.0295	46.46	1.95	24.50	27.09
C 6H 3F2N5	183.12	183.0356	39.35	1.65	20.75	38.24
C 6H 4F1N1	109.10	109.0328	66.05	3.70	17.41	12.84
C 6H 4F1N3	137.12	137.0389	52.56	2.94	13.86	30.65
C 6H 4F1N5	165.13	165.0451	43.64	2.44	11.51	42.41
C 6H 4F2N2	142.11	142.0343	50.71	2.84	26.74	19.71
C 6H 4F2N4	170.12	170.0404	42.36	2.37	22.33	32.93
C 6H 4F2N6	198.14	198.0465	36.37	2.03	19.18	42.42
C 6H 5F1N2	124.12	124.0437	58.06	4.06	15.31	22.57
C 6H 5F1N4	152.13	152.0498	47.37	3.31	12.49	36.83
C 6H 5F1N6	180.15	180.0560	40.00	2.80	10.55	46.65
C 6H 5F2N1	129.11	129.0390	55.82	3.90	29.43	10.85
C 6H 5F2N3	157.12	157.0451	45.87	3.21	24.18	26.74
C 6H 5F2N5	185.14	185.0513	38.93	2.72	20.52	37.83
C 6H 6F1N1	111.12	111.0484	64.86	5.44	17.10	12.61
C 6H 6F1N3	139.13	139.0546	51.80	4.35	13.65	30.20
C 6H 6F1N5	167.15	167.0607	43.12	3.62	11.37	41.90
C 6H 6F2N2	144.12	144.0499	50.00	4.20	26.36	19.44
C 6H 6F2N4	172.14	172.0560	41.87	3.51	22.07	32.55
C 6H 6F2N6	200.15	200.0622	36.01	3.02	18.98	41.99
C 6H 7F1N2	126.13	126.0593	57.13	5.59	15.06	22.21
C 6H 7F1N4	154.15	154.0655	46.75	4.58	12.32	36.35
C 6H 7F1N6	182.16	182.0716	39.56	3.87	10.43	46.14
C 6H 7F2N1	131.13	131.0547	54.96	5.38	28.98	10.68
C 6H 7F2N3	159.14	159.0608	45.29	4.43	23.88	26.40
C 6H 7F2N5	187.15	187.0669	38.51	3.77	20.30	37.42
C 6H 8F1N1	113.14	113.0641	63.70	7.13	16.79	12.38

C 6H 8F1N3	141.15	141.0702	51.06	5.71	13.46	29.77
C 6H 8F1N5	169.16	169.0764	42.60	4.77	11.23	41.40
C 6H 8F2N2	146.14	146.0655	49.31	5.52	26.00	19.17
C 6H 8F2N4	174.15	174.0717	41.38	4.63	21.82	32.17
C 6H 8F2N6	202.17	202.0778	35.65	3.99	18.79	41.57
C 6H 9F1N2	128.15	128.0750	56.24	7.08	14.83	21.86
C 6H 9F1N4	156.16	156.0811	46.15	5.81	12.17	35.88
C 6H 9F1N6	184.18	184.0873	39.13	4.93	10.32	45.63
C 6H 9F2N1	133.14	133.0703	54.13	6.81	28.54	10.52
C 6H 9F2N3	161.16	161.0764	44.72	5.63	23.58	26.07
C 6H 9F2N5	189.17	189.0826	38.10	4.80	20.09	37.02
C 6H10F1N1	115.15	115.0797	62.58	8.75	16.50	12.16
C 6H10F1N3	143.17	143.0859	50.34	7.04	13.27	29.35
C 6H10F1N5	171.18	171.0920	42.10	5.89	11.10	40.91
C 6H10F2N2	148.16	148.0812	48.64	6.80	25.65	18.91
C 6H10F2N4	176.17	176.0873	40.91	5.72	21.57	31.80
C 6H10F2N6	204.18	204.0935	35.30	4.94	18.61	41.16
C 6H11F1N2	130.17	130.0906	55.37	8.52	14.60	21.52
C 6H11F1N4	158.18	158.0968	45.56	7.01	12.01	35.42
C 6H11F1N6	186.19	186.1029	38.71	5.95	10.20	45.14
C 6H11F2N1	135.16	135.0859	53.32	8.20	28.11	10.36
C 6H11F2N3	163.17	163.0921	44.17	6.80	23.29	25.75
C 6H11F2N5	191.18	191.0982	37.69	5.80	19.87	36.63
C 6H12F1N1	117.17	117.0954	61.51	10.32	16.21	11.95
C 6H12F1N3	145.18	145.1015	49.64	8.33	13.09	28.94
C 6H12F1N5	173.19	173.1077	41.61	6.98	10.97	40.44
C 6H12F2N2	150.17	150.0968	47.99	8.05	25.30	18.65
C 6H12F2N4	178.19	178.1030	40.44	6.79	21.32	31.44
C 6H12F2N6	206.20	206.1091	34.95	5.87	18.43	40.76
C 6H13F1N2	132.18	132.1063	54.52	9.91	14.37	21.19
C 6H13F1N4	160.20	160.1124	44.99	8.18	11.86	34.97
C 6H13F1N6	188.21	188.1186	38.29	6.96	10.09	44.65
C 6H13F2N1	137.17	137.1016	52.54	9.55	27.70	10.21
C 6H13F2N3	165.19	165.1077	43.63	7.93	23.00	25.44
C 6H13F2N5	193.20	193.1139	37.30	6.78	19.67	36.25
C 6H14F1N1	119.18	119.1110	60.47	11.84	15.94	11.75
C 6H14F1N3	147.20	147.1172	48.96	9.59	12.91	28.55
C 6H14F1N5	175.21	175.1233	41.13	8.05	10.84	39.97
C 6H14F2N2	152.19	152.1125	47.35	9.27	24.97	18.41
C 6H14F2N4	180.20	180.1186	39.99	7.83	21.09	31.09
C 6H14F2N6	208.22	208.1248	34.61	6.78	18.25	40.36
C 6H15F1N2	134.20	134.1219	53.70	11.27	14.16	20.87
C 6H15F1N4	162.21	162.1281	44.43	9.32	11.71	34.54
C 6H15F1N6	190.23	190.1342	37.89	7.95	9.99	44.18
C 6H15F2N3	167.20	167.1234	43.10	9.04	22.72	25.13
C 6H15F2N5	195.22	195.1295	36.92	7.75	19.46	35.87
C 6H16F1N3	149.21	149.1328	48.30	10.81	12.73	28.16
C 6H16F1N5	177.23	177.1390	40.66	9.10	10.72	39.52
C 6H16F2N4	182.22	182.1343	39.55	8.85	20.85	30.75
C 6H16F2N6	210.23	210.1404	34.28	7.67	18.07	39.98
C 6H17F1N4	164.23	164.1437	43.88	10.43	11.57	34.12
C 6H17F1N6	192.24	192.1499	37.49	8.91	9.88	43.72
C 6H17F2N5	197.23	197.1452	36.54	8.69	19.26	35.51
C 6H18F1N5	179.24	179.1546	40.21	10.12	10.60	39.07
C 6H18F2N6	212.25	212.1561	33.95	8.55	17.90	39.60
C 6H19F1N6	194.26	194.1655	37.10	9.86	9.78	43.26
C 7H 3F1N2	134.11	134.0280	62.69	2.25	14.17	20.89
C 7H 3F1N4	162.13	162.0342	51.86	1.87	11.72	34.56
C 7H 3F1N6	190.14	190.0403	44.22	1.59	9.99	44.20
C 7H 3F2N1	139.11	139.0234	60.44	2.17	27.32	10.07

C 7H 3F2N3	167.12	167.0295	50.31	1.81	22.74	25.14
C 7H 3F2N5	195.13	195.0356	43.09	1.55	19.47	35.89
C 7H 4F1N1	121.12	121.0328	69.42	3.33	15.69	11.56
C 7H 4F1N3	149.13	149.0389	56.38	2.70	12.74	28.18
C 7H 4F1N5	177.14	177.0451	47.46	2.28	10.72	39.54
C 7H 4F2N2	154.12	154.0343	54.55	2.62	24.65	18.18
C 7H 4F2N4	182.13	182.0404	46.16	2.21	20.86	30.76
C 7H 4F2N6	210.15	210.0465	40.01	1.92	18.08	39.99
C 7H 5F1N2	136.13	136.0437	61.76	3.70	13.96	20.58
C 7H 5F1N4	164.14	164.0498	51.22	3.07	11.57	34.13
C 7H 5F1N6	192.16	192.0560	43.75	2.62	9.89	43.74
C 7H 5F2N1	141.12	141.0390	59.58	3.57	26.92	9.93
C 7H 5F2N3	169.13	169.0451	49.71	2.98	22.47	24.84
C 7H 5F2N5	197.15	197.0513	42.65	2.56	19.27	35.52
C 7H 6F1N1	123.13	123.0484	68.28	4.91	15.43	11.38
C 7H 6F1N3	151.14	151.0546	55.63	4.00	12.57	27.80
C 7H 6F1N5	179.16	179.0607	46.93	3.38	10.60	39.09
C 7H 6F2N2	156.14	156.0499	53.85	3.87	24.34	17.94
C 7H 6F2N4	184.15	184.0560	45.66	3.28	20.63	30.42
C 7H 6F2N6	212.16	212.0622	39.63	2.85	17.91	39.61
C 7H 7F1N2	138.15	138.0593	60.86	5.11	13.75	20.28
C 7H 7F1N4	166.16	166.0655	50.60	4.25	11.43	33.72
C 7H 7F1N6	194.17	194.0716	43.30	3.63	9.78	43.28
C 7H 7F2N1	143.14	143.0547	58.74	4.93	26.55	9.79
C 7H 7F2N3	171.15	171.0608	49.13	4.12	22.20	24.55
C 7H 7F2N5	199.16	199.0669	42.22	3.54	19.08	35.16
C 7H 8F1N1	125.15	125.0641	67.18	6.44	15.18	11.19
C 7H 8F1N3	153.16	153.0702	54.90	5.26	12.40	27.44
C 7H 8F1N5	181.17	181.0764	46.41	4.45	10.49	38.66
C 7H 8F2N2	158.15	158.0655	53.16	5.10	24.03	17.71
C 7H 8F2N4	186.17	186.0717	45.16	4.33	20.41	30.10
C 7H 8F2N6	214.18	214.0778	39.26	3.76	17.74	39.24
C 7H 9F1N2	140.16	140.0750	59.99	6.47	13.55	19.99
C 7H 9F1N4	168.17	168.0811	49.99	5.39	11.30	33.31
C 7H 9F1N6	196.19	196.0873	42.86	4.62	9.68	42.84
C 7H 9F2N1	145.15	145.0703	57.92	6.25	26.18	9.65
C 7H 9F2N3	173.17	173.0764	48.55	5.24	21.94	24.27
C 7H 9F2N5	201.18	201.0826	41.79	4.51	18.89	34.81
C 7H10F1N1	127.16	127.0797	66.12	7.93	14.94	11.01
C 7H10F1N3	155.18	155.0859	54.18	6.50	12.24	27.08
C 7H10F1N5	183.19	183.0920	45.90	5.50	10.37	38.23
C 7H10F2N2	160.17	160.0812	52.49	6.29	23.72	17.49
C 7H10F2N4	188.18	188.0873	44.68	5.36	20.19	29.77
C 7H10F2N6	216.19	216.0935	38.89	4.66	17.58	38.87
C 7H11F1N2	142.18	142.0906	59.14	7.80	13.36	19.70
C 7H11F1N4	170.19	170.0968	49.40	6.51	11.16	32.92
C 7H11F1N6	198.20	198.1029	42.42	5.59	9.59	42.40
C 7H11F2N1	147.17	147.0859	57.13	7.53	25.82	9.52
C 7H11F2N3	175.18	175.0921	47.99	6.33	21.69	23.99
C 7H11F2N5	203.20	203.0982	41.38	5.46	18.70	34.47
C 7H12F1N1	129.18	129.0954	65.09	9.36	14.71	10.84
C 7H12F1N3	157.19	157.1015	53.49	7.69	12.09	26.73
C 7H12F1N5	185.21	185.1077	45.40	6.53	10.26	37.81
C 7H12F2N2	162.18	162.0968	51.84	7.46	23.43	17.27
C 7H12F2N4	190.20	190.1030	44.21	6.36	19.98	29.46
C 7H12F2N6	218.21	218.1091	38.53	5.54	17.41	38.51
C 7H13F1N2	144.19	144.1063	58.31	9.09	13.18	19.43
C 7H13F1N4	172.21	172.1124	48.82	7.61	11.03	32.53
C 7H13F1N6	200.22	200.1186	41.99	6.54	9.49	41.97
C 7H13F2N1	149.19	149.1016	56.36	8.78	25.47	9.39

C 7H13F2N3	177.20	177.1077	47.45	7.39	21.44	23.71
C 7H13F2N5	205.21	205.1139	40.97	6.39	18.52	34.13
C 7H14F1N1	131.19	131.1110	64.09	10.76	14.48	10.68
C 7H14F1N3	159.21	159.1172	52.81	8.86	11.93	26.39
C 7H14F1N5	187.22	187.1233	44.91	7.54	10.15	37.41
C 7H14F2N2	164.20	164.1125	51.20	8.59	23.14	17.06
C 7H14F2N4	192.21	192.1186	43.74	7.34	19.77	29.15
C 7H14F2N6	220.23	220.1248	38.18	6.41	17.25	38.16
C 7H15F1N2	146.21	146.1219	57.51	10.34	12.99	19.16
C 7H15F1N4	174.22	174.1281	48.26	8.68	10.90	32.16
C 7H15F1N6	202.24	202.1342	41.57	7.48	9.39	41.56
C 7H15F2N1	151.20	151.1172	55.61	10.00	25.13	9.26
C 7H15F2N3	179.21	179.1234	46.91	8.44	21.20	23.45
C 7H15F2N5	207.23	207.1295	40.57	7.30	18.34	33.80
C 7H16F1N1	133.21	133.1267	63.12	12.11	14.26	10.51
C 7H16F1N3	161.22	161.1328	52.15	10.00	11.78	26.06
C 7H16F1N5	189.24	189.1390	44.43	8.52	10.04	37.01
C 7H16F2N2	166.22	166.1281	50.58	9.70	22.86	16.85
C 7H16F2N4	194.23	194.1343	43.29	8.30	19.56	28.85
C 7H16F2N6	222.24	222.1404	37.83	7.26	17.10	37.81
C 7H17F1N2	148.23	148.1376	56.72	11.56	12.82	18.90
C 7H17F1N4	176.24	176.1437	47.71	9.72	10.78	31.79
C 7H17F1N6	204.25	204.1499	41.16	8.39	9.30	41.15
C 7H17F2N3	181.23	181.1350	46.39	9.46	20.97	23.19
C 7H17F2N5	209.24	209.1452	40.18	8.19	18.16	33.47
C 7H18F1N3	163.24	163.1485	51.51	11.11	11.64	25.74
C 7H18F1N5	191.25	191.1546	43.96	9.49	9.93	36.62
C 7H18F2N4	196.25	196.1499	42.84	9.25	19.36	28.55
C 7H18F2N6	224.26	224.1561	37.49	8.09	16.94	37.47
C 7H19F1N4	178.25	178.1594	47.17	10.74	10.66	31.43
C 7H19F1N6	206.27	206.1655	40.76	9.28	9.21	40.74
C 7H19F2N5	211.26	211.1608	39.80	9.07	17.99	33.15
C 7H20F1N5	193.27	193.1703	43.50	10.43	9.83	36.24
C 7H20F2N6	226.27	226.1717	37.16	8.91	16.79	37.14
C 7H21F1N6	208.28	208.1812	40.37	10.16	9.12	40.35
C 8H 4F1N1	133.13	133.0328	72.18	3.03	14.27	10.52
C 8H 4F1N3	161.14	161.0389	59.63	2.50	11.79	26.08
C 8H 4F1N5	189.15	189.0451	50.80	2.13	10.04	37.02
C 8H 4F2N2	166.13	166.0343	57.84	2.43	22.87	16.86
C 8H 4F2N4	194.14	194.0404	49.49	2.08	19.57	28.86
C 8H 4F2N6	222.16	222.0465	43.25	1.81	17.10	37.83
C 8H 5F1N2	148.14	148.0437	64.86	3.40	12.82	18.91
C 8H 5F1N4	176.15	176.0498	54.55	2.86	10.79	31.81
C 8H 5F1N6	204.17	204.0560	47.06	2.47	9.31	41.16
C 8H 5F2N1	153.13	153.0390	62.75	3.29	24.81	9.15
C 8H 5F2N3	181.15	181.0451	53.05	2.78	20.98	23.20
C 8H 5F2N5	209.16	209.0513	45.94	2.41	18.17	33.48
C 8H 6F1N1	135.14	135.0484	71.10	4.48	14.06	10.36
C 8H 6F1N3	163.16	163.0546	58.89	3.71	11.64	25.75
C 8H 6F1N5	191.17	191.0607	50.26	3.16	9.94	36.63
C 8H 6F2N2	168.15	168.0499	57.15	3.60	22.60	16.66
C 8H 6F2N4	196.16	196.0560	48.98	3.08	19.37	28.56
C 8H 6F2N6	224.17	224.0622	42.86	2.70	16.95	37.49
C 8H 7F1N2	150.16	150.0593	63.99	4.70	12.65	18.66
C 8H 7F1N4	178.17	178.0655	53.93	3.96	10.66	31.45
C 8H 7F1N6	206.18	206.0716	46.60	3.42	9.21	40.76
C 8H 7F2N1	155.15	155.0547	61.93	4.55	24.49	9.03
C 8H 7F2N3	183.16	183.0608	52.46	3.85	20.74	22.94
C 8H 7F2N5	211.18	211.0669	45.50	3.34	17.99	33.16
C 8H 8F1N1	137.16	137.0641	70.06	5.88	13.85	10.21

C 8H 8F1N3	165.17	165.0702	58.18	4.88	11.50	25.44
C 8H 8F1N5	193.18	193.0764	49.74	4.17	9.83	36.25
C 8H 8F2N2	170.16	170.0655	56.47	4.74	22.33	16.46
C 8H 8F2N4	198.18	198.0717	48.49	4.07	19.17	28.27
C 8H 8F2N6	226.19	226.0778	42.48	3.57	16.80	37.15
C 8H 9F1N2	152.17	152.0750	63.14	5.96	12.48	18.41
C 8H 9F1N4	180.19	180.0811	53.33	5.03	10.54	31.09
C 8H 9F1N6	208.20	208.0873	46.15	4.36	9.13	40.37
C 8H 9F2N1	157.16	157.0703	61.14	5.77	24.18	8.91
C 8H 9F2N3	185.18	185.0764	51.89	4.90	20.52	22.69
C 8H 9F2N5	213.19	213.0826	45.07	4.26	17.82	32.85
C 8H10F1N1	139.17	139.0797	69.04	7.24	13.65	10.06
C 8H10F1N3	167.19	167.0859	57.47	6.03	11.36	25.13
C 8H10F1N5	195.20	195.0920	49.23	5.16	9.73	35.88
C 8H10F2N2	172.18	172.0812	55.81	5.85	22.07	16.27
C 8H10F2N4	200.19	200.0873	48.00	5.04	18.98	27.99
C 8H10F2N6	228.21	228.0935	42.11	4.42	16.65	36.83
C 8H11F1N2	154.19	154.0906	62.32	7.19	12.32	18.17
C 8H11F1N4	182.20	182.0968	52.74	6.09	10.43	30.75
C 8H11F1N6	210.22	210.1029	45.71	5.27	9.04	39.98
C 8H11F2N1	159.18	159.0859	60.36	6.97	23.87	8.80
C 8H11F2N3	187.19	187.0921	51.33	5.92	20.30	22.45
C 8H11F2N5	215.21	215.0982	44.65	5.15	17.66	32.54
C 8H12F1N1	141.19	141.0954	68.06	8.57	13.46	9.92
C 8H12F1N3	169.20	169.1015	56.79	7.15	11.23	24.83
C 8H12F1N5	197.22	197.1077	48.72	6.13	9.63	35.51
C 8H12F2N2	174.20	174.0968	55.16	6.94	21.81	16.08
C 8H12F2N4	202.21	202.1030	47.52	5.98	18.79	27.71
C 8H12F2N6	230.22	230.1091	41.74	5.25	16.50	36.50
C 8H13F1N2	156.20	156.1063	61.51	8.39	12.16	17.93
C 8H13F1N4	184.22	184.1124	52.16	7.11	10.31	30.41
C 8H13F1N6	212.23	212.1186	45.28	6.17	8.95	39.60
C 8H13F2N1	161.20	161.1016	59.61	8.13	23.57	8.69
C 8H13F2N3	189.21	189.1077	50.78	6.93	20.08	22.21
C 8H13F2N5	217.22	217.1139	44.24	6.03	17.49	32.24
C 8H14F1N1	143.21	143.1110	67.10	9.85	13.27	9.78
C 8H14F1N3	171.22	171.1172	56.12	8.24	11.10	24.54
C 8H14F1N5	199.23	199.1233	48.23	7.08	9.54	35.15
C 8H14F2N2	176.21	176.1125	54.53	8.01	21.56	15.90
C 8H14F2N4	204.22	204.1186	47.05	6.91	18.61	27.43
C 8H14F2N6	232.24	232.1248	41.38	6.08	16.36	36.19
C 8H15F1N2	158.22	158.1219	60.73	9.56	12.01	17.71
C 8H15F1N4	186.23	186.1281	51.60	8.12	10.20	30.08
C 8H15F1N6	214.25	214.1342	44.85	7.06	8.87	39.23
C 8H15F2N1	163.21	163.1172	58.87	9.26	23.28	8.58
C 8H15F2N3	191.23	191.1234	50.25	7.91	19.87	21.97
C 8H15F2N5	219.24	219.1295	43.83	6.90	17.33	31.94
C 8H16F1N1	145.22	145.1267	66.17	11.11	13.08	9.65
C 8H16F1N3	173.24	173.1328	55.47	9.31	10.97	24.26
C 8H16F1N5	201.25	201.1390	47.75	8.01	9.44	34.80
C 8H16F2N2	178.23	178.1281	53.91	9.05	21.32	15.72
C 8H16F2N4	206.24	206.1343	46.59	7.82	18.42	27.17
C 8H16F2N6	234.25	234.1404	41.02	6.88	16.22	35.88
C 8H17F1N2	160.24	160.1376	59.97	10.69	11.86	17.48
C 8H17F1N4	188.25	188.1437	51.04	9.10	10.09	29.76
C 8H17F1N6	216.26	216.1499	44.43	7.92	8.78	38.86
C 8H17F2N1	165.23	165.1329	58.16	10.37	23.00	8.48
C 8H17F2N3	193.24	193.1390	49.72	8.87	19.66	21.74
C 8H17F2N5	221.25	221.1452	43.43	7.74	17.17	31.65
C 8H18F1N1	147.24	147.1423	65.26	12.32	12.90	9.51

C 8H18F1N3	175.25	175.1485	54.83	10.35	10.84	23.98
C 8H18F1N5	203.26	203.1546	47.27	8.93	9.35	34.45
C 8H18F2N2	180.24	180.1438	53.31	10.07	21.08	15.54
C 8H18F2N4	208.26	208.1499	46.14	8.71	18.25	26.90
C 8H18F2N6	236.27	236.1561	40.67	7.68	16.08	35.57
C 8H19F1N2	162.25	162.1532	59.22	11.80	11.71	17.27
C 8H19F1N4	190.27	190.1594	50.50	10.07	9.99	29.45
C 8H19F1N6	218.28	218.1655	44.02	8.77	8.70	38.50
C 8H19F2N3	195.26	195.1547	49.21	9.81	19.46	21.52
C 8H19F2N5	223.27	223.1608	43.04	8.58	17.02	31.37
C 8H20F1N3	177.27	177.1641	54.21	11.37	10.72	23.70
C 8H20F1N5	205.28	205.1703	46.81	9.82	9.25	34.12
C 8H20F2N4	210.27	210.1656	45.70	9.59	18.07	26.64
C 8H20F2N6	238.29	238.1717	40.33	8.46	15.95	35.27
C 8H21F1N4	192.28	192.1750	49.97	11.01	9.88	29.14
C 8H21F1N6	220.30	220.1812	43.62	9.61	8.62	38.15
C 8H21F2N5	225.29	225.1765	42.65	9.40	16.87	31.09
C 8H22F1N5	207.30	207.1859	46.35	10.70	9.16	33.78
C 8H22F2N6	240.30	240.1874	39.99	9.23	15.81	34.97
C 8H23F1N6	222.31	222.1968	43.22	10.43	8.55	37.80
C 9H 4F1N1	145.14	145.0328	74.48	2.78	13.09	9.65
C 9H 4F1N3	173.15	173.0389	62.43	2.33	10.97	24.27
C 9H 4F1N5	201.16	201.0451	53.74	2.00	9.44	34.81
C 9H 4F2N2	178.14	178.0343	60.68	2.26	21.33	15.73
C 9H 4F2N4	206.16	206.0404	52.44	1.96	18.43	27.18
C 9H 4F2N6	234.17	234.0465	46.16	1.72	16.23	35.89
C 9H 5F1N2	160.15	160.0437	67.50	3.15	11.86	17.49
C 9H 5F1N4	188.17	188.0498	57.45	2.68	10.10	29.78
C 9H 5F1N6	216.18	216.0560	50.01	2.33	8.79	38.88
C 9H 5F2N1	165.14	165.0390	65.46	3.05	23.01	8.48
C 9H 5F2N3	193.16	193.0451	55.96	2.61	19.67	21.75
C 9H 5F2N5	221.17	221.0513	48.88	2.28	17.18	31.66
C 9H 6F1N1	147.15	147.0484	73.46	4.11	12.91	9.52
C 9H 6F1N3	175.17	175.0546	61.71	3.45	10.85	23.99
C 9H 6F1N5	203.18	203.0607	53.20	2.98	9.35	34.47
C 9H 6F2N2	180.16	180.0499	60.00	3.36	21.09	15.55
C 9H 6F2N4	208.17	208.0560	51.93	2.91	18.25	26.91
C 9H 6F2N6	236.19	236.0622	45.77	2.56	16.09	35.58
C 9H 7F1N2	162.17	162.0593	66.66	4.35	11.72	17.27
C 9H 7F1N4	190.18	190.0655	56.84	3.71	9.99	29.46
C 9H 7F1N6	218.19	218.0716	49.54	3.23	8.71	38.52
C 9H 7F2N1	167.16	167.0547	64.67	4.22	22.73	8.38
C 9H 7F2N3	195.17	195.0608	55.39	3.62	19.47	21.53
C 9H 7F2N5	223.19	223.0669	48.43	3.16	17.02	31.38
C 9H 8F1N1	149.17	149.0641	72.47	5.41	12.74	9.39
C 9H 8F1N3	177.18	177.0702	61.01	4.55	10.72	23.72
C 9H 8F1N5	205.20	205.0764	52.68	3.93	9.26	34.13
C 9H 8F2N2	182.17	182.0655	59.34	4.43	20.86	15.38
C 9H 8F2N4	210.19	210.0717	51.43	3.84	18.08	26.66
C 9H 8F2N6	238.20	238.0778	45.38	3.39	15.95	35.28
C 9H 9F1N2	164.18	164.0750	65.84	5.53	11.57	17.06
C 9H 9F1N4	192.20	192.0811	56.24	4.72	9.88	29.15
C 9H 9F1N6	220.21	220.0873	49.09	4.12	8.63	38.16
C 9H 9F2N1	169.18	169.0703	63.90	5.36	22.46	8.28
C 9H 9F2N3	197.19	197.0764	54.82	4.60	19.27	21.31
C 9H 9F2N5	225.20	225.0826	48.00	4.03	16.87	31.10
C 9H10F1N1	151.19	151.0797	71.50	6.67	12.57	9.26
C 9H10F1N3	179.20	179.0859	60.32	5.62	10.60	23.45
C 9H10F1N5	207.21	207.0920	52.17	4.86	9.17	33.80
C 9H10F2N2	184.19	184.0812	58.69	5.47	20.63	15.21

C 9H10F2N4	212.20	212.0873	50.94	4.75	17.91	26.40
C 9H10F2N6	240.22	240.0935	45.00	4.20	15.82	34.99
C 9H11F1N2	166.20	166.0906	65.04	6.67	11.43	16.86
C 9H11F1N4	194.21	194.0968	55.66	5.71	9.78	28.85
C 9H11F1N6	222.23	222.1029	48.64	4.99	8.55	37.82
C 9H11F2N1	171.19	171.0859	63.15	6.48	22.20	8.18
C 9H11F2N3	199.20	199.0921	54.27	5.57	19.07	21.09
C 9H11F2N5	227.22	227.0982	47.58	4.88	16.72	30.82
C 9H12F1N1	153.20	153.0954	70.56	7.90	12.40	9.14
C 9H12F1N3	181.21	181.1015	59.65	6.67	10.48	23.19
C 9H12F1N5	209.23	209.1077	51.67	5.78	9.08	33.47
C 9H12F2N2	186.21	186.0968	58.05	6.50	20.41	15.04
C 9H12F2N4	214.22	214.1030	50.46	5.65	17.74	26.15
C 9H12F2N6	242.23	242.1091	44.63	4.99	15.69	34.69
C 9H13F1N2	168.22	168.1063	64.26	7.79	11.29	16.65
C 9H13F1N4	196.23	196.1124	55.09	6.68	9.68	28.55
C 9H13F1N6	224.24	224.1186	48.21	5.84	8.47	37.48
C 9H13F2N1	173.21	173.1016	62.41	7.57	21.94	8.09
C 9H13F2N3	201.22	201.1077	53.72	6.51	18.88	20.88
C 9H13F2N5	229.23	229.1139	47.16	5.72	16.58	30.55
C 9H14F1N1	155.22	155.1110	69.64	9.09	12.24	9.02
C 9H14F1N3	183.23	183.1172	59.00	7.70	10.37	22.93
C 9H14F1N5	211.24	211.1233	51.17	6.68	8.99	33.15
C 9H14F2N2	188.22	188.1125	57.43	7.50	20.19	14.88
C 9H14F2N4	216.24	216.1186	49.99	6.53	17.57	25.91
C 9H14F2N6	244.25	244.1248	44.26	5.78	15.56	34.41
C 9H15F1N2	170.23	170.1219	63.50	8.88	11.16	16.46
C 9H15F1N4	198.25	198.1281	54.53	7.63	9.58	28.26
C 9H15F1N6	226.26	226.1342	47.78	6.68	8.40	37.14
C 9H15F2N1	175.22	175.1172	61.69	8.63	21.68	7.99
C 9H15F2N3	203.24	203.1234	53.19	7.44	18.70	20.68
C 9H15F2N5	231.25	231.1295	46.75	6.54	16.43	30.28
C 9H16F1N1	157.23	157.1267	68.75	10.26	12.08	8.91
C 9H16F1N3	185.25	185.1328	58.35	8.71	10.26	22.68
C 9H16F1N5	213.26	213.1390	50.69	7.56	8.91	32.84
C 9H16F2N2	190.24	190.1281	56.82	8.48	19.97	14.73
C 9H16F2N4	218.25	218.1343	49.53	7.39	17.41	25.67
C 9H16F2N6	246.26	246.1404	43.90	6.55	15.43	34.13
C 9H17F1N2	172.25	172.1376	62.76	9.95	11.03	16.26
C 9H17F1N4	200.26	200.1437	53.98	8.56	9.49	27.98
C 9H17F1N6	228.27	228.1499	47.36	7.51	8.32	36.82
C 9H17F2N1	177.24	177.1329	60.99	9.67	21.44	7.90
C 9H17F2N3	205.25	205.1390	52.67	8.35	18.51	20.47
C 9H17F2N5	233.27	233.1452	46.34	7.35	16.29	30.02
C 9H18F1N1	159.25	159.1423	67.88	11.39	11.93	8.80
C 9H18F1N3	187.26	187.1485	57.73	9.69	10.15	22.44
C 9H18F1N5	215.28	215.1546	50.21	8.43	8.83	32.53
C 9H18F2N2	192.25	192.1438	56.23	9.44	19.76	14.57
C 9H18F2N4	220.27	220.1499	49.08	8.24	17.25	25.44
C 9H18F2N6	248.28	248.1561	43.54	7.31	15.30	33.85
C 9H19F1N2	174.26	174.1532	62.03	10.99	10.90	16.08
C 9H19F1N4	202.28	202.1594	53.44	9.47	9.39	27.70
C 9H19F1N6	230.29	230.1655	46.94	8.32	8.25	36.49
C 9H19F2N1	179.26	179.1485	60.31	10.68	21.20	7.81
C 9H19F2N3	207.27	207.1547	52.15	9.24	18.33	20.27
C 9H19F2N5	235.28	235.1608	45.94	8.14	16.15	29.77
C 9H20F1N1	161.26	161.1580	67.03	12.50	11.78	8.69
C 9H20F1N3	189.28	189.1641	57.11	10.65	10.04	22.20
C 9H20F1N5	217.29	217.1703	49.75	9.28	8.74	32.23
C 9H20F2N2	194.27	194.1594	55.64	10.38	19.56	14.42

C 9H20F2N4	222.28	222.1656	48.63	9.07	17.09	25.21
C 9H20F2N6	250.30	250.1717	43.19	8.05	15.18	33.58
C 9H21F1N2	176.28	176.1689	61.32	12.01	10.78	15.89
C 9H21F1N4	204.29	204.1750	52.91	10.36	9.30	27.42
C 9H21F1N6	232.31	232.1812	46.53	9.11	8.18	36.18
C 9H21F2N3	209.28	209.1703	51.65	10.11	18.16	20.08
C 9H21F2N5	237.30	237.1765	45.55	8.92	16.01	29.51
C 9H22F1N3	191.29	191.1798	56.51	11.59	9.93	21.97
C 9H22F1N5	219.31	219.1859	49.29	10.11	8.66	31.93
C 9H22F2N4	224.30	224.1812	48.19	9.89	16.94	24.98
C 9H22F2N6	252.31	252.1874	42.84	8.79	15.06	33.31
C 9H23F1N4	206.31	206.1907	52.40	11.24	9.21	27.16
C 9H23F1N6	234.32	234.1968	46.13	9.89	8.11	35.87
C 9H23F2N5	239.31	239.1921	45.17	9.69	15.88	29.26
C 9H24F1N5	221.32	221.2016	48.84	10.93	8.58	31.64
C 9H24F2N6	254.33	254.2030	42.50	9.51	14.94	33.04
C 9H25F1N6	236.34	236.2125	45.74	10.66	8.04	35.56
C10H 5F1N2	172.16	172.0437	69.77	2.93	11.04	16.27
C10H 5F1N4	200.18	200.0498	60.00	2.52	9.49	27.99
C10H 5F1N6	228.19	228.0560	52.64	2.21	8.33	36.83
C10H 5F2N1	177.15	177.0390	67.80	2.84	21.45	7.91
C10H 5F2N3	205.17	205.0451	58.54	2.46	18.52	20.48
C10H 5F2N5	233.18	233.0513	51.51	2.16	16.29	30.03
C10H 6F1N1	159.16	159.0484	75.46	3.80	11.94	8.80
C10H 6F1N3	187.18	187.0546	64.17	3.23	10.15	22.45
C10H 6F1N5	215.19	215.0607	55.82	2.81	8.83	32.54
C10H 6F2N2	192.17	192.0499	62.50	3.15	19.77	14.58
C10H 6F2N4	220.18	220.0560	54.55	2.75	17.26	25.45
C10H 6F2N6	248.20	248.0622	48.39	2.44	15.31	33.86
C10H 7F1N2	174.18	174.0593	68.96	4.05	10.91	16.08
C10H 7F1N4	202.19	202.0655	59.40	3.49	9.40	27.71
C10H 7F1N6	230.21	230.0716	52.18	3.06	8.25	36.51
C10H 7F2N1	179.17	179.0547	67.04	3.94	21.21	7.82
C10H 7F2N3	207.18	207.0608	57.97	3.41	18.34	20.28
C10H 7F2N5	235.20	235.0669	51.07	3.00	16.16	29.78
C10H 8F1N1	161.18	161.0641	74.52	5.00	11.79	8.69
C10H 8F1N3	189.19	189.0702	63.49	4.26	10.04	22.21
C10H 8F1N5	217.21	217.0764	55.30	3.71	8.75	32.24
C10H 8F2N2	194.19	194.0655	61.85	4.15	19.57	14.43
C10H 8F2N4	222.20	222.0717	54.06	3.63	17.10	25.21
C10H 8F2N6	250.21	250.0778	48.00	3.22	15.19	33.59
C10H 9F1N2	176.20	176.0750	68.17	5.15	10.78	15.90
C10H 9F1N4	204.21	204.0811	58.82	4.44	9.30	27.44
C10H 9F1N6	232.22	232.0873	51.72	3.91	8.18	36.19
C10H 9F2N1	181.19	181.0703	66.29	5.01	20.97	7.73
C10H 9F2N3	209.20	209.0764	57.41	4.34	18.16	20.09
C10H 9F2N5	237.21	237.0826	50.63	3.82	16.02	29.52
C10H10F1N1	163.20	163.0797	73.60	6.18	11.64	8.58
C10H10F1N3	191.21	191.0859	62.82	5.27	9.94	21.98
C10H10F1N5	219.22	219.0920	54.79	4.60	8.67	31.95
C10H10F2N2	196.20	196.0812	61.22	5.14	19.37	14.28
C10H10F2N4	224.21	224.0873	53.57	4.50	16.95	24.99
C10H10F2N6	252.23	252.0935	47.62	4.00	15.06	33.32
C10H11F1N2	178.21	178.0906	67.40	6.22	10.66	15.72
C10H11F1N4	206.22	206.0968	58.24	5.38	9.21	27.17
C10H11F1N6	234.24	234.1029	51.28	4.73	8.11	35.88
C10H11F2N1	183.20	183.0859	65.56	6.05	20.74	7.65
C10H11F2N3	211.22	211.0921	56.87	5.25	17.99	19.89
C10H11F2N5	239.23	239.0982	50.21	4.63	15.88	29.27
C10H12F1N1	165.21	165.0954	72.70	7.32	11.50	8.48
C10H12F1N3	193.23	193.1015	62.16	6.26	9.83	21.75
C10H12F1N5	221.24	221.1077	54.29	5.47	8.59	31.66
C10H12F2N2	198.22	198.0968	60.60	6.10	19.17	14.13
C10H12F2N4	226.23	226.1030	53.09	5.35	16.80	24.77
C10H12F2N6	254.24	254.1091	47.24	4.76	14.95	33.05
C10H13F1N2	180.23	180.1063	66.64	7.27	10.54	15.54
C10H13F1N4	208.24	208.1124	57.68	6.29	9.12	26.90
C10H13F1N6	236.25	236.1186	50.84	5.55	8.04	35.57
C10H13F2N1	185.22	185.1016	64.85	7.07	20.51	7.56
C10H13F2N3	213.23	213.1077	56.33	6.15	17.82	19.71
C10H13F2N5	241.25	241.1139	49.79	5.43	15.75	29.03
C10H14F1N1	167.23	167.1110	71.82	8.44	11.36	8.38
C10H14F1N3	195.24	195.1172	61.52	7.23	9.73	21.52
C10H14F1N5	223.25	223.1233	53.80	6.32	8.51	31.37
C10H14F2N2	200.23	200.1125	59.99	7.05	18.98	13.99
C10H14F2N4	228.25	228.1186	52.62	6.18	16.65	24.55
C10H14F2N6	256.26	256.1248	46.87	5.51	14.83	32.79
C10H15F1N2	182.24	182.1219	65.91	8.30	10.42	15.37
C10H15F1N4	210.26	210.1281	57.13	7.19	9.04	26.65
C10H15F1N6	238.27	238.1342	50.41	6.35	7.97	35.27
C10H15F2N1	187.23	187.1172	64.15	8.08	20.29	7.48
C10H15F2N3	215.25	215.1234	55.80	7.02	17.65	19.52
C10H15F2N5	243.26	243.1295	49.38	6.22	15.62	28.79
C10H16F1N1	169.24	169.1267	70.97	9.53	11.23	8.28
C10H16F1N3	197.26	197.1328	60.89	8.18	9.63	21.30
C10H16F1N5	225.27	225.1390	53.32	7.16	8.43	31.09
C10H16F2N2	202.25	202.1281	59.39	7.97	18.79	13.85
C10H16F2N4	230.26	230.1343	52.16	7.00	16.50	24.33
C10H16F2N6	258.28	258.1404	46.51	6.24	14.71	32.54
C10H17F1N2	184.26	184.1376	65.19	9.30	10.31	15.20
C10H17F1N4	212.27	212.1437	56.58	8.07	8.95	26.39
C10H17F1N6	240.29	240.1499	49.99	7.13	7.91	34.98
C10H17F2N1	189.25	189.1329	63.47	9.05	20.08	7.40
C10H17F2N3	217.26	217.1390	55.28	7.89	17.49	19.34
C10H17F2N5	245.28	245.1452	48.97	6.99	15.49	28.55
C10H18F1N1	171.26	171.1423	70.13	10.59	11.09	8.18
C10H18F1N3	199.27	199.1485	60.27	9.10	9.53	21.09
C10H18F1N5	227.29	227.1546	52.85	7.98	8.36	30.81
C10H18F2N2	204.27	204.1438	58.80	8.88	18.60	13.71
C10H18F2N4	232.28	232.1499	51.71	7.81	16.36	24.12
C10H18F2N6	260.29	260.1561	46.14	6.97	14.60	32.29
C10H19F1N2	186.27	186.1532	64.48	10.28	10.20	15.04
C10H19F1N4	214.29	214.1594	56.05	8.94	8.87	26.15
C10H19F1N6	242.30	242.1655	49.57	7.90	7.84	34.68
C10H19F2N1	191.27	191.1485	62.80	10.01	19.87	7.32
C10H19F2N3	219.28	219.1547	54.78	8.73	17.33	19.16
C10H19F2N5	247.29	247.1608	48.57	7.74	15.37	28.32
C10H20F1N1	173.28	173.1580	69.32	11.63	10.96	8.08
C10H20F1N3	201.29	201.1641	59.67	10.02	9.44	20.88
C10H20F1N5	229.30	229.1703	52.38	8.79	8.29	30.54
C10H20F2N2	206.28	206.1594	58.23	9.77	18.42	13.58
C10H20F2N4	234.29	234.1656	51.27	8.60	16.22	23.91
C10H20F2N6	262.31	262.1717	45.79	7.69	14.49	32.04
C10H21F1N2	188.29	188.1689	63.79	11.24	10.09	14.88
C10H21F1N4	216.30	216.1750	55.53	9.79	8.78	25.90
C10H21F1N6	244.32	244.1812	49.16	8.66	7.78	34.40
C10H21F2N1	193.28	193.1642	62.14	10.95	19.66	7.25
C10H21F2N3	221.30	221.1703	54.28	9.57	17.17	18.99
C10H21F2N5	249.31	249.1765	48.18	8.49	15.24	28.09
C10H22F1N1	175.29	175.1736	68.52	12.65	10.84	7.99

C10H22F1N3	203.31	203.1798	59.08	10.91	9.34	20.67
C10H22F1N5	231.32	231.1859	51.92	9.59	P.21	30.28
C10H22F2N2	208.30	208.1751	57.66	10.65	18.24	13.45
C10H22F2N4	236.31	236.1812	50.83	9.38	16.08	23.71
C10H22F2N6	264.32	264.1874	45.44	8.39	14.38	31.79
C10H23F1N2	190.31	190.1845	63.11	12.18	9.98	14.72
C10H23F1N4	218.32	218.1907	55.02	10.62	8.70	25.66
C10H23F1N6	246.33	246.1968	48.76	9.41	7.71	34.12
C10H23F2N3	223.31	223.1860	53.79	10.38	17.02	18.82
C10H23F2N5	251.33	251.1921	47.79	9.22	15.12	27.87
C10H24F1N3	205.32	205.1954	58.50	11.78	9.25	20.47
C10H24F1N5	233.33	233.2016	51.48	10.37	8.14	30.01
C10H24F2N4	238.33	238.1969	50.40	10.15	15.94	23.51
C10H24F2N6	266.34	266.2030	45.10	9.08	14.27	31.55
C10H25F1N4	220.34	220.2063	54.51	11.44	8.62	25.43
C10H25F1N6	248.35	248.2125	48.36	10.15	7.65	33.84
C10H25F2N5	253.34	253.2078	47.41	9.95	15.00	27.64
C10H26F1N5	235.35	235.2172	51.04	11.14	8.07	29.76
C10H26F2N6	268.36	268.2187	44.76	9.77	14.16	31.32
C10H27F1N6	250.37	250.2281	47.97	10.87	7.59	33.57
C11H 5F1N2	184.17	184.0437	71.74	2.74	10.32	15.21
C11H 5F1N4	212.19	212.0498	62.27	2.38	8.95	26.40
C11H 5F1N6	240.20	240.0560	55.01	2.10	7.91	34.99
C11H 5F2N1	189.17	189.0390	69.84	2.66	20.09	7.40
C11H 5F2N3	217.18	217.0451	60.84	2.32	17.50	19.35
C11H 5F2N5	245.19	245.0513	53.89	2.06	15.50	28.56
C11H 6F1N1	171.18	171.0484	77.19	3.53	11.10	8.18
C11H 6F1N3	199.19	199.0546	66.33	3.04	9.54	21.10
C11H 6F1N5	227.20	227.0607	58.15	2.66	8.36	30.82
C11H 6F2N2	204.18	204.0499	64.71	2.96	18.61	13.72
C11H 6F2N4	232.19	232.0560	56.90	2.60	16.36	24.13
C11H 6F2N6	260.21	260.0622	50.78	2.32	14.60	32.30
C11H 7F1N2	186.19	186.0593	70.96	3.79	10.20	15.05
C11H 7F1N4	214.20	214.0655	61.68	3.29	8.87	26.16
C11H 7F1N6	242.22	242.0716	54.55	2.91	7.84	34.70
C11H 7F2N1	191.18	191.0547	69.11	3.69	19.87	7.33
C11H 7F2N3	219.20	219.0608	60.28	3.22	17.33	19.17
C11H 7F2N5	247.21	247.0669	53.45	2.85	15.37	28.33
C11H 8F1N1	173.19	173.0641	76.29	4.66	10.97	8.09
C11H 8F1N3	201.20	201.0702	65.67	4.01	9.44	20.88
C11H 8F1N5	229.22	229.0764	57.64	3.52	8.29	30.55
C11H 8F2N2	206.20	206.0655	64.08	3.91	18.43	13.59
C11H 8F2N4	234.21	234.0717	56.41	3.44	16.22	23.92
C11H 8F2N6	262.22	262.0778	50.39	3.08	14.49	32.05
C11H 9F1N2	188.21	188.0750	70.20	4.82	10.09	14.88
C11H 9F1N4	216.22	216.0811	61.11	4.20	8.79	25.91
C11H 9F1N6	244.23	244.0873	54.10	3.71	7.78	34.41
C11H 9F2N1	193.20	193.0703	68.39	4.70	19.67	7.25
C11H 9F2N3	221.21	221.0764	59.73	4.10	17.18	19.00
C11H 9F2N5	249.22	249.0826	53.01	3.64	15.25	28.10
C11H10F1N1	175.21	175.0797	75.41	5.75	10.84	7.99
C11H10F1N3	203.22	203.0859	65.01	4.96	9.35	20.68
C11H10F1N5	231.23	231.0920	57.14	4.36	8.22	30.29
C11H10F2N2	208.21	208.0812	63.46	4.84	18.25	13.45
C11H10F2N4	236.23	236.0873	55.93	4.27	16.08	23.72
C11H10F2N6	264.24	264.0935	50.00	3.81	14.38	31.80
C11H11F1N2	190.22	190.0906	69.46	5.83	9.99	14.73
C11H11F1N4	218.24	218.0968	60.54	5.08	8.71	25.67
C11H11F1N6	246.25	246.1029	53.65	4.50	7.72	34.13
C11H11F2N1	195.21	195.0859	67.68	5.68	19.46	7.18
C11H11F2N3	223.23	223.0921	59.19	4.97	17.02	18.82
C11H11F2N5	251.24	251.0982	52.59	4.41	15.12	27.88
C11H12F1N1	177.22	177.0954	74.55	6.83	10.72	7.90
C11H12F1N3	205.24	205.1015	64.38	5.89	9.26	20.47
C11H12F1N5	233.25	233.1077	56.64	5.19	8.15	30.03
C11H12F2N2	210.23	210.0968	62.85	5.75	18.07	13.33
C11H12F2N4	238.24	238.1030	55.46	5.08	15.95	23.52
C11H12F2N6	266.26	266.1091	49.62	4.54	14.27	31.56
C11H13F1N2	192.24	192.1063	68.73	6.82	9.88	14.57
C11H13F1N4	220.25	220.1124	59.99	5.95	8.63	25.44
C11H13F1N6	248.26	248.1186	53.22	5.28	7.65	33.85
C11H13F2N1	197.23	197.1016	66.99	6.64	19.27	7.10
C11H13F2N3	225.24	225.1077	58.66	5.82	16.87	18.66
C11H13F2N5	253.26	253.1139	52.17	5.17	15.00	27.65
C11H14F1N1	179.24	179.1110	73.71	7.87	10.60	7.81
C11H14F1N3	207.25	207.1172	63.75	6.81	9.17	20.27
C11H14F1N5	235.27	235.1233	56.16	6.00	8.08	29.77
C11H14F2N2	212.24	212.1125	62.25	6.65	17.90	13.20
C11H14F2N4	240.26	240.1186	54.99	5.87	15.82	23.32
C11H14F2N6	268.27	268.1248	49.25	5.26	14.16	31.33
C11H15F1N2	194.25	194.1219	68.02	7.78	9.78	14.42
C11H15F1N4	222.27	222.1281	59.44	6.80	8.55	25.21
C11H15F1N6	250.28	250.1342	52.79	6.04	7.59	33.58
C11H15F2N1	199.25	199.1172	66.31	7.59	19.07	7.03
C11H15F2N3	227.26	227.1234	58.14	6.65	16.72	18.49
C11H15F2N5	255.27	255.1295	51.76	5.92	14.88	27.43
C11H16F1N1	181.26	181.1267	72.89	8.90	10.48	7.73
C11H16F1N3	209.27	209.1328	63.14	7.71	9.08	20.08
C11H16F1N5	237.28	237.1390	55.68	6.80	8.01	29.51
C11H16F2N2	214.26	214.1281	61.66	7.53	17.73	13.07
C11H16F2N4	242.27	242.1343	54.53	6.66	15.68	23.13
C11H16F2N6	270.29	270.1404	48.88	5.97	14.06	31.09
C11H17F1N2	196.27	196.1376	67.32	8.73	9.68	14.27
C11H17F1N4	224.28	224.1437	58.91	7.64	8.47	24.98
C11H17F1N6	252.30	252.1499	52.37	6.79	7.53	33.31
C11H17F2N1	201.26	201.1329	65.65	8.51	18.88	6.96
C11H17F2N3	229.28	229.1390	57.63	7.47	16.57	18.33
C11H17F2N5	257.29	257.1452	51.35	6.66	14.77	27.22
C11H18F1N1	183.27	183.1423	72.09	9.90	10.37	7.64
C11H18F1N3	211.28	211.1485	62.53	8.59	8.99	19.89
C11H18F1N5	239.30	239.1546	55.21	7.58	7.94	29.27
C11H18F2N2	216.28	216.1438	61.09	8.39	17.57	12.95
C11H18F2N4	244.29	244.1499	54.08	7.43	15.55	22.93
C11H18F2N6	272.30	272.1561	48.52	6.66	13.95	30.86
C11H19F1N2	198.29	198.1532	66.63	9.66	9.58	14.13
C11H19F1N4	226.30	226.1594	58.38	8.46	8.40	24.76
C11H19F1N6	254.31	254.1655	51.95	7.53	7.47	33.05
C11H19F2N1	203.28	203.1485	65.00	9.42	18.69	6.89
C11H19F2N3	231.29	231.1547	57.12	8.28	16.43	18.17
C11H19F2N5	259.30	259.1608	50.95	7.39	14.65	27.01
C11H20F1N1	185.29	185.1580	71.31	10.88	10.25	7.56
C11H20F1N3	213.30	213.1641	61.94	9.45	8.91	19.70
C11H20F1N5	241.31	241.1703	54.75	8.35	7.87	29.02
C11H20F2N2	218.29	218.1594	60.53	9.24	17.41	12.83
C11H20F2N4	246.31	246.1656	53.64	8.18	15.43	22.75
C11H20F2N6	274.32	274.1717	48.16	7.35	13.85	30.64
C11H21F1N2	200.30	200.1689	65.96	10.57	9.48	13.99
C11H21F1N4	228.32	228.1750	57.87	9.27	8.32	24.54
C11H21F1N6	256.33	256.1812	51.54	8.26	7.41	32.79
C11H21F2N1	205.29	205.1642	64.36	10.31	18.51	6.82

C11H21F2N3	233.31	233.1703	56.63	9.07	16.29	18.01
C11H21F2N5	261.32	261.1765	50.56	8.10	14.54	26.80
C11H22F1N1	187.30	187.1736	70.54	11.84	10.14	7.48
C11H22F1N3	215.32	215.1798	61.36	10.30	8.82	19.52
C11H22F1N5	243.33	243.1859	54.30	9.11	7.81	28.78
C11H22F2N2	220.31	220.1751	59.97	10.07	17.25	12.72
C11H22F2N4	248.32	248.1812	53.21	8.93	15.30	22.56
C11H22F2N6	276.33	276.1874	47.81	8.02	13.75	30.41
C11H23F1N2	202.32	202.1845	65.30	11.46	9.39	13.85
C11H23F1N4	230.33	230.1907	57.36	10.07	8.25	24.32
C11H23F1N6	258.34	258.1968	51.14	8.97	7.35	32.53
C11H23F2N1	207.31	207.1798	63.73	11.18	18.33	6.76
C11H23F2N3	235.32	235.1860	56.15	9.85	16.15	17.86
C11H23F2N5	263.34	263.1921	50.17	8.80	14.43	26.59
C11H24F1N1	189.32	189.1893	69.79	12.78	10.04	7.40
C11H24F1N3	217.33	217.1954	60.79	11.13	8.74	19.33
C11H24F1N5	245.35	245.2016	53.85	9.86	7.74	28.54
C11H24F2N2	222.32	222.1907	59.43	10.88	17.09	12.60
C11H24F2N4	250.34	250.1969	52.78	9.66	15.18	22.38
C11H24F2N6	278.35	278.2030	47.47	8.69	13.65	30.19
C11H25F1N2	204.33	204.2002	64.66	12.33	9.30	13.71
C11H25F1N4	232.35	232.2063	56.86	10.85	8.18	24.11
C11H25F1N6	260.36	260.2125	50.75	9.68	7.30	32.28
C11H25F2N3	237.34	237.2016	55.67	10.62	16.01	17.70
C11H25F2N5	265.35	265.2078	49.79	9.50	14.32	26.39
C11H26F1N3	219.35	219.2111	60.23	11.95	8.66	19.16
C11H26F1N5	247.36	247.2172	53.41	10.59	7.68	28.31
C11H26F2N4	252.35	252.2125	52.36	10.39	15.06	22.20
C11H26F2N6	280.37	280.2187	47.12	9.35	13.55	29.98
C11H27F1N4	234.36	234.2220	56.38	11.61	8.11	23.91
C11H27F1N6	262.38	262.2281	50.36	10.37	7.24	32.03
C11H27F2N5	267.37	267.2234	49.42	10.18	14.21	26.19
C11H28F1N5	249.38	249.2329	52.98	11.32	7.62	28.08
C11H28F2N6	282.38	282.2343	46.79	9.99	13.46	29.76
C11H29F1N6	264.39	264.2438	49.97	11.06	7.19	31.79
C12H 6F1N1	183.19	183.0484	78.68	3.30	10.37	7.65
C12H 6F1N3	211.20	211.0546	68.25	2.86	9.00	19.90
C12H 6F1N5	239.21	239.0607	60.25	2.53	7.94	29.28
C12H 6F2N2	216.19	216.0499	66.67	2.80	17.58	12.96
C12H 6F2N4	244.21	244.0560	59.02	2.48	15.56	22.94
C12H 6F2N6	272.22	272.0622	52.95	2.22	13.96	30.87
C12H 7F1N2	198.20	198.0593	72.72	3.56	9.59	14.13
C12H 7F1N4	226.21	226.0655	63.72	3.12	8.40	24.77
C12H 7F1N6	254.23	254.0716	56.69	2.78	7.47	33.06
C12H 7F2N1	203.19	203.0547	70.93	3.47	18.70	6.89
C12H 7F2N3	231.21	231.0608	62.34	3.05	16.43	18.17
C12H 7F2N5	259.22	259.0669	55.60	2.72	14.66	27.02
C12H 8F1N1	185.20	185.0641	77.82	4.35	10.26	7.56
C12H 8F1N3	213.22	213.0702	67.60	3.78	8.91	19.71
C12H 8F1N5	241.23	241.0764	59.75	3.34	7.88	29.03
C12H 8F2N2	218.21	218.0655	66.05	3.70	17.41	12.84
C12H 8F2N4	246.22	246.0717	58.54	3.28	15.43	22.75
C12H 8F2N6	274.23	274.0778	52.56	2.94	13.86	30.65
C12H 9F1N2	200.22	200.0750	71.99	4.53	9.49	13.99
C12H 9F1N4	228.23	228.0811	63.15	3.97	8.32	24.55
C12H 9F1N6	256.24	256.0873	56.25	3.54	7.41	32.80
C12H 9F2N1	205.21	205.0703	70.24	4.42	18.52	6.83
C12H 9F2N3	233.22	233.0764	61.80	3.89	16.29	18.02
C12H 9F2N5	261.24	261.0826	55.17	3.47	14.55	26.81
C12H10F1N1	187.22	187.0797	76.99	5.38	10.15	7.48

C12H10F1N3	215.23	215.0859	66.97	4.68	8.83	19.52
C12H10F1N5	243.25	243.0920	59.25	4.14	7.81	28.79
C12H10F2N2	220.22	220.0812	65.45	4.58	17.25	12.72
C12H10F2N4	248.24	248.0873	58.06	4.06	15.31	22.57
C12H10F2N6	276.25	276.0935	52.18	3.65	13.75	30.42
C12H11F1N2	202.23	202.0906	71.27	5.48	9.39	13.85
C12H11F1N4	230.25	230.0968	62.60	4.82	8.25	24.33
C12H11F1N6	258.26	258.1029	55.81	4.29	7.36	32.54
C12H11F2N1	207.22	207.0859	69.55	5.35	18.34	6.76
C12H11F2N3	235.24	235.0921	61.27	4.71	16.15	17.86
C12H11F2N5	263.25	263.0982	54.75	4.21	14.43	26.60
C12H12F1N1	189.23	189.0954	76.17	6.39	10.04	7.40
C12H12F1N3	217.25	217.1015	66.35	5.57	8.75	19.34
C12H12F1N5	245.26	245.1077	58.77	4.93	7.75	28.55
C12H12F2N2	222.24	222.0968	64.86	5.44	17.10	12.61
C12H12F2N4	250.25	250.1030	57.60	4.83	15.18	22.39
C12H12F2N6	278.27	278.1091	51.80	4.35	13.65	30.20
C12H13F1N2	204.25	204.1063	70.57	6.42	9.30	13.72
C12H13F1N4	232.26	232.1124	62.06	5.64	8.18	24.12
C12H13F1N6	260.28	260.1186	55.38	5.03	7.30	32.29
C12H13F2N1	209.24	209.1016	68.88	6.26	18.16	6.69
C12H13F2N3	237.25	237.1077	60.75	5.52	16.02	17.71
C12H13F2N5	265.27	265.1139	54.34	4.94	14.32	26.40
C12H14F1N1	191.25	191.1110	75.36	7.38	9.93	7.32
C12H14F1N3	219.26	219.1172	65.74	6.44	8.66	19.16
C12H14F1N5	247.28	247.1233	58.29	5.71	7.68	28.32
C12H14F2N2	224.26	224.1125	64.27	6.29	16.94	12.49
C12H14F2N4	252.27	252.1186	57.13	5.59	15.06	22.21
C12H14F2N6	280.28	280.1248	51.42	5.03	13.56	29.98
C12H15F1N2	206.27	206.1219	69.88	7.33	9.21	13.58
C12H15F1N4	234.28	234.1281	61.52	6.45	8.11	23.91
C12H15F1N6	262.29	262.1342	54.95	5.76	7.24	32.04
C12H15F2N1	211.26	211.1172	68.23	7.16	17.99	6.63
C12H15F2N3	239.27	239.1234	60.24	6.32	15.88	17.56
C12H15F2N5	267.28	267.1295	53.93	5.66	14.22	26.20
C12H16F1N1	193.27	193.1267	74.58	8.34	9.83	7.25
C12H16F1N3	221.28	221.1328	65.14	7.29	8.59	18.99
C12H16F1N5	249.29	249.1390	57.82	6.47	7.62	28.09
C12H16F2N2	226.27	226.1281	63.70	7.13	16.79	12.38
C12H16F2N4	254.28	254.1343	56.68	6.34	14.94	22.03
C12H16F2N6	282.30	282.1404	51.06	5.71	13.46	29.77
C12H17F1N2	208.28	208.1376	69.20	8.23	9.12	13.45
C12H17F1N4	236.29	236.1437	61.00	7.25	8.04	23.71
C12H17F1N6	264.31	264.1499	54.53	6.48	7.19	31.80
C12H17F2N1	213.27	213.1329	67.58	8.03	17.82	6.57
C12H17F2N3	241.29	241.1390	59.74	7.10	15.75	17.42
C12H17F2N5	269.30	269.1452	53.52	6.36	14.11	26.01
C12H18F1N1	195.28	195.1423	73.81	9.29	9.73	7.17
C12H18F1N3	223.30	223.1485	64.55	8.13	8.51	18.82
C12H18F1N5	251.31	251.1546	57.35	7.22	7.56	27.87
C12H18F2N2	228.29	228.1438	63.14	7.95	16.64	12.27
C12H18F2N4	256.30	256.1499	56.24	7.08	14.83	21.86
C12H18F2N6	284.31	284.1561	50.70	6.38	13.36	29.56
C12H19F1N2	210.30	210.1532	68.54	9.11	9.03	13.32
C12H19F1N4	238.31	238.1594	60.48	8.04	7.97	23.51
C12H19F1N6	266.32	266.1655	54.12	7.19	7.13	31.56
C12H19F2N1	215.29	215.1485	66.95	8.90	17.65	6.51
C12H19F2N3	243.30	243.1547	59.24	7.87	15.62	17.27
C12H19F2N5	271.32	271.1608	53.12	7.06	14.00	25.81
C12H20F1N1	197.30	197.1580	73.05	10.22	9.63	7.10

C12H20F1N3	225.31	225.1641	63.97	8.95	8.43	18.65
C12H20F1N5	253.33	253.1703	56.90	7.96	7.50	27.65
C12H20F2N2	230.30	230.1594	62.58	8.75	16.50	12.16
C12H20F2N4	258.32	258.1656	55.80	7.80	14.71	21.69
C12H20F2N6	286.33	286.1717	50.34	7.04	13.27	29.35
C12H21F1N2	212.31	212.1689	67.89	9.97	8.95	13.19
C12H21F1N4	240.33	240.1750	59.97	8.81	7.91	23.31
C12H21F1N6	268.34	268.1812	53.71	7.89	7.08	31.32
C12H21F2N1	217.30	217.1642	66.33	9.74	17.49	6.45
C12H21F2N3	245.32	245.1703	58.75	8.63	15.49	17.13
C12H21F2N5	273.33	273.1765	52.73	7.74	13.90	25.62
C12H22F1N1	199.31	199.1736	72.31	11.13	9.53	7.03
C12H22F1N3	227.33	227.1798	63.40	9.75	8.36	18.48
C12H22F1N5	255.34	255.1859	56.45	8.68	7.44	27.43
C12H22F2N2	232.32	232.1751	62.04	9.55	16.36	12.06
C12H22F2N4	260.33	260.1812	55.37	8.52	14.60	21.52
C12H22F2N6	288.35	288.1874	49.99	7.69	13.18	29.15
C12H23F1N2	214.33	214.1845	67.25	10.82	8.86	13.07
C12H23F1N4	242.34	242.1907	59.48	9.57	7.84	23.12
C12H23F1N6	270.36	270.1968	53.31	8.58	7.03	31.09
C12H23F2N1	219.32	219.1798	65.72	10.57	17.32	6.39
C12H23F2N3	247.33	247.1860	58.27	9.37	15.36	16.99
C12H23F2N5	275.35	275.1921	52.35	8.42	13.80	25.43
C12H24F1N1	201.33	201.1893	71.59	12.02	9.44	6.96
C12H24F1N3	229.34	229.1954	62.85	10.55	8.28	18.32
C12H24F1N5	257.36	257.2016	56.01	9.40	7.38	27.21
C12H24F2N2	234.34	234.1907	61.51	10.32	16.21	11.95
C12H24F2N4	262.35	262.1969	54.94	9.22	14.48	21.36
C12H24F2N6	290.36	290.2030	49.64	8.33	13.09	28.94
C12H25F1N2	216.34	216.2002	66.62	11.65	8.78	12.95
C12H25F1N4	244.36	244.2063	58.98	10.31	7.77	22.93
C12H25F1N6	272.37	272.2125	52.92	9.25	6.98	30.85
C12H25F2N1	221.34	221.1955	65.12	11.39	17.17	6.33
C12H25F2N3	249.35	249.2016	57.80	10.11	15.24	16.85
C12H25F2N5	277.36	277.2078	51.97	9.09	13.70	25.25
C12H26F1N1	203.35	203.2049	70.88	12.89	9.34	6.89
C12H26F1N3	231.36	231.2111	62.30	11.33	8.21	18.16
C12H26F1N5	259.37	259.2172	55.57	10.10	7.32	27.00
C12H26F2N2	236.35	236.2064	60.98	11.09	16.08	11.85
C12H26F2N4	264.36	264.2125	54.52	9.91	14.37	21.19
C12H26F2N6	292.38	292.2187	49.30	8.96	13.00	28.74
C12H27F1N2	218.36	218.2158	66.01	12.46	8.70	12.83
C12H27F1N4	246.37	246.2220	58.50	11.05	7.71	22.74
C12H27F1N6	274.39	274.2281	52.53	9.92	6.92	30.63
C12H27F2N3	251.37	251.2173	57.34	10.83	15.12	16.72
C12H27F2N5	279.38	279.2234	51.59	9.74	13.60	25.07
C12H28F1N3	233.38	233.2267	61.76	12.09	8.14	18.01
C12H28F1N5	261.39	261.2329	55.14	10.80	7.27	26.79
C12H28F2N4	266.38	266.2282	54.11	10.60	14.26	21.03
C12H28F2N6	294.39	294.2343	48.96	9.59	12.91	28.55
C12H29F1N4	248.39	248.2376	58.03	11.77	7.65	22.56
C12H29F1N6	276.40	276.2438	52.15	10.58	6.87	30.40
C12H29F2N5	281.40	281.2391	51.22	10.39	13.50	24.89
C12H30F1N5	263.40	263.2485	54.72	11.48	7.21	26.59
C12H30F2N6	296.41	296.2500	48.63	10.20	12.82	28.35
C12H31F1N6	278.42	278.2594	51.77	11.22	6.82	30.18
C13H 6F1N1	195.20	195.0484	79.99	3.10	9.73	7.18
C13H 6F1N3	223.21	223.0546	69.95	2.71	8.51	18.83
C13H 6F1N5	251.22	251.0607	62.15	2.41	7.56	27.88
C13H 6F2N2	228.20	228.0499	68.42	2.65	16.65	12.28

C13H 6F2N4	256.22	256.0560	60.94	2.36	14.83	21.87
C13H 6F2N6	284.23	284.0622	54.94	2.13	13.37	29.57
C13H 7F1N2	210.21	210.0593	74.28	3.36	9.04	13.33
C13H 7F1N4	238.23	238.0655	65.54	2.96	7.97	23.52
C13H 7F1N6	266.24	266.0716	58.65	2.65	7.14	31.57
C13H 7F2N1	215.20	215.0547	72.56	3.28	17.66	6.51
C13H 7F2N3	243.22	243.0608	64.20	2.90	15.62	17.28
C13H 7F2N5	271.23	271.0669	57.57	2.60	14.01	25.82
C13H 8F1N1	197.21	197.0641	79.18	4.09	9.63	7.10
C13H 8F1N3	225.23	225.0702	69.33	3.58	8.44	18.66
C13H 8F1N5	253.24	253.0764	61.66	3.18	7.50	27.65
C13H 8F2N2	230.22	230.0655	67.82	3.50	16.50	12.17
C13H 8F2N4	258.23	258.0717	60.47	3.12	14.71	21.70
C13H 8F2N6	286.25	286.0778	54.55	2.82	13.27	29.36
C13H 9F1N2	212.23	212.0750	73.57	4.27	8.95	13.20
C13H 9F1N4	240.24	240.0811	64.99	3.78	7.91	23.32
C13H 9F1N6	268.26	268.0873	58.21	3.38	7.08	31.33
C13H 9F2N1	217.22	217.0703	71.88	4.18	17.49	6.45
C13H 9F2N3	245.23	245.0764	63.67	3.70	15.49	17.13
C13H 9F2N5	273.25	273.0826	57.14	3.32	13.91	25.63
C13H10F1N1	199.23	199.0797	78.37	5.06	9.54	7.03
C13H10F1N3	227.24	227.0859	68.71	4.44	8.36	18.49
C13H10F1N5	255.26	255.0920	61.17	3.95	7.44	27.44
C13H10F2N2	232.23	232.0812	67.24	4.34	16.36	12.06
C13H10F2N4	260.25	260.0873	60.00	3.87	14.60	21.53
C13H10F2N6	288.26	288.0935	54.17	3.50	13.18	29.15
C13H11F1N2	214.24	214.0906	72.88	5.18	8.87	13.08
C13H11F1N4	242.26	242.0968	64.45	4.58	7.84	23.13
C13H11F1N6	270.27	270.1029	57.77	4.10	7.03	31.09
C13H11F2N1	219.24	219.0859	71.22	5.06	17.33	6.39
C13H11F2N3	247.25	247.0921	63.15	4.48	15.37	17.00
C13H11F2N5	275.26	275.0982	56.73	4.03	13.80	25.44
C13H12F1N1	201.25	201.0954	77.59	6.01	9.44	6.96
C13H12F1N3	229.26	229.1015	68.11	5.28	8.29	18.33
C13H12F1N5	257.27	257.1077	60.69	4.70	7.38	27.22
C13H12F2N2	234.25	234.0968	66.66	5.16	16.22	11.96
C13H12F2N4	262.26	262.1030	59.54	4.61	14.49	21.36
C13H12F2N6	290.28	290.1091	53.79	4.17	13.09	28.95
C13H13F1N2	216.26	216.1063	72.20	6.06	8.78	12.95
C13H13F1N4	244.27	244.1124	63.92	5.36	7.78	22.94
C13H13F1N6	272.29	272.1186	57.35	4.81	6.98	30.86
C13H13F2N1	221.25	221.1016	70.57	5.92	17.17	6.33
C13H13F2N3	249.27	249.1077	62.64	5.26	15.24	16.86
C13H13F2N5	277.28	277.1139	56.31	4.73	13.70	25.26
C13H14F1N1	203.26	203.1110	76.82	6.94	9.35	6.89
C13H14F1N3	231.28	231.1172	67.51	6.10	8.21	18.17
C13H14F1N5	259.29	259.1233	60.22	5.44	7.33	27.01
C13H14F2N2	236.27	236.1125	66.09	5.97	16.08	11.86
C13H14F2N4	264.28	264.1186	59.08	5.34	14.38	21.20
C13H14F2N6	292.29	292.1248	53.42	4.83	13.00	28.75
C13H15F1N2	218.28	218.1219	71.54	6.93	8.70	12.83
C13H15F1N4	246.29	246.1281	63.40	6.14	7.71	22.75
C13H15F1N6	274.30	274.1342	56.92	5.51	6.93	30.64
C13H15F2N1	223.27	223.1172	69.94	6.77	17.02	6.27
C13H15F2N3	251.28	251.1234	62.14	6.02	15.12	16.72
C13H15F2N5	279.29	279.1295	55.91	5.41	13.60	25.08
C13H16F1N1	205.28	205.1267	76.07	7.86	9.25	6.82
C13H16F1N3	233.29	233 1328	66.93	6.91	8.14	18.01
C13H16F1N5	261.30	261.1390	59.76	6.17	7.27	26.80
C13H16F2N2	238.28	238.1281	65.53	6.77	15.95	11.76

C13H16F2N4	266.30	266.1343	58.64	6.06	14.27	21.04
C13H16F2N6	294.31	294.1404	53.05	5.48	12.91	28.56
C13H17F1N2	220.29	220.1376	70.88	7.78	8.62	12.72
C13H17F1N4	248.31	248.1437	62.88	6.90	7.65	22.56
C13H17F1N6	276.32	276.1499	56.51	6.20	6.88	30.41
C13H17F2N1	225.28	225.1329	69.31	7.61	16.87	6.22
C13H17F2N3	253.30	253.1390	61.64	6.76	15.00	16.59
C13H17F2N5	281.31	281.1452	55.51	6.09	13.51	24.90
C13H18F1N1	207.29	207.1423	75.33	8.75	9.16	6.76
C13H18F1N3	235.31	235.1485	66.36	7.71	8.07	17.86
C13H18F1N5	263.32	263.1546	59.30	6.89	7.21	26.60
C13H18F2N2	240.30	240.1438	64.98	7.55	15.81	11.66
C13H18F2N4	268.31	268.1499	58.20	6.76	14.16	20.88
C13H18F2N6	296.33	296.1561	52.69	6.12	12.82	28.36
C13H19F1N2	222.31	222.1532	70.24	8.61	8.55	12.60
C13H19F1N4	250.32	250.1594	62.38	7.65	7.59	22.38
C13H19F1N6	278.33	278.1655	56.10	6.88	6.83	30.19
C13H19F2N1	227.30	227.1485	68.70	8.43	16.72	6.16
C13H19F2N3	255.31	255.1547	61.16	7.50	14.88	16.46
C13H19F2N5	283.33	283.1608	55.11	6.76	13.41	24.72
C13H20F1N1	209.31	209.1580	74.60	9.63	9.08	6.69
C13H20F1N3	237.32	237.1641	65.79	8.49	8.01	17.71
C13H20F1N5	265.34	265.1703	58.85	7.60	7.16	26.39
C13H20F2N2	242.31	242.1594	64.44	8.32	15.68	11.56
C13H20F2N4	270.33	270.1656	57.76	7.46	14.06	20.73
C13H20F2N6	298.34	298.1717	52.34	6.76	12.74	28.17
C13H21F1N2	224.32	224.1689	69.61	9.44	8.47	12.49
C13H21F1N4	252.34	252.1750	61.88	8.39	7.53	22.20
C13H21F1N6	280.35	280.1812	55.70	7.55	6.78	29.98
C13H21F2N1	229.32	229.1642	68.09	9.23	16.57	6.11
C13H21F2N3	257.33	257.1703	60.68	8.23	14.77	16.33
C13H21F2N5	285.34	285.1765	54.72	7.42	13.32	24.54
C13H22F1N1	211.33	211.1736	73.89	10.49	8.99	6.63
C13H22F1N3	239.34	239.1798	65.24	9.27	7.94	17.56
C13H22F1N5	267.35	267.1859	58.40	8.29	7.11	26.20
C13H22F2N2	244.33	244.1751	63.91	9.08	15.55	11.47
C13H22F2N4	272.34	272.1812	57.33	8.14	13.95	20.57
C13H22F2N6	300.36	300.1874	51.99	7.38	12.65	27.98
C13H23F1N2	226.34	226.1845	68.99	10.24	8.39	12.38
C13H23F1N4	254.35	254.1907	61.39	9.11	7.47	22.03
C13H23F1N6	282.37	282.1968	55.30	8.21	6.73	29.76
C13H23F2N1	231.33	231.1798	67.50	10.02	16.43	6.05
C13H23F2N3	259.35	259.1860	60.21	8.94	14.65	16.20
C13H23F2N5	287.36	287.1921	54.34	8.07	13.22	24.37
C13H24F1N1	213.34	213.1893	73.19	11.34	8.91	6.57
C13H24F1N3	241.35	241.1954	64.70	10.02	7.87	17.41
C13H24F1N5	269.37	269.2016	57.97	8.98	7.05	26.00
C13H24F2N2	246.35	246.1907	63.38	9.82	15.42	11.37
C13H24F2N4	274.36	274.1969	56.91	8.82	13.85	20.42
C13H24F2N6	302.37	302.2030	51.64	8.00	12.57	27.79
C13H25F1N2	228.36	228.2002	68.38	11.04	8.32	12.27
C13H25F1N4	256.37	256.2063	60.91	9.83	7.41	21.85
C13H25F1N6	284.38	284.2125	54.91	8.86	6.68	29.55
C13H25F2N1	233.35	233.1955	66.92	10.80	16.28	6.00
C13H25F2N3	261.36	261.2016	59.74	9.64	14.54	16.08
C13H25F2N5	289.37	289.2078	53.96	8.71	13.13	24.20
C13H26F1N1	215.36	215.2049	72.51	12.17	8.82	6.50
C13H26F1N3	243.37	243.2111	64.16	10.77	7.81	17.27
C13H26F1N5	271.38	271.2172	57.54	9.66	7.00	25.81
C13H26F2N2	248.36	248.2064	62.87	10.55	15.30	11.28
C13H26F2N4	276.38	276.2125	56.50	9.48	13.75	20.27
C13H26F2N6	304.39	304.2187	51.30	8.61	12.48	27.61
C13H27F1N2	230.37	230.2158	67.78	11.81	8.25	12.16
C13H27F1N4	258.39	258.2220	60.43	10.53	7.35	21.68
C13H27F1N6	286.40	286.2281	54.52	9.50	6.63	29.34
C13H27F2N1	235.36	235.2111	66.34	11.56	16.14	5.95
C13H27F2N3	263.38	263.2173	59.29	10.33	14.43	15.95
C13H27F2N5	291.39	291.2234	53.59	9.34	13.04	24.03
C13H28F1N1	217.37	217.2206	71.83	12.98	8.74	6.44
C13H28F1N3	245.39	245.2267	63.63	11.50	7.74	17.12
C13H28F1N5	273.40	273.2329	57.11	10.32	6.95	25.62
C13H28F2N2	250.38	250.2220	62.36	11.27	15.18	11.19
C13H28F2N4	278.39	278.2282	56.09	10.14	13.65	20.13
C13H28F2N6	306.41	306.2343	50.96	9.21	12.40	27.43
C13H29F1N2	232.39	232.2315	67.19	12.58	8.18	12.05
C13H29F1N4	260.40	260.2376	59.96	11.23	7.30	21.52
C13H29F1N6	288.41	288.2438	54.14	10.14	6.59	29.14
C13H29F2N3	265.39	265.2329	58.84	11.01	14.32	15.83
C13H29F2N5	293.41	293.2391	53.22	9.96	12.95	23.87
C13H30F1N3	247.40	247.2424	63.11	12.22	7.68	16.98
C13H30F1N5	275.42	275.2485	56.69	10.98	6.90	25.43
C13H30F2N4	280.41	280.2438	55.68	10.78	13.55	19.98
C13H30F2N6	308.42	308.2500	50.63	9.80	12.32	27.25
C13H31F1N4	262.42	262.2533	59.50	11.91	7.24	21.35
C13H31F1N6	290.43	290.2594	53.76	10.76	6.54	28.94
C13H31F2N5	295.42	295.2547	52.85	10.58	12.86	23.71
C13H32F1N5	277.43	277.2642	56.28	11.63	6.85	25.24
C13H32F2N6	310.44	310.2656	50.30	10.39	12.24	27.07
C13H33F1N6	292.45	292.2751	53.39	11.37	6.50	28.74
C14H 7F1N2	222.22	222.0593	75.67	3.18	8.55	12.61
C14H 7F1N4	250.24	250.0655	67.20	2.82	7.59	22.39
C14H 7F1N6	278.25	278.0716	60.43	2.54	6.83	30.20
C14H 7F2N1	227.22	227.0547	74.01	3.11	16.72	6.16
C14H 7F2N3	255.23	255.0608	65.88	2.76	14.89	16.46
C14H 7F2N5	283.24	283.0669	59.37	2.49	13.41	24.73
C14H 8F1N1	209.22	209.0641	80.37	3.85	9.08	6.69
C14H 8F1N3	237.24	237.0702	70.88	3.40	8.01	17.71
C14H 8F1N5	265.25	265.0764	63.39	3.04	7.16	26.40
C14H 8F2N2	242.23	242.0655	69.42	3.33	15.69	11.56
C14H 8F2N4	270.24	270.0717	62.22	2.98	14.06	20.73
C14H 8F2N6	298.26	298.0778	56.38	2.70	12.74	28.18
C14H 9F1N2	224.24	224.0750	74.99	4.05	8.47	12.49
C14H 9F1N4	252.25	252.0811	66.66	3.60	7.53	22.21
C14H 9F1N6	280.27	280.0873	60.00	3.24	6.78	29.99
C14H 9F2N1	229.23	229.0703	73.36	3.96	16.58	6.11
C14H 9F2N3	257.24	257.0764	65.37	3.53	14.77	16.33
C14H 9F2N5	285.26	285.0826	58.95	3.18	13.32	24.55
C14H10F1N1	211.24	211.0797	79.60	4.77	8.99	6.63
C14H10F1N3	239.25	239.0859	70.28	4.21	7.94	17.56
C14H10F1N5	267.27	267.0920	62.92	3.77	7.11	26.20
C14H10F2N2	244.25	244.0812	68.85	4.13	15.56	11.47
C14H10F2N4	272.26	272.0873	61.76	3.70	13.96	20.58
C14H10F2N6	300.27	300.0935	56.00	3.36	12.65	27.99
C14H11F1N2	226.26	226.0906	74.32	4.90	8.40	12.38
C14H11F1N4	254.27	254.0968	66.13	4.36	7.47	22.03
C14H11F1N6	282.28	282.1029	59.57	3.93	6.73	29.77
C14H11F2N1	231.25	231.0859	72.72	4.79	16.43	6.06
C14H11F2N3	259.26	259.0921	64.86	4.28	14.66	16.21
C14H11F2N5	287.27	287.0982	58.54	3.86	13.23	24.38
C14H12F1N1	213.26	213.0954	78.85	5.67	8.91	6.57

C14H12F1N3	241.27	241.1015	69.70	5.01	7.87	17.42
C14H12F1N5	269.28	269.1077	62.45	4.49	7.06	26.01
C14H12F2N2	246.26	246.0968	68.28	4.91	15.43	11.38
C14H12F2N4	274.28	274.1030	61.31	4.41	13.85	20.43
C14H12F2N6	302.29	302.1091	55.63	4.00	12.57	27.80
C14H13F1N2	228.27	228.1063	73.66	5.74	8.32	12.27
C14H13F1N4	256.28	256.1124	65.61	5.11	7.41	21.86
C14H13F1N6	284.30	284.1186	59.15	4.61	6.68	29.56
C14H13F2N1	233.26	233.1016	72.09	5.62	16.29	6.00
C14H13F2N3	261.28	261.1077	64.36	5.02	14.54	16.08
C14H13F2N5	289.29	289.1139	58.13	4.53	13.13	24.21
C14H14F1N1	215.27	215.1110	78.11	6.56	8.83	6.51
C14H14F1N3	243.29	243.1172	69.12	5.80	7.81	17.27
C14H14F1N5	271.30	271.1233	61.98	5.20	7.00	25.81
C14H14F2N2	248.28	248.1125	67.73	5.68	15.30	11.28
C14H14F2N4	276.29	276.1186	60.86	5.11	13.75	20.28
C14H14F2N6	304.30	304.1248	55.26	4.64	12.49	27.62
C14H15F1N2	230.29	230.1219	73.02	6.57	8.25	12.16
C14H15F1N4	258.30	258.1281	65.10	5.85	7.36	21.69
C14H15F1N6	286.31	286.1342	58.73	5.28	6.64	29.35
C14H15F2N1	235.28	235.1172	71.47	6.43	16.15	5.95
C14H15F2N3	263.29	263.1234	63.87	5.74	14.43	15.96
C14H15F2N5	291.31	291.1295	57.72	5.19	13.04	24.04
C14H16F1N1	217.29	217.1267	77.39	7.42	8.74	6.45
C14H16F1N3	245.30	245.1328	68.55	6.57	7.74	17.13
C14H16F1N5	273.32	273.1390	61.52	5.90	6.95	25.62
C14H16F2N2	250.29	250.1281	67.18	6.44	15.18	11.19
C14H16F2N4	278.31	278.1343	60.42	5.79	13.65	20.13
C14H16F2N6	306.32	306.1404	54.90	5.26	12.40	27.44
C14H17F1N2	232.30	232.1376	72.39	7.38	8.18	12.06
C14H17F1N4	260.32	260.1437	64.60	6.58	7.30	21.52
C14H17F1N6	288.33	288.1499	58.32	5.94	6.59	29.15
C14H17F2N1	237.30	237.1329	70.86	7.22	16.01	5.90
C14H17F2N3	265.31	265.1390	63.38	6.46	14.32	15.84
C14H17F2N5	293.32	293.1452	57.33	5.84	12.95	23.88
C14H18F1N1	219.30	219.1423	76.68	8.27	8.66	6.39
C14H18F1N3	247.32	247.1485	67.99	7.34	7.68	16.99
C14H18F1N5	275.33	275.1546	61.07	6.59	6.90	25.44
C14H18F2N2	252.31	252.1438	66.65	7.19	15.06	11.10
C14H18F2N4	280.32	280.1499	59.99	6.47	13.55	19.99
C14H18F2N6	308.34	308.1561	54.54	5.88	12.32	27.26
C14H19F1N2	234.32	234.1532	71.76	8.17	8.11	11.96
C14H19F1N4	262.33	262.1594	64.10	7.30	7.24	21.36
C14H19F1N6	290.35	290.1655	57.92	6.60	6.54	28.94
C14H19F2N1	239.31	239.1485	70.27	8.00	15.88	5.85
C14H19F2N3	267.32	267.1547	62.90	7.16	14.21	15.72
C14H19F2N5	295.34	295.1608	56.94	6.48	12.87	23.71
C14H20F1N1	221.32	221.1580	75.98	9.11	8.58	6.33
C14H20F1N3	249.33	249.1641	67.44	8.09	7.62	16.85
C14H20F1N5	277.35	277.1703	60.63	7.27	6.85	25.25
C14H20F2N2	254.33	254.1594	66.12	7.93	14.94	11.01
C14H20F2N4	282.34	282.1656	59.56	7.14	13.46	19.84
C14H20F2N6	310.35	310.1717	54.18	6.50	12.24	27.08
C14H21F1N2	236.34	236.1689	71.15	8.96	8.04	11.85
C14H21F1N4	264.35	264.1750	63.61	8.01	7.19	21.19
C14H21F1N6	292.36	292.1812	57.52	7.24	6.50	28.75
C14H21F2N1	241.33	241.1642	69.68	8.77	15.74	5.80
C14H21F2N3	269.34	269.1703	62.43	7.86	14.11	15.60
C14H21F2N5	297.35	297.1765	56.55	7.12	12.78	23.55
C14H22F1N1	223.34	223.1736	75.29	9.93	8.51	6.27
C14H22F1N3	251.35	251.1798	66.90	8.82	7.56	16.72
C14H22F1N5	279.36	279.1859	60.19	7.94	6.80	25.07
C14H22F2N2	256.34	256.1751	65.60	8.65	14.82	10.93
C14H22F2N4	284.36	284.1812	59.14	7.80	13.36	19.70
C14H22F2N6	312.37	312.1874	53.83	7.10	12.16	26.90
C14H23F1N2	238.35	238.1845	70.55	9.73	7.97	11.75
C14H23F1N4	266.36	266.1907	63.13	8.70	7.13	21.03
C14H23F1N6	294.38	294.1968	57.12	7.88	6.45	28.55
C14H23F2N1	243.34	243.1798	69.10	9.53	15.61	5.76
C14H23F2N3	271.36	271.1860	61.97	8.54	14.00	15.49
C14H23F2N5	299.37	299.1921	56.17	7.74	12.69	23.39
C14H24F1N1	225.35	225.1893	74.62	10.73	8.43	6.22
C14H24F1N3	253.37	253.1954	66.37	9.55	7.50	16.58
C14H24F1N5	281.38	281.2016	59.76	8.60	6.75	24.89
C14H24F2N2	258.36	258.1907	65.09	9.36	14.71	10.84
C14H24F2N4	286.37	286.1969	58.72	8.45	13.27	19.56
C14H24F2N6	314.38	314.2030	53.49	7.69	12.09	26.73
C14H25F1N2	240.37	240.2002	69.96	10.48	7.90	11.65
C14H25F1N4	268.38	268.2063	62.66	9.39	7.08	20.88
C14H25F1N6	296.39	296.2125	56.73	8.50	6.41	28.35
C14H25F2N1	245.36	245.1955	68.53	10.27	15.49	5.71
C14H25F2N3	273.37	273.2016	61.51	9.22	13.90	15.37
C14H25F2N5	301.39	301.2078	55.79	8.36	12.61	23.24
C14H26F1N1	227.37	227.2049	73.96	11.53	8.36	6.16
C14H26F1N3	255.38	255.2111	65.84	10.26	7.44	16.45
C14H26F1N5	283.40	283.2172	59.34	9.25	6.70	24.71
C14H26F2N2	260.37	260.2064	64.58	10.07	14.59	10.76
C14H26F2N4	288.39	288.2125	58.31	9.09	13.18	19.43
C14H26F2N6	316.40	316.2187	53.15	8.28	12.01	26.56
C14H27F1N2	242.38	242.2158	69.38	11.23	7.84	11.56
C14H27F1N4	270.40	270.2220	62.19	10.06	7.03	20.72
C14H27F1N6	298.41	298.2281	56.35	9.12	6.37	28.16
C14H27F2N1	247.37	247.2111	67.98	11.00	15.36	5.66
C14H27F2N3	275.39	275.2173	61.06	9.88	13.80	15.26
C14H27F2N5	303.40	303.2234	55.42	8.97	12.52	23.08
C14H28F1N1	229.38	229.2206	73.31	12.30	8.28	6.11
C14H28F1N3	257.40	257.2267	65.33	10.96	7.38	16.32
C14H28F1N5	285.41	285.2329	58.92	9.89	6.66	24.54
C14H28F2N2	262.39	262.2220	64.09	10.76	14.48	10.68
C14H28F2N4	290.40	290.2282	57.90	9.72	13.08	19.29
C14H28F2N6	318.42	318.2343	52.81	8.86	11.93	26.39
C14H29F1N2	244.40	244.2315	68.80	11.96	7.77	11.46
C14H29F1N4	272.41	272.2376	61.73	10.73	6.97	20.57
C14H29F1N6	300.43	300.2438	55.97	9.73	6.32	27.97
C14H29F2N1	249.39	249.2268	67.43	11.72	15.24	5.62
C14H29F2N3	277.40	277.2329	60.62	10.54	13.70	15.15
C14H29F2N5	305.42	305.2391	55.06	9.57	12.44	22.93
C14H30F1N1	231.40	231.2362	72.67	13.07	8.21	6.05
C14H30F1N3	259.41	259.2424	64.82	11.66	7.32	16.20
C14H30F1N5	287.43	287.2485	58.50	10.52	6.61	24.37
C14H30F2N2	264.41	264.2377	63.60	11.44	14.37	10.59
C14H30F2N4	292.42	292.2438	57.51	10.34	12.99	19.16
C14H30F2N6	320.43	320.2500	52.48	9.44	11.86	26.23
C14H31F1N2	246.41	246.2471	68.24	12.68	7.71	11.37
C14H31F1N4	274.43	274.2533	61.28	11.39	6.92	20.42
C14H31F1N6	302.44	302.2594	55.60	10.33	6.28	27.79
C14H31F2N3	279.42	279.2486	60.18	11.18	13.60	15.04
C14H31F2N5	307.43	307.2547	54.70	10.16	12.36	22.78
C14H32F1N3	261.43	261.2580	64.32	12.34	7.27	16.07
C14H32F1N5	289.44	289.2642	58.10	11.14	6.56	24.20

C14H32F2N4	294.43	294.2595	57.11	10.95	12.90	19.03
C14H32F2N6	322.45	322.2656	52.15	10.00	11.78	26.06
C14H33F1N4	276.44	276.2689	60.83	12.03	6.87	20.27
C14H33F1N6	304.46	304.2751	55.23	10.93	6.24	27.60
C14H33F2N5	309.45	309.2704	54.34	10.75	12.28	22.63
C14H34F1N5	291.46	291.2798	57.69	11.76	6.52	24.03
C14H34F2N6	324.46	324.2813	51.83	10.56	11.71	25.90
C14H35F1N6	306.47	306.2907	54.87	11.51	6.20	27.42
C15H 7F1N2	234.23	234.0593	76.92	3.01	8.11	11.96
C15H 7F1N4	262.25	262.0655	68.70	2.69	7.24	21.36
C15H 7F1N6	290.26	290.0716	62.07	2.43	6.55	28.95
C15H 7F2N1	239.23	239.0547	75.31	2.95	15.88	5.85
C15H 7F2N3	267.24	267.0608	67.42	2.64	14.22	15.72
C15H 7F2N5	295.25	295.0669	61.02	2.39	12.87	23.72
C15H 8F1N1	221.24	221.0641	81.44	3.64	8.59	6.33
C15H 8F1N3	249.25	249.0702	72.28	3.24	7.62	16.86
C15H 8F1N5	277.26	277.0764	64.98	2.91	6.85	25.26
C15H 8F2N2	254.24	254.0655	70.86	3.17	14.95	11.02
C15H 8F2N4	282.25	282.0717	63.83	2.86	13.46	19.85
C15H 8F2N6	310.27	310.0778	58.07	2.60	12.25	27.09
C15H 9F1N2	236.25	236.0750	76.26	3.84	8.04	11.86
C15H 9F1N4	264.26	264.0811	68.18	3.43	7.19	21.20
C15H 9F1N6	292.28	292.0873	61.64	3.10	6.50	28.75
C15H 9F2N1	241.24	241.0703	74.68	3.76	15.75	5.81
C15H 9F2N3	269.26	269.0764	66.91	3.37	14.11	15.61
C15H 9F2N5	297.27	297.0826	60.61	3.05	12.78	23.56
C15H10F1N1	223.25	223.0797	80.70	4.51	8.51	6.27
C15H10F1N3	251.27	251.0859	71.70	4.01	7.56	16.72
C15H10F1N5	279.28	279.0920	64.51	3.61	6.80	25.08
C15H10F2N2	256.26	256.0812	70.31	3.93	14.83	10.93
C15H10F2N4	284.27	284.0873	63.38	3.55	13.37	19.71
C15H10F2N6	312.28	312.0935	57.69	3.23	12.17	26.91
C15H11F1N2	238.27	238.0906	75.62	4.65	7.97	11.76
C15H11F1N4	266.28	266.0968	67.66	4.16	7.13	21.04
C15H11F1N6	294.29	294.1029	61.22	3.77	6.46	28.56
C15H11F2N1	243.26	243.0859	74.06	4.56	15.62	5.76
C15H11F2N3	271.27	271.0921	66.42	4.09	14.01	15.49
C15H11F2N5	299.29	299.0982	60.20	3.70	12.70	23.40
C15H12F1N1	225.27	225.0954	79.98	5.37	8.43	6.22
C15H12F1N3	253.28	253.1015	71.13	4.78	7.50	16.59
C15H12F1N5	281.29	281.1077	64.05	4.30	6.75	24.90
C15H12F2N2	258.27	258.0968	69.76	4.68	14.71	10.85
C15H12F2N4	286.29	286.1030	62.93	4.23	13.27	19.57
C15H12F2N6	314.30	314.1091	57.32	3.85	12.09	26.74
C15H13F1N2	240.28	240.1063	74.98	5.45	7.91	11.66
C15H13F1N4	268.30	268.1124	67.15	4.88	7.08	20.88
C15H13F1N6	296.31	296.1186	60.80	4.42	6.41	28.36
C15H13F2N1	245.27	245.1016	73.46	5.34	15.49	5.71
C15H13F2N3	273.29	273.1077	65.93	4.79	13.90	15.38
C15H13F2N5	301.30	301.1139	59.80	4.35	12.61	23.24
C15H14F1N1	227.28	227.1110	79.27	6.21	8.36	6.16
C15H14F1N3	255.30	255.1172	70.57	5.53	7.44	16.46
C15H14F1N5	283.31	283.1233	63.59	4.98	6.71	24.72
C15H14F2N2	260.29	260.1125	69.22	5.42	14.60	10.76
C15H14F2N4	288.30	288.1186	62.49	4.89	13.18	19.43
C15H14F2N6	316.32	316.1248	56.96	4.46	12.01	26.57
C15H15F1N2	242.30	242.1219	74.36	6.24	7.84	11.56
C15H15F1N4	270.31	270.1281	66.65	5.59	7.03	20.73
C15H15F1N6	298.33	298.1342	60.39	5.07	6.37	28.17
C15H15F2N1	247.29	247.1172	72.86	6.11	15.37	5.66

C15H15F2N3	275.30	275.1234	65.44	5.49	13.80	15.26
C15H15F2N5	303.32	303.1295	59.40	4.98	12.53	23.09
C15H16F1N1	229.30	229.1267	78.57	7.03	8.29	6.11
C15H16F1N3	257.31	257.1328	70.02	6.27	7.38	16.33
C15H16F1N5	285.33	285.1390	63.14	5.65	6.66	24.55
C15H16F2N2	262.30	262.1281	68.69	6.15	14.49	10.68
C15H16F2N4	290.32	290.1343	62.06	5.56	13.09	19.30
C15H16F2N6	318.33	318.1404	56.60	5.07	11.94	26.40
C15H17F1N2	244.31	244.1376	73.74	7.01	7.78	11.47
C15H17F1N4	272.33	272.1437	66.16	6.29	6.98	20.57
C15H17F1N6	300.34	300.1499	59.99	5.71	6.33	27.98
C15H17F2N1	249.31	249.1329	72.27	6.87	15.24	5.62
C15H17F2N3	277.32	277.1390	64.97	6.18	13.70	15.15
C15H17F2N5	305.33	305.1452	59.01	5.61	12.44	22.94
C15H18F1N1	231.32	231.1423	77.89	7.84	8.21	6.06
C15H18F1N3	259.33	259.1485	69.47	7.00	7.33	16.20
C15H18F1N5	287.34	287.1546	62.70	6.31	6.61	24.37
C15H18F2N2	264.32	264.1438	68.16	6.86	14.38	10.60
C15H18F2N4	292.33	292.1499	61.63	6.21	13.00	19.17
C15H18F2N6	320.35	320.1561	56.24	5.66	11.86	26.23
C15H19F1N2	246.33	246.1532	73.14	7.77	7.71	11.37
C15H19F1N4	274.34	274.1594	65.67	6.98	6.93	20.42
C15H19F1N6	302.36	302.1655	59.59	6.33	6.28	27.79
C15H19F2N1	251.32	251.1485	71.69	7.62	15.12	5.57
C15H19F2N3	279.34	279.1547	64.50	6.86	13.60	15.04
C15H19F2N5	307.35	307.1608	58.62	6.23	12.36	22.79
C15H20F1N1	233.33	233.1580	77.22	8.64	8.14	6.00
C15H20F1N3	261.35	261.1641	68.94	7.71	7.27	16.08
C15H20F1N5	289.36	289.1703	62.26	6.97	6.57	24.20
C15H20F2N2	266.34	266.1594	67.65	7.57	14.27	10.52
C15H20F2N4	294.35	294.1656	61.21	6.85	12.91	19.03
C15H20F2N6	322.36	322.1717	55.89	6.25	11.79	26.07
C15H21F1N2	248.35	248.1689	72.55	8.52	7.65	11.28
C15H21F1N4	276.36	276.1750	65.19	7.66	6.87	20.27
C15H21F1N6	304.37	304.1812	59.19	6.95	6.24	27.61
C15H21F2N1	253.34	253.1642	71.12	8.36	15.00	5.53
C15H21F2N3	281.35	281.1703	64.04	7.52	13.51	14.94
C15H21F2N5	309.36	309.1765	58.24	6.84	12.28	22.64
C15H22F1N1	235.35	235.1736	76.55	9.42	8.07	5.95
C15H22F1N3	263.36	263.1798	68.41	8.42	7.21	15.96
C15H22F1N5	291.37	291.1859	61.83	7.61	6.52	24.04
C15H22F2N2	268.35	268.1751	67.14	8.26	14.16	10.44
C15H22F2N4	296.37	296.1812	60.79	7.48	12.82	18.90
C15H22F2N6	324.38	324.1874	55.54	6.84	11.71	25.91
C15H23F1N2	250.36	250.1845	71.96	9.26	7.59	11.19
C15H23F1N4	278.38	278.1907	64.72	8.33	6.82	20.13
C15H23F1N6	306.39	306.1968	58.80	7.57	6.20	27.43
C15H23F2N1	255.35	255.1798	70.56	9.08	14.88	5.49
C15H23F2N3	283.37	283.1860	63.58	8.18	13.41	14.83
C15H23F2N5	311.38	311.1921	57.86	7.45	12.20	22.49
C15H24F1N1	237.36	237.1893	75.90	10.19	8.00	5.90
C15H24F1N3	265.38	265.1954	67.89	9.12	7.16	15.83
C15H24F1N5	293.39	293.2016	61.41	8.25	6.48	23.87
C15H24F2N2	270.37	270.1907	66.64	8.95	14.05	10.36
C15H24F2N4	298.38	298.1969	60.38	8.11	12.73	18.78
C15H24F2N6	326.40	326.2030	55.20	7.41	11.64	25.75
C15H25F1N2	252.38	252.2002	71.39	9.98	7.53	11.10
C15H25F1N4	280.39	280.2063	64.26	8.99	6.78	19.98
C15H25F1N6	308.41	308.2125	58.42	8.17	6.16	27.25
C15H25F2N1	257.37	257.1955	70.00	9.79	14.76	5.44

C15H25F2N3	285.38	285.2016	63.13	8.83	13.31	14.72
C15H25F2N5	313.40	313.2078	57.49	8.04	12.12	22.35
C15H26F1N1	239.38	239.2049	75.26	10.95	7.94	5.85
C15H26F1N3	267.39	267.2111	67.38	9.80	7.11	15.71
C15H26F1N5	295.41	295.2172	60.99	8.87	6.43	23.71
C15H26F2N2	272.38	272.2064	66.14	9.62	13.95	10.28
C15H26F2N4	300.40	300.2125	59.98	8.72	12.65	18.65
C15H26F2N6	328.41	328.2187	54.86	7.98	11.57	25.59
C15H27F1N2	254.39	254.2158	70.82	10.70	7.47	11.01
C15H27F1N4	282.41	282.2220	63.80	9.64	6.73	19.84
C15H27F1N6	310.42	310.2281	58.04	8.77	6.12	27.07
C15H27F2N1	259.39	259.2111	69.46	10.49	14.65	5.40
C15H27F2N3	287.40	287.2173	62.69	9.47	13.22	14.62
C15H27F2N5	315.41	315.2234	57.12	8.63	12.05	22.20
C15H28F1N1	241.40	241.2206	74.64	11.69	7.87	5.80
C15H28F1N3	269.41	269.2267	66.88	10.48	7.05	15.60
C15H28F1N5	297.42	297.2329	60.58	9.49	6.39	23.55
C15H28F2N2	274.40	274.2220	65.66	10.29	13.85	10.21
C15H28F2N4	302.41	302.2282	59.58	9.33	12.56	18.53
C15H28F2N6	330.43	330.2343	54.53	8.54	11.50	25.43
C15H29F1N2	256.41	256.2315	70.27	11.40	7.41	10.93
C15H29F1N4	284.42	284.2376	63.34	10.28	6.68	19.70
C15H29F1N6	312.44	312.2438	57.67	9.36	6.08	26.90
C15H29F2N1	261.40	261.2268	68.92	11.18	14.54	5.36
C15H29F2N3	289.42	289.2329	62.25	10.10	13.13	14.52
C15H29F2N5	317.43	317.2391	56.76	9.21	11.97	22.06
C15H30F1N1	243.41	243.2362	74.02	12.42	7.81	5.75
C15H30F1N3	271.42	271.2424	66.38	11.14	7.00	15.48
C15H30F1N5	299.44	299.2485	60.17	10.10	6.34	23.39
C15H30F2N2	276.42	276.2377	65.18	10.94	13.75	10.13
C15H30F2N4	304.43	304.2438	59.18	9.93	12.48	18.40
C15H30F2N6	332.44	332.2500	54.19	9.10	11.43	25.28
C15H31F1N2	258.43	258.2471	69.72	12.09	7.35	10.84
C15H31F1N4	286.44	286.2533	62.90	10.91	6.63	19.56
C15H31F1N6	314.45	314.2594	57.30	9.94	6.04	26.73
C15H31F2N1	263.42	263.2424	68.40	11.86	14.42	5.32
C15H31F2N3	291.43	291.2486	61.82	10.72	13.04	14.42
C15H31F2N5	319.44	319.2547	56.40	9.78	11.89	21.92
C15H32F1N1	245.43	245.2519	73.41	13.14	7.74	5.71
C15H32F1N3	273.44	273.2580	65.89	11.80	6.95	15.37
C15H32F1N5	301.45	301.2642	59.77	10.70	6.30	23.23
C15H32F2N2	278.43	278.2533	64.71	11.58	13.65	10.06
C15H32F2N4	306.45	306.2595	58.79	10.53	12.40	18.28
C15H32F2N6	334.46	334.2656	53.87	9.64	11.36	25.13
C15H33F1N2	260.44	260.2628	69.18	12.77	7.29	10.76
C15H33F1N4	288.46	288.2689	62.46	11.53	6.59	19.42
C15H33F1N6	316.47	316.2751	56.93	10.51	6.00	26.56
C15H33F2N3	293.45	293.2642	61.40	11.34	12.95	14.32
C15H33F2N5	321.46	321.2704	56.05	10.35	11.82	21.79
C15H34F1N3	275.46	275.2737	65.41	12.44	6.90	15.25
C15H34F1N5	303.47	303.2798	59.37	11.29	6.26	23.08
C15H34F2N4	308.46	308.2751	58.41	11.11	12.32	18.16
C15H34F2N6	336.48	336.2813	53.55	10.19	11.29	24.98
C15H35F1N4	290.47	290.2846	62.03	12.15	6.54	19.29
C15H35F1N6	318.48	318.2907	56.57	11.08	5.97	26.39
C15H35F2N5	323.48	323.2860	55.70	10.91	11.75	21.65
C15H36F1N5	305.49	305.2955	58.98	11.88	6.22	22.93
C15H36F2N6	338.49	338.2969	53.23	10.72	11.23	24.83
C15H37F1N6	320.50	320.3064	56.21	11.64	5.93	26.22
C16H 8F1N1	233.25	233.0641	82.39	3.46	8.15	6.01

C16H 8F1N3	261.26	261.0702	73.56	3.09	7.27	16.08
C16H 8F1N5	289.27	289.0764	66.43	2.79	6.57	24.21
C16H 8F2N2	266.25	266.0655	72.18	3.03	14.27	10.52
C16H 8F2N4	294.27	294.0717	65.31	2.74	12.91	19.04
C16H 8F2N6	322.28	322.0778	59.63	2.50	11.79	26.08
C16H 9F1N2	248.26	248.0750	77.41	3.65	7.65	11.28
C16H 9F1N4	276.28	276.0811	69.56	3.28	6.88	20.28
C16H 9F1N6	304.29	304.0873	63.16	2.98	6.24	27.62
C16H 9F2N1	253.25	253.0703	75.88	3.58	15.00	5.53
C16H 9F2N3	281.27	281.0764	68.33	3.23	13.51	14.94
C16H 9F2N5	309.28	309.0826	62.14	2.93	12.29	22.64
C16H10F1N1	235.26	235.0797	81.69	4.28	8.08	5.95
C16H10F1N3	263.28	263.0859	72.99	3.83	7.22	15.96
C16H10F1N5	291.29	291.0920	65.97	3.46	6.52	24.04
C16H10F2N2	268.27	268.0812	71.64	3.76	14.16	10.44
C16H10F2N4	296.28	296.0873	64.86	3.40	12.82	18.91
C16H10F2N6	324.30	324.0935	59.26	3.11	11.72	25.91
C16H11F1N2	250.28	250.0906	76.79	4.43	7.59	11.19
C16H11F1N4	278.29	278.0968	69.06	3.98	6.83	20.13
C16H11F1N6	306.30	306.1029	62.74	3.62	6.20	27.44
C16H11F2N1	255.27	255.0859	75.28	4.34	14.88	5.49
C16H11F2N3	283.28	283.0921	67.84	3.91	13.41	14.83
C16H11F2N5	311.30	311.0982	61.73	3.56	12.21	22.50
C16H12F1N1	237.28	237.0954	80.99	5.10	8.01	5.90
C16H12F1N3	265.29	265.1015	72.44	4.56	7.16	15.84
C16H12F1N5	293.31	293.1077	65.52	4.12	6.48	23.88
C16H12F2N2	270.28	270.0968	71.10	4.48	14.06	10.36
C16H12F2N4	298.30	298.1030	64.43	4.05	12.74	18.78
C16H12F2N6	326.31	326.1091	58.89	3.71	11.64	25.75
C16H13F1N2	252.29	252.1063	76.17	5.19	7.53	11.10
C16H13F1N4	280.31	280.1124	68.56	4.67	6.78	19.99
C16H13F1N6	308.32	308.1186	62.33	4.25	6.16	27.26
C16H13F2N1	257.29	257.1016	74.69	5.09	14.77	5.44
C16H13F2N3	285.30	285.1077	67.36	4.59	13.32	14.73
C16H13F2N5	313.31	313.1139	61.34	4.18	12.13	22.35
C16H14F1N1	239.30	239.1110	80.31	5.90	7.94	5.85
C16H14F1N3	267.31	267.1172	71.89	5.28	7.11	15.72
C16H14F1N5	295.32	295.1233	65.07	4.78	6.43	23.71
C16H14F2N2	272.30	272.1125	70.58	5.18	13.95	10.29
C16H14F2N4	300.31	300.1186	63.99	4.70	12.65	18.66
C16H14F2N6	328.33	328.1248	58.53	4.30	11.57	25.60
C16H15F1N2	254.31	254.1219	75.57	5.95	7.47	11.02
C16H15F1N4	282.32	282.1281	68.07	5.36	6.73	19.84
C16H15F1N6	310.34	310.1342	61.93	4.87	6.12	27.08
C16H15F2N1	259.30	259.1172	74.11	5.83	14.65	5.40
C16H15F2N3	287.31	287.1234	66.89	5.26	13.22	14.63
C16H15F2N5	315.33	315.1295	60.95	4.79	12.05	22.21
C16H16F1N1	241.31	241.1267	79.64	6.68	7.87	5.80
C16H16F1N3	269.32	269.1328	71.36	5.99	7.05	15.60
C16H16F1N5	297.34	297.1390	64.63	5.42	6.39	23.55
C16H16F2N2	274.32	274.1281	70.06	5.88	13.85	10.21
C16H16F2N4	302.33	302.1343	63.57	5.33	12.57	18.53
C16H16F2N6	330.34	330.1404	58.18	4.88	11.50	25.44
C16H17F1N2	256.33	256.1376	74.97	6.69	7.41	10.93
C16H17F1N4	284.34	284.1437	67.59	6.03	6.68	19.70
C16H17F1N6	312.35	312.1499	61.53	5.49	6.08	26.91
C16H17F2N1	261.32	261.1329	73.54	6.56	14.54	5.36
C16H17F2N3	289.33	289.1390	66.42	5.92	13.13	14.52
C16H17F2N5	317.34	317.1452	60.56	5.40	11.97	22.07
C16H18F1N1	243.33	243.1423	78.98	7.46	7.81	5.76

C16H18F1N3	271.34	271.1485	70.83	6.69	7.00	15.49	C16H28F1N3	281.42	281.2267	68.29	10.03	6.75	14.93
C16H18F1N5	299.35	299.1546	64.20	6.06	6.35	23.39	C16H28F1N5	309.43	309.2329	62.11	9.12	6.14	22.63
C16H18F2N2	276.33	276.1438	69.55	6.57	13.75	10.14	C16H28F2N2	286.41	286.2220	67.10	9.85	13.27	9.78
C16H18F2N4	304.35	304.1499	63.14	5.96	12.48	18.41	C16H28F2N4	314.43	314.2282	61.12	8.98	12.08	17.82
C16H18F2N6	332.36	332.1561	57.82	5.46	11.43	25.29	C16H28F2N6	342.44	342.2343	56.12	8.24	11.10	24.54
C16H19F1N2	258.34	258.1532	74.39	7.41	7.35	10.84	C16H29F1N2	268.42	268.2315	71.60	10.89	7.08	10.44
C16H19F1N4	286.36	286.1594	67.11	6.69	6.63	19.57	C16H29F1N4	296.43	296.2376	64.83	9.86	6.41	18.90
C16H19F1N6	314.37	314.1655	61.13	6.09	6.04	26.73	C16H29F1N6	324.45	324.2438	59.23	9.01	5.86	25.90
C16H19F2N1	263.33	263.1485	72.98	7.27	14.43	5.32	C16H29F2N1	273.41	273.2268	70.29	10.69	13.90	5.12
C16H19F2N3	291.35	291.1547	65.96	6.57	13.04	14.42	C16H29F2N3	301.43	301.2329	63.76	9.70	12.61	13.94
C16H19F2N5	319.36	319.1608	60.18	6.00	11.90	21.93	C16H29F2N5	329.44	329.2391	58.33	8.87	11.53	21.26
C16H20F1N1	245.34	245.1580	78.33	8.22	7.74	5.71	C16H30F1N1	255.42	255.2362	75.24	11.84	7.44	5.48
C16H20F1N3	273.36	273.1641	70.30	7.37	6.95	15.37	C16H30F1N3	283.44	283.2424	67.80	10.67	6.70	14.83
C16H20F1N5	301.37	301.1703	63.77	6.69	6.30	23.24	C16H30F1N5	311.45	311.2485	61.70	9.71	6.10	22.49
C16H20F2N2	278.35	278.1594	69.04	7.24	13.65	10.06	C16H30F2N2	288.43	288.2377	66.63	10.48	13.17	9.71
C16H20F2N4	306.36	306.1656	62.73	6.58	12.40	18.29	C16H30F2N4	316.44	316.2438	60.73	9.56	12.01	17.71
C16H20F2N6	334.37	334.1717	57.47	6.03	11.36	25.13	C16H30F2N6	344.45	344.2500	55.79	8.78	11.03	24.40
C16H21F1N2	260.36	260.1689	73.81	8.13	7.30	10.76	C16H31F1N2	270.44	270.2471	71.06	11.55	7.03	10.36
C16H21F1N4	288.37	288.1750	66.64	7.34	6.59	19.43	C16H31F1N4	298.45	298.2533	64.39	10.47	6.37	18.77
C16H21F1N6	316.38	316.1812	60.74	6.69	6.00	26.56	C16H31F1N6	326.46	326.2594	58.87	9.57	5.82	25.74
C16H21F2N1	265.35	265.1642	72.42	7.98	14.32	5.28	C16H31F2N1	275.43	275.2424	69.77	11.34	13.80	5.09
C16H21F2N3	293.36	293.1703	65.51	7.22	12.95	14.32	C16H31F2N3	303.44	303.2486	63.33	10.30	12.52	13.85
C16H21F2N5	321.38	321.1765	59.80	6.59	11.82	21.79	C16H31F2N5	331.46	331.2547	57.98	9.43	11.46	21.13
C16H22F1N1	247.36	247.1736	77.69	8.96	7.68	5.66	C16H32F1N1	257.44	257.2519	74.65	12.53	7.38	5.44
C16H22F1N3	275.37	275.1798	69.79	8.05	6.90	15.26	C16H32F1N3	285.45	285.2580	67.32	11.30	6.66	14.72
C16H22F1N5	303.39	303.1859	63.34	7.31	6.26	23.08	C16H32F1N5	313.47	313.2642	61.31	10.29	6.06	22.34
C16H22F2N2	280.36	280.1751	68.55	7.91	13.55	9.99	C16H32F2N2	290.44	290.2533	66.17	11.11	13.08	9.65
C16H22F2N4	308.38	308.1812	62.32	7.19	12.32	18.17	C16H32F2N4	318.46	318.2595	60.35	10.13	11.93	17.59
C16H22F2N6	336.39	336.1874	57.13	6.59	11.30	24.98	C16H32F2N6	346.47	346.2656	55.47	9.31	10.97	24.26
C16H23F1N2	262.37	262.1845	73.25	8.84	7.24	10.68	C16H33F1N2	272.45	272.2628	70.54	12.21	6.97	10.28
C16H23F1N4	290.39	290.1907	66.18	7.98	6.54	19.29	C16H33F1N4	300.47	300.2689	63.96	11.07	6.32	18.65
C16H23F1N6	318.40	318.1968	60.36	7.28	5.97	26.39	C16H33F1N6	328.48	328.2751	58.51	10.13	5.78	25.58
C16H23F2N1	267.37	267.1798	71.88	8.67	14.21	5.24	C16H33F2N1	277.44	277.2581	69.27	11.99	13.70	5.05
C16H23F2N3	295.38	295.1860	65.06	7.85	12.86	14.23	C16H33F2N3	305.46	305.2642	62.91	10.89	12.44	13.76
C16H23F2N5	323.39	323.1921	59.43	7.17	11.75	21.66	C16H33F2N5	333.47	333.2704	57.63	9.97	11.39	21.00
C16H24F1N1	249.37	249.1893	77.06	9.70	7.62	5.62	C16H34F1N1	259.45	259.2675	74.07	13.21	7.32	5.40
C16H24F1N3	277.39	277.1954	69.28	8.72	6.85	15.15	C16H34F1N3	287.47	287.2737	66.85	11.92	6.61	14.62
C16H24F1N5	305.40	305.2016	62.93	7.92	6.22	22.93	C16H34F1N5	315.48	315.2798	60.92	10.86	6.02	22.20
C16H24F2N2	282.38	282.1907	68.06	8.57	13.46	9.92	C16H34F2N2	292.46	292.2690	65.71	11.72	12.99	9.58
C16H24F2N4	310.39	310.1969	61.91	7.79	12.24	18.05	C16H34F2N4	320.47	320.2751	59.97	10.69	11.86	17.48
C16H24F2N6	338.41	338.2030	56.79	7.15	11.23	24.83	C16H34F2N6	348.49	348.2813	55.15	9.83	10.90	24.12
C16H25F1N2	264.39	264.2002	72.69	9.53	7.19	10.60	C16H35F1N2	274.47	274.2784	70.02	12.85	6.92	10.21
C16H25F1N4	292.40	292.2063	65.72	8.62	6.50	19.16	C16H35F1N4	302.48	302.2846	63.53	11.66	6.28	18.52
C16H25F1N6	320.42	320.2125	59.98	7.86	5.93	26.23	C16H35F1N6	330.50	330.2907	58.15	10.67	5.75	25.43
C16H25F2N1	269.38	269.1955	71.34	9.35	14.11	5.20	C16H35F2N3	307.47	307.2799	62.50	11.47	12.36	13.67
C16H25F2N3	297.39	297.2016	64.62	8.47	12.78	14.13	C16H35F2N5	335.49	335.2860	57.28	10.52	11.33	20.88
C16H25F2N5	325.41	325.2078	59.06	7.74	11.68	21.52	C16H36F1N3	289.48	289.2893	66.39	12.54	6.56	14.52
C16H26F1N1	251.39	251.2049	76.45	10.42	7.56	5.57	C16H36F1N5	317.50	317.2955	60.53	11.43	5.98	22.06
C16H26F1N3	279.40	279.2111	68.78	9.38	6.80	15.04	C16H36F2N4	322.49	322.2908	59.59	11.25	11.78	17.37
C16H26F1N5	307.42	307.2172	62.51	8.52	6.18	22.78	C16H36F2N6	350.50	350.2969	54.83	10.35	10.84	23.98
C16H26F2N2	284.40	284.2064	67.57	9.22	13.36	9.85	C16H37F1N4	304.50	304.3002	63.11	12.25	6.24	18.40
C16H26F2N4	312.41	312.2125	61.51	8.39	12.16	17.93	C16H37F1N6	332.51	332.3064	57.80	11.22	5.71	25.27
C16H26F2N6	340.42	340.2187	56.45	7.70	11.16	24.69	C16H37F2N5	337.50	337.3017	56.94	11.05	11.26	20.75
C16H27F1N2	266.41	266.2158	72.14	10.22	7.13	10.52	C16H38F1N5	319.51	319.3111	60.15	11.99	5.95	21.92
C16H27F1N4	294.42	294.2220	65.27	9.24	6.45	19.03	C16H38F2N6	352.52	352.3126	54.52	10.87	10.78	23.84
C16H27F1N6	322.43	322.2281	59.60	8.44	5.89	26.06	C16H39F1N6	334.53	334.3220	57.45	11.75	5.68	25.12
C16H27F2N1	271.40	271.2111	70.81	10.03	14.00	5.16	C17H 8F1N1	245.26	245.0641	83.25	3.29	7.75	5.71
C16H27F2N3	299.41	299.2173	66.19	9.09	12.69	14.03	C17H 8F1N3	273.27	273.0702	74.72	2.95	6.95	15.38
C16H27F2N5	327.42	327.2234	58.69	8.31	11.60	21.39	C17H 8F1N5	301.29	301.0764	67.77	2.68	6.31	23.24
C16H28F1N1	253.41	253.2206	75.84	11.14	7.50	5.53	C17H 8F2N2	278.26	278.0655	73.38	2.90	13.65	10.07

C17H 8F2N4	306.28	306.0717	66.67	2.63	12.41	18.29	C17H18F2N4	316.36	316.1499	64.54	5.74	12.01	17.71
C17H 8F2N6	334.29	334.0778	61.08	2.41	11.37	25.14	C17H18F2N6	344.37	344.1561	59.29	5.27	11.03	24.40
C17H 9F1N2	260.27	260.0750	78.45	3.49	7.30	10.76	C17H19F1N2	270.35	270.1532	75.53	7.08	7.03	10.36
C17H 9F1N4	288.29	288.0811	70.83	3.15	6.59	19.43	C17H19F1N4	298.37	298.1594	68.44	6.42	6.37	18.78
C17H 9F1N6	316.30	316.0873	64.56	2.87	6.01	26.57	C17H19F1N6	326.38	326.1655	62.56	5.87	5.82	25.75
C17H 9F2N1	265.26	265.0703	76.98	3.42	14.32	5.28	C17H19F2N1	275.34	275.1485	74.16	6.96	13.80	5.09
C17H 9F2N3	293.28	293.0764	69.62	3.09	12.96	14.33	C17H19F2N3	303.36	303.1547	67.31	6.31	12.53	13.85
C17H 9F2N5	321.29	321.0826	63.55	2.82	11.83	21.80	C17H19F2N5	331.37	331.1608	61.62	5.78	11.47	21.13
C17H10F1N1	247.27	247.0797	82.58	4.08	7.68	5.66	C17H20F1N1	257.35	257.1580	79.34	7.83	7.38	5.44
C17H10F1N3	275.29	275.0859	74.17	3.66	6.90	15.26	C17H20F1N3	285.37	285.1641	71.55	7.06	6.66	14.72
C17H10F1N5	303.30	303.0920	67.32	3.32	6.26	23.09	C17H20F1N5	313.38	313.1703	65.16	6.43	6.06	22.35
C17H10F2N2	280.28	280.0812	72.85	3.60	13.56	9.99	C17H20F2N2	290.36	290.1594	70.32	6.94	13.09	9.65
C17H10F2N4	308.29	308.0873	66.23	3.27	12.32	18.17	C17H20F2N4	318.37	318.1656	64.14	6.33	11.93	17.60
C17H10F2N6	336.31	336.0935	60.72	3.00	11.30	24.99	C17H20F2N6	346.39	346.1717	58.95	5.82	10.97	24.26
C17H11F1N2	262.29	262.0906	77.85	4.23	7.24	10.68	C17H21F1N2	272.37	272.1689	74.97	7.77	6.98	10.29
C17H11F1N4	290.30	290.0968	70.34	3.82	6.54	19.30	C17H21F1N4	300.38	300.1750	67.98	7.05	6.32	18.65
C17H11F1N6	318.32	318.1029	64.15	3.48	5.97	26.40	C17H21F1N6	328.40	328.1812	62.18	6.45	5.79	25.59
C17H11F2N1	267.28	267.0859	76.40	4.15	14.22	5.24	C17H21F2N1	277.36	277.1642	73.62	7.63	13.70	5.05
C17H11F2N3	295.29	295.0921	69.15	3.75	12.87	14.23	C17H21F2N3	305.37	305.1703	66.87	6.93	12.44	13.76
C17H11F2N5	323.31	323.0982	63.16	3.43	11.75	21.66	C17H21F2N5	333.39	333.1765	61.25	6.35	11.40	21.01
C17H12F1N1	249.29	249.0954	81.91	4.85	7.62	5.62	C17H22F1N1	259.37	259.1736	78.73	8.55	7.32	5.40
C17H12F1N3	277.30	277.1015	73.63	4.36	6.85	15.15	C17H22F1N3	287.38	287.1798	71.05	7.72	6.61	14.62
C17H12F1N5	305.32	305.1077	66.88	3.96	6.22	22.94	C17H22F1N5	315.40	315.1859	64.74	7.03	6.02	22.20
C17H12F2N2	282.30	282.0968	72.33	4.28	13.46	9.92	C17H22F2N2	292.38	292.1751	69.84	7.58	13.00	9.58
C17H12F2N4	310.31	310.1030	65.80	3.90	12.24	18.06	C17H22F2N4	320.39	320.1812	63.73	6.92	11.86	17.49
C17H12F2N6	338.32	338.1091	60.35	3.58	11.23	24.84	C17H22F2N6	348.40	348.1874	58.61	6.36	10.91	24.12
C17H13F1N2	264.30	264.1063	77.26	4.96	7.19	10.60	C17H23F1N2	274.38	274.1845	74.42	8.45	6.92	10.21
C17H13F1N4	292.32	292.1124	69.85	4.48	6.50	19.17	C17H23F1N4	302.40	302.1907	67.52	7.67	6.28	18.53
C17H13F1N6	320.33	320.1186	63.74	4.09	5.93	26.24	C17H23F1N6	330.41	330.1968	61.80	7.02	5.75	25.44
C17H13F2N1	269.30	269.1016	75.82	4.87	14.11	5.20	C17H23F2N1	279.38	279.1798	73.09	8.30	13.60	5.01
C17H13F2N3	297.31	297.1077	68.68	4.41	12.78	14.13	C17H23F2N3	307.39	307.1860	66.43	7.54	12.36	13.67
C17H13F2N5	325.32	325.1139	62.77	4.03	11.68	21.53	C17H23F2N5	335.40	335.1921	60.88	6.91	11.33	20.88
C17H14F1N1	251.31	251.1110	81.25	5.62	7.56	5.57	C17H24F1N1	261.39	261.1893	78.12	9.26	7.27	5.36
C17H14F1N3	279.32	279.1172	73.10	5.05	6.80	15.04	C17H24F1N3	289.40	289.1954	70.56	8.36	6.56	14.52
C17H14F1N5	307.33	307.1233	66.44	4.59	6.18	22.79	C17H24F1N5	317.41	317.2016	64.33	7.62	5.99	22.06
C17H14F2N2	284.31	284.1125	71.82	4.96	13.36	9.85	C17H24F2N2	294.39	294.1907	69.36	8.22	12.91	9.52
C17H14F2N4	312.32	312.1186	65.38	4.52	12.17	17.94	C17H24F2N4	322.40	322.1969	63.33	7.50	11.79	17.38
C17H14F2N6	340.34	340.1248	60.00	4.15	11.16	24.69	C17H24F2N6	350.42	350.2030	58.27	6.90	10.84	23.98
C17H15F1N2	266.32	266.1219	76.67	5.68	7.13	10.52	C17H25F1N2	276.40	276.2002	73.87	9.12	6.87	10.14
C17H15F1N4	294.33	294.1281	69.37	5.14	6.45	19.04	C17H25F1N4	304.41	304.2063	67.08	8.28	6.24	18.40
C17H15F1N6	322.35	322.1342	63.34	4.69	5.89	26.07	C17H25F1N6	332.43	332.2125	61.42	7.58	5.72	25.28
C17H15F2N1	271.31	271.1172	75.26	5.57	14.00	5.16	C17H25F2N1	281.39	281.1955	72.56	8.96	13.50	4.98
C17H15F2N3	299.33	299.1234	68.22	5.05	12.69	14.04	C17H25F2N3	309.41	309.2016	65.99	8.14	12.28	13.58
C17H15F2N5	327.34	327.1295	62.38	4.62	11.61	21.39	C17H25F2N5	337.42	337.2078	60.52	7.47	11.26	20.76
C17H16F1N1	253.32	253.1267	80.60	6.37	7.50	5.53	C17H26F1N1	263.40	263.2049	77.52	9.95	7.21	5.32
C17H16F1N3	281.34	281.1328	72.58	5.73	6.75	14.94	C17H26F1N3	291.42	291.2111	70.07	8.99	6.52	14.42
C17H16F1N5	309.35	309.1390	66.01	5.21	6.14	22.64	C17H26F1N5	319.43	319.2172	63.92	8.20	5.95	21.92
C17H16F2N2	286.33	286.1281	71.31	5.63	13.27	9.78	C17H26F2N2	296.41	296.2064	68.89	8.84	12.82	9.45
C17H16F2N4	314.34	314.1343	64.96	5.13	12.09	17.82	C17H26F2N4	324.42	324.2125	62.94	8.08	11.71	17.27
C17H16F2N6	342.35	342.1404	59.64	4.71	11.10	24.55	C17H26F2N6	352.43	352.2187	57.94	7.44	10.78	23.85
C17H17F1N2	268.34	268.1376	76.09	6.39	7.08	10.44	C17H27F1N2	278.42	278.2158	73.34	9.77	6.82	10.06
C17H17F1N4	296.35	296.1437	68.90	5.78	6.41	18.91	C17H27F1N4	306.43	306.2220	66.63	8.88	6.20	18.28
C17H17F1N6	324.36	324.1499	62.95	5.28	5.86	25.91	C17H27F1N6	334.44	334.2281	61.05	8.14	5.68	25.13
C17H17F2N1	273.33	273.1329	74.70	6.27	13.90	5.12	C17H27F2N1	283.41	283.2111	72.05	9.60	13.41	4.94
C17H17F2N3	301.34	301.1390	67.76	5.69	12.61	13.94	C17H27F2N3	311.42	311.2173	65.57	8.74	12.20	13.49
C17H17F2N5	329.36	329.1452	62.00	5.20	11.54	21.26	C17H27F2N5	339.44	339.2234	60.16	8.02	11.19	20.63
C17H18F1N1	255.34	255.1423	79.97	7.11	7.44	5.49	C17H28F1N1	265.42	265.2206	76.93	10.63	7.16	5.28
C17H18F1N3	283.35	283.1485	72.06	6.40	6.70	14.83	C17H28F1N3	293.43	293.2267	69.59	9.62	6.47	14.32
C17H18F1N5	311.36	311.1546	65.58	5.83	6.10	22.49	C17H28F1N5	321.44	321.2329	63.52	8.78	5.91	21.79
C17H18F2N2	288.34	288.1438	70.81	6.29	13.18	9.72	C17H28F2N2	298.42	298.2220	68.42	9.46	12.73	9.39

C17H28F2N4	326.44	326.2282	62.55	8.65	11.64	17.16
C17H28F2N6	354.45	354.2343	57.61	7.96	10.72	23.71
C17H29F1N2	280.43	280.2315	72.81	10.42	6.77	9.99
C17H29F1N4	308.45	308.2376	66.20	9.48	6.16	18.16
C17H29F1N6	336.46	336.2438	60.69	8.69	5.65	24.98
C17H29F2N1	285.42	285.2268	71.54	10.24	13.31	4.91
C17H29F2N3	313.44	313.2329	65.15	9.33	12.12	13.41
C17H29F2N5	341.45	341.2391	59.80	8.56	11.13	20.51
C17H30F1N1	267.43	267.2362	76.35	11.31	7.10	5.24
C17H30F1N3	295.45	295.2424	69.11	10.24	6.43	14.22
C17H30F1N5	323.46	323.2485	63.13	9.35	5.87	21.65
C17H30F2N2	300.44	300.2377	67.96	10.06	12.65	9.32
C17H30F2N4	328.45	328.2438	62.17	9.21	11.57	17.06
C17H30F2N6	356.47	356.2500	57.28	8.48	10.66	23.58
C17H31F1N2	282.45	282.2471	72.29	11.06	6.73	9.92
C17H31F1N4	310.46	310.2533	65.77	10.06	6.12	18.05
C17H31F1N6	338.48	338.2594	60.33	9.23	5.61	24.83
C17H31F2N1	287.44	287.2424	71.04	10.87	13.22	4.87
C17H31F2N3	315.45	315.2486	64.73	9.91	12.05	13.32
C17H31F2N5	343.47	343.2547	59.45	9.10	11.06	20.39
C17H32F1N1	269.45	269.2519	75.78	11.97	7.05	5.20
C17H32F1N3	297.46	297.2580	68.64	10.84	6.39	14.13
C17H32F1N5	325.48	325.2642	62.74	9.91	5.84	21.52
C17H32F2N2	302.45	302.2533	67.51	10.66	12.56	9.26
C17H32F2N4	330.47	330.2595	61.79	9.76	11.50	16.95
C17H32F2N6	358.48	358.2656	56.96	9.00	10.60	23.44
C17H33F1N2	284.46	284.2628	71.78	11.69	6.68	9.85
C17H33F1N4	312.48	312.2689	65.35	10.64	6.08	17.93
C17H33F1N6	340.49	340.2751	59.97	9.77	5.58	24.68
C17H33F2N1	289.46	289.2581	70.54	11.49	13.13	4.84
C17H33F2N3	317.47	317.2642	64.32	10.48	11.97	13.24
C17H33F2N5	345.48	345.2704	59.10	9.63	11.00	20.27
C17H34F1N1	271.47	271.2675	75.22	12.62	7.00	5.16
C17H34F1N3	299.48	299.2737	68.18	11.44	6.34	14.03
C17H34F1N5	327.49	327.2798	62.35	10.46	5.80	21.38
C17H34F2N2	304.47	304.2690	67.06	11.26	12.48	9.20
C17H34F2N4	332.48	332.2751	61.41	10.31	11.43	16.85
C17H34F2N6	360.50	360.2813	56.64	9.51	10.54	23.31
C17H35F1N2	286.48	286.2784	71.28	12.31	6.63	9.78
C17H35F1N4	314.49	314.2846	64.93	11.22	6.04	17.81
C17H35F1N6	342.51	342.2907	59.62	10.30	5.55	24.54
C17H35F2N1	291.47	291.2737	70.05	12.10	13.04	4.81
C17H35F2N3	319.49	319.2799	63.91	11.04	11.89	13.15
C17H35F2N5	347.50	347.2860	58.76	10.15	10.93	20.15
C17H36F1N1	273.48	273.2832	74.66	13.27	6.95	5.12
C17H36F1N3	301.49	301.2893	67.73	12.04	6.30	13.94
C17H36F1N5	329.51	329.2955	61.97	11.01	5.77	21.25
C17H36F2N2	306.49	306.2846	66.62	11.84	12.40	9.14
C17H36F2N4	334.50	334.2908	61.04	10.85	11.36	16.75
C17H36F2N6	362.51	362.2969	56.33	10.01	10.48	23.18
C17H37F1N2	288.50	288.2941	70.78	12.93	6.59	9.71
C17H37F1N4	316.51	316.3002	64.51	11.78	6.00	17.70
C17H37F1N6	344.52	344.3064	59.27	10.83	5.51	24.39
C17H37F2N3	321.50	321.2955	63.51	11.60	11.82	13.07
C17H37F2N5	349.51	349.3017	58.42	10.67	10.87	20.04
C17H38F1N3	303.51	303.3050	67.28	12.62	6.26	13.84
C17H38F1N5	331.52	331.3111	61.59	11.55	5.73	21.12
C17H38F2N4	336.52	336.3064	60.68	11.38	11.29	16.65
C17H38F2N6	364.53	364.3126	56.01	10.51	10.42	23.05
C17H39F1N4	318.53	318.3159	64.10	12.34	5.96	17.59
C17H39F1N6	346.54	346.3220	58.92	11.34	5.48	24.25
C17H39F2N5	351.53	351.3173	58.09	11.18	10.81	19.92
C17H40F1N5	333.54	333.3268	61.22	12.09	5.70	21.00
C17H40F2N6	366.55	366.3282	55.71	11.00	10.37	22.93
C17H41F1N6	348.55	348.3377	58.58	11.86	5.45	24.11
C18H 9F1N2	272.28	272.0750	79.40	3.33	6.98	10.29
C18H 9F1N4	300.30	300.0811	72.00	3.02	6.33	18.66
C18H 9F1N6	328.31	328.0873	65.85	2.76	5.79	25.60
C18H 9F2N1	277.28	277.0703	77.97	3.27	13.70	5.05
C18H 9F2N3	305.29	305.0764	70.82	2.97	12.45	13.76
C18H 9F2N5	333.30	333.0826	64.87	2.72	11.40	21.01
C18H10F1N1	259.29	259.0797	83.38	3.89	7.33	5.40
C18H10F1N3	287.30	287.0859	75.25	3.51	6.61	14.63
C18H10F1N5	315.31	315.0920	68.57	3.20	6.03	22.21
C18H10F2N2	292.29	292.0812	73.97	3.45	13.00	9.58
C18H10F2N4	320.30	320.0873	67.50	3.15	11.86	17.49
C18H10F2N6	348.32	348.0935	62.07	2.89	10.91	24.13
C18H11F1N2	274.30	274.0906	78.82	4.04	6.93	10.21
C18H11F1N4	302.31	302.0968	71.52	3.67	6.28	18.53
C18H11F1N6	330.33	330.1029	65.45	3.36	5.75	25.44
C18H11F2N1	279.29	279.0859	77.41	3.97	13.60	5.02
C18H11F2N3	307.31	307.0921	70.35	3.61	12.36	13.67
C18H11F2N5	335.32	335.0982	64.48	3.31	11.33	20.89
C18H12F1N1	261.30	261.0954	82.74	4.63	7.27	5.36
C18H12F1N3	289.31	289.1015	74.73	4.18	6.57	14.52
C18H12F1N5	317.33	317.1077	68.13	3.81	5.99	22.07
C18H12F2N2	294.31	294.0968	73.46	4.11	12.91	9.52
C18H12F2N4	322.32	322.1030	67.08	3.75	11.79	17.38
C18H12F2N6	350.33	350.1091	61.71	3.45	10.85	23.99
C18H13F1N2	276.32	276.1063	78.24	4.74	6.88	10.14
C18H13F1N4	304.33	304.1124	71.04	4.31	6.24	18.41
C18H13F1N6	332.34	332.1186	65.05	3.94	5.72	25.29
C18H13F2N1	281.31	281.1016	76.86	4.66	13.51	4.98
C18H13F2N3	309.32	309.1077	69.90	4.24	12.28	13.58
C18H13F2N5	337.33	337.1139	64.09	3.88	11.26	20.76
C18H14F1N1	263.32	263.1110	82.11	5.36	7.22	5.32
C18H14F1N3	291.33	291.1172	74.21	4.84	6.52	14.42
C18H14F1N5	319.34	319.1233	67.70	4.42	5.95	21.93
C18H14F2N2	296.32	296.1125	72.96	4.76	12.82	9.45
C18H14F2N4	324.34	324.1186	66.66	4.35	11.72	17.27
C18H14F2N6	352.35	352.1248	61.36	4.00	10.78	23.85
C18H15F1N2	278.33	278.1219	77.68	5.43	6.83	10.06
C18H15F1N4	306.35	306.1281	70.57	4.94	6.20	18.29
C18H15F1N6	334.36	334.1342	64.66	4.52	5.68	25.13
C18H15F2N1	283.32	283.1172	76.31	5.34	13.41	4.94
C18H15F2N3	311.34	311.1234	69.44	4.86	12.20	13.50
C18H15F2N5	339.35	339.1295	63.71	4.46	11.20	20.64
C18H16F1N1	265.33	265.1267	81.48	6.08	7.16	5.28
C18H16F1N3	293.35	293.1328	73.70	5.50	6.48	14.32
C18H16F1N5	321.36	321.1390	67.28	5.02	5.91	21.79
C18H16F2N2	298.34	298.1281	72.47	5.41	12.74	9.39
C18H16F2N4	326.35	326.1343	66.25	4.94	11.64	17.17
C18H16F2N6	354.37	354.1404	61.01	4.55	10.72	23.72
C18H17F1N2	280.35	280.1376	77.12	6.11	6.78	9.99
C18H17F1N4	308.36	308.1437	70.11	5.56	6.16	18.17
C18H17F1N6	336.37	336.1499	64.27	5.09	5.65	24.98
C18H17F2N1	285.34	285.1329	75.77	6.01	13.32	4.91
C18H17F2N3	313.35	313.1390	69.00	5.47	12.13	13.41
C18H17F2N5	341.37	341.1452	63.33	5.02	11.13	20.52
C18H18F1N1	267.35	267.1423	80.87	6.79	7.11	5.24

C18H18F1N3	295.36	295.1485	73.20	6.14	6.43	14.23
C18H18F1N5	323.38	323.1546	66.86	5.61	5.88	21.66
C18H18F2N2	300.35	300.1438	71.98	6.04	12.65	9.33
C18H18F2N4	328.37	328.1499	65.84	5.53	11.57	17.06
C18H18F2N6	356.38	356.1561	60.67	5.09	10.66	23.58
C18H19F1N2	282.36	282.1532	76.57	6.78	6.73	9.92
C18H19F1N4	310.38	310.1594	69.66	6.17	6.12	18.05
C18H19F1N6	338.39	338.1655	63.89	5.66	5.61	24.84
C18H19F2N1	287.36	287.1485	75.24	6.66	13.22	4.87
C18H19F2N3	315.37	315.1547	68.55	6.07	12.05	13.32
C18H19F2N5	343.38	343.1608	62.96	5.58	11.07	20.40
C18H20F1N1	269.37	269.1580	80.26	7.48	7.05	5.20
C18H20F1N3	297.38	297.1641	72.70	6.78	6.39	14.13
C18H20F1N5	325.39	325.1703	66.44	6.20	5.84	21.52
C18H20F2N2	302.37	302.1594	71.50	6.67	12.57	9.26
C18H20F2N4	330.38	330.1656	65.44	6.10	11.50	16.96
C18H20F2N6	358.40	358.1717	60.32	5.62	10.60	23.45
C18H21F1N2	284.38	284.1689	76.03	7.44	6.68	9.85
C18H21F1N4	312.39	312.1750	69.21	6.78	6.08	17.93
C18H21F1N6	340.41	340.1812	63.51	6.22	5.58	24.69
C18H21F2N1	289.37	289.1642	74.71	7.31	13.13	4.84
C18H21F2N3	317.38	317.1703	68.12	6.67	11.97	13.24
C18H21F2N5	345.40	345.1765	62.59	6.13	11.00	20.28
C18H22F1N1	271.38	271.1736	79.67	8.17	7.00	5.16
C18H22F1N3	299.39	299.1798	72.21	7.41	6.35	14.04
C18H22F1N5	327.41	327.1859	66.03	6.77	5.80	21.39
C18H22F2N2	304.39	304.1751	71.03	7.29	12.48	9.20
C18H22F2N4	332.40	332.1812	65.04	6.67	11.43	16.86
C18H22F2N6	360.41	360.1874	59.99	6.15	10.54	23.32
C18H23F1N2	286.40	286.1845	75.49	8.09	6.63	9.78
C18H23F1N4	314.41	314.1907	68.76	7.37	6.04	17.82
C18H23F1N6	342.42	342.1968	63.14	6.77	5.55	24.54
C18H23F2N1	291.39	291.1798	74.20	7.96	13.04	4.81
C18H23F2N3	319.40	319.1860	67.69	7.26	11.90	13.16
C18H23F2N5	347.41	347.1921	62.23	6.67	10.94	20.16
C18H24F1N1	273.40	273.1893	79.08	8.85	6.95	5.12
C18H24F1N3	301.41	301.1954	71.73	8.03	6.30	13.94
C18H24F1N5	329.42	329.2016	65.63	7.34	5.77	21.26
C18H24F2N2	306.40	306.1907	70.56	7.90	12.40	9.14
C18H24F2N4	334.42	334.1969	64.65	7.23	11.36	16.75
C18H24F2N6	362.43	362.2030	59.65	6.67	10.48	23.19
C18H25F1N2	288.41	288.2002	74.96	8.74	6.59	9.71
C18H25F1N4	316.43	316.2063	68.33	7.96	6.00	17.71
C18H25F1N6	344.44	344.2125	62.77	7.32	5.52	24.40
C18H25F2N1	293.40	293.1955	73.69	8.59	12.95	4.77
C18H25F2N3	321.42	321.2016	67.26	7.84	11.82	13.07
C18H25F2N5	349.43	349.2078	61.87	7.21	10.87	20.04
C18H26F1N1	275.41	275.2049	78.50	9.52	6.90	5.09
C18H26F1N3	303.43	303.2111	71.25	8.64	6.26	13.85
C18H26F1N5	331.44	331.2172	65.23	7.91	5.73	21.13
C18H26F2N2	308.42	308.2064	70.10	8.50	12.32	9.08
C18H26F2N4	336.43	336.2125	64.26	7.79	11.29	16.65
C18H26F2N6	364.44	364.2187	59.32	7.19	10.43	23.06
C18H27F1N2	290.43	290.2158	74.44	9.37	6.54	9.65
C18H27F1N4	318.44	318.2220	67.89	8.55	5.97	17.59
C18H27F1N6	346.45	346.2281	62.40	7.86	5.48	24.26
C18H27F2N1	295.42	295.2111	73.18	9.21	12.86	4.74
C18H27F2N3	323.43	323.2173	66.85	8.41	11.75	12.99
C18H27F2N5	351.45	351.2234	61.52	7.74	10.81	19.93
C18H28F1N1	277.43	277.2206	77.93	10.17	6.85	5.05
C18H28F1N3	305.44	305.2267	70.78	9.24	6.22	13.76
C18H28F1N5	333.46	333.2329	64.84	8.46	5.70	21.00
C18H28F2N2	310.43	310.2220	69.64	9.09	12.24	9.02
C18H28F2N4	338.45	338.2282	63.88	8.34	11.23	16.55
C18H28F2N6	366.46	366.2343	59.00	7.70	10.37	22.93
C18H29F1N2	292.44	292.2315	73.93	10.00	6.50	9.58
C18H29F1N4	320.46	320.2376	67.47	9.12	5.93	17.48
C18H29F1N6	348.47	348.2438	62.04	8.39	5.45	24.12
C18H29F2N1	297.44	297.2268	72.69	9.83	12.77	4.71
C18H29F2N3	325.45	325.2329	66.43	8.98	11.68	12.91
C18H29F2N5	353.46	353.2391	61.17	8.27	10.75	19.81
C18H30F1N1	279.44	279.2362	77.37	10.82	6.80	5.01
C18H30F1N3	307.46	307.2424	70.32	9.84	6.18	13.67
C18H30F1N5	335.47	335.2485	64.45	9.01	5.66	20.88
C18H30F2N2	312.45	312.2377	69.20	9.68	12.16	8.97
C18H30F2N4	340.46	340.2438	63.50	8.88	11.16	16.46
C18H30F2N6	368.48	368.2500	58.67	8.21	10.31	22.81
C18H31F1N2	294.46	294.2471	73.42	10.61	6.45	9.51
C18H31F1N4	322.47	322.2533	67.04	9.69	5.89	17.37
C18H31F1N6	350.49	350.2594	61.69	8.92	5.42	23.98
C18H31F2N1	299.45	299.2424	72.20	10.43	12.69	4.68
C18H31F2N3	327.46	327.2486	66.02	9.54	11.60	12.83
C18H31F2N5	355.48	355.2547	60.82	8.79	10.69	19.70
C18H32F1N1	281.46	281.2519	76.81	11.46	6.75	4.98
C18H32F1N3	309.47	309.2580	69.86	10.42	6.14	13.58
C18H32F1N5	337.49	337.2642	64.06	9.56	5.63	20.75
C18H32F2N2	314.47	314.2533	68.75	10.26	12.08	8.91
C18H32F2N4	342.48	342.2595	63.13	9.42	11.09	16.36
C18H32F2N6	370.49	370.2656	58.35	8.71	10.26	22.68
C18H33F1N2	296.48	296.2628	72.92	11.22	6.41	9.45
C18H33F1N4	324.49	324.2689	66.63	10.25	5.85	17.27
C18H33F1N6	352.50	352.2751	61.33	9.44	5.39	23.84
C18H33F2N1	301.47	301.2581	71.72	11.03	12.60	4.65
C18H33F2N3	329.48	329.2642	65.62	10.10	11.53	12.75
C18H33F2N5	357.49	357.2704	60.48	9.30	10.63	19.59
C18H34F1N1	283.48	283.2675	76.27	12.09	6.70	4.94
C18H34F1N3	311.49	311.2737	69.41	11.00	6.10	13.49
C18H34F1N5	339.50	339.2798	63.68	10.09	5.60	20.63
C18H34F2N2	316.48	316.2690	68.31	10.83	12.01	8.85
C18H34F2N4	344.50	344.2751	62.76	9.95	11.03	16.26
C18H34F2N6	372.51	372.2813	58.04	9.20	10.20	22.56
C18H35F1N2	298.49	298.2784	72.43	11.82	6.36	9.38
C18H35F1N4	326.50	326.2846	66.22	10.81	5.82	17.16
C18H35F1N6	354.52	354.2907	60.98	9.95	5.36	23.71
C18H35F2N1	303.48	303.2737	71.24	11.62	12.52	4.62
C18H35F2N3	331.50	331.2799	65.22	10.64	11.46	12.68
C18H35F2N5	359.51	359.2860	60.14	9.81	10.57	19.48
C18H36F1N1	285.49	285.2832	75.73	12.71	6.65	4.91
C18H36F1N3	313.51	313.2893	68.96	11.57	6.06	13.40
C18H36F1N5	341.52	341.2955	63.31	10.63	5.56	20.51
C18H36F2N2	318.50	318.2846	67.88	11.39	11.93	8.80
C18H36F2N4	346.51	346.2908	62.39	10.47	10.97	16.17
C18H36F2N6	374.52	374.2969	57.73	9.69	10.15	22.44
C18H37F1N2	300.51	300.2941	71.95	12.41	6.32	9.32
C18H37F1N4	328.52	328.3002	65.81	11.35	5.78	17.05
C18H37F1N6	356.53	356.3064	60.64	10.46	5.33	23.57
C18H37F2N1	305.50	305.2894	70.77	12.21	12.44	4.58
C18H37F2N3	333.51	333.2955	64.83	11.18	11.39	12.60
C18H37F2N5	361.53	361.3017	59.80	10.32	10.51	19.37
C18H38F1N1	287.51	287.2988	75.20	13.32	6.61	4.87

C18H38F1N3	315.52	315.3050	68.52	12.14	6.02	13.32	C19H15F2N3	323.35	323.1234	70.58	4.68	11.75	13.00
C18H38F1N5	343.54	343.3111	62.93	11.15	5.53	20.39	C19H15F2N5	351.36	351.1295	64.95	4.30	10.81	19.93
C18H38F2N2	320.51	320.3003	67.45	11.95	11.85	8.74	C19H16F1N1	277.34	277.1267	82.28	5.81	6.85	5.05
C18H38F2N4	348.53	348.3064	62.03	10.99	10.90	16.08	C19H16F1N3	305.36	305.1328	74.74	5.28	6.22	13.76
C18H38F2N6	376.54	376.3126	57.42	10.17	10.09	22.32	C19H16F1N5	333.37	333.1390	68.46	4.84	5.70	21.01
C18H39F1N2	302.52	302.3097	71.47	12.99	6.28	9.26	C19H16F2N2	310.35	310.1281	73.53	5.20	12.24	9.03
C18H39F1N4	330.54	330.3159	65.41	11.89	5.75	16.95	C19H16F2N4	338.36	338.1343	67.45	4.77	11.23	16.56
C18H39F1N6	358.55	358.3220	60.30	10.96	5.30	23.44	C19H16F2N6	366.38	366.1404	62.29	4.40	10.37	22.94
C18H39F2N3	335.53	335.3112	64.44	11.72	11.32	12.52	C19H17F1N2	292.36	292.1376	78.06	5.86	6.50	9.58
C18H39F2N5	363.54	363.3173	59.47	10.81	10.45	19.26	C19H17F1N4	320.37	320.1437	71.23	5.35	5.93	17.49
C18H40F1N3	317.54	317.3206	68.09	12.70	5.98	13.23	C19H17F1N6	348.39	348.1499	65.51	4.92	5.45	24.12
C18H40F1N5	345.55	345.3268	62.57	11.67	5.50	20.27	C19H17F2N1	297.35	297.1329	76.75	5.76	12.78	4.71
C18H40F2N4	350.54	350.3221	61.68	11.50	10.84	15.98	C19H17F2N3	325.36	325.1390	70.14	5.27	11.68	12.91
C18H40F2N6	378.56	378.3282	57.11	10.65	10.04	22.20	C19H17F2N5	353.38	353.1452	64.58	4.85	10.75	19.82
C18H41F1N4	332.55	332.3315	65.01	12.43	5.71	16.85	C19H18F1N1	279.36	279.1423	81.69	6.49	6.80	5.01
C18H41F1N6	360.57	360.3377	59.96	11.46	5.27	23.31	C19H18F1N3	307.37	307.1485	74.25	5.90	6.18	13.67
C18H41F2N5	365.56	365.3330	59.14	11.31	10.39	19.16	C19H18F1N5	335.39	335.1546	68.04	5.41	5.66	20.88
C18H42F1N5	347.57	347.3424	62.20	12.18	5.47	20.15	C19H18F2N2	312.37	312.1438	73.06	5.81	12.16	8.97
C18H42F2N6	380.57	380.3439	56.81	11.12	9.98	22.08	C19H18F2N4	340.38	340.1499	67.05	5.33	11.16	16.46
C18H43F1N6	362.58	362.3533	59.63	11.95	5.24	23.18	C19H18F2N6	368.39	368.1561	61.95	4.93	10.31	22.81
C19H 9F1N2	284.30	284.0750	80.27	3.19	6.68	9.85	C19H19F1N2	294.38	294.1532	77.52	6.51	6.45	9.52
C19H 9F1N4	312.31	312.0811	73.07	2.90	6.08	17.94	C19H19F1N4	322.39	322.1594	70.79	5.94	5.89	17.38
C19H 9F1N6	340.32	340.0873	67.06	2.67	5.58	24.69	C19H19F1N6	350.40	350.1655	65.13	5.47	5.42	23.98
C19H 9F2N1	289.29	289.0703	78.89	3.14	13.13	4.84	C19H19F2N1	299.37	299.1485	76.23	6.40	12.69	4.68
C19H 9F2N3	317.30	317.0764	71.92	2.86	11.98	13.24	C19H19F2N3	327.38	327.1547	69.71	5.85	11.61	12.84
C19H 9F2N5	345.31	345.0826	66.09	2.63	11.00	20.28	C19H19F2N5	355.39	355.1608	64.21	5.39	10.69	19.71
C19H10F1N1	271.30	271.0797	84.12	3.72	7.00	5.16	C19H20F1N1	281.38	281.1580	81.11	7.16	6.75	4.98
C19H10F1N3	299.31	299.0859	76.25	3.37	6.35	14.04	C19H20F1N3	309.39	309.1641	73.76	6.52	6.14	13.58
C19H10F1N5	327.32	327.0920	69.72	3.08	5.80	21.40	C19H20F1N5	337.40	337.1703	67.64	5.97	5.63	20.76
C19H10F2N2	304.30	304.0812	75.00	3.31	12.49	9.21	C19H20F2N2	314.38	314.1594	72.59	6.41	12.09	8.91
C19H10F2N4	332.32	332.0873	68.67	3.03	11.43	16.86	C19H20F2N4	342.39	342.1656	66.65	5.89	11.10	16.36
C19H10F2N6	360.33	360.0935	63.33	2.80	10.55	23.32	C19H20F2N6	370.41	370.1717	61.61	5.44	10.26	22.69
C19H11F1N2	286.31	286.0906	79.71	3.87	6.64	9.78	C19H21F1N2	296.39	296.1689	77.00	7.14	6.41	9.45
C19H11F1N4	314.32	314.0968	72.60	3.53	6.04	17.82	C19H21F1N4	324.40	324.1750	70.35	6.52	5.86	17.27
C19H11F1N6	342.34	342.1029	66.66	3.24	5.55	24.55	C19H21F1N6	352.42	352.1812	64.76	6.01	5.39	23.85
C19H11F2N1	291.30	291.0859	78.34	3.81	13.04	4.81	C19H21F2N1	301.38	301.1642	75.72	7.02	12.61	4.65
C19H11F2N3	319.32	319.0921	71.47	3.47	11.90	13.16	C19H21F2N3	329.40	329.1703	69.28	6.43	11.54	12.76
C19H11F2N5	347.33	347.0982	65.70	3.19	10.94	20.16	C19H21F2N5	357.41	357.1765	63.85	5.92	10.63	19.59
C19H12F1N1	273.31	273.0954	83.50	4.43	6.95	5.12	C19H22F1N1	283.39	283.1736	80.53	7.82	6.70	4.94
C19H12F1N3	301.33	301.1015	75.74	4.01	6.30	13.95	C19H22F1N3	311.41	311.1798	73.28	7.12	6.10	13.49
C19H12F1N5	329.34	329.1077	69.25	3.67	5.77	21.26	C19H22F1N5	339.42	339.1859	67.24	6.53	5.60	20.63
C19H12F2N2	306.32	306.0968	74.50	3.95	12.40	9.15	C19H22F2N2	316.40	316.1751	72.13	7.01	12.01	8.85
C19H12F2N4	334.33	334.1030	68.26	3.62	11.37	16.76	C19H22F2N4	344.41	344.1812	66.26	6.44	11.03	16.27
C19H12F2N6	362.34	362.1091	62.98	3.34	10.49	23.19	C19H22F2N6	372.42	372.1874	61.28	5.95	10.20	22.57
C19H13F1N2	288.33	288.1063	79.15	4.54	6.59	9.72	C19H23F1N2	298.41	298.1845	76.48	7.77	6.37	9.39
C19H13F1N4	316.34	316.1124	72.14	4.14	6.01	17.71	C19H23F1N4	326.42	326.1907	69.91	7.10	5.82	17.16
C19H13F1N6	344.35	344.1186	66.27	3.81	5.52	24.41	C19H23F1N6	354.43	354.1968	64.39	6.54	5.36	23.71
C19H13F2N1	293.32	293.1016	77.80	4.47	12.95	4.78	C19H23F2N1	303.40	303.1798	75.22	7.64	12.52	4.62
C19H13F2N3	321.33	321.1077	71.02	4.08	11.82	13.08	C19H23F2N3	331.41	331.1860	68.86	7.00	11.47	12.68
C19H13F2N5	349.35	349.1139	65.33	3.75	10.88	20.05	C19H23F2N5	359.43	359.1921	63.49	6.45	10.57	19.48
C19H14F1N1	275.33	275.1110	82.89	5.13	6.90	5.09	C19H24F1N1	285.41	285.1893	79.96	8.48	6.66	4.91
C19H14F1N3	303.34	303.1172	75.23	4.65	6.26	13.85	C19H24F1N3	313.42	313.1954	72.81	7.72	6.06	13.41
C19H14F1N5	331.36	331.1233	68.87	4.26	5.73	21.14	C19H24F1N5	341.44	341.2016	66.84	7.09	5.56	20.51
C19H14F2N2	308.33	308.1125	74.01	4.58	12.32	9.09	C19H24F2N2	318.41	318.1907	71.67	7.60	11.93	8.80
C19H14F2N4	336.35	336.1186	67.85	4.20	11.30	16.66	C19H24F2N4	346.43	346.1969	65.88	6.98	10.97	16.17
C19H14F2N6	364.36	364.1248	62.63	3.87	10.43	23.07	C19H24F2N6	374.44	374.2030	60.95	6.46	10.15	22.44
C19H15F1N2	290.34	290.1219	78.60	5.21	6.54	9.65	C19H25F1N2	300.42	300.2002	75.96	8.39	6.32	9.32
C19H15F1N4	318.36	318.1281	71.68	4.75	5.97	17.60	C19H25F1N4	328.44	328.2063	69.48	7.67	5.78	17.06
C19H15F1N6	346.37	346.1342	65.89	4.37	5.49	24.26	C19H25F1N6	356.45	356.2125	64.02	7.07	5.33	23.58
C19H15F2N1	295.33	295.1172	77.27	5.12	12.87	4.74	C19H25F2N1	305.41	305.1955	74.72	8.25	12.44	4.59

C19H25F2N3	333.43	333.2016	68.44	7.56	11.40	12.60	C19H35F2N3	343.51	343.2799	66.44	10.27	11.06	12.23
C19H25F2N5	361.44	361.2078	63.14	6.97	10.51	19.38	C19H35F2N5	371.52	371.2860	61.43	9.50	10.23	18.85
C19H26F1N1	287.42	287.2049	79.40	9.12	6.61	4.87	C19H36F1N1	297.50	297.2832	76.71	12.20	6.39	4.71
C19H26F1N3	315.44	315.2111	72.35	8.31	6.02	13.32	C19H36F1N3	325.52	325.2893	70.11	11.15	5.84	12.91
C19H26F1N5	343.45	343.2172	66.45	7.63	5.53	20.39	C19H36F1N5	353.53	353.2955	64.55	10.26	5.37	19.81
C19H26F2N2	320.43	320.2064	71.22	8.18	11.86	8.74	C19H36F2N2	330.51	330.2846	69.05	10.98	11.50	8.48
C19H26F2N4	348.44	348.2125	65.49	7.52	10.90	16.08	C19H36F2N4	358.52	358.2908	63.65	10.12	10.60	15.63
C19H26F2N6	376.46	376.2187	60.62	6.96	10.09	22.32	C19H36F2N6	386.54	386.2969	59.04	9.39	9.83	21.74
C19H27F1N2	302.44	302.2158	75.46	9.00	6.28	9.26	C19H37F1N2	312.52	312.2941	73.02	11.93	6.08	8.96
C19H27F1N4	330.45	330.2220	69.06	8.24	5.75	16.95	C19H37F1N4	340.53	340.3002	67.02	10.95	5.58	16.45
C19H27F1N6	358.47	358.2281	63.66	7.59	5.30	23.44	C19H37F1N6	368.55	368.3064	61.92	10.12	5.15	22.80
C19H27F2N1	307.43	307.2111	74.23	8.85	12.36	4.56	C19H37F2N1	317.51	317.2894	71.88	11.75	11.97	4.41
C19H27F2N3	335.44	335.2173	68.03	8.11	11.33	12.53	C19H37F2N3	345.52	345.2955	66.05	10.79	11.00	12.16
C19H27F2N5	363.46	363.2234	62.79	7.49	10.45	19.27	C19H37F2N5	373.54	373.3017	61.09	9.98	10.17	18.75
C19H28F1N1	289.44	289.2206	78.85	9.75	6.56	4.84	C19H38F1N1	299.52	299.2988	76.19	12.79	6.34	4.68
C19H28F1N3	317.45	317.2267	71.89	8.89	5.98	13.24	C19H38F1N3	327.53	327.3050	69.68	11.69	5.80	12.83
C19H28F1N5	345.47	345.2329	66.06	8.17	5.50	20.27	C19H38F1N5	355.55	355.3111	64.19	10.77	5.34	19.70
C19H28F2N2	322.45	322.2220	70.78	8.75	11.78	8.69	C19H38F2N2	332.52	332.3003	68.63	11.52	11.43	8.42
C19H28F2N4	350.46	350.2282	65.12	8.05	10.84	15.99	C19H38F2N4	360.54	360.3064	63.30	10.62	10.54	15.54
C19H28F2N6	378.47	378.2343	60.30	7.46	10.04	22.21	C19H38F2N6	388.55	388.3126	58.73	9.86	9.78	21.63
C19H29F1N2	304.45	304.2315	74.96	9.60	6.24	9.20	C19H39F1N2	314.53	314.3097	72.56	12.50	6.04	8.91
C19H29F1N4	332.47	332.2376	68.64	8.79	5.71	16.85	C19H39F1N4	342.55	342.3159	66.62	11.48	5.55	16.36
C19H29F1N6	360.48	360.2438	63.31	8.11	5.27	23.31	C19H39F1N6	370.56	370.3220	61.59	10.61	5.13	22.68
C19H29F2N1	309.45	309.2268	73.75	9.45	12.28	4.53	C19H39F2N1	319.53	319.3050	71.42	12.30	11.89	4.38
C19H29F2N3	337.46	337.2329	67.63	8.66	11.26	12.45	C19H39F2N3	347.54	347.3112	65.66	11.31	10.93	12.09
C19H29F2N5	365.47	365.2391	62.44	8.00	10.40	19.16	C19H39F2N5	375.55	375.3173	60.77	10.47	10.12	18.65
C19H30F1N1	291.46	291.2362	78.30	10.38	6.52	4.81	C19H40F1N1	301.54	301.3145	75.68	13.37	6.30	4.65
C19H30F1N3	319.47	319.2424	71.43	9.47	5.95	13.15	C19H40F1N3	329.55	329.3206	69.25	12.23	5.76	12.75
C19H30F1N5	347.48	347.2485	65.68	8.70	5.47	20.15	C19H40F1N5	357.56	357.3268	63.82	11.28	5.31	19.59
C19H30F2N2	324.46	324.2377	70.34	9.32	11.71	8.63	C19H40F2N2	334.54	334.3159	68.22	12.05	11.36	8.37
C19H30F2N4	352.47	352.2438	64.75	8.58	10.78	15.90	C19H40F2N4	362.55	362.3221	62.95	11.12	10.48	15.45
C19H30F2N6	380.49	380.2500	59.98	7.95	9.99	22.09	C19H40F2N6	390.57	390.3282	58.43	10.32	9.73	21.52
C19H31F1N2	306.47	306.2471	74.46	10.20	6.20	9.14	C19H41F1N2	316.55	316.3254	72.09	13.06	6.00	8.85
C19H31F1N4	334.48	334.2533	68.23	9.34	5.68	16.75	C19H41F1N4	344.56	344.3315	66.23	11.99	5.51	16.26
C19H31F1N6	362.50	362.2594	62.96	8.62	5.24	23.18	C19H41F1N6	372.58	372.3377	61.25	11.09	5.10	22.56
C19H31F2N1	311.46	311.2424	73.27	10.03	12.20	4.50	C19H41F2N3	349.56	349.3268	65.29	11.82	10.87	12.02
C19H31F2N3	339.48	339.2486	67.22	9.20	11.19	12.38	C19H41F2N5	377.57	377.3330	60.44	10.95	10.06	18.55
C19H31F2N5	367.49	367.2547	62.10	8.50	10.34	19.06	C19H42F1N3	331.57	331.3363	68.83	12.77	5.73	12.67
C19H32F1N1	293.47	293.2519	77.76	10.99	6.47	4.77	C19H42F1N5	359.58	359.3424	63.47	11.77	5.28	19.48
C19H32F1N3	321.49	321.2580	70.99	10.03	5.91	13.07	C19H42F2N4	364.57	364.3377	62.60	11.61	10.42	15.37
C19H32F1N5	349.50	349.2642	65.30	9.23	5.44	20.04	C19H42F2N6	392.58	392.3439	58.13	10.78	9.68	21.41
C19H32F2N2	326.48	326.2533	69.90	9.88	11.64	8.58	C19H43F1N4	346.58	346.3472	65.85	12.51	5.48	16.17
C19H32F2N4	354.49	354.2595	64.38	9.10	10.72	15.80	C19H43F1N6	374.59	374.3533	60.92	11.57	5.07	22.44
C19H32F2N6	382.50	382.2656	59.66	8.43	9.93	21.97	C19H43F2N5	379.58	379.3486	60.12	11.42	10.01	18.45
C19H33F1N2	308.49	308.2628	73.98	10.78	6.16	9.08	C19H44F1N5	361.59	361.3581	63.11	12.27	5.25	19.37
C19H33F1N4	336.50	336.2689	67.82	9.88	5.65	16.65	C19H44F2N6	394.60	394.3595	57.83	11.24	9.63	21.30
C19H33F1N6	364.51	364.2751	62.61	9.13	5.21	23.06	C19H45F1N6	376.61	376.3690	60.60	12.04	5.04	22.31
C19H33F2N1	313.48	313.2581	72.80	10.61	12.12	4.47	C20H10F1N1	283.31	283.0797	84.79	3.56	6.71	4.94
C19H33F2N3	341.49	341.2642	66.83	9.74	11.13	12.30	C20H10F1N3	311.32	311.0859	77.16	3.24	6.10	13.50
C19H33F2N5	369.51	369.2704	61.76	9.00	10.28	18.95	C20H10F1N5	339.33	339.0920	70.79	2.97	5.60	20.64
C19H34F1N1	295.49	295.2675	77.23	11.60	6.43	4.74	C20H10F2N2	316.31	316.0812	75.94	3.19	12.01	8.86
C19H34F1N3	323.50	323.2737	70.54	10.59	5.87	12.99	C20H10F2N4	344.33	344.0873	69.77	2.93	11.04	16.27
C19H34F1N5	351.51	351.2798	64.92	9.75	5.40	19.92	C20H10F2N6	372.34	372.0935	64.52	2.71	10.20	22.57
C19H34F2N2	328.49	328.2690	69.47	10.43	11.57	8.53	C20H11F1N2	298.32	298.0906	80.52	3.72	6.37	9.39
C19H34F2N4	356.51	356.2751	64.01	9.61	10.66	15.72	C20H11F1N4	326.34	326.0968	73.61	3.40	5.82	17.17
C19H34F2N6	384.52	384.2813	59.35	8.91	9.88	21.86	C20H11F1N6	354.35	354.1029	67.79	3.13	5.36	23.72
C19H35F1N2	310.50	310.2784	73.50	11.36	6.12	9.02	C20H11F2N1	303.31	303.0859	79.20	3.66	12.53	4.62
C19H35F1N4	338.52	338.2846	67.42	10.42	5.61	16.55	C20H11F2N3	331.33	331.0921	72.50	3.35	11.47	12.68
C19H35F1N6	366.53	366.2907	62.26	9.63	5.18	22.93	C20H11F2N5	359.34	359.0982	66.85	3.09	10.57	19.49
C19H35F2N1	315.49	315.2737	72.33	11.18	12.04	4.44	C20H12F1N1	285.32	285.0954	84.19	4.24	6.66	4.91

C20H12F1N3	313.34	313.1015	76.67	3.86	6.06	13.41
C20H12F1N5	341.35	341.1077	70.37	3.54	5.57	20.52
C20H12F2N2	318.33	318.0968	75.46	3.80	11.94	8.80
C20H12F2N4	346.34	346.1030	69.36	3.49	10.97	16.18
C20H12F2N6	374.36	374.1091	64.17	3.23	10.15	22.45
C20H13F1N2	300.34	300.1063	79.98	4.36	6.33	9.33
C20H13F1N4	328.35	328.1124	73.16	3.99	5.79	17.06
C20H13F1N6	356.37	356.1186	67.41	3.68	5.33	23.58
C20H13F2N1	305.33	305.1016	78.68	4.29	12.44	4.59
C20H13F2N3	333.34	333.1077	72.06	3.93	11.40	12.61
C20H13F2N5	361.36	361.1139	66.48	3.63	10.52	19.38
C20H14F1N1	287.34	287.1110	83.60	4.91	6.61	4.87
C20H14F1N3	315.35	315.1172	76.18	4.47	6.02	13.32
C20H14F1N5	343.37	343.1233	69.96	4.11	5.53	20.40
C20H14F2N2	320.34	320.1125	74.95	4.41	11.86	8.74
C20H14F2N4	348.36	348.1186	68.96	4.05	10.91	16.08
C20H14F2N6	376.37	376.1248	63.83	3.75	10.10	22.33
C20H15F1N2	302.35	302.1219	79.45	5.00	6.28	9.27
C20H15F1N4	330.37	330.1281	72.71	4.58	5.75	16.96
C20H15F1N6	358.38	358.1342	67.03	4.22	5.30	23.45
C20H15F2N1	307.35	307.1172	78.16	4.92	12.36	4.56
C20H15F2N3	335.36	335.1234	71.63	4.51	11.33	12.53
C20H15F2N5	363.37	363.1295	66.11	4.16	10.46	19.27
C20H16F1N1	289.36	289.1267	83.02	5.57	6.57	4.84
C20H16F1N3	317.37	317.1328	75.69	5.08	5.99	13.24
C20H16F1N5	345.38	345.1390	69.55	4.67	5.50	20.28
C20H16F2N2	322.36	322.1281	74.52	5.00	11.79	8.69
C20H16F2N4	350.37	350.1343	68.56	4.60	10.84	15.99
C20H16F2N6	378.39	378.1404	63.49	4.26	10.04	22.21
C20H17F1N2	304.37	304.1376	78.92	5.63	6.24	9.20
C20H17F1N4	332.38	332.1437	72.27	5.16	5.72	16.86
C20H17F1N6	360.40	360.1499	66.66	4.75	5.27	23.32
C20H17F2N1	309.36	309.1329	77.65	5.54	12.28	4.53
C20H17F2N3	337.38	337.1390	71.20	5.08	11.26	12.45
C20H17F2N5	365.39	365.1452	65.74	4.69	10.40	19.17
C20H18F1N1	291.37	291.1423	82.45	6.23	6.52	4.81
C20H18F1N3	319.38	319.1485	75.21	5.68	5.95	13.16
C20H18F1N5	347.40	347.1546	69.15	5.22	5.47	20.16
C20H18F2N2	324.38	324.1438	74.06	5.59	11.71	8.64
C20H18F2N4	352.39	352.1499	68.17	5.15	10.78	15.90
C20H18F2N6	380.40	380.1561	63.15	4.77	9.99	22.09
C20H19F1N2	306.39	306.1532	78.41	6.25	6.20	9.14
C20H19F1N4	334.40	334.1594	71.84	5.73	5.68	16.75
C20H19F1N6	362.41	362.1655	66.28	5.28	5.24	23.19
C20H19F2N1	311.38	311.1485	77.15	6.15	12.20	4.50
C20H19F2N3	339.39	339.1547	70.78	5.64	11.20	12.38
C20H19F2N5	367.40	367.1608	65.38	5.21	10.34	19.06
C20H20F1N1	293.39	293.1580	81.88	6.87	6.48	4.77
C20H20F1N3	321.40	321.1641	74.74	6.27	5.91	13.07
C20H20F1N5	349.41	349.1703	68.75	5.77	5.44	20.04
C20H20F2N2	326.39	326.1594	73.60	6.18	11.64	8.58
C20H20F2N4	354.41	354.1656	67.78	5.69	10.72	15.81
C20H20F2N6	382.42	382.1717	62.82	5.27	9.94	21.98
C20H21F1N2	308.40	308.1689	77.89	6.86	6.16	9.08
C20H21F1N4	336.42	336.1750	71.41	6.29	5.65	16.65
C20H21F1N6	364.43	364.1812	65.92	5.81	5.21	23.06
C20H21F2N1	313.39	313.1642	76.65	6.75	12.12	4.47
C20H21F2N3	341.41	341.1703	70.36	6.20	11.13	12.31
C20H21F2N5	369.42	369.1765	65.03	5.73	10.29	18.96
C20H22F1N1	295.40	295.1736	81.32	7.51	6.43	4.74
C20H22F1N3	323.42	323.1798	74.28	6.86	5.87	12.99
C20H22F1N5	351.43	351.1859	68.36	6.31	5.41	19.93
C20H22F2N2	328.41	328.1751	73.15	6.75	11.57	8.53
C20H22F2N4	356.42	356.1812	67.40	6.22	10.66	15.72
C20H22F2N6	384.44	384.1874	62.49	5.77	9.88	21.86
C20H23F1N2	310.42	310.1845	77.39	7.47	6.12	9.02
C20H23F1N4	338.43	338.1907	70.98	6.85	5.61	16.55
C20H23F1N6	366.44	366.1968	65.56	6.33	5.18	22.93
C20H23F2N1	315.41	315.1798	76.16	7.35	12.05	4.44
C20H23F2N3	343.42	343.1860	69.95	6.75	11.06	12.24
C20H23F2N5	371.44	371.1921	64.67	6.24	10.23	18.85
C20H24F1N1	297.42	297.1893	80.77	8.13	6.39	4.71
C20H24F1N3	325.43	325.1954	73.82	7.43	5.84	12.91
C20H24F1N5	353.45	353.2016	67.97	6.84	5.38	19.81
C20H24F2N2	330.42	330.1907	72.70	7.32	11.50	8.48
C20H24F2N4	358.44	358.1969	67.02	6.75	10.60	15.63
C20H24F2N6	386.45	386.2030	62.16	6.26	9.83	21.75
C20H25F1N2	312.43	312.2002	76.89	8.07	6.08	8.97
C20H25F1N4	340.45	340.2063	70.56	7.40	5.58	16.46
C20H25F1N6	368.46	368.2125	65.20	6.84	5.16	22.81
C20H25F2N1	317.43	317.1955	75.68	7.94	11.97	4.41
C20H25F2N3	345.44	345.2016	69.54	7.29	11.00	12.16
C20H25F2N5	373.45	373.2078	64.32	6.75	10.17	18.75
C20H26F1N1	299.44	299.2049	80.23	8.75	6.34	4.68
C20H26F1N3	327.45	327.2111	73.36	8.00	5.80	12.83
C20H26F1N5	355.46	355.2172	67.58	7.37	5.34	19.70
C20H26F2N2	332.44	332.2064	72.26	7.88	11.43	8.43
C20H26F2N4	360.45	360.2125	66.64	7.27	10.54	15.54
C20H26F2N6	388.47	388.2187	61.84	6.75	9.78	21.63
C20H27F1N2	314.45	314.2158	76.39	8.65	6.04	8.91
C20H27F1N4	342.46	342.2220	70.15	7.95	5.55	16.36
C20H27F1N6	370.48	370.2281	64.84	7.35	5.13	22.68
C20H27F2N1	319.44	319.2111	75.20	8.52	11.89	4.38
C20H27F2N3	347.46	347.2173	69.14	7.83	10.94	12.09
C20H27F2N5	375.47	375.2234	63.98	7.25	10.12	18.65
C20H28F1N1	301.45	301.2206	79.69	9.36	6.30	4.65
C20H28F1N3	329.46	329.2267	72.91	8.57	5.77	12.75
C20H28F1N5	357.48	357.2329	67.20	7.90	5.31	19.59
C20H28F2N2	334.46	334.2220	71.82	8.44	11.36	8.38
C20H28F2N4	362.47	362.2282	66.27	7.79	10.48	15.46
C20H28F2N6	390.48	390.2343	61.52	7.23	9.73	21.52
C20H29F1N2	316.47	316.2315	75.91	9.24	6.00	8.85
C20H29F1N4	344.48	344.2376	69.74	8.49	5.52	16.26
C20H29F1N6	372.49	372.2438	64.49	7.85	5.10	22.56
C20H29F2N1	321.46	321.2268	74.73	9.09	11.82	4.36
C20H29F2N3	349.47	349.2329	68.74	8.36	10.87	12.02
C20H29F2N5	377.48	377.2391	63.64	7.74	10.07	18.55
C20H30F1N1	303.47	303.2362	79.16	9.96	6.26	4.62
C20H30F1N3	331.48	331.2424	72.47	9.12	5.73	12.68
C20H30F1N5	359.49	359.2485	66.82	8.41	5.28	19.48
C20H30F2N2	336.47	336.2377	71.39	8.99	11.29	8.33
C20H30F2N4	364.49	364.2438	65.91	8.30	10.42	15.37
C20H30F2N6	392.50	392.2500	61.20	7.70	9.68	21.41
C20H31F1N2	318.48	318.2471	75.43	9.81	5.97	8.80
C20H31F1N4	346.50	346.2533	69.33	9.02	5.48	16.17
C20H31F1N6	374.51	374.2594	64.14	8.34	5.07	22.44
C20H31F2N1	323.47	323.2424	74.26	9.66	11.75	4.33
C20H31F2N3	351.49	351.2486	68.34	8.89	10.81	11.95
C20H31F2N5	379.50	379.2547	63.30	8.23	10.01	18.45
C20H32F1N1	305.48	305.2519	78.64	10.56	6.22	4.59

C20H32F1N3	333.50	333.2580	72.03	9.67	5.70	12.60
C20H32F1N5	361.51	361.2642	66.45	8.92	5.26	19.37
C20H32F2N2	338.49	338.2533	70.97	9.53	11.23	8.28
C20H32F2N4	366.50	366.2595	65.54	8.80	10.37	15.29
C20H32F2N6	394.52	394.2656	60.89	8.18	9.63	21.30
C20H33F1N2	320.50	320.2628	74.95	10.38	5.93	8.74
C20H33F1N4	348.51	348.2689	68.93	9.54	5.45	16.08
C20H33F1N6	376.52	376.2751	63.80	8.83	5.05	22.32
C20H33F2N1	325.49	325.2581	73.80	10.22	11.67	4.30
C20H33F2N3	353.50	353.2642	67.96	9.41	10.75	11.89
C20H33F2N5	381.52	381.2704	62.97	8.72	9.96	18.36
C20H34F1N1	307.50	307.2675	78.12	11.15	6.18	4.56
C20H34F1N3	335.51	335.2737	71.60	10.21	5.66	12.52
C20H34F1N5	363.53	363.2798	66.08	9.43	5.23	19.27
C20H34F2N2	340.50	340.2690	70.55	10.06	11.16	8.23
C20H34F2N4	368.52	368.2751	65.19	9.30	10.31	15.20
C20H34F2N6	396.53	396.2813	60.58	8.64	9.58	21.19
C20H35F1N2	322.51	322.2784	74.48	10.94	5.89	8.69
C20H35F1N4	350.53	350.2846	68.53	10.06	5.42	15.98
C20H35F1N6	378.54	378.2907	63.46	9.32	5.02	22.20
C20H35F2N1	327.51	327.2737	73.35	10.77	11.60	4.28
C20H35F2N3	355.52	355.2799	67.57	9.92	10.69	11.82
C20H35F2N5	383.53	383.2860	62.63	9.20	9.91	18.26
C20H36F1N1	309.52	309.2832	77.61	11.72	6.14	4.53
C20H36F1N3	337.53	337.2893	71.17	10.75	5.63	12.45
C20H36F1N5	365.54	365.2955	65.72	9.93	5.20	19.16
C20H36F2N2	342.52	342.2846	70.13	10.59	11.09	8.18
C20H36F2N4	370.53	370.2908	64.83	9.79	10.25	15.12
C20H36F2N6	398.55	398.2969	60.27	9.10	9.53	21.09
C20H37F1N2	324.53	324.2941	74.02	11.49	5.85	8.63
C20H37F1N4	352.54	352.3002	68.14	10.58	5.39	15.89
C20H37F1N6	380.56	380.3064	63.12	9.80	4.99	22.08
C20H37F2N1	329.52	329.2894	72.90	11.32	11.53	4.25
C20H37F2N3	357.53	357.2955	67.19	10.43	10.63	11.75
C20H37F2N5	385.55	385.3017	62.31	9.67	9.86	18.16
C20H38F1N1	311.53	311.2988	77.11	12.30	6.10	4.50
C20H38F1N3	339.54	339.3050	70.75	11.28	5.60	12.38
C20H38F1N5	367.56	367.3111	65.36	10.42	5.17	19.05
C20H38F2N2	344.54	344.3003	69.72	11.12	11.03	8.13
C20H38F2N4	372.55	372.3064	64.48	10.28	10.20	15.04
C20H38F2N6	400.56	400.3126	59.97	9.56	9.49	20.98
C20H39F1N2	326.55	326.3097	73.56	12.04	5.82	8.58
C20H39F1N4	354.56	354.3159	67.75	11.09	5.36	15.80
C20H39F1N6	382.57	382.3220	62.79	10.28	4.97	21.97
C20H39F2N1	331.54	331.3050	72.46	11.86	11.46	4.22
C20H39F2N3	359.55	359.3112	66.81	10.93	10.57	11.69
C20H39F2N5	387.56	387.3173	61.98	10.14	9.80	18.07
C20H40F1N1	313.55	313.3145	76.61	12.86	6.06	4.47
C20H40F1N3	341.56	341.3206	70.33	11.80	5.56	12.30
C20H40F1N5	369.57	369.3268	65.00	10.91	5.14	18.95
C20H40F2N2	346.55	346.3159	69.32	11.63	10.96	8.08
C20H40F2N4	374.57	374.3221	64.13	10.76	10.14	14.96
C20H40F2N6	402.58	402.3282	59.67	10.02	9.44	20.88
C20H41F1N2	328.56	328.3254	73.11	12.58	5.78	8.53
C20H41F1N4	356.57	356.3315	67.37	11.59	5.33	15.71
C20H41F1N6	384.59	384.3377	62.46	10.75	4.94	21.85
C20H41F2N1	333.55	333.3207	72.02	12.39	11.39	4.20
C20H41F2N3	361.57	361.3268	66.44	11.43	10.51	11.62
C20H41F2N5	389.58	389.3330	61.66	10.61	9.75	17.98
C20H42F1N1	315.56	315.3301	76.13	13.42	6.02	4.44
C20H42F1N3	343.58	343.3363	69.92	12.32	5.53	12.23
C20H42F1N5	371.59	371.3424	64.65	11.39	5.11	18.85
C20H42F2N2	348.57	348.3316	68.92	12.15	10.90	8.04
C20H42F2N4	376.58	376.3377	63.79	11.24	10.09	14.88
C20H42F2N6	404.59	404.3439	59.37	10.46	9.39	20.77
C20H43F1N2	330.58	330.3410	72.67	13.11	5.75	8.47
C20H43F1N4	358.59	358.3472	66.99	12.09	5.30	15.62
C20H43F1N6	386.60	386.3533	62.14	11.21	4.91	21.74
C20H43F2N3	363.58	363.3425	66.07	11.92	10.45	11.56
C20H43F2N5	391.60	391.3486	61.34	11.07	9.70	17.88
C20H44F1N3	345.59	345.3519	69.51	12.83	5.50	12.16
C20H44F1N5	373.61	373.3581	64.30	11.87	5.09	18.75
C20H44F2N4	378.60	378.3534	63.45	11.71	10.04	14.80
C20H44F2N6	406.61	406.3595	59.08	10.91	9.34	20.67
C20H45F1N4	360.61	360.3628	66.62	12.58	5.27	15.54
C20H45F1N6	388.62	388.3690	61.81	11.67	4.89	21.63
C20H45F2N5	393.61	393.3643	61.03	11.52	9.65	17.79
C20H46F1N5	375.62	375.3737	63.95	12.34	5.06	18.64
C20H46F2N6	408.63	408.3752	58.79	11.35	9.30	20.57
C20H47F1N6	390.64	390.3846	61.50	12.13	4.86	21.51
C21H10F1N1	295.32	295.0797	85.41	3.41	6.43	4.74
C21H10F1N3	323.33	323.0859	78.01	3.12	5.88	13.00
C21H10F1N5	351.35	351.0920	71.79	2.87	5.41	19.93
C21H10F2N2	328.32	328.0812	76.82	3.07	11.57	8.53
C21H10F2N4	356.34	356.0873	70.79	2.83	10.66	15.72
C21H10F2N6	384.35	384.0935	65.63	2.62	9.89	21.87
C21H11F1N2	310.33	310.0906	81.28	3.57	6.12	9.03
C21H11F1N4	338.35	338.0968	74.55	3.28	5.62	16.56
C21H11F1N6	366.36	366.1029	68.85	3.03	5.19	22.94
C21H11F2N1	315.33	315.0859	79.99	3.52	12.05	4.44
C21H11F2N3	343.34	343.0921	73.47	3.23	11.07	12.24
C21H11F2N5	371.35	371.0982	67.92	2.99	10.23	18.86
C21H12F1N1	297.33	297.0954	84.83	4.07	6.39	4.71
C21H12F1N3	325.35	325.1015	77.53	3.72	5.84	12.92
C21H12F1N5	353.36	353.1077	71.38	3.42	5.38	19.82
C21H12F2N2	330.34	330.0968	76.36	3.66	11.50	8.48
C21H12F2N4	358.35	358.1030	70.39	3.38	10.60	15.63
C21H12F2N6	386.37	386.1091	65.28	3.13	9.83	21.75
C21H13F1N2	312.35	312.1063	80.75	4.20	6.08	8.97
C21H13F1N4	340.36	340.1124	74.11	3.85	5.58	16.46
C21H13F1N6	368.38	368.1186	68.47	3.56	5.16	22.81
C21H13F2N1	317.34	317.1016	79.48	4.13	11.97	4.41
C21H13F2N3	345.35	345.1077	73.04	3.79	11.00	12.17
C21H13F2N5	373.37	373.1139	67.56	3.51	10.18	18.76
C21H14F1N1	299.35	299.1110	84.26	4.71	6.35	4.68
C21H14F1N3	327.36	327.1172	77.05	4.31	5.80	12.84
C21H14F1N5	355.38	355.1233	70.98	3.97	5.35	19.71
C21H14F2N2	332.36	332.1125	75.89	4.25	11.43	8.43
C21H14F2N4	360.37	360.1186	69.99	3.92	10.54	15.55
C21H14F2N6	388.38	388.1248	64.94	3.63	9.78	21.64
C21H15F1N2	314.37	314.1219	80.24	4.81	6.04	8.91
C21H15F1N4	342.38	342.1281	73.67	4.42	5.55	16.36
C21H15F1N6	370.39	370.1342	68.10	4.08	5.13	22.69
C21H15F2N1	319.36	319.1172	78.98	4.73	11.90	4.39
C21H15F2N3	347.37	347.1234	72.61	4.35	10.94	12.10
C21H15F2N5	375.38	375.1295	67.19	4.03	10.12	18.66
C21H16F1N1	301.37	301.1267	83.70	5.35	6.30	4.65
C21H16F1N3	329.38	329.1328	76.58	4.90	5.77	12.76
C21H16F1N5	357.39	357.1390	70.58	4.51	5.32	19.60
C21H16F2N2	334.37	334.1281	75.44	4.82	11.36	8.38

10/21/66 218

C21H16F2N4	362.39	362.1343	69.60	4.45	10.49	15.46
C21H16F2N6	390.40	390.1404	64.61	4.13	9.73	21.53
C21H17F1N2	316.38	316.1376	79.72	5.42	6.00	8.85
C21H17F1N4	344.39	344.1437	73.24	4.98	5.52	16.27
C21H17F1N6	372.41	372.1499	67.73	4.60	5.10	22.57
C21H17F2N1	321.37	321.1329	78.49	5.33	11.82	4.36
C21H17F2N3	349.39	349.1390	72.19	4.90	10.88	12.03
C21H17F2N5	377.40	377.1452	66.83	4.54	10.07	18.56
C21H18F1N1	303.38	303.1423	83.14	5.98	6.26	4.62
C21H18F1N3	331.40	331.1485	76.11	5.47	5.73	12.68
C21H18F1N5	359.41	359.1546	70.18	5.05	5.29	19.49
C21H18F2N2	336.39	336.1438	74.98	5.39	11.30	8.33
C21H18F2N4	364.40	364.1499	69.22	4.98	10.43	15.38
C21H18F2N6	392.41	392.1561	64.28	4.62	9.68	21.42
C21H19F1N2	318.40	318.1532	79.22	6.01	5.97	8.80
C21H19F1N4	346.41	346.1594	72.81	5.53	5.48	16.17
C21H19F1N6	374.42	374.1655	67.37	5.11	5.07	22.45
C21H19F2N1	323.39	323.1485	78.00	5.92	11.75	4.33
C21H19F2N3	351.40	351.1547	71.78	5.45	10.81	11.96
C21H19F2N5	379.42	379.1608	66.48	5.05	10.01	18.46
C21H20F1N1	305.40	305.1580	82.59	6.60	6.22	4.59
C21H20F1N3	333.41	333.1641	75.65	6.05	5.70	12.60
C21H20F1N5	361.43	361.1703	69.79	5.58	5.26	19.38
C21H20F2N2	338.40	338.1594	74.54	5.96	11.23	8.28
C21H20F2N4	366.42	366.1656	68.84	5.50	10.37	15.29
C21H20F2N6	394.43	394.1717	63.95	5.11	9.63	21.31
C21H21F1N2	320.41	320.1689	78.72	6.61	5.93	8.74
C21H21F1N4	348.43	348.1750	72.39	6.08	5.45	16.08
C21H21F1N6	376.44	376.1812	67.01	5.62	5.05	22.32
C21H21F2N1	325.41	325.1642	77.51	6.50	11.68	4.30
C21H21F2N3	353.42	353.1703	71.37	5.99	10.75	11.89
C21H21F2N5	381.43	381.1765	66.13	5.55	9.96	18.36
C21H22F1N1	307.41	307.1736	82.05	7.21	6.18	4.56
C21H22F1N3	335.43	335.1798	75.20	6.61	5.66	12.53
C21H22F1N5	363.44	363.1859	69.40	6.10	5.23	19.27
C21H22F2N2	340.42	340.1751	74.10	6.51	11.16	8.23
C21H22F2N4	368.43	368.1812	68.46	6.02	10.31	15.21
C21H22F2N6	396.45	396.1874	63.62	5.59	9.58	21.20
C21H23F1N2	322.43	322.1845	78.23	7.19	5.89	8.69
C21H23F1N4	350.44	350.1907	71.98	6.62	5.42	15.99
C21H23F1N6	378.46	378.1968	66.65	6.13	5.02	22.21
C21H23F2N1	327.42	327.1798	77.04	7.08	11.60	4.28
C21H23F2N3	355.43	355.1860	70.97	6.52	10.69	11.82
C21H23F2N5	383.45	383.1921	65.78	6.05	9.91	18.26
C21H24F1N1	309.43	309.1893	81.52	7.82	6.14	4.53
C21H24F1N3	337.44	337.1954	74.75	7.17	5.63	12.45
C21H24F1N5	365.46	365.2016	69.02	6.62	5.20	19.16
C21H24F2N2	342.44	342.1907	73.66	7.06	11.10	8.18
C21H24F2N4	370.45	370.1969	68.09	6.53	10.26	15.12
C21H24F2N6	398.46	398.2030	63.30	6.07	9.54	21.09
C21H25F1N2	324.45	324.2002	77.74	7.77	5.86	8.63
C21H25F1N4	352.46	352.2063	71.56	7.15	5.39	15.90
C21H25F1N6	380.47	380.2125	66.30	6.62	4.99	22.09
C21H25F2N1	329.44	329.1955	76.57	7.65	11.53	4.25
C21H25F2N3	357.45	357.2016	70.56	7.05	10.63	11.76
C21H25F2N5	385.46	385.2078	65.44	6.54	9.86	18.17
C21H26F1N1	311.45	311.2049	80.99	8.41	6.10	4.50
C21H26F1N3	339.46	339.2111	74.30	7.72	5.60	12.38
C21H26F1N5	367.47	367.2172	68.64	7.13	5.17	19.06
C21H26F2N2	344.45	344.2064	73.23	7.61	11.03	8.13
C21H26F2N4	372.46	372.2125	67.72	7.04	10.20	15.04
C21H26F2N6	400.48	400.2187	62.98	6.54	9.49	20.98
C21H27F1N2	326.46	326.2158	77.26	8.34	5.82	8.58
C21H27F1N4	354.47	354.2220	71.16	7.68	5.36	15.81
C21H27F1N6	382.49	382.2281	65.95	7.12	4.97	21.97
C21H27F2N1	331.45	331.2111	76.10	8.21	11.46	4.23
C21H27F2N3	359.47	359.2173	70.17	7.57	10.57	11.69
C21H27F2N5	387.48	387.2234	65.10	7.02	9.81	18.07
C21H28F1N1	313.46	313.2206	80.47	9.00	6.06	4.47
C21H28F1N3	341.48	341.2267	73.87	8.27	5.56	12.31
C21H28F1N5	369.49	369.2329	68.27	7.64	5.14	18.95
C21H28F2N2	346.47	346.2220	72.80	8.15	10.97	8.09
C21H28F2N4	374.48	374.2282	67.36	7.54	10.15	14.96
C21H28F2N6	402.49	402.2343	62.67	7.01	9.44	20.88
C21H29F1N2	328.48	328.2315	76.79	8.90	5.78	8.53
C21H29F1N4	356.49	356.2376	70.75	8.20	5.33	15.72
C21H29F1N6	384.50	384.2438	65.60	7.60	4.94	21.86
C21H29F2N1	333.47	333.2268	75.64	8.77	11.39	4.20
C21H29F2N3	361.48	361.2329	69.78	8.09	10.51	11.62
C21H29F2N5	389.50	389.2391	64.76	7.50	9.76	17.98
C21H30F1N1	315.48	315.2362	79.95	9.59	6.02	4.44
C21H30F1N3	343.49	343.2424	73.43	8.80	5.53	12.23
C21H30F1N5	371.51	371.2485	67.90	8.14	5.11	18.85
C21H30F2N2	348.48	348.2377	72.38	8.68	10.90	8.04
C21H30F2N4	376.50	376.2438	67.00	8.03	10.09	14.88
C21H30F2N6	404.51	404.2500	62.36	7.48	9.39	20.78
C21H31F1N2	330.49	330.2471	76.32	9.45	5.75	8.48
C21H31F1N4	358.51	358.2533	70.36	8.72	5.30	15.63
C21H31F1N6	386.52	386.2594	65.26	8.08	4.92	21.74
C21H31F2N1	335.48	335.2424	75.18	9.31	11.33	4.18
C21H31F2N3	363.50	363.2486	69.39	8.60	10.45	11.56
C21H31F2N5	391.51	391.2547	64.43	7.98	9.71	17.89
C21H32F1N1	317.49	317.2519	79.45	10.16	5.98	4.41
C21H32F1N3	345.51	345.2580	73.00	9.34	5.50	12.16
C21H32F1N5	373.52	373.2642	67.53	8.64	5.09	18.75
C21H32F2N2	350.50	350.2533	71.96	9.20	10.84	7.99
C21H32F2N4	378.51	378.2595	66.64	8.52	10.04	14.80
C21H32F2N6	406.53	406.2656	62.05	7.93	9.35	20.67
C21H33F1N2	332.51	332.2628	75.86	10.00	5.71	8.42
C21H33F1N4	360.52	360.2689	69.96	9.23	5.27	15.54
C21H33F1N6	388.54	388.2751	64.92	8.56	4.89	21.63
C21H33F2N1	337.50	337.2581	74.74	9.86	11.26	4.15
C21H33F2N3	365.51	365.2642	69.01	9.10	10.40	11.50
C21H33F2N5	393.53	393.2704	64.10	8.45	9.66	17.80
C21H34F1N1	319.51	319.2675	78.94	10.73	5.95	4.38
C21H34F1N3	347.52	347.2737	72.58	9.86	5.47	12.09
C21H34F1N5	375.54	375.2798	67.17	9.13	5.06	18.65
C21H34F2N2	352.52	352.2690	71.55	9.72	10.78	7.95
C21H34F2N4	380.53	380.2751	66.29	9.01	9.99	14.72
C21H34F2N6	408.54	408.2813	61.74	8.39	9.30	20.57
C21H35F1N2	334.52	334.2784	75.40	10.55	5.68	8.37
C21H35F1N4	362.54	362.2846	69.57	9.73	5.24	15.45
C21H35F1N6	390.55	390.2907	64.58	9.03	4.86	21.52
C21H35F2N1	339.52	339.2737	74.29	10.39	11.19	4.13
C21H35F2N3	367.53	367.2799	68.63	9.60	10.34	11.43
C21H35F2N5	395.54	395.2860	63.77	8.92	9.61	17.71
C21H36F1N1	321.53	321.2832	78.45	11.29	5.91	4.36
C21H36F1N3	349.54	349.2893	72.16	10.38	5.44	12.02
C21H36F1N5	377.55	377.2955	66.81	9.61	5.03	18.55
C21H36F2N2	354.53	354.2846	71.15	10.24	10.72	7.90

10/21/66 219

C21H36F2N4	382.54	382.2908	65.94	9.49	9.93	14.65
C21H36F2N6	410.56	410.2969	61.44	8.84	9.25	20.47
C21H37F1N2	336.54	336.2941	74.95	11.08	5.65	8.32
C21H37F1N4	364.55	364.3002	69.19	10.23	5.21	15.37
C21H37F1N6	392.57	392.3064	64.25	9.50	4.84	21.41
C21H37F2N1	341.53	341.2894	73.85	10.92	11.13	4.10
C21H37F2N3	369.55	369.2955	68.26	10.09	10.28	11.37
C21H37F2N5	397.56	397.3017	63.45	9.38	9.56	17.62
C21H38F1N1	323.54	323.2988	77.96	11.84	5.87	4.33
C21H38F1N3	351.56	351.3050	71.75	10.90	5.40	11.95
C21H38F1N5	379.57	379.3111	66.45	10.09	5.01	18.45
C21H38F2N2	356.55	356.3003	70.74	10.74	10.66	7.86
C21H38F2N4	384.56	384.3064	65.59	9.96	9.88	14.57
C21H38F2N6	412.57	412.3126	61.14	9.28	9.21	20.37
C21H39F1N2	338.56	338.3097	74.50	11.61	5.61	8.27
C21H39F1N4	366.57	366.3159	68.81	10.72	5.18	15.28
C21H39F1N6	394.58	394.3220	63.92	9.96	4.81	21.30
C21H39F2N1	343.55	343.3050	73.42	11.44	11.06	4.08
C21H39F2N3	371.56	371.3112	67.88	10.58	10.23	11.31
C21H39F2N5	399.58	399.3173	63.13	9.84	9.51	17.53
C21H40F1N1	325.56	325.3145	77.48	12.38	5.84	4.30
C21H40F1N3	353.57	353.3206	71.34	11.40	5.37	11.88
C21H40F1N5	381.58	381.3268	66.10	10.57	4.98	18.35
C21H40F2N2	358.56	358.3159	70.35	11.24	10.60	7.81
C21H40F2N4	386.58	386.3221	65.25	10.43	9.83	14.49
C21H40F2N6	414.59	414.3282	60.84	9.72	9.16	20.27
C21H41F1N2	340.57	340.3254	74.06	12.13	5.58	8.23
C21H41F1N4	368.59	368.3315	68.43	11.21	5.15	15.20
C21H41F1N6	396.60	396.3377	63.60	10.42	4.79	21.19
C21H41F2N1	345.56	345.3207	72.99	11.96	11.00	4.05
C21H41F2N3	373.58	373.3268	67.52	11.06	10.17	11.25
C21H41F2N5	401.59	401.3330	62.81	10.29	9.46	17.44
C21H42F1N1	327.57	327.3301	77.00	12.92	5.80	4.28
C21H42F1N3	355.59	355.3363	70.93	11.91	5.34	11.82
C21H42F1N5	383.60	383.3424	65.75	11.04	4.95	18.26
C21H42F2N2	360.58	360.3316	69.95	11.74	10.54	7.77
C21H42F2N4	388.59	388.3377	64.91	10.89	9.78	14.42
C21H42F2N6	416.61	416.3439	60.55	10.16	9.12	20.17
C21H43F1N2	342.59	342.3410	73.63	12.65	5.55	8.18
C21H43F1N4	370.60	370.3472	68.06	11.70	5.13	15.12
C21H43F1N6	398.62	398.3533	63.28	10.87	4.77	21.08
C21H43F2N1	347.58	347.3363	72.57	12.47	10.93	4.03
C21H43F2N3	375.59	375.3425	67.16	11.54	10.12	11.19
C21H43F2N5	403.61	403.3486	62.49	10.74	9.41	17.35
C21H44F1N1	329.59	329.3458	76.53	13.46	5.76	4.25
C21H44F1N3	357.60	357.3519	70.53	12.40	5.31	11.75
C21H44F1N5	385.62	385.3581	65.41	11.50	4.93	18.16
C21H44F2N2	362.60	362.3472	69.56	12.23	10.48	7.73
C21H44F2N4	390.61	390.3534	64.57	11.35	9.73	14.34
C21H44F2N6	418.62	418.3595	60.25	10.59	9.08	20.08
C21H45F1N2	344.60	344.3567	73.20	13.16	5.51	8.13
C21H45F1N4	372.62	372.3628	67.69	12.17	5.10	15.04
C21H45F1N6	400.63	400.3690	62.96	11.32	4.74	20.98
C21H45F2N3	377.61	377.3581	66.80	12.01	10.06	11.13
C21H45F2N5	405.62	405.3643	62.18	11.18	9.37	17.27
C21H46F1N3	359.62	359.3676	70.14	12.89	5.28	11.68
C21H46F1N5	387.63	387.3737	65.07	11.96	4.90	18.07
C21H46F2N4	392.62	392.3690	64.24	11.81	9.68	14.27
C21H46F2N6	420.64	420.3752	59.96	11.02	9.03	19.98
C21H47F1N4	374.63	374.3785	67.33	12.65	5.07	14.96
C21H47F1N6	402.65	402.3846	62.64	11.77	4.72	20.87
C21H47F2N5	407.64	407.3799	61.88	11.62	9.32	17.18
C21H48F1N5	389.65	389.3894	64.73	12.42	4.88	17.97
C21H48F2N6	422.65	422.3908	59.68	11.45	8.99	19.88
C21H49F1N6	404.66	404.4003	62.33	12.21	4.69	20.77
C22H11F1N2	322.34	322.0906	81.98	3.44	5.89	8.69
C22H11F1N4	350.36	350.0968	75.42	3.16	5.42	15.99
C22H11F1N6	378.37	378.1029	69.84	2.93	5.02	22.21
C22H11F2N1	327.34	327.0859	80.73	3.39	11.61	4.28
C22H11F2N3	355.35	355.0921	74.36	3.12	10.69	11.82
C22H11F2N5	383.36	383.0982	68.93	2.89	9.91	18.27
C22H12F1N1	309.35	309.0954	85.42	3.91	6.14	4.53
C22H12F1N3	337.36	337.1015	78.33	3.59	5.63	12.46
C22H12F1N5	365.37	365.1077	72.32	3.31	5.20	19.17
C22H12F2N2	342.35	342.0968	77.19	3.53	11.10	8.18
C22H12F2N4	370.36	370.1030	71.35	3.27	10.26	15.13
C22H12F2N6	398.38	398.1091	66.33	3.04	9.54	21.10
C22H13F1N2	324.36	324.1063	81.47	4.04	5.86	8.64
C22H13F1N4	352.37	352.1124	74.99	3.72	5.39	15.90
C22H13F1N6	380.39	380.1186	69.47	3.44	4.99	22.09
C22H13F2N1	329.35	329.1016	80.23	3.98	11.54	4.25
C22H13F2N3	357.37	357.1077	73.94	3.67	10.63	11.76
C22H13F2N5	385.38	385.1139	68.57	3.40	9.86	18.17
C22H14F1N1	311.36	311.1110	84.87	4.53	6.10	4.50
C22H14F1N3	339.38	339.1172	77.86	4.16	5.60	12.38
C22H14F1N5	367.39	367.1233	71.93	3.84	5.17	19.06
C22H14F2N2	344.37	344.1125	76.73	4.10	11.03	8.13
C22H14F2N4	372.38	372.1186	70.96	3.79	10.20	15.05
C22H14F2N6	400.39	400.1248	66.00	3.52	9.49	20.99
C22H15F1N2	326.38	326.1219	80.96	4.63	5.82	8.58
C22H15F1N4	354.39	354.1281	74.56	4.27	5.36	15.81
C22H15F1N6	382.40	382.1342	69.10	3.95	4.97	21.98
C22H15F2N1	331.37	331.1172	79.74	4.56	11.47	4.23
C22H15F2N3	359.38	359.1234	73.53	4.21	10.57	11.69
C22H15F2N5	387.40	387.1295	68.21	3.90	9.81	18.08
C22H16F1N1	313.38	313.1267	84.32	5.15	6.06	4.47
C22H16F1N3	341.39	341.1328	77.40	4.72	5.56	12.31
C22H16F1N5	369.40	369.1390	71.53	4.37	5.14	18.96
C22H16F2N2	346.38	346.1281	76.29	4.66	10.97	8.09
C22H16F2N4	374.40	374.1343	70.58	4.31	10.15	14.96
C22H16F2N6	402.41	402.1404	65.67	4.01	9.44	20.88
C22H17F1N2	328.39	328.1376	80.47	5.22	5.79	8.53
C22H17F1N4	356.41	356.1437	74.14	4.81	5.33	15.72
C22H17F1N6	384.42	384.1499	68.74	4.46	4.94	21.86
C22H17F2N1	333.38	333.1329	79.26	5.14	11.40	4.20
C22H17F2N3	361.40	361.1390	73.12	4.74	10.51	11.63
C22H17F2N5	389.41	389.1452	67.86	4.40	9.76	17.98
C22H18F1N1	315.39	315.1423	83.78	5.75	6.02	4.44
C22H18F1N3	343.41	343.1485	76.95	5.28	5.53	12.24
C22H18F1N5	371.42	371.1546	71.14	4.88	5.12	18.86
C22H18F2N2	348.40	348.1438	75.85	5.21	10.91	8.04
C22H18F2N4	376.41	376.1499	70.20	4.82	10.09	14.88
C22H18F2N6	404.43	404.1561	65.34	4.49	9.40	20.78
C22H19F1N2	330.41	330.1532	79.98	5.80	5.75	8.48
C22H19F1N4	358.42	358.1594	73.72	5.34	5.30	15.63
C22H19F1N6	386.44	386.1655	68.38	4.96	4.92	21.75
C22H19F2N1	335.40	335.1485	78.79	5.71	11.33	4.18
C22H19F2N3	363.41	363.1547	72.71	5.27	10.46	11.56
C22H19F2N5	391.43	391.1608	67.51	4.89	9.71	17.89
C22H20F1N1	317.41	317.1580	83.25	6.35	5.99	4.41

C22H20F1N3	345.42	345.1641	76.50	5.84	5.50	12.16	C22H30F1N3	355.50	355.2424	74.33	8.51	5.34	11.82
C22H20F1N5	373.44	373.1703	70.76	5.40	5.09	18.75	C22H30F1N5	383.52	383.2485	68.90	7.88	4.95	18.26
C22H20F2N2	350.41	350.1594	75.41	5.75	10.84	7.99	C22H30F2N2	360.49	360.2377	73.30	8.39	10.54	7.77
C22H20F2N4	378.43	378.1656	69.83	5.33	10.04	14.81	C22H30F2N4	388.51	388.2438	68.02	7.78	9.78	14.42
C22H20F2N6	406.44	406.1717	65.01	4.96	9.35	20.68	C22H30F2N6	416.52	416.2500	63.44	7.26	9.12	20.18
C22H21F1N2	332.42	332.1689	79.49	6.37	5.72	8.43	C22H31F1N2	342.50	342.2471	77.15	9.12	5.55	8.18
C22H21F1N4	360.44	360.1750	73.31	5.87	5.27	15.54	C22H31F1N4	370.52	370.2533	71.32	8.43	5.13	15.12
C22H21F1N6	388.45	388.1812	68.03	5.45	4.89	21.63	C22H31F1N6	398.53	398.2594	66.30	7.84	4.77	21.09
C22H21F2N1	337.42	337.1642	78.31	6.27	11.26	4.15	C22H31F2N1	347.50	347.2424	76.04	8.99	10.93	4.03
C22H21F2N3	365.43	365.1703	72.31	5.79	10.40	11.50	C22H31F2N3	375.51	375.2486	70.37	8.32	10.12	11.19
C22H21F2N5	393.44	393.1765	67.16	5.38	9.66	17.80	C22H31F2N5	403.52	403.2547	65.48	7.74	9.42	17.36
C22H22F1N1	319.43	319.1736	82.73	6.94	5.95	4.38	C22H32F1N1	329.51	329.2519	80.19	9.79	5.77	4.25
C22H22F1N3	347.44	347.1798	76.06	6.38	5.47	12.09	C22H32F1N3	357.52	357.2580	73.91	9.02	5.31	11.75
C22H22F1N5	375.45	375.1859	70.38	5.91	5.06	18.65	C22H32F1N5	385.53	385.2642	68.54	8.37	4.93	18.17
C22H22F2N2	352.43	352.1751	74.98	6.29	10.78	7.95	C22H32F2N2	362.51	362.2533	72.89	8.90	10.48	7.73
C22H22F2N4	380.44	380.1812	69.46	5.83	9.99	14.73	C22H32F2N4	390.52	390.2595	67.66	8.26	9.73	14.35
C22H22F2N6	408.46	408.1874	64.69	5.43	9.30	20.58	C22H32F2N6	418.54	418.2656	63.14	7.71	9.08	20.08
C22H23F1N2	334.44	334.1845	79.01	6.93	5.68	8.38	C22H33F1N2	344.52	344.2628	76.70	9.65	5.51	8.13
C22H23F1N4	362.45	362.1907	72.90	6.40	5.24	15.46	C22H33F1N4	372.53	372.2689	70.93	8.93	5.10	15.04
C22H23F1N6	390.47	390.1968	67.67	5.94	4.87	21.52	C22H33F1N6	400.55	400.2751	65.97	8.30	4.74	20.98
C22H23F2N1	339.43	339.1798	77.85	6.83	11.19	4.13	C22H33F2N1	349.51	349.2581	75.60	9.52	10.87	4.01
C22H23F2N3	367.45	367.1860	71.91	6.31	10.34	11.44	C22H33F2N3	377.53	377.2642	69.99	8.81	10.06	11.13
C22H23F2N5	395.46	395.1921	66.82	5.86	9.61	17.71	C22H33F2N5	405.54	405.2704	65.16	8.20	9.37	17.27
C22H24F1N1	321.44	321.1893	82.21	7.53	5.91	4.36	C22H34F1N1	331.52	331.2675	79.71	10.34	5.73	4.22
C22H24F1N3	349.46	349.1954	75.62	6.92	5.44	12.02	C22H34F1N3	359.53	359.2737	73.50	9.53	5.28	11.69
C22H24F1N5	377.47	377.2016	70.00	6.41	5.03	18.55	C22H34F1N5	387.55	387.2798	68.18	8.84	4.90	18.07
C22H24F2N2	354.45	354.1907	74.55	6.83	10.72	7.90	C22H34F2N2	364.53	364.2690	72.49	9.40	10.42	7.68
C22H24F2N4	382.46	382.1969	69.09	6.33	9.93	14.65	C22H34F2N4	392.54	392.2751	67.32	8.73	9.68	14.27
C22H24F2N6	410.47	410.2030	64.38	5.89	9.26	20.47	C22H34F2N6	420.55	420.2813	62.83	8.15	9.03	19.98
C22H25F1N2	336.46	336.2002	78.54	7.49	5.65	8.33	C22H35F1N2	346.54	346.2784	76.25	10.18	5.48	8.08
C22H25F1N4	364.47	364.2063	72.50	6.91	5.21	15.37	C22H35F1N4	374.55	374.2846	70.55	9.42	5.07	14.96
C22H25F1N6	392.48	392.2125	67.33	6.42	4.84	21.41	C22H35F1N6	402.56	402.2907	65.64	8.76	4.72	20.88
C22H25F2N1	341.45	341.1955	77.39	7.38	11.13	4.10	C22H35F2N1	351.53	351.2737	75.17	10.04	10.81	3.98
C22H25F2N3	369.46	369.2016	71.52	6.82	10.28	11.37	C22H35F2N3	379.54	379.2799	69.62	9.30	10.01	11.07
C22H25F2N5	397.47	397.2078	66.48	6.34	9.56	17.62	C22H35F2N5	407.55	407.2860	64.84	8.66	9.32	17.18
C22H26F1N1	323.46	323.2049	81.69	8.10	5.87	4.33	C22H36F1N1	333.54	333.2832	79.23	10.88	5.70	4.20
C22H26F1N3	351.47	351.2111	75.18	7.46	5.41	11.96	C22H36F1N3	361.55	361.2893	73.09	10.04	5.25	11.62
C22H26F1N5	379.48	379.2172	69.63	6.91	5.01	18.45	C22H36F1N5	389.56	389.2955	67.83	9.31	4.88	17.98
C22H26F2N2	356.46	356.2064	74.13	7.35	10.66	7.86	C22H36F2N2	366.54	366.2846	72.09	9.90	10.37	7.64
C22H26F2N4	384.48	384.2125	68.73	6.82	9.88	14.57	C22H36F2N4	394.56	394.2908	66.97	9.20	9.63	14.20
C22H26F2N6	412.49	412.2187	64.06	6.35	9.21	20.37	C22H36F2N6	422.57	422.2969	62.53	8.59	8.99	19.89
C22H27F1N2	338.47	338.2158	78.07	8.04	5.61	8.28	C22H37F1N2	348.55	348.2941	75.81	10.70	5.45	8.04
C22H27F1N4	366.49	366.2220	72.10	7.43	5.18	15.29	C22H37F1N4	376.57	376.3002	70.17	9.90	5.05	14.88
C22H27F1N6	394.50	394.2281	66.98	6.90	4.82	21.30	C22H37F1N6	404.58	404.3064	65.31	9.22	4.70	20.77
C22H27F2N1	343.46	343.2111	76.94	7.92	11.06	4.08	C22H37F2N1	353.54	353.2894	74.74	10.55	10.75	3.96
C22H27F2N3	371.48	371.2173	71.13	7.33	10.23	11.31	C22H37F2N3	381.56	381.2955	69.25	9.77	9.96	11.01
C22H27F2N5	399.49	399.2234	66.15	6.81	9.51	17.53	C22H37F2N5	409.57	409.3017	64.52	9.11	9.28	17.10
C22H28F1N1	325.47	325.2206	81.19	8.67	5.84	4.30	C22H38F1N1	335.55	335.2988	78.75	11.41	5.66	4.17
C22H28F1N3	353.49	353.2267	74.75	7.98	5.37	11.89	C22H38F1N3	363.57	363.3050	72.68	10.54	5.23	11.56
C22H28F1N5	381.50	381.2329	69.26	7.40	4.98	18.36	C22H38F1N5	391.58	391.3111	67.48	9.78	4.85	17.88
C22H28F2N2	358.48	358.2220	73.71	7.87	10.60	7.81	C22H38F2N2	368.56	368.3003	71.70	10.39	10.31	7.60
C22H28F2N4	386.49	386.2282	68.37	7.30	9.83	14.50	C22H38F2N4	396.57	396.3064	66.63	9.66	9.58	14.13
C22H28F2N6	414.51	414.2343	63.75	6.81	9.17	20.27	C22H38F2N6	424.59	424.3126	62.24	9.02	8.95	19.79
C22H29F1N2	340.49	340.2315	77.61	8.59	5.58	8.23	C22H39F1N2	350.57	350.3097	75.38	11.21	5.42	7.99
C22H29F1N4	368.50	368.2376	71.71	7.93	5.16	15.20	C22H39F1N4	378.58	378.3159	69.80	10.38	5.02	14.80
C22H29F1N6	396.52	396.2438	66.64	7.37	4.79	21.19	C22H39F1N6	406.59	406.3220	64.99	9.67	4.67	20.67
C22H29F2N1	345.48	345.2268	76.49	8.46	11.00	4.05	C22H39F2N1	355.56	355.3050	74.32	11.06	10.69	3.94
C22H29F2N3	373.49	373.2329	70.75	7.83	10.17	11.25	C22H39F2N3	383.57	383.3112	68.89	10.25	9.91	10.95
C22H29F2N5	401.51	401.2391	65.81	7.28	9.46	17.44	C22H39F2N5	411.59	411.3173	64.20	9.55	9.23	17.02
C22H30F1N1	327.49	327.2362	80.69	9.23	5.80	4.28	C22H40F1N1	337.57	337.3145	78.28	11.94	5.63	4.15

C22H40F1N3	365.58	365.3206	72.28	11.03	5.20	11.49	C23H11F2N3	367.36	367.0921	75.20	3.02	10.34	11.44
C22H40F1N5	393.60	393.3268	67.14	10.24	4.83	17.79	C23H11F2N5	395.37	395.0982	69.87	2.80	9.61	17.71
C22H40F2N2	370.57	370.3159	71.31	10.88	10.25	7.56	C23H12F1N1	321.36	321.0954	85.97	3.76	5.91	4.36
C22H40F2N4	398.59	398.3221	66.30	10.12	9.53	14.06	C23H12F1N3	349.37	349.1015	79.07	3.46	5.44	12.03
C22H40F2N6	426.60	426.3282	61.94	9.45	8.91	19.70	C23H12F1N5	377.38	377.1077	73.20	3.21	5.03	18.56
C22H41F1N2	352.58	352.3254	74.95	11.72	5.39	7.95	C23H12F2N2	354.36	354.0968	77.96	3.41	10.72	7.91
C22H41F1N4	380.60	380.3315	69.43	10.86	4.99	14.72	C23H12F2N4	382.38	382.1030	72.25	3.16	9.94	14.65
C22H41F1N6	408.61	408.3377	64.67	10.11	4.65	20.57	C23H12F2N6	410.39	410.1091	67.32	2.95	9.26	20.48
C22H41F2N1	357.58	357.3207	73.90	11.56	10.63	3.92	C23H13F1N2	336.37	336.1063	82.13	3.90	5.65	8.33
C22H41F2N3	385.59	385.3268	68.53	10.72	9.85	10.90	C23H13F1N4	364.39	364.1124	75.81	3.60	5.21	15.38
C22H41F2N5	413.60	413.3330	63.89	9.99	9.19	16.93	C23H13F1N6	392.40	392.1186	70.40	3.34	4.84	21.42
C22H42F1N1	339.59	339.3301	77.81	12.47	5.59	4.12	C23H13F2N1	341.36	341.1016	80.93	3.84	11.13	4.10
C22H42F1N3	367.60	367.3363	71.88	11.52	5.17	11.43	C23H13F2N3	369.38	369.1077	74.79	3.55	10.29	11.38
C22H42F1N5	395.61	395.3424	66.79	10.70	4.80	17.70	C23H13F2N5	397.39	397.1139	69.52	3.30	9.56	17.62
C22H42F2N2	372.59	372.3316	70.92	11.36	10.20	7.52	C23H14F1N1	323.37	323.1110	85.43	4.36	5.88	4.33
C22H42F2N4	400.60	400.3377	65.96	10.57	9.48	13.99	C23H14F1N3	351.39	351.1172	78.62	4.02	5.41	11.96
C22H42F2N6	428.62	428.3439	61.65	9.88	8.86	19.61	C23H14F1N5	379.40	379.1233	72.81	3.72	5.01	18.46
C22H43F1N2	354.60	354.3410	74.52	12.22	5.36	7.90	C23H14F2N2	356.38	356.1125	77.52	3.96	10.66	7.86
C22H43F1N4	382.61	382.3472	69.06	11.33	4.97	14.64	C23H14F2N4	384.39	384.1186	71.87	3.67	9.88	14.58
C22H43F1N6	410.63	410.3533	64.35	10.56	4.63	20.47	C23H14F2N6	412.41	412.1248	66.99	3.42	9.21	20.38
C22H43F2N1	359.59	359.3363	73.48	12.05	10.57	3.90	C23H15F1N2	338.39	338.1219	81.64	4.47	5.61	8.28
C22H43F2N3	387.60	387.3425	68.17	11.18	9.80	10.84	C23H15F1N4	366.40	366.1281	75.40	4.13	5.19	15.29
C22H43F2N5	415.62	415.3486	63.58	10.43	9.14	16.85	C23H15F1N6	394.41	394.1342	70.04	3.83	4.82	21.31
C22H44F1N1	341.60	341.3458	77.35	12.98	5.56	4.10	C23H15F2N1	343.38	343.1172	80.45	4.40	11.07	4.08
C22H44F1N3	369.61	369.3519	71.49	12.00	5.14	11.37	C23H15F2N3	371.39	371.1234	74.38	4.07	10.23	11.31
C22H44F1N5	397.63	397.3581	66.46	11.15	4.78	17.61	C23H15F2N5	399.41	399.1295	69.17	3.79	9.51	17.53
C22H44F2N2	374.61	374.3472	70.54	11.84	10.14	7.48	C23H16F1N1	325.39	325.1267	84.90	4.96	5.84	4.30
C22H44F2N4	402.62	402.3534	65.63	11.02	9.44	13.92	C23H16F1N3	353.40	353.1328	78.17	4.56	5.38	11.89
C22H44F2N6	430.63	430.3595	61.36	10.30	8.82	19.52	C23H16F1N5	381.42	381.1390	72.43	4.23	4.98	18.36
C22H45F1N2	356.62	356.3567	74.10	12.72	5.33	7.86	C23H16F2N2	358.39	358.1281	77.08	4.50	10.60	7.82
C22H45F1N4	384.63	384.3628	68.70	11.79	4.94	14.57	C23H16F2N4	386.41	386.1343	71.49	4.17	9.83	14.50
C22H45F1N6	412.64	412.3690	64.04	10.99	4.60	20.37	C23H16F2N6	414.42	414.1404	66.66	3.89	9.17	20.28
C22H45F2N1	361.61	361.3520	73.08	12.54	10.51	3.87	C23H17F1N2	340.40	340.1376	81.16	5.03	5.58	8.23
C22H45F2N3	389.62	389.3581	67.82	11.64	9.75	10.78	C23H17F1N4	368.42	368.1437	74.98	4.65	5.16	15.21
C22H45F2N5	417.63	417.3643	63.27	10.86	9.10	16.77	C23H17F1N6	396.43	396.1499	69.69	4.32	4.79	21.20
C22H46F1N1	343.62	343.3614	76.90	13.49	5.53	4.08	C23H17F2N1	345.40	345.1329	79.98	4.96	11.00	4.06
C22H46F1N3	371.63	371.3676	71.10	12.48	5.11	11.31	C23H17F2N3	373.41	373.1390	73.98	4.59	10.18	11.25
C22H46F1N5	399.64	399.3737	66.12	11.60	4.75	17.52	C23H17F2N5	401.42	401.1452	68.82	4.27	9.47	17.45
C22H46F2N2	376.62	376.3629	70.16	12.31	10.09	7.44	C23H18F1N1	327.41	327.1423	84.38	5.54	5.80	4.28
C22H46F2N4	404.64	404.3690	65.30	11.46	9.39	13.85	C23H18F1N3	355.42	355.1485	77.73	5.10	5.35	11.82
C22H46F2N6	432.65	432.3752	61.08	10.72	8.78	19.42	C23H18F1N5	383.43	383.1546	72.05	4.73	4.95	18.26
C22H47F1N2	358.63	358.3723	73.68	13.21	5.30	7.81	C23H18F2N2	360.41	360.1438	76.65	5.03	10.54	7.77
C22H47F1N4	386.65	386.3785	68.34	12.25	4.91	14.49	C23H18F2N4	388.42	388.1499	71.12	4.67	9.78	14.42
C22H47F1N6	414.66	414.3846	63.73	11.42	4.58	20.27	C23H18F2N6	416.44	416.1561	66.34	4.36	9.12	20.18
C22H47F2N3	391.64	391.3738	67.47	12.10	9.70	10.73	C23H19F1N2	342.42	342.1532	80.68	5.59	5.55	8.18
C22H47F2N5	419.65	419.3799	62.97	11.29	9.05	16.69	C23H19F1N4	370.43	370.1594	74.58	5.17	5.13	15.12
C22H48F1N3	373.65	373.3832	70.72	12.95	5.08	11.25	C23H19F1N6	398.45	398.1655	69.33	4.81	4.77	21.09
C22H48F1N5	401.66	401.3894	65.79	12.05	4.73	17.44	C23H19F2N1	347.41	347.1485	79.52	5.51	10.94	4.03
C22H48F2N4	406.65	406.3847	64.98	11.90	9.34	13.78	C23H19F2N3	375.42	375.1547	73.59	5.10	10.12	11.19
C22H48F2N6	434.66	434.3908	60.79	11.13	8.74	19.33	C23H19F2N5	403.44	403.1608	68.48	4.75	9.42	17.36
C22H49F1N4	388.66	388.3941	67.99	12.71	4.89	14.42	C23H20F1N1	329.42	329.1580	83.86	6.12	5.77	4.25
C22H49F1N6	416.67	416.4003	63.42	11.85	4.56	20.17	C23H20F1N3	357.43	357.1641	77.29	5.64	5.32	11.76
C22H49F2N5	421.67	421.3956	62.67	11.71	9.01	16.61	C23H20F1N5	385.45	385.1703	71.67	5.23	4.93	18.17
C22H50F1N5	403.68	403.4050	65.46	12.48	4.71	17.35	C23H20F2N2	362.43	362.1594	76.22	5.56	10.48	7.73
C22H50F2N6	436.68	436.4065	60.51	11.54	8.70	19.25	C23H20F2N4	390.44	390.1656	70.76	5.16	9.73	14.35
C22H51F1N6	418.69	418.4159	63.11	12.28	4.54	20.07	C23H20F2N6	418.45	418.1717	66.02	4.82	9.08	20.08
C23H11F1N2	334.36	334.0906	82.62	3.32	5.68	8.38	C23H21F1N2	344.44	344.1689	80.21	6.15	5.52	8.13
C23H11F1N4	362.37	362.0968	76.24	3.06	5.24	15.46	C23H21F1N4	372.45	372.1750	74.17	5.68	5.10	15.04
C23H11F1N6	390.38	390.1029	70.77	2.84	4.87	21.53	C23H21F1N6	400.46	400.1812	68.98	5.29	4.74	20.99
C23H11F2N1	339.35	339.0859	81.41	3.27	11.20	4.13	C23H21F2N1	349.43	349.1642	79.06	6.06	10.87	4.01

C23H21F2N3	377.44	377.1703	73.19	5.61	10.07	11.13
C23H21F2N5	405.45	405.1765	68.14	5.22	9.37	17.27
C23H22F1N1	331.44	331.1736	83.35	6.69	5.73	4.23
C23H22F1N3	359.45	359.1798	76.86	6.17	5.29	11.69
C23H22F1N5	387.46	387.1859	71.30	5.72	4.90	18.07
C23H22F2N2	364.44	364.1751	75.80	6.08	10.43	7.69
C23H22F2N4	392.46	392.1812	70.39	5.65	9.68	14.28
C23H22F2N6	420.47	420.1874	65.70	5.27	9.04	19.99
C23H23F1N2	346.45	346.1845	79.74	6.69	5.48	8.09
C23H23F1N4	374.46	374.1907	73.77	6.19	5.07	14.96
C23H23F1N6	402.48	402.1968	68.64	5.76	4.72	20.88
C23H23F2N1	351.44	351.1798	78.61	6.60	10.81	3.99
C23H23F2N3	379.46	379.1860	72.80	6.11	10.01	11.07
C23H23F2N5	407.47	407.1921	67.80	5.69	9.33	17.19
C23H24F1N1	333.45	333.1893	82.85	7.25	5.70	4.20
C23H24F1N3	361.47	361.1954	76.43	6.69	5.26	11.62
C23H24F1N5	389.48	389.2016	70.93	6.21	4.88	17.98
C23H24F2N2	366.46	366.1907	75.39	6.60	10.37	7.64
C23H24F2N4	394.47	394.1969	70.03	6.13	9.63	14.20
C23H24F2N6	422.48	422.2030	65.39	5.73	8.99	19.89
C23H25F1N2	348.47	348.2002	79.28	7.23	5.45	8.04
C23H25F1N4	376.48	376.2063	73.38	6.69	5.05	14.88
C23H25F1N6	404.49	404.2125	68.30	6.23	4.70	20.78
C23H25F2N1	353.46	353.1955	78.16	7.13	10.75	3.96
C23H25F2N3	381.47	381.2016	72.42	6.61	9.96	11.02
C23H25F2N5	409.49	409.2078	67.46	6.15	9.28	17.10
C23H26F1N1	335.47	335.2049	82.35	7.81	5.66	4.18
C23H26F1N3	363.48	363.2111	76.00	7.21	5.23	11.56
C23H26F1N5	391.50	391.2172	70.56	6.69	4.85	17.89
C23H26F2N2	368.47	368.2064	74.97	7.11	10.31	7.60
C23H26F2N4	396.49	396.2125	69.68	6.61	9.58	14.13
C23H26F2N6	424.50	424.2187	65.08	6.17	8.95	19.80
C23H27F1N2	350.48	350.2158	78.82	7.77	5.42	7.99
C23H27F1N4	378.50	378.2220	72.99	7.19	5.02	14.80
C23H27F1N6	406.51	406.2281	67.96	6.69	4.67	20.67
C23H27F2N1	355.48	355.2111	77.71	7.66	10.69	3.94
C23H27F2N3	383.49	383.2173	72.04	7.10	9.91	10.96
C23H27F2N5	411.50	411.2234	67.13	6.61	9.23	17.02
C23H28F1N1	337.48	337.2206	81.86	8.36	5.63	4.15
C23H28F1N3	365.50	365.2267	75.58	7.72	5.20	11.50
C23H28F1N5	393.51	393.2329	70.20	7.17	4.83	17.80
C23H28F2N2	370.49	370.2220	74.57	7.62	10.26	7.56
C23H28F2N4	398.50	398.2282	69.32	7.08	9.53	14.06
C23H28F2N6	426.52	426.2343	64.77	6.62	8.91	19.70
C23H29F1N2	352.50	352.2315	78.37	8.29	5.39	7.95
C23H29F1N4	380.51	380.2376	72.60	7.68	4.99	14.72
C23H29F1N6	408.53	408.2438	67.62	7.16	4.65	20.57
C23H29F2N1	357.49	357.2268	77.28	8.18	10.63	3.92
C23H29F2N3	385.50	385.2329	71.66	7.58	9.86	10.90
C23H29F2N5	413.52	413.2391	66.81	7.07	9.19	16.94
C23H30F1N1	339.50	339.2362	81.37	8.91	5.60	4.13
C23H30F1N3	367.51	367.2424	75.17	8.23	5.17	11.43
C23H30F1N5	395.53	395.2485	69.85	7.65	4.80	17.71
C23H30F2N2	372.51	372.2377	74.16	8.12	10.20	7.52
C23H30F2N4	400.52	400.2438	68.97	7.55	9.49	13.99
C23H30F2N6	428.53	428.2500	64.47	7.06	8.87	19.61
C23H31F1N2	354.52	354.2471	77.93	8.81	5.36	7.90
C23H31F1N4	382.53	382.2533	72.22	8.17	4.97	14.65
C23H31F1N6	410.54	410.2594	67.29	7.61	4.63	20.47
C23H31F2N1	359.51	359.2424	76.84	8.69	10.57	3.90

C23H31F2N3	387.52	387.2486	71.29	8.06	9.81	10.84
C23H31F2N5	415.53	415.2547	66.48	7.52	9.14	16.85
C23H32F1N1	341.52	341.2519	80.89	9.44	5.56	4.10
C23H32F1N3	369.53	369.2580	74.76	8.73	5.14	11.37
C23H32F1N5	397.54	397.2642	69.49	8.11	4.78	17.62
C23H32F2N2	374.52	374.2533	73.76	8.61	10.15	7.48
C23H32F2N4	402.54	402.2595	68.63	8.01	9.44	13.92
C23H32F2N6	430.55	430.2656	64.16	7.49	8.83	19.52
C23H33F1N2	356.53	356.2628	77.48	9.33	5.33	7.86
C23H33F1N4	384.54	384.2689	71.84	8.65	4.94	14.57
C23H33F1N6	412.56	412.2751	66.96	8.06	4.61	20.37
C23H33F2N1	361.52	361.2581	76.41	9.20	10.51	3.87
C23H33F2N3	389.54	389.2642	70.92	8.54	9.75	10.79
C23H33F2N5	417.55	417.2704	66.16	7.97	9.10	16.77
C23H34F1N1	343.53	343.2675	80.42	9.98	5.53	4.08
C23H34F1N3	371.55	371.2737	74.35	9.22	5.11	11.31
C23H34F1N5	399.56	399.2798	69.14	8.58	4.75	17.53
C23H34F2N2	376.54	376.2690	73.37	9.10	10.09	7.44
C23H34F2N4	404.55	404.2751	68.29	8.47	9.39	13.85
C23H34F2N6	432.56	432.2813	63.86	7.92	8.78	19.43
C23H35F1N2	358.55	358.2784	77.05	9.84	5.30	7.81
C23H35F1N4	386.56	386.2846	71.47	9.13	4.91	14.49
C23H35F1N6	414.57	414.2907	66.64	8.51	4.58	20.27
C23H35F2N1	363.54	363.2737	75.99	9.70	10.45	3.85
C23H35F2N3	391.55	391.2799	70.55	9.01	9.70	10.73
C23H35F2N5	419.57	419.2860	65.84	8.41	9.06	16.69
C23H36F1N1	345.55	345.2832	79.95	10.50	5.50	4.05
C23H36F1N3	373.56	373.2893	73.95	9.71	5.09	11.25
C23H36F1N5	401.58	401.2955	68.79	9.04	4.73	17.44
C23H36F2N2	378.55	378.2846	72.98	9.59	10.04	7.40
C23H36F2N4	406.57	406.2908	67.95	8.93	9.35	13.78
C23H36F2N6	434.58	434.2969	63.57	8.35	8.74	19.34
C23H37F1N2	360.56	360.2941	76.62	10.34	5.27	7.77
C23H37F1N4	388.58	388.3002	71.09	9.60	4.89	14.42
C23H37F1N6	416.59	416.3064	66.31	8.95	4.56	20.17
C23H37F2N1	365.55	365.2894	75.57	10.20	10.39	3.83
C23H37F2N3	393.57	393.2955	70.19	9.48	9.65	10.68
C23H37F2N5	421.58	421.3017	65.53	8.85	9.01	16.61
C23H38F1N1	347.56	347.2988	79.48	11.02	5.47	4.03
C23H38F1N3	375.58	375.3050	73.56	10.20	5.06	11.19
C23H38F1N5	403.59	403.3111	68.45	9.49	4.71	17.35
C23H38F2N2	380.57	380.3003	72.59	10.06	9.98	7.36
C23H38F2N4	408.58	408.3064	67.61	9.37	9.30	13.71
C23H38F2N6	436.60	436.3126	63.28	8.77	8.70	19.25
C23H39F1N2	362.58	362.3097	76.19	10.84	5.24	7.73
C23H39F1N4	390.59	390.3159	70.73	10.06	4.86	14.34
C23H39F1N6	418.61	418.3220	65.99	9.39	4.54	20.08
C23H39F2N1	367.57	367.3050	75.16	10.69	10.34	3.81
C23H39F2N3	395.58	395.3112	69.84	9.94	9.61	10.62
C23H39F2N5	423.60	423.3173	65.22	9.28	8.97	16.53
C23H40F1N1	349.58	349.3145	79.03	11.53	5.43	4.01
C23H40F1N3	377.59	377.3206	73.16	10.68	5.03	11.13
C23H40F1N5	405.61	405.3268	68.11	9.94	4.68	17.27
C23H40F2N2	382.59	382.3159	72.21	10.54	9.93	7.32
C23H40F2N4	410.60	410.3221	67.28	9.82	9.25	13.65
C23H40F2N6	438.61	438.3282	62.98	9.19	8.66	19.16
C23H41F1N2	364.60	364.3254	75.77	11.33	5.21	7.68
C23H41F1N4	392.61	392.3315	70.36	10.53	4.84	14.27
C23H41F1N6	420.62	420.3377	65.68	9.83	4.52	19.98
C23H41F2N1	369.59	369.3207	74.75	11.18	10.28	3.79

C23H41F2N3	397.60	397.3268	69.48	10.39	9.56	10.57
C23H41F2N5	425.61	425.3330	64.91	9.71	8.93	16.45
C23H42F1N1	351.60	351.3301	78.57	12.04	5.40	3.98
C23H42F1N3	379.61	379.3363	72.77	11.15	5.00	11.07
C23H42F1N5	407.62	407.3424	67.77	10.39	4.66	17.18
C23H42F2N2	384.60	384.3316	71.83	11.01	9.88	7.28
C23H42F2N4	412.61	412.3377	66.95	10.26	9.21	13.58
C23H42F2N6	440.63	440.3439	62.70	9.61	8.62	19.07
C23H43F1N2	366.61	366.3410	75.35	11.82	5.18	7.64
C23H43F1N4	394.62	394.3472	70.00	10.98	4.81	14.20
C23H43F1N6	422.64	422.3533	65.36	10.26	4.50	19.88
C23H43F2N1	371.60	371.3363	74.34	11.66	10.23	3.77
C23H43F2N3	399.62	399.3425	69.13	10.85	9.51	10.52
C23H43F2N5	427.63	427.3486	64.60	10.14	8.89	16.38
C23H44F1N1	353.61	353.3458	78.12	12.54	5.37	3.96
C23H44F1N3	381.63	381.3519	72.39	11.62	4.98	11.01
C23H44F1N5	409.64	409.3581	67.44	10.83	4.64	17.10
C23H44F2N2	386.62	386.3472	71.45	11.47	9.83	7.25
C23H44F2N4	414.63	414.3534	66.63	10.70	9.16	13.51
C23H44F2N6	442.64	442.3595	62.41	10.02	8.58	18.99
C23H45F1N2	368.63	368.3567	74.94	12.30	5.15	7.60
C23H45F1N4	396.64	396.3628	69.65	11.44	4.79	14.13
C23H45F1N6	424.65	424.3690	65.05	10.68	4.47	19.79
C23H45F2N1	373.62	373.3520	73.94	12.14	10.17	3.75
C23H45F2N3	401.63	401.3581	68.78	11.29	9.46	10.46
C23H45F2N5	429.65	429.3643	64.30	10.56	8.84	16.30
C23H46F1N1	355.63	355.3614	77.68	13.04	5.34	3.94
C23H46F1N3	383.64	383.3676	72.01	12.09	4.95	10.95
C23H46F1N5	411.65	411.3737	67.11	11.26	4.62	17.01
C23H46F2N2	388.63	388.3629	71.08	11.93	9.78	7.21
C23H46F2N4	416.65	416.3690	66.30	11.13	9.12	13.45
C23H46F2N6	444.66	444.3752	62.13	10.43	8.55	18.90
C23H47F1N2	370.64	370.3723	74.53	12.78	5.13	7.56
C23H47F1N4	398.66	398.3785	69.30	11.88	4.77	14.05
C23H47F1N6	426.67	426.3846	64.75	11.10	4.45	19.70
C23H47F2N1	375.63	375.3676	73.54	12.61	10.12	3.73
C23H47F2N3	403.65	403.3738	68.44	11.74	9.41	10.41
C23H47F2N5	431.66	431.3799	64.00	10.97	8.80	16.22
C23H48F1N1	357.64	357.3771	77.24	13.53	5.31	3.92
C23H48F1N3	385.66	385.3832	71.63	12.55	4.93	10.90
C23H48F1N5	413.67	413.3894	66.78	11.70	4.59	16.93
C23H48F2N2	390.65	390.3785	70.72	12.39	9.73	7.17
C23H48F2N4	418.66	418.3847	65.99	11.56	9.08	13.38
C23H48F2N6	446.68	446.3908	61.85	10.83	8.51	18.81
C23H49F1N2	372.66	372.3880	74.13	13.25	5.10	7.52
C23H49F1N4	400.67	400.3941	68.95	12.33	4.74	13.98
C23H49F1N6	428.69	428.4003	64.44	11.52	4.43	19.60
C23H49F2N3	405.66	405.3894	68.10	12.18	9.37	10.36
C23H49F2N5	433.68	433.3956	63.70	11.39	8.76	16.15
C23H50F1N3	387.67	387.3989	71.26	13.00	4.90	10.84
C23H50F1N5	415.69	415.4050	66.46	12.12	4.57	16.85
C23H50F2N4	420.68	420.4003	65.67	11.98	9.03	13.32
C23H50F2N6	448.69	448.4065	61.57	11.23	8.47	18.73
C23H51F1N4	402.69	402.4098	68.60	12.77	4.72	13.91
C23H51F1N6	430.70	430.4159	64.14	11.94	4.41	19.51
C23H51F2N5	435.69	435.4112	63.41	11.80	8.72	16.07
C23H52F1N5	417.70	417.4207	66.14	12.55	4.55	16.77
C23H52F2N6	450.71	450.4221	61.29	11.63	8.43	18.65
C23H53F1N6	432.72	432.4315	63.84	12.35	4.39	19.42
C24H12F1N1	333.37	333.0954	86.47	3.63	5.70	4.20

C24H12F1N3	361.38	361.1015	79.77	3.35	5.26	11.63
C24H12F1N5	389.40	389.1077	74.03	3.11	4.88	17.99
C24H12F2N2	366.37	366.0968	78.68	3.30	10.37	7.65
C24H12F2N4	394.39	394.1030	73.09	3.07	9.63	14.21
C24H12F2N6	422.40	422.1091	68.25	2.86	9.00	19.90
C24H13F1N2	348.38	348.1063	82.74	3.76	5.45	8.04
C24H13F1N4	376.40	376.1124	76.59	3.48	5.05	14.89
C24H13F1N6	404.41	404.1186	71.28	3.24	4.70	20.78
C24H13F2N1	353.37	353.1016	81.58	3.71	10.75	3.96
C24H13F2N3	381.39	381.1077	75.58	3.44	9.96	11.02
C24H13F2N5	409.40	409.1139	70.41	3.20	9.28	17.11
C24H14F1N1	335.38	335.1110	85.95	4.21	5.66	4.18
C24H14F1N3	363.40	363.1172	79.33	3.88	5.23	11.56
C24H14F1N5	391:41	391.1233	73.65	3.61	4.85	17.89
C24H14F2N2	368.39	368.1125	78.25	3.83	10.31	7.60
C24H14F2N4	396.40	396.1186	72.72	3.56	9.59	14.13
C24H14F2N6	424.42	424.1248	67.92	3.32	8.95	19.80
C24H15F1N2	350.40	350.1219	82.27	4.31	5.42	7.99
C24H15F1N4	378.41	378.1281	76.18	4.00	5.02	14.81
C24H15F1N6	406.43	406.1342	70.93	3.72	4.67	20.68
C24H15F2N1	355.39	355.1172	81.11	4.25	10.69	3.94
C24H15F2N3	383.40	383.1234	75.19	3.94	9.91	10.96
C24H15F2N5	411.42	411.1295	70.07	3.67	9.24	17.02
C24H16F1N1	337.40	337.1267	85.44	4.78	5.63	4.15
C24H16F1N3	365.41	365.1328	78.89	4.41	5.20	11.50
C24H16F1N5	393.43	393.1390	73.27	4.10	4.83	17.80
C24H16F2N2	370.41	370.1281	77.82	4.35	10.26	7.56
C24H16F2N4	398.42	398.1343	72.35	4.05	9.54	14.06
C24H16F2N6	426.43	426.1404	67.60	3.78	8.91	19.71
C24H17F1N2	352.41	352.1376	81.80	4.86	5.39	7.95
C24H17F1N4	380.43	380.1437	75.77	4.50	4.99	14.73
C24H17F1N6	408.44	408.1499	70.58	4.20	4.65	20.58
C24H17F2N1	357.41	357.1329	80.66	4.79	10.63	3.92
C24H17F2N3	385.42	385.1390	74.79	4.45	9.86	10.90
C24H17F2N5	413.43	413.1452	69.73	4.14	9.19	16.94
C24H18F1N1	339.42	339.1423	84.93	5.35	5.60	4.13
C24H18F1N3	367.43	367.1485	78.46	4.94	5.17	11.44
C24H18F1N5	395.44	395.1546	72.90	4.59	4.80	17.71
C24H18F2N2	372.42	372.1438	77.40	4.87	10.20	7.52
C24H18F2N4	400.43	400.1499	71.99	4.53	9.49	13.99
C24H18F2N6	428.45	428.1561	67.28	4.23	8.87	19.62
C24H19F1N2	354.43	354.1532	81.33	5.40	5.36	7.90
C24H19F1N4	382.44	382.1594	75.38	5.01	4.97	14.65
C24H19F1N6	410.46	410.1655	70.23	4.67	4.63	20.47
C24H19F2N1	359.42	359.1485	80.20	5.33	10.57	3.90
C24H19F2N3	387.44	387.1547	74.40	4.94	9.81	10.85
C24H19F2N5	415.45	415.1608	69.39	4.61	9.15	16.86
C24H20F1N1	341.43	341.1580	84.43	5.90	5.56	4.10
C24H20F1N3	369.45	369.1641	78.03	5.46	5.14	11.37
C24H20F1N5	397.46	397.1703	72.53	5.07	4.78	17.62
C24H20F2N2	374.44	374.1594	76.99	5.38	10.15	7.48
C24H20F2N4	402.45	402.1656	71.63	5.01	9.44	13.92
C24H20F2N6	430.46	430.1717	66.97	4.68	8.83	19.52
C24H21F1N2	356.45	356.1689	80.87	5.94	5.33	7.86
C24H21F1N4	384.46	384.1750	74.98	5.51	4.94	14.57
C24H21F1N6	412.47	412.1812	69.89	5.13	4.61	20.37
C24H21F2N1	361.44	361.1642	79.76	5.86	10.51	3.88
C24H21F2N3	389.45	389.1703	74.02	5.44	9.76	10.79
C24H21F2N5	417.47	417.1765	69.05	5.07	9.10	16.78
C24H22F1N1	343.45	343.1736	83.93	6.46	5.53	4.08

C24H22F1N3	371.46	371.1798	77.60	5.97	5.11	11.31
C24H22F1N5	399.47	399.1859	72.16	5.55	4.76	17.53
C24H22F2N2	376.45	376.1751	76.57	5.89	10.09	7.44
C24H22F2N4	404.47	404.1812	71.27	5.48	9.39	13.85
C24H22F2N6	432.48	432.1874	66.65	5.13	8.79	19.43
C24H23F1N2	358.46	358.1845	80.42	6.47	5.30	7.81
C24H23F1N4	386.48	386.1907	74.59	6.00	4.92	14.50
C24H23F1N6	414.49	414.1968	69.55	5.59	4.58	20.28
C24H23F2N1	363.45	363.1798	79.31	6.38	10.45	3.85
C24H23F2N3	391.47	391.1860	73.64	5.92	9.71	10.73
C24H23F2N5	419.48	419.1921	68.72	5.53	9.06	16.70
C24H24F1N1	345.46	345.1893	83.44	7.00	5.50	4.05
C24H24F1N3	373.48	373.1954	77.18	6.48	5.09	11.25
C24H24F1N5	401.49	401.2016	71.80	6.03	4.73	17.44
C24H24F2N2	378.47	378.1907	76.17	6.39	10.04	7.40
C24H24F2N4	406.48	406.1969	70.92	5.95	9.35	13.78
C24H24F2N6	434.50	434.2030	66.35	5.57	8.75	19.34
C24H25F1N2	360.48	360.2002	79.97	6.99	5.27	7.77
C24H25F1N4	388.49	388.2063	74.20	6.49	4.89	14.42
C24H25F1N6	416.51	416.2125	69.21	6.05	4.56	20.18
C24H25F2N1	365.47	365.1955	78.88	6.90	10.40	3.83
C24H25F2N3	393.48	393.2016	73.26	6.40	9.66	10.68
C24H25F2N5	421.50	421.2078	68.39	5.98	9.01	16.62
C24H26F1N1	347.48	347.2049	82.96	7.54	5.47	4.03
C24H26F1N3	375.49	375.2111	76.77	6.98	5.06	11.19
C24H26F1N5	403.51	403.2172	71.44	6.49	4.71	17.36
C24H26F2N2	380.49	380.2064	75.76	6.89	9.99	7.36
C24H26F2N4	408.50	408.2125	70.57	6.42	9.30	13.72
C24H26F2N6	436.51	436.2187	66.04	6.00	8.70	19.25
C24H27F1N2	362.49	362.2158	79.52	7.51	5.24	7.73
C24H27F1N4	390.51	390.2220	73.82	6.97	4.87	14.35
C24H27F1N6	418.52	418.2281	68.88	6.50	4.54	20.08
C24H27F2N1	367.49	367.2111	78.44	7.41	10.34	3.81
C24H27F2N3	395.50	395.2173	72.89	6.88	9.61	10.62
C24H27F2N5	423.51	423.2234	68.07	6.43	8.97	16.54
C24H28F1N1	349.50	349.2206	82.48	8.08	5.44	4.01
C24H28F1N3	377.51	377.2267	76.36	7.48	5.03	11.13
C24H28F1N5	405.52	405.2329	71.09	6.96	4.68	17.27
C24H28F2N2	382.50	382.2220	75.36	7.38	9.93	7.32
C24H28F2N4	410.51	410.2282	70.22	6.88	9.26	13.65
C24H28F2N6	438.53	438.2343	65.74	6.44	8.66	19.16
C24H29F1N2	364.51	364.2315	79.08	8.02	5.21	7.69
C24H29F1N4	392.52	392.2376	73.44	7.45	4.84	14.27
C24H29F1N6	420.54	420.2438	68.55	6.95	4.52	19.98
C24H29F2N1	369.50	369.2268	78.02	7.91	10.28	3.79
C24H29F2N3	397.52	397.2329	72.52	7.35	9.56	10.57
C24H29F2N5	425.53	425.2391	67.74	6.87	8.93	16.46
C24H30F1N1	351.51	351.2362	82.01	8.60	5.40	3.98
C24H30F1N3	379.53	379.2424	75.95	7.97	5.01	11.07
C24H30F1N5	407.54	407.2485	70.73	7.42	4.66	17.18
C24H30F2N2	384.52	384.2377	74.97	7.86	9.88	7.29
C24H30F2N4	412.53	412.2438	69.88	7.33	9.21	13.58
C24H30F2N6	440.54	440.2500	65.43	6.86	8.62	19.08
C24H31F1N2	366.53	366.2471	78.65	8.53	5.18	7.64
C24H31F1N4	394.54	394.2533	73.06	7.92	4.82	14.20
C24H31F1N6	422.55	422.2594	68.22	7.39	4.50	19.89
C24H31F2N1	371.52	371.2424	77.59	8.41	10.23	3.77
C24H31F2N3	399.53	399.2486	72.15	7.82	9.51	10.52
C24H31F2N5	427.54	427.2547	67.42	7.31	8.89	16.38
C24H32F1N1	353.53	353.2519	81.54	9.12	5.37	3.96
C24H32F1N3	381.54	381.2580	75.55	8.45	4.98	11.01
C24H32F1N5	409.55	409.2642	70.39	7.88	4.64	17.10
C24H32F2N2	386.53	386.2533	74.58	8.34	9.83	7.25
C24H32F2N4	414.55	414.2595	69.54	7.78	9.17	13.52
C24H32F2N6	442.56	442.2656	65.14	7.29	8.59	18.99
C24H33F1N2	368.54	368.2628	78.22	9.03	5.16	7.60
C24H33F1N4	396.56	396.2689	72.69	8.39	4.79	14.13
C24H33F1N6	424.57	424.2751	67.90	7.83	4.47	19.79
C24H33F2N1	373.53	373.2581	77.17	8.90	10.17	3.75
C24H33F2N3	401.55	401.2642	71.79	8.28	9.46	10.46
C24H33F2N5	429.56	429.2704	67.11	7.74	8.85	16.30
C24H34F1N1	355.54	355.2675	81.08	9.64	5.34	3.94
C24H34F1N3	383.56	383.2737	75.16	8.94	4.95	10.96
C24H34F1N5	411.57	411.2798	70.04	8.33	4.62	17.02
C24H34F2N2	388.55	388.2690	74.19	8.82	9.78	7.21
C24H34F2N4	416.56	416.2751	69.20	8.23	9.12	13.45
C24H34F2N6	444.58	444.2813	64.84	7.71	8.55	18.90
C24H35F1N2	370.56	370.2784	77.79	9.52	5.13	7.56
C24H35F1N4	398.57	398.2846	72.33	8.85	4.77	14.06
C24H35F1N6	426.59	426.2907	67.58	8.27	4.45	19.70
C24H35F2N1	375.55	375.2737	76.76	9.39	10.12	3.73
C24H35F2N3	403.56	403.2799	71.43	8.74	9.42	10.41
C24H35F2N5	431.58	431.2860	66.79	8.17	8.80	16.23
C24H36F1N1	357.56	357.2832	80.62	10.15	5.31	3.92
C24H36F1N3	385.57	385.2893	74.76	9.41	4.93	10.90
C24H36F1N5	413.59	413.2955	69.70	8.77	4.59	16.93
C24H36F2N2	390.56	390.2846	73.81	9.29	9.73	7.17
C24H36F2N4	418.58	418.2908	68.87	8.67	9.08	13.39
C24H36F2N6	446.59	446.2969	64.55	8.13	8.51	18.82
C24H37F1N2	372.57	372.2941	77.37	10.01	5.10	7.52
C24H37F1N4	400.59	400.3002	71.96	9.31	4.74	13.99
C24H37F1N6	428.60	428.3064	67.26	8.70	4.43	19.61
C24H37F2N1	377.57	377.2894	76.35	9.88	10.06	3.71
C24H37F2N3	405.58	405.2955	71.08	9.20	9.37	10.36
C24H37F2N5	433.59	433.3017	66.48	8.60	8.76	16.15
C24H38F1N1	359.58	359.2988	80.17	10.65	5.28	3.90
C24H38F1N3	387.59	387.3050	74.37	9.88	4.90	10.84
C24H38F1N5	415.60	415.3111	69.36	9.22	4.57	16.85
C24H38F2N2	392.58	392.3003	73.43	9.76	9.68	7.14
C24H38F2N4	420.59	420.3064	68.54	9.11	9.03	13.32
C24H38F2N6	448.61	448.3126	64.26	8.54	8.47	18.73
C24H39F1N2	374.59	374.3097	76.96	10.49	5.07	7.48
C24H39F1N4	402.60	402.3159	71.60	9.76	4.72	13.92
C24H39F1N6	430.62	430.3220	66.94	9.13	4.41	19.52
C24H39F2N1	379.58	379.3050	75.94	10.36	10.01	3.69
C24H39F2N3	407.60	407.3112	70.72	9.64	9.32	10.31
C24H39F2N5	435.61	435.3173	66.18	9.02	8.72	16.08
C24H40F1N1	361.59	361.3145	79.72	11.15	5.25	3.87
C24H40F1N3	389.60	389.3206	73.99	10.35	4.88	10.79
C24H40F1N5	417.62	417.3268	69.03	9.65	4.55	16.77
C24H40F2N2	394.60	394.3159	73.05	10.22	9.63	7.10
C24H40F2N4	422.61	422.3221	68.21	9.54	8.99	13.26
C24H40F2N6	450.62	450.3282	63.97	8.95	8.43	18.65
C24H41F1N2	376.61	376.3254	76.54	10.97	5.04	7.44
C24H41F1N4	404.62	404.3315	71.24	10.21	4.70	13.85
C24H41F1N6	432.63	432.3377	66.63	9.55	4.39	19.43
C24H41F2N1	381.60	381.3207	75.54	10.83	9.96	3.67
C24H41F2N3	409.61	409.3268	70.38	10.09	9.28	10.26
C24H41F2N5	437.62	437.3330	65.87	9.44	8.68	16.00
C24H42F1N1	363.61	363.3301	79.28	11.64	5.22	3.85

C24H42F1N3	391.62	391.3363	73.61	10.81	4.85	10.73
C24H42F1N5	419.63	419.3424	68.69	10.09	4.53	16.69
C24H42F2N2	396.61	396.3316	72.68	10.67	9.58	7.06
C24H42F2N4	424.63	424.3377	67.89	9.97	8.95	13.19
C24H42F2N6	452.64	452.3439	63.69	9.35	8.39	18.57
C24H43F1N2	378.62	378.3410	76.14	11.45	5.02	7.40
C24H43F1N4	406.64	406.3472	70.89	10.66	4.67	13.78
C24H43F1N6	434.65	434.3533	66.32	9.97	4.37	19.34
C24H43F2N1	383.61	383.3363	75.15	11.30	9.90	3.65
C24H43F2N3	411.63	411.3425	70.03	10.53	9.23	10.21
C24H43F2N5	439.64	439.3486	65.57	9.86	8.64	15.93
C24H44F1N1	365.62	365.3458	78.84	12.13	5.20	3.83
C24H44F1N3	393.64	393.3519	73.23	11.27	4.83	10.67
C24H44F1N5	421.65	421.3581	68.37	10.52	4.51	16.61
C24H44F2N2	398.63	398.3472	72.31	11.13	9.53	7.03
C24H44F2N4	426.64	426.3534	67.57	10.40	8.91	13.13
C24H44F2N6	454.66	454.3595	63.40	9.75	8.36	18.48
C24H45F1N2	380.64	380.3567	75.73	11.92	4.99	7.36
C24H45F1N4	408.65	408.3628	70.54	11.10	4.65	13.71
C24H45F1N6	436.66	436.3690	66.02	10.39	4.35	19.25
C24H45F2N1	385.63	385.3520	74.75	11.76	9.85	3.63
C24H45F2N3	413.64	413.3581	69.69	10.97	9.19	10.16
C24H45F2N5	441.66	441.3643	65.27	10.27	8.60	15.86
C24H46F1N1	367.64	367.3614	78.41	12.61	5.17	3.81
C24H46F1N3	395.65	395.3676	72.86	11.72	4.80	10.62
C24H46F1N5	423.67	423.3737	68.04	10.94	4.48	16.53
C24H46F2N2	400.64	400.3629	71.95	11.57	9.48	6.99
C24H46F2N4	428.66	428.3690	67.25	10.82	8.86	13.07
C24H46F2N6	456.67	456.3752	63.12	10.15	8.32	18.40
C24H47F1N2	382.65	382.3723	75.33	12.38	4.96	7.32
C24H47F1N4	410.67	410.3785	70.19	11.54	4.63	13.64
C24H47F1N6	438.68	438.3846	65.71	10.80	4.33	19.16
C24H47F2N1	387.65	387.3676	74.36	12.22	9.80	3.61
C24H47F2N3	415.66	415.3738	69.35	11.40	9.14	10.11
C24H47F2N5	443.67	443.3799	64.97	10.68	8.56	15.78
C24H48F1N1	369.66	369.3771	77.98	13.09	5.14	3.79
C24H48F1N3	397.67	397.3832	72.49	12.17	4.78	10.57
C24H48F1N5	425.68	425.3894	67.72	11.37	4.46	16.45
C24H48F2N2	402.66	402.3785	71.59	12.02	9.44	6.96
C24H48F2N4	430.67	430.3847	66.93	11.23	8.82	13.01
C24H48F2N6	458.69	458.3908	62.85	10.55	8.28	18.32
C24H49F1N2	384.67	384.3880	74.94	12.84	4.94	7.28
C24H49F1N4	412.68	412.3941	69.85	11.97	4.60	13.58
C24H49F1N6	440.70	440.4003	65.41	11.21	4.31	19.07
C24H49F2N1	389.66	389.3833	73.98	12.68	9.75	3.59
C24H49F2N3	417.68	417.3894	69.02	11.83	9.10	10.06
C24H49F2N5	445.69	445.3956	64.68	11.08	8.53	15.71
C24H50F1N1	371.67	371.3927	77.56	13.56	5.11	3.77
C24H50F1N3	399.68	399.3989	72.12	12.61	4.75	10.51
C24H50F1N5	427.70	427.4050	67.40	11.78	4.44	16.37
C24H50F2N2	404.68	404.3942	71.23	12.45	9.39	6.92
C24H50F2N4	432.69	432.4003	66.62	11.65	8.78	12.95
C24H50F2N6	460.70	460.4065	62.57	10.94	8.25	18.24
C24H51F1N2	386.69	386.4036	74.55	13.29	4.91	7.24
C24H51F1N4	414.70	414.4098	69.51	12.40	4.58	13.51
C24H51F1N6	442.71	442.4159	65.11	11.61	4.29	18.98
C24H51F2N3	419.69	419.4051	68.69	12.25	9.05	10.01
C24H51F2N5	447.70	447.4112	64.39	11.48	8.49	15.64
C24H52F1N3	401.70	401.4145	71.76	13.05	4.73	10.46
C24H52F1N5	429.71	429.4207	67.08	12.20	4.42	16.30
C24H52F2N4	434.71	434.4160	66.31	12.06	8.74	12.89
C24H52F2N6	462.72	462.4221	62.30	11.33	8.21	18.16
C24H53F1N4	416.72	416.4254	69.18	12.82	4.56	13.44
C24H53F1N6	444.73	444.4315	64.82	12.01	4.27	18.90
C24H53F2N5	449.72	449.4269	64.10	11.88	8.45	15.57
C24H54F1N5	431.73	431.4363	66.77	12.61	4.40	16.22
C24H54F2N6	464.73	464.4378	62.03	11.71	8.18	18.08
C24H55F1N6	446.74	446.4472	64.53	12.41	4.25	18.81
C25H12F1N1	345.38	345.0954	86.94	3.50	5.50	4.06
C25H12F1N3	373.39	373.1015	80.42	3.24	5.09	11.25
C25H12F1N5	401.41	401.1077	74.81	3.01	4.73	17.45
C25H12F2N2	378.38	378.0968	79.36	3.20	10.04	7.40
C25H12F2N4	406.40	406.1030	73.89	2.98	9.35	13.79
C25H12F2N6	434.41	434.1091	69.12	2.78	8.75	19.35
C25H13F1N2	360.39	360.1063	83.32	3.64	5.27	7.77
C25H13F1N4	388.41	388.1124	77.31	3.37	4.89	14.42
C25H13F1N6	416.42	416.1186	72.11	3.15	4.56	20.18
C25H13F2N1	365.39	365.1016	82.18	3.59	10.40	3.83
C25H13F2N3	393.40	393.1077	76.33	3.33	9.66	10.68
C25H13F2N5	421.41	421.1139	71.26	3.11	9.02	16.62
C25H14F1N1	347.40	347.1110	86.44	4.06	5.47	4.03
C25H14F1N3	375.41	375.1172	79.99	3.76	5.06	11.19
C25H14F1N5	403.42	403.1233	74.43	3.50	4.71	17.36
C25H14F2N2	380.40	380.1125	78.94	3.71	9.99	7.36
C25H14F2N4	408.41	408.1186	73.52	3.46	9.30	13.72
C25H14F2N6	436.43	436.1248	68.80	3.23	8.71	19.26
C25H15F1N2	362.41	362.1219	82.86	4.17	5.24	7.73
C25H15F1N4	390.42	390.1281	76.91	3.87	4.87	14.35
C25H15F1N6	418.44	418.1342	71.76	3.61	4.54	20.08
C25H15F2N1	367.40	367.1172	81.73	4.12	10.34	3.81
C25H15F2N3	395.42	395.1234	75.94	3.82	9.61	10.63
C25H15F2N5	423.43	423.1295	70.92	3.57	8.97	16.54
C25H16F1N1	349.41	349.1267	85.94	4.62	5.44	4.01
C25H16F1N3	377.42	377.1328	79.56	4.27	5.03	11.13
C25H16F1N5	405.44	405.1390	74.06	3.98	4.69	17.27
C25H16F2N2	382.42	382.1281	78.52	4.22	9.94	7.33
C25H16F2N4	410.43	410.1343	73.16	3.93	9.26	13.65
C25H16F2N6	438.44	438.1404	68.49	3.68	8.67	19.17
C25H17F1N2	364.43	364.1376	82.40	4.70	5.21	7.69
C25H17F1N4	392.44	392.1437	76.52	4.37	4.84	14.28
C25H17F1N6	420.45	420.1499	71.42	4.08	4.52	19.99
C25H17F2N1	369.42	369.1329	81.28	4.64	10.29	3.79
C25H17F2N3	397.43	397.1390	75.55	4.31	9.56	10.57
C25H17F2N5	425.44	425.1452	70.58	4.03	8.93	16.46
C25H18F1N1	351.43	351.1423	85.45	5.16	5.41	3.99
C25H18F1N3	379.44	379.1485	79.14	4.78	5.01	11.07
C25H18F1N5	407.45	407.1546	73.70	4.45	4.66	17.19
C25H18F2N2	384.43	384.1438	78.11	4.72	9.88	7.29
C25H18F2N4	412.45	412.1499	72.80	4.40	9.21	13.58
C25H18F2N6	440.46	440.1561	68.17	4.12	8.63	19.08
C25H19F1N2	366.44	366.1532	81.94	5.23	5.18	7.64
C25H19F1N4	394.46	394.1594	76.12	4.86	4.82	14.20
C25H19F1N6	422.47	422.1655	71.08	4.53	4.50	19.89
C25H19F2N1	371.43	371.1485	80.84	5.16	10.23	3.77
C25H19F2N3	399.45	399.1547	75.17	4.79	9.51	10.52
C25H19F2N5	427.46	427.1608	70.25	4.48	8.89	16.38
C25H20F1N1	353.44	353.1580	84.96	5.70	5.38	3.96
C25H20F1N3	381.46	381.1641	78.72	5.28	4.98	11.02
C25H20F1N5	409.47	409.1703	73.33	4.92	4.64	17.10
C25H20F2N2	386.45	386.1594	77.70	5.22	9.83	7.25

C25H20F2N4	414.46	414.1656	72.45	4.86	9.17	13.52
C25H20F2N6	442.48	442.1717	67.86	4.56	8.59	18.99
C25H21F1N2	368.46	368.1689	81.50	5.74	5.16	7.60
C25H21F1N4	396.47	396.1750	75.74	5.34	4.79	14.13
C25H21F1N6	424.48	424.1812	70.74	4.99	4.48	19.80
C25H21F2N1	373.45	373.1642	80.41	5.67	10.17	3.75
C25H21F2N3	401.46	401.1703	74.80	5.27	9.46	10.47
C25H21F2N5	429.48	429.1765	69.92	4.93	8.85	16.31
C25H22F1N1	355.46	355.1736	84.48	6.24	5.34	3.94
C25H22F1N3	383.47	383.1798	78.31	5.78	4.95	10.96
C25H22F1N5	411.49	411.1859	72.97	5.39	4.62	17.02
C25H22F2N2	388.46	388.1751	77.30	5.71	9.78	7.21
C25H22F2N4	416.48	416.1812	72.10	5.32	9.12	13.45
C25H22F2N6	444.49	444.1874	67.56	4.99	8.55	18.91
C25H23F1N2	370.47	370.1845	81.05	6.26	5.13	7.56
C25H23F1N4	398.49	398.1907	75.35	5.82	4.77	14.06
C25H23F1N6	426.50	426.1968	70.41	5.44	4.45	19.70
C25H23F2N1	375.47	375.1798	79.98	6.17	10.12	3.73
C25H23F2N3	403.48	403.1860	74.42	5.75	9.42	10.41
C25H23F2N5	431.49	431.1921	69.59	5.37	8.81	16.23
C25H24F1N1	357.48	357.1893	84.00	6.77	5.31	3.92
C25H24F1N3	385.49	385.1954	77.90	6.28	4.93	10.90
C25H24F1N5	413.50	413.2016	72.62	5.85	4.59	16.94
C25H24F2N2	390.48	390.1907	76.90	6.20	9.73	7.17
C25H24F2N4	418.49	418.1969	71.75	5.78	9.08	13.39
C25H24F2N6	446.51	446.2030	67.25	5.42	8.51	18.82
C25H25F1N2	372.49	372.2002	80.61	6.77	5.10	7.52
C25H25F1N4	400.50	400.2063	74.98	6.29	4.74	13.99
C25H25F1N6	428.52	428.2125	70.07	5.88	4.43	19.61
C25H25F2N1	377.48	377.1955	79.55	6.68	10.07	3.71
C25H25F2N3	405.49	405.2016	74.05	6.21	9.37	10.36
C25H25F2N5	433.51	433.2078	69.27	5.81	8.76	16.16
C25H26F1N1	359.49	359.2049	83.53	7.29	5.28	3.90
C25H26F1N3	387.50	387.2111	77.49	6.76	4.90	10.84
C25H26F1N5	415.52	415.2172	72.27	6.31	4.57	16.85
C25H26F2N2	392.50	392.2064	76.50	6.68	9.68	7.14
C25H26F2N4	420.51	420.2125	71.41	6.23	9.04	13.32
C25H26F2N6	448.52	448.2187	66.95	5.84	8.47	18.74
C25H27F1N2	374.51	374.2158	80.18	7.27	5.07	7.48
C25H27F1N4	402.52	402.2220	74.60	6.76	4.72	13.92
C25H27F1N6	430.53	430.2281	69.75	6.32	4.41	19.52
C25H27F2N1	379.50	379.2111	79.13	7.17	10.01	3.69
C25H27F2N3	407.51	407.2173	73.69	6.68	9.32	10.31
C25H27F2N5	435.52	435.2234	68.95	6.25	8.72	16.08
C25H28F1N1	361.51	361.2206	83.06	7.81	5.26	3.87
C25H28F1N3	389.52	389.2267	77.09	7.25	4.88	10.79
C25H28F1N5	417.53	417.2329	71.92	6.76	4.55	16.77
C25H28F2N2	394.51	394.2220	76.11	7.15	9.63	7.10
C25H28F2N4	422.53	422.2282	71.07	6.68	8.99	13.26
C25H28F2N6	450.54	450.2343	66.65	6.26	8.43	18.65
C25H29F1N2	376.52	376.2315	79.75	7.76	5.05	7.44
C25H29F1N4	404.54	404.2376	74.23	7.23	4.70	13.85
C25H29F1N6	432.55	432.2438	69.42	6.76	4.39	19.43
C25H29F2N1	381.51	381.2268	78.71	7.66	9.96	3.67
C25H29F2N3	409.53	409.2329	73.32	7.14	9.28	10.26
C25H29F2N5	437.54	437.2391	68.63	6.68	8.68	16.01
C25H30F1N1	363.52	363.2362	82.60	8.32	5.23	3.85
C25H30F1N3	391.54	391.2424	76.69	7.72	4.85	10.73
C25H30F1N5	419.55	419.2485	71.57	7.21	4.53	16.69
C25H30F2N2	396.53	396.2377	75.73	7.63	9.58	7.06
C25H30F2N4	424.54	424.2438	70.73	7.12	8.95	13.20
C25H30F2N6	452.55	452.2500	66.35	6.68	8.40	18.57
C25H31F1N2	378.54	378.2471	79.33	8.25	5.02	7.40
C25H31F1N4	406.55	406.2533	73.86	7.69	4.67	13.78
C25H31F1N6	434.56	434.2594	69.10	7.19	4.37	19.34
C25H31F2N1	383.53	383.2424	78.29	8.15	9.91	3.65
C25H31F2N3	411.54	411.2486	72.96	7.59	9.23	10.21
C25H31F2N5	439.56	439.2547	68.31	7.11	8.64	15.93
C25H32F1N1	365.54	365.2519	82.15	8.82	5.20	3.83
C25H32F1N3	393.55	393.2580	76.30	8.20	4.83	10.68
C25H32F1N5	421.57	421.2642	71.23	7.65	4.51	16.61
C25H32F2N2	398.54	398.2533	75.34	8.09	9.53	7.03
C25H32F2N4	426.56	426.2595	70.40	7.56	8.91	13.13
C25H32F2N6	454.57	454.2656	66.06	7.10	8.36	18.49
C25H33F1N2	380.55	380.2628	78.91	8.74	4.99	7.36
C25H33F1N4	408.57	408.2689	73.50	8.14	4.65	13.71
C25H33F1N6	436.58	436.2751	68.78	7.62	4.35	19.25
C25H33F2N1	385.55	385.2581	77.88	8.63	9.86	3.63
C25H33F2N3	413.56	413.2642	72.61	8.04	9.19	10.16
C25H33F2N5	441.57	441.2704	68.00	7.53	8.60	15.86
C25H34F1N1	367.55	367.2675	81.70	9.32	5.17	3.81
C25H34F1N3	395.57	395.2737	75.91	8.66	4.80	10.62
C25H34F1N5	423.58	423.2798	70.89	8.09	4.49	16.53
C25H34F2N2	400.56	400.2690	74.96	8.56	9.49	6.99
C25H34F2N4	428.57	428.2751	70.06	8.00	8.87	13.07
C25H34F2N6	456.59	456.2813	65.77	7.51	8.32	18.41
C25H35F1N2	382.57	382.2784	78.49	9.22	4.97	7.32
C25H35F1N4	410.58	410.2846	73.13	8.59	4.63	13.65
C25H35F1N6	438.60	438.2907	68.46	8.04	4.33	19.16
C25H35F2N1	387.56	387.2737	77.48	9.10	9.80	3.61
C25H35F2N3	415.57	415.2799	72.26	8.49	9.14	10.11
C25H35F2N5	443.59	443.2860	67.69	7.95	8.57	15.79
C25H36F1N1	369.57	369.2832	81.25	9.82	5.14	3.79
C25H36F1N3	397.58	397.2893	75.53	9.13	4.78	10.57
C25H36F1N5	425.60	425.2955	70.55	8.53	4.46	16.46
C25H36F2N2	402.58	402.2846	74.59	9.01	9.44	6.96
C25H36F2N4	430.59	430.2908	69.74	8.43	8.82	13.01
C25H36F2N6	458.60	458.2969	65.48	7.91	8.29	18.33
C25H37F1N2	384.59	384.2941	78.08	9.70	4.94	7.28
C25H37F1N4	412.60	412.3002	72.78	9.04	4.60	13.58
C25H37F1N6	440.61	440.3064	68.15	8.46	4.31	19.07
C25H37F2N1	389.58	389.2894	77.08	9.57	9.75	3.60
C25H37F2N3	417.59	417.2955	71.91	8.93	9.10	10.06
C25H37F2N5	445.60	445.3017	67.39	8.37	8.53	15.72
C25H38F1N1	371.59	371.2988	80.81	10.31	5.11	3.77
C25H38F1N3	399.60	399.3050	75.14	9.59	4.75	10.52
C25H38F1N5	427.61	427.3111	70.22	8.96	4.44	16.38
C25H38F2N2	404.59	404.3003	74.22	9.47	9.39	6.92
C25H38F2N4	432.61	432.3064	69.41	8.85	8.78	12.95
C25H38F2N6	460.62	460.3126	65.19	8.32	8.25	18.25
C25H39F1N2	386.60	386.3097	77.67	10.17	4.91	7.25
C25H39F1N4	414.61	414.3159	72.42	9.48	4.58	13.51
C25H39F1N6	442.63	442.3220	67.84	8.88	4.29	18.99
C25H39F2N1	391.59	391.3050	76.68	10.04	9.70	3.58
C25H39F2N3	419.61	419.3112	71.56	9.37	9.06	10.01
C25H39F2N5	447.62	447.3173	67.08	8.78	8.49	15.65
C25H40F1N1	373.60	373.3145	80.37	10.79	5.09	3.75
C25H40F1N3	401.62	401.3206	74.77	10.04	4.73	10.46
C25H40F1N5	429.63	429.3268	69.89	9.38	4.42	16.30
C25H40F2N2	406.61	406.3159	73.85	9.92	9.34	6.89

C25H40F2N4	434.62	434.3221	69.09	9.28	8.74	12.89
C25H40F2N6	462.63	462.3282	64.91	8.72	8.21	18.17
C25H41F1N2	388.62	388.3254	77.27	10.63	4.89	7.21
C25H41F1N4	416.63	416.3315	72.07	9.92	4.56	13.45
C25H41F1N6	444.64	444.3377	67.53	9.29	4.27	18.90
C25H41F2N1	393.61	393.3207	76.29	10.50	9.65	3.56
C25H41F2N3	421.62	421.3268	71.22	9.80	9.01	9.97
C25H41F2N5	449.64	449.3330	66.78	9.19	8.45	15.58
C25H42F1N1	375.62	375.3301	79.94	11.27	5.06	3.73
C25H42F1N3	403.63	403.3363	74.39	10.49	4.71	10.41
C25H42F1N5	431.65	431.3424	69.57	9.81	4.40	16.22
C25H42F2N2	408.62	408.3316	73.49	10.36	9.30	6.86
C25H42F2N4	436.64	436.3377	68.77	9.70	8.70	12.83
C25H42F2N6	464.65	464.3439	64.62	9.11	8.18	18.09
C25H43F1N2	390.63	390.3410	76.87	11.10	4.86	7.17
C25H43F1N4	418.65	418.3472	71.73	10.35	4.54	13.38
C25H43F1N6	446.66	446.3533	67.23	9.70	4.25	18.82
C25H43F2N1	395.62	395.3363	75.90	10.96	9.60	3.54
C25H43F2N3	423.64	423.3425	70.88	10.23	8.97	9.92
C25H43F2N5	451.65	451.3486	66.48	9.60	8.41	15.51
C25H44F1N1	377.63	377.3458	79.52	11.74	5.03	3.71
C25H44F1N3	405.65	405.3519	74.02	10.93	4.68	10.36
C25H44F1N5	433.66	433.3581	69.24	10.23	4.38	16.15
C25H44F2N2	410.64	410.3472	73.12	10.80	9.25	6.82
C25H44F2N4	438.65	438.3534	68.45	10.11	8.66	12.77
C25H44F2N6	466.67	466.3595	64.35	9.50	8.14	18.01
C25H45F1N2	392.65	392.3567	76.48	11.55	4.84	7.13
C25H45F1N4	420.66	420.3628	71.38	10.78	4.52	13.32
C25H45F1N6	448.68	448.3690	66.93	10.11	4.23	18.73
C25H45F2N1	397.64	397.3520	75.52	11.41	9.56	3.52
C25H45F2N3	425.65	425.3581	70.55	10.66	8.93	9.87
C25H45F2N5	453.67	453.3643	66.19	10.00	8.38	15.44
C25H46F1N1	379.65	379.3614	79.09	12.21	5.00	3.69
C25H46F1N3	407.66	407.3676	73.66	11.37	4.66	10.31
C25H46F1N5	435.68	435.3737	68.92	10.64	4.36	16.07
C25H46F2N2	412.66	412.3629	72.77	11.24	9.21	6.79
C25H46F2N4	440.67	440.3690	68.14	10.52	8.62	12.71
C25H46F2N6	468.68	468.3752	64.07	9.89	8.11	17.93
C25H47F1N2	394.67	394.3723	76.08	12.00	4.81	7.10
C25H47F1N4	422.68	422.3785	71.04	11.21	4.49	13.26
C25H47F1N6	450.69	450.3846	66.63	10.51	4.22	18.65
C25H47F2N1	399.66	399.3676	75.13	11.85	9.51	3.50
C25H47F2N3	427.67	427.3738	70.21	11.08	8.88	9.83
C25H47F2N5	455.68	455.3799	65.90	10.40	8.34	15.37
C25H48F1N1	381.67	381.3771	78.68	12.68	4.98	3.67
C25H48F1N3	409.68	409.3832	73.30	11.81	4.64	10.26
C25H48F1N5	437.69	437.3894	68.60	11.05	4.34	16.00
C25H48F2N2	414.67	414.3785	72.41	11.67	9.16	6.76
C25H48F2N4	442.68	442.3847	67.83	10.93	8.58	12.66
C25H48F2N6	470.70	470.3908	63.79	10.28	8.07	17.85
C25H49F1N2	396.68	396.3880	75.70	12.45	4.79	7.06
C25H49F1N4	424.69	424.3941	70.70	11.63	4.47	13.19
C25H49F1N6	452.71	452.4003	66.33	10.91	4.20	18.56
C25H49F2N1	401.67	401.3833	74.76	12.30	9.46	3.49
C25H49F2N3	429.69	429.3894	69.88	11.49	8.84	9.78
C25H49F2N5	457.70	457.3956	65.61	10.79	8.30	15.30
C25H50F1N1	383.68	383.3927	78.26	13.14	4.95	3.65
C25H50F1N3	411.70	411.3989	72.94	12.24	4.61	10.21
C25H50F1N5	439.71	439.4050	68.29	11.46	4.32	15.93
C25H50F2N2	416.69	416.3942	72.06	12.10	9.12	6.72
C25H50F2N4	444.70	444.4003	67.52	11.33	8.54	12.60
C25H50F2N6	472.71	472.4065	63.52	10.66	8.04	17.78
C25H51F1N2	398.70	398.4036	75.32	12.89	4.77	7.03
C25H51F1N4	426.71	426.4098	70.37	12.05	4.45	13.13
C25H51F1N6	454.72	454.4159	66.04	11.30	4.18	18.48
C25H51F2N1	403.69	403.3989	74.38	12.73	9.41	3.47
C25H51F2N3	431.70	431.4051	69.56	11.91	8.80	9.73
C25H51F2N5	459.72	459.4112	65.32	11.18	8.27	15.23
C25H52F1N1	385.70	385.4084	77.85	13.59	4.93	3.63
C25H52F1N3	413.71	413.4145	72.58	12.67	4.59	10.16
C25H52F1N5	441.73	441.4207	67.98	11.87	4.30	15.85
C25H52F2N2	418.70	418.4098	71.72	12.52	9.07	6.69
C25H52F2N4	446.72	446.4160	67.22	11.73	8.51	12.54
C25H52F2N6	474.73	474.4221	63.25	11.04	8.00	17.70
C25H53F1N2	400.71	400.4193	74.94	13.33	4.74	6.99
C25H53F1N4	428.73	428.4254	70.04	12.46	4.43	13.07
C25H53F1N6	456.74	456.4315	65.74	11.70	4.16	18.40
C25H53F2N3	433.72	433.4207	69.23	12.32	8.76	9.69
C25H53F2N5	461.73	461.4269	65.03	11.57	8.23	15.17
C25H54F1N3	415.73	415.4302	72.23	13.09	4.57	10.11
C25H54F1N5	443.74	443.4363	67.67	12.27	4.28	15.78
C25H54F2N4	448.73	448.4316	66.92	12.13	8.47	12.49
C25H54F2N6	476.75	476.4378	62.99	11.42	7.97	17.63
C25H55F1N4	430.74	430.4411	69.71	12.87	4.41	13.01
C25H55F1N6	458.76	458.4472	65.46	12.08	4.14	18.32
C25H55F2N5	463.75	463.4425	64.75	11.95	8.19	15.10
C25H56F1N5	445.76	445.4519	67.36	12.66	4.26	15.71
C25H56F2N6	478.76	478.4534	62.72	11.79	7.94	17.55
C25H57F1N6	460.77	460.4628	65.17	12.47	4.12	18.24
C26H13F1N2	372.41	372.1063	83.86	3.52	5.10	7.52
C26H13F1N4	400.42	400.1124	77.99	3.27	4.74	13.99
C26H13F1N6	428.43	428.1186	72.89	3.06	4.43	19.62
C26H13F2N1	377.40	377.1016	82.75	3.47	10.07	3.71
C26H13F2N3	405.41	405.1077	77.03	3.23	9.37	10.36
C26H13F2N5	433.42	433.1139	72.05	3.02	8.77	16.16
C26H14F1N1	359.41	359.1110	86.89	3.93	5.29	3.90
C26H14F1N3	387.42	387.1172	80.61	3.64	4.90	10.85
C26H14F1N5	415.43	415.1233	75.17	3.40	4.57	16.86
C26H14F2N2	392.41	392.1125	79.58	3.60	9.68	7.14
C26H14F2N4	420.43	420.1186	74.28	3.36	9.04	13.33
C26H14F2N6	448.44	448.1248	69.64	3.15	8.47	18.74
C26H15F1N2	374.42	374.1219	83.41	4.04	5.07	7.48
C26H15F1N4	402.43	402.1281	77.60	3.76	4.72	13.92
C26H15F1N6	430.45	430.1342	72.55	3.51	4.41	19.52
C26H15F2N1	379.41	379.1172	82.31	3.98	10.01	3.69
C26H15F2N3	407.43	407.1234	76.65	3.71	9.33	10.31
C26H15F2N5	435.44	435.1295	71.72	3.47	8.73	16.08
C26H16F1N1	361.42	361.1267	86.41	4.46	5.26	3.88
C26H16F1N3	389.44	389.1328	80.19	4.14	4.88	10.79
C26H16F1N5	417.45	417.1390	74.81	3.86	4.55	16.78
C26H16F2N2	394.43	394.1281	79.18	4.09	9.63	7.10
C26H16F2N4	422.44	422.1343	73.93	3.82	8.99	13.26
C26H16F2N6	450.45	450.1404	69.33	3.58	8.44	18.66
C26H17F1N2	376.44	376.1376	82.96	4.55	5.05	7.44
C26H17F1N4	404.45	404.1437	77.21	4.24	4.70	13.85
C26H17F1N6	432.46	432.1499	72.21	3.96	4.39	19.43
C26H17F2N1	381.43	381.1329	81.87	4.49	9.96	3.67
C26H17F2N3	409.44	409.1390	76.27	4.19	9.28	10.26
C26H17F2N5	437.45	437.1452	71.39	3.92	8.69	16.01
C26H18F1N1	363.44	363.1423	85.93	4.99	5.23	3.85

C26H18F1N3	391.45	391.1485	79.78	4.63	4.85	10.73
C26H18F1N5	419.47	419.1546	74.45	4.33	4.53	16.70
C26H18F2N2	396.44	396.1438	78.77	4.58	9.58	7.07
C26H18F2N4	424.46	424.1499	73.57	4.27	8.95	13.20
C26H18F2N6	452.47	452.1561	69.02	4.01	8.40	18.57
C26H19F1N2	378.45	378.1532	82.52	5.06	5.02	7.40
C26H19F1N4	406.47	406.1594	76.83	4.71	4.67	13.78
C26H19F1N6	434.48	434.1655	71.88	4.41	4.37	19.34
C26H19F2N1	383.44	383.1485	81.44	4.99	9.91	3.65
C26H19F2N3	411.46	411.1547	75.90	4.65	9.23	10.21
C26H19F2N5	439.47	439.1608	71.06	4.36	8.65	15.94
C26H20F1N1	365.45	365.1580	85.45	5.52	5.20	3.83
C26H20F1N3	393.47	393.1641	79.37	5.12	4.83	10.68
C26H20F1N5	421.48	421.1703	74.09	4.78	4.51	16.62
C26H20F2N2	398.46	398.1594	78.37	5.06	9.54	7.03
C26H20F2N4	426.47	426.1656	73.23	4.73	8.91	13.14
C26H20F2N6	454.49	454.1717	68.71	4.44	8.36	18.49
C26H21F1N2	380.47	380.1689	82.08	5.56	4.99	7.36
C26H21F1N4	408.48	408.1750	76.45	5.18	4.65	13.72
C26H21F1N6	436.50	436.1812	71.54	4.85	4.35	19.25
C26H21F2N1	385.46	385.1642	81.02	5.49	9.86	3.63
C26H21F2N3	413.47	413.1703	75.53	5.12	9.19	10.16
C26H21F2N5	441.49	441.1765	70.74	4.79	8.61	15.86
C26H22F1N1	367.47	367.1736	84.98	6.03	5.17	3.81
C26H22F1N3	395.48	395.1798	78.96	5.61	4.80	10.62
C26H22F1N5	423.50	423.1859	73.74	5.24	4.49	16.54
C26H22F2N2	400.48	400.1751	77.98	5.54	9.49	7.00
C26H22F2N4	428.49	428.1812	72.88	5.18	8.87	13.08
C26H22F2N6	456.50	456.1874	68.41	4.86	8.32	18.41
C26H23F1N2	382.49	382.1845	81.65	6.06	4.97	7.32
C26H23F1N4	410.50	410.1907	76.08	5.65	4.63	13.65
C26H23F1N6	438.51	438.1968	71.22	5.29	4.33	19.16
C26H23F2N1	387.48	387.1798	80.60	5.98	9.81	3.61
C26H23F2N3	415.49	415.1860	75.16	5.58	9.15	10.11
C26H23F2N5	443.50	443.1921	70.41	5.23	8.57	15.79
C26H24F1N1	369.49	369.1893	84.52	6.55	5.14	3.79
C26H24F1N3	397.50	397.1954	78.56	6.09	4.78	10.57
C26H24F1N5	425.51	425.2016	73.39	5.69	4.46	16.46
C26H24F2N2	402.49	402.1907	77.59	6.01	9.44	6.96
C26H24F2N4	430.50	430.1969	72.54	5.62	8.83	13.01
C26H24F2N6	458.52	458.2030	68.11	5.28	8.29	18.33
C26H25F1N2	384.50	384.2002	81.22	6.55	4.94	7.29
C26H25F1N4	412.51	412.2063	75.70	6.11	4.61	13.58
C26H25F1N6	440.53	440.2125	70.89	5.72	4.31	19.08
C26H25F2N1	389.49	389.1955	80.18	6.47	9.76	3.60
C26H25F2N3	417.51	417.2016	74.80	6.04	9.10	10.06
C26H25F2N5	445.52	445.2078	70.10	5.66	8.53	15.72
C26H26F1N1	371.50	371.2049	84.06	7.05	5.11	3.77
C26H26F1N3	399.52	399.2111	78.17	6.56	4.76	10.52
C26H26F1N5	427.53	427.2172	73.05	6.13	4.44	16.38
C26H26F2N2	404.51	404.2064	77.20	6.48	9.39	6.93
C26H26F2N4	432.52	432.2125	72.20	6.06	8.78	12.95
C26H26F2N6	460.53	460.2187	67.81	5.69	8.25	18.25
C26H27F1N2	386.52	386.2158	80.80	7.04	4.92	7.25
C26H27F1N4	414.53	414.2220	75.34	6.57	4.58	13.52
C26H27F1N6	442.54	442.2281	70.57	6.15	4.29	18.99
C26H27F2N1	391.51	391.2111	79.77	6.95	9.71	3.58
C26H27F2N3	419.52	419.2173	74.44	6.49	9.06	10.02
C26H27F2N5	447.54	447.2234	69.78	6.08	8.49	15.65
C26H28F1N1	373.52	373.2206	83.61	7.56	5.09	3.75
C26H28F1N3	401.53	401.2267	77.77	7.03	4.73	10.46
C26H28F1N5	429.54	429.2329	72.70	6.57	4.42	16.30
C26H28F2N2	406.52	406.2220	76.82	6.94	9.35	6.89
C26H28F2N4	434.54	434.2282	71.87	6.50	8.74	12.89
C26H28F2N6	462.55	462.2343	67.51	6.10	8.21	18.17
C26H29F1N2	388.53	388.2315	80.38	7.52	4.89	7.21
C26H29F1N4	416.55	416.2376	74.97	7.02	4.56	13.45
C26H29F1N6	444.56	444.2438	70.25	6.58	4.27	18.90
C26H29F2N1	393.52	393.2268	79.36	7.43	9.66	3.56
C26H29F2N3	421.54	421.2329	74.08	6.93	9.01	9.97
C26H29F2N5	449.55	449.2391	69.47	6.50	8.45	15.58
C26H30F1N1	375.53	375.2362	83.16	8.05	5.06	3.73
C26H30F1N3	403.55	403.2424	77.39	7.49	4.71	10.41
C26H30F1N5	431.56	431.2485	72.36	7.01	4.40	16.23
C26H30F2N2	408.54	408.2377	76.44	7.40	9.30	6.86
C26H30F2N4	436.55	436.2438	71.54	6.93	8.70	12.83
C26H30F2N6	464.57	464.2500	67.22	6.51	8.18	18.09
C26H31F1N2	390.55	390.2471	79.96	8.00	4.86	7.17
C26H31F1N4	418.56	418.2533	74.61	7.47	4.54	13.39
C26H31F1N6	446.58	446.2594	69.93	7.00	4.25	18.82
C26H31F2N1	395.54	395.2424	78.95	7.90	9.61	3.54
C26H31F2N3	423.55	423.2486	73.73	7.38	8.97	9.92
C26H31F2N5	451.57	451.2547	69.16	6.92	8.41	15.51
C26H32F1N1	377.55	377.2519	82.71	8.54	5.03	3.71
C26H32F1N3	405.56	405.2580	77.00	7.95	4.68	10.36
C26H32F1N5	433.58	433.2642	72.03	7.44	4.38	16.15
C26H32F2N2	410.56	410.2533	76.07	7.86	9.25	6.82
C26H32F2N4	438.57	438.2595	71.21	7.35	8.66	12.77
C26H32F2N6	466.58	466.2656	66.93	6.91	8.14	18.01
C26H33F1N2	392.56	392.2628	79.55	8.47	4.84	7.14
C26H33F1N4	420.58	420.2689	74.25	7.91	4.52	13.32
C26H33F1N6	448.59	448.2751	69.62	7.41	4.24	18.73
C26H33F2N1	397.56	397.2581	78.55	8.37	9.56	3.52
C26H33F2N3	425.57	425.2642	73.38	7.82	8.93	9.87
C26H33F2N5	453.58	453.2704	68.85	7.33	8.38	15.44
C26H34F1N1	379.57	379.2675	82.28	9.03	5.01	3.69
C26H34F1N3	407.58	407.2737	76.62	8.41	4.66	10.31
C26H34F1N5	435.59	435.2798	71.69	7.87	4.36	16.08
C26H34F2N2	412.57	412.2690	75.69	8.31	9.21	6.79
C26H34F2N4	440.58	440.2751	70.88	7.78	8.62	12.72
C26H34F2N6	468.60	468.2813	66.64	7.31	8.11	17.93
C26H35F1N2	394.58	394.2784	79.14	8.94	4.81	7.10
C26H35F1N4	422.59	422.2846	73.90	8.35	4.50	13.26
C26H35F1N6	450.61	450.2907	69.30	7.83	4.22	18.65
C26H35F2N1	399.57	399.2737	78.16	8.83	9.51	3.51
C26H35F2N3	427.59	427.2799	73.04	8.25	8.89	9.83
C26H35F2N5	455.60	455.2860	68.54	7.74	8.34	15.37
C26H36F1N1	381.58	381.2832	81.84	9.51	4.98	3.67
C26H36F1N3	409.60	409.2893	76.24	8.86	4.64	10.26
C26H36F1N5	437.61	437.2955	71.36	8.29	4.34	16.00
C26H36F2N2	414.59	414.2846	75.33	8.75	9.16	6.76
C26H36F2N4	442.60	442.2908	70.56	8.20	8.58	12.66
C26H36F2N6	470.61	470.2969	66.36	7.71	8.07	17.86
C26H37F1N2	396.60	396.2941	78.74	9.40	4.79	7.06
C26H37F1N4	424.61	424.3002	73.55	8.78	4.47	13.19
C26H37F1N6	452.62	452.3064	69.00	8.24	4.20	18.57
C26H37F2N1	401.59	401.2894	77.76	9.29	9.46	3.49
C26H37F2N3	429.60	429.2955	72.69	8.68	8.84	9.78
C26H37F2N5	457.62	457.3017	68.24	8.15	8.30	15.30
C26H38F1N1	383.60	383.2988	81.41	9.99	4.95	3.65

10/21/66 229

C26H38F1N3	411.61	411.3050	75.87	9.31	4.62	10.21
C26H38F1N5	439.62	439.3111	71.04	8.71	4.32	15.93
C26H38F2N2	416.60	416.3003	74.96	9.19	9.12	6.72
C26H38F2N4	444.62	444.3064	70.24	8.61	8.55	12.60
C26H38F2N6	472.63	472.3126	66.07	8.10	8.04	17.78
C26H39F1N2	398.61	398.3097	78.34	9.86	4.77	7.03
C26H39F1N4	426.63	426.3159	73.20	9.21	4.45	13.13
C26H39F1N6	454.64	454.3220	68.69	8.65	4.18	18.49
C26H39F2N1	403.60	403.3050	77.38	9.74	9.41	3.47
C26H39F2N3	431.62	431.3112	72.35	9.11	8.80	9.74
C26H39F2N5	459.63	459.3173	67.94	8.55	8.27	15.24
C26H40F1N1	385.61	385.3145	80.99	10.46	4.93	3.63
C26H40F1N3	413.63	413.3206	75.50	9.75	4.59	10.16
C26H40F1N5	441.64	441.3268	70.71	9.13	4.30	15.86
C26H40F2N2	418.62	418.3159	74.60	9.63	9.08	6.69
C26H40F2N4	446.63	446.3221	69.92	9.03	8.51	12.54
C26H40F2N6	474.65	474.3282	65.79	8.49	8.01	17.71
C26H41F1N2	400.63	400.3254	77.95	10.32	4.74	6.99
C26H41F1N4	428.64	428.3315	72.86	9.64	4.43	13.07
C26H41F1N6	456.66	456.3377	68.39	9.05	4.16	18.40
C26H41F2N1	405.62	405.3207	76.99	10.19	9.37	3.45
C26H41F2N3	433.63	433.3268	72.02	9.53	8.76	9.69
C26H41F2N5	461.65	461.3330	67.65	8.95	8.23	15.17
C26H42F1N1	387.63	387.3301	80.56	10.92	4.90	3.61
C26H42F1N3	415.64	415.3363	75.13	10.19	4.57	10.11
C26H42F1N5	443.66	443.3424	70.39	9.54	4.28	15.79
C26H42F2N2	420.63	420.3316	74.24	10.06	9.03	6.66
C26H42F2N4	448.65	448.3377	69.61	9.44	8.47	12.49
C26H42F2N6	476.66	476.3439	65.52	8.88	7.97	17.63
C26H43F1N2	402.64	402.3410	77.56	10.76	4.72	6.96
C26H43F1N4	430.66	430.3472	72.51	10.06	4.41	13.01
C26H43F1N6	458.67	458.3533	68.09	9.45	4.14	18.32
C26H43F2N1	407.64	407.3363	76.61	10.63	9.32	3.44
C26H43F2N3	435.65	435.3425	71.68	9.95	8.72	9.65
C26H43F2N5	463.66	463.3486	67.35	9.35	8.19	15.10
C26H44F1N1	389.65	389.3458	80.15	11.38	4.88	3.59
C26H44F1N3	417.66	417.3519	74.77	10.62	4.55	10.06
C26H44F1N5	445.67	445.3581	70.07	9.95	4.26	15.71
C26H44F2N2	422.65	422.3472	73.89	10.49	8.99	6.63
C26H44F2N4	450.66	450.3534	69.30	9.84	8.43	12.43
C26H44F2N6	478.68	478.3595	65.24	9.27	7.94	17.56
C26H45F1N2	404.66	404.3567	77.17	11.21	4.69	6.92
C26H45F1N4	432.67	432.3628	72.18	10.48	4.39	12.95
C26H45F1N6	460.69	460.3690	67.79	9.85	4.12	18.24
C26H45F2N1	409.65	409.3520	76.23	11.07	9.28	3.42
C26H45F2N3	437.67	437.3581	71.35	10.36	8.68	9.60
C26H45F2N5	465.68	465.3643	67.06	9.74	8.16	15.04
C26H46F1N1	391.66	391.3614	79.73	11.84	4.85	3.58
C26H46F1N3	419.68	419.3676	74.41	11.05	4.53	10.01
C26H46F1N5	447.69	447.3737	69.76	10.36	4.24	15.64
C26H46F2N2	424.67	424.3629	73.54	10.92	8.95	6.60
C26H46F2N4	452.68	452.3690	68.99	10.24	8.39	12.38
C26H46F2N6	480.69	480.3752	64.97	9.65	7.90	17.48
C26H47F1N2	406.68	406.3723	76.79	11.65	4.67	6.89
C26H47F1N4	434.69	434.3785	71.84	10.90	4.37	12.89
C26H47F1N6	462.70	462.3846	67.49	10.24	4.11	18.16
C26H47F2N1	411.67	411.3676	75.86	11.51	9.23	3.40
C26H47F2N3	439.68	439.3738	71.03	10.77	8.64	9.56
C26H47F2N5	467.69	467.3799	66.77	10.13	8.12	14.97
C26H48F1N1	393.68	393.3771	79.33	12.29	4.83	3.56
C26H48F1N3	421.69	421.3832	74.06	11.47	4.51	9.96
C26H48F1N5	449.70	449.3894	69.44	10.76	4.22	15.57
C26H48F2N2	426.68	426.3785	73.19	11.34	8.91	6.57
C26H48F2N4	454.70	454.3847	68.68	10.64	8.36	12.32
C26H48F2N6	482.71	482.3908	64.70	10.02	7.87	17.41
C26H49F1N2	408.69	408.3880	76.41	12.09	4.65	6.85
C26H49F1N4	436.71	436.3941	71.51	11.31	4.35	12.83
C26H49F1N6	464.72	464.4003	67.20	10.63	4.09	18.08
C26H49F2N1	413.68	413.3833	75.49	11.94	9.18	3.39
C26H49F2N3	441.70	441.3894	70.70	11.18	8.60	9.51
C26H49F2N5	469.71	469.3956	66.49	10.52	8.09	14.91
C26H50F1N1	395.69	395.3927	78.92	12.74	4.80	3.54
C26H50F1N3	423.71	423.3989	73.70	11.89	4.48	9.92
C26H50F1N5	451.72	451.4050	69.13	11.16	4.21	15.50
C26H50F2N2	428.70	428.3942	72.85	11.76	8.86	6.53
C26H50F2N4	456.71	456.4003	68.38	11.04	8.32	12.27
C26H50F2N6	484.73	484.4065	64.43	10.40	7.84	17.34
C26H51F1N2	410.71	410.4036	76.04	12.52	4.63	6.82
C26H51F1N4	438.72	438.4098	71.18	11.72	4.33	12.77
C26H51F1N6	466.73	466.4159	66.91	11.01	4.07	18.01
C26H51F2N1	415.70	415.3989	75.12	12.37	9.14	3.37
C26H51F2N3	443.71	443.4051	70.38	11.59	8.56	9.47
C26H51F2N5	471.73	471.4112	66.20	10.90	8.05	14.85
C26H52F1N1	397.71	397.4084	78.52	13.18	4.78	3.52
C26H52F1N3	425.72	425.4145	73.36	12.31	4.46	9.87
C26H52F1N5	453.74	453.4207	68.83	11.55	4.19	15.43
C26H52F2N2	430.71	430.4098	72.51	12.17	8.82	6.50
C26H52F2N4	458.73	458.4160	68.08	11.43	8.28	12.21
C26H52F2N6	486.74	486.4221	64.16	10.77	7.81	17.27
C26H53F1N2	412.72	412.4193	75.67	12.94	4.60	6.79
C26H53F1N4	440.74	440.4254	70.86	12.12	4.31	12.71
C26H53F1N6	468.75	468.4315	66.62	11.40	4.05	17.93
C26H53F2N1	417.72	417.4146	74.76	12.79	9.10	3.35
C26H53F2N3	445.73	445.4207	70.06	11.99	8.52	9.43
C26H53F2N5	473.74	473.4269	65.92	11.28	8.02	14.78
C26H54F1N1	399.73	399.4240	78.13	13.62	4.75	3.50
C26H54F1N3	427.74	427.4302	73.01	12.73	4.44	9.82
C26H54F1N5	455.75	455.4363	68.52	11.94	4.17	15.37
C26H54F2N2	432.73	432.4255	72.17	12.58	8.78	6.47
C26H54F2N4	460.74	460.4316	67.78	11.81	8.25	12.16
C26H54F2N6	488.76	488.4378	63.89	11.14	7.77	17.19
C26H55F1N2	414.74	414.4349	75.30	13.37	4.58	6.75
C26H55F1N4	442.75	442.4411	70.53	12.52	4.29	12.65
C26H55F1N6	470.77	470.4472	66.34	11.78	4.04	17.85
C26H55F2N3	447.75	447.4364	69.75	12.38	8.49	9.38
C26H55F2N5	475.76	475.4425	65.64	11.65	7.99	14.72
C26H56F1N3	429.75	429.4458	72.67	13.13	4.42	9.78
C26H56F1N5	457.77	457.4519	68.22	12.33	4.15	15.30
C26H56F2N4	462.76	462.4473	67.48	12.20	8.21	12.11
C26H56F2N6	490.77	490.4534	63.63	11.50	7.74	17.12
C26H57F1N4	444.77	444.4567	70.21	12.92	4.27	12.60
C26H57F1N6	472.78	472.4628	66.05	12.15	4.02	17.78
C26H57F2N5	477.77	477.4582	65.36	12.03	7.95	14.66
C26H58F1N5	459.78	459.4676	67.92	12.72	4.13	15.23
C26H58F2N6	492.79	492.4691	63.37	11.86	7.71	17.05
C26H59F1N6	474.80	474.4785	65.77	12.53	4.00	17.70
C27H13F1N2	384.42	384.1063	84.36	3.41	4.94	7.29
C27H13F1N4	412.43	412.1124	78.63	3.18	4.61	13.58
C27H13F1N6	440.44	440.1186	73.63	2.98	4.31	19.08
C27H13F2N1	389.41	389.1016	83.28	3.37	9.76	3.60

C27H13F2N3	417.42	417.1077	77.69	3.14	9.10	10.07
C27H13F2N5	445.43	445.1139	72.81	2.94	8.53	15.72
C27H14F1N1	371.42	371.1110	87.31	3.80	5.12	3.77
C27H14F1N3	399.43	399.1172	81.19	3.53	4.76	10.52
C27H14F1N5	427.44	427.1233	75.87	3.30	4.44	16.38
C27H14F2N2	404.42	404.1125	80.19	3.49	9.40	6.93
C27H14F2N4	432.44	432.1186	74.99	3.26	8.79	12.96
C27H14F2N6	460.45	460.1248	70.43	3.06	8.25	18.25
C27H15F1N2	386.43	386.1219	83.92	3.91	4.92	7.25
C27H15F1N4	414.45	414.1281	78.25	3.65	4.58	13.52
C27H15F1N6	442.46	442.1342	73.30	3.42	4.29	18.99
C27H15F2N1	391.42	391.1172	82.85	3.86	9.71	3.58
C27H15F2N3	419.44	419.1234	77.32	3.60	9.06	10.02
C27H15F2N5	447.45	447.1295	72.48	3.38	8.49	15.65
C27H16F1N1	373.43	373.1267	86.84	4.32	5.09	3.75
C27H16F1N3	401.45	401.1328	80.78	4.02	4.73	10.47
C27H16F1N5	429.46	429.1390	75.51	3.76	4.42	16.31
C27H16F2N2	406.44	406.1281	79.79	3.97	9.35	6.89
C27H16F2N4	434.45	434.1343	74.65	3.71	8.75	12.90
C27H16F2N6	462.47	462.1404	70.12	3.49	8.22	18.17
C27H17F1N2	388.45	388.1376	83.49	4.41	4.89	7.21
C27H17F1N4	416.46	416.1437	77.87	4.11	4.56	13.45
C27H17F1N6	444.48	444.1499	72.96	3.86	4.27	18.91
C27H17F2N1	393.44	393.1329	82.43	4.36	9.66	3.56
C27H17F2N3	421.45	421.1390	76.95	4.07	9.02	9.97
C27H17F2N5	449.47	449.1452	72.15	3.81	8.45	15.58
C27H18F1N1	375.45	375.1423	86.38	4.83	5.06	3.73
C27H18F1N3	403.46	403.1485	80.38	4.50	4.71	10.41
C27H18F1N5	431.48	431.1546	75.16	4.20	4.40	16.23
C27H18F2N2	408.45	408.1438	79.40	4.44	9.30	6.86
C27H18F2N4	436.47	436.1499	74.30	4.16	8.71	12.84
C27H18F2N6	464.48	464.1561	69.82	3.91	8.18	18.09
C27H19F1N2	390.46	390.1532	83.06	4.90	4.87	7.17
C27H19F1N4	418.48	418.1594	77.50	4.58	4.54	13.39
C27H19F1N6	446.49	446.1655	72.63	4.29	4.26	18.82
C27H19F2N1	395.46	395.1485	82.01	4.84	9.61	3.54
C27H19F2N3	423.47	423.1547	76.58	4.52	8.97	9.92
C27H19F2N5	451.48	451.1608	71.83	4.24	8.42	15.51
C27H20F1N1	377.47	377.1580	85.92	5.34	5.03	3.71
C27H20F1N3	405.48	405.1641	79.98	4.97	4.69	10.36
C27H20F1N5	433.49	433.1703	74.81	4.65	4.38	16.16
C27H20F2N2	410.47	410.1594	79.01	4.91	9.26	6.82
C27H20F2N4	438.48	438.1656	73.96	4.60	8.67	12.78
C27H20F2N6	466.50	466.1717	69.52	4.32	8.15	18.02
C27H21F1N2	392.48	392.1689	82.63	5.39	4.84	7.14
C27H21F1N4	420.49	420.1750	77.12	5.03	4.52	13.32
C27H21F1N6	448.51	448.1812	72.31	4.72	4.24	18.74
C27H21F2N1	397.47	397.1642	81.59	5.33	9.56	3.52
C27H21F2N3	425.49	425.1703	76.22	4.97	8.93	9.88
C27H21F2N5	453.50	453.1765	71.51	4.67	8.38	15.44
C27H22F1N1	379.48	379.1736	85.46	5.84	5.01	3.69
C27H22F1N3	407.49	407.1798	79.58	5.44	4.66	10.31
C27H22F1N5	435.51	435.1859	74.46	5.09	4.36	16.08
C27H22F2N2	412.49	412.1751	78.62	5.38	9.21	6.79
C27H22F2N4	440.50	440.1812	73.62	5.03	8.63	12.72
C27H22F2N6	468.51	468.1874	69.22	4.73	8.11	17.94
C27H23F1N2	394.50	394.1845	82.21	5.88	4.82	7.10
C27H23F1N4	422.51	422.1907	76.76	5.49	4.50	13.26
C27H23F1N6	450.52	450.1968	71.98	5.15	4.22	18.65
C27H23F2N1	399.49	399.1798	81.18	5.80	9.51	3.51

C27H23F2N3	427.50	427.1860	75.86	5.42	8.89	9.83
C27H23F2N5	455.51	455.1921	71.19	5.09	8.34	15.37
C27H24F1N1	381.50	381.1893	85.01	6.34	4.98	3.67
C27H24F1N3	409.51	409.1954	79.19	5.91	4.64	10.26
C27H24F1N5	437.52	437.2016	74.12	5.53	4.34	16.01
C27H24F2N2	414.50	414.1907	78.24	5.84	9.17	6.76
C27H24F2N4	442.52	442.1969	73.29	5.47	8.59	12.66
C27H24F2N6	470.53	470.2030	68.92	5.14	8.08	17.86
C27H25F1N2	396.51	396.2002	81.79	6.36	4.79	7.06
C27H25F1N4	424.53	424.2063	76.39	5.94	4.48	13.20
C27H25F1N6	452.54	452.2125	71.66	5.57	4.20	18.57
C27H25F2N1	401.50	401.1955	80.77	6.28	9.46	3.49
C27H25F2N3	429.52	429.2016	75.50	5.87	8.85	9.78
C27H25F2N5	457.53	457.2078	70.88	5.51	8.30	15.31
C27H26F1N1	383.51	383.2049	84.56	6.83	4.95	3.65
C27H26F1N3	411.53	411.2111	78.80	6.37	4.62	10.21
C27H26F1N5	439.54	439.2172	73.78	5.96	4.32	15.93
C27H26F2N2	416.52	416.2064	77.86	6.29	9.12	6.73
C27H26F2N4	444.53	444.2125	72.95	5.90	8.55	12.60
C27H26F2N6	472.55	472.2187	68.63	5.55	8.04	17.78
C27H27F1N2	398.53	398.2158	81.37	6.83	4.77	7.03
C27H27F1N4	426.54	426.2220	76.03	6.38	4.45	13.14
C27H27F1N6	454.55	454.2281	71.34	5.99	4.18	18.49
C27H27F2N1	403.52	403.2111	80.37	6.74	9.42	3.47
C27H27F2N3	431.53	431.2173	75.15	6.31	8.81	9.74
C27H27F2N5	459.55	459.2234	70.57	5.92	8.27	15.24
C27H28F1N1	385.53	385.2206	84.12	7.32	4.93	3.63
C27H28F1N3	413.54	413.2267	78.42	6.82	4.59	10.16
C27H28F1N5	441.56	441.2329	73.45	6.39	4.30	15.86
C27H28F2N2	418.53	418.2220	77.48	6.74	9.08	6.69
C27H28F2N4	446.55	446.2282	72.62	6.32	8.51	12.55
C27H28F2N6	474.56	474.2343	68.34	5.95	8.01	17.71
C27H29F1N2	400.54	400.2315	80.97	7.30	4.74	6.99
C27H29F1N4	428.56	428.2376	75.67	6.82	4.43	13.07
C27H29F1N6	456.57	456.2438	71.03	6.40	4.16	18.41
C27H29F2N1	405.54	405.2268	79.97	7.21	9.37	3.45
C27H29F2N3	433.55	433.2329	74.80	6.74	8.76	9.69
C27H29F2N5	461.56	461.2391	70.26	6.33	8.23	15.17
C27H30F1N1	387.55	387.2362	83.68	7.80	4.90	3.61
C27H30F1N3	415.56	415.2424	78.04	7.28	4.57	10.11
C27H30F1N5	443.57	443.2485	73.11	6.82	4.28	15.79
C27H30F2N2	420.55	420.2377	77.11	7.19	9.04	6.66
C27H30F2N4	448.56	448.2438	72.30	6.74	8.47	12.49
C27H30F2N6	476.58	476.2500	68.05	6.35	7.97	17.63
C27H31F1N2	402.56	402.2471	80.56	7.76	4.72	6.96
C27H31F1N4	430.57	430.2533	75.32	7.26	4.41	13.01
C27H31F1N6	458.59	458.2594	70.72	6.81	4.14	18.33
C27H31F2N1	407.55	407.2424	79.57	7.67	9.32	3.44
C27H31F2N3	435.57	435.2486	74.46	7.17	8.72	9.65
C27H31F2N5	463.58	463.2547	69.96	6.74	8.20	15.11
C27H32F1N1	389.56	389.2519	83.25	8.28	4.88	3.60
C27H32F1N3	417.57	417.2580	77.66	7.72	4.55	10.06
C27H32F1N5	445.59	445.2642	72.78	7.24	4.26	15.72
C27H32F2N2	422.57	422.2533	76.75	7.63	8.99	6.63
C27H32F2N4	450.58	450.2595	71.97	7.16	8.43	12.43
C27H32F2N6	478.59	478.2656	67.76	6.74	7.94	17.56
C27H33F1N2	404.58	404.2628	80.16	8.22	4.70	6.92
C27H33F1N4	432.59	432.2689	74.97	7.69	4.39	12.95
C27H33F1N6	460.60	460.2751	70.41	7.22	4.12	18.25
C27H33F2N1	409.57	409.2581	79.18	8.12	9.28	3.42

C27H33F2N3	437.58	437.2642	74.11	7.60	8.68	9.60
C27H33F2N5	465.59	465.2704	69.65	7.14	8.16	15.04
C27H34F1N1	391.58	391.2675	82.82	8.75	4.85	3.58
C27H34F1N3	419.59	419.2737	77.29	8.17	4.53	10.01
C27H34F1N5	447.60	447.2798	72.45	7.66	4.24	15.65
C27H34F2N2	424.58	424.2690	76.38	8.07	8.95	6.60
C27H34F2N4	452.60	452.2751	71.65	7.57	8.40	12.38
C27H34F2N6	480.61	480.2813	67.48	7.13	7.91	17.49
C27H35F1N2	406.59	406.2784	79.76	8.68	4.67	6.89
C27H35F1N4	434.61	434.2846	74.62	8.12	4.37	12.89
C27H35F1N6	462.62	462.2907	70.10	7.63	4.11	18.17
C27H35F2N1	411.58	411.2737	78.79	8.57	9.23	3.40
C27H35F2N3	439.60	439.2799	73.77	8.03	8.64	9.56
C27H35F2N5	467.61	467.2860	69.35	7.54	8.13	14.98
C27H36F1N1	393.59	393.2832	82.40	9.22	4.83	3.56
C27H36F1N3	421.61	421.2893	76.92	8.61	4.51	9.97
C27H36F1N5	449.62	449.2955	72.13	8.07	4.23	15.58
C27H36F2N2	426.60	426.2846	76.02	8.51	8.91	6.57
C27H36F2N4	454.61	454.2908	71.34	7.98	8.36	12.32
C27H36F2N6	482.62	482.2969	67.20	7.52	7.87	17.41
C27H37F1N2	408.61	408.2941	79.37	9.13	4.65	6.86
C27H37F1N4	436.62	436.3002	74.28	8.54	4.35	12.83
C27H37F1N6	464.63	464.3064	69.80	8.03	4.09	18.09
C27H37F2N1	413.60	413.2894	78.41	9.02	9.19	3.39
C27H37F2N3	441.61	441.2955	73.44	8.45	8.60	9.52
C27H37F2N5	469.63	469.3017	69.06	7.94	8.09	14.91
C27H38F1N1	395.61	395.2988	81.98	9.68	4.80	3.54
C27H38F1N3	423.62	423.3050	76.55	9.04	4.48	9.92
C27H38F1N5	451.64	451.3111	71.81	8.48	4.21	15.51
C27H38F2N2	428.61	428.3003	75.66	8.94	8.87	6.54
C27H38F2N4	456.63	456.3064	71.02	8.39	8.32	12.27
C27H38F2N6	484.64	484.3126	66.92	7.90	7.84	17.34
C27H39F1N2	410.62	410.3097	78.98	9.57	4.63	6.82
C27H39F1N4	438.64	438.3159	73.93	8.96	4.33	12.77
C27H39F1N6	466.65	466.3220	69.50	8.42	4.07	18.01
C27H39F2N1	415.62	415.3050	78.03	9.46	9.14	3.37
C27H39F2N3	443.63	443.3112	73.10	8.86	8.56	9.47
C27H39F2N5	471.64	471.3173	68.76	8.33	8.06	14.85
C27H40F1N1	397.62	397.3145	81.56	10.14	4.78	3.52
C27H40F1N3	425.64	425.3206	76.19	9.47	4.46	9.87
C27H40F1N5	453.65	453.3268	71.49	8.89	4.19	15.44
C27H40F2N2	430.63	430.3159	75.31	9.36	8.82	6.51
C27H40F2N4	458.64	458.3221	70.71	8.79	8.28	12.22
C27H40F2N6	486.66	486.3282	66.64	8.28	7.81	17.27
C27H41F1N2	412.64	412.3254	78.59	10.02	4.60	6.79
C27H41F1N4	440.65	440.3315	73.60	9.38	4.31	12.71
C27H41F1N6	468.67	468.3377	69.20	8.82	4.05	17.93
C27H41F2N1	417.63	417.3207	77.65	9.90	9.10	3.35
C27H41F2N3	445.64	445.3268	72.77	9.27	8.53	9.43
C27H41F2N5	473.66	473.3330	68.47	8.73	8.02	14.79
C27H42F1N1	399.64	399.3301	81.15	10.59	4.75	3.50
C27H42F1N3	427.65	427.3363	75.83	9.90	4.44	9.83
C27H42F1N5	455.67	455.3424	71.17	9.29	4.17	15.37
C27H42F2N2	432.65	432.3316	74.96	9.79	8.78	6.47
C27H42F2N4	460.66	460.3377	70.40	9.19	8.25	12.16
C27H42F2N6	488.67	488.3439	66.36	8.66	7.78	17.20
C27H43F1N2	414.66	414.3410	78.21	10.45	4.58	6.76
C27H43F1N4	442.67	442.3472	73.26	9.79	4.29	12.66
C27H43F1N6	470.68	470.3533	68.90	9.21	4.04	17.85
C27H43F2N1	419.65	419.3363	77.28	10.33	9.05	3.34

C27H43F2N3	447.66	447.3425	72.44	9.68	8.49	9.39
C27H43F2N5	475.67	475.3486	68.18	9.11	7.99	14.72
C27H44F1N1	401.66	401.3458	80.74	11.04	4.73	3.49
C27H44F1N3	429.67	429.3519	75.48	10.32	4.42	9.78
C27H44F1N5	457.68	457.3581	70.86	9.69	4.15	15.30
C27H44F2N2	434.66	434.3472	74.61	10.20	8.74	6.44
C27H44F2N4	462.68	462.3534	70.09	9.59	8.21	12.11
C27H44F2N6	490.69	490.3595	66.09	9.04	7.74	17.13
C27H45F1N2	416.67	416.3567	77.83	10.89	4.56	6.72
C27H45F1N4	444.68	444.3628	72.93	10.20	4.27	12.60
C27H45F1N6	472.70	472.3690	68.61	9.60	4.02	17.78
C27H45F2N1	421.66	421.3520	76.91	10.76	9.01	3.32
C27H45F2N3	449.68	449.3581	72.12	10.09	8.45	9.34
C27H45F2N5	477.69	477.3643	67.89	9.50	7.95	14.66
C27H46F1N1	403.67	403.3614	80.34	11.49	4.71	3.47
C27H46F1N3	431.69	431.3676	75.12	10.74	4.40	9.73
C27H46F1N5	459.70	459.3737	70.55	10.09	4.13	15.23
C27H46F2N2	436.68	436.3629	74.27	10.62	8.70	6.42
C27H46F2N4	464.69	464.3690	69.79	9.98	8.18	12.06
C27H46F2N6	492.70	492.3752	65.82	9.41	7.71	17.06
C27H47F1N2	418.69	418.3723	77.46	11.32	4.54	6.69
C27H47F1N4	446.70	446.3785	72.60	10.61	4.25	12.54
C27H47F1N6	474.71	474.3846	68.32	9.98	4.00	17.70
C27H47F2N1	423.68	423.3676	76.54	11.18	8.97	3.31
C27H47F2N3	451.69	451.3738	71.80	10.49	8.41	9.30
C27H47F2N5	479.71	479.3799	67.60	9.88	7.92	14.60
C27H48F1N1	405.69	405.3771	79.94	11.93	4.68	3.45
C27H48F1N3	433.70	433.3832	74.78	11.16	4.38	9.69
C27H48F1N5	461.72	461.3894	70.24	10.48	4.11	15.17
C27H48F2N2	438.69	438.3785	73.92	11.03	8.66	6.39
C27H48F2N4	466.71	466.3847	69.49	10.37	8.14	12.00
C27H48F2N6	494.72	494.3908	65.55	9.78	7.68	16.99
C27H49F1N2	420.70	420.3880	77.09	11.74	4.52	6.66
C27H49F1N4	448.72	448.3941	72.27	11.01	4.23	12.49
C27H49F1N6	476.73	476.4003	68.03	10.36	3.99	17.63
C27H49F2N1	425.70	425.3833	76.18	11.60	8.93	3.29
C27H49F2N3	453.71	453.3894	71.48	10.89	8.37	9.26
C27H49F2N5	481.72	481.3956	67.32	10.25	7.89	14.54
C27H50F1N1	407.70	407.3927	79.54	12.36	4.66	3.44
C27H50F1N3	435.72	435.3989	74.43	11.57	4.36	9.64
C27H50F1N5	463.73	463.4050	69.93	10.87	4.10	15.10
C27H50F2N2	440.71	440.3942	73.59	11.44	8.62	6.36
C27H50F2N4	468.72	468.4003	69.19	10.75	8.11	11.95
C27H50F2N6	496.74	496.4065	65.29	10.15	7.65	16.92
C27H51F1N2	422.72	422.4036	76.72	12.16	4.49	6.63
C27H51F1N4	450.73	450.4098	71.95	11.41	4.22	12.43
C27H51F1N6	478.75	478.4159	67.74	10.74	3.97	17.55
C27H51F2N1	427.71	427.3989	75.82	12.02	8.88	3.27
C27H51F2N3	455.72	455.4051	71.16	11.28	8.34	9.22
C27H51F2N5	483.74	483.4112	67.04	10.63	7.85	14.48
C27H52F1N1	409.72	409.4084	79.15	12.79	4.64	3.42
C27H52F1N3	437.73	437.4145	74.09	11.97	4.34	9.60
C27H52F1N5	465.75	465.4207	69.63	11.25	4.08	15.04
C27H52F2N2	442.73	442.4098	73.25	11.84	8.58	6.33
C27H52F2N4	470.74	470.4160	68.89	11.13	8.07	11.90
C27H52F2N6	498.75	498.4221	65.02	10.51	7.62	16.85
C27H53F1N2	424.74	424.4193	76.35	12.58	4.47	6.60
C27H53F1N4	452.75	452.4254	71.63	11.80	4.20	12.37
C27H53F1N6	480.76	480.4315	67.46	11.11	3.95	17.48
C27H53F2N1	429.73	429.4146	75.47	12.43	8.84	3.26

C27H53F2N3	457.74	457.4207	70.85	11.67	8.30	9.18
C27H53F2N5	485.75	485.4269	66.76	11.00	7.82	14.42
C27H54F1N1	411.74	411.4240	78.76	13.22	4.61	3.40
C27H54F1N3	439.75	439.4302	73.75	12.38	4.32	9.56
C27H54F1N5	467.76	467.4363	69.33	11.64	4.06	14.97
C27H54F2N2	444.74	444.4255	72.92	12.24	8.54	6.30
C27H54F2N4	472.76	472.4316	68.60	11.51	8.04	11.85
C27H54F2N6	500.77	500.4378	64.76	10.87	7.59	16.78
C27H55F1N2	426.75	426.4349	75.99	12.99	4.45	6.56
C27H55F1N4	454.76	454.4411	71.31	12.19	4.18	12.32
C27H55F1N6	482.78	482.4472	67.17	11.48	3.94	17.41
C27H55F2N1	431.74	431.4302	75.11	12.84	8.80	3.24
C27H55F2N3	459.76	459.4364	70.54	12.06	8.26	9.14
C27H55F2N5	487.77	487.4425	66.49	11.37	7.79	14.36
C27H56F1N1	413.75	413.4397	78.38	13.64	4.59	3.39
C27H56F1N3	441.77	441.4458	73.41	12.78	4.30	9.51
C27H56F1N5	469.78	469.4519	69.03	12.02	4.04	14.91
C27H56F2N2	446.76	446.4411	72.59	12.63	8.51	6.27
C27H56F2N4	474.77	474.4473	68.31	11.89	8.00	11.80
C27H56F2N6	502.78	502.4534	64.50	11.23	7.56	16.71
C27H57F1N2	428.77	428.4506	75.64	13.40	4.43	6.53
C27H57F1N4	456.78	456.4567	71.00	12.58	4.16	12.27
C27H57F1N6	484.79	484.4628	66.85	11.85	3.92	17.34
C27H57F2N3	461.77	461.4520	70.23	12.44	8.23	9.10
C27H57F2N5	489.79	489.4582	66.21	11.73	7.76	14.30
C27H58F1N3	443.78	443.4615	73.08	13.17	4.28	9.47
C27H58F1N5	471.80	471.4676	68.74	12.39	4.03	14.84
C27H58F2N4	476.79	476.4629	68.02	12.26	7.97	11.75
C27H58F2N6	504.80	504.4691	64.24	11.58	7.53	16.65
C27H59F1N4	458.80	458.4723	70.69	12.96	4.14	12.21
C27H59F1N6	486.81	486.4785	66.62	12.22	3.90	17.26
C27H59F2N5	491.80	491.4738	65.94	12.09	7.73	14.24
C27H60F1N5	473.81	473.4832	68.45	12.76	4.01	14.78
C27H60F2N6	506.82	506.4847	63.99	11.93	7.50	16.58
C27H61F1N6	488.83	488.4941	66.34	12.58	3.89	17.19
C28H14F1N1	383.43	383.1110	87.71	3.68	4.95	3.65
C28H14F1N3	411.44	411.1172	81.74	3.43	4.62	10.21
C28H14F1N5	439.46	439.1233	76.53	3.21	4.32	15.94
C28H14F2N2	416.43	416.1125	80.76	3.39	9.12	6.73
C28H14F2N4	444.45	444.1186	75.67	3.18	8.55	12.61
C28H14F2N6	472.46	472.1248	71.18	2.99	8.04	17.79
C28H15F1N2	398.44	398.1219	84.41	3.79	4.77	7.03
C28H15F1N4	426.46	426.1281	78.86	3.55	4.45	13.14
C28H15F1N6	454.47	454.1342	74.00	3.33	4.18	18.49
C28H15F2N1	403.44	403.1172	83.36	3.75	9.42	3.47
C28H15F2N3	431.45	431.1234	77.95	3.50	8.81	9.74
C28H15F2N5	459.46	459.1295	73.20	3.29	8.27	15.24
C28H16F1N1	385.44	385.1267	87.25	4.18	4.93	3.63
C28H16F1N3	413.46	413.1328	81.34	3.90	4.59	10.16
C28H16F1N5	441.47	441.1390	76.18	3.65	4.30	15.86
C28H16F2N2	418.45	418.1281	80.37	3.85	9.08	6.69
C28H16F2N4	446.46	446.1343	75.33	3.61	8.51	12.55
C28H16F2N6	474.48	474.1404	70.88	3.40	8.01	17.71
C28H17F1N2	400.46	400.1376	83.98	4.28	4.74	7.00
C28H17F1N4	428.47	428.1437	78.49	4.00	4.43	13.08
C28H17F1N6	456.49	456.1499	73.67	3.75	4.16	18.41
C28H17F2N1	405.45	405.1329	82.95	4.23	9.37	3.45
C28H17F2N3	433.46	433.1390	77.59	3.95	8.77	9.69
C28H17F2N5	461.48	461.1452	72.88	3.71	8.23	15.18
C28H18F1N1	387.46	387.1423	86.80	4.68	4.90	3.61

C28H18F1N3	415.47	415.1485	80.95	4.37	4.57	10.11
C28H18F1N5	443.49	443.1546	75.83	4.09	4.28	15.79
C28H18F2N2	420.47	420.1438	79.99	4.32	9.04	6.66
C28H18F2N4	448.48	448.1499	74.99	4.05	8.47	12.49
C28H18F2N6	476.49	476.1561	70.58	3.81	7.97	17.64
C28H19F1N2	402.48	402.1532	83.56	4.76	4.72	6.96
C28H19F1N4	430.49	430.1594	78.12	4.45	4.41	13.01
C28H19F1N6	458.50	458.1655	73.35	4.18	4.14	18.33
C28H19F2N1	407.47	407.1485	82.54	4.70	9.33	3.44
C28H19F2N3	435.48	435.1547	77.23	4.40	8.73	9.65
C28H19F2N5	463.49	463.1608	72.56	4.13	8.20	15.11
C28H20F1N1	389.48	389.1580	86.35	5.18	4.88	3.60
C28H20F1N3	417.49	417.1641	80.56	4.83	4.55	10.06
C28H20F1N5	445.50	445.1703	75.49	4.53	4.26	15.72
C28H20F2N2	422.48	422.1594	79.60	4.77	8.99	6.63
C28H20F2N4	450.50	450.1656	74.65	4.47	8.43	12.44
C28H20F2N6	478.51	478.1717	70.28	4.21	7.94	17.56
C28H21F1N2	404.49	404.1689	83.14	5.23	4.70	6.93
C28H21F1N4	432.50	432.1750	77.76	4.89	4.39	12.95
C28H21F1N6	460.52	460.1812	73.03	4.60	4.13	18.25
C28H21F2N1	409.48	409.1642	82.13	5.17	9.28	3.42
C28H21F2N3	437.50	437.1703	76.87	4.84	8.69	9.60
C28H21F2N5	465.51	465.1765	72.25	4.55	8.16	15.04
C28H22F1N1	391.49	391.1736	85.91	5.66	4.85	3.58
C28H22F1N3	419.51	419.1798	80.17	5.29	4.53	10.02
C28H22F1N5	447.52	447.1859	75.15	4.96	4.25	15.65
C28H22F2N2	424.50	424.1751	79.23	5.22	8.95	6.60
C28H22F2N4	452.51	452.1812	74.32	4.90	8.40	12.38
C28H22F2N6	480.52	480.1874	69.99	4.61	7.91	17.49
C28H23F1N2	406.51	406.1845	82.73	5.70	4.67	6.89
C28H23F1N4	434.52	434.1907	77.40	5.34	4.37	12.89
C28H23F1N6	462.53	462.1968	72.71	5.01	4.11	18.17
C28H23F2N1	411.50	411.1798	81.73	5.63	9.23	3.40
C28H23F2N3	439.51	439.1860	76.52	5.27	8.65	9.56
C28H23F2N5	467.53	467.1921	71.93	4.96	8.13	14.98
C28H24F1N1	393.51	393.1893	85.47	6.15	4.83	3.56
C28H24F1N3	421.52	421.1954	79.79	5.74	4.51	9.97
C28H24F1N5	449.54	449.2016	74.81	5.38	4.23	15.58
C28H24F2N2	426.51	426.1907	78.85	5.67	8.91	6.57
C28H24F2N4	454.53	454.1969	73.99	5.32	8.36	12.33
C28H24F2N6	482.54	482.2030	69.70	5.01	7.87	17.42
C28H25F1N2	408.52	408.2002	82.32	6.17	4.65	6.86
C28H25F1N4	436.54	436.2063	77.04	5.77	4.35	12.83
C28H25F1N6	464.55	464.2125	72.40	5.42	4.09	18.09
C28H25F2N1	413.51	413.1955	81.33	6.09	9.19	3.39
C28H25F2N3	441.53	441.2016	76.17	5.71	8.61	9.52
C28H25F2N5	469.54	469.2078	71.63	5.37	8.09	14.92
C28H26F1N1	395.52	395.2049	85.03	6.63	4.80	3.54
C28H26F1N3	423.54	423.2111	79.41	6.19	4.49	9.92
C28H26F1N5	451.55	451.2172	74.48	5.80	4.21	15.51
C28H26F2N2	428.53	428.2064	78.48	6.12	8.87	6.54
C28H26F2N4	456.54	456.2125	73.66	5.74	8.32	12.27
C28H26F2N6	484.56	484.2187	69.41	5.41	7.84	17.34
C28H27F1N2	410.54	410.2158	81.92	6.63	4.63	6.82
C28H27F1N4	438.55	438.2220	76.69	6.21	4.33	12.78
C28H27F1N6	466.57	466.2281	72.08	5.83	4.07	18.01
C28H27F2N1	415.53	415.2111	80.94	6.55	9.14	3.37
C28H27F2N3	443.54	443.2173	75.82	6.14	8.57	9.47
C28H27F2N5	471.56	471.2234	71.32	5.77	8.06	14.85
C28H28F1N1	397.54	397.2206	84.60	7.10	4.78	3.52

C28H28F1N3	425.55	425.2267	79.03	6.63	4.46	9.87
C28H28F1N5	453.57	453.2329	74.15	6.22	4.19	15.44
C28H28F2N2	430.55	430.2220	78.11	6.56	8.83	6.51
C28H28F2N4	458.56	458.2282	73.34	6.15	8.29	12.22
C28H28F2N6	486.57	486.2343	69.12	5.80	7.81	17.27
C28H29F1N2	412.56	412.2315	81.52	7.09	4.61	6.79
C28H29F1N4	440.57	440.2376	76.34	6.63	4.31	12.72
C28H29F1N6	468.58	468.2438	71.77	6.24	4.05	17.94
C28H29F2N1	417.55	417.2268	80.54	7.00	9.10	3.35
C28H29F2N3	445.56	445.2329	75.48	6.56	8.53	9.43
C28H29F2N5	473.57	473.2391	71.02	6.17	8.02	14.79
C28H30F1N1	399.56	399.2362	84.17	7.57	4.75	3.51
C28H30F1N3	427.57	427.2424	78.66	7.07	4.44	9.83
C28H30F1N5	455.58	455.2485	73.82	6.64	4.17	15.37
C28H30F2N2	432.56	432.2377	77.75	6.99	8.78	6.48
C28H30F2N4	460.57	460.2438	73.02	6.57	8.25	12.16
C28H30F2N6	488.59	488.2500	68.83	6.19	7.78	17.20
C28H31F1N2	414.57	414.2471	81.12	7.54	4.58	6.76
C28H31F1N4	442.58	442.2533	75.99	7.06	4.29	12.66
C28H31F1N6	470.60	470.2594	71.46	6.64	4.04	17.86
C28H31F2N1	419.56	419.2424	80.16	7.45	9.06	3.34
C28H31F2N3	447.58	447.2486	75.14	6.98	8.49	9.39
C28H31F2N5	475.59	475.2547	70.71	6.57	7.99	14.73
C28H32F1N1	401.57	401.2519	83.75	8.03	4.73	3.49
C28H32F1N3	429.59	429.2580	78.29	7.51	4.42	9.78
C28H32F1N5	457.60	457.2642	73.49	7.05	4.15	15.30
C28H32F2N2	434.58	434.2533	77.39	7.42	8.74	6.45
C28H32F2N4	462.59	462.2595	72.70	6.97	8.21	12.11
C28H32F2N6	490.60	490.2656	68.55	6.57	7.74	17.13
C28H33F1N2	416.59	416.2628	80.73	7.98	4.56	6.72
C28H33F1N4	444.60	444.2689	75.64	7.48	4.27	12.60
C28H33F1N6	472.61	472.2751	71.16	7.04	4.02	17.78
C28H33F2N1	421.58	421.2581	79.77	7.89	9.01	3.32
C28H33F2N3	449.59	449.2642	74.80	7.40	8.45	9.35
C28H33F2N5	477.61	477.2704	70.42	6.96	7.96	14.66
C28H34F1N1	403.59	403.2675	83.33	8.49	4.71	3.47
C28H34F1N3	431.60	431.2737	77.92	7.94	4.40	9.74
C28H34F1N5	459.62	459.2798	73.17	7.46	4.13	15.24
C28H34F2N2	436.59	436.2690	77.03	7.85	8.70	6.42
C28H34F2N4	464.61	464.2751	72.39	7.38	8.18	12.06
C28H34F2N6	492.62	492.2813	68.27	6.96	7.71	17.06
C28H35F1N2	418.60	418.2784	80.34	8.43	4.54	6.69
C28H35F1N4	446.62	446.2846	75.30	7.90	4.25	12.54
C28H35F1N6	474.63	474.2907	70.86	7.43	4.00	17.71
C28H35F2N1	423.59	423.2737	79.39	8.33	8.97	3.31
C28H35F2N3	451.61	451.2799	74.47	7.81	8.41	9.30
C28H35F2N5	479.62	479.2860	70.12	7.36	7.92	14.60
C28H36F1N1	405.60	405.2832	82.92	8.95	4.68	3.45
C28H36F1N3	433.62	433.2893	77.56	8.37	4.38	9.69
C28H36F1N5	461.63	461.2955	72.85	7.86	4.12	15.17
C28H36F2N2	438.61	438.2846	76.68	8.27	8.66	6.39
C28H36F2N4	466.62	466.2908	72.07	7.78	8.14	12.01
C28H36F2N6	494.64	494.2969	67.99	7.34	7.68	16.99
C28H37F1N2	420.62	420.2941	79.96	8.87	4.52	6.66
C28H37F1N4	448.63	448.3002	74.96	8.31	4.23	12.49
C28H37F1N6	476.65	476.3064	70.56	7.82	3.99	17.63
C28H37F2N1	425.61	425.2894	79.02	8.76	8.93	3.29
C28H37F2N3	453.62	453.2955	74.14	8.22	8.38	9.26
C28H37F2N5	481.64	481.3017	69.83	7.74	7.89	14.54
C28H38F1N1	407.62	407.2988	82.51	9.40	4.66	3.44

C28H38F1N3	435.63	435.3050	77.20	8.79	4.36	9.65
C28H38F1N5	463.65	463.3111	72.54	8.26	4.10	15.10
C28H38F2N2	440.63	440.3003	76.33	8.69	8.62	6.36
C28H38F2N4	468.64	468.3064	71.76	8.17	8.11	11.96
C28H38F2N6	496.65	496.3126	67.72	7.71	7.65	16.92
C28H39F1N2	422.63	422.3097	79.58	9.30	4.50	6.63
C28H39F1N4	450.65	450.3159	74.63	8.72	4.22	12.43
C28H39F1N6	478.66	478.3220	70.26	8.21	3.97	17.56
C28H39F2N1	427.63	427.3050	78.65	9.19	8.89	3.28
C28H39F2N3	455.64	455.3112	73.81	8.63	8.34	9.22
C28H39F2N5	483.65	483.3173	69.54	8.13	7.86	14.48
C28H40F1N1	409.64	409.3145	82.10	9.84	4.64	3.42
C28H40F1N3	437.65	437.3206	76.85	9.21	4.34	9.60
C28H40F1N5	465.66	465.3268	72.22	8.66	4.08	15.04
C28H40F2N2	442.64	442.3159	75.98	9.11	8.58	6.33
C28H40F2N4	470.65	470.3221	71.46	8.57	8.07	11.90
C28H40F2N6	498.67	498.3282	67.44	8.09	7.62	16.85
C28H41F1N2	424.65	424.3254	79.20	9.73	4.47	6.60
C28H41F1N4	452.66	452.3315	74.30	9.13	4.20	12.38
C28H41F1N6	480.68	480.3377	69.97	8.60	3.95	17.48
C28H41F2N1	429.64	429.3207	78.28	9.62	8.84	3.26
C28H41F2N3	457.66	457.3268	73.49	9.03	8.30	9.18
C28H41F2N5	485.67	485.3330	69.25	8.51	7.82	14.42
C28H42F1N1	411.65	411.3301	81.70	10.28	4.62	3.40
C28H42F1N3	439.67	439.3363	76.49	9.63	4.32	9.56
C28H42F1N5	467.68	467.3424	71.91	9.05	4.06	14.97
C28H42F2N2	444.66	444.3316	75.63	9.52	8.55	6.30
C28H42F2N4	472.67	472.3377	71.15	8.96	8.04	11.85
C28H42F2N6	500.68	500.3439	67.17	8.46	7.59	16.79
C28H43F1N2	426.67	426.3410	78.82	10.16	4.45	6.57
C28H43F1N4	454.68	454.3472	73.97	9.53	4.18	12.32
C28H43F1N6	482.69	482.3533	69.67	8.98	3.94	17.41
C28H43F2N1	431.66	431.3363	77.91	10.04	8.80	3.24
C28H43F2N3	459.67	459.3425	73.16	9.43	8.27	9.14
C28H43F2N5	487.69	487.3486	68.96	8.89	7.79	14.36
C28H44F1N1	413.67	413.3458	81.30	10.72	4.59	3.39
C28H44F1N3	441.68	441.3519	76.14	10.04	4.30	9.51
C28H44F1N5	469.69	469.3581	71.60	9.44	4.04	14.91
C28H44F2N2	446.67	446.3472	75.29	9.93	8.51	6.27
C28H44F2N4	474.69	474.3534	70.85	9.34	8.00	11.80
C28H44F2N6	502.70	502.3595	66.90	8.82	7.56	16.72
C28H45F1N2	428.68	428.3567	78.45	10.58	4.43	6.53
C28H45F1N4	456.70	456.3628	73.64	9.93	4.16	12.27
C28H45F1N6	484.71	484.3690	69.38	9.36	3.92	17.34
C28H45F2N1	433.67	433.3520	77.55	10.46	8.76	3.23
C28H45F2N3	461.69	461.3581	72.84	9.82	8.23	9.10
C28H45F2N5	489.70	489.3643	68.68	9.26	7.76	14.30
C28H46F1N1	415.68	415.3614	80.91	11.15	4.57	3.37
C28H46F1N3	443.70	443.3676	75.80	10.45	4.28	9.47
C28H46F1N5	471.71	471.3737	71.30	9.83	4.03	14.85
C28H46F2N2	448.69	448.3629	74.95	10.33	8.47	6.24
C28H46F2N4	476.70	476.3690	70.55	9.73	7.97	11.75
C28H46F2N6	504.72	504.3752	66.63	9.19	7.53	16.65
C28H47F1N2	430.70	430.3723	78.09	11.00	4.41	6.50
C28H47F1N4	458.71	458.3785	73.32	10.33	4.14	12.21
C28H47F1N6	486.73	486.3846	69.10	9.73	3.90	17.27
C28H47F2N1	435.69	435.3676	77.19	10.87	8.72	3.21
C28H47F2N3	463.70	463.3738	72.53	10.22	8.19	9.06
C28H47F2N5	491.72	491.3799	68.40	9.63	7.73	14.24
C28H48F1N1	417.70	417.3771	80.52	11.58	4.55	3.35

C28H48F1N3	445.71	445.3832	75.45	10.86	4.26	9.43
C28H48F1N5	473.73	473.3894	70.99	10.21	4.01	14.78
C28H48F2N2	450.70	450.3785	74.62	10.73	8.43	6.22
C28H48F2N4	478.72	478.3847	70.25	10.11	7.94	11.70
C28H48F2N6	506.73	506.3908	66.37	9.55	7.50	16.58
C28H49F1N2	432.71	432.3880	77.72	11.41	4.39	6.47
C28H49F1N4	460.73	460.3941	73.00	10.72	4.12	12.16
C28H49F1N6	488.74	488.4003	68.81	10.11	3.89	17.20
C28H49F2N1	437.71	437.3833	76.84	11.28	8.68	3.20
C28H49F2N3	465.72	465.3894	72.21	10.61	8.16	9.02
C28H49F2N5	493.73	493.3956	68.12	10.00	7.70	14.18
C28H50F1N1	419.72	419.3927	80.13	12.01	4.53	3.34
C28H50F1N3	447.73	447.3989	75.12	11.26	4.24	9.39
C28H50F1N5	475.74	475.4050	70.69	10.59	3.99	14.72
C28H50F2N2	452.72	452.3942	74.29	11.13	8.39	6.19
C28H50F2N4	480.73	480.4003	69.96	10.48	7.90	11.65
C28H50F2N6	508.75	508.4065	66.11	9.91	7.47	16.52
C28H51F1N2	434.73	434.4036	77.36	11.82	4.37	6.44
C28H51F1N4	462.74	462.4098	72.68	11.11	4.11	12.11
C28H51F1N6	490.76	490.4159	68.53	10.47	3.87	17.12
C28H51F2N1	439.72	439.3989	76.48	11.69	8.64	3.19
C28H51F2N3	467.74	467.4051	71.90	10.99	8.12	8.98
C28H51F2N5	495.75	495.4112	67.84	10.37	7.66	14.13
C28H52F1N1	421.73	421.4084	79.75	12.43	4.50	3.32
C28H52F1N3	449.75	449.4145	74.78	11.65	4.22	9.34
C28H52F1N5	477.76	477.4207	70.39	10.97	3.98	14.66
C28H52F2N2	454.74	454.4098	73.96	11.53	8.36	6.16
C28H52F2N4	482.75	482.4160	69.67	10.86	7.87	11.61
C28H52F2N6	510.76	510.4221	65.84	10.26	7.44	16.45
C28H53F1N2	436.75	436.4193	77.00	12.23	4.35	6.41
C28H53F1N4	464.76	464.4254	72.36	11.49	4.09	12.06
C28H53F1N6	492.77	492.4315	68.25	10.84	3.86	17.05
C28H53F2N1	441.74	441.4146	76.13	12.09	8.60	3.17
C28H53F2N3	469.75	469.4207	71.59	11.37	8.09	8.95
C28H53F2N5	497.76	497.4269	67.56	10.73	7.63	14.07
C28H54F1N1	423.75	423.4240	79.37	12.84	4.48	3.31
C28H54F1N3	451.76	451.4302	74.44	12.05	4.21	9.30
C28H54F1N5	479.77	479.4363	70.10	11.34	3.96	14.60
C28H54F2N2	456.75	456.4255	73.63	11.92	8.32	6.13
C28H54F2N4	484.77	484.4316	69.38	11.23	7.84	11.56
C28H54F2N6	512.78	512.4378	65.59	10.61	7.41	16.39
C28H55F1N2	438.76	438.4349	76.65	12.64	4.33	6.38
C28H55F1N4	466.78	466.4411	72.05	11.88	4.07	12.00
C28H55F1N6	494.79	494.4472	67.97	11.20	3.84	16.99
C28H55F2N1	443.75	443.4302	75.79	12.49	8.56	3.16
C28H55F2N3	471.77	471.4364	71.29	11.75	8.05	8.91
C28H55F2N5	499.78	499.4425	67.29	11.09	7.60	14.01
C28H56F1N1	425.76	425.4397	78.99	13.26	4.46	3.29
C28H56F1N3	453.78	453.4458	74.11	12.44	4.19	9.26
C28H56F1N5	481.79	481.4519	69.80	11.72	3.94	14.54
C28H56F2N2	458.77	458.4411	73.31	12.30	8.28	6.11
C28H56F2N4	486.78	486.4473	69.09	11.60	7.81	11.51
C28H56F2N6	514.80	514.4534	65.33	10.96	7.38	16.32
C28H57F1N2	440.78	440.4506	76.30	13.03	4.31	6.36
C28H57F1N4	468.79	468.4567	71.74	12.26	4.05	11.95
C28H57F1N6	496.81	496.4628	67.69	11.56	3.82	16.92
C28H57F2N1	445.77	445.4459	75.45	12.89	8.52	3.14
C28H57F2N3	473.78	473.4520	70.98	12.13	8.02	8.87
C28H57F2N5	501.80	501.4582	67.02	11.45	7.57	13.96
C28H58F1N1	427.78	427.4553	78.62	13.67	4.44	3.27

C28H58F1N3	455.79	455.4615	73.79	12.83	4.17	9.22
C28H58F1N5	483.81	483.4676	69.51	12.08	3.93	14.48
C28H58F2N2	460.78	460.4568	72.99	12.69	8.25	6.08
C28H58F2N4	488.80	488.4629	68.80	11.96	7.77	11.46
C28H58F2N6	516.81	516.4691	65.07	11.31	7.35	16.26
C28H59F1N2	442.79	442.4662	75.95	13.43	4.29	6.33
C28H59F1N4	470.81	470.4723	71.43	12.63	4.04	11.90
C28H59F1N6	498.82	498.4785	67.42	11.92	3.81	16.85
C28H59F2N3	475.80	475.4677	70.68	12.50	7.99	8.83
C28H59F2N5	503.81	503.4738	66.75	11.80	7.54	13.90
C28H60F1N3	457.81	457.4771	73.46	13.21	4.15	9.18
C28H60F1N5	485.82	485.4832	69.23	12.45	3.91	14.42
C28H60F2N4	490.81	490.4786	68.52	12.32	7.74	11.42
C28H60F2N6	518.83	518.4847	64.82	11.66	7.32	16.20
C28H61F1N4	472.82	472.4880	71.13	13.00	4.02	11.85
C28H61F1N6	500.84	500.4941	67.15	12.28	3.79	16.78
C28H61F2N5	505.83	505.4895	66.49	12.16	7.51	13.85
C28H62F1N5	487.84	487.4989	68.94	12.81	3.89	14.36
C28H62F2N6	520.84	520.5004	64.57	12.00	7.30	16.14
C28H63F1N6	502.85	502.5098	66.88	12.63	3.78	16.71
C29H14F1N1	395.44	395.1110	88.08	3.57	4.80	3.54
C29H14F1N3	423.45	423.1172	82.26	3.33	4.49	9.92
C29H14F1N5	451.47	451.1233	77.15	3.13	4.21	15.51
C29H14F2N2	428.45	428.1125	81.30	3.29	8.87	6.54
C29H14F2N4	456.46	456.1186	76.31	3.09	8.32	12.27
C29H14F2N6	484.47	484.1248	71.90	2.91	7.84	17.35
C29H15F1N2	410.45	410.1219	84.86	3.68	4.63	6.82
C29H15F1N4	438.47	438.1281	79.44	3.45	4.33	12.78
C29H15F1N6	466.48	466.1342	74.67	3.24	4.07	18.02
C29H15F2N1	415.45	415.1172	83.84	3.64	9.15	3.37
C29H15F2N3	443.46	443.1234	78.55	3.41	8.57	9.48
C29H15F2N5	471.47	471.1295	73.88	3.21	8.06	14.85
C29H16F1N1	397.46	397.1267	87.64	4.06	4.78	3.52
C29H16F1N3	425.47	425.1328	81.87	3.79	4.47	9.88
C29H16F1N5	453.48	453.1390	76.81	3.56	4.19	15.44
C29H16F2N2	430.46	430.1281	80.92	3.75	8.83	6.51
C29H16F2N4	458.47	458.1343	75.97	3.52	8.29	12.22
C29H16F2N6	486.49	486.1404	71.60	3.32	7.81	17.27
C29H17F1N2	412.47	412.1376	84.45	4.15	4.61	6.79
C29H17F1N4	440.48	440.1437	79.08	3.89	4.31	12.72
C29H17F1N6	468.50	468.1499	74.35	3.66	4.06	17.94
C29H17F2N1	417.46	417.1329	83.44	4.10	9.10	3.36
C29H17F2N3	445.48	445.1390	78.19	3.85	8.53	9.43
C29H17F2N5	473.49	473.1452	73.57	3.62	8.02	14.79
C29H18F1N1	399.47	399.1423	87.20	4.54	4.76	3.51
C29H18F1N3	427.49	427.1485	81.48	4.24	4.44	9.83
C29H18F1N5	455.50	455.1546	76.47	3.98	4.17	15.38
C29H18F2N2	432.48	432.1438	80.54	4.20	8.79	6.48
C29H18F2N4	460.49	460.1499	75.64	3.94	8.25	12.17
C29H18F2N6	488.50	488.1561	71.30	3.71	7.78	17.20
C29H19F1N2	414.49	414.1532	84.04	4.62	4.58	6.76
C29H19F1N4	442.50	442.1594	78.72	4.33	4.29	12.66
C29H19F1N6	470.51	470.1655	74.03	4.07	4.04	17.86
C29H19F2N1	419.48	419.1485	83.04	4.57	9.06	3.34
C29H19F2N3	447.49	447.1547	77.84	4.28	8.49	9.39
C29H19F2N5	475.51	475.1608	73.25	4.03	7.99	14.73
C29H20F1N1	401.49	401.1580	86.76	5.02	4.73	3.49
C29H20F1N3	429.50	429.1641	81.10	4.69	4.42	9.78
C29H20F1N5	457.51	457.1703	76.13	4.41	4.15	15.31
C29H20F2N2	434.49	434.1594	80.17	4.64	8.75	6.45

C29H20F2N4	462.51	462.1656	75.31	4.36	8.22	12.11
C29H20F2N6	490.52	490.1717	71.01	4.11	7.75	17.13
C29H21F1N2	416.50	416.1689	83.63	5.08	4.56	6.73
C29H21F1N4	444.52	444.1750	78.36	4.76	4.27	12.60
C29H21F1N6	472.53	472.1812	73.71	4.48	4.02	17.79
C29H21F2N1	421.49	421.1642	82.64	5.02	9.01	3.32
C29H21F2N3	449.51	449.1703	77.49	4.71	8.45	9.35
C29H21F2N5	477.52	477.1765	72.94	4.43	7.96	14.67
C29H22F1N1	403.50	403.1736	86.32	5.50	4.71	3.47
C29H22F1N3	431.52	431.1798	80.72	5.14	4.40	9.74
C29H22F1N5	459.53	459.1859	75.80	4.83	4.13	15.24
C29H22F2N2	436.51	436.1751	79.80	5.08	8.70	6.42
C29H22F2N4	464.52	464.1812	74.99	4.77	8.18	12.06
C29H22F2N6	492.54	492.1874	70.72	4.50	7.71	17.06
C29H23F1N2	418.52	418.1845	83.23	5.54	4.54	6.69
C29H23F1N4	446.53	446.1907	78.01	5.19	4.25	12.55
C29H23F1N6	474.55	474.1968	73.40	4.89	4.00	17.71
C29H23F2N1	423.51	423.1798	82.25	5.47	8.97	3.31
C29H23F2N3	451.52	451.1860	77.14	5.13	8.42	9.31
C29H23F2N5	479.54	479.1921	72.64	4.83	7.92	14.60
C29H24F1N1	405.52	405.1893	85.90	5.97	4.68	3.45
C29H24F1N3	433.53	433.1954	80.35	5.58	4.38	9.69
C29H24F1N5	461.55	461.2016	75.47	5.24	4.12	15.17
C29H24F2N2	438.52	438.1907	79.43	5.52	8.66	6.39
C29H24F2N4	466.54	466.1969	74.66	5.19	8.14	12.01
C29H24F2N6	494.55	494.2030	70.43	4.89	7.68	16.99
C29H25F1N2	420.53	420.2002	82.83	5.99	4.52	6.66
C29H25F1N4	448.55	448.2063	77.66	5.62	4.24	12.49
C29H25F1N6	476.56	476.2125	73.09	5.29	3.99	17.63
C29H25F2N1	425.53	425.1955	81.86	5.92	8.93	3.29
C29H25F2N3	453.54	453.2016	76.80	5.56	8.38	9.26
C29H25F2N5	481.55	481.2078	72.33	5.23	7.89	14.54
C29H26F1N1	407.54	407.2049	85.47	6.43	4.66	3.44
C29H26F1N3	435.55	435.2111	79.97	6.02	4.36	9.65
C29H26F1N5	463.56	463.2172	75.14	5.65	4.10	15.11
C29H26F2N2	440.54	440.2064	79.07	5.95	8.63	6.36
C29H26F2N4	468.55	468.2125	74.34	5.59	8.11	11.96
C29H26F2N6	496.57	496.2187	70.15	5.28	7.65	16.92
C29H27F1N2	422.55	422.2158	82.43	6.44	4.50	6.63
C29H27F1N4	450.56	450.2220	77.31	6.04	4.22	12.43
C29H27F1N6	478.58	478.2281	72.78	5.69	3.97	17.56
C29H27F2N1	427.54	427.2111	81.47	6.37	8.89	3.28
C29H27F2N3	455.56	455.2173	76.46	5.97	8.34	9.22
C29H27F2N5	483.57	483.2234	72.03	5.63	7.86	14.48
C29H28F1N1	409.55	409.2206	85.05	6.89	4.64	3.42
C29H28F1N3	437.57	437.2267	79.60	6.45	4.34	9.60
C29H28F1N5	465.58	465.2329	74.82	6.06	4.08	15.04
C29H28F2N2	442.56	442.2220	78.71	6.38	8.59	6.33
C29H28F2N4	470.57	470.2282	74.02	6.00	8.07	11.91
C29H28F2N6	498.58	498.2343	69.86	5.66	7.62	16.86
C29H29F1N2	424.57	424.2315	82.04	6.88	4.47	6.60
C29H29F1N4	452.58	452.2376	76.96	6.46	4.20	12.38
C29H29F1N6	480.59	480.2438	72.48	6.08	3.95	17.49
C29H29F2N1	429.56	429.2268	81.09	6.80	8.85	3.26
C29H29F2N3	457.57	457.2329	76.12	6.39	8.30	9.18
C29H29F2N5	485.58	485.2391	71.73	6.02	7.82	14.42
C29H30F1N1	411.57	411.2362	84.63	7.35	4.62	3.40
C29H30F1N3	439.58	439.2424	79.24	6.88	4.32	9.56
C29H30F1N5	467.59	467.2485	74.49	6.47	4.06	14.98
C29H30F2N2	444.57	444.2377	78.35	6.80	8.55	6.30

C29H30F2N4	472.59	472.2438	73.71	6.40	8.04	11.86
C29H30F2N6	500.60	500.2500	69.58	6.04	7.59	16.79
C29H31F1N2	426.58	426.2471	81.65	7.32	4.45	6.57
C29H31F1N4	454.60	454.2533	76.62	6.87	4.18	12.32
C29H31F1N6	482.61	482.2594	72.18	6.47	3.94	17.41
C29H31F2N1	431.57	431.2424	80.71	7.24	8.80	3.25
C29H31F2N3	459.59	459.2486	75.79	6.80	8.27	9.14
C29H31F2N5	487.60	487.2547	71.44	6.41	7.79	14.36
C29H32F1N1	413.58	413.2519	84.22	7.80	4.59	3.39
C29H32F1N3	441.60	441.2580	78.88	7.30	4.30	9.52
C29H32F1N5	469.61	469.2642	74.17	6.87	4.05	14.91
C29H32F2N2	446.59	446.2533	78.00	7.22	8.51	6.27
C29H32F2N4	474.60	474.2595	73.39	6.80	8.01	11.81
C29H32F2N6	502.62	502.2656	69.30	6.42	7.56	16.72
C29H33F1N2	428.60	428.2628	81.27	7.76	4.43	6.54
C29H33F1N4	456.61	456.2689	76.28	7.28	4.16	12.27
C29H33F1N6	484.62	484.2751	71.87	6.86	3.92	17.34
C29H33F2N1	433.59	433.2581	80.33	7.67	8.76	3.23
C29H33F2N3	461.60	461.2642	75.46	7.21	8.23	9.10
C29H33F2N5	489.62	489.2704	71.14	6.79	7.76	14.30
C29H34F1N1	415.60	415.2675	83.81	8.25	4.57	3.37
C29H34F1N3	443.61	443.2737	78.52	7.73	4.28	9.47
C29H34F1N5	471.63	471.2798	73.86	7.27	4.03	14.85
C29H34F2N2	448.60	448.2690	77.65	7.64	8.47	6.24
C29H34F2N4	476.62	476.2751	73.08	7.19	7.97	11.76
C29H34F2N6	504.63	504.2813	69.03	6.79	7.53	16.65
C29H35F1N2	430.61	430.2784	80.89	8.19	4.41	6.51
C29H35F1N4	458.63	458.2846	75.95	7.69	4.14	12.22
C29H35F1N6	486.64	486.2907	71.58	7.25	3.90	17.27
C29H35F2N1	435.61	435.2737	79.96	8.10	8.72	3.22
C29H35F2N3	463.62	463.2799	75.13	7.61	8.20	9.06
C29H35F2N5	491.63	491.2860	70.85	7.18	7.73	14.25
C29H36F1N1	417.62	417.2832	83.41	8.69	4.55	3.35
C29H36F1N3	445.63	445.2893	78.16	8.14	4.26	9.43
C29H36F1N5	473.64	473.2955	73.54	7.66	4.01	14.79
C29H36F2N2	450.62	450.2846	77.30	8.05	8.43	6.22
C29H36F2N4	478.63	478.2908	72.77	7.58	7.94	11.71
C29H36F2N6	506.65	506.2969	68.75	7.16	7.50	16.59
C29H37F1N2	432.63	432.2941	80.51	8.62	4.39	6.48
C29H37F1N4	460.64	460.3002	75.62	8.10	4.12	12.16
C29H37F1N6	488.66	488.3064	71.28	7.63	3.89	17.20
C29H37F2N1	437.62	437.2894	79.59	8.52	8.68	3.20
C29H37F2N3	465.64	465.2955	74.81	8.01	8.16	9.02
C29H37F2N5	493.65	493.3017	70.56	7.55	7.70	14.19
C29H38F1N1	419.63	419.2988	83.01	9.13	4.53	3.34
C29H38F1N3	447.64	447.3050	77.81	8.56	4.24	9.39
C29H38F1N5	475.66	475.3111	73.23	8.05	3.99	14.72
C29H38F2N2	452.64	452.3003	76.95	8.46	8.39	6.19
C29H38F2N4	480.65	480.3064	72.47	7.97	7.91	11.66
C29H38F2N6	508.66	508.3126	68.48	7.53	7.47	16.52
C29H39F1N2	434.65	434.3097	80.14	9.04	4.37	6.45
C29H39F1N4	462.66	462.3159	75.29	8.50	4.11	12.11
C29H39F1N6	490.67	490.3220	70.99	8.01	3.87	17.13
C29H39F2N1	439.64	439.3050	79.23	8.94	8.64	3.19
C29H39F2N3	467.64	467.3112	74.48	8.41	8.13	8.99
C29H39F2N5	495.66	495.3173	70.27	7.93	7.67	14.13
C29H40F1N1	421.65	421.3145	82.61	9.56	4.51	3.32
C29H40F1N3	449.66	449.3206	77.46	8.97	4.23	9.34
C29H40F1N5	477.67	477.3268	72.92	8.44	3.98	14.66
C29H40F2N2	454.65	454.3159	76.61	8.87	8.36	6.16

C29H40F2N4	482.67	482.3221	72.17	8.35	7.87	11.61
C29H40F2N6	510.68	510.3282	68.21	7.90	7.44	16.46
C29H41F1N2	436.66	436.3254	79.77	9.46	4.35	6.42
C29H41F1N4	464.68	464.3315	74.96	8.89	4.09	12.06
C29H41F1N6	492.69	492.3377	70.70	8.39	3.86	17.06
C29H41F2N1	441.65	441.3207	78.87	9.36	8.60	3.17
C29H41F2N3	469.67	469.3268	74.16	8.80	8.09	8.95
C29H41F2N5	497.68	497.3330	69.99	8.30	7.63	14.07
C29H42F1N1	423.66	423.3301	82.22	9.99	4.48	3.31
C29H42F1N3	451.68	451.3363	77.12	9.37	4.21	9.30
C29H42F1N5	479.69	479.3424	72.61	8.83	3.96	14.60
C29H42F2N2	456.67	456.3316	76.27	9.27	8.32	6.13
C29H42F2N4	484.68	484.3377	71.87	8.73	7.84	11.56
C29H42F2N6	512.70	512.3439	67.94	8.26	7.41	16.39
C29H43F1N2	438.68	438.3410	79.40	9.88	4.33	6.39
C29H43F1N4	466.69	466.3472	74.64	9.29	4.07	12.01
C29H43F1N6	494.70	494.3533	70.41	8.76	3.84	16.99
C29H43F2N1	443.67	443.3363	78.51	9.77	8.56	3.16
C29H43F2N3	471.68	471.3425	73.85	9.19	8.06	8.91
C29H43F2N5	499.70	499.3486	69.71	8.67	7.60	14.02
C29H44F1N1	425.68	425.3458	81.83	10.42	4.46	3.29
C29H44F1N3	453.69	453.3519	76.78	9.78	4.19	9.26
C29H44F1N5	481.71	481.3581	72.31	9.21	3.94	14.54
C29H44F2N2	458.68	458.3472	75.94	9.67	8.28	6.11
C29H44F2N4	486.70	486.3534	71.57	9.11	7.81	11.51
C29H44F2N6	514.71	514.3595	67.67	8.62	7.38	16.33
C29H45F1N2	440.69	440.3567	79.04	10.29	4.31	6.36
C29H45F1N4	468.71	468.3628	74.32	9.68	4.05	11.95
C29H45F1N6	496.72	496.3690	70.12	9.13	3.82	16.92
C29H45F2N1	445.69	445.3520	78.15	10.18	8.53	3.14
C29H45F2N3	473.70	473.3581	73.53	9.58	8.02	8.87
C29H45F2N5	501.71	501.3643	69.43	9.04	7.57	13.96
C29H46F1N1	427.70	427.3614	81.44	10.84	4.44	3.27
C29H46F1N3	455.71	455.3676	76.44	10.17	4.17	9.22
C29H46F1N5	483.72	483.3737	72.01	9.59	3.93	14.48
C29H46F2N2	460.70	460.3629	75.61	10.06	8.25	6.08
C29H46F2N4	488.71	488.3690	71.27	9.49	7.77	11.46
C29H46F2N6	516.73	516.3752	67.41	8.97	7.35	16.26
C29H47F1N2	442.71	442.3723	78.68	10.70	4.29	6.33
C29H47F1N4	470.72	470.3785	74.00	10.06	4.04	11.90
C29H47F1N6	498.74	498.3846	69.84	9.50	3.81	16.85
C29H47F2N1	447.70	447.3676	77.80	10.58	8.49	3.13
C29H47F2N3	475.71	475.3738	73.22	9.96	7.99	8.83
C29H47F2N5	503.73	503.3799	69.15	9.40	7.54	13.90
C29H48F1N1	429.71	429.3771	81.06	11.26	4.42	3.26
C29H48F1N3	457.72	457.3832	76.10	10.57	4.15	9.18
C29H48F1N5	485.74	485.3894	71.71	9.96	3.91	14.42
C29H48F2N2	462.72	462.3785	75.28	10.46	8.21	6.05
C29H48F2N4	490.73	490.3847	70.98	9.86	7.74	11.42
C29H48F2N6	518.74	518.3908	67.15	9.33	7.32	16.20
C29H49F1N2	444.73	444.3880	78.32	11.11	4.27	6.30
C29H49F1N4	472.74	472.3941	73.68	10.45	4.02	11.85
C29H49F1N6	500.75	500.4003	69.56	9.86	3.79	16.78
C29H49F2N1	449.72	449.3833	77.45	10.98	8.45	3.11
C29H49F2N3	477.73	477.3894	72.91	10.34	7.95	8.80
C29H49F2N5	505.74	505.3956	68.87	9.77	7.51	13.85
C29H50F1N1	431.73	431.3927	80.68	11.67	4.40	3.24
C29H50F1N3	459.74	459.3989	75.77	10.96	4.13	9.14
C29H50F1N5	487.75	487.4050	71.41	10.33	3.90	14.36
C29H50F2N2	464.73	464.3942	74.95	10.84	8.18	6.03

C29H50F2N4	492.75	492.4003	70.69	10.23	7.71	11.37
C29H50F2N6	520.76	520.4065	66.89	9.68	7.30	16.14
C29H51F1N2	446.74	446.4036	77.97	11.51	4.25	6.27
C29H51F1N4	474.76	474.4098	73.37	10.83	4.00	11.80
C29H51F1N6	502.77	502.4159	69.28	10.22	3.78	16.72
C29H51F2N1	451.73	451.3989	77.11	11.38	8.41	3.10
C29H51F2N3	479.75	479.4051	72.61	10.72	7.92	8.76
C29H51F2N5	507.76	507.4112	68.60	10.12	7.48	13.79
C29H52F1N1	433.74	433.4084	80.31	12.08	4.38	3.23
C29H52F1N3	461.76	461.4145	75.43	11.35	4.11	9.10
C29H52F1N5	489.77	489.4207	71.12	10.70	3.88	14.30
C29H52F2N2	466.75	466.4098	74.63	11.23	8.14	6.00
C29H52F2N4	494.76	494.4160	70.40	10.59	7.68	11.32
C29H52F2N6	522.77	522.4221	66.63	10.03	7.27	16.08
C29H53F1N2	448.76	448.4193	77.62	11.90	4.23	6.24
C29H53F1N4	476.77	476.4254	73.06	11.21	3.98	11.75
C29H53F1N6	504.78	504.4315	69.00	10.58	3.76	16.65
C29H53F2N1	453.75	453.4146	76.77	11.77	8.37	3.09
C29H53F2N3	481.76	481.4207	72.30	11.09	7.89	8.72
C29H53F2N5	509.78	509.4269	68.33	10.48	7.45	13.74
C29H54F1N1	435.76	435.4240	79.93	12.49	4.36	3.21
C29H54F1N3	463.77	463.4302	75.11	11.74	4.10	9.06
C29H54F1N5	491.79	491.4363	70.83	11.07	3.86	14.24
C29H54F2N2	468.76	468.4255	74.31	11.61	8.11	5.98
C29H54F2N4	496.78	496.4316	70.12	10.96	7.65	11.28
C29H54F2N6	524.79	524.4378	66.37	10.37	7.24	16.01
C29H55F1N2	450.77	450.4349	77.27	12.30	4.21	6.21
C29H55F1N4	478.79	478.4411	72.75	11.58	3.97	11.70
C29H55F1N6	506.80	506.4472	68.73	10.94	3.75	16.58
C29H55F2N1	455.77	455.4302	76.43	12.16	8.34	3.07
C29H55F2N3	483.78	483.4364	72.00	11.46	7.85	8.69
C29H55F2N5	511.79	511.4425	68.06	10.83	7.42	13.68
C29H56F1N1	437.77	437.4397	79.57	12.89	4.34	3.20
C29H56F1N3	465.79	465.4458	74.78	12.12	4.08	9.02
C29H56F1N5	493.80	493.4519	70.54	11.43	3.85	14.18
C29H56F2N2	470.78	470.4411	73.99	11.99	8.07	5.95
C29H56F2N4	498.79	498.4473	69.83	11.32	7.62	11.23
C29H56F2N6	526.81	526.4534	66.12	10.71	7.21	15.95
C29H57F1N2	452.79	452.4506	76.93	12.69	4.20	6.19
C29H57F1N4	480.80	480.4567	72.45	11.95	3.95	11.65
C29H57F1N6	508.82	508.4628	68.46	11.29	3.73	16.52
C29H57F2N1	457.78	457.4459	76.09	12.55	8.30	3.06
C29H57F2N3	485.79	485.4520	71.70	11.83	7.82	8.65
C29H57F2N5	513.81	513.4582	67.79	11.18	7.40	13.63
C29H58F1N1	439.79	439.4553	79.20	13.29	4.32	3.18
C29H58F1N3	467.80	467.4615	74.46	12.50	4.06	8.98
C29H58F1N5	495.82	495.4676	70.25	11.79	3.83	14.12
C29H58F2N2	472.80	472.4568	73.67	12.37	8.04	5.93
C29H58F2N4	500.81	500.4629	69.55	11.67	7.59	11.19
C29H58F2N6	528.82	528.4691	65.87	11.06	7.19	15.89
C29H59F1N2	454.81	454.4662	76.59	13.08	4.18	6.16
C29H59F1N4	482.82	482.4723	72.14	12.32	3.93	11.60
C29H59F1N6	510.83	510.4785	68.19	11.64	3.72	16.45
C29H59F2N1	459.80	459.4615	75.76	12.93	8.26	3.05
C29H59F2N3	487.81	487.4677	71.41	12.19	7.79	8.61
C29H59F2N5	515.82	515.4738	67.53	11.53	7.37	13.58
C29H60F1N1	441.81	441.4710	78.84	13.69	4.30	3.17
C29H60F1N3	469.82	469.4771	74.14	12.87	4.04	8.94
C29H60F1N5	497.83	497.4832	69.97	12.15	3.82	14.07
C29H60F2N2	474.81	474.4724	73.36	12.74	8.00	5.90

10/21/66 237

C29H60F2N4	502.83	502.4786	69.27	12.03	7.56	11.14
C29H60F2N6	530.84	530.4847	65.62	11.39	7.16	15.83
C29H61F1N2	456.82	456.4819	76.25	13.46	4.16	6.13
C29H61F1N4	484.83	484.4880	71.84	12.68	3.92	11.56
C29H61F1N6	512.85	512.4941	67.92	11.99	3.70	16.39
C29H61F2N3	489.83	489.4833	71.11	12.55	7.76	8.58
C29H61F2N5	517.84	517.4895	67.26	11.87	7.34	13.52
C29H62F1N3	471.84	471.4927	73.82	13.24	4.03	8.91
C29H62F1N5	499.85	499.4989	69.69	12.50	3.80	14.01
C29H62F2N4	504.84	504.4942	69.00	12.38	7.53	11.10
C29H62F2N6	532.85	532.5004	65.37	11.73	7.13	15.77
C29H63F1N4	486.85	486.5036	71.55	13.04	3.90	11.51
C29H63F1N6	514.86	514.5098	67.65	12.33	3.69	16.32
C29H63F2N5	519.86	519.5051	67.00	12.22	7.31	13.47
C29H64F1N5	501.87	501.5145	69.41	12.85	3.79	13.95
C29H64F2N6	534.87	534.5160	65.12	12.06	7.10	15.71
C29H65F1N6	516.88	516.5254	67.39	12.68	3.68	16.26
C30H15F1N2	422.47	422.1219	85.29	3.58	4.50	6.63
C30H15F1N4	450.48	450.1281	79.99	3.36	4.22	12.44
C30H15F1N6	478.49	478.1342	75.31	3.16	3.97	17.56
C30H15F2N1	427.46	427.1172	84.30	3.54	8.89	3.28
C30H15F2N3	455.47	455.1234	79.11	3.32	8.34	9.23
C30H15F2N5	483.48	483.1295	74.53	3.13	7.86	14.49
C30H16F1N1	409.47	409.1267	88.00	3.94	4.64	3.42
C30H16F1N3	437.48	437.1328	82.37	3.69	4.34	9.61
C30H16F1N5	465.49	465.1390	77.41	3.46	4.08	15.04
C30H16F2N2	442.47	442.1281	81.44	3.64	8.59	6.33
C30H16F2N4	470.49	470.1343	76.59	3.43	8.08	11.91
C30H16F2N6	498.50	498.1404	72.28	3.24	7.62	16.86
C30H17F1N2	424.48	424.1376	84.89	4.04	4.48	6.60
C30H17F1N4	452.50	452.1437	79.63	3.79	4.20	12.38
C30H17F1N6	480.51	480.1499	74.99	3.57	3.95	17.49
C30H17F2N1	429.47	429.1329	83.90	3.99	8.85	3.26
C30H17F2N3	457.49	457.1390	78.76	3.75	8.31	9.18
C30H17F2N5	485.50	485.1452	74.22	3.53	7.83	14.43
C30H18F1N1	411.48	411.1423	87.57	4.41	4.62	3.40
C30H18F1N3	439.50	439.1485	81.99	4.13	4.32	9.56
C30H18F1N5	467.51	467.1546	77.08	3.88	4.06	14.98
C30H18F2N2	444.49	444.1438	81.07	4.08	8.55	6.30
C30H18F2N4	472.50	472.1499	76.26	3.84	8.04	11.86
C30H18F2N6	500.51	500.1561	71.99	3.62	7.59	16.79
C30H19F1N2	426.50	426.1532	84.49	4.49	4.45	6.57
C30H19F1N4	454.51	454.1594	79.28	4.21	4.18	12.33
C30H19F1N6	482.52	482.1655	74.68	3.97	3.94	17.42
C30H19F2N1	431.49	431.1485	83.51	4.44	8.81	3.25
C30H19F2N3	459.50	459.1547	78.42	4.17	8.27	9.14
C30H19F2N5	487.52	487.1608	73.91	3.93	7.79	14.37
C30H20F1N1	413.50	413.1580	87.14	4.88	4.59	3.39
C30H20F1N3	441.51	441.1641	81.61	4.57	4.30	9.52
C30H20F1N5	469.53	469.1703	76.74	4.29	4.05	14.92
C30H20F2N2	446.50	446.1594	80.70	4.51	8.51	6.27
C30H20F2N4	474.52	474.1656	75.94	4.25	8.01	11.81
C30H20F2N6	502.53	502.1717	71.70	4.01	7.56	16.72
C30H21F1N2	428.51	428.1689	84.09	4.94	4.43	6.54
C30H21F1N4	456.53	456.1750	78.93	4.64	4.16	12.27
C30H21F1N6	484.54	484.1812	74.37	4.37	3.92	17.34
C30H21F2N1	433.51	433.1642	83.12	4.88	8.77	3.23
C30H21F2N3	461.52	461.1703	78.08	4.59	8.23	9.10
C30H21F2N5	489.53	489.1765	73.61	4.32	7.76	14.31
C30H22F1N1	415.51	415.1736	86.72	5.34	4.57	3.37

C30H22F1N3	443.53	443.1798	81.24	5.00	4.28	9.47
C30H22F1N5	471.54	471.1859	76.42	4.70	4.03	14.85
C30H22F2N2	448.52	448.1751	80.34	4.94	8.47	6.25
C30H22F2N4	476.53	476.1812	75.62	4.65	7.97	11.76
C30H22F2N6	504.55	504.1874	71.42	4.40	7.53	16.66
C30H23F1N2	430.53	430.1845	83.70	5.38	4.41	6.51
C30H23F1N4	458.54	458.1907	78.58	5.06	4.14	12.22
C30H23F1N6	486.56	486.1968	74.06	4.76	3.90	17.27
C30H23F2N1	435.52	435.1798	82.74	5.32	8.72	3.22
C30H23F2N3	463.53	463.1860	77.74	5.00	8.20	9.07
C30H23F2N5	491.55	491.1921	73.31	4.72	7.73	14.25
C30H24F1N1	417.53	417.1893	86.30	5.79	4.55	3.35
C30H24F1N3	445.54	445.1954	80.88	5.43	4.26	9.43
C30H24F1N5	473.56	473.2016	76.09	5.11	4.01	14.79
C30H24F2N2	450.54	450.1907	79.98	5.37	8.43	6.22
C30H24F2N4	478.55	478.1969	75.30	5.06	7.94	11.71
C30H24F2N6	506.56	506.2030	71.13	4.78	7.50	16.59
C30H25F1N2	432.55	432.2002	83.31	5.83	4.39	6.48
C30H25F1N4	460.56	460.2063	78.24	5.47	4.13	12.16
C30H25F1N6	488.57	488.2125	73.75	5.16	3.89	17.20
C30H25F2N1	437.54	437.1955	82.36	5.76	8.68	3.20
C30H25F2N3	465.55	465.2016	77.40	5.41	8.16	9.03
C30H25F2N5	493.56	493.2078	73.01	5.11	7.70	14.19
C30H26F1N1	419.55	419.2049	85.89	6.25	4.53	3.34
C30H26F1N3	447.56	447.2111	80.51	5.86	4.24	9.39
C30H26F1N5	475.57	475.2172	75.77	5.51	3.99	14.73
C30H26F2N2	452.55	452.2064	79.62	5.79	8.40	6.19
C30H26F2N4	480.57	480.2125	74.98	5.45	7.91	11.66
C30H26F2N6	508.58	508.2187	70.85	5.15	7.47	16.52
C30H27F1N2	434.56	434.2158	82.92	6.26	4.37	6.45
C30H27F1N4	462.57	462.2220	77.90	5.88	4.11	12.11
C30H27F1N6	490.59	490.2281	73.45	5.55	3.87	17.13
C30H27F2N1	439.55	439.2111	81.98	6.19	8.64	3.19
C30H27F2N3	467.57	467.2173	77.07	5.82	8.13	8.99
C30H27F2N5	495.58	495.2234	72.71	5.49	7.67	14.13
C30H28F1N1	421.56	421.2206	85.48	6.69	4.51	3.32
C30H28F1N3	449.58	449.2267	80.15	6.28	4.23	9.35
C30H28F1N5	477.59	477.2329	75.45	5.91	3.98	14.66
C30H28F2N2	454.57	454.2220	79.27	6.21	8.36	6.16
C30H28F2N4	482.58	482.2282	74.67	5.85	7.87	11.61
C30H28F2N6	510.59	510.2343	70.57	5.53	7.44	16.46
C30H29F1N2	436.58	436.2315	82.54	6.70	4.35	6.42
C30H29F1N4	464.59	464.2376	77.56	6.29	4.09	12.06
C30H29F1N6	492.60	492.2438	73.15	5.93	3.86	17.06
C30H29F2N1	441.57	441.2268	81.60	6.62	8.60	3.17
C30H29F2N3	469.58	469.2329	76.74	6.22	8.09	8.95
C30H29F2N5	497.60	497.2391	72.42	5.87	7.64	14.07
C30H30F1N1	423.58	423.2362	85.07	7.14	4.49	3.31
C30H30F1N3	451.59	451.2424	79.79	6.70	4.21	9.30
C30H30F1N5	479.61	479.2485	75.13	6.30	3.96	14.60
C30H30F2N2	456.58	456.2377	78.92	6.62	8.32	6.14
C30H30F2N4	484.60	484.2438	74.36	6.24	7.84	11.56
C30H30F2N6	512.61	512.2500	70.29	5.90	7.41	16.39
C30H31F1N2	438.59	438.2471	82.16	7.12	4.33	6.39
C30H31F1N4	466.61	466.2533	77.22	6.70	4.07	12.01
C30H31F1N6	494.62	494.2594	72.85	6.32	3.84	16.99
C30H31F2N1	443.59	443.2424	81.23	7.04	8.57	3.16
C30H31F2N3	471.60	471.2486	76.41	6.63	8.06	8.91
C30H31F2N5	499.61	499.2547	72.12	6.25	7.61	14.02
C30H32F1N1	425.59	425.2519	84.67	7.58	4.46	3.29

C30H32F1N3	453.61	453.2580	79.44	7.11	4.19	9.26
C30H32F1N5	481.62	481.2642	74.82	6.70	3.94	14.54
C30H32F2N2	458.60	458.2533	78.57	7.03	8.29	6.11
C30H32F2N4	486.61	486.2595	74.05	6.63	7.81	11.51
C30H32F2N6	514.63	514.2656	70.02	6.27	7.38	16.33
C30H33F1N2	440.61	440.2628	81.78	7.55	4.31	6.36
C30H33F1N4	468.62	468.2689	76.89	7.10	4.05	11.96
C30H33F1N6	496.64	496.2751	72.56	6.70	3.83	16.92
C30H33F2N1	445.60	445.2581	80.86	7.46	8.53	3.14
C30H33F2N3	473.61	473.2642	76.08	7.02	8.02	8.87
C30H33F2N5	501.63	501.2704	71.83	6.63	7.57	13.96
C30H34F1N1	427.61	427.2675	84.27	8.01	4.44	3.28
C30H34F1N3	455.62	455.2737	79.09	7.52	4.17	9.22
C30H34F1N5	483.64	483.2798	74.51	7.09	3.93	14.48
C30H34F2N2	460.62	460.2690	78.23	7.44	8.25	6.08
C30H34F2N4	488.63	488.2751	73.74	7.01	7.78	11.47
C30H34F2N6	516.64	516.2813	69.75	6.63	7.35	16.27
C30H35F1N2	442.63	442.2784	81.41	7.97	4.29	6.33
C30H35F1N4	470.64	470.2846	76.56	7.50	4.04	11.90
C30H35F1N6	498.65	498.2907	72.26	7.07	3.81	16.85
C30H35F2N1	447.62	447.2737	80.50	7.88	8.49	3.13
C30H35F2N3	475.63	475.2799	75.76	7.42	7.99	8.83
C30H35F2N5	503.64	503.2860	71.55	7.00	7.54	13.91
C30H36F1N1	429.63	429.2832	83.87	8.45	4.42	3.26
C30H36F1N3	457.64	457.2893	78.74	7.93	4.15	9.18
C30H36F1N5	485.65	485.2955	74.20	7.47	3.91	14.42
C30H36F2N2	462.63	462.2846	77.89	7.84	8.21	6.06
C30H36F2N4	490.65	490.2908	73.44	7.40	7.74	11.42
C30H36F2N6	518.66	518.2969	69.47	7.00	7.33	16.20
C30H37F1N2	444.64	444.2941	81.04	8.39	4.27	6.30
C30H37F1N4	472.65	472.3002	76.24	7.89	4.02	11.85
C30H37F1N6	500.67	500.3064	71.97	7.45	3.79	16.79
C30H37F2N1	449.63	449.2894	80.14	8.29	8.45	3.12
C30H37F2N3	477.65	477.2955	75.44	7.81	7.96	8.80
C30H37F2N5	505.66	505.3017	71.26	7.38	7.51	13.85
C30H38F1N1	431.64	431.2988	83.48	8.87	4.40	3.24
C30H38F1N3	459.66	459.3050	78.39	8.33	4.13	9.14
C30H38F1N5	487.67	487.3111	73.89	7.85	3.90	14.36
C30H38F2N2	464.65	464.3003	77.55	8.24	8.18	6.03
C30H38F2N4	492.66	492.3064	73.14	7.77	7.71	11.37
C30H38F2N6	520.67	520.3126	69.21	7.36	7.30	16.14
C30H39F1N2	446.66	446.3097	80.67	8.80	4.25	6.27
C30H39F1N4	474.67	474.3159	75.91	8.28	4.00	11.80
C30H39F1N6	502.68	502.3220	71.68	7.82	3.78	16.72
C30H39F2N1	451.65	451.3050	79.78	8.70	8.41	3.10
C30H39F2N3	479.66	479.3112	75.12	8.20	7.92	8.76
C30H39F2N5	507.68	507.3173	70.98	7.74	7.48	13.79
C30H40F1N1	433.66	433.3145	83.09	9.30	4.38	3.23
C30H40F1N3	461.67	461.3206	78.05	8.73	4.12	9.10
C30H40F1N5	489.69	489.3268	73.58	8.23	3.88	14.30
C30H40F2N2	466.66	466.3159	77.22	8.64	8.14	6.00
C30H40F2N4	494.68	494.3221	72.84	8.15	7.68	11.33
C30H40F2N6	522.69	522.3282	68.94	7.71	7.27	16.08
C30H41F1N2	448.67	448.3254	80.31	9.21	4.23	6.24
C30H41F1N4	476.69	476.3315	75.59	8.67	3.99	11.75
C30H41F1N6	504.70	504.3377	71.40	8.19	3.76	16.65
C30H41F2N1	453.66	453.3207	79.43	9.11	8.38	3.09
C30H41F2N3	481.68	481.3268	74.81	8.58	7.89	8.72
C30H41F2N5	509.69	509.3330	70.70	8.11	7.45	13.74
C30H42F1N1	435.67	435.3301	82.71	9.72	4.36	3.21
C30H42F1N3	463.69	463.3363	77.71	9.13	4.10	9.06
C30H42F1N5	491.70	491.3424	73.28	8.61	3.86	14.24
C30H42F2N2	468.68	468.3316	76.88	9.03	8.11	5.98
C30H42F2N4	496.69	496.3377	72.55	8.52	7.65	11.28
C30H42F2N6	524.71	524.3439	68.67	8.07	7.24	16.02
C30H43F1N2	450.69	450.3410	79.95	9.62	4.22	6.22
C30H43F1N4	478.70	478.3472	75.27	9.05	3.97	11.70
C30H43F1N6	506.72	506.3533	71.11	8.55	3.75	16.59
C30H43F2N1	455.68	455.3363	79.08	9.51	8.34	3.07
C30H43F2N3	483.69	483.3425	74.50	8.96	7.86	8.69
C30H43F2N5	511.71	511.3486	70.42	8.47	7.43	13.69
C30H44F1N1	437.69	437.3458	82.33	10.13	4.34	3.20
C30H44F1N3	465.70	465.3519	77.37	9.52	4.08	9.02
C30H44F1N5	493.72	493.3581	72.98	8.98	3.85	14.18
C30H44F2N2	470.70	470.3472	76.55	9.42	8.07	5.95
C30H44F2N4	498.71	498.3534	72.25	8.89	7.62	11.23
C30H44F2N6	526.72	526.3595	68.41	8.42	7.21	15.96
C30H45F1N2	452.70	452.3567	79.60	10.02	4.20	6.19
C30H45F1N4	480.72	480.3628	74.96	9.44	3.95	11.65
C30H45F1N6	508.73	508.3690	70.83	8.92	3.73	16.52
C30H45F2N1	457.70	457.3520	78.73	9.91	8.30	3.06
C30H45F2N3	485.71	485.3581	74.19	9.34	7.82	8.65
C30H45F2N5	513.72	513.3643	70.14	8.83	7.40	13.63
C30H46F1N1	439.71	439.3614	81.95	10.54	4.32	3.19
C30H46F1N3	467.72	467.3676	77.04	9.91	4.06	8.98
C30H46F1N5	495.73	495.3737	72.69	9.35	3.83	14.13
C30H46F2N2	472.71	472.3629	76.23	9.81	8.04	5.93
C30H46F2N4	500.72	500.3690	71.96	9.26	7.59	11.19
C30H46F2N6	528.74	528.3752	68.15	8.77	7.19	15.89
C30H47F1N2	454.72	454.3723	79.24	10.42	4.18	6.16
C30H47F1N4	482.73	482.3785	74.64	9.81	3.94	11.61
C30H47F1N6	510.75	510.3846	70.55	9.28	3.72	16.45
C30H47F2N1	459.71	459.3676	78.38	10.31	8.27	3.05
C30H47F2N3	487.73	487.3738	73.88	9.71	7.79	8.62
C30H47F2N5	515.74	515.3799	69.87	9.19	7.37	13.58
C30H48F1N1	441.72	441.3771	81.57	10.95	4.30	3.17
C30H48F1N3	469.74	469.3832	76.71	10.30	4.04	8.95
C30H48F1N5	497.75	497.3894	72.39	9.72	3.82	14.07
C30H48F2N2	474.73	474.3785	75.90	10.19	8.00	5.90
C30H48F2N4	502.74	502.3847	71.67	9.62	7.56	11.14
C30H48F2N6	530.75	530.3908	67.89	9.12	7.16	15.83
C30H49F1N2	456.74	456.3880	78.89	10.81	4.16	6.13
C30H49F1N4	484.75	484.3941	74.33	10.19	3.92	11.56
C30H49F1N6	512.76	512.4003	70.27	9.63	3.71	16.39
C30H49F2N1	461.73	461.3833	78.04	10.70	8.23	3.03
C30H49F2N3	489.74	489.3894	73.58	10.09	7.76	8.58
C30H49F2N5	517.76	517.3956	69.60	9.54	7.34	13.53
C30H50F1N1	443.74	443.3927	81.20	11.36	4.28	3.16
C30H50F1N3	471.75	471.3989	76.38	10.68	4.03	8.91
C30H50F1N5	499.76	499.4050	72.10	10.08	3.80	14.01
C30H50F2N2	476.74	476.3942	75.58	10.57	7.97	5.88
C30H50F2N4	504.76	504.4003	71.39	9.98	7.53	11.10
C30H50F2N6	532.77	532.4065	67.63	9.46	7.13	15.77
C30H51F1N2	458.75	458.4036	78.55	11.21	4.14	6.11
C30H51F1N4	486.77	486.4098	74.03	10.56	3.90	11.51
C30H51F1N6	514.78	514.4159	70.00	9.99	3.69	16.33
C30H51F2N1	463.74	463.3989	77.70	11.09	8.19	3.02
C30H51F2N3	491.76	491.4051	73.27	10.45	7.73	8.54
C30H51F2N5	519.77	519.4112	69.33	9.89	7.31	13.47
C30H52F1N1	445.75	445.4084	80.84	11.76	4.26	3.14

C30H52F1N3	473.77	473.4145	76.06	11.06	4.01	8.87
C30H52F1N5	501.78	501.4207	71.81	10.45	3.79	13.96
C30H52F2N2	478.76	478.4098	75.26	10.95	7.94	5.85
C30H52F2N4	506.77	506.4160	71.10	10.34	7.50	11.06
C30H52F2N6	534.79	534.4221	67.38	9.80	7.11	15.71
C30H53F1N2	460.77	460.4193	78.20	11.59	4.12	6.08
C30H53F1N4	488.78	488.4254	73.72	10.93	3.89	11.46
C30H53F1N6	516.80	516.4315	69.72	10.34	3.68	16.26
C30H53F2N1	465.76	465.4146	77.36	11.47	8.16	3.01
C30H53F2N3	493.77	493.4207	72.98	10.82	7.70	8.51
C30H53F2N5	521.79	521.4269	69.06	10.24	7.28	13.42
C30H54F1N1	447.77	447.4240	80.47	12.16	4.24	3.13
C30H54F1N3	475.78	475.4302	75.73	11.44	3.99	8.83
C30H54F1N5	503.80	503.4363	71.52	10.80	3.77	13.90
C30H54F2N2	480.78	480.4255	74.95	11.32	7.90	5.83
C30H54F2N4	508.79	508.4316	70.82	10.70	7.47	11.01
C30H54F2N6	536.80	536.4378	67.13	10.14	7.08	15.66
C30H55F1N2	462.78	462.4349	77.86	11.98	4.11	6.05
C30H55F1N4	490.80	490.4411	73.42	11.30	3.87	11.42
C30H55F1N6	518.81	518.4472	69.45	10.69	3.66	16.20
C30H55F2N1	467.78	467.4302	77.03	11.85	8.12	2.99
C30H55F2N3	495.79	495.4364	72.68	11.18	7.66	8.48
C30H55F2N5	523.80	523.4425	68.79	10.58	7.25	13.37
C30H56F1N1	449.79	449.4397	80.11	12.55	4.22	3.11
C30H56F1N3	477.80	477.4458	75.42	11.81	3.98	8.79
C30H56F1N5	505.81	505.4519	71.24	11.16	3.76	13.85
C30H56F2N2	482.79	482.4411	74.64	11.69	7.87	5.80
C30H56F2N4	510.80	510.4473	70.54	11.05	7.44	10.97
C30H56F2N6	538.82	538.4534	66.88	10.48	7.05	15.60
C30H57F1N2	464.80	464.4506	77.52	12.36	4.09	6.03
C30H57F1N4	492.81	492.4567	73.12	11.66	3.86	11.37
C30H57F1N6	520.83	520.4628	69.19	11.03	3.65	16.14
C30H57F2N1	469.79	469.4459	76.70	12.23	8.09	2.98
C30H57F2N3	497.81	497.4520	72.38	11.54	7.63	8.44
C30H57F2N5	525.82	525.4582	68.53	10.93	7.23	13.32
C30H58F1N1	451.80	451.4553	79.75	12.94	4.21	3.10
C30H58F1N3	479.82	479.4615	75.10	12.18	3.96	8.76
C30H58F1N5	507.83	507.4676	70.96	11.51	3.74	13.79
C30H58F2N2	484.81	484.4568	74.33	12.06	7.84	5.78
C30H58F2N4	512.82	512.4629	70.27	11.40	7.41	10.93
C30H58F2N6	540.83	540.4691	66.63	10.81	7.03	15.54
C30H59F1N2	466.82	466.4662	77.19	12.74	4.07	6.00
C30H59F1N4	494.83	494.4723	72.82	12.02	3.84	11.32
C30H59F1N6	522.84	522.4785	68.92	11.37	3.63	16.07
C30H59F2N1	471.81	471.4615	76.37	12.60	8.05	2.97
C30H59F2N3	499.82	499.4677	72.09	11.90	7.60	8.41
C30H59F2N5	527.84	527.4738	68.27	11.27	7.20	13.27
C30H60F1N1	453.82	453.4710	79.40	13.33	4.19	3.09
C30H60F1N3	481.83	481.4771	74.78	12.55	3.94	8.72
C30H60F1N5	509.84	509.4832	70.68	11.86	3.73	13.74
C30H60F2N2	486.82	486.4724	74.02	12.42	7.81	5.75
C30H60F2N4	514.84	514.4786	69.99	11.75	7.38	10.88
C30H60F2N6	542.85	542.4847	66.38	11.14	7.00	15.48
C30H61F1N2	468.83	468.4819	76.86	13.11	4.05	5.98
C30H61F1N4	496.85	496.4880	72.52	12.38	3.82	11.28
C30H61F1N6	524.86	524.4941	68.65	11.71	3.62	16.01
C30H61F2N1	473.82	473.4772	76.05	12.98	8.02	2.96
C30H61F2N3	501.84	501.4833	71.80	12.25	7.57	8.37
C30H61F2N5	529.85	529.4895	68.01	11.60	7.17	13.22
C30H62F1N1	455.83	455.4866	79.05	13.71	4.17	3.07
C30H62F1N3	483.85	483.4927	74.47	12.92	3.93	8.68
C30H62F1N5	511.86	511.4989	70.40	12.21	3.71	13.68
C30H62F2N2	488.84	488.4881	73.71	12.78	7.77	5.73
C30H62F2N4	516.85	516.4942	69.72	12.09	7.35	10.84
C30H62F2N6	544.87	544.5004	66.13	11.47	6.97	15.42
C30H63F1N2	470.85	470.4975	76.53	13.49	4.03	5.95
C30H63F1N4	498.86	498.5036	72.23	12.73	3.81	11.23
C30H63F1N6	526.88	526.5098	68.39	12.05	3.61	15.95
C30H63F2N3	503.85	503.4990	71.52	12.60	7.54	8.34
C30H63F2N5	531.87	531.5051	67.75	11.94	7.14	13.17
C30H64F1N3	485.86	485.5084	74.16	13.28	3.91	8.65
C30H64F1N5	513.88	513.5145	70.12	12.55	3.70	13.63
C30H64F2N4	518.87	518.5099	69.45	12.43	7.32	10.80
C30H64F2N6	546.88	546.5160	65.89	11.80	6.95	15.37
C30H65F1N4	500.88	500.5193	71.94	13.08	3.79	11.19
C30H65F1N6	528.89	528.5254	68.13	12.39	3.59	15.89
C30H65F2N5	533.88	533.5208	67.49	12.27	7.12	13.12
C30H66F1N5	515.89	515.5302	69.85	12.90	3.68	13.58
C30H66F2N6	548.90	548.5317	65.65	12.12	6.92	15.31
C30H67F1N6	530.91	530.5411	67.87	12.72	3.58	15.83

6. CARBON, HYDROGEN, FLUORINE AND OXYGEN

6. KOHLENSTOFF, WASSERSTOFF, FLUOR UND SAUERSTOFF

6. CARBONE, HYDROGENE, FLUOR ET OXYGENE

P

MOLECULAR WEIGHTS AND PERCENTAGE COMPOSITIONS OF COMPOUNDS
CONTAINING CARBON, HYDROGEN, FLUORINE
AND OXYGEN

FORMULA	MOL. WEIGHT	ISOMOL. WEIGHT	PERC. C	PERC. H	PERC. F	PERC. O
C 1H 0F201	66.01	65.9917	18.20	0	57.56	24.24
C 1H 0F202	82.01	81.9866	14.65	0	46.33	39.02
C 1H 0F203	98.01	97.9815	12.26	0	38.77	48.97
C 1H 0F204	114.01	113.9765	10.54	0	33.33	56.14
C 1H 0F205	130.00	129.9714	9.24	0	29.23	61.53
C 1H 0F206	146.00	145.9663	8.23	0	26.02	65.75
C 1H 0F207	162.00	161.9612	7.41	0	23.45	69.13
C 1H 0F208	178.00	177.9561	6.75	0	21.35	71.91
C 1H 0F209	194.00	193.9510	6.19	0	19.59	74.22
C 1H 1F101	48.02	48.0011	25.01	2.10	39.57	33.32
C 1H 1F102	64.02	63.9961	18.76	1.57	29.68	49.99
C 1H 1F103	80.02	79.9910	15.01	1.26	23.74	59.99
C 1H 1F104	96.02	95.9859	12.51	1.05	19.79	66.65
C 1H 1F105	112.01	111.9808	10.72	.90	16.96	71.42
C 1H 1F106	128.01	127.9757	9.38	.79	14.84	74.99
C 1H 1F107	144.01	143.9706	8.34	.70	13.19	77.77
C 1H 1F108	160.01	159.9655	7.51	.63	11.87	79.99
C 1H 1F109	176.01	175.9605	6.82	.57	10.79	81.81
C 1H 2F201	68.02	68.0074	17.66	2.96	55.86	23.52
C 1H 2F202	84.02	84.0023	14.30	2.40	45.22	38.08
C 1H 2F203	100.02	99.9972	12.01	2.02	37.99	47.99
C 1H 2F204	116.02	115.9921	10.35	1.74	32.75	55.16
C 1H 2F205	132.02	131.9870	9.10	1.53	28.78	60.59
C 1H 2F206	148.02	147.9819	8.11	1.36	25.67	64.85
C 1H 2F207	164.02	163.9769	7.32	1.23	23.17	68.28
C 1H 2F208	180.02	179.9718	6.67	1.12	21.11	71.10
C 1H 2F209	196.02	195.9667	6.13	1.03	19.38	73.46
C 1H 3F101	50.03	50.0168	24.01	6.04	37.97	31.98
C 1H 3F102	66.03	66.0117	18.19	4.58	28.77	48.46
C 1H 3F103	82.03	82.0066	14.64	3.69	23.16	58.51
C 1H 3F104	98.03	98.0015	12.25	3.08	19.38	65.28
C 1H 3F105	114.03	113.9964	10.53	2.65	16.66	70.15
C 1H 3F106	130.03	129.9914	9.24	2.33	14.61	73.83
C 1H 3F107	146.03	145.9863	8.23	2.07	13.01	76.69
C 1H 3F108	162.03	161.9812	7.41	1.87	11.73	79.00
C 1H 3F109	178.03	177.9761	6.75	1.70	10.67	80.88
C 2H 1F101	60.03	60.0011	40.02	1.68	31.65	26.65
C 2H 1F102	76.03	75.9961	31.60	1.33	24.99	42.09
C 2H 1F103	92.03	91.9910	26.10	1.10	20.64	52.16
C 2H 1F104	108.03	107.9859	22.24	.93	17.59	59.24
C 2H 1F105	124.03	123.9808	19.37	.81	15.32	64.50
C 2H 1F106	140.03	139.9757	17.16	.72	13.57	68.56
C 2H 1F107	156.02	155.9706	15.40	.65	12.18	71.78
C 2H 1F108	172.02	171.9655	13.96	.59	11.04	74.41
C 2H 1F109	188.02	187.9605	12.78	.54	10.10	76.58
C 2H 2F201	80.03	80.0074	30.01	2.52	47.48	19.99
C 2H 2F202	96.03	96.0023	25.01	2.10	39.57	33.32
C 2H 2F203	112.03	111.9972	21.44	1.80	33.92	42.84
C 2H 2F204	128.03	127.9921	18.76	1.57	29.68	49.99
C 2H 2F205	144.03	143.9870	16.68	1.40	26.38	55.54
C 2H 2F206	160.03	159.9819	15.01	1.26	23.74	59.99
C 2H 2F207	176.03	175.9769	13.65	1.15	21.59	63.62
C 2H 2F208	192.03	191.9718	12.51	1.05	19.79	66.65
C 2H 2F209	208.03	207.9667	11.55	.97	18.27	69.22
C 2H 3F101	62.04	62.0168	38.72	4.87	30.62	25.79
C 2H 3F102	78.04	78.0117	30.78	3.87	24.34	41.00
C 2H 3F103	94.04	94.0066	25.54	3.22	20.20	51.04
C 2H 3F104	110.04	110.0015	21.83	2.75	17.26	58.16
C 2H 3F105	126.04	125.9964	19.06	2.40	15.07	63.47
C 2H 3F106	142.04	141.9914	16.91	2.13	13.38	67.58
C 2H 3F107	158.04	157.9863	15.20	1.91	12.02	70.87
C 2H 3F108	174.04	173.9812	13.80	1.74	10.92	73.54
C 2H 3F109	190.04	189.9761	12.64	1.59	10.00	75.77
C 2H 4F201	82.05	82.0230	29.28	4.91	46.31	19.50
C 2H 4F202	98.05	98.0179	24.50	4.11	38.75	32.64
C 2H 4F203	114.05	114.0128	21.06	3.54	33.32	42.09
C 2H 4F204	130.05	130.0078	18.47	3.10	29.22	49.21
C 2H 4F205	146.05	146.0027	16.45	2.76	26.02	54.77
C 2H 4F206	162.05	161.9976	14.82	2.49	23.45	59.24
C 2H 4F207	178.05	177.9925	13.49	2.26	21.34	62.90
C 2H 4F208	194.05	193.9874	12.38	2.08	19.58	65.96
C 2H 4F209	210.05	209.9823	11.44	1.92	18.09	68.55
C 2H 5F101	64.06	64.0324	37.50	7.87	29.66	24.98
C 2H 5F102	80.06	80.0274	30.01	6.30	23.73	39.97
C 2H 5F103	96.06	96.0223	25.01	5.25	19.78	49.97
C 2H 5F104	112.06	112.0172	21.44	4.50	16.95	57.11
C 2H 5F105	128.06	128.0121	18.76	3.94	14.84	62.47
C 2H 5F106	144.06	144.0070	16.68	3.50	13.19	66.64
C 2H 5F107	160.06	160.0019	15.01	3.15	11.87	69.97
C 2H 5F108	176.06	175.9968	13.64	2.86	10.79	72.70
C 2H 5F109	192.06	191.9918	12.51	2.62	9.89	74.98
C 3H 1F101	72.04	72.0011	50.02	1.40	26.37	22.21
C 3H 1F102	88.04	87.9961	40.93	1.14	21.58	36.35
C 3H 1F103	104.04	103.9910	34.63	.97	18.26	46.14
C 3H 1F104	120.04	119.9859	30.02	.84	15.83	53.31
C 3H 1F105	136.04	135.9808	26.49	.74	13.97	58.81
C 3H 1F106	152.04	151.9757	23.70	.66	12.50	63.14
C 3H 1F107	168.04	167.9706	21.44	.60	11.31	66.65
C 3H 1F108	184.04	183.9655	19.58	.55	10.32	69.55
C 3H 1F109	200.03	199.9605	18.01	.50	9.50	71.98
C 3H 2F201	92.05	92.0074	39.15	2.19	41.28	17.38
C 3H 2F202	108.04	108.0023	33.35	1.87	35.17	29.62
C 3H 2F203	124.04	123.9972	29.05	1.63	30.63	38.69
C 3H 2F204	140.04	139.9921	25.73	1.44	27.13	45.70
C 3H 2F205	156.04	155.9870	23.09	1.29	24.35	51.27
C 3H 2F206	172.04	171.9819	20.94	1.17	22.09	55.80
C 3H 2F207	188.04	187.9769	19.16	1.07	20.21	59.56
C 3H 2F208	204.04	203.9718	17.66	.99	18.62	62.73
C 3H 2F209	220.04	219.9667	16.38	.92	17.27	65.44
C 3H 3F101	74.06	74.0168	48.66	4.08	25.65	21.60

126

C 3H 3F102	90.05	90.0117	40.01	3.36	21.10	35.53
C 3H 3F103	106.05	106.0066	33.98	2.85	17.91	45.26
C 3H 3F104	122.05	122.0015	29.52	2.48	15.57	52.43
C 3H 3F105	138.05	137.9964	26.10	2.19	13.76	57.95
C 3H 3F106	154.05	153.9914	23.39	1.96	12.33	62.31
C 3H 3F107	170.05	169.9863	21.19	1.78	11.17	65.86
C 3H 3F108	186.05	185.9812	19.37	1.63	10.21	68.80
C 3H 3F109	202.05	201.9761	17.83	1.50	9.40	71.27
C 3H 4F201	94.06	94.0230	38.31	4.29	40.40	17.01
C 3H 4F202	110.06	110.0179	32.74	3.66	34.52	29.07
C 3H 4F203	126.06	126.0128	28.58	3.20	30.14	38.08
C 3H 4F204	142.06	142.0078	25.36	2.84	26.75	45.05
C 3H 4F205	158.06	158.0027	22.80	2.55	24.04	50.61
C 3H 4F206	174.06	173.9976	20.70	2.32	21.83	55.15
C 3H 4F207	190.06	189.9925	18.96	2.12	19.99	58.93
C 3H 4F208	206.06	205.9874	17.49	1.96	18.44	62.12
C 3H 4F209	222.06	221.9823	16.23	1.82	17.11	64.85
C 3H 5F101	76.07	76.0324	47.37	6.63	24.97	21.03
C 3H 5F102	92.07	92.0274	39.14	5.47	20.63	34.75
C 3H 5F103	108.07	108.0223	33.34	4.66	17.58	44.41
C 3H 5F104	124.07	124.0172	29.04	4.06	15.31	51.58
C 3H 5F105	140.07	140.0121	25.73	3.60	13.56	57.11
C 3H 5F106	156.07	156.0070	23.09	3.23	12.17	61.51
C 3H 5F107	172.07	172.0019	20.94	2.93	11.04	65.09
C 3H 5F108	188.07	187.9968	19.16	2.68	10.10	68.06
C 3H 5F109	204.07	203.9918	17.66	2.47	9.31	70.56
C 3H 6F201	96.08	96.0387	37.50	6.29	39.55	16.65
C 3H 6F202	112.08	112.0336	32.15	5.40	33.90	28.55
C 3H 6F203	128.08	128.0285	28.13	4.72	29.67	37.48
C 3H 6F204	144.08	144.0234	25.01	4.20	26.37	44.42
C 3H 6F205	160.08	160.0183	22.51	3.78	23.74	49.97
C 3H 6F206	176.07	176.0132	20.46	3.43	21.58	54.52
C 3H 6F207	192.07	192.0082	18.76	3.15	19.78	58.31
C 3H 6F208	208.07	208.0031	17.32	2.91	18.26	61.51
C 3H 6F209	224.07	223.9980	16.08	2.70	16.96	64.26
C 3H 7F101	78.09	78.0481	46.15	9.04	24.33	20.49
C 3H 7F102	94.09	94.0430	38.30	7.50	20.19	34.01
C 3H 7F103	110.09	110.0379	32.73	6.41	17.26	43.60
C 3H 7F104	126.09	126.0328	28.58	5.60	15.07	50.76
C 3H 7F105	142.08	142.0277	25.36	4.97	13.37	56.30
C 3H 7F106	158.08	158.0227	22.79	4.46	12.02	60.72
C 3H 7F107	174.08	174.0176	20.70	4.05	10.91	64.33
C 3H 7F108	190.08	190.0125	18.96	3.71	9.99	67.34
C 3H 7F109	206.08	206.0074	17.48	3.42	9.22	69.87
C 4H 2F201	104.06	104.0074	46.17	1.94	36.52	15.38
C 4H 2F202	120.06	120.0023	40.02	1.68	31.65	26.65
C 4H 2F203	136.06	135.9972	35.31	1.48	27.93	35.28
C 4H 2F204	152.05	151.9921	31.60	1.33	24.99	42.09
C 4H 2F205	168.05	167.9870	28.59	1.20	22.61	47.60
C 4H 2F206	184.05	183.9819	26.10	1.10	20.64	52.16
C 4H 2F207	200.05	199.9769	24.02	1.01	18.99	55.98
C 4H 2F208	216.05	215.9718	22.24	.93	17.59	59.24
C 4H 2F209	232.05	231.9667	20.70	.87	16.37	62.05
C 4H 3F101	86.07	86.0168	55.82	3.51	22.07	18.59
C 4H 3F102	102.07	102.0117	47.07	2.96	18.61	31.35
C 4H 3F103	118.07	118.0066	40.69	2.56	16.09	40.65
C 4H 3F104	134.06	134.0015	35.84	2.26	14.17	47.74
C 4H 3F105	150.06	149.9964	32.02	2.02	12.66	53.31
C 4H 3F106	166.06	165.9914	28.93	1.82	11.44	57.81
C 4H 3F107	182.06	181.9863	26.39	1.66	10.44	61.51
C 4H 3F108	198.06	197.9812	24.26	1.53	9.59	64.62
C 4H 3F109	214.06	213.9761	22.44	1.41	8.88	67.27
C 4H 4F201	106.07	106.0230	45.29	3.80	35.82	15.08
C 4H 4F202	122.07	122.0179	39.36	3.30	31.13	26.21
C 4H 4F203	138.07	138.0128	34.80	2.92	27.52	34.76
C 4H 4F204	154.07	154.0078	31.18	2.62	24.66	41.54
C 4H 4F205	170.07	170.0027	28.25	2.37	22.34	47.04
C 4H 4F206	186.07	185.9976	25.82	2.17	20.42	51.59
C 4H 4F207	202.07	201.9925	23.78	2.00	18.80	55.42
C 4H 4F208	218.07	217.9874	22.03	1.85	17.42	58.69
C 4H 4F209	234.07	233.9823	20.53	1.72	16.23	61.52
C 4H 5F101	88.08	88.0324	54.55	5.72	21.57	18.16
C 4H 5F102	104.08	104.0274	46.16	4.84	18.25	30.74
C 4H 5F103	120.08	120.0223	40.01	4.20	15.82	39.97
C 4H 5F104	136.08	136.0172	35.31	3.70	13.96	47.03
C 4H 5F105	152.08	152.0121	31.59	3.31	12.49	52.60
C 4H 5F106	168.08	168.0070	28.58	3.00	11.30	57.11
C 4H 5F107	184.08	184.0019	26.10	2.74	10.32	60.84
C 4H 5F108	200.08	199.9968	24.01	2.52	9.50	63.97
C 4H 5F109	216.08	215.9918	22.23	2.33	8.79	66.64
C 4H 6F201	108.09	108.0387	44.45	5.60	35.15	14.80
C 4H 6F202	124.09	124.0336	38.72	4.87	30.62	25.79
C 4H 6F203	140.09	140.0285	34.30	4.32	27.12	34.26
C 4H 6F204	156.09	156.0234	30.78	3.87	24.34	41.00
C 4H 6F205	172.09	172.0183	27.92	3.51	22.08	46.49
C 4H 6F206	188.09	188.0132	25.54	3.22	20.20	51.04
C 4H 6F207	204.09	204.0082	23.54	2.96	18.62	54.88
C 4H 6F208	220.08	220.0031	21.83	2.75	17.26	58.16
C 4H 6F209	236.08	235.9980	20.35	2.56	16.09	60.99
C 4H 7F101	90.10	90.0481	53.32	7.83	21.09	17.76
C 4H 7F102	106.10	106.0430	45.28	6.65	17.91	30.16
C 4H 7F103	122.10	122.0379	39.35	5.78	15.56	39.31
C 4H 7F104	138.10	138.0328	34.79	5.11	13.76	46.34
C 4H 7F105	154.10	154.0277	31.18	4.58	12.33	51.91
C 4H 7F106	170.10	170.0227	28.25	4.15	11.17	56.44
C 4H 7F107	186.09	186.0176	25.82	3.79	10.21	60.18
C 4H 7F108	202.09	202.0125	23.77	3.49	9.40	63.33
C 4H 7F109	218.09	218.0074	22.03	3.24	8.71	66.02
C 4H 8F201	110.10	110.0543	43.64	7.32	34.51	14.53
C 4H 8F202	126.10	126.0492	38.10	6.39	30.13	25.37
C 4H 8F203	142.10	142.0441	33.81	5.67	26.74	33.78
C 4H 8F204	158.10	158.0391	30.39	5.10	24.03	40.48
C 4H 8F205	174.10	174.0340	27.60	4.63	21.82	45.95
C 4H 8F206	190.10	190.0289	25.27	4.24	19.99	50.50
C 4H 8F207	206.10	206.0238	23.31	3.91	18.44	54.34
C 4H 8F208	222.10	222.0187	21.63	3.63	17.11	57.63
C 4H 8F209	238.10	238.0136	20.18	3.39	15.96	60.48
C 4H 9F101	92.11	92.0637	52.16	9.85	20.62	17.37
C 4H 9F102	108.11	108.0587	44.44	8.39	17.57	29.60
C 4H 9F103	124.11	124.0536	38.71	7.31	15.31	38.67
C 4H 9F104	140.11	140.0485	34.29	6.47	13.56	45.68
C 4H 9F105	156.11	156.0434	30.78	5.81	12.17	51.24
C 4H 9F106	172.11	172.0383	27.91	5.27	11.04	55.78
C 4H 9F107	188.11	188.0332	25.54	4.82	10.10	59.54
C 4H 9F108	204.11	204.0281	23.54	4.44	9.31	62.71
C 4H 9F109	220.11	220.0231	21.83	4.12	8.63	65.42
C 5H 2F201	116.07	116.0074	51.74	1.74	32.74	13.78
C 5H 2F202	132.07	132.0023	45.47	1.53	28.77	24.23
C 5H 2F203	148.07	147.9972	40.56	1.36	25.66	32.42
C 5H 2F204	164.07	163.9921	36.60	1.23	23.16	39.01

C 5H 2F205	180.07	179.9870	33.35	1.12	21.10	44.43
C 5H 2F206	196.06	195.9819	30.63	1.03	19.38	48.96
C 5H 2F207	212.06	211.9769	28.32	.95	17.92	52.81
C 5H 2F208	228.06	227.9718	26.33	.88	16.66	56.12
C 5H 2F209	244.06	243.9667	24.61	.83	15.57	59.00
C 5H 3F101	98.08	98.0168	61.23	3.08	19.37	16.31
C 5H 3F102	114.08	114.0117	52.64	2.65	16.65	28.05
C 5H 3F103	130.08	130.0066	46.17	2.32	14.61	36.90
C 5H 3F104	146.08	146.0015	41.11	2.07	13.01	43.81
C 5H 3F105	162.08	161.9964	37.05	1.87	11.72	49.36
C 5H 3F106	178.07	177.9914	33.73	1.70	10.67	53.91
C 5H 3F107	194.07	193.9863	30.94	1.56	9.79	57.71
C 5H 3F108	210.07	209.9812	28.59	1.44	9.04	60.93
C 5H 3F109	226.07	225.9761	26.56	1.34	8.40	63.69
C 5H 4F201	118.08	118.0230	50.86	3.41	32.18	13.55
C 5H 4F202	134.08	134.0179	44.79	3.01	28.34	23.86
C 5H 4F203	150.08	150.0128	40.02	2.69	25.32	31.98
C 5H 4F204	166.08	166.0078	36.16	2.43	22.88	38.53
C 5H 4F205	182.08	182.0027	32.98	2.21	20.87	43.93
C 5H 4F206	198.08	197.9976	30.32	2.04	19.18	48.46
C 5H 4F207	214.08	213.9925	28.05	1.88	17.75	52.31
C 5H 4F208	230.08	229.9874	26.10	1.75	16.51	55.63
C 5H 4F209	246.08	245.9823	24.41	1.64	15.44	58.52
C 5H 5F101	100.09	100.0324	60.00	5.04	18.98	15.98
C 5H 5F102	116.09	116.0274	51.73	4.34	16.36	27.56
C 5H 5F103	132.09	132.0223	45.47	3.82	14.38	36.34
C 5H 5F104	148.09	148.0172	40.55	3.40	12.83	43.21
C 5H 5F105	164.09	164.0121	36.60	3.07	11.58	48.75
C 5H 5F106	180.09	180.0070	33.35	2.80	10.55	53.30
C 5H 5F107	196.09	196.0019	30.63	2.57	9.69	57.11
C 5H 5F108	212.09	211.9968	28.32	2.38	8.96	60.35
C 5H 5F109	228.09	227.9918	26.33	2.21	8.33	63.13
C 5H 6F201	120.10	120.0387	50.00	5.04	31.64	13.32
C 5H 6F202	136.10	136.0336	44.13	4.44	27.92	23.51
C 5H 6F203	152.10	152.0285	39.48	3.98	24.98	31.56
C 5H 6F204	168.10	168.0234	35.73	3.60	22.60	38.07
C 5H 6F205	184.10	184.0183	32.62	3.29	20.64	43.45
C 5H 6F206	200.10	200.0132	30.01	3.02	18.99	47.97
C 5H 6F207	216.10	216.0082	27.79	2.80	17.58	51.83
C 5H 6F208	232.10	232.0031	25.88	2.61	16.37	55.15
C 5H 6F209	248.09	247.9980	24.21	2.44	15.32	58.04
C 5H 7F101	102.11	102.0481	58.82	6.91	18.61	15.67
C 5H 7F102	118.11	118.0430	50.85	5.97	16.09	27.09
C 5H 7F103	134.11	134.0379	44.78	5.26	14.17	35.79
C 5H 7F104	150.11	150.0328	40.01	4.70	12.66	42.63
C 5H 7F105	166.11	166.0277	36.15	4.25	11.44	48.16
C 5H 7F106	182.11	182.0227	32.98	3.87	10.43	52.71
C 5H 7F107	198.11	198.0176	30.31	3.56	9.59	56.53
C 5H 7F108	214.11	214.0125	28.05	3.30	8.87	59.78
C 5H 7F109	230.10	230.0074	26.10	3.07	8.26	62.58
C 5H 8F201	122.12	122.0543	49.18	6.60	31.12	13.10
C 5H 8F202	138.12	138.0492	43.48	5.84	27.51	23.17
C 5H 8F203	154.11	154.0441	38.97	5.23	24.65	31.14
C 5H 8F204	170.11	170.0391	35.30	4.74	22.34	37.62
C 5H 8F205	186.11	186.0340	32.27	4.33	20.42	42.98
C 5H 8F206	202.11	202.0289	29.71	3.99	18.80	47.50
C 5H 8F207	218.11	218.0238	27.53	3.70	17.42	51.35
C 5H 8F208	234.11	234.0187	25.65	3.44	16.23	54.67
C 5H 8F209	250.11	250.0136	24.01	3.22	15.19	57.57
C 5H 9F101	104.13	104.0637	57.68	8.71	18.25	15.37
C 5H 9F102	120.12	120.0587	49.99	7.55	15.82	26.64
C 5H 9F103	136.12	136.0536	44.12	6.66	13.96	35.26
C 5H 9F104	152.12	152.0485	39.48	5.96	12.49	42.07
C 5H 9F105	168.12	168.0434	35.72	5.40	11.30	47.58
C 5H 9F106	184.12	184.0383	32.62	4.93	10.32	52.14
C 5H 9F107	200.12	200.0332	30.01	4.53	9.49	55.96
C 5H 9F108	216.12	216.0281	27.79	4.20	8.79	59.22
C 5H 9F109	232.12	232.0231	25.87	3.91	8.18	62.03
C 5H10F201	124.13	124.0700	48.38	8.12	30.61	12.89
C 5H10F202	140.13	140.0649	42.86	7.19	27.12	22.83
C 5H10F203	156.13	156.0598	38.47	6.46	24.34	30.74
C 5H10F204	172.13	172.0547	34.89	5.86	22.07	37.18
C 5H10F205	188.13	188.0496	31.92	5.36	20.20	42.52
C 5H10F206	204.13	204.0445	29.42	4.94	18.61	47.03
C 5H10F207	220.13	220.0394	27.28	4.58	17.26	50.88
C 5H10F208	236.13	236.0344	25.43	4.27	16.09	54.21
C 5H10F209	252.13	252.0293	23.82	4.00	15.07	57.11
C 5H11F101	106.14	106.0794	56.58	10.45	17.90	15.07
C 5H11F102	122.14	122.0743	49.17	9.08	15.55	26.20
C 5H11F103	138.14	138.0692	43.47	8.03	13.75	34.75
C 5H11F104	154.14	154.0641	38.96	7.19	12.33	41.52
C 5H11F105	170.14	170.0590	35.30	6.52	11.17	47.02
C 5H11F106	186.14	186.0540	32.26	5.96	10.21	51.57
C 5H11F107	202.14	202.0489	29.71	5.49	9.40	55.41
C 5H11F108	218.14	218.0438	27.53	5.08	8.71	58.68
C 5H11F109	234.14	234.0387	25.65	4.74	8.11	61.50
C 6H 3F101	110.09	110.0168	65.46	2.75	17.26	14.53
C 6H 3F102	126.09	126.0117	57.16	2.40	15.07	25.38
C 6H 3F103	142.09	142.0066	50.72	2.13	13.37	33.78
C 6H 3F104	158.09	158.0015	45.59	1.91	12.02	40.48
C 6H 3F105	174.09	173.9964	41.40	1.74	10.91	45.95
C 6H 3F106	190.09	189.9914	37.91	1.59	9.99	50.50
C 6H 3F107	206.09	205.9863	34.97	1.47	9.22	54.34
C 6H 3F108	222.08	221.9812	32.45	1.36	8.55	57.63
C 6H 3F109	238.08	237.9761	30.27	1.27	7.98	60.48
C 6H 4F201	130.09	130.0230	55.40	3.10	29.21	12.30
C 6H 4F202	146.09	146.0179	49.33	2.76	26.01	21.90
C 6H 4F203	162.09	162.0128	44.46	2.49	23.44	29.61
C 6H 4F204	178.09	178.0078	40.47	2.26	21.34	35.93
C 6H 4F205	194.09	194.0027	37.13	2.08	19.58	41.22
C 6H 4F206	210.09	209.9976	34.30	1.92	18.09	45.69
C 6H 4F207	226.09	225.9925	31.88	1.78	16.81	49.54
C 6H 4F208	242.09	241.9874	29.77	1.67	15.70	52.87
C 6H 4F209	258.09	257.9823	27.92	1.56	14.72	55.79
C 6H 5F101	112.10	112.0324	64.29	4.50	16.95	14.27
C 6H 5F102	128.10	128.0274	56.26	3.93	14.83	24.98
C 6H 5F103	144.10	144.0223	50.01	3.50	13.18	33.31
C 6H 5F104	160.10	160.0172	45.01	3.15	11.87	39.97
C 6H 5F105	176.10	176.0121	40.92	2.86	10.79	45.43
C 6H 5F106	192.10	192.0070	37.52	2.62	9.89	49.97
C 6H 5F107	208.10	208.0019	34.63	2.42	9.13	53.82
C 6H 5F108	224.10	223.9968	32.16	2.25	8.48	57.12
C 6H 5F109	240.10	239.9918	30.02	2.10	7.91	59.97
C 6H 6F201	132.11	132.0387	54.55	4.58	28.76	12.11
C 6H 6F202	148.11	148.0336	48.66	4.08	25.65	21.60
C 6H 6F203	164.11	164.0285	43.91	3.69	23.15	29.25
C 6H 6F204	180.11	180.0234	40.01	3.36	21.10	35.53
C 6H 6F205	196.11	196.0183	36.75	3.08	19.38	40.79
C 6H 6F206	212.11	212.0132	33.98	2.85	17.91	45.26
C 6H 6F207	228.11	228.0082	31.59	2.65	16.66	49.10

C 6H 6F208	244.11	244.0031	29.52	2.48	15.57	52.43
C 6H 6F209	260.11	259.9980	27.71	2.33	14.61	55.36
C 6H 7F101	114.12	114.0481	63.15	6.18	16.65	14.02
C 6H 7F102	130.12	130.0430	55.38	5.42	14.60	24.59
C 6H 7F103	146.12	146.0379	49.32	4.83	13.00	32.85
C 6H 7F104	162.12	162.0328	44.45	4.35	11.72	39.48
C 6H 7F105	178.12	178.0277	40.46	3.96	10.67	44.91
C 6H 7F106	194.12	194.0227	37.13	3.63	9.79	49.45
C 6H 7F107	210.12	210.0176	34.30	3.36	9.04	53.30
C 6H 7F108	226.12	226.0125	31.87	3.12	8.40	56.61
C 6H 7F109	242.12	242.0074	29.77	2.91	7.85	59.47
C 6H 8F201	134.13	134.0543	53.73	6.01	28.33	11.93
C 6H 8F202	150.13	150.0492	48.00	5.37	25.31	21.31
C 6H 8F203	166.13	166.0441	43.38	4.85	22.87	28.89
C 6H 8F204	182.13	182.0391	39.57	4.43	20.86	35.14
C 6H 8F205	198.12	198.0340	36.37	4.07	19.18	40.38
C 6H 8F206	214.12	214.0289	33.66	3.77	17.75	44.83
C 6H 8F207	230.12	230.0238	31.32	3.50	16.51	48.67
C 6H 8F208	246.12	246.0187	29.28	3.28	15.44	52.00
C 6H 8F209	262.12	262.0136	27.49	3.08	14.50	54.93
C 6H 9F101	116.14	116.0637	62.05	7.81	16.36	13.78
C 6H 9F102	132.14	132.0587	54.54	6.87	14.38	24.22
C 6H 9F103	148.14	148.0536	48.65	6.12	12.83	32.40
C 6H 9F104	164.13	164.0485	43.91	5.53	11.57	38.99
C 6H 9F105	180.13	180.0434	40.01	5.04	10.55	44.41
C 6H 9F106	196.13	196.0383	36.74	4.63	9.69	48.94
C 6H 9F107	212.13	212.0332	33.97	4.28	8.96	52.80
C 6H 9F108	228.13	228.0281	31.59	3.98	8.33	56.11
C 6H 9F109	244.13	244.0231	29.52	3.72	7.78	58.98
C 6H10F201	136.14	136.0700	52.93	7.40	27.91	11.75
C 6H10F202	152.14	152.0649	47.37	6.63	24.97	21.03
C 6H10F203	168.14	168.0598	42.86	5.99	22.60	28.55
C 6H10F204	184.14	184.0547	39.14	5.47	20.63	34.75
C 6H10F205	200.14	200.0496	36.01	5.04	18.99	39.97
C 6H10F206	216.14	216.0445	33.34	4.66	17.58	44.41
C 6H10F207	232.14	232.0394	31.04	4.34	16.37	48.25
C 6H10F208	248.14	248.0344	29.04	4.06	15.31	51.58
C 6H10F209	264.14	264.0293	27.28	3.82	14.39	54.51
C 6H11F101	118.15	118.0794	60.99	9.38	16.08	13.54
C 6H11F102	134.15	134.0743	53.72	8.27	14.16	23.85
C 6H11F103	150.15	150.0692	48.00	7.38	12.65	31.97
C 6H11F104	166.15	166.0641	43.37	6.67	11.43	38.52
C 6H11F105	182.15	182.0590	39.56	6.09	10.43	43.92
C 6H11F106	198.15	198.0540	36.37	5.60	9.59	48.45
C 6H11F107	214.15	214.0489	33.65	5.18	8.87	52.30
C 6H11F108	230.15	230.0438	31.31	4.82	8.25	55.61
C 6H11F109	246.15	246.0387	29.28	4.50	7.72	58.50
C 6H12F201	138.16	138.0856	52.16	8.75	27.50	11.58
C 6H12F202	154.16	154.0805	46.75	7.85	24.65	20.76
C 6H12F203	170.16	170.0754	42.35	7.11	22.33	28.21
C 6H12F204	186.16	186.0704	38.71	6.50	20.41	34.38
C 6H12F205	202.16	202.0653	35.65	5.98	18.80	39.57
C 6H12F206	218.16	218.0602	33.03	5.54	17.42	44.00
C 6H12F207	234.16	234.0551	30.78	5.17	16.23	47.83
C 6H12F208	250.15	250.0500	28.81	4.84	15.19	51.17
C 6H12F209	266.15	266.0449	27.08	4.54	14.28	54.10
C 6H13F101	120.17	120.0950	59.97	10.90	15.81	13.31
C 6H13F102	136.17	136.0900	52.93	9.62	13.95	23.50
C 6H13F103	152.17	152.0849	47.36	8.61	12.49	31.54
C 6H13F104	168.17	168.0798	42.85	7.79	11.30	38.06
C 6H13F105	184.17	184.0747	39.13	7.12	10.32	43.44
C 6H13F106	200.17	200.0696	36.00	6.55	9.49	47.96
C 6H13F107	216.16	216.0645	33.34	6.06	8.79	51.81
C 6H13F108	232.16	232.0594	31.04	5.64	8.18	55.13
C 6H13F109	248.16	248.0543	29.04	5.28	7.66	58.02
C 7H 3F101	122.10	122.0168	68.86	2.48	15.56	13.10
C 7H 3F102	138.10	138.0117	60.88	2.19	13.76	23.17
C 7H 3F103	154.10	154.0066	54.56	1.96	12.33	31.15
C 7H 3F104	170.10	170.0015	49.43	1.78	11.17	37.62
C 7H 3F105	186.10	185.9964	45.18	1.62	10.21	42.99
C 7H 3F106	202.10	201.9914	41.60	1.50	9.40	47.50
C 7H 3F107	218.10	217.9863	38.55	1.39	8.71	51.35
C 7H 3F108	234.10	233.9812	35.92	1.29	8.12	54.68
C 7H 3F109	250.09	249.9761	33.62	1.21	7.60	57.58
C 7H 4F201	142.11	142.0230	59.17	2.84	26.74	11.26
C 7H 4F202	158.11	158.0179	53.18	2.55	24.03	20.24
C 7H 4F203	174.10	174.0128	48.29	2.32	21.82	27.57
C 7H 4F204	190.10	190.0078	44.23	2.12	19.99	33.66
C 7H 4F205	206.10	206.0027	40.79	1.96	18.44	38.81
C 7H 4F206	222.10	221.9976	37.86	1.82	17.11	43.22
C 7H 4F207	238.10	237.9925	35.31	1.69	15.96	47.04
C 7H 4F208	254.10	253.9874	33.09	1.59	14.95	50.37
C 7H 4F209	270.10	269.9823	31.13	1.49	14.07	53.31
C 7H 5F101	124.12	124.0324	67.74	4.06	15.31	12.89
C 7H 5F102	140.12	140.0274	60.01	3.60	13.56	22.84
C 7H 5F103	156.11	156.0223	53.86	3.23	12.17	30.75
C 7H 5F104	172.11	172.0172	48.85	2.93	11.04	37.18
C 7H 5F105	188.11	188.0121	44.70	2.68	10.10	42.53
C 7H 5F106	204.11	204.0070	41.19	2.47	9.31	47.03
C 7H 5F107	220.11	220.0019	38.20	2.29	8.63	50.88
C 7H 5F108	236.11	235.9968	35.61	2.13	8.05	54.21
C 7H 5F109	252.11	251.9918	33.35	2.00	7.54	57.12
C 7H 6F201	144.12	144.0387	58.34	4.20	26.36	11.10
C 7H 6F202	160.12	160.0336	52.51	3.78	23.73	19.98
C 7H 6F203	176.12	176.0285	47.74	3.43	21.57	27.25
C 7H 6F204	192.12	192.0234	43.76	3.15	19.78	33.31
C 7H 6F205	208.12	208.0183	40.40	2.91	18.26	38.44
C 7H 6F206	224.12	224.0132	37.51	2.70	16.95	42.83
C 7H 6F207	240.12	240.0082	35.02	2.52	15.82	46.64
C 7H 6F208	256.12	256.0031	32.83	2.36	14.84	49.98
C 7H 6F209	272.12	271.9980	30.90	2.22	13.96	52.92
C 7H 7F101	126.13	126.0481	66.66	5.59	15.06	12.68
C 7H 7F102	142.13	142.0430	59.16	4.96	13.37	22.51
C 7H 7F103	158.13	158.0379	53.17	4.46	12.01	30.35
C 7H 7F104	174.13	174.0328	48.28	4.05	10.91	36.75
C 7H 7F105	190.13	190.0277	44.22	3.71	9.99	42.08
C 7H 7F106	206.13	206.0227	40.79	3.42	9.22	46.57
C 7H 7F107	222.13	222.0176	37.85	3.18	8.55	50.42
C 7H 7F108	238.13	238.0125	35.31	2.96	7.98	53.75
C 7H 7F109	254.13	254.0074	33.09	2.78	7.48	56.66
C 7H 8F201	146.14	146.0543	57.53	5.52	26.00	10.95
C 7H 8F202	162.14	162.0492	51.86	4.97	23.43	19.74
C 7H 8F203	178.14	178.0441	47.20	4.53	21.33	26.94
C 7H 8F204	194.14	194.0391	43.31	4.15	19.57	32.97
C 7H 8F205	210.14	210.0340	40.01	3.84	18.08	38.07
C 7H 8F206	226.14	226.0289	37.18	3.57	16.80	42.45
C 7H 8F207	242.13	242.0238	34.72	3.33	15.69	46.25
C 7H 8F208	258.13	258.0187	32.57	3.12	14.72	49.58
C 7H 8F209	274.13	274.0136	30.67	2.94	13.86	52.53
C 7H 9F101	128.15	128.0637	65.61	7.08	14.83	12.49

C 7H 9F102	144.15	144.0587	58.33	6.29	13.18	22.20
C 7H 9F103	160.15	160.0536	52.50	5.66	11.86	29.97
C 7H 9F104	176.15	176.0485	47.73	5.15	10.79	36.33
C 7H 9F105	192.15	192.0434	43.76	4.72	9.89	41.63
C 7H 9F106	208.14	208.0383	40.39	4.36	9.13	46.12
C 7H 9F107	224.14	224.0332	37.51	4.05	8.48	49.97
C 7H 9F108	240.14	240.0281	35.01	3.78	7.91	53.30
C 7H 9F109	256.14	256.0231	32.82	3.54	7.42	56.22
C 7H10F201	148.15	148.0700	56.75	6.80	25.65	10.80
C 7H10F202	164.15	164.0649	51.22	6.14	23.15	19.49
C 7H10F203	180.15	180.0598	46.67	5.60	21.09	26.64
C 7H10F204	196.15	196.0547	42.86	5.14	19.37	32.63
C 7H10F205	212.15	212.0496	39.63	4.75	17.91	37.71
C 7H10F206	228.15	228.0445	36.85	4.42	16.65	42.08
C 7H10F207	244.15	244.0394	34.44	4.13	15.56	45.87
C 7H10F208	260.15	260.0344	32.32	3.87	14.61	49.20
C 7H10F209	276.15	276.0293	30.45	3.65	13.76	52.14
C 7H11F101	130.16	130.0794	64.59	8.52	14.60	12.29
C 7H11F102	146.16	146.0743	57.52	7.59	13.00	21.89
C 7H11F103	162.16	162.0692	51.85	6.84	11.72	29.60
C 7H11F104	178.16	178.0641	47.19	6.22	10.66	35.92
C 7H11F105	194.16	194.0590	43.30	5.71	9.78	41.20
C 7H11F106	210.16	210.0540	40.01	5.28	9.04	45.68
C 7H11F107	226.16	226.0489	37.18	4.90	8.40	49.52
C 7H11F108	242.16	242.0438	34.72	4.58	7.85	52.86
C 7H11F109	258.16	258.0387	32.57	4.29	7.36	55.78
C 7H12F201	150.17	150.0856	55.99	8.05	25.30	10.65
C 7H12F202	166.17	166.0805	50.60	7.28	22.87	19.26
C 7H12F203	182.17	182.0754	46.15	6.64	20.86	26.35
C 7H12F204	198.17	198.0704	42.43	6.10	19.17	32.29
C 7H12F205	214.17	214.0653	39.26	5.65	17.74	37.35
C 7H12F206	230.17	230.0602	36.53	5.26	16.51	41.71
C 7H12F207	246.17	246.0551	34.15	4.91	15.44	45.50
C 7H12F208	262.17	262.0500	32.07	4.61	14.49	48.82
C 7H12F209	278.17	278.0449	30.23	4.35	13.66	51.77
C 7H13F101	132.18	132.0950	63.61	9.91	14.37	12.10
C 7H13F102	148.18	148.0900	56.74	8.84	12.82	21.59
C 7H13F103	164.18	164.0849	51.21	7.98	11.57	29.24
C 7H13F104	180.18	180.0798	46.66	7.27	10.54	35.52
C 7H13F105	196.18	196.0747	42.86	6.68	9.68	40.78
C 7H13F106	212.18	212.0696	39.63	6.18	8.95	45.24
C 7H13F107	228.18	228.0645	36.85	5.74	8.33	49.08
C 7H13F108	244.18	244.0594	34.43	5.37	7.78	52.42
C 7H13F109	260.17	260.0543	32.32	5.04	7.30	55.35
C 7H14F201	152.19	152.1013	55.25	9.27	24.97	10.51
C 7H14F202	168.19	168.0962	49.99	8.39	22.59	19.03
C 7H14F203	184.18	184.0911	45.65	7.66	20.63	26.06
C 7H14F204	200.18	200.0860	42.00	7.05	18.98	31.97
C 7H14F205	216.18	216.0809	38.89	6.53	17.58	37.00
C 7H14F206	232.18	232.0758	36.21	6.08	16.37	41.35
C 7H14F207	248.18	248.0707	33.88	5.69	15.31	45.13
C 7H14F208	264.18	264.0657	31.83	5.34	14.38	48.45
C 7H14F209	280.18	280.0606	30.01	5.04	13.56	51.39
C 7H15F101	134.20	134.1107	62.65	11.27	14.16	11.92
C 7H15F102	150.19	150.1056	55.98	10.07	12.65	21.30
C 7H15F103	166.19	166.1005	50.59	9.10	11.43	28.88
C 7H15F104	182.19	182.0954	46.15	8.30	10.43	35.13
C 7H15F105	198.19	198.0903	42.42	7.63	9.59	40.36
C 7H15F106	214.19	214.0853	39.25	7.06	8.87	44.82
C 7H15F107	230.19	230.0802	36.53	6.57	8.25	48.65

C 7H15F108	246.19	246.0751	34.15	6.14	7.72	51.99
C 7H15F109	262.19	262.0700	32.07	5.77	7.25	54.92
C 8H 4F201	154.12	154.0230	62.35	2.62	24.65	10.38
C 8H 4F202	170.12	170.0179	56.48	2.37	22.34	18.81
C 8H 4F203	186.12	186.0128	51.63	2.17	20.42	25.79
C 8H 4F204	202.12	202.0078	47.54	1.99	18.80	31.66
C 8H 4F205	218.11	218.0027	44.05	1.85	17.42	36.68
C 8H 4F206	234.11	233.9976	41.04	1.72	16.23	41.00
C 8H 4F207	250.11	249.9925	38.42	1.61	15.19	44.78
C 8H 4F208	266.11	265.9874	36.11	1.52	14.28	48.10
C 8H 4F209	282.11	281.9823	34.06	1.43	13.47	51.04
C 8H 5F101	136.13	136.0324	70.59	3.70	13.96	11.75
C 8H 5F102	152.13	152.0274	63.16	3.31	12.49	21.03
C 8H 5F103	168.13	168.0223	57.15	3.00	11.30	28.55
C 8H 5F104	184.13	184.0172	52.19	2.74	10.32	34.76
C 8H 5F105	200.12	200.0121	48.01	2.52	9.49	39.97
C 8H 5F106	216.12	216.0070	44.46	2.33	8.79	44.42
C 8H 5F107	232.12	232.0019	41.40	2.17	8.18	48.25
C 8H 5F108	248.12	247.9968	38.73	2.03	7.66	51.59
C 8H 5F109	264.12	263.9918	36.38	1.91	7.19	54.52
C 8H 6F201	156.13	156.0387	61.54	3.87	24.34	10.25
C 8H 6F202	172.13	172.0336	55.82	3.51	22.07	18.59
C 8H 6F203	188.13	188.0285	51.08	3.21	20.20	25.51
C 8H 6F204	204.13	204.0234	47.07	2.96	18.61	31.35
C 8H 6F205	220.13	220.0183	43.65	2.75	17.26	36.34
C 8H 6F206	236.13	236.0132	40.69	2.56	16.09	40.65
C 8H 6F207	252.13	252.0082	38.11	2.40	15.07	44.42
C 8H 6F208	268.13	268.0031	35.84	2.26	14.17	47.74
C 8H 6F209	284.13	283.9980	33.82	2.13	13.37	50.68
C 8H 7F101	138.14	138.0481	69.56	5.11	13.75	11.58
C 8H 7F102	154.14	154.0430	62.34	4.58	12.33	20.76
C 8H 7F103	170.14	170.0379	56.48	4.15	11.17	28.21
C 8H 7F104	186.14	186.0328	51.62	3.79	10.21	34.38
C 8H 7F105	202.14	202.0277	47.54	3.49	9.40	39.57
C 8H 7F106	218.14	218.0227	44.05	3.23	8.71	44.01
C 8H 7F107	234.14	234.0176	41.04	3.01	8.11	47.83
C 8H 7F108	250.14	250.0125	38.41	2.82	7.60	51.17
C 8H 7F109	266.14	266.0074	36.11	2.65	7.14	54.11
C 8H 8F201	158.15	158.0543	60.76	5.10	24.03	10.12
C 8H 8F202	174.15	174.0492	55.18	4.63	21.82	18.37
C 8H 8F203	190.15	190.0441	50.53	4.24	19.98	25.24
C 8H 8F204	206.15	206.0391	46.61	3.91	18.43	31.04
C 8H 8F205	222.15	222.0340	43.25	3.63	17.10	36.01
C 8H 8F206	238.15	238.0289	40.35	3.39	15.96	40.31
C 8H 8F207	254.15	254.0238	37.81	3.17	14.95	44.07
C 8H 8F208	270.14	270.0187	35.57	2.98	14.07	47.38
C 8H 8F209	286.14	286.0136	33.58	2.82	13.28	50.32
C 8H 9F101	140.16	140.0637	68.56	6.47	13.55	11.42
C 8H 9F102	156.16	156.0587	61.53	5.81	12.17	20.49
C 8H 9F103	172.16	172.0536	55.81	5.27	11.04	27.88
C 8H 9F104	188.16	188.0485	51.07	4.82	10.10	34.01
C 8H 9F105	204.16	204.0434	47.07	4.44	9.31	39.18
C 8H 9F106	220.16	220.0383	43.65	4.12	8.63	43.60
C 8H 9F107	236.16	236.0332	40.69	3.84	8.04	47.42
C 8H 9F108	252.15	252.0281	38.11	3.60	7.53	50.76
C 8H 9F109	268.15	268.0231	35.83	3.38	7.08	53.70
C 8H10F201	160.17	160.0700	59.99	6.29	23.72	9.99
C 8H10F202	176.16	176.0649	54.55	5.72	21.57	18.16
C 8H10F203	192.16	192.0598	50.00	5.25	19.77	24.98
C 8H10F204	208.16	208.0547	46.16	4.84	18.25	30.74

C 8H10F205	224.16	224.0496	42.87	4.50	16.95	35.69
C 8H10F206	240.16	240.0445	40.01	4.20	15.82	39.97
C 8H10F207	256.16	256.0394	37.51	3.93	14.83	43.72
C 8H10F208	272.16	272.0344	35.31	3.70	13.96	47.03
C 8H10F209	288.16	288.0293	33.35	3.50	13.19	49.97
C 8H11F101	142.17	142.0794	67.59	7.80	13.36	11.25
C 8H11F102	158.17	158.0743	60.75	7.01	12.01	20.23
C 8H11F103	174.17	174.0692	55.17	6.37	10.91	27.56
C 8H11F104	190.17	190.0641	50.53	5.83	9.99	33.65
C 8H11F105	206.17	206.0590	46.61	5.38	9.21	38.80
C 8H11F106	222.17	222.0540	43.25	4.99	8.55	43.21
C 8H11F107	238.17	238.0489	40.34	4.66	7.98	47.02
C 8H11F108	254.17	254.0438	37.81	4.36	7.47	50.36
C 8H11F109	270.17	270.0387	35.57	4.10	7.03	53.30
C 8H12F201	162.18	162.0856	59.25	7.46	23.43	9.87
C 8H12F202	178.18	178.0805	53.93	6.79	21.32	17.96
C 8H12F203	194.18	194.0754	49.48	6.23	19.57	24.72
C 8H12F204	210.18	210.0704	45.72	5.75	18.08	30.45
C 8H12F205	226.18	226.0653	42.48	5.35	16.80	35.37
C 8H12F206	242.18	242.0602	39.68	4.99	15.69	39.64
C 8H12F207	258.18	258.0551	37.22	4.69	14.72	43.38
C 8H12F208	274.18	274.0500	35.05	4.41	13.86	46.68
C 8H12F209	290.18	290.0449	33.11	4.17	13.09	49.62
C 8H13F101	144.19	144.0950	66.64	9.09	13.18	11.10
C 8H13F102	160.19	160.0900	59.98	8.18	11.86	19.98
C 8H13F103	176.19	176.0849	54.54	7.44	10.78	27.24
C 8H13F104	192.19	192.0798	50.00	6.82	9.89	33.30
C 8H13F105	208.19	208.0747	46.15	6.29	9.13	38.43
C 8H13F106	224.19	224.0696	42.86	5.84	8.47	42.82
C 8H13F107	240.19	240.0645	40.01	5.46	7.91	46.63
C 8H13F108	256.19	256.0594	37.51	5.11	7.42	49.96
C 8H13F109	272.19	272.0543	35.30	4.81	6.98	52.90
C 8H14F201	164.20	164.1013	58.52	8.59	23.14	9.74
C 8H14F202	180.20	180.0962	53.32	7.83	21.09	17.76
C 8H14F203	196.20	196.0911	48.98	7.19	19.37	24.46
C 8H14F204	212.20	212.0860	45.28	6.65	17.91	30.16
C 8H14F205	228.19	228.0809	42.11	6.18	16.65	35.06
C 8H14F206	244.19	244.0758	39.35	5.78	15.56	39.31
C 8H14F207	260.19	260.0707	36.93	5.42	14.60	43.04
C 8H14F208	276.19	276.0657	34.79	5.11	13.76	46.34
C 8H14F209	292.19	292.0606	32.89	4.83	13.00	49.28
C 8H15F101	146.21	146.1107	65.72	10.34	12.99	10.94
C 8H15F102	162.21	162.1056	59.24	9.32	11.71	19.73
C 8H15F103	178.21	178.1005	53.92	8.48	10.66	26.93
C 8H15F104	194.20	194.0954	49.48	7.79	9.78	32.95
C 8H15F105	210.20	210.0903	45.71	7.19	9.04	38.06
C 8H15F106	226.20	226.0853	42.48	6.68	8.40	42.44
C 8H15F107	242.20	242.0802	39.67	6.24	7.84	46.24
C 8H15F108	258.20	258.0751	37.21	5.86	7.36	49.57
C 8H15F109	274.20	274.0700	35.04	5.51	6.93	52.51
C 8H16F201	166.21	166.1169	57.81	9.70	22.86	9.63
C 8H16F202	182.21	182.1118	52.73	8.85	20.85	17.56
C 8H16F203	198.21	198.1067	48.48	8.14	19.17	24.22
C 8H16F204	214.21	214.1017	44.86	7.53	17.74	29.88
C 8H16F205	230.21	230.0966	41.74	7.01	16.51	34.75
C 8H16F206	246.21	246.0915	39.03	6.55	15.43	38.99
C 8H16F207	262.21	262.0864	36.65	6.15	14.49	42.71
C 8H16F208	278.21	278.0813	34.54	5.80	13.66	46.01
C 8H16F209	294.21	294.0762	32.66	5.48	12.91	48.94
C 8H17F101	148.22	148.1263	64.83	11.56	12.82	10.79

C 8H17F102	164.22	164.1212	58.51	10.43	11.57	19.49
C 8H17F103	180.22	180.1162	53.32	9.51	10.54	26.63
C 8H17F104	196.22	196.1111	48.97	8.73	9.68	32.62
C 8H17F105	212.22	212.1060	45.28	8.07	8.95	37.70
C 8H17F106	228.22	228.1009	42.10	7.51	8.32	42.06
C 8H17F107	244.22	244.0958	39.35	7.02	7.78	45.86
C 8H17F108	260.22	260.0907	36.93	6.59	7.30	49.19
C 8H17F109	276.22	276.0856	34.79	6.20	6.88	52.13
C 9H 4F201	166.13	166.0230	65.07	2.43	22.87	9.63
C 9H 4F202	182.13	182.0179	59.35	2.21	20.86	17.57
C 9H 4F203	198.13	198.0128	54.56	2.03	19.18	24.23
C 9H 4F204	214.13	214.0078	50.48	1.88	17.75	29.89
C 9H 4F205	230.13	230.0027	46.97	1.75	16.51	34.76
C 9H 4F206	246.13	245.9976	43.92	1.64	15.44	39.00
C 9H 4F207	262.12	261.9925	41.24	1.54	14.50	42.73
C 9H 4F208	278.12	277.9874	38.87	1.45	13.66	46.02
C 9H 4F209	294.12	293.9823	36.75	1.37	12.92	48.96
C 9H 5F101	148.14	148.0324	72.97	3.40	12.82	10.80
C 9H 5F102	164.14	164.0274	65.86	3.07	11.57	19.50
C 9H 5F103	180.14	180.0223	60.01	2.80	10.55	26.65
C 9H 5F104	196.14	196.0172	55.11	2.57	9.69	32.63
C 9H 5F105	212.14	212.0121	50.96	2.38	8.96	37.71
C 9H 5F106	228.14	228.0070	47.38	2.21	8.33	42.08
C 9H 5F107	244.13	244.0019	44.28	2.06	7.78	45.87
C 9H 5F108	260.13	259.9968	41.56	1.94	7.30	49.20
C 9H 5F109	276.13	275.9918	39.15	1.83	6.88	52.15
C 9H 6F201	168.14	168.0387	64.29	3.60	22.60	9.52
C 9H 6F202	184.14	184.0336	58.70	3.28	20.63	17.38
C 9H 6F203	200.14	200.0285	54.01	3.02	18.98	23.98
C 9H 6F204	216.14	216.0234	50.01	2.80	17.58	29.61
C 9H 6F205	232.14	232.0183	46.57	2.61	16.37	34.46
C 9H 6F206	248.14	248.0132	43.56	2.44	15.31	38.69
C 9H 6F207	264.14	264.0082	40.93	2.29	14.39	42.40
C 9H 6F208	280.14	280.0031	38.59	2.16	13.56	45.69
C 9H 6F209	296.14	295.9980	36.50	2.04	12.83	48.62
C 9H 7F101	150.15	150.0481	71.99	4.70	12.65	10.66
C 9H 7F102	166.15	166.0430	65.06	4.25	11.43	19.26
C 9H 7F103	182.15	182.0379	59.35	3.87	10.43	26.35
C 9H 7F104	198.15	198.0328	54.55	3.56	9.59	32.30
C 9H 7F105	214.15	214.0277	50.48	3.29	8.87	37.36
C 9H 7F106	230.15	230.0227	46.97	3.07	8.25	41.71
C 9H 7F107	246.15	246.0176	43.92	2.87	7.72	45.50
C 9H 7F108	262.15	262.0125	41.24	2.69	7.25	48.83
C 9H 7F109	278.15	278.0074	38.86	2.54	6.83	51.77
C 9H 8F201	170.16	170.0543	63.53	4.74	22.33	9.40
C 9H 8F202	186.16	186.0492	58.07	4.33	20.41	17.19
C 9H 8F203	202.16	202.0441	53.47	3.99	18.80	23.74
C 9H 8F204	218.16	218.0391	49.55	3.70	17.42	29.34
C 9H 8F205	234.16	234.0340	46.17	3.44	16.23	34.16
C 9H 8F206	250.16	250.0289	43.21	3.22	15.19	38.37
C 9H 8F207	266.16	266.0238	40.62	3.03	14.28	42.08
C 9H 8F208	282.16	282.0187	38.31	2.86	13.47	45.36
C 9H 8F209	298.16	298.0136	36.26	2.70	12.74	48.30
C 9H 9F101	152.17	152.0637	71.04	5.96	12.48	10.51
C 9H 9F102	168.17	168.0587	64.28	5.39	11.30	19.03
C 9H 9F103	184.17	184.0536	58.70	4.93	10.32	26.06
C 9H 9F104	200.17	200.0485	54.00	4.53	9.49	31.97
C 9H 9F105	216.17	216.0434	50.01	4.20	8.79	37.01
C 9H 9F106	232.17	232.0383	46.56	3.91	8.18	41.35
C 9H 9F107	248.17	248.0332	43.56	3.66	7.66	45.13

C 9H 9F108	264.17	264.0281	40.92	3.43	7.19	48.45
C 9H 9F109	280.17	280.0231	38.58	3.24	6.78	51.40
C 9H10F201	172.18	172.0700	62.78	5.85	22.07	9.29
C 9H10F202	188.18	188.0649	57.45	5.36	20.19	17.00
C 9H10F203	204.18	204.0598	52.94	4.94	18.61	23.51
C 9H10F204	220.17	220.0547	49.10	4.58	17.26	29.07
C 9H10F205	236.17	236.0496	45.77	4.27	16.09	33.87
C 9H10F206	252.17	252.0445	42.87	4.00	15.07	38.07
C 9H10F207	268.17	268.0394	40.31	3.76	14.17	41.76
C 9H10F208	284.17	284.0344	38.04	3.55	13.37	45.04
C 9H10F209	300.17	300.0293	36.01	3.36	12.66	47.97
C 9H11F101	154.19	154.0794	70.11	7.19	12.32	10.38
C 9H11F102	170.19	170.0743	63.52	6.52	11.16	18.80
C 9H11F103	186.18	186.0692	58.06	5.96	10.20	25.78
C 9H11F104	202.18	202.0641	53.47	5.48	9.40	31.65
C 9H11F105	218.18	218.0590	49.55	5.08	8.71	36.67
C 9H11F106	234.18	234.0540	46.16	4.73	8.11	40.99
C 9H11F107	250.18	250.0489	43.21	4.43	7.59	44.77
C 9H11F108	266.18	266.0438	40.61	4.17	7.14	48.09
C 9H11F109	282.18	282.0387	38.31	3.93	6.73	51.03
C 9H12F201	174.19	174.0856	62.06	6.94	21.81	9.18
C 9H12F202	190.19	190.0805	56.84	6.36	19.98	16.82
C 9H12F203	206.19	206.0754	52.43	5.87	18.43	23.28
C 9H12F204	222.19	222.0704	48.65	5.44	17.10	28.80
C 9H12F205	238.19	238.0653	45.38	5.08	15.95	33.59
C 9H12F206	254.19	254.0602	42.53	4.76	14.95	37.77
C 9H12F207	270.19	270.0551	40.01	4.48	14.06	41.45
C 9H12F208	286.19	286.0500	37.77	4.23	13.28	44.72
C 9H12F209	302.19	302.0449	35.77	4.00	12.57	47.65
C 9H13F101	156.20	156.0950	69.21	8.39	12.16	10.24
C 9H13F102	172.20	172.0900	62.78	7.61	11.03	18.58
C 9H13F103	188.20	188.0849	57.44	6.96	10.09	25.50
C 9H13F104	204.20	204.0798	52.94	6.42	9.30	31.34
C 9H13F105	220.20	220.0747	49.09	5.95	8.63	36.33
C 9H13F106	236.20	236.0696	45.77	5.55	8.04	40.64
C 9H13F107	252.20	252.0645	42.86	5.20	7.53	44.41
C 9H13F108	268.20	268.0594	40.31	4.89	7.08	47.72
C 9H13F109	284.20	284.0543	38.04	4.61	6.68	50.67
C 9H14F201	176.21	176.1013	61.35	8.01	21.56	9.08
C 9H14F202	192.21	192.0962	56.24	7.34	19.77	16.65
C 9H14F203	208.21	208.0911	51.92	6.78	18.25	23.05
C 9H14F204	224.21	224.0860	48.21	6.29	16.95	28.54
C 9H14F205	240.21	240.0809	45.00	5.87	15.82	33.30
C 9H14F206	256.21	256.0758	42.19	5.51	14.83	37.47
C 9H14F207	272.20	272.0707	39.71	5.18	13.96	41.14
C 9H14F208	288.20	288.0657	37.51	4.90	13.18	44.41
C 9H14F209	304.20	304.0606	35.54	4.64	12.49	47.33
C 9H15F101	158.22	158.1107	68.32	9.56	12.01	10.11
C 9H15F102	174.22	174.1056	62.05	8.68	10.91	18.37
C 9H15F103	190.22	190.1005	56.83	7.95	9.99	25.23
C 9H15F104	206.22	206.0954	52.42	7.33	9.21	31.03
C 9H15F105	222.22	222.0903	48.65	6.80	8.55	36.00
C 9H15F106	238.21	238.0853	45.38	6.35	7.98	40.30
C 9H15F107	254.21	254.0802	42.52	5.95	7.47	44.06
C 9H15F108	270.21	270.0751	40.01	5.60	7.03	47.37
C 9H15F109	286.21	286.0700	37.77	5.28	6.64	50.31
C 9H16F201	178.22	178.1169	60.65	9.05	21.32	8.98
C 9H16F202	194.22	194.1118	55.66	8.30	19.56	16.48
C 9H16F203	210.22	210.1067	51.42	7.67	18.07	22.83
C 9H16F204	226.22	226.1017	47.79	7.13	16.80	28.29

C 9H16F205	242.22	242.0966	44.63	6.66	15.69	33.03
C 9H16F206	258.22	258.0915	41.86	6.25	14.71	37.18
C 9H16F207	274.22	274.0864	39.42	5.88	13.86	40.84
C 9H16F208	290.22	290.0813	37.25	5.56	13.09	44.10
C 9H16F209	306.22	306.0762	35.30	5.27	12.41	47.02
C 9H17F101	160.23	160.1263	67.46	10.69	11.86	9.99
C 9H17F102	176.23	176.1212	61.34	9.72	10.78	18.16
C 9H17F103	192.23	192.1162	56.23	8.91	9.88	24.97
C 9H17F104	208.23	208.1111	51.91	8.23	9.12	30.73
C 9H17F105	224.23	224.1060	48.21	7.64	8.47	35.68
C 9H17F106	240.23	240.1009	45.00	7.13	7.91	39.96
C 9H17F107	256.23	256.0958	42.19	6.69	7.41	43.71
C 9H17F108	272.23	272.0907	39.71	6.29	6.98	47.02
C 9H17F109	288.23	288.0856	37.51	5.95	6.59	49.96
C 9H18F201	180.24	180.1326	59.98	10.07	21.08	8.88
C 9H18F202	196.24	196.1275	55.09	9.25	19.36	16.31
C 9H18F203	212.24	212.1224	50.93	8.55	17.90	22.62
C 9H18F204	228.24	228.1173	47.36	7.95	16.65	28.04
C 9H18F205	244.24	244.1122	44.26	7.43	15.56	32.75
C 9H18F206	260.24	260.1071	41.54	6.97	14.60	36.89
C 9H18F207	276.24	276.1020	39.13	6.57	13.76	40.54
C 9H18F208	292.24	292.0970	36.99	6.21	13.00	43.80
C 9H18F209	308.24	308.0919	35.07	5.89	12.33	46.72
C 9H19F101	162.25	162.1420	66.63	11.80	11.71	9.86
C 9H19F102	178.25	178.1369	60.65	10.74	10.66	17.95
C 9H19F103	194.25	194.1318	55.65	9.86	9.78	24.71
C 9H19F104	210.25	210.1267	51.42	9.11	9.04	30.44
C 9H19F105	226.25	226.1216	47.78	8.46	8.40	35.36
C 9H19F106	242.25	242.1166	44.62	7.91	7.84	39.63
C 9H19F107	258.25	258.1115	41.86	7.42	7.36	43.37
C 9H19F108	274.25	274.1064	39.42	6.98	6.93	46.67
C 9H19F109	290.24	290.1013	37.24	6.60	6.55	49.61
C10H 5F101	160.15	160.0324	75.00	3.15	11.86	9.99
C10H 5F102	176.15	176.0274	68.19	2.86	10.79	18.17
C10H 5F103	192.15	192.0223	62.51	2.62	9.89	24.98
C10H 5F104	208.15	208.0172	57.71	2.42	9.13	30.75
C10H 5F105	224.15	224.0121	53.59	2.25	8.48	35.69
C10H 5F106	240.15	240.0070	50.02	2.10	7.91	39.97
C10H 5F107	256.15	256.0019	46.89	1.97	7.42	43.72
C10H 5F108	272.14	271.9968	44.14	1.85	6.98	47.03
C10H 5F109	288.14	287.9918	41.68	1.75	6.59	49.97
C10H 6F201	180.16	180.0387	66.67	3.36	21.09	8.88
C10H 6F202	196.15	196.0336	61.23	3.08	19.37	16.31
C10H 6F203	212.15	212.0285	56.62	2.85	17.91	22.62
C10H 6F204	228.15	228.0234	52.64	2.65	16.65	28.05
C10H 6F205	244.15	244.0183	49.20	2.48	15.56	32.77
C10H 6F206	260.15	260.0132	46.17	2.32	14.61	36.90
C10H 6F207	276.15	276.0082	43.49	2.19	13.76	40.56
C10H 6F208	292.15	292.0031	41.11	2.07	13.01	43.81
C10H 6F209	308.15	307.9980	38.98	1.96	12.33	46.73
C10H 7F101	162.17	162.0481	74.07	4.35	11.72	9.87
C10H 7F102	178.16	178.0430	67.42	3.96	10.66	17.96
C10H 7F103	194.16	194.0379	61.86	3.63	9.78	24.72
C10H 7F104	210.16	210.0328	57.15	3.36	9.04	30.45
C10H 7F105	226.16	226.0277	53.11	3.12	8.40	35.37
C10H 7F106	242.16	242.0227	49.60	2.91	7.85	39.64
C10H 7F107	258.16	258.0176	46.53	2.73	7.36	43.38
C10H 7F108	274.16	274.0125	43.81	2.57	6.93	46.69
C10H 7F109	290.16	290.0074	41.39	2.43	6.55	49.63
C10H 8F201	182.17	182.0543	65.93	4.43	20.86	8.78

C10H 8F202	198.17	198.0492	60.61	4.07	19.17	16.15
C10H 8F203	214.17	214.0441	56.08	3.77	17.74	22.41
C10H 8F204	230.17	230.0391	52.18	3.50	16.51	27.80
C10H 8F205	246.17	246.0340	48.79	3.28	15.44	32.50
C10H 8F206	262.17	262.0289	45.81	3.08	14.49	36.62
C10H 8F207	278.17	278.0238	43.18	2.90	13.66	40.26
C10H 8F208	294.17	294.0187	40.83	2.74	12.92	43.51
C10H 8F209	310.17	310.0136	38.72	2.60	12.25	46.42
C10H 9F101	164.18	164.0637	73.16	5.53	11.57	9.74
C10H 9F102	180.18	180.0587	66.66	5.03	10.54	17.76
C10H 9F103	196.18	196.0536	61.23	4.62	9.68	24.47
C10H 9F104	212.18	212.0485	56.61	4.28	8.95	30.16
C10H 9F105	228.18	228.0434	52.64	3.98	8.33	35.06
C10H 9F106	244.18	244.0383	49.19	3.72	7.78	39.31
C10H 9F107	260.18	260.0332	46.17	3.49	7.30	43.05
C10H 9F108	276.18	276.0281	43.49	3.28	6.88	46.35
C10H 9F109	292.18	292.0231	41.11	3.10	6.50	49.28
C10H10F201	184.19	184.0700	65.21	5.47	20.63	8.69
C10H10F202	200.19	200.0649	60.00	5.04	18.98	15.98
C10H10F203	216.19	216.0598	55.56	4.66	17.58	22.20
C10H10F204	232.19	232.0547	51.73	4.34	16.36	27.56
C10H10F205	248.19	248.0496	48.40	4.06	15.31	32.23
C10H10F206	264.18	264.0445	45.47	3.82	14.38	36.34
C10H10F207	280.18	280.0394	42.87	3.60	13.56	39.97
C10H10F208	296.18	296.0344	40.55	3.40	12.83	43.21
C10H10F209	312.18	312.0293	38.47	3.23	12.17	46.13
C10H11F101	166.20	166.0794	72.27	6.67	11.43	9.63
C10H11F102	182.20	182.0743	65.92	6.09	10.43	17.56
C10H11F103	198.20	198.0692	60.60	5.59	9.59	24.22
C10H11F104	214.20	214.0641	56.08	5.18	8.87	29.88
C10H11F105	230.19	230.0590	52.18	4.82	8.25	34.75
C10H11F106	246.19	246.0540	48.79	4.50	7.72	38.99
C10H11F107	262.19	262.0489	45.81	4.23	7.25	42.71
C10H11F108	278.19	278.0438	43.18	3.99	6.83	46.01
C10H11F109	294.19	294.0387	40.83	3.77	6.46	48.95
C10H12F201	186.20	186.0856	64.51	6.50	20.41	8.59
C10H12F202	202.20	202.0805	59.40	5.98	18.79	15.83
C10H12F203	218.20	218.0754	55.05	5.54	17.41	22.00
C10H12F204	234.20	234.0704	51.29	5.16	16.22	27.33
C10H12F205	250.20	250.0653	48.01	4.83	15.19	31.97
C10H12F206	266.20	266.0602	45.12	4.54	14.27	36.06
C10H12F207	282.20	282.0551	42.56	4.29	13.46	39.69
C10H12F208	298.20	298.0500	40.28	4.06	12.74	42.92
C10H12F209	314.20	314.0449	38.23	3.85	12.09	45.83
C10H13F101	168.21	168.0950	71.40	7.79	11.29	9.51
C10H13F102	184.21	184.0900	65.20	7.11	10.31	17.37
C10H13F103	200.21	200.0849	59.99	6.54	9.49	23.97
C10H13F104	216.21	216.0798	55.55	6.06	8.79	29.60
C10H13F105	232.21	232.0747	51.73	5.64	8.18	34.45
C10H13F106	248.21	248.0696	48.39	5.28	7.65	38.68
C10H13F107	264.21	264.0645	45.46	4.96	7.19	42.39
C10H13F108	280.21	280.0594	42.87	4.68	6.78	45.68
C10H13F109	296.21	296.0543	40.55	4.42	6.41	48.61
C10H14F201	188.22	188.1013	63.81	7.50	20.19	8.50
C10H14F202	204.22	204.0962	58.82	6.91	18.61	15.67
C10H14F203	220.22	220.0911	54.54	6.41	17.25	21.80
C10H14F204	236.22	236.0860	50.85	5.97	16.09	27.09
C10H14F205	252.22	252.0809	47.62	5.60	15.07	31.72
C10H14F206	268.22	268.0758	44.78	5.26	14.17	35.79
C10H14F207	284.22	284.0707	42.26	4.97	13.37	39.41

C10H14F208	300.22	300.0657	40.01	4.70	12.66	42.63
C10H14F209	316.21	316.0606	37.98	4.46	12.02	45.54
C10H15F101	170.23	170.1107	70.56	8.88	11.16	9.40
C10H15F102	186.23	186.1056	64.50	8.12	10.20	17.18
C10H15F103	202.23	202.1005	59.39	7.48	9.39	23.73
C10H15F104	218.23	218.0954	55.04	6.93	8.71	29.33
C10H15F105	234.23	234.0903	51.28	6.46	8.11	34.15
C10H15F106	250.23	250.0853	48.00	6.04	7.59	38.36
C10H15F107	266.23	266.0802	45.12	5.68	7.14	42.07
C10H15F108	282.22	282.0751	42.56	5.36	6.73	45.35
C10H15F109	298.22	298.0700	40.28	5.07	6.37	48.28
C10H16F201	190.24	190.1169	63.14	8.48	19.97	8.41
C10H16F202	206.23	206.1118	58.24	7.82	18.42	15.52
C10H16F203	222.23	222.1067	54.05	7.26	17.10	21.60
C10H16F204	238.23	238.1017	50.42	6.77	15.95	26.86
C10H16F205	254.23	254.0966	47.24	6.34	14.95	31.47
C10H16F206	270.23	270.0915	44.45	5.97	14.06	35.52
C10H16F207	286.23	286.0864	41.96	5.63	13.27	39.13
C10H16F208	302.23	302.0813	39.74	5.34	12.57	42.35
C10H16F209	318.23	318.0762	37.74	5.07	11.94	45.25
C10H17F101	172.24	172.1263	69.73	9.95	11.03	9.29
C10H17F102	188.24	188.1212	63.81	9.10	10.09	17.00
C10H17F103	204.24	204.1162	58.81	8.39	9.30	23.50
C10H17F104	220.24	220.1111	54.54	7.78	8.63	29.06
C10H17F105	236.24	236.1060	50.84	7.25	8.04	33.86
C10H17F106	252.24	252.1009	47.62	6.79	7.53	38.06
C10H17F107	268.24	268.0958	44.78	6.39	7.08	41.75
C10H17F108	284.24	284.0907	42.26	6.03	6.68	45.03
C10H17F109	300.24	300.0856	40.01	5.71	6.33	47.96
C10H18F201	192.25	192.1326	62.48	9.44	19.76	8.32
C10H18F202	208.25	208.1275	57.68	8.71	18.25	15.37
C10H18F203	224.25	224.1224	53.56	8.09	16.94	21.40
C10H18F204	240.25	240.1173	49.99	7.55	15.82	26.64
C10H18F205	256.25	256.1122	46.87	7.08	14.83	31.22
C10H18F206	272.25	272.1071	44.12	6.66	13.96	35.26
C10H18F207	288.25	288.1020	41.67	6.29	13.18	38.85
C10H18F208	304.25	304.0970	39.48	5.96	12.49	42.07
C10H18F209	320.25	320.0919	37.51	5.67	11.86	44.96
C10H19F101	174.26	174.1420	68.93	10.99	10.90	9.18
C10H19F102	190.26	190.1369	63.13	10.07	9.99	16.82
C10H19F103	206.26	206.1318	58.23	9.29	9.21	23.27
C10H19F104	222.26	222.1267	54.04	8.62	8.55	28.79
C10H19F105	238.26	238.1216	50.41	8.04	7.97	33.58
C10H19F106	254.26	254.1166	47.24	7.53	7.47	37.76
C10H19F107	270.26	270.1115	44.44	7.09	7.03	41.44
C10H19F108	286.26	286.1064	41.96	6.69	6.64	44.71
C10H19F109	302.26	302.1013	39.74	6.34	6.29	47.64
C10H20F201	194.27	194.1482	61.83	10.38	19.56	8.24
C10H20F202	210.27	210.1431	57.12	9.59	18.07	15.22
C10H20F203	226.27	226.1380	53.08	8.91	16.79	21.21
C10H20F204	242.27	242.1330	49.58	8.32	15.68	26.42
C10H20F205	258.26	258.1279	46.51	7.81	14.71	30.97
C10H20F206	274.26	274.1228	43.79	7.35	13.85	35.00
C10H20F207	290.26	290.1177	41.38	6.95	13.09	38.58
C10H20F208	306.26	306.1126	39.22	6.58	12.41	41.79
C10H20F209	322.26	322.1075	37.27	6.26	11.79	44.68
C10H21F101	176.28	176.1576	68.14	12.01	10.78	9.08
C10H21F102	192.28	192.1525	62.47	11.01	9.88	16.64
C10H21F103	208.28	208.1475	57.67	10.16	9.12	23.05
C10H21F104	224.27	224.1424	53.56	9.44	8.47	28.54

C10H21F105	240.27	240.1373	49.99	8.81	7.91	33.29
C10H21F106	256.27	256.1322	46.87	8.26	7.41	37.46
C10H21F107	272.27	272.1271	44.11	7.77	6.98	41.13
C10H21F108	288.27	288.1220	41.67	7.34	6.59	44.40
C10H21F109	304.27	304.1169	39.48	6.96	6.24	47.32
C11H 5F101	172.16	172.0324	76.74	2.93	11.04	9.29
C11H 5F102	188.16	188.0274	70.22	2.68	10.10	17.01
C11H 5F103	204.16	204.0223	64.72	2.47	9.31	23.51
C11H 5F104	220.16	220.0172	60.01	2.29	8.63	29.07
C11H 5F105	236.16	236.0121	55.95	2.13	8.04	33.87
C11H 5F106	252.16	252.0070	52.40	2.00	7.53	38.07
C11H 5F107	268.16	268.0019	49.27	1.88	7.08	41.77
C11H 5F108	284.16	283.9968	46.50	1.77	6.69	45.04
C11H 5F109	300.16	299.9918	44.02	1.68	6.33	47.97
C11H 6F201	192.17	192.0387	68.75	3.15	19.77	8.33
C11H 6F202	208.17	208.0336	63.47	2.91	18.25	15.37
C11H 6F203	224.17	224.0285	58.94	2.70	16.95	21.41
C11H 6F204	240.16	240.0234	55.01	2.52	15.82	26.65
C11H 6F205	256.16	256.0183	51.58	2.36	14.83	31.23
C11H 6F206	272.16	272.0132	48.55	2.22	13.96	35.27
C11H 6F207	288.16	288.0082	45.85	2.10	13.19	38.87
C11H 6F208	304.16	304.0031	43.44	1.99	12.49	42.08
C11H 6F209	320.16	319.9980	41.27	1.89	11.87	44.98
C11H 7F101	174.18	174.0481	75.86	4.05	10.91	9.19
C11H 7F102	190.18	190.0430	69.47	3.71	9.99	16.83
C11H 7F103	206.18	206.0379	64.08	3.42	9.21	23.28
C11H 7F104	222.17	222.0328	59.47	3.18	8.55	28.81
C11H 7F105	238.17	238.0277	55.47	2.96	7.98	33.59
C11H 7F106	254.17	254.0227	51.98	2.78	7.47	37.77
C11H 7F107	270.17	270.0176	48.90	2.61	7.03	41.45
C11H 7F108	286.17	286.0125	46.17	2.47	6.64	44.73
C11H 7F109	302.17	302.0074	43.72	2.34	6.29	47.65
C11H 8F201	194.18	194.0543	68.04	4.15	19.57	8.24
C11H 8F202	210.18	210.0492	62.86	3.84	18.08	15.22
C11H 8F203	226.18	226.0441	58.41	3.57	16.80	21.22
C11H 8F204	242.18	242.0391	54.56	3.33	15.69	26.43
C11H 8F205	258.18	258.0340	51.17	3.12	14.72	30.98
C11H 8F206	274.18	274.0289	48.19	2.94	13.86	35.01
C11H 8F207	290.18	290.0238	45.53	2.78	13.09	38.60
C11H 8F208	306.18	306.0187	43.15	2.63	12.41	41.80
C11H 8F209	322.18	322.0136	41.01	2.50	11.79	44.69
C11H 9F101	176.19	176.0637	74.99	5.15	10.78	9.08
C11H 9F102	192.19	192.0587	68.75	4.72	9.89	16.65
C11H 9F103	208.19	208.0536	63.46	4.36	9.13	23.05
C11H 9F104	224.19	224.0485	58.93	4.05	8.47	28.55
C11H 9F105	240.19	240.0434	55.01	3.78	7.91	33.31
C11H 9F106	256.19	256.0383	51.57	3.54	7.42	37.47
C11H 9F107	272.19	272.0332	48.54	3.33	6.98	41.15
C11H 9F108	288.19	288.0281	45.85	3.15	6.59	44.41
C11H 9F109	304.19	304.0231	43.43	2.98	6.25	47.34
C11H10F201	196.20	196.0700	67.34	5.14	19.37	8.15
C11H10F202	212.20	212.0649	62.26	4.75	17.91	15.08
C11H10F203	228.20	228.0598	57.90	4.42	16.65	21.03
C11H10F204	244.20	244.0547	54.10	4.13	15.56	26.21
C11H10F205	260.20	260.0496	50.78	3.87	14.60	30.74
C11H10F206	276.20	276.0445	47.84	3.65	13.76	34.76
C11H10F207	292.19	292.0394	45.22	3.45	13.00	38.33
C11H10F208	308.19	308.0344	42.87	3.27	12.33	41.53
C11H10F209	324.19	324.0293	40.75	3.11	11.72	44.42
C11H11F101	178.21	178.0794	74.14	6.22	10.66	8.98

C11H11F102	194.21	194.0743	68.03	5.71	9.78	16.48
C11H11F103	210.21	210.0692	62.85	5.27	9.04	22.83
C11H11F104	226.21	226.0641	58.41	4.90	8.40	28.29
C11H11F105	242.21	242.0590	54.55	4.58	7.84	33.03
C11H11F106	258.21	258.0540	51.17	4.29	7.36	37.18
C11H11F107	274.20	274.0489	48.18	4.04	6.93	40.84
C11H11F108	290.20	290.0438	45.53	3.82	6.55	44.11
C11H11F109	306.20	306.0387	43.15	3.62	6.20	47.03
C11H12F201	198.21	198.0856	66.66	6.10	19.17	8.07
C11H12F202	214.21	214.0805	61.68	5.65	17.74	14.94
C11H12F203	230.21	230.0754	57.39	5.25	16.51	20.85
C11H12F204	246.21	246.0704	53.66	4.91	15.43	25.99
C11H12F205	262.21	262.0653	50.39	4.61	14.49	30.51
C11H12F206	278.21	278.0602	47.49	4.35	13.66	34.50
C11H12F207	294.21	294.0551	44.91	4.11	12.91	38.07
C11H12F208	310.21	310.0500	42.59	3.90	12.25	41.26
C11H12F209	326.21	326.0449	40.50	3.71	11.65	44.14
C11H13F101	180.22	180.0950	73.31	7.27	10.54	8.88
C11H13F102	196.22	196.0900	67.33	6.68	9.68	16.31
C11H13F103	212.22	212.0849	62.26	6.17	8.95	22.62
C11H13F104	228.22	228.0798	57.89	5.74	8.32	28.04
C11H13F105	244.22	244.0747	54.10	5.37	7.78	32.76
C11H13F106	260.22	260.0696	50.77	5.04	7.30	36.89
C11H13F107	276.22	276.0645	47.83	4.74	6.88	40.55
C11H13F108	292.22	292.0594	45.21	4.48	6.50	43.80
C11H13F109	308.22	308.0543	42.87	4.25	6.16	46.72
C11H14F201	200.23	200.1013	65.99	7.05	18.98	7.99
C11H14F202	216.23	216.0962	61.10	6.53	17.57	14.80
C11H14F203	232.23	232.0911	56.89	6.08	16.36	20.67
C11H14F204	248.23	248.0860	53.23	5.68	15.31	25.78
C11H14F205	264.23	264.0809	50.00	5.34	14.38	30.28
C11H14F206	280.23	280.0758	47.15	5.04	13.56	34.26
C11H14F207	296.23	296.0707	44.60	4.76	12.83	37.81
C11H14F208	312.23	312.0657	42.32	4.52	12.17	40.99
C11H14F209	328.23	328.0606	40.25	4.30	11.58	43.87
C11H15F101	182.24	182.1107	72.50	8.30	10.42	8.78
C11H15F102	198.24	198.1056	66.65	7.63	9.58	16.14
C11H15F103	214.24	214.1005	61.67	7.06	8.87	22.40
C11H15F104	230.24	230.0954	57.39	6.57	8.25	27.80
C11H15F105	246.24	246.0903	53.66	6.14	7.72	32.49
C11H15F106	262.24	262.0853	50.38	5.77	7.24	36.61
C11H15F107	278.24	278.0802	47.49	5.43	6.83	40.25
C11H15F108	294.24	294.0751	44.90	5.14	6.46	43.50
C11H15F109	310.24	310.0700	42.59	4.87	6.12	46.41
C11H16F201	202.25	202.1169	65.33	7.97	18.79	7.91
C11H16F202	218.25	218.1118	60.54	7.39	17.41	14.66
C11H16F203	234.25	234.1067	56.40	6.88	16.22	20.49
C11H16F204	250.24	250.1017	52.80	6.44	15.18	25.57
C11H16F205	266.24	266.0966	49.62	6.06	14.27	30.05
C11H16F206	282.24	282.0915	46.81	5.71	13.46	34.01
C11H16F207	298.24	298.0864	44.30	5.41	12.74	37.55
C11H16F208	314.24	314.0813	42.04	5.13	12.09	40.73
C11H16F209	330.24	330.0762	40.01	4.88	11.51	43.60
C11H17F101	184.26	184.1263	71.71	9.30	10.31	8.68
C11H17F102	200.26	200.1212	65.98	8.56	9.49	15.98
C11H17F103	216.25	216.1162	61.10	7.92	8.79	22.20
C11H17F104	232.25	232.1111	56.89	7.38	8.18	27.55
C11H17F105	248.25	248.1060	53.22	6.90	7.65	32.22
C11H17F106	264.25	264.1009	50.00	6.48	7.19	36.33
C11H17F107	280.25	280.0958	47.14	6.11	6.78	39.96

134

C11H17F108	296.25	296.0907	44.60	5.78	6.41	43.20
C11H17F109	312.25	312.0856	42.31	5.49	6.08	46.11
C11H18F201	204.26	204.1326	64.68	8.88	18.60	7.83
C11H18F202	220.26	220.1275	59.98	8.24	17.25	14.53
C11H18F203	236.26	236.1224	55.92	7.68	16.08	20.32
C11H18F204	252.26	252.1173	52.38	7.19	15.06	25.37
C11H18F205	268.26	268.1122	49.25	6.76	14.16	29.82
C11H18F206	284.26	284.1071	46.48	6.38	13.37	33.77
C11H18F207	300.26	300.1020	44.00	6.04	12.65	37.30
C11H18F208	316.26	316.0970	41.78	5.74	12.01	40.47
C11H18F209	332.26	332.0919	39.77	5.46	11.44	43.34
C11H19F101	186.27	186.1420	70.93	10.28	10.20	8.59
C11H19F102	202.27	202.1369	65.32	9.47	9.39	15.82
C11H19F103	218.27	218.1318	60.53	8.77	8.70	21.99
C11H19F104	234.27	234.1267	56.40	8.17	8.11	27.32
C11H19F105	250.27	250.1216	52.79	7.65	7.59	31.96
C11H19F106	266.27	266.1166	49.62	7.19	7.14	36.05
C11H19F107	282.27	282.1115	46.81	6.78	6.73	39.68
C11H19F108	298.27	298.1064	44.30	6.42	6.37	42.91
C11H19F109	314.27	314.1013	42.04	6.09	6.05	45.82
C11H20F201	206.28	206.1482	64.05	9.77	18.42	7.76
C11H20F202	222.28	222.1431	59.44	9.07	17.09	14.40
C11H20F203	238.28	238.1380	55.45	8.46	15.95	20.14
C11H20F204	254.28	254.1330	51.96	7.93	14.94	25.17
C11H20F205	270.28	270.1279	48.88	7.46	14.06	29.60
C11H20F206	286.28	286.1228	46.15	7.04	13.27	33.53
C11H20F207	302.27	302.1177	43.71	6.67	12.57	37.05
C11H20F208	318.27	318.1126	41.51	6.33	11.94	40.22
C11H20F209	334.27	334.1075	39.53	6.03	11.37	43.08
C11H21F101	188.29	188.1576	70.17	11.24	10.09	8.50
C11H21F102	204.29	204.1525	64.67	10.36	9.30	15.66
C11H21F103	220.29	220.1475	59.98	9.61	8.62	21.79
C11H21F104	236.29	236.1424	55.92	8.96	8.04	27.08
C11H21F105	252.29	252.1373	52.37	8.39	7.53	31.71
C11H21F106	268.28	268.1322	49.25	7.89	7.08	35.78
C11H21F107	284.28	284.1271	46.48	7.45	6.68	39.40
C11H21F108	300.28	300.1220	44.00	7.05	6.33	42.62
C11H21F109	316.28	316.1169	41.77	6.69	6.01	45.53
C11H22F201	208.29	208.1639	63.43	10.65	18.24	7.68
C11H22F202	224.29	224.1588	58.91	9.89	16.94	14.27
C11H22F203	240.29	240.1537	54.98	9.23	15.81	19.97
C11H22F204	256.29	256.1486	51.55	8.65	14.83	24.97
C11H22F205	272.29	272.1435	48.52	8.14	13.95	29.38
C11H22F206	288.29	288.1384	45.83	7.69	13.18	33.30
C11H22F207	304.29	304.1333	43.42	7.29	12.49	36.81
C11H22F208	320.29	320.1283	41.25	6.92	11.86	39.96
C11H22F209	336.29	336.1232	39.29	6.59	11.30	42.82
C11H23F101	190.30	190.1733	69.43	12.18	9.98	8.41
C11H23F102	206.30	206.1682	64.04	11.24	9.21	15.51
C11H23F103	222.30	222.1631	59.43	10.43	8.55	21.59
C11H23F104	238.30	238.1580	55.44	9.73	7.97	26.86
C11H23F105	254.30	254.1529	51.96	9.12	7.47	31.46
C11H23F106	270.30	270.1479	48.88	8.58	7.03	35.51
C11H23F107	286.30	286.1428	46.15	8.10	6.64	39.12
C11H23F108	302.30	302.1377	43.71	7.67	6.28	42.34
C11H23F109	318.30	318.1326	41.51	7.28	5.97	45.24
C12H 6F201	204.18	204.0387	70.59	2.96	18.61	7.84
C12H 6F202	220.18	220.0336	65.46	2.75	17.26	14.53
C12H 6F203	236.18	236.0285	61.03	2.56	16.09	20.32
C12H 6F204	252.18	252.0234	57.16	2.40	15.07	25.38

C12H 6F205	268.18	268.0183	53.75	2.26	14.17	29.83
C12H 6F206	284.17	284.0132	50.72	2.13	13.37	33.78
C12H 6F207	300.17	300.0082	48.02	2.01	12.66	37.31
C12H 6F208	316.17	316.0031	45.59	1.91	12.02	40.48
C12H 6F209	332.17	331.9980	43.39	1.82	11.44	43.35
C12H 7F101	186.19	186.0481	77.41	3.79	10.20	8.59
C12H 7F102	202.19	202.0430	71.29	3.49	9.40	15.83
C12H 7F103	218.19	218.0379	66.06	3.23	8.71	22.00
C12H 7F104	234.19	234.0328	61.55	3.01	8.11	27.33
C12H 7F105	250.18	250.0277	57.61	2.82	7.59	31.98
C12H 7F106	266.18	266.0227	54.15	2.65	7.14	36.06
C12H 7F107	282.18	282.0176	51.08	2.50	6.73	39.69
C12H 7F108	298.18	298.0125	48.34	2.37	6.37	42.93
C12H 7F109	314.18	314.0074	45.88	2.25	6.05	45.83
C12H 8F201	206.19	206.0543	69.90	3.91	18.43	7.76
C12H 8F202	222.19	222.0492	64.87	3.63	17.10	14.40
C12H 8F203	238.19	238.0441	60.51	3.39	15.95	20.15
C12H 8F204	254.19	254.0391	56.70	3.17	14.95	25.18
C12H 8F205	270.19	270.0340	53.35	2.98	14.06	29.61
C12H 8F206	286.19	286.0289	50.36	2.82	13.28	33.54
C12H 8F207	302.19	302.0238	47.70	2.67	12.57	37.06
C12H 8F208	318.19	318.0187	45.30	2.53	11.94	40.23
C12H 8F209	334.19	334.0136	43.13	2.41	11.37	43.09
C12H 9F101	188.20	188.0637	76.58	4.82	10.09	8.50
C12H 9F102	204.20	204.0587	70.58	4.44	9.30	15.67
C12H 9F103	220.20	220.0536	65.46	4.12	8.63	21.80
C12H 9F104	236.20	236.0485	61.02	3.84	8.04	27.09
C12H 9F105	252.20	252.0434	57.15	3.60	7.53	31.72
C12H 9F106	268.20	268.0383	53.74	3.38	7.08	35.79
C12H 9F107	284.20	284.0332	50.72	3.19	6.68	39.41
C12H 9F108	300.20	300.0281	48.01	3.02	6.33	42.64
C12H 9F109	316.20	316.0231	45.58	2.87	6.01	45.54
C12H10F201	208.21	208.0700	69.23	4.84	18.25	7.68
C12H10F202	224.21	224.0649	64.29	4.50	16.95	14.27
C12H10F203	240.21	240.0598	60.00	4.20	15.82	19.98
C12H10F204	256.21	256.0547	56.26	3.93	14.83	24.98
C12H10F205	272.21	272.0496	52.95	3.70	13.96	29.39
C12H10F206	288.21	288.0445	50.01	3.50	13.18	33.31
C12H10F207	304.21	304.0394	47.38	3.31	12.49	36.82
C12H10F208	320.21	320.0344	45.01	3.15	11.87	39.97
C12H10F209	336.20	336.0293	42.87	3.00	11.30	42.83
C12H11F101	190.22	190.0794	75.77	5.83	9.99	8.41
C12H11F102	206.22	206.0743	69.89	5.38	9.21	15.52
C12H11F103	222.22	222.0692	64.86	4.99	8.55	21.60
C12H11F104	238.22	238.0641	60.51	4.65	7.98	26.87
C12H11F105	254.22	254.0590	56.70	4.36	7.47	31.47
C12H11F106	270.22	270.0540	53.34	4.10	7.03	35.53
C12H11F107	286.22	286.0489	50.36	3.87	6.64	39.13
C12H11F108	302.22	302.0438	47.69	3.67	6.29	42.35
C12H11F109	318.21	318.0387	45.29	3.48	5.97	45.25
C12H12F201	210.23	210.0856	68.56	5.75	18.07	7.61
C12H12F202	226.23	226.0805	63.71	5.35	16.80	14.14
C12H12F203	242.22	242.0754	59.50	4.99	15.69	19.82
C12H12F204	258.22	258.0704	55.82	4.68	14.71	24.78
C12H12F205	274.22	274.0653	52.56	4.41	13.86	29.17
C12H12F206	290.22	290.0602	49.66	4.17	13.09	33.08
C12H12F207	306.22	306.0551	47.07	3.95	12.41	36.57
C12H12F208	322.22	322.0500	44.73	3.75	11.79	39.72
C12H12F209	338.22	338.0449	42.62	3.58	11.23	42.57
C12H13F101	192.24	192.0950	74.98	6.82	9.88	8.32

135

C12H13F102	208.23	208.0900	69.22	6.29	9.12	15.37
C12H13F103	224.23	224.0849	64.28	5.84	8.47	21.41
C12H13F104	240.23	240.0798	60.00	5.45	7.91	26.64
C12H13F105	256.23	256.0747	56.25	5.11	7.41	31.22
C12H13F106	272.23	272.0696	52.95	4.81	6.98	35.26
C12H13F107	288.23	288.0645	50.01	4.55	6.59	38.86
C12H13F108	304.23	304.0594	47.38	4.31	6.24	42.07
C12H13F109	320.23	320.0543	45.01	4.09	5.93	44.97
C12H14F201	212.24	212.1013	67.91	6.65	17.90	7.54
C12H14F202	228.24	228.0962	63.15	6.18	16.65	14.02
C12H14F203	244.24	244.0911	59.01	5.78	15.56	19.65
C12H14F204	260.24	260.0860	55.38	5.42	14.60	24.59
C12H14F205	276.24	276.0809	52.18	5.11	13.76	28.96
C12H14F206	292.24	292.0758	49.32	4.83	13.00	32.85
C12H14F207	308.24	308.0707	46.76	4.58	12.33	36.33
C12H14F208	324.24	324.0657	44.45	4.35	11.72	39.48
C12H14F209	340.24	340.0606	42.36	4.15	11.17	42.32
C12H15F101	194.25	194.1107	74.20	7.78	9.78	8.24
C12H15F102	210.25	210.1056	68.55	7.19	9.04	15.22
C12H15F103	226.25	226.1005	63.71	6.68	8.40	21.21
C12H15F104	242.25	242.0954	59.50	6.24	7.84	26.42
C12H15F105	258.25	258.0903	55.81	5.85	7.36	30.98
C12H15F106	274.25	274.0853	52.56	5.51	6.93	35.00
C12H15F107	290.25	290.0802	49.66	5.21	6.55	38.59
C12H15F108	306.25	306.0751	47.06	4.94	6.20	41.79
C12H15F109	322.25	322.0700	44.73	4.69	5.90	44.68
C12H16F201	214.26	214.1169	67.27	7.53	17.73	7.47
C12H16F202	230.26	230.1118	62.60	7.00	16.50	13.90
C12H16F203	246.26	246.1067	58.53	6.55	15.43	19.49
C12H16F204	262.26	262.1017	54.96	6.15	14.49	24.40
C12H16F205	278.26	278.0966	51.80	5.80	13.66	28.75
C12H16F206	294.25	294.0915	48.98	5.48	12.91	32.62
C12H16F207	310.25	310.0864	46.46	5.20	12.25	36.10
C12H16F208	326.25	326.0813	44.18	4.94	11.65	39.23
C12H16F209	342.25	342.0762	42.11	4.71	11.10	42.07
C12H17F101	196.27	196.1263	73.44	8.73	9.68	8.15
C12H17F102	212.27	212.1212	67.90	8.07	8.95	15.07
C12H17F103	228.27	228.1162	63.14	7.51	8.32	21.03
C12H17F104	244.27	244.1111	59.01	7.02	7.78	26.20
C12H17F105	260.26	260.1060	55.38	6.58	7.30	30.74
C12H17F106	276.26	276.1009	52.17	6.20	6.88	34.75
C12H17F107	292.26	292.0958	49.32	5.86	6.50	38.32
C12H17F108	308.26	308.0907	46.76	5.56	6.16	41.52
C12H17F109	324.26	324.0856	44.45	5.28	5.86	44.41
C12H18F201	216.27	216.1326	66.64	8.39	17.57	7.40
C12H18F202	232.27	232.1275	62.05	7.81	16.36	13.78
C12H18F203	248.27	248.1224	58.05	7.31	15.30	19.33
C12H18F204	264.27	264.1173	54.54	6.87	14.38	24.22
C12H18F205	280.27	280.1122	51.43	6.47	13.56	28.54
C12H18F206	296.27	296.1071	48.65	6.12	12.83	32.40
C12H18F207	312.27	312.1020	46.16	5.81	12.17	35.87
C12H18F208	328.27	328.0970	43.91	5.53	11.57	38.99
C12H18F209	344.27	344.0919	41.87	5.27	11.04	41.83
C12H19F101	198.28	198.1420	72.69	9.66	9.58	8.07
C12H19F102	214.28	214.1369	67.26	8.94	8.87	14.93
C12H19F103	230.28	230.1318	62.59	8.32	8.25	20.84
C12H19F104	246.28	246.1267	58.52	7.78	7.71	25.99
C12H19F105	262.28	262.1216	54.95	7.30	7.24	30.50
C12H19F106	278.28	278.1166	51.79	6.88	6.83	34.50
C12H19F107	294.28	294.1115	48.98	6.51	6.46	38.06
C12H19F108	310.28	310.1064	46.45	6.17	6.12	41.25
C12H19F109	326.28	326.1013	44.18	5.87	5.82	44.13
C12H20F201	218.29	218.1482	66.03	9.24	17.41	7.33
C12H20F202	234.29	234.1431	61.52	8.60	16.22	13.66
C12H20F203	250.29	250.1380	57.59	8.05	15.18	19.18
C12H20F204	266.29	266.1330	54.13	7.57	14.27	24.03
C12H20F205	282.29	282.1279	51.06	7.14	13.46	28.34
C12H20F206	298.29	298.1228	48.32	6.76	12.74	32.18
C12H20F207	314.29	314.1177	45.86	6.41	12.09	35.64
C12H20F208	330.29	330.1126	43.64	6.10	11.50	38.75
C12H20F209	346.28	346.1075	41.62	5.82	10.97	41.58
C12H21F101	200.30	200.1576	71.96	10.57	9.49	7.99
C12H21F102	216.30	216.1525	66.64	9.79	8.78	14.79
C12H21F103	232.30	232.1475	62.05	9.11	8.18	20.66
C12H21F104	248.30	248.1424	58.05	8.53	7.65	25.77
C12H21F105	264.30	264.1373	54.53	8.01	7.19	30.27
C12H21F106	280.30	280.1322	51.42	7.55	6.78	34.25
C12H21F107	296.30	296.1271	48.65	7.14	6.41	37.80
C12H21F108	312.29	312.1220	46.15	6.78	6.08	40.99
C12H21F109	328.29	328.1169	43.90	6.45	5.79	43.86
C12H22F201	220.31	220.1639	65.42	10.07	17.25	7.26
C12H22F202	236.30	236.1588	60.99	9.38	16.08	13.54
C12H22F203	252.30	252.1537	57.13	8.79	15.06	19.02
C12H22F204	268.30	268.1486	53.72	8.27	14.16	23.85
C12H22F205	284.30	284.1435	50.70	7.80	13.36	28.14
C12H22F206	300.30	300.1384	48.00	7.38	12.65	31.97
C12H22F207	316.30	316.1333	45.57	7.01	12.01	35.41
C12H22F208	332.30	332.1283	43.37	6.67	11.43	38.52
C12H22F209	348.30	348.1232	41.38	6.37	10.91	41.34
C12H23F101	202.31	202.1733	71.24	11.46	9.39	7.91
C12H23F102	218.31	218.1682	66.02	10.62	8.70	14.66
C12H23F103	234.31	234.1631	61.51	9.89	8.11	20.48
C12H23F104	250.31	250.1580	57.58	9.26	7.59	25.57
C12H23F105	266.31	266.1529	54.12	8.71	7.13	30.04
C12H23F106	282.31	282.1479	51.05	8.21	6.73	34.00
C12H23F107	298.31	298.1428	48.32	7.77	6.37	37.54
C12H23F108	314.31	314.1377	45.86	7.38	6.04	40.72
C12H23F109	330.31	330.1326	43.64	7.02	5.75	43.59
C12H24F201	222.32	222.1795	64.83	10.88	17.09	7.20
C12H24F202	238.32	238.1744	60.48	10.15	15.94	13.43
C12H24F203	254.32	254.1693	56.67	9.51	14.94	18.87
C12H24F204	270.32	270.1643	53.32	8.95	14.06	23.67
C12H24F205	286.32	286.1592	50.34	8.45	13.27	27.94
C12H24F206	302.32	302.1541	47.68	8.00	12.57	31.75
C12H24F207	318.32	318.1490	45.28	7.60	11.94	35.18
C12H24F208	334.32	334.1439	43.11	7.24	11.37	38.29
C12H24F209	350.32	350.1388	41.14	6.91	10.85	41.10
C12H25F101	204.33	204.1889	70.54	12.33	9.30	7.83
C12H25F102	220.33	220.1838	65.42	11.44	8.62	14.52
C12H25F103	236.33	236.1788	60.99	10.66	8.04	20.31
C12H25F104	252.33	252.1737	57.12	9.99	7.53	25.36
C12H25F105	268.33	268.1686	53.72	9.39	7.08	29.81
C12H25F106	284.33	284.1635	50.69	8.86	6.68	33.76
C12H25F107	300.33	300.1584	47.99	8.39	6.33	37.29
C12H25F108	316.33	316.1533	45.56	7.97	6.01	40.46
C12H25F109	332.33	332.1482	43.37	7.58	5.72	43.33
C13H 6F201	216.19	216.0387	72.23	2.80	17.58	7.40
C13H 6F202	232.19	232.0336	67.25	2.60	16.36	13.78
C13H 6F203	248.19	248.0285	62.91	2.44	15.31	19.34
C13H 6F204	264.19	264.0234	59.10	2.29	14.38	24.22

136

C13H 6F205	280.19	280.0183	55.73	2.16	13.56	28.55
C13H 6F206	296.19	296.0132	52.72	2.04	12.83	32.41
C13H 6F207	312.19	312.0082	50.02	1.94	12.17	35.87
C13H 6F208	328.18	328.0031	47.58	1.84	11.58	39.00
C13H 6F209	344.18	343.9980	45.37	1.76	11.04	41.84
C13H 7F101	198.20	198.0481	78.78	3.56	9.59	8.07
C13H 7F102	214.20	214.0430	72.90	3.29	8.87	14.94
C13H 7F103	230.20	230.0379	67.83	3.07	8.25	20.85
C13H 7F104	246.20	246.0328	63.42	2.87	7.72	25.99
C13H 7F105	262.20	262.0277	59.55	2.69	7.25	30.51
C13H 7F106	278.20	278.0227	56.13	2.54	6.83	34.51
C13H 7F107	294.19	294.0176	53.08	2.40	6.46	38.07
C13H 7F108	310.19	310.0125	50.34	2.27	6.12	41.26
C13H 7F109	326.19	326.0074	47.87	2.16	5.82	44.14
C13H 8F201	218.20	218.0543	71.56	3.70	17.41	7.33
C13H 8F202	234.20	234.0492	66.67	3.44	16.22	13.66
C13H 8F203	250.20	250.0441	62.41	3.22	15.19	19.18
C13H 8F204	266.20	266.0391	58.66	3.03	14.27	24.04
C13H 8F205	282.20	282.0340	55.33	2.86	13.46	28.35
C13H 8F206	298.20	298.0289	52.36	2.70	12.74	32.19
C13H 8F207	314.20	314.0238	49.70	2.57	12.09	35.64
C13H 8F208	330.20	330.0187	47.29	2.44	11.51	38.76
C13H 8F209	346.20	346.0136	45.10	2.33	10.98	41.59
C13H 9F101	200.21	200.0637	77.99	4.53	9.49	7.99
C13H 9F102	216.21	216.0587	72.22	4.20	8.79	14.80
C13H 9F103	232.21	232.0536	67.24	3.91	8.18	20.67
C13H 9F104	248.21	248.0485	62.91	3.65	7.65	25.78
C13H 9F105	264.21	264.0434	59.10	3.43	7.19	30.28
C13H 9F106	280.21	280.0383	55.72	3.24	6.78	34.26
C13H 9F107	296.21	296.0332	52.71	3.06	6.41	37.81
C13H 9F108	312.21	312.0281	50.01	2.91	6.09	41.00
C13H 9F109	328.21	328.0231	47.57	2.76	5.79	43.87
C13H10F201	220.22	220.0700	70.90	4.58	17.25	7.27
C13H10F202	236.22	236.0649	66.10	4.27	16.09	13.55
C13H10F203	252.22	252.0598	61.91	4.00	15.06	19.03
C13H10F204	268.22	268.0547	58.22	3.76	14.17	23.86
C13H10F205	284.22	284.0496	54.94	3.55	13.37	28.15
C13H10F206	300.22	300.0445	52.01	3.36	12.66	31.98
C13H10F207	316.22	316.0394	49.38	3.19	12.02	35.42
C13H10F208	332.22	332.0344	47.00	3.03	11.44	38.53
C13H10F209	348.22	348.0293	44.84	2.89	10.91	41.35
C13H11F101	202.23	202.0794	77.21	5.48	9.39	7.91
C13H11F102	218.23	218.0743	71.55	5.08	8.71	14.66
C13H11F103	234.23	234.0692	66.66	4.73	8.11	20.49
C13H11F104	250.23	250.0641	62.40	4.43	7.59	25.58
C13H11F105	266.23	266.0590	58.65	4.16	7.14	30.05
C13H11F106	282.23	282.0540	55.33	3.93	6.73	34.01
C13H11F107	298.23	298.0489	52.36	3.72	6.37	37.55
C13H11F108	314.23	314.0438	49.69	3.53	6.05	40.73
C13H11F109	330.23	330.0387	47.28	3.36	5.75	43.60
C13H12F201	222.24	222.0856	70.26	5.44	17.10	7.20
C13H12F202	238.24	238.0805	65.54	5.08	15.95	13.43
C13H12F203	254.24	254.0754	61.42	4.76	14.95	18.88
C13H12F204	270.23	270.0704	57.78	4.48	14.06	23.68
C13H12F205	286.23	286.0653	54.55	4.23	13.27	27.95
C13H12F206	302.23	302.0602	51.66	4.00	12.57	31.76
C13H12F207	318.23	318.0551	49.07	3.80	11.94	35.19
C13H12F208	334.23	334.0500	46.72	3.62	11.37	38.30
C13H12F209	350.23	350.0449	44.58	3.45	10.85	41.11
C13H13F101	204.25	204.0950	76.45	6.42	9.30	7.83
C13H13F102	220.25	220.0900	70.90	5.95	8.63	14.53
C13H13F103	236.25	236.0849	66.09	5.55	8.04	20.32
C13H13F104	252.24	252.0798	61.90	5.19	7.53	25.37
C13H13F105	268.24	268.0747	58.21	4.88	7.08	29.82
C13H13F106	284.24	284.0696	54.93	4.61	6.68	33.77
C13H13F107	300.24	300.0645	52.01	4.36	6.33	37.30
C13H13F108	316.24	316.0594	49.38	4.14	6.01	40.47
C13H13F109	332.24	332.0543	47.00	3.94	5.72	43.34
C13H14F201	224.25	224.1013	69.63	6.29	16.94	7.13
C13H14F202	240.25	240.0962	64.99	5.87	15.82	13.32
C13H14F203	256.25	256.0911	60.93	5.51	14.83	18.73
C13H14F204	272.25	272.0860	57.35	5.18	13.96	23.51
C13H14F205	288.25	288.0809	54.17	4.90	13.18	27.75
C13H14F206	304.25	304.0758	51.32	4.64	12.49	31.55
C13H14F207	320.25	320.0707	48.76	4.41	11.86	34.97
C13H14F208	336.25	336.0657	46.44	4.20	11.30	38.07
C13H14F209	352.25	352.0606	44.33	4.01	10.79	40.88
C13H15F101	206.26	206.1107	75.70	7.33	9.21	7.76
C13H15F102	222.26	222.1056	70.25	6.80	8.55	14.40
C13H15F103	238.26	238.1005	65.54	6.35	7.97	20.15
C13H15F104	254.26	254.0954	61.41	5.95	7.47	25.17
C13H15F105	270.26	270.0903	57.78	5.59	7.03	29.60
C13H15F106	286.26	286.0853	54.55	5.28	6.64	33.53
C13H15F107	302.26	302.0802	51.66	5.00	6.29	37.05
C13H15F108	318.26	318.0751	49.06	4.75	5.97	40.22
C13H15F109	334.26	334.0700	46.71	4.52	5.68	43.08
C13H16F201	226.27	226.1169	69.01	7.13	16.79	7.07
C13H16F202	242.27	242.1118	64.45	6.66	15.68	13.21
C13H16F203	258.27	258.1067	60.46	6.24	14.71	18.58
C13H16F204	274.27	274.1017	56.93	5.88	13.85	23.33
C13H16F205	290.27	290.0966	53.79	5.56	13.09	27.56
C13H16F206	306.27	306.0915	50.98	5.27	12.41	31.34
C13H16F207	322.27	322.0864	48.45	5.00	11.79	34.75
C13H16F208	338.26	338.0813	46.16	4.77	11.23	37.84
C13H16F209	354.26	354.0762	44.08	4.55	10.73	40.65
C13H17F101	208.28	208.1263	74.97	8.23	9.12	7.68
C13H17F102	224.28	224.1212	69.62	7.64	8.47	14.27
C13H17F103	240.28	240.1162	64.99	7.13	7.91	19.98
C13H17F104	256.28	256.1111	60.93	6.69	7.41	24.97
C13H17F105	272.28	272.1060	57.35	6.29	6.98	29.38
C13H17F106	288.28	288.1009	54.17	5.94	6.59	33.30
C13H17F107	304.27	304.0958	51.32	5.63	6.24	36.81
C13H17F108	320.27	320.0907	48.75	5.35	5.93	39.96
C13H17F109	336.27	336.0856	46.43	5.10	5.65	42.82
C13H18F201	228.28	228.1326	68.40	7.95	16.64	7.01
C13H18F202	244.28	244.1275	63.92	7.43	15.55	13.10
C13H18F203	260.28	260.1224	59.99	6.97	14.60	18.44
C13H18F204	276.28	276.1173	56.52	6.57	13.75	23.16
C13H18F205	292.28	292.1122	53.42	6.21	13.00	27.37
C13H18F206	308.28	308.1071	50.65	5.89	12.33	31.14
C13H18F207	324.28	324.1020	48.15	5.59	11.72	34.54
C13H18F208	340.28	340.0970	45.89	5.33	11.17	37.61
C13H18F209	356.28	356.0919	43.83	5.09	10.66	40.42
C13H19F101	210.29	210.1420	74.25	9.11	9.03	7.61
C13H19F102	226.29	226.1369	69.00	8.46	8.40	14.14
C13H19F103	242.29	242.1318	64.44	7.90	7.84	19.81
C13H19F104	258.29	258.1267	60.45	7.41	7.56	24.78
C13H19F105	274.29	274.1216	56.93	6.98	6.93	29.16
C13H19F106	290.29	290.1166	53.79	6.60	6.54	33.07
C13H19F107	306.29	306.1115	50.98	6.25	6.20	36.57

C13H19F108	322.29	322.1064	48.45	5.94	5.89	39.71
C13H19F109	338.29	338.1013	46.16	5.66	5.62	42.57
C13H20F201	230.30	230.1482	67.80	8.75	16.50	6.95
C13H20F202	246.30	246.1431	63.40	8.18	15.43	12.99
C13H20F203	262.30	262.1380	59.53	7.69	14.49	18.30
C13H20F204	278.30	278.1330	56.11	7.24	13.65	23.00
C13H20F205	294.30	294.1279	53.06	6.85	12.91	27.18
C13H20F206	310.30	310.1228	50.32	6.50	12.25	30.94
C13H20F207	326.30	326.1177	47.85	6.18	11.64	34.32
C13H20F208	342.30	342.1126	45.62	5.89	11.10	37.39
C13H20F209	358.30	358.1075	43.58	5.63	10.60	40.19
C13H21F101	212.31	212.1576	73.55	9.97	8.95	7.54
C13H21F102	228.31	228.1525	68.39	9.27	8.32	14.02
C13H21F103	244.31	244.1475	63.91	8.66	7.78	19.65
C13H21F104	260.31	260.1424	59.98	8.13	7.30	24.59
C13H21F105	276.31	276.1373	56.51	7.66	6.88	28.95
C13H21F106	292.31	292.1322	53.42	7.24	6.50	32.84
C13H21F107	308.31	308.1271	50.65	6.87	6.16	36.33
C13H21F108	324.31	324.1220	48.15	6.53	5.86	39.47
C13H21F109	340.31	340.1169	45.88	6.22	5.58	42.31
C13H22F201	232.32	232.1639	67.21	9.55	16.36	6.89
C13H22F202	248.32	248.1588	62.88	8.93	15.30	12.89
C13H22F203	264.32	264.1537	59.08	8.39	14.38	18.16
C13H22F204	280.31	280.1486	55.70	7.91	13.56	22.83
C13H22F205	296.31	296.1435	52.70	7.48	12.82	27.00
C13H22F206	312.31	312.1384	50.00	7.10	12.17	30.74
C13H22F207	328.31	328.1333	47.56	6.75	11.57	34.11
C13H22F208	344.31	344.1283	45.35	6.44	11.04	37.17
C13H22F209	360.31	360.1232	43.34	6.15	10.55	39.96
C13H23F101	214.33	214.1733	72.85	10.82	8.86	7.46
C13H23F102	230.33	230.1682	67.79	10.07	8.25	13.89
C13H23F103	246.32	246.1631	63.39	9.41	7.71	19.49
C13H23F104	262.32	262.1580	59.52	8.84	7.24	24.40
C13H23F105	278.32	278.1529	56.10	8.33	6.83	28.74
C13H23F106	294.32	294.1479	53.05	7.88	6.45	32.62
C13H23F107	310.32	310.1428	50.32	7.47	6.12	36.09
C13H23F108	326.32	326.1377	47.85	7.10	5.82	39.22
C13H23F109	342.32	342.1326	45.61	6.77	5.55	42.06
C13H24F201	234.33	234.1795	66.63	10.32	16.21	6.83
C13H24F202	250.33	250.1744	62.38	9.66	15.18	12.78
C13H24F203	266.33	266.1693	58.63	9.08	14.27	18.02
C13H24F204	282.33	282.1643	55.31	8.57	13.46	22.67
C13H24F205	298.33	298.1592	52.34	8.11	12.74	26.81
C13H24F206	314.33	314.1541	49.68	7.70	12.09	30.54
C13H24F207	330.33	330.1490	47.27	7.32	11.50	33.90
C13H24F208	346.33	346.1439	45.09	6.99	10.97	36.96
C13H24F209	362.33	362.1388	43.09	6.68	10.49	39.74
C13H25F101	216.34	216.1889	72.18	11.65	8.78	7.40
C13H25F102	232.34	232.1838	67.20	10.85	8.18	13.77
C13H25F103	248.34	248.1788	62.88	10.15	7.65	19.33
C13H25F104	264.34	264.1737	59.07	9.53	7.19	24.21
C13H25F105	280.34	280.1686	55.70	8.99	6.78	28.54
C13H25F106	296.34	296.1635	52.69	8.50	6.41	32.39
C13H25F107	312.34	312.1584	49.99	8.07	6.08	35.86
C13H25F108	328.34	328.1533	47.56	7.67	5.79	38.98
C13H25F109	344.34	344.1482	45.35	7.32	5.52	41.82
C13H26F201	236.35	236.1952	66.07	11.09	16.08	6.77
C13H26F202	252.35	252.1901	61.88	10.39	15.06	12.68
C13H26F203	268.35	268.1850	58.19	9.77	14.16	17.89
C13H26F204	284.35	284.1799	54.91	9.22	13.36	22.51
C13H26F205	300.35	300.1748	51.99	8.73	12.65	26.63
C13H26F206	316.35	316.1697	49.36	8.28	12.01	30.35
C13H26F207	332.34	332.1646	46.98	7.89	11.43	33.70
C13H26F208	348.34	348.1596	44.82	7.52	10.91	36.74
C13H26F209	364.34	364.1545	42.86	7.19	10.43	39.52
C13H27F101	218.36	218.2046	71.51	12.46	8.70	7.33
C13H27F102	234.36	234.1995	66.63	11.61	8.11	13.65
C13H27F103	250.36	250.1944	62.37	10.87	7.59	19.17
C13H27F104	266.36	266.1893	58.62	10.22	7.13	24.03
C13H27F105	282.36	282.1842	55.30	9.64	6.73	28.33
C13H27F106	298.35	298.1792	52.34	9.12	6.37	32.18
C13H27F107	314.35	314.1741	49.67	8.66	6.04	35.63
C13H27F108	330.35	330.1690	47.27	8.24	5.75	38.74
C13H27F109	346.35	346.1639	45.08	7.86	5.49	41.57
C14H 7F101	210.21	210.0481	79.99	3.36	9.04	7.61
C14H 7F102	226.21	226.0430	74.34	3.12	8.40	14.15
C14H 7F103	242.21	242.0379	69.43	2.91	7.84	19.82
C14H 7F104	258.21	258.0328	65.12	2.73	7.36	24.79
C14H 7F105	274.21	274.0277	61.32	2.57	6.93	29.17
C14H 7F106	290.21	290.0227	57.94	2.43	6.55	33.08
C14H 7F107	306.21	306.0176	54.92	2.30	6.20	36.58
C14H 7F108	322.21	322.0125	52.19	2.19	5.90	39.72
C14H 7F109	338.20	338.0074	49.72	2.09	5.62	42.58
C14H 8F201	230.22	230.0543	73.04	3.50	16.50	6.95
C14H 8F202	246.22	246.0492	68.30	3.28	15.43	13.00
C14H 8F203	262.21	262.0441	64.13	3.08	14.49	18.30
C14H 8F204	278.21	278.0391	60.44	2.90	13.66	23.00
C14H 8F205	294.21	294.0340	57.15	2.74	12.91	27.19
C14H 8F206	310.21	310.0289	54.21	2.60	12.25	30.95
C14H 8F207	326.21	326.0238	51.55	2.47	11.65	34.33
C14H 8F208	342.21	342.0187	49.14	2.36	11.10	37.40
C14H 8F209	358.21	358.0136	46.94	2.25	10.61	40.20
C14H 9F101	212.23	212.0637	79.23	4.27	8.95	7.54
C14H 9F102	228.23	228.0587	73.68	3.97	8.32	14.02
C14H 9F103	244.22	244.0536	68.85	3.71	7.78	19.65
C14H 9F104	260.22	260.0485	64.62	3.49	7.30	24.59
C14H 9F105	276.22	276.0434	60.88	3.28	6.88	28.96
C14H 9F106	292.22	292.0383	57.54	3.10	6.50	32.85
C14H 9F107	308.22	308.0332	54.56	2.94	6.16	36.34
C14H 9F108	324.22	324.0281	51.86	2.80	5.86	39.48
C14H 9F109	340.22	340.0231	49.43	2.67	5.58	42.32
C14H10F201	232.23	232.0700	72.41	4.34	16.36	6.89
C14H10F202	248.23	248.0649	67.74	4.06	15.31	12.89
C14H10F203	264.23	264.0598	63.64	3.81	14.38	18.17
C14H10F204	280.23	280.0547	60.01	3.60	13.56	22.84
C14H10F205	296.23	296.0496	56.77	3.40	12.83	27.01
C14H10F206	312.23	312.0445	53.86	3.23	12.17	30.75
C14H10F207	328.23	328.0394	51.23	3.07	11.58	34.12
C14H10F208	344.23	344.0344	48.85	2.93	11.04	37.18
C14H10F209	360.23	360.0293	46.68	2.80	10.55	39.97
C14H11F101	214.24	214.0794	78.49	5.18	8.87	7.47
C14H11F102	230.24	230.0743	73.03	4.82	8.25	13.90
C14H11F103	246.24	246.0692	68.29	4.50	7.72	19.49
C14H11F104	262.24	262.0641	64.12	4.23	7.24	24.40
C14H11F105	278.24	278.0590	60.44	3.98	6.83	28.75
C14H11F106	294.24	294.0540	57.15	3.77	6.46	32.63
C14H11F107	310.24	310.0489	54.20	3.57	6.12	36.10
C14H11F108	326.24	326.0438	51.54	3.40	5.82	39.23
C14H11F109	342.24	342.0387	49.13	3.24	5.55	42.07
C14H12F201	234.25	234.0856	71.79	5.16	16.22	6.83

138

C14H12F202	250.25	250.0805	67.20	4.83	15.18	12.79
C14H12F203	266.25	266.0754	63.16	4.54	14.27	18.03
C14H12F204	282.25	282.0704	59.58	4.29	13.46	22.67
C14H12F205	298.25	298.0653	56.38	4.06	12.74	26.82
C14H12F206	314.24	314.0602	53.51	3.85	12.09	30.55
C14H12F207	330.24	330.0551	50.92	3.66	11.51	33.91
C14H12F208	346.24	346.0500	48.57	3.49	10.97	36.97
C14H12F209	362.24	362.0449	46.42	3.34	10.49	39.75
C14H13F101	216.26	216.0950	77.76	6.06	8.79	7.40
C14H13F102	232.26	232.0900	72.40	5.64	8.18	13.78
C14H13F103	248.26	248.0849	67.73	5.28	7.65	19.33
C14H13F104	264.26	264.0798	63.63	4.96	7.19	24.22
C14H13F105	280.26	280.0747	60.00	4.68	6.78	28.54
C14H13F106	296.25	296.0696	56.76	4.42	6.41	32.40
C14H13F107	312.25	312.0645	53.85	4.20	6.08	35.87
C14H13F108	328.25	328.0594	51.23	3.99	5.79	38.99
C14H13F109	344.25	344.0543	48.85	3.81	5.52	41.83
C14H14F201	236.26	236.1013	71.17	5.97	16.08	6.77
C14H14F202	252.26	252.0962	66.66	5.59	15.06	12.68
C14H14F203	268.26	268.0911	62.68	5.26	14.16	17.89
C14H14F204	284.26	284.0860	59.16	4.96	13.37	22.51
C14H14F205	300.26	300.0809	56.00	4.70	12.65	26.64
C14H14F206	316.26	316.0758	53.17	4.46	12.01	30.35
C14H14F207	332.26	332.0707	50.61	4.25	11.44	33.71
C14H14F208	348.26	348.0657	48.28	4.05	10.91	36.75
C14H14F209	364.26	364.0606	46.16	3.87	10.43	39.53
C14H15F101	218.27	218.1107	77.04	6.93	8.70	7.33
C14H15F102	234.27	234.1056	71.78	6.45	8.11	13.66
C14H15F103	250.27	250.1005	67.19	6.04	7.59	19.18
C14H15F104	266.27	266.0954	63.15	5.68	7.13	24.03
C14H15F105	282.27	282.0903	59.57	5.36	6.73	28.34
C14H15F106	298.27	298.0853	56.38	5.07	6.37	32.18
C14H15F107	314.27	314.0802	53.51	4.81	6.05	35.64
C14H15F108	330.27	330.0751	50.91	4.58	5.75	38.75
C14H15F109	346.27	346.0700	48.56	4.37	5.49	41.58
C14H16F201	238.28	238.1169	70.57	6.77	15.95	6.71
C14H16F202	254.28	254.1118	66.13	6.34	14.94	12.58
C14H16F203	270.28	270.1067	62.22	5.97	14.06	17.76
C14H16F204	286.28	286.1017	58.74	5.63	13.27	22.36
C14H16F205	302.28	302.0966	55.63	5.34	12.57	26.46
C14H16F206	318.28	318.0915	52.83	5.07	11.94	30.16
C14H16F207	334.28	334.0864	50.30	4.82	11.37	33.50
C14H16F208	350.28	350.0813	48.01	4.60	10.85	36.54
C14H16F209	366.28	366.0762	45.91	4.40	10.37	39.31
C14H17F101	220.29	220.1263	76.33	7.78	8.62	7.26
C14H17F102	236.29	236.1212	71.17	7.25	8.04	13.54
C14H17F103	252.29	252.1162	66.65	6.79	7.53	19.03
C14H17F104	268.29	268.1111	62.68	6.39	7.08	23.85
C14H17F105	284.29	284.1060	59.15	6.03	6.68	28.14
C14H17F106	300.29	300.1009	56.00	5.71	6.33	31.97
C14H17F107	316.29	316.0958	53.17	5.42	6.01	35.41
C14H17F108	332.29	332.0907	50.61	5.16	5.72	38.52
C14H17F109	348.28	348.0856	48.28	4.92	5.45	41.34
C14H18F201	240.30	240.1326	69.98	7.55	15.81	6.66
C14H18F202	256.30	256.1275	65.61	7.08	14.83	12.49
C14H18F203	272.29	272.1224	61.76	6.66	13.95	17.63
C14H18F204	288.29	288.1173	58.33	6.29	13.18	22.20
C14H18F205	304.29	304.1122	55.26	5.96	12.49	26.29
C14H18F206	320.29	320.1071	52.50	5.66	11.86	29.97
C14H18F207	336.29	336.1020	50.00	5.40	11.30	33.30
C14H18F208	352.29	352.0970	47.73	5.15	10.79	36.33
C14H18F209	368.29	368.0919	45.66	4.93	10.32	39.10
C14H19F101	222.31	222.1420	75.64	8.61	8.55	7.20
C14H19F102	238.30	238.1369	70.56	8.04	7.97	13.43
C14H19F103	254.30	254.1318	66.12	7.53	7.47	18.87
C14H19F104	270.30	270.1267	62.21	7.09	7.03	23.68
C14H19F105	286.30	286.1216	58.73	6.69	6.64	27.94
C14H19F106	302.30	302.1166	55.63	6.34	6.28	31.76
C14H19F107	318.30	318.1115	52.83	6.02	5.97	35.19
C14H19F108	334.30	334.1064	50.30	5.73	5.68	38.29
C14H19F109	350.30	350.1013	48.00	5.47	5.42	41.11
C14H20F201	242.31	242.1482	69.40	8.32	15.68	6.60
C14H20F202	258.31	258.1431	65.10	7.80	14.71	12.39
C14H20F203	274.31	274.1380	61.30	7.35	13.85	17.50
C14H20F204	290.31	290.1330	57.92	6.94	13.09	22.04
C14H20F205	306.31	306.1279	54.90	6.58	12.40	26.12
C14H20F206	322.31	322.1228	52.17	6.25	11.79	29.78
C14H20F207	338.31	338.1177	49.71	5.96	11.23	33.10
C14H20F208	354.31	354.1126	47.46	5.69	10.72	36.13
C14H20F209	370.31	370.1075	45.41	5.44	10.26	38.89
C14H21F101	224.32	224.1576	74.96	9.44	8.47	7.13
C14H21F102	240.32	240.1525	69.97	8.81	7.91	13.32
C14H21F103	256.32	256.1475	65.60	8.26	7.41	18.73
C14H21F104	272.32	272.1424	61.75	7.77	6.98	23.50
C14H21F105	288.32	288.1373	58.32	7.34	6.59	27.75
C14H21F106	304.32	304.1322	55.26	6.96	6.24	31.54
C14H21F107	320.32	320.1271	52.50	6.61	5.93	34.96
C14H21F108	336.32	336.1220	50.00	6.29	5.65	38.06
C14H21F109	352.32	352.1169	47.73	6.01	5.39	40.87
C14H22F201	244.33	244.1639	68.82	9.08	15.55	6.55
C14H22F202	260.33	260.1588	64.59	8.52	14.60	12.29
C14H22F203	276.33	276.1537	60.85	8.03	13.75	17.37
C14H22F204	292.33	292.1486	57.52	7.59	13.00	21.89
C14H22F205	308.33	308.1435	54.54	7.19	12.32	25.95
C14H22F206	324.32	324.1384	51.85	6.84	11.72	29.60
C14H22F207	340.32	340.1333	49.41	6.52	11.16	32.91
C14H22F208	356.32	356.1283	47.19	6.22	10.66	35.92
C14H22F209	372.32	372.1232	45.16	5.96	10.21	38.67
C14H23F101	226.34	226.1733	74.29	10.24	8.39	7.07
C14H23F102	242.34	242.1682	69.39	9.57	7.84	13.20
C14H23F103	258.34	258.1631	65.09	8.97	7.35	18.58
C14H23F104	274.34	274.1580	61.30	8.45	6.93	23.33
C14H23F105	290.33	290.1529	57.92	7.99	6.54	27.55
C14H23F106	306.33	306.1479	54.89	7.57	6.20	31.34
C14H23F107	322.33	322.1428	52.17	7.19	5.89	34.75
C14H23F108	338.33	338.1377	49.70	6.85	5.62	37.83
C14H23F109	354.33	354.1326	47.46	6.54	5.36	40.64
C14H24F201	246.34	246.1795	68.26	9.82	15.42	6.49
C14H24F202	262.34	262.1744	64.10	9.22	14.48	12.20
C14H24F203	278.34	278.1693	60.41	8.69	13.65	17.24
C14H24F204	294.34	294.1643	57.13	8.22	12.91	21.74
C14H24F205	310.34	310.1592	54.18	7.80	12.24	25.78
C14H24F206	326.34	326.1541	51.53	7.41	11.64	29.42
C14H24F207	342.34	342.1490	49.12	7.07	11.10	32.71
C14H24F208	358.34	358.1439	46.93	6.75	10.60	35.72
C14H24F209	374.34	374.1388	44.92	6.46	10.15	38.47
C14H25F101	228.35	228.1889	73.64	11.04	8.32	7.01
C14H25F102	244.35	244.1838	68.82	10.31	7.77	13.10
C14H25F103	260.35	260.1788	64.59	9.68	7.30	18.44
C14H25F104	276.35	276.1737	60.85	9.12	6.87	23.16

C14H25F105	292.35	292.1686	57.52	8.62	6.50	27.36
C14H25F106	308.35	308.1635	54.53	8.17	6.16	31.13
C14H25F107	324.35	324.1584	51.84	7.77	5.86	34.53
C14H25F108	340.35	340.1533	49.41	7.40	5.58	37.61
C14H25F109	356.35	356.1482	47.15	7.07	5.33	40.41
C14H26F201	248.36	248.1952	67.71	10.55	15.30	6.44
C14H26F202	264.36	264.1901	63.61	9.91	14.37	12.10
C14H26F203	280.36	280.1850	59.98	9.35	13.55	17.12
C14H26F204	296.36	296.1799	56.74	8.84	12.82	21.59
C14H26F205	312.36	312.1748	53.83	8.39	12.16	25.61
C14H26F206	328.36	328.1697	51.21	7.98	11.57	29.24
C14H26F207	344.36	344.1646	48.83	7.61	11.03	32.52
C14H26F208	360.36	360.1596	46.66	7.27	10.54	35.52
C14H26F209	376.35	376.1545	44.68	6.96	10.10	38.26
C14H27F101	230.37	230.2046	72.99	11.81	8.25	6.95
C14H27F102	246.37	246.1995	68.25	11.05	7.71	12.99
C14H27F103	262.37	262.1944	64.09	10.37	7.24	18.29
C14H27F104	278.37	278.1893	60.41	9.78	6.82	22.99
C14H27F105	294.37	294.1842	57.12	9.25	6.45	27.18
C14H27F106	310.37	310.1792	54.18	8.77	6.12	30.93
C14H27F107	326.37	326.1741	51.52	8.34	5.82	34.32
C14H27F108	342.36	342.1690	49.12	7.95	5.55	37.39
C14H27F109	358.36	358.1639	46.92	7.59	5.30	40.18
C14H28F201	250.38	250.2108	67.16	11.27	15.18	6.39
C14H28F202	266.37	266.2057	63.13	10.60	14.26	12.01
C14H28F203	282.37	282.2006	59.55	9.99	13.46	17.00
C14H28F204	298.37	298.1955	56.36	9.46	12.73	21.45
C14H28F205	314.37	314.1905	53.49	8.98	12.09	25.45
C14H28F206	330.37	330.1854	50.90	8.54	11.50	29.06
C14H28F207	346.37	346.1803	48.55	8.15	10.97	32.33
C14H28F208	362.37	362.1752	46.40	7.79	10.49	35.32
C14H28F209	378.37	378.1701	44.44	7.46	10.04	38.06
C14H29F101	232.39	232.2202	72.36	12.58	8.18	6.88
C14H29F102	248.38	248.2151	67.70	11.77	7.65	12.88
C14H29F103	264.38	264.2101	63.60	11.06	7.19	18.15
C14H29F104	280.38	280.2050	59.97	10.43	6.78	22.83
C14H29F105	296.38	296.1999	56.74	9.86	6.41	26.99
C14H29F106	312.38	312.1948	53.83	9.36	6.08	30.73
C14H29F107	328.38	328.1897	51.21	8.90	5.79	34.11
C14H29F108	344.38	344.1846	48.83	8.49	5.52	37.17
C14H29F109	360.38	360.1795	46.66	8.11	5.27	39.96
C15H 7F101	222.22	222.0481	81.08	3.18	8.55	7.20
C15H 7F102	238.22	238.0430	75.63	2.96	7.98	13.43
C15H 7F103	254.22	254.0379	70.87	2.78	7.47	18.88
C15H 7F104	270.22	270.0328	66.67	2.61	7.03	23.68
C15H 7F105	286.22	286.0277	62.95	2.47	6.64	27.95
C15H 7F106	302.22	302.0227	59.62	2.33	6.29	31.76
C15H 7F107	318.22	318.0176	56.62	2.22	5.97	35.19
C15H 7F108	334.22	334.0125	53.91	2.11	5.68	38.30
C15H 7F109	350.22	350.0074	51.44	2.01	5.42	41.12
C15H 8F201	242.23	242.0543	74.38	3.33	15.69	6.61
C15H 8F202	258.23	258.0492	69.77	3.12	14.71	12.39
C15H 8F203	274.23	274.0441	65.70	2.94	13.86	17.50
C15H 8F204	290.23	290.0391	62.08	2.78	13.09	22.05
C15H 8F205	306.22	306.0340	58.83	2.63	12.41	26.12
C15H 8F206	322.22	322.0289	55.91	2.50	11.79	29.79
C15H 8F207	338.22	338.0238	53.27	2.38	11.23	33.11
C15H 8F208	354.22	354.0187	50.86	2.28	10.73	36.13
C15H 8F209	370.22	370.0136	48.66	2.18	10.26	38.89
C15H 9F101	224.24	224.0637	80.35	4.05	8.47	7.14
C15H 9F102	240.24	240.0587	75.00	3.78	7.91	13.32
C15H 9F103	256.24	256.0536	70.31	3.54	7.41	18.73
C15H 9F104	272.23	272.0485	66.18	3.33	6.98	23.51
C15H 9F105	288.23	288.0434	62.51	3.15	6.59	27.75
C15H 9F106	304.23	304.0383	59.22	2.98	6.24	31.55
C15H 9F107	320.23	320.0332	56.26	2.83	5.93	34.97
C15H 9F108	336.23	336.0281	53.58	2.70	5.65	38.07
C15H 9F109	352.23	352.0231	51.15	2.58	5.39	40.88
C15H10F201	244.24	244.0700	73.77	4.13	15.56	6.55
C15H10F202	260.24	260.0649	69.23	3.87	14.60	12.30
C15H10F203	276.24	276.0598	65.22	3.65	13.75	17.38
C15H10F204	292.24	292.0547	61.65	3.45	13.00	21.90
C15H10F205	308.24	308.0496	58.45	3.27	12.33	25.95
C15H10F206	324.24	324.0445	55.57	3.11	11.72	29.61
C15H10F207	340.24	340.0394	52.95	2.96	11.17	32.92
C15H10F208	356.24	356.0344	50.57	2.83	10.67	35.93
C15H10F209	372.24	372.0293	48.40	2.71	10.21	38.68
C15H11F101	226.25	226.0794	79.63	4.90	8.40	7.07
C15H11F102	242.25	242.0743	74.37	4.58	7.84	13.21
C15H11F103	258.25	258.0692	69.76	4.29	7.36	18.59
C15H11F104	274.25	274.0641	65.69	4.04	6.93	23.34
C15H11F105	290.25	290.0590	62.07	3.82	6.55	27.56
C15H11F106	306.25	306.0540	58.83	3.62	6.20	31.35
C15H11F107	322.25	322.0489	55.91	3.44	5.90	34.75
C15H11F108	338.25	338.0438	53.26	3.28	5.62	37.84
C15H11F109	354.25	354.0387	50.86	3.13	5.36	40.65
C15H12F201	246.26	246.0856	73.16	4.91	15.43	6.50
C15H12F202	262.26	262.0805	68.70	4.61	14.49	12.20
C15H12F203	278.26	278.0754	64.75	4.35	13.66	17.25
C15H12F204	294.26	294.0704	61.23	4.11	12.91	21.75
C15H12F205	310.26	310.0653	58.07	3.90	12.25	25.78
C15H12F206	326.26	326.0602	55.22	3.71	11.65	29.42
C15H12F207	342.26	342.0551	52.64	3.53	11.10	32.72
C15H12F208	358.25	358.0500	50.29	3.38	10.61	35.73
C15H12F209	374.25	374.0449	48.14	3.23	10.15	38.48
C15H13F101	228.27	228.0950	78.93	5.74	8.32	7.01
C15H13F102	244.27	244.0900	73.76	5.36	7.78	13.10
C15H13F103	260.27	260.0849	69.22	5.03	7.30	18.44
C15H13F104	276.27	276.0798	65.21	4.74	6.88	23.17
C15H13F105	292.27	292.0747	61.64	4.48	6.50	27.37
C15H13F106	308.27	308.0696	58.45	4.25	6.16	31.14
C15H13F107	324.27	324.0645	55.56	4.04	5.86	34.54
C15H13F108	340.26	340.0594	52.95	3.85	5.58	37.62
C15H13F109	356.26	356.0543	50.57	3.68	5.33	40.42
C15H14F201	248.28	248.1013	72.57	5.68	15.30	6.44
C15H14F202	264.27	264.0962	68.17	5.34	14.38	12.11
C15H14F203	280.27	280.0911	64.28	5.03	13.56	17.13
C15H14F204	296.27	296.0860	60.81	4.76	12.82	21.60
C15H14F205	312.27	312.0809	57.70	4.52	12.17	25.62
C15H14F206	328.27	328.0758	54.88	4.30	11.57	29.24
C15H14F207	344.27	344.0707	52.33	4.10	11.04	32.53
C15H14F208	360.27	360.0657	50.01	3.92	10.55	35.53
C15H14F209	376.27	376.0606	47.88	3.75	10.10	38.27
C15H15F101	230.28	230.1107	78.24	6.57	8.25	6.95
C15H15F102	246.28	246.1056	73.15	6.14	7.71	12.99
C15H15F103	262.28	262.1005	68.69	5.76	7.24	18.30
C15H15F104	278.28	278.0954	64.74	5.43	6.83	23.00
C15H15F105	294.28	294.0903	61.22	5.14	6.46	27.18
C15H15F106	310.28	310.0853	58.07	4.87	6.12	30.94
C15H15F107	326.28	326.0802	55.22	4.63	5.82	34.32

C15H15F108	342.28	342.0751	52.64	4.42	5.55	37.39
C15H15F109	358.28	358.0700	50.29	4.22	5.30	40.19
C15H16F201	250.29	250.1169	71.98	6.44	15.18	6.39
C15H16F202	266.29	266.1118	67.66	6.06	14.27	12.02
C15H16F203	282.29	282.1067	63.82	5.71	13.46	17.00
C15H16F204	298.29	298.1017	60.40	5.41	12.74	21.45
C15H16F205	314.29	314.0966	57.33	5.13	12.09	25.45
C15H16F206	330.29	330.0915	54.55	4.88	11.50	29.06
C15H16F207	346.29	346.0864	52.03	4.66	10.97	32.34
C15H16F208	362.29	362.0813	49.73	4.45	10.49	35.33
C15H16F209	378.29	378.0762	47.63	4.26	10.04	38.06
C15H17F101	232.30	232.1263	77.56	7.38	8.18	6.89
C15H17F102	248.30	248.1212	72.56	6.90	7.65	12.89
C15H17F103	264.30	264.1162	68.17	6.48	7.19	18.16
C15H17F104	280.30	280.1111	64.28	6.11	6.78	22.83
C15H17F105	296.30	296.1060	60.81	5.78	6.41	27.00
C15H17F106	312.30	312.1009	57.69	5.49	6.08	30.74
C15H17F107	328.30	328.0958	54.88	5.22	5.79	34.11
C15H17F108	344.30	344.0907	52.33	4.98	5.52	37.18
C15H17F109	360.30	360.0856	50.01	4.76	5.27	39.97
C15H18F201	252.31	252.1326	71.41	7.19	15.06	6.34
C15H18F202	268.31	268.1275	67.15	6.76	14.16	11.93
C15H18F203	284.31	284.1224	63.37	6.38	13.36	16.88
C15H18F204	300.31	300.1173	59.99	6.04	12.65	21.31
C15H18F205	316.30	316.1122	56.96	5.74	12.01	25.29
C15H18F206	332.30	332.1071	54.22	5.46	11.43	28.89
C15H18F207	348.30	348.1020	51.73	5.21	10.91	32.15
C15H18F208	364.30	364.0970	49.46	4.98	10.43	35.13
C15H18F209	380.30	380.0919	47.37	4.77	9.99	37.86
C15H19F101	234.32	234.1420	76.89	8.17	8.11	6.83
C15H19F102	250.32	250.1369	71.98	7.65	7.59	12.78
C15H19F103	266.32	266.1318	67.65	7.19	7.13	18.02
C15H19F104	282.31	282.1267	63.82	6.78	6.73	22.67
C15H19F105	298.31	298.1216	60.40	6.42	6.37	26.82
C15H19F106	314.31	314.1166	57.32	6.09	6.04	30.54
C15H19F107	330.31	330.1115	54.54	5.80	5.75	33.91
C15H19F108	346.31	346.1064	52.02	5.53	5.49	36.96
C15H19F109	362.31	362.1013	49.73	5.29	5.24	39.74
C15H20F201	254.32	254.1482	70.84	7.93	14.94	6.29
C15H20F202	270.32	270.1431	66.65	7.46	14.06	11.84
C15H20F203	286.32	286.1380	62.92	7.04	13.27	16.76
C15H20F204	302.32	302.1330	59.59	6.67	12.57	21.17
C15H20F205	318.32	318.1279	56.60	6.33	11.94	25.13
C15H20F206	334.32	334.1228	53.89	6.03	11.37	28.71
C15H20F207	350.32	350.1177	51.43	5.75	10.85	31.97
C15H20F208	366.32	366.1126	49.18	5.50	10.37	34.94
C15H20F209	382.32	382.1075	47.12	5.27	9.94	37.66
C15H21F101	236.33	236.1576	76.23	8.96	8.04	6.77
C15H21F102	252.33	252.1525	71.40	8.39	7.53	12.68
C15H21F103	268.33	268.1475	67.14	7.89	7.08	17.89
C15H21F104	284.33	284.1424	63.37	7.44	6.68	22.51
C15H21F105	300.33	300.1373	59.99	7.05	6.33	26.64
C15H21F106	316.33	316.1322	56.96	6.69	6.01	30.35
C15H21F107	332.33	332.1271	54.21	6.37	5.72	33.70
C15H21F108	348.33	348.1220	51.72	6.08	5.45	36.75
C15H21F109	364.33	364.1169	49.45	5.81	5.21	39.52
C15H22F201	256.34	256.1639	70.28	8.65	14.82	6.24
C15H22F202	272.34	272.1588	66.16	8.14	13.95	11.75
C15H22F203	288.34	288.1537	62.48	7.69	13.18	16.65
C15H22F204	304.34	304.1486	59.20	7.29	12.49	21.03
C15H22F205	320.34	320.1435	56.24	6.92	11.86	24.97
C15H22F206	336.34	336.1384	53.57	6.59	11.30	28.54
C15H22F207	352.34	352.1333	51.14	6.29	10.78	31.79
C15H22F208	368.33	368.1283	48.91	6.02	10.32	34.75
C15H22F209	384.33	384.1232	46.88	5.77	9.89	37.47
C15H23F101	238.35	238.1733	75.59	9.73	7.97	6.71
C15H23F102	254.35	254.1682	70.84	9.11	7.47	12.58
C15H23F103	270.35	270.1631	66.64	8.58	7.03	17.75
C15H23F104	286.35	286.1580	62.92	8.10	6.63	22.35
C15H23F105	302.35	302.1529	59.59	7.67	6.28	26.46
C15H23F106	318.35	318.1479	56.59	7.28	5.97	30.15
C15H23F107	334.34	334.1428	53.89	6.93	5.68	33.50
C15H23F108	350.34	350.1377	51.43	6.62	5.42	36.53
C15H23F109	366.34	366.1326	49.18	6.33	5.19	39.31
C15H24F201	258.35	258.1795	69.74	9.36	14.71	6.19
C15H24F202	274.35	274.1744	65.67	8.82	13.85	11.66
C15H24F203	290.35	290.1693	62.05	8.33	13.09	16.53
C15H24F204	306.35	306.1643	58.81	7.90	12.40	20.89
C15H24F205	322.35	322.1592	55.89	7.50	11.79	24.82
C15H24F206	338.35	338.1541	53.25	7.15	11.23	28.37
C15H24F207	354.35	354.1490	50.84	6.83	10.72	31.61
C15H24F208	370.35	370.1439	48.65	6.53	10.26	34.56
C15H24F209	386.35	386.1388	46.63	6.26	9.83	37.27
C15H25F101	240.36	240.1889	74.96	10.48	7.90	6.66
C15H25F102	256.36	256.1838	70.28	9.83	7.41	12.48
C15H25F103	272.36	272.1788	66.15	9.25	6.98	17.62
C15H25F104	288.36	288.1737	62.48	8.74	6.59	22.19
C15H25F105	304.36	304.1686	59.20	8.28	6.24	26.28
C15H25F106	320.36	320.1635	56.24	7.87	5.93	29.97
C15H25F107	336.36	336.1584	53.56	7.49	5.65	33.30
C15H25F108	352.36	352.1533	51.13	7.15	5.39	36.33
C15H25F109	368.36	368.1482	48.91	6.84	5.16	39.09
C15H26F201	260.37	260.1952	69.20	10.07	14.59	6.14
C15H26F202	276.37	276.1901	65.19	9.48	13.75	11.58
C15H26F203	292.37	292.1850	61.62	8.96	13.00	16.42
C15H26F204	308.37	308.1799	58.43	8.50	12.32	20.75
C15H26F205	324.37	324.1748	55.54	8.08	11.71	24.66
C15H26F206	340.37	340.1697	52.93	7.70	11.16	28.20
C15H26F207	356.37	356.1646	50.56	7.35	10.66	31.43
C15H26F208	372.37	372.1596	48.38	7.04	10.20	34.37
C15H26F209	388.37	388.1545	46.39	6.75	9.78	37.08
C15H27F101	242.38	242.2046	74.33	11.23	7.84	6.60
C15H27F102	258.38	258.1995	69.73	10.53	7.35	12.38
C15H27F103	274.38	274.1944	65.66	9.92	6.92	17.49
C15H27F104	290.38	290.1893	62.05	9.37	6.54	22.04
C15H27F105	306.38	306.1842	58.81	8.88	6.20	26.11
C15H27F106	322.38	322.1792	55.89	8.44	5.89	29.78
C15H27F107	338.38	338.1741	53.24	8.04	5.61	33.10
C15H27F108	354.38	354.1690	50.84	7.68	5.36	36.12
C15H27F109	370.38	370.1639	48.64	7.35	5.13	38.88
C15H28F201	262.39	262.2108	68.66	10.76	14.48	6.10
C15H28F202	278.39	278.2057	64.72	10.14	13.65	11.49
C15H28F203	294.39	294.2006	61.20	9.59	12.91	16.30
C15H28F204	310.38	310.1955	58.05	9.09	12.24	20.62
C15H28F205	326.38	326.1905	55.20	8.65	11.64	24.51
C15H28F206	342.38	342.1854	52.62	8.24	11.10	28.04
C15H28F207	358.38	358.1803	50.27	7.88	10.60	31.25
C15H28F208	374.38	374.1752	48.12	7.54	10.15	34.19
C15H28F209	390.38	390.1701	46.15	7.23	9.73	36.89
C15H29F101	244.40	244.2202	73.72	11.96	7.77	6.55

C15H29F102	260.40	260.2151	69.19	11.23	7.30	12.29
C15H29F103	276.39	276.2101	65.18	10.58	6.87	17.37
C15H29F104	292.39	292.2050	61.62	10.00	6.50	21.89
C15H29F105	308.39	308.1999	58.42	9.48	6.16	25.94
C15H29F106	324.39	324.1948	55.54	9.01	5.86	29.59
C15H29F107	340.39	340.1897	52.93	8.59	5.58	32.90
C15H29F108	356.39	356.1846	50.55	8.20	5.33	35.91
C15H29F109	372.39	372.1795	48.38	7.85	5.10	38.67
C15H30F201	264.40	264.2265	68.14	11.44	14.37	6.05
C15H30F202	280.40	280.2214	64.25	10.78	13.55	11.41
C15H30F203	296.40	296.2163	60.78	10.20	12.82	16.19
C15H30F204	312.40	312.2112	57.67	9.68	12.16	20.49
C15H30F205	328.40	328.2061	54.86	9.21	11.57	24.36
C15H30F206	344.40	344.2010	52.31	8.78	11.03	27.87
C15H30F207	360.40	360.1959	49.99	8.39	10.54	31.08
C15H30F208	376.40	376.1909	47.87	8.03	10.09	34.01
C15H30F209	392.40	392.1858	45.91	7.71	9.68	36.70
C15H31F101	246.41	246.2359	73.12	12.68	7.71	6.49
C15H31F102	262.41	262.2308	68.66	11.91	7.24	12.19
C15H31F103	278.41	278.2257	64.71	11.22	6.82	17.24
C15H31F104	294.41	294.2206	61.20	10.61	6.45	21.74
C15H31F105	310.41	310.2155	58.04	10.07	6.12	25.77
C15H31F106	326.41	326.2104	55.20	9.57	5.82	29.41
C15H31F107	342.41	342.2054	52.62	9.13	5.55	32.71
C15H31F108	358.41	358.2003	50.27	8.72	5.30	35.71
C15H31F109	374.41	374.1952	48.12	8.35	5.07	38.46
C16H 8F201	254.24	254.0543	75.59	3.17	14.95	6.29
C16H 8F202	270.24	270.0492	71.11	2.98	14.06	11.84
C16H 8F203	286.24	286.0441	67.14	2.82	13.27	16.77
C16H 8F204	302.24	302.0391	63.59	2.67	12.57	21.17
C16H 8F205	318.24	318.0340	60.39	2.53	11.94	25.14
C16H 8F206	334.24	334.0289	57.50	2.41	11.37	28.72
C16H 8F207	350.23	350.0238	54.87	2.30	10.85	31.98
C16H 8F208	366.23	366.0187	52.47	2.20	10.38	34.95
C16H 8F209	382.23	382.0136	50.28	2.11	9.94	37.67
C16H 9F101	236.25	236.0637	81.35	3.84	8.04	6.77
C16H 9F102	252.25	252.0587	76.19	3.60	7.53	12.69
C16H 9F103	268.25	268.0536	71.64	3.38	7.08	17.89
C16H 9F104	284.25	284.0485	67.61	3.19	6.68	22.51
C16H 9F105	300.25	300.0434	64.01	3.02	6.33	26.64
C16H 9F106	316.24	316.0383	60.77	2.87	6.01	30.36
C16H 9F107	332.24	332.0332	57.84	2.73	5.72	33.71
C16H 9F108	348.24	348.0281	55.19	2.60	5.46	36.75
C16H 9F109	364.24	364.0231	52.76	2.49	5.22	39.53
C16H10F201	256.25	256.0700	75.00	3.93	14.83	6.24
C16H10F202	272.25	272.0649	70.59	3.70	13.96	11.75
C16H10F203	288.25	288.0598	66.67	3.50	13.18	16.65
C16H10F204	304.25	304.0547	63.16	3.31	12.49	21.03
C16H10F205	320.25	320.0496	60.01	3.15	11.86	24.98
C16H10F206	336.25	336.0445	57.15	3.00	11.30	28.55
C16H10F207	352.25	352.0394	54.56	2.86	10.79	31.79
C16H10F208	368.25	368.0344	52.19	2.74	10.32	34.76
C16H10F209	384.25	384.0293	50.01	2.62	9.89	37.47
C16H11F101	238.26	238.0794	80.66	4.65	7.97	6.71
C16H11F102	254.26	254.0743	75.58	4.36	7.47	12.58
C16H11F103	270.26	270.0692	71.11	4.10	7.03	17.76
C16H11F104	286.26	286.0641	67.13	3.87	6.64	22.36
C16H11F105	302.26	302.0590	63.58	3.67	6.29	26.47
C16H11F106	318.26	318.0540	60.38	3.48	5.97	30.16
C16H11F107	334.26	334.0489	57.49	3.32	5.68	33.51
C16H11F108	350.26	350.0438	54.87	3.17	5.42	36.54
C16H11F109	366.26	366.0387	52.47	3.03	5.19	39.31
C16H12F201	258.27	258.0856	74.41	4.68	14.71	6.19
C16H12F202	274.27	274.0805	70.07	4.41	13.85	11.67
C16H12F203	290.27	290.0754	66.21	4.17	13.09	16.54
C16H12F204	306.27	306.0704	62.75	3.95	12.41	20.90
C16H12F205	322.27	322.0653	59.63	3.75	11.79	24.82
C16H12F206	338.27	338.0602	56.81	3.58	11.23	28.38
C16H12F207	354.27	354.0551	54.25	3.41	10.73	31.61
C16H12F208	370.27	370.0500	51.90	3.27	10.26	34.57
C16H12F209	386.27	386.0449	49.75	3.13	9.84	37.28
C16H13F101	240.28	240.0950	79.98	5.45	7.91	6.66
C16H13F102	256.28	256.0900	74.99	5.11	7.41	12.49
C16H13F103	272.28	272.0849	70.58	4.81	6.98	17.63
C16H13F104	288.28	288.0798	66.66	4.55	6.59	22.20
C16H13F105	304.28	304.0747	63.16	4.31	6.24	26.29
C16H13F106	320.28	320.0696	60.00	4.09	5.93	29.97
C16H13F107	336.28	336.0645	57.15	3.90	5.65	33.30
C16H13F108	352.28	352.0594	54.55	3.72	5.39	36.33
C16H13F109	368.28	368.0543	52.18	3.56	5.16	39.10
C16H14F201	260.29	260.1013	73.83	5.42	14.60	6.15
C16H14F202	276.29	276.0962	69.56	5.11	13.75	11.58
C16H14F203	292.28	292.0911	65.75	4.83	13.00	16.42
C16H14F204	308.28	308.0860	62.34	4.58	12.33	20.76
C16H14F205	324.28	324.0809	59.26	4.35	11.72	24.67
C16H14F206	340.28	340.0758	56.48	4.15	11.17	28.21
C16H14F207	356.28	356.0707	53.94	3.96	10.66	31.43
C16H14F208	372.28	372.0657	51.62	3.79	10.21	34.38
C16H14F209	388.28	388.0606	49.49	3.63	9.79	37.09
C16H15F101	242.30	242.1107	79.32	6.24	7.84	6.60
C16H15F102	258.30	258.1056	74.40	5.85	7.36	12.39
C16H15F103	274.29	274.1005	70.06	5.51	6.93	17.50
C16H15F104	290.29	290.0954	66.20	5.21	6.54	22.05
C16H15F105	306.29	306.0903	62.74	4.94	6.20	26.12
C16H15F106	322.29	322.0853	59.63	4.69	5.89	29.79
C16H15F107	338.29	338.0802	56.81	4.47	5.62	33.11
C16H15F108	354.29	354.0751	54.24	4.27	5.36	36.13
C16H15F109	370.29	370.0700	51.90	4.08	5.13	38.89
C16H16F201	262.30	262.1169	73.27	6.15	14.49	6.10
C16H16F202	278.30	278.1118	69.05	5.79	13.65	11.50
C16H16F203	294.30	294.1067	65.30	5.48	12.91	16.31
C16H16F204	310.30	310.1017	61.93	5.20	12.25	20.62
C16H16F205	326.30	326.0966	58.90	4.94	11.64	24.52
C16H16F206	342.30	342.0915	56.14	4.71	11.10	28.04
C16H16F207	358.30	358.0864	53.64	4.50	10.60	31.26
C16H16F208	374.30	374.0813	51.34	4.31	10.15	34.20
C16H16F209	390.30	390.0762	49.24	4.13	9.74	36.89
C16H17F101	244.31	244.1263	78.66	7.01	7.78	6.55
C16H17F102	260.31	260.1212	73.83	6.58	7.30	12.29
C16H17F103	276.31	276.1162	69.55	6.20	6.88	17.37
C16H17F104	292.31	292.1111	65.74	5.86	6.50	21.89
C16H17F105	308.31	308.1060	62.33	5.56	6.16	25.95
C16H17F106	324.31	324.1009	59.26	5.28	5.86	29.60
C16H17F107	340.31	340.0958	56.47	5.04	5.58	32.91
C16H17F108	356.31	356.0907	53.94	4.81	5.33	35.92
C16H17F109	372.31	372.0856	51.62	4.60	5.10	38.68
C16H18F201	264.32	264.1326	72.71	6.86	14.38	6.05
C16H18F202	280.32	280.1275	68.56	6.47	13.55	11.42
C16H18F203	296.32	296.1224	64.86	6.12	12.82	16.20
C16H18F204	312.32	312.1173	61.53	5.81	12.17	20.49

C16H18F205	328.32	328.1122	58.53	5.53	11.57	24.37
C16H18F206	344.32	344.1071	55.81	5.27	11.04	27.88
C16H18F207	360.31	360.1020	53.34	5.04	10.55	31.08
C16H18F208	376.31	376.0970	51.07	4.82	10.10	34.01
C16H18F209	392.31	392.0919	48.99	4.62	9.69	36.70
C16H19F101	246.33	246.1420	78.02	7.77	7.71	6.50
C16H19F102	262.33	262.1369	73.26	7.30	7.24	12.20
C16H19F103	278.33	278.1318	69.05	6.88	6.83	17.25
C16H19F104	294.33	294.1267	65.29	6.51	6.45	21.74
C16H19F105	310.33	310.1216	61.93	6.17	6.12	25.78
C16H19F106	326.32	326.1166	58.89	5.87	5.82	29.42
C16H19F107	342.32	342.1115	56.14	5.59	5.55	32.72
C16H19F108	358.32	358.1064	53.63	5.34	5.30	35.72
C16H19F109	374.32	374.1013	51.34	5.12	5.08	38.47
C16H20F201	266.33	266.1482	72.16	7.57	14.27	6.01
C16H20F202	282.33	282.1431	68.07	7.14	13.46	11.33
C16H20F203	298.33	298.1380	64.42	6.76	12.74	16.09
C16H20F204	314.33	314.1330	61.14	6.41	12.09	20.36
C16H20F205	330.33	330.1279	58.18	6.10	11.50	24.22
C16H20F206	346.33	346.1228	55.49	5.82	10.97	27.72
C16H20F207	362.33	362.1177	53.04	5.56	10.49	30.91
C16H20F208	378.33	378.1126	50.80	5.33	10.04	33.83
C16H20F209	394.33	394.1075	48.74	5.11	9.64	36.52
C16H21F101	248.34	248.1576	77.38	8.52	7.65	6.44
C16H21F102	264.34	264.1525	72.70	8.01	7.19	12.11
C16H21F103	280.34	280.1475	68.55	7.55	6.78	17.12
C16H21F104	296.34	296.1424	64.85	7.14	6.41	21.60
C16H21F105	312.34	312.1373	61.53	6.78	6.08	25.61
C16H21F106	328.34	328.1322	58.53	6.45	5.79	29.24
C16H21F107	344.34	344.1271	55.81	6.15	5.52	32.52
C16H21F108	360.34	360.1220	53.33	5.87	5.27	35.52
C16H21F109	376.34	376.1169	51.07	5.62	5.05	38.26
C16H22F201	268.35	268.1639	71.61	8.26	14.16	5.96
C16H22F202	284.35	284.1588	67.59	7.80	13.36	11.25
C16H22F203	300.35	300.1537	63.99	7.38	12.65	15.98
C16H22F204	316.35	316.1486	60.75	7.01	12.01	20.23
C16H22F205	332.35	332.1435	57.82	6.67	11.43	24.07
C16H22F206	348.35	348.1384	55.17	6.37	10.91	27.56
C16H22F207	364.35	364.1333	52.75	6.09	10.43	30.74
C16H22F208	380.35	380.1283	50.53	5.83	9.99	33.65
C16H22F209	396.35	396.1232	48.49	5.59	9.59	36.33
C16H23F101	250.36	250.1733	76.76	9.26	7.59	6.39
C16H23F102	266.36	266.1682	72.15	8.70	7.13	12.01
C16H23F103	282.36	282.1631	68.06	8.21	6.73	17.00
C16H23F104	298.36	298.1580	64.41	7.77	6.37	21.45
C16H23F105	314.36	314.1529	61.13	7.37	6.04	25.45
C16H23F106	330.36	330.1479	58.17	7.02	5.75	29.06
C16H23F107	346.36	346.1428	55.49	6.69	5.49	32.34
C16H23F108	362.36	362.1377	53.04	6.40	5.24	35.32
C16H23F109	378.35	378.1326	50.79	6.13	5.02	38.06
C16H24F201	270.37	270.1795	71.08	8.95	14.05	5.92
C16H24F202	286.37	286.1744	67.11	8.45	13.27	11.17
C16H24F203	302.36	302.1693	63.56	8.00	12.57	15.87
C16H24F204	318.36	318.1643	60.36	7.60	11.94	20.10
C16H24F205	334.36	334.1592	57.48	7.24	11.36	23.93
C16H24F206	350.36	350.1541	54.85	6.90	10.84	27.40
C16H24F207	366.36	366.1490	52.46	6.60	10.37	30.57
C16H24F208	382.36	382.1439	50.26	6.33	9.94	33.47
C16H24F209	398.36	398.1388	48.24	6.07	9.54	36.15
C16H25F101	252.38	252.1889	76.15	9.98	7.53	6.34
C16H25F102	268.37	268.1838	71.61	9.39	7.08	11.92
C16H25F103	284.37	284.1788	67.58	8.86	6.68	16.88
C16H25F104	300.37	300.1737	63.98	8.39	6.32	21.31
C16H25F105	316.37	316.1686	60.74	7.97	6.01	25.29
C16H25F106	332.37	332.1635	57.82	7.58	5.72	28.88
C16H25F107	348.37	348.1584	55.16	7.23	5.45	32.15
C16H25F108	364.37	364.1533	52.74	6.92	5.21	35.13
C16H25F109	380.37	380.1482	50.52	6.62	4.99	37.86
C16H26F201	272.38	272.1952	70.55	9.62	13.95	5.87
C16H26F202	288.38	288.1901	66.64	9.09	13.18	11.10
C16H26F203	304.38	304.1850	63.14	8.61	12.48	15.77
C16H26F204	320.38	320.1799	59.98	8.18	11.86	19.98
C16H26F205	336.38	336.1748	57.13	7.79	11.30	23.78
C16H26F206	352.38	352.1697	54.54	7.44	10.78	27.24
C16H26F207	368.38	368.1646	52.17	7.11	10.31	30.40
C16H26F208	384.38	384.1596	50.00	6.82	9.89	33.30
C16H26F209	400.38	400.1545	48.00	6.55	9.49	35.96
C16H27F101	254.39	254.2046	75.54	10.70	7.47	6.29
C16H27F102	270.39	270.1995	71.07	10.07	7.03	11.83
C16H27F103	286.39	286.1944	67.10	9.50	6.63	16.76
C16H27F104	302.39	302.1893	63.55	9.00	6.28	21.16
C16H27F105	318.39	318.1842	60.36	8.55	5.97	25.13
C16H27F106	334.39	334.1792	57.47	8.14	5.68	28.71
C16H27F107	350.39	350.1741	54.85	7.77	5.42	31.96
C16H27F108	366.39	366.1690	52.45	7.43	5.19	34.93
C16H27F109	382.39	382.1639	50.26	7.12	4.97	37.66
C16H28F201	274.40	274.2108	70.04	10.29	13.85	5.83
C16H28F202	290.40	290.2057	66.18	9.72	13.08	11.02
C16H28F203	306.40	306.2006	62.72	9.21	12.40	15.67
C16H28F204	322.40	322.1955	59.61	8.75	11.79	19.85
C16H28F205	338.40	338.1905	56.79	8.34	11.23	23.64
C16H28F206	354.39	354.1854	54.23	7.96	10.72	27.09
C16H28F207	370.39	370.1803	51.88	7.62	10.26	30.24
C16H28F208	386.39	386.1752	49.74	7.30	9.83	33.13
C16H28F209	402.39	402.1701	47.76	7.01	9.44	35.78
C16H29F101	256.41	256.2202	74.95	11.40	7.41	6.24
C16H29F102	272.41	272.2151	70.55	10.73	6.97	11.75
C16H29F103	288.41	288.2101	66.63	10.14	6.59	16.64
C16H29F104	304.41	304.2050	63.13	9.60	6.24	21.02
C16H29F105	320.40	320.1999	59.98	9.12	5.93	24.97
C16H29F106	336.40	336.1948	57.13	8.69	5.65	28.54
C16H29F107	352.40	352.1897	54.53	8.29	5.39	31.78
C16H29F108	368.40	368.1846	52.17	7.93	5.16	34.74
C16H29F109	384.40	384.1795	49.99	7.60	4.94	37.46
C16H30F201	276.41	276.2265	69.53	10.94	13.75	5.79
C16H30F202	292.41	292.2214	65.72	10.34	12.99	10.94
C16H30F203	308.41	308.2163	62.31	9.80	12.32	15.56
C16H30F204	324.41	324.2112	59.24	9.32	11.71	19.73
C16H30F205	340.41	340.2061	56.45	8.88	11.16	23.50
C16H30F206	356.41	356.2010	53.92	8.48	10.66	26.93
C16H30F207	372.41	372.1959	51.60	8.12	10.20	30.07
C16H30F208	388.41	388.1909	49.48	7.79	9.78	32.95
C16H30F209	404.41	404.1858	47.52	7.48	9.40	35.61
C16H31F101	258.42	258.2359	74.37	12.09	7.35	6.19
C16H31F102	274.42	274.2308	70.03	11.39	6.92	11.66
C16H31F103	290.42	290.2257	66.17	10.76	6.54	16.53
C16H31F104	306.42	306.2206	62.72	10.20	6.20	20.89
C16H31F105	322.42	322.2155	59.60	9.69	5.89	24.81
C16H31F106	338.42	338.2104	56.79	9.23	5.61	28.37
C16H31F107	354.42	354.2054	54.22	8.82	5.36	31.60

C16H31F108	370.42	370.2003	51.88	8.44	5.13	34.55
C16H31F109	386.42	386.1952	49.73	8.09	4.92	37.26
C16H32F201	278.43	278.2421	69.02	11.58	13.65	5.75
C16H32F202	294.43	294.2370	65.27	10.96	12.91	10.87
C16H32F203	310.43	310.2319	61.91	10.39	12.24	15.46
C16H32F204	326.43	326.2268	58.87	9.88	11.64	19.61
C16H32F205	342.43	342.2218	56.12	9.42	11.10	23.36
C16H32F206	358.43	358.2167	53.62	9.00	10.60	26.78
C16H32F207	374.43	374.2116	51.33	8.61	10.15	29.91
C16H32F208	390.43	390.2065	49.22	8.26	9.73	32.78
C16H32F209	406.42	406.2014	47.29	7.94	9.35	35.43
C16H33F101	260.44	260.2515	73.79	12.77	7.29	6.14
C16H33F102	276.44	276.2464	69.52	12.03	6.87	11.58
C16H33F103	292.44	292.2414	65.72	11.37	6.50	16.41
C16H33F104	308.44	308.2363	62.31	10.78	6.16	20.75
C16H33F105	324.44	324.2312	59.23	10.25	5.86	24.66
C16H33F106	340.44	340.2261	56.45	9.77	5.58	28.20
C16H33F107	356.44	356.2210	53.92	9.33	5.33	31.42
C16H33F108	372.44	372.2159	51.60	8.93	5.10	34.37
C16H33F109	388.43	388.2108	49.48	8.56	4.89	37.07
C17H 8F201	266.25	266.0543	76.69	3.03	14.27	6.01
C17H 8F202	282.25	282.0492	72.34	2.86	13.46	11.34
C17H 8F203	298.25	298.0441	68.46	2.70	12.74	16.09
C17H 8F204	314.25	314.0391	64.98	2.57	12.09	20.37
C17H 8F205	330.25	330.0340	61.83	2.44	11.51	24.22
C17H 8F206	346.25	346.0289	58.97	2.33	10.97	27.72
C17H 8F207	362.25	362.0238	56.37	2.23	10.49	30.92
C17H 8F208	378.25	378.0187	53.98	2.13	10.05	33.84
C17H 8F209	394.24	394.0136	51.79	2.05	9.64	36.52
C17H 9F101	248.26	248.0637	82.25	3.65	7.65	6.44
C17H 9F102	264.26	264.0587	77.27	3.43	7.19	12.11
C17H 9F103	280.26	280.0536	72.86	3.24	6.78	17.13
C17H 9F104	296.26	296.0485	68.92	3.06	6.41	21.60
C17H 9F105	312.26	312.0434	65.39	2.91	6.08	25.62
C17H 9F106	328.26	328.0383	62.20	2.76	5.79	29.24
C17H 9F107	344.26	344.0332	59.31	2.64	5.52	32.53
C17H 9F108	360.25	360.0281	56.68	2.52	5.27	35.53
C17H 9F109	376.25	376.0231	54.27	2.41	5.05	38.27
C17H10F201	268.27	268.0700	76.11	3.76	14.16	5.96
C17H10F202	284.26	284.0649	71.83	3.55	13.37	11.26
C17H10F203	300.26	300.0598	68.00	3.36	12.65	15.99
C17H10F204	316.26	316.0547	64.56	3.19	12.01	20.24
C17H10F205	332.26	332.0496	61.45	3.03	11.44	24.08
C17H10F206	348.26	348.0445	58.63	2.89	10.91	27.56
C17H10F207	364.26	364.0394	56.06	2.77	10.43	30.75
C17H10F208	380.26	380.0344	53.70	2.65	9.99	33.66
C17H10F209	396.26	396.0293	51.53	2.54	9.59	36.34
C17H11F101	250.28	250.0794	81.59	4.43	7.59	6.39
C17H11F102	266.27	266.0743	76.68	4.16	7.13	12.02
C17H11F103	282.27	282.0692	72.34	3.93	6.73	17.00
C17H11F104	298.27	298.0641	68.46	3.72	6.37	21.46
C17H11F105	314.27	314.0590	64.97	3.53	6.05	25.45
C17H11F106	330.27	330.0540	61.82	3.36	5.75	29.07
C17H11F107	346.27	346.0489	58.97	3.20	5.49	32.34
C17H11F108	362.27	362.0438	56.36	3.06	5.24	35.33
C17H11F109	378.27	378.0387	53.98	2.93	5.02	38.07
C17H12F201	270.28	270.0856	75.55	4.48	14.06	5.92
C17H12F202	286.28	286.0805	71.32	4.23	13.27	11.18
C17H12F203	302.28	302.0754	67.55	4.00	12.57	15.88
C17H12F204	318.28	318.0704	64.15	3.80	11.94	20.11

C17H12F205	334.28	334.0653	61.08	3.62	11.37	23.93
C17H12F206	350.28	350.0602	58.29	3.45	10.85	27.41
C17H12F207	366.28	366.0551	55.75	3.30	10.37	30.58
C17H12F208	382.28	382.0500	53.41	3.16	9.94	33.48
C17H12F209	398.28	398.0449	51.27	3.04	9.54	36.15
C17H13F101	252.29	252.0950	80.93	5.19	7.53	6.34
C17H13F102	268.29	268.0900	76.11	4.88	7.08	11.93
C17H13F103	284.29	284.0849	71.82	4.61	6.68	16.88
C17H13F104	300.29	300.0798	68.00	4.36	6.33	21.31
C17H13F105	316.29	316.0747	64.56	4.14	6.01	25.29
C17H13F106	332.29	332.0696	61.45	3.94	5.72	28.89
C17H13F107	348.29	348.0645	58.63	3.76	5.45	32.16
C17H13F108	364.29	364.0594	56.05	3.60	5.22	35.14
C17H13F109	380.29	380.0543	53.69	3.45	5.00	37.86
C17H14F201	272.30	272.1013	74.99	5.18	13.95	5.88
C17H14F202	288.30	288.0962	70.83	4.89	13.18	11.10
C17H14F203	304.30	304.0911	67.10	4.64	12.49	15.77
C17H14F204	320.30	320.0860	63.75	4.41	11.86	19.98
C17H14F205	336.29	336.0809	60.72	4.20	11.30	23.79
C17H14F206	352.29	352.0758	57.96	4.01	10.79	27.25
C17H14F207	368.29	368.0707	55.44	3.83	10.32	30.41
C17H14F208	384.29	384.0657	53.13	3.67	9.89	33.31
C17H14F209	400.29	400.0606	51.01	3.53	9.49	35.97
C17H15F101	254.31	254.1107	80.29	5.95	7.47	6.29
C17H15F102	270.31	270.1056	75.54	5.59	7.03	11.84
C17H15F103	286.31	286.1005	71.32	5.28	6.64	16.76
C17H15F104	302.31	302.0954	67.54	5.00	6.28	21.17
C17H15F105	318.30	318.0903	64.15	4.75	5.97	25.13
C17H15F106	334.30	334.0853	61.08	4.52	5.68	28.72
C17H15F107	350.30	350.0802	58.29	4.32	5.42	31.97
C17H15F108	366.30	366.0751	55.74	4.13	5.19	34.94
C17H15F109	382.30	382.0700	53.41	3.95	4.97	37.67
C17H16F201	274.31	274.1169	74.44	5.88	13.85	5.83
C17H16F202	290.31	290.1118	70.33	5.56	13.09	11.02
C17H16F203	306.31	306.1067	66.66	5.27	12.40	15.67
C17H16F204	322.31	322.1017	63.35	5.00	11.79	19.86
C17H16F205	338.31	338.0966	60.36	4.77	11.23	23.65
C17H16F206	354.31	354.0915	57.63	4.55	10.72	27.09
C17H16F207	370.31	370.0864	55.14	4.36	10.26	30.24
C17H16F208	386.31	386.0813	52.86	4.17	9.84	33.13
C17H16F209	402.31	402.0762	50.75	4.01	9.44	35.79
C17H17F101	256.32	256.1263	79.66	6.69	7.41	6.24
C17H17F102	272.32	272.1212	74.98	6.29	6.98	11.75
C17H17F103	288.32	288.1162	70.82	5.94	6.59	16.65
C17H17F104	304.32	304.1111	67.10	5.63	6.24	21.03
C17H17F105	320.32	320.1060	63.75	5.35	5.93	24.97
C17H17F106	336.32	336.1009	60.71	5.09	5.65	28.54
C17H17F107	352.32	352.0958	57.96	4.86	5.39	31.79
C17H17F108	368.32	368.0907	55.44	4.65	5.16	34.75
C17H17F109	384.32	384.0856	53.13	4.46	4.94	37.47
C17H18F201	276.33	276.1326	73.89	6.57	13.75	5.79
C17H18F202	292.33	292.1275	69.85	6.21	13.00	10.95
C17H18F203	308.33	308.1224	66.22	5.88	12.32	15.57
C17H18F204	324.33	324.1173	62.96	5.59	11.72	19.73
C17H18F205	340.33	340.1122	60.00	5.33	11.16	23.51
C17H18F206	356.33	356.1071	57.30	5.09	10.66	26.94
C17H18F207	372.33	372.1020	54.84	4.87	10.21	30.08
C17H18F208	388.33	388.0970	52.58	4.67	9.78	32.96
C17H18F209	404.32	404.0919	50.50	4.49	9.40	35.61
C17H19F101	258.34	258.1420	79.04	7.41	7.35	6.19

C17H19F102	274.34	274.1369	74.43	6.98	6.93	11.66
C17H19F103	290.34	290.1318	70.33	6.60	6.54	16.53
C17H19F104	306.34	306.1267	66.66	6.25	6.20	20.89
C17H19F105	322.34	322.1216	63.35	5.94	5.89	24.82
C17H19F106	338.34	338.1166	60.35	5.66	5.62	28.37
C17H19F107	354.34	354.1115	57.63	5.40	5.36	31.61
C17H19F108	370.33	370.1064	55.14	5.17	5.13	34.56
C17H19F109	386.33	386.1013	52.85	4.96	4.92	37.27
C17H20F201	278.35	278.1482	73.36	7.24	13.65	5.75
C17H20F202	294.34	294.1431	69.37	6.85	12.91	10.87
C17H20F203	310.34	310.1380	65.79	6.50	12.24	15.47
C17H20F204	326.34	326.1330	62.57	6.18	11.64	19.61
C17H20F205	342.34	342.1279	59.64	5.89	11.10	23.37
C17H20F206	358.34	358.1228	56.98	5.63	10.60	26.79
C17H20F207	374.34	374.1177	54.55	5.39	10.15	29.92
C17H20F208	390.34	390.1126	52.31	5.16	9.73	32.79
C17H20F209	406.34	406.1075	50.25	4.96	9.35	35.44
C17H21F101	260.35	260.1576	78.43	8.13	7.30	6.15
C17H21F102	276.35	276.1525	73.89	7.66	6.87	11.58
C17H21F103	292.35	292.1475	69.84	7.24	6.50	16.42
C17H21F104	308.35	308.1424	66.22	6.86	6.16	20.75
C17H21F105	324.35	324.1373	62.95	6.53	5.86	24.66
C17H21F106	340.35	340.1322	59.99	6.22	5.58	28.21
C17H21F107	356.35	356.1271	57.30	5.94	5.33	31.43
C17H21F108	372.35	372.1220	54.84	5.68	5.10	34.37
C17H21F109	388.35	388.1169	52.58	5.45	4.89	37.08
C17H22F201	280.36	280.1639	72.83	7.91	13.55	5.71
C17H22F202	296.36	296.1588	68.90	7.48	12.82	10.80
C17H22F203	312.36	312.1537	65.37	7.10	12.16	15.37
C17H22F204	328.36	328.1486	62.18	6.75	11.57	19.49
C17H22F205	344.36	344.1435	59.30	6.44	11.03	23.23
C17H22F206	360.36	360.1384	56.66	6.15	10.54	26.64
C17H22F207	376.36	376.1333	54.25	5.89	10.10	29.76
C17H22F208	392.36	392.1283	52.04	5.65	9.68	32.62
C17H22F209	408.36	408.1232	50.00	5.43	9.30	35.26
C17H23F101	262.37	262.1733	77.82	8.84	7.24	6.10
C17H23F102	278.37	278.1682	73.35	8.33	6.82	11.50
C17H23F103	294.37	294.1631	69.37	7.88	6.45	16.31
C17H23F104	310.37	310.1580	65.79	7.47	6.12	20.62
C17H23F105	326.37	326.1529	62.56	7.10	5.82	24.51
C17H23F106	342.37	342.1479	59.64	6.77	5.55	28.04
C17H23F107	358.37	358.1428	56.98	6.47	5.30	31.25
C17H23F108	374.37	374.1377	54.54	6.19	5.07	34.19
C17H23F109	390.37	390.1326	52.31	5.94	4.87	36.89
C17H24F201	282.38	282.1795	72.31	8.57	13.46	5.67
C17H24F202	298.38	298.1744	68.43	8.11	12.73	10.72
C17H24F203	314.38	314.1693	64.95	7.70	12.09	15.27
C17H24F204	330.38	330.1643	61.81	7.32	11.50	19.37
C17H24F205	346.37	346.1592	58.95	6.98	10.97	23.10
C17H24F206	362.37	362.1541	56.35	6.68	10.49	26.49
C17H24F207	378.37	378.1490	53.97	6.39	10.04	29.60
C17H24F208	394.37	394.1439	51.78	6.13	9.63	32.46
C17H24F209	410.37	410.1388	49.76	5.89	9.26	35.09
C17H25F101	264.39	264.1889	77.23	9.53	7.19	6.05
C17H25F102	280.39	280.1838	72.82	8.99	6.78	11.41
C17H25F103	296.39	296.1788	68.89	8.50	6.41	16.19
C17H25F104	312.38	312.1737	65.36	8.07	6.08	20.49
C17H25F105	328.38	328.1686	62.18	7.67	5.79	24.36
C17H25F106	344.38	344.1635	59.29	7.32	5.52	27.87
C17H25F107	360.38	360.1584	56.66	6.99	5.27	31.08

C17H25F108	376.38	376.1533	54.25	6.70	5.05	34.01
C17H25F109	392.38	392.1482	52.04	6.42	4.84	36.70
C17H26F201	284.39	284.1952	71.80	9.22	13.36	5.63
C17H26F202	300.39	300.1901	67.97	8.72	12.65	10.65
C17H26F203	316.39	316.1850	64.54	8.28	12.01	15.17
C17H26F204	332.39	332.1799	61.43	7.88	11.43	19.25
C17H26F205	348.39	348.1748	58.61	7.52	10.91	22.96
C17H26F206	364.39	364.1697	56.04	7.19	10.43	26.34
C17H26F207	380.39	380.1646	53.68	6.89	9.99	29.44
C17H26F208	396.39	396.1596	51.51	6.61	9.59	32.29
C17H26F209	412.39	412.1545	49.51	6.35	9.21	34.92
C17H27F101	266.40	266.2046	76.65	10.22	7.13	6.01
C17H27F102	282.40	282.1995	72.30	9.64	6.73	11.33
C17H27F103	298.40	298.1944	68.43	9.12	6.37	16.09
C17H27F104	314.40	314.1893	64.95	8.66	6.04	20.36
C17H27F105	330.40	330.1842	61.80	8.24	5.75	24.21
C17H27F106	346.40	346.1792	58.95	7.86	5.48	27.71
C17H27F107	362.40	362.1741	56.34	7.51	5.24	30.90
C17H27F108	378.40	378.1690	53.96	7.19	5.02	33.83
C17H27F109	394.40	394.1639	51.77	6.90	4.82	36.51
C17H28F201	286.41	286.2108	71.29	9.85	13.27	5.59
C17H28F202	302.41	302.2057	67.52	9.33	12.56	10.58
C17H28F203	318.41	318.2006	64.13	8.86	11.93	15.07
C17H28F204	334.41	334.1955	61.06	8.44	11.36	19.14
C17H28F205	350.41	350.1905	58.27	8.05	10.84	22.83
C17H28F206	366.41	366.1854	55.73	7.70	10.37	26.20
C17H28F207	382.41	382.1803	53.40	7.38	9.94	29.29
C17H28F208	398.40	398.1752	51.25	7.08	9.54	32.13
C17H28F209	414.40	414.1701	49.27	6.81	9.17	34.75
C17H29F101	268.42	268.2202	76.07	10.89	7.08	5.96
C17H29F102	284.42	284.2151	71.79	10.28	6.68	11.25
C17H29F103	300.42	300.2101	67.97	9.73	6.32	15.98
C17H29F104	316.42	316.2050	64.53	9.24	6.00	20.23
C17H29F105	332.42	332.1999	61.43	8.79	5.72	24.07
C17H29F106	348.42	348.1948	58.61	8.39	5.45	27.55
C17H29F107	364.41	364.1897	56.03	8.02	5.21	30.73
C17H29F108	380.41	380.1846	53.68	7.68	4.99	33.65
C17H29F109	396.41	396.1795	51.51	7.37	4.79	36.32
C17H30F201	288.42	288.2265	70.79	10.48	13.17	5.55
C17H30F202	304.42	304.2214	67.07	9.93	12.48	10.51
C17H30F203	320.42	320.2163	63.72	9.44	11.86	14.98
C17H30F204	336.42	336.2112	60.69	8.99	11.29	19.02
C17H30F205	352.42	352.2061	57.94	8.58	10.78	22.70
C17H30F206	368.42	368.2010	55.42	8.21	10.31	26.06
C17H30F207	384.42	384.1959	53.12	7.87	9.88	29.13
C17H30F208	400.42	400.1909	50.99	7.55	9.49	31.97
C17H30F209	416.42	416.1858	49.03	7.26	9.12	34.58
C17H31F101	270.43	270.2359	75.50	11.55	7.03	5.92
C17H31F102	286.43	286.2308	71.29	10.91	6.63	11.17
C17H31F103	302.43	302.2257	67.52	10.33	6.28	15.87
C17H31F104	318.43	318.2206	64.12	9.81	5.97	20.10
C17H31F105	334.43	334.2155	61.06	9.34	5.68	23.92
C17H31F106	350.43	350.2104	58.27	8.92	5.42	27.39
C17H31F107	366.43	366.2054	55.72	8.53	5.18	30.56
C17H31F108	382.43	382.2003	53.39	8.17	4.97	33.47
C17H31F109	398.43	398.1952	51.25	7.84	4.77	36.14
C17H32F201	290.44	290.2421	70.30	11.11	13.08	5.51
C17H32F202	306.44	306.2370	66.63	10.53	12.40	10.44
C17H32F203	322.44	322.2319	63.33	10.00	11.78	14.89
C17H32F204	338.44	338.2268	60.33	9.53	11.23	18.91

C17H32F205	354.44	354.2218	57.61	9.10	10.72	22.57
C17H32F206	370.44	370.2167	55.12	8.71	10.26	25.91
C17H32F207	386.44	386.2116	52.84	8.35	9.83	28.98
C17H32F208	402.44	402.2065	50.74	8.01	9.44	31.81
C17H32F209	418.44	418.2014	48.80	7.71	9.08	34.41
C17H33F101	272.45	272.2515	74.95	12.21	6.97	5.87
C17H33F102	288.45	288.2464	70.75	11.53	6.59	11.09
C17H33F103	304.45	304.2414	67.07	10.93	6.24	15.77
C17H33F104	320.45	320.2363	63.72	10.38	5.93	19.97
C17H33F105	336.45	336.2312	60.69	9.89	5.65	23.78
C17H33F106	352.45	352.2261	57.93	9.44	5.39	27.24
C17H33F107	368.45	368.2210	55.42	9.03	5.16	30.40
C17H33F108	384.45	384.2159	53.11	8.65	4.94	33.29
C17H33F109	400.45	400.2108	50.99	8.31	4.74	35.96
C17H34F201	292.46	292.2578	69.82	11.72	12.99	5.47
C17H34F202	308.46	308.2527	66.20	11.11	12.32	10.37
C17H34F203	324.46	324.2476	62.93	10.56	11.71	14.79
C17H34F204	340.45	340.2425	59.98	10.07	11.16	18.80
C17H34F205	356.45	356.2374	57.28	9.61	10.66	22.44
C17H34F206	372.45	372.2323	54.82	9.20	10.20	25.77
C17H34F207	388.45	388.2272	52.56	8.82	9.78	28.83
C17H34F208	404.45	404.2222	50.49	8.47	9.39	31.65
C17H34F209	420.45	420.2171	48.56	8.15	9.04	34.25
C17H35F101	274.47	274.2672	74.40	12.85	6.92	5.83
C17H35F102	290.47	290.2621	70.30	12.15	6.54	11.02
C17H35F103	306.47	306.2570	66.63	11.51	6.20	15.66
C17H35F104	322.46	322.2519	63.32	10.94	5.89	19.85
C17H35F105	338.46	338.2468	60.33	10.42	5.61	23.64
C17H35F106	354.46	354.2417	57.61	9.95	5.36	27.08
C17H35F107	370.46	370.2367	55.12	9.52	5.13	30.23
C17H35F108	386.46	386.2316	52.84	9.13	4.92	33.12
C17H35F109	402.46	402.2265	50.74	8.77	4.72	35.78
C18H 9F101	260.27	260.0637	83.07	3.49	7.30	6.15
C18H 9F102	276.27	276.0587	78.26	3.28	6.88	11.58
C18H 9F103	292.27	292.0536	73.97	3.10	6.50	16.42
C18H 9F104	308.27	308.0485	70.13	2.94	6.16	20.76
C18H 9F105	324.27	324.0434	66.67	2.80	5.86	24.67
C18H 9F106	340.27	340.0383	63.54	2.67	5.58	28.21
C18H 9F107	356.27	356.0332	60.69	2.55	5.33	31.44
C18H 9F108	372.27	372.0281	58.08	2.44	5.10	34.38
C18H 9F109	388.27	388.0231	55.68	2.34	4.89	37.09
C18H10F201	280.28	280.0700	77.14	3.60	13.56	5.71
C18H10F202	296.28	296.0649	72.97	3.40	12.82	10.80
C18H10F203	312.28	312.0598	69.23	3.23	12.17	15.37
C18H10F204	328.27	328.0547	65.86	3.07	11.57	19.50
C18H10F205	344.27	344.0496	62.80	2.93	11.04	23.24
C18H10F206	360.27	360.0445	60.01	2.80	10.55	26.65
C18H10F207	376.27	376.0394	57.46	2.68	10.10	29.76
C18H10F208	392.27	392.0344	55.11	2.57	9.69	32.63
C18H10F209	408.27	408.0293	52.96	2.47	9.31	35.27
C18H11F101	262.29	262.0794	82.43	4.23	7.24	6.10
C18H11F102	278.29	278.0743	77.69	3.98	6.83	11.50
C18H11F103	294.28	294.0692	73.47	3.77	6.46	16.31
C18H11F104	310.28	310.0641	69.68	3.57	6.12	20.63
C18H11F105	326.28	326.0590	66.26	3.40	5.82	24.52
C18H11F106	342.28	342.0540	63.16	3.24	5.55	28.05
C18H11F107	358.28	358.0489	60.34	3.09	5.30	31.26
C18H11F108	374.28	374.0438	57.76	2.96	5.08	34.20
C18H11F109	390.28	390.0387	55.40	2.84	4.87	36.90
C18H12F201	282.29	282.0856	76.59	4.28	13.46	5.67

C18H12F202	298.29	298.0805	72.48	4.05	12.74	10.73
C18H12F203	314.29	314.0754	68.79	3.85	12.09	15.27
C18H12F204	330.29	330.0704	65.46	3.66	11.50	19.38
C18H12F205	346.29	346.0653	62.43	3.49	10.97	23.10
C18H12F206	362.29	362.0602	59.68	3.34	10.49	26.50
C18H12F207	378.29	378.0551	57.15	3.20	10.04	29.61
C18H12F208	394.29	394.0500	54.83	3.07	9.64	32.46
C18H12F209	410.29	410.0449	52.69	2.95	9.26	35.10
C18H13F101	264.30	264.0950	81.80	4.96	7.19	6.05
C18H13F102	280.30	280.0900	77.13	4.67	6.78	11.42
C18H13F103	296.30	296.0849	72.97	4.42	6.41	16.20
C18H13F104	312.30	312.0798	69.23	4.20	6.08	20.49
C18H13F105	328.30	328.0747	65.85	3.99	5.79	24.37
C18H13F106	344.30	344.0696	62.79	3.81	5.52	27.88
C18H13F107	360.30	360.0645	60.01	3.64	5.27	31.08
C18H13F108	376.30	376.0594	57.45	3.48	5.05	34.01
C18H13F109	392.30	392.0543	55.11	3.34	4.84	36.71
C18H14F201	284.31	284.1013	76.04	4.96	13.36	5.63
C18H14F202	300.31	300.0962	71.99	4.70	12.65	10.66
C18H14F203	316.31	316.0911	68.35	4.46	12.01	15.17
C18H14F204	332.31	332.0860	65.06	4.25	11.43	19.26
C18H14F205	348.31	348.0809	62.07	4.05	10.91	22.97
C18H14F206	364.31	364.0758	59.35	3.87	10.43	26.35
C18H14F207	380.30	380.0707	56.85	3.71	9.99	29.45
C18H14F208	396.30	396.0657	54.55	3.56	9.59	32.30
C18H14F209	412.30	412.0606	52.44	3.42	9.22	34.92
C18H15F101	266.32	266.1107	81.18	5.68	7.13	6.01
C18H15F102	282.32	282.1056	76.58	5.36	6.73	11.33
C18H15F103	298.32	298.1005	72.47	5.07	6.37	16.09
C18H15F104	314.32	314.0954	68.78	4.81	6.04	20.36
C18H15F105	330.32	330.0903	65.45	4.58	5.75	24.22
C18H15F106	346.32	346.0853	62.43	4.37	5.49	27.72
C18H15F107	362.31	362.0802	59.67	4.17	5.24	30.91
C18H15F108	378.31	378.0751	57.15	4.00	5.02	33.83
C18H15F109	394.31	394.0700	54.83	3.83	4.82	36.52
C18H16F201	286.32	286.1169	75.51	5.63	13.27	5.59
C18H16F202	302.32	302.1118	71.51	5.33	12.57	10.58
C18H16F203	318.32	318.1067	67.92	5.07	11.94	15.08
C18H16F204	334.32	334.1017	64.67	4.82	11.37	19.14
C18H16F205	350.32	350.0966	61.71	4.60	10.85	22.84
C18H16F206	366.32	366.0915	59.02	4.40	10.37	26.21
C18H16F207	382.32	382.0864	56.55	4.22	9.94	29.29
C18H16F208	398.32	398.0813	54.28	4.05	9.54	32.13
C18H16F209	414.32	414.0762	52.18	3.89	9.17	34.75
C18H17F101	268.33	268.1263	80.57	6.39	7.08	5.96
C18H17F102	284.33	284.1212	76.04	6.03	6.68	11.25
C18H17F103	300.33	300.1162	71.99	5.71	6.33	15.98
C18H17F104	316.33	316.1111	68.35	5.42	6.01	20.23
C18H17F105	332.33	332.1060	65.06	5.16	5.72	24.07
C18H17F106	348.33	348.1009	62.07	4.92	5.45	27.56
C18H17F107	364.33	364.0958	59.34	4.70	5.21	30.74
C18H17F108	380.33	380.0907	56.85	4.51	5.00	33.65
C18H17F109	396.33	396.0856	54.55	4.32	4.79	36.33
C18H18F201	288.34	288.1326	74.98	6.29	13.18	5.55
C18H18F202	304.34	304.1275	71.04	5.96	12.48	10.51
C18H18F203	320.34	320.1224	67.49	5.66	11.86	14.98
C18H18F204	336.34	336.1173	64.28	5.39	11.30	19.03
C18H18F205	352.34	352.1122	61.36	5.15	10.78	22.70
C18H18F206	368.34	368.1071	58.70	4.93	10.32	26.06
C18H18F207	384.34	384.1020	56.25	4.72	9.89	29.14

C18H18F208	400.34	400.0970	54.00	4.53	9.49	31.97
C18H18F209	416.34	416.0919	51.93	4.36	9.13	34.59
C18H19F101	270.35	270.1420	79.97	7.08	7.03	5.92
C18H19F102	286.35	286.1369	75.50	6.69	6.63	11.17
C18H19F103	302.35	302.1318	71.51	6.33	6.28	15.88
C18H19F104	318.35	318.1267	67.91	6.02	5.97	20.10
C18H19F105	334.35	334.1216	64.66	5.73	5.68	23.93
C18H19F106	350.35	350.1166	61.71	5.47	5.42	27.40
C18H19F107	366.35	366.1115	59.02	5.23	5.19	30.57
C18H19F108	382.35	382.1064	56.55	5.01	4.97	33.48
C18H19F109	398.35	398.1013	54.27	4.81	4.77	36.15
C18H20F201	290.36	290.1482	74.46	6.94	13.09	5.51
C18H20F202	306.36	306.1431	70.57	6.58	12.40	10.44
C18H20F203	322.36	322.1380	67.07	6.25	11.79	14.89
C18H20F204	338.35	338.1330	63.90	5.96	11.23	18.91
C18H20F205	354.35	354.1279	61.01	5.69	10.72	22.58
C18H20F206	370.35	370.1228	58.38	5.44	10.26	25.92
C18H20F207	386.35	386.1177	55.96	5.22	9.83	28.99
C18H20F208	402.35	402.1126	53.73	5.01	9.44	31.81
C18H20F209	418.35	418.1075	51.68	4.82	9.08	34.42
C18H21F101	272.37	272.1576	79.38	7.77	6.98	5.87
C18H21F102	288.37	288.1525	74.97	7.34	6.59	11.10
C18H21F103	304.36	304.1475	71.03	6.95	6.24	15.77
C18H21F104	320.36	320.1424	67.49	6.61	5.93	19.98
C18H21F105	336.36	336.1373	64.28	6.29	5.65	23.78
C18H21F106	352.36	352.1322	61.36	6.01	5.39	27.24
C18H21F107	368.36	368.1271	58.69	5.75	5.16	30.40
C18H21F108	384.36	384.1220	56.25	5.51	4.94	33.30
C18H21F109	400.36	400.1169	54.00	5.29	4.75	35.97
C18H22F201	292.37	292.1639	73.95	7.58	13.00	5.47
C18H22F202	308.37	308.1588	70.11	7.19	12.32	10.38
C18H22F203	324.37	324.1537	66.65	6.84	11.71	14.80
C18H22F204	340.37	340.1486	63.52	6.52	11.16	18.80
C18H22F205	356.37	356.1435	60.67	6.22	10.66	22.45
C18H22F206	372.37	372.1384	58.06	5.96	10.20	25.78
C18H22F207	388.37	388.1333	55.67	5.71	9.78	28.84
C18H22F208	404.37	404.1283	53.47	5.48	9.40	31.65
C18H22F209	420.37	420.1232	51.43	5.28	9.04	34.25
C18H23F101	274.38	274.1733	78.80	8.45	6.92	5.83
C18H23F102	290.38	290.1682	74.45	7.98	6.54	11.02
C18H23F103	306.38	306.1631	70.57	7.57	6.20	15.67
C18H23F104	322.38	322.1580	67.06	7.19	5.89	19.85
C18H23F105	338.38	338.1529	63.89	6.85	5.61	23.64
C18H23F106	354.38	354.1479	61.01	6.54	5.36	27.09
C18H23F107	370.38	370.1428	58.37	6.26	5.13	30.24
C18H23F108	386.38	386.1377	55.96	6.00	4.92	33.13
C18H23F109	402.38	402.1326	53.73	5.76	4.72	35.79
C18H24F201	294.39	294.1795	73.44	8.22	12.91	5.43
C18H24F202	310.39	310.1744	69.66	7.79	12.24	10.31
C18H24F203	326.39	326.1693	66.24	7.41	11.64	14.71
C18H24F204	342.39	342.1643	63.15	7.07	11.10	18.69
C18H24F205	358.39	358.1592	60.33	6.75	10.60	22.32
C18H24F206	374.39	374.1541	57.75	6.46	10.15	25.64
C18H24F207	390.38	390.1490	55.38	6.20	9.73	28.69
C18H24F208	406.38	406.1439	53.20	5.95	9.35	31.50
C18H24F209	422.38	422.1388	51.19	5.73	9.00	34.09
C18H25F101	276.40	276.1889	78.22	9.12	6.87	5.79
C18H25F102	292.40	292.1838	73.94	8.62	6.50	10.94
C18H25F103	308.40	308.1788	70.10	8.17	6.16	15.56
C18H25F104	324.40	324.1737	66.65	7.77	5.86	19.73

C18H25F105	340.40	340.1686	63.51	7.40	5.58	23.50
C18H25F106	356.39	356.1635	60.66	7.07	5.33	26.94
C18H25F107	372.39	372.1584	58.06	6.77	5.10	30.07
C18H25F108	388.39	388.1533	55.67	6.49	4.89	32.96
C18H25F109	404.39	404.1482	53.46	6.23	4.70	35.61
C18H26F201	296.40	296.1952	72.94	8.84	12.82	5.40
C18H26F202	312.40	312.1901	69.21	8.39	12.16	10.24
C18H26F203	328.40	328.1850	65.83	7.98	11.57	14.62
C18H26F204	344.40	344.1799	62.78	7.61	11.03	18.58
C18H26F205	360.40	360.1748	59.99	7.27	10.54	22.20
C18H26F206	376.40	376.1697	57.44	6.96	10.09	25.50
C18H26F207	392.40	392.1646	55.10	6.68	9.68	28.54
C18H26F208	408.40	408.1596	52.94	6.42	9.30	31.34
C18H26F209	424.40	424.1545	50.94	6.18	8.95	33.93
C18H27F101	278.41	278.2046	77.65	9.78	6.82	5.75
C18H27F102	294.41	294.1995	73.43	9.24	6.45	10.87
C18H27F103	310.41	310.1944	69.65	8.77	6.12	15.46
C18H27F104	326.41	326.1893	66.24	8.34	5.82	19.61
C18H27F105	342.41	342.1842	63.14	7.95	5.55	23.36
C18H27F106	358.41	358.1792	60.32	7.59	5.30	26.78
C18H27F107	374.41	374.1741	57.74	7.27	5.07	29.91
C18H27F108	390.41	390.1690	55.38	6.97	4.87	32.78
C18H27F109	406.41	406.1639	53.20	6.70	4.67	35.43
C18H28F201	298.42	298.2108	72.45	9.46	12.73	5.36
C18H28F202	314.42	314.2057	68.76	8.98	12.08	10.18
C18H28F203	330.42	330.2006	65.43	8.54	11.50	14.53
C18H28F204	346.42	346.1955	62.41	8.15	10.97	18.47
C18H28F205	362.42	362.1905	59.66	7.79	10.48	22.07
C18H28F206	378.42	378.1854	57.13	7.46	10.04	25.37
C18H28F207	394.42	394.1803	54.82	7.16	9.63	28.40
C18H28F208	410.42	410.1752	52.68	6.88	9.26	31.19
C18H28F209	426.42	426.1701	50.70	6.62	8.91	33.77
C18H29F101	280.43	280.2202	77.10	10.42	6.77	5.71
C18H29F102	296.43	296.2151	72.94	9.86	6.41	10.79
C18H29F103	312.43	312.2101	69.20	9.36	6.08	15.36
C18H29F104	328.43	328.2050	65.83	8.90	5.78	19.49
C18H29F105	344.43	344.1999	62.77	8.49	5.52	23.23
C18H29F106	360.43	360.1948	59.98	8.11	5.27	26.63
C18H29F107	376.43	376.1897	57.44	7.77	5.05	29.75
C18H29F108	392.43	392.1846	55.09	7.45	4.84	32.62
C18H29F109	408.42	408.1795	52.94	7.16	4.65	35.26
C18H30F201	300.44	300.2265	71.96	10.07	12.65	5.33
C18H30F202	316.44	316.2214	68.32	9.56	12.01	10.11
C18H30F203	332.43	332.2163	65.04	9.10	11.43	14.44
C18H30F204	348.43	348.2112	62.05	8.68	10.91	18.37
C18H30F205	364.43	364.2061	59.33	8.30	10.43	21.95
C18H30F206	380.43	380.2010	56.83	7.95	9.99	25.23
C18H30F207	396.43	396.1959	54.54	7.63	9.58	28.25
C18H30F208	412.43	412.1909	52.42	7.33	9.21	31.03
C18H30F209	428.43	428.1858	50.46	7.06	8.87	33.61
C18H31F101	282.45	282.2359	76.55	11.06	6.73	5.66
C18H31F102	298.44	298.2308	72.44	10.47	6.37	10.72
C18H31F103	314.44	314.2257	68.76	9.94	6.04	15.26
C18H31F104	330.44	330.2206	65.43	9.46	5.75	19.37
C18H31F105	346.44	346.2155	62.41	9.02	5.48	23.09
C18H31F106	362.44	362.2104	59.65	8.62	5.24	26.49
C18H31F107	378.44	378.2054	57.13	8.26	5.02	29.59
C18H31F108	394.44	394.2003	54.81	7.92	4.82	32.45
C18H31F109	410.44	410.1952	52.68	7.61	4.63	35.08
C18H32F201	302.45	302.2421	71.48	10.66	12.56	5.29

147

C18H32F202	318.45	318.2370	67.89	10.13	11.93	10.05
C18H32F203	334.45	334.2319	64.64	9.64	11.36	14.35
C18H32F204	350.45	350.2268	61.69	9.20	10.84	18.26
C18H32F205	366.45	366.2218	59.00	8.80	10.37	21.83
C18H32F206	382.45	382.2167	56.53	8.43	9.94	25.10
C18H32F207	398.45	398.2116	54.26	8.10	9.54	28.11
C18H32F208	414.45	414.2065	52.17	7.78	9.17	30.88
C18H32F209	430.45	430.2014	50.23	7.49	8.83	33.45
C18H33F101	284.46	284.2515	76.00	11.69	6.68	5.62
C18H33F102	300.46	300.2464	71.96	11.07	6.32	10.65
C18H33F103	316.46	316.2414	68.32	10.51	6.00	15.17
C18H33F104	332.46	332.2363	65.03	10.01	5.71	19.25
C18H33F105	348.46	348.2312	62.04	9.55	5.45	22.96
C18H33F106	364.46	364.2261	59.32	9.13	5.21	26.34
C18H33F107	380.46	380.2210	56.83	8.74	4.99	29.44
C18H33F108	396.46	396.2159	54.53	8.39	4.79	32.28
C18H33F109	412.46	412.2108	52.42	8.06	4.61	34.91
C18H34F201	304.47	304.2578	71.01	11.26	12.48	5.25
C18H34F202	320.47	320.2527	67.46	10.69	11.86	9.99
C18H34F203	336.47	336.2476	64.26	10.19	11.29	14.27
C18H34F204	352.47	352.2425	61.34	9.72	10.78	18.16
C18H34F205	368.47	368.2374	58.68	9.30	10.31	21.71
C18H34F206	384.46	384.2323	56.23	8.91	9.88	24.97
C18H34F207	400.46	400.2272	53.99	8.56	9.49	27.97
C18H34F208	416.46	416.2222	51.91	8.23	9.12	30.73
C18H34F209	432.46	432.2171	49.99	7.92	8.79	33.30
C18H35F101	286.48	286.2672	75.47	12.31	6.63	5.58
C18H35F102	302.48	302.2621	71.48	11.66	6.28	10.58
C18H35F103	318.48	318.2570	67.89	11.08	5.97	15.07
C18H35F104	334.48	334.2519	64.64	10.55	5.68	19.13
C18H35F105	350.48	350.2468	61.69	10.07	5.42	22.83
C18H35F106	366.47	366.2417	58.99	9.63	5.18	26.19
C18H35F107	382.47	382.2367	56.53	9.22	4.97	29.28
C18H35F108	398.47	398.2316	54.26	8.85	4.77	32.12
C18H35F109	414.47	414.2265	52.16	8.51	4.58	34.74
C18H36F201	306.48	306.2734	70.54	11.84	12.40	5.22
C18H36F202	322.48	322.2683	67.04	11.25	11.78	9.92
C18H36F203	338.48	338.2632	63.87	10.72	11.23	14.18
C18H36F204	354.48	354.2581	60.99	10.24	10.72	18.05
C18H36F205	370.48	370.2531	58.36	9.79	10.26	21.59
C18H36F206	386.48	386.2480	55.94	9.39	9.83	24.84
C18H36F207	402.48	402.2429	53.72	9.02	9.44	27.83
C18H36F208	418.48	418.2378	51.66	8.67	9.08	30.59
C18H36F209	434.48	434.2327	49.76	8.35	8.75	33.14
C18H37F101	288.49	288.2828	74.94	12.93	6.59	5.55
C18H37F102	304.49	304.2777	71.00	12.25	6.24	10.51
C18H37F103	320.49	320.2727	67.46	11.64	5.93	14.98
C18H37F104	336.49	336.2676	64.25	11.08	5.65	19.02
C18H37F105	352.49	352.2625	61.34	10.58	5.39	22.69
C18H37F106	368.49	368.2574	58.67	10.12	5.16	26.05
C18H37F107	384.49	384.2523	56.23	9.70	4.94	29.13
C18H37F108	400.49	400.2472	53.98	9.31	4.74	31.96
C18H37F109	416.49	416.2421	51.91	8.95	4.56	34.57
C19H 9F101	272.28	272.0637	83.81	3.33	6.98	5.88
C19H 9F102	288.28	288.0587	79.16	3.15	6.59	11.10
C19H 9F103	304.28	304.0536	75.00	2.98	6.24	15.77
C19H 9F104	320.28	320.0485	71.25	2.83	5.93	19.98
C19H 9F105	336.28	336.0434	67.86	2.70	5.65	23.79
C19H 9F106	352.28	352.0383	64.78	2.58	5.39	27.25
C19H 9F107	368.28	368.0332	61.97	2.46	5.16	30.41
C19H 9F108	384.28	384.0281	59.39	2.36	4.94	33.31
C19H 9F109	400.28	400.0231	57.01	2.27	4.75	35.97
C19H10F201	292.29	292.0700	78.08	3.45	13.00	5.47
C19H10F202	308.29	308.0649	74.03	3.27	12.33	10.38
C19H10F203	324.29	324.0598	70.37	3.11	11.72	14.80
C19H10F204	340.29	340.0547	67.06	2.96	11.17	18.81
C19H10F205	356.29	356.0496	64.05	2.83	10.66	22.45
C19H10F206	372.28	372.0445	61.30	2.71	10.21	25.79
C19H10F207	388.28	388.0394	58.77	2.60	9.79	28.84
C19H10F208	404.28	404.0344	56.45	2.49	9.40	31.66
C19H10F209	420.28	420.0293	54.30	2.40	9.04	34.26
C19H11F101	274.30	274.0794	83.20	4.04	6.93	5.83
C19H11F102	290.30	290.0743	78.61	3.82	6.54	11.02
C19H11F103	306.30	306.0692	74.51	3.62	6.20	15.67
C19H11F104	322.30	322.0641	70.81	3.44	5.89	19.86
C19H11F105	338.29	338.0590	67.46	3.28	5.62	23.65
C19H11F106	354.29	354.0540	64.41	3.13	5.36	27.10
C19H11F107	370.29	370.0489	61.63	2.99	5.13	30.25
C19H11F108	386.29	386.0438	59.08	2.87	4.92	33.13
C19H11F109	402.29	402.0387	56.73	2.76	4.72	35.79
C19H12F201	294.30	294.0856	77.54	4.11	12.91	5.44
C19H12F202	310.30	310.0805	73.54	3.90	12.25	10.31
C19H12F203	326.30	326.0754	69.94	3.71	11.64	14.71
C19H12F204	342.30	342.0704	66.67	3.53	11.10	18.70
C19H12F205	358.30	358.0653	63.69	3.38	10.60	22.33
C19H12F206	374.30	374.0602	60.97	3.23	10.15	25.65
C19H12F207	390.30	390.0551	58.47	3.10	9.74	28.69
C19H12F208	406.30	406.0500	56.17	2.98	9.35	31.50
C19H12F209	422.30	422.0449	54.04	2.86	9.00	34.10
C19H13F101	276.31	276.0950	82.59	4.74	6.88	5.79
C19H13F102	292.31	292.0900	78.07	4.48	6.50	10.95
C19H13F103	308.31	308.0849	74.02	4.25	6.16	15.57
C19H13F104	324.31	324.0798	70.37	4.04	5.86	19.73
C19H13F105	340.31	340.0747	67.06	3.85	5.58	23.51
C19H13F106	356.31	356.0696	64.05	3.68	5.33	26.94
C19H13F107	372.31	372.0645	61.30	3.52	5.10	30.08
C19H13F108	388.31	388.0594	58.77	3.37	4.89	32.96
C19H13F109	404.31	404.0543	56.44	3.24	4.70	35.62
C19H14F201	296.32	296.1013	77.02	4.76	12.82	5.40
C19H14F202	312.32	312.0962	73.07	4.52	12.17	10.25
C19H14F203	328.32	328.0911	69.51	4.30	11.57	14.62
C19H14F204	344.32	344.0860	66.28	4.10	11.04	18.59
C19H14F205	360.32	360.0809	63.34	3.92	10.55	22.20
C19H14F206	376.32	376.0758	60.64	3.75	10.10	25.51
C19H14F207	392.32	392.0707	58.17	3.60	9.69	28.55
C19H14F208	408.32	408.0657	55.89	3.46	9.31	31.35
C19H14F209	424.31	424.0606	53.78	3.33	8.95	33.94
C19H15F101	278.33	278.1107	81.99	5.43	6.83	5.75
C19H15F102	294.33	294.1056	77.54	5.14	6.45	10.87
C19H15F103	310.33	310.1005	73.54	4.87	6.12	15.47
C19H15F104	326.33	326.0954	69.93	4.63	5.82	19.61
C19H15F105	342.33	342.0903	66.66	4.42	5.55	23.37
C19H15F106	358.33	358.0853	63.69	4.22	5.30	26.79
C19H15F107	374.33	374.0802	60.97	4.04	5.08	29.92
C19H15F108	390.32	390.0751	58.47	3.87	4.87	32.79
C19H15F109	406.32	406.0700	56.16	3.72	4.68	35.44
C19H16F201	298.34	298.1169	76.50	5.41	12.74	5.36
C19H16F202	314.33	314.1118	72.60	5.13	12.09	10.18
C19H16F203	330.33	330.1067	69.09	4.88	11.50	14.53
C19H16F204	346.33	346.1017	65.89	4.66	10.97	18.48

148

C19H16F205	362.33	362.0966	62.98	4.45	10.49	22.08
C19H16F206	378.33	378.0915	60.32	4.26	10.04	25.37
C19H16F207	394.33	394.0864	57.87	4.09	9.64	28.40
C19H16F208	410.33	410.0813	55.62	3.93	9.26	31.19
C19H16F209	426.33	426.0762	53.53	3.78	8.91	33.78
C19H17F101	280.35	280.1263	81.40	6.11	6.78	5.71
C19H17F102	296.34	296.1212	77.01	5.78	6.41	10.80
C19H17F103	312.34	312.1162	73.06	5.49	6.08	15.37
C19H17F104	328.34	328.1111	69.50	5.22	5.79	19.49
C19H17F105	344.34	344.1060	66.27	4.98	5.52	23.23
C19H17F106	360.34	360.1009	63.33	4.76	5.27	26.64
C19H17F107	376.34	376.0958	60.64	4.55	5.05	29.76
C19H17F108	392.34	392.0907	58.17	4.37	4.84	32.62
C19H17F109	408.34	408.0856	55.89	4.20	4.65	35.26
C19H18F201	300.35	300.1326	75.98	6.04	12.65	5.33
C19H18F202	316.35	316.1275	72.14	5.74	12.01	10.11
C19H18F203	332.35	332.1224	68.67	5.46	11.43	14.44
C19H18F204	348.35	348.1173	65.51	5.21	10.91	18.37
C19H18F205	364.35	364.1122	62.64	4.98	10.43	21.96
C19H18F206	380.35	380.1071	60.00	4.77	9.99	25.24
C19H18F207	396.35	396.1020	57.58	4.58	9.59	28.26
C19H18F208	412.35	412.0970	55.34	4.40	9.21	31.04
C19H18F209	428.35	428.0919	53.28	4.24	8.87	33.62
C19H19F101	282.36	282.1420	80.82	6.78	6.73	5.67
C19H19F102	298.36	298.1369	76.49	6.42	6.37	10.72
C19H19F103	314.36	314.1318	72.60	6.09	6.04	15.27
C19H19F104	330.36	330.1267	69.08	5.80	5.75	19.37
C19H19F105	346.36	346.1216	65.89	5.53	5.49	23.10
C19H19F106	362.36	362.1166	62.98	5.29	5.24	26.49
C19H19F107	378.36	378.1115	60.32	5.06	5.02	29.60
C19H19F108	394.36	394.1064	57.87	4.86	4.82	32.46
C19H19F109	410.36	410.1013	55.61	4.67	4.63	35.09
C19H20F201	302.37	302.1482	75.48	6.67	12.57	5.29
C19H20F202	318.37	318.1431	71.68	6.33	11.93	10.05
C19H20F203	334.37	334.1380	68.25	6.03	11.36	14.35
C19H20F204	350.37	350.1330	65.14	5.75	10.84	18.27
C19H20F205	366.37	366.1279	62.29	5.50	10.37	21.84
C19H20F206	382.36	382.1228	59.68	5.27	9.94	25.11
C19H20F207	398.36	398.1177	57.29	5.06	9.54	28.11
C19H20F208	414.36	414.1126	55.08	4.87	9.17	30.89
C19H20F209	430.36	430.1075	53.03	4.68	8.83	33.46
C19H21F101	284.38	284.1576	80.25	7.44	6.68	5.63
C19H21F102	300.38	300.1525	75.98	7.05	6.32	10.65
C19H21F103	316.38	316.1475	72.13	6.69	6.01	15.17
C19H21F104	332.38	332.1424	68.66	6.37	5.72	19.25
C19H21F105	348.37	348.1373	65.51	6.08	5.45	22.96
C19H21F106	364.37	364.1322	62.63	5.81	5.21	26.35
C19H21F107	380.37	380.1271	60.00	5.56	4.99	29.44
C19H21F108	396.37	396.1220	57.58	5.34	4.79	32.29
C19H21F109	412.37	412.1169	55.34	5.13	4.61	34.92
C19H22F201	304.38	304.1639	74.98	7.29	12.48	5.26
C19H22F202	320.38	320.1588	71.23	6.92	11.86	9.99
C19H22F203	336.38	336.1537	67.84	6.59	11.30	14.27
C19H22F204	352.38	352.1486	64.76	6.29	10.78	18.16
C19H22F205	368.38	368.1435	61.95	6.02	10.31	21.72
C19H22F206	384.38	384.1384	59.37	5.77	9.89	24.97
C19H22F207	400.38	400.1333	57.00	5.54	9.49	27.97
C19H22F208	416.38	416.1283	54.81	5.33	9.13	30.74
C19H22F209	432.38	432.1232	52.78	5.13	8.79	33.30
C19H23F101	286.39	286.1733	79.68	8.09	6.63	5.59
C19H23F102	302.39	302.1682	75.47	7.67	6.28	10.58
C19H23F103	318.39	318.1631	71.68	7.28	5.97	15.08
C19H23F104	334.39	334.1580	68.25	6.93	5.68	19.14
C19H23F105	350.39	350.1529	65.13	6.62	5.42	22.83
C19H23F106	366.39	366.1479	62.29	6.33	5.19	26.20
C19H23F107	382.39	382.1428	59.68	6.06	4.97	29.29
C19H23F108	398.39	398.1377	57.28	5.82	4.77	32.13
C19H23F109	414.39	414.1326	55.07	5.59	4.58	34.75
C19H24F201	306.40	306.1795	74.48	7.90	12.40	5.22
C19H24F202	322.40	322.1744	70.79	7.50	11.79	9.93
C19H24F203	338.40	338.1693	67.44	7.15	11.23	14.18
C19H24F204	354.40	354.1643	64.39	6.83	10.72	18.06
C19H24F205	370.40	370.1592	61.61	6.53	10.26	21.60
C19H24F206	386.40	386.1541	59.06	6.26	9.83	24.84
C19H24F207	402.40	402.1490	56.71	6.01	9.44	27.83
C19H24F208	418.40	418.1439	54.54	5.78	9.08	30.59
C19H24F209	434.39	434.1388	52.54	5.57	8.75	33.15
C19H25F101	288.41	288.1889	79.13	8.74	6.59	5.55
C19H25F102	304.41	304.1838	74.97	8.28	6.24	10.51
C19H25F103	320.41	320.1788	71.23	7.86	5.93	14.98
C19H25F104	336.41	336.1737	67.84	7.49	5.65	19.02
C19H25F105	352.41	352.1686	64.76	7.15	5.39	22.70
C19H25F106	368.41	368.1635	61.95	6.84	5.16	26.06
C19H25F107	384.41	384.1584	59.37	6.56	4.94	29.13
C19H25F108	400.40	400.1533	57.00	6.29	4.74	31.97
C19H25F109	416.40	416.1482	54.81	6.05	4.56	34.58
C19H26F201	308.42	308.1952	73.99	8.50	12.32	5.19
C19H26F202	324.41	324.1901	70.35	8.08	11.71	9.86
C19H26F203	340.41	340.1850	67.04	7.70	11.16	14.10
C19H26F204	356.41	356.1799	64.03	7.35	10.66	17.96
C19H26F205	372.41	372.1748	61.28	7.04	10.20	21.48
C19H26F206	388.41	388.1697	58.76	6.75	9.78	24.72
C19H26F207	404.41	404.1646	56.43	6.48	9.40	27.69
C19H26F208	420.41	420.1596	54.28	6.23	9.04	30.45
C19H26F209	436.41	436.1545	52.29	6.01	8.71	33.00
C19H27F101	290.42	290.2046	78.58	9.37	6.54	5.51
C19H27F102	306.42	306.1995	74.48	8.88	6.20	10.44
C19H27F103	322.42	322.1944	70.78	8.44	5.89	14.89
C19H27F104	338.42	338.1893	67.43	8.04	5.61	18.91
C19H27F105	354.42	354.1842	64.39	7.68	5.36	22.57
C19H27F106	370.42	370.1792	61.61	7.35	5.13	25.92
C19H27F107	386.42	386.1741	59.06	7.04	4.92	28.98
C19H27F108	402.42	402.1690	56.71	6.76	4.72	31.81
C19H27F109	418.42	418.1639	54.54	6.50	4.54	34.41
C19H28F201	310.43	310.2108	73.51	9.09	12.24	5.15
C19H28F202	326.43	326.2057	69.91	8.65	11.64	9.80
C19H28F203	342.43	342.2006	66.64	8.24	11.10	14.02
C19H28F204	358.43	358.1955	63.67	7.87	10.60	17.86
C19H28F205	374.43	374.1905	60.95	7.54	10.15	21.37
C19H28F206	390.43	390.1854	58.45	7.23	9.73	24.59
C19H28F207	406.43	406.1803	56.15	6.94	9.35	27.56
C19H28F208	422.43	422.1752	54.02	6.68	8.99	30.30
C19H28F209	438.43	438.1701	52.05	6.44	8.67	32.84
C19H29F101	292.44	292.2202	78.04	10.00	6.50	5.47
C19H29F102	308.44	308.2151	73.99	9.48	6.16	10.37
C19H29F103	324.44	324.2101	70.34	9.01	5.86	14.79
C19H29F104	340.44	340.2050	67.03	8.59	5.58	18.80
C19H29F105	356.44	356.1999	64.03	8.20	5.33	22.44
C19H29F106	372.44	372.1948	61.28	7.85	5.10	25.78
C19H29F107	388.44	388.1897	58.75	7.53	4.89	28.83

C19H29F108	404.44	404.1846	56.43	7.23	4.70	31.65
C19H29F109	420.44	420.1795	54.28	6.95	4.52	34.25
C19H30F201	312.45	312.2265	73.04	9.68	12.16	5.12
C19H30F202	328.45	328.2214	69.48	9.21	11.57	9.74
C19H30F203	344.45	344.2163	66.25	8.78	11.03	13.93
C19H30F204	360.45	360.2112	63.31	8.39	10.54	17.76
C19H30F205	376.44	376.2061	60.62	8.03	10.09	21.25
C19H30F206	392.44	392.2010	58.15	7.71	9.68	24.46
C19H30F207	408.44	408.1959	55.87	7.40	9.30	27.42
C19H30F208	424.44	424.1909	53.77	7.12	8.95	30.16
C19H30F209	440.44	440.1858	51.81	6.87	8.63	32.69
C19H31F101	294.46	294.2359	77.50	10.61	6.45	5.43
C19H31F102	310.46	310.2308	73.51	10.06	6.12	10.31
C19H31F103	326.46	326.2257	69.91	9.57	5.82	14.70
C19H31F104	342.45	342.2206	66.64	9.12	5.55	18.69
C19H31F105	358.45	358.2155	63.67	8.72	5.30	22.32
C19H31F106	374.45	374.2104	60.95	8.34	5.07	25.64
C19H31F107	390.45	390.2054	58.45	8.00	4.87	28.68
C19H31F108	406.45	406.2003	56.15	7.69	4.67	31.49
C19H31F109	422.45	422.1952	54.02	7.40	4.50	34.09
C19H32F201	314.46	314.2421	72.57	10.26	12.08	5.09
C19H32F202	330.46	330.2370	69.06	9.76	11.50	9.68
C19H32F203	346.46	346.2319	65.87	9.31	10.97	13.85
C19H32F204	362.46	362.2268	62.96	8.90	10.48	17.66
C19H32F205	378.46	378.2218	60.30	8.52	10.04	21.14
C19H32F206	394.46	394.2167	57.85	8.18	9.63	24.34
C19H32F207	410.46	410.2116	55.60	7.86	9.26	27.29
C19H32F208	426.46	426.2065	53.51	7.56	8.91	30.01
C19H32F209	442.46	442.2014	51.58	7.29	8.59	32.54
C19H33F101	296.47	296.2515	76.98	11.22	6.41	5.40
C19H33F102	312.47	312.2464	73.03	10.65	6.08	10.24
C19H33F103	328.47	328.2414	69.48	10.13	5.78	14.61
C19H33F104	344.47	344.2363	66.25	9.66	5.52	18.58
C19H33F105	360.47	360.2312	63.31	9.23	5.27	22.19
C19H33F106	376.47	376.2261	60.62	8.84	5.05	25.50
C19H33F107	392.47	392.2210	58.15	8.48	4.84	28.54
C19H33F108	408.47	408.2159	55.87	8.14	4.65	31.34
C19H33F109	424.47	424.2108	53.76	7.84	4.48	33.92
C19H34F201	316.48	316.2578	72.11	10.83	12.01	5.06
C19H34F202	332.48	332.2527	68.64	10.31	11.43	9.62
C19H34F203	348.48	348.2476	65.49	9.83	10.90	13.77
C19H34F204	364.48	364.2425	62.61	9.40	10.43	17.56
C19H34F205	380.48	380.2374	59.98	9.01	9.99	21.03
C19H34F206	396.48	396.2323	57.56	8.64	9.58	24.21
C19H34F207	412.48	412.2272	55.33	8.31	9.21	27.15
C19H34F208	428.47	428.2222	53.26	8.00	8.87	29.87
C19H34F209	444.47	444.2171	51.34	7.71	8.55	32.40
C19H35F101	298.49	298.2672	76.46	11.82	6.36	5.36
C19H35F102	314.49	314.2621	72.57	11.22	6.04	10.17
C19H35F103	330.49	330.2570	69.05	10.67	5.75	14.52
C19H35F104	346.49	346.2519	65.86	10.18	5.48	18.47
C19H35F105	362.49	362.2468	62.96	9.73	5.24	22.07
C19H35F106	378.49	378.2417	60.30	9.32	5.02	25.36
C19H35F107	394.48	394.2367	57.85	8.94	4.82	28.39
C19H35F108	410.48	410.2316	55.60	8.59	4.63	31.18
C19H35F109	426.48	426.2265	53.51	8.27	4.45	33.76
C19H36F201	318.49	318.2734	71.65	11.39	11.93	5.02
C19H36F202	334.49	334.2683	68.23	10.85	11.36	9.57
C19H36F203	350.49	350.2632	65.11	10.35	10.84	13.69
C19H36F204	366.49	366.2581	62.27	9.90	10.37	17.46
C19H36F205	382.49	382.2531	59.66	9.49	9.93	20.91
C19H36F206	398.49	398.2480	57.27	9.11	9.54	24.09
C19H36F207	414.49	414.2429	55.06	8.75	9.17	27.02
C19H36F208	430.49	430.2378	53.01	8.43	8.83	29.73
C19H36F209	446.49	446.2327	51.11	8.13	8.51	32.25
C19H37F101	300.50	300.2828	75.94	12.41	6.32	5.32
C19H37F102	316.50	316.2777	72.10	11.78	6.00	10.11
C19H37F103	332.50	332.2727	68.63	11.22	5.71	14.44
C19H37F104	348.50	348.2676	65.48	10.70	5.45	18.36
C19H37F105	364.50	364.2625	62.61	10.23	5.21	21.95
C19H37F106	380.50	380.2574	59.98	9.80	4.99	25.23
C19H37F107	396.50	396.2523	57.56	9.41	4.79	28.25
C19H37F108	412.50	412.2472	55.32	9.04	4.61	31.03
C19H37F109	428.50	428.2421	53.26	8.70	4.43	33.60
C19H38F201	320.51	320.2891	71.20	11.95	11.86	4.99
C19H38F202	336.51	336.2840	67.82	11.38	11.29	9.51
C19H38F203	352.51	352.2789	64.74	10.87	10.78	13.62
C19H38F204	368.51	368.2738	61.93	10.39	10.31	17.37
C19H38F205	384.51	384.2687	59.35	9.96	9.88	20.81
C19H38F206	400.51	400.2636	56.98	9.56	9.49	23.97
C19H38F207	416.51	416.2585	54.79	9.20	9.12	26.89
C19H38F208	432.51	432.2535	52.76	8.86	8.79	29.59
C19H38F209	448.51	448.2484	50.88	8.54	8.47	32.11
C19H39F101	302.52	302.2985	75.44	12.99	6.28	5.29
C19H39F102	318.52	318.2934	71.65	12.34	5.96	10.05
C19H39F103	334.52	334.2883	68.22	11.75	5.68	14.35
C19H39F104	350.52	350.2832	65.11	11.22	5.42	18.26
C19H39F105	366.52	366.2781	62.26	10.73	5.18	21.83
C19H39F106	382.52	382.2730	59.66	10.28	4.97	25.10
C19H39F107	398.52	398.2680	57.27	9.86	4.77	28.10
C19H39F108	414.52	414.2629	55.05	9.48	4.58	30.88
C19H39F109	430.52	430.2578	53.01	9.13	4.41	33.45
C20H10F201	304.30	304.0700	78.94	3.31	12.49	5.26
C20H10F202	320.30	320.0649	75.00	3.15	11.86	9.99
C20H10F203	336.30	336.0598	71.43	3.00	11.30	14.27
C20H10F204	352.30	352.0547	68.19	2.86	10.79	18.17
C20H10F205	368.30	368.0496	65.23	2.74	10.32	21.72
C20H10F206	384.30	384.0445	62.51	2.62	9.89	24.98
C20H10F207	400.30	400.0394	60.01	2.52	9.49	27.98
C20H10F208	416.29	416.0344	57.71	2.42	9.13	30.75
C20H10F209	432.29	432.0293	55.57	2.33	8.79	33.31
C20H11F101	286.31	286.0794	83.90	3.87	6.64	5.59
C20H11F102	302.31	302.0743	79.46	3.67	6.28	10.58
C20H11F103	318.31	318.0692	75.47	3.48	5.97	15.08
C20H11F104	334.31	334.0641	71.86	3.32	5.68	19.14
C20H11F105	350.31	350.0590	68.58	3.17	5.42	22.84
C20H11F106	366.31	366.0540	65.58	3.03	5.19	26.21
C20H11F107	382.30	382.0489	62.84	2.90	4.97	29.29
C20H11F108	398.30	398.0438	60.31	2.78	4.77	32.14
C20H11F109	414.30	414.0387	57.98	2.68	4.59	34.76
C20H12F201	306.31	306.0856	78.42	3.95	12.40	5.22
C20H12F202	322.31	322.0805	74.53	3.75	11.79	9.93
C20H12F203	338.31	338.0754	71.01	3.58	11.23	14.19
C20H12F204	354.31	354.0704	67.80	3.41	10.72	18.06
C20H12F205	370.31	370.0653	64.87	3.27	10.26	21.60
C20H12F206	386.31	386.0602	62.18	3.13	9.84	24.85
C20H12F207	402.31	402.0551	59.71	3.01	9.44	27.84
C20H12F208	418.31	418.0500	57.43	2.89	9.08	30.60
C20H12F209	434.31	434.0449	55.31	2.79	8.75	33.15
C20H13F101	288.32	288.0950	83.32	4.54	6.59	5.55

C20H13F102	304.32	304.0900	78.94	4.31	6.24	10.51
C20H13F103	320.32	320.0849	74.99	4.09	5.93	14.98
C20H13F104	336.32	336.0798	71.43	3.90	5.65	19.03
C20H13F105	352.32	352.0747	68.18	3.72	5.39	22.71
C20H13F106	368.32	368.0696	65.22	3.56	5.16	26.06
C20H13F107	384.32	384.0645	62.51	3.41	4.94	29.14
C20H13F108	400.32	400.0594	60.01	3.27	4.75	31.97
C20H13F109	416.32	416.0543	57.70	3.15	4.56	34.59
C20H14F201	308.33	308.1013	77.91	4.58	12.32	5.19
C20H14F202	324.33	324.0962	74.07	4.35	11.72	9.87
C20H14F203	340.33	340.0911	70.59	4.15	11.16	14.10
C20H14F204	356.33	356.0860	67.42	3.96	10.66	17.96
C20H14F205	372.33	372.0809	64.52	3.79	10.21	21.49
C20H14F206	388.33	388.0758	61.86	3.63	9.78	24.72
C20H14F207	404.33	404.0707	59.41	3.49	9.40	27.70
C20H14F208	420.33	420.0657	57.15	3.36	9.04	30.45
C20H14F209	436.33	436.0606	55.06	3.23	8.71	33.00
C20H15F101	290.34	290.1107	82.74	5.21	6.54	5.51
C20H15F102	306.34	306.1056	78.42	4.94	6.20	10.45
C20H15F103	322.34	322.1005	74.52	4.69	5.89	14.89
C20H15F104	338.34	338.0954	71.00	4.47	5.62	18.92
C20H15F105	354.34	354.0903	67.79	4.27	5.36	22.58
C20H15F106	370.34	370.0853	64.87	4.08	5.13	25.92
C20H15F107	386.34	386.0802	62.18	3.91	4.92	28.99
C20H15F108	402.34	402.0751	59.71	3.76	4.72	31.81
C20H15F109	418.34	418.0700	57.42	3.61	4.54	34.42
C20H16F201	310.35	310.1169	77.40	5.20	12.24	5.16
C20H16F202	326.35	326.1118	73.61	4.94	11.64	9.81
C20H16F203	342.35	342.1067	70.17	4.71	11.10	14.02
C20H16F204	358.34	358.1017	67.04	4.50	10.60	17.86
C20H16F205	374.34	374.0966	64.17	4.31	10.15	21.37
C20H16F206	390.34	390.0915	61.54	4.13	9.73	24.59
C20H16F207	406.34	406.0864	59.12	3.97	9.35	27.56
C20H16F208	422.34	422.0813	56.88	3.82	9.00	30.31
C20H16F209	438.34	438.0762	54.80	3.68	8.67	32.85
C20H17F101	292.36	292.1263	82.17	5.86	6.50	5.47
C20H17F102	308.36	308.1212	77.90	5.56	6.16	10.38
C20H17F103	324.36	324.1162	74.06	5.28	5.86	14.80
C20H17F104	340.35	340.1111	70.58	5.03	5.58	18.80
C20H17F105	356.35	356.1060	67.41	4.81	5.33	22.45
C20H17F106	372.35	372.1009	64.51	4.60	5.10	25.78
C20H17F107	388.35	388.0958	61.86	4.41	4.89	28.84
C20H17F108	404.35	404.0907	59.41	4.24	4.70	31.65
C20H17F109	420.35	420.0856	57.15	4.08	4.52	34.26
C20H18F201	312.36	312.1326	76.91	5.81	12.16	5.12
C20H18F202	328.36	328.1275	73.16	5.53	11.57	9.74
C20H18F203	344.36	344.1224	69.76	5.27	11.03	13.94
C20H18F204	360.36	360.1173	66.66	5.03	10.54	17.76
C20H18F205	376.36	376.1122	63.83	4.82	10.10	21.26
C20H18F206	392.36	392.1071	61.23	4.62	9.68	24.47
C20H18F207	408.36	408.1020	58.83	4.44	9.30	27.43
C20H18F208	424.36	424.0970	56.61	4.28	8.95	30.16
C20H18F209	440.36	440.0919	54.55	4.12	8.63	32.70
C20H19F101	294.37	294.1420	81.61	6.51	6.45	5.44
C20H19F102	310.37	310.1369	77.40	6.17	6.12	10.31
C20H19F103	326.37	326.1318	73.60	5.87	5.82	14.71
C20H19F104	342.37	342.1267	70.16	5.59	5.55	18.69
C20H19F105	358.37	358.1216	67.03	5.34	5.30	22.32
C20H19F106	374.37	374.1166	64.17	5.12	5.07	25.64
C20H19F107	390.37	390.1115	61.54	4.91	4.87	28.69
C20H19F108	406.37	406.1064	59.11	4.71	4.68	31.50
C20H19F109	422.37	422.1013	56.88	4.53	4.50	34.09
C20H20F201	314.38	314.1482	76.41	6.41	12.09	5.09
C20H20F202	330.38	330.1431	72.71	6.10	11.50	9.69
C20H20F203	346.38	346.1380	69.35	5.82	10.97	13.86
C20H20F204	362.38	362.1330	66.29	5.56	10.49	17.66
C20H20F205	378.38	378.1279	63.49	5.33	10.04	21.14
C20H20F206	394.38	394.1228	60.91	5.11	9.63	24.34
C20H20F207	410.38	410.1177	58.54	4.91	9.26	27.29
C20H20F208	426.37	426.1126	56.34	4.73	8.91	30.02
C20H20F209	442.37	442.1075	54.30	4.56	8.59	32.55
C20H21F101	296.39	296.1576	81.05	7.14	6.41	5.40
C20H21F102	312.39	312.1525	76.90	6.78	6.08	10.24
C20H21F103	328.39	328.1475	73.15	6.45	5.79	14.62
C20H21F104	344.39	344.1424	69.75	6.15	5.52	18.58
C20H21F105	360.39	360.1373	66.66	5.87	5.27	22.20
C20H21F106	376.39	376.1322	63.82	5.62	5.05	25.50
C20H21F107	392.38	392.1271	61.22	5.39	4.84	28.54
C20H21F108	408.38	408.1220	58.82	5.18	4.65	31.34
C20H21F109	424.38	424.1169	56.61	4.99	4.48	33.93
C20H22F201	316.39	316.1639	75.93	7.01	12.01	5.06
C20H22F202	332.39	332.1588	72.27	6.67	11.43	9.63
C20H22F203	348.39	348.1537	68.95	6.37	10.91	13.78
C20H22F204	364.39	364.1486	65.92	6.09	10.43	17.56
C20H22F205	380.39	380.1435	63.15	5.83	9.99	21.03
C20H22F206	396.39	396.1384	60.60	5.59	9.59	24.22
C20H22F207	412.39	412.1333	58.25	5.38	9.21	27.16
C20H22F208	428.39	428.1283	56.08	5.18	8.87	29.88
C20H22F209	444.39	444.1232	54.06	4.99	8.55	32.40
C20H23F101	298.40	298.1733	80.50	7.77	6.37	5.36
C20H23F102	314.40	314.1682	76.41	7.37	6.04	10.18
C20H23F103	330.40	330.1631	72.71	7.02	5.75	14.53
C20H23F104	346.40	346.1580	69.35	6.69	5.48	18.47
C20H23F105	362.40	362.1529	66.29	6.40	5.24	22.07
C20H23F106	378.40	378.1479	63.48	6.13	5.02	25.37
C20H23F107	394.40	394.1428	60.91	5.88	4.82	28.40
C20H23F108	410.40	410.1377	58.53	5.65	4.63	31.19
C20H23F109	426.40	426.1326	56.34	5.44	4.46	33.77
C20H24F201	318.41	318.1795	75.44	7.60	11.93	5.02
C20H24F202	334.41	334.1744	71.83	7.23	11.36	9.57
C20H24F203	350.41	350.1693	68.55	6.90	10.84	13.70
C20H24F204	366.41	366.1643	65.56	6.60	10.37	17.47
C20H24F205	382.41	382.1592	62.82	6.33	9.94	20.92
C20H24F206	398.41	398.1541	60.30	6.07	9.54	24.10
C20H24F207	414.41	414.1490	57.97	5.84	9.17	27.03
C20H24F208	430.41	430.1439	55.81	5.62	8.83	29.74
C20H24F209	446.41	446.1388	53.81	5.42	8.51	32.26
C20H25F101	300.42	300.1889	79.96	8.39	6.32	5.33
C20H25F102	316.42	316.1838	75.92	7.96	6.00	10.11
C20H25F103	332.42	332.1788	72.27	7.58	5.72	14.44
C20H25F104	348.42	348.1737	68.95	7.23	5.45	18.37
C20H25F105	364.42	364.1686	65.92	6.91	5.21	21.95
C20H25F106	380.42	380.1635	63.15	6.62	4.99	25.23
C20H25F107	396.42	396.1584	60.60	6.36	4.79	28.25
C20H25F108	412.42'	412.1533	58.25	6.11	4.61	31.04
C20H25F109	428.42	428.1482	56.07	5.88	4.43	33.61
C20H26F201	320.43	320.1952	74.97	8.18	11.86	4.99
C20H26F202	336.43	336.1901	71.40	7.79	11.29	9.51
C20H26F203	352.43	352.1850	68.16	7.44	10.78	13.62
C20H26F204	368.42	368.1799	65.20	7.11	10.31	17.37

151

C20H26F205	384.42	384.1748	62.49	6.82	9.88	20.81
C20H26F206	400.42	400.1697	59.99	6.54	9.49	23.97
C20H26F207	416.42	416.1646	57.69	6.29	9.12	26.89
C20H26F208	432.42	432.1596	55.55	6.06	8.79	29.60
C20H26F209	448.42	448.1545	53.57	5.84	8.47	32.11
C20H27F101	302.44	302.2046	79.43	9.00	6.28	5.29
C20H27F102	318.44	318.1995	75.44	8.55	5.97	10.05
C20H27F103	334.43	334.1944	71.83	8.14	5.68	14.35
C20H27F104	350.43	350.1893	68.55	7.77	5.42	18.26
C20H27F105	366.43	366.1842	65.56	7.43	5.18	21.83
C20H27F106	382.43	382.1792	62.81	7.12	4.97	25.10
C20H27F107	398.43	398.1741	60.29	6.83	4.77	28.11
C20H27F108	414.43	414.1690	57.96	6.57	4.58	30.88
C20H27F109	430.43	430.1639	55.81	6.32	4.41	33.45
C20H28F201	322.44	322.2108	74.50	8.75	11.78	4.96
C20H28F202	338.44	338.2057	70.98	8.34	11.23	9.45
C20H28F203	354.44	354.2006	67.78	7.96	10.72	13.54
C20H28F204	370.44	370.1955	64.85	7.62	10.26	17.28
C20H28F205	386.44	386.1905	62.16	7.30	9.83	20.70
C20H28F206	402.44	402.1854	59.69	7.01	9.44	23.85
C20H28F207	418.44	418.1803	57.41	6.74	9.08	26.77
C20H28F208	434.44	434.1752	55.30	6.50	8.75	29.46
C20H28F209	450.44	450.1701	53.33	6.27	8.44	31.97
C20H29F101	304.45	304.2202	78.90	9.60	6.24	5.26
C20H29F102	320.45	320.2151	74.96	9.12	5.93	9.99
C20H29F103	336.45	336.2101	71.40	8.69	5.65	14.27
C20H29F104	352.45	352.2050	68.16	8.29	5.39	18.16
C20H29F105	368.45	368.1999	65.20	7.93	5.16	21.71
C20H29F106	384.45	384.1948	62.49	7.60	4.94	24.97
C20H29F107	400.45	400.1897	59.99	7.30	4.74	27.97
C20H29F108	416.45	416.1846	57.68	7.02	4.56	30.73
C20H29F109	432.45	432.1795	55.55	6.76	4.39	33.30
C20H30F201	324.46	324.2265	74.04	9.32	11.71	4.93
C20H30F202	340.46	340.2214	70.56	8.88	11.16	9.40
C20H30F203	356.46	356.2163	67.39	8.48	10.66	13.47
C20H30F204	372.46	372.2112	64.50	8.12	10.20	17.18
C20H30F205	388.46	388.2061	61.84	7.78	9.78	20.59
C20H30F206	404.46	404.2010	59.39	7.48	9.39	23.73
C20H30F207	420.45	420.1959	57.13	7.19	9.04	26.64
C20H30F208	436.45	436.1909	55.04	6.93	8.71	29.33
C20H30F209	452.45	452.1858	53.09	6.68	8.40	31.83
C20H31F101	306.47	306.2359	78.38	10.20	6.20	5.22
C20H31F102	322.47	322.2308	74.50	9.69	5.89	9.92
C20H31F103	338.47	338.2257	70.97	9.23	5.61	14.18
C20H31F104	354.47	354.2206	67.77	8.82	5.36	18.05
C20H31F105	370.47	370.2155	64.84	8.43	5.13	21.59
C20H31F106	386.46	386.2104	62.16	8.09	4.92	24.84
C20H31F107	402.46	402.2054	59.69	7.76	4.72	27.83
C20H31F108	418.46	418.2003	57.41	7.47	4.54	30.59
C20H31F109	434.46	434.1952	55.29	7.19	4.37	33.14
C20H32F201	326.47	326.2421	73.58	9.88	11.64	4.90
C20H32F202	342.47	342.2370	70.14	9.42	11.09	9.34
C20H32F203	358.47	358.2319	67.01	9.00	10.60	13.39
C20H32F204	374.47	374.2268	64.15	8.61	10.15	17.09
C20H32F205	390.47	390.2218	61.52	8.26	9.73	20.49
C20H32F206	406.47	406.2167	59.10	7.94	9.35	23.62
C20H32F207	422.47	422.2116	56.86	7.63	8.99	26.51
C20H32F208	438.47	438.2065	54.79	7.36	8.67	29.19
C20H32F209	454.47	454.2014	52.86	7.10	8.36	31.68
C20H33F101	308.48	308.2515	77.87	10.78	6.16	5.19

C20H33F102	324.48	324.2464	74.03	10.25	5.85	9.86
C20H33F103	340.48	340.2414	70.55	9.77	5.58	14.10
C20H33F104	356.48	356.2363	67.39	9.33	5.33	17.95
C20H33F105	372.48	372.2312	64.49	8.93	5.10	21.48
C20H33F106	388.48	388.2261	61.84	8.56	4.89	24.71
C20H33F107	404.48	404.2210	59.39	8.22	4.70	27.69
C20H33F108	420.48	420.2159	57.13	7.91	4.52	30.44
C20H33F109	436.48	436.2108	55.04	7.62	4.35	32.99
C20H34F201	328.49	328.2578	73.13	10.43	11.57	4.87
C20H34F202	344.49	344.2527	69.73	9.95	11.03	9.29
C20H34F203	360.49	360.2476	66.64	9.51	10.54	13.31
C20H34F204	376.49	376.2425	63.81	9.10	10.09	17.00
C20H34F205	392.49	392.2374	61.21	8.73	9.68	20.38
C20H34F206	408.49	408.2323	58.81	8.39	9.30	23.50
C20H34F207	424.49	424.2272	56.59	8.07	8.95	26.38
C20H34F208	440.49	440.2222	54.54	7.78	8.63	29.06
C20H34F209	456.49	456.2171	52.62	7.51	8.32	31.54
C20H35F101	310.50	310.2672	77.37	11.36	6.12	5.15
C20H35F102	326.50	326.2621	73.58	10.81	5.82	9.80
C20H35F103	342.50	342.2570	70.14	10.30	5.55	14.01
C20H35F104	358.50	358.2519	67.01	9.84	5.30	17.85
C20H35F105	374.50	374.2468	64.15	9.42	5.07	21.36
C20H35F106	390.50	390.2417	61.52	9.03	4.87	24.58
C20H35F107	406.50	406.2367	59.10	8.68	4.67	27.55
C20H35F108	422.50	422.2316	56.86	8.35	4.50	30.30
C20H35F109	438.49	438.2265	54.78	8.05	4.33	32.84
C20H36F201	330.51	330.2734	72.68	10.98	11.50	4.84
C20H36F202	346.51	346.2683	69.33	10.47	10.97	9.23
C20H36F203	362.50	362.2632	66.27	10.01	10.48	13.24
C20H36F204	378.50	378.2581	63.47	9.59	10.04	16.91
C20H36F205	394.50	394.2531	60.89	9.20	9.63	20.28
C20H36F206	410.50	410.2480	58.52	8.84	9.26	23.39
C20H36F207	426.50	426.2429	56.32	8.51	8.91	26.26
C20H36F208	442.50	442.2378	54.29	8.20	8.59	28.93
C20H36F209	458.50	458.2327	52.39	7.91	8.29	31.41
C20H37F101	312.52	312.2828	76.87	11.93	6.08	5.12
C20H37F102	328.52	328.2777	73.12	11.35	5.78	9.74
C20H37F103	344.51	344.2727	69.73	10.83	5.51	13.93
C20H37F104	360.51	360.2676	66.63	10.34	5.27	17.75
C20H37F105	376.51	376.2625	63.80	9.91	5.05	21.25
C20H37F106	392.51	392.2574	61.20	9.50	4.84	24.46
C20H37F107	408.51	408.2523	58.80	9.13	4.65	27.42
C20H37F108	424.51	424.2472	56.59	8.79	4.48	30.15
C20H37F109	440.51	440.2421	54.53	8.47	4.31	32.69
C20H38F201	332.52	332.2891	72.24	11.52	11.43	4.81
C20H38F202	348.52	348.2840	68.93	10.99	10.90	9.18
C20H38F203	364.52	364.2789	65.90	10.51	10.42	13.17
C20H38F204	380.52	380.2738	63.13	10.07	9.99	16.82
C20H38F205	396.52	396.2687	60.58	9.66	9.58	20.17
C20H38F206	412.52	412.2636	58.23	9.29	9.21	23.27
C20H38F207	428.52	428.2585	56.06	8.94	8.87	26.14
C20H38F208	444.52	444.2535	54.04	8.62	8.55	28.79
C20H38F209	460.52	460.2484	52.16	8.32	8.25	31.27
C20H39F101	314.53	314.2985	76.37	12.50	6.04	5.09
C20H39F102	330.53	330.2934	72.68	11.89	5.75	9.68
C20H39F103	346.53	346.2883	69.32	11.34	5.48	13.85
C20H39F104	362.53	362.2832	66.26	10.84	5.24	17.65
C20H39F105	378.53	378.2781	63.46	10.39	5.02	21.13
C20H39F106	394.53	394.2730	60.89	9.96	4.82	24.33
C20H39F107	410.53	410.2680	58.52	9.58	4.63	27.28

152

C20H39F108	426.53	426.2629	56.32	9.22	4.45	30.01
C20H39F109	442.53	442.2578	54.28	8.88	4.29	32.54
C20H40F201	334.54	334.3047	71.81	12.05	11.36	4.78
C20H40F202	350.54	350.2996	68.53	11.50	10.84	9.13
C20H40F203	366.54	366.2945	65.54	11.00	10.37	13.10
C20H40F204	382.54	382.2894	62.80	10.54	9.93	16.73
C20H40F205	398.54	398.2844	60.28	10.12	9.53	20.07
C20H40F206	414.53	414.2793	57.95	9.73	9.17	23.16
C20H40F207	430.53	430.2742	55.80	9.36	8.83	26.01
C20H40F208	446.53	446.2691	53.80	9.03	8.51	28.66
C20H40F209	462.53	462.2640	51.94	8.72	8.21	31.13
C20H41F101	316.55	316.3141	75.89	13.06	6.00	5.05
C20H41F102	332.55	332.3090	72.24	12.43	5.71	9.62
C20H41F103	348.55	348.3040	68.92	11.86	5.45	13.77
C20H41F104	364.55	364.2989	65.90	11.34	5.21	17.56
C20H41F105	380.55	380.2938	63.13	10.86	4.99	21.02
C20H41F106	396.54	396.2887	60.58	10.42	4.79	24.21
C20H41F107	412.54	412.2836	58.23	10.02	4.61	27.15
C20H41F108	428.54	428.2785	56.06	9.64	4.43	29.87
C20H41F109	444.54	444.2734	54.04	9.30	4.27	32.39
C21H10F201	316.31	316.0700	79.74	3.19	12.01	5.06
C21H10F202	332.31	332.0649	75.90	3.03	11.43	9.63
C21H10F203	348.31	348.0598	72.42	2.89	10.91	13.78
C21H10F204	364.31	364.0547	69.24	2.77	10.43	17.57
C21H10F205	380.31	380.0496	66.32	2.65	9.99	21.03
C21H10F206	396.31	396.0445	63.65	2.54	9.59	24.22
C21H10F207	412.31	412.0394	61.18	2.44	9.22	27.16
C21H10F208	428.31	428.0344	58.89	2.35	8.87	29.88
C21H10F209	444.31	444.0293	56.77	2.27	8.55	32.41
C21H11F101	298.32	298.0794	84.55	3.72	6.37	5.36
C21H11F102	314.32	314.0743	80.25	3.53	6.04	10.18
C21H11F103	330.32	330.0692	76.36	3.36	5.75	14.53
C21H11F104	346.32	346.0641	72.83	3.20	5.49	18.48
C21H11F105	362.32	362.0590	69.62	3.06	5.24	22.08
C21H11F106	378.32	378.0540	66.67	2.93	5.02	25.37
C21H11F107	394.32	394.0489	63.97	2.81	4.82	28.40
C21H11F108	410.32	410.0438	61.47	2.70	4.63	31.19
C21H11F109	426.31	426.0387	59.17	2.60	4.46	33.78
C21H12F201	318.33	318.0856	79.24	3.80	11.94	5.03
C21H12F202	334.33	334.0805	75.45	3.62	11.37	9.57
C21H12F203	350.32	350.0754	72.00	3.45	10.85	13.70
C21H12F204	366.32	366.0704	68.86	3.30	10.37	17.47
C21H12F205	382.32	382.0653	65.97	3.16	9.94	20.92
C21H12F206	398.32	398.0602	63.32	3.04	9.54	24.10
C21H12F207	414.32	414.0551	60.88	2.92	9.17	27.03
C21H12F208	430.32	430.0500	58.62	2.81	8.83	29.74
C21H12F209	446.32	446.0449	56.51	2.71	8.51	32.26
C21H13F101	300.34	300.0950	83.98	4.36	6.33	5.33
C21H13F102	316.33	316.0900	79.74	4.14	6.01	10.12
C21H13F103	332.33	332.0849	75.90	3.94	5.72	14.44
C21H13F104	348.33	348.0798	72.41	3.76	5.45	18.37
C21H13F105	364.33	364.0747	69.23	3.60	5.21	21.96
C21H13F106	380.33	380.0696	66.32	3.45	5.00	25.24
C21H13F107	396.33	396.0645	63.64	3.31	4.79	28.26
C21H13F108	412.33	412.0594	61.17	3.18	4.61	31.04
C21H13F109	428.33	428.0543	58.89	3.06	4.44	33.62
C21H14F201	320.34	320.1013	78.74	4.41	11.86	4.99
C21H14F202	336.34	336.0962	74.99	4.20	11.30	9.51
C21H14F203	352.34	352.0911	71.59	4.01	10.78	13.62
C21H14F204	368.34	368.0860	68.48	3.83	10.32	17.37

C21H14F205	384.34	384.0809	65.63	3.67	9.89	20.81
C21H14F206	400.34	400.0758	63.01	3.52	9.49	23.98
C21H14F207	416.34	416.0707	60.58	3.39	9.13	26.90
C21H14F208	432.34	432.0657	58.34	3.26	8.79	29.61
C21H14F209	448.34	448.0606	56.26	3.15	8.48	32.12
C21H15F101	302.35	302.1107	83.42	5.00	6.28	5.29
C21H15F102	318.35	318.1056	79.23	4.75	5.97	10.05
C21H15F103	334.35	334.1005	75.44	4.52	5.68	14.36
C21H15F104	350.35	350.0954	71.99	4.32	5.42	18.27
C21H15F105	366.35	366.0903	68.85	4.13	5.19	21.84
C21H15F106	382.35	382.0853	65.97	3.95	4.97	25.11
C21H15F107	398.35	398.0802	63.32	3.80	4.77	28.12
C21H15F108	414.35	414.0751	60.88	3.65	4.59	30.89
C21H15F109	430.35	430.0700	58.61	3.51	4.41	33.46
C21H16F201	322.36	322.1169	78.25	5.00	11.79	4.96
C21H16F202	338.36	338.1118	74.55	4.77	11.23	9.46
C21H16F203	354.36	354.1067	71.18	4.55	10.72	13.55
C21H16F204	370.36	370.1017	68.11	4.35	10.26	17.28
C21H16F205	386.36	386.0966	65.29	4.17	9.83	20.71
C21H16F206	402.35	402.0915	62.69	4.01	9.44	23.86
C21H16F207	418.35	418.0864	60.29	3.85	9.08	26.77
C21H16F208	434.35	434.0813	58.07	3.71	8.75	29.47
C21H16F209	450.35	450.0762	56.01	3.58	8.44	31.97
C21H17F101	304.37	304.1263	82.87	5.63	6.24	5.26
C21H17F102	320.37	320.1212	78.73	5.35	5.93	9.99
C21H17F103	336.37	336.1162	74.99	5.09	5.65	14.27
C21H17F104	352.37	352.1111	71.58	4.86	5.39	18.16
C21H17F105	368.37	368.1060	68.47	4.65	5.16	21.72
C21H17F106	384.36	384.1009	65.62	4.46	4.94	24.98
C21H17F107	400.36	400.0958	63.00	4.28	4.75	27.97
C21H17F108	416.36	416.0907	60.58	4.12	4.56	30.74
C21H17F109	432.36	432.0856	58.34	3.96	4.39	33.30
C21H18F201	324.37	324.1326	77.76	5.59	11.71	4.93
C21H18F202	340.37	340.1275	74.11	5.33	11.16	9.40
C21H18F203	356.37	356.1224	70.78	5.09	10.66	13.47
C21H18F204	372.37	372.1173	67.74	4.87	10.20	17.19
C21H18F205	388.37	388.1122	64.95	4.67	9.78	20.60
C21H18F206	404.37	404.1071	62.38	4.49	9.40	23.74
C21H18F207	420.37	420.1020	60.00	4.32	9.04	26.64
C21H18F208	436.37	436.0970	57.80	4.16	8.71	29.33
C21H18F209	452.37	452.0919	55.76	4.01	8.40	31.83
C21H19F101	306.38	306.1420	82.33	6.25	6.20	5.22
C21H19F102	322.38	322.1369	78.24	5.94	5.89	9.93
C21H19F103	338.38	338.1318	74.54	5.66	5.61	14.18
C21H19F104	354.38	354.1267	71.18	5.40	5.36	18.06
C21H19F105	370.38	370.1216	68.10	5.17	5.13	21.60
C21H19F106	386.38	386.1166	65.28	4.96	4.92	24.85
C21H19F107	402.38	402.1115	62.69	4.76	4.72	27.83
C21H19F108	418.38	418.1064	60.29	4.58	4.54	30.59
C21H19F109	434.38	434.1013	58.07	4.41	4.37	33.15
C21H20F201	326.39	326.1482	77.28	6.18	11.64	4.90
C21H20F202	342.39	342.1431	73.67	5.89	11.10	9.35
C21H20F203	358.39	358.1380	70.38	5.63	10.60	13.39
C21H20F204	374.39	374.1330	67.37	5.38	10.15	17.09
C21H20F205	390.39	390.1279	64.61	5.16	9.73	20.49
C21H20F206	406.39	406.1228	62.07	4.96	9.35	23.62
C21H20F207	422.39	422.1177	59.72	4.77	9.00	26.52
C21H20F208	438.39	438.1126	57.54	4.60	8.67	29.20
C21H20F209	454.38	454.1075	55.51	4.44	8.36	31.69
C21H21F101	308.40	308.1576	81.79	6.86	6.16	5.19

C21H21F102	324.40	324.1525	77.75	6.53	5.86	9.86
C21H21F103	340.40	340.1475	74.10	6.22	5.58	14.10
C21H21F104	356.40	356.1424	70.77	5.94	5.33	17.96
C21H21F105	372.40	372.1373	67.73	5.68	5.10	21.48
C21H21F106	388.40	388.1322	64.94	5.45	4.89	24.72
C21H21F107	404.40	404.1271	62.37	5.23	4.70	27.69
C21H21F108	420.40	420.1220	60.00	5.04	4.52	30.45
C21H21F109	436.39	436.1169	57.80	4.85	4.35	33.00
C21H22F201	328.41	328.1639	76.81	6.75	11.57	4.87
C21H22F202	344.41	344.1588	73.24	6.44	11.03	9.29
C21H22F203	360.40	360.1537	69.99	6.15	10.54	13.32
C21H22F204	376.40	376.1486	67.01	5.89	10.09	17.00
C21H22F205	392.40	392.1435	64.28	5.65	9.68	20.39
C21H22F206	408.40	408.1384	61.76	5.43	9.30	23.51
C21H22F207	424.40	424.1333	59.43	5.23	8.95	26.39
C21H22F208	440.40	440.1283	57.27	5.04	8.63	29.06
C21H22F209	456.40	456.1232	55.27	4.86	8.33	31.55
C21H23F101	310.42	310.1733	81.26	7.47	6.12	5.15
C21H23F102	326.41	326.1682	77.27	7.10	5.82	9.80
C21H23F103	342.41	342.1631	73.66	6.77	5.55	14.02
C21H23F104	358.41	358.1580	70.38	6.47	5.30	17.86
C21H23F105	374.41	374.1529	67.37	6.19	5.07	21.37
C21H23F106	390.41	390.1479	64.61	5.94	4.87	24.59
C21H23F107	406.41	406.1428	62.06	5.70	4.67	27.56
C21H23F108	422.41	422.1377	59.71	5.49	4.50	30.30
C21H23F109	438.41	438.1326	57.53	5.29	4.33	32.84
C21H24F201	330.42	330.1795	76.34	7.32	11.50	4.84
C21H24F202	346.42	346.1744	72.81	6.98	10.97	9.24
C21H24F203	362.42	362.1693	69.60	6.67	10.48	13.24
C21H24F204	378.42	378.1643	66.65	6.39	10.04	16.91
C21H24F205	394.42	394.1592	63.95	6.13	9.63	20.28
C21H24F206	410.42	410.1541	61.46	5.89	9.26	23.39
C21H24F207	426.42	426.1490	59.15	5.67	8.91	26.26
C21H24F208	442.42	442.1439	57.01	5.47	8.59	28.93
C21H24F209	458.42	458.1388	55.02	5.28	8.29	31.41
C21H25F101	312.43	312.1889	80.73	8.07	6.08	5.12
C21H25F102	328.43	328.1838	76.80	7.67	5.78	9.74
C21H25F103	344.43	344.1788	73.23	7.32	5.52	13.94
C21H25F104	360.43	360.1737	69.98	6.99	5.27	17.76
C21H25F105	376.43	376.1686	67.01	6.69	5.05	21.25
C21H25F106	392.43	392.1635	64.28	6.42	4.84	24.46
C21H25F107	408.43	408.1584	61.76	6.17	4.65	27.42
C21H25F108	424.43	424.1533	59.43	5.94	4.48	30.16
C21H25F109	440.43	440.1482	57.27	5.72	4.31	32.69
C21H26F201	332.44	332.1952	75.87	7.88	11.43	4.81
C21H26F202	348.44	348.1901	72.39	7.52	10.90	9.18
C21H26F203	364.44	364.1850	69.21	7.19	10.43	13.17
C21H26F204	380.44	380.1799	66.30	6.89	9.99	16.82
C21H26F205	396.44	396.1748	63.63	6.61	9.58	20.18
C21H26F206	412.43	412.1697	61.16	6.35	9.21	23.28
C21H26F207	428.43	428.1646	58.87	6.12	8.87	26.14
C21H26F208	444.43	444.1596	56.75	5.90	8.55	28.80
C21H26F209	460.43	460.1545	54.78	5.69	8.25	31.27
C21H27F101	314.45	314.2046	80.22	8.65	6.04	5.09
C21H27F102	330.45	330.1995	76.33	8.24	5.75	9.68
C21H27F103	346.45	346.1944	72.81	7.86	5.48	13.85
C21H27F104	362.45	362.1893	69.59	7.51	5.24	17.66
C21H27F105	378.44	378.1842	66.65	7.19	5.02	21.14
C21H27F106	394.44	394.1792	63.95	6.90	4.82	24.34
C21H27F107	410.44	410.1741	61.45	6.63	4.63	27.29
C21H27F108	426.44	426.1690	59.15	6.38	4.46	30.01
C21H27F109	442.44	442.1639	57.01	6.15	4.29	32.55
C21H28F201	334.45	334.2108	75.42	8.44	11.36	4.78
C21H28F202	350.45	350.2057	71.97	8.05	10.84	9.13
C21H28F203	366.45	366.2006	68.83	7.70	10.37	13.10
C21H28F204	382.45	382.1955	65.95	7.38	9.94	16.73
C21H28F205	398.45	398.1905	63.30	7.08	9.54	20.08
C21H28F206	414.45	414.1854	60.86	6.81	9.17	23.16
C21H28F207	430.45	430.1803	58.60	6.56	8.83	26.02
C21H28F208	446.45	446.1752	56.50	6.32	8.51	28.67
C21H28F209	462.45	462.1701	54.54	6.10	8.22	31.14
C21H29F101	316.46	316.2202	79.70	9.24	6.00	5.06
C21H29F102	332.46	332.2151	75.87	8.79	5.71	9.62
C21H29F103	348.46	348.2101	72.39	8.39	5.45	13.77
C21H29F104	364.46	364.2050	69.21	8.02	5.21	17.56
C21H29F105	380.46	380.1999	66.30	7.68	4.99	21.03
C21H29F106	396.46	396.1948	63.62	7.37	4.79	24.21
C21H29F107	412.46	412.1897	61.15	7.09	4.61	27.15
C21H29F108	428.46	428.1846	58.87	6.82	4.43	29.87
C21H29F109	444.46	444.1795	56.75	6.58	4.27	32.40
C21H30F201	336.47	336.2265	74.96	8.99	11.29	4.76
C21H30F202	352.47	352.2214	71.56	8.58	10.78	9.08
C21H30F203	368.47	368.2163	68.45	8.21	10.31	13.03
C21H30F204	384.47	384.2112	65.61	7.87	9.88	16.65
C21H30F205	400.47	400.2061	62.98	7.55	9.49	19.98
C21H30F206	416.47	416.2010	60.57	7.26	9.12	23.05
C21H30F207	432.47	432.1959	58.32	6.99	8.79	25.90
C21H30F208	448.47	448.1909	56.24	6.74	8.47	28.54
C21H30F209	464.46	464.1858	54.31	6.51	8.18	31.00
C21H31F101	318.48	318.2359	79.20	9.81	5.97	5.02
C21H31F102	334.48	334.2308	75.41	9.34	5.68	9.57
C21H31F103	350.48	350.2257	71.97	8.92	5.42	13.70
C21H31F104	366.48	366.2206	68.83	8.53	5.18	17.46
C21H31F105	382.48	382.2155	65.95	8.17	4.97	20.92
C21H31F106	398.48	398.2104	63.30	7.84	4.77	24.09
C21H31F107	414.48	414.2054	60.86	7.54	4.58	27.02
C21H31F108	430.47	430.2003	58.59	7.26	4.41	29.73
C21H31F109	446.47	446.1952	56.49	7.00	4.26	32.25
C21H32F201	338.49	338.2421	74.52	9.53	11.23	4.73
C21H32F202	354.48	354.2370	71.16	9.10	10.72	9.03
C21H32F203	370.48	370.2319	68.08	8.71	10.26	12.96
C21H32F204	386.48	386.2268	65.26	8.35	9.83	16.56
C21H32F205	402.48	402.2218	62.67	8.01	9.44	19.88
C21H32F206	418.48	418.2167	60.27	7.71	9.08	22.94
C21H32F207	434.48	434.2116	58.05	7.42	8.75	25.78
C21H32F208	450.48	450.2065	55.99	7.16	8.43	28.41
C21H32F209	466.48	466.2014	54.07	6.91	8.15	30.87
C21H33F101	320.49	320.2515	78.70	10.38	5.93	4.99
C21H33F102	336.49	336.2464	74.96	9.89	5.65	9.51
C21H33F103	352.49	352.2414	71.56	9.44	5.39	13.62
C21H33F104	368.49	368.2363	68.45	9.03	5.16	17.37
C21H33F105	384.49	384.2312	65.60	8.65	4.94	20.81
C21H33F106	400.49	400.2261	62.98	8.31	4.74	23.97
C21H33F107	416.49	416.2210	60.56	7.99	4.56	26.89
C21H33F108	432.49	432.2159	58.32	7.69	4.39	29.59
C21H33F109	448.49	448.2108	56.24	7.42	4.24	32.11
C21H34F201	340.50	340.2578	74.08	10.06	11.16	4.70
C21H34F202	356.50	356.2527	70.75	9.61	10.66	8.98
C21H34F203	372.50	372.2476	67.71	9.20	10.20	12.89
C21H34F204	388.50	388.2425	64.93	8.82	9.78	16.47

C21H34F205	404.50	404.2374	62.36	8.47	9.39	19.78
C21H34F206	420.50	420.2323	59.98	8.15	9.04	22.83
C21H34F207	436.50	436.2272	57.79	7.85	8.70	25.66
C21H34F208	452.50	452.2222	55.74	7.57	8.40	28.29
C21H34F209	468.50	468.2171	53.84	7.32	8.11	30.74
C21H35F101	322.51	322.2672	78.21	10.94	5.89	4.96
C21H35F102	338.51	338.2621	74.51	10.42	5.61	9.45
C21H35F103	354.51	354.2570	71.15	9.95	5.36	13.54
C21H35F104	370.51	370.2519	68.08	9.52	5.13	17.27
C21H35F105	386.51	386.2468	65.26	9.13	4.92	20.70
C21H35F106	402.51	402.2417	62.67	8.76	4.72	23.85
C21H35F107	418.51	418.2367	60.27	8.43	4.54	26.76
C21H35F108	434.51	434.2316	58.05	8.12	4.37	29.46
C21H35F109	450.51	450.2265	55.99	7.83	4.22	31.96
C21H36F201	342.52	342.2734	73.64	10.59	11.09	4.67
C21H36F202	358.52	358.2683	70.35	10.12	10.60	8.93
C21H36F203	374.52	374.2632	67.35	9.69	10.15	12.82
C21H36F204	390.52	390.2581	64.59	9.29	9.73	16.39
C21H36F205	406.51	406.2531	62.05	8.93	9.35	19.68
C21H36F206	422.51	422.2480	59.70	8.59	8.99	22.72
C21H36F207	438.51	438.2429	57.52	8.27	8.66	25.54
C21H36F208	454.51	454.2378	55.50	7.98	8.36	28.16
C21H36F209	470.51	470.2327	53.61	7.71	8.08	30.60
C21H37F101	324.53	324.2828	77.72	11.49	5.85	4.93
C21H37F102	340.53	340.2777	74.07	10.95	5.58	9.40
C21H37F103	356.53	356.2727	70.75	10.46	5.33	13.46
C21H37F104	372.53	372.2676	67.71	10.01	5.10	17.18
C21H37F105	388.52	388.2625	64.92	9.60	4.89	20.59
C21H37F106	404.52	404.2574	62.35	9.22	4.70	23.73
C21H37F107	420.52	420.2523	59.98	8.87	4.52	26.63
C21H37F108	436.52	436.2472	57.78	8.54	4.35	29.32
C21H37F109	452.52	452.2421	55.74	8.24	4.20	31.82
C21H38F201	344.53	344.2891	73.21	11.12	11.03	4.64
C21H38F202	360.53	360.2840	69.96	10.62	10.54	8.88
C21H38F203	376.53	376.2789	66.99	10.17	10.09	12.75
C21H38F204	392.53	392.2738	64.26	9.76	9.68	16.30
C21H38F205	408.53	408.2687	61.74	9.38	9.30	19.58
C21H38F206	424.53	424.2636	59.41	9.02	8.95	22.61
C21H38F207	440.53	440.2585	57.26	8.69	8.63	25.42
C21H38F208	456.53	456.2535	55.25	8.39	8.32	28.04
C21H38F209	472.53	472.2484	53.38	8.11	8.04	30.47
C21H39F101	326.54	326.2985	77.24	12.04	5.82	4.90
C21H39F102	342.54	342.2934	73.64	11.48	5.55	9.34
C21H39F103	358.54	358.2883	70.35	10.96	5.30	13.39
C21H39F104	374.54	374.2832	67.34	10.50	5.07	17.09
C21H39F105	390.54	390.2781	64.59	10.07	4.86	20.48
C21H39F106	406.54	406.2730	62.04	9.67	4.67	23.61
C21H39F107	422.54	422.2680	59.69	9.30	4.50	26.51
C21H39F108	438.54	438.2629	57.52	8.96	4.33	29.19
C21H39F109	454.54	454.2578	55.49	8.65	4.18	31.68
C21H40F201	346.55	346.3047	72.78	11.63	10.96	4.62
C21H40F202	362.55	362.2996	69.57	11.12	10.48	8.83
C21H40F203	378.55	378.2945	66.63	10.65	10.04	12.68
C21H40F204	394.55	394.2894	63.93	10.22	9.63	16.22
C21H40F205	410.55	410.2844	61.44	9.82	9.26	19.49
C21H40F206	426.55	426.2793	59.13	9.45	8.91	22.51
C21H40F207	442.55	442.2742	57.00	9.11	8.59	25.31
C21H40F208	458.54	458.2691	55.01	8.79	8.29	27.91
C21H40F209	474.54	474.2640	53.15	8.50	8.01	30.34
C21H41F101	328.56	328.3141	76.77	12.58	5.78	4.87
C21H41F102	344.56	344.3090	73.21	11.99	5.51	9.29
C21H41F103	360.56	360.3040	69.96	11.46	5.27	13.31
C21H41F104	376.56	376.2989	66.98	10.97	5.05	17.00
C21H41F105	392.56	392.2938	64.25	10.53	4.84	20.38
C21H41F106	408.56	408.2887	61.74	10.12	4.65	23.50
C21H41F107	424.56	424.2836	59.41	9.73	4.47	26.38
C21H41F108	440.55	440.2785	57.25	9.38	4.31	29.05
C21H41F109	456.55	456.2734	55.25	9.05	4.16	31.54
C21H42F201	348.57	348.3204	72.36	12.15	10.90	4.59
C21H42F202	364.56	364.3153	69.19	11.61	10.42	8.78
C21H42F203	380.56	380.3102	66.28	11.12	9.98	12.61
C21H42F204	396.56	396.3051	63.61	10.68	9.58	16.14
C21H42F205	412.56	412.3000	61.14	10.26	9.21	19.39
C21H42F206	428.56	428.2949	58.86	9.88	8.87	22.40
C21H42F207	444.56	444.2898	56.74	9.52	8.55	25.19
C21H42F208	460.56	460.2848	54.77	9.19	8.25	27.79
C21H42F209	476.56	476.2797	52.93	8.88	7.97	30.22
C21H43F101	330.57	330.3298	76.30	13.11	5.75	4.84
C21H43F102	346.57	346.3247	72.78	12.51	5.48	9.23
C21H43F103	362.57	362.3196	69.57	11.95	5.24	13.24
C21H43F104	378.57	378.3145	66.63	11.45	5.02	16.90
C21H43F105	394.57	394.3094	63.93	10.98	4.81	20.27
C21H43F106	410.57	410.3043	61.43	10.56	4.63	23.38
C21H43F107	426.57	426.2993	59.13	10.16	4.45	26.25
C21H43F108	442.57	442.2942	56.99	9.79	4.29	28.92
C21H43F109	458.57	458.2891	55.00	9.45	4.14	31.40
C22H11F101	310.33	310.0794	85.15	3.57	6.12	5.16
C22H11F102	326.33	326.0743	80.97	3.40	5.82	9.81
C22H11F103	342.33	342.0692	77.19	3.24	5.55	14.02
C22H11F104	358.33	358.0641	73.74	3.09	5.30	17.86
C22H11F105	374.33	374.0590	70.59	2.96	5.08	21.37
C22H11F106	390.33	390.0540	67.70	2.84	4.87	24.59
C22H11F107	406.33	406.0489	65.03	2.73	4.68	27.56
C22H11F108	422.33	422.0438	62.57	2.63	4.50	30.31
C22H11F109	438.33	438.0387	60.29	2.53	4.33	32.85
C22H12F201	330.34	330.0856	79.99	3.66	11.50	4.84
C22H12F202	346.34	346.0805	76.30	3.49	10.97	9.24
C22H12F203	362.34	362.0754	72.93	3.34	10.49	13.25
C22H12F204	378.34	378.0704	69.84	3.20	10.04	16.92
C22H12F205	394.33	394.0653	67.01	3.07	9.64	20.29
C22H12F206	410.33	410.0602	64.40	2.95	9.26	23.39
C22H12F207	426.33	426.0551	61.98	2.84	8.91	26.27
C22H12F208	442.33	442.0500	59.74	2.73	8.59	28.94
C22H12F209	458.33	458.0449	57.65	2.64	8.29	31.42
C22H13F101	312.35	312.0950	84.60	4.20	6.08	5.12
C22H13F102	328.35	328.0900	80.48	3.99	5.79	9.75
C22H13F103	344.35	344.0849	76.74	3.81	5.52	13.94
C22H13F104	360.34	360.0798	73.33	3.64	5.27	17.76
C22H13F105	376.34	376.0747	70.21	3.48	5.05	21.26
C22H13F106	392.34	392.0696	67.35	3.34	4.84	24.47
C22H13F107	408.34	408.0645	64.71	3.21	4.65	27.43
C22H13F108	424.34	424.0594	62.27	3.09	4.48	30.16
C22H13F109	440.34	440.0543	60.01	2.98	4.31	32.70
C22H14F201	332.35	332.1013	79.51	4.25	11.43	4.81
C22H14F202	348.35	348.0962	75.86	4.05	10.91	9.19
C22H14F203	364.35	364.0911	72.52	3.87	10.43	13.17
C22H14F204	380.35	380.0860	69.47	3.71	9.99	16.83
C22H14F205	396.35	396.0809	66.67	3.56	9.39	20.18
C22H14F206	412.35	412.0758	64.08	3.42	9.21	23.28
C22H14F207	428.35	428.0707	61.69	3.29	8.87	26.15

155

C22H14F208	444.35	444.0657	59.47	3.18	8.55	28.81
C22H14F209	460.35	460.0606	57.40	3.07	8.25	31.28
C22H15F101	314.36	314.1107	84.06	4.81	6.04	5.09
C22H15F102	330.36	330.1056	79.99	4.58	5.75	9.69
C22H15F103	346.36	346.1005	76.29	4.37	5.49	13.86
C22H15F104	362.36	362.0954	72.92	4.17	5.24	17.66
C22H15F105	378.36	378.0903	69.84	4.00	5.02	21.14
C22H15F106	394.36	394.0853	67.01	3.83	4.82	24.34
C22H15F107	410.36	410.0802	64.39	3.68	4.63	27.29
C22H15F108	426.36	426.0751	61.98	3.55	4.46	30.02
C22H15F109	442.36	442.0700	59.74	3.42	4.29	32.55
C22H16F201	334.37	334.1169	79.03	4.82	11.36	4.78
C22H16F202	350.37	350.1118	75.42	4.60	10.84	9.13
C22H16F203	366.37	366.1067	72.13	4.40	10.37	13.10
C22H16F204	382.37	382.1017	69.11	4.22	9.94	16.74
C22H16F205	398.37	398.0966	66.33	4.05	9.54	20.08
C22H16F206	414.37	414.0915	63.77	3.89	9.17	23.17
C22H16F207	430.37	430.0864	61.40	3.75	8.83	26.02
C22H16F208	446.36	446.0813	59.20	3.61	8.51	28.68
C22H16F209	462.36	462.0762	57.15	3.49	8.22	31.14
C22H17F101	316.38	316.1263	83.52	5.42	6.00	5.06
C22H17F102	332.38	332.1212	79.50	5.16	5.72	9.63
C22H17F103	348.38	348.1162	75.85	4.92	5.45	13.78
C22H17F104	364.38	364.1111	72.52	4.70	5.21	17.56
C22H17F105	380.38	380.1060	69.47	4.50	4.99	21.03
C22H17F106	396.38	396.1009	66.67	4.32	4.79	24.22
C22H17F107	412.37	412.0958	64.08	4.16	4.61	27.16
C22H17F108	428.37	428.0907	61.69	4.00	4.43	29.88
C22H17F109	444.37	444.0856	59.46	3.86	4.28	32.40
C22H18F201	336.38	336.1326	78.55	5.39	11.30	4.76
C22H18F202	352.38	352.1275	74.99	5.15	10.78	9.08
C22H18F203	368.38	368.1224	71.73	4.93	10.31	13.03
C22H18F204	384.38	384.1173	68.75	4.72	9.89	16.65
C22H18F205	400.38	400.1122	66.00	4.53	9.49	19.98
C22H18F206	416.38	416.1071	63.46	4.36	9.13	23.05
C22H18F207	432.38	432.1020	61.11	4.20	8.79	25.90
C22H18F208	448.38	448.0970	58.93	4.05	8.47	28.55
C22H18F209	464.38	464.0919	56.90	3.91	8.18	31.01
C22H19F101	318.39	318.1420	82.99	6.01	5.97	5.03
C22H19F102	334.39	334.1369	79.02	5.73	5.68	9.57
C22H19F103	350.39	350.1318	75.41	5.47	5.42	13.70
C22H19F104	366.39	366.1267	72.12	5.23	5.19	17.47
C22H19F105	382.39	382.1216	69.10	5.01	4.97	20.92
C22H19F106	398.39	398.1166	66.33	4.81	4.77	24.10
C22H19F107	414.39	414.1115	63.77	4.62	4.58	27.03
C22H19F108	430.39	430.1064	61.40	4.45	4.41	29.74
C22H19F109	446.39	446.1013	59.20	4.29	4.26	32.26
C22H20F201	338.40	338.1482	78.09	5.96	11.23	4.73
C22H20F202	354.40	354.1431	74.56	5.69	10.72	9.03
C22H20F203	370.40	370.1380	71.34	5.44	10.26	12.96
C22H20F204	386.40	386.1330	68.39	5.22	9.83	16.56
C22H20F205	402.40	402.1279	65.67	5.01	9.44	19.88
C22H20F206	418.40	418.1228	63.16	4.82	9.08	22.94
C22H20F207	434.40	434.1177	60.83	4.64	8.75	25.78
C22H20F208	450.40	450.1126	58.67	4.48	8.44	28.42
C22H20F209	466.40	466.1075	56.66	4.32	8.15	30.87
C22H21F101	320.41	320.1576	82.47	6.61	5.93	4.99
C22H21F102	336.41	336.1525	78.55	6.29	5.65	9.51
C22H21F103	352.41	352.1475	74.98	6.01	5.39	13.62
C22H21F104	368.41	368.1424	71.73	5.75	5.16	17.37
C22H21F105	384.41	384.1373	68.74	5.51	4.94	20.81
C22H21F106	400.41	400.1322	65.99	5.29	4.74	23.97
C22H21F107	416.41	416.1271	63.46	5.08	4.56	26.90
C22H21F108	432.41	432.1220	61.11	4.90	4.39	29.60
C22H21F109	448.41	448.1169	58.93	4.72	4.24	32.11
C22H22F201	340.42	340.1639	77.62	6.51	11.16	4.70
C22H22F202	356.42	356.1588	74.14	6.22	10.66	8.98
C22H22F203	372.42	372.1537	70.95	5.95	10.20	12.89
C22H22F204	388.42	388.1486	68.03	5.71	9.78	16.48
C22H22F205	404.41	404.1435	65.34	5.48	9.40	19.78
C22H22F206	420.41	420.1384	62.85	5.27	9.04	22.83
C22H22F207	436.41	436.1333	60.55	5.08	8.71	25.66
C22H22F208	452.41	452.1283	58.41	4.90	8.40	28.29
C22H22F209	468.41	468.1232	56.41	4.73	8.11	30.74
C22H23F101	322.43	322.1733	81.96	7.19	5.89	4.96
C22H23F102	338.43	338.1682	78.08	6.85	5.61	9.46
C22H23F103	354.43	354.1631	74.56	6.54	5.36	13.54
C22H23F104	370.42	370.1580	71.34	6.26	5.13	17.28
C22H23F105	386.42	386.1529	68.38	6.00	4.92	20.70
C22H23F106	402.42	402.1479	65.66	5.76	4.72	23.85
C22H23F107	418.42	418.1428	63.15	5.54	4.54	26.77
C22H23F108	434.42	434.1377	60.83	5.34	4.37	29.46
C22H23F109	450.42	450.1326	58.67	5.15	4.22	31.97
C22H24F201	342.43	342.1795	77.17	7.06	11.10	4.67
C22H24F202	358.43	358.1744	73.72	6.75	10.60	8.93
C22H24F203	374.43	374.1693	70.57	6.46	10.15	12.82
C22H24F204	390.43	390.1643	67.68	6.20	9.73	16.39
C22H24F205	406.43	406.1592	65.02	5.95	9.35	19.68
C22H24F206	422.43	422.1541	62.55	5.73	8.99	22.72
C22H24F207	438.43	438.1490	60.27	5.52	8.67	25.54
C22H24F208	454.43	454.1439	58.15	5.32	8.36	28.17
C22H24F209	470.43	470.1388	56.17	5.14	8.08	30.61
C22H25F101	324.44	324.1889	81.45	7.77	5.86	4.93
C22H25F102	340.44	340.1838	77.62	7.40	5.58	9.40
C22H25F103	356.44	356.1788	74.13	7.07	5.33	13.47
C22H25F104	372.44	372.1737	70.95	6.77	5.10	17.18
C22H25F105	388.44	388.1686	68.03	6.49	4.89	20.59
C22H25F106	404.44	404.1635	65.34	6.23	4.70	23.74
C22H25F107	420.44	420.1584	62.85	5.99	4.52	26.64
C22H25F108	436.44	436.1533	60.55	5.77	4.35	29.33
C22H25F109	452.44	452.1482	58.40	5.57	4.20	31.83
C22H26F201	344.45	344.1952	76.72	7.61	11.03	4.64
C22H26F202	360.45	360.1901	73.31	7.27	10.54	8.88
C22H26F203	376.45	376.1850	70.19	6.96	10.09	12.75
C22H26F204	392.45	392.1799	67.33	6.68	9.68	16.31
C22H26F205	408.45	408.1748	64.70	6.42	9.30	19.59
C22H26F206	424.45	424.1697	62.26	6.17	8.95	22.62
C22H26F207	440.45	440.1646	60.00	5.95	8.63	25.43
C22H26F208	456.44	456.1596	57.89	5.74	8.32	28.04
C22H26F209	472.44	472.1545	55.93	5.55	8.04	30.48
C22H27F101	326.46	326.2046	80.94	8.34	5.82	4.90
C22H27F102	342.46	342.1995	77.16	7.95	5.55	9.34
C22H27F103	358.46	358.1944	73.72	7.59	5.30	13.39
C22H27F104	374.46	374.1893	70.57	7.27	5.07	17.09
C22H27F105	390.46	390.1842	67.68	6.97	4.87	20.49
C22H27F106	406.46	406.1792	65.01	6.70	4.67	23.62
C22H27F107	422.45	422.1741	62.55	6.44	4.50	26.51
C22H27F108	438.45	438.1690	60.27	6.21	4.33	29.19
C22H27F109	454.45	454.1639	58.15	5.99	4.18	31.69
C22H28F201	346.46	346.2108	76.27	8.15	10.97	4.62

156

C22H28F202	362.46	362.2057	72.90	7.79	10.48	8.83
C22H28F203	378.46	378.2006	69.82	7.46	10.04	12.68
C22H28F204	394.46	394.1955	66.99	7.15	9.63	16.22
C22H28F205	410.46	410.1905	64.38	6.88	9.26	19.49
C22H28F206	426.46	426.1854	61.96	6.62	8.91	22.51
C22H28F207	442.46	442.1803	59.72	6.38	8.59	25.31
C22H28F208	458.46	458.1752	57.64	6.16	8.29	27.92
C22H28F209	474.46	474.1701	55.69	5.95	8.01	30.35
C22H29F101	328.47	328.2202	80.45	8.90	5.78	4.87
C22H29F102	344.47	344.2151	76.71	8.49	5.52	9.29
C22H29F103	360.47	360.2101	73.31	8.11	5.27	13.32
C22H29F104	376.47	376.2050	70.19	7.76	5.05	17.00
C22H29F105	392.47	392.1999	67.33	7.45	4.84	20.38
C22H29F106	408.47	408.1948	64.69	7.16	4.65	23.50
C22H29F107	424.47	424.1897	62.25	6.89	4.48	26.38
C22H29F108	440.47	440.1846	59.99	6.64	4.31	29.06
C22H29F109	456.47	456.1795	57.89	6.40	4.16	31.55
C22H30F201	348.48	348.2265	75.83	8.68	10.90	4.59
C22H30F202	364.48	364.2214	72.50	8.30	10.42	8.78
C22H30F203	380.48	380.2163	69.45	7.95	9.99	12.62
C22H30F204	396.48	396.2112	66.65	7.63	9.58	16.14
C22H30F205	412.48	412.2061	64.06	7.33	9.21	19.39
C22H30F206	428.48	428.2010	61.67	7.06	8.87	22.40
C22H30F207	444.48	444.1959	59.45	6.80	8.55	25.20
C22H30F208	460.48	460.1909	57.39	6.57	8.25	27.80
C22H30F209	476.48	476.1858	55.46	6.35	7.97	30.22
C22H31F101	330.49	330.2359	79.96	9.45	5.75	4.84
C22H31F102	346.49	346.2308	76.26	9.02	5.48	9.24
C22H31F103	362.49	362.2257	72.90	8.62	5.24	13.24
C22H31F104	378.49	378.2206	69.82	8.26	5.02	16.91
C22H31F105	394.49	394.2155	66.98	7.92	4.82	20.28
C22H31F106	410.49	410.2104	64.37	7.61	4.63	23.39
C22H31F107	426.49	426.2054	61.96	7.33	4.45	26.26
C22H31F108	442.49	442.2003	59.72	7.06	4.29	28.93
C22H31F109	458.49	458.1952	57.63	6.82	4.14	31.41
C22H32F201	350.50	350.2421	75.39	9.20	10.84	4.56
C22H32F202	366.50	366.2370	72.10	8.80	10.37	8.73
C22H32F203	382.50	382.2319	69.08	8.43	9.93	12.55
C22H32F204	398.49	398.2268	66.31	8.09	9.54	16.06
C22H32F205	414.49	414.2218	63.75	7.78	9.17	19.30
C22H32F206	430.49	430.2167	61.38	7.49	8.83	22.30
C22H32F207	446.49	446.2116	59.18	7.22	8.51	25.08
C22H32F208	462.49	462.2065	57.14	6.97	8.22	27.68
C22H32F209	478.49	478.2014	55.22	6.74	7.94	30.09
C22H33F101	332.51	332.2515	79.47	10.00	5.71	4.81
C22H33F102	348.51	348.2464	75.82	9.54	5.45	9.18
C22H33F103	364.50	364.2414	72.49	9.13	5.21	13.17
C22H33F104	380.50	380.2363	69.45	8.74	4.99	16.82
C22H33F105	396.50	396.2312	66.64	8.39	4.79	20.18
C22H33F106	412.50	412.2261	64.06	8.06	4.61	23.27
C22H33F107	428.50	428.2210	61.67	7.76	4.43	26.14
C22H33F108	444.50	444.2159	59.45	7.48	4.27	28.80
C22H33F109	460.50	460.2108	57.38	7.22	4.13	31.27
C22H34F201	352.51	352.2578	74.96	9.72	10.78	4.54
C22H34F202	368.51	368.2527	71.71	9.30	10.31	8.68
C22H34F203	384.51	384.2476	68.72	8.91	9.88	12.48
C22H34F204	400.51	400.2425	65.98	8.56	9.49	15.98
C22H34F205	416.51	416.2374	63.44	8.23	9.12	19.21
C22H34F206	432.51	432.2323	61.10	7.92	8.79	22.20
C22H34F207	448.51	448.2272	58.92	7.64	8.47	24.97
C22H34F208	464.51	464.2222	56.89	7.38	8.18	27.55
C22H34F209	480.51	480.2171	54.99	7.13	7.91	29.97
C22H35F101	334.52	334.2672	78.99	10.55	5.68	4.78
C22H35F102	350.52	350.2621	75.39	10.06	5.42	9.13
C22H35F103	366.52	366.2570	72.10	9.63	5.18	13.10
C22H35F104	382.52	382.2519	69.08	9.22	4.97	16.73
C22H35F105	398.52	398.2468	66.31	8.85	4.77	20.07
C22H35F106	414.52	414.2417	63.75	8.51	4.58	23.16
C22H35F107	430.52	430.2367	61.38	8.19	4.41	26.01
C22H35F108	446.52	446.2316	59.18	7.90	4.25	28.67
C22H35F109	462.52	462.2265	57.13	7.63	4.11	31.13
C22H36F201	354.53	354.2734	74.53	10.24	10.72	4.51
C22H36F202	370.53	370.2683	71.32	9.79	10.25	8.64
C22H36F203	386.53	386.2632	68.36	9.39	9.83	12.42
C22H36F204	402.53	402.2581	65.65	9.01	9.44	15.90
C22H36F205	418.53	418.2531	63.14	8.67	9.08	19.11
C22H36F206	434.53	434.2480	60.81	8.35	8.74	22.09
C22H36F207	450.52	450.2429	58.65	8.05	8.43	24.86
C22H36F208	466.52	466.2378	56.64	7.78	8.14	27.44
C22H36F209	482.52	482.2327	54.76	7.52	7.87	29.84
C22H37F101	336.54	336.2828	78.52	11.08	5.65	4.75
C22H37F102	352.54	352.2777	74.96	10.58	5.39	9.08
C22H37F103	368.54	368.2727	71.70	10.12	5.16	13.02
C22H37F104	384.54	384.2676	68.72	9.70	4.94	16.64
C22H37F105	400.54	400.2625	65.97	9.31	4.74	19.97
C22H37F106	416.53	416.2574	63.44	8.95	4.56	23.05
C22H37F107	432.53	432.2523	61.09	8.62	4.39	25.89
C22H37F108	448.53	448.2472	58.91	8.31	4.24	28.54
C22H37F109	464.53	464.2421	56.88	8.03	4.09	31.00
C22H38F201	356.54	356.2891	74.11	10.74	10.66	4.49
C22H38F202	372.54	372.2840	70.93	10.28	10.20	8.59
C22H38F203	388.54	388.2789	68.01	9.86	9.78	12.35
C22H38F204	404.54	404.2738	65.32	9.47	9.39	15.82
C22H38F205	420.54	420.2687	62.83	9.11	9.04	19.02
C22H38F206	436.54	436.2636	60.53	8.77	8.70	21.99
C22H38F207	452.54	452.2585	58.39	8.46	8.40	24.75
C22H38F208	468.54	468.2535	56.40	8.17	8.11	27.32
C22H38F209	484.54	484.2484	54.54	7.91	7.84	29.72
C22H39F101	338.55	338.2985	78.05	11.61	5.61	4.73
C22H39F102	354.55	354.2934	74.53	11.09	5.36	9.03
C22H39F103	370.55	370.2883	71.31	10.61	5.13	12.95
C22H39F104	386.55	386.2832	68.36	10.17	4.91	16.56
C22H39F105	402.55	402.2781	65.64	9.77	4.72	19.87
C22H39F106	418.55	418.2730	63.13	9.39	4.54	22.94
C22H39F107	434.55	434.2680	60.81	9.05	4.37	25.77
C22H39F108	450.55	450.2629	58.65	8.73	4.22	28.41
C22H39F109	466.55	466.2578	56.64	8.43	4.07	30.86
C22H40F201	358.56	358.3047	73.70	11.24	10.60	4.46
C22H40F202	374.56	374.2996	70.55	10.76	10.14	8.54
C22H40F203	390.56	390.2945	67.66	10.32	9.73	12.29
C22H40F204	406.56	406.2894	65.00	9.92	9.35	15.74
C22H40F205	422.56	422.2844	62.53	9.54	8.99	18.93
C22H40F206	438.56	438.2793	60.25	9.19	8.66	21.89
C22H40F207	454.56	454.2742	58.13	8.87	8.36	24.64
C22H40F208	470.56	470.2691	56.16	8.57	8.07	27.20
C22H40F209	486.56	486.2640	54.31	8.29	7.81	29.59
C22H41F101	340.57	340.3141	77.59	12.13	5.58	4.70
C22H41F102	356.57	356.3090	74.11	11.59	5.33	8.97
C22H41F103	372.57	372.3040	70.93	11.09	5.10	12.88
C22H41F104	388.57	388.2989	68.00	10.64	4.89	16.47

C22H41F105	404.57	404.2938	65.32	10.22	4.70	19.77
C22H41F106	420.57	420.2887	62.83	9.83	4.52	22.83
C22H41F107	436.57	436.2836	60.53	9.47	4.35	25.65
C22H41F108	452.57	452.2785	58.39	9.13	4.20	28.28
C22H41F109	468.57	468.2734	56.39	8.82	4.05	30.73
C22H42F201	360.58	360.3204	73.28	11.74	10.54	4.44
C22H42F202	376.58	376.3153	70.17	11.24	10.09	8.50
C22H42F203	392.58	392.3102	67.31	10.78	9.68	12.23
C22H42F204	408.57	408.3051	64.67	10.36	9.30	15.66
C22H42F205	424.57	424.3000	62.24	9.97	8.95	18.84
C22H42F206	440.57	440.2949	59.98	9.61	8.62	21.79
C22H42F207	456.57	456.2898	57.88	9.27	8.32	24.53
C22H42F208	472.57	472.2848	55.92	8.96	8.04	27.08
C22H42F209	488.57	488.2797	54.09	8.67	7.78	29.47
C22H43F101	342.59	342.3298	77.13	12.65	5.55	4.67
C22H43F102	358.59	358.3247	73.69	12.09	5.30	8.92
C22H43F103	374.58	374.3196	70.54	11.57	5.07	12.81
C22H43F104	390.58	390.3145	67.65	11.10	4.86	16.39
C22H43F105	406.58	406.3094	64.99	10.66	4.67	19.68
C22H43F106	422.58	422.3043	62.53	10.26	4.50	22.72
C22H43F107	438.58	438.2993	60.25	9.88	4.33	25.54
C22H43F108	454.58	454.2942	58.13	9.53	4.18	28.16
C22H43F109	470.58	470.2891	56.15	9.21	4.04	30.60
C22H44F201	362.59	362.3360	72.88	12.23	10.48	4.41
C22H44F202	378.59	378.3309	69.80	11.71	10.04	8.45
C22H44F203	394.59	394.3258	66.97	11.24	9.63	12.16
C22H44F204	410.59	410.3207	64.36	10.80	9.25	15.59
C22H44F205	426.59	426.3157	61.94	10.40	8.91	18.75
C22H44F206	442.59	442.3106	59.70	10.02	8.59	21.69
C22H44F207	458.59	458.3055	57.62	9.67	8.29	24.42
C22H44F208	474.59	474.3004	55.68	9.35	8.01	26.97
C22H44F209	490.59	490.2953	53.86	9.04	7.75	29.35
C22H45F101	344.60	344.3454	76.68	13.16	5.51	4.64
C22H45F102	360.60	360.3403	73.28	12.58	5.27	8.87
C22H45F103	376.60	376.3353	70.17	12.04	5.04	12.75
C22H45F104	392.60	392.3302	67.31	11.55	4.84	16.30
C22H45F105	408.60	408.3251	64.67	11.10	4.65	19.58
C22H45F106	424.60	424.3200	62.23	10.68	4.47	22.61
C22H45F107	440.60	440.3149	59.97	10.29	4.31	25.42
C22H45F108	456.60	456.3098	57.87	9.93	4.16	28.03
C22H45F109	472.60	472.3047	55.91	9.60	4.02	30.47
C23H11F101	322.34	322.0794	85.70	3.44	5.89	4.96
C23H11F102	338.34	338.0743	81.65	3.28	5.62	9.46
C23H11F103	354.34	354.0692	77.96	3.13	5.36	13.55
C23H11F104	370.34	370.0641	74.60	2.99	5.13	17.28
C23H11F105	386.34	386.0590	71.51	2.87	4.92	20.71
C23H11F106	402.34	402.0540	68.66	2.76	4.72	23.86
C23H11F107	418.34	418.0489	66.04	2.65	4.54	26.77
C23H11F108	434.34	434.0438	63.60	2.55	4.37	29.47
C23H11F109	450.34	450.0387	61.34	2.46	4.22	31.97
C23H12F201	342.35	342.0856	80.69	3.53	11.10	4.67
C23H12F202	358.35	358.0805	77.09	3.38	10.60	8.93
C23H12F203	374.35	374.0754	73.80	3.23	10.15	12.82
C23H12F204	390.35	390.0704	70.77	3.10	9.73	16.40
C23H12F205	406.35	406.0653	67.95	2.98	9.35	19.69
C23H12F206	422.35	422.0602	65.41	2.86	9.00	22.73
C23H12F207	438.34	438.0551	63.02	2.76	8.67	25.55
C23H12F208	454.34	454.0500	60.80	2.66	8.36	28.17
C23H12F209	470.34	470.0449	58.74	2.57	8.08	30.61
C23H13F101	324.36	324.0950	85.17	4.04	5.86	4.93

C23H13F102	340.36	340.0900	81.17	3.85	5.58	9.40
C23H13F103	356.36	356.0849	77.52	3.68	5.33	13.47
C23H13F104	372.36	372.0798	74.19	3.52	5.10	17.19
C23H13F105	388.36	388.0747	71.13	3.37	4.89	20.60
C23H13F106	404.35	404.0696	68.32	3.24	4.70	23.74
C23H13F107	420.35	420.0645	65.72	3.12	4.52	26.64
C23H13F108	436.35	436.0594	63.31	3.00	4.35	29.33
C23H13F109	452.35	452.0543	61.07	2.90	4.20	31.83
C23H14F201	344.36	344.1013	80.22	4.10	11.03	4.65
C23H14F202	360.36	360.0962	76.66	3.92	10.54	8.88
C23H14F203	376.36	376.0911	73.40	3.75	10.10	12.75
C23H14F204	392.36	392.0860	70.41	3.60	9.68	16.31
C23H14F205	408.36	408.0809	67.65	3.46	9.30	19.59
C23H14F206	424.36	424.0758	65.10	3.33	8.95	22.62
C23H14F207	440.36	440.0707	62.73	3.20	8.63	25.43
C23H14F208	456.36	456.0657	60.53	3.09	8.33	28.05
C23H14F209	472.36	472.0606	58.48	2.99	8.04	30.48
C23H15F101	326.37	326.1107	84.64	4.63	5.82	4.90
C23H15F102	342.37	342.1056	80.69	4.42	5.55	9.35
C23H15F103	358.37	358.1005	77.09	4.22	5.30	13.39
C23H15F104	374.37	374.0954	73.79	4.04	5.07	17.09
C23H15F105	390.37	390.0903	70.77	3.87	4.87	20.49
C23H15F106	406.37	406.0853	67.98	3.72	4.68	23.62
C23H15F107	422.37	422.0802	65.41	3.58	4.50	26.52
C23H15F108	438.37	438.0751	63.02	3.45	4.33	29.20
C23H15F109	454.37	454.0700	60.80	3.33	4.18	31.69
C23H16F201	346.38	346.1169	79.76	4.66	10.97	4.62
C23H16F202	362.38	362.1118	76.23	4.45	10.49	8.83
C23H16F203	378.38	378.1067	73.01	4.26	10.04	12.69
C23H16F204	394.38	394.1017	70.05	4.09	9.63	16.23
C23H16F205	410.38	410.0966	67.32	3.93	9.26	19.49
C23H16F206	426.38	426.0915	64.79	3.78	8.91	22.51
C23H16F207	442.38	442.0864	62.45	3.65	8.59	25.32
C23H16F208	458.38	458.0813	60.27	3.52	8.29	27.92
C23H16F209	474.38	474.0762	58.24	3.40	8.01	30.35
C23H17F101	328.39	328.1263	84.12	5.22	5.79	4.87
C23H17F102	344.39	344.1212	80.22	4.98	5.52	9.29
C23H17F103	360.39	360.1162	76.66	4.75	5.27	13.32
C23H17F104	376.39	376.1111	73.40	4.55	5.05	17.00
C23H17F105	392.39	392.1060	70.40	4.37	4.84	20.39
C23H17F106	408.39	408.1009	67.65	4.20	4.65	23.51
C23H17F107	424.39	424.0958	65.10	4.04	4.48	26.39
C23H17F108	440.39	440.0907	62.73	3.89	4.31	29.06
C23H17F109	456.38	456.0856	60.53	3.75	4.16	31.55
C23H18F201	348.40	348.1326	79.29	5.21	10.91	4.59
C23H18F202	364.40	364.1275	75.81	4.98	10.43	8.78
C23H18F203	380.39	380.1224	72.62	4.77	9.99	12.62
C23H18F204	396.39	396.1173	69.69	4.58	9.59	16.14
C23H18F205	412.39	412.1122	66.99	4.40	9.21	19.40
C23H18F206	428.39	428.1071	64.49	4.24	8.87	22.41
C23H18F207	444.39	444.1020	62.16	4.08	8.55	25.20
C23H18F208	460.39	460.0970	60.00	3.94	8.25	27.80
C23H18F209	476.39	476.0919	57.99	3.81	7.98	30.23
C23H19F101	330.41	330.1420	83.61	5.80	5.75	4.84
C23H19F102	346.41	346.1369	79.75	5.53	5.48	9.24
C23H19F103	362.40	362.1318	76.23	5.28	5.24	13.24
C23H19F104	378.40	378.1267	73.01	5.06	5.02	16.91
C23H19F105	394.40	394.1216	70.04	4.86	4.82	20.28
C23H19F106	410.40	410.1166	67.31	4.67	4.63	23.39
C23H19F107	426.40	426.1115	64.79	4.49	4.46	26.27

158

C23H19F108	442.40	442.1064	62.44	4.33	4.29	28.93
C23H19F109	458.40	458.1013	60.27	4.18	4.14	31.41
C23H20F201	350.41	350.1482	78.84	5.75	10.84	4.57
C23H20F202	366.41	366.1431	75.40	5.50	10.37	8.73
C23H20F203	382.41	382.1380	72.24	5.27	9.94	12.55
C23H20F204	398.41	398.1330	69.34	5.06	9.54	16.06
C23H20F205	414.41	414.1279	66.66	4.86	9.17	19.30
C23H20F206	430.41	430.1228	64.18	4.68	8.83	22.30
C23H20F207	446.41	446.1177	61.88	4.52	8.51	25.09
C23H20F208	462.41	462.1126	59.74	4.36	8.22	27.68
C23H20F209	478.41	478.1075	57.75	4.21	7.94	30.10
C23H21F101	332.42	332.1576	83.10	6.37	5.72	4.81
C23H21F102	348.42	348.1525	79.29	6.08	5.45	9.18
C23H21F103	364.42	364.1475	75.81	5.81	5.21	13.17
C23H21F104	380.42	380.1424	72.62	5.56	4.99	16.82
C23H21F105	396.42	396.1373	69.69	5.34	4.79	20.18
C23H21F106	412.42	412.1322	66.98	5.13	4.61	23.28
C23H21F107	428.42	428.1271	64.48	4.94	4.43	26.14
C23H21F108	444.42	444.1220	62.16	4.76	4.27	28.80
C23H21F109	460.42	460.1169	60.00	4.60	4.13	31.27
C23H22F201	352.43	352.1639	78.39	6.29	10.78	4.54
C23H22F202	368.43	368.1588	74.98	6.02	10.31	8.69
C23H22F203	384.43	384.1537	71.86	5.77	9.88	12.49
C23H22F204	400.43	400.1486	68.99	5.54	9.49	15.98
C23H22F205	416.43	416.1435	66.34	5.33	9.12	19.21
C23H22F206	432.42	432.1384	63.89	5.13	8.79	22.20
C23H22F207	448.42	448.1333	61.61	4.95	8.47	24.98
C23H22F208	464.42	464.1283	59.48	4.77	8.18	27.56
C23H22F209	480.42	480.1232	57.50	4.62	7.91	29.97
C23H23F101	334.44	334.1733	82.60	6.93	5.68	4.78
C23H23F102	350.44	350.1682	78.83	6.62	5.42	9.13
C23H23F103	366.44	366.1631	75.39	6.33	5.18	13.10
C23H23F104	382.44	382.1580	72.24	6.06	4.97	16.73
C23H23F105	398.44	398.1529	69.34	5.82	4.77	20.08
C23H23F106	414.43	414.1479	66.66	5.59	4.58	23.16
C23H23F107	430.43	430.1428	64.18	5.39	4.41	26.02
C23H23F108	446.43	446.1377	61.88	5.19	4.26	28.67
C23H23F109	462.43	462.1326	59.74	5.01	4.11	31.14
C23H24F201	354.44	354.1795	77.94	6.83	10.72	4.51
C23H24F202	370.44	370.1744	74.57	6.53	10.26	8.64
C23H24F203	386.44	386.1693	71.49	6.26	9.83	12.42
C23H24F204	402.44	402.1643	68.65	6.01	9.44	15.90
C23H24F205	418.44	418.1592	66.02	5.78	9.08	19.12
C23H24F206	434.44	434.1541	63.59	5.57	8.75	22.10
C23H24F207	450.44	450.1490	61.33	5.37	8.44	24.86
C23H24F208	466.44	466.1439	59.23	5.19	8.15	27.44
C23H24F209	482.44	482.1388	57.26	5.01	7.88	29.85
C23H25F101	336.45	336.1889	82.11	7.49	5.65	4.76
C23H25F102	352.45	352.1838	78.38	7.15	5.39	9.08
C23H25F103	368.45	368.1788	74.98	6.84	5.16	13.03
C23H25F104	384.45	384.1737	71.86	6.55	4.94	16.65
C23H25F105	400.45	400.1686	68.99	6.29	4.74	19.98
C23H25F106	416.45	416.1635	66.34	6.05	4.56	23.05
C23H25F107	432.45	432.1584	63.88	5.83	4.39	25.90
C23H25F108	448.45	448.1533	61.60	5.62	4.24	28.54
C23H25F109	464.45	464.1482	59.48	5.43	4.09	31.00
C23H26F201	356.46	356.1952	77.50	7.35	10.66	4.49
C23H26F202	372.46	372.1901	74.17	7.04	10.20	8.59
C23H26F203	388.46	388.1850	71.12	6.75	9.78	12.36
C23H26F204	404.46	404.1799	68.30	6.48	9.39	15.82

C23H26F205	420.46	420.1748	65.70	6.23	9.04	19.03
C23H26F206	436.46	436.1697	63.30	6.00	8.71	21.99
C23H26F207	452.46	452.1646	61.06	5.79	8.40	24.75
C23H26F208	468.46	468.1596	58.97	5.59	8.11	27.32
C23H26F209	484.46	484.1545	57.02	5.41	7.84	29.72
C23H27F101	338.47	338.2046	81.62	8.04	5.61	4.73
C23H27F102	354.47	354.1995	77.94	7.68	5.36	9.03
C23H27F103	370.47	370.1944	74.57	7.35	5.13	12.96
C23H27F104	386.47	386.1893	71.48	7.04	4.92	16.56
C23H27F105	402.47	402.1842	68.64	6.76	4.72	19.88
C23H27F106	418.47	418.1792	66.02	6.50	4.54	22.94
C23H27F107	434.47	434.1741	63.59	6.26	4.37	25.78
C23H27F108	450.47	450.1690	61.33	6.04	4.22	28.41
C23H27F109	466.46	466.1639	59.22	5.83	4.07	30.87
C23H28F201	358.48	358.2108	77.06	7.87	10.60	4.46
C23H28F202	374.48	374.2057	73.77	7.54	10.15	8.54
C23H28F203	390.47	390.2006	70.75	7.23	9.73	12.29
C23H28F204	406.47	406.1955	67.96	6.94	9.35	15.74
C23H28F205	422.47	422.1905	65.39	6.68	8.99	18.94
C23H28F206	438.47	438.1854	63.00	6.44	8.67	21.89
C23H28F207	454.47	454.1803	60.79	6.21	8.36	24.64
C23H28F208	470.47	470.1752	58.72	6.00	8.08	27.21
C23H28F209	486.47	486.1701	56.79	5.80	7.81	29.60
C23H29F101	340.49	340.2202	81.14	8.59	5.58	4.70
C23H29F102	356.48	356.2151	77.49	8.20	5.33	8.98
C23H29F103	372.48	372.2101	74.17	7.85	5.10	12.89
C23H29F104	388.48	388.2050	71.11	7.52	4.89	16.47
C23H29F105	404.48	404.1999	68.30	7.23	4.70	19.78
C23H29F106	420.48	420.1948	65.70	6.95	4.52	22.83
C23H29F107	436.48	436.1897	63.29	6.70	4.35	25.66
C23H29F108	452.48	452.1846	61.05	6.46	4.20	28.29
C23H29F109	468.48	468.1795	58.97	6.24	4.06	30.74
C23H30F201	360.49	360.2265	76.63	8.39	10.54	4.44
C23H30F202	376.49	376.2214	73.38	8.03	10.09	8.50
C23H30F203	392.49	392.2163	70.39	7.70	9.68	12.23
C23H30F204	408.49	408.2112	67.63	7.40	9.30	15.67
C23H30F205	424.49	424.2061	65.08	7.12	8.95	18.85
C23H30F206	440.49	440.2010	62.72	6.86	8.63	21.79
C23H30F207	456.49	456.1959	60.52	6.62	8.32	24.53
C23H30F208	472.49	472.1909	58.47	6.40	8.04	27.09
C23H30F209	488.49	488.1858	56.55	6.19	7.78	29.48
C23H31F101	342.50	342.2359	80.66	9.12	5.55	4.67
C23H31F102	358.50	358.2308	77.06	8.72	5.30	8.93
C23H31F103	374.50	374.2257	73.77	8.34	5.07	12.82
C23H31F104	390.50	390.2206	70.74	8.00	4.87	16.39
C23H31F105	406.50	406.2155	67.96	7.69	4.67	19.68
C23H31F106	422.50	422.2104	65.39	7.40	4.50	22.72
C23H31F107	438.50	438.2054	63.00	7.13	4.33	25.54
C23H31F108	454.50	454.2003	60.78	6.88	4.18	28.16
C23H31F109	470.50	470.1952	58.72	6.64	4.04	30.60
C23H32F201	362.51	362.2421	76.21	8.90	10.48	4.41
C23H32F202	378.51	378.2370	72.99	8.52	10.04	8.45
C23H32F203	394.51	394.2319	70.03	8.18	9.63	12.17
C23H32F204	410.51	410.2268	67.30	7.86	9.26	15.59
C23H32F205	426.51	426.2218	64.77	7.56	8.91	18.76
C23H32F206	442.50	442.2167	62.43	7.29	8.59	21.69
C23H32F207	458.50	458.2116	60.25	7.03	8.29	24.43
C23H32F208	474.50	474.2065	58.22	6.80	8.01	26.97
C23H32F209	490.50	490.2014	56.32	6.58	7.75	29.36
C23H33F101	344.52	344.2515	80.19	9.65	5.51	4.64

159

C23H33F102	360.52	360.2464	76.63	9.23	5.27	8.88
C23H33F103	376.52	376.2414	73.37	8.83	5.05	12.75
C23H33F104	392.52	392.2363	70.38	8.47	4.84	16.30
C23H33F105	408.51	408.2312	67.62	8.14	4.65	19.58
C23H33F106	424.51	424.2261	65.08	7.84	4.48	22.61
C23H33F107	440.51	440.2210	62.71	7.55	4.31	25.42
C23H33F108	456.51	456.2159	60.51	7.29	4.16	28.04
C23H33F109	472.51	472.2108	58.47	7.04	4.02	30.47
C23H34F201	364.52	364.2578	75.79	9.40	10.42	4.39
C23H34F202	380.52	380.2527	72.60	9.01	9.99	8.41
C23H34F203	396.52	396.2476	69.67	8.64	9.58	12.10
C23H34F204	412.52	412.2425	66.97	8.31	9.21	15.51
C23H34F205	428.52	428.2374	64.47	8.00	8.87	18.67
C23H34F206	444.52	444.2323	62.15	7.71	8.55	21.60
C23H34F207	460.52	460.2272	59.99	7.44	8.25	24.32
C23H34F208	476.52	476.2222	57.97	7.19	7.97	26.86
C23H34F209	492.52	492.2171	56.09	6.96	7.71	29.24
C23H35F101	346.53	346.2672	79.72	10.18	5.48	4.62
C23H35F102	362.53	362.2621	76.20	9.73	5.24	8.83
C23H35F103	378.53	378.2570	72.98	9.32	5.02	12.68
C23H35F104	394.53	394.2519	70.02	8.94	4.82	16.22
C23H35F105	410.53	410.2468	67.29	8.59	4.63	19.49
C23H35F106	426.53	426.2417	64.77	8.27	4.45	22.51
C23H35F107	442.53	442.2367	62.43	7.97	4.29	25.31
C23H35F108	458.53	458.2316	60.25	7.69	4.14	27.91
C23H35F109	474.53	474.2265	58.22	7.43	4.00	30.34
C23H36F201	366.54	366.2734	75.37	9.90	10.37	4.36
C23H36F202	382.54	382.2683	72.22	9.49	9.93	8.36
C23H36F203	398.54	398.2632	69.32	9.11	9.53	12.04
C23H36F204	414.54	414.2581	66.64	8.75	9.17	15.44
C23H36F205	430.54	430.2531	64.17	8.43	8.83	18.58
C23H36F206	446.54	446.2480	61.87	8.13	8.51	21.50
C23H36F207	462.54	462.2429	59.73	7.85	8.21	24.21
C23H36F208	478.54	478.2378	57.73	7.58	7.94	26.75
C23H36F209	494.53	494.2327	55.86	7.34	7.68	29.12
C23H37F101	348.55	348.2828	79.26	10.70	5.45	4.59
C23H37F102	364.55	364.2777	75.78	10.23	5.21	8.78
C23H37F103	380.55	380.2727	72.59	9.80	4.99	12.61
C23H37F104	396.55	396.2676	69.67	9.40	4.79	16.14
C23H37F105	412.55	412.2625	66.96	9.04	4.61	19.39
C23H37F106	428.55	428.2574	64.46	8.70	4.43	22.40
C23H37F107	444.55	444.2523	62.14	8.39	4.27	25.19
C23H37F108	460.54	460.2472	59.98	8.10	4.13	27.79
C23H37F109	476.54	476.2421	57.97	7.83	3.99	30.22
C23H38F201	368.56	368.2891	74.96	10.39	10.31	4.34
C23H38F202	384.55	384.2840	71.84	9.96	9.88	8.32
C23H38F203	400.55	400.2789	68.97	9.56	9.49	11.98
C23H38F204	416.55	416.2738	66.32	9.20	9.12	15.36
C23H38F205	432.55	432.2687	63.87	8.86	8.78	18.49
C23H38F206	448.55	448.2636	61.59	8.54	8.47	21.40
C23H38F207	464.55	464.2585	59.47	8.25	8.18	24.11
C23H38F208	480.55	480.2535	57.49	7.97	7.91	26.64
C23H38F209	496.55	496.2484	55.64	7.71	7.65	29.00
C23H39F101	350.57	350.2985	78.80	11.21	5.42	4.56
C23H39F102	366.56	366.2934	75.36	10.72	5.18	8.73
C23H39F103	382.56	382.2883	72.21	10.28	4.97	12.55
C23H39F104	398.56	398.2832	69.31	9.86	4.77	16.06
C23H39F105	414.56	414.2781	66.64	9.48	4.58	19.30
C23H39F106	430.56	430.2730	64.16	9.13	4.41	22.30
C23H39F107	446.56	446.2680	61.86	8.80	4.25	25.08
C23H39F108	462.56	462.2629	59.72	8.50	4.11	27.67
C23H39F109	478.56	478.2578	57.73	8.21	3.97	30.09
C23H40F201	370.57	370.3047	74.55	10.88	10.25	4.32
C23H40F202	386.57	386.2996	71.46	10.43	9.83	8.28
C23H40F203	402.57	402.2945	68.62	10.02	9.44	11.92
C23H40F204	418.57	418.2894	66.00	9.63	9.08	15.29
C23H40F205	434.57	434.2844	63.57	9.28	8.74	18.41
C23H40F206	450.57	450.2793	61.31	8.95	8.43	21.31
C23H40F207	466.57	466.2742	59.21	8.64	8.14	24.00
C23H40F208	482.57	482.2691	57.25	8.36	7.87	26.52
C23H40F209	498.57	498.2640	55.41	8.09	7.62	28.88
C23H41F101	352.58	352.3141	78.35	11.72	5.39	4.54
C23H41F102	368.58	368.3090	74.95	11.21	5.15	8.68
C23H41F103	384.58	384.3040	71.83	10.75	4.94	12.48
C23H41F104	400.58	400.2989	68.96	10.32	4.74	15.98
C23H41F105	416.58	416.2938	66.32	9.92	4.56	19.20
C23H41F106	432.58	432.2887	63.86	9.55	4.39	22.19
C23H41F107	448.58	448.2836	61.59	9.21	4.24	24.97
C23H41F108	464.58	464.2785	59.46	8.90	4.09	27.55
C23H41F109	480.58	480.2734	57.48	8.60	3.95	29.96
C23H42F201	372.59	372.3204	74.15	11.36	10.20	4.29
C23H42F202	388.59	388.3153	71.09	10.89	9.78	8.23
C23H42F203	404.59	404.3102	68.28	10.46	9.39	11.86
C23H42F204	420.59	420.3051	65.68	10.07	9.03	15.22
C23H42F205	436.58	436.3000	63.28	9.70	8.70	18.32
C23H42F206	452.58	452.2949	61.04	9.35	8.40	21.21
C23H42F207	468.58	468.2898	58.96	9.03	8.11	23.90
C23H42F208	484.58	484.2848	57.01	8.74	7.84	26.41
C23H42F209	500.58	500.2797	55.19	8.46	7.59	28.77
C23H43F101	354.60	354.3298	77.91	12.22	5.36	4.51
C23H43F102	370.60	370.3247	74.54	11.70	5.13	8.63
C23H43F103	386.60	386.3196	71.46	11.21	4.91	12.42
C23H43F104	402.60	402.3145	68.62	10.77	4.72	15.90
C23H43F105	418.59	418.3094	66.00	10.35	4.54	19.11
C23H43F106	434.59	434.3043	63.57	9.97	4.37	22.09
C23H43F107	450.59	450.2993	61.31	9.62	4.22	24.86
C23H43F108	466.59	466.2942	59.21	9.29	4.07	27.43
C23H43F109	482.59	482.2891	57.24	8.98	3.94	29.84
C23H44F201	374.60	374.3360	73.75	11.84	10.14	4.27
C23H44F202	390.60	390.3309	70.73	11.35	9.73	8.19
C23H44F203	406.60	406.3258	67.94	10.91	9.34	11.80
C23H44F204	422.60	422.3207	65.37	10.49	8.99	15.14
C23H44F205	438.60	438.3157	62.99	10.11	8.66	18.24
C23H44F206	454.60	454.3106	60.77	9.76	8.36	21.12
C23H44F207	470.60	470.3055	58.70	9.42	8.07	23.80
C23H44F208	486.60	486.3004	56.77	9.11	7.81	26.30
C23H44F209	502.60	502.2953	54.97	8.82	7.56	28.65
C23H45F101	356.61	356.3454	77.47	12.72	5.33	4.49
C23H45F102	372.61	372.3403	74.14	12.17	5.10	8.59
C23H45F103	388.61	388.3353	71.09	11.67	4.89	12.35
C23H45F104	404.61	404.3302	68.28	11.21	4.70	15.82
C23H45F105	420.61	420.3251	65.68	10.78	4.52	19.02
C23H45F106	436.61	436.3200	63.27	10.39	4.35	21.99
C23H45F107	452.61	452.3149	61.04	10.02	4.20	24.74
C23H45F108	468.61	468.3098	58.95	9.68	4.05	27.31
C23H45F109	484.61	484.3047	57.01	9.36	3.92	29.71
C23H46F201	376.62	376.3517	73.35	12.31	10.09	4.25
C23H46F202	392.62	392.3466	70.36	11.81	9.68	8.15
C23H46F203	408.62	408.3415	67.61	11.35	9.30	11.75
C23H46F204	424.62	424.3364	65.06	10.92	8.95	15.07

160

C23H46F205	440.62	440.3313	62.70	10.52	8.62	18.16
C23H46F206	456.62	456.3262	60.50	10.15	8.32	21.02
C23H46F207	472.62	472.3211	58.45	9.81	8.04	23.70
C23H46F208	488.62	488.3160	56.54	9.49	7.78	26.20
C23H46F209	504.61	504.3110	54.75	9.19	7.53	28.54
C23H47F101	358.63	358.3611	77.03	13.21	5.30	4.46
C23H47F102	374.63	374.3560	73.74	12.65	5.07	8.54
C23H47F103	390.63	390.3509	70.72	12.13	4.86	12.29
C23H47F104	406.63	406.3458	67.94	11.65	4.67	15.74
C23H47F105	422.63	422.3407	65.37	11.21	4.50	18.93
C23H47F106	438.63	438.3356	62.98	10.80	4.33	21.89
C23H47F107	454.63	454.3306	60.77	10.42	4.18	24.63
C23H47F108	470.62	470.3255	58.70	10.07	4.04	27.20
C23H47F109	486.62	486.3204	56.77	9.74	3.90	29.59
C24H12F201	354.36	354.0856	81.35	3.41	10.72	4.52
C24H12F202	370.36	370.0805	77.83	3.27	10.26	8.64
C24H12F203	386.36	386.0754	74.61	3.13	9.83	12.42
C24H12F204	402.36	402.0704	71.64	3.01	9.44	15.91
C24H12F205	418.36	418.0653	68.90	2.89	9.08	19.12
C24H12F206	434.36	434.0602	66.37	2.78	8.75	22.10
C24H12F207	450.36	450.0551	64.01	2.69	8.44	24.87
C24H12F208	466.36	466.0500	61.81	2.59	8.15	27.45
C24H12F209	482.35	482.0449	59.76	2.51	7.88	29.85
C24H13F101	336.37	336.0950	85.70	3.90	5.65	4.76
C24H13F102	352.37	352.0900	81.81	3.72	5.39	9.08
C24H13F103	368.37	368.0849	78.26	3.56	5.16	13.03
C24H13F104	384.37	384.0798	75.00	3.41	4.94	16.65
C24H13F105	400.37	400.0747	72.00	3.27	4.75	19.98
C24H13F106	416.37	416.0696	69.23	3.15	4.56	23.06
C24H13F107	432.37	432.0645	66.67	3.03	4.39	25.90
C24H13F108	448.36	448.0594	64.29	2.92	4.24	28.55
C24H13F109	464.36	464.0543	62.08	2.82	4.09	31.01
C24H14F201	356.38	356.1013	80.89	3.96	10.66	4.49
C24H14F202	372.37	372.0962	77.41	3.79	10.20	8.59
C24H14F203	388.37	388.0911	74.22	3.63	9.78	12.36
C24H14F204	404.37	404.0860	71.29	3.49	9.40	15.83
C24H14F205	420.37	420.0809	68.57	3.36	9.04	19.03
C24H14F206	436.37	436.0758	66.06	3.23	8.71	22.00
C24H14F207	452.37	452.0707	63.72	3.12	8.40	24.76
C24H14F208	468.37	468.0657	61.55	3.01	8.11	27.33
C24H14F209	484.37	484.0606	59.51	2.91	7.84	29.73
C24H15F101	338.38	338.1107	85.19	4.47	5.61	4.73
C24H15F102	354.38	354.1056	81.34	4.27	5.36	9.03
C24H15F103	370.38	370.1005	77.83	4.08	5.13	12.96
C24H15F104	386.38	386.0954	74.61	3.91	4.92	16.56
C24H15F105	402.38	402.0903	71.64	3.76	4.72	19.88
C24H15F106	418.38	418.0853	68.90	3.61	4.54	22.94
C24H15F107	434.38	434.0802	66.36	3.48	4.37	25.78
C24H15F108	450.38	450.0751	64.01	3.36	4.22	28.42
C24H15F109	466.38	466.0700	61.81	3.24	4.07	30.87
C24H16F201	358.39	358.1169	80.43	4.50	10.60	4.46
C24H16F202	374.39	374.1118	77.00	4.31	10.15	8.55
C24H16F203	390.39	390.1067	73.84	4.13	9.73	12.29
C24H16F204	406.39	406.1017	70.93	3.97	9.35	15.75
C24H16F205	422.39	422.0966	68.25	3.82	9.00	18.94
C24H16F206	438.39	438.0915	65.76	3.68	8.67	21.90
C24H16F207	454.39	454.0864	63.44	3.55	8.36	24.65
C24H16F208	470.39	470.0813	61.28	3.43	8.08	27.21
C24H16F209	486.39	486.0762	59.27	3.32	7.81	29.60
C24H17F101	340.40	340.1263	84.68	5.03	5.58	4.70
C24H17F102	356.40	356.1212	80.88	4.81	5.33	8.98
C24H17F103	372.40	372.1162	77.41	4.60	5.10	12.89
C24H17F104	388.40	388.1111	74.22	4.41	4.89	16.48
C24H17F105	404.40	404.1060	71.28	4.24	4.70	19.78
C24H17F106	420.40	420.1009	68.57	4.08	4.52	22.83
C24H17F107	436.40	436.0958	66.06	3.93	4.35	25.66
C24H17F108	452.40	452.0907	63.72	3.79	4.20	28.29
C24H17F109	468.40	468.0856	61.54	3.66	4.06	30.74
C24H18F201	360.41	360.1326	79.98	5.03	10.54	4.44
C24H18F202	376.41	376.1275	76.58	4.82	10.09	8.50
C24H18F203	392.41	392.1224	73.46	4.62	9.68	12.23
C24H18F204	408.41	408.1173	70.58	4.44	9.30	15.67
C24H18F205	424.40	424.1122	67.92	4.28	8.95	18.85
C24H18F206	440.40	440.1071	65.46	4.12	8.63	21.80
C24H18F207	456.40	456.1020	63.16	3.98	8.33	24.54
C24H18F208	472.40	472.0970	61.02	3.84	8.04	27.09
C24H18F209	488.40	488.0919	59.02	3.71	7.78	29.48
C24H19F101	342.42	342.1420	84.19	5.59	5.55	4.67
C24H19F102	358.42	358.1369	80.43	5.34	5.30	8.93
C24H19F103	374.42	374.1318	76.99	5.12	5.07	12.82
C24H19F104	390.42	390.1267	73.84	4.91	4.87	16.39
C24H19F105	406.41	406.1216	70.93	4.71	4.67	19.68
C24H19F106	422.41	422.1166	68.24	4.53	4.50	22.73
C24H19F107	438.41	438.1115	65.75	4.37	4.33	25.55
C24H19F108	454.41	454.1064	63.44	4.21	4.18	28.17
C24H19F109	470.41	470.1013	61.28	4.07	4.04	30.61
C24H20F201	362.42	362.1482	79.54	5.56	10.48	4.41
C24H20F202	378.42	378.1431	76.18	5.33	10.04	8.46
C24H20F203	394.42	394.1380	73.09	5.11	9.63	12.17
C24H20F204	410.42	410.1330	70.24	4.91	9.26	15.59
C24H20F205	426.42	426.1279	67.60	4.73	8.91	18.76
C24H20F206	442.42	442.1228	65.16	4.56	8.59	21.70
C24H20F207	458.42	458.1177	62.88	4.40	8.29	24.43
C24H20F208	474.42	474.1126	60.76	4.25	8.01	26.98
C24H20F209	490.42	490.1075	58.78	4.11	7.75	29.36
C24H21F101	344.43	344.1576	83.69	6.15	5.52	4.65
C24H21F102	360.43	360.1525	79.98	5.87	5.27	8.88
C24H21F103	376.43	376.1475	76.58	5.62	5.05	12.75
C24H21F104	392.43	392.1424	73.46	5.39	4.84	16.31
C24H21F105	408.43	408.1373	70.58	5.18	4.65	19.59
C24H21F106	424.43	424.1322	67.92	4.99	4.48	22.62
C24H21F107	440.43	440.1271	65.45	4.81	4.31	25.43
C24H21F108	456.43	456.1220	63.16	4.64	4.16	28.04
C24H21F109	472.43	472.1169	61.02	4.48	4.02	30.48
C24H22F201	364.44	364.1639	79.10	6.08	10.43	4.39
C24H22F202	380.44	380.1588	75.77	5.83	9.99	8.41
C24H22F203	396.44	396.1537	72.71	5.59	9.58	12.11
C24H22F204	412.44	412.1486	69.89	5.38	9.21	15.52
C24H22F205	428.44	428.1435	67.28	5.18	8.87	18.67
C24H22F206	444.44	444.1384	64.86	4.99	8.55	21.60
C24H22F207	460.44	460.1333	62.61	4.82	8.25	24.32
C24H22F208	476.43	476.1283	60.51	4.65	7.98	26.87
C24H22F209	492.43	492.1232	58.54	4.50	7.72	29.24
C24H23F101	346.45	346.1733	83.21	6.69	5.48	4.62
C24H23F102	362.45	362.1682	79.53	6.40	5.24	8.83
C24H23F103	378.45	378.1631	76.17	6.13	5.02	12.68
C24H23F104	394.45	394.1580	73.08	5.88	4.82	16.22
C24H23F105	410.45	410.1529	70.23	5.65	4.63	19.49
C24H23F106	426.45	426.1479	67.60	5.44	4.46	22.51
C24H23F107	442.45	442.1428	65.15	5.24	4.29	25.31

C24H23F108	458.44	458.1377	62.88	5.06	4.14	27.92	C24H30F205	436.50	436.2061	66.04	6.93	8.70	18.33
C24H23F109	474.44	474.1326	60.76	4.89	4.00	30.35	C24H30F206	452.50	452.2010	63.71	6.68	8.40	21.21
C24H24F201	366.46	366.1795	78.66	6.60	10.37	4.37	C24H30F207	468.50	468.1959	61.53	6.45	8.11	23.91
C24H24F202	382.45	382.1744	75.37	6.33	9.93	8.37	C24H30F208	484.50	484.1909	59.50	6.24	7.84	26.42
C24H24F203	398.45	398.1693	72.35	6.07	9.54	12.05	C24H30F209	500.50	500.1858	57.60	6.04	7.59	28.77
C24H24F204	414.45	414.1643	69.55	5.84	9.17	15.44	C24H31F101	354.51	354.2359	81.31	8.81	5.36	4.51
C24H24F205	430.45	430.1592	66.97	5.62	8.83	18.58	C24H31F102	370.51	370.2308	77.80	8.43	5.13	8.64
C24H24F206	446.45	446.1541	64.57	5.42	8.51	21.50	C24H31F103	386.51	386.2257	74.58	8.08	4.92	12.42
C24H24F207	462.45	462.1490	62.33	5.23	8.22	24.22	C24H31F104	402.51	402.2206	71.62	7.76	4.72	15.90
C24H24F208	478.45	478.1439	60.25	5.06	7.94	26.75	C24H31F105	418.51	418.2155	68.88	7.47	4.54	19.11
C24H24F209	494.45	494.1388	58.30	4.89	7.68	29.12	C24H31F106	434.51	434.2104	66.34	7.19	4.37	22.09
C24H25F101	348.46	348.1889	82.73	7.23	5.45	4.59	C24H31F107	450.51	450.2054	63.99	6.94	4.22	24.86
C24H25F102	364.46	364.1838	79.09	6.91	5.21	8.78	C24H31F108	466.51	466.2003	61.79	6.70	4.07	27.44
C24H25F103	380.46	380.1788	75.77	6.62	4.99	12.62	C24H31F109	482.51	482.1952	59.74	6.48	3.94	29.84
C24H25F104	396.46	396.1737	72.71	6.36	4.79	16.14	C24H32F201	374.52	374.2421	76.97	8.61	10.15	4.27
C24H25F105	412.46	412.1686	69.89	6.11	4.61	19.39	C24H32F202	390.52	390.2370	73.82	8.26	9.73	8.19
C24H25F106	428.46	428.1635	67.28	5.88	4.43	22.40	C24H32F203	406.52	406.2319	70.91	7.93	9.35	11.81
C24H25F107	444.46	444.1584	64.86	5.67	4.27	25.20	C24H32F204	422.52	422.2268	68.23	7.63	8.99	15.15
C24H25F108	460.46	460.1533	62.60	5.47	4.13	27.80	C24H32F205	438.52	438.2218	65.74	7.36	8.66	18.24
C24H25F109	476.46	476.1482	60.50	5.29	3.99	30.22	C24H32F206	454.52	454.2167	63.42	7.10	8.36	21.12
C24H26F201	368.47	368.1952	78.23	7.11	10.31	4.34	C24H32F207	470.52	470.2116	61.27	6.86	8.08	23.80
C24H26F202	384.47	384.1901	74.98	6.82	9.88	8.32	C24H32F208	486.51	486.2065	59.25	6.63	7.81	26.31
C24H26F203	400.47	400.1850	71.98	6.54	9.49	11.99	C24H32F209	502.51	502.2014	57.37	6.42	7.56	28.65
C24H26F204	416.47	416.1799	69.22	6.29	9.12	15.37	C24H33F101	356.53	356.2515	80.85	9.33	5.33	4.49
C24H26F205	432.47	432.1748	66.66	6.06	8.79	18.50	C24H33F102	372.53	372.2464	77.38	8.93	5.10	8.59
C24H26F206	448.47	448.1697	64.28	5.84	8.47	21.41	C24H33F103	388.53	388.2414	74.19	8.56	4.89	12.35
C24H26F207	464.47	464.1646	62.06	5.64	8.18	24.11	C24H33F104	404.53	404.2363	71.26	8.22	4.70	15.82
C24H26F208	480.47	480.1596	60.00	5.45	7.91	26.64	C24H33F105	420.53	420.2312	68.55	7.91	4.52	19.02
C24H26F209	496.47	496.1545	58.06	5.28	7.65	29.00	C24H33F106	436.53	436.2261	66.04	7.62	4.35	21.99
C24H27F101	350.48	350.2046	82.25	7.77	5.42	4.56	C24H33F107	452.52	452.2210	63.70	7.35	4.20	24.75
C24H27F102	366.48	366.1995	78.66	7.43	5.18	8.73	C24H33F108	468.52	468.2159	61.53	7.10	4.05	27.32
C24H27F103	382.48	382.1944	75.37	7.12	4.97	12.55	C24H33F109	484.52	484.2108	59.50	6.87	3.92	29.72
C24H27F104	398.48	398.1893	72.34	6.83	4.77	16.06	C24H34F201	376.53	376.2578	76.56	9.10	10.09	4.25
C24H27F105	414.48	414.1842	69.55	6.57	4.58	19.30	C24H34F202	392.53	392.2527	73.44	8.73	9.68	8.15
C24H27F106	430.48	430.1792	66.96	6.32	4.41	22.30	C24H34F203	408.53	408.2476	70.56	8.39	9.30	11.75
C24H27F107	446.48	446.1741	64.56	6.10	4.26	25.08	C24H34F204	424.53	424.2425	67.90	8.07	8.95	15.07
C24H27F108	462.48	462.1690	62.33	5.88	4.11	27.68	C24H34F205	440.53	440.2374	65.44	7.78	8.63	18.16
C24H27F109	478.48	478.1639	60.25	5.69	3.97	30.09	C24H34F206	456.53	456.2323	63.14	7.51	8.32	21.03
C24H28F201	370.49	370.2108	77.81	7.62	10.26	4.32	C24H34F207	472.53	472.2272	61.00	7.25	8.04	23.70
C24H28F202	386.49	386.2057	74.55	7.30	9.83	8.28	C24H34F208	488.53	488.2222	59.01	7.02	7.78	26.20
C24H28F203	402.49	402.2006	71.62	7.01	9.44	11.93	C24H34F209	504.53	504.2171	57.14	6.79	7.53	28.54
C24H28F204	418.49	418.1955	68.88	6.74	9.08	15.29	C24H35F101	358.54	358.2672	80.40	9.84	5.30	4.46
C24H28F205	434.48	434.1905	66.35	6.50	8.75	18.41	C24H35F102	374.54	374.2621	76.97	9.42	5.07	8.54
C24H28F206	450.48	450.1854	63.95	6.27	8.43	21.31	C24H35F103	390.54	390.2570	73.81	9.03	4.86	12.29
C24H28F207	466.48	466.1803	61.80	6.05	8.15	24.01	C24H35F104	406.54	406.2519	70.91	8.68	4.67	15.74
C24H28F208	482.48	482.1752	59.75	5.85	7.88	26.53	C24H35F105	422.54	422.2468	68.22	8.35	4.50	18.93
C24H28F209	498.48	498.1701	57.83	5.66	7.62	28.89	C24H35F106	438.54	438.2417	65.73	8.04	4.33	21.89
C24H29F101	352.50	352.2202	81.78	8.29	5.39	4.54	C24H35F107	454.54	454.2367	63.42	7.76	4.18	24.64
C24H29F102	368.50	368.2151	78.23	7.93	5.16	8.68	C24H35F108	470.54	470.2316	61.26	7.50	4.04	27.20
C24H29F103	384.50	384.2101	74.97	7.60	4.94	12.48	C24H35F109	486.54	486.2265	59.25	7.25	3.90	29.60
C24H29F104	400.49	400.2050	71.98	7.30	4.74	15.98	C24H36F201	378.55	378.2734	76.15	9.59	10.04	4.23
C24H29F105	416.49	416.1999	69.21	7.02	4.56	19.21	C24H36F202	394.55	394.2683	73.06	9.20	9.63	8.11
C24H29F106	432.49	432.1948	66.65	6.76	4.39	22.20	C24H36F203	410.55	410.2632	70.22	8.84	9.26	11.69
C24H29F107	448.49	448.1897	64.27	6.52	4.24	24.97	C24H36F204	426.55	426.2581	67.58	8.51	8.91	15.00
C24H29F108	464.49	464.1846	62.06	6.29	4.09	27.56	C24H36F205	442.55	442.2531	65.14	8.20	8.59	18.08
C24H29F109	480.49	480.1795	59.99	6.08	3.95	29.97	C24H36F206	458.55	458.2480	62.87	7.91	8.29	20.93
C24H30F201	372.50	372.2265	77.39	8.12	10.20	4.30	C24H36F207	474.55	474.2429	60.75	7.65	8.01	23.60
C24H30F202	388.50	388.2214	74.20	7.78	9.78	8.24	C24H36F208	490.55	490.2378	58.76	7.40	7.75	26.09
C24H30F203	404.50	404.2163	71.26	7.48	9.39	11.87	C24H36F209	506.55	506.2327	56.91	7.16	7.50	28.43
C24H30F204	420.50	420.2112	68.55	7.19	9.04	15.22	C24H37F101	360.56	360.2828	79.95	10.34	5.27	4.44

C24H37F102	376.56	376.2777	76.55	9.90	5.05	8.50	C24H43F108	478.60	478.2942	60.23	9.06	3.97	26.74
C24H37F103	392.56	392.2727	73.43	9.50	4.84	12.23	C24H43F109	494.60	494.2891	58.28	8.76	3.84	29.11
C24H37F104	408.56	408.2676	70.56	9.13	4.65	15.66	C24H44F201	386.61	386.3360	74.56	11.47	9.83	4.14
C24H37F105	424.56	424.2625	67.90	8.78	4.47	18.84	C24H44F202	402.61	402.3309	71.60	11.02	9.44	7.95
C24H37F106	440.56	440.2574	65.43	8.47	4.31	21.79	C24H44F203	418.61	418.3258	68.86	10.59	9.08	11.47
C24H37F107	456.56	456.2523	63.14	8.17	4.16	24.53	C24H44F204	434.61	434.3207	66.33	10.20	8.74	14.73
C24H37F108	472.56	472.2472	61.00	7.89	4.02	27.09	C24H44F205	450.61	450.3157	63.97	9.84	8.43	17.75
C24H37F109	488.56	488.2421	59.00	7.63	3.89	29.47	C24H44F206	466.61	466.3106	61.78	9.50	8.14	20.57
C24H38F201	380.57	380.2891	75.75	10.06	9.98	4.20	C24H44F207	482.61	482.3055	59.73	9.19	7.87	23.21
C24H38F202	396.57	396.2840	72.69	9.66	9.58	8.07	C24H44F208	498.61	498.3004	57.81	8.89	7.62	25.67
C24H38F203	412.57	412.2789	69.87	9.28	9.21	11.63	C24H44F209	514.61	514.2953	56.02	8.62	7.38	27.98
C24H38F204	428.56	428.2738	67.26	8.94	8.87	14.93	C24H45F101	368.62	368.3454	78.20	12.30	5.15	4.34
C24H38F205	444.56	444.2687	64.84	8.62	8.55	17.99	C24H45F102	384.62	384.3403	74.95	11.79	4.94	8.32
C24H38F206	460.56	460.2636	62.59	8.32	8.25	20.84	C24H45F103	400.62	400.3353	71.95	11.32	4.74	11.98
C24H38F207	476.56	476.2585	60.49	8.04	7.97	23.50	C24H45F104	416.62	416.3302	69.19	10.89	4.56	15.36
C24H38F208	492.56	492.2535	58.52	7.78	7.71	25.99	C24H45F105	432.62	432.3251	66.63	10.48	4.39	18.49
C24H38F209	508.56	508.2484	56.68	7.53	7.47	28.31	C24H45F106	448.62	448.3200	64.26	10.11	4.23	21.40
C24H39F101	362.58	362.2985	79.51	10.84	5.24	4.41	C24H45F107	464.62	464.3149	62.04	9.76	4.09	24.10
C24H39F102	378.58	378.2934	76.15	10.38	5.02	8.45	C24H45F108	480.62	480.3098	59.98	9.44	3.95	26.63
C24H39F103	394.58	394.2883	73.06	9.96	4.81	12.16	C24H45F109	496.62	496.3047	58.05	9.13	3.83	28.99
C24H39F104	410.57	410.2832	70.21	9.57	4.63	15.59	C24H46F201	388.63	388.3517	74.18	11.93	9.78	4.12
C24H39F105	426.57	426.2781	67.58	9.22	4.45	18.75	C24H46F202	404.63	404.3466	71.24	11.46	9.39	7.91
C24H39F106	442.57	442.2730	65.13	8.88	4.29	21.69	C24H46F203	420.63	420.3415	68.53	11.02	9.03	11.41
C24H39F107	458.57	458.2680	62.86	8.57	4.14	24.42	C24H46F204	436.63	436.3364	66.02	10.62	8.70	14.66
C24H39F108	474.57	474.2629	60.74	8.28	4.00	26.97	C24H46F205	452.63	452.3313	63.69	10.24	8.39	17.67
C24H39F109	490.57	490.2578	58.76	8.01	3.87	29.35	C24H46F206	468.63	468.3262	61.51	9.89	8.11	20.48
C24H40F201	382.58	382.3047	75.35	10.54	9.93	4.18	C24H46F207	484.63	484.3211	59.48	9.57	7.84	23.11
C24H40F202	398.58	398.2996	72.32	10.12	9.53	8.03	C24H46F208	500.63	500.3160	57.58	9.26	7.59	25.57
C24H40F203	414.58	414.2945	69.53	9.73	9.17	11.58	C24H46F209	516.63	516.3110	55.80	8.97	7.35	27.87
C24H40F204	430.58	430.2894	66.95	9.36	8.82	14.86	C24H47F101	370.64	370.3611	77.78	12.78	5.13	4.32
C24H40F205	446.58	446.2844	64.55	9.03	8.51	17.91	C24H47F102	386.64	386.3560	74.56	12.25	4.91	8.28
C24H40F206	462.58	462.2793	62.32	8.72	8.21	20.75	C24H47F103	402.64	402.3509	71.59	11.77	4.72	11.92
C24H40F207	478.58	478.2742	60.23	8.42	7.94	23.40	C24H47F104	418.64	418.3458	68.86	11.32	4.54	15.29
C24H40F208	494.58	494.2691	58.29	8.15	7.68	25.88	C24H47F105	434.64	434.3407	66.32	10.90	4.37	18.41
C24H40F209	510.58	510.2640	56.46	7.90	7.44	28.20	C24H47F106	450.64	450.3356	63.97	10.51	4.22	21.30
C24H41F101	364.59	364.3141	79.07	11.34	5.21	4.39	C24H47F107	466.64	466.3306	61.78	10.15	4.07	24.00
C24H41F102	380.59	380.3090	75.74	10.86	4.99	8.41	C24H47F108	482.64	482.3255	59.73	9.82	3.94	26.52
C24H41F103	396.59	396.3040	72.69	10.42	4.79	12.10	C24H47F109	498.64	498.3204	57.81	9.50	3.81	28.88
C24H41F104	412.59	412.2989	69.87	10.02	4.60	15.51	C24H48F201	390.65	390.3673	73.79	12.39	9.73	4.10
C24H41F105	428.59	428.2938	67.26	9.64	4.43	18.67	C24H48F202	406.65	406.3622	70.89	11.90	9.34	7.87
C24H41F106	444.59	444.2887	64.84	9.30	4.27	21.59	C24H48F203	422.65	422.3571	68.21	11.45	8.99	11.36
C24H41F107	460.59	460.2836	62.59	8.97	4.12	24.32	C24H48F204	438.64	438.3520	65.72	11.03	8.66	14.59
C24H41F108	476.59	476.2785	60.49	8.67	3.99	26.86	C24H48F205	454.64	454.3470	63.41	10.64	8.36	17.60
C24H41F109	492.59	492.2734	58.52	8.39	3.86	29.23	C24H48F206	470.64	470.3419	61.25	10.28	8.07	20.40
C24H42F201	384.60	384.3204	74.95	11.01	9.88	4.16	C24H48F207	486.64	486.3368	59.24	9.94	7.81	23.01
C24H42F202	400.60	400.3153	71.96	10.57	9.49	7.99	C24H48F208	502.64	502.3317	57.35	9.63	7.56	25.46
C24H42F203	416.60	416.3102	69.20	10.16	9.12	11.52	C24H48F209	518.64	518.3266	55.58	9.33	7.33	27.76
C24H42F204	432.60	432.3051	66.64	9.79	8.78	14.79	C24H49F101	372.66	372.3767	77.35	13.25	5.10	4.29
C24H42F205	448.60	448.3000	64.26	9.44	8.47	17.83	C24H49F102	388.66	388.3716	74.17	12.71	4.89	8.23
C24H42F206	464.60	464.2949	62.05	9.11	8.18	20.66	C24H49F103	404.65	404.3665	71.24	12.21	4.69	11.86
C24H42F207	480.59	480.2898	59.98	8.81	7.91	23.30	C24H49F104	420.65	420.3615	68.53	11.74	4.52	15.21
C24H42F208	496.59	496.2848	58.05	8.53	7.65	25.77	C24H49F105	436.65	436.3564	66.02	11.31	4.35	18.32
C24H42F209	512.59	512.2797	56.24	8.26	7.41	28.09	C24H49F106	452.65	452.3513	63.68	10.91	4.20	21.21
C24H43F101	366.61	366.3298	78.63	11.82	5.18	4.36	C24H49F107	468.65	468.3462	61.51	10.54	4.05	23.90
C24H43F102	382.61	382.3247	75.34	11.33	4.97	8.36	C24H49F108	484.65	484.3411	59.48	10.19	3.92	26.41
C24H43F103	398.61	398.3196	72.32	10.87	4.77	12.04	C24H49F109	500.65	500.3360	57.58	9.87	3.79	28.76
C24H43F104	414.61	414.3145	69.53	10.45	4.58	15.44	C25H12F201	366.37	366.0856	81.96	3.30	10.37	4.37
C24H43F105	430.61	430.3094	66.94	10.07	4.41	18.58	C25H12F202	382.37	382.0805	78.53	3.16	9.94	8.37
C24H43F106	446.61	446.3043	64.55	9.70	4.25	21.49	C25H12F203	398.37	398.0754	75.38	3.04	9.54	12.05
C24H43F107	462.60	462.2993	62.31	9.37	4.11	24.21	C25H12F204	414.37	414.0704	72.47	2.92	9.17	15.44

C25H12F205	430.37	430.0653	69.77	2.81	8.83	18.59
C25H12F206	446.37	446.0602	67.27	2.71	8.51	21.51
C25H12F207	462.37	462.0551	64.94	2.62	8.22	24.22
C25H12F208	478.37	478.0500	62.77	2.53	7.94	26.76
C25H12F209	494.37	494.0449	60.74	2.45	7.69	29.13
C25H13F101	348.38	348.0950	86.19	3.76	5.45	4.59
C25H13F102	364.38	364.0900	82.41	3.60	5.21	8.78
C25H13F103	380.38	380.0849	78.94	3.44	4.99	12.62
C25H13F104	396.38	396.0798	75.76	3.31	4.79	16.15
C25H13F105	412.38	412.0747	72.82	3.18	4.61	19.40
C25H13F106	428.38	428.0696	70.10	3.06	4.43	22.41
C25H13F107	444.38	444.0645	67.57	2.95	4.28	25.20
C25H13F108	460.38	460.0594	65.22	2.85	4.13	27.80
C25H13F109	476.38	476.0543	63.03	2.75	3.99	30.23
C25H14F201	368.39	368.1013	81.51	3.83	10.31	4.34
C25H14F202	384.39	384.0962	78.12	3.67	9.89	8.32
C25H14F203	400.39	400.0911	75.00	3.52	9.49	11.99
C25H14F204	416.38	416.0860	72.12	3.39	9.13	15.37
C25H14F205	432.38	432.0809	69.45	3.26	8.79	18.50
C25H14F206	448.38	448.0758	66.97	3.15	8.47	21.41
C25H14F207	464.38	464.0707	64.66	3.04	8.18	24.12
C25H14F208	480.38	480.0657	62.51	2.94	7.91	26.64
C25H14F209	496.38	496.0606	60.49	2.84	7.65	29.01
C25H15F101	350.40	350.1107	85.70	4.31	5.42	4.57
C25H15F102	366.40	366.1056	81.95	4.13	5.19	8.73
C25H15F103	382.39	382.1005	78.53	3.95	4.97	12.55
C25H15F104	398.39	398.0954	75.37	3.80	4.77	16.06
C25H15F105	414.39	414.0903	72.46	3.65	4.58	19.30
C25H15F106	430.39	430.0853	69.77	3.51	4.41	22.30
C25H15F107	446.39	446.0802	67.27	3.39	4.26	25.09
C25H15F108	462.39	462.0751	64.94	3.27	4.11	27.68
C25H15F109	478.39	478.0700	62.77	3.16	3.97	30.10
C25H16F201	370.40	370.1169	81.07	4.35	10.26	4.32
C25H16F202	386.40	386.1118	77.71	4.17	9.83	8.28
C25H16F203	402.40	402.1067	74.62	4.01	9.44	11.93
C25H16F204	418.40	418.1017	71.77	3.85	9.08	15.30
C25H16F205	434.40	434.0966	69.12	3.71	8.75	18.42
C25H16F206	450.40	450.0915	66.67	3.58	8.44	21.31
C25H16F207	466.40	466.0864	64.38	3.46	8.15	24.01
C25H16F208	482.40	482.0813	62.25	3.34	7.88	26.53
C25H16F209	498.40	498.0762	60.25	3.24	7.62	28.89
C25H17F101	352.41	352.1263	85.21	4.86	5.39	4.54
C25H17F102	368.41	368.1212	81.51	4.65	5.16	8.69
C25H17F103	384.41	384.1162	78.11	4.46	4.94	12.49
C25H17F104	400.41	400.1111	74.99	4.28	4.74	15.98
C25H17F105	416.41	416.1060	72.11	4.12	4.56	19.21
C25H17F106	432.41	432.1009	69.44	3.96	4.39	22.20
C25H17F107	448.41	448.0958	66.97	3.82	4.24	24.98
C25H17F108	464.41	464.0907	64.66	3.69	4.09	27.56
C25H17F109	480.41	480.0856	62.51	3.57	3.95	29.97
C25H18F201	372.42	372.1326	80.63	4.87	10.20	4.30
C25H18F202	388.42	388.1275	77.31	4.67	9.78	8.24
C25H18F203	404.42	404.1224	74.25	4.49	9.40	11.87
C25H18F204	420.42	420.1173	71.42	4.32	9.04	15.22
C25H18F205	436.42	436.1122	68.81	4.16	8.71	18.33
C25H18F206	452.42	452.1071	66.37	4.01	8.40	21.22
C25H18F207	468.41	468.1020	64.11	3.87	8.11	23.91
C25H18F208	484.41	484.0970	61.99	3.75	7.84	26.42
C25H18F209	500.41	500.0919	60.01	3.63	7.59	28.78
C25H19F101	354.43	354.1420	84.72	5.40	5.36	4.51
C25H19F102	370.43	370.1369	81.06	5.17	5.13	8.64
C25H19F103	386.43	386.1318	77.71	4.96	4.92	12.42
C25H19F104	402.43	402.1267	74.62	4.76	4.72	15.90
C25H19F105	418.43	418.1216	71.76	4.58	4.54	19.12
C25H19F106	434.42	434.1166	69.12	4.41	4.37	22.10
C25H19F107	450.42	450.1115	66.67	4.25	4.22	24.86
C25H19F108	466.42	466.1064	64.38	4.11	4.07	27.44
C25H19F109	482.42	482.1013	62.24	3.97	3.94	29.85
C25H20F201	374.43	374.1482	80.20	5.38	10.15	4.27
C25H20F202	390.43	390.1431	76.91	5.16	9.73	8.20
C25H20F203	406.43	406.1380	73.88	4.96	9.35	11.81
C25H20F204	422.43	422.1330	71.08	4.77	8.99	15.15
C25H20F205	438.43	438.1279	68.49	4.60	8.67	18.25
C25H20F206	454.43	454.1228	66.08	4.44	8.36	21.12
C25H20F207	470.43	470.1177	63.83	4.29	8.08	23.81
C25H20F208	486.43	486.1126	61.73	4.14	7.81	26.31
C25H20F209	502.43	502.1075	59.77	4.01	7.56	28.66
C25H21F101	356.44	356.1576	84.24	5.94	5.33	4.49
C25H21F102	372.44	372.1525	80.62	5.68	5.10	8.59
C25H21F103	388.44	388.1475	77.30	5.45	4.89	12.36
C25H21F104	404.44	404.1424	74.25	5.23	4.70	15.82
C25H21F105	420.44	420.1373	71.42	5.03	4.52	19.03
C25H21F106	436.44	436.1322	68.80	4.85	4.35	22.00
C25H21F107	452.44	452.1271	66.37	4.68	4.20	24.75
C25H21F108	468.44	468.1220	64.10	4.52	4.06	27.32
C25H21F109	484.44	484.1169	61.98	4.37	3.92	29.72
C25H22F201	376.45	376.1639	79.77	5.89	10.09	4.25
C25H22F202	392.45	392.1588	76.51	5.65	9.68	8.15
C25H22F203	408.45	408.1537	73.52	5.43	9.30	11.75
C25H22F204	424.45	424.1486	70.75	5.22	8.95	15.08
C25H22F205	440.45	440.1435	68.18	5.03	8.63	18.16
C25H22F206	456.45	456.1384	65.79	4.86	8.32	21.03
C25H22F207	472.45	472.1333	63.56	4.69	8.04	23.71
C25H22F208	488.45	488.1283	61.48	4.54	7.78	26.20
C25H22F209	504.45	504.1232	59.53	4.40	7.53	28.55
C25H23F101	358.46	358.1733	83.77	6.47	5.30	4.46
C25H23F102	374.46	374.1682	80.19	6.19	5.07	8.55
C25H23F103	390.46	390.1631	76.90	5.94	4.87	12.29
C25H23F104	406.46	406.1580	73.88	5.70	4.67	15.75
C25H23F105	422.46	422.1529	71.08	5.49	4.50	18.94
C25H23F106	438.46	438.1479	68.49	5.29	4.33	21.89
C25H23F107	454.46	454.1428	66.07	5.10	4.18	24.64
C25H23F108	470.46	470.1377	63.83	4.93	4.04	27.21
C25H23F109	486.46	486.1326	61.73	4.77	3.91	29.60
C25H24F201	378.47	378.1795	79.34	6.39	10.04	4.23
C25H24F202	394.47	394.1744	76.12	6.13	9.63	8.11
C25H24F203	410.47	410.1693	73.16	5.89	9.26	11.69
C25H24F204	426.46	426.1643	70.41	5.67	8.91	15.01
C25H24F205	442.46	442.1592	67.87	5.47	8.59	18.08
C25H24F206	458.46	458.1541	65.50	5.28	8.29	20.94
C25H24F207	474.46	474.1490	63.29	5.10	8.01	23.60
C25H24F208	490.46	490.1439	61.22	4.93	7.75	26.10
C25H24F209	506.46	506.1388	59.29	4.78	7.50	28.43
C25H25F101	360.48	360.1889	83.30	6.99	5.27	4.44
C25H25F102	376.48	376.1838	79.76	6.69	5.05	8.50
C25H25F103	392.47	392.1788	76.51	6.42	4.84	12.23
C25H25F104	408.47	408.1737	73.51	6.17	4.65	15.67
C25H25F105	424.47	424.1686	70.74	5.94	4.48	18.85
C25H25F106	440.47	440.1635	68.17	5.72	4.31	21.79
C25H25F107	456.47	456.1584	65.78	5.52	4.16	24.54

C25H25F108	472.47	472.1533	63.55	5.33	4.02	27.09
C25H25F109	488.47	488.1482	61.47	5.16	3.89	29.48
C25H26F201	380.48	380.1952	78.92	6.89	9.99	4.21
C25H26F202	396.48	396.1901	75.74	6.61	9.58	8.07
C25H26F203	412.48	412.1850	72.80	6.35	9.21	11.64
C25H26F204	428.48	428.1799	70.08	6.12	8.87	14.94
C25H26F205	444.48	444.1748	67.56	5.90	8.55	18.00
C25H26F206	460.48	460.1697	65.21	5.69	8.25	20.85
C25H26F207	476.48	476.1646	63.02	5.50	7.97	23.50
C25H26F208	492.48	492.1596	60.97	5.32	7.72	25.99
C25H26F209	508.48	508.1545	59.05	5.15	7.47	28.32
C25H27F101	362.49	362.2046	82.84	7.51	5.24	4.41
C25H27F102	378.49	378.1995	79.34	7.19	5.02	8.45
C25H27F103	394.49	394.1944	76.12	6.90	4.82	12.17
C25H27F104	410.49	410.1893	73.15	6.63	4.63	15.59
C25H27F105	426.49	426.1842	70.41	6.38	4.45	18.76
C25H27F106	442.49	442.1792	67.86	6.15	4.29	21.69
C25H27F107	458.49	458.1741	65.45	5.94	4.14	24.43
C25H27F108	474.49	474.1690	63.28	5.74	4.00	26.98
C25H27F109	490.49	490.1639	61.22	5.55	3.87	29.36
C25H28F201	382.50	382.2108	78.50	7.38	9.93	4.18
C25H28F202	398.50	398.2057	75.35	7.08	9.54	8.03
C25H28F203	414.50	414.2006	72.44	6.81	9.17	11.58
C25H28F204	430.50	430.1955	69.75	6.56	8.83	14.87
C25H28F205	446.50	446.1905	67.25	6.32	8.51	17.92
C25H28F206	462.50	462.1854	64.93	6.10	8.22	20.76
C25H28F207	478.49	478.1803	62.75	5.90	7.94	23.41
C25H28F208	494.49	494.1752	60.72	5.71	7.68	25.88
C25H28F209	510.49	510.1701	58.82	5.53	7.44	28.21
C25H29F101	364.51	364.2202	82.38	8.02	5.21	4.39
C25H29F102	380.51	380.2151	78.92	7.68	4.99	8.41
C25H29F103	396.51	396.2101	75.73	7.37	4.79	12.11
C25H29F104	412.51	412.2050	72.79	7.09	4.61	15.51
C25H29F105	428.51	428.1999	70.08	6.82	4.43	18.67
C25H29F106	444.50	444.1948	67.55	6.58	4.27	21.60
C25H29F107	460.50	460.1897	65.21	6.35	4.13	24.32
C25H29F108	476.50	476.1846	63.02	6.13	3.99	26.86
C25H29F109	492.50	492.1795	60.97	5.94	3.86	29.24
C25H30F201	384.51	384.2265	78.09	7.86	9.88	4.16
C25H30F202	400.51	400.2214	74.97	7.55	9.49	7.99
C25H30F203	416.51	416.2163	72.09	7.26	9.12	11.52
C25H30F204	432.51	432.2112	69.43	6.99	8.79	14.80
C25H30F205	448.51	448.2061	66.95	6.74	8.47	17.84
C25H30F206	464.51	464.2010	64.64	6.51	8.18	20.67
C25H30F207	480.51	480.1959	62.49	6.29	7.91	23.31
C25H30F208	496.51	496.1909	60.48	6.09	7.65	25.78
C25H30F209	512.51	512.1858	58.59	5.90	7.41	28.10
C25H31F101	366.52	366.2359	81.93	8.53	5.18	4.37
C25H31F102	382.52	382.2308	78.50	8.17	4.97	8.37
C25H31F103	398.52	398.2257	75.35	7.84	4.77	12.04
C25H31F104	414.52	414.2206	72.44	7.54	4.58	15.44
C25H31F105	430.52	430.2155	69.75	7.26	4.41	18.58
C25H31F106	446.52	446.2104	67.25	7.00	4.25	21.50
C25H31F107	462.52	462.2054	64.92	6.76	4.11	24.21
C25H31F108	478.52	478.2003	62.75	6.53	3.97	26.75
C25H31F109	494.52	494.1952	60.72	6.32	3.84	29.12
C25H32F201	386.53	386.2421	77.69	8.34	9.83	4.14
C25H32F202	402.53	402.2370	74.60	8.01	9.44	7.95
C25H32F203	418.53	418.2319	71.75	7.71	9.08	11.47
C25H32F204	434.53	434.2268	69.10	7.42	8.74	14.73
C25H32F205	450.53	450.2218	66.65	7.16	8.43	17.76
C25H32F206	466.53	466.2167	64.36	6.91	8.14	20.58
C25H32F207	482.53	482.2116	62.23	6.68	7.87	23.21
C25H32F208	498.53	498.2065	60.23	6.47	7.62	25.67
C25H32F209	514.53	514.2014	58.36	6.27	7.38	27.99
C25H33F101	368.54	368.2515	81.48	9.03	5.16	4.34
C25H33F102	384.54	384.2464	78.09	8.65	4.94	8.32
C25H33F103	400.54	400.2414	74.97	8.30	4.74	11.98
C25H33F104	416.54	416.2363	72.09	7.99	4.56	15.36
C25H33F105	432.54	432.2312	69.42	7.69	4.39	18.49
C25H33F106	448.54	448.2261	66.95	7.42	4.24	21.40
C25H33F107	464.54	464.2210	64.64	7.16	4.09	24.11
C25H33F108	480.54	480.2159	62.49	6.92	3.95	26.64
C25H33F109	496.53	496.2108	60.47	6.70	3.83	29.00
C25H34F201	388.55	388.2578	77.28	8.82	9.78	4.12
C25H34F202	404.55	404.2527	74.23	8.47	9.39	7.91
C25H34F203	420.54	420.2476	71.40	8.15	9.04	11.41
C25H34F204	436.54	436.2425	68.79	7.85	8.70	14.66
C25H34F205	452.54	452.2374	66.35	7.57	8.40	17.68
C25H34F206	468.54	468.2323	64.09	7.31	8.11	20.49
C25H34F207	484.54	484.2272	61.97	7.07	7.84	23.11
C25H34F208	500.54	500.2222	59.99	6.85	7.59	25.57
C25H34F209	516.54	516.2171	58.13	6.63	7.36	27.88
C25H35F101	370.56	370.2672	81.03	9.52	5.13	4.32
C25H35F102	386.55	386.2621	77.68	9.13	4.91	8.28
C25H35F103	402.55	402.2570	74.59	8.76	4.72	11.92
C25H35F104	418.55	418.2519	71.74	8.43	4.54	15.29
C25H35F105	434.55	434.2468	69.10	8.12	4.37	18.41
C25H35F106	450.55	450.2417	66.65	7.83	4.22	21.31
C25H35F107	466.55	466.2367	64.36	7.56	4.07	24.01
C25H35F108	482.55	482.2316	62.23	7.31	3.94	26.52
C25H35F109	498.55	498.2265	60.23	7.08	3.81	28.88
C25H36F201	390.56	390.2734	76.88	9.29	9.73	4.10
C25H36F202	406.56	406.2683	73.86	8.93	9.35	7.87
C25H36F203	422.56	422.2632	71.06	8.59	8.99	11.36
C25H36F204	438.56	438.2581	68.47	8.27	8.66	14.59
C25H36F205	454.56	454.2531	66.06	7.98	8.36	17.60
C25H36F206	470.56	470.2480	63.81	7.71	8.07	20.40
C25H36F207	486.56	486.2429	61.71	7.46	7.81	23.02
C25H36F208	502.56	502.2378	59.75	7.22	7.56	25.47
C25H36F209	518.56	518.2327	57.91	7.00	7.33	27.77
C25H37F101	372.57	372.2828	80.60	10.01	5.10	4.29
C25H37F102	388.57	388.2777	77.28	9.60	4.89	8.23
C25H37F103	404.57	404.2727	74.22	9.22	4.70	11.86
C25H37F104	420.57	420.2676	71.40	8.87	4.52	15.22
C25H37F105	436.57	436.2625	68.78	8.54	4.35	18.32
C25H37F106	452.57	452.2574	66.35	8.24	4.20	21.21
C25H37F107	468.57	468.2523	64.08	7.96	4.05	23.90
C25H37F108	484.57	484.2472	61.97	7.70	3.92	26.41
C25H37F109	500.57	500.2421	59.99	7.45	3.80	28.77
C25H38F201	392.58	392.2891	76.49	9.76	9.68	4.08
C25H38F202	408.58	408.2840	73.49	9.37	9.30	7.83
C25H38F203	424.58	424.2789	70.72	9.02	8.95	11.30
C25H38F204	440.58	440.2738	68.16	8.69	8.62	14.53
C25H38F205	456.58	456.2687	65.77	8.39	8.32	17.52
C25H38F206	472.57	472.2636	63.54	8.11	8.04	20.31
C25H38F207	488.57	488.2585	61.46	7.84	7.78	22.92
C25H38F208	504.57	504.2535	59.51	7.59	7.53	25.37
C25H38F209	520.57	520.2484	57.68	7.36	7.30	27.66
C25H39F101	374.59	374.2985	80.16	10.49	5.07	4.27

C25H39F102	390.59	390.2934	76.88	10.06	4.86	8.19
C25H39F103	406.59	406.2883	73.85	9.67	4.67	11.81
C25H39F104	422.59	422.2832	71.06	9.30	4.50	15.14
C25H39F105	438.58	438.2781	68.47	8.96	4.33	18.24
C25H39F106	454.58	454.2730	66.06	8.65	4.18	21.12
C25H39F107	470.58	470.2680	63.81	8.35	4.04	23.80
C25H39F108	486.58	486.2629	61.71	8.08	3.90	26.30
C25H39F109	502.58	502.2578	59.75	7.82	3.78	28.65
C25H40F201	394.59	394.3047	76.10	10.22	9.63	4.05
C25H40F202	410.59	410.2996	73.13	9.82	9.25	7.79
C25H40F203	426.59	426.2945	70.39	9.45	8.91	11.25
C25H40F204	442.59	442.2894	67.85	9.11	8.59	14.46
C25H40F205	458.59	458.2844	65.48	8.79	8.29	17.44
C25H40F206	474.59	474.2793	63.27	8.50	8.01	20.23
C25H40F207	490.59	490.2742	61.21	8.22	7.75	22.83
C25H40F208	506.59	506.2691	59.27	7.96	7.50	25.27
C25H40F209	522.59	522.2640	57.46	7.72	7.27	27.55
C25H41F101	376.60	376.3141	79.73	10.97	5.04	4.25
C25H41F102	392.60	392.3090	76.48	10.53	4.84	8.15
C25H41F103	408.60	408.3040	73.49	10.11	4.65	11.75
C25H41F104	424.60	424.2989	70.72	9.73	4.47	15.07
C25H41F105	440.60	440.2938	68.15	9.38	4.31	18.16
C25H41F106	456.60	456.2887	65.76	9.05	4.16	21.02
C25H41F107	472.60	472.2836	63.54	8.74	4.02	23.70
C25H41F108	488.60	488.2785	61.46	8.46	3.89	26.20
C25H41F109	504.60	504.2734	59.51	8.19	3.77	28.54
C25H42F201	396.61	396.3204	75.71	10.67	9.58	4.03
C25H42F202	412.61	412.3153	72.78	10.26	9.21	7.76
C25H42F203	428.61	428.3102	70.06	9.88	8.87	11.20
C25H42F204	444.61	444.3051	67.54	9.52	8.55	14.39
C25H42F205	460.61	460.3000	65.19	9.19	8.25	17.37
C25H42F206	476.61	476.2949	63.00	8.88	7.97	20.14
C25H42F207	492.61	492.2898	60.96	8.59	7.71	22.74
C25H42F208	508.61	508.2848	59.04	8.32	7.47	25.17
C25H42F209	524.60	524.2797	57.24	8.07	7.24	27.45
C25H43F101	378.62	378.3298	79.31	11.45	5.02	4.23
C25H43F102	394.62	394.3247	76.09	10.98	4.81	8.11
C25H43F103	410.62	410.3196	73.13	10.56	4.63	11.69
C25H43F104	426.62	426.3145	70.39	10.16	4.45	15.00
C25H43F105	442.62	442.3094	67.84	9.79	4.29	18.07
C25H43F106	458.62	458.3043	65.47	9.45	4.14	20.93
C25H43F107	474.62	474.2993	63.27	9.13	4.00	23.60
C25H43F108	490.62	490.2942	61.20	8.83	3.87	26.09
C25H43F109	506.61	506.2891	59.27	8.56	3.75	28.42
C25H44F201	398.63	398.3360	75.33	11.13	9.53	4.01
C25H44F202	414.63	414.3309	72.42	10.70	9.16	7.72
C25H44F203	430.62	430.3258	69.73	10.30	8.82	11.15
C25H44F204	446.62	446.3207	67.23	9.93	8.51	14.33
C25H44F205	462.62	462.3157	64.91	9.59	8.21	17.29
C25H44F206	478.62	478.3106	62.74	9.27	7.94	20.06
C25H44F207	494.62	494.3055	60.71	8.97	7.68	22.64
C25H44F208	510.62	510.3004	58.81	8.69	7.44	25.07
C25H44F209	526.62	526.2953	57.02	8.42	7.22	27.34
C25H45F101	380.64	380.3454	78.89	11.92	4.99	4.20
C25H45F102	396.63	396.3403	75.71	11.44	4.79	8.07
C25H45F103	412.63	412.3353	72.77	10.99	4.60	11.63
C25H45F104	428.63	428.3302	70.05	10.58	4.43	14.93
C25H45F105	444.63	444.3251	67.53	10.20	4.27	17.99
C25H45F106	460.63	460.3200	65.19	9.85	4.12	20.84
C25H45F107	476.63	476.3149	63.00	9.52	3.99	23.50
C25H45F108	492.63	492.3098	60.95	9.21	3.86	25.98
C25H45F109	508.63	508.3047	59.04	8.92	3.74	28.31
C25H46F201	400.64	400.3517	74.95	11.57	9.48	3.99
C25H46F202	416.64	416.3466	72.07	11.13	9.12	7.68
C25H46F203	432.64	432.3415	69.41	10.72	8.78	11.09
C25H46F204	448.64	448.3364	66.93	10.33	8.47	14.26
C25H46F205	464.64	464.3313	64.63	9.98	8.18	17.22
C25H46F206	480.64	480.3262	62.47	9.65	7.91	19.97
C25H46F207	496.64	496.3211	60.46	9.34	7.65	22.55
C25H46F208	512.64	512.3160	58.58	9.04	7.41	24.97
C25H46F209	528.64	528.3110	56.80	8.77	7.19	27.24
C25H47F101	382.65	382.3611	78.47	12.38	4.96	4.18
C25H47F102	398.65	398.3560	75.32	11.88	4.77	8.03
C25H47F103	414.65	414.3509	72.42	11.43	4.58	11.58
C25H47F104	430.65	430.3458	69.73	11.00	4.41	14.86
C25H47F105	446.65	446.3407	67.23	10.61	4.25	17.91
C25H47F106	462.65	462.3356	64.90	10.24	4.11	20.75
C25H47F107	478.65	478.3306	62.73	9.90	3.97	23.40
C25H47F108	494.65	494.3255	60.71	9.58	3.84	25.88
C25H47F109	510.65	510.3204	58.80	9.28	3.72	28.20
C25H48F201	402.66	402.3673	74.57	12.02	9.44	3.97
C25H48F202	418.66	418.3622	71.72	11.56	9.08	7.64
C25H48F203	434.66	434.3571	69.08	11.13	8.74	11.04
C25H48F204	450.66	450.3520	66.63	10.74	8.43	14.20
C25H48F205	466.66	466.3470	64.35	10.37	8.14	17.14
C25H48F206	482.65	482.3419	62.21	10.02	7.87	19.89
C25H48F207	498.65	498.3368	60.22	9.70	7.62	22.46
C25H48F208	514.65	514.3317	58.35	9.40	7.38	24.87
C25H48F209	530.65	530.3266	56.59	9.12	7.16	27.14
C25H49F101	384.67	384.3767	78.06	12.84	4.94	4.16
C25H49F102	400.67	400.3716	74.94	12.33	4.74	7.99
C25H49F103	416.67	416.3665	72.07	11.85	4.56	11.52
C25H49F104	432.67	432.3615	69.40	11.42	4.39	14.79
C25H49F105	448.66	448.3564	66.93	11.01	4.23	17.83
C25H49F106	464.66	464.3513	64.62	10.63	4.09	20.66
C25H49F107	480.66	480.3462	62.47	10.28	3.95	23.30
C25H49F108	496.66	496.3411	60.46	9.94	3.83	25.77
C25H49F109	512.66	512.3360	58.57	9.63	3.71	28.09
C25H50F201	404.67	404.3829	74.20	12.45	9.39	3.95
C25H50F202	420.67	420.3779	71.38	11.98	9.03	7.61
C25H50F203	436.67	436.3728	68.77	11.54	8.70	10.99
C25H50F204	452.67	452.3677	66.33	11.13	8.39	14.14
C25H50F205	468.67	468.3626	64.07	10.75	8.11	17.07
C25H50F206	484.67	484.3575	61.96	10.40	7.84	19.81
C25H50F207	500.67	500.3524	59.98	10.07	7.59	22.37
C25H50F208	516.67	516.3473	58.12	9.75	7.35	24.77
C25H50F209	532.67	532.3423	56.37	9.46	7.13	27.03
C25H51F101	386.68	386.3924	77.66	13.29	4.91	4.14
C25H51F102	402.68	402.3873	74.57	12.77	4.72	7.95
C25H51F103	418.68	418.3822	71.72	12.28	4.54	11.46
C25H51F104	434.68	434.3771	69.08	11.83	4.37	14.72
C25H51F105	450.68	450.3720	66.63	11.41	4.22	17.75
C25H51F106	466.68	466.3669	64.34	11.02	4.07	20.57
C25H51F107	482.68	482.3619	62.21	10.65	3.94	23.20
C25H51F108	498.68	498.3568	60.21	10.31	3.81	25.67
C25H51F109	514.68	514.3517	58.34	9.99	3.69	27.98
C26H13F101	360.39	360.0950	86.65	3.64	5.27	4.44
C26H13F102	376.39	376.0900	82.97	3.48	5.05	8.50
C26H13F103	392.39	392.0849	79.59	3.34	4.84	12.23
C26H13F104	408.39	408.0798	76.47	3.21	4.65	15.67

166

C26H13F105	424.39	424.0747	73.59	3.09	4.48	18.85
C26H13F106	440.39	440.0696	70.91	2.98	4.31	21.80
C26H13F107	456.39	456.0645	68.43	2.87	4.16	24.54
C26H13F108	472.39	472.0594	66.11	2.77	4.02	27.10
C26H13F109	488.39	488.0543	63.94	2.68	3.89	29.48
C26H14F201	380.40	380.1013	82.10	3.71	9.99	4.21
C26H14F202	396.40	396.0962	78.78	3.56	9.59	8.07
C26H14F203	412.40	412.0911	75.73	3.42	9.21	11.64
C26H14F204	428.40	428.0860	72.90	3.29	8.87	14.94
C26H14F205	444.40	444.0809	70.27	3.18	8.55	18.00
C26H14F206	460.39	460.0758	67.83	3.07	8.25	20.85
C26H14F207	476.39	476.0707	65.55	2.96	7.98	23.51
C26H14F208	492.39	492.0657	63.42	2.87	7.72	25.99
C26H14F209	508.39	508.0606	61.43	2.78	7.47	28.32
C26H15F101	362.41	362.1107	86.17	4.17	5.24	4.41
C26H15F102	378.41	378.1056	82.53	4.00	5.02	8.46
C26H15F103	394.41	394.1005	79.18	3.83	4.82	12.17
C26H15F104	410.41	410.0954	76.09	3.68	4.63	15.59
C26H15F105	426.40	426.0903	73.24	3.55	4.46	18.76
C26H15F106	442.40	442.0853	70.59	3.42	4.29	21.70
C26H15F107	458.40	458.0802	68.13	3.30	4.14	24.43
C26H15F108	474.40	474.0751	65.83	3.19	4.00	26.98
C26H15F109	490.40	490.0700	63.68	3.08	3.87	29.36
C26H16F201	382.41	382.1169	81.66	4.22	9.94	4.18
C26H16F202	398.41	398.1118	78.38	4.05	9.54	8.03
C26H16F203	414.41	414.1067	75.36	3.89	9.17	11.58
C26H16F204	430.41	430.1017	72.56	3.75	8.83	14.87
C26H16F205	446.41	446.0966	69.96	3.61	8.51	17.92
C26H16F206	462.41	462.0915	67.54	3.49	8.22	20.76
C26H16F207	478.41	478.0864	65.28	3.37	7.94	23.41
C26H16F208	494.41	494.0813	63.16	3.26	7.69	25.89
C26H16F209	510.41	510.0762	61.18	3.16	7.44	28.21
C26H17F101	364.42	364.1263	85.69	4.70	5.21	4.39
C26H17F102	380.42	380.1212	82.09	4.50	4.99	8.41
C26H17F103	396.42	396.1162	78.78	4.32	4.79	12.11
C26H17F104	412.42	412.1111	75.72	4.15	4.61	15.52
C26H17F105	428.42	428.1060	72.89	4.00	4.43	18.67
C26H17F106	444.42	444.1009	70.27	3.86	4.27	21.60
C26H17F107	460.42	460.0958	67.83	3.72	4.13	24.32
C26H17F108	476.42	476.0907	65.55	3.60	3.99	26.87
C26H17F109	492.42	492.0856	63.42	3.48	3.86	29.24
C26H18F201	384.43	384.1326	81.23	4.72	9.88	4.16
C26H18F202	400.43	400.1275	77.99	4.53	9.49	7.99
C26H18F203	416.43	416.1224	74.99	4.36	9.12	11.53
C26H18F204	432.43	432.1173	72.22	4.20	8.79	14.80
C26H18F205	448.43	448.1122	69.64	4.05	8.47	17.84
C26H18F206	464.43	464.1071	67.24	3.91	8.18	20.67
C26H18F207	480.43	480.1020	65.00	3.78	7.91	23.31
C26H18F208	496.43	496.0970	62.91	3.65	7.65	25.78
C26H18F209	512.42	512.0919	60.94	3.54	7.42	28.10
C26H19F101	366.44	366.1420	85.22	5.23	5.18	4.37
C26H19F102	382.44	382.1369	81.66	5.01	4.97	8.37
C26H19F103	398.44	398.1318	78.38	4.81	4.77	12.05
C26H19F104	414.44	414.1267	75.35	4.62	4.58	15.44
C26H19F105	430.44	430.1216	72.55	4.45	4.41	18.59
C26H19F106	446.44	446.1166	69.95	4.29	4.26	21.50
C26H19F107	462.44	462.1115	67.53	4.14	4.11	24.22
C26H19F108	478.43	478.1064	65.27	4.00	3.97	26.75
C26H19F109	494.43	494.1013	63.16	3.87	3.84	29.12
C26H20F201	386.45	386.1482	80.81	5.22	9.83	4.14
C26H20F202	402.44	402.1431	77.60	5.01	9.44	7.95
C26H20F203	418.44	418.1380	74.63	4.82	9.08	11.47
C26H20F204	434.44	434.1330	71.88	4.64	8.75	14.73
C26H20F205	450.44	450.1279	69.33	4.48	8.44	17.76
C26H20F206	466.44	466.1228	66.95	4.32	8.15	20.58
C26H20F207	482.44	482.1177	64.73	4.18	7.88	23.21
C26H20F208	498.44	498.1126	62.65	4.04	7.62	25.68
C26H20F209	514.44	514.1075	60.70	3.92	7.39	27.99
C26H21F101	368.46	368.1576	84.76	5.74	5.16	4.34
C26H21F102	384.45	384.1525	81.23	5.51	4.94	8.32
C26H21F103	400.45	400.1475	77.98	5.29	4.74	11.99
C26H21F104	416.45	416.1424	74.99	5.08	4.56	15.37
C26H21F105	432.45	432.1373	72.21	4.89	4.39	18.50
C26H21F106	448.45	448.1322	69.64	4.72	4.24	21.41
C26H21F107	464.45	464.1271	67.24	4.56	4.09	24.11
C26H21F108	480.45	480.1220	65.00	4.41	3.95	26.64
C26H21F109	496.45	496.1169	62.90	4.26	3.83	29.00
C26H22F201	388.46	388.1639	80.39	5.71	9.78	4.12
C26H22F202	404.46	404.1588	77.21	5.48	9.39	7.91
C26H22F203	420.46	420.1537	74.27	5.27	9.04	11.42
C26H22F204	436.46	436.1486	71.55	5.08	8.71	14.66
C26H22F205	452.46	452.1435	69.02	4.90	8.40	17.68
C26H22F206	468.46	468.1384	66.66	4.73	8.11	20.49
C26H22F207	484.46	484.1333	64.46	4.58	7.84	23.12
C26H22F208	500.46	500.1283	62.40	4.43	7.59	25.58
C26H22F209	516.46	516.1232	60.47	4.29	7.36	27.88
C26H23F101	370.47	370.1733	84.30	6.26	5.13	4.32
C26H23F102	386.47	386.1682	80.81	6.00	4.92	8.28
C26H23F103	402.47	402.1631	77.59	5.76	4.72	11.93
C26H23F104	418.47	418.1580	74.63	5.54	4.54	15.29
C26H23F105	434.47	434.1529	71.88	5.34	4.37	18.41
C26H23F106	450.47	450.1479	69.33	5.15	4.22	21.31
C26H23F107	466.47	466.1428	66.95	4.97	4.07	24.01
C26H23F108	482.47	482.1377	64.73	4.81	3.94	26.53
C26H23F109	498.47	498.1326	62.65	4.65	3.81	28.89
C26H24F201	390.48	390.1795	79.98	6.20	9.73	4.10
C26H24F202	406.48	406.1744	76.83	5.95	9.35	7.87
C26H24F203	422.48	422.1693	73.92	5.73	8.99	11.36
C26H24F204	438.48	438.1643	71.22	5.52	8.67	14.60
C26H24F205	454.47	454.1592	68.71	5.32	8.36	17.60
C26H24F206	470.47	470.1541	66.38	5.14	8.08	20.40
C26H24F207	486.47	486.1490	64.19	4.97	7.81	23.02
C26H24F208	502.47	502.1439	62.15	4.81	7.56	25.47
C26H24F209	518.47	518.1388	60.23	4.67	7.33	27.77
C26H25F101	372.49	372.1889	83.84	6.77	5.10	4.30
C26H25F102	388.49	388.1838	80.39	6.49	4.89	8.24
C26H25F103	404.49	404.1788	77.21	6.23	4.70	11.87
C26H25F104	420.49	420.1737	74.27	5.99	4.52	15.22
C26H25F105	436.48	436.1686	71.55	5.77	4.35	18.33
C26H25F106	452.48	452.1635	69.02	5.57	4.20	21.22
C26H25F107	468.48	468.1584	66.66	5.38	4.06	23.91
C26H25F108	484.48	484.1533	64.46	5.20	3.92	26.42
C26H25F109	500.48	500.1482	62.40	5.03	3.80	28.77
C26H26F201	392.49	392.1952	79.57	6.68	9.68	4.08
C26H26F202	408.49	408.1901	76.45	6.42	9.30	7.83
C26H26F203	424.49	424.1850	73.57	6.17	8.95	11.31
C26H26F204	440.49	440.1799	70.90	5.95	8.63	14.53
C26H26F205	456.49	456.1748	68.41	5.74	8.32	17.52
C26H26F206	472.49	472.1697	66.09	5.55	8.04	20.32
C26H26F207	488.49	488.1646	63.93	5.36	7.78	22.93

167

C26H26F208	504.49	504.1596	61.90	5.19	7.53	25.37
C26H26F209	520.49	520.1545	60.00	5.04	7.30	27.67
C26H27F101	374.50	374.2046	83.39	7.27	5.07	4.27
C26H27F102	390.50	390.1995	79.97	6.97	4.87	8.19
C26H27F103	406.50	406.1944	76.82	6.69	4.67	11.81
C26H27F104	422.50	422.1893	73.91	6.44	4.50	15.15
C26H27F105	438.50	438.1842	71.22	6.21	4.33	18.24
C26H27F106	454.50	454.1792	68.71	5.99	4.18	21.12
C26H27F107	470.50	470.1741	66.37	5.78	4.04	23.80
C26H27F108	486.50	486.1690	64.19	5.59	3.91	26.31
C26H27F109	502.50	502.1639	62.15	5.42	3.78	28.66
C26H28F201	394.51	394.2108	79.16	7.15	9.63	4.06
C26H28F202	410.51	410.2057	76.07	6.88	9.26	7.79
C26H28F203	426.51	426.2006	73.22	6.62	8.91	11.25
C26H28F204	442.51	442.1955	70.57	6.38	8.59	14.46
C26H28F205	458.51	458.1905	68.11	6.16	8.29	17.45
C26H28F206	474.51	474.1854	65.81	5.95	8.01	20.23
C26H28F207	490.51	490.1803	63.67	5.75	7.75	22.83
C26H28F208	506.51	506.1752	61.66	5.57	7.50	25.27
C26H28F209	522.50	522.1701	59.77	5.40	7.27	27.56
C26H29F101	376.52	376.2202	82.94	7.76	5.05	4.25
C26H29F102	392.52	392.2151	79.56	7.45	4.84	8.15
C26H29F103	408.52	408.2101	76.44	7.16	4.65	11.75
C26H29F104	424.52	424.2050	73.56	6.89	4.48	15.08
C26H29F105	440.52	440.1999	70.89	6.64	4.31	18.16
C26H29F106	456.52	456.1948	68.41	6.40	4.16	21.03
C26H29F107	472.52	472.1897	66.09	6.19	4.02	23.70
C26H29F108	488.51	488.1846	63.93	5.98	3.89	26.20
C26H29F109	504.51	504.1795	61.90	5.79	3.77	28.54
C26H30F201	396.53	396.2265	78.76	7.63	9.58	4.03
C26H30F202	412.52	412.2214	75.70	7.33	9.21	7.76
C26H30F203	428.52	428.2163	72.88	7.06	8.87	11.20
C26H30F204	444.52	444.2112	70.25	6.80	8.55	14.40
C26H30F205	460.52	460.2061	67.81	6.57	8.25	17.37
C26H30F206	476.52	476.2010	65.54	6.35	7.97	20.15
C26H30F207	492.52	492.1959	63.41	6.14	7.71	22.74
C26H30F208	508.52	508.1909	61.41	5.95	7.47	25.17
C26H30F209	524.52	524.1858	59.54	5.77	7.24	27.45
C26H31F101	378.53	378.2359	82.50	8.25	5.02	4.23
C26H31F102	394.53	394.2308	79.15	7.92	4.82	8.11
C26H31F103	410.53	410.2257	76.07	7.61	4.63	11.69
C26H31F104	426.53	426.2206	73.22	7.33	4.45	15.00
C26H31F105	442.53	442.2155	70.57	7.06	4.29	18.08
C26H31F106	458.53	458.2104	68.11	6.81	4.14	20.94
C26H31F107	474.53	474.2054	65.81	6.58	4.00	23.60
C26H31F108	490.53	490.2003	63.66	6.37	3.87	26.09
C26H31F109	506.53	506.1952	61.65	6.17	3.75	28.43
C26H32F201	398.54	398.2421	78.36	8.09	9.53	4.01
C26H32F202	414.54	414.2370	75.33	7.78	9.17	7.72
C26H32F203	430.54	430.2319	72.53	7.49	8.83	11.15
C26H32F204	446.54	446.2268	69.94	7.22	8.51	14.33
C26H32F205	462.54	462.2218	67.52	6.97	8.21	17.30
C26H32F206	478.54	478.2167	65.26	6.74	7.94	20.06
C26H32F207	494.54	494.2116	63.15	6.52	7.68	22.65
C26H32F208	510.54	510.2065	61.17	6.32	7.44	25.07
C26H32F209	526.54	526.2014	59.31	6.13	7.22	27.35
C26H33F101	380.55	380.2515	82.06	8.74	4.99	4.20
C26H33F102	396.55	396.2464	78.75	8.39	4.79	8.07
C26H33F103	412.55	412.2414	75.70	8.06	4.61	11.63
C26H33F104	428.55	428.2363	72.87	7.76	4.43	14.93
C26H33F105	444.55	444.2312	70.25	7.48	4.27	18.00
C26H33F106	460.55	460.2261	67.81	7.22	4.13	20.84
C26H33F107	476.55	476.2210	65.53	6.98	3.99	23.50
C26H33F108	492.55	492.2159	63.40	6.75	3.86	25.99
C26H33F109	508.55	508.2108	61.41	6.54	3.74	28.31
C26H34F201	400.56	400.2578	77.96	8.56	9.49	3.99
C26H34F202	416.56	416.2527	74.97	8.23	9.12	7.68
C26H34F203	432.56	432.2476	72.20	7.92	8.78	11.10
C26H34F204	448.56	448.2425	69.62	7.64	8.47	14.27
C26H34F205	464.55	464.2374	67.22	7.38	8.18	17.22
C26H34F206	480.55	480.2323	64.99	7.13	7.91	19.98
C26H34F207	496.55	496.2272	62.89	6.90	7.65	22.55
C26H34F208	512.55	512.2222	60.93	6.69	7.41	24.97
C26H34F209	528.55	528.2171	59.08	6.48	7.19	27.24
C26H35F101	382.57	382.2672	81.63	9.22	4.97	4.18
C26H35F102	398.57	398.2621	78.35	8.85	4.77	8.03
C26H35F103	414.57	414.2570	75.33	8.51	4.58	11.58
C26H35F104	430.56	430.2519	72.53	8.19	4.41	14.86
C26H35F105	446.56	446.2468	69.93	7.90	4.25	17.91
C26H35F106	462.56	462.2417	67.51	7.63	4.11	20.75
C26H35F107	478.56	478.2367	65.26	7.37	3.97	23.40
C26H35F108	494.56	494.2316	63.14	7.13	3.84	25.88
C26H35F109	510.56	510.2265	61.17	6.91	3.72	28.20
C26H36F201	402.57	402.2734	77.57	9.01	9.44	3.97
C26H36F202	418.57	418.2683	74.61	8.67	9.08	7.64
C26H36F203	434.57	434.2632	71.86	8.35	8.74	11.04
C26H36F204	450.57	450.2581	69.31	8.05	8.43	14.20
C26H36F205	466.57	466.2531	66.93	7.78	8.14	17.15
C26H36F206	482.57	482.2480	64.71	7.52	7.87	19.89
C26H36F207	498.57	498.2429	62.64	7.28	7.62	22.46
C26H36F208	514.57	514.2378	60.69	7.05	7.38	24.87
C26H36F209	530.57	530.2327	58.86	6.84	7.16	27.14
C26H37F101	384.58	384.2828	81.20	9.70	4.94	4.16
C26H37F102	400.58	400.2777	77.96	9.31	4.74	7.99
C26H37F103	416.58	416.2727	74.96	8.95	4.56	11.52
C26H37F104	432.58	432.2676	72.19	8.62	4.39	14.79
C26H37F105	448.58	448.2625	69.62	8.31	4.24	17.83
C26H37F106	464.58	464.2574	67.22	8.03	4.09	20.66
C26H37F107	480.58	480.2523	64.98	7.76	3.95	23.30
C26H37F108	496.58	496.2472	62.89	7.51	3.83	25.78
C26H37F109	512.58	512.2421	60.93	7.28	3.71	28.09
C26H38F201	404.59	404.2891	77.19	9.47	9.39	3.95
C26H38F202	420.59	420.2840	74.25	9.11	9.03	7.61
C26H38F203	436.59	436.2789	71.53	8.77	8.70	10.99
C26H38F204	452.59	452.2738	69.00	8.46	8.40	14.14
C26H38F205	468.59	468.2687	66.65	8.17	8.11	17.07
C26H38F206	484.59	484.2636	64.44	7.90	7.84	19.81
C26H38F207	500.59	500.2585	62.38	7.65	7.59	22.37
C26H38F208	516.58	516.2535	60.45	7.41	7.36	24.78
C26H38F209	532.58	532.2484	58.64	7.19	7.13	27.04
C26H39F101	386.60	386.2985	80.78	10.17	4.91	4.14
C26H39F102	402.60	402.2934	77.57	9.76	4.72	7.95
C26H39F103	418.60	418.2883	74.60	9.39	4.54	11.47
C26H39F104	434.60	434.2832	71.86	9.05	4.37	14.73
C26H39F105	450.60	450.2781	69.31	8.72	4.22	17.75
C26H39F106	466.60	466.2730	66.93	8.43	4.07	20.57
C26H39F107	482.59	482.2680	64.71	8.15	3.94	23.21
C26H39F108	498.59	498.2629	62.63	7.88	3.81	25.67
C26H39F109	514.59	514.2578	60.69	7.64	3.69	27.98
C26H40F201	406.60	406.3047	76.80	9.92	9.34	3.93

168

C26H40F202	422.60	422.2996	73.90	9.54	8.99	7.57
C26H40F203	438.60	438.2945	71.20	9.19	8.66	10.94
C26H40F204	454.60	454.2894	68.70	8.87	8.36	14.08
C26H40F205	470.60	470.2844	66.36	8.57	8.07	17.00
C26H40F206	486.60	486.2793	64.18	8.29	7.81	19.73
C26H40F207	502.60	502.2742	62.13	8.02	7.56	22.28
C26H40F208	518.60	518.2691	60.22	7.77	7.33	24.68
C26H40F209	534.60	534.2640	58.42	7.54	7.11	26.94
C26H41F101	388.61	388.3141	80.36	10.63	4.89	4.12
C26H41F102	404.61	404.3090	77.18	10.21	4.70	7.91
C26H41F103	420.61	420.3040	74.25	9.83	4.52	11.41
C26H41F104	436.61	436.2989	71.53	9.47	4.35	14.66
C26H41F105	452.61	452.2938	69.00	9.13	4.20	17.67
C26H41F106	468.61	468.2887	66.64	8.82	4.05	20.49
C26H41F107	484.61	484.2836	64.44	8.53	3.92	23.11
C26H41F108	500.61	500.2785	62.38	8.26	3.80	25.57
C26H41F109	516.61	516.2734	60.45	8.00	3.68	27.87
C26H42F201	408.62	408.3204	76.43	10.36	9.30	3.92
C26H42F202	424.62	424.3153	73.55	9.97	8.95	7.54
C26H42F203	440.62	440.3102	70.88	9.61	8.62	10.89
C26H42F204	456.62	456.3051	68.39	9.27	8.32	14.02
C26H42F205	472.62	472.3000	66.08	8.96	8.04	16.93
C26H42F206	488.62	488.2949	63.91	8.66	7.78	19.65
C26H42F207	504.62	504.2898	61.89	8.39	7.53	22.19
C26H42F208	520.62	520.2848	59.98	8.13	7.30	24.59
C26H42F209	536.62	536.2797	58.20	7.89	7.08	26.83
C26H43F101	390.63	390.3298	79.95	11.10	4.86	4.10
C26H43F102	406.63	406.3247	76.80	10.66	4.67	7.87
C26H43F103	422.63	422.3196	73.89	10.26	4.50	11.36
C26H43F104	438.63	438.3145	71.20	9.88	4.33	14.59
C26H43F105	454.63	454.3094	68.69	9.53	4.18	17.60
C26H43F106	470.63	470.3043	66.36	9.21	4.04	20.40
C26H43F107	486.63	486.2993	64.17	8.91	3.90	23.01
C26H43F108	502.63	502.2942	62.13	8.62	3.78	25.47
C26H43F109	518.63	518.2891	60.21	8.36	3.66	27.76
C26H44F201	410.64	410.3360	76.05	10.80	9.25	3.90
C26H44F202	426.64	426.3309	73.20	10.40	8.91	7.50
C26H44F203	442.64	442.3258	70.55	10.02	8.58	10.84
C26H44F204	458.63	458.3207	68.09	9.67	8.28	13.95
C26H44F205	474.63	474.3157	65.80	9.34	8.01	16.85
C26H44F206	490.63	490.3106	63.65	9.04	7.74	19.57
C26H44F207	506.63	506.3055	61.64	8.75	7.50	22.11
C26H44F208	522.63	522.3004	59.75	8.49	7.27	24.49
C26H44F209	538.63	538.2953	57.98	8.23	7.05	26.73
C26H45F101	392.65	392.3454	79.53	11.55	4.84	4.07
C26H45F102	408.65	408.3403	76.42	11.10	4.65	7.83
C26H45F103	424.65	424.3353	73.54	10.68	4.47	11.30
C26H45F104	440.64	440.3302	70.87	10.29	4.31	14.52
C26H45F105	456.64	456.3251	68.39	9.93	4.16	17.52
C26H45F106	472.64	472.3200	66.07	9.60	4.02	20.31
C26H45F107	488.64	488.3149	63.91	9.28	3.89	22.92
C26H45F108	504.64	504.3098	61.88	8.99	3.76	25.36
C26H45F109	520.64	520.3047	59.98	8.71	3.65	27.66
C26H46F201	412.65	412.3517	75.68	11.24	9.21	3.88
C26H46F202	428.65	428.3466	72.85	10.82	8.86	7.46
C26H46F203	444.65	444.3415	70.23	10.43	8.55	10.79
C26H46F204	460.65	460.3364	67.79	10.07	8.25	13.89
C26H46F205	476.65	476.3313	65.52	9.73	7.97	16.78
C26H46F206	492.65	492.3262	63.39	9.41	7.71	19.49
C26H46F207	508.65	508.3211	61.40	9.12	7.47	22.02
C26H46F208	524.65	524.3160	59.52	8.84	7.24	24.40
C26H46F209	540.65	540.3110	57.76	8.58	7.03	26.63
C26H47F101	394.66	394.3611	79.13	12.00	4.81	4.05
C26H47F102	410.66	410.3560	76.05	11.54	4.63	7.79
C26H47F103	426.66	426.3509	73.19	11.10	4.45	11.25
C26H47F104	442.66	442.3458	70.55	10.70	4.29	14.46
C26H47F105	458.66	458.3407	68.09	10.33	4.14	17.44
C26H47F106	474.66	474.3356	65.79	9.98	4.00	20.22
C26H47F107	490.66	490.3306	63.65	9.66	3.87	22.83
C26H47F108	506.66	506.3255	61.64	9.35	3.75	25.26
C26H47F109	522.66	522.3204	59.75	9.06	3.63	27.55
C26H48F201	414.67	414.3673	75.31	11.67	9.16	3.86
C26H48F202	430.67	430.3622	72.51	11.23	8.82	7.43
C26H48F203	446.67	446.3571	69.92	10.83	8.51	10.75
C26H48F204	462.67	462.3520	67.50	10.46	8.21	13.83
C26H48F205	478.67	478.3470	65.24	10.11	7.94	16.71
C26H48F206	494.67	494.3419	63.13	9.78	7.68	19.41
C26H48F207	510.67	510.3368	61.15	9.47	7.44	21.93
C26H48F208	526.66	526.3317	59.30	9.19	7.21	24.30
C26H48F209	542.66	542.3266	57.55	8.92	7.00	26.53
C26H49F101	396.68	396.3767	78.73	12.45	4.79	4.03
C26H49F102	412.68	412.3716	75.67	11.97	4.60	7.75
C26H49F103	428.68	428.3665	72.85	11.52	4.43	11.20
C26H49F104	444.68	444.3615	70.23	11.11	4.27	14.39
C26H49F105	460.68	460.3564	67.79	10.72	4.12	17.37
C26H49F106	476.68	476.3513	65.51	10.36	3.99	20.14
C26H49F107	492.67	492.3462	63.39	10.02	3.86	22.73
C26H49F108	508.67	508.3411	61.39	9.71	3.73	25.1?
C26H49F109	524.67	524.3360	59.52	9.41	3.62	27.44
C26H50F201	416.68	416.3829	74.95	12.10	9.12	3.84
C26H50F202	432.68	432.3779	72.18	11.65	8.78	7.40
C26H50F203	448.68	448.3728	69.60	11.23	8.47	10.70
C26H50F204	464.68	464.3677	67.20	10.85	8.18	13.77
C26H50F205	480.68	480.3626	64.97	10.48	7.90	16.64
C26H50F206	496.68	496.3575	62.88	10.15	7.65	19.33
C26H50F207	512.68	512.3524	60.91	9.83	7.41	21.85
C26H50F208	528.68	528.3473	59.07	9.53	7.19	24.21
C26H50F209	544.68	544.3423	57.33	9.25	6.98	26.44
C26H51F101	398.69	398.3924	78.33	12.89	4.77	4.01
C26H51F102	414.69	414.3873	75.31	12.40	4.58	7.72
C26H51F103	430.69	430.3822	72.51	11.94	4.41	11.14
C26H51F104	446.69	446.3771	69.91	11.51	4.25	14.33
C26H51F105	462.69	462.3720	67.49	11.11	4.11	17.29
C26H51F106	478.69	478.3669	65.24	10.74	3.97	20.05
C26H51F107	494.69	494.3619	63.13	10.39	3.84	22.64
C26H51F108	510.69	510.3568	61.15	10.07	3.72	25.06
C26H51F109	526.69	526.3517	59.29	9.76	3.61	27.34
C26H52F201	418.70	418.3986	74.59	12.52	9.07	3.82
C26H52F202	434.70	434.3935	71.84	12.06	8.74	7.36
C26H52F203	450.70	450.3884	69.29	11.63	8.43	10.65
C26H52F204	466.70	466.3833	66.91	11.23	8.14	13.71
C26H52F205	482.70	482.3783	64.70	10.86	7.87	16.57
C26H52F206	498.70	498.3732	62.62	10.51	7.62	19.25
C26H52F207	514.70	514.3681	60.67	10.18	7.38	21.76
C26H52F208	530.70	530.3630	58.85	9.88	7.16	24.12
C26H52F209	546.70	546.3579	57.12	9.59	6.95	26.34
C26H53F101	400.71	400.4080	77.93	13.33	4.74	3.99
C26H53F102	416.71	416.4029	74.94	12.82	4.56	7.68
C26H53F103	432.71	432.3978	72.17	12.35	4.39	11.09
C26H53F104	448.71	448.3928	69.60	11.91	4.23	14.26

C26H53F105	464.71	464.3877	67.20	11.50	4.09	17.21
C26H53F106	480.71	480.3826	64.96	11.11	3.95	19.97
C26H53F107	496.71	496.3775	62.87	10.76	3.82	22.55
C26H53F108	512.71	512.3724	60.91	10.42	3.71	24.96
C26H53F109	528.71	528.3673	59.07	10.10	3.59	27.24
C27H13F101	372.40	372.0950	87.08	3.52	5.10	4.30
C27H13F102	388.40	388.0900	83.50	3.37	4.89	8.24
C27H13F103	404.40	404.0849	80.19	3.24	4.70	11.87
C27H13F104	420.40	420.0798	77.14	3.12	4.52	15.22
C27H13F105	436.40	436.0747	74.31	3.00	4.35	18.33
C27H13F106	452.40	452.0696	71.68	2.90	4.20	21.22
C27H13F107	468.40	468.0645	69.24	2.80	4.06	23.91
C27H13F108	484.40	484.0594	66.95	2.71	3.92	26.42
C27H13F109	500.40	500.0543	64.81	2.62	3.80	28.78
C27H14F201	392.41	392.1013	82.64	3.60	9.68	4.08
C27H14F202	408.41	408.0962	79.41	3.46	9.30	7.84
C27H14F203	424.41	424.0911	76.41	3.33	8.95	11.31
C27H14F204	440.41	440.0860	73.64	3.20	8.63	14.53
C27H14F205	456.41	456.0809	71.06	3.09	8.33	17.53
C27H14F206	472.41	472.0758	68.65	2.99	8.04	20.32
C27H14F207	488.41	488.0707	66.40	2.89	7.78	22.93
C27H14F208	504.40	504.0657	64.29	2.80	7.53	25.38
C27H14F209	520.40	520.0606	62.32	2.71	7.30	27.67
C27H15F101	374.42	374.1107	86.61	4.04	5.07	4.27
C27H15F102	390.42	390.1056	83.07	3.87	4.87	8.20
C27H15F103	406.42	406.1005	79.80	3.72	4.67	11.81
C27H15F104	422.42	422.0954	76.77	3.58	4.50	15.15
C27H15F105	438.42	438.0903	73.97	3.45	4.33	18.25
C27H15F106	454.42	454.0853	71.37	3.33	4.18	21.13
C27H15F107	470.41	470.0802	68.94	3.21	4.04	23.81
C27H15F108	486.41	486.0751	66.67	3.11	3.91	26.31
C27H15F109	502.41	502.0700	64.55	3.01	3.78	28.66
C27H16F201	394.42	394.1169	82.22	4.09	9.63	4.06
C27H16F202	410.42	410.1118	79.02	3.93	9.26	7.80
C27H16F203	426.42	426.1067	76.05	3.78	8.91	11.26
C27H16F204	442.42	442.1017	73.30	3.65	8.59	14.47
C27H16F205	458.42	458.0966	70.74	3.52	8.29	17.45
C27H16F206	474.42	474.0915	68.36	3.40	8.01	20.23
C27H16F207	490.42	490.0864	66.13	3.29	7.75	22.84
C27H16F208	506.42	506.0813	64.04	3.18	7.50	25.27
C27H16F209	522.42	522.0762	62.08	3.09	7.27	27.56
C27H17F101	376.43	376.1263	86.15	4.55	5.05	4.25
C27H17F102	392.43	392.1212	82.64	4.37	4.84	8.15
C27H17F103	408.43	408.1162	79.40	4.20	4.65	11.75
C27H17F104	424.43	424.1111	76.41	4.04	4.48	15.08
C27H17F105	440.43	440.1060	73.63	3.89	4.31	18.16
C27H17F106	456.43	456.1009	71.05	3.75	4.16	21.03
C27H17F107	472.43	472.0958	68.65	3.63	4.02	23.71
C27H17F108	488.43	488.0907	66.40	3.51	3.89	26.21
C27H17F109	504.43	504.0856	64.29	3.40	3.77	28.55
C27H18F201	396.44	396.1326	81.80	4.58	9.58	4.04
C27H18F202	412.44	412.1275	78.63	4.40	9.21	7.76
C27H18F203	428.44	428.1224	75.69	4.23	8.87	11.20
C27H18F204	444.44	444.1173	72.97	4.08	8.55	14.40
C27H18F205	460.44	460.1122	70.43	3.94	8.25	17.37
C27H18F206	476.44	476.1071	68.07	3.81	7.98	20.15
C27H18F207	492.44	492.1020	65.86	3.68	7.72	22.74
C27H18F208	508.44	508.0970	63.78	3.57	7.47	25.17
C27H18F209	524.44	524.0919	61.84	3.46	7.25	27.46
C27H19F101	378.45	378.1420	85.69	5.06	5.02	4.23
C27H19F102	394.45	394.1369	82.22	4.86	4.82	8.11
C27H19F103	410.45	410.1318	79.01	4.67	4.63	11.69
C27H19F104	426.45	426.1267	76.05	4.49	4.46	15.01
C27H19F105	442.45	442.1216	73.30	4.33	4.29	18.08
C27H19F106	458.45	458.1166	70.74	4.18	4.14	20.94
C27H19F107	474.45	474.1115	68.35	4.04	4.00	23.61
C27H19F108	490.45	490.1064	66.12	3.90	3.87	26.10
C27H19F109	506.45	506.1013	64.03	3.78	3.75	28.43
C27H20F201	398.46	398.1482	81.39	5.06	9.54	4.02
C27H20F202	414.46	414.1431	78.25	4.86	9.17	7.72
C27H20F203	430.46	430.1380	75.34	4.68	8.83	11.15
C27H20F204	446.45	446.1330	72.64	4.52	8.51	14.33
C27H20F205	462.45	462.1279	70.13	4.36	8.22	17.30
C27H20F206	478.45	478.1228	67.78	4.21	7.94	20.06
C27H20F207	494.45	494.1177	65.59	4.08	7.68	22.65
C27H20F208	510.45	510.1126	63.53	3.95	7.44	25.07
C27H20F209	526.45	526.1075	61.60	3.83	7.22	27.35
C27H21F101	380.47	380.1576	85.24	5.56	4.99	4.21
C27H21F102	396.47	396.1525	81.80	5.34	4.79	8.07
C27H21F103	412.47	412.1475	78.63	5.13	4.61	11.64
C27H21F104	428.46	428.1424	75.69	4.94	4.43	14.94
C27H21F105	444.46	444.1373	72.96	4.76	4.27	18.00
C27H21F106	460.46	460.1322	70.43	4.60	4.13	20.85
C27H21F107	476.46	476.1271	68.06	4.44	3.99	23.51
C27H21F108	492.46	492.1220	65.85	4.30	3.86	25.99
C27H21F109	508.46	508.1169	63.78	4.16	3.74	28.32
C27H22F201	400.47	400.1639	80.98	5.54	9.49	4.00
C27H22F202	416.47	416.1588	77.87	5.32	9.12	7.68
C27H22F203	432.47	432.1537	74.99	5.13	8.79	11.10
C27H22F204	448.47	448.1486	72.31	4.94	8.47	14.27
C27H22F205	464.47	464.1435	69.82	4.77	8.18	17.22
C27H22F206	480.47	480.1384	67.50	4.62	7.91	19.98
C27H22F207	496.47	496.1333	65.32	4.47	7.65	22.56
C27H22F208	512.47	512.1283	63.28	4.33	7.41	24.98
C27H22F209	528.47	528.1232	61.37	4.20	7.19	27.25
C27H23F101	382.48	382.1733	84.79	6.06	4.97	4.18
C27H23F102	398.48	398.1682	81.38	5.82	4.77	8.03
C27H23F103	414.48	414.1631	78.24	5.59	4.58	11.58
C27H23F104	430.48	430.1580	75.33	5.39	4.41	14.87
C27H23F105	446.48	446.1529	72.64	5.19	4.26	17.92
C27H23F106	462.48	462.1479	70.12	5.01	4.11	20.76
C27H23F107	478.48	478.1428	67.78	4.85	3.97	23.41
C27H23F108	494.48	494.1377	65.58	4.69	3.84	25.88
C27H23F109	510.48	510.1326	63.53	4.54	3.72	28.21
C27H24F201	402.49	402.1795	80.57	6.01	9.44	3.98
C27H24F202	418.49	418.1744	77.49	5.78	9.08	7.65
C27H24F203	434.49	434.1693	74.64	5.57	8.75	11.05
C27H24F204	450.49	450.1643	71.99	5.37	8.43	14.21
C27H24F205	466.49	466.1592	69.52	5.19	8.15	17.15
C27H24F206	482.49	482.1541	67.21	5.01	7.88	19.90
C27H24F207	498.48	498.1490	65.06	4.85	7.62	22.47
C27H24F208	514.48	514.1439	63.03	4.70	7.39	24.88
C27H24F209	530.48	530.1388	61.13	4.56	7.16	27.14
C27H25F101	384.50	384.1889	84.34	6.55	4.94	4.16
C27H25F102	400.50	400.1838	80.97	6.29	4.74	7.99
C27H25F103	416.50	416.1788	77.86	6.05	4.56	11.52
C27H25F104	432.50	432.1737	74.98	5.83	4.39	14.80
C27H25F105	448.50	448.1686	72.31	5.62	4.24	17.84
C27H25F106	464.50	464.1635	69.82	5.43	4.09	20.67
C27H25F107	480.49	480.1584	67.49	5.24	3.95	23.31

C27H25F108	496.49	496.1533	65.32	5.08	3.83	25.78
C27H25F109	512.49	512.1482	63.28	4.92	3.71	28.10
C27H26F201	404.50	404.1952	80.17	6.48	9.39	3.96
C27H26F202	420.50	420.1901	77.12	6.23	9.04	7.61
C27H26F203	436.50	436.1850	74.30	6.00	8.70	11.00
C27H26F204	452.50	452.1799	71.67	5.79	8.40	14.14
C27H26F205	468.50	468.1748	69.22	5.59	8.11	17.08
C27H26F206	484.50	484.1697	66.93	5.41	7.84	19.81
C27H26F207	500.50	500.1646	64.80	5.24	7.59	22.38
C27H26F208	516.50	516.1596	62.79	5.07	7.36	24.78
C27H26F209	532.50	532.1545	60.90	4.92	7.14	27.04
C27H27F101	386.51	386.2046	83.90	7.04	4.92	4.14
C27H27F102	402.51	402.1995	80.57	6.76	4.72	7.95
C27H27F103	418.51	418.1944	77.49	6.50	4.54	11.47
C27H27F104	434.51	434.1893	74.64	6.26	4.37	14.73
C27H27F105	450.51	450.1842	71.99	6.04	4.22	17.76
C27H27F106	466.51	466.1792	69.52	5.83	4.07	20.58
C27H27F107	482.51	482.1741	67.21	5.64	3.94	23.21
C27H27F108	498.51	498.1690	65.05	5.46	3.81	25.68
C27H27F109	514.51	514.1639	63.03	5.29	3.69	27.99
C27H28F201	406.52	406.2108	79.77	6.94	9.35	3.94
C27H28F202	422.52	422.2057	76.75	6.68	8.99	7.57
C27H28F203	438.52	438.2006	73.95	6.44	8.66	10.95
C27H28F204	454.52	454.1955	71.35	6.21	8.36	14.08
C27H28F205	470.52	470.1905	68.92	6.00	8.08	17.00
C27H28F206	486.52	486.1854	66.66	5.80	7.81	19.73
C27H28F207	502.52	502.1803	64.54	5.62	7.56	22.29
C27H28F208	518.52	518.1752	62.54	5.44	7.33	24.68
C27H28F209	534.52	534.1701	60.67	5.28	7.11	26.94
C27H29F101	388.53	388.2202	83.47	7.52	4.89	4.12
C27H29F102	404.53	404.2151	80.17	7.23	4.70	7.91
C27H29F103	420.53	420.2101	77.12	6.95	4.52	11.41
C27H29F104	436.53	436.2050	74.29	6.70	4.35	14.66
C27H29F105	452.53	452.1999	71.66	6.46	4.20	17.68
C27H29F106	468.53	468.1948	69.22	6.24	4.05	20.49
C27H29F107	484.53	484.1897	66.93	6.03	3.92	23.11
C27H29F108	500.53	500.1846	64.79	5.84	3.80	25.57
C27H29F109	516.53	516.1795	62.79	5.66	3.68	27.88
C27H30F201	408.54	408.2265	79.38	7.40	9.30	3.92
C27H30F202	424.54	424.2214	76.39	7.12	8.95	7.54
C27H30F203	440.54	440.2163	73.62	6.86	8.63	10.90
C27H30F204	456.53	456.2112	71.04	6.62	8.32	14.02
C27H30F205	472.53	472.2061	68.63	6.40	8.04	16.93
C27H30F206	488.53	488.2010	66.38	6.19	7.78	19.65
C27H30F207	504.53	504.1959	64.28	5.99	7.53	22.20
C27H30F208	520.53	520.1909	62.30	5.81	7.30	24.59
C27H30F209	536.53	536.1858	60.44	5.64	7.08	26.84
C27H31F101	390.55	390.2359	83.04	8.00	4.86	4.10
C27H31F102	406.55	406.2308	79.77	7.69	4.67	7.87
C27H31F103	422.54	422.2257	76.75	7.39	4.50	11.36
C27H31F104	438.54	438.2206	73.95	7.13	4.33	14.59
C27H31F105	454.54	454.2155	71.35	6.87	4.18	17.60
C27H31F106	470.54	470.2104	68.92	6.64	4.04	20.40
C27H31F107	486.54	486.2054	66.65	6.42	3.90	23.02
C27H31F108	502.54	502.2003	64.53	6.22	3.78	25.47
C27H31F109	518.54	518.1952	62.54	6.03	3.66	27.77
C27H32F201	410.55	410.2421	78.99	7.86	9.26	3.90
C27H32F202	426.55	426.2370	76.03	7.56	8.91	7.50
C27H32F203	442.55	442.2319	73.28	7.29	8.59	10.85
C27H32F204	458.55	458.2268	70.72	7.03	8.29	13.96
C27H32F205	474.55	474.2218	68.34	6.80	8.01	16.86
C27H32F206	490.55	490.2167	66.11	6.58	7.75	19.57
C27H32F207	506.55	506.2116	64.02	6.37	7.50	22.11
C27H32F208	522.55	522.2065	62.06	6.17	7.27	24.49
C27H32F209	538.55	538.2014	60.22	5.99	7.06	26.74
C27H33F101	392.56	392.2515	82.61	8.47	4.84	4.08
C27H33F102	408.56	408.2464	79.38	8.14	4.65	7.83
C27H33F103	424.56	424.2414	76.39	7.83	4.47	11.31
C27H33F104	440.56	440.2363	73.61	7.55	4.31	14.53
C27H33F105	456.56	456.2312	71.03	7.29	4.16	17.52
C27H33F106	472.56	472.2261	68.63	7.04	4.02	20.31
C27H33F107	488.56	488.2210	66.38	6.81	3.89	22.92
C27H33F108	504.56	504.2159	64.27	6.59	3.77	25.37
C27H33F109	520.56	520.2108	62.30	6.39	3.65	27.66
C27H34F201	412.57	412.2578	78.61	8.31	9.21	3.88
C27H34F202	428.57	428.2527	75.67	8.00	8.87	7.47
C27H34F203	444.57	444.2476	72.95	7.71	8.55	10.80
C27H34F204	460.57	460.2425	70.41	7.44	8.25	13.90
C27H34F205	476.57	476.2374	68.05	7.19	7.97	16.79
C27H34F206	492.57	492.2323	65.84	6.96	7.71	19.49
C27H34F207	508.56	508.2272	63.77	6.74	7.47	22.02
C27H34F208	524.56	524.2222	61.82	6.53	7.24	24.40
C27H34F209	540.56	540.2171	59.99	6.34	7.03	26.64
C27H35F101	394.58	394.2672	82.19	8.94	4.81	4.05
C27H35F102	410.58	410.2621	78.99	8.59	4.63	7.79
C27H35F103	426.58	426.2570	76.02	8.27	4.45	11.25
C27H35F104	442.58	442.2519	73.28	7.97	4.29	14.46
C27H35F105	458.58	458.2468	70.72	7.69	4.14	17.44
C27H35F106	474.57	474.2417	68.34	7.43	4.00	20.23
C27H35F107	490.57	490.2367	66.11	7.19	3.87	22.83
C27H35F108	506.57	506.2316	64.02	6.96	3.75	25.27
C27H35F109	522.57	522.2265	62.06	6.75	3.64	27.55
C27H36F201	414.58	414.2734	78.22	8.75	9.17	3.86
C27H36F202	430.58	430.2683	75.32	8.43	8.82	7.43
C27H36F203	446.58	446.2632	72.62	8.13	8.51	10.75
C27H36F204	462.58	462.2581	70.11	7.84	8.21	13.83
C27H36F205	478.58	478.2531	67.76	7.58	7.94	16.72
C27H36F206	494.58	494.2480	65.57	7.34	7.68	19.41
C27H36F207	510.58	510.2429	63.52	7.11	7.44	21.93
C27H36F208	526.58	526.2378	61.59	6.89	7.22	24.31
C27H36F209	542.58	542.2327	59.77	6.69	7.00	26.54
C27H37F101	396.59	396.2828	81.77	9.40	4.79	4.03
C27H37F102	412.59	412.2777	78.60	9.04	4.60	7.76
C27H37F103	428.59	428.2727	75.67	8.70	4.43	11.20
C27H37F104	444.59	444.2676	72.94	8.39	4.27	14.39
C27H37F105	460.59	460.2625	70.41	8.10	4.12	17.37
C27H37F106	476.59	476.2574	68.05	7.83	3.99	20.14
C27H37F107	492.59	492.2523	65.84	7.57	3.86	22.74
C27H37F108	508.59	508.2472	63.76	7.33	3.74	25.17
C27H37F109	524.59	524.2421	61.82	7.11	3.62	27.45
C27H38F201	416.60	416.2891	77.84	9.19	9.12	3.84
C27H38F202	432.60	432.2840	74.97	8.85	8.78	7.40
C27H38F203	448.60	448.2789	72.29	8.54	8.47	10.70
C27H38F204	464.60	464.2738	69.80	8.24	8.18	13.77
C27H38F205	480.60	480.2687	67.48	7.97	7.91	16.65
C27H38F206	496.60	496.2636	65.30	7.71	7.65	19.33
C27H38F207	512.60	512.2585	63.27	7.47	7.41	21.85
C27H38F208	528.60	528.2535	61.35	7.25	7.19	24.21
C27H38F209	544.60	544.2484	59.55	7.03	6.98	26.44
C27H39F101	398.61	398.2985	81.36	9.86	4.77	4.01

171

C27H39F102	414.61	414.2934	78.22	9.48	4.58	7.72
C27H39F103	430.61	430.2883	75.31	9.13	4.41	11.15
C27H39F104	446.61	446.2832	72.61	8.80	4.25	14.33
C27H39F105	462.61	462.2781	70.10	8.50	4.11	17.29
C27H39F106	478.61	478.2730	67.76	8.21	3.97	20.06
C27H39F107	494.61	494.2680	65.57	7.95	3.84	22.64
C27H39F108	510.61	510.2629	63.51	7.70	3.72	25.07
C27H39F109	526.60	526.2578	61.58	7.46	3.61	27.34
C27H40F201	418.62	418.3047	77.47	9.63	9.08	3.82
C27H40F202	434.62	434.2996	74.62	9.28	8.74	7.36
C27H40F203	450.61	450.2945	71.97	8.95	8.43	10.65
C27H40F204	466.61	466.2894	69.50	8.64	8.14	13.72
C27H40F205	482.61	482.2844	67.20	8.35	7.87	16.58
C27H40F206	498.61	498.2793	65.04	8.09	7.62	19.25
C27H40F207	514.61	514.2742	63.02	7.83	7.38	21.76
C27H40F208	530.61	530.2691	61.12	7.60	7.16	24.12
C27H40F209	546.61	546.2640	59.33	7.38	6.95	26.34
C27H41F101	400.63	400.3141	80.95	10.32	4.74	3.99
C27H41F102	416.63	416.3090	77.84	9.92	4.56	7.68
C27H41F103	432.62	432.3040	74.96	9.55	4.39	11.09
C27H41F104	448.62	448.2989	72.29	9.21	4.23	14.27
C27H41F105	464.62	464.2938	69.80	8.89	4.09	17.22
C27H41F106	480.62	480.2887	67.48	8.60	3.95	19.97
C27H41F107	496.62	496.2836	65.30	8.32	3.83	22.55
C27H41F108	512.62	512.2785	63.26	8.06	3.71	24.97
C27H41F109	528.62	528.2734	61.35	7.82	3.59	27.24
C27H42F201	420.63	420.3204	77.10	10.06	9.03	3.80
C27H42F202	436.63	436.3153	74.27	9.70	8.70	7.33
C27H42F203	452.63	452.3102	71.65	9.35	8.39	10.60
C27H42F204	468.63	468.3051	69.20	9.03	8.11	13.66
C27H42F205	484.63	484.3000	66.92	8.74	7.84	16.51
C27H42F206	500.63	500.2949	64.78	8.46	7.59	19.18
C27H42F207	516.63	516.2898	62.77	8.19	7.35	21.68
C27H42F208	532.63	532.2848	60.89	7.95	7.13	24.03
C27H42F209	548.63	548.2797	59.11	7.72	6.93	26.25
C27H43F101	402.64	402.3298	80.54	10.76	4.72	3.97
C27H43F102	418.64	418.3247	77.47	10.35	4.54	7.64
C27H43F103	434.64	434.3196	74.61	9.97	4.37	11.04
C27H43F104	450.64	450.3145	71.96	9.62	4.22	14.20
C27H43F105	466.64	466.3094	69.50	9.29	4.07	17.14
C27H43F106	482.64	482.3043	67.19	8.98	3.94	19.89
C27H43F107	498.64	498.2993	65.04	8.69	3.81	22.46
C27H43F108	514.64	514.2942	63.02	8.42	3.69	24.87
C27H43F109	530.64	530.2891	61.12	8.17	3.58	27.14
C27H44F201	422.65	422.3360	76.73	10.49	8.99	3.79
C27H44F202	438.65	438.3309	73.93	10.11	8.66	7.29
C27H44F203	454.65	454.3258	71.33	9.75	8.36	10.56
C27H44F204	470.65	470.3207	68.91	9.42	8.07	13.60
C27H44F205	486.65	486.3157	66.64	9.11	7.81	16.44
C27H44F206	502.64	502.3106	64.52	8.82	7.56	19.10
C27H44F207	518.64	518.3055	62.53	8.55	7.33	21.59
C27H44F208	534.64	534.3004	60.66	8.30	7.11	23.94
C27H44F209	550.64	550.2953	58.89	8.05	6.90	26.15
C27H45F101	404.66	404.3454	80.14	11.21	4.69	3.95
C27H45F102	420.66	420.3403	77.09	10.78	4.52	7.61
C27H45F103	436.66	436.3353	74.27	10.39	4.35	10.99
C27H45F104	452.66	452.3302	71.64	10.02	4.20	14.14
C27H45F105	468.66	468.3251	69.20	9.68	4.05	17.07
C27H45F106	484.65	484.3200	66.91	9.36	3.92	19.81
C27H45F107	500.65	500.3149	64.78	9.06	3.79	22.37
C27H45F108	516.65	516.3098	62.77	8.78	3.68	24.77
C27H45F109	532.65	532.3047	60.88	8.52	3.57	27.03
C27H46F201	424.66	424.3517	76.37	10.92	8.95	3.77
C27H46F202	440.66	440.3466	73.59	10.52	8.62	7.26
C27H46F203	456.66	456.3415	71.02	10.15	8.32	10.51
C27H46F204	472.66	472.3364	68.61	9.81	8.04	13.54
C27H46F205	488.66	488.3313	66.37	9.49	7.78	16.37
C27H46F206	504.66	504.3262	64.26	9.19	7.53	19.02
C27H46F207	520.66	520.3211	62.29	8.91	7.30	21.51
C27H46F208	536.66	536.3160	60.43	8.64	7.08	23.85
C27H46F209	552.66	552.3110	58.68	8.39	6.88	26.05
C27H47F101	406.67	406.3611	79.74	11.65	4.67	3.93
C27H47F102	422.67	422.3560	76.73	11.21	4.49	7.57
C27H47F103	438.67	438.3509	73.93	10.80	4.33	10.94
C27H47F104	454.67	454.3458	71.33	10.42	4.18	14.08
C27H47F105	470.67	470.3407	68.90	10.07	4.04	17.00
C27H47F106	486.67	486.3356	66.64	9.73	3.90	19.73
C27H47F107	502.67	502.3306	64.52	9.42	3.78	22.28
C27H47F108	518.67	518.3255	62.53	9.13	3.66	24.68
C27H47F109	534.67	534.3204	60.65	8.86	3.55	26.93
C27H48F201	426.68	426.3673	76.01	11.34	8.91	3.75
C27H48F202	442.68	442.3622	73.26	10.93	8.58	7.23
C27H48F203	458.68	458.3571	70.70	10.55	8.28	10.46
C27H48F204	474.68	474.3520	68.32	10.19	8.00	13.48
C27H48F205	490.68	490.3470	66.09	9.86	7.74	16.30
C27H48F206	506.68	506.3419	64.01	9.55	7.50	18.95
C27H48F207	522.68	522.3368	62.05	9.26	7.27	21.43
C27H48F208	538.68	538.3317	60.20	8.98	7.05	23.76
C27H48F209	554.68	554.3266	58.47	8.72	6.85	25.96
C27H49F101	408.69	408.3767	79.35	12.09	4.65	3.91
C27H49F102	424.69	424.3716	76.36	11.63	4.47	7.53
C27H49F103	440.69	440.3665	73.59	11.21	4.31	10.89
C27H49F104	456.69	456.3615	71.01	10.81	4.16	14.01
C27H49F105	472.69	472.3564	68.61	10.45	4.02	16.92
C27H49F106	488.69	488.3513	66.36	10.11	3.89	19.64
C27H49F107	504.69	504.3462	64.26	9.79	3.76	22.19
C27H49F108	520.69	520.3411	62.28	9.49	3.65	24.58
C27H49F109	536.68	536.3360	60.43	9.20	3.54	26.83
C27H50F201	428.70	428.3829	75.65	11.76	8.86	3.73
C27H50F202	444.70	444.3779	72.93	11.33	8.54	7.20
C27H50F203	460.69	460.3728	70.39	10.94	8.25	10.42
C27H50F204	476.69	476.3677	68.03	10.57	7.97	13.43
C27H50F205	492.69	492.3626	65.82	10.23	7.71	16.24
C27H50F206	508.69	508.3575	63.75	9.91	7.47	18.87
C27H50F207	524.69	524.3524	61.81	9.61	7.24	21.35
C27H50F208	540.69	540.3473	59.98	9.32	7.03	23.67
C27H50F209	556.69	556.3423	58.26	9.05	6.83	25.87
C27H51F101	410.71	410.3924	78.96	12.52	4.63	3.90
C27H51F102	426.70	426.3873	76.00	12.05	4.45	7.50
C27H51F103	442.70	442.3822	73.25	11.61	4.29	10.84
C27H51F104	458.70	458.3771	70.70	11.21	4.14	13.95
C27H51F105	474.70	474.3720	68.32	10.83	4.00	16.85
C27H51F106	490.70	490.3669	66.09	10.48	3.87	19.56
C27H51F107	506.70	506.3619	64.00	10.15	3.75	22.10
C27H51F108	522.70	522.3568	62.04	9.83	3.63	24.49
C27H51F109	538.70	538.3517	60.20	9.54	3.53	26.73
C27H52F201	430.71	430.3986	75.29	12.17	8.82	3.71
C27H52F202	446.71	446.3935	72.60	11.73	8.51	7.16
C27H52F203	462.71	462.3884	70.09	11.33	8.21	10.37
C27H52F204	478.71	478.3833	67.74	10.95	7.94	13.37

172

C27H52F205	494.71	494.3783	65.55	10.59	7.68	16.17
C27H52F206	510.71	510.3732	63.50	10.26	7.44	18.80
C27H52F207	526.71	526.3681	61.57	9.95	7.21	21.26
C27H52F208	542.71	542.3630	59.76	9.66	7.00	23.58
C27H52F209	558.71	558.3579	58.04	9.38	6.80	25.77
C27H53F101	412.72	412.4080	78.58	12.94	4.60	3.88
C27H53F102	428.72	428.4029	75.64	12.46	4.43	7.46
C27H53F103	444.72	444.3978	72.92	12.01	4.27	10.79
C27H53F104	460.72	460.3928	70.39	11.60	4.12	13.89
C27H53F105	476.72	476.3877	68.03	11.21	3.99	16.78
C27H53F106	492.72	492.3826	65.82	10.84	3.86	19.48
C27H53F107	508.72	508.3775	63.75	10.50	3.73	22.02
C27H53F108	524.72	524.3724	61.80	10.18	3.62	24.39
C27H53F109	540.72	540.3673	59.98	9.88	3.51	26.63
C27H54F201	432.73	432.4142	74.94	12.58	8.78	3.70
C27H54F202	448.73	448.4092	72.27	12.13	8.47	7.13
C27H54F203	464.73	464.4041	69.78	11.71	8.18	10.33
C27H54F204	480.73	480.3990	67.46	11.32	7.90	13.31
C27H54F205	496.73	496.3939	65.29	10.96	7.65	16.10
C27H54F206	512.72	512.3888	63.25	10.62	7.41	18.72
C27H54F207	528.72	528.3837	61.34	10.29	7.19	21.18
C27H54F208	544.72	544.3786	59.53	9.99	6.98	23.50
C27H54F209	560.72	560.3736	57.84	9.71	6.78	25.68
C27H55F101	414.74	414.4237	78.19	13.37	4.58	3.86
C27H55F102	430.74	430.4186	75.29	12.87	4.41	7.43
C27H55F103	446.74	446.4135	72.59	12.41	4.25	10.74
C27H55F104	462.74	462.4084	70.08	11.98	4.11	13.83
C27H55F105	478.73	478.4033	67.74	11.58	3.97	16.71
C27H55F106	494.73	494.3982	65.55	11.21	3.84	19.40
C27H55F107	510.73	510.3932	63.50	10.85	3.72	21.93
C27H55F108	526.73	526.3881	61.57	10.52	3.61	24.30
C27H55F109	542.73	542.3830	59.75	10.21	3.50	26.53
C28H14F201	404.42	404.1013	83.16	3.49	9.40	3.96
C28H14F202	420.42	420.0962	79.99	3.36	9.04	7.61
C28H14F203	436.42	436.0911	77.06	3.23	8.71	11.00
C28H14F204	452.42	452.0860	74.34	3.12	8.40	14.15
C28H14F205	468.42	468.0809	71.80	3.01	8.11	17.08
C28H14F206	484.42	484.0758	69.43	2.91	7.84	19.82
C28H14F207	500.42	500.0707	67.21	2.82	7.59	22.38
C28H14F208	516.42	516.0657	65.12	2.73	7.36	24.79
C28H14F209	532.42	532.0606	63.17	2.65	7.14	27.05
C28H15F101	386.43	386.1107	87.03	3.91	4.92	4.14
C28H15F102	402.43	402.1056	83.57	3.76	4.72	7.95
C28H15F103	418.43	418.1005	80.38	3.61	4.54	11.47
C28H15F104	434.43	434.0954	77.41	3.48	4.37	14.73
C28H15F105	450.43	450.0903	74.67	3.36	4.22	17.76
C28H15F106	466.43	466.0853	72.10	3.24	4.07	20.58
C28H15F107	482.43	482.0802	69.71	3.13	3.94	23.22
C28H15F108	498.43	498.0751	67.47	3.03	3.81	25.68
C28H15F109	514.42	514.0700	65.38	2.94	3.69	27.99
C28H16F201	406.44	406.1169	82.75	3.97	9.35	3.94
C28H16F202	422.44	422.1118	79.61	3.82	8.99	7.57
C28H16F203	438.43	438.1067	76.71	3.68	8.67	10.95
C28H16F204	454.43	454.1017	74.01	3.55	8.36	14.08
C28H16F205	470.43	470.0966	71.49	3.43	8.08	17.00
C28H16F206	486.43	486.0915	69.14	3.32	7.81	19.73
C28H16F207	502.43	502.0864	66.94	3.21	7.56	22.29
C28H16F208	518.43	518.0813	64.87	3.11	7.33	24.69
C28H16F209	534.43	534.0762	62.93	3.02	7.11	26.94
C28H17F101	388.45	388.1263	86.58	4.41	4.89	4.12
C28H17F102	404.44	404.1212	83.15	4.24	4.70	7.91
C28H17F103	420.44	420.1162	79.99	4.08	4.52	11.42
C28H17F104	436.44	436.1111	77.06	3.93	4.35	14.66
C28H17F105	452.44	452.1060	74.33	3.79	4.20	17.68
C28H17F106	468.44	468.1009	71.79	3.66	4.06	20.49
C28H17F107	484.44	484.0958	69.42	3.54	3.92	23.12
C28H17F108	500.44	500.0907	67.20	3.42	3.80	25.58
C28H17F109	516.44	516.0856	65.12	3.32	3.68	27.88
C28H18F201	408.45	408.1326	82.34	4.44	9.30	3.92
C28H18F202	424.45	424.1275	79.23	4.27	8.95	7.54
C28H18F203	440.45	440.1224	76.36	4.12	8.63	10.90
C28H18F204	456.45	456.1173	73.68	3.97	8.32	14.02
C28H18F205	472.45	472.1122	71.18	3.84	8.04	16.93
C28H18F206	488.45	488.1071	68.85	3.71	7.78	19.65
C28H18F207	504.45	504.1020	66.67	3.60	7.53	22.20
C28H18F208	520.45	520.0970	64.62	3.49	7.30	24.59
C28H18F209	536.45	536.0919	62.69	3.38	7.08	26.84
C28H19F101	390.46	390.1420	86.13	4.90	4.87	4.10
C28H19F102	406.46	406.1369	82.74	4.71	4.67	7.87
C28H19F103	422.46	422.1318	79.61	4.53	4.50	11.36
C28H19F104	438.46	438.1267	76.70	4.37	4.33	14.60
C28H19F105	454.46	454.1216	74.00	4.21	4.18	17.60
C28H19F106	470.46	470.1166	71.49	4.07	4.04	20.40
C28H19F107	486.46	486.1115	69.13	3.94	3.91	23.02
C28H19F108	502.46	502.1064	66.93	3.81	3.78	25.47
C28H19F109	518.46	518.1013	64.87	3.69	3.66	27.77
C28H20F201	410.47	410.1482	81.93	4.91	9.26	3.90
C28H20F202	426.47	426.1431	78.86	4.73	8.91	7.50
C28H20F203	442.47	442.1380	76.01	4.56	8.59	10.85
C28H20F204	458.47	458.1330	73.36	4.40	8.29	13.96
C28H20F205	474.47	474.1279	70.88	4.25	8.01	16.86
C28H20F206	490.46	490.1228	68.57	4.11	7.75	19.57
C28H20F207	506.46	506.1177	66.40	3.98	7.50	22.11
C28H20F208	522.46	522.1126	64.37	3.86	7.27	24.50
C28H20F209	538.46	538.1075	62.46	3.74	7.06	26.74
C28H21F101	392.48	392.1576	85.69	5.39	4.84	4.08
C28H21F102	408.48	408.1525	82.33	5.18	4.65	7.83
C28H21F103	424.48	424.1475	79.23	4.99	4.48	11.31
C28H21F104	440.48	440.1424	76.35	4.81	4.31	14.53
C28H21F105	456.47	456.1373	73.68	4.64	4.16	17.52
C28H21F106	472.47	472.1322	71.18	4.48	4.02	20.32
C28H21F107	488.47	488.1271	68.85	4.33	3.89	22.93
C28H21F108	504.47	504.1220	66.67	4.20	3.77	25.37
C28H21F109	520.47	520.1169	64.62	4.07	3.65	27.67
C28H22F201	412.48	412.1639	81.53	5.38	9.21	3.88
C28H22F202	428.48	428.1588	78.49	5.18	8.87	7.47
C28H22F203	444.48	444.1537	75.66	4.99	8.55	10.80
C28H22F204	460.48	460.1486	73.03	4.82	8.25	13.90
C28H22F205	476.48	476.1435	70.58	4.65	7.97	16.79
C28H22F206	492.48	492.1384	68.29	4.50	7.72	19.49
C28H22F207	508.48	508.1333	66.14	4.36	7.47	22.03
C28H22F208	524.48	524.1283	64.12	4.23	7.24	24.40
C28H22F209	540.48	540.1232	62.22	4.10	7.03	26.64
C28H23F101	394.49	394.1733	85.25	5.88	4.82	4.06
C28H23F102	410.49	410.1682	81.93	5.65	4.63	7.80
C28H23F103	426.49	426.1631	78.86	5.44	4.45	11.25
C28H23F104	442.49	442.1580	76.00	5.24	4.29	14.46
C28H23F105	458.49	458.1529	73.35	5.06	4.14	17.45
C28H23F106	474.49	474.1479	70.88	4.89	4.00	20.23
C28H23F107	490.49	490.1428	68.57	4.73	3.87	22.83

173

C28H23F108	506.49	506.1377	66.40	4.58	3.75	25.27
C28H23F109	522.49	522.1326	64.37	4.44	3.64	27.56
C28H24F201	414.50	414.1795	81.14	5.84	9.17	3.86
C28H24F202	430.50	430.1744	78.12	5.62	8.83	7.43
C28H24F203	446.50	446.1693	75.32	5.42	8.51	10.75
C28H24F204	462.50	462.1643	72.72	5.23	8.22	13.84
C28H24F205	478.50	478.1592	70.29	5.06	7.94	16.72
C28H24F206	494.50	494.1541	68.01	4.89	7.68	19.41
C28H24F207	510.50	510.1490	65.88	4.74	7.44	21.94
C28H24F208	526.50	526.1439	63.88	4.59	7.22	24.31
C28H24F209	542.49	542.1388	61.99	4.46	7.00	26.54
C28H25F101	396.51	396.1889	84.82	6.36	4.79	4.04
C28H25F102	412.51	412.1838	81.53	6.11	4.61	7.76
C28H25F103	428.51	428.1788	78.48	5.88	4.43	11.20
C28H25F104	444.51	444.1737	75.66	5.67	4.27	14.40
C28H25F105	460.51	460.1686	73.03	5.47	4.13	17.37
C28H25F106	476.51	476.1635	70.58	5.29	3.99	20.15
C28H25F107	492.51	492.1584	68.29	5.12	3.86	22.74
C28H25F108	508.51	508.1533	66.14	4.96	3.74	25.17
C28H25F109	524.50	524.1482	64.12	4.80	3.62	27.45
C28H26F201	416.52	416.1952	80.74	6.29	9.12	3.84
C28H26F202	432.52	432.1901	77.76	6.06	8.79	7.40
C28H26F203	448.51	448.1850	74.98	5.84	8.47	10.70
C28H26F204	464.51	464.1799	72.40	5.64	8.18	13.78
C28H26F205	480.51	480.1748	69.99	5.45	7.91	16.65
C28H26F206	496.51	496.1697	67.73	5.28	7.65	19.33
C28H26F207	512.51	512.1646	65.62	5.11	7.41	21.85
C28H26F208	528.51	528.1596	63.63	4.96	7.19	24.22
C28H26F209	544.51	544.1545	61.76	4.81	6.98	26.44
C28H27F101	398.53	398.2046	84.39	6.83	4.77	4.01
C28H27F102	414.52	414.1995	81.13	6.57	4.58	7.72
C28H27F103	430.52	430.1944	78.12	6.32	4.41	11.15
C28H27F104	446.52	446.1893	75.32	6.09	4.25	14.33
C28H27F105	462.52	462.1842	72.71	5.88	4.11	17.30
C28H27F106	478.52	478.1792	70.28	5.69	3.97	20.06
C28H27F107	494.52	494.1741	68.01	5.50	3.84	22.65
C28H27F108	510.52	510.1690	65.88	5.33	3.72	25.07
C28H27F109	526.52	526.1639	63.87	5.17	3.61	27.35
C28H28F201	418.53	418.2108	80.36	6.74	9.08	3.82
C28H28F202	434.53	434.2057	77.40	6.50	8.74	7.36
C28H28F203	450.53	450.2006	74.65	6.26	8.43	10.65
C28H28F204	466.53	466.1955	72.09	6.05	8.14	13.72
C28H28F205	482.53	482.1905	69.70	5.85	7.87	16.58
C28H28F206	498.53	498.1854	67.46	5.66	7.62	19.26
C28H28F207	514.53	514.1803	65.36	5.49	7.38	21.77
C28H28F208	530.53	530.1752	63.39	5.32	7.16	24.13
C28H28F209	546.53	546.1701	61.54	5.16	6.95	26.35
C28H29F101	400.54	400.2202	83.96	7.30	4.74	3.99
C28H29F102	416.54	416.2151	80.74	7.02	4.56	7.68
C28H29F103	432.54	432.2101	77.75	6.76	4.39	11.10
C28H29F104	448.54	448.2050	74.98	6.52	4.24	14.27
C28H29F105	464.54	464.1999	72.40	6.29	4.09	17.22
C28H29F106	480.54	480.1948	69.99	6.08	3.95	19.98
C28H29F107	496.54	496.1897	67.73	5.89	3.83	22.56
C28H29F108	512.54	512.1846	65.62	5.70	3.71	24.97
C28H29F109	528.54	528.1795	63.63	5.53	3.59	27.24
C28H30F201	420.55	420.2265	79.97	7.19	9.04	3.80
C28H30F202	436.55	436.2214	77.04	6.93	8.70	7.33
C28H30F203	452.55	452.2163	74.32	6.68	8.40	10.61
C28H30F204	468.55	468.2112	71.78	6.45	8.11	13.66

C28H30F205	484.55	484.2061	69.41	6.24	7.84	16.51
C28H30F206	500.54	500.2010	67.19	6.04	7.59	19.18
C28H30F207	516.54	516.1959	65.11	5.85	7.36	21.68
C28H30F208	532.54	532.1909	63.15	5.68	7.13	24.03
C28H30F209	548.54	548.1858	61.31	5.51	6.93	26.25
C28H31F101	402.56	402.2359	83.54	7.76	4.72	3.97
C28H31F102	418.56	418.2308	80.35	7.47	4.54	7.65
C28H31F103	434.56	434.2257	77.39	7.19	4.37	11.05
C28H31F104	450.56	450.2206	74.64	6.94	4.22	14.20
C28H31F105	466.55	466.2155	72.08	6.70	4.07	17.15
C28H31F106	482.55	482.2104	69.69	6.48	3.94	19.89
C28H31F107	498.55	498.2054	67.46	6.27	3.81	22.46
C28H31F108	514.55	514.2003	65.36	6.07	3.69	24.88
C28H31F109	530.55	530.1952	63.39	5.89	3.58	27.14
C28H32F201	422.56	422.2421	79.59	7.63	8.99	3.79
C28H32F202	438.56	438.2370	76.69	7.35	8.66	7.30
C28H32F203	454.56	454.2319	73.99	7.10	8.36	10.56
C28H32F204	470.56	470.2268	71.47	6.85	8.07	13.60
C28H32F205	486.56	486.2218	69.12	6.63	7.81	16.44
C28H32F206	502.56	502.2167	66.92	6.42	7.56	19.10
C28H32F207	518.56	518.2116	64.86	6.22	7.33	21.60
C28H32F208	534.56	534.2065	62.91	6.03	7.11	23.94
C28H32F209	550.56	550.2014	61.09	5.86	6.90	26.15
C28H33F101	404.57	404.2515	83.13	8.22	4.70	3.95
C28H33F102	420.57	420.2464	79.97	7.91	4.52	7.61
C28H33F103	436.57	436.2414	77.03	7.62	4.35	10.99
C28H33F104	452.57	452.2363	74.31	7.35	4.20	14.14
C28H33F105	468.57	468.2312	71.77	7.10	4.05	17.07
C28H33F106	484.57	484.2261	69.40	6.86	3.92	19.81
C28H33F107	500.57	500.2210	67.19	6.65	3.80	22.37
C28H33F108	516.57	516.2159	65.11	6.44	3.68	24.78
C28H33F109	532.57	532.2108	63.15	6.25	3.57	27.04
C28H34F201	424.58	424.2578	79.21	8.07	8.95	3.77
C28H34F202	440.58	440.2527	76.33	7.78	8.62	7.26
C28H34F203	456.58	456.2476	73.66	7.51	8.32	10.51
C28H34F204	472.58	472.2425	71.17	7.25	8.04	13.54
C28H34F205	488.58	488.2374	68.84	7.01	7.78	16.37
C28H34F206	504.58	504.2323	66.65	6.79	7.53	19.03
C28H34F207	520.58	520.2272	64.60	6.58	7.30	21.51
C28H34F208	536.58	536.2222	62.68	6.39	7.08	23.85
C28H34F209	552.57	552.2171	60.86	6.20	6.88	26.06
C28H35F101	406.59	406.2672	82.72	8.68	4.67	3.94
C28H35F102	422.59	422.2621	79.58	8.35	4.50	7.57
C28H35F103	438.59	438.2570	76.68	8.04	4.33	10.94
C28H35F104	454.59	454.2519	73.98	7.76	4.18	14.08
C28H35F105	470.59	470.2468	71.47	7.50	4.04	17.00
C28H35F106	486.59	486.2417	69.12	7.25	3.90	19.73
C28H35F107	502.59	502.2367	66.92	7.02	3.78	22.28
C28H35F108	518.58	518.2316	64.85	6.80	3.66	24.68
C28H35F109	534.58	534.2265	62.91	6.60	3.55	26.94
C28H36F201	426.60	426.2734	78.84	8.51	8.91	3.75
C28H36F202	442.59	442.2683	75.99	8.20	8.59	7.23
C28H36F203	458.59	458.2632	73.34	7.91	8.29	10.47
C28H36F204	474.59	474.2581	70.86	7.65	8.01	13.48
C28H36F205	490.59	490.2531	68.55	7.40	7.75	16.31
C28H36F206	506.59	506.2480	66.39	7.16	7.50	18.95
C28H36F207	522.59	522.2429	64.35	6.94	7.27	21.43
C28H36F208	538.59	538.2378	62.44	6.74	7.05	23.76
C28H36F209	554.59	554.2327	60.64	6.54	6.85	25.96
C28H37F101	408.60	408.2828	82.31	9.13	4.65	3.92

174

C28H37F102	424.60	424.2777	79.21	8.78	4.47	7.54
C28H37F103	440.60	440.2727	76.33	8.46	4.31	10.89
C28H37F104	456.60	456.2676	73.66	8.17	4.16	14.02
C28H37F105	472.60	472.2625	71.16	7.89	4.02	16.93
C28H37F106	488.60	488.2574	68.83	7.63	3.89	19.65
C28H37F107	504.60	504.2523	66.65	7.39	3.77	22.19
C28H37F108	520.60	520.2472	64.60	7.16	3.65	24.59
C28H37F109	536.60	536.2421	62.67	6.95	3.54	26.83
C28H38F201	428.61	428.2891	78.47	8.94	8.87	3.73
C28H38F202	444.61	444.2840	75.64	8.61	8.55	7.20
C28H38F203	460.61	460.2789	73.01	8.32	8.25	10.42
C28H38F204	476.61	476.2738	70.56	8.04	7.97	13.43
C28H38F205	492.61	492.2687	68.27	7.78	7.71	16.24
C28H38F206	508.61	508.2636	66.12	7.53	7.47	18.87
C28H38F207	524.61	524.2585	64.11	7.30	7.24	21.35
C28H38F208	540.61	540.2535	62.21	7.09	7.03	23.68
C28H38F209	556.61	556.2484	60.42	6.88	6.83	25.87
C28H39F101	410.62	410.2985	81.90	9.57	4.63	3.90
C28H39F102	426.62	426.2934	78.83	9.21	4.45	7.50
C28H39F103	442.62	442.2883	75.98	8.88	4.29	10.84
C28H39F104	458.62	458.2832	73.33	8.57	4.14	13.95
C28H39F105	474.62	474.2781	70.86	8.28	4.00	16.86
C28H39F106	490.62	490.2730	68.55	8.01	3.87	19.57
C28H39F107	506.62	506.2680	66.38	7.76	3.75	22.11
C28H39F108	522.62	522.2629	64.35	7.52	3.64	24.49
C28H39F109	538.62	538.2578	62.44	7.30	3.53	26.73
C28H40F201	430.63	430.3047	78.10	9.36	8.82	3.72
C28H40F202	446.63	446.2996	75.30	9.03	8.51	7.16
C28H40F203	462.63	462.2945	72.70	8.72	8.21	10.38
C28H40F204	478.63	478.2894	70.27	8.42	7.94	13.37
C28H40F205	494.62	494.2844	67.99	8.15	7.68	16.17
C28H40F206	510.62	510.2793	65.86	7.90	7.44	18.80
C28H40F207	526.62	526.2742	63.86	7.66	7.22	21.27
C28H40F208	542.62	542.2691	61.98	7.43	7.00	23.59
C28H40F209	558.62	558.2640	60.20	7.22	6.80	25.78
C28H41F101	412.64	412.3141	81.50	10.02	4.60	3.88
C28H41F102	428.64	428.3090	78.46	9.64	4.43	7.47
C28H41F103	444.64	444.3040	75.64	9.29	4.27	10.79
C28H41F104	460.63	460.2989	73.01	8.97	4.12	13.89
C28H41F105	476.63	476.2938	70.56	8.67	3.99	16.78
C28H41F106	492.63	492.2887	68.27	8.39	3.86	19.49
C28H41F107	508.63	508.2836	66.12	8.13	3.74	22.02
C28H41F108	524.63	524.2785	64.10	7.88	3.62	24.40
C28H41F109	540.63	540.2734	62.21	7.64	3.51	26.63
C28H42F201	432.64	432.3204	77.73	9.79	8.78	3.70
C28H42F202	448.64	448.3153	74.96	9.44	8.47	7.13
C28H42F203	464.64	464.3102	72.38	9.11	8.18	10.33
C28H42F204	480.64	480.3051	69.97	8.81	7.91	13.32
C28H42F205	496.64	496.3000	67.72	8.52	7.65	16.11
C28H42F206	512.64	512.2949	65.60	8.26	7.41	18.73
C28H42F207	528.64	528.2898	63.62	8.01	7.19	21.19
C28H42F208	544.64	544.2848	61.75	7.77	6.98	23.50
C28H42F209	560.64	560.2797	59.99	7.55	6.78	25.68
C28H43F101	414.65	414.3298	81.11	10.45	4.58	3.86
C28H43F102	430.65	430.3247	78.09	10.06	4.41	7.43
C28H43F103	446.65	446.3196	75.30	9.70	4.25	10.75
C28H43F104	462.65	462.3145	72.69	9.37	4.11	13.83
C28H43F105	478.65	478.3094	70.26	9.06	3.97	16.71
C28H43F106	494.65	494.3043	67.99	8.76	3.84	19.41
C28H43F107	510.65	510.2993	65.86	8.49	3.72	21.93

C28H43F108	526.65	526.2942	63.86	8.23	3.61	24.30
C28H43F109	542.65	542.2891	61.98	7.99	3.50	26.54
C28H44F201	434.66	434.3360	77.37	10.20	8.74	3.68
C28H44F202	450.66	450.3309	74.63	9.84	8.43	7.10
C28H44F203	466.66	466.3258	72.07	9.50	8.14	10.29
C28H44F204	482.66	482.3207	69.68	9.19	7.87	13.26
C28H44F205	498.66	498.3157	67.44	8.89	7.62	16.04
C28H44F206	514.66	514.3106	65.35	8.62	7.38	18.65
C28H44F207	530.66	530.3055	63.38	8.36	7.16	21.11
C28H44F208	546.65	546.3004	61.52	8.11	6.95	23.41
C28H44F209	562.65	562.2953	59.77	7.88	6.75	25.59
C28H45F101	416.67	416.3454	80.71	10.89	4.56	3.84
C28H45F102	432.67	432.3403	77.73	10.48	4.39	7.40
C28H45F103	448.67	448.3353	74.96	10.11	4.23	10.70
C28H45F104	464.67	464.3302	72.38	9.76	4.09	13.77
C28H45F105	480.67	480.3251	69.97	9.44	3.95	16.64
C28H45F106	496.67	496.3200	67.71	9.13	3.83	19.33
C28H45F107	512.67	512.3149	65.60	8.85	3.71	21.85
C28H45F108	528.66	528.3098	63.62	8.58	3.59	24.21
C28H45F109	544.66	544.3047	61.75	8.33	3.49	26.44
C28H46F201	436.68	436.3517	77.02	10.62	8.70	3.66
C28H46F202	452.67	452.3466	74.29	10.24	8.39	7.07
C28H46F203	468.67	468.3415	71.76	9.89	8.11	10.24
C28H46F204	484.67	484.3364	69.39	9.57	7.84	13.20
C28H46F205	500.67	500.3313	67.17	9.26	7.59	15.98
C28H46F206	516.67	516.3262	65.09	8.97	7.35	18.58
C28H46F207	532.67	532.3211	63.14	8.70	7.13	21.03
C28H46F208	548.67	548.3160	61.30	8.45	6.93	23.33
C28H46F209	564.67	564.3110	59.56	8.21	6.73	25.50
C28H47F101	418.68	418.3611	80.33	11.32	4.54	3.82
C28H47F102	434.68	434.3560	77.37	10.90	4.37	7.36
C28H47F103	450.68	450.3509	74.62	10.51	4.22	10.65
C28H47F104	466.68	466.3458	72.06	10.15	4.07	13.71
C28H47F105	482.68	482.3407	69.68	9.81	3.94	16.57
C28H47F106	498.68	498.3356	67.44	9.50	3.81	19.25
C28H47F107	514.68	514.3306	65.34	9.20	3.69	21.76
C28H47F108	530.68	530.3255	63.37	8.93	3.58	24.12
C28H47F109	546.68	546.3204	61.52	8.67	3.48	26.34
C28H48F201	438.69	438.3673	76.66	11.03	8.66	3.65
C28H48F202	454.69	454.3622	73.97	10.64	8.36	7.04
C28H48F203	470.69	470.3571	71.45	10.28	8.07	10.20
C28H48F204	486.69	486.3520	69.10	9.94	7.81	13.15
C28H48F205	502.69	502.3470	66.90	9.62	7.56	15.91
C28H48F206	518.69	518.3419	64.84	9.33	7.33	18.51
C28H48F207	534.69	534.3368	62.90	9.05	7.11	20.95
C28H48F208	550.69	550.3317	61.07	8.79	6.90	23.24
C28H48F209	566.69	566.3266	59.35	8.54	6.71	25.41
C28H49F101	420.70	420.3767	79.94	11.74	4.52	3.80
C28H49F102	436.70	436.3716	77.01	11.31	4.35	7.33
C28H49F103	452.70	452.3665	74.29	10.91	4.20	10.60
C28H49F104	468.70	468.3615	71.75	10.54	4.05	13.65
C28H49F105	484.70	484.3564	69.39	10.19	3.92	16.50
C28H49F106	500.70	500.3513	67.17	9.86	3.79	19.17
C28H49F107	516.70	516.3462	65.09	9.56	3.68	21.68
C28H49F108	532.70	532.3411	63.13	9.27	3.57	24.03
C28H49F109	548.70	548.3360	61.29	9.00	3.46	26.24
C28H50F201	440.71	440.3829	76.31	11.44	8.62	3.63
C28H50F202	456.71	456.3779	73.64	11.04	8.32	7.01
C28H50F203	472.71	472.3728	71.15	10.66	8.04	10.15
C28H50F204	488.71	488.3677	68.82	10.31	7.77	13.10

175

C28H50F205	504.70	504.3626	66.64	9.99	7.53	15.85
C28H50F206	520.70	520.3575	64.59	9.68	7.30	18.44
C28H50F207	536.70	536.3524	62.66	9.39	7.08	20.87
C28H50F208	552.70	552.3473	60.85	9.12	6.87	23.16
C28H50F209	568.70	568.3423	59.14	8.86	6.68	25.32
C28H51F101	422.72	422.3924	79.56	12.16	4.49	3.78
C28H51F102	438.72	438.3873	76.66	11.72	4.33	7.29
C28H51F103	454.72	454.3822	73.96	11.31	4.18	10.56
C28H51F104	470.71	470.3771	71.45	10.92	4.04	13.60
C28H51F105	486.71	486.3720	69.10	10.56	3.90	16.44
C28H51F106	502.71	502.3669	66.90	10.23	3.78	19.10
C28H51F107	518.71	518.3619	64.84	9.91	3.66	21.59
C28H51F108	534.71	534.3568	62.90	9.61	3.55	23.94
C28H51F109	550.71	550.3517	61.07	9.33	3.45	26.15
C28H52F201	442.72	442.3986	75.96	11.84	8.58	3.61
C28H52F202	458.72	458.3935	73.31	11.43	8.28	6.98
C28H52F203	474.72	474.3884	70.84	11.04	8.00	10.11
C28H52F204	490.72	490.3833	68.53	10.68	7.74	13.04
C28H52F205	506.72	506.3783	66.37	10.34	7.50	15.79
C28H52F206	522.72	522.3732	64.34	10.03	7.27	18.36
C28H52F207	538.72	538.3681	62.43	9.73	7.05	20.79
C28H52F208	554.72	554.3630	60.63	9.45	6.85	23.07
C28H52F209	570.72	570.3579	58.93	9.18	6.66	25.23
C28H53F101	424.73	424.4080	79.18	12.58	4.47	3.77
C28H53F102	440.73	440.4029	76.31	12.12	4.31	7.26
C28H53F103	456.73	456.3978	73.63	11.70	4.16	10.51
C28H53F104	472.73	472.3928	71.14	11.30	4.02	13.54
C28H53F105	488.73	488.3877	68.81	10.93	3.89	16.37
C28H53F106	504.73	504.3826	66.63	10.58	3.76	19.02
C28H53F107	520.73	520.3775	64.58	10.26	3.65	21.51
C28H53F108	536.73	536.3724	62.66	9.95	3.54	23.85
C28H53F109	552.73	552.3673	60.85	9.67	3.44	26.05
C28H54F201	444.74	444.4142	75.62	12.24	8.54	3.60
C28H54F202	460.74	460.4092	72.99	11.81	8.25	6.95
C28H54F203	476.74	476.4041	70.54	11.42	7.97	10.07
C28H54F204	492.74	492.3990	68.25	11.05	7.71	12.99
C28H54F205	508.74	508.3939	66.11	10.70	7.47	15.72
C28H54F206	524.74	524.3888	64.09	10.37	7.24	18.29
C28H54F207	540.74	540.3837	62.20	10.07	7.03	20.71
C28H54F208	556.73	556.3786	60.41	9.78	6.82	22.99
C28H54F209	572.73	572.3736	58.72	9.50	6.63	25.14
C28H55F101	426.75	426.4237	78.81	12.99	4.45	3.75
C28H55F102	442.75	442.4186	75.96	12.52	4.29	7.23
C28H55F103	458.75	458.4135	73.31	12.08	4.14	10.46
C28H55F104	474.75	474.4084	70.84	11.68	4.00	13.48
C28H55F105	490.75	490.4033	68.53	11.30	3.87	16.30
C28H55F106	506.75	506.3982	66.37	10.94	3.75	18.94
C28H55F107	522.74	522.3932	64.34	10.61	3.63	21.42
C28H55F108	538.74	538.3881	62.43	10.29	3.53	23.76
C28H55F109	554.74	554.3830	60.62	9.99	3.42	25.96
C28H56F201	446.75	446.4299	75.28	12.63	8.51	3.58
C28H56F202	462.75	462.4248	72.68	12.20	8.21	6.91
C28H56F203	478.75	478.4197	70.25	11.79	7.94	10.03
C28H56F204	494.75	494.4146	67.98	11.41	7.68	12.94
C28H56F205	510.75	510.4096	65.85	11.05	7.44	15.66
C28H56F206	526.75	526.4045	63.85	10.72	7.21	18.22
C28H56F207	542.75	542.3994	61.96	10.40	7.00	20.63
C28H56F208	558.75	558.3943	60.19	10.10	6.80	22.91
C28H56F209	574.75	574.3892	58.51	9.82	6.61	25.05
C28H57F101	428.76	428.4393	78.44	13.40	4.43	3.73
C28H57F102	444.76	444.4342	75.62	12.92	4.27	7.19
C28H57F103	460.76	460.4291	72.99	12.47	4.12	10.42
C28H57F104	476.76	476.4241	70.54	12.05	3.98	13.42
C28H57F105	492.76	492.4190	68.25	11.66	3.86	16.23
C28H57F106	508.76	508.4139	66.10	11.29	3.73	18.87
C28H57F107	524.76	524.4088	64.09	10.95	3.62	21.34
C28H57F108	540.76	540.4037	62.19	10.62	3.51	23.67
C28H57F109	556.76	556.3986	60.41	10.32	3.41	25.86
C29H14F201	416.43	416.1013	83.64	3.39	9.12	3.84
C29H14F202	432.43	432.0962	80.55	3.26	8.79	7.40
C29H14F203	448.43	448.0911	77.68	3.15	8.47	10.70
C29H14F204	464.43	464.0860	75.00	3.04	8.18	13.78
C29H14F205	480.43	480.0809	72.50	2.94	7.91	16.65
C29H14F206	496.43	496.0758	70.17	2.84	7.65	19.34
C29H14F207	512.43	512.0707	67.98	2.75	7.42	21.86
C29H14F208	528.43	528.0657	65.92	2.67	7.19	24.22
C29H14F209	544.43	544.0606	63.98	2.59	6.98	26.45
C29H15F101	398.44	398.1107	87.42	3.79	4.77	4.02
C29H15F102	414.44	414.1056	84.05	3.65	4.58	7.72
C29H15F103	430.44	430.1005	80.92	3.51	4.41	11.15
C29H15F104	446.44	446.0954	78.02	3.39	4.26	14.34
C29H15F105	462.44	462.0903	75.32	3.27	4.11	17.30
C29H15F106	478.44	478.0853	72.80	3.16	3.97	20.06
C29H15F107	494.44	494.0802	70.45	3.06	3.84	22.65
C29H15F108	510.44	510.0751	68.24	2.96	3.72	25.08
C29H15F109	526.44	526.0700	66.17	2.87	3.61	27.35
C29H16F201	418.45	418.1169	83.24	3.85	9.08	3.82
C29H16F202	434.45	434.1118	80.18	3.71	8.75	7.37
C29H16F203	450.45	450.1067	77.33	3.58	8.44	10.66
C29H16F204	466.45	466.1017	74.68	3.46	8.15	13.72
C29H16F205	482.44	482.0966	72.20	3.34	7.88	16.58
C29H16F206	498.44	498.0915	69.88	3.24	7.62	19.26
C29H16F207	514.44	514.0864	67.71	3.13	7.39	21.77
C29H16F208	530.44	530.0813	65.67	3.04	7.16	24.13
C29H16F209	546.44	546.0762	63.74	2.95	6.95	26.35
C29H17F101	400.46	400.1263	86.98	4.28	4.74	4.00
C29H17F102	416.46	416.1212	83.64	4.11	4.56	7.68
C29H17F103	432.46	432.1162	80.55	3.96	4.39	11.10
C29H17F104	448.45	448.1111	77.67	3.82	4.24	14.27
C29H17F105	464.45	464.1060	75.00	3.69	4.09	17.22
C29H17F106	480.45	480.1009	72.50	3.57	3.95	19.98
C29H17F107	496.45	496.0958	70.16	3.45	3.83	22.56
C29H17F108	512.45	512.0907	67.97	3.34	3.71	24.98
C29H17F109	528.45	528.0856	65.91	3.24	3.60	27.25
C29H18F201	420.46	420.1326	82.84	4.32	9.04	3.81
C29H18F202	436.46	436.1275	79.81	4.16	8.71	7.33
C29H18F203	452.46	452.1224	76.98	4.01	8.40	10.61
C29H18F204	468.46	468.1173	74.35	3.87	8.11	13.66
C29H18F205	484.46	484.1122	71.90	3.75	7.84	16.51
C29H18F206	500.46	500.1071	69.60	3.63	7.59	19.18
C29H18F207	516.46	516.1020	67.44	3.51	7.36	21.69
C29H18F208	532.46	532.0970	65.42	3.41	7.14	24.04
C29H18F209	548.46	548.0919	63.51	3.31	6.93	26.25
C29H19F101	402.47	402.1420	86.55	4.76	4.72	3.98
C29H19F102	418.47	418.1369	83.24	4.58	4.54	7.65
C29H19F103	434.47	434.1318	80.17	4.41	4.37	11.05
C29H19F104	450.47	450.1267	77.32	4.25	4.22	14.21
C29H19F105	466.47	466.1216	74.67	4.11	4.07	17.15
C29H19F106	482.47	482.1166	72.20	3.97	3.94	19.90
C29H19F107	498.47	498.1115	69.88	3.84	3.81	22.47

176

C29H19F108	514.47	514.1064	67.71	3.72	3.69	24.88
C29H19F109	530.47	530.1013	65.66	3.61	3.58	27.14
C29H20F201	422.48	422.1482	82.45	4.77	8.99	3.79
C29H20F202	438.48	438.1431	79.44	4.60	8.67	7.30
C29H20F203	454.48	454.1380	76.64	4.44	8.36	10.56
C29H20F204	470.48	470.1330	74.04	4.28	8.08	13.60
C29H20F205	486.48	486.1279	71.60	4.14	7.81	16.44
C29H20F206	502.48	502.1228	69.32	4.01	7.56	19.10
C29H20F207	518.48	518.1177	67.18	3.89	7.33	21.60
C29H20F208	534.47	534.1126	65.17	3.77	7.11	23.95
C29H20F209	550.47	550.1075	63.28	3.66	6.90	26.16
C29H21F101	404.49	404.1576	86.11	5.23	4.70	3.96
C29H21F102	420.49	420.1525	82.84	5.03	4.52	7.61
C29H21F103	436.49	436.1475	79.80	4.85	4.35	11.00
C29H21F104	452.49	452.1424	76.98	4.68	4.20	14.14
C29H21F105	468.49	468.1373	74.35	4.52	4.06	17.08
C29H21F106	484.49	484.1322	71.90	4.37	3.92	19.81
C29H21F107	500.48	500.1271	69.60	4.23	3.80	22.38
C29H21F108	516.48	516.1220	67.44	4.10	3.68	24.78
C29H21F109	532.48	532.1169	65.41	3.98	3.57	27.04
C29H22F201	424.49	424.1639	82.06	5.22	8.95	3.77
C29H22F202	440.49	440.1588	79.08	5.03	8.63	7.26
C29H22F203	456.49	456.1537	76.30	4.86	8.32	10.51
C29H22F204	472.49	472.1486	73.72	4.69	8.04	13.54
C29H22F205	488.49	488.1435	71.31	4.54	7.78	16.38
C29H22F206	504.49	504.1384	69.04	4.40	7.53	19.03
C29H22F207	520.49	520.1333	66.92	4.26	7.30	21.52
C29H22F208	536.49	536.1283	64.93	4.13	7.08	23.86
C29H22F209	552.49	552.1232	63.05	4.01	6.88	26.06
C29H23F101	406.50	406.1733	85.69	5.70	4.67	3.94
C29H23F102	422.50	422.1682	82.44	5.49	4.50	7.57
C29H23F103	438.50	438.1631	79.43	5.29	4.33	10.95
C29H23F104	454.50	454.1580	76.64	5.10	4.18	14.08
C29H23F105	470.50	470.1529	74.03	4.93	4.04	17.00
C29H23F106	486.50	486.1479	71.60	4.77	3.91	19.73
C29H23F107	502.50	502.1428	69.32	4.61	3.78	22.29
C29H23F108	518.50	518.1377	67.18	4.47	3.66	24.69
C29H23F109	534.50	534.1326	65.17	4.34	3.55	26.94
C29H24F201	426.51	426.1795	81.67	5.67	8.91	3.75
C29H24F202	442.51	442.1744	78.72	5.47	8.59	7.23
C29H24F203	458.51	458.1693	75.97	5.28	8.29	10.47
C29H24F204	474.51	474.1643	73.41	5.10	8.01	13.49
C29H24F205	490.51	490.1592	71.01	4.93	7.75	16.31
C29H24F206	506.51	506.1541	68.77	4.78	7.50	18.95
C29H24F207	522.51	522.1490	66.66	4.63	7.27	21.43
C29H24F208	538.51	538.1439	64.68	4.49	7.06	23.77
C29H24F209	554.51	554.1388	62.82	4.36	6.85	25.97
C29H25F101	408.52	408.1889	85.26	6.17	4.65	3.92
C29H25F102	424.52	424.1838	82.05	5.94	4.48	7.54
C29H25F103	440.52	440.1788	79.07	5.72	4.31	10.90
C29H25F104	456.52	456.1737	76.30	5.52	4.16	14.02
C29H25F105	472.52	472.1686	73.72	5.33	4.02	16.93
C29H25F106	488.52	488.1635	71.30	5.16	3.89	19.65
C29H25F107	504.52	504.1584	69.04	4.99	3.77	22.20
C29H25F108	520.52	520.1533	66.92	4.84	3.65	24.59
C29H25F109	536.52	536.1482	64.92	4.70	3.54	26.84
C29H26F201	428.53	428.1952	81.28	6.12	8.87	3.73
C29H26F202	444.53	444.1901	78.36	5.90	8.55	7.20
C29H26F203	460.53	460.1850	75.64	5.69	8.25	10.42
C29H26F204	476.52	476.1799	73.10	5.50	7.97	13.43
C29H26F205	492.52	492.1748	70.72	5.32	7.71	16.24
C29H26F206	508.52	508.1697	68.50	5.15	7.47	18.88
C29H26F207	524.52	524.1646	66.41	5.00	7.24	21.35
C29H26F208	540.52	540.1596	64.44	4.85	7.03	23.68
C29H26F209	556.52	556.1545	62.59	4.71	6.83	25.87
C29H27F101	410.54	410.2046	84.85	6.63	4.63	3.90
C29H27F102	426.54	426.1995	81.66	6.38	4.45	7.50
C29H27F103	442.54	442.1944	78.71	6.15	4.29	10.85
C29H27F104	458.53	458.1893	75.96	5.94	4.14	13.96
C29H27F105	474.53	474.1842	73.40	5.74	4.00	16.86
C29H27F106	490.53	490.1792	71.01	5.55	3.87	19.57
C29H27F107	506.53	506.1741	68.77	5.37	3.75	22.11
C29H27F108	522.53	522.1690	66.66	5.21	3.64	24.50
C29H27F109	538.53	538.1639	64.68	5.05	3.53	26.74
C29H28F201	430.54	430.2108	80.90	6.56	8.83	3.72
C29H28F202	446.54	446.2057	78.00	6.32	8.51	7.17
C29H28F203	462.54	462.2006	75.31	6.10	8.21	10.38
C29H28F204	478.54	478.1955	72.79	5.90	7.94	13.37
C29H28F205	494.54	494.1905	70.43	5.71	7.68	16.18
C29H28F206	510.54	510.1854	68.23	5.53	7.44	18.80
C29H28F207	526.54	526.1803	66.15	5.36	7.22	21.27
C29H28F208	542.54	542.1752	64.20	5.20	7.00	23.59
C29H28F209	558.54	558.1701	62.36	5.05	6.80	25.78
C29H29F101	412.55	412.2202	84.43	7.09	4.61	3.88
C29H29F102	428.55	428.2151	81.28	6.82	4.43	7.47
C29H29F103	444.55	444.2101	78.35	6.58	4.27	10.80
C29H29F104	460.55	460.2050	75.63	6.35	4.13	13.90
C29H29F105	476.55	476.1999	73.09	6.13	3.99	16.79
C29H29F106	492.55	492.1948	70.72	5.93	3.86	19.49
C29H29F107	508.55	508.1897	68.49	5.75	3.74	22.02
C29H29F108	524.55	524.1846	66.40	5.57	3.62	24.40
C29H29F109	540.55	540.1795	64.44	5.41	3.51	26.64
C29H30F201	432.56	432.2265	80.53	6.99	8.78	3.70
C29H30F202	448.56	448.2214	77.65	6.74	8.47	7.13
C29H30F203	464.56	464.2163	74.98	6.51	8.18	10.33
C29H30F204	480.56	480.2112	72.48	6.29	7.91	13.32
C29H30F205	496.56	496.2061	70.15	6.09	7.65	16.11
C29H30F206	512.56	512.2010	67.96	5.90	7.41	18.73
C29H30F207	528.56	528.1959	65.90	5.72	7.19	21.19
C29H30F208	544.55	544.1909	63.96	5.55	6.98	23.50
C29H30F209	560.55	560.1858	62.14	5.39	6.78	25.69
C29H31F101	414.57	414.2359	84.02	7.54	4.58	3.86
C29H31F102	430.57	430.2308	80.90	7.26	4.41	7.43
C29H31F103	446.57	446.2257	78.00	7.00	4.25	10.75
C29H31F104	462.57	462.2206	75.30	6.76	4.11	13.84
C29H31F105	478.57	478.2155	72.78	6.53	3.97	16.72
C29H31F106	494.57	494.2104	70.43	6.32	3.84	19.41
C29H31F107	510.56	510.2054	68.22	6.12	3.72	21.94
C29H31F108	526.56	526.2003	66.15	5.93	3.61	24.31
C29H31F109	542.56	542.1952	64.20	5.76	3.50	26.54
C29H32F201	434.57	434.2421	80.15	7.42	8.74	3.68
C29H32F202	450.57	450.2370	77.31	7.16	8.43	7.10
C29H32F203	466.57	466.2319	74.66	6.91	8.14	10.29
C29H32F204	482.57	482.2268	72.18	6.68	7.87	13.26
C29H32F205	498.57	498.2218	69.86	6.47	7.62	16.05
C29H32F206	514.57	514.2167	67.69	6.27	7.38	18.66
C29H32F207	530.57	530.2116	65.65	6.08	7.16	21.11
C29H32F208	546.57	546.2065	63.73	5.90	6.95	23.42
C29H32F209	562.57	562.2014	61.92	5.73	6.75	25.60
C29H33F101	416.58	416.2515	83.61	7.98	4.56	3.84

C29H33F102	432.58	432.2464	80.52	7.69	4.39	7.40
C29H33F103	448.58	448.2414	77.65	7.42	4.24	10.70
C29H33F104	464.58	464.2363	74.98	7.16	4.09	13.78
C29H33F105	480.58	480.2312	72.48	6.92	3.95	16.65
C29H33F106	496.58	496.2261	70.14	6.70	3.83	19.33
C29H33F107	512.58	512.2210	67.95	6.49	3.71	21.85
C29H33F108	528.58	528.2159	65.90	6.29	3.59	24.21
C29H33F109	544.58	544.2108	63.96	6.11	3.49	26.44
C29H34F201	436.59	436.2578	79.78	7.85	8.70	3.66
C29H34F202	452.59	452.2527	76.96	7.57	8.40	7.07
C29H34F203	468.59	468.2476	74.33	7.31	8.11	10.24
C29H34F204	484.59	484.2425	71.88	7.07	7.84	13.21
C29H34F205	500.59	500.2374	69.58	6.85	7.59	15.98
C29H34F206	516.59	516.2323	67.43	6.63	7.36	18.58
C29H34F207	532.59	532.2272	65.40	6.43	7.13	21.03
C29H34F208	548.59	548.2222	63.49	6.25	6.93	23.33
C29H34F209	564.59	564.2171	61.70	6.07	6.73	25.50
C29H35F101	418.60	418.2672	83.21	8.43	4.54	3.82
C29H35F102	434.60	434.2621	80.15	8.12	4.37	7.36
C29H35F103	450.60	450.2570	77.30	7.83	4.22	10.65
C29H35F104	466.60	466.2519	74.65	7.56	4.07	13.72
C29H35F105	482.60	482.2468	72.18	7.31	3.94	16.58
C29H35F106	498.60	498.2417	69.86	7.08	3.81	19.25
C29H35F107	514.60	514.2367	67.69	6.86	3.69	21.76
C29H35F108	530.60	530.2316	65.65	6.65	3.58	24.12
C29H35F109	546.60	546.2265	63.73	6.45	3.48	26.34
C29H36F201	438.61	438.2734	79.42	8.27	8.66	3.65
C29H36F202	454.61	454.2683	76.62	7.98	8.36	7.04
C29H36F203	470.61	470.2632	74.02	7.71	8.07	10.20
C29H36F204	486.60	486.2581	71.58	7.46	7.81	13.15
C29H36F205	502.60	502.2531	69.30	7.22	7.56	15.92
C29H36F206	518.60	518.2480	67.17	7.00	7.33	18.51
C29H36F207	534.60	534.2429	65.16	6.79	7.11	20.95
C29H36F208	550.60	550.2378	63.26	6.59	6.90	23.25
C29H36F209	566.60	566.2327	61.48	6.40	6.71	25.41
C29H37F101	420.62	420.2828	82.81	8.87	4.52	3.80
C29H37F102	436.62	436.2777	79.78	8.54	4.35	7.33
C29H37F103	452.61	452.2727	76.96	8.24	4.20	10.60
C29H37F104	468.61	468.2676	74.33	7.96	4.05	13.66
C29H37F105	484.61	484.2625	71.88	7.70	3.92	16.51
C29H37F106	500.61	500.2574	69.58	7.45	3.80	19.18
C29H37F107	516.61	516.2523	67.42	7.22	3.68	21.68
C29H37F108	532.61	532.2472	65.40	7.00	3.57	24.03
C29H37F109	548.61	548.2421	63.49	6.80	3.46	26.25
C29H38F201	440.62	440.2891	79.05	8.69	8.62	3.63
C29H38F202	456.62	456.2840	76.28	8.39	8.32	7.01
C29H38F203	472.62	472.2789	73.70	8.10	8.04	10.16
C29H38F204	488.62	488.2738	71.29	7.84	7.78	13.10
C29H38F205	504.62	504.2687	69.03	7.59	7.53	15.85
C29H38F206	520.62	520.2636	66.91	7.36	7.30	18.44
C29H38F207	536.62	536.2585	64.91	7.14	7.08	20.87
C29H38F208	552.62	552.2535	63.03	6.93	6.88	23.16
C29H38F209	568.62	568.2484	61.26	6.74	6.68	25.32
C29H39F101	422.63	422.2985	82.42	9.30	4.50	3.79
C29H39F102	438.63	438.2934	79.41	8.96	4.33	7.30
C29H39F103	454.63	454.2883	76.62	8.65	4.18	10.56
C29H39F104	470.63	470.2832	74.01	8.35	4.04	13.60
C29H39F105	486.63	486.2781	71.58	8.08	3.90	16.44
C29H39F106	502.63	502.2730	69.30	7.82	3.78	19.10
C29H39F107	518.63	518.2680	67.16	7.58	3.66	21.59
C29H39F108	534.63	534.2629	65.15	7.35	3.55	23.94
C29H39F109	550.63	550.2578	63.26	7.14	3.45	26.15
C29H40F201	442.64	442.3047	78.69	9.11	8.58	3.61
C29H40F202	458.64	458.2996	75.95	8.79	8.28	6.98
C29H40F203	474.64	474.2945	73.39	8.49	8.01	10.11
C29H40F204	490.64	490.2894	70.99	8.22	7.74	13.04
C29H40F205	506.64	506.2844	68.75	7.96	7.50	15.79
C29H40F206	522.64	522.2793	66.65	7.71	7.27	18.37
C29H40F207	538.63	538.2742	64.67	7.49	7.05	20.79
C29H40F208	554.63	554.2691	62.80	7.27	6.85	23.08
C29H40F209	570.63	570.2640	61.04	7.07	6.66	25.23
C29H41F101	424.65	424.3141	82.03	9.73	4.47	3.77
C29H41F102	440.65	440.3090	79.05	9.38	4.31	7.26
C29H41F103	456.65	456.3040	76.28	9.05	4.16	10.51
C29H41F104	472.65	472.2989	73.70	8.74	4.02	13.54
C29H41F105	488.65	488.2938	71.28	8.46	3.89	16.37
C29H41F106	504.64	504.2887	69.02	8.19	3.76	19.02
C29H41F107	520.64	520.2836	66.90	7.94	3.65	21.51
C29H41F108	536.64	536.2785	64.91	7.70	3.54	23.85
C29H41F109	552.64	552.2734	63.03	7.48	3.44	26.06
C29H42F201	444.65	444.3204	78.34	9.52	8.55	3.60
C29H42F202	460.65	460.3153	75.62	9.19	8.25	6.95
C29H42F203	476.65	476.3102	73.08	8.88	7.97	10.07
C29H42F204	492.65	492.3051	70.70	8.59	7.71	12.99
C29H42F205	508.65	508.3000	68.48	8.32	7.47	15.73
C29H42F206	524.65	524.2949	66.39	8.07	7.24	18.30
C29H42F207	540.65	540.2898	64.43	7.83	7.03	20.72
C29H42F208	556.65	556.2848	62.57	7.61	6.83	22.99
C29H42F209	572.65	572.2797	60.83	7.39	6.64	25.15
C29H43F101	426.66	426.3298	81.64	10.16	4.45	3.75
C29H43F102	442.66	442.3247	78.69	9.79	4.29	7.23
C29H43F103	458.66	458.3196	75.94	9.45	4.14	10.46
C29H43F104	474.66	474.3145	73.38	9.13	4.00	13.48
C29H43F105	490.66	490.3094	70.99	8.83	3.87	16.30
C29H43F106	506.66	506.3043	68.75	8.55	3.75	18.95
C29H43F107	522.66	522.2993	66.64	8.29	3.63	21.43
C29H43F108	538.66	538.2942	64.66	8.05	3.53	23.76
C29H43F109	554.66	554.2891	62.80	7.81	3.43	25.96
C29H44F201	446.67	446.3360	77.98	9.93	8.51	3.58
C29H44F202	462.67	462.3309	75.29	9.59	8.21	6.92
C29H44F203	478.67	478.3258	72.77	9.27	7.94	10.03
C29H44F204	494.67	494.3207	70.42	8.97	7.68	12.94
C29H44F205	510.67	510.3157	68.21	8.68	7.44	15.67
C29H44F206	526.67	526.3106	66.14	8.42	7.21	18.23
C29H44F207	542.67	542.3055	64.19	8.17	7.00	20.64
C29H44F208	558.67	558.3004	62.35	7.94	6.80	22.91
C29H44F209	574.67	574.2953	60.61	7.72	6.61	25.06
C29H45F101	428.68	428.3454	81.25	10.58	4.43	3.73
C29H45F102	444.68	444.3403	78.33	10.20	4.27	7.20
C29H45F103	460.68	460.3353	75.61	9.85	4.12	10.42
C29H45F104	476.68	476.3302	73.07	9.52	3.99	13.43
C29H45F105	492.68	492.3251	70.70	9.21	3.86	16.24
C29H45F106	508.68	508.3200	68.48	8.92	3.73	18.87
C29H45F107	524.68	524.3149	66.39	8.65	3.62	21.35
C29H45F108	540.68	540.3098	64.42	8.39	3.51	23.67
C29H45F109	556.68	556.3047	62.57	8.15	3.41	25.87
C29H46F201	448.69	448.3517	77.63	10.33	8.47	3.57
C29H46F202	464.69	464.3466	74.96	9.98	8.18	6.89
C29H46F203	480.68	480.3415	72.46	9.65	7.90	9.99
C29H46F204	496.68	496.3364	70.13	9.34	7.65	12.88

C29H46F205	512.68	512.3313	67.94	9.04	7.41	15.60
C29H46F206	528.68	528.3262	65.89	8.77	7.19	18.16
C29H46F207	544.68	544.3211	63.95	8.51	6.98	20.56
C29H46F208	560.68	560.3160	62.12	8.27	6.78	22.83
C29H46F209	576.68	576.3110	60.40	8.04	6.59	24.97
C29H47F101	430.70	430.3611	80.87	11.00	4.41	3.71
C29H47F102	446.70	446.3560	77.98	10.61	4.25	7.16
C29H47F103	462.69	462.3509	75.28	10.24	4.11	10.37
C29H47F104	478.69	478.3458	72.77	9.90	3.97	13.37
C29H47F105	494.69	494.3407	70.41	9.58	3.84	16.17
C29H47F106	510.69	510.3356	68.21	9.28	3.72	18.80
C29H47F107	526.69	526.3306	66.13	8.99	3.61	21.26
C29H47F108	542.69	542.3255	64.18	8.73	3.50	23.59
C29H47F109	558.69	558.3204	62.35	8.48	3.40	25.77
C29H48F201	450.70	450.3673	77.28	10.73	8.43	3.55
C29H48F202	466.70	466.3622	74.64	10.37	8.14	6.86
C29H48F203	482.70	482.3571	72.16	10.02	7.87	9.94
C29H48F204	498.70	498.3520	69.85	9.70	7.62	12.83
C29H48F205	514.70	514.3470	67.68	9.40	7.38	15.54
C29H48F206	530.70	530.3419	65.63	9.12	7.16	18.09
C29H48F207	546.70	546.3368	63.71	8.85	6.95	20.49
C29H48F208	562.70	562.3317	61.90	8.60	6.75	22.75
C29H48F209	578.70	578.3266	60.19	8.36	6.57	24.88
C29H49F101	432.71	432.3767	80.50	11.41	4.39	3.70
C29H49F102	448.71	448.3716	77.63	11.01	4.23	7.13
C29H49F103	464.71	464.3665	74.95	10.63	4.09	10.33
C29H49F104	480.71	480.3615	72.46	10.27	3.95	13.31
C29H49F105	496.71	496.3564	70.13	9.94	3.82	16.11
C29H49F106	512.71	512.3513	67.94	9.63	3.71	18.72
C29H49F107	528.71	528.3462	65.88	9.34	3.59	21.18
C29H49F108	544.71	544.3411	63.95	9.07	3.49	23.50
C29H49F109	560.71	560.3360	62.12	8.81	3.39	25.68
C29H50F201	452.72	452.3829	76.94	11.13	8.39	3.53
C29H50F202	468.72	468.3779	74.31	10.75	8.11	6.83
C29H50F203	484.72	484.3728	71.86	10.40	7.84	9.90
C29H50F204	500.72	500.3677	69.57	10.07	7.59	12.78
C29H50F205	516.72	516.3626	67.41	9.75	7.35	15.48
C29H50F206	532.72	532.3575	65.39	9.46	7.13	18.02
C29H50F207	548.71	548.3524	63.48	9.18	6.92	20.41
C29H50F208	564.71	564.3473	61.68	8.92	6.73	22.67
C29H50F209	580.71	580.3423	59.98	8.68	6.54	24.80
C29H51F101	434.73	434.3924	80.12	11.82	4.37	3.68
C29H51F102	450.73	450.3873	77.28	11.41	4.22	7.10
C29H51F103	466.73	466.3822	74.63	11.01	4.07	10.28
C29H51F104	482.73	482.3771	72.16	10.65	3.94	13.26
C29H51F105	498.73	498.3720	69.84	10.31	3.81	16.04
C29H51F106	514.72	514.3669	67.67	9.99	3.69	18.65
C29H51F107	530.72	530.3619	65.63	9.69	3.58	21.10
C29H51F108	546.72	546.3568	63.71	9.40	3.47	23.41
C29H51F109	562.72	562.3517	61.90	9.14	3.38	25.59
C29H52F201	454.73	454.3986	76.60	11.53	8.36	3.52
C29H52F202	470.73	470.3935	74.00	11.13	8.07	6.80
C29H52F203	486.73	486.3884	71.56	10.77	7.81	9.85
C29H52F204	502.73	502.3833	69.29	10.43	7.56	12.73
C29H52F205	518.73	518.3783	67.15	10.10	7.32	15.42
C29H52F206	534.73	534.3732	65.14	9.80	7.11	17.95
C29H52F207	550.73	550.3681	63.25	9.52	6.90	20.34
C29H52F208	566.73	566.3630	61.46	9.25	6.70	22.58
C29H52F209	582.73	582.3579	59.77	8.99	6.52	24.71
C29H53F101	436.74	436.4080	79.75	12.23	4.35	3.66
C29H53F102	452.74	452.4029	76.94	11.80	4.20	7.07
C29H53F103	468.74	468.3978	74.31	11.40	4.05	10.24
C29H53F104	484.74	484.3928	71.86	11.02	3.92	13.20
C29H53F105	500.74	500.3877	69.56	10.67	3.79	15.98
C29H53F106	516.74	516.3826	67.41	10.34	3.68	18.58
C29H53F107	532.74	532.3775	65.38	10.03	3.57	21.02
C29H53F108	548.74	548.3724	63.48	9.74	3.46	23.33
C29H53F109	564.74	564.3673	61.68	9.46	3.36	25.50
C29H54F201	456.75	456.4142	76.26	11.92	8.32	3.50
C29H54F202	472.75	472.4092	73.68	11.51	8.04	6.77
C29H54F203	488.75	488.4041	71.27	11.14	7.77	9.82
C29H54F204	504.75	504.3990	69.01	10.78	7.53	12.68
C29H54F205	520.75	520.3939	66.89	10.45	7.30	15.36
C29H54F206	536.75	536.3888	64.90	10.14	7.08	17.88
C29H54F207	552.75	552.3837	63.02	9.85	6.87	20.26
C29H54F208	568.75	568.3786	61.24	9.57	6.68	22.50
C29H54F209	584.75	584.3736	59.57	9.31	6.50	24.63
C29H55F101	438.76	438.4237	79.39	12.64	4.33	3.65
C29H55F102	454.76	454.4186	76.60	12.19	4.18	7.04
C29H55F103	470.76	470.4135	73.99	11.78	4.04	10.20
C29H55F104	486.76	486.4084	71.56	11.39	3.90	13.15
C29H55F105	502.76	502.4033	69.28	11.03	3.78	15.91
C29H55F106	518.76	518.3982	67.15	10.69	3.66	18.51
C29H55F107	534.76	534.3932	65.14	10.37	3.55	20.94
C29H55F108	550.76	550.3881	63.24	10.07	3.45	23.24
C29H55F109	566.75	566.3830	61.46	9.78	3.35	25.41
C29H56F201	458.77	458.4299	75.93	12.30	8.28	3.49
C29H56F202	474.77	474.4248	73.37	11.89	8.00	6.74
C29H56F203	490.76	490.4197	70.98	11.50	7.74	9.78
C29H56F204	506.76	506.4146	68.73	11.14	7.50	12.63
C29H56F205	522.76	522.4096	66.63	10.80	7.27	15.30
C29H56F206	538.76	538.4045	64.65	10.48	7.05	17.82
C29H56F207	554.76	554.3994	62.79	10.17	6.85	20.19
C29H56F208	570.76	570.3943	61.03	9.89	6.66	22.43
C29H56F209	586.76	586.3892	59.36	9.62	6.48	24.54
C29H57F101	440.78	440.4393	79.03	13.03	4.31	3.63
C29H57F102	456.77	456.4342	76.26	12.58	4.16	7.01
C29H57F103	472.77	472.4291	73.68	12.15	4.02	10.15
C29H57F104	488.77	488.4241	71.26	11.75	3.89	13.09
C29H57F105	504.77	504.4190	69.01	11.38	3.76	15.85
C29H57F106	520.77	520.4139	66.89	11.03	3.65	18.43
C29H57F107	536.77	536.4088	64.89	10.70	3.54	20.86
C29H57F108	552.77	552.4037	63.01	10.39	3.44	23.16
C29H57F109	568.77	568.3986	61.24	10.10	3.34	25.32
C29H58F201	460.78	460.4455	75.59	12.69	8.25	3.47
C29H58F202	476.78	476.4405	73.06	12.26	7.97	6.71
C29H58F203	492.78	492.4354	70.69	11.86	7.71	9.74
C29H58F204	508.78	508.4303	68.46	11.49	7.47	12.58
C29H58F205	524.78	524.4252	66.38	11.14	7.24	15.24
C29H58F206	540.78	540.4201	64.41	10.81	7.03	17.75
C29H58F207	556.78	556.4150	62.56	10.50	6.82	20.11
C29H58F208	572.78	572.4099	60.81	10.21	6.63	22.35
C29H58F209	588.78	588.4049	59.16	9.93	6.45	24.46
C29H59F101	442.79	442.4550	78.67	13.43	4.29	3.61
C29H59F102	458.79	458.4499	75.92	12.96	4.14	6.97
C29H59F103	474.79	474.4448	73.36	12.53	4.00	10.11
C29H59F104	490.79	490.4397	70.97	12.12	3.87	13.04
C29H59F105	506.79	506.4346	68.73	11.73	3.75	15.79
C29H59F106	522.79	522.4295	66.63	11.38	3.63	18.36
C29H59F107	538.79	538.4245	64.65	11.04	3.53	20.79

179

C29H59F108	554.79	554.4194	62.79	10.72	3.42	23.07
C29H59F109	570.79	570.4143	61.03	10.42	3.33	25.23
C30H15F101	410.45	410.1107	87.79	3.68	4.63	3.90
C30H15F102	426.45	426.1056	84.50	3.55	4.45	7.50
C30H15F103	442.45	442.1005	81.44	3.42	4.29	10.85
C30H15F104	458.45	458.0954	78.60	3.30	4.14	13.96
C30H15F105	474.45	474.0903	75.95	3.19	4.00	16.86
C30H15F106	490.45	490.0853	73.47	3.08	3.87	19.57
C30H15F107	506.45	506.0802	71.15	2.99	3.75	22.11
C30H15F108	522.45	522.0751	68.97	2.89	3.64	24.50
C30H15F109	538.45	538.0700	66.92	2.81	3.53	26.74
C30H16F201	430.46	430.1169	83.71	3.75	8.83	3.72
C30H16F202	446.46	446.1118	80.71	3.61	8.51	7.17
C30H16F203	462.46	462.1067	77.92	3.49	8.22	10.38
C30H16F204	478.46	478.1017	75.31	3.37	7.94	13.38
C30H16F205	494.46	494.0966	72.87	3.26	7.68	16.18
C30H16F206	510.46	510.0915	70.59	3.16	7.44	18.81
C30H16F207	526.45	526.0864	68.45	3.06	7.22	21.27
C30H16F208	542.45	542.0813	66.43	2.97	7.00	23.60
C30H16F209	558.45	558.0762	64.52	2.89	6.80	25.78
C30H17F101	412.47	412.1263	87.36	4.15	4.61	3.88
C30H17F102	428.47	428.1212	84.10	4.00	4.43	7.47
C30H17F103	444.47	444.1162	81.07	3.86	4.27	10.80
C30H17F104	460.47	460.1111	78.25	3.72	4.13	13.90
C30H17F105	476.47	476.1060	75.63	3.60	3.99	16.79
C30H17F106	492.46	492.1009	73.17	3.48	3.86	19.49
C30H17F107	508.46	508.0958	70.87	3.37	3.74	22.03
C30H17F108	524.46	524.0907	68.71	3.27	3.62	24.40
C30H17F109	540.46	540.0856	66.67	3.17	3.52	26.64
C30H18F201	432.47	432.1326	83.32	4.20	8.79	3.70
C30H18F202	448.47	448.1275	80.35	4.05	8.47	7.14
C30H18F203	464.47	464.1224	77.58	3.91	8.18	10.33
C30H18F204	480.47	480.1173	75.00	3.78	7.91	13.32
C30H18F205	496.47	496.1122	72.58	3.65	7.65	16.11
C30H18F206	512.47	512.1071	70.31	3.54	7.41	18.73
C30H18F207	528.47	528.1020	68.18	3.43	7.19	21.19
C30H18F208	544.47	544.0970	66.18	3.33	6.98	23.51
C30H18F209	560.47	560.0919	64.29	3.24	6.78	25.69
C30H19F101	414.48	414.1420	86.94	4.62	4.58	3.86
C30H19F102	430.48	430.1369	83.70	4.45	4.41	7.43
C30H19F103	446.48	446.1318	80.71	4.29	4.26	10.75
C30H19F104	462.48	462.1267	77.91	4.14	4.11	13.84
C30H19F105	478.48	478.1216	75.31	4.00	3.97	16.72
C30H19F106	494.48	494.1166	72.87	3.87	3.84	19.41
C30H19F107	510.48	510.1115	70.59	3.75	3.72	21.94
C30H19F108	526.48	526.1064	68.44	3.64	3.61	24.31
C30H19F109	542.48	542.1013	66.42	3.53	3.50	26.54
C30H20F201	434.49	434.1482	82.93	4.64	8.75	3.68
C30H20F202	450.49	450.1431	79.99	4.47	8.43	7.10
C30H20F203	466.49	466.1380	77.24	4.32	8.15	10.29
C30H20F204	482.49	482.1330	74.68	4.18	7.88	13.26
C30H20F205	498.49	498.1279	72.29	4.04	7.62	16.05
C30H20F206	514.49	514.1228	70.04	3.92	7.39	18.66
C30H20F207	530.49	530.1177	67.93	3.80	7.16	21.11
C30H20F208	546.49	546.1126	65.94	3.69	6.95	23.42
C30H20F209	562.49	562.1075	64.06	3.58	6.76	25.60
C30H21F101	416.50	416.1576	86.51	5.08	4.56	3.84
C30H21F102	432.50	432.1525	83.31	4.89	4.39	7.40
C30H21F103	448.50	448.1475	80.34	4.72	4.24	10.70
C30H21F104	464.50	464.1424	77.58	4.56	4.09	13.78
C30H21F105	480.50	480.1373	74.99	4.41	3.95	16.65
C30H21F106	496.50	496.1322	72.58	4.26	3.83	19.33
C30H21F107	512.50	512.1271	70.31	4.13	3.71	21.85
C30H21F108	528.50	528.1220	68.18	4.01	3.59	24.22
C30H21F109	544.49	544.1169	66.18	3.89	3.49	26.45
C30H22F201	436.51	436.1639	82.55	5.08	8.70	3.67
C30H22F202	452.51	452.1588	79.63	4.90	8.40	7.07
C30H22F203	468.50	468.1537	76.91	4.73	8.11	10.24
C30H22F204	484.50	484.1486	74.37	4.58	7.84	13.21
C30H22F205	500.50	500.1435	71.99	4.43	7.59	15.98
C30H22F206	516.50	516.1384	69.76	4.29	7.36	18.59
C30H22F207	532.50	532.1333	67.67	4.16	7.14	21.03
C30H22F208	548.50	548.1283	65.69	4.04	6.93	23.34
C30H22F209	564.50	564.1232	63.83	3.93	6.73	25.51
C30H23F101	418.52	418.1733	86.10	5.54	4.54	3.82
C30H23F102	434.52	434.1682	82.93	5.34	4.37	7.36
C30H23F103	450.51	450.1631	79.98	5.15	4.22	10.65
C30H23F104	466.51	466.1580	77.24	4.97	4.07	13.72
C30H23F105	482.51	482.1529	74.68	4.80	3.94	16.58
C30H23F106	498.51	498.1479	72.28	4.65	3.81	19.26
C30H23F107	514.51	514.1428	70.03	4.51	3.69	21.77
C30H23F108	530.51	530.1377	67.92	4.37	3.58	24.13
C30H23F109	546.51	546.1326	65.93	4.24	3.48	26.35
C30H24F201	438.52	438.1795	82.17	5.52	8.66	3.65
C30H24F202	454.52	454.1744	79.28	5.32	8.36	7.04
C30H24F203	470.52	470.1693	76.58	5.14	8.08	10.20
C30H24F204	486.52	486.1643	74.06	4.97	7.81	13.15
C30H24F205	502.52	502.1592	71.71	4.81	7.56	15.92
C30H24F206	518.52	518.1541	69.49	4.67	7.33	18.51
C30H24F207	534.52	534.1490	67.41	4.53	7.11	20.95
C30H24F208	550.52	550.1439	65.45	4.39	6.90	23.25
C30H24F209	566.52	566.1388	63.61	4.27	6.71	25.42
C30H25F101	420.53	420.1889	85.69	5.99	4.52	3.80
C30H25F102	436.53	436.1838	82.55	5.77	4.35	7.33
C30H25F103	452.53	452.1788	79.63	5.57	4.20	10.61
C30H25F104	468.53	468.1737	76.91	5.38	4.05	13.66
C30H25F105	484.53	484.1686	74.37	5.20	3.92	16.51
C30H25F106	500.53	500.1635	71.99	5.03	3.80	19.18
C30H25F107	516.53	516.1584	69.76	4.88	3.68	21.68
C30H25F108	532.53	532.1533	67.66	4.73	3.57	24.04
C30H25F109	548.53	548.1482	65.69	4.59	3.46	26.25
C30H26F201	440.54	440.1952	81.79	5.95	8.63	3.63
C30H26F202	456.54	456.1901	78.93	5.74	8.32	7.01
C30H26F203	472.54	472.1850	76.26	5.55	8.04	10.16
C30H26F204	488.54	488.1799	73.76	5.36	7.78	13.10
C30H26F205	504.54	504.1748	71.42	5.19	7.53	15.86
C30H26F206	520.53	520.1697	69.22	5.03	7.30	18.44
C30H26F207	536.53	536.1646	67.16	4.88	7.08	20.87
C30H26F208	552.53	552.1596	65.21	4.74	6.88	23.17
C30H26F209	568.53	568.1545	63.38	4.61	6.68	25.33
C30H27F101	422.55	422.2046	85.28	6.44	4.50	3.79
C30H27F102	438.55	438.1995	82.17	6.21	4.33	7.30
C30H27F103	454.55	454.1944	79.27	5.99	4.18	10.56
C30H27F104	470.55	470.1893	76.58	5.78	4.04	13.60
C30H27F105	486.55	486.1842	74.06	5.59	3.90	16.44
C30H27F106	502.54	502.1792	71.70	5.42	3.78	19.10
C30H27F107	518.54	518.1741	69.49	5.25	3.66	21.60
C30H27F108	534.54	534.1690	67.41	5.09	3.55	23.94
C30H27F109	550.54	550.1639	65.45	4.94	3.45	26.16
C30H28F201	442.55	442.2108	81.42	6.38	8.59	3.62

180

C30H28F202	458.55	458.2057	78.58	6.15	8.29	6.98
C30H28F203	474.55	474.2006	75.93	5.95	8.01	10.11
C30H28F204	490.55	490.1955	73.45	5.75	7.75	13.05
C30H28F205	506.55	506.1905	71.13	5.57	7.50	15.79
C30H28F206	522.55	522.1854	68.96	5.40	7.27	18.37
C30H28F207	538.55	538.1803	66.91	5.24	7.06	20.80
C30H28F208	554.55	554.1752	64.98	5.09	6.85	23.08
C30H28F209	570.55	570.1701	63.16	4.95	6.66	25.24
C30H29F101	424.56	424.2202	84.87	6.88	4.47	3.77
C30H29F102	440.56	440.2151	81.79	6.63	4.31	7.26
C30H29F103	456.56	456.2101	78.92	6.40	4.16	10.51
C30H29F104	472.56	472.2050	76.25	6.19	4.02	13.54
C30H29F105	488.56	488.1999	73.75	5.98	3.89	16.37
C30H29F106	504.56	504.1948	71.42	5.79	3.77	19.03
C30H29F107	520.56	520.1897	69.22	5.62	3.65	21.51
C30H29F108	536.56	536.1846	67.16	5.45	3.54	23.85
C30H29F109	552.56	552.1795	65.21	5.29	3.44	26.06
C30H30F201	444.57	444.2265	81.05	6.80	8.55	3.60
C30H30F202	460.57	460.2214	78.24	6.57	8.25	6.95
C30H30F203	476.57	476.2163	75.61	6.35	7.97	10.07
C30H30F204	492.57	492.2112	73.15	6.14	7.71	12.99
C30H30F205	508.57	508.2061	70.85	5.95	7.47	15.73
C30H30F206	524.57	524.2010	68.69	5.76	7.24	18.30
C30H30F207	540.57	540.1959	66.66	5.59	7.03	20.72
C30H30F208	556.57	556.1909	64.74	5.43	6.83	23.00
C30H30F209	572.56	572.1858	62.93	5.28	6.64	25.15
C30H31F101	426.58	426.2359	84.47	7.33	4.45	3.75
C30H31F102	442.58	442.2308	81.42	7.06	4.29	7.23
C30H31F103	458.58	458.2257	78.58	6.81	4.14	10.47
C30H31F104	474.58	474.2206	75.93	6.58	4.00	13.49
C30H31F105	490.58	490.2155	73.45	6.37	3.87	16.31
C30H31F106	506.58	506.2104	71.13	6.17	3.75	18.95
C30H31F107	522.58	522.2054	68.95	5.98	3.64	21.43
C30H31F108	538.58	538.2003	66.91	5.80	3.53	23.77
C30H31F109	554.57	554.1952	64.97	5.63	3.43	25.96
C30H32F201	446.59	446.2421	80.69	7.22	8.51	3.58
C30H32F202	462.59	462.2370	77.90	6.97	8.21	6.92
C30H32F203	478.58	478.2319	75.29	6.74	7.94	10.03
C30H32F204	494.58	494.2268	72.86	6.52	7.68	12.94
C30H32F205	510.58	510.2218	70.57	6.32	7.44	15.67
C30H32F206	526.58	526.2167	68.43	6.13	7.22	18.23
C30H32F207	542.58	542.2116	66.41	5.94	7.00	20.64
C30H32F208	558.58	558.2065	64.51	5.77	6.80	22.91
C30H32F209	574.58	574.2014	62.71	5.61	6.61	25.06
C30H33F101	428.60	428.2515	84.07	7.76	4.43	3.73
C30H33F102	444.59	444.2464	81.05	7.48	4.27	7.20
C30H33F103	460.59	460.2414	78.23	7.22	4.12	10.42
C30H33F104	476.59	476.2363	75.61	6.98	3.99	13.43
C30H33F105	492.59	492.2312	73.15	6.75	3.86	16.24
C30H33F106	508.59	508.2261	70.85	6.54	3.74	18.87
C30H33F107	524.59	524.2210	68.69	6.34	3.62	21.35
C30H33F108	540.59	540.2159	66.66	6.15	3.51	23.68
C30H33F109	556.59	556.2108	64.74	5.98	3.41	25.87
C30H34F201	448.60	448.2578	80.32	7.64	8.47	3.57
C30H34F202	464.60	464.2527	77.56	7.38	8.18	6.89
C30H34F203	480.60	480.2476	74.98	7.13	7.91	9.99
C30H34F204	496.60	496.2425	72.56	6.90	7.65	12.89
C30H34F205	512.60	512.2374	70.30	6.69	7.41	15.61
C30H34F206	528.60	528.2323	68.17	6.48	7.19	18.16
C30H34F207	544.60	544.2272	66.17	6.29	6.98	20.56
C30H34F208	560.60	560.2222	64.28	6.11	6.78	22.83
C30H34F209	576.60	576.2171	62.49	5.94	6.59	24.97
C30H35F101	430.61	430.2672	83.68	8.19	4.41	3.72
C30H35F102	446.61	446.2621	80.68	7.90	4.25	7.16
C30H35F103	462.61	462.2570	77.89	7.63	4.11	10.38
C30H35F104	478.61	478.2519	75.29	7.37	3.97	13.37
C30H35F105	494.61	494.2468	72.85	7.13	3.84	16.17
C30H35F106	510.61	510.2417	70.57	6.91	3.72	18.80
C30H35F107	526.61	526.2367	68.43	6.70	3.61	21.27
C30H35F108	542.61	542.2316	66.41	6.50	3.50	23.59
C30H35F109	558.61	558.2265	64.51	6.32	3.40	25.78
C30H36F201	450.62	450.2734	79.96	8.05	8.43	3.55
C30H36F202	466.62	466.2683	77.22	7.78	8.14	6.86
C30H36F203	482.62	482.2632	74.66	7.52	7.87	9.95
C30H36F204	498.62	498.2581	72.27	7.28	7.62	12.84
C30H36F205	514.62	514.2531	70.02	7.05	7.38	15.55
C30H36F206	530.61	530.2480	67.91	6.84	7.16	18.09
C30H36F207	546.61	546.2429	65.92	6.64	6.95	20.49
C30H36F208	562.61	562.2378	64.05	6.45	6.75	22.75
C30H36F209	578.61	578.2327	62.28	6.27	6.57	24.89
C30H37F101	432.63	432.2828	83.29	8.62	4.39	3.70
C30H37F102	448.63	448.2777	80.32	8.31	4.23	7.13
C30H37F103	464.63	464.2727	77.55	8.03	4.09	10.33
C30H37F104	480.63	480.2676	74.97	7.76	3.95	13.32
C30H37F105	496.62	496.2625	72.56	7.51	3.83	16.11
C30H37F106	512.62	512.2574	70.29	7.28	3.71	18.73
C30H37F107	528.62	528.2523	68.16	7.06	3.59	21.19
C30H37F108	544.62	544.2472	66.16	6.85	3.49	23.50
C30H37F109	560.62	560.2421	64.27	6.65	3.39	25.68
C30H38F201	452.63	452.2891	79.61	8.46	8.39	3.53
C30H38F202	468.63	468.2840	76.89	8.17	8.11	6.83
C30H38F203	484.63	484.2789	74.35	7.90	7.84	9.90
C30H38F204	500.63	500.2738	71.98	7.65	7.59	12.78
C30H38F205	516.63	516.2687	69.75	7.41	7.35	15.48
C30H38F206	532.63	532.2636	67.65	7.19	7.13	18.02
C30H38F207	548.63	548.2585	65.68	6.98	6.93	20.41
C30H38F208	564.63	564.2535	63.82	6.78	6.73	22.67
C30H38F209	580.63	580.2484	62.06	6.60	6.54	24.80
C30H39F101	434.64	434.2985	82.90	9.04	4.37	3.68
C30H39F102	450.64	450.2934	79.96	8.72	4.22	7.10
C30H39F103	466.64	466.2883	77.22	8.42	4.07	10.29
C30H39F104	482.64	482.2832	74.66	8.14	3.94	13.26
C30H39F105	498.64	498.2781	72.26	7.88	3.81	16.04
C30H39F106	514.64	514.2730	70.02	7.64	3.69	18.65
C30H39F107	530.64	530.2680	67.91	7.41	3.58	21.11
C30H39F108	546.64	546.2629	65.92	7.19	3.48	23.41
C30H39F109	562.64	562.2578	64.04	6.99	3.38	25.59
C30H40F201	454.65	454.3047	79.26	8.87	8.36	3.52
C30H40F202	470.65	470.2996	76.56	8.57	8.07	6.80
C30H40F203	486.65	486.2945	74.04	8.28	7.81	9.86
C30H40F204	502.65	502.2894	71.69	8.02	7.56	12.73
C30H40F205	518.65	518.2844	69.48	7.77	7.33	15.42
C30H40F206	534.65	534.2793	67.40	7.54	7.11	17.96
C30H40F207	550.65	550.2742	65.44	7.32	6.90	20.34
C30H40F208	566.65	566.2691	63.59	7.12	6.71	22.59
C30H40F209	582.64	582.2640	61.84	6.92	6.52	24.71
C30H41F101	436.66	436.3141	82.52	9.46	4.35	3.66
C30H41F102	452.66	452.3090	79.60	9.13	4.20	7.07
C30H41F103	468.66	468.3040	76.89	8.82	4.05	10.24
C30H41F104	484.66	484.2989	74.35	8.53	3.92	13.20

C30H41F105	500.66	500.2938	71.97	8.25	3.79	15.98
C30H41F106	516.66	516.2887	69.74	8.00	3.68	18.58
C30H41F107	532.66	532.2836	67.65	7.76	3.57	21.03
C30H41F108	548.65	548.2785	65.68	7.53	3.46	23.33
C30H41F109	564.65	564.2734	63.82	7.32	3.36	25.50
C30H42F201	456.67	456.3204	78.91	9.27	8.32	3.50
C30H42F202	472.66	472.3153	76.23	8.96	8.04	6.77
C30H42F203	488.66	488.3102	73.74	8.66	7.78	9.82
C30H42F204	504.66	504.3051	71.40	8.39	7.53	12.68
C30H42F205	520.66	520.3000	69.21	8.13	7.30	15.36
C30H42F206	536.66	536.2949	67.14	7.89	7.08	17.89
C30H42F207	552.66	552.2898	65.20	7.66	6.88	20.26
C30H42F208	568.66	568.2848	63.37	7.44	6.68	22.51
C30H42F209	584.66	584.2797	61.63	7.24	6.50	24.63
C30H43F101	438.68	438.3298	82.14	9.88	4.33	3.65
C30H43F102	454.67	454.3247	79.25	9.53	4.18	7.04
C30H43F103	470.67	470.3196	76.56	9.21	4.04	10.20
C30H43F104	486.67	486.3145	74.04	8.91	3.90	13.15
C30H43F105	502.67	502.3094	71.68	8.62	3.78	15.91
C30H43F106	518.67	518.3043	69.47	8.36	3.66	18.51
C30H43F107	534.67	534.2993	67.39	8.11	3.55	20.95
C30H43F108	550.67	550.2942	65.44	7.87	3.45	23.24
C30H43F109	566.67	566.2891	63.59	7.65	3.35	25.41
C30H44F201	458.68	458.3360	78.56	9.67	8.28	3.49
C30H44F202	474.68	474.3309	75.91	9.34	8.00	6.74
C30H44F203	490.68	490.3258	73.44	9.04	7.74	9.78
C30H44F204	506.68	506.3207	71.12	8.75	7.50	12.63
C30H44F205	522.68	522.3157	68.94	8.49	7.27	15.31
C30H44F206	538.68	538.3106	66.89	8.23	7.05	17.82
C30H44F207	554.68	554.3055	64.96	8.00	6.85	20.19
C30H44F208	570.68	570.3004	63.14	7.77	6.66	22.43
C30H44F209	586.68	586.2953	61.42	7.56	6.48	24.54
C30H45F101	440.69	440.3454	81.77	10.29	4.31	3.63
C30H45F102	456.69	456.3403	78.90	9.93	4.16	7.01
C30H45F103	472.69	472.3353	76.23	9.60	4.02	10.15
C30H45F104	488.69	488.3302	73.73	9.28	3.89	13.10
C30H45F105	504.69	504.3251	71.40	8.99	3.76	15.85
C30H45F106	520.69	520.3200	69.20	8.71	3.65	18.44
C30H45F107	536.69	536.3149	67.14	8.45	3.54	20.87
C30H45F108	552.69	552.3098	65.20	8.21	3.44	23.16
C30H45F109	568.69	568.3047	63.36	7.98	3.34	25.32
C30H46F201	460.70	460.3517	78.22	10.06	8.25	3.47
C30H46F202	476.70	476.3466	75.59	9.73	7.97	6.71
C30H46F203	492.70	492.3415	73.14	9.41	7.71	9.74
C30H46F204	508.70	508.3364	70.84	9.11	7.47	12.58
C30H46F205	524.69	524.3313	68.68	8.84	7.24	15.25
C30H46F206	540.69	540.3262	66.64	8.58	7.03	17.75
C30H46F207	556.69	556.3211	64.73	8.33	6.83	20.12
C30H46F208	572.69	572.3160	62.92	8.10	6.63	22.35
C30H46F209	588.69	588.3110	61.21	7.88	6.45	24.46
C30H47F101	442.71	442.3611	81.39	10.70	4.29	3.61
C30H47F102	458.71	458.3560	78.55	10.33	4.14	6.98
C30H47F103	474.71	474.3509	75.91	9.98	4.00	10.11
C30H47F104	490.71	490.3458	73.43	9.65	3.87	13.04
C30H47F105	506.70	506.3407	71.11	9.35	3.75	15.79
C30H47F106	522.70	522.3356	68.94	9.06	3.63	18.37
C30H47F107	538.70	538.3306	66.89	8.79	3.53	20.79
C30H47F108	554.70	554.3255	64.96	8.54	3.42	23.07
C30H47F109	570.70	570.3204	63.14	8.30	3.33	25.23
C30H48F201	462.71	462.3673	77.87	10.46	8.21	3.46
C30H48F202	478.71	478.3622	75.27	10.11	7.94	6.68
C30H48F203	494.71	494.3571	72.84	9.78	7.68	9.70
C30H48F204	510.71	510.3520	70.56	9.47	7.44	12.53
C30H48F205	526.71	526.3470	68.41	9.19	7.21	15.19
C30H48F206	542.71	542.3419	66.40	8.91	7.00	17.69
C30H48F207	558.71	558.3368	64.49	8.66	6.80	20.05
C30H48F208	574.71	574.3317	62.70	8.42	6.61	22.27
C30H48F209	590.71	590.3266	61.00	8.19	6.43	24.38
C30H49F101	444.72	444.3767	81.02	11.11	4.27	3.60
C30H49F102	460.72	460.3716	78.21	10.72	4.12	6.95
C30H49F103	476.72	476.3665	75.59	10.36	3.99	10.07
C30H49F104	492.72	492.3615	73.13	10.02	3.86	12.99
C30H49F105	508.72	508.3564	70.83	9.71	3.73	15.73
C30H49F106	524.72	524.3513	68.67	9.41	3.62	18.29
C30H49F107	540.72	540.3462	66.64	9.13	3.51	20.71
C30H49F108	556.72	556.3411	64.72	8.87	3.41	22.99
C30H49F109	572.72	572.3360	62.92	8.62	3.32	25.14
C30H50F201	464.73	464.3829	77.54	10.84	8.18	3.44
C30H50F202	480.73	480.3779	74.96	10.48	7.90	6.66
C30H50F203	496.73	496.3728	72.54	10.15	7.65	9.66
C30H50F204	512.73	512.3677	70.28	9.83	7.41	12.48
C30H50F205	528.73	528.3626	68.15	9.53	7.19	15.13
C30H50F206	544.73	544.3575	66.15	9.25	6.98	17.62
C30H50F207	560.73	560.3524	64.26	8.99	6.78	19.97
C30H50F208	576.72	576.3473	62.48	8.74	6.59	22.19
C30H50F209	592.72	592.3423	60.79	8.50	6.41	24.29
C30H51F101	446.74	446.3924	80.66	11.51	4.25	3.58
C30H51F102	462.74	462.3873	77.87	11.11	4.11	6.92
C30H51F103	478.74	478.3822	75.27	10.74	3.97	10.03
C30H51F104	494.74	494.3771	72.83	10.39	3.84	12.94
C30H51F105	510.74	510.3720	70.55	10.07	3.72	15.66
C30H51F106	526.74	526.3669	68.41	9.76	3.61	18.22
C30H51F107	542.74	542.3619	66.39	9.47	3.50	20.64
C30H51F108	558.73	558.3568	64.49	9.20	3.40	22.91
C30H51F109	574.73	574.3517	62.70	8.94	3.31	25.05
C30H52F201	466.75	466.3986	77.20	11.23	8.14	3.43
C30H52F202	482.74	482.3935	74.64	10.86	7.87	6.63
C30H52F203	498.74	498.3884	72.25	10.51	7.62	9.62
C30H52F204	514.74	514.3833	70.00	10.18	7.38	12.43
C30H52F205	530.74	530.3783	67.89	9.88	7.16	15.07
C30H52F206	546.74	546.3732	65.91	9.59	6.95	17.56
C30H52F207	562.74	562.3681	64.03	9.31	6.75	19.90
C30H52F208	578.74	578.3630	62.26	9.06	6.57	22.12
C30H52F209	594.74	594.3579	60.59	8.81	6.39	24.21
C30H53F101	448.75	448.4080	80.30	11.90	4.23	3.57
C30H53F102	464.75	464.4029	77.53	11.49	4.09	6.89
C30H53F103	480.75	480.3978	74.95	11.11	3.95	9.98
C30H53F104	496.75	496.3928	72.54	10.75	3.82	12.88
C30H53F105	512.75	512.3877	70.27	10.42	3.71	15.60
C30H53F106	528.75	528.3826	68.15	10.10	3.59	18.16
C30H53F107	544.75	544.3775	66.15	9.81	3.49	20.56
C30H53F108	560.75	560.3724	64.26	9.53	3.39	22.83
C30H53F109	576.75	576.3673	62.48	9.26	3.29	24.97
C30H54F201	468.76	468.4142	76.87	11.61	8.11	3.41
C30H54F202	484.76	484.4092	74.33	11.23	7.84	6.60
C30H54F203	500.76	500.4041	71.96	10.87	7.59	9.59
C30H54F204	516.76	516.3990	69.73	10.53	7.35	12.38
C30H54F205	532.76	532.3939	67.64	10.22	7.13	15.02
C30H54F206	548.76	548.3888	65.66	9.92	6.92	17.49
C30H54F207	564.76	564.3837	63.80	9.64	6.73	19.83

C30H54F208	580.76	580.3786	62.05	9.37	6.54	22.04
C30H54F209	596.76	596.3736	60.38	9.12	6.37	24.13
C30H55F101	450.77	450.4237	79.94	12.30	4.21	3.55
C30H55F102	466.77	466.4186	77.20	11.88	4.07	6.86
C30H55F103	482.77	482.4135	74.64	11.48	3.94	9.94
C30H55F104	498.77	498.4084	72.24	11.12	3.81	12.83
C30H55F105	514.77	514.4033	70.00	10.77	3.69	15.54
C30H55F106	530.77	530.3982	67.89	10.44	3.58	18.09
C30H55F107	546.77	546.3932	65.90	10.14	3.47	20.48
C30H55F108	562.77	562.3881	64.03	9.85	3.38	22.74
C30H55F109	578.77	578.3830	62.26	9.58	3.28	24.88
C30H56F201	470.78	470.4299	76.54	11.99	8.07	3.40
C30H56F202	486.78	486.4248	74.02	11.60	7.81	6.57
C30H56F203	502.78	502.4197	71.67	11.23	7.56	9.55
C30H56F204	518.78	518.4146	69.46	10.88	7.32	12.34
C30H56F205	534.77	534.4096	67.38	10.56	7.11	14.96
C30H56F206	550.77	550.4045	65.42	10.25	6.90	17.43
C30H56F207	566.77	566.3994	63.58	9.96	6.70	19.76
C30H56F208	582.77	582.3943	61.83	9.69	6.52	21.96
C30H56F209	598.77	598.3892	60.18	9.43	6.35	24.05
C30H57F101	452.79	452.4393	79.58	12.69	4.20	3.53
C30H57F102	468.79	468.4342	76.87	12.26	4.05	6.83
C30H57F103	484.79	484.4291	74.33	11.85	3.92	9.90
C30H57F104	500.78	500.4241	71.95	11.47	3.79	12.78
C30H57F105	516.78	516.4190	69.73	11.12	3.68	15.48
C30H57F106	532.78	532.4139	67.63	10.78	3.57	18.02
C30H57F107	548.78	548.4088	65.66	10.47	3.46	20.41
C30H57F108	564.78	564.4037	63.80	10.17	3.36	22.66
C30H57F109	580.78	580.3986	62.04	9.89	3.27	24.79
C30H58F201	472.79	472.4455	76.21	12.37	8.04	3.38
C30H58F202	488.79	488.4405	73.72	11.96	7.77	6.55
C30H58F203	504.79	504.4354	71.38	11.58	7.53	9.51
C30H58F204	520.79	520.4303	69.19	11.23	7.30	12.29
C30H58F205	536.79	536.4252	67.13	10.89	7.08	14.90
C30H58F206	552.79	552.4201	65.18	10.58	6.87	17.37
C30H58F207	568.79	568.4150	63.35	10.28	6.68	19.69
C30H58F208	584.79	584.4099	61.62	10.00	6.50	21.89
C30H58F209	600.79	600.4049	59.98	9.73	6.32	23.97
C30H59F101	454.80	454.4550	79.23	13.08	4.18	3.52
C30H59F102	470.80	470.4499	76.54	12.63	4.04	6.80
C30H59F103	486.80	486.4448	74.02	12.22	3.90	9.86
C30H59F104	502.80	502.4397	71.67	11.83	3.78	12.73
C30H59F105	518.80	518.4346	69.46	11.46	3.66	15.42
C30H59F106	534.80	534.4295	67.38	11.12	3.55	17.95
C30H59F107	550.80	550.4245	65.42	10.80	3.45	20.33
C30H59F108	566.80	566.4194	63.57	10.49	3.35	22.58
C30H59F109	582.80	582.4143	61.83	10.20	3.26	24.71
C30H60F201	474.81	474.4612	75.89	12.74	8.00	3.37
C30H60F202	490.81	490.4561	73.42	12.32	7.74	6.52
C30H60F203	506.81	506.4510	71.10	11.93	7.50	9.47
C30H60F204	522.81	522.4459	68.92	11.57	7.27	12.24
C30H60F205	538.81	538.4409	66.88	11.22	7.05	14.85
C30H60F206	554.81	554.4358	64.95	10.90	6.85	17.30
C30H60F207	570.81	570.4307	63.13	10.60	6.66	19.62
C30H60F208	586.80	586.4256	61.41	10.31	6.48	21.81
C30H60F209	602.80	602.4205	59.78	10.03	6.30	23.89
C30H61F101	456.82	456.4706	78.88	13.46	4.16	3.50
C30H61F102	472.82	472.4655	76.21	13.00	4.02	6.77
C30H61F103	488.82	488.4604	73.72	12.58	3.89	9.82
C30H61F104	504.82	504.4554	71.38	12.18	3.76	12.68
C30H61F105	520.82	520.4503	69.19	11.81	3.65	15.36
C30H61F106	536.82	536.4452	67.12	11.45	3.54	17.88
C30H61F107	552.81	552.4401	65.18	11.12	3.44	20.26
C30H61F108	568.81	568.4350	63.35	10.81	3.34	22.50
C30H61F109	584.81	584.4299	61.62	10.51	3.25	24.62

7. CARBON, HYDROGEN, IODINE AND NITROGEN

7. KOHLENSTOFF, WASSERSTOFF, JOD UND STICKSTOFF

7. CARBONE, HYDROGENE, IODE ET AZOTE

R

FORMULA	MOL. WEIGHT	ISOMOL. WEIGHT	PERC. C	PERC. H	PERC. I	PERC. N
C 1H 0I1N1	152.92	152.9077	7.85	0	82.99	9.16
C 1H 0I1N3	180.94	180.9139	6.64	0	70.14	23.22
C 1H 0I1N5	208.95	208.9200	5.75	0	60.73	33.52
C 1H 0I2N2	293.83	293.8155	4.09	0	86.38	9.53
C 1H 0I2N4	321.85	321.8216	3.73	0	78.86	17.41
C 1H 0I2N6	349.86	349.8278	3.43	0	72.55	24.02
C 1H 1I1N2	167.94	167.9186	7.15	.60	75.57	16.68
C 1H 1I1N4	195.95	195.9248	6.13	.51	64.76	28.59
C 1H 1I1N6	223.96	223.9309	5.36	.45	56.66	37.52
C 1H 1I2N1	280.83	280.8202	4.28	.36	90.38	4.99
C 1H 1I2N3	308.85	308.8264	3.89	.33	82.18	13.61
C 1H 1I2N5	336.86	336.8325	3.57	.30	75.35	20.79
C 1H 2I1N1	154.94	154.9234	7.75	1.30	81.91	9.04
C 1H 2I1N3	182.95	182.9295	6.57	1.10	69.37	22.97
C 1H 2I1N5	210.96	210.9357	5.69	.96	60.15	33.20
C 1H 2I2N2	295.85	295.8311	4.06	.68	85.79	9.47
C 1H 2I2N4	323.86	323.8373	3.71	.62	78.37	17.30
C 1H 2I2N6	351.88	351.8434	3.41	.57	72.13	23.88
C 1H 3I1N2	169.95	169.9343	7.07	1.78	74.67	16.48
C 1H 3I1N4	197.97	197.9404	6.07	1.53	64.10	28.30
C 1H 3I1N6	225.98	225.9466	5.32	1.34	56.16	37.19
C 1H 3I2N1	282.85	282.8359	4.25	1.07	89.73	4.95
C 1H 3I2N3	310.86	310.8420	3.86	.97	81.65	13.52
C 1H 3I2N5	338.88	338.8482	3.54	.89	74.90	20.67
C 1H 4I1N1	156.95	156.9390	7.65	2.57	80.85	8.92
C 1H 4I1N3	184.97	184.9452	6.49	2.18	68.61	22.72
C 1H 4I1N5	212.98	212.9513	5.64	1.89	59.58	32.88
C 1H 4I2N2	297.87	297.8468	4.03	1.35	85.21	9.40
C 1H 4I2N4	325.88	325.8529	3.69	1.24	77.88	17.19
C 1H 4I2N6	353.89	353.8591	3.39	1.14	71.72	23.75
C 1H 5I1N2	171.97	171.9499	6.98	2.93	73.80	16.29
C 1H 5I1N4	199.98	199.9561	6.01	2.52	63.46	28.02
C 1H 5I1N6	228.00	227.9622	5.27	2.21	55.66	36.86
C 1H 5I2N3	312.88	312.8577	3.84	1.61	81.12	13.43
C 1H 5I2N5	340.89	340.8638	3.52	1.48	74.45	20.54
C 1H 6I1N3	186.98	186.9608	6.42	3.23	67.87	22.47
C 1H 6I1N5	215.00	214.9670	5.59	2.81	59.03	32.57
C 1H 6I2N4	327.89	327.8686	3.66	1.84	77.41	17.09
C 1H 6I2N6	355.91	355.8747	3.37	1.70	71.31	23.61
C 1H 7I1N4	202.00	201.9717	5.95	3.49	62.82	27.74
C 1H 7I1N6	230.01	229.9779	5.22	3.07	55.17	36.54
C 1H 7I2N5	342.91	342.8795	3.50	2.06	74.02	20.42
C 1H 8I1N5	217.01	216.9826	5.53	3.72	58.48	32.27
C 1H 8I2N6	357.92	357.8904	3.36	2.25	70.91	23.48
C 1H 9I1N6	232.03	231.9935	5.18	3.91	54.69	36.22
C 2H 1I1N2	179.95	179.9186	13.35	.56	70.52	15.57
C 2H 1I1N4	207.96	207.9248	11.55	.48	61.02	26.94
C 2H 1I1N6	235.97	235.9309	10.18	.43	53.78	35.61
C 2H 1I2N1	292.85	292.8202	8.20	.34	86.67	4.78
C 2H 1I2N3	320.86	320.8264	7.49	.31	79.10	13.10

FORMULA	MOL. WEIGHT	ISOMOL. WEIGHT	PERC. C	PERC. H	PERC. I	PERC. N
C 2H 1I2N5	348.87	348.8325	6.89	.29	72.75	20.07
C 2H 2I1N1	166.95	166.9234	14.39	1.21	76.01	8.39
C 2H 2I1N3	194.96	194.9295	12.32	1.03	65.09	21.55
C 2H 2I1N5	222.98	222.9357	10.77	.90	56.91	31.41
C 2H 2I2N2	307.86	307.8311	7.80	.65	82.44	9.10
C 2H 2I2N4	335.87	335.8373	7.15	.60	75.57	16.68
C 2H 2I2N6	363.89	363.8434	6.60	.55	69.75	23.10
C 2H 3I1N2	181.96	181.9343	13.20	1.66	69.74	15.40
C 2H 3I1N4	209.98	209.9404	11.44	1.44	60.44	26.68
C 2H 3I1N6	237.99	237.9466	10.09	1.27	53.32	35.31
C 2H 3I2N1	294.86	294.8359	8.15	1.03	86.08	4.75
C 2H 3I2N3	322.88	322.8420	7.44	.94	78.61	13.01
C 2H 3I2N5	350.89	350.8482	6.85	.86	72.33	19.96
C 2H 4I1N1	168.97	168.9390	14.22	2.39	75.11	8.29
C 2H 4I1N3	196.98	196.9452	12.20	2.05	64.43	21.33
C 2H 4I1N5	224.99	224.9513	10.68	1.79	56.40	31.13
C 2H 4I2N2	309.88	309.8468	7.75	1.30	81.91	9.04
C 2H 4I2N4	337.89	337.8529	7.11	1.19	75.12	16.58
C 2H 4I2N6	365.90	365.8591	6.57	1.10	69.37	22.97
C 2H 5I1N2	183.98	183.9499	13.06	2.74	68.98	15.23
C 2H 5I1N4	211.99	211.9561	11.33	2.38	59.86	26.43
C 2H 5I1N6	240.01	239.9622	10.01	2.10	52.88	35.02
C 2H 5I2N1	296.88	296.8515	8.09	1.70	85.49	4.72
C 2H 5I2N3	324.89	324.8577	7.39	1.55	78.12	12.93
C 2H 5I2N5	352.90	352.8638	6.81	1.43	71.92	19.84
C 2H 6I1N1	170.98	170.9547	14.05	3.54	74.22	8.19
C 2H 6I1N3	198.99	198.9608	12.07	3.04	63.77	21.12
C 2H 6I1N5	227.01	226.9670	10.58	2.66	55.90	30.85
C 2H 6I2N2	311.89	311.8624	7.70	1.94	81.38	8.98
C 2H 6I2N4	339.91	339.8686	7.07	1.78	74.67	16.48
C 2H 6I2N6	367.92	367.8747	6.53	1.64	68.98	22.84
C 2H 7I1N2	186.00	185.9656	12.92	3.79	68.23	15.06
C 2H 7I1N4	214.01	213.9717	11.22	3.30	59.30	26.18
C 2H 7I1N6	242.02	241.9779	9.93	2.92	52.43	34.72
C 2H 7I2N3	326.91	326.8733	7.35	2.16	77.64	12.85
C 2H 7I2N5	354.92	354.8795	6.77	1.99	71.51	19.73
C 2H 8I1N3	201.01	200.9765	11.95	4.01	63.13	20.90
C 2H 8I1N5	229.02	228.9826	10.49	3.52	55.41	30.58
C 2H 8I2N4	341.92	341.8842	7.03	2.36	74.23	16.39
C 2H 8I2N6	369.94	369.8904	6.49	2.18	68.61	22.72
C 2H 9I1N4	216.03	215.9874	11.12	4.20	58.75	25.94
C 2H 9I1N6	244.04	243.9935	9.84	3.72	52.00	34.44
C 2H 9I2N5	356.94	356.8951	6.73	2.54	71.11	19.62
C 2H10I1N5	231.04	230.9983	10.40	4.36	54.93	30.31
C 2H10I2N6	371.95	371.9060	6.46	2.71	68.24	22.59
C 2H11I1N6	246.05	246.0092	9.76	4.51	51.58	34.16
C 3H 1I1N2	191.96	191.9186	18.77	.53	66.11	14.59
C 3H 1I1N4	219.97	219.9248	16.38	.46	57.69	25.47
C 3H 1I1N6	247.99	247.9309	14.53	.41	51.17	33.89
C 3H 1I2N1	304.86	304.8202	11.82	.33	83.26	4.59

010

FORMULA	MOL. WEIGHT	ISOMOL. WEIGHT	PERC. C	PERC. H	PERC. I	PERC. N
C 3H 1I2N3	332.87	332.8264	10.83	.30	76.25	12.62
C 3H 1I2N5	360.88	360.8325	9.98	.28	70.33	19.41
C 3H 2I1N1	178.96	178.9234	20.13	1.13	70.91	7.83
C 3H 2I1N3	206.97	206.9295	17.41	.97	61.31	20.30
C 3H 2I1N5	234.99	234.9357	15.33	.86	54.00	29.80
C 3H 2I2N2	319.87	319.8311	11.26	.63	79.35	8.76
C 3H 2I2N4	347.88	347.8373	10.36	.58	72.96	16.10
C 3H 2I2N6	375.90	375.8434	9.59	.54	67.52	22.36
C 3H 3I1N2	193.98	193.9343	18.58	1.56	65.42	14.44
C 3H 3I1N4	221.99	221.9404	16.23	1.36	57.17	25.24
C 3H 3I1N6	250.00	249.9466	14.41	1.21	50.76	33.62
C 3H 3I2N1	306.87	306.8359	11.74	.99	82.71	4.56
C 3H 3I2N3	334.89	334.8420	10.76	.90	75.79	12.55
C 3H 3I2N5	362.90	362.8482	9.93	.83	69.94	19.30
C 3H 4I1N1	180.98	180.9390	19.91	2.23	70.12	7.74
C 3H 4I1N3	208.99	208.9452	17.24	1.93	60.72	20.11
C 3H 4I1N5	237.00	236.9513	15.20	1.70	53.55	29.55
C 3H 4I2N2	321.89	321.8468	11.19	1.25	78.85	8.70
C 3H 4I2N4	349.90	349.8529	10.30	1.15	72.54	16.01
C 3H 4I2N6	377.91	377.8591	9.53	1.07	67.16	22.24
C 3H 5I1N2	195.99	195.9499	18.39	2.57	64.75	14.29
C 3H 5I1N4	224.00	223.9561	16.09	2.25	56.65	25.01
C 3H 5I1N6	252.02	251.9622	14.30	2.00	50.36	33.35
C 3H 5I2N1	308.89	308.8515	11.67	1.63	82.17	4.53
C 3H 5I2N3	336.90	336.8577	10.70	1.50	75.34	12.47
C 3H 5I2N5	364.92	364.8638	9.87	1.38	69.55	19.19
C 3H 6I1N1	182.99	182.9547	19.69	3.30	69.35	7.65
C 3H 6I1N3	211.01	210.9608	17.08	2.87	60.14	19.91
C 3H 6I1N5	239.02	238.9670	15.08	2.53	53.09	29.30
C 3H 6I2N2	323.90	323.8624	11.12	1.87	78.36	8.65
C 3H 6I2N4	351.92	351.8686	10.24	1.72	72.12	15.92
C 3H 6I2N6	379.93	379.8747	9.48	1.59	66.80	22.12
C 3H 7I1N2	198.01	197.9656	18.20	3.56	64.09	14.15
C 3H 7I1N4	226.02	225.9717	15.94	3.12	56.15	24.79
C 3H 7I1N6	254.03	253.9779	14.18	2.78	49.96	33.08
C 3H 7I2N1	310.90	310.8672	11.59	2.27	81.64	4.51
C 3H 7I2N3	338.92	338.8733	10.63	2.08	74.89	12.40
C 3H 7I2N5	366.93	366.8795	9.82	1.92	69.17	19.09
C 3H 8I1N1	185.01	184.9703	19.48	4.36	68.59	7.57
C 3H 8I1N3	213.02	212.9765	16.92	3.79	59.57	19.73
C 3H 8I1N5	241.04	240.9826	14.95	3.35	52.65	29.06
C 3H 8I2N2	325.92	325.8781	11.06	2.47	77.87	8.60
C 3H 8I2N4	353.93	353.8842	10.18	2.28	71.71	15.83
C 3H 8I2N6	381.95	381.8904	9.43	2.11	66.45	22.00
C 3H 9I1N2	200.02	199.9812	18.01	4.54	63.44	14.01
C 3H 9I1N4	228.04	227.9874	15.80	3.98	55.65	24.57
C 3H 9I1N6	256.05	255.9935	14.07	3.54	49.56	32.82
C 3H 9I2N3	340.93	340.8890	10.57	2.66	74.45	12.32
C 3H 9I2N5	368.95	368.8951	9.77	2.46	68.79	18.98
C 3H10I1N3	215.04	214.9921	16.76	4.69	59.01	19.54
C 3H10I1N5	243.05	242.9983	14.83	4.15	52.21	28.81
C 3H10I2N4	355.95	355.8999	10.12	2.83	71.30	15.74
C 3H10I2N6	383.96	383.9060	9.38	2.63	66.10	21.89
C 3H11I1N4	230.05	230.0030	15.66	4.82	55.16	24.35
C 3H11I1N6	258.07	258.0092	13.96	4.30	49.18	32.57
C 3H11 2N5	370.96	370.9108	9.71	2.99	68.42	18.88
C 3H12I1N5	245.07	245.0139	14.70	4.94	51.78	28.58
C 3H12I2N6	385.98	385.9217	9.34	3.13	65.76	21.77
C 3H13I1N6	260.08	260.0248	13.85	5.04	48.79	32.31
C 4H 2I1N1	190.97	190.9234	25.16	1.06	66.45	7.33

FORMULA	MOL. WEIGHT	ISOMOL. WEIGHT	PERC. C	PERC. H	PERC. I	PERC. N
C 4H 2I1N3	218.99	218.9295	21.94	.92	57.95	19.19
C 4H 2I1N5	247.00	246.9357	19.45	.82	51.38	28.35
C 4H 2I2N2	331.88	331.8311	14.48	.61	76.48	8.44
C 4H 2I2N4	359.90	359.8373	13.35	.56	70.52	15.57
C 4H 2I2N6	387.91	387.8434	12.39	.52	65.43	21.66
C 4H 3I1N2	205.99	205.9343	23.32	1.47	61.61	13.60
C 4H 3I1N4	234.00	233.9404	20.53	1.29	54.23	23.94
C 4H 3I1N6	262.01	261.9466	18.34	1.15	48.43	32.07
C 4H 3I2N1	318.88	318.8359	15.07	.95	79.59	4.39
C 4H 3I2N3	346.90	346.8420	13.85	.87	73.17	12.11
C 4H 3I2N5	374.91	374.8482	12.81	.81	67.70	18.68
C 4H 4I1N1	192.99	192.9390	24.90	2.09	65.76	7.26
C 4H 4I1N3	221.00	220.9452	21.74	1.82	57.42	19.01
C 4H 4I1N5	249.01	248.9513	19.29	1.62	50.96	28.12
C 4H 4I2N2	333.90	333.8468	14.39	1.21	76.01	8.39
C 4H 4I2N4	361.91	361.8529	13.28	1.11	70.13	15.48
C 4H 4I2N6	389.93	389.8591	12.32	1.03	65.09	21.55
C 4H 5I1N2	208.00	207.9499	23.10	2.42	61.01	13.47
C 4H 5I1N4	236.02	235.9561	20.36	2.14	53.77	23.74
C 4H 5I1N6	264.03	263.9622	18.20	1.91	48.06	31.83
C 4H 5I2N1	320.90	320.8515	14.97	1.57	79.09	4.36
C 4H 5I2N3	348.91	348.8577	13.77	1.44	72.74	12.04
C 4H 5I2N5	376.93	376.8638	12.75	1.34	67.34	18.58
C 4H 6I1N1	195.00	194.9547	24.64	3.10	65.08	7.18
C 4H 6I1N3	223.02	222.9608	21.54	2.71	56.90	18.84
C 4H 6I1N5	251.03	250.9670	19.14	2.41	50.55	27.90
C 4H 6I2N2	335.91	335.8624	14.30	1.80	75.56	8.34
C 4H 6I2N4	363.93	363.8686	13.20	1.66	69.74	15.40
C 4H 6I2N6	391.94	391.8747	12.26	1.54	64.76	21.44
C 4H 7I1N2	210.02	209.9656	22.88	3.36	60.43	13.34
C 4H 7I1N4	238.03	237.9717	20.18	2.96	53.31	23.54
C 4H 7I1N6	266.04	265.9779	18.06	2.65	47.70	31.59
C 4H 7I2N1	322.92	322.8672	14.88	2.19	78.60	4.34
C 4H 7I2N3	350.93	350.8733	13.69	2.01	72.32	11.97
C 4H 7I2N5	378.94	378.8795	12.68	1.86	66.98	18.48
C 4H 8I1N1	197.02	196.9703	24.39	4.09	64.41	7.11
C 4H 8I1N3	225.03	224.9765	21.35	3.58	56.39	18.67
C 4H 8I1N5	253.05	252.9826	18.99	3.19	50.15	27.68
C 4H 8I2N2	337.93	337.8781	14.22	2.39	75.11	8.29
C 4H 8I2N4	365.94	365.8842	13.13	2.20	69.36	15.31
C 4H 8I2N6	393.96	393.8904	12.20	2.05	64.43	21.33
C 4H 9I1N2	212.03	211.9812	22.66	4.28	59.85	13.21
C 4H 9I1N4	240.05	239.9874	20.01	3.78	52.87	23.34
C 4H 9I1N6	268.06	267.9935	17.92	3.38	47.34	31.35
C 4H 9I2N1	324.93	324.8828	14.79	2.79	78.11	4.31
C 4H 9I2N3	352.95	352.8890	13.61	2.57	71.91	11.91
C 4H 9I2N5	380.96	380.8951	12.61	2.38	66.62	18.38
C 4H10I1N1	199.04	198.9860	24.14	5.06	63.76	7.04
C 4H10I1N3	227.05	226.9921	21.16	4.44	55.89	18.51
C 4H10I1N5	255.06	254.9983	18.84	3.95	49.75	27.46
C 4H10I2N2	339.95	339.8937	14.13	2.97	74.66	8.24
C 4H10I2N4	367.96	367.8999	13.06	2.74	68.98	15.23
C 4H10I2N6	395.97	395.9060	12.13	2.55	64.10	21.22
C 4H11I1N2	214.05	213.9969	22.45	5.18	59.29	13.09
C 4H11I1N4	242.06	242.0030	19.85	4.58	52.43	23.15
C 4H11I1N6	270.08	270.0092	17.79	4.11	46.99	31.12
C 4H11I2N3	354.96	354.9046	13.54	3.12	71.50	11.84
C 4H11I2N5	382.97	382.9108	12.55	2.90	66.27	18.29
C 4H12I1N3	229.06	229.0078	20.97	5.28	55.40	18.34
C 4H12I1N5	257.08	257.0139	18.69	4.71	49.36	27.24

011

C 4H12I2N4	369.98	369.9155	12.99	3.27	68.60	15.14
C 4H12I2N6	397.99	397.9217	12.07	3.04	63.77	21.12
C 4H13I1N4	244.08	244.0187	19.68	5.37	51.99	22.95
C 4H13I1N6	272.09	272.0248	17.66	4.82	46.64	30.89
C 4H13I2N5	384.99	384.9264	12.48	3.40	65.93	18.19
C 4H14I1N5	259.09	259.0296	18.54	5.45	48.98	27.03
C 4H14I2N6	400.01	399.9373	12.01	3.53	63.45	21.01
C 4H15I1N6	274.11	274.0405	17.53	5.52	46.30	30.66
C 5H 2I1N1	202.98	202.9234	29.59	.99	62.52	6.90
C 5H 2I1N3	231.00	230.9295	26.00	.87	54.94	18.19
C 5H 2I1N5	259.01	258.9357	23.19	.78	49.00	27.04
C 5H 2I2N2	343.89	343.8311	17.46	.59	73.80	8.15
C 5H 2I2N4	371.91	371.8373	16.15	.54	68.25	15.06
C 5H 2I2N6	399.92	399.8434	15.02	.50	63.46	21.01
C 5H 3I1N2	218.00	217.9343	27.55	1.39	58.21	12.85
C 5H 3I1N4	246.01	245.9404	24.41	1.23	51.58	22.77
C 5H 3I1N6	274.02	273.9466	21.92	1.10	46.31	30.67
C 5H 3I2N1	330.90	330.8359	18.15	.91	76.70	4.23
C 5H 3I2N3	358.91	358.8420	16.73	.84	70.72	11.71
C 5H 3I2N5	386.92	386.8482	15.52	.78	65.60	18.10
C 5H 4I1N1	205.00	204.9390	29.30	1.97	61.90	6.83
C 5H 4I1N3	233.01	232.9452	25.77	1.73	54.46	18.03
C 5H 4I1N5	261.03	260.9513	23.01	1.54	48.62	26.83
C 5H 4I2N2	345.91	345.8468	17.36	1.17	73.37	8.10
C 5H 4I2N4	373.92	373.8529	16.06	1.08	67.88	14.98
C 5H 4I2N6	401.94	401.8591	14.94	1.00	63.15	20.91
C 5H 5I1N2	220.01	219.9499	27.30	2.29	57.68	12.73
C 5H 5I1N4	248.03	247.9561	24.21	2.03	51.17	22.59
C 5H 5I1N6	276.04	275.9622	21.76	1.83	45.97	30.44
C 5H 5I2N1	332.91	332.8515	18.04	1.51	76.24	4.21
C 5H 5I2N3	360.92	360.8577	16.64	1.40	70.32	11.64
C 5H 5I2N5	388.94	388.8638	15.44	1.30	65.26	18.01
C 5H 6I1N1	207.01	206.9547	29.01	2.92	61.30	6.77
C 5H 6I1N3	235.03	234.9608	25.55	2.57	54.00	17.88
C 5H 6I1N5	263.04	262.9670	22.83	2.30	48.25	26.62
C 5H 6I2N2	347.93	347.8624	17.26	1.74	72.95	8.05
C 5H 6I2N4	375.94	375.8686	15.97	1.61	67.51	14.90
C 5H 6I2N6	403.95	403.8747	14.87	1.50	62.83	20.80
C 5H 7I1N2	222.03	221.9656	27.05	3.18	57.16	12.62
C 5H 7I1N4	250.04	249.9717	24.02	2.82	50.75	22.41
C 5H 7I1N6	278.06	277.9779	21.60	2.54	45.64	30.22
C 5H 7I2N1	334.93	334.8672	17.93	2.11	75.78	4.18
C 5H 7I2N3	362.94	362.8733	16.55	1.94	69.93	11.58
C 5H 7I2N5	390.95	390.8795	15.36	1.80	64.92	17.91
C 5H 8I1N1	209.03	208.9703	28.73	3.86	60.71	6.70
C 5H 8I1N3	237.04	236.9765	25.34	3.40	53.54	17.73
C 5H 8I1N5	265.06	264.9826	22.66	3.04	47.88	26.42
C 5H 8I2N2	349.94	349.8781	17.16	2.30	72.53	8.01
C 5H 8I2N4	377.96	377.8842	15.89	2.13	67.15	14.82
C 5H 8I2N6	405.97	405.8904	14.79	1.99	62.52	20.70
C 5H 9I1N2	224.05	223.9812	26.81	4.05	56.64	12.50
C 5H 9I1N4	252.06	251.9874	23.83	3.60	50.35	22.23
C 5H 9I1N6	280.07	279.9935	21.44	3.24	45.31	30.01
C 5H 9I2N1	336.94	336.8828	17.82	2.69	75.33	4.16
C 5H 9I2N3	364.96	364.8890	16.46	2.49	69.54	11.51
C 5H 9I2N5	392.97	392.8951	15.28	2.31	64.59	17.82
C 5H10I1N1	211.05	210.9860	28.46	4.78	60.13	6.64
C 5H10I1N3	239.06	238.9921	25.12	4.22	53.08	17.58
C 5H10I1N5	267.07	266.9983	22.49	3.77	47.52	26.22
C 5H10I2N2	351.96	351.8937	17.06	2.86	72.11	7.96
C 5H10I2N4	379.97	379.8999	15.81	2.65	66.80	14.75
C 5H10I2N6	407.98	407.9060	14.72	2.47	62.21	20.60
C 5H11I1N2	226.06	225.9969	26.57	4.90	56.14	12.39
C 5H11I1N4	254.07	254.0030	23.64	4.36	49.95	22.05
C 5H11I1N6	282.09	282.0092	21.29	3.93	44.99	29.79
C 5H11I2N1	338.96	338.8985	17.72	3.27	74.88	4.13
C 5H11I2N3	366.97	366.9046	16.37	3.02	69.16	11.45
C 5H11I2N5	394.99	394.9108	15.20	2.81	64.26	17.73
C 5H12I1N1	213.06	213.0016	28.19	5.68	59.56	6.57
C 5H12I1N3	241.08	241.0078	24.91	5.02	52.64	17.43
C 5H12I1N5	269.09	269.0139	22.32	4.50	47.16	26.03
C 5H12I2N2	353.97	353.9094	16.97	3.42	71.70	7.91
C 5H12I2N4	381.99	381.9155	15.72	3.17	66.44	14.67
C 5H12I2N6	410.00	409.9217	14.65	2.95	61.90	20.50
C 5H13I1N2	228.08	228.0125	26.33	5.75	55.64	12.28
C 5H13I1N4	256.09	256.0187	23.45	5.12	49.55	21.88
C 5H13I1N6	284.10	284.0248	21.14	4.61	44.67	29.58
C 5H13I2N3	368.99	368.9203	16.28	3.55	68.79	11.39
C 5H13I2N5	397.00	396.9264	15.13	3.30	63.93	17.64
C 5H14I1N3	243.09	243.0234	24.70	5.81	52.20	17.29
C 5H14I1N5	271.11	271.0296	22.15	5.21	46.81	25.83
C 5H14I2N4	384.00	383.9312	15.64	3.67	66.10	14.59
C 5H14I2N6	412.02	411.9373	14.58	3.43	61.60	20.40
C 5H15I1N4	258.11	258.0343	23.27	5.86	49.17	21.71
C 5H15I1N6	286.12	286.0405	20.99	5.28	44.35	29.37
C 5H15I2N5	399.02	398.9421	15.05	3.79	63.61	17.55
C 5H16I1N5	273.12	273.0452	21.99	5.90	46.46	25.64
C 5H16I2N6	414.03	413.9530	14.51	3.90	61.30	20.30
C 5H17I1N6	288.14	288.0561	20.84	5.95	44.04	29.17
C 6H 3I1N2	230.01	229.9343	31.33	1.31	55.17	12.18
C 6H 3I1N4	258.02	257.9404	27.93	1.17	49.18	21.71
C 6H 3I1N6	286.04	285.9466	25.20	1.06	44.37	29.38
C 6H 3I2N1	342.91	342.8359	21.02	.88	74.02	4.08
C 6H 3I2N3	370.92	370.8420	19.43	.82	68.43	11.33
C 6H 3I2N5	398.93	398.8482	18.06	.76	63.62	17.56
C 6H 4I1N1	217.01	216.9390	33.21	1.86	58.48	6.45
C 6H 4I1N3	245.02	244.9452	29.41	1.65	51.79	17.15
C 6H 4I1N5	273.04	272.9513	26.39	1.48	46.48	25.65
C 6H 4I2N2	357.92	357.8468	20.13	1.13	70.91	7.83
C 6H 4I2N4	385.93	385.8529	18.67	1.04	65.76	14.52
C 6H 4I2N6	413.95	413.8591	17.41	.97	61.31	20.30
C 6H 5I1N2	232.02	231.9499	31.06	2.17	54.69	12.07
C 6H 5I1N4	260.04	259.9561	27.71	1.94	48.80	21.55
C 6H 5I1N6	288.05	287.9622	25.02	1.75	44.06	29.18
C 6H 5I2N1	344.92	344.8515	20.89	1.46	73.58	4.06
C 6H 5I2N3	372.94	372.8577	19.32	1.35	68.06	11.27
C 6H 5I2N5	400.95	400.8638	17.97	1.26	63.30	17.47
C 6H 6I1N1	219.03	218.9547	32.90	2.76	57.94	6.39
C 6H 6I1N3	247.04	246.9608	29.17	2.45	51.37	17.01
C 6H 6I1N5	275.05	274.9670	26.20	2.20	46.14	25.46
C 6H 6I2N2	359.94	359.8624	20.02	1.68	70.51	7.78
C 6H 6I2N4	387.95	387.8686	18.58	1.56	65.42	14.44
C 6H 6I2N6	415.96	415.8747	17.33	1.45	61.02	20.20
C 6H 7I1N2	234.04	233.9656	30.79	3.01	54.22	11.97
C 6H 7I1N4	262.05	261.9717	27.50	2.69	48.43	21.38
C 6H 7I1N6	290.07	289.9779	24.84	2.43	43.75	28.97
C 6H 7I2N1	346.94	346.8672	20.77	2.03	73.16	4.04
C 6H 7I2N3	374.95	374.8733	19.22	1.88	67.69	11.21
C 6H 7I2N5	402.96	402.8795	17.88	1.75	62.99	17.38
C 6H 8I1N1	221.04	220.9703	32.60	3.65	57.41	6.34

012

C 6H 8I1N3	249.06	248.9765	28.94	3.24	50.95	16.87
C 6H 8I1N5	277.07	276.9826	26.01	2.91	45.80	25.28
C 6H 8I2N2	361.95	361.8781	19.91	2.23	70.12	7.74
C 6H 8I2N4	389.97	389.8842	18.48	2.07	65.08	14.37
C 6H 8I2N6	417.98	417.8904	17.24	1.93	60.72	20.11
C 6H 9I1N2	236.06	235.9812	30.53	3.84	53.76	11.87
C 6H 9I1N4	264.07	263.9874	27.29	3.44	48.06	21.22
C 6H 9I1N6	292.08	291.9935	24.67	3.11	43.45	28.77
C 6H 9I2N1	348.95	348.8828	20.65	2.60	72.73	4.01
C 6H 9I2N3	376.97	376.8890	19.12	2.41	67.33	11.15
C 6H 9I2N5	404.98	404.8951	17.80	2.24	62.67	17.29
C 6H10I1N1	223.06	222.9860	32.31	4.52	56.89	6.28
C 6H10I1N3	251.07	250.9921	28.70	4.01	50.55	16.74
C 6H10I1N5	279.08	278.9983	25.82	3.61	45.47	25.09
C 6H10I2N2	363.97	363.8937	19.80	2.77	69.73	7.70
C 6H10I2N4	391.98	391.8999	18.39	2.57	64.75	14.29
C 6H10I2N6	420.00	419.9060	17.16	2.40	60.43	20.01
C 6H11I1N2	238.07	237.9969	30.27	4.66	53.30	11.77
C 6H11I1N4	266.09	266.0030	27.08	4.17	47.69	21.06
C 6H11I1N6	294.10	294.0092	24.50	3.77	43.15	28.58
C 6H11I2N1	350.97	350.8985	20.53	3.16	72.32	3.99
C 6H11I2N3	378.98	378.9046	19.02	2.93	66.97	11.09
C 6H11I2N5	407.00	406.9108	17.71	2.72	62.36	17.21
C 6H12I1N1	225.07	225.0016	32.02	5.37	56.38	6.22
C 6H12I1N3	253.09	253.0078	28.48	4.78	50.14	16.60
C 6H12I1N5	281.10	281.0139	25.64	4.30	45.15	24.91
C 6H12I2N2	365.98	365.9094	19.69	3.30	69.35	7.65
C 6H12I2N4	394.00	393.9155	18.29	3.07	64.42	14.22
C 6H12I2N6	422.01	421.9217	17.08	2.87	60.14	19.91
C 6H13I1N2	240.09	240.0125	30.02	5.46	52.86	11.67
C 6H13I1N4	268.10	268.0187	26.88	4.89	47.33	20.90
C 6H13I1N6	296.12	296.0248	24.34	4.43	42.86	28.38
C 6H13I2N1	352.99	352.9141	20.42	3.71	71.90	3.97
C 6H13I2N3	381.00	380.9203	18.92	3.44	66.62	11.03
C 6H13I2N5	409.01	408.9264	17.62	3.20	62.05	17.12
C 6H14I1N1	227.09	227.0173	31.74	6.21	55.88	6.17
C 6H14I1N3	255.10	255.0234	28.25	5.53	49.75	16.47
C 6H14I1N5	283.12	283.0296	25.45	4.98	44.82	24.74
C 6H14I2N2	368.00	367.9250	19.58	3.83	68.97	7.61
C 6H14I2N4	396.01	395.9312	18.20	3.56	64.09	14.15
C 6H14I2N6	424.03	423.9373	17.00	3.33	59.86	19.82
C 6H15I1N2	242.10	242.0282	29.77	6.25	52.42	11.57
C 6H15I1N4	270.12	270.0343	26.68	5.60	46.98	20.74
C 6H15I1N6	298.13	298.0405	24.17	5.07	42.57	28.19
C 6H15I2N3	383.02	382.9359	18.82	3.95	66.27	10.97
C 6H15I2N5	411.03	410.9421	17.53	3.68	61.75	17.04
C 6H16I1N3	257.12	257.0391	28.03	6.27	49.36	16.34
C 6H16I1N5	285.13	285.0452	25.27	5.66	44.51	24.56
C 6H16I2N4	398.03	397.9468	18.11	4.05	63.77	14.08
C 6H16I2N6	426.04	425.9530	16.92	3.79	59.57	19.73
C 6H17I1N4	272.13	272.0500	26.48	6.30	46.63	20.59
C 6H17I1N6	300.15	300.0561	24.01	5.71	42.28	28.00
C 6H17I2N5	413.04	412.9577	17.45	4.15	61.45	16.96
C 6H18I1N5	287.15	287.0609	25.10	6.32	44.19	24.39
C 6H18I2N6	428.06	427.9686	16.84	4.24	59.29	19.63
C 6H19I1N6	302.16	302.0718	23.85	6.34	42.00	27.81
C 7H 3I1N2	242.02	241.9343	34.74	1.25	52.44	11.57
C 7H 3I1N4	270.03	269.9404	31.14	1.12	47.00	20.75
C 7H 3I1N6	298.05	297.9466	28.21	1.01	42.58	28.20
C 7H 3I2N1	354.92	354.8359	23.69	.85	71.51	3.95
C 7H 3I2N3	382.93	382.8420	21.96	.79	66.28	10.97
C 7H 3I2N5	410.94	410.8482	20.46	.74	61.76	17.04
C 7H 4I1N1	229.02	228.9390	36.71	1.76	55.41	6.12
C 7H 4I1N3	257.03	256.9452	32.71	1.57	49.37	16.35
C 7H 4I1N5	285.05	284.9513	29.50	1.41	44.52	24.57
C 7H 4I2N2	369.93	369.8468	22.73	1.09	68.61	7.57
C 7H 4I2N4	397.95	397.8529	21.13	1.01	63.78	14.08
C 7H 4I2N6	425.96	425.8591	19.74	.95	59.59	19.73
C 7H 5I1N2	244.04	243.9499	34.45	2.07	52.00	11.48
C 7H 5I1N4	272.05	271.9561	30.91	1.85	46.65	20.59
C 7H 5I1N6	300.06	299.9622	28.02	1.68	42.29	28.01
C 7H 5I2N1	356.93	356.8515	23.56	1.41	71.11	3.92
C 7H 5I2N3	384.95	384.8577	21.84	1.31	65.93	10.92
C 7H 5I2N5	412.96	412.8638	20.36	1.22	61.46	16.96
C 7H 6I1N1	231.04	230.9547	36.39	2.62	54.93	6.06
C 7H 6I1N3	259.05	258.9608	32.46	2.33	48.99	16.22
C 7H 6I1N5	287.06	286.9670	29.29	2.11	44.21	24.40
C 7H 6I2N2	371.95	371.8624	22.60	1.63	68.24	7.53
C 7H 6I2N4	399.96	399.8686	21.02	1.51	63.46	14.01
C 7H 6I2N6	427.97	427.8747	19.65	1.41	59.30	19.64
C 7H 7I1N2	246.05	245.9656	34.17	2.87	51.58	11.39
C 7H 7I1N4	274.07	273.9717	30.68	2.57	46.30	20.44
C 7H 7I1N6	302.08	301.9779	27.83	2.34	42.01	27.82
C 7H 7I2N1	358.95	358.8672	23.42	1.97	70.71	3.90
C 7H 7I2N3	386.96	386.8733	21.73	1.82	65.59	10.86
C 7H 7I2N5	414.98	414.8795	20.26	1.70	61.16	16.88
C 7H 8I1N1	233.05	232.9703	36.08	3.46	54.45	6.01
C 7H 8I1N3	261.07	260.9765	32.21	3.09	48.61	16.10
C 7H 8I1N5	289.08	288.9826	29.08	2.79	43.90	24.23
C 7H 8I2N2	373.96	373.8781	22.48	2.16	67.87	7.49
C 7H 8I2N4	401.98	401.8842	20.92	2.01	63.14	13.94
C 7H 8I2N6	429.99	429.8904	19.55	1.88	59.03	19.54
C 7H 9I1N2	248.07	247.9812	33.89	3.66	51.16	11.29
C 7H 9I1N4	276.08	275.9874	30.45	3.29	45.97	20.29
C 7H 9I1N6	304.09	303.9935	27.65	2.98	41.73	27.64
C 7H 9I2N1	360.97	360.8828	23.29	2.51	70.31	3.88
C 7H 9I2N3	388.98	388.8890	21.62	2.33	65.25	10.80
C 7H 9I2N5	416.99	416.8951	20.16	2.18	60.87	16.79
C 7H10I1N1	235.07	234.9860	35.77	4.29	53.99	5.96
C 7H10I1N3	263.08	262.9921	31.96	3.83	48.24	15.97
C 7H10I1N5	291.10	290.9983	28.88	3.46	43.60	24.06
C 7H10I2N2	375.98	375.8937	22.36	2.68	67.51	7.45
C 7H10I2N4	403.99	403.8999	20.81	2.50	62.82	13.87
C 7H10I2N6	432.01	431.9060	19.46	2.33	58.75	19.45
C 7H11I1N2	250.08	249.9969	33.62	4.43	50.74	11.20
C 7H11I1N4	278.10	278.0030	30.23	3.99	45.63	20.15
C 7H11I1N6	306.11	306.0092	27.47	3.62	41.46	27.45
C 7H11I2N1	362.98	362.8985	23.16	3.05	69.92	3.86
C 7H11I2N3	390.99	390.9046	21.50	2.84	64.91	10.75
C 7H11I2N5	419.01	418.9108	20.07	2.65	60.57	16.71
C 7H12I1N1	237.08	237.0016	35.46	5.10	53.53	5.91
C 7H12I1N3	265.10	265.0078	31.72	4.56	47.87	15.85
C 7H12I1N5	293.11	293.0139	28.68	4.13	43.30	23.89
C 7H12I2N2	378.00	377.9094	22.24	3.20	67.15	7.41
C 7H12I2N4	406.01	405.9155	20.71	2.98	62.51	13.80
C 7H12I2N6	434.02	433.9217	19.37	2.79	58.48	19.36
C 7H13I1N2	252.10	252.0125	33.35	5.20	50.34	11.11
C 7H13I1N4	280.11	280.0187	30.02	4.68	45.30	20.00
C 7H13I1N6	308.13	308.0248	27.29	4.25	41.19	27.27
C 7H13I2N1	365.00	364.9141	23.04	3.59	69.54	3.84

C 7H13I2N3	393.01	392.9203	21.39	3.33	64.58	10.69
C 7H13I2N5	421.02	420.9264	19.97	3.11	60.28	16.63
C 7H14I1N1	239.10	239.0173	35.16	5.90	53.08	5.86
C 7H14I1N3	267.11	267.0234	31.48	5.28	47.51	15.73
C 7H14I1N5	295.13	295.0296	28.49	4.78	43.00	23.73
C 7H14I2N2	380.01	379.9250	22.13	3.71	66.79	7.37
C 7H14I2N4	408.03	407.9312	20.61	3.46	62.20	13.73
C 7H14I2N6	436.04	435.9373	19.28	3.24	58.21	19.27
C 7H15I1N2	254.12	254.0282	33.09	5.95	49.94	11.02
C 7H15I1N4	282.13	282.0343	29.80	5.36	44.98	19.86
C 7H15I1N6	310.14	310.0405	27.11	4.88	40.92	27.10
C 7H15I2N1	367.01	366.9298	22.91	4.12	69.16	3.82
C 7H15I2N3	395.03	394.9359	21.28	3.83	64.25	10.64
C 7H15I2N5	423.04	422.9421	19.87	3.57	60.00	16.55
C 7H16I1N1	241.12	241.0329	34.87	6.69	52.63	5.81
C 7H16I1N3	269.13	269.0391	31.24	5.99	47.15	15.61
C 7H16I1N5	297.14	297.0452	28.30	5.43	42.71	23.57
C 7H16I2N2	382.03	381.9407	22.01	4.22	66.44	7.33
C 7H16I2N4	410.04	409.9468	20.50	3.93	61.90	13.66
C 7H16I2N6	438.05	437.9530	19.19	3.68	57.94	19.18
C 7H17I1N2	256.13	256.0438	32.83	6.69	49.55	10.94
C 7H17I1N4	284.14	284.0500	29.59	6.03	44.66	19.72
C 7H17I1N6	312.16	312.0561	26.93	5.49	40.65	26.92
C 7H17I2N3	397.04	396.9516	21.18	4.32	63.92	10.58
C 7H17I2N5	425.06	424.9577	19.78	4.03	59.71	16.48
C 7H18I1N3	271.15	271.0547	31.01	6.69	46.80	15.50
C 7H18I1N5	299.16	299.0609	28.10	6.06	42.42	23.41
C 7H18I2N4	412.06	411.9625	20.40	4.40	61.60	13.60
C 7H18I2N6	440.07	439.9686	19.11	4.12	57.67	19.10
C 7H19I1N4	286.16	286.0656	29.38	6.69	44.35	19.58
C 7H19I1N6	314.17	314.0718	26.76	6.10	40.39	26.75
C 7H19I2N5	427.07	426.9734	19.69	4.48	59.43	16.40
C 7H20I1N5	301.18	301.0765	27.92	6.69	42.14	23.25
C 7H20I2N6	442.09	441.9843	19.02	4.56	57.41	19.01
C 7H21I1N6	316.19	316.0874	26.59	6.69	40.14	26.58
C 8H 4I1N1	241.03	240.9390	39.87	1.67	52.65	5.81
C 8H 4I1N3	269.05	268.9452	35.71	1.50	47.17	15.62
C 8H 4I1N5	297.06	296.9513	32.35	1.36	42.72	23.58
C 8H 4I2N2	381.94	381.8468	25.16	1.06	66.45	7.33
C 8H 4I2N4	409.96	409.8529	23.44	.98	61.91	13.67
C 8H 4I2N6	437.97	437.8591	21.94	.92	57.95	19.19
C 8H 5I1N2	256.05	255.9499	37.53	1.97	49.56	10.94
C 8H 5I1N4	284.06	283.9561	33.83	1.77	44.68	19.72
C 8H 5I1N6	312.07	311.9622	30.79	1.61	40.66	26.93
C 8H 5I2N1	368.94	368.8515	26.04	1.37	68.79	3.80
C 8H 5I2N3	396.96	396.8577	24.21	1.27	63.94	10.59
C 8H 5I2N5	424.97	424.8638	22.61	1.19	59.72	16.48
C 8H 6I1N1	243.05	242.9547	39.54	2.49	52.21	5.76
C 8H 6I1N3	271.06	270.9608	35.45	2.23	46.82	15.50
C 8H 6I1N5	299.07	298.9670	32.13	2.02	42.43	23.42
C 8H 6I2N2	383.96	383.8624	25.03	1.58	66.10	7.30
C 8H 6I2N4	411.97	411.8686	23.32	1.47	61.61	13.60
C 8H 6I2N6	439.99	439.8747	21.84	1.37	57.69	19.10
C 8H 7I1N2	258.06	257.9656	37.23	2.73	49.18	10.86
C 8H 7I1N4	286.08	285.9717	33.59	2.47	44.36	19.58
C 8H 7I1N6	314.09	313.9779	30.59	2.25	40.40	26.76
C 8H 7I2N1	370.96	370.8672	25.90	1.90	68.42	3.78
C 8H 7I2N3	398.97	398.8733	24.08	1.77	63.62	10.53
C 8H 7I2N5	426.99	426.8795	22.50	1.65	59.44	16.40
C 8H 8I1N1	245.06	244.9703	39.21	3.29	51.78	5.72
C 8H 8I1N3	273.08	272.9765	35.19	2.95	46.47	15.39
C 8H 8I1N5	301.09	300.9826	31.91	2.68	42.15	23.26
C 8H 8I2N2	385.98	385.8781	24.90	2.09	65.76	7.26
C 8H 8I2N4	413.99	413.8842	23.21	1.95	61.31	13.53
C 8H 8I2N6	442.00	441.8904	21.74	1.82	57.42	19.01
C 8H 9I1N2	260.08	259.9812	36.95	3.49	48.79	10.77
C 8H 9I1N4	288.09	287.9874	33.35	3.15	44.05	19.45
C 8H 9I1N6	316.11	315.9935	30.40	2.87	40.15	26.59
C 8H 9I2N1	372.98	372.8828	25.76	2.43	68.05	3.76
C 8H 9I2N3	400.99	400.8890	23.96	2.26	63.30	10.48
C 8H 9I2N5	429.00	428.8951	22.40	2.11	59.16	16.32
C 8H10I1N1	247.08	246.9860	38.89	4.08	51.36	5.67
C 8H10I1N3	275.09	274.9921	34.93	3.66	46.13	15.27
C 8H10I1N5	303.11	302.9983	31.70	3.33	41.87	23.11
C 8H10I2N2	387.99	387.8937	24.77	2.60	65.42	7.22
C 8H10I2N4	416.00	415.8999	23.10	2.42	61.01	13.47
C 8H10I2N6	444.02	443.9060	21.64	2.27	57.16	18.93
C 8H11I1N2	262.09	261.9969	36.66	4.23	48.42	10.69
C 8H11I1N4	290.11	290.0030	33.12	3.82	43.74	19.31
C 8H11I1N6	318.12	318.0092	30.21	3.49	39.89	26.42
C 8H11I2N1	374.99	374.8985	25.62	2.96	67.68	3.74
C 8H11I2N3	403.01	402.9046	23.84	2.75	62.98	10.43
C 8H11I2N5	431.02	430.9108	22.29	2.57	58.89	16.25
C 8H12I1N1	249.10	249.0016	38.58	4.86	50.95	5.62
C 8H12I1N3	277.11	277.0078	34.68	4.36	45.80	15.16
C 8H12I1N5	305.12	305.0139	31.49	3.96	41.59	22.95
C 8H12I2N2	390.01	389.9094	24.64	3.10	65.08	7.18
C 8H12I2N4	418.02	417.9155	22.99	2.89	60.72	13.40
C 8H12I2N6	446.03	445.9217	21.54	2.71	56.90	18.84
C 8H13I1N2	264.11	264.0125	36.38	4.96	48.05	10.61
C 8H13I1N4	292.12	292.0187	32.89	4.49	43.44	19.18
C 8H13I1N6	320.14	320.0248	30.01	4.09	39.64	26.25
C 8H13I2N1	377.01	376.9141	25.49	3.48	67.32	3.72
C 8H13I2N3	405.02	404.9203	23.72	3.24	62.67	10.37
C 8H13I2N5	433.04	432.9264	22.19	3.03	58.61	16.17
C 8H14I1N1	251.11	251.0173	38.27	5.62	50.54	5.58
C 8H14I1N3	279.13	279.0234	34.43	5.06	45.47	15.05
C 8H14I1N5	307.14	307.0296	31.29	4.59	41.32	22.80
C 8H14I2N2	392.02	391.9250	24.51	3.60	64.74	7.15
C 8H14I2N4	420.04	419.9312	22.88	3.36	60.43	13.34
C 8H14I2N6	448.05	447.9373	21.45	3.15	56.65	18.76
C 8H15I1N2	266.13	266.0282	36.11	5.68	47.69	10.53
C 8H15I1N4	294.14	294.0343	32.67	5.14	43.14	19.05
C 8H15I1N6	322.15	322.0405	29.83	4.69	39.39	26.09
C 8H15I2N1	379.02	378.9298	25.35	3.99	66.96	3.70
C 8H15I2N3	407.04	406.9359	23.61	3.71	62.36	10.32
C 8H15I2N5	435.05	434.9421	22.09	3.48	58.34	16.10
C 8H16I1N1	253.13	253.0329	37.96	6.37	50.13	5.53
C 8H16I1N3	281.14	281.0391	34.18	5.74	45.14	14.95
C 8H16I1N5	309.15	309.0452	31.08	5.22	41.05	22.65
C 8H16I2N2	394.04	393.9407	24.39	4.09	64.41	7.11
C 8H16I2N4	422.05	421.9468	22.77	3.82	60.14	13.27
C 8H16I2N6	450.07	449.9530	21.35	3.58	56.39	18.67
C 8H17I1N2	268.14	268.0438	35.84	6.39	47.33	10.45
C 8H17I1N4	296.16	296.0500	32.45	5.79	42.85	18.92
C 8H17I1N6	324.17	324.0561	29.64	5.29	39.15	25.92
C 8H17I2N1	381.04	380.9454	25.22	4.50	66.61	3.68
C 8H17I2N3	409.05	408.9516	23.49	4.19	62.05	10.27
C 8H17I2N5	437.07	436.9577	21.99	3.92	58.07	16.02
C 8H18I1N1	255.14	255.0486	37.66	7.11	49.74	5.49

C 8H18I1N3	283.16	283.0547	33.93	6.41	44.82	14.84
C 8H18I1N5	311.17	311.0609	30.88	5.83	40.78	22.51
C 8H18I2N2	396.05	395.9563	24.26	4.58	64.08	7.07
C 8H18I2N4	424.07	423.9625	22.66	4.28	59.85	13.21
C 8H18I2N6	452.08	451.9686	21.25	4.01	56.14	18.59
C 8H19I1N2	270.16	270.0595	35.57	7.09	46.97	10.37
C 8H19I1N4	298.17	298.0656	32.23	6.42	42.56	18.79
C 8H19I1N6	326.19	326.0718	29.46	5.87	38.91	25.76
C 8H19I2N3	411.07	410.9672	23.38	4.66	61.74	10.22
C 8H19I2N5	439.08	438.9734	21.88	4.36	57.80	15.95
C 8H20I1N3	285.17	285.0704	33.70	7.07	44.50	14.73
C 8H20I1N5	313.19	313.0765	30.68	6.44	40.52	22.36
C 8H20I2N4	426.08	425.9781	22.55	4.73	59.57	13.15
C 8H20I2N6	454.10	453.9843	21.16	4.44	55.89	18.51
C 8H21I1N4	300.19	300.0813	32.01	7.05	42.28	18.66
C 8H21I1N6	328.20	328.0874	29.28	6.45	38.67	25.61
C 8H21I2N5	441.10	440.9890	21.78	4.80	57.54	15.88
C 8H22I1N5	315.20	315.0922	30.48	7.04	40.26	22.22
C 8H22I2N6	456.11	455.9999	21.07	4.86	55.65	18.43
C 8H23I1N6	330.22	330.1031	29.10	7.02	38.43	25.45
C 9H 4I1N1	253.04	252.9390	42.72	1.59	50.15	5.54
C 9H 4I1N3	281.06	280.9452	38.46	1.43	45.15	14.95
C 9H 4I1N5	309.07	308.9513	34.98	1.30	41.06	22.66
C 9H 4I2N2	393.95	393.8468	27.44	1.02	64.43	7.11
C 9H 4I2N4	421.97	421.8529	25.62	.96	60.15	13.28
C 9H 4I2N6	449.98	449.8591	24.02	.90	56.40	18.68
C 9H 5I1N2	268.06	267.9499	40.33	1.88	47.34	10.45
C 9H 5I1N4	296.07	295.9561	36.51	1.70	42.86	18.92
C 9H 5I1N6	324.08	323.9622	33.36	1.56	39.16	25.93
C 9H 5I2N1	380.96	380.8515	28.38	1.32	66.62	3.68
C 9H 5I2N3	408.97	408.8577	26.43	1.23	62.06	10.27
C 9H 5I2N5	436.98	436.8638	24.74	1.15	58.08	16.03
C 9H 6I1N1	255.06	254.9547	42.38	2.37	49.75	5.49
C 9H 6I1N3	283.07	282.9608	38.19	2.14	44.83	14.84
C 9H 6I1N5	311.09	310.9670	34.75	1.94	40.79	22.51
C 9H 6I2N2	395.97	395.8624	27.30	1.53	64.10	7.07
C 9H 6I2N4	423.98	423.8686	25.50	1.43	59.86	13.21
C 9H 6I2N6	452.00	451.8747	23.92	1.34	56.15	18.59
C 9H 7I1N2	270.07	269.9656	40.03	2.61	46.99	10.37
C 9H 7I1N4	298.09	297.9717	36.26	2.37	42.57	18.80
C 9H 7I1N6	326.10	325.9779	33.15	2.16	38.92	25.77
C 9H 7I2N1	382.97	382.8672	28.23	1.84	66.27	3.66
C 9H 7I2N3	410.99	410.8733	26.30	1.72	61.76	10.22
C 9H 7I2N5	439.00	438.8795	24.62	1.61	57.82	15.95
C 9H 8I1N1	257.08	256.9703	42.05	3.14	49.36	5.45
C 9H 8I1N3	285.09	284.9765	37.92	2.83	44.51	14.74
C 9H 8I1N5	313.10	312.9826	34.53	2.58	40.53	22.37
C 9H 8I2N2	397.99	397.8781	27.16	2.03	63.77	7.04
C 9H 8I2N4	426.00	425.8842	25.38	1.89	59.58	13.15
C 9H 8I2N6	454.01	453.8904	23.81	1.78	55.90	18.51
C 9H 9I1N2	272.09	271.9812	39.73	3.33	46.64	10.30
C 9H 9I1N4	300.10	299.9874	36.02	3.02	42.29	18.67
C 9H 9I1N6	328.12	327.9935	32.95	2.76	38.68	25.61
C 9H 9I2N1	384.99	384.8828	28.08	2.36	65.93	3.64
C 9H 9I2N3	413.00	412.8890	26.17	2.20	61.45	10.17
C 9H 9I2N5	441.01	440.8951	24.51	2.06	57.55	15.88
C 9H10I1N1	259.09	258.9860	41.72	3.89	48.98	5.41
C 9H10I1N3	287.10	286.9921	37.65	3.51	44.20	14.64
C 9H10I1N5	315.12	314.9983	34.30	3.20	40.27	22.22
C 9H10I2N2	400.00	399.8937	27.02	2.52	63.45	7.00
C 9H10I2N4	428.02	427.8999	25.26	2.35	59.30	13.09
C 9H10I2N6	456.03	455.9060	23.70	2.21	55.66	18.43
C 9H11I1N2	274.11	273.9969	39.44	4.05	46.30	10.22
C 9H11I1N4	302.12	302.0030	35.78	3.67	42.00	18.54
C 9H11I1N6	330.13	330.0092	32.74	3.36	38.44	25.46
C 9H11I2N1	387.00	386.8985	27.93	2.87	65.58	3.62
C 9H11I2N3	415.02	414.9046	26.05	2.67	61.16	10.12
C 9H11I2N5	443.03	442.9108	24.40	2.50	57.29	15.81
C 9H12I1N1	261.11	261.0016	41.40	4.63	48.60	5.36
C 9H12I1N3	289.12	289.0078	37.39	4.18	43.89	14.53
C 9H12I1N5	317.13	317.0139	34.09	3.81	40.02	22.08
C 9H12I2N2	402.02	401.9094	26.89	3.01	63.13	6.97
C 9H12I2N4	430.03	429.9155	25.14	2.81	59.02	13.03
C 9H12I2N6	458.04	457.9217	23.60	2.64	55.41	18.35
C 9H13I1N2	276.12	276.0125	39.15	4.75	45.96	10.15
C 9H13I1N4	304.14	304.0187	35.54	4.31	41.73	18.42
C 9H13I1N6	332.15	332.0248	32.55	3.95	38.21	25.30
C 9H13I2N1	389.02	388.9141	27.79	3.37	65.24	3.60
C 9H13I2N3	417.03	416.9203	25.92	3.14	60.86	10.08
C 9H13I2N5	445.05	444.9264	24.29	2.94	57.03	15.74
C 9H14I1N1	263.12	263.0173	41.08	5.36	48.23	5.32
C 9H14I1N3	291.14	291.0234	37.13	4.85	43.59	14.43
C 9H14I1N5	319.15	319.0296	33.87	4.42	39.76	21.94
C 9H14I2N2	404.03	403.9250	26.76	3.49	62.82	6.93
C 9H14I2N4	432.05	431.9312	25.02	3.27	58.75	12.97
C 9H14I2N6	460.06	459.9373	23.50	3.07	55.17	18.27
C 9H15I1N2	278.14	278.0282	38.87	5.44	45.63	10.07
C 9H15I1N4	306.15	306.0343	35.31	4.94	41.45	18.30
C 9H15I1N6	334.16	334.0405	32.35	4.52	37.98	25.15
C 9H15I2N1	391.04	390.9298	27.64	3.87	64.91	3.58
C 9H15I2N3	419.05	418.9359	25.80	3.61	60.57	10.03
C 9H15I2N5	447.06	446.9421	24.18	3.38	56.77	15.67
C 9H16I1N1	265.14	265.0329	40.77	6.08	47.86	5.28
C 9H16I1N3	293.15	293.0391	36.88	5.50	43.29	14.33
C 9H16I1N5	321.17	321.0452	33.66	5.02	39.51	21.81
C 9H16I2N2	406.05	405.9407	26.62	3.97	62.51	6.90
C 9H16I2N4	434.06	433.9468	24.90	3.72	58.47	12.91
C 9H16I2N6	462.08	461.9530	23.39	3.49	54.93	18.19
C 9H17I1N2	280.15	280.0438	38.59	6.12	45.30	10.00
C 9H17I1N4	308.17	308.0500	35.08	5.56	41.18	18.18
C 9H17I1N6	336.18	336.0561	32.16	5.10	37.75	25.00
C 9H17I2N1	393.05	392.9454	27.50	4.36	64.57	3.56
C 9H17I2N3	421.06	420.9516	25.67	4.07	60.28	9.98
C 9H17I2N5	449.08	448.9577	24.07	3.82	56.52	15.59
C 9H18I1N1	267.15	267.0486	40.46	6.79	47.50	5.24
C 9H18I1N3	295.17	295.0547	36.62	6.15	42.99	14.24
C 9H18I1N5	323.18	323.0609	33.45	5.61	39.27	21.67
C 9H18I2N2	408.07	407.9563	26.49	4.45	62.20	6.86
C 9H18I2N4	436.08	435.9625	24.79	4.16	58.20	12.85
C 9H18I2N6	464.09	463.9686	23.29	3.91	54.69	18.11
C 9H19I1N2	282.17	282.0595	38.31	6.79	44.97	9.93
C 9H19I1N4	310.18	310.0656	34.85	6.17	40.91	18.06
C 9H19I1N6	338.20	338.0718	31.96	5.66	37.52	24.85
C 9H19I2N1	395.07	394.9611	27.36	4.85	64.24	3.55
C 9H19I2N3	423.08	422.9672	25.55	4.53	59.99	9.93
C 9H19I2N5	451.09	450.9734	23.96	4.25	56.27	15.53
C 9H20I1N1	269.17	269.0642	40.16	7.49	47.15	5.20
C 9H20I1N3	297.18	297.0704	36.37	6.78	42.70	14.14
C 9H20I1N5	325.20	325.0765	33.24	6.20	39.02	21.54
C 9H20I2N2	410.08	409.9720	26.36	4.92	61.89	6.83

015

C 9H20I2N4	438.10	437.9781	24.68	4.60	57.93	12.79
C 9H20I2N6	466.11	465.9843	23.19	4.33	54.45	18.03
C 9H21I1N2	284.19	284.0751	38.04	7.45	44.66	9.86
C 9H21I1N4	312.20	312.0813	34.63	6.78	40.65	17.95
C 9H21I1N6	340.21	340.0874	31.77	6.22	37.30	24.70
C 9H21I2N3	425.10	424.9829	25.43	4.98	59.71	9.88
C 9H21I2N5	453.11	452.9890	23.86	4.67	56.01	15.46
C 9H22I1N3	299.20	299.0860	36.13	7.41	42.41	14.04
C 9H22I1N5	327.21	327.0922	33.04	6.78	38.78	21.40
C 9H22I2N4	440.11	439.9938	24.56	5.04	57.67	12.73
C 9H22I2N6	468.12	467.9999	23.09	4.74	54.22	17.95
C 9H23I1N4	314.21	314.0969	34.40	7.38	40.39	17.83
C 9H23I1N6	342.23	342.1031	31.59	6.77	37.08	24.56
C 9H23I2N5	455.13	455.0047	23.75	5.09	55.77	15.39
C 9H24I1N5	329.23	329.1078	32.83	7.35	38.55	21.27
C 9H24I2N6	470.14	470.0156	22.99	5.15	53.99	17.88
C 9H25I1N6	344.24	344.1187	31.40	7.32	36.86	24.41
C10H 5I1N2	280.07	279.9499	42.89	1.80	45.31	10.00
C10H 5I1N4	308.08	307.9561	38.99	1.64	41.19	18.19
C10H 5I1N6	336.10	335.9622	35.74	1.50	37.76	25.00
C10H 5I2N1	392.97	392.8515	30.57	1.28	64.59	3.56
C10H 5I2N3	420.98	420.8577	28.53	1.20	60.29	9.98
C10H 5I2N5	448.99	448.8638	26.75	1.12	56.53	15.60
C10H 6I1N1	267.07	266.9547	44.97	2.26	47.52	5.24
C10H 6I1N3	295.08	294.9608	40.70	2.05	43.01	14.24
C10H 6I1N5	323.10	322.9670	37.18	1.87	39.28	21.68
C10H 6I2N2	407.98	407.8624	29.44	1.48	62.21	6.87
C10H 6I2N4	435.99	435.8686	27.55	1.39	58.21	12.85
C10H 6I2N6	464.01	463.8747	25.89	1.30	54.70	18.11
C10H 7I1N2	282.09	281.9656	42.58	2.50	44.99	9.93
C10H 7I1N4	310.10	309.9717	38.73	2.28	40.92	18.07
C10H 7I1N6	338.11	337.9779	35.52	2.09	37.53	24.86
C10H 7I2N1	394.98	394.8672	30.41	1.79	64.26	3.55
C10H 7I2N3	423.00	422.8733	28.40	1.67	60.00	9.93
C10H 7I2N5	451.01	450.8795	26.63	1.56	56.28	15.53
C10H 8I1N1	269.09	268.9703	44.64	3.00	47.16	5.21
C10H 8I1N3	297.10	296.9765	40.43	2.71	42.71	14.14
C10H 8I1N5	325.11	324.9826	36.94	2.48	39.03	21.54
C10H 8I2N2	410.00	409.8781	29.30	1.97	61.90	6.83
C10H 8I2N4	438.01	437.8842	27.42	1.84	57.95	12.79
C10H 8I2N6	466.02	465.8904	25.77	1.73	54.46	18.03
C10H 9I1N2	284.10	283.9812	42.28	3.19	44.67	9.86
C10H 9I1N4	312.11	311.9874	38.48	2.91	40.66	17.95
C10H 9I1N6	340.13	339.9935	35.31	2.67	37.31	24.71
C10H 9I2N1	397.00	396.8828	30.25	2.29	63.93	3.53
C10H 9I2N3	425.01	424.8890	28.26	2.13	59.72	9.89
C10H 9I2N5	453.03	452.8951	26.51	2.00	56.03	15.46
C10H10I1N1	271.10	270.9860	44.30	3.72	46.81	5.17
C10H10I1N3	299.12	298.9921	40.16	3.37	42.43	14.05
C10H10I1N5	327.13	326.9983	36.72	3.08	38.79	21.41
C10H10I2N2	412.01	411.8937	29.15	2.45	61.60	6.80
C10H10I2N4	440.03	439.8999	27.30	2.29	57.68	12.73
C10H10I2N6	468.04	467.9060	25.66	2.15	54.23	17.96
C10H11I1N2	286.12	285.9969	41.98	3.88	44.35	9.79
C10H11I1N4	314.13	314.0030	38.24	3.53	40.40	17.84
C10H11I1N6	342.14	342.0092	35.11	3.24	37.09	24.56
C10H11I2N1	399.01	398.8985	30.10	2.78	63.61	3.51
C10H11I2N3	427.03	426.9046	28.13	2.60	59.44	9.84
C10H11I2N5	455.04	454.9108	26.40	2.44	55.78	15.39
C10H12I1N1	273.12	273.0016	43.98	4.43	46.47	5.13
C10H12I1N3	301.13	301.0078	39.89	4.02	42.14	13.95
C10H12I1N5	329.15	329.0139	36.49	3.67	38.56	21.28
C10H12I2N2	414.03	413.9094	29.01	2.92	61.30	6.77
C10H12I2N4	442.04	441.9155	27.17	2.74	57.42	12.67
C10H12I2N6	470.06	469.9217	25.55	2.57	54.00	17.88
C10H13I1N2	288.13	288.0125	41.69	4.55	44.04	9.72
C10H13I1N4	316.15	316.0187	37.99	4.14	40.14	17.72
C10H13I1N6	344.16	344.0248	34.90	3.81	36.87	24.42
C10H13I2N1	401.03	400.9141	29.95	3.27	63.29	3.49
C10H13I2N3	429.04	428.9203	28.00	3.05	59.16	9.79
C10H13I2N5	457.06	456.9264	26.28	2.87	55.53	15.32
C10H14I1N1	275.13	275.0173	43.66	5.13	46.12	5.09
C10H14I1N3	303.15	303.0234	39.62	4.66	41.86	13.86
C10H14I1N5	331.16	331.0296	36.27	4.26	38.32	21.15
C10H14I2N2	416.05	415.9250	28.87	3.39	61.01	6.73
C10H14I2N4	444.06	443.9312	27.05	3.18	57.16	12.62
C10H14I2N6	472.07	471.9373	25.44	2.99	53.76	17.80
C10H15I1N2	290.15	290.0282	41.40	5.21	43.74	9.65
C10H15I1N4	318.16	318.0343	37.75	4.75	39.89	17.61
C10H15I1N6	346.18	346.0405	34.70	4.37	36.66	24.28
C10H15I2N1	403.05	402.9298	29.80	3.75	62.97	3.48
C10H15I2N3	431.06	430.9359	27.86	3.51	58.88	9.75
C10H15I2N5	459.07	458.9421	26.16	3.29	55.29	15.26
C10H16I1N1	277.15	277.0329	43.34	5.82	45.79	5.05
C10H16I1N3	305.16	305.0391	39.36	5.28	41.59	13.77
C10H16I1N5	333.18	333.0452	36.05	4.84	38.09	21.02
C10H16I2N2	418.06	417.9407	28.73	3.86	60.71	6.70
C10H16I2N4	446.07	445.9468	26.93	3.62	56.90	12.56
C10H16I2N6	474.09	473.9530	25.34	3.40	53.54	17.73
C10H17I1N2	292.16	292.0438	41.11	5.87	43.44	9.59
C10H17I1N4	320.18	320.0500	37.51	5.35	39.64	17.50
C10H17I1N6	348.19	348.0561	34.50	4.92	36.45	24.14
C10H17I2N1	405.06	404.9454	29.65	4.23	62.66	3.46
C10H17I2N3	433.08	432.9516	27.73	3.96	58.61	9.70
C10H17I2N5	461.09	460.9577	26.05	3.72	55.05	15.19
C10H18I1N1	279.17	279.0486	43.03	6.50	45.46	5.02
C10H18I1N3	307.18	307.0547	39.10	5.91	41.31	13.68
C10H18I1N5	335.19	335.0609	35.83	5.41	37.86	20.89
C10H18I2N2	420.08	419.9563	28.59	4.32	60.42	6.67
C10H18I2N4	448.09	447.9625	26.81	4.05	56.64	12.50
C10H18I2N6	476.10	475.9686	25.23	3.81	53.31	17.65
C10H19I1N2	294.18	294.0595	40.83	6.51	43.14	9.52
C10H19I1N4	322.19	322.0656	37.28	5.94	39.39	17.39
C10H19I1N6	350.21	350.0718	34.30	5.47	36.24	24.00
C10H19I2N1	407.08	406.9611	29.51	4.70	62.35	3.44
C10H19I2N3	435.09	434.9672	27.61	4.40	58.33	9.66
C10H19I2N5	463.11	462.9734	25.94	4.14	54.81	15.12
C10H20I1N1	281.18	281.0642	42.72	7.17	45.13	4.98
C10H20I1N3	309.20	309.0704	38.85	6.52	41.04	13.59
C10H20I1N5	337.21	337.0765	35.62	5.98	37.63	20.77
C10H20I2N2	422.09	421.9720	28.46	4.78	60.13	6.64
C10H20I2N4	450.11	449.9781	26.69	4.48	56.39	12.45
C10H20I2N6	478.12	477.9843	25.12	4.22	53.08	17.58
C10H21I1N2	296.20	296.0751	40.55	7.15	42.84	9.46
C10H21I1N4	324.21	324.0813	37.05	6.53	39.14	17.28
C10H21I1N6	352.22	352.0874	34.10	6.01	36.03	23.86
C10H21I2N1	409.09	408.9767	29.36	5.17	62.04	3.42
C10H21I2N3	437.11	436.9829	27.48	4.84	58.07	9.61
C10H21I2N5	465.12	464.9890	25.82	4.55	54.57	15.06
C10H22I1N1	283.20	283.0799	42.41	7.83	44.81	4.95

016

C10H22I1N3	311.21	311.0860	38.59	7.13	40.78	13.50
C10H22I1N5	339.22	339.0922	35.41	6.54	37.41	20.65
C10H22I2N2	424.11	423.9876	28.32	5.23	59.85	6.61
C10H22I2N4	452.12	451.9938	26.57	4.90	56.14	12.39
C10H22I2N6	480.14	479.9999	25.02	4.62	52.86	17.50
C10H23I1N2	298.21	298.0908	40.28	7.77	42.56	9.39
C10H23I1N4	326.23	326.0969	36.82	7.11	38.90	17.17
C10H23I1N6	354.24	354.1031	33.91	6.54	35.82	23.72
C10H23I2N3	439.12	438.9985	27.35	5.28	57.80	9.57
C10H23I2N5	467.14	467.0047	25.71	4.96	54.33	14.99
C10H24I1N3	313.23	313.1017	38.35	7.72	40.52	13.42
C10H24I1N5	341.24	341.1078	35.20	7.09	37.19	20.52
C10H24I2N4	454.14	454.0094	26.45	5.33	55.89	12.34
C10H24I2N6	482.15	482.0156	24.91	5.02	52.64	17.43
C10H25I1N4	328.24	328.1126	36.59	7.68	38.66	17.07
C10H25I1N6	356.26	356.1187	33.72	7.07	35.62	23.59
C10H25I2N5	469.15	469.0203	25.60	5.37	54.10	14.93
C10H26I1N5	343.26	343.1235	34.99	7.63	36.97	20.40
C10H26I2N6	484.17	484.0312	24.81	5.41	52.42	17.36
C10H27I1N6	358.27	358.1344	33.53	7.60	35.42	23.46
C11H 5I1N2	292.08	291.9499	45.24	1.73	43.45	9.59
C11H 5I1N4	320.09	319.9561	41.28	1.57	39.65	17.50
C11H 5I1N6	348.11	347.9622	37.95	1.45	36.46	24.14
C11H 5I2N1	404.98	404.8515	32.62	1.24	62.67	3.46
C11H 5I2N3	432.99	432.8577	30.51	1.16	58.62	9.70
C11H 5I2N5	461.00	460.8638	28.66	1.09	55.06	15.19
C11H 6I1N1	279.08	278.9547	47.34	2.17	45.47	5.02
C11H 6I1N3	307.09	306.9608	43.02	1.97	41.32	13.68
C11H 6I1N5	335.11	334.9670	39.43	1.80	37.87	20.90
C11H 6I2N2	419.99	419.8624	31.46	1.44	60.43	6.67
C11H 6I2N4	448.01	447.8686	29.49	1.35	56.65	12.51
C11H 6I2N6	476.02	475.8747	27.76	1.27	53.32	17.65
C11H 7I1N2	294.10	293.9656	44.92	2.40	43.15	9.53
C11H 7I1N4	322.11	321.9717	41.02	2.19	39.40	17.39
C11H 7I1N6	350.12	349.9779	37.74	2.02	36.25	24.00
C11H 7I2N1	406.99	406.8672	32.46	1.73	62.36	3.44
C11H 7I2N3	435.01	434.8733	30.37	1.62	58.35	9.66
C11H 7I2N5	463.02	462.8795	28.53	1.52	54.82	15.13
C11H 8I1N1	281.10	280.9703	47.00	2.87	45.15	4.98
C11H 8I1N3	309.11	308.9765	42.74	2.61	41.05	13.59
C11H 8I1N5	337.12	336.9826	39.19	2.39	37.64	20.77
C11H 8I2N2	422.01	421.8781	31.31	1.91	60.14	6.64
C11H 8I2N4	450.02	449.8842	29.36	1.79	56.40	12.45
C11H 8I2N6	478.04	477.8904	27.64	1.69	53.09	17.58
C11H 9I1N2	296.11	295.9812	44.62	3.06	42.86	9.46
C11H 9I1N4	324.13	323.9874	40.76	2.80	39.15	17.29
C11H 9I1N6	352.14	351.9935	37.52	2.58	36.04	23.87
C11H 9I2N1	409.01	408.8828	32.30	2.22	62.05	3.42
C11H 9I2N3	437.02	436.8890	30.23	2.08	58.08	9.62
C11H 9I2N5	465.04	464.8951	28.41	1.95	54.58	15.06
C11H10I1N1	283.11	282.9860	46.67	3.56	44.82	4.95
C11H10I1N3	311.13	310.9921	42.47	3.24	40.79	13.51
C11H10I1N5	339.14	338.9983	38.96	2.97	37.42	20.65
C11H10I2N2	424.02	423.8937	31.16	2.38	59.86	6.61
C11H10I2N4	452.04	451.8999	29.23	2.23	56.15	12.39
C11H10I2N6	480.05	479.9060	27.52	2.10	52.87	17.51
C11H11I1N2	298.13	297.9969	44.32	3.72	42.57	9.40
C11H11I1N4	326.14	326.0030	40.51	3.40	38.91	17.18
C11H11I1N6	354.15	354.0092	37.31	3.13	35.83	23.73
C11H11I2N1	411.03	410.8985	32.14	2.70	61.75	3.41
C11H11I2N3	439.04	438.9046	30.09	2.53	57.81	9.57
C11H11I2N5	467.05	466.9108	28.29	2.37	54.34	14.99
C11H12I1N1	285.13	285.0016	46.34	4.24	44.51	4.91
C11H12I1N3	313.14	313.0078	42.19	3.86	40.53	13.42
C11H12I1N5	341.16	341.0139	38.73	3.55	37.20	20.53
C11H12I2N2	426.04	425.9094	31.01	2.84	59.57	6.58
C11H12I2N4	454.05	453.9155	29.10	2.66	55.90	12.34
C11H12I2N6	482.07	481.9217	27.41	2.51	52.65	17.43
C11H13I1N2	300.14	300.0125	44.02	4.37	42.28	9.33
C11H13I1N4	328.16	328.0187	40.26	3.99	38.67	17.07
C11H13I1N6	356.17	356.0248	37.10	3.68	35.63	23.60
C11H13I2N1	413.04	412.9141	31.99	3.17	61.45	3.39
C11H13I2N3	441.06	440.9203	29.96	2.97	57.55	9.53
C11H13I2N5	469.07	468.9264	28.17	2.79	54.11	14.93
C11H14I1N1	287.15	287.0173	46.01	4.91	44.20	4.88
C11H14I1N3	315.16	315.0234	41.92	4.48	40.27	13.33
C11H14I1N5	343.17	343.0296	38.50	4.11	36.98	20.41
C11H14I2N2	428.06	427.9250	30.87	3.30	59.29	6.54
C11H14I2N4	456.07	455.9312	28.97	3.09	55.65	12.28
C11H14I2N6	484.08	483.9373	27.29	2.92	52.43	17.36
C11H15I1N2	302.16	302.0282	43.73	5.00	42.00	9.27
C11H15I1N4	330.17	330.0343	40.02	4.58	38.44	16.97
C11H15I1N6	358.19	358.0405	36.89	4.22	35.43	23.46
C11H15I2N1	415.06	414.9298	31.83	3.64	61.15	3.37
C11H15I2N3	443.07	442.9359	29.82	3.41	57.28	9.48
C11H15I2N5	471.08	470.9421	28.05	3.21	53.88	14.87
C11H16I1N1	289.16	289.0329	45.69	5.58	43.89	4.84
C11H16I1N3	317.17	317.0391	41.66	5.08	40.01	13.25
C11H16I1N5	345.19	345.0452	38.28	4.67	36.76	20.29
C11H16I2N2	430.07	429.9407	30.72	3.75	59.02	6.51
C11H16I2N4	458.09	457.9468	28.84	3.52	55.41	12.23
C11H16I2N6	486.10	485.9530	27.18	3.32	52.21	17.29
C11H17I1N2	304.18	304.0438	43.44	5.63	41.72	9.21
C11H17I1N4	332.19	332.0500	39.77	5.16	38.20	16.87
C11H17I1N6	360.20	360.0561	36.68	4.76	35.23	23.33
C11H17I2N1	417.07	416.9454	31.68	4.11	60.85	3.36
C11H17I2N3	445.09	444.9516	29.68	3.85	57.02	9.44
C11H17I2N5	473.10	472.9577	27.93	3.62	53.65	14.80
C11H18I1N1	291.18	291.0486	45.38	6.23	43.58	4.81
C11H18I1N3	319.19	319.0547	41.39	5.68	39.76	13.16
C11H18I1N5	347.20	347.0609	38.05	5.23	36.55	20.17
C11H18I2N2	432.09	431.9563	30.58	4.20	58.74	6.48
C11H18I2N4	460.10	459.9625	28.72	3.94	55.16	12.18
C11H18I2N6	488.12	487.9686	27.07	3.72	52.00	17.22
C11H19I1N2	306.19	306.0595	43.15	6.25	41.45	9.15
C11H19I1N4	334.21	334.0656	39.53	5.73	37.97	16.76
C11H19I1N6	362.22	362.0718	36.48	5.29	35.04	23.20
C11H19I2N1	419.09	418.9611	31.53	4.57	60.56	3.34
C11H19I2N3	447.10	446.9672	29.55	4.28	56.77	9.40
C11H19I2N5	475.12	474.9734	27.81	4.03	53.42	14.74
C11H20I1N1	293.19	293.0642	45.06	6.88	43.28	4.78
C11H20I1N3	321.21	321.0704	41.13	6.28	39.51	13.08
C11H20I1N5	349.22	349.0765	37.83	5.77	36.34	20.05
C11H20I2N2	434.10	433.9720	30.44	4.64	58.47	6.45
C11H20I2N4	462.12	461.9781	28.59	4.36	54.92	12.12
C11H20I2N6	490.13	489.9843	26.96	4.11	51.78	17.15
C11H21I1N2	308.21	308.0751	42.87	6.87	41.17	9.09
C11H21I1N4	336.22	336.0813	39.30	6.30	37.74	16.66
C11H21I1N6	364.23	364.0874	36.27	5.81	34.84	23.07
C11H21I2N1	421.11	420.9767	31.38	5.03	60.27	3.33

C11H21I2N3	449.12	448.9829	29.42	4.71	56.51	9.36
C11H21I2N5	477.13	476.9890	27.69	4.44	53.19	14.68
C11H22I1N1	295.21	295.0799	44.76	7.51	42.99	4.74
C11H22I1N3	323.22	323.0860	40.88	6.86	39.26	13.00
C11H22I1N5	351.24	351.0922	37.62	6.31	36.13	19.94
C11H22I2N2	436.12	435.9876	30.30	5.08	58.20	6.42
C11H22I2N4	464.13	463.9938	28.47	4.78	54.68	12.07
C11H22I2N6	492.15	491.9999	26.85	4.51	51.57	17.08
C11H23I1N2	310.22	310.0908	42.59	7.47	40.91	9.03
C11H23I1N4	338.24	338.0969	39.06	6.85	37.52	16.56
C11H23I1N6	366.25	366.1031	36.07	6.33	34.65	22.95
C11H23I2N1	423.12	422.9924	31.23	5.48	59.98	3.31
C11H23I2N3	451.13	450.9985	29.29	5.14	56.26	9.31
C11H23I2N5	479.15	479.0047	27.57	4.84	52.97	14.62
C11H24I1N1	297.23	297.0955	44.45	8.14	42.70	4.71
C11H24I1N3	325.24	325.1017	40.62	7.44	39.02	12.92
C11H24I1N5	353.25	353.1078	37.40	6.85	35.92	19.83
C11H24I2N2	438.14	438.0033	30.16	5.52	57.93	6.39
C11H24I2N4	466.15	466.0094	28.34	5.19	54.45	12.02
C11H24I2N6	494.16	494.0156	26.74	4.90	51.36	17.01
C11H25I1N2	312.24	312.1064	42.31	8.07	40.64	8.97
C11H25I1N4	340.25	340.1126	38.83	7.41	37.30	16.47
C11H25I1N6	368.27	368.1187	35.88	6.84	34.46	22.82
C11H25I2N3	453.15	453.0142	29.16	5.56	56.01	9.27
C11H25I2N5	481.16	481.0203	27.46	5.24	52.75	14.56
C11H26I1N3	327.25	327.1173	40.37	8.01	38.78	12.84
C11H26I1N5	355.27	355.1235	37.19	7.38	35.72	19.71
C11H26I2N4	468.17	468.0251	28.22	5.60	54.21	11.97
C11H26I2N6	496.18	496.0312	26.63	5.28	51.15	16.94
C11H27I1N4	342.27	342.1282	38.60	7.95	37.08	16.37
C11H27I1N6	370.28	370.1344	35.68	7.35	34.27	22.70
C11H27I2N5	483.18	483.0360	27.34	5.63	52.53	14.49
C11H28I1N5	357.28	357.1391	36.98	7.90	35.52	19.60
C11H28I2N6	498.19	498.0469	26.52	5.67	50.95	16.87
C11H29I1N6	372.30	372.1500	35.49	7.85	34.09	22.57
C12H 6I1N1	291.09	290.9547	49.51	2.08	43.60	4.81
C12H 6I1N3	319.11	318.9608	45.17	1.90	39.77	13.17
C12H 6I1N5	347.12	346.9670	41.52	1.74	36.56	20.18
C12H 6I2N2	432.00	431.8624	33.36	1.40	58.75	6.48
C12H 6I2N4	460.02	459.8686	31.33	1.31	55.17	12.18
C12H 6I2N6	488.03	487.8747	29.53	1.24	52.01	17.22
C12H 7I1N2	306.11	305.9656	47.09	2.31	41.46	9.15
C12H 7I1N4	334.12	333.9717	43.14	2.11	37.98	16.77
C12H 7I1N6	362.13	361.9779	39.80	1.95	35.04	23.21
C12H 7I2N1	419.01	418.8672	34.40	1.68	60.57	3.34
C12H 7I2N3	447.02	446.8733	32.24	1.58	56.78	9.40
C12H 7I2N5	475.03	474.8795	30.34	1.49	53.43	14.74
C12H 8I1N1	293.11	292.9703	49.17	2.75	43.30	4.78
C12H 8I1N3	321.12	320.9765	44.88	2.51	39.52	13.09
C12H 8I1N5	349.14	348.9826	41.28	2.31	36.35	20.06
C12H 8I2N2	434.02	433.8781	33.21	1.86	58.48	6.45
C12H 8I2N4	462.03	461.8842	31.20	1.75	54.93	12.13
C12H 8I2N6	490.05	489.8904	29.41	1.65	51.79	17.15
C12H 9I1N2	308.12	307.9812	46.78	2.94	41.19	9.09
C12H 9I1N4	336.14	335.9874	42.88	2.70	37.75	16.67
C12H 9I1N6	364.15	363.9935	39.58	2.49	34.85	23.08
C12H 9I2N1	421.02	420.8828	34.23	2.15	60.28	3.33
C12H 9I2N3	449.03	448.8890	32.10	2.02	56.52	9.36
C12H 9I2N5	477.05	476.8951	30.21	1.90	53.20	14.68
C12H10I1N1	295.12	294.9860	48.84	3.42	43.00	4.75

C12H10I1N3	323.14	322.9921	44.60	3.12	39.27	13.00
C12H10I1N5	351.15	350.9983	41.05	2.87	36.14	19.94
C12H10I2N2	436.04	435.8937	33.06	2.31	58.21	6.42
C12H10I2N4	464.05	463.8999	31.06	2.17	54.69	12.07
C12H10I2N6	492.06	491.9060	29.29	2.05	51.58	17.08
C12H11I1N2	310.14	309.9969	46.47	3.58	40.92	9.03
C12H11I1N4	338.15	338.0030	42.62	3.28	37.53	16.57
C12H11I1N6	366.17	366.0092	39.36	3.03	34.66	22.95
C12H11I2N1	423.04	422.8985	34.07	2.62	60.00	3.31
C12H11I2N3	451.05	450.9046	31.96	2.46	56.27	9.32
C12H11I2N5	479.06	478.9108	30.09	2.31	52.98	14.62
C12H12I1N1	297.14	297.0016	48.51	4.07	42.71	4.71
C12H12I1N3	325.15	325.0078	44.33	3.72	39.03	12.92
C12H12I1N5	353.17	353.0139	40.81	3.42	35.93	19.83
C12H12I2N2	438.05	437.9094	32.90	2.76	57.94	6.39
C12H12I2N4	466.07	465.9155	30.93	2.60	54.46	12.02
C12H12I2N6	494.08	493.9217	29.17	2.45	51.37	17.01
C12H13I1N2	312.16	312.0125	46.17	4.20	40.65	8.97
C12H13I1N4	340.17	340.0187	42.37	3.85	37.31	16.47
C12H13I1N6	368.18	368.0248	39.15	3.56	34.47	22.83
C12H13I2N1	425.05	424.9141	33.91	3.08	59.71	3.30
C12H13I2N3	453.07	452.9203	31.81	2.89	56.02	9.27
C12H13I2N5	481.08	480.9264	29.96	2.72	52.76	14.56
C12H14I1N1	299.16	299.0173	48.18	4.72	42.42	4.68
C12H14I1N3	327.17	327.0234	44.05	4.31	38.79	12.84
C12H14I1N5	355.18	355.0296	40.58	3.97	35.73	19.72
C12H14I2N2	440.07	439.9250	32.75	3.21	57.67	6.37
C12H14I2N4	468.08	467.9312	30.79	3.01	54.22	11.97
C12H14I2N6	496.09	495.9373	29.05	2.84	51.16	16.94
C12H15I1N2	314.17	314.0282	45.88	4.81	40.39	8.92
C12H15I1N4	342.18	342.0343	42.12	4.42	37.09	16.37
C12H15I1N6	370.20	370.0405	38.93	4.08	34.28	22.70
C12H15I2N1	427.07	426.9298	33.75	3.54	59.43	3.28
C12H15I2N3	455.08	454.9359	31.67	3.32	55.77	9.23
C12H15I2N5	483.10	482.9421	29.84	3.13	52.54	14.50
C12H16I1N1	301.17	301.0329	47.86	5.35	42.14	4.65
C12H16I1N3	329.19	329.0391	43.78	4.90	38.55	12.76
C12H16I1N5	357.20	357.0452	40.35	4.51	35.53	19.61
C12H16I2N2	442.08	441.9407	32.60	3.65	57.41	6.34
C12H16I2N4	470.10	469.9468	30.66	3.43	53.99	11.92
C12H16I2N6	498.11	497.9530	28.94	3.24	50.95	16.87
C12H17I1N2	316.19	316.0438	45.58	5.42	40.14	8.86
C12H17I1N4	344.20	344.0500	41.87	4.98	36.87	16.28
C12H17I1N6	372.21	372.0561	38.72	4.60	34.09	22.58
C12H17I2N1	429.08	428.9454	33.59	3.99	59.15	3.26
C12H17I2N3	457.10	456.9516	31.53	3.75	55.53	9.19
C12H17I2N5	485.11	484.9577	29.71	3.53	52.32	14.44
C12H18I1N1	303.19	303.0486	47.54	5.98	41.86	4.62
C12H18I1N3	331.20	331.0547	43.52	5.48	38.32	12.69
C12H18I1N5	359.22	359.0609	40.12	5.05	35.33	19.50
C12H18I2N2	444.10	443.9563	32.46	4.09	57.15	6.31
C12H18I2N4	472.11	471.9625	30.53	3.84	53.76	11.87
C12H18I2N6	500.13	499.9686	28.82	3.63	50.75	16.80
C12H19I1N2	318.20	318.0595	45.30	6.02	39.88	8.80
C12H19I1N4	346.22	346.0656	41.63	5.53	36.65	16.18
C12H19I1N6	374.23	374.0718	38.51	5.12	33.91	22.46
C12H19I2N1	431.10	430.9611	33.43	4.44	58.87	3.25
C12H19I2N3	459.11	458.9672	31.39	4.17	55.28	9.15
C12H19I2N5	487.13	486.9734	29.59	3.93	52.10	14.38
C12H20I1N1	305.20	305.0642	47.23	6.61	41.58	4.59

018

C12H20I1N3	333.22	333.0704	43.26	6.05	38.08	12.61
C12H20I1N5	361.23	361.0765	39.90	5.58	35.13	19.39
C12H20I2N2	446.12	445.9720	32.31	4.52	56.89	6.28
C12H20I2N4	474.13	473.9781	30.40	4.25	53.53	11.82
C12H20I2N6	502.14	501.9843	28.70	4.01	50.55	16.74
C12H21I1N2	320.22	320.0751	45.01	6.61	39.63	8.75
C12H21I1N4	348.23	348.0813	41.39	6.08	36.44	16.09
C12H21I1N6	376.25	376.0874	38.31	5.63	33.73	22.34
C12H21I2N1	433.12	432.9767	33.28	4.89	58.60	3.23
C12H21I2N3	461.13	460.9829	31.26	4.59	55.04	9.11
C12H21I2N5	489.14	488.9890	29.47	4.33	51.89	14.32
C12H22I1N1	307.22	307.0799	46.92	7.22	41.31	4.56
C12H22I1N3	335.23	335.0860	43.00	6.61	37.86	12.53
C12H22I1N5	363.25	363.0922	39.68	6.10	34.94	19.28
C12H22I2N2	448.13	447.9876	32.16	4.95	56.64	6.25
C12H22I2N4	476.14	475.9938	30.27	4.66	53.30	11.77
C12H22I2N6	504.16	503.9999	28.59	4.40	50.34	16.67
C12H23I1N2	322.23	322.0908	44.73	7.19	39.38	8.69
C12H23I1N4	350.25	350.0969	41.15	6.62	36.23	16.00
C12H23I1N6	378.26	378.1031	38.10	6.13	33.55	22.22
C12H23I2N1	435.13	434.9924	33.12	5.33	58.33	3.22
C12H23I2N3	463.15	462.9985	31.12	5.01	54.80	9.07
C12H23I2N5	491.16	491.0047	29.35	4.72	51.68	14.26
C12H24I1N1	309.24	309.0955	46.61	7.82	41.04	4.53
C12H24I1N3	337.25	337.1017	42.74	7.17	37.63	12.46
C12H24I1N5	365.26	365.1078	39.46	6.62	34.74	19.17
C12H24I2N2	450.15	450.0033	32.02	5.37	56.38	6.22
C12H24I2N4	478.16	478.0094	30.14	5.06	53.08	11.72
C12H24I2N6	506.17	506.0156	28.48	4.78	50.14	16.60
C12H25I1N2	324.25	324.1064	44.45	7.77	39.14	8.64
C12H25I1N4	352.26	352.1126	40.92	7.15	36.03	15.90
C12H25I1N6	380.28	380.1187	37.90	6.63	33.37	22.10
C12H25I2N1	437.15	437.0080	32.97	5.76	58.06	3.20
C12H25I2N3	465.16	465.0142	30.99	5.42	54.56	9.03
C12H25I2N5	493.18	493.0203	29.23	5.11	51.46	14.20
C12H26I1N1	311.25	311.1112	46.31	8.42	40.77	4.50
C12H26I1N3	339.27	339.1173	42.48	7.72	37.41	12.39
C12H26I1N5	367.28	367.1235	39.24	7.14	34.55	19.07
C12H26I2N2	452.16	452.0189	31.88	5.80	56.13	6.20
C12H26I2N4	480.18	480.0251	30.02	5.46	52.86	11.67
C12H26I2N6	508.19	508.0312	28.36	5.16	49.94	16.54
C12H27I1N2	326.27	326.1221	44.18	8.34	38.90	8.59
C12H27I1N4	354.28	354.1282	40.68	7.68	35.82	15.81
C12H27I1N6	382.29	382.1344	37.70	7.12	33.20	21.98
C12H27I2N3	467.18	467.0298	30.85	5.83	54.33	8.99
C12H27I2N5	495.19	495.0360	29.11	5.50	51.25	14.14
C12H28I1N3	341.28	341.1330	42.23	8.27	37.18	12.31
C12H28I1N5	369.29	369.1391	39.03	7.64	34.36	18.96
C12H28I2N4	482.19	482.0407	29.89	5.85	52.64	11.62
C12H28I2N6	510.21	510.0469	28.25	5.53	49.75	16.47
C12H29I1N4	356.30	356.1439	40.45	8.20	35.62	15.72
C12H29I1N6	384.31	384.1500	37.50	7.61	33.02	21.87
C12H29I2N5	497.21	497.0516	28.99	5.88	51.05	14.09
C12H30I1N5	371.31	371.1548	38.82	8.14	34.18	18.86
C12H30I2N6	512.22	512.0625	28.14	5.90	49.55	16.41
C12H31I1N6	386.33	386.1657	37.31	8.09	32.85	21.75
C13H 6I1N1	303.10	302.9547	51.52	2.00	41.87	4.62
C13H 6I1N3	331.12	330.9608	47.16	1.83	38.33	12.69
C13H 6I1N5	359.13	358.9670	43.48	1.68	35.34	19.50
C13H 6I2N2	444.01	443.8624	35.17	1.36	57.16	6.31

C13H 6I2N4	472.03	471.8686	33.08	1.28	53.77	11.87
C13H 6I2N6	500.04	499.8747	31.23	1.21	50.76	16.81
C13H 7I1N2	318.12	317.9656	49.08	2.22	39.89	8.81
C13H 7I1N4	346.13	345.9717	45.11	2.04	36.66	16.19
C13H 7I1N6	374.15	373.9779	41.73	1.89	33.92	22.46
C13H 7I2N1	431.02	430.8672	36.23	1.64	58.89	3.25
C13H 7I2N3	459.03	458.8733	34.02	1.54	55.29	9.15
C13H 7I2N5	487.04	486.8795	32.06	1.45	52.11	14.38
C13H 8I1N1	305.12	304.9703	51.17	2.64	41.59	4.59
C13H 8I1N3	333.13	332.9765	46.87	2.42	38.09	12.61
C13H 8I1N5	361.15	360.9826	43.24	2.23	35.14	19.39
C13H 8I2N2	446.03	445.8781	35.01	1.81	56.90	6.28
C13H 8I2N4	474.04	473.8842	32.94	1.70	53.54	11.82
C13H 8I2N6	502.06	501.8904	31.10	1.61	50.55	16.74
C13H 9I1N2	320.13	319.9812	48.77	2.83	39.64	8.75
C13H 9I1N4	348.15	347.9874	44.85	2.61	36.45	16.09
C13H 9I1N6	376.16	375.9935	41.51	2.41	33.74	22.34
C13H 9I2N1	433.03	432.8828	36.06	2.09	58.61	3.23
C13H 9I2N3	461.05	460.8890	33.87	1.97	55.05	9.11
C13H 9I2N5	489.06	488.8951	31.93	1.85	51.90	14.32
C13H10I1N1	307.14	306.9860	50.84	3.28	41.32	4.56
C13H10I1N3	335.15	334.9921	46.59	3.01	37.87	12.54
C13H10I1N5	363.16	362.9983	43.00	2.78	34.94	19.28
C13H10I2N2	448.05	447.8937	34.85	2.25	56.65	6.25
C13H10I2N4	476.06	475.8999	32.80	2.12	53.31	11.77
C13H10I2N6	504.07	503.9060	30.98	2.00	50.35	16.67
C13H11I1N2	322.15	321.9969	48.47	3.44	39.39	8.70
C13H11I1N4	350.16	350.0030	44.59	3.17	36.24	16.00
C13H11I1N6	378.18	378.0092	41.29	2.93	33.56	22.22
C13H11I2N1	435.05	434.8985	35.89	2.55	58.34	3.22
C13H11I2N3	463.06	462.9046	33.72	2.39	54.81	9.07
C13H11I2N5	491.07	490.9108	31.80	2.26	51.68	14.26
C13H12I1N1	309.15	309.0016	50.51	3.91	41.05	4.53
C13H12I1N3	337.17	337.0078	46.31	3.59	37.64	12.46
C13H12I1N5	365.18	365.0139	42.76	3.31	34.75	19.18
C13H12I2N2	450.06	449.9094	34.69	2.69	56.39	6.22
C13H12I2N4	478.08	477.9155	32.66	2.53	53.09	11.72
C13H12I2N6	506.09	505.9217	30.85	2.39	50.15	16.61
C13H13I1N2	324.17	324.0125	48.17	4.04	39.15	8.64
C13H13I1N4	352.18	352.0187	44.34	3.72	36.03	15.91
C13H13I1N6	380.19	380.0248	41.07	3.45	33.38	22.10
C13H13I2N1	437.06	436.9141	35.73	3.00	58.07	3.20
C13H13I2N3	465.08	464.9203	33.57	2.82	54.57	9.04
C13H13I2N5	493.09	492.9264	31.67	2.66	51.47	14.20
C13H14I1N1	311.17	311.0173	50.18	4.54	40.78	4.50
C13H14I1N3	339.18	339.0234	46.04	4.16	37.41	12.39
C13H14I1N5	367.19	367.0296	42.52	3.84	34.56	19.07
C13H14I2N2	452.08	451.9250	34.54	3.12	56.14	6.20
C13H14I2N4	480.09	479.9312	32.52	2.94	52.87	11.67
C13H14I2N6	508.11	507.9373	30.73	2.78	49.95	16.54
C13H15I1N2	326.18	326.0282	47.87	4.64	38.91	8.59
C13H15I1N4	354.20	354.0343	44.08	4.27	35.83	15.82
C13H15I1N6	382.21	382.0405	40.85	3.96	33.20	21.99
C13H15I2N1	439.08	438.9298	35.56	3.44	57.80	3.19
C13H15I2N3	467.09	466.9359	33.43	3.24	54.34	9.00
C13H15I2N5	495.11	494.9421	31.54	3.05	51.26	14.15
C13H16I1N1	313.18	313.0329	49.86	5.15	40.52	4.47
C13H16I1N3	341.20	341.0391	45.76	4.73	37.19	12.32
C13H16I1N5	369.21	369.0452	42.29	4.37	34.37	18.97
C13H16I2N2	454.09	453.9407	34.39	3.55	55.89	6.17

019

C13H16I2N4	482.11	481.9468	32.39	3.35	52.65	11.62
C13H16I2N6	510.12	509.9530	30.61	3.16	49.75	16.47
C13H17I1N2	328.20	328.0438	47.58	5.22	38.67	8.54
C13H17I1N4	356.21	356.0500	43.83	4.81	35.63	15.73
C13H17I1N6	384.23	384.0561	40.64	4.46	33.03	21.87
C13H17I2N1	441.10	440.9454	35.40	3.88	57.54	3.18
C13H17I2N3	469.11	468.9516	33.29	3.65	54.10	8.96
C13H17I2N5	497.12	496.9577	31.41	3.45	51.06	14.09
C13H18I1N1	315.20	315.0486	49.54	5.76	40.26	4.44
C13H18I1N3	343.21	343.0547	45.50	5.29	36.98	12.24
C13H18I1N5	371.23	371.0609	42.06	4.89	34.19	18.87
C13H18I2N2	456.11	455.9563	34.23	3.98	55.65	6.14
C13H18I2N4	484.12	483.9625	32.25	3.75	52.43	11.57
C13H18I2N6	512.14	511.9686	30.45	3.54	49.56	16.41
C13H19I1N2	330.21	330.0595	47.29	5.80	38.43	8.48
C13H19I1N4	358.23	358.0656	43.59	5.35	35.43	15.64
C13H19I1N6	386.24	386.0718	40.43	4.96	32.86	21.76
C13H19I2N1	443.11	442.9611	35.24	4.32	57.28	3.16
C13H19I2N3	471.13	470.9672	33.14	4.07	53.87	8.92
C13H19I2N5	499.14	498.9734	31.28	3.84	50.85	14.03
C13H20I1N1	317.22	317.0642	49.22	6.36	40.01	4.42
C13H20I1N3	345.23	345.0704	45.23	5.84	36.76	12.17
C13H20I1N5	373.24	373.0765	41.83	5.40	34.00	18.76
C13H20I2N2	458.13	457.9720	34.08	4.40	55.40	6.11
C13H20I2N4	486.14	485.9781	32.12	4.15	52.21	11.52
C13H20I2N6	514.15	513.9843	30.37	3.92	49.36	16.35
C13H21I1N2	332.23	332.0751	47.00	6.37	38.20	8.43
C13H21I1N4	360.24	360.0813	43.34	5.88	35.23	15.55
C13H21I1N6	388.26	388.0874	40.22	5.45	32.69	21.65
C13H21I2N1	445.13	444.9767	35.08	4.76	57.02	3.15
C13H21I2N3	473.14	472.9829	33.00	4.47	53.64	8.88
C13H21I2N5	501.15	500.9890	31.16	4.22	50.64	13.97
C13H22I1N1	319.23	319.0799	48.91	6.95	39.75	4.39
C13H22I1N3	347.24	347.0860	44.97	6.39	36.55	12.10
C13H22I1N5	375.26	375.0922	41.61	5.91	33.82	18.66
C13H22I2N2	460.14	459.9876	33.93	4.82	55.16	6.09
C13H22I2N4	488.16	487.9938	31.99	4.54	51.99	11.48
C13H22I2N6	516.17	515.9999	30.25	4.30	49.17	16.28
C13H23I1N2	334.25	334.0908	46.72	6.94	37.97	8.38
C13H23I1N4	362.26	362.0969	43.10	6.40	35.03	15.47
C13H23I1N6	390.27	390.1031	40.01	5.94	32.52	21.53
C13H23I2N1	447.14	446.9924	34.92	5.18	56.76	3.13
C13H23I2N3	475.16	474.9985	32.86	4.88	53.42	8.84
C13H23I2N5	503.17	503.0047	31.03	4.61	50.44	13.92
C13H24I1N1	321.25	321.0955	48.61	7.53	39.50	4.36
C13H24I1N3	349.26	349.1017	44.71	6.93	36.34	12.03
C13H24I1N5	377.27	377.1078	41.39	6.41	33.64	18.56
C13H24I2N2	462.16	462.0033	33.79	5.23	54.92	6.06
C13H24I2N4	490.17	490.0094	31.86	4.94	51.78	11.43
C13H24I2N6	518.19	518.0156	30.13	4.67	48.98	16.22
C13H25I1N2	336.26	336.1064	46.44	7.49	37.74	8.33
C13H25I1N4	364.28	364.1126	42.86	6.92	34.84	15.38
C13H25I1N6	392.29	392.1187	39.80	6.42	32.35	21.42
C13H25I2N1	449.16	449.0080	34.76	5.61	56.51	3.12
C13H25I2N3	477.17	477.0142	32.72	5.28	53.19	8.81
C13H25I2N5	505.19	505.0203	30.91	4.99	50.24	13.86
C13H26I1N1	323.26	323.1112	48.30	8.11	39.26	4.33
C13H26I1N3	351.28	351.1173	44.45	7.46	36.13	11.96
C13H26I1N5	379.29	379.1235	41.17	6.91	33.46	18.46
C13H26I2N2	464.17	464.0189	33.64	5.65	54.68	6.04
C13H26I2N4	492.19	492.0251	31.72	5.32	51.57	11.38
C13H26I2N6	520.20	520.0312	30.02	5.04	48.79	16.16
C13H27I1N2	338.28	338.1221	46.16	8.05	37.51	8.28
C13H27I1N4	366.29	366.1282	42.63	7.43	34.65	15.30
C13H27I1N6	394.30	394.1344	39.60	6.90	32.18	21.31
C13H27I2N1	451.18	451.0237	34.61	6.03	56.25	3.10
C13H27I2N3	479.19	479.0298	32.59	5.68	52.97	8.77
C13H27I2N5	507.20	507.0360	30.79	5.37	50.04	13.81
C13H28I1N1	325.28	325.1268	48.00	8.68	39.01	4.31
C13H28I1N3	353.29	353.1330	44.20	7.99	35.92	11.89
C13H28I1N5	381.31	381.1391	40.95	7.40	33.28	18.37
C13H28I2N2	466.19	466.0346	33.49	6.05	54.44	6.01
C13H28I2N4	494.20	494.0407	31.60	5.71	51.36	11.34
C13H28I2N6	522.22	522.0469	29.90	5.40	48.60	16.09
C13H29I1N2	340.29	340.1377	45.89	8.59	37.29	8.23
C13H29I1N4	368.31	368.1439	42.40	7.94	34.46	15.21
C13H29I1N6	396.32	396.1500	39.40	7.38	32.02	21.21
C13H29I2N3	481.20	481.0455	32.45	6.07	52.74	8.73
C13H29I2N5	509.22	509.0516	30.66	5.74	49.84	13.75
C13H30I1N3	355.31	355.1486	43.95	8.51	35.72	11.83
C13H30I1N5	383.32	383.1548	40.73	7.89	33.11	18.27
C13H30I2N4	496.22	496.0564	31.47	6.09	51.15	11.29
C13H30I2N6	524.23	524.0625	29.79	5.77	48.42	16.03
C13H31I1N4	370.32	370.1595	42.16	8.44	34.27	15.13
C13H31I1N6	398.34	398.1657	39.20	7.84	31.86	21.10
C13H31I2N5	511.23	511.0673	30.54	6.11	49.65	13.70
C13H32I1N5	385.34	385.1704	40.52	8.37	32.93	18.17
C13H32I2N6	526.25	526.0782	29.67	6.13	48.23	15.97
C13H33I1N6	400.35	400.1813	39.00	8.31	31.70	20.99
C14H 7I1N2	330.13	329.9656	50.94	2.14	38.44	8.49
C14H 7I1N4	358.14	357.9717	46.95	1.97	35.43	15.64
C14H 7I1N6	386.16	385.9779	43.55	1.83	32.86	21.76
C14H 7I2N1	443.03	442.8672	37.96	1.59	57.29	3.16
C14H 7I2N3	471.04	470.8733	35.70	1.50	53.88	8.92
C14H 7I2N5	499.05	498.8795	33.69	1.41	50.86	14.03
C14H 8I1N1	317.13	316.9703	53.02	2.54	40.02	4.42
C14H 8I1N3	345.14	344.9765	48.72	2.34	36.77	12.17
C14H 8I1N5	373.16	372.9826	45.06	2.16	34.01	18.77
C14H 8I2N2	458.04	457.8781	36.71	1.76	55.41	6.12
C14H 8I2N4	486.06	485.8842	34.60	1.66	52.22	11.53
C14H 8I2N6	514.07	513.8904	32.71	1.57	49.37	16.35
C14H 9I1N2	332.15	331.9812	50.63	2.73	38.21	8.43
C14H 9I1N4	360.16	359.9874	46.69	2.52	35.24	15.56
C14H 9I1N6	388.17	387.9935	43.32	2.34	32.69	21.65
C14H 9I2N1	445.04	444.8828	37.78	2.04	57.03	3.15
C14H 9I2N3	473.06	472.8890	35.55	1.92	53.65	8.88
C14H 9I2N5	501.07	500.8951	33.56	1.81	50.65	13.98
C14H10I1N1	319.15	318.9860	52.69	3.16	39.76	4.39
C14H10I1N3	347.16	346.9921	48.44	2.90	36.55	12.10
C14H10I1N5	375.17	374.9983	44.82	2.69	33.83	18.67
C14H10I2N2	460.06	459.8937	36.55	2.19	55.17	6.09
C14H10I2N4	488.07	487.8999	34.45	2.07	52.00	11.48
C14H10I2N6	516.08	515.9060	32.58	1.95	49.18	16.28
C14H11I1N2	334.16	333.9969	50.32	3.32	37.98	8.38
C14H11I1N4	362.17	362.0030	46.43	3.06	35.04	15.47
C14H11I1N6	390.19	390.0092	43.10	2.84	32.52	21.54
C14H11I2N1	447.06	446.8985	37.61	2.48	56.77	3.13
C14H11I2N3	475.07	474.9046	35.40	2.33	53.43	8.84
C14H11I2N5	503.09	502.9108	33.42	2.20	50.45	13.92
C14H12I1N1	321.16	321.0016	52.36	3.77	39.51	4.36

020

C14H12I1N3	349.18	349.0078	48.16	3.46	36.34	12.03
C14H12I1N5	377.19	377.0139	44.58	3.21	33.64	18.57
C14H12I2N2	462.07	461.9094	36.39	2.62	54.93	6.06
C14H12I2N4	490.09	489.9155	34.31	2.47	51.79	11.43
C14H12I2N6	518.10	517.9217	32.46	2.33	48.99	16.22
C14H13I1N2	336.18	336.0125	50.02	3.90	37.75	8.33
C14H13I1N4	364.19	364.0187	46.17	3.60	34.85	15.38
C14H13I1N6	392.20	392.0248	42.87	3.34	32.36	21.43
C14H13I2N1	449.08	448.9141	37.44	2.92	56.52	3.12
C14H13I2N3	477.09	476.9203	35.25	2.75	53.20	8.81
C14H13I2N5	505.10	504.9264	33.29	2.59	50.25	13.87
C14H14I1N1	323.18	323.0173	52.03	4.37	39.27	4.33
C14H14I1N3	351.19	351.0234	47.88	4.02	36.14	11.96
C14H14I1N5	379.21	379.0296	44.34	3.72	33.47	18.47
C14H14I2N2	464.09	463.9250	36.23	3.04	54.69	6.04
C14H14I2N4	492.10	491.9312	34.17	2.87	51.58	11.39
C14H14I2N6	520.12	519.9373	32.33	2.71	48.80	16.16
C14H15I1N2	338.19	338.0282	49.72	4.47	37.52	8.28
C14H15I1N4	366.21	366.0343	45.92	4.13	34.65	15.30
C14H15I1N6	394.22	394.0405	42.66	3.84	32.19	21.32
C14H15I2N1	451.09	450.9298	37.28	3.35	56.27	3.11
C14H15I2N3	479.10	478.9359	35.10	3.16	52.98	8.77
C14H15I2N5	507.12	506.9421	33.16	2.98	50.05	13.81
C14H16I1N1	325.19	325.0329	51.71	4.96	39.02	4.31
C14H16I1N3	353.21	353.0391	47.61	4.57	35.93	11.90
C14H16I1N5	381.22	381.0452	44.11	4.23	33.29	18.37
C14H16I2N2	466.11	465.9407	36.08	3.46	54.45	6.01
C14H16I2N4	494.12	493.9468	34.03	3.26	51.37	11.34
C14H16I2N6	522.13	521.9530	32.21	3.09	48.61	16.10
C14H17I1N2	340.21	340.0438	49.43	5.04	37.30	8.23
C14H17I1N4	368.22	368.0500	45.67	4.65	34.46	15.22
C14H17I1N6	396.24	396.0561	42.44	4.32	32.03	21.21
C14H17I2N1	453.11	452.9454	37.11	3.78	56.02	3.09
C14H17I2N3	481.12	480.9516	34.95	3.56	52.75	8.73
C14H17I2N5	509.13	508.9577	33.03	3.37	49.85	13.76
C14H18I1N1	327.21	327.0486	51.39	5.54	38.78	4.28
C14H18I1N3	355.22	355.0547	47.34	5.11	35.73	11.83
C14H18I1N5	383.24	383.0609	43.88	4.73	33.11	18.27
C14H18I2N2	468.12	467.9563	35.92	3.88	54.22	5.98
C14H18I2N4	496.14	495.9625	33.89	3.66	51.16	11.29
C14H18I2N6	524.15	523.9686	32.08	3.46	48.42	16.03
C14H19I1N2	342.23	342.0595	49.14	5.60	37.08	8.19
C14H19I1N4	370.24	370.0656	45.42	5.17	34.28	15.13
C14H19I1N6	398.25	398.0718	42.22	4.81	31.87	21.10
C14H19I2N1	455.12	454.9611	36.95	4.21	55.77	3.08
C14H19I2N3	483.14	482.9672	34.81	3.96	52.53	8.70
C14H19I2N5	511.15	510.9734	32.90	3.75	49.65	13.70
C14H20I1N1	329.23	329.0642	51.08	6.12	38.55	4.25
C14H20I1N3	357.24	357.0704	47.07	5.64	35.52	11.76
C14H20I1N5	385.25	385.0765	43.65	5.23	32.94	18.18
C14H20I2N2	470.14	469.9720	35.77	4.29	53.99	5.96
C14H20I2N4	498.15	497.9781	33.76	4.05	50.95	11.25
C14H20I2N6	526.16	525.9843	31.96	3.83	48.24	15.97
C14H21I1N2	344.24	344.0751	48.85	6.15	36.86	8.14
C14H21I1N4	372.25	372.0813	45.17	5.69	34.09	15.05
C14H21I1N6	400.27	400.0874	42.01	5.29	31.70	21.00
C14H21I2N1	457.14	456.9767	36.78	4.63	55.52	3.06
C14H21I2N3	485.15	484.9829	34.66	4.36	52.32	8.66
C14H21I2N5	513.17	512.9890	32.77	4.12	49.46	13.65
C14H22I1N1	331.24	331.0799	50.77	6.69	38.31	4.23
C14H22I1N3	359.26	359.0860	46.81	6.17	35.32	11.70
C14H22I1N5	387.27	387.0922	43.42	5.73	32.77	18.08
C14H22I2N2	472.15	471.9876	35.61	4.70	53.76	5.93
C14H22I2N4	500.17	499.9938	33.62	4.43	50.74	11.20
C14H22I2N6	528.18	527.9999	31.84	4.20	48.05	15.91
C14H23I1N2	346.26	346.0908	48.56	6.70	36.65	8.09
C14H23I1N4	374.27	374.0969	44.93	6.19	33.91	14.97
C14H23I1N6	402.28	402.1031	41.80	5.76	31.55	20.89
C14H23I2N1	459.15	458.9924	36.62	5.05	55.28	3.05
C14H23I2N3	487.17	486.9985	34.52	4.76	52.10	8.63
C14H23I2N5	515.18	515.0047	32.64	4.50	49.27	13.59
C14H24I1N1	333.26	333.0955	50.46	7.26	38.08	4.20
C14H24I1N3	361.27	361.1017	46.55	6.70	35.13	11.63
C14H24I1N5	389.29	389.1078	43.20	6.21	32.60	17.99
C14H24I2N2	474.17	474.0033	35.46	5.10	53.53	5.91
C14H24I2N4	502.18	502.0094	33.49	4.82	50.54	11.16
C14H24I2N6	530.20	530.0156	31.72	4.56	47.87	15.85
C14H25I1N2	348.27	348.1064	48.28	7.24	36.44	8.04
C14H25I1N4	376.29	376.1126	44.69	6.70	33.73	14.89
C14H25I1N6	404.30	404.1187	41.59	6.23	31.39	20.79
C14H25I2N1	461.17	461.0080	36.46	5.46	55.04	3.04
C14H25I2N3	489.18	489.0142	34.37	5.15	51.88	8.59
C14H25I2N5	517.20	517.0203	32.51	4.87	49.07	13.54
C14H26I1N1	335.27	335.1112	50.15	7.82	37.85	4.18
C14H26I1N3	363.29	363.1173	46.29	7.21	34.93	11.57
C14H26I1N5	391.30	391.1235	42.97	6.70	32.43	17.90
C14H26I2N2	476.19	476.0189	35.31	5.50	53.30	5.88
C14H26I2N4	504.20	504.0251	33.35	5.20	50.34	11.11
C14H26I2N6	532.21	532.0312	31.60	4.92	47.69	15.79
C14H27I1N2	350.29	350.1221	48.00	7.77	36.23	8.00
C14H27I1N4	378.30	378.1282	44.45	7.19	33.55	14.81
C14H27I1N6	406.32	406.1344	41.39	6.70	31.23	20.68
C14H27I2N1	463.19	463.0237	36.30	5.88	54.80	3.02
C14H27I2N3	491.20	491.0298	34.23	5.54	51.67	8.55
C14H27I2N5	519.21	519.0360	32.39	5.24	48.88	13.49
C14H28I1N1	337.29	337.1268	49.85	8.37	37.62	4.15
C14H28I1N3	365.30	365.1330	46.03	7.73	34.74	11.50
C14H28I1N5	393.32	393.1391	42.75	7.18	32.27	17.81
C14H28I2N2	478.20	478.0346	35.16	5.90	53.08	5.86
C14H28I2N4	506.21	506.0407	33.22	5.58	50.14	11.07
C14H28I2N6	534.23	534.0469	31.48	5.28	47.51	15.73
C14H29I1N2	352.31	352.1377	47.73	8.30	36.02	7.95
C14H29I1N4	380.32	380.1439	44.21	7.69	33.37	14.73
C14H29I1N6	408.33	408.1500	41.18	7.16	31.08	20.58
C14H29I2N1	465.20	465.0393	36.15	6.28	54.56	3.01
C14H29I2N3	493.22	493.0455	34.09	5.93	51.46	8.52
C14H29I2N5	521.23	521.0516	32.26	5.61	48.69	13.44
C14H30I1N1	339.31	339.1425	49.56	8.91	37.40	4.13
C14H30I1N3	367.32	367.1486	45.78	8.23	34.55	11.44
C14H30I1N5	395.33	395.1548	42.54	7.65	32.10	17.72
C14H30I2N2	480.22	480.0502	35.02	6.30	52.85	5.83
C14H30I2N4	508.23	508.0564	33.09	5.95	49.94	11.02
C14H30I2N6	536.24	536.0625	31.36	5.64	47.33	15.67
C14H31I1N2	354.32	354.1534	47.46	8.82	35.82	7.91
C14H31I1N4	382.33	382.1595	43.98	8.17	33.19	14.65
C14H31I1N6	410.35	410.1657	40.98	7.61	30.93	20.48
C14H31I2N3	495.23	495.0611	33.96	6.31	51.25	8.48
C14H31I2N5	523.25	523.0673	32.14	5.97	48.51	13.38
C14H32I1N3	369.34	369.1643	45.53	8.73	34.36	11.38
C14H32I1N5	397.35	397.1704	42.32	8.12	31.94	17.63

C14H32I2N4	510.25	510.0720	32.96	6.32	49.74	10.98	
C14H32I2N6	538.26	538.0782	31.24	5.99	47.15	15.61	
C14H33I1N4	384.35	384.1752	43.75	8.65	33.02	14.58	
C14H33I1N6	412.36	412.1813	40.78	8.07	30.77	20.38	
C14H33I2N5	525.26	525.0829	32.01	6.33	48.32	13.33	
C14H34I1N5	399.36	399.1861	42.11	8.58	31.78	17.54	
C14H34I2N6	540.28	540.0938	31.12	6.34	46.98	15.56	
C14H35I1N6	414.38	414.1970	40.58	8.51	30.63	20.28	
C15H 7I1N2	342.14	341.9656	52.66	2.06	37.09	8.19	
C15H 7I1N4	370.15	369.9717	48.67	1.91	34.28	15.14	
C15H 7I1N6	398.17	397.9779	45.25	1.77	31.87	21.11	
C15H 7I2N1	455.04	454.8672	39.59	1.55	55.78	3.08	
C15H 7I2N3	483.05	482.8733	37.30	1.46	52.54	8.70	
C15H 7I2N5	511.07	510.8795	35.25	1.38	49.66	13.70	
C15H 8I1N1	329.14	328.9703	54.74	2.45	38.56	4.26	
C15H 8I1N3	357.16	356.9765	50.45	2.26	35.53	11.77	
C15H 8I1N5	385.17	384.9826	46.78	2.09	32.95	18.18	
C15H 8I2N2	470.05	469.8781	38.33	1.72	54.00	5.96	
C15H 8I2N4	498.07	497.8842	36.17	1.62	50.96	11.25	
C15H 8I2N6	526.08	525.8904	34.25	1.53	48.25	15.97	
C15H 9I1N2	344.16	343.9812	52.35	2.64	36.87	8.14	
C15H 9I1N4	372.17	371.9874	48.41	2.44	34.10	15.05	
C15H 9I1N6	400.18	399.9935	45.02	2.27	31.71	21.00	
C15H 9I2N1	457.05	456.8828	39.42	1.98	55.53	3.06	
C15H 9I2N3	485.07	484.8890	37.14	1.87	52.32	8.66	
C15H 9I2N5	513.08	512.8951	35.11	1.77	49.47	13.65	
C15H10I1N1	331.16	330.9860	54.41	3.04	38.32	4.23	
C15H10I1N3	359.17	358.9921	50.16	2.81	35.33	11.70	
C15H10I1N5	387.18	386.9983	46.53	2.60	32.78	18.09	
C15H10I2N2	472.07	471.8937	38.17	2.14	53.77	5.93	
C15H10I2N4	500.08	499.8999	36.03	2.02	50.75	11.20	
C15H10I2N6	528.10	527.9060	34.12	1.91	48.06	15.91	
C15H11I1N2	346.17	345.9969	52.05	3.20	36.66	8.09	
C15H11I1N4	374.19	374.0030	48.15	2.96	33.91	14.97	
C15H11I1N6	402.20	402.0092	44.80	2.76	31.55	20.90	
C15H11I2N1	459.07	458.8985	39.25	2.42	55.29	3.05	
C15H11I2N3	487.08	486.9046	36.99	2.28	52.11	8.63	
C15H11I2N5	515.10	514.9108	34.98	2.15	49.27	13.60	
C15H12I1N1	333.17	333.0016	54.08	3.63	38.09	4.20	
C15H12I1N3	361.19	361.0078	49.88	3.35	35.14	11.63	
C15H12I1N5	389.20	389.0139	46.29	3.11	32.61	17.99	
C15H12I2N2	474.09	473.9094	38.00	2.55	53.54	5.91	
C15H12I2N4	502.10	501.9155	35.88	2.41	50.55	11.16	
C15H12I2N6	530.11	529.9217	33.99	2.28	47.88	15.85	
C15H13I1N2	348.19	348.0125	51.74	3.76	36.45	8.05	
C15H13I1N4	376.20	376.0187	47.89	3.48	33.73	14.89	
C15H13I1N6	404.22	404.0248	44.57	3.24	31.40	20.79	
C15H13I2N1	461.09	460.9141	39.07	2.84	55.05	3.04	
C15H13I2N3	489.10	488.9203	36.84	2.68	51.89	8.59	
C15H13I2N5	517.11	516.9264	34.84	2.53	49.08	13.54	
C15H14I1N1	335.19	335.0173	53.75	4.21	37.86	4.18	
C15H14I1N3	363.20	363.0234	49.61	3.89	34.94	11.57	
C15H14I1N5	391.22	391.0296	46.05	3.61	32.44	17.90	
C15H14I2N2	476.10	475.9250	37.84	2.96	53.31	5.88	
C15H14I2N4	504.11	503.9312	35.74	2.80	50.35	11.11	
C15H14I2N6	532.13	531.9373	33.86	2.65	47.70	15.79	
C15H15I1N2	350.20	350.0282	51.45	4.32	36.24	8.00	
C15H15I1N4	378.22	378.0343	47.64	4.00	33.55	14.81	
C15H15I1N6	406.23	406.0405	44.35	3.72	31.24	20.69	
C15H15I2N1	463.10	462.9298	38.90	3.26	54.81	3.02	
C15H15I2N3	491.12	490.9359	36.69	3.08	51.68	8.56	
C15H15I2N5	519.13	518.9421	34.71	2.91	48.89	13.49	
C15H16I1N1	337.21	337.0329	53.43	4.78	37.63	4.15	
C15H16I1N3	365.22	365.0391	49.33	4.42	34.75	11.51	
C15H16I1N5	393.23	393.0452	45.82	4.10	32.27	17.81	
C15H16I2N2	478.12	477.9407	37.68	3.37	53.09	5.86	
C15H16I2N4	506.13	505.9468	35.60	3.19	50.15	11.07	
C15H16I2N6	534.14	533.9530	33.73	3.02	47.52	15.73	
C15H17I1N2	352.22	352.0438	51.15	4.86	36.03	7.95	
C15H17I1N4	380.23	380.0500	47.38	4.51	33.38	14.73	
C15H17I1N6	408.25	408.0561	44.13	4.20	31.09	20.59	
C15H17I2N1	465.12	464.9454	38.74	3.68	54.57	3.01	
C15H17I2N3	493.13	492.9516	36.54	3.47	51.47	8.52	
C15H17I2N5	521.15	520.9577	34.57	3.29	48.70	13.44	
C15H18I1N1	339.22	339.0486	53.11	5.35	37.41	4.13	
C15H18I1N3	367.24	367.0547	49.06	4.94	34.56	11.44	
C15H18I1N5	395.25	395.0609	45.58	4.59	32.11	17.72	
C15H18I2N2	480.13	479.9563	37.52	3.78	52.86	5.83	
C15H18I2N4	508.15	507.9625	35.46	3.57	49.95	11.03	
C15H18I2N6	536.16	535.9686	33.60	3.38	47.34	15.67	
C15H19I1N2	354.24	354.0595	50.86	5.41	35.82	7.91	
C15H19I1N4	382.25	382.0656	47.13	5.01	33.20	14.66	
C15H19I1N6	410.26	410.0718	43.92	4.67	30.93	20.48	
C15H19I2N1	467.13	466.9611	38.57	4.10	54.33	3.00	
C15H19I2N3	495.15	494.9672	36.39	3.87	51.26	8.49	
C15H19I2N5	523.16	522.9734	34.44	3.66	48.51	13.39	
C15H20I1N1	341.24	341.0642	52.80	5.91	37.19	4.10	
C15H20I1N3	369.25	369.0704	48.79	5.46	34.37	11.38	
C15H20I1N5	397.26	397.0765	45.35	5.07	31.94	17.63	
C15H20I2N2	482.15	481.9720	37.37	4.18	52.64	5.81	
C15H20I2N4	510.16	509.9781	35.32	3.95	49.75	10.98	
C15H20I2N6	538.18	537.9843	33.48	3.75	47.16	15.62	
C15H21I1N2	356.25	356.0751	50.57	5.94	35.62	7.86	
C15H21I1N4	384.27	384.0813	46.89	5.51	33.03	14.58	
C15H21I1N6	412.28	412.0874	43.70	5.13	30.78	20.38	
C15H21I2N1	469.15	468.9767	38.40	4.51	54.10	2.99	
C15H21I2N3	497.16	496.9829	36.24	4.26	51.05	8.45	
C15H21I2N5	525.18	524.9890	34.31	4.03	48.33	13.34	
C15H22I1N1	343.25	343.0799	52.49	6.46	36.97	4.08	
C15H22I1N3	371.27	371.0860	48.53	5.97	34.18	11.32	
C15H22I1N5	399.28	399.0922	45.12	5.55	31.78	17.54	
C15H22I2N2	484.16	483.9876	37.21	4.58	52.42	5.79	
C15H22I2N4	512.18	511.9938	35.18	4.33	49.55	10.94	
C15H22I2N6	540.19	539.9999	33.35	4.11	46.98	15.56	
C15H23I1N2	358.27	358.0908	50.29	6.47	35.42	7.82	
C15H23I1N4	386.28	386.0969	46.64	6.00	32.85	14.50	
C15H23I1N6	414.30	414.1031	43.49	5.60	30.63	20.29	
C15H23I2N1	471.17	470.9924	38.24	4.92	53.87	2.97	
C15H23I2N3	499.18	498.9985	36.09	4.64	50.85	8.42	
C15H23I2N5	527.19	527.0047	34.17	4.40	48.14	13.28	
C15H24I1N1	345.27	345.0955	52.18	7.01	36.76	4.06	
C15H24I1N3	373.28	373.1017	48.27	6.48	34.00	11.26	
C15H24I1N5	401.30	401.1078	44.90	6.03	31.62	17.45	
C15H24I2N2	486.18	486.0033	37.06	4.98	52.20	5.76	
C15H24I2N4	514.19	514.0094	35.04	4.70	49.36	10.90	
C15H24I2N6	542.21	542.0156	33.23	4.46	46.81	15.50	
C15H25I1N2	360.28	360.1064	50.01	6.99	35.22	7.78	
C15H25I1N4	388.30	388.1126	46.40	6.49	32.68	14.43	
C15H25I1N6	416.31	416.1187	43.28	6.05	30.48	20.19	
C15H25I2N1	473.18	473.0080	38.08	5.33	53.54	2.96	

022

C15H25I2N3	501.20	501.0142	35.95	5.03	50.64	8.38
C15H25I2N5	529.21	529.0203	34.04	4.76	47.96	13.23
C15H26I1N1	347.29	347.1112	51.88	7.55	36.54	4.03
C15H26I1N3	375.30	375.1173	48.01	6.98	33.81	11.20
C15H26I1N5	403.31	403.1235	44.67	6.50	31.47	17.36
C15H26I2N2	488.20	488.0189	36.90	5.37	51.99	5.74
C15H26I2N4	516.21	516.0251	34.90	5.08	49.17	10.85
C15H26I2N6	544.22	544.0312	33.11	4.82	46.64	15.44
C15H27I1N2	362.30	362.1221	49.73	7.51	35.03	7.73
C15H27I1N4	390.31	390.1282	46.16	6.97	32.51	14.35
C15H27I1N6	418.33	418.1344	43.07	6.51	30.34	20.09
C15H27I2N1	475.20	475.0237	37.91	5.73	53.41	2.95
C15H27I2N3	503.21	503.0298	35.80	5.41	50.44	8.35
C15H27I2N5	531.22	531.0360	33.92	5.12	47.78	13.18
C15H28I1N1	349.30	349.1268	51.58	8.08	36.33	4.01
C15H28I1N3	377.31	377.1330	47.75	7.48	33.63	11.14
C15H28I1N5	405.33	405.1391	44.45	6.96	31.31	17.28
C15H28I2N2	490.21	490.0346	36.75	5.76	51.78	5.71
C15H28I2N4	518.23	518.0407	34.77	5.45	48.98	10.81
C15H28I2N6	546.24	546.0469	32.98	5.17	46.46	15.39
C15H29I1N2	364.32	364.1377	49.45	8.02	34.83	7.69
C15H29I1N4	392.33	392.1439	45.92	7.45	32.35	14.28
C15H29I1N6	420.34	420.1500	42.86	6.95	30.19	19.99
C15H29I2N1	477.21	477.0393	37.75	6.13	53.19	2.94
C15H29I2N3	505.23	505.0455	35.66	5.79	50.24	8.32
C15H29I2N5	533.24	533.0516	33.79	5.48	47.60	13.13
C15H30I1N1	351.32	351.1425	51.28	8.61	36.12	3.99
C15H30I1N3	379.33	379.1486	47.50	7.97	33.45	11.08
C15H30I1N5	407.34	407.1548	44.23	7.42	31.15	17.19
C15H30I2N2	492.23	492.0502	36.60	6.14	51.56	5.69
C15H30I2N4	520.24	520.0564	34.63	5.81	48.79	10.77
C15H30I2N6	548.26	548.0625	32.86	5.52	46.29	15.33
C15H31I1N2	366.33	366.1534	49.18	8.53	34.64	7.65
C15H31I1N4	394.35	394.1595	45.69	7.92	32.18	14.21
C15H31I1N6	422.36	422.1657	42.66	7.40	30.05	19.90
C15H31I2N1	479.23	479.0550	37.60	6.52	52.96	2.92
C15H31I2N3	507.24	507.0611	35.52	6.16	50.04	8.28
C15H31I2N5	535.26	535.0673	33.66	5.84	47.42	13.08
C15H32I1N1	353.33	353.1581	50.99	9.13	35.92	3.96
C15H32I1N3	381.35	381.1643	47.24	8.46	33.28	11.02
C15H32I1N5	409.36	409.1704	44.01	7.88	31.00	17.11
C15H32I2N2	494.24	494.0659	36.45	6.53	51.35	5.67
C15H32I2N4	522.26	522.0720	34.50	6.18	48.60	10.73
C15H32I2N6	550.27	550.0782	32.74	5.86	46.12	15.27
C15H33I1N2	368.35	368.1690	48.91	9.03	34.45	7.61
C15H33I1N4	396.36	396.1752	45.46	8.39	32.02	14.14
C15H33I1N6	424.37	424.1813	42.45	7.84	29.90	19.80
C15H33I2N3	509.26	509.0768	35.38	6.53	49.84	8.25
C15H33I2N5	537.27	537.0829	33.53	6.19	47.24	13.04
C15H34I1N3	383.36	383.1799	47.00	8.94	33.10	10.96
C15H34I1N5	411.38	411.1861	43.80	8.33	30.85	17.02
C15H34I2N4	524.27	524.0877	34.37	6.54	48.41	10.69
C15H34I2N6	552.29	552.0938	32.62	6.21	45.96	15.22
C15H35I1N4	398.38	398.1908	45.23	8.86	31.86	14.06
C15H35I1N6	426.39	426.1970	42.25	8.27	29.76	19.71
C15H35I2N5	539.29	539.0985	33.41	6.54	47.06	12.99
C15H36I1N5	413.39	413.2017	43.58	8.78	30.70	16.94
C15H36I2N6	554.30	554.1094	32.50	6.55	45.79	15.16
C15H37I1N6	428.41	428.2126	42.06	8.71	29.62	19.62
C16H 8I1N1	341.15	340.9703	56.33	2.36	37.20	4.11
C16H 8I1N3	369.17	368.9765	52.06	2.18	34.38	11.38
C16H 8I1N5	397.18	396.9826	48.39	2.03	31.95	17.63
C16H 8I2N2	482.06	481.8781	39.87	1.67	52.65	5.81
C16H 8I2N4	510.08	509.8842	37.68	1.58	49.76	10.98
C16H 8I2N6	538.09	537.8904	35.71	1.50	47.17	15.62
C16H 9I1N2	356.17	355.9812	53.96	2.55	35.63	7.87
C16H 9I1N4	384.18	383.9874	50.02	2.36	33.03	14.58
C16H 9I1N6	412.19	411.9935	46.62	2.20	30.79	20.39
C16H 9I2N1	469.07	468.8828	40.97	1.93	54.11	2.99
C16H 9I2N3	497.08	496.8890	38.66	1.83	51.06	8.45
C16H 9I2N5	525.09	524.8951	36.60	1.73	48.34	13.34
C16H10I1N1	343.17	342.9860	56.00	2.94	36.98	4.08
C16H10I1N3	371.18	370.9921	51.77	2.72	34.19	11.32
C16H10I1N5	399.20	398.9983	48.14	2.53	31.79	17.54
C16H10I2N2	484.08	483.8937	39.70	2.08	52.43	5.79
C16H10I2N4	512.09	511.8999	37.53	1.97	49.56	10.94
C16H10I2N6	540.11	539.9060	35.58	1.87	46.99	15.56
C16H11I1N2	358.18	357.9969	53.65	3.10	35.43	7.82
C16H11I1N4	386.20	386.0030	49.76	2.87	32.86	14.51
C16H11I1N6	414.21	414.0092	46.40	2.68	30.64	20.29
C16H11I2N1	471.08	470.8985	40.80	2.35	53.88	2.97
C16H11I2N3	499.09	498.9046	38.51	2.22	50.85	8.42
C16H11I2N5	527.11	526.9108	36.46	2.10	48.15	13.29
C16H12I1N1	345.19	345.0016	55.67	3.50	36.76	4.06
C16H12I1N3	373.20	373.0078	51.49	3.24	34.00	11.26
C16H12I1N5	401.21	401.0139	47.90	3.01	31.63	17.46
C16H12I2N2	486.10	485.9094	39.54	2.49	52.21	5.76
C16H12I2N4	514.11	513.9155	37.38	2.35	49.37	10.90
C16H12I2N6	542.12	541.9217	35.45	2.23	46.82	15.50
C16H13I1N2	360.20	360.0125	53.35	3.64	35.23	7.78
C16H13I1N4	388.21	388.0187	49.50	3.38	32.69	14.43
C16H13I1N6	416.23	416.0248	46.17	3.15	30.49	20.19
C16H13I2N1	473.10	472.9141	40.62	2.77	53.65	2.96
C16H13I2N3	501.11	500.9203	38.35	2.61	50.65	8.39
C16H13I2N5	529.12	528.9264	36.32	2.48	47.97	13.24
C16H14I1N1	347.20	347.0173	55.35	4.06	36.55	4.03
C16H14I1N3	375.21	375.0234	51.22	3.76	33.82	11.20
C16H14I1N5	403.23	403.0296	47.66	3.50	31.47	17.37
C16H14I2N2	488.11	487.9250	39.37	2.89	52.00	5.74
C16H14I2N4	516.13	515.9312	37.23	2.73	49.18	10.86
C16H14I2N6	544.14	543.9373	35.32	2.59	46.64	15.44
C16H15I1N2	362.22	362.0282	53.06	4.17	35.04	7.73
C16H15I1N4	390.23	390.0343	49.25	3.87	32.52	14.36
C16H15I1N6	418.24	418.0405	45.95	3.62	30.34	20.09
C16H15I2N1	475.11	474.9298	40.45	3.18	53.42	2.95
C16H15I2N3	503.13	502.9359	38.20	3.01	50.45	8.35
C16H15I2N5	531.14	530.9421	36.18	2.85	47.79	13.19
C16H16I1N1	349.22	349.0329	55.03	4.62	36.34	4.01
C16H16I1N3	377.23	377.0391	50.94	4.28	33.64	11.14
C16H16I1N5	405.24	405.0452	47.42	3.98	31.32	17.28
C16H16I2N2	490.13	489.9407	39.21	3.29	51.78	5.72
C16H16I2N4	518.14	517.9468	37.09	3.11	48.98	10.81
C16H16I2N6	546.15	545.9530	35.19	2.95	46.47	15.39
C16H17I1N2	364.23	364.0438	52.76	4.70	34.84	7.69
C16H17I1N4	392.25	392.0500	48.99	4.37	32.35	14.28
C16H17I1N6	420.26	420.0561	45.73	4.08	30.20	20.00
C16H17I2N1	477.13	476.9454	40.28	3.59	53.19	2.94
C16H17I2N3	505.14	504.9516	38.04	3.39	50.24	8.32
C16H17I2N5	533.16	532.9577	36.05	3.21	47.60	13.14
C16H18I1N1	351.23	351.0486	54.72	5.17	36.13	3.99

023

C16H18I1N3	379.25	379.0547	50.67	4.78	33.46	11.08
C16H18I1N5	407.26	407.0609	47.19	4.46	31.16	17.20
C16H18I2N2	492.14	491.9563	39.05	3.69	51.57	5.69
C16H18I2N4	520.16	519.9625	36.95	3.49	48.79	10.77
C16H18I2N6	548.17	547.9686	35.06	3.31	46.30	15.33
C16H19I1N2	366.25	366.0595	52.47	5.23	34.65	7.65
C16H19I1N4	394.26	394.0656	48.74	4.86	32.19	14.21
C16H19I1N6	422.27	422.0718	45.51	4.54	30.05	19.90
C16H19I2N1	479.15	478.9611	40.11	4.00	52.97	2.92
C16H19I2N3	507.16	506.9672	37.89	3.78	50.05	8.29
C16H19I2N5	535.17	534.9734	35.91	3.58	47.43	13.09
C16H20I1N1	353.25	353.0642	54.40	5.71	35.92	3.97
C16H20I1N3	381.26	381.0704	50.41	5.29	33.29	11.02
C16H20I1N5	409.28	409.0765	46.96	4.93	31.01	17.11
C16H20I2N2	494.16	493.9720	38.89	4.08	51.36	5.67
C16H20I2N4	522.17	521.9781	36.80	3.86	48.61	10.73
C16H20I2N6	550.19	549.9843	34.93	3.66	46.13	15.27
C16H21I1N2	368.26	368.0751	52.19	5.75	34.46	7.61
C16H21I1N4	396.28	396.0813	48.50	5.34	32.02	14.14
C16H21I1N6	424.29	424.0874	45.29	4.99	29.91	19.81
C16H21I2N1	481.16	480.9767	39.94	4.40	52.75	2.91
C16H21I2N3	509.17	508.9829	37.74	4.16	49.85	8.25
C16H21I2N5	537.19	536.9890	35.77	3.94	47.25	13.04
C16H22I1N1	355.26	355.0799	54.09	6.24	35.72	3.94
C16H22I1N3	383.28	383.0860	50.14	5.79	33.11	10.96
C16H22I1N5	411.29	411.0922	46.73	5.39	30.86	17.03
C16H22I2N2	496.18	495.9876	38.73	4.47	51.15	5.65
C16H22I2N4	524.19	523.9938	36.66	4.23	48.42	10.69
C16H22I2N6	552.20	551.9999	34.80	4.02	45.96	15.22
C16H23I1N2	370.28	370.0908	51.90	6.26	34.27	7.57
C16H23I1N4	398.29	398.0969	48.25	5.82	31.86	14.07
C16H23I1N6	426.31	426.1031	45.08	5.44	29.77	19.71
C16H23I2N1	483.18	482.9924	39.77	4.80	52.53	2.90
C16H23I2N3	511.19	510.9985	37.59	4.54	49.65	8.22
C16H23I2N5	539.20	539.0047	35.64	4.30	47.07	12.99
C16H24I1N1	357.28	357.0955	53.79	6.77	35.52	3.92
C16H24I1N3	385.29	385.1017	49.88	6.28	32.94	10.91
C16H24I1N5	413.31	413.1078	46.50	5.85	30.70	16.94
C16H24I2N2	498.19	498.0033	38.58	4.86	50.95	5.62
C16H24I2N4	526.21	526.0094	36.52	4.60	48.23	10.65
C16H24I2N6	554.22	554.0156	34.68	4.36	45.80	15.16
C16H25I1N2	372.30	372.1064	51.62	6.77	34.09	7.52
C16H25I1N4	400.31	400.1126	48.01	6.29	31.70	14.00
C16H25I1N6	428.32	428.1187	44.87	5.88	29.63	19.62
C16H25I2N1	485.19	485.0080	39.61	5.19	52.31	2.89
C16H25I2N3	513.21	513.0142	37.45	4.91	49.46	8.19
C16H25I2N5	541.22	541.0203	35.51	4.66	46.90	12.94
C16H26I1N1	359.30	359.1112	53.49	7.29	35.32	3.90
C16H26I1N3	387.31	387.1173	49.62	6.77	32.77	10.85
C16H26I1N5	415.32	415.1235	46.27	6.31	30.56	16.86
C16H26I2N2	500.21	500.0189	38.42	5.24	50.74	5.60
C16H26I2N4	528.22	528.0251	36.38	4.96	48.05	10.61
C16H26I2N6	556.23	556.0312	34.55	4.71	45.63	15.11
C16H27I1N2	374.31	374.1221	51.34	7.27	33.90	7.48
C16H27I1N4	402.32	402.1282	47.77	6.76	31.54	13.93
C16H27I1N6	430.34	430.1344	44.66	6.32	29.49	19.53
C16H27I2N1	487.21	487.0237	39.44	5.59	52.09	2.87
C16H27I2N3	515.22	515.0298	37.30	5.28	49.26	8.16
C16H27I2N5	543.24	543.0360	35.38	5.01	46.72	12.89
C16H28I1N1	361.31	361.1268	53.19	7.81	35.12	3.88
C16H28I1N3	389.33	389.1330	49.36	7.25	32.60	10.79
C16H28I1N5	417.34	417.1391	46.05	6.76	30.41	16.78
C16H28I2N2	502.22	502.0346	38.27	5.62	50.54	5.58
C16H28I2N4	530.24	530.0407	36.24	5.32	47.87	10.57
C16H28I2N6	558.25	558.0469	34.43	5.06	45.47	15.05
C16H29I1N2	376.33	376.1377	51.07	7.77	33.72	7.44
C16H29I1N4	404.34	404.1439	47.53	7.23	31.39	13.86
C16H29I1N6	432.35	432.1500	44.45	6.76	29.35	19.44
C16H29I2N1	489.23	489.0393	39.28	5.97	51.88	2.86
C16H29I2N3	517.24	517.0455	37.15	5.65	49.07	8.12
C16H29I2N5	545.25	545.0516	35.25	5.36	46.55	12.84
C16H30I1N1	363.33	363.1425	52.89	8.32	34.93	3.86
C16H30I1N3	391.34	391.1486	49.11	7.73	32.43	10.74
C16H30I1N5	419.36	419.1548	45.83	7.21	30.26	16.70
C16H30I2N2	504.24	504.0502	38.11	6.00	50.33	5.56
C16H30I2N4	532.25	532.0564	36.11	5.68	47.69	10.53
C16H30I2N6	560.27	560.0625	34.30	5.40	45.30	15.00
C16H31I1N2	378.34	378.1534	50.79	8.26	33.54	7.40
C16H31I1N4	406.36	406.1595	47.29	7.69	31.23	13.79
C16H31I1N6	434.37	434.1657	44.24	7.19	29.22	19.35
C16H31I2N1	491.24	491.0550	39.12	6.36	51.67	2.85
C16H31I2N3	519.25	519.0611	37.01	6.02	48.88	8.09
C16H31I2N5	547.27	547.0673	35.12	5.71	46.38	12.80
C16H32I1N1	365.34	365.1581	52.60	8.83	34.74	3.83
C16H32I1N3	393.36	393.1643	48.86	8.20	32.26	10.68
C16H32I1N5	421.37	421.1704	45.61	7.65	30.12	16.62
C16H32I2N2	506.26	506.0659	37.96	6.37	50.13	5.53
C16H32I2N4	534.27	534.0720	35.97	6.04	47.51	10.49
C16H32I2N6	562.28	562.0782	34.18	5.74	45.14	14.95
C16H33I1N2	380.36	380.1690	50.53	8.75	33.36	7.36
C16H33I1N4	408.37	408.1752	47.06	8.15	31.08	13.72
C16H33I1N6	436.39	436.1813	44.04	7.62	29.08	19.26
C16H33I2N1	493.26	493.0706	38.96	6.74	51.46	2.84
C16H33I2N3	521.27	521.0768	36.87	6.38	48.69	8.06
C16H33I2N5	549.28	549.0829	34.99	6.06	46.21	12.75
C16H34I1N1	367.36	367.1738	52.31	9.33	34.54	3.81
C16H34I1N3	395.37	395.1799	48.61	8.67	32.10	10.63
C16H34I1N5	423.39	423.1861	45.39	8.09	29.97	16.54
C16H34I2N2	508.27	508.0815	37.81	6.74	49.94	5.51
C16H34I2N4	536.28	536.0877	35.84	6.39	47.33	10.45
C16H34I2N6	564.30	564.0938	34.06	6.07	44.98	14.89
C16H35I1N2	382.38	382.1847	50.26	9.23	33.19	7.33
C16H35I1N4	410.39	410.1908	46.83	8.60	30.92	13.65
C16H35I1N6	438.40	438.1970	43.84	8.05	28.95	19.17
C16H35I2N3	523.29	523.0924	36.73	6.74	48.50	8.03
C16H35I2N5	551.30	551.0985	34.86	6.40	46.04	12.70
C16H36I1N3	397.39	397.1956	48.36	9.13	31.93	10.57
C16H36I1N5	425.40	425.2017	45.18	8.53	29.83	16.46
C16H36I2N4	538.30	538.1033	35.70	6.74	47.15	10.41
C16H36I2N6	566.31	566.1094	33.93	6.41	44.82	14.84
C16H37I1N4	412.40	412.2065	46.60	9.04	30.77	13.59
C16H37I1N6	440.42	440.2126	43.64	8.47	28.81	19.08
C16H37I2N5	553.32	553.1142	34.73	6.74	45.87	12.66
C16H38I1N5	427.42	427.2174	44.96	8.96	29.69	16.39
C16H38I2N6	568.33	568.1251	33.81	6.74	44.66	14.79
C16H39I1N6	442.43	442.2283	43.44	8.89	28.68	18.99
C17H 8I1N1	353.16	352.9703	57.82	2.28	35.93	3.97
C17H 8I1N3	381.18	380.9765	53.57	2.12	33.29	11.02
C17H 8I1N5	409.19	408.9826	49.90	1.97	31.01	17.12
C17H 8I2N2	494.08	493.8781	41.33	1.63	51.37	5.67

024

C17H 8I2N4	522.09	521.8842	39.11	1.54	48.61	10.73
C17H 8I2N6	550.10	549.8904	37.12	1.47	46.14	15.28
C17H 9I1N2	368.18	367.9812	55.46	2.46	34.47	7.61
C17H 9I1N4	396.19	395.9874	51.54	2.29	32.03	14.14
C17H 9I1N6	424.21	423.9935	48.13	2.14	29.92	19.81
C17H 9I2N1	481.08	480.8828	42.44	1.89	52.76	2.91
C17H 9I2N3	509.09	508.8890	40.11	1.78	49.86	8.25
C17H 9I2N5	537.10	536.8951	38.02	1.69	47.26	13.04
C17H10I1N1	355.18	354.9860	57.49	2.84	35.73	3.94
C17H10I1N3	383.19	382.9921	53.29	2.63	33.12	10.97
C17H10I1N5	411.21	410.9983	49.66	2.45	30.86	17.03
C17H10I2N2	496.09	495.8937	41.16	2.03	51.16	5.65
C17H10I2N4	524.10	523.8999	38.96	1.92	48.43	10.69
C17H10I2N6	552.12	551.9060	36.98	1.83	45.97	15.22
C17H11I1N2	370.20	369.9969	55.16	3.00	34.28	7.57
C17H11I1N4	398.21	398.0030	51.28	2.78	31.87	14.07
C17H11I1N6	426.22	426.0092	47.91	2.60	29.77	19.72
C17H11I2N1	483.09	482.8985	42.27	2.30	52.54	2.90
C17H11I2N3	511.11	510.9046	39.95	2.17	49.66	8.22
C17H11I2N5	539.12	538.9108	37.87	2.06	47.08	12.99
C17H12I1N1	357.20	357.0016	57.16	3.39	35.53	3.92
C17H12I1N3	385.21	385.0078	53.01	3.14	32.94	10.91
C17H12I1N5	413.22	413.0139	49.41	2.93	30.71	16.95
C17H12I2N2	498.11	497.9094	40.99	2.43	50.95	5.62
C17H12I2N4	526.12	525.9155	38.81	2.30	48.24	10.65
C17H12I2N6	554.13	553.9217	36.85	2.18	45.80	15.17
C17H13I1N2	372.21	372.0125	54.86	3.52	34.09	7.53
C17H13I1N4	400.22	400.0187	51.02	3.27	31.71	14.00
C17H13I1N6	428.24	428.0248	47.68	3.06	29.63	19.62
C17H13I2N1	485.11	484.9141	42.09	2.70	52.32	2.89
C17H13I2N3	513.12	512.9203	39.79	2.55	49.46	8.19
C17H13I2N5	541.14	540.9264	37.73	2.42	46.90	12.94
C17H14I1N1	359.21	359.0173	56.84	3.93	35.33	3.90
C17H14I1N3	387.23	387.0234	52.73	3.64	32.77	10.85
C17H14I1N5	415.24	415.0296	49.17	3.40	30.56	16.87
C17H14I2N2	500.12	499.9250	40.83	2.82	50.75	5.60
C17H14I2N4	528.14	527.9312	38.66	2.67	48.06	10.61
C17H14I2N6	556.15	555.9373	36.71	2.54	45.64	15.11
C17H15I1N2	374.23	374.0282	54.56	4.04	33.91	7.49
C17H15I1N4	402.24	402.0343	50.76	3.76	31.55	13.93
C17H15I1N6	430.25	430.0405	47.46	3.51	29.50	19.53
C17H15I2N1	487.12	486.9298	41.92	3.10	52.10	2.88
C17H15I2N3	515.14	514.9359	39.64	2.94	49.27	8.16
C17H15I2N5	543.15	542.9421	37.59	2.78	46.73	12.89
C17H16I1N1	361.23	361.0329	56.53	4.46	35.13	3.88
C17H16I1N3	389.24	389.0391	52.46	4.14	32.60	10.80
C17H16I1N5	417.25	417.0452	48.94	3.87	30.41	16.78
C17H16I2N2	502.14	501.9407	40.66	3.21	50.55	5.58
C17H16I2N4	530.15	529.9468	38.52	3.04	47.87	10.57
C17H16I2N6	558.17	557.9530	36.58	2.89	45.47	15.06
C17H17I1N2	376.24	376.0438	54.27	4.55	33.73	7.45
C17H17I1N4	404.26	404.0500	50.51	4.24	31.39	13.86
C17H17I1N6	432.27	432.0561	47.24	3.96	29.36	19.44
C17H17I2N1	489.14	488.9454	41.74	3.50	51.89	2.86
C17H17I2N3	517.15	516.9516	39.48	3.31	49.08	8.13
C17H17I2N5	545.17	544.9577	37.45	3.14	46.56	12.85
C17H18I1N1	363.24	363.0486	56.21	4.99	34.94	3.86
C17H18I1N3	391.26	391.0547	52.19	4.64	32.44	10.74
C17H18I1N5	419.27	419.0609	48.70	4.33	30.27	16.70
C17H18I2N2	504.16	503.9563	40.50	3.60	50.34	5.56
C17H18I2N4	532.17	531.9625	38.37	3.41	47.69	10.53
C17H18I2N6	560.18	559.9686	36.45	3.24	45.31	15.00
C17H19I1N2	378.26	378.0595	53.98	5.06	33.55	7.41
C17H19I1N4	406.27	406.0656	50.26	4.71	31.24	13.79
C17H19I1N6	434.29	434.0718	47.02	4.41	29.22	19.35
C17H19I2N1	491.16	490.9611	41.57	3.90	51.68	2.85
C17H19I2N3	519.17	518.9672	39.33	3.69	48.89	8.09
C17H19I2N5	547.18	546.9734	37.32	3.50	46.38	12.80
C17H20I1N1	365.26	365.0642	55.90	5.52	34.74	3.83
C17H20I1N3	393.27	393.0704	51.92	5.13	32.27	10.68
C17H20I1N5	421.29	421.0765	48.47	4.79	30.12	16.62
C17H20I2N2	506.17	505.9720	40.34	3.98	50.14	5.53
C17H20I2N4	534.18	533.9781	38.22	3.77	47.51	10.49
C17H20I2N6	562.20	561.9843	36.32	3.59	45.15	14.95
C17H21I1N2	380.27	380.0751	53.70	5.57	33.37	7.37
C17H21I1N4	408.29	408.0813	50.01	5.18	31.08	13.72
C17H21I1N6	436.30	436.0874	46.80	4.85	29.09	19.26
C17H21I2N1	493.17	492.9767	41.40	4.29	51.46	2.84
C17H21I2N3	521.19	520.9829	39.18	4.06	48.70	8.06
C17H21I2N5	549.20	548.9890	37.18	3.85	46.21	12.75
C17H22I1N1	367.28	367.0799	55.60	6.04	34.55	3.81
C17H22I1N3	395.29	395.0860	51.66	5.61	32.10	10.63
C17H22I1N5	423.30	423.0922	48.24	5.24	29.98	16.54
C17H22I2N2	508.19	507.9876	40.18	4.36	49.94	5.51
C17H22I2N4	536.20	535.9938	38.08	4.14	47.33	10.45
C17H22I2N6	564.21	563.9999	36.19	3.93	44.98	14.90
C17H23I1N2	382.29	382.0908	53.41	6.06	33.20	7.33
C17H23I1N4	410.30	410.0969	49.77	5.65	30.93	13.65
C17H23I1N6	438.32	438.1031	46.58	5.29	28.95	19.17
C17H23I2N1	495.19	494.9924	41.23	4.68	51.26	2.83
C17H23I2N3	523.20	522.9985	39.03	4.43	48.51	8.03
C17H23I2N5	551.22	551.0047	37.04	4.21	46.05	12.71
C17H24I1N1	369.29	369.0955	55.29	6.55	34.36	3.79
C17H24I1N3	397.31	397.1017	51.39	6.09	31.94	10.58
C17H24I1N5	425.32	425.1078	48.01	5.69	29.84	16.47
C17H24I2N2	510.20	510.0033	40.02	4.74	49.75	5.49
C17H24I2N4	538.22	538.0094	37.94	4.49	47.16	10.41
C17H24I2N6	566.23	566.0156	36.06	4.27	44.82	14.84
C17H25I1N2	384.31	384.1064	53.13	6.56	33.02	7.29
C17H25I1N4	412.32	412.1126	49.52	6.11	30.78	13.59
C17H25I1N6	440.33	440.1187	46.37	5.72	28.82	19.09
C17H25I2N1	497.20	497.0080	41.07	5.07	51.05	2.82
C17H25I2N3	525.22	525.0142	38.88	4.80	48.32	8.00
C17H25I2N5	553.23	553.0203	36.91	4.55	45.88	12.66
C17H26I1N1	371.31	371.1112	54.99	7.06	34.18	3.77
C17H26I1N3	399.32	399.1173	51.13	6.56	31.78	10.52
C17H26I1N5	427.33	427.1235	47.78	6.13	29.70	16.39
C17H26I2N2	512.22	512.0189	39.86	5.12	49.55	5.47
C17H26I2N4	540.23	540.0251	37.80	4.85	46.98	10.37
C17H26I2N6	568.25	568.0312	35.93	4.61	44.67	14.79
C17H27I1N2	386.32	386.1221	52.85	7.04	32.85	7.25
C17H27I1N4	414.34	414.1282	49.28	6.57	30.63	13.52
C17H27I1N6	442.35	442.1344	46.16	6.15	28.69	19.00
C17H27I2N1	499.22	499.0237	40.90	5.45	50.84	2.81
C17H27I2N3	527.23	527.0298	38.73	5.16	48.14	7.97
C17H27I2N5	555.25	555.0360	36.77	4.90	45.71	12.61
C17H28I1N1	373.32	373.1268	54.70	7.56	33.99	3.75
C17H28I1N3	401.34	401.1330	50.88	7.03	31.62	10.47
C17H28I1N5	429.35	429.1391	47.56	6.57	29.56	16.31
C17H28I2N2	514.23	514.0346	39.71	5.49	49.36	5.45

C17H28I2N4	542.25	542.0407	37.66	5.20	46.81	10.33
C17H28I2N6	570.26	570.0469	35.81	4.95	44.51	14.74
C17H29I1N2	388.34	388.1377	52.58	7.53	32.68	7.21
C17H29I1N4	416.35	416.1439	49.04	7.02	30.48	13.46
C17H29I1N6	444.37	444.1500	45.95	6.58	28.56	18.91
C17H29I2N1	501.24	501.0393	40.74	5.83	50.64	2.79
C17H29I2N3	529.25	529.0455	38.58	5.52	47.96	7.94
C17H29I2N5	557.26	557.0516	36.64	5.25	45.55	12.57
C17H30I1N1	375.34	375.1425	54.40	8.06	33.81	3.73
C17H30I1N3	403.35	403.1486	50.62	7.50	31.46	10.42
C17H30I1N5	431.37	431.1548	47.34	7.01	29.42	16.24
C17H30I2N2	516.25	516.0502	39.55	5.86	49.16	5.43
C17H30I2N4	544.26	544.0564	37.52	5.56	46.63	10.29
C17H30I2N6	572.28	572.0625	35.68	5.28	44.35	14.69
C17H31I1N2	390.35	390.1534	52.31	8.00	32.51	7.18
C17H31I1N4	418.37	418.1595	48.81	7.47	30.33	13.39
C17H31I1N6	446.38	446.1657	45.74	7.00	28.43	18.83
C17H31I2N1	503.25	503.0550	40.57	6.21	50.43	2.78
C17H31I2N3	531.27	531.0611	38.43	5.88	47.77	7.91
C17H31I2N5	559.28	559.0673	36.51	5.59	45.38	12.52
C17H32I1N1	377.36	377.1581	54.11	8.55	33.63	3.71
C17H32I1N3	405.37	405.1643	50.37	7.96	31.31	10.37
C17H32I1N5	433.38	433.1704	47.12	7.44	29.28	16.16
C17H32I2N2	518.27	518.0659	39.40	6.22	48.97	5.41
C17H32I2N4	546.28	546.0720	37.38	5.90	46.46	10.26
C17H32I2N6	574.29	574.0782	35.55	5.62	44.19	14.63
C17H33I1N2	392.37	392.1690	52.04	8.48	32.34	7.14
C17H33I1N4	420.38	420.1752	48.57	7.91	30.19	13.33
C17H33I1N6	448.40	448.1813	45.54	7.42	28.30	18.74
C17H33I2N1	505.27	505.0706	40.41	6.58	50.23	2.77
C17H33I2N3	533.28	533.0768	38.29	6.24	47.59	7.88
C17H33I2N5	561.29	561.0829	36.38	5.93	45.22	12.48
C17H34I1N1	379.37	379.1738	53.82	9.03	33.45	3.69
C17H34I1N3	407.39	407.1799	50.12	8.41	31.15	10.31
C17H34I1N5	435.40	435.1861	46.90	7.87	29.15	16.08
C17H34I2N2	520.28	520.0815	39.25	6.59	48.78	5.38
C17H34I2N4	548.30	548.0877	37.24	6.25	46.29	10.22
C17H34I2N6	576.31	576.0938	35.43	5.95	44.04	14.58
C17H35I1N2	394.39	394.1847	51.77	8.95	32.18	7.10
C17H35I1N4	422.40	422.1908	48.34	8.35	30.04	13.26
C17H35I1N6	450.41	450.1970	45.33	7.83	28.18	18.66
C17H35I2N1	507.28	507.0863	40.25	6.95	50.03	2.76
C17H35I2N3	535.30	535.0924	38.15	6.59	47.41	7.85
C17H35I2N5	563.31	563.0985	36.25	6.26	45.06	12.43
C17H36I1N1	381.39	381.1894	53.54	9.51	33.27	3.67
C17H36I1N3	409.40	409.1956	49.88	8.86	31.00	10.26
C17H36I1N5	437.41	437.2017	46.68	8.30	29.01	16.01
C17H36I2N2	522.30	522.0972	39.09	6.95	48.59	5.36
C17H36I2N4	550.31	550.1033	37.10	6.59	46.12	10.18
C17H36I2N6	578.33	578.1094	35.31	6.27	43.89	14.53
C17H37I1N2	396.40	396.2003	51.51	9.41	32.01	7.07
C17H37I1N4	424.42	424.2065	48.11	8.79	29.90	13.20
C17H37I1N6	452.43	452.2126	45.13	8.24	28.05	18.58
C17H37I2N3	537.31	537.1081	38.00	6.94	47.24	7.82
C17H37I2N5	565.33	565.1142	36.12	6.60	44.90	12.39
C17H38I1N3	411.42	411.2112	49.63	9.31	30.85	10.21
C17H38I1N5	439.43	439.2174	46.47	8.72	28.88	15.94
C17H38I2N4	552.33	552.1189	36.97	6.93	45.95	10.14
C17H38I2N6	580.34	580.1251	35.18	6.60	43.73	14.48
C17H39I1N4	426.43	426.2221	47.88	9.22	29.76	13.14
C17H39I1N6	454.44	454.2283	44.93	8.65	27.93	18.49
C17H39I2N5	567.34	567.1298	35.99	6.93	44.74	12.34
C17H40I1N5	441.45	441.2330	46.25	9.13	28.75	15.86
C17H40I2N6	582.36	582.1407	35.06	6.92	43.58	14.43
C17H41I1N6	456.46	456.2439	44.73	9.05	27.80	18.41
C18H 9I1N2	380.19	379.9812	56.87	2.39	33.38	7.37
C18H 9I1N4	408.20	407.9874	52.96	2.22	31.09	13.73
C18H 9I1N6	436.22	435.9935	49.56	2.08	29.09	19.27
C18H 9I2N1	493.09	492.8828	43.85	1.84	51.47	2.84
C18H 9I2N3	521.10	520.8890	41.49	1.74	43.71	8.06
C18H 9I2N5	549.11	548.8951	39.37	1.65	45.22	12.75
C18H10I1N1	367.19	366.9860	58.88	2.75	34.56	3.81
C18H10I1N3	395.20	394.9921	54.71	2.55	32.11	10.63
C18H10I1N5	423.22	422.9983	51.08	2.38	29.99	16.55
C18H10I2N2	508.10	507.8937	42.55	1.98	49.95	5.51
C18H10I2N4	536.12	535.8999	40.33	1.88	47.34	10.45
C18H10I2N6	564.13	563.9060	38.32	1.79	44.99	14.90
C18H11I1N2	382.21	381.9969	56.57	2.90	33.20	7.33
C18H11I1N4	410.22	410.0030	52.70	2.70	30.94	13.66
C18H11I1N6	438.23	438.0092	49.33	2.53	28.96	19.18
C18H11I2N1	495.10	494.8985	43.67	2.24	51.26	2.83
C18H11I2N3	523.12	522.9046	41.33	2.12	48.52	8.03
C18H11I2N5	551.13	550.9108	39.23	2.01	46.05	12.71
C18H12I1N1	369.21	369.0016	58.56	3.28	34.37	3.79
C18H12I1N3	397.22	397.0078	54.43	3.05	31.95	10.58
C18H12I1N5	425.23	425.0139	50.84	2.84	29.84	16.47
C18H12I2N2	510.12	509.9094	42.38	2.37	49.75	5.49
C18H12I2N4	538.13	537.9155	40.18	2.25	47.16	10.41
C18H12I2N6	566.15	565.9217	38.19	2.14	44.83	14.84
C18H13I1N2	384.22	384.0125	56.27	3.41	33.03	7.29
C18H13I1N4	412.24	412.0187	52.45	3.18	30.78	13.59
C18H13I1N6	440.25	440.0248	49.11	2.98	28.83	19.09
C18H13I2N1	497.12	496.9141	43.49	2.64	51.06	2.82
C18H13I2N3	525.13	524.9203	41.17	2.50	48.33	8.00
C18H13I2N5	553.15	552.9264	39.09	2.37	45.88	12.66
C18H14I1N1	371.22	371.0173	58.24	3.80	34.19	3.77
C18H14I1N3	399.24	399.0234	54.15	3.53	31.79	10.53
C18H14I1N5	427.25	427.0296	50.60	3.30	29.70	16.39
C18H14I2N2	512.13	511.9250	42.22	2.76	49.56	5.47
C18H14I2N4	540.15	539.9312	40.03	2.61	46.99	10.37
C18H14I2N6	568.16	567.9373	38.05	2.48	44.67	14.79
C18H15I1N2	386.24	386.0282	55.98	3.91	32.86	7.25
C18H15I1N4	414.25	414.0343	52.19	3.65	30.63	13.52
C18H15I1N6	442.26	442.0405	48.88	3.42	28.69	19.00
C18H15I2N1	499.14	498.9298	43.32	3.03	50.85	2.81
C18H15I2N3	527.15	526.9359	41.01	2.87	48.15	7.97
C18H15I2N5	555.16	554.9421	38.94	2.72	45.72	12.61
C18H16I1N1	373.24	373.0329	57.93	4.32	34.00	3.75
C18H16I1N3	401.25	401.0391	53.88	4.02	31.63	10.47
C18H16I1N5	429.27	429.0452	50.37	3.76	29.56	16.31
C18H16I2N2	514.15	513.9407	42.05	3.14	49.36	5.45
C18H16I2N4	542.16	541.9468	39.88	2.97	46.81	10.33
C18H16I2N6	570.18	569.9530	37.92	2.83	44.51	14.74
C18H17I1N2	388.25	388.0438	55.69	4.41	32.69	7.22
C18H17I1N4	416.27	416.0500	51.94	4.12	30.49	13.46
C18H17I1N6	444.28	444.0561	48.66	3.86	28.56	18.92
C18H17I2N1	501.15	500.9454	43.14	3.42	50.65	2.79
C18H17I2N3	529.17	528.9516	40.86	3.24	47.96	7.94
C18H17I2N5	557.18	556.9577	38.80	3.08	45.55	12.57
C18H18I1N1	375.26	375.0486	57.61	4.83	33.82	3.73

026

C18H18I1N3	403.27	403.0547	53.61	4.50	31.47	10.42
C18H18I1N5	431.28	431.0609	50.13	4.21	29.42	16.24
C18H18I2N2	516.17	515.9563	41.89	3.52	49.17	5.43
C18H18I2N4	544.18	543.9625	39.73	3.33	46.64	10.30
C18H18I2N6	572.19	571.9686	37.78	3.17	44.36	14.69
C18H19I1N2	390.27	390.0595	55.40	4.91	32.52	7.18
C18H19I1N4	418.28	418.0656	51.69	4.58	30.34	13.39
C18H19I1N6	446.30	446.0718	48.44	4.29	28.43	18.83
C18H19I2N1	503.17	502.9611	42.97	3.81	50.44	2.78
C18H19I2N3	531.18	530.9672	40.70	3.61	47.78	7.91
C18H19I2N5	559.19	558.9734	38.66	3.42	45.39	12.52
C18H20I1N1	377.27	377.0642	57.31	5.34	33.64	3.71
C18H20I1N3	405.28	405.0704	53.35	4.97	31.31	10.37
C18H20I1N5	433.30	433.0765	49.90	4.65	29.29	16.16
C18H20I2N2	518.18	517.9720	41.72	3.89	48.98	5.41
C18H20I2N4	546.20	545.9781	39.58	3.69	46.47	10.26
C18H20I2N6	574.21	573.9843	37.65	3.51	44.20	14.64
C18H21I1N2	392.29	392.0751	55.11	5.40	32.35	7.14
C18H21I1N4	420.30	420.0813	51.44	5.04	30.19	13.33
C18H21I1N6	448.31	448.0874	48.23	4.72	28.31	18.75
C18H21I2N1	505.18	504.9767	42.80	4.19	50.24	2.77
C18H21I2N3	533.20	532.9829	40.55	3.97	47.60	7.88
C18H21I2N5	561.21	560.9890	38.52	3.77	45.23	12.48
C18H22I1N1	379.29	379.0799	57.00	5.85	33.46	3.69
C18H22I1N3	407.30	407.0860	53.08	5.44	31.16	10.32
C18H22I1N5	435.31	435.0922	49.67	5.09	29.15	16.09
C18H22I2N2	520.20	519.9876	41.56	4.26	48.79	5.39
C18H22I2N4	548.21	547.9938	39.44	4.05	46.30	10.22
C18H22I2N6	576.23	575.9999	37.52	3.85	44.05	14.58
C18H23I1N2	394.30	394.0908	54.83	5.88	32.18	7.10
C18H23I1N4	422.32	422.0969	51.19	5.49	30.05	13.27
C18H23I1N6	450.33	450.1031	48.01	5.15	28.18	18.66
C18H23I2N1	507.20	506.9924	42.63	4.57	50.04	2.76
C18H23I2N3	535.21	534.9985	40.40	4.33	47.42	7.85
C18H23I2N5	563.23	563.0047	38.39	4.12	45.06	12.43
C18H24I1N1	381.30	381.0955	56.70	6.34	33.28	3.67
C18H24I1N3	409.32	409.1017	52.82	5.91	31.00	10.27
C18H24I1N5	437.33	437.1078	49.44	5.53	29.02	16.01
C18H24I2N2	522.21	522.0033	41.40	4.63	48.60	5.36
C18H24I2N4	550.23	550.0094	39.29	4.40	46.13	10.18
C18H24I2N6	578.24	578.0156	37.39	4.18	43.89	14.53
C18H25I1N2	396.32	396.1064	54.55	6.36	32.02	7.07
C18H25I1N4	424.33	424.1126	50.95	5.94	29.91	13.20
C18H25I1N6	452.34	452.1187	47.80	5.57	28.05	18.58
C18H25I2N1	509.22	509.0080	42.46	4.95	49.84	2.75
C18H25I2N3	537.23	537.0142	40.24	4.69	47.24	7.82
C18H25I2N5	565.24	565.0203	38.25	4.46	44.90	12.39
C18H26I1N1	383.32	383.1112	56.40	6.84	33.11	3.65
C18H26I1N3	411.33	411.1173	52.56	6.37	30.85	10.22
C18H26I1N5	439.35	439.1235	49.21	5.97	28.88	15.94
C18H26I2N2	524.23	524.0189	41.24	5.00	48.42	5.34
C18H26I2N4	552.24	552.0251	39.15	4.75	45.96	10.15
C18H26I2N6	580.26	580.0312	37.26	4.52	43.74	14.48
C18H27I1N2	398.33	398.1221	54.28	6.83	31.86	7.03
C18H27I1N4	426.35	426.1282	50.71	6.38	29.77	13.14
C18H27I1N6	454.36	454.1344	47.58	5.99	27.93	18.50
C18H27I2N1	511.23	511.0237	42.29	5.32	49.65	2.74
C18H27I2N3	539.24	539.0298	40.09	5.05	47.07	7.79
C18H27I2N5	567.26	567.0360	38.11	4.80	44.74	12.35
C18H28I1N1	385.33	385.1268	56.11	7.32	32.93	3.63
C18H28I1N3	413.35	413.1330	52.30	6.83	30.70	10.17
C18H28I1N5	441.36	441.1391	48.98	6.39	28.75	15.87
C18H28I2N2	526.25	526.0346	41.08	5.36	48.23	5.32
C18H28I2N4	554.26	554.0407	39.01	5.09	45.79	10.11
C18H28I2N6	582.27	582.0469	37.13	4.85	43.59	14.43
C18H29I1N2	400.35	400.1377	54.00	7.30	31.70	7.00
C18H29I1N4	428.36	428.1439	50.47	6.82	29.63	13.08
C18H29I1N6	456.38	456.1500	47.37	6.41	27.81	18.41
C18H29I2N1	513.25	513.0393	42.12	5.70	49.45	2.73
C18H29I2N3	541.26	541.0455	39.94	5.40	46.89	7.76
C18H29I2N5	569.27	569.0516	37.98	5.13	44.58	12.30
C18H30I1N1	387.35	387.1425	55.82	7.81	32.76	3.62
C18H30I1N3	415.36	415.1486	52.05	7.28	30.55	10.12
C18H30I1N5	443.38	443.1548	48.76	6.82	28.62	15.80
C18H30I2N2	528.26	528.0502	40.93	5.72	48.05	5.30
C18H30I2N4	556.28	556.0564	38.87	5.44	45.63	10.07
C18H30I2N6	584.29	584.0625	37.00	5.18	43.44	14.38
C18H31I1N2	402.37	402.1534	53.73	7.77	31.54	6.96
C18H31I1N4	430.38	430.1595	50.23	7.26	29.49	13.02
C18H31I1N6	458.39	458.1657	47.16	6.82	27.68	18.33
C18H31I2N1	515.26	515.0550	41.96	6.06	49.26	2.72
C18H31I2N3	543.28	543.0611	39.80	5.75	46.72	7.73
C18H31I2N5	571.29	571.0673	37.84	5.47	44.43	12.26
C18H32I1N1	389.37	389.1581	55.53	8.28	32.59	3.60
C18H32I1N3	417.38	417.1643	51.80	7.73	30.40	10.07
C18H32I1N5	445.39	445.1704	48.54	7.24	28.49	15.72
C18H32I2N2	530.28	530.0659	40.77	6.08	47.86	5.28
C18H32I2N4	558.29	558.0720	38.73	5.78	45.46	10.04
C18H32I2N6	586.30	586.0782	36.88	5.50	43.29	14.33
C18H33I1N2	404.38	404.1690	53.46	8.23	31.38	6.93
C18H33I1N4	432.39	432.1752	50.00	7.69	29.35	12.96
C18H33I1N6	460.41	460.1813	46.96	7.22	27.56	18.25
C18H33I2N1	517.28	517.0706	41.80	6.43	49.07	2.71
C18H33I2N3	545.29	545.0768	39.65	6.10	46.55	7.71
C18H33I2N5	573.31	573.0829	37.71	5.80	44.27	12.22
C18H34I1N1	391.38	391.1738	55.24	8.76	32.42	3.58
C18H34I1N3	419.40	419.1799	51.55	8.17	30.26	10.02
C18H34I1N5	447.41	447.1861	48.32	7.66	28.36	15.65
C18H34I2N2	532.29	532.0815	40.62	6.44	47.68	5.26
C18H34I2N4	560.31	560.0877	38.59	6.12	45.30	10.00
C18H34I2N6	588.32	588.0938	36.75	5.83	43.14	14.28
C18H35I1N2	406.40	406.1847	53.20	8.68	31.23	6.89
C18H35I1N4	434.41	434.1908	49.77	8.12	29.21	12.90
C18H35I1N6	462.42	462.1970	46.75	7.63	27.44	18.17
C18H35I2N1	519.30	519.0863	41.63	6.79	48.88	2.70
C18H35I2N3	547.31	547.0924	39.50	6.45	46.37	7.68
C18H35I2N5	575.32	575.0985	37.58	6.13	44.12	12.17
C18H36I1N1	393.40	393.1894	54.96	9.22	32.26	3.56
C18H36I1N3	421.41	421.1956	51.30	8.61	30.11	9.97
C18H36I1N5	449.43	449.2017	48.11	8.07	28.24	15.58
C18H36I2N2	534.31	534.0972	40.46	6.79	47.50	5.24
C18H36I2N4	562.32	562.1033	38.45	6.45	45.14	9.96
C18H36I2N6	590.34	590.1094	36.62	6.15	42.99	14.24
C18H37I1N2	408.41	408.2003	52.94	9.13	31.07	6.86
C18H37I1N4	436.43	436.2065	49.54	8.55	29.08	12.84
C18H37I1N6	464.44	464.2126	46.55	8.03	27.32	18.09
C18H37I2N1	521.31	521.1019	41.47	7.15	48.69	2.69
C18H37I2N3	549.32	549.1081	39.36	6.79	46.20	7.65
C18H37I2N5	577.34	577.1142	37.45	6.46	43.96	12.13
C18H38I1N1	395.41	395.2051	54.68	9.69	32.09	3.54

027

C18H38I1N3	423.43	423.2112	51.06	9.05	29.97	9.92
C18H38I1N5	451.44	451.2174	47.89	8.48	28.11	15.51
C18H38I2N2	536.33	536.1128	40.31	7.14	47.32	5.22
C18H38I2N4	564.34	564.1189	38.31	6.79	44.97	9.93
C18H38I2N6	592.35	592.1251	36.50	6.47	42.85	14.19
C18H39I1N2	410.43	410.2160	52.68	9.58	30.92	6.83
C18H39I1N4	438.44	438.2221	49.31	8.97	28.94	12.78
C18H39I1N6	466.46	466.2283	46.35	8.43	27.21	18.02
C18H39I2N3	551.34	551.1237	39.21	7.13	46.03	7.62
C18H39I2N5	579.35	579.1298	37.32	6.79	43.81	12.09
C18H40I1N3	425.44	425.2269	50.82	9.48	29.83	9.88
C18H40I1N5	453.46	453.2330	47.68	8.89	27.99	15.44
C18H40I2N4	566.36	566.1346	38.17	7.12	44.81	9.89
C18H40I2N6	594.37	594.1407	36.37	6.78	42.70	14.14
C18H41I1N4	440.46	440.2378	49.09	9.38	28.81	12.72
C18H41I1N6	468.47	468.2439	46.15	8.82	27.09	17.94
C18H41I2N5	581.37	581.1455	37.19	7.11	43.66	12.05
C18H42I1N5	455.47	455.2487	47.47	9.29	27.86	15.38
C18H42I2N6	596.38	596.1564	36.25	7.10	42.56	14.09
C18H43I1N6	470.49	470.2596	45.95	9.21	26.97	17.86
C19H 9I1N2	392.20	391.9812	58.19	2.31	32.36	7.14
C19H 9I1N4	420.21	419.9874	54.31	2.16	30.20	13.33
C19H 9I1N6	448.23	447.9935	50.91	2.02	28.31	18.75
C19H 9I2N1	505.10	504.8828	45.18	1.80	50.25	2.77
C19H 9I2N3	533.11	532.8890	42.81	1.70	47.61	7.88
C19H 9I2N5	561.13	560.8951	40.67	1.62	45.23	12.48
C19H10I1N1	379.20	378.9860	60.18	2.66	33.47	3.69
C19H10I1N3	407.22	406.9921	56.04	2.48	31.16	10.32
C19H10I1N5	435.23	434.9983	52.43	2.32	29.16	16.09
C19H10I2N2	520.11	519.8937	43.88	1.94	48.80	5.39
C19H10I2N4	548.13	547.8999	41.63	1.84	46.30	10.22
C19H10I2N6	576.14	575.9060	39.61	1.75	44.05	14.59
C19H11I1N2	394.22	393.9969	57.89	2.81	32.19	7.11
C19H11I1N4	422.23	422.0030	54.05	2.63	30.06	13.27
C19H11I1N6	450.24	450.0092	50.69	2.46	28.19	18.67
C19H11I2N1	507.12	506.8985	45.00	2.19	50.05	2.76
C19H11I2N3	535.13	534.9046	42.65	2.07	47.43	7.85
C19H11I2N5	563.14	562.9108	40.52	1.97	45.07	12.44
C19H12I1N1	381.22	381.0016	59.86	3.17	33.29	3.67
C19H12I1N3	409.23	409.0078	55.77	2.96	31.01	10.27
C19H12I1N5	437.25	437.0139	52.19	2.77	29.02	16.02
C19H12I2N2	522.13	521.9094	43.71	2.32	48.61	5.37
C19H12I2N4	550.14	549.9155	41.48	2.20	46.14	10.18
C19H12I2N6	578.16	577.9217	39.47	2.09	43.90	14.54
C19H13I1N2	396.23	396.0125	57.60	3.31	32.03	7.07
C19H13I1N4	424.25	424.0187	53.79	3.09	29.91	13.21
C19H13I1N6	452.26	452.0248	50.46	2.90	28.06	18.58
C19H13I2N1	509.13	508.9141	44.82	2.57	49.85	2.75
C19H13I2N3	537.14	536.9203	42.49	2.44	47.25	7.82
C19H13I2N5	565.16	564.9264	40.38	2.32	44.91	12.39
C19H14I1N1	383.23	383.0173	59.55	3.68	33.11	3.65
C19H14I1N3	411.25	411.0234	55.49	3.43	30.86	10.22
C19H14I1N5	439.26	439.0296	51.95	3.21	28.89	15.94
C19H14I2N2	524.15	523.9250	43.54	2.69	48.42	5.34
C19H14I2N4	552.16	551.9312	41.33	2.56	45.97	10.15
C19H14I2N6	580.17	579.9373	39.34	2.43	43.75	14.49
C19H15I1N2	398.25	398.0282	57.30	3.80	31.87	7.03
C19H15I1N4	426.26	426.0343	53.54	3.55	29.77	13.14
C19H15I1N6	454.28	454.0405	50.24	3.33	27.94	18.50
C19H15I2N1	511.15	510.9298	44.65	2.96	49.65	2.74
C19H15I2N3	539.16	538.9359	42.33	2.80	47.07	7.79
C19H15I2N5	567.17	566.9421	40.24	2.67	44.75	12.35
C19H16I1N1	385.25	385.0329	59.24	4.19	32.94	3.64
C19H16I1N3	413.26	413.0391	55.22	3.90	30.71	10.17
C19H16I1N5	441.28	441.0452	51.72	3.65	28.76	15.87
C19H16I2N2	526.16	525.9407	43.37	3.07	48.24	5.32
C19H16I2N4	554.17	553.9468	41.18	2.91	45.80	10.11
C19H16I2N6	582.19	581.9530	39.20	2.77	43.60	14.44
C19H17I1N2	400.27	400.0438	57.02	4.28	31.71	7.00
C19H17I1N4	428.28	428.0500	53.29	4.00	29.63	13.08
C19H17I1N6	456.29	456.0561	50.01	3.76	27.81	18.42
C19H17I2N1	513.16	512.9454	44.47	3.34	49.46	2.73
C19H17I2N3	541.18	540.9516	42.17	3.17	46.90	7.76
C19H17I2N5	569.19	568.9577	40.09	3.01	44.59	12.30
C19H18I1N1	387.27	387.0486	58.93	4.69	32.77	3.62
C19H18I1N3	415.28	415.0547	54.95	4.37	30.56	10.12
C19H18I1N5	443.29	443.0609	51.48	4.09	28.63	15.80
C19H18I2N2	528.18	527.9563	43.21	3.44	48.05	5.30
C19H18I2N4	556.19	555.9625	41.03	3.26	45.63	10.07
C19H18I2N6	584.20	583.9686	39.06	3.11	43.45	14.39
C19H19I1N2	402.28	402.0595	56.73	4.76	31.55	6.96
C19H19I1N4	430.29	430.0656	53.04	4.45	29.49	13.02
C19H19I1N6	458.31	458.0718	49.79	4.18	27.69	18.34
C19H19I2N1	515.18	514.9611	44.30	3.72	49.27	2.72
C19H19I2N3	543.19	542.9672	42.01	3.53	46.73	7.74
C19H19I2N5	571.21	570.9734	39.95	3.35	44.43	12.26
C19H20I1N1	389.28	389.0642	58.62	5.18	32.60	3.60
C19H20I1N3	417.30	417.0704	54.69	4.83	30.41	10.07
C19H20I1N5	445.31	445.0765	51.25	4.53	28.50	15.73
C19H20I2N2	530.19	529.9720	43.04	3.80	47.87	5.28
C19H20I2N4	558.21	557.9781	40.88	3.61	45.47	10.04
C19H20I2N6	586.22	585.9843	38.93	3.44	43.30	14.34
C19H21I1N2	404.30	404.0751	56.45	5.24	31.39	6.93
C19H21I1N4	432.31	432.0813	52.79	4.90	29.35	12.96
C19H21I1N6	460.32	460.0874	49.58	4.60	27.57	18.26
C19H21I2N1	517.19	516.9767	44.12	4.09	49.07	2.71
C19H21I2N3	545.21	544.9829	41.86	3.88	46.55	7.71
C19H21I2N5	573.22	572.9890	39.81	3.69	44.28	12.22
C19H22I1N1	391.30	391.0799	58.32	5.67	32.43	3.58
C19H22I1N3	419.31	419.0860	54.43	5.29	30.26	10.02
C19H22I1N5	447.33	447.0922	51.02	4.96	28.37	15.66
C19H22I2N2	532.21	531.9876	42.88	4.17	47.69	5.26
C19H22I2N4	560.22	559.9938	40.74	3.96	45.30	10.00
C19H22I2N6	588.24	587.9999	38.80	3.77	43.15	14.29
C19H23I1N2	406.31	406.0908	56.17	5.71	31.23	6.89
C19H23I1N4	434.33	434.0969	52.54	5.34	29.22	12.90
C19H23I1N6	462.34	462.1031	49.36	5.01	27.45	18.18
C19H23I2N1	519.21	518.9924	43.95	4.47	48.88	2.70
C19H23I2N3	547.22	546.9985	41.70	4.24	46.38	7.68
C19H23I2N5	575.24	575.0047	39.67	4.03	44.12	12.17
C19H24I1N1	393.31	393.0955	58.02	6.15	32.27	3.56
C19H24I1N3	421.33	421.1017	54.16	5.74	30.12	9.97
C19H24I1N5	449.34	449.1078	50.79	5.38	28.24	15.59
C19H24I2N2	534.23	534.0033	42.72	4.53	47.51	5.24
C19H24I2N4	562.24	562.0094	40.59	4.30	45.14	9.96
C19H24I2N6	590.25	590.0156	38.66	4.10	43.00	14.24
C19H25I1N2	408.33	408.1064	55.89	6.17	31.08	6.86
C19H25I1N4	436.34	436.1126	52.30	5.78	29.08	12.84
C19H25I1N6	464.36	464.1187	49.15	5.43	27.33	18.10
C19H25I2N1	521.23	521.0080	43.78	4.83	48.69	2.69

028

C19H25I2N3	549.24	549.0142	41.55	4.59	46.21	7.65
C19H25I2N5	577.25	577.0203	39.53	4.37	43.97	12.13
C19H26I1N1	395.33	395.1112	57.73	6.63	32.10	3.54
C19H26I1N3	423.34	423.1173	53.91	6.19	29.98	9.93
C19H26I1N5	451.36	451.1235	50.56	5.81	28.12	15.52
C19H26I2N2	536.24	536.0189	42.56	4.89	47.33	5.22
C19H26I2N4	564.25	564.0251	40.44	4.64	44.98	9.93
C19H26I2N6	592.27	592.0312	38.53	4.42	42.85	14.19
C19H27I1N2	410.34	410.1221	55.61	6.63	30.93	6.83
C19H27I1N4	438.36	438.1282	52.06	6.21	28.95	12.78
C19H27I1N6	466.37	466.1344	48.93	5.84	27.21	18.02
C19H27I2N1	523.24	523.0237	43.61	5.20	48.51	2.68
C19H27I2N3	551.26	551.0298	41.40	4.94	46.04	7.62
C19H27I2N5	579.27	579.0360	39.40	4.70	43.82	12.09
C19H28I1N1	397.35	397.1268	57.43	7.10	31.94	3.53
C19H28I1N3	425.36	425.1330	53.65	6.64	29.83	9.88
C19H28I1N5	453.37	453.1391	50.34	6.23	27.99	15.45
C19H28I2N2	538.26	538.0346	42.40	5.24	47.15	5.20
C19H28I2N4	566.27	566.0407	40.30	4.98	44.82	9.89
C19H28I2N6	594.28	594.0469	38.40	4.75	42.71	14.14
C19H29I1N2	412.36	412.1377	55.34	7.09	30.78	6.79
C19H29I1N4	440.37	440.1439	51.82	6.64	28.82	12.72
C19H29I1N6	468.39	468.1500	48.72	6.24	27.09	17.94
C19H29I2N1	525.26	525.0393	43.45	5.57	48.32	2.67
C19H29I2N3	553.27	553.0455	41.25	5.28	45.87	7.59
C19H29I2N5	581.29	581.0516	39.26	5.03	43.66	12.05
C19H30I1N1	399.36	399.1425	57.14	7.57	31.78	3.51
C19H30I1N3	427.38	427.1486	53.40	7.08	29.69	9.83
C19H30I1N5	455.39	455.1548	50.11	6.64	27.87	15.38
C19H30I2N2	540.27	540.0502	42.24	5.60	46.98	5.19
C19H30I2N4	568.29	568.0564	40.16	5.32	44.66	9.86
C19H30I2N6	596.30	596.0625	38.27	5.07	42.56	14.09
C19H31I1N2	414.38	414.1534	55.07	7.54	30.63	6.76
C19H31I1N4	442.39	442.1595	51.59	7.06	28.69	12.66
C19H31I1N6	470.40	470.1657	48.51	6.64	26.98	17.87
C19H31I2N1	527.27	527.0550	43.28	5.93	48.14	2.66
C19H31I2N3	555.29	555.0611	41.10	5.63	45.71	7.57
C19H31I2N5	583.30	583.0673	39.12	5.36	43.51	12.01
C19H32I1N1	401.38	401.1581	56.86	8.04	31.62	3.49
C19H32I1N3	429.39	429.1643	53.15	7.51	29.55	9.79
C19H32I1N5	457.40	457.1704	49.89	7.05	27.74	15.31
C19H32I2N2	542.29	542.0659	42.08	5.95	46.80	5.17
C19H32I2N4	570.30	570.0720	40.02	5.66	44.50	9.82
C19H32I2N6	598.32	598.0781	38.14	5.39	42.42	14.05
C19H33I1N2	416.39	416.1690	54.81	7.99	30.48	6.73
C19H33I1N4	444.41	444.1752	51.35	7.48	28.56	12.61
C19H33I1N6	472.42	472.1813	48.31	7.04	26.86	17.79
C19H33I2N1	529.29	529.0706	43.12	6.28	47.95	2.65
C19H33I2N3	557.30	557.0768	40.95	5.97	45.54	7.54
C19H33I2N5	585.32	585.0829	38.99	5.68	43.36	11.97
C19H34I1N1	403.39	403.1738	56.57	8.50	31.46	3.47
C19H34I1N3	431.41	431.1799	52.90	7.94	29.42	9.74
C19H34I1N5	459.42	459.1861	49.67	7.46	27.62	15.24
C19H34I2N2	544.31	544.0815	41.93	6.30	46.63	5.15
C19H34I2N4	572.32	572.0877	39.87	5.99	44.35	9.79
C19H34I2N6	600.33	600.0938	38.01	5.71	42.28	14.00
C19H35I1N2	418.41	418.1847	54.54	8.43	30.33	6.70
C19H35I1N4	446.42	446.1908	51.12	7.90	28.43	12.55
C19H35I1N6	474.44	474.1970	48.10	7.44	26.75	17.71
C19H35I2N1	531.31	531.0863	42.95	6.64	47.77	2.64
C19H35I2N3	559.32	559.0924	40.80	6.31	45.38	7.51
C19H35I2N5	587.33	587.0985	38.86	6.01	43.21	11.92
C19H36I1N1	405.41	405.1894	56.29	8.95	31.30	3.45
C19H36I1N3	433.42	433.1956	52.65	8.37	29.28	9.69
C19H36I1N5	461.44	461.2017	49.46	7.86	27.50	15.18
C19H36I2N2	546.32	546.0972	41.77	6.64	46.46	5.13
C19H36I2N4	574.33	574.1033	39.74	6.32	44.19	9.76
C19H36I2N6	602.35	602.1094	37.89	6.02	42.14	13.95
C19H37I1N2	420.42	420.2003	54.28	8.87	30.18	6.66
C19H37I1N4	448.44	448.2065	50.89	8.32	28.30	12.49
C19H37I1N6	476.45	476.2126	47.90	7.83	26.64	17.64
C19H37I2N1	533.32	533.1019	42.79	6.99	47.59	2.63
C19H37I2N3	561.34	561.1081	40.66	6.64	45.22	7.49
C19H37I2N5	589.35	589.1142	38.72	6.33	43.07	11.88
C19H38I1N1	407.43	407.2051	56.01	9.40	31.15	3.44
C19H38I1N3	435.45	435.2112	52.41	8.80	29.14	9.65
C19H38I1N5	463.45	463.2174	49.24	8.26	27.38	15.11
C19H38I2N2	548.34	548.1128	41.62	6.99	46.29	5.11
C19H38I2N4	576.35	576.1189	39.60	6.65	44.04	9.72
C19H38I2N6	604.36	604.1251	37.76	6.34	42.00	13.91
C19H39I1N2	422.44	422.2160	54.02	9.31	30.04	6.63
C19H39I1N4	450.45	450.2221	50.66	8.73	28.17	12.44
C19H39I1N6	478.47	478.2283	47.70	8.22	26.52	17.56
C19H39I2N1	535.34	535.1176	42.63	7.34	47.41	2.62
C19H39I2N3	563.35	563.1237	40.51	6.98	45.05	7.46
C19H39I2N5	591.36	591.1298	38.59	6.65	42.92	11.84
C19H40I1N1	409.44	409.2207	55.74	9.85	30.99	3.42
C19H40I1N3	437.46	437.2269	52.17	9.22	29.01	9.61
C19H40I1N5	465.47	465.2330	49.03	8.66	27.26	15.05
C19H40I2N2	550.35	550.1285	41.47	7.33	46.12	5.09
C19H40I2N4	578.37	578.1346	39.46	6.97	43.88	9.69
C19H40I2N6	606.38	606.1407	37.64	6.65	41.86	13.86
C19H41I1N2	424.46	424.2316	53.77	9.74	29.90	6.60
C19H41I1N4	452.47	452.2378	50.44	9.13	28.05	12.38
C19H41I1N6	480.48	480.2439	47.50	8.60	26.41	17.49
C19H41I2N3	565.37	565.1393	40.37	7.31	44.89	7.43
C19H41I2N5	593.38	593.1455	38.46	6.96	42.77	11.80
C19H42I1N3	439.47	439.2425	51.93	9.63	28.88	9.56
C19H42I1N5	467.48	467.2487	48.82	9.06	27.15	14.98
C19H42I2N4	580.38	580.1502	39.32	7.29	43.73	9.65
C19H42I2N6	608.40	608.1564	37.51	6.96	41.72	13.81
C19H43I1N4	454.49	454.2534	50.21	9.54	27.92	12.33
C19H43I1N6	482.50	482.2596	47.30	8.98	26.30	17.42
C19H43I2N5	595.40	595.1611	38.33	7.28	42.63	11.76
C19H44I1N5	469.50	469.2643	48.61	9.45	27.03	14.92
C19H44I2N6	610.41	610.1720	37.39	7.27	41.58	13.77
C19H45I1N6	484.52	484.2752	47.10	9.36	26.19	17.35
C20H10I1N1	391.21	390.9860	61.40	2.58	32.44	3.58
C20H10I1N3	419.23	418.9921	57.30	2.40	30.27	10.02
C20H10I1N5	447.24	446.9983	53.71	2.25	28.37	15.66
C20H10I2N2	532.12	531.8937	45.14	1.89	47.70	5.26
C20H10I2N4	560.14	559.8999	42.89	1.80	45.31	10.00
C20H10I2N6	588.15	587.9060	40.84	1.71	43.15	14.29
C20H11I1N2	406.23	405.9969	59.13	2.73	31.24	6.90
C20H11I1N4	434.24	434.0030	55.35	2.55	29.22	12.90
C20H11I1N6	462.26	462.0092	51.97	2.40	27.45	18.18
C20H11I2N1	519.13	518.8985	46.27	2.14	48.89	2.70
C20H11I2N3	547.14	546.9046	43.91	2.03	46.39	7.68
C20H11I2N5	575.15	574.9108	41.77	1.93	44.13	12.18
C20H12I1N1	393.23	393.0016	61.09	3.08	32.27	3.56

C20H12I1N3	421.24	421.0078	57.03	2.87	30.13	9.98
C20H12I1N5	449.26	449.0139	53.47	2.69	28.25	15.59
C20H12I2N2	534.14	533.9094	44.97	2.26	47.52	5.24
C20H12I2N4	562.15	561.9155	42.73	2.15	45.15	9.97
C20H12I2N6	590.17	589.9217	40.70	2.05	43.01	14.24
C20H13I1N2	408.24	408.0125	58.84	3.21	31.09	6.86
C20H13I1N4	436.26	436.0187	55.06	3.00	29.09	12.84
C20H13I1N6	464.27	464.0248	51.74	2.82	27.33	18.10
C20H13I2N1	521.14	520.9141	46.10	2.51	48.70	2.69
C20H13I2N3	549.16	548.9203	43.74	2.39	46.22	7.65
C20H13I2N5	577.17	576.9264	41.62	2.27	43.97	12.13
C20H14I1N1	395.25	395.0173	60.78	3.57	32.11	3.54
C20H14I1N3	423.26	423.0234	56.76	3.33	29.98	9.93
C20H14I1N5	451.27	451.0296	53.23	3.13	28.12	15.52
C20H14I2N2	536.16	535.9250	44.80	2.63	47.34	5.22
C20H14I2N4	564.17	563.9312	42.58	2.50	44.99	9.93
C20H14I2N6	592.18	591.9373	40.57	2.38	42.86	14.19
C20H15I1N2	410.26	410.0282	58.55	3.69	30.93	6.83
C20H15I1N4	438.27	438.0343	54.81	3.45	28.96	12.78
C20H15I1N6	466.29	466.0405	51.52	3.24	27.22	18.02
C20H15I2N1	523.16	522.9298	45.92	2.89	48.51	2.68
C20H15I2N3	551.17	550.9359	43.58	2.74	46.05	7.62
C20H15I2N5	579.18	578.9421	41.48	2.61	43.82	12.09
C20H16I1N1	397.26	397.0329	60.47	4.06	31.94	3.53
C20H16I1N3	425.28	425.0391	56.49	3.79	29.84	9.88
C20H16I1N5	453.29	453.0452	53.00	3.56	28.00	15.45
C20H16I2N2	538.17	537.9407	44.64	3.00	47.16	5.21
C20H16I2N4	566.19	565.9468	42.43	2.85	44.83	9.90
C20H16I2N6	594.20	593.9530	40.43	2.71	42.71	14.14
C20H17I1N2	412.28	412.0438	58.27	4.16	30.78	6.79
C20H17I1N4	440.29	440.0500	54.56	3.89	28.82	12.72
C20H17I1N6	468.30	468.0561	51.30	3.66	27.10	17.95
C20H17I2N1	525.17	524.9454	45.74	3.26	48.33	2.67
C20H17I2N3	553.19	552.9516	43.43	3.10	45.88	7.60
C20H17I2N5	581.20	580.9577	41.33	2.95	43.67	12.05
C20H18I1N1	399.28	399.0486	60.16	4.54	31.78	3.51
C20H18I1N3	427.29	427.0547	56.22	4.25	29.70	9.83
C20H18I1N5	455.30	455.0609	52.76	3.98	27.87	15.38
C20H18I2N2	540.19	539.9563	44.47	3.36	46.99	5.19
C20H18I2N4	568.20	567.9625	42.28	3.19	44.67	9.86
C20H18I2N6	596.22	595.9686	40.29	3.04	42.57	14.10
C20H19I1N2	414.29	414.0595	57.98	4.62	30.63	6.76
C20H19I1N4	442.31	442.0656	54.31	4.33	28.69	12.67
C20H19I1N6	470.32	470.0718	51.08	4.07	26.98	17.87
C20H19I2N1	527.19	526.9611	45.57	3.63	48.14	2.66
C20H19I2N3	555.20	554.9672	43.27	3.45	45.71	7.57
C20H19I2N5	583.22	582.9734	41.19	3.28	43.52	12.01
C20H20I1N1	401.29	401.0642	59.86	5.02	31.62	3.49
C20H20I1N3	429.31	429.0704	55.96	4.70	29.56	9.79
C20H20I1N5	457.32	457.0765	52.53	4.41	27.75	15.31
C20H20I2N2	542.20	541.9720	44.30	3.72	46.81	5.17
C20H20I2N4	570.22	569.9781	42.13	3.54	44.51	9.83
C20H20I2N6	598.23	597.9843	40.16	3.37	42.43	14.05
C20H21I1N2	416.31	416.0751	57.70	5.08	30.48	6.73
C20H21I1N4	444.32	444.0813	54.07	4.76	28.56	12.61
C20H21I1N6	472.33	472.0874	50.86	4.48	26.87	17.79
C20H21I2N1	529.21	528.9767	45.39	4.00	47.96	2.65
C20H21I2N3	557.22	556.9829	43.11	3.80	45.55	7.54
C20H21I2N5	585.23	584.9890	41.05	3.62	43.37	11.97
C20H22I1N1	403.31	403.0799	59.56	5.50	31.47	3.47
C20H22I1N3	431.32	431.0860	55.69	5.14	29.42	9.74
C20H22I1N5	459.34	459.0922	52.30	4.83	27.63	15.25
C20H22I2N2	544.22	543.9876	44.14	4.07	46.64	5.15
C20H22I2N4	572.23	571.9938	41.98	3.88	44.35	9.79
C20H22I2N6	600.25	599.9999	40.02	3.69	42.28	14.00
C20H23I1N2	418.32	418.0908	57.43	5.54	30.34	6.70
C20H23I1N4	446.34	446.0969	53.82	5.19	28.43	12.55
C20H23I1N6	474.35	474.1031	50.64	4.89	26.75	17.72
C20H23I2N1	531.22	530.9924	45.22	4.36	47.78	2.64
C20H23I2N3	559.24	558.9985	42.96	4.15	45.38	7.51
C20H23I2N5	587.25	587.0047	40.91	3.95	43.22	11.93
C20H24I1N1	405.33	405.0955	59.27	5.97	31.31	3.46
C20H24I1N3	433.34	433.1017	55.44	5.58	29.29	9.70
C20H24I1N5	461.35	461.1078	52.07	5.24	27.51	15.18
C20H24I2N2	546.24	546.0033	43.98	4.43	46.47	5.13
C20H24I2N4	574.25	574.0094	41.83	4.21	44.20	9.76
C20H24I2N6	602.26	602.0156	39.89	4.02	42.14	13.95
C20H25I1N2	420.34	420.1064	57.15	5.99	30.19	6.66
C20H25I1N4	448.35	448.1126	53.58	5.62	28.30	12.50
C20H25I1N6	476.37	476.1187	50.43	5.29	26.64	17.64
C20H25I2N1	533.24	533.0080	45.05	4.73	47.60	2.63
C20H25I2N3	561.25	561.0142	42.80	4.49	45.22	7.49
C20H25I2N5	589.26	589.0203	40.77	4.28	43.07	11.88
C20H26I1N1	407.34	407.1112	58.97	6.43	31.15	3.44
C20H26I1N3	435.35	435.1173	55.18	6.02	29.15	9.65
C20H26I1N5	463.37	463.1235	51.84	5.66	27.39	15.11
C20H26I2N2	548.25	548.0189	43.82	4.78	46.29	5.11
C20H26I2N4	576.27	576.0251	41.69	4.55	44.04	9.72
C20H26I2N6	604.28	604.0312	39.75	4.34	42.00	13.91
C20H27I1N2	422.36	422.1221	56.88	6.44	30.05	6.63
C20H27I1N4	450.37	450.1282	53.34	6.04	28.18	12.44
C20H27I1N6	478.38	478.1344	50.22	5.69	26.53	17.57
C20H27I2N1	535.25	535.0237	44.88	5.08	47.42	2.62
C20H27I2N3	563.27	563.0298	42.65	4.83	45.06	7.46
C20H27I2N5	591.28	591.0360	40.63	4.60	42.93	11.84
C20H28I1N1	409.36	409.1268	58.68	6.89	31.00	3.42
C20H28I1N3	437.37	437.1330	54.92	6.45	29.02	9.61
C20H28I1N5	465.38	465.1391	51.62	6.06	27.27	15.05
C20H28I2N2	550.27	550.0346	43.66	5.13	46.12	5.09
C20H28I2N4	578.28	578.0407	41.54	4.88	43.89	9.69
C20H28I2N6	606.30	606.0469	39.62	4.66	41.86	13.86
C20H29I1N2	424.37	424.1377	56.61	6.89	29.90	6.60
C20H29I1N4	452.39	452.1439	53.10	6.46	28.05	12.38
C20H29I1N6	480.40	480.1500	50.00	6.08	26.42	17.49
C20H29I2N1	537.27	537.0393	44.71	5.44	47.24	2.61
C20H29I2N3	565.28	565.0455	42.50	5.17	44.90	7.43
C20H29I2N5	593.30	593.0516	40.49	4.93	42.78	11.80
C20H30I1N1	411.37	411.1425	58.40	7.35	30.85	3.40
C20H30I1N3	439.39	439.1486	54.67	6.88	28.88	9.56
C20H30I1N5	467.40	467.1548	51.40	6.47	27.15	14.98
C20H30I2N2	552.28	552.0502	43.50	5.48	45.96	5.07
C20H30I2N4	580.30	580.0564	41.40	5.21	43.74	9.65
C20H30I2N6	608.31	608.0625	39.49	4.97	41.72	13.82
C20H31I1N2	426.39	426.1534	56.34	7.33	29.76	6.57
C20H31I1N4	454.40	454.1595	52.87	6.88	27.93	12.33
C20H31I1N6	482.41	482.1657	49.80	6.48	26.31	17.42
C20H31I2N1	539.29	539.0550	44.54	5.79	47.06	2.60
C20H31I2N3	567.30	567.0611	42.35	5.51	44.74	7.41
C20H31I2N5	595.31	595.0673	40.35	5.25	42.63	11.76
C20H32I1N1	413.39	413.1581	58.11	7.80	30.70	3.39

030

C20H32I1N3	441.40	441.1643	54.42	7.31	28.75	9.52
C20H32I1N5	469.42	469.1704	51.17	6.87	27.03	14.92
C20H32I2N2	554.30	554.0659	43.34	5.82	45.79	5.05
C20H32I2N4	582.31	582.0720	41.25	5.54	43.59	9.62
C20H32I2N6	610.33	610.0781	39.36	5.28	41.59	13.77
C20H33I1N2	428.40	428.1690	56.07	7.76	29.62	6.54
C20H33I1N4	456.42	456.1752	52.63	7.29	27.80	12.28
C20H33I1N6	484.43	484.1813	49.59	6.87	26.20	17.35
C20H33I2N1	541.30	541.0706	44.38	6.15	46.89	2.59
C20H33I2N3	569.31	569.0768	42.20	5.84	44.58	7.38
C20H33I2N5	597.33	597.0829	40.22	5.57	42.49	11.72
C20H34I1N1	415.41	415.1738	57.83	8.25	30.55	3.37
C20H34I1N3	443.42	443.1799	54.18	7.73	28.62	9.48
C20H34I1N5	471.43	471.1861	50.96	7.27	26.92	14.86
C20H34I2N2	556.32	556.0815	43.18	6.16	45.62	5.04
C20H34I2N4	584.33	584.0877	41.11	5.87	43.44	9.59
C20H34I2N6	612.34	612.0938	39.23	5.60	41.45	13.72
C20H35I1N2	430.42	430.1847	55.81	8.20	29.48	6.51
C20H35I1N4	458.43	458.1908	52.40	7.70	27.68	12.22
C20H35I1N6	486.45	486.1970	49.38	7.25	26.09	17.28
C20H35I2N1	543.32	543.0863	44.21	6.49	46.71	2.58
C20H35I2N3	571.33	571.0924	42.05	6.17	44.42	7.35
C20H35I2N5	599.34	599.0985	40.08	5.89	42.35	11.69
C20H36I1N1	417.42	417.1894	57.55	8.69	30.40	3.36
C20H36I1N3	445.43	445.1956	53.93	8.15	28.49	9.43
C20H36I1N5	473.45	473.2017	50.74	7.66	26.80	14.79
C20H36I2N2	558.33	558.0972	43.03	6.50	45.46	5.02
C20H36I2N4	586.35	586.1033	40.97	6.19	43.29	9.56
C20H36I2N6	614.36	614.1094	39.10	5.91	41.31	13.68
C20H37I1N2	432.44	432.2003	55.55	8.62	29.35	6.48
C20H37I1N4	460.45	460.2065	52.17	8.10	27.56	12.17
C20H37I1N6	488.46	488.2126	49.18	7.64	25.98	17.21
C20H37I2N1	545.33	545.1019	44.05	6.84	46.54	2.57
C20H37I2N3	573.35	573.1081	41.90	6.50	44.27	7.33
C20H37I2N5	601.36	601.1142	39.95	6.20	42.21	11.65
C20H38I1N1	419.44	419.2051	57.27	9.13	30.26	3.34
C20H38I1N3	447.45	447.2112	53.69	8.56	28.36	9.39
C20H38I1N5	475.46	475.2174	50.52	8.06	26.69	14.73
C20H38I2N2	560.35	560.1128	42.87	6.84	45.29	5.00
C20H38I2N4	588.36	588.1189	40.83	6.51	43.14	9.52
C20H38I2N6	616.37	616.1251	38.97	6.21	41.18	13.63
C20H39I1N2	434.45	434.2160	55.29	9.05	29.21	6.45
C20H39I1N4	462.47	462.2221	51.94	8.50	27.44	12.11
C20H39I1N6	490.48	490.2283	48.98	8.01	25.87	17.13
C20H39I2N1	547.35	547.1176	43.89	7.18	46.37	2.56
C20H39I2N3	575.36	575.1237	41.75	6.83	44.11	7.30
C20H39I2N5	603.38	603.1298	39.81	6.52	42.06	11.61
C20H40I1N1	421.45	421.2207	57.00	9.57	30.11	3.32
C20H40I1N3	449.47	449.2269	53.45	8.97	28.23	9.35
C20H40I1N5	477.48	477.2330	50.31	8.44	26.58	14.67
C20H40I2N2	562.36	562.1285	42.72	7.17	45.13	4.98
C20H40I2N4	590.38	590.1346	40.69	6.83	42.99	9.49
C20H40I2N6	618.39	618.1407	38.85	6.52	41.04	13.59
C20H41I1N2	436.47	436.2316	55.04	9.47	29.08	6.42
C20H41I1N4	464.48	464.2378	51.72	8.90	27.32	12.06
C20H41I1N6	492.49	492.2439	48.78	8.39	25.77	17.06
C20H41I2N1	549.37	549.1332	43.73	7.52	46.20	2.55
C20H41I2N3	577.38	577.1393	41.61	7.16	43.96	7.28
C20H41I2N5	605.39	605.1455	39.68	6.83	41.92	11.57
C20H42I1N1	423.47	423.2364	56.73	10.00	29.97	3.31
C20H42I1N3	451.48	451.2425	53.21	9.38	28.11	9.31
C20H42I1N5	479.50	479.2487	50.10	8.83	26.47	14.61
C20H42I2N2	564.38	564.1441	42.56	7.50	44.97	4.96
C20H42I2N4	592.39	592.1502	40.55	7.15	42.84	9.46
C20H42I2N6	620.41	620.1564	38.72	6.82	40.91	13.55
C20H43I1N2	438.48	438.2473	54.78	9.88	28.94	6.39
C20H43I1N4	466.50	466.2534	51.50	9.29	27.20	12.01
C20H43I1N6	494.51	494.2596	48.58	8.76	25.66	16.99
C20H43I2N3	579.39	579.1550	41.46	7.48	43.81	7.25
C20H43I2N5	607.41	607.1611	39.55	7.14	41.79	11.53
C20H44I1N3	453.50	453.2582	52.97	9.78	27.98	9.27
C20H44I1N5	481.51	481.2643	49.89	9.21	26.36	14.54
C20H44I2N4	594.41	594.1659	40.41	7.46	42.70	9.43
C20H44I2N6	622.42	622.1720	38.59	7.13	40.78	13.50
C20H45I1N4	468.51	468.2691	51.27	9.68	27.09	11.96
C20H45I1N6	496.53	496.2752	48.38	9.14	25.56	16.93
C20H45I2N5	609.42	609.1768	39.42	7.44	41.65	11.49
C20H46I1N5	483.53	483.2800	49.68	9.59	26.25	14.48
C20H46I2N6	624.44	624.1877	38.47	7.43	40.65	13.46
C20H47I1N6	498.54	498.2909	48.19	9.50	25.46	16.86
C21H10I1N1	403.22	402.9860	62.55	2.50	31.47	3.47
C21H10I1N3	431.24	430.9921	58.49	2.34	29.43	9.74
C21H10I1N5	459.25	458.9983	54.92	2.19	27.63	15.25
C21H10I2N2	544.14	543.8937	46.35	1.85	46.64	5.15
C21H10I2N4	572.15	571.8999	44.09	1.76	44.36	9.79
C21H10I2N6	600.16	599.9060	42.03	1.68	42.29	14.00
C21H11I1N2	418.24	417.9969	60.31	2.65	30.34	6.70
C21H11I1N4	446.25	446.0030	56.52	2.48	28.44	12.55
C21H11I1N6	474.27	474.0092	53.18	2.34	26.76	17.72
C21H11I2N1	531.14	530.8985	47.49	2.09	47.79	2.64
C21H11I2N3	559.15	558.9046	45.11	1.98	45.39	7.51
C21H11I2N5	587.16	586.9108	42.96	1.89	43.23	11.93
C21H12I1N1	405.24	405.0016	62.24	2.98	31.32	3.46
C21H12I1N3	433.25	433.0078	58.22	2.79	29.29	9.70
C21H12I1N5	461.27	461.0139	54.68	2.62	27.51	15.18
C21H12I2N2	546.15	545.9094	46.18	2.21	46.47	5.13
C21H12I2N4	574.17	573.9155	43.93	2.11	44.20	9.76
C21H12I2N6	602.18	601.9217	41.89	2.01	42.15	13.96
C21H13I1N2	420.26	420.0125	60.02	3.12	30.20	6.67
C21H13I1N4	448.27	448.0187	56.27	2.92	28.31	12.50
C21H13I1N6	476.28	476.0248	52.96	2.75	26.64	17.65
C21H13I2N1	533.15	532.9141	47.31	2.46	47.61	2.63
C21H13I2N3	561.17	560.9203	44.95	2.34	45.23	7.49
C21H13I2N5	589.18	588.9264	42.81	2.22	43.08	11.89
C21H14I1N1	407.26	407.0173	61.93	3.47	31.16	3.44
C21H14I1N3	435.27	435.0234	57.95	3.24	29.16	9.65
C21H14I1N5	463.28	463.0296	54.44	3.05	27.39	15.12
C21H14I2N2	548.17	547.9250	46.01	2.57	46.30	5.11
C21H14I2N4	576.18	575.9312	43.78	2.45	44.05	9.72
C21H14I2N6	604.19	603.9373	41.75	2.34	42.01	13.91
C21H15I1N2	422.27	422.0282	59.73	3.58	30.05	6.63
C21H15I1N4	450.28	450.0343	56.02	3.36	28.18	12.44
C21H15I1N6	478.30	478.0405	52.74	3.16	26.53	17.57
C21H15I2N1	535.17	534.9298	47.13	2.83	47.43	2.62
C21H15I2N3	563.18	562.9359	44.79	2.68	45.07	7.46
C21H15I2N5	591.20	590.9421	42.67	2.56	42.93	11.85
C21H16I1N1	409.27	409.0329	61.63	3.94	31.01	3.42
C21H16I1N3	437.29	437.0391	57.68	3.69	29.02	9.61
C21H16I1N5	465.30	465.0452	54.21	3.47	27.27	15.05
C21H16I2N2	550.18	549.9407	45.85	2.93	46.13	5.09

031

C21H16I2N4	578.20	577.9468	43.62	2.79	43.90	9.69	C21H26I2N4	588.28	588.0251	42.88	4.45	43.14	9.52
C21H16I2N6	606.21	605.9530	41.61	2.66	41.87	13.86	C21H26I2N6	616.29	616.0312	40.93	4.25	41.18	13.64
C21H17I1N2	424.29	424.0438	59.45	4.04	29.91	6.60	C21H27I1N2	434.37	434.1221	58.07	6.27	29.22	6.45
C21H17I1N4	452.30	452.0500	55.77	3.79	28.06	12.39	C21H27I1N4	462.38	462.1282	54.55	5.89	27.45	12.12
C21H17I1N6	480.31	480.0561	52.51	3.57	26.42	17.50	C21H27I1N6	490.39	490.1344	51.44	5.55	25.88	17.14
C21H17I2N1	537.19	536.9454	46.95	3.19	47.25	2.61	C21H27I2N1	547.26	547.0237	46.09	4.97	46.38	2.56
C21H17I2N3	565.20	564.9516	44.63	3.03	44.91	7.43	C21H27I2N3	575.28	575.0298	43.85	4.73	44.12	7.30
C21H17I2N5	593.21	592.9577	42.52	2.89	42.79	11.81	C21H27I2N5	603.29	603.0360	41.81	4.51	42.07	11.61
C21H18I1N1	411.29	411.0486	61.33	4.41	30.86	3.41	C21H28I1N1	421.37	421.1268	59.86	6.70	30.12	3.32
C21H18I1N3	439.30	439.0547	57.42	4.13	28.89	9.57	C21H28I1N3	449.38	449.1330	56.13	6.28	28.24	9.35
C21H18I1N5	467.32	467.0609	53.98	3.88	27.16	14.99	C21H28I1N5	477.40	477.1391	52.84	5.91	26.58	14.67
C21H18I2N2	552.20	551.9563	45.68	3.29	45.96	5.07	C21H28I2N2	562.28	562.0346	44.86	5.02	45.14	4.98
C21H18I2N4	580.21	579.9625	43.47	3.13	43.74	9.66	C21H28I2N4	590.29	590.0407	42.73	4.78	43.00	9.49
C21H18I2N6	608.23	607.9686	41.47	2.98	41.73	13.82	C21H28I2N6	618.31	618.0469	40.79	4.56	41.05	13.59
C21H19I1N2	426.30	426.0595	59.17	4.49	29.77	6.57	C21H29I1N2	436.38	436.1377	57.80	6.70	29.08	6.42
C21H19I1N4	454.32	454.0656	55.52	4.22	27.93	12.33	C21H29I1N4	464.40	464.1439	54.31	6.29	27.33	12.06
C21H19I1N6	482.33	482.0718	52.29	3.97	26.31	17.42	C21H29I1N6	492.41	492.1500	51.22	5.94	25.77	17.07
C21H19I2N1	539.20	538.9611	46.78	3.55	47.07	2.60	C21H29I2N1	549.28	549.0393	45.92	5.32	46.21	2.55
C21H19I2N3	567.21	566.9672	44.47	3.38	44.75	7.41	C21H29I2N3	577.29	577.0455	43.69	5.06	43.97	7.28
C21H19I2N5	595.23	594.9734	42.38	3.22	42.64	11.77	C21H29I2N5	605.31	605.0516	41.67	4.83	41.93	11.57
C21H20I1N1	413.30	413.0642	61.03	4.88	30.70	3.39	C21H30I1N1	423.38	423.1425	59.58	7.14	29.97	3.31
C21H20I1N3	441.32	441.0704	57.15	4.57	28.76	9.52	C21H30I1N3	451.40	451.1486	55.88	6.70	28.11	9.31
C21H20I1N5	469.33	469.0765	53.74	4.30	27.04	14.92	C21H30I1N5	479.41	479.1548	52.61	6.31	26.47	14.61
C21H20I2N2	554.22	553.9720	45.51	3.64	45.80	5.05	C21H30I2N2	564.30	564.0502	44.70	5.36	44.98	4.96
C21H20I2N4	582.23	581.9781	43.32	3.46	43.59	9.62	C21H30I2N4	592.31	592.0564	42.58	5.11	42.85	9.46
C21H20I2N6	610.24	609.9843	41.33	3.30	41.59	13.77	C21H30I2N6	620.32	620.0625	40.66	4.87	40.92	13.55
C21H21I1N2	428.32	428.0751	58.89	4.94	29.63	6.54	C21H31I1N2	438.40	438.1534	57.54	7.13	28.95	6.39
C21H21I1N4	456.33	456.0813	55.27	4.64	27.81	12.28	C21H31I1N4	466.41	466.1595	54.08	6.70	27.21	12.01
C21H21I1N6	484.35	484.0874	52.08	4.37	26.20	17.35	C21H31I1N6	494.43	494.1657	51.02	6.32	25.67	17.00
C21H21I2N1	541.22	540.9767	46.60	3.91	46.90	2.59	C21H31I2N1	551.30	551.0550	45.75	5.67	46.04	2.54
C21H21I2N3	569.23	568.9829	44.31	3.72	44.59	7.38	C21H31I2N3	579.31	579.0611	43.54	5.39	43.81	7.25
C21H21I2N5	597.24	596.9890	42.23	3.54	42.50	11.73	C21H31I2N5	607.32	607.0673	41.53	5.15	41.79	11.53
C21H22I1N1	415.32	415.0799	60.73	5.34	30.56	3.37	C21H32I1N1	425.40	425.1581	59.29	7.58	29.83	3.29
C21H22I1N3	443.33	443.0860	56.89	5.00	28.63	9.48	C21H32I1N3	453.41	453.1643	55.63	7.11	27.99	9.27
C21H22I1N5	471.35	471.0922	53.51	4.70	26.92	14.86	C21H32I1N5	481.43	481.1704	52.39	6.70	26.36	14.55
C21H22I2N2	556.23	555.9876	45.35	3.99	45.63	5.04	C21H32I2N2	566.31	566.0659	44.54	5.70	44.82	4.95
C21H22I2N4	584.25	583.9938	43.17	3.80	43.44	9.59	C21H32I2N4	594.32	594.0720	42.44	5.43	42.71	9.43
C21H22I2N6	612.26	611.9999	41.20	3.62	41.45	13.73	C21H32I2N6	622.34	622.0781	40.53	5.18	40.78	13.50
C21H23I1N2	430.34	430.0908	58.61	5.39	29.49	6.51	C21H33I1N2	440.41	440.1690	57.27	7.55	28.81	6.36
C21H23I1N4	458.35	458.0969	55.03	5.06	27.69	12.22	C21H33I1N4	468.43	468.1752	53.85	7.10	27.09	11.96
C21H23I1N6	486.36	486.1031	51.86	4.77	26.09	17.28	C21H33I1N6	496.44	496.1813	50.81	6.70	25.56	16.93
C21H23I2N1	543.23	542.9924	46.43	4.27	46.72	2.58	C21H33I2N1	553.31	553.0706	45.59	6.01	45.87	2.53
C21H23I2N3	571.25	570.9985	44.16	4.06	44.43	7.36	C21H33I2N3	581.33	581.0768	43.39	5.72	43.66	7.23
C21H23I2N5	599.26	599.0047	42.09	3.87	42.35	11.69	C21H33I2N5	609.34	609.0829	41.39	5.46	41.65	11.49
C21H24I1N1	417.34	417.0955	60.44	5.80	30.41	3.36	C21H34I1N1	427.42	427.1738	59.01	8.02	29.69	3.28
C21H24I1N3	445.35	445.1017	56.64	5.43	28.50	9.44	C21H34I1N3	455.43	455.1799	55.38	7.52	27.86	9.23
C21H24I1N5	473.36	473.1078	53.29	5.11	26.81	14.79	C21H34I1N5	483.44	483.1861	52.17	7.09	26.25	14.49
C21H24I2N2	558.25	558.0033	45.18	4.33	45.47	5.02	C21H34I2N2	568.33	568.0815	44.38	6.03	44.66	4.93
C21H24I2N4	586.26	586.0094	43.02	4.13	43.29	9.56	C21H34I2N4	596.34	596.0877	42.30	5.75	42.56	9.40
C21H24I2N6	614.27	614.0156	41.06	3.94	41.32	13.68	C21H34I2N6	624.35	624.0938	40.40	5.49	40.65	13.46
C21H25I1N2	432.35	432.1064	58.34	5.83	29.35	6.48	C21H35I1N2	442.43	442.1847	57.01	7.97	28.68	6.33
C21H25I1N4	460.36	460.1126	54.79	5.47	27.57	12.17	C21H35I1N4	470.44	470.1908	53.62	7.50	26.98	11.91
C21H25I1N6	488.38	488.1187	51.65	5.16	25.98	17.21	C21H35I1N6	498.46	498.1970	50.60	7.08	25.46	16.86
C21H25I2N1	545.25	545.0080	46.26	4.62	46.55	2.57	C21H35I2N1	555.33	555.0863	45.42	6.35	45.70	2.52
C21H25I2N3	573.26	573.0142	44.00	4.40	44.27	7.33	C21H35I2N3	583.34	583.0924	43.24	6.05	43.51	7.20
C21H25I2N5	601.28	601.0203	41.95	4.19	42.21	11.65	C21H35I2N5	611.36	611.0985	41.26	5.77	41.52	11.46
C21H26I1N1	419.35	419.1112	60.15	6.25	30.26	3.34	C21H36I1N1	429.43	429.1894	58.74	8.45	29.55	3.26
C21H26I1N3	447.37	447.1173	56.38	5.86	28.37	9.39	C21H36I1N3	457.45	457.1956	55.14	7.93	27.74	9.19
C21H26I1N5	475.38	475.1235	53.06	5.51	26.70	14.73	C21H36I1N5	485.46	485.2017	51.96	7.47	26.14	14.43
C21H26I2N2	560.26	560.0189	45.02	4.68	45.30	5.00	C21H36I2N2	570.34	570.0972	44.22	6.36	44.50	4.91

032

C21H36I2N4	598.36	598.1033	42.15	6.06	42.42	9.36	C21H47I1N6	510.55	510.2909	49.40	9.28	24.86	16.46
C21H36I2N6	626.37	626.1094	40.27	5.79	40.52	13.42	C21H47I2N5	623.45	623.1924	40.46	7.60	40.71	11.23
C21H37I1N2	444.45	444.2003	56.75	8.39	28.55	6.30	C21H48I1N5	497.55	497.2956	50.69	9.72	25.51	14.08
C21H37I1N4	472.46	472.2065	53.39	7.89	26.86	11.86	C21H48I2N6	638.47	638.2033	39.51	7.58	39.75	13.16
C21H37I1N6	500.47	500.2126	50.40	7.45	25.36	16.79	C21H49I1N6	512.57	512.3065	49.21	9.64	24.76	16.40
C21H37I2N1	557.34	557.1019	45.26	6.69	45.54	2.51	C22H11I1N2	430.25	429.9969	61.42	2.58	29.50	6.51
C21H37I2N3	585.36	585.1081	43.09	6.37	43.36	7.18	C22H11I1N4	458.26	458.0030	57.66	2.42	27.69	12.23
C21H37I2N5	613.37	613.1142	41.12	6.08	41.38	11.42	C22H11I1N6	486.28	486.0092	54.34	2.28	26.10	17.28
C21H38I1N1	431.45	431.2051	58.46	8.88	29.41	3.25	C22H11I2N1	543.15	542.8985	48.65	2.04	46.73	2.58
C21H38I1N3	459.46	459.2112	54.90	8.34	27.62	9.15	C22H11I2N3	571.16	570.9046	46.26	1.94	44.44	7.36
C21H38I1N5	487.47	487.2174	51.74	7.86	26.03	14.37	C22H11I2N5	599.18	598.9108	44.10	1.85	42.36	11.69
C21H38I2N2	572.36	572.1128	44.07	6.69	44.34	4.89	C22H12I1N1	417.25	417.0016	63.33	2.90	30.41	3.36
C21H38I2N4	600.37	600.1189	42.01	6.38	42.28	9.33	C22H12I1N3	445.27	445.0078	59.35	2.72	28.50	9.44
C21H38I2N6	628.39	628.1251	40.14	6.10	40.39	13.37	C22H12I1N5	473.28	473.0139	55.83	2.56	26.81	14.80
C21H39I1N2	446.46	446.2160	56.50	8.80	28.42	6.27	C22H12I2N2	558.16	557.9094	47.34	2.17	45.47	5.02
C21H39I1N4	474.48	474.2221	53.16	8.29	26.75	11.81	C22H12I2N4	586.18	585.9155	45.08	2.06	43.30	9.56
C21H39I1N6	502.49	502.2283	50.20	7.82	25.26	16.72	C22H12I2N6	614.19	613.9217	43.02	1.97	41.32	13.68
C21H39I2N1	559.36	559.1176	45.09	7.03	45.37	2.50	C22H13I1N2	432.27	432.0125	61.13	3.03	29.36	6.48
C21H39I2N3	587.37	587.1237	42.94	6.69	43.21	7.15	C22H13I1N4	460.28	460.0187	57.41	2.85	27.57	12.17
C21H39I2N5	615.39	615.1298	40.99	6.39	41.24	11.38	C22H13I1N6	488.29	488.0248	54.12	2.68	25.99	17.21
C21H40I1N1	433.46	433.2207	58.19	9.30	29.28	3.23	C22H13I2N1	545.16	544.9141	48.47	2.40	46.56	2.57
C21H40I1N3	461.48	461.2269	54.66	8.74	27.50	9.11	C22H13I2N3	573.18	572.9203	46.10	2.29	44.28	7.33
C21H40I1N5	489.49	489.2330	51.53	8.24	25.93	14.31	C22H13I2N5	601.19	600.9264	43.95	2.18	42.22	11.65
C21H40I2N2	574.38	574.1285	43.91	7.02	44.19	4.88	C22H14I1N1	419.27	419.0173	63.03	3.37	30.27	3.34
C21H40I2N4	602.39	602.1346	41.87	6.69	42.13	9.30	C22H14I1N3	447.28	447.0234	59.08	3.15	28.37	9.39
C21H40I2N6	630.40	630.1407	40.01	6.40	40.26	13.33	C22H14I1N5	475.29	475.0296	55.60	2.97	26.70	14.73
C21H41I1N2	448.48	448.2316	56.24	9.21	28.30	6.25	C22H14I2N2	560.18	559.9250	47.17	2.52	45.31	5.00
C21H41I1N4	476.49	476.2378	52.94	8.67	26.63	11.76	C22H14I2N4	588.19	587.9312	44.92	2.40	43.15	9.53
C21H41I1N6	504.51	504.2439	50.00	8.19	25.15	16.66	C22H14I2N6	616.21	615.9373	42.88	2.29	41.19	13.64
C21H41I2N1	561.38	561.1332	44.93	7.36	45.21	2.50	C22H15I1N2	434.28	434.0282	60.85	3.48	29.22	6.45
C21H41I2N3	589.39	589.1393	42.80	7.01	43.06	7.13	C22H15I1N4	462.30	462.0343	57.16	3.27	27.45	12.12
C21H41I2N5	617.40	617.1455	40.85	6.69	41.11	11.34	C22H15I1N6	490.31	490.0405	53.89	3.08	25.88	17.14
C21H42I1N1	435.48	435.2364	57.92	9.72	29.14	3.22	C22H15I2N1	547.18	546.9298	48.29	2.76	46.38	2.56
C21H42I1N3	463.49	463.2425	54.42	9.13	27.38	9.07	C22H15I2N3	575.19	574.9359	45.94	2.63	44.13	7.31
C21H42I1N5	491.51	491.2487	51.32	8.61	25.82	14.25	C22H15I2N5	603.21	602.9421	43.81	2.51	42.08	11.61
C21H42I2N2	576.39	576.1441	43.76	7.34	44.03	4.86	C22H16I1N1	421.28	421.0329	62.72	3.83	30.12	3.32
C21H42I2N4	604.40	604.1502	41.73	7.00	41.99	9.27	C22H16I1N3	449.30	449.0391	58.81	3.59	28.25	9.35
C21H42I2N6	632.42	632.1564	39.88	6.69	40.13	13.29	C22H16I1N5	477.31	477.0452	55.36	3.38	26.59	14.67
C21H43I1N2	450.49	450.2473	55.99	9.62	28.17	6.22	C22H16I2N2	562.20	561.9407	47.00	2.87	45.15	4.98
C21H43I1N4	478.51	478.2534	52.71	9.06	26.52	11.71	C22H16I2N4	590.21	589.9468	44.77	2.73	43.00	9.49
C21H43I1N6	506.52	506.2596	49.80	8.56	25.05	16.59	C22H16I2N6	618.22	617.9530	42.74	2.61	41.05	13.59
C21H43I2N1	563.39	563.1489	44.77	7.59	45.05	2.49	C22H17I1N2	436.30	436.0438	60.57	3.93	29.09	6.42
C21H43I2N3	591.41	591.1550	42.65	7.33	42.92	7.11	C22H17I1N4	464.31	464.0500	56.91	3.69	27.33	12.07
C21H43I2N5	619.42	619.1611	40.72	7.00	40.98	11.31	C22H17I1N6	492.33	492.0561	53.67	3.48	25.78	17.07
C21H44I1N1	437.50	437.2520	57.65	10.14	29.01	3.20	C22H17I2N1	549.20	548.9454	48.11	3.12	46.21	2.55
C21H44I1N3	465.51	465.2582	54.18	9.53	27.26	9.03	C22H17I2N3	577.21	576.9516	45.78	2.97	43.97	7.28
C21H44I1N5	493.52	493.2643	51.11	8.99	25.71	14.19	C22H17I2N5	605.22	604.9577	43.66	2.83	41.94	11.57
C21H44I2N2	578.41	578.1597	43.61	7.67	43.88	4.84	C22H18I1N1	423.30	423.0486	62.43	4.29	29.98	3.31
C21H44I2N4	606.42	606.1659	41.59	7.31	41.85	9.24	C22H18I1N3	451.31	451.0547	58.55	4.02	28.12	9.31
C21H44I2N6	634.43	634.1720	39.76	6.99	40.01	13.25	C22H18I1N5	479.33	479.0609	55.13	3.79	26.48	14.61
C21H45I1N2	452.51	452.2629	55.74	10.02	28.04	6.19	C22H18I2N2	564.21	563.9563	46.83	3.22	44.98	4.97
C21H45I1N4	480.52	480.2691	52.49	9.44	26.41	11.66	C22H18I2N4	592.22	591.9625	44.62	3.06	42.86	9.46
C21H45I1N6	508.54	508.2752	49.60	8.92	24.95	16.53	C22H18I2N6	620.24	619.9686	42.60	2.93	40.92	13.55
C21H45I2N3	593.42	593.1706	42.51	7.64	42.77	7.08	C22H19I1N2	438.31	438.0595	60.29	4.37	28.95	6.39
C21H45I2N5	621.44	621.1768	40.59	7.30	40.84	11.27	C22H19I1N4	466.33	466.0656	56.67	4.11	27.21	12.01
C21H46I1N3	467.53	467.2738	53.95	9.92	27.14	8.99	C22H19I1N6	494.34	494.0718	53.45	3.87	25.67	17.00
C21H46I1N5	495.54	495.2800	50.90	9.36	25.61	14.13	C22H19I2N1	551.21	550.9611	47.94	3.47	46.05	2.54
C21H46I2N4	608.44	608.1815	41.46	7.62	41.71	9.21	C22H19I2N3	579.23	578.9672	45.62	3.31	43.82	7.25
C21H46I2N6	636.45	636.1877	39.63	7.29	39.88	13.20	C22H19I2N5	607.24	606.9734	43.52	3.15	41.80	11.53
C21H47I1N4	482.54	482.2847	52.27	9.82	26.30	11.61	C22H20I1N1	425.32	425.0642	62.13	4.74	29.84	3.29

C22H20I1N3	453.33	453.0704	58.29	4.45	27.99	9.27
C22H20I1N5	481.34	481.0765	54.90	4.19	26.36	14.55
C22H20I2N2	566.23	565.9720	46.67	3.56	44.82	4.95
C22H20I2N4	594.24	593.9781	44.47	3.39	42.71	9.43
C22H20I2N6	622.25	621.9843	42.47	3.24	40.79	13.51
C22H21I1N2	440.33	440.0751	60.01	4.81	28.82	6.36
C22H21I1N4	468.34	468.0813	56.42	4.52	27.10	11.96
C22H21I1N6	496.36	496.0874	53.24	4.26	25.57	16.93
C22H21I2N1	553.23	552.9767	47.76	3.83	45.88	2.53
C22H21I2N3	581.24	580.9829	45.46	3.64	43.67	7.23
C22H21I2N5	609.25	608.9890	43.37	3.47	41.66	11.49
C22H22I1N1	427.33	427.0799	61.84	5.19	29.70	3.28
C22H22I1N3	455.35	455.0860	58.03	4.87	27.87	9.23
C22H22I1N5	483.36	483.0922	54.67	4.59	26.25	14.49
C22H22I2N2	568.24	567.9876	46.50	3.90	44.67	4.93
C22H22I2N4	596.26	595.9938	44.32	3.72	42.57	9.40
C22H22I2N6	624.27	623.9999	42.33	3.55	40.66	13.46
C22H23I1N2	442.35	442.0908	59.74	5.24	28.69	6.33
C22H23I1N4	470.36	470.0969	56.18	4.93	26.98	11.91
C22H23I1N6	498.37	498.1031	53.02	4.65	25.46	16.86
C22H23I2N1	555.24	554.9924	47.59	4.18	45.71	2.52
C22H23I2N3	583.26	582.9985	45.31	3.97	43.52	7.20
C22H23I2N5	611.27	611.0047	43.23	3.79	41.52	11.46
C22H24I1N1	429.35	429.0955	61.55	5.63	29.56	3.26
C22H24I1N3	457.36	457.1017	57.78	5.29	27.75	9.19
C22H24I1N5	485.37	485.1078	54.44	4.98	26.15	14.43
C22H24I2N2	570.26	570.0033	46.34	4.24	44.51	4.91
C22H24I2N4	598.27	598.0094	44.17	4.04	42.42	9.36
C22H24I2N6	626.29	626.0156	42.19	3.86	40.53	13.42
C22H25I1N2	444.36	444.1064	59.47	5.67	28.56	6.30
C22H25I1N4	472.38	472.1126	55.94	5.33	26.87	11.86
C22H25I1N6	500.39	500.1187	52.81	5.04	25.36	16.79
C22H25I2N1	557.26	557.0080	47.42	4.52	45.55	2.51
C22H25I2N3	585.27	585.0142	45.15	4.31	43.37	7.18
C22H25I2N5	613.29	613.0203	43.09	4.11	41.39	11.42
C22H26I1N1	431.36	431.1112	61.26	6.08	29.42	3.25
C22H26I1N3	459.38	459.1173	57.52	5.70	27.63	9.15
C22H26I1N5	487.39	487.1235	54.22	5.38	26.04	14.37
C22H26I2N2	572.27	572.0189	46.17	4.58	44.35	4.90
C22H26I2N4	600.29	600.0251	44.02	4.37	42.28	9.33
C22H26I2N6	628.30	628.0312	42.06	4.17	40.40	13.38
C22H27I1N2	446.38	446.1221	59.20	6.10	28.43	6.28
C22H27I1N4	474.39	474.1282	55.70	5.74	26.75	11.81
C22H27I1N6	502.41	502.1344	52.60	5.42	25.26	16.73
C22H27I2N1	559.28	559.0237	47.25	4.87	45.38	2.50
C22H27I2N3	587.29	587.0298	44.99	4.63	43.22	7.15
C22H27I2N5	615.30	615.0360	42.95	4.42	41.25	11.38
C22H28I1N1	433.38	433.1268	60.97	6.51	29.28	3.23
C22H28I1N3	461.39	461.1330	57.27	6.12	27.50	9.11
C22H28I1N5	489.41	489.1391	53.99	5.77	25.93	14.31
C22H28I2N2	574.29	574.0346	46.01	4.91	44.20	4.88
C22H28I2N4	602.30	602.0407	43.87	4.69	42.14	9.30
C22H28I2N6	630.32	630.0469	41.92	4.48	40.27	13.33
C22H29I1N2	448.39	448.1377	58.93	6.52	28.30	6.25
C22H29I1N4	476.41	476.1439	55.47	6.14	26.64	11.76
C22H29I1N6	504.42	504.1500	52.39	5.79	25.16	16.66
C22H29I2N1	561.29	561.0393	47.08	5.21	45.22	2.50
C22H29I2N3	589.31	589.0455	44.84	4.96	43.07	7.13
C22H29I2N5	617.32	617.0516	42.81	4.74	41.11	11.34
C22H30I1N1	435.40	435.1425	60.69	6.95	29.15	3.22
C22H30I1N3	463.41	463.1486	57.02	6.53	27.38	9.07
C22H30I1N5	491.42	491.1548	53.77	6.15	25.82	14.25
C22H30I2N2	576.31	576.0502	45.85	5.25	44.04	4.86
C22H30I2N4	604.32	604.0564	43.73	5.00	42.00	9.27
C22H30I2N6	632.33	632.0625	41.79	4.78	40.14	13.29
C22H31I1N2	450.41	450.1534	58.67	6.94	28.18	6.22
C22H31I1N4	478.42	478.1595	55.23	6.53	26.53	11.71
C22H31I1N6	506.44	506.1657	52.18	6.17	25.06	16.59
C22H31I2N1	563.31	563.0550	46.91	5.55	45.06	2.49
C22H31I2N3	591.32	591.0611	44.69	5.28	42.92	7.11
C22H31I2N5	619.33	619.0673	42.67	5.05	40.98	11.31
C22H32I1N1	437.41	437.1581	60.41	7.37	29.01	3.20
C22H32I1N3	465.42	465.1643	56.78	6.93	27.27	9.03
C22H32I1N5	493.44	493.1704	53.55	6.54	25.72	14.19
C22H32I2N2	578.32	578.0659	45.69	5.58	43.89	4.84
C22H32I2N4	606.34	606.0720	43.58	5.32	41.86	9.24
C22H32I2N6	634.35	634.0781	41.66	5.08	40.01	13.25
C22H33I1N2	452.43	452.1690	58.41	7.35	28.05	6.19
C22H33I1N4	480.44	480.1752	55.00	6.92	26.41	11.66
C22H33I1N6	508.45	508.1813	51.97	6.54	24.96	16.53
C22H33I2N1	565.32	565.0706	46.74	5.88	44.90	2.48
C22H33I2N3	593.34	593.0768	44.54	5.61	42.78	7.08
C22H33I2N5	621.35	621.0829	42.53	5.35	40.85	11.27
C22H34I1N1	439.43	439.1738	60.13	7.80	28.88	3.19
C22H34I1N3	467.44	467.1799	56.53	7.33	27.15	8.99
C22H34I1N5	495.45	495.1861	53.33	6.92	25.61	14.14
C22H34I2N2	580.34	580.0815	45.53	5.91	43.73	4.83
C22H34I2N4	608.35	608.0877	43.44	5.63	41.72	9.21
C22H34I2N6	636.37	636.0938	41.52	5.39	39.88	13.21
C22H35I1N2	454.44	454.1847	58.15	7.76	27.93	6.16
C22H35I1N4	482.46	482.1908	54.77	7.31	26.30	11.61
C22H35I1N6	510.47	510.1970	51.77	6.91	24.86	16.46
C22H35I2N1	567.34	567.0863	46.58	6.22	44.74	2.47
C22H35I2N3	595.35	595.0924	44.38	5.93	42.63	7.06
C22H35I2N5	623.37	623.0985	42.39	5.66	40.72	11.23
C22H36I1N1	441.44	441.1894	59.86	8.22	28.75	3.17
C22H36I1N3	469.46	469.1956	56.29	7.73	27.03	8.95
C22H36I1N5	497.47	497.2017	53.12	7.29	25.51	14.08
C22H36I2N2	582.35	582.0972	45.38	6.23	43.58	4.81
C22H36I2N4	610.37	610.1033	43.29	5.95	41.58	9.18
C22H36I2N6	638.38	638.1094	41.39	5.68	39.76	13.16
C22H37I1N2	456.46	456.2003	57.89	8.17	27.80	6.14
C22H37I1N4	484.47	484.2065	54.54	7.70	26.19	11.56
C22H37I1N6	512.48	512.2126	51.56	7.28	24.76	16.40
C22H37I2N1	569.36	569.1019	46.41	6.55	44.58	2.46
C22H37I2N3	597.37	597.1081	44.23	6.24	42.49	7.03
C22H37I2N5	625.38	625.1142	42.25	5.96	40.58	11.20
C22H38I1N1	443.46	443.2051	59.59	8.64	28.62	3.16
C22H38I1N3	471.47	471.2112	56.05	8.12	26.92	8.91
C22H38I1N5	499.49	499.2174	52.90	7.67	25.41	14.02
C22H38I2N2	584.37	584.1128	45.22	6.55	43.43	4.79
C22H38I2N4	612.38	612.1189	43.15	6.25	41.45	9.15
C22H38I2N6	640.40	640.1251	41.26	5.98	39.63	13.12
C22H39I1N2	458.47	458.2160	57.64	8.57	27.68	6.11
C22H39I1N4	486.49	486.2221	54.32	8.08	26.09	11.52
C22H39I1N6	514.50	514.2283	51.36	7.64	24.67	16.33
C22H39I2N1	571.37	571.1176	46.25	6.88	44.42	2.45
C22H39I2N3	599.39	599.1237	44.09	6.56	42.34	7.01
C22H39I2N5	627.40	627.1298	42.12	6.27	40.45	11.16
C22H40I1N1	445.48	445.2207	59.32	9.05	28.49	3.14

034

C22H40I1N3	473.49	473.2269	55.81	8.52	26.80	8.87
C22H40I1N5	501.50	501.2330	52.69	8.04	25.30	13.96
C22H40I2N2	586.39	586.1285	45.06	6.88	43.28	4.78
C22H40I2N4	614.40	614.1346	43.01	6.56	41.31	9.12
C22H40I2N6	642.41	642.1407	41.13	6.28	39.51	13.08
C22H41I1N2	460.49	460.2316	57.38	8.97	27.56	6.08
C22H41I1N4	488.50	488.2378	54.09	8.46	25.98	11.47
C22H41I1N6	516.52	516.2439	51.16	8.00	24.57	16.27
C22H41I2N1	573.39	573.1332	46.08	7.21	44.26	2.44
C22H41I2N3	601.40	601.1394	43.94	6.87	42.20	6.99
C22H41I2N5	629.41	629.1455	41.98	6.57	40.32	11.13
C22H42I1N1	447.49	447.2364	59.05	9.46	28.36	3.13
C22H42I1N3	475.50	475.2425	55.57	8.90	26.69	8.84
C22H42I1N5	503.52	503.2487	52.48	8.41	25.20	13.91
C22H42I2N2	588.40	588.1441	44.91	7.19	43.14	4.76
C22H42I2N4	616.42	616.1502	42.87	6.87	41.17	9.09
C22H42I2N6	644.43	644.1564	41.00	6.57	39.39	13.04
C22H43I1N2	462.51	462.2473	57.13	9.37	27.44	6.06
C22H43I1N4	490.52	490.2534	53.87	8.84	25.87	11.42
C22H43I1N6	518.53	518.2596	50.96	8.36	24.47	16.21
C22H43I2N1	575.40	575.1489	45.92	7.53	44.11	2.43
C22H43I2N3	603.42	603.1550	43.79	7.18	42.06	6.96
C22H43I2N5	631.43	631.1611	41.85	6.86	40.20	11.09
C22H44I1N1	449.51	449.2520	58.79	9.87	28.23	3.12
C22H44I1N3	477.52	477.2582	55.34	9.29	26.58	8.80
C22H44I1N5	505.53	505.2643	52.27	8.77	25.10	13.85
C22H44I2N2	590.42	590.1597	44.76	7.51	42.99	4.74
C22H44I2N4	618.43	618.1659	42.73	7.17	41.04	9.06
C22H44I2N6	646.44	646.1720	40.88	6.86	39.26	13.00
C22H45I1N2	464.52	464.2629	56.89	9.76	27.32	6.03
C22H45I1N4	492.54	492.2691	53.65	9.21	25.77	11.38
C22H45I1N6	520.55	520.2752	50.76	8.71	24.38	16.14
C22H45I2N1	577.42	577.1645	45.76	7.86	43.96	2.43
C22H45I2N3	605.43	605.1706	43.65	7.49	41.92	6.94
C22H45I2N5	633.45	633.1768	41.72	7.16	40.07	11.06
C22H46I1N1	451.52	451.2677	58.52	10.27	28.11	3.10
C22H46I1N3	479.54	479.2738	55.10	9.67	26.46	8.76
C22H46I1N5	507.55	507.2800	52.06	9.14	25.00	13.80
C22H46I2N2	592.43	592.1754	44.60	7.83	42.84	4.73
C22H46I2N4	620.45	620.1815	42.59	7.47	40.91	9.03
C22H46I2N6	648.46	648.1877	40.75	7.15	39.14	12.96
C22H47I1N2	466.54	466.2786	56.64	10.15	27.20	6.00
C22H47I1N4	494.55	494.2847	53.43	9.58	25.66	11.33
C22H47I1N6	522.56	522.2909	50.57	9.07	24.28	16.08
C22H47I2N3	607.45	607.1863	43.50	7.80	41.78	6.92
C22H47I2N5	635.46	635.1924	41.58	7.46	39.94	11.02
C22H48I1N3	481.55	481.2895	54.87	10.05	26.35	8.73
C22H48I1N5	509.57	509.2956	51.86	9.49	24.90	13.74
C22H48I2N4	622.46	622.1972	42.45	7.77	40.77	9.00
C22H48I2N6	650.48	650.2033	40.62	7.44	39.02	12.92
C22H49I1N4	496.57	496.3004	53.21	9.95	25.56	11.28
C22H49I1N6	524.58	524.3065	50.37	9.42	24.19	16.02
C22H49I2N5	637.48	637.2081	41.45	7.75	39.81	10.99
C22H50I1N5	511.58	511.3113	51.65	9.85	24.81	13.69
C22H50I2N6	652.49	652.2190	40.50	7.72	38.90	12.88
C22H51I1N6	526.60	526.3222	50.18	9.76	24.10	15.96
C23H11I1N2	442.26	441.9969	62.46	2.51	28.69	6.33
C23H11I1N4	470.28	470.0030	58.74	2.36	26.99	11.91
C23H11I1N6	498.29	498.0092	55.44	2.23	25.47	16.87
C23H11I2N1	555.16	554.8985	49.76	2.00	45.72	2.52
C23H11I2N3	583.17	582.9046	47.37	1.90	43.52	7.21
C23H11I2N5	611.19	610.9108	45.20	1.81	41.53	11.46
C23H12I1N1	429.26	429.0016	64.36	2.82	29.56	3.26
C23H12I1N3	457.28	457.0078	60.41	2.65	27.75	9.19
C23H12I1N5	485.29	485.0139	56.93	2.49	26.15	14.43
C23H12I2N2	570.17	569.9094	48.45	2.12	44.51	4.91
C23H12I2N4	598.19	597.9155	46.18	2.02	42.43	9.37
C23H12I2N6	626.20	625.9217	44.12	1.93	40.53	13.42
C23H13I1N2	444.28	444.0125	62.18	2.95	28.56	6.31
C23H13I1N4	472.29	472.0187	58.49	2.77	26.87	11.86
C23H13I1N6	500.30	500.0248	55.22	2.62	25.37	16.80
C23H13I2N1	557.18	556.9141	49.58	2.35	45.55	2.51
C23H13I2N3	585.19	584.9203	47.21	2.24	43.37	7.18
C23H13I2N5	613.20	612.9264	45.05	2.14	41.39	11.42
C23H14I1N1	431.28	431.0173	64.06	3.27	29.43	3.25
C23H14I1N3	459.29	459.0234	60.15	3.07	27.63	9.15
C23H14I1N5	487.31	487.0296	56.69	2.90	26.04	14.37
C23H14I2N2	572.19	571.9250	48.28	2.47	44.36	4.90
C23H14I2N4	600.20	599.9312	46.03	2.35	42.29	9.33
C23H14I2N6	628.22	627.9373	43.97	2.25	40.40	13.38
C23H15I1N2	446.29	446.0282	61.90	3.39	28.44	6.28
C23H15I1N4	474.31	474.0343	58.24	3.19	26.76	11.81
C23H15I1N6	502.32	502.0405	55.00	3.01	25.26	16.73
C23H15I2N1	559.19	558.9298	49.40	2.70	45.39	2.50
C23H15I2N3	587.20	586.9359	47.05	2.57	43.22	7.16
C23H15I2N5	615.22	614.9421	44.90	2.46	41.26	11.38
C23H16I1N1	433.30	433.0329	63.76	3.72	29.29	3.23
C23H16I1N3	461.31	461.0391	59.89	3.50	27.51	9.11
C23H16I1N5	489.32	489.0452	56.46	3.30	25.93	14.31
C23H16I2N2	574.21	573.9407	48.11	2.81	44.20	4.88
C23H16I2N4	602.22	601.9468	45.87	2.68	42.15	9.30
C23H16I2N6	630.23	629.9530	43.83	2.56	40.27	13.33
C23H17I1N2	448.31	448.0438	61.62	3.82	28.31	6.25
C23H17I1N4	476.32	476.0500	58.00	3.60	26.64	11.76
C23H17I1N6	504.34	504.0561	54.78	3.40	25.16	16.66
C23H17I2N1	561.21	560.9454	49.23	3.05	45.23	2.50
C23H17I2N3	589.22	588.9516	46.89	2.91	43.08	7.13
C23H17I2N5	617.23	616.9577	44.76	2.78	41.12	11.35
C23H18I1N1	435.31	435.0486	63.46	4.17	29.15	3.22
C23H18I1N3	463.32	463.0547	59.62	3.92	27.39	9.07
C23H18I1N5	491.34	491.0609	56.23	3.69	25.83	14.25
C23H18I2N2	576.22	575.9563	47.94	3.15	44.05	4.86
C23H18I2N4	604.24	603.9625	45.72	3.00	42.00	9.27
C23H18I2N6	632.25	631.9686	43.69	2.87	40.14	13.29
C23H19I1N2	450.33	450.0595	61.35	4.25	28.18	6.22
C23H19I1N4	478.34	478.0656	57.75	4.00	26.53	11.71
C23H19I1N6	506.35	506.0718	54.56	3.78	25.06	16.60
C23H19I2N1	563.22	562.9611	49.05	3.40	45.06	2.49
C23H19I2N3	591.24	590.9672	46.73	3.24	42.93	7.11
C23H19I2N5	619.25	618.9734	44.61	3.09	40.99	11.31
C23H20I1N1	437.33	437.0642	63.17	4.61	29.02	3.20
C23H20I1N3	465.34	465.0704	59.37	4.33	27.27	9.03
C23H20I1N5	493.35	493.0765	56.00	4.09	25.72	14.20
C23H20I2N2	578.24	577.9720	47.78	3.49	43.89	4.84
C23H20I2N4	606.25	605.9781	45.57	3.33	41.87	9.24
C23H20I2N6	634.26	633.9843	43.56	3.18	40.02	13.25
C23H21I1N2	452.34	452.0751	61.07	4.68	28.05	6.19
C23H21I1N4	480.36	480.0813	57.51	4.41	26.42	11.66
C23H21I1N6	508.37	508.0874	54.34	4.16	24.96	16.53
C23H21I2N1	565.24	564.9767	48.87	3.74	44.90	2.48

035

C23H21I2N3	593.25	592.9829	46.57	3.57	42.78	7.08
C23H21I2N5	621.27	620.9890	44.47	3.41	40.85	11.27
C23H22I1N1	439.34	439.0799	62.88	5.05	28.89	3.19
C23H22I1N3	467.36	467.0860	59.11	4.74	27.15	8.99
C23H22I1N5	495.37	495.0922	55.77	4.48	25.62	14.14
C23H22I2N2	580.25	579.9876	47.61	3.82	43.74	4.83
C23H22I2N4	608.27	607.9938	45.42	3.65	41.73	9.21
C23H22I2N6	636.28	635.9999	43.42	3.49	39.89	13.21
C23H23I1N2	454.36	454.0908	60.80	5.10	27.93	6.17
C23H23I1N4	482.37	482.0969	57.27	4.81	26.31	11.61
C23H23I1N6	510.38	510.1031	54.13	4.54	24.86	16.47
C23H23I2N1	567.26	566.9924	48.70	4.09	44.74	2.47
C23H23I2N3	595.27	594.9985	46.41	3.89	42.64	7.06
C23H23I2N5	623.28	623.0047	44.32	3.72	40.72	11.24
C23H24I1N1	441.36	441.0955	62.59	5.48	28.75	3.17
C23H24I1N3	469.37	469.1017	58.86	5.15	27.04	8.95
C23H24I1N5	497.39	497.1078	55.54	4.86	25.51	14.08
C23H24I2N2	582.27	582.0033	47.44	4.15	43.59	4.81
C23H24I2N4	610.28	610.0094	45.27	3.96	41.59	9.18
C23H24I2N6	638.30	638.0156	43.28	3.79	39.76	13.17
C23H25I1N2	456.37	456.1064	60.53	5.52	27.81	6.14
C23H25I1N4	484.39	484.1126	57.03	5.20	26.20	11.57
C23H25I1N6	512.40	512.1187	53.91	4.92	24.77	16.40
C23H25I2N1	569.27	569.0080	48.53	4.43	44.58	2.46
C23H25I2N3	597.28	597.0142	46.25	4.22	42.49	7.04
C23H25I2N5	625.30	625.0203	44.18	4.03	40.59	11.20
C23H26I1N1	443.37	443.1112	62.31	5.91	28.62	3.16
C23H26I1N3	471.39	471.1173	58.60	5.56	26.92	8.91
C23H26I1N5	499.40	499.1235	55.32	5.25	25.41	14.02
C23H26I2N2	584.29	584.0189	47.28	4.49	43.44	4.79
C23H26I2N4	612.30	612.0251	45.12	4.28	41.45	9.15
C23H26I2N6	640.31	640.0312	43.14	4.09	39.64	13.12
C23H27I1N2	458.39	458.1221	60.27	5.94	27.68	6.11
C23H27I1N4	486.40	486.1282	56.80	5.60	26.09	11.52
C23H27I1N6	514.42	514.1344	53.70	5.29	24.67	16.34
C23H27I2N1	571.29	571.0237	48.36	4.76	44.43	2.45
C23H27I2N3	599.30	599.0298	46.10	4.54	42.35	7.01
C23H27I2N5	627.31	627.0360	44.04	4.34	40.46	11.16
C23H28I1N1	445.39	445.1268	62.03	6.34	28.49	3.14
C23H28I1N3	473.40	473.1330	58.36	5.96	26.81	8.88
C23H28I1N5	501.42	501.1391	55.10	5.63	25.31	13.97
C23H28I2N2	586.30	586.0346	47.12	4.81	43.29	4.78
C23H28I2N4	614.32	614.0407	44.97	4.59	41.32	9.12
C23H28I2N6	642.33	642.0469	43.01	4.39	39.51	13.08
C23H29I1N2	460.41	460.1377	60.00	6.35	27.56	6.08
C23H29I1N4	488.42	488.1439	56.56	5.98	25.98	11.47
C23H29I1N6	516.43	516.1500	53.49	5.66	24.57	16.27
C23H29I2N1	573.30	573.0393	48.19	5.10	44.27	2.44
C23H29I2N3	601.32	601.0455	45.94	4.86	42.21	6.99
C23H29I2N5	629.33	629.0516	43.90	4.64	40.33	11.13
C23H30I1N1	447.41	447.1425	61.75	6.76	28.36	3.13
C23H30I1N3	475.42	475.1486	58.11	6.36	26.69	8.84
C23H30I1N5	503.43	503.1548	54.87	6.01	25.21	13.91
C23H30I2N2	588.32	588.0502	46.96	5.14	43.14	4.76
C23H30I2N4	616.33	616.0564	44.82	4.91	41.18	9.09
C23H30I2N6	644.34	644.0625	42.87	4.69	39.39	13.04
C23H31I1N2	462.42	462.1534	59.74	6.76	27.44	6.06
C23H31I1N4	490.43	490.1595	56.33	6.37	25.88	11.42
C23H31I1N6	518.45	518.1657	53.29	6.03	24.48	16.21
C23H31I2N1	575.32	575.0550	48.02	5.43	44.12	2.43

C23H31I2N3	603.33	603.0611	45.79	5.18	42.07	6.96
C23H31I2N5	631.35	631.0673	43.76	4.95	40.20	11.09
C23H32I1N1	449.42	449.1581	61.47	7.18	28.24	3.12
C23H32I1N3	477.44	477.1643	57.86	6.76	26.58	8.80
C23H32I1N5	505.45	505.1704	54.66	6.38	25.11	13.86
C23H32I2N2	590.33	590.0659	46.80	5.46	42.99	4.75
C23H32I2N4	618.35	618.0720	44.68	5.22	41.05	9.06
C23H32I2N6	646.36	646.0781	42.74	4.99	39.27	13.00
C23H33I1N2	464.44	464.1690	59.48	7.16	27.32	6.03
C23H33I1N4	492.45	492.1752	56.10	6.75	25.77	11.38
C23H33I1N6	520.46	520.1813	53.08	6.39	24.38	16.15
C23H33I2N1	577.33	577.0706	47.85	5.76	43.96	2.43
C23H33I2N3	605.35	605.0768	45.64	5.49	41.93	6.94
C23H33I2N5	633.36	633.0829	43.62	5.25	40.07	11.06
C23H34I1N1	451.44	451.1738	61.19	7.59	28.11	3.10
C23H34I1N3	479.45	479.1799	57.62	7.15	26.47	8.76
C23H34I1N5	507.47	507.1861	54.44	6.75	25.01	13.80
C23H34I2N2	592.35	592.0815	46.64	5.79	42.85	4.73
C23H34I2N4	620.36	620.0877	44.53	5.52	40.91	9.03
C23H34I2N6	648.38	648.0938	42.61	5.29	39.15	12.96
C23H35I1N2	466.45	466.1847	59.22	7.56	27.21	6.01
C23H35I1N4	494.47	494.1908	55.87	7.13	25.66	11.33
C23H35I1N6	522.48	522.1970	52.87	6.75	24.29	16.08
C23H35I2N1	579.35	579.0863	47.68	6.09	43.81	2.42
C23H35I2N3	607.36	607.0924	45.48	5.81	41.79	6.92
C23H35I2N5	635.38	635.0985	43.48	5.55	39.95	11.02
C23H36I1N1	453.45	453.1894	60.92	8.00	27.99	3.09
C23H36I1N3	481.47	481.1956	57.38	7.54	26.36	8.73
C23H36I1N5	509.48	509.2017	54.22	7.12	24.91	13.75
C23H36I2N2	594.37	594.0972	46.48	6.11	42.70	4.71
C23H36I2N4	622.38	622.1033	44.39	5.83	40.78	9.00
C23H36I2N6	650.39	650.1094	42.48	5.58	39.02	12.92
C23H37I1N2	468.47	468.2003	58.97	7.96	27.09	5.98
C23H37I1N4	496.48	496.2065	55.64	7.51	25.56	11.28
C23H37I1N6	524.50	524.2126	52.67	7.11	24.20	16.02
C23H37I2N1	581.37	581.1019	47.52	6.42	43.66	2.41
C23H37I2N3	609.38	609.1081	45.33	6.12	41.65	6.90
C23H37I2N5	637.39	637.1142	43.34	5.85	39.82	10.99
C23H38I1N1	455.47	455.2051	60.65	8.41	27.86	3.08
C23H38I1N3	483.48	483.2112	57.14	7.92	26.25	8.69
C23H38I1N5	511.50	511.2174	54.01	7.49	24.81	13.69
C23H38I2N2	596.38	596.1128	46.32	6.42	42.56	4.70
C23H38I2N4	624.39	624.1189	44.24	6.13	40.65	8.97
C23H38I2N6	652.41	652.1251	42.34	5.87	38.90	12.88
C23H39I1N2	470.49	470.2160	58.72	8.36	26.97	5.95
C23H39I1N4	498.50	498.2221	55.42	7.89	25.46	11.24
C23H39I1N6	526.51	526.2283	52.47	7.47	24.10	15.96
C23H39I2N1	583.38	583.1176	47.35	6.74	43.51	2.40
C23H39I2N3	611.40	611.1237	45.18	6.43	41.51	6.87
C23H39I2N5	639.41	639.1298	43.20	6.15	39.69	10.95
C23H40I1N1	457.49	457.2207	60.39	8.81	27.74	3.06
C23H40I1N3	485.50	485.2269	56.90	8.30	26.14	8.66
C23H40I1N5	513.51	513.2330	53.80	7.85	24.71	13.64
C23H40I2N2	598.40	598.1285	46.17	6.74	42.41	4.68
C23H40I2N4	626.41	626.1346	44.10	6.44	40.52	8.94
C23H40I2N6	654.42	654.1407	42.21	6.16	38.78	12.84
C23H41I1N2	472.50	472.2316	58.47	8.75	26.86	5.93
C23H41I1N4	500.51	500.2378	55.19	8.26	25.35	11.19
C23H41I1N6	528.53	528.2439	52.27	7.82	24.01	15.90
C23H41I2N1	585.40	585.1332	47.19	7.06	43.36	2.39

036

C23H41I2N3	613.41	613.1394	45.04	6.74	41.38	6.85
C23H41I2N5	641.43	641.1455	43.07	6.44	39.57	10.92
C23H42I1N1	459.50	459.2364	60.12	9.21	27.62	3.05
C23H42I1N3	487.52	487.2425	56.67	8.68	26.03	8.62
C23H42I1N5	515.53	515.2487	53.59	8.21	24.62	13.58
C23H42I2N2	600.41	600.1441	46.01	7.05	42.27	4.67
C23H42I2N4	628.43	628.1502	43.96	6.74	40.39	8.92
C23H42I2N6	656.44	656.1564	42.08	6.45	38.66	12.80
C23H43I1N2	474.52	474.2473	58.22	9.13	26.74	5.90
C23H43I1N4	502.53	502.2534	54.97	8.62	25.25	11.15
C23H43I1N6	530.54	530.2596	52.07	8.17	23.92	15.84
C23H43I2N1	587.41	587.1489	47.03	7.38	43.21	2.38
C23H43I2N3	615.43	615.1550	44.89	7.04	41.24	6.83
C23H43I2N5	643.44	643.1611	42.93	6.74	39.45	10.88
C23H44I1N1	461.52	461.2520	59.86	9.61	27.50	3.03
C23H44I1N3	489.53	489.2582	56.43	9.06	25.92	8.58
C23H44I1N5	517.55	517.2643	53.38	8.57	24.52	13.53
C23H44I2N2	602.43	602.1597	45.86	7.36	42.13	4.65
C23H44I2N4	630.44	630.1659	43.82	7.03	40.26	8.89
C23H44I2N6	658.46	658.1720	41.96	6.74	38.55	12.76
C23H45I1N2	476.53	476.2629	57.97	9.52	26.63	5.88
C23H45I1N4	504.55	504.2691	54.75	8.99	25.15	11.10
C23H45I1N6	532.56	532.2752	51.87	8.52	23.83	15.78
C23H45I2N1	589.43	589.1645	46.87	7.70	43.06	2.38
C23H45I2N3	617.44	617.1706	44.74	7.35	41.11	6.81
C23H45I2N5	645.46	645.1768	42.80	7.03	39.32	10.85
C23H46I1N1	463.53	463.2677	59.60	10.00	27.38	3.02
C23H46I1N3	491.55	491.2738	56.20	9.43	25.82	8.55
C23H46I1N5	519.56	519.2800	53.17	8.92	24.43	13.48
C23H46I2N2	604.45	604.1754	45.70	7.67	41.99	4.63
C23H46I2N4	632.46	632.1815	43.68	7.33	40.13	8.86
C23H46I2N6	660.47	660.1877	41.83	7.02	38.43	12.72
C23H47I1N2	478.55	478.2786	57.73	9.90	26.52	5.85
C23H47I1N4	506.56	506.2847	54.54	9.35	25.05	11.06
C23H47I1N6	534.58	534.2909	51.68	8.86	23.74	15.72
C23H47I2N1	591.45	591.1802	46.71	8.01	42.91	2.37
C23H47I2N3	619.46	619.1863	44.60	7.65	40.97	6.78
C23H47I2N5	647.47	647.1924	42.67	7.32	39.20	10.82
C23H48I1N1	465.55	465.2833	59.34	10.39	27.26	3.01
C23H48I1N3	493.56	493.2895	55.97	9.80	25.71	8.51
C23H48I1N5	521.58	521.2956	52.97	9.28	24.33	13.43
C23H48I2N2	606.46	606.1910	45.55	7.98	41.85	4.62
C23H48I2N4	634.47	634.1972	43.54	7.63	40.00	8.83
C23H48I2N6	662.49	662.2033	41.70	7.30	38.31	12.69
C23H49I1N2	480.56	480.2942	57.49	10.28	26.41	5.83
C23H49I1N4	508.58	508.3004	54.32	9.71	24.95	11.02
C23H49I1N6	536.59	536.3065	51.48	9.20	23.65	15.66
C23H49I2N3	621.48	621.2019	44.45	7.95	40.84	6.76
C23H49I2N5	649.49	649.2081	42.53	7.60	39.08	10.78
C23H50I1N3	495.58	495.3051	55.74	10.17	25.61	8.48
C23H50I1N5	523.59	523.3113	52.76	9.63	24.24	13.38
C23H50I2N4	636.49	636.2128	43.40	7.92	39.88	8.80
C23H50I2N6	664.50	664.2190	41.57	7.58	38.20	12.65
C23H51I1N4	510.59	510.3160	54.10	10.07	24.85	10.97
C23H51I1N6	538.61	538.3222	51.29	9.54	23.56	15.60
C23H51I2N5	651.51	651.2237	42.40	7.89	38.96	10.75
C23H52I1N5	525.61	525.3269	52.56	9.97	24.14	13.32
C23H52I2N6	666.52	666.2346	41.45	7.86	38.08	12.61
C23H53I1N6	540.62	540.3378	51.10	9.88	23.47	15.55
C24H12I1N1	441.27	441.0016	65.33	2.74	28.76	3.17

C24H12I1N3	469.29	469.0078	61.43	2.58	27.04	8.95
C24H12I1N5	497.30	497.0139	57.97	2.43	25.52	14.08
C24H12I2N2	582.19	581.9094	49.51	2.08	43.60	4.81
C24H12I2N4	610.20	609.9155	47.24	1.98	41.59	9.18
C24H12I2N6	638.21	637.9217	45.17	1.90	39.77	13.17
C24H13I1N2	456.29	456.0125	63.18	2.87	27.81	6.14
C24H13I1N4	484.30	484.0187	59.52	2.71	26.20	11.57
C24H13I1N6	512.32	512.0248	56.27	2.56	24.77	16.40
C24H13I2N1	569.19	568.9141	50.65	2.30	44.59	2.46
C24H13I2N3	597.20	596.9203	48.27	2.19	42.50	7.04
C24H13I2N5	625.21	624.9264	46.11	2.10	40.60	11.20
C24H14I1N1	443.29	443.0173	65.03	3.18	28.63	3.16
C24H14I1N3	471.30	471.0234	61.16	2.99	26.93	8.92
C24H14I1N5	499.32	499.0296	57.73	2.83	25.42	14.03
C24H14I2N2	584.20	583.9250	49.34	2.42	43.45	4.80
C24H14I2N4	612.21	611.9312	47.09	2.31	41.46	9.15
C24H14I2N6	640.23	639.9373	45.03	2.20	39.64	13.13
C24H15I1N2	458.30	458.0282	62.90	3.30	27.69	6.11
C24H15I1N4	486.32	486.0343	59.28	3.11	26.09	11.52
C24H15I1N6	514.33	514.0405	56.05	2.94	24.67	16.34
C24H15I2N1	571.20	570.9298	50.47	2.65	44.43	2.45
C24H15I2N3	599.22	598.9359	48.11	2.52	42.36	7.01
C24H15I2N5	627.23	626.9421	45.96	2.41	40.47	11.17
C24H16I1N1	445.31	445.0329	64.73	3.62	28.50	3.15
C24H16I1N3	473.32	473.0391	60.90	3.41	26.81	8.88
C24H16I1N5	501.33	501.0452	57.50	3.22	25.31	13.97
C24H16I2N2	586.22	585.9407	49.17	2.75	43.30	4.78
C24H16I2N4	614.23	613.9468	46.93	2.63	41.32	9.12
C24H16I2N6	642.24	641.9530	44.88	2.51	39.52	13.09
C24H17I1N2	460.32	460.0438	62.62	3.72	27.57	6.09
C24H17I1N4	488.33	488.0500	59.03	3.51	25.99	11.47
C24H17I1N6	516.35	516.0561	55.83	3.32	24.58	16.28
C24H17I2N1	573.22	572.9454	50.29	2.99	44.28	2.44
C24H17I2N3	601.23	600.9516	47.95	2.85	42.21	6.99
C24H17I2N5	629.25	628.9577	45.81	2.72	40.34	11.13
C24H18I1N1	447.32	447.0486	64.44	4.06	28.37	3.13
C24H18I1N3	475.34	475.0547	60.65	3.82	26.70	8.84
C24H18I1N5	503.35	503.0609	57.27	3.60	25.21	13.91
C24H18I2N2	588.23	587.9563	49.01	3.08	43.15	4.76
C24H18I2N4	616.25	615.9625	46.78	2.94	41.19	9.09
C24H18I2N6	644.26	643.9686	44.74	2.82	39.40	13.04
C24H19I1N2	462.34	462.0595	62.35	4.14	27.45	6.06
C24H19I1N4	490.35	490.0656	58.79	3.91	25.88	11.43
C24H19I1N6	518.36	518.0718	55.61	3.69	24.48	16.21
C24H19I2N1	575.23	574.9611	50.11	3.33	44.12	2.43
C24H19I2N3	603.25	602.9672	47.79	3.17	42.07	6.97
C24H19I2N5	631.26	630.9734	45.67	3.03	40.21	11.09
C24H20I1N1	449.34	449.0642	64.15	4.49	28.24	3.12
C24H20I1N3	477.35	477.0704	60.39	4.22	26.59	8.80
C24H20I1N5	505.36	505.0765	57.04	3.99	25.11	13.86
C24H20I2N2	590.25	589.9720	48.84	3.42	43.00	4.75
C24H20I2N4	618.26	617.9781	46.63	3.26	41.05	9.06
C24H20I2N6	646.28	645.9843	44.60	3.12	39.27	13.00
C24H21I1N2	464.35	464.0751	62.08	4.56	27.33	6.03
C24H21I1N4	492.37	492.0813	58.55	4.30	25.77	11.38
C24H21I1N6	520.38	520.0874	55.40	4.07	24.39	16.15
C24H21I2N1	577.25	576.9767	49.94	3.67	43.97	2.43
C24H21I2N3	605.26	604.9829	47.63	3.50	41.93	6.94
C24H21I2N5	633.28	632.9890	45.52	3.34	40.08	11.06
C24H22I1N1	451.35	451.0799	63.87	4.91	28.12	3.10

037

C24H22I1N3	479.37	479.0860	60.13	4.63	26.47	8.77
C24H22I1N5	507.38	507.0922	56.81	4.37	25.01	13.80
C24H22I2N2	592.27	591.9876	48.67	3.74	42.85	4.73
C24H22I2N4	620.28	619.9938	46.47	3.58	40.92	9.03
C24H22I2N6	648.29	647.9999	44.47	3.42	39.15	12.96
C24H23I1N2	466.37	466.0908	61.81	4.97	27.21	6.01
C24H23I1N4	494.38	494.0969	58.31	4.69	25.67	11.33
C24H23I1N6	522.40	522.1031	55.18	4.44	24.29	16.09
C24H23I2N1	579.27	578.9924	49.76	4.00	43.82	2.42
C24H23I2N3	607.28	606.9985	47.47	3.82	41.79	6.92
C24H23I2N5	635.29	635.0047	45.38	3.65	39.95	11.02
C24H24I1N1	453.37	453.0955	63.58	5.34	27.99	3.09
C24H24I1N3	481.38	481.1017	59.88	5.03	26.36	8.73
C24H24I1N5	509.40	509.1078	56.59	4.75	24.91	13.75
C24H24I2N2	594.28	594.0033	48.51	4.07	42.71	4.71
C24H24I2N4	622.29	622.0094	46.32	3.89	40.79	9.00
C24H24I2N6	650.31	650.0156	44.33	3.72	39.03	12.92
C24H25I1N2	468.38	468.1064	61.55	5.38	27.09	5.98
C24H25I1N4	496.40	496.1126	58.07	5.08	25.57	11.29
C24H25I1N6	524.41	524.1187	54.97	4.81	24.20	16.03
C24H25I2N1	581.28	581.0080	49.59	4.34	43.66	2.41
C24H25I2N3	609.30	609.0142	47.31	4.14	41.66	6.90
C24H25I2N5	637.31	637.0203	45.23	3.95	39.83	10.99
C24H26I1N1	455.39	455.1112	63.30	5.75	27.87	3.08
C24H26I1N3	483.40	483.1173	59.63	5.42	26.25	8.69
C24H26I1N5	511.41	511.1235	56.37	5.12	24.81	13.69
C24H26I2N2	596.30	596.0189	48.34	4.39	42.56	4.70
C24H26I2N4	624.31	624.0251	46.17	4.20	40.65	8.97
C24H26I2N6	652.32	652.0312	44.19	4.02	38.91	12.88
C24H27I1N2	470.40	470.1221	61.28	5.79	26.98	5.96
C24H27I1N4	498.41	498.1282	57.84	5.46	25.46	11.24
C24H27I1N6	526.43	526.1344	54.76	5.17	24.11	15.96
C24H27I2N1	583.30	583.0237	49.42	4.67	43.51	2.40
C24H27I2N3	611.31	611.0298	47.16	4.45	41.52	6.87
C24H27I2N5	639.33	639.0360	45.09	4.26	39.70	10.95
C24H28I1N1	457.40	457.1268	63.02	6.17	27.74	3.06
C24H28I1N3	485.42	485.1330	59.39	5.81	26.14	8.66
C24H28I1N5	513.43	513.1391	56.15	5.50	24.72	13.64
C24H28I2N2	598.31	598.0346	48.18	4.72	42.42	4.68
C24H28I2N4	626.33	626.0407	46.03	4.51	40.52	8.95
C24H28I2N6	654.34	654.0469	44.05	4.31	38.79	12.84
C24H29I1N2	472.42	472.1377	61.02	6.19	26.86	5.93
C24H29I1N4	500.43	500.1439	57.60	5.84	25.36	11.20
C24H29I1N6	528.44	528.1500	54.55	5.53	24.01	15.90
C24H29I2N1	585.31	585.0393	49.25	4.99	43.36	2.39
C24H29I2N3	613.33	613.0455	47.00	4.77	41.38	6.85
C24H29I2N5	641.34	641.0516	44.95	4.56	39.57	10.92
C24H30I1N1	459.42	459.1425	62.75	6.58	27.62	3.05
C24H30I1N3	487.43	487.1486	59.14	6.20	26.04	8.62
C24H30I1N5	515.44	515.1548	55.93	5.87	24.62	13.59
C24H30I2N2	600.33	600.0502	48.02	5.04	42.28	4.67
C24H30I2N4	628.34	628.0564	45.88	4.81	40.39	8.92
C24H30I2N6	656.36	656.0625	43.92	4.61	38.67	12.80
C24H31I1N2	474.43	474.1534	60.76	6.59	26.75	5.90
C24H31I1N4	502.45	502.1595	57.37	6.22	25.26	11.15
C24H31I1N6	530.46	530.1657	54.34	5.89	23.92	15.84
C24H31I2N1	587.33	587.0550	49.08	5.32	43.21	2.38
C24H31I2N3	615.34	615.0611	46.85	5.08	41.25	6.83
C24H31I2N5	643.36	643.0673	44.81	4.86	39.45	10.89
C24H32I1N1	461.43	461.1581	62.47	6.99	27.50	3.04
C24H32I1N3	489.45	489.1643	58.90	6.59	25.93	8.59
C24H32I1N5	517.46	517.1704	55.71	6.23	24.52	13.53
C24H32I2N2	602.34	602.0659	47.86	5.35	42.14	4.65
C24H32I2N4	630.36	630.0720	45.73	5.12	40.26	8.89
C24H32I2N6	658.37	658.0781	43.78	4.90	38.55	12.76
C24H33I1N2	476.45	476.1690	60.50	6.98	26.64	5.88
C24H33I1N4	504.46	504.1752	57.14	6.59	25.16	11.11
C24H33I1N6	532.48	532.1813	54.14	6.25	23.83	15.78
C24H33I2N1	589.35	589.0706	48.91	5.64	43.07	2.38
C24H33I2N3	617.36	617.0768	46.69	5.39	41.11	6.81
C24H33I2N5	645.37	645.0829	44.67	5.15	39.33	10.85
C24H34I1N1	463.45	463.1738	62.20	7.39	27.38	3.02
C24H34I1N3	491.46	491.1799	58.65	6.97	25.82	8.55
C24H34I1N5	519.48	519.1861	55.49	6.60	24.43	13.48
C24H34I2N2	604.36	604.0815	47.70	5.67	42.00	4.64
C24H34I2N4	632.37	632.0877	45.58	5.42	40.14	8.86
C24H34I2N6	660.39	660.0938	43.65	5.19	38.43	12.73
C24H35I1N2	478.46	478.1847	60.25	7.37	26.52	5.85
C24H35I1N4	506.48	506.1908	56.92	6.97	25.06	11.06
C24H35I1N6	534.49	534.1970	53.93	6.60	23.74	15.72
C24H35I2N1	591.36	591.0863	48.75	5.97	42.92	2.37
C24H35I2N3	619.38	619.0924	46.54	5.70	40.98	6.78
C24H35I2N5	647.39	647.0985	44.53	5.45	39.21	10.82
C24H36I1N1	465.47	465.1894	61.93	7.80	27.26	3.01
C24H36I1N3	493.48	493.1956	58.42	7.35	25.72	8.52
C24H36I1N5	521.49	521.2017	55.28	6.96	24.33	13.43
C24H36I2N2	606.38	606.0972	47.54	5.98	41.86	4.62
C24H36I2N4	634.39	634.1033	45.44	5.72	40.01	8.83
C24H36I2N6	662.40	662.1094	43.52	5.48	38.32	12.69
C24H37I1N2	480.48	480.2003	60.00	7.76	26.41	5.83
C24H37I1N4	508.49	508.2065	56.69	7.33	24.96	11.02
C24H37I1N6	536.51	536.2126	53.73	6.95	23.65	15.66
C24H37I2N1	593.38	593.1019	48.58	6.29	42.77	2.36
C24H37I2N3	621.39	621.1081	46.39	6.00	40.85	6.76
C24H37I2N5	649.40	649.1142	44.39	5.74	39.08	10.78
C24H38I1N1	467.48	467.2051	61.66	8.19	27.15	3.00
C24H38I1N3	495.49	495.2112	58.18	7.73	25.61	8.48
C24H38I1N5	523.51	523.2174	55.06	7.32	24.24	13.38
C24H38I2N2	608.39	608.1128	47.38	6.30	41.72	4.60
C24H38I2N4	636.41	636.1189	45.30	6.02	39.88	8.80
C24H38I2N6	664.42	664.1251	43.39	5.76	38.20	12.65
C24H39I1N2	482.50	482.2160	59.75	8.15	26.30	5.81
C24H39I1N4	510.51	510.2221	56.47	7.70	24.86	10.97
C24H39I1N6	538.52	538.2283	53.53	7.30	23.57	15.61
C24H39I2N1	595.39	595.1176	48.42	6.60	42.63	2.35
C24H39I2N3	623.41	623.1237	46.24	6.31	40.71	6.74
C24H39I2N5	651.42	651.1298	44.25	6.03	38.96	10.75
C24H40I1N1	469.50	469.2207	61.40	8.59	27.03	2.98
C24H40I1N3	497.51	497.2269	57.94	8.10	25.51	8.45
C24H40I1N5	525.52	525.2330	54.85	7.67	24.15	13.33
C24H40I2N2	610.41	610.1285	47.23	6.61	41.58	4.59
C24H40I2N4	638.42	638.1346	45.15	6.32	39.76	8.78
C24H40I2N6	666.44	666.1407	43.26	6.05	38.08	12.61
C24H41I1N2	484.51	484.2316	59.50	8.53	26.19	5.78
C24H41I1N4	512.53	512.2378	56.24	8.06	24.76	10.93
C24H41I1N6	540.54	540.2439	53.33	7.65	23.48	15.55
C24H41I2N1	597.41	597.1332	48.25	6.92	42.48	2.34
C24H41I2N3	625.42	625.1394	46.09	6.61	40.58	6.72
C24H41I2N5	653.44	653.1455	44.12	6.32	38.84	10.72
C24H42I1N1	471.51	471.2364	61.14	8.98	26.91	2.97

038

C24H42I1N3	499.53	499.2425	57.71	8.47	25.40	8.41
C24H42I1N5	527.54	527.2487	54.64	8.02	24.06	13.28
C24H42I2N2	612.42	612.1441	47.07	6.91	41.44	4.57
C24H42I2N4	640.44	640.1502	45.01	6.61	39.63	8.75
C24H42I2N6	668.45	668.1564	43.12	6.33	37.97	12.57
C24H43I1N2	486.53	486.2473	59.25	8.91	26.08	5.76
C24H43I1N4	514.54	514.2534	56.02	8.42	24.66	10.89
C24H43I1N6	542.55	542.2596	53.13	7.99	23.39	15.49
C24H43I2N1	599.43	599.1489	48.09	7.23	42.34	2.34
C24H43I2N3	627.44	627.1550	45.94	6.91	40.45	6.70
C24H43I2N5	655.45	655.1611	43.98	6.61	38.72	10.68
C24H44I1N1	473.53	473.2520	60.88	9.37	26.80	2.96
C24H44I1N3	501.54	501.2582	57.48	8.84	25.30	8.38
C24H44I1N5	529.56	529.2643	54.44	8.38	23.96	13.22
C24H44I2N2	614.44	614.1597	46.92	7.22	41.31	4.56
C24H44I2N4	642.45	642.1659	44.87	6.90	39.51	8.72
C24H44I2N6	670.47	670.1720	43.00	6.61	37.86	12.53
C24H45I1N2	488.54	488.2629	59.01	9.28	25.98	5.73
C24H45I1N4	516.56	516.2691	55.81	8.78	24.57	10.85
C24H45I1N6	544.57	544.2752	52.93	8.33	23.30	15.43
C24H45I2N1	601.44	601.1645	47.93	7.54	42.20	2.33
C24H45I2N3	629.46	629.1706	45.80	7.21	40.32	6.68
C24H45I2N5	657.47	657.1768	43.85	6.90	38.60	10.65
C24H46I1N1	475.55	475.2677	60.62	9.75	26.69	2.95
C24H46I1N3	503.56	503.2738	57.25	9.21	25.20	8.34
C24H46I1N5	531.57	531.2800	54.23	8.72	23.87	13.17
C24H46I2N2	616.46	616.1754	46.76	7.52	41.17	4.54
C24H46I2N4	644.47	644.1815	44.73	7.19	39.38	8.69
C24H46I2N6	672.48	672.1877	42.87	6.89	37.74	12.50
C24H47I1N2	490.56	490.2786	58.76	9.66	25.87	5.71
C24H47I1N4	518.57	518.2847	55.59	9.14	24.47	10.80
C24H47I1N6	546.59	546.2909	52.74	8.67	23.22	15.38
C24H47I2N1	603.46	603.1802	47.77	7.85	42.06	2.32
C24H47I2N3	631.47	631.1863	45.65	7.50	40.19	6.65
C24H47I2N5	659.48	659.1924	43.71	7.18	38.49	10.62
C24H48I1N1	477.56	477.2833	60.36	10.13	26.57	2.93
C24H48I1N3	505.57	505.2895	57.02	9.57	25.10	8.31
C24H48I1N5	533.59	533.2956	54.02	9.07	23.78	13.13
C24H48I2N2	618.47	618.1910	46.61	7.82	41.04	4.53
C24H48I2N4	646.49	646.1972	44.59	7.48	39.26	8.67
C24H48I2N6	674.50	674.2033	42.74	7.17	37.63	12.46
C24H49I1N2	492.58	492.2942	58.52	10.03	25.76	5.69
C24H49I1N4	520.59	520.3004	55.37	9.49	24.38	10.76
C24H49I1N6	548.60	548.3065	52.55	9.00	23.13	15.32
C24H49I2N1	605.47	605.1958	47.61	8.16	41.92	2.31
C24H49I2N3	633.49	633.2019	45.50	7.80	40.07	6.63
C24H49I2N5	661.50	661.2081	43.58	7.47	38.37	10.59
C24H50I1N1	479.58	479.2990	60.11	10.51	26.46	2.92
C24H50I1N3	507.59	507.3051	56.79	9.93	25.00	8.28
C24H50I1N5	535.60	535.3113	53.82	9.41	23.69	13.08
C24H50I2N2	620.49	620.2067	46.46	8.12	40.90	4.51
C24H50I2N4	648.50	648.2128	44.45	7.77	39.14	8.64
C24H50I2N6	676.52	676.2190	42.61	7.45	37.52	12.42
C24H51I1N2	494.59	494.3099	58.28	10.39	25.66	5.66
C24H51I1N4	522.61	522.3160	55.16	9.84	24.28	10.72
C24H51I1N6	550.62	550.3222	52.35	9.34	23.05	15.26
C24H51I2N3	635.50	635.2176	45.36	8.09	39.94	6.61
C24H51I2N5	663.52	663.2237	43.45	7.75	38.25	10.55
C24H52I1N3	509.61	509.3208	56.57	10.29	24.90	8.25
C24H52I1N5	537.62	537.3269	53.62	9.75	23.60	13.03
C24H52I2N4	650.52	650.2285	44.31	8.06	39.02	8.61
C24H52I2N6	678.53	678.2346	42.48	7.72	37.41	12.39
C24H53I1N4	524.62	524.3317	54.95	10.18	24.19	10.68
C24H53I1N6	552.63	552.3378	52.16	9.67	22.96	15.21
C24H53I2N5	665.53	665.2394	43.31	8.03	38.14	10.52
C24H54I1N5	539.64	539.3426	53.42	10.09	23.52	12.98
C24H54I2N6	680.55	680.2503	42.36	8.00	37.29	12.35
C24H55I1N6	554.65	554.3535	51.97	10.00	22.88	15.15
C25H12I1N1	453.29	453.0016	66.24	2.67	28.00	3.09
C25H12I1N3	481.30	481.0078	62.39	2.51	26.37	8.73
C25H12I1N5	509.31	509.0139	58.96	2.37	24.92	13.75
C25H12I2N2	594.20	593.9094	50.54	2.04	42.71	4.71
C25H12I2N4	622.21	621.9155	48.26	1.94	40.79	9.00
C25H12I2N6	650.22	649.9217	46.18	1.86	39.03	12.92
C25H13I1N2	468.30	468.0125	64.12	2.80	27.10	5.98
C25H13I1N4	496.31	496.0187	60.50	2.64	25.57	11.29
C25H13I1N6	524.33	524.0248	57.27	2.50	24.20	16.03
C25H13I2N1	581.20	580.9141	51.67	2.25	43.67	2.41
C25H13I2N3	609.21	608.9203	49.29	2.15	41.66	6.90
C25H13I2N5	637.22	636.9264	47.12	2.06	39.83	10.99
C25H14I1N1	455.30	455.0173	65.95	3.10	27.87	3.08
C25H14I1N3	483.31	483.0234	62.13	2.92	26.26	8.69
C25H14I1N5	511.33	511.0296	58.73	2.76	24.82	13.70
C25H14I2N2	596.21	595.9250	50.36	2.37	42.57	4.70
C25H14I2N4	624.23	623.9312	48.10	2.26	40.66	8.98
C25H14I2N6	652.24	651.9373	46.04	2.16	38.91	12.88
C25H15I1N2	470.32	470.0282	63.85	3.21	26.98	5.96
C25H15I1N4	498.33	498.0343	60.26	3.03	25.47	11.24
C25H15I1N6	526.34	526.0405	57.05	2.87	24.11	15.97
C25H15I2N1	583.21	582.9298	51.49	2.59	43.52	2.40
C25H15I2N3	611.23	610.9359	49.13	2.47	41.52	6.87
C25H15I2N5	639.24	638.9421	46.97	2.37	39.70	10.96
C25H16I1N1	457.32	457.0329	65.66	3.53	27.75	3.06
C25H16I1N3	485.33	485.0391	61.87	3.32	26.15	8.66
C25H16I1N5	513.34	513.0452	58.49	3.14	24.72	13.64
C25H16I2N2	598.23	597.9407	50.19	2.70	42.43	4.68
C25H16I2N4	626.24	625.9468	47.95	2.58	40.53	8.95
C25H16I2N6	654.26	653.9530	45.90	2.47	38.79	12.85
C25H17I1N2	472.33	472.0438	63.57	3.63	26.87	5.93
C25H17I1N4	500.35	500.0500	60.01	3.42	25.36	11.20
C25H17I1N6	528.36	528.0561	56.83	3.24	24.02	15.91
C25H17I2N1	585.23	584.9454	51.31	2.93	43.37	2.39
C25H17I2N3	613.24	612.9516	48.97	2.79	41.39	6.85
C25H17I2N5	641.26	640.9577	46.83	2.67	39.58	10.92
C25H18I1N1	459.33	459.0486	65.37	3.95	27.63	3.05
C25H18I1N3	487.35	487.0547	61.62	3.72	26.04	8.62
C25H18I1N5	515.36	515.0609	58.27	3.52	24.62	13.59
C25H18I2N2	600.24	599.9563	50.03	3.02	42.28	4.67
C25H18I2N4	628.26	627.9625	47.80	2.89	40.40	8.92
C25H18I2N6	656.27	655.9686	45.76	2.76	38.67	12.81
C25H19I1N2	474.35	474.0595	63.30	4.04	26.75	5.91
C25H19I1N4	502.36	502.0656	59.77	3.81	25.26	11.15
C25H19I1N6	530.37	530.0718	56.62	3.61	23.93	15.85
C25H19I2N1	587.25	586.9611	51.13	3.26	43.22	2.39
C25H19I2N3	615.26	614.9672	48.81	3.11	41.25	6.83
C25H19I2N5	643.27	642.9734	46.68	2.98	39.46	10.89
C25H20I1N1	461.35	461.0642	65.09	4.37	27.51	3.04
C25H20I1N3	489.36	489.0704	61.36	4.12	25.93	8.59
C25H20I1N5	517.38	517.0765	58.04	3.90	24.53	13.54
C25H20I2N2	602.26	601.9720	49.86	3.35	42.14	4.65

039

C25H20I2N4	630.27	629.9781	47.64	3.20	40.27	8.89
C25H20I2N6	658.29	657.9843	45.62	3.06	38.56	12.77
C25H21I1N2	476.36	476.0751	63.04	4.44	26.64	5.88
C25H21I1N4	504.38	504.0813	59.53	4.20	25.16	11.11
C25H21I1N6	532.39	532.0874	56.40	3.98	23.84	15.79
C25H21I2N1	589.26	588.9767	50.96	3.59	43.07	2.38
C25H21I2N3	617.28	616.9829	48.65	3.43	41.12	6.81
C25H21I2N5	645.29	644.9890	46.53	3.28	39.33	10.85
C25H22I1N1	463.37	463.0799	64.80	4.79	27.39	3.02
C25H22I1N3	491.38	491.0860	61.11	4.51	25.83	8.55
C25H22I1N5	519.39	519.0922	57.81	4.27	24.43	13.48
C25H22I2N2	604.28	603.9876	49.69	3.67	42.00	4.64
C25H22I2N4	632.29	631.9938	47.49	3.51	40.14	8.86
C25H22I2N6	660.30	659.9999	45.48	3.36	38.44	12.73
C25H23I1N2	478.38	478.0908	62.77	4.85	26.53	5.86
C25H23I1N4	506.39	506.0969	59.30	4.58	25.06	11.06
C25H23I1N6	534.41	534.1031	56.19	4.34	23.75	15.73
C25H23I2N1	591.28	590.9924	50.78	3.92	42.93	2.37
C25H23I2N3	619.29	618.9985	48.49	3.74	40.98	6.79
C25H23I2N5	647.30	647.0047	46.39	3.58	39.21	10.82
C25H24I1N1	465.38	465.0955	64.52	5.20	27.27	3.01
C25H24I1N3	493.39	493.1017	60.86	4.90	25.72	8.52
C25H24I1N5	521.41	521.1078	57.59	4.64	24.34	13.43
C25H24I2N2	606.29	606.0033	49.53	3.99	41.86	4.62
C25H24I2N4	634.31	634.0094	47.34	3.81	40.01	8.83
C25H24I2N6	662.32	662.0156	45.34	3.65	38.32	12.69
C25H25I1N2	480.40	480.1064	62.51	5.25	26.42	5.83
C25H25I1N4	508.41	508.1126	59.06	4.96	24.96	11.02
C25H25I1N6	536.42	536.1187	55.98	4.70	23.66	15.67
C25H25I2N1	593.29	593.0080	50.61	4.25	42.78	2.36
C25H25I2N3	621.31	621.0142	48.33	4.06	40.85	6.76
C25H25I2N5	649.32	649.0203	46.25	3.88	39.09	10.79
C25H26I1N1	467.40	467.1112	64.24	5.61	27.15	3.00
C25H26I1N3	495.41	495.1173	60.61	5.29	25.62	8.48
C25H26I1N5	523.42	523.1235	57.37	5.01	24.25	13.38
C25H26I2N2	608.31	608.0189	49.36	4.31	41.72	4.61
C25H26I2N4	636.32	636.0251	47.19	4.12	39.89	8.80
C25H26I2N6	664.33	664.0312	45.20	3.94	38.20	12.65
C25H27I1N2	482.41	482.1221	62.25	5.64	26.31	5.81
C25H27I1N4	510.43	510.1282	58.83	5.33	24.86	10.98
C25H27I1N6	538.44	538.1344	55.77	5.05	23.57	15.61
C25H27I2N1	595.31	595.0237	50.44	4.57	42.63	2.35
C25H27I2N3	623.32	623.0298	48.17	4.37	40.72	6.74
C25H27I2N5	651.34	651.0360	46.10	4.18	38.97	10.75
C25H28I1N1	469.41	469.1268	63.97	6.01	27.03	2.98
C25H28I1N3	497.43	497.1330	60.37	5.67	25.51	8.45
C25H28I1N5	525.44	525.1391	57.15	5.37	24.15	13.33
C25H28I2N2	610.32	610.0346	49.20	4.62	41.59	4.59
C25H28I2N4	638.34	638.0407	47.04	4.42	39.76	8.78
C25H28I2N6	666.35	666.0469	45.06	4.24	38.09	12.61
C25H29I1N2	484.43	484.1377	61.99	6.03	26.20	5.78
C25H29I1N4	512.44	512.1439	58.60	5.70	24.76	10.93
C25H29I1N6	540.45	540.1500	55.56	5.41	23.48	15.55
C25H29I2N1	597.33	597.0393	50.27	4.89	42.49	2.34
C25H29I2N3	625.34	625.0455	48.02	4.67	40.59	6.72
C25H29I2N5	653.35	653.0516	45.96	4.47	38.85	10.72
C25H30I1N1	471.43	471.1425	63.70	6.41	26.92	2.97
C25H30I1N3	499.44	499.1486	60.12	6.05	25.41	8.41
C25H30I1N5	527.46	527.1548	56.93	5.73	24.06	13.28
C25H30I2N2	612.34	612.0502	49.04	4.94	41.45	4.57
C25H30I2N4	640.35	640.0564	46.89	4.72	39.64	8.75
C25H30I2N6	668.37	668.0625	44.93	4.52	37.97	12.57
C25H31I1N2	486.44	486.1534	61.73	6.42	26.09	5.76
C25H31I1N4	514.46	514.1595	58.37	6.07	24.67	10.89
C25H31I1N6	542.47	542.1657	55.35	5.76	23.39	15.49
C25H31I2N1	599.34	599.0550	50.10	5.21	42.35	2.34
C25H31I2N3	627.35	627.0611	47.86	4.98	40.46	6.70
C25H31I2N5	655.37	655.0673	45.82	4.77	38.73	10.69
C25H32I1N1	473.44	473.1581	63.42	6.81	26.80	2.96
C25H32I1N3	501.46	501.1643	59.88	6.43	25.31	8.38
C25H32I1N5	529.47	529.1704	56.71	6.09	23.97	13.23
C25H32I2N2	614.36	614.0659	48.88	5.25	41.31	4.56
C25H32I2N4	642.37	642.0720	46.75	5.02	39.51	8.72
C25H32I2N6	670.38	670.0781	44.79	4.81	37.86	12.54
C25H33I1N2	488.46	488.1690	61.47	6.81	25.98	5.74
C25H33I1N4	516.47	516.1752	58.14	6.44	24.57	10.85
C25H33I1N6	544.49	544.1813	55.15	6.11	23.31	15.43
C25H33I2N1	601.36	601.0706	49.93	5.53	42.21	2.33
C25H33I2N3	629.37	629.0768	47.71	5.29	40.33	6.68
C25H33I2N5	657.38	657.0829	45.68	5.06	38.61	10.65
C25H34I1N1	475.46	475.1738	63.16	7.21	26.69	2.95
C25H34I1N3	503.47	503.1799	59.64	6.81	25.21	8.35
C25H34I1N5	531.49	531.1861	56.50	6.45	23.88	13.18
C25H34I2N2	616.37	616.0815	48.72	5.56	41.18	4.54
C25H34I2N4	644.39	644.0877	46.60	5.32	39.39	8.69
C25H34I2N6	672.40	672.0938	44.66	5.10	37.75	12.50
C25H35I1N2	490.48	490.1847	61.22	7.19	25.87	5.71
C25H35I1N4	518.49	518.1908	57.91	6.80	24.48	10.81
C25H35I1N6	546.50	546.1970	54.95	6.46	23.22	15.38
C25H35I2N1	603.37	603.0863	49.77	5.85	42.06	2.32
C25H35I2N3	631.39	631.0924	47.56	5.59	40.20	6.66
C25H35I2N5	659.40	659.0985	45.54	5.35	38.49	10.62
C25H36I1N1	477.48	477.1894	62.89	7.60	26.58	2.93
C25H36I1N3	505.49	505.1956	59.40	7.18	25.11	8.31
C25H36I1N5	533.50	533.2017	56.28	6.80	23.79	13.13
C25H36I2N2	618.39	618.0972	48.56	5.87	41.04	4.53
C25H36I2N4	646.40	646.1033	46.45	5.61	39.26	8.67
C25H36I2N6	674.41	674.1094	44.52	5.38	37.63	12.46
C25H37I1N2	492.49	492.2003	60.97	7.57	25.77	5.69
C25H37I1N4	520.50	520.2065	57.69	7.17	24.38	10.76
C25H37I1N6	548.52	548.2126	54.74	6.80	23.14	15.32
C25H37I2N1	605.39	605.1019	49.60	6.16	41.92	2.31
C25H37I2N3	633.40	633.1081	47.41	5.89	40.07	6.63
C25H37I2N5	661.42	661.1142	45.40	5.64	38.37	10.59
C25H38I1N1	479.49	479.2051	62.62	7.99	26.47	2.92
C25H38I1N3	507.51	507.2112	59.17	7.55	25.01	8.28
C25H38I1N5	535.52	535.2174	56.07	7.15	23.70	13.08
C25H38I2N2	620.40	620.1128	48.40	6.17	40.91	4.52
C25H38I2N4	648.42	648.1189	46.31	5.91	39.14	8.64
C25H38I2N6	676.43	676.1251	44.39	5.66	37.52	12.42
C25H39I1N2	494.51	494.2160	60.72	7.95	25.66	5.66
C25H39I1N4	522.52	522.2221	57.47	7.52	24.29	10.72
C25H39I1N6	550.53	550.2283	54.54	7.14	23.05	15.27
C25H39I2N1	607.41	607.1176	49.44	6.47	41.79	2.31
C25H39I2N3	635.42	635.1237	47.26	6.19	39.94	6.61
C25H39I2N5	663.43	663.1298	45.26	5.93	38.26	10.56
C25H40I1N1	481.51	481.2207	62.36	8.37	26.36	2.91
C25H40I1N3	509.52	509.2269	58.93	7.91	24.91	8.25
C25H40I1N5	537.54	537.2330	55.86	7.50	23.61	13.03
C25H40I2N2	622.42	622.1285	48.24	6.48	40.78	4.50

040

C25H40I2N4	650.43	650.1346	46.17	6.20	39.02	8.61
C25H40I2N6	678.45	678.1407	44.26	5.94	37.41	12.39
C25H41I1N2	496.52	496.2316	60.48	8.32	25.56	5.64
C25H41I1N4	524.54	524.2378	57.25	7.88	24.19	10.68
C25H41I1N6	552.55	552.2439	54.34	7.48	22.97	15.21
C25H41I2N1	609.42	609.1332	49.27	6.78	41.65	2.30
C25H41I2N3	637.43	637.1394	47.11	6.48	39.82	6.59
C25H41I2N5	665.45	665.1455	45.12	6.21	38.14	10.52
C25H42I1N1	483.52	483.2364	62.10	8.76	26.25	2.90
C25H42I1N3	511.54	511.2425	58.70	8.28	24.81	8.21
C25H42I1N5	539.55	539.2487	55.65	7.85	23.52	12.98
C25H42I2N2	624.44	624.1441	48.09	6.78	40.65	4.49
C25H42I2N4	652.45	652.1502	46.02	6.49	38.90	8.59
C25H42I2N6	680.46	680.1564	44.13	6.22	37.30	12.35
C25H43I1N2	498.54	498.2473	60.23	8.69	25.46	5.62
C25H43I1N4	526.55	526.2534	57.03	8.23	24.10	10.64
C25H43I1N6	554.57	554.2596	54.15	7.82	22.88	15.15
C25H43I2N1	611.44	611.1489	49.11	7.09	41.51	2.29
C25H43I2N3	639.45	639.1550	46.96	6.78	39.69	6.57
C25H43I2N5	667.46	667.1611	44.99	6.49	38.03	10.49
C25H44I1N1	485.54	485.2520	61.84	9.13	26.14	2.88
C25H44I1N3	513.55	513.2582	58.47	8.64	24.71	8.18
C25H44I1N5	541.57	541.2643	55.45	8.19	23.43	12.93
C25H44I2N2	626.45	626.1597	47.93	7.08	40.52	4.47
C25H44I2N4	654.47	654.1659	45.88	6.78	38.78	8.56
C25H44I2N6	682.48	682.1720	44.00	6.50	37.19	12.31
C25H45I1N2	500.56	500.2629	59.99	9.06	25.35	5.60
C25H45I1N4	528.57	528.2691	56.81	8.58	24.01	10.60
C25H45I1N6	556.58	556.2752	53.95	8.15	22.80	15.10
C25H45I2N1	613.45	613.1645	48.95	7.39	41.37	2.28
C25H45I2N3	641.47	641.1706	46.81	7.07	39.57	6.55
C25H45I2N5	669.48	669.1768	44.85	6.78	37.91	10.46
C25H46I1N1	487.56	487.2677	61.59	9.51	26.03	2.87
C25H46I1N3	515.57	515.2738	58.24	8.99	24.61	8.15
C25H46I1N5	543.58	543.2800	55.24	8.53	23.35	12.88
C25H46I2N2	628.47	628.1754	47.78	7.38	40.39	4.46
C25H46I2N4	656.48	656.1815	45.74	7.06	38.66	8.53
C25H46I2N6	684.49	684.1877	43.87	6.77	37.08	12.28
C25H47I1N2	502.57	502.2786	59.75	9.43	25.25	5.57
C25H47I1N4	530.58	530.2847	56.59	8.93	23.92	10.56
C25H47I1N6	558.60	558.2909	53.76	8.48	22.72	15.04
C25H47I2N1	615.47	615.1802	48.79	7.70	41.24	2.28
C25H47I2N3	643.48	643.1863	46.66	7.36	39.44	6.53
C25H47I2N5	671.50	671.1924	44.72	7.06	37.80	10.43
C25H48I1N1	489.57	489.2833	61.33	9.88	25.92	2.86
C25H48I1N3	517.59	517.2895	58.02	9.35	24.52	8.12
C25H48I1N5	545.60	545.2956	55.04	8.87	23.26	12.84
C25H48I2N2	630.48	630.1910	47.63	7.67	40.26	4.44
C25H48I2N4	658.50	658.1972	45.60	7.35	38.54	8.51
C25H48I2N6	686.51	686.2033	43.74	7.05	36.97	12.24
C25H49I1N2	504.59	504.2942	59.51	9.79	25.15	5.55
C25H49I1N4	532.60	532.3004	56.38	9.27	23.83	10.52
C25H49I1N6	560.61	560.3065	53.56	8.81	22.64	14.99
C25H49I2N1	617.48	617.1958	48.63	8.00	41.10	2.27
C25H49I2N3	645.50	645.2019	46.52	7.65	39.32	6.51
C25H49I2N5	673.51	673.2081	44.58	7.33	37.68	10.40
C25H50I1N1	491.59	491.2990	61.08	10.25	25.82	2.85
C25H50I1N3	519.60	519.3051	57.79	9.70	24.42	8.09
C25H50I1N5	547.62	547.3113	54.83	9.20	23.17	12.79
C25H50I2N2	632.50	632.2067	47.47	7.97	40.13	4.43
C25H50I2N4	660.51	660.2128	45.46	7.63	38.43	8.48
C25H50I2N6	688.53	688.2190	43.61	7.32	36.86	12.21
C25H51I1N2	506.60	506.3099	59.27	10.15	25.05	5.53
C25H51I1N4	534.62	534.3160	56.17	9.62	23.74	10.48
C25H51I1N6	562.63	562.3222	53.37	9.14	22.56	14.94
C25H51I2N1	619.50	619.2114	48.47	8.30	40.97	2.26
C25H51I2N3	647.51	647.2176	46.37	7.94	39.20	6.49
C25H51I2N5	675.53	675.2237	44.45	7.61	37.57	10.37
C25H52I1N1	493.60	493.3146	60.83	10.62	25.71	2.84
C25H52I1N3	521.62	521.3208	57.57	10.05	24.33	8.06
C25H52I1N5	549.63	549.3269	54.63	9.54	23.09	12.74
C25H52I2N2	634.52	634.2223	47.32	8.26	40.00	4.41
C25H52I2N4	662.53	662.2285	45.32	7.91	38.31	8.46
C25H52I2N6	690.54	690.2346	43.48	7.59	36.76	12.17
C25H53I1N2	508.62	508.3255	59.04	10.50	24.95	5.51
C25H53I1N4	536.63	536.3317	55.96	9.96	23.65	10.44
C25H53I1N6	564.65	564.3378	53.18	9.46	22.48	14.88
C25H53I2N3	649.53	649.2332	46.23	8.22	39.08	6.47
C25H53I2N5	677.54	677.2394	44.32	7.88	37.46	10.34
C25H54I1N3	523.63	523.3364	57.35	10.39	24.24	8.02
C25H54I1N5	551.65	551.3426	54.43	9.87	23.00	12.70
C25H54I2N4	664.54	664.2441	45.19	8.19	38.19	8.43
C25H54I2N6	692.56	692.2503	43.36	7.86	36.65	12.13
C25H55I1N4	538.65	538.3473	55.75	10.29	23.56	10.40
C25H55I1N6	566.66	566.3535	52.99	9.78	22.40	14.83
C25H55I2N5	679.56	679.2550	44.19	8.16	37.35	10.31
C25H56I1N5	553.66	553.3582	54.23	10.20	22.92	12.65
C25H56I2N6	694.57	694.2659	43.23	8.13	36.54	12.10
C25H57I1N6	568.68	568.3691	52.80	10.10	22.32	14.78
C26H13I1N2	480.31	480.0125	65.02	2.73	26.42	5.83
C26H13I1N4	508.32	508.0187	61.44	2.58	24.97	11.02
C26H13I1N6	536.34	536.0248	58.23	2.44	23.66	15.67
C26H13I2N1	593.21	592.9141	52.64	2.21	42.79	2.36
C26H13I2N3	621.22	620.9203	50.27	2.11	40.86	6.76
C26H13I2N5	649.24	648.9264	48.10	2.02	39.09	10.79
C26H14I1N1	467.31	467.0173	66.83	3.02	27.16	3.00
C26H14I1N3	495.33	495.0234	63.05	2.85	25.62	8.48
C26H14I1N5	523.34	523.0296	59.67	2.70	24.25	13.38
C26H14I2N2	608.22	607.9250	51.34	2.32	41.73	4.61
C26H14I2N4	636.24	635.9312	49.08	2.22	39.89	8.81
C26H14I2N6	664.25	663.9373	47.01	2.12	38.21	12.65
C26H15I1N2	482.33	482.0282	64.75	3.13	26.31	5.81
C26H15I1N4	510.34	510.0343	61.19	2.96	24.87	10.98
C26H15I1N6	538.35	538.0405	58.01	2.81	23.57	15.61
C26H15I2N1	595.22	594.9298	52.47	2.54	42.64	2.35
C26H15I2N3	623.24	622.9359	50.11	2.43	40.72	6.74
C26H15I2N5	651.25	650.9421	47.95	2.32	38.97	10.75
C26H16I1N1	469.33	469.0329	66.54	3.44	27.04	2.98
C26H16I1N3	497.34	497.0391	62.79	3.24	25.52	8.45
C26H16I1N5	525.36	525.0452	59.44	3.07	24.16	13.33
C26H16I2N2	610.24	609.9407	51.17	2.64	41.59	4.59
C26H16I2N4	638.25	637.9468	48.93	2.53	39.77	8.78
C26H16I2N6	666.27	665.9530	46.87	2.42	38.09	12.61
C26H17I1N2	484.34	484.0438	64.48	3.54	26.20	5.78
C26H17I1N4	512.36	512.0500	60.95	3.34	24.77	10.94
C26H17I1N6	540.37	540.0561	57.79	3.17	23.48	15.55
C26H17I2N1	597.24	596.9454	52.29	2.87	42.50	2.35
C26H17I2N3	625.25	624.9516	49.95	2.74	40.59	6.72
C26H17I2N5	653.27	652.9577	47.80	2.62	38.85	10.72
C26H18I1N1	471.34	471.0486	66.26	3.85	26.92	2.97

041

C26H18I1N3	499.36	499.0547	62.54	3.63	25.41	8.41
C26H18I1N5	527.37	527.0609	59.22	3.44	24.06	13.28
C26H18I2N2	612.26	611.9563	51.01	2.96	41.45	4.58
C26H18I2N4	640.27	639.9625	48.77	2.83	39.64	8.75
C26H18I2N6	668.28	667.9686	46.73	2.71	37.98	12.58
C26H19I1N2	486.36	486.0595	64.21	3.94	26.09	5.76
C26H19I1N4	514.37	514.0656	60.71	3.72	24.67	10.89
C26H19I1N6	542.39	542.0718	57.58	3.53	23.40	15.49
C26H19I2N1	599.26	598.9611	52.11	3.20	42.35	2.34
C26H19I2N3	627.27	626.9672	49.79	3.05	40.46	6.70
C26H19I2N5	655.28	654.9734	47.66	2.92	38.73	10.69
C26H20I1N1	473.36	473.0642	65.97	4.26	26.81	2.96
C26H20I1N3	501.37	501.0704	62.29	4.02	25.31	8.38
C26H20I1N5	529.39	529.0765	58.99	3.81	23.97	13.23
C26H20I2N2	614.27	613.9720	50.84	3.28	41.32	4.56
C26H20I2N4	642.28	641.9781	48.62	3.14	39.52	8.72
C26H20I2N6	670.30	669.9843	46.59	3.01	37.87	12.54
C26H21I1N2	488.38	488.0751	63.94	4.33	25.99	5.74
C26H21I1N4	516.39	516.0813	60.48	4.10	24.58	10.85
C26H21I1N6	544.40	544.0874	57.36	3.89	23.31	15.44
C26H21I2N1	601.27	600.9767	51.94	3.52	42.21	2.33
C26H21I2N3	629.29	628.9829	49.63	3.36	40.33	6.68
C26H21I2N5	657.30	656.9890	47.51	3.22	38.61	10.65
C26H22I1N1	475.38	475.0799	65.69	4.66	26.70	2.95
C26H22I1N3	503.39	503.0860	62.04	4.41	25.21	8.35
C26H22I1N5	531.40	531.0922	58.77	4.17	23.88	13.18
C26H22I2N2	616.29	615.9876	50.67	3.60	41.18	4.55
C26H22I2N4	644.30	643.9938	48.47	3.44	39.39	8.70
C26H22I2N6	672.31	671.9999	46.45	3.30	37.75	12.50
C26H23I1N2	490.39	490.0908	63.68	4.73	25.88	5.71
C26H23I1N4	518.40	518.0969	60.24	4.47	24.48	10.81
C26H23I1N6	546.42	546.1031	57.15	4.24	23.22	15.38
C26H23I2N1	603.29	602.9924	51.76	3.84	42.07	2.32
C26H23I2N3	631.30	630.9985	49.47	3.67	40.20	6.66
C26H23I2N5	659.32	659.0047	47.37	3.52	38.50	10.62
C26H24I1N1	477.39	477.0955	65.42	5.07	26.58	2.93
C26H24I1N3	505.41	505.1017	61.79	4.79	25.11	8.31
C26H24I1N5	533.42	533.1078	58.54	4.54	23.79	13.13
C26H24I2N2	618.30	618.0033	50.51	3.91	41.05	4.53
C26H24I2N4	646.32	646.0094	48.32	3.74	39.27	8.67
C26H24I2N6	674.33	674.0156	46.31	3.59	37.64	12.46
C26H25I1N2	492.41	492.1064	63.42	5.12	25.77	5.69
C26H25I1N4	520.42	520.1126	60.01	4.84	24.38	10.77
C26H25I1N6	548.43	548.1187	56.94	4.59	23.14	15.32
C26H25I2N1	605.30	605.0080	51.59	4.16	41.93	2.31
C26H25I2N3	633.32	633.0142	49.31	3.98	40.08	6.63
C26H25I2N5	661.33	661.0203	47.22	3.81	38.38	10.59
C26H26I1N1	479.41	479.1112	65.14	5.47	26.47	2.92
C26H26I1N3	507.42	507.1173	61.54	5.16	25.01	8.28
C26H26I1N5	535.44	535.1235	58.32	4.89	23.70	13.08
C26H26I2N2	620.32	620.0189	50.34	4.22	40.92	4.52
C26H26I2N4	648.33	648.0251	48.17	4.04	39.15	8.64
C26H26I2N6	676.35	676.0312	46.17	3.87	37.53	12.43
C26H27I1N2	494.42	494.1221	63.16	5.50	25.67	5.67
C26H27I1N4	522.44	522.1282	59.78	5.21	24.29	10.72
C26H27I1N6	550.45	550.1344	56.73	4.94	23.05	15.27
C26H27I2N1	607.32	607.0237	51.42	4.48	41.79	2.31
C26H27I2N3	635.33	635.0298	49.15	4.28	39.95	6.61
C26H27I2N5	663.35	663.0360	47.08	4.10	38.26	10.56
C26H28I1N1	481.42	481.1268	64.87	5.86	26.36	2.91
C26H28I1N3	509.44	509.1330	61.30	5.54	24.91	8.25
C26H28I1N5	537.45	537.1391	58.11	5.25	23.61	13.03
C26H28I2N2	622.34	622.0346	50.18	4.54	40.78	4.50
C26H28I2N4	650.35	650.0407	48.02	4.34	39.03	8.61
C26H28I2N6	678.36	678.0469	46.04	4.16	37.41	12.39
C26H29I1N2	496.44	496.1377	62.91	5.89	25.56	5.64
C26H29I1N4	524.45	524.1439	59.55	5.57	24.20	10.68
C26H29I1N6	552.47	552.1500	56.53	5.29	22.97	15.21
C26H29I2N1	609.34	609.0393	51.25	4.80	41.65	2.30
C26H29I2N3	637.35	637.0455	49.00	4.59	39.82	6.59
C26H29I2N5	665.36	665.0516	46.94	4.39	38.15	10.53
C26H30I1N1	483.44	483.1425	64.60	6.25	26.25	2.90
C26H30I1N3	511.45	511.1486	61.06	5.91	24.81	8.22
C26H30I1N5	539.47	539.1548	57.89	5.61	23.52	12.98
C26H30I2N2	624.35	624.0502	50.02	4.84	40.65	4.49
C26H30I2N4	652.36	652.0564	47.87	4.64	38.91	8.59
C26H30I2N6	680.38	680.0625	45.90	4.44	37.30	12.35
C26H31I1N2	498.45	498.1534	62.65	6.27	25.46	5.62
C26H31I1N4	526.47	526.1595	59.32	5.94	24.10	10.64
C26H31I1N6	554.48	554.1657	56.32	5.64	22.89	15.16
C26H31I2N1	611.35	611.0550	51.08	5.11	41.52	2.29
C26H31I2N3	639.37	639.0611	48.84	4.89	39.70	6.57
C26H31I2N5	667.38	667.0673	46.79	4.68	38.03	10.49
C26H32I1N1	485.46	485.1581	64.33	6.64	26.14	2.89
C26H32I1N3	513.47	513.1643	60.82	6.28	24.72	8.18
C26H32I1N5	541.48	541.1704	57.67	5.96	23.44	12.93
C26H32I2N2	626.37	626.0659	49.86	5.15	40.52	4.47
C26H32I2N4	654.38	654.0720	47.72	4.93	38.79	8.56
C26H32I2N6	682.39	682.0781	45.76	4.73	37.19	12.32
C26H33I1N2	500.47	500.1690	62.40	6.65	25.36	5.60
C26H33I1N4	528.48	528.1752	59.09	6.29	24.01	10.60
C26H33I1N6	556.50	556.1813	56.12	5.98	22.80	15.10
C26H33I2N1	613.37	613.0706	50.91	5.42	41.38	2.28
C26H33I2N3	641.38	641.0768	48.69	5.19	39.57	6.55
C26H33I2N5	669.40	669.0829	46.65	4.97	37.92	10.46
C26H34I1N1	487.47	487.1738	64.06	7.03	26.03	2.87
C26H34I1N3	515.49	515.1799	60.58	6.65	24.62	8.15
C26H34I1N5	543.50	543.1861	57.46	6.31	23.35	12.89
C26H34I2N2	628.38	628.0815	49.70	5.45	40.39	4.46
C26H34I2N4	656.40	656.0877	47.58	5.22	38.67	8.54
C26H34I2N6	684.41	684.0938	45.63	5.01	37.08	12.28
C26H35I1N2	502.49	502.1847	62.15	7.02	25.26	5.57
C26H35I1N4	530.50	530.1908	58.87	6.65	23.92	10.56
C26H35I1N6	558.51	558.1970	55.91	6.32	22.72	15.05
C26H35I2N1	615.38	615.0863	50.75	5.73	41.24	2.28
C26H35I2N3	643.40	643.0924	48.54	5.48	39.45	6.53
C26H35I2N5	671.41	671.0985	46.51	5.25	37.80	10.43
C26H36I1N1	489.49	489.1894	63.80	7.41	25.93	2.86
C26H36I1N3	517.50	517.1956	60.35	7.01	24.52	8.12
C26H36I1N5	545.51	545.2017	57.25	6.65	23.26	12.84
C26H36I2N2	630.40	630.0972	49.54	5.76	40.26	4.44
C26H36I2N4	658.41	658.1033	47.43	5.51	38.55	8.51
C26H36I2N6	686.43	686.1094	45.50	5.29	36.98	12.24
C26H37I1N2	504.50	504.2003	61.90	7.39	25.15	5.55
C26H37I1N4	532.52	532.2065	58.64	7.00	23.83	10.52
C26H37I1N6	560.53	560.2126	55.71	6.65	22.64	14.99
C26H37I2N1	617.40	617.1019	50.58	6.04	41.11	2.27
C26H37I2N3	645.41	645.1081	48.39	5.78	39.32	6.51
C26H37I2N5	673.43	673.1142	46.37	5.54	37.69	10.40
C26H38I1N1	491.50	491.2051	63.54	7.79	25.82	2.85

042

C26H38I1N3	519.52	519.2112	60.11	7.37	24.43	8.09
C26H38I1N5	547.53	547.2174	57.04	7.00	23.18	12.79
C26H38I2N2	632.41	632.1128	49.38	6.06	40.13	4.43
C26H38I2N4	660.43	660.1189	47.29	5.80	38.43	8.48
C26H38I2N6	688.44	688.1251	45.36	5.56	36.87	12.21
C26H39I1N2	506.52	506.2160	61.65	7.76	25.05	5.53
C26H39I1N4	534.53	534.2221	58.42	7.35	23.74	10.48
C26H39I1N6	562.55	562.2283	55.51	6.99	22.56	14.94
C26H39I2N1	619.42	619.1176	50.42	6.35	40.98	2.26
C26H39I2N3	647.43	647.1237	48.24	6.07	39.20	6.49
C26H39I2N5	675.44	675.1298	46.23	5.82	37.58	10.37
C26H40I1N1	493.52	493.2207	63.28	8.17	25.71	2.84
C26H40I1N3	521.53	521.2269	59.88	7.73	24.33	8.06
C26H40I1N5	549.55	549.2330	56.83	7.34	23.09	12.74
C26H40I2N2	634.43	634.1285	49.22	6.36	40.01	4.42
C26H40I2N4	662.44	662.1346	47.14	6.09	38.31	8.46
C26H40I2N6	690.46	690.1407	45.23	5.84	36.76	12.17
C26H41I1N2	508.53	508.2316	61.41	8.13	24.95	5.51
C26H41I1N4	536.55	536.2378	58.20	7.70	23.65	10.44
C26H41I1N6	564.56	564.2439	55.32	7.32	22.48	14.89
C26H41I2N1	621.43	621.1332	50.25	6.65	40.84	2.25
C26H41I2N3	649.45	649.1394	48.09	6.36	39.08	6.47
C26H41I2N5	677.46	677.1455	46.10	6.10	37.46	10.34
C26H42I1N1	495.54	495.2364	63.02	8.54	25.61	2.83
C26H42I1N3	523.55	523.2425	59.65	8.09	24.24	8.03
C26H42I1N5	551.56	551.2487	56.62	7.68	23.01	12.70
C26H42I2N2	636.45	636.1441	49.07	6.65	39.88	4.40
C26H42I2N4	664.46	664.1502	47.00	6.37	38.20	8.43
C26H42I2N6	692.47	692.1564	45.10	6.11	36.65	12.14
C26H43I1N2	510.55	510.2473	61.17	8.49	24.86	5.49
C26H43I1N4	538.56	538.2534	57.99	8.05	23.56	10.40
C26H43I1N6	566.58	566.2596	55.12	7.65	22.40	14.83
C26H43I2N1	623.45	623.1489	50.09	6.95	40.71	2.25
C26H43I2N3	651.46	651.1550	47.94	6.65	38.96	6.45
C26H43I2N5	679.47	679.1611	45.96	6.38	37.35	10.31
C26H44I1N1	497.55	497.2520	62.77	8.91	25.51	2.82
C26H44I1N3	525.57	525.2582	59.42	8.44	24.15	8.00
C26H44I1N5	553.58	553.2643	56.41	8.01	22.92	12.65
C26H44I2N2	638.46	638.1597	48.91	6.95	39.75	4.39
C26H44I2N4	666.48	666.1659	46.86	6.65	38.08	8.41
C26H44I2N6	694.49	694.1720	44.97	6.39	36.55	12.10
C26H45I1N2	512.57	512.2629	60.93	8.85	24.76	5.47
C26H45I1N4	540.58	540.2691	57.77	8.39	23.48	10.36
C26H45I1N6	568.59	568.2752	54.92	7.98	22.32	14.78
C26H45I2N1	625.46	625.1645	49.93	7.25	40.58	2.24
C26H45I2N3	653.48	653.1706	47.79	6.94	38.84	6.43
C26H45I2N5	681.49	681.1768	45.82	6.66	37.24	10.28
C26H46I1N1	499.57	499.2677	62.51	9.28	25.40	2.80
C26H46I1N3	527.58	527.2738	59.19	8.79	24.05	7.96
C26H46I1N5	555.59	555.2800	56.21	8.35	22.84	12.61
C26H46I2N2	640.48	640.1754	48.76	7.24	39.63	4.37
C26H46I2N4	668.49	668.1815	46.72	6.94	37.97	8.38
C26H46I2N6	696.51	696.1877	44.84	6.66	36.44	12.07
C26H47I1N2	514.58	514.2786	60.69	9.21	24.66	5.44
C26H47I1N4	542.60	542.2847	57.55	8.73	23.39	10.33
C26H47I1N6	570.61	570.2909	54.73	8.30	22.24	14.73
C26H47I2N1	627.48	627.1802	49.77	7.55	40.45	2.23
C26H47I2N3	655.49	655.1863	47.64	7.23	38.72	6.41
C26H47I2N5	683.51	683.1924	45.69	6.93	37.13	10.25
C26H48I1N1	501.58	501.2833	62.26	9.65	25.30	2.79
C26H48I1N3	529.60	529.2895	58.97	9.14	23.96	7.93
C26H48I1N5	557.61	557.2956	56.01	8.68	22.76	12.56
C26H48I2N2	642.49	642.1910	48.61	7.53	39.50	4.36
C26H48I2N4	670.51	670.1972	46.58	7.22	37.85	8.36
C26H48I2N6	698.52	698.2033	44.71	6.93	36.34	12.03
C26H49I1N2	516.60	516.2942	60.45	9.56	24.57	5.42
C26H49I1N4	544.61	544.3004	57.34	9.07	23.30	10.29
C26H49I1N6	572.63	572.3065	54.54	8.63	22.16	14.68
C26H49I2N1	629.50	629.1958	49.61	7.85	40.32	2.23
C26H49I2N3	657.51	657.2019	47.50	7.51	38.60	6.39
C26H49I2N5	685.52	685.2081	45.56	7.20	37.02	10.22
C26H50I1N1	503.60	503.2990	62.01	10.01	25.20	2.78
C26H50I1N3	531.61	531.3051	58.74	9.48	23.87	7.90
C26H50I1N5	559.63	559.3113	55.80	9.01	22.68	12.51
C26H50I2N2	644.51	644.2067	48.45	7.82	39.38	4.35
C26H50I2N4	672.52	672.2128	46.44	7.49	37.74	8.33
C26H50I2N6	700.54	700.2190	44.58	7.19	36.23	12.00
C26H51I1N2	518.61	518.3099	60.22	9.91	24.47	5.40
C26H51I1N4	546.63	546.3160	57.13	9.40	23.22	10.25
C26H51I1N6	574.64	574.3222	54.35	8.95	22.08	14.62
C26H51I2N1	631.51	631.2114	49.45	8.14	40.19	2.22
C26H51I2N3	659.53	659.2176	47.35	7.79	38.48	6.37
C26H51I2N5	687.54	687.2237	45.42	7.48	36.92	10.19
C26H52I1N1	505.62	505.3146	61.76	10.37	25.10	2.77
C26H52I1N3	533.63	533.3208	58.52	9.82	23.78	7.87
C26H52I1N5	561.64	561.3269	55.60	9.33	22.60	12.47
C26H52I2N2	646.53	646.2223	48.30	8.11	39.26	4.33
C26H52I2N4	674.54	674.2285	46.30	7.77	37.63	8.31
C26H52I2N6	702.55	702.2346	44.45	7.46	36.13	11.96
C26H53I1N2	520.63	520.3255	59.98	10.26	24.38	5.38
C26H53I1N4	548.64	548.3317	56.92	9.74	23.13	10.21
C26H53I1N6	576.66	576.3378	54.16	9.26	22.01	14.57
C26H53I2N1	633.53	633.2271	49.29	8.43	40.06	2.21
C26H53I2N3	661.54	661.2332	47.21	8.08	38.37	6.35
C26H53I2N5	689.55	689.2394	45.29	7.75	36.81	10.16
C26H54I1N1	507.63	507.3303	61.52	10.72	25.00	2.76
C26H54I1N3	535.64	535.3364	58.30	10.16	23.69	7.84
C26H54I1N5	563.66	563.3426	55.40	9.66	22.51	12.42
C26H54I2N2	648.54	648.2380	48.15	8.39	39.14	4.32
C26H54I2N4	676.56	676.2441	46.16	8.05	37.51	8.28
C26H54I2N6	704.57	704.2503	44.32	7.73	36.02	11.93
C26H55I1N2	522.65	522.3412	59.75	10.61	24.28	5.36
C26H55I1N4	550.66	550.3473	56.71	10.07	23.05	10.17
C26H55I1N6	578.67	578.3535	53.97	9.58	21.93	14.52
C26H55I2N3	663.56	663.2489	47.06	8.35	38.25	6.33
C26H55I2N5	691.57	691.2550	45.16	8.02	36.70	10.13
C26H56I1N3	537.66	537.3521	58.08	10.50	23.60	7.82
C26H56I1N5	565.67	565.3582	55.21	9.98	22.43	12.38
C26H56I2N4	678.57	678.2598	46.02	8.32	37.40	8.26
C26H56I2N6	706.59	706.2659	44.20	7.99	35.92	11.89
C26H57I1N4	552.68	552.3630	56.51	10.40	22.96	10.14
C26H57I1N6	580.69	580.3691	53.78	9.89	21.85	14.47
C26H57I2N5	693.59	693.2707	45.03	8.28	36.59	10.10
C26H58I1N5	567.69	567.3739	55.01	10.30	22.35	12.34
C26H58I2N6	708.60	708.2816	44.07	8.25	35.82	11.86
C26H59I1N6	582.70	582.3848	53.59	10.21	21.78	14.42
C27H13I1N2	492.32	492.0125	65.87	2.66	25.78	5.69
C27H13I1N4	520.34	520.0187	62.33	2.52	24.39	10.77
C27H13I1N6	548.35	548.0248	59.14	2.39	23.14	15.33
C27H13I2N1	605.22	604.9141	53.58	2.17	41.94	2.31

043

C27H13I2N3	633.23	632.9203	51.21	2.07	40.08	6.64
C27H13I2N5	661.25	660.9264	49.04	1.98	38.38	10.59
C27H14I1N1	479.32	479.0173	67.66	2.94	26.48	2.92
C27H14I1N3	507.34	507.0234	63.92	2.78	25.01	8.28
C27H14I1N5	535.35	535.0296	60.58	2.64	23.70	13.08
C27H14I2N2	620.23	619.9250	52.29	2.28	40.92	4.52
C27H14I2N4	648.25	647.9312	50.03	2.18	39.15	8.64
C27H14I2N6	676.26	675.9373	47.95	2.09	37.53	12.43
C27H15I1N2	494.34	494.0282	65.60	3.06	25.67	5.67
C27H15I1N4	522.35	522.0343	62.08	2.89	24.29	10.73
C27H15I1N6	550.37	550.0405	58.92	2.75	23.06	15.27
C27H15I2N1	607.24	606.9298	53.41	2.49	41.80	2.31
C27H15I2N3	635.25	634.9359	51.05	2.38	39.95	6.61
C27H15I2N5	663.26	662.9421	48.89	2.28	38.27	10.56
C27H16I1N1	481.34	481.0329	67.37	3.35	26.36	2.91
C27H16I1N3	509.35	509.0391	63.67	3.17	24.91	8.25
C27H16I1N5	537.37	537.0452	60.35	3.00	23.62	13.03
C27H16I2N2	622.25	621.9407	52.12	2.59	40.79	4.50
C27H16I2N4	650.26	649.9468	49.87	2.48	39.03	8.62
C27H16I2N6	678.28	677.9530	47.81	2.38	37.42	12.39
C27H17I1N2	496.35	496.0438	65.34	3.45	25.57	5.64
C27H17I1N4	524.37	524.0500	61.85	3.27	24.20	10.68
C27H17I1N6	552.38	552.0561	58.71	3.10	22.97	15.21
C27H17I2N1	609.25	608.9454	53.23	2.81	41.66	2.30
C27H17I2N3	637.27	636.9516	50.89	2.69	39.83	6.59
C27H17I2N5	665.28	664.9577	48.75	2.58	38.15	10.53
C27H18I1N1	483.36	483.0486	67.09	3.75	26.25	2.90
C27H18I1N3	511.37	511.0547	63.42	3.55	24.82	8.22
C27H18I1N5	539.38	539.0609	60.12	3.36	23.53	12.98
C27H18I2N2	624.27	623.9563	51.95	2.91	40.66	4.49
C27H18I2N4	652.28	651.9625	49.72	2.78	38.91	8.59
C27H18I2N6	680.29	679.9686	47.67	2.67	37.31	12.35
C27H19I1N2	498.37	498.0595	65.07	3.84	25.46	5.62
C27H19I1N4	526.38	526.0656	61.61	3.64	24.11	10.64
C27H19I1N6	554.40	554.0718	58.50	3.45	22.89	15.16
C27H19I2N1	611.27	610.9611	53.05	3.13	41.52	2.29
C27H19I2N3	639.28	638.9672	50.73	3.00	39.70	6.57
C27H19I2N5	667.29	666.9734	48.60	2.87	38.04	10.50
C27H20I1N1	485.37	485.0642	66.82	4.15	26.15	2.89
C27H20I1N3	513.38	513.0704	63.17	3.93	24.72	8.18
C27H20I1N5	541.40	541.0765	59.90	3.72	23.44	12.94
C27H20I2N2	626.28	625.9720	51.78	3.22	40.53	4.47
C27H20I2N4	654.30	653.9781	49.56	3.08	38.79	8.56
C27H20I2N6	682.31	681.9843	47.53	2.95	37.20	12.32
C27H21I1N2	500.39	500.0751	64.81	4.23	25.36	5.60
C27H21I1N4	528.40	528.0813	61.37	4.01	24.02	10.60
C27H21I1N6	556.41	556.0874	58.28	3.80	22.81	15.10
C27H21I2N1	613.28	612.9767	52.88	3.45	41.39	2.28
C27H21I2N3	641.30	640.9829	50.57	3.30	39.58	6.55
C27H21I2N5	669.31	668.9890	48.45	3.16	37.92	10.46
C27H22I1N1	487.39	487.0799	66.54	4.55	26.04	2.87
C27H22I1N3	515.40	515.0860	62.92	4.30	24.62	8.15
C27H22I1N5	543.41	543.0922	59.68	4.08	23.35	12.89
C27H22I2N2	628.30	627.9876	51.62	3.53	40.40	4.46
C27H22I2N4	656.31	655.9938	49.41	3.38	38.67	8.54
C27H22I2N6	684.33	683.9999	47.39	3.24	37.09	12.28
C27H23I1N2	502.40	502.0908	64.55	4.61	25.26	5.58
C27H23I1N4	530.42	530.0969	61.14	4.37	23.93	10.56
C27H23I1N6	558.43	558.1031	58.07	4.15	22.73	15.05
C27H23I2N1	615.30	614.9924	52.71	3.77	41.25	2.28
C27H23I2N3	643.31	642.9985	50.41	3.60	39.45	6.53
C27H23I2N5	671.33	671.0047	48.31	3.45	37.81	10.43
C27H24I1N1	489.40	489.0955	66.26	4.94	25.93	2.86
C27H24I1N3	517.42	517.1017	62.68	4.68	24.53	8.12
C27H24I1N5	545.43	545.1078	59.46	4.44	23.27	12.84
C27H24I2N2	630.31	630.0033	51.45	3.84	40.27	4.44
C27H24I2N4	658.33	658.0094	49.26	3.67	38.55	8.51
C27H24I2N6	686.34	686.0156	47.25	3.52	36.98	12.24
C27H25I1N2	504.42	504.1064	64.29	5.00	25.16	5.55
C27H25I1N4	532.43	532.1126	60.91	4.73	23.83	10.52
C27H25I1N6	560.44	560.1187	57.86	4.50	22.64	15.00
C27H25I2N1	617.32	617.0080	52.53	4.08	41.11	2.27
C27H25I2N3	645.33	645.0142	50.25	3.90	39.33	6.51
C27H25I2N5	673.34	673.0203	48.16	3.74	37.69	10.40
C27H26I1N1	491.42	491.1112	65.99	5.33	25.82	2.85
C27H26I1N3	519.43	519.1173	62.43	5.05	24.43	8.09
C27H26I1N5	547.45	547.1235	59.24	4.79	23.18	12.79
C27H26I2N2	632.33	632.0189	51.29	4.14	40.14	4.43
C27H26I2N4	660.34	660.0251	49.11	3.97	38.44	8.48
C27H26I2N6	688.36	688.0312	47.11	3.81	36.87	12.21
C27H27I1N2	506.43	506.1221	64.04	5.37	25.06	5.53
C27H27I1N4	534.45	534.1282	60.68	5.09	23.74	10.48
C27H27I1N6	562.46	562.1344	57.66	4.84	22.56	14.94
C27H27I2N1	619.33	619.0237	52.36	4.39	40.98	2.26
C27H27I2N3	647.35	647.0298	50.10	4.20	39.21	6.49
C27H27I2N5	675.36	675.0360	48.02	4.03	37.58	10.37
C27H28I1N1	493.44	493.1268	65.72	5.72	25.72	2.84
C27H28I1N3	521.45	521.1330	62.19	5.41	24.34	8.06
C27H28I1N5	549.46	549.1391	59.02	5.14	23.10	12.75
C27H28I2N2	634.35	634.0346	51.12	4.45	40.01	4.42
C27H28I2N4	662.36	662.0407	48.96	4.26	38.32	8.46
C27H28I2N6	690.37	690.0469	46.97	4.09	36.76	12.17
C27H29I1N2	508.45	508.1377	63.78	5.75	24.96	5.51
C27H29I1N4	536.46	536.1439	60.45	5.45	23.66	10.44
C27H29I1N6	564.48	564.1500	57.45	5.18	22.48	14.89
C27H29I2N1	621.35	621.0393	52.19	4.70	40.85	2.25
C27H29I2N3	649.36	649.0455	49.94	4.50	39.09	6.47
C27H29I2N5	677.37	677.0516	47.88	4.32	37.47	10.34
C27H30I1N1	495.45	495.1425	65.46	6.10	25.61	2.83
C27H30I1N3	523.46	523.1486	61.95	5.78	24.24	8.03
C27H30I1N5	551.48	551.1548	58.81	5.48	23.01	12.70
C27H30I2N2	636.36	636.0502	50.96	4.75	39.88	4.40
C27H30I2N4	664.38	664.0564	48.81	4.55	38.20	8.43
C27H30I2N6	692.39	692.0625	46.84	4.37	36.66	12.14
C27H31I1N2	510.47	510.1534	63.53	6.12	24.86	5.49
C27H31I1N4	538.48	538.1595	60.23	5.80	23.57	10.40
C27H31I1N6	566.49	566.1657	57.25	5.52	22.40	14.84
C27H31I2N1	623.36	623.0550	52.02	5.01	40.72	2.25
C27H31I2N3	651.38	651.0611	49.79	4.80	38.96	6.45
C27H31I2N5	679.39	679.0673	47.73	4.60	37.36	10.31
C27H32I1N1	497.47	497.1581	65.19	6.48	25.51	2.82
C27H32I1N3	525.48	525.1643	61.72	6.14	24.15	8.00
C27H32I1N5	553.49	553.1704	58.59	5.83	22.93	12.65
C27H32I2N2	638.38	638.0659	50.80	5.05	39.76	4.39
C27H32I2N4	666.39	666.0720	48.67	4.84	38.09	8.41
C27H32I2N6	694.41	694.0781	46.70	4.64	36.55	12.10
C27H33I1N2	512.48	512.1690	63.28	6.49	24.76	5.47
C27H33I1N4	540.50	540.1752	60.00	6.15	23.48	10.37
C27H33I1N6	568.51	568.1813	57.04	5.85	22.32	14.78
C27H33I2N1	625.38	625.0706	51.86	5.32	40.58	2.24

044

C27H33I2N3	653.39	653.0768	49.63	5.09	38.84	6.43
C27H33I2N5	681.41	681.0829	47.59	4.88	37.25	10.28
C27H34I1N1	499.48	499.1738	64.93	6.86	25.41	2.80
C27H34I1N3	527.50	527.1799	61.48	6.50	24.06	7.97
C27H34I1N5	555.51	555.1861	58.38	6.17	22.84	12.61
C27H34I2N2	640.39	640.0815	50.64	5.35	39.63	4.37
C27H34I2N4	668.41	668.0877	48.52	5.13	37.97	8.38
C27H34I2N6	696.42	696.0938	46.57	4.92	36.44	12.07
C27H35I1N2	514.50	514.1847	63.03	6.86	24.67	5.44
C27H35I1N4	542.51	542.1908	59.78	6.50	23.39	10.33
C27H35I1N6	570.52	570.1970	56.84	6.18	22.24	14.73
C27H35I2N1	627.40	627.0863	51.69	5.62	40.45	2.23
C27H35I2N3	655.41	655.0924	49.48	5.38	38.73	6.41
C27H35I2N5	683.42	683.0985	47.45	5.16	37.14	10.25
C27H36I1N1	501.50	501.1894	64.67	7.24	25.31	2.79
C27H36I1N3	529.51	529.1956	61.25	6.85	23.97	7.94
C27H36I1N5	557.53	557.2017	58.17	6.51	22.76	12.56
C27H36I2N2	642.41	642.0972	50.48	5.65	39.51	4.36
C27H36I2N4	670.42	670.1033	48.37	5.41	37.86	8.36
C27H36I2N6	698.44	698.1094	46.43	5.20	36.34	12.03
C27H37I1N2	516.51	516.2003	62.79	7.22	24.57	5.42
C27H37I1N4	544.53	544.2065	59.56	6.85	23.31	10.29
C27H37I1N6	572.54	572.2126	56.64	6.51	22.17	14.68
C27H37I2N1	629.41	629.1019	51.52	5.93	40.32	2.23
C27H37I2N3	657.42	657.1081	49.33	5.67	38.61	6.39
C27H37I2N5	685.44	685.1142	47.31	5.44	37.03	10.22
C27H38I1N1	503.52	503.2051	64.41	7.61	25.20	2.78
C27H38I1N3	531.53	531.2112	61.01	7.21	23.88	7.91
C27H38I1N5	559.54	559.2174	57.96	6.85	22.68	12.52
C27H38I2N2	644.43	644.1128	50.32	5.94	39.39	4.35
C27H38I2N4	672.44	672.1189	48.23	5.70	37.74	8.33
C27H38I2N6	700.45	700.1251	46.30	5.47	36.23	12.00
C27H39I1N2	518.53	518.2160	62.54	7.58	24.47	5.40
C27H39I1N4	546.54	546.2221	59.34	7.19	23.22	10.25
C27H39I1N6	574.56	574.2283	56.44	6.84	22.09	14.63
C27H39I2N1	631.43	631.1176	51.36	6.23	40.20	2.22
C27H39I2N3	659.44	659.1237	49.18	5.96	38.49	6.37
C27H39I2N5	687.45	687.1298	47.17	5.72	36.92	10.19
C27H40I1N1	505.53	505.2207	64.15	7.98	25.10	2.77
C27H40I1N3	533.54	533.2269	60.78	7.56	23.79	7.88
C27H40I1N5	561.56	561.2330	57.75	7.18	22.60	12.47
C27H40I2N2	646.44	646.1285	50.17	6.24	39.26	4.33
C27H40I2N4	674.46	674.1346	48.08	5.98	37.63	8.31
C27H40I2N6	702.47	702.1407	46.17	5.74	36.13	11.96
C27H41I1N2	520.55	520.2316	62.30	7.94	24.38	5.38
C27H41I1N4	548.56	548.2378	59.12	7.53	23.13	10.21
C27H41I1N6	576.57	576.2439	56.25	7.17	22.01	14.58
C27H41I2N1	633.44	633.1332	51.20	6.52	40.07	2.21
C27H41I2N3	661.46	661.1394	49.03	6.25	38.37	6.35
C27H41I2N5	689.47	689.1455	47.04	5.99	36.81	10.16
C27H42I1N1	507.55	507.2364	63.90	8.34	25.00	2.76
C27H42I1N3	535.56	535.2425	60.55	7.90	23.70	7.85
C27H42I1N5	563.57	563.2487	57.54	7.51	22.52	12.43
C27H42I2N2	648.46	648.1441	50.01	6.53	39.14	4.32
C27H42I2N4	676.47	676.1502	47.94	6.26	37.52	8.28
C27H42I2N6	704.48	704.1564	46.03	6.01	36.03	11.93
C27H43I1N2	522.56	522.2473	62.06	8.29	24.29	5.36
C27H43I1N4	550.57	550.2534	58.90	7.87	23.05	10.18
C27H43I1N6	578.59	578.2596	56.05	7.49	21.93	14.53
C27H43I2N1	635.46	635.1489	51.03	6.82	39.94	2.20
C27H43I2N3	663.47	663.1550	48.88	6.53	38.25	6.33
C27H43I2N5	691.49	691.1611	46.90	6.27	36.70	10.13
C27H44I1N1	509.56	509.2520	63.64	8.70	24.90	2.75
C27H44I1N3	537.58	537.2582	60.33	8.25	23.61	7.82
C27H44I1N5	565.59	565.2643	57.34	7.84	22.44	12.38
C27H44I2N2	650.47	650.1597	49.86	6.82	39.02	4.31
C27H44I2N4	678.49	678.1659	47.80	6.54	37.41	8.26
C27H44I2N6	706.50	706.1720	45.90	6.28	35.92	11.90
C27H45I1N2	524.58	524.2629	61.82	8.65	24.19	5.34
C27H45I1N4	552.59	552.2691	58.69	8.21	22.97	10.14
C27H45I1N6	580.60	580.2752	55.86	7.81	21.86	14.47
C27H45I2N1	637.48	637.1645	50.87	7.12	39.81	2.20
C27H45I2N3	665.49	665.1706	48.73	6.82	38.14	6.31
C27H45I2N5	693.50	693.1768	46.76	6.54	36.60	10.10
C27H46I1N1	511.58	511.2677	63.39	9.06	24.81	2.74
C27H46I1N3	539.59	539.2738	60.10	8.59	23.52	7.79
C27H46I1N5	567.61	567.2800	57.13	8.17	22.36	12.34
C27H46I2N2	652.49	652.1754	49.70	7.11	38.90	4.29
C27H46I2N4	680.50	680.1815	47.66	6.81	37.30	8.23
C27H46I2N6	708.52	708.1877	45.77	6.54	35.82	11.86
C27H47I1N2	526.59	526.2786	61.58	9.00	24.10	5.32
C27H47I1N4	554.61	554.2847	58.47	8.54	22.88	10.10
C27H47I1N6	582.62	582.2909	55.66	8.13	21.78	14.42
C27H47I2N1	639.49	639.1802	50.71	7.41	39.69	2.19
C27H47I2N3	667.50	667.1863	48.58	7.10	38.02	6.30
C27H47I2N5	695.52	695.1924	46.63	6.81	36.49	10.07
C27H48I1N1	513.59	513.2833	63.14	9.42	24.71	2.73
C27H48I1N3	541.61	541.2895	59.88	8.93	23.43	7.76
C27H48I1N5	569.62	569.2956	56.93	8.49	22.28	12.29
C27H48I2N2	654.51	654.1910	49.55	7.39	38.78	4.28
C27H48I2N4	682.52	682.1972	47.52	7.09	37.19	8.21
C27H48I2N6	710.53	710.2033	45.64	6.81	35.72	11.83
C27H49I1N2	528.61	528.2942	61.35	9.34	24.01	5.30
C27H49I1N4	556.62	556.3004	58.26	8.87	22.80	10.07
C27H49I1N6	584.64	584.3065	55.47	8.45	21.71	14.37
C27H49I2N1	641.51	641.1958	50.55	7.70	39.56	2.18
C27H49I2N3	669.52	669.2019	48.44	7.38	37.91	6.28
C27H49I2N5	697.53	697.2081	46.49	7.08	36.39	10.04
C27H50I1N1	515.61	515.2990	62.90	9.77	24.61	2.72
C27H50I1N3	543.62	543.3051	59.66	9.27	23.34	7.73
C27H50I1N5	571.64	571.3113	56.73	8.82	22.20	12.25
C27H50I2N2	656.52	656.2067	49.40	7.68	38.66	4.27
C27H50I2N4	684.54	684.2128	47.38	7.36	37.08	8.18
C27H50I2N6	712.55	712.2190	45.51	7.07	35.62	11.79
C27H51I1N2	530.63	530.3099	61.12	9.69	23.92	5.28
C27H51I1N4	558.64	558.3160	58.05	9.20	22.72	10.03
C27H51I1N6	586.65	586.3222	55.28	8.76	21.63	14.33
C27H51I2N1	643.52	643.2114	50.39	7.99	39.44	2.18
C27H51I2N3	671.54	671.2176	48.29	7.66	37.80	6.26
C27H51I2N5	699.55	699.2237	46.36	7.35	36.28	10.01
C27H52I1N1	517.63	517.3146	62.65	10.13	24.52	2.71
C27H52I1N3	545.64	545.3208	59.43	9.61	23.26	7.70
C27H52I1N5	573.65	573.3269	56.53	9.14	22.12	12.21
C27H52I2N2	658.54	658.2223	49.25	7.96	38.54	4.25
C27H52I2N4	686.55	686.2285	47.24	7.63	36.97	8,16
C27H52I2N6	714.56	714.2346	45.38	7.34	35.52	11.76
C27H53I1N2	532.64	532.3255	60.89	10.03	23.83	5.26
C27H53I1N4	560.65	560.3317	57.84	9.53	22.64	9.99
C27H53I1N6	588.67	588.3378	55.09	9.08	21.56	14.28
C27H53I2N1	645.54	645.2271	50.24	8.28	39.32	2.17

C27H53I2N3	673.55	673.2332	48.15	7.93	37.68	6.24
C27H53I2N5	701.57	701.2394	46.23	7.61	36.18	9.98
C27H54I1N1	519.64	519.3303	62.41	10.47	24.42	2.70
C27H54I1N3	547.66	547.3364	59.22	9.94	23.17	7.67
C27H54I1N5	575.67	575.3426	56.33	9.46	22.04	12.17
C27H54I2N2	660.55	660.2380	49.10	8.24	38.42	4.24
C27H54I2N4	688.57	688.2441	47.10	7.90	36.86	8.14
C27H54I2N6	716.58	716.2503	45.26	7.60	35.42	11.73
C27H55I1N2	534.66	534.3412	60.66	10.37	23.74	5.24
C27H55I1N4	562.67	562.3473	57.64	9.85	22.55	9.96
C27H55I1N6	590.68	590.3535	54.90	9.39	21.48	14.23
C27H55I2N1	647.55	647.2427	50.08	8.56	39.19	2.16
C27H55I2N3	675.57	675.2489	48.00	8.21	37.57	6.22
C27H55I2N5	703.58	703.2550	46.09	7.88	36.07	9.95
C27H56I1N1	521.66	521.3459	62.17	10.82	24.33	2.69
C27H56I1N3	549.67	549.3521	59.00	10.27	23.09	7.64
C27H56I1N5	577.69	577.3582	56.14	9.77	21.97	12.12
C27H56I2N2	662.57	662.2536	48.95	8.52	38.31	4.23
C27H56I2N4	690.58	690.2598	46.96	8.17	36.75	8.11
C27H56I2N6	718.60	718.2659	45.13	7.86	35.32	11.70
C27H57I1N2	536.67	536.3568	60.43	10.71	23.65	5.22
C27H57I1N4	564.69	564.3630	57.43	10.17	22.47	9.92
C27H57I1N6	592.70	592.3691	54.72	9.69	21.41	14.18
C27H57I2N3	677.58	677.2645	47.86	8.48	37.46	6.20
C27H57I2N5	705.60	705.2707	45.96	8.14	35.97	9.93
C27H58I1N3	551.69	551.3677	58.78	10.60	23.00	7.62
C27H58I1N5	579.70	579.3739	55.94	10.08	21.89	12.08
C27H58I2N4	692.60	692.2754	46.82	8.44	36.65	8.09
C27H58I2N6	720.61	720.2816	45.00	8.11	35.22	11.66
C27H59I1N4	566.70	566.3786	57.23	10.49	22.39	9.89
C27H59I1N6	594.72	594.3848	54.53	10.00	21.34	14.13
C27H59I2N5	707.61	707.2863	45.83	8.40	35.87	9.90
C27H60I1N5	581.72	581.3895	55.75	10.40	21.82	12.04
C27H60I2N6	722.63	722.2972	44.88	8.37	35.12	11.63
C27H61I1N6	596.73	596.4004	54.35	10.30	21.27	14.08
C28H14I1N1	491.33	491.0173	68.45	2.87	25.83	2.85
C28H14I1N3	519.35	519.0234	64.76	2.72	24.44	8.09
C28H14I1N5	547.36	547.0296	61.44	2.58	23.18	12.79
C28H14I2N2	632.25	631.9250	53.19	2.23	40.14	4.43
C28H14I2N4	660.26	659.9312	50.94	2.14	38.44	8.49
C28H14I2N6	688.27	687.9373	48.86	2.05	36.88	12.21
C28H15I1N2	506.35	506.0282	66.42	2.99	25.06	5.53
C28H15I1N4	534.36	534.0343	62.94	2.83	23.75	10.48
C28H15I1N6	562.38	562.0405	59.80	2.69	22.57	14.94
C28H15I2N1	619.25	618.9298	54.31	2.44	40.99	2.26
C28H15I2N3	647.26	646.9359	51.96	2.34	39.21	6.49
C28H15I2N5	675.27	674.9421	49.80	2.24	37.59	10.37
C28H16I1N1	493.35	493.0329	68.17	3.27	25.72	2.84
C28H16I1N3	521.36	521.0391	64.51	3.09	24.34	8.06
C28H16I1N5	549.38	549.0452	61.22	2.94	23.10	12.75
C28H16I2N2	634.26	633.9407	53.02	2.54	40.02	4.42
C28H16I2N4	662.28	661.9468	50.78	2.44	38.32	8.46
C28H16I2N6	690.29	689.9530	48.72	2.34	36.77	12.17
C28H17I1N2	508.37	508.0438	66.16	3.37	24.96	5.51
C28H17I1N4	536.38	536.0500	62.70	3.19	23.66	10.45
C28H17I1N6	564.39	564.0561	59.59	3.04	22.49	14.89
C28H17I2N1	621.26	620.9454	54.13	2.76	40.85	2.25
C28H17I2N3	649.28	648.9516	51.80	2.64	39.09	6.47
C28H17I2N5	677.29	676.9577	49.66	2.53	37.47	10.34
C28H18I1N1	495.37	495.0486	67.89	3.66	25.62	2.83
C28H18I1N3	523.38	523.0547	64.26	3.47	24.25	8.03
C28H18I1N5	551.39	551.0609	60.99	3.29	23.02	12.70
C28H18I2N2	636.28	635.9563	52.86	2.85	39.89	4.40
C28H18I2N4	664.29	663.9625	50.63	2.73	38.21	8.43
C28H18I2N6	692.30	691.9686	48.58	2.62	36.66	12.14
C28H19I1N2	510.38	510.0595	65.89	3.75	24.86	5.49
C28H19I1N4	538.39	538.0656	62.47	3.56	23.57	10.41
C28H19I1N6	566.41	566.0718	59.38	3.38	22.41	14.84
C28H19I2N1	623.28	622.9611	53.96	3.07	40.72	2.25
C28H19I2N3	651.29	650.9672	51.64	2.94	38.97	6.45
C28H19I2N5	679.31	678.9734	49.51	2.82	37.36	10.31
C28H20I1N1	497.38	497.0642	67.62	4.05	25.51	2.82
C28H20I1N3	525.40	525.0704	64.01	3.84	24.15	8.00
C28H20I1N5	553.41	553.0765	60.77	3.64	22.93	12.65
C28H20I2N2	638.29	637.9720	52.69	3.16	39.76	4.39
C28H20I2N4	666.31	665.9781	50.47	3.03	38.09	8.41
C28H20I2N6	694.32	693.9843	48.44	2.90	36.55	12.10
C28H21I1N2	512.40	512.0751	65.64	4.13	24.77	5.47
C28H21I1N4	540.41	540.0813	62.23	3.92	23.48	10.37
C28H21I1N6	568.42	568.0874	59.17	3.72	22.33	14.78
C28H21I2N1	625.30	624.9767	53.78	3.39	40.59	2.24
C28H21I2N3	653.31	652.9829	51.48	3.24	38.85	6.43
C28H21I2N5	681.32	680.9890	49.36	3.11	37.25	10.28
C28H22I1N1	499.40	499.0799	67.34	4.44	25.41	2.80
C28H22I1N3	527.41	527.0860	63.77	4.20	24.06	7.97
C28H22I1N5	555.43	555.0922	60.55	3.99	22.85	12.61
C28H22I2N2	640.31	639.9876	52.52	3.46	39.64	4.37
C28H22I2N4	668.32	667.9938	50.32	3.32	37.98	8.38
C28H22I2N6	696.34	695.9999	48.30	3.18	36.45	12.07
C28H23I1N2	514.41	514.0908	65.38	4.51	24.67	5.45
C28H23I1N4	542.43	542.0969	62.00	4.27	23.40	10.33
C28H23I1N6	570.44	570.1031	58.96	4.06	22.25	14.73
C28H23I2N1	627.31	626.9924	53.61	3.70	40.46	2.23
C28H23I2N3	655.32	654.9985	51.32	3.54	38.73	6.41
C28H23I2N5	683.34	683.0047	49.22	3.39	37.14	10.25
C28H24I1N1	501.41	501.0955	67.07	4.82	25.31	2.79
C28H24I1N3	529.43	529.1017	63.52	4.57	23.97	7.94
C28H24I1N5	557.44	557.1078	60.33	4.34	22.77	12.56
C28H24I2N2	642.33	642.0033	52.36	3.77	39.51	4.36
C28H24I2N4	670.34	670.0094	50.17	3.61	37.86	8.36
C28H24I2N6	698.35	698.0156	48.16	3.46	36.34	12.03
C28H25I1N2	516.43	516.1064	65.12	4.88	24.57	5.42
C28H25I1N4	544.44	544.1126	61.77	4.63	23.31	10.29
C28H25I1N6	572.46	572.1187	58.75	4.40	22.17	14.68
C28H25I2N1	629.33	629.0080	53.44	4.00	40.33	2.23
C28H25I2N3	657.34	657.0142	51.16	3.83	38.61	6.39
C28H25I2N5	685.35	685.0203	49.07	3.68	37.03	10.22
C28H26I1N1	503.43	503.1112	66.80	5.21	25.21	2.78
C28H26I1N3	531.44	531.1173	63.28	4.93	23.88	7.91
C28H26I1N5	559.46	559.1235	60.11	4.68	22.68	12.52
C28H26I2N2	644.34	644.0189	52.19	4.07	39.39	4.35
C28H26I2N4	672.36	672.0251	50.02	3.90	37.75	8.33
C28H26I2N6	700.37	700.0312	48.02	3.74	36.24	12.00
C28H27I1N2	518.45	518.1221	64.87	5.25	24.48	5.40
C28H27I1N4	546.46	546.1282	61.54	4.98	23.22	10.25
C28H27I1N6	574.47	574.1344	58.54	4.74	22.09	14.63
C28H27I2N1	631.34	631.0237	53.27	4.31	40.20	2.22
C28H27I2N3	659.36	659.0298	51.01	4.13	38.49	6.37
C28H27I2N5	687.37	687.0360	48.93	3.96	36.92	10.19
C28H28I1N1	505.45	505.1268	66.54	5.58	25.11	2.77

C28H28I1N3	533.46	533.1330	63.04	5.29	23.79	7.88
C28H28I1N5	561.47	561.1391	59.90	5.03	22.60	12.47
C28H28I2N2	646.36	646.0346	52.03	4.37	39.27	4.33
C28H28I2N4	674.37	674.0407	49.87	4.19	37.64	8.31
C28H28I2N6	702.38	702.0469	47.88	4.02	36.14	11.96
C28H29I1N2	520.46	520.1377	64.62	5.62	24.38	5.38
C28H29I1N4	548.47	548.1439	61.32	5.33	23.14	10.22
C28H29I1N6	576.49	576.1500	58.34	5.07	22.01	14.58
C28H29I2N1	633.36	633.0393	53.10	4.62	40.07	2.21
C28H29I2N3	661.37	661.0455	50.85	4.42	38.38	6.35
C28H29I2N5	689.39	689.0516	48.78	4.24	36.82	10.16
C28H30I1N1	507.46	507.1425	66.27	5.96	25.01	2.76
C28H30I1N3	535.48	535.1486	62.81	5.65	23.70	7.85
C28H30I1N5	563.49	563.1548	59.68	5.37	22.52	12.43
C28H30I2N2	648.37	648.0502	51.87	4.66	39.15	4.32
C28H30I2N4	676.39	676.0564	49.72	4.47	37.52	8.28
C28H30I2N6	704.40	704.0625	47.74	4.29	36.03	11.93
C28H31I1N2	522.48	522.1534	64.37	5.98	24.29	5.36
C28H31I1N4	550.49	550.1595	61.09	5.68	23.05	10.18
C28H31I1N6	578.50	578.1657	58.13	5.40	21.94	14.53
C28H31I2N1	635.37	635.0550	52.93	4.92	39.95	2.20
C28H31I2N3	663.39	663.0611	50.70	4.71	38.26	6.33
C28H31I2N5	691.40	691.0673	48.64	4.52	36.71	10.13
C28H32I1N1	509.48	509.1581	66.01	6.33	24.91	2.75
C28H32I1N3	537.49	537.1643	62.57	6.00	23.61	7.82
C28H32I1N5	565.51	565.1704	59.47	5.70	22.44	12.38
C28H32I2N2	650.39	650.0659	51.71	4.96	39.02	4.31
C28H32I2N4	678.40	678.0720	49.57	4.75	37.41	8.26
C28H32I2N6	706.42	706.0781	47.61	4.57	35.93	11.90
C28H33I1N2	524.49	524.1690	64.12	6.34	24.20	5.34
C28H33I1N4	552.51	552.1752	60.87	6.02	22.97	10.14
C28H33I1N6	580.52	580.1813	57.93	5.73	21.86	14.48
C28H33I2N1	637.39	637.0706	52.76	5.22	39.82	2.20
C28H33I2N3	665.40	665.0768	50.54	5.00	38.14	6.31
C28H33I2N5	693.42	693.0829	48.50	4.80	36.60	10.10
C28H34I1N1	511.49	511.1738	65.75	6.70	24.81	2.74
C28H34I1N3	539.51	539.1799	62.34	6.35	23.52	7.79
C28H34I1N5	567.52	567.1861	59.26	6.04	22.36	12.34
C28H34I2N2	652.41	652.0815	51.55	5.25	38.90	4.29
C28H34I2N4	680.42	680.0877	49.43	5.04	37.30	8.23
C28H34I2N6	708.43	708.0938	47.47	4.84	35.83	11.86
C28H35I1N2	526.51	526.1847	63.88	6.70	24.10	5.32
C28H35I1N4	554.52	554.1908	60.65	6.36	22.89	10.10
C28H35I1N6	582.54	582.1970	57.73	6.06	21.78	14.43
C28H35I2N1	639.41	639.0863	52.60	5.52	39.69	2.19
C28H35I2N3	667.42	667.0924	50.39	5.29	38.03	6.30
C28H35I2N5	695.43	695.0985	48.36	5.07	36.50	10.07
C28H36I1N1	513.51	513.1894	65.49	7.07	24.71	2.73
C28H36I1N3	541.52	541.1956	62.10	6.70	23.43	7.76
C28H36I1N5	569.54	569.2017	59.05	6.37	22.28	12.30
C28H36I2N2	654.42	654.0972	51.39	5.54	38.78	4.28
C28H36I2N4	682.43	682.1033	49.28	5.32	37.19	8.21
C28H36I2N6	710.45	710.1094	47.34	5.11	35.73	11.83
C28H37I1N2	528.52	528.2003	63.63	7.06	24.01	5.30
C28H37I1N4	556.54	556.2065	60.43	6.70	22.80	10.07
C28H37I1N6	584.55	584.2126	57.53	6.38	21.71	14.38
C28H37I2N1	641.42	641.1019	52.43	5.81	39.57	2.18
C28H37I2N3	669.44	669.1081	50.24	5.57	37.91	6.28
C28H37I2N5	697.45	697.1142	48.22	5.35	36.39	10.04
C28H38I1N1	515.53	515.2051	65.24	7.43	24.62	2.72
C28H38I1N3	543.54	543.2112	61.87	7.05	23.35	7.73
C28H38I1N5	571.55	571.2174	58.84	6.70	22.20	12.25
C28H38I2N2	656.44	656.1128	51.23	5.83	38.66	4.27
C28H38I2N4	684.45	684.1189	49.14	5.60	37.08	8.19
C28H38I2N6	712.46	712.1251	47.20	5.38	35.62	11.80
C28H39I1N2	530.54	530.2160	63.39	7.41	23.92	5.28
C28H39I1N4	558.55	558.2221	60.21	7.04	22.72	10.03
C28H39I1N6	586.57	586.2283	57.34	6.70	21.64	14.33
C28H39I2N1	643.44	643.1176	52.27	6.11	39.45	2.18
C28H39I2N3	671.45	671.1237	50.09	5.85	37.80	6.26
C28H39I2N5	699.47	699.1298	48.08	5.62	36.29	10.01
C28H40I1N1	517.54	517.2207	64.98	7.79	24.52	2.71
C28H40I1N3	545.56	545.2269	61.65	7.39	23.26	7.70
C28H40I1N5	573.57	573.2330	58.64	7.03	22.13	12.21
C28H40I2N2	658.45	658.1285	51.08	6.12	38.55	4.25
C28H40I2N4	686.47	686.1346	48.99	5.87	36.97	8.16
C28H40I2N6	714.48	714.1407	47.07	5.64	35.52	11.76
C28H41I1N2	532.56	532.2316	63.15	7.76	23.83	5.26
C28H41I1N4	560.57	560.2378	59.99	7.37	22.64	9.99
C28H41I1N6	588.58	588.2439	57.14	7.02	21.56	14.28
C28H41I2N1	645.45	645.1332	52.10	6.40	39.32	2.17
C28H41I2N3	673.47	673.1394	49.94	6.14	37.69	6.24
C28H41I2N5	701.48	701.1455	47.94	5.89	36.18	9.98
C28H42I1N1	519.56	519.2364	64.73	8.15	24.43	2.70
C28H42I1N3	547.57	547.2425	61.42	7.73	23.18	7.67
C28H42I1N5	575.58	575.2487	58.43	7.36	22.05	12.17
C28H42I2N2	660.47	660.1441	50.92	6.41	38.43	4.24
C28H42I2N4	688.48	688.1502	48.85	6.15	36.86	8.14
C28H42I2N6	716.50	716.1564	46.94	5.91	35.42	11.73
C28H43I1N2	534.57	534.2473	62.91	8.11	23.74	5.24
C28H43I1N4	562.59	562.2534	59.78	7.70	22.56	9.96
C28H43I1N6	590.60	590.2596	56.94	7.34	21.49	14.23
C28H43I2N1	647.47	647.1489	51.94	6.69	39.20	2.16
C28H43I2N3	675.48	675.1550	49.79	6.42	37.57	6.22
C28H43I2N5	703.50	703.1611	47.81	6.16	36.08	9.96
C28H44I1N1	521.57	521.2520	64.48	8.50	24.33	2.69
C28H44I1N3	549.59	549.2582	61.19	8.07	23.09	7.65
C28H44I1N5	577.60	577.2643	58.23	7.68	21.97	12.12
C28H44I2N2	662.49	662.1597	50.77	6.69	38.31	4.23
C28H44I2N4	690.50	690.1659	48.71	6.42	36.76	8.11
C28H44I2N6	718.51	718.1720	46.81	6.17	35.32	11.70
C28H45I1N2	536.59	536.2629	62.68	8.45	23.65	5.22
C28H45I1N4	564.60	564.2691	59.57	8.03	22.48	9.92
C28H45I1N6	592.62	592.2752	56.75	7.65	21.41	14.18
C28H45I2N1	649.49	649.1645	51.78	6.98	39.08	2.16
C28H45I2N3	677.50	677.1706	49.64	6.70	37.46	6.20
C28H45I2N5	705.51	705.1768	47.67	6.43	35.98	9.93
C28H46I1N1	523.59	523.2677	64.23	8.86	24.24	2.68
C28H46I1N3	551.60	551.2738	60.97	8.41	23.01	7.62
C28H46I1N5	579.62	579.2800	58.02	8.00	21.89	12.08
C28H46I2N2	664.50	664.1754	50.61	6.98	38.20	4.22
C28H46I2N4	692.51	692.1815	48.56	6.70	36.65	8.09
C28H46I2N6	720.53	720.1877	46.68	6.44	35.23	11.66
C28H47I1N2	538.60	538.2786	62.44	8.80	23.56	5.20
C28H47I1N4	566.62	566.2847	59.35	8.36	22.40	9.89
C28H47I1N6	594.63	594.2909	56.56	7.97	21.34	14.13
C28H47I2N1	651.50	651.1802	51.62	7.27	38.96	2.15
C28H47I2N3	679.52	679.1863	49.49	6.97	37.35	6.18
C28H47I2N5	707.53	707.1924	47.53	6.70	35.87	9.90
C28H48I1N1	525.61	525.2833	63.99	9.21	24.14	2.66

047

C28H48I1N3	553.62	553.2895	60.75	8.74	22.92	7.59
C28H48I1N5	581.63	581.2956	57.82	8.32	21.82	12.04
C28H48I2N2	666.52	666.1910	50.46	7.26	38.08	4.20
C28H48I2N4	694.53	694.1972	48.42	6.97	36.54	8.07
C28H48I2N6	722.54	722.2033	46.55	6.70	35.13	11.63
C28H49I1N2	540.62	540.2942	62.21	9.14	23.47	5.18
C28H49I1N4	568.63	568.3004	59.14	8.69	22.32	9.85
C28H49I1N6	596.65	596.3065	56.37	8.28	21.27	14.09
C28H49I2N1	653.52	653.1958	51.46	7.56	38.84	2.14
C28H49I2N3	681.53	681.2019	49.35	7.25	37.24	6.17
C28H49I2N5	709.55	709.2081	47.40	6.96	35.77	9.87
C28H50I1N1	527.62	527.2990	63.74	9.55	24.05	2.65
C28H50I1N3	555.64	555.3051	60.53	9.07	22.84	7.56
C28H50I1N5	583.65	583.3113	57.62	8.64	21.74	12.00
C28H50I2N2	668.53	668.2067	50.31	7.54	37.97	4.19
C28H50I2N4	696.55	696.2128	48.28	7.24	36.44	8.04
C28H50I2N6	724.56	724.2190	46.42	6.96	35.03	11.60
C28H51I1N2	542.64	542.3099	61.98	9.47	23.39	5.16
C28H51I1N4	570.65	570.3160	58.93	9.01	22.24	9.82
C28H51I1N6	598.66	598.3222	56.18	8.59	21.20	14.04
C28H51I2N1	655.53	655.2114	51.30	7.84	38.72	2.14
C28H51I2N3	683.55	683.2176	49.20	7.52	37.13	6.15
C28H51I2N5	711.56	711.2237	47.26	7.22	35.67	9.84
C28H52I1N1	529.64	529.3146	63.50	9.90	23.96	2.64
C28H52I1N3	557.65	557.3208	60.31	9.40	22.76	7.54
C28H52I1N5	585.66	585.3269	57.42	8.95	21.67	11.96
C28H52I2N2	670.55	670.2223	50.15	7.82	37.85	4.18
C28H52I2N4	698.56	698.2285	48.14	7.50	36.33	8.02
C28H52I2N6	726.58	726.2346	46.29	7.21	34.93	11.57
C28H53I1N2	544.65	544.3255	61.75	9.81	23.30	5.14
C28H53I1N4	572.67	572.3317	58.73	9.33	22.16	9.78
C28H53I1N6	600.68	600.3378	55.99	8.89	21.13	13.99
C28H53I2N1	657.55	657.2271	51.15	8.12	38.60	2.13
C28H53I2N3	685.56	685.2332	49.06	7.79	37.02	6.13
C28H53I2N5	713.58	713.2394	47.13	7.49	35.57	9.81
C28H54I1N1	531.65	531.3303	63.26	10.24	23.87	2.63
C28H54I1N3	559.67	559.3364	60.09	9.73	22.67	7.51
C28H54I1N5	587.68	587.3426	57.23	9.26	21.59	11.92
C28H54I2N2	672.56	672.2380	50.00	8.09	37.74	4.17
C28H54I2N4	700.58	700.2441	48.00	7.77	36.23	8.00
C28H54I2N6	728.59	728.2503	46.16	7.47	34.84	11.53
C28H55I1N2	546.67	546.3412	61.52	10.14	23.21	5.12
C28H55I1N4	574.68	574.3473	58.52	9.65	22.08	9.75
C28H55I1N6	602.70	602.3535	55.80	9.20	21.06	13.94
C28H55I2N1	659.57	659.2427	50.99	8.41	38.48	2.12
C28H55I2N3	687.58	687.2489	48.91	8.06	36.91	6.11
C28H55I2N5	715.59	715.2550	47.00	7.75	35.47	9.79
C28H56I1N1	533.67	533.3459	63.02	10.58	23.78	2.62
C28H56I1N3	561.68	561.3521	59.88	10.05	22.59	7.48
C28H56I1N5	589.70	589.3582	57.03	9.57	21.52	11.88
C28H56I2N2	674.58	674.2536	49.85	8.37	37.62	4.15
C28H56I2N4	702.59	702.2598	47.87	8.03	36.12	7.97
C28H56I2N6	730.61	730.2659	46.03	7.73	34.74	11.50
C28H57I1N2	548.68	548.3568	61.29	10.47	23.13	5.11
C28H57I1N4	576.70	576.3630	58.32	9.96	22.01	9.72
C28H57I1N6	604.71	604.3691	55.62	9.50	20.99	13.90
C28H57I2N1	661.58	661.2584	50.83	8.68	38.36	2.12
C28H57I2N3	689.60	689.2645	48.77	8.33	36.81	6.09
C28H57I2N5	717.61	717.2707	46.87	8.01	35.37	9.76
C28H58I1N1	535.69	535.3616	62.78	10.91	23.69	2.61
C28H58I1N3	563.70	563.3677	59.66	10.37	22.51	7.45
C28H58I1N5	591.71	591.3739	56.84	9.88	21.45	11.84
C28H58I2N2	676.60	676.2693	49.71	8.64	37.51	4.14
C28H58I2N4	704.61	704.2754	47.73	8.30	36.02	7.95
C28H58I2N6	732.62	732.2816	45.91	7.98	34.64	11.47
C28H59I1N2	550.70	550.3725	61.07	10.80	23.04	5.09
C28H59I1N4	578.71	578.3786	58.11	10.28	21.93	9.68
C28H59I1N6	606.73	606.3848	55.43	9.80	20.92	13.85
C28H59I2N3	691.61	691.2802	48.63	8.60	36.70	6.08
C28H59I2N5	719.62	719.2863	46.73	8.26	35.27	9.73
C28H60I1N3	565.71	565.3834	59.45	10.69	22.43	7.43
C28H60I1N5	593.73	593.3895	56.64	10.19	21.37	11.80
C28H60I2N4	706.63	706.2911	47.59	8.56	35.92	7.93
C28H60I2N6	734.64	734.2972	45.78	8.23	34.55	11.44
C28H61I1N4	580.73	580.3943	57.91	10.59	21.85	9.65
C28H61I1N6	608.74	608.4004	55.25	10.10	20.85	13.81
C28H61I2N5	721.64	721.3020	46.60	8.52	35.17	9.70
C28H62I1N5	595.74	595.4052	56.45	10.49	21.30	11.76
C28H62I2N6	736.66	736.3129	45.65	8.48	34.45	11.41
C28H63I1N6	610.76	610.4161	55.06	10.40	20.78	13.76
C29H14I1N1	503.35	503.0173	69.20	2.80	25.21	2.78
C29H14I1N3	531.36	531.0234	65.55	2.66	23.88	7.91
C29H14I1N5	559.37	559.0296	62.27	2.52	22.69	12.52
C29H14I2N2	644.26	643.9250	54.07	2.19	39.40	4.35
C29H14I2N4	672.27	671.9312	51.81	2.10	37.75	8.33
C29H14I2N6	700.28	699.9373	49.74	2.02	36.24	12.00
C29H15I1N2	518.36	518.0282	67.20	2.92	24.48	5.40
C29H15I1N4	546.37	546.0343	63.75	2.77	23.23	10.25
C29H15I1N6	574.39	574.0405	60.64	2.63	22.09	14.63
C29H15I2N1	631.26	630.9298	55.18	2.40	40.21	2.22
C29H15I2N3	659.27	658.9359	52.83	2.29	38.50	6.37
C29H15I2N5	687.29	686.9421	50.68	2.20	36.93	10.19
C29H16I1N1	505.36	505.0329	68.93	3.19	25.11	2.77
C29H16I1N3	533.38	533.0391	65.31	3.02	23.79	7.88
C29H16I1N5	561.39	561.0452	62.05	2.87	22.61	12.48
C29H16I2N2	646.27	645.9407	53.90	2.50	39.27	4.33
C29H16I2N4	674.29	673.9468	51.66	2.39	37.64	8.31
C29H16I2N6	702.30	701.9530	49.60	2.30	36.14	11.97
C29H17I1N2	520.38	520.0438	66.94	3.29	24.39	5.38
C29H17I1N4	548.39	548.0500	63.52	3.12	23.14	10.22
C29H17I1N6	576.40	576.0561	60.43	2.97	22.02	14.58
C29H17I2N1	633.27	632.9454	55.00	2.71	40.08	2.21
C29H17I2N3	661.29	660.9516	52.67	2.59	38.38	6.35
C29H17I2N5	689.30	688.9577	50.53	2.49	36.82	10.16
C29H18I1N1	507.38	507.0486	68.65	3.58	25.01	2.76
C29H18I1N3	535.39	535.0547	65.06	3.39	23.70	7.85
C29H18I1N5	563.40	563.0609	61.82	3.22	22.52	12.43
C29H18I2N2	648.29	647.9563	53.73	2.80	39.15	4.32
C29H18I2N4	676.30	675.9625	51.50	2.68	37.53	8.28
C29H18I2N6	704.32	703.9686	49.46	2.58	36.04	11.93
C29H19I1N2	522.39	522.0595	66.68	3.67	24.29	5.36
C29H19I1N4	550.41	550.0656	63.28	3.48	23.06	10.18
C29H19I1N6	578.42	578.0718	60.22	3.31	21.94	14.53
C29H19I2N1	635.29	634.9611	54.83	3.01	39.95	2.20
C29H19I2N3	663.30	662.9672	52.51	2.89	38.26	6.33
C29H19I2N5	691.32	690.9734	50.39	2.77	36.71	10.13
C29H20I1N1	509.39	509.0642	68.38	3.96	24.91	2.75
C29H20I1N3	537.41	537.0704	64.82	3.75	23.61	7.82
C29H20I1N5	565.42	565.0765	61.60	3.57	22.44	12.39
C29H20I2N2	650.30	649.9720	53.56	3.10	39.03	4.31

048

C29H20I2N4	678.32	677.9781	51.35	2.97	37.42	8.26
C29H20I2N6	706.33	705.9843	49.31	2.85	35.93	11.90
C29H21I1N2	524.41	524.0751	66.42	4.04	24.20	5.34
C29H21I1N4	552.42	552.0813	63.05	3.83	22.97	10.14
C29H21I1N6	580.44	580.0874	60.01	3.65	21.86	14.48
C29H21I2N1	637.31	636.9767	54.66	3.32	39.83	2.20
C29H21I2N3	665.32	664.9829	52.35	3.18	38.15	6.32
C29H21I2N5	693.33	692.9890	50.24	3.05	36.61	10.10
C29H22I1N1	511.41	511.0799	68.11	4.34	24.81	2.74
C29H22I1N3	539.42	539.0860	64.57	4.11	23.53	7.79
C29H22I1N5	567.44	567.0922	61.39	3.91	22.36	12.34
C29H22I2N2	652.32	651.9876	53.40	3.40	38.91	4.29
C29H22I2N4	680.33	679.9938	51.20	3.26	37.31	8.24
C29H22I2N6	708.35	707.9999	49.17	3.13	35.83	11.86
C29H23I1N2	526.42	526.0908	66.17	4.40	24.11	5.32
C29H23I1N4	554.44	554.0969	62.82	4.18	22.89	10.11
C29H23I1N6	582.45	582.1031	59.80	3.98	21.79	14.43
C29H23I2N1	639.32	638.9924	54.48	3.63	39.70	2.19
C29H23I2N3	667.34	666.9985	52.20	3.47	38.03	6.30
C29H23I2N5	695.35	695.0047	50.09	3.33	36.50	10.07
C29H24I1N1	513.43	513.0955	67.84	4.71	24.72	2.73
C29H24I1N3	541.44	541.1017	64.33	4.47	23.44	7.76
C29H24I1N5	569.45	569.1078	61.17	4.25	22.29	12.30
C29H24I2N2	654.34	654.0033	53.23	3.70	38.79	4.28
C29H24I2N4	682.35	682.0094	51.05	3.55	37.20	8.21
C29H24I2N6	710.36	710.0156	49.03	3.41	35.73	11.83
C29H25I1N2	528.44	528.1064	65.92	4.77	24.01	5.30
C29H25I1N4	556.45	556.1126	62.60	4.53	22.81	10.07
C29H25I1N6	584.47	584.1187	59.60	4.31	21.71	14.38
C29H25I2N1	641.34	641.0080	54.31	3.93	39.57	2.18
C29H25I2N3	669.35	669.0142	52.04	3.76	37.92	6.28
C29H25I2N5	697.36	697.0203	49.95	3.61	36.40	10.04
C29H26I1N1	515.44	515.1112	67.58	5.08	24.62	2.72
C29H26I1N3	543.46	543.1173	64.09	4.82	23.35	7.73
C29H26I1N5	571.47	571.1235	60.95	4.59	22.21	12.26
C29H26I2N2	656.35	656.0189	53.07	3.99	38.67	4.27
C29H26I2N4	684.37	684.0251	50.90	3.83	37.09	8.19
C29H26I2N6	712.38	712.0312	48.90	3.68	35.63	11.80
C29H27I1N2	530.46	530.1221	65.66	5.13	23.92	5.28
C29H27I1N4	558.47	558.1282	62.37	4.87	22.72	10.03
C29H27I1N6	586.48	586.1344	59.39	4.64	21.64	14.33
C29H27I2N1	643.35	643.0237	54.14	4.23	39.45	2.18
C29H27I2N3	671.37	671.0298	51.88	4.05	37.80	6.26
C29H27I2N5	699.38	699.0360	49.80	3.89	36.29	10.01
C29H28I1N1	517.46	517.1268	67.31	5.45	24.52	2.71
C29H28I1N3	545.47	545.1330	63.86	5.17	23.27	7.70
C29H28I1N5	573.48	573.1391	60.74	4.92	22.13	12.21
C29H28I2N2	658.37	658.0346	52.91	4.29	38.55	4.25
C29H28I2N4	686.38	686.0407	50.75	4.11	36.98	8.16
C29H28I2N6	714.40	714.0469	48.76	3.95	35.53	11.76
C29H29I1N2	532.47	532.1377	65.42	5.49	23.83	5.26
C29H29I1N4	560.49	560.1439	62.15	5.22	22.64	10.00
C29H29I1N6	588.50	588.1500	59.19	4.97	21.56	14.28
C29H29I2N1	645.37	645.0393	53.97	4.53	39.33	2.17
C29H29I2N3	673.38	673.0455	51.73	4.34	37.69	6.24
C29H29I2N5	701.40	701.0516	49.66	4.17	36.19	9.98
C29H30I1N1	519.47	519.1425	67.05	5.82	24.43	2.70
C29H30I1N3	547.49	547.1486	63.62	5.52	23.18	7.68
C29H30I1N5	575.50	575.1548	60.53	5.25	22.05	12.17
C29H30I2N2	660.38	660.0502	52.75	4.58	38.43	4.24
C29H30I2N4	688.40	688.0564	50.60	4.39	36.87	8.14
C29H30I2N6	716.41	716.0625	48.62	4.22	35.43	11.73
C29H31I1N2	534.49	534.1534	65.17	5.85	23.74	5.24
C29H31I1N4	562.50	562.1595	61.92	5.56	22.56	9.96
C29H31I1N6	590.52	590.1657	58.99	5.29	21.49	14.23
C29H31I2N1	647.39	647.0550	53.80	4.83	39.21	2.16
C29H31I2N3	675.40	675.0611	51.57	4.63	37.58	6.22
C29H31I2N5	703.41	703.0673	49.52	4.44	36.08	9.96
C29H32I1N1	521.49	521.1581	66.79	6.19	24.33	2.69
C29H32I1N3	549.50	549.1643	63.39	5.87	23.09	7.65
C29H32I1N5	577.52	577.1704	60.31	5.59	21.97	12.13
C29H32I2N2	662.40	662.0659	52.58	4.87	38.32	4.23
C29H32I2N4	690.41	690.0720	50.45	4.67	36.76	8.11
C29H32I2N6	718.43	718.0781	48.48	4.49	35.33	11.70
C29H33I1N2	536.50	536.1690	64.92	6.20	23.65	5.22
C29H33I1N4	564.52	564.1752	61.70	5.89	22.48	9.92
C29H33I1N6	592.53	592.1813	58.79	5.61	21.42	14.18
C29H33I2N1	649.40	649.0706	53.64	5.12	39.08	2.16
C29H33I2N3	677.42	677.0768	51.42	4.91	37.47	6.20
C29H33I2N5	705.43	705.0829	49.38	4.72	35.98	9.93
C29H34I1N1	523.51	523.1738	66.54	6.55	24.24	2.68
C29H34I1N3	551.52	551.1799	63.16	6.21	23.01	7.62
C29H34I1N5	579.53	579.1861	60.10	5.91	21.90	12.08
C29H34I2N2	664.42	664.0815	52.43	5.16	38.20	4.22
C29H34I2N4	692.43	692.0877	50.30	4.95	36.65	8.09
C29H34I2N6	720.44	720.0938	48.35	4.76	35.23	11.67
C29H35I1N2	538.52	538.1847	64.68	6.55	23.57	5.20
C29H35I1N4	566.53	566.1908	61.48	6.23	22.40	9.89
C29H35I1N6	594.55	594.1970	58.59	5.93	21.34	14.14
C29H35I2N1	651.42	651.0863	53.47	5.42	38.96	2.15
C29H35I2N3	679.43	679.0924	51.27	5.19	37.36	6.18
C29H35I2N5	707.44	707.0985	49.24	4.99	35.88	9.90
C29H36I1N1	525.52	525.1894	66.28	6.90	24.15	2.67
C29H36I1N3	553.53	553.1956	62.93	6.56	22.93	7.59
C29H36I1N5	581.55	581.2017	59.90	6.24	21.82	12.04
C29H36I2N2	666.43	666.0972	52.27	5.44	38.08	4.20
C29H36I2N4	694.45	694.1033	50.16	5.23	36.55	8.07
C29H36I2N6	722.46	722.1094	48.21	5.02	35.13	11.63
C29H37I1N2	540.54	540.2003	64.44	6.90	23.48	5.18
C29H37I1N4	568.55	568.2065	61.27	6.56	22.32	9.85
C29H37I1N6	596.56	596.2126	58.39	6.25	21.27	14.09
C29H37I2N1	653.43	653.1019	53.31	5.71	38.84	2.14
C29H37I2N3	681.45	681.1081	51.12	5.47	37.25	6.17
C29H37I2N5	709.46	709.1142	49.10	5.26	35.77	9.87
C29H38I1N1	527.54	527.2051	66.03	7.26	24.06	2.66
C29H38I1N3	555.55	555.2112	62.70	6.89	22.84	7.56
C29H38I1N5	583.56	583.2174	59.69	6.56	21.75	12.00
C29H38I2N2	668.45	668.1128	52.11	5.73	37.97	4.19
C29H38I2N4	696.46	696.1189	50.01	5.50	36.44	8.04
C29H38I2N6	724.48	724.1251	48.08	5.29	35.03	11.60
C29H39I1N2	542.55	542.2160	64.20	7.25	23.39	5.16
C29H39I1N4	570.57	570.2221	61.05	6.89	22.24	9.82
C29H39I1N6	598.58	598.2283	58.19	6.57	21.20	14.04
C29H39I2N1	655.45	655.1176	53.14	6.00	38.72	2.14
C29H39I2N3	683.46	683.1237	50.96	5.75	37.14	6.15
C29H39I2N5	711.48	711.1298	48.96	5.53	35.67	9.84
C29H40I1N1	529.55	529.2207	65.78	7.61	23.96	2.65
C29H40I1N3	557.57	557.2269	62.47	7.23	22.76	7.54
C29H40I1N5	585.58	585.2330	59.48	6.89	21.67	11.96
C29H40I2N2	670.46	670.1285	51.95	6.01	37.86	4.18

C29H40I2N4	698.48	698.1346	49.87	5.77	36.34	8.02
C29H40I2N6	726.49	726.1407	47.95	5.55	34.94	11.57
C29H41I1N2	544.57	544.2316	63.96	7.59	23.30	5.14
C29H41I1N4	572.58	572.2378	60.83	7.22	22.16	9.78
C29H41I1N6	600.59	600.2439	58.00	6.88	21.13	13.99
C29H41I2N1	657.47	657.1332	52.98	6.29	38.60	2.13
C29H41I2N3	685.48	685.1394	50.81	6.03	37.03	6.13
C29H41I2N5	713.49	713.1455	48.82	5.79	35.57	9.82
C29H42I1N1	531.57	531.2364	65.53	7.96	23.87	2.63
C29H42I1N3	559.58	559.2425	62.25	7.57	22.68	7.51
C29H42I1N5	587.60	587.2487	59.28	7.20	21.60	11.92
C29H42I2N2	672.48	672.1441	51.80	6.30	37.74	4.17
C29H42I2N4	700.49	700.1502	49.73	6.04	36.23	8.00
C29H42I2N6	728.51	728.1564	47.81	5.81	34.84	11.54
C29H43I1N2	546.58	546.2473	63.73	7.93	23.22	5.13
C29H43I1N4	574.60	574.2534	60.62	7.54	22.09	9.75
C29H43I1N6	602.61	602.2596	57.80	7.19	21.06	13.95
C29H43I2N1	659.48	659.1489	52.82	6.57	38.49	2.12
C29H43I2N3	687.49	687.1550	50.67	6.30	36.92	6.11
C29H43I2N5	715.51	715.1611	48.68	6.06	35.47	9.79
C29H44I1N1	533.59	533.2520	65.28	8.31	23.78	2.63
C29H44I1N3	561.60	561.2582	62.02	7.90	22.60	7.48
C29H44I1N5	589.61	589.2643	59.08	7.52	21.52	11.88
C29H44I2N2	674.50	674.1597	51.64	6.58	37.63	4.15
C29H44I2N4	702.51	702.1659	49.58	6.31	36.13	7.98
C29H44I2N6	730.52	730.1720	47.68	6.07	34.74	11.50
C29H45I1N2	548.60	548.2629	63.49	8.27	23.13	5.11
C29H45I1N4	576.61	576.2691	60.41	7.87	22.01	9.72
C29H45I1N6	604.63	604.2752	57.61	7.50	20.99	13.90
C29H45I2N1	661.50	661.1645	52.66	6.86	38.37	2.12
C29H45I2N3	689.51	689.1706	50.52	6.58	36.81	6.09
C29H45I2N5	717.52	717.1768	48.55	6.32	35.37	9.76
C29H46I1N1	535.60	535.2677	65.03	8.66	23.69	2.62
C29H46I1N3	563.61	563.2738	61.80	8.23	22.52	7.46
C29H46I1N5	591.63	591.2800	58.88	7.84	21.45	11.84
C29H46I2N2	676.51	676.1754	51.49	6.85	37.52	4.14
C29H46I2N4	704.53	704.1815	49.44	6.58	36.03	7.95
C29H46I2N6	732.54	732.1877	47.55	6.33	34.65	11.47
C29H47I1N2	550.62	550.2786	63.26	8.60	23.05	5.09
C29H47I1N4	578.63	578.2847	60.20	8.19	21.93	9.68
C29H47I1N6	606.64	606.2909	57.42	7.81	20.92	13.85
C29H47I2N1	663.51	663.1802	52.50	7.14	38.25	2.11
C29H47I2N3	691.53	691.1863	50.37	6.85	36.70	6.08
C29H47I2N5	719.54	719.1924	48.41	6.58	35.27	9.73
C29H48I1N1	537.62	537.2833	64.79	9.00	23.60	2.61
C29H48I1N3	565.63	565.2895	61.58	8.55	22.44	7.43
C29H48I1N5	593.64	593.2956	58.68	8.15	21.38	11.80
C29H48I2N2	678.53	678.1910	51.34	7.13	37.41	4.13
C29H48I2N4	706.54	706.1972	49.30	6.85	35.92	7.93
C29H48I2N6	734.55	734.2033	47.42	6.59	34.55	11.44
C29H49I1N2	552.63	552.2942	63.03	8.94	22.96	5.07
C29H49I1N4	580.65	580.3004	59.99	8.51	21.86	9.65
C29H49I1N6	608.66	608.3065	57.23	8.11	20.85	13.81
C29H49I2N1	665.53	665.1958	52.34	7.42	38.14	2.10
C29H49I2N3	693.54	693.2019	50.22	7.12	36.60	6.06
C29H49I2N5	721.56	721.2081	48.27	6.85	35.18	9.71
C29H50I1N1	539.63	539.2990	64.55	9.34	23.52	2.60
C29H50I1N3	567.65	567.3051	61.36	8.88	22.36	7.40
C29H50I1N5	595.66	595.3113	58.48	8.46	21.30	11.76
C29H50I2N2	680.54	680.2067	51.18	7.41	37.29	4.12
C29H50I2N4	708.56	708.2128	49.16	7.11	35.82	7.91
C29H50I2N6	736.57	736.2190	47.29	6.84	34.46	11.41
C29H51I1N2	554.65	554.3099	62.80	9.27	22.88	5.05
C29H51I1N4	582.66	582.3160	59.78	8.82	21.78	9.62
C29H51I1N6	610.67	610.3222	57.04	8.42	20.78	13.76
C29H51I2N1	667.55	667.2114	52.18	7.70	38.02	2.10
C29H51I2N3	695.56	695.2176	50.08	7.39	36.49	6.04
C29H51I2N5	723.57	723.2237	48.14	7.10	35.08	9.68
C29H52I1N1	541.65	541.3146	64.31	9.68	23.43	2.59
C29H52I1N3	569.66	569.3208	61.15	9.20	22.28	7.38
C29H52I1N5	597.68	597.3269	58.28	8.77	21.23	11.72
C29H52I2N2	682.56	682.2223	51.03	7.68	37.18	4.10
C29H52I2N4	710.57	710.2285	49.02	7.38	35.72	7.88
C29H52I2N6	738.59	738.2346	47.16	7.10	34.36	11.38
C29H53I1N2	556.66	556.3255	62.57	9.60	22.80	5.03
C29H53I1N4	584.68	584.3317	59.58	9.14	21.71	9.58
C29H53I1N6	612.69	612.3378	56.85	8.72	20.71	13.72
C29H53I2N1	669.56	669.2271	52.02	7.98	37.91	2.09
C29H53I2N3	697.57	697.2332	49.93	7.66	36.38	6.02
C29H53I2N5	725.59	725.2394	48.01	7.36	34.98	9.65
C29H54I1N1	543.66	543.3303	64.07	10.01	23.34	2.58
C29H54I1N3	571.68	571.3364	60.93	9.52	22.20	7.35
C29H54I1N5	599.69	599.3426	58.08	9.08	21.16	11.68
C29H54I2N2	684.58	684.2380	50.88	7.95	37.08	4.09
C29H54I2N4	712.59	712.2441	48.88	7.64	35.62	7.86
C29H54I2N6	740.60	740.2503	47.03	7.35	34.27	11.35
C29H55I1N2	558.68	558.3412	62.35	9.92	22.72	5.01
C29H55I1N4	586.69	586.3473	59.37	9.45	21.63	9.55
C29H55I1N6	614.71	614.3535	56.67	9.02	20.64	13.67
C29H55I2N1	671.58	671.2427	51.87	8.25	37.79	2.09
C29H55I2N3	699.59	699.2489	49.79	7.92	36.28	6.01
C29H55I2N5	727.60	727.2550	47.87	7.62	34.88	9.63
C29H56I1N1	545.68	545.3459	63.83	10.34	23.26	2.57
C29H56I1N3	573.69	573.3521	60.72	9.84	22.12	7.32
C29H56I1N5	601.71	601.3582	57.89	9.38	21.09	11.64
C29H56I2N2	686.59	686.2536	50.73	8.22	36.97	4.08
C29H56I2N4	714.61	714.2598	48.74	7.90	35.52	7.84
C29H56I2N6	742.62	742.2659	46.90	7.60	34.18	11.32
C29H57I1N2	560.70	560.3568	62.12	10.25	22.63	5.00
C29H57I1N4	588.71	588.3630	59.17	9.76	21.56	9.52
C29H57I1N6	616.72	616.3691	56.48	9.32	20.58	13.63
C29H57I2N1	673.59	673.2584	51.71	8.53	37.68	2.08
C29H57I2N3	701.61	701.2645	49.65	8.19	36.18	5.99
C29H57I2N5	729.62	729.2707	47.74	7.87	34.79	9.60
C29H58I1N1	547.70	547.3616	63.60	10.67	23.17	2.56
C29H58I1N3	575.71	575.3677	60.50	10.15	22.04	7.30
C29H58I1N5	603.72	603.3739	57.70	9.68	21.02	11.60
C29H58I2N2	688.61	688.2693	50.58	8.49	36.86	4.07
C29H58I2N4	716.62	716.2754	48.61	8.16	35.42	7.82
C29H58I2N6	744.63	744.2816	46.78	7.85	34.09	11.29
C29H59I1N2	562.71	562.3725	61.90	10.57	22.55	4.98
C29H59I1N4	590.72	590.3786	58.97	10.07	21.48	9.48
C29H59I1N6	618.74	618.3848	56.30	9.61	20.51	13.58
C29H59I2N1	675.61	675.2740	51.56	8.80	37.57	2.07
C29H59I2N3	703.62	703.2802	49.50	8.45	36.07	5.97
C29H59I2N5	731.64	731.2863	47.61	8.13	34.69	9.57
C29H60I1N1	549.71	549.3772	63.36	11.00	23.09	2.55
C29H60I1N3	577.73	577.3834	60.29	10.47	21.97	7.27
C29H60I1N5	605.74	605.3895	57.50	9.98	20.95	11.56
C29H60I2N2	690.62	690.2849	50.44	8.76	36.75	4.06

050

C29H60I2N4	718.64	718.2911	48.47	8.42	35.32	7.80
C29H60I2N6	746.65	746.2972	46.65	8.10	33.99	11.26
C29H61I1N2	564.73	564.3881	61.68	10.89	22.47	4.96
C29H61I1N4	592.74	592.3943	58.76	10.37	21.41	9.45
C29H61I1N6	620.75	620.4004	56.11	9.91	20.44	13.54
C29H61I2N3	705.64	705.2958	49.36	8.71	35.97	5.95
C29H61I2N5	733.65	733.3020	47.48	8.38	34.60	9.55
C29H62I1N3	579.74	579.3990	60.08	10.78	21.89	7.25
C29H62I1N5	607.76	607.4052	57.31	10.28	20.88	11.52
C29H62I2N4	720.65	720.3067	48.33	8.67	35.22	7.77
C29H62I2N6	748.67	748.3129	46.53	8.35	33.90	11.23
C29H63I1N4	594.76	594.4099	58.57	10.68	21.34	9.42
C29H63I1N6	622.77	622.4161	55.93	10.20	20.38	13.49
C29H63I2N5	735.67	735.3176	47.35	8.63	34.50	9.52
C29H64I1N5	609.77	609.4208	57.12	10.58	20.81	11.49
C29H64I2N6	750.68	750.3285	46.40	8.59	33.81	11.20
C29H65I1N6	624.79	624.4317	55.75	10.49	20.31	13.45
C30H15I1N2	530.37	530.0282	67.94	2.85	23.93	5.28
C30H15I1N4	558.39	558.0343	64.53	2.71	22.73	10.03
C30H15I1N6	586.40	586.0405	61.45	2.58	21.64	14.33
C30H15I2N1	643.27	642.9298	56.02	2.35	39.46	2.18
C30H15I2N3	671.28	670.9359	53.68	2.25	37.81	6.26
C30H15I2N5	699.30	698.9421	51.53	2.16	36.29	10.01
C30H16I1N1	517.37	517.0329	69.65	3.12	24.53	2.71
C30H16I1N3	545.39	545.0391	66.07	2.96	23.27	7.70
C30H16I1N5	573.40	573.0452	62.84	2.81	22.13	12.21
C30H16I2N2	658.28	657.9407	54.74	2.45	38.56	4.26
C30H16I2N4	686.30	685.9468	52.50	2.35	36.98	8.16
C30H16I2N6	714.31	713.9530	50.45	2.26	35.53	11.77
C30H17I1N2	532.39	532.0438	67.68	3.22	23.84	5.26
C30H17I1N4	560.40	560.0500	64.30	3.06	22.65	10.00
C30H17I1N6	588.41	588.0561	61.24	2.91	21.57	14.28
C30H17I2N1	645.29	644.9454	55.84	2.66	39.33	2.17
C30H17I2N3	673.30	672.9516	53.52	2.55	37.70	6.24
C30H17I2N5	701.31	700.9577	51.38	2.44	36.19	9.99
C30H18I1N1	519.39	519.0486	69.38	3.49	24.43	2.70
C30H18I1N3	547.40	547.0547	65.83	3.31	23.18	7.68
C30H18I1N5	575.42	575.0609	62.62	3.15	22.05	12.17
C30H18I2N2	660.30	659.9563	54.57	2.75	38.44	4.24
C30H18I2N4	688.31	687.9625	52.35	2.64	36.87	8.14
C30H18I2N6	716.33	715.9686	50.30	2.53	35.43	11.73
C30H19I1N2	534.40	534.0595	67.43	3.58	23.75	5.24
C30H19I1N4	562.42	562.0656	64.07	3.41	22.56	9.96
C30H19I1N6	590.43	590.0718	61.03	3.24	21.49	14.23
C30H19I2N1	647.30	646.9611	55.67	2.96	39.21	2.16
C30H19I2N3	675.31	674.9672	53.36	2.84	37.58	6.22
C30H19I2N5	703.33	702.9734	51.23	2.72	36.09	9.96
C30H20I1N1	521.41	521.0642	69.11	3.87	24.34	2.69
C30H20I1N3	549.42	549.0704	65.58	3.67	23.10	7.65
C30H20I1N5	577.43	577.0765	62.40	3.49	21.98	12.13
C30H20I2N2	662.32	661.9720	54.41	3.04	38.32	4.23
C30H20I2N4	690.33	689.9781	52.20	2.92	36.77	8.12
C30H20I2N6	718.34	717.9843	50.16	2.81	35.33	11.70
C30H21I1N2	536.42	536.0751	67.17	3.95	23.66	5.22
C30H21I1N4	564.43	564.0813	63.84	3.75	22.48	9.93
C30H21I1N6	592.45	592.0874	60.82	3.57	21.42	14.19
C30H21I2N1	649.32	648.9767	55.49	3.26	39.09	2.16
C30H21I2N3	677.33	676.9829	53.20	3.13	37.47	6.20
C30H21I2N5	705.34	704.9890	51.09	3.00	35.98	9.93
C30H22I1N1	523.42	523.0799	68.84	4.24	24.25	2.68
C30H22I1N3	551.43	551.0860	65.34	4.02	23.01	7.62
C30H22I1N5	579.45	579.0922	62.19	3.83	21.90	12.09
C30H22I2N2	664.33	663.9876	54.24	3.34	38.21	4.22
C30H22I2N4	692.35	691.9938	52.05	3.20	36.66	8.09
C30H22I2N6	720.36	719.9999	50.02	3.08	35.23	11.67
C30H23I1N2	538.44	538.0908	66.92	4.31	23.57	5.20
C30H23I1N4	566.45	566.0969	63.61	4.09	22.40	9.89
C30H23I1N6	594.46	594.1031	60.62	3.90	21.35	14.14
C30H23I2N1	651.33	650.9924	55.32	3.56	38.97	2.15
C30H23I2N3	679.35	678.9985	53.04	3.41	37.36	6.19
C30H23I2N5	707.36	707.0047	50.94	3.28	35.88	9.90
C30H24I1N1	525.44	525.0955	68.58	4.60	24.15	2.67
C30H24I1N3	553.45	553.1017	65.11	4.37	22.93	7.59
C30H24I1N5	581.46	581.1078	61.97	4.16	21.82	12.04
C30H24I2N2	666.35	666.0033	54.08	3.63	38.09	4.20
C30H24I2N4	694.36	694.0094	51.89	3.48	36.55	8.07
C30H24I2N6	722.37	722.0156	49.88	3.35	35.14	11.63
C30H25I1N2	540.45	540.1064	66.67	4.66	23.48	5.18
C30H25I1N4	568.46	568.1126	63.39	4.43	22.32	9.86
C30H25I1N6	596.48	596.1187	60.41	4.22	21.28	14.09
C30H25I2N1	653.35	653.0080	55.15	3.86	38.85	2.14
C30H25I2N3	681.36	681.0142	52.88	3.70	37.25	6.17
C30H25I2N5	709.38	709.0203	50.80	3.55	35.78	9.87
C30H26I1N1	527.45	527.1112	68.32	4.97	24.06	2.66
C30H26I1N3	555.47	555.1173	64.87	4.72	22.85	7.56
C30H26I1N5	583.48	583.1235	61.76	4.49	21.75	12.00
C30H26I2N2	668.36	668.0189	53.91	3.92	37.97	4.19
C30H26I2N4	696.38	696.0251	51.74	3.76	36.45	8.05
C30H26I2N6	724.39	724.0312	49.74	3.62	35.04	11.60
C30H27I1N2	542.47	542.1221	66.43	5.02	23.39	5.16
C30H27I1N4	570.48	570.1282	63.16	4.77	22.25	9.82
C30H27I1N6	598.49	598.1344	60.21	4.55	21.20	14.04
C30H27I2N1	655.37	655.0237	54.98	4.15	38.73	2.14
C30H27I2N3	683.38	683.0298	52.73	3.98	37.14	6.15
C30H27I2N5	711.39	711.0360	50.65	3.83	35.68	9.84
C30H28I1N1	529.47	529.1268	68.06	5.33	23.97	2.65
C30H28I1N3	557.48	557.1330	64.64	5.06	22.76	7.54
C30H28I1N5	585.50	585.1391	61.54	4.82	21.67	11.96
C30H28I2N2	670.38	670.0346	53.75	4.21	37.86	4.18
C30H28I2N4	698.39	698.0407	51.59	4.04	36.34	8.02
C30H28I2N6	726.41	726.0469	49.61	3.89	34.94	11.57
C30H29I1N2	544.48	544.1377	66.18	5.37	23.31	5.14
C30H29I1N4	572.50	572.1439	62.94	5.11	22.17	9.79
C30H29I1N6	600.51	600.1500	60.00	4.87	21.13	13.99
C30H29I2N1	657.38	657.0393	54.81	4.45	38.61	2.13
C30H29I2N3	685.39	685.0455	52.57	4.26	37.03	6.13
C30H29I2N5	713.41	713.0516	50.51	4.10	35.58	9.82
C30H30I1N1	531.48	531.1425	67.80	5.69	23.88	2.64
C30H30I1N3	559.50	559.1486	64.40	5.40	22.68	7.51
C30H30I1N5	587.51	587.1548	61.33	5.15	21.60	11.92
C30H30I2N2	672.40	672.0502	53.59	4.50	37.75	4.17
C30H30I2N4	700.41	700.0564	51.45	4.32	36.24	8.00
C30H30I2N6	728.42	728.0625	49.47	4.15	34.84	11.54
C30H31I1N2	546.50	546.1534	65.94	5.72	23.22	5.13
C30H31I1N4	574.51	574.1595	62.72	5.44	22.09	9.75
C30H31I1N6	602.53	602.1657	59.80	5.19	21.06	13.95
C30H31I2N1	659.40	659.0550	54.65	4.74	38.49	2.12
C30H31I2N3	687.41	687.0611	52.42	4.55	36.92	6.11
C30H31I2N5	715.42	715.0673	50.37	4.37	35.48	9.79
C30H32I1N1	533.50	533.1581	67.54	6.05	23.79	2.63

051

C30H32I1N3	561.51	561.1643	64.17	5.74	22.60	7.48
C30H32I1N5	589.53	589.1704	61.12	5.47	21.53	11.88
C30H32I2N2	674.41	674.0659	53.43	4.78	37.63	4.15
C30H32I2N4	702.43	702.0720	51.30	4.59	36.13	7.98
C30H32I2N6	730.44	730.0781	49.33	4.42	34.75	11.51
C30H33I1N2	548.52	548.1690	65.69	6.06	23.14	5.11
C30H33I1N4	576.53	576.1752	62.50	5.77	22.01	9.72
C30H33I1N6	604.54	604.1813	59.60	5.50	20.99	13.90
C30H33I2N1	661.41	661.0706	54.48	5.03	38.37	2.12
C30H33I2N3	689.43	689.0768	52.27	4.82	36.81	6.09
C30H33I2N5	717.44	717.0829	50.23	4.64	35.38	9.76
C30H34I1N1	535.52	535.1738	67.29	6.40	23.70	2.62
C30H34I1N3	563.53	563.1799	63.94	6.08	22.52	7.46
C30H34I1N5	591.54	591.1861	60.91	5.79	21.45	11.84
C30H34I2N2	676.43	676.0815	53.27	5.07	37.52	4.14
C30H34I2N4	704.44	704.0877	51.15	4.86	36.03	7.95
C30H34I2N6	732.45	732.0938	49.20	4.68	34.65	11.47
C30H35I1N2	550.53	550.1847	65.45	6.41	23.05	5.09
C30H35I1N4	578.54	578.1908	62.28	6.10	21.94	9.68
C30H35I1N6	606.56	606.1970	59.41	5.82	20.92	13.86
C30H35I2N1	663.43	663.0863	54.31	5.32	38.26	2.11
C30H35I2N3	691.44	691.0924	52.11	5.10	36.71	6.08
C30H35I2N5	719.46	719.0985	50.08	4.90	35.28	9.73
C30H36I1N1	537.53	537.1894	67.03	6.75	23.61	2.61
C30H36I1N3	565.55	565.1956	63.71	6.42	22.44	7.43
C30H36I1N5	593.56	593.2017	60.71	6.11	21.38	11.80
C30H36I2N2	678.44	678.0972	53.11	5.35	37.41	4.13
C30H36I2N4	706.46	706.1033	51.01	5.14	35.93	7.93
C30H36I2N6	734.47	734.1094	49.06	4.94	34.56	11.44
C30H37I1N2	552.55	552.2003	65.21	6.75	22.97	5.07
C30H37I1N4	580.56	580.2065	62.07	6.42	21.86	9.65
C30H37I1N6	608.57	608.2126	59.21	6.13	20.85	13.81
C30H37I2N1	665.44	665.1019	54.15	5.60	38.14	2.10
C30H37I2N3	693.46	693.1081	51.96	5.38	36.60	6.06
C30H37I2N5	721.47	721.1142	49.94	5.17	35.18	9.71
C30H38I1N1	539.55	539.2051	66.78	7.10	23.52	2.60
C30H38I1N3	567.56	567.2112	63.49	6.75	22.36	7.40
C30H38I1N5	595.58	595.2174	60.50	6.43	21.31	11.76
C30H38I2N2	680.46	680.1128	52.95	5.63	37.30	4.12
C30H38I2N4	708.47	708.1189	50.86	5.41	35.82	7.91
C30H38I2N6	736.49	736.1251	48.93	5.20	34.46	11.41
C30H39I1N2	554.56	554.2160	64.98	7.09	22.88	5.05
C30H39I1N4	582.58	582.2221	61.85	6.75	21.78	9.62
C30H39I1N6	610.59	610.2283	59.01	6.44	20.78	13.76
C30H39I2N1	667.46	667.1176	53.99	5.89	38.03	2.10
C30H39I2N3	695.47	695.1237	51.81	5.65	36.49	6.04
C30H39I2N5	723.49	723.1298	49.81	5.43	35.08	9.68
C30H40I1N1	541.56	541.2207	66.54	7.44	23.43	2.59
C30H40I1N3	569.58	569.2269	63.26	7.08	22.28	7.38
C30H40I1N5	597.59	597.2330	60.30	6.75	21.24	11.72
C30H40I2N2	682.48	682.1285	52.80	5.91	37.19	4.10
C30H40I2N4	710.49	710.1346	50.72	5.67	35.72	7.89
C30H40I2N6	738.50	738.1407	48.79	5.46	34.37	11.38
C30H41I1N2	556.58	556.2316	64.74	7.43	22.80	5.03
C30H41I1N4	584.59	584.2378	61.64	7.07	21.71	9.58
C30H41I1N6	612.61	612.2439	58.82	6.75	20.72	13.72
C30H41I2N1	669.48	669.1332	53.82	6.17	37.91	2.09
C30H41I2N3	697.49	697.1394	51.66	5.93	36.39	6.02
C30H41I2N5	725.50	725.1455	49.67	5.70	34.98	9.65
C30H42I1N1	543.58	543.2364	66.29	7.79	23.35	2.58
C30H42I1N3	571.59	571.2425	63.04	7.41	22.20	7.35
C30H42I1N5	599.61	599.2487	60.10	7.06	21.16	11.68
C30H42I2N2	684.49	684.1441	52.64	6.18	37.08	4.09
C30H42I2N4	712.50	712.1502	50.57	5.94	35.62	7.86
C30H42I2N6	740.52	740.1564	48.66	5.72	34.27	11.35
C30H43I1N2	558.60	558.2473	64.51	7.76	22.72	5.01
C30H43I1N4	586.61	586.2534	61.43	7.39	21.63	9.55
C30H43I1N6	614.62	614.2596	58.63	7.05	20.65	13.67
C30H43I2N1	671.49	671.1489	53.66	6.45	37.80	2.09
C30H43I2N3	699.51	699.1550	51.51	6.20	36.28	6.01
C30H43I2N5	727.52	727.1611	49.53	5.96	34.89	9.63
C30H44I1N1	545.60	545.2520	66.04	8.13	23.26	2.57
C30H44I1N3	573.61	573.2582	62.82	7.73	22.12	7.33
C30H44I1N5	601.62	601.2643	59.89	7.37	21.09	11.64
C30H44I2N2	686.51	686.1597	52.49	6.46	36.97	4.08
C30H44I2N4	714.52	714.1659	50.43	6.21	35.52	7.84
C30H44I2N6	742.53	742.1720	48.53	5.97	34.18	11.32
C30H45I1N2	560.61	560.2629	64.28	8.09	22.64	5.00
C30H45I1N4	588.62	588.2691	61.22	7.71	21.56	9.52
C30H45I1N6	616.64	616.2752	58.44	7.36	20.58	13.63
C30H45I2N1	673.51	673.1645	53.50	6.73	37.68	2.08
C30H45I2N3	701.52	701.1706	51.36	6.47	36.18	5.99
C30H45I2N5	729.54	729.1768	49.39	6.22	34.79	9.60
C30H46I1N1	547.61	547.2677	65.80	8.47	23.17	2.56
C30H46I1N3	575.63	575.2738	62.60	8.05	22.05	7.30
C30H46I1N5	603.64	603.2800	59.69	7.68	21.02	11.60
C30H46I2N2	688.52	688.1754	52.33	6.73	36.86	4.07
C30H46I2N4	716.54	716.1815	50.29	6.47	35.42	7.82
C30H46I2N6	744.55	744.1877	48.40	6.23	34.09	11.29
C30H47I1N2	562.63	562.2786	64.05	8.42	22.56	4.98
C30H47I1N4	590.64	590.2847	61.01	8.02	21.49	9.49
C30H47I1N6	618.65	618.2909	58.24	7.66	20.51	13.58
C30H47I2N1	675.52	675.1802	53.34	7.01	37.57	2.07
C30H47I2N3	703.54	703.1863	51.22	6.73	36.08	5.97
C30H47I2N5	731.55	731.1924	49.26	6.48	34.69	9.57
C30H48I1N1	549.63	549.2833	65.56	8.80	23.09	2.55
C30H48I1N3	577.64	577.2895	62.38	8.38	21.97	7.27
C30H48I1N5	605.65	605.2956	59.50	7.99	20.95	11.56
C30H48I2N2	690.54	690.1910	52.18	7.01	36.76	4.06
C30H48I2N4	718.55	718.1972	50.15	6.73	35.32	7.80
C30H48I2N6	746.57	746.2033	48.27	6.48	34.00	11.26
C30H49I1N2	564.64	564.2942	63.82	8.75	22.48	4.96
C30H49I1N4	592.66	592.3004	60.80	8.33	21.41	9.45
C30H49I1N6	620.67	620.3065	58.06	7.96	20.45	13.54
C30H49I2N1	677.54	677.1958	53.18	7.29	37.46	2.07
C30H49I2N3	705.55	705.2019	51.07	7.00	35.97	5.96
C30H49I2N5	733.57	733.2081	49.12	6.73	34.60	9.55
C30H50I1N1	551.64	551.2990	65.32	9.14	23.00	2.54
C30H50I1N3	579.66	579.3051	62.16	8.69	21.89	7.25
C30H50I1N5	607.67	607.3113	59.30	8.29	20.88	11.52
C30H50I2N2	692.56	692.2067	52.03	7.28	36.65	4.04
C30H50I2N4	720.57	720.2128	50.01	6.99	35.22	7.78
C30H50I2N6	748.58	748.2190	48.14	6.73	33.91	11.23
C30H51I1N2	566.66	566.3099	63.59	9.07	22.40	4.94
C30H51I1N4	594.67	594.3160	60.59	8.64	21.34	9.42
C30H51I1N6	622.69	622.3222	57.87	8.26	20.38	13.50
C30H51I2N1	679.56	679.2114	53.02	7.56	37.35	2.06
C30H51I2N3	707.57	707.2176	50.93	7.27	35.87	5.94
C30H51I2N5	735.58	735.2237	48.99	6.99	34.50	9.52
C30H52I1N1	553.66	553.3146	65.08	9.47	22.92	2.53

052

C30H52I1N3	581.67	581.3208	61.95	9.01	21.82	7.22
C30H52I1N5	609.69	609.3269	59.10	8.60	20.81	11.49
C30H52I2N2	694.57	694.2223	51.88	7.55	36.54	4.03
C30H52I2N4	722.58	722.2285	49.87	7.25	35.13	7.75
C30H52I2N6	750.60	750.2346	48.01	6.98	33.81	11.20
C30H53I1N2	568.67	568.3255	63.36	9.39	22.32	4.93
C30H53I1N4	596.69	596.3317	60.39	8.95	21.27	9.39
C30H53I1N6	624.70	624.3378	57.68	8.55	20.31	13.45
C30H53I2N1	681.57	681.2271	52.87	7.84	37.24	2.06
C30H53I2N3	709.59	709.2332	50.78	7.53	35.77	5.92
C30H53I2N5	737.60	737.2394	48.85	7.24	34.41	9.49
C30H54I1N1	555.68	555.3303	64.85	9.80	22.84	2.52
C30H54I1N3	583.69	583.3364	61.73	9.33	21.74	7.20
C30H54I1N5	611.70	611.3426	58.91	8.90	20.75	11.45
C30H54I2N2	696.59	696.2380	51.73	7.81	36.44	4.02
C30H54I2N4	724.60	724.2441	49.73	7.51	35.03	7.73
C30H54I2N6	752.61	752.2503	47.88	7.23	33.72	11.17
C30H55I1N2	570.69	570.3412	63.14	9.71	22.24	4.91
C30H55I1N4	598.70	598.3473	60.19	9.26	21.20	9.36
C30H55I1N6	626.72	626.3535	57.50	8.85	20.25	13.41
C30H55I2N1	683.59	683.2427	52.71	8.11	37.13	2.05
C30H55I2N3	711.60	711.2489	50.64	7.79	35.67	5.91
C30H55I2N5	739.62	739.2550	48.72	7.50	34.32	9.47
C30H56I1N1	557.69	557.3459	64.61	10.12	22.76	2.51
C30H56I1N3	585.71	585.3521	61.52	9.64	21.67	7.17
C30H56I1N5	613.72	613.3582	58.71	9.20	20.68	11.41
C30H56I2N2	698.60	698.2536	51.58	8.08	36.33	4.01
C30H56I2N4	726.62	726.2598	49.59	7.77	34.93	7.71
C30H56I2N6	754.63	754.2659	47.75	7.48	33.63	11.14
C30H57I1N2	572.71	572.3568	62.92	10.03	22.16	4.89
C30H57I1N4	600.72	600.3630	59.98	9.56	21.13	9.33
C30H57I1N6	628.73	628.3691	57.31	9.14	20.18	13.37
C30H57I2N1	685.60	685.2584	52.56	8.38	37.02	2.04
C30H57I2N3	713.62	713.2645	50.49	8.05	35.57	5.89
C30H57I2N5	741.63	741.2707	48.59	7.75	34.22	9.44
C30H58I1N1	559.71	559.3616	64.38	10.45	22.67	2.50
C30H58I1N3	587.72	587.3677	61.31	9.95	21.59	7.15
C30H58I1N5	615.73	615.3739	58.52	9.49	20.61	11.37
C30H58I2N2	700.62	700.2693	51.43	8.34	36.23	4.00
C30H58I2N4	728.63	728.2754	49.45	8.02	34.83	7.69
C30H58I2N6	756.65	756.2816	47.62	7.73	33.54	11.11
C30H59I1N2	574.72	574.3725	62.70	10.35	22.08	4.87
C30H59I1N4	602.74	602.3786	59.78	9.87	21.05	9.30
C30H59I1N6	630.75	630.3848	57.13	9.43	20.12	13.32
C30H59I2N1	687.62	687.2740	52.40	8.65	36.91	2.04
C30H59I2N3	715.63	715.2802	50.35	8.31	35.47	5.87
C30H59I2N5	743.65	743.2863	48.46	8.00	34.13	9.42
C30H60I1N1	561.72	561.3772	64.15	10.77	22.59	2.49
C30H60I1N3	589.74	589.3834	61.10	10.26	21.52	7.13
C30H60I1N5	617.75	617.3895	58.33	9.79	20.54	11.34
C30H60I2N2	702.63	702.2849	51.28	8.61	36.12	3.99
C30H60I2N4	730.65	730.2911	49.32	8.28	34.74	7.67
C30H60I2N6	758.66	758.2972	47.50	7.97	33.45	11.08
C30H61I1N2	576.74	576.3881	62.48	10.66	22.00	4.86
C30H61I1N4	604.75	604.3943	59.58	10.17	20.98	9.26
C30H61I1N6	632.77	632.4004	56.95	9.72	20.06	13.28
C30H61I2N1	689.64	689.2897	52.25	8.92	36.80	2.03
C30H61I2N3	717.65	717.2958	50.21	8.57	35.37	5.86
C30H61I2N5	745.66	745.3020	48.32	8.25	34.04	9.39
C30H62I1N1	563.74	563.3929	63.92	11.09	22.51	2.48
C30H62I1N3	591.75	591.3990	60.89	10.56	21.45	7.10
C30H62I1N5	619.77	619.4052	58.14	10.08	20.48	11.30
C30H62I2N2	704.65	704.3006	51.14	8.87	36.02	3.98
C30H62I2N4	732.66	732.3067	49.18	8.53	34.64	7.65
C30H62I2N6	760.68	760.3129	47.37	8.22	33.37	11.05
C30H63I1N2	578.75	578.4038	62.26	10.97	21.93	4.84
C30H63I1N4	606.77	606.4099	59.39	10.47	20.91	9.23
C30H63I1N6	634.78	634.4161	56.77	10.00	19.99	13.24
C30H63I2N3	719.67	719.3115	50.07	8.82	35.27	5.84
C30H63I2N5	747.68	747.3176	48.19	8.49	33.95	9.37
C30H64I1N3	593.77	593.4147	60.69	10.86	21.37	7.08
C30H64I1N5	621.78	621.4208	57.95	10.38	20.41	11.26
C30H64I2N4	734.68	734.3224	49.05	8.78	34.55	7.63
C30H64I2N6	762.69	762.3285	47.24	8.46	33.28	11.02
C30H65I1N4	608.78	608.4256	59.19	10.76	20.85	9.20
C30H65I1N6	636.80	636.4317	56.59	10.29	19.93	13.20
C30H65I2N5	749.69	749.3333	48.06	8.74	33.85	9.34
C30H66I1N5	623.80	623.4365	57.76	10.66	20.34	11.23
C30H66I2N6	764.71	764.3442	47.12	8.70	33.19	10.99
C30H67I1N6	638.81	638.4474	56.41	10.57	19.87	13.16

8. CARBON, HYDROGEN, IODINE AND OXYGEN

8. KOHLENSTOFF, WASSERSTOFF, JOD UND SAUERSTOFF

8. CARBONE, HYDROGENE, IODE ET OXYGENE

MOLECULAR WEIGHTS AND PERCENTAGE COMPOSITIONS OF COMPOUNDS CONTAINING CARBON, HYDROGEN, IODINE AND OXYGEN

FORMULA	MOL. WEIGHT	ISOMOL. WEIGHT	PERC. C	PERC. H	PERC. I	PERC. O
C 1H 0I201	281.82	281.8042	4.26	0	90.06	5.68
C 1H 0I202	297.82	297.7991	4.03	0	85.22	10.74
C 1H 0I203	313.82	313.7941	3.83	0	80.88	15.29
C 1H 0I204	329.82	329.7890	3.64	0	76.95	19.40
C 1H 0I205	345.82	345.7839	3.47	0	73.39	23.13
C 1H 0I206	361.82	361.7788	3.32	0	70.15	26.53
C 1H 0I207	377.82	377.7737	3.18	0	67.18	29.64
C 1H 0I208	393.82	393.7686	3.05	0	64.45	32.50
C 1H 0I209	409.81	409.7635	2.93	0	61.93	35.14
C 1H 1I101	155.92	155.9074	7.70	.65	81.39	10.26
C 1H 1I102	171.92	171.9023	6.99	.59	73.81	18.61
C 1H 1I103	187.92	187.8972	6.39	.54	67.53	25.54
C 1H 1I104	203.92	203.8921	5.89	.49	62.23	31.38
C 1H 1I105	219.92	219.8871	5.46	.46	57.70	36.38
C 1H 1I106	235.92	235.8820	5.09	.43	53.79	40.69
C 1H 1I107	251.92	251.8769	4.77	.40	50.38	44.46
C 1H 1I108	267.92	267.8718	4.48	.38	47.37	47.77
C 1H 1I109	283.92	283.8667	4.23	.36	44.70	50.72
C 1H 2I201	283.84	283.8199	4.23	.71	89.42	5.64
C 1H 2I202	299.83	299.8148	4.01	.67	84.65	10.67
C 1H 2I203	315.83	315.8097	3.80	.64	80.36	15.20
C 1H 2I204	331.83	331.8046	3.62	.61	76.49	19.29
C 1H 2I205	347.83	347.7995	3.45	.58	72.97	23.00
C 1H 2I206	363.83	363.7945	3.30	.55	69.76	26.38
C 1H 2I207	379.83	379.7894	3.16	.53	66.82	29.49
C 1H 2I208	395.83	395.7843	3.03	.51	64.12	32.34
C 1H 2I209	411.83	411.7792	2.92	.49	61.63	34.96
C 1H 3I101	157.94	157.9230	7.60	1.91	80.35	10.13
C 1H 3I102	173.94	173.9180	6.91	1.74	72.96	18.40
C 1H 3I103	189.94	189.9129	6.32	1.59	66.81	25.27
C 1H 3I104	205.94	205.9078	5.83	1.47	61.62	31.08
C 1H 3I105	221.94	221.9027	5.41	1.36	57.18	36.05
C 1H 3I106	237.94	237.8976	5.05	1.27	53.34	40.35
C 1H 3I107	253.94	253.8925	4.73	1.19	49.98	44.10
C 1H 3I108	269.93	269.8874	4.45	1.12	47.01	47.42
C 1H 3I109	285.93	285.8824	4.20	1.06	44.38	50.36
C 2H 1I101	167.93	167.9074	14.30	.60	75.57	9.53
C 2H 1I102	183.93	183.9023	13.06	.55	68.99	17.40
C 2H 1I103	199.93	199.8972	12.02	.50	63.47	24.01
C 2H 1I104	215.93	215.8921	11.12	.47	58.77	29.64
C 2H 1I105	231.93	231.8871	10.36	.43	54.72	34.49
C 2H 1I106	247.93	247.8820	9.69	.41	51.19	38.72
C 2H 1I107	263.93	263.8769	9.10	.38	48.08	42.43
C 2H 1I108	279.93	279.8718	8.58	.36	45.33	45.72
C 2H 1I109	295.93	295.8667	8.12	.34	42.88	48.66
C 2H 2I201	295.85	295.8199	8.12	.68	85.79	5.41
C 2H 2I202	311.85	311.8148	7.70	.65	81.39	10.26
C 2H 2I203	327.85	327.8097	7.33	.61	77.42	14.64
C 2H 2I204	343.84	343.8046	6.99	.59	73.81	18.61
C 2H 2I205	359.84	359.7995	6.68	.56	70.53	22.23
C 2H 2I206	375.84	375.7945	6.39	.54	67.53	25.54
C 2H 2I207	391.84	391.7894	6.13	.51	64.77	28.58
C 2H 2I208	407.84	407.7843	5.89	.49	62.23	31.38
C 2H 2I209	423.84	423.7792	5.67	.48	59.88	33.97
C 2H 3I101	169.95	169.9230	14.13	1.78	74.67	9.41
C 2H 3I102	185.95	185.9180	12.92	1.63	68.25	17.21
C 2H 3I103	201.95	201.9129	11.90	1.50	62.84	23.77
C 2H 3I104	217.95	217.9078	11.02	1.39	58.23	29.36
C 2H 3I105	233.95	233.9027	10.27	1.29	54.24	34.19
C 2H 3I106	249.95	249.8976	9.61	1.21	50.77	38.41
C 2H 3I107	265.95	265.8925	9.03	1.14	47.72	42.11
C 2H 3I108	281.95	281.8874	8.52	1.07	45.01	45.40
C 2H 3I109	297.95	297.8824	8.06	1.01	42.59	48.33
C 2H 4I201	297.86	297.8355	8.06	1.35	85.21	5.37
C 2H 4I202	313.86	313.8304	7.65	1.28	80.87	10.20
C 2H 4I203	329.86	329.8254	7.28	1.22	76.94	14.55
C 2H 4I204	345.86	345.8203	6.95	1.17	73.38	18.50
C 2H 4I205	361.86	361.8152	6.64	1.11	70.14	22.11
C 2H 4I206	377.86	377.8101	6.36	1.07	67.17	25.41
C 2H 4I207	393.86	393.8050	6.10	1.02	64.44	28.44
C 2H 4I208	409.86	409.7999	5.86	.98	61.93	31.23
C 2H 4I209	425.86	425.7948	5.64	.95	59.60	33.81
C 2H 5I101	171.97	171.9387	13.97	2.93	73.80	9.30
C 2H 5I102	187.97	187.9336	12.78	2.68	67.51	17.02
C 2H 5I103	203.96	203.9285	11.78	2.47	62.22	23.53
C 2H 5I104	219.96	219.9234	10.92	2.29	57.69	29.09
C 2H 5I105	235.96	235.9184	10.18	2.14	53.78	33.90
C 2H 5I106	251.96	251.9133	9.53	2.00	50.37	38.10
C 2H 5I107	267.96	267.9082	8.96	1.88	47.36	41.80
C 2H 5I108	283.96	283.9031	8.46	1.77	44.69	45.07
C 2H 5I109	299.96	299.8980	8.01	1.68	42.31	48.00
C 3H 1I101	179.95	179.9074	20.02	.56	70.52	8.89
C 3H 1I102	195.94	195.9023	18.39	.51	64.77	16.33
C 3H 1I103	211.94	211.8972	17.00	.48	59.88	22.65
C 3H 1I104	227.94	227.8921	15.81	.44	55.67	28.08
C 3H 1I105	243.94	243.8871	14.77	.41	52.02	32.79
C 3H 1I106	259.94	259.8820	13.86	.39	48.82	36.93
C 3H 1I107	275.94	275.8769	13.06	.37	45.99	40.59
C 3H 1I108	291.94	291.8718	12.34	.35	43.47	43.84
C 3H 1I109	307.94	307.8667	11.70	.33	41.21	46.76
C 3H 2I201	307.86	307.8199	11.70	.65	82.44	5.20
C 3H 2I202	323.86	323.8148	11.13	.62	78.37	9.88
C 3H 2I203	339.86	339.8097	10.60	.59	74.68	14.12
C 3H 2I204	355.86	355.8046	10.13	.57	71.32	17.98
C 3H 2I205	371.86	371.7995	9.69	.54	68.25	21.51
C 3H 2I206	387.85	387.7945	9.29	.52	65.44	24.75
C 3H 2I207	403.85	403.7894	8.92	.50	62.85	27.73
C 3H 2I208	419.85	419.7843	8.58	.48	60.45	30.49
C 3H 2I209	435.85	435.7792	8.27	.46	58.23	33.04
C 3H 3I101	181.96	181.9230	19.80	1.66	69.74	8.79

184

FORMULA	MOL. WEIGHT	ISOMOL. WEIGHT	PERC. C	PERC. H	PERC. I	PERC. O
C 3H 3I102	197.96	197.9180	18.20	1.53	64.11	16.16
C 3H 3I103	213.96	213.9129	16.84	1.41	59.31	22.43
C 3H 3I104	229.96	229.9078	15.67	1.31	55.19	27.83
C 3H 3I105	245.96	245.9027	14.65	1.23	51.60	32.52
C 3H 3I106	261.96	261.8976	13.76	1.15	48.44	36.65
C 3H 3I107	277.96	277.8925	12.96	1.09	45.66	40.29
C 3H 3I108	293.96	293.8874	12.26	1.03	43.17	43.54
C 3H 3I109	309.96	309.8824	11.63	.98	40.94	46.46
C 3H 4I201	309.87	309.8355	11.63	1.30	81.91	5.16
C 3H 4I202	325.87	325.8304	11.06	1.24	77.89	9.82
C 3H 4I203	341.87	341.8254	10.54	1.18	74.24	14.04
C 3H 4I204	357.87	357.8203	10.07	1.13	70.92	17.88
C 3H 4I205	373.87	373.8152	9.64	1.08	67.89	21.40
C 3H 4I206	389.87	389.8101	9.24	1.03	65.10	24.62
C 3H 4I207	405.87	405.8050	8.88	.99	62.53	27.59
C 3H 4I208	421.87	421.7999	8.54	.96	60.16	30.34
C 3H 4I209	437.87	437.7948	8.23	.92	57.96	32.89
C 3H 5I101	183.98	183.9387	19.59	2.74	68.98	8.70
C 3H 5I102	199.98	199.9336	18.02	2.52	63.46	16.00
C 3H 5I103	215.98	215.9285	16.68	2.33	58.76	22.22
C 3H 5I104	231.98	231.9234	15.53	2.17	54.71	27.59
C 3H 5I105	247.97	247.9184	14.53	2.03	51.18	32.26
C 3H 5I106	263.97	263.9133	13.65	1.91	48.07	36.37
C 3H 5I107	279.97	279.9082	12.87	1.80	45.33	40.00
C 3H 5I108	295.97	295.9031	12.17	1.70	42.88	43.25
C 3H 5I109	311.97	311.8980	11.55	1.62	40.68	46.16
C 3H 6I201	311.89	311.8512	11.55	1.94	81.38	5.13
C 3H 6I202	327.89	327.8461	10.99	1.84	77.41	9.76
C 3H 6I203	343.89	343.8410	10.48	1.76	73.81	13.96
C 3H 6I204	359.89	359.8359	10.01	1.68	70.52	17.78
C 3H 6I205	375.89	375.8308	9.59	1.61	67.52	21.28
C 3H 6I206	391.89	391.8258	9.19	1.54	64.77	24.50
C 3H 6I207	407.89	407.8207	8.83	1.48	62.23	27.46
C 3H 6I208	423.89	423.8156	8.50	1.43	59.88	30.20
C 3H 6I209	439.88	439.8105	8.19	1.37	57.70	32.73
C 3H 7I101	185.99	185.9543	19.37	3.79	68.23	8.60
C 3H 7I102	201.99	201.9493	17.84	3.49	62.83	15.84
C 3H 7I103	217.99	217.9442	16.53	3.24	58.22	22.02
C 3H 7I104	233.99	233.9391	15.40	3.02	54.23	27.35
C 3H 7I105	249.99	249.9340	14.41	2.82	50.76	32.00
C 3H 7I106	265.99	265.9289	13.55	2.65	47.71	36.09
C 3H 7I107	281.99	281.9238	12.78	2.50	45.00	39.72
C 3H 7I108	297.99	297.9187	12.09	2.37	42.59	42.95
C 3H 7I109	313.99	313.9137	11.48	2.25	40.42	45.86
C 4H 2I201	319.87	319.8199	15.02	.63	79.35	5.00
C 4H 2I202	335.87	335.8148	14.30	.60	75.57	9.53
C 4H 2I203	351.87	351.8097	13.65	.57	72.13	13.64
C 4H 2I204	367.87	367.8046	13.06	.55	68.99	17.40
C 4H 2I205	383.87	383.7995	12.52	.53	66.12	20.84
C 4H 2I206	399.87	399.7945	12.02	.50	63.47	24.01
C 4H 2I207	415.87	415.7894	11.55	.48	61.03	26.93
C 4H 2I208	431.86	431.7843	11.12	.47	58.77	29.64
C 4H 2I209	447.86	447.7792	10.73	.45	56.67	32.15
C 4H 3I101	193.97	193.9230	24.77	1.56	65.42	8.25
C 4H 3I102	209.97	209.9180	22.88	1.44	60.44	15.24
C 4H 3I103	225.97	225.9129	21.26	1.34	56.16	21.24
C 4H 3I104	241.97	241.9078	19.86	1.25	52.45	26.45
C 4H 3I105	257.97	257.9027	18.62	1.17	49.19	31.01
C 4H 3I106	273.97	273.8976	17.54	1.10	46.32	35.04
C 4H 3I107	289.97	289.8925	16.57	1.04	43.76	38.62
C 4H 3I108	305.97	305.8874	15.70	.99	41.48	41.83
C 4H 3I109	321.97	321.8824	14.92	.94	39.42	44.72
C 4H 4I201	321.88	321.8355	14.93	1.25	78.85	4.97
C 4H 4I202	337.88	337.8304	14.22	1.19	75.12	9.47
C 4H 4I203	353.88	353.8254	13.58	1.14	71.72	13.56
C 4H 4I204	369.88	369.8203	12.99	1.09	68.62	17.30
C 4H 4I205	385.88	385.8152	12.45	1.04	65.77	20.73
C 4H 4I206	401.88	401.8101	11.95	1.00	63.16	23.89
C 4H 4I207	417.88	417.8050	11.50	.96	60.74	26.80
C 4H 4I208	433.88	433.7999	11.07	.93	58.50	29.50
C 4H 4I209	449.88	449.7948	10.68	.90	56.42	32.01
C 4H 5I101	195.99	195.9387	24.51	2.57	64.75	8.16
C 4H 5I102	211.99	211.9336	22.66	2.38	59.86	15.09
C 4H 5I103	227.99	227.9285	21.07	2.21	55.66	21.05
C 4H 5I104	243.99	243.9234	19.69	2.07	52.01	26.23
C 4H 5I105	259.99	259.9184	18.48	1.94	48.81	30.77
C 4H 5I106	275.99	275.9133	17.41	1.83	45.98	34.78
C 4H 5I107	291.98	291.9082	16.45	1.73	43.46	38.36
C 4H 5I108	307.98	307.9031	15.60	1.64	41.20	41.56
C 4H 5I109	323.98	323.8980	14.83	1.56	39.17	44.45
C 4H 6I201	323.90	323.8512	14.83	1.87	78.36	4.94
C 4H 6I202	339.90	339.8461	14.13	1.78	74.67	9.41
C 4H 6I203	355.90	355.8410	13.50	1.70	71.31	13.49
C 4H 6I204	371.90	371.8359	12.92	1.63	68.25	17.21
C 4H 6I205	387.90	387.8308	12.39	1.56	65.43	20.62
C 4H 6I206	403.90	403.8258	11.90	1.50	62.84	23.77
C 4H 6I207	419.90	419.8207	11.44	1.44	60.45	26.67
C 4H 6I208	435.90	435.8156	11.02	1.39	58.23	29.36
C 4H 6I209	451.90	451.8105	10.63	1.34	56.17	31.86
C 4H 7I101	198.00	197.9543	24.26	3.56	64.09	8.08
C 4H 7I102	214.00	213.9493	22.45	3.30	59.30	14.95
C 4H 7I103	230.00	229.9442	20.89	3.07	55.18	20.87
C 4H 7I104	246.00	245.9391	19.53	2.87	51.59	26.02
C 4H 7I105	262.00	261.9340	18.34	2.69	48.44	30.53
C 4H 7I106	278.00	277.9289	17.28	2.54	45.65	34.53
C 4H 7I107	294.00	293.9238	16.34	2.40	43.16	38.09
C 4H 7I108	310.00	309.9187	15.50	2.28	40.94	41.29
C 4H 7I109	326.00	325.9137	14.74	2.16	38.93	44.17
C 4H 8I201	325.92	325.8668	14.74	2.47	77.88	4.91
C 4H 8I202	341.92	341.8617	14.05	2.36	74.23	9.36
C 4H 8I203	357.92	357.8567	13.42	2.25	70.91	13.41
C 4H 8I204	373.91	373.8516	12.85	2.16	67.88	17.12
C 4H 8I205	389.91	389.8465	12.32	2.07	65.09	20.52
C 4H 8I206	405.91	405.8414	11.84	1.99	62.53	23.65
C 4H 8I207	421.91	421.8363	11.39	1.91	60.16	26.54
C 4H 8I208	437.91	437.8312	10.97	1.84	57.96	29.23
C 4H 8I209	453.91	453.8261	10.58	1.78	55.92	31.72
C 4H 9I101	200.02	199.9700	24.02	4.54	63.45	8.00
C 4H 9I102	216.02	215.9649	22.24	4.20	58.75	14.81
C 4H 9I103	232.02	231.9598	20.71	3.91	54.70	20.69
C 4H 9I104	248.02	247.9547	19.37	3.66	51.17	25.80
C 4H 9I105	264.02	263.9497	18.20	3.44	48.07	30.30
C 4H 9I106	280.02	279.9446	17.16	3.24	45.32	34.28
C 4H 9I107	296.02	295.9395	16.23	3.06	42.87	37.83
C 4H 9I108	312.02	311.9344	15.40	2.91	40.67	41.02
C 4H 9I109	328.02	327.9293	14.65	2.77	38.69	43.90
C 5H 2I201	331.88	331.8199	18.10	.61	76.48	4.82
C 5H 2I202	347.88	347.8148	17.26	.58	72.96	9.20
C 5H 2I203	363.88	363.8097	16.50	.55	69.75	13.19
C 5H 2I204	379.88	379.8046	15.81	.53	66.81	16.85

185

C 5H 2I205	395.88	395.7995	15.17	.51	64.11	20.21
C 5H 2I206	411.88	411.7945	14.58	.49	61.62	23.31
C 5H 2I207	427.88	427.7894	14.04	.47	59.32	26.17
C 5H 2I208	443.88	443.7843	13.53	.45	57.18	28.84
C 5H 2I209	459.88	459.7792	13.06	.44	55.19	31.31
C 5H 3I101	205.98	205.9230	29.16	1.47	61.61	7.77
C 5H 3I102	221.98	221.9180	27.05	1.36	57.17	14.41
C 5H 3I103	237.98	237.9129	25.24	1.27	53.33	20.17
C 5H 3I104	253.98	253.9078	23.65	1.19	49.97	25.20
C 5H 3I105	269.98	269.9027	22.24	1.12	47.00	29.63
C 5H 3I106	285.98	285.8976	21.00	1.06	44.38	33.57
C 5H 3I107	301.98	301.8925	19.89	1.00	42.02	37.09
C 5H 3I108	317.98	317.8874	18.89	.95	39.91	40.25
C 5H 3I109	333.98	333.8824	17.98	.91	38.00	43.11
C 5H 4I201	333.90	333.8355	17.99	1.21	76.01	4.79
C 5H 4I202	349.90	349.8304	17.16	1.15	72.54	9.15
C 5H 4I203	365.89	365.8254	16.41	1.10	69.37	13.12
C 5H 4I204	381.89	381.8203	15.73	1.06	66.46	16.76
C 5H 4I205	397.89	397.8152	15.09	1.01	63.79	20.11
C 5H 4I206	413.89	413.8101	14.51	.97	61.32	23.19
C 5H 4I207	429.89	429.8050	13.97	.94	59.04	26.05
C 5H 4I208	445.89	445.7999	13.47	.90	56.92	28.71
C 5H 4I209	461.89	461.7948	13.00	.87	54.95	31.18
C 5H 5I101	208.00	207.9387	28.87	2.42	61.01	7.69
C 5H 5I102	224.00	223.9336	26.81	2.25	56.65	14.29
C 5H 5I103	240.00	239.9285	25.02	2.10	52.88	20.00
C 5H 5I104	256.00	255.9234	23.46	1.97	49.57	25.00
C 5H 5I105	272.00	271.9184	22.08	1.85	46.66	29.41
C 5H 5I106	288.00	287.9133	20.85	1.75	44.06	33.33
C 5H 5I107	304.00	303.9082	19.76	1.66	41.75	36.84
C 5H 5I108	320.00	319.9031	18.77	1.57	39.66	40.00
C 5H 5I109	335.99	335.8980	17.87	1.50	37.77	42.86
C 5H 6I201	335.91	335.8512	17.88	1.80	75.56	4.76
C 5H 6I202	351.91	351.8461	17.07	1.72	72.12	9.09
C 5H 6I203	367.91	367.8410	16.32	1.64	68.99	13.05
C 5H 6I204	383.91	383.8359	15.64	1.58	66.11	16.67
C 5H 6I205	399.91	399.8308	15.02	1.51	63.47	20.00
C 5H 6I206	415.91	415.8258	14.44	1.45	61.03	23.08
C 5H 6I207	431.91	431.8207	13.90	1.40	58.76	25.93
C 5H 6I208	447.91	447.8156	13.41	1.35	56.67	28.58
C 5H 6I209	463.91	463.8105	12.95	1.30	54.71	31.04
C 5H 7I101	210.02	209.9543	28.60	3.36	60.43	7.62
C 5H 7I102	226.01	225.9493	26.57	3.12	56.15	14.16
C 5H 7I103	242.01	241.9442	24.81	2.92	52.44	19.83
C 5H 7I104	258.01	257.9391	23.28	2.73	49.19	24.80
C 5H 7I105	274.01	273.9340	21.92	2.57	46.31	29.19
C 5H 7I106	290.01	289.9289	20.71	2.43	43.76	33.10
C 5H 7I107	306.01	305.9238	19.63	2.31	41.47	36.60
C 5H 7I108	322.01	321.9187	18.65	2.19	39.41	39.75
C 5H 7I109	338.01	337.9137	17.77	2.09	37.54	42.60
C 5H 8I201	337.93	337.8668	17.77	2.39	75.11	4.73
C 5H 8I202	353.93	353.8617	16.97	2.28	71.71	9.04
C 5H 8I203	369.93	369.8567	16.23	2.18	68.61	12.98
C 5H 8I204	385.93	385.8516	15.56	2.09	65.77	16.58
C 5H 8I205	401.93	401.8465	14.94	2.01	63.15	19.90
C 5H 8I206	417.92	417.8414	14.37	1.93	60.73	22.97
C 5H 8I207	433.92	433.8363	13.84	1.86	58.49	25.81
C 5H 8I208	449.92	449.8312	13.35	1.79	56.41	28.45
C 5H 8I209	465.92	465.8261	12.89	1.73	54.47	30.91
C 5H 9I101	212.03	211.9700	28.32	4.28	59.85	7.55
C 5H 9I102	228.03	227.9649	26.34	3.98	55.65	14.03
C 5H 9I103	244.03	243.9598	24.61	3.72	52.00	19.67
C 5H 9I104	260.03	259.9547	23.10	3.49	48.80	24.61
C 5H 9I105	276.03	275.9497	21.76	3.29	45.98	28.98
C 5H 9I106	292.03	291.9446	20.57	3.11	43.46	32.87
C 5H 9I107	308.03	307.9395	19.50	2.95	41.20	36.36
C 5H 9I108	324.03	323.9344	18.53	2.80	39.16	39.50
C 5H 9I109	340.03	339.9293	17.66	2.67	37.32	42.35
C 5H10I201	339.94	339.8825	17.67	2.97	74.66	4.71
C 5H10I202	355.94	355.8774	16.87	2.83	71.31	8.99
C 5H10I203	371.94	371.8723	16.15	2.71	68.24	12.90
C 5H10I204	387.94	387.8672	15.48	2.60	65.42	16.50
C 5H10I205	403.94	403.8621	14.87	2.50	62.83	19.80
C 5H10I206	419.94	419.8571	14.30	2.40	60.44	22.86
C 5H10I207	435.94	435.8520	13.78	2.31	58.22	25.69
C 5H10I208	451.94	451.8469	13.29	2.23	56.16	28.32
C 5H10I209	467.94	467.8418	12.83	2.15	54.24	30.77
C 5H11I101	214.05	213.9856	28.06	5.18	59.29	7.47
C 5H11I102	230.05	229.9806	26.11	4.82	55.16	13.91
C 5H11I103	246.05	245.9755	24.41	4.51	51.58	19.51
C 5H11I104	262.05	261.9704	22.92	4.23	48.43	24.42
C 5H11I105	278.04	277.9653	21.60	3.99	45.64	28.77
C 5H11I106	294.04	293.9602	20.42	3.77	43.16	32.65
C 5H11I107	310.04	309.9551	19.37	3.58	40.93	36.12
C 5H11I108	326.04	325.9500	18.42	3.40	38.92	39.26
C 5H11I109	342.04	341.9450	17.56	3.24	37.10	42.10
C 6H 3I101	217.99	217.9230	33.06	1.39	58.21	7.34
C 6H 3I102	233.99	233.9180	30.80	1.29	54.23	13.68
C 6H 3I103	249.99	249.9129	28.83	1.21	50.76	19.20
C 6H 3I104	265.99	265.9078	27.09	1.14	47.71	24.06
C 6H 3I105	281.99	281.9027	25.56	1.07	45.00	28.37
C 6H 3I106	297.99	297.8976	24.18	1.01	42.59	32.21
C 6H 3I107	313.99	313.8925	22.95	.96	40.42	35.67
C 6H 3I108	329.99	329.8874	21.84	.92	38.46	38.79
C 6H 3I109	345.99	345.8824	20.83	.87	36.68	41.62
C 6H 4I201	345.91	345.8355	20.83	1.17	73.37	4.63
C 6H 4I202	361.91	361.8304	19.91	1.11	70.13	8.84
C 6H 4I203	377.91	377.8254	19.07	1.07	67.16	12.70
C 6H 4I204	393.91	393.8203	18.30	1.02	64.43	16.25
C 6H 4I205	409.90	409.8152	17.58	.98	61.92	19.52
C 6H 4I206	425.90	425.8101	16.92	.95	59.59	22.54
C 6H 4I207	441.90	441.8050	16.31	.91	57.44	25.34
C 6H 4I208	457.90	457.7999	15.74	.88	55.43	27.95
C 6H 4I209	473.90	473.7948	15.21	.85	53.56	30.38
C 6H 5I101	220.01	219.9387	32.76	2.29	57.68	7.27
C 6H 5I102	236.01	235.9336	30.54	2.14	53.77	13.56
C 6H 5I103	252.01	251.9285	28.60	2.00	50.36	19.05
C 6H 5I104	268.01	267.9234	26.89	1.88	47.35	23.88
C 6H 5I105	284.01	283.9184	25.37	1.77	44.68	28.17
C 6H 5I106	300.01	299.9133	24.02	1.68	42.30	32.00
C 6H 5I107	316.01	315.9082	22.81	1.59	40.16	35.44
C 6H 5I108	332.01	331.9031	21.71	1.52	38.22	38.55
C 6H 5I109	348.01	347.8980	20.71	1.45	36.47	41.38
C 6H 6I201	347.92	347.8512	20.71	1.74	72.95	4.60
C 6H 6I202	363.92	363.8461	19.80	1.66	69.74	8.79
C 6H 6I203	379.92	379.8410	18.97	1.59	66.81	12.63
C 6H 6I204	395.92	395.8359	18.20	1.53	64.11	16.16
C 6H 6I205	411.92	411.8308	17.50	1.47	61.62	19.42
C 6H 6I206	427.92	427.8258	16.84	1.41	59.31	22.43
C 6H 6I207	443.92	443.8207	16.23	1.36	57.17	25.23

186

C 6H 6I208	459.92	459.8156	15.67	1.31	55.19	27.83
C 6H 6I209	475.92	475.8105	15.14	1.27	53.33	30.26
C 6H 7I101	222.03	221.9543	32.46	3.18	57.16	7.21
C 6H 7I102	238.03	237.9493	30.28	2.96	53.32	13.44
C 6H 7I103	254.03	253.9442	28.37	2.78	49.96	18.90
C 6H 7I104	270.02	269.9391	26.69	2.61	47.00	23.70
C 6H 7I105	286.02	285.9340	25.20	2.47	44.37	27.97
C 6H 7I106	302.02	301.9289	23.86	2.34	42.02	31.78
C 6H 7I107	318.02	317.9238	22.66	2.22	39.90	35.22
C 6H 7I108	334.02	333.9187	21.58	2.11	37.99	38.32
C 6H 7I109	350.02	349.9137	20.59	2.02	36.26	41.14
C 6H 8I201	349.94	349.8668	20.59	2.30	72.53	4.57
C 6H 8I202	365.94	365.8617	19.69	2.20	69.36	8.74
C 6H 8I203	381.94	381.8567	18.87	2.11	66.45	12.57
C 6H 8I204	397.94	397.8516	18.11	2.03	63.78	16.08
C 6H 8I205	413.94	413.8465	17.41	1.95	61.32	19.33
C 6H 8I206	429.94	429.8414	16.76	1.88	59.03	22.33
C 6H 8I207	445.94	445.8363	16.16	1.81	56.92	25.11
C 6H 8I208	461.93	461.8312	15.60	1.75	54.94	27.71
C 6H 8I209	477.93	477.8261	15.08	1.69	53.11	30.13
C 6H 9I101	224.04	223.9700	32.17	4.05	56.64	7.14
C 6H 9I102	240.04	239.9649	30.02	3.78	52.87	13.33
C 6H 9I103	256.04	255.9598	28.15	3.54	49.56	18.75
C 6H 9I104	272.04	271.9547	26.49	3.33	46.65	23.53
C 6H 9I105	288.04	287.9497	25.02	3.15	44.06	27.77
C 6H 9I106	304.04	303.9446	23.70	2.98	41.74	31.57
C 6H 9I107	320.04	319.9395	22.52	2.83	39.65	34.99
C 6H 9I108	336.04	335.9344	21.45	2.70	37.76	38.09
C 6H 9I109	352.04	351.9293	20.47	2.58	36.05	40.90
C 6H10I201	351.95	351.8825	20.48	2.86	72.11	4.55
C 6H10I202	367.95	367.8774	19.59	2.74	68.98	8.70
C 6H10I203	383.95	383.8723	18.77	2.63	66.10	12.50
C 6H10I204	399.95	399.8672	18.02	2.52	63.46	16.00
C 6H10I205	415.95	415.8621	17.33	2.42	61.02	19.23
C 6H10I206	431.95	431.8571	16.68	2.33	58.76	22.22
C 6H10I207	447.95	447.8520	16.09	2.25	56.66	25.00
C 6H10I208	463.95	463.8469	15.53	2.17	54.71	27.59
C 6H10I209	479.95	479.8418	15.02	2.10	52.88	30.00
C 6H11I101	226.06	225.9856	31.88	4.90	56.14	7.08
C 6H11I102	242.06	241.9806	29.77	4.58	52.43	13.22
C 6H11I103	258.06	257.9755	27.93	4.30	49.18	18.60
C 6H11I104	274.06	273.9704	26.30	4.05	46.31	23.35
C 6H11I105	290.06	289.9653	24.85	3.82	43.75	27.58
C 6H11I106	306.06	305.9602	23.55	3.62	41.46	31.37
C 6H11I107	322.05	321.9551	22.38	3.44	39.40	34.78
C 6H11I108	338.05	337.9500	21.32	3.28	37.54	37.86
C 6H11I109	354.05	353.9450	20.35	3.13	35.84	40.67
C 6H12I201	353.97	353.8981	20.36	3.42	71.70	4.52
C 6H12I202	369.97	369.8930	19.48	3.27	68.60	8.65
C 6H12I203	385.97	385.8880	18.67	3.13	65.76	12.44
C 6H12I204	401.97	401.8829	17.93	3.01	63.14	15.92
C 6H12I205	417.97	417.8778	17.24	2.89	60.72	19.14
C 6H12I206	433.97	433.8727	16.61	2.79	58.49	22.12
C 6H12I207	449.97	449.8676	16.02	2.69	56.41	24.89
C 6H12I208	465.97	465.8625	15.47	2.60	54.47	27.47
C 6H12I209	481.97	481.8574	14.95	2.51	52.66	29.88
C 6H13I101	228.07	228.0013	31.60	5.75	55.64	7.01
C 6H13I102	244.07	243.9962	29.53	5.37	51.99	13.11
C 6H13I103	260.07	259.9911	27.71	5.04	48.80	18.46
C 6H13I104	276.07	275.9860	26.10	4.75	45.97	23.18
C 6H13I105	292.07	291.9810	24.67	4.49	43.45	27.39
C 6H13I106	308.07	307.9759	23.39	4.25	41.19	31.16
C 6H13I107	324.07	323.9708	22.24	4.04	39.16	34.56
C 6H13I108	340.07	339.9657	21.19	3.85	37.32	37.64
C 6H13I109	356.07	355.9606	20.24	3.68	35.64	40.44
C 7H 3I101	230.01	229.9230	36.55	1.31	55.17	6.96
C 7H 3I102	246.01	245.9180	34.18	1.23	51.59	13.01
C 7H 3I103	262.00	261.9129	32.09	1.15	48.44	18.32
C 7H 3I104	278.00	277.9078	30.24	1.09	45.65	23.02
C 7H 3I105	294.00	293.9027	28.60	1.03	43.16	27.21
C 7H 3I106	310.00	309.8976	27.12	.98	40.94	30.97
C 7H 3I107	326.00	325.8925	25.79	.93	38.93	34.35
C 7H 3I108	342.00	341.8874	24.58	.88	37.11	37.43
C 7H 3I109	358.00	357.8824	23.49	.84	35.45	40.22
C 7H 4I201	357.92	357.8355	23.49	1.13	70.91	4.47
C 7H 4I202	373.92	373.8304	22.49	1.08	67.88	8.56
C 7H 4I203	389.92	389.8254	21.56	1.03	65.09	12.31
C 7H 4I204	405.92	405.8203	20.71	.99	62.53	15.77
C 7H 4I205	421.92	421.8152	19.93	.96	60.16	18.96
C 7H 4I206	437.92	437.8101	19.20	.92	57.96	21.92
C 7H 4I207	453.91	453.8050	18.52	.89	55.92	24.67
C 7H 4I208	469.91	469.7999	17.89	.86	54.01	27.24
C 7H 4I209	485.91	485.7948	17.30	.83	52.23	29.63
C 7H 5I101	232.02	231.9387	36.24	2.17	54.70	6.90
C 7H 5I102	248.02	247.9336	33.90	2.03	51.17	12.90
C 7H 5I103	264.02	263.9285	31.85	1.91	48.07	18.18
C 7H 5I104	280.02	279.9234	30.03	1.80	45.32	22.85
C 7H 5I105	296.02	295.9184	28.40	1.70	42.87	27.02
C 7H 5I106	312.02	311.9133	26.95	1.62	40.67	30.77
C 7H 5I107	328.02	327.9082	25.63	1.54	38.69	34.14
C 7H 5I108	344.02	343.9031	24.44	1.46	36.89	37.21
C 7H 5I109	360.02	359.8980	23.35	1.40	35.25	40.00
C 7H 6I201	359.93	359.8512	23.36	1.68	70.52	4.45
C 7H 6I202	375.93	375.8461	22.37	1.61	67.51	8.51
C 7H 6I203	391.93	391.8410	21.45	1.54	64.76	12.25
C 7H 6I204	407.93	407.8359	20.61	1.48	62.22	15.69
C 7H 6I205	423.93	423.8308	19.83	1.43	59.87	18.87
C 7H 6I206	439.93	439.8258	19.11	1.37	57.69	21.82
C 7H 6I207	455.93	455.8207	18.44	1.33	55.67	24.56
C 7H 6I208	471.93	471.8156	17.82	1.28	53.78	27.12
C 7H 6I209	487.93	487.8105	17.23	1.24	52.02	29.51
C 7H 7I101	234.04	233.9543	35.93	3.01	54.22	6.84
C 7H 7I102	250.04	249.9493	33.63	2.82	50.75	12.80
C 7H 7I103	266.04	265.9442	31.60	2.65	47.70	18.04
C 7H 7I104	282.04	281.9391	29.81	2.50	45.00	22.69
C 7H 7I105	298.04	297.9340	28.21	2.37	42.58	26.84
C 7H 7I106	314.03	313.9289	26.77	2.25	40.41	30.57
C 7H 7I107	330.03	329.9238	25.48	2.14	38.45	33.93
C 7H 7I108	346.03	345.9187	24.30	2.04	36.67	36.99
C 7H 7I109	362.03	361.9137	23.22	1.95	35.05	39.77
C 7H 8I201	361.95	361.8668	23.23	2.23	70.12	4.42
C 7H 8I202	377.95	377.8617	22.25	2.13	67.15	8.47
C 7H 8I203	393.95	393.8567	21.34	2.05	64.43	12.18
C 7H 8I204	409.95	409.8516	20.51	1.97	61.91	15.61
C 7H 8I205	425.95	425.8465	19.74	1.89	59.59	18.78
C 7H 8I206	441.95	441.8414	19.02	1.82	57.43	21.72
C 7H 8I207	457.95	457.8363	18.36	1.76	55.42	24.46
C 7H 8I208	473.95	473.8312	17.74	1.70	53.55	27.01
C 7H 8I209	489.95	489.8261	17.16	1.65	51.80	29.39
C 7H 9I101	236.05	235.9700	35.62	3.84	53.76	6.78

C 7H 9I102	252.05	251.9649	33.36	3.60	50.35	12.70
C 7H 9I103	268.05	267.9598	31.37	3.38	47.34	17.91
C 7H 9I104	284.05	283.9547	29.60	3.19	44.68	22.53
C 7H 9I105	300.05	299.9497	28.02	3.02	42.29	26.66
C 7H 9I106	316.05	315.9446	26.60	2.87	40.15	30.37
C 7H 9I107	332.05	331.9395	25.32	2.73	38.22	33.73
C 7H 9I108	348.05	347.9344	24.16	2.61	36.46	36.78
C 7H 9I109	364.05	363.9293	23.10	2.49	34.86	39.55
C 7H10I201	363.97	363.8825	23.10	2.77	69.73	4.40
C 7H10I202	379.97	379.8774	22.13	2.65	66.80	8.42
C 7H10I203	395.96	395.8723	21.23	2.55	64.10	12.12
C 7H10I204	411.96	411.8672	20.41	2.45	61.61	15.53
C 7H10I205	427.96	427.8621	19.65	2.36	59.31	18.69
C 7H10I206	443.96	443.8571	18.94	2.27	57.17	21.62
C 7H10I207	459.96	459.8520	18.28	2.19	55.18	24.35
C 7H10I208	475.96	475.8469	17.66	2.12	53.33	26.89
C 7H10I209	491.96	491.8418	17.09	2.05	51.59	29.27
C 7H11I101	238.07	237.9856	35.32	4.66	53.31	6.72
C 7H11I102	254.07	253.9806	33.09	4.36	49.95	12.59
C 7H11I103	270.07	269.9755	31.13	4.11	46.99	17.77
C 7H11I104	286.07	285.9704	29.39	3.88	44.36	22.37
C 7H11I105	302.07	301.9653	27.83	3.67	42.01	26.48
C 7H11I106	318.07	317.9602	26.43	3.49	39.90	30.18
C 7H11I107	334.07	333.9551	25.17	3.32	37.99	33.53
C 7H11I108	350.07	349.9500	24.02	3.17	36.25	36.56
C 7H11I109	366.06	365.9450	22.97	3.03	34.67	39.34
C 7H12I201	365.98	365.8981	22.97	3.30	69.35	4.37
C 7H12I202	381.98	381.8930	22.01	3.17	66.45	8.38
C 7H12I203	397.98	397.8880	21.13	3.04	63.77	12.06
C 7H12I204	413.98	413.8829	20.31	2.92	61.31	15.46
C 7H12I205	429.98	429.8778	19.55	2.81	59.03	18.60
C 7H12I206	445.98	445.8727	18.85	2.71	56.91	21.52
C 7H12I207	461.98	461.8676	18.20	2.62	54.94	24.24
C 7H12I208	477.98	477.8625	17.59	2.53	53.10	26.78
C 7H12I209	493.98	493.8574	17.02	2.45	51.38	29.15
C 7H13I101	240.09	240.0013	35.02	5.46	52.86	6.66
C 7H13I102	256.08	255.9962	32.83	5.12	49.56	12.50
C 7H13I103	272.08	271.9911	30.90	4.82	46.64	17.64
C 7H13I104	288.08	287.9860	29.19	4.55	44.05	22.21
C 7H13I105	304.08	303.9810	27.65	4.31	41.73	26.31
C 7H13I106	320.08	319.9759	26.27	4.09	39.65	29.99
C 7H13I107	336.08	335.9708	25.02	3.90	37.76	33.32
C 7H13I108	352.08	351.9657	23.88	3.72	36.04	36.35
C 7H13I109	368.08	367.9606	22.84	3.56	34.48	39.12
C 7H14I201	368.00	367.9138	22.85	3.83	68.97	4.35
C 7H14I202	384.00	383.9087	21.90	3.67	66.10	8.33
C 7H14I203	400.00	399.9036	21.02	3.53	63.45	12.00
C 7H14I204	416.00	415.8985	20.21	3.39	61.01	15.38
C 7H14I205	432.00	431.8934	19.46	3.27	58.75	18.52
C 7H14I206	447.99	447.8883	18.77	3.15	56.65	21.43
C 7H14I207	463.99	463.8833	18.12	3.04	54.70	24.14
C 7H14I208	479.99	479.8782	17.52	2.94	52.88	26.67
C 7H14I209	495.99	495.8731	16.95	2.85	51.17	29.03
C 7H15I101	242.10	242.0169	34.73	6.25	52.42	6.61
C 7H15I102	258.10	258.0119	32.58	5.86	49.17	12.40
C 7H15I103	274.10	274.0068	30.67	5.52	46.30	17.51
C 7H15I104	290.10	290.0017	28.98	5.21	43.75	22.06
C 7H15I105	306.10	305.9966	27.47	4.94	41.46	26.13
C 7H15I106	322.10	321.9915	26.10	4.69	39.40	29.80
C 7H15I107	338.10	337.9864	24.87	4.47	37.53	33.13
C 7H15I108	354.10	353.9813	23.74	4.27	35.84	36.15
C 7H15I109	370.10	369.9763	22.72	4.09	34.29	38.91
C 8H 4I201	369.93	369.8355	25.98	1.09	68.61	4.32
C 8H 4I202	385.93	385.8304	24.90	1.04	65.77	8.29
C 8H 4I203	401.93	401.8254	23.91	1.00	63.15	11.94
C 8H 4I204	417.93	417.8203	22.99	.96	60.73	15.31
C 8H 4I205	433.93	433.8152	22.14	.93	58.49	18.44
C 8H 4I206	449.93	449.8101	21.36	.90	56.41	21.34
C 8H 4I207	465.93	465.8050	20.62	.87	54.47	24.04
C 8H 4I208	481.93	481.7999	19.94	.84	52.67	26.56
C 8H 4I209	497.92	497.7948	19.30	.81	50.97	28.92
C 8H 5I101	244.03	243.9387	39.38	2.07	52.00	6.56
C 8H 5I102	260.03	259.9336	36.95	1.94	48.80	12.31
C 8H 5I103	276.03	275.9285	34.81	1.83	45.97	17.39
C 8H 5I104	292.03	291.9234	32.90	1.73	43.46	21.91
C 8H 5I105	308.03	307.9184	31.19	1.64	41.20	25.97
C 8H 5I106	324.03	323.9133	29.65	1.56	39.16	29.63
C 8H 5I107	340.03	339.9082	28.26	1.48	37.32	32.94
C 8H 5I108	356.03	355.9031	26.99	1.42	35.64	35.95
C 8H 5I109	372.03	371.8980	25.83	1.35	34.11	38.71
C 8H 6I201	371.95	371.8512	25.83	1.63	68.24	4.30
C 8H 6I202	387.94	387.8461	24.77	1.56	65.42	8.25
C 8H 6I203	403.94	403.8410	23.79	1.50	62.83	11.88
C 8H 6I204	419.94	419.8359	22.88	1.44	60.44	15.24
C 8H 6I205	435.94	435.8308	22.04	1.39	58.22	18.35
C 8H 6I206	451.94	451.8258	21.26	1.34	56.16	21.24
C 8H 6I207	467.94	467.8207	20.53	1.29	54.24	23.93
C 8H 6I208	483.94	483.8156	19.86	1.25	52.45	26.45
C 8H 6I209	499.94	499.8105	19.22	1.21	50.77	28.80
C 8H 7I101	246.05	245.9543	39.05	2.87	51.58	6.50
C 8H 7I102	262.05	261.9493	36.67	2.69	48.43	12.21
C 8H 7I103	278.05	277.9442	34.56	2.54	45.64	17.26
C 8H 7I104	294.05	293.9391	32.68	2.40	43.16	21.76
C 8H 7I105	310.05	309.9340	30.99	2.28	40.93	25.80
C 8H 7I106	326.05	325.9289	29.47	2.16	38.92	29.44
C 8H 7I107	342.05	341.9238	28.09	2.06	37.10	32.74
C 8H 7I108	358.04	357.9187	26.84	1.97	35.44	35.75
C 8H 7I109	374.04	373.9137	25.69	1.89	33.93	38.50
C 8H 8I201	373.96	373.8668	25.69	2.16	67.87	4.28
C 8H 8I202	389.96	389.8617	24.64	2.07	65.09	8.21
C 8H 8I203	405.96	405.8567	23.67	1.99	62.52	11.82
C 8H 8I204	421.96	421.8516	22.77	1.91	60.15	15.17
C 8H 8I205	437.96	437.8465	21.94	1.84	57.95	18.27
C 8H 8I206	453.96	453.8414	21.17	1.78	55.91	21.15
C 8H 8I207	469.96	469.8363	20.45	1.72	54.01	23.83
C 8H 8I208	485.96	485.8312	19.77	1.66	52.23	26.34
C 8H 8I209	501.96	501.8261	19.14	1.61	50.56	28.69
C 8H 9I101	248.06	247.9700	38.74	3.66	51.16	6.45
C 8H 9I102	264.06	263.9649	36.39	3.44	48.06	12.12
C 8H 9I103	280.06	279.9598	34.31	3.24	45.31	17.14
C 8H 9I104	296.06	295.9547	32.46	3.06	42.86	21.62
C 8H 9I105	312.06	311.9497	30.79	2.91	40.67	25.63
C 8H 9I106	328.06	327.9446	29.29	2.77	38.68	29.26
C 8H 9I107	344.06	343.9395	27.93	2.64	36.88	32.55
C 8H 9I108	360.06	359.9344	26.69	2.52	35.25	35.55
C 8H 9I109	376.06	375.9293	25.55	2.41	33.75	38.29
C 8H10I201	375.98	375.8825	25.56	2.68	67.51	4.26
C 8H10I202	391.98	391.8774	24.51	2.57	64.75	8.16
C 8H10I203	407.98	407.8723	23.55	2.47	62.21	11.76
C 8H10I204	423.98	423.8672	22.66	2.38	59.86	15.09

188

C 8H10I205	439.97	439.8621	21.84	2.29	57.69	18.18
C 8H10I206	455.97	455.8571	21.07	2.21	55.66	21.05
C 8H10I207	471.97	471.8520	20.36	2.14	53.78	23.73
C 8H10I208	487.97	487.8469	19.69	2.07	52.01	26.23
C 8H10I209	503.97	503.8418	19.07	2.00	50.36	28.57
C 8H11I101	250.08	249.9856	38.42	4.43	50.75	6.40
C 8H11I102	266.08	265.9806	36.11	4.17	47.69	12.03
C 8H11I103	282.08	281.9755	34.06	3.93	44.99	17.02
C 8H11I104	298.08	297.9704	32.24	3.72	42.57	21.47
C 8H11I105	314.08	313.9653	30.59	3.53	40.41	25.47
C 8H11I106	330.08	329.9602	29.11	3.36	38.45	29.08
C 8H11I107	346.08	345.9551	27.77	3.20	36.67	32.36
C 8H11I108	362.08	361.9500	26.54	3.06	35.05	35.35
C 8H11I109	378.08	377.9450	25.42	2.93	33.57	38.09
C 8H12I201	377.99	377.8981	25.42	3.20	67.15	4.23
C 8H12I202	393.99	393.8930	24.39	3.07	64.42	8.12
C 8H12I203	409.99	409.8880	23.44	2.95	61.91	11.71
C 8H12I204	425.99	425.8829	22.56	2.84	59.58	15.02
C 8H12I205	441.99	441.8778	21.74	2.74	57.42	18.10
C 8H12I206	457.99	457.8727	20.98	2.64	55.42	20.96
C 8H12I207	473.99	473.8676	20.27	2.55	53.55	23.63
C 8H12I208	489.99	489.8625	19.61	2.47	51.80	26.12
C 8H12I209	505.99	505.8574	18.99	2.39	50.16	28.46
C 8H13I101	252.10	252.0013	38.12	5.20	50.34	6.35
C 8H13I102	268.10	267.9962	35.84	4.89	47.34	11.94
C 8H13I103	284.10	283.9911	33.82	4.61	44.67	16.90
C 8H13I104	300.09	299.9860	32.02	4.37	42.29	21.33
C 8H13I105	316.09	315.9810	30.40	4.15	40.15	25.31
C 8H13I106	332.09	331.9759	28.93	3.95	38.21	28.91
C 8H13I107	348.09	347.9708	27.60	3.76	36.46	32.17
C 8H13I108	364.09	363.9657	26.39	3.60	34.85	35.15
C 8H13I109	380.09	379.9606	25.28	3.45	33.39	37.88
C 8H14I201	380.01	379.9138	25.29	3.71	66.79	4.21
C 8H14I202	396.01	395.9087	24.26	3.56	64.09	8.08
C 8H14I203	412.01	411.9036	23.32	3.43	61.60	11.65
C 8H14I204	428.01	427.8985	22.45	3.30	59.30	14.95
C 8H14I205	444.01	443.8934	21.64	3.18	57.16	18.02
C 8H14I206	460.01	459.8883	20.89	3.07	55.18	20.87
C 8H14I207	476.01	475.8833	20.19	2.96	53.32	23.53
C 8H14I208	492.00	491.8782	19.53	2.87	51.59	26.02
C 8H14I209	508.00	507.8731	18.92	2.78	49.96	28.35
C 8H15I101	254.11	254.0169	37.81	5.95	49.94	6.30
C 8H15I102	270.11	270.0119	35.57	5.60	46.98	11.85
C 8H15I103	286.11	286.0068	33.58	5.28	44.35	16.78
C 8H15I104	302.11	302.0017	31.81	5.00	42.01	21.18
C 8H15I105	318.11	317.9966	30.21	4.75	39.89	25.15
C 8H15I106	334.11	333.9915	28.76	4.53	37.98	28.73
C 8H15I107	350.11	349.9864	27.45	4.32	36.25	31.99
C 8H15I108	366.11	365.9813	26.25	4.13	34.66	34.96
C 8H15I109	382.11	381.9763	25.15	3.96	33.21	37.68
C 8H16I201	382.02	381.9294	25.15	4.22	66.44	4.19
C 8H16I202	398.02	397.9243	24.14	4.05	63.77	8.04
C 8H16I203	414.02	413.9193	23.21	3.90	61.30	11.59
C 8H16I204	430.02	429.9142	22.35	3.75	59.02	14.88
C 8H16I205	446.02	445.9091	21.54	3.62	56.90	17.94
C 8H16I206	462.02	461.9040	20.80	3.49	54.93	20.78
C 8H16I207	478.02	477.8989	20.10	3.37	53.10	23.43
C 8H16I208	494.02	493.8938	19.45	3.26	51.38	25.91
C 8H16I209	510.02	509.8887	18.84	3.16	49.76	28.23
C 8H17I101	256.13	256.0326	37.52	6.69	49.55	6.25
C 8H17I102	272.13	272.0275	35.31	6.30	46.63	11.76
C 8H17I103	288.13	288.0224	33.35	5.95	44.04	16.66
C 8H17I104	304.13	304.0173	31.60	5.63	41.73	21.04
C 8H17I105	320.13	320.0122	30.02	5.35	39.64	24.99
C 8H17I106	336.13	336.0072	28.59	5.10	37.76	28.56
C 8H17I107	352.12	352.0021	27.29	4.87	36.04	31.81
C 8H17I108	368.12	367.9970	26.10	4.65	34.47	34.77
C 8H17I109	384.12	383.9919	25.02	4.46	33.04	37.49
C 9H 4I201	381.94	381.8355	28.30	1.06	66.45	4.19
C 9H 4I202	397.94	397.8304	27.16	1.01	63.78	8.04
C 9H 4I203	413.94	413.8254	26.12	.97	61.32	11.60
C 9H 4I204	429.94	429.8203	25.14	.94	59.03	14.89
C 9H 4I205	445.94	445.8152	24.24	.90	56.92	17.94
C 9H 4I206	461.94	461.8101	23.40	.87	54.94	20.78
C 9H 4I207	477.94	477.8050	22.62	.84	53.11	23.43
C 9H 4I208	493.94	493.7999	21.89	.82	51.38	25.91
C 9H 4I209	509.94	509.7948	21.20	.79	49.77	28.24
C 9H 5I101	256.04	255.9387	42.22	1.97	49.56	6.25
C 9H 5I102	272.04	271.9336	39.74	1.85	46.65	11.76
C 9H 5I103	288.04	287.9285	37.53	1.75	44.06	16.66
C 9H 5I104	304.04	303.9234	35.55	1.66	41.74	21.05
C 9H 5I105	320.04	319.9184	33.78	1.57	39.65	25.00
C 9H 5I106	336.04	335.9133	32.17	1.50	37.76	28.57
C 9H 5I107	352.04	351.9082	30.71	1.43	36.05	31.81
C 9H 5I108	368.04	367.9031	29.37	1.37	34.48	34.78
C 9H 5I109	384.04	383.8980	28.15	1.31	33.04	37.49
C 9H 6I201	383.96	383.8512	28.15	1.58	66.10	4.17
C 9H 6I202	399.96	399.8461	27.03	1.51	63.46	8.00
C 9H 6I203	415.96	415.8410	25.99	1.45	61.02	11.54
C 9H 6I204	431.95	431.8359	25.03	1.40	58.76	14.82
C 9H 6I205	447.95	447.8308	24.13	1.35	56.66	17.86
C 9H 6I206	463.95	463.8258	23.30	1.30	54.71	20.69
C 9H 6I207	479.95	479.8207	22.52	1.26	52.88	23.33
C 9H 6I208	495.95	495.8156	21.80	1.22	51.18	25.81
C 9H 6I209	511.95	511.8105	21.12	1.18	49.58	28.13
C 9H 7I101	258.06	257.9543	41.89	2.73	49.18	6.20
C 9H 7I102	274.06	273.9493	39.44	2.57	46.31	11.68
C 9H 7I103	290.06	289.9442	37.27	2.43	43.75	16.55
C 9H 7I104	306.06	305.9391	35.32	2.31	41.46	20.91
C 9H 7I105	322.06	321.9340	33.57	2.19	39.40	24.84
C 9H 7I106	338.06	337.9289	31.98	2.09	37.54	28.40
C 9H 7I107	354.06	353.9238	30.53	1.99	35.84	31.63
C 9H 7I108	370.06	369.9187	29.21	1.91	34.29	34.59
C 9H 7I109	386.06	385.9137	28.00	1.83	32.87	37.30
C 9H 8I201	385.97	385.8668	28.01	2.09	65.76	4.15
C 9H 8I202	401.97	401.8617	26.89	2.01	63.14	7.96
C 9H 8I203	417.97	417.8567	25.86	1.93	60.72	11.48
C 9H 8I204	433.97	433.8516	24.91	1.86	58.49	14.75
C 9H 8I205	449.97	449.8465	24.02	1.79	56.41	17.78
C 9H 8I206	465.97	465.8414	23.20	1.73	54.47	20.60
C 9H 8I207	481.97	481.8363	22.43	1.67	52.66	23.24
C 9H 8I208	497.97	497.8312	21.71	1.62	50.97	25.70
C 9H 8I209	513.97	513.8261	21.03	1.57	49.38	28.02
C 9H 9I101	260.08	259.9700	41.56	3.49	48.80	6.15
C 9H 9I102	276.08	275.9649	39.16	3.29	45.97	11.59
C 9H 9I103	292.07	291.9598	37.01	3.11	43.45	16.43
C 9H 9I104	308.07	307.9547	35.09	2.94	41.19	20.77
C 9H 9I105	324.07	323.9497	33.36	2.80	39.16	24.68
C 9H 9I106	340.07	339.9446	31.79	2.67	37.32	28.23
C 9H 9I107	356.07	355.9395	30.36	2.55	35.64	31.45

C 9H 9I108	372.07	371.9344	29.05	2.44	34.11	34.40
C 9H 9I109	388.07	387.9293	27.86	2.34	32.70	37.11
C 9H10I201	387.99	387.8825	27.86	2.60	65.42	4.12
C 9H10I202	403.99	403.8774	26.76	2.50	62.83	7.92
C 9H10I203	419.99	419.8723	25.74	2.40	60.43	11.43
C 9H10I204	435.99	435.8672	24.79	2.31	58.21	14.68
C 9H10I205	451.99	451.8621	23.92	2.23	56.15	17.70
C 9H10I206	467.99	467.8571	23.10	2.15	54.23	20.51
C 9H10I207	483.98	483.8520	22.34	2.08	52.44	23.14
C 9H10I208	499.98	499.8469	21.62	2.02	50.76	25.60
C 9H10I209	515.98	515.8418	20.95	1.95	49.19	27.91
C 9H11I101	262.09	261.9856	41.25	4.23	48.42	6.10
C 9H11I102	278.09	277.9806	38.87	3.99	45.63	11.51
C 9H11I103	294.09	293.9755	36.76	3.77	43.15	16.32
C 9H11I104	310.09	309.9704	34.86	3.58	40.93	20.64
C 9H11I105	326.09	325.9653	33.15	3.40	38.92	24.53
C 9H11I106	342.09	341.9602	31.60	3.24	37.10	28.06
C 9H11I107	358.09	357.9551	30.19	3.10	35.44	31.28
C 9H11I108	374.09	373.9500	28.90	2.96	33.92	34.22
C 9H11I109	390.09	389.9450	27.71	2.84	32.53	36.91
C 9H12I201	390.00	389.8981	27.72	3.10	65.08	4.10
C 9H12I202	406.00	405.8930	26.63	2.98	62.51	7.88
C 9H12I203	422.00	421.8880	25.62	2.87	60.14	11.37
C 9H12I204	438.00	437.8829	24.68	2.76	57.95	14.61
C 9H12I205	454.00	453.8778	23.81	2.66	55.90	17.62
C 9H12I206	470.00	469.8727	23.00	2.57	54.00	20.42
C 9H12I207	486.00	485.8676	22.24	2.49	52.22	23.04
C 9H12I208	502.00	501.8625	21.53	2.41	50.56	25.50
C 9H12I209	518.00	517.8574	20.87	2.34	49.00	27.80
C 9H13I101	264.11	264.0013	40.93	4.96	48.05	6.06
C 9H13I102	280.11	279.9962	38.59	4.68	45.31	11.42
C 9H13I103	296.11	295.9911	36.51	4.43	42.86	16.21
C 9H13I104	312.11	311.9860	34.64	4.20	40.66	20.51
C 9H13I105	328.11	327.9810	32.95	3.99	38.68	24.38
C 9H13I106	344.10	343.9759	31.41	3.81	36.88	27.90
C 9H13I107	360.10	359.9708	30.02	3.64	35.24	31.10
C 9H13I108	376.10	375.9657	28.74	3.48	33.74	34.03
C 9H13I109	392.10	391.9606	27.57	3.34	32.37	36.72
C 9H14I201	392.02	391.9138	27.58	3.60	64.74	4.08
C 9H14I202	408.02	407.9087	26.49	3.46	62.21	7.84
C 9H14I203	424.02	423.9036	25.49	3.33	59.86	11.32
C 9H14I204	440.02	439.8985	24.57	3.21	57.68	14.54
C 9H14I205	456.02	455.8934	23.71	3.09	55.66	17.54
C 9H14I206	472.02	471.8883	22.90	2.99	53.77	20.34
C 9H14I207	488.02	487.8833	22.15	2.89	52.01	22.95
C 9H14I208	504.02	503.8782	21.45	2.80	50.36	25.40
C 9H14I209	520.02	519.8731	20.79	2.71	48.81	27.69
C 9H15I101	266.12	266.0169	40.62	5.68	47.69	6.01
C 9H15I102	282.12	282.0119	38.32	5.36	44.98	11.34
C 9H15I103	298.12	298.0068	36.26	5.07	42.57	16.10
C 9H15I104	314.12	314.0017	34.41	4.81	40.40	20.37
C 9H15I105	330.12	329.9966	32.75	4.58	38.44	24.23
C 9H15I106	346.12	345.9915	31.23	4.37	36.66	27.73
C 9H15I107	362.12	361.9864	29.85	4.18	35.04	30.93
C 9H15I108	378.12	377.9813	28.59	4.00	33.56	33.85
C 9H15I109	394.12	393.9763	27.43	3.84	32.20	36.54
C 9H16I201	394.04	393.9294	27.43	4.09	64.41	4.06
C 9H16I202	410.04	409.9243	26.36	3.93	61.90	7.80
C 9H16I203	426.03	425.9193	25.37	3.79	59.57	11.27
C 9H16I204	442.03	441.9142	24.46	3.65	57.42	14.48
C 9H16I205	458.03	457.9091	23.60	3.52	55.41	17.47
C 9H16I206	474.03	473.9040	22.80	3.40	53.54	20.25
C 9H16I207	490.03	489.8989	22.06	3.29	51.79	22.85
C 9H16I208	506.03	505.8938	21.36	3.19	50.16	25.29
C 9H16I209	522.03	521.8887	20.71	3.09	48.62	27.58
C 9H17I101	268.14	268.0326	40.31	6.39	47.33	5.97
C 9H17I102	284.14	284.0275	38.04	6.03	44.66	11.26
C 9H17I103	300.14	300.0224	36.02	5.71	42.28	15.99
C 9H17I104	316.14	316.0173	34.19	5.42	40.14	20.24
C 9H17I105	332.14	332.0122	32.55	5.16	38.21	24.09
C 9H17I106	348.14	348.0072	31.05	4.92	36.45	27.57
C 9H17I107	364.14	364.0021	29.69	4.71	34.85	30.76
C 9H17I108	380.14	379.9970	28.44	4.51	33.38	33.67
C 9H17I109	396.13	395.9919	27.29	4.33	32.04	36.35
C 9H18I201	396.05	395.9451	27.29	4.58	64.08	4.04
C 9H18I202	412.05	411.9400	26.23	4.40	61.60	7.77
C 9H18I203	428.05	427.9349	25.25	4.24	59.29	11.21
C 9H18I204	444.05	443.9298	24.34	4.09	57.16	14.41
C 9H18I205	460.05	459.9247	23.50	3.94	55.17	17.39
C 9H18I206	476.05	475.9196	22.71	3.81	53.32	20.17
C 9H18I207	492.05	491.9146	21.97	3.69	51.58	22.76
C 9H18I208	508.05	507.9095	21.28	3.57	49.96	25.19
C 9H18I209	524.05	523.9044	20.63	3.46	48.43	27.48
C 9H19I101	270.16	270.0482	40.01	7.09	46.97	5.92
C 9H19I102	286.15	286.0432	37.78	6.69	44.35	11.18
C 9H19I103	302.15	302.0381	35.78	6.34	42.00	15.89
C 9H19I104	318.15	318.0330	33.98	6.02	39.89	20.12
C 9H19I105	334.15	334.0279	32.35	5.73	37.98	23.94
C 9H19I106	350.15	350.0228	30.87	5.47	36.24	27.42
C 9H19I107	366.15	366.0177	29.52	5.23	34.66	30.59
C 9H19I108	382.15	382.0126	28.29	5.01	33.21	33.49
C 9H19I109	398.15	398.0076	27.15	4.81	31.87	36.17
C10H 5I101	268.06	267.9387	44.81	1.88	47.34	5.97
C10H 5I102	284.05	283.9336	42.28	1.77	44.68	11.27
C10H 5I103	300.05	299.9285	40.03	1.68	42.29	16.00
C10H 5I104	316.05	315.9234	38.00	1.59	40.15	20.25
C10H 5I105	332.05	331.9184	36.17	1.52	38.22	24.09
C10H 5I106	348.05	347.9133	34.51	1.45	36.46	27.58
C10H 5I107	364.05	363.9082	32.99	1.38	34.86	30.76
C10H 5I108	380.05	379.9031	31.60	1.33	33.39	33.68
C10H 5I109	396.05	395.8980	30.33	1.27	32.04	36.36
C10H 6I201	395.97	395.8512	30.33	1.53	64.10	4.04
C10H 6I202	411.97	411.8461	29.16	1.47	61.61	7.77
C10H 6I203	427.97	427.8410	28.07	1.41	59.31	11.22
C10H 6I204	443.97	443.8359	27.05	1.36	57.17	14.41
C10H 6I205	459.97	459.8308	26.11	1.31	55.18	17.39
C10H 6I206	475.96	475.8258	25.24	1.27	53.33	20.17
C10H 6I207	491.96	491.8207	24.41	1.23	51.59	22.77
C10H 6I208	507.96	507.8156	23.65	1.19	49.97	25.20
C10H 6I209	523.96	523.8105	22.92	1.15	48.44	27.48
C10H 7I101	270.07	269.9543	44.47	2.61	46.99	5.92
C10H 7I102	286.07	285.9493	41.99	2.47	44.36	11.19
C10H 7I103	302.07	301.9442	39.76	2.34	42.01	15.89
C10H 7I104	318.07	317.9391	37.76	2.22	39.90	20.12
C10H 7I105	334.07	333.9340	35.95	2.11	37.99	23.95
C10H 7I106	350.07	349.9289	34.31	2.02	36.25	27.42
C10H 7I107	366.07	365.9238	32.81	1.93	34.67	30.59
C10H 7I108	382.07	381.9187	31.44	1.85	33.22	33.50
C10H 7I109	398.07	397.9137	30.17	1.77	31.88	36.17
C10H 8I201	397.98	397.8668	30.18	2.03	63.77	4.02

C10H 8I202	413.98	413.8617	29.01	1.95	61.31	7.73
C10H 8I203	429.98	429.8567	27.93	1.88	59.03	11.16
C10H 8I204	445.98	445.8516	26.93	1.81	56.91	14.35
C10H 8I205	461.98	461.8465	26.00	1.75	54.94	17.32
C10H 8I206	477.98	477.8414	25.13	1.69	53.10	20.08
C10H 8I207	493.98	493.8363	24.32	1.63	51.38	22.67
C10H 8I208	509.98	509.8312	23.55	1.58	49.77	25.10
C10H 8I209	525.98	525.8261	22.84	1.53	48.25	27.38
C10H 9I101	272.09	271.9700	44.14	3.33	46.64	5.88
C10H 9I102	288.09	287.9649	41.69	3.15	44.05	11.11
C10H 9I103	304.09	303.9598	39.50	2.98	41.73	15.78
C10H 9I104	320.09	319.9547	37.52	2.83	39.65	19.99
C10H 9I105	336.08	335.9497	35.74	2.70	37.76	23.80
C10H 9I106	352.08	351.9446	34.11	2.58	36.04	27.27
C10H 9I107	368.08	367.9395	32.63	2.46	34.48	30.43
C10H 9I108	384.08	383.9344	31.27	2.36	33.04	33.32
C10H 9I109	400.08	399.9293	30.02	2.27	31.72	35.99
C10H10I201	400.00	399.8825	30.03	2.52	63.45	4.00
C10H10I202	416.00	415.8774	28.87	2.42	61.01	7.69
C10H10I203	432.00	431.8723	27.80	2.33	58.75	11.11
C10H10I204	448.00	447.8672	26.81	2.25	56.65	14.29
C10H10I205	464.00	463.8621	25.89	2.17	54.70	17.24
C10H10I206	480.00	479.8571	25.02	2.10	52.88	20.00
C10H10I207	496.00	495.8520	24.22	2.03	51.17	22.58
C10H10I208	512.00	511.8469	23.46	1.97	49.57	25.00
C10H10I209	527.99	527.8418	22.75	1.91	48.07	27.27
C10H11I101	274.10	273.9856	43.82	4.05	46.30	5.84
C10H11I102	290.10	289.9806	41.40	3.82	43.74	11.03
C10H11I103	306.10	305.9755	39.24	3.62	41.46	15.68
C10H11I104	322.10	321.9704	37.29	3.44	39.40	19.87
C10H11I105	338.10	337.9653	35.53	3.28	37.53	23.66
C10H11I106	354.10	353.9602	33.92	3.13	35.84	27.11
C10H11I107	370.10	369.9551	32.45	3.00	34.29	30.26
C10H11I108	386.10	385.9500	31.11	2.87	32.87	33.15
C10H11I109	402.10	401.9450	29.87	2.76	31.56	35.81
C10H12I201	402.02	401.8981	29.88	3.01	63.13	3.98
C10H12I202	418.01	417.8930	28.73	2.89	60.72	7.65
C10H12I203	434.01	433.8880	27.67	2.79	58.48	11.06
C10H12I204	450.01	449.8829	26.69	2.69	56.40	14.22
C10H12I205	466.01	465.8778	25.77	2.60	54.46	17.17
C10H12I206	482.01	481.8727	24.92	2.51	52.66	19.92
C10H12I207	498.01	497.8676	24.12	2.43	50.96	22.49
C10H12I208	514.01	513.8625	23.37	2.35	49.38	24.90
C10H12I209	530.01	529.8574	22.66	2.28	47.89	27.17
C10H13I101	276.12	276.0013	43.50	4.75	45.96	5.79
C10H13I102	292.12	291.9962	41.12	4.49	43.44	10.95
C10H13I103	308.12	307.9911	38.98	4.25	41.19	15.58
C10H13I104	324.12	323.9860	37.06	4.04	39.15	19.75
C10H13I105	340.12	339.9810	35.31	3.85	37.31	23.52
C10H13I106	356.12	355.9759	33.73	3.68	35.64	26.96
C10H13I107	372.12	371.9708	32.28	3.52	34.10	30.10
C10H13I108	388.11	387.9657	30.95	3.38	32.70	32.98
C10H13I109	404.11	403.9606	29.72	3.24	31.40	35.63
C10H14I201	404.03	403.9138	29.73	3.49	62.82	3.96
C10H14I202	420.03	419.9087	28.60	3.36	60.43	7.62
C10H14I203	436.03	435.9036	27.55	3.24	58.21	11.01
C10H14I204	452.03	451.8985	26.57	3.12	56.15	14.16
C10H14I205	468.03	467.8934	25.66	3.02	54.23	17.09
C10H14I206	484.03	483.8883	24.81	2.92	52.44	19.83
C10H14I207	500.03	499.8833	24.02	2.82	50.76	22.40
C10H14I208	516.03	515.8782	23.28	2.73	49.19	24.80
C10H14I209	532.03	531.8731	22.58	2.65	47.71	27.07
C10H15I101	278.13	278.0169	43.18	5.44	45.63	5.75
C10H15I102	294.13	294.0119	40.84	5.14	43.15	10.88
C10H15I103	310.13	310.0068	38.73	4.88	40.92	15.48
C10H15I104	326.13	326.0017	36.83	4.64	38.91	19.62
C10H15I105	342.13	341.9966	35.11	4.42	37.09	23.38
C10H15I106	358.13	357.9915	33.54	4.22	35.44	26.80
C10H15I107	374.13	373.9864	32.10	4.04	33.92	29.93
C10H15I108	390.13	389.9813	30.79	3.88	32.53	32.81
C10H15I109	406.13	405.9763	29.57	3.72	31.25	35.46
C10H16I201	406.05	405.9294	29.58	3.97	62.51	3.94
C10H16I202	422.05	421.9243	28.46	3.82	60.14	7.58
C10H16I203	438.05	437.9193	27.42	3.68	57.94	10.96
C10H16I204	454.05	453.9142	26.45	3.55	55.90	14.09
C10H16I205	470.04	469.9091	25.55	3.43	54.00	17.02
C10H16I206	486.04	485.9040	24.71	3.32	52.22	19.75
C10H16I207	502.04	501.8989	23.92	3.21	50.56	22.31
C10H16I208	518.04	517.8938	23.19	3.11	48.99	24.71
C10H16I209	534.04	533.8887	22.49	3.02	47.53	26.96
C10H17I101	280.15	280.0326	42.87	6.12	45.30	5.71
C10H17I102	296.15	296.0275	40.56	5.79	42.85	10.80
C10H17I103	312.15	312.0224	38.48	5.49	40.65	15.38
C10H17I104	328.15	328.0173	36.60	5.22	38.67	19.50
C10H17I105	344.15	344.0122	34.90	4.98	36.87	23.24
C10H17I106	360.15	360.0072	33.35	4.76	35.24	26.65
C10H17I107	376.15	376.0021	31.93	4.56	33.74	29.77
C10H17I108	392.15	391.9970	30.63	4.37	32.36	32.64
C10H17I109	408.15	407.9919	29.43	4.20	31.09	35.28
C10H18I201	408.06	407.9451	29.43	4.45	62.20	3.92
C10H18I202	424.06	423.9400	28.32	4.28	59.85	7.55
C10H18I203	440.06	439.9349	27.29	4.12	57.68	10.91
C10H18I204	456.06	455.9298	26.34	3.98	55.65	14.03
C10H18I205	472.06	471.9247	25.44	3.84	53.77	16.95
C10H18I206	488.06	487.9196	24.61	3.72	52.00	19.67
C10H18I207	504.06	503.9146	23.83	3.60	50.35	22.22
C10H18I208	520.06	519.9095	23.10	3.49	48.80	24.61
C10H18I209	536.06	535.9044	22.41	3.38	47.35	26.86
C10H19I101	282.17	282.0482	42.57	6.79	44.97	5.67
C10H19I102	298.17	298.0432	40.28	6.42	42.56	10.73
C10H19I103	314.17	314.0381	38.23	6.10	40.39	15.28
C10H19I104	330.16	330.0330	36.38	5.80	38.44	19.38
C10H19I105	346.16	346.0279	34.70	5.53	36.66	23.11
C10H19I106	362.16	362.0228	33.16	5.29	35.04	26.51
C10H19I107	378.16	378.0177	31.76	5.06	33.56	29.62
C10H19I108	394.16	394.0126	30.47	4.86	32.20	32.47
C10H19I109	410.16	410.0076	29.28	4.67	30.94	35.11
C10H20I201	410.08	409.9607	29.29	4.92	61.89	3.90
C10H20I202	426.08	425.9556	28.19	4.73	59.57	7.51
C10H20I203	442.08	441.9506	27.17	4.56	57.41	10.86
C10H20I204	458.08	457.9455	26.22	4.40	55.41	13.97
C10H20I205	474.08	473.9404	25.34	4.25	53.54	16.87
C10H20I206	490.08	489.9353	24.51	4.11	51.79	19.59
C10H20I207	506.08	505.9302	23.73	3.98	50.15	22.13
C10H20I208	522.07	521.9251	23.01	3.86	48.62	24.52
C10H20I209	538.07	537.9200	22.32	3.75	47.17	26.76
C10H21I101	284.18	284.0639	42.27	7.45	44.66	5.63
C10H21I102	300.18	300.0588	40.01	7.05	42.28	10.66
C10H21I103	316.18	316.0537	37.99	6.69	40.14	15.18
C10H21I104	332.18	332.0486	36.16	6.37	38.20	19.27

C10H21I105	348.18	348.0435	34.50	6.08	36.45	22.98
C10H21I106	364.18	364.0385	32.98	5.81	34.85	26.36
C10H21I107	380.18	380.0334	31.59	5.57	33.38	29.46
C10H21I108	396.18	396.0283	30.32	5.34	32.03	32.31
C10H21I109	412.18	412.0232	29.14	5.14	30.79	34.94
C11H 5I101	280.07	279.9387	47.18	1.80	45.31	5.71
C11H 5I102	296.07	295.9336	44.63	1.70	42.86	10.81
C11H 5I103	312.07	311.9285	42.34	1.61	40.67	15.38
C11H 5I104	328.06	327.9234	40.27	1.54	38.68	19.51
C11H 5I105	344.06	343.9184	38.40	1.46	36.88	23.25
C11H 5I106	360.06	359.9133	36.69	1.40	35.25	26.66
C11H 5I107	376.06	375.9082	35.13	1.34	33.75	29.78
C11H 5I108	392.06	391.9031	33.70	1.29	32.37	32.65
C11H 5I109	408.06	407.8980	32.38	1.24	31.10	35.29
C11H 6I201	407.98	407.8512	32.38	1.48	62.21	3.92
C11H 6I202	423.98	423.8461	31.16	1.43	59.86	7.55
C11H 6I203	439.98	439.8410	30.03	1.37	57.69	10.91
C11H 6I204	455.98	455.8359	28.98	1.33	55.66	14.04
C11H 6I205	471.98	471.8308	27.99	1.28	53.78	16.95
C11H 6I206	487.98	487.8258	27.08	1.24	52.01	19.67
C11H 6I207	503.98	503.8207	26.22	1.20	50.36	22.22
C11H 6I208	519.97	519.8156	25.41	1.16	48.81	24.62
C11H 6I209	535.97	535.8105	24.65	1.13	47.35	26.87
C11H 7I101	282.08	281.9543	46.84	2.50	44.99	5.67
C11H 7I102	298.08	297.9493	44.32	2.37	42.57	10.73
C11H 7I103	314.08	313.9442	42.07	2.25	40.40	15.28
C11H 7I104	330.08	329.9391	40.03	2.14	38.45	19.39
C11H 7I105	346.08	345.9340	38.18	2.04	36.67	23.12
C11H 7I106	362.08	361.9289	36.49	1.95	35.05	26.51
C11H 7I107	378.08	377.9238	34.95	1.87	33.57	29.62
C11H 7I108	394.08	393.9187	33.53	1.79	32.20	32.48
C11H 7I109	410.08	409.9137	32.22	1.72	30.95	35.11
C11H 8I201	409.99	409.8668	32.23	1.97	61.91	3.90
C11H 8I202	425.99	425.8617	31.02	1.89	59.58	7.51
C11H 8I203	441.99	441.8567	29.89	1.82	57.42	10.86
C11H 8I204	457.99	457.8516	28.85	1.76	55.42	13.97
C11H 8I205	473.99	473.8465	27.87	1.70	53.55	16.88
C11H 8I206	489.99	489.8414	26.96	1.65	51.80	19.59
C11H 8I207	505.99	505.8363	26.11	1.59	50.16	22.13
C11H 8I208	521.99	521.8312	25.31	1.54	48.62	24.52
C11H 8I209	537.99	537.8261	24.56	1.50	47.18	26.77
C11H 9I101	284.10	283.9700	46.51	3.19	44.67	5.63
C11H 9I102	300.10	299.9649	44.03	3.02	42.29	10.66
C11H 9I103	316.10	315.9598	41.80	2.87	40.15	15.18
C11H 9I104	332.10	331.9547	39.78	2.73	38.21	19.27
C11H 9I105	348.10	347.9497	37.96	2.61	36.46	22.98
C11H 9I106	364.10	363.9446	36.29	2.49	34.85	26.37
C11H 9I107	380.09	379.9395	34.76	2.39	33.39	29.47
C11H 9I108	396.09	395.9344	33.36	2.29	32.04	32.31
C11H 9I109	412.09	411.9293	32.06	2.20	30.80	34.94
C11H10I201	412.01	411.8825	32.07	2.45	61.60	3.88
C11H10I202	428.01	427.8774	30.87	2.36	59.30	7.48
C11H10I203	444.01	443.8723	29.76	2.27	57.16	10.81
C11H10I204	460.01	459.8672	28.72	2.19	55.17	13.91
C11H10I205	476.01	475.8621	27.76	2.12	53.32	16.81
C11H10I206	492.01	491.8571	26.85	2.05	51.59	19.51
C11H10I207	508.01	507.8520	26.01	1.98	49.96	22.05
C11H10I208	524.01	523.8469	25.21	1.92	48.44	24.43
C11H10I209	540.01	539.8418	24.47	1.87	47.00	26.67
C11H11I101	286.11	285.9856	46.18	3.88	44.35	5.59
C11H11I102	302.11	301.9806	43.73	3.67	42.01	10.59
C11H11I103	318.11	317.9755	41.53	3.49	39.89	15.09
C11H11I104	334.11	333.9704	39.54	3.32	37.98	19.15
C11H11I105	350.11	349.9653	37.74	3.17	36.25	22.85
C11H11I106	366.11	365.9602	36.09	3.03	34.66	26.22
C11H11I107	382.11	381.9551	34.58	2.90	33.21	29.31
C11H11I108	398.11	397.9500	33.19	2.79	31.88	32.15
C11H11I109	414.11	413.9450	31.91	2.68	30.65	34.77
C11H12I201	414.03	413.8981	31.91	2.92	61.30	3.86
C11H12I202	430.03	429.8930	30.72	2.81	59.02	7.44
C11H12I203	446.03	445.8880	29.62	2.71	56.90	10.76
C11H12I204	462.02	461.8829	28.60	2.62	54.93	13.85
C11H12I205	478.02	477.8778	27.64	2.53	53.10	16.73
C11H12I206	494.02	493.8727	26.74	2.45	51.38	19.43
C11H12I207	510.02	509.8676	25.91	2.37	49.76	21.96
C11H12I208	526.02	525.8625	25.12	2.30	48.25	24.33
C11H12I209	542.02	541.8574	24.38	2.23	46.83	26.57
C11H13I101	288.13	288.0013	45.86	4.55	44.04	5.55
C11H13I102	304.13	303.9962	43.44	4.31	41.73	10.52
C11H13I103	320.13	319.9911	41.27	4.09	39.64	14.99
C11H13I104	336.13	335.9860	39.31	3.90	37.75	19.04
C11H13I105	352.13	351.9810	37.52	3.72	36.04	22.72
C11H13I106	368.13	367.9759	35.89	3.56	34.47	26.08
C11H13I107	384.13	383.9708	34.40	3.41	33.04	29.16
C11H13I108	400.13	399.9657	33.02	3.27	31.72	31.99
C11H13I109	416.13	415.9606	31.75	3.15	30.50	34.60
C11H14I201	416.04	415.9138	31.76	3.39	61.01	3.85
C11H14I202	432.04	431.9087	30.58	3.27	58.75	7.41
C11H14I203	448.04	447.9036	29.49	3.15	56.65	10.71
C11H14I204	464.04	463.8985	28.47	3.04	54.70	13.79
C11H14I205	480.04	479.8934	27.52	2.94	52.87	16.66
C11H14I206	496.04	495.8883	26.64	2.84	51.17	19.35
C11H14I207	512.04	511.8833	25.80	2.76	49.57	21.87
C11H14I208	528.04	527.8782	25.02	2.67	48.07	24.24
C11H14I209	544.04	543.8731	24.29	2.59	46.65	26.47
C11H15I101	290.15	290.0169	45.54	5.21	43.74	5.51
C11H15I102	306.15	306.0119	43.16	4.94	41.45	10.45
C11H15I103	322.14	322.0068	41.01	4.69	39.39	14.90
C11H15I104	338.14	338.0017	39.07	4.47	37.53	18.93
C11H15I105	354.14	353.9966	37.31	4.27	35.83	22.59
C11H15I106	370.14	369.9915	35.70	4.08	34.29	25.93
C11H15I107	386.14	385.9864	34.22	3.92	32.86	29.00
C11H15I108	402.14	401.9813	32.85	3.76	31.56	31.83
C11H15I109	418.14	417.9763	31.60	3.62	30.35	34.44
C11H16I201	418.06	417.9294	31.60	3.86	60.71	3.83
C11H16I202	434.06	433.9243	30.44	3.72	58.47	7.37
C11H16I203	450.06	449.9193	29.36	3.58	56.39	10.66
C11H16I204	466.06	465.9142	28.35	3.46	54.46	13.73
C11H16I205	482.06	481.9091	27.41	3.35	52.65	16.59
C11H16I206	498.06	497.9040	26.53	3.24	50.96	19.27
C11H16I207	514.05	513.8989	25.70	3.14	49.37	21.79
C11H16I208	530.05	529.8938	24.93	3.04	47.88	24.15
C11H16I209	546.05	545.8887	24.20	2.95	46.48	26.37
C11H17I101	292.16	292.0326	45.22	5.87	43.44	5.48
C11H17I102	308.16	308.0275	42.87	5.56	41.18	10.38
C11H17I103	324.16	324.0224	40.76	5.29	39.15	14.81
C11H17I104	340.16	340.0173	38.84	5.04	37.31	18.81
C11H17I105	356.16	356.0122	37.10	4.81	35.63	22.46
C11H17I106	372.16	372.0072	35.50	4.60	34.10	25.79
C11H17I107	388.16	388.0021	34.04	4.41	32.69	28.85

19

C11H17I108	404.16	403.9970	32.69	4.24	31.40	31.67
C11H17I109	420.16	419.9919	31.45	4.08	30.20	34.27
C11H18I201	420.07	419.9451	31.45	4.32	60.42	3.81
C11H18I202	436.07	435.9400	30.30	4.16	58.20	7.34
C11H18I203	452.07	451.9349	29.23	4.01	56.14	10.62
C11H18I204	468.07	467.9298	28.23	3.88	54.22	13.67
C11H18I205	484.07	483.9247	27.29	3.75	52.43	16.53
C11H18I206	500.07	499.9196	26.42	3.63	50.75	19.20
C11H18I207	516.07	515.9146	25.60	3.52	49.18	21.70
C11H18I208	532.07	531.9095	24.83	3.41	47.70	24.06
C11H18I209	548.07	547.9044	24.11	3.31	46.31	26.27
C11H19I101	294.18	294.0482	44.91	6.51	43.14	5.44
C11H19I102	310.18	310.0432	42.60	6.17	40.91	10.32
C11H19I103	326.18	326.0381	40.51	5.87	38.91	14.72
C11H19I104	342.18	342.0330	38.61	5.60	37.09	18.70
C11H19I105	358.18	358.0279	36.89	5.35	35.43	22.33
C11H19I106	374.17	374.0228	35.31	5.12	33.92	25.66
C11H19I107	390.17	390.0177	33.86	4.91	32.53	28.70
C11H19I108	406.17	406.0126	32.53	4.72	31.24	31.51
C11H19I109	422.17	422.0076	31.30	4.54	30.06	34.11
C11H20I201	422.09	421.9607	31.30	4.78	60.13	3.79
C11H20I202	438.09	437.9556	30.16	4.60	57.94	7.30
C11H20I203	454.09	453.9506	29.10	4.44	55.89	10.57
C11H20I204	470.09	469.9455	28.11	4.29	53.99	13.61
C11H20I205	486.09	485.9404	27.18	4.15	52.21	16.46
C11H20I206	502.09	501.9353	26.31	4.02	50.55	19.12
C11H20I207	518.09	517.9302	25.50	3.89	48.99	21.62
C11H20I208	534.09	533.9251	24.74	3.77	47.52	23.97
C11H20I209	550.09	549.9200	24.02	3.66	46.14	26.18
C11H21I101	296.19	296.0639	44.61	7.15	42.85	5.40
C11H21I102	312.19	312.0588	42.32	6.78	40.65	10.25
C11H21I103	328.19	328.0537	40.26	6.45	38.67	14.63
C11H21I104	344.19	344.0486	38.39	6.15	36.87	18.59
C11H21I105	360.19	360.0435	36.68	5.88	35.23	22.21
C11H21I106	376.19	376.0385	35.12	5.63	33.73	25.52
C11H21I107	392.19	392.0334	33.69	5.40	32.36	28.56
C11H21I108	408.19	408.0283	32.37	5.19	31.09	31.36
C11H21I109	424.19	424.0232	31.15	4.99	29.92	33.95
C11H22I201	424.11	423.9764	31.15	5.23	59.85	3.77
C11H22I202	440.11	439.9713	30.02	5.04	57.67	7.27
C11H22I203	456.10	455.9662	28.97	4.86	55.65	10.52
C11H22I204	472.10	471.9611	27.99	4.70	53.76	13.56
C11H22I205	488.10	487.9560	27.07	4.54	52.00	16.39
C11H22I206	504.10	503.9509	26.21	4.40	50.35	19.04
C11H22I207	520.10	519.9459	25.40	4.26	48.80	21.53
C11H22I208	536.10	535.9408	24.65	4.14	47.34	23.88
C11H22I209	552.10	551.9357	23.93	4.02	45.97	26.08
C11H23I101	298.21	298.0795	44.31	7.77	42.56	5.37
C11H23I102	314.21	314.0745	42.05	7.38	40.39	10.18
C11H23I103	330.21	330.0694	40.01	7.02	38.43	14.54
C11H23I104	346.21	346.0643	38.16	6.70	36.66	18.49
C11H23I105	362.21	362.0592	36.48	6.40	35.04	22.09
C11H23I106	378.21	378.0541	34.93	6.13	33.55	25.38
C11H23I107	394.21	394.0490	33.52	5.88	32.19	28.41
C11H23I108	410.21	410.0439	32.21	5.65	30.94	31.20
C11H23I109	426.20	426.0389	31.00	5.44	29.78	33.79
C12H 6I201	419.99	419.8512	34.32	1.44	60.43	3.81
C12H 6I202	435.99	435.8461	33.06	1.39	58.21	7.34
C12H 6I203	451.99	451.8410	31.89	1.34	56.15	10.62
C12H 6I204	467.99	467.8359	30.80	1.29	54.23	13.68
C12H 6I205	483.99	483.8308	29.78	1.25	52.44	16.53
C12H 6I206	499.99	499.8258	28.83	1.21	50.76	19.20
C12H 6I207	515.99	515.8207	27.93	1.17	49.19	21.71
C12H 6I208	531.99	531.8156	27.09	1.14	47.71	24.06
C12H 6I209	547.99	547.8105	26.30	1.10	46.32	26.28
C12H 7I101	294.09	293.9543	49.01	2.40	43.15	5.44
C12H 7I102	310.09	309.9493	46.48	2.28	40.92	10.32
C12H 7I103	326.09	325.9442	44.20	2.16	38.92	14.72
C12H 7I104	342.09	341.9391	42.13	2.06	37.10	18.71
C12H 7I105	358.09	357.9340	40.25	1.97	35.44	22.34
C12H 7I106	374.09	373.9289	38.53	1.89	33.92	25.66
C12H 7I107	390.09	389.9238	36.95	1.81	32.53	28.71
C12H 7I108	406.09	405.9187	35.49	1.74	31.25	31.52
C12H 7I109	422.09	421.9137	34.15	1.67	30.07	34.11
C12H 8I201	422.01	421.8668	34.15	1.91	60.14	3.79
C12H 8I202	438.01	437.8617	32.91	1.84	57.95	7.31
C12H 8I203	454.00	453.8567	31.75	1.78	55.90	10.57
C12H 8I204	470.00	469.8516	30.67	1.72	54.00	13.62
C12H 8I205	486.00	485.8465	29.66	1.66	52.22	16.46
C12H 8I206	502.00	501.8414	28.71	1.61	50.56	19.12
C12H 8I207	518.00	517.8363	27.82	1.56	49.00	21.62
C12H 8I208	534.00	533.8312	26.99	1.51	47.53	23.97
C12H 8I209	550.00	549.8261	26.21	1.47	46.15	26.18
C12H 9I101	296.11	295.9700	48.68	3.06	42.86	5.40
C12H 9I102	312.11	311.9649	46.18	2.91	40.66	10.25
C12H 9I103	328.11	327.9598	43.93	2.76	38.68	14.63
C12H 9I104	344.11	343.9547	41.89	2.64	36.88	18.60
C12H 9I105	360.11	359.9497	40.03	2.52	35.24	22.21
C12H 9I106	376.11	375.9446	38.32	2.41	33.74	25.52
C12H 9I107	392.11	391.9395	36.76	2.31	32.36	28.56
C12H 9I108	408.11	407.9344	35.32	2.22	31.10	31.36
C12H 9I109	424.10	423.9293	33.99	2.14	29.92	33.95
C12H10I201	424.02	423.8825	33.99	2.38	59.86	3.77
C12H10I202	440.02	439.8774	32.76	2.29	57.68	7.27
C12H10I203	456.02	455.8723	31.61	2.21	55.66	10.53
C12H10I204	472.02	471.8672	30.54	2.14	53.77	13.56
C12H10I205	488.02	487.8621	29.53	2.07	52.01	16.39
C12H10I206	504.02	503.8571	28.60	2.00	50.36	19.05
C12H10I207	520.02	519.8520	27.72	1.94	48.81	21.54
C12H10I208	536.02	535.8469	26.89	1.88	47.35	23.88
C12H10I209	552.02	551.8418	26.11	1.83	45.98	26.09
C12H11I101	298.13	297.9856	48.35	3.72	42.57	5.37
C12H11I102	314.12	313.9806	45.88	3.53	40.40	10.19
C12H11I103	330.12	329.9755	43.66	3.36	38.44	14.54
C12H11I104	346.12	345.9704	41.64	3.20	36.66	18.49
C12H11I105	362.12	361.9653	39.80	3.06	35.04	22.09
C12H11I106	378.12	377.9602	38.12	2.93	33.56	25.39
C12H11I107	394.12	393.9551	36.57	2.81	32.20	28.42
C12H11I108	410.12	409.9500	35.14	2.70	30.94	31.21
C12H11I109	426.12	425.9450	33.82	2.60	29.78	33.79
C12H12I201	426.04	425.8981	33.83	2.84	59.57	3.76
C12H12I202	442.04	441.8930	32.61	2.74	57.42	7.24
C12H12I203	458.04	457.8880	31.47	2.64	55.41	10.48
C12H12I204	474.04	473.8829	30.41	2.55	53.54	13.50
C12H12I205	490.04	489.8778	29.41	2.47	51.79	16.32
C12H12I206	506.03	505.8727	28.48	2.39	50.16	18.97
C12H12I207	522.03	521.8676	27.61	2.32	48.62	21.45
C12H12I208	538.03	537.8625	26.79	2.25	47.17	23.79
C12H12I209	554.03	553.8574	26.02	2.18	45.81	25.99
C12H13I101	300.14	300.0013	48.02	4.37	42.28	5.33

193

C12H13I102	316.14	315.9962	45.59	4.14	40.14	10.12
C12H13I103	332.14	331.9911	43.40	3.95	38.21	14.45
C12H13I104	348.14	347.9860	41.40	3.76	36.45	18.38
C12H13I105	364.14	363.9810	39.58	3.60	34.85	21.97
C12H13I106	380.14	379.9759	37.92	3.45	33.38	25.25
C12H13I107	396.14	395.9708	36.38	3.31	32.04	28.27
C12H13I108	412.14	411.9657	34.97	3.18	30.79	31.06
C12H13I109	428.14	427.9606	33.67	3.06	29.64	33.63
C12H14I201	428.05	427.9138	33.67	3.30	59.29	3.74
C12H14I202	444.05	443.9087	32.46	3.18	57.16	7.21
C12H14I203	460.05	459.9036	31.33	3.07	55.17	10.43
C12H14I204	476.05	475.8985	30.28	2.96	53.32	13.44
C12H14I205	492.05	491.8934	29.29	2.87	51.58	16.26
C12H14I206	508.05	507.8883	28.37	2.78	49.96	18.90
C12H14I207	524.05	523.8833	27.50	2.69	48.43	21.37
C12H14I208	540.05	539.8782	26.69	2.61	47.00	23.70
C12H14I209	556.05	555.8731	25.92	2.54	45.65	25.90
C12H15I101	302.16	302.0169	47.70	5.00	42.00	5.30
C12H15I102	318.16	318.0119	45.30	4.75	39.89	10.06
C12H15I103	334.16	334.0068	43.13	4.52	37.98	14.36
C12H15I104	350.16	350.0017	41.16	4.32	36.24	18.28
C12H15I105	366.15	365.9966	39.36	4.13	34.66	21.85
C12H15I106	382.15	381.9915	37.72	3.96	33.21	25.12
C12H15I107	398.15	397.9864	36.20	3.80	31.87	28.13
C12H15I108	414.15	413.9813	34.80	3.65	30.64	30.91
C12H15I109	430.15	429.9763	33.51	3.51	29.50	33.48
C12H16I201	430.07	429.9294	33.51	3.75	59.02	3.72
C12H16I202	446.07	445.9243	32.31	3.62	56.90	7.17
C12H16I203	462.07	461.9193	31.19	3.49	54.93	10.39
C12H16I204	478.07	477.9142	30.15	3.37	53.09	13.39
C12H16I205	494.07	493.9091	29.17	3.26	51.37	16.19
C12H16I206	510.07	509.9040	28.26	3.16	49.76	18.82
C12H16I207	526.07	525.8989	27.40	3.07	48.25	21.29
C12H16I208	542.07	541.8938	26.59	2.98	46.82	23.61
C12H16I209	558.06	557.8887	25.83	2.89	45.48	25.80
C12H17I101	304.17	304.0326	47.39	5.63	41.72	5.26
C12H17I102	320.17	320.0275	45.02	5.35	39.64	9.99
C12H17I103	336.17	336.0224	42.88	5.10	37.75	14.28
C12H17I104	352.17	352.0173	40.93	4.87	36.03	18.17
C12H17I105	368.17	368.0122	39.15	4.65	34.47	21.73
C12H17I106	384.17	384.0072	37.52	4.46	33.03	24.99
C12H17I107	400.17	400.0021	36.02	4.28	31.71	27.99
C12H17I108	416.17	415.9970	34.63	4.12	30.49	30.76
C12H17I109	432.17	431.9919	33.35	3.97	29.36	33.32
C12H18I201	432.09	431.9451	33.36	4.20	58.74	3.70
C12H18I202	448.08	447.9400	32.17	4.05	56.64	7.14
C12H18I203	464.08	463.9349	31.06	3.91	54.69	10.34
C12H18I204	480.08	479.9298	30.02	3.78	52.87	13.33
C12H18I205	496.08	495.9247	29.05	3.66	51.16	16.13
C12H18I206	512.08	511.9196	28.15	3.54	49.56	18.75
C12H18I207	528.08	527.9146	27.29	3.44	48.06	21.21
C12H18I208	544.08	543.9095	26.49	3.33	46.65	23.53
C12H18I209	560.08	559.9044	25.73	3.24	45.32	25.71
C12H19I101	306.19	306.0482	47.07	6.25	41.45	5.23
C12H19I102	322.19	322.0432	44.74	5.94	39.39	9.93
C12H19I103	338.19	338.0381	42.62	5.66	37.52	14.19
C12H19I104	354.19	354.0330	40.69	5.41	35.83	18.07
C12H19I105	370.19	370.0279	38.94	5.17	34.28	21.61
C12H19I106	386.19	386.0228	37.32	4.96	32.86	24.86
C12H19I107	402.19	402.0177	35.84	4.76	31.55	27.85

C12H19I108	418.18	418.0126	34.47	4.58	30.35	30.61
C12H19I109	434.18	434.0076	33.20	4.41	29.23	33.16
C12H20I201	434.10	433.9607	33.20	4.64	58.47	3.69
C12H20I202	450.10	449.9556	32.02	4.48	56.39	7.11
C12H20I203	466.10	465.9506	30.92	4.33	54.45	10.30
C12H20I204	482.10	481.9455	29.90	4.18	52.65	13.27
C12H20I205	498.10	497.9404	28.94	4.05	50.96	16.06
C12H20I206	514.10	513.9353	28.04	3.92	49.37	18.67
C12H20I207	530.10	529.9302	27.19	3.80	47.88	21.13
C12H20I208	546.10	545.9251	26.39	3.69	46.48	23.44
C12H20I209	562.10	561.9200	25.64	3.59	45.15	25.62
C12H21I101	308.20	308.0639	46.77	6.87	41.18	5.19
C12H21I102	324.20	324.0588	44.46	6.53	39.14	9.87
C12H21I103	340.20	340.0537	42.37	6.22	37.30	14.11
C12H21I104	356.20	356.0486	40.46	5.94	35.63	17.97
C12H21I105	372.20	372.0435	38.72	5.69	34.10	21.49
C12H21I106	388.20	388.0385	37.13	5.45	32.69	24.73
C12H21I107	404.20	404.0334	35.66	5.24	31.40	27.71
C12H21I108	420.20	420.0283	34.30	5.04	30.20	30.46
C12H21I109	436.20	436.0232	33.04	4.85	29.09	33.01
C12H22I201	436.12	435.9764	33.05	5.08	58.20	3.67
C12H22I202	452.12	451.9713	31.88	4.90	56.14	7.08
C12H22I203	468.12	467.9662	30.79	4.74	54.22	10.25
C12H22I204	484.12	483.9611	29.77	4.58	52.43	13.22
C12H22I205	500.11	499.9560	28.82	4.43	50.75	16.00
C12H22I206	516.11	515.9509	27.93	4.30	49.18	18.60
C12H22I207	532.11	531.9459	27.09	4.17	47.70	21.05
C12H22I208	548.11	547.9408	26.30	4.05	46.31	23.35
C12H22I209	564.11	563.9357	25.55	3.93	44.99	25.53
C12H23I101	310.22	310.0795	46.46	7.47	40.91	5.16
C12H23I102	326.22	326.0745	44.18	7.11	38.90	9.81
C12H23I103	342.22	342.0694	42.12	6.77	37.08	14.03
C12H23I104	358.22	358.0643	40.24	6.47	35.43	17.87
C12H23I105	374.22	374.0592	38.52	6.20	33.91	21.38
C12H23I106	390.22	390.0541	36.94	5.94	32.52	24.60
C12H23I107	406.22	406.0490	35.48	5.71	31.24	27.57
C12H23I108	422.22	422.0439	34.14	5.49	30.06	30.32
C12H23I109	438.22	438.0389	32.89	5.29	28.96	32.86
C12H24I201	438.13	437.9920	32.90	5.52	57.93	3.65
C12H24I202	454.13	453.9869	31.74	5.33	55.89	7.05
C12H24I203	470.13	469.9819	30.66	5.15	53.99	10.21
C12H24I204	486.13	485.9768	29.65	4.98	52.21	13.16
C12H24I205	502.13	501.9717	28.70	4.82	50.55	15.93
C12H24I206	518.13	517.9666	27.82	4.67	48.99	18.53
C12H24I207	534.13	533.9615	26.98	4.53	47.52	20.97
C12H24I208	550.13	549.9564	26.20	4.40	46.14	23.27
C12H24I209	566.13	565.9513	25.46	4.27	44.83	25.43
C12H25I101	312.24	312.0952	46.16	8.07	40.64	5.12
C12H25I102	328.24	328.0901	43.91	7.68	38.66	9.75
C12H25I103	344.24	344.0850	41.87	7.32	36.87	13.94
C12H25I104	360.24	360.0799	40.01	7.00	35.23	17.77
C12H25I105	376.23	376.0748	38.31	6.70	33.73	21.26
C12H25I106	392.23	392.0698	36.75	6.42	32.35	24.47
C12H25I107	408.23	408.0647	35.31	6.17	31.09	27.43
C12H25I108	424.23	424.0596	33.98	5.94	29.91	30.17
C12H25I109	440.23	440.0545	32.74	5.72	28.83	32.71
C13H 6I201	432.00	431.8512	36.14	1.40	58.75	3.70
C13H 6I202	448.00	447.8461	34.85	1.35	56.65	7.14
C13H 6I203	464.00	463.8410	33.65	1.30	54.70	10.34
C13H 6I204	480.00	479.8359	32.53	1.26	52.88	13.33

194

C13H 6I205	496.00	495.8308	31.48	1.22	51.17	16.13
C13H 6I206	512.00	511.8258	30.50	1.18	49.57	18.75
C13H 6I207	528.00	527.8207	29.57	1.15	48.07	21.21
C13H 6I208	544.00	543.8156	28.70	1.11	46.66	23.53
C13H 6I209	560.00	559.8105	27.88	1.08	45.32	25.71
C13H 7I101	306.10	305.9543	51.01	2.31	41.46	5.23
C13H 7I102	322.10	321.9493	48.48	2.19	39.40	9.93
C13H 7I103	338.10	337.9442	46.18	2.09	37.53	14.20
C13H 7I104	354.10	353.9391	44.10	1.99	35.84	18.07
C13H 7I105	370.10	369.9340	42.19	1.91	34.29	21.61
C13H 7I106	386.10	385.9289	40.44	1.83	32.87	24.86
C13H 7I107	402.10	401.9238	38.83	1.75	31.56	27.85
C13H 7I108	418.10	417.9187	37.35	1.69	30.35	30.61
C13H 7I109	434.10	433.9137	35.97	1.63	29.23	33.17
C13H 8I201	434.02	433.8668	35.98	1.86	58.48	3.69
C13H 8I202	450.02	449.8617	34.70	1.79	56.40	7.11
C13H 8I203	466.02	465.8567	33.51	1.73	54.46	10.30
C13H 8I204	482.02	481.8516	32.39	1.67	52.66	13.28
C13H 8I205	498.01	497.8465	31.35	1.62	50.96	16.06
C13H 8I206	514.01	513.8414	30.38	1.57	49.38	18.68
C13H 8I207	530.01	529.8363	29.46	1.52	47.89	21.13
C13H 8I208	546.01	545.8312	28.60	1.48	46.48	23.44
C13H 8I209	562.01	561.8261	27.78	1.43	45.16	25.62
C13H 9I101	308.12	307.9700	50.68	2.94	41.19	5.19
C13H 9I102	324.12	323.9649	48.18	2.80	39.15	9.87
C13H 9I103	340.12	339.9598	45.91	2.67	37.31	14.11
C13H 9I104	356.12	355.9547	43.85	2.55	35.64	17.97
C13H 9I105	372.12	371.9497	41.96	2.44	34.10	21.50
C13H 9I106	388.12	387.9446	40.23	2.34	32.70	24.73
C13H 9I107	404.12	403.9395	38.64	2.24	31.40	27.71
C13H 9I108	420.12	419.9344	37.17	2.16	30.21	30.47
C13H 9I109	436.12	435.9293	35.80	2.08	29.10	33.02
C13H10I201	436.03	435.8825	35.81	2.31	58.21	3.67
C13H10I202	452.03	451.8774	34.54	2.23	56.15	7.08
C13H10I203	468.03	467.8723	33.36	2.15	54.23	10.26
C13H10I204	484.03	483.8672	32.26	2.08	52.44	13.22
C13H10I205	500.03	499.8621	31.23	2.02	50.76	16.00
C13H10I206	516.03	515.8571	30.26	1.95	49.18	18.60
C13H10I207	532.03	531.8520	29.35	1.89	47.71	21.05
C13H10I208	548.03	547.8469	28.49	1.84	46.31	23.36
C13H10I209	564.03	563.8418	27.68	1.79	45.00	25.53
C13H11I101	310.14	309.9856	50.35	3.58	40.92	5.16
C13H11I102	326.14	325.9806	47.88	3.40	38.91	9.81
C13H11I103	342.14	341.9755	45.64	3.24	37.09	14.03
C13H11I104	358.13	357.9704	43.60	3.10	35.43	17.87
C13H11I105	374.13	373.9653	41.74	2.96	33.92	21.38
C13H11I106	390.13	389.9602	40.02	2.84	32.53	24.61
C13H11I107	406.13	405.9551	38.45	2.73	31.25	27.58
C13H11I108	422.13	421.9500	36.99	2.63	30.06	30.32
C13H11I109	438.13	437.9450	35.64	2.53	28.96	32.87
C13H12I201	438.05	437.8981	35.65	2.76	57.94	3.65
C13H12I202	454.05	453.8930	34.39	2.66	55.90	7.05
C13H12I203	470.05	469.8880	33.22	2.57	54.00	10.21
C13H12I204	486.05	485.8829	32.13	2.49	52.22	13.17
C13H12I205	502.05	501.8778	31.10	2.41	50.55	15.93
C13H12I206	518.05	517.8727	30.14	2.33	48.99	18.53
C13H12I207	534.05	533.8676	29.24	2.26	47.53	20.97
C13H12I208	550.04	549.8625	28.39	2.20	46.14	23.27
C13H12I209	566.04	565.8574	27.59	2.14	44.84	25.44
C13H13I101	312.15	312.0013	50.02	4.20	40.65	5.13

C13H13I102	328.15	327.9962	47.58	3.99	38.67	9.75
C13H13I103	344.15	343.9911	45.37	3.81	36.87	13.95
C13H13I104	360.15	359.9860	43.36	3.64	35.24	17.77
C13H13I105	376.15	375.9810	41.51	3.48	33.74	21.27
C13H13I106	392.15	391.9759	39.82	3.34	32.36	24.48
C13H13I107	408.15	407.9708	38.26	3.21	31.09	27.44
C13H13I108	424.15	423.9657	36.81	3.09	29.92	30.18
C13H13I109	440.15	439.9606	35.48	2.98	28.83	32.72
C13H14I201	440.06	439.9138	35.48	3.21	57.68	3.64
C13H14I202	456.06	455.9087	34.24	3.09	55.65	7.02
C13H14I203	472.06	471.9036	33.08	2.99	53.77	10.17
C13H14I204	488.06	487.8985	31.99	2.89	52.00	13.11
C13H14I205	504.06	503.8934	30.98	2.80	50.35	15.87
C13H14I206	520.06	519.8883	30.02	2.71	48.80	18.46
C13H14I207	536.06	535.8833	29.13	2.63	47.35	20.89
C13H14I208	552.06	551.8782	28.28	2.56	45.97	23.18
C13H14I209	568.06	567.8731	27.49	2.48	44.68	25.35
C13H15I101	314.17	314.0169	49.70	4.81	40.39	5.09
C13H15I102	330.17	330.0119	47.29	4.58	38.44	9.69
C13H15I103	346.17	346.0068	45.11	4.37	36.66	13.87
C13H15I104	362.17	362.0017	43.11	4.17	35.04	17.67
C13H15I105	378.17	377.9966	41.29	4.00	33.56	21.15
C13H15I106	394.17	393.9915	39.61	3.84	32.20	24.35
C13H15I107	410.16	409.9864	38.07	3.69	30.94	27.31
C13H15I108	426.16	425.9813	36.64	3.55	29.78	30.03
C13H15I109	442.16	441.9763	35.31	3.42	28.70	32.57
C13H16I201	442.08	441.9294	35.32	3.65	57.41	3.62
C13H16I202	458.08	457.9243	34.09	3.52	55.41	6.99
C13H16I203	474.08	473.9193	32.94	3.40	53.54	10.12
C13H16I204	490.08	489.9142	31.86	3.29	51.79	13.06
C13H16I205	506.08	505.9091	30.85	3.19	50.15	15.81
C13H16I206	522.08	521.9040	29.91	3.09	48.62	18.39
C13H16I207	538.08	537.8989	29.02	3.00	47.17	20.81
C13H16I208	554.08	553.8938	28.18	2.91	45.81	23.10
C13H16I209	570.08	569.8887	27.39	2.83	44.52	25.26
C13H17I101	316.18	316.0326	49.38	5.42	40.14	5.06
C13H17I102	332.18	332.0275	47.01	5.16	38.20	9.63
C13H17I103	348.18	348.0224	44.85	4.92	36.45	13.79
C13H17I104	364.18	364.0173	42.88	4.71	34.85	17.57
C13H17I105	380.18	380.0122	41.07	4.51	33.38	21.04
C13H17I106	396.18	396.0072	39.41	4.33	32.03	24.23
C13H17I107	412.18	412.0021	37.88	4.16	30.79	27.17
C13H17I108	428.18	427.9970	36.47	4.00	29.64	29.89
C13H17I109	444.18	443.9919	35.15	3.86	28.57	32.42
C13H18I201	444.10	443.9451	35.16	4.09	57.15	3.60
C13H18I202	460.10	459.9400	33.94	3.94	55.16	6.95
C13H18I203	476.10	475.9349	32.80	3.81	53.31	10.08
C13H18I204	492.09	491.9298	31.73	3.69	51.58	13.01
C13H18I205	508.09	507.9247	30.73	3.57	49.95	15.74
C13H18I206	524.09	523.9196	29.79	3.46	48.43	18.32
C13H18I207	540.09	539.9146	28.91	3.36	46.99	20.74
C13H18I208	556.09	555.9095	28.08	3.26	45.64	23.02
C13H18I209	572.09	571.9044	27.29	3.17	44.37	25.17
C13H19I101	318.20	318.0482	49.07	6.02	39.88	5.03
C13H19I102	334.20	334.0432	46.72	5.73	37.97	9.57
C13H19I103	350.20	350.0381	44.59	5.47	36.24	13.71
C13H19I104	366.20	366.0330	42.64	5.23	34.65	17.48
C13H19I105	382.20	382.0279	40.85	5.01	33.20	20.93
C13H19I106	398.20	398.0228	39.21	4.81	31.87	24.11
C13H19I107	414.20	414.0177	37.70	4.62	30.64	27.04

195

C13H19I108	430.20	430.0126	36.30	4.45	29.50	29.75	C13H26I205	516.16	515.9873	30.25	5.08	49.17	15.50
C13H19I109	446.20	446.0076	34.95	4.29	28.44	32.27	C13H26I206	532.16	531.9822	29.34	4.92	47.69	18.04
C13H20I201	446.11	445.9607	35.00	4.52	56.89	3.59	C13H26I207	548.16	547.9772	28.49	4.78	46.30	20.43
C13H20I202	462.11	461.9556	33.79	4.36	54.92	6.92	C13H26I208	564.16	563.9721	27.68	4.65	44.99	22.69
C13H20I203	478.11	477.9506	32.66	4.22	53.09	10.04	C13H26I209	580.16	579.9670	26.91	4.52	43.75	24.82
C13H20I204	494.11	493.9455	31.60	4.08	51.37	12.95	C13H27I101	326.26	326.1108	47.86	8.34	38.90	4.90
C13H20I205	510.11	509.9404	30.61	3.95	49.76	15.68	C13H27I102	342.26	342.1058	45.62	7.95	37.08	9.35
C13H20I206	526.11	525.9353	29.68	3.83	48.24	18.25	C13H27I103	358.26	358.1007	43.58	7.60	35.42	13.40
C13H20I207	542.11	541.9302	28.80	3.72	46.82	20.66	C13H27I104	374.26	374.0956	41.72	7.27	33.91	17.10
C13H20I208	558.11	557.9251	27.98	3.61	45.48	22.93	C13H27I105	390.26	390.0905	40.01	6.97	32.52	20.50
C13H20I209	574.11	573.9200	27.20	3.51	44.21	25.08	C13H27I106	406.26	406.0854	38.43	6.70	31.24	23.63
C13H21I101	320.22	320.0639	48.76	6.61	39.63	5.00	C13H27I107	422.26	422.0803	36.98	6.45	30.05	26.52
C13H21I102	336.22	336.0588	46.44	6.30	37.74	9.52	C13H27I108	438.26	438.0752	35.63	6.21	28.96	29.21
C13H21I103	352.21	352.0537	44.33	6.01	36.03	13.63	C13H27I109	454.26	454.0702	34.37	5.99	27.94	31.70
C13H21I104	368.21	368.0486	42.41	5.75	34.46	17.38	C14H 7I101	318.12	317.9543	52.86	2.22	39.89	5.03
C13H21I105	384.21	384.0435	40.64	5.51	33.03	20.82	C14H 7I102	334.12	333.9493	50.33	2.11	37.98	9.58
C13H21I106	400.21	400.0385	39.02	5.29	31.71	23.99	C14H 7I103	350.11	349.9442	48.03	2.02	36.25	13.71
C13H21I107	416.21	416.0334	37.52	5.09	30.49	26.91	C14H 7I104	366.11	365.9391	45.93	1.93	34.66	17.48
C13H21I108	432.21	432.0283	36.13	4.90	29.36	29.61	C14H 7I105	382.11	381.9340	44.01	1.85	33.21	20.94
C13H21I109	448.21	448.0232	34.84	4.72	28.31	32.13	C14H 7I106	398.11	397.9289	42.24	1.77	31.88	24.11
C13H22I201	448.13	447.9764	34.84	4.95	56.64	3.57	C14H 7I107	414.11	413.9238	40.61	1.70	30.64	27.04
C13H22I202	464.13	463.9713	33.64	4.78	54.69	6.89	C14H 7I108	430.11	429.9187	39.10	1.64	29.51	29.76
C13H22I203	480.13	479.9662	32.52	4.62	52.86	10.00	C14H 7I109	446.11	445.9137	37.69	1.58	28.45	32.28
C13H22I204	496.13	495.9611	31.47	4.47	51.16	12.90	C14H 8I201	446.03	445.8668	37.70	1.81	56.90	3.59
C13H22I205	512.13	511.9560	30.49	4.33	49.56	15.62	C14H 8I202	462.03	461.8617	36.40	1.75	54.93	6.93
C13H22I206	528.13	527.9509	29.57	4.20	48.06	18.18	C14H 8I203	478.03	477.8567	35.18	1.69	53.10	10.04
C13H22I207	544.12	543.9459	28.70	4.08	46.65	20.58	C14H 8I204	494.03	493.8516	34.04	1.63	51.38	12.95
C13H22I208	560.12	559.9408	27.88	3.96	45.31	22.85	C14H 8I205	510.03	509.8465	32.97	1.58	49.76	15.68
C13H22I209	576.12	575.9357	27.10	3.85	44.05	24.99	C14H 8I206	526.03	525.8414	31.97	1.53	48.25	18.25
C13H23I101	322.23	322.0795	48.46	7.19	39.38	4.97	C14H 8I207	542.02	541.8363	31.02	1.49	46.83	20.66
C13H23I102	338.23	338.0745	46.17	6.85	37.52	9.46	C14H 8I208	558.02	557.8312	30.13	1.45	45.48	22.94
C13H23I103	354.23	354.0694	44.08	6.54	35.83	13.55	C14H 8I209	574.02	573.8261	29.29	1.40	44.22	25.09
C13H23I104	370.23	370.0643	42.18	6.26	34.28	17.29	C14H 9I101	320.13	319.9700	52.53	2.83	39.64	5.00
C13H23I105	386.23	386.0592	40.43	6.00	32.86	20.71	C14H 9I102	336.13	335.9649	50.03	2.70	37.75	9.52
C13H23I106	402.23	402.0541	38.82	5.76	31.55	23.87	C14H 9I103	352.13	351.9598	47.75	2.58	36.04	13.63
C13H23I107	418.23	418.0490	37.33	5.54	30.34	26.78	C14H 9I104	368.13	367.9547	45.68	2.46	34.47	17.38
C13H23I108	434.23	434.0439	35.96	5.34	29.23	29.48	C14H 9I105	384.13	383.9497	43.78	2.36	33.04	20.83
C13H23I109	450.23	450.0389	34.68	5.15	28.19	31.98	C14H 9I106	400.13	399.9446	42.03	2.27	31.72	23.99
C13H24I201	450.14	449.9920	34.69	5.37	56.38	3.55	C14H 9I107	416.13	415.9395	40.41	2.18	30.50	26.91
C13H24I202	466.14	465.9869	33.50	5.19	54.45	6.86	C14H 9I108	432.13	431.9344	38.91	2.10	29.37	29.62
C13H24I203	482.14	481.9819	32.39	5.02	52.64	9.96	C14H 9I109	448.13	447.9293	37.52	2.02	28.32	32.13
C13H24I204	498.14	497.9768	31.35	4.86	50.95	12.85	C14H10I201	448.04	447.8825	37.53	2.25	56.65	3.57
C13H24I205	514.14	513.9717	30.37	4.71	49.37	15.56	C14H10I202	464.04	463.8774	36.24	2.17	54.70	6.90
C13H24I206	530.14	529.9666	29.45	4.56	47.88	18.11	C14H10I203	480.04	479.8723	35.03	2.10	52.87	10.00
C13H24I207	546.14	545.9615	28.59	4.43	46.47	20.51	C14H10I204	496.04	495.8672	33.90	2.03	51.17	12.90
C13H24I208	562.14	561.9564	27.78	4.30	45.15	22.77	C14H10I205	512.04	511.8621	32.84	1.97	49.57	15.62
C13H24I209	578.14	577.9513	27.01	4.18	43.90	24.91	C14H10I206	528.04	527.8571	31.85	1.91	48.07	18.18
C13H25I101	324.25	324.0952	48.16	7.77	39.14	4.93	C14H10I207	544.04	543.8520	30.91	1.85	46.65	20.59
C13H25I102	340.25	340.0901	45.89	7.41	37.30	9.40	C14H10I208	560.04	559.8469	30.03	1.80	45.32	22.85
C13H25I103	356.25	356.0850	43.83	7.07	35.62	13.47	C14H10I209	576.04	575.8418	29.19	1.75	44.06	25.00
C13H25I104	372.25	372.0799	41.95	6.77	34.09	17.19	C14H11I101	322.15	321.9856	52.20	3.44	39.39	4.97
C13H25I105	388.25	388.0748	40.22	6.49	32.69	20.60	C14H11I102	338.15	337.9806	49.73	3.28	37.53	9.46
C13H25I106	404.24	404.0698	38.63	6.23	31.39	23.75	C14H11I103	354.15	353.9755	47.48	3.13	35.83	13.55
C13H25I107	420.24	420.0647	37.16	6.00	30.20	26.65	C14H11I104	370.15	369.9704	45.43	3.00	34.28	17.29
C13H25I108	436.24	436.0596	35.79	5.78	29.09	29.34	C14H11I105	386.15	385.9653	43.55	2.87	32.86	20.72
C13H25I109	452.24	452.0545	34.53	5.57	28.06	31.84	C14H11I106	402.14	401.9602	41.81	2.76	31.56	23.87
C13H26I201	452.16	452.0077	34.53	5.80	56.13	3.54	C14H11I107	418.14	417.9551	40.21	2.65	30.35	26.78
C13H26I202	468.16	468.0026	33.35	5.60	54.21	6.84	C14H11I108	434.14	433.9500	38.73	2.55	29.23	29.48
C13H26I203	484.16	483.9975	32.25	5.41	52.42	9.91	C14H11I109	450.14	449.9450	37.36	2.46	28.19	31.99
C13H26I204	500.16	499.9924	31.22	5.24	50.75	12.80	C14H12I201	450.06	449.8981	37.36	2.69	56.39	3.55

196

C14H12I202	466.06	465.8930	36.08	2.60	54.46	6.87	C14H18I208	568.10	567.9095	29.60	3.19	44.68	22.53
C14H12I203	482.06	481.8880	34.88	2.51	52.65	9.96	C14H18I209	584.10	583.9044	28.79	3.11	43.45	24.65
C14H12I204	498.06	497.8829	33.76	2.43	50.96	12.85	C14H19I101	330.21	330.0482	50.92	5.80	38.43	4.85
C14H12I205	514.06	513.8778	32.71	2.35	49.37	15.56	C14H19I102	346.21	346.0432	48.57	5.53	36.66	9.24
C14H12I206	530.06	529.8727	31.72	2.28	47.88	18.11	C14H19I103	362.21	362.0381	46.43	5.29	35.04	13.25
C14H12I207	546.06	545.8676	30.79	2.22	46.48	20.51	C14H19I104	378.21	378.0330	44.46	5.06	33.55	16.92
C14H12I208	562.06	561.8625	29.92	2.15	45.16	22.77	C14H19I105	394.21	394.0279	42.66	4.86	32.19	20.29
C14H12I209	578.06	577.8574	29.09	2.09	43.91	24.91	C14H19I106	410.21	410.0228	40.99	4.67	30.94	23.40
C14H13I101	324.16	324.0013	51.87	4.04	39.15	4.94	C14H19I107	426.21	426.0177	39.45	4.49	29.78	26.28
C14H13I102	340.16	339.9962	49.43	3.85	37.31	9.41	C14H19I108	442.21	442.0126	38.03	4.33	28.70	28.94
C14H13I103	356.16	355.9911	47.21	3.68	35.63	13.48	C14H19I109	458.21	458.0076	36.70	4.18	27.70	31.43
C14H13I104	372.16	371.9860	45.18	3.52	34.10	17.20	C14H20I201	458.12	457.9607	36.71	4.40	55.40	3.49
C14H13I105	388.16	387.9810	43.32	3.38	32.69	20.61	C14H20I202	474.12	473.9556	35.47	4.25	53.53	6.75
C14H13I106	404.16	403.9759	41.61	3.24	31.40	23.75	C14H20I203	490.12	489.9506	34.31	4.11	51.78	9.79
C14H13I107	420.16	419.9708	40.02	3.12	30.20	26.66	C14H20I204	506.12	505.9455	33.22	3.98	50.15	12.64
C14H13I108	436.16	435.9657	38.55	3.00	29.10	29.35	C14H20I205	522.12	521.9404	32.21	3.86	48.61	15.32
C14H13I109	452.16	451.9606	37.19	2.90	28.07	31.85	C14H20I206	538.12	537.9353	31.25	3.75	47.17	17.84
C14H14I201	452.08	451.9138	37.20	3.12	56.14	3.54	C14H20I207	554.12	553.9302	30.35	3.64	45.80	20.21
C14H14I202	468.08	467.9087	35.93	3.01	54.22	6.84	C14H20I208	570.12	569.9251	29.49	3.54	44.52	22.45
C14H14I203	484.07	483.9036	34.74	2.92	52.43	9.92	C14H20I209	586.12	585.9200	28.69	3.44	43.30	24.57
C14H14I204	500.07	499.8985	33.63	2.82	50.75	12.80	C14H21I101	332.23	332.0639	50.61	6.37	38.20	4.82
C14H14I205	516.07	515.8934	32.58	2.73	49.18	15.50	C14H21I102	348.23	348.0588	48.29	6.08	36.44	9.19
C14H14I206	532.07	531.8883	31.60	2.65	47.70	18.04	C14H21I103	364.23	364.0537	46.17	5.81	34.84	13.18
C14H14I207	548.07	547.8833	30.68	2.57	46.31	20.43	C14H21I104	380.23	380.0486	44.23	5.57	33.38	16.83
C14H14I208	564.07	563.8782	29.81	2.50	45.00	22.69	C14H21I105	396.22	396.0435	42.44	5.34	32.03	20.19
C14H14I209	580.07	579.8731	28.99	2.43	43.75	24.82	C14H21I106	412.22	412.0385	40.79	5.13	30.79	23.29
C14H15I101	326.18	326.0169	51.55	4.64	38.91	4.91	C14H21I107	428.22	428.0334	39.27	4.94	29.64	26.15
C14H15I102	342.18	342.0119	49.14	4.42	37.09	9.35	C14H21I108	444.22	444.0283	37.85	4.77	28.57	28.81
C14H15I103	358.18	358.0068	46.95	4.22	35.43	13.40	C14H21I109	460.22	460.0232	36.54	4.60	27.57	31.29
C14H15I104	374.18	374.0017	44.94	4.04	33.92	17.10	C14H22I201	460.14	459.9764	36.54	4.82	55.16	3.48
C14H15I105	390.18	389.9966	43.10	3.88	32.52	20.50	C14H22I202	476.14	475.9713	35.32	4.66	53.31	6.72
C14H15I106	406.18	405.9915	41.40	3.72	31.24	23.63	C14H22I203	492.14	491.9662	34.17	4.51	51.57	9.75
C14H15I107	422.18	421.9864	39.83	3.58	30.06	26.53	C14H22I204	508.14	507.9611	33.09	4.36	49.95	12.59
C14H15I108	438.18	437.9813	38.38	3.45	28.96	29.21	C14H22I205	524.14	523.9560	32.08	4.23	48.42	15.26
C14H15I109	454.17	453.9763	37.02	3.33	27.94	31.70	C14H22I206	540.14	539.9509	31.13	4.11	46.99	17.77
C14H16I201	454.09	453.9294	37.03	3.55	55.89	3.52	C14H22I207	556.14	555.9459	30.24	3.99	45.64	20.14
C14H16I202	470.09	469.9243	35.77	3.43	53.99	6.81	C14H22I208	572.14	571.9408	29.39	3.88	44.36	22.37
C14H16I203	486.09	485.9193	34.59	3.32	52.21	9.87	C14H22I209	588.13	587.9357	28.59	3.77	43.15	24.48
C14H16I204	502.09	501.9142	33.49	3.21	50.55	12.75	C14H23I101	334.24	334.0795	50.31	6.94	37.97	4.79
C14H16I205	518.09	517.9091	32.46	3.11	48.99	15.44	C14H23I102	350.24	350.0745	48.01	6.62	36.23	9.14
C14H16I206	534.09	533.9040	31.48	3.02	47.52	17.97	C14H23I103	366.24	366.0694	45.91	6.33	34.65	13.11
C14H16I207	550.09	549.8989	30.57	2.93	46.14	20.36	C14H23I104	382.24	382.0643	43.99	6.07	33.20	16.74
C14H16I208	566.09	565.8938	29.70	2.85	44.84	22.61	C14H23I105	398.24	398.0592	42.22	5.82	31.87	20.09
C14H16I209	582.09	581.8887	28.89	2.77	43.60	24.74	C14H23I106	414.24	414.0541	40.59	5.60	30.64	23.17
C14H17I101	328.20	328.0326	51.24	5.22	38.67	4.87	C14H23I107	430.24	430.0490	39.08	5.39	29.50	26.03
C14H17I102	344.19	344.0275	48.85	4.98	36.87	9.30	C14H23I108	446.24	446.0439	37.68	5.20	28.44	28.68
C14H17I103	360.19	360.0224	46.68	4.76	35.23	13.33	C14H23I109	462.24	462.0389	36.38	5.02	27.45	31.15
C14H17I104	376.19	376.0173	44.70	4.55	33.73	17.01	C14H24I201	462.16	461.9920	36.39	5.23	54.92	3.46
C14H17I105	392.19	392.0122	42.88	4.37	32.36	20.40	C14H24I202	478.15	477.9869	35.17	5.06	53.08	6.69
C14H17I106	408.19	408.0072	41.20	4.20	31.09	23.52	C14H24I203	494.15	493.9819	34.03	4.90	51.36	9.71
C14H17I107	424.19	424.0021	39.64	4.04	29.92	26.40	C14H24I204	510.15	509.9768	32.96	4.74	49.75	12.54
C14H17I108	440.19	439.9970	38.20	3.89	28.83	29.08	C14H24I205	526.15	525.9717	31.96	4.60	48.24	15.20
C14H17I109	456.19	455.9919	36.86	3.76	27.82	31.56	C14H24I206	542.15	541.9666	31.02	4.46	46.82	17.71
C14H18I201	456.11	455.9451	36.87	3.98	55.65	3.51	C14H24I207	558.15	557.9615	30.13	4.33	45.47	20.07
C14H18I202	472.11	471.9400	35.62	3.84	53.76	6.78	C14H24I208	574.15	573.9564	29.29	4.21	44.21	22.29
C14H18I203	488.11	487.9349	34.45	3.72	52.00	9.83	C14H24I209	590.15	589.9513	28.49	4.10	43.01	24.40
C14H18I204	504.11	503.9298	33.36	3.60	50.35	12.70	C14H25I101	336.26	336.0952	50.01	7.49	37.74	4.76
C14H18I205	520.11	519.9247	32.33	3.49	48.80	15.38	C14H25I102	352.26	352.0901	47.74	7.15	36.03	9.08
C14H18I206	536.10	535.9196	31.37	3.38	47.34	17.91	C14H25I103	368.26	368.0850	45.66	6.84	34.46	13.03
C14H18I207	552.10	551.9146	30.46	3.29	45.97	20.29	C14H25I104	384.26	384.0799	43.76	6.56	33.03	16.65

197

C14H25I105	400.26	400.0748	42.01	6.30	31.71	19.99
C14H25I106	416.26	416.0698	40.40	6.05	30.49	23.06
C14H25I107	432.26	432.0647	38.90	5.83	29.36	25.91
C14H25I108	448.25	448.0596	37.51	5.62	28.31	28.55
C14H25I109	464.25	464.0545	36.22	5.43	27.34	31.02
C14H26I201	464.17	464.0077	36.23	5.65	54.68	3.45
C14H26I202	480.17	480.0026	35.02	5.46	52.86	6.66
C14H26I203	496.17	495.9975	33.89	5.28	51.15	9.67
C14H26I204	512.17	511.9924	32.83	5.12	49.56	12.50
C14H26I205	528.17	527.9873	31.84	4.96	48.05	15.15
C14H26I206	544.17	543.9822	30.90	4.82	46.64	17.64
C14H26I207	560.17	559.9772	30.02	4.68	45.31	19.99
C14H26I208	576.17	575.9721	29.19	4.55	44.05	22.21
C14H26I209	592.17	591.9670	28.40	4.43	42.86	24.32
C14H27I101	338.28	338.1108	49.71	8.05	37.52	4.73
C14H27I102	354.27	354.1058	47.46	7.68	35.82	9.03
C14H27I103	370.27	370.1007	45.41	7.35	34.27	12.96
C14H27I104	386.27	386.0956	43.53	7.05	32.85	16.57
C14H27I105	402.27	402.0905	41.80	6.77	31.55	19.89
C14H27I106	418.27	418.0854	40.20	6.51	30.34	22.95
C14H27I107	434.27	434.0803	38.72	6.27	29.22	25.79
C14H27I108	450.27	450.0752	37.35	6.04	28.18	28.43
C14H27I109	466.27	466.0702	36.06	5.84	27.22	30.88
C14H28I201	466.19	466.0233	36.07	6.05	54.44	3.43
C14H28I202	482.19	482.0182	34.87	5.85	52.64	6.64
C14H28I203	498.19	498.0132	33.75	5.67	50.95	9.63
C14H28I204	514.19	514.0081	32.70	5.49	49.36	12.45
C14H28I205	530.19	530.0030	31.72	5.32	47.87	15.09
C14H28I206	546.18	545.9979	30.79	5.17	46.47	17.58
C14H28I207	562.18	561.9928	29.91	5.02	45.15	19.92
C14H28I208	578.18	577.9877	29.08	4.88	43.90	22.14
C14H28I209	594.18	593.9826	28.30	4.75	42.72	24.23
C14H29I101	340.29	340.1265	49.42	8.59	37.29	4.70
C14H29I102	356.29	356.1214	47.20	8.20	35.62	8.98
C14H29I103	372.29	372.1163	45.17	7.85	34.09	12.89
C14H29I104	388.29	388.1112	43.31	7.53	32.68	16.48
C14H29I105	404.29	404.1061	41.59	7.23	31.39	19.79
C14H29I106	420.29	420.1011	40.01	6.96	30.19	22.84
C14H29I107	436.29	436.0960	38.54	6.70	29.09	25.67
C14H29I108	452.29	452.0909	37.18	6.46	28.06	28.30
C14H29I109	468.29	468.0858	35.91	6.24	27.10	30.75
C15H 7I101	330.13	329.9543	54.58	2.14	38.44	4.85
C15H 7I102	346.13	345.9493	52.05	2.04	36.66	9.24
C15H 7I103	362.13	361.9442	49.75	1.95	35.04	13.25
C15H 7I104	378.13	377.9391	47.65	1.87	33.56	16.92
C15H 7I105	394.12	393.9340	45.71	1.79	32.20	20.30
C15H 7I106	410.12	409.9289	43.93	1.72	30.94	23.41
C15H 7I107	426.12	425.9238	42.28	1.66	29.78	26.28
C15H 7I108	442.12	441.9187	40.75	1.60	28.70	28.95
C15H 7I109	458.12	457.9137	39.33	1.54	27.70	31.43
C15H 8I201	458.04	457.8668	39.33	1.76	55.41	3.49
C15H 8I202	474.04	473.8617	38.01	1.70	53.54	6.75
C15H 8I203	490.04	489.8567	36.77	1.65	51.79	9.79
C15H 8I204	506.04	505.8516	35.60	1.59	50.16	12.65
C15H 8I205	522.04	521.8465	34.51	1.54	48.62	15.32
C15H 8I206	538.04	537.8414	33.49	1.50	47.17	17.84
C15H 8I207	554.04	553.8363	32.52	1.46	45.81	20.21
C15H 8I208	570.04	569.8312	31.61	1.41	44.53	22.45
C15H 8I209	586.03	585.8261	30.74	1.38	43.31	24.57
C15H 9I101	332.14	331.9700	54.24	2.73	38.21	4.82
C15H 9I102	348.14	347.9649	51.75	2.61	36.45	9.19
C15H 9I103	364.14	363.9598	49.48	2.49	34.85	13.18
C15H 9I104	380.14	379.9547	47.39	2.39	33.38	16.84
C15H 9I105	396.14	395.9497	45.48	2.29	32.04	20.19
C15H 9I106	412.14	411.9446	43.72	2.20	30.79	23.29
C15H 9I107	428.14	427.9395	42.08	2.12	29.64	26.16
C15H 9I108	444.14	443.9344	40.57	2.04	28.57	28.82
C15H 9I109	460.14	459.9293	39.16	1.97	27.58	31.29
C15H10I201	460.06	459.8825	39.16	2.19	55.17	3.48
C15H10I202	476.05	475.8774	37.85	2.12	53.32	6.72
C15H10I203	492.05	491.8723	36.62	2.05	51.58	9.75
C15H10I204	508.05	507.8672	35.46	1.98	49.96	12.60
C15H10I205	524.05	523.8621	34.38	1.92	48.43	15.27
C15H10I206	540.05	539.8571	33.36	1.87	47.00	17.78
C15H10I207	556.05	555.8520	32.40	1.81	45.64	20.14
C15H10I208	572.05	571.8469	31.49	1.76	44.37	22.37
C15H10I209	588.05	587.8418	30.64	1.71	43.16	24.49
C15H11I101	334.16	333.9856	53.92	3.32	37.98	4.79
C15H11I102	350.16	349.9806	51.45	3.17	36.24	9.14
C15H11I103	366.16	365.9755	49.20	3.03	34.66	13.11
C15H11I104	382.16	381.9704	47.14	2.90	33.21	16.75
C15H11I105	398.16	397.9653	45.25	2.78	31.87	20.09
C15H11I106	414.16	413.9602	43.50	2.68	30.64	23.18
C15H11I107	430.16	429.9551	41.88	2.58	29.50	26.04
C15H11I108	446.15	445.9500	40.38	2.49	28.44	28.69
C15H11I109	462.15	461.9450	38.98	2.40	27.46	31.16
C15H12I201	462.07	461.8981	38.99	2.62	54.93	3.46
C15H12I202	478.07	477.8930	37.69	2.53	53.09	6.69
C15H12I203	494.07	493.8880	36.47	2.45	51.37	9.71
C15H12I204	510.07	509.8829	35.32	2.37	49.76	12.55
C15H12I205	526.07	525.8778	34.25	2.30	48.25	15.21
C15H12I206	542.07	541.8727	33.24	2.23	46.82	17.71
C15H12I207	558.07	557.8676	32.28	2.17	45.48	20.07
C15H12I208	574.07	573.8625	31.38	2.11	44.21	22.30
C15H12I209	590.07	589.8574	30.53	2.05	43.01	24.40
C15H13I101	336.17	336.0013	53.59	3.90	37.75	4.76
C15H13I102	352.17	351.9962	51.16	3.72	36.03	9.09
C15H13I103	368.17	367.9911	48.94	3.56	34.47	13.04
C15H13I104	384.17	383.9860	46.90	3.41	33.03	16.66
C15H13I105	400.17	399.9810	45.02	3.27	31.71	19.99
C15H13I106	416.17	415.9759	43.29	3.15	30.49	23.07
C15H13I107	432.17	431.9708	41.69	3.03	29.36	25.91
C15H13I108	448.17	447.9657	40.20	2.92	28.32	28.56
C15H13I109	464.17	463.9606	38.81	2.82	27.34	31.02
C15H14I201	464.09	463.9138	38.82	3.04	54.69	3.45
C15H14I202	480.09	479.9087	37.53	2.94	52.87	6.67
C15H14I203	496.09	495.9036	36.32	2.84	51.16	9.68
C15H14I204	512.09	511.8985	35.18	2.76	49.56	12.50
C15H14I205	528.08	527.8934	34.12	2.67	48.06	15.15
C15H14I206	544.08	543.8883	33.11	2.59	46.65	17.64
C15H14I207	560.08	559.8833	32.17	2.52	45.32	20.00
C15H14I208	576.08	575.8782	31.27	2.45	44.06	22.22
C15H14I209	592.08	591.8731	30.43	2.38	42.87	24.32
C15H15I101	338.19	338.0169	53.27	4.47	37.52	4.73
C15H15I102	354.19	354.0119	50.87	4.27	35.83	9.03
C15H15I103	370.19	370.0068	48.67	4.08	34.28	12.97
C15H15I104	386.19	386.0017	46.65	3.92	32.86	16.57
C15H15I105	402.19	401.9966	44.80	3.76	31.55	19.89
C15H15I106	418.19	417.9915	43.08	3.62	30.35	22.96
C15H15I107	434.19	433.9864	41.50	3.48	29.23	25.79

198

C15H15I108	450.19	449.9813	40.02	3.36	28.19	28.43
C15H15I109	466.19	465.9763	38.65	3.24	27.22	30.89
C15H16I201	466.10	465.9294	38.65	3.46	54.45	3.43
C15H16I202	482.10	481.9243	37.37	3.35	52.65	6.64
C15H16I203	498.10	497.9193	36.17	3.24	50.96	9.64
C15H16I204	514.10	513.9142	35.05	3.14	49.37	12.45
C15H16I205	530.10	529.9091	33.99	3.04	47.88	15.09
C15H16I206	546.10	545.9040	32.99	2.95	46.48	17.58
C15H16I207	562.10	561.8989	32.05	2.87	45.15	19.92
C15H16I208	578.10	577.8938	31.17	2.79	43.90	22.14
C15H16I209	594.10	593.8887	30.33	2.71	42.72	24.24
C15H17I101	340.21	340.0326	52.96	5.04	37.30	4.70
C15H17I102	356.21	356.0275	50.58	4.81	35.63	8.98
C15H17I103	372.21	372.0224	48.41	4.60	34.10	12.90
C15H17I104	388.20	388.0173	46.41	4.41	32.69	16.49
C15H17I105	404.20	404.0122	44.57	4.24	31.40	19.79
C15H17I106	420.20	420.0072	42.88	4.08	30.20	22.85
C15H17I107	436.20	436.0021	41.30	3.93	29.09	25.68
C15H17I108	452.20	451.9970	39.84	3.79	28.06	28.30
C15H17I109	468.20	467.9919	38.48	3.66	27.10	30.75
C15H18I201	468.12	467.9451	38.49	3.88	54.22	3.42
C15H18I202	484.12	483.9400	37.22	3.75	52.43	6.61
C15H18I203	500.12	499.9349	36.02	3.63	50.75	9.60
C15H18I204	516.12	515.9298	34.91	3.52	49.18	12.40
C15H18I205	532.12	531.9247	33.86	3.41	47.70	15.03
C15H18I206	548.12	547.9196	32.87	3.31	46.31	17.51
C15H18I207	564.12	563.9146	31.94	3.22	44.99	19.85
C15H18I208	580.11	579.9095	31.06	3.13	43.75	22.06
C15H18I209	596.11	595.9044	30.22	3.04	42.58	24.16
C15H19I101	342.22	342.0482	52.65	5.60	37.08	4.68
C15H19I102	358.22	358.0432	50.29	5.35	35.43	8.93
C15H19I103	374.22	374.0381	48.14	5.12	33.91	12.83
C15H19I104	390.22	390.0330	46.17	4.91	32.52	16.40
C15H19I105	406.22	406.0279	44.35	4.71	31.24	19.69
C15H19I106	422.22	422.0228	42.67	4.54	30.06	22.74
C15H19I107	438.22	438.0177	41.11	4.37	28.96	25.56
C15H19I108	454.22	454.0126	39.67	4.22	27.94	28.18
C15H19I109	470.22	470.0076	38.32	4.07	26.99	30.62
C15H20I201	470.13	469.9607	38.32	4.29	53.99	3.40
C15H20I202	486.13	485.9556	37.06	4.15	52.21	6.58
C15H20I203	502.13	501.9506	35.88	4.01	50.55	9.56
C15H20I204	518.13	517.9455	34.77	3.89	48.99	12.35
C15H20I205	534.13	533.9404	33.73	3.77	47.52	14.98
C15H20I206	550.13	549.9353	32.75	3.66	46.14	17.45
C15H20I207	566.13	565.9302	31.82	3.56	44.83	19.78
C15H20I208	582.13	581.9251	30.95	3.46	43.60	21.99
C15H20I209	598.13	597.9200	30.12	3.37	42.43	24.07
C15H21I101	344.24	344.0639	52.34	6.15	36.87	4.65
C15H21I102	360.24	360.0588	50.01	5.88	35.23	8.88
C15H21I103	376.24	376.0537	47.89	5.63	33.73	12.76
C15H21I104	392.24	392.0486	45.93	5.40	32.35	16.32
C15H21I105	408.24	408.0435	44.13	5.19	31.09	19.60
C15H21I106	424.24	424.0385	42.47	4.99	29.91	22.63
C15H21I107	440.23	440.0334	40.93	4.81	28.83	25.44
C15H21I108	456.23	456.0283	39.49	4.64	27.82	28.05
C15H21I109	472.23	472.0232	38.15	4.48	26.87	30.49
C15H22I201	472.15	471.9764	38.16	4.70	53.76	3.39
C15H22I202	488.15	487.9713	36.91	4.54	51.99	6.56
C15H22I203	504.15	503.9662	35.74	4.40	50.34	9.52
C15H22I204	520.15	519.9611	34.64	4.26	48.80	12.30
C15H22I205	536.15	535.9560	33.60	4.14	47.34	14.92
C15H22I206	552.15	551.9509	32.63	4.02	45.97	17.39
C15H22I207	568.15	567.9459	31.71	3.90	44.67	19.71
C15H22I208	584.15	583.9408	30.84	3.80	43.45	21.91
C15H22I209	600.15	599.9357	30.02	3.69	42.29	23.99
C15H23I101	346.25	346.0795	52.03	6.70	36.65	4.62
C15H23I102	362.25	362.0745	49.74	6.40	35.03	8.83
C15H23I103	378.25	378.0694	47.63	6.13	33.55	12.69
C15H23I104	394.25	394.0643	45.70	5.88	32.19	16.23
C15H23I105	410.25	410.0592	43.92	5.65	30.93	19.50
C15H23I106	426.25	426.0541	42.27	5.44	29.77	22.52
C15H23I107	442.25	442.0490	40.74	5.24	28.70	25.32
C15H23I108	458.25	458.0439	39.32	5.06	27.69	27.93
C15H23I109	474.25	474.0389	37.99	4.89	26.76	30.36
C15H24I201	474.17	473.9920	38.00	5.10	53.53	3.37
C15H24I202	490.17	489.9869	36.76	4.94	51.78	6.53
C15H24I203	506.17	505.9819	35.59	4.78	50.14	9.48
C15H24I204	522.16	521.9768	34.50	4.63	48.61	12.26
C15H24I205	538.16	537.9717	33.48	4.50	47.16	14.86
C15H24I206	554.16	553.9666	32.51	4.37	45.80	17.32
C15H24I207	570.16	569.9615	31.60	4.24	44.52	19.64
C15H24I208	586.16	585.9564	30.74	4.13	43.30	21.84
C15H24I209	602.16	601.9513	29.92	4.02	42.15	23.91
C15H25I101	348.27	348.0952	51.73	7.24	36.44	4.59
C15H25I102	364.27	364.0901	49.46	6.92	34.84	8.78
C15H25I103	380.27	380.0850	47.38	6.63	33.37	12.62
C15H25I104	396.27	396.0799	45.47	6.36	32.02	16.15
C15H25I105	412.27	412.0748	43.70	6.11	30.78	19.40
C15H25I106	428.27	428.0698	42.07	5.88	29.63	22.42
C15H25I107	444.27	444.0647	40.55	5.67	28.56	25.21
C15H25I108	460.27	460.0596	39.14	5.47	27.57	27.81
C15H25I109	476.27	476.0545	37.83	5.29	26.65	30.23
C15H26I201	476.18	476.0077	37.84	5.50	53.30	3.36
C15H26I202	492.18	492.0026	36.61	5.32	51.57	6.50
C15H26I203	508.18	507.9975	35.45	5.16	49.94	9.45
C15H26I204	524.18	523.9924	34.37	5.00	48.42	12.21
C15H26I205	540.18	539.9873	33.35	4.85	46.99	14.81
C15H26I206	556.18	555.9822	32.39	4.71	45.63	17.26
C15H26I207	572.18	571.9772	31.49	4.58	44.36	19.57
C15H26I208	588.18	587.9721	30.63	4.46	43.15	21.76
C15H26I209	604.18	603.9670	29.82	4.34	42.01	23.83
C15H27I101	350.29	350.1108	51.43	7.77	36.23	4.57
C15H27I102	366.29	366.1058	49.19	7.43	34.65	8.74
C15H27I103	382.29	382.1007	47.13	7.12	33.20	12.56
C15H27I104	398.28	398.0956	45.24	6.83	31.86	16.07
C15H27I105	414.28	414.0905	43.49	6.57	30.63	19.31
C15H27I106	430.28	430.0854	41.87	6.32	29.49	22.31
C15H27I107	446.28	446.0803	40.37	6.10	28.44	25.10
C15H27I108	462.28	462.0752	38.97	5.89	27.45	27.69
C15H27I109	478.28	478.0702	37.67	5.69	26.53	30.11
C15H28I201	478.20	478.0233	37.68	5.90	53.08	3.35
C15H28I202	494.20	494.0182	36.46	5.71	51.36	6.47
C15H28I203	510.20	510.0132	35.31	5.53	49.75	9.41
C15H28I204	526.20	526.0081	34.24	5.36	48.23	12.16
C15H28I205	542.20	542.0030	33.23	5.21	46.81	14.75
C15H28I206	558.20	557.9979	32.28	5.06	45.47	17.20
C15H28I207	574.20	573.9928	31.38	4.92	44.20	19.50
C15H28I208	590.19	589.9877	30.53	4.78	43.00	21.69
C15H28I209	606.19	605.9826	29.72	4.66	41.87	23.75
C15H29I101	352.30	352.1265	51.14	8.30	36.02	4.54

C15H29I102	368.30	368.1214	48.92	7.94	34.46	8.69
C15H29I103	384.30	384.1163	46.88	7.61	33.02	12.49
C15H29I104	400.30	400.1112	45.01	7.30	31.70	15.99
C15H29I105	416.30	416.1061	43.28	7.02	30.48	19.22
C15H29I106	432.30	432.1011	41.68	6.76	29.36	22.21
C15H29I107	448.30	448.0960	40.19	6.52	28.31	24.98
C15H29I108	464.30	464.0909	38.80	6.30	27.33	27.57
C15H29I109	480.30	480.0858	37.51	6.09	26.42	29.98
C15H30I201	480.21	480.0390	37.52	6.30	52.85	3.33
C15H30I202	496.21	496.0339	36.31	6.09	51.15	6.45
C15H30I203	512.21	512.0288	35.17	5.90	49.55	9.37
C15H30I204	528.21	528.0237	34.11	5.72	48.05	12.12
C15H30I205	544.21	544.0186	33.11	5.56	46.64	14.70
C15H30I206	560.21	560.0135	32.16	5.40	45.31	17.14
C15H30I207	576.21	576.0085	31.27	5.25	44.05	19.44
C15H30I208	592.21	592.0034	30.42	5.11	42.86	21.61
C15H30I209	608.21	607.9983	29.62	4.97	41.73	23.68
C15H31I101	354.32	354.1421	50.85	8.82	35.82	4.52
C15H31I102	370.32	370.1371	48.65	8.44	34.27	8.64
C15H31I103	386.32	386.1320	46.64	8.09	32.85	12.42
C15H31I104	402.32	402.1269	44.78	7.77	31.54	15.91
C15H31I105	418.32	418.1218	43.07	7.47	30.34	19.12
C15H31I106	434.32	434.1167	41.48	7.19	29.22	22.10
C15H31I107	450.31	450.1116	40.01	6.94	28.18	24.87
C15H31I108	466.31	466.1065	38.64	6.70	27.21	27.45
C15H31I109	482.31	482.1014	37.35	6.48	26.31	29.85
C16H 8I201	470.05	469.8668	40.88	1.72	54.00	3.40
C16H 8I202	486.05	485.8617	39.54	1.66	52.22	6.58
C16H 8I203	502.05	501.8567	38.28	1.61	50.55	9.56
C16H 8I204	518.05	517.8516	37.10	1.56	48.99	12.35
C16H 8I205	534.05	533.8465	35.99	1.51	47.53	14.98
C16H 8I206	550.05	549.8414	34.94	1.47	46.14	17.45
C16H 8I207	566.05	565.8363	33.95	1.42	44.84	19.79
C16H 8I208	582.05	581.8312	33.02	1.39	43.61	21.99
C16H 8I209	598.05	597.8261	32.13	1.35	42.44	24.08
C16H 9I101	344.15	343.9700	55.84	2.64	36.87	4.65
C16H 9I102	360.15	359.9649	53.36	2.52	35.24	8.88
C16H 9I103	376.15	375.9598	51.09	2.41	33.74	12.76
C16H 9I104	392.15	391.9547	49.01	2.31	32.36	16.32
C16H 9I105	408.15	407.9497	47.09	2.22	31.09	19.60
C16H 9I106	424.15	423.9446	45.31	2.14	29.92	22.63
C16H 9I107	440.15	439.9395	43.66	2.06	28.83	25.44
C16H 9I108	456.15	455.9344	42.13	1.99	27.82	28.06
C16H 9I109	472.15	471.9293	40.70	1.92	26.88	30.50
C16H10I201	472.07	471.8825	40.71	2.14	53.77	3.39
C16H10I202	488.07	487.8774	39.38	2.07	52.00	6.56
C16H10I203	504.07	503.8723	38.13	2.00	50.35	9.52
C16H10I204	520.06	519.8672	36.95	1.94	48.80	12.31
C16H10I205	536.06	535.8621	35.85	1.88	47.35	14.92
C16H10I206	552.06	551.8571	34.81	1.83	45.97	17.39
C16H10I207	568.06	567.8520	33.83	1.77	44.68	19.72
C16H10I208	584.06	583.8469	32.90	1.73	43.46	21.91
C16H10I209	600.06	599.8418	32.03	1.68	42.30	24.00
C16H11I101	346.17	345.9856	55.52	3.20	36.66	4.62
C16H11I102	362.17	361.9806	53.06	3.06	35.04	8.84
C16H11I103	378.17	377.9755	50.82	2.93	33.56	12.69
C16H11I104	394.17	393.9704	48.76	2.81	32.20	16.24
C16H11I105	410.17	409.9653	46.85	2.70	30.94	19.50
C16H11I106	426.17	425.9602	45.09	2.60	29.78	22.53
C16H11I107	442.17	441.9551	43.46	2.51	28.70	25.33

C16H11I108	458.17	457.9500	41.95	2.42	27.70	27.94
C16H11I109	474.17	473.9450	40.53	2.34	26.76	30.37
C16H12I201	474.08	473.8981	40.54	2.55	53.54	3.37
C16H12I202	490.08	489.8930	39.21	2.47	51.79	6.53
C16H12I203	506.08	505.8880	37.97	2.39	50.15	9.48
C16H12I204	522.08	521.8829	36.81	2.32	48.61	12.26
C16H12I205	538.08	537.8778	35.72	2.25	47.17	14.87
C16H12I206	554.08	553.8727	34.68	2.18	45.81	17.33
C16H12I207	570.08	569.8676	33.71	2.12	44.52	19.65
C16H12I208	586.08	585.8625	32.79	2.06	43.31	21.84
C16H12I209	602.08	601.8574	31.92	2.01	42.16	23.92
C16H13I101	348.19	348.0013	55.19	3.76	36.45	4.60
C16H13I102	364.19	363.9962	52.77	3.60	34.85	8.79
C16H13I103	380.18	379.9911	50.55	3.45	33.38	12.62
C16H13I104	396.18	395.9860	48.51	3.31	32.03	16.15
C16H13I105	412.18	411.9810	46.62	3.18	30.79	19.41
C16H13I106	428.18	427.9759	44.88	3.06	29.64	22.42
C16H13I107	444.18	443.9708	43.27	2.95	28.57	25.21
C16H13I108	460.18	459.9657	41.76	2.85	27.58	27.81
C16H13I109	476.18	475.9606	40.36	2.75	26.65	30.24
C16H14I201	476.10	475.9138	40.37	2.96	53.31	3.36
C16H14I202	492.10	491.9087	39.05	2.87	51.58	6.50
C16H14I203	508.10	507.9036	37.82	2.78	49.95	9.45
C16H14I204	524.10	523.8985	36.67	2.69	48.43	12.21
C16H14I205	540.10	539.8934	35.58	2.61	46.99	14.81
C16H14I206	556.10	555.8883	34.56	2.54	45.64	17.26
C16H14I207	572.09	571.8833	33.59	2.47	44.36	19.58
C16H14I208	588.09	587.8782	32.68	2.40	43.16	21.76
C16H14I209	604.09	603.8731	31.81	2.34	42.01	23.84
C16H15I101	350.20	350.0169	54.88	4.32	36.24	4.57
C16H15I102	366.20	366.0119	52.48	4.13	34.65	8.74
C16H15I103	382.20	382.0068	50.28	3.96	33.20	12.56
C16H15I104	398.20	398.0017	48.26	3.80	31.87	16.07
C16H15I105	414.20	413.9966	46.40	3.65	30.64	19.31
C16H15I106	430.20	429.9915	44.67	3.51	29.50	22.31
C16H15I107	446.20	445.9864	43.07	3.39	28.44	25.10
C16H15I108	462.20	461.9813	41.58	3.27	27.46	27.69
C16H15I109	478.20	477.9763	40.19	3.16	26.54	30.11
C16H16I201	478.11	477.9294	40.20	3.37	53.09	3.35
C16H16I202	494.11	493.9243	38.89	3.26	51.37	6.48
C16H16I203	510.11	509.9193	37.67	3.16	49.76	9.41
C16H16I204	526.11	525.9142	36.53	3.07	48.24	12.16
C16H16I205	542.11	541.9091	35.45	2.97	46.82	14.76
C16H16I206	558.11	557.9040	34.43	2.89	45.48	17.20
C16H16I207	574.11	573.8989	33.47	2.81	44.21	19.51
C16H16I208	590.11	589.8938	32.57	2.73	43.01	21.69
C16H16I209	606.11	605.8887	31.71	2.66	41.88	23.76
C16H17I101	352.22	352.0326	54.56	4.87	36.03	4.54
C16H17I102	368.22	368.0275	52.19	4.65	34.46	8.69
C16H17I103	384.22	384.0224	50.02	4.46	33.03	12.49
C16H17I104	400.22	400.0173	48.02	4.28	31.71	15.99
C16H17I105	416.22	416.0122	46.17	4.12	30.49	19.22
C16H17I106	432.21	432.0072	44.46	3.96	29.36	22.21
C16H17I107	448.21	448.0021	42.88	3.82	28.31	24.99
C16H17I108	464.21	463.9970	41.40	3.69	27.34	27.57
C16H17I109	480.21	479.9919	40.02	3.57	26.43	29.99
C16H18I201	480.13	479.9451	40.03	3.78	52.86	3.33
C16H18I202	496.13	495.9400	38.74	3.66	51.16	6.45
C16H18I203	512.13	511.9349	37.53	3.54	49.56	9.37
C16H18I204	528.13	527.9298	36.39	3.44	48.06	12.12

C16H18I205	544.13	543.9247	35.32	3.33	46.65	14.70
C16H18I206	560.13	559.9196	34.31	3.24	45.31	17.14
C16H18I207	576.13	575.9146	33.36	3.15	44.05	19.44
C16H18I208	592.13	591.9095	32.46	3.06	42.86	21.62
C16H18I209	608.13	607.9044	31.60	2.98	41.74	23.68
C16H19I101	354.23	354.0482	54.25	5.41	35.83	4.52
C16H19I102	370.23	370.0432	51.91	5.17	34.28	8.64
C16H19I103	386.23	386.0381	49.76	4.96	32.86	12.43
C16H19I104	402.23	402.0330	47.78	4.76	31.55	15.91
C16H19I105	418.23	418.0279	45.95	4.58	30.34	19.13
C16H19I106	434.23	434.0228	44.26	4.41	29.23	22.11
C16H19I107	450.23	450.0177	42.68	4.25	28.19	24.88
C16H19I108	466.23	466.0126	41.22	4.11	27.22	27.45
C16H19I109	482.23	482.0076	39.85	3.97	26.32	29.86
C16H20I201	482.15	481.9607	39.86	4.18	52.64	3.32
C16H20I202	498.15	497.9556	38.58	4.05	50.95	6.42
C16H20I203	514.14	513.9506	37.38	3.92	49.37	9.34
C16H20I204	530.14	529.9455	36.25	3.80	47.88	12.07
C16H20I205	546.14	545.9404	35.19	3.69	46.47	14.65
C16H20I206	562.14	561.9353	34.19	3.59	45.15	17.08
C16H20I207	578.14	577.9302	33.24	3.49	43.90	19.37
C16H20I208	594.14	593.9251	32.35	3.39	42.72	21.54
C16H20I209	610.14	609.9200	31.50	3.30	41.60	23.60
C16H21I101	356.25	356.0639	53.94	5.94	35.62	4.49
C16H21I102	372.25	372.0588	51.63	5.69	34.09	8.60
C16H21I103	388.25	388.0537	49.50	5.45	32.69	12.36
C16H21I104	404.25	404.0486	47.54	5.24	31.39	15.83
C16H21I105	420.25	420.0435	45.73	5.04	30.20	19.04
C16H21I106	436.25	436.0385	44.05	4.85	29.09	22.01
C16H21I107	452.25	452.0334	42.49	4.68	28.06	24.76
C16H21I108	468.25	468.0283	41.04	4.52	27.10	27.34
C16H21I109	484.24	484.0232	39.69	4.37	26.21	29.74
C16H22I201	484.16	483.9764	39.69	4.58	52.42	3.30
C16H22I202	500.16	499.9713	38.42	4.43	50.75	6.40
C16H22I203	516.16	515.9662	37.23	4.30	49.17	9.30
C16H22I204	532.16	531.9611	36.11	4.17	47.69	12.03
C16H22I205	548.16	547.9560	35.06	4.05	46.30	14.59
C16H22I206	564.16	563.9509	34.06	3.93	44.99	17.02
C16H22I207	580.16	579.9459	33.13	3.82	43.75	19.30
C16H22I208	596.16	595.9408	32.24	3.72	42.57	21.47
C16H22I209	612.16	611.9357	31.39	3.62	41.46	23.52
C16H23I101	358.27	358.0795	53.64	6.47	35.42	4.47
C16H23I102	374.26	374.0745	51.35	6.19	33.91	8.55
C16H23I103	390.26	390.0694	49.24	5.94	32.52	12.30
C16H23I104	406.26	406.0643	47.30	5.71	31.24	15.75
C16H23I105	422.26	422.0592	45.51	5.49	30.05	18.94
C16H23I106	438.26	438.0541	43.85	5.29	28.96	21.90
C16H23I107	454.26	454.0490	42.31	5.10	27.94	24.65
C16H23I108	470.26	470.0439	40.87	4.93	26.99	27.22
C16H23I109	486.26	486.0389	39.52	4.77	26.10	29.61
C16H24I201	486.18	485.9920	39.53	4.98	52.20	3.29
C16H24I202	502.18	501.9869	38.27	4.82	50.54	6.37
C16H24I203	518.18	517.9819	37.09	4.67	48.98	9.26
C16H24I204	534.18	533.9768	35.98	4.53	47.51	11.98
C16H24I205	550.18	549.9717	34.93	4.40	46.13	14.54
C16H24I206	566.17	565.9666	33.94	4.27	44.83	16.96
C16H24I207	582.17	581.9615	33.01	4.16	43.60	19.24
C16H24I208	598.17	597.9564	32.13	4.04	42.43	21.40
C16H24I209	614.17	613.9513	31.29	3.94	41.33	23.45
C16H25I101	360.28	360.0952	53.34	6.99	35.22	4.44

C16H25I102	376.28	376.0901	51.07	6.70	33.73	8.50
C16H25I103	392.28	392.0850	48.99	6.42	32.35	12.24
C16H25I104	408.28	408.0799	47.07	6.17	31.08	15.67
C16H25I105	424.28	424.0748	45.30	5.94	29.91	18.85
C16H25I106	440.28	440.0698	43.65	5.72	28.82	21.80
C16H25I107	456.28	456.0647	42.12	5.52	27.81	24.55
C16H25I108	472.28	472.0596	40.69	5.34	26.87	27.10
C16H25I109	488.28	488.0545	39.36	5.16	25.99	29.49
C16H26I201	488.19	488.0077	39.37	5.37	51.99	3.28
C16H26I202	504.19	504.0026	38.12	5.20	50.34	6.35
C16H26I203	520.19	519.9975	36.94	5.04	48.79	9.23
C16H26I204	536.19	535.9924	35.84	4.89	47.34	11.94
C16H26I205	552.19	551.9873	34.80	4.75	45.96	14.49
C16H26I206	568.19	567.9822	33.82	4.61	44.67	16.90
C16H26I207	584.19	583.9772	32.90	4.49	43.45	19.17
C16H26I208	600.19	599.9721	32.02	4.37	42.29	21.33
C16H26I209	616.19	615.9670	31.19	4.25	41.19	23.37
C16H27I101	362.30	362.1108	53.04	7.51	35.03	4.42
C16H27I102	378.30	378.1058	50.80	7.19	33.55	8.46
C16H27I103	394.30	394.1007	48.74	6.90	32.19	12.17
C16H27I104	410.30	410.0956	46.84	6.63	30.93	15.60
C16H27I105	426.29	426.0905	45.08	6.38	29.77	18.77
C16H27I106	442.29	442.0854	43.45	6.15	28.69	21.70
C16H27I107	458.29	458.0803	41.93	5.94	27.69	24.44
C16H27I108	474.29	474.0752	40.52	5.74	26.76	26.99
C16H27I109	490.29	490.0702	39.20	5.55	25.88	29.37
C16H28I201	490.21	490.0233	39.20	5.76	51.78	3.26
C16H28I202	506.21	506.0182	37.96	5.58	50.14	6.32
C16H28I203	522.21	522.0132	36.80	5.40	48.60	9.19
C16H28I204	538.21	538.0081	35.71	5.24	47.16	11.89
C16H28I205	554.21	554.0030	34.68	5.09	45.80	14.43
C16H28I206	570.21	569.9979	33.70	4.95	44.51	16.84
C16H28I207	586.21	585.9928	32.78	4.81	43.30	19.11
C16H28I208	602.21	601.9877	31.91	4.69	42.15	21.25
C16H28I209	618.20	617.9826	31.09	4.57	41.06	23.29
C16H29I101	364.31	364.1265	52.75	8.02	34.83	4.39
C16H29I102	380.31	380.1214	50.53	7.69	33.37	8.41
C16H29I103	396.31	396.1163	48.49	7.38	32.02	12.11
C16H29I104	412.31	412.1112	46.61	7.09	30.78	15.52
C16H29I105	428.31	428.1061	44.87	6.82	29.63	18.68
C16H29I106	444.31	444.1011	43.25	6.58	28.56	21.61
C16H29I107	460.31	460.0960	41.75	6.35	27.57	24.33
C16H29I108	476.31	476.0909	40.35	6.14	26.64	26.87
C16H29I109	492.31	492.0858	39.04	5.94	25.78	29.25
C16H30I201	492.23	492.0390	39.04	6.14	51.56	3.25
C16H30I202	508.23	508.0339	37.81	5.95	49.94	6.30
C16H30I203	524.22	524.0288	36.66	5.77	48.42	9.16
C16H30I204	540.22	540.0237	35.57	5.60	46.98	11.85
C16H30I205	556.22	556.0186	34.55	5.44	45.63	14.38
C16H30I206	572.22	572.0135	33.58	5.28	44.35	16.78
C16H30I207	588.22	588.0085	32.67	5.14	43.15	19.04
C16H30I208	604.22	604.0034	31.81	5.00	42.01	21.18
C16H30I209	620.22	619.9983	30.99	4.88	40.92	23.22
C16H31I101	366.33	366.1421	52.46	8.53	34.64	4.37
C16H31I102	382.33	382.1371	50.27	8.17	33.19	8.37
C16H31I103	398.33	398.1320	48.25	7.84	31.86	12.05
C16H31I104	414.33	414.1269	46.38	7.54	30.63	15.45
C16H31I105	430.33	430.1218	44.66	7.26	29.49	18.59
C16H31I106	446.33	446.1167	43.06	7.00	28.43	21.51
C16H31I107	462.33	462.1116	41.57	6.76	27.45	24.22

201

C16H31I108	478.33	478.1065	40.18	6.53	26.53	26.76
C16H31I109	494.32	494.1014	38.88	6.32	25.67	29.13
C16H32I201	494.24	494.0546	38.88	6.53	51.35	3.24
C16H32I202	510.24	510.0495	37.66	6.32	49.74	6.27
C16H32I203	526.24	526.0444	36.52	6.13	48.23	9.12
C16H32I204	542.24	542.0394	35.44	5.95	46.81	11.80
C16H32I205	558.24	558.0343	34.43	5.78	45.47	14.33
C16H32I206	574.24	574.0292	33.47	5.62	44.20	16.72
C16H32I207	590.24	590.0241	32.56	5.46	43.00	18.97
C16H32I208	606.24	606.0190	31.70	5.32	41.87	21.11
C16H32I209	622.24	622.0139	30.89	5.18	40.79	23.14
C16H33I101	368.35	368.1578	52.17	9.03	34.45	4.34
C16H33I102	384.34	384.1527	50.00	8.65	33.02	8.33
C16H33I103	400.34	400.1476	48.00	8.31	31.70	11.99
C16H33I104	416.34	416.1425	46.16	7.99	30.48	15.37
C16H33I105	432.34	432.1374	44.45	7.69	29.35	18.50
C16H33I106	448.34	448.1324	42.86	7.42	28.31	21.41
C16H33I107	464.34	464.1273	41.39	7.16	27.33	24.12
C16H33I108	480.34	480.1222	40.01	6.92	26.42	26.65
C16H33I109	496.34	496.1171	38.72	6.70	25.57	29.01
C17H 8I201	482.06	481.8668	42.36	1.67	52.65	3.32
C17H 8I202	498.06	497.8617	41.00	1.62	50.96	6.42
C17H 8I203	514.06	513.8567	39.72	1.57	49.37	9.34
C17H 8I204	530.06	529.8516	38.52	1.52	47.88	12.07
C17H 8I205	546.06	545.8465	37.39	1.48	46.48	14.65
C17H 8I206	562.06	561.8414	36.33	1.43	45.16	17.08
C17H 8I207	578.06	577.8363	35.32	1.39	43.91	19.37
C17H 8I208	594.06	593.8312	34.37	1.36	42.72	21.55
C17H 8I209	610.06	609.8261	33.47	1.32	41.60	23.60
C17H 9I101	356.17	355.9700	57.33	2.55	35.63	4.49
C17H 9I102	372.16	371.9649	54.87	2.44	34.10	8.60
C17H 9I103	388.16	387.9598	52.60	2.34	32.69	12.37
C17H 9I104	404.16	403.9547	50.52	2.24	31.40	15.83
C17H 9I105	420.16	419.9497	48.60	2.16	30.20	19.04
C17H 9I106	436.16	435.9446	46.82	2.08	29.10	22.01
C17H 9I107	452.16	451.9395	45.16	2.01	28.07	24.77
C17H 9I108	468.16	467.9344	43.62	1.94	27.11	27.34
C17H 9I109	484.16	483.9293	42.17	1.87	26.21	29.74
C17H10I201	484.08	483.8825	42.18	2.08	52.43	3.31
C17H10I202	500.08	499.8774	40.83	2.02	50.75	6.40
C17H10I203	516.08	515.8723	39.57	1.95	49.18	9.30
C17H10I204	532.08	531.8672	38.38	1.89	47.70	12.03
C17H10I205	548.08	547.8621	37.26	1.84	46.31	14.60
C17H10I206	564.07	563.8571	36.20	1.79	45.00	17.02
C17H10I207	580.07	579.8520	35.20	1.74	43.75	19.31
C17H10I208	596.07	595.8469	34.26	1.69	42.58	21.47
C17H10I209	612.07	611.8418	33.36	1.65	41.47	23.53
C17H11I101	358.18	357.9856	57.01	3.10	35.43	4.47
C17H11I102	374.18	373.9806	54.57	2.96	33.92	8.55
C17H11I103	390.18	389.9755	52.33	2.84	32.52	12.30
C17H11I104	406.18	405.9704	50.27	2.73	31.24	15.76
C17H11I105	422.18	421.9653	48.37	2.63	30.06	18.95
C17H11I106	438.18	437.9602	46.60	2.53	28.96	21.91
C17H11I107	454.18	453.9551	44.96	2.44	27.94	24.66
C17H11I108	470.18	469.9500	43.43	2.36	26.99	27.22
C17H11I109	486.18	485.9450	42.00	2.28	26.10	29.62
C17H12I201	486.09	485.8981	42.01	2.49	52.21	3.29
C17H12I202	502.09	501.8930	40.67	2.41	50.55	6.37
C17H12I203	518.09	517.8880	39.41	2.33	48.99	9.26
C17H12I204	534.09	533.8829	38.23	2.26	47.52	11.98
C17H12I205	550.09	549.8778	37.12	2.20	46.14	14.54
C17H12I206	566.09	565.8727	36.07	2.14	44.84	16.96
C17H12I207	582.09	581.8676	35.08	2.08	43.60	19.24
C17H12I208	598.09	597.8625	34.14	2.02	42.44	21.40
C17H12I209	614.09	613.8574	33.25	1.97	41.33	23.45
C17H13I101	360.20	360.0013	56.69	3.64	35.23	4.44
C17H13I102	376.20	375.9962	54.28	3.48	33.73	8.51
C17H13I103	392.20	391.9911	52.06	3.34	32.36	12.24
C17H13I104	408.20	407.9860	50.02	3.21	31.09	15.68
C17H13I105	424.19	423.9810	48.14	3.09	29.92	18.86
C17H13I106	440.19	439.9759	46.39	2.98	28.83	21.81
C17H13I107	456.19	455.9708	44.76	2.87	27.82	24.55
C17H13I108	472.19	471.9657	43.24	2.78	26.88	27.11
C17H13I109	488.19	487.9606	41.83	2.68	25.99	29.50
C17H14I201	488.11	487.9138	41.83	2.89	52.00	3.28
C17H14I202	504.11	503.9087	40.51	2.80	50.35	6.35
C17H14I203	520.11	519.9036	39.26	2.71	48.80	9.23
C17H14I204	536.11	535.8985	38.09	2.63	47.34	11.94
C17H14I205	552.11	551.8934	36.98	2.56	45.97	14.49
C17H14I206	568.11	567.8883	35.94	2.48	44.68	16.90
C17H14I207	584.11	583.8833	34.96	2.42	43.45	19.17
C17H14I208	600.11	599.8782	34.03	2.35	42.29	21.33
C17H14I209	616.10	615.8731	33.14	2.29	41.20	23.37
C17H15I101	362.21	362.0169	56.37	4.17	35.04	4.42
C17H15I102	378.21	378.0119	53.99	4.00	33.55	8.46
C17H15I103	394.21	394.0068	51.80	3.84	32.19	12.18
C17H15I104	410.21	410.0017	49.78	3.69	30.94	15.60
C17H15I105	426.21	425.9966	47.91	3.55	29.78	18.77
C17H15I106	442.21	441.9915	46.17	3.42	28.70	21.71
C17H15I107	458.21	457.9864	44.56	3.30	27.70	24.44
C17H15I108	474.21	473.9813	43.06	3.19	26.76	26.99
C17H15I109	490.21	489.9763	41.65	3.08	25.89	29.37
C17H16I201	490.13	489.9294	41.66	3.29	51.78	3.26
C17H16I202	506.12	505.9243	40.34	3.19	50.15	6.32
C17H16I203	522.12	521.9193	39.11	3.09	48.61	9.19
C17H16I204	538.12	537.9142	37.94	3.00	47.17	11.89
C17H16I205	554.12	553.9091	36.85	2.91	45.80	14.44
C17H16I206	570.12	569.9040	35.82	2.83	44.52	16.84
C17H16I207	586.12	585.8989	34.84	2.75	43.30	19.11
C17H16I208	602.12	601.8938	33.91	2.68	42.15	21.26
C17H16I209	618.12	617.8887	33.03	2.61	41.06	23.30
C17H17I101	364.23	364.0326	56.06	4.70	34.84	4.39
C17H17I102	380.23	380.0275	53.70	4.51	33.38	8.42
C17H17I103	396.23	396.0224	51.53	4.32	32.03	12.11
C17H17I104	412.23	412.0173	49.53	4.16	30.79	15.52
C17H17I105	428.23	428.0122	47.68	4.00	29.63	18.68
C17H17I106	444.23	444.0072	45.97	3.86	28.57	21.61
C17H17I107	460.23	460.0021	44.37	3.72	27.57	24.33
C17H17I108	476.22	475.9970	42.88	3.60	26.65	26.88
C17H17I109	492.22	491.9919	41.48	3.48	25.78	29.25
C17H18I201	492.14	491.9451	41.49	3.69	51.57	3.25
C17H18I202	508.14	507.9400	40.18	3.57	49.95	6.30
C17H18I203	524.14	523.9349	38.96	3.46	48.42	9.16
C17H18I204	540.14	539.9298	37.80	3.36	46.99	11.85
C17H18I205	556.14	555.9247	36.72	3.26	45.64	14.38
C17H18I206	572.14	571.9196	35.69	3.17	44.36	16.78
C17H18I207	588.14	587.9146	34.72	3.08	43.15	19.04
C17H18I208	604.14	603.9095	33.80	3.00	42.01	21.19
C17H18I209	620.14	619.9044	32.93	2.93	40.93	23.22
C17H19I101	366.24	366.0482	55.75	5.23	34.65	4.37

202

C17H19I102	382.24	382.0432	53.42	5.01	33.20	8.37
C17H19I103	398.24	398.0381	51.27	4.81	31.87	12.05
C17H19I104	414.24	414.0330	49.29	4.62	30.64	15.45
C17H19I105	430.24	430.0279	47.46	4.45	29.50	18.59
C17H19I106	446.24	446.0228	45.76	4.29	28.44	21.51
C17H19I107	462.24	462.0177	44.17	4.14	27.45	24.23
C17H19I108	478.24	478.0126	42.70	4.00	26.54	26.76
C17H19I109	494.24	494.0076	41.31	3.87	25.68	29.13
C17H20I201	494.16	493.9607	41.32	4.08	51.36	3.24
C17H20I202	510.16	509.9556	40.02	3.95	49.75	6.27
C17H20I203	526.16	525.9506	38.81	3.83	48.24	9.12
C17H20I204	542.16	541.9455	37.66	3.72	46.81	11.80
C17H20I205	558.15	557.9404	36.58	3.61	45.47	14.33
C17H20I206	574.15	573.9353	35.56	3.51	44.21	16.72
C17H20I207	590.15	589.9302	34.60	3.42	43.01	18.98
C17H20I208	606.15	605.9251	33.69	3.33	41.87	21.12
C17H20I209	622.15	621.9200	32.82	3.24	40.80	23.14
C17H21I101	368.26	368.0639	55.45	5.75	34.46	4.34
C17H21I102	384.26	384.0588	53.14	5.51	33.03	8.33
C17H21I103	400.26	400.0537	51.01	5.29	31.71	11.99
C17H21I104	416.26	416.0486	49.05	5.09	30.49	15.37
C17H21I105	432.26	432.0435	47.24	4.90	29.36	18.51
C17H21I106	448.26	448.0385	45.55	4.72	28.31	21.42
C17H21I107	464.26	464.0334	43.98	4.56	27.33	24.12
C17H21I108	480.26	480.0283	42.52	4.41	26.42	26.65
C17H21I109	496.26	496.0232	41.15	4.27	25.57	29.02
C17H22I201	496.17	495.9764	41.15	4.47	51.15	3.22
C17H22I202	512.17	511.9713	39.87	4.33	49.56	6.25
C17H22I203	528.17	527.9662	38.66	4.20	48.05	9.09
C17H22I204	544.17	543.9611	37.52	4.08	46.64	11.76
C17H22I205	560.17	559.9560	36.45	3.96	45.31	14.28
C17H22I206	576.17	575.9509	35.44	3.85	44.05	16.66
C17H22I207	592.17	591.9459	34.48	3.74	42.86	18.91
C17H22I208	608.17	607.9408	33.57	3.65	41.73	21.05
C17H22I209	624.17	623.9357	32.71	3.55	40.66	23.07
C17H23I101	370.28	370.0795	55.15	6.26	34.27	4.32
C17H23I102	386.28	386.0745	52.86	6.00	32.85	8.28
C17H23I103	402.28	402.0694	50.76	5.76	31.55	11.93
C17H23I104	418.27	418.0643	48.82	5.54	30.34	15.30
C17H23I105	434.27	434.0592	47.02	5.34	29.22	18.42
C17H23I106	450.27	450.0541	45.35	5.15	28.18	21.32
C17H23I107	466.27	466.0490	43.79	4.97	27.22	24.02
C17H23I108	482.27	482.0439	42.34	4.81	26.31	26.54
C17H23I109	498.27	498.0389	40.98	4.65	25.47	28.90
C17H24I201	498.19	497.9920	40.99	4.86	50.95	3.21
C17H24I202	514.19	513.9869	39.71	4.70	49.36	6.22
C17H24I203	530.19	529.9819	38.51	4.56	47.87	9.05
C17H24I204	546.19	545.9768	37.38	4.43	46.47	11.72
C17H24I205	562.19	561.9717	36.32	4.30	45.15	14.23
C17H24I206	578.19	577.9666	35.32	4.18	43.90	16.60
C17H24I207	594.19	593.9615	34.36	4.07	42.72	18.85
C17H24I208	610.18	609.9564	33.46	3.96	41.60	20.98
C17H24I209	626.18	625.9513	32.61	3.86	40.53	23.00
C17H25I101	372.29	372.0952	54.85	6.77	34.09	4.30
C17H25I102	388.29	388.0901	52.59	6.49	32.68	8.24
C17H25I103	404.29	404.0850	50.51	6.23	31.39	11.87
C17H25I104	420.29	420.0799	48.58	6.00	30.19	15.23
C17H25I105	436.29	436.0748	46.80	5.78	29.09	18.34
C17H25I106	452.29	452.0698	45.15	5.57	28.06	21.22
C17H25I107	468.29	468.0647	43.60	5.38	27.10	23.92
C17H25I108	484.29	484.0596	42.16	5.20	26.20	26.43
C17H25I109	500.29	500.0545	40.81	5.04	25.37	28.78
C17H26I201	500.20	500.0077	40.82	5.24	50.74	3.20
C17H26I202	516.20	516.0026	39.56	5.08	49.17	6.20
C17H26I203	532.20	531.9975	38.37	4.92	47.69	9.02
C17H26I204	548.20	547.9924	37.25	4.78	46.30	11.67
C17H26I205	564.20	563.9873	36.19	4.65	44.99	14.18
C17H26I206	580.20	579.9822	35.19	4.52	43.74	16.55
C17H26I207	596.20	595.9772	34.25	4.40	42.57	18.78
C17H26I208	612.20	611.9721	33.35	4.28	41.46	20.91
C17H26I209	628.20	627.9670	32.50	4.17	40.40	22.92
C17H27I101	374.31	374.1108	54.55	7.27	33.90	4.27
C17H27I102	390.31	390.1058	52.31	6.97	32.51	8.20
C17H27I103	406.31	406.1007	50.25	6.70	31.23	11.81
C17H27I104	422.31	422.0956	48.35	6.44	30.05	15.15
C17H27I105	438.31	438.0905	46.59	6.21	28.95	18.25
C17H27I106	454.31	454.0854	44.95	5.99	27.93	21.13
C17H27I107	470.30	470.0803	43.42	5.79	26.98	23.81
C17H27I108	486.30	486.0752	41.99	5.60	26.10	26.32
C17H27I109	502.30	502.0702	40.65	5.42	25.26	28.67
C17H28I201	502.22	502.0233	40.66	5.62	50.54	3.19
C17H28I202	518.22	518.0182	39.40	5.45	48.98	6.17
C17H28I203	534.22	534.0132	38.22	5.28	47.51	8.98
C17H28I204	550.22	550.0081	37.11	5.13	46.13	11.63
C17H28I205	566.22	566.0030	36.06	4.98	44.83	14.13
C17H28I206	582.22	581.9979	35.07	4.85	43.59	16.49
C17H28I207	598.22	597.9928	34.13	4.72	42.43	18.72
C17H28I208	614.22	613.9877	33.24	4.59	41.32	20.84
C17H28I209	630.22	629.9826	32.40	4.48	40.27	22.85
C17H29I101	376.32	376.1265	54.26	7.77	33.72	4.25
C17H29I102	392.32	392.1214	52.05	7.45	32.35	8.16
C17H29I103	408.32	408.1163	50.01	7.16	31.08	11.75
C17H29I104	424.32	424.1112	48.12	6.89	29.91	15.08
C17H29I105	440.32	440.1061	46.37	6.64	28.82	18.17
C17H29I106	456.32	456.1011	44.75	6.41	27.81	21.04
C17H29I107	472.32	472.0960	43.23	6.19	26.87	23.71
C17H29I108	488.32	488.0909	41.81	5.99	25.99	26.21
C17H29I109	504.32	504.0858	40.49	5.80	25.16	28.55
C17H30I201	504.24	504.0390	40.49	6.00	50.34	3.17
C17H30I202	520.24	520.0339	39.25	5.81	48.79	6.15
C17H30I203	536.24	536.0288	38.08	5.64	47.33	8.95
C17H30I204	552.24	552.0237	36.98	5.48	45.96	11.59
C17H30I205	568.23	568.0186	35.93	5.32	44.67	14.08
C17H30I206	584.23	584.0135	34.95	5.18	43.44	16.43
C17H30I207	600.23	600.0085	34.02	5.04	42.29	18.66
C17H30I208	616.23	616.0034	33.14	4.91	41.19	20.77
C17H30I209	632.23	631.9983	32.30	4.78	40.14	22.78
C17H31I101	378.34	378.1421	53.97	8.26	33.54	4.23
C17H31I102	394.34	394.1371	51.78	7.92	32.18	8.11
C17H31I103	410.34	410.1320	49.76	7.61	30.93	11.70
C17H31I104	426.34	426.1269	47.89	7.33	29.77	15.01
C17H31I105	442.34	442.1218	46.16	7.06	28.69	18.09
C17H31I106	458.34	458.1167	44.55	6.82	27.69	20.94
C17H31I107	474.34	474.1116	43.05	6.59	26.75	23.61
C17H31I108	490.34	490.1065	41.64	6.37	25.88	26.10
C17H31I109	506.34	506.1014	40.33	6.17	25.06	28.44
C17H32I201	506.25	506.0546	40.33	6.37	50.13	3.16
C17H32I202	522.25	522.0495	39.10	6.18	48.60	6.13
C17H32I203	538.25	538.0444	37.94	5.99	47.15	8.92
C17H32I204	554.25	554.0394	36.84	5.82	45.79	11.55

C17H32I205	570.25	570.0343	35.81	5.66	44.51	14.03
C17H32I206	586.25	586.0292	34.83	5.50	43.29	16.37
C17H32I207	602.25	602.0241	33.90	5.36	42.14	18.60
C17H32I208	618.25	618.0190	33.03	5.22	41.05	20.70
C17H32I209	634.25	634.0139	32.19	5.09	40.02	22.70
C17H33I101	380.36	380.1578	53.68	8.75	33.36	4.21
C17H33I102	396.36	396.1527	51.52	8.39	32.02	8.07
C17H33I103	412.36	412.1476	49.52	8.07	30.78	11.64
C17H33I104	428.35	428.1425	47.67	7.77	29.63	14.94
C17H33I105	444.35	444.1374	45.95	7.49	28.56	18.00
C17H33I106	460.35	460.1324	44.35	7.23	27.57	20.85
C17H33I107	476.35	476.1273	42.87	6.98	26.64	23.51
C17H33I108	492.35	492.1222	41.47	6.76	25.78	26.00
C17H33I109	508.35	508.1171	40.17	6.54	24.96	28.33
C17H34I201	508.27	508.0703	40.17	6.74	49.94	3.15
C17H34I202	524.27	524.0652	38.95	6.54	48.41	6.10
C17H34I203	540.27	540.0601	37.79	6.34	46.98	8.88
C17H34I204	556.27	556.0550	36.71	6.16	45.63	11.50
C17H34I205	572.27	572.0499	35.68	5.99	44.35	13.98
C17H34I206	588.27	588.0448	34.71	5.83	43.15	16.32
C17H34I207	604.27	604.0398	33.79	5.67	42.00	18.53
C17H34I208	620.26	620.0347	32.92	5.53	40.92	20.64
C17H34I209	636.26	636.0296	32.09	5.39	39.89	22.63
C17H35I101	382.37	382.1734	53.40	9.23	33.19	4.18
C17H35I102	398.37	398.1683	51.26	8.86	31.86	8.03
C17H35I103	414.37	414.1633	49.28	8.51	30.63	11.58
C17H35I104	430.37	430.1582	47.45	8.20	29.49	14.87
C17H35I105	446.37	446.1531	45.74	7.90	28.43	17.92
C17H35I106	462.37	462.1480	44.16	7.63	27.45	20.76
C17H35I107	478.37	478.1429	42.68	7.37	26.53	23.41
C17H35I108	494.37	494.1378	41.30	7.14	25.67	25.89
C17H35I109	510.37	510.1327	40.01	6.91	24.87	28.21
C18H 9I101	368.18	367.9700	58.72	2.46	34.47	4.35
C18H 9I102	384.18	383.9649	56.28	2.36	33.03	8.33
C18H 9I103	400.18	399.9598	54.03	2.27	31.71	11.99
C18H 9I104	416.17	415.9547	51.95	2.18	30.49	15.38
C18H 9I105	432.17	431.9497	50.03	2.10	29.36	18.51
C18H 9I106	448.17	447.9446	48.24	2.02	28.32	21.42
C18H 9I107	464.17	463.9395	46.58	1.95	27.34	24.13
C18H 9I108	480.17	479.9344	45.03	1.89	26.43	26.66
C18H 9I109	496.17	495.9293	43.57	1.83	25.58	29.02
C18H10I201	496.09	495.8825	43.58	2.03	51.16	3.23
C18H10I202	512.09	511.8774	42.22	1.97	49.56	6.25
C18H10I203	528.09	527.8723	40.94	1.91	48.06	9.09
C18H10I204	544.09	543.8672	39.74	1.85	46.65	11.76
C18H10I205	560.09	559.8621	38.60	1.80	45.32	14.28
C18H10I206	576.09	575.8571	37.53	1.75	44.06	16.66
C18H10I207	592.09	591.8520	36.52	1.70	42.87	18.92
C18H10I208	608.08	607.8469	35.55	1.66	41.74	21.05
C18H10I209	624.08	623.8418	34.64	1.62	40.67	23.07
C18H11I101	370.19	369.9856	58.40	3.00	34.28	4.32
C18H11I102	386.19	385.9806	55.98	2.87	32.86	8.29
C18H11I103	402.19	401.9755	53.76	2.76	31.55	11.93
C18H11I104	418.19	417.9704	51.70	2.65	30.35	15.30
C18H11I105	434.19	433.9653	49.79	2.55	29.23	18.42
C18H11I106	450.19	449.9602	48.02	2.46	28.19	21.32
C18H11I107	466.19	465.9551	46.38	2.38	27.22	24.02
C18H11I108	482.19	481.9500	44.84	2.30	26.32	26.54
C18H11I109	498.19	497.9450	43.40	2.23	25.47	28.90
C18H12I201	498.10	497.8981	43.40	2.43	50.95	3.21

C18H12I202	514.10	513.8930	42.05	2.35	49.37	6.22
C18H12I203	530.10	529.8880	40.78	2.28	47.88	9.05
C18H12I204	546.10	545.8829	39.59	2.21	46.48	11.72
C18H12I205	562.10	561.8778	38.46	2.15	45.15	14.23
C18H12I206	578.10	577.8727	37.40	2.09	43.90	16.61
C18H12I207	594.10	593.8676	36.39	2.04	42.72	18.85
C18H12I208	610.10	609.8625	35.44	1.98	41.60	20.98
C18H12I209	626.10	625.8574	34.53	1.93	40.54	23.00
C18H13I101	372.21	372.0013	58.09	3.52	34.10	4.30
C18H13I102	388.21	387.9962	55.69	3.38	32.69	8.24
C18H13I103	404.21	403.9911	53.49	3.24	31.40	11.87
C18H13I104	420.21	419.9860	51.45	3.12	30.20	15.23
C18H13I105	436.21	435.9810	49.56	3.00	29.09	18.34
C18H13I106	452.21	451.9759	47.81	2.90	28.06	21.23
C18H13I107	468.20	467.9708	46.18	2.80	27.10	23.92
C18H13I108	484.20	483.9657	44.65	2.71	26.21	26.43
C18H13I109	500.20	499.9606	43.22	2.62	25.37	28.79
C18H14I201	500.12	499.9138	43.23	2.82	50.75	3.20
C18H14I202	516.12	515.9087	41.89	2.73	49.18	6.20
C18H14I203	532.12	531.9036	40.63	2.65	47.70	9.02
C18H14I204	548.12	547.8985	39.44	2.57	46.31	11.68
C18H14I205	564.12	563.8934	38.33	2.50	44.99	14.18
C18H14I206	580.12	579.8883	37.27	2.43	43.75	16.55
C18H14I207	596.12	595.8833	36.27	2.37	42.58	18.79
C18H14I208	612.12	611.8782	35.32	2.31	41.46	20.91
C18H14I209	628.12	627.8731	34.42	2.25	40.41	22.92
C18H15I101	374.22	374.0169	57.77	4.04	33.91	4.28
C18H15I102	390.22	390.0119	55.40	3.87	32.52	8.20
C18H15I103	406.22	406.0068	53.22	3.72	31.24	11.82
C18H15I104	422.22	422.0017	51.21	3.58	30.06	15.16
C18H15I105	438.22	437.9966	49.34	3.45	28.96	18.25
C18H15I106	454.22	453.9915	47.60	3.33	27.94	21.13
C18H15I107	470.22	469.9864	45.98	3.22	26.99	23.82
C18H15I108	486.22	485.9813	44.47	3.11	26.10	26.32
C18H15I109	502.22	501.9763	43.05	3.01	25.27	28.67
C18H16I201	502.14	501.9294	43.06	3.21	50.55	3.19
C18H16I202	518.14	517.9243	41.73	3.11	48.98	6.18
C18H16I203	534.14	533.9193	40.48	3.02	47.52	8.99
C18H16I204	550.13	549.9142	39.30	2.93	46.14	11.63
C18H16I205	566.13	565.9091	38.19	2.85	44.83	14.13
C18H16I206	582.13	581.9040	37.14	2.77	43.60	16.49
C18H16I207	598.13	597.8989	36.15	2.70	42.43	18.72
C18H16I208	614.13	613.8938	35.20	2.63	41.33	20.84
C18H16I209	630.13	629.8887	34.31	2.56	40.28	22.85
C18H17I101	376.24	376.0326	57.46	4.55	33.73	4.25
C18H17I102	392.24	392.0275	55.12	4.37	32.35	8.16
C18H17I103	408.24	408.0224	52.96	4.20	31.09	11.76
C18H17I104	424.24	424.0173	50.96	4.04	29.91	15.09
C18H17I105	440.24	440.0122	49.11	3.89	28.83	18.17
C18H17I106	456.24	456.0072	47.39	3.76	27.82	21.04
C18H17I107	472.24	472.0021	45.78	3.63	26.87	23.72
C18H17I108	488.24	487.9970	44.28	3.51	25.99	26.22
C18H17I109	504.24	503.9919	42.88	3.40	25.17	28.56
C18H18I201	504.15	503.9451	42.88	3.60	50.34	3.17
C18H18I202	520.15	519.9400	41.56	3.49	48.80	6.15
C18H18I203	536.15	535.9349	40.32	3.38	47.34	8.95
C18H18I204	552.15	551.9298	39.16	3.29	45.97	11.59
C18H18I205	568.15	567.9247	38.05	3.19	44.67	14.08
C18H18I206	584.15	583.9196	37.01	3.11	43.45	16.43
C18H18I207	600.15	599.9146	36.02	3.02	42.29	18.66

C18H18I208	616.15	615.9095	35.09	2.94	41.19	20.77
C18H18I209	632.15	631.9044	34.20	2.87	40.15	22.78
C18H19I101	378.26	378.0482	57.16	5.06	33.55	4.23
C18H19I102	394.26	394.0432	54.84	4.86	32.19	8.12
C18H19I103	410.25	410.0381	52.70	4.67	30.93	11.70
C18H19I104	426.25	426.0330	50.72	4.49	29.77	15.01
C18H19I105	442.25	442.0279	48.89	4.33	28.69	18.09
C18H19I106	458.25	458.0228	47.18	4.18	27.69	20.95
C18H19I107	474.25	474.0177	45.59	4.04	26.76	23.62
C18H19I108	490.25	490.0126	44.10	3.91	25.89	26.11
C18H19I109	506.25	506.0076	42.71	3.78	25.07	28.44
C18H20I201	506.17	505.9607	42.71	3.98	50.14	3.16
C18H20I202	522.17	521.9556	41.40	3.86	48.61	6.13
C18H20I203	538.17	537.9506	40.17	3.75	47.16	8.92
C18H20I204	554.17	553.9455	39.01	3.64	45.80	11.55
C18H20I205	570.17	569.9404	37.92	3.54	44.51	14.03
C18H20I206	586.17	585.9353	36.88	3.44	43.30	16.38
C18H20I207	602.16	601.9302	35.90	3.35	42.15	18.60
C18H20I208	618.16	617.9251	34.97	3.26	41.06	20.71
C18H20I209	634.16	633.9200	34.09	3.18	40.02	22.71
C18H21I101	380.27	380.0639	56.85	5.57	33.37	4.21
C18H21I102	396.27	396.0588	54.56	5.34	32.02	8.07
C18H21I103	412.27	412.0537	52.44	5.13	30.78	11.64
C18H21I104	428.27	428.0486	50.48	4.94	29.63	14.94
C18H21I105	444.27	444.0435	48.66	4.76	28.56	18.01
C18H21I106	460.27	460.0385	46.97	4.60	27.57	20.86
C18H21I107	476.27	476.0334	45.39	4.44	26.65	23.52
C18H21I108	492.27	492.0283	43.92	4.30	25.78	26.00
C18H21I109	508.27	508.0232	42.54	4.16	24.97	28.33
C18H22I201	508.18	507.9764	42.54	4.36	49.94	3.15
C18H22I202	524.18	523.9713	41.25	4.23	48.42	6.10
C18H22I203	540.18	539.9662	40.02	4.11	46.99	8.89
C18H22I204	556.18	555.9611	38.87	3.99	45.63	11.51
C18H22I205	572.18	571.9560	37.79	3.88	44.36	13.98
C18H22I206	588.18	587.9509	36.76	3.77	43.15	16.32
C18H22I207	604.18	603.9459	35.78	3.67	42.01	18.54
C18H22I208	620.18	619.9408	34.86	3.58	40.93	20.64
C18H22I209	636.18	635.9357	33.98	3.49	39.90	22.63
C18H23I101	382.29	382.0795	56.55	6.06	33.20	4.19
C18H23I102	398.29	398.0745	54.28	5.82	31.86	8.03
C18H23I103	414.29	414.0694	52.19	5.60	30.63	11.59
C18H23I104	430.29	430.0643	50.25	5.39	29.49	14.87
C18H23I105	446.29	446.0592	48.44	5.19	28.44	17.93
C18H23I106	462.28	462.0541	46.77	5.01	27.45	20.77
C18H23I107	478.28	478.0490	45.20	4.85	26.53	23.42
C18H23I108	494.28	494.0439	43.74	4.69	25.67	25.90
C18H23I109	510.28	510.0389	42.37	4.54	24.87	28.22
C18H24I201	510.20	509.9920	42.38	4.74	49.75	3.14
C18H24I202	526.20	525.9869	41.09	4.60	48.23	6.08
C18H24I203	542.20	541.9819	39.87	4.46	46.81	8.85
C18H24I204	558.20	557.9768	38.73	4.33	45.47	11.47
C18H24I205	574.20	573.9717	37.65	4.21	44.20	13.93
C18H24I206	590.20	589.9666	36.63	4.10	43.00	16.27
C18H24I207	606.20	605.9615	35.67	3.99	41.87	18.48
C18H24I208	622.20	621.9564	34.75	3.89	40.79	20.57
C18H24I209	638.20	637.9513	33.88	3.79	39.77	22.56
C18H25I101	384.30	384.0952	56.26	6.56	33.02	4.16
C18H25I102	400.30	400.0901	54.01	6.30	31.70	7.99
C18H25I103	416.30	416.0850	51.93	6.05	30.48	11.53
C18H25I104	432.30	432.0799	50.01	5.83	29.36	14.80

C18H25I105	448.30	448.0748	48.23	5.62	28.31	17.84
C18H25I106	464.30	464.0698	46.56	5.43	27.33	20.68
C18H25I107	480.30	480.0647	45.01	5.25	26.42	23.32
C18H25I108	496.30	496.0596	43.56	5.08	25.57	25.79
C18H25I109	512.30	512.0545	42.20	4.92	24.77	28.11
C18H26I201	512.22	512.0077	42.21	5.12	49.55	3.12
C18H26I202	528.22	528.0026	40.93	4.96	48.05	6.06
C18H26I203	544.21	543.9975	39.73	4.82	46.64	8.82
C18H26I204	560.21	559.9924	38.59	4.68	45.31	11.42
C18H26I205	576.21	575.9873	37.52	4.55	44.05	13.88
C18H26I206	592.21	591.9822	36.51	4.43	42.86	16.21
C18H26I207	608.21	607.9772	35.55	4.31	41.73	18.41
C18H26I208	624.21	623.9721	34.64	4.20	40.66	20.51
C18H26I209	640.21	639.9670	33.77	4.09	39.64	22.49
C18H27I101	386.32	386.1108	55.96	7.04	32.85	4.14
C18H27I102	402.32	402.1058	53.74	6.76	31.54	7.95
C18H27I103	418.32	418.1007	51.68	6.51	30.34	11.47
C18H27I104	434.32	434.0956	49.78	6.27	29.22	14.74
C18H27I105	450.32	450.0905	48.01	6.04	28.18	17.76
C18H27I106	466.32	466.0854	46.36	5.84	27.21	20.59
C18H27I107	482.32	482.0803	44.83	5.64	26.31	23.22
C18H27I108	498.32	498.0752	43.39	5.46	25.47	25.69
C18H27I109	514.31	514.0702	42.04	5.29	24.67	28.00
C18H28I201	514.23	514.0233	42.04	5.49	49.36	3.11
C18H28I202	530.23	530.0182	40.77	5.32	47.87	6.03
C18H28I203	546.23	546.0132	39.58	5.17	46.47	8.79
C18H28I204	562.23	562.0081	38.45	5.02	45.14	11.38
C18H28I205	578.23	578.0030	37.39	4.88	43.89	13.83
C18H28I206	594.23	593.9979	36.38	4.75	42.71	16.15
C18H28I207	610.23	609.9928	35.43	4.63	41.59	18.35
C18H28I208	626.23	625.9877	34.52	4.51	40.53	20.44
C18H28I209	642.23	641.9826	33.66	4.39	39.52	22.42
C18H29I101	388.34	388.1265	55.67	7.53	32.68	4.12
C18H29I102	404.34	404.1214	53.47	7.23	31.39	7.91
C18H29I103	420.33	420.1163	51.44	6.95	30.19	11.42
C18H29I104	436.33	436.1112	49.55	6.70	29.08	14.67
C18H29I105	452.33	452.1061	47.80	6.46	28.06	17.69
C18H29I106	468.33	468.1011	46.16	6.24	27.10	20.50
C18H29I107	484.33	484.0960	44.64	6.04	26.20	23.12
C18H29I108	500.33	500.0909	43.21	5.84	25.36	25.58
C18H29I109	516.33	516.0858	41.87	5.66	24.58	27.89
C18H30I201	516.25	516.0390	41.88	5.86	49.16	3.10
C18H30I202	532.25	532.0339	40.62	5.68	47.69	6.01
C18H30I203	548.25	548.0288	39.43	5.52	46.29	8.75
C18H30I204	564.25	564.0237	38.32	5.36	44.98	11.34
C18H30I205	580.25	580.0186	37.26	5.21	43.74	13.79
C18H30I206	596.25	596.0135	36.26	5.07	42.57	16.10
C18H30I207	612.24	612.0085	35.31	4.94	41.46	18.29
C18H30I208	628.24	628.0034	34.41	4.81	40.40	20.37
C18H30I209	644.24	643.9983	33.56	4.69	39.40	22.35
C18H31I101	390.35	390.1421	55.39	8.00	32.51	4.10
C18H31I102	406.35	406.1371	53.21	7.69	31.23	7.87
C18H31I103	422.35	422.1320	51.19	7.40	30.05	11.36
C18H31I104	438.35	438.1269	49.32	7.13	28.95	14.60
C18H31I105	454.35	454.1218	47.58	6.88	27.93	17.61
C18H31I106	470.35	470.1167	45.97	6.64	26.98	20.41
C18H31I107	486.35	486.1116	44.45	6.42	26.09	23.03
C18H31I108	502.35	502.1065	43.04	6.22	25.26	25.48
C18H31I109	518.35	518.1014	41.71	6.03	24.48	27.78
C18H32I201	518.26	518.0546	41.72	6.22	48.97	3.09

C18H32I202	534.26	534.0495	40.47	6.04	47.51	5.99
C18H32I203	550.26	550.0444	39.29	5.86	46.13	8.72
C18H32I204	566.26	566.0394	38.18	5.70	44.82	11.30
C18H32I205	582.26	582.0343	37.13	5.54	43.59	13.74
C18H32I206	598.26	598.0292	36.14	5.39	42.42	16.05
C18H32I207	614.26	614.0241	35.20	5.25	41.32	18.23
C18H32I208	630.26	630.0190	34.30	5.12	40.27	20.31
C18H32I209	646.26	646.0139	33.45	4.99	39.27	22.28
C18H33I101	392.37	392.1578	55.10	8.48	32.34	4.08
C18H33I102	408.37	408.1527	52.94	8.15	31.08	7.84
C18H33I103	424.37	424.1476	50.95	7.84	29.90	11.31
C18H33I104	440.37	440.1425	49.10	7.55	28.82	14.53
C18H33I105	456.37	456.1374	47.37	7.29	27.81	17.53
C18H33I106	472.36	472.1324	45.77	7.04	26.87	20.32
C18H33I107	488.36	488.1273	44.27	6.81	25.99	22.93
C18H33I108	504.36	504.1222	42.87	6.60	25.16	25.38
C18H33I109	520.36	520.1171	41.55	6.39	24.39	27.67
C18H34I201	520.28	520.0703	41.55	6.59	48.78	3.08
C18H34I202	536.28	536.0652	40.31	6.39	47.33	5.97
C18H34I203	552.28	552.0601	39.15	6.21	45.96	8.69
C18H34I204	568.28	568.0550	38.04	6.03	44.66	11.26
C18H34I205	584.28	584.0499	37.00	5.87	43.44	13.69
C18H34I206	600.28	600.0448	36.02	5.71	42.28	15.99
C18H34I207	616.28	616.0398	35.08	5.56	41.18	18.17
C18H34I208	632.28	632.0347	34.19	5.42	40.14	20.24
C18H34I209	648.28	648.0296	33.35	5.29	39.15	22.21
C18H35I101	394.38	394.1734	54.82	8.95	32.18	4.06
C18H35I102	410.38	410.1683	52.68	8.60	30.92	7.80
C18H35I103	426.38	426.1633	50.71	8.27	29.76	11.26
C18H35I104	442.38	442.1582	48.87	7.97	28.69	14.47
C18H35I105	458.38	458.1531	47.17	7.70	27.69	17.45
C18H35I106	474.38	474.1480	45.58	7.44	26.75	20.24
C18H35I107	490.38	490.1429	44.09	7.19	25.88	22.84
C18H35I108	506.38	506.1378	42.70	6.97	25.06	25.28
C18H35I109	522.38	522.1327	41.39	6.75	24.29	27.57
C18H36I201	522.30	522.0859	41.39	6.95	48.59	3.06
C18H36I202	538.30	538.0808	40.16	6.74	47.15	5.94
C18H36I203	554.29	554.0757	39.00	6.55	45.79	8.66
C18H36I204	570.29	570.0707	37.91	6.36	44.50	11.22
C18H36I205	586.29	586.0656	36.88	6.19	43.29	13.64
C18H36I206	602.29	602.0605	35.90	6.02	42.14	15.94
C18H36I207	618.29	618.0554	34.97	5.87	41.05	18.11
C18H36I208	634.29	634.0503	34.09	5.72	40.01	20.18
C18H36I209	650.29	650.0452	33.25	5.58	39.03	22.14
C18H37I101	396.40	396.1891	54.54	9.41	32.01	4.04
C18H37I102	412.40	412.1840	52.43	9.04	30.77	7.76
C18H37I103	428.40	428.1789	50.47	8.71	29.62	11.20
C18H37I104	444.40	444.1738	48.65	8.39	28.56	14.40
C18H37I105	460.40	460.1687	46.96	8.10	27.56	17.38
C18H37I106	476.40	476.1637	45.38	7.83	26.64	20.15
C18H37I107	492.40	492.1586	43.91	7.57	25.77	22.75
C18H37I108	508.40	508.1535	42.53	7.34	24.96	25.18
C18H37I109	524.39	524.1484	41.23	7.11	24.20	27.46
C19H 9I101	380.19	379.9700	60.03	2.39	33.38	4.21
C19H 9I102	396.19	395.9649	57.60	2.29	32.03	8.08
C19H 9I103	412.19	411.9598	55.37	2.20	30.79	11.64
C19H 9I104	428.19	427.9547	53.30	2.12	29.64	14.95
C19H 9I105	444.18	443.9497	51.38	2.04	28.57	18.01
C19H 9I106	460.18	459.9446	49.59	1.97	27.58	20.86
C19H 9I107	476.18	475.9395	47.93	1.91	26.65	23.52
C19H 9I108	492.18	491.9344	46.37	1.84	25.78	26.01
C19H 9I109	508.18	507.9293	44.91	1.79	24.97	28.34
C19H10I201	508.10	507.8825	44.91	1.98	49.95	3.15
C19H10I202	524.10	523.8774	43.54	1.92	48.43	6.11
C19H10I203	540.10	539.8723	42.25	1.87	46.99	8.89
C19H10I204	556.10	555.8672	41.04	1.81	45.64	11.51
C19H10I205	572.10	571.8621	39.89	1.76	44.36	13.98
C19H10I206	588.10	587.8571	38.81	1.71	43.16	16.32
C19H10I207	604.10	603.8520	37.78	1.67	42.01	18.54
C19H10I208	620.10	619.8469	36.80	1.63	40.93	20.64
C19H10I209	636.09	635.8418	35.88	1.58	39.90	22.64
C19H11I101	382.20	381.9856	59.71	2.90	33.20	4.19
C19H11I102	398.20	397.9806	57.31	2.78	31.87	8.04
C19H11I103	414.20	413.9755	55.10	2.68	30.64	11.59
C19H11I104	430.20	429.9704	53.05	2.58	29.50	14.88
C19H11I105	446.20	445.9653	51.15	2.48	28.44	17.93
C19H11I106	462.20	461.9602	49.38	2.40	27.46	20.77
C19H11I107	478.20	477.9551	47.72	2.32	26.54	23.42
C19H11I108	494.20	493.9500	46.18	2.24	25.68	25.90
C19H11I109	510.20	509.9450	44.73	2.17	24.87	28.22
C19H12I201	510.12	509.8981	44.74	2.37	49.76	3.14
C19H12I202	526.12	525.8930	43.38	2.30	48.24	6.08
C19H12I203	542.11	541.8880	42.10	2.23	46.82	8.85
C19H12I204	558.11	557.8829	40.89	2.17	45.48	11.47
C19H12I205	574.11	573.8778	39.75	2.11	44.21	13.93
C19H12I206	590.11	589.8727	38.67	2.05	43.01	16.27
C19H12I207	606.11	605.8676	37.65	2.00	41.87	18.48
C19H12I208	622.11	621.8625	36.68	1.94	40.80	20.57
C19H12I209	638.11	637.8574	35.76	1.90	39.78	22.57
C19H13I101	384.22	384.0013	59.40	3.41	33.03	4.16
C19H13I102	400.22	399.9962	57.02	3.27	31.71	8.00
C19H13I103	416.22	415.9911	54.83	3.15	30.49	11.53
C19H13I104	432.22	431.9860	52.80	3.03	29.36	14.81
C19H13I105	448.22	447.9810	50.92	2.92	28.31	17.85
C19H13I106	464.22	463.9759	49.16	2.82	27.34	20.68
C19H13I107	480.22	479.9708	47.52	2.73	26.43	23.32
C19H13I108	496.22	495.9657	45.99	2.64	25.57	25.79
C19H13I109	512.21	511.9606	44.55	2.56	24.78	28.11
C19H14I201	512.13	511.9138	44.56	2.76	49.56	3.12
C19H14I202	528.13	527.9087	43.21	2.67	48.06	6.06
C19H14I203	544.13	543.9036	41.94	2.59	46.64	8.82
C19H14I204	560.13	559.8985	40.74	2.52	45.31	11.43
C19H14I205	576.13	575.8934	39.61	2.45	44.05	13.89
C19H14I206	592.13	591.8883	38.54	2.38	42.86	16.21
C19H14I207	608.13	607.8833	37.53	2.32	41.74	18.42
C19H14I208	624.13	623.8782	36.56	2.26	40.67	20.51
C19H14I209	640.13	639.8731	35.65	2.20	39.65	22.49
C19H15I101	386.24	386.0169	59.09	3.91	32.86	4.14
C19H15I102	402.23	402.0119	56.74	3.76	31.55	7.96
C19H15I103	418.23	418.0068	54.57	3.62	30.34	11.48
C19H15I104	434.23	434.0017	52.56	3.48	29.22	14.74
C19H15I105	450.23	449.9966	50.69	3.36	28.19	17.77
C19H15I106	466.23	465.9915	48.95	3.24	27.22	20.59
C19H15I107	482.23	481.9864	47.32	3.14	26.32	23.22
C19H15I108	498.23	497.9813	45.80	3.03	25.47	25.69
C19H15I109	514.23	513.9763	44.38	2.94	24.68	28.00
C19H16I201	514.15	513.9294	44.39	3.14	49.36	3.11
C19H16I202	530.15	529.9243	43.05	3.04	47.88	6.04
C19H16I203	546.15	545.9193	41.79	2.95	46.47	8.79
C19H16I204	562.15	561.9142	40.60	2.87	45.15	11.38

C19H16I205	578.15	577.9091	39.47	2.79	43.90	13.84
C19H16I206	594.14	593.9040	38.41	2.71	42.72	16.16
C19H16I207	610.14	609.8989	37.40	2.64	41.60	18.36
C19H16I208	626.14	625.8938	36.45	2.58	40.54	20.44
C19H16I209	642.14	641.8887	35.54	2.51	39.53	22.42
C19H17I101	388.25	388.0326	58.78	4.41	32.69	4.12
C19H17I102	404.25	404.0275	56.45	4.24	31.39	7.92
C19H17I103	420.25	420.0224	54.30	4.08	30.20	11.42
C19H17I104	436.25	436.0173	52.31	3.93	29.09	14.67
C19H17I105	452.25	452.0122	50.46	3.79	28.06	17.69
C19H17I106	468.25	468.0072	48.74	3.66	27.10	20.50
C19H17I107	484.25	484.0021	47.13	3.54	26.21	23.13
C19H17I108	500.25	499.9970	45.62	3.43	25.37	25.59
C19H17I109	516.25	515.9919	44.21	3.32	24.58	27.89
C19H18I201	516.16	515.9451	44.21	3.52	49.17	3.10
C19H18I202	532.16	531.9400	42.88	3.41	47.69	6.01
C19H18I203	548.16	547.9349	41.63	3.31	46.30	8.76
C19H18I204	564.16	563.9298	40.45	3.22	44.99	11.34
C19H18I205	580.16	579.9247	39.34	3.13	43.75	13.79
C19H18I206	596.16	595.9196	38.28	3.04	42.57	16.10
C19H18I207	612.16	611.9146	37.28	2.96	41.46	18.30
C19H18I208	628.16	627.9095	36.33	2.89	40.41	20.38
C19H18I209	644.16	643.9044	35.43	2.82	39.40	22.35
C19H19I101	390.27	390.0482	58.48	4.91	32.52	4.10
C19H19I102	406.27	406.0432	56.17	4.71	31.24	7.88
C19H19I103	422.27	422.0381	54.04	4.54	30.05	11.37
C19H19I104	438.27	438.0330	52.07	4.37	28.96	14.60
C19H19I105	454.26	454.0279	50.24	4.22	27.94	17.61
C19H19I106	470.26	470.0228	48.53	4.07	26.99	20.41
C19H19I107	486.26	486.0177	46.93	3.94	26.10	23.03
C19H19I108	502.26	502.0126	45.44	3.81	25.27	25.48
C19H19I109	518.26	518.0076	44.03	3.70	24.49	27.78
C19H20I201	518.18	517.9607	44.04	3.89	48.98	3.09
C19H20I202	534.18	533.9556	42.72	3.77	47.51	5.99
C19H20I203	550.18	549.9506	41.48	3.66	46.13	8.72
C19H20I204	566.18	565.9455	40.31	3.56	44.83	11.30
C19H20I205	582.18	581.9404	39.20	3.46	43.60	13.74
C19H20I206	598.18	597.9353	38.15	3.37	42.43	16.05
C19H20I207	614.18	613.9302	37.16	3.28	41.33	18.24
C19H20I208	630.18	629.9251	36.21	3.20	40.28	20.31
C19H20I209	646.17	645.9200	35.32	3.12	39.28	22.28
C19H21I101	392.28	392.0639	58.18	5.40	32.35	4.08
C19H21I102	408.28	408.0588	55.90	5.18	31.08	7.84
C19H21I103	424.28	424.0537	53.79	4.99	29.91	11.31
C19H21I104	440.28	440.0486	51.83	4.81	28.82	14.54
C19H21I105	456.28	456.0435	50.02	4.64	27.81	17.53
C19H21I106	472.28	472.0385	48.32	4.48	26.87	20.33
C19H21I107	488.28	488.0334	46.74	4.34	25.99	22.94
C19H21I108	504.28	504.0283	45.26	4.20	25.17	25.38
C19H21I109	520.28	520.0232	43.86	4.07	24.39	27.68
C19H22I201	520.20	519.9764	43.87	4.26	48.79	3.08
C19H22I202	536.19	535.9713	42.56	4.14	47.34	5.97
C19H22I203	552.19	551.9662	41.33	4.02	45.96	8.69
C19H22I204	568.19	567.9611	40.16	3.90	44.67	11.26
C19H22I205	584.19	583.9560	39.06	3.80	43.45	13.69
C19H22I206	600.19	599.9509	38.02	3.69	42.29	15.99
C19H22I207	616.19	615.9459	37.04	3.60	41.19	18.18
C19H22I208	632.19	631.9408	36.10	3.51	40.15	20.25
C19H22I209	648.19	647.9357	35.21	3.42	39.16	22.21
C19H23I101	394.30	394.0795	57.88	5.88	32.18	4.06
C19H23I102	410.30	410.0745	55.62	5.65	30.93	7.80
C19H23I103	426.30	426.0694	53.53	5.44	29.77	11.26
C19H23I104	442.30	442.0643	51.60	5.24	28.69	14.47
C19H23I105	458.30	458.0592	49.80	5.06	27.69	17.46
C19H23I106	474.30	474.0541	48.12	4.89	26.76	20.24
C19H23I107	490.30	490.0490	46.55	4.73	25.88	22.84
C19H23I108	506.29	506.0439	45.07	4.58	25.07	25.28
C19H23I109	522.29	522.0389	43.69	4.44	24.30	27.57
C19H24I201	522.21	521.9920	43.70	4.63	48.60	3.06
C19H24I202	538.21	537.9869	42.40	4.49	47.16	5.95
C19H24I203	554.21	553.9819	41.18	4.37	45.80	8.66
C19H24I204	570.21	569.9768	40.02	4.24	44.51	11.22
C19H24I205	586.21	585.9717	38.93	4.13	43.30	13.65
C19H24I206	602.21	601.9666	37.90	4.02	42.15	15.94
C19H24I207	618.21	617.9615	36.92	3.91	41.06	18.12
C19H24I208	634.21	633.9564	35.98	3.81	40.02	20.18
C19H24I209	650.21	649.9513	35.10	3.72	39.04	22.15
C19H25I101	396.31	396.0952	57.58	6.36	32.02	4.04
C19H25I102	412.31	412.0901	55.35	6.11	30.78	7.76
C19H25I103	428.31	428.0850	53.28	5.88	29.63	11.21
C19H25I104	444.31	444.0799	51.36	5.67	28.56	14.40
C19H25I105	460.31	460.0748	49.58	5.47	27.57	17.38
C19H25I106	476.31	476.0698	47.91	5.29	26.64	20.15
C19H25I107	492.31	492.0647	46.36	5.12	25.78	22.75
C19H25I108	508.31	508.0596	44.90	4.96	24.97	25.18
C19H25I109	524.31	524.0545	43.53	4.81	24.20	27.46
C19H26I201	524.23	524.0077	43.53	5.00	48.42	3.05
C19H26I202	540.23	540.0026	42.24	4.85	46.98	5.92
C19H26I203	556.23	555.9975	41.03	4.71	45.63	8.63
C19H26I204	572.23	571.9924	39.88	4.58	44.35	11.18
C19H26I205	588.22	587.9873	38.80	4.46	43.15	13.60
C19H26I206	604.22	603.9822	37.77	4.34	42.01	15.89
C19H26I207	620.22	619.9772	36.80	4.23	40.92	18.06
C19H26I208	636.22	635.9721	35.87	4.12	39.89	20.12
C19H26I209	652.22	651.9670	34.99	4.02	38.91	22.08
C19H27I101	398.33	398.1108	57.29	6.83	31.86	4.02
C19H27I102	414.33	414.1058	55.08	6.57	30.63	7.72
C19H27I103	430.33	430.1007	53.03	6.32	29.49	11.15
C19H27I104	446.33	446.0956	51.13	6.10	28.43	14.34
C19H27I105	462.33	462.0905	49.36	5.89	27.45	17.30
C19H27I106	478.33	478.0854	47.71	5.69	26.53	20.07
C19H27I107	494.33	494.0803	46.17	5.51	25.67	22.66
C19H27I108	510.33	510.0752	44.72	5.33	24.87	25.08
C19H27I109	526.33	526.0702	43.36	5.17	24.11	27.36
C19H28I201	526.24	526.0233	43.37	5.36	48.23	3.04
C19H28I202	542.24	542.0182	42.09	5.20	46.81	5.90
C19H28I203	558.24	558.0132	40.88	5.06	45.47	8.60
C19H28I204	574.24	574.0081	39.74	4.91	44.20	11.14
C19H28I205	590.24	590.0030	38.66	4.78	43.00	13.55
C19H28I206	606.24	605.9979	37.64	4.66	41.87	15.83
C19H28I207	622.24	621.9928	36.68	4.54	40.79	18.00
C19H28I208	638.24	637.9877	35.76	4.42	39.77	20.05
C19H28I209	654.24	653.9826	34.88	4.31	38.79	22.01
C19H29I101	400.35	400.1265	57.00	7.30	31.70	4.00
C19H29I102	416.35	416.1214	54.81	7.02	30.48	7.69
C19H29I103	432.35	432.1163	52.78	6.76	29.35	11.10
C19H29I104	448.34	448.1112	50.90	6.52	28.31	14.27
C19H29I105	464.34	464.1061	49.15	6.30	27.33	17.23
C19H29I106	480.34	480.1011	47.51	6.09	26.42	19.98
C19H29I107	496.34	496.0960	45.98	5.89	25.57	22.56

C19H29I108	512.34	512.0909	44.54	5.71	24.77	24.98
C19H29I109	528.34	528.0858	43.19	5.53	24.02	27.25
C19H30I201	528.26	528.0390	43.20	5.72	48.05	3.03
C19H30I202	544.26	544.0339	41.93	5.56	46.63	5.88
C19H30I203	560.26	560.0288	40.73	5.40	45.30	8.57
C19H30I204	576.26	576.0237	39.60	5.25	44.04	11.11
C19H30I205	592.26	592.0186	38.53	5.11	42.85	13.51
C19H30I206	608.26	608.0135	37.52	4.97	41.73	15.78
C19H30I207	624.26	624.0085	36.56	4.84	40.66	17.94
C19H30I208	640.25	640.0034	35.64	4.72	39.64	19.99
C19H30I209	656.25	655.9983	34.77	4.61	38.68	21.94
C19H31I101	402.36	402.1421	56.72	7.77	31.54	3.98
C19H31I102	418.36	418.1371	54.55	7.47	30.33	7.65
C19H31I103	434.36	434.1320	52.54	7.19	29.22	11.05
C19H31I104	450.36	450.1269	50.67	6.94	28.18	14.21
C19H31I105	466.36	466.1218	48.93	6.70	27.21	17.15
C19H31I106	482.36	482.1167	47.31	6.48	26.31	19.90
C19H31I107	498.36	498.1116	45.79	6.27	25.46	22.47
C19H31I108	514.36	514.1065	44.37	6.07	24.67	24.88
C19H31I109	530.36	530.1014	43.03	5.89	23.93	27.15
C19H32I201	530.28	530.0546	43.04	6.08	47.86	3.02
C19H32I202	546.27	546.0495	41.78	5.90	46.46	5.86
C19H32I203	562.27	562.0444	40.59	5.74	45.14	8.54
C19H32I204	578.27	578.0394	39.46	5.58	43.89	11.07
C19H32I205	594.27	594.0343	38.40	5.43	42.71	13.46
C19H32I206	610.27	610.0292	37.40	5.29	41.59	15.73
C19H32I207	626.27	626.0241	36.44	5.15	40.53	17.88
C19H32I208	642.27	642.0190	35.53	5.02	39.52	19.93
C19H32I209	658.27	658.0139	34.67	4.90	38.56	21.87
C19H33I101	404.38	404.1578	56.44	8.23	31.38	3.96
C19H33I102	420.38	420.1527	54.29	7.91	30.19	7.61
C19H33I103	436.38	436.1476	52.30	7.62	29.08	11.00
C19H33I104	452.38	452.1425	50.45	7.35	28.05	14.15
C19H33I105	468.38	468.1374	48.72	7.10	27.09	17.08
C19H33I106	484.38	484.1324	47.11	6.87	26.20	19.82
C19H33I107	500.38	500.1273	45.61	6.65	25.36	22.38
C19H33I108	516.37	516.1222	44.20	6.44	24.58	24.79
C19H33I109	532.37	532.1171	42.87	6.25	23.84	27.05
C19H34I201	532.29	532.0703	42.87	6.44	47.68	3.01
C19H34I202	548.29	548.0652	41.62	6.25	46.29	5.84
C19H34I203	564.29	564.0601	40.44	6.07	44.98	8.51
C19H34I204	580.29	580.0550	39.33	5.91	43.74	11.03
C19H34I205	596.29	596.0499	38.27	5.75	42.56	13.42
C19H34I206	612.29	612.0448	37.27	5.60	41.45	15.68
C19H34I207	628.29	628.0398	36.32	5.45	40.40	17.83
C19H34I208	644.29	644.0347	35.42	5.32	39.39	19.87
C19H34I209	660.29	660.0296	34.56	5.19	38.44	21.81
C19H35I101	406.39	406.1734	56.16	8.68	31.23	3.94
C19H35I102	422.39	422.1683	54.03	8.35	30.04	7.58
C19H35I103	438.39	438.1633	52.06	8.05	28.95	10.95
C19H35I104	454.39	454.1582	50.22	7.76	27.93	14.08
C19H35I105	470.39	470.1531	48.52	7.50	26.98	17.01
C19H35I106	486.39	486.1480	46.92	7.25	26.09	19.74
C19H35I107	502.39	502.1429	45.43	7.02	25.26	22.29
C19H35I108	518.39	518.1378	44.02	6.81	24.48	24.69
C19H35I109	534.39	534.1327	42.71	6.60	23.75	26.95
C19H36I201	534.31	534.0859	42.71	6.79	47.50	2.99
C19H36I202	550.31	550.0808	41.47	6.59	46.12	5.81
C19H36I203	566.31	566.0757	40.30	6.41	44.82	8.48
C19H36I204	582.31	582.0707	39.19	6.23	43.59	10.99

C19H36I205	598.30	598.0656	38.14	6.06	42.42	13.37
C19H36I206	614.30	614.0605	37.15	5.91	41.32	15.63
C19H36I207	630.30	630.0554	36.21	5.76	40.27	17.77
C19H36I208	646.30	646.0503	35.31	5.61	39.27	19.80
C19H36I209	662.30	662.0452	34.46	5.48	38.32	21.74
C19H37I101	408.41	408.1891	55.88	9.13	31.07	3.92
C19H37I102	424.41	424.1840	53.77	8.79	29.90	7.54
C19H37I103	440.41	440.1789	51.82	8.47	28.82	10.90
C19H37I104	456.41	456.1738	50.00	8.17	27.80	14.02
C19H37I105	472.41	472.1687	48.31	7.89	26.86	16.93
C19H37I106	488.41	488.1637	46.73	7.64	25.98	19.65
C19H37I107	504.41	504.1586	45.24	7.39	25.16	22.20
C19H37I108	520.41	520.1535	43.85	7.17	24.39	24.60
C19H37I109	536.41	536.1484	42.54	6.95	23.66	26.84
C19H38I201	536.32	536.1016	42.55	7.14	47.32	2.98
C19H38I202	552.32	552.0965	41.32	6.93	45.95	5.79
C19H38I203	568.32	568.0914	40.16	6.74	44.66	8.45
C19H38I204	584.32	584.0863	39.06	6.56	43.44	10.95
C19H38I205	600.32	600.0812	38.02	6.38	42.28	13.33
C19H38I206	616.32	616.0761	37.03	6.21	41.18	15.58
C19H38I207	632.32	632.0711	36.09	6.06	40.14	17.71
C19H38I208	648.32	648.0660	35.20	5.91	39.15	19.74
C19H38I209	664.32	664.0609	34.35	5.77	38.21	21.68
C19H39I101	410.43	410.2047	55.60	9.58	30.92	3.90
C19H39I102	426.43	426.1996	53.52	9.22	29.76	7.50
C19H39I103	442.43	442.1946	51.58	8.89	28.68	10.85
C19H39I104	458.42	458.1895	49.78	8.58	27.68	13.96
C19H39I105	474.42	474.1844	48.10	8.29	26.75	16.86
C19H39I106	490.42	490.1793	46.53	8.02	25.88	19.57
C19H39I107	506.42	506.1742	45.06	7.76	25.06	22.12
C19H39I108	522.42	522.1691	43.68	7.52	24.29	24.50
C19H39I109	538.42	538.1640	42.39	7.30	23.57	26.74
C20H10I201	520.11	519.8825	46.19	1.94	48.80	3.08
C20H10I202	536.11	535.8774	44.81	1.88	47.34	5.97
C20H10I203	552.11	551.8723	43.51	1.83	45.97	8.69
C20H10I204	568.11	567.8672	42.28	1.77	44.68	11.27
C20H10I205	584.11	583.8621	41.13	1.73	43.45	13.70
C20H10I206	600.11	599.8571	40.03	1.68	42.29	16.00
C20H10I207	616.11	615.8520	38.99	1.64	41.20	18.18
C20H10I208	632.11	631.8469	38.00	1.59	40.15	20.25
C20H10I209	648.11	647.8418	37.07	1.56	39.16	22.22
C20H11I101	394.21	393.9856	60.94	2.81	32.19	4.06
C20H11I102	410.21	409.9806	58.56	2.70	30.94	7.80
C20H11I103	426.21	425.9755	56.36	2.60	29.77	11.26
C20H11I104	442.21	441.9704	54.32	2.51	28.70	14.47
C20H11I105	458.21	457.9653	52.43	2.42	27.70	17.46
C20H11I106	474.21	473.9602	50.66	2.34	26.76	20.24
C20H11I107	490.21	489.9551	49.00	2.26	25.89	22.85
C20H11I108	506.21	505.9500	47.46	2.19	25.07	25.28
C20H11I109	522.21	521.9450	46.00	2.12	24.30	27.57
C20H12I201	522.13	521.8981	46.01	2.32	48.61	3.06
C20H12I202	538.13	537.8930	44.64	2.25	47.17	5.95
C20H12I203	554.13	553.8880	43.35	2.18	45.80	8.66
C20H12I204	570.13	569.8829	42.14	2.12	44.52	11.23
C20H12I205	586.12	585.8778	40.98	2.06	43.30	13.65
C20H12I206	602.12	601.8727	39.90	2.01	42.15	15.94
C20H12I207	618.12	617.8676	38.86	1.96	41.06	18.12
C20H12I208	634.12	633.8625	37.88	1.91	40.03	20.18
C20H12I209	650.12	649.8574	36.95	1.86	39.04	22.15
C20H13I101	396.23	396.0013	60.63	3.31	32.03	4.04

C20H13I102	412.23	411.9962	58.27	3.18	30.78	7.76
C20H13I103	428.23	427.9911	56.10	3.06	29.63	11.21
C20H13I104	444.23	443.9860	54.08	2.95	28.57	14.41
C20H13I105	460.23	459.9810	52.20	2.85	27.57	17.38
C20H13I106	476.23	475.9759	50.44	2.75	26.65	20.16
C20H13I107	492.23	491.9708	48.80	2.66	25.78	22.75
C20H13I108	508.23	507.9657	47.27	2.58	24.97	25.18
C20H13I109	524.23	523.9606	45.82	2.50	24.21	27.47
C20H14I201	524.14	523.9138	45.83	2.69	48.42	3.05
C20H14I202	540.14	539.9087	44.47	2.61	46.99	5.92
C20H14I203	556.14	555.9036	43.19	2.54	45.64	8.63
C20H14I204	572.14	571.8985	41.99	2.47	44.36	11.19
C20H14I205	588.14	587.8934	40.84	2.40	43.15	13.60
C20H14I206	604.14	603.8883	39.76	2.34	42.01	15.89
C20H14I207	620.14	619.8833	38.74	2.28	40.93	18.06
C20H14I208	636.14	635.8782	37.76	2.22	39.90	20.12
C20H14I209	652.14	651.8731	36.84	2.16	38.92	22.08
C20H15I101	398.25	398.0169	60.32	3.80	31.87	4.02
C20H15I102	414.25	414.0119	57.99	3.65	30.64	7.72
C20H15I103	430.25	430.0068	55.83	3.51	29.50	11.16
C20H15I104	446.24	446.0017	53.83	3.39	28.44	14.34
C20H15I105	462.24	461.9966	51.97	3.27	27.45	17.31
C20H15I106	478.24	477.9915	50.23	3.16	26.54	20.07
C20H15I107	494.24	493.9864	48.60	3.06	25.68	22.66
C20H15I108	510.24	509.9813	47.08	2.96	24.87	25.09
C20H15I109	526.24	525.9763	45.65	2.87	24.12	27.36
C20H16I201	526.16	525.9294	45.66	3.07	48.24	3.04
C20H16I202	542.16	541.9243	44.31	2.97	46.81	5.90
C20H16I203	558.16	557.9193	43.04	2.89	45.47	8.60
C20H16I204	574.16	573.9142	41.84	2.81	44.21	11.15
C20H16I205	590.16	589.9091	40.70	2.73	43.01	13.56
C20H16I206	606.16	605.9040	39.63	2.66	41.87	15.84
C20H16I207	622.16	621.8989	38.61	2.59	40.80	18.00
C20H16I208	638.15	637.8938	37.64	2.53	39.77	20.06
C20H16I209	654.15	653.8887	36.72	2.47	38.80	22.01
C20H17I101	400.26	400.0326	60.02	4.28	31.71	4.00
C20H17I102	416.26	416.0275	57.71	4.12	30.49	7.69
C20H17I103	432.26	432.0224	55.57	3.96	29.36	11.10
C20H17I104	448.26	448.0173	53.59	3.82	28.31	14.28
C20H17I105	464.26	464.0122	51.74	3.69	27.33	17.23
C20H17I106	480.26	480.0072	50.02	3.57	26.42	19.99
C20H17I107	496.26	496.0021	48.41	3.45	25.57	22.57
C20H17I108	512.26	511.9970	46.89	3.35	24.77	24.99
C20H17I109	528.26	527.9919	45.47	3.24	24.02	27.26
C20H18I201	528.17	527.9451	45.48	3.44	48.05	3.03
C20H18I202	544.17	543.9400	44.14	3.33	46.64	5.88
C20H18I203	560.17	559.9349	42.88	3.24	45.31	8.57
C20H18I204	576.17	575.9298	41.69	3.15	44.05	11.11
C20H18I205	592.17	591.9247	40.57	3.06	42.86	13.51
C20H18I206	608.17	607.9196	39.50	2.98	41.73	15.78
C20H18I207	624.17	623.9146	38.49	2.91	40.66	17.94
C20H18I208	640.17	639.9095	37.52	2.83	39.65	19.99
C20H18I209	656.17	655.9044	36.61	2.77	38.68	21.94
C20H19I101	402.28	402.0482	59.72	4.76	31.55	3.98
C20H19I102	418.28	418.0432	57.43	4.58	30.34	7.65
C20H19I103	434.28	434.0381	55.32	4.41	29.22	11.05
C20H19I104	450.28	450.0330	53.35	4.25	28.18	14.21
C20H19I105	466.28	466.0279	51.52	4.11	27.22	17.16
C20H19I106	482.28	482.0228	49.81	3.97	26.31	19.90
C20H19I107	498.27	498.0177	48.21	3.84	25.47	22.48

C20H19I108	514.27	514.0126	46.71	3.72	24.68	24.89
C20H19I109	530.27	530.0076	45.30	3.61	23.93	27.15
C20H20I201	530.19	529.9607	45.31	3.80	47.87	3.02
C20H20I202	546.19	545.9556	43.98	3.69	46.47	5.86
C20H20I203	562.19	561.9506	42.73	3.59	45.15	8.54
C20H20I204	578.19	577.9455	41.55	3.49	43.90	11.07
C20H20I205	594.19	593.9404	40.43	3.39	42.72	13.46
C20H20I206	610.19	609.9353	39.37	3.30	41.60	15.73
C20H20I207	626.19	625.9302	38.36	3.22	40.53	17.89
C20H20I208	642.19	641.9251	37.41	3.14	39.52	19.93
C20H20I209	658.19	657.9200	36.50	3.06	38.56	21.88
C20H21I101	404.29	404.0639	59.42	5.24	31.39	3.96
C20H21I102	420.29	420.0588	57.16	5.04	30.19	7.61
C20H21I103	436.29	436.0537	55.06	4.85	29.09	11.00
C20H21I104	452.29	452.0486	53.11	4.68	28.06	14.15
C20H21I105	468.29	468.0435	51.30	4.52	27.10	17.08
C20H21I106	484.29	484.0385	49.60	4.37	26.20	19.82
C20H21I107	500.29	500.0334	48.02	4.23	25.37	22.39
C20H21I108	516.29	516.0283	46.53	4.10	24.58	24.79
C20H21I109	532.29	532.0232	45.13	3.98	23.84	27.05
C20H22I201	532.21	531.9764	45.14	4.17	47.69	3.01
C20H22I202	548.21	547.9713	43.82	4.05	46.30	5.84
C20H22I203	564.21	563.9662	42.58	3.93	44.99	8.51
C20H22I204	580.20	579.9611	41.40	3.82	43.74	11.03
C20H22I205	596.20	595.9560	40.29	3.72	42.57	13.42
C20H22I206	612.20	611.9509	39.24	3.62	41.46	15.68
C20H22I207	628.20	627.9459	38.24	3.53	40.40	17.83
C20H22I208	644.20	643.9408	37.29	3.44	39.40	19.87
C20H22I209	660.20	659.9357	36.39	3.36	38.44	21.81
C20H23I101	406.31	406.0795	59.12	5.71	31.23	3.94
C20H23I102	422.31	422.0745	56.88	5.49	30.05	7.58
C20H23I103	438.31	438.0694	54.81	5.29	28.95	10.95
C20H23I104	454.31	454.0643	52.88	5.10	27.93	14.09
C20H23I105	470.31	470.0592	51.08	4.93	26.98	17.01
C20H23I106	486.31	486.0541	49.40	4.77	26.10	19.74
C20H23I107	502.31	502.0490	47.82	4.62	25.26	22.30
C20H23I108	518.31	518.0439	46.35	4.47	24.48	24.69
C20H23I109	534.31	534.0389	44.96	4.34	23.75	26.95
C20H24I201	534.22	533.9920	44.97	4.53	47.51	2.99
C20H24I202	550.22	549.9869	43.66	4.40	46.13	5.82
C20H24I203	566.22	565.9819	42.43	4.27	44.83	8.48
C20H24I204	582.22	581.9768	41.26	4.16	43.59	10.99
C20H24I205	598.22	597.9717	40.16	4.04	42.43	13.37
C20H24I206	614.22	613.9666	39.11	3.94	41.32	15.63
C20H24I207	630.22	629.9615	38.12	3.84	40.27	17.77
C20H24I208	646.22	645.9564	37.17	3.74	39.28	19.81
C20H24I209	662.22	661.9513	36.28	3.65	38.33	21.74
C20H25I101	408.33	408.0952	58.83	6.17	31.08	3.92
C20H25I102	424.33	424.0901	56.61	5.94	29.91	7.54
C20H25I103	440.32	440.0850	54.56	5.72	28.82	10.90
C20H25I104	456.32	456.0799	52.64	5.52	27.81	14.02
C20H25I105	472.32	472.0748	50.86	5.34	26.87	16.94
C20H25I106	488.32	488.0698	49.19	5.16	25.99	19.66
C20H25I107	504.32	504.0647	47.63	5.00	25.16	22.21
C20H25I108	520.32	520.0596	46.17	4.84	24.39	24.60
C20H25I109	536.32	536.0545	44.79	4.70	23.66	26.85
C20H26I201	536.24	536.0077	44.80	4.89	47.33	2.98
C20H26I202	552.24	552.0026	43.50	4.75	45.96	5.79
C20H26I203	568.24	567.9975	42.28	4.61	44.67	8.45
C20H26I204	584.24	583.9924	41.12	4.49	43.44	10.95

C20H26I2O5	600.24	599.9873	40.02	4.37	42.28	13.33
C20H26I2O6	616.24	615.9822	38.98	4.25	41.19	15.58
C20H26I2O7	632.23	631.9772	38.00	4.15	40.14	17.71
C20H26I2O8	648.23	647.9721	37.06	4.04	39.15	19.75
C20H26I2O9	664.23	663.9670	36.17	3.95	38.21	21.68
C20H27I1O1	410.34	410.1108	58.54	6.63	30.93	3.90
C20H27I1O2	426.34	426.1058	56.35	6.38	29.77	7.51
C20H27I1O3	442.34	442.1007	54.31	6.15	28.69	10.85
C20H27I1O4	458.34	458.0956	52.41	5.94	27.69	13.96
C20H27I1O5	474.34	474.0905	50.64	5.74	26.75	16.86
C20H27I1O6	490.34	490.0854	48.99	5.55	25.88	19.58
C20H27I1O7	506.34	506.0803	47.44	5.37	25.06	22.12
C20H27I1O8	522.34	522.0752	45.99	5.21	24.30	24.50
C20H27I1O9	538.34	538.0702	44.62	5.06	23.57	26.75
C20H28I2O1	538.25	538.0233	44.63	5.24	47.15	2.97
C20H28I2O2	554.25	554.0182	43.34	5.09	45.79	5.77
C20H28I2O3	570.25	570.0132	42.13	4.95	44.51	8.42
C20H28I2O4	586.25	586.0081	40.98	4.81	43.29	10.92
C20H28I2O5	602.25	602.0030	39.89	4.69	42.14	13.28
C20H28I2O6	618.25	617.9979	38.86	4.56	41.05	15.53
C20H28I2O7	634.25	633.9928	37.88	4.45	40.02	17.66
C20H28I2O8	650.25	649.9877	36.94	4.34	39.03	19.68
C20H28I2O9	666.25	665.9826	36.06	4.24	38.10	21.61
C20H29I1O1	412.36	412.1265	58.26	7.09	30.78	3.88
C20H29I1O2	428.36	428.1214	56.08	6.82	29.63	7.47
C20H29I1O3	444.36	444.1163	54.06	6.58	28.56	10.80
C20H29I1O4	460.36	460.1112	52.18	6.35	27.57	13.90
C20H29I1O5	476.36	476.1061	50.43	6.14	26.64	16.79
C20H29I1O6	492.35	492.1011	48.79	5.94	25.77	19.50
C20H29I1O7	508.35	508.0960	47.26	5.75	24.96	22.03
C20H29I1O8	524.35	524.0909	45.81	5.57	24.20	24.41
C20H29I1O9	540.35	540.0858	44.46	5.41	23.49	26.65
C20H30I2O1	540.27	540.0390	44.46	5.60	46.98	2.96
C20H30I2O2	556.27	556.0339	43.18	5.44	45.63	5.75
C20H30I2O3	572.27	572.0288	41.98	5.28	44.35	8.39
C20H30I2O4	588.27	588.0237	40.84	5.14	43.15	10.88
C20H30I2O5	604.27	604.0186	39.75	5.00	42.00	13.24
C20H30I2O6	620.27	620.0135	38.73	4.88	40.92	15.48
C20H30I2O7	636.27	636.0085	37.76	4.75	39.89	17.60
C20H30I2O8	652.27	652.0034	36.83	4.64	38.91	19.62
C20H30I2O9	668.27	667.9983	35.95	4.53	37.98	21.55
C20H31I1O1	414.37	414.1421	57.97	7.54	30.63	3.86
C20H31I1O2	430.37	430.1371	55.82	7.26	29.49	7.44
C20H31I1O3	446.37	446.1320	53.82	7.00	28.43	10.75
C20H31I1O4	462.37	462.1269	51.95	6.76	27.45	13.84
C20H31I1O5	478.37	478.1218	50.22	6.53	26.53	16.72
C20H31I1O6	494.37	494.1167	48.59	6.32	25.67	19.42
C20H31I1O7	510.37	510.1116	47.07	6.12	24.87	21.94
C20H31I1O8	526.37	526.1065	45.64	5.94	24.11	24.32
C20H31I1O9	542.37	542.1014	44.29	5.76	23.40	26.55
C20H32I2O1	542.29	542.0546	44.30	5.95	46.80	2.95
C20H32I2O2	558.29	558.0495	43.03	5.78	45.46	5.73
C20H32I2O3	574.29	574.0444	41.83	5.62	44.20	8.36
C20H32I2O4	590.28	590.0394	40.70	5.46	43.00	10.84
C20H32I2O5	606.28	606.0343	39.62	5.32	41.86	13.19
C20H32I2O6	622.28	622.0292	38.60	5.18	40.79	15.43
C20H32I2O7	638.28	638.0241	37.64	5.05	39.76	17.55
C20H32I2O8	654.28	654.0190	36.72	4.93	38.79	19.56
C20H32I2O9	670.28	670.0139	35.84	4.81	37.87	21.48
C20H33I1O1	416.39	416.1578	57.69	7.99	30.48	3.84
C20H33I1O2	432.39	432.1527	55.56	7.69	29.35	7.40
C20H33I1O3	448.39	448.1476	53.57	7.42	28.30	10.70
C20H33I1O4	464.39	464.1425	51.73	7.16	27.33	13.78
C20H33I1O5	480.39	480.1374	50.01	6.92	26.42	16.65
C20H33I1O6	496.39	496.1324	48.39	6.70	25.57	19.34
C20H33I1O7	512.39	512.1273	46.88	6.49	24.77	21.86
C20H33I1O8	528.39	528.1222	45.46	6.30	24.02	24.22
C20H33I1O9	544.39	544.1171	44.13	6.11	23.31	26.45
C20H34I2O1	544.30	544.0703	44.13	6.30	46.63	2.94
C20H34I2O2	560.30	560.0652	42.87	6.12	45.30	5.71
C20H34I2O3	576.30	576.0601	41.68	5.95	44.04	8.33
C20H34I2O4	592.30	592.0550	40.56	5.79	42.85	10.80
C20H34I2O5	608.30	608.0499	39.49	5.63	41.72	13.15
C20H34I2O6	624.30	624.0448	38.48	5.49	40.65	15.38
C20H34I2O7	640.30	640.0398	37.52	5.35	39.64	17.49
C20H34I2O8	656.30	656.0347	36.60	5.22	38.67	19.50
C20H34I2O9	672.30	672.0296	35.73	5.10	37.75	21.42
C20H35I1O1	418.41	418.1734	57.41	8.43	30.33	3.82
C20H35I1O2	434.41	434.1683	55.30	8.12	29.21	7.37
C20H35I1O3	450.40	450.1633	53.33	7.83	28.18	10.66
C20H35I1O4	466.40	466.1582	51.51	7.56	27.21	13.72
C20H35I1O5	482.40	482.1531	49.80	7.31	26.31	16.58
C20H35I1O6	498.40	498.1480	48.20	7.08	25.46	19.26
C20H35I1O7	514.40	514.1429	46.70	6.86	24.67	21.77
C20H35I1O8	530.40	530.1378	45.29	6.65	23.93	24.13
C20H35I1O9	546.40	546.1327	43.96	6.46	23.23	26.35
C20H36I2O1	546.32	546.0859	43.97	6.64	46.46	2.93
C20H36I2O2	562.32	562.0808	42.72	6.45	45.14	5.69
C20H36I2O3	578.32	578.0757	41.54	6.27	43.89	8.30
C20H36I2O4	594.32	594.0707	40.42	6.11	42.71	10.77
C20H36I2O5	610.32	610.0656	39.36	5.95	41.59	13.11
C20H36I2O6	626.32	626.0605	38.35	5.79	40.52	15.33
C20H36I2O7	642.31	642.0554	37.40	5.65	39.51	17.44
C20H36I2O8	658.31	658.0503	36.49	5.51	38.55	19.44
C20H36I2O9	674.31	674.0452	35.62	5.38	37.64	21.35
C20H37I1O1	420.42	420.1891	57.14	8.87	30.19	3.81
C20H37I1O2	436.42	436.1840	55.04	8.55	29.08	7.33
C20H37I1O3	452.42	452.1789	53.10	8.24	28.05	10.61
C20H37I1O4	468.42	468.1738	51.28	7.96	27.09	13.66
C20H37I1O5	484.42	484.1687	49.59	7.70	26.20	16.51
C20H37I1O6	500.42	500.1637	48.00	7.45	25.36	19.18
C20H37I1O7	516.42	516.1586	46.52	7.22	24.57	21.69
C20H37I1O8	532.42	532.1535	45.12	7.00	23.84	24.04
C20H37I1O9	548.42	548.1484	43.80	6.80	23.14	26.26
C20H38I2O1	548.33	548.1016	43.81	6.99	46.29	2.92
C20H38I2O2	564.33	564.0965	42.57	6.79	44.97	5.67
C20H38I2O3	580.33	580.0914	41.39	6.60	43.74	8.27
C20H38I2O4	596.33	596.0863	40.28	6.42	42.56	10.73
C20H38I2O5	612.33	612.0812	39.23	6.26	41.45	13.06
C20H38I2O6	628.33	628.0761	38.23	6.10	40.39	15.28
C20H38I2O7	644.33	644.0711	37.28	5.94	39.39	17.38
C20H38I2O8	660.33	660.0660	36.38	5.80	38.44	19.38
C20H38I2O9	676.33	676.0609	35.52	5.66	37.53	21.29
C20H39I1O1	422.44	422.2047	56.87	9.31	30.04	3.79
C20H39I1O2	438.44	438.1996	54.79	8.97	28.94	7.30
C20H39I1O3	454.44	454.1946	52.86	8.65	27.93	10.56
C20H39I1O4	470.44	470.1895	51.06	8.36	26.98	13.60
C20H39I1O5	486.44	486.1844	49.38	8.08	26.09	16.45
C20H39I1O6	502.43	502.1793	47.81	7.82	25.26	19.11
C20H39I1O7	518.43	518.1742	46.34	7.58	24.48	21.60

C20H39I1O8	534.43	534.1691	44.95	7.36	23.75	23.95
C20H39I1O9	550.43	550.1640	43.64	7.14	23.06	26.16
C20H40I2O1	550.35	550.1172	43.65	7.33	46.12	2.91
C20H40I2O2	566.35	566.1121	42.42	7.12	44.81	5.65
C20H40I2O3	582.35	582.1070	41.25	6.92	43.58	8.24
C20H40I2O4	598.35	598.1020	40.15	6.74	42.42	10.70
C20H40I2O5	614.35	614.0969	39.10	6.56	41.31	13.02
C20H40I2O6	630.35	630.0918	38.11	6.40	40.26	15.23
C20H40I2O7	646.35	646.0867	37.17	6.24	39.27	17.33
C20H40I2O8	662.35	662.0816	36.27	6.09	38.32	19.32
C20H40I2O9	678.35	678.0765	35.41	5.94	37.42	21.23
C20H41I1O1	424.45	424.2204	56.60	9.74	29.90	3.77
C20H41I1O2	440.45	440.2153	54.54	9.38	28.81	7.26
C20H41I1O3	456.45	456.2102	52.63	9.05	27.80	10.52
C20H41I1O4	472.45	472.2051	50.85	8.75	26.86	13.55
C20H41I1O5	488.45	488.2000	49.18	8.46	25.98	16.38
C20H41I1O6	504.45	504.1950	47.62	8.19	25.16	19.03
C20H41I1O7	520.45	520.1899	46.16	7.94	24.38	21.52
C20H41I1O8	536.45	536.1848	44.78	7.70	23.66	23.86
C20H41I1O9	552.45	552.1797	43.48	7.48	22.97	26.06
C21H10I2O1	532.12	531.8825	47.40	1.89	47.70	3.01
C21H10I2O2	548.12	547.8774	46.02	1.84	46.31	5.84
C21H10I2O3	564.12	563.8723	44.71	1.79	44.99	8.51
C21H10I2O4	580.12	579.8672	43.48	1.74	43.75	11.03
C21H10I2O5	596.12	595.8621	42.31	1.69	42.58	13.42
C21H10I2O6	612.12	611.8571	41.21	1.65	41.46	15.68
C21H10I2O7	628.12	627.8520	40.16	1.60	40.41	17.83
C21H10I2O8	644.12	643.8469	39.16	1.56	39.40	19.87
C21H10I2O9	660.12	659.8418	38.21	1.53	38.45	21.81
C21H11I1O1	406.23	405.9856	62.09	2.73	31.24	3.94
C21H11I1O2	422.23	421.9806	59.74	2.63	30.06	7.58
C21H11I1O3	438.22	437.9755	57.56	2.53	28.96	10.95
C21H11I1O4	454.22	453.9704	55.53	2.44	27.94	14.09
C21H11I1O5	470.22	469.9653	53.64	2.36	26.99	17.01
C21H11I1O6	486.22	485.9602	51.88	2.28	26.10	19.74
C21H11I1O7	502.22	501.9551	50.22	2.21	25.27	22.30
C21H11I1O8	518.22	517.9500	48.67	2.14	24.49	24.70
C21H11I1O9	534.22	533.9450	47.22	2.08	23.76	26.95
C21H12I2O1	534.14	533.8981	47.22	2.26	47.52	3.00
C21H12I2O2	550.14	549.8930	45.85	2.20	46.14	5.82
C21H12I2O3	566.14	565.8880	44.55	2.14	44.83	8.48
C21H12I2O4	582.14	581.8829	43.33	2.08	43.60	10.99
C21H12I2O5	598.14	597.8778	42.17	2.02	42.43	13.37
C21H12I2O6	614.13	613.8727	41.07	1.97	41.33	15.63
C21H12I2O7	630.13	629.8676	40.03	1.92	40.28	17.77
C21H12I2O8	646.13	645.8625	39.04	1.87	39.28	19.81
C21H12I2O9	662.13	661.8574	38.09	1.83	38.33	21.75
C21H13I1O1	408.24	408.0013	61.79	3.21	31.09	3.92
C21H13I1O2	424.24	423.9962	59.46	3.09	29.91	7.54
C21H13I1O3	440.24	439.9911	57.29	2.98	28.83	10.90
C21H13I1O4	456.24	455.9860	55.29	2.87	27.82	14.03
C21H13I1O5	472.24	471.9810	53.41	2.77	26.87	16.94
C21H13I1O6	488.24	487.9759	51.66	2.68	25.99	19.66
C21H13I1O7	504.24	503.9708	50.02	2.60	25.17	22.21
C21H13I1O8	520.24	519.9657	48.48	2.52	24.39	24.60
C21H13I1O9	536.24	535.9606	47.04	2.44	23.67	26.85
C21H14I2O1	536.15	535.9138	47.05	2.63	47.34	2.98
C21H14I2O2	552.15	551.9087	45.68	2.56	45.97	5.80
C21H14I2O3	568.15	567.9036	44.40	2.48	44.67	8.45
C21H14I2O4	584.15	583.8985	43.18	2.42	43.45	10.96
C21H14I2O5	600.15	599.8934	42.03	2.35	42.29	13.33
C21H14I2O6	616.15	615.8883	40.94	2.29	41.19	15.58
C21H14I2O7	632.15	631.8833	39.90	2.23	40.15	17.72
C21H14I2O8	648.15	647.8782	38.92	2.18	39.16	19.75
C21H14I2O9	664.15	663.8731	37.98	2.12	38.22	21.68
C21H15I1O1	410.26	410.0169	61.48	3.69	30.93	3.90
C21H15I1O2	426.26	426.0119	59.17	3.55	29.77	7.51
C21H15I1O3	442.26	442.0068	57.03	3.42	28.69	10.85
C21H15I1O4	458.26	458.0017	55.04	3.30	27.69	13.97
C21H15I1O5	474.26	473.9966	53.19	3.19	26.76	16.87
C21H15I1O6	490.25	489.9915	51.45	3.08	25.89	19.58
C21H15I1O7	506.25	505.9864	49.82	2.99	25.07	22.12
C21H15I1O8	522.25	521.9813	48.30	2.90	24.30	24.51
C21H15I1O9	538.25	537.9763	46.86	2.81	23.58	26.75
C21H16I2O1	538.17	537.9294	46.87	3.00	47.16	2.97
C21H16I2O2	554.17	553.9243	45.52	2.91	45.80	5.77
C21H16I2O3	570.17	569.9193	44.24	2.83	44.51	8.42
C21H16I2O4	586.17	585.9142	43.03	2.75	43.30	10.92
C21H16I2O5	602.17	601.9091	41.89	2.68	42.15	13.28
C21H16I2O6	618.17	617.9040	40.80	2.61	41.06	15.53
C21H16I2O7	634.17	633.8989	39.77	2.54	40.02	17.66
C21H16I2O8	650.17	649.8938	38.80	2.48	39.04	19.69
C21H16I2O9	666.17	665.8887	37.86	2.42	38.10	21.62
C21H17I1O1	412.27	412.0326	61.18	4.16	30.78	3.88
C21H17I1O2	428.27	428.0275	58.90	4.00	29.63	7.47
C21H17I1O3	444.27	444.0224	56.77	3.86	28.56	10.80
C21H17I1O4	460.27	460.0173	54.80	3.72	27.57	13.90
C21H17I1O5	476.27	476.0122	52.96	3.60	26.65	16.80
C21H17I1O6	492.27	492.0072	51.24	3.48	25.78	19.50
C21H17I1O7	508.27	508.0021	49.63	3.37	24.97	22.03
C21H17I1O8	524.27	523.9970	48.11	3.27	24.21	24.41
C21H17I1O9	540.27	539.9919	46.69	3.17	23.49	26.65
C21H18I2O1	540.19	539.9451	46.69	3.36	46.99	2.96
C21H18I2O2	556.19	555.9400	45.35	3.26	45.63	5.75
C21H18I2O3	572.18	571.9349	44.08	3.17	44.36	8.39
C21H18I2O4	588.18	587.9298	42.88	3.08	43.15	10.88
C21H18I2O5	604.18	603.9247	41.75	3.00	42.01	13.24
C21H18I2O6	620.18	619.9196	40.67	2.93	40.92	15.48
C21H18I2O7	636.18	635.9146	39.65	2.85	39.90	17.60
C21H18I2O8	652.18	651.9095	38.68	2.78	38.92	19.63
C21H18I2O9	668.18	667.9044	37.75	2.72	37.99	21.55
C21H19I1O1	414.29	414.0482	60.88	4.62	30.63	3.86
C21H19I1O2	430.29	430.0432	58.62	4.45	29.49	7.44
C21H19I1O3	446.29	446.0381	56.52	4.29	28.44	10.75
C21H19I1O4	462.29	462.0330	54.56	4.14	27.45	13.84
C21H19I1O5	478.29	478.0279	52.74	4.00	26.53	16.73
C21H19I1O6	494.29	494.0228	51.03	3.87	25.67	19.42
C21H19I1O7	510.29	510.0177	49.43	3.75	24.87	21.95
C21H19I1O8	526.29	526.0126	47.93	3.64	24.11	24.32
C21H19I1O9	542.28	542.0076	46.51	3.53	23.40	26.55
C21H20I2O1	542.20	541.9607	46.52	3.72	46.81	2.95
C21H20I2O2	558.20	557.9556	45.19	3.61	45.47	5.73
C21H20I2O3	574.20	573.9506	43.93	3.51	44.20	8.36
C21H20I2O4	590.20	589.9455	42.74	3.42	43.00	10.84
C21H20I2O5	606.20	605.9404	41.61	3.33	41.87	13.20
C21H20I2O6	622.20	621.9353	40.54	3.24	40.79	15.43
C21H20I2O7	638.20	637.9302	39.52	3.16	39.77	17.55
C21H20I2O8	654.20	653.9251	38.56	3.08	38.80	19.57
C21H20I2O9	670.20	669.9200	37.64	3.01	37.87	21.49
C21H21I1O1	416.21	416.0639	60.59	5.08	30.48	3.84

C21H21I102	432.30	432.0588	58.35	4.90	29.36	7.40
C21H21I103	448.30	448.0537	56.26	4.72	28.31	10.71
C21H21I104	464.30	464.0486	54.33	4.56	27.33	13.78
C21H21I105	480.30	480.0435	52.52	4.41	26.42	16.66
C21H21I106	496.30	496.0385	50.82	4.27	25.57	19.34
C21H21I107	512.30	512.0334	49.24	4.13	24.77	21.86
C21H21I108	528.30	528.0283	47.74	4.01	24.02	24.23
C21H21I109	544.30	544.0232	46.34	3.89	23.32	26.45
C21H22I201	544.22	543.9764	46.35	4.07	46.64	2.94
C21H22I202	560.22	559.9713	45.02	3.96	45.31	5.71
C21H22I203	576.22	575.9662	43.77	3.85	44.05	8.33
C21H22I204	592.22	591.9611	42.59	3.74	42.86	10.81
C21H22I205	608.22	607.9560	41.47	3.65	41.73	13.15
C21H22I206	624.21	623.9509	40.41	3.55	40.66	15.38
C21H22I207	640.21	639.9459	39.40	3.46	39.64	17.49
C21H22I208	656.21	655.9408	38.44	3.38	38.68	19.51
C21H22I209	672.21	671.9357	37.52	3.30	37.76	21.42
C21H23I101	418.32	418.0795	60.30	5.54	30.34	3.82
C21H23I102	434.32	434.0745	58.08	5.34	29.22	7.37
C21H23I103	450.32	450.0694	56.01	5.15	28.18	10.66
C21H23I104	466.32	466.0643	54.09	4.97	27.21	13.72
C21H23I105	482.32	482.0592	52.30	4.81	26.31	16.59
C21H23I106	498.32	498.0541	50.62	4.65	25.47	19.26
C21H23I107	514.32	514.0490	49.04	4.51	24.67	21.78
C21H23I108	530.32	530.0439	47.56	4.37	23.93	24.14
C21H23I109	546.32	546.0389	46.17	4.24	23.23	26.36
C21H24I201	546.23	545.9920	46.18	4.43	46.47	2.93
C21H24I202	562.23	561.9869	44.86	4.30	45.14	5.69
C21H24I203	578.23	577.9819	43.62	4.18	43.89	8.30
C21H24I204	594.23	593.9768	42.45	4.07	42.71	10.77
C21H24I205	610.23	609.9717	41.33	3.96	41.59	13.11
C21H24I206	626.23	625.9666	40.28	3.86	40.53	15.33
C21H24I207	642.23	641.9615	39.27	3.77	39.52	17.44
C21H24I208	658.23	657.9564	38.32	3.68	38.56	19.45
C21H24I209	674.23	673.9513	37.41	3.59	37.64	21.36
C21H25I101	420.34	420.0952	60.01	6.00	30.19	3.81
C21H25I102	436.34	436.0901	57.81	5.78	29.08	7.33
C21H25I103	452.34	452.0850	55.76	5.57	28.06	10.61
C21H25I104	468.34	468.0799	53.86	5.38	27.10	13.66
C21H25I105	484.33	484.0748	52.08	5.20	26.20	16.52
C21H25I106	500.33	500.0698	50.41	5.04	25.36	19.19
C21H25I107	516.33	516.0647	48.85	4.88	24.58	21.69
C21H25I108	532.33	532.0596	47.38	4.73	23.84	24.04
C21H25I109	548.33	548.0545	46.00	4.60	23.14	26.26
C21H26I201	548.25	548.0077	46.01	4.78	46.29	2.92
C21H26I202	564.25	564.0026	44.70	4.64	44.98	5.67
C21H26I203	580.25	579.9975	43.47	4.52	43.74	8.27
C21H26I204	596.25	595.9924	42.30	4.40	42.57	10.73
C21H26I205	612.25	611.9873	41.20	4.28	41.46	13.07
C21H26I206	628.25	627.9822	40.15	4.17	40.40	15.28
C21H26I207	644.25	643.9772	39.15	4.07	39.40	17.38
C21H26I208	660.25	659.9721	38.20	3.97	38.44	19.39
C21H26I209	676.24	675.9670	37.30	3.88	37.53	21.29
C21H27I101	422.35	422.1108	59.72	6.44	30.05	3.79
C21H27I102	438.35	438.1058	57.54	6.21	28.95	7.30
C21H27I103	454.35	454.1007	55.52	5.99	27.93	10.56
C21H27I104	470.35	470.0956	53.63	5.79	26.98	13.61
C21H27I105	486.35	486.0905	51.86	5.60	26.09	16.45
C21H27I106	502.35	502.0854	50.21	5.42	25.26	19.11
C21H27I107	518.35	518.0803	48.66	5.25	24.48	21.61

C21H27I108	534.35	534.0752	47.20	5.09	23.75	23.95
C21H27I109	550.35	550.0702	45.83	4.95	23.06	26.16
C21H28I201	550.27	550.0233	45.84	5.13	46.12	2.91
C21H28I202	566.26	566.0182	44.54	4.98	44.82	5.65
C21H28I203	582.26	582.0132	43.32	4.85	43.59	8.24
C21H28I204	598.26	598.0081	42.16	4.72	42.42	10.70
C21H28I205	614.26	614.0030	41.06	4.59	41.32	13.02
C21H28I206	630.26	629.9979	40.02	4.48	40.27	15.23
C21H28I207	646.26	645.9928	39.03	4.37	39.27	17.33
C21H28I208	662.26	661.9877	38.09	4.26	38.32	19.33
C21H28I209	678.26	677.9826	37.19	4.16	37.42	21.23
C21H29I101	424.37	424.1265	59.44	6.89	29.90	3.77
C21H29I102	440.37	440.1214	57.28	6.64	28.82	7.27
C21H29I103	456.37	456.1163	55.27	6.41	27.81	10.52
C21H29I104	472.37	472.1112	53.40	6.19	26.87	13.55
C21H29I105	488.37	488.1061	51.65	5.99	25.99	16.38
C21H29I106	504.37	504.1011	50.01	5.80	25.16	19.03
C21H29I107	520.37	520.0960	48.47	5.62	24.39	21.52
C21H29I108	536.36	536.0909	47.03	5.45	23.66	23.86
C21H29I109	552.36	552.0858	45.66	5.29	22.97	26.07
C21H30I201	552.28	552.0390	45.67	5.48	45.96	2.90
C21H30I202	568.28	568.0339	44.39	5.32	44.66	5.63
C21H30I203	584.28	584.0288	43.17	5.18	43.44	8.21
C21H30I204	600.28	600.0237	42.02	5.04	42.28	10.66
C21H30I205	616.28	616.0186	40.93	4.91	41.18	12.98
C21H30I206	632.28	632.0135	39.89	4.78	40.14	15.18
C21H30I207	648.28	648.0085	38.91	4.66	39.15	17.28
C21H30I208	664.28	664.0034	37.97	4.55	38.21	19.27
C21H30I209	680.28	679.9983	37.08	4.45	37.31	21.17
C21H31I101	426.39	426.1421	59.16	7.33	29.76	3.75
C21H31I102	442.38	442.1371	57.02	7.06	28.69	7.23
C21H31I103	458.38	458.1320	55.03	6.82	27.69	10.47
C21H31I104	474.38	474.1269	53.17	6.59	26.75	13.49
C21H31I105	490.38	490.1218	51.44	6.37	25.88	16.31
C21H31I106	506.38	506.1167	49.81	6.17	25.06	18.96
C21H31I107	522.38	522.1116	48.29	5.98	24.29	21.44
C21H31I108	538.38	538.1065	46.85	5.80	23.57	23.77
C21H31I109	554.38	554.1014	45.50	5.64	22.89	25.97
C21H32I201	554.30	554.0546	45.51	5.82	45.79	2.89
C21H32I202	570.30	570.0495	44.23	5.66	44.50	5.61
C21H32I203	586.30	586.0444	43.02	5.50	43.29	8.19
C21H32I204	602.30	602.0394	41.88	5.36	42.14	10.63
C21H32I205	618.29	618.0343	40.80	5.22	41.05	12.94
C21H32I206	634.29	634.0292	39.77	5.09	40.01	15.13
C21H32I207	650.29	650.0241	38.79	4.96	39.03	17.22
C21H32I208	666.29	666.0190	37.86	4.84	38.09	19.21
C21H32I209	682.29	682.0139	36.97	4.73	37.20	21.10
C21H33I101	428.40	428.1578	58.88	7.76	29.62	3.73
C21H33I102	444.40	444.1527	56.76	7.48	28.56	7.20
C21H33I103	460.40	460.1476	54.79	7.22	27.56	10.43
C21H33I104	476.40	476.1425	52.95	6.98	26.64	13.43
C21H33I105	492.40	492.1374	51.23	6.76	25.77	16.25
C21H33I106	508.40	508.1324	49.61	6.54	24.96	18.88
C21H33I107	524.40	524.1273	48.10	6.34	24.20	21.36
C21H33I108	540.40	540.1222	46.68	6.16	23.48	23.69
C21H33I109	556.40	556.1171	45.33	5.98	22.81	25.88
C21H34I201	556.31	556.0703	45.34	6.16	45.62	2.88
C21H34I202	572.31	572.0652	44.07	5.99	44.35	5.59
C21H34I203	588.31	588.0601	42.87	5.83	43.14	8.16
C21H34I204	604.31	604.0550	41.74	5.67	42.00	10.59

212

C21H34I205	620.31	620.0499	40.66	5.52	40.92	12.90
C21H34I206	636.31	636.0448	39.64	5.39	39.89	15.09
C21H34I207	652.31	652.0398	38.67	5.25	38.91	17.17
C21H34I208	668.31	668.0347	37.74	5.13	37.98	19.15
C21H34I209	684.31	684.0296	36.86	5.01	37.09	21.04
C21H35I101	430.42	430.1734	58.60	8.20	29.48	3.72
C21H35I102	446.42	446.1683	56.50	7.90	28.43	7.17
C21H35I103	462.42	462.1633	54.55	7.63	27.44	10.38
C21H35I104	478.42	478.1582	52.72	7.37	26.53	13.38
C21H35I105	494.41	494.1531	51.02	7.14	25.67	16.18
C21H35I106	510.41	510.1480	49.42	6.91	24.86	18.81
C21H35I107	526.41	526.1429	47.92	6.70	24.11	21.28
C21H35I108	542.41	542.1378	46.50	6.50	23.40	23.60
C21H35I109	558.41	558.1327	45.17	6.32	22.73	25.79
C21H36I201	558.33	558.0859	45.18	6.50	45.46	2.87
C21H36I202	574.33	574.0808	43.92	6.32	44.19	5.57
C21H36I203	590.33	590.0757	42.73	6.15	42.99	8.13
C21H36I204	606.33	606.0707	41.60	5.98	41.86	10.55
C21H36I205	622.33	622.0656	40.53	5.83	40.78	12.85
C21H36I206	638.33	638.0605	39.51	5.68	39.76	15.04
C21H36I207	654.33	654.0554	38.55	5.55	38.79	17.12
C21H36I208	670.33	670.0503	37.63	5.41	37.86	19.09
C21H36I209	686.32	686.0452	36.75	5.29	36.98	20.98
C21H37I101	432.43	432.1891	58.33	8.62	29.35	3.70
C21H37I102	448.43	448.1840	56.25	8.32	28.30	7.14
C21H37I103	464.43	464.1789	54.31	8.03	27.32	10.33
C21H37I104	480.43	480.1738	52.50	7.76	26.41	13.32
C21H37I105	496.43	496.1687	50.81	7.51	25.56	16.11
C21H37I106	512.43	512.1637	49.22	7.28	24.77	18.73
C21H37I107	528.43	528.1586	47.73	7.06	24.02	21.19
C21H37I108	544.43	544.1535	46.33	6.85	23.31	23.51
C21H37I109	560.43	560.1484	45.01	6.65	22.64	25.69
C21H38I201	560.35	560.1016	45.01	6.84	45.30	2.86
C21H38I202	576.34	576.0965	43.76	6.65	44.04	5.55
C21H38I203	592.34	592.0914	42.58	6.47	42.85	8.10
C21H38I204	608.34	608.0863	41.46	6.30	41.72	10.52
C21H38I205	624.34	624.0812	40.40	6.13	40.65	12.81
C21H38I206	640.34	640.0761	39.39	5.98	39.64	14.99
C21H38I207	656.34	656.0711	38.43	5.84	38.67	17.06
C21H38I208	672.34	672.0660	37.52	5.70	37.75	19.04
C21H38I209	688.34	688.0609	36.64	5.56	36.87	20.92
C21H39I101	434.45	434.2047	58.06	9.05	29.21	3.68
C21H39I102	450.45	450.1996	56.00	8.73	28.17	7.10
C21H39I103	466.45	466.1946	54.08	8.43	27.21	10.29
C21H39I104	482.45	482.1895	52.28	8.15	26.30	13.27
C21H39I105	498.45	498.1844	50.60	7.89	25.46	16.05
C21H39I106	514.45	514.1793	49.03	7.64	24.67	18.66
C21H39I107	530.45	530.1742	47.55	7.41	23.92	21.11
C21H39I108	546.44	546.1691	46.16	7.19	23.22	23.42
C21H39I109	562.44	562.1640	44.85	6.99	22.56	25.60
C21H40I201	562.36	562.1172	44.85	7.17	45.13	2.85
C21H40I202	578.36	578.1121	43.61	6.97	43.88	5.53
C21H40I203	594.36	594.1070	42.44	6.78	42.70	8.08
C21H40I204	610.36	610.1020	41.33	6.61	41.58	10.49
C21H40I205	626.36	626.0969	40.27	6.44	40.52	12.77
C21H40I206	642.36	642.0918	39.27	6.28	39.51	14.94
C21H40I207	658.36	658.0867	38.31	6.12	38.55	17.01
C21H40I208	674.36	674.0816	37.40	5.98	37.64	18.98
C21H40I209	690.36	690.0765	36.54	5.84	36.76	20.86
C21H41I101	436.46	436.2204	57.79	9.47	29.08	3.67

C21H41I102	452.46	452.2153	55.75	9.13	28.05	7.07
C21H41I103	468.46	468.2102	53.84	8.82	27.09	10.25
C21H41I104	484.46	484.2051	52.06	8.53	26.19	13.21
C21H41I105	500.46	500.2000	50.40	8.26	25.36	15.98
C21H41I106	516.46	516.1950	48.84	8.00	24.57	18.59
C21H41I107	532.46	532.1899	47.37	7.76	23.83	21.03
C21H41I108	548.46	548.1848	45.99	7.54	23.14	23.34
C21H41I109	564.46	564.1797	44.69	7.32	22.48	25.51
C21H42I201	564.38	564.1329	44.69	7.50	44.97	2.83
C21H42I202	580.38	580.1278	43.46	7.29	43.73	5.51
C21H42I203	596.38	596.1227	42.29	7.10	42.56	8.05
C21H42I204	612.38	612.1176	41.19	6.91	41.45	10.45
C21H42I205	628.37	628.1125	40.14	6.74	40.39	12.73
C21H42I206	644.37	644.1074	39.14	6.57	39.39	14.90
C21H42I207	660.37	660.1024	38.20	6.41	38.43	16.96
C21H42I208	676.37	676.0973	37.29	6.26	37.52	18.92
C21H42I209	692.37	692.0922	36.43	6.11	36.66	20.80
C21H43I101	438.48	438.2360	57.52	9.88	28.94	3.65
C21H43I102	454.48	454.2309	55.50	9.54	27.92	7.04
C21H43I103	470.48	470.2259	53.61	9.21	26.97	10.20
C21H43I104	486.48	486.2208	51.85	8.91	26.09	13.16
C21H43I105	502.48	502.2157	50.20	8.63	25.26	15.92
C21H43I106	518.48	518.2106	48.65	8.36	24.48	18.52
C21H43I107	534.48	534.2055	47.19	8.11	23.74	20.95
C21H43I108	550.48	550.2004	45.82	7.87	23.05	23.25
C21H43I109	566.48	566.1953	44.53	7.65	22.40	25.42
C22H11I101	418.24	417.9856	63.18	2.65	30.34	3.83
C22H11I102	434.24	433.9806	60.85	2.55	29.22	7.37
C22H11I103	450.24	449.9755	58.69	2.46	28.19	10.66
C22H11I104	466.23	465.9704	56.68	2.38	27.22	13.73
C22H11I105	482.23	481.9653	54.80	2.30	26.32	16.59
C22H11I106	498.23	497.9602	53.04	2.23	25.47	19.27
C22H11I107	514.23	513.9551	51.39	2.16	24.68	21.78
C22H11I108	530.23	529.9500	49.84	2.09	23.93	24.14
C22H11I109	546.23	545.9450	48.38	2.03	23.23	26.36
C22H12I201	546.15	545.8981	48.38	2.21	46.47	2.93
C22H12I202	562.15	561.8930	47.01	2.15	45.15	5.69
C22H12I203	578.15	577.8880	45.71	2.09	43.90	8.30
C22H12I204	594.15	593.8829	44.47	2.04	42.72	10.77
C22H12I205	610.15	609.8778	43.31	1.98	41.60	13.11
C22H12I206	626.15	625.8727	42.20	1.93	40.54	15.33
C22H12I207	642.15	641.8676	41.15	1.88	39.53	17.44
C22H12I208	658.14	657.8625	40.15	1.84	38.56	19.45
C22H12I209	674.14	673.8574	39.20	1.79	37.65	21.36
C22H13I101	420.25	420.0013	62.88	3.12	30.20	3.81
C22H13I102	436.25	435.9962	60.57	3.00	29.09	7.33
C22H13I103	452.25	451.9911	58.43	2.90	28.06	10.61
C22H13I104	468.25	467.9860	56.43	2.80	27.10	13.67
C22H13I105	484.25	483.9810	54.57	2.71	26.21	16.52
C22H13I106	500.25	499.9759	52.82	2.62	25.37	19.19
C22H13I107	516.25	515.9708	51.19	2.54	24.58	21.69
C22H13I108	532.25	531.9657	49.65	2.46	23.84	24.05
C22H13I109	548.25	547.9606	48.20	2.39	23.15	26.26
C22H14I201	548.17	547.9138	48.21	2.57	46.30	2.92
C22H14I202	564.16	563.9087	46.84	2.50	44.99	5.67
C22H14I203	580.16	579.9036	45.55	2.43	43.75	8.27
C22H14I204	596.16	595.8985	44.32	2.37	42.57	10.73
C22H14I205	612.16	611.8934	43.17	2.31	41.46	13.07
C22H14I206	628.16	627.8883	42.07	2.25	40.40	15.28
C22H14I207	644.16	643.8833	41.02	2.19	39.40	17.39

C22H14I2O8	660.16	659.8782	40.03	2.14	38.45	19.39
C22H14I2O9	676.16	675.8731	39.08	2.09	37.54	21.30
C22H15I1O1	422.27	422.0169	62.58	3.58	30.05	3.79
C22H15I1O2	438.27	438.0119	60.29	3.45	28.96	7.30
C22H15I1O3	454.27	454.0068	58.17	3.33	27.94	10.57
C22H15I1O4	470.27	470.0017	56.19	3.22	26.99	13.61
C22H15I1O5	486.27	485.9966	54.34	3.11	26.10	16.45
C22H15I1O6	502.27	501.9915	52.61	3.01	25.27	19.11
C22H15I1O7	518.27	517.9864	50.99	2.92	24.49	21.61
C22H15I1O8	534.26	533.9813	49.46	2.83	23.75	23.96
C22H15I1O9	550.26	549.9763	48.02	2.75	23.06	26.17
C22H16I2O1	550.18	549.9294	48.03	2.93	46.13	2.91
C22H16I2O2	566.18	565.9243	46.67	2.85	44.83	5.65
C22H16I2O3	582.18	581.9193	45.39	2.77	43.60	8.24
C22H16I2O4	598.18	597.9142	44.17	2.70	42.43	10.70
C22H16I2O5	614.18	613.9091	43.02	2.63	41.32	13.03
C22H16I2O6	630.18	629.9040	41.93	2.56	40.28	15.23
C22H16I2O7	646.18	645.8989	40.89	2.50	39.28	17.33
C22H16I2O8	662.18	661.8938	39.91	2.44	38.33	19.33
C22H16I2O9	678.18	677.8887	38.96	2.38	37.43	21.23
C22H17I1O1	424.28	424.0326	62.28	4.04	29.91	3.77
C22H17I1O2	440.28	440.0275	60.02	3.89	28.82	7.27
C22H17I1O3	456.28	456.0224	57.91	3.76	27.81	10.52
C22H17I1O4	472.28	472.0173	55.95	3.63	26.87	13.55
C22H17I1O5	488.28	488.0122	54.12	3.51	25.99	16.38
C22H17I1O6	504.28	504.0072	52.40	3.40	25.17	19.04
C22H17I1O7	520.28	520.0021	50.79	3.29	24.39	21.53
C22H17I1O8	536.28	535.9970	49.27	3.20	23.66	23.87
C22H17I1O9	552.28	551.9919	47.85	3.10	22.98	26.07
C22H18I2O1	552.20	551.9451	47.85	3.29	45.96	2.90
C22H18I2O2	568.20	567.9400	46.51	3.19	44.67	5.63
C22H18I2O3	584.20	583.9349	45.23	3.11	43.45	8.22
C22H18I2O4	600.20	599.9298	44.03	3.02	42.29	10.66
C22H18I2O5	616.19	615.9247	42.88	2.94	41.19	12.98
C22H18I2O6	632.19	631.9196	41.80	2.87	40.15	15.18
C22H18I2O7	648.19	647.9146	40.77	2.80	39.16	17.28
C22H18I2O8	664.19	663.9095	39.78	2.73	38.21	19.27
C22H18I2O9	680.19	679.9044	38.85	2.67	37.31	21.17
C22H19I1O1	426.30	426.0482	61.99	4.49	29.77	3.75
C22H19I1O2	442.30	442.0432	59.74	4.33	28.69	7.23
C22H19I1O3	458.30	458.0381	57.66	4.18	27.69	10.47
C22H19I1O4	474.30	474.0330	55.71	4.04	26.76	13.49
C22H19I1O5	490.30	490.0279	53.89	3.91	25.88	16.32
C22H19I1O6	506.30	506.0228	52.19	3.78	25.07	18.96
C22H19I1O7	522.30	522.0177	50.59	3.67	24.30	21.44
C22H19I1O8	538.30	538.0126	49.09	3.56	23.58	23.78
C22H19I1O9	554.30	554.0076	47.67	3.46	22.89	25.98
C22H20I2O1	554.21	553.9607	47.68	3.64	45.80	2.89
C22H20I2O2	570.21	569.9556	46.34	3.54	44.51	5.61
C22H20I2O3	586.21	585.9506	45.08	3.44	43.30	8.19
C22H20I2O4	602.21	601.9455	43.88	3.35	42.15	10.63
C22H20I2O5	618.21	617.9404	42.74	3.26	41.06	12.94
C22H20I2O6	634.21	633.9353	41.67	3.18	40.02	15.14
C22H20I2O7	650.21	649.9302	40.64	3.10	39.03	17.22
C22H20I2O8	666.21	665.9251	39.66	3.03	38.10	19.21
C22H20I2O9	682.21	681.9200	38.73	2.96	37.20	21.11
C22H21I1O1	428.32	428.0639	61.69	4.94	29.63	3.74
C22H21I1O2	444.32	444.0588	59.47	4.76	28.56	7.20
C22H21I1O3	460.32	460.0537	57.41	4.60	27.57	10.43
C22H21I1O4	476.31	476.0486	55.48	4.44	26.64	13.44
C22H21I1O5	492.31	492.0435	53.67	4.30	25.78	16.25
C22H21I1O6	508.31	508.0385	51.98	4.16	24.97	18.89
C22H21I1O7	524.31	524.0334	50.40	4.04	24.20	21.36
C22H21I1O8	540.31	540.0283	48.91	3.92	23.49	23.69
C22H21I1O9	556.31	556.0232	47.50	3.80	22.81	25.88
C22H22I2O1	556.23	555.9764	47.51	3.99	45.63	2.88
C22H22I2O2	572.23	571.9713	46.18	3.88	44.35	5.59
C22H22I2O3	588.23	587.9662	44.92	3.77	43.15	8.16
C22H22I2O4	604.23	603.9611	43.73	3.67	42.01	10.59
C22H22I2O5	620.23	619.9560	42.60	3.58	40.92	12.90
C22H22I2O6	636.23	635.9509	41.53	3.49	39.89	15.09
C22H22I2O7	652.23	651.9459	40.51	3.40	38.91	17.17
C22H22I2O8	668.22	667.9408	39.54	3.32	37.98	19.15
C22H22I2O9	684.22	683.9357	38.62	3.24	37.09	21.04
C22H23I1O1	430.33	430.0795	61.40	5.39	29.49	3.72
C22H23I1O2	446.33	446.0745	59.20	5.19	28.43	7.17
C22H23I1O3	462.33	462.0694	57.15	5.01	27.45	10.38
C22H23I1O4	478.33	478.0643	55.24	4.85	26.53	13.38
C22H23I1O5	494.33	494.0592	53.46	4.69	25.67	16.18
C22H23I1O6	510.33	510.0541	51.78	4.54	24.87	18.81
C22H23I1O7	526.33	526.0490	50.21	4.40	24.11	21.28
C22H23I1O8	542.33	542.0439	48.72	4.27	23.40	23.60
C22H23I1O9	558.33	558.0389	47.33	4.15	22.73	25.79
C22H24I2O1	558.24	557.9920	47.34	4.33	45.47	2.87
C22H24I2O2	574.24	573.9869	46.02	4.21	44.20	5.57
C22H24I2O3	590.24	589.9819	44.77	4.10	43.00	8.13
C22H24I2O4	606.24	605.9768	43.59	3.99	41.87	10.56
C22H24I2O5	622.24	621.9717	42.47	3.89	40.79	12.86
C22H24I2O6	638.24	637.9666	41.40	3.79	39.77	15.04
C22H24I2O7	654.24	653.9615	40.39	3.70	38.79	17.12
C22H24I2O8	670.24	669.9564	39.43	3.61	37.87	19.10
C22H24I2O9	686.24	685.9513	38.51	3.53	36.99	20.98
C22H25I1O1	432.35	432.0952	61.12	5.83	29.35	3.70
C22H25I1O2	448.35	448.0901	58.94	5.62	28.30	7.14
C22H25I1O3	464.35	464.0850	56.91	5.43	27.33	10.34
C22H25I1O4	480.35	480.0799	55.01	5.25	26.42	13.32
C22H25I1O5	496.35	496.0748	53.24	5.08	25.57	16.12
C22H25I1O6	512.35	512.0698	51.58	4.92	24.77	18.74
C22H25I1O7	528.34	528.0647	50.01	4.77	24.02	21.20
C22H25I1O8	544.34	544.0596	48.54	4.63	23.31	23.51
C22H25I1O9	560.34	560.0545	47.16	4.50	22.65	25.70
C22H26I2O1	560.26	560.0077	47.16	4.68	45.30	2.86
C22H26I2O2	576.26	576.0026	45.86	4.55	44.04	5.55
C22H26I2O3	592.26	591.9975	44.62	4.42	42.85	8.10
C22H26I2O4	608.26	607.9924	43.44	4.31	41.73	10.52
C22H26I2O5	624.26	623.9873	42.33	4.20	40.66	12.81
C22H26I2O6	640.26	639.9822	41.27	4.09	39.64	14.99
C22H26I2O7	656.26	655.9772	40.27	3.99	38.68	17.07
C22H26I2O8	672.26	671.9721	39.31	3.90	37.75	19.04
C22H26I2O9	688.26	687.9670	38.39	3.81	36.88	20.92
C22H27I1O1	434.36	434.1108	60.83	6.27	29.22	3.68
C22H27I1O2	450.36	450.1058	58.67	6.04	28.18	7.11
C22H27I1O3	466.36	466.1007	56.66	5.84	27.21	10.29
C22H27I1O4	482.36	482.0956	54.78	5.64	26.31	13.27
C22H27I1O5	498.36	498.0905	53.02	5.46	25.46	16.05
C22H27I1O6	514.36	514.0854	51.37	5.29	24.67	18.66
C22H27I1O7	530.36	530.0803	49.82	5.13	23.93	21.12
C22H27I1O8	546.36	546.0752	48.36	4.98	23.23	23.43
C22H27I1O9	562.36	562.0702	46.99	4.84	22.57	25.61
C22H28I2O1	562.28	562.0233	47.00	5.02	45.14	2.85

C22H28I2O2	578.28	578.0182	45.70	4.88	43.89	5.53
C22H28I2O3	594.28	594.0132	44.47	4.75	42.71	8.08
C22H28I2O4	610.27	610.0081	43.30	4.62	41.59	10.49
C22H28I2O5	626.27	626.0030	42.19	4.51	40.53	12.77
C22H28I2O6	642.27	641.9979	41.14	4.39	39.52	14.95
C22H28I2O7	658.27	657.9928	40.14	4.29	38.56	17.01
C22H28I2O8	674.27	673.9877	39.19	4.19	37.64	18.98
C22H28I2O9	690.27	689.9826	38.28	4.09	36.77	20.86
C22H29I1O1	436.38	436.1265	60.55	6.70	29.08	3.67
C22H29I1O2	452.38	452.1214	58.41	6.46	28.05	7.07
C22H29I1O3	468.38	468.1163	56.42	6.24	27.09	10.25
C22H29I1O4	484.38	484.1112	54.55	6.03	26.20	13.21
C22H29I1O5	500.38	500.1061	52.81	5.84	25.36	15.99
C22H29I1O6	516.38	516.1011	51.17	5.66	24.58	18.59
C22H29I1O7	532.38	532.0960	49.64	5.49	23.84	21.04
C22H29I1O8	548.38	548.0909	48.19	5.33	23.14	23.34
C22H29I1O9	564.38	564.0858	46.82	5.18	22.49	25.51
C22H30I2O1	564.29	564.0390	46.83	5.36	44.98	2.84
C22H30I2O2	580.29	580.0339	45.54	5.21	43.74	5.51
C22H30I2O3	596.29	596.0288	44.31	5.07	42.56	8.05
C22H30I2O4	612.29	612.0237	43.16	4.94	41.45	10.45
C22H30I2O5	628.29	628.0186	42.06	4.81	40.40	12.73
C22H30I2O6	644.29	644.0135	41.01	4.69	39.39	14.90
C22H30I2O7	660.29	660.0085	40.02	4.58	38.44	16.96
C22H30I2O8	676.29	676.0034	39.07	4.47	37.53	18.93
C22H30I2O9	692.29	691.9983	38.17	4.37	36.66	20.80
C22H31I1O1	438.40	438.1421	60.28	7.13	28.95	3.65
C22H31I1O2	454.40	454.1371	58.15	6.88	27.93	7.04
C22H31I1O3	470.39	470.1320	56.18	6.64	26.98	10.20
C22H31I1O4	486.39	486.1269	54.33	6.42	26.09	13.16
C22H31I1O5	502.39	502.1218	52.60	6.22	25.26	15.92
C22H31I1O6	518.39	518.1167	50.97	6.03	24.48	18.52
C22H31I1O7	534.39	534.1116	49.45	5.85	23.75	20.96
C22H31I1O8	550.39	550.1065	48.01	5.68	23.06	23.26
C22H31I1O9	566.39	566.1014	46.65	5.52	22.41	25.42
C22H32I2O1	566.31	566.0546	46.66	5.70	44.82	2.83
C22H32I2O2	582.31	582.0495	45.38	5.54	43.59	5.50
C22H32I2O3	598.31	598.0444	44.17	5.39	42.42	8.02
C22H32I2O4	614.31	614.0394	43.02	5.25	41.32	10.42
C22H32I2O5	630.31	630.0343	41.92	5.12	40.27	12.69
C22H32I2O6	646.31	646.0292	40.89	4.99	39.27	14.85
C22H32I2O7	662.30	662.0241	39.90	4.87	38.32	16.91
C22H32I2O8	678.30	678.0190	38.96	4.76	37.42	18.87
C22H32I2O9	694.30	694.0139	38.06	4.65	36.56	20.74
C22H33I1O1	440.41	440.1578	60.00	7.55	28.81	3.63
C22H33I1O2	456.41	456.1527	57.90	7.29	27.80	7.01
C22H33I1O3	472.41	472.1476	55.94	7.04	26.86	10.16
C22H33I1O4	488.41	488.1425	54.10	6.81	25.98	13.10
C22H33I1O5	504.41	504.1374	52.39	6.59	25.16	15.86
C22H33I1O6	520.41	520.1324	50.78	6.39	24.39	18.45
C22H33I1O7	536.41	536.1273	49.26	6.20	23.66	20.88
C22H33I1O8	552.41	552.1222	47.84	6.02	22.97	23.17
C22H33I1O9	568.41	568.1171	46.49	5.85	22.33	25.33
C22H34I2O1	568.32	568.0703	46.50	6.03	44.66	2.82
C22H34I2O2	584.32	584.0652	45.22	5.87	43.44	5.48
C22H34I2O3	600.32	600.0601	44.02	5.71	42.28	8.00
C22H34I2O4	616.32	616.0550	42.87	5.56	41.18	10.38
C22H34I2O5	632.32	632.0499	41.79	5.42	40.14	12.65
C22H34I2O6	648.32	648.0448	40.76	5.29	39.15	14.81
C22H34I2O7	664.32	664.0398	39.78	5.16	38.21	16.86
C22H34I2O8	680.32	680.0347	38.84	5.04	37.31	18.81
C22H34I2O9	696.32	696.0296	37.95	4.92	36.45	20.68
C22H35I1O1	442.43	442.1734	59.73	7.97	28.68	3.62
C22H35I1O2	458.43	458.1683	57.64	7.70	27.68	6.98
C22H35I1O3	474.43	474.1633	55.70	7.44	26.75	10.12
C22H35I1O4	490.43	490.1582	53.88	7.19	25.88	13.05
C22H35I1O5	506.43	506.1531	52.18	6.97	25.06	15.80
C22H35I1O6	522.43	522.1480	50.58	6.75	24.29	18.38
C22H35I1O7	538.42	538.1429	49.08	6.55	23.57	20.80
C22H35I1O8	554.42	554.1378	47.66	6.36	22.89	23.09
C22H35I1O9	570.42	570.1327	46.32	6.18	22.25	25.24
C22H36I2O1	570.34	570.0859	46.33	6.36	44.50	2.81
C22H36I2O2	586.34	586.0808	45.07	6.19	43.29	5.46
C22H36I2O3	602.34	602.0757	43.87	6.02	42.14	7.97
C22H36I2O4	618.34	618.0707	42.73	5.87	41.05	10.35
C22H36I2O5	634.34	634.0656	41.66	5.72	40.01	12.61
C22H36I2O6	650.34	650.0605	40.63	5.58	39.03	14.76
C22H36I2O7	666.34	666.0554	39.66	5.45	38.09	16.81
C22H36I2O8	682.34	682.0503	38.73	5.32	37.20	18.76
C22H36I2O9	698.34	698.0452	37.84	5.20	36.34	20.62
C22H37I1O1	444.44	444.1891	59.46	8.39	28.55	3.60
C22H37I1O2	460.44	460.1840	57.39	8.10	27.56	6.95
C22H37I1O3	476.44	476.1789	55.46	7.83	26.64	10.07
C22H37I1O4	492.44	492.1738	53.66	7.57	25.77	13.00
C22H37I1O5	508.44	508.1687	51.97	7.34	24.96	15.73
C22H37I1O6	524.44	524.1637	50.39	7.11	24.20	18.30
C22H37I1O7	540.44	540.1586	48.89	6.90	23.48	20.72
C22H37I1O8	556.44	556.1535	47.49	6.70	22.81	23.00
C22H37I1O9	572.44	572.1484	46.16	6.52	22.17	25.15
C22H38I2O1	572.36	572.1016	46.17	6.69	44.34	2.80
C22H38I2O2	588.36	588.0965	44.91	6.51	43.14	5.44
C22H38I2O3	604.36	604.0914	43.72	6.34	42.00	7.94
C22H38I2O4	620.35	620.0863	42.60	6.17	40.91	10.32
C22H38I2O5	636.35	636.0812	41.52	6.02	39.88	12.57
C22H38I2O6	652.35	652.0761	40.51	5.87	38.91	14.72
C22H38I2O7	668.35	668.0711	39.54	5.73	37.98	16.76
C22H38I2O8	684.35	684.0660	38.61	5.60	37.09	18.70
C22H38I2O9	700.35	700.0609	37.73	5.47	36.24	20.56
C22H39I1O1	446.46	446.2047	59.19	8.81	28.42	3.58
C22H39I1O2	462.46	462.1996	57.14	8.50	27.44	6.92
C22H39I1O3	478.46	478.1946	55.23	8.22	26.52	10.03
C22H39I1O4	494.46	494.1895	53.44	7.95	25.67	12.94
C22H39I1O5	510.46	510.1844	51.77	7.70	24.86	15.67
C22H39I1O6	526.46	526.1793	50.19	7.47	24.11	18.23
C22H39I1O7	542.46	542.1742	48.71	7.25	23.39	20.65
C22H39I1O8	558.46	558.1691	47.32	7.04	22.72	22.92
C22H39I1O9	574.46	574.1640	46.00	6.84	22.09	25.07
C22H40I2O1	574.37	574.1172	46.01	7.02	44.19	2.79
C22H40I2O2	590.37	590.1121	44.76	6.83	42.99	5.42
C22H40I2O3	606.37	606.1070	43.58	6.65	41.86	7.92
C22H40I2O4	622.37	622.1020	42.46	6.48	40.78	10.28
C22H40I2O5	638.37	638.0969	41.39	6.32	39.76	12.53
C22H40I2O6	654.37	654.0918	40.38	6.16	38.79	14.67
C22H40I2O7	670.37	670.0867	39.42	6.01	37.86	16.71
C22H40I2O8	686.37	686.0816	38.50	5.87	36.98	18.65
C22H40I2O9	702.37	702.0765	37.62	5.74	36.14	20.50
C22H41I1O1	448.48	448.2204	58.92	9.21	28.30	3.57
C22H41I1O2	464.48	464.2153	56.89	8.90	27.32	6.89
C22H41I1O3	480.47	480.2102	55.00	8.60	26.41	9.99
C22H41I1O4	496.47	496.2051	53.22	8.32	25.56	12.89

C22H41I105	512.47	512.2000	51.56	8.06	24.76	15.61
C22H41I106	528.47	528.1950	50.00	7.82	24.01	18.16
C22H41I107	544.47	544.1899	48.53	7.59	23.31	20.57
C22H41I108	560.47	560.1848	47.15	7.37	22.64	22.84
C22H41I109	576.47	576.1797	45.84	7.17	22.01	24.98
C22H42I201	576.39	576.1329	45.85	7.34	44.03	2.78
C22H42I202	592.39	592.1278	44.61	7.15	42.85	5.40
C22H42I203	608.39	608.1227	43.43	6.96	41.72	7.89
C22H42I204	624.39	624.1176	42.32	6.78	40.65	10.25
C22H42I205	640.39	640.1125	41.26	6.61	39.63	12.49
C22H42I206	656.39	656.1074	40.26	6.45	38.67	14.63
C22H42I207	672.38	672.1024	39.30	6.30	37.75	16.66
C22H42I208	688.38	688.0973	38.39	6.15	36.87	18.59
C22H42I209	704.38	704.0922	37.51	6.01	36.03	20.44
C22H43I101	450.49	450.2360	58.66	9.62	28.17	3.55
C22H43I102	466.49	466.2309	56.65	9.29	27.20	6.86
C22H43I103	482.49	482.2259	54.77	8.98	26.30	9.95
C22H43I104	498.49	498.2208	53.01	8.69	25.46	12.84
C22H43I105	514.49	514.2157	51.36	8.42	24.67	15.55
C22H43I106	530.49	530.2106	49.81	8.17	23.92	18.10
C22H43I107	546.49	546.2055	48.35	7.93	23.22	20.49
C22H43I108	562.49	562.2004	46.98	7.71	22.56	22.76
C22H43I109	578.49	578.1953	45.68	7.49	21.94	24.89
C22H44I201	578.40	578.1485	45.69	7.67	43.88	2.77
C22H44I202	594.40	594.1434	44.46	7.46	42.70	5.38
C22H44I203	610.40	610.1383	43.29	7.27	41.58	7.86
C22H44I204	626.40	626.1333	42.18	7.08	40.52	10.22
C22H44I205	642.40	642.1282	41.13	6.90	39.51	12.45
C22H44I206	658.40	658.1231	40.13	6.74	38.55	14.58
C22H44I207	674.40	674.1180	39.18	6.58	37.63	16.61
C22H44I208	690.40	690.1129	38.27	6.42	36.76	18.54
C22H44I209	706.40	706.1078	37.41	6.28	35.93	20.38
C22H45I101	452.51	452.2517	58.40	10.02	28.04	3.54
C22H45I102	468.51	468.2466	56.40	9.68	27.09	6.83
C22H45I103	484.51	484.2415	54.54	9.36	26.19	9.91
C22H45I104	500.51	500.2364	52.80	9.06	25.36	12.79
C22H45I105	516.51	516.2313	51.16	8.78	24.57	15.49
C22H45I106	532.50	532.2263	49.62	8.52	23.83	18.03
C22H45I107	548.50	548.2212	48.18	8.27	23.14	20.42
C22H45I108	564.50	564.2161	46.81	8.04	22.48	22.67
C22H45I109	580.50	580.2110	45.52	7.81	21.86	24.81
C23H11I101	430.25	429.9856	64.21	2.58	29.50	3.72
C23H11I102	446.25	445.9806	61.91	2.48	28.44	7.17
C23H11I103	462.25	461.9755	59.76	2.40	27.45	10.38
C23H11I104	478.25	477.9704	57.76	2.32	26.54	13.38
C23H11I105	494.25	493.9653	55.89	2.24	25.68	16.19
C23H11I106	510.24	509.9602	54.14	2.17	24.87	18.81
C23H11I107	526.24	525.9551	52.50	2.11	24.12	21.28
C23H11I108	542.24	541.9500	50.95	2.04	23.40	23.60
C23H11I109	558.24	557.9450	49.49	1.99	22.73	25.79
C23H12I201	558.16	557.8981	49.49	2.17	45.47	2.87
C23H12I202	574.16	573.8930	48.11	2.11	44.21	5.57
C23H12I203	590.16	589.8880	46.81	2.05	43.01	8.13
C23H12I204	606.16	605.8829	45.57	2.00	41.87	10.56
C23H12I205	622.16	621.8778	44.40	1.94	40.79	12.86
C23H12I206	638.16	637.8727	43.29	1.90	39.77	15.04
C23H12I207	654.16	653.8676	42.23	1.85	38.80	17.12
C23H12I208	670.16	669.8625	41.22	1.80	37.87	19.10
C23H12I209	686.16	685.8574	40.26	1.76	36.99	20.99
C23H13I101	432.26	432.0013	63.91	3.03	29.36	3.70
C23H13I102	448.26	447.9962	61.63	2.92	28.31	7.14
C23H13I103	464.26	463.9911	59.50	2.82	27.33	10.34
C23H13I104	480.26	479.9860	57.52	2.73	26.42	13.33
C23H13I105	496.26	495.9810	55.67	2.64	25.57	16.12
C23H13I106	512.26	511.9759	53.93	2.56	24.77	18.74
C23H13I107	528.26	527.9708	52.30	2.48	24.02	21.20
C23H13I108	544.26	543.9657	50.76	2.41	23.32	23.52
C23H13I109	560.26	559.9606	49.31	2.34	22.65	25.70
C23H14I201	560.18	559.9138	49.32	2.52	45.31	2.86
C23H14I202	576.18	575.9087	47.95	2.45	44.05	5.55
C23H14I203	592.18	591.9036	46.65	2.38	42.86	8.11
C23H14I204	608.17	607.8985	45.42	2.32	41.73	10.52
C23H14I205	624.17	623.8934	44.26	2.26	40.66	12.82
C23H14I206	640.17	639.8883	43.15	2.20	39.65	15.00
C23H14I207	656.17	655.8833	42.10	2.15	38.68	17.07
C23H14I208	672.17	671.8782	41.10	2.10	37.76	19.04
C23H14I209	688.17	687.8731	40.14	2.05	36.88	20.92
C23H15I101	434.28	434.0169	63.61	3.48	29.22	3.68
C23H15I102	450.28	450.0119	61.35	3.36	28.18	7.11
C23H15I103	466.28	466.0068	59.25	3.24	27.22	10.29
C23H15I104	482.28	482.0017	57.28	3.14	26.31	13.27
C23H15I105	498.28	497.9966	55.44	3.03	25.47	16.05
C23H15I106	514.28	513.9915	53.72	2.94	24.68	18.67
C23H15I107	530.28	529.9864	52.10	2.85	23.93	21.12
C23H15I108	546.28	545.9813	50.57	2.77	23.23	23.43
C23H15I109	562.28	561.9763	49.13	2.69	22.57	25.61
C23H16I201	562.19	561.9294	49.14	2.87	45.15	2.85
C23H16I202	578.19	577.9243	47.78	2.79	43.90	5.53
C23H16I203	594.19	593.9193	46.49	2.71	42.72	8.08
C23H16I204	610.19	609.9142	45.27	2.64	41.60	10.49
C23H16I205	626.19	625.9091	44.12	2.58	40.53	12.78
C23H16I206	642.19	641.9040	43.02	2.51	39.52	14.95
C23H16I207	658.19	657.8989	41.97	2.45	38.56	17.02
C23H16I208	674.19	673.8938	40.98	2.39	37.65	18.99
C23H16I209	690.19	689.8887	40.03	2.34	36.77	20.86
C23H17I101	436.30	436.0326	63.32	3.93	29.09	3.67
C23H17I102	452.30	452.0275	61.08	3.79	28.06	7.07
C23H17I103	468.29	468.0224	58.99	3.66	27.10	10.25
C23H17I104	484.29	484.0173	57.04	3.54	26.20	13.21
C23H17I105	500.29	500.0122	55.22	3.43	25.37	15.99
C23H17I106	516.29	516.0072	53.51	3.32	24.58	18.59
C23H17I107	532.29	532.0021	51.90	3.22	23.84	21.04
C23H17I108	548.29	547.9970	50.38	3.13	23.15	23.34
C23H17I109	564.29	563.9919	48.96	3.04	22.49	25.52
C23H18I201	564.21	563.9451	48.96	3.22	44.98	2.84
C23H18I202	580.21	579.9400	47.61	3.13	43.74	5.52
C23H18I203	596.21	595.9349	46.34	3.04	42.57	8.05
C23H18I204	612.21	611.9298	45.12	2.96	41.46	10.45
C23H18I205	628.21	627.9247	43.98	2.89	40.40	12.73
C23H18I206	644.21	643.9196	42.88	2.82	39.40	14.90
C23H18I207	660.20	659.9146	41.84	2.75	38.44	16.96
C23H18I208	676.20	675.9095	40.85	2.68	37.53	18.93
C23H18I209	692.20	691.9044	39.91	2.62	36.67	20.80
C23H19I101	438.31	438.0482	63.03	4.37	28.95	3.65
C23H19I102	454.31	454.0432	60.81	4.22	27.93	7.04
C23H19I103	470.31	470.0381	58.74	4.07	26.98	10.21
C23H19I104	486.31	486.0330	56.81	3.94	26.10	13.16
C23H19I105	502.31	502.0279	55.00	3.81	25.26	15.93
C23H19I106	518.31	518.0228	53.30	3.69	24.48	18.52
C23H19I107	534.31	534.0177	51.70	3.58	23.75	20.96

C23H19I108	550.31	550.0126	50.20	3.48	23.06	23.26
C23H19I109	566.31	566.0076	48.78	3.38	22.41	25.43
C23H20I201	566.22	565.9607	48.79	3.56	44.82	2.83
C23H20I202	582.22	581.9556	47.45	3.46	43.59	5.50
C23H20I203	598.22	597.9506	46.18	3.37	42.43	8.02
C23H20I204	614.22	613.9455	44.98	3.28	41.32	10.42
C23H20I205	630.22	629.9404	43.83	3.20	40.27	12.69
C23H20I206	646.22	645.9353	42.75	3.12	39.28	14.86
C23H20I207	662.22	661.9302	41.72	3.04	38.33	16.91
C23H20I208	678.22	677.9251	40.73	2.97	37.42	18.87
C23H20I209	694.22	693.9200	39.79	2.90	36.56	20.74
C23H21I101	440.33	440.0639	62.74	4.81	28.82	3.63
C23H21I102	456.33	456.0588	60.54	4.64	27.81	7.01
C23H21I103	472.33	472.0537	58.49	4.48	26.87	10.16
C23H21I104	488.33	488.0486	56.57	4.33	25.99	13.11
C23H21I105	504.33	504.0435	54.78	4.20	25.16	15.86
C23H21I106	520.32	520.0385	53.09	4.07	24.39	18.45
C23H21I107	536.32	536.0334	51.51	3.95	23.66	20.88
C23H21I108	552.32	552.0283	50.02	3.83	22.98	23.17
C23H21I109	568.32	568.0232	48.61	3.72	22.33	25.34
C23H22I201	568.24	567.9764	48.62	3.90	44.67	2.82
C23H22I202	584.24	583.9713	47.28	3.80	43.44	5.48
C23H22I203	600.24	599.9662	46.02	3.69	42.28	8.00
C23H22I204	616.24	615.9611	44.83	3.60	41.19	10.39
C23H22I205	632.24	631.9560	43.70	3.51	40.14	12.65
C23H22I206	648.24	647.9509	42.62	3.42	39.15	14.81
C23H22I207	664.24	663.9459	41.59	3.34	38.21	16.86
C23H22I208	680.24	679.9408	40.61	3.26	37.31	18.82
C23H22I209	696.24	695.9357	39.68	3.19	36.45	20.68
C23H23I101	442.34	442.0795	62.45	5.24	28.69	3.62
C23H23I102	458.34	458.0745	60.27	5.06	27.69	6.98
C23H23I103	474.34	474.0694	58.24	4.89	26.75	10.12
C23H23I104	490.34	490.0643	56.34	4.73	25.88	13.05
C23H23I105	506.34	506.0592	54.56	4.58	25.06	15.80
C23H23I106	522.34	522.0541	52.89	4.44	24.30	18.38
C23H23I107	538.34	538.0490	51.32	4.31	23.57	20.80
C23H23I108	554.34	554.0439	49.84	4.18	22.89	23.09
C23H23I109	570.34	570.0389	48.44	4.06	22.25	25.25
C23H24I201	570.26	569.9920	48.44	4.24	44.51	2.81
C23H24I202	586.26	585.9869	47.12	4.13	43.29	5.46
C23H24I203	602.25	601.9819	45.87	4.02	42.14	7.97
C23H24I204	618.25	617.9768	44.68	3.91	41.05	10.35
C23H24I205	634.25	633.9717	43.56	3.81	40.02	12.61
C23H24I206	650.25	649.9666	42.48	3.72	39.03	14.76
C23H24I207	666.25	665.9615	41.46	3.63	38.09	16.81
C23H24I208	682.25	681.9564	40.49	3.55	37.20	18.76
C23H24I209	698.25	697.9513	39.56	3.46	36.35	20.62
C23H25I101	444.36	444.0952	62.17	5.67	28.56	3.60
C23H25I102	460.36	460.0901	60.01	5.47	27.57	6.95
C23H25I103	476.36	476.0850	57.99	5.29	26.64	10.08
C23H25I104	492.36	492.0799	56.11	5.12	25.77	13.00
C23H25I105	508.36	508.0748	54.34	4.96	24.96	15.74
C23H25I106	524.36	524.0698	52.68	4.81	24.20	18.31
C23H25I107	540.36	540.0647	51.12	4.66	23.49	20.73
C23H25I108	556.36	556.0596	49.65	4.53	22.81	23.01
C23H25I109	572.35	572.0545	48.27	4.40	22.17	25.16
C23H26I201	572.27	572.0077	48.27	4.58	44.35	2.80
C23H26I202	588.27	588.0026	46.96	4.45	43.14	5.44
C23H26I203	604.27	603.9975	45.72	4.34	42.00	7.94
C23H26I204	620.27	619.9924	44.54	4.23	40.92	10.32
C23H26I205	636.27	635.9873	43.42	4.12	39.89	12.57
C23H26I206	652.27	651.9822	42.35	4.02	38.91	14.72
C23H26I207	668.27	667.9772	41.34	3.92	37.98	16.76
C23H26I208	684.27	683.9721	40.37	3.83	37.09	18.71
C23H26I209	700.27	699.9670	39.45	3.74	36.24	20.56
C23H27I101	446.38	446.1108	61.89	6.10	28.43	3.58
C23H27I102	462.37	462.1058	59.75	5.89	27.45	6.92
C23H27I103	478.37	478.1007	57.75	5.69	26.53	10.03
C23H27I104	494.37	494.0956	55.88	5.50	25.67	12.95
C23H27I105	510.37	510.0905	54.13	5.33	24.87	15.67
C23H27I106	526.37	526.0854	52.48	5.17	24.11	18.24
C23H27I107	542.37	542.0803	50.93	5.02	23.40	20.65
C23H27I108	558.37	558.0752	49.48	4.87	22.73	22.92
C23H27I109	574.37	574.0702	48.10	4.74	22.09	25.07
C23H28I201	574.29	574.0233	48.10	4.91	44.20	2.79
C23H28I202	590.29	590.0182	46.80	4.78	43.00	5.42
C23H28I203	606.29	606.0132	45.57	4.66	41.86	7.92
C23H28I204	622.29	622.0081	44.39	4.54	40.79	10.28
C23H28I205	638.29	638.0030	43.28	4.42	39.76	12.53
C23H28I206	654.28	653.9979	42.22	4.31	38.79	14.67
C23H28I207	670.28	669.9928	41.21	4.21	37.87	16.71
C23H28I208	686.28	685.9877	40.25	4.11	36.98	18.65
C23H28I209	702.28	701.9826	39.34	4.02	36.14	20.50
C23H29I101	448.39	448.1265	61.61	6.52	28.30	3.57
C23H29I102	464.39	464.1214	59.49	6.29	27.33	6.89
C23H29I103	480.39	480.1163	57.51	6.08	26.42	9.99
C23H29I104	496.39	496.1112	55.65	5.89	25.57	12.89
C23H29I105	512.39	512.1061	53.92	5.70	24.77	15.61
C23H29I106	528.39	528.1011	52.28	5.53	24.02	18.17
C23H29I107	544.39	544.0960	50.75	5.37	23.31	20.57
C23H29I108	560.39	560.0909	49.30	5.22	22.65	22.84
C23H29I109	576.39	576.0858	47.93	5.07	22.02	24.98
C23H30I201	576.30	576.0390	47.94	5.25	44.04	2.78
C23H30I202	592.30	592.0339	46.64	5.11	42.85	5.40
C23H30I203	608.30	608.0288	45.41	4.97	41.72	7.89
C23H30I204	624.30	624.0237	44.25	4.84	40.65	10.25
C23H30I205	640.30	640.0186	43.14	4.72	39.64	12.49
C23H30I206	656.30	656.0135	42.09	4.61	38.67	14.63
C23H30I207	672.30	672.0085	41.09	4.50	37.75	16.66
C23H30I208	688.30	688.0034	40.14	4.39	36.87	18.60
C23H30I209	704.30	703.9983	39.22	4.29	36.04	20.45
C23H31I101	450.41	450.1421	61.33	6.94	28.18	3.55
C23H31I102	466.41	466.1371	59.23	6.70	27.21	6.86
C23H31I103	482.41	482.1320	57.27	6.48	26.31	9.95
C23H31I104	498.41	498.1269	55.43	6.27	25.46	12.84
C23H31I105	514.40	514.1218	53.70	6.07	24.67	15.55
C23H31I106	530.40	530.1167	52.08	5.89	23.93	18.10
C23H31I107	546.40	546.1116	50.56	5.72	23.23	20.50
C23H31I108	562.40	562.1065	49.12	5.56	22.56	22.76
C23H31I109	578.40	578.1014	47.76	5.40	21.94	24.90
C23H32I201	578.32	578.0546	47.77	5.58	43.89	2.77
C23H32I202	594.32	594.0495	46.48	5.43	42.71	5.38
C23H32I203	610.32	610.0444	45.26	5.28	41.59	7.86
C23H32I204	626.32	626.0394	44.11	5.15	40.52	10.22
C23H32I205	642.32	642.0343	43.01	5.02	39.51	12.45
C23H32I206	658.32	658.0292	41.96	4.90	38.55	14.58
C23H32I207	674.32	674.0241	40.97	4.78	37.64	16.61
C23H32I208	690.32	690.0190	40.02	4.67	36.77	18.54
C23H32I209	706.31	706.0139	39.11	4.57	35.93	20.39
C23H33I101	452.42	452.1578	61.06	7.35	28.05	3.54

C23H33I102	468.42	468.1527	58.98	7.10	27.09	6.83
C23H33I103	484.42	484.1476	57.03	6.87	26.20	9.91
C23H33I104	500.42	500.1425	55.20	6.65	25.36	12.79
C23H33I105	516.42	516.1374	53.49	6.44	24.57	15.49
C23H33I106	532.42	532.1324	51.89	6.25	23.84	18.03
C23H33I107	548.42	548.1273	50.37	6.07	23.14	20.42
C23H33I108	564.42	564.1222	48.95	5.89	22.48	22.68
C23H33I109	580.42	580.1171	47.60	5.73	21.86	24.81
C23H34I201	580.34	580.0703	47.60	5.91	43.73	2.76
C23H34I202	596.34	596.0652	46.33	5.75	42.56	5.37
C23H34I203	612.33	612.0601	45.12	5.60	41.45	7.84
C23H34I204	628.33	628.0550	43.97	5.45	40.39	10.19
C23H34I205	644.33	644.0499	42.87	5.32	39.39	12.42
C23H34I206	660.33	660.0448	41.84	5.19	38.44	14.54
C23H34I207	676.33	676.0398	40.85	5.07	37.53	16.56
C23H34I208	692.33	692.0347	39.90	4.95	36.66	18.49
C23H34I209	708.33	708.0296	39.00	4.84	35.83	20.33
C23H35I101	454.44	454.1734	60.79	7.76	27.93	3.52
C23H35I102	470.44	470.1683	58.72	7.50	26.98	6.80
C23H35I103	486.44	486.1633	56.79	7.25	26.09	9.87
C23H35I104	502.44	502.1582	54.98	7.02	25.26	12.74
C23H35I105	518.44	518.1531	53.29	6.80	24.48	15.43
C23H35I106	534.44	534.1480	51.69	6.60	23.75	17.96
C23H35I107	550.44	550.1429	50.19	6.41	23.06	20.35
C23H35I108	566.44	566.1378	48.77	6.23	22.40	22.60
C23H35I109	582.43	582.1327	47.43	6.06	21.79	24.72
C23H36I201	582.35	582.0859	47.44	6.23	43.58	2.75
C23H36I202	598.35	598.0808	46.17	6.06	42.42	5.35
C23H36I203	614.35	614.0757	44.97	5.91	41.31	7.81
C23H36I204	630.35	630.0707	43.83	5.76	40.26	10.15
C23H36I205	646.35	646.0656	42.74	5.61	39.27	12.38
C23H36I206	662.35	662.0605	41.71	5.48	38.32	14.49
C23H36I207	678.35	678.0554	40.72	5.35	37.42	16.51
C23H36I208	694.35	694.0503	39.79	5.23	36.55	18.43
C23H36I209	710.35	710.0452	38.89	5.11	35.73	20.27
C23H37I101	456.46	456.1891	60.52	8.17	27.80	3.51
C23H37I102	472.45	472.1840	58.47	7.89	26.86	6.77
C23H37I103	488.45	488.1789	56.56	7.64	25.98	9.83
C23H37I104	504.45	504.1738	54.76	7.39	25.16	12.69
C23H37I105	520.45	520.1687	53.08	7.17	24.38	15.37
C23H37I106	536.45	536.1637	51.50	6.95	23.66	17.89
C23H37I107	552.45	552.1586	50.01	6.75	22.97	20.27
C23H37I108	568.45	568.1535	48.60	6.56	22.32	22.52
C23H37I109	584.45	584.1484	47.27	6.38	21.71	24.64
C23H38I201	584.37	584.1016	47.27	6.55	43.43	2.74
C23H38I202	600.37	600.0965	46.01	6.38	42.28	5.33
C23H38I203	616.37	616.0914	44.82	6.21	41.18	7.79
C23H38I204	632.37	632.0863	43.69	6.06	40.14	10.12
C23H38I205	648.37	648.0812	42.61	5.91	39.15	12.34
C23H38I206	664.36	664.0761	41.58	5.77	38.20	14.45
C23H38I207	680.36	680.0711	40.60	5.63	37.30	16.46
C23H38I208	696.36	696.0660	39.67	5.50	36.45	18.38
C23H38I209	712.36	712.0609	38.78	5.38	35.63	20.21
C23H39I101	458.47	458.2047	60.26	8.57	27.68	3.49
C23H39I102	474.47	474.1996	58.22	8.29	26.75	6.74
C23H39I103	490.47	490.1946	56.32	8.01	25.87	9.79
C23H39I104	506.47	506.1895	54.55	7.76	25.06	12.64
C23H39I105	522.47	522.1844	52.88	7.52	24.29	15.31
C23H39I106	538.47	538.1793	51.30	7.30	23.57	17.83
C23H39I107	554.47	554.1742	49.82	7.09	22.89	20.20
C23H39I108	570.47	570.1691	48.43	6.89	22.25	22.44
C23H39I109	586.47	586.1640	47.11	6.70	21.64	24.55
C23H40I201	586.38	586.1172	47.11	6.88	43.28	2.73
C23H40I202	602.38	602.1121	45.86	6.69	42.13	5.31
C23H40I203	618.38	618.1070	44.67	6.52	41.04	7.76
C23H40I204	634.38	634.1020	43.55	6.36	40.01	10.09
C23H40I205	650.38	650.0969	42.48	6.20	39.02	12.30
C23H40I206	666.38	666.0918	41.46	6.05	38.09	14.41
C23H40I207	682.38	682.0867	40.48	5.91	37.19	16.41
C23H40I208	698.38	698.0816	39.56	5.77	36.34	18.33
C23H40I209	714.38	714.0765	38.67	5.64	35.53	20.16
C23H41I101	460.49	460.2204	59.99	8.97	27.56	3.47
C23H41I102	476.49	476.2153	57.98	8.67	26.63	6.72
C23H41I103	492.49	492.2102	56.09	8.39	25.77	9.75
C23H41I104	508.49	508.2051	54.33	8.13	24.96	12.59
C23H41I105	524.48	524.2000	52.67	7.88	24.20	15.25
C23H41I106	540.48	540.1950	51.11	7.65	23.48	17.76
C23H41I107	556.48	556.1899	49.64	7.43	22.80	20.13
C23H41I108	572.48	572.1848	48.26	7.22	22.17	22.36
C23H41I109	588.48	588.1797	46.94	7.02	21.56	24.47
C23H42I201	588.40	588.1329	46.95	7.19	43.14	2.72
C23H42I202	604.40	604.1278	45.71	7.00	41.99	5.29
C23H42I203	620.40	620.1227	44.53	6.82	40.91	7.74
C23H42I204	636.40	636.1176	43.41	6.65	39.88	10.06
C23H42I205	652.40	652.1125	42.34	6.49	38.90	12.26
C23H42I206	668.40	668.1074	41.33	6.33	37.97	14.36
C23H42I207	684.40	684.1024	40.37	6.19	37.09	16.36
C23H42I208	700.40	700.0973	39.44	6.04	36.24	18.27
C23H42I209	716.39	716.0922	38.56	5.91	35.43	20.10
C23H43I101	462.50	462.2360	59.73	9.37	27.44	3.46
C23H43I102	478.50	478.2309	57.73	9.06	26.52	6.69
C23H43I103	494.50	494.2259	55.87	8.76	25.66	9.71
C23H43I104	510.50	510.2208	54.11	8.49	24.86	12.54
C23H43I105	526.50	526.2157	52.47	8.23	24.10	15.19
C23H43I106	542.50	542.2106	50.92	7.99	23.39	17.70
C23H43I107	558.50	558.2055	49.46	7.76	22.72	20.05
C23H43I108	574.50	574.2004	48.09	7.54	22.09	22.28
C23H43I109	590.50	590.1953	46.78	7.34	21.49	24.39
C23H44I201	590.42	590.1485	46.79	7.51	42.99	2.71
C23H44I202	606.41	606.1434	45.56	7.31	41.85	5.28
C23H44I203	622.41	622.1383	44.38	7.13	40.78	7.71
C23H44I204	638.41	638.1333	43.27	6.95	39.76	10.02
C23H44I205	654.41	654.1282	42.21	6.78	38.78	12.22
C23H44I206	670.41	670.1231	41.21	6.62	37.86	14.32
C23H44I207	686.41	686.1180	40.25	6.46	36.98	16.32
C23H44I208	702.41	702.1129	39.33	6.31	36.13	18.22
C23H44I209	718.41	718.1078	38.45	6.17	35.33	20.04
C23H45I101	464.52	464.2517	59.47	9.76	27.32	3.44
C23H45I102	480.52	480.2466	57.49	9.44	26.41	6.66
C23H45I103	496.52	496.2415	55.64	9.14	25.56	9.67
C23H45I104	512.52	512.2364	53.90	8.85	24.76	12.49
C23H45I105	528.52	528.2313	52.27	8.58	24.01	15.14
C23H45I106	544.52	544.2263	50.73	8.33	23.31	17.63
C23H45I107	560.52	560.2212	49.29	8.09	22.64	19.98
C23H45I108	576.51	576.2161	47.92	7.87	22.01	22.20
C23H45I109	592.51	592.2110	46.62	7.66	21.42	24.30
C23H46I201	592.43	592.1642	46.63	7.83	42.84	2.70
C23H46I202	608.43	608.1591	45.40	7.62	41.72	5.26
C23H46I203	624.43	624.1540	44.24	7.43	40.65	7.69
C23H46I204	640.43	640.1489	43.14	7.24	39.63	9.99

C23H46I205	656.43	656.1438	42.08	7.06	38.67	12.19
C23H46I206	672.43	672.1387	41.08	6.90	37.75	14.28
C23H46I207	688.43	688.1337	40.13	6.74	36.87	16.27
C23H46I208	704.43	704.1286	39.22	6.58	36.03	18.17
C23H46I209	720.43	720.1235	38.35	6.44	35.23	19.99
C23H47I101	466.53	466.2673	59.21	10.15	27.20	3.43
C23H47I102	482.53	482.2622	57.25	9.82	26.30	6.63
C23H47I103	498.53	498.2572	55.41	9.50	25.46	9.63
C23H47I104	514.53	514.2521	53.69	9.21	24.66	12.44
C23H47I105	530.53	530.2470	52.07	8.93	23.92	15.08
C23H47I106	546.53	546.2419	50.55	8.67	23.22	17.56
C23H47I107	562.53	562.2368	49.11	8.42	22.56	19.91
C23H47I108	578.53	578.2317	47.75	8.19	21.94	22.12
C23H47I109	594.53	594.2266	46.47	7.97	21.35	24.22
C24H12I201	570.17	569.8981	50.56	2.12	44.51	2.81
C24H12I202	586.17	585.8930	49.18	2.06	43.30	5.46
C24H12I203	602.17	601.8880	47.87	2.01	42.15	7.97
C24H12I204	618.17	617.8829	46.63	1.96	41.06	10.35
C24H12I205	634.17	633.8778	45.46	1.91	40.02	12.61
C24H12I206	650.17	649.8727	44.34	1.86	39.04	14.76
C24H12I207	666.17	665.8676	43.27	1.82	38.10	16.81
C24H12I208	682.17	681.8625	42.26	1.77	37.21	18.76
C24H12I209	698.17	697.8574	41.29	1.73	36.35	20.62
C24H13I101	444.28	444.0013	64.88	2.95	28.56	3.60
C24H13I102	460.27	459.9962	62.63	2.85	27.57	6.95
C24H13I103	476.27	475.9911	60.53	2.75	26.65	10.08
C24H13I104	492.27	491.9860	58.56	2.66	25.78	13.00
C24H13I105	508.27	507.9810	56.72	2.58	24.97	15.74
C24H13I106	524.27	523.9759	54.98	2.50	24.21	18.31
C24H13I107	540.27	539.9708	53.36	2.43	23.49	20.73
C24H13I108	556.27	555.9657	51.82	2.36	22.81	23.01
C24H13I109	572.27	571.9606	50.37	2.29	22.18	25.16
C24H14I201	572.19	571.9138	50.38	2.47	44.36	2.80
C24H14I202	588.19	587.9087	49.01	2.40	43.15	5.44
C24H14I203	604.19	603.9036	47.71	2.34	42.01	7.94
C24H14I204	620.19	619.8985	46.48	2.28	40.92	10.32
C24H14I205	636.18	635.8934	45.31	2.22	39.90	12.57
C24H14I206	652.18	651.8883	44.20	2.16	38.92	14.72
C24H14I207	668.18	667.8833	43.14	2.11	37.98	16.76
C24H14I208	684.18	683.8782	42.13	2.06	37.10	18.71
C24H14I209	700.18	699.8731	41.17	2.02	36.25	20.57
C24H15I101	446.29	446.0169	64.59	3.39	28.44	3.58
C24H15I102	462.29	462.0119	62.36	3.27	27.45	6.92
C24H15I103	478.29	478.0068	60.27	3.16	26.53	10.04
C24H15I104	494.29	494.0017	58.32	3.06	25.67	12.95
C24H15I105	510.29	509.9966	56.49	2.96	24.87	15.68
C24H15I106	526.29	525.9915	54.77	2.87	24.11	18.24
C24H15I107	542.29	541.9864	53.16	2.79	23.40	20.65
C24H15I108	558.29	557.9813	51.63	2.71	22.73	22.93
C24H15I109	574.29	573.9763	50.20	2.63	22.10	25.07
C24H16I201	574.20	573.9294	50.20	2.81	44.20	2.79
C24H16I202	590.20	589.9243	48.84	2.73	43.00	5.42
C24H16I203	606.20	605.9193	47.55	2.66	41.87	7.92
C24H16I204	622.20	621.9142	46.33	2.59	40.79	10.29
C24H16I205	638.20	637.9091	45.17	2.53	39.77	12.53
C24H16I206	654.20	653.9040	44.06	2.47	38.80	14.67
C24H16I207	670.20	669.8989	43.01	2.41	37.87	16.71
C24H16I208	686.20	685.8938	42.01	2.35	36.99	18.65
C24H16I209	702.20	701.8887	41.05	2.30	36.14	20.51
C24H17I101	448.31	448.0326	64.30	3.82	28.31	3.57
C24H17I102	464.31	464.0275	62.09	3.69	27.33	6.89
C24H17I103	480.31	480.0224	60.02	3.57	26.42	9.99
C24H17I104	496.31	496.0173	58.08	3.45	25.57	12.89
C24H17I105	512.30	512.0122	56.27	3.34	24.77	15.62
C24H17I106	528.30	528.0072	54.56	3.24	24.02	18.17
C24H17I107	544.30	544.0021	52.96	3.15	23.32	20.58
C24H17I108	560.30	559.9970	51.45	3.06	22.65	22.84
C24H17I109	576.30	575.9919	50.02	2.97	22.02	24.99
C24H18I201	576.22	575.9451	50.03	3.15	44.05	2.78
C24H18I202	592.22	591.9400	48.68	3.06	42.86	5.40
C24H18I203	608.22	607.9349	47.40	2.98	41.73	7.89
C24H18I204	624.22	623.9298	46.18	2.91	40.66	10.25
C24H18I205	640.22	639.9247	45.03	2.83	39.64	12.50
C24H18I206	656.22	655.9196	43.93	2.76	38.68	14.63
C24H18I207	672.22	671.9146	42.88	2.70	37.76	16.66
C24H18I208	688.22	687.9095	41.89	2.64	36.88	18.60
C24H18I209	704.21	703.9044	40.93	2.58	36.04	20.45
C24H19I101	450.32	450.0482	64.01	4.25	28.18	3.55
C24H19I102	466.32	466.0432	61.82	4.11	27.21	6.86
C24H19I103	482.32	482.0381	59.77	3.97	26.31	9.95
C24H19I104	498.32	498.0330	57.85	3.84	25.47	12.84
C24H19I105	514.32	514.0279	56.05	3.72	24.67	15.55
C24H19I106	530.32	530.0228	54.36	3.61	23.93	18.10
C24H19I107	546.32	546.0177	52.77	3.51	23.23	20.50
C24H19I108	562.32	562.0126	51.26	3.41	22.57	22.76
C24H19I109	578.32	578.0076	49.85	3.31	21.94	24.90
C24H20I201	578.24	577.9607	49.85	3.49	43.89	2.77
C24H20I202	594.23	593.9556	48.51	3.39	42.71	5.38
C24H20I203	610.23	609.9506	47.24	3.30	41.59	7.87
C24H20I204	626.23	625.9455	46.03	3.22	40.53	10.22
C24H20I205	642.23	641.9404	44.89	3.14	39.52	12.46
C24H20I206	658.23	657.9353	43.79	3.06	38.56	14.58
C24H20I207	674.23	673.9302	42.75	2.99	37.64	16.61
C24H20I208	690.23	689.9251	41.76	2.92	36.77	18.54
C24H20I209	706.23	705.9200	40.82	2.85	35.94	20.39
C24H21I101	452.34	452.0639	63.73	4.68	28.06	3.54
C24H21I102	468.34	468.0588	61.55	4.52	27.10	6.83
C24H21I103	484.34	484.0537	59.52	4.37	26.20	9.91
C24H21I104	500.34	500.0486	57.61	4.23	25.36	12.79
C24H21I105	516.34	516.0435	55.83	4.10	24.58	15.49
C24H21I106	532.34	532.0385	54.15	3.98	23.84	18.03
C24H21I107	548.34	548.0334	52.57	3.86	23.14	20.42
C24H21I108	564.33	564.0283	51.08	3.75	22.49	22.68
C24H21I109	580.33	580.0232	49.67	3.65	21.87	24.81
C24H22I201	580.25	579.9764	49.68	3.82	43.74	2.76
C24H22I202	596.25	595.9713	48.35	3.72	42.57	5.37
C24H22I203	612.25	611.9662	47.08	3.62	41.46	7.84
C24H22I204	628.25	627.9611	45.88	3.53	40.40	10.19
C24H22I205	644.25	643.9560	44.74	3.44	39.40	12.42
C24H22I206	660.25	659.9509	43.66	3.36	38.44	14.54
C24H22I207	676.25	675.9459	42.63	3.28	37.53	16.56
C24H22I208	692.25	691.9408	41.64	3.20	36.66	18.49
C24H22I209	708.25	707.9357	40.70	3.13	35.84	20.33
C24H23I101	454.35	454.0795	63.45	5.10	27.93	3.52
C24H23I102	470.35	470.0745	61.29	4.93	26.98	6.80
C24H23I103	486.35	486.0694	59.27	4.77	26.09	9.87
C24H23I104	502.35	502.0643	57.38	4.61	25.26	12.74
C24H23I105	518.35	518.0592	55.61	4.47	24.48	15.43
C24H23I106	534.35	534.0541	53.95	4.34	23.75	17.97
C24H23I107	550.35	550.0490	52.38	4.21	23.06	20.35

C24H23I108	566.35	566.0439	50.90	4.09	22.41	22.60
C24H23I109	582.35	582.0389	49.50	3.98	21.79	24.73
C24H24I201	582.27	581.9920	49.51	4.15	43.59	2.75
C24H24I202	598.27	597.9869	48.18	4.04	42.42	5.35
C24H24I203	614.27	613.9819	46.93	3.94	41.32	7.81
C24H24I204	630.27	629.9768	45.74	3.84	40.27	10.15
C24H24I205	646.26	645.9717	44.61	3.74	39.27	12.38
C24H24I206	662.26	661.9666	43.53	3.65	38.32	14.50
C24H24I207	678.26	677.9615	42.50	3.57	37.42	16.51
C24H24I208	694.26	693.9564	41.52	3.48	36.56	18.44
C24H24I209	710.26	709.9513	40.59	3.41	35.73	20.27
C24H25I101	456.37	456.0952	63.17	5.52	27.81	3.51
C24H25I102	472.37	472.0901	61.03	5.33	26.87	6.77
C24H25I103	488.37	488.0850	59.03	5.16	25.99	9.83
C24H25I104	504.37	504.0799	57.15	5.00	25.16	12.69
C24H25I105	520.37	520.0748	55.40	4.84	24.39	15.37
C24H25I106	536.37	536.0698	53.74	4.70	23.66	17.90
C24H25I107	552.37	552.0647	52.19	4.56	22.97	20.28
C24H25I108	568.37	568.0596	50.72	4.43	22.33	22.52
C24H25I109	584.37	584.0545	49.33	4.31	21.72	24.64
C24H26I201	584.28	584.0077	49.34	4.49	43.44	2.74
C24H26I202	600.28	600.0026	48.02	4.37	42.28	5.33
C24H26I203	616.28	615.9975	46.78	4.25	41.18	7.79
C24H26I204	632.28	631.9924	45.59	4.14	40.14	10.12
C24H26I205	648.28	647.9873	44.47	4.04	39.15	12.34
C24H26I206	664.28	663.9822	43.40	3.95	38.21	14.45
C24H26I207	680.28	679.9772	42.37	3.85	37.31	16.46
C24H26I208	696.28	695.9721	41.40	3.76	36.45	18.38
C24H26I209	712.28	711.9670	40.47	3.68	35.63	20.22
C24H27I101	458.39	458.1108	62.89	5.94	27.69	3.49
C24H27I102	474.39	474.1058	60.77	5.74	26.75	6.75
C24H27I103	490.39	490.1007	58.78	5.55	25.88	9.79
C24H27I104	506.38	506.0956	56.93	5.37	25.06	12.64
C24H27I105	522.38	522.0905	55.18	5.21	24.29	15.31
C24H27I106	538.38	538.0854	53.54	5.05	23.57	17.83
C24H27I107	554.38	554.0803	52.00	4.91	22.89	20.20
C24H27I108	570.38	570.0752	50.54	4.77	22.25	22.44
C24H27I109	586.38	586.0702	49.16	4.64	21.64	24.56
C24H28I201	586.30	586.0233	49.17	4.81	43.29	2.73
C24H28I202	602.30	602.0182	47.86	4.69	42.14	5.31
C24H28I203	618.30	618.0132	46.62	4.56	41.05	7.76
C24H28I204	634.30	634.0081	45.45	4.45	40.01	10.09
C24H28I205	650.30	650.0030	44.33	4.34	39.03	12.30
C24H28I206	666.30	665.9979	43.26	4.24	38.09	14.41
C24H28I207	682.30	681.9928	42.25	4.14	37.20	16.41
C24H28I208	698.29	697.9877	41.28	4.04	36.35	18.33
C24H28I209	714.29	713.9826	40.36	3.95	35.53	20.16
C24H29I101	460.40	460.1265	62.61	6.35	27.56	3.48
C24H29I102	476.40	476.1214	60.51	6.14	26.64	6.72
C24H29I103	492.40	492.1163	58.54	5.94	25.77	9.75
C24H29I104	508.40	508.1112	56.70	5.75	24.96	12.59
C24H29I105	524.40	524.1061	54.97	5.57	24.20	15.25
C24H29I106	540.40	540.1011	53.34	5.41	23.48	17.76
C24H29I107	556.40	556.0960	51.81	5.25	22.81	20.13
C24H29I108	572.40	572.0909	50.36	5.11	22.17	22.36
C24H29I109	588.40	588.0858	48.99	4.97	21.57	24.47
C24H30I201	588.31	588.0390	49.00	5.14	43.14	2.72
C24H30I202	604.31	604.0339	47.70	5.00	42.00	5.30
C24H30I203	620.31	620.0288	46.47	4.87	40.92	7.74
C24H30I204	636.31	636.0237	45.30	4.75	39.89	10.06

C24H30I205	652.31	652.0186	44.19	4.64	38.91	12.26
C24H30I206	668.31	668.0135	43.13	4.52	37.98	14.36
C24H30I207	684.31	684.0085	42.13	4.42	37.09	16.37
C24H30I208	700.31	700.0034	41.16	4.32	36.24	18.28
C24H30I209	716.31	715.9983	40.24	4.22	35.43	20.10
C24H31I101	462.42	462.1421	62.34	6.76	27.44	3.46
C24H31I102	478.42	478.1371	60.25	6.53	26.53	6.69
C24H31I103	494.42	494.1320	58.30	6.32	25.67	9.71
C24H31I104	510.42	510.1269	56.48	6.12	24.86	12.54
C24H31I105	526.42	526.1218	54.76	5.94	24.11	15.20
C24H31I106	542.42	542.1167	53.15	5.76	23.40	17.70
C24H31I107	558.41	558.1116	51.62	5.60	22.73	20.06
C24H31I108	574.41	574.1065	50.18	5.44	22.09	22.28
C24H31I109	590.41	590.1014	48.82	5.29	21.49	24.39
C24H32I201	590.33	590.0546	48.83	5.46	42.99	2.71
C24H32I202	606.33	606.0495	47.54	5.32	41.86	5.28
C24H32I203	622.33	622.0444	46.32	5.18	40.78	7.71
C24H32I204	638.33	638.0394	45.16	5.05	39.76	10.03
C24H32I205	654.33	654.0343	44.06	4.93	38.79	12.23
C24H32I206	670.33	670.0292	43.00	4.81	37.86	14.32
C24H32I207	686.33	686.0241	42.00	4.70	36.98	16.32
C24H32I208	702.33	702.0190	41.04	4.59	36.14	18.22
C24H32I209	718.33	718.0139	40.13	4.49	35.33	20.05
C24H33I101	464.43	464.1578	62.07	7.16	27.32	3.44
C24H33I102	480.43	480.1527	60.00	6.92	26.41	6.66
C24H33I103	496.43	496.1476	58.07	6.70	25.56	9.67
C24H33I104	512.43	512.1425	56.25	6.49	24.77	12.49
C24H33I105	528.43	528.1374	54.55	6.29	24.02	15.14
C24H33I106	544.43	544.1324	52.95	6.11	23.31	17.63
C24H33I107	560.43	560.1273	51.44	5.94	22.64	19.98
C24H33I108	576.43	576.1222	50.01	5.77	22.02	22.20
C24H33I109	592.43	592.1171	48.66	5.61	21.42	24.31
C24H34I201	592.35	592.0703	48.67	5.79	42.85	2.70
C24H34I202	608.35	608.0652	47.39	5.63	41.72	5.26
C24H34I203	624.35	624.0601	46.17	5.49	40.65	7.69
C24H34I204	640.34	640.0550	45.02	5.35	39.64	9.99
C24H34I205	656.34	656.0499	43.92	5.22	38.67	12.19
C24H34I206	672.34	672.0448	42.88	5.10	37.75	14.28
C24H34I207	688.34	688.0398	41.88	4.98	36.87	16.27
C24H34I208	704.34	704.0347	40.93	4.87	36.03	18.17
C24H34I209	720.34	720.0296	40.02	4.76	35.23	19.99
C24H35I101	466.45	466.1734	61.80	7.56	27.21	3.43
C24H35I102	482.45	482.1683	59.75	7.31	26.30	6.63
C24H35I103	498.45	498.1633	57.83	7.08	25.46	9.63
C24H35I104	514.45	514.1582	56.03	6.86	24.67	12.44
C24H35I105	530.45	530.1531	54.34	6.65	23.92	15.08
C24H35I106	546.45	546.1480	52.75	6.46	23.22	17.57
C24H35I107	562.45	562.1429	51.25	6.27	22.56	19.91
C24H35I108	578.45	578.1378	49.83	6.10	21.94	22.13
C24H35I109	594.45	594.1327	48.49	5.93	21.35	24.22
C24H36I201	594.36	594.0859	48.50	6.11	42.70	2.69
C24H36I202	610.36	610.0808	47.23	5.95	41.58	5.24
C24H36I203	626.36	626.0757	46.02	5.79	40.52	7.66
C24H36I204	642.36	642.0707	44.88	5.65	39.51	9.96
C24H36I205	658.36	658.0656	43.79	5.51	38.55	12.15
C24H36I206	674.36	674.0605	42.75	5.38	37.64	14.24
C24H36I207	690.36	690.0554	41.76	5.26	36.76	16.22
C24H36I208	706.36	706.0503	40.81	5.14	35.93	18.12
C24H36I209	722.36	722.0452	39.91	5.02	35.14	19.93
C24H37I101	468.47	468.1891	61.53	7.96	27.09	3.42

C24H37I102	484.47	484.1840	59.50	7.70	26.19	6.60
C24H37I103	500.47	500.1789	57.60	7.45	25.36	9.59
C24H37I104	516.46	516.1738	55.82	7.22	24.57	12.39
C24H37I105	532.46	532.1687	54.14	7.00	23.83	15.02
C24H37I106	548.46	548.1637	52.56	6.80	23.14	17.50
C24H37I107	564.46	564.1586	51.07	6.61	22.48	19.84
C24H37I108	580.46	580.1535	49.66	6.43	21.86	22.05
C24H37I109	596.46	596.1484	48.33	6.25	21.28	24.14
C24H38I201	596.38	596.1016	48.34	6.42	42.56	2.68
C24H38I202	612.38	612.0965	47.07	6.25	41.45	5.23
C24H38I203	628.38	628.0914	45.87	6.10	40.39	7.64
C24H38I204	644.38	644.0863	44.74	5.94	39.39	9.93
C24H38I205	660.38	660.0812	43.65	5.80	38.43	12.11
C24H38I206	676.38	676.0761	42.62	5.66	37.52	14.19
C24H38I207	692.38	692.0711	41.63	5.53	36.66	16.18
C24H38I208	708.37	708.0660	40.69	5.41	35.83	18.07
C24H38I209	724.37	724.0609	39.80	5.29	35.04	19.88
C24H39I101	470.48	470.2047	61.27	8.36	26.97	3.40
C24H39I102	486.48	486.1996	59.26	8.08	26.09	6.58
C24H39I103	502.48	502.1946	57.37	7.82	25.26	9.55
C24H39I104	518.48	518.1895	55.60	7.58	24.48	12.34
C24H39I105	534.48	534.1844	53.93	7.35	23.74	14.97
C24H39I106	550.48	550.1793	52.37	7.14	23.05	17.44
C24H39I107	566.48	566.1742	50.89	6.94	22.40	19.77
C24H39I108	582.48	582.1691	49.49	6.75	21.79	21.97
C24H39I109	598.48	598.1640	48.17	6.57	21.20	24.06
C24H40I201	598.39	598.1172	48.17	6.74	42.41	2.67
C24H40I202	614.39	614.1121	46.92	6.56	41.31	5.21
C24H40I203	630.39	630.1070	45.73	6.40	40.26	7.61
C24H40I204	646.39	646.1020	44.60	6.24	39.27	9.90
C24H40I205	662.39	662.0969	43.52	6.09	38.32	12.08
C24H40I206	678.39	678.0918	42.49	5.94	37.41	14.15
C24H40I207	694.39	694.0867	41.51	5.81	36.55	16.13
C24H40I208	710.39	710.0816	40.58	5.68	35.73	18.02
C24H40I209	726.39	726.0765	39.68	5.55	34.94	19.82
C24H41I101	472.50	472.2204	61.01	8.75	26.86	3.39
C24H41I102	488.50	488.2153	59.01	8.46	25.98	6.55
C24H41I103	504.50	504.2102	57.14	8.19	25.15	9.51
C24H41I104	520.50	520.2051	55.38	7.94	24.38	12.30
C24H41I105	536.50	536.2000	53.73	7.70	23.65	14.91
C24H41I106	552.50	552.1950	52.18	7.48	22.97	17.38
C24H41I107	568.49	568.1899	50.71	7.27	22.32	19.70
C24H41I108	584.49	584.1848	49.32	7.07	21.71	21.90
C24H41I109	600.49	600.1797	48.01	6.88	21.13	23.98
C24H42I201	600.41	600.1329	48.01	7.05	42.27	2.66
C24H42I202	616.41	616.1278	46.77	6.87	41.18	5.19
C24H42I203	632.41	632.1227	45.58	6.69	40.13	7.59
C24H42I204	648.41	648.1176	44.46	6.53	39.14	9.87
C24H42I205	664.41	664.1125	43.39	6.37	38.20	12.04
C24H42I206	680.41	680.1074	42.37	6.22	37.30	14.11
C24H42I207	696.41	696.1024	41.39	6.08	36.45	16.08
C24H42I208	712.41	712.0973	40.46	5.94	35.63	17.97
C24H42I209	728.41	728.0922	39.58	5.81	34.84	19.77
C24H43I101	474.51	474.2360	60.75	9.13	26.74	3.37
C24H43I102	490.51	490.2309	58.77	8.84	25.87	6.52
C24H43I103	506.51	506.2259	56.91	8.56	25.05	9.48
C24H43I104	522.51	522.2208	55.17	8.30	24.29	12.25
C24H43I105	538.51	538.2157	53.53	8.05	23.57	14.86
C24H43I106	554.51	554.2106	51.99	7.82	22.89	17.31
C24H43I107	570.51	570.2055	50.53	7.60	22.24	19.63

C24H43I108	586.51	586.2004	49.15	7.39	21.64	21.82
C24H43I109	602.51	602.1953	47.84	7.19	21.06	23.90
C24H44I201	602.43	602.1485	47.85	7.36	42.13	2.66
C24H44I202	618.43	618.1434	46.61	7.17	41.04	5.17
C24H44I203	634.43	634.1383	45.44	6.99	40.01	7.57
C24H44I204	650.42	650.1333	44.32	6.82	39.02	9.84
C24H44I205	666.42	666.1282	43.26	6.66	38.09	12.00
C24H44I206	682.42	682.1231	42.24	6.50	37.19	14.07
C24H44I207	698.42	698.1180	41.27	6.35	36.34	16.04
C24H44I208	714.42	714.1129	40.35	6.21	35.53	17.92
C24H44I209	730.42	730.1078	39.47	6.07	34.75	19.71
C24H45I101	476.53	476.2517	60.49	9.52	26.63	3.36
C24H45I102	492.53	492.2466	58.53	9.21	25.77	6.50
C24H45I103	508.53	508.2415	56.69	8.92	24.96	9.44
C24H45I104	524.53	524.2364	54.96	8.65	24.19	12.20
C24H45I105	540.53	540.2313	53.33	8.39	23.48	14.80
C24H45I106	556.53	556.2263	51.80	8.15	22.80	17.25
C24H45I107	572.53	572.2212	50.35	7.92	22.17	19.56
C24H45I108	588.53	588.2161	48.98	7.71	21.56	21.75
C24H45I109	604.53	604.2110	47.68	7.50	20.99	23.82
C24H46I201	604.44	604.1642	47.69	7.67	41.99	2.65
C24H46I202	620.44	620.1591	46.46	7.47	40.91	5.16
C24H46I203	636.44	636.1540	45.29	7.29	39.88	7.54
C24H46I204	652.44	652.1489	44.18	7.11	38.90	9.81
C24H46I205	668.44	668.1438	43.13	6.94	37.97	11.97
C24H46I206	684.44	684.1387	42.12	6.77	37.08	14.03
C24H46I207	700.44	700.1337	41.16	6.62	36.24	15.99
C24H46I208	716.44	716.1286	40.24	6.47	35.43	17.87
C24H46I209	732.44	732.1235	39.36	6.33	34.65	19.66
C24H47I101	478.55	478.2673	60.24	9.90	26.52	3.34
C24H47I102	494.55	494.2622	58.29	9.58	25.66	6.47
C24H47I103	510.54	510.2572	56.46	9.28	24.86	9.40
C24H47I104	526.54	526.2521	54.75	9.00	24.10	12.15
C24H47I105	542.54	542.2470	53.13	8.73	23.39	14.74
C24H47I106	558.54	558.2419	51.61	8.48	22.72	17.19
C24H47I107	574.54	574.2368	50.17	8.25	22.09	19.49
C24H47I108	590.54	590.2317	48.81	8.02	21.49	21.67
C24H47I109	606.54	606.2266	47.53	7.81	20.92	23.74
C24H48I201	606.46	606.1798	47.53	7.98	41.85	2.64
C24H48I202	622.46	622.1747	46.31	7.77	40.78	5.14
C24H48I203	638.46	638.1696	45.15	7.58	39.75	7.52
C24H48I204	654.46	654.1646	44.05	7.39	38.78	9.78
C24H48I205	670.46	670.1595	43.00	7.22	37.86	11.93
C24H48I206	686.46	686.1544	41.99	7.05	36.97	13.98
C24H48I207	702.45	702.1493	41.04	6.89	36.13	15.94
C24H48I208	718.45	718.1442	40.12	6.73	35.33	17.82
C24H48I209	734.45	734.1391	39.25	6.59	34.56	19.61
C24H49I101	480.56	480.2830	59.99	10.28	26.41	3.33
C24H49I102	496.56	496.2779	58.05	9.95	25.56	6.44
C24H49I103	512.56	512.2728	56.24	9.64	24.76	9.36
C24H49I104	528.56	528.2677	54.54	9.34	24.01	12.11
C24H49I105	544.56	544.2626	52.94	9.07	23.30	14.69
C24H49I106	560.56	560.2576	51.43	8.81	22.64	17.13
C24H49I107	576.56	576.2525	50.00	8.57	22.01	19.42
C24H49I108	592.56	592.2474	48.65	8.34	21.42	21.60
C24H49I109	608.56	608.2423	47.37	8.12	20.85	23.66
C25H12I201	582.18	581.8981	51.58	2.08	43.60	2.75
C25H12I202	598.18	597.8930	50.20	2.02	42.43	5.35
C25H12I203	614.18	613.8880	48.89	1.97	41.32	7.81
C25H12I204	630.18	629.8829	47.65	1.92	40.28	10.16

221

C25H12I205	646.18	645.8778	46.47	1.87	39.28	12.38
C25H12I206	662.18	661.8727	45.35	1.83	38.33	14.50
C25H12I207	678.18	677.8676	44.28	1.78	37.43	16.51
C25H12I208	694.18	693.8625	43.26	1.74	36.56	18.44
C25H12I209	710.18	709.8574	42.28	1.70	35.74	20.28
C25H13I101	456.29	456.0013	65.81	2.87	27.81	3.51
C25H13I102	472.29	471.9962	63.58	2.77	26.87	6.78
C25H13I103	488.28	487.9911	61.50	2.68	25.99	9.83
C25H13I104	504.28	503.9860	59.55	2.60	25.17	12.69
C25H13I105	520.28	519.9810	57.71	2.52	24.39	15.38
C25H13I106	536.28	535.9759	55.99	2.44	23.66	17.90
C25H13I107	552.28	551.9708	54.37	2.37	22.98	20.28
C25H13I108	568.28	567.9657	52.84	2.31	22.33	22.52
C25H13I109	584.28	583.9606	51.39	2.24	21.72	24.64
C25H14I201	584.20	583.9138	51.40	2.42	43.45	2.74
C25H14I202	600.20	599.9087	50.03	2.35	42.29	5.33
C25H14I203	616.20	615.9036	48.73	2.29	41.19	7.79
C25H14I204	632.20	631.8985	47.50	2.23	40.15	10.12
C25H14I205	648.20	647.8934	46.33	2.18	39.16	12.34
C25H14I206	664.20	663.8883	45.21	2.12	38.21	14.45
C25H14I207	680.19	679.8833	44.15	2.07	37.31	16.47
C25H14I208	696.19	695.8782	43.13	2.03	36.46	18.38
C25H14I209	712.19	711.8731	42.16	1.98	35.64	20.22
C25H15I101	458.30	458.0169	65.52	3.30	27.69	3.49
C25H15I102	474.30	474.0119	63.31	3.19	26.76	6.75
C25H15I103	490.30	490.0068	61.24	3.08	25.88	9.79
C25H15I104	506.30	506.0017	59.31	2.99	25.07	12.64
C25H15I105	522.30	521.9966	57.49	2.89	24.30	15.32
C25H15I106	538.30	537.9915	55.78	2.81	23.58	17.83
C25H15I107	554.30	553.9864	54.17	2.73	22.89	20.20
C25H15I108	570.30	569.9813	52.65	2.65	22.25	22.44
C25H15I109	586.30	585.9763	51.22	2.58	21.65	24.56
C25H16I201	586.21	585.9294	51.22	2.75	43.30	2.73
C25H16I202	602.21	601.9243	49.86	2.68	42.15	5.31
C25H16I203	618.21	617.9193	48.57	2.61	41.06	7.76
C25H16I204	634.21	633.9142	47.35	2.54	40.02	10.09
C25H16I205	650.21	649.9091	46.18	2.48	39.03	12.30
C25H16I206	666.21	665.9040	45.07	2.42	38.10	14.41
C25H16I207	682.21	681.8989	44.02	2.36	37.20	16.42
C25H16I208	698.21	697.8938	43.01	2.31	36.35	18.33
C25H16I209	714.21	713.8887	42.04	2.26	35.54	20.16
C25H17I101	460.32	460.0326	65.23	3.72	27.57	3.48
C25H17I102	476.32	476.0275	63.04	3.60	26.64	6.72
C25H17I103	492.32	492.0224	60.99	3.48	25.78	9.75
C25H17I104	508.32	508.0173	59.07	3.37	24.97	12.59
C25H17I105	524.32	524.0122	57.27	3.27	24.20	15.26
C25H17I106	540.32	540.0072	55.57	3.17	23.49	17.77
C25H17I107	556.31	556.0021	53.98	3.08	22.81	20.13
C25H17I108	572.31	571.9970	52.47	2.99	22.17	22.36
C25H17I109	588.31	587.9919	51.04	2.91	21.57	24.48
C25H18I201	588.23	587.9451	51.05	3.08	43.15	2.72
C25H18I202	604.23	603.9400	49.70	3.00	42.01	5.30
C25H18I203	620.23	619.9349	48.41	2.93	40.92	7.74
C25H18I204	636.23	635.9298	47.20	2.85	39.89	10.06
C25H18I205	652.23	651.9247	46.04	2.78	38.91	12.27
C25H18I206	668.23	667.9196	44.94	2.72	37.98	14.37
C25H18I207	684.23	683.9146	43.89	2.65	37.09	16.37
C25H18I208	700.23	699.9095	42.88	2.59	36.25	18.28
C25H18I209	716.23	715.9044	41.93	2.53	35.44	20.10
C25H19I101	462.33	462.0482	64.95	4.14	27.45	3.46
C25H19I102	478.33	478.0432	62.78	4.00	26.53	6.69
C25H19I103	494.33	494.0381	60.74	3.87	25.67	9.71
C25H19I104	510.33	510.0330	58.84	3.75	24.87	12.54
C25H19I105	526.33	526.0279	57.05	3.64	24.11	15.20
C25H19I106	542.33	542.0228	55.37	3.53	23.40	17.70
C25H19I107	558.33	558.0177	53.78	3.43	22.73	20.06
C25H19I108	574.33	574.0126	52.28	3.33	22.10	22.29
C25H19I109	590.33	590.0076	50.87	3.24	21.50	24.39
C25H20I201	590.25	589.9607	50.87	3.42	43.00	2.71
C25H20I202	606.25	605.9556	49.53	3.33	41.87	5.28
C25H20I203	622.25	621.9506	48.26	3.24	40.79	7.71
C25H20I204	638.24	637.9455	47.05	3.16	39.77	10.03
C25H20I205	654.24	653.9404	45.90	3.08	38.79	12.23
C25H20I206	670.24	669.9353	44.80	3.01	37.87	14.32
C25H20I207	686.24	685.9302	43.76	2.94	36.99	16.32
C25H20I208	702.24	701.9251	42.76	2.87	36.14	18.23
C25H20I209	718.24	717.9200	41.81	2.81	35.34	20.05
C25H21I101	464.35	464.0639	64.67	4.56	27.33	3.45
C25H21I102	480.35	480.0588	62.51	4.41	26.42	6.66
C25H21I103	496.35	496.0537	60.50	4.26	25.57	9.67
C25H21I104	512.35	512.0486	58.61	4.13	24.77	12.49
C25H21I105	528.35	528.0435	56.83	4.01	24.02	15.14
C25H21I106	544.35	544.0385	55.16	3.89	23.31	17.64
C25H21I107	560.35	560.0334	53.59	3.78	22.65	19.99
C25H21I108	576.35	576.0283	52.10	3.67	22.02	22.21
C25H21I109	592.35	592.0232	50.69	3.57	21.42	24.31
C25H22I201	592.26	591.9764	50.70	3.74	42.85	2.70
C25H22I202	608.26	607.9713	49.37	3.65	41.73	5.26
C25H22I203	624.26	623.9662	48.10	3.55	40.66	7.69
C25H22I204	640.26	639.9611	46.90	3.46	39.64	10.00
C25H22I205	656.26	655.9560	45.76	3.38	38.68	12.19
C25H22I206	672.26	671.9509	44.67	3.30	37.75	14.28
C25H22I207	688.26	687.9459	43.63	3.22	36.88	16.27
C25H22I208	704.26	703.9408	42.64	3.15	36.04	18.17
C25H22I209	720.26	719.9357	41.69	3.08	35.24	19.99
C25H23I101	466.37	466.0795	64.39	4.97	27.21	3.43
C25H23I102	482.37	482.0745	62.25	4.81	26.31	6.63
C25H23I103	498.36	498.0694	60.25	4.65	25.46	9.63
C25H23I104	514.36	514.0643	58.38	4.51	24.67	12.44
C25H23I105	530.36	530.0592	56.62	4.37	23.93	15.08
C25H23I106	546.36	546.0541	54.96	4.24	23.23	17.57
C25H23I107	562.36	562.0490	53.40	4.12	22.57	19.92
C25H23I108	578.36	578.0439	51.92	4.01	21.94	22.13
C25H23I109	594.36	594.0389	50.52	3.90	21.35	24.23
C25H24I201	594.28	593.9920	50.53	4.07	42.71	2.69
C25H24I202	610.28	609.9869	49.20	3.96	41.59	5.24
C25H24I203	626.28	625.9819	47.95	3.86	40.53	7.66
C25H24I204	642.28	641.9768	46.75	3.77	39.52	9.96
C25H24I205	658.28	657.9717	45.62	3.67	38.56	12.15
C25H24I206	674.28	673.9666	44.53	3.59	37.64	14.24
C25H24I207	690.27	689.9615	43.50	3.50	36.77	16.22
C25H24I208	706.27	705.9564	42.52	3.43	35.94	18.12
C25H24I209	722.27	721.9513	41.57	3.35	35.14	19.94
C25H25I101	468.38	468.0952	64.11	5.38	27.09	3.42
C25H25I102	484.38	484.0901	61.99	5.20	26.20	6.61
C25H25I103	500.38	500.0850	60.01	5.04	25.36	9.59
C25H25I104	516.38	516.0799	58.15	4.88	24.58	12.39
C25H25I105	532.38	532.0748	56.40	4.73	23.84	15.03
C25H25I106	548.38	548.0698	54.76	4.60	23.14	17.51
C25H25I107	564.38	564.0647	53.21	4.46	22.49	19.84

222

C25H25I108	580.38	580.0596	51.74	4.34	21.87	22.05
C25H25I109	596.38	596.0545	50.35	4.23	21.28	24.14
C25H26I201	596.29	596.0077	50.36	4.40	42.56	2.68
C25H26I202	612.29	612.0026	49.04	4.28	41.45	5.23
C25H26I203	628.29	627.9975	47.79	4.17	40.40	7.64
C25H26I204	644.29	643.9924	46.61	4.07	39.39	9.93
C25H26I205	660.29	659.9873	45.48	3.97	38.44	12.12
C25H26I206	676.29	675.9822	44.40	3.88	37.53	14.19
C25H26I207	692.29	691.9772	43.37	3.79	36.66	16.18
C25H26I208	708.29	707.9721	42.39	3.70	35.83	18.07
C25H26I209	724.29	723.9670	41.46	3.62	35.04	19.88
C25H27I101	470.40	470.1108	63.84	5.79	26.98	3.40
C25H27I102	486.40	486.1058	61.74	5.60	26.09	6.58
C25H27I103	502.40	502.1007	59.77	5.42	25.26	9.55
C25H27I104	518.40	518.0956	57.92	5.25	24.48	12.35
C25H27I105	534.40	534.0905	56.19	5.09	23.75	14.97
C25H27I106	550.39	550.0854	54.56	4.94	23.06	17.44
C25H27I107	566.39	566.0803	53.02	4.80	22.41	19.77
C25H27I108	582.39	582.0752	51.56	4.67	21.79	21.98
C25H27I109	598.39	598.0702	50.18	4.55	21.21	24.06
C25H28I201	598.31	598.0233	50.19	4.72	42.42	2.67
C25H28I202	614.31	614.0182	48.88	4.59	41.32	5.21
C25H28I203	630.31	630.0132	47.64	4.48	40.27	7.62
C25H28I204	646.31	646.0081	46.46	4.37	39.27	9.90
C25H28I205	662.31	662.0030	45.34	4.26	38.32	12.08
C25H28I206	678.31	677.9979	44.27	4.16	37.42	14.15
C25H28I207	694.31	693.9928	43.25	4.06	36.56	16.13
C25H28I208	710.31	709.9877	42.27	3.97	35.73	18.02
C25H28I209	726.31	725.9826	41.34	3.89	34.95	19.83
C25H29I101	472.41	472.1265	63.56	6.19	26.86	3.39
C25H29I102	488.41	488.1214	61.48	5.98	25.98	6.55
C25H29I103	504.41	504.1163	59.53	5.80	25.16	9.52
C25H29I104	520.41	520.1112	57.70	5.62	24.39	12.30
C25H29I105	536.41	536.1061	55.98	5.45	23.66	14.91
C25H29I106	552.41	552.1011	54.36	5.29	22.97	17.38
C25H29I107	568.41	568.0960	52.83	5.14	22.33	19.70
C25H29I108	584.41	584.0909	51.38	5.00	21.71	21.90
C25H29I109	600.41	600.0858	50.01	4.87	21.14	23.98
C25H30I201	600.33	600.0390	50.02	5.04	42.28	2.67
C25H30I202	616.33	616.0339	48.72	4.91	41.18	5.19
C25H30I203	632.32	632.0288	47.49	4.78	40.14	7.59
C25H30I204	648.32	648.0237	46.32	4.66	39.15	9.87
C25H30I205	664.32	664.0186	45.20	4.55	38.21	12.04
C25H30I206	680.32	680.0135	44.14	4.44	37.31	14.11
C25H30I207	696.32	696.0085	43.12	4.34	36.45	16.08
C25H30I208	712.32	712.0034	42.15	4.25	35.63	17.97
C25H30I209	728.32	727.9983	41.23	4.15	34.85	19.77
C25H31I101	474.43	474.1421	63.29	6.59	26.75	3.37
C25H31I102	490.43	490.1371	61.23	6.37	25.88	6.52
C25H31I103	506.43	506.1320	59.29	6.17	25.06	9.48
C25H31I104	522.43	522.1269	57.48	5.98	24.29	12.25
C25H31I105	538.43	538.1218	55.77	5.80	23.57	14.86
C25H31I106	554.43	554.1167	54.16	5.64	22.89	17.31
C25H31I107	570.43	570.1116	52.64	5.48	22.25	19.63
C25H31I108	586.43	586.1065	51.20	5.33	21.64	21.83
C25H31I109	602.42	602.1014	49.85	5.19	21.07	23.90
C25H32I201	602.34	602.0546	49.85	5.35	42.14	2.66
C25H32I202	618.34	618.0495	48.56	5.22	41.05	5.17
C25H32I203	634.34	634.0444	47.34	5.08	40.01	7.57
C25H32I204	650.34	650.0394	46.17	4.96	39.03	9.84
C25H32I205	666.34	666.0343	45.06	4.84	38.09	12.01
C25H32I206	682.34	682.0292	44.01	4.73	37.20	14.07
C25H32I207	698.34	698.0241	43.00	4.62	36.34	16.04
C25H32I208	714.34	714.0190	42.04	4.52	35.53	17.92
C25H32I209	730.34	730.0139	41.12	4.42	34.75	19.72
C25H33I101	476.45	476.1578	63.02	6.98	26.64	3.36
C25H33I102	492.44	492.1527	60.98	6.75	25.77	6.50
C25H33I103	508.44	508.1476	59.06	6.54	24.96	9.44
C25H33I104	524.44	524.1425	57.26	6.34	24.20	12.20
C25H33I105	540.44	540.1374	55.56	6.15	23.48	14.80
C25H33I106	556.44	556.1324	53.96	5.98	22.81	17.25
C25H33I107	572.44	572.1273	52.46	5.81	22.17	19.56
C25H33I108	588.44	588.1222	51.03	5.65	21.57	21.75
C25H33I109	604.44	604.1171	49.68	5.50	21.00	23.82
C25H34I201	604.36	604.0703	49.69	5.67	42.00	2.65
C25H34I202	620.36	620.0652	48.40	5.52	40.91	5.16
C25H34I203	636.36	636.0601	47.19	5.39	39.88	7.54
C25H34I204	652.36	652.0550	46.03	5.25	38.91	9.81
C25H34I205	668.36	668.0499	44.93	5.13	37.98	11.97
C25H34I206	684.35	684.0448	43.88	5.01	37.09	14.03
C25H34I207	700.35	700.0398	42.88	4.89	36.24	15.99
C25H34I208	716.35	716.0347	41.92	4.78	35.43	17.87
C25H34I209	732.35	732.0296	41.00	4.68	34.66	19.66
C25H35I101	478.46	478.1734	62.76	7.37	26.52	3.34
C25H35I102	494.46	494.1683	60.73	7.13	25.67	6.47
C25H35I103	510.46	510.1633	58.83	6.91	24.86	9.40
C25H35I104	526.46	526.1582	57.04	6.70	24.11	12.16
C25H35I105	542.46	542.1531	55.36	6.50	23.39	14.75
C25H35I106	558.46	558.1480	53.77	6.32	22.72	17.19
C25H35I107	574.46	574.1429	52.27	6.14	22.09	19.50
C25H35I108	590.46	590.1378	50.86	5.97	21.49	21.68
C25H35I109	606.46	606.1327	49.51	5.82	20.93	23.74
C25H36I201	606.37	606.0859	49.52	5.98	41.86	2.64
C25H36I202	622.37	622.0808	48.25	5.83	40.78	5.14
C25H36I203	638.37	638.0757	47.04	5.68	39.76	7.52
C25H36I204	654.37	654.0707	45.89	5.55	38.79	9.78
C25H36I205	670.37	670.0656	44.79	5.41	37.86	11.93
C25H36I206	686.37	686.0605	43.75	5.29	36.98	13.99
C25H36I207	702.37	702.0554	42.75	5.17	36.14	15.95
C25H36I208	718.37	718.0503	41.80	5.05	35.33	17.82
C25H36I209	734.37	734.0452	40.89	4.94	34.56	19.61
C25H37I101	480.48	480.1891	62.50	7.76	26.41	3.33
C25H37I102	496.48	496.1840	60.48	7.51	25.56	6.45
C25H37I103	512.48	512.1789	58.59	7.28	24.76	9.37
C25H37I104	528.48	528.1738	56.82	7.06	24.01	12.11
C25H37I105	544.48	544.1687	55.15	6.85	23.31	14.69
C25H37I106	560.47	560.1637	53.58	6.65	22.64	17.13
C25H37I107	576.47	576.1586	52.09	6.47	22.01	19.43
C25H37I108	592.47	592.1535	50.68	6.29	21.42	21.60
C25H37I109	608.47	608.1484	49.35	6.13	20.86	23.66
C25H38I201	608.39	608.1016	49.36	6.30	41.72	2.63
C25H38I202	624.39	624.0965	48.09	6.13	40.65	5.12
C25H38I203	640.39	640.0914	46.89	5.98	39.63	7.50
C25H38I204	656.39	656.0863	45.75	5.84	38.67	9.75
C25H38I205	672.39	672.0812	44.66	5.70	37.75	11.90
C25H38I206	688.39	688.0761	43.62	5.56	36.87	13.95
C25H38I207	704.39	704.0711	42.63	5.44	36.03	15.90
C25H38I208	720.39	720.0660	41.68	5.32	35.23	17.77
C25H38I209	736.39	736.0609	40.78	5.20	34.47	19.55
C25H39I101	482.49	482.2047	62.23	8.15	26.30	3.32

C25H39I102	498.49	498.1996	60.24	7.89	25.46	6.42
C25H39I103	514.49	514.1946	58.36	7.64	24.67	9.33
C25H39I104	530.49	530.1895	56.60	7.41	23.92	12.06
C25H39I105	546.49	546.1844	54.95	7.19	23.22	14.64
C25H39I106	562.49	562.1793	53.38	6.99	22.56	17.07
C25H39I107	578.49	578.1742	51.91	6.80	21.94	19.36
C25H39I108	594.49	594.1691	50.51	6.61	21.35	21.53
C25H39I109	610.49	610.1640	49.19	6.44	20.79	23.59
C25H40I201	610.41	610.1172	49.19	6.61	41.58	2.62
C25H40I202	626.41	626.1121	47.94	6.44	40.52	5.11
C25H40I203	642.40	642.1070	46.74	6.28	39.51	7.47
C25H40I204	658.40	658.1020	45.61	6.12	38.55	9.72
C25H40I205	674.40	674.0969	44.53	5.98	37.63	11.86
C25H40I206	690.40	690.0918	43.49	5.84	36.76	13.90
C25H40I207	706.40	706.0867	42.51	5.71	35.93	15.85
C25H40I208	722.40	722.0816	41.57	5.58	35.13	17.72
C25H40I209	738.40	738.0765	40.67	5.46	34.37	19.50
C25H41I101	484.51	484.2204	61.98	8.53	26.19	3.30
C25H41I102	500.51	500.2153	59.95	8.26	25.36	6.39
C25H41I103	516.51	516.2102	58.14	8.00	24.57	9.29
C25H41I104	532.51	532.2051	56.39	7.76	23.83	12.02
C25H41I105	548.51	548.2000	54.74	7.53	23.14	14.58
C25H41I106	564.51	564.1950	53.19	7.32	22.48	17.01
C25H41I107	580.51	580.1899	51.73	7.12	21.86	19.29
C25H41I108	596.51	596.1848	50.34	6.93	21.27	21.46
C25H41I109	612.50	612.1797	49.02	6.75	20.72	23.51
C25H42I201	612.42	612.1329	49.03	6.91	41.44	2.61
C25H42I202	628.42	628.1278	47.78	6.74	40.39	5.09
C25H42I203	644.42	644.1227	46.60	6.57	39.39	7.45
C25H42I204	660.42	660.1176	45.47	6.41	38.43	9.69
C25H42I205	676.42	676.1125	44.39	6.26	37.52	11.83
C25H42I206	692.42	692.1074	43.37	6.11	36.66	13.86
C25H42I207	708.42	708.1024	42.39	5.98	35.83	15.81
C25H42I208	724.42	724.0973	41.45	5.84	35.04	17.67
C25H42I209	740.42	740.0922	40.56	5.72	34.28	19.45
C25H43I101	486.53	486.2360	61.72	8.91	26.08	3.29
C25H43I102	502.52	502.2309	59.75	8.62	25.25	6.37
C25H43I103	518.52	518.2259	57.91	8.36	24.47	9.26
C25H43I104	534.52	534.2208	56.18	8.11	23.74	11.97
C25H43I105	550.52	550.2157	54.54	7.87	23.05	14.53
C25H43I106	566.52	566.2106	53.00	7.65	22.40	16.94
C25H43I107	582.52	582.2055	51.55	7.44	21.79	19.23
C25H43I108	598.52	598.2004	50.17	7.24	21.20	21.39
C25H43I109	614.52	614.1953	48.86	7.05	20.65	23.43
C25H44I201	614.44	614.1485	48.87	7.22	41.31	2.60
C25H44I202	630.44	630.1434	47.63	7.03	40.26	5.08
C25H44I203	646.44	646.1383	46.45	6.86	39.26	7.43
C25H44I204	662.44	662.1333	45.33	6.70	38.31	9.66
C25H44I205	678.44	678.1282	44.26	6.54	37.41	11.79
C25H44I206	694.43	694.1231	43.24	6.39	36.55	13.82
C25H44I207	710.43	710.1180	42.27	6.24	35.73	15.76
C25H44I208	726.43	726.1129	41.34	6.11	34.94	17.62
C25H44I209	742.43	742.1078	40.45	5.97	34.19	19.39
C25H45I101	488.54	488.2517	61.46	9.28	25.98	3.27
C25H45I102	504.54	504.2466	59.52	8.99	25.15	6.34
C25H45I103	520.54	520.2415	57.69	8.71	24.38	9.22
C25H45I104	536.54	536.2364	55.97	8.45	23.65	11.93
C25H45I105	552.54	552.2313	54.35	8.21	22.97	14.48
C25H45I106	568.54	568.2263	52.82	7.98	22.32	16.88
C25H45I107	584.54	584.2212	51.37	7.76	21.71	19.16
C25H45I108	600.54	600.2161	50.00	7.55	21.13	21.31
C25H45I109	616.54	616.2110	48.70	7.36	20.58	23.36
C25H46I201	616.45	616.1642	48.71	7.52	41.17	2.60
C25H46I202	632.45	632.1591	47.48	7.33	40.13	5.06
C25H46I203	648.45	648.1540	46.31	7.15	39.14	7.40
C25H46I204	664.45	664.1489	45.19	6.98	38.20	9.63
C25H46I205	680.45	680.1438	44.13	6.81	37.30	11.76
C25H46I206	696.45	696.1387	43.12	6.66	36.44	13.78
C25H46I207	712.45	712.1337	42.15	6.51	35.62	15.72
C25H46I208	728.45	728.1286	41.22	6.37	34.84	17.57
C25H46I209	744.45	744.1235	40.34	6.23	34.09	19.34
C25H47I101	490.56	490.2673	61.21	9.66	25.87	3.26
C25H47I102	506.56	506.2622	59.28	9.35	25.05	6.32
C25H47I103	522.56	522.2572	57.46	9.07	24.29	9.19
C25H47I104	538.56	538.2521	55.76	8.80	23.56	11.88
C25H47I105	554.55	554.2470	54.15	8.54	22.88	14.43
C25H47I106	570.55	570.2419	52.63	8.30	22.24	16.83
C25H47I107	586.55	586.2368	51.19	8.08	21.64	19.09
C25H47I108	602.55	602.2317	49.83	7.86	21.06	21.24
C25H47I109	618.55	618.2266	48.55	7.66	20.52	23.28
C25H48I201	618.47	618.1798	48.55	7.82	41.04	2.59
C25H48I202	634.47	634.1747	47.33	7.63	40.00	5.04
C25H48I203	650.47	650.1696	46.16	7.44	39.02	7.38
C25H48I204	666.47	666.1646	45.06	7.26	38.08	9.60
C25H48I205	682.47	682.1595	44.00	7.09	37.19	11.72
C25H48I206	698.47	698.1544	42.99	6.93	36.34	13.74
C25H48I207	714.47	714.1493	42.03	6.77	35.52	15.68
C25H48I208	730.47	730.1442	41.11	6.62	34.75	17.52
C25H48I209	746.46	746.1391	40.23	6.48	34.00	19.29
C25H49I101	492.57	492.2830	60.96	10.03	25.76	3.25
C25H49I102	508.57	508.2779	59.04	9.71	24.95	6.29
C25H49I103	524.57	524.2728	57.24	9.42	24.19	9.15
C25H49I104	540.57	540.2677	55.55	9.14	23.48	11.84
C25H49I105	556.57	556.2626	53.95	8.87	22.80	14.37
C25H49I106	572.57	572.2576	52.44	8.63	22.16	16.77
C25H49I107	588.57	588.2525	51.02	8.39	21.56	19.03
C25H49I108	604.57	604.2474	49.67	8.17	20.99	21.17
C25H49I109	620.57	620.2423	48.39	7.96	20.45	23.20
C25H50I201	620.49	620.1955	48.39	8.12	40.90	2.58
C25H50I202	636.48	636.1904	47.18	7.92	39.88	5.03
C25H50I203	652.48	652.1853	46.02	7.72	38.90	7.36
C25H50I204	668.48	668.1802	44.92	7.54	37.97	9.57
C25H50I205	684.48	684.1751	43.87	7.36	37.08	11.69
C25H50I206	700.48	700.1700	42.87	7.19	36.23	13.70
C25H50I207	716.48	716.1649	41.91	7.03	35.42	15.63
C25H50I208	732.48	732.1599	40.99	6.88	34.65	17.47
C25H50I209	748.48	748.1548	40.12	6.73	33.91	19.24
C25H51I101	494.59	494.2986	60.71	10.39	25.66	3.23
C25H51I102	510.59	510.2935	58.81	10.07	24.85	6.27
C25H51I103	526.59	526.2885	57.02	9.76	24.10	9.11
C25H51I104	542.59	542.2834	55.34	9.47	23.39	11.79
C25H51I105	558.59	558.2783	53.76	9.20	22.72	14.32
C25H51I106	574.59	574.2732	52.26	8.95	22.09	16.71
C25H51I107	590.59	590.2681	50.84	8.70	21.49	18.96
C25H51I108	606.58	606.2630	49.50	8.47	20.92	21.10
C25H51I109	622.58	622.2579	48.23	8.26	20.38	23.13
C26H13I101	468.30	468.0013	66.69	2.80	27.10	3.42
C26H13I102	484.30	483.9962	64.48	2.71	26.20	6.61
C26H13I103	500.30	499.9911	62.42	2.62	25.37	9.59
C26H13I104	516.30	515.9860	60.49	2.54	24.58	12.40

224

C26H13I105	532.29	531.9810	58.67	2.46	23.84	15.03
C26H13I106	548.29	547.9759	56.96	2.39	23.15	17.51
C26H13I107	564.29	563.9708	55.34	2.32	22.49	19.85
C26H13I108	580.29	579.9657	53.82	2.26	21.87	22.06
C26H13I109	596.29	595.9606	52.37	2.20	21.28	24.15
C26H14I201	596.21	595.9138	52.38	2.37	42.57	2.68
C26H14I202	612.21	611.9087	51.01	2.31	41.46	5.23
C26H14I203	628.21	627.9036	49.71	2.25	40.40	7.64
C26H14I204	644.21	643.8985	48.48	2.19	39.40	9.93
C26H14I205	660.21	659.8934	47.30	2.14	38.44	12.12
C26H14I206	676.21	675.8883	46.18	2.09	37.53	14.20
C26H14I207	692.21	691.8833	45.12	2.04	36.67	16.18
C26H14I208	708.21	707.8782	44.10	1.99	35.84	18.07
C26H14I209	724.20	723.8731	43.12	1.95	35.05	19.88
C26H15I101	470.31	470.0169	66.40	3.21	26.98	3.40
C26H15I102	486.31	486.0119	64.22	3.11	26.10	6.58
C26H15I103	502.31	502.0068	62.17	3.01	25.26	9.56
C26H15I104	518.31	518.0017	60.25	2.92	24.48	12.35
C26H15I105	534.31	533.9966	58.45	2.83	23.75	14.97
C26H15I106	550.31	549.9915	56.75	2.75	23.06	17.44
C26H15I107	566.31	565.9864	55.14	2.67	22.41	19.78
C26H15I108	582.31	581.9813	53.63	2.60	21.79	21.98
C26H15I109	598.31	597.9763	52.20	2.53	21.21	24.07
C26H16I201	598.23	597.9294	52.20	2.70	42.43	2.67
C26H16I202	614.23	613.9243	50.84	2.63	41.32	5.21
C26H16I203	630.22	629.9193	49.55	2.56	40.27	7.62
C26H16I204	646.22	645.9142	48.33	2.50	39.28	9.90
C26H16I205	662.22	661.9091	47.16	2.44	38.33	12.08
C26H16I206	678.22	677.9040	46.05	2.38	37.42	14.15
C26H16I207	694.22	693.8989	44.98	2.32	36.56	16.13
C26H16I208	710.22	709.8938	43.97	2.27	35.74	18.02
C26H16I209	726.22	725.8887	43.00	2.22	34.95	19.83
C26H17I101	472.33	472.0326	66.12	3.63	26.87	3.39
C26H17I102	488.33	488.0275	63.95	3.51	25.99	6.55
C26H17I103	504.33	504.0224	61.92	3.40	25.16	9.52
C26H17I104	520.33	520.0173	60.02	3.29	24.39	12.30
C26H17I105	536.33	536.0122	58.23	3.19	23.66	14.92
C26H17I106	552.33	552.0072	56.54	3.10	22.98	17.38
C26H17I107	568.33	568.0021	54.95	3.02	22.33	19.71
C26H17I108	584.32	583.9970	53.44	2.93	21.72	21.90
C26H17I109	600.32	599.9919	52.02	2.85	21.14	23.99
C26H18I201	600.24	599.9451	52.03	3.02	42.28	2.67
C26H18I202	616.24	615.9400	50.68	2.94	41.19	5.19
C26H18I203	632.24	631.9349	49.39	2.87	40.14	7.59
C26H18I204	648.24	647.9298	48.18	2.80	39.15	9.87
C26H18I205	664.24	663.9247	47.01	2.73	38.21	12.04
C26H18I206	680.24	679.9196	45.91	2.67	37.31	14.11
C26H18I207	696.24	695.9146	44.85	2.61	36.45	16.09
C26H18I208	712.24	711.9095	43.85	2.55	35.64	17.97
C26H18I209	728.24	727.9044	42.88	2.49	34.85	19.77
C26H19I101	474.35	474.0482	65.84	4.04	26.75	3.37
C26H19I102	490.34	490.0432	63.69	3.91	25.88	6.53
C26H19I103	506.34	506.0381	61.68	3.78	25.06	9.48
C26H19I104	522.34	522.0330	59.79	3.67	24.30	12.25
C26H19I105	538.34	538.0279	58.01	3.56	23.57	14.86
C26H19I106	554.34	554.0228	56.34	3.45	22.89	17.32
C26H19I107	570.34	570.0177	54.75	3.36	22.25	19.64
C26H19I108	586.34	586.0126	53.26	3.27	21.64	21.83
C26H19I109	602.34	602.0076	51.85	3.18	21.07	23.91
C26H20I201	602.26	601.9607	51.85	3.35	42.14	2.66
C26H20I202	618.26	617.9556	50.51	3.26	41.05	5.18
C26H20I203	634.26	633.9506	49.24	3.18	40.02	7.57
C26H20I204	650.26	649.9455	48.03	3.10	39.03	9.84
C26H20I205	666.26	665.9404	46.87	3.03	38.09	12.01
C26H20I206	682.25	681.9353	45.77	2.95	37.20	14.07
C26H20I207	698.25	697.9302	44.72	2.89	36.35	16.04
C26H20I208	714.25	713.9251	43.72	2.82	35.53	17.92
C26H20I209	730.25	729.9200	42.76	2.76	34.76	19.72
C26H21I101	476.36	476.0639	65.56	4.44	26.64	3.36
C26H21I102	492.36	492.0588	63.43	4.30	25.77	6.50
C26H21I103	508.36	508.0537	61.43	4.16	24.96	9.44
C26H21I104	524.36	524.0486	59.56	4.04	24.20	12.20
C26H21I105	540.36	540.0435	57.79	3.92	23.49	14.80
C26H21I106	556.36	556.0385	56.13	3.80	22.81	17.25
C26H21I107	572.36	572.0334	54.56	3.70	22.17	19.57
C26H21I108	588.36	588.0283	53.08	3.60	21.57	21.75
C26H21I109	604.36	604.0232	51.67	3.50	21.00	23.83
C26H22I201	604.27	603.9764	51.68	3.67	42.00	2.65
C26H22I202	620.27	619.9713	50.35	3.58	40.92	5.16
C26H22I203	636.27	635.9662	49.08	3.49	39.89	7.54
C26H22I204	652.27	651.9611	47.88	3.40	38.91	9.81
C26H22I205	668.27	667.9560	46.73	3.32	37.98	11.97
C26H22I206	684.27	683.9509	45.64	3.24	37.09	14.03
C26H22I207	700.27	699.9459	44.60	3.17	36.24	15.99
C26H22I208	716.27	715.9408	43.60	3.10	35.43	17.87
C26H22I209	732.27	731.9357	42.65	3.03	34.66	19.66
C26H23I101	478.38	478.0795	65.28	4.85	26.53	3.34
C26H23I102	494.38	494.0745	63.17	4.69	25.67	6.47
C26H23I103	510.38	510.0694	61.19	4.54	24.86	9.40
C26H23I104	526.38	526.0643	59.33	4.40	24.11	12.16
C26H23I105	542.37	542.0592	57.58	4.27	23.40	14.75
C26H23I106	558.37	558.0541	55.93	4.15	22.73	17.19
C26H23I107	574.37	574.0490	54.37	4.04	22.09	19.50
C26H23I108	590.37	590.0439	52.90	3.93	21.50	21.68
C26H23I109	606.37	606.0389	51.50	3.82	20.93	23.75
C26H24I201	606.29	605.9920	51.51	3.99	41.86	2.64
C26H24I202	622.29	621.9869	50.18	3.89	40.79	5.14
C26H24I203	638.29	637.9819	48.93	3.79	39.76	7.52
C26H24I204	654.29	653.9768	47.73	3.70	38.79	9.78
C26H24I205	670.29	669.9717	46.59	3.61	37.87	11.93
C26H24I206	686.29	685.9666	45.50	3.52	36.98	13.99
C26H24I207	702.29	701.9615	44.47	3.44	36.14	15.95
C26H24I208	718.29	717.9564	43.48	3.37	35.34	17.82
C26H24I209	734.28	733.9513	42.53	3.29	34.57	19.61
C26H25I101	480.39	480.0952	65.01	5.25	26.42	3.33
C26H25I102	496.39	496.0901	62.91	5.08	25.57	6.45
C26H25I103	512.39	512.0850	60.95	4.92	24.77	9.37
C26H25I104	528.39	528.0799	59.10	4.77	24.02	12.11
C26H25I105	544.39	544.0748	57.37	4.63	23.31	14.69
C26H25I106	560.39	560.0698	55.73	4.50	22.65	17.13
C26H25I107	576.39	576.0647	54.18	4.37	22.02	19.43
C26H25I108	592.39	592.0596	52.72	4.25	21.42	21.61
C26H25I109	608.39	608.0545	51.33	4.14	20.86	23.67
C26H26I201	608.31	608.0077	51.34	4.31	41.72	2.63
C26H26I202	624.30	624.0026	50.02	4.20	40.65	5.13
C26H26I203	640.30	639.9975	48.77	4.09	39.64	7.50
C26H26I204	656.30	655.9924	47.58	3.99	38.67	9.75
C26H26I205	672.30	671.9873	46.45	3.90	37.75	11.90
C26H26I206	688.30	687.9822	45.37	3.81	36.87	13.95
C26H26I207	704.30	703.9772	44.34	3.72	36.04	15.90

225

C26H26I208	720.30	719.9721	43.36	3.64	35.24	17.77
C26H26I209	736.30	735.9670	42.41	3.56	34.47	19.56
C26H27I101	482.41	482.1108	64.74	5.64	26.31	3.32
C26H27I102	498.41	498.1058	62.66	5.46	25.46	6.42
C26H27I103	514.41	514.1007	60.71	5.29	24.67	9.33
C26H27I104	530.41	530.0956	58.88	5.13	23.93	12.07
C26H27I105	546.41	546.0905	57.15	4.98	23.23	14.64
C26H27I106	562.41	562.0854	55.53	4.84	22.56	17.07
C26H27I107	578.41	578.0803	53.99	4.71	21.94	19.36
C26H27I108	594.40	594.0752	52.54	4.58	21.35	21.53
C26H27I109	610.40	610.0702	51.16	4.46	20.79	23.59
C26H28I201	610.32	610.0233	51.17	4.62	41.59	2.62
C26H28I202	626.32	626.0182	49.86	4.51	40.52	5.11
C26H28I203	642.32	642.0132	48.62	4.39	39.51	7.47
C26H28I204	658.32	658.0081	47.44	4.29	38.55	9.72
C26H28I205	674.32	674.0030	46.31	4.19	37.64	11.86
C26H28I206	690.32	689.9979	45.24	4.09	36.77	13.91
C26H28I207	706.32	705.9928	44.21	4.00	35.93	15.86
C26H28I208	722.32	721.9877	43.23	3.91	35.14	17.72
C26H28I209	738.32	737.9826	42.30	3.82	34.38	19.50
C26H29I101	484.42	484.1265	64.47	6.03	26.20	3.30
C26H29I102	500.42	500.1214	62.41	5.84	25.36	6.39
C26H29I103	516.42	516.1163	60.47	5.66	24.57	9.29
C26H29I104	532.42	532.1112	58.65	5.49	23.84	12.02
C26H29I105	548.42	548.1061	56.94	5.33	23.14	14.59
C26H29I106	564.42	564.1011	55.33	5.18	22.48	17.01
C26H29I107	580.42	580.0960	53.80	5.04	21.86	19.30
C26H29I108	596.42	596.0909	52.36	4.90	21.28	21.46
C26H29I109	612.42	612.0858	50.99	4.77	20.72	23.51
C26H30I201	612.34	612.0390	51.00	4.94	41.45	2.61
C26H30I202	628.34	628.0339	49.70	4.81	40.39	5.09
C26H30I203	644.34	644.0288	48.47	4.69	39.39	7.45
C26H30I204	660.34	660.0237	47.29	4.58	38.44	9.69
C26H30I205	676.33	676.0186	46.17	4.47	37.53	11.83
C26H30I206	692.33	692.0135	45.11	4.37	36.66	13.87
C26H30I207	708.33	708.0085	44.09	4.27	35.83	15.81
C26H30I208	724.33	724.0034	43.11	4.17	35.04	17.67
C26H30I209	740.33	739.9983	42.18	4.08	34.28	19.45
C26H31I101	486.44	486.1421	64.20	6.42	26.09	3.29
C26H31I102	502.44	502.1371	62.15	6.22	25.26	6.37
C26H31I103	518.44	518.1320	60.24	6.03	24.48	9.26
C26H31I104	534.44	534.1269	58.43	5.85	23.75	11.97
C26H31I105	550.44	550.1218	56.73	5.68	23.06	14.53
C26H31I106	566.44	566.1167	55.13	5.52	22.40	16.95
C26H31I107	582.44	582.1116	53.62	5.36	21.79	19.23
C26H31I108	598.44	598.1065	52.18	5.22	21.21	21.39
C26H31I109	614.44	614.1014	50.83	5.09	20.65	23.44
C26H32I201	614.35	614.0546	50.83	5.25	41.31	2.60
C26H32I202	630.35	630.0495	49.54	5.12	40.26	5.08
C26H32I203	646.35	646.0444	48.32	4.99	39.27	7.43
C26H32I204	662.35	662.0394	47.15	4.87	38.32	9.66
C26H32I205	678.35	678.0343	46.04	4.75	37.42	11.79
C26H32I206	694.35	694.0292	44.98	4.65	36.55	13.83
C26H32I207	710.35	710.0241	43.96	4.54	35.73	15.77
C26H32I208	726.35	726.0190	42.99	4.44	34.94	17.62
C26H32I209	742.35	742.0139	42.07	4.35	34.19	19.40
C26H33I101	488.46	488.1578	63.93	6.81	25.98	3.28
C26H33I102	504.46	504.1527	61.91	6.59	25.16	6.34
C26H33I103	520.46	520.1476	60.00	6.39	24.38	9.22
C26H33I104	536.45	536.1425	58.21	6.20	23.66	11.93
C26H33I105	552.45	552.1374	56.53	6.02	22.97	14.48
C26H33I106	568.45	568.1324	54.94	5.85	22.32	16.89
C26H33I107	584.45	584.1273	53.43	5.69	21.71	19.16
C26H33I108	600.45	600.1222	52.01	5.54	21.13	21.32
C26H33I109	616.45	616.1171	50.66	5.40	20.59	23.36
C26H34I201	616.37	616.0703	50.67	5.56	41.18	2.60
C26H34I202	632.37	632.0652	49.38	5.42	40.14	5.06
C26H34I203	648.37	648.0601	48.17	5.29	39.15	7.40
C26H34I204	664.37	664.0550	47.01	5.16	38.20	9.63
C26H34I205	680.37	680.0499	45.90	5.04	37.30	11.76
C26H34I206	696.37	696.0448	44.85	4.92	36.45	13.79
C26H34I207	712.37	712.0398	43.84	4.81	35.63	15.72
C26H34I208	728.36	728.0347	42.88	4.71	34.85	17.57
C26H34I209	744.36	744.0296	41.95	4.60	34.10	19.34
C26H35I101	490.47	490.1734	63.67	7.19	25.87	3.26
C26H35I102	506.47	506.1683	61.66	6.97	25.06	6.32
C26H35I103	522.47	522.1633	59.77	6.75	24.29	9.19
C26H35I104	538.47	538.1582	58.00	6.55	23.57	11.89
C26H35I105	554.47	554.1531	56.32	6.36	22.89	14.43
C26H35I106	570.47	570.1480	54.74	6.18	22.25	16.83
C26H35I107	586.47	586.1429	53.25	6.02	21.64	19.10
C26H35I108	602.47	602.1378	51.84	5.86	21.06	21.25
C26H35I109	618.47	618.1327	50.49	5.70	20.52	23.28
C26H36I201	618.39	618.0859	50.50	5.87	41.04	2.59
C26H36I202	634.38	634.0808	49.23	5.72	40.01	5.04
C26H36I203	650.38	650.0757	48.02	5.58	39.02	7.38
C26H36I204	666.38	666.0707	46.86	5.45	38.09	9.60
C26H36I205	682.38	682.0656	45.76	5.32	37.19	11.72
C26H36I206	698.38	698.0605	44.72	5.20	36.34	13.75
C26H36I207	714.38	714.0554	43.71	5.08	35.53	15.68
C26H36I208	730.38	730.0503	42.76	4.97	34.75	17.52
C26H36I209	746.38	746.0452	41.84	4.86	34.01	19.29
C26H37I101	492.49	492.1891	63.41	7.57	25.77	3.25
C26H37I102	508.49	508.1840	61.42	7.33	24.96	6.29
C26H37I103	524.49	524.1789	59.54	7.11	24.20	9.15
C26H37I104	540.49	540.1738	57.78	6.90	23.48	11.84
C26H37I105	556.49	556.1687	56.12	6.70	22.80	14.38
C26H37I106	572.49	572.1637	54.55	6.51	22.17	16.77
C26H37I107	588.48	588.1586	53.07	6.34	21.56	19.03
C26H37I108	604.48	604.1535	51.66	6.17	20.99	21.17
C26H37I109	620.48	620.1484	50.33	6.01	20.45	23.21
C26H38I201	620.40	620.1016	50.34	6.17	40.91	2.58
C26H38I202	636.40	636.0965	49.07	6.02	39.88	5.03
C26H38I203	652.40	652.0914	47.87	5.87	38.90	7.36
C26H38I204	668.40	668.0863	46.72	5.73	37.97	9.57
C26H38I205	684.40	684.0812	45.63	5.60	37.08	11.69
C26H38I206	700.40	700.0761	44.59	5.47	36.24	13.71
C26H38I207	716.40	716.0711	43.59	5.35	35.43	15.63
C26H38I208	732.40	732.0660	42.64	5.23	34.65	17.48
C26H38I209	748.40	748.0609	41.73	5.12	33.91	19.24
C26H39I101	494.50	494.2047	63.15	7.95	25.66	3.24
C26H39I102	510.50	510.1996	61.17	7.70	24.86	6.27
C26H39I103	526.50	526.1946	59.31	7.47	24.10	9.12
C26H39I104	542.50	542.1895	57.56	7.25	23.39	11.80
C26H39I105	558.50	558.1844	55.92	7.04	22.72	14.32
C26H39I106	574.50	574.1793	54.36	6.84	22.09	16.71
C26H39I107	590.50	590.1742	52.89	6.66	21.49	18.97
C26H39I108	606.50	606.1691	51.49	6.48	20.92	21.10
C26H39I109	622.50	622.1640	50.17	6.31	20.39	23.13
C26H40I201	622.42	622.1172	50.17	6.48	40.78	2.57

226

C26H40I202	638.42	638.1121	48.92	6.32	39.76	5.01
C26H40I203	654.42	654.1070	47.72	6.16	38.78	7.33
C26H40I204	670.42	670.1020	46.58	6.01	37.86	9.55
C26H40I205	686.41	686.0969	45.50	5.87	36.98	11.65
C26H40I206	702.41	702.0918	44.46	5.74	36.13	13.67
C26H40I207	718.41	718.0867	43.47	5.61	35.33	15.59
C26H40I208	734.41	734.0816	42.52	5.49	34.56	17.43
C26H40I209	750.41	750.0765	41.62	5.37	33.82	19.19
C26H41I101	496.52	496.2204	62.90	8.32	25.56	3.22
C26H41I102	512.52	512.2153	60.93	8.06	24.76	6.24
C26H41I103	528.52	528.2102	59.09	7.82	24.01	9.08
C26H41I104	544.52	544.2051	57.35	7.59	23.31	11.75
C26H41I105	560.52	560.2000	55.71	7.37	22.64	14.27
C26H41I106	576.52	576.1950	54.17	7.17	22.01	16.65
C26H41I107	592.52	592.1899	52.71	6.97	21.42	18.90
C26H41I108	608.52	608.1848	51.32	6.79	20.85	21.03
C26H41I109	624.52	624.1797	50.01	6.62	20.32	23.06
C26H42I201	624.43	624.1329	50.01	6.78	40.65	2.56
C26H42I202	640.43	640.1278	48.76	6.61	39.63	5.00
C26H42I203	656.43	656.1227	47.57	6.45	38.66	7.31
C26H42I204	672.43	672.1176	46.44	6.30	37.74	9.52
C26H42I205	688.43	688.1125	45.36	6.15	36.87	11.62
C26H42I206	704.43	704.1074	44.33	6.01	36.03	13.63
C26H42I207	720.43	720.1024	43.35	5.88	35.23	15.55
C26H42I208	736.43	736.0973	42.41	5.75	34.46	17.38
C26H42I209	752.43	752.0922	41.50	5.63	33.73	19.14
C26H43I101	498.54	498.2360	62.64	8.69	25.46	3.21
C26H43I102	514.54	514.2309	60.69	8.42	24.66	6.22
C26H43I103	530.54	530.2259	58.86	8.17	23.92	9.05
C26H43I104	546.53	546.2208	57.14	7.93	23.22	11.71
C26H43I105	562.53	562.2157	55.51	7.70	22.56	14.22
C26H43I106	578.53	578.2106	53.98	7.49	21.94	16.59
C26H43I107	594.53	594.2055	52.53	7.29	21.35	18.84
C26H43I108	610.53	610.2004	51.15	7.10	20.79	20.96
C26H43I109	626.53	626.1953	49.84	6.92	20.26	22.98
C26H44I201	626.45	626.1485	49.85	7.08	40.52	2.55
C26H44I202	642.45	642.1434	48.61	6.90	39.51	4.98
C26H44I203	658.45	658.1383	47.43	6.74	38.55	7.29
C26H44I204	674.45	674.1333	46.30	6.58	37.63	9.49
C26H44I205	690.45	690.1282	45.23	6.42	36.76	11.59
C26H44I206	706.45	706.1231	44.21	6.28	35.93	13.59
C26H44I207	722.45	722.1180	43.23	6.14	35.13	15.50
C26H44I208	738.44	738.1129	42.29	6.01	34.37	17.33
C26H44I209	754.44	754.1078	41.39	5.88	33.64	19.09
C26H45I101	500.55	500.2517	62.39	9.06	25.35	3.20
C26H45I102	516.55	516.2466	60.46	8.78	24.57	6.19
C26H45I103	532.55	532.2415	58.64	8.52	23.83	9.01
C26H45I104	548.55	548.2364	56.93	8.27	23.13	11.67
C26H45I105	564.55	564.2313	55.32	8.03	22.48	14.17
C26H45I106	580.55	580.2263	53.79	7.81	21.86	16.54
C26H45I107	596.55	596.2212	52.35	7.60	21.27	18.77
C26H45I108	612.55	612.2161	50.98	7.40	20.72	20.90
C26H45I109	628.55	628.2110	49.68	7.22	20.19	22.91
C26H46I201	628.46	628.1642	49.69	7.38	40.39	2.55
C26H46I202	644.46	644.1591	48.46	7.19	39.38	4.97
C26H46I203	660.46	660.1540	47.28	7.02	38.43	7.27
C26H46I204	676.46	676.1489	46.17	6.85	37.52	9.46
C26H46I205	692.46	692.1438	45.10	6.70	36.65	11.55
C26H46I206	708.46	708.1387	44.08	6.54	35.83	13.55
C26H46I207	724.46	724.1337	43.11	6.40	35.03	15.46
C26H46I208	740.46	740.1286	42.18	6.26	34.28	17.29
C26H46I209	756.46	756.1235	41.28	6.13	33.55	19.04
C26H47I101	502.57	502.2673	62.14	9.43	25.25	3.18
C26H47I102	518.57	518.2622	60.22	9.14	24.47	6.17
C26H47I103	534.57	534.2572	58.42	8.86	23.74	8.98
C26H47I104	550.57	550.2521	56.72	8.60	23.05	11.62
C26H47I105	566.57	566.2470	55.12	8.36	22.40	14.12
C26H47I106	582.57	582.2419	53.61	8.13	21.78	16.48
C26H47I107	598.56	598.2368	52.17	7.91	21.20	18.71
C26H47I108	614.56	614.2317	50.81	7.71	20.65	20.83
C26H47I109	630.56	630.2266	49.53	7.51	20.13	22.84
C26H48I201	630.48	630.1798	49.53	7.67	40.26	2.54
C26H48I202	646.48	646.1747	48.31	7.48	39.26	4.95
C26H48I203	662.48	662.1696	47.14	7.30	38.31	7.25
C26H48I204	678.48	678.1646	46.03	7.13	37.41	9.43
C26H48I205	694.48	694.1595	44.97	6.97	36.55	11.52
C26H48I206	710.48	710.1544	43.95	6.81	35.72	13.51
C26H48I207	726.48	726.1493	42.99	6.66	34.94	15.42
C26H48I208	742.48	742.1442	42.06	6.52	34.18	17.24
C26H48I209	758.48	758.1391	41.17	6.38	33.46	18.98
C26H49I101	504.58	504.2830	61.89	9.79	25.15	3.17
C26H49I102	520.58	520.2779	59.99	9.49	24.38	6.15
C26H49I103	536.58	536.2728	58.20	9.20	23.65	8.95
C26H49I104	552.58	552.2677	56.51	8.94	22.97	11.58
C26H49I105	568.58	568.2626	54.92	8.69	22.32	14.07
C26H49I106	584.58	584.2576	53.42	8.45	21.71	16.42
C26H49I107	600.58	600.2525	52.00	8.22	21.13	18.65
C26H49I108	616.58	616.2474	50.65	8.01	20.58	20.76
C26H49I109	632.58	632.2423	49.37	7.81	20.06	22.76
C26H50I201	632.50	632.1955	49.37	7.97	40.13	2.53
C26H50I202	648.50	648.1904	48.16	7.77	39.14	4.93
C26H50I203	664.50	664.1853	47.00	7.58	38.20	7.22
C26H50I204	680.49	680.1802	45.89	7.41	37.30	9.40
C26H50I205	696.49	696.1751	44.84	7.24	36.44	11.49
C26H50I206	712.49	712.1700	43.83	7.07	35.62	13.47
C26H50I207	728.49	728.1649	42.87	6.92	34.84	15.37
C26H50I208	744.49	744.1599	41.95	6.77	34.09	17.19
C26H50I209	760.49	760.1548	41.06	6.63	33.37	18.93
C26H51I101	506.60	506.2986	61.64	10.15	25.05	3.16
C26H51I102	522.60	522.2935	59.76	9.84	24.28	6.12
C26H51I103	538.60	538.2885	57.98	9.54	23.56	8.91
C26H51I104	554.60	554.2834	56.31	9.27	22.88	11.54
C26H51I105	570.60	570.2783	54.73	9.01	22.24	14.02
C26H51I106	586.60	586.2732	53.24	8.76	21.63	16.36
C26H51I107	602.60	602.2681	51.82	8.53	21.06	18.59
C26H51I108	618.60	618.2630	50.48	8.31	20.51	20.69
C26H51I109	634.60	634.2579	49.21	8.10	20.00	22.69
C26H52I201	634.51	634.2111	49.22	8.26	40.00	2.52
C26H52I202	650.51	650.2060	48.01	8.06	39.02	4.92
C26H52I203	666.51	666.2009	46.85	7.86	38.08	7.20
C26H52I204	682.51	682.1959	45.76	7.68	37.19	9.38
C26H52I205	698.51	698.1908	44.71	7.50	36.34	11.45
C26H52I206	714.51	714.1857	43.71	7.34	35.52	13.44
C26H52I207	730.51	730.1806	42.75	7.18	34.74	15.33
C26H52I208	746.51	746.1755	41.83	7.02	34.00	17.15
C26H52I209	762.51	762.1704	40.96	6.87	33.29	18.88
C26H53I101	508.62	508.3143	61.40	10.50	24.95	3.15
C26H53I102	524.62	524.3092	59.53	10.18	24.19	6.10
C26H53I103	540.61	540.3041	57.77	9.88	23.47	8.88
C26H53I104	556.61	556.2990	56.11	9.60	22.80	11.50

C26H53I105	572.61	572.2939	54.54	9.33	22.16	13.97
C26H53I106	588.61	588.2888	53.06	9.08	21.56	16.31
C26H53I107	604.61	604.2838	51.65	8.84	20.99	18.52
C26H53I108	620.61	620.2787	50.32	8.61	20.45	20.62
C26H53I109	636.61	636.2736	49.06	8.39	19.93	22.62
C27H13I101	480.31	480.0013	67.52	2.73	26.42	3.33
C27H13I102	496.31	495.9962	65.34	2.64	25.57	6.45
C27H13I103	512.31	511.9911	63.30	2.56	24.77	9.37
C27H13I104	528.31	527.9860	61.39	2.48	24.02	12.11
C27H13I105	544.31	543.9810	59.58	2.41	23.31	14.70
C27H13I106	560.31	559.9759	57.88	2.34	22.65	17.13
C27H13I107	576.30	575.9708	56.27	2.27	22.02	19.43
C27H13I108	592.30	591.9657	54.75	2.21	21.43	21.61
C27H13I109	608.30	607.9606	53.31	2.15	20.86	23.67
C27H14I201	608.22	607.9138	53.32	2.32	41.73	2.63
C27H14I202	624.22	623.9087	51.95	2.26	40.66	5.13
C27H14I203	640.22	639.9036	50.65	2.20	39.64	7.50
C27H14I204	656.22	655.8985	49.42	2.15	38.68	9.75
C27H14I205	672.22	671.8934	48.24	2.10	37.76	11.90
C27H14I206	688.22	687.8883	47.12	2.05	36.88	13.95
C27H14I207	704.22	703.8833	46.05	2.00	36.04	15.90
C27H14I208	720.22	719.8782	45.03	1.96	35.24	17.77
C27H14I209	736.22	735.8731	44.05	1.92	34.47	19.56
C27H15I101	482.32	482.0169	67.24	3.13	26.31	3.32
C27H15I102	498.32	498.0119	65.08	3.03	25.47	6.42
C27H15I103	514.32	514.0068	63.05	2.94	24.67	9.33
C27H15I104	530.32	530.0017	61.15	2.85	23.93	12.07
C27H15I105	546.32	545.9966	59.36	2.77	23.23	14.64
C27H15I106	562.32	561.9915	57.67	2.69	22.57	17.07
C27H15I107	578.32	577.9864	56.08	2.61	21.94	19.37
C27H15I108	594.32	593.9813	54.57	2.54	21.35	21.54
C27H15I109	610.32	609.9763	53.14	2.48	20.79	23.59
C27H16I201	610.24	609.9294	53.14	2.64	41.59	2.62
C27H16I202	626.24	625.9243	51.79	2.58	40.53	5.11
C27H16I203	642.24	641.9193	50.50	2.51	39.52	7.47
C27H16I204	658.23	657.9142	49.27	2.45	38.56	9.72
C27H16I205	674.23	673.9091	48.10	2.39	37.64	11.86
C27H16I206	690.23	689.9040	46.98	2.34	36.77	13.91
C27H16I207	706.23	705.8989	45.92	2.28	35.94	15.86
C27H16I208	722.23	721.8938	44.90	2.23	35.14	17.72
C27H16I209	738.23	737.8887	43.93	2.18	34.38	19.51
C27H17I101	484.34	484.0326	66.96	3.54	26.20	3.30
C27H17I102	500.34	500.0275	64.82	3.42	25.36	6.40
C27H17I103	516.34	516.0224	62.81	3.32	24.58	9.30
C27H17I104	532.34	532.0173	60.92	3.22	23.84	12.02
C27H17I105	548.34	548.0122	59.14	3.12	23.14	14.59
C27H17I106	564.34	564.0072	57.47	3.04	22.49	17.01
C27H17I107	580.34	580.0021	55.88	2.95	21.87	19.30
C27H17I108	596.34	595.9970	54.38	2.87	21.28	21.46
C27H17I109	612.34	611.9919	52.96	2.80	20.72	23.52
C27H18I201	612.25	611.9451	52.97	2.96	41.45	2.61
C27H18I202	628.25	627.9400	51.62	2.89	40.40	5.09
C27H18I203	644.25	643.9349	50.34	2.82	39.40	7.45
C27H18I204	660.25	659.9298	49.12	2.75	38.44	9.69
C27H18I205	676.25	675.9247	47.96	2.68	37.53	11.83
C27H18I206	692.25	691.9196	46.85	2.62	36.66	13.87
C27H18I207	708.25	707.9146	45.79	2.56	35.84	15.81
C27H18I208	724.25	723.9095	44.78	2.51	35.04	17.67
C27H18I209	740.25	739.9044	43.81	2.45	34.29	19.45
C27H19I101	486.36	486.0482	66.68	3.94	26.09	3.29

C27H19I102	502.36	502.0432	64.56	3.81	25.26	6.37
C27H19I103	518.36	518.0381	62.56	3.69	24.48	9.26
C27H19I104	534.35	534.0330	60.69	3.58	23.75	11.98
C27H19I105	550.35	550.0279	58.93	3.48	23.06	14.54
C27H19I106	566.35	566.0228	57.26	3.38	22.41	16.95
C27H19I107	582.35	582.0177	55.69	3.29	21.79	19.23
C27H19I108	598.35	598.0126	54.20	3.20	21.21	21.39
C27H19I109	614.35	614.0076	52.79	3.12	20.66	23.44
C27H20I201	614.27	613.9607	52.79	3.28	41.32	2.60
C27H20I202	630.27	629.9556	51.45	3.20	40.27	5.08
C27H20I203	646.27	645.9506	50.18	3.12	39.27	7.43
C27H20I204	662.27	661.9455	48.97	3.04	38.32	9.66
C27H20I205	678.27	677.9404	47.81	2.97	37.42	11.79
C27H20I206	694.27	693.9353	46.71	2.90	36.56	13.83
C27H20I207	710.27	709.9302	45.66	2.84	35.73	15.77
C27H20I208	726.26	725.9251	44.65	2.78	34.95	17.62
C27H20I209	742.26	741.9200	43.69	2.72	34.19	19.40
C27H21I101	488.37	488.0639	66.40	4.33	25.99	3.28
C27H21I102	504.37	504.0588	64.30	4.20	25.16	6.34
C27H21I103	520.37	520.0537	62.32	4.07	24.39	9.22
C27H21I104	536.37	536.0486	60.46	3.95	23.66	11.93
C27H21I105	552.37	552.0435	58.71	3.83	22.97	14.48
C27H21I106	568.37	568.0385	57.06	3.72	22.33	16.89
C27H21I107	584.37	584.0334	55.50	3.62	21.72	19.17
C27H21I108	600.37	600.0283	54.02	3.53	21.14	21.32
C27H21I109	616.37	616.0232	52.61	3.43	20.59	23.36
C27H22I201	616.28	615.9764	52.62	3.60	41.18	2.60
C27H22I202	632.28	631.9713	51.29	3.51	40.14	5.06
C27H22I203	648.28	647.9662	50.02	3.42	39.15	7.40
C27H22I204	664.28	663.9611	48.82	3.34	38.21	9.63
C27H22I205	680.28	679.9560	47.67	3.26	37.31	11.76
C27H22I206	696.28	695.9509	46.58	3.18	36.45	13.79
C27H22I207	712.28	711.9459	45.53	3.11	35.63	15.72
C27H22I208	728.28	727.9408	44.53	3.04	34.85	17.57
C27H22I209	744.28	743.9357	43.57	2.98	34.10	19.35
C27H23I101	490.39	490.0795	66.13	4.73	25.88	3.26
C27H23I102	506.39	506.0745	64.04	4.58	25.06	6.32
C27H23I103	522.39	522.0694	62.08	4.44	24.29	9.19
C27H23I104	538.39	538.0643	60.24	4.31	23.57	11.89
C27H23I105	554.39	554.0592	58.50	4.18	22.89	14.43
C27H23I106	570.39	570.0541	56.86	4.06	22.25	16.83
C27H23I107	586.38	586.0490	55.31	3.95	21.64	19.10
C27H23I108	602.38	602.0439	53.84	3.85	21.07	21.25
C27H23I109	618.38	618.0389	52.44	3.75	20.52	23.29
C27H24I201	618.30	617.9920	52.45	3.91	41.05	2.59
C27H24I202	634.30	633.9869	51.13	3.81	40.01	5.04
C27H24I203	650.30	649.9819	49.87	3.72	39.03	7.38
C27H24I204	666.30	665.9768	48.67	3.63	38.09	9.60
C27H24I205	682.30	681.9717	47.53	3.55	37.20	11.72
C27H24I206	698.30	697.9666	46.44	3.46	36.35	13.75
C27H24I207	714.30	713.9615	45.40	3.39	35.53	15.68
C27H24I208	730.30	729.9564	44.41	3.31	34.75	17.53
C27H24I209	746.30	745.9513	43.45	3.24	34.01	19.29
C27H25I101	492.40	492.0952	65.86	5.12	25.77	3.25
C27H25I102	508.40	508.0901	63.79	4.96	24.96	6.29
C27H25I103	524.40	524.0850	61.84	4.81	24.20	9.15
C27H25I104	540.40	540.0799	60.01	4.66	23.48	11.84
C27H25I105	556.40	556.0748	58.29	4.53	22.81	14.38
C27H25I106	572.40	572.0698	56.66	4.40	22.17	16.77
C27H25I107	588.40	588.0647	55.12	4.28	21.57	19.03

228

C27H25I108	604.40	604.0596	53.66	4.17	21.00	21.18
C27H25I109	620.40	620.0545	52.27	4.06	20.46	23.21
C27H26I201	620.32	620.0077	52.28	4.22	40.92	2.58
C27H26I202	636.32	636.0026	50.97	4.12	39.89	5.03
C27H26I203	652.32	651.9975	49.72	4.02	38.91	7.36
C27H26I204	668.31	667.9924	48.53	3.92	37.98	9.58
C27H26I205	684.31	683.9873	47.39	3.83	37.09	11.69
C27H26I206	700.31	699.9822	46.31	3.74	36.24	13.71
C27H26I207	716.31	715.9772	45.27	3.66	35.43	15.64
C27H26I208	732.31	731.9721	44.28	3.58	34.66	17.48
C27H26I209	748.31	747.9670	43.34	3.50	33.92	19.24
C27H27I101	494.42	494.1108	65.59	5.50	25.67	3.24
C27H27I102	510.42	510.1058	63.54	5.33	24.86	6.27
C27H27I103	526.42	526.1007	61.61	5.17	24.11	9.12
C27H27I104	542.42	542.0956	59.79	5.02	23.40	11.80
C27H27I105	558.42	558.0905	58.08	4.87	22.73	14.33
C27H27I106	574.42	574.0854	56.46	4.74	22.09	16.71
C27H27I107	590.42	590.0803	54.93	4.61	21.49	18.97
C27H27I108	606.42	606.0752	53.48	4.49	20.93	21.11
C27H27I109	622.42	622.0702	52.10	4.37	20.39	23.13
C27H28I201	622.33	622.0233	52.11	4.54	40.78	2.57
C27H28I202	638.33	638.0182	50.80	4.42	39.76	5.01
C27H28I203	654.33	654.0132	49.56	4.31	38.79	7.34
C27H28I204	670.33	670.0081	48.38	4.21	37.86	9.55
C27H28I205	686.33	686.0030	47.25	4.11	36.98	11.66
C27H28I206	702.33	701.9979	46.18	4.02	36.14	13.67
C27H28I207	718.33	717.9928	45.15	3.93	35.33	15.59
C27H28I208	734.33	733.9877	44.16	3.84	34.56	17.43
C27H28I209	750.33	749.9826	43.22	3.76	33.83	19.19
C27H29I101	496.44	496.1265	65.33	5.89	25.56	3.22
C27H29I102	512.44	512.1214	63.29	5.70	24.76	6.24
C27H29I103	528.43	528.1163	61.37	5.53	24.02	9.08
C27H29I104	544.43	544.1112	59.57	5.37	23.31	11.75
C27H29I105	560.43	560.1061	57.87	5.22	22.64	14.27
C27H29I106	576.43	576.1011	56.26	5.07	22.02	16.65
C27H29I107	592.43	592.0960	54.74	4.93	21.42	18.90
C27H29I108	608.43	608.0909	53.30	4.80	20.86	21.04
C27H29I109	624.43	624.0858	51.94	4.68	20.32	23.06
C27H30I201	624.35	624.0390	51.94	4.84	40.65	2.56
C27H30I202	640.35	640.0339	50.64	4.72	39.64	5.00
C27H30I203	656.35	656.0288	49.41	4.61	38.67	7.31
C27H30I204	672.35	672.0237	48.23	4.50	37.75	9.52
C27H30I205	688.35	688.0186	47.11	4.39	36.87	11.62
C27H30I206	704.35	704.0135	46.04	4.29	36.03	13.63
C27H30I207	720.34	720.0085	45.02	4.20	35.23	15.55
C27H30I208	736.34	736.0034	44.04	4.11	34.47	17.38
C27H30I209	752.34	751.9983	43.11	4.02	33.74	19.14
C27H31I101	498.45	498.1421	65.06	6.27	25.46	3.21
C27H31I102	514.45	514.1371	63.04	6.07	24.67	6.22
C27H31I103	530.45	530.1320	61.14	5.89	23.92	9.05
C27H31I104	546.45	546.1269	59.35	5.72	23.22	11.71
C27H31I105	562.45	562.1218	57.66	5.56	22.56	14.22
C27H31I106	578.45	578.1167	56.06	5.40	21.94	16.60
C27H31I107	594.45	594.1116	54.55	5.26	21.35	18.84
C27H31I108	610.45	610.1065	53.13	5.12	20.79	20.97
C27H31I109	626.45	626.1014	51.77	4.99	20.26	22.99
C27H32I201	626.36	626.0546	51.78	5.15	40.52	2.55
C27H32I202	642.36	642.0495	50.49	5.02	39.51	4.98
C27H32I203	658.36	658.0444	49.26	4.90	38.55	7.29
C27H32I204	674.36	674.0394	48.09	4.78	37.64	9.49

C27H32I205	690.36	690.0343	46.98	4.67	36.76	11.59
C27H32I206	706.36	706.0292	45.91	4.57	35.93	13.59
C27H32I207	722.36	722.0241	44.89	4.47	35.14	15.50
C27H32I208	738.36	738.0190	43.92	4.37	34.37	17.34
C27H32I209	754.36	754.0139	42.99	4.28	33.65	19.09
C27H33I101	500.47	500.1578	64.80	6.65	25.36	3.20
C27H33I102	516.47	516.1527	62.79	6.44	24.57	6.20
C27H33I103	532.47	532.1476	60.91	6.25	23.83	9.01
C27H33I104	548.47	548.1425	59.13	6.06	23.14	11.67
C27H33I105	564.47	564.1374	57.45	5.89	22.48	14.17
C27H33I106	580.46	580.1324	55.87	5.73	21.86	16.54
C27H33I107	596.46	596.1273	54.37	5.58	21.28	18.78
C27H33I108	612.46	612.1222	52.95	5.43	20.72	20.90
C27H33I109	628.46	628.1171	51.60	5.29	20.19	22.91
C27H34I201	628.38	628.0703	51.61	5.45	40.39	2.55
C27H34I202	644.38	644.0652	50.33	5.32	39.39	4.97
C27H34I203	660.38	660.0601	49.11	5.19	38.43	7.27
C27H34I204	676.38	676.0550	47.95	5.07	37.52	9.46
C27H34I205	692.38	692.0499	46.84	4.95	36.66	11.55
C27H34I206	708.38	708.0448	45.78	4.84	35.83	13.55
C27H34I207	724.38	724.0398	44.77	4.73	35.04	15.46
C27H34I208	740.38	740.0347	43.80	4.63	34.28	17.29
C27H34I209	756.38	756.0296	42.88	4.53	33.56	19.04
C27H35I101	502.48	502.1734	64.54	7.02	25.26	3.18
C27H35I102	518.48	518.1683	62.55	6.80	24.48	6.17
C27H35I103	534.48	534.1633	60.68	6.60	23.74	8.98
C27H35I104	550.48	550.1582	58.91	6.41	23.05	11.63
C27H35I105	566.48	566.1531	57.25	6.23	22.40	14.12
C27H35I106	582.48	582.1480	55.68	6.06	21.79	16.48
C27H35I107	598.48	598.1429	54.19	5.89	21.20	18.71
C27H35I108	614.48	614.1378	52.78	5.74	20.65	20.83
C27H35I109	630.48	630.1327	51.44	5.60	20.13	22.84
C27H36I201	630.40	630.0859	51.44	5.76	40.26	2.54
C27H36I202	646.40	646.0808	50.17	5.61	39.27	4.95
C27H36I203	662.39	662.0757	48.96	5.48	38.32	7.25
C27H36I204	678.39	678.0707	47.80	5.35	37.41	9.43
C27H36I205	694.39	694.0656	46.70	5.23	36.55	11.52
C27H36I206	710.39	710.0605	45.65	5.11	35.73	13.51
C27H36I207	726.39	726.0554	44.65	5.00	34.94	15.42
C27H36I208	742.39	742.0503	43.68	4.89	34.19	17.24
C27H36I209	758.39	758.0452	42.76	4.78	33.47	18.99
C27H37I101	504.50	504.1891	64.28	7.39	25.15	3.17
C27H37I102	520.50	520.1840	62.31	7.17	24.38	6.15
C27H37I103	536.50	536.1789	60.45	6.95	23.65	8.95
C27H37I104	552.50	552.1738	58.70	6.75	22.97	11.58
C27H37I105	568.50	568.1687	57.05	6.56	22.32	14.07
C27H37I106	584.50	584.1637	55.48	6.38	21.71	16.42
C27H37I107	600.50	600.1586	54.01	6.21	21.13	18.65
C27H37I108	616.50	616.1535	52.60	6.05	20.58	20.76
C27H37I109	632.49	632.1484	51.27	5.90	20.06	22.77
C27H38I201	632.41	632.1016	51.28	6.06	40.13	2.53
C27H38I202	648.41	648.0965	50.01	5.91	39.14	4.93
C27H38I203	664.41	664.0914	48.81	5.76	38.20	7.22
C27H38I204	680.41	680.0863	47.66	5.63	37.30	9.41
C27H38I205	696.41	696.0812	46.57	5.50	36.45	11.49
C27H38I206	712.41	712.0761	45.52	5.38	35.63	13.47
C27H38I207	728.41	728.0711	44.52	5.26	34.84	15.38
C27H38I208	744.41	744.0660	43.56	5.15	34.10	17.19
C27H38I209	760.41	760.0609	42.65	5.04	33.38	18.94
C27H39I101	506.52	506.2047	64.03	7.76	25.05	3.16

u

C27H39I102	522.52	522.1996	62.07	7.52	24.29	6.12
C27H39I103	538.51	538.1946	60.22	7.30	23.57	8.91
C27H39I104	554.51	554.1895	58.48	7.09	22.89	11.54
C27H39I105	570.51	570.1844	56.84	6.89	22.24	14.02
C27H39I106	586.51	586.1793	55.29	6.70	21.64	16.37
C27H39I107	602.51	602.1742	53.82	6.52	21.06	18.59
C27H39I108	618.51	618.1691	52.43	6.36	20.52	20.69
C27H39I109	634.51	634.1640	51.11	6.20	20.00	22.69
C27H40I201	634.43	634.1172	51.12	6.36	40.01	2.52
C27H40I202	650.43	650.1121	49.86	6.20	39.02	4.92
C27H40I203	666.43	666.1070	48.66	6.05	38.09	7.20
C27H40I204	682.43	682.1020	47.52	5.91	37.19	9.38
C27H40I205	698.43	698.0969	46.43	5.77	36.34	11.45
C27H40I206	714.43	714.0918	45.39	5.64	35.53	13.44
C27H40I207	730.42	730.0867	44.40	5.52	34.75	15.33
C27H40I208	746.42	746.0816	43.45	5.40	34.00	17.15
C27H40I209	762.42	762.0765	42.54	5.29	33.29	18.89
C27H41I101	508.53	508.2204	63.77	8.13	24.96	3.15
C27H41I102	524.53	524.2153	61.83	7.88	24.19	6.10
C27H41I103	540.53	540.2102	60.00	7.65	23.48	8.88
C27H41I104	556.53	556.2051	58.27	7.43	22.80	11.50
C27H41I105	572.53	572.2000	56.64	7.22	22.17	13.97
C27H41I106	588.53	588.1950	55.10	7.02	21.56	16.31
C27H41I107	604.53	604.1899	53.65	6.84	20.99	18.53
C27H41I108	620.53	620.1848	52.26	6.66	20.45	20.63
C27H41I109	636.53	636.1797	50.95	6.49	19.94	22.62
C27H42I201	636.44	636.1329	50.96	6.65	39.88	2.51
C27H42I202	652.44	652.1278	49.71	6.49	38.90	4.90
C27H42I203	668.44	668.1227	48.52	6.33	37.97	7.18
C27H42I204	684.44	684.1176	47.38	6.19	37.08	9.35
C27H42I205	700.44	700.1125	46.30	6.04	36.24	11.42
C27H42I206	716.44	716.1074	45.27	5.91	35.43	13.40
C27H42I207	732.44	732.1024	44.28	5.78	34.65	15.29
C27H42I208	748.44	748.0973	43.33	5.66	33.91	17.10
C27H42I209	764.44	764.0922	42.42	5.54	33.20	18.84
C27H43I101	510.55	510.2360	63.52	8.49	24.86	3.13
C27H43I102	526.55	526.2309	61.59	8.23	24.10	6.08
C27H43I103	542.55	542.2259	59.77	7.99	23.39	8.85
C27H43I104	558.55	558.2208	58.06	7.76	22.72	11.46
C27H43I105	574.55	574.2157	56.44	7.54	22.09	13.92
C27H43I106	590.54	590.2106	54.92	7.34	21.49	16.26
C27H43I107	606.54	606.2055	53.47	7.15	20.92	18.46
C27H43I108	622.54	622.2004	52.09	6.96	20.38	20.56
C27H43I109	638.54	638.1953	50.79	6.79	19.87	22.55
C27H44I201	638.46	638.1485	50.79	6.95	39.75	2.51
C27H44I202	654.46	654.1434	49.55	6.78	38.78	4.89
C27H44I203	670.46	670.1383	48.37	6.61	37.86	7.16
C27H44I204	686.46	686.1333	47.24	6.46	36.97	9.32
C27H44I205	702.46	702.1282	46.17	6.31	36.13	11.39
C27H44I206	718.46	718.1231	45.14	6.17	35.33	13.36
C27H44I207	734.46	734.1180	44.16	6.04	34.56	15.25
C27H44I208	750.46	750.1129	43.21	5.91	33.82	17.06
C27H44I209	766.46	766.1078	42.31	5.79	33.11	18.79
C27H45I101	512.56	512.2517	63.27	8.85	24.76	3.12
C27H45I102	528.56	528.2466	61.36	8.58	24.01	6.05
C27H45I103	544.56	544.2415	59.55	8.33	23.30	8.81
C27H45I104	560.56	560.2364	57.85	8.09	22.64	11.42
C27H45I105	576.56	576.2313	56.25	7.87	22.01	13.87
C27H45I106	592.56	592.2263	54.73	7.65	21.42	16.20
C27H45I107	608.56	608.2212	53.29	7.45	20.85	18.40
C27H45I108	624.56	624.2161	51.92	7.26	20.32	20.49
C27H45I109	640.56	640.2110	50.63	7.08	19.81	22.48
C27H46I201	640.48	640.1642	50.63	7.24	39.63	2.50
C27H46I202	656.48	656.1591	49.40	7.06	38.66	4.87
C27H46I203	672.47	672.1540	48.23	6.89	37.74	7.14
C27H46I204	688.47	688.1489	47.10	6.73	36.87	9.30
C27H46I205	704.47	704.1438	46.03	6.58	36.03	11.36
C27H46I206	720.47	720.1387	45.01	6.44	35.23	13.32
C27H46I207	736.47	736.1337	44.03	6.30	34.46	15.21
C27H46I208	752.47	752.1286	43.10	6.16	33.73	17.01
C27H46I209	768.47	768.1235	42.20	6.03	33.03	18.74
C27H47I101	514.58	514.2673	63.02	9.21	24.66	3.11
C27H47I102	530.58	530.2622	61.12	8.93	23.92	6.03
C27H47I103	546.58	546.2572	59.33	8.67	23.22	8.78
C27H47I104	562.58	562.2521	57.65	8.42	22.56	11.38
C27H47I105	578.58	578.2470	56.05	8.19	21.93	13.83
C27H47I106	594.58	594.2419	54.54	7.97	21.34	16.15
C27H47I107	610.58	610.2368	53.11	7.76	20.78	18.34
C27H47I108	626.58	626.2317	51.76	7.56	20.25	20.43
C27H47I109	642.57	642.2266	50.47	7.37	19.75	22.41
C27H48I201	642.49	642.1798	50.48	7.53	39.50	2.49
C27H48I202	658.49	658.1747	49.25	7.35	38.54	4.86
C27H48I203	674.49	674.1696	48.08	7.17	37.63	7.12
C27H48I204	690.49	690.1646	46.97	7.01	36.76	9.27
C27H48I205	706.49	706.1595	45.90	6.85	35.93	11.32
C27H48I206	722.49	722.1544	44.89	6.70	35.13	13.29
C27H48I207	738.49	738.1493	43.91	6.55	34.37	15.17
C27H48I208	754.49	754.1442	42.98	6.41	33.64	16.96
C27H48I209	770.49	770.1391	42.09	6.28	32.94	18.69
C27H49I101	516.60	516.2830	62.78	9.56	24.57	3.10
C27H49I102	532.59	532.2779	60.89	9.27	23.83	6.01
C27H49I103	548.59	548.2728	59.11	9.00	23.13	8.75
C27H49I104	564.59	564.2677	57.44	8.75	22.48	11.34
C27H49I105	580.59	580.2626	55.86	8.51	21.86	13.78
C27H49I106	596.59	596.2576	54.36	8.28	21.27	16.09
C27H49I107	612.59	612.2525	52.94	8.06	20.72	18.28
C27H49I108	628.59	628.2474	51.59	7.86	20.19	20.36
C27H49I109	644.59	644.2423	50.31	7.66	19.69	22.34
C27H50I201	644.51	644.1955	50.32	7.82	39.38	2.48
C27H50I202	660.51	660.1904	49.10	7.63	38.43	4.84
C27H50I203	676.51	676.1853	47.94	7.45	37.52	7.10
C27H50I204	692.51	692.1802	46.83	7.28	36.65	9.24
C27H50I205	708.51	708.1751	45.77	7.11	35.82	11.29
C27H50I206	724.50	724.1700	44.76	6.96	35.03	13.25
C27H50I207	740.50	740.1649	43.79	6.81	34.28	15.12
C27H50I208	756.50	756.1599	42.87	6.66	33.55	16.92
C27H50I209	772.50	772.1548	41.98	6.52	32.86	18.64
C27H51I101	518.61	518.2986	62.53	9.91	24.47	3.09
C27H51I102	534.61	534.2935	60.66	9.62	23.74	5.99
C27H51I103	550.61	550.2885	58.90	9.34	23.05	8.72
C27H51I104	566.61	566.2834	57.24	9.07	22.40	11.29
C27H51I105	582.61	582.2783	55.66	8.82	21.78	13.73
C27H51I106	598.61	598.2732	54.18	8.59	21.20	16.04
C27H51I107	614.61	614.2681	52.77	8.36	20.65	18.22
C27H51I108	630.61	630.2630	51.43	8.15	20.12	20.30
C27H51I109	646.61	646.2579	50.15	7.95	19.63	22.27
C27H52I201	646.52	646.2111	50.16	8.11	39.26	2.47
C27H52I202	662.52	662.2060	48.95	7.91	38.31	4.83
C27H52I203	678.52	678.2009	47.80	7.72	37.41	7.07
C27H52I204	694.52	694.1959	46.69	7.55	36.54	9.21

C27H52I205	710.52	710.1908	45.64	7.38	35.72	11.26
C27H52I206	726.52	726.1857	44.64	7.21	34.93	13.21
C27H52I207	742.52	742.1806	43.68	7.06	34.18	15.08
C27H52I208	758.52	758.1755	42.75	6.91	33.46	16.87
C27H52I209	774.52	774.1704	41.87	6.77	32.77	18.59
C27H53I101	520.63	520.3143	62.29	10.26	24.38	3.07
C27H53I102	536.63	536.3092	60.43	9.96	23.65	5.96
C27H53I103	552.63	552.3041	58.68	9.67	22.96	8.69
C27H53I104	568.63	568.2990	57.03	9.40	22.32	11.25
C27H53I105	584.62	584.2939	55.47	9.14	21.71	13.68
C27H53I106	600.62	600.2888	53.99	8.89	21.13	15.98
C27H53I107	616.62	616.2838	52.59	8.66	20.58	18.16
C27H53I108	632.62	632.2787	51.26	8.44	20.06	20.23
C27H53I109	648.62	648.2736	50.00	8.24	19.57	22.20
C27H54I201	648.54	648.2268	50.00	8.39	39.14	2.47
C27H54I202	664.54	664.2217	48.80	8.19	38.19	4.82
C27H54I203	680.54	680.2166	47.65	8.00	37.30	7.05
C27H54I204	696.54	696.2115	46.56	7.81	36.44	9.19
C27H54I205	712.54	712.2064	45.51	7.64	35.62	11.23
C27H54I206	728.54	728.2013	44.51	7.47	34.84	13.18
C27H54I207	744.54	744.1962	43.56	7.31	34.09	15.04
C27H54I208	760.54	760.1912	42.64	7.16	33.37	16.83
C27H54I209	776.53	776.1861	41.76	7.01	32.68	18.54
C27H55I101	522.64	522.3299	62.05	10.61	24.28	3.06
C27H55I102	538.64	538.3248	60.21	10.29	23.56	5.94
C27H55I103	554.64	554.3198	58.47	10.00	22.88	8.65
C27H55I104	570.64	570.3147	56.83	9.72	22.24	11.22
C27H55I105	586.64	586.3096	55.28	9.45	21.63	13.64
C27H55I106	602.64	602.3045	53.81	9.20	21.06	15.93
C27H55I107	618.64	618.2994	52.42	8.96	20.51	18.10
C27H55I108	634.64	634.2943	51.10	8.74	20.00	20.17
C27H55I109	650.64	650.2892	49.84	8.52	19.50	22.13
C28H14I201	620.23	619.9138	54.22	2.28	40.92	2.58
C28H14I202	636.23	635.9087	52.86	2.22	39.89	5.03
C28H14I203	652.23	651.9036	51.56	2.16	38.91	7.36
C28H14I204	668.23	667.8985	50.33	2.11	37.98	9.58
C28H14I205	684.23	683.8934	49.15	2.06	37.09	11.69
C28H14I206	700.23	699.8883	48.03	2.02	36.25	13.71
C28H14I207	716.23	715.8833	46.96	1.97	35.44	15.64
C28H14I208	732.23	731.8782	45.93	1.93	34.66	17.48
C28H14I209	748.23	747.8731	44.95	1.89	33.92	19.24
C28H15I101	494.34	494.0169	68.03	3.06	25.67	3.24
C28H15I102	510.33	510.0119	65.90	2.96	24.87	6.27
C28H15I103	526.33	526.0068	63.90	2.87	24.11	9.12
C28H15I104	542.33	542.0017	62.01	2.79	23.40	11.80
C28H15I105	558.33	557.9966	60.24	2.71	22.73	14.33
C28H15I106	574.33	573.9915	58.56	2.63	22.10	16.71
C28H15I107	590.33	589.9864	56.97	2.56	21.50	18.97
C28H15I108	606.33	605.9813	55.47	2.49	20.93	21.11
C28H15I109	622.33	621.9763	54.04	2.43	20.39	23.14
C28H16I201	622.25	621.9294	54.05	2.59	40.79	2.57
C28H16I202	638.25	637.9243	52.69	2.53	39.77	5.01
C28H16I203	654.25	653.9193	51.40	2.47	38.79	7.34
C28H16I204	670.25	669.9142	50.18	2.41	37.87	9.55
C28H16I205	686.25	685.9091	49.01	2.35	36.99	11.66
C28H16I206	702.24	701.9040	47.89	2.30	36.14	13.67
C28H16I207	718.24	717.8989	46.82	2.25	35.34	15.59
C28H16I208	734.24	733.8938	45.80	2.20	34.57	17.43
C28H16I209	750.24	749.8887	44.83	2.15	33.83	19.19
C28H17I101	496.35	496.0326	67.76	3.45	25.57	3.22
C28H17I102	512.35	512.0275	65.64	3.34	24.77	6.25
C28H17I103	528.35	528.0224	63.65	3.24	24.02	9.08
C28H17I104	544.35	544.0173	61.78	3.15	23.31	11.76
C28H17I105	560.35	560.0122	60.02	3.06	22.65	14.28
C28H17I106	576.35	576.0072	58.35	2.97	22.02	16.66
C28H17I107	592.35	592.0021	56.78	2.89	21.42	18.91
C28H17I108	608.35	607.9970	55.28	2.82	20.86	21.04
C28H17I109	624.35	623.9919	53.87	2.74	20.33	23.06
C28H18I201	624.26	623.9451	53.87	2.91	40.66	2.56
C28H18I202	640.26	639.9400	52.53	2.83	39.64	5.00
C28H18I203	656.26	655.9349	51.25	2.76	38.67	7.31
C28H18I204	672.26	671.9298	50.03	2.70	37.75	9.52
C28H18I205	688.26	687.9247	48.86	2.64	36.88	11.62
C28H18I206	704.26	703.9196	47.75	2.58	36.04	13.63
C28H18I207	720.26	719.9146	46.69	2.52	35.24	15.55
C28H18I208	736.26	735.9095	45.68	2.46	34.47	17.38
C28H18I209	752.26	751.9044	44.71	2.41	33.74	19.14
C28H19I101	498.37	498.0482	67.48	3.84	25.46	3.21
C28H19I102	514.37	514.0432	65.38	3.72	24.67	6.22
C28H19I103	530.37	530.0381	63.41	3.61	23.93	9.05
C28H19I104	546.37	546.0330	61.55	3.51	23.23	11.71
C28H19I105	562.37	562.0279	59.80	3.41	22.57	14.23
C28H19I106	578.36	578.0228	58.15	3.31	21.94	16.60
C28H19I107	594.36	594.0177	56.58	3.22	21.35	18.84
C28H19I108	610.36	610.0126	55.10	3.14	20.79	20.97
C28H19I109	626.36	626.0076	53.69	3.06	20.26	22.99
C28H20I201	626.28	625.9607	53.70	3.22	40.53	2.55
C28H20I202	642.28	641.9556	52.36	3.14	39.52	4.98
C28H20I203	658.28	657.9506	51.09	3.06	38.56	7.29
C28H20I204	674.28	673.9455	49.88	2.99	37.64	9.49
C28H20I205	690.28	689.9404	48.72	2.92	36.77	11.59
C28H20I206	706.28	705.9353	47.62	2.85	35.94	13.59
C28H20I207	722.28	721.9302	46.56	2.79	35.14	15.51
C28H20I208	738.28	737.9251	45.55	2.73	34.38	17.34
C28H20I209	754.27	753.9200	44.59	2.67	33.65	19.09
C28H21I101	500.38	500.0639	67.21	4.23	25.36	3.20
C28H21I102	516.38	516.0588	65.13	4.10	24.58	6.20
C28H21I103	532.38	532.0537	63.17	3.98	23.84	9.02
C28H21I104	548.38	548.0486	61.33	3.86	23.14	11.67
C28H21I105	564.38	564.0435	59.59	3.75	22.49	14.17
C28H21I106	580.38	580.0385	57.95	3.65	21.87	16.54
C28H21I107	596.38	596.0334	56.39	3.55	21.28	18.78
C28H21I108	612.38	612.0283	54.92	3.46	20.72	20.90
C28H21I109	628.38	628.0232	53.52	3.37	20.20	22.92
C28H22I201	628.30	627.9764	53.53	3.53	40.40	2.55
C28H22I202	644.30	643.9713	52.20	3.44	39.39	4.97
C28H22I203	660.29	659.9662	50.93	3.36	38.44	7.27
C28H22I204	676.29	675.9611	49.73	3.28	37.53	9.46
C28H22I205	692.29	691.9560	48.58	3.20	36.66	11.56
C28H22I206	708.29	707.9509	47.48	3.13	35.83	13.55
C28H22I207	724.29	723.9459	46.43	3.06	35.04	15.46
C28H22I208	740.29	739.9408	45.43	3.00	34.28	17.29
C28H22I209	756.29	755.9357	44.47	2.93	33.56	19.04
C28H23I101	502.40	502.0795	66.94	4.61	25.26	3.18
C28H23I102	518.40	518.0745	64.88	4.47	24.48	6.17
C28H23I103	534.40	534.0694	62.93	4.34	23.75	8.98
C28H23I104	550.40	550.0643	61.10	4.21	23.06	11.63
C28H23I105	566.40	566.0592	59.38	4.09	22.41	14.12
C28H23I106	582.40	582.0541	57.75	3.98	21.79	16.48
C28H23I107	598.40	598.0490	56.20	3.87	21.21	18.72

C28H23I108	614.40	614.0439	54.74	3.77	20.66	20.83
C28H23I109	630.39	630.0389	53.35	3.68	20.13	22.84
C28H24I201	630.31	629.9920	53.36	3.84	40.27	2.54
C28H24I202	646.31	645.9869	52.04	3.74	39.27	4.95
C28H24I203	662.31	661.9819	50.78	3.65	38.32	7.25
C28H24I204	678.31	677.9768	49.58	3.57	37.42	9.43
C28H24I205	694.31	693.9717	48.44	3.48	36.56	11.52
C28H24I206	710.31	709.9666	47.35	3.41	35.73	13.51
C28H24I207	726.31	725.9615	46.30	3.33	34.95	15.42
C28H24I208	742.31	741.9564	45.31	3.26	34.19	17.24
C28H24I209	758.31	757.9513	44.35	3.19	33.47	18.99
C28H25I101	504.42	504.0952	66.67	5.00	25.16	3.17
C28H25I102	520.41	520.0901	64.62	4.84	24.39	6.15
C28H25I103	536.41	536.0850	62.70	4.70	23.66	8.95
C28H25I104	552.41	552.0799	60.88	4.56	22.97	11.59
C28H25I105	568.41	568.0748	59.17	4.43	22.33	14.07
C28H25I106	584.41	584.0698	57.55	4.31	21.71	16.43
C28H25I107	600.41	600.0647	56.01	4.20	21.14	18.65
C28H25I108	616.41	616.0596	54.56	4.09	20.59	20.76
C28H25I109	632.41	632.0545	53.18	3.98	20.07	22.77
C28H26I201	632.33	632.0077	53.19	4.14	40.14	2.53
C28H26I202	648.33	648.0026	51.87	4.04	39.15	4.94
C28H26I203	664.33	663.9975	50.62	3.94	38.21	7.23
C28H26I204	680.33	679.9924	49.43	3.85	37.31	9.41
C28H26I205	696.33	695.9873	48.30	3.76	36.45	11.49
C28H26I206	712.32	711.9822	47.21	3.68	35.63	13.48
C28H26I207	728.32	727.9772	46.18	3.60	34.85	15.38
C28H26I208	744.32	743.9721	45.18	3.52	34.10	17.20
C28H26I209	760.32	759.9670	44.23	3.45	33.38	18.94
C28H27I101	506.43	506.1108	66.41	5.37	25.06	3.16
C28H27I102	522.43	522.1058	64.37	5.21	24.29	6.12
C28H27I103	538.43	538.1007	62.46	5.05	23.57	8.91
C28H27I104	554.43	554.0956	60.66	4.91	22.89	11.54
C28H27I105	570.43	570.0905	58.96	4.77	22.25	14.02
C28H27I106	586.43	586.0854	57.35	4.64	21.64	16.37
C28H27I107	602.43	602.0803	55.83	4.52	21.07	18.59
C28H27I108	618.43	618.0752	54.38	4.40	20.52	20.70
C28H27I109	634.43	634.0702	53.01	4.29	20.00	22.70
C28H28I201	634.34	634.0233	53.02	4.45	40.01	2.52
C28H28I202	650.34	650.0182	51.71	4.34	39.03	4.92
C28H28I203	666.34	666.0132	50.47	4.24	38.09	7.20
C28H28I204	682.34	682.0081	49.29	4.14	37.20	9.38
C28H28I205	698.34	698.0030	48.16	4.04	36.34	11.46
C28H28I206	714.34	713.9979	47.08	3.95	35.53	13.44
C28H28I207	730.34	729.9928	46.05	3.86	34.75	15.33
C28H28I208	746.34	745.9877	45.06	3.78	34.01	17.15
C28H28I209	762.34	761.9826	44.12	3.70	33.29	18.89
C28H29I101	508.45	508.1265	66.14	5.75	24.96	3.15
C28H29I102	524.45	524.1214	64.13	5.57	24.20	6.10
C28H29I103	540.45	540.1163	62.23	5.41	23.48	8.88
C28H29I104	556.45	556.1112	60.44	5.25	22.81	11.50
C28H29I105	572.44	572.1061	58.75	5.11	22.17	13.97
C28H29I106	588.44	588.1011	57.15	4.97	21.57	16.31
C28H29I107	604.44	604.0960	55.64	4.84	21.00	18.53
C28H29I108	620.44	620.0909	54.21	4.71	20.45	20.63
C28H29I109	636.44	636.0858	52.84	4.59	19.94	22.62
C28H30I201	636.36	636.0390	52.85	4.75	39.88	2.51
C28H30I202	652.36	652.0339	51.55	4.64	38.91	4.91
C28H30I203	668.36	668.0288	50.32	4.52	37.97	7.18
C28H30I204	684.36	684.0237	49.14	4.42	37.09	9.35
C28H30I205	700.36	700.0186	48.02	4.32	36.24	11.42
C28H30I206	716.36	716.0135	46.95	4.22	35.43	13.40
C28H30I207	732.36	732.0085	45.92	4.13	34.66	15.29
C28H30I208	748.36	748.0034	44.94	4.04	33.92	17.10
C28H30I209	764.35	763.9983	44.00	3.96	33.21	18.84
C28H31I101	510.46	510.1421	65.88	6.12	24.86	3.13
C28H31I102	526.46	526.1371	63.88	5.94	24.11	6.08
C28H31I103	542.46	542.1320	62.00	5.76	23.39	8.85
C28H31I104	558.46	558.1269	60.22	5.60	22.72	11.46
C28H31I105	574.46	574.1218	58.54	5.44	22.09	13.93
C28H31I106	590.46	590.1167	56.96	5.29	21.49	16.26
C28H31I107	606.46	606.1116	55.46	5.15	20.93	18.47
C28H31I108	622.46	622.1065	54.03	5.02	20.39	20.56
C28H31I109	638.46	638.1014	52.68	4.89	19.88	22.55
C28H32I201	638.38	638.0546	52.68	5.05	39.76	2.51
C28H32I202	654.37	654.0495	51.39	4.93	38.79	4.89
C28H32I203	670.37	670.0444	50.17	4.81	37.86	7.16
C28H32I204	686.37	686.0394	49.00	4.70	36.98	9.32
C28H32I205	702.37	702.0343	47.88	4.59	36.14	11.39
C28H32I206	718.37	718.0292	46.82	4.49	35.33	13.36
C28H32I207	734.37	734.0241	45.80	4.39	34.56	15.25
C28H32I208	750.37	750.0190	44.82	4.30	33.82	17.06
C28H32I209	766.37	766.0139	43.88	4.21	33.12	18.79
C28H33I101	512.48	512.1578	65.62	6.49	24.76	3.12
C28H33I102	528.48	528.1527	63.64	6.29	24.01	6.05
C28H33I103	544.48	544.1476	61.77	6.11	23.31	8.82
C28H33I104	560.48	560.1425	60.00	5.93	22.64	11.42
C28H33I105	576.48	576.1374	58.34	5.77	22.01	13.88
C28H33I106	592.48	592.1324	56.76	5.61	21.42	16.20
C28H33I107	608.48	608.1273	55.27	5.47	20.86	18.41
C28H33I108	624.47	624.1222	53.86	5.33	20.32	20.50
C28H33I109	640.47	640.1171	52.51	5.19	19.81	22.48
C28H34I201	640.39	640.0703	52.52	5.35	39.63	2.50
C28H34I202	656.39	656.0652	51.24	5.22	38.67	4.87
C28H34I203	672.39	672.0601	50.02	5.10	37.75	7.14
C28H34I204	688.39	688.0550	48.85	4.98	36.87	9.30
C28H34I205	704.39	704.0499	47.75	4.87	36.03	11.36
C28H34I206	720.39	720.0448	46.68	4.76	35.23	13.33
C28H34I207	736.39	736.0398	45.67	4.65	34.47	15.21
C28H34I208	752.39	752.0347	44.70	4.55	33.73	17.01
C28H34I209	768.39	768.0296	43.77	4.46	33.03	18.74
C28H35I101	514.49	514.1734	65.37	6.86	24.67	3.11
C28H35I102	530.49	530.1683	63.40	6.65	23.92	6.03
C28H35I103	546.49	546.1633	61.54	6.46	23.22	8.78
C28H35I104	562.49	562.1582	59.79	6.27	22.56	11.38
C28H35I105	578.49	578.1531	58.14	6.10	21.94	13.83
C28H35I106	594.49	594.1480	56.57	5.93	21.35	16.15
C28H35I107	610.49	610.1429	55.09	5.78	20.79	18.35
C28H35I108	626.49	626.1378	53.68	5.63	20.26	20.43
C28H35I109	642.49	642.1327	52.35	5.49	19.75	22.41
C28H36I201	642.41	642.0859	52.35	5.65	39.51	2.49
C28H36I202	658.41	658.0808	51.08	5.51	38.55	4.86
C28H36I203	674.41	674.0757	49.87	5.38	37.63	7.12
C28H36I204	690.41	690.0707	48.71	5.26	36.76	9.27
C28H36I205	706.40	706.0656	47.61	5.14	35.93	11.32
C28H36I206	722.40	722.0605	46.55	5.02	35.13	13.29
C28H36I207	738.40	738.0554	45.55	4.91	34.37	15.17
C28H36I208	754.40	754.0503	44.58	4.81	33.64	16.97
C28H36I209	770.40	770.0452	43.65	4.71	32.94	18.69
C28H37I101	516.51	516.1891	65.11	7.22	24.57	3.10

232

C28H37I102	532.51	532.1840	63.16	7.00	23.83	6.01
C28H37I103	548.51	548.1789	61.31	6.80	23.14	8.75
C28H37I104	564.51	564.1738	59.58	6.61	22.48	11.34
C28H37I105	580.51	580.1687	57.93	6.42	21.86	13.78
C28H37I106	596.51	596.1637	56.38	6.25	21.27	16.09
C28H37I107	612.51	612.1586	54.91	6.09	20.72	18.28
C28H37I108	628.51	628.1535	53.51	5.93	20.19	20.36
C28H37I109	644.51	644.1484	52.18	5.79	19.69	22.34
C28H38I201	644.42	644.1016	52.19	5.94	39.39	2.48
C28H38I202	660.42	660.0965	50.92	5.80	38.43	4.85
C28H38I203	676.42	676.0914	49.72	5.66	37.52	7.10
C28H38I204	692.42	692.0863	48.57	5.53	36.66	9.24
C28H38I205	708.42	708.0812	47.47	5.41	35.83	11.29
C28H38I206	724.42	724.0761	46.43	5.29	35.04	13.25
C28H38I207	740.42	740.0711	45.42	5.17	34.28	15.13
C28H38I208	756.42	756.0660	44.46	5.06	33.55	16.92
C28H38I209	772.42	772.0609	43.54	4.96	32.86	18.64
C28H39I101	518.53	518.2047	64.86	7.58	24.47	3.09
C28H39I102	534.53	534.1996	62.92	7.35	23.74	5.99
C28H39I103	550.53	550.1946	61.09	7.14	23.05	8.72
C28H39I104	566.53	566.1895	59.36	6.94	22.40	11.30
C28H39I105	582.52	582.1844	57.73	6.75	21.79	13.73
C28H39I106	598.52	598.1793	56.19	6.57	21.20	16.04
C28H39I107	614.52	614.1742	54.73	6.40	20.65	18.22
C28H39I108	630.52	630.1691	53.34	6.23	20.13	20.30
C28H39I109	646.52	646.1640	52.02	6.08	19.63	22.27
C28H40I201	646.44	646.1172	52.03	6.24	39.26	2.48
C28H40I202	662.44	662.1121	50.77	6.09	38.31	4.83
C28H40I203	678.44	678.1070	49.57	5.94	37.41	7.07
C28H40I204	694.44	694.1020	48.43	5.81	36.55	9.22
C28H40I205	710.44	710.0969	47.34	5.68	35.73	11.26
C28H40I206	726.44	726.0918	46.30	5.55	34.94	13.21
C28H40I207	742.44	742.0867	45.30	5.43	34.19	15.08
C28H40I208	758.44	758.0816	44.34	5.32	33.46	16.88
C28H40I209	774.43	774.0765	43.43	5.21	32.77	18.59
C28H41I101	520.54	520.2204	64.61	7.94	24.38	3.07
C28H41I102	536.54	536.2153	62.68	7.70	23.65	5.96
C28H41I103	552.54	552.2102	60.87	7.48	22.97	8.69
C28H41I104	568.54	568.2051	59.15	7.27	22.32	11.26
C28H41I105	584.54	584.2000	57.53	7.07	21.71	13.69
C28H41I106	600.54	600.1950	56.00	6.88	21.13	15.99
C28H41I107	616.54	616.1899	54.55	6.70	20.58	18.17
C28H41I108	632.54	632.1848	53.17	6.53	20.06	20.24
C28H41I109	648.54	648.1797	51.86	6.37	19.57	22.20
C28H42I201	648.46	648.1329	51.86	6.53	39.14	2.47
C28H42I202	664.45	664.1278	50.61	6.37	38.20	4.82
C28H42I203	680.45	680.1227	49.42	6.22	37.30	7.05
C28H42I204	696.45	696.1176	48.29	6.08	36.44	9.19
C28H42I205	712.45	712.1125	47.20	5.94	35.62	11.23
C28H42I206	728.45	728.1074	46.17	5.81	34.84	13.18
C28H42I207	744.45	744.1024	45.18	5.69	34.09	15.04
C28H42I208	760.45	760.0973	44.23	5.57	33.38	16.83
C28H42I209	776.45	776.0922	43.31	5.45	32.69	18.55
C28H43I101	522.56	522.2360	64.36	8.29	24.29	3.06
C28H43I102	538.56	538.2309	62.45	8.05	23.56	5.94
C28H43I103	554.56	554.2259	60.65	7.82	22.88	8.66
C28H43I104	570.56	570.2208	58.94	7.60	22.24	11.22
C28H43I105	586.56	586.2157	57.34	7.39	21.64	13.64
C28H43I106	602.56	602.2106	55.81	7.19	21.06	15.93
C28H43I107	618.56	618.2055	54.37	7.01	20.52	18.11
C28H43I108	634.55	634.2004	53.00	6.83	20.00	20.17
C28H43I109	650.55	650.1953	51.70	6.66	19.51	22.13
C28H44I201	650.47	650.1485	51.70	6.82	39.02	2.46
C28H44I202	666.47	666.1434	50.46	6.65	38.08	4.80
C28H44I203	682.47	682.1383	49.28	6.50	37.19	7.03
C28H44I204	698.47	698.1333	48.15	6.35	36.34	9.16
C28H44I205	714.47	714.1282	47.07	6.21	35.52	11.20
C28H44I206	730.47	730.1231	46.04	6.07	34.75	13.14
C28H44I207	746.47	746.1180	45.05	5.94	34.00	15.00
C28H44I208	762.47	762.1129	44.11	5.82	33.29	16.79
C28H44I209	778.47	778.1078	43.20	5.70	32.60	18.50
C28H45I101	524.57	524.2517	64.11	8.65	24.19	3.05
C28H45I102	540.57	540.2466	62.21	8.39	23.48	5.92
C28H45I103	556.57	556.2415	60.43	8.15	22.80	8.62
C28H45I104	572.57	572.2364	58.74	7.92	22.16	11.18
C28H45I105	588.57	588.2313	57.14	7.71	21.56	13.59
C28H45I106	604.57	604.2263	55.63	7.50	20.99	15.88
C28H45I107	620.57	620.2212	54.19	7.31	20.45	18.05
C28H45I108	636.57	636.2161	52.83	7.13	19.94	20.11
C28H45I109	652.57	652.2110	51.54	6.95	19.45	22.07
C28H46I201	652.49	652.1642	51.54	7.11	38.90	2.45
C28H46I202	668.49	668.1591	50.31	6.94	37.97	4.79
C28H46I203	684.49	684.1540	49.13	6.77	37.08	7.01
C28H46I204	700.49	700.1489	48.01	6.62	36.23	9.14
C28H46I205	716.48	716.1438	46.94	6.47	35.42	11.17
C28H46I206	732.48	732.1387	45.91	6.33	34.65	13.11
C28H46I207	748.48	748.1337	44.93	6.19	33.91	14.96
C28H46I208	764.48	764.1286	43.99	6.07	33.20	16.74
C28H46I209	780.48	780.1235	43.09	5.94	32.52	18.45
C28H47I101	526.59	526.2673	63.87	9.00	24.10	3.04
C28H47I102	542.59	542.2622	61.98	8.73	23.39	5.90
C28H47I103	558.59	558.2572	60.21	8.48	22.72	8.59
C28H47I104	574.59	574.2521	58.53	8.24	22.09	11.14
C28H47I105	590.59	590.2470	56.95	8.02	21.49	13.55
C28H47I106	606.59	606.2419	55.44	7.81	20.92	15.83
C28H47I107	622.59	622.2368	54.02	7.61	20.38	17.99
C28H47I108	638.59	638.2317	52.67	7.42	19.87	20.04
C28H47I109	654.59	654.2266	51.38	7.24	19.39	22.00
C28H48I201	654.50	654.1798	51.38	7.39	38.78	2.44
C28H48I202	670.50	670.1747	50.16	7.22	37.85	4.77
C28H48I203	686.50	686.1696	48.99	7.05	36.97	6.99
C28H48I204	702.50	702.1646	47.87	6.89	36.13	9.11
C28H48I205	718.50	718.1595	46.81	6.73	35.32	11.13
C28H48I206	734.50	734.1544	45.79	6.59	34.56	13.07
C28H48I207	750.50	750.1493	44.81	6.45	33.82	14.92
C28H48I208	766.50	766.1442	43.88	6.31	33.11	16.70
C28H48I209	782.50	782.1391	42.98	6.18	32.44	18.40
C28H49I101	528.61	528.2830	63.62	9.34	24.01	3.03
C28H49I102	544.61	544.2779	61.75	9.07	23.30	5.88
C28H49I103	560.61	560.2728	59.99	8.81	22.64	8.56
C28H49I104	576.60	576.2677	58.33	8.57	22.01	11.10
C28H49I105	592.60	592.2626	56.75	8.33	21.41	13.50
C28H49I106	608.60	608.2576	55.26	8.12	20.85	15.77
C28H49I107	624.60	624.2525	53.84	7.91	20.32	17.93
C28H49I108	640.60	640.2474	52.50	7.71	19.81	19.98
C28H49I109	656.60	656.2423	51.22	7.52	19.33	21.93
C28H50I201	656.52	656.1955	51.23	7.68	38.66	2.44
C28H50I202	672.52	672.1904	50.01	7.49	37.74	4.76
C28H50I203	688.52	688.1853	48.85	7.32	36.86	6.97
C28H50I204	704.52	704.1802	47.74	7.15	36.03	9.08

233

C28H50I205	720.52	720.1751	46.68	6.99	35.23	11.10
C28H50I206	736.52	736.1700	45.66	6.84	34.46	13.03
C28H50I207	752.52	752.1649	44.69	6.70	33.73	14.88
C28H50I208	768.51	768.1599	43.76	6.56	33.03	16.65
C28H50I209	784.51	784.1548	42.87	6.42	32.35	18.35
C28H51I101	530.62	530.2986	63.38	9.69	23.92	3.02
C28H51I102	546.62	546.2935	61.53	9.40	23.22	5.85
C28H51I103	562.62	562.2885	59.78	9.14	22.56	8.53
C28H51I104	578.62	578.2834	58.12	8.88	21.93	11.06
C28H51I105	594.62	594.2783	56.56	8.65	21.34	13.45
C28H51I106	610.62	610.2732	55.08	8.42	20.78	15.72
C28H51I107	626.62	626.2681	53.67	8.20	20.25	17.87
C28H51I108	642.62	642.2630	52.33	8.00	19.75	19.92
C28H51I109	658.62	658.2579	51.06	7.81	19.27	21.86
C28H52I201	658.53	658.2111	51.07	7.96	38.54	2.43
C28H52I202	674.53	674.2060	49.86	7.77	37.63	4.74
C28H52I203	690.53	690.2009	48.70	7.59	36.76	6.95
C28H52I204	706.53	706.1959	47.60	7.42	35.92	9.06
C28H52I205	722.53	722.1908	46.55	7.25	35.13	11.07
C28H52I206	738.53	738.1857	45.54	7.10	34.37	13.00
C28H52I207	754.53	754.1806	44.57	6.95	33.64	14.84
C28H52I208	770.53	770.1755	43.65	6.80	32.94	16.61
C28H52I209	786.53	786.1704	42.76	6.66	32.27	18.31
C28H53I101	532.64	532.3143	63.14	10.03	23.83	3.00
C28H53I102	548.64	548.3092	61.30	9.74	23.13	5.83
C28H53I103	564.64	564.3041	59.56	9.46	22.48	8.50
C28H53I104	580.64	580.2990	57.92	9.20	21.86	11.02
C28H53I105	596.64	596.2939	56.37	8.95	21.27	13.41
C28H53I106	612.64	612.2888	54.90	8.72	20.71	15.67
C28H53I107	628.63	628.2838	53.50	8.50	20.19	17.82
C28H53I108	644.63	644.2787	52.17	8.29	19.69	19.86
C28H53I109	660.63	660.2736	50.91	8.09	19.21	21.80
C28H54I201	660.55	660.2268	50.91	8.24	38.42	2.42
C28H54I202	676.55	676.2217	49.71	8.05	37.52	4.73
C28H54I203	692.55	692.2166	48.56	7.86	36.65	6.93
C28H54I204	708.55	708.2115	47.46	7.68	35.82	9.03
C28H54I205	724.55	724.2064	46.42	7.51	35.03	11.04
C28H54I206	740.55	740.2013	45.41	7.35	34.27	12.96
C28H54I207	756.55	756.1962	44.45	7.19	33.55	14.80
C28H54I208	772.55	772.1912	43.53	7.05	32.85	16.57
C28H54I209	788.55	788.1861	42.65	6.90	32.19	18.26
C28H55I101	534.65	534.3299	62.90	10.37	23.74	2.99
C28H55I102	550.65	550.3248	61.08	10.07	23.05	5.81
C28H55I103	566.65	566.3198	59.35	9.78	22.40	8.47
C28H55I104	582.65	582.3147	57.72	9.51	21.78	10.98
C28H55I105	598.65	598.3096	56.18	9.26	21.20	13.36
C28H55I106	614.65	614.3045	54.72	9.02	20.65	15.62
C28H55I107	630.65	630.2994	53.33	8.79	20.12	17.76
C28H55I108	646.65	646.2943	52.01	8.57	19.62	19.79
C28H55I109	662.65	662.2892	50.75	8.37	19.15	21.73
C28H56I201	662.57	662.2424	50.76	8.52	38.31	2.41
C28H56I202	678.57	678.2373	49.56	8.32	37.40	4.72
C28H56I203	694.57	694.2322	48.42	8.13	36.54	6.91
C28H56I204	710.56	710.2272	47.33	7.94	35.72	9.01
C28H56I205	726.56	726.2221	46.29	7.77	34.93	11.01
C28H56I206	742.56	742.2170	45.29	7.60	34.18	12.93
C28H56I207	758.56	758.2119	44.34	7.44	33.46	14.76
C28H56I208	774.56	774.2068	43.42	7.29	32.77	16.52
C28H56I209	790.56	790.2017	42.54	7.14	32.10	18.21
C28H57I101	536.67	536.3456	62.67	10.71	23.65	2.98
C28H57I102	552.67	552.3405	60.85	10.40	22.96	5.79
C28H57I103	568.67	568.3354	59.14	10.10	22.32	8.44
C28H57I104	584.67	584.3303	57.52	9.83	21.71	10.95
C28H57I105	600.67	600.3252	55.99	9.57	21.13	13.32
C28H57I106	616.67	616.3201	54.54	9.32	20.58	15.57
C28H57I107	632.67	632.3151	53.16	9.08	20.06	17.70
C28H57I108	648.67	648.3100	51.85	8.86	19.56	19.73
C28H57I109	664.67	664.3049	50.60	8.64	19.09	21.66
C29H14I201	632.24	631.9138	55.09	2.23	40.14	2.53
C29H14I202	648.24	647.9087	53.73	2.18	39.15	4.94
C29H14I203	664.24	663.9036	52.44	2.12	38.21	7.23
C29H14I204	680.24	679.8985	51.21	2.07	37.31	9.41
C29H14I205	696.24	695.8934	50.03	2.03	36.45	11.49
C29H14I206	712.24	711.8883	48.91	1.98	35.64	13.48
C29H14I207	728.24	727.8833	47.83	1.94	34.85	15.38
C29H14I208	744.24	743.8782	46.80	1.90	34.10	17.20
C29H14I209	760.24	759.8731	45.82	1.86	33.39	18.94
C29H15I101	506.35	506.0169	68.79	2.99	25.06	3.16
C29H15I102	522.35	522.0119	66.68	2.89	24.30	6.13
C29H15I103	538.35	538.0068	64.70	2.81	23.57	8.92
C29H15I104	554.34	554.0017	62.84	2.73	22.89	11.54
C29H15I105	570.34	569.9966	61.07	2.65	22.25	14.03
C29H15I106	586.34	585.9915	59.41	2.58	21.64	16.37
C29H15I107	602.34	601.9864	57.83	2.51	21.07	18.59
C29H15I108	618.34	617.9813	56.33	2.45	20.52	20.70
C29H15I109	634.34	633.9763	54.91	2.38	20.01	22.70
C29H16I201	634.26	633.9294	54.92	2.54	40.02	2.52
C29H16I202	650.26	649.9243	53.57	2.48	39.03	4.92
C29H16I203	666.26	665.9193	52.28	2.42	38.09	7.20
C29H16I204	682.26	681.9142	51.05	2.36	37.20	9.38
C29H16I205	698.26	697.9091	49.88	2.31	36.35	11.46
C29H16I206	714.26	713.9040	48.77	2.26	35.53	13.44
C29H16I207	730.26	729.8989	47.70	2.21	34.76	15.34
C29H16I208	746.25	745.8938	46.68	2.16	34.01	17.15
C29H16I209	762.25	761.8887	45.70	2.12	33.30	18.89
C29H17I101	508.36	508.0326	68.52	3.37	24.96	3.15
C29H17I102	524.36	524.0275	66.43	3.27	24.20	6.10
C29H17I103	540.36	540.0224	64.46	3.17	23.49	8.88
C29H17I104	556.36	556.0173	62.61	3.08	22.81	11.50
C29H17I105	572.36	572.0122	60.86	2.99	22.17	13.98
C29H17I106	588.36	588.0072	59.20	2.91	21.57	16.32
C29H17I107	604.36	604.0021	57.64	2.84	21.00	18.53
C29H17I108	620.36	619.9970	56.15	2.76	20.46	20.63
C29H17I109	636.36	635.9919	54.74	2.69	19.94	22.63
C29H18I201	636.28	635.9451	54.74	2.85	39.89	2.51
C29H18I202	652.27	651.9400	53.40	2.78	38.91	4.91
C29H18I203	668.27	667.9349	52.12	2.71	37.98	7.18
C29H18I204	684.27	683.9298	50.90	2.65	37.09	9.35
C29H18I205	700.27	699.9247	49.74	2.59	36.24	11.42
C29H18I206	716.27	715.9196	48.63	2.53	35.43	13.40
C29H18I207	732.27	731.9146	47.57	2.48	34.66	15.29
C29H18I208	748.27	747.9095	46.55	2.42	33.92	17.11
C29H18I209	764.27	763.9044	45.58	2.37	33.21	18.84
C29H19I101	510.38	510.0482	68.25	3.75	24.86	3.13
C29H19I102	526.38	526.0432	66.17	3.64	24.11	6.08
C29H19I103	542.38	542.0381	64.22	3.53	23.40	8.85
C29H19I104	558.38	558.0330	62.38	3.43	22.73	11.46
C29H19I105	574.38	574.0279	60.64	3.33	22.09	13.93
C29H19I106	590.38	590.0228	59.00	3.24	21.50	16.26
C29H19I107	606.37	606.0177	57.44	3.16	20.93	18.47

234

C29H19I108	622.37	622.0126	55.97	3.08	20.39	20.57
C29H19I109	638.37	638.0076	54.56	3.00	19.88	22.56
C29H20I201	638.29	637.9607	54.57	3.16	39.76	2.51
C29H20I202	654.29	653.9556	53.24	3.08	38.79	4.89
C29H20I203	670.29	669.9506	51.97	3.01	37.87	7.16
C29H20I204	686.29	685.9455	50.75	2.94	36.98	9.33
C29H20I205	702.29	701.9404	49.60	2.87	36.14	11.39
C29H20I206	718.29	717.9353	48.49	2.81	35.34	13.36
C29H20I207	734.29	733.9302	47.44	2.75	34.57	15.25
C29H20I208	750.29	749.9251	46.43	2.69	33.83	17.06
C29H20I209	766.29	765.9200	45.46	2.63	33.12	18.79
C29H21I101	512.39	512.0639	67.98	4.13	24.77	3.12
C29H21I102	528.39	528.0588	65.92	4.01	24.02	6.06
C29H21I103	544.39	544.0537	63.98	3.89	23.31	8.82
C29H21I104	560.39	560.0486	62.16	3.78	22.65	11.42
C29H21I105	576.39	576.0435	60.43	3.67	22.02	13.88
C29H21I106	592.39	592.0385	58.80	3.57	21.42	16.20
C29H21I107	608.39	608.0334	57.25	3.48	20.86	18.41
C29H21I108	624.39	624.0283	55.79	3.39	20.32	20.50
C29H21I109	640.39	640.0232	54.39	3.31	19.82	22.49
C29H22I201	640.31	639.9764	54.40	3.46	39.64	2.50
C29H22I202	656.31	655.9713	53.07	3.38	38.67	4.88
C29H22I203	672.31	671.9662	51.81	3.30	37.75	7.14
C29H22I204	688.31	687.9611	50.61	3.22	36.87	9.30
C29H22I205	704.30	703.9560	49.46	3.15	36.04	11.36
C29H22I206	720.30	719.9509	48.36	3.08	35.24	13.33
C29H22I207	736.30	735.9459	47.31	3.01	34.47	15.21
C29H22I208	752.30	751.9408	46.30	2.95	33.74	17.01
C29H22I209	768.30	767.9357	45.34	2.89	33.04	18.74
C29H23I101	514.41	514.0795	67.71	4.51	24.67	3.11
C29H23I102	530.41	530.0745	65.67	4.37	23.93	6.03
C29H23I103	546.41	546.0694	63.75	4.24	23.23	8.78
C29H23I104	562.41	562.0643	61.93	4.12	22.56	11.38
C29H23I105	578.41	578.0592	60.22	4.01	21.94	13.83
C29H23I106	594.41	594.0541	58.60	3.90	21.35	16.15
C29H23I107	610.41	610.0490	57.06	3.80	20.79	18.35
C29H23I108	626.41	626.0439	55.61	3.70	20.26	20.43
C29H23I109	642.41	642.0389	54.22	3.61	19.75	22.41
C29H24I201	642.32	641.9920	54.23	3.77	39.51	2.49
C29H24I202	658.32	657.9869	52.91	3.67	38.55	4.86
C29H24I203	674.32	673.9819	51.66	3.59	37.64	7.12
C29H24I204	690.32	689.9768	50.46	3.50	36.77	9.27
C29H24I205	706.32	705.9717	49.32	3.42	35.93	11.33
C29H24I206	722.32	721.9666	48.22	3.35	35.14	13.29
C29H24I207	738.32	737.9615	47.18	3.28	34.38	15.17
C29H24I208	754.32	753.9564	46.18	3.21	33.65	16.97
C29H24I209	770.32	769.9513	45.22	3.14	32.95	18.69
C29H25I101	516.43	516.0952	67.45	4.88	24.57	3.10
C29H25I102	532.43	532.0901	65.42	4.73	23.84	6.01
C29H25I103	548.43	548.0850	63.51	4.59	23.14	8.75
C29H25I104	564.42	564.0799	61.71	4.46	22.48	11.34
C29H25I105	580.42	580.0748	60.01	4.34	21.86	13.78
C29H25I106	596.42	596.0698	58.40	4.23	21.28	16.10
C29H25I107	612.42	612.0647	56.88	4.11	20.72	18.29
C29H25I108	628.42	628.0596	55.43	4.01	20.19	20.37
C29H25I109	644.42	644.0545	54.05	3.91	19.69	22.34
C29H26I201	644.34	644.0077	54.06	4.07	39.39	2.48
C29H26I202	660.34	660.0026	52.75	3.97	38.44	4.85
C29H26I203	676.34	675.9975	51.50	3.87	37.53	7.10
C29H26I204	692.34	691.9924	50.31	3.79	36.66	9.24
C29H26I205	708.34	707.9873	49.17	3.70	35.83	11.29
C29H26I206	724.34	723.9822	48.09	3.62	35.04	13.25
C29H26I207	740.34	739.9772	47.05	3.54	34.28	15.13
C29H26I208	756.33	755.9721	46.05	3.47	33.56	16.92
C29H26I209	772.33	771.9670	45.10	3.39	32.86	18.64
C29H27I101	518.44	518.1108	67.19	5.25	24.48	3.09
C29H27I102	534.44	534.1058	65.18	5.09	23.75	5.99
C29H27I103	550.44	550.1007	63.28	4.94	23.06	8.72
C29H27I104	566.44	566.0956	61.49	4.80	22.40	11.30
C29H27I105	582.44	582.0905	59.80	4.67	21.79	13.73
C29H27I106	598.44	598.0854	58.21	4.55	21.21	16.04
C29H27I107	614.44	614.0803	56.69	4.43	20.65	18.23
C29H27I108	630.44	630.0752	55.25	4.32	20.13	20.30
C29H27I109	646.44	646.0702	53.88	4.21	19.63	22.28
C29H28I201	646.35	646.0233	53.89	4.37	39.27	2.48
C29H28I202	662.35	662.0182	52.59	4.26	38.32	4.83
C29H28I203	678.35	678.0132	51.35	4.16	37.42	7.08
C29H28I204	694.35	694.0081	50.17	4.06	36.55	9.22
C29H28I205	710.35	710.0030	49.04	3.97	35.73	11.26
C29H28I206	726.35	725.9979	47.96	3.89	34.94	13.22
C29H28I207	742.35	741.9928	46.92	3.80	34.19	15.09
C29H28I208	758.35	757.9877	45.93	3.72	33.47	16.88
C29H28I209	774.35	773.9826	44.98	3.64	32.78	18.60
C29H29I101	520.46	520.1265	66.93	5.62	24.38	3.07
C29H29I102	536.46	536.1214	64.93	5.45	23.66	5.96
C29H29I103	552.46	552.1163	63.05	5.29	22.97	8.69
C29H29I104	568.46	568.1112	61.28	5.14	22.32	11.26
C29H29I105	584.46	584.1061	59.60	5.00	21.71	13.69
C29H29I106	600.46	600.1011	58.01	4.87	21.13	15.99
C29H29I107	616.45	616.0960	56.50	4.74	20.59	18.17
C29H29I108	632.45	632.0909	55.07	4.62	20.07	20.24
C29H29I109	648.45	648.0858	53.72	4.51	19.57	22.21
C29H30I201	648.37	648.0390	53.72	4.66	39.15	2.47
C29H30I202	664.37	664.0339	52.43	4.55	38.20	4.82
C29H30I203	680.37	680.0288	51.20	4.44	37.30	7.05
C29H30I204	696.37	696.0237	50.02	4.34	36.45	9.19
C29H30I205	712.37	712.0186	48.90	4.24	35.63	11.23
C29H30I206	728.37	728.0135	47.82	4.15	34.85	13.18
C29H30I207	744.37	744.0085	46.79	4.06	34.10	15.05
C29H30I208	760.37	760.0034	45.81	3.98	33.38	16.83
C29H30I209	776.37	775.9983	44.87	3.89	32.69	18.55
C29H31I101	522.47	522.1421	66.67	5.98	24.29	3.06
C29H31I102	538.47	538.1371	64.69	5.80	23.57	5.94
C29H31I103	554.47	554.1320	62.82	5.64	22.89	8.66
C29H31I104	570.47	570.1269	61.06	5.48	22.25	11.22
C29H31I105	586.47	586.1218	59.39	5.33	21.64	13.64
C29H31I106	602.47	602.1167	57.82	5.19	21.06	15.93
C29H31I107	618.47	618.1116	56.32	5.05	20.52	18.11
C29H31I108	634.47	634.1065	54.90	4.92	20.00	20.17
C29H31I109	650.47	650.1014	53.55	4.80	19.51	22.14
C29H32I201	650.39	650.0546	53.56	4.96	39.02	2.46
C29H32I202	666.39	666.0495	52.27	4.84	38.09	4.80
C29H32I203	682.39	682.0444	51.04	4.73	37.19	7.03
C29H32I204	698.38	698.0394	49.88	4.62	36.34	9.16
C29H32I205	714.38	714.0343	48.76	4.52	35.53	11.20
C29H32I206	730.38	730.0292	47.69	4.42	34.75	13.14
C29H32I207	746.38	746.0241	46.67	4.32	34.01	15.01
C29H32I208	762.38	762.0190	45.69	4.23	33.29	16.79
C29H32I209	778.38	778.0139	44.75	4.14	32.61	18.50
C29H33I101	524.49	524.1578	66.41	6.34	24.20	3.05

C29H33I102	540.49	540.1527	64.45	6.15	23.48	5.92
C29H33I103	556.49	556.1476	62.59	5.98	22.80	8.63
C29H33I104	572.49	572.1425	60.84	5.81	22.17	11.18
C29H33I105	588.49	588.1374	59.19	5.65	21.56	13.59
C29H33I106	604.49	604.1324	57.62	5.50	20.99	15.88
C29H33I107	620.49	620.1273	56.14	5.36	20.45	18.05
C29H33I108	636.49	636.1222	54.73	5.23	19.94	20.11
C29H33I109	652.49	652.1171	53.38	5.10	19.45	22.07
C29H34I201	652.40	652.0703	53.39	5.25	38.90	2.45
C29H34I202	668.40	668.0652	52.11	5.13	37.97	4.79
C29H34I203	684.40	684.0601	50.89	5.01	37.08	7.01
C29H34I204	700.40	700.0550	49.73	4.89	36.24	9.14
C29H34I205	716.40	716.0499	48.62	4.78	35.43	11.17
C29H34I206	732.40	732.0448	47.56	4.68	34.65	13.11
C29H34I207	748.40	748.0398	46.54	4.58	33.91	14.96
C29H34I208	764.40	764.0347	45.57	4.48	33.20	16.74
C29H34I209	780.40	780.0296	44.63	4.39	32.52	18.45
C29H35I101	526.51	526.1734	66.16	6.70	24.10	3.04
C29H35I102	542.51	542.1683	64.21	6.50	23.39	5.90
C29H35I103	558.50	558.1633	62.37	6.32	22.72	8.59
C29H35I104	574.50	574.1582	60.63	6.14	22.09	11.14
C29H35I105	590.50	590.1531	58.99	5.97	21.49	13.55
C29H35I106	606.50	606.1480	57.43	5.82	20.92	15.83
C29H35I107	622.50	622.1429	55.96	5.67	20.39	17.99
C29H35I108	638.50	638.1378	54.55	5.53	19.88	20.05
C29H35I109	654.50	654.1327	53.22	5.39	19.39	22.00
C29H36I201	654.42	654.0859	53.23	5.54	38.78	2.44
C29H36I202	670.42	670.0808	51.96	5.41	37.86	4.77
C29H36I203	686.42	686.0757	50.75	5.29	36.98	6.99
C29H36I204	702.42	702.0707	49.59	5.17	36.13	9.11
C29H36I205	718.42	718.0656	48.48	5.05	35.33	11.14
C29H36I206	734.42	734.0605	47.43	4.94	34.56	13.07
C29H36I207	750.41	750.0554	46.42	4.84	33.82	14.92
C29H36I208	766.41	766.0503	45.45	4.73	33.12	16.70
C29H36I209	782.41	782.0452	44.52	4.64	32.44	18.40
C29H37I101	528.52	528.1891	65.91	7.06	24.01	3.03
C29H37I102	544.52	544.1840	63.97	6.85	23.31	5.88
C29H37I103	560.52	560.1789	62.14	6.65	22.64	8.56
C29H37I104	576.52	576.1738	60.42	6.47	22.01	11.10
C29H37I105	592.52	592.1687	58.79	6.29	21.42	13.50
C29H37I106	608.52	608.1637	57.24	6.13	20.85	15.78
C29H37I107	624.52	624.1586	55.77	5.97	20.32	17.93
C29H37I108	640.52	640.1535	54.38	5.82	19.81	19.98
C29H37I109	656.52	656.1484	53.06	5.68	19.33	21.93
C29H38I201	656.43	656.1016	53.06	5.83	38.66	2.44
C29H38I202	672.43	672.0965	51.80	5.70	37.74	4.76
C29H38I203	688.43	688.0914	50.60	5.56	36.87	6.97
C29H38I204	704.43	704.0863	49.45	5.44	36.03	9.08
C29H38I205	720.43	720.0812	48.35	5.32	35.23	11.10
C29H38I206	736.43	736.0761	47.30	5.20	34.46	13.04
C29H38I207	752.43	752.0711	46.29	5.09	33.73	14.88
C29H38I208	768.43	768.0660	45.33	4.98	33.03	16.66
C29H38I209	784.43	784.0609	44.40	4.88	32.36	18.36
C29H39I101	530.54	530.2047	65.65	7.41	23.92	3.02
C29H39I102	546.54	546.1996	63.73	7.19	23.22	5.85
C29H39I103	562.54	562.1946	61.92	6.99	22.56	8.53
C29H39I104	578.54	578.1895	60.21	6.79	21.94	11.06
C29H39I105	594.54	594.1844	58.59	6.61	21.35	13.46
C29H39I106	610.53	610.1793	57.05	6.44	20.79	15.72
C29H39I107	626.53	626.1742	55.60	6.27	20.25	17.88

C29H39I108	642.53	642.1691	54.21	6.12	19.75	19.92
C29H39I109	658.53	658.1640	52.89	5.97	19.27	21.87
C29H40I201	658.45	658.1172	52.90	6.12	38.55	2.43
C29H40I202	674.45	674.1121	51.65	5.98	37.63	4.74
C29H40I203	690.45	690.1070	50.45	5.84	36.76	6.95
C29H40I204	706.45	706.1020	49.31	5.71	35.93	9.06
C29H40I205	722.45	722.0969	48.21	5.58	35.13	11.07
C29H40I206	738.45	738.0918	47.17	5.46	34.37	13.00
C29H40I207	754.45	754.0867	46.17	5.34	33.64	14.84
C29H40I208	770.45	770.0816	45.21	5.23	32.94	16.61
C29H40I209	786.45	786.0765	44.29	5.13	32.27	18.31
C29H41I101	532.55	532.2204	65.41	7.76	23.83	3.00
C29H41I102	548.55	548.2153	63.50	7.53	23.13	5.83
C29H41I103	564.55	564.2102	61.70	7.32	22.48	8.50
C29H41I104	580.55	580.2051	60.00	7.12	21.86	11.02
C29H41I105	596.55	596.2000	58.39	6.93	21.27	13.41
C29H41I106	612.55	612.1950	56.86	6.75	20.72	15.67
C29H41I107	628.55	628.1899	55.42	6.57	20.19	17.82
C29H41I108	644.55	644.1848	54.04	6.41	19.69	19.86
C29H41I109	660.55	660.1797	52.73	6.26	19.21	21.80
C29H42I201	660.47	660.1329	52.74	6.41	38.43	2.42
C29H42I202	676.47	676.1278	51.49	6.26	37.52	4.73
C29H42I203	692.47	692.1227	50.30	6.11	36.65	6.93
C29H42I204	708.46	708.1176	49.17	5.98	35.83	9.03
C29H42I205	724.46	724.1125	48.08	5.84	35.03	11.04
C29H42I206	740.46	740.1074	47.04	5.72	34.28	12.96
C29H42I207	756.46	756.1024	46.05	5.60	33.55	14.81
C29H42I208	772.46	772.0973	45.09	5.48	32.86	16.57
C29H42I209	788.46	788.0922	44.18	5.37	32.19	18.26
C29H43I101	534.57	534.2360	65.16	8.11	23.74	2.99
C29H43I102	550.57	550.2309	63.27	7.87	23.05	5.81
C29H43I103	566.57	566.2259	61.48	7.65	22.40	8.47
C29H43I104	582.57	582.2208	59.79	7.44	21.78	10.99
C29H43I105	598.57	598.2157	58.19	7.24	21.20	13.36
C29H43I106	614.57	614.2106	56.68	7.05	20.65	15.62
C29H43I107	630.57	630.2055	55.24	6.87	20.13	17.76
C29H43I108	646.57	646.2004	53.87	6.70	19.63	19.80
C29H43I109	662.57	662.1953	52.57	6.54	19.15	21.73
C29H44I201	662.48	662.1485	52.58	6.69	38.31	2.42
C29H44I202	678.48	678.1434	51.34	6.54	37.41	4.72
C29H44I203	694.48	694.1383	50.16	6.39	36.55	6.91
C29H44I204	710.48	710.1333	49.03	6.24	35.72	9.01
C29H44I205	726.48	726.1282	47.95	6.10	34.94	11.01
C29H44I206	742.48	742.1231	46.91	5.97	34.18	12.93
C29H44I207	758.48	758.1180	45.92	5.85	33.46	14.77
C29H44I208	774.48	774.1129	44.98	5.73	32.77	16.53
C29H44I209	790.48	790.1078	44.06	5.61	32.11	18.22
C29H45I101	536.59	536.2517	64.91	8.45	23.65	2.98
C29H45I102	552.59	552.2466	63.04	8.21	22.97	5.79
C29H45I103	568.58	568.2415	61.26	7.98	22.32	8.44
C29H45I104	584.58	584.2364	59.58	7.76	21.71	10.95
C29H45I105	600.58	600.2313	58.00	7.55	21.13	13.32
C29H45I106	616.58	616.2263	56.49	7.36	20.58	15.57
C29H45I107	632.58	632.2212	55.06	7.17	20.06	17.70
C29H45I108	648.58	648.2161	53.71	6.99	19.57	19.73
C29H45I109	664.58	664.2110	52.41	6.83	19.10	21.67
C29H46I201	664.50	664.1642	52.42	6.98	38.20	2.41
C29H46I202	680.50	680.1591	51.19	6.81	37.30	4.70
C29H46I203	696.50	696.1540	50.01	6.66	36.44	6.89
C29H46I204	712.50	712.1489	48.89	6.51	35.62	8.98

C29H46I205	728.50	728.1438	47.81	6.36	34.84	10.98
C29H46I206	744.50	744.1387	46.79	6.23	34.09	12.89
C29H46I207	760.49	760.1337	45.80	6.10	33.37	14.73
C29H46I208	776.49	776.1286	44.86	5.97	32.69	16.48
C29H46I209	792.49	792.1235	43.95	5.85	32.03	18.17
C29H47I101	538.60	538.2673	64.67	8.80	23.56	2.97
C29H47I102	554.60	554.2622	62.81	8.54	22.88	5.77
C29H47I103	570.60	570.2572	61.05	8.30	22.24	8.41
C29H47I104	586.60	586.2521	59.38	8.08	21.63	10.91
C29H47I105	602.60	602.2470	57.80	7.86	21.06	13.28
C29H47I106	618.60	618.2419	56.31	7.66	20.51	15.52
C29H47I107	634.60	634.2368	54.89	7.47	20.00	17.65
C29H47I108	650.60	650.2317	53.54	7.28	19.51	19.67
C29H47I109	666.60	666.2266	52.25	7.11	19.04	21.60
C29H48I201	666.51	666.1798	52.26	7.26	38.08	2.40
C29H48I202	682.51	682.1747	51.04	7.09	37.19	4.69
C29H48I203	698.51	698.1696	49.87	6.93	36.34	6.87
C29H48I204	714.51	714.1646	48.75	6.77	35.52	8.96
C29H48I205	730.51	730.1595	47.68	6.62	34.74	10.95
C29H48I206	746.51	746.1544	46.66	6.48	34.00	12.86
C29H48I207	762.51	762.1493	45.68	6.35	33.29	14.69
C29H48I208	778.51	778.1442	44.74	6.21	32.60	16.44
C29H48I209	794.51	794.1391	43.84	6.09	31.95	18.12
C29H49I101	540.62	540.2830	64.43	9.14	23.47	2.96
C29H49I102	556.62	556.2779	62.58	8.87	22.80	5.75
C29H49I103	572.62	572.2728	60.83	8.63	22.16	8.38
C29H49I104	588.62	588.2677	59.18	8.39	21.56	10.87
C29H49I105	604.62	604.2626	57.61	8.17	20.99	13.23
C29H49I106	620.61	620.2576	56.13	7.96	20.45	15.47
C29H49I107	636.61	636.2525	54.71	7.76	19.93	17.59
C29H49I108	652.61	652.2474	53.37	7.57	19.45	19.61
C29H49I109	668.61	668.2423	52.10	7.39	18.98	21.54
C29H50I201	668.53	668.1955	52.10	7.54	37.97	2.39
C29H50I202	684.53	684.1904	50.89	7.36	37.08	4.67
C29H50I203	700.53	700.1853	49.72	7.19	36.23	6.85
C29H50I204	716.53	716.1802	48.61	7.03	35.42	8.93
C29H50I205	732.53	732.1751	47.55	6.88	34.65	10.92
C29H50I206	748.53	748.1700	46.53	6.73	33.91	12.82
C29H50I207	764.53	764.1649	45.56	6.59	33.20	14.65
C29H50I208	780.53	780.1599	44.63	6.46	32.52	16.40
C29H50I209	796.53	796.1548	43.73	6.33	31.86	18.08
C29H51I101	542.63	542.2986	64.19	9.47	23.39	2.95
C29H51I102	558.63	558.2935	62.35	9.20	22.72	5.73
C29H51I103	574.63	574.2885	60.62	8.95	22.08	8.35
C29H51I104	590.63	590.2834	58.97	8.70	21.49	10.84
C29H51I105	606.63	606.2783	57.42	8.47	20.92	13.19
C29H51I106	622.63	622.2732	55.94	8.26	20.38	15.42
C29H51I107	638.63	638.2681	54.54	8.05	19.87	17.54
C29H51I108	654.63	654.2630	53.21	7.85	19.39	19.55
C29H51I109	670.63	670.2579	51.94	7.67	18.92	21.47
C29H52I201	670.55	670.2111	51.95	7.82	37.85	2.39
C29H52I202	686.55	686.2060	50.74	7.63	36.97	4.66
C29H52I203	702.54	702.2009	49.58	7.46	36.13	6.83
C29H52I204	718.54	718.1959	48.48	7.29	35.32	8.91
C29H52I205	734.54	734.1908	47.42	7.14	34.55	10.89
C29H52I206	750.54	750.1857	46.41	6.98	33.82	12.79
C29H52I207	766.54	766.1806	45.44	6.84	33.11	14.61
C29H52I208	782.54	782.1755	44.51	6.70	32.43	16.36
C29H52I209	798.54	798.1704	43.62	6.56	31.78	18.03
C29H53I101	544.65	544.3143	63.95	9.81	23.30	2.94

C29H53I102	560.65	560.3092	62.13	9.53	22.64	5.71
C29H53I103	576.65	576.3041	60.40	9.26	22.01	8.32
C29H53I104	592.65	592.2990	58.77	9.01	21.41	10.80
C29H53I105	608.65	608.2939	57.23	8.78	20.85	13.14
C29H53I106	624.65	624.2888	55.76	8.55	20.32	15.37
C29H53I107	640.65	640.2838	54.37	8.34	19.81	17.48
C29H53I108	656.65	656.2787	53.05	8.14	19.33	19.49
C29H53I109	672.64	672.2736	51.78	7.94	18.87	21.41
C29H54I201	672.56	672.2268	51.79	8.09	37.74	2.38
C29H54I202	688.56	688.2217	50.59	7.90	36.86	4.65
C29H54I203	704.56	704.2166	49.44	7.73	36.02	6.81
C29H54I204	720.56	720.2115	48.34	7.55	35.22	8.88
C29H54I205	736.56	736.2064	47.29	7.39	34.46	10.86
C29H54I206	752.56	752.2013	46.29	7.23	33.73	12.76
C29H54I207	768.56	768.1962	45.32	7.08	33.02	14.57
C29H54I208	784.56	784.1912	44.40	6.94	32.35	16.31
C29H54I209	800.56	800.1861	43.51	6.80	31.70	17.99
C29H55I101	546.67	546.3299	63.72	10.14	23.21	2.93
C29H55I102	562.66	562.3248	61.91	9.85	22.55	5.69
C29H55I103	578.66	578.3198	60.19	9.58	21.93	8.29
C29H55I104	594.66	594.3147	58.57	9.32	21.34	10.76
C29H55I105	610.66	610.3096	57.04	9.08	20.78	13.10
C29H55I106	626.66	626.3045	55.58	8.85	20.25	15.32
C29H55I107	642.66	642.2994	54.20	8.63	19.75	17.43
C29H55I108	658.66	658.2943	52.88	8.42	19.27	19.43
C29H55I109	674.66	674.2892	51.63	8.22	18.81	21.34
C29H56I201	674.58	674.2424	51.64	8.37	37.62	2.37
C29H56I202	690.58	690.2373	50.44	8.17	36.75	4.63
C29H56I203	706.58	706.2322	49.30	7.99	35.92	6.79
C29H56I204	722.58	722.2272	48.21	7.81	35.13	8.86
C29H56I205	738.58	738.2221	47.16	7.64	34.36	10.83
C29H56I206	754.57	754.2170	46.16	7.48	33.64	12.72
C29H56I207	770.57	770.2119	45.20	7.33	32.94	14.53
C29H56I208	786.57	786.2068	44.28	7.18	32.27	16.27
C29H56I209	802.57	802.2017	43.40	7.03	31.62	17.94
C29H57I101	548.68	548.3456	63.48	10.47	23.13	2.92
C29H57I102	564.68	564.3405	61.68	10.17	22.47	5.67
C29H57I103	580.68	580.3354	59.99	9.89	21.85	8.27
C29H57I104	596.68	596.3303	58.38	9.63	21.27	10.73
C29H57I105	612.68	612.3252	56.85	9.38	20.71	13.06
C29H57I106	628.68	628.3201	55.41	9.14	20.19	15.27
C29H57I107	644.68	644.3151	54.03	8.91	19.68	17.37
C29H57I108	660.68	660.3100	52.72	8.70	19.21	19.37
C29H57I109	676.68	676.3049	51.48	8.49	18.75	21.28
C29H58I201	676.59	676.2581	51.48	8.64	37.51	2.36
C29H58I202	692.59	692.2530	50.29	8.44	36.65	4.62
C29H58I203	708.59	708.2479	49.16	8.25	35.82	6.77
C29H58I204	724.59	724.2428	48.07	8.07	35.03	8.83
C29H58I205	740.59	740.2377	47.03	7.89	34.27	10.80
C29H58I206	756.59	756.2326	46.04	7.73	33.55	12.69
C29H58I207	772.59	772.2275	45.09	7.57	32.85	14.50
C29H58I208	788.59	788.2225	44.17	7.41	32.19	16.23
C29H58I209	804.59	804.2174	43.29	7.27	31.55	17.90
C29H59I101	550.70	550.3612	63.25	10.80	23.04	2.91
C29H59I102	566.70	566.3561	61.47	10.49	22.39	5.65
C29H59I103	582.70	582.3511	59.78	10.21	21.78	8.24
C29H59I104	598.70	598.3460	58.18	9.93	21.20	10.69
C29H59I105	614.69	614.3409	56.67	9.67	20.65	13.01
C29H59I106	630.69	630.3358	55.23	9.43	20.12	15.22
C29H59I107	646.69	646.3307	53.86	9.20	19.62	17.32

C29H59I108	662.69	662.3256	52.56	8.97	19.15	19.31
C29H59I109	678.69	678.3205	51.32	8.76	18.70	21.22
C30H15I101	518.36	518.0169	69.51	2.92	24.48	3.09
C30H15I102	534.36	534.0119	67.43	2.83	23.75	5.99
C30H15I103	550.36	550.0068	65.47	2.75	23.06	8.72
C30H15I104	566.36	566.0017	63.62	2.67	22.41	11.30
C30H15I105	582.36	581.9966	61.88	2.60	21.79	13.74
C30H15I106	598.35	597.9915	60.22	2.53	21.21	16.04
C30H15I107	614.35	613.9864	58.65	2.46	20.66	18.23
C30H15I108	630.35	629.9813	57.16	2.40	20.13	20.31
C30H15I109	646.35	645.9763	55.75	2.34	19.63	22.28
C30H16I201	646.27	645.9294	55.76	2.50	39.27	2.48
C30H16I202	662.27	661.9243	54.41	2.44	38.32	4.83
C30H16I203	678.27	677.9193	53.13	2.38	37.42	7.08
C30H16I204	694.27	693.9142	51.90	2.32	36.56	9.22
C30H16I205	710.27	709.9091	50.73	2.27	35.73	11.26
C30H16I206	726.27	725.9040	49.61	2.22	34.95	13.22
C30H16I207	742.27	741.8989	48.55	2.17	34.19	15.09
C30H16I208	758.27	757.8938	47.52	2.13	33.47	16.88
C30H16I209	774.27	773.8887	46.54	2.08	32.78	18.60
C30H17I101	520.37	520.0326	69.25	3.29	24.39	3.07
C30H17I102	536.37	536.0275	67.18	3.19	23.66	5.97
C30H17I103	552.37	552.0224	65.23	3.10	22.97	8.69
C30H17I104	568.37	568.0173	63.40	3.01	22.33	11.26
C30H17I105	584.37	584.0122	61.66	2.93	21.72	13.69
C30H17I106	600.37	600.0072	60.02	2.85	21.14	15.99
C30H17I107	616.37	616.0021	58.46	2.78	20.59	18.17
C30H17I108	632.37	631.9970	56.98	2.71	20.07	20.24
C30H17I109	648.37	647.9919	55.58	2.64	19.57	22.21
C30H18I201	648.29	647.9451	55.58	2.80	39.15	2.47
C30H18I202	664.29	663.9400	54.24	2.73	38.21	4.82
C30H18I203	680.28	679.9349	52.97	2.67	37.31	7.06
C30H18I204	696.28	695.9298	51.75	2.61	36.45	9.19
C30H18I205	712.28	711.9247	50.59	2.55	35.63	11.23
C30H18I206	728.28	727.9196	49.48	2.49	34.85	13.18
C30H18I207	744.28	743.9146	48.41	2.44	34.10	15.05
C30H18I208	760.28	759.9095	47.39	2.39	33.38	16.84
C30H18I209	776.28	775.9044	46.42	2.34	32.70	18.55
C30H19I101	522.39	522.0482	68.98	3.67	24.29	3.06
C30H19I102	538.39	538.0432	66.93	3.56	23.57	5.94
C30H19I103	554.39	554.0381	65.00	3.45	22.89	8.66
C30H19I104	570.39	570.0330	63.17	3.36	22.25	11.22
C30H19I105	586.39	586.0279	61.45	3.27	21.64	13.64
C30H19I106	602.39	602.0228	59.82	3.18	21.07	15.94
C30H19I107	618.39	618.0177	58.27	3.10	20.52	18.11
C30H19I108	634.39	634.0126	56.80	3.02	20.00	20.18
C30H19I109	650.38	650.0076	55.40	2.94	19.51	22.14
C30H20I201	650.30	649.9607	55.41	3.10	39.03	2.46
C30H20I202	666.30	665.9556	54.08	3.03	38.09	4.80
C30H20I203	682.30	681.9506	52.81	2.95	37.20	7.03
C30H20I204	698.30	697.9455	51.60	2.89	36.35	9.16
C30H20I205	714.30	713.9404	50.45	2.82	35.53	11.20
C30H20I206	730.30	729.9353	49.34	2.76	34.75	13.14
C30H20I207	746.30	745.9302	48.28	2.70	34.01	15.01
C30H20I208	762.30	761.9251	47.27	2.64	33.30	16.79
C30H20I209	778.30	777.9200	46.30	2.59	32.61	18.50
C30H21I101	524.41	524.0639	68.71	4.04	24.20	3.05
C30H21I102	540.41	540.0588	66.68	3.92	23.48	5.92
C30H21I103	556.40	556.0537	64.76	3.80	22.81	8.63
C30H21I104	572.40	572.0486	62.95	3.70	22.17	11.18
C30H21I105	588.40	588.0435	61.24	3.60	21.57	13.60
C30H21I106	604.40	604.0385	59.62	3.50	21.00	15.88
C30H21I107	620.40	620.0334	58.08	3.41	20.46	18.05
C30H21I108	636.40	636.0283	56.62	3.33	19.94	20.11
C30H21I109	652.40	652.0232	55.23	3.24	19.45	22.07
C30H22I201	652.32	651.9764	55.24	3.40	38.91	2.45
C30H22I202	668.32	667.9713	53.92	3.32	37.98	4.79
C30H22I203	684.32	683.9662	52.66	3.24	37.09	7.01
C30H22I204	700.32	699.9611	51.45	3.17	36.24	9.14
C30H22I205	716.32	715.9560	50.30	3.10	35.43	11.17
C30H22I206	732.32	731.9509	49.20	3.03	34.66	13.11
C30H22I207	748.31	747.9459	48.15	2.96	33.92	14.97
C30H22I208	764.31	763.9408	47.14	2.90	33.21	16.75
C30H22I209	780.31	779.9357	46.18	2.84	32.53	18.45
C30H23I101	526.42	526.0795	68.45	4.40	24.11	3.04
C30H23I102	542.42	542.0745	66.43	4.27	23.40	5.90
C30H23I103	558.42	558.0694	64.53	4.15	22.73	8.60
C30H23I104	574.42	574.0643	62.73	4.04	22.09	11.14
C30H23I105	590.42	590.0592	61.03	3.93	21.49	13.55
C30H23I106	606.42	606.0541	59.42	3.82	20.93	15.83
C30H23I107	622.42	622.0490	57.89	3.72	20.39	17.99
C30H23I108	638.42	638.0439	56.44	3.63	19.88	20.05
C30H23I109	654.42	654.0389	55.06	3.54	19.39	22.00
C30H24I201	654.33	653.9920	55.07	3.70	38.79	2.45
C30H24I202	670.33	669.9869	53.75	3.61	37.86	4.77
C30H24I203	686.33	685.9819	52.50	3.52	36.98	6.99
C30H24I204	702.33	701.9768	51.31	3.44	36.14	9.11
C30H24I205	718.33	717.9717	50.16	3.37	35.33	11.14
C30H24I206	734.33	733.9666	49.07	3.29	34.56	13.07
C30H24I207	750.33	749.9615	48.02	3.22	33.83	14.93
C30H24I208	766.33	765.9564	47.02	3.16	33.12	16.70
C30H24I209	782.33	781.9513	46.06	3.09	32.44	18.41
C30H25I101	528.44	528.0952	68.19	4.77	24.02	3.03
C30H25I102	544.44	544.0901	66.18	4.63	23.31	5.88
C30H25I103	560.44	560.0850	64.30	4.50	22.64	8.56
C30H25I104	576.44	576.0799	62.51	4.37	22.02	11.10
C30H25I105	592.44	592.0748	60.82	4.25	21.42	13.50
C30H25I106	608.43	608.0698	59.22	4.14	20.86	15.78
C30H25I107	624.43	624.0647	57.71	4.04	20.32	17.94
C30H25I108	640.43	640.0596	56.26	3.93	19.82	19.99
C30H25I109	656.43	656.0545	54.89	3.84	19.33	21.94
C30H26I201	656.35	656.0077	54.90	3.99	38.67	2.44
C30H26I202	672.35	672.0026	53.59	3.90	37.75	4.76
C30H26I203	688.35	687.9975	52.35	3.81	36.87	6.97
C30H26I204	704.35	703.9924	51.16	3.72	36.03	9.09
C30H26I205	720.35	719.9873	50.02	3.64	35.23	11.11
C30H26I206	736.35	735.9822	48.94	3.56	34.47	13.04
C30H26I207	752.35	751.9772	47.89	3.48	33.74	14.89
C30H26I208	768.35	767.9721	46.90	3.41	33.03	16.66
C30H26I209	784.35	783.9670	45.94	3.34	32.36	18.36
C30H27I101	530.45	530.1108	67.93	5.13	23.92	3.02
C30H27I102	546.45	546.1058	65.94	4.98	23.22	5.86
C30H27I103	562.45	562.1007	64.06	4.84	22.56	8.53
C30H27I104	578.45	578.0956	62.29	4.70	21.94	11.06
C30H27I105	594.45	594.0905	60.62	4.58	21.35	13.46
C30H27I106	610.45	610.0854	59.03	4.46	20.79	15.73
C30H27I107	626.45	626.0803	57.52	4.34	20.26	17.88
C30H27I108	642.45	642.0752	56.09	4.24	19.75	19.92
C30H27I109	658.45	658.0702	54.72	4.13	19.27	21.87
C30H28I201	658.37	658.0233	54.73	4.29	38.55	2.43

C30H28I202	674.37	674.0182	53.43	4.19	37.64	4.75
C30H28I203	690.36	690.0132	52.19	4.09	36.76	6.95
C30H28I204	706.36	706.0081	51.01	4.00	35.93	9.06
C30H28I205	722.36	722.0030	49.88	3.91	35.14	11.07
C30H28I206	738.36	737.9979	48.80	3.82	34.37	13.00
C30H28I207	754.36	753.9928	47.77	3.74	33.65	14.85
C30H28I208	770.36	769.9877	46.77	3.66	32.95	16.61
C30H28I209	786.36	785.9826	45.82	3.59	32.28	18.31
C30H29I101	532.47	532.1265	67.67	5.49	23.83	3.00
C30H29I102	548.47	548.1214	65.70	5.33	23.14	5.83
C30H29I103	564.47	564.1163	63.84	5.18	22.48	8.50
C30H29I104	580.47	580.1112	62.08	5.04	21.86	11.03
C30H29I105	596.47	596.1061	60.41	4.90	21.28	13.41
C30H29I106	612.47	612.1011	58.83	4.77	20.72	15.67
C30H29I107	628.47	628.0960	57.34	4.65	20.19	17.82
C30H29I108	644.47	644.0909	55.91	4.54	19.69	19.86
C30H29I109	660.46	660.0858	54.56	4.43	19.21	21.80
C30H30I201	660.38	660.0390	54.56	4.58	38.43	2.42
C30H30I202	676.38	676.0339	53.27	4.47	37.52	4.73
C30H30I203	692.38	692.0288	52.04	4.37	36.66	6.93
C30H30I204	708.38	708.0237	50.87	4.27	35.83	9.03
C30H30I205	724.38	724.0186	49.74	4.17	35.04	11.04
C30H30I206	740.38	740.0135	48.67	4.08	34.28	12.97
C30H30I207	756.38	756.0085	47.64	4.00	33.56	14.81
C30H30I208	772.38	772.0034	46.65	3.92	32.86	16.57
C30H30I209	788.38	787.9983	45.71	3.84	32.19	18.26
C30H31I101	534.49	534.1421	67.42	5.85	23.74	2.99
C30H31I102	550.48	550.1371	65.46	5.68	23.05	5.81
C30H31I103	566.48	566.1320	63.61	5.52	22.40	8.47
C30H31I104	582.48	582.1269	61.86	5.36	21.79	10.99
C30H31I105	598.48	598.1218	60.21	5.22	21.20	13.37
C30H31I106	614.48	614.1167	58.64	5.09	20.65	15.62
C30H31I107	630.48	630.1116	57.15	4.96	20.13	17.76
C30H31I108	646.48	646.1065	55.74	4.83	19.63	19.80
C30H31I109	662.48	662.1014	54.39	4.72	19.16	21.74
C30H32I201	662.40	662.0546	54.40	4.87	38.32	2.42
C30H32I202	678.40	678.0495	53.12	4.75	37.41	4.72
C30H32I203	694.40	694.0444	51.89	4.65	36.55	6.91
C30H32I204	710.40	710.0394	50.72	4.54	35.73	9.01
C30H32I205	726.40	726.0343	49.61	4.44	34.94	11.01
C30H32I206	742.39	742.0292	48.54	4.34	34.19	12.93
C30H32I207	758.39	758.0241	47.51	4.25	33.47	14.77
C30H32I208	774.39	774.0190	46.53	4.17	32.78	16.53
C30H32I209	790.39	790.0139	45.59	4.08	32.11	18.22
C30H33I101	536.50	536.1578	67.16	6.20	23.65	2.98
C30H33I102	552.50	552.1527	65.22	6.02	22.97	5.79
C30H33I103	568.50	568.1476	63.38	5.85	22.32	8.44
C30H33I104	584.50	584.1425	61.65	5.69	21.71	10.95
C30H33I105	600.50	600.1374	60.01	5.54	21.13	13.32
C30H33I106	616.50	616.1324	58.45	5.40	20.58	15.57
C30H33I107	632.50	632.1273	56.97	5.26	20.06	17.71
C30H33I108	648.50	648.1222	55.56	5.13	19.57	19.74
C30H33I109	664.50	664.1171	54.23	5.01	19.10	21.67
C30H34I201	664.41	664.0703	54.23	5.16	38.20	2.41
C30H34I202	680.41	680.0652	52.96	5.04	37.30	4.70
C30H34I203	696.41	696.0601	51.74	4.92	36.45	6.89
C30H34I204	712.41	712.0550	50.58	4.81	35.63	8.98
C30H34I205	728.41	728.0499	49.47	4.70	34.84	10.98
C30H34I206	744.41	744.0448	48.41	4.60	34.10	12.90
C30H34I207	760.41	760.0398	47.39	4.51	33.38	14.73
C30H34I208	776.41	776.0347	46.41	4.41	32.69	16.49
C30H34I209	792.41	792.0296	45.47	4.32	32.03	18.17
C30H35I101	538.52	538.1734	66.91	6.55	23.57	2.97
C30H35I102	554.52	554.1683	64.98	6.36	22.89	5.77
C30H35I103	570.52	570.1633	63.16	6.18	22.24	8.41
C30H35I104	586.52	586.1582	61.44	6.02	21.64	10.91
C30H35I105	602.51	602.1531	59.81	5.86	21.06	13.28
C30H35I106	618.51	618.1480	58.26	5.70	20.52	15.52
C30H35I107	634.51	634.1429	56.79	5.56	20.00	17.65
C30H35I108	650.51	650.1378	55.39	5.42	19.51	19.68
C30H35I109	666.51	666.1327	54.06	5.29	19.04	21.60
C30H36I201	666.43	666.0859	54.07	5.44	38.08	2.40
C30H36I202	682.43	682.0808	52.80	5.32	37.19	4.69
C30H36I203	698.43	698.0757	51.59	5.20	36.34	6.87
C30H36I204	714.43	714.0707	50.44	5.08	35.53	8.96
C30H36I205	730.43	730.0656	49.33	4.97	34.75	10.95
C30H36I206	746.43	746.0605	48.27	4.86	34.00	12.86
C30H36I207	762.43	762.0554	47.26	4.76	33.29	14.69
C30H36I208	778.43	778.0503	46.29	4.66	32.61	16.44
C30H36I209	794.42	794.0452	45.36	4.57	31.95	18.13
C30H37I101	540.53	540.1891	66.66	6.90	23.48	2.96
C30H37I102	556.53	556.1840	64.75	6.70	22.80	5.75
C30H37I103	572.53	572.1789	62.94	6.51	22.17	8.38
C30H37I104	588.53	588.1738	61.23	6.34	21.56	10.87
C30H37I105	604.53	604.1687	59.61	6.17	20.99	13.23
C30H37I106	620.53	620.1637	58.07	6.01	20.45	15.47
C30H37I107	636.53	636.1586	56.61	5.86	19.94	17.59
C30H37I108	652.53	652.1535	55.22	5.72	19.45	19.62
C30H37I109	668.53	668.1484	53.90	5.58	18.98	21.54
C30H38I201	668.45	668.1016	53.91	5.73	37.97	2.39
C30H38I202	684.44	684.0965	52.65	5.60	37.08	4.68
C30H38I203	700.44	700.0914	51.44	5.47	36.24	6.85
C30H38I204	716.44	716.0863	50.29	5.35	35.43	8.93
C30H38I205	732.44	732.0812	49.20	5.23	34.65	10.92
C30H38I206	748.44	748.0761	48.14	5.12	33.91	12.83
C30H38I207	764.44	764.0711	47.14	5.01	33.20	14.65
C30H38I208	780.44	780.0660	46.17	4.91	32.52	16.40
C30H38I209	796.44	796.0609	45.24	4.81	31.87	18.08
C30H39I101	542.55	542.2047	66.42	7.25	23.39	2.95
C30H39I102	558.55	558.1996	64.51	7.04	22.72	5.73
C30H39I103	574.55	574.1946	62.72	6.84	22.09	8.35
C30H39I104	590.55	590.1895	61.02	6.66	21.49	10.84
C30H39I105	606.55	606.1844	59.41	6.48	20.92	13.19
C30H39I106	622.55	622.1793	57.88	6.31	20.38	15.42
C30H39I107	638.55	638.1742	56.43	6.16	19.87	17.54
C30H39I108	654.54	654.1691	55.05	6.01	19.39	19.55
C30H39I109	670.54	670.1640	53.74	5.86	18.93	21.47
C30H40I201	670.46	670.1172	53.74	6.01	37.86	2.39
C30H40I202	686.46	686.1121	52.49	5.87	36.97	4.66
C30H40I203	702.46	702.1070	51.30	5.74	36.13	6.83
C30H40I204	718.46	718.1020	50.15	5.61	35.33	8.91
C30H40I205	734.46	734.0969	49.06	5.49	34.56	10.89
C30H40I206	750.46	750.0918	48.02	5.37	33.82	12.79
C30H40I207	766.46	766.0867	47.01	5.26	33.11	14.61
C30H40I208	782.46	782.0816	46.05	5.15	32.44	16.36
C30H40I209	798.46	798.0765	45.13	5.05	31.79	18.03
C30H41I101	544.57	544.2204	66.17	7.59	23.30	2.94
C30H41I102	560.56	560.2153	64.28	7.37	22.64	5.71
C30H41I103	576.56	576.2102	62.50	7.17	22.01	8.32
C30H41I104	592.56	592.2051	60.81	6.97	21.42	10.80

C30H41I105	608.56	608.2000	59.21	6.79	20.85	13.15
C30H41I106	624.56	624.1950	57.69	6.62	20.32	15.37
C30H41I107	640.56	640.1899	56.25	6.45	19.81	17.48
C30H41I108	656.56	656.1848	54.88	6.29	19.33	19.49
C30H41I109	672.56	672.1797	53.58	6.14	18.87	21.41
C30H42I201	672.48	672.1329	53.58	6.30	37.74	2.38
C30H42I202	688.48	688.1278	52.34	6.15	36.87	4.65
C30H42I203	704.48	704.1227	51.15	6.01	36.03	6.81
C30H42I204	720.48	720.1176	50.01	5.88	35.23	8.88
C30H42I205	736.48	736.1125	48.93	5.75	34.46	10.86
C30H42I206	752.47	752.1074	47.89	5.63	33.73	12.76
C30H42I207	768.47	768.1024	46.89	5.51	33.03	14.57
C30H42I208	784.47	784.0973	45.93	5.40	32.35	16.32
C30H42I209	800.47	800.0922	45.02	5.29	31.71	17.99
C30H43I101	546.58	546.2360	65.93	7.93	23.22	2.93
C30H43I102	562.58	562.2309	64.05	7.70	22.56	5.69
C30H43I103	578.58	578.2259	62.28	7.49	21.93	8.30
C30H43I104	594.58	594.2208	60.60	7.29	21.34	10.76
C30H43I105	610.58	610.2157	59.02	7.10	20.78	13.10
C30H43I106	626.58	626.2106	57.51	6.92	20.25	15.32
C30H43I107	642.58	642.2055	56.08	6.75	19.75	17.43
C30H43I108	658.58	658.2004	54.71	6.58	19.27	19.44
C30H43I109	674.58	674.1953	53.42	6.43	18.81	21.35
C30H44I201	674.49	674.1485	53.42	6.58	37.63	2.37
C30H44I202	690.49	690.1434	52.19	6.42	36.76	4.63
C30H44I203	706.49	706.1383	51.00	6.28	35.93	6.79
C30H44I204	722.49	722.1333	49.87	6.14	35.13	8.86
C30H44I205	738.49	738.1282	48.79	6.01	34.37	10.83
C30H44I206	754.49	754.1231	47.76	5.88	33.64	12.72
C30H44I207	770.49	770.1180	46.77	5.76	32.94	14.54
C30H44I208	786.49	786.1129	45.82	5.64	32.27	16.27
C30H44I209	802.49	802.1078	44.90	5.53	31.63	17.94
C30H45I101	548.60	548.2517	65.68	8.27	23.13	2.92
C30H45I102	564.60	564.2466	63.82	8.03	22.48	5.67
C30H45I103	580.60	580.2415	62.06	7.81	21.86	8.27
C30H45I104	596.60	596.2364	60.40	7.60	21.27	10.73
C30H45I105	612.59	612.2313	58.82	7.40	20.72	13.06
C30H45I106	628.59	628.2263	57.32	7.22	20.19	15.27
C30H45I107	644.59	644.2212	55.90	7.04	19.69	17.37
C30H45I108	660.59	660.2161	54.55	6.87	19.21	19.38
C30H45I109	676.59	676.2110	53.26	6.70	18.76	21.28
C30H46I201	676.51	676.1642	53.26	6.85	37.52	2.36
C30H46I202	692.51	692.1591	52.03	6.70	36.65	4.62
C30H46I203	708.51	708.1540	50.86	6.54	35.82	6.77
C30H46I204	724.51	724.1489	49.74	6.40	35.03	8.83
C30H46I205	740.51	740.1438	48.66	6.26	34.28	10.80
C30H46I206	756.51	756.1387	47.63	6.13	33.55	12.69
C30H46I207	772.51	772.1337	46.64	6.00	32.86	14.50
C30H46I208	788.51	788.1286	45.70	5.88	32.19	16.23
C30H46I209	804.50	804.1235	44.79	5.76	31.55	17.90
C30H47I101	550.61	550.2673	65.44	8.60	23.05	2.91
C30H47I102	566.61	566.2622	63.59	8.36	22.40	5.65
C30H47I103	582.61	582.2572	61.85	8.13	21.78	8.24
C30H47I104	598.61	598.2521	60.20	7.91	21.20	10.69
C30H47I105	614.61	614.2470	58.63	7.71	20.65	13.02
C30H47I106	630.61	630.2419	57.14	7.51	20.12	15.22
C30H47I107	646.61	646.2368	55.73	7.33	19.63	17.32
C30H47I108	662.61	662.2317	54.38	7.15	19.15	19.32
C30H47I109	678.61	678.2266	53.10	6.98	18.70	21.22
C30H48I201	678.53	678.1798	53.11	7.13	37.41	2.36
C30H48I202	694.52	694.1747	51.88	6.97	36.54	4.61
C30H48I203	710.52	710.1696	50.71	6.81	35.72	6.76
C30H48I204	726.52	726.1646	49.60	6.66	34.93	8.81
C30H48I205	742.52	742.1595	48.53	6.52	34.18	10.77
C30H48I206	758.52	758.1544	47.50	6.38	33.46	12.66
C30H48I207	774.52	774.1493	46.52	6.25	32.77	14.46
C30H48I208	790.52	790.1442	45.58	6.12	32.11	16.19
C30H48I209	806.52	806.1391	44.68	6.00	31.47	17.85
C30H49I101	552.63	552.2830	65.20	8.94	22.96	2.90
C30H49I102	568.63	568.2779	63.37	8.69	22.32	5.63
C30H49I103	584.63	584.2728	61.63	8.45	21.71	8.21
C30H49I104	600.63	600.2677	59.99	8.22	21.13	10.66
C30H49I105	616.63	616.2626	58.44	8.01	20.58	12.97
C30H49I106	632.63	632.2576	56.96	7.81	20.06	15.17
C30H49I107	648.63	648.2525	55.55	7.61	19.57	17.27
C30H49I108	664.62	664.2474	54.22	7.43	19.09	19.26
C30H49I109	680.62	680.2423	52.94	7.26	18.65	21.16
C30H50I201	680.54	680.1955	52.95	7.41	37.30	2.35
C30H50I202	696.54	696.1904	51.73	7.24	36.44	4.59
C30H50I203	712.54	712.1853	50.57	7.07	35.62	6.74
C30H50I204	728.54	728.1802	49.46	6.92	34.84	8.78
C30H50I205	744.54	744.1751	48.40	6.77	34.09	10.74
C30H50I206	760.54	760.1700	47.38	6.63	33.37	12.62
C30H50I207	776.54	776.1649	46.40	6.49	32.68	14.42
C30H50I208	792.54	792.1599	45.47	6.36	32.02	16.15
C30H50I209	808.54	808.1548	44.57	6.23	31.39	17.81
C30H51I101	554.64	554.2986	64.97	9.27	22.88	2.88
C30H51I102	570.64	570.2935	63.15	9.01	22.24	5.61
C30H51I103	586.64	586.2885	61.42	8.76	21.63	8.18
C30H51I104	602.64	602.2834	59.79	8.53	21.06	10.62
C30H51I105	618.64	618.2783	58.25	8.31	20.51	12.93
C30H51I106	634.64	634.2732	56.78	8.10	20.00	15.13
C30H51I107	650.64	650.2681	55.38	7.90	19.50	17.21
C30H51I108	666.64	666.2630	54.05	7.71	19.04	19.20
C30H51I109	682.64	682.2579	52.79	7.53	18.59	21.09
C30H52I201	682.56	682.2111	52.79	7.68	37.18	2.34
C30H52I202	698.56	698.2060	51.58	7.50	36.33	4.58
C30H52I203	714.56	714.2009	50.43	7.34	35.52	6.72
C30H52I204	730.56	730.1959	49.32	7.17	34.74	8.76
C30H52I205	746.55	746.1908	48.27	7.02	34.00	10.72
C30H52I206	762.55	762.1857	47.25	6.87	33.28	12.59
C30H52I207	778.55	778.1806	46.28	6.73	32.60	14.39
C30H52I208	794.55	794.1755	45.35	6.60	31.94	16.11
C30H52I209	810.55	810.1704	44.46	6.47	31.31	17.76
C30H53I101	556.66	556.3143	64.73	9.60	22.80	2.87
C30H53I102	572.66	572.3092	62.92	9.33	22.16	5.59
C30H53I103	588.66	588.3041	61.21	9.08	21.56	8.15
C30H53I104	604.66	604.2990	59.59	8.84	20.99	10.58
C30H53I105	620.66	620.2939	58.06	8.61	20.45	12.89
C30H53I106	636.66	636.2888	56.60	8.39	19.93	15.08
C30H53I107	652.66	652.2838	55.21	8.19	19.44	17.16
C30H53I108	668.66	668.2787	53.89	7.99	18.98	19.14
C30H53I109	684.66	684.2736	52.63	7.80	18.54	21.03
C30H54I201	684.57	684.2268	52.64	7.95	37.08	2.34
C30H54I202	700.57	700.2217	51.43	7.77	36.23	4.57
C30H54I203	716.57	716.2166	50.29	7.60	35.42	6.70
C30H54I204	732.57	732.2115	49.19	7.43	34.65	8.74
C30H54I205	748.57	748.2064	48.14	7.27	33.91	10.69
C30H54I206	764.57	764.2013	47.13	7.12	33.20	12.56
C30H54I207	780.57	780.1962	46.16	6.97	32.52	14.35

C30H54I208	796.57	796.1912	45.24	6.83	31.86	16.07
C30H54I209	812.57	812.1861	44.35	6.70	31.24	17.72
C30H55I101	558.68	558.3299	64.50	9.92	22.72	2.86
C30H55I102	574.68	574.3248	62.70	9.65	22.08	5.57
C30H55I103	590.68	590.3198	61.00	9.39	21.48	8.13
C30H55I104	606.67	606.3147	59.39	9.14	20.92	10.55
C30H55I105	622.67	622.3096	57.87	8.90	20.38	12.85
C30H55I106	638.67	638.3045	56.42	8.68	19.87	15.03
C30H55I107	654.67	654.2994	55.04	8.47	19.38	17.11
C30H55I108	670.67	670.2943	53.73	8.27	18.92	19.08
C30H55I109	686.67	686.2892	52.48	8.07	18.48	20.97
C30H56I201	686.59	686.2424	52.48	8.22	36.97	2.33
C30H56I202	702.59	702.2373	51.29	8.03	36.12	4.55
C30H56I203	718.59	718.2322	50.14	7.86	35.32	6.68
C30H56I204	734.59	734.2272	49.05	7.68	34.55	8.71
C30H56I205	750.59	750.2221	48.01	7.52	33.81	10.66
C30H56I206	766.59	766.2170	47.01	7.36	33.11	12.52
C30H56I207	782.59	782.2119	46.04	7.21	32.43	14.31
C30H56I208	798.58	798.2068	45.12	7.07	31.78	16.03
C30H56I209	814.58	814.2017	44.24	6.93	31.16	17.68
C30H57I101	560.69	560.3456	64.27	10.25	22.63	2.85
C30H57I102	576.69	576.3405	62.48	9.96	22.01	5.55
C30H57I103	592.69	592.3354	60.80	9.69	21.41	8.10
C30H57I104	608.69	608.3303	59.20	9.44	20.85	10.51
C30H57I105	624.69	624.3252	57.68	9.20	20.31	12.81
C30H57I106	640.69	640.3201	56.24	8.97	19.81	14.98
C30H57I107	656.69	656.3151	54.87	8.75	19.32	17.05
C30H57I108	672.69	672.3100	53.57	8.54	18.87	19.03
C30H57I109	688.69	688.3049	52.32	8.34	18.43	20.91
C30H58I201	688.60	688.2581	52.33	8.49	36.86	2.32
C30H58I202	704.60	704.2530	51.14	8.30	36.02	4.54
C30H58I203	720.60	720.2479	50.00	8.11	35.22	6.66
C30H58I204	736.60	736.2428	48.92	7.94	34.46	8.69
C30H58I205	752.60	752.2377	47.88	7.77	33.72	10.63
C30H58I206	768.60	768.2326	46.88	7.61	33.02	12.49
C30H58I207	784.60	784.2275	45.93	7.45	32.35	14.27
C30H58I208	800.60	800.2225	45.01	7.30	31.70	15.99
C30H58I209	816.60	816.2174	44.13	7.16	31.08	17.63
C30H59I101	562.71	562.3612	64.04	10.57	22.55	2.84
C30H59I102	578.71	578.3561	62.27	10.28	21.93	5.53
C30H59I103	594.71	594.3511	60.59	10.00	21.34	8.07
C30H59I104	610.71	610.3460	59.00	9.74	20.78	10.48
C30H59I105	626.71	626.3409	57.50	9.49	20.25	12.76
C30H59I106	642.71	642.3358	56.07	9.25	19.75	14.94
C30H59I107	658.70	658.3307	54.70	9.03	19.27	17.00
C30H59I108	674.70	674.3256	53.41	8.81	18.81	18.97
C30H59I109	690.70	690.3205	52.17	8.61	18.37	20.85
C30H60I201	690.62	690.2737	52.18	8.76	36.75	2.32
C30H60I202	706.62	706.2686	50.99	8.56	35.92	4.53
C30H60I203	722.62	722.2635	49.87	8.37	35.12	6.64
C30H60I204	738.62	738.2585	48.78	8.19	34.36	8.66
C30H60I205	754.62	754.2534	47.75	8.01	33.63	10.60
C30H60I206	770.62	770.2483	46.76	7.85	32.94	12.46
C30H60I207	786.62	786.2432	45.81	7.69	32.27	14.24
C30H60I208	802.62	802.2381	44.89	7.54	31.62	15.95
C30H60I209	818.62	818.2330	44.02	7.39	31.00	17.59
C30H61I101	564.72	564.3769	63.81	10.89	22.47	2.83
C30H61I102	580.72	580.3718	62.05	10.59	21.85	5.51
C30H61I103	596.72	596.3667	60.39	10.30	21.27	8.04
C30H61I104	612.72	612.3616	58.81	10.03	20.71	10.44
C30H61I105	628.72	628.3565	57.31	9.78	20.18	12.72
C30H61I106	644.72	644.3514	55.89	9.54	19.68	14.89
C30H61I107	660.72	660.3464	54.54	9.31	19.21	16.95
C30H61I108	676.72	676.3413	53.25	9.09	18.75	18.91
C30H61I109	692.72	692.3362	52.02	8.88	18.32	20.79

9. CARBON, HYDROGEN, NITROGEN AND OXYGEN

9. KOHLENSTOFF, WASSERSTOFF, STICKSTOFF UND SAUERSTOFF

9. CARBONE, HYDROGENE, AZOTE ET OXYGENE

MOLECULAR WEIGHTS AND PERCENTAGE COMPOSITIONS OF COMPOUNDS
CONTAINING CARBON, HYDROGEN, NITROGEN
AND OXYGEN

FORMULA	MOL. WEIGHT	ISOMOL. WEIGHT	PERC. C	PERC. H	PERC. N	PERC. O
C 1H 0N201	56.02	56.0011	21.44	0	50.00	28.56
C 1H 0N202	72.02	71.9960	16.68	0	38.89	44.43
C 1H 0N203	88.02	87.9909	13.65	0	31.83	54.53
C 1H 0N204	104.02	103.9858	11.55	0	26.93	61.52
C 1H 0N205	120.02	119.9807	10.01	0	23.34	66.65
C 1H 0N206	136.02	135.9756	8.83	0	20.59	70.57
C 1H 0N207	152.02	151.9705	7.90	0	18.43	73.67
C 1H 0N208	168.02	167.9655	7.15	0	16.67	76.18
C 1H 0N209	184.02	183.9604	6.53	0	15.22	78.25
C 1H 0N401	84.04	84.0072	14.29	0	66.67	19.04
C 1H 0N402	100.04	100.0021	12.01	0	56.01	31.99
C 1H 0N403	116.04	115.9970	10.35	0	48.28	41.36
C 1H 0N404	132.04	131.9920	9.10	0	42.43	48.47
C 1H 0N405	148.03	147.9869	8.11	0	37.85	54.04
C 1H 0N406	164.03	163.9818	7.32	0	34.16	58.52
C 1H 0N407	180.03	179.9767	6.67	0	31.12	62.21
C 1H 0N408	196.03	195.9716	6.13	0	28.58	65.29
C 1H 0N409	212.03	211.9665	5.66	0	26.42	67.91
C 1H 0N601	112.05	112.0134	10.72	0	75.00	14.28
C 1H 0N602	128.05	128.0083	9.38	0	65.63	24.99
C 1H 0N603	144.05	144.0032	8.34	0	58.34	33.32
C 1H 0N604	160.05	159.9981	7.50	0	52.51	39.99
C 1H 0N605	176.05	175.9930	6.82	0	47.74	45.44
C 1H 0N606	192.05	191.9879	6.25	0	43.76	49.99
C 1H 0N607	208.05	207.9828	5.77	0	40.39	53.83
C 1H 0N608	224.05	223.9778	5.36	0	37.51	57.13
C 1H 0N609	240.05	239.9727	5.00	0	35.01	59.99
C 1H 1N101	43.03	43.0058	27.92	2.34	32.55	37.19
C 1H 1N102	59.02	59.0007	20.35	1.71	23.73	54.21
C 1H 1N103	75.02	74.9956	16.01	1.34	18.67	63.98
C 1H 1N104	91.02	90.9906	13.20	1.11	15.39	70.31
C 1H 1N105	107.02	106.9855	11.22	.94	13.09	74.75
C 1H 1N106	123.02	122.9804	9.76	.82	11.39	78.03
C 1H 1N107	139.02	138.9753	8.64	.73	10.08	80.56
C 1H 1N108	155.02	154.9702	7.75	.65	9.04	82.57
C 1H 1N109	171.02	170.9651	7.02	.59	8.19	84.20
C 1H 1N301	71.04	71.0120	16.91	1.42	59.15	22.52
C 1H 1N302	87.04	87.0069	13.80	1.16	48.28	36.76
C 1H 1N303	103.04	103.0018	11.66	.98	40.78	46.58
C 1H 1N304	119.04	118.9967	10.09	.85	35.30	53.76
C 1H 1N305	135.04	134.9916	8.89	.75	31.12	59.24
C 1H 1N306	151.04	150.9865	7.95	.67	27.82	63.56
C 1H 1N307	167.04	166.9814	7.19	.60	25.16	67.05
C 1H 1N308	183.03	182.9764	6.56	.55	22.96	69.93
C 1H 1N309	199.03	198.9713	6.03	.51	21.11	72.35
C 1H 1N501	99.05	99.0181	12.13	1.02	70.70	16.15
C 1H 1N502	115.05	115.0130	10.44	.88	60.87	27.81
C 1H 1N503	131.05	131.0079	9.17	.77	53.44	36.63
C 1H 1N504	147.05	147.0028	8.17	.69	47.63	43.52
C 1H 1N505	163.05	162.9978	7.37	.62	42.95	49.06
C 1H 1N506	179.05	178.9927	6.71	.56	39.11	53.61
C 1H 1N507	195.05	194.9876	6.16	.52	35.91	57.42
C 1H 1N508	211.05	210.9825	5.69	.48	33.18	60.65
C 1H 1N509	227.05	226.9774	5.29	.44	30.85	63.42
C 1H 2N201	58.04	58.0167	20.69	3.47	48.27	27.57
C 1H 2N202	74.04	74.0116	16.22	2.72	37.84	43.22
C 1H 2N203	90.04	90.0065	13.34	2.24	31.11	53.31
C 1H 2N204	106.04	106.0015	11.33	1.90	26.42	60.35
C 1H 2N205	122.04	121.9964	9.84	1.65	22.95	65.55
C 1H 2N206	138.04	137.9913	8.70	1.46	20.29	69.54
C 1H 2N207	154.04	153.9862	7.80	1.31	18.19	72.71
C 1H 2N208	170.04	169.9811	7.06	1.19	16.48	75.28
C 1H 2N209	186.04	185.9760	6.46	1.08	15.06	77.40
C 1H 2N401	86.05	86.0229	13.96	2.34	65.11	18.59
C 1H 2N402	102.05	102.0178	11.77	1.98	54.90	31.36
C 1H 2N403	118.05	118.0127	10.17	1.71	47.46	40.66
C 1H 2N404	134.05	134.0076	8.96	1.50	41.79	47.74
C 1H 2N405	150.05	150.0025	8.00	1.34	37.34	53.31
C 1H 2N406	166.05	165.9974	7.23	1.21	33.74	57.81
C 1H 2N407	182.05	181.9923	6.60	1.11	30.78	61.52
C 1H 2N408	198.05	197.9873	6.06	1.02	28.29	64.63
C 1H 2N409	214.05	213.9822	5.61	.94	26.17	67.27
C 1H 2N601	114.07	114.0290	10.53	1.77	73.68	14.03
C 1H 2N602	130.07	130.0239	9.23	1.55	64.61	24.60
C 1H 2N603	146.07	146.0188	8.22	1.38	57.54	32.86
C 1H 2N604	162.06	162.0137	7.41	1.24	51.86	39.49
C 1H 2N605	178.06	178.0087	6.75	1.13	47.20	44.93
C 1H 2N606	194.06	194.0036	6.19	1.04	43.31	49.47
C 1H 2N607	210.06	209.9985	5.72	.96	40.01	53.32
C 1H 2N608	226.06	225.9934	5.31	.89	37.18	56.62
C 1H 2N609	242.06	241.9883	4.96	.83	34.72	59.49
C 1H 3N101	45.04	45.0215	26.67	6.71	31.10	35.52
C 1H 3N102	61.04	61.0164	19.68	4.95	22.95	52.42
C 1H 3N103	77.04	77.0113	15.59	3.93	18.18	62.30
C 1H 3N104	93.04	93.0062	12.91	3.25	15.05	68.79
C 1H 3N105	109.04	109.0011	11.02	2.77	12.85	73.37
C 1H 3N106	125.04	124.9960	9.61	2.42	11.20	76.77
C 1H 3N107	141.04	140.9909	8.52	2.14	9.93	79.41
C 1H 3N108	157.04	156.9859	7.65	1.93	8.92	81.51
C 1H 3N109	173.04	172.9808	6.94	1.75	8.09	83.22
C 1H 3N301	73.05	73.0276	16.44	4.14	57.52	21.90
C 1H 3N302	89.05	89.0225	13.49	3.40	47.18	35.93
C 1H 3N303	105.05	105.0174	11.43	2.88	40.00	45.69
C 1H 3N304	121.05	121.0124	9.92	2.50	34.71	52.87
C 1H 3N305	137.05	137.0073	8.76	2.21	30.66	58.37
C 1H 3N306	153.05	153.0022	7.85	1.98	27.45	62.72
C 1H 3N307	169.05	168.9971	7.11	1.79	24.86	66.25
C 1H 3N308	185.05	184.9920	6.49	1.63	22.71	69.17
C 1H 3N309	201.05	200.9869	5.97	1.50	20.90	71.62
C 1H 3N501	101.07	101.0338	11.88	2.99	69.29	15.83

017

FORMULA	MOL. WEIGHT	ISOMOL. WEIGHT	PERC. C	PERC. H	PERC. N	PERC. O
C 1H 3N502	117.07	117.0287	10.26	2.58	59.82	27.33
C 1H 3N503	133.07	133.0236	9.03	2.27	52.63	36.07
C 1H 3N504	149.07	149.0185	8.06	2.03	46.98	42.93
C 1H 3N505	165.07	165.0134	7.28	1.83	42.43	48.46
C 1H 3N506	181.06	181.0083	6.63	1.67	38.68	53.02
C 1H 3N507	197.06	197.0032	6.10	1.53	35.54	56.83
C 1H 3N508	213.06	212.9982	5.64	1.42	32.87	60.07
C 1H 3N509	229.06	228.9931	5.24	1.32	30.57	62.86
C 1H 4N201	60.06	60.0324	20.00	6.71	46.65	26.64
C 1H 4N202	76.06	76.0273	15.79	5.30	36.83	42.07
C 1H 4N203	92.05	92.0222	13.05	4.38	30.43	52.14
C 1H 4N204	108.05	108.0171	11.12	3.73	25.93	59.23
C 1H 4N205	124.05	124.0120	9.68	3.25	22.58	64.49
C 1H 4N206	140.05	140.0069	8.58	2.88	20.00	68.54
C 1H 4N207	156.05	156.0018	7.70	2.58	17.95	71.77
C 1H 4N208	172.05	171.9968	6.98	2.34	16.28	74.39
C 1H 4N209	188.05	187.9917	6.39	2.14	14.90	76.57
C 1H 4N401	88.07	88.0385	13.64	4.58	63.62	18.17
C 1H 4N402	104.07	104.0334	11.54	3.87	53.84	30.75
C 1H 4N403	120.07	120.0283	10.00	3.36	46.66	39.98
C 1H 4N404	136.07	136.0232	8.83	2.96	41.18	47.03
C 1H 4N405	152.07	152.0182	7.90	2.65	36.84	52.61
C 1H 4N406	168.07	168.0131	7.15	2.40	33.34	57.12
C 1H 4N407	184.07	184.0080	6.53	2.19	30.44	60.85
C 1H 4N408	200.07	200.0029	6.00	2.02	28.00	63.98
C 1H 4N409	216.06	215.9978	5.56	1.87	25.93	66.64
C 1H 4N601	116.08	116.0447	10.35	3.47	72.40	13.78
C 1H 4N602	132.08	132.0396	9.09	3.05	63.63	24.23
C 1H 4N603	148.08	148.0345	8.11	2.72	56.75	32.41
C 1H 4N604	164.08	164.0294	7.32	2.46	51.22	39.00
C 1H 4N605	180.08	180.0243	6.67	2.24	46.67	44.42
C 1H 4N606	196.08	196.0192	6.13	2.06	42.86	48.96
C 1H 4N607	212.08	212.0141	5.66	1.90	39.63	52.81
C 1H 4N608	228.08	228.0091	5.27	1.77	36.85	56.12
C 1H 4N609	244.08	244.0040	4.92	1.65	34.43	59.00
C 1H 5N101	47.06	47.0371	25.52	10.71	29.77	34.00
C 1H 5N102	63.06	63.0320	19.05	7.99	22.21	50.75
C 1H 5N103	79.06	79.0269	15.19	6.38	17.72	60.71
C 1H 5N104	95.06	95.0219	12.64	5.30	14.74	67.33
C 1H 5N105	111.05	111.0168	10.82	4.54	12.61	72.03
C 1H 5N106	127.05	127.0117	9.45	3.97	11.02	75.56
C 1H 5N107	143.05	143.0066	8.40	3.52	9.79	78.29
C 1H 5N108	159.05	159.0015	7.55	3.17	8.81	80.47
C 1H 5N109	175.05	174.9964	6.86	2.88	8.00	82.26
C 1H 5N301	75.07	75.0433	16.00	6.71	55.97	21.31
C 1H 5N302	91.07	91.0382	13.19	5.53	46.14	35.14
C 1H 5N303	107.07	107.0331	11.22	4.71	39.25	44.83
C 1H 5N304	123.07	123.0280	9.76	4.10	34.14	52.00
C 1H 5N305	139.07	139.0229	8.64	3.62	30.22	57.52
C 1H 5N306	155.07	155.0178	7.75	3.25	27.10	61.91
C 1H 5N307	171.07	171.0127	7.02	2.95	24.56	65.47
C 1H 5N308	187.07	187.0077	6.42	2.69	22.46	68.42
C 1H 5N309	203.07	203.0026	5.91	2.48	20.69	70.91
C 1H 5N501	103.08	103.0494	11.65	4.89	67.94	15.52
C 1H 5N502	119.08	119.0443	10.09	4.23	58.81	26.87
C 1H 5N503	135.08	135.0392	8.89	3.73	51.84	35.53
C 1H 5N504	151.08	151.0341	7.95	3.34	46.35	42.36
C 1H 5N505	167.08	167.0291	7.19	3.02	41.92	47.88
C 1H 5N506	183.08	183.0240	6.56	2.75	38.25	52.43
C 1H 5N507	199.08	199.0189	6.03	2.53	35.18	56.26
C 1H 5N508	215.08	215.0138	5.58	2.34	32.56	59.51
C 1H 5N509	231.08	231.0087	5.20	2.18	30.31	62.31
C 1H 6N201	62.07	62.0480	19.35	9.74	45.13	25.78
C 1H 6N202	78.07	78.0429	15.38	7.75	35.88	40.99
C 1H 6N203	94.07	94.0378	12.77	6.43	29.78	51.02
C 1H 6N204	110.07	110.0328	10.91	5.49	25.45	58.14
C 1H 6N205	126.07	126.0277	9.53	4.80	22.22	63.45
C 1H 6N206	142.07	142.0226	8.45	4.26	19.72	67.57
C 1H 6N207	158.07	158.0175	7.60	3.83	17.72	70.85
C 1H 6N208	174.07	174.0124	6.90	3.47	16.09	73.53
C 1H 6N209	190.07	190.0073	6.32	3.18	14.74	75.76
C 1H 6N401	90.09	90.0542	13.33	6.71	62.19	17.76
C 1H 6N402	106.08	106.0491	11.32	5.70	52.81	30.16
C 1H 6N403	122.08	122.0440	9.84	4.95	45.89	39.32
C 1H 6N404	138.08	138.0389	8.70	4.38	40.57	46.35
C 1H 6N405	154.08	154.0338	7.80	3.93	36.36	51.92
C 1H 6N406	170.08	170.0287	7.06	3.56	32.94	56.44
C 1H 6N407	186.08	186.0236	6.45	3.25	30.11	60.19
C 1H 6N408	202.08	202.0186	5.94	2.99	27.72	63.34
C 1H 6N409	218.08	218.0135	5.51	2.77	25.69	66.03
C 1H 6N601	118.10	118.0603	10.17	5.12	71.16	13.55
C 1H 6N602	134.10	134.0552	8.96	4.51	62.67	23.86
C 1H 6N603	150.10	150.0501	8.00	4.03	55.99	31.98
C 1H 6N604	166.10	166.0450	7.23	3.64	50.60	38.53
C 1H 6N605	182.10	182.0400	6.60	3.32	46.15	43.93
C 1H 6N606	198.10	198.0349	6.06	3.05	42.42	48.46
C 1H 6N607	214.09	214.0298	5.61	2.82	39.25	52.31
C 1H 6N608	230.09	230.0247	5.22	2.63	36.52	55.63
C 1H 6N609	246.09	246.0196	4.88	2.46	34.15	58.51
C 1H 7N301	77.09	77.0589	15.58	9.15	54.51	20.76
C 1H 7N302	93.09	93.0538	12.90	7.58	45.14	34.38
C 1H 7N303	109.09	109.0487	11.01	6.47	38.52	44.00
C 1H 7N304	125.08	125.0436	9.60	5.64	33.59	51.16
C 1H 7N305	141.08	141.0386	8.51	5.00	29.78	56.70
C 1H 7N306	157.08	157.0335	7.65	4.49	26.75	61.11
C 1H 7N307	173.08	173.0284	6.94	4.08	24.28	64.71
C 1H 7N308	189.08	189.0233	6.35	3.73	22.22	67.69
C 1H 7N309	205.08	205.0182	5.86	3.44	20.49	70.21
C 1H 7N501	105.10	105.0651	11.43	6.71	66.64	15.22
C 1H 7N502	121.10	121.0600	9.92	5.83	57.83	26.42
C 1H 7N503	137.10	137.0549	8.76	5.15	51.08	35.01
C 1H 7N504	153.10	153.0498	7.85	4.61	45.74	41.80
C 1H 7N505	169.10	169.0447	7.10	4.17	41.42	47.31
C 1H 7N506	185.10	185.0396	6.49	3.81	37.84	51.86
C 1H 7N507	201.10	201.0345	5.97	3.51	34.83	55.69
C 1H 7N508	217.10	217.0295	5.53	3.25	32.26	58.96
C 1H 7N509	233.10	233.0244	5.15	3.03	30.05	61.78
C 1H 8N401	92.10	92.0698	13.04	8.76	60.83	17.37
C 1H 8N402	108.10	108.0647	11.11	7.46	51.83	29.60
C 1H 8N403	124.10	124.0596	9.68	6.50	45.15	38.68
C 1H 8N404	140.10	140.0545	8.57	5.76	39.99	45.68
C 1H 8N405	156.10	156.0495	7.69	5.17	35.89	51.25
C 1H 8N406	172.10	172.0444	6.98	4.69	32.56	55.78
C 1H 8N407	188.10	188.0393	6.39	4.29	29.79	59.54
C 1H 8N408	204.10	204.0342	5.89	3.95	27.45	62.71
C 1H 8N409	220.10	220.0291	5.46	3.66	25.46	65.42
C 1H 8N601	120.11	120.0760	10.00	6.71	69.97	13.32
C 1H 8N602	136.11	136.0709	8.82	5.92	61.74	23.51
C 1H 8N603	152.11	152.0658	7.90	5.30	55.25	31.55
C 1H 8N604	168.11	168.0607	7.14	4.80	49.99	38.07

018

C 1H 8N605	184.11	184.0556	6.52	4.38	45.65	43.45
C 1H 8N606	200.11	200.0505	6.00	4.03	42.00	47.97
C 1H 8N607	216.11	216.0454	5.56	3.73	38.89	51.82
C 1H 8N608	232.11	232.0404	5.17	3.47	36.21	55.14
C 1H 8N609	248.11	248.0353	4.84	3.25	33.87	58.04
C 1H 9N501	107.12	107.0807	11.21	8.47	65.38	14.94
C 1H 9N502	123.12	123.0756	9.76	7.37	56.88	25.99
C 1H 9N503	139.11	139.0705	8.63	6.52	50.34	34.50
C 1H 9N504	155.11	155.0654	7.74	5.85	45.15	41.26
C 1H 9N505	171.11	171.0604	7.02	5.30	40.93	46.75
C 1H 9N506	187.11	187.0553	6.42	4.85	37.43	51.30
C 1H 9N507	203.11	203.0502	5.91	4.47	34.48	55.14
C 1H 9N508	219.11	219.0451	5.48	4.14	31.96	58.42
C 1H 9N509	235.11	235.0400	5.11	3.86	29.79	61.25
C 1H10N601	122.13	122.0916	9.83	8.25	68.81	13.10
C 1H10N602	138.13	138.0865	8.70	7.30	60.84	23.17
C 1H10N603	154.13	154.0814	7.79	6.54	54.53	31.14
C 1H10N604	170.13	170.0763	7.06	5.92	49.40	37.62
C 1H10N605	186.13	186.0713	6.45	5.42	45.15	42.98
C 1H10N606	202.13	202.0662	5.94	4.99	41.58	47.49
C 1H10N607	218.13	218.0611	5.51	4.62	38.53	51.34
C 1H10N608	234.13	234.0560	5.13	4.31	35.90	54.67
C 1H10N609	250.13	250.0509	4.80	4.03	33.60	57.57
C 2H 1N101	55.04	55.0058	43.65	1.83	25.45	29.07
C 2H 1N102	71.04	71.0007	33.82	1.42	19.72	45.05
C 2H 1N103	87.04	86.9956	27.60	1.16	16.09	55.15
C 2H 1N104	103.03	102.9906	23.31	.98	13.59	62.11
C 2H 1N105	119.03	118.9855	20.18	.85	11.77	67.21
C 2H 1N106	135.03	134.9804	17.79	.75	10.37	71.09
C 2H 1N107	151.03	150.9753	15.91	.67	9.27	74.15
C 2H 1N108	167.03	166.9702	14.38	.60	8.39	76.63
C 2H 1N109	183.03	182.9651	13.12	.55	7.65	78.67
C 2H 1N301	83.05	83.0120	28.93	1.21	50.60	19.26
C 2H 1N302	99.05	99.0069	24.25	1.02	42.42	32.31
C 2H 1N303	115.05	115.0018	20.88	.88	36.52	41.72
C 2H 1N304	131.05	130.9967	18.33	.77	32.06	48.84
C 2H 1N305	147.05	146.9916	16.34	.69	28.58	54.40
C 2H 1N306	163.05	162.9865	14.73	.62	25.77	58.88
C 2H 1N307	179.05	178.9814	13.42	.56	23.47	62.55
C 2H 1N308	195.05	194.9764	12.32	.52	21.54	65.62
C 2H 1N309	211.04	210.9713	11.38	.48	19.91	68.23
C 2H 1N501	111.06	111.0181	21.63	.91	63.06	14.41
C 2H 1N502	127.06	127.0130	18.91	.79	55.12	25.18
C 2H 1N503	143.06	143.0079	16.79	.70	48.95	33.55
C 2H 1N504	159.06	159.0028	15.10	.63	44.03	40.23
C 2H 1N505	175.06	174.9978	13.72	.58	40.01	45.70
C 2H 1N506	191.06	190.9927	12.57	.53	36.66	50.24
C 2H 1N507	207.06	206.9876	11.60	.49	33.82	54.09
C 2H 1N508	223.06	222.9825	10.77	.45	31.40	57.38
C 2H 1N509	239.06	238.9774	10.05	.42	29.30	60.23
C 2H 2N201	70.05	70.0167	34.29	2.88	39.99	22.84
C 2H 2N202	86.05	86.0116	27.92	2.34	32.55	37.19
C 2H 2N203	102.05	102.0065	23.54	1.98	27.45	47.03
C 2H 2N204	118.05	118.0015	20.35	1.71	23.73	54.21
C 2H 2N205	134.05	133.9964	17.92	1.50	20.90	59.68
C 2H 2N206	150.05	149.9913	16.01	1.34	18.67	63.98
C 2H 2N207	166.05	165.9862	14.47	1.21	16.87	67.45
C 2H 2N208	182.05	181.9811	13.20	1.11	15.39	70.31
C 2H 2N209	198.05	197.9760	12.13	1.02	14.14	72.71
C 2H 2N401	98.06	98.0229	24.50	2.06	57.13	16.32

C 2H 2N402	114.06	114.0178	21.06	1.77	49.12	28.05
C 2H 2N403	130.06	130.0127	18.47	1.55	43.08	36.90
C 2H 2N404	146.06	146.0076	16.45	1.38	38.36	43.82
C 2H 2N405	162.06	162.0025	14.82	1.24	34.57	49.36
C 2H 2N406	178.06	177.9974	13.49	1.13	31.46	53.91
C 2H 2N407	194.06	193.9923	12.38	1.04	28.87	57.71
C 2H 2N408	210.06	209.9873	11.44	.96	26.67	60.93
C 2H 2N409	226.06	225.9822	10.63	.89	24.78	63.70
C 2H 2N601	126.08	126.0290	19.05	1.60	66.66	12.69
C 2H 2N602	142.08	142.0239	16.91	1.42	59.15	22.52
C 2H 2N603	158.08	158.0188	15.20	1.28	53.16	30.36
C 2H 2N604	174.08	174.0137	13.80	1.16	48.28	36.76
C 2H 2N605	190.08	190.0087	12.64	1.06	44.21	42.09
C 2H 2N606	206.07	206.0036	11.66	.98	40.78	46.58
C 2H 2N607	222.07	221.9985	10.82	.91	37.84	50.43
C 2H 2N608	238.07	237.9934	10.09	.85	35.30	53.76
C 2H 2N609	254.07	253.9883	9.45	.79	33.08	56.67
C 2H 3N101	57.05	57.0215	42.11	5.30	24.55	28.04
C 2H 3N102	73.05	73.0164	32.88	4.14	19.17	43.80
C 2H 3N103	89.05	89.0113	26.98	3.40	15.73	53.90
C 2H 3N104	105.05	105.0062	22.87	2.88	13.33	60.92
C 2H 3N105	121.05	121.0011	19.84	2.50	11.57	66.09
C 2H 3N106	137.05	136.9960	17.53	2.21	10.22	70.05
C 2H 3N107	153.05	152.9909	15.70	1.98	9.15	73.18
C 2H 3N108	169.05	168.9859	14.21	1.79	8.29	75.72
C 2H 3N109	185.05	184.9808	12.98	1.63	7.57	77.81
C 2H 3N301	85.07	85.0276	28.24	3.55	49.40	18.81
C 2H 3N302	101.07	101.0225	23.77	2.99	41.58	31.66
C 2H 3N303	117.06	117.0174	20.52	2.58	35.89	41.00
C 2H 3N304	133.06	133.0124	18.05	2.27	31.58	48.10
C 2H 3N305	149.06	149.0073	16.12	2.03	28.19	53.67
C 2H 3N306	165.06	165.0022	14.55	1.83	25.46	58.16
C 2H 3N307	181.06	180.9971	13.27	1.67	23.21	61.85
C 2H 3N308	197.06	196.9920	12.19	1.53	21.32	64.95
C 2H 3N309	213.06	212.9869	11.27	1.42	19.72	67.58
C 2H 3N501	113.08	113.0338	21.24	2.67	61.93	14.15
C 2H 3N502	129.08	129.0287	18.61	2.34	54.26	24.79
C 2H 3N503	145.08	145.0236	16.56	2.08	48.27	33.08
C 2H 3N504	161.08	161.0185	14.91	1.88	43.48	39.73
C 2H 3N505	177.08	177.0134	13.57	1.71	39.55	45.18
C 2H 3N506	193.08	193.0083	12.44	1.57	36.27	49.72
C 2H 3N507	209.08	209.0032	11.49	1.45	33.50	53.57
C 2H 3N508	225.07	224.9982	10.67	1.34	31.12	56.87
C 2H 3N509	241.07	240.9931	9.96	1.25	29.05	59.73
C 2H 4N201	72.07	72.0324	33.33	5.59	38.87	22.20
C 2H 4N202	88.07	88.0273	27.28	4.58	31.81	36.33
C 2H 4N203	104.07	104.0222	23.08	3.87	26.92	46.12
C 2H 4N204	120.07	120.0171	20.01	3.36	23.33	53.30
C 2H 4N205	136.06	136.0120	17.66	2.96	20.59	58.79
C 2H 4N206	152.06	152.0069	15.80	2.65	18.42	63.13
C 2H 4N207	168.06	168.0018	14.29	2.40	16.67	66.64
C 2H 4N208	184.06	183.9968	13.05	2.19	15.22	69.54
C 2H 4N209	200.06	199.9917	12.01	2.02	14.00	71.97
C 2H 4N401	100.08	100.0385	24.00	4.03	55.98	15.99
C 2H 4N402	116.08	116.0334	20.69	3.47	48.27	27.57
C 2H 4N403	132.08	132.0283	18.19	3.05	42.42	36.34
C 2H 4N404	148.08	148.0232	16.22	2.72	37.84	43.22
C 2H 4N405	164.08	164.0182	14.64	2.46	34.15	48.76
C 2H 4N406	180.08	180.0131	13.34	2.24	31.11	53.31
C 2H 4N407	196.08	196.0080	12.25	2.06	28.57	57.12

019

C 2H 4N408	212.08	212.0029	11.33	1.90	26.42	60.35
C 2H 4N409	228.08	227.9978	10.53	1.77	24.57	63.13
C 2H 4N601	128.09	128.0447	18.75	3.15	65.61	12.49
C 2H 4N602	144.09	144.0396	16.67	2.80	58.32	22.21
C 2H 4N603	160.09	160.0345	15.01	2.52	52.49	29.98
C 2H 4N604	176.09	176.0294	13.64	2.29	47.73	36.34
C 2H 4N605	192.09	192.0243	12.51	2.10	43.75	41.65
C 2H 4N606	208.09	208.0192	11.54	1.94	40.39	46.13
C 2H 4N607	224.09	224.0141	10.72	1.80	37.50	49.98
C 2H 4N608	240.09	240.0091	10.01	1.68	35.00	53.31
C 2H 4N609	256.09	256.0040	9.38	1.57	32.82	56.23
C 2H 5N101	59.07	59.0371	40.67	8.53	23.71	27.09
C 2H 5N102	75.07	75.0320	32.00	6.71	18.66	42.63
C 2H 5N103	91.07	91.0269	26.38	5.53	15.38	52.71
C 2H 5N104	107.07	107.0219	22.44	4.71	13.08	59.77
C 2H 5N105	123.07	123.0168	19.52	4.10	11.38	65.00
C 2H 5N106	139.07	139.0117	17.27	3.62	10.07	69.03
C 2H 5N107	155.06	155.0066	15.49	3.25	9.03	72.23
C 2H 5N108	171.06	171.0015	14.04	2.95	8.19	74.82
C 2H 5N109	187.06	186.9964	12.84	2.69	7.49	76.98
C 2H 5N301	87.08	87.0433	27.59	5.79	48.25	18.37
C 2H 5N302	103.08	103.0382	23.30	4.89	40.76	31.04
C 2H 5N303	119.08	119.0331	20.17	4.23	35.29	40.31
C 2H 5N304	135.08	135.0280	17.78	3.73	31.11	47.38
C 2H 5N305	151.08	151.0229	15.90	3.34	27.81	52.95
C 2H 5N306	167.08	167.0178	14.38	3.02	25.15	57.46
C 2H 5N307	183.08	183.0127	13.12	2.75	22.95	61.17
C 2H 5N308	199.08	199.0077	12.07	2.53	21.11	64.29
C 2H 5N309	215.08	215.0026	11.17	2.34	19.54	66.95
C 2H 5N501	115.10	115.0494	20.87	4.38	60.85	13.90
C 2H 5N502	131.09	131.0443	18.32	3.84	53.42	24.41
C 2H 5N503	147.09	147.0392	16.33	3.43	47.61	32.63
C 2H 5N504	163.09	163.0341	14.73	3.09	42.94	39.24
C 2H 5N505	179.09	179.0291	13.41	2.81	39.10	44.67
C 2H 5N506	195.09	195.0240	12.31	2.58	35.90	49.21
C 2H 5N507	211.09	211.0189	11.38	2.39	33.18	53.06
C 2H 5N508	227.09	227.0138	10.58	2.22	30.84	56.36
C 2H 5N509	243.09	243.0087	9.88	2.07	28.81	59.24
C 2H 6N201	74.08	74.0480	32.43	8.16	37.81	21.60
C 2H 6N202	90.08	90.0429	26.67	6.71	31.10	35.52
C 2H 6N203	106.08	106.0378	22.65	5.70	26.41	45.25
C 2H 6N204	122.08	122.0328	19.68	4.95	22.95	52.42
C 2H 6N205	138.08	138.0277	17.40	4.38	20.29	57.94
C 2H 6N206	154.08	154.0226	15.59	3.93	18.18	62.30
C 2H 6N207	170.08	170.0175	14.12	3.56	16.47	65.85
C 2H 6N208	186.08	186.0124	12.91	3.25	15.05	68.79
C 2H 6N209	202.08	202.0073	11.89	2.99	13.86	71.26
C 2H 6N401	102.10	102.0542	23.53	5.92	54.88	15.67
C 2H 6N402	118.10	118.0491	20.34	5.12	47.44	27.10
C 2H 6N403	134.10	134.0440	17.91	4.51	41.78	35.79
C 2H 6N404	150.09	150.0389	16.00	4.03	37.33	42.64
C 2H 6N405	166.09	166.0338	14.46	3.64	33.73	48.16
C 2H 6N406	182.09	182.0287	13.19	3.32	30.77	52.72
C 2H 6N407	198.09	198.0236	12.13	3.05	28.28	56.54
C 2H 6N408	214.09	214.0186	11.22	2.82	26.17	59.79
C 2H 6N409	230.09	230.0135	10.44	2.63	24.35	62.58
C 2H 6N601	130.11	130.0603	18.46	4.65	64.59	12.30
C 2H 6N602	146.11	146.0552	16.44	4.14	57.52	21.90
C 2H 6N603	162.11	162.0501	14.82	3.73	51.84	29.61
C 2H 6N604	178.11	178.0450	13.49	3.40	47.18	35.93

C 2H 6N605	194.11	194.0400	12.38	3.12	43.30	41.21
C 2H 6N606	210.11	210.0349	11.43	2.88	40.00	45.69
C 2H 6N607	226.11	226.0298	10.62	2.67	37.17	49.53
C 2H 6N608	242.11	242.0247	9.92	2.50	34.71	52.87
C 2H 6N609	258.10	258.0196	9.31	2.34	32.56	55.79
C 2H 7N101	61.08	61.0528	39.33	11.55	22.93	26.19
C 2H 7N102	77.08	77.0477	31.16	9.15	18.17	41.51
C 2H 7N103	93.08	93.0426	25.81	7.58	15.05	51.56
C 2H 7N104	109.08	109.0375	22.02	6.47	12.84	58.67
C 2H 7N105	125.08	125.0324	19.21	5.64	11.20	63.96
C 2H 7N106	141.08	141.0273	17.03	5.00	9.93	68.04
C 2H 7N107	157.08	157.0222	15.29	4.49	8.92	71.30
C 2H 7N108	173.08	173.0172	13.88	4.08	8.09	73.95
C 2H 7N109	189.08	189.0121	12.70	3.73	7.41	76.16
C 2H 7N301	89.10	89.0589	26.96	7.92	47.16	17.96
C 2H 7N302	105.10	105.0538	22.86	6.71	39.98	30.45
C 2H 7N303	121.10	121.0487	19.84	5.83	34.70	39.64
C 2H 7N304	137.10	137.0436	17.52	5.15	30.65	46.68
C 2H 7N305	153.10	153.0386	15.69	4.61	27.45	52.25
C 2H 7N306	169.09	169.0335	14.21	4.17	24.85	56.77
C 2H 7N307	185.09	185.0284	12.98	3.81	22.70	60.51
C 2H 7N308	201.09	201.0233	11.95	3.51	20.90	63.65
C 2H 7N309	217.09	217.0182	11.07	3.25	19.36	66.33
C 2H 7N501	117.11	117.0651	20.51	6.02	59.80	13.66
C 2H 7N502	133.11	133.0600	18.05	5.30	52.61	24.04
C 2H 7N503	149.11	149.0549	16.11	4.73	46.97	32.19
C 2H 7N504	165.11	165.0498	14.55	4.27	42.42	38.76
C 2H 7N505	181.11	181.0447	13.26	3.90	38.67	44.17
C 2H 7N506	197.11	197.0396	12.19	3.58	35.53	48.70
C 2H 7N507	213.11	213.0345	11.27	3.31	32.86	52.55
C 2H 7N508	229.11	229.0295	10.49	3.08	30.57	55.87
C 2H 7N509	245.11	245.0244	9.80	2.88	28.57	58.75
C 2H 8N201	76.10	76.0637	31.57	10.60	36.81	21.02
C 2H 8N202	92.10	92.0586	26.08	8.76	30.42	34.74
C 2H 8N203	108.10	108.0535	22.22	7.46	25.91	44.40
C 2H 8N204	124.10	124.0484	19.36	6.50	22.57	51.57
C 2H 8N205	140.10	140.0433	17.15	5.76	20.00	57.10
C 2H 8N206	156.10	156.0382	15.39	5.17	17.95	61.50
C 2H 8N207	172.10	172.0331	13.96	4.69	16.28	65.08
C 2H 8N208	188.09	188.0281	12.77	4.29	14.89	68.05
C 2H 8N209	204.09	204.0230	11.77	3.95	13.73	70.55
C 2H 8N401	104.11	104.0698	23.07	7.75	53.81	15.37
C 2H 8N402	120.11	120.0647	20.00	6.71	46.65	26.64
C 2H 8N403	136.11	136.0596	17.65	5.92	41.16	35.26
C 2H 8N404	152.11	152.0545	15.79	5.30	36.83	42.07
C 2H 8N405	168.11	168.0495	14.29	4.80	33.33	47.59
C 2H 8N406	184.11	184.0444	13.05	4.38	30.43	52.14
C 2H 8N407	200.11	200.0393	12.00	4.03	28.00	55.97
C 2H 8N408	216.11	216.0342	11.12	3.73	25.93	59.23
C 2H 8N409	232.11	232.0291	10.35	3.47	24.14	62.04
C 2H 8N601	132.13	132.0760	18.18	6.10	63.61	12.11
C 2H 8N602	148.13	148.0709	16.22	5.44	56.74	21.60
C 2H 8N603	164.12	164.0658	14.64	4.91	51.21	29.25
C 2H 8N604	180.12	180.0607	13.34	4.48	46.66	35.53
C 2H 8N605	196.12	196.0556	12.25	4.11	42.85	40.79
C 2H 8N606	212.12	212.0505	11.32	3.80	39.62	45.26
C 2H 8N607	228.12	228.0454	10.53	3.53	36.84	49.09
C 2H 8N608	244.12	244.0404	9.84	3.30	34.43	52.43
C 2H 8N609	260.12	260.0353	9.24	3.10	32.31	55.36
C 2H 9N301	91.11	91.0746	26.37	9.96	46.12	17.56

C 2H 9N302	107.11	107.0695	22.43	8.47	39.23	29.87
C 2H 9N303	123.11	123.0644	19.51	7.37	34.13	38.99
C 2H 9N304	139.11	139.0593	17.27	6.52	30.21	46.00
C 2H 9N305	155.11	155.0542	15.45	5.85	27.09	51.57
C 2H 9N306	171.11	171.0491	14.04	5.30	24.56	56.10
C 2H 9N307	187.11	187.0440	12.84	4.85	22.46	59.86
C 2H 9N308	203.11	203.0390	11.83	4.47	20.69	63.02
C 2H 9N309	219.11	219.0339	10.96	4.14	19.18	65.72
C 2H 9N501	119.13	119.0807	20.17	7.62	58.79	13.43
C 2H 9N502	135.13	135.0756	17.78	6.71	51.83	23.68
C 2H 9N503	151.13	151.0705	15.90	6.00	46.34	31.76
C 2H 9N504	167.13	167.0654	14.37	5.43	41.90	38.29
C 2H 9N505	183.12	183.0604	13.12	4.95	38.24	43.68
C 2H 9N506	199.12	199.0553	12.06	4.56	35.17	48.21
C 2H 9N507	215.12	215.0502	11.17	4.22	32.56	52.06
C 2H 9N508	231.12	231.0451	10.39	3.93	30.30	55.38
C 2H 9N509	247.12	247.0400	9.72	3.67	28.34	58.27
C 2H10N401	106.13	106.0855	22.64	9.50	52.79	15.08
C 2H10N402	122.13	122.0804	19.67	8.25	45.88	26.20
C 2H10N403	138.13	138.0753	17.39	7.30	40.56	34.75
C 2H10N404	154.13	154.0702	15.59	6.54	36.35	41.52
C 2H10N405	170.13	170.0651	14.12	5.92	32.93	47.02
C 2H10N406	186.13	186.0600	12.91	5.42	30.10	51.58
C 2H10N407	202.12	202.0549	11.88	4.99	27.72	55.41
C 2H10N408	218.12	218.0499	11.01	4.62	25.69	58.68
C 2H10N409	234.12	234.0448	10.26	4.31	23.93	61.50
C 2H10N601	134.14	134.0916	17.91	7.51	62.65	11.93
C 2H10N602	150.14	150.0865	16.00	6.71	55.97	21.31
C 2H10N603	166.14	166.0814	14.46	6.07	50.58	28.89
C 2H10N604	182.14	182.0763	13.19	5.53	46.14	35.14
C 2H10N605	198.14	198.0713	12.12	5.09	42.41	40.37
C 2H10N606	214.14	214.0662	11.22	4.71	39.25	44.83
C 2H10N607	230.14	230.0611	10.44	4.38	36.52	48.66
C 2H10N608	246.14	246.0560	9.76	4.10	34.14	52.00
C 2H10N609	262.14	262.0509	9.16	3.85	32.06	54.93
C 2H11N501	121.14	121.0964	19.83	9.15	57.81	13.21
C 2H11N502	137.14	137.0913	17.52	8.08	51.07	23.33
C 2H11N503	153.14	153.0862	15.69	7.24	45.73	31.34
C 2H11N504	169.14	169.0811	14.20	6.56	41.41	37.84
C 2H11N505	185.14	185.0760	12.98	5.99	37.83	43.21
C 2H11N506	201.14	201.0709	11.94	5.51	34.82	47.73
C 2H11N507	217.14	217.0658	11.06	5.11	32.25	51.58
C 2H11N508	233.14	233.0608	10.30	4.76	30.04	54.90
C 2H11N509	249.14	249.0557	9.64	4.45	28.11	57.80
C 2H12N601	136.16	136.1073	17.64	8.88	61.72	11.75
C 2H12N602	152.16	152.1022	15.79	7.95	55.23	21.03
C 2H12N603	168.16	168.0971	14.29	7.19	49.98	28.54
C 2H12N604	184.16	184.0920	13.04	6.57	45.64	34.75
C 2H12N605	200.16	200.0869	12.00	6.04	41.99	39.97
C 2H12N606	216.15	216.0818	11.11	5.60	38.88	44.41
C 2H12N607	232.15	232.0767	10.35	5.21	36.20	48.24
C 2H12N608	248.15	248.0717	9.68	4.87	33.87	51.58
C 2H12N609	264.15	264.0666	9.09	4.58	31.82	54.51
C 3H 1N101	67.05	67.0058	53.74	1.50	20.89	23.86
C 3H 1N102	83.05	83.0007	43.39	1.21	16.87	38.53
C 3H 1N103	99.05	98.9956	36.38	1.02	14.14	48.46
C 3H 1N104	115.05	114.9906	31.32	.88	12.17	55.63
C 3H 1N105	131.05	130.9855	27.50	.77	10.69	61.05
C 3H 1N106	147.04	146.9804	24.51	.69	9.53	65.28
C 3H 1N107	163.04	162.9753	22.10	.62	8.59	68.69
C 3H 1N108	179.04	178.9702	20.13	.56	7.82	71.49
C 3H 1N109	195.04	194.9651	18.47	.52	7.18	73.83
C 3H 1N301	95.06	95.0120	37.91	1.06	44.20	16.83
C 3H 1N302	111.06	111.0069	32.44	.91	37.84	28.81
C 3H 1N303	127.06	127.0018	28.36	.79	33.07	37.78
C 3H 1N304	143.06	142.9967	25.19	.70	29.37	44.74
C 3H 1N305	159.06	158.9916	22.65	.63	26.42	50.29
C 3H 1N306	175.06	174.9865	20.58	.58	24.00	54.84
C 3H 1N307	191.06	190.9814	18.86	.53	21.99	58.62
C 3H 1N308	207.06	206.9764	17.40	.49	20.29	61.82
C 3H 1N309	223.06	222.9713	16.15	.45	18.84	64.56
C 3H 1N501	123.07	123.0181	29.28	.82	56.90	13.00
C 3H 1N502	139.07	139.0130	25.91	.72	50.36	23.01
C 3H 1N503	155.07	155.0079	23.24	.65	45.16	30.95
C 3H 1N504	171.07	171.0028	21.06	.59	40.94	37.41
C 3H 1N505	187.07	186.9978	19.26	.54	37.44	42.76
C 3H 1N506	203.07	202.9927	17.74	.50	34.49	47.27
C 3H 1N507	219.07	218.9876	16.45	.46	31.97	51.12
C 3H 1N508	235.07	234.9825	15.33	.43	29.79	54.45
C 3H 1N509	251.07	250.9774	14.35	.40	27.89	57.35
C 3H 2N201	82.06	82.0167	43.91	2.46	34.14	19.50
C 3H 2N202	98.06	98.0116	36.75	2.06	28.57	32.63
C 3H 2N203	114.06	114.0065	31.59	1.77	24.56	42.08
C 3H 2N204	130.06	130.0015	27.71	1.55	21.54	49.21
C 3H 2N205	146.06	145.9964	24.67	1.38	19.18	54.77
C 3H 2N206	162.06	161.9913	22.23	1.24	17.29	59.24
C 3H 2N207	178.06	177.9862	20.24	1.13	15.73	62.90
C 3H 2N208	194.06	193.9811	18.57	1.04	14.44	65.96
C 3H 2N209	210.06	209.9760	17.15	.96	13.34	68.55
C 3H 2N401	110.08	110.0229	32.74	1.83	50.90	14.53
C 3H 2N402	126.07	126.0178	28.58	1.60	44.44	25.38
C 3H 2N403	142.07	142.0127	25.36	1.42	39.43	33.78
C 3H 2N404	158.07	158.0076	22.80	1.28	35.44	40.49
C 3H 2N405	174.07	174.0025	20.70	1.16	32.19	45.96
C 3H 2N406	190.07	189.9974	18.96	1.06	29.48	50.51
C 3H 2N407	206.07	205.9923	17.49	.98	27.19	54.35
C 3H 2N408	222.07	221.9873	16.23	.91	25.23	57.64
C 3H 2N409	238.07	237.9822	15.14	.85	23.53	60.48
C 3H 2N601	138.09	138.0290	26.09	1.46	60.86	11.59
C 3H 2N602	154.09	154.0239	23.38	1.31	54.54	20.77
C 3H 2N603	170.09	170.0188	21.19	1.19	49.41	28.22
C 3H 2N604	186.09	186.0137	19.36	1.08	45.16	34.39
C 3H 2N605	202.09	202.0087	17.83	1.00	41.59	39.59
C 3H 2N606	218.09	218.0036	16.52	.92	38.54	44.02
C 3H 2N607	234.09	233.9985	15.39	.86	35.90	47.84
C 3H 2N608	250.08	249.9934	14.41	.81	33.60	51.18
C 3H 2N609	266.08	265.9883	13.54	.76	31.58	54.12
C 3H 3N101	69.06	69.0215	52.17	4.38	20.28	23.17
C 3H 3N102	85.06	85.0164	42.36	3.55	16.47	37.62
C 3H 3N103	101.06	101.0113	35.65	2.99	13.86	47.49
C 3H 3N104	117.06	117.0062	30.78	2.58	11.97	54.67
C 3H 3N105	133.06	133.0011	27.08	2.27	10.53	60.12
C 3H 3N106	149.06	148.9960	24.17	2.03	9.40	64.40
C 3H 3N107	165.06	164.9909	21.83	1.83	8.49	67.85
C 3H 3N108	181.06	180.9859	19.90	1.67	7.74	70.69
C 3H 3N109	197.06	196.9808	18.29	1.53	7.11	73.07
C 3H 3N301	97.08	97.0276	37.12	3.11	43.29	16.48
C 3H 3N302	113.08	113.0225	31.87	2.67	37.16	28.30
C 3H 3N303	129.08	129.0174	27.92	2.34	32.55	37.19
C 3H 3N304	145.08	145.0124	24.84	2.08	28.96	44.11

021

C 3H 3N305	161.07	161.0073	22.37	1.88	26.09	49.66
C 3H 3N306	177.07	177.0022	20.35	1.71	23.73	54.21
C 3H 3N307	193.07	192.9971	18.66	1.57	21.76	58.01
C 3H 3N308	209.07	208.9920	17.23	1.45	20.10	61.22
C 3H 3N309	225.07	224.9869	16.01	1.34	18.67	63.98
C 3H 3N501	125.09	125.0338	28.81	2.42	55.99	12.79
C 3H 3N502	141.09	141.0287	25.54	2.14	49.64	22.68
C 3H 3N503	157.09	157.0236	22.94	1.92	44.58	30.55
C 3H 3N504	173.09	173.0185	20.82	1.75	40.46	36.97
C 3H 3N505	189.09	189.0134	19.06	1.60	37.04	42.31
C 3H 3N506	205.09	205.0083	17.57	1.47	34.15	46.81
C 3H 3N507	221.09	221.0032	16.30	1.37	31.68	50.66
C 3H 3N508	237.09	236.9982	15.20	1.28	29.54	53.99
C 3H 3N509	253.09	252.9931	14.24	1.19	27.67	56.90
C 3H 4N201	84.08	84.0324	42.86	4.80	33.32	19.03
C 3H 4N202	100.08	100.0273	36.01	4.03	27.99	31.97
C 3H 4N203	116.08	116.0222	31.04	3.47	24.13	41.35
C 3H 4N204	132.08	132.0171	27.28	3.05	21.21	48.46
C 3H 4N205	148.08	148.0120	24.33	2.72	18.92	54.02
C 3H 4N206	164.08	164.0069	21.96	2.46	17.07	58.51
C 3H 4N207	180.07	180.0018	20.01	2.24	15.56	62.19
C 3H 4N208	196.07	195.9968	18.38	2.06	14.29	65.28
C 3H 4N209	212.07	211.9917	16.99	1.90	13.21	67.90
C 3H 4N401	112.09	112.0385	32.15	3.60	49.98	14.27
C 3H 4N402	128.09	128.0334	28.13	3.15	43.74	24.98
C 3H 4N403	144.09	144.0283	25.01	2.80	38.88	33.31
C 3H 4N404	160.09	160.0232	22.51	2.52	35.00	39.98
C 3H 4N405	176.09	176.0182	20.46	2.29	31.82	45.43
C 3H 4N406	192.09	192.0131	18.76	2.10	29.17	49.98
C 3H 4N407	208.09	208.0080	17.32	1.94	26.92	53.82
C 3H 4N408	224.09	224.0029	16.08	1.80	25.00	57.12
C 3H 4N409	240.09	239.9978	15.01	1.68	23.34	59.98
C 3H 4N601	140.10	140.0447	25.72	2.88	59.98	11.42
C 3H 4N602	156.10	156.0396	23.08	2.58	53.84	20.50
C 3H 4N603	172.10	172.0345	20.94	2.34	48.83	27.89
C 3H 4N604	188.10	188.0294	19.16	2.14	44.68	34.02
C 3H 4N605	204.10	204.0243	17.65	1.98	41.18	39.19
C 3H 4N606	220.10	220.0192	16.37	1.83	38.18	43.61
C 3H 4N607	236.10	236.0141	15.26	1.71	35.59	47.44
C 3H 4N608	252.10	252.0091	14.29	1.60	33.34	50.77
C 3H 4N609	268.10	268.0040	13.44	1.50	31.35	53.71
C 3H 5N101	71.08	71.0371	50.69	7.09	19.71	22.51
C 3H 5N102	87.08	87.0320	41.38	5.79	16.09	36.75
C 3H 5N103	103.08	103.0269	34.96	4.89	13.59	46.56
C 3H 5N104	119.08	119.0219	30.26	4.23	11.76	53.74
C 3H 5N105	135.08	135.0168	26.68	3.73	10.37	59.22
C 3H 5N106	151.08	151.0117	23.85	3.34	9.27	63.54
C 3H 5N107	167.08	167.0066	21.57	3.02	8.38	67.03
C 3H 5N108	183.08	183.0015	19.68	2.75	7.65	69.91
C 3H 5N109	199.07	198.9964	18.10	2.53	7.04	72.33
C 3H 5N301	99.09	99.0433	36.36	5.09	42.40	16.15
C 3H 5N302	115.09	115.0382	31.31	4.38	36.51	27.80
C 3H 5N303	131.09	131.0331	27.49	3.84	32.05	36.61
C 3H 5N304	147.09	147.0280	24.50	3.43	28.57	43.51
C 3H 5N305	163.09	163.0229	22.09	3.09	25.76	49.05
C 3H 5N306	179.09	179.0178	20.12	2.81	23.46	53.60
C 3H 5N307	195.09	195.0127	18.47	2.58	21.54	57.41
C 3H 5N308	211.09	211.0077	17.07	2.39	19.91	60.64
C 3H 5N309	227.09	227.0026	15.87	2.22	18.50	63.41
C 3H 5N501	127.11	127.0494	28.35	3.97	55.10	12.59
C 3H 5N502	143.11	143.0443	25.18	3.52	48.94	22.36
C 3H 5N503	159.11	159.0392	22.65	3.17	44.02	30.17
C 3H 5N504	175.10	175.0341	20.58	2.88	40.00	36.55
C 3H 5N505	191.10	191.0291	18.86	2.64	36.65	41.86
C 3H 5N506	207.10	207.0240	17.40	2.43	33.82	46.35
C 3H 5N507	223.10	223.0189	16.15	2.26	31.39	50.20
C 3H 5N508	239.10	239.0138	15.07	2.11	29.29	53.53
C 3H 5N509	255.10	255.0087	14.13	1.98	27.45	56.45
C 3H 6N201	86.09	86.0480	41.85	7.02	32.54	18.58
C 3H 6N202	102.09	102.0429	35.29	5.92	27.44	31.34
C 3H 6N203	118.09	118.0378	30.51	5.12	23.72	40.64
C 3H 6N204	134.09	134.0328	26.87	4.51	20.89	47.73
C 3H 6N205	150.09	150.0277	24.01	4.03	18.66	53.30
C 3H 6N206	166.09	166.0226	21.69	3.64	16.87	57.80
C 3H 6N207	182.09	182.0175	19.79	3.32	15.38	61.51
C 3H 6N208	198.09	198.0124	18.19	3.05	14.14	64.61
C 3H 6N209	214.09	214.0073	16.83	2.82	13.08	67.26
C 3H 6N401	114.11	114.0542	31.58	5.30	49.10	14.02
C 3H 6N402	130.11	130.0491	27.70	4.65	43.06	24.59
C 3H 6N403	146.11	146.0440	24.66	4.14	38.35	32.85
C 3H 6N404	162.11	162.0389	22.23	3.73	34.56	39.48
C 3H 6N405	178.11	178.0338	20.23	3.40	31.46	44.92
C 3H 6N406	194.10	194.0287	18.56	3.12	28.86	49.46
C 3H 6N407	210.10	210.0236	17.15	2.88	26.67	53.30
C 3H 6N408	226.10	226.0186	15.94	2.67	24.78	56.61
C 3H 6N409	242.10	242.0135	14.88	2.50	23.14	59.48
C 3H 6N601	142.12	142.0603	25.35	4.26	59.13	11.26
C 3H 6N602	158.12	158.0552	22.79	3.82	53.15	20.24
C 3H 6N603	174.12	174.0501	20.69	3.47	48.27	27.57
C 3H 6N604	190.12	190.0450	18.95	3.18	44.20	33.66
C 3H 6N605	206.12	206.0400	17.48	2.93	40.77	38.81
C 3H 6N606	222.12	222.0349	16.22	2.72	37.84	43.22
C 3H 6N607	238.12	238.0298	15.13	2.54	35.29	47.03
C 3H 6N608	254.12	254.0247	14.18	2.38	33.07	50.37
C 3H 6N609	270.12	270.0196	13.34	2.24	31.11	53.31
C 3H 7N101	73.10	73.0528	49.30	9.65	19.16	21.89
C 3H 7N102	89.09	89.0477	40.44	7.92	15.72	35.92
C 3H 7N103	105.09	105.0426	34.29	6.71	13.33	45.67
C 3H 7N104	121.09	121.0375	29.76	5.83	11.57	52.85
C 3H 7N105	137.09	137.0324	26.28	5.15	10.22	58.35
C 3H 7N106	153.09	153.0273	23.54	4.61	9.15	62.70
C 3H 7N107	169.09	169.0222	21.31	4.17	8.28	66.23
C 3H 7N108	185.09	185.0172	19.47	3.81	7.57	69.15
C 3H 7N109	201.09	201.0121	17.92	3.51	6.97	71.61
C 3H 7N301	101.11	101.0589	35.64	6.98	41.56	15.82
C 3H 7N302	117.11	117.0538	30.77	6.03	35.88	27.32
C 3H 7N303	133.11	133.0487	27.07	5.30	31.57	36.06
C 3H 7N304	149.11	149.0436	24.17	4.73	28.18	42.92
C 3H 7N305	165.11	165.0386	21.82	4.27	25.45	48.45
C 3H 7N306	181.11	181.0335	19.90	3.90	23.20	53.01
C 3H 7N307	197.11	197.0284	18.28	3.58	21.32	56.82
C 3H 7N308	213.10	213.0233	16.91	3.31	19.72	60.06
C 3H 7N309	229.10	229.0182	15.73	3.08	18.34	62.85
C 3H 7N501	129.12	129.0651	27.91	5.46	54.24	12.39
C 3H 7N502	145.12	145.0600	24.83	4.86	48.26	22.05
C 3H 7N503	161.12	161.0549	22.36	4.38	43.47	29.79
C 3H 7N504	177.12	177.0498	20.34	3.98	39.54	36.13
C 3H 7N505	193.12	193.0447	18.66	3.65	36.26	41.42
C 3H 7N506	209.12	209.0396	17.23	3.37	33.49	45.91
C 3H 7N507	225.12	225.0345	16.01	3.13	31.11	49.75

022

C 3H 7N508	241.12	241.0295	14.94	2.93	29.05	53.08
C 3H 7N509	257.12	257.0244	14.01	2.74	27.24	56.00
C 3H 8N201	88.11	88.0637	40.90	9.15	31.79	18.16
C 3H 8N202	104.11	104.0586	34.61	7.75	26.91	30.74
C 3H 8N203	120.11	120.0535	30.00	6.71	23.32	39.96
C 3H 8N204	136.11	136.0484	26.47	5.92	20.58	47.02
C 3H 8N205	152.11	152.0433	23.69	5.30	18.42	52.59
C 3H 8N206	168.11	168.0382	21.43	4.80	16.66	57.10
C 3H 8N207	184.11	184.0331	19.57	4.38	15.22	60.83
C 3H 8N208	200.11	200.0281	18.01	4.03	14.00	63.96
C 3H 8N209	216.11	216.0230	16.67	3.73	12.96	66.63
C 3H 8N401	116.12	116.0698	31.03	6.94	48.25	13.78
C 3H 8N402	132.12	132.0647	27.27	6.10	42.41	24.22
C 3H 8N403	148.12	148.0596	24.33	5.44	37.82	32.40
C 3H 8N404	164.12	164.0545	21.96	4.91	34.14	38.99
C 3H 8N405	180.12	180.0495	20.01	4.48	31.11	44.41
C 3H 8N406	196.12	196.0444	18.37	4.11	28.57	48.95
C 3H 8N407	212.12	212.0393	16.99	3.80	26.41	52.80
C 3H 8N408	228.12	228.0342	15.80	3.53	24.56	56.11
C 3H 8N409	244.12	244.0291	14.76	3.30	22.95	58.99
C 3H 8N601	144.14	144.0760	25.00	5.59	58.31	11.10
C 3H 8N602	160.14	160.0709	22.50	5.04	52.48	19.98
C 3H 8N603	176.14	176.0658	20.46	4.58	47.71	27.25
C 3H 8N604	192.14	192.0607	18.75	4.20	43.74	33.31
C 3H 8N605	208.13	208.0556	17.31	3.87	40.38	38.44
C 3H 8N606	224.13	224.0505	16.08	3.60	37.50	42.83
C 3H 8N607	240.13	240.0454	15.01	3.36	35.00	46.64
C 3H 8N608	256.13	256.0404	14.07	3.15	32.81	49.97
C 3H 8N609	272.13	272.0353	13.24	2.96	30.88	52.91
C 3H 9N101	75.11	75.0684	47.97	12.08	18.65	21.30
C 3H 9N102	91.11	91.0633	39.55	9.96	15.37	35.12
C 3H 9N103	107.11	107.0582	33.64	8.47	13.08	44.81
C 3H 9N104	123.11	123.0532	29.27	7.37	11.38	51.98
C 3H 9N105	139.11	139.0481	25.90	6.52	10.07	57.51
C 3H 9N106	155.11	155.0430	23.23	5.85	9.03	61.89
C 3H 9N107	171.11	171.0379	21.06	5.30	8.19	65.45
C 3H 9N108	187.11	187.0328	19.26	4.85	7.49	68.41
C 3H 9N109	203.11	203.0277	17.74	4.47	6.90	70.90
C 3H 9N301	103.12	103.0746	34.94	8.80	40.75	15.51
C 3H 9N302	119.12	119.0695	30.25	7.62	35.27	26.86
C 3H 9N303	135.12	135.0644	26.67	6.71	31.10	35.52
C 3H 9N304	151.12	151.0593	23.84	6.00	27.81	42.35
C 3H 9N305	167.12	167.0542	21.56	5.43	25.14	47.87
C 3H 9N306	183.12	183.0491	19.68	4.95	22.95	52.42
C 3H 9N307	199.12	199.0440	18.10	4.56	21.10	56.25
C 3H 9N308	215.12	215.0390	16.75	4.22	19.53	59.50
C 3H 9N309	231.12	231.0339	15.59	3.93	18.18	62.30
C 3H 9N501	131.14	131.0807	27.48	6.92	53.40	12.20
C 3H 9N502	147.14	147.0756	24.49	6.17	47.60	21.75
C 3H 9N503	163.14	163.0705	22.09	5.56	42.93	29.42
C 3H 9N504	179.14	179.0654	20.12	5.06	39.10	35.73
C 3H 9N505	195.14	195.0604	18.47	4.65	35.89	41.00
C 3H 9N506	211.14	211.0553	17.07	4.30	33.17	45.47
C 3H 9N507	227.13	227.0502	15.86	3.99	30.83	49.31
C 3H 9N508	243.13	243.0451	14.82	3.73	28.80	52.64
C 3H 9N509	259.13	259.0400	13.91	3.50	27.03	55.57
C 3H10N201	90.13	90.0793	39.98	11.18	31.08	17.75
C 3H10N202	106.13	106.0742	33.95	9.50	26.40	30.15
C 3H10N203	122.12	122.0691	29.51	8.25	22.94	39.30
C 3H10N204	138.12	138.0641	26.09	7.30	20.28	46.33
C 3H10N205	154.12	154.0590	23.38	6.54	18.18	51.90
C 3H10N206	170.12	170.0539	21.18	5.92	16.47	56.43
C 3H10N207	186.12	186.0488	19.36	5.42	15.05	60.17
C 3H10N208	202.12	202.0437	17.83	4.99	13.86	63.33
C 3H10N209	218.12	218.0386	16.52	4.62	12.84	66.02
C 3H10N401	118.14	118.0855	30.50	8.53	47.42	13.54
C 3H10N402	134.14	134.0804	26.86	7.51	41.77	23.86
C 3H10N403	150.14	150.0753	24.00	6.71	37.32	31.97
C 3H10N404	166.14	166.0702	21.69	6.07	33.72	38.52
C 3H10N405	182.14	182.0651	19.78	5.53	30.76	43.92
C 3H10N406	198.14	198.0600	18.19	5.09	28.28	48.45
C 3H10N407	214.14	214.0549	16.83	4.71	26.16	52.30
C 3H10N408	230.14	230.0499	15.66	4.38	24.35	55.62
C 3H10N409	246.13	246.0448	14.64	4.10	22.76	58.50
C 3H10N601	146.15	146.0916	24.65	6.90	57.50	10.95
C 3H10N602	162.15	162.0865	22.22	6.22	51.83	19.73
C 3H10N603	178.15	178.0814	20.23	5.66	47.17	26.94
C 3H10N604	194.15	194.0763	18.56	5.19	43.29	32.96
C 3H10N605	210.15	210.0713	17.15	4.80	39.99	38.07
C 3H10N606	226.15	226.0662	15.93	4.46	37.16	42.45
C 3H10N607	242.15	242.0611	14.88	4.16	34.71	46.25
C 3H10N608	258.15	258.0560	13.96	3.90	32.55	49.58
C 3H10N609	274.15	274.0509	13.14	3.68	30.66	52.52
C 3H11N301	105.14	105.0902	34.27	10.55	39.97	15.22
C 3H11N302	121.14	121.0851	29.75	9.15	34.69	26.41
C 3H11N303	137.14	137.0800	26.28	8.08	30.64	35.00
C 3H11N304	153.14	153.0749	23.53	7.24	27.44	41.79
C 3H11N305	169.14	169.0699	21.30	6.56	24.84	47.30
C 3H11N306	185.14	185.0648	19.46	5.99	22.70	51.85
C 3H11N307	201.14	201.0597	17.91	5.51	20.89	55.68
C 3H11N308	217.14	217.0546	16.59	5.11	19.35	58.95
C 3H11N309	233.14	233.0495	15.46	4.76	18.02	61.76
C 3H11N501	133.15	133.0964	27.06	8.33	52.60	12.02
C 3H11N502	149.15	149.0913	24.16	7.43	46.95	21.45
C 3H11N503	165.15	165.0862	21.82	6.71	42.41	29.06
C 3H11N504	181.15	181.0811	19.89	6.12	38.66	35.33
C 3H11N505	197.15	197.0760	18.28	5.62	35.52	40.58
C 3H11N506	213.15	213.0709	16.91	5.20	32.86	45.04
C 3H11N507	229.15	229.0658	15.72	4.84	30.56	48.87
C 3H11N508	245.15	245.0608	14.70	4.52	28.57	52.21
C 3H11N509	261.15	261.0557	13.80	4.25	26.82	55.14
C 3H12N401	120.16	120.1011	29.99	10.07	46.63	13.32
C 3H12N402	136.15	136.0960	26.47	8.88	41.15	23.50
C 3H12N403	152.15	152.0909	23.68	7.95	36.82	31.55
C 3H12N404	168.15	168.0858	21.43	7.19	33.32	38.06
C 3H12N405	184.15	184.0808	19.57	6.57	30.42	43.44
C 3H12N406	200.15	200.0757	18.00	6.04	27.99	47.96
C 3H12N407	216.15	216.0706	16.67	5.60	25.92	51.81
C 3H12N408	232.15	232.0655	15.52	5.21	24.13	55.13
C 3H12N409	248.15	248.0604	14.52	4.87	22.58	58.03
C 3H12N601	148.17	148.1073	24.32	8.16	56.72	10.80
C 3H12N602	164.17	164.1022	21.95	7.37	51.19	19.49
C 3H12N603	180.17	180.0971	20.00	6.71	46.65	26.64
C 3H12N604	196.17	196.0920	18.37	6.17	42.84	32.62
C 3H12N605	212.17	212.0869	16.98	5.70	39.61	37.70
C 3H12N606	228.17	228.0818	15.79	5.30	36.83	42.07
C 3H12N607	244.17	244.0767	14.76	4.95	34.42	45.87
C 3H12N608	260.16	260.0717	13.85	4.65	32.30	49.20
C 3H12N609	276.16	276.0666	13.05	4.38	30.43	52.14
C 3H13N501	135.17	135.1120	26.66	9.69	51.81	11.84

023

C 3H13N502	151.17	151.1069	23.84	8.67	46.33	21.17
C 3H13N503	167.17	167.1018	21.56	7.84	41.89	28.71
C 3H13N504	183.17	183.0967	19.67	7.15	38.23	34.94
C 3H13N505	199.17	199.0917	18.09	6.58	35.16	40.17
C 3H13N506	215.17	215.0866	16.75	6.09	32.55	44.61
C 3H13N507	231.17	231.0815	15.59	5.67	30.30	48.45
C 3H13N508	247.17	247.0764	14.58	5.30	28.33	51.79
C 3H13N509	263.17	263.0713	13.69	4.98	26.61	54.72
C 3H14N601	150.18	150.1229	23.99	9.40	55.96	10.65
C 3H14N602	166.18	166.1178	21.68	8.49	50.57	19.26
C 3H14N603	182.18	182.1127	19.78	7.75	46.13	26.35
C 3H14N604	198.18	198.1076	18.18	7.12	42.41	32.29
C 3H14N605	214.18	214.1026	16.82	6.59	39.24	37.35
C 3H14N606	230.18	230.0975	15.65	6.13	36.51	41.70
C 3H14N607	246.18	246.0924	14.64	5.73	34.14	45.49
C 3H14N608	262.18	262.0873	13.74	5.38	32.05	48.82
C 3H14N609	278.18	278.0822	12.95	5.07	30.21	51.76
C 4H 2N201	94.07	94.0167	51.07	2.14	29.78	17.01
C 4H 2N202	110.07	110.0116	43.65	1.83	25.45	29.07
C 4H 2N203	126.07	126.0065	38.11	1.60	22.22	38.07
C 4H 2N204	142.07	142.0015	33.82	1.42	19.72	45.05
C 4H 2N205	158.07	157.9964	30.39	1.28	17.72	50.61
C 4H 2N206	174.07	173.9913	27.60	1.16	16.09	55.15
C 4H 2N207	190.07	189.9862	25.28	1.06	14.74	58.92
C 4H 2N208	206.07	205.9811	23.31	.98	13.59	62.11
C 4H 2N209	222.07	221.9760	21.64	.91	12.61	64.84
C 4H 2N401	122.09	122.0229	39.35	1.65	45.89	13.10
C 4H 2N402	138.09	138.0178	34.79	1.46	40.57	23.17
C 4H 2N403	154.09	154.0127	31.18	1.31	36.36	31.15
C 4H 2N404	170.08	170.0076	28.25	1.19	32.94	37.63
C 4H 2N405	186.08	186.0025	25.82	1.08	30.11	42.99
C 4H 2N406	202.08	201.9974	23.77	1.00	27.72	47.50
C 4H 2N407	218.08	217.9923	22.03	.92	25.69	51.35
C 4H 2N408	234.08	233.9873	20.52	.86	23.93	54.68
C 4H 2N409	250.08	249.9822	19.21	.81	22.40	57.58
C 4H 2N601	150.10	150.0290	32.01	1.34	55.99	10.66
C 4H 2N602	166.10	166.0239	28.93	1.21	50.60	19.26
C 4H 2N603	182.10	182.0188	26.38	1.11	46.15	26.36
C 4H 2N604	198.10	198.0137	24.25	1.02	42.42	32.31
C 4H 2N605	214.10	214.0087	22.44	.94	39.25	37.36
C 4H 2N606	230.10	230.0036	20.88	.88	36.52	41.72
C 4H 2N607	246.10	245.9985	19.52	.82	34.15	45.51
C 4H 2N608	262.10	261.9934	18.33	.77	32.06	48.84
C 4H 2N609	278.10	277.9883	17.28	.72	30.22	51.78
C 4H 3N101	81.07	81.0215	59.26	3.73	17.28	19.73
C 4H 3N102	97.07	97.0164	49.49	3.12	14.43	32.96
C 4H 3N103	113.07	113.0113	42.49	2.67	12.39	42.45
C 4H 3N104	129.07	129.0062	37.22	2.34	10.85	49.58
C 4H 3N105	145.07	145.0011	33.12	2.08	9.65	55.14
C 4H 3N106	161.07	160.9960	29.83	1.88	8.70	59.60
C 4H 3N107	177.07	176.9909	27.13	1.71	7.91	63.25
C 4H 3N108	193.07	192.9859	24.88	1.57	7.25	66.29
C 4H 3N109	209.07	208.9808	22.98	1.45	6.70	68.87
C 4H 3N301	109.09	109.0276	44.04	2.77	38.52	14.67
C 4H 3N302	125.09	125.0225	38.41	2.42	33.59	25.58
C 4H 3N303	141.09	141.0174	34.05	2.14	29.78	34.02
C 4H 3N304	157.09	157.0124	30.58	1.93	26.75	40.74
C 4H 3N305	173.09	173.0073	27.76	1.75	24.28	46.22
C 4H 3N306	189.09	189.0022	25.41	1.60	22.22	50.77
C 4H 3N307	205.08	204.9971	23.43	1.47	20.49	54.61
C 4H 3N308	221.08	220.9920	21.73	1.37	19.01	57.89
C 4H 3N309	237.08	236.9869	20.26	1.28	17.72	60.74
C 4H 3N501	137.10	137.0338	35.04	2.21	51.08	11.67
C 4H 3N502	153.10	153.0287	31.38	1.98	45.74	20.90
C 4H 3N503	169.10	169.0236	28.41	1.79	41.42	28.38
C 4H 3N504	185.10	185.0185	25.96	1.63	37.84	34.57
C 4H 3N505	201.10	201.0134	23.89	1.50	34.83	39.78
C 4H 3N506	217.10	217.0083	22.13	1.39	32.26	44.22
C 4H 3N507	233.10	233.0032	20.61	1.30	30.04	48.05
C 4H 3N508	249.10	248.9982	19.29	1.21	28.11	51.38
C 4H 3N509	265.10	264.9931	18.12	1.14	26.42	54.32
C 4H 4N201	96.09	96.0324	50.00	4.20	29.15	16.65
C 4H 4N202	112.09	112.0273	42.86	3.60	24.99	28.55
C 4H 4N203	128.09	128.0222	37.51	3.15	21.87	37.47
C 4H 4N204	144.09	144.0171	33.34	2.80	19.44	44.42
C 4H 4N205	160.09	160.0120	30.01	2.52	17.50	49.97
C 4H 4N206	176.09	176.0069	27.28	2.29	15.91	54.52
C 4H 4N207	192.09	192.0018	25.01	2.10	14.58	58.31
C 4H 4N208	208.09	207.9968	23.09	1.94	13.46	61.51
C 4H 4N209	224.08	223.9917	21.44	1.80	12.50	64.26
C 4H 4N401	124.10	124.0385	38.71	3.25	45.15	12.89
C 4H 4N402	140.10	140.0334	34.29	2.88	39.99	22.84
C 4H 4N403	156.10	156.0283	30.78	2.58	35.89	30.75
C 4H 4N404	172.10	172.0232	27.92	2.34	32.55	37.19
C 4H 4N405	188.10	188.0182	25.54	2.14	29.79	42.53
C 4H 4N406	204.10	204.0131	23.54	1.98	27.45	47.03
C 4H 4N407	220.10	220.0080	21.83	1.83	25.46	50.88
C 4H 4N408	236.10	236.0029	20.35	1.71	23.73	54.21
C 4H 4N409	252.10	251.9978	19.06	1.60	22.22	57.12
C 4H 4N601	152.12	152.0447	31.58	2.65	55.25	10.52
C 4H 4N602	168.12	168.0396	28.58	2.40	49.99	19.03
C 4H 4N603	184.11	184.0345	26.09	2.19	45.65	26.07
C 4H 4N604	200.11	200.0294	24.01	2.01	42.00	31.98
C 4H 4N605	216.11	216.0243	22.23	1.87	38.89	37.02
C 4H 4N606	232.11	232.0192	20.70	1.74	36.21	41.36
C 4H 4N607	248.11	248.0141	19.36	1.63	33.87	45.14
C 4H 4N608	264.11	264.0091	18.19	1.53	31.82	48.46
C 4H 4N609	280.11	280.0040	17.15	1.44	30.00	51.41
C 4H 5N101	83.09	83.0371	57.82	6.07	16.86	19.26
C 4H 5N102	99.09	99.0320	48.49	5.09	14.14	32.29
C 4H 5N103	115.09	115.0269	41.75	4.38	12.17	41.71
C 4H 5N104	131.09	131.0219	36.65	3.84	10.68	48.82
C 4H 5N105	147.09	147.0168	32.66	3.43	9.52	54.39
C 4H 5N106	163.09	163.0117	29.46	3.09	8.59	58.86
C 4H 5N107	179.09	179.0066	26.83	2.81	7.82	62.54
C 4H 5N108	195.09	195.0015	24.63	2.58	7.18	65.61
C 4H 5N109	211.09	210.9964	22.76	2.39	6.64	68.22
C 4H 5N301	111.10	111.0433	43.24	4.54	37.82	14.40
C 4H 5N302	127.10	127.0382	37.80	3.97	33.06	25.18
C 4H 5N303	143.10	143.0331	33.57	3.52	29.36	33.54
C 4H 5N304	159.10	159.0280	30.20	3.17	26.41	40.22
C 4H 5N305	175.10	175.0229	27.44	2.88	24.00	45.69
C 4H 5N306	191.10	191.0178	25.14	2.64	21.99	50.23
C 4H 5N307	207.10	207.0127	23.20	2.43	20.29	54.08
C 4H 5N308	223.10	223.0077	21.54	2.26	18.83	57.37
C 4H 5N309	239.10	239.0026	20.09	2.11	17.57	60.22
C 4H 5N501	139.12	139.0494	34.54	3.62	50.34	11.50
C 4H 5N502	155.12	155.0443	30.97	3.25	45.15	20.63
C 4H 5N503	171.12	171.0392	28.08	2.95	40.93	28.05
C 4H 5N504	187.12	187.0341	25.68	2.69	37.43	34.20

C 4H 5N505	203.11	203.0291	23.65	2.48	34.48	39.39
C 4H 5N506	219.11	219.0240	21.93	2.30	31.96	43.81
C 4H 5N507	235.11	235.0189	20.43	2.14	29.79	47.63
C 4H 5N508	251.11	251.0138	19.13	2.01	27.89	50.97
C 4H 5N509	267.11	267.0087	17.99	1.89	26.22	53.91
C 4H 6N201	98.11	98.0480	48.97	6.16	28.55	16.31
C 4H 6N202	114.10	114.0429	42.11	5.30	24.55	28.04
C 4H 6N203	130.10	130.0378	36.93	4.65	21.53	36.89
C 4H 6N204	146.10	146.0328	32.88	4.14	19.17	43.80
C 4H 6N205	162.10	162.0277	29.64	3.73	17.28	49.35
C 4H 6N206	178.10	178.0226	26.98	3.40	15.73	53.90
C 4H 6N207	194.10	194.0175	24.75	3.12	14.43	57.70
C 4H 6N208	210.10	210.0124	22.87	2.88	13.33	60.92
C 4H 6N209	226.10	226.0073	21.25	2.67	12.39	63.69
C 4H 6N401	126.12	126.0542	38.09	4.80	44.42	12.69
C 4H 6N402	142.12	142.0491	33.81	4.26	39.42	22.52
C 4H 6N403	158.12	158.0440	30.39	3.82	35.43	30.36
C 4H 6N404	174.12	174.0389	27.59	3.47	32.18	36.76
C 4H 6N405	190.12	190.0338	25.27	3.18	29.47	42.08
C 4H 6N406	206.12	206.0287	23.31	2.93	27.18	46.57
C 4H 6N407	222.12	222.0236	21.63	2.72	25.22	50.42
C 4H 6N408	238.11	238.0186	20.18	2.54	23.53	53.75
C 4H 6N409	254.11	254.0135	18.91	2.38	22.05	56.67
C 4H 6N601	154.13	154.0603	31.17	3.92	54.52	10.38
C 4H 6N602	170.13	170.0552	28.24	3.55	49.40	18.81
C 4H 6N603	186.13	186.0501	25.81	3.25	45.15	25.79
C 4H 6N604	202.13	202.0450	23.77	2.99	41.58	31.66
C 4H 6N605	218.13	218.0400	22.03	2.77	38.53	36.67
C 4H 6N606	234.13	234.0349	20.52	2.58	35.89	41.00
C 4H 6N607	250.13	250.0298	19.21	2.42	33.60	44.78
C 4H 6N608	266.13	266.0247	18.05	2.27	31.58	48.10
C 4H 6N609	282.13	282.0196	17.03	2.14	29.79	51.04
C 4H 7N101	85.11	85.0528	56.45	8.29	16.46	18.80
C 4H 7N102	101.11	101.0477	47.52	6.98	13.85	31.65
C 4H 7N103	117.11	117.0426	41.03	6.03	11.96	40.99
C 4H 7N104	133.10	133.0375	36.10	5.30	10.52	48.08
C 4H 7N105	149.10	149.0324	32.22	4.73	9.39	53.65
C 4H 7N106	165.10	165.0273	29.10	4.27	8.48	58.14
C 4H 7N107	181.10	181.0222	26.53	3.90	7.73	61.84
C 4H 7N108	197.10	197.0172	24.38	3.58	7.11	64.94
C 4H 7N109	213.10	213.0121	22.55	3.31	6.57	67.57
C 4H 7N301	113.12	113.0589	42.47	6.24	37.15	14.14
C 4H 7N302	129.12	129.0538	37.21	5.46	32.54	24.78
C 4H 7N303	145.12	145.0487	33.11	4.86	28.96	33.08
C 4H 7N304	161.12	161.0436	29.82	4.38	26.08	39.72
C 4H 7N305	177.12	177.0386	27.13	3.98	23.72	45.17
C 4H 7N306	193.12	193.0335	24.88	3.65	21.76	49.71
C 4H 7N307	209.12	209.0284	22.98	3.37	20.09	53.56
C 4H 7N308	225.12	225.0233	21.34	3.13	18.67	56.86
C 4H 7N309	241.12	241.0182	19.93	2.93	17.43	59.72
C 4H 7N501	141.13	141.0651	34.04	5.00	49.62	11.34
C 4H 7N502	157.13	157.0600	30.58	4.49	44.57	20.36
C 4H 7N503	173.13	173.0549	27.75	4.08	40.45	27.72
C 4H 7N504	189.13	189.0498	25.40	3.73	37.03	33.84
C 4H 7N505	205.13	205.0447	23.42	3.44	34.14	39.00
C 4H 7N506	221.13	221.0396	21.73	3.19	31.67	43.41
C 4H 7N507	237.13	237.0345	20.26	2.98	29.53	47.23
C 4H 7N508	253.13	253.0295	18.98	2.79	27.67	50.57
C 4H 7N509	269.13	269.0244	17.85	2.62	26.02	53.50
C 4H 8N201	100.12	100.0637	47.99	8.05	27.98	15.98
C 4H 8N202	116.12	116.0586	41.37	6.94	24.12	27.56
C 4H 8N203	132.12	132.0535	36.36	6.10	21.20	36.33
C 4H 8N204	148.12	148.0484	32.44	5.44	18.91	43.21
C 4H 8N205	164.12	164.0433	29.27	4.91	17.07	48.74
C 4H 8N206	180.12	180.0382	26.67	4.48	15.55	53.30
C 4H 8N207	196.12	196.0331	24.50	4.11	14.28	57.11
C 4H 8N208	212.12	212.0281	22.65	3.80	13.21	60.34
C 4H 8N209	228.12	228.0230	21.06	3.53	12.28	63.12
C 4H 8N401	128.13	128.0698	37.50	6.29	43.72	12.49
C 4H 8N402	144.13	144.0647	33.33	5.59	38.87	22.20
C 4H 8N403	160.13	160.0596	30.00	5.04	34.99	29.97
C 4H 8N404	176.13	176.0545	27.28	4.58	31.81	36.33
C 4H 8N405	192.13	192.0495	25.01	4.20	29.16	41.64
C 4H 8N406	208.13	208.0444	23.08	3.87	26.92	46.12
C 4H 8N407	224.13	224.0393	21.44	3.60	25.00	49.97
C 4H 8N408	240.13	240.0342	20.01	3.36	23.33	53.30
C 4H 8N409	256.13	256.0291	18.76	3.15	21.87	56.22
C 4H 8N601	156.15	156.0760	30.77	5.16	53.82	10.25
C 4H 8N602	172.15	172.0709	27.91	4.68	48.82	18.59
C 4H 8N603	188.15	188.0658	25.54	4.29	44.67	25.51
C 4H 8N604	204.15	204.0607	23.53	3.95	41.17	31.35
C 4H 8N605	220.15	220.0556	21.82	3.66	38.17	36.34
C 4H 8N606	236.14	236.0505	20.35	3.41	35.59	40.65
C 4H 8N607	252.14	252.0454	19.05	3.20	33.33	44.42
C 4H 8N608	268.14	268.0404	17.92	3.01	31.34	47.73
C 4H 8N609	284.14	284.0353	16.91	2.84	29.58	50.68
C 4H 9N101	87.12	87.0684	55.15	10.41	16.08	18.36
C 4H 9N102	103.12	103.0633	46.59	8.80	13.58	31.03
C 4H 9N103	119.12	119.0582	40.33	7.62	11.76	40.29
C 4H 9N104	135.12	135.0532	35.56	6.71	10.37	47.36
C 4H 9N105	151.12	151.0481	31.79	6.00	9.27	52.94
C 4H 9N106	167.12	167.0430	28.75	5.43	8.38	57.44
C 4H 9N107	183.12	183.0379	26.24	4.95	7.65	61.16
C 4H 9N108	199.12	199.0328	24.13	4.56	7.03	64.28
C 4H 9N109	215.12	215.0277	22.33	4.22	6.51	66.94
C 4H 9N301	115.14	115.0746	41.73	7.88	36.50	13.90
C 4H 9N302	131.14	131.0695	36.64	6.92	32.04	24.40
C 4H 9N303	147.13	147.0644	32.65	6.17	28.56	32.62
C 4H 9N304	163.13	163.0593	29.45	5.56	25.76	39.23
C 4H 9N305	179.13	179.0542	26.82	5.06	23.46	44.66
C 4H 9N306	195.13	195.0491	24.62	4.65	21.53	49.20
C 4H 9N307	211.13	211.0440	22.76	4.30	19.90	53.05
C 4H 9N308	227.13	227.0390	21.15	3.99	18.50	56.35
C 4H 9N309	243.13	243.0339	19.76	3.73	17.28	59.23
C 4H 9N501	143.15	143.0807	33.56	6.34	48.92	11.18
C 4H 9N502	159.15	159.0756	30.19	5.70	44.01	20.11
C 4H 9N503	175.15	175.0705	27.43	5.18	39.99	27.40
C 4H 9N504	191.15	191.0654	25.13	4.75	36.64	33.48
C 4H 9N505	207.15	207.0604	23.19	4.38	33.81	38.62
C 4H 9N506	223.15	223.0553	21.53	4.07	31.38	43.02
C 4H 9N507	239.15	239.0502	20.09	3.79	29.28	46.83
C 4H 9N508	255.15	255.0451	18.83	3.56	27.45	50.17
C 4H 9N509	271.14	271.0400	17.72	3.35	25.83	53.11
C 4H10N201	102.14	102.0793	47.04	9.87	27.43	15.66
C 4H10N202	118.14	118.0742	40.67	8.53	23.71	27.09
C 4H10N203	134.14	134.0691	35.82	7.51	20.88	35.78
C 4H10N204	150.14	150.0641	32.00	6.71	18.66	42.63
C 4H10N205	166.13	166.0590	28.92	6.07	16.86	48.15
C 4H10N206	182.13	182.0539	26.38	5.53	15.38	52.71
C 4H10N207	198.13	198.0488	24.25	5.09	14.14	56.53

025

C 4H10N208	214.13	214.0437	22.44	4.71	13.08	59.77
C 4H10N209	230.13	230.0386	20.88	4.38	12.17	62.57
C 4H10N401	130.15	130.0855	36.91	7.74	43.05	12.29
C 4H10N402	146.15	146.0804	32.87	6.90	38.34	21.89
C 4H10N403	162.15	162.0753	29.63	6.22	34.55	29.60
C 4H10N404	178.15	178.0702	26.97	5.66	31.45	35.92
C 4H10N405	194.15	194.0651	24.75	5.19	28.86	41.20
C 4H10N406	210.15	210.0600	22.86	4.80	26.66	45.68
C 4H10N407	226.15	226.0549	21.24	4.46	24.77	49.52
C 4H10N408	242.15	242.0499	19.84	4.16	23.14	52.86
C 4H10N409	258.15	258.0448	18.61	3.90	21.70	55.78
C 4H10N601	158.16	158.0916	30.38	6.37	53.13	10.12
C 4H10N602	174.16	174.0865	27.59	5.79	48.25	18.37
C 4H10N603	190.16	190.0814	25.26	5.30	44.19	25.24
C 4H10N604	206.16	206.0763	23.30	4.89	40.76	31.04
C 4H10N605	222.16	222.0713	21.63	4.54	37.83	36.01
C 4H10N606	238.16	238.0662	20.17	4.23	35.29	40.31
C 4H10N607	254.16	254.0611	18.90	3.97	33.07	44.07
C 4H10N608	270.16	270.0560	17.78	3.73	31.11	47.38
C 4H10N609	286.16	286.0509	16.79	3.52	29.37	50.32
C 4H11N101	89.14	89.0841	53.90	12.44	15.71	17.95
C 4H11N102	105.14	105.0790	45.70	10.55	13.32	30.44
C 4H11N103	121.14	121.0739	39.66	9.15	11.56	39.62
C 4H11N104	137.14	137.0688	35.03	8.09	10.21	46.67
C 4H11N105	153.14	153.0637	31.37	7.24	9.15	52.24
C 4H11N106	169.14	169.0586	28.41	6.56	8.28	56.76
C 4H11N107	185.13	185.0535	25.95	5.99	7.57	60.49
C 4H11N108	201.13	201.0485	23.89	5.51	6.96	63.64
C 4H11N109	217.13	217.0434	22.13	5.11	6.45	66.32
C 4H11N301	117.15	117.0902	41.01	9.46	35.87	13.66
C 4H11N302	133.15	133.0851	36.08	8.33	31.56	24.03
C 4H11N303	149.15	149.0800	32.21	7.43	28.17	32.18
C 4H11N304	165.15	165.0749	29.09	6.71	25.44	38.75
C 4H11N305	181.15	181.0699	26.52	6.12	23.20	44.16
C 4H11N306	197.15	197.0648	24.37	5.62	21.31	48.69
C 4H11N307	213.15	213.0597	22.54	5.20	19.71	52.54
C 4H11N308	229.15	229.0546	20.97	4.84	18.34	55.86
C 4H11N309	245.15	245.0495	19.60	4.52	17.14	58.74
C 4H11N501	145.17	145.0964	33.10	7.64	48.24	11.02
C 4H11N502	161.16	161.0913	29.81	6.88	43.45	19.85
C 4H11N503	177.16	177.0862	27.12	6.26	39.53	27.09
C 4H11N504	193.16	193.0811	24.87	5.74	36.26	33.13
C 4H11N505	209.16	209.0760	22.97	5.30	33.48	38.25
C 4H11N506	225.16	225.0709	21.34	4.92	31.10	42.63
C 4H11N507	241.16	241.0658	19.92	4.60	29.04	46.44
C 4H11N508	257.16	257.0608	18.68	4.31	27.23	49.77
C 4H11N509	273.16	273.0557	17.59	4.06	25.64	52.71
C 4H12N201	104.15	104.0950	46.13	11.61	26.90	15.36
C 4H12N202	120.15	120.0899	39.99	10.07	23.31	26.63
C 4H12N203	136.15	136.0848	35.29	8.88	20.58	35.25
C 4H12N204	152.15	152.0797	31.58	7.95	18.41	42.06
C 4H12N205	168.15	168.0746	28.57	7.19	16.66	47.57
C 4H12N206	184.15	184.0695	26.09	6.57	15.21	52.13
C 4H12N207	200.15	200.0644	24.00	6.04	14.00	55.96
C 4H12N208	216.15	216.0594	22.23	5.60	12.96	59.22
C 4H12N209	232.15	232.0543	20.70	5.21	12.07	62.03
C 4H12N401	132.17	132.1011	36.35	9.15	42.39	12.11
C 4H12N402	148.17	148.0960	32.43	8.16	37.81	21.60
C 4H12N403	164.17	164.0909	29.27	7.37	34.13	29.24
C 4H12N404	180.16	180.0858	26.67	6.71	31.10	35.52
C 4H12N405	196.16	196.0808	24.49	6.17	28.56	40.78
C 4H12N406	212.16	212.0757	22.65	5.70	26.41	45.25
C 4H12N407	228.16	228.0706	21.06	5.30	24.56	49.09
C 4H12N408	244.16	244.0655	19.68	4.95	22.95	52.42
C 4H12N409	260.16	260.0604	18.47	4.65	21.54	55.35
C 4H12N601	160.18	160.1073	29.99	7.55	52.47	9.99
C 4H12N602	176.18	176.1022	27.27	6.87	47.70	18.16
C 4H12N603	192.18	192.0971	25.00	6.29	43.73	24.98
C 4H12N604	208.18	208.0920	23.08	5.81	40.37	30.74
C 4H12N605	224.18	224.0869	21.43	5.40	37.49	35.68
C 4H12N606	240.18	240.0818	20.00	5.04	34.99	39.97
C 4H12N607	256.18	256.0767	18.75	4.72	32.81	43.72
C 4H12N608	272.18	272.0717	17.65	4.44	30.88	47.03
C 4H12N609	288.18	288.0666	16.67	4.20	29.16	49.97
C 4H13N301	119.17	119.1059	40.32	11.00	35.26	13.43
C 4H13N302	135.17	135.1008	35.54	9.69	31.09	23.67
C 4H13N303	151.17	151.0957	31.78	8.67	27.80	31.75
C 4H13N304	167.17	167.0906	28.74	7.84	25.14	38.28
C 4H13N305	183.17	183.0855	26.23	7.15	22.94	43.67
C 4H13N306	199.16	199.0804	24.12	6.58	21.10	48.20
C 4H13N307	215.16	215.0753	22.33	6.09	19.53	52.05
C 4H13N308	231.16	231.0703	20.78	5.67	18.18	55.37
C 4H13N309	247.16	247.0652	19.44	5.30	17.00	58.26
C 4H13N501	147.18	147.1120	32.64	8.90	47.58	10.87
C 4H13N502	163.18	163.1069	29.44	8.03	42.92	19.61
C 4H13N503	179.18	179.1018	26.81	7.31	39.09	26.79
C 4H13N504	195.18	195.0967	24.62	6.71	35.88	32.79
C 4H13N505	211.18	211.0917	22.75	6.20	33.16	37.88
C 4H13N506	227.18	227.0866	21.15	5.77	30.83	42.26
C 4H13N507	243.18	243.0815	19.76	5.39	28.80	46.06
C 4H13N508	259.18	259.0764	18.54	5.06	27.02	49.39
C 4H13N509	275.18	275.0713	17.46	4.76	25.45	52.33
C 4H14N401	134.18	134.1168	35.81	10.52	41.75	11.92
C 4H14N402	150.18	150.1117	31.99	9.40	37.31	21.31
C 4H14N403	166.18	166.1066	28.91	8.49	33.71	28.88
C 4H14N404	182.18	182.1015	26.37	7.75	30.75	35.13
C 4H14N405	198.18	198.0964	24.24	7.12	28.27	40.37
C 4H14N406	214.18	214.0913	22.43	6.59	26.16	44.82
C 4H14N407	230.18	230.0862	20.87	6.13	24.34	48.66
C 4H14N408	246.18	246.0812	19.52	5.73	22.76	51.99
C 4H14N409	262.18	262.0761	18.33	5.38	21.37	54.92
C 4H14N601	162.20	162.1229	29.62	8.70	51.81	9.86
C 4H14N602	178.20	178.1178	26.96	7.92	47.16	17.96
C 4H14N603	194.19	194.1127	24.74	7.27	43.28	24.72
C 4H14N604	210.19	210.1076	22.86	6.71	39.98	30.45
C 4H14N605	226.19	226.1026	21.24	6.24	37.15	35.37
C 4H14N606	242.19	242.0975	19.84	5.83	34.70	39.64
C 4H14N607	258.19	258.0924	18.61	5.47	32.55	43.38
C 4H14N608	274.19	274.0873	17.52	5.15	30.65	46.68
C 4H14N609	290.19	290.0822	16.56	4.86	28.96	49.62
C 4H15N501	149.20	149.1277	32.20	10.13	46.94	10.72
C 4H15N502	165.20	165.1226	29.08	9.15	42.39	19.37
C 4H15N503	181.20	181.1175	26.52	8.34	38.65	26.49
C 4H15N504	197.20	197.1124	24.36	7.67	35.51	32.45
C 4H15N505	213.19	213.1073	22.54	7.09	32.85	37.52
C 4H15N506	229.19	229.1022	20.96	6.60	30.56	41.88
C 4H15N507	245.19	245.0971	19.59	6.17	28.56	45.68
C 4H15N508	261.19	261.0921	18.39	5.79	26.81	49.00
C 4H15N509	277.19	277.0870	17.33	5.45	25.27	51.95
C 4H16N601	164.21	164.1386	29.26	9.82	51.18	9.74

026

C 4H16N6O2	180.21	180.1335	26.66	8.95	46.63	17.76
C 4H16N6O3	196.21	196.1284	24.49	8.22	42.83	24.46
C 4H16N6O4	212.21	212.1233	22.64	7.60	39.60	30.16
C 4H16N6O5	228.21	228.1182	21.05	7.07	36.83	35.05
C 4H16N6O6	244.21	244.1131	19.67	6.60	34.41	39.31
C 4H16N6O7	260.21	260.1080	18.46	6.20	32.30	43.04
C 4H16N6O8	276.21	276.1029	17.39	5.84	30.43	46.34
C 4H16N6O9	292.21	292.0979	16.44	5.52	28.76	49.28
C 5H 2N2O1	106.08	106.0167	56.61	1.90	26.41	15.08
C 5H 2N2O2	122.08	122.0116	49.19	1.65	22.95	26.21
C 5H 2N2O3	138.08	138.0065	43.49	1.46	20.29	34.76
C 5H 2N2O4	154.08	154.0015	38.98	1.31	18.18	41.53
C 5H 2N2O5	170.08	169.9964	35.31	1.19	16.47	47.03
C 5H 2N2O6	186.08	185.9913	32.27	1.08	15.05	51.59
C 5H 2N2O7	202.08	201.9862	29.72	1.00	13.86	55.42
C 5H 2N2O8	218.08	217.9811	27.54	.92	12.85	58.69
C 5H 2N2O9	234.08	233.9760	25.66	.86	11.97	61.52
C 5H 2N4O1	134.10	134.0229	44.79	1.50	41.78	11.93
C 5H 2N4O2	150.10	150.0178	40.01	1.34	37.33	21.32
C 5H 2N4O3	166.10	166.0127	36.16	1.21	33.73	28.90
C 5H 2N4O4	182.10	182.0076	32.98	1.11	30.77	35.14
C 5H 2N4O5	198.10	198.0025	30.32	1.02	28.28	40.38
C 5H 2N4O6	214.09	213.9974	28.05	.94	26.17	44.84
C 5H 2N4O7	230.09	229.9923	26.10	.88	24.35	48.67
C 5H 2N4O8	246.09	245.9873	24.40	.82	22.77	52.01
C 5H 2N4O9	262.09	261.9822	22.91	.77	21.38	54.94
C 5H 2N6O1	162.11	162.0290	37.05	1.24	51.84	9.87
C 5H 2N6O2	178.11	178.0239	33.72	1.13	47.18	17.97
C 5H 2N6O3	194.11	194.0188	30.94	1.04	43.30	24.73
C 5H 2N6O4	210.11	210.0137	28.58	.96	40.00	30.46
C 5H 2N6O5	226.11	226.0087	26.56	.89	37.17	35.38
C 5H 2N6O6	242.11	242.0036	24.81	.83	34.71	39.65
C 5H 2N6O7	258.11	257.9985	23.27	.78	32.56	43.39
C 5H 2N6O8	274.11	273.9934	21.91	.74	30.66	46.70
C 5H 2N6O9	290.11	289.9883	20.70	.69	28.97	49.64
C 5H 3N1O1	93.09	93.0215	64.52	3.25	15.05	17.19
C 5H 3N1O2	109.09	109.0164	55.05	2.77	12.84	29.33
C 5H 3N1O3	125.08	125.0113	48.01	2.42	11.20	38.37
C 5H 3N1O4	141.08	141.0062	42.57	2.14	9.93	45.36
C 5H 3N1O5	157.08	157.0011	38.23	1.93	8.92	50.93
C 5H 3N1O6	173.08	172.9960	34.70	1.75	8.09	55.46
C 5H 3N1O7	189.08	188.9909	31.76	1.60	7.41	59.23
C 5H 3N1O8	205.08	204.9859	29.28	1.47	6.83	62.41
C 5H 3N1O9	221.08	220.9808	27.16	1.37	6.34	65.13
C 5H 3N3O1	121.10	121.0276	49.59	2.50	34.70	13.21
C 5H 3N3O2	137.10	137.0225	43.80	2.21	30.65	23.34
C 5H 3N3O3	153.10	153.0174	39.23	1.98	27.45	31.35
C 5H 3N3O4	169.10	169.0124	35.52	1.79	24.85	37.85
C 5H 3N3O5	185.10	185.0073	32.45	1.63	22.70	43.22
C 5H 3N3O6	201.10	201.0022	29.86	1.50	20.90	47.74
C 5H 3N3O7	217.10	216.9971	27.66	1.39	19.36	51.59
C 5H 3N3O8	233.09	232.9920	25.76	1.30	18.03	54.91
C 5H 3N3O9	249.09	248.9869	24.11	1.21	16.87	57.81
C 5H 3N5O1	149.11	149.0338	40.28	2.03	46.97	10.73
C 5H 3N5O2	165.11	165.0287	36.37	1.83	42.42	19.38
C 5H 3N5O3	181.11	181.0236	33.16	1.67	38.67	26.50
C 5H 3N5O4	197.11	197.0185	30.47	1.53	35.53	32.47
C 5H 3N5O5	213.11	213.0134	28.18	1.42	32.86	37.54
C 5H 3N5O6	229.11	229.0083	26.21	1.32	30.57	41.90
C 5H 3N5O7	245.11	245.0032	24.50	1.23	28.57	45.69

C 5H 3N5O8	261.11	260.9982	23.00	1.16	26.82	49.02
C 5H 3N5O9	277.11	276.9931	21.67	1.09	25.27	51.96
C 5H 4N2O1	108.10	108.0324	55.56	3.73	25.91	14.80
C 5H 4N2O2	124.10	124.0273	48.39	3.25	22.57	25.78
C 5H 4N2O3	140.10	140.0222	42.87	2.88	20.00	34.26
C 5H 4N2O4	156.10	156.0171	38.47	2.58	17.95	41.00
C 5H 4N2O5	172.10	172.0120	34.90	2.34	16.28	46.48
C 5H 4N2O6	188.10	188.0069	31.93	2.14	14.89	51.04
C 5H 4N2O7	204.10	204.0018	29.43	1.98	13.73	54.87
C 5H 4N2O8	220.10	219.9968	27.29	1.83	12.73	58.15
C 5H 4N2O9	236.10	235.9917	25.44	1.71	11.87	60.99
C 5H 4N4O1	136.11	136.0385	44.12	2.96	41.16	11.75
C 5H 4N4O2	152.11	152.0334	39.48	2.65	36.83	21.04
C 5H 4N4O3	168.11	168.0283	35.72	2.40	33.33	28.55
C 5H 4N4O4	184.11	184.0232	32.62	2.19	30.43	34.76
C 5H 4N4O5	200.11	200.0182	30.01	2.01	28.00	39.98
C 5H 4N4O6	216.11	216.0131	27.79	1.87	25.93	44.42
C 5H 4N4O7	232.11	232.0080	25.87	1.74	24.14	48.25
C 5H 4N4O8	248.11	248.0029	24.21	1.63	22.58	51.59
C 5H 4N4O9	264.11	263.9978	22.74	1.53	21.21	54.52
C 5H 4N6O1	164.13	164.0447	36.59	2.46	51.20	9.75
C 5H 4N6O2	180.13	180.0396	33.34	2.24	46.66	17.76
C 5H 4N6O3	196.13	196.0345	30.62	2.06	42.85	24.47
C 5H 4N6O4	212.13	212.0294	28.31	1.90	39.62	30.17
C 5H 4N6O5	228.12	228.0243	26.33	1.77	36.84	35.07
C 5H 4N6O6	244.12	244.0192	24.60	1.65	34.43	39.32
C 5H 4N6O7	260.12	260.0141	23.09	1.55	32.31	43.05
C 5H 4N6O8	276.12	276.0091	21.75	1.46	30.44	46.35
C 5H 4N6O9	292.12	292.0040	20.56	1.38	28.77	49.29
C 5H 5N1O1	95.10	95.0371	63.15	5.30	14.73	16.82
C 5H 5N1O2	111.10	111.0320	54.06	4.54	12.61	28.80
C 5H 5N1O3	127.10	127.0269	47.25	3.97	11.02	37.76
C 5H 5N1O4	143.10	143.0219	41.97	3.52	9.79	44.72
C 5H 5N1O5	159.10	159.0168	37.75	3.17	8.80	50.28
C 5H 5N1O6	175.10	175.0117	34.30	2.88	8.00	54.82
C 5H 5N1O7	191.10	191.0066	31.43	2.64	7.33	58.61
C 5H 5N1O8	207.10	207.0015	29.00	2.43	6.76	61.80
C 5H 5N1O9	223.10	222.9964	26.92	2.26	6.28	64.54
C 5H 5N3O1	123.12	123.0433	48.78	4.09	34.13	13.00
C 5H 5N3O2	139.11	139.0382	43.17	3.62	30.21	23.00
C 5H 5N3O3	155.11	155.0331	38.72	3.25	27.09	30.94
C 5H 5N3O4	171.11	171.0280	35.10	2.95	24.56	37.40
C 5H 5N3O5	187.11	187.0229	32.10	2.69	22.46	42.75
C 5H 5N3O6	203.11	203.0178	29.57	2.48	20.69	47.26
C 5H 5N3O7	219.11	219.0127	27.41	2.30	19.18	51.11
C 5H 5N3O8	235.11	235.0077	25.54	2.14	17.87	54.44
C 5H 5N3O9	251.11	251.0026	23.92	2.01	16.73	57.34
C 5H 5N5O1	151.13	151.0494	39.74	3.33	46.34	10.59
C 5H 5N5O2	167.13	167.0443	35.93	3.02	41.90	19.15
C 5H 5N5O3	183.13	183.0392	32.79	2.75	38.24	26.21
C 5H 5N5O4	199.13	199.0341	30.16	2.53	35.17	32.14
C 5H 5N5O5	215.13	215.0291	27.92	2.34	32.55	37.19
C 5H 5N5O6	231.13	231.0240	25.98	2.18	30.30	41.53
C 5H 5N5O7	247.12	247.0189	24.30	2.04	28.34	45.32
C 5H 5N5O8	263.12	263.0138	22.82	1.92	26.62	48.64
C 5H 5N5O9	279.12	279.0087	21.52	1.81	25.09	51.59
C 5H 6N2O1	110.12	110.0480	54.54	5.49	25.44	14.53
C 5H 6N2O2	126.12	126.0429	47.62	4.80	22.21	25.37
C 5H 6N2O3	142.12	142.0378	42.26	4.26	19.71	33.77
C 5H 6N2O4	158.11	158.0328	37.98	3.82	17.72	40.48

027

C 5H 6N2O5	174.11	174.0277	34.49	3.47	16.09	45.95
C 5H 6N2O6	190.11	190.0226	31.59	3.18	14.74	50.49
C 5H 6N2O7	206.11	206.0175	29.14	2.93	13.59	54.34
C 5H 6N2O8	222.11	222.0124	27.04	2.72	12.61	57.63
C 5H 6N2O9	238.11	238.0073	25.22	2.54	11.76	60.47
C 5H 6N4O1	138.13	138.0542	43.48	4.38	40.56	11.58
C 5H 6N4O2	154.13	154.0491	38.96	3.92	36.35	20.76
C 5H 6N4O3	170.13	170.0440	35.30	3.55	32.93	28.21
C 5H 6N4O4	186.13	186.0389	32.27	3.25	30.10	34.38
C 5H 6N4O5	202.13	202.0338	29.71	2.99	27.72	39.58
C 5H 6N4O6	218.13	218.0287	27.53	2.77	25.69	44.01
C 5H 6N4O7	234.13	234.0236	25.65	2.58	23.93	47.84
C 5H 6N4O8	250.13	250.0186	24.01	2.42	22.40	51.17
C 5H 6N4O9	266.12	266.0135	22.57	2.27	21.05	54.11
C 5H 6N6O1	166.14	166.0603	36.15	3.64	50.58	9.63
C 5H 6N6O2	182.14	182.0552	32.97	3.32	46.14	17.57
C 5H 6N6O3	198.14	198.0501	30.31	3.05	42.41	24.22
C 5H 6N6O4	214.14	214.0450	28.04	2.82	39.25	29.89
C 5H 6N6O5	230.14	230.0400	26.10	2.63	36.52	34.76
C 5H 6N6O6	246.14	246.0349	24.40	2.46	34.14	39.00
C 5H 6N6O7	262.14	262.0298	22.91	2.31	32.06	42.72
C 5H 6N6O8	278.14	278.0247	21.59	2.17	30.22	46.02
C 5H 6N6O9	294.14	294.0196	20.42	2.06	28.57	48.95
C 5H 7N1O1	97.12	97.0528	61.84	7.27	14.42	16.47
C 5H 7N1O2	113.12	113.0477	53.09	6.24	12.38	28.29
C 5H 7N1O3	129.12	129.0426	46.51	5.46	10.85	37.17
C 5H 7N1O4	145.12	145.0375	41.38	4.86	9.65	44.10
C 5H 7N1O5	161.12	161.0324	37.28	4.38	8.69	49.65
C 5H 7N1O6	177.11	177.0273	33.91	3.98	7.91	54.20
C 5H 7N1O7	193.11	193.0222	31.10	3.65	7.25	57.99
C 5H 7N1O8	209.11	209.0172	28.72	3.37	6.70	61.21
C 5H 7N1O9	225.11	225.0121	26.68	3.13	6.22	63.97
C 5H 7N3O1	125.13	125.0589	47.99	5.64	33.58	12.79
C 5H 7N3O2	141.13	141.0538	42.55	5.00	29.77	22.67
C 5H 7N3O3	157.13	157.0487	38.22	4.49	26.74	30.55
C 5H 7N3O4	173.13	173.0437	34.69	4.08	24.27	36.97
C 5H 7N3O5	189.13	189.0386	31.75	3.73	22.22	42.30
C 5H 7N3O6	205.13	205.0335	29.28	3.44	20.48	46.80
C 5H 7N3O7	221.13	221.0284	27.16	3.19	19.00	50.65
C 5H 7N3O8	237.13	237.0233	25.33	2.98	17.72	53.98
C 5H 7N3O9	253.13	253.0182	23.73	2.79	16.60	56.89
C 5H 7N5O1	153.14	153.0651	39.22	4.61	45.73	10.45
C 5H 7N5O2	169.14	169.0600	35.51	4.17	41.40	18.92
C 5H 7N5O3	185.14	185.0549	32.44	3.81	37.83	25.92
C 5H 7N5O4	201.14	201.0498	29.86	3.51	34.82	31.82
C 5H 7N5O5	217.14	217.0447	27.66	3.25	32.25	36.84
C 5H 7N5O6	233.14	233.0396	25.76	3.03	30.04	41.18
C 5H 7N5O7	249.14	249.0345	24.11	2.83	28.11	44.95
C 5H 7N5O8	265.14	265.0295	22.65	2.66	26.41	48.27
C 5H 7N5O9	281.14	281.0244	21.36	2.51	24.91	51.22
C 5H 8N2O1	112.13	112.0637	53.56	7.19	24.98	14.27
C 5H 8N2O2	128.13	128.0586	46.87	6.29	21.86	24.97
C 5H 8N2O3	144.13	144.0535	41.67	5.59	19.44	33.30
C 5H 8N2O4	160.13	160.0484	37.50	5.04	17.49	39.97
C 5H 8N2O5	176.13	176.0433	34.10	4.58	15.90	45.42
C 5H 8N2O6	192.13	192.0382	31.26	4.20	14.58	49.96
C 5H 8N2O7	208.13	208.0331	28.86	3.87	13.46	53.81
C 5H 8N2O8	224.13	224.0281	26.80	3.60	12.50	57.11
C 5H 8N2O9	240.13	240.0230	25.01	3.36	11.67	59.97
C 5H 8N4O1	140.15	140.0698	42.85	5.75	39.98	11.42

C 5H 8N4O2	156.15	156.0647	38.46	5.16	35.88	20.49
C 5H 8N4O3	172.14	172.0596	34.89	4.68	32.55	27.88
C 5H 8N4O4	188.14	188.0545	31.92	4.29	29.78	34.02
C 5H 8N4O5	204.14	204.0495	29.42	3.95	27.44	39.19
C 5H 8N4O6	220.14	220.0444	27.28	3.66	25.45	43.61
C 5H 8N4O7	236.14	236.0393	25.43	3.41	23.73	47.43
C 5H 8N4O8	252.14	252.0342	23.82	3.20	22.22	50.76
C 5H 8N4O9	268.14	268.0291	22.40	3.01	20.89	53.70
C 5H 8N6O1	168.16	168.0760	35.71	4.80	49.98	9.51
C 5H 8N6O2	184.16	184.0709	32.61	4.38	45.63	17.38
C 5H 8N6O3	200.16	200.0658	30.00	4.03	41.99	23.98
C 5H 8N6O4	216.16	216.0607	27.78	3.73	38.88	29.61
C 5H 8N6O5	232.16	232.0556	25.87	3.47	36.20	34.46
C 5H 8N6O6	248.16	248.0505	24.20	3.25	33.87	38.68
C 5H 8N6O7	264.16	264.0454	22.73	3.05	31.81	42.40
C 5H 8N6O8	280.15	280.0404	21.44	2.88	30.00	45.69
C 5H 8N6O9	296.15	296.0353	20.28	2.72	28.38	48.62
C 5H 9N1O1	99.13	99.0684	60.58	9.15	14.13	16.14
C 5H 9N1O2	115.13	115.0633	52.16	7.88	12.17	27.79
C 5H 9N1O3	131.13	131.0582	45.80	6.92	10.68	36.60
C 5H 9N1O4	147.13	147.0532	40.82	6.17	9.52	43.50
C 5H 9N1O5	163.13	163.0481	36.81	5.56	8.59	49.04
C 5H 9N1O6	179.13	179.0430	33.53	5.06	7.82	53.59
C 5H 9N1O7	195.13	195.0379	30.78	4.65	7.18	57.40
C 5H 9N1O8	211.13	211.0328	28.44	4.30	6.63	60.62
C 5H 9N1O9	227.13	227.0277	26.44	3.99	6.17	63.40
C 5H 9N3O1	127.15	127.0746	47.23	7.13	33.05	12.58
C 5H 9N3O2	143.15	143.0695	41.95	6.34	29.35	22.35
C 5H 9N3O3	159.15	159.0644	37.74	5.70	26.40	30.16
C 5H 9N3O4	175.15	175.0593	34.29	5.18	23.99	36.54
C 5H 9N3O5	191.14	191.0542	31.42	4.75	21.98	41.85
C 5H 9N3O6	207.14	207.0491	28.99	4.38	20.29	46.34
C 5H 9N3O7	223.14	223.0440	26.91	4.07	18.83	50.19
C 5H 9N3O8	239.14	239.0390	25.11	3.79	17.57	53.52
C 5H 9N3O9	255.14	255.0339	23.54	3.56	16.47	56.44
C 5H 9N5O1	155.16	155.0807	38.71	5.85	45.14	10.31
C 5H 9N5O2	171.16	171.0756	35.09	5.30	40.92	18.70
C 5H 9N5O3	187.16	187.0705	32.09	4.85	37.42	25.65
C 5H 9N5O4	203.16	203.0654	29.56	4.47	34.47	31.50
C 5H 9N5O5	219.16	219.0604	27.40	4.14	31.96	36.50
C 5H 9N5O6	235.16	235.0553	25.54	3.86	29.78	40.82
C 5H 9N5O7	251.16	251.0502	23.91	3.61	27.88	44.59
C 5H 9N5O8	267.16	267.0451	22.48	3.40	26.21	47.91
C 5H 9N5O9	283.16	283.0400	21.21	3.20	24.73	50.85
C 5H10N2O1	114.15	114.0793	52.61	8.83	24.54	14.02
C 5H10N2O2	130.15	130.0742	46.14	7.74	21.52	24.59
C 5H10N2O3	146.15	146.0691	41.09	6.90	19.17	32.84
C 5H10N2O4	162.15	162.0641	37.04	6.22	17.28	39.47
C 5H10N2O5	178.15	178.0590	33.71	5.66	15.72	44.91
C 5H10N2O6	194.15	194.0539	30.93	5.19	14.43	49.45
C 5H10N2O7	210.14	210.0488	28.58	4.80	13.33	53.29
C 5H10N2O8	226.14	226.0437	26.56	4.46	12.39	56.60
C 5H10N2O9	242.14	242.0386	24.80	4.16	11.57	59.47
C 5H10N4O1	142.16	142.0855	42.24	7.09	39.41	11.25
C 5H10N4O2	158.16	158.0804	37.97	6.37	35.42	20.23
C 5H10N4O3	174.16	174.0753	34.48	5.79	32.17	27.56
C 5H10N4O4	190.16	190.0702	31.58	5.30	29.46	33.65
C 5H10N4O5	206.16	206.0651	29.13	4.89	27.18	38.80
C 5H10N4O6	222.16	222.0600	27.03	4.54	25.22	43.21
C 5H10N4O7	238.16	238.0549	25.22	4.23	23.53	47.03

C 5H10N408	254.16	254.0499	23.63	3.97	22.04	50.36
C 5H10N409	270.16	270.0448	22.23	3.73	20.74	53.30
C 5H10N601	170.18	170.0916	35.29	5.92	49.38	9.40
C 5H10N602	186.17	186.0865	32.26	5.41	45.14	17.19
C 5H10N603	202.17	202.0814	29.71	4.99	41.57	23.74
C 5H10N604	218.17	218.0763	27.53	4.62	38.52	29.33
C 5H10N605	234.17	234.0713	25.65	4.30	35.89	34.16
C 5H10N606	250.17	250.0662	24.01	4.03	33.59	38.37
C 5H10N607	266.17	266.0611	22.56	3.79	31.57	42.08
C 5H10N608	282.17	282.0560	21.28	3.57	29.78	45.36
C 5H10N609	298.17	298.0509	20.14	3.38	28.19	48.29
C 5H11N101	101.15	101.0841	59.37	10.96	13.85	15.82
C 5H11N102	117.15	117.0790	51.26	9.46	11.96	27.31
C 5H11N103	133.15	133.0739	45.10	8.33	10.52	36.05
C 5H11N104	149.15	149.0688	40.27	7.43	9.39	42.91
C 5H11N105	165.15	165.0637	36.36	6.71	8.48	48.44
C 5H11N106	181.15	181.0586	33.15	6.12	7.73	52.99
C 5H11N107	197.15	197.0535	30.46	5.62	7.10	56.81
C 5H11N108	213.15	213.0485	28.18	5.20	6.57	60.05
C 5H11N109	229.14	229.0434	26.21	4.84	6.11	62.84
C 5H11N301	129.16	129.0902	46.50	8.58	32.53	12.39
C 5H11N302	145.16	145.0851	41.37	7.64	28.95	22.04
C 5H11N303	161.16	161.0800	37.26	6.88	26.07	29.78
C 5H11N304	177.16	177.0749	33.90	6.26	23.72	36.12
C 5H11N305	193.16	193.0699	31.09	5.74	21.75	41.41
C 5H11N306	209.16	209.0648	28.71	5.30	20.09	45.90
C 5H11N307	225.16	225.0597	26.67	4.92	18.66	49.74
C 5H11N308	241.16	241.0546	24.90	4.60	17.42	53.08
C 5H11N309	257.16	257.0495	23.35	4.31	16.34	55.99
C 5H11N501	157.18	157.0964	38.21	7.05	44.56	10.18
C 5H11N502	173.18	173.0913	34.68	6.40	40.44	18.48
C 5H11N503	189.18	189.0862	31.75	5.86	37.02	25.37
C 5H11N504	205.17	205.0811	29.27	5.40	34.13	31.19
C 5H11N505	221.17	221.0760	27.15	5.01	31.66	36.17
C 5H11N506	237.17	237.0709	25.32	4.67	29.53	40.48
C 5H11N507	253.17	253.0658	23.72	4.38	27.66	44.24
C 5H11N508	269.17	269.0608	22.31	4.12	26.02	47.55
C 5H11N509	285.17	285.0557	21.06	3.89	24.56	50.49
C 5H12N201	116.16	116.0950	51.70	10.41	24.12	13.77
C 5H12N202	132.16	132.0899	45.44	9.15	21.20	24.21
C 5H12N203	148.16	148.0848	40.53	8.16	18.91	32.40
C 5H12N204	164.16	164.0797	36.58	7.37	17.06	38.98
C 5H12N205	180.16	180.0746	33.33	6.71	15.55	44.40
C 5H12N206	196.16	196.0695	30.62	6.17	14.28	48.94
C 5H12N207	212.16	212.0644	28.31	5.70	13.20	52.79
C 5H12N208	228.16	228.0594	26.32	5.30	12.28	56.10
C 5H12N209	244.16	244.0543	24.60	4.95	11.47	58.98
C 5H12N401	144.18	144.1011	41.65	8.39	38.86	11.10
C 5H12N402	160.18	160.0960	37.49	7.55	34.98	19.98
C 5H12N403	176.18	176.0909	34.09	6.87	31.80	27.24
C 5H12N404	192.18	192.0858	31.25	6.29	29.15	33.30
C 5H12N405	208.18	208.0808	28.85	5.81	26.91	38.43
C 5H12N406	224.17	224.0757	26.79	5.40	24.99	42.82
C 5H12N407	240.17	240.0706	25.01	5.04	23.33	46.63
C 5H12N408	256.17	256.0655	23.44	4.72	21.87	49.96
C 5H12N409	272.17	272.0604	22.07	4.44	20.59	52.91
C 5H12N601	172.19	172.1073	34.88	7.02	48.81	9.29
C 5H12N602	188.19	188.1022	31.91	6.43	44.66	17.00
C 5H12N603	204.19	204.0971	29.41	5.92	41.16	23.51
C 5H12N604	220.19	220.0920	27.27	5.49	38.17	29.06

C 5H12N605	236.19	236.0869	25.43	5.12	35.58	33.87
C 5H12N606	252.19	252.0818	23.81	4.80	33.32	38.07
C 5H12N607	268.19	268.0767	22.39	4.51	31.34	41.76
C 5H12N608	284.19	284.0717	21.13	4.26	29.57	45.04
C 5H12N609	300.19	300.0666	20.01	4.03	28.00	47.97
C 5H13N101	103.17	103.0997	58.21	12.70	13.58	15.51
C 5H13N102	119.16	119.0946	50.40	11.00	11.75	26.85
C 5H13N103	135.16	135.0895	44.43	9.69	10.36	35.51
C 5H13N104	151.16	151.0844	39.73	8.67	9.27	42.34
C 5H13N105	167.16	167.0794	35.93	7.84	8.38	47.86
C 5H13N106	183.16	183.0743	32.79	7.15	7.65	52.41
C 5H13N107	199.16	199.0692	30.15	6.58	7.03	56.23
C 5H13N108	215.16	215.0641	27.91	6.09	6.51	59.49
C 5H13N109	231.16	231.0590	25.98	5.67	6.06	62.29
C 5H13N301	131.18	131.1059	45.78	9.99	32.03	12.20
C 5H13N302	147.18	147.1008	40.80	8.90	28.55	21.74
C 5H13N303	163.18	163.0957	36.80	8.03	25.75	29.41
C 5H13N304	179.18	179.0906	33.52	7.31	23.45	35.72
C 5H13N305	195.18	195.0855	30.77	6.71	21.53	40.99
C 5H13N306	211.18	211.0804	28.44	6.21	19.90	45.46
C 5H13N307	227.18	227.0753	26.44	5.77	18.50	49.30
C 5H13N308	243.17	243.0703	24.70	5.39	17.28	52.64
C 5H13N309	259.17	259.0652	23.17	5.06	16.21	55.56
C 5H13N501	159.19	159.1120	37.73	8.23	43.99	10.05
C 5H13N502	175.19	175.1069	34.28	7.48	39.98	18.27
C 5H13N503	191.19	191.1018	31.41	6.85	36.63	25.10
C 5H13N504	207.19	207.0967	28.99	6.32	33.80	30.89
C 5H13N505	223.19	223.0917	26.91	5.87	31.38	35.84
C 5H13N506	239.19	239.0866	25.11	5.48	29.28	40.13
C 5H13N507	255.19	255.0815	23.53	5.13	27.44	43.89
C 5H13N508	271.19	271.0764	22.15	4.83	25.82	47.20
C 5H13N509	287.19	287.0713	20.91	4.56	24.39	50.14
C 5H14N201	118.18	118.1106	50.82	11.94	23.70	13.54
C 5H14N202	134.18	134.1055	44.76	10.52	20.88	23.85
C 5H14N203	150.18	150.1004	39.99	9.40	18.65	31.96
C 5H14N204	166.18	166.0953	36.14	8.49	16.86	38.51
C 5H14N205	182.18	182.0903	32.97	7.75	15.38	43.91
C 5H14N206	198.18	198.0852	30.30	7.12	14.14	48.44
C 5H14N207	214.18	214.0801	28.04	6.59	13.08	52.29
C 5H14N208	230.18	230.0750	26.09	6.13	12.17	55.61
C 5H14N209	246.18	246.0699	24.40	5.73	11.38	58.49
C 5H14N401	146.19	146.1168	41.08	9.65	38.32	10.94
C 5H14N402	162.19	162.1117	37.03	8.70	34.54	19.73
C 5H14N403	178.19	178.1066	33.70	7.92	31.44	26.94
C 5H14N404	194.19	194.1015	30.93	7.27	28.85	32.96
C 5H14N405	210.19	210.0964	28.57	6.71	26.66	38.06
C 5H14N406	226.19	226.0913	26.55	6.24	24.77	42.44
C 5H14N407	242.19	242.0862	24.80	5.83	23.13	46.24
C 5H14N408	258.19	258.0812	23.26	5.47	21.70	49.57
C 5H14N409	274.19	274.0761	21.90	5.15	20.43	52.52
C 5H14N601	174.21	174.1229	34.47	8.10	48.24	9.18
C 5H14N602	190.21	190.1178	31.57	7.42	44.18	16.82
C 5H14N603	206.21	206.1127	29.12	6.84	40.76	23.28
C 5H14N604	222.21	222.1076	27.03	6.35	37.82	28.80
C 5H14N605	238.20	238.1026	25.21	5.92	35.28	33.58
C 5H14N606	254.20	254.0975	23.63	5.55	33.06	37.76
C 5H14N607	270.20	270.0924	22.23	5.22	31.10	41.45
C 5H14N608	286.20	286.0873	20.98	4.93	29.36	44.72
C 5H14N609	302.20	302.0822	19.87	4.67	27.81	47.65
C 5H15N301	133.19	133.1215	45.09	11.35	31.55	12.01

029

C 5H15N302	149.19	149.1164	40.25	10.13	28.16	21.45
C 5H15N303	165.19	165.1113	36.35	9.15	25.44	29.06
C 5H15N304	181.19	181.1062	33.14	8.34	23.19	35.32
C 5H15N305	197.19	197.1012	30.46	7.67	21.31	40.57
C 5H15N306	213.19	213.0961	28.17	7.09	19.71	45.03
C 5H15N307	229.19	229.0910	26.20	6.60	18.33	48.87
C 5H15N308	245.19	245.0859	24.49	6.17	17.14	52.20
C 5H15N309	261.19	261.0808	22.99	5.79	16.09	55.13
C 5H15N501	161.21	161.1277	37.25	9.38	43.44	9.92
C 5H15N502	177.21	177.1226	33.89	8.53	39.52	18.06
C 5H15N503	193.21	193.1175	31.08	7.83	36.25	24.84
C 5H15N504	209.21	209.1124	28.71	7.23	33.48	30.59
C 5H15N505	225.21	225.1073	26.67	6.71	31.10	35.52
C 5H15N506	241.21	241.1022	24.90	6.27	29.03	39.80
C 5H15N507	257.20	257.0971	23.35	5.88	27.23	43.54
C 5H15N508	273.20	273.0921	21.98	5.53	25.63	46.85
C 5H15N509	289.20	289.0870	20.77	5.23	24.22	49.79
C 5H16N401	148.21	148.1324	40.52	10.88	37.80	10.80
C 5H16N402	164.21	164.1273	36.57	9.82	34.12	19.49
C 5H16N403	180.21	180.1222	33.33	8.95	31.09	26.63
C 5H16N404	196.21	196.1171	30.61	8.22	28.55	32.62
C 5H16N405	212.21	212.1121	28.30	7.60	26.40	37.70
C 5H16N406	228.21	228.1070	26.32	7.07	24.55	42.07
C 5H16N407	244.21	244.1019	24.59	6.60	22.94	45.86
C 5H16N408	260.21	260.0968	23.08	6.20	21.53	49.19
C 5H16N409	276.20	276.0917	21.74	5.84	20.28	52.13
C 5H16N601	176.22	176.1386	34.08	9.15	47.69	9.08
C 5H16N602	192.22	192.1335	31.24	8.39	43.72	16.65
C 5H16N603	208.22	208.1284	28.84	7.75	40.36	23.05
C 5H16N604	224.22	224.1233	26.78	7.19	37.48	28.54
C 5H16N605	240.22	240.1182	25.00	6.71	34.98	33.30
C 5H16N606	256.22	256.1131	23.44	6.29	32.80	37.47
C 5H16N607	272.22	272.1080	22.06	5.92	30.87	41.14
C 5H16N608	288.22	288.1029	20.84	5.60	29.16	44.41
C 5H16N609	304.22	304.0979	19.74	5.30	27.62	47.33
C 5H17N501	163.22	163.1433	36.79	10.50	42.91	9.80
C 5H17N502	179.22	179.1382	33.51	9.56	39.08	17.85
C 5H17N503	195.22	195.1331	30.76	8.78	35.87	24.59
C 5H17N504	211.22	211.1280	28.43	8.11	33.16	30.30
C 5H17N505	227.22	227.1230	26.43	7.54	30.82	35.21
C 5H17N506	243.22	243.1179	24.69	7.05	28.79	39.47
C 5H17N507	259.22	259.1128	23.17	6.61	27.02	43.20
C 5H17N508	275.22	275.1077	21.82	6.23	25.45	46.51
C 5H17N509	291.22	291.1026	20.62	5.88	24.05	49.45
C 5H18N601	178.24	178.1542	33.69	10.18	47.15	8.98
C 5H18N602	194.24	194.1491	30.92	9.34	43.27	16.47
C 5H18N603	210.24	210.1440	28.57	8.63	39.97	22.83
C 5H18N604	226.24	226.1389	26.55	8.02	37.15	28.29
C 5H18N605	242.24	242.1339	24.79	7.49	34.69	33.02
C 5H18N606	258.24	258.1288	23.26	7.03	32.54	37.17
C 5H18N607	274.24	274.1237	21.90	6.62	30.65	40.84
C 5H18N608	290.23	290.1186	20.69	6.25	28.96	44.10
C 5H18N609	306.23	306.1135	19.61	5.92	27.44	47.02
C 6H 3N101	105.10	105.0215	68.57	2.88	13.33	15.22
C 6H 3N102	121.10	121.0164	59.51	2.50	11.57	26.42
C 6H 3N103	137.10	137.0113	52.57	2.21	10.22	35.01
C 6H 3N104	153.10	153.0062	47.07	1.98	9.15	41.80
C 6H 3N105	169.09	169.0011	42.62	1.79	8.28	47.31
C 6H 3N106	185.09	184.9960	38.94	1.63	7.57	51.86
C 6H 3N107	201.09	200.9909	35.84	1.50	6.97	55.69

C 6H 3N108	217.09	216.9859	33.20	1.39	6.45	58.96
C 6H 3N109	233.09	232.9808	30.92	1.30	6.01	61.78
C 6H 3N301	133.11	133.0276	54.14	2.27	31.57	12.02
C 6H 3N302	149.11	149.0225	48.33	2.03	28.18	21.46
C 6H 3N303	165.11	165.0174	43.65	1.83	25.45	29.07
C 6H 3N304	181.11	181.0124	39.79	1.67	23.20	35.34
C 6H 3N305	197.11	197.0073	36.56	1.53	21.32	40.59
C 6H 3N306	213.11	213.0022	33.82	1.42	19.72	45.05
C 6H 3N307	229.11	228.9971	31.46	1.32	18.34	48.88
C 6H 3N308	245.11	244.9920	29.40	1.23	17.14	52.22
C 6H 3N309	261.11	260.9869	27.60	1.16	16.09	55.15
C 6H 3N501	161.12	161.0338	44.73	1.88	43.47	9.93
C 6H 3N502	177.12	177.0287	40.69	1.71	39.54	18.07
C 6H 3N503	193.12	193.0236	37.32	1.57	36.26	24.85
C 6H 3N504	209.12	209.0185	34.46	1.45	33.49	30.60
C 6H 3N505	225.12	225.0134	32.01	1.34	31.11	35.54
C 6H 3N506	241.12	241.0083	29.89	1.25	29.04	39.81
C 6H 3N507	257.12	257.0032	28.03	1.18	27.24	43.56
C 6H 3N508	273.12	272.9982	26.39	1.11	25.64	46.86
C 6H 3N509	289.12	288.9931	24.93	1.05	24.22	49.80
C 6H 4N201	120.11	120.0324	60.00	3.36	23.32	13.32
C 6H 4N202	136.11	136.0273	52.95	2.96	20.58	23.51
C 6H 4N203	152.11	152.0222	47.38	2.65	18.42	31.55
C 6H 4N204	168.11	168.0171	42.87	2.40	16.66	38.07
C 6H 4N205	184.11	184.0120	39.14	2.19	15.22	43.45
C 6H 4N206	200.11	200.0069	36.01	2.01	14.00	47.97
C 6H 4N207	216.11	216.0018	33.35	1.87	12.96	51.82
C 6H 4N208	232.11	231.9968	31.05	1.74	12.07	55.14
C 6H 4N209	248.11	247.9917	29.05	1.63	11.29	58.04
C 6H 4N401	148.12	148.0385	48.65	2.72	37.82	10.80
C 6H 4N402	164.12	164.0334	43.91	2.46	34.14	19.50
C 6H 4N403	180.12	180.0283	40.01	2.24	31.10	26.65
C 6H 4N404	196.12	196.0233	36.75	2.06	28.57	32.63
C 6H 4N405	212.12	212.0182	33.97	1.90	26.41	37.71
C 6H 4N406	228.12	228.0131	31.59	1.77	24.56	42.08
C 6H 4N407	244.12	244.0080	29.52	1.65	22.95	45.88
C 6H 4N408	260.12	260.0029	27.71	1.55	21.54	49.21
C 6H 4N409	276.12	275.9978	26.10	1.46	20.29	52.15
C 6H 4N601	176.14	176.0447	40.91	2.29	47.71	9.08
C 6H 4N602	192.14	192.0396	37.51	2.10	43.74	16.65
C 6H 4N603	208.14	208.0345	34.62	1.94	40.38	23.06
C 6H 4N604	224.14	224.0294	32.15	1.80	37.50	28.55
C 6H 4N605	240.14	240.0243	30.01	1.68	35.00	33.31
C 6H 4N606	256.14	256.0192	28.14	1.57	32.81	37.48
C 6H 4N607	272.13	272.0141	26.48	1.48	30.88	41.15
C 6H 4N608	288.13	288.0091	25.01	1.40	29.17	44.42
C 6H 4N609	304.13	304.0040	23.70	1.33	27.63	47.35
C 6H 5N101	107.11	107.0371	67.28	4.71	13.08	14.94
C 6H 5N102	123.11	123.0320	58.54	4.09	11.38	25.99
C 6H 5N103	139.11	139.0269	51.81	3.62	10.07	34.50
C 6H 5N104	155.11	155.0219	46.46	3.25	9.03	41.26
C 6H 5N105	171.11	171.0168	42.12	2.95	8.19	46.75
C 6H 5N106	187.11	187.0117	38.52	2.69	7.49	51.30
C 6H 5N107	203.11	203.0066	35.48	2.48	6.90	55.14
C 6H 5N108	219.11	219.0015	32.89	2.30	6.39	58.42
C 6H 5N109	235.11	234.9964	30.65	2.14	5.96	61.25
C 6H 5N301	135.13	135.0433	53.33	3.73	31.10	11.84
C 6H 5N302	151.13	151.0382	47.69	3.33	27.80	21.17
C 6H 5N303	167.13	167.0331	43.12	3.02	25.14	28.72
C 6H 5N304	183.12	183.0280	39.35	2.75	22.95	34.95

030

C 6H 5N305	199.12	199.0229	36.19	2.53	21.10	40.17
C 6H 5N306	215.12	215.0178	33.50	2.34	19.53	44.62
C 6H 5N307	231.12	231.0127	31.18	2.18	18.18	48.46
C 6H 5N308	247.12	247.0077	29.16	2.04	17.00	51.79
C 6H 5N309	263.12	263.0026	27.39	1.92	15.97	54.73
C 6H 5N501	163.14	163.0494	44.17	3.09	42.93	9.81
C 6H 5N502	179.14	179.0443	40.23	2.81	39.09	17.86
C 6H 5N503	195.14	195.0392	36.93	2.58	35.89	24.60
C 6H 5N504	211.14	211.0341	34.13	2.39	33.17	30.31
C 6H 5N505	227.14	227.0291	31.73	2.22	30.83	35.22
C 6H 5N506	243.14	243.0240	29.64	2.07	28.80	39.48
C 6H 5N507	259.14	259.0189	27.81	1.94	27.03	43.22
C 6H 5N508	275.14	275.0138	26.19	1.83	25.45	46.52
C 6H 5N509	291.13	291.0087	24.75	1.73	24.06	49.46
C 6H 6N201	122.13	122.0480	59.01	4.95	22.94	13.10
C 6H 6N202	138.13	138.0429	52.17	4.38	20.28	23.17
C 6H 6N203	154.13	154.0378	46.76	3.92	18.18	31.14
C 6H 6N204	170.13	170.0328	42.36	3.55	16.47	37.62
C 6H 6N205	186.13	186.0277	38.72	3.25	15.05	42.98
C 6H 6N206	202.12	202.0226	35.65	2.99	13.86	47.49
C 6H 6N207	218.12	218.0175	33.04	2.77	12.84	51.35
C 6H 6N208	234.12	234.0124	30.78	2.58	11.97	54.67
C 6H 6N209	250.12	250.0073	28.81	2.42	11.20	57.57
C 6H 6N401	150.14	150.0542	48.00	4.03	37.32	10.66
C 6H 6N402	166.14	166.0491	43.38	3.64	33.72	19.26
C 6H 6N403	182.14	182.0440	39.57	3.32	30.76	26.35
C 6H 6N404	198.14	198.0389	36.37	3.05	28.28	32.30
C 6H 6N405	214.14	214.0328	33.65	2.82	26.16	37.36
C 6H 6N406	230.14	230.0287	31.31	2.63	24.34	41.71
C 6H 6N407	246.14	246.0236	29.28	2.46	22.76	45.50
C 6H 6N408	262.14	262.0186	27.49	2.31	21.37	48.83
C 6H 6N409	278.14	278.0135	25.91	2.17	20.14	51.77
C 6H 6N601	178.15	178.0603	40.45	3.39	47.17	8.98
C 6H 6N602	194.15	194.0552	37.12	3.11	43.29	16.48
C 6H 6N603	210.15	210.0501	34.29	2.88	39.99	22.84
C 6H 6N604	226.15	226.0450	31.87	2.67	37.16	28.30
C 6H 6N605	242.15	242.0400	29.76	2.50	34.71	33.04
C 6H 6N606	258.15	258.0349	27.92	2.34	32.55	37.19
C 6H 6N607	274.15	274.0298	26.29	2.21	30.65	40.85
C 6H 6N608	290.15	290.0247	24.84	2.08	28.96	44.11
C 6H 6N609	306.15	306.0196	23.54	1.98	27.45	47.03
C 6H 7N101	109.13	109.0528	66.04	6.47	12.84	14.66
C 6H 7N102	125.13	125.0477	57.59	5.64	11.19	25.57
C 6H 7N103	141.13	141.0426	51.07	5.00	9.92	34.01
C 6H 7N104	157.13	157.0375	45.87	4.49	8.91	40.73
C 6H 7N105	173.13	173.0324	41.63	4.08	8.09	46.21
C 6H 7N106	189.13	189.0273	38.11	3.73	7.41	50.76
C 6H 7N107	205.13	205.0222	35.13	3.44	6.83	54.60
C 6H 7N108	221.12	221.0172	32.59	3.19	6.33	57.88
C 6H 7N109	237.12	237.0121	30.39	2.98	5.91	60.73
C 6H 7N301	137.14	137.0589	52.55	5.14	30.64	11.67
C 6H 7N302	153.14	153.0538	47.06	4.61	27.44	20.89
C 6H 7N303	169.14	169.0487	42.61	4.17	24.84	28.38
C 6H 7N304	185.14	185.0437	38.93	3.81	22.70	34.57
C 6H 7N305	201.14	201.0386	35.83	3.51	20.89	39.77
C 6H 7N306	217.14	217.0335	33.19	3.25	19.35	44.21
C 6H 7N307	233.14	233.0284	30.91	3.03	18.02	48.04
C 6H 7N308	249.14	249.0233	28.93	2.83	16.87	51.38
C 6H 7N309	265.14	265.0182	27.18	2.66	15.85	54.31
C 6H 7N501	165.16	165.0651	43.64	4.27	42.40	9.69
C 6H 7N502	181.15	181.0600	39.78	3.89	38.66	17.66
C 6H 7N503	197.15	197.0549	36.55	3.58	35.52	24.35
C 6H 7N504	213.15	213.0498	33.81	3.31	32.86	30.02
C 6H 7N505	229.15	229.0447	31.45	3.08	30.56	34.91
C 6H 7N506	245.15	245.0396	29.40	2.88	28.57	39.16
C 6H 7N507	261.15	261.0345	27.60	2.70	26.82	42.89
C 6H 7N508	277.15	277.0295	26.00	2.55	25.27	46.18
C 6H 7N509	293.15	293.0244	24.58	2.41	23.89	49.12
C 6H 8N201	124.14	124.0637	58.05	6.50	22.57	12.89
C 6H 8N202	140.14	140.0586	51.42	5.75	19.99	22.83
C 6H 8N203	156.14	156.0535	46.15	5.16	17.94	30.74
C 6H 8N204	172.14	172.0484	41.86	4.68	16.27	37.18
C 6H 8N205	188.14	188.0433	38.30	4.29	14.89	42.52
C 6H 8N206	204.14	204.0382	35.30	3.95	13.72	47.02
C 6H 8N207	220.14	220.0331	32.74	3.66	12.73	50.87
C 6H 8N208	236.14	236.0281	30.52	3.41	11.86	54.20
C 6H 8N209	252.14	252.0230	28.58	3.20	11.11	57.11
C 6H 8N401	152.16	152.0698	47.36	5.30	36.82	10.52
C 6H 8N402	168.16	168.0647	42.86	4.80	33.32	19.03
C 6H 8N403	184.16	184.0596	39.13	4.38	30.42	26.06
C 6H 8N404	200.16	200.0545	36.01	4.03	27.99	31.97
C 6H 8N405	216.15	216.0495	33.34	3.73	25.92	37.01
C 6H 8N406	232.15	232.0444	31.04	3.47	24.13	41.35
C 6H 8N407	248.15	248.0393	29.04	3.25	22.58	45.13
C 6H 8N408	264.15	264.0342	27.28	3.05	21.21	48.46
C 6H 8N409	280.15	280.0291	25.72	2.88	20.00	51.40
C 6H 8N601	180.17	180.0760	40.00	4.48	46.64	8.88
C 6H 8N602	196.17	196.0709	36.74	4.11	42.84	16.31
C 6H 8N603	212.17	212.0658	33.97	3.80	39.61	22.62
C 6H 8N604	228.17	228.0607	31.58	3.53	36.83	28.05
C 6H 8N605	244.17	244.0556	29.52	3.30	34.42	32.76
C 6H 8N606	260.17	260.0505	27.70	3.10	32.30	36.90
C 6H 8N607	276.17	276.0454	26.10	2.92	30.43	40.55
C 6H 8N608	292.17	292.0404	24.67	2.76	28.76	43.81
C 6H 8N609	308.17	308.0353	23.39	2.62	27.27	46.73
C 6H 9N101	111.14	111.0684	64.84	8.16	12.60	14.40
C 6H 9N102	127.14	127.0633	56.68	7.13	11.02	25.17
C 6H 9N103	143.14	143.0582	50.35	6.34	9.79	33.53
C 6H 9N104	159.14	159.0532	45.28	5.70	8.80	40.21
C 6H 9N105	175.14	175.0481	41.15	5.18	8.00	45.68
C 6H 9N106	191.14	191.0430	37.70	4.75	7.33	50.22
C 6H 9N107	207.14	207.0379	34.79	4.38	6.76	54.07
C 6H 9N108	223.14	223.0328	32.30	4.07	6.28	57.36
C 6H 9N109	239.14	239.0277	30.14	3.79	5.86	60.21
C 6H 9N301	139.16	139.0746	51.79	6.52	30.20	11.50
C 6H 9N302	155.16	155.0695	46.45	5.85	27.08	20.62
C 6H 9N303	171.16	171.0644	42.11	5.30	24.55	28.04
C 6H 9N304	187.16	187.0593	38.51	4.85	22.45	34.19
C 6H 9N305	203.16	203.0542	35.47	4.47	20.68	39.38
C 6H 9N306	219.16	219.0491	32.88	4.14	19.17	43.80
C 6H 9N307	235.15	235.0440	30.65	3.86	17.87	47.63
C 6H 9N308	251.15	251.0390	28.69	3.61	16.73	50.96
C 6H 9N309	267.15	267.0339	26.98	3.40	15.73	53.90
C 6H 9N501	167.17	167.0807	43.11	5.43	41.89	9.57
C 6H 9N502	183.17	183.0756	39.34	4.95	38.23	17.47
C 6H 9N503	199.17	199.0705	36.18	4.55	35.16	24.10
C 6H 9N504	215.17	215.0654	33.49	4.22	32.55	29.74
C 6H 9N505	231.17	231.0604	31.17	3.92	30.30	34.61
C 6H 9N506	247.17	247.0553	29.16	3.67	28.33	38.84
C 6H 9N507	263.17	263.0502	27.38	3.45	26.61	42.56

031

C 6H 9N508	279.17	279.0451	25.81	3.25	25.09	45.85
C 6H 9N509	295.17	295.0400	24.42	3.07	23.73	48.78
C 6H10N201	126.16	126.0793	57.12	7.99	22.20	12.68
C 6H10N202	142.16	142.0742	50.69	7.09	19.71	22.51
C 6H10N203	158.16	158.0691	45.57	6.37	17.71	30.35
C 6H10N204	174.16	174.0641	41.38	5.79	16.09	36.75
C 6H10N205	190.16	190.0590	37.90	5.30	14.73	42.07
C 6H10N206	206.16	206.0539	34.96	4.89	13.59	46.56
C 6H10N207	222.16	222.0488	32.44	4.54	12.61	50.41
C 6H10N208	238.16	238.0437	30.26	4.23	11.76	53.74
C 6H10N209	254.15	254.0386	28.36	3.97	11.02	56.66
C 6H10N401	154.17	154.0855	46.74	6.54	36.34	10.38
C 6H10N402	170.17	170.0804	42.35	5.92	32.92	18.80
C 6H10N403	186.17	186.0753	38.71	5.41	30.09	25.78
C 6H10N404	202.17	202.0702	35.65	4.99	27.71	31.66
C 6H10N405	218.17	218.0651	33.03	4.62	25.68	36.67
C 6H10N406	234.17	234.0600	30.78	4.30	23.93	40.99
C 6H10N407	250.17	250.0549	28.81	4.03	22.40	44.77
C 6H10N408	266.17	266.0499	27.08	3.79	21.05	48.09
C 6H10N409	282.17	262.0448	25.54	3.57	19.86	51.03
C 6H10N601	182.19	182.0916	39.56	5.53	46.13	8.78
C 6H10N602	198.19	198.0865	36.36	5.09	42.40	16.15
C 6H10N603	214.18	214.0814	33.65	4.71	39.24	22.41
C 6H10N604	230.18	230.0763	31.31	4.38	36.51	27.80
C 6H10N605	246.18	246.0713	29.27	4.09	34.14	32.49
C 6H10N606	262.18	262.0662	27.49	3.84	32.05	36.61
C 6H10N607	278.18	278.0611	25.91	3.62	30.21	40.26
C 6H10N608	294.18	294.0560	24.50	3.43	28.57	43.51
C 6H10N609	310.18	310.0509	23.23	3.25	27.09	46.42
C 6H11N101	113.16	113.0841	63.69	9.80	12.38	14.14
C 6H11N102	129.16	129.0790	55.80	8.58	10.84	24.77
C 6H11N103	145.16	145.0739	49.65	7.64	9.65	33.07
C 6H11N104	161.16	161.0688	44.72	6.88	8.69	39.71
C 6H11N105	177.16	177.0637	40.68	6.26	7.91	45.16
C 6H11N106	193.16	193.0586	37.31	5.74	7.25	49.70
C 6H11N107	209.16	209.0535	34.46	5.30	6.70	53.55
C 6H11N108	225.16	225.0485	32.01	4.92	6.22	56.85
C 6H11N109	241.16	241.0434	29.88	4.60	5.81	59.71
C 6H11N301	141.17	141.0902	51.05	7.85	29.76	11.33
C 6H11N302	157.17	157.0851	45.85	7.05	26.73	20.36
C 6H11N303	173.17	173.0800	41.62	6.40	24.26	27.72
C 6H11N304	189.17	189.0749	38.10	5.86	22.21	33.83
C 6H11N305	205.17	205.0699	35.13	5.40	20.48	38.99
C 6H11N306	221.17	221.0648	32.58	5.01	19.00	43.40
C 6H11N307	237.17	237.0597	30.39	4.67	17.72	47.22
C 6H11N308	253.17	253.0546	28.47	4.38	16.60	50.56
C 6H11N309	269.17	269.0495	26.77	4.12	15.61	53.50
C 6H11N501	169.19	169.0964	42.60	6.55	41.39	9.46
C 6H11N502	185.19	185.0913	38.92	5.99	37.82	17.28
C 6H11N503	201.19	201.0862	35.82	5.51	34.81	23.86
C 6H11N504	217.19	217.0811	33.18	5.11	32.25	29.47
C 6H11N505	233.19	233.0760	30.91	4.75	30.03	34.31
C 6H11N506	249.18	249.0709	28.92	4.45	28.11	38.52
C 6H11N507	265.18	265.0658	27.18	4.18	26.41	42.23
C 6H11N508	281.18	281.0608	25.63	3.94	24.91	45.52
C 6H11N509	297.18	297.0557	24.25	3.73	23.57	48.45
C 6H12N201	128.18	128.0950	56.23	9.44	21.86	12.48
C 6H12N202	144.17	144.0899	49.99	8.39	19.43	22.19
C 6H12N203	160.17	160.0848	44.99	7.55	17.49	29.97
C 6H12N204	176.17	176.0797	40.91	6.87	15.90	36.33
C 6H12N205	192.17	192.0746	37.50	6.29	14.58	41.63
C 6H12N206	208.17	208.0695	34.62	5.81	13.46	46.11
C 6H12N207	224.17	224.0644	32.15	5.40	12.50	49.96
C 6H12N208	240.17	240.0594	30.01	5.04	11.66	53.29
C 6H12N209	256.17	256.0543	28.13	4.72	10.94	56.21
C 6H12N401	156.19	156.1011	46.14	7.74	35.87	10.24
C 6H12N402	172.19	172.0960	41.85	7.02	32.54	18.58
C 6H12N403	188.19	188.0909	38.30	6.43	29.77	25.51
C 6H12N404	204.19	204.0858	35.29	5.92	27.44	31.34
C 6H12N405	220.19	220.0808	32.73	5.49	25.45	36.33
C 6H12N406	236.19	236.0757	30.51	5.12	23.72	40.64
C 6H12N407	252.19	252.0706	28.58	4.80	22.22	44.41
C 6H12N408	268.18	268.0655	26.87	4.51	20.89	47.73
C 6H12N409	284.18	284.0604	25.36	4.26	19.71	50.67
C 6H12N601	184.20	184.1073	39.12	6.57	45.62	8.69
C 6H12N602	200.20	200.1022	36.00	6.04	41.98	15.98
C 6H12N603	216.20	216.0971	33.33	5.59	38.87	22.20
C 6H12N604	232.20	232.0920	31.04	5.21	36.19	27.56
C 6H12N605	248.20	248.0869	29.04	4.87	33.86	32.23
C 6H12N606	264.20	264.0818	27.28	4.58	31.81	36.33
C 6H12N607	280.20	280.0767	25.72	4.32	29.99	39.97
C 6H12N608	296.20	296.0717	24.33	4.08	28.37	43.21
C 6H12N609	312.20	312.0666	23.08	3.87	26.92	46.12
C 6H13N101	115.18	115.0997	62.57	11.38	12.16	13.89
C 6H13N102	131.18	131.0946	54.94	9.99	10.68	24.39
C 6H13N103	147.18	147.0895	48.97	8.90	9.52	32.61
C 6H13N104	163.17	163.0844	44.17	8.03	8.58	39.22
C 6H13N105	179.17	179.0794	40.22	7.31	7.82	44.65
C 6H13N106	195.17	195.0743	36.92	6.71	7.18	49.19
C 6H13N107	211.17	211.0692	34.13	6.21	6.63	53.04
C 6H13N108	227.17	227.0641	31.72	5.77	6.17	56.34
C 6H13N109	243.17	243.0590	29.64	5.39	5.76	59.22
C 6H13N301	143.19	143.1059	50.33	9.15	29.35	11.17
C 6H13N302	159.19	159.1008	45.27	8.23	26.40	20.10
C 6H13N303	175.19	175.0957	41.14	7.48	23.99	27.40
C 6H13N304	191.19	191.0906	37.69	6.85	21.98	33.47
C 6H13N305	207.19	207.0855	34.78	6.32	20.28	38.61
C 6H13N306	223.19	223.0804	32.29	5.87	18.83	43.01
C 6H13N307	239.19	239.0753	30.13	5.48	17.57	46.82
C 6H13N308	255.19	255.0703	28.24	5.13	16.47	50.16
C 6H13N309	271.19	271.0652	26.57	4.83	15.49	53.10
C 6H13N501	171.20	171.1120	42.09	7.65	40.91	9.35
C 6H13N502	187.20	187.1069	38.50	7.00	37.41	17.09
C 6H13N503	203.20	203.1018	35.47	6.45	34.46	23.62
C 6H13N504	219.20	219.0967	32.88	5.98	31.95	29.20
C 6H13N505	235.20	235.0917	30.64	5.57	29.78	34.01
C 6H13N506	251.20	251.0866	28.69	5.22	27.88	38.22
C 6H13N507	267.20	267.0815	26.97	4.90	26.21	41.91
C 6H13N508	283.20	283.0764	25.45	4.63	24.73	45.20
C 6H13N509	299.20	299.0713	24.09	4.38	23.41	48.13
C 6H14N201	130.19	130.1106	55.35	10.84	21.52	12.29
C 6H14N202	146.19	146.1055	49.30	9.65	19.16	21.89
C 6H14N203	162.19	162.1004	44.43	8.70	17.27	29.59
C 6H14N204	178.19	178.0953	40.44	7.92	15.72	35.92
C 6H14N205	194.19	194.0903	37.11	7.27	14.43	41.20
C 6H14N206	210.19	210.0852	34.29	6.71	13.33	45.67
C 6H14N207	226.19	226.0801	31.86	6.24	12.39	49.51
C 6H14N208	242.19	242.0750	29.76	5.83	11.57	52.85
C 6H14N209	258.19	258.0699	27.91	5.47	10.85	55.77
C 6H14N401	158.20	158.1168	45.55	8.92	35.41	10.11

C 6H14N4O2	174.20	174.1117	41.37	8.10	32.16	18.37
C 6H14N4O3	190.20	190.1066	37.89	7.42	29.46	25.24
C 6H14N4O4	206.20	206.1015	34.95	6.84	27.17	31.04
C 6H14N4O5	222.20	222.0964	32.43	6.35	25.21	36.00
C 6H14N4O6	238.20	238.0913	30.25	5.92	23.52	40.30
C 6H14N4O7	254.20	254.0862	28.35	5.55	22.04	44.06
C 6H14N4O8	270.20	270.0812	26.67	5.22	20.74	47.37
C 6H14N4O9	286.20	286.0761	25.18	4.93	19.58	50.31
C 6H14N6O1	186.22	186.1229	38.70	7.58	45.13	8.59
C 6H14N6O2	202.22	202.1178	35.64	6.98	41.56	15.82
C 6H14N6O3	218.22	218.1127	33.03	6.47	38.51	22.00
C 6H14N6O4	234.22	234.1076	30.77	6.03	35.88	27.32
C 6H14N6O5	250.22	250.1026	28.80	5.64	33.59	31.97
C 6H14N6O6	266.22	266.0975	27.07	5.30	31.57	36.06
C 6H14N6O7	282.21	282.0924	25.54	5.00	29.78	39.68
C 6H14N6O8	298.21	298.0873	24.17	4.73	28.18	42.92
C 6H14N6O9	314.21	314.0822	22.94	4.49	26.75	45.83
C 6H15N1O1	117.19	117.1154	61.49	12.90	11.95	13.65
C 6H15N1O2	133.19	133.1103	54.11	11.35	10.52	24.02
C 6H15N1O3	149.19	149.1052	48.31	10.13	9.39	32.17
C 6H15N1O4	165.19	165.1001	43.63	9.15	8.48	38.74
C 6H15N1O5	181.19	181.0950	39.77	8.34	7.73	44.15
C 6H15N1O6	197.19	197.0899	36.55	7.67	7.10	48.68
C 6H15N1O7	213.19	213.0848	33.80	7.09	6.57	52.53
C 6H15N1O8	229.19	229.0798	31.44	6.60	6.11	55.85
C 6H15N1O9	245.19	245.0747	29.39	6.17	5.71	58.73
C 6H15N3O1	145.21	145.1215	49.63	10.41	28.94	11.02
C 6H15N3O2	161.21	161.1164	44.71	9.38	26.07	19.85
C 6H15N3O3	177.20	177.1113	40.67	8.53	23.71	27.09
C 6H15N3O4	193.20	193.1062	37.30	7.83	21.75	33.12
C 6H15N3O5	209.20	209.1012	34.45	7.23	20.09	38.24
C 6H15N3O6	225.20	225.0961	32.00	6.71	18.66	42.63
C 6H15N3O7	241.20	241.0910	29.88	6.27	17.42	46.43
C 6H15N3O8	257.20	257.0859	28.02	5.88	16.34	49.76
C 6H15N3O9	273.20	273.0808	26.38	5.53	15.38	52.71
C 6H15N5O1	173.22	173.1277	41.60	8.73	40.43	9.24
C 6H15N5O2	189.22	189.1226	38.09	7.99	37.01	16.91
C 6H15N5O3	205.22	205.1175	35.12	7.37	34.13	23.39
C 6H15N5O4	221.22	221.1124	32.58	6.83	31.66	28.93
C 6H15N5O5	237.22	237.1073	30.38	6.37	29.52	33.72
C 6H15N5O6	253.22	253.1022	28.46	5.97	27.66	37.91
C 6H15N5O7	269.22	269.0971	26.77	5.62	26.01	41.60
C 6H15N5O8	285.22	285.0921	25.27	5.30	24.55	44.88
C 6H15N5O9	301.21	301.0870	23.93	5.02	23.25	47.80
C 6H16N2O1	132.21	132.1263	54.51	12.20	21.19	12.10
C 6H16N2O2	148.21	148.1212	48.63	10.88	18.90	21.59
C 6H16N2O3	164.21	164.1161	43.89	9.82	17.06	29.23
C 6H16N2O4	180.21	180.1110	39.99	8.95	15.55	35.51
C 6H16N2O5	196.20	196.1059	36.73	8.22	14.28	40.77
C 6H16N2O6	212.20	212.1008	33.96	7.60	13.20	45.24
C 6H16N2O7	228.20	228.0957	31.58	7.07	12.28	49.08
C 6H16N2O8	244.20	244.0907	29.51	6.60	11.47	52.41
C 6H16N2O9	260.20	260.0856	27.70	6.20	10.77	55.34
C 6H16N4O1	160.22	160.1324	44.98	10.07	34.97	9.99
C 6H16N4O2	176.22	176.1273	40.90	9.15	31.79	18.16
C 6H16N4O3	192.22	192.1222	37.49	8.39	29.15	24.97
C 6H16N4O4	208.22	208.1171	34.61	7.75	26.91	30.74
C 6H16N4O5	224.22	224.1121	32.14	7.19	24.99	35.68
C 6H16N4O6	240.22	240.1070	30.00	6.71	23.32	39.96
C 6H16N4O7	256.22	256.1019	28.13	6.29	21.87	43.71

C 6H16N4O8	272.22	272.0968	26.47	5.92	20.58	47.02
C 6H16N4O9	288.22	288.0917	25.00	5.60	19.44	49.96
C 6H16N6O1	188.23	188.1386	38.29	8.57	44.65	8.50
C 6H16N6O2	204.23	204.1335	35.29	7.90	41.15	15.67
C 6H16N6O3	220.23	220.1284	32.72	7.32	38.16	21.79
C 6H16N6O4	236.23	236.1233	30.51	6.83	35.58	27.09
C 6H16N6O5	252.23	252.1182	28.57	6.39	33.32	31.72
C 6H16N6O6	268.23	268.1131	26.87	6.01	31.33	35.79
C 6H16N6O7	284.23	284.1080	25.36	5.67	29.57	39.40
C 6H16N6O8	300.23	300.1029	24.00	5.37	27.99	42.63
C 6H16N6O9	316.23	316.0979	22.79	5.10	26.58	45.53
C 6H17N3O1	147.22	147.1372	48.95	11.64	28.54	10.87
C 6H17N3O2	163.22	163.1321	44.15	10.50	25.74	19.60
C 6H17N3O3	179.22	179.1270	40.21	9.56	23.45	26.78
C 6H17N3O4	195.22	195.1219	36.92	8.78	21.52	32.78
C 6H17N3O5	211.22	211.1168	34.12	8.11	19.89	37.87
C 6H17N3O6	227.22	227.1117	31.72	7.54	18.49	42.25
C 6H17N3O7	243.22	243.1066	29.63	7.05	17.28	46.05
C 6H17N3O8	259.22	259.1016	27.80	6.61	16.21	49.38
C 6H17N3O9	275.22	275.0965	26.19	6.23	15.27	52.32
C 6H17N5O1	175.24	175.1433	41.13	9.78	39.97	9.13
C 6H17N5O2	191.23	191.1382	37.69	8.96	36.62	16.73
C 6H17N5O3	207.23	207.1331	34.78	8.27	33.79	23.16
C 6H17N5O4	223.23	223.1280	32.28	7.68	31.37	28.67
C 6H17N5O5	239.23	239.1230	30.12	7.16	29.27	33.44
C 6H17N5O6	255.23	255.1179	28.24	6.71	27.44	37.61
C 6H17N5O7	271.23	271.1128	26.57	6.32	25.82	41.29
C 6H17N5O8	287.23	287.1077	25.09	5.97	24.38	44.56
C 6H17N5O9	303.23	303.1026	23.77	5.65	23.10	47.49
C 6H18N4O1	162.24	162.1481	44.42	11.18	34.53	9.86
C 6H18N4O2	178.24	178.1430	40.43	10.18	31.43	17.95
C 6H18N4O3	194.24	194.1379	37.10	9.34	28.84	24.71
C 6H18N4O4	210.23	210.1328	34.28	8.63	26.65	30.44
C 6H18N4O5	226.23	226.1277	31.86	8.02	24.76	35.36
C 6H18N4O6	242.23	242.1226	29.75	7.49	23.13	39.63
C 6H18N4O7	258.23	258.1175	27.91	7.03	21.70	43.37
C 6H18N4O8	274.23	274.1125	26.28	6.62	20.43	46.67
C 6H18N4O9	290.23	290.1074	24.83	6.25	19.30	49.61
C 6H18N6O1	190.25	190.1542	37.88	9.54	44.17	8.41
C 6H18N6O2	206.25	206.1491	34.94	8.80	40.75	15.51
C 6H18N6O3	222.25	222.1440	32.43	8.16	37.81	21.60
C 6H18N6O4	238.25	238.1389	30.25	7.62	35.27	26.86
C 6H18N6O5	254.25	254.1339	28.35	7.14	33.05	31.46
C 6H18N6O6	270.25	270.1288	26.67	6.71	31.10	35.52
C 6H18N6O7	286.25	286.1237	25.18	6.34	29.36	39.13
C 6H18N6O8	302.25	302.1186	23.84	6.00	27.81	42.35
C 6H18N6O9	318.25	318.1135	22.65	5.70	26.41	45.25
C 6H19N5O1	177.25	177.1590	40.66	10.80	39.51	9.03
C 6H19N5O2	193.25	193.1539	37.29	9.91	36.24	16.56
C 6H19N5O3	209.25	209.1488	34.44	9.15	33.47	22.94
C 6H19N5O4	225.25	225.1437	31.99	8.50	31.09	28.41
C 6H19N5O5	241.25	241.1386	29.87	7.94	29.03	33.16
C 6H19N5O6	257.25	257.1335	28.01	7.44	27.22	37.32
C 6H19N5O7	273.25	273.1284	26.37	7.01	25.63	40.99
C 6H19N5O8	289.25	289.1233	24.92	6.62	24.21	44.25
C 6H19N5O9	305.25	305.1183	23.61	6.27	22.94	47.17
C 6H20N6O1	192.27	192.1698	37.48	10.49	43.71	8.32
C 6H20N6O2	208.27	208.1648	34.60	9.68	40.35	15.36
C 6H20N6O3	224.26	224.1597	32.13	8.99	37.47	21.40
C 6H20N6O4	240.26	240.1546	29.99	8.39	34.98	26.64

033

C 6H20N6O5	256.26	256.1495	28.12	7.87	32.79	31.22
C 6H20N6O6	272.26	272.1444	26.47	7.40	30.87	35.26
C 6H20N6O7	288.26	288.1393	25.00	6.99	29.15	38.85
C 6H20N6O8	304.26	304.1342	23.69	6.63	27.62	42.07
C 6H20N6O9	320.26	320.1292	22.50	6.29	26.24	44.96
C 7H 3N1O1	117.11	117.0215	71.80	2.58	11.96	13.66
C 7H 3N1O2	133.11	133.0164	63.17	2.27	10.52	24.04
C 7H 3N1O3	149.11	149.0113	56.39	2.03	9.39	32.19
C 7H 3N1O4	165.11	165.0062	50.92	1.83	8.48	38.76
C 7H 3N1O5	181.11	181.0011	46.42	1.67	7.73	44.17
C 7H 3N1O6	197.11	196.9960	42.66	1.53	7.11	48.70
C 7H 3N1O7	213.10	212.9909	39.45	1.42	6.57	52.55
C 7H 3N1O8	229.10	228.9859	36.70	1.32	6.11	55.87
C 7H 3N1O9	245.10	244.9808	34.30	1.23	5.71	58.75
C 7H 3N3O1	145.12	145.0276	57.94	2.08	28.96	11.02
C 7H 3N3O2	161.12	161.0225	52.18	1.88	26.08	19.86
C 7H 3N3O3	177.12	177.0174	47.47	1.71	23.72	27.10
C 7H 3N3O4	193.12	193.0124	43.54	1.57	21.76	33.14
C 7H 3N3O5	209.12	209.0073	40.21	1.45	20.09	38.25
C 7H 3N3O6	225.12	225.0022	37.35	1.34	18.67	42.64
C 7H 3N3O7	241.12	240.9971	34.87	1.25	17.43	46.45
C 7H 3N3O8	257.12	256.9920	32.70	1.18	16.34	49.78
C 7H 3N3O9	273.12	272.9869	30.78	1.11	15.39	52.72
C 7H 3N5O1	173.13	173.0338	48.56	1.75	40.45	9.24
C 7H 3N5O2	189.13	189.0287	44.45	1.60	37.03	16.92
C 7H 3N5O3	205.13	205.0236	40.99	1.47	34.14	23.40
C 7H 3N5O4	221.13	221.0185	38.02	1.37	31.67	28.94
C 7H 3N5O5	237.13	237.0134	35.46	1.28	29.53	33.74
C 7H 3N5O6	253.13	253.0083	33.22	1.19	27.67	37.92
C 7H 3N5O7	269.13	269.0032	31.24	1.12	26.02	41.61
C 7H 3N5O8	285.13	284.9982	29.49	1.06	24.56	44.89
C 7H 3N5O9	301.13	300.9931	27.92	1.00	23.26	47.82
C 7H 4N2O1	132.12	132.0324	63.64	3.05	21.20	12.11
C 7H 4N2O2	148.12	148.0273	56.76	2.72	18.91	21.60
C 7H 4N2O3	164.12	164.0222	51.23	2.46	17.07	29.25
C 7H 4N2O4	180.12	180.0171	46.68	2.24	15.55	35.53
C 7H 4N2O5	196.12	196.0120	42.87	2.06	14.28	40.79
C 7H 4N2O6	212.12	212.0069	39.64	1.90	13.21	45.26
C 7H 4N2O7	228.12	228.0018	36.86	1.77	12.28	49.10
C 7H 4N2O8	244.12	243.9968	34.44	1.65	11.48	52.43
C 7H 4N2O9	260.12	259.9917	32.32	1.55	10.77	55.36
C 7H 4N4O1	160.14	160.0385	52.50	2.52	34.99	9.99
C 7H 4N4O2	176.14	176.0334	47.73	2.29	31.81	18.17
C 7H 4N4O3	192.13	192.0283	43.76	2.10	29.16	24.98
C 7H 4N4O4	208.13	208.0233	40.40	1.94	26.92	30.75
C 7H 4N4O5	224.13	224.0182	37.51	1.80	25.00	35.69
C 7H 4N4O6	240.13	240.0131	35.01	1.68	23.33	39.98
C 7H 4N4O7	256.13	256.0080	32.83	1.57	21.87	43.73
C 7H 4N4O8	272.13	272.0029	30.90	1.48	20.59	47.03
C 7H 4N4O9	288.13	287.9978	29.18	1.40	19.44	49.98
C 7H 4N6O1	188.15	188.0447	44.69	2.14	44.67	8.50
C 7H 4N6O2	204.15	204.0396	41.18	1.97	41.17	15.67
C 7H 4N6O3	220.15	220.0345	38.19	1.83	38.17	21.80
C 7H 4N6O4	236.15	236.0294	35.60	1.71	35.59	27.10
C 7H 4N6O5	252.15	252.0243	33.34	1.60	33.33	31.73
C 7H 4N6O6	268.15	268.0192	31.36	1.50	31.34	35.80
C 7H 4N6O7	284.15	284.0141	29.59	1.42	29.58	39.41
C 7H 4N6O8	300.15	300.0091	28.01	1.34	28.00	42.64
C 7H 4N6O9	316.14	316.0040	26.59	1.28	26.58	45.55
C 7H 5N1O1	119.12	119.0371	70.58	4.23	11.76	13.43

C 7H 5N1O2	135.12	135.0320	62.22	3.73	10.37	23.68
C 7H 5N1O3	151.12	151.0269	55.64	3.33	9.27	31.76
C 7H 5N1O4	167.12	167.0219	50.31	3.02	8.38	38.29
C 7H 5N1O5	183.12	183.0168	45.91	2.75	7.65	43.69
C 7H 5N1O6	199.12	199.0117	42.22	2.53	7.03	48.21
C 7H 5N1O7	215.12	215.0066	39.08	2.34	6.51	52.06
C 7H 5N1O8	231.12	231.0015	36.38	2.18	6.06	55.38
C 7H 5N1O9	247.12	246.9964	34.02	2.04	5.67	58.27
C 7H 5N3O1	147.14	147.0433	57.14	3.43	28.56	10.87
C 7H 5N3O2	163.14	163.0382	51.54	3.09	25.76	19.61
C 7H 5N3O3	179.14	179.0331	46.94	2.81	23.46	26.79
C 7H 5N3O4	195.14	195.0280	43.09	2.58	21.53	32.80
C 7H 5N3O5	211.14	211.0229	39.82	2.39	19.90	37.89
C 7H 5N3O6	227.13	227.0178	37.02	2.22	18.50	42.26
C 7H 5N3O7	243.13	243.0127	34.58	2.07	17.28	46.06
C 7H 5N3O8	259.13	259.0077	32.45	1.94	16.22	49.39
C 7H 5N3O9	275.13	275.0026	30.56	1.83	15.27	52.34
C 7H 5N5O1	175.15	175.0494	48.00	2.88	39.98	9.13
C 7H 5N5O2	191.15	191.0443	43.99	2.64	36.64	16.74
C 7H 5N5O3	207.15	207.0392	40.59	2.43	33.81	23.17
C 7H 5N5O4	223.15	223.0341	37.68	2.26	31.38	28.68
C 7H 5N5O5	239.15	239.0291	35.16	2.11	29.28	33.45
C 7H 5N5O6	255.15	255.0240	32.95	1.98	27.45	37.62
C 7H 5N5O7	271.15	271.0189	31.01	1.86	25.83	41.30
C 7H 5N5O8	287.15	287.0138	29.28	1.76	24.39	44.57
C 7H 5N5O9	303.15	303.0087	27.74	1.66	23.10	47.50
C 7H 6N2O1	134.14	134.0480	62.68	4.51	20.88	11.93
C 7H 6N2O2	150.14	150.0429	56.00	4.03	18.66	21.31
C 7H 6N2O3	166.14	166.0378	50.61	3.64	16.86	28.89
C 7H 6N2O4	182.14	182.0328	46.16	3.32	15.38	35.14
C 7H 6N2O5	198.14	198.0277	42.43	3.05	14.14	40.37
C 7H 6N2O6	214.14	214.0226	39.26	2.82	13.08	44.83
C 7H 6N2O7	230.14	230.0175	36.53	2.63	12.17	48.67
C 7H 6N2O8	246.13	246.0124	34.16	2.46	11.38	52.00
C 7H 6N2O9	262.13	262.0073	32.07	2.31	10.69	54.93
C 7H 6N4O1	162.15	162.0542	51.85	3.73	34.55	9.87
C 7H 6N4O2	178.15	178.0491	47.19	3.39	31.45	17.96
C 7H 6N4O3	194.15	194.0440	43.31	3.12	28.86	24.72
C 7H 6N4O4	210.15	210.0389	40.01	2.88	26.66	30.45
C 7H 6N4O5	226.15	226.0338	37.18	2.67	24.77	35.37
C 7H 6N4O6	242.15	242.0287	34.72	2.50	23.14	39.64
C 7H 6N4O7	258.15	258.0236	32.57	2.34	21.70	43.38
C 7H 6N4O8	274.15	274.0186	30.67	2.21	20.44	46.69
C 7H 6N4O9	290.15	290.0135	28.98	2.08	19.31	49.63
C 7H 6N6O1	190.17	190.0603	44.21	3.18	44.19	8.41
C 7H 6N6O2	206.16	206.0552	40.78	2.93	40.76	15.52
C 7H 6N6O3	222.16	222.0501	37.84	2.72	37.83	21.60
C 7H 6N6O4	238.16	238.0450	35.30	2.54	35.29	26.87
C 7H 6N6O5	254.16	254.0400	33.08	2.38	33.07	31.47
C 7H 6N6O6	270.16	270.0349	31.12	2.24	31.11	35.53
C 7H 6N6O7	286.16	286.0298	29.38	2.11	29.37	39.14
C 7H 6N6O8	302.16	302.0247	27.83	2.00	27.81	42.36
C 7H 6N6O9	318.16	318.0196	26.43	1.90	26.41	45.26
C 7H 7N1O1	121.14	121.0528	69.41	5.82	11.56	13.21
C 7H 7N1O2	137.14	137.0477	61.31	5.14	10.21	23.33
C 7H 7N1O3	153.14	153.0426	54.90	4.61	9.15	31.34
C 7H 7N1O4	169.14	169.0375	49.71	4.17	8.28	37.84
C 7H 7N1O5	185.14	185.0324	45.41	3.81	7.57	43.21
C 7H 7N1O6	201.14	201.0273	41.80	3.51	6.96	47.73
C 7H 7N1O7	217.14	217.0222	38.72	3.25	6.45	51.58

034

C 7H 7N108	233.14	233.0172	36.06	3.03	6.01	54.90
C 7H 7N109	249.14	249.0121	33.75	2.83	5.62	57.80
C 7H 7N301	149.15	149.0589	56.37	4.73	28.17	10.73
C 7H 7N302	165.15	165.0538	50.91	4.27	25.44	19.38
C 7H 7N303	181.15	181.0487	46.41	3.89	23.20	26.50
C 7H 7N304	197.15	197.0437	42.65	3.58	21.31	32.46
C 7H 7N305	213.15	213.0386	39.45	3.31	19.71	37.53
C 7H 7N306	229.15	229.0335	36.69	3.08	18.34	41.89
C 7H 7N307	245.15	245.0284	34.30	2.88	17.14	45.68
C 7H 7N308	261.15	261.0233	32.20	2.70	16.09	49.01
C 7H 7N309	277.15	277.0182	30.34	2.55	15.16	51.96
C 7H 7N501	177.17	177.0651	47.46	3.98	39.53	9.03
C 7H 7N502	193.17	193.0600	43.53	3.65	36.26	16.57
C 7H 7N503	209.17	209.0549	40.20	3.37	33.48	22.95
C 7H 7N504	225.16	225.0498	37.34	3.13	31.10	28.42
C 7H 7N505	241.16	241.0447	34.86	2.93	29.04	33.17
C 7H 7N506	257.16	257.0396	32.69	2.74	27.23	37.33
C 7H 7N507	273.16	273.0345	30.78	2.58	25.64	41.00
C 7H 7N508	289.16	289.0295	29.08	2.44	24.22	44.26
C 7H 7N509	305.16	305.0244	27.55	2.31	22.95	47.19
C 7H 8N201	136.15	136.0637	61.75	5.92	20.57	11.75
C 7H 8N202	152.15	152.0586	55.26	5.30	18.41	21.03
C 7H 8N203	168.15	168.0535	50.00	4.80	16.66	28.54
C 7H 8N204	184.15	184.0484	45.66	4.38	15.21	34.75
C 7H 8N205	200.15	200.0433	42.01	4.03	14.00	39.97
C 7H 8N206	216.15	216.0382	38.90	3.73	12.96	44.41
C 7H 8N207	232.15	232.0331	36.22	3.47	12.07	48.24
C 7H 8N208	248.15	248.0281	33.88	3.25	11.29	51.58
C 7H 8N209	264.15	264.0230	31.83	3.05	10.61	54.51
C 7H 8N401	164.17	164.0698	51.21	4.91	34.13	9.75
C 7H 8N402	180.17	180.0647	46.67	4.48	31.10	17.76
C 7H 8N403	196.17	196.0596	42.86	4.11	28.56	24.47
C 7H 8N404	212.17	212.0545	39.63	3.80	26.41	30.16
C 7H 8N405	228.17	228.0495	36.85	3.53	24.56	35.06
C 7H 8N406	244.17	244.0444	34.43	3.30	22.95	39.32
C 7H 8N407	260.16	260.0393	32.32	3.10	21.54	43.05
C 7H 8N408	276.16	276.0342	30.44	2.92	20.29	46.35
C 7H 8N409	292.16	292.0291	28.78	2.76	19.18	49.29
C 7H 8N601	192.18	192.0760	43.75	4.20	43.73	8.33
C 7H 8N602	208.18	208.0709	40.39	3.87	40.37	15.37
C 7H 8N603	224.18	224.0658	37.50	3.60	37.49	21.41
C 7H 8N604	240.18	240.0607	35.01	3.36	34.99	26.65
C 7H 8N605	256.18	256.0556	32.82	3.15	32.81	31.23
C 7H 8N606	272.18	272.0505	30.89	2.96	30.88	35.27
C 7H 8N607	288.18	288.0454	29.18	2.80	29.16	38.86
C 7H 8N608	304.18	304.0404	27.64	2.65	27.63	42.08
C 7H 8N609	320.18	320.0353	26.26	2.52	26.25	44.97
C 7H 9N101	123.16	123.0684	68.27	7.37	11.37	12.99
C 7H 9N102	139.16	139.0633	60.42	6.52	10.07	23.00
C 7H 9N103	155.15	155.0582	54.19	5.85	9.03	30.94
C 7H 9N104	171.15	171.0532	49.12	5.30	8.18	37.39
C 7H 9N105	187.15	187.0481	44.92	4.85	7.48	42.74
C 7H 9N106	203.15	203.0430	41.39	4.47	6.89	47.25
C 7H 9N107	219.15	219.0379	38.37	4.14	6.39	51.10
C 7H 9N108	235.15	235.0328	35.75	3.86	5.96	54.43
C 7H 9N109	251.15	251.0277	33.48	3.61	5.58	57.33
C 7H 9N301	151.17	151.0746	55.62	6.00	27.80	10.58
C 7H 9N302	167.17	167.0695	50.30	5.43	25.14	19.14
C 7H 9N303	183.17	183.0644	45.90	4.95	22.94	26.20
C 7H 9N304	199.17	199.0593	42.21	4.55	21.10	32.13
C 7H 9N305	215.17	215.0542	39.08	4.22	19.53	37.18
C 7H 9N306	231.17	231.0491	36.37	3.92	18.18	41.53
C 7H 9N307	247.17	247.0440	34.02	3.67	17.00	45.31
C 7H 9N308	263.17	263.0390	31.95	3.45	15.97	48.64
C 7H 9N309	279.16	279.0339	30.12	3.25	15.05	51.58
C 7H 9N501	179.18	179.0807	46.92	5.06	39.08	8.93
C 7H 9N502	195.18	195.0756	43.08	4.65	35.88	16.39
C 7H 9N503	211.18	211.0705	39.81	4.30	33.16	22.73
C 7H 9N504	227.18	227.0654	37.01	3.99	30.83	28.17
C 7H 9N505	243.18	243.0604	34.57	3.73	28.80	32.90
C 7H 9N506	259.18	259.0553	32.44	3.50	27.02	37.04
C 7H 9N507	275.18	275.0502	30.55	3.30	25.45	40.70
C 7H 9N508	291.18	291.0451	28.88	3.12	24.05	43.96
C 7H 9N509	307.18	307.0400	27.37	2.95	22.80	46.88
C 7H10N201	138.17	138.0793	60.85	7.30	20.27	11.58
C 7H10N202	154.17	154.0742	54.54	6.54	18.17	20.76
C 7H10N203	170.17	170.0691	49.41	5.92	16.46	28.21
C 7H10N204	186.17	186.0641	45.16	5.41	15.05	34.38
C 7H10N205	202.17	202.0590	41.59	4.99	13.86	39.57
C 7H10N206	218.17	218.0539	38.54	4.62	12.84	44.00
C 7H10N207	234.17	234.0488	35.91	4.30	11.96	47.83
C 7H10N208	250.17	250.0437	33.61	4.03	11.20	51.16
C 7H10N209	266.17	266.0386	31.59	3.79	10.52	54.10
C 7H10N401	166.18	166.0855	50.59	6.07	33.71	9.63
C 7H10N402	182.18	182.0804	46.15	5.53	30.75	17.56
C 7H10N403	198.18	198.0753	42.42	5.09	28.27	24.22
C 7H10N404	214.18	214.0702	39.26	4.71	26.16	29.88
C 7H10N405	230.18	230.0651	36.53	4.38	24.34	34.75
C 7H10N406	246.18	246.0600	34.15	4.09	22.76	38.99
C 7H10N407	262.18	262.0549	32.07	3.84	21.37	42.72
C 7H10N408	278.18	278.0499	30.22	3.62	20.14	46.01
C 7H10N409	294.18	294.0448	28.58	3.43	19.05	48.95
C 7H10N601	194.20	194.0916	43.30	5.19	43.28	8.24
C 7H10N602	210.20	210.0865	40.00	4.80	39.98	15.22
C 7H10N603	226.20	226.0814	37.17	4.46	37.15	21.22
C 7H10N604	242.20	242.0763	34.71	4.16	34.70	26.42
C 7H10N605	258.19	258.0713	32.56	3.90	32.55	30.98
C 7H10N606	274.19	274.0662	30.66	3.68	30.65	35.01
C 7H10N607	290.19	290.0611	28.97	3.47	28.96	38.59
C 7H10N608	306.19	306.0560	27.46	3.29	27.45	41.80
C 7H10N609	322.19	322.0509	26.10	3.13	26.08	44.69
C 7H11N101	125.17	125.0841	67.17	8.86	11.19	12.78
C 7H11N102	141.17	141.0790	59.56	7.85	9.92	22.67
C 7H11N103	157.17	157.0739	53.49	7.05	8.91	30.54
C 7H11N104	173.17	173.0688	48.55	6.40	8.09	36.96
C 7H11N105	189.17	189.0637	44.45	5.86	7.40	42.29
C 7H11N106	205.17	205.0586	40.98	5.40	6.83	46.79
C 7H11N107	221.17	221.0535	38.02	5.01	6.33	50.64
C 7H11N108	237.17	237.0485	35.45	4.68	5.91	53.97
C 7H11N109	253.17	253.0434	33.21	4.38	5.53	56.88
C 7H11N301	153.19	153.0902	54.89	7.24	27.43	10.44
C 7H11N302	169.18	169.0851	49.70	6.55	24.84	18.91
C 7H11N303	185.18	185.0800	45.40	5.99	22.69	25.92
C 7H11N304	201.18	201.0749	41.79	5.51	20.89	31.81
C 7H11N305	217.18	217.0699	38.71	5.11	19.35	36.83
C 7H11N306	233.18	233.0648	36.06	4.75	18.02	41.17
C 7H11N307	249.18	249.0597	33.74	4.45	16.86	44.95
C 7H11N308	265.18	265.0546	31.71	4.18	15.85	48.27
C 7H11N309	281.18	281.0495	29.90	3.94	14.94	51.21
C 7H11N501	181.20	181.0964	46.40	6.12	38.65	8.83

035

C 7H11N502	197.20	197.0913	42.64	5.62	35.51	16.23
C 7H11N503	213.20	213.0862	39.44	5.20	32.85	22.51
C 7H11N504	229.20	229.0811	36.68	4.84	30.56	27.92
C 7H11N505	245.20	245.0760	34.29	4.52	28.56	32.63
C 7H11N506	261.20	261.0709	32.19	4.24	26.81	36.75
C 7H11N507	277.20	277.0658	30.33	4.00	25.27	40.40
C 7H11N508	293.19	293.0608	28.68	3.78	23.89	43.66
C 7H11N509	309.19	309.0557	27.19	3.59	22.65	46.57
C 7H12N201	140.19	140.0950	59.98	8.63	19.98	11.41
C 7H12N202	156.19	156.0899	53.83	7.74	17.94	20.49
C 7H12N203	172.19	172.0848	48.83	7.02	16.27	27.88
C 7H12N204	188.18	188.0797	44.68	6.43	14.89	34.01
C 7H12N205	204.18	204.0746	41.18	5.92	13.72	39.18
C 7H12N206	220.18	220.0695	38.19	5.49	12.72	43.60
C 7H12N207	236.18	236.0644	35.60	5.12	11.86	47.42
C 7H12N208	252.18	252.0594	33.34	4.80	11.11	50.76
C 7H12N209	268.18	268.0543	31.35	4.51	10.45	53.69
C 7H12N401	168.20	168.1011	49.99	7.19	33.31	9.51
C 7H12N402	184.20	184.0960	45.65	6.57	30.42	17.37
C 7H12N403	200.20	200.0909	42.00	6.04	27.99	23.98
C 7H12N404	216.20	216.0858	38.89	5.59	25.91	29.60
C 7H12N405	232.20	232.0808	36.21	5.21	24.13	34.45
C 7H12N406	248.20	248.0757	33.88	4.87	22.57	38.68
C 7H12N407	264.20	264.0706	31.82	4.58	21.21	42.39
C 7H12N408	280.20	280.0655	30.01	4.32	20.00	45.68
C 7H12N409	296.20	296.0604	28.39	4.08	18.92	48.61
C 7H12N601	196.21	196.1073	42.85	6.16	42.83	8.15
C 7H12N602	212.21	212.1022	39.62	5.70	39.60	15.08
C 7H12N603	228.21	228.0971	36.84	5.30	36.83	21.03
C 7H12N604	244.21	244.0920	34.43	4.95	34.41	26.21
C 7H12N605	260.21	260.0869	32.31	4.65	32.30	30.74
C 7H12N606	276.21	276.0818	30.44	4.38	30.43	34.75
C 7H12N607	292.21	292.0767	28.77	4.14	28.76	38.33
C 7H12N608	308.21	308.0717	27.28	3.92	27.27	41.53
C 7H12N609	324.21	324.0666	25.93	3.73	25.92	44.41
C 7H13N101	127.19	127.0997	66.11	10.30	11.01	12.58
C 7H13N102	143.19	143.0946	58.72	9.15	9.78	22.35
C 7H13N103	159.19	159.0895	52.82	8.23	8.80	30.15
C 7H13N104	175.19	175.0844	47.99	7.48	8.00	36.53
C 7H13N105	191.19	191.0794	43.98	6.85	7.33	41.84
C 7H13N106	207.18	207.0743	40.58	6.32	6.76	46.33
C 7H13N107	223.18	223.0692	37.67	5.87	6.28	50.18
C 7H13N108	239.18	239.0641	35.15	5.48	5.86	53.51
C 7H13N109	255.18	255.0590	32.95	5.13	5.49	56.43
C 7H13N301	155.20	155.1059	54.17	8.44	27.07	10.31
C 7H13N302	171.20	171.1008	49.11	7.65	24.54	18.69
C 7H13N303	187.20	187.0957	44.91	7.00	22.45	25.64
C 7H13N304	203.20	203.0906	41.38	6.45	20.68	31.49
C 7H13N305	219.20	219.0855	38.36	5.98	19.17	36.50
C 7H13N306	235.20	235.0804	35.75	5.57	17.87	40.82
C 7H13N307	251.20	251.0753	33.47	5.22	16.73	44.58
C 7H13N308	267.20	267.0703	31.47	4.90	15.73	47.90
C 7H13N309	283.20	283.0652	29.69	4.63	14.84	50.85
C 7H13N501	183.21	183.1120	45.89	7.15	38.22	8.73
C 7H13N502	199.21	199.1069	42.20	6.58	35.15	16.06
C 7H13N503	215.21	215.1018	39.07	6.09	32.54	22.30
C 7H13N504	231.21	231.0967	36.36	5.67	30.29	27.68
C 7H13N505	247.21	247.0917	34.01	5.30	28.33	32.36
C 7H13N506	263.21	263.0866	31.94	4.98	26.61	36.47
C 7H13N507	279.21	279.0815	30.11	4.69	25.08	40.11
C 7H13N508	295.21	295.0764	28.48	4.44	23.72	43.36
C 7H13N509	311.21	311.0713	27.02	4.21	22.50	46.27
C 7H14N201	142.20	142.1106	59.13	9.92	19.70	11.25
C 7H14N202	158.20	158.1055	53.15	8.92	17.71	20.23
C 7H14N203	174.20	174.1004	48.26	8.10	16.08	27.55
C 7H14N204	190.20	190.0953	44.20	7.42	14.73	33.65
C 7H14N205	206.20	206.0903	40.77	6.84	13.59	38.80
C 7H14N206	222.20	222.0852	37.84	6.35	12.61	43.20
C 7H14N207	238.20	238.0801	35.30	5.92	11.76	47.02
C 7H14N208	254.20	254.0750	33.08	5.55	11.02	50.35
C 7H14N209	270.20	270.0699	31.12	5.22	10.37	53.29
C 7H14N401	170.22	170.1168	49.39	8.29	32.92	9.40
C 7H14N402	186.22	186.1117	45.15	7.58	30.09	17.18
C 7H14N403	202.21	202.1066	41.58	6.98	27.71	23.74
C 7H14N404	218.21	218.1015	38.53	6.47	25.68	29.33
C 7H14N405	234.21	234.0964	35.90	6.03	23.92	34.16
C 7H14N406	250.21	250.0913	33.60	5.64	22.39	38.37
C 7H14N407	266.21	266.0862	31.58	5.30	21.05	42.07
C 7H14N408	282.21	282.0812	29.79	5.00	19.85	45.35
C 7H14N409	298.21	298.0761	28.19	4.73	18.79	48.29
C 7H14N601	198.23	198.1229	42.41	7.12	42.40	8.07
C 7H14N602	214.23	214.1178	39.25	6.59	39.23	14.94
C 7H14N603	230.23	230.1127	36.52	6.13	36.50	20.85
C 7H14N604	246.23	246.1076	34.15	5.73	34.13	25.99
C 7H14N605	262.23	262.1026	32.06	5.38	32.05	30.51
C 7H14N606	278.23	278.0975	30.22	5.07	30.21	34.50
C 7H14N607	294.23	294.0924	28.58	4.80	28.56	38.06
C 7H14N608	310.23	310.0873	27.10	4.55	27.09	41.26
C 7H14N609	326.22	326.0822	25.77	4.33	25.76	44.14
C 7H15N101	129.20	129.1154	65.07	11.70	10.84	12.38
C 7H15N102	145.20	145.1103	57.90	10.41	9.65	22.04
C 7H15N103	161.20	161.1052	52.16	9.38	8.69	29.78
C 7H15N104	177.20	177.1001	47.45	8.53	7.90	36.12
C 7H15N105	193.20	193.0950	43.52	7.83	7.25	41.41
C 7H15N106	209.20	209.0899	40.19	7.23	6.70	45.89
C 7H15N107	225.20	225.0848	37.33	6.71	6.22	49.73
C 7H15N108	241.20	241.0798	34.86	6.27	5.81	53.07
C 7H15N109	257.20	257.0747	32.69	5.88	5.45	55.99
C 7H15N301	157.22	157.1215	53.48	9.62	26.73	10.18
C 7H15N302	173.22	173.1164	48.54	8.73	24.26	18.47
C 7H15N303	189.22	189.1113	44.43	7.99	22.21	25.37
C 7H15N304	205.22	205.1062	40.97	7.37	20.48	31.19
C 7H15N305	221.21	221.1012	38.01	6.83	19.00	36.16
C 7H15N306	237.21	237.0961	35.44	6.37	17.71	40.47
C 7H15N307	253.21	253.0910	33.20	5.97	16.59	44.23
C 7H15N308	269.21	269.0859	31.23	5.62	15.61	47.54
C 7H15N309	285.21	285.0808	29.48	5.30	14.73	50.49
C 7H15N501	185.23	185.1277	45.39	8.16	37.81	8.64
C 7H15N502	201.23	201.1226	41.78	7.51	34.80	15.90
C 7H15N503	217.23	217.1175	38.70	6.96	32.24	22.10
C 7H15N504	233.23	233.1124	36.05	6.48	30.03	27.44
C 7H15N505	249.23	249.1073	33.74	6.07	28.10	32.10
C 7H15N506	265.23	265.1022	31.70	5.70	26.41	36.19
C 7H15N507	281.23	281.0971	29.90	5.38	24.90	39.82
C 7H15N508	297.23	297.0921	28.29	5.09	23.56	43.06
C 7H15N509	313.23	313.0870	26.84	4.83	22.36	45.97
C 7H16N201	144.22	144.1263	58.30	11.18	19.42	11.09
C 7H16N202	160.22	160.1212	52.48	10.07	17.48	19.97
C 7H16N203	176.22	176.1161	47.71	9.15	15.90	27.24
C 7H16N204	192.22	192.1110	43.74	8.39	14.57	33.29

C 7H16N205	208.22	208.1059	40.38	7.75	13.45	38.42
C 7H16N206	224.22	224.1008	37.50	7.19	12.49	42.81
C 7H16N207	240.21	240.0957	35.00	6.71	11.66	46.62
C 7H16N208	256.21	256.0907	32.82	6.29	10.93	49.96
C 7H16N209	272.21	272.0856	30.89	5.92	10.29	52.90
C 7H16N401	172.23	172.1324	48.82	9.36	32.53	9.29
C 7H16N402	188.23	188.1273	44.67	8.57	29.76	17.00
C 7H16N403	204.23	204.1222	41.17	7.90	27.43	23.50
C 7H16N404	220.23	220.1171	38.18	7.32	25.44	29.06
C 7H16N405	236.23	236.1121	35.59	6.83	23.72	33.86
C 7H16N406	252.23	252.1070	33.33	6.39	22.21	38.06
C 7H16N407	268.23	268.1019	31.35	6.01	20.89	41.75
C 7H16N408	284.23	284.0968	29.58	5.67	19.71	45.03
C 7H16N409	300.23	300.0917	28.00	5.37	18.66	47.96
C 7H16N601	200.25	200.1386	41.99	8.05	41.97	7.99
C 7H16N602	216.24	216.1335	38.88	7.46	38.86	14.80
C 7H16N603	232.24	232.1284	36.20	6.94	36.19	20.67
C 7H16N604	248.24	248.1233	33.87	6.50	33.85	25.78
C 7H16N605	264.24	264.1182	31.82	6.10	31.80	30.27
C 7H16N606	280.24	280.1131	30.00	5.75	29.99	34.25
C 7H16N607	296.24	296.1080	28.38	5.44	28.37	37.81
C 7H16N608	312.24	312.1029	26.93	5.17	26.92	40.99
C 7H16N609	328.24	328.0979	25.61	4.91	25.60	43.87
C 7H17N101	131.22	131.1310	64.07	13.06	10.67	12.19
C 7H17N102	147.22	147.1259	57.11	11.64	9.51	21.74
C 7H17N103	163.22	163.1208	51.51	10.50	8.58	29.41
C 7H17N104	179.22	179.1157	46.91	9.56	7.82	35.71
C 7H17N105	195.22	195.1107	43.07	8.78	7.17	40.98
C 7H17N106	211.22	211.1056	39.81	8.11	6.63	45.45
C 7H17N107	227.22	227.1005	37.00	7.54	6.16	49.29
C 7H17N108	243.22	243.0954	34.57	7.05	5.76	52.63
C 7H17N109	259.21	259.0903	32.44	6.61	5.40	55.55
C 7H17N301	159.23	159.1372	52.80	10.76	26.39	10.05
C 7H17N302	175.23	175.1321	47.98	9.78	23.98	18.26
C 7H17N303	191.23	191.1270	43.97	8.96	21.97	25.10
C 7H17N304	207.23	207.1219	40.57	8.27	20.28	30.88
C 7H17N305	223.23	223.1168	37.66	7.68	18.82	35.84
C 7H17N306	239.23	239.1117	35.15	7.16	17.56	40.13
C 7H17N307	255.23	255.1066	32.94	6.71	16.46	43.88
C 7H17N308	271.23	271.1016	31.00	6.32	15.49	47.19
C 7H17N309	287.23	287.0965	29.27	5.97	14.63	50.13
C 7H17N501	187.25	187.1433	44.90	9.15	37.40	8.54
C 7H17N502	203.25	203.1382	41.37	8.43	34.46	15.74
C 7H17N503	219.25	219.1331	38.35	7.82	31.94	21.89
C 7H17N504	235.24	235.1280	35.74	7.28	29.77	27.20
C 7H17N505	251.24	251.1230	33.46	6.82	27.87	31.84
C 7H17N506	267.24	267.1179	31.46	6.41	26.21	35.92
C 7H17N507	283.24	283.1128	29.68	6.05	24.73	39.54
C 7H17N508	299.24	299.1077	28.10	5.73	23.40	42.77
C 7H17N509	315.24	315.1026	26.67	5.44	22.22	45.68
C 7H18N201	146.23	146.1419	57.50	12.41	19.16	10.94
C 7H18N202	162.23	162.1368	51.83	11.18	17.27	19.72
C 7H18N203	178.23	178.1317	47.17	10.18	15.72	26.93
C 7H18N204	194.23	194.1266	43.29	9.34	14.42	32.95
C 7H18N205	210.23	210.1216	39.99	8.63	13.32	38.05
C 7H18N206	226.23	226.1165	37.16	8.02	12.38	42.43
C 7H18N207	242.23	242.1114	34.71	7.49	11.56	46.24
C 7H18N208	258.23	258.1063	32.56	7.03	10.85	49.57
C 7H18N209	274.23	274.1012	30.66	6.62	10.22	52.51
C 7H18N401	174.25	174.1481	48.25	10.41	32.15	9.18
C 7H18N402	190.25	190.1430	44.19	9.54	29.45	16.82
C 7H18N403	206.25	206.1379	40.77	8.80	27.16	23.27
C 7H18N404	222.25	222.1328	37.83	8.16	25.21	28.80
C 7H18N405	238.25	238.1277	35.29	7.62	23.52	33.58
C 7H18N406	254.24	254.1226	33.07	7.14	22.04	37.76
C 7H18N407	270.24	270.1175	31.11	6.71	20.73	41.44
C 7H18N408	286.24	286.1125	29.37	6.34	19.57	44.72
C 7H18N409	302.24	302.1074	27.82	6.00	18.54	47.64
C 7H18N601	202.26	202.1542	41.57	8.97	41.55	7.91
C 7H18N602	218.26	218.1491	38.52	8.31	38.50	14.66
C 7H18N603	234.26	234.1440	35.89	7.75	35.87	20.49
C 7H18N604	250.26	250.1389	33.60	7.25	33.58	25.57
C 7H18N605	266.26	266.1339	31.58	6.81	31.56	30.04
C 7H18N606	282.26	282.1288	29.79	6.43	29.77	34.01
C 7H18N607	298.26	298.1237	28.19	6.08	28.18	37.55
C 7H18N608	314.26	314.1186	26.75	5.77	26.74	40.73
C 7H18N609	330.26	330.1135	25.46	5.49	25.45	43.60
C 7H19N301	161.25	161.1528	52.14	11.88	26.06	9.92
C 7H19N302	177.25	177.1477	47.44	10.80	23.71	18.05
C 7H19N303	193.25	193.1426	43.51	9.91	21.74	24.84
C 7H19N304	209.25	209.1375	40.18	9.15	20.08	30.58
C 7H19N305	225.25	225.1325	37.33	8.50	18.66	35.52
C 7H19N306	241.25	241.1274	34.85	7.94	17.42	39.79
C 7H19N307	257.25	257.1223	32.68	7.44	16.33	43.54
C 7H19N308	273.24	273.1172	30.77	7.01	15.38	46.84
C 7H19N309	289.24	289.1121	29.07	6.62	14.53	49.78
C 7H19N501	189.26	189.1590	44.42	10.12	37.00	8.45
C 7H19N502	205.26	205.1539	40.96	9.33	34.12	15.59
C 7H19N503	221.26	221.1488	38.00	8.66	31.65	21.69
C 7H19N504	237.26	237.1437	35.44	8.07	29.52	26.97
C 7H19N505	253.26	253.1386	33.20	7.56	27.65	31.59
C 7H19N506	269.26	269.1335	31.23	7.11	26.01	35.65
C 7H19N507	285.26	285.1284	29.47	6.71	24.55	39.26
C 7H19N508	301.26	301.1233	27.91	6.36	23.25	42.49
C 7H19N509	317.26	317.1183	26.50	6.04	22.07	45.39
C 7H20N401	176.26	176.1637	47.70	11.44	31.79	9.08
C 7H20N402	192.26	192.1586	43.73	10.49	29.14	16.64
C 7H20N403	208.26	208.1535	40.37	9.68	26.90	23.05
C 7H20N404	224.26	224.1484	37.49	8.99	24.98	28.54
C 7H20N405	240.26	240.1434	34.99	8.39	23.32	33.30
C 7H20N406	256.26	256.1383	32.81	7.87	21.86	37.46
C 7H20N407	272.26	272.1332	30.88	7.40	20.58	41.14
C 7H20N408	288.26	288.1281	29.17	6.99	19.44	44.40
C 7H20N409	304.26	304.1230	27.63	6.63	18.41	47.33
C 7H20N601	204.28	204.1698	41.16	9.87	41.14	7.83
C 7H20N602	220.28	220.1648	38.17	9.15	38.15	14.53
C 7H20N603	236.28	236.1597	35.58	8.53	35.57	20.31
C 7H20N604	252.28	252.1546	33.33	7.99	33.31	25.37
C 7H20N605	268.27	268.1495	31.34	7.51	31.33	29.82
C 7H20N606	284.27	284.1444	29.58	7.09	29.56	33.77
C 7H20N607	300.27	300.1393	28.00	6.71	27.99	37.30
C 7H20N608	316.27	316.1342	26.58	6.37	26.57	40.47
C 7H20N609	332.27	332.1292	25.30	6.07	25.29	43.34
C 7H21N501	191.28	191.1746	43.96	11.07	36.61	8.36
C 7H21N502	207.28	207.1695	40.56	10.21	33.79	15.44
C 7H21N503	223.28	223.1644	37.66	9.48	31.37	21.50
C 7H21N504	239.28	239.1593	35.14	8.85	29.27	26.75
C 7H21N505	255.28	255.1543	32.94	8.29	27.43	31.34
C 7H21N506	271.28	271.1492	30.99	7.80	25.82	35.39
C 7H21N507	287.27	287.1441	29.27	7.37	24.38	38.99

C 7H21N508	303.27	303.1390	27.72	6.98	23.09	42.20
C 7H21N509	319.27	319.1339	26.33	6.63	21.94	45.10
C 7H22N601	206.29	206.1855	40.76	10.75	40.74	7.76
C 7H22N602	222.29	222.1804	37.82	9.98	37.81	14.39
C 7H22N603	238.29	238.1753	35.28	9.31	35.27	20.14
C 7H22N604	254.29	254.1702	33.06	8.72	33.05	25.17
C 7H22N605	270.29	270.1652	31.11	8.20	31.09	29.60
C 7H22N606	286.29	286.1601	29.37	7.75	29.35	33.53
C 7H22N607	302.29	302.1550	27.81	7.34	27.80	37.05
C 7H22N608	318.29	318.1499	26.42	6.97	26.40	40.21
C 7H22N609	334.29	334.1448	25.15	6.63	25.14	43.07
C 8H 4N201	144.13	144.0324	66.67	2.80	19.44	11.10
C 8H 4N202	160.13	160.0273	60.01	2.52	17.49	19.98
C 8H 4N203	176.13	176.0222	54.56	2.29	15.90	27.25
C 8H 4N204	192.13	192.0171	50.01	2.10	14.58	33.31
C 8H 4N205	208.13	208.0120	46.17	1.94	13.46	38.44
C 8H 4N206	224.13	224.0069	42.87	1.80	12.50	42.83
C 8H 4N207	240.13	240.0018	40.02	1.68	11.67	46.64
C 8H 4N208	256.13	255.9968	37.52	1.57	10.94	49.97
C 8H 4N209	272.13	271.9917	35.31	1.48	10.29	52.91
C 8H 4N401	172.15	172.0385	55.82	2.34	32.55	9.29
C 8H 4N402	188.15	188.0334	51.07	2.14	29.78	17.01
C 8H 4N403	204.15	204.0283	47.07	1.97	27.44	23.51
C 8H 4N404	220.15	220.0233	43.65	1.83	25.45	29.07
C 8H 4N405	236.14	236.0182	40.69	1.71	23.73	33.88
C 8H 4N406	252.14	252.0131	38.11	1.60	22.22	38.07
C 8H 4N407	268.14	268.0080	35.83	1.50	20.89	41.77
C 8H 4N408	284.14	284.0029	33.82	1.42	19.72	45.05
C 8H 4N409	300.14	299.9978	32.01	1.34	18.67	47.98
C 8H 4N601	200.16	200.0447	48.01	2.01	41.99	7.99
C 8H 4N602	216.16	216.0396	44.45	1.87	38.88	14.80
C 8H 4N603	232.16	232.0345	41.39	1.74	36.20	20.67
C 8H 4N604	248.16	248.0294	38.72	1.62	33.87	25.79
C 8H 4N605	264.16	264.0243	36.38	1.53	31.81	30.28
C 8H 4N606	280.16	280.0192	34.30	1.44	30.00	34.27
C 8H 4N607	296.16	296.0141	32.45	1.36	28.38	37.82
C 8H 4N608	312.16	312.0091	30.78	1.29	26.92	41.00
C 8H 4N609	328.16	328.0040	29.28	1.23	25.61	43.88
C 8H 5N101	131.14	131.0371	73.27	3.84	10.68	12.20
C 8H 5N102	147.13	147.0320	65.31	3.43	9.52	21.75
C 8H 5N103	163.13	163.0269	58.90	3.09	8.59	29.42
C 8H 5N104	179.13	179.0219	53.64	2.81	7.82	35.73
C 8H 5N105	195.13	195.0168	49.24	2.58	7.18	41.00
C 8H 5N106	211.13	211.0117	45.51	2.39	6.63	45.47
C 8H 5N107	227.13	227.0066	42.31	2.22	6.17	49.31
C 8H 5N108	243.13	243.0015	39.52	2.07	5.76	52.64
C 8H 5N109	259.13	258.9964	37.08	1.94	5.41	55.57
C 8H 5N301	159.15	159.0433	60.38	3.17	26.40	10.05
C 8H 5N302	175.15	175.0382	54.86	2.88	23.99	18.27
C 8H 5N303	191.15	191.0331	50.27	2.64	21.98	25.11
C 8H 5N304	207.15	207.0280	46.39	2.43	20.29	30.89
C 8H 5N305	223.15	223.0229	43.06	2.26	18.83	35.85
C 8H 5N306	239.15	239.0178	40.18	2.11	17.57	40.14
C 8H 5N307	255.14	255.0127	37.66	1.98	16.47	43.89
C 8H 5N308	271.14	271.0077	35.44	1.86	15.50	47.21
C 8H 5N309	287.14	287.0026	33.46	1.76	14.63	50.15
C 8H 5N501	187.16	187.0494	51.34	2.69	37.42	8.55
C 8H 5N502	203.16	203.0443	47.30	2.48	34.47	15.75
C 8H 5N503	219.16	219.0392	43.84	2.30	31.96	21.90
C 8H 5N504	235.16	235.0341	40.86	2.14	29.78	27.21
C 8H 5N505	251.16	251.0291	38.26	2.01	27.88	31.85
C 8H 5N506	267.16	267.0240	35.97	1.89	26.21	35.93
C 8H 5N507	283.16	283.0189	33.93	1.78	24.73	39.55
C 8H 5N508	299.16	299.0138	32.12	1.68	23.41	42.79
C 8H 5N509	315.16	315.0087	30.49	1.60	22.22	45.69
C 8H 6N201	146.15	146.0480	65.75	4.14	19.17	10.95
C 8H 6N202	162.15	162.0429	59.26	3.73	17.28	19.73
C 8H 6N203	178.15	178.0378	53.94	3.39	15.72	26.94
C 8H 6N204	194.15	194.0328	49.49	3.12	14.43	32.96
C 8H 6N205	210.15	210.0277	45.72	2.88	13.33	38.07
C 8H 6N206	226.15	226.0226	42.49	2.67	12.39	42.45
C 8H 6N207	242.15	242.0175	39.68	2.50	11.57	46.25
C 8H 6N208	258.15	258.0124	37.22	2.34	10.85	49.58
C 8H 6N209	274.15	274.0073	35.05	2.21	10.22	52.52
C 8H 6N401	174.16	174.0542	55.17	3.47	32.17	9.19
C 8H 6N402	190.16	190.0491	50.53	3.18	29.46	16.83
C 8H 6N403	206.16	206.0440	46.61	2.93	27.18	23.28
C 8H 6N404	222.16	222.0389	43.25	2.72	25.22	28.81
C 8H 6N405	238.16	238.0338	40.35	2.54	23.52	33.59
C 8H 6N406	254.16	254.0287	37.81	2.38	22.04	37.77
C 8H 6N407	270.16	270.0236	35.57	2.24	20.74	41.46
C 8H 6N408	286.16	286.0186	33.58	2.11	19.58	44.73
C 8H 6N409	302.16	302.0135	31.80	2.00	18.54	47.66
C 8H 6N601	202.18	202.0603	47.53	2.99	41.57	7.91
C 8H 6N602	218.18	218.0552	44.04	2.77	38.52	14.67
C 8H 6N603	234.18	234.0501	41.03	2.58	35.89	20.50
C 8H 6N604	250.17	250.0450	38.41	2.42	33.59	25.58
C 8H 6N605	266.17	266.0400	36.10	2.27	31.57	30.05
C 8H 6N606	282.17	282.0349	34.05	2.14	29.78	34.02
C 8H 6N607	298.17	298.0298	32.23	2.03	28.19	37.56
C 8H 6N608	314.17	314.0247	30.58	1.93	26.75	40.74
C 8H 6N609	330.17	330.0196	29.10	1.83	25.45	43.61
C 8H 7N101	133.15	133.0528	72.17	5.30	10.52	12.02
C 8H 7N102	149.15	149.0477	64.42	4.73	9.39	21.45
C 8H 7N103	165.15	165.0426	58.18	4.27	8.48	29.06
C 8H 7N104	181.15	181.0375	53.04	3.90	7.73	35.33
C 8H 7N105	197.15	197.0324	48.74	3.58	7.10	40.58
C 8H 7N106	213.15	213.0273	45.08	3.31	6.57	45.04
C 8H 7N107	229.15	229.0222	41.93	3.08	6.11	48.87
C 8H 7N108	245.15	245.0172	39.20	2.88	5.71	52.21
C 8H 7N109	261.15	261.0121	36.80	2.70	5.36	55.14
C 8H 7N301	161.16	161.0589	59.62	4.38	26.07	9.93
C 8H 7N302	177.16	177.0538	54.24	3.98	23.72	18.06
C 8H 7N303	193.16	193.0487	49.75	3.65	21.75	24.85
C 8H 7N304	209.16	209.0437	45.94	3.37	20.09	30.60
C 8H 7N305	225.16	225.0386	42.68	3.13	18.66	35.53
C 8H 7N306	241.16	241.0335	39.84	2.93	17.42	39.81
C 8H 7N307	257.16	257.0284	37.37	2.74	16.34	43.55
C 8H 7N308	273.16	273.0233	35.18	2.58	15.38	46.86
C 8H 7N309	289.16	289.0182	33.23	2.44	14.53	49.80
C 8H 7N501	189.18	189.0651	50.79	3.73	37.02	8.46
C 8H 7N502	205.18	205.0600	46.83	3.44	34.13	15.60
C 8H 7N503	221.18	221.0549	43.44	3.19	31.66	21.70
C 8H 7N504	237.18	237.0498	40.51	2.97	29.53	26.98
C 8H 7N505	253.18	253.0447	37.95	2.79	27.66	31.60
C 8H 7N506	269.17	269.0396	35.70	2.62	26.02	35.66
C 8H 7N507	285.17	285.0345	33.69	2.47	24.56	39.27
C 8H 7N508	301.17	301.0295	31.90	2.34	23.25	42.50
C 8H 7N509	317.17	317.0244	30.30	2.22	22.08	45.40
C 8H 8N201	148.17	148.0637	64.85	5.44	18.91	10.80

038

C 8H 8N202	164.17	164.0586	58.53	4.91	17.06	19.49
C 8H 8N203	180.16	180.0535	53.33	4.48	15.55	26.64
C 8H 8N204	196.16	196.0484	48.98	4.11	14.28	32.62
C 8H 8N205	212.16	212.0433	45.29	3.80	13.20	37.71
C 8H 8N206	228.16	228.0382	42.11	3.53	12.28	42.07
C 8H 8N207	244.16	244.0331	39.35	3.30	11.47	45.87
C 8H 8N208	260.16	260.0281	36.93	3.10	10.77	49.20
C 8H 8N209	276.16	276.0230	34.79	2.92	10.14	52.14
C 8H 8N401	176.18	176.0698	54.54	4.58	31.80	9.08
C 8H 8N402	192.18	192.0647	50.00	4.20	29.15	16.65
C 8H 8N403	208.18	208.0596	46.16	3.87	26.91	23.06
C 8H 8N404	224.18	224.0545	42.86	3.60	24.99	28.55
C 8H 8N405	240.18	240.0495	40.01	3.36	23.33	33.31
C 8H 8N406	256.18	256.0444	37.51	3.15	21.87	37.47
C 8H 8N407	272.18	272.0393	35.30	2.96	20.58	41.15
C 8H 8N408	288.17	288.0342	33.34	2.80	19.44	44.42
C 8H 8N409	304.17	304.0291	31.59	2.65	18.42	47.34
C 8H 8N601	204.19	204.0760	47.06	3.95	41.16	7.84
C 8H 8N602	220.19	220.0709	43.64	3.66	38.17	14.53
C 8H 8N603	236.19	236.0658	40.68	3.41	35.58	20.32
C 8H 8N604	252.19	252.0607	38.10	3.20	33.32	25.38
C 8H 8N605	268.19	268.0556	35.83	3.01	31.34	29.83
C 8H 8N606	284.19	284.0505	33.81	2.84	29.57	33.78
C 8H 8N607	300.19	300.0454	32.01	2.69	28.00	37.31
C 8H 8N608	316.19	316.0404	30.39	2.55	26.58	40.48
C 8H 8N609	332.19	332.0353	28.93	2.43	25.30	43.35
C 8H 9N101	135.17	135.0684	71.09	6.71	10.36	11.84
C 8H 9N102	151.17	151.0633	63.57	6.00	9.27	21.17
C 8H 9N103	167.17	167.0582	57.48	5.43	8.38	28.71
C 8H 9N104	183.17	183.0532	52.46	4.95	7.65	34.94
C 8H 9N105	199.16	199.0481	48.25	4.55	7.03	40.17
C 8H 9N106	215.16	215.0430	44.66	4.22	6.51	44.62
C 8H 9N107	231.16	231.0379	41.57	3.92	6.06	48.45
C 8H 9N108	247.16	247.0328	38.88	3.67	5.67	51.79
C 8H 9N109	263.16	263.0277	36.51	3.45	5.32	54.72
C 8H 9N301	163.18	163.0746	58.89	5.56	25.75	9.80
C 8H 9N302	179.18	179.0695	53.63	5.06	23.45	17.86
C 8H 9N303	195.18	195.0644	49.23	4.65	21.53	24.59
C 8H 9N304	211.18	211.0593	45.50	4.30	19.90	30.30
C 8H 9N305	227.18	227.0542	42.30	3.99	18.50	35.21
C 8H 9N306	243.18	243.0491	39.51	3.73	17.28	39.48
C 8H 9N307	259.18	259.0440	37.07	3.50	16.21	43.21
C 8H 9N308	275.18	275.0390	34.92	3.30	15.27	46.51
C 8H 9N309	291.18	291.0339	33.00	3.12	14.43	49.45
C 8H 9N501	191.19	191.0807	50.26	4.74	36.63	8.37
C 8H 9N502	207.19	207.0756	46.38	4.38	33.80	15.44
C 8H 9N503	223.19	223.0705	43.05	4.06	31.38	21.51
C 8H 9N504	239.19	239.0654	40.17	3.79	29.28	26.76
C 8H 9N505	255.19	255.0604	37.65	3.55	27.44	31.35
C 8H 9N506	271.19	271.0553	35.43	3.35	25.82	35.40
C 8H 9N507	287.19	287.0502	33.46	3.16	24.39	39.00
C 8H 9N508	303.19	303.0451	31.69	2.99	23.10	42.22
C 8H 9N509	319.19	319.0400	30.10	2.84	21.94	45.11
C 8H10N201	150.18	150.0793	63.98	6.71	18.65	10.65
C 8H10N202	166.18	166.0742	57.82	6.07	16.86	19.26
C 8H10N203	182.18	182.0691	52.74	5.53	15.38	26.35
C 8H10N204	198.18	198.0641	48.49	5.09	14.14	32.29
C 8H10N205	214.18	214.0590	44.86	4.71	13.08	37.35
C 8H10N206	230.18	230.0539	41.75	4.38	12.17	41.71
C 8H10N207	246.18	246.0488	39.03	4.09	11.38	45.49

C 8H10N208	262.18	262.0437	36.65	3.84	10.68	48.82
C 8H10N209	278.18	278.0386	34.54	3.62	10.07	51.76
C 8H10N401	178.20	178.0855	53.92	5.66	31.44	8.98
C 8H10N402	194.19	194.0804	49.48	5.19	28.85	16.48
C 8H10N403	210.19	210.0753	45.71	4.80	26.65	22.84
C 8H10N404	226.19	226.0702	42.48	4.46	24.77	28.29
C 8H10N405	242.19	242.0651	39.67	4.16	23.13	33.03
C 8H10N406	258.19	258.0600	37.22	3.90	21.70	37.18
C 8H10N407	274.19	274.0549	35.04	3.68	20.43	40.85
C 8H10N408	290.19	290.0499	33.11	3.47	19.31	44.11
C 8H10N409	306.19	306.0448	31.38	3.29	18.30	47.03
C 8H10N601	206.21	206.0916	46.60	4.89	40.75	7.76
C 8H10N602	222.21	222.0865	43.24	4.54	37.82	14.40
C 8H10N603	238.21	238.0814	40.34	4.23	35.28	20.15
C 8H10N604	254.21	254.0763	37.80	3.97	33.06	25.18
C 8H10N605	270.21	270.0713	35.56	3.73	31.10	29.61
C 8H10N606	286.21	286.0662	33.57	3.52	29.36	33.54
C 8H10N607	302.20	302.0611	31.80	3.34	27.81	37.06
C 8H10N608	318.20	318.0560	30.20	3.17	26.41	40.22
C 8H10N609	334.20	334.0509	28.75	3.02	25.15	43.09
C 8H11N101	137.18	137.0841	70.04	8.08	10.21	11.66
C 8H11N102	153.18	153.0790	62.73	7.24	9.14	20.89
C 8H11N103	169.18	169.0739	56.80	6.55	8.28	28.37
C 8H11N104	185.18	185.0688	51.89	5.99	7.56	34.56
C 8H11N105	201.18	201.0637	47.76	5.51	6.96	39.76
C 8H11N106	217.18	217.0586	44.24	5.11	6.45	44.20
C 8H11N107	233.18	233.0535	41.21	4.75	6.01	48.03
C 8H11N108	249.18	249.0485	38.56	4.45	5.62	51.37
C 8H11N109	265.18	265.0434	36.24	4.18	5.28	54.30
C 8H11N301	165.20	165.0902	58.17	6.71	25.44	9.69
C 8H11N302	181.20	181.0851	53.03	6.12	23.19	17.66
C 8H11N303	197.20	197.0800	48.73	5.62	21.31	24.34
C 8H11N304	213.19	213.0749	45.07	5.20	19.71	30.02
C 8H11N305	229.19	229.0699	41.92	4.84	18.33	34.90
C 8H11N306	245.19	245.0648	39.19	4.52	17.14	39.15
C 8H11N307	261.19	261.0597	36.79	4.25	16.09	42.88
C 8H11N308	277.19	277.0546	34.67	4.00	15.16	46.18
C 8H11N309	293.19	293.0495	32.77	3.78	14.33	49.11
C 8H11N501	193.21	193.0964	49.73	5.74	36.25	8.28
C 8H11N502	209.21	209.0913	45.93	5.30	33.48	15.30
C 8H11N503	225.21	225.0862	42.67	4.92	31.10	21.31
C 8H11N504	241.21	241.0811	39.84	4.60	29.03	26.53
C 8H11N505	257.21	257.0760	37.36	4.31	27.23	31.10
C 8H11N506	273.21	273.0709	35.17	4.06	25.63	35.14
C 8H11N507	289.21	289.0658	33.23	3.83	24.22	38.73
C 8H11N508	305.21	305.0608	31.48	3.63	22.95	41.94
C 8H11N509	321.20	321.0557	29.92	3.45	21.80	44.83
C 8H12N201	152.20	152.0950	63.13	7.95	18.41	10.51
C 8H12N202	168.20	168.0899	57.13	7.19	16.66	19.02
C 8H12N203	184.20	184.0848	52.17	6.57	15.21	26.06
C 8H12N204	200.20	200.0797	48.00	6.04	13.99	31.97
C 8H12N205	216.20	216.0746	44.45	5.59	12.96	37.00
C 8H12N206	232.19	232.0695	41.38	5.21	12.06	41.34
C 8H12N207	248.19	248.0644	38.72	4.87	11.29	45.12
C 8H12N208	264.19	264.0594	36.37	4.58	10.60	48.45
C 8H12N209	280.19	280.0543	34.29	4.32	10.00	51.39
C 8H12N401	180.21	180.1011	53.32	6.71	31.09	8.88
C 8H12N402	196.21	196.0960	48.97	6.16	28.55	16.31
C 8H12N403	212.21	212.0909	45.28	5.70	26.40	22.62
C 8H12N404	228.21	228.0858	42.11	5.30	24.55	28.04

039

C 8H12N405	244.21	244.0808	39.35	4.95	22.94	32.76
C 8H12N406	260.21	260.0757	36.93	4.65	21.53	36.89
C 8H12N407	276.21	276.0706	34.79	4.38	20.28	40.55
C 8H12N408	292.21	292.0655	32.88	4.14	19.17	43.80
C 8H12N409	308.21	308.0604	31.18	3.92	18.18	46.72
C 8H12N601	208.22	208.1073	46.15	5.81	40.36	7.68
C 8H12N602	224.22	224.1022	42.85	5.39	37.48	14.27
C 8H12N603	240.22	240.0971	40.00	5.04	34.98	19.98
C 8H12N604	256.22	256.0920	37.50	4.72	32.80	24.98
C 8H12N605	272.22	272.0869	35.30	4.44	30.87	29.39
C 8H12N606	288.22	288.0818	33.34	4.20	29.16	33.31
C 8H12N607	304.22	304.0767	31.59	3.98	27.62	36.81
C 8H12N608	320.22	320.0717	30.01	3.78	26.24	39.97
C 8H12N609	336.22	336.0666	28.58	3.60	25.00	42.83
C 8H13N101	139.20	139.0997	69.03	9.41	10.06	11.49
C 8H13N102	155.20	155.0946	61.91	8.44	9.03	20.62
C 8H13N103	171.20	171.0895	56.13	7.65	8.18	28.04
C 8H13N104	187.20	187.0844	51.33	7.00	7.48	34.19
C 8H13N105	203.20	203.0794	47.29	6.45	6.89	39.37
C 8H13N106	219.20	219.0743	43.84	5.98	6.39	43.79
C 8H13N107	235.20	235.0692	40.86	5.57	5.96	47.62
C 8H13N108	251.19	251.0641	38.25	5.22	5.58	50.95
C 8H13N109	267.19	267.0590	35.96	4.90	5.24	53.89
C 8H13N301	167.21	167.1059	57.47	7.84	25.13	9.57
C 8H13N302	183.21	183.1008	52.45	7.15	22.94	17.47
C 8H13N303	199.21	199.0957	48.23	6.58	21.09	24.09
C 8H13N304	215.21	215.0906	44.65	6.09	19.53	29.74
C 8H13N305	231.21	231.0855	41.56	5.67	18.17	34.60
C 8H13N306	247.21	247.0804	38.87	5.30	17.00	38.83
C 8H13N307	263.21	263.0753	36.51	4.98	15.96	42.55
C 8H13N308	279.21	279.0703	34.41	4.69	15.05	45.84
C 8H13N309	295.21	295.0652	32.55	4.44	14.23	48.78
C 8H13N501	195.23	195.1120	49.22	6.71	35.87	8.20
C 8H13N502	211.23	211.1069	45.49	6.20	33.16	15.15
C 8H13N503	227.22	227.1018	42.29	5.77	30.82	21.12
C 8H13N504	243.22	243.0967	39.51	5.39	28.79	26.31
C 8H13N505	259.22	259.0917	37.07	5.05	27.02	30.86
C 8H13N506	275.22	275.0866	34.91	4.76	25.45	34.88
C 8H13N507	291.22	291.0815	33.00	4.50	24.05	38.46
C 8H13N508	307.22	307.0764	31.28	4.27	22.80	41.66
C 8H13N509	323.22	323.0713	29.73	4.05	21.67	44.55
C 8H14N201	154.21	154.1106	62.31	9.15	18.17	10.37
C 8H14N202	170.21	170.1055	56.45	8.29	16.46	18.80
C 8H14N203	186.21	186.1004	51.60	7.58	15.04	25.78
C 8H14N204	202.21	202.0953	47.52	6.98	13.85	31.65
C 8H14N205	218.21	218.0903	44.03	6.47	12.84	36.66
C 8H14N206	234.21	234.0852	41.03	6.03	11.96	40.99
C 8H14N207	250.21	250.0801	38.40	5.64	11.20	44.76
C 8H14N208	266.21	266.0750	36.10	5.30	10.52	48.08
C 8H14N209	282.21	282.0699	34.05	5.00	9.93	51.02
C 8H14N401	182.23	182.1168	52.73	7.74	30.75	8.78
C 8H14N402	198.23	198.1117	48.47	7.12	28.26	16.14
C 8H14N403	214.23	214.1066	44.85	6.59	26.15	22.41
C 8H14N404	230.23	230.1015	41.74	6.13	24.34	27.80
C 8H14N405	246.22	246.0964	39.03	5.73	22.75	32.49
C 8H14N406	262.22	262.0913	36.64	5.38	21.37	36.61
C 8H14N407	278.22	278.0862	34.54	5.07	20.14	40.25
C 8H14N408	294.22	294.0812	32.66	4.80	19.04	43.50
C 8H14N409	310.22	310.0761	30.97	4.55	18.06	46.42
C 8H14N601	210.24	210.1229	45.70	6.71	39.97	7.61

C 8H14N602	226.24	226.1178	42.47	6.24	37.15	14.14
C 8H14N603	242.24	242.1127	39.67	5.83	34.69	19.81
C 8H14N604	258.24	258.1076	37.21	5.46	32.54	24.78
C 8H14N605	274.24	274.1026	35.04	5.15	30.64	29.17
C 8H14N606	290.24	290.0975	33.11	4.86	28.96	33.08
C 8H14N607	306.24	306.0924	31.38	4.61	27.44	36.57
C 8H14N608	322.24	322.0873	29.82	4.38	26.08	39.72
C 8H14N609	338.24	338.0822	28.41	4.17	24.85	42.57
C 8H15N101	141.21	141.1154	68.04	10.71	9.92	11.33
C 8H15N102	157.21	157.1103	61.12	9.62	8.91	20.35
C 8H15N103	173.21	173.1052	55.47	8.73	8.09	27.71
C 8H15N104	189.21	189.1001	50.78	7.99	7.40	33.82
C 8H15N105	205.21	205.0950	46.82	7.37	6.83	38.98
C 8H15N106	221.21	221.0899	43.44	6.83	6.33	43.40
C 8H15N107	237.21	237.0848	40.51	6.37	5.90	47.21
C 8H15N108	253.21	253.0798	37.95	5.97	5.53	50.55
C 8H15N109	269.21	269.0747	35.69	5.62	5.20	53.49
C 8H15N301	169.23	169.1215	56.78	8.93	24.83	9.45
C 8H15N302	185.23	185.1164	51.88	8.16	22.69	17.28
C 8H15N303	201.23	201.1113	47.75	7.51	20.88	23.85
C 8H15N304	217.23	217.1062	44.23	6.96	19.34	29.46
C 8H15N305	233.23	233.1012	41.20	6.48	18.02	34.30
C 8H15N306	249.23	249.0961	38.56	6.07	16.86	38.52
C 8H15N307	265.22	265.0910	36.23	5.70	15.84	42.23
C 8H15N308	281.22	281.0859	34.17	5.38	14.94	45.51
C 8H15N309	297.22	297.0808	32.33	5.09	14.14	48.45
C 8H15N501	197.24	197.1277	48.72	7.67	35.51	8.11
C 8H15N502	213.24	213.1226	45.06	7.09	32.84	15.01
C 8H15N503	229.24	229.1175	41.92	6.60	30.55	20.94
C 8H15N504	245.24	245.1124	39.18	6.17	28.56	26.10
C 8H15N505	261.24	261.1073	36.78	5.79	26.81	30.62
C 8H15N506	277.24	277.1022	34.66	5.45	25.26	34.63
C 8H15N507	293.24	293.0971	32.77	5.16	23.88	38.19
C 8H15N508	309.24	309.0921	31.07	4.89	22.65	41.39
C 8H15N509	325.24	325.0870	29.54	4.65	21.53	44.27
C 8H16N201	156.23	156.1263	61.51	10.32	17.93	10.24
C 8H16N202	172.23	172.1212	55.79	9.36	16.27	18.58
C 8H16N203	188.23	188.1161	51.05	8.57	14.88	25.50
C 8H16N204	204.23	204.1110	47.05	7.90	13.72	31.34
C 8H16N205	220.23	220.1059	43.63	7.32	12.72	36.32
C 8H16N206	236.23	236.1008	40.68	6.83	11.86	40.64
C 8H16N207	252.23	252.0957	38.10	6.39	11.11	44.40
C 8H16N208	268.23	268.0907	35.82	6.01	10.44	47.72
C 8H16N209	284.22	284.0856	33.81	5.67	9.86	50.66
C 8H16N401	184.24	184.1324	52.15	8.75	30.41	8.68
C 8H16N402	200.24	200.1273	47.99	8.05	27.98	15.98
C 8H16N403	216.24	216.1222	44.44	7.46	25.91	22.20
C 8H16N404	232.24	232.1171	41.37	6.94	24.12	27.56
C 8H16N405	248.24	248.1121	38.71	6.50	22.57	32.23
C 8H16N406	264.24	264.1070	36.36	6.10	21.20	36.33
C 8H16N407	280.24	280.1019	34.29	5.75	19.99	39.96
C 8H16N408	296.24	296.0968	32.44	5.44	18.91	43.21
C 8H16N409	312.24	312.0917	30.77	5.17	17.94	46.12
C 8H16N601	212.26	212.1386	45.27	7.60	39.59	7.54
C 8H16N602	228.26	228.1335	42.10	7.07	36.82	14.02
C 8H16N603	244.26	244.1284	39.34	6.60	34.41	19.65
C 8H16N604	260.25	260.1233	36.92	6.20	32.29	24.59
C 8H16N605	276.25	276.1182	34.78	5.84	30.42	28.96
C 8H16N606	292.25	292.1131	32.88	5.52	28.76	32.85
C 8H16N607	308.25	308.1080	31.17	5.23	27.26	36.33

C 8H16N608	324.25	324.1029	29.63	4.97	25.92	39.47
C 8H16N609	340.25	340.0979	28.24	4.74	24.70	42.32
C 8H17N101	143.23	143.1310	67.09	11.96	9.78	11.17
C 8H17N102	159.23	159.1259	60.35	10.76	8.80	20.10
C 8H17N103	175.23	175.1208	54.84	9.78	7.99	27.39
C 8H17N104	191.23	191.1157	50.25	8.96	7.32	33.47
C 8H17N105	207.23	207.1107	46.37	8.27	6.76	38.60
C 8H17N106	223.23	223.1056	43.05	7.68	6.27	43.00
C 8H17N107	239.23	239.1005	40.17	7.16	5.85	46.82
C 8H17N108	255.23	255.0954	37.65	6.71	5.49	50.15
C 8H17N109	271.23	271.0903	35.43	6.32	5.16	53.09
C 8H17N301	171.24	171.1372	56.11	10.01	24.54	9.34
C 8H17N302	187.24	187.1321	51.32	9.15	22.44	17.09
C 8H17N303	203.24	203.1270	47.28	8.43	20.67	23.62
C 8H17N304	219.24	219.1219	43.83	7.82	19.17	29.19
C 8H17N305	235.24	235.1168	40.85	7.28	17.86	34.01
C 8H17N306	251.24	251.1117	38.25	6.82	16.73	38.21
C 8H17N307	267.24	267.1066	35.96	6.41	15.72	41.91
C 8H17N308	283.24	283.1016	33.93	6.05	14.84	45.19
C 8H17N309	299.24	299.0965	32.11	5.73	14.04	48.12
C 8H17N501	199.26	199.1433	48.22	8.60	35.15	8.03
C 8H17N502	215.26	215.1382	44.64	7.96	32.53	14.87
C 8H17N503	231.26	231.1331	41.55	7.41	30.28	20.76
C 8H17N504	247.26	247.1280	38.86	6.93	28.32	25.88
C 8H17N505	263.26	263.1230	36.50	6.51	26.60	30.39
C 8H17N506	279.25	279.1179	34.41	6.14	25.08	34.38
C 8H17N507	295.25	295.1128	32.54	5.80	23.72	37.93
C 8H17N508	311.25	311.1077	30.87	5.51	22.50	41.12
C 8H17N509	327.25	327.1026	29.36	5.24	21.40	44.00
C 8H18N201	158.25	158.1419	60.72	11.47	17.70	10.11
C 8H18N202	174.24	174.1368	55.15	10.41	16.08	18.36
C 8H18N203	190.24	190.1317	50.51	9.54	14.72	25.23
C 8H18N204	206.24	206.1266	46.59	8.80	13.58	31.03
C 8H18N205	222.24	222.1216	43.24	8.16	12.60	36.00
C 8H18N206	238.24	238.1165	40.33	7.62	11.76	40.29
C 8H18N207	254.24	254.1114	37.79	7.14	11.02	44.05
C 8H18N208	270.24	270.1063	35.56	6.71	10.37	47.36
C 8H18N209	286.24	286.1012	33.57	6.34	9.79	50.31
C 8H18N401	186.26	186.1481	51.59	9.74	30.08	8.59
C 8H18N402	202.26	202.1430	47.51	8.97	27.70	15.82
C 8H18N403	218.26	218.1379	44.03	8.31	25.67	21.99
C 8H18N404	234.26	234.1328	41.02	7.75	23.92	27.32
C 8H18N405	250.26	250.1277	38.40	7.25	22.39	31.97
C 8H18N406	266.26	266.1226	36.09	6.81	21.04	36.05
C 8H18N407	282.26	282.1175	34.04	6.43	19.85	39.68
C 8H18N408	298.25	298.1125	32.22	6.08	18.78	42.91
C 8H18N409	314.25	314.1074	30.58	5.77	17.83	45.82
C 8H18N601	214.27	214.1542	44.84	8.47	39.22	7.47
C 8H18N602	230.27	230.1491	41.73	7.88	36.50	13.90
C 8H18N603	246.27	246.1440	39.02	7.37	34.13	19.49
C 8H18N604	262.27	262.1389	36.64	6.92	32.04	24.40
C 8H18N605	278.27	278.1339	34.53	6.52	30.20	28.75
C 8H18N606	294.27	294.1288	32.65	6.17	28.56	32.62
C 8H18N607	310.27	310.1237	30.97	5.85	27.09	36.10
C 8H18N608	326.27	326.1186	29.45	5.56	25.76	39.23
C 8H18N609	342.27	342.1135	28.07	5.30	24.55	42.07
C 8H19N101	145.25	145.1467	66.16	13.19	9.64	11.02
C 8H19N102	161.25	161.1416	59.59	11.88	8.69	19.84
C 8H19N103	177.25	177.1365	54.21	10.81	7.90	27.08
C 8H19N104	193.24	193.1314	49.72	9.91	7.25	33.12

C 8H19N105	209.24	209.1263	45.92	9.15	6.69	38.23
C 8H19N106	225.24	225.1212	42.66	8.50	6.22	42.62
C 8H19N107	241.24	241.1161	39.83	7.94	5.81	46.42
C 8H19N108	257.24	257.1111	37.35	7.44	5.44	49.76
C 8H19N109	273.24	273.1060	35.17	7.01	5.13	52.70
C 8H19N301	173.26	173.1528	55.46	11.05	24.25	9.23
C 8H19N302	189.26	189.1477	50.77	10.12	22.20	16.91
C 8H19N303	205.26	205.1426	46.81	9.33	20.47	23.38
C 8H19N304	221.26	221.1375	43.43	8.66	18.99	28.92
C 8H19N305	237.26	237.1325	40.50	8.07	17.71	33.72
C 8H19N306	253.26	253.1274	37.94	7.56	16.59	37.90
C 8H19N307	269.26	269.1223	35.69	7.11	15.61	41.59
C 8H19N308	285.26	285.1172	33.69	6.71	14.73	44.87
C 8H19N309	301.26	301.1121	31.90	6.36	13.95	47.80
C 8H19N501	201.27	201.1590	47.74	9.52	34.80	7.95
C 8H19N502	217.27	217.1539	44.23	8.81	32.23	14.73
C 8H19N503	233.27	233.1488	41.19	8.21	30.02	20.58
C 8H19N504	249.27	249.1437	38.55	7.68	28.10	25.67
C 8H19N505	265.27	265.1386	36.22	7.22	26.40	30.16
C 8H19N506	281.27	281.1335	34.16	6.81	24.90	34.13
C 8H19N507	297.27	297.1284	32.32	6.44	23.56	37.67
C 8H19N508	313.27	313.1233	30.67	6.11	22.36	40.86
C 8H19N509	329.27	329.1183	29.18	5.82	21.27	43.73
C 8H20N201	160.26	160.1576	59.96	12.58	17.48	9.98
C 8H20N202	176.26	176.1525	54.52	11.44	15.89	18.15
C 8H20N203	192.26	192.1474	49.98	10.49	14.57	24.97
C 8H20N204	208.26	208.1423	46.14	9.68	13.45	30.73
C 8H20N205	224.26	224.1372	42.85	8.99	12.49	35.67
C 8H20N206	240.26	240.1321	39.99	8.39	11.66	39.96
C 8H20N207	256.26	256.1270	37.50	7.87	10.93	43.70
C 8H20N208	272.26	272.1220	35.29	7.40	10.29	47.01
C 8H20N209	288.26	288.1169	33.33	6.99	9.72	49.95
C 8H20N401	188.27	188.1637	51.04	10.71	29.76	8.50
C 8H20N402	204.27	204.1586	47.04	9.87	27.43	15.66
C 8H20N403	220.27	220.1535	43.62	9.15	25.44	21.79
C 8H20N404	236.27	236.1484	40.67	8.53	23.71	27.09
C 8H20N405	252.27	252.1434	38.09	7.99	22.21	31.71
C 8H20N406	268.27	268.1383	35.82	7.51	20.88	35.78
C 8H20N407	284.27	284.1332	33.80	7.09	19.71	39.40
C 8H20N408	300.27	300.1281	32.00	6.71	18.66	42.63
C 8H20N409	316.27	316.1230	30.38	6.37	17.71	45.53
C 8H20N601	216.29	216.1698	44.43	9.32	38.86	7.40
C 8H20N602	232.29	232.1648	41.37	8.68	36.18	13.78
C 8H20N603	248.29	248.1597	38.70	8.12	33.85	19.33
C 8H20N604	264.29	264.1546	36.36	7.63	31.80	24.22
C 8H20N605	280.29	280.1495	34.28	7.19	29.98	28.54
C 8H20N606	296.29	296.1444	32.43	6.80	28.36	32.40
C 8H20N607	312.28	312.1393	30.77	6.46	26.91	35.86
C 8H20N608	328.28	328.1342	29.27	6.14	25.60	38.99
C 8H20N609	344.28	344.1292	27.91	5.86	24.41	41.82
C 8H21N301	175.28	175.1685	54.82	12.08	23.97	9.13
C 8H21N302	191.28	191.1634	50.24	11.07	21.97	16.73
C 8H21N303	207.27	207.1583	46.36	10.21	20.27	23.16
C 8H21N304	223.27	223.1532	43.04	9.48	18.82	28.66
C 8H21N305	239.27	239.1481	40.16	8.85	17.56	33.43
C 8H21N306	255.27	255.1430	37.64	8.29	16.46	37.61
C 8H21N307	271.27	271.1379	35.42	7.80	15.49	41.29
C 8H21N308	287.27	287.1329	33.45	7.37	14.63	44.56
C 8H21N309	303.27	303.1278	31.68	6.98	13.86	47.48
C 8H21N501	203.29	203.1746	47.27	10.41	34.45	7.87

041

C 8H21N502	219.29	219.1695	43.82	9.65	31.94	14.59
C 8H21N503	235.29	235.1644	40.84	9.00	29.76	20.40
C 8H21N504	251.29	251.1593	38.24	8.42	27.87	25.47
C 8H21N505	267.29	267.1543	35.95	7.92	26.20	29.93
C 8H21N506	283.29	283.1492	33.92	7.47	24.72	33.89
C 8H21N507	299.29	299.1441	32.11	7.07	23.40	37.42
C 8H21N508	315.29	315.1390	30.48	6.71	22.21	40.60
C 8H21N509	331.28	331.1339	29.01	6.39	21.14	43.47
C 8H22N401	190.29	190.1794	50.50	11.65	29.44	8.41
C 8H22N402	206.29	206.1743	46.58	10.75	27.16	15.51
C 8H22N403	222.29	222.1692	43.23	9.98	25.20	21.59
C 8H22N404	238.29	238.1641	40.32	9.31	23.51	26.86
C 8H22N405	254.29	254.1590	37.79	8.72	22.03	31.46
C 8H22N406	270.29	270.1539	35.55	8.20	20.73	35.52
C 8H22N407	286.29	286.1488	33.56	7.75	19.57	39.12
C 8H22N408	302.29	302.1437	31.79	7.34	18.53	42.34
C 8H22N409	318.29	318.1387	30.19	6.97	17.60	45.24
C 8H22N601	218.30	218.1855	44.02	10.16	38.50	7.33
C 8H22N602	234.30	234.1804	41.01	9.46	35.87	13.66
C 8H22N603	250.30	250.1753	38.39	8.86	33.58	19.18
C 8H22N604	266.30	266.1702	36.08	8.33	31.56	24.03
C 8H22N605	282.30	282.1652	34.04	7.86	29.77	28.34
C 8H22N606	298.30	298.1601	32.21	7.43	28.17	32.18
C 8H22N607	314.30	314.1550	30.57	7.06	26.74	35.63
C 8H22N608	330.30	330.1499	29.09	6.71	25.44	38.75
C 8H22N609	346.30	346.1448	27.75	6.40	24.27	41.58
C 8H23N501	205.31	205.1902	46.80	11.29	34.11	7.79
C 8H23N502	221.30	221.1852	43.42	10.48	31.65	14.46
C 8H23N503	237.30	237.1801	40.49	9.77	29.51	20.23
C 8H23N504	253.30	253.1750	37.93	9.15	27.65	25.27
C 8H23N505	269.30	269.1699	35.68	8.61	26.01	29.71
C 8H23N506	285.30	285.1648	33.68	8.13	24.55	33.65
C 8H23N507	301.30	301.1597	31.89	7.69	23.24	37.17
C 8H23N508	317.30	317.1546	30.28	7.31	22.07	40.34
C 8H23N509	333.30	333.1496	28.83	6.96	21.01	43.20
C 8H24N601	220.32	220.2011	43.61	10.98	38.14	7.26
C 8H24N602	236.32	236.1961	40.66	10.24	35.56	13.54
C 8H24N603	252.32	252.1910	38.08	9.59	33.31	19.02
C 8H24N604	268.32	268.1859	35.81	9.02	31.32	23.85
C 8H24N605	284.32	284.1808	33.80	8.51	29.56	28.14
C 8H24N606	300.32	300.1757	32.00	8.06	27.98	31.97
C 8H24N607	316.32	316.1706	30.38	7.65	26.57	35.41
C 8H24N608	332.32	332.1655	28.92	7.28	25.29	38.52
C 8H24N609	348.32	348.1605	27.59	6.95	24.13	41.34
C 9H 4N201	156.15	156.0324	69.23	2.58	17.94	10.25
C 9H 4N202	172.14	172.0273	62.80	2.34	16.27	18.59
C 9H 4N203	188.14	188.0222	57.46	2.14	14.89	25.51
C 9H 4N204	204.14	204.0171	52.95	1.98	13.72	31.35
C 9H 4N205	220.14	220.0120	49.10	1.83	12.73	36.34
C 9H 4N206	236.14	236.0069	45.78	1.71	11.86	40.65
C 9H 4N207	252.14	252.0018	42.87	1.60	11.11	44.42
C 9H 4N208	268.14	267.9968	40.31	1.50	10.45	47.73
C 9H 4N209	284.14	283.9917	38.04	1.42	9.86	50.68
C 9H 4N401	184.16	184.0385	58.70	2.19	30.42	8.69
C 9H 4N402	200.16	200.0334	54.01	2.01	27.99	15.99
C 9H 4N403	216.16	216.0283	50.01	1.87	25.92	22.21
C 9H 4N404	232.16	232.0233	46.56	1.74	24.13	27.57
C 9H 4N405	248.16	248.0182	43.56	1.62	22.58	32.24
C 9H 4N406	264.16	264.0131	40.92	1.53	21.21	36.34
C 9H 4N407	280.15	280.0080	38.59	1.44	20.00	39.98

C 9H 4N408	296.15	296.0029	36.50	1.36	18.92	43.22
C 9H 4N409	312.15	311.9978	34.63	1.29	17.95	46.13
C 9H 4N601	212.17	212.0447	50.95	1.90	39.61	7.54
C 9H 4N602	228.17	228.0396	47.38	1.77	36.83	14.02
C 9H 4N603	244.17	244.0345	44.27	1.65	34.42	19.66
C 9H 4N604	260.17	260.0294	41.55	1.55	32.30	24.60
C 9H 4N605	276.17	276.0243	39.14	1.46	30.43	28.97
C 9H 4N606	292.17	292.0192	37.00	1.38	28.76	32.86
C 9H 4N607	308.17	308.0141	35.08	1.31	27.27	36.34
C 9H 4N608	324.17	324.0091	33.35	1.24	25.92	39.48
C 9H 4N609	340.17	340.0040	31.78	1.19	24.71	42.33
C 9H 5N101	143.15	143.0371	75.52	3.52	9.78	11.18
C 9H 5N102	159.15	159.0320	67.93	3.17	8.80	20.11
C 9H 5N103	175.15	175.0269	61.72	2.88	8.00	27.40
C 9H 5N104	191.14	191.0219	56.55	2.64	7.33	33.48
C 9H 5N105	207.14	207.0168	52.19	2.43	6.76	38.62
C 9H 5N106	223.14	223.0117	48.44	2.26	6.28	43.02
C 9H 5N107	239.14	239.0066	45.20	2.11	5.86	46.83
C 9H 5N108	255.14	255.0015	42.37	1.98	5.49	50.17
C 9H 5N109	271.14	270.9964	39.87	1.86	5.17	53.11
C 9H 5N301	171.16	171.0433	63.16	2.94	24.55	9.35
C 9H 5N302	187.16	187.0382	57.76	2.69	22.45	17.10
C 9H 5N303	203.16	203.0331	53.21	2.48	20.68	23.63
C 9H 5N304	219.16	219.0280	49.33	2.30	19.17	29.20
C 9H 5N305	235.16	235.0229	45.97	2.14	17.87	34.02
C 9H 5N306	251.16	251.0178	43.04	2.01	16.73	38.22
C 9H 5N307	267.16	267.0127	40.46	1.89	15.73	41.92
C 9H 5N308	283.16	283.0077	38.18	1.78	14.84	45.20
C 9H 5N309	299.15	299.0026	36.14	1.68	14.05	48.13
C 9H 5N501	199.17	199.0494	54.27	2.53	35.16	8.03
C 9H 5N502	215.17	215.0443	50.24	2.34	32.55	14.87
C 9H 5N503	231.17	231.0392	46.76	2.18	30.29	20.76
C 9H 5N504	247.17	247.0341	43.73	2.04	28.33	25.89
C 9H 5N505	263.17	263.0291	41.08	1.92	26.61	30.40
C 9H 5N506	279.17	279.0240	38.72	1.81	25.09	34.39
C 9H 5N507	295.17	295.0189	36.62	1.71	23.73	37.94
C 9H 5N508	311.17	311.0138	34.74	1.62	22.51	41.13
C 9H 5N509	327.17	327.0087	33.04	1.54	21.41	44.01
C 9H 6N201	158.16	158.0480	68.35	3.82	17.71	10.12
C 9H 6N202	174.16	174.0429	62.07	3.47	16.08	18.37
C 9H 6N203	190.16	190.0378	56.85	3.18	14.73	25.24
C 9H 6N204	206.16	206.0328	52.44	2.93	13.59	31.04
C 9H 6N205	222.16	222.0277	48.66	2.72	12.61	36.01
C 9H 6N206	238.16	238.0226	45.39	2.54	11.76	40.31
C 9H 6N207	254.16	254.0175	42.53	2.38	11.02	44.07
C 9H 6N208	270.16	270.0124	40.01	2.24	10.37	47.38
C 9H 6N209	286.16	286.0073	37.78	2.11	9.79	50.32
C 9H 6N401	186.17	186.0542	58.06	3.25	30.09	8.59
C 9H 6N402	202.17	202.0491	53.47	2.99	27.71	15.83
C 9H 6N403	218.17	218.0440	49.55	2.77	25.68	22.00
C 9H 6N404	234.17	234.0389	46.16	2.58	23.93	27.33
C 9H 6N405	250.17	250.0338	43.21	2.42	22.40	31.98
C 9H 6N406	266.17	266.0287	40.61	2.27	21.05	36.07
C 9H 6N407	282.17	282.0236	38.31	2.14	19.86	39.69
C 9H 6N408	298.17	298.0186	36.25	2.03	18.79	42.93
C 9H 6N409	314.17	314.0135	34.41	1.93	17.83	45.83
C 9H 6N601	214.19	214.0603	50.47	2.82	39.24	7.47
C 9H 6N602	230.19	230.0552	46.96	2.63	36.51	13.90
C 9H 6N603	246.19	246.0501	43.91	2.46	34.14	19.50
C 9H 6N604	262.19	262.0450	41.23	2.31	32.05	24.41

042

C 9H 6N6O5	278.19	278.0400	38.86	2.17	30.21	28.76
C 9H 6N6O6	294.18	294.0349	36.75	2.06	28.57	32.63
C 9H 6N6O7	310.18	310.0298	34.85	1.95	27.09	36.11
C 9H 6N6O8	326.18	326.0247	33.14	1.85	25.76	39.24
C 9H 6N6O9	342.18	342.0196	31.59	1.77	24.56	42.08
C 9H 7N1O1	145.16	145.0528	74.47	4.86	9.65	11.02
C 9H 7N1O2	161.16	161.0477	67.08	4.38	8.69	19.86
C 9H 7N1O3	177.16	177.0426	61.02	3.98	7.91	27.09
C 9H 7N1O4	193.16	193.0375	55.96	3.65	7.25	33.13
C 9H 7N1O5	209.16	209.0324	51.68	3.37	6.70	38.25
C 9H 7N1O6	225.16	225.0273	48.01	3.13	6.22	42.63
C 9H 7N1O7	241.16	241.0222	44.83	2.93	5.81	46.44
C 9H 7N1O8	257.16	257.0172	42.04	2.74	5.45	49.77
C 9H 7N1O9	273.16	273.0121	39.57	2.58	5.13	52.71
C 9H 7N3O1	173.18	173.0589	62.42	4.07	24.26	9.24
C 9H 7N3O2	189.18	189.0538	57.14	3.73	22.21	16.91
C 9H 7N3O3	205.17	205.0487	52.69	3.44	20.48	23.39
C 9H 7N3O4	221.17	221.0437	48.88	3.19	19.00	28.94
C 9H 7N3O5	237.17	237.0386	45.58	2.97	17.72	33.73
C 9H 7N3O6	253.17	253.0335	42.70	2.79	16.60	37.92
C 9H 7N3O7	269.17	269.0284	40.16	2.62	15.61	41.61
C 9H 7N3O8	285.17	285.0233	37.91	2.47	14.74	44.88
C 9H 7N3O9	301.17	301.0182	35.89	2.34	13.95	47.81
C 9H 7N5O1	201.19	201.0651	53.73	3.51	34.81	7.95
C 9H 7N5O2	217.19	217.0600	49.77	3.25	32.25	14.73
C 9H 7N5O3	233.19	233.0549	46.36	3.03	30.03	20.58
C 9H 7N5O4	249.19	249.0498	43.38	2.83	28.10	25.68
C 9H 7N5O5	265.19	265.0447	40.76	2.66	26.41	30.17
C 9H 7N5O6	281.19	281.0396	38.44	2.51	24.91	34.14
C 9H 7N5O7	297.19	297.0345	36.37	2.37	23.57	37.69
C 9H 7N5O8	313.18	313.0295	34.52	2.25	22.36	40.87
C 9H 7N5O9	329.18	329.0244	32.84	2.14	21.27	43.74
C 9H 8N2O1	160.18	160.0637	67.49	5.03	17.49	9.99
C 9H 8N2O2	176.18	176.0586	61.36	4.58	15.90	18.16
C 9H 8N2O3	192.18	192.0535	56.25	4.20	14.58	24.98
C 9H 8N2O4	208.18	208.0484	51.93	3.87	13.46	30.74
C 9H 8N2O5	224.17	224.0433	48.22	3.60	12.50	35.69
C 9H 8N2O6	240.17	240.0382	45.01	3.36	11.66	39.97
C 9H 8N2O7	256.17	256.0331	42.20	3.15	10.94	43.72
C 9H 8N2O8	272.17	272.0281	39.72	2.96	10.29	47.03
C 9H 8N2O9	288.17	288.0230	37.51	2.80	9.72	49.97
C 9H 8N4O1	188.19	188.0698	57.44	4.28	29.77	8.50
C 9H 8N4O2	204.19	204.0647	52.94	3.95	27.44	15.67
C 9H 8N4O3	220.19	220.0596	49.09	3.66	25.44	21.80
C 9H 8N4O4	236.19	236.0545	45.77	3.41	23.72	27.10
C 9H 8N4O5	252.19	252.0495	42.87	3.20	22.22	31.72
C 9H 8N4O6	268.19	268.0444	40.31	3.01	20.89	35.79
C 9H 8N4O7	284.19	284.0393	38.04	2.84	19.71	39.41
C 9H 8N4O8	300.19	300.0342	36.01	2.69	18.66	42.64
C 9H 8N4O9	316.19	316.0291	34.19	2.55	17.72	45.54
C 9H 8N6O1	216.20	216.0760	50.00	3.73	38.87	7.40
C 9H 8N6O2	232.20	232.0709	46.55	3.47	36.19	13.78
C 9H 8N6O3	248.20	248.0658	43.55	3.25	33.86	19.34
C 9H 8N6O4	264.20	264.0607	40.92	3.05	31.81	24.22
C 9H 8N6O5	280.20	280.0556	38.58	2.88	29.99	28.55
C 9H 8N6O6	296.20	296.0505	36.50	2.72	28.37	32.41
C 9H 8N6O7	312.20	312.0454	34.63	2.58	26.92	35.87
C 9H 8N6O8	328.20	328.0404	32.94	2.46	25.61	39.00
C 9H 8N6O9	344.20	344.0353	31.41	2.34	24.42	41.83
C 9H 9N1O1	147.18	147.0684	73.45	6.16	9.52	10.87
C 9H 9N1O2	163.18	163.0633	66.25	5.56	8.58	19.61
C 9H 9N1O3	179.18	179.0582	60.33	5.06	7.82	26.79
C 9H 9N1O4	195.18	195.0532	55.39	4.65	7.18	32.79
C 9H 9N1O5	211.18	211.0481	51.19	4.30	6.63	37.88
C 9H 9N1O6	227.18	227.0430	47.58	3.99	6.17	42.26
C 9H 9N1O7	243.17	243.0379	44.45	3.73	5.76	46.06
C 9H 9N1O8	259.17	259.0328	41.71	3.50	5.40	49.39
C 9H 9N1O9	275.17	275.0277	39.28	3.30	5.09	52.33
C 9H 9N3O1	175.19	175.0746	61.70	5.18	23.99	9.13
C 9H 9N3O2	191.19	191.0695	56.54	4.74	21.98	16.74
C 9H 9N3O3	207.19	207.0644	52.17	4.38	20.28	23.17
C 9H 9N3O4	223.19	223.0593	48.43	4.06	18.83	28.67
C 9H 9N3O5	239.19	239.0542	45.19	3.79	17.57	33.45
C 9H 9N3O6	255.19	255.0491	42.36	3.55	16.47	37.62
C 9H 9N3O7	271.19	271.0440	39.86	3.35	15.49	41.30
C 9H 9N3O8	287.19	287.0390	37.64	3.16	14.63	44.57
C 9H 9N3O9	303.19	303.0339	35.65	2.99	13.86	47.49
C 9H 9N5O1	203.20	203.0807	53.20	4.46	34.46	7.87
C 9H 9N5O2	219.20	219.0756	49.31	4.14	31.95	14.60
C 9H 9N5O3	235.20	235.0705	45.96	3.86	29.78	20.41
C 9H 9N5O4	251.20	251.0654	43.03	3.61	27.88	25.48
C 9H 9N5O5	267.20	267.0604	40.46	3.40	26.21	29.94
C 9H 9N5O6	283.20	283.0553	38.17	3.20	24.73	33.90
C 9H 9N5O7	299.20	299.0502	36.13	3.03	23.41	37.43
C 9H 9N5O8	315.20	315.0451	34.30	2.88	22.22	40.61
C 9H 9N5O9	331.20	331.0400	32.64	2.74	21.15	43.48
C 9H10N2O1	162.19	162.0793	66.65	6.21	17.27	9.86
C 9H10N2O2	178.19	178.0742	60.67	5.66	15.72	17.96
C 9H10N2O3	194.19	194.0691	55.67	5.19	14.43	24.72
C 9H10N2O4	210.19	210.0641	51.43	4.80	13.33	30.45
C 9H10N2O5	226.19	226.0590	47.79	4.46	12.38	35.37
C 9H10N2O6	242.19	242.0539	44.63	4.16	11.57	39.64
C 9H10N2O7	258.19	258.0488	41.87	3.90	10.85	43.38
C 9H10N2O8	274.19	274.0437	39.43	3.68	10.22	46.68
C 9H10N2O9	290.19	290.0386	37.25	3.47	9.65	49.62
C 9H10N4O1	190.21	190.0855	56.83	5.30	29.46	8.41
C 9H10N4O2	206.21	206.0804	52.42	4.89	27.17	15.52
C 9H10N4O3	222.21	222.0753	48.65	4.54	25.21	21.60
C 9H10N4O4	238.20	238.0702	45.38	4.23	23.52	26.87
C 9H10N4O5	254.20	254.0651	42.53	3.97	22.04	31.47
C 9H10N4O6	270.20	270.0600	40.01	3.73	20.74	35.53
C 9H10N4O7	286.20	286.0549	37.77	3.52	19.58	39.13
C 9H10N4O8	302.20	302.0499	35.77	3.34	18.54	42.35
C 9H10N4O9	318.20	318.0448	33.97	3.17	17.61	45.25
C 9H10N6O1	218.22	218.0916	49.54	4.62	38.51	7.33
C 9H10N6O2	234.22	234.0865	46.15	4.30	35.88	13.66
C 9H10N6O3	250.22	250.0814	43.20	4.03	33.59	19.18
C 9H10N6O4	266.22	266.0763	40.61	3.79	31.57	24.04
C 9H10N6O5	282.22	282.0713	38.30	3.57	29.78	28.35
C 9H10N6O6	298.22	298.0662	36.25	3.38	28.18	32.19
C 9H10N6O7	314.22	314.0611	34.40	3.21	26.75	35.64
C 9H10N6O8	330.22	330.0560	32.74	3.05	25.45	38.76
C 9H10N6O9	346.21	346.0509	31.22	2.91	24.27	41.59
C 9H11N1O1	149.19	149.0841	72.46	7.43	9.39	10.72
C 9H11N1O2	165.19	165.0790	65.44	6.71	8.48	19.37
C 9H11N1O3	181.19	181.0739	59.66	6.12	7.73	26.49
C 9H11N1O4	197.19	197.0688	54.82	5.62	7.10	32.45
C 9H11N1O5	213.19	213.0637	50.71	5.20	6.57	37.52
C 9H11N1O6	229.19	229.0586	47.17	4.84	6.11	41.88
C 9H11N1O7	245.19	245.0535	44.09	4.52	5.71	45.68

043

C 9H11N1O8	261.19	261.0485	41.39	4.25	5.36	49.00
C 9H11N1O9	277.19	277.0434	39.00	4.00	5.05	51.95
C 9H11N3O1	177.21	177.0902	61.00	6.26	23.71	9.03
C 9H11N3O2	193.21	193.0851	55.95	5.74	21.75	16.56
C 9H11N3O3	209.21	209.0800	51.67	5.30	20.09	22.94
C 9H11N3O4	225.21	225.0749	48.00	4.92	18.66	28.42
C 9H11N3O5	241.21	241.0699	44.82	4.60	17.42	33.17
C 9H11N3O6	257.20	257.0648	42.03	4.31	16.34	37.32
C 9H11N3O7	273.20	273.0597	39.57	4.06	15.38	40.99
C 9H11N3O8	289.20	289.0546	37.38	3.83	14.53	44.26
C 9H11N3O9	305.20	305.0495	35.42	3.63	13.77	47.18
C 9H11N5O1	205.22	205.0964	52.68	5.40	34.13	7.80
C 9H11N5O2	221.22	221.0913	48.87	5.01	31.66	14.46
C 9H11N5O3	237.22	237.0862	45.57	4.67	29.52	20.23
C 9H11N5O4	253.22	253.0811	42.69	4.38	27.66	25.27
C 9H11N5O5	269.22	269.0760	40.15	4.12	26.01	29.71
C 9H11N5O6	285.22	285.0709	37.90	3.89	24.55	33.66
C 9H11N5O7	301.22	301.0658	35.89	3.68	23.25	37.18
C 9H11N5O8	317.22	317.0608	34.08	3.50	22.08	40.35
C 9H11N5O9	333.22	333.0557	32.44	3.33	21.02	43.21
C 9H12N2O1	164.21	164.0950	65.83	7.37	17.06	9.74
C 9H12N2O2	180.21	180.0899	59.99	6.71	15.55	17.76
C 9H12N2O3	196.21	196.0848	55.09	6.16	14.28	24.46
C 9H12N2O4	212.21	212.0797	50.94	5.70	13.20	30.16
C 9H12N2O5	228.21	228.0746	47.37	5.30	12.28	35.05
C 9H12N2O6	244.21	244.0695	44.27	4.95	11.47	39.31
C 9H12N2O7	260.21	260.0644	41.54	4.65	10.77	43.04
C 9H12N2O8	276.20	276.0594	39.14	4.38	10.14	46.34
C 9H12N2O9	292.20	292.0543	36.99	4.14	9.59	49.28
C 9H12N4O1	192.22	192.1011	56.24	6.29	29.15	8.32
C 9H12N4O2	208.22	208.0960	51.92	5.81	26.91	15.37
C 9H12N4O3	224.22	224.0909	48.21	5.39	24.99	21.41
C 9H12N4O4	240.22	240.0858	45.00	5.04	23.32	26.64
C 9H12N4O5	256.22	256.0808	42.19	4.72	21.87	31.22
C 9H12N4O6	272.22	272.0757	39.71	4.44	20.58	35.26
C 9H12N4O7	288.22	288.0706	37.51	4.20	19.44	38.86
C 9H12N4O8	304.22	304.0655	35.53	3.98	18.42	42.07
C 9H12N4O9	320.22	320.0604	33.76	3.78	17.50	44.97
C 9H12N6O1	220.24	220.1073	49.08	5.49	38.16	7.26
C 9H12N6O2	236.23	236.1022	45.76	5.12	35.57	13.55
C 9H12N6O3	252.23	252.0971	42.86	4.80	33.32	19.03
C 9H12N6O4	268.23	268.0920	40.30	4.51	31.33	23.86
C 9H12N6O5	284.23	284.0869	38.03	4.26	29.57	28.14
C 9H12N6O6	300.23	300.0818	36.01	4.03	27.99	31.97
C 9H12N6O7	316.23	316.0767	34.18	3.82	26.58	35.42
C 9H12N6O8	332.23	332.0717	32.54	3.64	25.30	38.53
C 9H12N6O9	348.23	348.0666	31.04	3.47	24.13	41.35
C 9H13N1O1	151.21	151.0997	71.49	8.67	9.26	10.58
C 9H13N1O2	167.21	167.0946	64.65	7.84	8.38	19.14
C 9H13N1O3	183.21	183.0895	59.00	7.15	7.65	26.20
C 9H13N1O4	199.21	199.0845	54.26	6.58	7.03	32.13
C 9H13N1O5	215.21	215.0794	50.23	6.09	6.51	37.17
C 9H13N1O6	231.21	231.0743	46.75	5.67	6.06	41.52
C 9H13N1O7	247.21	247.0692	43.73	5.30	5.67	45.30
C 9H13N1O8	263.21	263.0641	41.07	4.98	5.32	48.63
C 9H13N1O9	279.21	279.0590	38.72	4.69	5.02	51.57
C 9H13N3O1	179.22	179.1059	60.32	7.31	23.45	8.93
C 9H13N3O2	195.22	195.1008	55.37	6.71	21.52	16.39
C 9H13N3O3	211.22	211.0957	51.18	6.20	19.89	22.72
C 9H13N3O4	227.22	227.0906	47.57	5.77	18.49	28.17
C 9H13N3O5	243.22	243.0855	44.45	5.39	17.28	32.89
C 9H13N3O6	259.22	259.0804	41.70	5.06	16.21	37.03
C 9H13N3O7	275.22	275.0753	39.28	4.76	15.27	40.69
C 9H13N3O8	291.22	291.0703	37.12	4.50	14.43	43.95
C 9H13N3O9	307.22	307.0652	35.19	4.27	13.68	46.87
C 9H13N5O1	207.24	207.1120	52.16	6.32	33.79	7.72
C 9H13N5O2	223.24	223.1069	48.42	5.87	31.37	14.33
C 9H13N5O3	239.24	239.1018	45.19	5.48	29.27	20.06
C 9H13N5O4	255.24	255.0967	42.35	5.13	27.44	25.07
C 9H13N5O5	271.23	271.0917	39.85	4.83	25.82	29.49
C 9H13N5O6	287.23	287.0866	37.63	4.56	24.38	33.42
C 9H13N5O7	303.23	303.0815	35.65	4.32	23.10	36.93
C 9H13N5O8	319.23	319.0764	33.86	4.10	21.94	40.09
C 9H13N5O9	335.23	335.0713	32.25	3.91	20.89	42.95
C 9H14N2O1	166.22	166.1106	65.03	8.49	16.85	9.63
C 9H14N2O2	182.22	182.1055	59.32	7.74	15.37	17.56
C 9H14N2O3	198.22	198.1004	54.53	7.12	14.13	24.21
C 9H14N2O4	214.22	214.0953	50.46	6.59	13.08	29.87
C 9H14N2O5	230.22	230.0903	46.95	6.13	12.17	34.75
C 9H14N2O6	246.22	246.0852	43.90	5.73	11.38	38.99
C 9H14N2O7	262.22	262.0801	41.22	5.38	10.68	42.71
C 9H14N2O8	278.22	278.0750	38.85	5.07	10.07	46.00
C 9H14N2O9	294.22	294.0699	36.74	4.80	9.52	48.94
C 9H14N4O1	194.24	194.1168	55.65	7.27	28.84	8.24
C 9H14N4O2	210.24	210.1117	51.42	6.71	26.65	15.22
C 9H14N4O3	226.24	226.1066	47.78	6.24	24.76	21.22
C 9H14N4O4	242.24	242.1015	44.63	5.83	23.13	26.42
C 9H14N4O5	258.24	258.0964	41.86	5.46	21.70	30.98
C 9H14N4O6	274.24	274.0913	39.42	5.15	20.43	35.01
C 9H14N4O7	290.23	290.0862	37.25	4.86	19.30	38.59
C 9H14N4O8	306.23	306.0812	35.30	4.61	18.30	41.80
C 9H14N4O9	322.23	322.0761	33.55	4.38	17.39	44.69
C 9H14N6O1	222.25	222.1229	48.64	6.35	37.81	7.20
C 9H14N6O2	238.25	238.1178	45.37	5.92	35.27	13.43
C 9H14N6O3	254.25	254.1127	42.52	5.55	33.05	18.88
C 9H14N6O4	270.25	270.1076	40.00	5.22	31.10	23.68
C 9H14N6O5	286.25	286.1026	37.76	4.93	29.36	27.95
C 9H14N6O6	302.25	302.0975	35.77	4.67	27.80	31.76
C 9H14N6O7	318.25	318.0924	33.97	4.43	26.41	35.19
C 9H14N6O8	334.25	334.0873	32.34	4.22	25.14	38.29
C 9H14N6O9	350.25	350.0822	30.86	4.03	23.99	41.11
C 9H15N1O1	153.23	153.1154	70.55	9.87	9.14	10.44
C 9H15N1O2	169.23	169.1103	63.88	8.93	8.28	18.91
C 9H15N1O3	185.22	185.1052	58.36	8.16	7.56	25.91
C 9H15N1O4	201.22	201.1001	53.72	7.51	6.96	31.80
C 9H15N1O5	217.22	217.0950	49.76	6.96	6.45	36.83
C 9H15N1O6	233.22	233.0899	46.35	6.48	6.01	41.16
C 9H15N1O7	249.22	249.0848	43.38	6.07	5.62	44.94
C 9H15N1O8	265.22	265.0798	40.76	5.70	5.28	48.26
C 9H15N1O9	281.22	281.0747	38.44	5.38	4.98	51.20
C 9H15N3O1	181.24	181.1215	59.65	8.34	23.18	8.83
C 9H15N3O2	197.24	197.1164	54.81	7.67	21.30	16.22
C 9H15N3O3	213.24	213.1113	50.69	7.09	19.71	22.51
C 9H15N3O4	229.24	229.1062	47.16	6.60	18.33	27.92
C 9H15N3O5	245.24	245.1012	44.08	6.17	17.13	32.62
C 9H15N3O6	261.24	261.0961	41.38	5.79	16.09	36.75
C 9H15N3O7	277.24	277.0910	38.99	5.45	15.16	40.40
C 9H15N3O8	293.24	293.0859	36.86	5.16	14.33	43.65
C 9H15N3O9	309.23	309.0808	34.96	4.89	13.59	46.56
C 9H15N5O1	209.25	209.1277	51.66	7.23	33.47	7.65

C 9H15N5O2	225.25	225.1226	47.99	6.71	31.09	14.21
C 9H15N5O3	241.25	241.1175	44.81	6.27	29.03	19.90
C 9H15N5O4	257.25	257.1124	42.02	5.88	27.22	24.88
C 9H15N5O5	273.25	273.1073	39.56	5.53	25.63	29.28
C 9H15N5O6	289.25	289.1022	37.37	5.23	24.21	33.19
C 9H15N5O7	305.25	305.0971	35.41	4.95	22.94	36.69
C 9H15N5O8	321.25	321.0921	33.65	4.71	21.80	39.84
C 9H15N5O9	337.25	337.0870	32.05	4.48	20.77	42.70
C 9H16N2O1	168.24	168.1263	64.25	9.59	16.65	9.51
C 9H16N2O2	184.24	184.1212	58.67	8.75	15.20	17.37
C 9H16N2O3	200.24	200.1161	53.99	8.05	13.99	23.97
C 9H16N2O4	216.24	216.1110	49.99	7.46	12.95	29.60
C 9H16N2O5	232.24	232.1059	46.55	6.94	12.06	34.45
C 9H16N2O6	248.24	248.1008	43.55	6.50	11.28	38.67
C 9H16N2O7	264.24	264.0957	40.91	6.10	10.60	42.38
C 9H16N2O8	280.24	280.0907	38.57	5.75	10.00	45.67
C 9H16N2O9	296.24	296.0856	36.49	5.44	9.46	48.61
C 9H16N4O1	196.25	196.1324	55.08	8.22	28.55	8.15
C 9H16N4O2	212.25	212.1273	50.93	7.60	26.40	15.08
C 9H16N4O3	228.25	228.1222	47.36	7.07	24.55	21.03
C 9H16N4O4	244.25	244.1171	44.26	6.60	22.94	26.20
C 9H16N4O5	260.25	260.1121	41.54	6.20	21.53	30.74
C 9H16N4O6	276.25	276.1070	39.13	5.84	20.28	34.75
C 9H16N4O7	292.25	292.1019	36.99	5.52	19.17	38.32
C 9H16N4O8	308.25	308.0968	35.07	5.23	18.18	41.52
C 9H16N4O9	324.25	324.0917	33.34	4.97	17.28	44.41
C 9H16N6O1	224.27	224.1386	48.20	7.19	37.47	7.13
C 9H16N6O2	240.27	240.1335	44.99	6.71	34.98	13.32
C 9H16N6O3	256.27	256.1284	42.18	6.29	32.79	18.73
C 9H16N6O4	272.27	272.1233	39.70	5.92	30.87	23.51
C 9H16N6O5	288.27	288.1182	37.50	5.59	29.15	27.75
C 9H16N6O6	304.26	304.1131	35.53	5.30	27.62	31.55
C 9H16N6O7	320.26	320.1080	33.75	5.04	26.24	34.97
C 9H16N6O8	336.26	336.1029	32.15	4.80	24.99	38.06
C 9H16N6O9	352.26	352.0979	30.69	4.58	23.86	40.88
C 9H17N1O1	155.24	155.1310	69.63	11.04	9.02	10.31
C 9H17N1O2	171.24	171.1259	63.13	10.01	8.18	18.69
C 9H17N1O3	187.24	187.1208	57.73	9.15	7.48	25.63
C 9H17N1O4	203.24	203.1157	53.19	8.43	6.89	31.49
C 9H17N1O5	219.24	219.1107	49.31	7.82	6.39	36.49
C 9H17N1O6	235.24	235.1056	45.95	7.28	5.95	40.81
C 9H17N1O7	251.24	251.1005	43.03	6.82	5.58	44.58
C 9H17N1O8	267.24	267.0954	40.45	6.41	5.24	47.90
C 9H17N1O9	283.24	283.0903	38.17	6.05	4.95	50.84
C 9H17N3O1	183.26	183.1372	58.99	9.35	22.93	8.73
C 9H17N3O2	199.25	199.1321	54.25	8.60	21.09	16.06
C 9H17N3O3	215.25	215.1270	50.22	7.96	19.52	22.30
C 9H17N3O4	231.25	231.1219	46.75	7.41	18.17	27.67
C 9H17N3O5	247.25	247.1168	43.72	6.93	16.99	32.35
C 9H17N3O6	263.25	263.1117	41.06	6.51	15.96	36.47
C 9H17N3O7	279.25	279.1066	38.71	6.14	15.05	40.11
C 9H17N3O8	295.25	295.1016	36.61	5.80	14.23	43.35
C 9H17N3O9	311.25	311.0965	34.73	5.51	13.50	46.26
C 9H17N5O1	211.27	211.1433	51.17	8.11	33.15	7.57
C 9H17N5O2	227.27	227.1382	47.57	7.54	30.82	14.08
C 9H17N5O3	243.27	243.1331	44.44	7.04	28.79	19.73
C 9H17N5O4	259.27	259.1280	41.69	6.61	27.01	24.68
C 9H17N5O5	275.27	275.1230	39.27	6.23	25.44	29.06
C 9H17N5O6	291.27	291.1179	37.11	5.88	24.04	32.96
C 9H17N5O7	307.27	307.1128	35.18	5.58	22.79	36.45
C 9H17N5O8	323.26	323.1077	33.44	5.30	21.66	39.59
C 9H17N5O9	339.26	339.1026	31.86	5.05	20.64	42.44
C 9H18N2O1	170.26	170.1419	63.49	10.66	16.45	9.40
C 9H18N2O2	186.26	186.1368	58.04	9.74	15.04	17.18
C 9H18N2O3	202.26	202.1317	53.45	8.97	13.85	23.73
C 9H18N2O4	218.25	218.1266	49.53	8.31	12.84	29.32
C 9H18N2O5	234.25	234.1216	46.15	7.75	11.96	34.15
C 9H18N2O6	250.25	250.1165	43.20	7.25	11.19	38.36
C 9H18N2O7	266.25	266.1114	40.60	6.81	10.52	42.06
C 9H18N2O8	282.25	282.1063	38.30	6.43	9.92	45.35
C 9H18N2O9	298.25	298.1012	36.24	6.08	9.39	48.28
C 9H18N4O1	198.27	198.1481	54.52	9.15	28.26	8.07
C 9H18N4O2	214.27	214.1430	50.45	8.47	26.15	14.93
C 9H18N4O3	230.27	230.1379	46.95	7.88	24.33	20.84
C 9H18N4O4	246.27	246.1328	43.90	7.37	22.75	25.99
C 9H18N4O5	262.27	262.1277	41.22	6.92	21.36	30.50
C 9H18N4O6	278.27	278.1226	38.85	6.52	20.13	34.50
C 9H18N4O7	294.27	294.1175	36.74	6.17	19.04	38.06
C 9H18N4O8	310.27	310.1125	34.84	5.85	18.06	41.25
C 9H18N4O9	326.27	326.1074	33.13	5.56	17.17	44.13
C 9H18N6O1	226.28	226.1542	47.77	8.02	37.14	7.07
C 9H18N6O2	242.28	242.1491	44.62	7.49	34.69	13.21
C 9H18N6O3	258.28	258.1440	41.85	7.02	32.54	18.58
C 9H18N6O4	274.28	274.1389	39.41	6.61	30.64	23.33
C 9H18N6O5	290.28	290.1339	37.24	6.25	28.95	27.56
C 9H18N6O6	306.28	306.1288	35.29	5.92	27.44	31.34
C 9H18N6O7	322.28	322.1237	33.54	5.63	26.08	34.75
C 9H18N6O8	338.28	338.1186	31.96	5.36	24.84	37.84
C 9H18N6O9	354.28	354.1135	30.51	5.12	23.72	40.64
C 9H19N1O1	157.26	157.1467	68.74	12.18	8.91	10.17
C 9H19N1O2	173.26	173.1416	62.39	11.05	8.08	18.47
C 9H19N1O3	189.26	189.1365	57.12	10.12	7.40	25.36
C 9H19N1O4	205.26	205.1314	52.67	9.33	6.82	31.18
C 9H19N1O5	221.26	221.1263	48.86	8.66	6.33	36.16
C 9H19N1O6	237.25	237.1212	45.56	8.07	5.90	40.46
C 9H19N1O7	253.25	253.1161	42.68	7.56	5.53	44.22
C 9H19N1O8	269.25	269.1111	40.15	7.11	5.20	47.54
C 9H19N1O9	285.25	285.1060	37.90	6.71	4.91	50.48
C 9H19N3O1	185.27	185.1528	58.35	10.34	22.68	8.64
C 9H19N3O2	201.27	201.1477	53.71	9.52	20.88	15.90
C 9H19N3O3	217.27	217.1426	49.75	8.81	19.34	22.09
C 9H19N3O4	233.27	233.1375	46.34	8.21	18.01	27.44
C 9H19N3O5	249.27	249.1325	43.37	7.68	16.86	32.09
C 9H19N3O6	265.27	265.1274	40.75	7.22	15.84	36.19
C 9H19N3O7	281.27	281.1223	38.43	6.81	14.94	39.82
C 9H19N3O8	297.27	297.1172	36.36	6.44	14.14	43.06
C 9H19N3O9	313.27	313.1121	34.51	6.11	13.41	45.97
C 9H19N5O1	213.28	213.1590	50.68	8.98	32.84	7.50
C 9H19N5O2	229.28	229.1539	47.15	8.35	30.54	13.96
C 9H19N5O3	245.28	245.1488	44.07	7.81	28.55	19.57
C 9H19N5O4	261.28	261.1437	41.37	7.33	26.80	24.49
C 9H19N5O5	277.28	277.1386	38.99	6.91	25.26	28.85
C 9H19N5O6	293.28	293.1335	36.86	6.53	23.88	32.73
C 9H19N5O7	309.28	309.1284	34.95	6.19	22.64	36.21
C 9H19N5O8	325.28	325.1233	33.23	5.89	21.53	39.35
C 9H19N5O9	341.28	341.1183	31.67	5.61	20.52	42.19
C 9H20N2O1	172.27	172.1576	62.75	11.70	16.26	9.29
C 9H20N2O2	188.27	188.1525	57.42	10.71	14.88	17.00
C 9H20N2O3	204.27	204.1474	52.92	9.87	13.71	23.50
C 9H20N2O4	220.27	220.1423	49.08	9.15	12.72	29.05

C 9H20N2O5	236.27	236.1372	45.75	8.53	11.86	33.86
C 9H20N2O6	252.27	252.1321	42.85	7.99	11.10	38.05
C 9H20N2O7	268.27	268.1270	40.30	7.51	10.44	41.75
C 9H20N2O8	284.27	284.1220	38.03	7.09	9.85	45.03
C 9H20N2O9	300.27	300.1169	36.00	6.71	9.33	47.96
C 9H20N4O1	200.29	200.1637	53.97	10.07	27.97	7.99
C 9H20N4O2	216.29	216.1586	49.98	9.32	25.90	14.79
C 9H20N4O3	232.28	232.1535	46.54	8.68	24.12	20.66
C 9H20N4O4	248.28	248.1484	43.54	8.12	22.57	25.78
C 9H20N4O5	264.28	264.1434	40.90	7.63	21.20	30.27
C 9H20N4O6	280.28	280.1383	38.57	7.19	19.99	34.25
C 9H20N4O7	296.28	296.1332	36.49	6.80	18.91	37.80
C 9H20N4O8	312.28	312.1281	34.62	6.46	17.94	40.99
C 9H20N4O9	328.28	328.1230	32.93	6.14	17.07	43.86
C 9H20N6O1	228.30	228.1698	47.35	8.83	36.81	7.01
C 9H20N6O2	244.30	244.1648	44.25	8.25	34.40	13.10
C 9H20N6O3	260.30	260.1597	41.53	7.74	32.29	18.44
C 9H20N6O4	276.30	276.1546	39.12	7.30	30.42	23.16
C 9H20N6O5	292.30	292.1495	36.98	6.90	28.75	27.37
C 9H20N6O6	308.30	308.1444	35.06	6.54	27.26	31.14
C 9H20N6O7	324.30	324.1393	33.33	6.22	25.91	34.54
C 9H20N6O8	340.30	340.1342	31.77	5.92	24.70	37.61
C 9H20N6O9	356.29	356.1292	30.34	5.66	23.59	40.41
C 9H21N1O1	159.27	159.1623	67.87	13.29	8.79	10.05
C 9H21N1O2	175.27	175.1572	61.68	12.08	7.99	18.26
C 9H21N1O3	191.27	191.1521	56.52	11.07	7.32	25.09
C 9H21N1O4	207.27	207.1470	52.15	10.21	6.76	30.88
C 9H21N1O5	223.27	223.1420	48.42	9.48	6.27	35.83
C 9H21N1O6	239.27	239.1369	45.18	8.85	5.85	40.12
C 9H21N1O7	255.27	255.1318	42.35	8.29	5.49	43.87
C 9H21N1O8	271.27	271.1267	39.85	7.80	5.16	47.18
C 9H21N1O9	287.27	287.1216	37.63	7.37	4.88	50.13
C 9H21N3O1	187.29	187.1685	57.72	11.30	22.44	8.54
C 9H21N3O2	203.29	203.1634	53.18	10.41	20.67	15.74
C 9H21N3O3	219.29	219.1583	49.30	9.65	19.16	21.89
C 9H21N3O4	235.29	235.1532	45.94	9.00	17.86	27.20
C 9H21N3O5	251.28	251.1481	43.02	8.42	16.72	31.84
C 9H21N3O6	267.28	267.1430	40.44	7.92	15.72	35.92
C 9H21N3O7	283.28	283.1379	38.16	7.47	14.83	39.53
C 9H21N3O8	299.28	299.1329	36.12	7.07	14.04	42.77
C 9H21N3O9	315.28	315.1278	34.29	6.71	13.33	45.67
C 9H21N5O1	215.30	215.1746	50.21	9.83	32.53	7.43
C 9H21N5O2	231.30	231.1695	46.74	9.15	30.28	13.83
C 9H21N5O3	247.30	247.1644	43.71	8.56	28.32	19.41
C 9H21N5O4	263.30	263.1593	41.06	8.04	26.60	24.31
C 9H21N5O5	279.30	279.1543	38.70	7.58	25.07	28.64
C 9H21N5O6	295.30	295.1492	36.61	7.17	23.72	32.51
C 9H21N5O7	311.30	311.1441	34.73	6.80	22.50	35.98
C 9H21N5O8	327.30	327.1390	33.03	6.47	21.40	39.11
C 9H21N5O9	343.30	343.1339	31.49	6.17	20.40	41.94
C 9H22N2O1	174.29	174.1732	62.02	12.72	16.07	9.18
C 9H22N2O2	190.29	190.1681	56.81	11.65	14.72	16.82
C 9H22N2O3	206.29	206.1630	52.40	10.75	13.58	23.27
C 9H22N2O4	222.29	222.1579	48.63	9.98	12.60	28.79
C 9H22N2O5	238.29	238.1529	45.37	9.31	11.76	33.57
C 9H22N2O6	254.29	254.1478	42.51	8.72	11.02	37.75
C 9H22N2O7	270.28	270.1427	39.99	8.20	10.36	41.44
C 9H22N2O8	286.28	286.1376	37.76	7.75	9.79	44.71
C 9H22N2O9	302.28	302.1325	35.76	7.34	9.27	47.64
C 9H22N4O1	202.30	202.1794	53.44	10.96	27.69	7.91
C 9H22N4O2	218.30	218.1743	49.52	10.16	25.66	14.66
C 9H22N4O3	234.30	234.1692	46.14	9.46	23.91	20.49
C 9H22N4O4	250.30	250.1641	43.19	8.86	22.38	25.57
C 9H22N4O5	266.30	266.1590	40.59	8.33	21.04	30.04
C 9H22N4O6	282.30	282.1539	38.29	7.86	19.85	34.01
C 9H22N4O7	298.30	298.1488	36.24	7.43	18.78	37.54
C 9H22N4O8	314.30	314.1437	34.39	7.06	17.83	40.72
C 9H22N4O9	330.30	330.1387	32.73	6.71	16.96	43.60
C 9H22N6O1	230.32	230.1855	46.94	9.63	36.49	6.95
C 9H22N6O2	246.31	246.1804	43.89	9.00	34.12	12.99
C 9H22N6O3	262.31	262.1753	41.21	8.45	32.04	18.30
C 9H22N6O4	278.31	278.1702	38.84	7.97	30.20	22.99
C 9H22N6O5	294.31	294.1652	36.73	7.53	28.55	27.18
C 9H22N6O6	310.31	310.1601	34.84	7.15	27.08	30.94
C 9H22N6O7	326.31	326.1550	33.13	6.80	25.75	34.32
C 9H22N6O8	342.31	342.1499	31.58	6.48	24.55	37.39
C 9H22N6O9	358.31	358.1448	30.17	6.19	23.45	40.19
C 9H23N3O1	189.30	189.1841	57.10	12.25	22.20	8.45
C 9H23N3O2	205.30	205.1790	52.65	11.29	20.47	15.59
C 9H23N3O3	221.30	221.1739	48.85	10.48	18.99	21.69
C 9H23N3O4	237.30	237.1688	45.55	9.77	17.71	26.97
C 9H23N3O5	253.30	253.1638	42.68	9.15	16.59	31.58
C 9H23N3O6	269.30	269.1587	40.14	8.61	15.60	35.65
C 9H23N3O7	285.30	285.1536	37.89	8.13	14.73	39.26
C 9H23N3O8	301.30	301.1485	35.88	7.69	13.95	42.48
C 9H23N3O9	317.30	317.1434	34.07	7.31	13.24	45.38
C 9H23N5O1	217.32	217.1902	49.74	10.67	32.23	7.36
C 9H23N5O2	233.32	233.1852	46.33	9.94	30.02	13.71
C 9H23N5O3	249.32	249.1801	43.36	9.30	28.09	19.25
C 9H23N5O4	265.31	265.1750	40.74	8.74	26.40	24.12
C 9H23N5O5	281.31	281.1699	38.43	8.24	24.90	28.44
C 9H23N5O6	297.31	297.1648	36.36	7.80	23.56	32.29
C 9H23N5O7	313.31	313.1597	34.50	7.40	22.35	35.75
C 9H23N5O8	329.31	329.1546	32.83	7.04	21.27	38.87
C 9H23N5O9	345.31	345.1496	31.31	6.71	20.28	41.70
C 9H24N4O1	204.32	204.1950	52.91	11.84	27.42	7.83
C 9H24N4O2	220.32	220.1899	49.07	10.98	25.43	14.52
C 9H24N4O3	236.32	236.1848	45.74	10.24	23.71	20.31
C 9H24N4O4	252.32	252.1797	42.84	9.59	22.21	25.36
C 9H24N4O5	268.32	268.1747	40.29	9.02	20.88	29.81
C 9H24N4O6	284.31	284.1696	38.02	8.51	19.71	33.76
C 9H24N4O7	300.31	300.1645	36.00	8.06	18.66	37.29
C 9H24N4O8	316.31	316.1594	34.18	7.65	17.71	40.46
C 9H24N4O9	332.31	332.1543	32.53	7.28	16.86	43.33
C 9H24N6O1	232.33	232.2011	46.53	10.41	36.17	6.89
C 9H24N6O2	248.33	248.1961	43.53	9.74	33.84	12.89
C 9H24N6O3	264.33	264.1910	40.90	9.15	31.79	18.16
C 9H24N6O4	280.33	280.1859	38.56	8.63	29.98	22.83
C 9H24N6O5	296.33	296.1808	36.48	8.16	28.36	27.00
C 9H24N6O6	312.33	312.1757	34.61	7.75	26.91	30.74
C 9H24N6O7	328.33	328.1706	32.92	7.37	25.60	34.11
C 9H24N6O8	344.33	344.1655	31.39	7.03	24.41	37.17
C 9H24N6O9	360.33	360.1605	30.00	6.71	23.32	39.96
C 9H25N5O1	219.33	219.2059	49.29	11.49	31.93	7.29
C 9H25N5O2	235.33	235.2008	45.94	10.71	29.76	13.60
C 9H25N5O3	251.33	251.1957	43.01	10.03	27.87	19.10
C 9H25N5O4	267.33	267.1906	40.44	9.43	26.20	23.94
C 9H25N5O5	283.33	283.1856	38.15	8.89	24.72	28.23
C 9H25N5O6	299.33	299.1805	36.11	8.42	23.40	32.07
C 9H25N5O7	315.33	315.1754	34.28	7.99	22.21	35.52

046

C 9H25N508	331.33	331.1703	32.63	7.61	21.14	38.63
C 9H25N509	347.33	347.1652	31.12	7.26	20.16	41.46
C 9H26N601	234.35	234.2168	46.13	11.18	35.86	6.83
C 9H26N602	250.35	250.2117	43.18	10.47	33.57	12.78
C 9H26N603	266.35	266.2066	40.59	9.84	31.55	18.02
C 9H26N604	282.35	282.2015	38.29	9.28	29.77	22.67
C 9H26N605	298.34	298.1965	36.23	8.78	28.17	26.81
C 9H26N606	314.34	314.1914	34.39	8.34	26.74	30.54
C 9H26N607	330.34	330.1863	32.72	7.93	25.44	33.90
C 9H26N608	346.34	346.1812	31.21	7.57	24.27	36.96
C 9H26N609	362.34	362.1761	29.83	7.23	23.19	39.74
C10H 5N101	155.16	155.0371	77.41	3.25	9.03	10.31
C10H 5N102	171.16	171.0320	70.18	2.94	8.18	18.70
C10H 5N103	187.16	187.0269	64.18	2.69	7.48	25.65
C10H 5N104	203.16	203.0219	59.12	2.48	6.89	31.50
C10H 5N105	219.16	219.0168	54.81	2.30	6.39	36.50
C10H 5N106	235.15	235.0117	51.08	2.14	5.96	40.82
C10H 5N107	251.15	251.0066	47.82	2.01	5.58	44.59
C10H 5N108	267.15	267.0015	44.96	1.89	5.24	47.91
C10H 5N109	283.15	282.9964	42.42	1.78	4.95	50.85
C10H 5N301	183.17	183.0433	65.57	2.75	22.94	8.73
C10H 5N302	199.17	199.0382	60.31	2.53	21.10	16.07
C10H 5N303	215.17	215.0331	55.82	2.34	19.53	22.31
C10H 5N304	231.17	231.0280	51.96	2.18	18.18	27.68
C10H 5N305	247.17	247.0229	48.59	2.04	17.00	32.37
C10H 5N306	263.17	263.0178	45.64	1.92	15.97	36.48
C10H 5N307	279.17	279.0127	43.02	1.81	15.05	40.12
C10H 5N308	295.17	295.0077	40.69	1.71	14.24	43.36
C10H 5N309	311.17	311.0026	38.60	1.62	13.50	46.28
C10H 5N501	211.18	211.0494	56.88	2.39	33.16	7.58
C10H 5N502	227.18	227.0443	52.87	2.22	30.83	14.08
C10H 5N503	243.18	243.0392	49.39	2.07	28.80	19.74
C10H 5N504	259.18	259.0341	46.34	1.94	27.02	24.69
C10H 5N505	275.18	275.0291	43.65	1.83	25.45	29.07
C10H 5N506	291.18	291.0240	41.25	1.73	24.05	32.97
C10H 5N507	307.18	307.0189	39.10	1.64	22.80	36.46
C10H 5N508	323.18	323.0138	37.17	1.56	21.67	39.60
C10H 5N509	339.18	339.0087	35.41	1.49	20.65	42.45
C10H 6N201	170.17	170.0480	70.58	3.55	16.46	9.40
C10H 6N202	186.17	186.0429	64.52	3.25	15.05	17.19
C10H 6N203	202.17	202.0378	59.41	2.99	13.86	23.74
C10H 6N204	218.17	218.0328	55.05	2.77	12.84	29.33
C10H 6N205	234.17	234.0277	51.29	2.58	11.96	34.16
C10H 6N206	250.17	250.0226	48.01	2.42	11.20	38.37
C10H 6N207	266.17	266.0175	45.13	2.27	10.52	42.08
C10H 6N208	282.17	282.0124	42.57	2.14	9.93	45.36
C10H 6N209	298.17	298.0073	40.28	2.03	9.40	48.29
C10H 6N401	198.19	198.0542	60.61	3.05	28.27	8.07
C10H 6N402	214.18	214.0491	56.08	2.82	26.16	14.94
C10H 6N403	230.18	230.0440	52.18	2.63	24.34	20.85
C10H 6N404	246.18	246.0389	48.79	2.46	22.76	26.00
C10H 6N405	262.18	262.0338	45.81	2.31	21.37	30.51
C10H 6N406	278.18	278.0287	43.18	2.17	20.14	34.51
C10H 6N407	294.18	294.0236	40.83	2.06	19.04	38.07
C10H 6N408	310.18	310.0186	38.72	1.95	18.06	41.26
C10H 6N409	326.18	326.0135	36.82	1.85	17.18	44.15
C10H 6N601	226.20	226.0603	53.10	2.67	37.15	7.07
C10H 6N602	242.20	242.0552	49.59	2.50	34.70	13.21
C10H 6N603	258.20	258.0501	46.52	2.34	32.55	18.59
C10H 6N604	274.20	274.0450	43.80	2.21	30.65	23.34
C10H 6N605	290.20	290.0400	41.39	2.08	28.96	27.57
C10H 6N606	306.20	306.0349	39.23	1.98	27.45	31.35
C10H 6N607	322.20	322.0298	37.28	1.88	26.08	34.76
C10H 6N608	338.19	338.0247	35.52	1.79	24.85	37.85
C10H 6N609	354.19	354.0196	33.91	1.71	23.73	40.65
C10H 7N101	157.17	157.0528	76.42	4.49	8.91	10.18
C10H 7N102	173.17	173.0477	69.36	4.07	8.09	18.48
C10H 7N103	189.17	189.0426	63.49	3.73	7.40	25.37
C10H 7N104	205.17	205.0375	58.54	3.44	6.83	31.19
C10H 7N105	221.17	221.0324	54.31	3.19	6.33	36.17
C10H 7N106	237.17	237.0273	50.64	2.97	5.91	40.48
C10H 7N107	253.17	253.0222	47.44	2.79	5.53	44.24
C10H 7N108	269.17	269.0172	44.62	2.62	5.20	47.55
C10H 7N109	285.17	285.0121	42.12	2.47	4.91	50.49
C10H 7N301	185.19	185.0589	64.86	3.81	22.69	8.64
C10H 7N302	201.19	201.0538	59.70	3.51	20.89	15.91
C10H 7N303	217.19	217.0487	55.30	3.25	19.35	22.10
C10H 7N304	233.18	233.0437	51.51	3.03	18.02	27.44
C10H 7N305	249.18	249.0386	48.20	2.83	16.86	32.10
C10H 7N306	265.18	265.0335	45.29	2.66	15.85	36.20
C10H 7N307	281.18	281.0284	42.72	2.51	14.94	39.83
C10H 7N308	297.18	297.0233	40.42	2.37	14.14	43.07
C10H 7N309	313.18	313.0182	38.35	2.25	13.42	45.98
C10H 7N501	213.20	213.0651	56.34	3.31	32.85	7.50
C10H 7N502	229.20	229.0600	52.40	3.08	30.56	13.96
C10H 7N503	245.20	245.0549	48.99	2.88	28.56	19.58
C10H 7N504	261.20	261.0498	45.98	2.70	26.81	24.50
C10H 7N505	277.20	277.0447	43.33	2.55	25.26	28.86
C10H 7N506	293.20	293.0396	40.97	2.41	23.89	32.74
C10H 7N507	309.20	309.0345	38.85	2.28	22.65	36.22
C10H 7N508	325.20	325.0295	36.94	2.17	21.54	39.36
C10H 7N509	341.20	341.0244	35.20	2.07	20.53	42.20
C10H 8N201	172.19	172.0637	69.76	4.68	16.27	9.29
C10H 8N202	188.19	188.0586	63.83	4.28	14.89	17.00
C10H 8N203	204.19	204.0535	58.82	3.95	13.72	23.51
C10H 8N204	220.19	220.0484	54.55	3.66	12.72	29.07
C10H 8N205	236.19	236.0433	50.85	3.41	11.86	33.87
C10H 8N206	252.19	252.0382	47.63	3.20	11.11	38.07
C10H 8N207	268.18	268.0331	44.79	3.01	10.45	41.76
C10H 8N208	284.18	284.0281	42.27	2.84	9.86	45.04
C10H 8N209	300.18	300.0230	40.01	2.69	9.33	47.97
C10H 8N401	200.20	200.0698	60.00	4.03	27.99	7.99
C10H 8N402	216.20	216.0647	55.56	3.73	25.91	14.80
C10H 8N403	232.20	232.0596	51.73	3.47	24.13	20.67
C10H 8N404	248.20	248.0545	48.39	3.25	22.57	25.78
C10H 8N405	264.20	264.0495	45.46	3.05	21.21	30.28
C10H 8N406	280.20	280.0444	42.87	2.88	20.00	34.26
C10H 8N407	296.20	296.0393	40.55	2.72	18.92	37.81
C10H 8N408	312.20	312.0342	38.47	2.58	17.95	41.00
C10H 8N409	328.20	328.0291	36.60	2.46	17.07	43.87
C10H 8N601	228.21	228.0760	52.63	3.53	36.83	7.01
C10H 8N602	244.21	244.0709	49.18	3.30	34.41	13.10
C10H 8N603	260.21	260.0658	46.16	3.10	32.30	18.45
C10H 8N604	276.21	276.0607	43.49	2.92	30.43	23.17
C10H 8N605	292.21	292.0556	41.10	2.76	28.76	27.38
C10H 8N606	308.21	308.0505	38.97	2.62	27.27	31.15
C10H 8N607	324.21	324.0454	37.05	2.49	25.92	34.54
C10H 8N608	340.21	340.0404	35.31	2.37	24.70	37.62
C10H 8N609	356.21	356.0353	33.72	2.26	23.59	40.42
C10H 9N101	159.19	159.0684	75.45	5.70	8.80	10.05

047

C10H 9N102	175.19	175.0633	68.56	5.18	8.00	18.27
C10H 9N103	191.19	191.0582	62.82	4.74	7.33	25.11
C10H 9N104	207.19	207.0532	57.97	4.38	6.76	30.89
C10H 9N105	223.19	223.0481	53.82	4.06	6.28	35.84
C10H 9N106	239.19	239.0430	50.22	3.79	5.86	40.13
C10H 9N107	255.19	255.0379	47.07	3.55	5.49	43.89
C10H 9N108	271.19	271.0328	44.29	3.35	5.16	47.20
C10H 9N109	287.18	287.0277	41.82	3.16	4.88	50.14
C10H 9N301	187.20	187.0746	64.16	4.85	22.45	8.55
C10H 9N302	203.20	203.0695	59.11	4.46	20.68	15.75
C10H 9N303	219.20	219.0644	54.80	4.14	19.17	21.90
C10H 9N304	235.20	235.0593	51.07	3.86	17.87	27.21
C10H 9N305	251.20	251.0542	47.82	3.61	16.73	31.85
C10H 9N306	267.20	267.0491	44.95	3.40	15.73	35.93
C10H 9N307	283.20	283.0440	42.41	3.20	14.84	39.55
C10H 9N308	299.20	299.0390	40.14	3.03	14.04	42.78
C10H 9N309	315.20	315.0339	38.11	2.88	13.33	45.68
C10H 9N501	215.22	215.0807	55.81	4.22	32.54	7.43
C10H 9N502	231.22	231.0756	51.95	3.92	30.29	13.84
C10H 9N503	247.21	247.0705	48.59	3.67	28.33	19.42
C10H 9N504	263.21	263.0654	45.63	3.45	26.61	24.31
C10H 9N505	279.21	279.0604	43.02	3.25	25.08	28.65
C10H 9N506	295.21	295.0553	40.69	3.07	23.72	32.52
C10H 9N507	311.21	311.0502	38.59	2.91	22.50	35.99
C10H 9N508	327.21	327.0451	36.71	2.77	21.40	39.12
C10H 9N509	343.21	343.0400	35.00	2.64	20.41	41.96
C10H10N201	174.20	174.0793	68.95	5.79	16.08	9.18
C10H10N202	190.20	190.0742	63.15	5.30	14.73	16.82
C10H10N203	206.20	206.0691	58.25	4.89	13.59	23.28
C10H10N204	222.20	222.0641	54.06	4.54	12.61	28.80
C10H10N205	238.20	238.0590	50.42	4.23	11.76	33.58
C10H10N206	254.20	254.0539	47.25	3.97	11.02	37.76
C10H10N207	270.20	270.0488	44.45	3.73	10.37	41.45
C10H10N208	286.20	286.0437	41.97	3.52	9.79	44.72
C10H10N209	302.20	302.0386	39.75	3.34	9.27	47.65
C10H10N401	202.22	202.0855	59.40	4.98	27.71	7.91
C10H10N402	218.22	218.0804	55.04	4.62	25.67	14.66
C10H10N403	234.22	234.0753	51.28	4.30	23.92	20.49
C10H10N404	250.22	250.0702	48.00	4.03	22.39	25.58
C10H10N405	266.22	266.0651	45.12	3.79	21.05	30.05
C10H10N406	282.21	282.0600	42.56	3.57	19.85	34.02
C10H10N407	298.21	298.0549	40.28	3.38	18.79	37.56
C10H10N408	314.21	314.0499	38.23	3.21	17.83	40.74
C10H10N409	330.21	330.0448	36.37	3.05	16.97	43.61
C10H10N601	230.23	230.0916	52.17	4.38	36.50	6.95
C10H10N602	246.23	246.0865	48.78	4.09	34.13	13.00
C10H10N603	262.23	262.0814	45.80	3.84	32.05	18.30
C10H10N604	278.23	278.0763	43.17	3.62	30.21	23.00
C10H10N605	294.23	294.0713	40.82	3.43	28.56	27.19
C10H10N606	310.23	310.0662	38.72	3.25	27.09	30.94
C10H10N607	326.23	326.0611	36.82	3.09	25.76	34.33
C10H10N608	342.23	342.0560	35.10	2.95	24.56	37.40
C10H10N609	358.23	358.0509	33.53	2.81	23.46	40.20
C10H11N101	161.21	161.0841	74.51	6.88	8.69	9.92
C10H11N102	177.20	177.0790	67.78	6.26	7.90	18.06
C10H11N103	193.20	193.0739	62.17	5.74	7.25	24.84
C10H11N104	209.20	209.0688	57.41	5.30	6.70	30.59
C10H11N105	225.20	225.0637	53.33	4.92	6.22	35.52
C10H11N106	241.20	241.0586	49.80	4.60	5.81	39.80
C10H11N107	257.20	257.0535	46.70	4.31	5.45	43.54
C10H11N108	273.20	273.0485	43.96	4.06	5.13	46.85
C10H11N109	289.20	289.0434	41.53	3.83	4.84	49.79
C10H11N301	189.22	189.0902	63.48	5.86	22.21	8.46
C10H11N302	205.22	205.0851	58.53	5.40	20.48	15.59
C10H11N303	221.22	221.0800	54.30	5.01	18.99	21.70
C10H11N304	237.22	237.0749	50.63	4.67	17.71	26.98
C10H11N305	253.22	253.0699	47.43	4.38	16.59	31.59
C10H11N306	269.22	269.0648	44.62	4.12	15.61	35.66
C10H11N307	285.22	285.0597	42.11	3.89	14.73	39.27
C10H11N308	301.21	301.0546	39.88	3.68	13.95	42.49
C10H11N309	317.21	317.0495	37.86	3.50	13.25	45.39
C10H11N501	217.23	217.0964	55.29	5.10	32.24	7.37
C10H11N502	233.23	233.0913	51.50	4.75	30.03	13.72
C10H11N503	249.23	249.0862	48.19	4.45	28.10	19.26
C10H11N504	265.23	265.0811	45.29	4.18	26.40	24.13
C10H11N505	281.23	281.0760	42.71	3.94	24.90	28.45
C10H11N506	297.23	297.0709	40.41	3.73	23.56	32.30
C10H11N507	313.23	313.0658	38.35	3.54	22.36	35.76
C10H11N508	329.23	329.0608	36.48	3.37	21.27	38.88
C10H11N509	345.23	345.0557	34.79	3.21	20.29	41.71
C10H12N201	176.22	176.0950	68.16	6.86	15.90	9.08
C10H12N202	192.22	192.0899	62.49	6.29	14.57	16.65
C10H12N203	208.22	208.0848	57.69	5.81	13.45	23.05
C10H12N204	224.22	224.0797	53.57	5.39	12.49	28.54
C10H12N205	240.22	240.0746	50.00	5.04	11.66	33.30
C10H12N206	256.22	256.0695	46.88	4.72	10.93	37.47
C10H12N207	272.22	272.0644	44.12	4.44	10.29	41.14
C10H12N208	288.22	288.0594	41.67	4.20	9.72	44.41
C10H12N209	304.22	304.0543	39.48	3.98	9.21	47.33
C10H12N401	204.23	204.1011	58.81	5.92	27.43	7.83
C10H12N402	220.23	220.0960	54.54	5.49	25.44	14.53
C10H12N403	236.23	236.0909	50.84	5.12	23.72	20.32
C10H12N404	252.23	252.0858	47.62	4.80	22.21	25.37
C10H12N405	268.23	268.0808	44.78	4.51	20.89	29.82
C10H12N406	284.23	284.0757	42.26	4.26	19.71	33.77
C10H12N407	300.23	300.0706	40.01	4.03	18.66	37.30
C10H12N408	316.23	316.0655	37.98	3.82	17.72	40.48
C10H12N409	332.23	332.0604	36.15	3.64	16.86	43.34
C10H12N601	232.25	232.1073	51.72	5.21	36.19	6.89
C10H12N602	248.25	248.1022	48.38	4.87	33.85	12.89
C10H12N603	264.25	264.0971	45.45	4.58	31.80	18.16
C10H12N604	280.24	280.0920	42.86	4.32	29.99	22.84
C10H12N605	296.24	296.0869	40.54	4.08	28.37	27.00
C10H12N606	312.24	312.0818	38.47	3.87	26.91	30.74
C10H12N607	328.24	328.0767	36.59	3.68	25.60	34.12
C10H12N608	344.24	344.0717	34.89	3.51	24.41	37.18
C10H12N609	360.24	360.0666	33.34	3.36	23.33	39.97
C10H13N101	163.22	163.0997	73.59	8.03	8.58	9.80
C10H13N102	179.22	179.0946	67.02	7.31	7.82	17.85
C10H13N103	195.22	195.0895	61.53	6.71	7.17	24.59
C10H13N104	211.22	211.0845	56.87	6.20	6.63	30.30
C10H13N105	227.22	227.0794	52.86	5.77	6.16	35.21
C10H13N106	243.22	243.0743	49.38	5.39	5.76	39.47
C10H13N107	259.22	259.0692	46.34	5.06	5.40	43.21
C10H13N108	275.22	275.0641	43.64	4.76	5.09	46.51
C10H13N109	291.22	291.0590	41.24	4.50	4.81	49.45
C10H13N301	191.23	191.1059	62.81	6.85	21.97	8.37
C10H13N302	207.23	207.1008	57.96	6.32	20.28	15.44
C10H13N303	223.23	223.0957	53.81	5.87	18.82	21.50
C10H13N304	239.23	239.0906	50.21	5.48	17.56	26.75

C10H13N305	255.23	255.0855	47.06	5.13	16.46	31.34
C10H13N306	271.23	271.0804	44.28	4.83	15.49	35.39
C10H13N307	287.23	287.0753	41.82	4.56	14.63	38.99
C10H13N308	303.23	303.0703	39.61	4.32	13.86	42.21
C10H13N309	319.23	319.0652	37.63	4.10	13.16	45.11
C10H13N501	219.25	219.1120	54.78	5.98	31.94	7.30
C10H13N502	235.25	235.1069	51.06	5.57	29.77	13.60
C10H13N503	251.25	251.1018	47.81	5.22	27.87	19.10
C10H13N504	267.25	267.0967	44.94	4.90	26.21	23.95
C10H13N505	283.25	283.0917	42.41	4.63	24.73	28.24
C10H13N506	299.25	299.0866	40.14	4.38	23.40	32.08
C10H13N507	315.24	315.0815	38.10	4.16	22.22	35.53
C10H13N508	331.24	331.0764	36.26	3.96	21.14	38.64
C10H13N509	347.24	347.0713	34.59	3.77	20.17	41.47
C10H14N201	178.24	178.1106	67.39	7.92	15.72	8.98
C10H14N202	194.24	194.1055	61.84	7.27	14.42	16.47
C10H14N203	210.23	210.1004	57.13	6.71	13.32	22.83
C10H14N204	226.23	226.0953	53.09	6.24	12.38	28.29
C10H14N205	242.23	242.0903	49.59	5.83	11.56	33.02
C10H14N206	258.23	258.0852	46.51	5.46	10.85	37.17
C10H14N207	274.23	274.0801	43.80	5.15	10.22	40.84
C10H14N208	290.23	290.0750	41.38	4.86	9.65	44.10
C10H14N209	306.23	306.0699	39.22	4.61	9.15	47.02
C10H14N401	206.25	206.1168	58.24	6.84	27.16	7.76
C10H14N402	222.25	222.1117	54.04	6.35	25.21	14.40
C10H14N403	238.25	238.1066	50.41	5.92	23.52	20.15
C10H14N404	254.25	254.1015	47.24	5.55	22.04	25.17
C10H14N405	270.25	270.0964	44.45	5.22	20.73	29.60
C10H14N406	286.25	286.0913	41.96	4.93	19.57	33.54
C10H14N407	302.25	302.0862	39.74	4.67	18.54	37.05
C10H14N408	318.25	318.0812	37.74	4.43	17.60	40.22
C10H14N409	334.24	334.0761	35.94	4.22	16.76	43.08
C10H14N601	234.26	234.1229	51.27	6.02	35.87	6.83
C10H14N602	250.26	250.1178	47.99	5.64	33.58	12.79
C10H14N603	266.26	266.1127	45.11	5.30	31.56	18.03
C10H14N604	282.26	282.1076	42.55	5.00	29.77	22.67
C10H14N605	298.26	298.1026	40.27	4.73	28.18	26.82
C10H14N606	314.26	314.0975	38.22	4.49	26.74	30.55
C10H14N607	330.26	330.0924	36.37	4.27	25.45	33.91
C10H14N608	346.26	346.0873	34.69	4.08	24.27	36.97
C10H14N609	362.26	362.0822	33.16	3.90	23.20	39.75
C10H15N101	165.24	165.1154	72.69	9.15	8.48	9.68
C10H15N102	181.24	181.1103	66.27	8.34	7.73	17.66
C10H15N103	197.24	197.1052	60.90	7.67	7.10	24.34
C10H15N104	213.24	213.1001	56.33	7.09	6.57	30.01
C10H15N105	229.23	229.0950	52.40	6.60	6.11	34.90
C10H15N106	245.23	245.0899	48.98	6.17	5.71	39.14
C10H15N107	261.23	261.0848	45.98	5.79	5.36	42.87
C10H15N108	277.23	277.0798	43.33	5.45	5.05	46.17
C10H15N109	293.23	293.0747	40.96	5.16	4.78	49.11
C10H15N301	193.25	193.1215	62.15	7.82	21.74	8.28
C10H15N302	209.25	209.1164	57.40	7.23	20.08	15.29
C10H15N303	225.25	225.1113	53.32	6.71	18.65	21.31
C10H15N304	241.25	241.1062	49.79	6.27	17.42	26.53
C10H15N305	257.25	257.1012	46.69	5.88	16.33	31.10
C10H15N306	273.25	273.0961	43.96	5.53	15.38	35.13
C10H15N307	289.25	289.0910	41.53	5.23	14.53	38.72
C10H15N308	305.25	305.0859	39.35	4.95	13.77	41.93
C10H15N309	321.25	321.0808	37.39	4.71	13.08	44.82
C10H15N501	221.26	221.1277	54.28	6.83	31.65	7.23

C10H15N502	237.26	237.1226	50.62	6.37	29.52	13.49
C10H15N503	253.26	253.1175	47.43	5.97	27.65	18.95
C10H15N504	269.26	269.1124	44.61	5.62	26.01	23.77
C10H15N505	285.26	285.1073	42.11	5.30	24.55	28.04
C10H15N506	301.26	301.1022	39.87	5.02	23.25	31.86
C10H15N507	317.26	317.0971	37.86	4.77	22.07	35.30
C10H15N508	333.26	333.0921	36.04	4.54	21.01	38.41
C10H15N509	349.26	349.0870	34.39	4.33	20.05	41.23
C10H16N201	180.25	180.1263	66.64	8.95	15.54	8.88
C10H16N202	196.25	196.1212	61.20	8.22	14.27	16.31
C10H16N203	212.25	212.1161	56.59	7.60	13.20	22.61
C10H16N204	228.25	228.1110	52.62	7.07	12.27	28.04
C10H16N205	244.25	244.1059	49.18	6.60	11.47	32.75
C10H16N206	260.25	260.1008	46.15	6.20	10.76	36.89
C10H16N207	276.25	276.0957	43.48	5.84	10.14	40.54
C10H16N208	292.25	292.0907	41.10	5.52	9.59	43.80
C10H16N209	308.25	308.0856	38.97	5.23	9.09	46.71
C10H16N401	208.27	208.1324	57.67	7.74	26.90	7.68
C10H16N402	224.26	224.1273	53.56	7.19	24.98	14.27
C10H16N403	240.26	240.1222	49.99	6.71	23.32	19.98
C10H16N404	256.26	256.1171	46.87	6.29	21.86	24.97
C10H16N405	272.26	272.1121	44.12	5.92	20.58	29.38
C10H16N406	288.26	288.1070	41.67	5.59	19.44	33.30
C10H16N407	304.26	304.1019	39.48	5.30	18.41	36.81
C10H16N408	320.26	320.0968	37.50	5.04	17.49	39.97
C10H16N409	336.26	336.0917	35.72	4.80	16.66	42.82
C10H16N601	236.28	236.1386	50.83	6.83	35.57	6.77
C10H16N602	252.28	252.1335	47.61	6.39	33.31	12.68
C10H16N603	268.28	268.1284	44.77	6.01	31.33	17.89
C10H16N604	284.28	284.1233	42.25	5.67	29.56	22.51
C10H16N605	300.28	300.1182	40.00	5.37	27.99	26.64
C10H16N606	316.28	316.1131	37.98	5.10	26.57	30.35
C10H16N607	332.28	332.1080	36.15	4.85	25.29	33.71
C10H16N608	348.27	348.1029	34.49	4.63	24.13	36.75
C10H16N609	364.27	364.0979	32.97	4.43	23.07	39.53
C10H17N101	167.25	167.1310	71.81	10.25	8.37	9.57
C10H17N102	183.25	183.1259	65.54	9.35	7.64	17.46
C10H17N103	199.25	199.1208	60.28	8.60	7.03	24.09
C10H17N104	215.25	215.1157	55.80	7.96	6.51	29.73
C10H17N105	231.25	231.1107	51.94	7.41	6.06	34.59
C10H17N106	247.25	247.1056	48.58	6.93	5.66	38.83
C10H17N107	263.25	263.1005	45.63	6.51	5.32	42.54
C10H17N108	279.25	279.0954	43.01	6.14	5.02	45.84
C10H17N109	295.25	295.0903	40.68	5.80	4.74	48.77
C10H17N301	195.27	195.1372	61.51	8.78	21.52	8.19
C10H17N302	211.27	211.1321	56.85	8.11	19.89	15.15
C10H17N303	227.27	227.1270	52.85	7.54	18.49	21.12
C10H17N304	243.26	243.1219	49.37	7.04	17.27	26.31
C10H17N305	259.26	259.1168	46.33	6.61	16.21	30.86
C10H17N306	275.26	275.1117	43.64	6.23	15.27	34.87
C10H17N307	291.26	291.1066	41.24	5.88	14.43	38.45
C10H17N308	307.26	307.1016	39.09	5.58	13.68	41.66
C10H17N309	323.26	323.0965	37.16	5.30	13.00	44.54
C10H17N501	223.28	223.1433	53.79	7.67	31.37	7.17
C10H17N502	239.28	239.1382	50.20	7.16	29.27	13.37
C10H17N503	255.28	255.1331	47.05	6.71	27.43	18.80
C10H17N504	271.28	271.1280	44.28	6.32	25.82	23.59
C10H17N505	287.28	287.1230	41.81	5.96	24.38	27.85
C10H17N506	303.28	303.1179	39.60	5.65	23.09	31.65
C10H17N507	319.28	319.1128	37.62	5.37	21.94	35.08

C10H17N508	335.28	335.1077	35.82	5.11	20.89	38.18
C10H17N509	351.28	351.1026	34.19	4.88	19.94	40.99
C10H18N201	182.27	182.1419	65.90	9.95	15.37	8.78
C10H18N202	198.27	198.1368	60.58	9.15	14.13	16.14
C10H18N203	214.27	214.1317	56.06	8.47	13.07	22.40
C10H18N204	230.27	230.1266	52.16	7.88	12.17	27.79
C10H18N205	246.27	246.1216	48.77	7.37	11.38	32.48
C10H18N206	262.26	262.1165	45.80	6.92	10.68	36.60
C10H18N207	278.26	278.1114	43.16	6.52	10.07	40.25
C10H18N208	294.26	294.1063	40.82	6.17	9.52	43.50
C10H18N209	310.26	310.1012	38.71	5.85	9.03	46.41
C10H18N401	210.28	210.1481	57.12	8.63	26.64	7.61
C10H18N402	226.28	226.1430	53.08	8.02	24.76	14.14
C10H18N403	242.28	242.1379	49.58	7.49	23.12	19.81
C10H18N404	258.28	258.1328	46.50	7.02	21.69	24.78
C10H18N405	274.28	274.1277	43.79	6.61	20.43	29.17
C10H18N406	290.28	290.1226	41.38	6.25	19.30	33.07
C10H18N407	306.28	306.1175	39.22	5.92	18.29	36.57
C10H18N408	322.28	322.1125	37.27	5.63	17.38	39.72
C10H18N409	338.28	338.1074	35.51	5.36	16.56	42.57
C10H18N601	238.29	238.1542	50.40	7.61	35.27	6.71
C10H18N602	254.29	254.1491	47.23	7.13	33.05	12.58
C10H18N603	270.29	270.1440	44.44	6.71	31.09	17.76
C10H18N604	286.29	286.1389	41.95	6.34	29.35	22.35
C10H18N605	302.29	302.1339	39.73	6.00	27.80	26.46
C10H18N606	318.29	318.1288	37.74	5.70	26.40	30.16
C10H18N607	334.29	334.1237	35.93	5.43	25.14	33.50
C10H18N608	350.29	350.1186	34.29	5.18	23.99	36.54
C10H18N609	366.29	366.1135	32.79	4.95	22.94	39.31
C10H19N101	169.27	169.1467	70.96	11.31	8.27	9.45
C10H19N102	185.27	185.1416	64.83	10.34	7.56	17.27
C10H19N103	201.27	201.1365	59.68	9.52	6.96	23.85
C10H19N104	217.27	217.1314	55.28	8.81	6.45	29.46
C10H19N105	233.27	233.1263	51.49	8.21	6.00	34.29
C10H19N106	249.27	249.1212	48.19	7.68	5.62	38.51
C10H19N107	265.27	265.1161	45.28	7.22	5.28	42.22
C10H19N108	281.26	281.1111	42.70	6.81	4.98	45.51
C10H19N109	297.26	297.1060	40.41	6.44	4.71	48.44
C10H19N301	197.28	197.1528	60.88	9.71	21.30	8.11
C10H19N302	213.28	213.1477	56.32	8.98	19.70	15.00
C10H19N303	229.28	229.1426	52.39	8.35	18.33	20.93
C10H19N304	245.28	245.1375	48.97	7.81	17.13	26.09
C10H19N305	261.28	261.1325	45.97	7.33	16.08	30.62
C10H19N306	277.28	277.1274	43.32	6.91	15.15	34.62
C10H19N307	293.28	293.1223	40.95	6.53	14.33	38.19
C10H19N308	309.28	309.1172	38.84	6.19	13.59	41.39
C10H19N309	325.28	325.1121	36.93	5.89	12.92	44.27
C10H19N501	225.30	225.1590	53.31	8.50	31.09	7.10
C10H19N502	241.30	241.1539	49.78	7.94	29.02	13.26
C10H19N503	257.29	257.1488	46.68	7.44	27.22	18.65
C10H19N504	273.29	273.1437	43.95	7.01	25.63	23.42
C10H19N505	289.29	289.1386	41.52	6.62	24.21	27.65
C10H19N506	305.29	305.1335	39.34	6.27	22.94	31.44
C10H19N507	321.29	321.1284	37.38	5.96	21.80	34.86
C10H19N508	337.29	337.1233	35.61	5.68	20.76	37.95
C10H19N509	353.29	353.1183	34.00	5.42	19.82	40.76
C10H20N201	184.28	184.1576	65.18	10.94	15.20	8.68
C10H20N202	200.28	200.1525	59.97	10.07	13.99	15.98
C10H20N203	216.28	216.1474	55.53	9.32	12.95	22.19
C10H20N204	232.28	232.1423	51.71	8.68	12.06	27.55

C10H20N205	248.28	248.1372	48.38	8.12	11.28	32.22
C10H20N206	264.28	264.1321	45.45	7.63	10.60	36.32
C10H20N207	280.28	280.1270	42.85	7.19	9.99	39.96
C10H20N208	296.28	296.1220	40.54	6.80	9.46	43.20
C10H20N209	312.28	312.1169	38.46	6.46	8.97	46.11
C10H20N401	212.30	212.1637	56.58	9.50	26.39	7.54
C10H20N402	228.30	228.1586	52.61	8.83	24.54	14.02
C10H20N403	244.30	244.1535	49.17	8.25	22.93	19.65
C10H20N404	260.30	260.1484	46.14	7.74	21.52	24.59
C10H20N405	276.29	276.1434	43.47	7.30	20.28	28.95
C10H20N406	292.29	292.1383	41.09	6.90	19.17	32.84
C10H20N407	308.29	308.1332	38.96	6.54	18.17	36.33
C10H20N408	324.29	324.1281	37.04	6.22	17.28	39.47
C10H20N409	340.29	340.1230	35.30	5.92	16.46	42.31
C10H20N601	240.31	240.1698	49.98	8.39	34.97	6.66
C10H20N602	256.31	256.1648	46.86	7.87	32.79	12.48
C10H20N603	272.31	272.1597	44.11	7.40	30.86	17.63
C10H20N604	288.31	288.1546	41.66	6.99	29.15	22.20
C10H20N605	304.31	304.1495	39.47	6.62	27.62	26.29
C10H20N606	320.31	320.1444	37.50	6.29	26.24	29.97
C10H20N607	336.31	336.1393	35.71	5.99	24.99	33.30
C10H20N608	352.31	352.1342	34.09	5.72	23.85	36.33
C10H20N609	368.31	368.1292	32.61	5.47	22.82	39.10
C10H21N101	171.28	171.1623	70.12	12.36	8.18	9.34
C10H21N102	187.28	187.1572	64.13	11.30	7.48	17.09
C10H21N103	203.28	203.1521	59.09	10.41	6.89	23.61
C10H21N104	219.28	219.1470	54.77	9.65	6.39	29.18
C10H21N105	235.28	235.1420	51.05	9.00	5.95	34.00
C10H21N106	251.28	251.1369	47.80	8.42	5.57	38.20
C10H21N107	267.28	267.1318	44.94	7.92	5.24	41.90
C10H21N108	283.28	283.1267	42.40	7.47	4.94	45.18
C10H21N109	299.28	299.1216	40.13	7.07	4.68	48.11
C10H21N301	199.30	199.1685	60.27	10.62	21.08	8.03
C10H21N302	215.30	215.1634	55.79	9.83	19.52	14.86
C10H21N303	231.30	231.1583	51.93	9.15	18.17	20.75
C10H21N304	247.30	247.1532	48.57	8.56	16.99	25.88
C10H21N305	263.30	263.1481	45.62	8.04	15.96	30.38
C10H21N306	279.30	279.1430	43.01	7.58	15.05	34.37
C10H21N307	295.29	295.1379	40.68	7.17	14.23	37.93
C10H21N308	311.29	311.1329	38.58	6.80	13.50	41.12
C10H21N309	327.29	327.1278	36.70	6.47	12.84	44.00
C10H21N501	227.31	227.1746	52.84	9.31	30.81	7.04
C10H21N502	243.31	243.1695	49.37	8.70	28.78	13.15
C10H21N503	259.31	259.1644	46.32	8.16	27.01	18.51
C10H21N504	275.31	275.1593	43.63	7.69	25.44	23.25
C10H21N505	291.31	291.1543	41.23	7.27	24.04	27.46
C10H21N506	307.31	307.1492	39.08	6.89	22.79	31.24
C10H21N507	323.31	323.1441	37.15	6.55	21.66	34.64
C10H21N508	339.31	339.1390	35.40	6.24	20.64	37.72
C10H21N509	355.31	355.1339	33.80	5.96	19.71	40.53
C10H22N201	186.30	186.1732	64.47	11.90	15.04	8.59
C10H22N202	202.30	202.1681	59.37	10.96	13.85	15.82
C10H22N203	218.30	218.1630	55.02	10.16	12.83	21.99
C10H22N204	234.30	234.1579	51.26	9.46	11.96	27.31
C10H22N205	250.30	250.1529	47.99	8.86	11.19	31.96
C10H22N206	266.30	266.1478	45.10	8.33	10.52	36.05
C10H22N207	282.30	282.1427	42.55	7.86	9.92	39.67
C10H22N208	298.30	298.1376	40.27	7.43	9.39	42.91
C10H22N209	314.29	314.1325	38.22	7.06	8.91	45.82
C10H22N401	214.31	214.1794	56.04	10.35	26.14	7.47

050

C10H22N4O2	230.31	230.1743	52.15	9.63	24.33	13.89
C10H22N4O3	246.31	246.1692	48.76	9.00	22.75	19.49
C10H22N4O4	262.31	262.1641	45.79	8.45	21.36	24.40
C10H22N4O5	278.31	278.1590	43.16	7.97	20.13	28.74
C10H22N4O6	294.31	294.1539	40.81	7.53	19.04	32.62
C10H22N4O7	310.31	310.1488	38.71	7.15	18.06	36.09
C10H22N4O8	326.31	326.1437	36.81	6.80	17.17	39.23
C10H22N4O9	342.31	342.1387	35.09	6.48	16.37	42.07
C10H22N6O1	242.33	242.1855	49.57	9.15	34.68	6.60
C10H22N6O2	258.33	258.1804	46.50	8.58	32.53	12.39
C10H22N6O3	274.33	274.1753	43.78	8.08	30.64	17.50
C10H22N6O4	290.32	290.1702	41.37	7.64	28.95	22.04
C10H22N6O5	306.32	306.1652	39.21	7.24	27.44	26.12
C10H22N6O6	322.32	322.1601	37.26	6.88	26.07	29.78
C10H22N6O7	338.32	338.1550	35.50	6.55	24.84	33.10
C10H22N6O8	354.32	354.1499	33.90	6.26	23.72	36.12
C10H22N6O9	370.32	370.1448	32.43	5.99	22.69	38.88
C10H23N1O1	173.30	173.1780	69.31	13.38	8.08	9.23
C10H23N1O2	189.30	189.1729	63.45	12.25	7.40	16.90
C10H23N1O3	205.30	205.1678	58.51	11.29	6.82	23.38
C10H23N1O4	221.30	221.1627	54.28	10.48	6.33	28.92
C10H23N1O5	237.30	237.1576	50.62	9.77	5.90	33.71
C10H23N1O6	253.30	253.1525	47.42	9.15	5.53	37.90
C10H23N1O7	269.30	269.1474	44.60	8.61	5.20	41.59
C10H23N1O8	285.30	285.1424	42.10	8.13	4.91	44.86
C10H23N1O9	301.30	301.1373	39.86	7.69	4.65	47.79
C10H23N3O1	201.31	201.1841	59.66	11.52	20.87	7.95
C10H23N3O2	217.31	217.1790	55.27	10.67	19.34	14.72
C10H23N3O3	233.31	233.1739	51.48	9.94	18.01	20.57
C10H23N3O4	249.31	249.1688	48.18	9.30	16.85	25.67
C10H23N3O5	265.31	265.1638	45.27	8.74	15.84	30.15
C10H23N3O6	281.31	281.1587	42.70	8.24	14.94	34.12
C10H23N3O7	297.31	297.1536	40.40	7.80	14.13	37.67
C10H23N3O8	313.31	313.1485	38.34	7.40	13.41	40.85
C10H23N3O9	329.31	329.1434	36.47	7.04	12.76	43.73
C10H23N5O1	229.33	229.1902	52.38	10.11	30.54	6.98
C10H23N5O2	245.33	245.1852	48.96	9.45	28.55	13.04
C10H23N5O3	261.33	261.1801	45.96	8.87	26.80	18.37
C10H23N5O4	277.33	277.1750	43.31	8.36	25.25	23.08
C10H23N5O5	293.33	293.1699	40.95	7.90	23.88	27.27
C10H23N5O6	309.32	309.1648	38.83	7.49	22.64	31.03
C10H23N5O7	325.32	325.1597	36.92	7.13	21.53	34.43
C10H23N5O8	341.32	341.1546	35.19	6.79	20.52	37.50
C10H23N5O9	357.32	357.1496	33.61	6.49	19.60	40.30
C10H24N2O1	188.32	188.1889	63.78	12.85	14.88	8.50
C10H24N2O2	204.31	204.1838	58.79	11.84	13.71	15.66
C10H24N2O3	220.31	220.1787	54.52	10.98	12.72	21.79
C10H24N2O4	236.31	236.1736	50.83	10.24	11.85	27.08
C10H24N2O5	252.31	252.1685	47.60	9.59	11.10	31.71
C10H24N2O6	268.31	268.1634	44.77	9.02	10.44	35.78
C10H24N2O7	284.31	284.1583	42.25	8.51	9.85	39.39
C10H24N2O8	300.31	300.1533	40.00	8.06	9.33	42.62
C10H24N2O9	316.31	316.1482	37.97	7.65	8.86	45.52
C10H24N4O1	216.33	216.1950	55.52	11.18	25.90	7.40
C10H24N4O2	232.33	232.1899	51.70	10.41	24.12	13.77
C10H24N4O3	248.33	248.1848	48.37	9.74	22.56	19.33
C10H24N4O4	264.33	264.1797	45.44	9.15	21.20	24.21
C10H24N4O5	280.33	280.1747	42.85	8.63	19.99	28.54
C10H24N4O6	296.33	296.1696	40.53	8.16	18.91	32.40
C10H24N4O7	312.33	312.1645	38.46	7.75	17.94	35.86
C10H24N4O8	328.32	328.1594	36.58	7.37	17.06	38.98
C10H24N4O9	344.32	344.1543	34.88	7.03	16.27	41.82
C10H24N6O1	244.34	244.2011	49.16	9.90	34.39	6.55
C10H24N6O2	260.34	260.1961	46.14	9.29	32.28	12.29
C10H24N6O3	276.34	276.1910	43.46	8.75	30.41	17.37
C10H24N6O4	292.34	292.1859	41.09	8.28	28.75	21.89
C10H24N6O5	308.34	308.1808	38.95	7.85	27.26	25.94
C10H24N6O6	324.34	324.1757	37.03	7.46	25.91	29.60
C10H24N6O7	340.34	340.1706	35.29	7.11	24.69	32.91
C10H24N6O8	356.34	356.1655	33.71	6.79	23.58	35.92
C10H24N6O9	372.34	372.1605	32.26	6.50	22.57	38.67
C10H25N3O1	203.33	203.1998	59.07	12.39	20.67	7.87
C10H25N3O2	219.33	219.1947	54.76	11.49	19.16	14.59
C10H25N3O3	235.33	235.1896	51.04	10.71	17.86	20.40
C10H25N3O4	251.33	251.1845	47.79	10.03	16.72	25.46
C10H25N3O5	267.33	267.1794	44.93	9.43	15.72	29.92
C10H25N3O6	283.33	283.1743	42.39	8.89	14.83	33.88
C10H25N3O7	299.33	299.1692	40.13	8.42	14.04	37.42
C10H25N3O8	315.33	315.1641	38.09	7.99	13.33	40.59
C10H25N3O9	331.33	331.1591	36.25	7.61	12.68	43.46
C10H25N5O1	231.34	231.2059	51.92	10.89	30.27	6.92
C10H25N5O2	247.34	247.2008	48.56	10.19	28.31	12.94
C10H25N5O3	263.34	263.1957	45.61	9.57	26.59	18.23
C10H25N5O4	279.34	279.1906	43.00	9.02	25.07	22.91
C10H25N5O5	295.34	295.1856	40.67	8.53	23.71	27.09
C10H25N5O6	311.34	311.1805	38.58	8.09	22.49	30.83
C10H25N5O7	327.34	327.1754	36.69	7.70	21.39	34.21
C10H25N5O8	343.34	343.1703	34.98	7.34	20.40	37.28
C10H25N5O9	359.34	359.1652	33.43	7.01	19.49	40.07
C10H26N4O1	218.34	218.2106	55.01	12.00	25.66	7.33
C10H26N4O2	234.34	234.2056	51.25	11.18	23.91	13.65
C10H26N4O3	250.34	250.2005	47.98	10.47	22.38	19.17
C10H26N4O4	266.34	266.1954	45.10	9.84	21.04	24.03
C10H26N4O5	282.34	282.1903	42.54	9.28	19.84	28.33
C10H26N4O6	298.34	298.1852	40.26	8.78	18.78	32.18
C10H26N4O7	314.34	314.1801	38.21	8.34	17.82	35.63
C10H26N4O8	330.34	330.1750	36.36	7.93	16.96	38.75
C10H26N4O9	346.34	346.1700	34.68	7.57	16.18	41.58
C10H26N6O1	246.36	246.2168	48.75	10.64	34.11	6.49
C10H26N6O2	262.36	262.2117	45.78	9.99	32.03	12.20
C10H26N6O3	278.36	278.2066	43.15	9.41	30.19	17.24
C10H26N6O4	294.36	294.2015	40.80	8.90	28.55	21.74
C10H26N6O5	310.36	310.1965	38.70	8.44	27.08	25.78
C10H26N6O6	326.36	326.1914	36.80	8.03	25.75	29.41
C10H26N6O7	342.35	342.1863	35.08	7.65	24.55	32.71
C10H26N6O8	358.35	358.1812	33.52	7.31	23.45	35.72
C10H26N6O9	374.35	374.1761	32.09	7.00	22.45	38.46
C10H27N5O1	233.36	233.2215	51.47	11.66	30.01	6.86
C10H27N5O2	249.36	249.2165	48.17	10.91	28.09	12.83
C10H27N5O3	265.36	265.2114	45.26	10.26	26.39	18.09
C10H27N5O4	281.36	281.2063	42.69	9.67	24.89	22.75
C10H27N5O5	297.36	297.2012	40.39	9.15	23.55	26.90
C10H27N5O6	313.36	313.1961	38.33	8.69	22.35	30.63
C10H27N5O7	329.36	329.1910	36.47	8.26	21.26	34.00
C10H27N5O8	345.36	345.1859	34.78	7.88	20.28	37.06
C10H27N5O9	361.35	361.1809	33.24	7.53	19.38	39.85
C10H28N6O1	248.37	248.2324	48.36	11.36	33.84	6.44
C10H28N6O2	264.37	264.2274	45.43	10.68	31.79	12.10
C10H28N6O3	280.37	280.2223	42.84	10.07	29.97	17.12
C10H28N6O4	296.37	296.2172	40.53	9.52	28.36	21.59

051

C10H28N6O5	312.37	312.2121	38.45	9.04	26.90	25.61
C10H28N6O6	328.37	328.2070	36.58	8.59	25.59	29.23
C10H28N6O7	344.37	344.2019	34.88	8.20	24.40	32.52
C10H28N6O8	360.37	360.1968	33.33	7.83	23.32	35.52
C10H28N6O9	376.37	376.1918	31.91	7.50	22.33	38.26
C11H 5N1O1	167.17	167.0371	79.04	3.01	8.38	9.57
C11H 5N1O2	183.17	183.0320	72.13	2.75	7.65	17.47
C11H 5N1O3	199.17	199.0269	66.34	2.53	7.03	24.10
C11H 5N1O4	215.17	215.0219	61.40	2.34	6.51	29.74
C11H 5N1O5	231.17	231.0168	57.15	2.18	6.06	34.61
C11H 5N1O6	247.17	247.0117	53.46	2.04	5.67	38.84
C11H 5N1O7	263.16	263.0066	50.21	1.92	5.32	42.56
C11H 5N1O8	279.16	279.0015	47.33	1.81	5.02	45.85
C11H 5N1O9	295.16	294.9964	44.76	1.71	4.75	48.78
C11H 5N3O1	195.18	195.0433	67.69	2.58	21.53	8.20
C11H 5N3O2	211.18	211.0382	62.56	2.39	19.90	15.15
C11H 5N3O3	227.18	227.0331	58.16	2.22	18.50	21.13
C11H 5N3O4	243.18	243.0280	54.33	2.07	17.28	26.32
C11H 5N3O5	259.18	259.0229	50.98	1.94	16.21	30.87
C11H 5N3O6	275.18	275.0178	48.01	1.83	15.27	34.89
C11H 5N3O7	291.18	291.0127	45.38	1.73	14.43	38.46
C11H 5N3O8	307.18	307.0077	43.01	1.64	13.68	41.67
C11H 5N3O9	323.18	323.0026	40.88	1.56	13.00	44.56
C11H 5N5O1	223.20	223.0494	59.20	2.26	31.38	7.17
C11H 5N5O2	239.19	239.0443	55.24	2.11	29.28	13.38
C11H 5N5O3	255.19	255.0392	51.77	1.97	27.44	18.81
C11H 5N5O4	271.19	271.0341	48.72	1.86	25.82	23.60
C11H 5N5O5	287.19	287.0291	46.00	1.75	24.39	27.85
C11H 5N5O6	303.19	303.0240	43.58	1.66	23.10	31.66
C11H 5N5O7	319.19	319.0189	41.39	1.58	21.94	35.09
C11H 5N5O8	335.19	335.0138	39.42	1.50	20.89	38.19
C11H 5N5O9	351.19	351.0087	37.62	1.44	19.94	41.00
C11H 6N2O1	182.18	182.0480	72.52	3.32	15.38	8.78
C11H 6N2O2	198.18	198.0429	66.67	3.05	14.14	16.15
C11H 6N2O3	214.18	214.0378	61.69	2.82	13.08	22.41
C11H 6N2O4	230.18	230.0328	57.40	2.63	12.17	27.80
C11H 6N2O5	246.18	246.0277	53.67	2.46	11.38	32.50
C11H 6N2O6	262.18	262.0226	50.39	2.31	10.68	36.61
C11H 6N2O7	278.18	278.0175	47.50	2.17	10.07	40.26
C11H 6N2O8	294.18	294.0124	44.91	2.06	9.52	43.51
C11H 6N2O9	310.18	310.0073	42.60	1.95	9.03	46.42
C11H 6N4O1	210.20	210.0542	62.86	2.88	26.65	7.61
C11H 6N4O2	226.20	226.0491	58.41	2.67	24.77	14.15
C11H 6N4O3	242.20	242.0440	54.55	2.50	23.13	19.82
C11H 6N4O4	258.19	258.0389	51.17	2.34	21.70	24.79
C11H 6N4O5	274.19	274.0338	48.19	2.21	20.43	29.18
C11H 6N4O6	290.19	290.0287	45.53	2.08	19.31	33.08
C11H 6N4O7	306.19	306.0236	43.15	1.98	18.30	36.58
C11H 6N4O8	322.19	322.0186	41.01	1.88	17.39	39.73
C11H 6N4O9	338.19	338.0135	39.07	1.79	16.57	42.58
C11H 6N6O1	238.21	238.0603	55.46	2.54	35.28	6.72
C11H 6N6O2	254.21	254.0552	51.97	2.38	33.06	12.59
C11H 6N6O3	270.21	270.0501	48.90	2.24	31.10	17.76
C11H 6N6O4	286.21	286.0450	46.16	2.11	29.36	22.36
C11H 6N6O5	302.21	302.0400	43.72	2.00	27.81	26.47
C11H 6N6O6	318.21	318.0349	41.52	1.90	26.41	30.17
C11H 6N6O7	334.21	334.0298	39.53	1.81	25.15	33.51
C11H 6N6O8	350.21	350.0247	37.73	1.73	24.00	36.55
C11H 6N6O9	366.21	366.0196	36.08	1.65	22.95	39.32
C11H 7N1O1	169.18	169.0528	78.09	4.17	8.28	9.46
C11H 7N1O2	185.18	185.0477	71.35	3.81	7.56	17.28
C11H 7N1O3	201.18	201.0426	65.67	3.51	6.96	23.86
C11H 7N1O4	217.18	217.0375	60.83	3.25	6.45	29.47
C11H 7N1O5	233.18	233.0324	56.66	3.03	6.01	34.31
C11H 7N1O6	249.18	249.0273	53.02	2.83	5.62	38.52
C11H 7N1O7	265.18	265.0222	49.82	2.66	5.28	42.23
C11H 7N1O8	281.18	281.0172	46.99	2.51	4.98	45.52
C11H 7N1O9	297.18	297.0121	44.46	2.37	4.71	48.45
C11H 7N3O1	197.20	197.0589	67.00	3.58	21.31	8.11
C11H 7N3O2	213.20	213.0538	61.97	3.31	19.71	15.01
C11H 7N3O3	229.20	229.0487	57.65	3.08	18.33	20.94
C11H 7N3O4	245.20	245.0437	53.88	2.88	17.14	26.10
C11H 7N3O5	261.20	261.0386	50.58	2.70	16.09	30.63
C11H 7N3O6	277.19	277.0335	47.66	2.55	15.16	34.63
C11H 7N3O7	293.19	293.0284	45.06	2.41	14.33	38.20
C11H 7N3O8	309.19	309.0233	42.73	2.28	13.59	41.40
C11H 7N3O9	325.19	325.0182	40.63	2.17	12.92	44.28
C11H 7N5O1	225.21	225.0651	58.67	3.13	31.10	7.10
C11H 7N5O2	241.21	241.0600	54.77	2.93	29.03	13.27
C11H 7N5O3	257.21	257.0549	51.37	2.74	27.23	18.66
C11H 7N5O4	273.21	273.0498	48.36	2.58	25.63	23.42
C11H 7N5O5	289.21	289.0447	45.68	2.44	24.22	27.66
C11H 7N5O6	305.21	305.0396	43.29	2.31	22.95	31.45
C11H 7N5O7	321.21	321.0345	41.13	2.20	21.80	34.87
C11H 7N5O8	337.21	337.0295	39.18	2.09	20.77	37.96
C11H 7N5O9	353.21	353.0244	37.41	2.00	19.83	40.77
C11H 8N2O1	184.20	184.0637	71.73	4.38	15.21	8.69
C11H 8N2O2	200.20	200.0586	66.00	4.03	13.99	15.98
C11H 8N2O3	216.20	216.0535	61.11	3.73	12.96	22.20
C11H 8N2O4	232.20	232.0484	56.90	3.47	12.06	27.56
C11H 8N2O5	248.20	248.0433	53.23	3.25	11.29	32.23
C11H 8N2O6	264.20	264.0382	50.01	3.05	10.60	36.34
C11H 8N2O7	280.20	280.0331	47.15	2.88	10.00	39.97
C11H 8N2O8	296.20	296.0281	44.61	2.72	9.46	43.21
C11H 8N2O9	312.19	312.0230	42.32	2.58	8.97	46.12
C11H 8N4O1	212.21	212.0698	62.26	3.80	26.40	7.54
C11H 8N4O2	228.21	228.0647	57.89	3.53	24.55	14.02
C11H 8N4O3	244.21	244.0596	54.10	3.30	22.94	19.65
C11H 8N4O4	260.21	260.0545	50.78	3.10	21.53	24.59
C11H 8N4O5	276.21	276.0495	47.83	2.92	20.28	28.96
C11H 8N4O6	292.21	292.0444	45.22	2.76	19.17	32.85
C11H 8N4O7	308.21	308.0393	42.87	2.62	18.18	36.34
C11H 8N4O8	324.21	324.0342	40.75	2.49	17.28	39.48
C11H 8N4O9	340.21	340.0291	38.84	2.37	16.47	42.33
C11H 8N6O1	240.23	240.0760	55.00	3.36	34.98	6.66
C11H 8N6O2	256.23	256.0709	51.57	3.15	32.80	12.49
C11H 8N6O3	272.22	272.0658	48.53	2.96	30.87	17.63
C11H 8N6O4	288.22	288.0607	45.84	2.80	29.16	22.20
C11H 8N6O5	304.22	304.0556	43.43	2.65	27.62	26.30
C11H 8N6O6	320.22	320.0505	41.26	2.52	26.24	29.98
C11H 8N6O7	336.22	336.0454	39.30	2.40	25.00	33.31
C11H 8N6O8	352.22	352.0404	37.51	2.29	23.86	36.34
C11H 8N6O9	368.22	368.0353	35.88	2.19	22.82	39.11
C11H 9N1O1	171.20	171.0684	77.17	5.30	8.18	9.35
C11H 9N1O2	187.20	187.0633	70.58	4.85	7.48	17.09
C11H 9N1O3	203.20	203.0582	65.02	4.46	6.89	23.62
C11H 9N1O4	219.20	219.0532	60.28	4.14	6.39	29.20
C11H 9N1O5	235.20	235.0481	56.18	3.86	5.96	34.01
C11H 9N1O6	251.20	251.0430	52.60	3.61	5.58	38.22
C11H 9N1O7	267.20	267.0379	49.45	3.40	5.24	41.92

C11H 9N108	283.20	283.0328	46.65	3.20	4.95	45.20
C11H 9N109	299.20	299.0277	44.16	3.03	4.68	48.13
C11H 9N301	199.21	199.0746	66.32	4.55	21.09	8.03
C11H 9N302	215.21	215.0695	61.39	4.22	19.52	14.87
C11H 9N303	231.21	231.0644	57.14	3.92	18.17	20.76
C11H 9N304	247.21	247.0593	53.45	3.67	17.00	25.89
C11H 9N305	263.21	263.0542	50.20	3.45	15.96	30.39
C11H 9N306	279.21	279.0491	47.32	3.25	15.05	34.38
C11H 9N307	295.21	295.0440	44.76	3.07	14.23	37.94
C11H 9N308	311.21	311.0390	42.45	2.91	13.50	41.13
C11H 9N309	327.21	327.0339	40.38	2.77	12.84	44.01
C11H 9N501	227.23	227.0807	58.15	3.99	30.82	7.04
C11H 9N502	243.23	243.0756	54.32	3.73	28.79	13.16
C11H 9N503	259.23	259.0705	50.97	3.50	27.02	18.52
C11H 9N504	275.23	275.0654	48.01	3.30	25.45	23.25
C11H 9N505	291.22	291.0604	45.37	3.12	24.05	27.47
C11H 9N506	307.22	307.0553	43.01	2.95	22.80	31.25
C11H 9N507	323.22	323.0502	40.88	2.81	21.67	34.65
C11H 9N508	339.22	339.0451	38.95	2.67	20.65	37.73
C11H 9N509	355.22	355.0400	37.19	2.55	19.72	40.54
C11H10N201	186.22	186.0793	70.95	5.41	15.04	8.59
C11H10N202	202.21	202.0742	65.34	4.98	13.85	15.82
C11H10N203	218.21	218.0691	60.55	4.62	12.84	22.00
C11H10N204	234.21	234.0641	56.41	4.30	11.96	27.32
C11H10N205	250.21	250.0590	52.80	4.03	11.20	31.97
C11H10N206	266.21	266.0539	49.63	3.79	10.52	36.06
C11H10N207	282.21	282.0488	46.82	3.57	9.93	39.69
C11H10N208	298.21	298.0437	44.31	3.38	9.39	42.92
C11H10N209	314.21	314.0386	42.05	3.21	8.92	45.83
C11H10N401	214.23	214.0855	61.67	4.71	26.15	7.47
C11H10N402	230.23	230.0804	57.39	4.38	24.34	13.90
C11H10N403	246.23	246.0753	53.66	4.09	22.75	19.49
C11H10N404	262.23	262.0702	50.38	3.84	21.37	24.41
C11H10N405	278.23	278.0651	47.49	3.62	20.14	28.75
C11H10N406	294.23	294.0600	44.91	3.43	19.04	32.63
C11H10N407	310.22	310.0549	42.59	3.25	18.06	36.10
C11H10N408	326.22	326.0499	40.50	3.09	17.17	39.24
C11H10N409	342.22	342.0448	38.61	2.95	16.37	42.08
C11H10N601	242.24	242.0916	54.54	4.16	34.69	6.60
C11H10N602	258.24	258.0865	51.16	3.90	32.54	12.39
C11H10N603	274.24	274.0814	48.18	3.68	30.64	17.50
C11H10N604	290.24	290.0763	45.52	3.47	28.96	22.05
C11H10N605	306.24	306.0713	43.14	3.29	27.44	26.12
C11H10N606	322.24	322.0662	41.00	3.13	26.08	29.79
C11H10N607	338.24	338.0611	39.06	2.98	24.85	33.11
C11H10N608	354.24	354.0560	37.30	2.85	23.72	36.13
C11H10N609	370.24	370.0509	35.69	2.72	22.70	38.89
C11H11N101	173.22	173.0841	76.28	6.40	8.09	9.24
C11H11N102	189.22	189.0790	69.83	5.86	7.40	16.91
C11H11N103	205.22	205.0739	64.38	5.40	6.83	23.39
C11H11N104	221.21	221.0688	59.73	5.01	6.33	28.93
C11H11N105	237.21	237.0637	55.70	4.67	5.90	33.72
C11H11N106	253.21	253.0586	52.18	4.38	5.53	37.91
C11H11N107	269.21	269.0535	49.08	4.12	5.20	41.60
C11H11N108	285.21	285.0485	46.32	3.89	4.91	44.88
C11H11N109	301.21	301.0434	43.86	3.68	4.65	47.81
C11H11N301	201.23	201.0902	65.66	5.51	20.88	7.95
C11H11N302	217.23	217.0851	60.82	5.10	19.34	14.73
C11H11N303	233.23	233.0800	56.65	4.75	18.02	20.58
C11H11N304	249.23	249.0749	53.01	4.45	16.86	25.68
C11H11N305	265.23	265.0699	49.81	4.18	15.84	30.16
C11H11N306	281.23	281.0648	46.98	3.94	14.94	34.13
C11H11N307	297.23	297.0597	44.45	3.73	14.14	37.68
C11H11N308	313.23	313.0546	42.18	3.54	13.42	40.86
C11H11N309	329.23	329.0495	40.13	3.37	12.76	43.74
C11H11N501	229.24	229.0964	57.63	4.84	30.55	6.98
C11H11N502	245.24	245.0913	53.87	4.52	28.56	13.05
C11H11N503	261.24	261.0862	50.57	4.24	26.81	18.37
C11H11N504	277.24	277.0811	47.66	4.00	25.26	23.08
C11H11N505	293.24	293.0760	45.06	3.78	23.88	27.28
C11H11N506	309.24	309.0709	42.72	3.59	22.65	31.04
C11H11N507	325.24	325.0658	40.62	3.41	21.53	34.43
C11H11N508	341.24	341.0608	38.72	3.25	20.52	37.51
C11H11N509	357.24	357.0557	36.98	3.10	19.60	40.31
C11H12N201	188.23	188.0950	70.19	6.43	14.88	8.50
C11H12N202	204.23	204.0899	64.69	5.92	13.72	15.67
C11H12N203	220.23	220.0848	59.99	5.49	12.72	21.79
C11H12N204	236.23	236.0797	55.93	5.12	11.86	27.09
C11H12N205	252.23	252.0746	52.38	4.80	11.11	31.72
C11H12N206	268.23	268.0695	49.26	4.51	10.44	35.79
C11H12N207	284.23	284.0644	46.48	4.26	9.86	39.40
C11H12N208	300.23	300.0594	44.01	4.03	9.33	42.63
C11H12N209	316.23	316.0543	41.78	3.82	8.86	45.54
C11H12N401	216.24	216.1011	61.10	5.59	25.91	7.40
C11H12N402	232.24	232.0960	56.89	5.21	24.12	13.78
C11H12N403	248.24	248.0909	53.22	4.87	22.57	19.34
C11H12N404	264.24	264.0858	50.00	4.58	21.20	24.22
C11H12N405	280.24	280.0808	47.15	4.32	19.99	28.55
C11H12N406	296.24	296.0757	44.60	4.08	18.91	32.40
C11H12N407	312.24	312.0706	42.31	3.87	17.94	35.87
C11H12N408	328.24	328.0655	40.25	3.68	17.07	38.99
C11H12N409	344.24	344.0604	38.38	3.51	16.28	41.83
C11H12N601	244.26	244.1073	54.09	4.95	34.41	6.55
C11H12N602	260.26	260.1022	50.77	4.65	32.29	12.30
C11H12N603	276.26	276.0971	47.83	4.38	30.42	17.37
C11H12N604	292.26	292.0920	45.21	4.14	28.76	21.90
C11H12N605	308.26	308.0869	42.86	3.92	27.26	25.95
C11H12N606	324.25	324.0818	40.75	3.73	25.92	29.61
C11H12N607	340.25	340.0767	38.83	3.55	24.70	32.92
C11H12N608	356.25	356.0717	37.09	3.40	23.59	35.93
C11H12N609	372.25	372.0666	35.49	3.25	22.58	38.68
C11H13N101	175.23	175.0997	75.40	7.48	7.99	9.13
C11H13N102	191.23	191.0946	69.09	6.85	7.32	16.73
C11H13N103	207.23	207.0895	63.76	6.32	6.76	23.16
C11H13N104	223.23	223.0845	59.19	5.87	6.27	28.67
C11H13N105	239.23	239.0794	55.23	5.48	5.85	33.44
C11H13N106	255.23	255.0743	51.77	5.13	5.49	37.61
C11H13N107	271.23	271.0692	48.71	4.83	5.16	41.29
C11H13N108	287.23	287.0641	46.00	4.56	4.88	44.56
C11H13N109	303.23	303.0590	43.57	4.32	4.62	47.49
C11H13N301	203.25	203.1059	65.01	6.45	20.67	7.87
C11H13N302	219.25	219.1008	60.26	5.98	19.17	14.59
C11H13N303	235.24	235.0957	56.16	5.57	17.86	20.40
C11H13N304	251.24	251.0906	52.59	5.22	16.72	25.47
C11H13N305	267.24	267.0855	49.44	4.90	15.72	29.93
C11H13N306	283.24	283.0804	46.65	4.63	14.84	33.89
C11H13N307	299.24	299.0753	44.15	4.38	14.04	37.43
C11H13N308	315.24	315.0703	41.91	4.16	13.33	40.60
C11H13N309	331.24	331.0652	39.89	3.96	12.69	43.47
C11H13N501	231.26	231.1120	57.13	5.67	30.28	6.92

053

C11H13N502	247.26	247.1069	53.44	5.30	28.32	12.94
C11H13N503	263.26	263.1018	50.19	4.98	26.60	18.23
C11H13N504	279.26	279.0967	47.31	4.69	25.08	22.92
C11H13N505	295.26	295.0917	44.75	4.44	23.72	27.09
C11H13N506	311.26	311.0866	42.45	4.21	22.50	30.84
C11H13N507	327.26	327.0815	40.37	4.00	21.40	34.22
C11H13N508	343.25	343.0764	38.49	3.82	20.40	37.29
C11H13N509	359.25	359.0713	36.78	3.65	19.49	40.08
C11H14N201	190.25	190.1106	69.45	7.42	14.72	8.41
C11H14N202	206.25	206.1055	64.06	6.84	13.58	15.51
C11H14N203	222.25	222.1004	59.45	6.35	12.60	21.60
C11H14N204	238.25	238.0953	55.46	5.92	11.76	26.86
C11H14N205	254.24	254.0903	51.97	5.55	11.02	31.46
C11H14N206	270.24	270.0852	48.89	5.22	10.37	35.52
C11H14N207	286.24	286.0801	46.16	4.93	9.79	39.13
C11H14N208	302.24	302.0750	43.71	4.67	9.27	42.35
C11H14N209	318.24	318.0699	41.52	4.43	8.80	45.25
C11H14N401	218.26	218.1168	60.53	6.47	25.67	7.33
C11H14N402	234.26	234.1117	56.40	6.02	23.92	13.66
C11H14N403	250.26	250.1066	52.79	5.64	22.39	19.18
C11H14N404	266.26	266.1015	49.62	5.30	21.04	24.04
C11H14N405	282.26	282.0964	46.81	5.00	19.85	28.34
C11H14N406	298.26	298.0913	44.30	4.73	18.78	32.19
C11H14N407	314.26	314.0862	42.04	4.49	17.83	35.64
C11H14N408	330.26	330.0812	40.01	4.27	16.96	38.76
C11H14N409	346.26	346.0761	38.16	4.08	16.18	41.59
C11H14N601	246.27	246.1229	53.65	5.73	34.12	6.50
C11H14N602	262.27	262.1178	50.38	5.38	32.04	12.20
C11H14N603	278.27	278.1127	47.48	5.07	30.20	17.25
C11H14N604	294.27	294.1076	44.90	4.80	28.56	21.75
C11H14N605	310.27	310.1026	42.58	4.55	27.09	25.78
C11H14N606	326.27	326.0975	40.49	4.33	25.76	29.42
C11H14N607	342.27	342.0924	38.60	4.12	24.55	32.72
C11H14N608	358.27	358.0873	36.88	3.94	23.46	35.73
C11H14N609	374.27	374.0822	35.30	3.77	22.45	38.47
C11H15N101	177.25	177.1154	74.54	8.53	7.90	9.03
C11H15N102	193.25	193.1103	68.37	7.82	7.25	16.56
C11H15N103	209.25	209.1052	63.14	7.23	6.69	22.94
C11H15N104	225.25	225.1001	58.66	6.71	6.22	28.41
C11H15N105	241.25	241.0950	54.77	6.27	5.81	33.16
C11H15N106	257.25	257.0899	51.36	5.88	5.44	37.32
C11H15N107	273.24	273.0848	48.35	5.53	5.13	40.99
C11H15N108	289.24	289.0798	45.68	5.23	4.84	44.25
C11H15N109	305.24	305.0747	43.28	4.95	4.59	47.17
C11H15N301	205.26	205.1215	64.37	7.37	20.47	7.79
C11H15N302	221.26	221.1164	59.71	6.83	18.99	14.46
C11H15N303	237.26	237.1113	55.69	6.37	17.71	20.23
C11H15N304	253.26	253.1062	52.17	5.97	16.59	25.27
C11H15N305	269.26	269.1012	49.07	5.62	15.61	29.71
C11H15N306	285.26	285.0961	46.32	5.30	14.73	33.65
C11H15N307	301.26	301.0910	43.86	5.02	13.95	37.18
C11H15N308	317.26	317.0859	41.65	4.77	13.24	40.34
C11H15N309	333.26	333.0808	39.65	4.54	12.61	43.21
C11H15N501	233.28	233.1277	56.64	6.48	30.02	6.86
C11H15N502	249.27	249.1226	53.00	6.07	28.09	12.84
C11H15N503	265.27	265.1175	49.81	5.70	26.40	18.09
C11H15N504	281.27	281.1124	46.97	5.38	24.90	22.75
C11H15N505	297.27	297.1073	44.44	5.09	23.56	26.91
C11H15N506	313.27	313.1022	42.18	4.83	22.36	30.64
C11H15N507	329.27	329.0971	40.13	4.59	21.27	34.01
C11H15N508	345.27	345.0921	38.27	4.38	20.28	37.07
C11H15N509	361.27	361.0870	36.57	4.19	19.39	39.86
C11H16N201	192.26	192.1263	68.72	8.39	14.57	8.32
C11H16N202	208.26	208.1212	63.44	7.74	13.45	15.36
C11H16N203	224.26	224.1161	58.91	7.19	12.49	21.40
C11H16N204	240.26	240.1110	54.99	6.71	11.66	26.64
C11H16N205	256.26	256.1059	51.56	6.29	10.93	31.22
C11H16N206	272.26	272.1008	48.53	5.92	10.29	35.26
C11H16N207	288.26	288.0957	45.83	5.59	9.72	38.85
C11H16N208	304.26	304.0907	43.42	5.30	9.21	42.07
C11H16N209	320.26	320.0856	41.26	5.04	8.75	44.96
C11H16N401	220.28	220.1324	59.98	7.32	25.43	7.26
C11H16N402	236.28	236.1273	55.92	6.83	23.71	13.54
C11H16N403	252.28	252.1222	52.37	6.39	22.21	19.03
C11H16N404	268.27	268.1171	49.25	6.01	20.88	23.86
C11H16N405	284.27	284.1121	46.48	5.67	19.71	28.14
C11H16N406	300.27	300.1070	44.00	5.37	18.66	31.97
C11H16N407	316.27	316.1019	41.77	5.10	17.71	35.41
C11H16N408	332.27	332.0968	39.76	4.85	16.86	38.52
C11H16N409	348.27	348.0917	37.94	4.63	16.09	41.35
C11H16N601	248.29	248.1386	53.21	6.50	33.85	6.44
C11H16N602	264.29	264.1335	49.99	6.10	31.80	12.11
C11H16N603	280.29	280.1284	47.14	5.75	29.98	17.12
C11H16N604	296.29	296.1233	44.59	5.44	28.36	21.60
C11H16N605	312.29	312.1182	42.31	5.16	26.91	25.62
C11H16N606	328.29	328.1131	40.25	4.91	25.60	29.24
C11H16N607	344.29	344.1080	38.38	4.68	24.41	32.53
C11H16N608	360.29	360.1029	36.67	4.48	23.33	35.53
C11H16N609	376.28	376.0979	35.11	4.29	22.33	38.27
C11H17N101	179.26	179.1310	73.70	9.56	7.81	8.93
C11H17N102	195.26	195.1259	67.66	8.78	7.17	16.39
C11H17N103	211.26	211.1208	62.54	8.11	6.63	22.72
C11H17N104	227.26	227.1157	58.14	7.54	6.16	28.16
C11H17N105	243.26	243.1107	54.31	7.04	5.76	32.89
C11H17N106	259.26	259.1056	50.96	6.61	5.40	37.03
C11H17N107	275.26	275.1005	48.00	6.23	5.09	40.69
C11H17N108	291.26	291.0954	45.36	5.88	4.81	43.95
C11H17N109	307.26	307.0903	43.00	5.58	4.56	46.86
C11H17N301	207.28	207.1372	63.74	8.27	20.27	7.72
C11H17N302	223.28	223.1321	59.17	7.67	18.82	14.33
C11H17N303	239.28	239.1270	55.22	7.16	17.56	20.06
C11H17N304	255.28	255.1219	51.76	6.71	16.46	25.07
C11H17N305	271.28	271.1168	48.70	6.32	15.49	29.49
C11H17N306	287.27	287.1117	45.99	5.96	14.63	33.42
C11H17N307	303.27	303.1066	43.57	5.65	13.86	36.93
C11H17N308	319.27	319.1016	41.38	5.37	13.16	40.09
C11H17N309	335.27	335.0965	39.41	5.11	12.53	42.95
C11H17N501	235.29	235.1433	56.15	7.28	29.76	6.80
C11H17N502	251.29	251.1382	52.58	6.82	27.87	12.73
C11H17N503	267.29	267.1331	49.43	6.41	26.20	17.96
C11H17N504	283.29	283.1280	46.64	6.05	24.72	22.59
C11H17N505	299.29	299.1230	44.15	5.73	23.40	26.73
C11H17N506	315.29	315.1179	41.91	5.43	22.21	30.45
C11H17N507	331.29	331.1128	39.88	5.17	21.14	33.81
C11H17N508	347.29	347.1077	38.04	4.93	20.17	36.86
C11H17N509	363.29	363.1026	36.37	4.72	19.28	39.64
C11H18N201	194.28	194.1419	68.01	9.34	14.42	8.24
C11H18N202	210.28	210.1368	62.83	8.63	13.32	15.22
C11H18N203	226.28	226.1317	58.39	8.02	12.38	21.21
C11H18N204	242.28	242.1266	54.53	7.49	11.56	26.42

054

C11H18N205	258.28	258.1216	51.16	7.02	10.85	30.97
C11H18N206	274.28	274.1165	48.17	6.62	10.21	35.00
C11H18N207	290.28	290.1114	45.52	6.25	9.65	38.58
C11H18N208	306.27	306.1063	43.14	5.92	9.15	41.79
C11H18N209	322.27	322.1012	41.00	5.63	8.69	44.68
C11H18N401	222.29	222.1481	59.44	8.16	25.20	7.20
C11H18N402	238.29	238.1430	55.45	7.61	23.51	13.43
C11H18N403	254.29	254.1379	51.96	7.13	22.03	18.88
C11H18N404	270.29	270.1328	48.88	6.71	20.73	23.68
C11H18N405	286.29	286.1277	46.15	6.34	19.57	27.94
C11H18N406	302.29	302.1226	43.71	6.00	18.53	31.76
C11H18N407	318.29	318.1175	41.51	5.70	17.60	35.19
C11H18N408	334.29	334.1125	39.52	5.43	16.76	38.29
C11H18N409	350.29	350.1074	37.72	5.18	15.99	41.11
C11H18N601	250.31	250.1542	52.78	7.25	33.58	6.39
C11H18N602	266.31	266.1491	49.61	6.81	31.56	12.02
C11H18N603	282.30	282.1440	46.80	6.43	29.77	17.00
C11H18N604	298.30	298.1389	44.29	6.08	28.17	21.45
C11H18N605	314.30	314.1339	42.04	5.77	26.74	25.45
C11H18N606	330.30	330.1288	40.00	5.49	25.44	29.06
C11H18N607	346.30	346.1237	38.15	5.24	24.27	32.34
C11H18N608	362.30	362.1186	36.47	5.01	23.20	35.33
C11H18N609	378.30	378.1135	34.93	4.80	22.22	38.06
C11H19N101	181.28	181.1467	72.88	10.56	7.73	8.83
C11H19N102	197.28	197.1416	66.97	9.71	7.10	16.22
C11H19N103	213.28	213.1365	61.95	8.98	6.57	22.50
C11H19N104	229.28	229.1314	57.63	8.35	6.11	27.91
C11H19N105	245.28	245.1263	53.87	7.81	5.71	32.61
C11H19N106	261.28	261.1212	50.57	7.33	5.36	36.74
C11H19N107	277.28	277.1161	47.65	6.91	5.05	40.39
C11H19N108	293.28	293.1111	45.05	6.53	4.78	43.64
C11H19N109	309.28	309.1060	42.72	6.19	4.53	46.56
C11H19N301	209.29	209.1528	63.13	9.15	20.08	7.64
C11H19N302	225.29	225.1477	58.64	8.50	18.65	14.20
C11H19N303	241.29	241.1426	54.76	7.94	17.41	19.89
C11H19N304	257.29	257.1375	51.35	7.44	16.33	24.87
C11H19N305	273.29	273.1325	48.35	7.01	15.38	29.27
C11H19N306	289.29	289.1274	45.67	6.62	14.53	33.18
C11H19N307	305.29	305.1223	43.28	6.27	13.76	36.69
C11H19N308	321.29	321.1172	41.12	5.96	13.08	39.84
C11H19N309	337.29	337.1121	39.17	5.68	12.46	42.69
C11H19N501	237.31	237.1590	55.68	8.07	29.51	6.74
C11H19N502	253.31	253.1539	52.16	7.56	27.65	12.63
C11H19N503	269.31	269.1488	49.06	7.11	26.01	17.82
C11H19N504	285.31	285.1437	46.31	6.71	24.55	22.43
C11H19N505	301.30	301.1386	43.85	6.36	23.24	26.55
C11H19N506	317.30	317.1335	41.64	6.04	22.07	30.25
C11H19N507	333.30	333.1284	39.64	5.75	21.01	33.60
C11H19N508	349.30	349.1233	37.82	5.48	20.05	36.64
C11H19N509	365.30	365.1183	36.17	5.24	19.17	39.42
C11H20N201	196.29	196.1576	67.31	10.27	14.27	8.15
C11H20N202	212.29	212.1525	62.24	9.50	13.20	15.07
C11H20N203	228.29	228.1474	57.87	8.83	12.27	21.02
C11H20N204	244.29	244.1423	54.08	8.25	11.47	26.20
C11H20N205	260.29	260.1372	50.76	7.74	10.76	30.73
C11H20N206	276.29	276.1321	47.82	7.30	10.14	34.74
C11H20N207	292.29	292.1270	45.20	6.90	9.58	38.32
C11H20N208	308.29	308.1220	42.86	6.54	9.09	41.52
C11H20N209	324.29	324.1169	40.74	6.22	8.64	44.40
C11H20N401	224.31	224.1637	58.90	8.99	24.98	7.13
C11H20N402	240.31	240.1586	54.98	8.39	23.31	13.32
C11H20N403	256.31	256.1535	51.55	7.87	21.86	18.73
C11H20N404	272.31	272.1484	48.52	7.40	20.57	23.50
C11H20N405	288.31	288.1434	45.83	6.99	19.43	27.75
C11H20N406	304.31	304.1383	43.42	6.62	18.41	31.55
C11H20N407	320.30	320.1332	41.25	6.29	17.49	34.97
C11H20N408	336.30	336.1281	39.29	5.99	16.66	38.06
C11H20N409	352.30	352.1230	37.50	5.72	15.90	40.87
C11H20N601	252.32	252.1698	52.36	7.99	33.31	6.34
C11H20N602	268.32	268.1648	49.24	7.51	31.32	11.93
C11H20N603	284.32	284.1597	46.47	7.09	29.56	16.88
C11H20N604	300.32	300.1546	43.99	6.71	27.98	21.31
C11H20N605	316.32	316.1495	41.77	6.37	26.57	25.29
C11H20N606	332.32	332.1444	39.76	6.07	25.29	28.89
C11H20N607	348.32	348.1393	37.93	5.79	24.13	32.15
C11H20N608	364.32	364.1342	36.27	5.53	23.07	35.13
C11H20N609	380.32	380.1292	34.74	5.30	22.10	37.86
C11H21N101	183.30	183.1623	72.08	11.55	7.64	8.73
C11H21N102	199.30	199.1572	66.29	10.62	7.03	16.06
C11H21N103	215.29	215.1521	61.37	9.83	6.51	22.29
C11H21N104	231.29	231.1470	57.12	9.15	6.06	27.67
C11H21N105	247.29	247.1420	53.43	8.56	5.66	32.35
C11H21N106	263.29	263.1369	50.18	8.04	5.32	36.46
C11H21N107	279.29	279.1318	47.31	7.58	5.02	40.10
C11H21N108	295.29	295.1267	44.74	7.17	4.74	43.35
C11H21N109	311.29	311.1216	42.44	6.80	4.50	46.26
C11H21N301	211.31	211.1685	62.53	10.02	19.89	7.57
C11H21N302	227.31	227.1634	58.12	9.31	18.49	14.08
C11H21N303	243.31	243.1583	54.30	8.70	17.27	19.73
C11H21N304	259.31	259.1532	50.95	8.16	16.20	24.68
C11H21N305	275.31	275.1481	47.99	7.69	15.26	29.06
C11H21N306	291.31	291.1430	45.36	7.27	14.42	32.95
C11H21N307	307.31	307.1379	42.99	6.89	13.67	36.44
C11H21N308	323.31	323.1329	40.87	6.55	13.00	39.59
C11H21N309	339.30	339.1278	38.94	6.24	12.38	42.44
C11H21N501	239.32	239.1746	55.21	8.84	29.26	6.69
C11H21N502	255.32	255.1695	51.75	8.29	27.43	12.53
C11H21N503	271.32	271.1644	48.70	7.80	25.81	17.69
C11H21N504	287.32	287.1593	45.98	7.37	24.37	22.27
C11H21N505	303.32	303.1543	43.56	6.98	23.09	26.37
C11H21N506	319.32	319.1492	41.38	6.63	21.93	30.06
C11H21N507	335.32	335.1441	39.40	6.31	20.89	33.40
C11H21N508	351.32	351.1390	37.61	6.03	19.93	36.43
C11H21N509	367.32	367.1339	35.97	5.76	19.07	39.20
C11H22N201	198.31	198.1732	66.62	11.18	14.13	8.07
C11H22N202	214.31	214.1681	61.65	10.35	13.07	14.93
C11H22N203	230.31	230.1630	57.37	9.63	12.16	20.84
C11H22N204	246.31	246.1579	53.64	9.00	11.37	25.98
C11H22N205	262.31	262.1529	50.37	8.45	10.68	30.50
C11H22N206	278.31	278.1478	47.47	7.97	10.07	34.49
C11H22N207	294.31	294.1427	44.89	7.53	9.52	38.05
C11H22N208	310.31	310.1376	42.58	7.15	9.03	41.25
C11H22N209	326.31	326.1325	40.49	6.80	8.59	44.13
C11H22N401	226.32	226.1794	58.38	9.80	24.76	7.07
C11H22N402	242.32	242.1743	54.52	9.15	23.12	13.20
C11H22N403	258.32	258.1692	51.15	8.58	21.69	18.58
C11H22N404	274.32	274.1641	48.16	8.08	20.42	23.33
C11H22N405	290.32	290.1590	45.51	7.64	19.30	27.55
C11H22N406	306.32	306.1539	43.13	7.24	18.29	31.34
C11H22N407	322.32	322.1488	40.99	6.88	17.38	34.75

055

C11H22N408	338.32	338.1437	39.05	6.55	16.56	37.83
C11H22N409	354.32	354.1387	37.29	6.26	15.81	40.64
C11H22N601	254.34	254.1855	51.95	8.72	33.04	6.29
C11H22N602	270.34	270.1804	48.87	8.20	31.09	11.84
C11H22N603	286.34	286.1753	46.14	7.74	29.35	16.76
C11H22N604	302.34	302.1702	43.70	7.33	27.80	21.17
C11H22N605	318.34	318.1652	41.50	6.97	26.40	25.13
C11H22N606	334.33	334.1601	39.52	6.63	25.14	28.71
C11H22N607	350.33	350.1550	37.71	6.33	23.99	31.97
C11H22N608	366.33	366.1499	36.07	6.05	22.94	34.94
C11H22N609	382.33	382.1448	34.56	5.80	21.98	37.66
C11H23N101	185.31	185.1780	71.30	12.51	7.56	8.63
C11H23N102	201.31	201.1729	65.63	11.52	6.96	15.90
C11H23N103	217.31	217.1678	60.80	10.67	6.45	22.09
C11H23N104	233.31	233.1627	56.63	9.94	6.00	27.43
C11H23N105	249.31	249.1576	53.00	9.30	5.62	32.09
C11H23N106	265.31	265.1525	49.80	8.74	5.28	36.18
C11H23N107	281.31	281.1474	46.97	8.24	4.98	39.81
C11H23N108	297.31	297.1424	44.44	7.80	4.71	43.05
C11H23N109	313.31	313.1373	42.17	7.40	4.47	45.96
C11H23N301	213.33	213.1841	61.93	10.87	19.70	7.50
C11H23N302	229.32	229.1790	57.61	10.11	18.32	13.95
C11H23N303	245.32	245.1739	53.86	9.45	17.13	19.57
C11H23N304	261.32	261.1688	50.56	8.87	16.08	24.49
C11H23N305	277.32	277.1638	47.64	8.36	15.15	28.85
C11H23N306	293.32	293.1587	45.04	7.90	14.33	32.73
C11H23N307	309.32	309.1536	42.71	7.49	13.58	36.21
C11H23N308	325.32	325.1485	40.61	7.13	12.92	39.34
C11H23N309	341.32	341.1434	38.71	6.79	12.31	42.19
C11H23N501	241.34	241.1902	54.75	9.61	29.02	6.63
C11H23N502	257.34	257.1852	51.34	9.01	27.21	12.43
C11H23N503	273.34	273.1801	48.34	8.48	25.62	17.56
C11H23N504	289.34	289.1750	45.66	8.01	24.20	22.12
C11H23N505	305.34	305.1699	43.27	7.59	22.94	26.20
C11H23N506	321.34	321.1648	41.12	7.21	21.79	29.87
C11H23N507	337.34	337.1597	39.17	6.87	20.76	33.20
C11H23N508	353.33	353.1546	37.39	6.56	19.82	36.22
C11H23N509	369.33	369.1496	35.77	6.28	18.96	38.99
C11H24N201	200.33	200.1889	65.95	12.08	13.98	7.99
C11H24N202	216.33	216.1838	61.08	11.18	12.95	14.79
C11H24N203	232.33	232.1787	56.87	10.41	12.06	20.66
C11H24N204	248.32	248.1736	53.21	9.74	11.28	25.77
C11H24N205	264.32	264.1685	49.99	9.15	10.60	30.26
C11H24N206	280.32	280.1634	47.13	8.63	9.99	34.24
C11H24N207	296.32	296.1583	44.59	8.16	9.45	37.80
C11H24N208	312.32	312.1533	42.30	7.75	8.97	40.98
C11H24N209	328.32	328.1482	40.24	7.37	8.53	43.86
C11H24N401	228.34	228.1950	57.86	10.59	24.54	7.01
C11H24N402	244.34	244.1899	54.07	9.90	22.93	13.10
C11H24N403	260.34	260.1848	50.75	9.29	21.52	18.44
C11H24N404	276.34	276.1797	47.81	8.75	20.27	23.16
C11H24N405	292.34	292.1747	45.20	8.28	19.17	27.36
C11H24N406	308.34	308.1696	42.85	7.85	18.17	31.13
C11H24N407	324.34	324.1645	40.74	7.46	17.27	34.53
C11H24N408	340.34	340.1594	38.82	7.11	16.46	37.61
C11H24N409	356.34	356.1543	37.08	6.79	15.72	40.41
C11H24N601	256.35	256.2011	51.54	9.44	32.78	6.24
C11H24N602	272.35	272.1961	48.51	8.88	30.86	11.75
C11H24N603	288.35	288.1910	45.82	8.39	29.14	16.65
C11H24N604	304.35	304.1859	43.41	7.95	27.61	21.03
C11H24N605	320.35	320.1808	41.24	7.55	26.23	24.97
C11H24N606	336.35	336.1757	39.28	7.19	24.99	28.54
C11H24N607	352.35	352.1706	37.50	6.87	23.85	31.79
C11H24N608	368.35	368.1655	35.87	6.57	22.82	34.75
C11H24N609	384.35	384.1605	34.38	6.29	21.87	37.46
C11H25N101	187.33	187.1936	70.53	13.45	7.48	8.54
C11H25N102	203.33	203.1885	64.98	12.39	6.89	15.74
C11H25N103	219.33	219.1834	60.24	11.49	6.39	21.88
C11H25N104	235.33	235.1783	56.14	10.71	5.95	27.20
C11H25N105	251.33	251.1733	52.57	10.03	5.57	31.83
C11H25N106	267.33	267.1682	49.42	9.43	5.24	35.91
C11H25N107	283.32	283.1631	46.63	8.89	4.94	39.53
C11H25N108	299.32	299.1580	44.14	8.42	4.68	42.76
C11H25N109	315.32	315.1529	41.90	7.99	4.44	45.67
C11H25N301	215.34	215.1998	61.35	11.70	19.51	7.43
C11H25N302	231.34	231.1947	57.11	10.89	18.16	13.83
C11H25N303	247.34	247.1896	53.42	10.19	16.99	19.41
C11H25N304	263.34	263.1845	50.17	9.57	15.96	24.30
C11H25N305	279.34	279.1794	47.30	9.02	15.04	28.64
C11H25N306	295.34	295.1743	44.74	8.53	14.23	32.50
C11H25N307	311.34	311.1692	42.44	8.09	13.50	35.97
C11H25N308	327.34	327.1641	40.36	7.70	12.84	39.10
C11H25N309	343.34	343.1591	38.48	7.34	12.24	41.94
C11H25N501	243.35	243.2059	54.29	10.35	28.78	6.57
C11H25N502	259.35	259.2008	50.94	9.72	27.00	12.34
C11H25N503	275.35	275.1957	47.98	9.15	25.43	17.43
C11H25N504	291.35	291.1906	45.35	8.65	24.04	21.97
C11H25N505	307.35	307.1856	42.99	8.20	22.79	26.03
C11H25N506	323.35	323.1805	40.86	7.79	21.66	29.69
C11H25N507	339.35	339.1754	38.93	7.43	20.64	33.00
C11H25N508	355.35	355.1703	37.18	7.09	19.71	36.02
C11H25N509	371.35	371.1652	35.58	6.79	18.86	38.78
C11H26N201	202.34	202.2045	65.30	12.95	13.84	7.91
C11H26N202	218.34	218.1994	60.51	12.00	12.83	14.66
C11H26N203	234.34	234.1943	56.38	11.18	11.95	20.48
C11H26N204	250.34	250.1892	52.78	10.47	11.19	25.56
C11H26N205	266.34	266.1842	49.61	9.84	10.52	30.04
C11H26N206	282.34	282.1791	46.80	9.28	9.92	34.00
C11H26N207	298.34	298.1740	44.29	8.78	9.39	37.54
C11H26N208	314.34	314.1689	42.03	8.34	8.91	40.72
C11H26N209	330.34	330.1638	40.00	7.93	8.48	43.59
C11H26N401	230.36	230.2106	57.36	11.38	24.32	6.95
C11H26N402	246.36	246.2056	53.63	10.64	22.74	12.99
C11H26N403	262.35	262.2005	50.36	9.99	21.36	18.30
C11H26N404	278.35	278.1954	47.47	9.42	20.13	22.99
C11H26N405	294.35	294.1903	44.89	8.90	19.03	27.18
C11H26N406	310.35	310.1852	42.57	8.44	18.05	30.93
C11H26N407	326.35	326.1801	40.48	8.03	17.17	34.32
C11H26N408	342.35	342.1750	38.59	7.66	16.37	37.39
C11H26N409	358.35	358.1700	36.87	7.31	15.63	40.18
C11H26N601	258.37	258.2168	51.14	10.14	32.53	6.19
C11H26N602	274.37	274.2117	48.16	9.55	30.63	11.66
C11H26N603	290.37	290.2066	45.50	9.03	28.94	16.53
C11H26N604	306.37	306.2015	43.13	8.55	27.43	20.89
C11H26N605	322.37	322.1965	40.99	8.13	26.07	24.82
C11H26N606	338.37	338.1914	39.05	7.75	24.84	28.37
C11H26N607	354.37	354.1863	37.28	7.40	23.72	31.60
C11H26N608	370.37	370.1812	35.67	7.08	22.69	34.56
C11H26N609	386.36	386.1761	34.20	6.78	21.75	37.27
C11H27N301	217.36	217.2154	60.79	12.52	19.33	7.36

056

C11H27N302	233.36	233.2103	56.62	11.66	18.01	13.71
C11H27N303	249.36	249.2052	52.99	10.91	16.85	19.25
C11H27N304	265.36	265.2001	49.79	10.26	15.84	24.12
C11H27N305	281.35	281.1951	46.96	9.67	14.93	28.43
C11H27N306	297.35	297.1900	44.43	9.15	14.13	32.28
C11H27N307	313.35	313.1849	42.16	8.69	13.41	35.74
C11H27N308	329.35	329.1798	40.12	8.26	12.76	38.86
C11H27N309	345.35	345.1747	38.26	7.88	12.17	41.69
C11H27N501	245.37	245.2215	53.85	11.09	28.54	6.52
C11H27N502	261.37	261.2165	50.55	10.41	26.79	12.24
C11H27N503	277.37	277.2114	47.63	9.81	25.25	17.30
C11H27N504	293.37	293.2063	45.04	9.28	23.87	21.81
C11H27N505	309.37	309.2012	42.71	8.80	22.64	25.86
C11H27N506	325.37	325.1961	40.61	8.36	21.52	29.50
C11H27N507	341.37	341.1910	38.70	7.97	20.52	32.81
C11H27N508	357.37	357.1859	36.97	7.62	19.60	35.82
C11H27N509	373.37	373.1809	35.39	7.29	18.76	38.57
C11H28N401	232.37	232.2263	56.86	12.15	24.11	6.89
C11H28N402	248.37	248.2212	53.20	11.36	22.56	12.88
C11H28N403	264.37	264.2161	49.98	10.68	21.19	18.16
C11H28N404	280.37	280.2110	47.12	10.07	19.98	22.83
C11H28N405	296.37	296.2060	44.58	9.52	18.90	26.99
C11H28N406	312.37	312.2009	42.30	9.04	17.94	30.73
C11H28N407	328.37	328.1958	40.24	8.59	17.06	34.11
C11H28N408	344.37	344.1907	38.37	8.20	16.27	37.17
C11H28N409	360.37	360.1856	36.66	7.83	15.55	39.96
C11H28N601	260.39	260.2324	50.74	10.84	32.28	6.14
C11H28N602	276.38	276.2274	47.80	10.21	30.41	11.58
C11H28N603	292.38	292.2223	45.19	9.65	28.74	16.42
C11H28N604	308.38	308.2172	42.84	9.15	27.25	20.75
C11H28N605	324.38	324.2121	40.73	8.70	25.91	24.66
C11H28N606	340.38	340.2070	38.82	8.29	24.69	28.20
C11H28N607	356.38	356.2019	37.07	7.92	23.58	31.43
C11H28N608	372.38	372.1968	35.48	7.58	22.57	34.37
C11H28N609	388.38	388.1918	34.02	7.27	21.64	37.08
C11H29N501	247.39	247.2372	53.41	11.82	28.31	6.47
C11H29N502	263.39	263.2321	50.16	11.10	26.59	12.15
C11H29N503	279.39	279.2270	47.29	10.46	25.07	17.18
C11H29N504	295.38	295.2219	44.73	9.90	23.71	21.67
C11H29N505	311.38	311.2169	42.43	9.39	22.49	25.69
C11H29N506	327.38	327.2118	40.36	8.93	21.39	29.32
C11H29N507	343.38	343.2067	38.48	8.51	20.40	32.62
C11H29N508	359.38	359.2016	36.76	8.13	19.49	35.62
C11H29N509	375.38	375.1965	35.20	7.79	18.66	38.36
C11H30N601	262.40	262.2481	50.35	11.52	32.03	6.10
C11H30N602	278.40	278.2430	47.46	10.86	30.19	11.49
C11H30N603	294.40	294.2379	44.88	10.27	28.55	16.30
C11H30N604	310.40	310.2328	42.57	9.74	27.07	20.62
C11H30N605	326.40	326.2278	40.48	9.26	25.75	24.51
C11H30N606	342.40	342.2227	38.59	8.83	24.54	28.04
C11H30N607	358.40	358.2176	36.86	8.44	23.45	31.25
C11H30N608	374.40	374.2125	35.29	8.08	22.45	34.19
C11H30N609	390.40	390.2074	33.84	7.75	21.53	36.88
C12H 6N201	194.19	194.0480	74.22	3.11	14.43	8.24
C12H 6N202	210.19	210.0429	68.57	2.88	13.33	15.22
C12H 6N203	226.19	226.0378	63.72	2.67	12.38	21.22
C12H 6N204	242.19	242.0328	59.51	2.50	11.57	26.42
C12H 6N205	258.19	258.0277	55.82	2.34	10.85	30.98
C12H 6N206	274.19	274.0226	52.57	2.21	10.22	35.01
C12H 6N207	290.19	290.0175	49.67	2.08	9.65	38.59
C12H 6N208	306.19	306.0124	47.07	1.98	9.15	41.80
C12H 6N209	322.19	322.0073	44.74	1.88	8.69	44.69
C12H 6N401	222.21	222.0542	64.86	2.72	25.21	7.20
C12H 6N402	238.21	238.0491	60.51	2.54	23.52	13.43
C12H 6N403	254.21	254.0440	56.70	2.38	22.04	18.88
C12H 6N404	270.21	270.0389	53.34	2.24	20.73	23.68
C12H 6N405	286.21	286.0338	50.36	2.11	19.58	27.95
C12H 6N406	302.20	302.0287	47.69	2.00	18.54	31.77
C12H 6N407	318.20	318.0236	45.30	1.90	17.61	35.20
C12H 6N408	334.20	334.0186	43.13	1.81	16.76	38.30
C12H 6N409	350.20	350.0135	41.16	1.73	16.00	41.12
C12H 6N601	250.22	250.0603	57.60	2.42	33.59	6.39
C12H 6N602	266.22	266.0552	54.14	2.27	31.57	12.02
C12H 6N603	282.22	282.0501	51.07	2.14	29.78	17.01
C12H 6N604	298.22	298.0450	48.33	2.03	28.18	21.46
C12H 6N605	314.22	314.0400	45.87	1.92	26.75	25.46
C12H 6N606	330.22	330.0349	43.65	1.83	25.45	29.07
C12H 6N607	346.22	346.0298	41.63	1.75	24.27	32.35
C12H 6N608	362.22	362.0247	39.79	1.67	23.20	35.34
C12H 6N609	378.22	378.0196	38.11	1.60	22.22	38.07
C12H 7N101	181.20	181.0528	79.55	3.89	7.73	8.83
C12H 7N102	197.20	197.0477	73.09	3.58	7.10	16.23
C12H 7N103	213.19	213.0426	67.61	3.31	6.57	22.51
C12H 7N104	229.19	229.0375	62.89	3.08	6.11	27.92
C12H 7N105	245.19	245.0324	58.78	2.88	5.71	32.63
C12H 7N106	261.19	261.0273	55.18	2.70	5.36	36.75
C12H 7N107	277.19	277.0222	52.00	2.55	5.05	40.40
C12H 7N108	293.19	293.0172	49.16	2.41	4.78	43.66
C12H 7N109	309.19	309.0121	46.62	2.28	4.53	46.57
C12H 7N301	209.21	209.0589	68.89	3.37	20.09	7.65
C12H 7N302	225.21	225.0538	64.00	3.13	18.66	14.21
C12H 7N303	241.21	241.0487	59.76	2.93	17.42	19.90
C12H 7N304	257.21	257.0437	56.04	2.74	16.34	24.88
C12H 7N305	273.21	273.0386	52.76	2.58	15.38	29.28
C12H 7N306	289.21	289.0335	49.84	2.44	14.53	33.19
C12H 7N307	305.21	305.0284	47.23	2.31	13.77	36.70
C12H 7N308	321.20	321.0233	44.87	2.20	13.08	39.85
C12H 7N309	337.20	337.0182	42.74	2.09	12.46	42.70
C12H 7N501	237.22	237.0651	60.76	2.97	29.52	6.74
C12H 7N502	253.22	253.0600	56.92	2.79	27.66	12.64
C12H 7N503	269.22	269.0549	53.54	2.62	26.01	17.83
C12H 7N504	285.22	285.0498	50.53	2.47	24.55	22.44
C12H 7N505	301.22	301.0447	47.85	2.34	23.25	26.56
C12H 7N506	317.22	317.0396	45.44	2.22	22.08	30.26
C12H 7N507	333.22	333.0345	43.25	2.12	21.02	33.61
C12H 7N508	349.22	349.0295	41.27	2.02	20.05	36.65
C12H 7N509	365.22	365.0244	39.47	1.93	19.18	39.43
C12H 8N201	196.21	196.0637	73.46	4.11	14.28	8.15
C12H 8N202	212.21	212.0586	67.92	3.80	13.20	15.08
C12H 8N203	228.21	228.0535	63.16	3.53	12.28	21.03
C12H 8N204	244.21	244.0484	59.02	3.30	11.47	26.21
C12H 8N205	260.21	260.0433	55.39	3.10	10.77	30.74
C12H 8N206	276.21	276.0382	52.18	2.92	10.14	34.76
C12H 8N207	292.21	292.0331	49.33	2.76	9.59	38.33
C12H 8N208	308.21	308.0281	46.77	2.62	9.09	41.53
C12H 8N209	324.21	324.0230	44.46	2.49	8.64	44.41
C12H 8N401	224.22	224.0698	64.28	3.60	24.99	7.14
C12H 8N402	240.22	240.0647	60.00	3.36	23.32	13.32
C12H 8N403	256.22	256.0596	56.25	3.15	21.87	18.73
C12H 8N404	272.22	272.0545	52.95	2.96	20.58	23.51

057

C12H 8N405	288.22	288.0495	50.01	2.80	19.44	27.76
C12H 8N406	304.22	304.0444	47.38	2.65	18.42	31.55
C12H 8N407	320.22	320.0393	45.01	2.52	17.50	34.97
C12H 8N408	336.22	336.0342	42.87	2.40	16.66	38.07
C12H 8N409	352.22	352.0291	40.92	2.29	15.91	40.88
C12H 8N601	252.24	252.0760	57.14	3.20	33.32	6.34
C12H 8N602	268.24	268.0709	53.73	3.01	31.33	11.93
C12H 8N603	284.24	284.0658	50.71	2.84	29.57	16.89
C12H 8N604	300.24	300.0607	48.01	2.69	27.99	21.32
C12H 8N605	316.23	316.0556	45.58	2.55	26.58	25.30
C12H 8N606	332.23	332.0505	43.38	2.43	25.30	28.89
C12H 8N607	348.23	348.0454	41.39	2.32	24.13	32.16
C12H 8N608	364.23	364.0404	39.57	2.21	23.07	35.14
C12H 8N609	380.23	380.0353	37.91	2.12	22.10	37.87
C12H 9N101	183.21	183.0684	78.67	4.95	7.65	8.73
C12H 9N102	199.21	199.0633	72.35	4.55	7.03	16.06
C12H 9N103	215.21	215.0582	66.97	4.22	6.51	22.30
C12H 9N104	231.21	231.0532	62.34	3.92	6.06	27.68
C12H 9N105	247.21	247.0481	58.30	3.67	5.67	32.36
C12H 9N106	263.21	263.0430	54.76	3.45	5.32	36.47
C12H 9N107	279.21	279.0379	51.62	3.25	5.02	40.11
C12H 9N108	295.21	295.0328	48.82	3.07	4.74	43.36
C12H 9N109	311.21	311.0277	46.31	2.92	4.50	46.27
C12H 9N301	211.23	211.0746	68.24	4.29	19.89	7.57
C12H 9N302	227.22	227.0695	63.43	3.99	18.49	14.08
C12H 9N303	243.22	243.0644	59.26	3.73	17.28	19.73
C12H 9N304	259.22	259.0593	55.60	3.50	16.21	24.69
C12H 9N305	275.22	275.0542	52.37	3.30	15.27	29.07
C12H 9N306	291.22	291.0491	49.49	3.12	14.43	32.96
C12H 9N307	307.22	307.0440	46.92	2.95	13.68	36.45
C12H 9N308	323.22	323.0390	44.59	2.81	13.00	39.60
C12H 9N309	339.22	339.0339	42.49	2.67	12.39	42.45
C12H 9N501	239.24	239.0807	60.25	3.79	29.27	6.69
C12H 9N502	255.24	255.0756	56.47	3.55	27.44	12.54
C12H 9N503	271.24	271.0705	53.14	3.34	25.82	17.70
C12H 9N504	287.24	287.0654	50.18	3.16	24.38	22.28
C12H 9N505	303.24	303.0604	47.53	2.99	23.10	26.38
C12H 9N506	319.24	319.0553	45.15	2.84	21.94	30.07
C12H 9N507	335.23	335.0502	42.99	2.71	20.89	33.41
C12H 9N508	351.23	351.0451	41.04	2.58	19.94	36.44
C12H 9N509	367.23	367.0400	39.25	2.47	19.07	39.21
C12H10N201	198.23	198.0793	72.71	5.08	14.13	8.07
C12H10N202	214.23	214.0742	67.28	4.71	13.08	14.94
C12H10N203	230.23	230.0691	62.61	4.38	12.17	20.85
C12H10N204	246.22	246.0641	58.54	4.09	11.38	25.99
C12H10N205	262.22	262.0590	54.97	3.84	10.68	30.51
C12H10N206	278.22	278.0539	51.81	3.62	10.07	34.50
C12H10N207	294.22	294.0488	48.99	3.43	9.52	38.06
C12H10N208	310.22	310.0437	46.46	3.25	9.03	41.26
C12H10N209	326.22	326.0386	44.18	3.09	8.59	44.14
C12H10N401	226.24	226.0855	63.71	4.46	24.76	7.07
C12H10N402	242.24	242.0804	59.50	4.16	23.13	13.21
C12H10N403	258.24	258.0753	55.81	3.90	21.70	18.59
C12H10N404	274.24	274.0702	52.56	3.68	20.43	23.34
C12H10N405	290.24	290.0651	49.66	3.47	19.30	27.56
C12H10N406	306.24	306.0600	47.07	3.29	18.30	31.35
C12H10N407	322.24	322.0549	44.73	3.13	17.39	34.76
C12H10N408	338.24	338.0499	42.61	2.98	16.56	37.84
C12H10N409	354.23	354.0448	40.69	2.85	15.82	40.65
C12H10N601	254.25	254.0916	56.69	3.96	33.05	6.29
C12H10N602	270.25	270.0865	53.33	3.73	31.10	11.84
C12H10N603	286.25	286.0814	50.35	3.52	29.36	16.77
C12H10N604	302.25	302.0763	47.69	3.33	27.80	21.17
C12H10N605	318.25	318.0713	45.29	3.17	26.41	25.14
C12H10N606	334.25	334.0662	43.12	3.02	25.14	28.72
C12H10N607	350.25	350.0611	41.15	2.88	23.99	31.98
C12H10N608	366.25	366.0560	39.35	2.75	22.95	34.95
C12H10N609	382.25	382.0509	37.71	2.64	21.99	37.67
C12H11N101	185.23	185.0841	77.81	5.99	7.56	8.64
C12H11N102	201.23	201.0790	71.63	5.51	6.96	15.90
C12H11N103	217.23	217.0739	66.35	5.10	6.45	22.10
C12H11N104	233.23	233.0688	61.80	4.75	6.01	27.44
C12H11N105	249.23	249.0637	57.83	4.45	5.62	32.10
C12H11N106	265.22	265.0586	54.34	4.18	5.28	36.19
C12H11N107	281.22	281.0535	51.25	3.94	4.98	39.82
C12H11N108	297.22	297.0485	48.49	3.73	4.71	43.06
C12H11N109	313.22	313.0434	46.02	3.54	4.47	45.97
C12H11N301	213.24	213.0902	67.59	5.20	19.71	7.50
C12H11N302	229.24	229.0851	62.87	4.84	18.33	13.96
C12H11N303	245.24	245.0800	58.77	4.52	17.13	19.57
C12H11N304	261.24	261.0749	55.17	4.24	16.08	24.50
C12H11N305	277.24	277.0699	51.99	4.00	15.16	28.85
C12H11N306	293.24	293.0648	49.15	3.78	14.33	32.74
C12H11N307	309.24	309.0597	46.61	3.59	13.59	36.22
C12H11N308	325.24	325.0546	44.32	3.41	12.92	39.35
C12H11N309	341.24	341.0495	42.24	3.25	12.31	42.20
C12H11N501	241.25	241.0964	59.74	4.60	29.03	6.63
C12H11N502	257.25	257.0913	56.03	4.31	27.22	12.44
C12H11N503	273.25	273.0862	52.75	4.06	25.63	17.57
C12H11N504	289.25	289.0811	49.83	3.83	24.21	22.13
C12H11N505	305.25	305.0760	47.22	3.63	22.94	26.21
C12H11N506	321.25	321.0709	44.87	3.45	21.80	29.88
C12H11N507	337.25	337.0658	42.74	3.29	20.77	33.21
C12H11N508	353.25	353.0608	40.80	3.14	19.83	36.23
C12H11N509	369.25	369.0557	39.03	3.00	18.97	39.00
C12H12N201	200.24	200.0950	71.98	6.04	13.99	7.99
C12H12N202	216.24	216.0899	66.65	5.59	12.95	14.80
C12H12N203	232.24	232.0848	62.06	5.21	12.06	20.67
C12H12N204	248.24	248.0797	58.06	4.87	11.28	25.78
C12H12N205	264.24	264.0746	54.55	4.58	10.60	30.27
C12H12N206	280.24	280.0695	51.43	4.32	10.00	34.26
C12H12N207	296.24	296.0644	48.65	4.08	9.46	37.81
C12H12N208	312.24	312.0594	46.16	3.87	8.97	40.99
C12H12N209	328.24	328.0543	43.91	3.69	8.53	43.87
C12H12N401	228.26	228.1011	63.15	5.30	24.55	7.01
C12H12N402	244.26	244.0960	59.01	4.95	22.94	13.10
C12H12N403	260.25	260.0909	55.38	4.65	21.53	18.44
C12H12N404	276.25	276.0858	52.17	4.38	20.28	23.17
C12H12N405	292.25	292.0808	49.32	4.14	19.17	27.37
C12H12N406	308.25	308.0757	46.76	3.92	18.18	31.14
C12H12N407	324.25	324.0706	44.45	3.73	17.28	34.54
C12H12N408	340.25	340.0655	42.36	3.55	16.47	37.62
C12H12N409	356.25	356.0604	40.46	3.40	15.73	40.42
C12H12N601	256.27	256.1073	56.24	4.72	32.79	6.24
C12H12N602	272.27	272.1022	52.94	4.44	30.87	11.75
C12H12N603	288.27	288.0971	50.00	4.20	29.15	16.65
C12H12N604	304.27	304.0920	47.37	3.98	27.62	21.03
C12H12N605	320.27	320.0869	45.00	3.78	26.24	24.98
C12H12N606	336.27	336.0818	42.86	3.60	24.99	28.55
C12H12N607	352.27	352.0767	40.92	3.43	23.86	31.79

058

C12H12N608	368.26	368.0717	39.14	3.28	22.82	34.76
C12H12N609	384.26	384.0666	37.51	3.15	21.87	37.47
C12H13N101	187.24	187.0997	76.98	7.00	7.48	8.54
C12H13N102	203.24	203.0946	70.92	6.45	6.89	15.74
C12H13N103	219.24	219.0895	65.74	5.98	6.39	21.89
C12H13N104	235.24	235.0845	61.27	5.57	5.95	27.21
C12H13N105	251.24	251.0794	57.37	5.22	5.58	31.84
C12H13N106	267.24	267.0743	53.93	4.90	5.24	35.92
C12H13N107	283.24	283.0692	50.89	4.63	4.95	39.54
C12H13N108	299.24	299.0641	48.17	4.38	4.68	42.77
C12H13N109	315.24	315.0590	45.72	4.16	4.44	45.68
C12H13N301	215.26	215.1059	66.96	6.09	19.52	7.43
C12H13N302	231.26	231.1008	62.33	5.67	18.17	13.84
C12H13N303	247.26	247.0957	58.29	5.30	16.99	19.41
C12H13N304	263.26	263.0906	54.75	4.98	15.96	24.31
C12H13N305	279.25	279.0855	51.61	4.69	15.05	28.65
C12H13N306	295.25	295.0804	48.82	4.44	14.23	32.51
C12H13N307	311.25	311.0753	46.31	4.21	13.50	35.98
C12H13N308	327.25	327.0703	44.04	4.00	12.84	39.11
C12H13N309	343.25	343.0652	41.99	3.82	12.24	41.95
C12H13N501	243.27	243.1120	59.25	5.39	28.79	6.58
C12H13N502	259.27	259.1069	55.59	5.05	27.01	12.34
C12H13N503	275.27	275.1018	52.36	4.76	25.44	17.44
C12H13N504	291.27	291.0967	49.48	4.50	24.04	21.97
C12H13N505	307.27	307.0917	46.91	4.26	22.79	26.03
C12H13N506	323.27	323.0866	44.59	4.05	21.66	29.70
C12H13N507	339.27	339.0815	42.48	3.86	20.64	33.01
C12H13N508	355.27	355.0764	40.57	3.69	19.71	36.03
C12H13N509	371.27	371.0713	38.82	3.53	18.86	38.78
C12H14N201	202.26	202.1106	71.26	6.98	13.85	7.91
C12H14N202	218.26	218.1055	66.04	6.47	12.84	14.66
C12H14N203	234.26	234.1004	61.53	6.02	11.96	20.49
C12H14N204	250.26	250.0953	57.59	5.64	11.19	25.57
C12H14N205	266.26	266.0903	54.13	5.30	10.52	30.05
C12H14N206	282.26	282.0852	51.07	5.00	9.92	34.01
C12H14N207	298.25	298.0801	48.33	4.73	9.39	37.55
C12H14N208	314.25	314.0750	45.87	4.49	8.91	40.73
C12H14N209	330.25	330.0699	43.64	4.27	8.48	43.60
C12H14N401	230.27	230.1168	62.59	6.13	24.33	6.95
C12H14N402	246.27	246.1117	58.53	5.73	22.75	12.99
C12H14N403	262.27	262.1066	54.96	5.38	21.36	18.30
C12H14N404	278.27	278.1015	51.80	5.07	20.13	23.00
C12H14N405	294.27	294.0964	48.98	4.80	19.04	27.18
C12H14N406	310.27	310.0913	46.45	4.55	18.06	30.94
C12H14N407	326.27	326.0862	44.18	4.33	17.17	34.33
C12H14N408	342.27	342.0812	42.11	4.12	16.37	37.40
C12H14N409	358.27	358.0761	40.23	3.94	15.64	40.19
C12H14N601	258.28	258.1229	55.80	5.46	32.54	6.19
C12H14N602	274.28	274.1178	52.55	5.14	30.64	11.67
C12H14N603	290.28	290.1127	49.65	4.86	28.95	16.53
C12H14N604	306.28	306.1076	47.06	4.61	27.44	20.89
C12H14N605	322.28	322.1026	44.72	4.38	26.08	24.82
C12H14N606	338.28	338.0975	42.61	4.17	24.84	28.38
C12H14N607	354.28	354.0924	40.68	3.98	23.72	31.61
C12H14N608	370.28	370.0873	38.93	3.81	22.70	34.57
C12H14N609	386.28	386.0822	37.31	3.65	21.76	37.28
C12H15N101	189.26	189.1154	76.16	7.99	7.40	8.45
C12H15N102	205.26	205.1103	70.22	7.37	6.82	15.59
C12H15N103	221.26	221.1052	65.14	6.83	6.33	21.69
C12H15N104	237.26	237.1001	60.75	6.37	5.90	26.97

C12H15N105	253.26	253.0950	56.91	5.97	5.53	31.59
C12H15N106	269.26	269.0899	53.53	5.62	5.20	35.65
C12H15N107	285.26	285.0848	50.53	5.30	4.91	39.26
C12H15N108	301.26	301.0798	47.84	5.02	4.65	42.49
C12H15N109	317.25	317.0747	45.43	4.77	4.41	45.39
C12H15N301	217.27	217.1215	66.34	6.96	19.34	7.36
C12H15N302	233.27	233.1164	61.79	6.48	18.01	13.72
C12H15N303	249.27	249.1113	57.82	6.07	16.86	19.26
C12H15N304	265.27	265.1062	54.33	5.70	15.84	24.13
C12H15N305	281.27	281.1012	51.24	5.38	14.94	28.44
C12H15N306	297.27	297.0961	48.49	5.09	14.14	32.29
C12H15N307	313.27	313.0910	46.01	4.83	13.41	35.75
C12H15N308	329.27	329.0859	43.77	4.59	12.76	38.87
C12H15N309	345.27	345.0808	41.75	4.38	12.17	41.71
C12H15N501	245.29	245.1277	58.76	6.16	28.55	6.52
C12H15N502	261.29	261.1226	55.16	5.79	26.80	12.25
C12H15N503	277.29	277.1175	51.98	5.45	25.26	17.31
C12H15N504	293.28	293.1124	49.14	5.16	23.88	21.82
C12H15N505	309.28	309.1073	46.60	4.89	22.64	25.87
C12H15N506	325.28	325.1022	44.31	4.65	21.53	29.51
C12H15N507	341.28	341.0971	42.23	4.43	20.52	32.82
C12H15N508	357.28	357.0921	40.34	4.23	19.60	35.82
C12H15N509	373.28	373.0870	38.61	4.05	18.76	38.58
C12H16N201	204.27	204.1263	70.56	7.90	13.71	7.83
C12H16N202	220.27	220.1212	65.43	7.32	12.72	14.53
C12H16N203	236.27	236.1161	61.00	6.83	11.86	20.31
C12H16N204	252.27	252.1110	57.13	6.39	11.10	25.37
C12H16N205	268.27	268.1059	53.73	6.01	10.44	29.82
C12H16N206	284.27	284.1008	50.70	5.67	9.85	33.77
C12H16N207	300.27	300.0957	48.00	5.37	9.33	37.30
C12H16N208	316.27	316.0907	45.57	5.10	8.86	40.47
C12H16N209	332.27	332.0856	43.38	4.85	8.43	43.34
C12H16N401	232.29	232.1324	62.05	6.94	24.12	6.89
C12H16N402	248.29	248.1273	58.05	6.50	22.57	12.89
C12H16N403	264.29	264.1222	54.54	6.10	21.20	18.16
C12H16N404	280.29	280.1171	51.42	5.75	19.99	22.83
C12H16N405	296.29	296.1121	48.65	5.44	18.91	27.00
C12H16N406	312.28	312.1070	46.15	5.16	17.94	30.74
C12H16N407	328.28	328.1019	43.91	4.91	17.07	34.12
C12H16N408	344.28	344.0968	41.86	4.68	16.27	37.18
C12H16N409	360.28	360.0917	40.01	4.48	15.55	39.97
C12H16N601	260.30	260.1386	55.37	6.20	32.29	6.15
C12H16N602	276.30	276.1335	52.17	5.84	30.42	11.58
C12H16N603	292.30	292.1284	49.31	5.52	28.75	16.42
C12H16N604	308.30	308.1233	46.75	5.23	27.26	20.76
C12H16N605	324.30	324.1182	44.44	4.97	25.91	24.67
C12H16N606	340.30	340.1131	42.36	4.74	24.70	28.21
C12H16N607	356.30	356.1080	40.45	4.53	23.59	31.43
C12H16N608	372.30	372.1029	38.71	4.33	22.57	34.38
C12H16N609	388.30	388.0979	37.12	4.15	21.64	37.08
C12H17N101	191.28	191.1310	75.35	8.96	7.32	8.36
C12H17N102	207.27	207.1259	69.54	8.27	6.76	15.44
C12H17N103	223.27	223.1208	64.55	7.67	6.27	21.50
C12H17N104	239.27	239.1157	60.24	7.16	5.85	26.75
C12H17N105	255.27	255.1107	56.46	6.71	5.49	31.34
C12H17N106	271.27	271.1056	53.13	6.32	5.16	35.39
C12H17N107	287.27	287.1005	50.17	5.96	4.88	38.99
C12H17N108	303.27	303.0954	47.53	5.65	4.62	42.20
C12H17N109	319.27	319.0903	45.14	5.37	4.39	45.10
C12H17N301	219.29	219.1372	65.73	7.81	19.16	7.30

059

C12H17N302	235.29	235.1321	61.26	7.28	17.86	13.60
C12H17N303	251.29	251.1270	57.36	6.82	16.72	19.10
C12H17N304	267.29	267.1219	53.92	6.41	15.72	23.94
C12H17N305	283.29	283.1168	50.88	6.05	14.83	28.24
C12H17N306	299.29	299.1117	48.16	5.73	14.04	32.08
C12H17N307	315.29	315.1066	45.72	5.43	13.33	35.52
C12H17N308	331.28	331.1016	43.51	5.17	12.68	38.64
C12H17N309	347.28	347.0965	41.50	4.93	12.10	41.46
C12H17N501	247.30	247.1433	58.28	6.93	28.32	6.47
C12H17N502	263.30	263.1382	54.74	6.51	26.60	12.15
C12H17N503	279.30	279.1331	51.61	6.14	25.07	17.19
C12H17N504	295.30	295.1280	48.81	5.80	23.72	21.67
C12H17N505	311.30	311.1230	46.30	5.50	22.50	25.70
C12H17N506	327.30	327.1179	44.04	5.24	21.40	29.33
C12H17N507	343.30	343.1128	41.98	4.99	20.40	32.62
C12H17N508	359.30	359.1077	40.12	4.77	19.49	35.62
C12H17N509	375.30	375.1026	38.41	4.57	18.66	38.37
C12H18N201	206.29	206.1419	69.87	8.80	13.58	7.76
C12H18N202	222.29	222.1368	64.84	8.16	12.60	14.40
C12H18N203	238.29	238.1317	60.49	7.61	11.76	20.14
C12H18N204	254.29	254.1266	56.68	7.13	11.02	25.17
C12H18N205	270.29	270.1216	53.33	6.71	10.36	29.60
C12H18N206	286.29	286.1165	50.35	6.34	9.79	33.53
C12H18N207	302.29	302.1114	47.68	6.00	9.27	37.05
C12H18N208	318.29	318.1063	45.28	5.70	8.80	40.21
C12H18N209	334.29	334.1012	43.12	5.43	8.38	43.08
C12H18N401	234.30	234.1481	61.52	7.74	23.91	6.83
C12H18N402	250.30	250.1430	57.58	7.25	22.38	12.78
C12H18N403	266.30	266.1379	54.12	6.81	21.04	18.02
C12H18N404	282.30	282.1328	51.06	6.43	19.85	22.67
C12H18N405	298.30	298.1277	48.32	6.08	18.78	26.82
C12H18N406	314.30	314.1226	45.86	5.77	17.83	30.54
C12H18N407	330.30	330.1175	43.64	5.49	16.96	33.91
C12H18N408	346.30	346.1125	41.62	5.24	16.18	36.96
C12H18N409	362.30	362.1074	39.78	5.01	15.46	39.74
C12H18N601	262.32	262.1542	54.95	6.92	32.04	6.10
C12H18N602	278.32	278.1491	51.79	6.52	30.20	11.50
C12H18N603	294.32	294.1440	48.97	6.16	28.55	16.31
C12H18N604	310.32	310.1389	46.45	5.85	27.08	20.62
C12H18N605	326.31	326.1339	44.17	5.56	25.75	24.52
C12H18N606	342.31	342.1288	42.11	5.30	24.55	28.04
C12H18N607	358.31	358.1237	40.23	5.06	23.45	31.26
C12H18N608	374.31	374.1186	38.51	4.85	22.45	34.19
C12H18N609	390.31	390.1135	36.93	4.65	21.53	36.89
C12H19N101	193.29	193.1467	74.57	9.91	7.25	8.28
C12H19N102	209.29	209.1416	68.87	9.15	6.69	15.29
C12H19N103	225.29	225.1365	63.98	8.50	6.22	21.31
C12H19N104	241.29	241.1314	59.73	7.94	5.80	26.52
C12H19N105	257.29	257.1263	56.02	7.44	5.44	31.09
C12H19N106	273.29	273.1212	52.74	7.01	5.13	35.13
C12H19N107	289.29	289.1161	49.82	6.62	4.84	38.71
C12H19N108	305.29	305.1111	47.21	6.27	4.59	41.93
C12H19N109	321.29	321.1060	44.86	5.96	4.36	44.82
C12H19N301	221.30	221.1528	65.13	8.65	18.99	7.23
C12H19N302	237.30	237.1477	60.74	8.07	17.71	13.48
C12H19N303	253.30	253.1426	56.90	7.56	16.59	18.95
C12H19N304	269.30	269.1375	53.52	7.11	15.60	23.76
C12H19N305	285.30	285.1325	50.52	6.71	14.73	28.04
C12H19N306	301.30	301.1274	47.84	6.36	13.95	31.86
C12H19N307	317.30	317.1223	45.42	6.04	13.24	35.30

C12H19N308	333.30	333.1172	43.24	5.75	12.61	38.40
C12H19N309	349.30	349.1121	41.26	5.48	12.03	41.22
C12H19N501	249.32	249.1590	57.81	7.68	28.09	6.42
C12H19N502	265.32	265.1539	54.33	7.22	26.40	12.06
C12H19N503	281.32	281.1488	51.24	6.81	24.89	17.06
C12H19N504	297.32	297.1437	48.48	6.44	23.56	21.53
C12H19N505	313.32	313.1386	46.00	6.11	22.35	25.53
C12H19N506	329.32	329.1335	43.77	5.82	21.27	29.15
C12H19N507	345.31	345.1284	41.74	5.55	20.28	32.43
C12H19N508	361.31	361.1233	39.89	5.30	19.38	35.42
C12H19N509	377.31	377.1183	38.20	5.08	18.56	38.16
C12H20N201	208.31	208.1576	69.19	9.68	13.45	7.68
C12H20N202	224.31	224.1525	64.26	8.99	12.49	14.27
C12H20N203	240.30	240.1474	59.98	8.39	11.66	19.97
C12H20N204	256.30	256.1423	56.24	7.87	10.93	24.97
C12H20N205	272.30	272.1372	52.93	7.40	10.29	29.38
C12H20N206	288.30	288.1321	49.99	6.99	9.72	33.30
C12H20N207	304.30	304.1270	47.37	6.62	9.21	36.80
C12H20N208	320.30	320.1220	45.00	6.29	8.75	39.96
C12H20N209	336.30	336.1169	42.86	5.99	8.33	42.82
C12H20N401	236.32	236.1637	60.99	8.53	23.71	6.77
C12H20N402	252.32	252.1586	57.12	7.99	22.20	12.68
C12H20N403	268.32	268.1535	53.72	7.51	20.88	17.89
C12H20N404	284.32	284.1484	50.69	7.09	19.71	22.51
C12H20N405	300.32	300.1434	47.99	6.71	18.66	26.64
C12H20N406	316.32	316.1383	45.57	6.37	17.71	30.35
C12H20N407	332.32	332.1332	43.37	6.07	16.86	33.70
C12H20N408	348.32	348.1281	41.38	5.79	16.09	36.75
C12H20N409	364.31	364.1230	39.56	5.53	15.38	39.52
C12H20N601	264.33	264.1698	54.53	7.63	31.79	6.05
C12H20N602	280.33	280.1648	51.42	7.19	29.98	11.41
C12H20N603	296.33	296.1597	48.64	6.80	28.36	16.20
C12H20N604	312.33	312.1546	46.15	6.45	26.91	20.49
C12H20N605	328.33	328.1495	43.90	6.14	25.60	24.36
C12H20N606	344.33	344.1444	41.86	5.85	24.41	27.88
C12H20N607	360.33	360.1393	40.00	5.59	23.32	31.08
C12H20N608	376.33	376.1342	38.30	5.36	22.33	34.01
C12H20N609	392.33	392.1292	36.74	5.14	21.42	36.70
C12H21N101	195.31	195.1623	73.80	10.84	7.17	8.19
C12H21N102	211.31	211.1572	68.21	10.02	6.63	15.14
C12H21N103	227.31	227.1521	63.41	9.31	6.16	21.12
C12H21N104	243.31	243.1470	59.24	8.70	5.76	26.30
C12H21N105	259.30	259.1420	55.58	8.16	5.40	30.85
C12H21N106	275.30	275.1369	52.35	7.69	5.09	34.87
C12H21N107	291.30	291.1318	49.48	7.27	4.81	38.45
C12H21N108	307.30	307.1267	46.90	6.89	4.56	41.65
C12H21N109	323.30	323.1216	44.58	6.55	4.33	44.54
C12H21N301	223.32	223.1685	64.54	9.48	18.82	7.16
C12H21N302	239.32	239.1634	60.23	8.84	17.56	13.37
C12H21N303	255.32	255.1583	56.45	8.29	16.46	18.80
C12H21N304	271.32	271.1532	53.12	7.80	15.49	23.59
C12H21N305	287.32	287.1481	50.17	7.37	14.62	27.84
C12H21N306	303.32	303.1430	47.52	6.98	13.85	31.65
C12H21N307	319.32	319.1379	45.14	6.63	13.16	35.07
C12H21N308	335.32	335.1329	42.98	6.31	12.53	38.17
C12H21N309	351.32	351.1278	41.03	6.03	11.96	40.99
C12H21N501	251.33	251.1746	57.35	8.42	27.86	6.37
C12H21N502	267.33	267.1695	53.92	7.92	26.20	11.97
C12H21N503	283.33	283.1644	50.87	7.47	24.72	16.94
C12H21N504	299.33	299.1593	48.15	7.07	23.40	21.38

C12H21N505	315.33	315.1543	45.71	6.71	22.21	25.37
C12H21N506	331.33	331.1492	43.50	6.39	21.14	28.97
C12H21N507	347.33	347.1441	41.50	6.09	20.16	32.24
C12H21N508	363.33	363.1390	39.67	5.83	19.28	35.23
C12H21N509	379.33	379.1339	38.00	5.58	18.46	37.96
C12H22N201	210.32	210.1732	68.53	10.54	13.32	7.61
C12H22N202	226.32	226.1681	63.69	9.80	12.38	14.14
C12H22N203	242.32	242.1630	59.48	9.15	11.56	19.81
C12H22N204	258.32	258.1579	55.80	8.58	10.84	24.77
C12H22N205	274.32	274.1529	52.54	8.08	10.21	29.16
C12H22N206	290.32	290.1478	49.65	7.64	9.65	33.07
C12H22N207	306.32	306.1427	47.05	7.24	9.15	36.56
C12H22N208	322.32	322.1376	44.72	6.88	8.69	39.71
C12H22N209	338.32	338.1325	42.60	6.55	8.28	42.56
C12H22N401	238.34	238.1794	60.48	9.30	23.51	6.71
C12H22N402	254.33	254.1743	56.67	8.72	22.03	12.58
C12H22N403	270.33	270.1692	53.32	8.20	20.73	17.76
C12H22N404	286.33	286.1641	50.34	7.74	19.57	22.35
C12H22N405	302.33	302.1590	47.67	7.33	18.53	26.46
C12H22N406	318.33	318.1539	45.28	6.97	17.60	30.16
C12H22N407	334.33	334.1488	43.11	6.63	16.76	33.50
C12H22N408	350.33	350.1437	41.14	6.33	15.99	36.54
C12H22N409	366.33	366.1387	39.35	6.05	15.29	39.31
C12H22N601	266.35	266.1855	54.11	8.33	31.55	6.01
C12H22N602	282.35	282.1804	51.05	7.85	29.76	11.33
C12H22N603	298.35	298.1753	48.31	7.43	28.17	16.09
C12H22N604	314.35	314.1702	45.85	7.05	26.73	20.36
C12H22N605	330.35	330.1652	43.63	6.71	25.44	24.22
C12H22N606	346.35	346.1601	41.62	6.40	24.26	27.72
C12H22N607	362.35	362.1550	39.78	6.12	23.19	30.91
C12H22N608	378.34	378.1499	38.10	5.86	22.21	33.83
C12H22N609	394.34	394.1448	36.55	5.62	21.31	36.51
C12H23N101	197.32	197.1780	73.04	11.75	7.10	8.11
C12H23N102	213.32	213.1729	67.57	10.87	6.57	15.00
C12H23N103	229.32	229.1678	62.85	10.11	6.11	20.93
C12H23N104	245.32	245.1627	58.75	9.45	5.71	26.09
C12H23N105	261.32	261.1576	55.16	8.87	5.36	30.61
C12H23N106	277.32	277.1525	51.97	8.36	5.05	34.62
C12H23N107	293.32	293.1474	49.14	7.90	4.78	38.18
C12H23N108	309.32	309.1424	46.60	7.49	4.53	41.38
C12H23N109	325.32	325.1373	44.31	7.13	4.31	44.26
C12H23N301	225.34	225.1841	63.96	10.29	18.65	7.10
C12H23N302	241.34	241.1790	59.72	9.61	17.41	13.26
C12H23N303	257.34	257.1739	56.01	9.01	16.33	18.65
C12H23N304	273.33	273.1688	52.73	8.48	15.37	23.41
C12H23N305	289.33	289.1638	49.82	8.01	14.52	27.65
C12H23N306	305.33	305.1587	47.21	7.59	13.76	31.44
C12H23N307	321.33	321.1536	44.85	7.21	13.08	34.85
C12H23N308	337.33	337.1485	42.73	6.87	12.46	37.94
C12H23N309	353.33	353.1434	40.79	6.56	11.89	40.75
C12H23N501	253.35	253.1902	56.89	9.15	27.64	6.32
C12H23N502	269.35	269.1852	53.51	8.61	26.00	11.88
C12H23N503	285.35	285.1801	50.51	8.12	24.54	16.82
C12H23N504	301.35	301.1750	47.83	7.69	23.24	21.24
C12H23N505	317.35	317.1699	45.42	7.31	22.07	25.21
C12H23N506	333.35	333.1648	43.24	6.95	21.01	28.80
C12H23N507	349.35	349.1597	41.26	6.64	20.05	32.06
C12H23N508	365.35	365.1546	39.45	6.35	19.17	35.03
C12H23N509	381.35	381.1496	37.80	6.08	18.36	37.76
C12H24N201	212.34	212.1889	67.88	11,39	13.19	7.53

C12H24N202	228.34	228.1838	63.12	10.59	12.27	14.01
C12H24N203	244.34	244.1787	58.99	9.90	11.47	19.64
C12H24N204	260.34	260.1736	55.36	9.29	10.76	24.58
C12H24N205	276.34	276.1685	52.16	8.75	10.14	28.95
C12H24N206	292.33	292.1634	49.30	8.28	9.58	32.84
C12H24N207	308.33	308.1583	46.75	7.85	9.09	36.32
C12H24N208	324.33	324.1533	44.44	7.46	8.64	39.46
C12H24N209	340.33	340.1482	42.35	7.11	8.23	42.31
C12H24N401	240.35	240.1950	59.97	10.06	23.31	6.66
C12H24N402	256.35	256.1899	56.23	9.44	21.86	12.48
C12H24N403	272.35	272.1848	52.92	8.88	20.57	17.62
C12H24N404	288.35	288.1797	49.99	8.39	19.43	22.19
C12H24N405	304.35	304.1747	47.36	7.95	18.41	26.28
C12H24N406	320.35	320.1696	44.99	7.55	17.49	29.97
C12H24N407	336.35	336.1645	42.85	7.19	16.66	33.30
C12H24N408	352.35	352.1594	40.91	6.87	15.90	36.33
C12H24N409	368.35	368.1543	39.13	6.57	15.21	39.09
C12H24N601	268.36	268.2011	53.71	9.01	31.32	5.96
C12H24N602	284.36	284.1961	50.69	8.51	29.55	11.25
C12H24N603	300.36	300.1910	47.99	8.05	27.98	15.98
C12H24N604	316.36	316.1859	45.56	7.65	26.56	20.23
C12H24N605	332.36	332.1808	43.37	7.28	25.29	24.07
C12H24N606	348.36	348.1757	41.37	6.94	24.12	27.56
C12H24N607	364.36	364.1706	39.56	6.64	23.07	30.74
C12H24N608	380.36	380.1655	37.89	6.36	22.09	33.65
C12H24N609	396.36	396.1605	36.36	6.10	21.20	36.33
C12H25N101	199.34	199.1936	72.31	12.64	7.03	8.03
C12H25N102	215.34	215.1885	66.93	11.70	6.50	14.86
C12H25N103	231.34	231.1834	62.30	10.89	6.05	20.75
C12H25N104	247.34	247.1783	58.27	10.19	5.66	25.87
C12H25N105	263.34	263.1733	54.73	9.57	5.32	30.38
C12H25N106	279.34	279.1682	51.60	9.02	5.01	34.37
C12H25N107	295.34	295.1631	48.80	8.53	4.74	37.92
C12H25N108	311.33	311.1580	46.30	8.09	4.50	41.11
C12H25N109	327.33	327.1529	44.03	7.70	4.28	43.99
C12H25N301	227.35	227.1998	63.40	11.08	18.48	7.04
C12H25N302	243.35	243.1947	59.23	10.36	17.27	13.15
C12H25N303	259.35	259.1896	55.57	9.72	16.20	18.51
C12H25N304	275.35	275.1845	52.35	9.15	15.26	23.24
C12H25N305	291.35	291.1794	49.47	8.65	14.42	27.46
C12H25N306	307.35	307.1743	46.90	8.20	13.67	31.23
C12H25N307	323.35	323.1692	44.58	7.79	13.00	34.64
C12H25N308	339.35	339.1641	42.47	7.43	12.38	37.72
C12H25N309	355.35	355.1591	40.56	7.09	11.83	40.52
C12H25N501	255.37	255.2059	56.44	9.87	27.42	6.27
C12H25N502	271.37	271.2008	53.11	9.29	25.81	11.79
C12H25N503	287.36	287.1957	50.16	8.77	24.37	16.70
C12H25N504	303.36	303.1906	47.51	8.31	23.09	21.10
C12H25N505	319.36	319.1856	45.13	7.89	21.93	25.05
C12H25N506	335.36	335.1805	42.98	7.51	20.88	28.62
C12H25N507	351.36	351.1754	41.02	7.17	19.93	31.87
C12H25N508	367.36	367.1703	39.23	6.86	19.06	34.84
C12H25N509	383.36	383.1652	37.60	6.57	18.27	37.56
C12H26N201	214.35	214.2045	67.24	12.23	13.07	7.46
C12H26N202	230.35	230.1994	62.57	11.38	12.16	13.89
C12H26N203	246.35	246.1943	58.51	10.64	11.37	19.48
C12H26N204	262.35	262.1892	54.94	9.99	10.68	24.39
C12H26N205	278.35	278.1842	51.78	9.42	10.06	28.74
C12H26N206	294.35	294.1791	48.97	8.90	9.52	32.61
C12H26N207	310.35	310.1740	46.44	8.44	9.03	36.09

061

C12H26N208	326.35	326.1689	44.17	8.03	8.58	39.22
C12H26N209	342.35	342.1638	42.10	7.66	8.18	42.06
C12H26N401	242.37	242.2106	59.47	10.81	23.12	6.60
C12H26N402	258.37	258.2056	55.79	10.14	21.68	12.39
C12H26N403	274.37	274.2005	52.53	9.55	20.42	17.49
C12H26N404	290.37	290.1954	49.64	9.03	19.30	22.04
C12H26N405	306.36	306.1903	47.05	8.55	18.29	26.11
C12H26N406	322.36	322.1852	44.71	8.13	17.38	29.78
C12H26N407	338.36	338.1801	42.60	7.75	16.56	33.10
C12H26N408	354.36	354.1750	40.67	7.40	15.81	36.12
C12H26N409	370.36	370.1700	38.92	7.08	15.13	38.88
C12H26N601	270.38	270.2168	53.31	9.69	31.08	5.92
C12H26N602	286.38	286.2117	50.33	9.15	29.35	11.17
C12H26N603	302.38	302.2066	47.67	8.67	27.79	15.87
C12H26N604	318.38	318.2015	45.27	8.23	26.40	20.10
C12H26N605	334.38	334.1965	43.11	7.84	25.13	23.92
C12H26N606	350.38	350.1914	41.14	7.48	23.99	27.40
C12H26N607	366.38	366.1863	39.34	7.15	22.94	30.57
C12H26N608	382.38	382.1812	37.69	6.85	21.98	33.47
C12H26N609	398.38	398.1761	36.18	6.58	21.10	36.15
C12H27N101	201.36	201.2093	71.58	13.52	6.96	7.95
C12H27N102	217.35	217.2042	66.31	12.52	6.44	14.72
C12H27N103	233.35	233.1991	61.77	11.66	6.00	20.57
C12H27N104	249.35	249.1940	57.80	10.91	5.62	25.67
C12H27N105	265.35	265.1889	54.32	10.26	5.28	30.15
C12H27N106	281.35	281.1838	51.23	9.67	4.98	34.12
C12H27N107	297.35	297.1787	48.47	9.15	4.71	37.66
C12H27N108	313.35	313.1737	46.00	8.69	4.47	40.85
C12H27N109	329.35	329.1686	43.76	8.26	4.25	43.72
C12H27N301	229.37	229.2154	62.84	11.87	18.32	6.98
C12H27N302	245.37	245.2103	58.74	11.09	17.13	13.04
C12H27N303	261.37	261.2052	55.15	10.41	16.08	18.36
C12H27N304	277.37	277.2001	51.97	9.81	15.15	23.07
C12H27N305	293.37	293.1951	49.13	9.28	14.32	27.27
C12H27N306	309.37	309.1900	46.59	8.80	13.58	31.03
C12H27N307	325.36	325.1849	44.30	8.36	12.91	34.42
C12H27N308	341.36	341.1798	42.22	7.97	12.31	37.50
C12H27N309	357.36	357.1747	40.33	7.62	11.76	40.29
C12H27N501	257.38	257.2215	56.00	10.57	27.21	6.22
C12H27N502	273.38	273.2165	52.72	9.96	25.62	11.70
C12H27N503	289.38	289.2114	49.81	9.40	24.20	16.59
C12H27N504	305.38	305.2063	47.20	8.91	22.93	20.96
C12H27N505	321.38	321.2012	44.85	8.47	21.79	24.89
C12H27N506	337.38	337.1961	42.72	8.07	20.76	28.45
C12H27N507	353.38	353.1910	40.79	7.70	19.82	31.69
C12H27N508	369.38	369.1859	39.02	7.37	18.96	34.65
C12H27N509	385.38	385.1809	37.40	7.06	18.17	37.36
C12H28N201	216.37	216.2202	66.61	13.04	12.95	7.39
C12H28N202	232.37	232.2151	62.03	12.15	12.06	13.77
C12H28N203	248.37	248.2100	58.03	11.36	11.28	19.33
C12H28N204	264.37	264.2049	54.52	10.68	10.60	24.21
C12H28N205	280.37	280.1998	51.41	10.07	9.99	28.53
C12H28N206	296.37	296.1947	48.63	9.52	9.45	32.39
C12H28N207	312.37	312.1896	46.14	9.04	8.97	35.85
C12H28N208	328.37	328.1845	43.89	8.60	8.53	38.98
C12H28N209	344.36	344.1795	41.85	8.20	8.13	41.81
C12H28N401	244.38	244.2263	58.98	11.55	22.93	6.55
C12H28N402	260.38	260.2212	55.35	10.84	21.52	12.29
C12H28N403	276.38	276.2161	52.15	10.21	20.27	17.37
C12H28N404	292.38	292.2110	49.30	9.65	19.16	21.89

C12H28N405	308.38	308.2060	46.74	9.15	18.17	25.94
C12H28N406	324.38	324.2009	44.43	8.70	17.27	29.59
C12H28N407	340.38	340.1958	42.35	8.29	16.46	32.90
C12H28N408	356.38	356.1907	40.44	7.92	15.72	35.92
C12H28N409	372.38	372.1856	38.71	7.58	15.05	38.67
C12H28N601	272.40	272.2324	52.91	10.36	30.85	5.87
C12H28N602	288.40	288.2274	49.98	9.79	29.14	11.10
C12H28N603	304.40	304.2223	47.35	9.27	27.61	15.77
C12H28N604	320.39	320.2172	44.99	8.81	26.23	19.97
C12H28N605	336.39	336.2121	42.85	8.39	24.98	23.78
C12H28N606	352.39	352.2070	40.90	8.01	23.85	27.24
C12H28N607	368.39	368.2019	39.13	7.66	22.81	30.40
C12H28N608	384.39	384.1968	37.50	7.34	21.86	33.30
C12H28N609	400.39	400.1918	36.00	7.05	20.99	35.96
C12H29N301	231.38	231.2310	62.29	12.63	18.16	6.91
C12H29N302	247.38	247.2260	58.26	11.82	16.99	12.93
C12H29N303	263.38	263.2209	54.72	11.10	15.95	18.22
C12H29N304	279.38	279.2158	51.59	10.46	15.04	22.91
C12H29N305	295.38	295.2107	48.80	9.90	14.23	27.08
C12H29N306	311.38	311.2056	46.29	9.39	13.49	30.83
C12H29N307	327.38	327.2005	44.03	8.93	12.84	34.21
C12H29N308	343.38	343.1954	41.97	8.51	12.24	37.28
C12H29N309	359.38	359.1904	40.11	8.13	11.69	40.07
C12H29N501	259.40	259.2372	55.56	11.27	27.00	6.17
C12H29N502	275.40	275.2321	52.34	10.61	25.43	11.62
C12H29N503	291.40	291.2270	49.46	10.03	24.03	16.47
C12H29N504	307.40	307.2219	46.89	9.51	22.78	20.82
C12H29N505	323.40	323.2169	44.57	9.04	21.66	24.74
C12H29N506	339.39	339.2118	42.47	8.61	20.63	28.28
C12H29N507	355.39	355.2067	40.56	8.22	19.71	31.51
C12H29N508	371.39	371.2016	38.81	7.87	18.86	34.46
C12H29N509	387.39	387.1965	37.21	7.55	18.08	37.17
C12H30N401	246.40	246.2419	58.50	12.27	22.74	6.49
C12H30N402	262.40	262.2369	54.93	11.52	21.35	12.19
C12H30N403	278.40	278.2318	51.77	10.86	20.12	17.24
C12H30N404	294.40	294.2267	48.96	10.27	19.03	21.74
C12H30N405	310.40	310.2216	46.44	9.74	18.05	25.77
C12H30N406	326.40	326.2165	44.16	9.26	17.17	29.41
C12H30N407	342.40	342.2114	42.10	8.83	16.36	32.71
C12H30N408	358.39	358.2063	40.22	8.44	15.63	35.71
C12H30N409	374.39	374.2013	38.50	8.08	14.96	38.46
C12H30N601	274.41	274.2481	52.52	11.02	30.63	5.83
C12H30N602	290.41	290.2430	49.63	10.41	28.94	11.02
C12H30N603	306.41	306.2379	47.04	9.87	27.43	15.66
C12H30N604	322.41	322.2328	44.71	9.38	26.07	19.85
C12H30N605	338.41	338.2278	42.59	8.94	24.83	23.64
C12H30N606	354.41	354.2227	40.67	8.53	23.71	27.09
C12H30N607	370.41	370.2176	38.91	8.16	22.69	30.24
C12H30N608	386.41	386.2125	37.30	7.83	21.75	33.12
C12H30N609	402.41	402.2074	35.82	7.51	20.88	35.78
C12H31N501	261.41	261.2528	55.14	11.95	26.79	6.12
C12H31N502	277.41	277.2478	51.96	11.26	25.25	11.53
C12H31N503	293.41	293.2427	49.12	10.65	23.87	16.36
C12H31N504	309.41	309.2376	46.58	10.10	22.63	20.68
C12H31N505	325.41	325.2325	44.29	9.60	21.52	24.58
C12H31N506	341.41	341.2274	42.22	9.15	20.51	28.12
C12H31N507	357.41	357.2223	40.33	8.74	19.59	31.34
C12H31N508	373.41	373.2172	38.60	8.37	18.76	34.28
C12H31N509	389.41	389.2122	37.01	8.02	17.98	36.98
C12H32N601	276.43	276.2637	52.14	11.67	30.40	5.79

062

C12H32N6O2	292.43	292.2587	49.29	11.03	28.74	10.94
C12H32N6O3	308.43	308.2536	46.73	10.46	27.25	15.56
C12H32N6O4	324.43	324.2485	44.43	9.94	25.90	19.73
C12H32N6O5	340.43	340.2434	42.34	9.47	24.69	23.50
C12H32N6O6	356.43	356.2383	40.44	9.05	23.58	26.93
C12H32N6O7	372.42	372.2332	38.70	8.66	22.57	30.07
C12H32N6O8	388.42	388.2281	37.11	8.30	21.64	32.95
C12H32N6O9	404.42	404.2231	35.64	7.98	20.78	35.60
C13H 6N2O1	206.21	206.0480	75.72	2.93	13.59	7.76
C13H 6N2O2	222.20	222.0429	70.27	2.72	12.61	14.40
C13H 6N2O3	238.20	238.0378	65.55	2.54	11.76	20.15
C13H 6N2O4	254.20	254.0328	61.43	2.38	11.02	25.18
C13H 6N2O5	270.20	270.0277	57.79	2.24	10.37	29.61
C13H 6N2O6	286.20	286.0226	54.56	2.11	9.79	33.54
C13H 6N2O7	302.20	302.0175	51.67	2.00	9.27	37.06
C13H 6N2O8	318.20	318.0124	49.07	1.90	8.80	40.22
C13H 6N2O9	334.20	334.0073	46.72	1.81	8.38	43.09
C13H 6N4O1	234.22	234.0542	66.67	2.58	23.92	6.83
C13H 6N4O2	250.22	250.0491	62.40	2.42	22.39	12.79
C13H 6N4O3	266.22	266.0440	58.65	2.27	21.05	18.03
C13H 6N4O4	282.22	282.0389	55.33	2.14	19.85	22.68
C13H 6N4O5	298.22	298.0338	52.36	2.03	18.79	26.83
C13H 6N4O6	314.22	314.0287	49.69	1.92	17.83	30.55
C13H 6N4O7	330.22	330.0236	47.29	1.83	16.97	33.92
C13H 6N4O8	346.21	346.0186	45.10	1.75	16.18	36.97
C13H 6N4O9	362.21	362.0135	43.11	1.67	15.47	39.75
C13H 6N6O1	262.23	262.0603	59.54	2.31	32.05	6.10
C13H 6N6O2	278.23	278.0552	56.12	2.17	30.21	11.50
C13H 6N6O3	294.23	294.0501	53.07	2.06	28.56	16.31
C13H 6N6O4	310.23	310.0450	50.33	1.95	27.09	20.63
C13H 6N6O5	326.23	326.0400	47.86	1.85	25.76	24.52
C13H 6N6O6	342.23	342.0349	45.63	1.77	24.56	28.05
C13H 6N6O7	358.23	358.0298	43.59	1.69	23.46	31.26
C13H 6N6O8	374.23	374.0247	41.72	1.62	22.46	34.20
C13H 6N6O9	390.23	390.0196	40.01	1.55	21.54	36.90
C13H 7N1O1	193.21	193.0528	80.82	3.65	7.25	8.28
C13H 7N1O2	209.21	209.0477	74.64	3.37	6.70	15.30
C13H 7N1O3	225.21	225.0426	69.33	3.13	6.22	21.31
C13H 7N1O4	241.21	241.0375	64.74	2.93	5.81	26.53
C13H 7N1O5	257.20	257.0324	60.71	2.74	5.45	31.10
C13H 7N1O6	273.20	273.0273	57.15	2.58	5.13	35.14
C13H 7N1O7	289.20	289.0222	53.99	2.44	4.84	38.73
C13H 7N1O8	305.20	305.0172	51.16	2.31	4.59	41.94
C13H 7N1O9	321.20	321.0121	48.61	2.20	4.36	44.83
C13H 7N3O1	221.22	221.0589	70.58	3.19	18.99	7.23
C13H 7N3O2	237.22	237.0538	65.82	2.97	17.71	13.49
C13H 7N3O3	253.22	253.0487	61.66	2.79	16.59	18.96
C13H 7N3O4	269.22	269.0437	58.00	2.62	15.61	23.77
C13H 7N3O5	285.22	285.0386	54.75	2.47	14.73	28.05
C13H 7N3O6	301.22	301.0335	51.84	2.34	13.95	31.87
C13H 7N3O7	317.22	317.0284	49.22	2.22	13.25	35.31
C13H 7N3O8	333.22	333.0233	46.86	2.12	12.61	38.41
C13H 7N3O9	349.22	349.0182	44.71	2.02	12.03	41.23
C13H 7N5O1	249.23	249.0651	62.65	2.83	28.10	6.42
C13H 7N5O2	265.23	265.0600	58.87	2.66	26.40	12.06
C13H 7N5O3	281.23	281.0549	55.52	2.51	24.90	17.07
C13H 7N5O4	297.23	297.0498	52.53	2.37	23.56	21.53
C13H 7N5O5	313.23	313.0447	49.85	2.25	22.36	25.54
C13H 7N5O6	329.23	329.0396	47.43	2.14	21.27	29.16
C13H 7N5O7	345.23	345.0345	45.23	2.04	20.29	32.44

C13H 7N5O8	361.23	361.0295	43.23	1.95	19.39	35.43
C13H 7N5O9	377.23	377.0244	41.39	1.87	18.57	38.17
C13H 8N2O1	208.22	208.0637	74.99	3.87	13.45	7.68
C13H 8N2O2	224.22	224.0586	69.64	3.60	12.49	14.27
C13H 8N2O3	240.22	240.0535	65.00	3.36	11.66	19.98
C13H 8N2O4	256.22	256.0484	60.94	3.15	10.93	24.98
C13H 8N2O5	272.22	272.0433	57.36	2.96	10.29	29.39
C13H 8N2O6	288.22	288.0382	54.18	2.80	9.72	33.31
C13H 8N2O7	304.22	304.0331	51.33	2.65	9.21	36.81
C13H 8N2O8	320.22	320.0281	48.76	2.52	8.75	39.97
C13H 8N2O9	336.22	336.0230	46.44	2.40	8.33	42.83
C13H 8N4O1	236.23	236.0698	66.10	3.41	23.72	6.77
C13H 8N4O2	252.23	252.0647	61.90	3.20	22.21	12.69
C13H 8N4O3	268.23	268.0596	58.21	3.01	20.89	17.89
C13H 8N4O4	284.23	284.0545	54.94	2.84	19.71	22.52
C13H 8N4O5	300.23	300.0495	52.01	2.69	18.66	26.65
C13H 8N4O6	316.23	316.0444	49.38	2.55	17.72	30.36
C13H 8N4O7	332.23	332.0393	47.00	2.43	16.86	33.71
C13H 8N4O8	348.23	348.0342	44.84	2.32	16.09	36.76
C13H 8N4O9	364.23	364.0291	42.87	2.21	15.38	39.53
C13H 8N6O1	264.25	264.0760	59.09	3.05	31.80	6.05
C13H 8N6O2	280.25	280.0709	55.72	2.88	29.99	11.42
C13H 8N6O3	296.25	296.0658	52.71	2.72	28.37	16.20
C13H 8N6O4	312.25	312.0607	50.01	2.58	26.91	20.50
C13H 8N6O5	328.25	328.0556	47.57	2.46	25.60	24.37
C13H 8N6O6	344.25	344.0505	45.36	2.34	24.41	27.89
C13H 8N6O7	360.24	360.0454	43.34	2.24	23.33	31.09
C13H 8N6O8	376.24	376.0404	41.50	2.14	22.34	34.02
C13H 8N6O9	392.24	392.0353	39.81	2.06	21.43	36.71
C13H 9N1O1	195.22	195.0684	79.98	4.65	7.17	8.20
C13H 9N1O2	211.22	211.0633	73.92	4.29	6.63	15.15
C13H 9N1O3	227.22	227.0582	68.72	3.99	6.16	21.12
C13H 9N1O4	243.22	243.0532	64.20	3.73	5.76	26.31
C13H 9N1O5	259.22	259.0481	60.24	3.50	5.40	30.86
C13H 9N1O6	275.22	275.0430	56.73	3.30	5.09	34.88
C13H 9N1O7	291.22	291.0379	53.62	3.12	4.81	38.46
C13H 9N1O8	307.22	307.0328	50.83	2.95	4.56	41.66
C13H 9N1O9	323.22	323.0277	48.31	2.81	4.33	44.55
C13H 9N3O1	223.24	223.0746	69.95	4.06	18.82	7.17
C13H 9N3O2	239.24	239.0695	65.27	3.79	17.56	13.38
C13H 9N3O3	255.23	255.0644	61.18	3.55	16.46	18.81
C13H 9N3O4	271.23	271.0593	57.57	3.34	15.49	23.59
C13H 9N3O5	287.23	287.0542	54.36	3.16	14.63	27.85
C13H 9N3O6	303.23	303.0491	51.49	2.99	13.86	31.66
C13H 9N3O7	319.23	319.0440	48.91	2.84	13.16	35.08
C13H 9N3O8	335.23	335.0390	46.58	2.71	12.53	38.18
C13H 9N3O9	351.23	351.0339	44.46	2.58	11.96	41.00
C13H 9N5O1	251.25	251.0807	62.15	3.61	27.87	6.37
C13H 9N5O2	267.25	267.0756	58.43	3.39	26.21	11.97
C13H 9N5O3	283.25	283.0705	55.13	3.20	24.73	16.95
C13H 9N5O4	299.25	299.0654	52.18	3.03	23.40	21.39
C13H 9N5O5	315.25	315.0604	49.53	2.88	22.22	25.38
C13H 9N5O6	331.25	331.0553	47.14	2.74	21.14	28.98
C13H 9N5O7	347.25	347.0502	44.97	2.61	20.17	32.25
C13H 9N5O8	363.25	363.0451	42.99	2.50	19.28	35.24
C13H 9N5O9	379.24	379.0400	41.17	2.39	18.47	37.97
C13H10N2O1	210.24	210.0793	74.27	4.79	13.32	7.61
C13H10N2O2	226.24	226.0742	69.02	4.46	12.38	14.14
C13H10N2O3	242.24	242.0691	64.46	4.16	11.56	19.81
C13H10N2O4	258.24	258.0641	60.47	3.90	10.85	24.78

063

C13H10N2O5	274.24	274.0590	56.94	3.68	10.22	29.17
C13H10N2O6	290.23	290.0539	53.80	3.47	9.65	33.08
C13H10N2O7	306.23	306.0488	50.99	3.29	9.15	36.57
C13H10N2O8	322.23	322.0437	48.46	3.13	8.69	39.72
C13H10N2O9	338.23	338.0386	46.16	2.98	8.28	42.57
C13H10N4O1	238.25	238.0855	65.54	4.23	23.52	6.72
C13H10N4O2	254.25	254.0804	61.41	3.96	22.04	12.59
C13H10N4O3	270.25	270.0753	57.78	3.73	20.73	17.76
C13H10N4O4	286.25	286.0702	54.55	3.52	19.57	22.36
C13H10N4O5	302.25	302.0651	51.66	3.33	18.54	26.47
C13H10N4O6	318.25	318.0600	49.06	3.17	17.60	30.16
C13H10N4O7	334.25	334.0549	46.72	3.02	16.76	33.51
C13H10N4O8	350.25	350.0499	44.58	2.88	16.00	36.54
C13H10N4O9	366.25	366.0448	42.63	2.75	15.30	39.32
C13H10N6O1	266.26	266.0916	58.64	3.79	31.56	6.01
C13H10N6O2	282.26	282.0865	55.32	3.57	29.77	11.34
C13H10N6O3	298.26	298.0814	52.35	3.38	28.18	16.09
C13H10N6O4	314.26	314.0763	49.69	3.21	26.74	20.36
C13H10N6O5	330.26	330.0713	47.28	3.05	25.45	24.22
C13H10N6O6	346.26	346.0662	45.09	2.91	24.27	27.72
C13H10N6O7	362.26	362.0611	43.10	2.78	23.20	30.92
C13H10N6O8	378.26	378.0560	41.28	2.66	22.22	33.84
C13H10N6O9	394.26	394.0509	39.60	2.56	21.32	36.52
C13H11N1O1	197.24	197.0841	79.17	5.62	7.10	8.11
C13H11N1O2	213.24	213.0790	73.23	5.20	6.57	15.01
C13H11N1O3	229.24	229.0739	68.11	4.84	6.11	20.94
C13H11N1O4	245.24	245.0688	63.67	4.52	5.71	26.10
C13H11N1O5	261.24	261.0637	59.77	4.24	5.36	30.62
C13H11N1O6	277.24	277.0586	56.32	4.00	5.05	34.63
C13H11N1O7	293.24	293.0535	53.25	3.78	4.78	38.19
C13H11N1O8	309.23	309.0485	50.49	3.59	4.53	41.39
C13H11N1O9	325.23	325.0434	48.01	3.41	4.31	44.27
C13H11N3O1	225.25	225.0902	69.32	4.92	18.65	7.10
C13H11N3O2	241.25	241.0851	64.72	4.60	17.42	13.26
C13H11N3O3	257.25	257.0800	60.70	4.31	16.33	18.66
C13H11N3O4	273.25	273.0749	57.14	4.06	15.38	23.42
C13H11N3O5	289.25	289.0699	53.98	3.83	14.53	27.66
C13H11N3O6	305.25	305.0648	51.15	3.63	13.77	31.45
C13H11N3O7	321.25	321.0597	48.61	3.45	13.08	34.86
C13H11N3O8	337.25	337.0546	46.30	3.29	12.46	37.95
C13H11N3O9	353.25	353.0495	44.20	3.14	11.90	40.76
C13H11N5O1	253.27	253.0964	61.65	4.38	27.65	6.32
C13H11N5O2	269.26	269.0913	57.99	4.12	26.01	11.88
C13H11N5O3	285.26	285.0862	54.74	3.89	24.55	16.83
C13H11N5O4	301.26	301.0811	51.83	3.68	23.25	21.24
C13H11N5O5	317.26	317.0760	49.22	3.49	22.07	25.21
C13H11N5O6	333.26	333.0709	46.85	3.33	21.01	28.81
C13H11N5O7	349.26	349.0658	44.71	3.17	20.05	32.07
C13H11N5O8	365.26	365.0608	42.75	3.04	19.17	35.04
C13H11N5O9	381.26	381.0557	40.95	2.91	18.37	37.77
C13H12N2O1	212.25	212.0950	73.57	5.70	13.20	7.54
C13H12N2O2	228.25	228.0899	68.41	5.30	12.27	14.02
C13H12N2O3	244.25	244.0848	63.93	4.95	11.47	19.65
C13H12N2O4	260.25	260.0797	60.00	4.65	10.76	24.59
C13H12N2O5	276.25	276.0746	56.52	4.38	10.14	28.96
C13H12N2O6	292.25	292.0695	53.43	4.14	9.59	32.85
C13H12N2O7	308.25	308.0644	50.66	3.92	9.09	36.33
C13H12N2O8	324.25	324.0594	48.16	3.73	8.64	39.47
C13H12N2O9	340.25	340.0543	45.89	3.55	8.23	42.32
C13H12N4O1	240.27	240.1011	64.99	5.03	23.32	6.66

C13H12N4O2	256.27	256.0960	60.93	4.72	21.86	12.49
C13H12N4O3	272.27	272.0909	57.35	4.44	20.58	17.63
C13H12N4O4	288.26	288.0858	54.17	4.20	19.44	22.20
C13H12N4O5	304.26	304.0808	51.32	3.98	18.41	26.29
C13H12N4O6	320.26	320.0757	48.76	3.78	17.49	29.97
C13H12N4O7	336.26	336.0706	46.44	3.60	16.66	33.31
C13H12N4O8	352.26	352.0655	44.33	3.43	15.90	36.34
C13H12N4O9	368.26	368.0604	42.40	3.28	15.21	39.10
C13H12N6O1	268.28	268.1073	58.20	4.51	31.33	5.96
C13H12N6O2	284.28	284.1022	54.93	4.25	29.56	11.26
C13H12N6O3	300.28	300.0971	52.00	4.03	27.99	15.98
C13H12N6O4	316.28	316.0920	49.37	3.82	26.57	20.23
C13H12N6O5	332.28	332.0869	46.99	3.64	25.29	24.08
C13H12N6O6	348.28	348.0818	44.83	3.47	24.13	27.56
C13H12N6O7	364.28	364.0767	42.86	3.32	23.07	30.74
C13H12N6O8	380.28	380.0717	41.06	3.18	22.10	33.66
C13H12N6O9	396.28	396.0666	39.40	3.05	21.21	36.34
C13H13N1O1	199.25	199.0997	78.36	6.58	7.03	8.03
C13H13N1O2	215.25	215.0946	72.54	6.09	6.51	14.87
C13H13N1O3	231.25	231.0895	67.52	5.67	6.06	20.76
C13H13N1O4	247.25	247.0845	63.15	5.30	5.66	25.88
C13H13N1O5	263.25	263.0794	59.31	4.98	5.32	30.39
C13H13N1O6	279.25	279.0743	55.92	4.69	5.02	34.38
C13H13N1O7	295.25	295.0692	52.89	4.44	4.74	37.93
C13H13N1O8	311.25	311.0641	50.17	4.21	4.50	41.12
C13H13N1O9	327.25	327.0590	47.71	4.00	4.28	44.00
C13H13N3O1	227.27	227.1059	68.71	5.77	18.49	7.04
C13H13N3O2	243.27	243.1008	64.19	5.39	17.27	13.15
C13H13N3O3	259.27	259.0957	60.23	5.05	16.21	18.51
C13H13N3O4	275.27	275.0906	56.73	4.76	15.27	23.25
C13H13N3O5	291.27	291.0855	53.61	4.50	14.43	27.47
C13H13N3O6	307.27	307.0804	50.82	4.26	13.68	31.24
C13H13N3O7	323.26	323.0753	48.30	4.05	13.00	34.65
C13H13N3O8	339.26	339.0703	46.02	3.86	12.39	37.73
C13H13N3O9	355.26	355.0652	43.95	3.69	11.83	40.53
C13H13N5O1	255.28	255.1120	61.17	5.13	27.43	6.27
C13H13N5O2	271.28	271.1069	57.56	4.83	25.82	11.80
C13H13N5O3	287.28	287.1018	54.35	4.56	24.38	16.71
C13H13N5O4	303.28	303.0967	51.49	4.32	23.09	21.10
C13H13N5O5	319.28	319.0917	48.91	4.10	21.93	25.06
C13H13N5O6	335.28	335.0866	46.57	3.91	20.89	28.63
C13H13N5O7	351.28	351.0815	44.45	3.73	19.94	31.88
C13H13N5O8	367.28	367.0764	42.51	3.57	19.07	34.85
C13H13N5O9	383.28	383.0713	40.74	3.42	18.27	37.57
C13H14N2O1	214.27	214.1106	72.87	6.59	13.07	7.47
C13H14N2O2	230.27	230.1055	67.81	6.13	12.17	13.90
C13H14N2O3	246.27	246.1004	63.40	5.73	11.38	19.49
C13H14N2O4	262.27	262.0953	59.54	5.38	10.68	24.40
C13H14N2O5	278.27	278.0903	56.11	5.07	10.07	28.75
C13H14N2O6	294.27	294.0852	53.06	4.80	9.52	32.62
C13H14N2O7	310.27	310.0801	50.33	4.55	9.03	36.10
C13H14N2O8	326.27	326.0750	47.86	4.33	8.59	39.23
C13H14N2O9	342.26	342.0699	45.62	4.12	8.18	42.07
C13H14N4O1	242.28	242.1168	64.45	5.82	23.12	6.60
C13H14N4O2	258.28	258.1117	60.46	5.46	21.69	12.39
C13H14N4O3	274.28	274.1066	56.93	5.14	20.43	17.50
C13H14N4O4	290.28	290.1015	53.79	4.86	19.30	22.05
C13H14N4O5	306.28	306.0964	50.98	4.61	18.29	26.12
C13H14N4O6	322.28	322.0913	48.45	4.38	17.38	29.79
C13H14N4O7	338.28	338.0862	46.16	4.17	16.56	33.11

C13H14N408	354.28	354.0812	44.07	3.98	15.81	36.13
C13H14N409	370.28	370.0761	42.17	3.81	15.13	38.89
C13H14N601	270.30	270.1229	57.77	5.22	31.09	5.92
C13H14N602	286.30	286.1178	54.54	4.93	29.35	11.18
C13H14N603	302.29	302.1127	51.65	4.67	27.80	15.88
C13H14N604	318.29	318.1076	49.06	4.43	26.40	20.11
C13H14N605	334.29	334.1026	46.71	4.22	25.14	23.93
C13H14N606	350.29	350.0975	44.58	4.03	23.99	27.40
C13H14N607	366.29	366.0924	42.63	3.85	22.94	30.58
C13H14N608	382.29	382.0873	40.84	3.69	21.98	33.48
C13H14N609	398.29	398.0822	39.20	3.54	21.10	36.15
C13H15N101	201.27	201.1154	77.58	7.51	6.96	7.95
C13H15N102	217.27	217.1103	71.87	6.96	6.45	14.73
C13H15N103	233.27	233.1052	66.94	6.48	6.00	20.58
C13H15N104	249.27	249.1001	62.64	6.07	5.62	25.67
C13H15N105	265.27	265.0950	58.86	5.70	5.28	30.16
C13H15N106	281.27	281.0899	55.51	5.38	4.98	34.13
C13H15N107	297.27	297.0848	52.53	5.09	4.71	37.68
C13H15N108	313.27	313.0798	49.84	4.83	4.47	40.86
C13H15N109	329.27	329.0747	47.42	4.59	4.25	43.73
C13H15N301	229.28	229.1215	68.10	6.59	18.33	6.98
C13H15N302	245.28	245.1164	63.66	6.16	17.13	13.05
C13H15N303	261.28	261.1113	59.76	5.79	16.08	18.37
C13H15N304	277.28	277.1062	56.31	5.45	15.15	23.08
C13H15N305	293.28	293.1012	53.24	5.16	14.33	27.28
C13H15N306	309.28	309.0961	50.49	4.89	13.59	31.04
C13H15N307	325.28	325.0910	48.00	4.65	12.92	34.43
C13H15N308	341.28	341.0859	45.75	4.43	12.31	37.50
C13H15N309	357.28	357.0808	43.70	4.23	11.76	40.30
C13H15N501	257.30	257.1277	60.69	5.88	27.22	6.22
C13H15N502	273.30	273.1226	57.13	5.53	25.63	11.71
C13H15N503	289.30	289.1175	53.97	5.23	24.21	16.59
C13H15N504	305.30	305.1124	51.15	4.95	22.94	20.96
C13H15N505	321.30	321.1073	48.60	4.71	21.80	24.90
C13H15N506	337.29	337.1022	46.29	4.48	20.76	28.46
C13H15N507	353.29	353.0971	44.20	4.28	19.82	31.70
C13H15N508	369.29	369.0921	42.28	4.09	18.96	34.66
C13H15N509	385.29	385.0870	40.53	3.92	18.18	37.37
C13H16N201	216.29	216.1263	72.19	7.46	12.95	7.40
C13H16N202	232.28	232.1212	67.22	6.94	12.06	13.78
C13H16N203	248.28	248.1161	62.89	6.50	11.28	19.33
C13H16N204	264.28	264.1110	59.08	6.10	10.60	24.22
C13H16N205	280.28	280.1059	55.71	5.75	9.99	28.54
C13H16N206	296.28	296.1008	52.70	5.44	9.45	32.40
C13H16N207	312.28	312.0957	50.00	5.16	8.97	35.86
C13H16N208	328.28	328.0907	47.56	4.91	8.53	38.99
C13H16N209	344.28	344.0856	45.35	4.68	8.14	41.82
C13H16N401	244.30	244.1324	63.92	6.60	22.93	6.55
C13H16N402	260.30	260.1273	59.99	6.20	21.52	12.29
C13H16N403	276.30	276.1222	56.51	5.84	20.28	17.37
C13H16N404	292.30	292.1171	53.42	5.52	19.17	21.89
C13H16N405	308.30	308.1121	50.65	5.23	18.17	25.95
C13H16N406	324.30	324.1070	48.15	4.97	17.28	29.60
C13H16N407	340.30	340.1019	45.89	4.74	16.46	32.91
C13H16N408	356.29	356.0968	43.82	4.53	15.72	35.92
C13H16N409	372.29	372.0917	41.94	4.33	15.05	38.68
C13H16N601	272.31	272.1386	57.34	5.92	30.86	5.88
C13H16N602	288.31	288.1335	54.16	5.59	29.15	11.10
C13H16N603	304.31	304.1284	51.31	5.30	27.62	15.77
C13H16N604	320.31	320.1233	48.75	5.03	26.24	19.98

C13H16N605	336.31	336.1182	46.43	4.80	24.99	23.79
C13H16N606	352.31	352.1131	44.32	4.58	23.85	27.25
C13H16N607	368.31	368.1080	42.40	4.38	22.82	30.41
C13H16N608	384.31	384.1029	40.63	4.20	21.87	33.31
C13H16N609	400.31	400.0979	39.01	4.03	20.99	35.97
C13H17N101	203.29	203.1310	76.81	8.43	6.89	7.87
C13H17N102	219.29	219.1259	71.21	7.81	6.39	14.59
C13H17N103	235.29	235.1208	66.36	7.28	5.95	20.40
C13H17N104	251.28	251.1157	62.14	6.82	5.57	25.47
C13H17N105	267.28	267.1107	58.42	6.41	5.24	29.93
C13H17N106	283.28	283.1056	55.12	6.05	4.94	33.89
C13H17N107	299.28	299.1005	52.17	5.73	4.68	37.42
C13H17N108	315.28	315.0954	49.53	5.43	4.44	40.60
C13H17N109	331.28	331.0903	47.13	5.17	4.23	43.47
C13H17N301	231.30	231.1372	67.51	7.41	18.17	6.92
C13H17N302	247.30	247.1321	63.14	6.93	16.99	12.94
C13H17N303	263.30	263.1270	59.30	6.51	15.96	18.23
C13H17N304	279.30	279.1219	55.91	6.14	15.04	22.91
C13H17N305	295.30	295.1168	52.88	5.80	14.23	27.09
C13H17N306	311.30	311.1117	50.16	5.50	13.50	30.84
C13H17N307	327.30	327.1066	47.71	5.24	12.84	34.22
C13H17N308	343.30	343.1016	45.48	4.99	12.24	37.28
C13H17N309	359.30	359.0965	43.46	4.77	11.70	40.08
C13H17N501	259.31	259.1433	60.21	6.61	27.01	6.17
C13H17N502	275.31	275.1382	56.72	6.22	25.44	11.62
C13H17N503	291.31	291.1331	53.60	5.88	24.04	16.48
C13H17N504	307.31	307.1280	50.81	5.58	22.79	20.82
C13H17N505	323.31	323.1230	48.30	5.30	21.66	24.74
C13H17N506	339.31	339.1179	46.02	5.05	20.64	28.29
C13H17N507	355.31	355.1128	43.95	4.82	19.71	31.52
C13H17N508	371.31	371.1077	42.05	4.61	18.86	34.47
C13H17N509	387.31	387.1026	40.32	4.42	18.08	37.18
C13H18N201	218.30	218.1419	71.53	8.31	12.83	7.33
C13H18N202	234.30	234.1368	66.64	7.74	11.96	13.66
C13H18N203	250.30	250.1317	62.38	7.25	11.19	19.18
C13H18N204	266.30	266.1266	58.64	6.81	10.52	24.03
C13H18N205	282.30	282.1216	55.31	6.43	9.92	28.34
C13H18N206	298.30	298.1165	52.35	6.08	9.39	32.18
C13H18N207	314.30	314.1114	49.68	5.77	8.91	35.63
C13H18N208	330.30	330.1063	47.27	5.49	8.48	38.75
C13H18N209	346.30	346.1012	45.09	5.24	8.09	41.58
C13H18N401	246.31	246.1481	63.39	7.37	22.75	6.50
C13H18N402	262.31	262.1430	59.53	6.92	21.36	12.20
C13H18N403	278.31	278.1379	56.10	6.52	20.13	17.25
C13H18N404	294.31	294.1328	53.05	6.16	19.04	21.74
C13H18N405	310.31	310.1277	50.32	5.85	18.05	25.78
C13H18N406	326.31	326.1226	47.85	5.56	17.17	29.42
C13H18N407	342.31	342.1175	45.61	5.30	16.37	32.72
C13H18N408	358.31	358.1125	43.58	5.06	15.64	35.72
C13H18N409	374.31	374.1074	41.72	4.85	14.97	38.47
C13H18N601	274.33	274.1542	56.92	6.61	30.63	5.83
C13H18N602	290.33	290.1491	53.78	6.25	28.95	11.02
C13H18N603	306.33	306.1440	50.97	5.92	27.43	15.67
C13H18N604	322.33	322.1389	48.44	5.63	26.07	19.85
C13H18N605	338.33	338.1339	46.15	5.36	24.84	23.64
C13H18N606	354.33	354.1288	44.07	5.12	23.72	27.09
C13H18N607	370.32	370.1237	42.16	4.90	22.69	30.24
C13H18N608	386.32	386.1186	40.42	4.70	21.75	33.13
C13H18N609	402.32	402.1135	38.81	4.51	20.89	35.79
C13H19N101	205.30	205.1467	76.06	9.33	6.82	7.79

065

C13H19N102	221.30	221.1416	70.56	8.65	6.33	14.46
C13H19N103	237.30	237.1365	65.80	8.07	5.90	20.23
C13H19N104	253.30	253.1314	61.64	7.56	5.53	25.27
C13H19N105	269.30	269.1263	57.98	7.11	5.20	29.71
C13H19N106	285.30	285.1212	54.73	6.71	4.91	33.65
C13H19N107	301.30	301.1161	51.82	6.36	4.65	37.17
C13H19N108	317.30	317.1111	49.21	6.04	4.41	40.34
C13H19N109	333.30	333.1060	46.85	5.75	4.20	43.20
C13H19N301	233.32	233.1528	66.92	8.21	18.01	6.86
C13H19N302	249.32	249.1477	62.63	7.68	16.85	12.83
C13H19N303	265.31	265.1426	58.85	7.22	15.84	18.09
C13H19N304	281.31	281.1375	55.51	6.81	14.94	22.75
C13H19N305	297.31	297.1325	52.52	6.44	14.13	26.91
C13H19N306	313.31	313.1274	49.84	6.11	13.41	30.64
C13H19N307	329.31	329.1223	47.42	5.82	12.76	34.01
C13H19N308	345.31	345.1172	45.22	5.55	12.17	37.07
C13H19N309	361.31	361.1121	43.22	5.30	11.63	39.85
C13H19N501	261.33	261.1590	59.75	7.33	26.80	6.12
C13H19N502	277.33	277.1539	56.30	6.91	25.25	11.54
C13H19N503	293.33	293.1488	53.23	6.53	23.88	16.36
C13H19N504	309.33	309.1437	50.48	6.19	22.64	20.69
C13H19N505	325.33	325.1386	48.00	5.89	21.53	24.59
C13H19N506	341.33	341.1335	45.75	5.61	20.52	28.12
C13H19N507	357.33	357.1284	43.70	5.36	19.60	31.34
C13H19N508	373.33	373.1233	41.83	5.13	18.76	34.29
C13H19N509	389.32	389.1183	40.11	4.92	17.99	36.99
C13H20N201	220.32	220.1576	70.87	9.15	12.72	7.26
C13H20N202	236.32	236.1525	66.07	8.53	11.85	13.54
C13H20N203	252.32	252.1474	61.88	7.99	11.10	19.02
C13H20N204	268.32	268.1423	58.19	7.51	10.44	23.85
C13H20N205	284.31	284.1372	54.92	7.09	9.85	28.14
C13H20N206	300.31	300.1321	51.99	6.71	9.33	31.97
C13H20N207	316.31	316.1270	49.36	6.37	8.86	35.41
C13H20N208	332.31	332.1220	46.99	6.07	8.43	38.52
C13H20N209	348.31	348.1169	44.83	5.79	8.04	41.34
C13H20N401	248.33	248.1637	62.88	8.12	22.56	6.44
C13H20N402	264.33	264.1586	59.07	7.63	21.20	12.11
C13H20N403	280.33	280.1535	55.70	7.19	19.99	17.12
C13H20N404	296.33	296.1484	52.69	6.80	18.91	21.60
C13H20N405	312.33	312.1434	49.99	6.45	17.94	25.61
C13H20N406	328.33	328.1383	47.56	6.14	17.06	29.24
C13H20N407	344.33	344.1332	45.35	5.85	16.27	32.53
C13H20N408	360.33	360.1281	43.33	5.59	15.55	35.52
C13H20N409	376.33	376.1230	41.49	5.36	14.89	38.26
C13H20N601	276.34	276.1698	56.50	7.30	30.41	5.79
C13H20N602	292.34	292.1648	53.41	6.90	28.75	10.95
C13H20N603	308.34	308.1597	50.64	6.54	27.26	15.57
C13H20N604	324.34	324.1546	48.14	6.22	25.91	19.73
C13H20N605	340.34	340.1495	45.88	5.92	24.69	23.50
C13H20N606	356.34	356.1444	43.82	5.66	23.58	26.94
C13H20N607	372.34	372.1393	41.94	5.41	22.57	30.08
C13H20N608	388.34	388.1342	40.21	5.19	21.64	32.96
C13H20N609	404.34	404.1292	38.62	4.99	20.78	35.61
C13H21N101	207.32	207.1623	75.32	10.21	6.76	7.72
C13H21N102	223.32	223.1572	69.92	9.48	6.27	14.33
C13H21N103	239.32	239.1521	65.25	8.84	5.85	20.06
C13H21N104	255.32	255.1470	61.16	8.29	5.49	25.07
C13H21N105	271.32	271.1420	57.55	7.80	5.16	29.48
C13H21N106	287.32	287.1369	54.35	7.37	4.88	33.41
C13H21N107	303.31	303.1318	51.48	6.98	4.62	36.92

C13H21N108	319.31	319.1267	48.90	6.63	4.39	40.08
C13H21N109	335.31	335.1216	46.57	6.31	4.18	42.94
C13H21N301	235.33	235.1685	66.35	8.99	17.86	6.80
C13H21N302	251.33	251.1634	62.13	8.42	16.72	12.73
C13H21N303	267.33	267.1583	58.41	7.92	15.72	17.95
C13H21N304	283.33	283.1532	55.11	7.47	14.83	22.59
C13H21N305	299.33	299.1481	52.16	7.07	14.04	26.73
C13H21N306	315.33	315.1430	49.52	6.71	13.33	30.44
C13H21N307	331.33	331.1379	47.13	6.39	12.68	33.80
C13H21N308	347.33	347.1329	44.96	6.09	12.10	36.85
C13H21N309	363.33	363.1278	42.98	5.83	11.57	39.63
C13H21N501	263.35	263.1746	59.29	8.04	26.59	6.08
C13H21N502	279.34	279.1695	55.90	7.58	25.07	11.45
C13H21N503	295.34	295.1644	52.87	7.17	23.71	16.25
C13H21N504	311.34	311.1593	50.15	6.80	22.49	20.56
C13H21N505	327.34	327.1543	47.70	6.47	21.39	24.44
C13H21N506	343.34	343.1492	45.48	6.17	20.40	27.96
C13H21N507	359.34	359.1441	43.45	5.89	19.49	31.17
C13H21N508	375.34	375.1390	41.60	5.64	18.66	34.10
C13H21N509	391.34	391.1339	39.90	5.41	17.90	36.80
C13H22N201	222.33	222.1732	70.23	9.97	12.60	7.20
C13H22N202	238.33	238.1681	65.52	9.30	11.75	13.43
C13H22N203	254.33	254.1630	61.39	8.72	11.01	18.87
C13H22N204	270.33	270.1579	57.76	8.20	10.36	23.67
C13H22N205	286.33	286.1529	54.53	7.74	9.78	27.94
C13H22N206	302.33	302.1478	51.65	7.33	9.27	31.75
C13H22N207	318.33	318.1427	49.05	6.97	8.80	35.18
C13H22N208	334.33	334.1376	46.70	6.63	8.38	38.28
C13H22N209	350.33	350.1325	44.57	6.33	8.00	41.10
C13H22N401	250.35	250.1794	62.37	8.86	22.38	6.39
C13H22N402	266.35	266.1743	58.62	8.33	21.04	12.01
C13H22N403	282.35	282.1692	55.30	7.85	19.84	17.00
C13H22N404	298.34	298.1641	52.34	7.43	18.78	21.45
C13H22N405	314.34	314.1590	49.67	7.05	17.82	25.45
C13H22N406	330.34	330.1539	47.27	6.71	16.96	29.06
C13H22N407	346.34	346.1488	45.08	6.40	16.18	32.34
C13H22N408	362.34	362.1437	43.09	6.12	15.46	35.32
C13H22N409	378.34	378.1387	41.27	5.86	14.81	38.06
C13H22N601	278.36	278.1855	56.09	7.97	30.19	5.75
C13H22N602	294.36	294.1804	53.05	7.53	28.55	10.87
C13H22N603	310.36	310.1753	50.31	7.15	27.08	15.47
C13H22N604	326.36	326.1702	47.84	6.79	25.75	19.61
C13H22N605	342.36	342.1652	45.61	6.48	24.55	23.37
C13H22N606	358.36	358.1601	43.57	6.19	23.45	26.79
C13H22N607	374.36	374.1550	41.71	5.92	22.45	29.92
C13H22N608	390.36	390.1499	40.00	5.68	21.53	32.79
C13H22N609	406.36	406.1448	38.43	5.46	20.68	35.44
C13H23N101	209.33	209.1780	74.59	11.07	6.69	7.64
C13H23N102	225.33	225.1729	69.29	10.29	6.22	14.20
C13H23N103	241.33	241.1678	64.70	9.61	5.80	19.89
C13H23N104	257.33	257.1627	60.68	9.01	5.44	24.87
C13H23N105	273.33	273.1576	57.13	8.48	5.12	29.27
C13H23N106	289.33	289.1525	53.97	8.01	4.84	33.18
C13H23N107	305.33	305.1474	51.14	7.59	4.59	36.68
C13H23N108	321.33	321.1424	48.59	7.21	4.36	39.83
C13H23N109	337.33	337.1373	46.29	6.87	4.15	42.69
C13H23N301	237.35	237.1841	65.79	9.77	17.70	6.74
C13H23N302	253.35	253.1790	61.63	9.15	16.59	12.63
C13H23N303	269.35	269.1739	57.97	8.61	15.60	17.82
C13H23N304	285.35	285.1688	54.72	8.12	14.73	22.43

066

C13H23N305	301.35	301.1638	51.82	7.69	13.94	26.55
C13H23N306	317.34	317.1587	49.20	7.31	13.24	30.25
C13H23N307	333.34	333.1536	46.84	6.95	12.61	33.60
C13H23N308	349.34	349.1485	44.70	6.64	12.03	36.64
C13H23N309	365.34	365.1434	42.74	6.35	11.50	39.41
C13H23N501	265.36	265.1902	58.84	8.74	26.39	6.03
C13H23N502	281.36	281.1852	55.50	8.24	24.89	11.37
C13H23N503	297.36	297.1801	52.51	7.80	23.55	16.14
C13H23N504	313.36	313.1750	49.83	7.40	22.35	20.42
C13H23N505	329.36	329.1699	47.41	7.04	21.26	24.29
C13H23N506	345.36	345.1648	45.21	6.71	20.28	27.80
C13H23N507	361.36	361.1597	43.21	6.42	19.38	30.99
C13H23N508	377.36	377.1546	41.38	6.14	18.56	33.92
C13H23N509	393.36	393.1496	39.70	5.89	17.80	36.61
C13H24N201	224.35	224.1889	69.60	10.78	12.49	7.13
C13H24N202	240.35	240.1838	64.97	10.07	11.66	13.31
C13H24N203	256.35	256.1787	60.91	9.44	10.93	18.72
C13H24N204	272.35	272.1736	57.33	8.88	10.29	23.50
C13H24N205	288.35	288.1685	54.15	8.39	9.72	27.74
C13H24N206	304.35	304.1634	51.31	7.95	9.20	31.54
C13H24N207	320.35	320.1583	48.74	7.55	8.74	34.96
C13H24N208	336.34	336.1533	46.42	7.19	8.33	38.05
C13H24N209	352.34	352.1482	44.32	6.87	7.95	40.87
C13H24N401	252.36	252.1950	61.87	9.59	22.20	6.34
C13H24N402	268.36	268.1899	58.18	9.01	20.88	11.92
C13H24N403	284.36	284.1848	54.91	8.51	19.70	16.88
C13H24N404	300.36	300.1797	51.99	8.05	18.65	21.31
C13H24N405	316.36	316.1747	49.36	7.65	17.71	25.29
C13H24N406	332.36	332.1696	46.98	7.28	16.86	28.88
C13H24N407	348.36	348.1645	44.82	6.94	16.08	32.15
C13H24N408	364.36	364.1594	42.85	6.64	15.38	35.13
C13H24N409	380.36	380.1543	41.05	6.36	14.73	37.86
C13H24N601	280.38	280.2011	55.69	8.63	29.97	5.71
C13H24N602	296.38	296.1961	52.68	8.16	28.36	10.80
C13H24N603	312.37	312.1910	49.99	7.74	26.90	15.37
C13H24N604	328.37	328.1859	47.55	7.37	25.59	19.49
C13H24N605	344.37	344.1808	45.34	7.02	24.40	23.23
C13H24N606	360.37	360.1757	43.33	6.71	23.32	26.64
C13H24N607	376.37	376.1706	41.49	6.43	22.33	29.76
C13H24N608	392.37	392.1655	39.80	6.17	21.42	32.62
C13H24N609	408.37	408.1605	38.24	5.92	20.58	35.26
C13H25N101	211.35	211.1936	73.88	11.92	6.63	7.57
C13H25N102	227.35	227.1885	68.68	11.08	6.16	14.07
C13H25N103	243.35	243.1834	64.16	10.36	5.76	19.72
C13H25N104	259.35	259.1783	60.21	9.72	5.40	24.68
C13H25N105	275.35	275.1733	56.71	9.15	5.09	29.05
C13H25N106	291.35	291.1682	53.59	8.65	4.81	32.95
C13H25N107	307.35	307.1631	50.80	8.20	4.56	36.44
C13H25N108	323.35	323.1580	48.29	7.79	4.33	39.58
C13H25N109	339.35	339.1529	46.01	7.43	4.13	42.43
C13H25N301	239.36	239.1998	65.23	10.53	17.55	6.68
C13H25N302	255.36	255.1947	61.15	9.87	16.46	12.53
C13H25N303	271.36	271.1896	57.54	9.29	15.48	17.69
C13H25N304	287.36	287.1845	54.34	8.77	14.62	22.27
C13H25N305	303.36	303.1794	51.47	8.31	13.85	26.37
C13H25N306	319.36	319.1743	48.89	7.89	13.16	30.06
C13H25N307	335.36	335.1692	46.56	7.51	12.53	33.40
C13H25N308	351.36	351.1641	44.44	7.17	11.96	36.43
C13H25N309	367.36	367.1591	42.50	6.86	11.44	39.20
C13H25N501	267.38	267.2059	58.40	9.42	26.19	5.98
C13H25N502	283.38	283.2008	55.10	8.89	24.71	11.29
C13H25N503	299.38	299.1957	52.16	8.42	23.39	16.03
C13H25N504	315.38	315.1906	49.51	7.99	22.21	20.29
C13H25N505	331.37	331.1856	47.12	7.60	21.13	24.14
C13H25N506	347.37	347.1805	44.95	7.25	20.16	27.63
C13H25N507	363.37	363.1754	42.97	6.93	19.27	30.82
C13H25N508	379.37	379.1703	41.16	6.64	18.46	33.74
C13H25N509	395.37	395.1652	39.49	6.37	17.71	36.42
C13H26N201	226.36	226.2045	68.98	11.58	12.38	7.07
C13H26N202	242.36	242.1994	64.43	10.81	11.56	13.20
C13H26N203	258.36	258.1943	60.44	10.14	10.84	18.58
C13H26N204	274.36	274.1892	56.91	9.55	10.21	23.33
C13H26N205	290.36	290.1842	53.78	9.03	9.65	27.55
C13H26N206	306.36	306.1791	50.97	8.55	9.14	31.33
C13H26N207	322.36	322.1740	48.44	8.13	8.69	34.74
C13H26N208	338.36	338.1689	46.15	7.75	8.28	37.83
C13H26N209	354.36	354.1638	44.06	7.40	7.91	40.64
C13H26N401	254.38	254.2106	61.38	10.30	22.02	6.29
C13H26N402	270.38	270.2056	57.75	9.69	20.72	11.83
C13H26N403	286.38	286.2005	54.52	9.15	19.56	16.76
C13H26N404	302.38	302.1954	51.64	8.67	18.53	21.16
C13H26N405	318.38	318.1903	49.04	8.23	17.60	25.13
C13H26N406	334.38	334.1852	46.70	7.84	16.76	28.71
C13H26N407	350.37	350.1801	44.57	7.48	15.99	31.96
C13H26N408	366.37	366.1750	42.62	7.15	15.29	34.94
C13H26N409	382.37	382.1700	40.84	6.85	14.65	37.66
C13H26N601	282.39	282.2168	55.29	9.28	29.76	5.67
C13H26N602	298.39	298.2117	52.33	8.78	28.16	10.72
C13H26N603	314.39	314.2066	49.67	8.34	26.73	15.27
C13H26N604	330.39	330.2015	47.26	7.93	25.44	19.37
C13H26N605	346.39	346.1965	45.08	7.57	24.26	23.09
C13H26N606	362.39	362.1914	43.09	7.23	23.19	26.49
C13H26N607	378.39	378.1863	41.27	6.93	22.21	29.60
C13H26N608	394.39	394.1812	39.59	6.65	21.31	32.45
C13H26N609	410.39	410.1761	38.05	6.39	20.48	35.09
C13H27N101	213.37	213.2093	73.18	12.76	6.56	7.50
C13H27N102	229.37	229.2042	68.08	11.87	6.11	13.95
C13H27N103	245.37	245.1991	63.64	11.09	5.71	19.56
C13H27N104	261.36	261.1940	59.74	10.41	5.36	24.49
C13H27N105	277.36	277.1889	56.30	9.81	5.05	28.84
C13H27N106	293.36	293.1838	53.23	9.28	4.77	32.72
C13H27N107	309.36	309.1787	50.47	8.80	4.53	36.20
C13H27N108	325.36	325.1737	47.99	8.36	4.30	39.34
C13H27N109	341.36	341.1686	45.74	7.97	4.10	42.18
C13H27N301	241.38	241.2154	64.69	11.27	17.41	6.63
C13H27N302	257.38	257.2103	60.67	10.57	16.33	12.43
C13H27N303	273.38	273.2052	57.12	9.96	15.37	17.56
C13H27N304	289.38	289.2001	53.96	9.40	14.52	22.12
C13H27N305	305.38	305.1951	51.13	8.91	13.76	26.20
C13H27N306	321.38	321.1900	48.59	8.47	13.08	29.87
C13H27N307	337.38	337.1849	46.28	8.07	12.45	33.20
C13H27N308	353.38	353.1798	44.19	7.70	11.89	36.22
C13H27N309	369.38	369.1747	42.27	7.37	11.38	38.98
C13H27N501	269.39	269.2215	57.96	10.10	26.00	5.94
C13H27N502	285.39	285.2165	54.71	9.54	24.54	11.21
C13H27N503	301.39	301.2114	51.81	9.03	23.24	15.93
C13H27N504	317.39	317.2063	49.20	8.57	22.07	20.16
C13H27N505	333.39	333.2012	46.84	8.16	21.01	23.99
C13H27N506	349.39	349.1961	44.69	7.79	20.04	27.48
C13H27N507	365.39	365.1910	42.73	7.45	19.17	30.65

067

C13H27N508	381.39	381.1859	40.94	7.14	18.36	33.56
C13H27N509	397.39	397.1809	39.29	6.85	17.62	36.24
C13H28N201	228.38	228.2202	68.37	12.36	12.27	7.01
C13H28N202	244.38	244.2151	63.89	11.55	11.46	13.09
C13H28N203	260.38	260.2100	59.97	10.84	10.76	18.43
C13H28N204	276.38	276.2049	56.50	10.21	10.14	23.16
C13H28N205	292.38	292.1998	53.41	9.65	9.58	27.36
C13H28N206	308.38	308.1947	50.63	9.15	9.08	31.13
C13H28N207	324.38	324.1896	48.14	8.70	8.64	34.53
C13H28N208	340.38	340.1845	45.87	8.29	8.23	37.60
C13H28N209	356.38	356.1795	43.81	7.92	7.86	40.41
C13H28N401	256.39	256.2263	60.90	11.01	21.85	6.24
C13H28N402	272.39	272.2212	57.32	10.36	20.57	11.75
C13H28N403	288.39	288.2161	54.14	9.79	19.43	16.64
C13H28N404	304.39	304.2110	51.30	9.27	18.41	21.02
C13H28N405	320.39	320.2060	48.74	8.81	17.49	24.97
C13H28N406	336.39	336.2009	46.42	8.39	16.66	28.54
C13H28N407	352.39	352.1958	44.31	8.01	15.90	31.78
C13H28N408	368.39	368.1907	42.39	7.66	15.21	34.74
C13H28N409	384.39	384.1856	40.62	7.34	14.58	37.46
C13H28N601	284.41	284.2324	54.90	9.92	29.55	5.63
C13H28N602	300.41	300.2274	51.98	9.39	27.98	10.65
C13H28N603	316.41	316.2223	49.35	8.92	26.56	15.17
C13H28N604	332.41	332.2172	46.97	8.49	25.28	19.25
C13H28N605	348.41	348.2121	44.82	8.10	24.12	22.96
C13H28N606	364.40	364.2070	42.85	7.75	23.06	26.34
C13H28N607	380.40	380.2019	41.05	7.42	22.09	29.44
C13H28N608	396.40	396.1968	39.39	7.12	21.20	32.29
C13H28N609	412.40	412.1918	37.86	6.84	20.38	34.92
C13H29N101	215.38	215.2249	72.50	13.57	6.50	7.43
C13H29N102	231.38	231.2198	67.48	12.63	6.05	13.83
C13H29N103	247.38	247.2147	63.12	11.82	5.66	19.40
C13H29N104	263.38	263.2096	59.28	11.10	5.32	24.30
C13H29N105	279.38	279.2046	55.89	10.46	5.01	28.63
C13H29N106	295.38	295.1995	52.86	9.90	4.74	32.50
C13H29N107	311.38	311.1944	50.15	9.39	4.50	35.97
C13H29N108	327.38	327.1893	47.70	8.93	4.28	39.10
C13H29N109	343.38	343.1842	45.47	8.51	4.08	41.93
C13H29N301	243.40	243.2310	64.15	12.01	17.26	6.57
C13H29N302	259.39	259.2260	60.20	11.27	16.20	12.34
C13H29N303	275.39	275.2209	56.70	10.61	15.26	17.43
C13H29N304	291.39	291.2158	53.59	10.03	14.42	21.96
C13H29N305	307.39	307.2107	50.80	9.51	13.67	26.02
C13H29N306	323.39	323.2056	48.28	9.04	12.99	29.68
C13H29N307	339.39	339.2005	46.01	8.61	12.38	33.00
C13H29N308	355.39	355.1954	43.94	8.23	11.82	36.02
C13H29N309	371.39	371.1904	42.04	7.87	11.31	38.77
C13H29N501	271.41	271.2372	57.53	10.77	25.80	5.89
C13H29N502	287.41	287.2321	54.33	10.17	24.37	11.13
C13H29N503	303.41	303.2270	51.46	9.63	23.08	15.82
C13H29N504	319.41	319.2219	48.89	9.15	21.93	20.04
C13H29N505	335.41	335.2169	46.55	8.72	20.88	23.85
C13H29N506	351.41	351.2118	44.43	8.32	19.93	27.32
C13H29N507	367.41	367.2067	42.50	7.96	19.06	30.48
C13H29N508	383.40	383.2016	40.73	7.62	18.27	33.38
C13H29N509	399.40	399.1965	39.09	7.32	17.53	36.05
C13H30N201	230.40	230.2358	67.77	13.12	12.16	6.94
C13H30N202	246.40	246.2307	63.37	12.27	11.37	12.99
C13H30N203	262.40	262.2256	59.51	11.52	10.68	18.29
C13H30N204	278.40	278.2205	56.09	10.86	10.06	22.99
C13H30N205	294.39	294.2155	53.04	10.27	9.52	27.17
C13H30N206	310.39	310.2104	50.31	9.74	9.03	30.93
C13H30N207	326.39	326.2053	47.84	9.26	8.58	34.31
C13H30N208	342.39	342.2002	45.60	8.83	8.18	37.38
C13H30N209	358.39	358.1951	43.57	8.44	7.82	40.18
C13H30N401	258.41	258.2419	60.43	11.70	21.68	6.19
C13H30N402	274.41	274.2369	56.90	11.02	20.42	11.66
C13H30N403	290.41	290.2318	53.77	10.41	19.29	16.53
C13H30N404	306.41	306.2267	50.96	9.87	18.29	20.89
C13H30N405	322.41	322.2216	48.43	9.38	17.38	24.81
C13H30N406	338.41	338.2165	46.14	8.94	16.56	28.37
C13H30N407	354.41	354.2114	44.06	8.53	15.81	31.60
C13H30N408	370.41	370.2063	42.16	8.16	15.13	34.56
C13H30N409	386.41	386.2013	40.41	7.83	14.50	37.27
C13H30N601	286.42	286.2481	54.52	10.56	29.34	5.59
C13H30N602	302.42	302.2430	51.63	10.00	27.79	10.58
C13H30N603	318.42	318.2379	49.04	9.50	26.39	15.07
C13H30N604	334.42	334.2328	46.69	9.04	25.13	19.14
C13H30N605	350.42	350.2278	44.56	8.63	23.98	22.83
C13H30N606	366.42	366.2227	42.61	8.25	22.94	26.20
C13H30N607	382.42	382.2176	40.83	7.91	21.98	29.29
C13H30N608	398.42	398.2125	39.19	7.59	21.09	32.13
C13H30N609	414.42	414.2074	37.68	7.30	20.28	34.75
C13H31N301	245.41	245.2467	63.63	12.73	17.12	6.52
C13H31N302	261.41	261.2416	59.73	11.95	16.07	12.24
C13H31N303	277.41	277.2365	56.29	11.26	15.15	17.30
C13H31N304	293.41	293.2314	53.22	10.65	14.32	21.81
C13H31N305	309.41	309.2264	50.47	10.10	13.58	25.85
C13H31N306	325.41	325.2213	47.98	9.60	12.91	29.50
C13H31N307	341.41	341.2162	45.74	9.15	12.31	32.80
C13H31N308	357.41	357.2111	43.69	8.74	11.76	35.81
C13H31N309	373.41	373.2060	41.82	8.37	11.25	38.56
C13H31N501	273.42	273.2528	57.11	11.43	25.61	5.85
C13H31N502	289.42	289.2478	53.95	10.80	24.20	11.06
C13H31N503	305.42	305.2427	51.12	10.23	22.93	15.72
C13H31N504	321.42	321.2376	48.58	9.72	21.79	19.91
C13H31N505	337.42	337.2325	46.28	9.26	20.76	23.71
C13H31N506	353.42	353.2274	44.18	8.84	19.82	27.16
C13H31N507	369.42	369.2223	42.27	8.46	18.96	30.32
C13H31N508	385.42	385.2172	40.51	8.11	18.17	33.21
C13H31N509	401.42	401.2122	38.90	7.78	17.45	35.87
C13H32N401	260.43	260.2576	59.96	12.39	21.51	6.14
C13H32N402	276.43	276.2525	56.49	11.67	20.27	11.58
C13H32N403	292.42	292.2474	53.40	11.03	19.16	16.41
C13H32N404	308.42	308.2423	50.63	10.46	18.17	20.75
C13H32N405	324.42	324.2373	48.13	9.94	17.27	24.66
C13H32N406	340.42	340.2322	45.87	9.47	16.46	28.20
C13H32N407	356.42	356.2271	43.81	9.05	15.72	31.42
C13H32N408	372.42	372.2220	41.93	8.66	15.04	34.37
C13H32N409	388.42	388.2169	40.20	8.30	14.42	37.07
C13H32N601	288.44	288.2637	54.13	11.18	29.14	5.55
C13H32N602	304.44	304.2587	51.29	10.59	27.60	10.51
C13H32N603	320.44	320.2536	48.73	10.07	26.23	14.98
C13H32N604	336.44	336.2485	46.41	9.59	24.98	19.02
C13H32N605	352.44	352.2434	44.30	9.15	23.85	22.70
C13H32N606	368.44	368.2383	42.38	8.75	22.81	26.06
C13H32N607	384.44	384.2332	40.62	8.39	21.86	29.13
C13H32N608	400.44	400.2281	38.99	8.05	20.99	31.96
C13H32N609	416.43	416.2231	37.50	7.75	20.18	34.58
C13H33N501	275.44	275.2685	56.69	12.08	25.43	5.81

C13H33N502	291.44	291.2634	53.58	11.41	24.03	10.98
C13H33N503	307.44	307.2583	50.79	10.82	22.78	15.61
C13H33N504	323.44	323.2532	48.28	10.28	21.65	19.79
C13H33N505	339.44	339.2482	46.00	9.80	20.63	23.57
C13H33N506	355.44	355.2431	43.93	9.36	19.70	27.01
C13H33N507	371.44	371.2380	42.04	8.96	18.85	30.15
C13H33N508	387.44	387.2329	40.30	8.59	18.08	33.04
C13H33N509	403.44	403.2278	38.70	8.24	17.36	35.69
C13H34N601	290.46	290.2794	53.76	11.80	28.93	5.51
C13H34N602	306.45	306.2743	50.95	11.18	27.42	10.44
C13H34N603	322.45	322.2692	48.42	10.63	26.06	14.89
C13H34N604	338.45	338.2641	46.13	10.13	24.83	18.91
C13H34N605	354.45	354.2590	44.05	9.67	23.71	22.57
C13H34N606	370.45	370.2540	42.15	9.25	22.69	25.91
C13H34N607	386.45	386.2489	40.40	8.87	21.75	28.98
C13H34N608	402.45	402.2438	38.80	8.52	20.88	31.80
C13H34N609	418.45	418.2387	37.32	8.19	20.08	34.41
C14H 7N101	205.22	205.0528	81.94	3.44	6.83	7.80
C14H 7N102	221.22	221.0477	76.01	3.19	6.33	14.46
C14H 7N103	237.22	237.0426	70.89	2.97	5.90	20.23
C14H 7N104	253.22	253.0375	66.41	2.79	5.53	25.27
C14H 7N105	269.22	269.0324	62.46	2.62	5.20	29.71
C14H 7N106	285.21	285.0273	58.96	2.47	4.91	33.66
C14H 7N107	301.21	301.0222	55.83	2.34	4.65	37.18
C14H 7N108	317.21	317.0172	53.01	2.22	4.42	40.35
C14H 7N109	333.21	333.0121	50.47	2.12	4.20	43.21
C14H 7N301	233.23	233.0589	72.10	3.03	18.02	6.86
C14H 7N302	249.23	249.0538	67.47	2.83	16.86	12.84
C14H 7N303	265.23	265.0487	63.40	2.66	15.84	18.10
C14H 7N304	281.23	281.0437	59.79	2.51	14.94	22.76
C14H 7N305	297.23	297.0386	56.57	2.37	14.14	26.91
C14H 7N306	313.23	313.0335	53.68	2.25	13.42	30.65
C14H 7N307	329.23	329.0284	51.08	2.14	12.76	34.02
C14H 7N308	345.23	345.0233	48.71	2.04	12.17	37.08
C14H 7N309	361.23	361.0182	46.55	1.95	11.63	39.86
C14H 7N501	261.24	261.0651	64.37	2.70	26.81	6.12
C14H 7N502	277.24	277.0600	60.65	2.54	25.26	11.54
C14H 7N503	293.24	293.0549	57.34	2.41	23.88	16.37
C14H 7N504	309.24	309.0498	54.38	2.28	22.65	20.69
C14H 7N505	325.24	325.0447	51.70	2.17	21.53	24.60
C14H 7N506	341.24	341.0396	49.28	2.07	20.52	28.13
C14H 7N507	357.24	357.0345	47.07	1.98	19.60	31.35
C14H 7N508	373.24	373.0295	45.05	1.89	18.76	34.29
C14H 7N509	389.24	389.0244	43.20	1.81	17.99	36.99
C14H 8N201	220.23	220.0637	76.35	3.66	12.72	7.26
C14H 8N202	236.23	236.0586	71.18	3.41	11.86	13.55
C14H 8N203	252.23	252.0535	66.67	3.20	11.11	19.03
C14H 8N204	268.23	268.0484	62.69	3.01	10.44	23.86
C14H 8N205	284.23	284.0433	59.16	2.84	9.86	28.15
C14H 8N206	300.23	300.0382	56.01	2.69	9.33	31.97
C14H 8N207	316.23	316.0331	53.18	2.55	8.86	35.42
C14H 8N208	332.23	332.0281	50.61	2.43	8.43	38.53
C14H 8N209	348.23	348.0230	48.29	2.32	8.04	41.35
C14H 8N401	248.25	248.0698	67.74	3.25	22.57	6.44
C14H 8N402	264.25	264.0647	63.64	3.05	21.20	12.11
C14H 8N403	280.24	280.0596	60.00	2.88	19.99	17.13
C14H 8N404	296.24	296.0545	56.76	2.72	18.91	21.60
C14H 8N405	312.24	312.0495	53.85	2.58	17.94	25.62
C14H 8N406	328.24	328.0444	51.23	2.46	17.07	29.25
C14H 8N407	344.24	344.0393	48.85	2.34	16.28	32.53
C14H 8N408	360.24	360.0342	46.68	2.24	15.55	35.53
C14H 8N409	376.24	376.0291	44.69	2.14	14.89	38.27
C14H 8N601	276.26	276.0760	60.87	2.92	30.42	5.79
C14H 8N602	292.26	292.0709	57.54	2.76	28.76	10.95
C14H 8N603	308.26	308.0658	54.55	2.62	27.26	15.57
C14H 8N604	324.26	324.0607	51.86	2.49	25.92	19.74
C14H 8N605	340.26	340.0556	49.42	2.37	24.70	23.51
C14H 8N606	356.26	356.0505	47.20	2.26	23.59	26.95
C14H 8N607	372.26	372.0454	45.17	2.17	22.58	30.09
C14H 8N608	388.26	388.0404	43.31	2.08	21.65	32.97
C14H 8N609	404.25	404.0353	41.60	1.99	20.79	35.62
C14H 9N101	207.23	207.0684	81.14	4.38	6.76	7.72
C14H 9N102	223.23	223.0633	75.33	4.06	6.27	14.33
C14H 9N103	239.23	239.0582	70.29	3.79	5.85	20.06
C14H 9N104	255.23	255.0532	65.88	3.55	5.49	25.07
C14H 9N105	271.23	271.0481	62.00	3.34	5.16	29.49
C14H 9N106	287.23	287.0430	58.54	3.16	4.88	33.42
C14H 9N107	303.23	303.0379	55.45	2.99	4.62	36.93
C14H 9N108	319.23	319.0328	52.68	2.84	4.39	40.10
C14H 9N109	335.23	335.0277	50.16	2.71	4.18	42.95
C14H 9N301	235.25	235.0746	71.48	3.86	17.86	6.80
C14H 9N302	251.25	251.0695	66.93	3.61	16.72	12.74
C14H 9N303	267.25	267.0644	62.92	3.39	15.72	17.96
C14H 9N304	283.25	283.0593	59.37	3.20	14.84	22.59
C14H 9N305	299.24	299.0542	56.19	3.03	14.04	26.73
C14H 9N306	315.24	315.0491	53.34	2.88	13.33	30.45
C14H 9N307	331.24	331.0440	50.77	2.74	12.69	33.81
C14H 9N308	347.24	347.0390	48.43	2.61	12.10	36.86
C14H 9N309	363.24	363.0339	46.29	2.50	11.57	39.64
C14H 9N501	263.26	263.0807	63.87	3.45	26.60	6.08
C14H 9N502	279.26	279.0756	60.21	3.25	25.08	11.46
C14H 9N503	295.26	295.0705	56.95	3.07	23.72	16.26
C14H 9N504	311.26	311.0654	54.02	2.91	22.50	20.56
C14H 9N505	327.26	327.0604	51.38	2.77	21.40	24.44
C14H 9N506	343.26	343.0553	48.99	2.64	20.40	27.97
C14H 9N507	359.26	359.0502	46.81	2.53	19.49	31.17
C14H 9N508	375.26	375.0451	44.81	2.42	18.66	34.11
C14H 9N509	391.26	391.0400	42.98	2.32	17.90	36.80
C14H10N201	222.25	222.0793	75.66	4.54	12.60	7.20
C14H10N202	238.25	238.0742	70.58	4.23	11.76	13.43
C14H10N203	254.25	254.0691	66.14	3.96	11.02	18.88
C14H10N204	270.25	270.0641	62.22	3.73	10.37	23.68
C14H10N205	286.25	286.0590	58.75	3.52	9.79	27.95
C14H10N206	302.25	302.0539	55.64	3.33	9.27	31.76
C14H10N207	318.24	318.0488	52.84	3.17	8.80	35.19
C14H10N208	334.24	334.0437	50.31	3.02	8.38	38.29
C14H10N209	350.24	350.0386	48.01	2.88	8.00	41.11
C14H10N401	250.26	250.0855	67.19	4.03	22.39	6.39
C14H10N402	266.26	266.0804	63.15	3.79	21.04	12.02
C14H10N403	282.26	282.0753	59.57	3.57	19.85	17.00
C14H10N404	298.26	298.0702	56.38	3.38	18.78	21.46
C14H10N405	314.26	314.0651	53.51	3.21	17.83	25.46
C14H10N406	330.26	330.0600	50.92	3.05	16.96	29.07
C14H10N407	346.26	346.0549	48.56	2.91	16.18	32.34
C14H10N408	362.26	362.0499	46.42	2.78	15.47	35.33
C14H10N409	378.26	378.0448	44.46	2.66	14.81	38.07
C14H10N601	278.28	278.0916	60.43	3.62	30.20	5.75
C14H10N602	294.27	294.0865	57.14	3.43	28.56	10.87
C14H10N603	310.27	310.0814	54.20	3.25	27.09	15.47
C14H10N604	326.27	326.0763	51.54	3.09	25.76	19.61

069

C14H10N605	342.27	342.0713	49.13	2.94	24.55	23.37
C14H10N606	358.27	358.0662	46.94	2.81	23.46	26.79
C14H10N607	374.27	374.0611	44.93	2.69	22.45	29.92
C14H10N608	390.27	390.0560	43.09	2.58	21.53	32.80
C14H10N609	406.27	406.0509	41.39	2.48	20.69	35.44
C14H11N101	209.25	209.0841	80.36	5.30	6.69	7.65
C14H11N102	225.25	225.0790	74.65	4.92	6.22	14.21
C14H11N103	241.25	241.0739	69.70	4.60	5.81	19.90
C14H11N104	257.25	257.0688	65.37	4.31	5.44	24.88
C14H11N105	273.25	273.0637	61.54	4.06	5.13	29.28
C14H11N106	289.25	289.0586	58.14	3.83	4.84	33.19
C14H11N107	305.25	305.0535	55.09	3.63	4.59	36.69
C14H11N108	321.25	321.0485	52.35	3.45	4.36	39.84
C14H11N109	337.25	337.0434	49.86	3.29	4.15	42.70
C14H11N301	237.26	237.0902	70.87	4.67	17.71	6.74
C14H11N302	253.26	253.0851	66.40	4.38	16.59	12.63
C14H11N303	269.26	269.0800	62.45	4.12	15.61	17.83
C14H11N304	285.26	285.0749	58.95	3.89	14.73	22.43
C14H11N305	301.26	301.0699	55.82	3.68	13.95	26.55
C14H11N306	317.26	317.0648	53.00	3.49	13.24	30.26
C14H11N307	333.26	333.0597	50.46	3.33	12.61	33.61
C14H11N308	349.26	349.0546	48.15	3.17	12.03	36.65
C14H11N309	365.26	365.0495	46.04	3.04	11.50	39.42
C14H11N501	265.28	265.0964	63.39	4.18	26.40	6.03
C14H11N502	281.28	281.0913	59.78	3.94	24.90	11.38
C14H11N503	297.28	297.0862	56.57	3.73	23.56	16.15
C14H11N504	313.27	313.0811	53.68	3.54	22.36	20.43
C14H11N505	329.27	329.0760	51.07	3.37	21.27	24.29
C14H11N506	345.27	345.0709	48.70	3.21	20.28	27.80
C14H11N507	361.27	361.0658	46.55	3.07	19.39	31.00
C14H11N508	377.27	377.0608	44.57	2.94	18.56	33.93
C14H11N509	393.27	393.0557	42.76	2.82	17.81	36.61
C14H12N201	224.26	224.0950	74.98	5.39	12.49	7.13
C14H12N202	240.26	240.0899	69.99	5.03	11.66	13.32
C14H12N203	256.26	256.0848	65.62	4.72	10.93	18.73
C14H12N204	272.26	272.0797	61.76	4.44	10.29	23.51
C14H12N205	288.26	288.0746	58.33	4.20	9.72	27.75
C14H12N206	304.26	304.0695	55.27	3.98	9.21	31.55
C14H12N207	320.26	320.0644	52.51	3.78	8.75	34.97
C14H12N208	336.26	336.0594	50.01	3.60	8.33	38.06
C14H12N209	352.26	352.0543	47.74	3.43	7.95	40.88
C14H12N401	252.28	252.1011	66.66	4.79	22.21	6.34
C14H12N402	268.28	268.0960	62.68	4.51	20.88	11.93
C14H12N403	284.28	284.0909	59.15	4.25	19.71	16.88
C14H12N404	300.28	300.0858	56.00	4.03	18.66	21.31
C14H12N405	316.28	316.0808	53.17	3.82	17.71	25.29
C14H12N406	332.27	332.0757	50.61	3.64	16.86	28.89
C14H12N407	348.27	348.0706	48.28	3.47	16.09	32.16
C14H12N408	364.27	364.0655	46.16	3.32	15.38	35.14
C14H12N409	380.27	380.0604	44.22	3.18	14.73	37.87
C14H12N601	280.29	280.1073	59.99	4.32	29.98	5.71
C14H12N602	296.29	296.1022	56.75	4.08	28.36	10.80
C14H12N603	312.29	312.0971	53.85	3.87	26.91	15.37
C14H12N604	328.29	328.0920	51.22	3.68	25.60	19.49
C14H12N605	344.29	344.0869	48.84	3.51	24.41	23.24
C14H12N606	360.29	360.0818	46.67	3.36	23.33	26.64
C14H12N607	376.29	376.0767	44.69	3.21	22.33	29.76
C14H12N608	392.29	392.0717	42.87	3.08	21.42	32.63
C14H12N609	408.29	408.0666	41.19	2.96	20.58	35.27
C14H13N101	211.27	211.0997	79.59	6.20	6.63	7.57
C14H13N102	227.27	227.0946	73.99	5.77	6.16	14.08
C14H13N103	243.26	243.0895	69.12	5.39	5.76	19.73
C14H13N104	259.26	259.0844	64.86	5.05	5.40	24.68
C14H13N105	275.26	275.0794	61.09	4.76	5.09	29.06
C14H13N106	291.26	291.0743	57.73	4.50	4.81	32.96
C14H13N107	307.26	307.0692	54.73	4.26	4.56	36.45
C14H13N108	323.26	323.0641	52.02	4.05	4.33	39.59
C14H13N109	339.26	339.0590	49.57	3.86	4.13	42.44
C14H13N301	239.28	239.1059	70.28	5.48	17.56	6.69
C14H13N302	255.28	255.1008	65.87	5.13	16.46	12.53
C14H13N303	271.28	271.0957	61.99	4.83	15.49	17.69
C14H13N304	287.28	287.0906	58.53	4.56	14.63	22.28
C14H13N305	303.28	303.0855	55.45	4.32	13.86	26.38
C14H13N306	319.28	319.0804	52.67	4.10	13.16	30.07
C14H13N307	335.28	335.0753	50.15	3.91	12.53	33.40
C14H13N308	351.28	351.0703	47.87	3.73	11.96	36.44
C14H13N309	367.27	367.0652	45.78	3.57	11.44	39.21
C14H13N501	267.29	267.1120	62.91	4.90	26.20	5.99
C14H13N502	283.29	283.1069	59.36	4.63	24.72	11.30
C14H13N503	299.29	299.1018	56.18	4.38	23.40	16.04
C14H13N504	315.29	315.0967	53.33	4.16	22.21	20.30
C14H13N505	331.29	331.0917	50.76	3.96	21.14	24.15
C14H13N506	347.29	347.0866	48.42	3.77	20.17	27.64
C14H13N507	363.29	363.0815	46.29	3.61	19.28	30.83
C14H13N508	379.29	379.0764	44.33	3.45	18.46	33.75
C14H13N509	395.29	395.0713	42.54	3.31	17.72	36.43
C14H14N201	226.28	226.1106	74.31	6.24	12.38	7.07
C14H14N202	242.28	242.1055	69.41	5.82	11.56	13.21
C14H14N203	258.28	258.1004	65.11	5.46	10.85	18.58
C14H14N204	274.28	274.0953	61.31	5.14	10.21	23.33
C14H14N205	290.28	290.0903	57.93	4.86	9.65	27.56
C14H14N206	306.28	306.0852	54.90	4.61	9.15	31.34
C14H14N207	322.28	322.0801	52.18	4.38	8.69	34.75
C14H14N208	338.28	338.0750	49.71	4.17	8.28	37.84
C14H14N209	354.28	354.0699	47.46	3.98	7.91	40.64
C14H14N401	254.29	254.1168	66.13	5.55	22.03	6.29
C14H14N402	270.29	270.1117	62.21	5.22	20.73	11.84
C14H14N403	286.29	286.1066	58.74	4.93	19.57	16.77
C14H14N404	302.29	302.1015	55.63	4.67	18.53	21.17
C14H14N405	318.29	318.0964	52.83	4.43	17.60	25.13
C14H14N406	334.29	334.0913	50.30	4.22	16.76	28.72
C14H14N407	350.29	350.0862	48.00	4.03	15.99	31.97
C14H14N408	366.29	366.0812	45.91	3.85	15.30	34.94
C14H14N409	382.29	382.0761	43.99	3.69	14.66	37.67
C14H14N601	282.31	282.1229	59.56	5.00	29.77	5.67
C14H14N602	298.31	298.1178	56.37	4.73	28.17	10.73
C14H14N603	314.31	314.1127	53.50	4.49	26.74	15.27
C14H14N604	330.31	330.1076	50.91	4.27	25.44	19.38
C14H14N605	346.30	346.1026	48.56	4.07	24.27	23.10
C14H14N606	362.30	362.0975	46.41	3.89	23.20	26.50
C14H14N607	378.30	378.0924	44.45	3.73	22.22	29.60
C14H14N608	394.30	394.0873	42.65	3.58	21.31	32.46
C14H14N609	410.30	410.0822	40.98	3.44	20.48	35.09
C14H15N101	213.28	213.1154	78.84	7.09	6.57	7.50
C14H15N102	229.28	229.1103	73.34	6.59	6.11	13.96
C14H15N103	245.28	245.1052	68.56	6.16	5.71	19.57
C14H15N104	261.28	261.1001	64.36	5.79	5.36	24.49
C14H15N105	277.28	277.0950	60.65	5.45	5.05	28.85
C14H15N106	293.28	293.0899	57.34	5.16	4.78	32.73
C14H15N107	309.28	309.0848	54.37	4.89	4.53	36.21

070

C14H15N108	325.28	325.0798	51.70	4.65	4.31	39.35
C14H15N109	341.28	341.0747	49.27	4.43	4.10	42.19
C14H15N301	241.30	241.1215	69.69	6.27	17.41	6.63
C14H15N302	257.29	257.1164	65.36	5.88	16.33	12.44
C14H15N303	273.29	273.1113	61.53	5.53	15.38	17.56
C14H15N304	289.29	289.1062	58.13	5.23	14.53	22.12
C14H15N305	305.29	305.1012	55.08	4.95	13.76	26.20
C14H15N306	321.29	321.0961	52.34	4.71	13.08	29.88
C14H15N307	337.29	337.0910	49.85	4.48	12.46	33.20
C14H15N308	353.29	353.0859	47.60	4.28	11.89	36.23
C14H15N309	369.29	369.0808	45.53	4.09	11.38	38.99
C14H15N501	269.31	269.1277	62.44	5.61	26.00	5.94
C14H15N502	285.31	285.1226	58.94	5.30	24.55	11.22
C14H15N503	301.31	301.1175	55.81	5.02	23.24	15.93
C14H15N504	317.31	317.1124	52.99	4.76	22.07	20.17
C14H15N505	333.31	333.1073	50.45	4.54	21.01	24.00
C14H15N506	349.31	349.1022	48.14	4.33	20.05	27.48
C14H15N507	365.30	365.0971	46.03	4.14	19.17	30.66
C14H15N508	381.30	381.0921	44.10	3.97	18.37	33.57
C14H15N509	397.30	397.0870	42.32	3.81	17.63	36.24
C14H16N201	228.30	228.1263	73.66	7.06	12.27	7.01
C14H16N202	244.30	244.1212	68.83	6.60	11.47	13.10
C14H16N203	260.30	260.1161	64.60	6.20	10.76	18.44
C14H16N204	276.29	276.1110	60.86	5.84	10.14	23.16
C14H16N205	292.29	292.1059	57.53	5.52	9.58	27.37
C14H16N206	308.29	308.1008	54.54	5.23	9.09	31.14
C14H16N207	324.29	324.0957	51.85	4.97	8.64	34.54
C14H16N208	340.29	340.0907	49.42	4.74	8.23	37.61
C14H16N209	356.29	356.0856	47.20	4.53	7.86	40.41
C14H16N401	256.31	256.1324	65.61	6.29	21.86	6.24
C14H16N402	272.31	272.1273	61.75	5.92	20.57	11.75
C14H16N403	288.31	288.1222	58.33	5.59	19.43	16.65
C14H16N404	304.31	304.1171	55.26	5.30	18.41	21.03
C14H16N405	320.31	320.1121	52.50	5.04	17.49	24.98
C14H16N406	336.31	336.1070	50.00	4.80	16.66	28.54
C14H16N407	352.31	352.1019	47.73	4.58	15.90	31.79
C14H16N408	368.31	368.0968	45.66	4.38	15.21	34.75
C14H16N409	384.31	384.0917	43.76	4.20	14.58	37.47
C14H16N601	284.32	284.1386	59.14	5.67	29.56	5.63
C14H16N602	300.32	300.1335	55.99	5.37	27.98	10.65
C14H16N603	316.32	316.1284	53.16	5.10	26.57	15.17
C14H16N604	332.32	332.1233	50.60	4.85	25.29	19.26
C14H16N605	348.32	348.1182	48.28	4.63	24.13	22.97
C14H16N606	364.32	364.1131	46.16	4.43	23.07	26.35
C14H16N607	380.32	380.1080	44.21	4.24	22.10	29.45
C14H16N608	396.32	396.1029	42.43	4.07	21.21	32.30
C14H16N609	412.32	412.0979	40.78	3.91	20.38	34.92
C14H17N101	215.30	215.1310	78.10	7.96	6.51	7.43
C14H17N102	231.30	231.1259	72.70	7.41	6.06	13.83
C14H17N103	247.30	247.1208	68.00	6.93	5.66	19.41
C14H17N104	263.30	263.1157	63.87	6.51	5.32	24.31
C14H17N105	279.30	279.1107	60.21	6.14	5.02	28.64
C14H17N106	295.29	295.1056	56.95	5.80	4.74	32.51
C14H17N107	311.29	311.1005	54.02	5.50	4.50	35.98
C14H17N108	327.29	327.0954	51.38	5.24	4.28	39.11
C14H17N109	343.29	343.0903	48.98	4.99	4.08	41.95
C14H17N301	243.31	243.1372	69.11	7.04	17.27	6.58
C14H17N302	259.31	259.1321	64.85	6.61	16.20	12.34
C14H17N303	275.31	275.1270	61.08	6.22	15.26	17.43
C14H17N304	291.31	291.1219	57.72	5.88	14.42	21.97
C14H17N305	307.31	307.1168	54.72	5.58	13.67	26.03
C14H17N306	323.31	323.1117	52.01	5.30	13.00	29.69
C14H17N307	339.31	339.1066	49.56	5.05	12.38	33.01
C14H17N308	355.31	355.1016	47.33	4.82	11.83	36.02
C14H17N309	371.31	371.0965	45.29	4.61	11.32	38.78
C14H17N501	271.32	271.1433	61.98	6.32	25.81	5.90
C14H17N502	287.32	287.1382	58.52	5.96	24.37	11.14
C14H17N503	303.32	303.1331	55.44	5.65	23.09	15.82
C14H17N504	319.32	319.1280	52.66	5.37	21.93	20.04
C14H17N505	335.32	335.1230	50.15	5.11	20.89	23.86
C14H17N506	351.32	351.1179	47.86	4.88	19.93	27.32
C14H17N507	367.32	367.1128	45.78	4.66	19.07	30.49
C14H17N508	383.32	383.1077	43.87	4.47	18.27	33.39
C14H17N509	399.32	399.1026	42.11	4.29	17.54	36.06
C14H18N201	230.31	230.1419	73.01	7.88	12.16	6.95
C14H18N202	246.31	246.1368	68.27	7.37	11.37	12.99
C14H18N203	262.31	262.1317	64.11	6.92	10.68	18.30
C14H18N204	278.31	278.1266	60.42	6.52	10.07	23.00
C14H18N205	294.31	294.1216	57.14	6.16	9.52	27.18
C14H18N206	310.31	310.1165	54.19	5.85	9.03	30.94
C14H18N207	326.31	326.1114	51.53	5.56	8.58	34.32
C14H18N208	342.31	342.1063	49.12	5.30	8.18	37.39
C14H18N209	358.31	358.1012	46.93	5.06	7.82	40.19
C14H18N401	258.33	258.1481	65.09	7.02	21.69	6.19
C14H18N402	274.33	274.1430	61.30	6.61	20.42	11.66
C14H18N403	290.32	290.1379	57.92	6.25	19.30	16.53
C14H18N404	306.32	306.1328	54.89	5.92	18.29	20.89
C14H18N405	322.32	322.1277	52.17	5.63	17.38	24.82
C14H18N406	338.32	338.1226	49.70	5.36	16.56	28.37
C14H18N407	354.32	354.1175	47.46	5.12	15.81	31.61
C14H18N408	370.32	370.1125	45.41	4.90	15.13	34.56
C14H18N409	386.32	386.1074	43.53	4.70	14.50	37.27
C14H18N601	286.34	286.1542	58.73	6.34	29.35	5.59
C14H18N602	302.34	302.1491	55.62	6.00	27.80	10.58
C14H18N603	318.34	318.1440	52.82	5.70	26.40	15.08
C14H18N604	334.34	334.1389	50.30	5.43	25.14	19.14
C14H18N605	350.34	350.1339	48.00	5.18	23.99	22.83
C14H18N606	366.34	366.1288	45.90	4.95	22.94	26.20
C14H18N607	382.34	382.1237	43.98	4.75	21.98	29.29
C14H18N608	398.33	398.1186	42.21	4.55	21.10	32.13
C14H18N609	414.33	414.1135	40.58	4.38	20.28	34.75
C14H19N101	217.31	217.1467	77.38	8.81	6.45	7.36
C14H19N102	233.31	233.1416	72.07	8.21	6.00	13.71
C14H19N103	249.31	249.1365	67.45	7.68	5.62	19.25
C14H19N104	265.31	265.1314	63.38	7.22	5.28	24.12
C14H19N105	281.31	281.1263	59.78	6.81	4.98	28.44
C14H19N106	297.31	297.1212	56.56	6.44	4.71	32.29
C14H19N107	313.31	313.1161	53.67	6.11	4.47	35.75
C14H19N108	329.31	329.1111	51.06	5.82	4.25	38.87
C14H19N109	345.31	345.1060	48.70	5.55	4.06	41.70
C14H19N301	245.33	245.1528	68.54	7.81	17.13	6.52
C14H19N302	261.33	261.1477	64.35	7.33	16.08	12.24
C14H19N303	277.33	277.1426	60.63	6.91	15.15	17.31
C14H19N304	293.33	293.1375	57.33	6.53	14.33	21.82
C14H19N305	309.32	309.1325	54.36	6.19	13.58	25.86
C14H19N306	325.32	325.1274	51.69	5.89	12.92	29.51
C14H19N307	341.32	341.1223	49.27	5.61	12.31	32.81
C14H19N308	357.32	357.1172	47.06	5.36	11.76	35.82
C14H19N309	373.32	373.1121	45.04	5.13	11.26	38.57
C14H19N501	273.34	273.1590	61.52	7.01	25.62	5.85

071

C14H19N502	289.34	289.1539	58.12	6.62	24.20	11.06
C14H19N503	305.34	305.1488	55.07	6.27	22.94	15.72
C14H19N504	321.34	321.1437	52.33	5.96	21.79	19.92
C14H19N505	337.34	337.1386	49.85	5.68	20.76	23.71
C14H19N506	353.34	353.1335	47.59	5.42	19.82	27.17
C14H19N507	369.34	369.1284	45.53	5.19	18.96	30.32
C14H19N508	385.34	385.1233	43.64	4.97	18.17	33.22
C14H19N509	401.34	401.1183	41.90	4.77	17.45	35.88
C14H20N201	232.33	232.1576	72.38	8.68	12.06	6.89
C14H20N202	248.33	248.1525	67.72	8.12	11.28	12.89
C14H20N203	264.33	264.1474	63.62	7.63	10.60	18.16
C14H20N204	280.33	280.1423	59.99	7.19	9.99	22.83
C14H20N205	296.33	296.1372	56.75	6.80	9.45	27.00
C14H20N206	312.33	312.1321	53.84	6.45	8.97	30.74
C14H20N207	328.32	328.1270	51.22	6.14	8.53	34.11
C14H20N208	344.32	344.1220	48.84	5.85	8.14	37.17
C14H20N209	360.32	360.1169	46.67	5.59	7.77	39.96
C14H20N401	260.34	260.1637	64.59	7.74	21.52	6.15
C14H20N402	276.34	276.1586	60.85	7.30	20.27	11.58
C14H20N403	292.34	292.1535	57.52	6.90	19.16	16.42
C14H20N404	308.34	308.1484	54.54	6.54	18.17	20.76
C14H20N405	324.34	324.1434	51.85	6.22	17.27	24.66
C14H20N406	340.34	340.1383	49.41	5.92	16.46	28.21
C14H20N407	356.34	356.1332	47.19	5.66	15.72	31.43
C14H20N408	372.34	372.1281	45.16	5.41	15.05	34.38
C14H20N409	388.34	388.1230	43.30	5.19	14.43	37.08
C14H20N601	288.36	288.1698	58.32	6.99	29.14	5.55
C14H20N602	304.35	304.1648	55.25	6.62	27.61	10.51
C14H20N603	320.35	320.1597	52.49	6.29	26.23	14.98
C14H20N604	336.35	336.1546	49.99	5.99	24.99	19.03
C14H20N605	352.35	352.1495	47.72	5.72	23.85	22.70
C14H20N606	368.35	368.1444	45.65	5.47	22.82	26.06
C14H20N607	384.35	384.1393	43.75	5.25	21.87	29.14
C14H20N608	400.35	400.1342	42.00	5.04	20.99	31.97
C14H20N609	416.35	416.1292	40.39	4.84	20.18	34.58
C14H21N101	219.33	219.1623	76.67	9.65	6.39	7.29
C14H21N102	235.33	235.1572	71.46	8.99	5.95	13.60
C14H21N103	251.33	251.1521	66.91	8.42	5.57	19.10
C14H21N104	267.33	267.1470	62.90	7.92	5.24	23.94
C14H21N105	283.33	283.1420	59.35	7.47	4.94	28.23
C14H21N106	299.33	299.1369	56.18	7.07	4.68	32.07
C14H21N107	315.33	315.1318	53.33	6.71	4.44	35.52
C14H21N108	331.33	331.1267	50.75	6.39	4.23	38.63
C14H21N109	347.32	347.1216	48.41	6.09	4.03	41.46
C14H21N301	247.34	247.1685	67.98	8.56	16.99	6.47
C14H21N302	263.34	263.1634	63.85	8.04	15.96	12.15
C14H21N303	279.34	279.1583	60.20	7.58	15.04	17.18
C14H21N304	295.34	295.1532	56.94	7.17	14.23	21.67
C14H21N305	311.34	311.1481	54.01	6.80	13.50	25.69
C14H21N306	327.34	327.1430	51.37	6.47	12.84	29.33
C14H21N307	343.34	343.1379	48.98	6.17	12.24	32.62
C14H21N308	359.34	359.1329	46.80	5.89	11.69	35.62
C14H21N309	375.34	375.1278	44.80	5.64	11.20	38.36
C14H21N501	275.36	275.1746	61.07	7.69	25.43	5.81
C14H21N502	291.36	291.1695	57.72	7.27	24.04	10.98
C14H21N503	307.36	307.1644	54.71	6.89	22.79	15.62
C14H21N504	323.35	323.1593	52.00	6.55	21.66	19.79
C14H21N505	339.35	339.1543	49.55	6.24	20.64	23.57
C14H21N506	355.35	355.1492	47.32	5.96	19.71	27.01
C14H21N507	371.35	371.1441	45.28	5.70	18.86	30.16
C14H21N508	387.35	387.1390	43.41	5.46	18.08	33.04
C14H21N509	403.35	403.1339	41.69	5.25	17.36	35.70
C14H22N201	234.34	234.1732	71.76	9.46	11.95	6.83
C14H22N202	250.34	250.1681	67.17	8.86	11.19	12.78
C14H22N203	266.34	266.1630	63.14	8.33	10.52	18.02
C14H22N204	282.34	282.1579	59.56	7.85	9.92	22.67
C14H22N205	298.34	298.1529	56.36	7.43	9.39	26.81
C14H22N206	314.34	314.1478	53.49	7.05	8.91	30.54
C14H22N207	330.34	330.1427	50.90	6.71	8.48	33.90
C14H22N208	346.34	346.1376	48.55	6.40	8.09	36.96
C14H22N209	362.34	362.1325	46.41	6.12	7.73	39.74
C14H22N401	262.36	262.1794	64.09	8.45	21.36	6.10
C14H22N402	278.36	278.1743	60.41	7.97	20.13	11.50
C14H22N403	294.36	294.1692	57.13	7.53	19.03	16.31
C14H22N404	310.36	310.1641	54.18	7.15	18.05	20.62
C14H22N405	326.36	326.1590	51.53	6.79	17.17	24.51
C14H22N406	342.35	342.1539	49.12	6.48	16.37	28.04
C14H22N407	358.35	358.1488	46.92	6.19	15.63	31.25
C14H22N408	374.35	374.1437	44.92	5.92	14.97	34.19
C14H22N409	390.35	390.1387	43.08	5.68	14.35	36.89
C14H22N601	290.37	290.1855	57.91	7.64	28.94	5.51
C14H22N602	306.37	306.1804	54.89	7.24	27.43	10.44
C14H22N603	322.37	322.1753	52.16	6.88	26.07	14.89
C14H22N604	338.37	338.1702	49.70	6.55	24.84	18.91
C14H22N605	354.37	354.1652	47.45	6.26	23.72	22.57
C14H22N606	370.37	370.1601	45.40	5.99	22.69	25.92
C14H22N607	386.37	386.1550	43.52	5.74	21.75	28.99
C14H22N608	402.37	402.1499	41.79	5.51	20.89	31.81
C14H22N609	418.37	418.1448	40.19	5.30	20.09	34.42
C14H23N101	221.35	221.1780	75.97	10.47	6.33	7.23
C14H23N102	237.34	237.1729	70.85	9.77	5.90	13.48
C14H23N103	253.34	253.1678	66.37	9.15	5.53	18.95
C14H23N104	269.34	269.1627	62.43	8.61	5.20	23.76
C14H23N105	285.34	285.1576	58.93	8.12	4.91	28.04
C14H23N106	301.34	301.1525	55.80	7.69	4.65	31.86
C14H23N107	317.34	317.1474	52.99	7.31	4.41	35.29
C14H23N108	333.34	333.1424	50.45	6.95	4.20	38.40
C14H23N109	349.34	349.1373	48.14	6.64	4.01	41.22
C14H23N301	249.36	249.1841	67.44	9.30	16.85	6.42
C14H23N302	265.36	265.1790	63.37	8.74	15.84	12.06
C14H23N303	281.36	281.1739	59.77	8.24	14.93	17.06
C14H23N304	297.36	297.1688	56.55	7.80	14.13	21.52
C14H23N305	313.36	313.1638	53.66	7.40	13.41	25.53
C14H23N306	329.36	329.1587	51.06	7.04	12.76	29.15
C14H23N307	345.36	345.1536	48.69	6.71	12.17	32.43
C14H23N308	361.35	361.1485	46.53	6.42	11.63	35.42
C14H23N309	377.35	377.1434	44.56	6.14	11.14	38.16
C14H23N501	277.37	277.1902	60.62	8.36	25.25	5.77
C14H23N502	293.37	293.1852	57.32	7.90	23.87	10.91
C14H23N503	309.37	309.1801	54.35	7.49	22.64	15.51
C14H23N504	325.37	325.1750	51.68	7.13	21.52	19.67
C14H23N505	341.37	341.1699	49.26	6.79	20.52	23.43
C14H23N506	357.37	357.1648	47.05	6.49	19.60	26.86
C14H23N507	373.37	373.1597	45.04	6.21	18.76	30.00
C14H23N508	389.37	389.1546	43.19	5.95	17.99	32.87
C14H23N509	405.37	405.1496	41.48	5.72	17.28	35.52
C14H24N201	236.36	236.1889	71.14	10.23	11.85	6.77
C14H24N202	252.36	252.1838	66.63	9.59	11.10	12.68
C14H24N203	268.36	268.1787	62.66	9.01	10.44	17.89
C14H24N204	284.36	284.1736	59.14	8.51	9.85	22.51

C14H24N2O5	300.36	300.1685	55.99	8.05	9.33	26.63
C14H24N2O6	316.36	316.1634	53.15	7.65	8.85	30.34
C14H24N2O7	332.36	332.1583	50.60	7.28	8.43	33.70
C14H24N2O8	348.36	348.1533	48.27	6.94	8.04	36.74
C14H24N2O9	364.36	364.1482	46.15	6.64	7.69	39.52
C14H24N4O1	264.37	264.1950	63.61	9.15	21.19	6.05
C14H24N4O2	280.37	280.1899	59.98	8.63	19.98	11.41
C14H24N4O3	296.37	296.1848	56.74	8.16	18.90	16.20
C14H24N4O4	312.37	312.1797	53.83	7.74	17.94	20.49
C14H24N4O5	328.37	328.1747	51.21	7.37	17.06	24.36
C14H24N4O6	344.37	344.1696	48.83	7.02	16.27	27.88
C14H24N4O7	360.37	360.1645	46.66	6.71	15.55	31.08
C14H24N4O8	376.37	376.1594	44.68	6.43	14.89	34.01
C14H24N4O9	392.37	392.1543	42.86	6.17	14.28	36.70
C14H24N6O1	292.39	292.2011	57.51	8.27	28.74	5.47
C14H24N6O2	308.39	308.1961	54.53	7.84	27.25	10.38
C14H24N6O3	324.39	324.1910	51.84	7.46	25.91	14.80
C14H24N6O4	340.39	340.1859	49.40	7.11	24.69	18.80
C14H24N6O5	356.38	356.1808	47.18	6.79	23.58	22.45
C14H24N6O6	372.38	372.1757	45.16	6.50	22.57	25.78
C14H24N6O7	388.38	388.1706	43.30	6.23	21.64	28.84
C14H24N6O8	404.38	404.1655	41.58	5.98	20.78	31.65
C14H24N6O9	420.38	420.1605	40.00	5.75	19.99	34.25
C14H25N1O1	223.36	223.1936	75.28	11.28	6.27	7.16
C14H25N1O2	239.36	239.1885	70.25	10.53	5.85	13.37
C14H25N1O3	255.36	255.1834	65.85	9.87	5.49	18.80
C14H25N1O4	271.36	271.1783	61.97	9.29	5.16	23.58
C14H25N1O5	287.36	287.1733	58.52	8.77	4.87	27.84
C14H25N1O6	303.36	303.1682	55.43	8.31	4.62	31.64
C14H25N1O7	319.36	319.1631	52.65	7.89	4.39	35.07
C14H25N1O8	335.36	335.1580	50.14	7.51	4.18	38.17
C14H25N1O9	351.36	351.1529	47.86	7.17	3.99	40.98
C14H25N3O1	251.37	251.1998	66.89	10.02	16.72	6.36
C14H25N3O2	267.37	267.1947	62.89	9.42	15.72	11.97
C14H25N3O3	283.37	283.1896	59.34	8.89	14.83	16.94
C14H25N3O4	299.37	299.1845	56.17	8.42	14.04	21.38
C14H25N3O5	315.37	315.1794	53.32	7.99	13.32	25.37
C14H25N3O6	331.37	331.1743	50.75	7.60	12.68	28.97
C14H25N3O7	347.37	347.1692	48.41	7.25	12.10	32.24
C14H25N3O8	363.37	363.1641	46.28	6.93	11.56	35.22
C14H25N3O9	379.37	379.1591	44.33	6.64	11.08	37.96
C14H25N5O1	279.39	279.2059	60.19	9.02	25.07	5.73
C14H25N5O2	295.39	295.2008	56.93	8.53	23.71	10.83
C14H25N5O3	311.39	311.1957	54.00	8.09	22.49	15.41
C14H25N5O4	327.39	327.1906	51.36	7.70	21.39	19.55
C14H25N5O5	343.39	343.1856	48.97	7.34	20.39	23.30
C14H25N5O6	359.39	359.1805	46.79	7.01	19.49	26.71
C14H25N5O7	375.38	375.1754	44.80	6.71	18.66	29.83
C14H25N5O8	391.38	391.1703	42.96	6.44	17.89	32.70
C14H25N5O9	407.38	407.1652	41.28	6.19	17.19	35.35
C14H26N2O1	238.38	238.2045	70.54	10.99	11.75	6.71
C14H26N2O2	254.38	254.1994	66.11	10.30	11.01	12.58
C14H26N2O3	270.37	270.1943	62.19	9.69	10.36	17.75
C14H26N2O4	286.37	286.1892	58.72	9.15	9.78	22.35
C14H26N2O5	302.37	302.1842	55.61	8.67	9.26	26.46
C14H26N2O6	318.37	318.1791	52.82	8.23	8.80	30.15
C14H26N2O7	334.37	334.1740	50.29	7.84	8.38	33.49
C14H26N2O8	350.37	350.1689	47.99	7.48	8.00	36.53
C14H26N2O9	366.37	366.1638	45.90	7.15	7.65	39.30
C14H26N4O1	266.39	266.2106	63.12	9.84	21.03	6.01
C14H26N4O2	282.39	282.2056	59.55	9.28	19.84	11.33
C14H26N4O3	298.39	298.2005	56.35	8.78	18.78	16.09
C14H26N4O4	314.39	314.1954	53.49	8.34	17.82	20.36
C14H26N4O5	330.39	330.1903	50.90	7.93	16.96	24.21
C14H26N4O6	346.39	346.1852	48.55	7.57	16.17	27.71
C14H26N4O7	362.39	362.1801	46.40	7.23	15.46	30.91
C14H26N4O8	378.39	378.1750	44.44	6.93	14.81	33.83
C14H26N4O9	394.38	394.1700	42.64	6.65	14.21	36.51
C14H26N6O1	294.40	294.2168	57.12	8.90	28.55	5.43
C14H26N6O2	310.40	310.2117	54.17	8.44	27.07	10.31
C14H26N6O3	326.40	326.2066	51.52	8.03	25.75	14.71
C14H26N6O4	342.40	342.2015	49.11	7.65	24.54	18.69
C14H26N6O5	358.40	358.1965	46.92	7.31	23.45	22.32
C14H26N6O6	374.40	374.1914	44.91	7.00	22.45	25.64
C14H26N6O7	390.40	390.1863	43.07	6.71	21.53	28.69
C14H26N6O8	406.40	406.1812	41.38	6.45	20.68	31.49
C14H26N6O9	422.40	422.1761	39.81	6.20	19.90	34.09
C14H27N1O1	225.38	225.2093	74.61	12.08	6.21	7.10
C14H27N1O2	241.38	241.2042	69.67	11.27	5.80	13.26
C14H27N1O3	257.38	257.1991	65.33	10.57	5.44	18.65
C14H27N1O4	273.38	273.1940	61.51	9.96	5.12	23.41
C14H27N1O5	289.37	289.1889	58.11	9.40	4.84	27.64
C14H27N1O6	305.37	305.1838	55.07	8.91	4.59	31.44
C14H27N1O7	321.37	321.1787	52.32	8.47	4.36	34.85
C14H27N1O8	337.37	337.1737	49.84	8.07	4.15	37.94
C14H27N1O9	353.37	353.1686	47.59	7.70	3.96	40.75
C14H27N3O1	253.39	253.2154	66.36	10.74	16.58	6.31
C14H27N3O2	269.39	269.2103	62.42	10.10	15.60	11.88
C14H27N3O3	285.39	285.2052	58.92	9.54	14.72	16.82
C14H27N3O4	301.39	301.2001	55.79	9.03	13.94	21.23
C14H27N3O5	317.39	317.1951	52.98	8.57	13.24	25.20
C14H27N3O6	333.39	333.1900	50.44	8.16	12.60	28.79
C14H27N3O7	349.39	349.1849	48.13	7.79	12.03	32.05
C14H27N3O8	365.39	365.1798	46.02	7.45	11.50	35.03
C14H27N3O9	381.39	381.1747	44.09	7.14	11.02	37.76
C14H27N5O1	281.40	281.2215	59.76	9.67	24.89	5.69
C14H27N5O2	297.40	297.2165	56.54	9.15	23.55	10.76
C14H27N5O3	313.40	313.2114	53.65	8.68	22.35	15.32
C14H27N5O4	329.40	329.2063	51.05	8.26	21.26	19.43
C14H27N5O5	345.40	345.2012	48.68	7.88	20.28	23.16
C14H27N5O6	361.40	361.1961	46.53	7.53	19.38	26.56
C14H27N5O7	377.40	377.1910	44.56	7.21	18.56	29.68
C14H27N5O8	393.40	393.1859	42.74	6.92	17.80	32.54
C14H27N5O9	409.40	409.1809	41.07	6.65	17.11	35.17
C14H28N2O1	240.39	240.2202	69.95	11.74	11.65	6.66
C14H28N2O2	256.39	256.2151	65.59	11.01	10.93	12.48
C14H28N2O3	272.39	272.2100	61.73	10.36	10.28	17.62
C14H28N2O4	288.39	288.2049	58.31	9.79	9.71	22.19
C14H28N2O5	304.39	304.1998	55.24	9.27	9.20	26.28
C14H28N2O6	320.39	320.1947	52.48	8.81	8.74	29.96
C14H28N2O7	336.39	336.1896	49.99	8.39	8.33	33.29
C14H28N2O8	352.39	352.1845	47.72	8.01	7.95	36.32
C14H28N2O9	368.39	368.1795	45.65	7.66	7.60	39.09
C14H28N4O1	268.41	268.2263	62.65	10.52	20.87	5.96
C14H28N4O2	284.40	284.2212	59.13	9.92	19.70	11.25
C14H28N4O3	300.40	300.2161	55.98	9.40	18.65	15.98
C14H28N4O4	316.40	316.2110	53.15	8.92	17.71	20.23
C14H28N4O5	332.40	332.2060	50.59	8.49	16.86	24.07
C14H28N4O6	348.40	348.2009	48.26	8.10	16.08	27.55
C14H28N4O7	364.40	364.1958	46.15	7.75	15.38	30.73

073

C14H28N4O8	380.40	380.1907	44.20	7.42	14.73	33.65
C14H28N4O9	396.40	396.1856	42.42	7.12	14.13	36.33
C14H28N6O1	296.42	296.2324	56.73	9.52	28.35	5.40
C14H28N6O2	312.42	312.2274	53.82	9.03	26.90	10.24
C14H28N6O3	328.42	328.2223	51.20	8.59	25.59	14.61
C14H28N6O4	344.42	344.2172	48.82	8.19	24.40	18.58
C14H28N6O5	360.42	360.2121	46.66	7.83	23.32	22.20
C14H28N6O6	376.42	376.2070	44.67	7.50	22.33	25.50
C14H28N6O7	392.42	392.2019	42.85	7.19	21.42	28.54
C14H28N6O8	408.41	408.1968	41.17	6.91	20.58	31.34
C14H28N6O9	424.41	424.1918	39.62	6.65	19.80	33.93
C14H29N1O1	227.39	227.2249	73.95	12.85	6.16	7.04
C14H29N1O2	243.39	243.2198	69.09	12.01	5.75	13.15
C14H29N1O3	259.39	259.2147	64.83	11.27	5.40	18.50
C14H29N1O4	275.39	275.2096	61.06	10.61	5.09	23.24
C14H29N1O5	291.39	291.2046	57.71	10.03	4.81	27.45
C14H29N1O6	307.39	307.1995	54.70	9.51	4.56	31.23
C14H29N1O7	323.39	323.1944	52.00	9.04	4.33	34.63
C14H29N1O8	339.39	339.1893	49.55	8.61	4.13	37.71
C14H29N1O9	355.39	355.1842	47.32	8.23	3.94	40.52
C14H29N3O1	255.41	255.2310	65.84	11.44	16.45	6.26
C14H29N3O2	271.41	271.2260	61.96	10.77	15.48	11.79
C14H29N3O3	287.41	287.2209	58.51	10.17	14.62	16.70
C14H29N3O4	303.40	303.2158	55.42	9.63	13.85	21.09
C14H29N3O5	319.40	319.2107	52.65	9.15	13.16	25.05
C14H29N3O6	335.40	335.2056	50.14	8.72	12.53	28.62
C14H29N3O7	351.40	351.2005	47.85	8.32	11.96	31.87
C14H29N3O8	367.40	367.1954	45.77	7.96	11.44	34.84
C14H29N3O9	383.40	383.1904	43.86	7.62	10.96	37.56
C14H29N5O1	283.42	283.2372	59.33	10.31	24.71	5.65
C14H29N5O2	299.42	299.2321	56.16	9.76	23.39	10.69
C14H29N5O3	315.42	315.2270	53.31	9.27	22.20	15.22
C14H29N5O4	331.42	331.2219	50.74	8.82	21.13	19.31
C14H29N5O5	347.42	347.2169	48.40	8.41	20.16	23.03
C14H29N5O6	363.42	363.2118	46.27	8.04	19.27	26.41
C14H29N5O7	379.42	379.2067	44.32	7.70	18.46	29.52
C14H29N5O8	395.42	395.2016	42.53	7.39	17.71	32.37
C14H29N5O9	411.42	411.1965	40.87	7.11	17.02	35.00
C14H30N2O1	242.41	242.2358	69.37	12.47	11.56	6.60
C14H30N2O2	258.41	258.2307	65.07	11.70	10.84	12.38
C14H30N2O3	274.41	274.2256	61.28	11.02	10.21	17.49
C14H30N2O4	290.41	290.2205	57.90	10.41	9.65	22.04
C14H30N2O5	306.41	306.2155	54.88	9.87	9.14	26.11
C14H30N2O6	322.41	322.2104	52.16	9.38	8.69	29.78
C14H30N2O7	338.40	338.2053	49.69	8.94	8.28	33.10
C14H30N2O8	354.40	354.2002	47.45	8.53	7.90	36.12
C14H30N2O9	370.40	370.1951	45.40	8.16	7.56	38.88
C14H30N4O1	270.42	270.2419	62.18	11.18	20.72	5.92
C14H30N4O2	286.42	286.2369	58.71	10.56	19.56	11.17
C14H30N4O3	302.42	302.2318	55.60	10.00	18.53	15.87
C14H30N4O4	318.42	318.2267	52.81	9.50	17.60	20.10
C14H30N4O5	334.42	334.2216	50.28	9.04	16.75	23.92
C14H30N4O6	350.42	350.2165	47.99	8.63	15.99	27.39
C14H30N4O7	366.42	366.2114	45.89	8.25	15.29	30.57
C14H30N4O8	382.42	382.2063	43.97	7.91	14.65	33.47
C14H30N4O9	398.42	398.2013	42.21	7.59	14.06	36.14
C14H30N6O1	298.43	298.2481	56.35	10.13	28.16	5.36
C14H30N6O2	314.43	314.2430	53.48	9.62	26.73	10.18
C14H30N6O3	330.43	330.2379	50.89	9.15	25.43	14.53
C14H30N6O4	346.43	346.2328	48.54	8.73	24.26	18.47
C14H30N6O5	362.43	362.2278	46.40	8.34	23.19	22.07
C14H30N6O6	378.43	378.2227	44.43	7.99	22.21	25.37
C14H30N6O7	394.43	394.2176	42.63	7.67	21.31	28.39
C14H30N6O8	410.43	410.2125	40.97	7.37	20.48	31.19
C14H30N6O9	426.43	426.2074	39.43	7.09	19.71	33.77
C14H31N1O1	229.41	229.2406	73.30	13.62	6.11	6.97
C14H31N1O2	245.41	245.2355	68.52	12.73	5.71	13.04
C14H31N1O3	261.41	261.2304	64.33	11.95	5.36	18.36
C14H31N1O4	277.41	277.2253	60.62	11.26	5.05	23.07
C14H31N1O5	293.41	293.2202	57.31	10.65	4.77	27.26
C14H31N1O6	309.41	309.2151	54.35	10.10	4.53	31.03
C14H31N1O7	325.41	325.2100	51.68	9.60	4.30	34.42
C14H31N1O8	341.41	341.2049	49.25	9.15	4.10	37.49
C14H31N1O9	357.40	357.1999	47.05	8.74	3.92	40.29
C14H31N3O1	257.42	257.2467	65.32	12.14	16.32	6.22
C14H31N3O2	273.42	273.2416	61.50	11.43	15.37	11.70
C14H31N3O3	289.42	289.2365	58.10	10.80	14.52	16.58
C14H31N3O4	305.42	305.2314	55.06	10.23	13.76	20.95
C14H31N3O5	321.42	321.2264	52.32	9.72	13.07	24.89
C14H31N3O6	337.42	337.2213	49.84	9.26	12.45	28.45
C14H31N3O7	353.42	353.2162	47.58	8.84	11.89	31.69
C14H31N3O8	369.42	369.2111	45.52	8.46	11.37	34.65
C14H31N3O9	385.42	385.2060	43.63	8.11	10.90	37.36
C14H31N5O1	285.44	285.2528	58.91	10.95	24.54	5.61
C14H31N5O2	301.44	301.2478	55.79	10.37	23.23	10.62
C14H31N5O3	317.43	317.2427	52.97	9.84	22.06	15.12
C14H31N5O4	333.43	333.2376	50.43	9.37	21.00	19.19
C14H31N5O5	349.43	349.2325	48.12	8.94	20.04	22.89
C14H31N5O6	365.43	365.2274	46.02	8.55	19.16	26.27
C14H31N5O7	381.43	381.2223	44.09	8.19	18.36	29.36
C14H31N5O8	397.43	397.2172	42.31	7.86	17.62	32.21
C14H31N5O9	413.43	413.2122	40.67	7.56	16.94	34.83
C14H32N2O1	244.42	244.2514	68.80	13.20	11.46	6.55
C14H32N2O2	260.42	260.2464	64.57	12.39	10.76	12.29
C14H32N2O3	276.42	276.2413	60.83	11.67	10.13	17.36
C14H32N2O4	292.42	292.2362	57.50	11.03	9.58	21.89
C14H32N2O5	308.42	308.2311	54.52	10.46	9.08	25.94
C14H32N2O6	324.42	324.2260	51.83	9.94	8.63	29.59
C14H32N2O7	340.42	340.2209	49.40	9.48	8.23	32.90
C14H32N2O8	356.42	356.2158	47.18	9.05	7.86	35.91
C14H32N2O9	372.42	372.2108	45.15	8.66	7.52	38.66
C14H32N4O1	272.44	272.2576	61.72	11.84	20.57	5.87
C14H32N4O2	288.44	288.2525	58.30	11.18	19.42	11.09
C14H32N4O3	304.44	304.2474	55.24	10.60	18.40	15.77
C14H32N4O4	320.44	320.2423	52.48	10.07	17.48	19.97
C14H32N4O5	336.43	336.2373	49.98	9.59	16.65	23.78
C14H32N4O6	352.43	352.2322	47.71	9.15	15.90	27.24
C14H32N4O7	368.43	368.2271	45.64	8.75	15.21	30.40
C14H32N4O8	384.43	384.2220	43.74	8.39	14.57	33.29
C14H32N4O9	400.43	400.2169	41.99	8.06	13.99	35.96
C14H32N6O1	300.45	300.2637	55.97	10.74	27.97	5.33
C14H32N6O2	316.45	316.2587	53.14	10.19	26.56	10.11
C14H32N6O3	332.45	332.2536	50.58	9.70	25.28	14.44
C14H32N6O4	348.45	348.2485	48.26	9.26	24.12	18.37
C14H32N6O5	364.45	364.2434	46.14	8.85	23.06	21.95
C14H32N6O6	380.45	380.2383	44.20	8.48	22.09	25.23
C14H32N6O7	396.45	396.2332	42.42	8.14	21.20	28.25
C14H32N6O8	412.45	412.2281	40.77	7.82	20.38	31.03
C14H32N6O9	428.45	428.2231	39.25	7.53	19.62	33.61
C14H33N3O1	259.44	259.2623	64.82	12.82	16.20	6.17

074

C14H33N302	275.44	275.2573	61.05	12.08	15.26	11.62
C14H33N303	291.44	291.2522	57.70	11.41	14.42	16.47
C14H33N304	307.44	307.2471	54.70	10.82	13.67	20.82
C14H33N305	323.44	323.2420	51.99	10.28	12.99	24.73
C14H33N306	339.44	339.2369	49.54	9.80	12.38	28.28
C14H33N307	355.44	355.2318	47.31	9.36	11.82	31.51
C14H33N308	371.43	371.2267	45.27	8.96	11.31	34.46
C14H33N309	387.43	387.2217	43.40	8.59	10.85	37.17
C14H33N501	287.45	287.2685	58.50	11.57	24.36	5.57
C14H33N502	303.45	303.2634	55.41	10.96	23.08	10.54
C14H33N503	319.45	319.2583	52.64	10.41	21.92	15.03
C14H33N504	335.45	335.2532	50.13	9.92	20.88	19.08
C14H33N505	351.45	351.2482	47.85	9.46	19.93	22.76
C14H33N506	367.45	367.2431	45.76	9.05	19.06	26.13
C14H33N507	383.45	383.2380	43.85	8.67	18.26	29.21
C14H33N508	399.45	399.2329	42.10	8.33	17.53	32.04
C14H33N509	415.45	415.2278	40.48	8.01	16.86	34.66
C14H34N401	274.45	274.2732	61.27	12.49	20.41	5.83
C14H34N402	290.45	290.2682	57.89	11.80	19.29	11.02
C14H34N403	306.45	306.2631	54.87	11.18	18.28	15.66
C14H34N404	322.45	322.2580	52.15	10.63	17.38	19.85
C14H34N405	338.45	338.2529	49.68	10.13	16.55	23.64
C14H34N406	354.45	354.2478	47.44	9.67	15.81	27.08
C14H34N407	370.45	370.2427	45.39	9.25	15.12	30.23
C14H34N408	386.45	386.2376	43.51	8.87	14.50	33.12
C14H34N409	402.45	402.2326	41.78	8.52	13.92	35.78
C14H34N601	302.47	302.2794	55.59	11.33	27.78	5.29
C14H34N602	318.47	318.2743	52.80	10.76	26.39	10.05
C14H34N603	334.47	334.2692	50.28	10.25	25.13	14.35
C14H34N604	350.46	350.2641	47.98	9.78	23.98	18.26
C14H34N605	366.46	366.2590	45.89	9.35	22.93	21.83
C14H34N606	382.46	382.2540	43.97	8.96	21.97	25.10
C14H34N607	398.46	398.2489	42.20	8.60	21.09	28.11
C14H34N608	414.46	414.2438	40.57	8.27	20.28	30.88
C14H34N609	430.46	430.2387	39.06	7.96	19.52	33.45
C14H35N501	289.47	289.2841	58.09	12.19	24.19	5.53
C14H35N502	305.47	305.2791	55.05	11.55	22.93	10.48
C14H35N503	321.47	321.2740	52.31	10.97	21.79	14.93
C14H35N504	337.47	337.2689	49.83	10.45	20.75	18.96
C14H35N505	353.47	353.2638	47.57	9.98	19.81	22.63
C14H35N506	369.46	369.2587	45.51	9.55	18.96	25.98
C14H35N507	385.46	385.2536	43.62	9.15	18.17	29.05
C14H35N508	401.46	401.2485	41.89	8.79	17.44	31.88
C14H35N509	417.46	417.2435	40.28	8.45	16.78	34.49
C14H36N601	304.48	304.2950	55.23	11.92	27.60	5.25
C14H36N602	320.48	320.2900	52.47	11.32	26.22	9.98
C14H36N603	336.48	336.2849	49.97	10.78	24.98	14.26
C14H36N604	352.48	352.2798	47.71	10.29	23.84	18.16
C14H36N605	368.48	368.2747	45.64	9.85	22.81	21.71
C14H36N606	384.48	384.2696	43.74	9.44	21.86	24.97
C14H36N607	400.48	400.2645	41.99	9.06	20.98	27.97
C14H36N608	416.48	416.2594	40.38	8.71	20.18	30.73
C14H36N609	432.48	432.2544	38.88	8.39	19.43	33.30
C15H 7N101	217.23	217.0528	82.94	3.25	6.45	7.37
C15H 7N102	233.23	233.0477	77.25	3.03	6.01	13.72
C15H 7N103	249.23	249.0426	72.29	2.83	5.62	19.26
C15H 7N104	265.23	265.0375	67.93	2.66	5.28	24.13
C15H 7N105	281.23	281.0324	64.06	2.51	4.98	28.45
C15H 7N106	297.23	297.0273	60.62	2.37	4.71	32.30
C15H 7N107	313.23	313.0222	57.52	2.25	4.47	35.76
C15H 7N108	329.22	329.0172	54.72	2.14	4.25	38.88
C15H 7N109	345.22	345.0121	52.19	2.04	4.06	41.71
C15H 7N301	245.24	245.0589	73.46	2.88	17.13	6.52
C15H 7N302	261.24	261.0538	68.97	2.70	16.08	12.25
C15H 7N303	277.24	277.0487	64.99	2.54	15.16	17.31
C15H 7N304	293.24	293.0437	61.44	2.41	14.33	21.82
C15H 7N305	309.24	309.0386	58.26	2.28	13.59	25.87
C15H 7N306	325.24	325.0335	55.40	2.17	12.92	29.52
C15H 7N307	341.24	341.0284	52.80	2.07	12.31	32.82
C15H 7N308	357.24	357.0233	50.43	1.98	11.76	35.83
C15H 7N309	373.24	373.0182	48.27	1.89	11.26	38.58
C15H 7N501	273.26	273.0651	65.93	2.58	25.63	5.86
C15H 7N502	289.26	289.0600	62.29	2.44	24.21	11.06
C15H 7N503	305.25	305.0549	59.02	2.31	22.94	15.72
C15H 7N504	321.25	321.0498	56.08	2.20	21.80	19.92
C15H 7N505	337.25	337.0447	53.42	2.09	20.77	23.72
C15H 7N506	353.25	353.0396	51.00	2.00	19.83	27.17
C15H 7N507	369.25	369.0345	48.79	1.91	18.97	30.33
C15H 7N508	385.25	385.0295	46.77	1.83	18.18	33.22
C15H 7N509	401.25	401.0244	44.90	1.76	17.45	35.89
C15H 8N201	232.24	232.0637	77.58	3.47	12.06	6.89
C15H 8N202	248.24	248.0586	72.58	3.25	11.28	12.89
C15H 8N203	264.24	264.0535	68.18	3.05	10.60	18.16
C15H 8N204	280.24	280.0484	64.29	2.88	10.00	22.84
C15H 8N205	296.24	296.0433	60.82	2.72	9.46	27.00
C15H 8N206	312.24	312.0382	57.70	2.58	8.97	30.74
C15H 8N207	328.24	328.0331	54.89	2.46	8.53	34.12
C15H 8N208	344.24	344.0281	52.34	2.34	8.14	37.18
C15H 8N209	360.24	360.0230	50.01	2.24	7.78	39.97
C15H 8N401	260.26	260.0698	69.23	3.10	21.53	6.15
C15H 8N402	276.26	276.0647	65.22	2.92	20.28	11.58
C15H 8N403	292.26	292.0596	61.65	2.76	19.17	16.42
C15H 8N404	308.26	308.0545	58.45	2.62	18.18	20.76
C15H 8N405	324.25	324.0495	55.56	2.49	17.28	24.67
C15H 8N406	340.25	340.0444	52.95	2.37	16.47	28.21
C15H 8N407	356.25	356.0393	50.57	2.26	15.73	31.44
C15H 8N408	372.25	372.0342	48.40	2.17	15.05	34.38
C15H 8N409	388.25	388.0291	46.40	2.08	14.43	37.09
C15H 8N601	288.27	288.0760	62.50	2.80	29.15	5.55
C15H 8N602	304.27	304.0709	59.21	2.65	27.62	10.52
C15H 8N603	320.27	320.0658	56.25	2.52	26.24	14.99
C15H 8N604	336.27	336.0607	53.58	2.40	24.99	19.03
C15H 8N605	352.27	352.0556	51.14	2.29	23.86	22.71
C15H 8N606	368.27	368.0505	48.92	2.19	22.82	26.07
C15H 8N607	384.27	384.0454	46.89	2.10	21.87	29.15
C15H 8N608	400.27	400.0404	45.01	2.01	21.00	31.98
C15H 8N609	416.27	416.0353	43.28	1.94	20.19	34.59
C15H 9N101	219.25	219.0684	82.18	4.14	6.39	7.30
C15H 9N102	235.24	235.0633	76.59	3.86	5.95	13.60
C15H 9N103	251.24	251.0582	71.71	3.61	5.57	19.10
C15H 9N104	267.24	267.0532	67.42	3.39	5.24	23.95
C15H 9N105	283.24	283.0481	63.61	3.20	4.95	28.24
C15H 9N106	299.24	299.0430	60.21	3.03	4.68	32.08
C15H 9N107	315.24	315.0379	57.15	2.88	4.44	35.53
C15H 9N108	331.24	331.0328	54.39	2.74	4.23	38.64
C15H 9N109	347.24	347.0277	51.89	2.61	4.03	41.47
C15H 9N301	247.26	247.0746	72.87	3.67	16.99	6.47
C15H 9N302	263.26	263.0695	68.44	3.45	15.96	12.15
C15H 9N303	279.26	279.0644	64.52	3.25	15.05	17.19
C15H 9N304	295.26	295.0593	61.02	3.07	14.23	21.68

075

C15H 9N305	311.26	311.0542	57.88	2.91	13.50	25.70
C15H 9N306	327.26	327.0491	55.05	2.77	12.84	29.33
C15H 9N307	343.25	343.0440	52.49	2.64	12.24	32.63
C15H 9N308	359.25	359.0390	50.15	2.53	11.70	35.63
C15H 9N309	375.25	375.0339	48.01	2.42	11.20	38.37
C15H 9N501	275.27	275.0807	65.45	3.30	25.44	5.81
C15H 9N502	291.27	291.0756	61.86	3.11	24.04	10.99
C15H 9N503	307.27	307.0705	58.63	2.95	22.79	15.62
C15H 9N504	323.27	323.0654	55.73	2.81	21.66	19.80
C15H 9N505	339.27	339.0604	53.10	2.67	20.64	23.58
C15H 9N506	355.27	355.0553	50.71	2.55	19.71	27.02
C15H 9N507	371.27	371.0502	48.53	2.44	18.86	30.17
C15H 9N508	387.27	387.0451	46.52	2.34	18.08	33.05
C15H 9N509	403.27	403.0400	44.68	2.25	17.37	35.71
C15H10N201	234.26	234.0793	76.91	4.30	11.96	6.83
C15H10N202	250.26	250.0742	71.99	4.03	11.19	12.79
C15H10N203	266.26	266.0691	67.67	3.79	10.52	18.03
C15H10N204	282.26	282.0641	63.83	3.57	9.92	22.67
C15H10N205	298.26	298.0590	60.41	3.38	9.39	26.82
C15H10N206	314.26	314.0539	57.33	3.21	8.91	30.55
C15H10N207	330.26	330.0488	54.55	3.05	8.48	33.91
C15H10N208	346.26	346.0437	52.03	2.91	8.09	36.97
C15H10N209	362.25	362.0386	49.73	2.78	7.73	39.75
C15H10N401	262.27	262.0855	68.69	3.84	21.36	6.10
C15H10N402	278.27	278.0804	64.74	3.62	20.13	11.50
C15H10N403	294.27	294.0753	61.22	3.43	19.04	16.31
C15H10N404	310.27	310.0702	58.07	3.25	18.06	20.63
C15H10N405	326.27	326.0651	55.22	3.09	17.17	24.52
C15H10N406	342.27	342.0600	52.64	2.94	16.37	28.05
C15H10N407	358.27	358.0549	50.29	2.81	15.64	31.26
C15H10N408	374.27	374.0499	48.14	2.69	14.97	34.20
C15H10N409	390.27	390.0448	46.16	2.58	14.36	36.90
C15H10N601	290.29	290.0916	62.07	3.47	28.95	5.51
C15H10N602	306.29	306.0865	58.82	3.29	27.44	10.45
C15H10N603	322.29	322.0814	55.90	3.13	26.08	14.89
C15H10N604	338.28	338.0763	53.26	2.98	24.84	18.92
C15H10N605	354.28	354.0713	50.85	2.85	23.72	22.58
C15H10N606	370.28	370.0662	48.66	2.72	22.70	25.93
C15H10N607	386.28	386.0611	46.64	2.61	21.76	28.99
C15H10N608	402.28	402.0560	44.79	2.51	20.89	31.82
C15H10N609	418.28	418.0509	43.07	2.41	20.09	34.43
C15H11N101	221.26	221.0841	81.43	5.01	6.33	7.23
C15H11N102	237.26	237.0790	75.94	4.67	5.90	13.49
C15H11N103	253.26	253.0739	71.14	4.38	5.53	18.95
C15H11N104	269.26	269.0688	66.91	4.12	5.20	23.77
C15H11N105	285.26	285.0637	63.16	3.89	4.91	28.04
C15H11N106	301.26	301.0586	59.80	3.68	4.65	31.87
C15H11N107	317.26	317.0535	56.79	3.49	4.41	35.30
C15H11N108	333.26	333.0485	54.06	3.33	4.20	38.41
C15H11N109	349.26	349.0434	51.59	3.17	4.01	41.23
C15H11N301	249.27	249.0902	72.28	4.45	16.86	6.42
C15H11N302	265.27	265.0851	67.92	4.18	15.84	12.06
C15H11N303	281.27	281.0800	64.05	3.94	14.94	17.06
C15H11N304	297.27	297.0749	60.61	3.73	14.14	21.53
C15H11N305	313.27	313.0699	57.51	3.54	13.41	25.54
C15H11N306	329.27	329.0648	54.72	3.37	12.76	29.15
C15H11N307	345.27	345.0597	52.18	3.21	12.17	32.44
C15H11N308	361.27	361.0546	49.87	3.07	11.63	35.43
C15H11N309	377.27	377.0495	47.76	2.94	11.14	38.17
C15H11N501	277.29	277.0964	64.97	4.00	25.26	5.77
C15H11N502	293.29	293.0913	61.43	3.78	23.88	10.91
C15H11N503	309.29	309.0862	58.25	3.58	22.64	15.52
C15H11N504	325.29	325.0811	55.39	3.41	21.53	19.67
C15H11N505	341.29	341.0760	52.79	3.25	20.52	23.44
C15H11N506	357.28	357.0709	50.43	3.10	19.60	26.87
C15H11N507	373.28	373.0658	48.27	2.97	18.76	30.00
C15H11N508	389.28	389.0608	46.28	2.85	17.99	32.88
C15H11N509	405.28	405.0557	44.45	2.74	17.28	35.53
C15H12N201	236.28	236.0950	76.25	5.12	11.86	6.77
C15H12N202	252.28	252.0899	71.42	4.79	11.10	12.68
C15H12N203	268.27	268.0848	67.16	4.51	10.44	17.89
C15H12N204	284.27	284.0797	63.38	4.25	9.85	22.51
C15H12N205	300.27	300.0746	60.00	4.03	9.33	26.64
C15H12N206	316.27	316.0695	56.97	3.82	8.86	30.35
C15H12N207	332.27	332.0644	54.22	3.64	8.43	33.71
C15H12N208	348.27	348.0594	51.73	3.47	8.04	36.75
C15H12N209	364.27	364.0543	49.46	3.32	7.69	39.53
C15H12N401	264.29	264.1011	68.17	4.58	21.20	6.05
C15H12N402	280.29	280.0960	64.28	4.32	19.99	11.42
C15H12N403	296.29	296.0909	60.81	4.08	18.91	16.20
C15H12N404	312.29	312.0858	57.69	3.87	17.94	20.49
C15H12N405	328.29	328.0808	54.88	3.68	17.07	24.37
C15H12N406	344.29	344.0757	52.33	3.51	16.27	27.88
C15H12N407	360.29	360.0706	50.01	3.36	15.55	31.09
C15H12N408	376.28	376.0655	47.88	3.21	14.89	34.02
C15H12N409	392.28	392.0604	45.93	3.08	14.28	36.71
C15H12N601	292.30	292.1073	61.64	4.14	28.75	5.47
C15H12N602	308.30	308.1022	58.44	3.92	27.26	10.38
C15H12N603	324.30	324.0971	55.56	3.73	25.91	14.80
C15H12N604	340.30	340.0920	52.94	3.55	24.70	18.81
C15H12N605	356.30	356.0869	50.57	3.39	23.59	22.45
C15H12N606	372.30	372.0818	48.39	3.25	22.57	25.78
C15H12N607	388.30	388.0767	46.40	3.12	21.64	28.84
C15H12N608	404.30	404.0717	44.56	2.99	20.79	31.66
C15H12N609	420.30	420.0666	42.87	2.88	20.00	34.26
C15H13N101	223.28	223.0997	80.69	5.87	6.27	7.17
C15H13N102	239.28	239.0946	75.30	5.48	5.85	13.37
C15H13N103	255.28	255.0895	70.58	5.13	5.49	18.80
C15H13N104	271.28	271.0844	66.41	4.83	5.16	23.59
C15H13N105	287.27	287.0794	62.72	4.56	4.88	27.85
C15H13N106	303.27	303.0743	59.41	4.32	4.62	31.65
C15H13N107	319.27	319.0692	56.43	4.10	4.39	35.08
C15H13N108	335.27	335.0641	53.74	3.91	4.18	38.18
C15H13N109	351.27	351.0590	51.29	3.73	3.99	40.99
C15H13N301	251.29	251.1059	71.70	5.21	16.72	6.37
C15H13N302	267.29	267.1008	67.41	4.90	15.72	11.97
C15H13N303	283.29	283.0957	63.60	4.63	14.83	16.94
C15H13N304	299.29	299.0906	60.20	4.38	14.04	21.38
C15H13N305	315.29	315.0855	57.14	4.16	13.33	25.37
C15H13N306	331.29	331.0804	54.38	3.96	12.68	28.98
C15H13N307	347.29	347.0753	51.88	3.77	12.10	32.25
C15H13N308	363.29	363.0703	49.59	3.61	11.57	35.23
C15H13N309	379.29	379.0652	47.50	3.45	11.08	37.96
C15H13N501	279.30	279.1120	64.51	4.69	25.07	5.73
C15H13N502	295.30	295.1069	61.01	4.44	23.72	10.84
C15H13N503	311.30	311.1018	57.88	4.21	22.50	15.42
C15H13N504	327.30	327.0967	55.05	4.00	21.40	19.55
C15H13N505	343.30	343.0917	52.48	3.82	20.40	23.30
C15H13N506	359.30	359.0866	50.14	3.65	19.49	26.72
C15H13N507	375.30	375.0815	48.01	3.49	18.66	29.84

C15H13N508	391.30	391.0764	46.04	3.35	17.90	32.71
C15H13N509	407.30	407.0713	44.23	3.22	17.19	35.35
C15H14N201	238.29	238.1106	75.61	5.92	11.76	6.71
C15H14N202	254.29	254.1055	70.85	5.55	11.02	12.58
C15H14N203	270.29	270.1004	66.66	5.22	10.36	17.76
C15H14N204	286.29	286.0953	62.93	4.93	9.78	22.35
C15H14N205	302.29	302.0903	59.60	4.67	9.27	26.46
C15H14N206	318.29	318.0852	56.60	4.43	8.80	30.16
C15H14N207	334.29	334.0801	53.90	4.22	8.38	33.50
C15H14N208	350.29	350.0750	51.43	4.03	8.00	36.54
C15H14N209	366.29	366.0699	49.19	3.85	7.65	39.31
C15H14N401	266.31	266.1168	67.65	5.30	21.04	6.01
C15H14N402	282.30	282.1117	63.82	5.00	19.85	11.33
C15H14N403	298.30	298.1066	60.40	4.73	18.78	16.09
C15H14N404	314.30	314.1015	57.32	4.49	17.83	20.36
C15H14N405	330.30	330.0964	54.55	4.27	16.96	24.22
C15H14N406	346.30	346.0913	52.03	4.07	16.18	27.72
C15H14N407	362.30	362.0862	49.73	3.89	15.46	30.91
C15H14N408	378.30	378.0812	47.63	3.73	14.81	33.83
C15H14N409	394.30	394.0761	45.69	3.58	14.21	36.52
C15H14N601	294.32	294.1229	61.22	4.79	28.55	5.44
C15H14N602	310.32	310.1178	58.06	4.55	27.08	10.31
C15H14N603	326.32	326.1127	55.21	4.32	25.75	14.71
C15H14N604	342.32	342.1076	52.63	4.12	24.55	18.70
C15H14N605	358.32	358.1026	50.28	3.94	23.45	22.33
C15H14N606	374.32	374.0975	48.13	3.77	22.45	25.65
C15H14N607	390.31	390.0924	46.16	3.62	21.53	28.69
C15H14N608	406.31	406.0873	44.34	3.47	20.68	31.50
C15H14N609	422.31	422.0822	42.66	3.34	19.90	34.10
C15H15N101	225.29	225.1154	79.97	6.71	6.22	7.10
C15H15N102	241.29	241.1103	74.67	6.27	5.80	13.26
C15H15N103	257.29	257.1052	70.02	5.88	5.44	18.66
C15H15N104	273.29	273.1001	65.93	5.53	5.13	23.42
C15H15N105	289.29	289.0950	62.28	5.23	4.84	27.65
C15H15N106	305.29	305.0899	59.02	4.95	4.59	31.44
C15H15N107	321.29	321.0848	56.08	4.71	4.36	34.86
C15H15N108	337.29	337.0798	53.42	4.48	4.15	37.95
C15H15N109	353.29	353.0747	51.00	4.28	3.96	40.76
C15H15N301	253.31	253.1215	71.13	5.97	16.59	6.32
C15H15N302	269.31	269.1164	66.90	5.61	15.60	11.88
C15H15N303	285.31	285.1113	63.15	5.30	14.73	16.82
C15H15N304	301.30	301.1062	59.80	5.02	13.95	21.24
C15H15N305	317.30	317.1012	56.78	4.77	13.24	25.21
C15H15N306	333.30	333.0961	54.06	4.54	12.61	28.80
C15H15N307	349.30	349.0910	51.58	4.33	12.03	32.06
C15H15N308	365.30	365.0859	49.32	4.14	11.50	35.04
C15H15N309	381.30	381.0808	47.25	3.97	11.02	37.76
C15H15N501	281.32	281.1277	64.04	5.37	24.89	5.69
C15H15N502	297.32	297.1226	60.60	5.09	23.55	10.76
C15H15N503	313.32	313.1175	57.50	4.83	22.35	15.32
C15H15N504	329.32	329.1124	54.71	4.59	21.27	19.43
C15H15N505	345.32	345.1073	52.17	4.38	20.28	23.17
C15H15N506	361.32	361.1022	49.86	4.18	19.38	26.57
C15H15N507	377.32	377.0971	47.75	4.01	18.56	29.68
C15H15N508	393.32	393.0921	45.81	3.84	17.81	32.54
C15H15N509	409.31	409.0870	44.02	3.69	17.11	35.18
C15H16N201	240.31	240.1263	74.97	6.71	11.66	6.66
C15H16N202	256.31	256.1212	70.29	6.29	10.93	12.48
C15H16N203	272.31	272.1161	66.16	5.92	10.29	17.63
C15H16N204	288.31	288.1110	62.49	5.59	9.72	22.20
C15H16N205	304.31	304.1059	59.21	5.30	9.21	26.29
C15H16N206	320.30	320.1008	56.25	5.04	8.75	29.97
C15H16N207	336.30	336.0957	53.57	4.80	8.33	33.30
C15H16N208	352.30	352.0907	51.14	4.58	7.95	36.33
C15H16N209	368.30	368.0856	48.92	4.38	7.61	39.10
C15H16N401	268.32	268.1324	67.15	6.01	20.88	5.96
C15H16N402	284.32	284.1273	63.37	5.67	19.71	11.25
C15H16N403	300.32	300.1222	59.99	5.37	18.66	15.98
C15H16N404	316.32	316.1171	56.96	5.10	17.71	20.23
C15H16N405	332.32	332.1121	54.22	4.85	16.86	24.07
C15H16N406	348.32	348.1070	51.72	4.63	16.08	27.56
C15H16N407	364.32	364.1019	49.45	4.43	15.38	30.74
C15H16N408	380.32	380.0968	47.37	4.24	14.73	33.65
C15H16N409	396.32	396.0917	45.46	4.07	14.14	36.33
C15H16N601	296.33	296.1386	60.80	5.44	28.36	5.40
C15H16N602	312.33	312.1335	57.68	5.16	26.91	10.25
C15H16N603	328.33	328.1284	54.87	4.91	25.60	14.62
C15H16N604	344.33	344.1233	52.32	4.68	24.41	18.59
C15H16N605	360.33	360.1182	50.00	4.48	23.32	22.20
C15H16N606	376.33	376.1131	47.87	4.29	22.33	25.51
C15H16N607	392.33	392.1080	45.92	4.11	21.42	28.55
C15H16N608	408.33	408.1029	44.12	3.95	20.58	31.35
C15H16N609	424.33	424.0979	42.46	3.80	19.81	33.93
C15H17N101	227.31	227.1310	79.26	7.54	6.16	7.04
C15H17N102	243.31	243.1259	74.05	7.04	5.76	13.15
C15H17N103	259.31	259.1208	69.48	6.61	5.40	18.51
C15H17N104	275.31	275.1157	65.44	6.22	5.09	23.25
C15H17N105	291.31	291.1107	61.85	5.88	4.81	27.46
C15H17N106	307.31	307.1056	58.63	5.58	4.56	31.24
C15H17N107	323.31	323.1005	55.73	5.30	4.33	34.64
C15H17N108	339.30	339.0954	53.10	5.05	4.13	37.72
C15H17N109	355.30	355.0903	50.71	4.82	3.94	40.53
C15H17N301	255.32	255.1372	70.56	6.71	16.46	6.27
C15H17N302	271.32	271.1321	66.40	6.32	15.49	11.79
C15H17N303	287.32	287.1270	62.71	5.96	14.62	16.71
C15H17N304	303.32	303.1219	59.40	5.65	13.85	21.10
C15H17N305	319.32	319.1168	56.42	5.37	13.16	25.05
C15H17N306	335.32	335.1117	53.73	5.11	12.53	28.63
C15H17N307	351.32	351.1066	51.28	4.88	11.96	31.88
C15H17N308	367.32	367.1016	49.05	4.67	11.44	34.85
C15H17N309	383.32	383.0965	47.00	4.47	10.96	37.57
C15H17N501	283.34	283.1433	63.59	6.05	24.72	5.65
C15H17N502	299.34	299.1382	60.19	5.72	23.40	10.69
C15H17N503	315.33	315.1331	57.14	5.43	22.21	15.22
C15H17N504	331.33	331.1280	54.38	5.17	21.14	19.32
C15H17N505	347.33	347.1230	51.87	4.93	20.16	23.03
C15H17N506	363.33	363.1179	49.59	4.72	19.28	26.42
C15H17N507	379.33	379.1128	47.50	4.52	18.46	29.52
C15H17N508	395.33	395.1077	45.57	4.33	17.72	32.38
C15H17N509	411.33	411.1026	43.80	4.17	17.03	35.01
C15H18N201	242.32	242.1419	74.35	7.49	11.56	6.60
C15H18N202	258.32	258.1368	69.74	7.02	10.84	12.39
C15H18N203	274.32	274.1317	65.68	6.61	10.21	17.50
C15H18N204	290.32	290.1266	62.06	6.25	9.65	22.04
C15H18N205	306.32	306.1216	58.82	5.92	9.15	26.12
C15H18N206	322.32	322.1165	55.90	5.63	8.69	29.78
C15H18N207	338.32	338.1114	53.25	5.36	8.28	33.10
C15H18N208	354.32	354.1063	50.85	5.12	7.91	36.12
C15H18N209	370.32	370.1012	48.65	4.90	7.56	38.88
C15H18N401	270.34	270.1481	66.65	6.71	20.72	5.92

C15H18N402	286.34	286.1430	62.92	6.34	19.57	11.18
C15H18N403	302.34	302.1379	59.59	6.00	18.53	15.88
C15H18N404	318.34	318.1328	56.60	5.70	17.60	20.10
C15H18N405	334.33	334.1277	53.89	5.43	16.76	23.93
C15H18N406	350.33	350.1226	51.43	5.18	15.99	27.40
C15H18N407	366.33	366.1175	49.18	4.95	15.29	30.57
C15H18N408	382.33	382.1125	47.12	4.75	14.65	33.48
C15H18N409	398.33	398.1074	45.23	4.55	14.07	36.15
C15H18N601	298.35	298.1542	60.39	6.08	28.17	5.36
C15H18N602	314.35	314.1491	57.31	5.77	26.73	10.18
C15H18N603	330.35	330.1440	54.54	5.49	25.44	14.53
C15H18N604	346.35	346.1389	52.02	5.24	24.26	18.48
C15H18N605	362.35	362.1339	49.72	5.01	23.19	22.08
C15H18N606	378.35	378.1288	47.62	4.80	22.21	25.37
C15H18N607	394.35	394.1237	45.69	4.60	21.31	28.40
C15H18N608	410.35	410.1186	43.91	4.42	20.48	31.19
C15H18N609	426.35	426.1135	42.26	4.26	19.71	33.77
C15H19N101	229.32	229.1467	78.56	8.35	6.11	6.98
C15H19N102	245.32	245.1416	73.44	7.81	5.71	13.04
C15H19N103	261.32	261.1365	68.94	7.33	5.36	18.37
C15H19N104	277.32	277.1314	64.97	6.91	5.05	23.08
C15H19N105	293.32	293.1263	61.42	6.53	4.78	27.27
C15H19N106	309.32	309.1212	58.25	6.19	4.53	31.03
C15H19N107	325.32	325.1161	55.38	5.89	4.31	34.43
C15H19N108	341.32	341.1111	52.79	5.61	4.10	37.50
C15H19N109	357.32	357.1060	50.42	5.36	3.92	40.30
C15H19N301	257.34	257.1528	70.01	7.44	16.33	6.22
C15H19N302	273.34	273.1477	65.91	7.01	15.37	11.71
C15H19N303	289.34	289.1426	62.27	6.62	14.52	16.59
C15H19N304	305.34	305.1375	59.01	6.27	13.76	20.96
C15H19N305	321.34	321.1325	56.07	5.96	13.08	24.90
C15H19N306	337.34	337.1274	53.41	5.68	12.46	28.46
C15H19N307	353.33	353.1223	50.99	5.42	11.89	31.70
C15H19N308	369.33	369.1172	48.78	5.19	11.38	34.66
C15H19N309	385.33	385.1121	46.76	4.97	10.90	37.37
C15H19N501	285.35	285.1590	63.14	6.71	24.54	5.61
C15H19N502	301.35	301.1539	59.79	6.36	23.24	10.62
C15H19N503	317.35	317.1488	56.77	6.03	22.07	15.12
C15H19N504	333.35	333.1437	54.05	5.75	21.01	19.20
C15H19N505	349.35	349.1386	51.57	5.48	20.05	22.90
C15H19N506	365.35	365.1335	49.31	5.24	19.17	26.28
C15H19N507	381.35	381.1284	47.24	5.02	18.36	29.37
C15H19N508	397.35	397.1233	45.34	4.82	17.63	32.21
C15H19N509	413.35	413.1183	43.59	4.63	16.94	34.84
C15H20N201	244.34	244.1576	73.74	8.25	11.46	6.55
C15H20N202	260.34	260.1525	69.20	7.74	10.76	12.29
C15H20N203	276.34	276.1474	65.20	7.30	10.14	17.37
C15H20N204	292.34	292.1423	61.63	6.90	9.58	21.89
C15H20N205	308.34	308.1372	58.43	6.54	9.09	25.94
C15H20N206	324.34	324.1321	55.55	6.22	8.64	29.60
C15H20N207	340.34	340.1270	52.94	5.92	8.23	32.91
C15H20N208	356.34	356.1220	50.56	5.66	7.86	35.92
C15H20N209	372.33	372.1169	48.39	5.41	7.52	38.67
C15H20N401	272.35	272.1637	66.15	7.40	20.57	5.87
C15H20N402	288.35	288.1586	62.48	6.99	19.43	11.10
C15H20N403	304.35	304.1535	59.20	6.62	18.41	15.77
C15H20N404	320.35	320.1484	56.24	6.29	17.49	19.98
C15H20N405	336.35	336.1434	53.57	5.99	16.66	23.78
C15H20N406	352.35	352.1383	51.13	5.72	15.90	27.24
C15H20N407	368.35	368.1332	48.91	5.47	15.21	30.40
C15H20N408	384.35	384.1281	46.88	5.25	14.58	33.30
C15H20N409	400.35	400.1230	45.00	5.04	13.99	35.97
C15H20N601	300.37	300.1698	59.98	6.71	27.98	5.33
C15H20N602	316.37	316.1648	56.95	6.37	26.56	10.11
C15H20N603	332.37	332.1597	54.21	6.07	25.29	14.44
C15H20N604	348.36	348.1546	51.72	5.79	24.12	18.37
C15H20N605	364.36	364.1495	49.45	5.53	23.06	21.96
C15H20N606	380.36	380.1444	47.37	5.30	22.09	25.24
C15H20N607	396.36	396.1393	45.46	5.09	21.20	28.26
C15H20N608	412.36	412.1342	43.69	4.89	20.38	31.04
C15H20N609	428.36	428.1292	42.06	4.71	19.62	33.62
C15H21N101	231.34	231.1623	77.88	9.15	6.05	6.92
C15H21N102	247.34	247.1572	72.84	8.56	5.66	12.94
C15H21N103	263.34	263.1521	68.42	8.04	5.32	18.23
C15H21N104	279.34	279.1470	64.50	7.58	5.01	22.91
C15H21N105	295.34	295.1420	61.00	7.17	4.74	27.09
C15H21N106	311.34	311.1369	57.87	6.80	4.50	30.83
C15H21N107	327.34	327.1318	55.04	6.47	4.28	34.21
C15H21N108	343.34	343.1267	52.48	6.17	4.08	37.28
C15H21N109	359.34	359.1216	50.14	5.89	3.90	40.07
C15H21N301	259.35	259.1685	69.47	8.16	16.20	6.17
C15H21N302	275.35	275.1634	65.43	7.69	15.26	11.62
C15H21N303	291.35	291.1583	61.84	7.27	14.42	16.47
C15H21N304	307.35	307.1532	58.62	6.89	13.67	20.82
C15H21N305	323.35	323.1481	55.72	6.55	13.00	24.74
C15H21N306	339.35	339.1430	53.09	6.24	12.38	28.29
C15H21N307	355.35	355.1379	50.70	5.96	11.82	31.52
C15H21N308	371.35	371.1329	48.52	5.70	11.32	34.47
C15H21N309	387.35	387.1278	46.51	5.46	10.85	37.17
C15H21N501	287.37	287.1746	62.70	7.37	24.37	5.57
C15H21N502	303.37	303.1695	59.39	6.98	23.09	10.55
C15H21N503	319.37	319.1644	56.41	6.63	21.93	15.03
C15H21N504	335.37	335.1593	53.72	6.31	20.88	19.08
C15H21N505	351.37	351.1543	51.28	6.02	19.93	22.77
C15H21N506	367.36	367.1492	49.04	5.76	19.06	26.13
C15H21N507	383.36	383.1441	47.00	5.52	18.27	29.21
C15H21N508	399.36	399.1390	45.11	5.30	17.54	32.05
C15H21N509	415.36	415.1339	43.38	5.10	16.86	34.67
C15H22N201	246.36	246.1732	73.13	9.00	11.37	6.49
C15H22N202	262.35	262.1681	68.67	8.45	10.68	12.20
C15H22N203	278.35	278.1630	64.73	7.97	10.06	17.24
C15H22N204	294.35	294.1579	61.21	7.53	9.52	21.74
C15H22N205	310.35	310.1529	58.05	7.15	9.03	25.78
C15H22N206	326.35	326.1478	55.21	6.79	8.58	29.41
C15H22N207	342.35	342.1427	52.63	6.48	8.18	32.71
C15H22N208	358.35	358.1376	50.28	6.19	7.82	35.72
C15H22N209	374.35	374.1325	48.13	5.92	7.48	38.47
C15H22N401	274.37	274.1794	65.67	8.08	20.42	5.83
C15H22N402	290.37	290.1743	62.05	7.64	19.30	11.02
C15H22N403	306.37	306.1692	58.81	7.24	18.29	15.67
C15H22N404	322.37	322.1641	55.89	6.88	17.38	19.85
C15H22N405	338.37	338.1590	53.25	6.55	16.56	23.64
C15H22N406	354.37	354.1539	50.84	6.26	15.81	27.09
C15H22N407	370.37	370.1488	48.65	5.99	15.13	30.24
C15H22N408	386.36	386.1437	46.63	5.74	14.50	33.13
C15H22N409	402.36	402.1387	44.78	5.51	13.92	35.79
C15H22N601	302.38	302.1855	59.58	7.33	27.79	5.29
C15H22N602	318.38	318.1804	56.59	6.97	26.40	10.05
C15H22N603	334.38	334.1753	53.88	6.63	25.13	14.35
C15H22N604	350.38	350.1702	51.42	6.33	23.99	18.27

078

C15H22N605	366.38	366.1652	49.17	6.05	22.94	21.83
C15H22N606	382.38	382.1601	47.12	5.80	21.98	25.11
C15H22N607	398.38	398.1550	45.23	5.57	21.10	28.11
C15H22N608	414.38	414.1499	43.48	5.35	20.28	30.89
C15H22N609	430.38	430.1448	41.86	5.15	19.53	33.46
C15H23N101	233.36	233.1780	77.21	9.93	6.00	6.86
C15H23N102	249.36	249.1729	72.25	9.30	5.62	12.83
C15H23N103	265.36	265.1678	67.90	8.74	5.28	18.09
C15H23N104	281.35	281.1627	64.04	8.24	4.98	22.75
C15H23N105	297.35	297.1576	60.59	7.80	4.71	26.90
C15H23N106	313.35	313.1525	57.50	7.40	4.47	30.64
C15H23N107	329.35	329.1474	54.70	7.04	4.25	34.00
C15H23N108	345.35	345.1424	52.17	6.71	4.06	37.06
C15H23N109	361.35	361.1373	49.86	6.42	3.88	39.85
C15H23N301	261.37	261.1841	68.93	8.87	16.08	6.12
C15H23N302	277.37	277.1790	64.96	8.36	15.15	11.54
C15H23N303	293.37	293.1739	61.41	7.90	14.32	16.36
C15H23N304	309.37	309.1688	58.24	7.49	13.58	20.69
C15H23N305	325.37	325.1638	55.37	7.13	12.91	24.59
C15H23N306	341.37	341.1587	52.78	6.79	12.31	28.12
C15H23N307	357.37	357.1536	50.42	6.49	11.76	31.34
C15H23N308	373.37	373.1485	48.25	6.21	11.25	34.28
C15H23N309	389.37	389.1434	46.27	5.95	10.79	36.98
C15H23N501	289.38	289.1902	62.26	8.01	24.20	5.53
C15H23N502	305.38	305.1852	59.00	7.59	22.93	10.48
C15H23N503	321.38	321.1801	56.06	7.21	21.79	14.93
C15H23N504	337.38	337.1750	53.40	6.87	20.76	18.97
C15H23N505	353.38	353.1699	50.98	6.56	19.82	22.64
C15H23N506	369.38	369.1648	48.78	6.28	18.96	25.99
C15H23N507	385.38	385.1597	46.75	6.02	18.17	29.06
C15H23N508	401.38	401.1546	44.89	5.78	17.45	31.89
C15H23N509	417.38	417.1496	43.17	5.55	16.78	34.50
C15H24N201	248.37	248.1889	72.54	9.74	11.28	6.44
C15H24N202	264.37	264.1838	68.15	9.15	10.60	12.10
C15H24N203	280.37	280.1787	64.26	8.63	9.99	17.12
C15H24N204	296.37	296.1736	60.79	8.16	9.45	21.59
C15H24N205	312.37	312.1685	57.68	7.74	8.97	25.61
C15H24N206	328.37	328.1634	54.87	7.37	8.53	29.23
C15H24N207	344.37	344.1583	52.32	7.02	8.13	32.52
C15H24N208	360.37	360.1533	50.00	6.71	7.77	35.52
C15H24N209	376.37	376.1482	47.87	6.43	7.44	38.26
C15H24N401	276.38	276.1950	65.19	8.75	20.27	5.79
C15H24N402	292.38	292.1899	61.62	8.27	19.16	10.94
C15H24N403	308.38	308.1848	58.42	7.84	18.17	15.56
C15H24N404	324.38	324.1797	55.54	7.46	17.27	19.73
C15H24N405	340.38	340.1747	52.93	7.11	16.46	23.50
C15H24N406	356.38	356.1696	50.55	6.79	15.72	26.94
C15H24N407	372.38	372.1645	48.38	6.50	15.05	30.08
C15H24N408	388.38	388.1594	46.39	6.23	14.43	32.96
C15H24N409	404.38	404.1543	44.55	5.98	13.85	35.61
C15H24N601	304.40	304.2011	59.19	7.95	27.61	5.26
C15H24N602	320.40	320.1961	56.23	7.55	26.23	9.99
C15H24N603	336.40	336.1910	53.56	7.19	24.98	14.27
C15H24N604	352.40	352.1859	51.13	6.86	23.85	18.16
C15H24N605	368.40	368.1808	48.91	6.57	22.81	21.71
C15H24N606	384.40	384.1757	46.87	6.29	21.86	24.97
C15H24N607	400.39	400.1706	45.00	6.04	20.99	27.97
C15H24N608	416.39	416.1655	43.27	5.81	20.18	30.74
C15H24N609	432.39	432.1605	41.67	5.59	19.44	33.30
C15H25N101	235.37	235.1936	76.55	10.71	5.95	6.80

C15H25N102	251.37	251.1885	71.67	10.02	5.57	12.73
C15H25N103	267.37	267.1834	67.38	9.42	5.24	17.95
C15H25N104	283.37	283.1783	63.58	8.89	4.94	22.58
C15H25N105	299.37	299.1733	60.18	8.42	4.68	26.72
C15H25N106	315.37	315.1682	57.13	7.99	4.44	30.44
C15H25N107	331.37	331.1631	54.37	7.60	4.23	33.80
C15H25N108	347.37	347.1580	51.87	7.25	4.03	36.85
C15H25N109	363.37	363.1529	49.58	6.93	3.85	39.63
C15H25N301	263.39	263.1998	68.40	9.57	15.95	6.07
C15H25N302	279.39	279.1947	64.49	9.02	15.04	11.45
C15H25N303	295.38	295.1896	60.99	8.53	14.23	16.25
C15H25N304	311.38	311.1845	57.86	8.09	13.49	20.55
C15H25N305	327.38	327.1794	55.03	7.70	12.84	24.44
C15H25N306	343.38	343.1743	52.47	7.34	12.24	27.96
C15H25N307	359.38	359.1692	50.13	7.01	11.69	31.16
C15H25N308	375.38	375.1641	48.00	6.71	11.19	34.10
C15H25N309	391.38	391.1591	46.03	6.44	10.74	36.79
C15H25N501	291.40	291.2059	61.83	8.65	24.03	5.49
C15H25N502	307.40	307.2008	58.61	8.20	22.78	10.41
C15H25N503	323.40	323.1957	55.71	7.79	21.66	14.84
C15H25N504	339.40	339.1906	53.08	7.42	20.63	18.86
C15H25N505	355.40	355.1856	50.69	7.09	19.71	22.51
C15H25N506	371.40	371.1805	48.51	6.79	18.86	25.85
C15H25N507	387.40	387.1754	46.51	6.50	18.08	28.91
C15H25N508	403.40	403.1703	44.66	6.25	17.36	31.73
C15H25N509	419.39	419.1652	42.96	6.01	16.70	34.33
C15H26N201	250.39	250.2045	71.96	10.47	11.19	6.39
C15H26N202	266.39	266.1994	67.63	9.84	10.52	12.01
C15H26N203	282.39	282.1943	63.80	9.28	9.92	17.00
C15H26N204	298.39	298.1892	60.38	8.78	9.39	21.45
C15H26N205	314.38	314.1842	57.31	8.34	8.91	25.45
C15H26N206	330.38	330.1791	54.53	7.93	8.48	29.06
C15H26N207	346.38	346.1740	52.01	7.57	8.09	32.33
C15H26N208	362.38	362.1689	49.72	7.23	7.73	35.32
C15H26N209	378.38	378.1638	47.62	6.93	7.40	38.06
C15H26N401	278.40	278.2106	64.72	9.41	20.12	5.75
C15H26N402	294.40	294.2056	61.20	8.90	19.03	10.87
C15H26N403	310.40	310.2005	58.04	8.44	18.05	15.46
C15H26N404	326.40	326.1954	55.20	8.03	17.17	19.61
C15H26N405	342.40	342.1903	52.62	7.65	16.36	23.36
C15H26N406	358.40	358.1852	50.27	7.31	15.63	26.78
C15H26N407	374.40	374.1801	48.12	7.00	14.96	29.91
C15H26N408	390.40	390.1750	46.15	6.71	14.35	32.79
C15H26N409	406.40	406.1700	44.33	6.45	13.79	35.43
C15H26N601	306.41	306.2168	58.80	8.55	27.43	5.22
C15H26N602	322.41	322.2117	55.88	8.13	26.07	9.92
C15H26N603	338.41	338.2066	53.24	7.74	24.83	14.18
C15H26N604	354.41	354.2015	50.84	7.39	23.71	18.06
C15H26N605	370.41	370.1965	48.64	7.08	22.69	21.60
C15H26N606	386.41	386.1914	46.63	6.78	21.75	24.84
C15H26N607	402.41	402.1863	44.77	6.51	20.88	27.83
C15H26N608	418.41	418.1812	43.06	6.26	20.09	30.59
C15H26N609	434.41	434.1761	41.47	6.03	19.35	33.15
C15H27N101	237.39	237.2093	75.90	11.46	5.90	6.74
C15H27N102	253.39	253.2042	71.10	10.74	5.53	12.63
C15H27N103	269.39	269.1991	66.88	10.10	5.20	17.82
C15H27N104	285.39	285.1940	63.13	9.54	4.91	22.42
C15H27N105	301.39	301.1889	59.78	9.03	4.65	26.54
C15H27N106	317.39	317.1838	56.77	8.57	4.41	30.25
C15H27N107	333.38	333.1787	54.04	8.16	4.20	33.59

079

C15H27N108	349.38	349.1737	51.57	7.79	4.01	36.63
C15H27N109	365.38	365.1686	49.31	7.45	3.83	39.41
C15H27N301	265.40	265.2154	67.88	10.25	15.83	6.03
C15H27N302	281.40	281.2103	64.03	9.67	14.93	11.37
C15H27N303	297.40	297.2052	60.58	9.15	14.13	16.14
C15H27N304	313.40	313.2001	57.49	8.68	13.41	20.42
C15H27N305	329.40	329.1951	54.70	8.26	12.76	24.29
C15H27N306	345.40	345.1900	52.16	7.88	12.17	27.79
C15H27N307	361.40	361.1849	49.85	7.53	11.63	30.99
C15H27N308	377.40	377.1798	47.74	7.21	11.13	33.92
C15H27N309	393.40	393.1747	45.80	6.92	10.68	36.60
C15H27N501	293.42	293.2215	61.40	9.28	23.87	5.45
C15H27N502	309.41	309.2165	58.23	8.80	22.63	10.34
C15H27N503	325.41	325.2114	55.37	8.36	21.52	14.75
C15H27N504	341.41	341.2063	52.77	7.97	20.51	18.74
C15H27N505	357.41	357.2012	50.41	7.61	19.59	22.38
C15H27N506	373.41	373.1961	48.25	7.29	18.76	25.71
C15H27N507	389.41	389.1910	46.27	6.99	17.98	28.76
C15H27N508	405.41	405.1859	44.44	6.71	17.27	31.57
C15H27N509	421.41	421.1809	42.75	6.46	16.62	34.17
C15H28N201	252.40	252.2202	71.38	11.18	11.10	6.34
C15H28N202	268.40	268.2151	67.13	10.52	10.44	11.92
C15H28N203	284.40	284.2100	63.35	9.92	9.85	16.88
C15H28N204	300.40	300.2049	59.98	9.40	9.33	21.30
C15H28N205	316.40	316.1998	56.94	8.92	8.85	25.28
C15H28N206	332.40	332.1947	54.20	8.49	8.43	28.88
C15H28N207	348.40	348.1896	51.71	8.10	8.04	32.15
C15H28N208	364.40	364.1845	49.44	7.75	7.69	35.13
C15H28N209	380.40	380.1795	47.36	7.42	7.36	37.85
C15H28N401	280.42	280.2263	64.25	10.06	19.98	5.71
C15H28N402	296.42	296.2212	60.78	9.52	18.90	10.80
C15H28N403	312.42	312.2161	57.67	9.03	17.93	15.36
C15H28N404	328.41	328.2110	54.86	8.59	17.06	19.49
C15H28N405	344.41	344.2060	52.31	8.19	16.27	23.23
C15H28N406	360.41	360.2009	49.99	7.83	15.55	26.64
C15H28N407	376.41	376.1958	47.86	7.50	14.88	29.75
C15H28N408	392.41	392.1907	45.91	7.19	14.28	32.62
C15H28N409	408.41	408.1856	44.11	6.91	13.72	35.26
C15H28N601	308.43	308.2324	58.41	9.15	27.25	5.19
C15H28N602	324.43	324.2274	55.53	8.70	25.90	9.86
C15H28N603	340.43	340.2223	52.92	8.29	24.69	14.10
C15H28N604	356.43	356.2172	50.55	7.92	23.58	17.96
C15H28N605	372.43	372.2121	48.38	7.58	22.57	21.48
C15H28N606	388.43	388.2070	46.38	7.27	21.64	24.71
C15H28N607	404.43	404.2019	44.55	6.98	20.78	27.69
C15H28N608	420.43	420.1968	42.85	6.71	19.99	30.44
C15H28N609	436.43	436.1918	41.28	6.47	19.26	32.99
C15H29N101	239.40	239.2249	75.26	12.21	5.85	6.68
C15H29N102	255.40	255.2198	70.54	11.45	5.48	12.53
C15H29N103	271.40	271.2147	66.38	10.77	5.16	17.69
C15H29N104	287.40	287.2096	62.69	10.17	4.87	22.27
C15H29N105	303.40	303.2046	59.38	9.63	4.62	26.37
C15H29N106	319.40	319.1995	56.41	9.15	4.39	30.06
C15H29N107	335.40	335.1944	53.72	8.72	4.18	33.39
C15H29N108	351.40	351.1893	51.27	8.32	3.99	36.42
C15H29N109	367.40	367.1842	49.04	7.96	3.81	39.19
C15H29N301	267.42	267.2310	67.37	10.93	15.71	5.98
C15H29N302	283.42	283.2260	63.57	10.31	14.83	11.29
C15H29N303	299.42	299.2209	60.17	9.76	14.03	16.03
C15H29N304	315.42	315.2158	57.12	9.27	13.32	20.29

C15H29N305	331.42	331.2107	54.36	8.82	12.68	24.14
C15H29N306	347.41	347.2056	51.86	8.41	12.10	27.63
C15H29N307	363.41	363.2005	49.58	8.04	11.56	30.82
C15H29N308	379.41	379.1954	47.49	7.70	11.08	33.73
C15H29N309	395.41	395.1904	45.56	7.39	10.63	36.42
C15H29N501	295.43	295.2372	60.98	9.89	23.71	5.42
C15H29N502	311.43	311.2321	57.85	9.39	22.49	10.27
C15H29N503	327.43	327.2270	55.02	8.93	21.39	14.66
C15H29N504	343.43	343.2219	52.46	8.51	20.39	18.63
C15H29N505	359.43	359.2169	50.13	8.13	19.48	22.26
C15H29N506	375.43	375.2118	47.99	7.79	18.65	25.57
C15H29N507	391.43	391.2067	46.03	7.47	17.89	28.61
C15H29N508	407.43	407.2016	44.22	7.17	17.19	31.42
C15H29N509	423.43	423.1965	42.55	6.90	16.54	34.01
C15H30N201	254.42	254.2358	70.82	11.89	11.01	6.29
C15H30N202	270.42	270.2307	66.63	11.18	10.36	11.83
C15H30N203	286.42	286.2256	62.90	10.56	9.78	16.76
C15H30N204	302.42	302.2205	59.58	10.00	9.26	21.16
C15H30N205	318.42	318.2155	56.58	9.50	8.80	25.12
C15H30N206	334.42	334.2104	53.88	9.04	8.38	28.71
C15H30N207	350.42	350.2053	51.42	8.63	7.99	31.96
C15H30N208	366.41	366.2002	49.17	8.25	7.65	34.93
C15H30N209	382.41	382.1951	47.11	7.91	7.33	37.65
C15H30N401	282.43	282.2419	63.79	10.71	19.84	5.66
C15H30N402	298.43	298.2369	60.37	10.13	18.77	10.72
C15H30N403	314.43	314.2318	57.30	9.62	17.82	15.27
C15H30N404	330.43	330.2267	54.52	9.15	16.96	19.37
C15H30N405	346.43	346.2216	52.01	8.73	16.17	23.09
C15H30N406	362.43	362.2165	49.71	8.34	15.46	26.49
C15H30N407	378.43	378.2114	47.61	7.99	14.81	29.59
C15H30N408	394.43	394.2063	45.68	7.67	14.20	32.45
C15H30N409	410.43	410.2013	43.90	7.37	13.65	35.08
C15H30N601	310.45	310.2481	58.03	9.74	27.07	5.15
C15H30N602	326.45	326.2430	55.19	9.26	25.74	9.80
C15H30N603	342.44	342.2379	52.61	8.83	24.54	14.02
C15H30N604	358.44	358.2328	50.26	8.44	23.45	17.85
C15H30N605	374.44	374.2278	48.12	8.08	22.44	21.36
C15H30N606	390.44	390.2227	46.14	7.74	21.52	24.59
C15H30N607	406.44	406.2176	44.33	7.44	20.68	27.56
C15H30N608	422.44	422.2125	42.65	7.16	19.89	30.30
C15H30N609	438.44	438.2074	41.09	6.90	19.17	32.84
C15H31N101	241.42	241.2406	74.63	12.94	5.80	6.63
C15H31N102	257.42	257.2355	69.99	12.14	5.44	12.43
C15H31N103	273.42	273.2304	65.89	11.43	5.12	17.55
C15H31N104	289.42	289.2253	62.25	10.80	4.84	22.11
C15H31N105	305.42	305.2202	58.99	10.23	4.59	26.19
C15H31N106	321.42	321.2151	56.05	9.72	4.36	29.87
C15H31N107	337.42	337.2100	53.40	9.26	4.15	33.19
C15H31N108	353.42	353.2049	50.98	8.84	3.96	36.22
C15H31N109	369.42	369.1999	48.77	8.46	3.79	38.98
C15H31N301	269.43	269.2467	66.87	11.60	15.60	5.94
C15H31N302	285.43	285.2416	63.12	10.95	14.72	11.21
C15H31N303	301.43	301.2365	59.77	10.37	13.94	15.92
C15H31N304	317.43	317.2314	56.76	9.84	13.24	20.16
C15H31N305	333.43	333.2264	54.03	9.37	12.60	23.99
C15H31N306	349.43	349.2213	51.56	8.94	12.03	27.47
C15H31N307	365.43	365.2162	49.30	8.55	11.50	30.65
C15H31N308	381.43	381.2111	47.23	8.19	11.02	33.56
C15H31N309	397.43	397.2060	45.33	7.86	10.57	36.23
C15H31N501	297.45	297.2528	60.57	10.51	23.54	5.38

C15H31N502	313.45	313.2478	57.48	9.97	22.34	10.21
C15H31N503	329.45	329.2427	54.69	9.48	21.26	14.57
C15H31N504	345.45	345.2376	52.16	9.05	20.27	18.53
C15H31N505	361.44	361.2325	49.85	8.65	19.38	22.13
C15H31N506	377.44	377.2274	47.73	8.28	18.55	25.43
C15H31N507	393.44	393.2223	45.79	7.94	17.80	28.47
C15H31N508	409.44	409.2172	44.00	7.63	17.10	31.26
C15H31N509	425.44	425.2122	42.35	7.34	16.46	33.85
C15H32N201	256.44	256.2514	70.26	12.58	10.92	6.24
C15H32N202	272.43	272.2464	66.13	11.84	10.28	11.75
C15H32N203	288.43	288.2413	62.46	11.18	9.71	16.64
C15H32N204	304.43	304.2362	59.18	10.60	9.20	21.02
C15H32N205	320.43	320.2311	56.23	10.07	8.74	24.97
C15H32N206	336.43	336.2260	53.55	9.59	8.33	28.53
C15H32N207	352.43	352.2209	51.12	9.15	7.95	31.78
C15H32N208	368.43	368.2158	48.90	8.75	7.60	34.74
C15H32N209	384.43	384.2108	46.87	8.39	7.29	37.46
C15H32N401	284.45	284.2576	63.34	11.34	19.70	5.62
C15H32N402	300.45	300.2525	59.97	10.74	18.65	10.65
C15H32N403	316.45	316.2474	56.93	10.19	17.70	15.17
C15H32N404	332.45	332.2423	54.19	9.70	16.85	19.25
C15H32N405	348.45	348.2373	51.71	9.26	16.08	22.96
C15H32N406	364.45	364.2322	49.44	8.85	15.37	26.34
C15H32N407	380.44	380.2271	47.36	8.48	14.73	29.44
C15H32N408	396.44	396.2220	45.45	8.14	14.13	32.29
C15H32N409	412.44	412.2169	43.68	7.82	13.58	34.91
C15H32N601	312.46	312.2637	57.66	10.32	26.90	5.12
C15H32N602	328.46	328.2587	54.85	9.82	25.59	9.74
C15H32N603	344.46	344.2536	52.30	9.36	24.40	13.93
C15H32N604	360.46	360.2485	49.98	8.95	23.31	17.75
C15H32N605	376.46	376.2434	47.86	8.57	22.32	21.25
C15H32N606	392.46	392.2383	45.91	8.22	21.41	24.46
C15H32N607	408.46	408.2332	44.11	7.90	20.57	27.42
C15H32N608	424.46	424.2281	42.45	7.60	19.80	30.15
C15H32N609	440.46	440.2231	40.90	7.32	19.08	32.69
C15H33N101	243.44	243.2562	74.01	13.66	5.75	6.57
C15H33N102	259.44	259.2511	69.45	12.82	5.40	12.33
C15H33N103	275.44	275.2460	65.41	12.08	5.09	17.43
C15H33N104	291.43	291.2409	61.82	11.41	4.81	21.96
C15H33N105	307.43	307.2359	58.60	10.82	4.56	26.02
C15H33N106	323.43	323.2308	55.70	10.28	4.33	29.68
C15H33N107	339.43	339.2257	53.08	9.80	4.13	32.99
C15H33N108	355.43	355.2206	50.69	9.36	3.94	36.01
C15H33N109	371.43	371.2155	48.51	8.96	3.77	38.77
C15H33N301	271.45	271.2623	66.37	12.25	15.48	5.89
C15H33N302	287.45	287.2573	62.68	11.57	14.62	11.13
C15H33N303	303.45	303.2522	59.37	10.96	13.85	15.82
C15H33N304	319.45	319.2471	56.40	10.41	13.15	20.03
C15H33N305	335.45	335.2420	53.71	9.92	12.53	23.85
C15H33N306	351.45	351.2369	51.26	9.46	11.96	27.31
C15H33N307	367.45	367.2318	49.03	9.05	11.44	30.48
C15H33N308	383.45	383.2267	46.99	8.67	10.96	33.38
C15H33N309	399.44	399.2217	45.10	8.33	10.52	36.05
C15H33N501	299.46	299.2685	60.16	11.11	23.39	5.34
C15H33N502	315.46	315.2634	57.11	10.54	22.20	10.14
C15H33N503	331.46	331.2583	54.36	10.04	21.13	14.48
C15H33N504	347.46	347.2532	51.85	9.57	20.16	18.42
C15H33N505	363.46	363.2482	49.57	9.15	19.27	22.01
C15H33N506	379.46	379.2431	47.48	8.77	18.46	25.30
C15H33N507	395.46	395.2380	45.56	8.41	17.71	28.32
C15H33N508	411.46	411.2329	43.79	8.08	17.02	31.11
C15H33N509	427.46	427.2278	42.15	7.78	16.38	33.69
C15H34N201	258.45	258.2671	69.71	13.26	10.84	6.19
C15H34N202	274.45	274.2620	65.65	12.49	10.21	11.66
C15H34N203	290.45	290.2569	62.03	11.80	9.64	16.53
C15H34N204	306.45	306.2518	58.79	11.18	9.14	20.88
C15H34N205	322.45	322.2468	55.87	10.63	8.69	24.81
C15H34N206	338.45	338.2417	53.23	10.13	8.28	28.36
C15H34N207	354.45	354.2366	50.83	9.67	7.90	31.60
C15H34N208	370.45	370.2315	48.64	9.25	7.56	34.55
C15H34N209	386.45	386.2264	46.62	8.87	7.25	37.26
C15H34N401	286.46	286.2732	62.89	11.96	19.56	5.59
C15H34N402	302.46	302.2682	59.57	11.33	18.52	10.58
C15H34N403	318.46	318.2631	56.57	10.76	17.59	15.07
C15H34N404	334.46	334.2580	53.87	10.25	16.75	19.13
C15H34N405	350.46	350.2529	51.41	9.78	15.99	22.83
C15H34N406	366.46	366.2478	49.16	9.35	15.29	26.20
C15H34N407	382.46	382.2427	47.11	8.96	14.65	29.28
C15H34N408	398.46	398.2376	45.22	8.60	14.06	32.12
C15H34N409	414.46	414.2326	43.47	8.27	13.52	34.74
C15H34N601	314.48	314.2794	57.29	10.90	26.72	5.09
C15H34N602	330.48	330.2743	54.52	10.37	25.43	9.68
C15H34N603	346.48	346.2692	52.00	9.89	24.26	13.85
C15H34N604	362.48	362.2641	49.70	9.45	23.19	17.66
C15H34N605	378.48	378.2590	47.60	9.06	22.20	21.14
C15H34N606	394.47	394.2540	45.67	8.69	21.30	24.34
C15H34N607	410.47	410.2489	43.89	8.35	20.47	27.28
C15H34N608	426.47	426.2438	42.25	8.04	19.71	30.01
C15H34N609	442.47	442.2387	40.72	7.75	18.99	32.54
C15H35N301	273.47	273.2780	65.88	12.90	15.37	5.85
C15H35N302	289.47	289.2729	62.24	12.19	14.52	11.05
C15H35N303	305.46	305.2678	58.98	11.55	13.76	15.71
C15H35N304	321.46	321.2627	56.05	10.97	13.07	19.91
C15H35N305	337.46	337.2577	53.39	10.45	12.45	23.71
C15H35N306	353.46	353.2526	50.97	9.98	11.89	27.16
C15H35N307	369.46	369.2475	48.76	9.55	11.37	30.31
C15H35N308	385.46	385.2424	46.74	9.15	10.90	33.21
C15H35N309	401.46	401.2373	44.88	8.79	10.47	35.87
C15H35N501	301.48	301.2841	59.76	11.70	23.23	5.31
C15H35N502	317.48	317.2791	56.75	11.11	22.06	10.08
C15H35N503	333.48	333.2740	54.03	10.58	21.00	14.39
C15H35N504	349.48	349.2689	51.55	10.09	20.04	18.31
C15H35N505	365.48	365.2638	49.30	9.65	19.16	21.89
C15H35N506	381.48	381.2587	47.23	9.25	18.36	25.16
C15H35N507	397.48	397.2536	45.33	8.88	17.62	28.18
C15H35N508	413.47	413.2485	43.57	8.53	16.94	30.96
C15H35N509	429.47	429.2435	41.95	8.21	16.31	33.53
C15H36N401	288.48	288.2889	62.45	12.58	19.42	5.55
C15H36N402	304.48	304.2838	59.17	11.92	18.40	10.51
C15H36N403	320.48	320.2787	56.22	11.32	17.48	14.98
C15H36N404	336.48	336.2736	53.54	10.78	16.65	19.02
C15H36N405	352.48	352.2686	51.11	10.29	15.90	22.70
C15H36N406	368.48	368.2635	48.90	9.85	15.20	26.05
C15H36N407	384.48	384.2584	46.86	9.44	14.57	29.13
C15H36N408	400.48	400.2533	44.99	9.06	13.99	31.96
C15H36N409	416.48	416.2482	43.26	8.71	13.45	34.57
C15H36N601	316.49	316.2950	56.93	11.47	26.55	5.06
C15H36N602	332.49	332.2900	54.19	10.91	25.28	9.62
C15H36N603	348.49	348.2849	51.70	10.41	24.12	13.77
C15H36N604	364.49	364.2798	49.43	9.96	23.06	17.56

C15H36N605	380.49	380.2747	47.35	9.54	22.09	21.02
C15H36N606	396.49	396.2696	45.44	9.15	21.20	24.21
C15H36N607	412.49	412.2645	43.68	8.80	20.37	27.15
C15H36N608	428.49	428.2594	42.05	8.47	19.61	29.87
C15H36N609	444.49	444.2544	40.53	8.16	18.91	32.40
C15H37N501	303.50	303.2998	59.36	12.29	23.08	5.27
C15H37N502	319.49	319.2947	56.39	11.67	21.92	10.02
C15H37N503	335.49	335.2896	53.70	11.12	20.87	14.31
C15H37N504	351.49	351.2845	51.26	10.61	19.92	18.21
C15H37N505	367.49	367.2794	49.03	10.15	19.06	21.77
C15H37N506	383.49	383.2744	46.98	9.73	18.26	25.03
C15H37N507	399.49	399.2693	45.10	9.34	17.53	28.03
C15H37N508	415.49	415.2642	43.36	8.98	16.86	30.81
C15H37N509	431.49	431.2591	41.75	8.64	16.23	33.37
C15H38N601	318.51	318.3107	56.57	12.03	26.39	5.02
C15H38N602	334.51	334.3056	53.86	11.45	25.12	9.57
C15H38N603	350.51	350.3005	51.40	10.93	23.98	13.69
C15H38N604	366.51	366.2954	49.16	10.45	22.93	17.46
C15H38N605	382.51	382.2903	47.10	10.01	21.97	20.91
C15H38N606	398.51	398.2853	45.21	9.61	21.09	24.09
C15H38N607	414.51	414.2802	43.47	9.24	20.27	27.02
C15H38N608	430.51	430.2751	41.85	8.90	19.52	29.73
C15H38N609	446.50	446.2700	40.35	8.58	18.82	32.25
C16H 8N201	244.25	244.0637	78.68	3.30	11.47	6.55
C16H 8N202	260.25	260.0586	73.84	3.10	10.76	12.30
C16H 8N203	276.25	276.0535	69.57	2.92	10.14	17.37
C16H 8N204	292.25	292.0484	65.76	2.76	9.59	21.90
C16H 8N205	308.25	308.0433	62.34	2.62	9.09	25.95
C16H 8N206	324.25	324.0382	59.27	2.49	8.64	29.61
C16H 8N207	340.25	340.0331	56.48	2.37	8.23	32.92
C16H 8N208	356.25	356.0281	53.94	2.26	7.86	35.93
C16H 8N209	372.25	372.0230	51.63	2.17	7.53	38.68
C16H 8N401	272.27	272.0698	70.58	2.96	20.58	5.88
C16H 8N402	288.27	288.0647	66.67	2.80	19.44	11.10
C16H 8N403	304.27	304.0596	63.16	2.65	18.41	15.78
C16H 8N404	320.27	320.0545	60.01	2.52	17.49	19.98
C16H 8N405	336.27	336.0495	57.15	2.40	16.66	23.79
C16H 8N406	352.27	352.0444	54.56	2.29	15.90	27.25
C16H 8N407	368.26	368.0393	52.18	2.19	15.21	30.41
C16H 8N408	384.26	384.0342	50.01	2.10	14.58	33.31
C16H 8N409	400.26	400.0291	48.01	2.01	14.00	35.97
C16H 8N601	300.28	300.0760	64.00	2.69	27.99	5.33
C16H 8N602	316.28	316.0709	60.76	2.55	26.57	10.12
C16H 8N603	332.28	332.0658	57.84	2.43	25.29	14.45
C16H 8N604	348.28	348.0607	55.18	2.32	24.13	18.38
C16H 8N605	364.28	364.0556	52.76	2.21	23.07	21.96
C16H 8N606	380.28	380.0505	50.54	2.12	22.10	25.24
C16H 8N607	396.28	396.0454	48.50	2.03	21.21	28.26
C16H 8N608	412.28	412.0404	46.61	1.96	20.38	31.05
C16H 8N609	428.23	428.0353	44.87	1.88	19.62	33.62
C16H 9N101	231.26	231.0684	83.10	3.92	6.06	6.92
C16H 9N102	247.26	247.0633	77.72	3.67	5.66	12.94
C16H 9N103	263.26	263.0582	73.00	3.45	5.32	18.23
C16H 9N104	279.25	279.0532	68.82	3.25	5.02	22.92
C16H 9N105	295.25	295.0481	65.09	3.07	4.74	27.09
C16H 9N106	311.25	311.0430	61.74	2.91	4.50	30.84
C16H 9N107	327.25	327.0379	58.72	2.77	4.28	34.22
C16H 9N108	343.25	343.0328	55.95	2.64	4.08	37.29
C16H 9N109	359.25	359.0277	53.49	2.53	3.90	40.08
C16H 9N301	259.27	259.0746	74.12	3.50	16.21	6.17
C16H 9N302	275.27	275.0695	69.81	3.30	15.27	11.62
C16H 9N303	291.27	291.0644	65.98	3.11	14.43	16.48
C16H 9N304	307.27	307.0593	62.54	2.95	13.68	20.83
C16H 9N305	323.27	323.0542	59.45	2.81	13.00	24.75
C16H 9N306	339.27	339.0491	56.65	2.67	12.39	28.30
C16H 9N307	355.27	355.0440	54.09	2.55	11.83	31.52
C16H 9N308	371.27	371.0390	51.76	2.44	11.32	34.48
C16H 9N309	387.26	387.0339	49.62	2.34	10.85	37.18
C16H 9N501	287.28	287.0807	66.90	3.16	24.38	5.57
C16H 9N502	303.28	303.0756	63.37	2.99	23.09	10.55
C16H 9N503	319.28	319.0705	60.19	2.84	21.93	15.03
C16H 9N504	335.28	335.0654	57.32	2.71	20.89	19.09
C16H 9N505	351.28	351.0604	54.71	2.58	19.94	22.77
C16H 9N506	367.28	367.0553	52.32	2.47	19.07	26.14
C16H 9N507	383.28	383.0502	50.14	2.37	18.27	29.22
C16H 9N508	399.28	399.0451	48.13	2.27	17.54	32.06
C16H 9N509	415.28	415.0400	46.28	2.18	16.86	34.67
C16H10N201	246.27	246.0793	78.04	4.09	11.38	6.50
C16H10N202	262.27	262.0742	73.27	3.84	10.68	12.20
C16H10N203	278.27	278.0691	69.06	3.62	10.07	17.25
C16H10N204	294.27	294.0641	65.31	3.43	9.52	21.75
C16H10N205	310.27	310.0590	61.94	3.25	9.03	25.78
C16H10N206	326.27	326.0539	58.90	3.09	8.59	29.42
C16H10N207	342.27	342.0488	56.15	2.94	8.18	32.72
C16H10N208	358.27	358.0437	53.64	2.81	7.82	35.73
C16H10N209	374.27	374.0386	51.35	2.69	7.48	38.47
C16H10N401	274.28	274.0855	70.07	3.67	20.43	5.83
C16H10N402	290.28	290.0804	66.20	3.47	19.30	11.02
C16H10N403	306.28	306.0753	62.75	3.29	18.29	15.67
C16H10N404	322.28	322.0702	59.63	3.13	17.38	19.86
C16H10N405	338.28	338.0651	56.81	2.98	16.56	23.65
C16H10N406	354.28	354.0600	54.24	2.85	15.81	27.10
C16H10N407	370.28	370.0549	51.90	2.72	15.13	30.25
C16H10N408	386.28	386.0499	49.75	2.61	14.50	33.14
C16H10N409	402.28	402.0448	47.77	2.51	13.93	35.79
C16H10N601	302.30	302.0916	63.57	3.33	27.80	5.29
C16H10N602	318.30	318.0865	60.38	3.17	26.40	10.05
C16H10N603	334.30	334.0814	57.49	3.02	25.14	14.36
C16H10N604	350.30	350.0763	54.86	2.88	23.99	18.27
C16H10N605	366.30	366.0713	52.47	2.75	22.94	21.84
C16H10N606	382.29	382.0662	50.27	2.64	21.98	25.11
C16H10N607	398.29	398.0611	48.25	2.53	21.10	28.12
C16H10N608	414.29	414.0560	46.39	2.43	20.29	30.89
C16H10N609	430.29	430.0509	44.66	2.34	19.53	33.46
C16H11N101	233.27	233.0841	82.38	4.75	6.00	6.86
C16H11N102	249.27	249.0790	77.10	4.45	5.62	12.84
C16H11N103	265.27	265.0739	72.45	4.18	5.28	18.09
C16H11N104	281.27	281.0688	68.33	3.94	4.98	22.75
C16H11N105	297.27	297.0637	64.65	3.73	4.71	26.91
C16H11N106	313.27	313.0586	61.35	3.54	4.47	30.64
C16H11N107	329.27	329.0535	58.37	3.37	4.25	34.01
C16H11N108	345.27	345.0485	55.66	3.21	4.06	37.07
C16H11N109	361.27	361.0434	53.20	3.07	3.88	39.86
C16H11N301	261.29	261.0902	73.55	4.24	16.08	6.12
C16H11N302	277.28	277.0851	69.31	4.00	15.15	11.54
C16H11N303	293.28	293.0800	65.53	3.78	14.33	16.37
C16H11N304	309.28	309.0749	62.14	3.58	13.59	20.69
C16H11N305	325.28	325.0699	59.08	3.41	12.92	24.59
C16H11N306	341.28	341.0648	56.31	3.25	12.31	28.13
C16H11N307	357.28	357.0597	53.79	3.10	11.76	31.35

082

C16H11N3O8	373.28	373.0546	51.48	2.97	11.26	34.29
C16H11N3O9	389.28	389.0495	49.37	2.85	10.79	36.99
C16H11N5O1	289.30	289.0964	66.43	3.83	24.21	5.53
C16H11N5O2	305.30	305.0913	62.95	3.63	22.94	10.48
C16H11N5O3	321.30	321.0862	59.81	3.45	21.80	14.94
C16H11N5O4	337.30	337.0811	56.98	3.29	20.76	18.97
C16H11N5O5	353.30	353.0760	54.40	3.14	19.82	22.64
C16H11N5O6	369.30	369.0709	52.04	3.00	18.96	25.99
C16H11N5O7	385.30	385.0658	49.88	2.88	18.18	29.07
C16H11N5O8	401.29	401.0608	47.89	2.76	17.45	31.90
C16H11N5O9	417.29	417.0557	46.05	2.66	16.78	34.51
C16H12N2O1	248.29	248.0950	77.40	4.87	11.28	6.44
C16H12N2O2	264.29	264.0899	72.72	4.58	10.60	12.11
C16H12N2O3	280.29	280.0848	68.57	4.32	9.99	17.12
C16H12N2O4	296.29	296.0797	64.86	4.08	9.45	21.60
C16H12N2O5	312.28	312.0746	61.54	3.87	8.97	25.62
C16H12N2O6	328.28	328.0695	58.54	3.68	8.53	29.24
C16H12N2O7	344.28	344.0644	55.82	3.51	8.14	32.53
C16H12N2O8	360.28	360.0594	53.34	3.36	7.78	35.53
C16H12N2O9	376.28	376.0543	51.07	3.21	7.44	38.27
C16H12N4O1	276.30	276.1011	69.55	4.38	20.28	5.79
C16H12N4O2	292.30	292.0960	65.75	4.14	19.17	10.95
C16H12N4O3	308.30	308.0909	62.34	3.92	18.17	15.57
C16H12N4O4	324.30	324.0858	59.26	3.73	17.28	19.73
C16H12N4O5	340.30	340.0808	56.47	3.55	16.46	23.51
C16H12N4O6	356.30	356.0757	53.94	3.39	15.72	26.94
C16H12N4O7	372.30	372.0706	51.62	3.25	15.05	30.08
C16H12N4O8	388.30	388.0655	49.49	3.12	14.43	32.96
C16H12N4O9	404.30	404.0604	47.53	2.99	13.86	35.62
C16H12N6O1	304.31	304.1073	63.15	3.97	27.62	5.26
C16H12N6O2	320.31	320.1022	60.00	3.78	26.24	9.99
C16H12N6O3	336.31	336.0971	57.14	3.60	24.99	14.27
C16H12N6O4	352.31	352.0920	54.55	3.43	23.85	18.17
C16H12N6O5	368.31	368.0869	52.18	3.28	22.82	21.72
C16H12N6O6	384.31	384.0818	50.01	3.15	21.87	24.98
C16H12N6O7	400.31	400.0767	48.01	3.02	20.99	27.98
C16H12N6O8	416.31	416.0717	46.16	2.91	20.19	30.75
C16H12N6O9	432.31	432.0666	44.45	2.80	19.44	33.31
C16H13N1O1	235.29	235.0997	81.68	5.57	5.95	6.80
C16H13N1O2	251.29	251.0946	76.48	5.21	5.57	12.73
C16H13N1O3	267.29	267.0895	71.90	4.90	5.24	17.96
C16H13N1O4	283.29	283.0844	67.84	4.63	4.94	22.59
C16H13N1O5	299.29	299.0794	64.21	4.38	4.68	26.73
C16H13N1O6	315.29	315.0743	60.95	4.16	4.44	30.45
C16H13N1O7	331.28	331.0692	58.01	3.96	4.23	33.81
C16H13N1O8	347.28	347.0641	55.34	3.77	4.03	36.86
C16H13N1O9	363.28	363.0590	52.90	3.61	3.86	39.64
C16H13N3O1	263.30	263.1059	72.99	4.98	15.96	6.08
C16H13N3O2	279.30	279.1008	68.81	4.69	15.04	11.46
C16H13N3O3	295.30	295.0957	65.08	4.44	14.23	16.25
C16H13N3O4	311.30	311.0906	61.73	4.21	13.50	20.56
C16H13N3O5	327.30	327.0855	58.72	4.00	12.84	24.44
C16H13N3O6	343.30	343.0804	55.98	3.82	12.24	27.96
C16H13N3O7	359.30	359.0753	53.49	3.65	11.70	31.17
C16H13N3O8	375.30	375.0703	51.21	3.49	11.20	34.11
C16H13N3O9	391.30	391.0652	49.11	3.35	10.74	36.80
C16H13N5O1	291.31	291.1120	65.97	4.50	24.04	5.49
C16H13N5O2	307.31	307.1069	62.53	4.26	22.79	10.41
C16H13N5O3	323.31	323.1018	59.44	4.05	21.66	14.85
C16H13N5O4	339.31	339.0967	56.64	3.86	20.64	18.86

C16H13N5O5	355.31	355.0917	54.09	3.69	19.71	22.51
C16H13N5O6	371.31	371.0866	51.76	3.53	18.86	25.85
C16H13N5O7	387.31	387.0815	49.62	3.38	18.08	28.92
C16H13N5O8	403.31	403.0764	47.65	3.25	17.36	31.74
C16H13N5O9	419.31	419.0713	45.83	3.13	16.70	34.34
C16H14N2O1	250.30	250.1106	76.78	5.64	11.19	6.39
C16H14N2O2	266.30	266.1055	72.17	5.30	10.52	12.02
C16H14N2O3	282.30	282.1004	68.08	5.00	9.92	17.00
C16H14N2O4	298.30	298.0953	64.42	4.73	9.39	21.45
C16H14N2O5	314.30	314.0903	61.14	4.49	8.91	25.45
C16H14N2O6	330.30	330.0852	58.18	4.27	8.48	29.06
C16H14N2O7	346.30	346.0801	55.49	4.07	8.09	32.34
C16H14N2O8	362.30	362.0750	53.04	3.90	7.73	35.33
C16H14N2O9	378.30	378.0699	50.80	3.73	7.41	38.06
C16H14N4O1	278.32	278.1168	69.05	5.07	20.13	5.75
C16H14N4O2	294.32	294.1117	65.30	4.79	19.04	10.87
C16H14N4O3	310.31	310.1066	61.93	4.55	18.05	15.47
C16H14N4O4	326.31	326.1015	58.89	4.32	17.17	19.61
C16H14N4O5	342.31	342.0964	56.14	4.12	16.37	23.37
C16H14N4O6	358.31	358.0913	53.63	3.94	15.64	26.79
C16H14N4O7	374.31	374.0862	51.34	3.77	14.97	29.92
C16H14N4O8	390.31	390.0812	49.24	3.62	14.35	32.79
C16H14N4O9	406.31	406.0761	47.30	3.47	13.79	35.44
C16H14N6O1	306.33	306.1229	62.74	4.61	27.43	5.22
C16H14N6O2	322.33	322.1178	59.62	4.38	26.07	9.93
C16H14N6O3	338.33	338.1127	56.80	4.17	24.84	14.19
C16H14N6O4	354.33	354.1076	54.24	3.98	23.72	18.06
C16H14N6O5	370.33	370.1026	51.89	3.81	22.69	21.60
C16H14N6O6	386.33	386.0975	49.75	3.65	21.75	24.85
C16H14N6O7	402.33	402.0924	47.77	3.51	20.89	27.84
C16H14N6O8	418.33	418.0873	45.94	3.37	20.09	30.60
C16H14N6O9	434.32	434.0822	44.25	3.25	19.35	33.15
C16H15N1O1	237.30	237.1154	80.98	6.37	5.90	6.74
C16H15N1O2	253.30	253.1103	75.87	5.97	5.53	12.63
C16H15N1O3	269.30	269.1052	71.36	5.61	5.20	17.82
C16H15N1O4	285.30	285.1001	67.36	5.30	4.91	22.43
C16H15N1O5	301.30	301.0950	63.78	5.02	4.65	26.55
C16H15N1O6	317.30	317.0899	60.57	4.77	4.41	30.25
C16H15N1O7	333.30	333.0848	57.66	4.54	4.20	33.60
C16H15N1O8	349.30	349.0798	55.02	4.33	4.01	36.64
C16H15N1O9	365.30	365.0747	52.61	4.14	3.83	39.42
C16H15N3O1	265.32	265.1215	72.43	5.70	15.84	6.03
C16H15N3O2	281.32	281.1164	68.31	5.37	14.94	11.37
C16H15N3O3	297.32	297.1113	64.64	5.09	14.13	16.14
C16H15N3O4	313.32	313.1062	61.34	4.83	13.41	20.43
C16H15N3O5	329.32	329.1012	58.36	4.59	12.76	24.29
C16H15N3O6	345.31	345.0961	55.65	4.38	12.17	27.80
C16H15N3O7	361.31	361.0910	53.19	4.18	11.63	31.00
C16H15N3O8	377.31	377.0859	50.93	4.01	11.14	33.92
C16H15N3O9	393.31	393.0808	48.86	3.84	10.68	36.61
C16H15N5O1	293.33	293.1277	65.52	5.15	23.88	5.45
C16H15N5O2	309.33	309.1226	62.13	4.89	22.64	10.34
C16H15N5O3	325.33	325.1175	59.07	4.65	21.53	14.75
C16H15N5O4	341.33	341.1124	56.30	4.43	20.52	18.75
C16H15N5O5	357.33	357.1073	53.78	4.23	19.60	22.39
C16H15N5O6	373.33	373.1022	51.48	4.05	18.76	25.71
C16H15N5O7	389.33	389.0971	49.36	3.88	17.99	28.77
C16H15N5O8	405.33	405.0921	47.41	3.73	17.28	31.58
C16H15N5O9	421.33	421.0870	45.61	3.59	16.62	34.18
C16H16N2O1	252.32	252.1263	76.16	6.39	11.10	6.34

083

C16H16N2O2	268.32	268.1212	71.62	6.01	10.44	11.93
C16H16N2O3	284.32	284.1161	67.59	5.67	9.85	16.88
C16H16N2O4	300.32	300.1110	63.99	5.37	9.33	21.31
C16H16N2O5	316.32	316.1059	60.76	5.10	8.86	25.29
C16H16N2O6	332.32	332.1008	57.83	4.85	8.43	28.89
C16H16N2O7	348.32	348.0957	55.17	4.63	8.04	32.15
C16H16N2O8	364.31	364.0907	52.75	4.43	7.69	35.13
C16H16N2O9	380.31	380.0856	50.53	4.24	7.37	37.86
C16H16N4O1	280.33	280.1324	68.55	5.75	19.99	5.71
C16H16N4O2	296.33	296.1273	64.85	5.44	18.91	10.80
C16H16N4O3	312.33	312.1222	61.53	5.16	17.94	15.37
C16H16N4O4	328.33	328.1171	58.53	4.91	17.06	19.49
C16H16N4O5	344.33	344.1121	55.81	4.68	16.27	23.23
C16H16N4O6	360.33	360.1070	53.33	4.48	15.55	26.64
C16H16N4O7	376.33	376.1019	51.07	4.29	14.89	29.76
C16H16N4O8	392.33	392.0968	48.98	4.11	14.28	32.62
C16H16N4O9	408.33	408.0917	47.06	3.95	13.72	35.26
C16H16N6O1	308.35	308.1386	62.33	5.23	27.26	5.19
C16H16N6O2	324.34	324.1335	59.25	4.97	25.91	9.87
C16H16N6O3	340.34	340.1284	56.47	4.74	24.69	14.10
C16H16N6O4	356.34	356.1233	53.93	4.53	23.58	17.96
C16H16N6O5	372.34	372.1182	51.61	4.33	22.57	21.48
C16H16N6O6	388.34	388.1131	49.49	4.15	21.64	24.72
C16H16N6O7	404.34	404.1080	47.53	3.99	20.78	27.70
C16H16N6O8	420.34	420.1029	45.72	3.84	19.99	30.45
C16H16N6O9	436.34	436.0979	44.04	3.70	19.26	33.00
C16H17N1O1	239.32	239.1310	80.30	7.16	5.85	6.69
C16H17N1O2	255.32	255.1259	75.27	6.71	5.49	12.53
C16H17N1O3	271.32	271.1208	70.83	6.32	5.16	17.69
C16H17N1O4	287.32	287.1157	66.89	5.96	4.87	22.27
C16H17N1O5	303.32	303.1107	63.36	5.65	4.62	26.37
C16H17N1O6	319.32	319.1056	60.18	5.37	4.39	30.06
C16H17N1O7	335.32	335.1005	57.31	5.11	4.18	33.40
C16H17N1O8	351.32	351.0954	54.70	4.88	3.99	36.43
C16H17N1O9	367.32	367.0903	52.32	4.67	3.81	39.20
C16H17N3O1	267.33	267.1372	71.89	6.41	15.72	5.98
C16H17N3O2	283.33	283.1321	67.83	6.05	14.83	11.29
C16H17N3O3	299.33	299.1270	64.20	5.72	14.04	16.04
C16H17N3O4	315.33	315.1219	60.94	5.43	13.33	20.30
C16H17N3O5	331.33	331.1168	58.00	5.17	12.68	24.14
C16H17N3O6	347.33	347.1117	55.33	4.93	12.10	27.64
C16H17N3O7	363.33	363.1066	52.89	4.72	11.57	30.82
C16H17N3O8	379.33	379.1016	50.66	4.52	11.08	33.74
C16H17N3O9	395.33	395.0965	48.61	4.33	10.63	36.42
C16H17N5O1	295.35	295.1433	65.07	5.80	23.71	5.42
C16H17N5O2	311.35	311.1382	61.72	5.50	22.49	10.28
C16H17N5O3	327.35	327.1331	58.71	5.23	21.39	14.66
C16H17N5O4	343.34	343.1280	55.97	4.99	20.40	18.64
C16H17N5O5	359.34	359.1230	53.48	4.77	19.49	22.26
C16H17N5O6	375.34	375.1179	51.20	4.57	18.66	25.58
C16H17N5O7	391.34	391.1128	49.11	4.38	17.90	28.62
C16H17N5O8	407.34	407.1077	47.18	4.21	17.19	31.42
C16H17N5O9	423.34	423.1026	45.40	4.05	16.54	34.01
C16H18N2O1	254.33	254.1419	75.56	7.13	11.01	6.29
C16H18N2O2	270.33	270.1368	71.09	6.71	10.36	11.84
C16H18N2O3	286.33	286.1317	67.12	6.34	9.78	16.76
C16H18N2O4	302.33	302.1266	63.57	6.00	9.27	21.17
C16H18N2O5	318.33	318.1216	60.37	5.70	8.80	25.13
C16H18N2O6	334.33	334.1165	57.48	5.43	8.38	28.71
C16H18N2O7	350.33	350.1114	54.86	5.18	8.00	31.97

C16H18N2O8	366.33	366.1063	52.46	4.95	7.65	34.94
C16H18N2O9	382.33	382.1012	50.27	4.75	7.33	37.66
C16H18N4O1	282.35	282.1481	68.06	6.43	19.84	5.67
C16H18N4O2	298.35	298.1430	64.41	6.08	18.78	10.73
C16H18N4O3	314.35	314.1379	61.14	5.77	17.82	15.27
C16H18N4O4	330.35	330.1328	58.17	5.49	16.96	19.37
C16H18N4O5	346.35	346.1277	55.49	5.24	16.18	23.10
C16H18N4O6	362.35	362.1226	53.04	5.01	15.46	26.49
C16H18N4O7	378.34	378.1175	50.79	4.80	14.81	29.60
C16H18N4O8	394.34	394.1125	48.73	4.60	14.21	32.46
C16H18N4O9	410.34	410.1074	46.83	4.42	13.65	35.09
C16H18N6O1	310.36	310.1542	61.92	5.85	27.08	5.16
C16H18N6O2	326.36	326.1491	58.89	5.56	25.75	9.80
C16H18N6O3	342.36	342.1440	56.13	5.30	24.55	14.02
C16H18N6O4	358.36	358.1389	53.63	5.06	23.45	17.86
C16H18N6O5	374.36	374.1339	51.34	4.85	22.45	21.37
C16H18N6O6	390.36	390.1288	49.23	4.65	21.53	24.59
C16H18N6O7	406.36	406.1237	47.29	4.46	20.68	27.56
C16H18N6O8	422.36	422.1186	45.50	4.30	19.90	30.30
C16H18N6O9	438.36	438.1135	43.84	4.14	19.17	32.85
C16H19N1O1	241.34	241.1467	79.63	7.94	5.80	6.63
C16H19N1O2	257.34	257.1416	74.68	7.44	5.44	12.43
C16H19N1O3	273.33	273.1365	70.31	7.01	5.12	17.56
C16H19N1O4	289.33	289.1314	66.42	6.62	4.84	22.12
C16H19N1O5	305.33	305.1263	62.94	6.27	4.59	26.20
C16H19N1O6	321.33	321.1212	59.81	5.96	4.36	29.87
C16H19N1O7	337.33	337.1161	56.97	5.68	4.15	33.20
C16H19N1O8	353.33	353.1111	54.39	5.42	3.96	36.23
C16H19N1O9	369.33	369.1060	52.03	5.19	3.79	38.99
C16H19N3O1	269.35	269.1528	71.35	7.11	15.60	5.94
C16H19N3O2	285.35	285.1477	67.35	6.71	14.73	11.21
C16H19N3O3	301.35	301.1426	63.77	6.36	13.94	15.93
C16H19N3O4	317.35	317.1375	60.56	6.03	13.24	20.17
C16H19N3O5	333.35	333.1325	57.65	5.75	12.61	24.00
C16H19N3O6	349.35	349.1274	55.01	5.48	12.03	27.48
C16H19N3O7	365.35	365.1223	52.60	5.24	11.50	30.65
C16H19N3O8	381.35	381.1172	50.39	5.02	11.02	33.56
C16H19N3O9	397.34	397.1121	48.37	4.82	10.58	36.24
C16H19N5O1	297.36	297.1590	64.63	6.44	23.55	5.38
C16H19N5O2	313.36	313.1539	61.33	6.11	22.35	10.21
C16H19N5O3	329.36	329.1488	58.35	5.81	21.26	14.57
C16H19N5O4	345.36	345.1437	55.65	5.55	20.28	18.53
C16H19N5O5	361.36	361.1386	53.18	5.30	19.38	22.14
C16H19N5O6	377.36	377.1335	50.93	5.08	18.56	25.44
C16H19N5O7	393.36	393.1284	48.86	4.87	17.80	28.47
C16H19N5O8	409.36	409.1233	46.95	4.68	17.11	31.27
C16H19N5O9	425.36	425.1183	45.18	4.50	16.46	33.85
C16H20N2O1	256.35	256.1576	74.97	7.86	10.93	6.24
C16H20N2O2	272.35	272.1525	70.56	7.40	10.29	11.75
C16H20N2O3	288.35	288.1474	66.65	6.99	9.72	16.65
C16H20N2O4	304.35	304.1423	63.14	6.62	9.20	21.03
C16H20N2O5	320.35	320.1372	59.99	6.29	8.74	24.97
C16H20N2O6	336.35	336.1321	57.14	5.99	8.33	28.54
C16H20N2O7	352.35	352.1270	54.54	5.72	7.95	31.79
C16H20N2O8	368.35	368.1220	52.17	5.47	7.61	34.75
C16H20N2O9	384.35	384.1169	50.00	5.25	7.29	37.46
C16H20N4O1	284.36	284.1637	67.58	7.09	19.70	5.63
C16H20N4O2	300.36	300.1586	63.98	6.71	18.65	10.65
C16H20N4O3	316.36	316.1535	60.75	6.37	17.71	15.17
C16H20N4O4	332.36	332.1484	57.82	6.07	16.86	19.26

084

C16H20N4O5	348.36	348.1434	55.17	5.79	16.08	22.96
C16H20N4O6	364.36	364.1383	52.74	5.53	15.38	26.35
C16H20N4O7	380.36	380.1332	50.53	5.30	14.73	29.44
C16H20N4O8	396.36	396.1281	48.49	5.09	14.14	32.29
C16H20N4O9	412.36	412.1230	46.60	4.89	13.59	34.92
C16H20N6O1	312.38	312.1698	61.52	6.45	26.90	5.12
C16H20N6O2	328.38	328.1648	58.52	6.14	25.59	9.74
C16H20N6O3	344.38	344.1597	55.80	5.85	24.40	13.94
C16H20N6O4	360.38	360.1546	53.33	5.59	23.32	17.76
C16H20N6O5	376.38	376.1495	51.06	5.36	22.33	21.25
C16H20N6O6	392.37	392.1444	48.98	5.14	21.42	24.47
C16H20N6O7	408.37	408.1393	47.06	4.94	20.58	27.42
C16H20N6O8	424.37	424.1342	45.29	4.75	19.80	30.16
C16H20N6O9	440.37	440.1292	43.64	4.58	19.08	32.70
C16H21N1O1	243.35	243.1623	78.97	8.70	5.76	6.57
C16H21N1O2	259.35	259.1572	74.10	8.16	5.40	12.34
C16H21N1O3	275.35	275.1521	69.79	7.69	5.09	17.43
C16H21N1O4	291.35	291.1470	65.96	7.27	4.81	21.97
C16H21N1O5	307.35	307.1420	62.53	6.89	4.56	26.03
C16H21N1O6	323.35	323.1369	59.43	6.55	4.33	29.69
C16H21N1O7	339.35	339.1318	56.63	6.24	4.13	33.00
C16H21N1O8	355.35	355.1267	54.08	5.96	3.94	36.02
C16H21N1O9	371.35	371.1216	51.75	5.70	3.77	38.78
C16H21N3O1	271.37	271.1685	70.82	7.80	15.48	5.90
C16H21N3O2	287.36	287.1634	66.88	7.37	14.62	11.14
C16H21N3O3	303.36	303.1583	63.35	6.98	13.85	15.82
C16H21N3O4	319.36	319.1532	60.18	6.63	13.16	20.04
C16H21N3O5	335.36	335.1481	57.30	6.31	12.53	23.85
C16H21N3O6	351.36	351.1430	54.70	6.02	11.96	27.32
C16H21N3O7	367.36	367.1379	52.31	5.76	11.44	30.49
C16H21N3O8	383.36	383.1329	50.13	5.52	10.96	33.39
C16H21N3O9	399.36	399.1278	48.12	5.30	10.52	36.06
C16H21N5O1	299.38	299.1746	64.19	7.07	23.39	5.34
C16H21N5O2	315.38	315.1695	60.94	6.71	22.21	10.15
C16H21N5O3	331.38	331.1644	57.99	6.39	21.13	14.48
C16H21N5O4	347.38	347.1593	55.32	6.09	20.16	18.42
C16H21N5O5	363.38	363.1543	52.89	5.83	19.27	22.01
C16H21N5O6	379.38	379.1492	50.66	5.58	18.46	25.30
C16H21N5O7	395.38	395.1441	48.61	5.35	17.71	28.33
C16H21N5O8	411.37	411.1390	46.72	5.15	17.02	31.11
C16H21N5O9	427.37	427.1339	44.97	4.95	16.39	33.69
C16H22N2O1	258.37	258.1732	74.38	8.58	10.84	6.19
C16H22N2O2	274.37	274.1681	70.04	8.08	10.21	11.66
C16H22N2O3	290.37	290.1630	66.19	7.64	9.65	16.53
C16H22N2O4	306.36	306.1579	62.73	7.24	9.14	20.89
C16H22N2O5	322.36	322.1529	59.62	6.88	8.69	24.82
C16H22N2O6	338.36	338.1478	56.80	6.55	8.28	28.37
C16H22N2O7	354.36	354.1427	54.23	6.26	7.91	31.60
C16H22N2O8	370.36	370.1376	51.89	5.99	7.56	34.56
C16H22N2O9	386.36	386.1325	49.74	5.74	7.25	37.27
C16H22N4O1	286.38	286.1794	67.11	7.74	19.56	5.59
C16H22N4O2	302.38	302.1743	63.56	7.33	18.53	10.58
C16H22N4O3	318.38	318.1692	60.36	6.97	17.60	15.08
C16H22N4O4	334.38	334.1641	57.47	6.63	16.76	19.14
C16H22N4O5	350.38	350.1590	54.85	6.33	15.99	22.83
C16H22N4O6	366.38	366.1539	52.45	6.05	15.29	26.20
C16H22N4O7	382.38	382.1488	50.26	5.80	14.65	29.29
C16H22N4O8	398.38	398.1437	48.24	5.57	14.06	32.13
C16H22N4O9	414.38	414.1387	46.38	5.35	13.52	34.75
C16H22N6O1	314.39	314.1855	61.13	7.05	26.73	5.09

C16H22N6O2	330.39	330.1804	58.17	6.71	25.44	9.69
C16H22N6O3	346.39	346.1753	55.48	6.40	24.26	13.86
C16H22N6O4	362.39	362.1702	53.03	6.12	23.19	17.66
C16H22N6O5	378.39	378.1652	50.79	5.86	22.21	21.14
C16H22N6O6	394.39	394.1601	48.73	5.62	21.31	24.34
C16H22N6O7	410.39	410.1550	46.83	5.40	20.48	27.29
C16H22N6O8	426.39	426.1499	45.07	5.20	19.71	30.02
C16H22N6O9	442.39	442.1448	43.44	5.01	19.00	32.55
C16H23N1O1	245.37	245.1780	78.32	9.45	5.71	6.52
C16H23N1O2	261.37	261.1729	73.53	8.87	5.36	12.24
C16H23N1O3	277.37	277.1678	69.29	8.36	5.05	17.30
C16H23N1O4	293.37	293.1627	65.51	7.90	4.77	21.81
C16H23N1O5	309.37	309.1576	62.12	7.49	4.53	25.86
C16H23N1O6	325.36	325.1525	59.07	7.13	4.30	29.50
C16H23N1O7	341.36	341.1474	56.30	6.79	4.10	32.81
C16H23N1O8	357.36	357.1424	53.78	6.49	3.92	35.82
C16H23N1O9	373.36	373.1373	51.47	6.21	3.75	38.57
C16H23N3O1	273.38	273.1841	70.30	8.48	15.37	5.85
C16H23N3O2	289.38	289.1790	66.41	8.01	14.52	11.06
C16H23N3O3	305.38	305.1739	62.93	7.59	13.76	15.72
C16H23N3O4	321.38	321.1688	59.80	7.21	13.07	19.91
C16H23N3O5	337.38	337.1638	56.96	6.87	12.45	23.71
C16H23N3O6	353.38	353.1587	54.38	6.56	11.89	27.17
C16H23N3O7	369.38	369.1536	52.03	6.28	11.38	30.32
C16H23N3O8	385.38	385.1485	49.87	6.02	10.90	33.21
C16H23N3O9	401.38	401.1434	47.88	5.78	10.47	35.88
C16H23N5O1	301.39	301.1902	63.76	7.69	23.24	5.31
C16H23N5O2	317.39	317.1852	60.55	7.30	22.07	10.08
C16H23N5O3	333.39	333.1801	57.64	6.95	21.01	14.40
C16H23N5O4	349.39	349.1750	55.00	6.64	20.04	18.32
C16H23N5O5	365.39	365.1699	52.60	6.34	19.17	21.89
C16H23N5O6	381.39	381.1648	50.39	6.08	18.36	25.17
C16H23N5O7	397.39	397.1597	48.36	5.83	17.62	28.18
C16H23N5O8	413.39	413.1546	46.49	5.61	16.94	30.96
C16H23N5O9	429.39	429.1496	44.76	5.40	16.31	33.53
C16H24N2O1	260.38	260.1889	73.81	9.29	10.76	6.14
C16H24N2O2	276.38	276.1838	69.53	8.75	10.14	11.58
C16H24N2O3	292.38	292.1787	65.73	8.27	9.58	16.42
C16H24N2O4	308.38	308.1736	62.32	7.84	9.08	20.75
C16H24N2O5	324.38	324.1685	59.24	7.46	8.64	24.66
C16H24N2O6	340.38	340.1634	56.46	7.11	8.23	28.20
C16H24N2O7	356.38	356.1583	53.93	6.79	7.86	31.43
C16H24N2O8	372.38	372.1533	51.61	6.50	7.52	34.37
C16H24N2O9	388.38	388.1482	49.48	6.23	7.21	37.08
C16H24N4O1	288.40	288.1950	66.64	8.39	19.43	5.55
C16H24N4O2	304.40	304.1899	63.13	7.95	18.41	10.51
C16H24N4O3	320.39	320.1848	59.98	7.55	17.49	14.98
C16H24N4O4	336.39	336.1797	57.13	7.19	16.66	19.02
C16H24N4O5	352.39	352.1747	54.54	6.86	15.90	22.70
C16H24N4O6	368.39	368.1696	52.17	6.57	15.21	26.06
C16H24N4O7	384.39	384.1645	50.00	6.29	14.58	29.14
C16H24N4O8	400.39	400.1594	48.00	6.04	13.99	31.97
C16H24N4O9	416.39	416.1543	46.15	5.81	13.46	34.58
C16H24N6O1	316.41	316.2011	60.74	7.65	26.56	5.06
C16H24N6O2	332.41	332.1961	57.81	7.28	25.28	9.63
C16H24N6O3	348.41	348.1910	55.16	6.94	24.12	13.78
C16H24N6O4	364.41	364.1859	52.74	6.64	23.06	17.56
C16H24N6O5	380.41	380.1808	50.52	6.36	22.09	21.03
C16H24N6O6	396.41	396.1757	48.48	6.10	21.20	24.22
C16H24N6O7	412.41	412.1706	46.60	5.87	20.38	27.16

085

C16H24N6O8	428.41	428.1655	44.86	5.65	19.62	29.88
C16H24N6O9	444.40	444.1605	43.24	5.44	18.91	32.40
C16H25N1O1	247.38	247.1936	77.68	10.19	5.66	6.47
C16H25N1O2	263.38	263.1885	72.97	9.57	5.32	12.15
C16H25N1O3	279.38	279.1834	68.79	9.02	5.01	17.18
C16H25N1O4	295.38	295.1783	65.06	8.53	4.74	21.67
C16H25N1O5	311.38	311.1733	61.72	8.09	4.50	25.69
C16H25N1O6	327.38	327.1682	58.70	7.70	4.28	29.32
C16H25N1O7	343.38	343.1631	55.97	7.34	4.08	32.62
C16H25N1O8	359.38	359.1580	53.48	7.01	3.90	35.62
C16H25N1O9	375.38	375.1529	51.20	6.71	3.73	38.36
C16H25N3O1	275.40	275.1998	69.78	9.15	15.26	5.81
C16H25N3O2	291.40	291.1947	65.95	8.65	14.42	10.98
C16H25N3O3	307.40	307.1896	62.52	8.20	13.67	15.61
C16H25N3O4	323.40	323.1845	59.43	7.79	12.99	19.79
C16H25N3O5	339.39	339.1794	56.62	7.42	12.38	23.57
C16H25N3O6	355.39	355.1743	54.07	7.09	11.82	27.01
C16H25N3O7	371.39	371.1692	51.75	6.79	11.31	30.16
C16H25N3O8	387.39	387.1641	49.61	6.50	10.85	33.04
C16H25N3O9	403.39	403.1591	47.64	6.25	10.42	35.70
C16H25N5O1	303.41	303.2059	63.34	8.31	23.08	5.27
C16H25N5O2	319.41	319.2008	60.17	7.89	21.93	10.02
C16H25N5O3	335.41	335.1957	57.30	7.51	20.88	14.31
C16H25N5O4	351.41	351.1906	54.69	7.17	19.93	18.21
C16H25N5O5	367.41	367.1856	52.31	6.86	19.06	21.77
C16H25N5O6	383.41	383.1805	50.12	6.57	18.27	25.04
C16H25N5O7	399.41	399.1754	48.12	6.31	17.53	28.04
C16H25N5O8	415.41	415.1703	46.26	6.07	16.86	30.81
C16H25N5O9	431.41	431.1652	44.55	5.84	16.23	33.38
C16H26N2O1	262.40	262.2045	73.24	9.99	10.68	6.10
C16H26N2O2	278.40	278.1994	69.03	9.41	10.06	11.49
C16H26N2O3	294.40	294.1943	65.28	8.90	9.52	16.30
C16H26N2O4	310.40	310.1892	61.91	8.44	9.03	20.62
C16H26N2O5	326.40	326.1842	58.88	8.03	8.58	24.51
C16H26N2O6	342.40	342.1791	56.13	7.65	8.18	28.04
C16H26N2O7	358.39	358.1740	53.62	7.31	7.82	31.25
C16H26N2O8	374.39	374.1689	51.33	7.00	7.48	34.19
C16H26N2O9	390.39	390.1638	49.23	6.71	7.18	36.88
C16H26N4O1	290.41	290.2106	66.17	9.02	19.29	5.51
C16H26N4O2	306.41	306.2056	62.72	8.55	18.28	10.44
C16H26N4O3	322.41	322.2005	59.61	8.13	17.38	14.89
C16H26N4O4	338.41	338.1954	56.79	7.74	16.56	18.91
C16H26N4O5	354.41	354.1903	54.22	7.39	15.81	22.57
C16H26N4O6	370.41	370.1852	51.88	7.08	15.13	25.92
C16H26N4O7	386.41	386.1801	49.73	6.78	14.50	28.98
C16H26N4O8	402.41	402.1750	47.76	6.51	13.92	31.81
C16H26N4O9	418.41	418.1700	45.93	6.26	13.39	34.41
C16H26N6O1	318.43	318.2168	60.35	8.23	26.39	5.02
C16H26N6O2	334.42	334.2117	57.47	7.84	25.13	9.57
C16H26N6O3	350.42	350.2066	54.84	7.48	23.98	13.70
C16H26N6O4	366.42	366.2015	52.45	7.15	22.94	17.47
C16H26N6O5	382.42	382.1965	50.25	6.85	21.98	20.92
C16H26N6O6	398.42	398.1914	48.23	6.58	21.09	24.09
C16H26N6O7	414.42	414.1863	46.37	6.32	20.28	27.02
C16H26N6O8	430.42	430.1812	44.65	6.09	19.53	29.74
C16H26N6O9	446.42	446.1761	43.05	5.87	18.83	32.26
C16H27N1O1	249.40	249.2093	77.06	10.91	5.62	6.42
C16H27N1O2	265.40	265.2042	72.41	10.25	5.28	12.06
C16H27N1O3	281.40	281.1991	68.29	9.67	4.98	17.06
C16H27N1O4	297.40	297.1940	64.62	9.15	4.71	21.52

C16H27N1O5	313.40	313.1889	61.32	8.68	4.47	25.53
C16H27N1O6	329.40	329.1838	58.34	8.26	4.25	29.14
C16H27N1O7	345.40	345.1787	55.64	7.88	4.06	32.43
C16H27N1O8	361.40	361.1737	53.18	7.53	3.88	35.42
C16H27N1O9	377.39	377.1686	50.92	7.21	3.71	38.15
C16H27N3O1	277.41	277.2154	69.28	9.81	15.15	5.77
C16H27N3O2	293.41	293.2103	65.50	9.28	14.32	10.91
C16H27N3O3	309.41	309.2052	62.11	8.80	13.58	15.51
C16H27N3O4	325.41	325.2001	59.06	8.36	12.91	19.67
C16H27N3O5	341.41	341.1951	56.29	7.97	12.31	23.43
C16H27N3O6	357.41	357.1900	53.77	7.61	11.76	26.86
C16H27N3O7	373.41	373.1849	51.47	7.29	11.25	29.99
C16H27N3O8	389.41	389.1798	49.35	6.99	10.79	32.87
C16H27N3O9	405.41	405.1747	47.40	6.71	10.36	35.52
C16H27N5O1	305.43	305.2215	62.92	8.91	22.93	5.24
C16H27N5O2	321.43	321.2165	59.79	8.47	21.79	9.96
C16H27N5O3	337.43	337.2114	56.95	8.07	20.76	14.22
C16H27N5O4	353.42	353.2063	54.38	7.70	19.82	18.11
C16H27N5O5	369.42	369.2012	52.02	7.37	18.96	21.65
C16H27N5O6	385.42	385.1961	49.86	7.06	18.17	24.91
C16H27N5O7	401.42	401.1910	47.87	6.78	17.45	27.90
C16H27N5O8	417.42	417.1859	46.04	6.52	16.78	30.66
C16H27N5O9	433.42	433.1809	44.34	6.28	16.16	33.22
C16H28N2O1	264.41	264.2202	72.68	10.67	10.59	6.05
C16H28N2O2	280.41	280.2151	68.53	10.06	9.99	11.41
C16H28N2O3	296.41	296.2100	64.83	9.52	9.45	16.19
C16H28N2O4	312.41	312.2049	61.51	9.03	8.97	20.48
C16H28N2O5	328.41	328.1998	58.52	8.59	8.53	24.36
C16H28N2O6	344.41	344.1947	55.80	8.19	8.13	27.87
C16H28N2O7	360.41	360.1896	53.32	7.83	7.77	31.07
C16H28N2O8	376.41	376.1845	51.06	7.50	7.44	34.00
C16H28N2O9	392.41	392.1795	48.97	7.19	7.14	36.69
C16H28N4O1	292.43	292.2263	65.72	9.65	19.16	5.47
C16H28N4O2	308.43	308.2212	62.31	9.15	18.17	10.37
C16H28N4O3	324.43	324.2161	59.24	8.70	17.27	14.79
C16H28N4O4	340.43	340.2110	56.45	8.29	16.46	18.80
C16H28N4O5	356.43	356.2060	53.92	7.92	15.72	22.44
C16H28N4O6	372.42	372.2009	51.60	7.58	15.04	25.78
C16H28N4O7	388.42	388.1958	49.48	7.27	14.42	28.83
C16H28N4O8	404.42	404.1907	47.52	6.98	13.85	31.65
C16H28N4O9	420.42	420.1856	45.71	6.71	13.33	34.25
C16H28N6O1	320.44	320.2324	59.97	8.81	26.23	4.99
C16H28N6O2	336.44	336.2274	57.12	8.39	24.98	9.51
C16H28N6O3	352.44	352.2223	54.53	8.01	23.85	13.62
C16H28N6O4	368.44	368.2172	52.16	7.66	22.81	17.37
C16H28N6O5	384.44	384.2121	49.99	7.34	21.86	20.81
C16H28N6O6	400.44	400.2070	47.99	7.05	20.99	23.97
C16H28N6O7	416.44	416.2019	46.15	6.78	20.18	26.89
C16H28N6O8	432.44	432.1968	44.44	6.53	19.43	29.60
C16H28N6O9	448.44	448.1918	42.86	6.29	18.74	32.11
C16H29N1O1	251.42	251.2249	76.44	11.63	5.57	6.36
C16H29N1O2	267.42	267.2198	71.87	10.93	5.24	11.97
C16H29N1O3	283.41	283.2147	67.81	10.31	4.94	16.94
C16H29N1O4	299.41	299.2096	64.18	9.76	4.68	21.37
C16H29N1O5	315.41	315.2046	60.93	9.27	4.44	25.36
C16H29N1O6	331.41	331.1995	57.99	8.82	4.23	28.97
C16H29N1O7	347.41	347.1944	55.32	8.41	4.03	32.24
C16H29N1O8	363.41	363.1893	52.88	8.04	3.85	35.22
C16H29N1O9	379.41	379.1842	50.65	7.70	3.69	37.95
C16H29N3O1	279.43	279.2310	68.78	10.46	15.04	5.73

086

C16H29N3O2	295.43	295.2260	65.05	9.89	14.22	10.83
C16H29N3O3	311.43	311.2209	61.71	9.39	13.49	15.41
C16H29N3O4	327.43	327.2158	58.69	8.93	12.83	19.55
C16H29N3O5	343.43	343.2107	55.96	8.51	12.24	23.29
C16H29N3O6	359.43	359.2056	53.47	8.13	11.69	26.71
C16H29N3O7	375.43	375.2005	51.19	7.79	11.19	29.83
C16H29N3O8	391.42	391.1954	49.10	7.47	10.74	32.70
C16H29N3O9	407.42	407.1904	47.17	7.17	10.31	35.34
C16H29N5O1	307.44	307.2372	62.51	9.51	22.78	5.20
C16H29N5O2	323.44	323.2321	59.42	9.04	21.65	9.89
C16H29N5O3	339.44	339.2270	56.62	8.61	20.63	14.14
C16H29N5O4	355.44	355.2219	54.07	8.22	19.70	18.01
C16H29N5O5	371.44	371.2169	51.74	7.87	18.85	21.54
C16H29N5O6	387.44	387.2118	49.60	7.54	18.08	24.78
C16H29N5O7	403.44	403.2067	47.64	7.25	17.36	27.76
C16H29N5O8	419.44	419.2016	45.82	6.97	16.70	30.52
C16H29N5O9	435.44	435.1965	44.13	6.71	16.08	33.07
C16H30N2O1	266.43	266.2358	72.13	11.35	10.51	6.01
C16H30N2O2	282.43	282.2307	68.04	10.71	9.92	11.33
C16H30N2O3	298.43	298.2256	64.40	10.13	9.39	16.08
C16H30N2O4	314.43	314.2205	61.12	9.62	8.91	20.35
C16H30N2O5	330.43	330.2155	58.16	9.15	8.48	24.21
C16H30N2O6	346.43	346.2104	55.47	8.73	8.09	27.71
C16H30N2O7	362.43	362.2053	53.03	8.34	7.73	30.90
C16H30N2O8	378.43	378.2002	50.78	7.99	7.40	33.82
C16H30N2O9	394.43	394.1951	48.72	7.67	7.10	36.51
C16H30N4O1	294.44	294.2419	65.27	10.27	19.03	5.43
C16H30N4O2	310.44	310.2369	61.90	9.74	18.05	10.31
C16H30N4O3	326.44	326.2318	58.87	9.26	17.16	14.70
C16H30N4O4	342.44	342.2267	56.12	8.83	16.36	18.69
C16H30N4O5	358.44	358.2216	53.62	8.44	15.63	22.32
C16H30N4O6	374.44	374.2165	51.32	8.08	14.96	25.64
C16H30N4O7	390.44	390.2114	49.22	7.74	14.35	28.68
C16H30N4O8	406.44	406.2063	47.28	7.44	13.78	31.49
C16H30N4O9	422.44	422.2013	45.49	7.16	13.26	34.09
C16H30N6O1	322.46	322.2481	59.60	9.38	26.06	4.96
C16H30N6O2	338.46	338.2430	56.78	8.93	24.83	9.45
C16H30N6O3	354.46	354.2379	54.22	8.53	23.71	13.54
C16H30N6O4	370.46	370.2328	51.88	8.16	22.69	17.28
C16H30N6O5	386.45	386.2278	49.73	7.82	21.75	20.70
C16H30N6O6	402.45	402.2227	47.75	7.51	20.88	23.85
C16H30N6O7	418.45	418.2176	45.93	7.23	20.08	26.76
C16H30N6O8	434.45	434.2125	44.23	6.96	19.34	29.46
C16H30N6O9	450.45	450.2074	42.66	6.71	18.66	31.97
C16H31N1O1	253.43	253.2406	75.83	12.33	5.53	6.31
C16H31N1O2	269.43	269.2355	71.33	11.60	5.20	11.88
C16H31N1O3	285.43	285.2304	67.33	10.95	4.91	16.82
C16H31N1O4	301.43	301.2253	63.76	10.37	4.65	21.23
C16H31N1O5	317.43	317.2202	60.54	9.84	4.41	25.20
C16H31N1O6	333.43	333.2151	57.64	9.37	4.20	28.79
C16H31N1O7	349.43	349.2100	55.00	8.94	4.01	32.05
C16H31N1O8	365.43	365.2049	52.59	8.55	3.83	35.03
C16H31N1O9	381.43	381.1999	50.38	8.19	3.67	37.75
C16H31N3O1	281.44	281.2467	68.28	11.10	14.93	5.68
C16H31N3O2	297.44	297.2416	64.61	10.51	14.13	10.76
C16H31N3O3	313.44	313.2365	61.31	9.97	13.41	15.31
C16H31N3O4	329.44	329.2314	58.33	9.48	12.75	19.43
C16H31N3O5	345.44	345.2264	55.63	9.05	12.16	23.16
C16H31N3O6	361.44	361.2213	53.17	8.65	11.63	26.56
C16H31N3O7	377.44	377.2162	50.92	8.28	11.13	29.67

C16H31N3O8	393.44	393.2111	48.85	7.94	10.68	32.53
C16H31N3O9	409.44	409.2060	46.94	7.63	10.26	35.17
C16H31N5O1	309.46	309.2528	62.10	10.10	22.63	5.17
C16H31N5O2	325.46	325.2478	59.05	9.60	21.52	9.83
C16H31N5O3	341.46	341.2427	56.28	9.15	20.51	14.06
C16H31N5O4	357.46	357.2376	53.76	8.74	19.59	17.90
C16H31N5O5	373.46	373.2325	51.46	8.37	18.75	21.42
C16H31N5O6	389.46	389.2274	49.35	8.02	17.98	24.65
C16H31N5O7	405.45	405.2223	47.40	7.71	17.27	27.62
C16H31N5O8	421.45	421.2172	45.60	7.41	16.62	30.37
C16H31N5O9	437.45	437.2122	43.93	7.14	16.01	32.92
C16H32N2O1	268.45	268.2514	71.59	12.02	10.44	5.96
C16H32N2O2	284.45	284.2464	67.56	11.34	9.85	11.25
C16H32N2O3	300.45	300.2413	63.96	10.74	9.32	15.98
C16H32N2O4	316.44	316.2362	60.73	10.19	8.85	20.22
C16H32N2O5	332.44	332.2311	57.81	9.70	8.43	24.06
C16H32N2O6	348.44	348.2260	55.15	9.26	8.04	27.55
C16H32N2O7	364.44	364.2209	52.73	8.85	7.69	30.73
C16H32N2O8	380.44	380.2158	50.51	8.48	7.36	33.64
C16H32N2O9	396.44	396.2108	48.48	8.14	7.07	36.32
C16H32N4O1	296.46	296.2576	64.82	10.88	18.90	5.40
C16H32N4O2	312.46	312.2525	61.51	10.32	17.93	10.24
C16H32N4O3	328.46	328.2474	58.51	9.82	17.06	14.61
C16H32N4O4	344.46	344.2423	55.79	9.36	16.27	18.58
C16H32N4O5	360.46	360.2373	53.32	8.95	15.54	22.19
C16H32N4O6	376.46	376.2322	51.05	8.57	14.88	25.50
C16H32N4O7	392.46	392.2271	48.97	8.22	14.28	28.54
C16H32N4O8	408.46	408.2220	47.05	7.90	13.72	31.34
C16H32N4O9	424.45	424.2169	45.28	7.60	13.20	33.92
C16H32N6O1	324.47	324.2637	59.23	9.94	25.90	4.93
C16H32N6O2	340.47	340.2587	56.44	9.47	24.68	9.40
C16H32N6O3	356.47	356.2536	53.91	9.05	23.58	13.46
C16H32N6O4	372.47	372.2485	51.60	8.66	22.56	17.18
C16H32N6O5	388.47	388.2434	49.47	8.30	21.63	20.59
C16H32N6O6	404.47	404.2383	47.51	7.97	20.78	23.73
C16H32N6O7	420.47	420.2332	45.71	7.67	19.99	26.64
C16H32N6O8	436.47	436.2281	44.03	7.39	19.25	29.33
C16H32N6O9	452.47	452.2231	42.47	7.13	18.57	31.82
C16H33N1O1	255.45	255.2562	75.23	13.02	5.48	6.26
C16H33N1O2	271.45	271.2511	70.80	12.25	5.16	11.79
C16H33N1O3	287.45	287.2460	66.86	11.57	4.87	16.70
C16H33N1O4	303.45	303.2409	63.33	10.96	4.62	21.09
C16H33N1O5	319.45	319.2359	60.16	10.41	4.38	25.04
C16H33N1O6	335.44	335.2308	57.29	9.92	4.18	28.62
C16H33N1O7	351.44	351.2257	54.68	9.46	3.99	31.87
C16H33N1O8	367.44	367.2206	52.30	9.05	3.81	34.83
C16H33N1O9	383.44	383.2155	50.12	8.67	3.65	37.55
C16H33N3O1	283.46	283.2623	67.80	11.73	14.82	5.64
C16H33N3O2	299.46	299.2573	64.17	11.11	14.03	10.69
C16H33N3O3	315.46	315.2522	60.92	10.54	13.32	15.22
C16H33N3O4	331.46	331.2471	57.98	10.04	12.68	19.31
C16H33N3O5	347.46	347.2420	55.31	9.57	12.09	23.02
C16H33N3O6	363.46	363.2369	52.88	9.15	11.56	26.41
C16H33N3O7	379.46	379.2318	50.65	8.77	11.07	29.51
C16H33N3O8	395.46	395.2267	48.60	8.41	10.63	32.37
C16H33N3O9	411.46	411.2217	46.71	8.08	10.21	35.00
C16H33N5O1	311.47	311.2685	61.70	10.68	22.48	5.14
C16H33N5O2	327.47	327.2634	58.69	10.16	21.39	9.77
C16H33N5O3	343.47	343.2583	55.95	9.68	20.39	13.97
C16H33N5O4	359.47	359.2532	53.46	9.25	19.48	17.80

087

C16H33N5O5	375.47	375.2482	51.18	8.86	18.65	21.31
C16H33N5O6	391.47	391.2431	49.09	8.50	17.89	24.52
C16H33N5O7	407.47	407.2380	47.16	8.16	17.19	27.49
C16H33N5O8	423.47	423.2329	45.38	7.85	16.54	30.23
C16H33N5O9	439.47	439.2278	43.73	7.57	15.94	32.77
C16H34N2O1	270.46	270.2671	71.06	12.67	10.36	5.92
C16H34N2O2	286.46	286.2620	67.05	11.96	9.78	11.17
C16H34N2O3	302.46	302.2569	63.54	11.33	9.26	15.87
C16H34N2O4	318.46	318.2518	60.35	10.76	8.80	20.10
C16H34N2O5	334.46	334.2468	57.46	10.25	8.38	23.92
C16H34N2O6	350.46	350.2417	54.84	9.78	7.99	27.39
C16H34N2O7	366.46	366.2366	52.44	9.35	7.64	30.56
C16H34N2O8	382.46	382.2315	50.25	8.96	7.32	33.47
C16H34N2O9	398.46	398.2264	48.23	8.60	7.03	36.14
C16H34N4O1	298.48	298.2732	64.39	11.48	18.77	5.36
C16H34N4O2	314.47	314.2682	61.11	10.90	17.82	10.18
C16H34N4O3	330.47	330.2631	58.15	10.37	16.95	14.52
C16H34N4O4	346.47	346.2580	55.47	9.89	16.17	18.47
C16H34N4O5	362.47	362.2529	53.02	9.45	15.46	22.07
C16H34N4O6	378.47	378.2478	50.78	9.06	14.80	25.36
C16H34N4O7	394.47	394.2427	48.72	8.69	14.20	28.39
C16H34N4O8	410.47	410.2376	46.82	8.35	13.65	31.18
C16H34N4O9	426.47	426.2326	45.06	8.04	13.14	33.76
C16H34N6O1	326.49	326.2794	58.86	10.50	25.74	4.90
C16H34N6O2	342.49	342.2743	56.11	10.01	24.54	9.34
C16H34N6O3	358.49	358.2692	53.61	9.56	23.44	13.39
C16H34N6O4	374.49	374.2641	51.32	9.15	22.44	17.09
C16H34N6O5	390.49	390.2590	49.22	8.78	21.52	20.49
C16H34N6O6	406.49	406.2540	47.28	8.43	20.67	23.62
C16H34N6O7	422.49	422.2489	45.49	8.11	19.89	26.51
C16H34N6O8	438.48	438.2438	43.83	7.82	19.17	29.19
C16H34N6O9	454.48	454.2387	42.28	7.54	18.49	31.68
C16H35N1O1	257.46	257.2718	74.64	13.70	5.44	6.21
C16H35N1O2	273.46	273.2668	70.28	12.90	5.12	11.70
C16H35N1O3	289.46	289.2617	66.39	12.19	4.84	16.58
C16H35N1O4	305.46	305.2566	62.91	11.55	4.59	20.95
C16H35N1O5	321.46	321.2515	59.78	10.97	4.36	24.89
C16H35N1O6	337.46	337.2464	56.95	10.45	4.15	28.45
C16H35N1O7	353.46	353.2413	54.37	9.98	3.96	31.69
C16H35N1O8	369.46	369.2362	52.02	9.55	3.79	34.64
C16H35N1O9	385.46	385.2312	49.86	9.15	3.63	37.36
C16H35N3O1	285.48	285.2780	67.32	12.36	14.72	5.60
C16H35N3O2	301.48	301.2729	63.75	11.70	13.94	10.61
C16H35N3O3	317.48	317.2678	60.53	11.11	13.24	15.12
C16H35N3O4	333.48	333.2627	57.63	10.58	12.60	19.19
C16H35N3O5	349.47	349.2577	54.99	10.09	12.02	22.89
C16H35N3O6	365.47	365.2526	52.58	9.65	11.50	26.27
C16H35N3O7	381.47	381.2475	50.38	9.25	11.02	29.36
C16H35N3O8	397.47	397.2424	48.35	8.88	10.57	32.20
C16H35N3O9	413.47	413.2373	46.48	8.53	10.16	34.83
C16H35N5O1	313.49	313.2841	61.30	11.25	22.34	5.10
C16H35N5O2	329.49	329.2791	58.33	10.71	21.26	9.71
C16H35N5O3	345.49	345.2740	55.63	10.21	20.27	13.89
C16H35N5O4	361.49	361.2689	53.16	9.76	19.37	17.70
C16H35N5O5	377.49	377.2638	50.91	9.35	18.55	21.19
C16H35N5O6	393.49	393.2587	48.84	8.97	17.80	24.40
C16H35N5O7	409.49	409.2536	46.93	8.62	17.10	27.35
C16H35N5O8	425.49	425.2485	45.17	8.29	16.46	30.08
C16H35N5O9	441.49	441.2435	43.53	7.99	15.86	32.62
C16H36N2O1	272.48	272.2827	70.53	13.32	10.28	5.87

C16H36N2O2	288.48	288.2777	66.62	12.58	9.71	11.09
C16H36N2O3	304.48	304.2726	63.12	11.92	9.20	15.76
C16H36N2O4	320.48	320.2675	59.97	11.32	8.74	19.97
C16H36N2O5	336.48	336.2624	57.12	10.78	8.33	23.77
C16H36N2O6	352.48	352.2573	54.52	10.29	7.95	27.23
C16H36N2O7	368.47	368.2522	52.16	9.85	7.60	30.39
C16H36N2O8	384.47	384.2471	49.98	9.44	7.29	33.29
C16H36N2O9	400.47	400.2421	47.99	9.06	7.00	35.96
C16H36N4O1	300.49	300.2889	63.95	12.08	18.65	5.32
C16H36N4O2	316.49	316.2838	60.72	11.47	17.70	10.11
C16H36N4O3	332.49	332.2787	57.80	10.91	16.85	14.44
C16H36N4O4	348.49	348.2736	55.15	10.41	16.08	18.36
C16H36N4O5	364.49	364.2686	52.73	9.96	15.37	21.95
C16H36N4O6	380.49	380.2635	50.51	9.54	14.72	25.23
C16H36N4O7	396.49	396.2584	48.47	9.15	14.13	28.25
C16H36N4O8	412.49	412.2533	46.59	8.80	13.58	31.03
C16H36N4O9	428.49	428.2482	44.85	8.47	13.08	33.61
C16H36N6O1	328.50	328.2950	58.50	11.05	25.58	4.87
C16H36N6O2	344.50	344.2900	55.78	10.53	24.39	9.29
C16H36N6O3	360.50	360.2849	53.31	10.07	23.31	13.31
C16H36N6O4	376.50	376.2798	51.04	9.64	22.32	17.00
C16H36N6O5	392.50	392.2747	48.96	9.25	21.41	20.38
C16H36N6O6	408.50	408.2696	47.04	8.88	20.57	23.50
C16H36N6O7	424.50	424.2645	45.27	8.55	19.80	26.38
C16H36N6O8	440.50	440.2594	43.63	8.24	19.08	29.06
C16H36N6O9	456.50	456.2544	42.10	7.95	18.41	31.54
C16H37N3O1	287.49	287.2936	66.85	12.97	14.62	5.57
C16H37N3O2	303.49	303.2886	63.32	12.29	13.85	10.54
C16H37N3O3	319.49	319.2835	60.15	11.67	13.15	15.02
C16H37N3O4	335.49	335.2784	57.28	11.12	12.52	19.08
C16H37N3O5	351.49	351.2733	54.68	10.61	11.95	22.76
C16H37N3O6	367.49	367.2682	52.29	10.15	11.43	26.12
C16H37N3O7	383.49	383.2631	50.11	9.73	10.96	29.20
C16H37N3O8	399.49	399.2580	48.11	9.34	10.52	32.04
C16H37N3O9	415.49	415.2530	46.25	8.98	10.11	34.66
C16H37N5O1	315.51	315.2998	60.91	11.82	22.20	5.07
C16H37N5O2	331.51	331.2947	57.97	11.25	21.13	9.65
C16H37N5O3	347.50	347.2896	55.30	10.73	20.15	13.81
C16H37N5O4	363.50	363.2845	52.87	10.26	19.27	17.61
C16H37N5O5	379.50	379.2794	50.64	9.83	18.45	21.08
C16H37N5O6	395.50	395.2744	48.59	9.43	17.71	24.27
C16H37N5O7	411.50	411.2693	46.70	9.06	17.02	27.22
C16H37N5O8	427.50	427.2642	44.95	8.72	16.38	29.94
C16H37N5O9	443.50	443.2591	43.33	8.41	15.79	32.47
C16H38N4O1	302.51	302.3045	63.53	12.66	18.52	5.29
C16H38N4O2	318.51	318.2995	60.34	12.03	17.59	10.05
C16H38N4O3	334.51	334.2944	57.45	11.45	16.75	14.35
C16H38N4O4	350.51	350.2893	54.83	10.93	15.98	18.26
C16H38N4O5	366.51	366.2842	52.44	10.45	15.29	21.83
C16H38N4O6	382.50	382.2791	50.24	10.01	14.65	25.10
C16H38N4O7	398.50	398.2740	48.22	9.61	14.06	28.10
C16H38N4O8	414.50	414.2689	46.36	9.24	13.52	30.88
C16H38N4O9	430.50	430.2639	44.64	8.90	13.01	33.45
C16H38N6O1	330.52	330.3107	58.14	11.59	25.43	4.84
C16H38N6O2	346.52	346.3056	55.46	11.05	24.25	9.23
C16H38N6O3	362.52	362.3005	53.01	10.57	23.18	13.24
C16H38N6O4	378.52	378.2954	50.77	10.12	22.20	16.91
C16H38N6O5	394.52	394.2903	48.71	9.71	21.30	20.28
C16H38N6O6	410.52	410.2853	46.81	9.33	20.47	23.38
C16H38N6O7	426.52	426.2802	45.06	8.98	19.70	26.26

C16H38N608	442.52	442.2751	43.43	8.66	18.99	28.92	C17H 9N305	335.28	335.0542	60.90	2.71	12.53	23.86
C16H38N609	458.52	458.2700	41.91	8.35	18.33	31.40	C17H 9N306	351.28	351.0491	58.13	2.58	11.96	27.33
C16H39N501	317.52	317.3154	60.52	12.38	22.06	5.04	C17H 9N307	367.28	367.0440	55.60	2.47	11.44	30.49
C16H39N502	333.52	333.3104	57.62	11.79	21.00	9.59	C17H 9N308	383.28	383.0390	53.27	2.37	10.96	33.39
C16H39N503	349.52	349.3053	54.98	11.25	20.04	13.73	C17H 9N309	399.28	399.0339	51.14	2.27	10.52	36.06
C16H39N504	365.52	365.3002	52.58	10.75	19.16	17.51	C17H 9N501	299.29	299.0807	68.22	3.03	23.40	5.35
C16H39N505	381.52	381.2951	50.37	10.30	18.36	20.97	C17H 9N502	315.29	315.0756	64.76	2.88	22.21	10.15
C16H39N506	397.52	397.2900	48.34	9.89	17.62	24.15	C17H 9N503	331.29	331.0705	61.63	2.74	21.14	14.49
C16H39N507	413.52	413.2849	46.47	9.51	16.94	27.08	C17H 9N504	347.29	347.0654	58.79	2.61	20.17	18.43
C16H39N508	429.52	429.2798	44.74	9.15	16.31	29.80	C17H 9N505	363.29	363.0604	56.21	2.50	19.28	22.02
C16H39N509	445.52	445.2748	43.14	8.82	15.72	32.32	C17H 9N506	379.29	379.0553	53.83	2.39	18.46	25.31
C16H40N601	332.54	332.3263	57.79	12.12	25.27	4.81	C17H 9N507	395.29	395.0502	51.66	2.29	17.72	28.33
C16H40N602	348.54	348.3213	55.14	11.57	24.11	9.18	C17H 9N508	411.29	411.0451	49.65	2.21	17.03	31.12
C16H40N603	364.54	364.3162	52.72	11.06	23.05	13.17	C17H 9N509	427.29	427.0400	47.79	2.12	16.39	33.70
C16H40N604	380.54	380.3111	50.50	10.60	22.08	16.82	C17H10N201	258.28	258.0793	79.06	3.90	10.85	6.19
C16H40N605	396.53	396.3060	48.46	10.17	21.19	20.17	C17H10N202	274.28	274.0742	74.45	3.67	10.21	11.67
C16H40N606	412.53	412.3009	46.58	9.77	20.37	23.27	C17H10N203	290.28	290.0691	70.34	3.47	9.65	16.54
C16H40N607	428.53	428.2958	44.85	9.41	19.61	26.13	C17H10N204	306.28	306.0641	66.67	3.29	9.15	20.90
C16H40N608	444.53	444.2907	43.23	9.07	18.91	28.79	C17H10N205	322.28	322.0590	63.36	3.13	8.69	24.82
C16H40N609	460.53	460.2857	41.73	8.75	18.25	31.27	C17H10N206	338.28	338.0539	60.36	2.98	8.28	28.38
C17H 8N201	256.27	256.0637	79.68	3.15	10.93	6.24	C17H10N207	354.28	354.0488	57.64	2.85	7.91	31.61
C17H 8N202	272.27	272.0586	75.00	2.96	10.29	11.75	C17H10N208	370.28	370.0437	55.14	2.72	7.57	34.57
C17H 8N203	288.26	288.0535	70.83	2.80	9.72	16.65	C17H10N209	386.28	386.0386	52.86	2.61	7.25	37.28
C17H 8N204	304.26	304.0484	67.11	2.65	9.21	21.03	C17H10N401	286.30	286.0855	71.32	3.52	19.57	5.59
C17H 8N205	320.26	320.0433	63.76	2.52	8.75	24.98	C17H10N402	302.29	302.0804	67.55	3.33	18.53	10.59
C17H 8N206	336.26	336.0382	60.72	2.40	8.33	28.55	C17H10N403	318.29	318.0753	64.15	3.17	17.60	15.08
C17H 8N207	352.26	352.0331	57.97	2.29	7.95	31.79	C17H10N404	334.29	334.0702	61.08	3.02	16.76	19.14
C17H 8N208	368.26	368.0281	55.45	2.19	7.61	34.76	C17H10N405	350.29	350.0651	58.29	2.88	15.99	22.84
C17H 8N209	384.26	384.0230	53.14	2.10	7.29	37.47	C17H10N406	366.29	366.0600	55.74	2.75	15.30	26.21
C17H 8N401	284.28	284.0698	71.83	2.84	19.71	5.63	C17H10N407	382.29	382.0549	53.41	2.64	14.66	29.30
C17H 8N402	300.28	300.0647	68.00	2.69	18.66	10.66	C17H10N408	398.29	398.0499	51.27	2.53	14.07	32.14
C17H 8N403	316.28	316.0596	64.56	2.55	17.71	15.18	C17H10N409	414.29	414.0448	49.29	2.43	13.52	34.76
C17H 8N404	332.28	332.0545	61.45	2.43	16.86	19.26	C17H10N601	314.31	314.0916	64.96	3.21	26.74	5.09
C17H 8N405	348.28	348.0495	58.63	2.32	16.09	22.97	C17H10N602	330.31	330.0865	61.82	3.05	25.44	9.69
C17H 8N406	364.28	364.0444	56.05	2.21	15.38	26.35	C17H10N603	346.31	346.0814	58.96	2.91	24.27	13.86
C17H 8N407	380.28	380.0393	53.70	2.12	14.73	29.45	C17H10N604	362.31	362.0763	56.36	2.78	23.20	17.66
C17H 8N408	396.28	396.0342	51.53	2.03	14.14	32.30	C17H10N605	378.31	378.0713	53.97	2.66	22.21	21.15
C17H 8N409	412.27	412.0291	49.53	1.96	13.59	34.93	C17H10N606	394.31	394.0662	51.78	2.56	21.31	24.35
C17H 8N601	312.29	312.0760	65.38	2.58	26.91	5.12	C17H10N607	410.31	410.0611	49.77	2.46	20.48	27.30
C17H 8N602	328.29	328.0709	62.20	2.46	25.60	9.75	C17H10N608	426.30	426.0560	47.90	2.36	19.71	30.02
C17H 8N603	344.29	344.0658	59.31	2.34	24.41	13.94	C17H10N609	442.30	442.0509	46.16	2.28	19.00	32.56
C17H 8N604	360.29	360.0607	56.67	2.24	23.33	17.76	C17H11N101	245.28	245.0841	83.25	4.52	5.71	6.52
C17H 8N605	376.29	376.0556	54.26	2.14	22.33	21.26	C17H11N102	261.28	261.0790	78.15	4.24	5.36	12.25
C17H 8N606	392.29	392.0505	52.05	2.06	21.42	24.47	C17H11N103	277.28	277.0739	73.64	4.00	5.05	17.31
C17H 8N607	408.29	408.0454	50.01	1.98	20.58	27.43	C17H11N104	293.28	293.0688	69.62	3.78	4.78	21.82
C17H 8N608	424.29	424.0404	48.13	1.90	19.81	30.17	C17H11N105	309.28	309.0637	66.02	3.58	4.53	25.87
C17H 8N609	440.29	440.0353	46.38	1.83	19.09	32.70	C17H11N106	325.28	325.0586	62.77	3.41	4.31	29.51
C17H 9N101	243.27	243.0684	83.94	3.73	5.76	6.58	C17H11N107	341.28	341.0535	59.83	3.25	4.10	32.82
C17H 9N102	259.27	259.0633	78.76	3.50	5.40	12.34	C17H11N108	357.28	357.0485	57.15	3.10	3.92	35.82
C17H 9N103	275.27	275.0582	74.18	3.30	5.09	17.44	C17H11N109	373.28	373.0434	54.70	2.97	3.75	38.58
C17H 9N104	291.27	291.0532	70.10	3.11	4.81	21.97	C17H11N301	273.30	273.0902	74.71	4.06	15.38	5.85
C17H 9N105	307.26	307.0481	66.45	2.95	4.56	26.04	C17H11N302	289.30	289.0851	70.58	3.83	14.52	11.06
C17H 9N106	323.26	323.0430	63.16	2.81	4.33	29.70	C17H11N303	305.30	305.0800	66.88	3.63	13.76	15.72
C17H 9N107	339.26	339.0379	60.19	2.67	4.13	33.01	C17H11N304	321.29	321.0749	63.55	3.45	13.08	19.92
C17H 9N108	355.26	355.0328	57.48	2.55	3.94	36.03	C17H11N305	337.29	337.0699	60.54	3.29	12.46	23.72
C17H 9N109	371.26	371.0277	55.00	2.44	3.77	38.79	C17H11N306	353.29	353.0648	57.80	3.14	11.89	27.17
C17H 9N301	271.28	271.0746	75.27	3.34	15.49	5.90	C17H11N307	369.29	369.0597	55.29	3.00	11.38	30.33
C17H 9N302	287.28	287.0695	71.08	3.16	14.63	11.14	C17H11N308	385.29	385.0546	53.00	2.88	10.91	33.22
C17H 9N303	303.28	303.0644	67.33	2.99	13.86	15.83	C17H11N309	401.29	401.0495	50.88	2.76	10.47	35.88
C17H 9N304	319.28	319.0593	63.95	2.84	13.16	20.04	C17H11N501	301.31	301.0964	67.77	3.68	23.24	5.31

C17H11N502	317.31	317.0913	64.35	3.49	22.07	10.08	C17H13N508	415.32	415.0764	49.16	3.16	16.86	30.82
C17H11N503	333.31	333.0862	61.26	3.33	21.01	14.40	C17H13N509	431.32	431.0713	47.34	3.04	16.24	33.38
C17H11N504	349.31	349.0811	58.46	3.17	20.05	18.32	C17H14N201	262.31	262.1106	77.84	5.38	10.68	6.10
C17H11N505	365.31	365.0760	55.90	3.04	19.17	21.90	C17H14N202	278.31	278.1055	73.37	5.07	10.07	11.50
C17H11N506	381.31	381.0709	53.55	2.91	18.37	25.18	C17H14N203	294.31	294.1004	69.38	4.79	9.52	16.31
C17H11N507	397.31	397.0658	51.39	2.79	17.63	28.19	C17H14N204	310.31	310.0953	65.80	4.55	9.03	20.62
C17H11N508	413.31	413.0608	49.40	2.68	16.94	30.97	C17H14N205	326.31	326.0903	62.58	4.32	8.58	24.52
C17H11N509	429.31	429.0557	47.56	2.58	16.31	33.54	C17H14N206	342.31	342.0852	59.65	4.12	8.18	28.04
C17H12N201	260.30	260.0950	78.44	4.65	10.76	6.15	C17H14N207	358.31	358.0801	56.99	3.94	7.82	31.26
C17H12N202	276.30	276.0899	73.90	4.38	10.14	11.58	C17H14N208	374.31	374.0750	54.55	3.77	7.48	34.19
C17H12N203	292.30	292.0848	69.86	4.14	9.58	16.42	C17H14N209	390.31	390.0699	52.31	3.62	7.18	36.89
C17H12N204	308.30	308.0797	66.23	3.92	9.09	20.76	C17H14N401	290.33	290.1168	70.33	4.86	19.30	5.51
C17H12N205	324.30	324.0746	62.96	3.73	8.64	24.67	C17H14N402	306.33	306.1117	66.66	4.61	18.29	10.45
C17H12N206	340.29	340.0695	60.00	3.55	8.23	28.21	C17H14N403	322.33	322.1066	63.35	4.38	17.38	14.89
C17H12N207	356.29	356.0644	57.31	3.39	7.86	31.43	C17H14N404	338.33	338.1015	60.35	4.17	16.56	18.92
C17H12N208	372.29	372.0594	54.85	3.25	7.52	34.38	C17H14N405	354.32	354.0964	57.63	3.98	15.81	22.58
C17H12N209	388.29	388.0543	52.59	3.12	7.21	37.08	C17H14N406	370.32	370.0913	55.14	3.81	15.13	25.92
C17H12N401	288.31	288.1011	70.82	4.20	19.43	5.55	C17H14N407	386.32	386.0862	52.85	3.65	14.50	28.99
C17H12N402	304.31	304.0960	67.10	3.97	18.41	10.52	C17H14N408	402.32	402.0812	50.75	3.51	13.93	31.81
C17H12N403	320.31	320.0909	63.75	3.78	17.49	14.98	C17H14N409	418.32	418.0761	48.81	3.37	13.39	34.42
C17H12N404	336.31	336.0858	60.71	3.60	16.66	19.03	C17H14N601	318.34	318.1229	64.14	4.43	26.40	5.03
C17H12N405	352.31	352.0808	57.96	3.43	15.90	22.71	C17H14N602	334.34	334.1178	61.07	4.22	25.14	9.57
C17H12N406	368.31	368.0757	55.44	3.28	15.21	26.06	C17H14N603	350.34	350.1127	58.28	4.03	23.99	13.70
C17H12N407	384.31	384.0706	53.13	3.15	14.58	29.14	C17H14N604	366.34	366.1076	55.74	3.85	22.94	17.47
C17H12N408	400.31	400.0655	51.01	3.02	14.00	31.97	C17H14N605	382.34	382.1026	53.41	3.69	21.98	20.92
C17H12N409	416.31	416.0604	49.05	2.91	13.46	34.59	C17H14N606	398.34	398.0975	51.26	3.54	21.10	24.10
C17H12N601	316.32	316.1073	64.55	3.82	26.57	5.06	C17H14N607	414.34	414.0924	49.28	3.41	20.28	27.03
C17H12N602	332.32	332.1022	61.44	3.64	25.29	9.63	C17H14N608	430.34	430.0873	47.45	3.28	19.53	29.74
C17H12N603	348.32	348.0971	58.62	3.47	24.13	13.78	C17H14N609	446.34	446.0822	45.75	3.16	18.83	32.26
C17H12N604	364.32	364.0920	56.05	3.32	23.07	17.57	C17H15N101	249.32	249.1154	81.90	6.06	5.62	6.42
C17H12N605	380.32	380.0869	53.69	3.18	22.10	21.03	C17H15N102	265.31	265.1103	76.96	5.70	5.28	12.06
C17H12N606	396.32	396.0818	51.52	3.05	21.21	24.22	C17H15N103	281.31	281.1052	72.58	5.37	4.98	17.06
C17H12N607	412.32	412.0767	49.52	2.93	20.38	27.16	C17H15N104	297.31	297.1001	68.68	5.09	4.71	21.53
C17H12N608	428.32	428.0717	47.67	2.82	19.62	29.88	C17H15N105	313.31	313.0950	65.17	4.83	4.47	25.53
C17H12N609	444.32	444.0666	45.96	2.72	18.91	32.41	C17H15N106	329.31	329.0899	62.00	4.59	4.25	29.15
C17H13N101	247.30	247.0997	82.57	5.30	5.66	6.47	C17H15N107	345.31	345.0848	59.13	4.38	4.06	32.43
C17H13N102	263.30	263.0946	77.55	4.98	5.32	12.15	C17H15N108	361.31	361.0798	56.51	4.18	3.88	35.43
C17H13N103	279.30	279.0895	73.11	4.69	5.01	17.19	C17H15N109	377.31	377.0747	54.12	4.01	3.71	38.16
C17H13N104	295.30	295.0844	69.15	4.44	4.74	21.67	C17H15N301	277.33	277.1215	73.63	5.45	15.15	5.77
C17H13N105	311.30	311.0794	65.59	4.21	4.50	25.70	C17H15N302	293.33	293.1164	69.61	5.15	14.33	10.91
C17H13N106	327.30	327.0743	62.39	4.00	4.28	29.33	C17H15N303	309.33	309.1113	66.01	4.89	13.58	15.52
C17H13N107	343.30	343.0692	59.48	3.82	4.08	32.62	C17H15N304	325.33	325.1062	62.76	4.65	12.92	19.67
C17H13N108	359.30	359.0641	56.83	3.65	3.90	35.62	C17H15N305	341.33	341.1012	59.82	4.43	12.31	23.44
C17H13N109	375.29	375.0590	54.41	3.49	3.73	38.37	C17H15N306	357.33	357.0961	57.14	4.23	11.76	26.87
C17H13N301	275.31	275.1059	74.17	4.76	15.26	5.81	C17H15N307	373.33	373.0910	54.69	4.05	11.26	30.00
C17H13N302	291.31	291.1008	70.09	4.50	14.42	10.98	C17H15N308	389.32	389.0859	52.45	3.88	10.79	32.88
C17H13N303	307.31	307.0957	66.44	4.26	13.67	15.62	C17H15N309	405.32	405.0808	50.38	3.73	10.37	35.53
C17H13N304	323.31	323.0906	63.16	4.05	13.00	19.79	C17H15N501	305.34	305.1277	66.87	4.95	22.94	5.24
C17H13N305	339.31	339.0855	60.18	3.86	12.38	23.58	C17H15N502	321.34	321.1226	63.54	4.71	21.79	9.96
C17H13N306	355.31	355.0804	57.47	3.69	11.83	27.02	C17H15N503	337.34	337.1175	60.53	4.48	20.76	14.23
C17H13N307	371.31	371.0753	54.99	3.53	11.32	30.16	C17H15N504	353.34	353.1124	57.79	4.28	19.82	18.11
C17H13N308	387.31	387.0703	52.72	3.38	10.85	33.05	C17H15N505	369.34	369.1073	55.29	4.09	18.96	21.66
C17H13N309	403.31	403.0652	50.63	3.25	10.42	35.70	C17H15N506	385.34	385.1022	52.99	3.92	18.17	24.91
C17H13N501	303.33	303.1120	67.32	4.32	23.09	5.27	C17H15N507	401.34	401.0971	50.88	3.77	17.45	27.91
C17H13N502	319.33	319.1069	63.94	4.10	21.93	10.02	C17H15N508	417.34	417.0921	48.93	3.62	16.78	30.67
C17H13N503	335.32	335.1018	60.89	3.91	20.89	14.31	C17H15N509	433.34	433.0870	47.12	3.49	16.16	33.23
C17H13N504	351.32	351.0967	58.12	3.73	19.93	18.22	C17H16N201	264.33	264.1263	77.25	6.10	10.60	6.05
C17H13N505	367.32	367.0917	55.59	3.57	19.07	21.78	C17H16N202	280.33	280.1212	72.84	5.75	9.99	11.41
C17H13N506	383.32	383.0866	53.27	3.42	18.27	25.04	C17H16N203	296.33	296.1161	68.91	5.44	9.45	16.20
C17H13N507	399.32	399.0815	51.13	3.28	17.54	28.05	C17H16N204	312.33	312.1110	65.38	5.16	8.97	20.49

C17H16N2O5	328.33	328.1059	62.19	4.91	8.53	24.37
C17H16N2O6	344.33	344.1008	59.30	4.68	8.14	27.88
C17H16N2O7	360.33	360.0957	56.67	4.48	7.77	31.08
C17H16N2O8	376.33	376.0907	54.26	4.29	7.44	34.01
C17H16N2O9	392.33	392.0856	52.05	4.11	7.14	36.70
C17H16N4O1	292.34	292.1324	69.85	5.52	19.16	5.47
C17H16N4O2	308.34	308.1273	66.22	5.23	18.17	10.38
C17H16N4O3	324.34	324.1222	62.95	4.97	17.27	14.80
C17H16N4O4	340.34	340.1171	60.00	4.74	16.46	18.80
C17H16N4O5	356.34	356.1121	57.30	4.53	15.72	22.45
C17H16N4O6	372.34	372.1070	54.84	4.33	15.05	25.78
C17H16N4O7	388.34	388.1019	52.58	4.15	14.43	28.84
C17H16N4O8	404.34	404.0968	50.50	3.99	13.86	31.66
C17H16N4O9	420.34	420.0917	48.58	3.84	13.33	34.26
C17H16N6O1	320.36	320.1386	63.74	5.03	26.23	4.99
C17H16N6O2	336.36	336.1335	60.71	4.79	24.99	9.51
C17H16N6O3	352.36	352.1284	57.95	4.58	23.85	13.62
C17H16N6O4	368.35	368.1233	55.43	4.38	22.82	17.37
C17H16N6O5	384.35	384.1182	53.13	4.20	21.87	20.81
C17H16N6O6	400.35	400.1131	51.00	4.03	20.99	23.98
C17H16N6O7	416.35	416.1080	49.04	3.87	20.18	26.90
C17H16N6O8	432.35	432.1029	47.23	3.73	19.44	29.60
C17H16N6O9	448.35	448.0979	45.54	3.60	18.74	32.12
C17H17N1O1	251.33	251.1310	81.24	6.82	5.57	6.37
C17H17N1O2	267.33	267.1259	76.38	6.41	5.24	11.97
C17H17N1O3	283.33	283.1208	72.07	6.05	4.94	16.94
C17H17N1O4	299.33	299.1157	68.22	5.72	4.68	21.38
C17H17N1O5	315.33	315.1107	64.75	5.43	4.44	25.37
C17H17N1O6	331.33	331.1056	61.63	5.17	4.23	28.97
C17H17N1O7	347.33	347.1005	58.79	4.93	4.03	32.25
C17H17N1O8	363.33	363.0954	56.20	4.72	3.86	35.23
C17H17N1O9	379.33	379.0903	53.83	4.52	3.69	37.96
C17H17N3O1	279.34	279.1372	73.10	6.13	15.04	5.73
C17H17N3O2	295.34	295.1321	69.14	5.80	14.23	10.83
C17H17N3O3	311.34	311.1270	65.58	5.50	13.50	15.42
C17H17N3O4	327.34	327.1219	62.38	5.23	12.84	19.55
C17H17N3O5	343.34	343.1168	59.47	4.99	12.24	23.30
C17H17N3O6	359.34	359.1117	56.82	4.77	11.69	26.71
C17H17N3O7	375.34	375.1066	54.40	4.57	11.20	29.84
C17H17N3O8	391.34	391.1016	52.18	4.38	10.74	32.71
C17H17N3O9	407.34	407.0965	50.13	4.21	10.32	35.35
C17H17N5O1	307.36	307.1433	66.43	5.58	22.79	5.21
C17H17N5O2	323.36	323.1382	63.15	5.30	21.66	9.90
C17H17N5O3	339.36	339.1331	60.17	5.05	20.64	14.14
C17H17N5O4	355.36	355.1280	57.46	4.82	19.71	18.01
C17H17N5O5	371.36	371.1230	54.98	4.61	18.86	21.54
C17H17N5O6	387.35	387.1179	52.71	4.42	18.08	24.78
C17H17N5O7	403.35	403.1128	50.62	4.25	17.36	27.77
C17H17N5O8	419.35	419.1077	48.69	4.09	16.70	30.52
C17H17N5O9	435.35	435.1026	46.90	3.94	16.09	33.08
C17H18N2O1	266.35	266.1419	76.66	6.81	10.52	6.01
C17H18N2O2	282.35	282.1368	72.32	6.43	9.92	11.33
C17H18N2O3	298.34	298.1317	68.44	6.08	9.39	16.09
C17H18N2O4	314.34	314.1266	64.96	5.77	8.91	20.36
C17H18N2O5	330.34	330.1216	61.81	5.49	8.48	24.22
C17H18N2O6	346.34	346.1165	58.96	5.24	8.09	27.72
C17H18N2O7	362.34	362.1114	56.35	5.01	7.73	30.91
C17H18N2O8	378.34	378.1063	53.97	4.80	7.40	33.83
C17H18N2O9	394.34	394.1012	51.78	4.60	7.10	36.52
C17H18N4O1	294.36	294.1481	69.37	6.16	19.03	5.44

C17H18N4O2	310.36	310.1430	65.79	5.85	18.05	10.31
C17H18N4O3	326.36	326.1379	62.57	5.56	17.17	14.71
C17H18N4O4	342.36	342.1328	59.64	5.30	16.37	18.69
C17H18N4O5	358.36	358.1277	56.98	5.06	15.63	22.32
C17H18N4O6	374.36	374.1226	54.54	4.85	14.97	25.64
C17H18N4O7	390.36	390.1175	52.31	4.65	14.35	28.69
C17H18N4O8	406.36	406.1125	50.25	4.46	13.79	31.50
C17H18N4O9	422.35	422.1074	48.35	4.30	13.27	34.09
C17H18N6O1	322.37	322.1542	63.34	5.63	26.07	4.96
C17H18N6O2	338.37	338.1491	60.34	5.36	24.84	9.46
C17H18N6O3	354.37	354.1440	57.62	5.12	23.72	13.54
C17H18N6O4	370.37	370.1389	55.13	4.90	22.69	17.28
C17H18N6O5	386.37	386.1339	52.85	4.70	21.75	20.70
C17H18N6O6	402.37	402.1288	50.75	4.51	20.89	23.86
C17H18N6O7	418.37	418.1237	48.81	4.34	20.09	26.77
C17H18N6O8	434.37	434.1186	47.01	4.18	19.35	29.47
C17H18N6O9	450.37	450.1135	45.34	4.03	18.66	31.97
C17H19N1O1	253.35	253.1467	80.60	7.56	5.53	6.32
C17H19N1O2	269.35	269.1416	75.81	7.11	5.20	11.88
C17H19N1O3	285.35	285.1365	71.56	6.71	4.91	16.82
C17H19N1O4	301.35	301.1314	67.76	6.36	4.65	21.24
C17H19N1O5	317.34	317.1263	64.34	6.03	4.41	25.21
C17H19N1O6	333.34	333.1212	61.25	5.75	4.20	28.80
C17H19N1O7	349.34	349.1161	58.45	5.48	4.01	32.06
C17H19N1O8	365.34	365.1111	55.89	5.24	3.83	35.03
C17H19N1O9	381.34	381.1060	53.54	5.02	3.67	37.76
C17H19N3O1	281.36	281.1528	72.57	6.81	14.93	5.69
C17H19N3O2	297.36	297.1477	68.67	6.44	14.13	10.76
C17H19N3O3	313.36	313.1426	65.16	6.11	13.41	15.32
C17H19N3O4	329.36	329.1375	62.00	5.81	12.76	19.43
C17H19N3O5	345.36	345.1325	59.12	5.55	12.17	23.16
C17H19N3O6	361.36	361.1274	56.51	5.30	11.63	26.57
C17H19N3O7	377.36	377.1223	54.11	5.08	11.14	29.68
C17H19N3O8	393.36	393.1172	51.91	4.87	10.68	32.54
C17H19N3O9	409.36	409.1121	49.88	4.68	10.26	35.18
C17H19N5O1	309.37	309.1590	66.00	6.19	22.64	5.17
C17H19N5O2	325.37	325.1539	62.76	5.89	21.52	9.83
C17H19N5O3	341.37	341.1488	59.81	5.61	20.52	14.06
C17H19N5O4	357.37	357.1437	57.14	5.36	19.60	17.91
C17H19N5O5	373.37	373.1386	54.69	5.13	18.76	21.43
C17H19N5O6	389.37	389.1335	52.44	4.92	17.99	24.65
C17H19N5O7	405.37	405.1284	50.37	4.72	17.28	27.63
C17H19N5O8	421.37	421.1233	48.46	4.55	16.62	30.38
C17H19N5O9	437.37	437.1183	46.69	4.38	16.01	32.92
C17H20N2O1	268.36	268.1576	76.09	7.51	10.44	5.96
C17H20N2O2	284.36	284.1525	71.81	7.09	9.85	11.25
C17H20N2O3	300.36	300.1474	67.98	6.71	9.33	15.98
C17H20N2O4	316.36	316.1423	64.54	6.37	8.85	20.23
C17H20N2O5	332.36	332.1372	61.44	6.07	8.43	24.07
C17H20N2O6	348.36	348.1321	58.61	5.79	8.04	27.56
C17H20N2O7	364.36	364.1270	56.04	5.53	7.69	30.74
C17H20N2O8	380.36	380.1220	53.68	5.30	7.37	33.65
C17H20N2O9	396.36	396.1169	51.52	5.09	7.07	36.33
C17H20N4O1	296.38	296.1637	68.90	6.80	18.90	5.40
C17H20N4O2	312.37	312.1586	65.37	6.45	17.94	10.24
C17H20N4O3	328.37	328.1535	62.18	6.14	17.06	14.62
C17H20N4O4	344.37	344.1484	59.29	5.85	16.27	18.58
C17H20N4O5	360.37	360.1434	56.66	5.59	15.55	22.20
C17H20N4O6	376.37	376.1383	54.25	5.36	14.89	25.51
C17H20N4O7	392.37	392.1332	52.04	5.14	14.28	28.54

091

C17H20N4O8	408.37	408.1281	50.00	4.94	13.72	31.34
C17H20N4O9	424.37	424.1230	48.12	4.75	13.20	33.93
C17H20N6O1	324.39	324.1698	62.95	6.21	25.91	4.93
C17H20N6O2	340.39	340.1648	59.99	5.92	24.69	9.40
C17H20N6O3	356.39	356.1597	57.29	5.66	23.58	13.47
C17H20N6O4	372.39	372.1546	54.83	5.41	22.57	17.19
C17H20N6O5	388.39	388.1495	52.57	5.19	21.64	20.60
C17H20N6O6	404.39	404.1444	50.49	4.99	20.78	23.74
C17H20N6O7	420.38	420.1393	48.57	4.80	19.99	26.64
C17H20N6O8	436.38	436.1342	46.79	4.62	19.26	29.33
C17H20N6O9	452.38	452.1292	45.14	4.46	18.58	31.83
C17H21N1O1	255.36	255.1623	79.96	8.29	5.49	6.27
C17H21N1O2	271.36	271.1572	75.25	7.80	5.16	11.79
C17H21N1O3	287.36	287.1521	71.06	7.37	4.87	16.70
C17H21N1O4	303.36	303.1470	67.31	6.98	4.62	21.10
C17H21N1O5	319.36	319.1420	63.94	6.63	4.39	25.05
C17H21N1O6	335.36	335.1369	60.89	6.31	4.18	28.62
C17H21N1O7	351.36	351.1318	58.11	6.02	3.99	31.87
C17H21N1O8	367.36	367.1267	55.58	5.76	3.81	34.84
C17H21N1O9	383.36	383.1216	53.26	5.52	3.65	37.56
C17H21N3O1	283.38	283.1685	72.06	7.47	14.83	5.65
C17H21N3O2	299.38	299.1634	68.21	7.07	14.04	10.69
C17H21N3O3	315.38	315.1583	64.74	6.71	13.32	15.22
C17H21N3O4	331.37	331.1532	61.62	6.39	12.68	19.31
C17H21N3O5	347.37	347.1481	58.78	6.09	12.10	23.03
C17H21N3O6	363.37	363.1430	56.19	5.83	11.56	26.42
C17H21N3O7	379.37	379.1379	53.82	5.58	11.08	29.52
C17H21N3O8	395.37	395.1329	51.64	5.35	10.63	32.37
C17H21N3O9	411.37	411.1278	49.64	5.15	10.21	35.00
C17H21N5O1	311.39	311.1746	65.57	6.80	22.49	5.14
C17H21N5O2	327.39	327.1695	62.37	6.47	21.39	9.77
C17H21N5O3	343.39	343.1644	59.46	6.16	20.39	13.98
C17H21N5O4	359.39	359.1593	56.82	5.89	19.49	17.81
C17H21N5O5	375.39	375.1543	54.39	5.64	18.66	21.31
C17H21N5O6	391.39	391.1492	52.17	5.41	17.89	24.53
C17H21N5O7	407.39	407.1441	50.12	5.20	17.19	27.49
C17H21N5O8	423.39	423.1390	48.23	5.00	16.54	30.23
C17H21N5O9	439.39	439.1339	46.47	4.82	15.94	32.77
C17H22N2O1	270.38	270.1732	75.52	8.20	10.36	5.92
C17H22N2O2	286.38	286.1681	71.30	7.74	9.78	11.17
C17H22N2O3	302.38	302.1630	67.53	7.33	9.26	15.87
C17H22N2O4	318.38	318.1579	64.13	6.97	8.80	20.10
C17H22N2O5	334.38	334.1529	61.07	6.63	8.38	23.92
C17H22N2O6	350.37	350.1478	58.28	6.33	8.00	27.40
C17H22N2O7	366.37	366.1427	55.73	6.05	7.65	30.57
C17H22N2O8	382.37	382.1376	53.40	5.80	7.33	33.47
C17H22N2O9	398.37	398.1325	51.26	5.57	7.03	36.15
C17H22N4O1	298.39	298.1794	68.43	7.43	18.78	5.36
C17H22N4O2	314.39	314.1743	64.95	7.05	17.82	10.18
C17H22N4O3	330.39	330.1692	61.80	6.71	16.96	14.53
C17H22N4O4	346.39	346.1641	58.95	6.40	16.17	18.48
C17H22N4O5	362.39	362.1590	56.35	6.12	15.46	22.07
C17H22N4O6	378.39	378.1539	53.96	5.86	14.81	25.37
C17H22N4O7	394.39	394.1488	51.77	5.62	14.21	28.40
C17H22N4O8	410.39	410.1437	49.76	5.40	13.65	31.19
C17H22N4O9	426.39	426.1387	47.89	5.20	13.14	33.77
C17H22N6O1	326.40	326.1855	62.56	6.79	25.75	4.90
C17H22N6O2	342.40	342.1804	59.63	6.48	24.54	9.35
C17H22N6O3	358.40	358.1753	56.97	6.19	23.45	13.39
C17H22N6O4	374.40	374.1702	54.54	5.92	22.45	17.09

C17H22N6O5	390.40	390.1652	52.30	5.68	21.53	20.49
C17H22N6O6	406.40	406.1601	50.24	5.46	20.68	23.62
C17H22N6O7	422.40	422.1550	48.34	5.25	19.90	26.51
C17H22N6O8	438.40	438.1499	46.58	5.06	19.17	29.20
C17H22N6O9	454.40	454.1448	44.94	4.88	18.49	31.69
C17H23N1O1	257.38	257.1780	79.33	9.01	5.44	6.22
C17H23N1O2	273.38	273.1729	74.69	8.48	5.12	11.70
C17H23N1O3	289.38	289.1678	70.56	8.01	4.84	16.59
C17H23N1O4	305.38	305.1627	66.86	7.59	4.59	20.96
C17H23N1O5	321.38	321.1576	63.54	7.21	4.36	24.89
C17H23N1O6	337.38	337.1525	60.52	6.87	4.15	28.45
C17H23N1O7	353.38	353.1474	57.78	6.56	3.96	31.69
C17H23N1O8	369.37	369.1424	55.28	6.28	3.79	34.65
C17H23N1O9	385.37	385.1373	52.98	6.02	3.63	37.36
C17H23N3O1	285.39	285.1841	71.55	8.12	14.72	5.61
C17H23N3O2	301.39	301.1790	67.75	7.69	13.94	10.62
C17H23N3O3	317.39	317.1739	64.33	7.30	13.24	15.12
C17H23N3O4	333.39	333.1688	61.25	6.95	12.60	19.20
C17H23N3O5	349.39	349.1638	58.44	6.64	12.03	22.90
C17H23N3O6	365.39	365.1587	55.88	6.34	11.50	26.27
C17H23N3O7	381.39	381.1536	53.54	6.08	11.02	29.37
C17H23N3O8	397.39	397.1485	51.38	5.83	10.57	32.21
C17H23N3O9	413.39	413.1434	49.39	5.61	10.16	34.83
C17H23N5O1	313.41	313.1902	65.15	7.40	22.35	5.11
C17H23N5O2	329.41	329.1852	61.99	7.04	21.26	9.71
C17H23N5O3	345.40	345.1801	59.12	6.71	20.28	13.90
C17H23N5O4	361.40	361.1750	56.50	6.41	19.38	17.71
C17H23N5O5	377.40	377.1699	54.10	6.14	18.56	21.20
C17H23N5O6	393.40	393.1648	51.90	5.89	17.80	24.40
C17H23N5O7	409.40	409.1597	49.88	5.66	17.11	27.36
C17H23N5O8	425.40	425.1546	48.00	5.45	16.46	30.09
C17H23N5O9	441.40	441.1496	46.26	5.25	15.87	32.62
C17H24N2O1	272.39	272.1889	74.96	8.88	10.28	5.87
C17H24N2O2	288.39	288.1838	70.80	8.39	9.71	11.10
C17H24N2O3	304.39	304.1787	67.08	7.95	9.20	15.77
C17H24N2O4	320.39	320.1736	63.73	7.55	8.74	19.97
C17H24N2O5	336.39	336.1685	60.70	7.19	8.33	23.78
C17H24N2O6	352.39	352.1634	57.94	6.86	7.95	27.24
C17H24N2O7	368.39	368.1583	55.43	6.57	7.60	30.40
C17H24N2O8	384.39	384.1533	53.12	6.29	7.29	33.30
C17H24N2O9	400.39	400.1482	51.00	6.04	7.00	35.96
C17H24N4O1	300.41	300.1950	67.97	8.05	18.65	5.33
C17H24N4O2	316.41	316.1899	64.53	7.65	17.71	10.11
C17H24N4O3	332.41	332.1848	61.43	7.28	16.85	14.44
C17H24N4O4	348.41	348.1797	58.61	6.94	16.08	18.37
C17H24N4O5	364.40	364.1747	56.03	6.64	15.37	21.95
C17H24N4O6	380.40	380.1696	53.68	6.36	14.73	25.24
C17H24N4O7	396.40	396.1645	51.51	6.10	14.13	28.25
C17H24N4O8	412.40	412.1594	49.51	5.87	13.59	31.04
C17H24N4O9	428.40	428.1543	47.66	5.65	13.08	33.61
C17H24N6O1	328.42	328.2011	62.17	7.37	25.59	4.87
C17H24N6O2	344.42	344.1961	59.29	7.02	24.40	9.29
C17H24N6O3	360.42	360.1910	56.65	6.71	23.32	13.32
C17H24N6O4	376.42	376.1859	54.25	6.43	22.33	17.00
C17H24N6O5	392.42	392.1808	52.03	6.16	21.42	20.39
C17H24N6O6	408.42	408.1757	50.00	5.92	20.58	23.50
C17H24N6O7	424.42	424.1706	48.11	5.70	19.80	26.39
C17H24N6O8	440.42	440.1655	46.36	5.49	19.08	29.06
C17H24N6O9	456.42	456.1605	44.74	5.30	18.41	31.55
C17H25N1O1	259.39	259.1936	78.72	9.71	5.40	6.17

C17H25N102	275.39	275.1885	74.14	9.15	5.09	11.62
C17H25N103	291.39	291.1834	70.07	8.65	4.81	16.47
C17H25N104	307.39	307.1783	66.43	8.20	4.56	20.82
C17H25N105	323.39	323.1733	63.14	7.79	4.33	24.74
C17H25N106	339.39	339.1682	60.16	7.42	4.13	28.28
C17H25N107	355.39	355.1631	57.45	7.09	3.94	31.51
C17H25N108	371.39	371.1580	54.98	6.79	3.77	34.46
C17H25N109	387.39	387.1529	52.71	6.50	3.62	37.17
C17H25N301	287.41	287.1998	71.05	8.77	14.62	5.57
C17H25N302	303.41	303.1947	67.30	8.31	13.85	10.55
C17H25N303	319.41	319.1896	63.93	7.89	13.16	15.03
C17H25N304	335.41	335.1845	60.88	7.51	12.53	19.08
C17H25N305	351.41	351.1794	58.11	7.17	11.96	22.76
C17H25N306	367.41	367.1743	55.58	6.86	11.44	26.13
C17H25N307	383.40	383.1692	53.26	6.57	10.96	29.21
C17H25N308	399.40	399.1641	51.12	6.31	10.52	32.05
C17H25N309	415.40	415.1591	49.15	6.07	10.12	34.66
C17H25N501	315.42	315.2059	64.74	7.99	22.20	5.07
C17H25N502	331.42	331.2008	61.61	7.60	21.13	9.66
C17H25N503	347.42	347.1957	58.77	7.25	20.16	13.82
C17H25N504	363.42	363.1906	56.19	6.93	19.27	17.61
C17H25N505	379.42	379.1856	53.82	6.64	18.46	21.08
C17H25N506	395.42	395.1805	51.64	6.37	17.71	24.28
C17H25N507	411.42	411.1754	49.63	6.12	17.02	27.22
C17H25N508	427.42	427.1703	47.77	5.90	16.39	29.95
C17H25N509	443.42	443.1652	46.05	5.68	15.79	32.47
C17H26N201	274.41	274.2045	74.41	9.55	10.21	5.83
C17H26N202	290.41	290.1994	70.31	9.02	9.65	11.02
C17H26N203	306.41	306.1943	66.64	8.55	9.14	15.66
C17H26N204	322.41	322.1892	63.33	8.13	8.69	19.85
C17H26N205	338.41	338.1842	60.34	7.74	8.28	23.64
C17H26N206	354.41	354.1791	57.61	7.39	7.90	27.09
C17H26N207	370.41	370.1740	55.13	7.08	7.56	30.24
C17H26N208	386.41	386.1689	52.84	6.78	7.25	33.12
C17H26N209	402.40	402.1638	50.74	6.51	6.96	35.78
C17H26N401	302.42	302.2106	67.52	8.67	18.53	5.29
C17H26N402	318.42	318.2056	64.13	8.23	17.60	10.05
C17H26N403	334.42	334.2005	61.06	7.84	16.75	14.35
C17H26N404	350.42	350.1954	58.27	7.48	15.99	18.26
C17H26N405	366.42	366.1903	55.73	7.15	15.29	21.83
C17H26N406	382.42	382.1852	53.39	6.85	14.65	25.10
C17H26N407	398.42	398.1801	51.25	6.58	14.06	28.11
C17H26N408	414.42	414.1750	49.27	6.32	13.52	30.89
C17H26N409	430.42	430.1700	47.44	6.09	13.02	33.45
C17H26N601	330.44	330.2168	61.79	7.93	25.43	4.84
C17H26N602	346.44	346.2117	58.94	7.56	24.26	9.24
C17H26N603	362.44	362.2066	56.34	7.23	23.19	13.24
C17H26N604	378.43	378.2015	53.96	6.93	22.21	16.91
C17H26N605	394.43	394.1965	51.77	6.64	21.31	20.28
C17H26N606	410.43	410.1914	49.75	6.39	20.48	23.39
C17H26N607	426.43	426.1863	47.88	6.15	19.71	26.26
C17H26N608	442.43	442.1812	46.15	5.92	19.00	28.93
C17H26N609	458.43	458.1761	44.54	5.72	18.33	31.41
C17H27N101	261.41	261.2093	78.11	10.41	5.36	6.12
C17H27N102	277.41	277.2042	73.61	9.81	5.05	11.53
C17H27N103	293.41	293.1991	69.59	9.28	4.77	16.36
C17H27N104	309.41	309.1940	65.99	8.80	4.53	20.68
C17H27N105	325.41	325.1889	62.75	8.36	4.30	24.58
C17H27N106	341.41	341.1838	59.81	7.97	4.10	28.12
C17H27N107	357.41	357.1787	57.13	7.61	3.92	31.34
C17H27N108	373.41	373.1737	54.68	7.29	3.75	34.28
C17H27N109	389.41	389.1686	52.44	6.99	3.60	36.98
C17H27N301	289.42	289.2154	70.55	9.40	14.52	5.53
C17H27N302	305.42	305.2103	66.85	8.91	13.76	10.48
C17H27N303	321.42	321.2052	63.53	8.47	13.07	14.93
C17H27N304	337.42	337.2001	60.51	8.07	12.45	18.97
C17H27N305	353.42	353.1951	57.78	7.70	11.89	22.63
C17H27N306	369.42	369.1900	55.27	7.37	11.37	25.99
C17H27N307	385.42	385.1849	52.98	7.06	10.90	29.06
C17H27N308	401.42	401.1798	50.87	6.78	10.47	31.89
C17H27N309	417.42	417.1747	48.92	6.52	10.07	34.50
C17H27N501	317.44	317.2215	64.32	8.57	22.06	5.04
C17H27N502	333.44	333.2165	61.24	8.16	21.00	9.60
C17H27N503	349.44	349.2114	58.43	7.79	20.04	13.74
C17H27N504	365.44	365.2063	55.88	7.45	19.16	17.51
C17H27N505	381.44	381.2012	53.53	7.13	18.36	20.97
C17H27N506	397.43	397.1961	51.38	6.85	17.62	24.15
C17H27N507	413.43	413.1910	49.39	6.58	16.94	27.09
C17H27N508	429.43	429.1859	47.55	6.34	16.31	29.81
C17H27N509	445.43	445.1809	45.84	6.11	15.72	32.33
C17H28N201	276.43	276.2202	73.87	10.21	10.13	5.79
C17H28N202	292.42	292.2151	69.83	9.65	9.58	10.94
C17H28N203	308.42	308.2100	66.20	9.15	9.08	15.56
C17H28N204	324.42	324.2049	62.94	8.70	8.63	19.73
C17H28N205	340.42	340.1998	59.98	8.29	8.23	23.50
C17H28N206	356.42	356.1947	57.29	7.92	7.86	26.93
C17H28N207	372.42	372.1896	54.83	7.58	7.52	30.07
C17H28N208	388.42	388.1845	52.57	7.27	7.21	32.95
C17H28N209	404.42	404.1795	50.49	6.98	6.93	35.61
C17H28N401	304.44	304.2263	67.07	9.27	18.40	5.26
C17H28N402	320.44	320.2212	63.72	8.81	17.48	9.99
C17H28N403	336.44	336.2161	60.69	8.39	16.65	14.27
C17H28N404	352.44	352.2110	57.94	8.01	15.90	18.16
C17H28N405	368.44	368.2060	55.42	7.66	15.21	21.71
C17H28N406	384.44	384.2009	53.11	7.34	14.57	24.97
C17H28N407	400.44	400.1958	50.99	7.05	13.99	27.97
C17H28N408	416.43	416.1907	49.03	6.78	13.45	30.74
C17H28N409	432.43	432.1856	47.22	6.53	12.96	33.30
C17H28N601	332.45	332.2324	61.42	8.49	25.28	4.81
C17H28N602	348.45	348.2274	58.60	8.10	24.12	9.18
C17H28N603	364.45	364.2223	56.03	7.74	23.06	13.17
C17H28N604	380.45	380.2172	53.67	7.42	22.09	16.82
C17H28N605	396.45	396.2121	51.50	7.12	21.20	20.18
C17H28N606	412.45	412.2070	49.51	6.84	20.38	23.27
C17H28N607	428.45	428.2019	47.66	6.59	19.61	26.14
C17H28N608	444.45	444.1968	45.94	6.35	18.91	28.80
C17H28N609	460.45	460.1918	44.35	6.13	18.25	31.27
C17H29N101	263.43	263.2249	77.51	11.10	5.32	6.07
C17H29N102	279.43	279.2198	73.07	10.46	5.01	11.45
C17H29N103	295.43	295.2147	69.12	9.89	4.74	16.25
C17H29N104	311.42	311.2096	65.57	9.39	4.50	20.55
C17H29N105	327.42	327.2046	62.36	8.93	4.28	24.43
C17H29N106	343.42	343.1995	59.46	8.51	4.08	27.95
C17H29N107	359.42	359.1944	56.81	8.13	3.90	31.16
C17H29N108	375.42	375.1893	54.39	7.79	3.73	34.09
C17H29N109	391.42	391.1842	52.17	7.47	3.58	36.79
C17H29N301	291.44	291.2310	70.06	10.03	14.42	5.49
C17H29N302	307.44	307.2260	66.42	9.51	13.67	10.41
C17H29N303	323.44	323.2209	63.13	9.04	12.99	14.84
C17H29N304	339.44	339.2158	60.16	8.61	12.38	18.85

093

C17H29N305	355.44	355.2107	57.45	8.22	11.82	22.51
C17H29N306	371.44	371.2056	54.97	7.87	11.31	25.84
C17H29N307	387.44	387.2005	52.70	7.54	10.85	28.91
C17H29N308	403.44	403.1954	50.61	7.25	10.42	31.73
C17H29N309	419.44	419.1904	48.68	6.97	10.02	34.33
C17H29N501	319.45	319.2372	63.92	9.15	21.92	5.01
C17H29N502	335.45	335.2321	60.87	8.71	20.88	9.54
C17H29N503	351.45	351.2270	58.10	8.32	19.93	13.66
C17H29N504	367.45	367.2219	55.57	7.96	19.06	17.42
C17H29N505	383.45	383.2169	53.25	7.62	18.26	20.86
C17H29N506	399.45	399.2118	51.12	7.32	17.53	24.03
C17H29N507	415.45	415.2067	49.15	7.04	16.86	26.96
C17H29N508	431.45	431.2016	47.33	6.78	16.23	29.67
C17H29N509	447.45	447.1965	45.63	6.53	15.65	32.18
C17H30N201	278.44	278.2358	73.33	10.86	10.06	5.75
C17H30N202	294.44	294.2307	69.35	10.27	9.51	10.87
C17H30N203	310.44	310.2256	65.77	9.74	9.02	15.46
C17H30N204	326.44	326.2205	62.55	9.26	8.58	19.60
C17H30N205	342.44	342.2155	59.63	8.83	8.18	23.36
C17H30N206	358.44	358.2104	56.97	8.44	7.82	26.78
C17H30N207	374.44	374.2053	54.53	8.08	7.48	29.91
C17H30N208	390.44	390.2002	52.30	7.74	7.17	32.78
C17H30N209	406.44	406.1951	50.24	7.44	6.89	35.43
C17H30N401	306.45	306.2419	66.63	9.87	18.28	5.22
C17H30N402	322.45	322.2369	63.32	9.38	17.38	9.92
C17H30N403	338.45	338.2318	60.33	8.93	16.55	14.18
C17H30N404	354.45	354.2267	57.61	8.53	15.81	18.06
C17H30N405	370.45	370.2216	55.12	8.16	15.12	21.59
C17H30N406	386.45	386.2165	52.84	7.82	14.50	24.84
C17H30N407	402.45	402.2114	50.74	7.51	13.92	27.83
C17H30N408	418.45	418.2063	48.80	7.23	13.39	30.59
C17H30N409	434.45	434.2013	47.00	6.96	12.90	33.14
C17H30N601	334.47	334.2481	61.05	9.04	25.13	4.78
C17H30N602	350.47	350.2430	58.26	8.63	23.98	9.13
C17H30N603	366.47	366.2379	55.72	8.25	22.93	13.10
C17H30N604	382.47	382.2328	53.39	7.91	21.97	16.73
C17H30N605	398.47	398.2278	51.24	7.59	21.09	20.08
C17H30N606	414.47	414.2227	49.27	7.30	20.28	23.16
C17H30N607	430.46	430.2176	47.43	7.02	19.52	26.02
C17H30N608	446.46	446.2125	45.73	6.77	18.82	28.67
C17H30N609	462.46	462.2074	44.15	6.54	18.17	31.14
C17H31N101	265.44	265.2406	76.92	11.77	5.28	6.03
C17H31N102	281.44	281.2355	72.55	11.10	4.98	11.37
C17H31N103	297.44	297.2304	68.65	10.51	4.71	16.14
C17H31N104	313.44	313.2253	65.14	9.97	4.47	20.42
C17H31N105	329.44	329.2202	61.98	9.48	4.25	24.28
C17H31N106	345.44	345.2151	59.11	9.05	4.05	27.79
C17H31N107	361.44	361.2100	56.49	8.65	3.88	30.99
C17H31N108	377.44	377.2049	54.10	8.28	3.71	33.91
C17H31N109	393.44	393.1999	51.90	7.94	3.56	36.60
C17H31N301	293.46	293.2467	69.58	10.65	14.32	5.45
C17H31N302	309.46	309.2416	65.98	10.10	13.58	10.34
C17H31N303	325.45	325.2365	62.74	9.60	12.91	14.75
C17H31N304	341.45	341.2314	59.80	9.15	12.31	18.74
C17H31N305	357.45	357.2264	57.12	8.74	11.76	22.38
C17H31N306	373.45	373.2213	54.68	8.37	11.25	25.71
C17H31N307	389.45	389.2162	52.43	8.02	10.79	28.76
C17H31N308	405.45	405.2111	50.36	7.71	10.36	31.57
C17H31N309	421.45	421.2060	48.45	7.41	9.97	34.17
C17H31N501	321.47	321.2528	63.52	9.72	21.79	4.98
C17H31N502	337.47	337.2478	60.51	9.26	20.75	9.48
C17H31N503	353.47	353.2427	57.77	8.84	19.81	13.58
C17H31N504	369.47	369.2376	55.27	8.46	18.96	17.32
C17H31N505	385.47	385.2325	52.97	8.11	18.17	20.75
C17H31N506	401.47	401.2274	50.86	7.78	17.44	23.91
C17H31N507	417.47	417.2223	48.91	7.48	16.78	26.83
C17H31N508	433.47	433.2172	47.11	7.21	16.16	29.53
C17H31N509	449.46	449.2122	45.43	6.95	15.58	32.04
C17H32N201	280.46	280.2514	72.81	11.50	9.99	5.70
C17H32N202	296.46	296.2464	68.88	10.88	9.45	10.79
C17H32N203	312.46	312.2413	65.35	10.32	8.97	15.36
C17H32N204	328.46	328.2362	62.17	9.82	8.53	19.48
C17H32N205	344.45	344.2311	59.28	9.36	8.13	23.22
C17H32N206	360.45	360.2260	56.65	8.95	7.77	26.63
C17H32N207	376.45	376.2209	54.24	8.57	7.44	29.75
C17H32N208	392.45	392.2158	52.03	8.22	7.14	32.61
C17H32N209	408.45	408.2108	49.99	7.90	6.86	35.25
C17H32N401	308.47	308.2576	66.19	10.46	18.16	5.19
C17H32N402	324.47	324.2525	62.93	9.94	17.27	9.86
C17H32N403	340.47	340.2474	59.97	9.47	16.46	14.10
C17H32N404	356.47	356.2423	57.28	9.05	15.72	17.95
C17H32N405	372.47	372.2373	54.82	8.66	15.04	21.48
C17H32N406	388.47	388.2322	52.56	8.30	14.42	24.71
C17H32N407	404.47	404.2271	50.48	7.97	13.85	27.69
C17H32N408	420.47	420.2220	48.56	7.67	13.32	30.44
C17H32N409	436.47	436.2169	46.78	7.39	12.84	32.99
C17H32N601	336.48	336.2637	60.68	9.59	24.98	4.75
C17H32N602	352.48	352.2587	57.93	9.15	23.84	9.08
C17H32N603	368.48	368.2536	55.41	8.75	22.81	13.03
C17H32N604	384.48	384.2485	53.11	8.39	21.86	16.65
C17H32N605	400.48	400.2434	50.99	8.05	20.98	19.98
C17H32N606	416.48	416.2383	49.03	7.74	20.18	23.05
C17H32N607	432.48	432.2332	47.21	7.46	19.43	25.90
C17H32N608	448.48	448.2281	45.53	7.19	18.74	28.54
C17H32N609	464.48	464.2231	43.96	6.94	18.09	31.00
C17H33N101	267.46	267.2562	76.34	12.44	5.24	5.98
C17H33N102	283.46	283.2511	72.04	11.73	4.94	11.29
C17H33N103	299.46	299.2460	68.19	11.11	4.68	16.03
C17H33N104	315.46	315.2409	64.73	10.54	4.44	20.29
C17H33N105	331.46	331.2359	61.60	10.04	4.23	24.14
C17H33N106	347.46	347.2308	58.77	9.57	4.03	27.63
C17H33N107	363.46	363.2257	56.18	9.15	3.85	30.81
C17H33N108	379.45	379.2206	53.81	8.77	3.69	33.73
C17H33N109	395.45	395.2155	51.63	8.41	3.54	36.41
C17H33N301	295.47	295.2623	69.11	11.26	14.22	5.41
C17H33N302	311.47	311.2573	65.56	10.68	13.49	10.27
C17H33N303	327.47	327.2522	62.35	10.16	12.83	14.66
C17H33N304	343.47	343.2471	59.45	9.68	12.23	18.63
C17H33N305	359.47	359.2420	56.80	9.25	11.69	22.25
C17H33N306	375.47	375.2369	54.38	8.86	11.19	25.57
C17H33N307	391.47	391.2318	52.16	8.50	10.73	28.61
C17H33N308	407.47	407.2267	50.11	8.16	10.31	31.41
C17H33N309	423.47	423.2217	48.22	7.85	9.92	34.00
C17H33N501	323.49	323.2685	63.12	10.28	21.65	4.95
C17H33N502	339.48	339.2634	60.15	9.80	20.63	9.43
C17H33N503	355.48	355.2583	57.44	9.36	19.70	13.50
C17H33N504	371.48	371.2532	54.97	8.95	18.85	17.23
C17H33N505	387.48	387.2482	52.70	8.58	18.07	20.65
C17H33N506	403.48	403.2431	50.61	8.24	17.36	23.79
C17H33N507	419.48	419.2380	48.68	7.93	16.70	26.70

094

C17H33N5O8	435.48	435.2329	46.89	7.64	16.08	29.39
C17H33N5O9	451.48	451.2278	45.23	7.37	15.51	31.89
C17H34N2O1	282.47	282.2671	72.29	12.13	9.92	5.66
C17H34N2O2	298.47	298.2620	68.41	11.48	9.39	10.72
C17H34N2O3	314.47	314.2569	64.93	10.90	8.91	15.26
C17H34N2O4	330.47	330.2518	61.79	10.37	8.48	19.37
C17H34N2O5	346.47	346.2468	58.93	9.89	8.09	23.09
C17H34N2O6	362.47	362.2417	56.33	9.45	7.73	26.48
C17H34N2O7	378.47	378.2366	53.95	9.06	7.40	29.59
C17H34N2O8	394.47	394.2315	51.76	8.69	7.10	32.45
C17H34N2O9	410.47	410.2264	49.75	8.35	6.82	35.08
C17H34N4O1	310.49	310.2732	65.76	11.04	18.04	5.15
C17H34N4O2	326.49	326.2682	62.54	10.50	17.16	9.80
C17H34N4O3	342.49	342.2631	59.62	10.01	16.36	14.01
C17H34N4O4	358.48	358.2580	56.96	9.56	15.63	17.85
C17H34N4O5	374.48	374.2529	54.53	9.15	14.96	21.36
C17H34N4O6	390.48	390.2478	52.29	8.78	14.35	24.58
C17H34N4O7	406.48	406.2427	50.23	8.43	13.78	27.55
C17H34N4O8	422.48	422.2376	48.33	8.11	13.26	30.30
C17H34N4O9	438.48	438.2326	46.57	7.82	12.78	32.84
C17H34N6O1	338.50	338.2794	60.32	10.12	24.83	4.73
C17H34N6O2	354.50	354.2743	57.60	9.67	23.71	9.03
C17H34N6O3	370.50	370.2692	55.11	9.25	22.68	12.96
C17H34N6O4	386.50	386.2641	52.83	8.87	21.74	16.56
C17H34N6O5	402.50	402.2590	50.73	8.51	20.88	19.88
C17H34N6O6	418.50	418.2540	48.79	8.19	20.08	22.94
C17H34N6O7	434.50	434.2489	46.99	7.89	19.34	25.78
C17H34N6O8	450.50	450.2438	45.33	7.61	18.66	28.41
C17H34N6O9	466.50	466.2387	43.77	7.35	18.02	30.87
C17H35N1O1	269.47	269.2718	75.77	13.09	5.20	5.94
C17H35N1O2	285.47	285.2668	71.53	12.36	4.91	11.21
C17H35N1O3	301.47	301.2617	67.73	11.70	4.65	15.92
C17H35N1O4	317.47	317.2566	64.32	11.11	4.41	20.16
C17H35N1O5	333.47	333.2515	61.23	10.58	4.20	23.99
C17H35N1O6	349.47	349.2464	58.43	10.09	4.01	27.47
C17H35N1O7	365.47	365.2413	55.87	9.65	3.83	30.64
C17H35N1O8	381.47	381.2362	53.53	9.25	3.67	33.55
C17H35N1O9	397.47	397.2312	51.37	8.88	3.52	36.23
C17H35N3O1	297.49	297.2780	68.64	11.86	14.12	5.38
C17H35N3O2	313.49	313.2729	65.13	11.25	13.40	10.21
C17H35N3O3	329.49	329.2678	61.97	10.71	12.75	14.57
C17H35N3O4	345.49	345.2627	59.10	10.21	12.16	18.52
C17H35N3O5	361.49	361.2577	56.49	9.76	11.62	22.13
C17H35N3O6	377.48	377.2526	54.09	9.35	11.13	25.43
C17H35N3O7	393.48	393.2475	51.89	8.97	10.68	28.46
C17H35N3O8	409.48	409.2424	49.87	8.62	10.26	31.26
C17H35N3O9	425.48	425.2373	47.99	8.29	9.88	33.84
C17H35N5O1	325.50	325.2841	62.73	10.84	21.52	4.92
C17H35N5O2	341.50	341.2791	59.79	10.33	20.51	9.37
C17H35N5O3	357.50	357.2740	57.12	9.87	19.59	13.43
C17H35N5O4	373.50	373.2689	54.67	9.45	18.75	17.13
C17H35N5O5	389.50	389.2638	52.42	9.06	17.98	20.54
C17H35N5O6	405.50	405.2587	50.36	8.70	17.27	23.67
C17H35N5O7	421.50	421.2536	48.44	8.37	16.62	26.57
C17H35N5O8	437.50	437.2485	46.67	8.06	16.01	29.26
C17H35N5O9	453.50	453.2435	45.03	7.78	15.44	31.75
C17H36N2O1	284.49	284.2827	71.77	12.76	9.85	5.62
C17H36N2O2	300.49	300.2777	67.95	12.08	9.32	10.65
C17H36N2O3	316.49	316.2726	64.52	11.47	8.85	15.17
C17H36N2O4	332.49	332.2675	61.41	10.91	8.43	19.25

C17H36N2O5	348.49	348.2624	58.59	10.41	8.04	22.96
C17H36N2O6	364.49	364.2573	56.02	9.96	7.69	26.34
C17H36N2O7	380.49	380.2522	53.67	9.54	7.36	29.43
C17H36N2O8	396.49	396.2471	51.50	9.15	7.07	32.28
C17H36N2O9	412.48	412.2421	49.50	8.80	6.79	34.91
C17H36N4O1	312.50	312.2889	65.34	11.61	17.93	5.12
C17H36N4O2	328.50	328.2838	62.16	11.05	17.06	9.74
C17H36N4O3	344.50	344.2787	59.27	10.53	16.26	13.93
C17H36N4O4	360.50	360.2736	56.64	10.07	15.54	17.75
C17H36N4O5	376.50	376.2686	54.23	9.64	14.88	21.25
C17H36N4O6	392.50	392.2635	52.02	9.25	14.27	24.46
C17H36N4O7	408.50	408.2584	49.99	8.88	13.72	27.42
C17H36N4O8	424.50	424.2533	48.10	8.55	13.20	30.15
C17H36N4O9	440.50	440.2482	46.35	8.24	12.72	32.69
C17H36N6O1	340.52	340.2950	59.96	10.66	24.68	4.70
C17H36N6O2	356.52	356.2900	57.27	10.18	23.57	8.98
C17H36N6O3	372.51	372.2849	54.81	9.74	22.56	12.88
C17H36N6O4	388.51	388.2798	52.56	9.34	21.63	16.47
C17H36N6O5	404.51	404.2747	50.48	8.97	20.78	19.78
C17H36N6O6	420.51	420.2696	48.56	8.63	19.99	22.83
C17H36N6O7	436.51	436.2645	46.78	8.31	19.25	25.66
C17H36N6O8	452.51	452.2594	45.12	8.02	18.57	28.29
C17H36N6O9	468.51	468.2544	43.58	7.75	17.94	30.73
C17H37N1O1	271.49	271.2875	75.21	13.74	5.16	5.89
C17H37N1O2	287.49	287.2824	71.02	12.97	4.87	11.13
C17H37N1O3	303.49	303.2773	67.28	12.29	4.62	15.82
C17H37N1O4	319.49	319.2722	63.91	11.67	4.38	20.03
C17H37N1O5	335.49	335.2672	60.86	11.12	4.18	23.84
C17H37N1O6	351.49	351.2621	58.09	10.61	3.98	27.31
C17H37N1O7	367.49	367.2570	55.56	10.15	3.81	30.48
C17H37N1O8	383.49	383.2519	53.25	9.73	3.65	33.38
C17H37N1O9	399.49	399.2468	51.11	9.34	3.51	36.04
C17H37N3O1	299.50	299.2936	68.18	12.45	14.03	5.34
C17H37N3O2	315.50	315.2886	64.72	11.82	13.32	10.14
C17H37N3O3	331.50	331.2835	61.60	11.25	12.68	14.48
C17H37N3O4	347.50	347.2784	58.76	10.73	12.09	18.42
C17H37N3O5	363.50	363.2733	56.17	10.26	11.56	22.01
C17H37N3O6	379.50	379.2682	53.80	9.83	11.07	25.30
C17H37N3O7	395.50	395.2631	51.63	9.43	10.62	28.32
C17H37N3O8	411.50	411.2580	49.62	9.06	10.21	31.10
C17H37N3O9	427.50	427.2530	47.76	8.72	9.83	33.68
C17H37N5O1	327.52	327.2998	62.34	11.39	21.38	4.89
C17H37N5O2	343.52	343.2947	59.44	10.86	20.39	9.32
C17H37N5O3	359.52	359.2896	56.80	10.37	19.48	13.35
C17H37N5O4	375.52	375.2845	54.38	9.93	18.65	17.04
C17H37N5O5	391.51	391.2794	52.15	9.53	17.89	20.43
C17H37N5O6	407.51	407.2744	50.11	9.15	17.19	23.56
C17H37N5O7	423.51	423.2693	48.21	8.81	16.54	26.44
C17H37N5O8	439.51	439.2642	46.46	8.49	15.93	29.12
C17H37N5O9	455.51	455.2591	44.83	8.19	15.37	31.61
C17H38N2O1	286.51	286.2984	71.27	13.37	9.78	5.58
C17H38N2O2	302.50	302.2933	67.50	12.66	9.26	10.58
C17H38N2O3	318.50	318.2882	64.11	12.03	8.80	15.07
C17H38N2O4	334.50	334.2831	61.04	11.45	8.37	19.13
C17H38N2O5	350.50	350.2781	58.26	10.93	7.99	22.82
C17H38N2O6	366.50	366.2730	55.71	10.45	7.64	26.19
C17H38N2O7	382.50	382.2679	53.38	10.01	7.32	29.28
C17H38N2O8	398.50	398.2628	51.24	9.61	7.03	32.12
C17H38N2O9	414.50	414.2577	49.26	9.24	6.76	34.74
C17H38N4O1	314.52	314.3045	64.92	12.18	17.81	5.09

095

C17H38N4O2	330.52	330.2995	61.78	11.59	16.95	9.68
C17H38N4O3	346.52	346.2944	58.93	11.05	16.17	13.85
C17H38N4O4	362.52	362.2893	56.33	10.57	15.45	17.65
C17H38N4O5	378.52	378.2842	53.94	10.12	14.80	21.13
C17H38N4O6	394.52	394.2791	51.76	9.71	14.20	24.33
C17H38N4O7	410.52	410.2740	49.74	9.33	13.65	27.28
C17H38N4O8	426.51	426.2689	47.87	8.98	13.14	30.01
C17H38N4O9	442.51	442.2639	46.14	8.66	12.66	32.54
C17H38N6O1	342.53	342.3107	59.61	11.18	24.53	4.67
C17H38N6O2	358.53	358.3056	56.95	10.68	23.44	8.92
C17H38N6O3	374.53	374.3005	54.52	10.23	22.44	12.82
C17H38N6O4	390.53	390.2954	52.29	9.81	21.52	16.39
C17H38N6O5	406.53	406.2903	50.23	9.42	20.67	19.68
C17H38N6O6	422.53	422.2853	48.33	9.07	19.89	22.72
C17H38N6O7	438.53	438.2802	46.56	8.73	19.16	25.54
C17H38N6O8	454.53	454.2751	44.92	8.43	18.49	28.16
C17H38N6O9	470.53	470.2700	43.40	8.14	17.86	30.60
C17H39N3O1	301.52	301.3093	67.72	13.04	13.94	5.31
C17H39N3O2	317.52	317.3042	64.31	12.38	13.23	10.08
C17H39N3O3	333.52	333.2991	61.22	11.79	12.60	14.39
C17H39N3O4	349.52	349.2940	58.42	11.25	12.02	18.31
C17H39N3O5	365.52	365.2890	55.86	10.75	11.50	21.89
C17H39N3O6	381.52	381.2839	53.52	10.30	11.01	25.16
C17H39N3O7	397.52	397.2788	51.37	9.89	10.57	28.17
C17H39N3O8	413.52	413.2737	49.38	9.51	10.16	30.95
C17H39N3O9	429.52	429.2686	47.54	9.15	9.78	33.52
C17H39N5O1	329.53	329.3154	61.96	11.93	21.25	4.86
C17H39N5O2	345.53	345.3104	59.09	11.38	20.27	9.26
C17H39N5O3	361.53	361.3053	56.48	10.87	19.37	13.28
C17H39N5O4	377.53	377.3002	54.09	10.41	18.55	16.95
C17H39N5O5	393.53	393.2951	51.89	9.99	17.80	20.33
C17H39N5O6	409.53	409.2900	49.86	9.60	17.10	23.44
C17H39N5O7	425.53	425.2849	47.98	9.24	16.46	26.32
C17H39N5O8	441.53	441.2798	46.25	8.90	15.86	28.99
C17H39N5O9	457.53	457.2748	44.63	8.59	15.31	31.47
C17H40N4O1	316.53	316.3202	64.51	12.74	17.70	5.05
C17H40N4O2	332.53	332.3151	61.40	12.12	16.85	9.62
C17H40N4O3	348.53	348.3100	58.59	11.57	16.08	13.77
C17H40N4O4	364.53	364.3049	56.01	11.06	15.37	17.56
C17H40N4O5	380.53	380.2998	53.66	10.60	14.72	21.02
C17H40N4O6	396.53	396.2948	51.49	10.17	14.13	24.21
C17H40N4O7	412.53	412.2897	49.50	9.77	13.58	27.15
C17H40N4O8	428.53	428.2846	47.65	9.41	13.07	29.87
C17H40N4O9	444.53	444.2795	45.93	9.07	12.60	32.39
C17H40N6O1	344.55	344.3263	59.26	11.70	24.39	4.64
C17H40N6O2	360.55	360.3213	56.63	11.18	23.31	8.88
C17H40N6O3	376.55	376.3162	54.23	10.71	22.32	12.75
C17H40N6O4	392.55	392.3111	52.02	10.27	21.41	16.30
C17H40N6O5	408.55	408.3060	49.98	9.87	20.57	19.58
C17H40N6O6	424.54	424.3009	48.10	9.50	19.80	22.61
C17H40N6O7	440.54	440.2958	46.35	9.15	19.08	25.42
C17H40N6O8	456.54	456.2907	44.73	8.83	18.41	28.04
C17H40N6O9	472.54	472.2857	43.21	8.53	17.78	30.47
C17H41N5O1	331.55	331.3311	61.59	12.46	21.12	4.83
C17H41N5O2	347.55	347.3260	58.75	11.89	20.15	9.21
C17H41N5O3	363.55	363.3209	56.17	11.37	19.26	13.20
C17H41N5O4	379.55	379.3158	53.80	10.89	18.45	16.86
C17H41N5O5	395.55	395.3107	51.62	10.45	17.71	20.22
C17H41N5O6	411.55	411.3057	49.62	10.04	17.02	23.33
C17H41N5O7	427.55	427.3006	47.76	9.67	16.38	26.20

C17H41N5O8	443.55	443.2955	46.04	9.32	15.79	28.86
C17H41N5O9	459.54	459.2904	44.43	8.99	15.24	31.33
C17H42N6O1	346.56	346.3420	58.92	12.22	24.25	4.62
C17H42N6O2	362.56	362.3369	56.32	11.68	23.18	8.83
C17H42N6O3	378.56	378.3318	53.94	11.18	22.20	12.68
C17H42N6O4	394.56	394.3267	51.75	10.73	21.30	16.22
C17H42N6O5	410.56	410.3216	49.73	10.31	20.47	19.48
C17H42N6O6	426.56	426.3166	47.87	9.92	19.70	22.50
C17H42N6O7	442.56	442.3115	46.14	9.57	18.99	25.31
C17H42N6O8	458.56	458.3064	44.53	9.23	18.33	27.91
C17H42N6O9	474.56	474.3013	43.03	8.92	17.71	30.34
C18H 9N1O1	255.28	255.0684	84.69	3.55	5.49	6.27
C18H 9N1O2	271.28	271.0633	79.70	3.34	5.16	11.80
C18H 9N1O3	287.28	287.0582	75.26	3.16	4.88	16.71
C18H 9N1O4	303.28	303.0532	71.29	2.99	4.62	21.10
C18H 9N1O5	319.28	319.0481	67.72	2.84	4.39	25.06
C18H 9N1O6	335.28	335.0430	64.48	2.71	4.18	28.63
C18H 9N1O7	351.27	351.0379	61.55	2.58	3.99	31.88
C18H 9N1O8	367.27	367.0328	58.87	2.47	3.81	34.85
C18H 9N1O9	383.27	383.0277	56.41	2.37	3.65	37.57
C18H 9N3O1	283.29	283.0746	76.32	3.20	14.83	5.65
C18H 9N3O2	299.29	299.0695	72.24	3.03	14.04	10.69
C18H 9N3O3	315.29	315.0644	68.57	2.88	13.33	15.22
C18H 9N3O4	331.29	331.0593	65.26	2.74	12.68	19.32
C18H 9N3O5	347.29	347.0542	62.25	2.61	12.10	23.03
C18H 9N3O6	363.29	363.0491	59.51	2.50	11.57	26.42
C18H 9N3O7	379.29	379.0440	57.00	2.39	11.08	29.53
C18H 9N3O8	395.29	395.0390	54.69	2.29	10.63	32.38
C18H 9N3O9	411.29	411.0339	52.57	2.21	10.22	35.01
C18H 9N5O1	311.31	311.0807	69.45	2.91	22.50	5.14
C18H 9N5O2	327.30	327.0756	66.05	2.77	21.40	9.78
C18H 9N5O3	343.30	343.0705	62.98	2.64	20.40	13.98
C18H 9N5O4	359.30	359.0654	60.17	2.52	19.49	17.81
C18H 9N5O5	375.30	375.0604	57.61	2.42	18.66	21.32
C18H 9N5O6	391.30	391.0553	55.25	2.32	17.90	24.53
C18H 9N5O7	407.30	407.0502	53.08	2.23	17.19	27.50
C18H 9N5O8	423.30	423.0451	51.07	2.14	16.54	30.24
C18H 9N5O9	439.30	439.0400	49.21	2.07	15.94	32.78
C18H10N2O1	270.29	270.0793	79.99	3.73	10.36	5.92
C18H10N2O2	286.29	286.0742	75.52	3.52	9.78	11.18
C18H10N2O3	302.29	302.0691	71.52	3.33	9.27	15.88
C18H10N2O4	318.29	318.0641	67.93	3.17	8.80	20.11
C18H10N2O5	334.29	334.0590	64.67	3.02	8.38	23.93
C18H10N2O6	350.29	350.0539	61.72	2.88	8.00	27.40
C18H10N2O7	366.29	366.0488	59.02	2.75	7.65	30.58
C18H10N2O8	382.29	382.0437	56.55	2.64	7.33	33.48
C18H10N2O9	398.29	398.0386	54.28	2.53	7.03	36.15
C18H10N4O1	298.31	298.0855	72.48	3.38	18.78	5.36
C18H10N4O2	314.31	314.0804	68.79	3.21	17.83	10.18
C18H10N4O3	330.31	330.0753	65.45	3.05	16.96	14.53
C18H10N4O4	346.30	346.0702	62.43	2.91	16.18	18.48
C18H10N4O5	362.30	362.0651	59.67	2.78	15.46	22.08
C18H10N4O6	378.30	378.0600	57.15	2.66	14.81	25.38
C18H10N4O7	394.30	394.0549	54.83	2.56	14.21	28.40
C18H10N4O8	410.30	410.0499	52.69	2.46	13.66	31.20
C18H10N4O9	426.30	426.0448	50.72	2.36	13.14	33.78
C18H10N6O1	326.32	326.0916	66.25	3.09	25.75	4.90
C18H10N6O2	342.32	342.0865	63.16	2.94	24.55	9.35
C18H10N6O3	358.32	358.0814	60.34	2.81	23.45	13.40
C18H10N6O4	374.32	374.0763	57.76	2.69	22.45	17.10

C18H10N605	390.32	390.0713	55.39	2.58	21.53	20.50
C18H10N606	406.32	406.0662	53.21	2.48	20.68	23.63
C18H10N607	422.32	422.0611	51.19	2.39	19.90	26.52
C18H10N608	438.32	438.0560	49.33	2.30	19.17	29.20
C18H10N609	454.32	454.0509	47.59	2.22	18.50	31.69
C18H11N101	257.29	257.0841	84.03	4.31	5.44	6.22
C18H11N102	273.29	273.0790	79.11	4.06	5.13	11.71
C18H11N103	289.29	289.0739	74.73	3.83	4.84	16.59
C18H11N104	305.29	305.0688	70.82	3.63	4.59	20.96
C18H11N105	321.29	321.0637	67.29	3.45	4.36	24.90
C18H11N106	337.29	337.0586	64.10	3.29	4.15	28.46
C18H11N107	353.29	353.0535	61.20	3.14	3.96	31.70
C18H11N108	369.29	369.0485	58.54	3.00	3.79	34.66
C18H11N109	385.29	385.0434	56.11	2.88	3.64	37.37
C18H11N301	285.31	285.0902	75.78	3.89	14.73	5.61
C18H11N302	301.31	301.0851	71.75	3.68	13.95	10.62
C18H11N303	317.31	317.0800	68.14	3.49	13.24	15.13
C18H11N304	333.31	333.0749	64.87	3.33	12.61	19.20
C18H11N305	349.31	349.0699	61.89	3.17	12.03	22.90
C18H11N306	365.30	365.0648	59.18	3.04	11.50	26.28
C18H11N307	381.30	381.0597	56.70	2.91	11.02	29.37
C18H11N308	397.30	397.0546	54.42	2.79	10.58	32.22
C18H11N309	413.30	413.0495	52.31	2.68	10.17	34.84
C18H11N501	313.32	313.0964	69.00	3.54	22.35	5.11
C18H11N502	329.32	329.0913	65.65	3.37	21.27	9.72
C18H11N503	345.32	345.0862	62.61	3.21	20.28	13.90
C18H11N504	361.32	361.0811	59.84	3.07	19.38	17.71
C18H11N505	377.32	377.0760	57.30	2.94	18.56	21.20
C18H11N506	393.32	393.0709	54.97	2.82	17.81	24.41
C18H11N507	409.32	409.0658	52.82	2.71	17.11	27.36
C18H11N508	425.32	425.0608	50.83	2.61	16.47	30.09
C18H11N509	441.32	441.0557	48.99	2.51	15.87	32.63
C18H12N201	272.31	272.0950	79.40	4.44	10.29	5.88
C18H12N202	288.31	288.0899	74.99	4.20	9.72	11.10
C18H12N203	304.31	304.0848	71.05	3.97	9.21	15.77
C18H12N204	320.31	320.0797	67.50	3.78	8.75	19.98
C18H12N205	336.31	336.0746	64.29	3.60	8.33	23.79
C18H12N206	352.31	352.0695	61.37	3.43	7.95	27.25
C18H12N207	368.31	368.0644	58.70	3.28	7.61	30.41
C18H12N208	384.30	384.0594	56.26	3.15	7.29	33.31
C18H12N209	400.30	400.0543	54.01	3.02	7.00	35.97
C18H12N401	300.32	300.1011	71.99	4.03	18.66	5.33
C18H12N402	316.32	316.0960	68.35	3.82	17.71	10.12
C18H12N403	332.32	332.0909	65.06	3.64	16.86	14.44
C18H12N404	348.32	348.0858	62.07	3.47	16.08	18.37
C18H12N405	364.32	364.0808	59.34	3.32	15.38	21.96
C18H12N406	380.32	380.0757	56.85	3.18	14.73	25.24
C18H12N407	396.32	396.0706	54.55	3.05	14.14	28.26
C18H12N408	412.32	412.0655	52.44	2.93	13.59	31.04
C18H12N409	428.32	428.0604	50.48	2.82	13.08	33.62
C18H12N601	328.34	328.1073	65.85	3.68	25.60	4.87
C18H12N602	344.34	344.1022	62.79	3.51	24.41	9.29
C18H12N603	360.33	360.0971	60.00	3.36	23.32	13.32
C18H12N604	376.33	376.0920	57.45	3.21	22.33	17.01
C18H12N605	392.33	392.0869	55.11	3.08	21.42	20.39
C18H12N606	408.33	408.0818	52.95	2.96	20.58	23.51
C18H12N607	424.33	424.0767	50.95	2.85	19.81	26.39
C18H12N608	440.33	440.0717	49.10	2.75	19.09	29.07
C18H12N609	456.33	456.0666	47.38	2.65	18.42	31.55
C18H13N101	259.31	259.0997	83.38	5.05	5.40	6.17

C18H13N102	275.31	275.0946	78.53	4.76	5.09	11.62
C18H13N103	291.31	291.0895	74.22	4.50	4.81	16.48
C18H13N104	307.31	307.0844	70.35	4.26	4.56	20.83
C18H13N105	323.31	323.0794	66.87	4.05	4.33	24.74
C18H13N106	339.31	339.0743	63.72	3.86	4.13	28.29
C18H13N107	355.31	355.0692	60.85	3.69	3.94	31.52
C18H13N108	371.31	371.0641	58.23	3.53	3.77	34.47
C18H13N109	387.31	387.0590	55.82	3.38	3.62	37.18
C18H13N301	287.32	287.1059	75.25	4.56	14.62	5.57
C18H13N302	303.32	303.1008	71.28	4.32	13.85	10.55
C18H13N303	319.32	319.0957	67.71	4.10	13.16	15.03
C18H13N304	335.32	335.0906	64.48	3.91	12.53	19.09
C18H13N305	351.32	351.0855	61.54	3.73	11.96	22.77
C18H13N306	367.32	367.0804	58.86	3.57	11.44	26.13
C18H13N307	383.32	383.0753	56.40	3.42	10.96	29.22
C18H13N308	399.32	399.0703	54.14	3.28	10.52	32.05
C18H13N309	415.32	415.0652	52.06	3.16	10.12	34.67
C18H13N501	315.34	315.1120	68.56	4.16	22.21	5.07
C18H13N502	331.34	331.1069	65.25	3.95	21.14	9.66
C18H13N503	347.34	347.1018	62.25	3.77	20.16	13.82
C18H13N504	363.34	363.0967	59.50	3.61	19.28	17.61
C18H13N505	379.33	379.0917	56.99	3.45	18.46	21.09
C18H13N506	395.33	395.0866	54.69	3.31	17.72	24.28
C18H13N507	411.33	411.0815	52.56	3.19	17.03	27.23
C18H13N508	427.33	427.0764	50.59	3.07	16.39	29.95
C18H13N509	443.33	443.0713	48.77	2.96	15.80	32.48
C18H14N201	274.33	274.1106	78.81	5.14	10.21	5.83
C18H14N202	290.32	290.1055	74.47	4.86	9.65	11.02
C18H14N203	306.32	306.1004	70.58	4.61	9.15	15.67
C18H14N204	322.32	322.0953	67.08	4.38	8.69	19.86
C18H14N205	338.32	338.0903	63.90	4.17	8.28	23.65
C18H14N206	354.32	354.0852	61.02	3.98	7.91	27.09
C18H14N207	370.32	370.0801	58.38	3.81	7.56	30.24
C18H14N208	386.32	386.0750	55.96	3.65	7.25	33.13
C18H14N209	402.32	402.0699	53.74	3.51	6.96	35.79
C18H14N401	302.34	302.1168	71.51	4.67	18.53	5.29
C18H14N402	318.34	318.1117	67.92	4.43	17.60	10.05
C18H14N403	334.34	334.1066	64.67	4.22	16.76	14.36
C18H14N404	350.34	350.1015	61.71	4.03	15.99	18.27
C18H14N405	366.34	366.0964	59.02	3.85	15.29	21.84
C18H14N406	382.34	382.0913	56.55	3.69	14.65	25.11
C18H14N407	398.33	398.0862	54.28	3.54	14.07	28.12
C18H14N408	414.33	414.0812	52.18	3.41	13.52	30.89
C18H14N409	430.33	430.0761	50.24	3.28	13.02	33.46
C18H14N601	330.35	330.1229	65.45	4.27	25.44	4.84
C18H14N602	346.35	346.1178	62.42	4.07	24.26	9.24
C18H14N603	362.35	362.1127	59.67	3.89	23.19	13.25
C18H14N604	378.35	378.1076	57.14	3.73	22.21	16.91
C18H14N605	394.35	394.1026	54.82	3.58	21.31	20.29
C18H14N606	410.35	410.0975	52.69	3.44	20.48	23.39
C18H14N607	426.35	426.0924	50.71	3.31	19.71	26.27
C18H14N608	442.35	442.0873	48.88	3.19	19.00	28.94
C18H14N609	458.35	458.0822	47.17	3.08	18.34	31.42
C18H15N101	261.33	261.1154	82.73	5.79	5.36	6.12
C18H15N102	277.33	277.1103	77.96	5.45	5.05	11.54
C18H15N103	293.33	293.1052	73.71	5.15	4.78	16.36
C18H15N104	309.32	309.1001	69.89	4.89	4.53	20.69
C18H15N105	325.32	325.0950	66.46	4.65	4.31	24.59
C18H15N106	341.32	341.0899	63.34	4.43	4.10	28.12
C18H15N107	357.32	357.0848	60.51	4.23	3.92	31.34

097

C18H15N108	373.32	373.0798	57.91	4.05	3.75	34.29
C18H15N109	389.32	389.0747	55.53	3.88	3.60	36.99
C18H15N301	289.34	289.1215	74.72	5.23	14.52	5.53
C18H15N302	305.34	305.1164	70.81	4.95	13.76	10.48
C18H15N303	321.34	321.1113	67.28	4.71	13.08	14.94
C18H15N304	337.34	337.1062	64.09	4.48	12.46	18.97
C18H15N305	353.34	353.1012	61.19	4.28	11.89	22.64
C18H15N306	369.34	369.0961	58.54	4.09	11.38	25.99
C18H15N307	385.34	385.0910	56.11	3.92	10.90	29.06
C18H15N308	401.34	401.0859	53.87	3.77	10.47	31.89
C18H15N309	417.33	417.0808	51.81	3.62	10.07	34.50
C18H15N501	317.35	317.1277	68.13	4.76	22.07	5.04
C18H15N502	333.35	333.1226	64.86	4.54	21.01	9.60
C18H15N503	349.35	349.1175	61.89	4.33	20.05	13.74
C18H15N504	365.35	365.1124	59.18	4.14	19.17	17.52
C18H15N505	381.35	381.1073	56.69	3.96	18.36	20.98
C18H15N506	397.35	397.1022	54.41	3.81	17.63	24.16
C18H15N507	413.35	413.0971	52.30	3.66	16.94	27.09
C18H15N508	429.35	429.0921	50.36	3.52	16.31	29.81
C18H15N509	445.35	445.0870	48.55	3.39	15.73	32.33
C18H16N201	276.34	276.1263	78.24	5.84	10.14	5.79
C18H16N202	292.34	292.1212	73.96	5.52	9.58	10.95
C18H16N203	308.34	308.1161	70.12	5.23	9.09	15.57
C18H16N204	324.34	324.1110	66.66	4.97	8.64	19.73
C18H16N205	340.34	340.1059	63.53	4.74	8.23	23.51
C18H16N206	356.34	356.1008	60.67	4.53	7.86	26.94
C18H16N207	372.34	372.0957	58.07	4.33	7.52	30.08
C18H16N208	388.34	388.0907	55.67	4.15	7.21	32.96
C18H16N209	404.34	404.0856	53.47	3.99	6.93	35.61
C18H16N401	304.35	304.1324	71.04	5.30	18.41	5.26
C18H16N402	320.35	320.1273	67.49	5.03	17.49	9.99
C18H16N403	336.35	336.1222	64.28	4.79	16.66	14.27
C18H16N404	352.35	352.1171	61.36	4.58	15.90	18.16
C18H16N405	368.35	368.1121	58.69	4.38	15.21	21.72
C18H16N406	384.35	384.1070	56.25	4.20	14.58	24.98
C18H16N407	400.35	400.1019	54.00	4.03	13.99	27.97
C18H16N408	416.35	416.0968	51.93	3.87	13.46	30.74
C18H16N409	432.35	432.0917	50.01	3.73	12.96	33.31
C18H16N601	332.37	332.1386	65.05	4.85	25.29	4.81
C18H16N602	348.37	348.1335	62.06	4.63	24.12	9.19
C18H16N603	364.37	364.1284	59.34	4.43	23.06	13.17
C18H16N604	380.37	380.1233	56.84	4.24	22.09	16.83
C18H16N605	396.37	396.1182	54.55	4.07	21.20	20.18
C18H16N606	412.36	412.1131	52.43	3.91	20.38	23.28
C18H16N607	428.36	428.1080	50.47	3.76	19.62	26.14
C18H16N608	444.36	444.1029	48.65	3.63	18.91	28.80
C18H16N609	460.36	460.0979	46.96	3.50	18.26	31.28
C18H17N101	263.34	263.1310	82.10	6.51	5.32	6.08
C18H17N102	279.34	279.1259	77.40	6.13	5.01	11.46
C18H17N103	295.34	295.1208	73.20	5.80	4.74	16.25
C18H17N104	311.34	311.1157	69.44	5.50	4.50	20.56
C18H17N105	327.34	327.1107	66.05	5.23	4.28	24.44
C18H17N106	343.34	343.1056	62.97	4.99	4.08	27.96
C18H17N107	359.34	359.1005	60.17	4.77	3.90	31.17
C18H17N108	375.34	375.0954	57.60	4.57	3.73	34.10
C18H17N109	391.34	391.0903	55.25	4.38	3.58	36.80
C18H17N301	291.36	291.1372	74.21	5.88	14.42	5.49
C18H17N302	307.36	307.1321	70.34	5.58	13.67	10.41
C18H17N303	323.35	323.1270	66.86	5.30	13.00	14.84
C18H17N304	339.35	339.1219	63.71	5.05	12.38	18.86

C18H17N305	355.35	355.1168	60.84	4.82	11.82	22.51
C18H17N306	371.35	371.1117	58.22	4.61	11.32	25.85
C18H17N307	387.35	387.1066	55.82	4.42	10.85	28.91
C18H17N308	403.35	403.1016	53.60	4.25	10.42	31.73
C18H17N309	419.35	419.0965	51.56	4.09	10.02	34.34
C18H17N501	319.37	319.1433	67.70	5.37	21.93	5.01
C18H17N502	335.37	335.1382	64.47	5.11	20.88	9.54
C18H17N503	351.37	351.1331	61.53	4.88	19.93	13.66
C18H17N504	367.37	367.1280	58.85	4.66	19.06	17.42
C18H17N505	383.37	383.1230	56.40	4.47	18.27	20.87
C18H17N506	399.37	399.1179	54.14	4.29	17.54	24.04
C18H17N507	415.37	415.1128	52.05	4.13	16.86	26.96
C18H17N508	431.36	431.1077	50.12	3.97	16.24	29.67
C18H17N509	447.36	447.1026	48.33	3.83	15.65	32.19
C18H18N201	278.36	278.1419	77.67	6.52	10.06	5.75
C18H18N202	294.36	294.1368	73.45	6.16	9.52	10.87
C18H18N203	310.36	310.1317	69.66	5.85	9.03	15.47
C18H18N204	326.36	326.1266	66.25	5.56	8.58	19.61
C18H18N205	342.35	342.1216	63.15	5.30	8.18	23.37
C18H18N206	358.35	358.1165	60.33	5.06	7.82	26.79
C18H18N207	374.35	374.1114	57.75	4.85	7.48	29.92
C18H18N208	390.35	390.1063	55.39	4.65	7.18	32.79
C18H18N209	406.35	406.1012	53.21	4.46	6.89	35.44
C18H18N401	306.37	306.1481	70.57	5.92	18.29	5.22
C18H18N402	322.37	322.1430	67.07	5.63	17.38	9.93
C18H18N403	338.37	338.1379	63.89	5.36	16.56	14.19
C18H18N404	354.37	354.1328	61.01	5.12	15.81	18.06
C18H18N405	370.37	370.1277	58.37	4.90	15.13	21.60
C18H18N406	386.37	386.1226	55.96	4.70	14.50	24.85
C18H18N407	402.37	402.1175	53.73	4.51	13.92	27.83
C18H18N408	418.37	418.1125	51.68	4.34	13.39	30.59
C18H18N409	434.37	434.1074	49.77	4.18	12.90	33.15
C18H18N601	334.38	334.1542	64.66	5.43	25.13	4.78
C18H18N602	350.38	350.1491	61.70	5.18	23.99	9.13
C18H18N603	366.38	366.1440	59.01	4.95	22.94	13.10
C18H18N604	382.38	382.1389	56.54	4.74	21.98	16.74
C18H18N605	398.38	398.1339	54.27	4.55	21.10	20.08
C18H18N606	414.38	414.1288	52.17	4.38	20.28	23.17
C18H18N607	430.38	430.1237	50.23	4.22	19.53	26.02
C18H18N608	446.38	446.1186	48.43	4.06	18.83	28.67
C18H18N609	462.38	462.1135	46.76	3.92	18.18	31.14
C18H19N101	265.36	265.1467	81.48	7.22	5.28	6.03
C18H19N102	281.36	281.1416	76.84	6.81	4.98	11.37
C18H19N103	297.36	297.1365	72.71	6.44	4.71	16.14
C18H19N104	313.36	313.1314	69.00	6.11	4.47	20.42
C18H19N105	329.36	329.1263	65.64	5.81	4.25	24.29
C18H19N106	345.36	345.1212	62.60	5.55	4.06	27.80
C18H19N107	361.35	361.1161	59.83	5.30	3.88	30.99
C18H19N108	377.35	377.1111	57.29	5.08	3.71	33.92
C18H19N109	393.35	393.1060	54.96	4.87	3.56	36.61
C18H19N301	293.37	293.1528	73.70	6.53	14.32	5.45
C18H19N302	309.37	309.1477	69.88	6.19	13.58	10.34
C18H19N303	325.37	325.1426	66.45	5.89	12.91	14.75
C18H19N304	341.37	341.1375	63.33	5.61	12.31	18.75
C18H19N305	357.37	357.1325	60.50	5.36	11.76	22.38
C18H19N306	373.37	373.1274	57.91	5.13	11.25	25.71
C18H19N307	389.37	389.1223	55.53	4.92	10.79	28.76
C18H19N308	405.37	405.1172	53.33	4.72	10.37	31.58
C18H19N309	421.37	421.1121	51.31	4.55	9.97	34.17
C18H19N501	321.39	321.1590	67.27	5.96	21.79	4.98

098

C18H19N5O2	337.38	337.1539	64.08	5.68	20.76	9.48	C18H21N5O8	435.40	435.1390	49.66	4.86	16.08	29.40
C18H19N5O3	353.38	353.1488	61.18	5.42	19.82	13.58	C18H21N5O9	451.40	451.1339	47.90	4.69	15.51	31.90
C18H19N5O4	369.38	369.1437	58.53	5.18	18.96	17.33	C18H22N2O1	282.39	282.1732	76.56	7.85	9.92	5.67
C18H19N5O5	385.38	385.1386	56.10	4.97	18.17	20.76	C18H22N2O2	298.39	298.1681	72.46	7.43	9.39	10.72
C18H19N5O6	401.38	401.1335	53.86	4.77	17.45	23.92	C18H22N2O3	314.39	314.1630	68.77	7.05	8.91	15.27
C18H19N5O7	417.38	417.1284	51.80	4.59	16.78	26.83	C18H22N2O4	330.39	330.1579	65.44	6.71	8.48	19.37
C18H19N5O8	433.38	433.1233	49.89	4.42	16.16	29.53	C18H22N2O5	346.39	346.1529	62.42	6.40	8.09	23.09
C18H19N5O9	449.38	449.1183	48.11	4.26	15.58	32.04	C18H22N2O6	362.39	362.1478	59.66	6.12	7.73	26.49
C18H20N2O1	280.37	280.1576	77.11	7.19	9.99	5.71	C18H22N2O7	378.39	378.1427	57.14	5.86	7.40	29.60
C18H20N2O2	296.37	296.1525	72.95	6.80	9.45	10.80	C18H22N2O8	394.38	394.1376	54.82	5.62	7.10	32.45
C18H20N2O3	312.37	312.1474	69.21	6.45	8.97	15.37	C18H22N2O9	410.38	410.1325	52.68	5.40	6.83	35.09
C18H20N2O4	328.37	328.1423	65.84	6.14	8.53	19.49	C18H22N4O1	310.40	310.1794	69.65	7.14	18.05	5.15
C18H20N2O5	344.37	344.1372	62.78	5.85	8.13	23.23	C18H22N4O2	326.40	326.1743	66.24	6.79	17.16	9.80
C18H20N2O6	360.37	360.1321	59.99	5.59	7.77	26.64	C18H22N4O3	342.40	342.1692	63.14	6.48	16.36	14.02
C18H20N2O7	376.37	376.1270	57.44	5.36	7.44	29.76	C18H22N4O4	358.40	358.1641	60.32	6.19	15.63	17.86
C18H20N2O8	392.37	392.1220	55.10	5.14	7.14	32.62	C18H22N4O5	374.40	374.1590	57.75	5.92	14.96	21.37
C18H20N2O9	408.37	408.1169	52.94	4.94	6.86	35.26	C18H22N4O6	390.40	390.1539	55.38	5.68	14.35	24.59
C18H20N4O1	308.39	308.1637	70.11	6.54	18.17	5.19	C18H22N4O7	406.40	406.1488	53.20	5.46	13.79	27.56
C18H20N4O2	324.39	324.1586	66.65	6.21	17.27	9.86	C18H22N4O8	422.40	422.1437	51.18	5.25	13.26	30.30
C18H20N4O3	340.39	340.1535	63.52	5.92	16.46	14.10	C18H22N4O9	438.40	438.1387	49.32	5.06	12.78	32.85
C18H20N4O4	356.38	356.1484	60.67	5.66	15.72	17.96	C18H22N6O1	338.42	338.1855	63.89	6.55	24.83	4.73
C18H20N4O5	372.38	372.1434	58.06	5.41	15.05	21.48	C18H22N6O2	354.42	354.1804	61.00	6.26	23.71	9.03
C18H20N4O6	388.38	388.1383	55.67	5.19	14.43	24.72	C18H22N6O3	370.41	370.1753	58.37	5.99	22.69	12.96
C18H20N4O7	404.38	404.1332	53.46	4.99	13.85	27.70	C18H22N6O4	386.41	386.1702	55.95	5.74	21.75	16.56
C18H20N4O8	420.38	420.1281	51.43	4.80	13.33	30.45	C18H22N6O5	402.41	402.1652	53.73	5.51	20.88	19.88
C18H20N4O9	436.38	436.1230	49.54	4.62	12.84	33.00	C18H22N6O6	418.41	418.1601	51.67	5.30	20.09	22.94
C18H20N6O1	336.40	336.1698	64.27	5.99	24.98	4.76	C18H22N6O7	434.41	434.1550	49.77	5.10	19.35	25.78
C18H20N6O2	352.40	352.1648	61.35	5.72	23.85	9.08	C18H22N6O8	450.41	450.1499	48.00	4.92	18.66	28.42
C18H20N6O3	368.40	368.1597	58.69	5.47	22.81	13.03	C18H22N6O9	466.41	466.1448	46.35	4.75	18.02	30.87
C18H20N6O4	384.40	384.1546	56.24	5.24	21.86	16.65	C18H23N1O1	269.39	269.1780	80.26	8.61	5.20	5.94
C18H20N6O5	400.40	400.1495	54.00	5.03	20.99	19.98	C18H23N1O2	285.39	285.1729	75.76	8.12	4.91	11.21
C18H20N6O6	416.40	416.1444	51.92	4.84	20.18	23.05	C18H23N1O3	301.39	301.1678	71.73	7.69	4.65	15.93
C18H20N6O7	432.40	432.1393	50.00	4.66	19.44	25.90	C18H23N1O4	317.39	317.1627	68.12	7.30	4.41	20.16
C18H20N6O8	448.40	448.1342	48.22	4.50	18.74	28.55	C18H23N1O5	333.39	333.1576	64.85	6.95	4.20	24.00
C18H20N6O9	464.39	464.1292	46.56	4.34	18.10	31.01	C18H23N1O6	349.39	349.1525	61.88	6.64	4.01	27.48
C18H21N1O1	267.37	267.1623	80.86	7.92	5.24	5.98	C18H23N1O7	365.39	365.1474	59.17	6.34	3.83	30.65
C18H21N1O2	283.37	283.1572	76.30	7.47	4.94	11.29	C18H23N1O8	381.39	381.1424	56.69	6.08	3.67	33.56
C18H21N1O3	299.37	299.1521	72.22	7.07	4.68	16.03	C18H23N1O9	397.39	397.1373	54.41	5.83	3.52	36.24
C18H21N1O4	315.37	315.1470	68.55	6.71	4.44	20.29	C18H23N3O1	297.40	297.1841	72.70	7.80	14.13	5.38
C18H21N1O5	331.37	331.1420	65.24	6.39	4.23	24.14	C18H23N3O2	313.40	313.1790	68.98	7.40	13.41	10.21
C18H21N1O6	347.37	347.1369	62.24	6.09	4.03	27.64	C18H23N3O3	329.40	329.1739	65.63	7.04	12.76	14.57
C18H21N1O7	363.37	363.1318	59.50	5.83	3.85	30.82	C18H23N3O4	345.40	345.1688	62.59	6.71	12.17	18.53
C18H21N1O8	379.37	379.1267	56.99	5.58	3.69	33.74	C18H23N3O5	361.40	361.1638	59.82	6.41	11.63	22.14
C18H21N1O9	395.37	395.1216	54.68	5.35	3.54	36.42	C18H23N3O6	377.40	377.1587	57.29	6.14	11.13	25.44
C18H21N3O1	295.39	295.1685	73.19	7.17	14.23	5.42	C18H23N3O7	393.40	393.1536	54.96	5.89	10.68	28.47
C18H21N3O2	311.39	311.1634	69.43	6.80	13.49	10.28	C18H23N3O8	409.40	409.1485	52.81	5.66	10.26	31.26
C18H21N3O3	327.39	327.1583	66.04	6.47	12.84	14.66	C18H23N3O9	425.40	425.1434	50.82	5.45	9.88	33.85
C18H21N3O4	343.39	343.1532	62.96	6.16	12.24	18.64	C18H23N5O1	325.42	325.1902	66.44	7.12	21.52	4.92
C18H21N3O5	359.39	359.1481	60.16	5.89	11.69	22.26	C18H23N5O2	341.42	341.1852	63.32	6.79	20.51	9.37
C18H21N3O6	375.38	375.1430	57.59	5.64	11.19	25.57	C18H23N5O3	357.42	357.1801	60.49	6.49	19.59	13.43
C18H21N3O7	391.38	391.1379	55.24	5.41	10.74	28.62	C18H23N5O4	373.42	373.1750	57.90	6.21	18.75	17.14
C18H21N3O8	407.38	407.1329	53.07	5.20	10.31	31.42	C18H23N5O5	389.41	389.1699	55.52	5.95	17.98	20.54
C18H21N3O9	423.38	423.1278	51.07	5.00	9.92	34.01	C18H23N5O6	405.41	405.1648	53.33	5.72	17.27	23.68
C18H21N5O1	323.40	323.1746	66.85	6.55	21.66	4.95	C18H23N5O7	421.41	421.1597	51.30	5.50	16.62	26.58
C18H21N5O2	339.40	339.1695	63.70	6.24	20.63	9.43	C18H23N5O8	437.41	437.1546	49.43	5.30	16.01	29.26
C18H21N5O3	355.40	355.1644	60.83	5.96	19.71	13.51	C18H23N5O9	453.41	453.1496	47.68	5.11	15.45	31.76
C18H21N5O4	371.40	371.1593	58.21	5.70	18.86	17.23	C18H24N2O1	284.40	284.1889	76.02	8.51	9.85	5.63
C18H21N5O5	387.40	387.1543	55.81	5.46	18.08	20.65	C18H24N2O2	300.40	300.1838	71.97	8.05	9.33	10.65
C18H21N5O6	403.40	403.1492	53.59	5.25	17.36	23.80	C18H24N2O3	316.40	316.1787	68.33	7.65	8.85	15.17
C18H21N5O7	419.40	419.1441	51.55	5.05	16.70	26.70	C18H24N2O4	332.40	332.1736	65.04	7.28	8.43	19.25

099

C18H24N2O5	348.40	348.1685	62.05	6.94	8.04	22.96	C18H26N4O2	330.43	330.2056	65.43	7.93	16.96	9.68
C18H24N2O6	364.40	364.1634	59.33	6.64	7.69	26.34	C18H26N4O3	346.43	346.2005	62.41	7.56	16.17	13.85
C18H24N2O7	380.40	380.1583	56.83	6.36	7.36	29.44	C18H26N4O4	362.43	362.1954	59.65	7.23	15.46	17.66
C18H24N2O8	396.40	396.1533	54.54	6.10	7.07	32.29	C18H26N4O5	378.43	378.1903	57.13	6.93	14.80	21.14
C18H24N2O9	412.40	412.1482	52.43	5.87	6.79	34.92	C18H26N4O6	394.43	394.1852	54.81	6.64	14.20	24.34
C18H24N4O1	312.42	312.1950	69.20	7.74	17.93	5.12	C18H26N4O7	410.43	410.1801	52.68	6.39	13.65	27.29
C18H24N4O2	328.42	328.1899	65.83	7.37	17.06	9.74	C18H26N4O8	426.43	426.1750	50.70	6.15	13.14	30.02
C18H24N4O3	344.42	344.1848	62.77	7.02	16.27	13.94	C18H26N4O9	442.43	442.1700	48.87	5.92	12.66	32.55
C18H24N4O4	360.42	360.1797	59.99	6.71	15.55	17.76	C18H26N6O1	342.45	342.2168	63.13	7.65	24.54	4.67
C18H24N4O5	376.42	376.1747	57.44	6.43	14.88	21.25	C18H26N6O2	358.45	358.2117	60.32	7.31	23.45	8.93
C18H24N4O6	392.42	392.1696	55.09	6.16	14.28	24.46	C18H26N6O3	374.45	374.2066	57.74	7.00	22.44	12.82
C18H24N4O7	408.41	408.1645	52.94	5.92	13.72	27.42	C18H26N6O4	390.45	390.2015	55.37	6.71	21.52	16.39
C18H24N4O8	424.41	424.1594	50.94	5.70	13.20	30.16	C18H26N6O5	406.45	406.1965	53.19	6.45	20.68	19.68
C18H24N4O9	440.41	440.1543	49.09	5.49	12.72	32.70	C18H26N6O6	422.44	422.1914	51.18	6.20	19.89	22.72
C18H24N6O1	340.43	340.2011	63.51	7.11	24.69	4.70	C18H26N6O7	438.44	438.1863	49.31	5.98	19.17	25.54
C18H24N6O2	356.43	356.1961	60.66	6.79	23.58	8.98	C18H26N6O8	454.44	454.1812	47.57	5.77	18.49	28.17
C18H24N6O3	372.43	372.1910	58.05	6.50	22.57	12.89	C18H26N6O9	470.44	470.1761	45.96	5.57	17.86	30.61
C18H24N6O4	388.43	388.1859	55.66	6.23	21.64	16.48	C18H27N1O1	273.42	273.2093	79.07	9.95	5.12	5.85
C18H24N6O5	404.43	404.1808	53.46	5.98	20.78	19.78	C18H27N1O2	289.42	289.2042	74.70	9.40	4.84	11.06
C18H24N6O6	420.43	420.1757	51.42	5.75	19.99	22.83	C18H27N1O3	305.42	305.1991	70.79	8.91	4.59	15.72
C18H24N6O7	436.43	436.1706	49.54	5.54	19.26	25.66	C18H27N1O4	321.42	321.1940	67.26	8.47	4.36	19.91
C18H24N6O8	452.43	452.1655	47.79	5.35	18.58	28.29	C18H27N1O5	337.42	337.1889	64.07	8.07	4.15	23.71
C18H24N6O9	468.43	468.1605	46.15	5.16	17.94	30.74	C18H27N1O6	353.42	353.1838	61.17	7.70	3.96	27.16
C18H25N1O1	271.41	271.1936	79.66	9.28	5.16	5.90	C18H27N1O7	369.42	369.1787	58.52	7.37	3.79	30.32
C18H25N1O2	287.41	287.1885	75.22	8.77	4.87	11.13	C18H27N1O8	385.42	385.1737	56.10	7.06	3.63	33.21
C18H25N1O3	303.40	303.1834	71.26	8.31	4.62	15.82	C18H27N1O9	401.42	401.1686	53.86	6.78	3.49	35.87
C18H25N1O4	319.40	319.1783	67.69	7.89	4.39	20.04	C18H27N3O1	301.44	301.2154	71.72	9.03	13.94	5.31
C18H25N1O5	335.40	335.1733	64.46	7.51	4.18	23.85	C18H27N3O2	317.43	317.2103	68.11	8.57	13.24	10.08
C18H25N1O6	351.40	351.1682	61.52	7.17	3.99	27.32	C18H27N3O3	333.43	333.2052	64.84	8.16	12.60	14.40
C18H25N1O7	367.40	367.1631	58.85	6.86	3.81	30.48	C18H27N3O4	349.43	349.2001	61.87	7.79	12.03	18.31
C18H25N1O8	383.40	383.1580	56.39	6.57	3.65	33.38	C18H27N3O5	365.43	365.1951	59.16	7.45	11.50	21.89
C18H25N1O9	399.40	399.1529	54.13	6.31	3.51	36.05	C18H27N3O6	381.43	381.1900	56.68	7.13	11.02	25.17
C18H25N3O1	299.42	299.1998	72.21	8.42	14.03	5.34	C18H27N3O7	397.43	397.1849	54.40	6.85	10.57	28.18
C18H25N3O2	315.42	315.1947	68.54	7.99	13.32	10.14	C18H27N3O8	413.43	413.1798	52.29	6.58	10.16	30.96
C18H25N3O3	331.42	331.1896	65.24	7.60	12.68	14.48	C18H27N3O9	429.43	429.1747	50.35	6.34	9.79	33.53
C18H25N3O4	347.42	347.1845	62.23	7.25	12.09	18.42	C18H27N5O1	329.45	329.2215	65.62	8.26	21.26	4.86
C18H25N3O5	363.42	363.1794	59.49	6.93	11.56	22.01	C18H27N5O2	345.45	345.2165	62.59	7.88	20.27	9.26
C18H25N3O6	379.42	379.1743	56.98	6.64	11.07	25.30	C18H27N5O3	361.45	361.2114	59.82	7.53	19.38	13.28
C18H25N3O7	395.42	395.1692	54.68	6.37	10.63	28.32	C18H27N5O4	377.45	377.2063	57.28	7.21	18.55	16.96
C18H25N3O8	411.42	411.1641	52.55	6.13	10.21	31.11	C18H27N5O5	393.45	393.2012	54.95	6.92	17.80	20.33
C18H25N3O9	427.41	427.1591	50.58	5.90	9.83	33.69	C18H27N5O6	409.45	409.1961	52.80	6.65	17.10	23.45
C18H25N5O1	327.43	327.2059	66.03	7.70	21.39	4.89	C18H27N5O7	425.45	425.1910	50.82	6.40	16.46	26.32
C18H25N5O2	343.43	343.2008	62.95	7.34	20.39	9.32	C18H27N5O8	441.44	441.1859	48.98	6.17	15.86	28.99
C18H25N5O3	359.43	359.1957	60.15	7.01	19.48	13.35	C18H27N5O9	457.44	457.1809	47.26	5.95	15.31	31.48
C18H25N5O4	375.43	375.1906	57.59	6.71	18.65	17.05	C18H28N2O1	288.44	288.2202	74.96	9.78	9.71	5.55
C18H25N5O5	391.43	391.1856	55.23	6.44	17.89	20.44	C18H28N2O2	304.44	304.2151	71.02	9.27	9.20	10.51
C18H25N5O6	407.43	407.1805	53.06	6.18	17.19	23.56	C18H28N2O3	320.44	320.2100	67.47	8.81	8.74	14.98
C18H25N5O7	423.43	423.1754	51.06	5.95	16.54	26.45	C18H28N2O4	336.43	336.2049	64.26	8.39	8.33	19.02
C18H25N5O8	439.43	439.1703	49.20	5.73	15.94	29.13	C18H28N2O5	352.43	352.1998	61.34	8.01	7.95	22.70
C18H25N5O9	455.43	455.1652	47.47	5.53	15.38	31.62	C18H28N2O6	368.43	368.1947	58.68	7.66	7.60	26.06
C18H26N2O1	286.42	286.2045	75.48	9.15	9.78	5.59	C18H28N2O7	384.43	384.1896	56.24	7.34	7.29	29.13
C18H26N2O2	302.42	302.1994	71.49	8.67	9.26	10.58	C18H28N2O8	400.43	400.1845	53.99	7.05	7.00	31.96
C18H26N2O3	318.42	318.1943	67.90	8.23	8.80	15.07	C18H28N2O9	416.43	416.1795	51.92	6.78	6.73	34.58
C18H26N2O4	334.42	334.1892	64.65	7.84	8.38	19.14	C18H28N4O1	316.45	316.2263	68.32	8.92	17.70	5.06
C18H26N2O5	350.42	350.1842	61.70	7.48	7.99	22.83	C18H28N4O2	332.45	332.2212	65.03	8.49	16.85	9.63
C18H26N2O6	366.42	366.1791	59.00	7.15	7.65	26.20	C18H28N4O3	348.45	348.2161	62.05	8.10	16.08	13.77
C18H26N2O7	382.42	382.1740	56.54	6.85	7.33	29.29	C18H28N4O4	364.45	364.2110	59.32	7.74	15.37	17.56
C18H26N2O8	398.42	398.1689	54.26	6.58	7.03	32.13	C18H28N4O5	380.45	380.2060	56.83	7.42	14.73	21.03
C18H26N2O9	414.42	414.1638	52.17	6.32	6.76	34.75	C18H28N4O6	396.45	396.2009	54.53	7.12	14.13	24.21
C18H26N4O1	314.43	314.2106	68.76	8.33	17.82	5.09	C18H28N4O7	412.45	412.1958	52.42	6.84	13.58	27.15

C18H28N408	428.45	428.1907	50.46	6.59	13.08	29.87
C18H28N409	444.45	444.1856	48.65	6.35	12.61	32.40
C18H28N601	344.46	344.2324	62.76	8.19	24.40	4.64
C18H28N602	360.46	360.2274	59.98	7.83	23.31	8.88
C18H28N603	376.46	376.2223	57.43	7.50	22.32	12.75
C18H28N604	392.46	392.2172	55.09	7.19	21.41	16.31
C18H28N605	408.46	408.2121	52.93	6.91	20.57	19.58
C18H28N606	424.46	424.2070	50.94	6.65	19.80	22.62
C18H28N607	440.46	440.2019	49.09	6.41	19.08	25.43
C18H28N608	456.46	456.1968	47.36	6.18	18.41	28.04
C18H28N609	472.46	472.1918	45.76	5.97	17.79	30.48
C18H29N101	275.44	275.2249	78.49	10.61	5.09	5.81
C18H29N102	291.44	291.2198	74.18	10.03	4.81	10.98
C18H29N103	307.44	307.2147	70.32	9.51	4.56	15.61
C18H29N104	323.44	323.2096	66.84	9.04	4.33	19.79
C18H29N105	339.44	339.2046	63.69	8.61	4.13	23.57
C18H29N106	355.43	355.1995	60.83	8.22	3.94	27.01
C18H29N107	371.43	371.1944	58.21	7.87	3.77	30.15
C18H29N108	387.43	387.1893	55.80	7.54	3.62	33.04
C18H29N109	403.43	403.1842	53.59	7.25	3.47	35.69
C18H29N301	303.45	303.2310	71.25	9.63	13.85	5.27
C18H29N302	319.45	319.2260	67.68	9.15	13.15	10.02
C18H29N303	335.45	335.2209	64.45	8.71	12.53	14.31
C18H29N304	351.45	351.2158	61.52	8.32	11.96	18.21
C18H29N305	367.45	367.2107	58.84	7.96	11.44	21.77
C18H29N306	383.45	383.2056	56.38	7.62	10.96	25.04
C18H29N307	399.45	399.2005	54.12	7.32	10.52	28.04
C18H29N308	415.45	415.1954	52.04	7.04	10.11	30.81
C18H29N309	431.45	431.1904	50.11	6.78	9.74	33.37
C18H29N501	331.46	331.2372	65.23	8.82	21.13	4.83
C18H29N502	347.46	347.2321	62.22	8.41	20.16	9.21
C18H29N503	363.46	363.2270	59.48	8.04	19.27	13.21
C18H29N504	379.46	379.2219	56.98	7.70	18.46	16.87
C18H29N505	395.46	395.2169	54.67	7.39	17.71	20.23
C18H29N506	411.46	411.2118	52.54	7.10	17.02	23.33
C18H29N507	427.46	427.2067	50.58	6.84	16.38	26.20
C18H29N508	443.46	443.2016	48.75	6.59	15.79	28.86
C18H29N509	459.46	459.1965	47.06	6.36	15.24	31.34
C18H30N201	290.45	290.2358	74.44	10.41	9.64	5.51
C18H30N202	306.45	306.2307	70.55	9.87	9.14	10.44
C18H30N203	322.45	322.2256	67.05	9.38	8.69	14.89
C18H30N204	338.45	338.2205	63.88	8.93	8.28	18.91
C18H30N205	354.45	354.2155	61.00	8.53	7.90	22.57
C18H30N206	370.45	370.2104	58.36	8.16	7.56	25.91
C18H30N207	386.45	386.2053	55.95	7.82	7.25	28.98
C18H30N208	402.45	402.2002	53.72	7.51	6.96	31.80
C18H30N209	418.45	418.1951	51.67	7.23	6.69	34.41
C18H30N401	318.47	318.2419	67.89	9.50	17.59	5.02
C18H30N402	334.47	334.2369	64.64	9.04	16.75	9.57
C18H30N403	350.46	350.2318	61.69	8.63	15.99	13.70
C18H30N404	366.46	366.2267	59.00	8.25	15.29	17.46
C18H30N405	382.46	382.2216	56.53	7.91	14.65	20.92
C18H30N406	398.46	398.2165	54.26	7.59	14.06	24.09
C18H30N407	414.46	414.2114	52.16	7.30	13.52	27.02
C18H30N408	430.46	430.2063	50.23	7.02	13.02	29.73
C18H30N409	446.46	446.2013	48.43	6.77	12.55	32.25
C18H30N601	346.48	346.2481	62.40	8.73	24.26	4.62
C18H30N602	362.48	362.2430	59.65	8.34	23.18	8.83
C18H30N603	378.48	378.2379	57.12	7.99	22.20	12.68
C18H30N604	394.48	394.2328	54.81	7.67	21.30	16.22

C18H30N605	410.48	410.2278	52.67	7.37	20.47	19.49
C18H30N606	426.48	426.2227	50.69	7.09	19.71	22.51
C18H30N607	442.48	442.2176	48.86	6.83	18.99	25.31
C18H30N608	458.48	458.2125	47.16	6.60	18.33	27.92
C18H30N609	474.47	474.2074	45.57	6.37	17.71	30.35
C18H31N101	277.45	277.2406	77.92	11.26	5.05	5.77
C18H31N102	293.45	293.2355	73.67	10.65	4.77	10.90
C18H31N103	309.45	309.2304	69.87	10.10	4.53	15.51
C18H31N104	325.45	325.2253	66.43	9.60	4.30	19.66
C18H31N105	341.45	341.2202	63.32	9.15	4.10	23.43
C18H31N106	357.45	357.2151	60.48	8.74	3.92	26.86
C18H31N107	373.45	373.2100	57.89	8.37	3.75	29.99
C18H31N108	389.45	389.2049	55.51	8.02	3.60	32.87
C18H31N109	405.45	405.1999	53.32	7.71	3.45	35.51
C18H31N301	305.47	305.2467	70.78	10.23	13.76	5.24
C18H31N302	321.47	321.2416	67.25	9.72	13.07	9.95
C18H31N303	337.47	337.2365	64.07	9.26	12.45	14.22
C18H31N304	353.47	353.2314	61.17	8.84	11.89	18.11
C18H31N305	369.46	369.2264	58.52	8.46	11.37	21.65
C18H31N306	385.46	385.2213	56.09	8.11	10.90	24.90
C18H31N307	401.46	401.2162	53.85	7.78	10.47	27.90
C18H31N308	417.46	417.2111	51.79	7.48	10.07	30.66
C18H31N309	433.46	433.2060	49.88	7.21	9.69	33.22
C18H31N501	333.48	333.2528	64.83	9.37	21.00	4.80
C18H31N502	349.48	349.2478	61.86	8.94	20.04	9.16
C18H31N503	365.48	365.2427	59.16	8.55	19.16	13.13
C18H31N504	381.48	381.2376	56.67	8.19	18.36	16.78
C18H31N505	397.48	397.2325	54.39	7.86	17.62	20.13
C18H31N506	413.48	413.2274	52.29	7.56	16.94	23.22
C18H31N507	429.48	429.2223	50.34	7.28	16.31	26.08
C18H31N508	445.48	445.2172	48.53	7.01	15.72	28.73
C18H31N509	461.48	461.2122	46.85	6.77	15.18	31.20
C18H32N201	292.47	292.2514	73.92	11.03	9.58	5.47
C18H32N202	308.47	308.2464	70.09	10.46	9.08	10.37
C18H32N203	324.47	324.2413	66.63	9.94	8.63	14.79
C18H32N204	340.47	340.2362	63.50	9.47	8.23	18.80
C18H32N205	356.47	356.2311	60.65	9.05	7.86	22.44
C18H32N206	372.47	372.2260	58.05	8.66	7.52	25.77
C18H32N207	388.46	388.2209	55.66	8.30	7.21	28.83
C18H32N208	404.46	404.2158	53.45	7.97	6.93	31.65
C18H32N209	420.46	420.2108	51.42	7.67	6.66	34.25
C18H32N401	320.48	320.2576	67.46	10.06	17.48	4.99
C18H32N402	336.48	336.2525	64.25	9.59	16.65	9.51
C18H32N403	352.48	352.2474	61.34	9.15	15.89	13.62
C18H32N404	368.48	368.2423	58.67	8.75	15.20	17.37
C18H32N405	384.48	384.2373	56.23	8.39	14.57	20.81
C18H32N406	400.48	400.2322	53.99	8.05	13.99	23.97
C18H32N407	416.48	416.2271	51.91	7.74	13.45	26.89
C18H32N408	432.48	432.2220	49.99	7.46	12.95	29.60
C18H32N409	448.48	448.2169	48.21	7.19	12.49	32.11
C18H32N601	348.50	348.2637	62.04	9.26	24.12	4.59
C18H32N602	364.49	364.2587	59.32	8.85	23.06	8.78
C18H32N603	380.49	380.2536	56.82	8.48	22.09	12.61
C18H32N604	396.49	396.2485	54.53	8.14	21.20	16.14
C18H32N605	412.49	412.2434	52.41	7.82	20.37	19.39
C18H32N606	428.49	428.2383	50.46	7.53	19.61	22.40
C18H32N607	444.49	444.2332	48.64	7.26	18.91	25.20
C18H32N608	460.49	460.2281	46.95	7.00	18.25	27.80
C18H32N609	476.49	476.2231	45.37	6.77	17.64	30.22
C18H33N101	279.47	279.2562	77.36	11.90	5.01	5.72

C18H33N102	295.47	295.2511	73.17	11.26	4.74	10.83
C18H33N103	311.47	311.2460	69.41	10.68	4.50	15.41
C18H33N104	327.47	327.2409	66.02	10.16	4.28	19.54
C18H33N105	343.47	343.2359	62.95	9.68	4.08	23.29
C18H33N106	359.47	359.2308	60.14	9.25	3.90	26.71
C18H33N107	375.47	375.2257	57.58	8.86	3.73	29.83
C18H33N108	391.47	391.2206	55.23	8.50	3.58	32.70
C18H33N109	407.47	407.2155	53.06	8.16	3.44	35.34
C18H33N301	307.48	307.2623	70.31	10.82	13.67	5.20
C18H33N302	323.48	323.2573	66.84	10.28	12.99	9.89
C18H33N303	339.48	339.2522	63.69	9.80	12.38	14.14
C18H33N304	355.48	355.2471	60.82	9.36	11.82	18.00
C18H33N305	371.48	371.2420	58.20	8.95	11.31	21.53
C18H33N306	387.48	387.2369	55.80	8.58	10.84	24.77
C18H33N307	403.48	403.2318	53.58	8.24	10.41	27.76
C18H33N308	419.48	419.2267	51.54	7.93	10.02	30.51
C18H33N309	435.48	435.2217	49.65	7.64	9.65	33.07
C18H33N501	335.50	335.2685	64.44	9.91	20.87	4.77
C18H33N502	351.50	351.2634	61.51	9.46	19.92	9.10
C18H33N503	367.50	367.2583	58.83	9.05	19.06	13.06
C18H33N504	383.49	383.2532	56.38	8.67	18.26	16.69
C18H33N505	399.49	399.2482	54.12	8.33	17.53	20.02
C18H33N506	415.49	415.2431	52.03	8.01	16.86	23.10
C18H33N507	431.49	431.2380	50.11	7.71	16.23	25.96
C18H33N508	447.49	447.2329	48.31	7.43	15.65	28.60
C18H33N509	463.49	463.2278	46.65	7.18	15.11	31.07
C18H34N201	294.48	294.2671	73.42	11.64	9.51	5.43
C18H34N202	310.48	310.2620	69.63	11.04	9.02	10.31
C18H34N203	326.48	326.2569	66.22	10.50	8.58	14.70
C18H34N204	342.48	342.2518	63.13	10.01	8.18	18.69
C18H34N205	358.48	358.2468	60.31	9.56	7.81	22.32
C18H34N206	374.48	374.2417	57.73	9.15	7.48	25.63
C18H34N207	390.48	390.2366	55.37	8.78	7.17	28.68
C18H34N208	406.48	406.2315	53.19	8.43	6.89	31.49
C18H34N209	422.48	422.2264	51.17	8.11	6.63	34.08
C18H34N401	322.50	322.2732	67.04	10.63	17.37	4.96
C18H34N402	338.50	338.2682	63.87	10.12	16.55	9.45
C18H34N403	354.50	354.2631	60.99	9.67	15.80	13.54
C18H34N404	370.50	370.2580	58.35	9.25	15.12	17.27
C18H34N405	386.50	386.2529	55.94	8.87	14.50	20.70
C18H34N406	402.49	402.2478	53.72	8.51	13.92	23.85
C18H34N407	418.49	418.2427	51.66	8.19	13.39	26.76
C18H34N408	434.49	434.2376	49.76	7.89	12.89	29.46
C18H34N409	450.49	450.2326	47.99	7.61	12.44	31.96
C18H34N601	350.51	350.2794	61.68	9.78	23.98	4.56
C18H34N602	366.51	366.2743	58.99	9.35	22.93	8.73
C18H34N603	382.51	382.2692	56.52	8.96	21.97	12.55
C18H34N604	398.51	398.2641	54.25	8.60	21.09	16.06
C18H34N605	414.51	414.2590	52.16	8.27	20.27	19.30
C18H34N606	430.51	430.2540	50.22	7.96	19.52	22.30
C18H34N607	446.51	446.2489	48.42	7.68	18.82	25.08
C18H34N608	462.51	462.2438	46.75	7.41	18.17	27.67
C18H34N609	478.51	478.2387	45.18	7.16	17.56	30.09
C18H35N101	281.49	281.2718	76.81	12.53	4.98	5.68
C18H35N102	297.49	297.2668	72.68	11.86	4.71	10.76
C18H35N103	313.48	313.2617	68.97	11.25	4.47	15.31
C18H35N104	329.48	329.2566	65.62	10.71	4.25	19.42
C18H35N105	345.48	345.2515	62.58	10.21	4.05	23.16
C18H35N106	361.48	361.2464	59.81	9.76	3.87	26.56
C18H35N107	377.48	377.2413	57.27	9.35	3.71	29.67

C18H35N108	393.48	393.2362	54.95	8.97	3.56	32.53
C18H35N109	409.48	409.2312	52.80	8.62	3.42	35.17
C18H35N301	309.50	309.2780	69.86	11.40	13.58	5.17
C18H35N302	325.50	325.2729	66.42	10.84	12.91	9.83
C18H35N303	341.50	341.2678	63.31	10.33	12.30	14.06
C18H35N304	357.50	357.2627	60.48	9.87	11.75	17.90
C18H35N305	373.50	373.2577	57.89	9.45	11.25	21.42
C18H35N306	389.50	389.2526	55.51	9.06	10.79	24.65
C18H35N307	405.50	405.2475	53.32	8.70	10.36	27.62
C18H35N308	421.49	421.2424	51.29	8.37	9.97	30.37
C18H35N309	437.49	437.2373	49.42	8.06	9.60	32.91
C18H35N501	337.51	337.2841	64.06	10.45	20.75	4.74
C18H35N502	353.51	353.2791	61.16	9.98	19.81	9.05
C18H35N503	369.51	369.2740	58.51	9.55	18.95	12.99
C18H35N504	385.51	385.2689	56.08	9.15	18.17	16.60
C18H35N505	401.51	401.2638	53.85	8.79	17.44	19.92
C18H35N506	417.51	417.2587	51.78	8.45	16.77	22.99
C18H35N507	433.51	433.2536	49.87	8.14	16.16	25.83
C18H35N508	449.51	449.2485	48.10	7.85	15.58	28.47
C18H35N509	465.51	465.2435	46.44	7.58	15.04	30.93
C18H36N201	296.50	296.2827	72.92	12.24	9.45	5.40
C18H36N202	312.50	312.2777	69.18	11.61	8.96	10.24
C18H36N203	328.50	328.2726	65.81	11.05	8.53	14.61
C18H36N204	344.50	344.2675	62.76	10.53	8.13	18.58
C18H36N205	360.50	360.2624	59.97	10.07	7.77	22.19
C18H36N206	376.50	376.2573	57.42	9.64	7.44	25.50
C18H36N207	392.50	392.2522	55.08	9.25	7.14	28.53
C18H36N208	408.50	408.2471	52.93	8.88	6.86	31.33
C18H36N209	424.50	424.2421	50.93	8.55	6.60	33.92
C18H36N401	324.51	324.2889	66.62	11.18	17.26	4.93
C18H36N402	340.51	340.2838	63.49	10.66	16.45	9.40
C18H36N403	356.51	356.2787	60.64	10.18	15.72	13.46
C18H36N404	372.51	372.2736	58.04	9.74	15.04	17.18
C18H36N405	388.51	388.2686	55.65	9.34	14.42	20.59
C18H36N406	404.51	404.2635	53.45	8.97	13.85	23.73
C18H36N407	420.51	420.2584	51.41	8.63	13.32	26.63
C18H36N408	436.51	436.2533	49.53	8.31	12.84	29.32
C18H36N409	452.51	452.2482	47.78	8.02	12.38	31.82
C18H36N601	352.53	352.2950	61.33	10.29	23.84	4.54
C18H36N602	368.53	368.2900	58.67	9.85	22.80	8.68
C18H36N603	384.53	384.2849	56.23	9.44	21.86	12.48
C18H36N604	400.53	400.2798	53.98	9.06	20.98	15.98
C18H36N605	416.52	416.2747	51.91	8.71	20.18	19.21
C18H36N606	432.52	432.2696	49.99	8.39	19.43	22.19
C18H36N607	448.52	448.2645	48.20	8.09	18.74	24.97
C18H36N608	464.52	464.2594	46.54	7.81	18.09	27.55
C18H36N609	480.52	480.2544	44.99	7.55	17.49	29.97
C18H37N101	283.50	283.2875	76.26	13.16	4.94	5.64
C18H37N102	299.50	299.2824	72.19	12.45	4.68	10.68
C18H37N103	315.50	315.2773	68.53	11.82	4.44	15.21
C18H37N104	331.50	331.2722	65.22	11.25	4.23	19.31
C18H37N105	347.50	347.2672	62.22	10.73	4.03	23.02
C18H37N106	363.50	363.2621	59.48	10.26	3.85	26.41
C18H37N107	379.50	379.2570	56.97	9.83	3.69	29.51
C18H37N108	395.50	395.2519	54.67	9.43	3.54	32.36
C18H37N109	411.50	411.2468	52.54	9.06	3.40	34.99
C18H37N301	311.52	311.2936	69.40	11.97	13.49	5.14
C18H37N302	327.51	327.2886	66.01	11.39	12.83	9.77
C18H37N303	343.51	343.2835	62.94	10.86	12.23	13.97
C18H37N304	359.51	359.2784	60.14	10.37	11.69	17.80

C18H37N3O5	375.51	375.2733	57.57	9.93	11.19	21.30
C18H37N3O6	391.51	391.2682	55.22	9.53	10.73	24.52
C18H37N3O7	407.51	407.2631	53.05	9.15	10.31	27.48
C18H37N3O8	423.51	423.2580	51.05	8.81	9.92	30.22
C18H37N3O9	439.51	439.2530	49.19	8.49	9.56	32.76
C18H37N5O1	339.53	339.2998	63.68	10.98	20.63	4.71
C18H37N5O2	355.53	355.2947	60.81	10.49	19.70	9.00
C18H37N5O3	371.53	371.2896	58.19	10.04	18.85	12.92
C18H37N5O4	387.53	387.2845	55.79	9.62	18.07	16.51
C18H37N5O5	403.53	403.2794	53.58	9.24	17.36	19.82
C18H37N5O6	419.53	419.2744	51.53	8.89	16.69	22.88
C18H37N5O7	435.52	435.2693	49.64	8.56	16.08	25.72
C18H37N5O8	451.52	451.2642	47.88	8.26	15.51	28.35
C18H37N5O9	467.52	467.2591	46.24	7.98	14.98	30.80
C18H38N2O1	298.52	298.2984	72.43	12.83	9.38	5.36
C18H38N2O2	314.52	314.2933	68.74	12.18	8.91	10.17
C18H38N2O3	330.52	330.2882	65.41	11.59	8.48	14.52
C18H38N2O4	346.51	346.2831	62.39	11.05	8.08	18.47
C18H38N2O5	362.51	362.2781	59.64	10.57	7.73	22.07
C18H38N2O6	378.51	378.2730	57.12	10.12	7.40	25.36
C18H38N2O7	394.51	394.2679	54.80	9.71	7.10	28.39
C18H38N2O8	410.51	410.2628	52.67	9.33	6.82	31.18
C18H38N2O9	426.51	426.2577	50.69	8.98	6.57	33.76
C18H38N4O1	326.53	326.3045	66.21	11.73	17.16	4.90
C18H38N4O2	342.53	342.2995	63.12	11.18	16.36	9.34
C18H38N4O3	358.53	358.2944	60.30	10.68	15.63	13.39
C18H38N4O4	374.53	374.2893	57.73	10.23	14.96	17.09
C18H38N4O5	390.53	390.2842	55.36	9.81	14.35	20.48
C18H38N4O6	406.53	406.2791	53.18	9.42	13.78	23.61
C18H38N4O7	422.53	422.2740	51.17	9.07	13.26	26.51
C18H38N4O8	438.53	438.2689	49.30	8.73	12.78	29.19
C18H38N4O9	454.52	454.2639	47.57	8.43	12.33	31.68
C18H38N6O1	354.54	354.3107	60.98	10.80	23.70	4.51
C18H38N6O2	370.54	370.3056	58.35	10.34	22.68	8.64
C18H38N6O3	386.54	386.3005	55.93	9.91	21.74	12.42
C18H38N6O4	402.54	402.2954	53.71	9.52	20.88	15.90
C18H38N6O5	418.54	418.2903	51.66	9.15	20.08	19.11
C18H38N6O6	434.54	434.2853	49.75	8.81	19.34	22.09
C18H38N6O7	450.54	450.2802	47.99	8.50	18.65	24.86
C18H38N6O8	466.54	466.2751	46.34	8.21	18.01	27.44
C18H38N6O9	482.54	482.2700	44.80	7.94	17.42	29.84
C18H39N1O1	285.52	285.3031	75.72	13.77	4.91	5.60
C18H39N1O2	301.52	301.2981	71.70	13.04	4.65	10.61
C18H39N1O3	317.52	317.2930	68.09	12.38	4.41	15.12
C18H39N1O4	333.52	333.2879	64.82	11.79	4.20	19.19
C18H39N1O5	349.52	349.2828	61.86	11.25	4.01	22.89
C18H39N1O6	365.51	365.2777	59.15	10.75	3.83	26.26
C18H39N1O7	381.51	381.2726	56.67	10.30	3.67	29.36
C18H39N1O8	397.51	397.2675	54.39	9.89	3.52	32.20
C18H39N1O9	413.51	413.2625	52.28	9.51	3.39	34.82
C18H39N3O1	313.53	313.3093	68.96	12.54	13.40	5.10
C18H39N3O2	329.53	329.3042	65.61	11.93	12.75	9.71
C18H39N3O3	345.53	345.2991	62.57	11.38	12.16	13.89
C18H39N3O4	361.53	361.2940	59.80	10.87	11.62	17.70
C18H39N3O5	377.53	377.2890	57.27	10.41	11.13	21.19
C18H39N3O6	393.53	393.2839	54.94	9.99	10.68	24.39
C18H39N3O7	409.53	409.2788	52.79	9.60	10.26	27.35
C18H39N3O8	425.53	425.2737	50.81	9.24	9.87	30.08
C18H39N3O9	441.53	441.2686	48.97	8.90	9.52	32.61
C18H39N5O1	341.54	341.3154	63.30	11.51	20.50	4.68
C18H39N5O2	357.54	357.3104	60.47	10.99	19.59	8.95
C18H39N5O3	373.54	373.3053	57.88	10.52	18.75	12.85
C18H39N5O4	389.54	389.3002	55.50	10.09	17.98	16.43
C18H39N5O5	405.54	405.2951	53.31	9.69	17.27	19.73
C18H39N5O6	421.54	421.2900	51.29	9.33	16.61	22.77
C18H39N5O7	437.54	437.2849	49.41	8.98	16.01	25.60
C18H39N5O8	453.54	453.2798	47.67	8.67	15.44	28.22
C18H39N5O9	469.54	469.2748	46.05	8.37	14.92	30.67
C18H40N2O1	300.53	300.3140	71.94	13.42	9.32	5.32
C18H40N2O2	316.53	316.3090	68.30	12.74	8.85	10.11
C18H40N2O3	332.53	332.3039	65.02	12.12	8.42	14.43
C18H40N2O4	348.53	348.2988	62.03	11.57	8.04	18.36
C18H40N2O5	364.53	364.2937	59.31	11.06	7.68	21.95
C18H40N2O6	380.53	380.2886	56.82	10.60	7.36	25.23
C18H40N2O7	396.53	396.2835	54.52	10.17	7.06	28.24
C18H40N2O8	412.53	412.2784	52.41	9.77	6.79	31.03
C18H40N2O9	428.53	428.2734	50.45	9.41	6.54	33.60
C18H40N4O1	328.55	328.3202	65.81	12.27	17.05	4.87
C18H40N4O2	344.55	344.3151	62.75	11.70	16.26	9.29
C18H40N4O3	360.54	360.3100	59.97	11.18	15.54	13.31
C18H40N4O4	376.54	376.3049	57.42	10.71	14.88	17.00
C18H40N4O5	392.54	392.2998	55.08	10.27	14.27	20.38
C18H40N4O6	408.54	408.2948	52.92	9.87	13.71	23.50
C18H40N4O7	424.54	424.2897	50.93	9.50	13.20	26.38
C18H40N4O8	440.54	440.2846	49.08	9.15	12.72	29.05
C18H40N4O9	456.54	456.2795	47.36	8.83	12.27	31.54
C18H40N6O1	356.56	356.3263	60.64	11.31	23.57	4.49
C18H40N6O2	372.56	372.3213	58.03	10.82	22.56	8.59
C18H40N6O3	388.56	388.3162	55.64	10.38	21.63	12.35
C18H40N6O4	404.56	404.3111	53.44	9.97	20.77	15.82
C18H40N6O5	420.56	420.3060	51.41	9.59	19.98	19.02
C18H40N6O6	436.56	436.3009	49.52	9.24	19.25	21.99
C18H40N6O7	452.56	452.2958	47.77	8.91	18.57	24.75
C18H40N6O8	468.55	468.2907	46.14	8.60	17.94	27.32
C18H40N6O9	484.55	484.2857	44.62	8.32	17.34	29.72
C18H41N3O1	315.55	315.3249	68.52	13.10	13.32	5.07
C18H41N3O2	331.55	331.3199	65.21	12.46	12.67	9.65
C18H41N3O3	347.55	347.3148	62.21	11.89	12.09	13.81
C18H41N3O4	363.55	363.3097	59.47	11.37	11.56	17.60
C18H41N3O5	379.54	379.3046	56.96	10.89	11.07	21.08
C18H41N3O6	395.54	395.2995	54.66	10.45	10.62	24.27
C18H41N3O7	411.54	411.2944	52.53	10.04	10.21	27.21
C18H41N3O8	427.54	427.2893	50.57	9.67	9.83	29.94
C18H41N3O9	443.54	443.2843	48.74	9.32	9.47	32.46
C18H41N5O1	343.56	343.3311	62.93	12.03	20.38	4.66
C18H41N5O2	359.56	359.3260	60.13	11.49	19.48	8.90
C18H41N5O3	375.56	375.3209	57.57	11.00	18.65	12.78
C18H41N5O4	391.56	391.3158	55.22	10.55	17.89	16.34
C18H41N5O5	407.56	407.3107	53.05	10.14	17.18	19.63
C18H41N5O6	423.56	423.3057	51.04	9.76	16.53	22.66
C18H41N5O7	439.56	439.3006	49.19	9.40	15.93	25.48
C18H41N5O8	455.56	455.2955	47.46	9.07	15.37	28.10
C18H41N5O9	471.56	471.2904	45.85	8.76	14.85	30.54
C18H42N4O1	330.56	330.3358	65.40	12.81	16.95	4.84
C18H42N4O2	346.56	346.3308	62.38	12.22	16.17	9.23
C18H42N4O3	362.56	362.3257	59.63	11.68	15.45	13.24
C18H42N4O4	378.56	378.3206	57.11	11.18	14.80	16.91
C18H42N4O5	394.56	394.3155	54.80	10.73	14.20	20.28
C18H42N4O6	410.56	410.3104	52.66	10.31	13.65	23.38
C18H42N4O7	426.56	426.3053	50.68	9.92	13.13	26.26

103

C18H42N4O8	442.56	442.3002	48.85	9.57	12.66	28.92
C18H42N4O9	458.56	458.2952	47.15	9.23	12.22	31.40
C18H42N6O1	358.58	358.3420	60.29	11.81	23.44	4.46
C18H42N6O2	374.57	374.3369	57.72	11.30	22.44	8.54
C18H42N6O3	390.57	390.3318	55.35	10.84	21.52	12.29
C18H42N6O4	406.57	406.3267	53.18	10.41	20.67	15.74
C18H42N6O5	422.57	422.3216	51.16	10.02	19.89	18.93
C18H42N6O6	438.57	438.3166	49.30	9.65	19.16	21.89
C18H42N6O7	454.57	454.3115	47.56	9.31	18.49	24.64
C18H42N6O8	470.57	470.3064	45.94	9.00	17.86	27.20
C18H42N6O9	486.57	486.3013	44.43	8.70	17.27	29.59
C18H43N5O1	345.58	345.3467	62.56	12.54	20.27	4.63
C18H43N5O2	361.58	361.3417	59.79	11.99	19.37	8.85
C18H43N5O3	377.58	377.3366	57.26	11.48	18.55	12.71
C18H43N5O4	393.57	393.3315	54.93	11.01	17.79	16.26
C18H43N5O5	409.57	409.3264	52.79	10.58	17.10	19.53
C18H43N5O6	425.57	425.3213	50.80	10.18	16.46	22.56
C18H43N5O7	441.57	441.3162	48.96	9.82	15.86	25.36
C18H43N5O8	457.57	457.3111	47.25	9.47	15.31	27.97
C18H43N5O9	473.57	473.3061	45.65	9.15	14.79	30.41
C18H44N6O1	360.59	360.3576	59.96	12.30	23.31	4.44
C18H44N6O2	376.59	376.3526	57.41	11.78	22.32	8.50
C18H44N6O3	392.59	392.3475	55.07	11.30	21.41	12.23
C18H44N6O4	408.59	408.3424	52.91	10.85	20.57	15.66
C18H44N6O5	424.59	424.3373	50.92	10.45	19.79	18.84
C18H44N6O6	440.59	440.3322	49.07	10.07	19.07	21.79
C18H44N6O7	456.59	456.3271	47.35	9.71	18.41	24.53
C18H44N6O8	472.59	472.3220	45.75	9.38	17.78	27.08
C18H44N6O9	488.59	488.3170	44.25	9.08	17.20	29.47
C19H 9N1O1	267.29	267.0684	85.38	3.39	5.24	5.99
C19H 9N1O2	283.29	283.0633	80.56	3.20	4.94	11.30
C19H 9N1O3	299.29	299.0582	76.25	3.03	4.68	16.04
C19H 9N1O4	315.29	315.0532	72.38	2.88	4.44	20.30
C19H 9N1O5	331.29	331.0481	68.89	2.74	4.23	24.15
C19H 9N1O6	347.29	347.0430	65.71	2.61	4.03	27.64
C19H 9N1O7	363.29	363.0379	62.82	2.50	3.86	30.83
C19H 9N1O8	379.29	379.0328	60.17	2.39	3.69	33.75
C19H 9N1O9	395.28	395.0277	57.73	2.29	3.54	36.43
C19H 9N3O1	295.30	295.0746	77.28	3.07	14.23	5.42
C19H 9N3O2	311.30	311.0695	73.31	2.91	13.50	10.28
C19H 9N3O3	327.30	327.0644	69.73	2.77	12.84	14.66
C19H 9N3O4	343.30	343.0593	66.48	2.64	12.24	18.64
C19H 9N3O5	359.30	359.0542	63.52	2.52	11.69	22.26
C19H 9N3O6	375.30	375.0491	60.81	2.42	11.20	25.58
C19H 9N3O7	391.30	391.0440	58.32	2.32	10.74	28.62
C19H 9N3O8	407.30	407.0390	56.03	2.23	10.32	31.43
C19H 9N3O9	423.30	423.0339	53.91	2.14	9.93	34.02
C19H 9N5O1	323.32	323.0807	70.58	2.81	21.66	4.95
C19H 9N5O2	339.32	339.0756	67.26	2.67	20.64	9.43
C19H 9N5O3	355.32	355.0705	64.23	2.55	19.71	13.51
C19H 9N5O4	371.31	371.0654	61.46	2.44	18.86	17.24
C19H 9N5O5	387.31	387.0604	58.92	2.34	18.08	20.65
C19H 9N5O6	403.31	403.0553	56.58	2.25	17.36	23.80
C19H 9N5O7	419.31	419.0502	54.43	2.16	16.70	26.71
C19H 9N5O8	435.31	435.0451	52.42	2.08	16.09	29.40
C19H 9N5O9	451.31	451.0400	50.57	2.01	15.52	31.91
C19H10N2O1	282.30	282.0793	80.84	3.57	9.92	5.67
C19H10N2O2	298.30	298.0742	76.50	3.38	9.39	10.73
C19H10N2O3	314.30	314.0691	72.61	3.21	8.91	15.27
C19H10N2O4	330.30	330.0640	69.09	3.05	8.48	19.38
C19H10N2O5	346.30	346.0590	65.90	2.91	8.09	23.10
C19H10N2O6	362.30	362.0539	62.99	2.78	7.73	26.50
C19H10N2O7	378.30	378.0488	60.33	2.66	7.41	29.60
C19H10N2O8	394.30	394.0437	57.88	2.56	7.10	32.46
C19H10N2O9	410.30	410.0386	55.62	2.46	6.83	35.09
C19H10N4O1	310.32	310.0855	73.54	3.25	18.05	5.16
C19H10N4O2	326.32	326.0804	69.94	3.09	17.17	9.81
C19H10N4O3	342.32	342.0753	66.67	2.94	16.37	14.02
C19H10N4O4	358.32	358.0702	63.69	2.81	15.64	17.86
C19H10N4O5	374.32	374.0651	60.97	2.69	14.97	21.37
C19H10N4O6	390.31	390.0600	58.47	2.58	14.35	24.59
C19H10N4O7	406.31	406.0549	56.17	2.48	13.79	27.56
C19H10N4O8	422.31	422.0499	54.04	2.39	13.27	30.31
C19H10N4O9	438.31	438.0448	52.07	2.30	12.78	32.85
C19H10N6O1	338.33	338.0916	67.45	2.98	24.84	4.73
C19H10N6O2	354.33	354.0865	64.41	2.84	23.72	9.03
C19H10N6O3	370.33	370.0814	61.62	2.72	22.69	12.96
C19H10N6O4	386.33	386.0763	59.07	2.61	21.75	16.57
C19H10N6O5	402.33	402.0713	56.72	2.51	20.89	19.88
C19H10N6O6	418.33	418.0662	54.55	2.41	20.09	22.95
C19H10N6O7	434.33	434.0611	52.54	2.32	19.35	25.79
C19H10N6O8	450.33	450.0560	50.68	2.24	18.66	28.42
C19H10N6O9	466.33	466.0509	48.94	2.16	18.02	30.88
C19H11N1O1	269.31	269.0841	84.74	4.12	5.20	5.94
C19H11N1O2	285.31	285.0790	79.99	3.89	4.91	11.22
C19H11N1O3	301.30	301.0739	75.74	3.68	4.65	15.93
C19H11N1O4	317.30	317.0688	71.92	3.49	4.41	20.17
C19H11N1O5	333.30	333.0637	68.47	3.33	4.20	24.00
C19H11N1O6	349.30	349.0586	65.33	3.17	4.01	27.48
C19H11N1O7	365.30	365.0535	62.47	3.04	3.83	30.66
C19H11N1O8	381.30	381.0485	59.85	2.91	3.67	33.57
C19H11N1O9	397.30	397.0434	57.44	2.79	3.53	36.24
C19H11N3O1	297.32	297.0902	76.76	3.73	14.13	5.38
C19H11N3O2	313.32	313.0851	72.84	3.54	13.41	10.21
C19H11N3O3	329.32	329.0800	69.30	3.37	12.76	14.58
C19H11N3O4	345.32	345.0749	66.09	3.21	12.17	18.53
C19H11N3O5	361.32	361.0699	63.16	3.07	11.63	22.14
C19H11N3O6	377.32	377.0648	60.48	2.94	11.14	25.44
C19H11N3O7	393.32	393.0597	58.02	2.82	10.68	28.47
C19H11N3O8	409.32	409.0546	55.75	2.71	10.27	31.27
C19H11N3O9	425.31	425.0495	53.66	2.61	9.88	33.86
C19H11N5O1	325.33	325.0964	70.15	3.41	21.53	4.92
C19H11N5O2	341.33	341.0913	66.86	3.25	20.52	9.37
C19H11N5O3	357.33	357.0862	63.87	3.10	19.60	13.43
C19H11N5O4	373.33	373.0811	61.13	2.97	18.76	17.14
C19H11N5O5	389.33	389.0760	58.62	2.85	17.99	20.55
C19H11N5O6	405.33	405.0709	56.30	2.74	17.28	23.68
C19H11N5O7	421.33	421.0658	54.16	2.63	16.62	26.58
C19H11N5O8	437.33	437.0608	52.18	2.54	16.01	29.27
C19H11N5O9	453.33	453.0557	50.34	2.45	15.45	31.76
C19H12N2O1	284.32	284.0950	80.27	4.25	9.85	5.63
C19H12N2O2	300.32	300.0899	75.99	4.03	9.33	10.65
C19H12N2O3	316.32	316.0848	72.15	3.82	8.86	15.17
C19H12N2O4	332.32	332.0797	68.67	3.64	8.43	19.26
C19H12N2O5	348.32	348.0746	65.52	3.47	8.04	22.97
C19H12N2O6	364.32	364.0695	62.64	3.32	7.69	26.35
C19H12N2O7	380.32	380.0644	60.01	3.18	7.37	29.45
C19H12N2O8	396.32	396.0594	57.58	3.05	7.07	32.30
C19H12N2O9	412.32	412.0543	55.35	2.93	6.79	34.92
C19H12N4O1	312.33	312.1011	73.07	3.87	17.94	5.12

C19H12N402	328.33	328.0960	69.51	3.68	17.06	9.75
C19H12N403	344.33	344.0909	66.28	3.51	16.27	13.94
C19H12N404	360.33	360.0858	63.33	3.36	15.55	17.76
C19H12N405	376.33	376.0808	60.64	3.21	14.89	21.26
C19H12N406	392.33	392.0757	58.17	3.08	14.28	24.47
C19H12N407	408.33	408.0706	55.89	2.96	13.72	27.43
C19H12N408	424.33	424.0655	53.78	2.85	13.20	30.16
C19H12N409	440.33	440.0604	51.83	2.75	12.72	32.70
C19H12N601	340.35	340.1073	67.05	3.55	24.69	4.70
C19H12N602	356.35	356.1022	64.04	3.39	23.58	8.98
C19H12N603	372.35	372.0971	61.29	3.25	22.57	12.89
C19H12N604	388.35	388.0920	58.77	3.11	21.64	16.48
C19H12N605	404.34	404.0869	56.44	2.99	20.78	19.78
C19H12N606	420.34	420.0818	54.29	2.88	19.99	22.84
C19H12N607	436.34	436.0767	52.30	2.77	19.26	25.67
C19H12N608	452.34	452.0717	50.45	2.67	18.58	28.30
C19H12N609	468.34	468.0666	48.73	2.58	17.94	30.75
C19H13N101	271.32	271.0997	84.11	4.83	5.16	5.90
C19H13N102	287.32	287.0946	79.43	4.56	4.87	11.14
C19H13N103	303.32	303.0895	75.24	4.32	4.62	15.82
C19H13N104	319.32	319.0844	71.47	4.10	4.39	20.04
C19H13N105	335.32	335.0794	68.06	3.91	4.18	23.86
C19H13N106	351.32	351.0743	64.96	3.73	3.99	27.32
C19H13N107	367.32	367.0692	62.13	3.57	3.81	30.49
C19H13N108	383.32	383.0641	59.54	3.42	3.65	33.39
C19H13N109	399.32	399.0590	57.15	3.28	3.51	36.06
C19H13N301	299.33	299.1059	76.24	4.38	14.04	5.34
C19H13N302	315.33	315.1008	72.37	4.16	13.33	10.15
C19H13N303	331.33	331.0957	68.88	3.95	12.68	14.49
C19H13N304	347.33	347.0906	65.70	3.77	12.10	18.43
C19H13N305	363.33	363.0855	62.81	3.61	11.57	22.02
C19H13N306	379.33	379.0804	60.16	3.45	11.08	25.31
C19H13N307	395.33	395.0753	57.73	3.31	10.63	28.33
C19H13N308	411.33	411.0703	55.48	3.19	10.22	31.12
C19H13N309	427.33	427.0652	53.40	3.07	9.83	33.70
C19H13N501	327.35	327.1120	69.72	4.00	21.39	4.89
C19H13N502	343.35	343.1069	66.47	3.82	20.40	9.32
C19H13N503	359.35	359.1018	63.51	3.65	19.49	13.36
C19H13N504	375.35	375.0967	60.80	3.49	18.66	17.05
C19H13N505	391.35	391.0917	58.31	3.35	17.90	20.44
C19H13N506	407.35	407.0866	56.02	3.22	17.19	23.57
C19H13N507	423.34	423.0815	53.91	3.10	16.54	26.45
C19H13N508	439.34	439.0764	51.94	2.98	15.94	29.13
C19H13N509	455.34	455.0713	50.12	2.88	15.38	31.62
C19H14N201	286.34	286.1106	79.70	4.93	9.78	5.59
C19H14N202	302.34	302.1055	75.48	4.67	9.27	10.58
C19H14N203	318.34	318.1004	71.69	4.43	8.80	15.08
C19H14N204	334.33	334.0953	68.26	4.22	8.38	19.14
C19H14N205	350.33	350.0903	65.14	4.03	8.00	22.83
C19H14N206	366.33	366.0852	62.30	3.85	7.65	26.20
C19H14N207	382.33	382.0801	59.69	3.69	7.33	29.29
C19H14N208	398.33	398.0750	57.29	3.54	7.03	32.13
C19H14N209	414.33	414.0699	55.08	3.41	6.76	34.75
C19H14N401	314.35	314.1168	72.60	4.49	17.82	5.09
C19H14N402	330.35	330.1117	69.08	4.27	16.96	9.69
C19H14N403	346.35	346.1066	65.89	4.07	16.18	13.86
C19H14N404	362.35	362.1015	62.98	3.89	15.46	17.66
C19H14N405	378.35	378.0964	60.32	3.73	14.81	21.14
C19H14N406	394.35	394.0913	57.87	3.58	14.21	24.34
C19H14N407	410.35	410.0862	55.61	3.44	13.65	27.29

C19H14N408	426.35	426.0812	53.53	3.31	13.14	30.02
C19H14N409	442.34	442.0761	51.59	3.19	12.67	32.55
C19H14N601	342.36	342.1229	66.66	4.12	24.55	4.67
C19H14N602	358.36	358.1178	63.68	3.94	23.45	8.93
C19H14N603	374.36	374.1127	60.96	3.77	22.45	12.82
C19H14N604	390.36	390.1076	58.46	3.62	21.53	16.39
C19H14N605	406.36	406.1026	56.16	3.47	20.68	19.69
C19H14N606	422.36	422.0975	54.03	3.34	19.90	22.73
C19H14N607	438.36	438.0924	52.06	3.22	19.17	25.55
C19H14N608	454.36	454.0873	50.23	3.11	18.50	28.17
C19H14N609	470.36	470.0822	48.52	3.00	17.87	30.61
C19H15N101	273.34	273.1154	83.49	5.53	5.12	5.85
C19H15N102	289.34	289.1103	78.87	5.23	4.84	11.06
C19H15N103	305.34	305.1052	74.74	4.95	4.59	15.72
C19H15N104	321.34	321.1001	71.02	4.71	4.36	19.92
C19H15N105	337.34	337.0950	67.65	4.48	4.15	23.71
C19H15N106	353.33	353.0899	64.59	4.28	3.96	27.17
C19H15N107	369.33	369.0848	61.79	4.09	3.79	30.32
C19H15N108	385.33	385.0798	59.22	3.92	3.63	33.22
C19H15N109	401.33	401.0747	56.86	3.77	3.49	35.88
C19H15N301	301.35	301.1215	75.73	5.02	13.94	5.31
C19H15N302	317.35	317.1164	71.91	4.76	13.24	10.08
C19H15N303	333.35	333.1113	68.46	4.54	12.61	14.40
C19H15N304	349.35	349.1062	65.32	4.33	12.03	18.32
C19H15N305	365.35	365.1012	62.46	4.14	11.50	21.90
C19H15N306	381.35	381.0961	59.84	3.96	11.02	25.17
C19H15N307	397.35	397.0910	57.43	3.81	10.58	28.19
C19H15N308	413.35	413.0859	55.21	3.66	10.17	30.97
C19H15N309	429.35	429.0808	53.15	3.52	9.79	33.54
C19H15N501	329.36	329.1277	69.29	4.59	21.26	4.86
C19H15N502	345.36	345.1226	66.08	4.38	20.28	9.27
C19H15N503	361.36	361.1175	63.15	4.18	19.38	13.28
C19H15N504	377.36	377.1124	60.48	4.01	18.56	16.96
C19H15N505	393.36	393.1073	58.02	3.84	17.80	20.34
C19H15N506	409.36	409.1022	55.75	3.69	17.11	23.45
C19H15N507	425.36	425.0971	53.65	3.55	16.46	26.33
C19H15N508	441.36	441.0921	51.71	3.43	15.87	29.00
C19H15N509	457.36	457.0870	49.90	3.31	15.31	31.48
C19H16N201	288.35	288.1263	79.14	5.59	9.71	5.55
C19H16N202	304.35	304.1212	74.98	5.30	9.20	10.51
C19H16N203	320.35	320.1161	71.24	5.03	8.74	14.98
C19H16N204	336.35	336.1110	67.85	4.79	8.33	19.03
C19H16N205	352.35	352.1059	64.77	4.58	7.95	22.70
C19H16N206	368.35	368.1008	61.96	4.38	7.61	26.06
C19H16N207	384.35	384.0957	59.38	4.20	7.29	29.14
C19H16N208	400.35	400.0907	57.00	4.03	7.00	31.97
C19H16N209	416.35	416.0856	54.81	3.87	6.73	34.59
C19H16N401	316.37	316.1324	72.14	5.10	17.71	5.06
C19H16N402	332.36	332.1273	68.66	4.85	16.86	9.63
C19H16N403	348.36	348.1222	65.51	4.63	16.08	13.78
C19H16N404	364.36	364.1171	62.63	4.43	15.38	17.56
C19H16N405	380.36	380.1121	60.00	4.24	14.73	21.03
C19H16N406	396.36	396.1070	57.58	4.07	14.14	24.22
C19H16N407	412.36	412.1019	55.34	3.91	13.59	27.16
C19H16N408	428.36	428.0968	53.28	3.76	13.08	29.88
C19H16N409	444.36	444.0917	51.36	3.63	12.61	32.40
C19H16N601	344.38	344.1386	66.27	4.68	24.40	4.65
C19H16N602	360.38	360.1335	63.33	4.48	23.32	8.88
C19H16N603	376.38	376.1284	60.63	4.28	22.33	12.75
C19H16N604	392.38	392.1233	58.16	4.11	21.42	16.31

C19H16N605	408.38	408.1182	55.88	3.95	20.58	19.59
C19H16N606	424.38	424.1131	53.78	3.80	19.80	22.62
C19H16N607	440.38	440.1080	51.82	3.66	19.08	25.43
C19H16N608	456.37	456.1029	50.01	3.53	18.41	28.05
C19H16N609	472.37	472.0979	48.31	3.41	17.79	30.48
C19H17N101	275.35	275.1310	82.88	6.22	5.09	5.81
C19H17N102	291.35	291.1259	78.33	5.88	4.81	10.98
C19H17N103	307.35	307.1208	74.25	5.58	4.56	15.62
C19H17N104	323.35	323.1157	70.58	5.30	4.33	19.79
C19H17N105	339.35	339.1107	67.25	5.05	4.13	23.57
C19H17N106	355.35	355.1056	64.22	4.82	3.94	27.01
C19H17N107	371.35	371.1005	61.45	4.61	3.77	30.16
C19H17N108	387.35	387.0954	58.92	4.42	3.62	33.04
C19H17N109	403.35	403.0903	56.58	4.25	3.47	35.70
C19H17N301	303.37	303.1372	75.23	5.65	13.85	5.27
C19H17N302	319.37	319.1321	71.46	5.37	13.16	10.02
C19H17N303	335.37	335.1270	68.05	5.11	12.53	14.31
C19H17N304	351.37	351.1219	64.95	4.88	11.96	18.21
C19H17N305	367.36	367.1168	62.12	4.66	11.44	21.78
C19H17N306	383.36	383.1117	59.53	4.47	10.96	25.04
C19H17N307	399.36	399.1066	57.14	4.29	10.52	28.04
C19H17N308	415.36	415.1016	54.94	4.13	10.12	30.82
C19H17N309	431.36	431.0965	52.90	3.97	9.74	33.38
C19H17N501	331.38	331.1433	68.87	5.17	21.13	4.83
C19H17N502	347.38	347.1382	65.70	4.93	20.16	9.21
C19H17N503	363.38	363.1331	62.80	4.72	19.27	13.21
C19H17N504	379.38	379.1280	60.15	4.52	18.46	16.87
C19H17N505	395.38	395.1230	57.72	4.33	17.71	20.23
C19H17N506	411.38	411.1179	55.48	4.17	17.02	23.34
C19H17N507	427.38	427.1128	53.40	4.01	16.39	26.21
C19H17N508	443.38	443.1077	51.47	3.86	15.80	28.87
C19H17N509	459.38	459.1026	49.68	3.73	15.25	31.35
C19H18N201	290.37	290.1419	78.59	6.25	9.65	5.51
C19H18N202	306.37	306.1368	74.49	5.92	9.14	10.44
C19H18N203	322.37	322.1317	70.79	5.63	8.69	14.89
C19H18N204	338.37	338.1266	67.45	5.36	8.28	18.91
C19H18N205	354.37	354.1216	64.40	5.12	7.91	22.57
C19H18N206	370.37	370.1165	61.62	4.90	7.56	25.92
C19H18N207	386.36	386.1114	59.07	4.70	7.25	28.99
C19H18N208	402.36	402.1063	56.72	4.51	6.96	31.81
C19H18N209	418.36	418.1012	54.55	4.34	6.70	34.42
C19H18N401	318.38	318.1481	71.68	5.70	17.60	5.03
C19H18N402	334.38	334.1430	68.25	5.43	16.76	9.57
C19H18N403	350.38	350.1379	65.13	5.18	15.99	13.70
C19H18N404	366.38	366.1328	62.29	4.95	15.29	17.47
C19H18N405	382.38	382.1277	59.68	4.74	14.65	20.92
C19H18N406	398.38	398.1226	57.29	4.55	14.06	24.10
C19H18N407	414.38	414.1175	55.07	4.38	13.52	27.03
C19H18N408	430.38	430.1125	53.03	4.22	13.02	29.74
C19H18N409	446.38	446.1074	51.13	4.06	12.55	32.26
C19H18N601	346.39	346.1542	65.88	5.24	24.26	4.62
C19H18N602	362.39	362.1491	62.97	5.01	23.19	8.83
C19H18N603	378.39	378.1440	60.31	4.79	22.21	12.68
C19H18N604	394.39	394.1389	57.86	4.60	21.31	16.23
C19H18N605	410.39	410.1339	55.61	4.42	20.48	19.49
C19H18N606	426.39	426.1288	53.52	4.26	19.71	22.51
C19H18N607	442.39	442.1237	51.59	4.10	19.00	25.32
C19H18N608	458.39	458.1186	49.79	3.96	18.33	27.92
C19H18N609	474.39	474.1135	48.11	3.82	17.72	30.35
C19H19N101	277.37	277.1467	82.28	6.90	5.05	5.77

C19H19N102	293.37	293.1416	77.79	6.53	4.77	10.91
C19H19N103	309.37	309.1365	73.77	6.19	4.53	15.51
C19H19N104	325.37	325.1314	70.14	5.89	4.30	19.67
C19H19N105	341.37	341.1263	66.85	5.61	4.10	23.43
C19H19N106	357.37	357.1212	63.86	5.36	3.92	26.86
C19H19N107	373.37	373.1161	61.12	5.13	3.75	30.00
C19H19N108	389.37	389.1111	58.61	4.92	3.60	32.87
C19H19N109	405.36	405.1060	56.30	4.72	3.46	35.52
C19H19N301	305.38	305.1528	74.73	6.27	13.76	5.24
C19H19N302	321.38	321.1477	71.01	5.96	13.07	9.96
C19H19N303	337.38	337.1426	67.64	5.68	12.45	14.23
C19H19N304	353.38	353.1375	64.58	5.42	11.89	18.11
C19H19N305	369.38	369.1325	61.78	5.18	11.38	21.66
C19H19N306	385.38	385.1274	59.22	4.97	10.90	24.91
C19H19N307	401.38	401.1223	56.86	4.77	10.47	27.90
C19H19N308	417.38	417.1172	54.68	4.59	10.07	30.67
C19H19N309	433.38	433.1121	52.66	4.42	9.70	33.23
C19H19N501	333.40	333.1590	68.45	5.74	21.01	4.80
C19H19N502	349.40	349.1539	65.32	5.48	20.04	9.16
C19H19N503	365.39	365.1488	62.46	5.24	19.17	13.14
C19H19N504	381.39	381.1437	59.84	5.02	18.36	16.78
C19H19N505	397.39	397.1386	57.43	4.82	17.62	20.13
C19H19N506	413.39	413.1335	55.20	4.63	16.94	23.22
C19H19N507	429.39	429.1284	53.15	4.46	16.31	26.08
C19H19N508	445.39	445.1233	51.24	4.30	15.72	28.74
C19H19N509	461.39	461.1183	49.46	4.15	15.18	31.21
C19H20N201	292.38	292.1576	78.05	6.89	9.58	5.47
C19H20N202	308.38	308.1525	74.00	6.54	9.08	10.38
C19H20N203	324.38	324.1474	70.35	6.21	8.64	14.80
C19H20N204	340.38	340.1423	67.05	5.92	8.23	18.80
C19H20N205	356.38	356.1372	64.04	5.66	7.86	22.45
C19H20N206	372.38	372.1321	61.28	5.41	7.52	25.78
C19H20N207	388.38	388.1270	58.76	5.19	7.21	28.84
C19H20N208	404.38	404.1220	56.44	4.99	6.93	31.65
C19H20N209	420.38	420.1169	54.29	4.80	6.66	34.25
C19H20N401	320.40	320.1637	71.23	6.29	17.49	4.99
C19H20N402	336.40	336.1586	67.84	5.99	16.65	9.51
C19H20N403	352.40	352.1535	64.76	5.72	15.90	13.62
C19H20N404	368.40	368.1484	61.95	5.47	15.21	17.37
C19H20N405	384.40	384.1434	59.37	5.24	14.58	20.81
C19H20N406	400.39	400.1383	57.00	5.03	13.99	23.98
C19H20N407	416.39	416.1332	54.81	4.84	13.46	26.90
C19H20N408	432.39	432.1281	52.78	4.66	12.96	29.60
C19H20N409	448.39	448.1230	50.90	4.50	12.50	32.11
C19H20N601	348.41	348.1698	65.50	5.79	24.12	4.59
C19H20N602	364.41	364.1648	62.62	5.53	23.06	8.78
C19H20N603	380.41	380.1597	59.99	5.30	22.09	12.62
C19H20N604	396.41	396.1546	57.57	5.09	21.20	16.14
C19H20N605	412.41	412.1495	55.34	4.89	20.38	19.40
C19H20N606	428.41	428.1444	53.27	4.71	19.62	22.41
C19H20N607	444.41	444.1393	51.35	4.54	18.91	25.20
C19H20N608	460.41	460.1342	49.57	4.38	18.25	27.80
C19H20N609	476.41	476.1292	47.90	4.23	17.64	30.23
C19H21N101	279.39	279.1623	81.68	7.58	5.01	5.73
C19H21N102	295.38	295.1572	77.26	7.17	4.74	10.83
C19H21N103	311.38	311.1521	73.29	6.80	4.50	15.41
C19H21N104	327.38	327.1470	69.71	6.47	4.28	19.55
C19H21N105	343.38	343.1420	66.46	6.16	4.08	23.30
C19H21N106	359.38	359.1369	63.50	5.89	3.90	26.71
C19H21N107	375.38	375.1318	60.79	5.64	3.73	29.84

106

C19H21N108	391.38	391.1267	58.31	5.41	3.58	32.70
C19H21N109	407.38	407.1216	56.02	5.20	3.44	35.35
C19H21N301	307.40	307.1685	74.24	6.89	13.67	5.20
C19H21N302	323.40	323.1634	70.57	6.55	12.99	9.89
C19H21N303	339.40	339.1583	67.24	6.24	12.38	14.14
C19H21N304	355.40	355.1532	64.21	5.96	11.82	18.01
C19H21N305	371.40	371.1481	61.45	5.70	11.31	21.54
C19H21N306	387.40	387.1430	58.91	5.46	10.85	24.78
C19H21N307	403.40	403.1379	56.57	5.25	10.42	27.76
C19H21N308	419.39	419.1329	54.41	5.05	10.02	30.52
C19H21N309	435.39	435.1278	52.42	4.86	9.65	33.07
C19H21N501	335.41	335.1746	68.04	6.31	20.88	4.77
C19H21N502	351.41	351.1695	64.94	6.02	19.93	9.11
C19H21N503	367.41	367.1644	62.11	5.76	19.06	13.06
C19H21N504	383.41	383.1593	59.52	5.52	18.27	16.69
C19H21N505	399.41	399.1543	57.14	5.30	17.53	20.03
C19H21N506	415.41	415.1492	54.94	5.10	16.86	23.11
C19H21N507	431.41	431.1441	52.90	4.91	16.23	25.96
C19H21N508	447.41	447.1390	51.01	4.73	15.65	28.61
C19H21N509	463.41	463.1339	49.25	4.57	15.11	31.07
C19H22N201	294.40	294.1732	77.52	7.53	9.52	5.43
C19H22N202	310.40	310.1681	73.52	7.14	9.02	10.31
C19H22N203	326.40	326.1630	69.92	6.79	8.58	14.71
C19H22N204	342.40	342.1579	66.65	6.48	8.18	18.69
C19H22N205	358.40	358.1529	63.68	6.19	7.82	22.32
C19H22N206	374.40	374.1478	60.95	5.92	7.48	25.64
C19H22N207	390.40	390.1427	58.46	5.68	7.18	28.69
C19H22N208	406.40	406.1376	56.16	5.46	6.89	31.50
C19H22N209	422.40	422.1325	54.03	5.25	6.63	34.09
C19H22N401	322.41	322.1794	70.78	6.88	17.38	4.96
C19H22N402	338.41	338.1743	67.44	6.55	16.56	9.46
C19H22N403	354.41	354.1692	64.39	6.26	15.81	13.54
C19H22N404	370.41	370.1641	61.61	5.99	15.13	17.28
C19H22N405	386.41	386.1590	59.06	5.74	14.50	20.70
C19H22N406	402.41	402.1539	56.71	5.51	13.92	23.86
C19H22N407	418.41	418.1488	54.54	5.30	13.39	26.77
C19H22N408	434.41	434.1437	52.53	5.10	12.90	29.46
C19H22N409	450.41	450.1387	50.67	4.92	12.44	31.97
C19H22N601	350.43	350.1855	65.12	6.33	23.98	4.57
C19H22N602	366.43	366.1804	62.28	6.05	22.94	8.73
C19H22N603	382.43	382.1753	59.67	5.80	21.98	12.55
C19H22N604	398.42	398.1702	57.28	5.57	21.09	16.06
C19H22N605	414.42	414.1652	55.07	5.35	20.28	19.30
C19H22N606	430.42	430.1601	53.02	5.15	19.52	22.30
C19H22N607	446.42	446.1550	51.12	4.97	18.83	25.09
C19H22N608	462.42	462.1499	49.35	4.80	18.17	27.68
C19H22N609	478.42	478.1448	47.70	4.64	17.57	30.10
C19H23N101	281.40	281.1780	81.10	8.24	4.98	5.69
C19H23N102	297.40	297.1729	76.74	7.80	4.71	10.76
C19H23N103	313.40	313.1678	72.82	7.40	4.47	15.32
C19H23N104	329.40	329.1627	69.28	7.04	4.25	19.43
C19H23N105	345.40	345.1576	66.07	6.71	4.06	23.16
C19H23N106	361.40	361.1525	63.15	6.41	3.88	26.56
C19H23N107	377.40	377.1474	60.47	6.14	3.71	29.68
C19H23N108	393.40	393.1424	58.01	5.89	3.56	32.54
C19H23N109	409.40	409.1373	55.74	5.66	3.42	35.17
C19H23N301	309.41	309.1841	73.76	7.49	13.58	5.17
C19H23N302	325.41	325.1790	70.13	7.12	12.91	9.83
C19H23N303	341.41	341.1739	66.84	6.79	12.31	14.06
C19H23N304	357.41	357.1688	63.85	6.49	11.76	17.91
C19H23N305	373.41	373.1638	61.12	6.21	11.25	21.42
C19H23N306	389.41	389.1587	58.60	5.95	10.79	24.65
C19H23N307	405.41	405.1536	56.29	5.72	10.36	27.63
C19H23N308	421.41	421.1485	54.15	5.50	9.97	30.37
C19H23N309	437.41	437.1434	52.17	5.30	9.61	32.92
C19H23N501	337.43	337.1902	67.63	6.87	20.76	4.74
C19H23N502	353.43	353.1852	64.57	6.56	19.82	9.05
C19H23N503	369.43	369.1801	61.77	6.28	18.96	12.99
C19H23N504	385.43	385.1750	59.21	6.01	18.17	16.60
C19H23N505	401.43	401.1699	56.85	5.78	17.45	19.93
C19H23N506	417.43	417.1648	54.67	5.55	16.78	23.00
C19H23N507	433.42	433.1597	52.65	5.35	16.16	25.84
C19H23N508	449.42	449.1546	50.78	5.16	15.58	28.48
C19H23N509	465.42	465.1496	49.03	4.98	15.05	30.94
C19H24N201	296.42	296.1889	76.99	8.16	9.45	5.40
C19H24N202	312.42	312.1838	73.05	7.74	8.97	10.24
C19H24N203	328.41	328.1787	69.49	7.37	8.53	14.62
C19H24N204	344.41	344.1736	66.26	7.02	8.13	18.58
C19H24N205	360.41	360.1685	63.32	6.71	7.77	22.20
C19H24N206	376.41	376.1634	60.63	6.43	7.44	25.50
C19H24N207	392.41	392.1583	58.16	6.16	7.14	28.54
C19H24N208	408.41	408.1533	55.88	5.92	6.86	31.34
C19H24N209	424.41	424.1482	53.77	5.70	6.60	33.93
C19H24N401	324.43	324.1950	70.34	7.46	17.27	4.93
C19H24N402	340.43	340.1899	67.04	7.11	16.46	9.40
C19H24N403	356.43	356.1848	64.03	6.79	15.72	13.47
C19H24N404	372.43	372.1797	61.28	6.50	15.04	17.18
C19H24N405	388.43	388.1747	58.75	6.23	14.42	20.60
C19H24N406	404.43	404.1696	56.43	5.98	13.85	23.74
C19H24N407	420.43	420.1645	54.28	5.75	13.33	26.64
C19H24N408	436.43	436.1594	52.29	5.54	12.84	29.33
C19H24N409	452.42	452.1543	50.44	5.35	12.38	31.83
C19H24N601	352.44	352.2011	64.75	6.86	23.85	4.54
C19H24N602	368.44	368.1961	61.94	6.57	22.81	8.68
C19H24N603	384.44	384.1910	59.36	6.29	21.86	12.49
C19H24N604	400.44	400.1859	56.99	6.04	20.99	15.98
C19H24N605	416.44	416.1808	54.80	5.81	20.18	19.21
C19H24N606	432.44	432.1757	52.77	5.59	19.43	22.20
C19H24N607	448.44	448.1706	50.89	5.39	18.74	24.97
C19H24N608	464.44	464.1655	49.14	5.21	18.10	27.56
C19H24N609	480.44	480.1605	47.50	5.04	17.49	29.97
C19H25N101	283.42	283.1936	80.52	8.89	4.94	5.65
C19H25N102	299.42	299.1885	76.22	8.42	4.68	10.69
C19H25N103	315.42	315.1834	72.35	7.99	4.44	15.22
C19H25N104	331.42	331.1783	68.86	7.60	4.23	19.31
C19H25N105	347.41	347.1733	65.69	7.25	4.03	23.03
C19H25N106	363.41	363.1682	62.80	6.93	3.85	26.42
C19H25N107	379.41	379.1631	60.15	6.64	3.69	29.52
C19H25N108	395.41	395.1580	57.71	6.37	3.54	32.37
C19H25N109	411.41	411.1529	55.47	6.13	3.40	35.00
C19H25N301	311.43	311.1998	73.28	8.09	13.49	5.14
C19H25N302	327.43	327.1947	69.70	7.70	12.83	9.77
C19H25N303	343.43	343.1896	66.45	7.34	12.24	13.98
C19H25N304	359.43	359.1845	63.49	7.01	11.69	17.81
C19H25N305	375.43	375.1794	60.79	6.71	11.19	21.31
C19H25N306	391.43	391.1743	58.30	6.44	10.74	24.52
C19H25N307	407.43	407.1692	56.01	6.18	10.31	27.49
C19H25N308	423.43	423.1641	53.90	5.95	9.92	30.23
C19H25N309	439.43	439.1591	51.93	5.73	9.56	32.77
C19H25N501	339.44	339.2059	67.23	7.42	20.63	4.71

107

C19H25N502	355.44	355.2008	64.20	7.09	19.70	9.00
C19H25N503	371.44	371.1957	61.44	6.78	18.85	12.92
C19H25N504	387.44	387.1906	58.90	6.50	18.08	16.52
C19H25N505	403.44	403.1856	56.57	6.25	17.36	19.83
C19H25N506	419.44	419.1805	54.41	6.01	16.70	22.89
C19H25N507	435.44	435.1754	52.41	5.79	16.08	25.72
C19H25N508	451.44	451.1703	50.55	5.58	15.51	28.35
C19H25N509	467.44	467.1652	48.82	5.39	14.98	30.80
C19H26N201	298.43	298.2045	76.47	8.78	9.39	5.36
C19H26N202	314.43	314.1994	72.58	8.33	8.91	10.18
C19H26N203	330.43	330.1943	69.06	7.93	8.48	14.53
C19H26N204	346.43	346.1892	65.88	7.56	8.09	18.47
C19H26N205	362.43	362.1842	62.97	7.23	7.73	22.07
C19H26N206	378.43	378.1791	60.31	6.93	7.40	25.37
C19H26N207	394.43	394.1740	57.86	6.64	7.10	28.39
C19H26N208	410.43	410.1689	55.60	6.39	6.83	31.19
C19H26N209	426.43	426.1638	53.52	6.15	6.57	33.77
C19H26N401	326.45	326.2106	69.91	8.03	17.16	4.90
C19H26N402	342.44	342.2056	66.64	7.65	16.36	9.34
C19H26N403	358.44	358.2005	63.67	7.31	15.63	13.39
C19H26N404	374.44	374.1954	60.95	7.00	14.96	17.09
C19H26N405	390.44	390.1903	58.45	6.71	14.35	20.49
C19H26N406	406.44	406.1852	56.15	6.45	13.78	23.62
C19H26N407	422.44	422.1801	54.02	6.20	13.26	26.51
C19H26N408	438.44	438.1750	52.05	5.98	12.78	29.19
C19H26N409	454.44	454.1700	50.22	5.77	12.33	31.69
C19H26N601	354.46	354.2168	64.38	7.39	23.71	4.51
C19H26N602	370.46	370.2117	61.60	7.07	22.69	8.64
C19H26N603	386.46	386.2066	59.05	6.78	21.75	12.42
C19H26N604	402.46	402.2015	56.70	6.51	20.88	15.90
C19H26N605	418.46	418.1965	54.54	6.26	20.08	19.12
C19H26N606	434.46	434.1914	52.53	6.03	19.34	22.10
C19H26N607	450.46	450.1863	50.66	5.82	18.66	24.86
C19H26N608	466.45	466.1812	48.92	5.62	18.02	27.44
C19H26N609	482.45	482.1761	47.30	5.43	17.42	29.85
C19H27N101	285.43	285.2093	79.95	9.53	4.91	5.61
C19H27N102	301.43	301.2042	75.71	9.03	4.65	10.62
C19H27N103	317.43	317.1991	71.89	8.57	4.41	15.12
C19H27N104	333.43	333.1940	68.44	8.16	4.20	19.19
C19H27N105	349.43	349.1889	65.31	7.79	4.01	22.89
C19H27N106	365.43	365.1838	62.45	7.45	3.83	26.27
C19H27N107	381.43	381.1787	59.83	7.14	3.67	29.36
C19H27N108	397.43	397.1737	57.42	6.85	3.52	32.21
C19H27N109	413.43	413.1686	55.20	6.58	3.39	34.83
C19H27N301	313.45	313.2154	72.81	8.68	13.41	5.10
C19H27N302	329.45	329.2103	69.27	8.26	12.75	9.71
C19H27N303	345.45	345.2052	66.06	7.88	12.16	13.89
C19H27N304	361.44	361.2001	63.14	7.53	11.63	17.71
C19H27N305	377.44	377.1951	60.46	7.21	11.13	21.19
C19H27N306	393.44	393.1900	58.00	6.92	10.68	24.40
C19H27N307	409.44	409.1849	55.74	6.65	10.26	27.35
C19H27N308	425.44	425.1798	53.64	6.40	9.88	30.09
C19H27N309	441.44	441.1747	51.70	6.17	9.52	32.62
C19H27N501	341.46	341.2215	66.83	7.97	20.51	4.69
C19H27N502	357.46	357.2165	63.84	7.61	19.59	8.95
C19H27N503	373.46	373.2114	61.11	7.29	18.75	12.85
C19H27N504	389.46	389.2063	58.60	6.99	17.98	16.43
C19H27N505	405.46	405.2012	56.29	6.71	17.27	19.73
C19H27N506	421.46	421.1961	54.15	6.46	16.62	22.78
C19H27N507	437.46	437.1910	52.17	6.22	16.01	25.60
C19H27N508	453.46	453.1859	50.33	6.00	15.44	28.23
C19H27N509	469.46	469.1809	48.61	5.80	14.92	30.67
C19H28N201	300.45	300.2202	75.96	9.39	9.32	5.33
C19H28N202	316.45	316.2151	72.12	8.92	8.85	10.11
C19H28N203	332.45	332.2100	68.65	8.49	8.43	14.44
C19H28N204	348.45	348.2049	65.49	8.10	8.04	18.37
C19H28N205	364.45	364.1998	62.62	7.74	7.69	21.95
C19H28N206	380.44	380.1947	59.99	7.42	7.36	25.23
C19H28N207	396.44	396.1896	57.56	7.12	7.07	28.25
C19H28N208	412.44	412.1845	55.33	6.84	6.79	31.03
C19H28N209	428.44	428.1795	53.27	6.59	6.54	33.61
C19H28N401	328.46	328.2263	69.48	8.59	17.06	4.87
C19H28N402	344.46	344.2212	66.25	8.19	16.27	9.29
C19H28N403	360.46	360.2161	63.31	7.83	15.54	13.32
C19H28N404	376.46	376.2110	60.62	7.50	14.88	17.00
C19H28N405	392.46	392.2060	58.15	7.19	14.28	20.38
C19H28N406	408.46	408.2009	55.87	6.91	13.72	23.50
C19H28N407	424.46	424.1958	53.77	6.65	13.20	26.39
C19H28N408	440.46	440.1907	51.81	6.41	12.72	29.06
C19H28N409	456.46	456.1856	50.00	6.18	12.27	31.55
C19H28N601	356.47	356.2324	64.02	7.92	23.58	4.49
C19H28N602	372.47	372.2274	61.27	7.58	22.56	8.59
C19H28N603	388.47	388.2223	58.75	7.27	21.63	12.36
C19H28N604	404.47	404.2172	56.42	6.98	20.78	15.82
C19H28N605	420.47	420.2121	54.28	6.71	19.99	19.03
C19H28N606	436.47	436.2070	52.29	6.47	19.25	21.99
C19H28N607	452.47	452.2019	50.44	6.24	18.57	24.75
C19H28N608	468.47	468.1968	48.71	6.02	17.94	27.32
C19H28N609	484.47	484.1918	47.11	5.83	17.35	29.72
C19H29N101	287.45	287.2249	79.39	10.17	4.87	5.57
C19H29N102	303.45	303.2198	75.21	9.63	4.62	10.55
C19H29N103	319.45	319.2147	71.44	9.15	4.38	15.03
C19H29N104	335.45	335.2096	68.03	8.71	4.18	19.08
C19H29N105	351.45	351.2046	64.93	8.32	3.99	22.76
C19H29N106	367.45	367.1995	62.11	7.96	3.81	26.13
C19H29N107	383.45	383.1944	59.52	7.62	3.65	29.21
C19H29N108	399.44	399.1893	57.13	7.32	3.51	32.04
C19H29N109	415.44	415.1842	54.93	7.04	3.37	34.66
C19H29N301	315.46	315.2310	72.34	9.27	13.32	5.07
C19H29N302	331.46	331.2260	68.85	8.82	12.68	9.65
C19H29N303	347.46	347.2209	65.68	8.41	12.09	13.81
C19H29N304	363.46	363.2158	62.79	8.04	11.56	17.61
C19H29N305	379.46	379.2107	60.14	7.70	11.07	21.08
C19H29N306	395.46	395.2056	57.71	7.39	10.63	24.27
C19H29N307	411.46	411.2005	55.46	7.10	10.21	27.22
C19H29N308	427.46	427.1954	53.39	6.84	9.83	29.94
C19H29N309	443.46	443.1904	51.46	6.59	9.48	32.47
C19H29N501	343.48	343.2372	66.44	8.51	20.39	4.66
C19H29N502	359.48	359.2321	63.48	8.13	19.48	8.90
C19H29N503	375.47	375.2270	60.78	7.79	18.65	12.78
C19H29N504	391.47	391.2219	58.30	7.47	17.89	16.35
C19H29N505	407.47	407.2169	56.01	7.17	17.19	19.63
C19H29N506	423.47	423.2118	53.89	6.90	16.54	22.67
C19H29N507	439.47	439.2067	51.93	6.65	15.94	25.48
C19H29N508	455.47	455.2016	50.10	6.42	15.38	28.10
C19H29N509	471.47	471.1965	48.40	6.20	14.85	30.54
C19H30N201	302.46	302.2358	75.45	10.00	9.26	5.29
C19H30N202	318.46	318.2307	71.66	9.50	8.80	10.05
C19H30N203	334.46	334.2256	68.23	9.04	8.38	14.35
C19H30N204	350.46	350.2205	65.12	8.63	7.99	18.26

C19H30N205	366.46	366.2155	62.27	8.25	7.64	21.83
C19H30N206	382.46	382.2104	59.67	7.91	7.32	25.10
C19H30N207	398.46	398.2053	57.27	7.59	7.03	28.11
C19H30N208	414.46	414.2002	55.06	7.30	6.76	30.88
C19H30N209	430.46	430.1951	53.02	7.02	6.51	33.45
C19H30N401	330.48	330.2419	69.06	9.15	16.95	4.84
C19H30N402	346.48	346.2369	65.87	8.73	16.17	9.24
C19H30N403	362.48	362.2318	62.96	8.34	15.46	13.24
C19H30N404	378.48	378.2267	60.30	7.99	14.80	16.91
C19H30N405	394.47	394.2216	57.85	7.67	14.20	20.28
C19H30N406	410.47	410.2165	55.60	7.37	13.65	23.39
C19H30N407	426.47	426.2114	53.51	7.09	13.14	26.26
C19H30N408	442.47	442.2063	51.58	6.83	12.66	28.93
C19H30N409	458.47	458.2013	49.78	6.60	12.22	31.41
C19H30N601	358.49	358.2481	63.66	8.44	23.44	4.46
C19H30N602	374.49	374.2430	60.94	8.07	22.44	8.54
C19H30N603	390.49	390.2379	58.44	7.74	21.52	12.29
C19H30N604	406.49	406.2328	56.14	7.44	20.67	15.74
C19H30N605	422.49	422.2278	54.02	7.16	19.89	18.93
C19H30N606	438.49	438.2227	52.05	6.90	19.17	21.89
C19H30N607	454.49	454.2176	50.21	6.65	18.49	24.64
C19H30N608	470.49	470.2125	48.51	6.43	17.86	27.20
C19H30N609	486.49	486.2074	46.91	6.22	17.27	29.60
C19H31N101	289.47	289.2406	78.84	10.79	4.84	5.53
C19H31N102	305.46	305.2355	74.71	10.23	4.59	10.48
C19H31N103	321.46	321.2304	70.99	9.72	4.36	14.93
C19H31N104	337.46	337.2253	67.63	9.26	4.15	18.96
C19H31N105	353.46	353.2202	64.56	8.84	3.96	22.63
C19H31N106	369.46	369.2151	61.77	8.46	3.79	25.98
C19H31N107	385.46	385.2100	59.20	8.11	3.63	29.05
C19H31N108	401.46	401.2049	56.85	7.78	3.49	31.88
C19H31N109	417.46	417.1999	54.67	7.49	3.36	34.49
C19H31N301	317.48	317.2467	71.88	9.84	13.24	5.04
C19H31N302	333.48	333.2416	68.43	9.37	12.60	9.60
C19H31N303	349.48	349.2365	65.30	8.94	12.02	13.73
C19H31N304	365.48	365.2314	62.44	8.55	11.50	17.51
C19H31N305	361.48	381.2264	59.82	8.19	11.02	20.97
C19H31N306	397.48	397.2213	57.42	7.86	10.57	24.15
C19H31N307	413.47	413.2162	55.19	7.56	10.16	27.09
C19H31N308	429.47	429.2111	53.14	7.28	9.78	29.80
C19H31N309	445.47	445.2060	51.23	7.01	9.43	32.32
C19H31N501	345.49	345.2528	66.05	9.04	20.27	4.63
C19H31N502	361.49	361.2478	63.13	8.64	19.37	8.85
C19H31N503	377.49	377.2427	60.45	8.28	18.55	12.72
C19H31N504	393.49	393.2376	58.00	7.94	17.80	16.26
C19H31N505	409.49	409.2325	55.73	7.63	17.10	19.54
C19H31N506	425.49	425.2274	53.64	7.34	16.46	22.56
C19H31N507	441.49	441.2223	51.69	7.08	15.86	25.37
C19H31N508	457.49	457.2172	49.88	6.83	15.31	27.98
C19H31N509	473.49	473.2122	48.20	6.60	14.79	30.41
C19H32N201	304.48	304.2514	74.95	10.59	9.20	5.25
C19H32N202	320.48	320.2464	71.21	10.06	8.74	9.98
C19H32N203	336.48	336.2413	67.82	9.59	8.33	14.26
C19H32N204	352.48	352.2362	64.75	9.15	7.95	18.16
C19H32N205	368.48	368.2311	61.93	8.75	7.60	21.71
C19H32N206	384.48	384.2260	59.36	8.39	7.29	24.97
C19H32N207	400.48	400.2209	56.99	8.05	7.00	27.97
C19H32N208	416.48	416.2158	54.80	7.74	6.73	30.73
C19H32N209	432.47	432.2108	52.77	7.46	6.48	33.30
C19H32N401	332.49	332.2576	68.64	9.70	16.85	4.81
C19H32N402	348.49	348.2525	65.49	9.26	16.08	9.18
C19H32N403	364.49	364.2474	62.61	8.85	15.37	13.17
C19H32N404	380.49	380.2423	59.98	8.48	14.72	16.82
C19H32N405	396.49	396.2373	57.56	8.14	14.13	20.18
C19H32N406	412.49	412.2322	55.33	7.82	13.58	23.27
C19H32N407	428.49	428.2271	53.26	7.53	13.08	26.14
C19H32N408	444.49	444.2220	51.34	7.26	12.60	28.80
C19H32N409	460.49	460.2169	49.56	7.00	12.17	31.27
C19H32N601	360.51	360.2637	63.30	8.95	23.31	4.44
C19H32N602	376.51	376.2587	60.61	8.57	22.32	8.50
C19H32N603	392.51	392.2536	58.14	8.22	21.41	12.23
C19H32N604	408.50	408.2485	55.87	7.90	20.57	15.67
C19H32N605	424.50	424.2434	53.76	7.60	19.80	18.84
C19H32N606	440.50	440.2383	51.81	7.32	19.08	21.79
C19H32N607	456.50	456.2332	49.99	7.07	18.41	24.53
C19H32N608	472.50	472.2281	48.30	6.83	17.79	27.09
C19H32N609	488.50	488.2231	46.72	6.60	17.20	29.48
C19H33N101	291.48	291.2562	78.29	11.41	4.81	5.49
C19H33N102	307.48	307.2511	74.22	10.82	4.56	10.41
C19H33N103	323.48	323.2460	70.55	10.28	4.33	14.84
C19H33N104	339.48	339.2409	67.22	9.80	4.13	18.85
C19H33N105	355.48	355.2359	64.20	9.36	3.94	22.50
C19H33N106	371.48	371.2308	61.43	8.95	3.77	25.84
C19H33N107	387.48	387.2257	58.90	8.58	3.61	28.90
C19H33N108	403.48	403.2206	56.56	8.24	3.47	31.72
C19H33N109	419.48	419.2155	54.40	7.93	3.34	34.33
C19H33N301	319.49	319.2623	71.43	10.41	13.15	5.01
C19H33N302	335.49	335.2573	68.02	9.91	12.52	9.54
C19H33N303	351.49	351.2522	64.93	9.46	11.95	13.66
C19H33N304	367.49	367.2471	62.10	9.05	11.43	17.41
C19H33N305	383.49	383.2420	59.51	8.67	10.96	20.86
C19H33N306	399.49	399.2369	57.13	8.33	10.52	24.03
C19H33N307	415.49	415.2318	54.93	8.01	10.11	26.96
C19H33N308	431.49	431.2267	52.89	7.71	9.74	29.66
C19H33N309	447.49	447.2217	51.00	7.43	9.39	32.18
C19H33N501	347.51	347.2685	65.67	9.57	20.15	4.60
C19H33N502	363.51	363.2634	62.78	9.15	19.27	8.80
C19H33N503	379.51	379.2583	60.13	8.76	18.45	12.65
C19H33N504	395.51	395.2532	57.70	8.41	17.71	16.18
C19H33N505	411.51	411.2482	55.46	8.08	17.02	19.44
C19H33N506	427.50	427.2431	53.38	7.78	16.38	22.46
C19H33N507	443.50	443.2380	51.46	7.50	15.79	25.25
C19H33N508	459.50	459.2329	49.66	7.24	15.24	27.86
C19H33N509	475.50	475.2278	47.99	7.00	14.73	30.28
C19H34N201	306.50	306.2671	74.46	11.18	9.14	5.22
C19H34N202	322.50	322.2620	70.76	10.63	8.69	9.92
C19H34N203	338.49	338.2569	67.42	10.12	8.28	14.18
C19H34N204	354.49	354.2518	64.38	9.67	7.90	18.05
C19H34N205	370.49	370.2468	61.60	9.25	7.56	21.59
C19H34N206	386.49	386.2417	59.05	8.87	7.25	24.84
C19H34N207	402.49	402.2366	56.70	8.51	6.96	27.83
C19H34N208	418.49	418.2315	54.53	8.19	6.69	30.58
C19H34N209	434.49	434.2264	52.52	7.89	6.45	33.14
C19H34N401	334.51	334.2732	68.22	10.25	16.75	4.78
C19H34N402	350.51	350.2682	65.11	9.78	15.98	9.13
C19H34N403	366.51	366.2631	62.27	9.35	15.29	13.10
C19H34N404	382.51	382.2580	59.66	8.96	14.65	16.73
C19H34N405	398.51	398.2529	57.27	8.60	14.06	20.07
C19H34N406	414.51	414.2478	55.06	8.27	13.52	23.16
C19H34N407	430.51	430.2427	53.01	7.96	13.01	26.01

C19H34N408	446.50	446.2376	51.11	7.68	12.55	28.67
C19H34N409	462.50	462.2326	49.34	7.41	12.11	31.13
C19H34N601	362.52	362.2794	62.95	9.45	23.18	4.41
C19H34N602	378.52	378.2743	60.29	9.05	22.20	8.45
C19H34N603	394.52	394.2692	57.85	8.69	21.30	12.17
C19H34N604	410.52	410.2641	55.59	8.35	20.47	15.59
C19H34N605	426.52	426.2590	53.51	8.04	19.70	18.76
C19H34N606	442.52	442.2540	51.57	7.74	18.99	21.69
C19H34N607	458.52	458.2489	49.77	7.47	18.33	24.43
C19H34N608	474.52	474.2438	48.09	7.22	17.71	26.97
C19H34N609	490.52	490.2387	46.52	6.99	17.13	29.36
C19H35N101	293.50	293.2718	77.76	12.02	4.77	5.45
C19H35N102	309.50	309.2668	73.74	11.40	4.53	10.34
C19H35N103	325.50	325.2617	70.11	10.84	4.30	14.75
C19H35N104	341.50	341.2566	66.83	10.33	4.10	18.74
C19H35N105	357.49	357.2515	63.84	9.87	3.92	22.38
C19H35N106	373.49	373.2464	61.10	9.45	3.75	25.70
C19H35N107	389.49	389.2413	58.59	9.06	3.60	28.75
C19H35N108	405.49	405.2362	56.28	8.70	3.45	31.57
C19H35N109	421.49	421.2312	54.14	8.37	3.32	34.16
C19H35N301	321.51	321.2780	70.98	10.97	13.07	4.98
C19H35N302	337.51	337.2729	67.62	10.45	12.45	9.48
C19H35N303	353.51	353.2678	64.56	9.98	11.89	13.58
C19H35N304	369.51	369.2627	61.76	9.55	11.37	17.32
C19H35N305	385.51	385.2577	59.20	9.15	10.90	20.75
C19H35N306	401.51	401.2526	56.84	8.79	10.47	23.91
C19H35N307	417.51	417.2475	54.66	8.45	10.06	26.82
C19H35N308	433.51	433.2424	52.64	8.14	9.69	29.53
C19H35N309	449.51	449.2373	50.77	7.85	9.35	32.03
C19H35N501	349.52	349.2841	65.29	10.09	20.04	4.58
C19H35N502	365.52	365.2791	62.43	9.65	19.16	8.75
C19H35N503	381.52	381.2740	59.82	9.25	18.36	12.58
C19H35N504	397.52	397.2689	57.41	8.87	17.62	16.10
C19H35N505	413.52	413.2638	55.19	8.53	16.94	19.35
C19H35N506	429.52	429.2587	53.13	8.21	16.31	22.35
C19H35N507	445.52	445.2536	51.22	7.92	15.72	25.14
C19H35N508	461.52	461.2485	49.45	7.64	15.17	27.73
C19H35N509	477.52	477.2435	47.79	7.39	14.67	30.15
C19H36N201	308.51	308.2827	73.97	11.76	9.08	5.19
C19H36N202	324.51	324.2777	70.32	11.18	8.63	9.86
C19H36N203	340.51	340.2726	67.02	10.66	8.23	14.10
C19H36N204	356.51	356.2675	64.01	10.18	7.86	17.95
C19H36N205	372.51	372.2624	61.26	9.74	7.52	21.48
C19H36N206	388.51	388.2573	58.74	9.34	7.21	24.71
C19H36N207	404.51	404.2522	56.42	8.97	6.93	27.69
C19H36N208	420.51	420.2471	54.27	8.63	6.66	30.44
C19H36N209	436.51	436.2421	52.28	8.31	6.42	32.99
C19H36N401	336.52	336.2889	67.81	10.78	16.65	4.75
C19H36N402	352.52	352.2838	64.74	10.29	15.89	9.08
C19H36N403	368.52	368.2787	61.93	9.85	15.20	13.02
C19H36N404	384.52	384.2736	59.35	9.44	14.57	16.64
C19H36N405	400.52	400.2686	56.98	9.06	13.99	19.97
C19H36N406	416.52	416.2635	54.79	8.71	13.45	23.05
C19H36N407	432.52	432.2584	52.76	8.39	12.95	25.89
C19H36N408	448.52	448.2533	50.88	8.09	12.49	28.54
C19H36N409	464.52	464.2482	49.13	7.81	12.06	31.00
C19H36N601	364.54	364.2950	62.60	9.95	23.05	4.39
C19H36N602	380.54	380.2900	59.97	9.54	22.08	8.41
C19H36N603	396.54	396.2849	57.55	9.15	21.19	12.10
C19H36N604	412.54	412.2798	55.32	8.80	20.37	15.51
C19H36N605	428.54	428.2747	53.25	8.47	19.61	18.67
C19H36N606	444.54	444.2696	51.34	8.16	18.91	21.59
C19H36N607	460.53	460.2645	49.55	7.88	18.25	24.32
C19H36N608	476.53	476.2594	47.89	7.61	17.64	26.86
C19H36N609	492.53	492.2544	46.33	7.37	17.06	29.24
C19H37N101	295.51	295.2875	77.23	12.62	4.74	5.41
C19H37N102	311.51	311.2824	73.26	11.97	4.50	10.27
C19H37N103	327.51	327.2773	69.68	11.39	4.28	14.66
C19H37N104	343.51	343.2722	66.44	10.86	4.08	18.63
C19H37N105	359.51	359.2672	63.48	10.37	3.90	22.25
C19H37N106	375.51	375.2621	60.77	9.93	3.73	25.56
C19H37N107	391.51	391.2570	58.29	9.53	3.58	28.61
C19H37N108	407.51	407.2519	56.00	9.15	3.44	31.41
C19H37N109	423.51	423.2468	53.89	8.81	3.31	34.00
C19H37N301	323.53	323.2936	70.54	11.53	12.99	4.95
C19H37N302	339.53	339.2886	67.21	10.98	12.38	9.42
C19H37N303	355.53	355.2835	64.19	10.49	11.82	13.50
C19H37N304	371.52	371.2784	61.43	10.04	11.31	17.23
C19H37N305	387.52	387.2733	58.89	9.62	10.84	20.64
C19H37N306	403.52	403.2682	56.55	9.24	10.41	23.79
C19H37N307	419.52	419.2631	54.40	8.89	10.02	26.70
C19H37N308	435.52	435.2580	52.40	8.56	9.65	29.39
C19H37N309	451.52	451.2530	50.54	8.26	9.31	31.89
C19H37N501	351.54	351.2998	64.92	10.61	19.92	4.55
C19H37N502	367.54	367.2947	62.09	10.15	19.05	8.71
C19H37N503	383.54	383.2896	59.50	9.72	18.26	12.51
C19H37N504	399.54	399.2845	57.12	9.33	17.53	16.02
C19H37N505	415.54	415.2794	54.92	8.98	16.85	19.25
C19H37N506	431.54	431.2744	52.88	8.64	16.23	22.25
C19H37N507	447.54	447.2693	50.99	8.33	15.65	25.02
C19H37N508	463.54	463.2642	49.23	8.05	15.11	27.61
C19H37N509	479.53	479.2591	47.59	7.78	14.60	30.03
C19H38N201	310.53	310.2984	73.49	12.33	9.02	5.15
C19H38N202	326.53	326.2933	69.89	11.73	8.58	9.80
C19H38N203	342.53	342.2882	66.63	11.18	8.18	14.01
C19H38N204	358.53	358.2831	63.65	10.68	7.81	17.85
C19H38N205	374.53	374.2781	60.93	10.23	7.48	21.36
C19H38N206	390.52	390.2730	58.44	9.81	7.17	24.58
C19H38N207	406.52	406.2679	56.14	9.42	6.89	27.55
C19H38N208	422.52	422.2628	54.01	9.07	6.63	30.29
C19H38N209	438.52	438.2577	52.04	8.73	6.39	32.84
C19H38N401	338.54	338.3045	67.41	11.31	16.55	4.73
C19H38N402	354.54	354.2995	64.37	10.80	15.80	9.03
C19H38N403	370.54	370.2944	61.59	10.34	15.12	12.95
C19H38N404	386.54	386.2893	59.04	9.91	14.49	16.56
C19H38N405	402.54	402.2842	56.69	9.52	13.92	19.87
C19H38N406	418.54	418.2791	54.53	9.15	13.39	22.94
C19H38N407	434.54	434.2740	52.52	8.81	12.89	25.77
C19H38N408	450.54	450.2689	50.65	8.50	12.44	28.41
C19H38N409	466.54	466.2639	48.92	8.21	12.01	30.86
C19H38N601	366.55	366.3107	62.26	10.45	22.93	4.36
C19H38N602	382.55	382.3056	59.65	10.01	21.97	8.36
C19H38N603	398.55	398.3005	57.26	9.61	21.09	12.04
C19H38N604	414.55	414.2954	55.05	9.24	20.27	15.44
C19H38N605	430.55	430.2903	53.00	8.90	19.52	18.58
C19H38N606	446.55	446.2853	51.11	8.58	18.82	21.50
C19H38N607	462.55	462.2802	49.34	8.28	18.17	24.21
C19H38N608	478.55	478.2751	47.69	8.00	17.56	26.75
C19H38N609	494.55	494.2700	46.15	7.75	16.99	29.12
C19H39N101	297.53	297.3031	76.70	13.21	4.71	5.38

C19H39N1O2	313.53	313.2981	72.79	12.54	4.47	10.21
C19H39N1O3	329.53	329.2930	69.25	11.93	4.25	14.57
C19H39N1O4	345.53	345.2879	66.05	11.38	4.05	18.52
C19H39N1O5	361.53	361.2828	63.12	10.87	3.87	22.13
C19H39N1O6	377.53	377.2777	60.45	10.41	3.71	25.43
C19H39N1O7	393.53	393.2726	57.99	9.99	3.56	28.46
C19H39N1O8	409.52	409.2675	55.73	9.60	3.42	31.25
C19H39N1O9	425.52	425.2625	53.63	9.24	3.29	33.84
C19H39N3O1	325.54	325.3093	70.10	12.08	12.91	4.91
C19H39N3O2	341.54	341.3042	66.82	11.51	12.30	9.37
C19H39N3O3	357.54	357.2991	63.83	10.99	11.75	13.42
C19H39N3O4	373.54	373.2940	61.09	10.52	11.25	17.13
C19H39N3O5	389.54	389.2890	58.58	10.09	10.79	20.54
C19H39N3O6	405.54	405.2839	56.27	9.69	10.36	23.67
C19H39N3O7	421.54	421.2788	54.14	9.33	9.97	26.57
C19H39N3O8	437.54	437.2737	52.16	8.98	9.60	29.25
C19H39N3O9	453.54	453.2686	50.32	8.67	9.26	31.75
C19H39N5O1	353.56	353.3154	64.55	11.12	19.81	4.53
C19H39N5O2	369.55	369.3104	61.75	10.64	18.95	8.66
C19H39N5O3	385.55	385.3053	59.19	10.20	18.16	12.45
C19H39N5O4	401.55	401.3002	56.83	9.79	17.44	15.94
C19H39N5O5	417.55	417.2951	54.65	9.41	16.77	19.16
C19H39N5O6	433.55	433.2900	52.64	9.07	16.15	22.14
C19H39N5O7	449.55	449.2849	50.76	8.74	15.58	24.91
C19H39N5O8	465.55	465.2798	49.02	8.44	15.04	27.49
C19H39N5O9	481.55	481.2748	47.39	8.16	14.54	29.90
C19H40N2O1	312.54	312.3140	73.02	12.90	8.96	5.12
C19H40N2O2	328.54	328.3090	69.46	12.27	8.53	9.74
C19H40N2O3	344.54	344.3039	66.24	11.70	8.13	13.93
C19H40N2O4	360.54	360.2988	63.30	11.18	7.77	17.75
C19H40N2O5	376.54	376.2937	60.61	10.71	7.44	21.25
C19H40N2O6	392.54	392.2886	58.14	10.27	7.14	24.46
C19H40N2O7	408.54	408.2835	55.86	9.87	6.86	27.41
C19H40N2O8	424.54	424.2784	53.76	9.50	6.60	30.15
C19H40N2O9	440.54	440.2734	51.80	9.15	6.36	32.69
C19H40N4O1	340.56	340.3202	67.01	11.84	16.45	4.70
C19H40N4O2	356.56	356.3151	64.00	11.31	15.71	8.97
C19H40N4O3	372.56	372.3100	61.26	10.82	15.04	12.88
C19H40N4O4	388.56	388.3049	58.73	10.38	14.42	16.47
C19H40N4O5	404.55	404.2998	56.41	9.97	13.85	19.77
C19H40N4O6	420.55	420.2948	54.26	9.59	13.32	22.83
C19H40N4O7	436.55	436.2897	52.28	9.24	12.83	25.65
C19H40N4O8	452.55	452.2846	50.43	8.91	12.38	28.28
C19H40N4O9	468.55	468.2795	48.71	8.60	11.96	30.73
C19H40N6O1	368.57	368.3263	61.92	10.94	22.80	4.34
C19H40N6O2	384.57	384.3213	59.34	10.48	21.85	8.32
C19H40N6O3	400.57	400.3162	56.97	10.07	20.98	11.98
C19H40N6O4	416.57	416.3111	54.78	9.68	20.17	15.36
C19H40N6O5	432.57	432.3060	52.76	9.32	19.43	18.49
C19H40N6O6	448.57	448.3009	50.88	8.99	18.74	21.40
C19H40N6O7	464.57	464.2958	49.12	8.68	18.09	24.11
C19H40N6O8	480.57	480.2907	47.49	8.39	17.49	26.63
C19H40N6O9	496.57	496.2857	45.96	8.12	16.92	29.00
C19H41N1O1	299.54	299.3188	76.19	13.80	4.68	5.34
C19H41N1O2	315.54	315.3137	72.32	13.10	4.44	10.14
C19H41N1O3	331.54	331.3086	68.83	12.46	4.22	14.48
C19H41N1O4	347.54	347.3035	65.66	11.89	4.03	18.41
C19H41N1O5	363.54	363.2985	62.77	11.37	3.85	22.00
C19H41N1O6	379.54	379.2934	60.13	10.89	3.69	25.29
C19H41N1O7	395.54	395.2883	57.70	10.45	3.54	28.31

C19H41N1O8	411.54	411.2832	55.45	10.04	3.40	31.10
C19H41N1O9	427.54	427.2781	53.38	9.67	3.28	33.68
C19H41N3O1	327.56	327.3249	69.67	12.62	12.83	4.88
C19H41N3O2	343.56	343.3199	66.43	12.03	12.23	9.31
C19H41N3O3	359.56	359.3148	63.47	11.49	11.69	13.35
C19H41N3O4	375.56	375.3097	60.77	11.00	11.19	17.04
C19H41N3O5	391.56	391.3046	58.28	10.55	10.73	20.43
C19H41N3O6	407.56	407.2995	56.00	10.14	10.31	23.55
C19H41N3O7	423.55	423.2944	53.88	9.76	9.92	26.44
C19H41N3O8	439.55	439.2893	51.92	9.40	9.56	29.12
C19H41N3O9	455.55	455.2843	50.10	9.07	9.22	31.61
C19H41N5O1	355.57	355.3311	64.18	11.62	19.70	4.50
C19H41N5O2	371.57	371.3260	61.42	11.12	18.85	8.61
C19H41N5O3	387.57	387.3209	58.88	10.66	18.07	12.38
C19H41N5O4	403.57	403.3158	56.55	10.24	17.35	15.86
C19H41N5O5	419.57	419.3107	54.39	9.85	16.69	19.07
C19H41N5O6	435.57	435.3057	52.39	9.49	16.08	22.04
C19H41N5O7	451.57	451.3006	50.54	9.15	15.51	24.80
C19H41N5O8	467.57	467.2955	48.81	8.84	14.98	27.37
C19H41N5O9	483.57	483.2904	47.19	8.55	14.48	29.78
C19H42N2O1	314.56	314.3297	72.55	13.46	8.91	5.09
C19H42N2O2	330.56	330.3246	69.04	12.81	8.47	9.68
C19H42N2O3	346.56	346.3195	65.85	12.22	8.08	13.85
C19H42N2O4	362.56	362.3144	62.94	11.68	7.73	17.65
C19H42N2O5	378.56	378.3094	60.28	11.18	7.40	21.13
C19H42N2O6	394.56	394.3043	57.84	10.73	7.10	24.33
C19H42N2O7	410.56	410.2992	55.59	10.31	6.82	27.28
C19H42N2O8	426.56	426.2941	53.50	9.92	6.57	30.01
C19H42N2O9	442.55	442.2890	51.57	9.57	6.33	32.54
C19H42N4O1	342.57	342.3358	66.62	12.36	16.35	4.67
C19H42N4O2	358.57	358.3308	63.64	11.81	15.62	8.92
C19H42N4O3	374.57	374.3257	60.93	11.30	14.96	12.81
C19H42N4O4	390.57	390.3206	58.43	10.84	14.34	16.39
C19H42N4O5	406.57	406.3155	56.13	10.41	13.78	19.68
C19H42N4O6	422.57	422.3104	54.01	10.02	13.26	22.72
C19H42N4O7	438.57	438.3053	52.04	9.65	12.77	25.54
C19H42N4O8	454.57	454.3002	50.20	9.31	12.33	28.16
C19H42N4O9	470.57	470.2952	48.50	9.00	11.91	30.60
C19H42N6O1	370.59	370.3420	61.58	11.42	22.68	4.32
C19H42N6O2	386.59	386.3369	59.03	10.95	21.74	8.28
C19H42N6O3	402.58	402.3318	56.69	10.52	20.88	11.92
C19H42N6O4	418.58	418.3267	54.52	10.11	20.08	15.29
C19H42N6O5	434.58	434.3216	52.51	9.74	19.34	18.41
C19H42N6O6	450.58	450.3166	50.65	9.40	18.65	21.30
C19H42N6O7	466.58	466.3115	48.91	9.07	18.01	24.00
C19H42N6O8	482.58	482.3064	47.29	8.77	17.41	26.52
C19H42N6O9	498.58	498.3013	45.77	8.49	16.86	28.88
C19H43N3O1	329.57	329.3406	69.24	13.15	12.75	4.85
C19H43N3O2	345.57	345.3355	66.04	12.54	12.16	9.26
C19H43N3O3	361.57	361.3304	63.12	11.99	11.62	13.27
C19H43N3O4	377.57	377.3253	60.44	11.48	11.13	16.95
C19H43N3O5	393.57	393.3202	57.98	11.01	10.68	20.33
C19H43N3O6	409.57	409.3152	55.72	10.58	10.26	23.44
C19H43N3O7	425.57	425.3101	53.62	10.18	9.87	26.32
C19H43N3O8	441.57	441.3050	51.68	9.82	9.52	28.99
C19H43N3O9	457.57	457.2999	49.87	9.47	9.18	31.47
C19H43N5O1	357.59	357.3467	63.82	12.12	19.58	4.47
C19H43N5O2	373.59	373.3417	61.09	11.60	18.75	8.57
C19H43N5O3	389.59	389.3366	58.58	11.13	17.98	12.32
C19H43N5O4	405.59	405.3315	56.27	10.69	17.27	15.78

C19H43N5O5	421.59	421.3264	54.13	10.28	16.61	18.98
C19H43N5O6	437.58	437.3213	52.15	9.90	16.00	21.94
C19H43N5O7	453.58	453.3162	50.31	9.56	15.44	24.69
C19H43N5O8	469.58	469.3111	48.60	9.23	14.91	27.26
C19H43N5O9	485.58	485.3061	47.00	8.93	14.42	29.65
C19H44N4O1	344.59	344.3515	66.23	12.87	16.26	4.64
C19H44N4O2	360.59	360.3464	63.29	12.30	15.54	8.87
C19H44N4O3	376.59	376.3413	60.60	11.78	14.88	12.75
C19H44N4O4	392.59	392.3362	58.13	11.30	14.27	16.30
C19H44N4O5	408.59	408.3311	55.85	10.85	13.71	19.58
C19H44N4O6	424.59	424.3261	53.75	10.45	13.20	22.61
C19H44N4O7	440.59	440.3210	51.80	10.07	12.72	25.42
C19H44N4O8	456.58	456.3159	49.98	9.71	12.27	28.03
C19H44N4O9	472.58	472.3108	48.29	9.38	11.86	30.47
C19H44N6O1	372.60	372.3576	61.25	11.90	22.55	4.29
C19H44N6O2	388.60	388.3526	58.73	11.41	21.63	8.23
C19H44N6O3	404.60	404.3475	56.40	10.96	20.77	11.86
C19H44N6O4	420.60	420.3424	54.26	10.54	19.98	15.22
C19H44N6O5	436.60	436.3373	52.27	10.16	19.25	18.32
C19H44N6O6	452.60	452.3322	50.42	9.80	18.57	21.21
C19H44N6O7	468.60	468.3271	48.70	9.46	17.93	23.90
C19H44N6O8	484.60	484.3220	47.09	9.15	17.34	26.41
C19H44N6O9	500.60	500.3170	45.59	8.86	16.79	28.76
C19H45N5O1	359.60	359.3624	63.46	12.61	19.48	4.45
C19H45N5O2	375.60	375.3573	60.76	12.08	18.65	8.52
C19H45N5O3	391.60	391.3522	58.28	11.58	17.88	12.26
C19H45N5O4	407.60	407.3471	55.99	11.13	17.18	15.70
C19H45N5O5	423.60	423.3420	53.87	10.71	16.53	18.88
C19H45N5O6	439.60	439.3370	51.91	10.32	15.93	21.84
C19H45N5O7	455.60	455.3319	50.09	9.96	15.37	24.58
C19H45N5O8	471.60	471.3268	48.39	9.62	14.85	27.14
C19H45N5O9	487.60	487.3217	46.80	9.30	14.36	29.53
C19H46N6O1	374.62	374.3733	60.92	12.38	22.43	4.27
C19H46N6O2	390.62	390.3682	58.42	11.87	21.51	8.19
C19H46N6O3	406.62	406.3631	56.12	11.40	20.67	11.80
C19H46N6O4	422.62	422.3580	54.00	10.97	19.89	15.14
C19H46N6O5	438.62	438.3529	52.03	10.57	19.16	18.24
C19H46N6O6	454.62	454.3479	50.20	10.20	18.49	21.12
C19H46N6O7	470.61	470.3428	48.49	9.85	17.86	23.80
C19H46N6O8	486.61	486.3377	46.90	9.53	17.27	26.30
C19H46N6O9	502.61	502.3326	45.41	9.23	16.72	28.65
C20H10N2O1	294.32	294.0793	81.62	3.42	9.52	5.44
C20H10N2O2	310.31	310.0742	77.41	3.25	9.03	10.31
C20H10N2O3	326.31	326.0691	73.62	3.09	8.58	14.71
C20H10N2O4	342.31	342.0640	70.18	2.94	8.18	18.70
C20H10N2O5	358.31	358.0590	67.04	2.81	7.82	22.33
C20H10N2O6	374.31	374.0539	64.18	2.69	7.48	25.65
C20H10N2O7	390.31	390.0488	61.55	2.58	7.18	28.69
C20H10N2O8	406.31	406.0437	59.12	2.48	6.89	31.50
C20H10N2O9	422.31	422.0386	56.88	2.39	6.63	34.10
C20H10N4O1	322.33	322.0855	74.53	3.13	17.38	4.96
C20H10N4O2	338.33	338.0804	71.00	2.98	16.56	9.46
C20H10N4O3	354.33	354.0753	67.80	2.84	15.81	13.55
C20H10N4O4	370.33	370.0702	64.87	2.72	15.13	17.28
C20H10N4O5	386.33	386.0651	62.18	2.61	14.50	20.71
C20H10N4O6	402.33	402.0600	59.71	2.51	13.93	23.86
C20H10N4O7	418.33	418.0549	57.42	2.41	13.39	26.77
C20H10N4O8	434.32	434.0499	55.31	2.32	12.90	29.47
C20H10N4O9	450.32	450.0448	53.34	2.24	12.44	31.98
C20H10N6O1	350.34	350.0916	68.57	2.88	23.99	4.57

C20H10N6O2	366.34	366.0865	65.57	2.75	22.94	8.73
C20H10N6O3	382.34	382.0814	62.83	2.64	21.98	12.55
C20H10N6O4	398.34	398.0763	60.31	2.53	21.10	16.07
C20H10N6O5	414.34	414.0713	57.98	2.43	20.28	19.31
C20H10N6O6	430.34	430.0662	55.82	2.34	19.53	22.31
C20H10N6O7	446.34	446.0611	53.82	2.26	18.83	25.09
C20H10N6O8	462.34	462.0560	51.96	2.18	18.18	27.68
C20H10N6O9	478.34	478.0509	50.22	2.11	17.57	30.10
C20H11N1O1	281.32	281.0841	85.39	3.94	4.98	5.69
C20H11N1O2	297.32	297.0790	80.80	3.73	4.71	10.76
C20H11N1O3	313.32	313.0739	76.67	3.54	4.47	15.32
C20H11N1O4	329.31	329.0688	72.95	3.37	4.25	19.43
C20H11N1O5	345.31	345.0637	69.57	3.21	4.06	23.17
C20H11N1O6	361.31	361.0586	66.49	3.07	3.88	26.57
C20H11N1O7	377.31	377.0535	63.67	2.94	3.71	29.68
C20H11N1O8	393.31	393.0485	61.08	2.82	3.56	32.54
C20H11N1O9	409.31	409.0434	58.69	2.71	3.42	35.18
C20H11N3O1	309.33	309.0902	77.66	3.58	13.58	5.17
C20H11N3O2	325.33	325.0851	73.84	3.41	12.92	9.84
C20H11N3O3	341.33	341.0800	70.38	3.25	12.31	14.06
C20H11N3O4	357.33	357.0749	67.23	3.10	11.76	17.91
C20H11N3O5	373.33	373.0699	64.35	2.97	11.26	21.43
C20H11N3O6	389.33	389.0648	61.70	2.85	10.79	24.66
C20H11N3O7	405.33	405.0597	59.27	2.74	10.37	27.63
C20H11N3O8	421.33	421.0546	57.02	2.63	9.97	30.38
C20H11N3O9	437.33	437.0495	54.93	2.54	9.61	32.93
C20H11N5O1	337.34	337.0964	71.21	3.29	20.76	4.74
C20H11N5O2	353.34	353.0913	67.99	3.14	19.82	9.06
C20H11N5O3	369.34	369.0862	65.04	3.00	18.96	13.00
C20H11N5O4	385.34	385.0811	62.34	2.88	18.17	16.61
C20H11N5O5	401.34	401.0760	59.86	2.76	17.45	19.93
C20H11N5O6	417.34	417.0709	57.56	2.66	16.78	23.00
C20H11N5O7	433.34	433.0658	55.44	2.56	16.16	25.84
C20H11N5O8	449.34	449.0608	53.46	2.47	15.59	28.49
C20H11N5O9	465.34	465.0557	51.62	2.38	15.05	30.94
C20H12N2O1	296.33	296.0950	81.07	4.08	9.45	5.40
C20H12N2O2	312.33	312.0899	76.91	3.87	8.97	10.25
C20H12N2O3	328.33	328.0848	73.17	3.68	8.53	14.62
C20H12N2O4	344.33	344.0797	69.77	3.51	8.14	18.59
C20H12N2O5	360.33	360.0746	66.67	3.36	7.77	22.20
C20H12N2O6	376.33	376.0695	63.83	3.21	7.44	25.51
C20H12N2O7	392.33	392.0644	61.23	3.08	7.14	28.55
C20H12N2O8	408.33	408.0594	58.83	2.96	6.86	31.35
C20H12N2O9	424.33	424.0543	56.61	2.85	6.60	33.93
C20H12N4O1	324.34	324.1011	74.06	3.73	17.27	4.93
C20H12N4O2	340.34	340.0960	70.58	3.55	16.46	9.40
C20H12N4O3	356.34	356.0909	67.41	3.39	15.72	13.47
C20H12N4O4	372.34	372.0858	64.52	3.25	15.05	17.19
C20H12N4O5	388.34	388.0808	61.86	3.11	14.43	20.60
C20H12N4O6	404.34	404.0757	59.41	2.99	13.86	23.74
C20H12N4O7	420.34	420.0706	57.15	2.88	13.33	26.64
C20H12N4O8	436.34	436.0655	55.05	2.77	12.84	29.33
C20H12N4O9	452.34	452.0604	53.11	2.67	12.39	31.83
C20H12N6O1	352.36	352.1073	68.18	3.43	23.85	4.54
C20H12N6O2	368.36	368.1022	65.21	3.28	22.81	8.69
C20H12N6O3	384.36	384.0971	62.50	3.15	21.87	12.49
C20H12N6O4	400.36	400.0920	60.00	3.02	20.99	15.99
C20H12N6O5	416.36	416.0869	57.70	2.91	20.18	19.21
C20H12N6O6	432.36	432.0818	55.56	2.80	19.44	22.20
C20H12N6O7	448.35	448.0767	53.58	2.70	18.74	24.98

C20H12N608	464.35	464.0717	51.73	2.60	18.10	27.56
C20H12N609	480.35	480.0666	50.01	2.52	17.50	29.98
C20H13N101	283.33	283.0997	84.78	4.62	4.94	5.65
C20H13N102	299.33	299.0946	80.25	4.38	4.68	10.69
C20H13N103	315.33	315.0895	76.18	4.16	4.44	15.22
C20H13N104	331.33	331.0844	72.50	3.95	4.23	19.32
C20H13N105	347.33	347.0794	69.16	3.77	4.03	23.03
C20H13N106	363.33	363.0743	66.12	3.61	3.86	26.42
C20H13N107	379.33	379.0692	63.33	3.45	3.69	29.52
C20H13N108	395.33	395.0641	60.77	3.31	3.54	32.38
C20H13N109	411.33	411.0590	58.40	3.19	3.41	35.01
C20H13N301	311.35	311.1059	77.16	4.21	13.50	5.14
C20H13N302	327.35	327.1008	73.39	4.00	12.84	9.78
C20H13N303	343.34	343.0957	69.97	3.82	12.24	13.98
C20H13N304	359.34	359.0906	66.85	3.65	11.69	17.81
C20H13N305	375.34	375.0855	64.00	3.49	11.20	21.31
C20H13N306	391.34	391.0804	61.38	3.35	10.74	24.53
C20H13N307	407.34	407.0753	58.97	3.22	10.32	27.49
C20H13N308	423.34	423.0703	56.74	3.10	9.93	30.23
C20H13N309	439.34	439.0652	54.68	2.98	9.56	32.78
C20H13N501	339.36	339.1120	70.79	3.86	20.64	4.71
C20H13N502	355.36	355.1069	67.60	3.69	19.71	9.00
C20H13N503	371.36	371.1018	64.69	3.53	18.86	12.93
C20H13N504	387.36	387.0967	62.02	3.38	18.08	16.52
C20H13N505	403.36	403.0917	59.56	3.25	17.36	19.83
C20H13N506	419.36	419.0866	57.28	3.12	16.70	22.89
C20H13N507	435.36	435.0815	55.18	3.01	16.09	25.73
C20H13N508	451.36	451.0764	53.22	2.90	15.52	28.36
C20H13N509	467.35	467.0713	51.40	2.80	14.99	30.81
C20H14N201	298.35	298.1106	80.52	4.73	9.39	5.36
C20H14N202	314.35	314.1055	76.42	4.49	8.91	10.18
C20H14N203	330.35	330.1004	72.72	4.27	8.48	14.53
C20H14N204	346.35	346.0953	69.36	4.07	8.09	18.48
C20H14N205	362.34	362.0903	66.30	3.89	7.73	22.08
C20H14N206	378.34	378.0852	63.49	3.73	7.40	25.37
C20H14N207	394.34	394.0801	60.92	3.58	7.10	28.40
C20H14N208	410.34	410.0750	58.54	3.44	6.83	31.19
C20H14N209	426.34	426.0699	56.35	3.31	6.57	33.77
C20H14N401	326.36	326.1168	73.61	4.32	17.17	4.90
C20H14N402	342.36	342.1117	70.17	4.12	16.36	9.35
C20H14N403	358.36	358.1066	67.03	3.94	15.63	13.39
C20H14N404	374.36	374.1015	64.17	3.77	14.97	17.10
C20H14N405	390.36	390.0964	61.54	3.62	14.35	20.49
C20H14N406	406.36	406.0913	59.12	3.47	13.79	23.62
C20H14N407	422.36	422.0862	56.88	3.34	13.27	26.52
C20H14N408	438.36	438.0812	54.80	3.22	12.78	29.20
C20H14N409	454.36	454.0761	52.87	3.11	12.33	31.69
C20H14N601	354.37	354.1229	67.79	3.98	23.72	4.51
C20H14N602	370.37	370.1178	64.86	3.81	22.69	8.64
C20H14N603	386.37	386.1127	62.17	3.65	21.75	12.42
C20H14N604	402.37	402.1076	59.70	3.51	20.89	15.91
C20H14N605	418.37	418.1026	57.42	3.37	20.09	19.12
C20H14N606	434.37	434.0975	55.30	3.25	19.35	22.10
C20H14N607	450.37	450.0924	53.34	3.13	18.66	24.87
C20H14N608	466.37	466.0873	51.51	3.03	18.02	27.44
C20H14N609	482.37	482.0822	49.80	2.93	17.42	29.85
C20H15N101	285.35	285.1154	84.19	5.30	4.91	5.61
C20H15N102	301.35	301.1103	79.72	5.02	4.65	10.62
C20H15N103	317.35	317.1052	75.70	4.76	4.41	15.12
C20H15N104	333.35	333.1001	72.06	4.54	4.20	19.20
C20H15N105	349.35	349.0950	68.76	4.33	4.01	22.90
C20H15N106	365.35	365.0899	65.75	4.14	3.83	26.28
C20H15N107	381.35	381.0848	62.99	3.96	3.67	29.37
C20H15N108	397.34	397.0798	60.46	3.81	3.53	32.21
C20H15N109	413.34	413.0747	58.12	3.66	3.39	34.84
C20H15N301	313.36	313.1215	76.66	4.82	13.41	5.11
C20H15N302	329.36	329.1164	72.94	4.59	12.76	9.72
C20H15N303	345.36	345.1113	69.56	4.38	12.17	13.90
C20H15N304	361.36	361.1062	66.48	4.18	11.63	17.71
C20H15N305	377.36	377.1012	63.66	4.01	11.14	21.20
C20H15N306	393.36	393.0961	61.07	3.84	10.68	24.40
C20H15N307	409.36	409.0910	58.68	3.69	10.26	27.36
C20H15N308	425.36	425.0859	56.48	3.55	9.88	30.09
C20H15N309	441.36	441.0808	54.43	3.43	9.52	32.63
C20H15N501	341.38	341.1277	70.37	4.43	20.52	4.69
C20H15N502	357.37	357.1226	67.22	4.23	19.60	8.95
C20H15N503	373.37	373.1175	64.34	4.05	18.76	12.86
C20H15N504	389.37	389.1124	61.69	3.88	17.99	16.44
C20H15N505	405.37	405.1073	59.26	3.73	17.28	19.73
C20H15N506	421.37	421.1022	57.01	3.59	16.62	22.78
C20H15N507	437.37	437.0971	54.92	3.46	16.01	25.61
C20H15N508	453.37	453.0921	52.99	3.33	15.45	28.23
C20H15N509	469.37	469.0870	51.18	3.22	14.92	30.68
C20H16N201	300.36	300.1263	79.98	5.37	9.33	5.33
C20H16N202	316.36	316.1212	75.93	5.10	8.85	10.11
C20H16N203	332.36	332.1161	72.28	4.85	8.43	14.44
C20H16N204	348.36	348.1110	68.96	4.63	8.04	18.37
C20H16N205	364.36	364.1059	65.93	4.43	7.69	21.96
C20H16N206	380.36	380.1008	63.16	4.24	7.36	25.24
C20H16N207	396.36	396.0957	60.61	4.07	7.07	28.26
C20H16N208	412.36	412.0907	58.26	3.91	6.79	31.04
C20H16N209	428.36	428.0856	56.08	3.76	6.54	33.62
C20H16N401	328.38	328.1324	73.15	4.91	17.06	4.87
C20H16N402	344.38	344.1273	69.76	4.68	16.27	9.29
C20H16N403	360.38	360.1222	66.66	4.48	15.55	13.32
C20H16N404	376.37	376.1171	63.83	4.28	14.89	17.00
C20H16N405	392.37	392.1121	61.22	4.11	14.28	20.39
C20H16N406	408.37	408.1070	58.82	3.95	13.72	23.51
C20H16N407	424.37	424.1019	56.61	3.80	13.20	26.39
C20H16N408	440.37	440.0968	54.55	3.66	12.72	29.07
C20H16N409	456.37	456.0917	52.64	3.53	12.28	31.55
C20H16N601	356.39	356.1386	67.40	4.53	23.58	4.49
C20H16N602	372.39	372.1335	64.51	4.33	22.57	8.59
C20H16N603	388.39	388.1284	61.85	4.15	21.64	12.36
C20H16N604	404.39	404.1233	59.40	3.99	20.78	15.83
C20H16N605	420.39	420.1182	57.14	3.84	19.99	19.03
C20H16N606	436.39	436.1131	55.05	3.70	19.26	22.00
C20H16N607	452.39	452.1080	53.10	3.56	18.58	24.76
C20H16N608	468.39	468.1029	51.29	3.44	17.94	27.33
C20H16N609	484.39	484.0979	49.59	3.33	17.35	29.73
C20H17N101	287.36	287.1310	83.60	5.96	4.87	5.57
C20H17N102	303.36	303.1259	79.19	5.65	4.62	10.55
C20H17N103	319.36	319.1208	75.22	5.37	4.39	15.03
C20H17N104	335.36	335.1157	71.63	5.11	4.18	19.08
C20H17N105	351.36	351.1107	68.37	4.88	3.99	22.77
C20H17N106	367.36	367.1056	65.39	4.66	3.81	26.13
C20H17N107	383.36	383.1005	62.66	4.47	3.65	29.21
C20H17N108	399.36	399.0954	60.15	4.29	3.51	32.05
C20H17N109	415.36	415.0903	57.83	4.13	3.37	34.67
C20H17N301	315.38	315.1372	76.17	5.43	13.32	5.07

C20H17N302	331.38	331.1321	72.49	5.17	12.68	9.66
C20H17N303	347.38	347.1270	69.15	4.93	12.10	13.82
C20H17N304	363.38	363.1219	66.11	4.72	11.56	17.61
C20H17N305	379.38	379.1168	63.32	4.52	11.08	21.09
C20H17N306	395.37	395.1117	60.76	4.33	10.63	24.28
C20H17N307	411.37	411.1066	58.40	4.17	10.21	27.22
C20H17N308	427.37	427.1016	56.21	4.01	9.83	29.95
C20H17N309	443.37	443.0965	54.18	3.86	9.48	32.48
C20H17N501	343.39	343.1433	69.96	4.99	20.39	4.66
C20H17N502	359.39	359.1382	66.84	4.77	19.49	8.90
C20H17N503	375.39	375.1331	63.99	4.56	18.66	12.79
C20H17N504	391.39	391.1280	61.38	4.38	17.89	16.35
C20H17N505	407.39	407.1230	58.97	4.21	17.19	19.64
C20H17N506	423.39	423.1179	56.74	4.05	16.54	22.67
C20H17N507	439.39	439.1128	54.67	3.90	15.94	25.49
C20H17N508	455.39	455.1077	52.75	3.76	15.38	28.11
C20H17N509	471.39	471.1026	50.96	3.64	14.86	30.55
C20H18N201	302.38	302.1419	79.44	6.00	9.26	5.29
C20H18N202	318.38	318.1368	75.45	5.70	8.80	10.05
C20H18N203	334.38	334.1317	71.84	5.43	8.38	14.35
C20H18N204	350.38	350.1266	68.56	5.18	8.00	18.27
C20H18N205	366.38	366.1216	65.57	4.95	7.65	21.83
C20H18N206	382.38	382.1165	62.82	4.74	7.33	25.11
C20H18N207	398.38	398.1114	60.30	4.55	7.03	28.11
C20H18N208	414.38	414.1063	57.97	4.38	6.76	30.89
C20H18N209	430.37	430.1012	55.82	4.22	6.51	33.46
C20H18N401	330.39	330.1481	72.71	5.49	16.96	4.84
C20H18N402	346.39	346.1430	69.35	5.24	16.17	9.24
C20H18N403	362.39	362.1379	66.29	5.01	15.46	13.24
C20H18N404	378.39	378.1328	63.49	4.79	14.81	16.91
C20H18N405	394.39	394.1277	60.91	4.60	14.21	20.28
C20H18N406	410.39	410.1226	58.54	4.42	13.65	23.39
C20H18N407	426.39	426.1175	56.34	4.26	13.14	26.27
C20H18N408	442.39	442.1125	54.30	4.10	12.66	28.93
C20H18N409	458.39	458.1074	52.41	3.96	12.22	31.41
C20H18N601	358.41	358.1542	67.03	5.06	23.45	4.46
C20H18N602	374.41	374.1491	64.16	4.85	22.45	8.55
C20H18N603	390.40	390.1440	61.53	4.65	21.53	12.29
C20H18N604	406.40	406.1389	59.11	4.46	20.68	15.75
C20H18N605	422.40	422.1339	56.87	4.30	19.90	18.94
C20H18N606	438.40	438.1288	54.80	4.14	19.17	21.90
C20H18N607	454.40	454.1237	52.87	3.99	18.49	24.65
C20H18N608	470.40	470.1186	51.07	3.86	17.87	27.21
C20H18N609	486.40	486.1135	49.39	3.73	17.28	29.60
C20H19N101	289.38	289.1467	83.01	6.62	4.84	5.53
C20H19N102	305.38	305.1416	78.66	6.27	4.59	10.48
C20H19N103	321.38	321.1365	74.75	5.96	4.36	14.94
C20H19N104	337.38	337.1314	71.20	5.68	4.15	18.97
C20H19N105	353.38	353.1263	67.98	5.42	3.96	22.64
C20H19N106	369.38	369.1212	65.03	5.18	3.79	25.99
C20H19N107	385.38	385.1161	62.33	4.97	3.63	29.06
C20H19N108	401.38	401.1111	59.85	4.77	3.49	31.89
C20H19N109	417.38	417.1060	57.56	4.59	3.36	34.50
C20H19N301	317.39	317.1528	75.69	6.03	13.24	5.04
C20H19N302	333.39	333.1477	72.05	5.74	12.60	9.60
C20H19N303	349.39	349.1426	68.75	5.48	12.03	13.74
C20H19N304	365.39	365.1375	65.74	5.24	11.50	17.51
C20H19N305	381.39	381.1325	62.99	5.02	11.02	20.98
C20H19N306	397.39	397.1274	60.45	4.82	10.57	24.16
C20H19N307	413.39	413.1223	58.11	4.63	10.16	27.09
C20H19N308	429.39	429.1172	55.95	4.46	9.79	29.81
C20H19N309	445.39	445.1121	53.94	4.30	9.43	32.33
C20H19N501	345.41	345.1590	69.55	5.54	20.28	4.63
C20H19N502	361.41	361.1539	66.47	5.30	19.38	8.85
C20H19N503	377.41	377.1488	63.65	5.07	18.56	12.72
C20H19N504	393.41	393.1437	61.06	4.87	17.80	16.27
C20H19N505	409.40	409.1386	58.68	4.68	17.11	19.54
C20H19N506	425.40	425.1335	56.47	4.50	16.46	22.57
C20H19N507	441.40	441.1284	54.42	4.34	15.87	25.37
C20H19N508	457.40	457.1233	52.52	4.19	15.31	27.98
C20H19N509	473.40	473.1183	50.74	4.05	14.79	30.42
C20H20N201	304.40	304.1576	78.92	6.62	9.20	5.26
C20H20N202	320.39	320.1525	74.98	6.29	8.74	9.99
C20H20N203	336.39	336.1474	71.41	5.99	8.33	14.27
C20H20N204	352.39	352.1423	68.17	5.72	7.95	18.16
C20H20N205	368.39	368.1372	65.21	5.47	7.60	21.72
C20H20N206	384.39	384.1321	62.49	5.24	7.29	24.97
C20H20N207	400.39	400.1270	60.00	5.03	7.00	27.97
C20H20N208	416.39	416.1220	57.69	4.84	6.73	30.74
C20H20N209	432.39	432.1169	55.56	4.66	6.48	33.30
C20H20N401	332.41	332.1637	72.27	6.06	16.85	4.81
C20H20N402	348.41	348.1586	68.95	5.79	16.08	9.18
C20H20N403	364.41	364.1535	65.92	5.53	15.37	13.17
C20H20N404	380.41	380.1484	63.15	5.30	14.73	16.82
C20H20N405	396.41	396.1434	60.60	5.09	14.13	20.18
C20H20N406	412.41	412.1383	58.25	4.89	13.59	23.28
C20H20N407	428.41	428.1332	56.07	4.71	13.08	26.14
C20H20N408	444.40	444.1281	54.06	4.54	12.61	28.80
C20H20N409	460.40	460.1230	52.18	4.38	12.17	31.28
C20H20N601	360.42	360.1698	66.65	5.59	23.32	4.44
C20H20N602	376.42	376.1648	63.82	5.36	22.33	8.50
C20H20N603	392.42	392.1597	61.22	5.14	21.42	12.23
C20H20N604	408.42	408.1546	58.82	4.94	20.58	15.67
C20H20N605	424.42	424.1495	56.60	4.75	19.80	18.85
C20H20N606	440.42	440.1444	54.54	4.58	19.08	21.80
C20H20N607	456.42	456.1393	52.63	4.42	18.41	24.54
C20H20N608	472.42	472.1342	50.85	4.27	17.79	27.09
C20H20N609	488.42	488.1292	49.18	4.13	17.21	29.48
C20H21N101	291.40	291.1623	82.44	7.26	4.81	5.49
C20H21N102	307.40	307.1572	78.15	6.89	4.56	10.41
C20H21N103	323.40	323.1521	74.28	6.55	4.33	14.84
C20H21N104	339.39	339.1470	70.78	6.24	4.13	18.86
C20H21N105	355.39	355.1420	67.59	5.96	3.94	22.51
C20H21N106	371.39	371.1369	64.68	5.70	3.77	25.85
C20H21N107	387.39	387.1318	62.01	5.46	3.62	28.91
C20H21N108	403.39	403.1267	59.55	5.25	3.47	31.73
C20H21N109	419.39	419.1216	57.28	5.05	3.34	34.33
C20H21N301	319.41	319.1685	75.21	6.63	13.16	5.01
C20H21N302	335.41	335.1634	71.62	6.31	12.53	9.54
C20H21N303	351.41	351.1583	68.36	6.02	11.96	13.66
C20H21N304	367.41	367.1532	65.38	5.76	11.44	17.42
C20H21N305	383.41	383.1481	62.65	5.52	10.96	20.86
C20H21N306	399.41	399.1430	60.14	5.30	10.52	24.03
C20H21N307	415.41	415.1379	57.83	5.10	10.12	26.96
C20H21N308	431.41	431.1329	55.68	4.91	9.74	29.67
C20H21N309	447.41	447.1278	53.69	4.73	9.39	32.18
C20H21N501	347.42	347.1746	69.14	6.09	20.16	4.61
C20H21N502	363.42	363.1695	66.10	5.82	19.27	8.80
C20H21N503	379.42	379.1644	63.31	5.58	18.46	12.65
C20H21N504	395.42	395.1593	60.75	5.35	17.71	16.18

C20H21N5O5	411.42	411.1543	58.39	5.14	17.02	19.44
C20H21N5O6	427.42	427.1492	56.20	4.95	16.39	22.46
C20H21N5O7	443.42	443.1441	54.18	4.77	15.79	25.26
C20H21N5O8	459.42	459.1390	52.29	4.61	15.24	27.86
C20H21N5O9	475.42	475.1339	50.53	4.45	14.73	30.29
C20H22N2O1	306.41	306.1732	78.40	7.24	9.14	5.22
C20H22N2O2	322.41	322.1681	74.51	6.88	8.69	9.92
C20H22N2O3	338.41	338.1630	70.99	6.55	8.28	14.18
C20H22N2O4	354.41	354.1579	67.78	6.26	7.90	18.06
C20H22N2O5	370.41	370.1529	64.85	5.99	7.56	21.60
C20H22N2O6	386.41	386.1478	62.17	5.74	7.25	24.84
C20H22N2O7	402.41	402.1427	59.70	5.51	6.96	27.83
C20H22N2O8	418.41	418.1376	57.41	5.30	6.70	30.59
C20H22N2O9	434.41	434.1325	55.30	5.10	6.45	33.15
C20H22N4O1	334.42	334.1794	71.83	6.63	16.75	4.78
C20H22N4O2	350.42	350.1743	68.55	6.33	15.99	9.13
C20H22N4O3	366.42	366.1692	65.56	6.05	15.29	13.10
C20H22N4O4	382.42	382.1641	62.82	5.80	14.65	16.73
C20H22N4O5	398.42	398.1590	60.29	5.57	14.06	20.08
C20H22N4O6	414.42	414.1539	57.97	5.35	13.52	23.16
C20H22N4O7	430.42	430.1488	55.81	5.15	13.02	26.02
C20H22N4O8	446.42	446.1437	53.81	4.97	12.55	28.67
C20H22N4O9	462.42	462.1387	51.95	4.80	12.12	31.14
C20H22N6O1	362.44	362.1855	66.28	6.12	23.19	4.41
C20H22N6O2	378.44	378.1804	63.48	5.86	22.21	8.46
C20H22N6O3	394.44	394.1753	60.90	5.62	21.31	12.17
C20H22N6O4	410.44	410.1702	58.53	5.40	20.48	15.59
C20H22N6O5	426.44	426.1652	56.33	5.20	19.71	18.76
C20H22N6O6	442.43	442.1601	54.30	5.01	18.99	21.70
C20H22N6O7	458.43	458.1550	52.40	4.84	18.33	24.43
C20H22N6O8	474.43	474.1499	50.63	4.67	17.71	26.98
C20H22N6O9	490.43	490.1448	48.98	4.52	17.14	29.36
C20H23N1O1	293.41	293.1780	81.87	7.90	4.77	5.45
C20H23N1O2	309.41	309.1729	77.64	7.49	4.53	10.34
C20H23N1O3	325.41	325.1678	73.82	7.12	4.30	14.75
C20H23N1O4	341.41	341.1627	70.36	6.79	4.10	18.75
C20H23N1O5	357.41	357.1576	67.21	6.49	3.92	22.38
C20H23N1O6	373.41	373.1525	64.33	6.21	3.75	25.71
C20H23N1O7	389.41	389.1474	61.69	5.95	3.60	28.76
C20H23N1O8	405.41	405.1424	59.25	5.72	3.45	31.57
C20H23N1O9	421.41	421.1373	57.00	5.50	3.32	34.17
C20H23N3O1	321.43	321.1841	74.74	7.21	13.07	4.98
C20H23N3O2	337.43	337.1790	71.19	6.87	12.45	9.48
C20H23N3O3	353.42	353.1739	67.97	6.56	11.89	13.58
C20H23N3O4	369.42	369.1688	65.03	6.28	11.37	17.32
C20H23N3O5	385.42	385.1638	62.33	6.02	10.90	20.76
C20H23N3O6	401.42	401.1587	59.84	5.78	10.47	23.91
C20H23N3O7	417.42	417.1536	57.55	5.55	10.07	26.83
C20H23N3O8	433.42	433.1485	55.42	5.35	9.69	29.53
C20H23N3O9	449.42	449.1434	53.45	5.16	9.35	32.04
C20H23N5O1	349.44	349.1902	68.75	6.63	20.04	4.58
C20H23N5O2	365.44	365.1852	65.74	6.34	19.16	8.76
C20H23N5O3	381.44	381.1801	62.98	6.08	18.36	12.58
C20H23N5O4	397.44	397.1750	60.44	5.83	17.62	16.10
C20H23N5O5	413.44	413.1699	58.10	5.61	16.94	19.35
C20H23N5O6	429.44	429.1648	55.94	5.40	16.31	22.35
C20H23N5O7	445.44	445.1597	53.93	5.20	15.72	25.14
C20H23N5O8	461.44	461.1546	52.06	5.02	15.18	27.74
C20H23N5O9	477.43	477.1496	50.32	4.86	14.67	30.16
C20H24N2O1	308.43	308.1889	77.89	7.84	9.08	5.19

C20H24N2O2	324.43	324.1838	74.05	7.46	8.63	9.86
C20H24N2O3	340.43	340.1787	70.57	7.11	8.23	14.10
C20H24N2O4	356.43	356.1736	67.40	6.79	7.86	17.96
C20H24N2O5	372.42	372.1685	64.50	6.50	7.52	21.48
C20H24N2O6	388.42	388.1634	61.85	6.23	7.21	24.71
C20H24N2O7	404.42	404.1583	59.40	5.98	6.93	27.69
C20H24N2O8	420.42	420.1533	57.14	5.75	6.66	30.44
C20H24N2O9	436.42	436.1482	55.04	5.54	6.42	32.99
C20H24N4O1	336.44	336.1950	71.40	7.19	16.65	4.76
C20H24N4O2	352.44	352.1899	68.16	6.86	15.90	9.08
C20H24N4O3	368.44	368.1848	65.20	6.57	15.21	13.03
C20H24N4O4	384.44	384.1797	62.49	6.29	14.57	16.65
C20H24N4O5	400.44	400.1747	59.99	6.04	13.99	19.98
C20H24N4O6	416.44	416.1696	57.69	5.81	13.45	23.05
C20H24N4O7	432.44	432.1645	55.55	5.59	12.96	25.90
C20H24N4O8	448.44	448.1594	53.57	5.39	12.49	28.54
C20H24N4O9	464.44	464.1543	51.72	5.21	12.06	31.00
C20H24N6O1	364.45	364.2011	65.91	6.64	23.06	4.39
C20H24N6O2	380.45	380.1961	63.14	6.36	22.09	8.41
C20H24N6O3	396.45	396.1910	60.59	6.10	21.20	12.11
C20H24N6O4	412.45	412.1859	58.24	5.87	20.38	15.52
C20H24N6O5	428.45	428.1808	56.07	5.65	19.61	18.67
C20H24N6O6	444.45	444.1757	54.05	5.44	18.91	21.60
C20H24N6O7	460.45	460.1706	52.17	5.25	18.25	24.32
C20H24N6O8	476.45	476.1655	50.42	5.08	17.64	26.86
C20H24N6O9	492.45	492.1605	48.78	4.91	17.07	29.24
C20H25N1O1	295.43	295.1936	81.31	8.53	4.74	5.42
C20H25N1O2	311.43	311.1885	77.14	8.09	4.50	10.27
C20H25N1O3	327.43	327.1834	73.37	7.70	4.28	14.66
C20H25N1O4	343.43	343.1783	69.95	7.34	4.08	18.64
C20H25N1O5	359.43	359.1733	66.84	7.01	3.90	22.26
C20H25N1O6	375.43	375.1682	63.99	6.71	3.73	25.57
C20H25N1O7	391.42	391.1631	61.37	6.44	3.58	28.61
C20H25N1O8	407.42	407.1580	58.96	6.19	3.44	31.42
C20H25N1O9	423.42	423.1529	56.73	5.95	3.31	34.01
C20H25N3O1	323.44	323.1998	74.27	7.79	12.99	4.95
C20H25N3O2	339.44	339.1947	70.77	7.42	12.38	9.43
C20H25N3O3	355.44	355.1896	67.58	7.09	11.82	13.50
C20H25N3O4	371.44	371.1845	64.67	6.78	11.31	17.23
C20H25N3O5	387.44	387.1794	62.00	6.50	10.85	20.65
C20H25N3O6	403.44	403.1743	59.54	6.25	10.42	23.79
C20H25N3O7	419.44	419.1692	57.27	6.01	10.02	26.70
C20H25N3O8	435.44	435.1641	55.17	5.79	9.65	29.39
C20H25N3O9	451.44	451.1591	53.21	5.58	9.31	31.90
C20H25N5O1	351.46	351.2059	68.35	7.17	19.93	4.55
C20H25N5O2	367.45	367.2008	65.37	6.86	19.06	8.71
C20H25N5O3	383.45	383.1957	62.65	6.57	18.26	12.52
C20H25N5O4	399.45	399.1906	60.14	6.31	17.53	16.02
C20H25N5O5	415.45	415.1856	57.82	6.07	16.86	19.26
C20H25N5O6	431.45	431.1805	55.68	5.84	16.23	22.25
C20H25N5O7	447.45	447.1754	53.69	5.63	15.65	25.03
C20H25N5O8	463.45	463.1703	51.83	5.44	15.11	27.62
C20H25N5O9	479.45	479.1652	50.10	5.26	14.61	30.03
C20H26N2O1	310.44	310.2045	77.38	8.44	9.02	5.15
C20H26N2O2	326.44	326.1994	73.59	8.03	8.58	9.80
C20H26N2O3	342.44	342.1943	70.15	7.65	8.18	14.02
C20H26N2O4	358.44	358.1892	67.02	7.31	7.82	17.85
C20H26N2O5	374.44	374.1842	64.16	7.00	7.48	21.36
C20H26N2O6	390.44	390.1791	61.53	6.71	7.17	24.59
C20H26N2O7	406.44	406.1740	59.10	6.45	6.89	27.56

C20H26N2O8	422.44	422.1689	56.87	6.20	6.63	30.30
C20H26N2O9	438.44	438.1638	54.79	5.98	6.39	32.84
C20H26N4O1	338.46	338.2106	70.98	7.74	16.55	4.73
C20H26N4O2	354.46	354.2056	67.77	7.39	15.81	9.03
C20H26N4O3	370.46	370.2005	64.85	7.07	15.12	12.96
C20H26N4O4	386.45	386.1954	62.16	6.78	14.50	16.56
C20H26N4O5	402.45	402.1903	59.69	6.51	13.92	19.88
C20H26N4O6	418.45	418.1852	57.41	6.26	13.39	22.94
C20H26N4O7	434.45	434.1801	55.29	6.03	12.90	25.78
C20H26N4O8	450.45	450.1750	53.33	5.82	12.44	28.41
C20H26N4O9	466.45	466.1700	51.50	5.62	12.01	30.87
C20H26N6O1	366.47	366.2168	65.55	7.15	22.93	4.37
C20H26N6O2	382.47	382.2117	62.81	6.85	21.97	8.37
C20H26N6O3	398.47	398.2066	60.29	6.58	21.09	12.05
C20H26N6O4	414.47	414.2015	57.96	6.32	20.28	15.44
C20H26N6O5	430.47	430.1965	55.81	6.09	19.52	18.58
C20H26N6O6	446.47	446.1914	53.81	5.87	18.82	21.50
C20H26N6O7	462.47	462.1863	51.94	5.67	18.17	24.22
C20H26N6O8	478.47	478.1812	50.21	5.48	17.56	26.75
C20H26N6O9	494.47	494.1761	48.58	5.30	17.00	29.12
C20H27N1O1	297.44	297.2093	80.76	9.15	4.71	5.38
C20H27N1O2	313.44	313.2042	76.64	8.68	4.47	10.21
C20H27N1O3	329.44	329.1991	72.92	8.26	4.25	14.57
C20H27N1O4	345.44	345.1940	69.54	7.88	4.05	18.53
C20H27N1O5	361.44	361.1889	66.46	7.53	3.88	22.13
C20H27N1O6	377.44	377.1838	63.65	7.21	3.71	25.43
C20H27N1O7	393.44	393.1787	61.06	6.92	3.56	28.47
C20H27N1O8	409.44	409.1737	58.67	6.65	3.42	31.26
C20H27N1O9	425.44	425.1686	56.46	6.40	3.29	33.85
C20H27N3O1	325.46	325.2154	73.81	8.36	12.91	4.92
C20H27N3O2	341.46	341.2103	70.35	7.97	12.31	9.37
C20H27N3O3	357.46	357.2052	67.20	7.61	11.76	13.43
C20H27N3O4	373.46	373.2001	64.32	7.29	11.25	17.14
C20H27N3O5	389.46	389.1951	61.68	6.99	10.79	20.54
C20H27N3O6	405.45	405.1900	59.25	6.71	10.36	23.68
C20H27N3O7	421.45	421.1849	57.00	6.46	9.97	26.57
C20H27N3O8	437.45	437.1798	54.91	6.22	9.61	29.26
C20H27N3O9	453.45	453.1747	52.98	6.00	9.27	31.76
C20H27N5O1	353.47	353.2215	67.96	7.70	19.81	4.53
C20H27N5O2	369.47	369.2165	65.02	7.37	18.96	8.66
C20H27N5O3	385.47	385.2114	62.32	7.06	18.17	12.45
C20H27N5O4	401.47	401.2063	59.84	6.78	17.44	15.94
C20H27N5O5	417.47	417.2012	57.54	6.52	16.78	19.16
C20H27N5O6	433.47	433.1961	55.42	6.28	16.16	22.15
C20H27N5O7	449.47	449.1910	53.45	6.05	15.58	24.92
C20H27N5O8	465.47	465.1859	51.61	5.85	15.05	27.50
C20H27N5O9	481.47	481.1809	49.89	5.65	14.55	29.91
C20H28N2O1	312.46	312.2202	76.88	9.03	8.97	5.12
C20H28N2O2	328.46	328.2151	73.14	8.59	8.53	9.74
C20H28N2O3	344.46	344.2100	69.74	8.19	8.13	13.93
C20H28N2O4	360.46	360.2049	66.64	7.83	7.77	17.75
C20H28N2O5	376.46	376.1998	63.81	7.50	7.44	21.25
C20H28N2O6	392.46	392.1947	61.21	7.19	7.14	24.46
C20H28N2O7	408.46	408.1896	58.81	6.91	6.86	27.42
C20H28N2O8	424.45	424.1845	56.60	6.65	6.60	30.16
C20H28N2O9	440.45	440.1795	54.54	6.41	6.36	32.69
C20H28N4O1	340.47	340.2263	70.56	8.29	16.46	4.70
C20H28N4O2	356.47	356.2212	67.39	7.92	15.72	8.98
C20H28N4O3	372.47	372.2161	64.49	7.58	15.04	12.89
C20H28N4O4	388.47	388.2110	61.84	7.27	14.42	16.47

C20H28N4O5	404.47	404.2060	59.39	6.98	13.85	19.78
C20H28N4O6	420.47	420.2009	57.13	6.71	13.32	22.83
C20H28N4O7	436.47	436.1958	55.04	6.47	12.84	25.66
C20H28N4O8	452.47	452.1907	53.09	6.24	12.38	28.29
C20H28N4O9	468.47	468.1856	51.28	6.02	11.96	30.74
C20H28N6O1	368.49	368.2324	65.19	7.66	22.81	4.34
C20H28N6O2	384.49	384.2274	62.48	7.34	21.86	8.32
C20H28N6O3	400.48	400.2223	59.98	7.05	20.98	11.99
C20H28N6O4	416.48	416.2172	57.68	6.78	20.18	15.37
C20H28N6O5	432.48	432.2121	55.55	6.53	19.43	18.50
C20H28N6O6	448.48	448.2070	53.56	6.29	18.74	21.40
C20H28N6O7	464.48	464.2019	51.72	6.08	18.09	24.11
C20H28N6O8	480.48	480.1968	50.00	5.87	17.49	26.64
C20H28N6O9	496.48	496.1918	48.39	5.68	16.93	29.00
C20H29N1O1	299.46	299.2249	80.22	9.76	4.68	5.34
C20H29N1O2	315.46	315.2198	76.15	9.27	4.44	10.14
C20H29N1O3	331.46	331.2147	72.47	8.82	4.23	14.48
C20H29N1O4	347.46	347.2096	69.14	8.41	4.03	18.42
C20H29N1O5	363.46	363.2046	66.09	8.04	3.85	22.01
C20H29N1O6	379.46	379.1995	63.31	7.70	3.69	25.30
C20H29N1O7	395.46	395.1944	60.75	7.39	3.54	28.32
C20H29N1O8	411.46	411.1893	58.38	7.10	3.40	31.11
C20H29N1O9	427.46	427.1842	56.20	6.84	3.28	33.69
C20H29N3O1	327.47	327.2310	73.36	8.93	12.83	4.89
C20H29N3O2	343.47	343.2260	69.94	8.51	12.23	9.32
C20H29N3O3	359.47	359.2209	66.83	8.13	11.69	13.35
C20H29N3O4	375.47	375.2158	63.98	7.79	11.19	17.04
C20H29N3O5	391.47	391.2107	61.36	7.47	10.73	20.43
C20H29N3O6	407.47	407.2056	58.95	7.17	10.31	23.56
C20H29N3O7	423.47	423.2005	56.73	6.90	9.92	26.45
C20H29N3O8	439.47	439.1954	54.66	6.65	9.56	29.12
C20H29N3O9	455.47	455.1904	52.74	6.42	9.23	31.61
C20H29N5O1	355.49	355.2372	67.58	8.22	19.70	4.50
C20H29N5O2	371.49	371.2321	64.67	7.87	18.85	8.61
C20H29N5O3	387.49	387.2270	62.00	7.54	18.07	12.39
C20H29N5O4	403.49	403.2219	59.54	7.24	17.36	15.86
C20H29N5O5	419.48	419.2169	57.27	6.97	16.70	19.07
C20H29N5O6	435.48	435.2118	55.16	6.71	16.08	22.04
C20H29N5O7	451.48	451.2067	53.21	6.47	15.51	24.81
C20H29N5O8	467.48	467.2016	51.39	6.25	14.98	27.38
C20H29N5O9	483.48	483.1965	49.69	6.05	14.49	29.78
C20H30N2O1	314.47	314.2358	76.39	9.62	8.91	5.09
C20H30N2O2	330.47	330.2307	72.69	9.15	8.48	9.68
C20H30N2O3	346.47	346.2256	69.33	8.73	8.09	13.85
C20H30N2O4	362.47	362.2205	66.27	8.34	7.73	17.66
C20H30N2O5	378.47	378.2155	63.47	7.99	7.40	21.14
C20H30N2O6	394.47	394.2104	60.90	7.67	7.10	24.34
C20H30N2O7	410.47	410.2053	58.52	7.37	6.82	27.28
C20H30N2O8	426.47	426.2002	56.33	7.09	6.57	30.01
C20H30N2O9	442.47	442.1951	54.29	6.83	6.33	32.54
C20H30N4O1	342.49	342.2419	70.14	8.83	16.36	4.67
C20H30N4O2	358.49	358.2369	67.01	8.44	15.63	8.93
C20H30N4O3	374.49	374.2318	64.15	8.07	14.96	12.82
C20H30N4O4	390.49	390.2267	61.52	7.74	14.35	16.39
C20H30N4O5	406.49	406.2216	59.10	7.44	13.78	19.68
C20H30N4O6	422.49	422.2165	56.86	7.16	13.26	22.72
C20H30N4O7	438.48	438.2114	54.78	6.90	12.78	25.54
C20H30N4O8	454.48	454.2063	52.86	6.65	12.33	28.16
C20H30N4O9	470.48	470.2013	51.06	6.43	11.91	30.61
C20H30N6O1	370.50	370.2481	64.84	8.16	22.68	4.32

C20H30N602	386.50	386.2430	62.15	7.82	21.74	8.28
C20H30N603	402.50	402.2379	59.68	7.51	20.88	11.93
C20H30N604	418.50	418.2328	57.40	7.23	20.08	15.29
C20H30N605	434.50	434.2278	55.29	6.96	19.34	18.41
C20H30N606	450.50	450.2227	53.32	6.71	18.65	21.31
C20H30N607	466.50	466.2176	51.49	6.48	18.02	24.01
C20H30N608	482.50	482.2125	49.79	6.27	17.42	26.53
C20H30N609	498.50	498.2074	48.19	6.07	16.86	28.89
C20H31N101	301.48	301.2406	79.68	10.36	4.65	5.31
C20H31N102	317.48	317.2355	75.67	9.84	4.41	10.08
C20H31N103	333.47	333.2304	72.04	9.37	4.20	14.39
C20H31N104	349.47	349.2253	68.74	8.94	4.01	18.31
C20H31N105	365.47	365.2202	65.73	8.55	3.83	21.89
C20H31N106	381.47	381.2151	62.97	8.19	3.67	25.16
C20H31N107	397.47	397.2100	60.44	7.86	3.52	28.18
C20H31N108	413.47	413.2049	58.10	7.56	3.39	30.96
C20H31N109	429.47	429.1999	55.93	7.28	3.26	33.53
C20H31N301	329.49	329.2467	72.91	9.48	12.75	4.86
C20H31N302	345.49	345.2416	69.53	9.04	12.16	9.26
C20H31N303	361.49	361.2365	66.45	8.64	11.62	13.28
C20H31N304	377.49	377.2314	63.64	8.28	11.13	16.95
C20H31N305	393.49	393.2264	61.05	7.94	10.68	20.33
C20H31N306	409.49	409.2213	58.66	7.63	10.26	23.44
C20H31N307	425.49	425.2162	56.46	7.34	9.88	26.32
C20H31N308	441.49	441.2111	54.41	7.08	9.52	28.99
C20H31N309	457.48	457.2060	52.51	6.83	9.19	31.48
C20H31N501	357.50	357.2528	67.19	8.74	19.59	4.48
C20H31N502	373.50	373.2478	64.32	8.37	18.75	8.57
C20H31N503	389.50	389.2427	61.67	8.02	17.98	12.32
C20H31N504	405.50	405.2376	59.24	7.71	17.27	15.78
C20H31N505	421.50	421.2325	56.99	7.41	16.62	18.98
C20H31N506	437.50	437.2274	54.91	7.14	16.01	21.94
C20H31N507	453.50	453.2223	52.97	6.89	15.44	24.70
C20H31N508	469.50	469.2172	51.17	6.66	14.92	27.26
C20H31N509	485.50	485.2122	49.48	6.44	14.43	29.66
C20H32N201	316.49	316.2514	75.90	10.19	8.85	5.06
C20H32N202	332.49	332.2464	72.25	9.70	8.43	9.62
C20H32N203	348.49	348.2413	68.93	9.26	8.04	13.77
C20H32N204	364.49	364.2362	65.91	8.85	7.69	17.56
C20H32N205	380.49	380.2311	63.14	8.48	7.36	21.02
C20H32N206	396.49	396.2260	60.59	8.14	7.07	24.21
C20H32N207	412.49	412.2209	58.24	7.82	6.79	27.15
C20H32N208	428.49	428.2158	56.06	7.53	6.54	29.87
C20H32N209	444.49	444.2108	54.05	7.26	6.30	32.40
C20H32N401	344.50	344.2576	69.73	9.36	16.26	4.64
C20H32N402	360.50	360.2525	66.64	8.95	15.54	8.88
C20H32N403	376.50	376.2474	63.80	8.57	14.88	12.75
C20H32N404	392.50	392.2423	61.20	8.22	14.27	16.31
C20H32N405	408.50	408.2373	58.81	7.90	13.72	19.58
C20H32N406	424.50	424.2322	56.59	7.60	13.20	22.61
C20H32N407	440.50	440.2271	54.53	7.32	12.72	25.42
C20H32N408	456.50	456.2220	52.62	7.07	12.27	28.04
C20H32N409	472.50	472.2169	50.84	6.83	11.86	30.48
C20H32N601	372.52	372.2637	64.49	8.66	22.56	4.29
C20H32N602	388.52	388.2587	61.83	8.30	21.63	8.24
C20H32N603	404.52	404.2536	59.39	7.97	20.78	11.87
C20H32N604	420.52	420.2485	57.13	7.67	19.99	15.22
C20H32N605	436.52	436.2434	55.03	7.39	19.25	18.33
C20H32N606	452.51	452.2383	53.09	7.13	18.57	21.21
C20H32N607	468.51	468.2332	51.27	6.88	17.94	23.90
C20H32N608	484.51	484.2281	49.58	6.66	17.35	26.42
C20H32N609	500.51	500.2231	48.00	6.44	16.79	28.77
C20H33N101	303.49	303.2562	79.15	10.96	4.62	5.27
C20H33N102	319.49	319.2511	75.19	10.41	4.38	10.02
C20H33N103	335.49	335.2460	71.60	9.91	4.17	14.31
C20H33N104	351.49	351.2409	68.34	9.46	3.98	18.21
C20H33N105	367.49	367.2359	65.37	9.05	3.81	21.77
C20H33N106	383.49	383.2308	62.64	8.67	3.65	25.03
C20H33N107	399.49	399.2257	60.13	8.33	3.51	28.03
C20H33N108	415.49	415.2206	57.82	8.01	3.37	30.81
C20H33N109	431.49	431.2155	55.67	7.71	3.25	33.37
C20H33N301	331.51	331.2623	72.46	10.03	12.68	4.83
C20H33N302	347.50	347.2573	69.13	9.57	12.09	9.21
C20H33N303	363.50	363.2522	66.09	9.15	11.56	13.20
C20H33N304	379.50	379.2471	63.30	8.76	11.07	16.86
C20H33N305	395.50	395.2420	60.74	8.41	10.62	20.23
C20H33N306	411.50	411.2369	58.38	8.08	10.21	23.33
C20H33N307	427.50	427.2318	56.19	7.78	9.83	26.20
C20H33N308	443.50	443.2267	54.17	7.50	9.47	28.86
C20H33N309	459.50	459.2217	52.28	7.24	9.14	31.34
C20H33N501	359.52	359.2685	66.82	9.25	19.48	4.45
C20H33N502	375.52	375.2634	63.97	8.86	18.65	8.52
C20H33N503	391.52	391.2583	61.36	8.50	17.89	12.26
C20H33N504	407.52	407.2532	58.95	8.16	17.19	15.70
C20H33N505	423.52	423.2482	56.72	7.85	16.54	18.89
C20H33N506	439.52	439.2431	54.66	7.57	15.93	21.84
C20H33N507	455.52	455.2380	52.74	7.30	15.37	24.59
C20H33N508	471.51	471.2329	50.95	7.05	14.85	27.15
C20H33N509	487.51	487.2278	49.28	6.82	14.37	29.54
C20H34N201	318.51	318.2671	75.42	10.76	8.80	5.02
C20H34N202	334.51	334.2620	71.81	10.25	8.37	9.57
C20H34N203	350.51	350.2569	68.54	9.78	7.99	13.69
C20H34N204	366.50	366.2518	65.54	9.35	7.64	17.46
C20H34N205	382.50	382.2468	62.80	8.96	7.32	20.91
C20H34N206	398.50	398.2417	60.28	8.60	7.03	24.09
C20H34N207	414.50	414.2366	57.95	8.27	6.76	27.02
C20H34N208	430.50	430.2315	55.80	7.96	6.51	29.73
C20H34N209	446.50	446.2264	53.80	7.68	6.27	32.25
C20H34N401	346.52	346.2732	69.32	9.89	16.17	4.62
C20H34N402	362.52	362.2682	66.26	9.45	15.45	8.83
C20H34N403	378.52	378.2631	63.46	9.05	14.80	12.68
C20H34N404	394.52	394.2580	60.89	8.69	14.20	16.22
C20H34N405	410.52	410.2529	58.52	8.35	13.65	19.49
C20H34N406	426.52	426.2478	56.32	8.04	13.14	22.51
C20H34N407	442.52	442.2427	54.29	7.74	12.66	25.31
C20H34N408	458.52	458.2376	52.39	7.47	12.22	27.92
C20H34N409	474.52	474.2326	50.62	7.22	11.81	30.35
C20H34N601	374.53	374.2794	64.14	9.15	22.44	4.27
C20H34N602	390.53	390.2743	61.51	8.78	21.52	8.19
C20H34N603	406.53	406.2692	59.09	8.43	20.67	11.81
C20H34N604	422.53	422.2641	56.85	8.11	19.89	15.15
C20H34N605	438.53	438.2590	54.78	7.81	19.16	18.24
C20H34N606	454.53	454.2540	52.85	7.54	18.49	21.12
C20H34N607	470.53	470.2489	51.05	7.28	17.86	23.80
C20H34N608	486.53	486.2438	49.37	7.04	17.27	26.31
C20H34N609	502.53	502.2387	47.80	6.82	16.72	28.65
C20H35N101	305.51	305.2718	78.63	11.55	4.58	5.24
C20H35N102	321.51	321.2668	74.72	10.97	4.36	9.95
C20H35N103	337.51	337.2617	71.18	10.45	4.15	14.22
C20H35N104	353.51	353.2566	67.95	9.98	3.96	18.10

C20H35N105	369.51	369.2515	65.01	9.55	3.79	21.65
C20H35N106	385.51	385.2464	62.31	9.15	3.63	24.90
C20H35N107	401.50	401.2413	59.83	8.79	3.49	27.89
C20H35N108	417.50	417.2362	57.54	8.45	3.35	30.66
C20H35N109	433.50	433.2312	55.41	8.14	3.23	33.22
C20H35N301	333.52	333.2780	72.03	10.58	12.60	4.80
C20H35N302	349.52	349.2729	68.73	10.09	12.02	9.16
C20H35N303	365.52	365.2678	65.72	9.65	11.50	13.13
C20H35N304	381.52	381.2627	62.96	9.25	11.01	16.77
C20H35N305	397.52	397.2577	60.43	8.87	10.57	20.12
C20H35N306	413.52	413.2526	58.09	8.53	10.16	23.21
C20H35N307	429.52	429.2475	55.93	8.21	9.78	26.07
C20H35N308	445.52	445.2424	53.92	7.92	9.43	28.73
C20H35N309	461.52	461.2373	52.05	7.64	9.10	31.20
C20H35N501	361.53	361.2841	66.45	9.76	19.37	4.43
C20H35N502	377.53	377.2791	63.63	9.34	18.55	8.48
C20H35N503	393.53	393.2740	61.04	8.96	17.80	12.20
C20H35N504	409.53	409.2689	58.66	8.61	17.10	15.63
C20H35N505	425.53	425.2638	56.45	8.29	16.46	18.80
C20H35N506	441.53	441.2587	54.41	7.99	15.86	21.74
C20H35N507	457.53	457.2536	52.50	7.71	15.31	24.48
C20H35N508	473.53	473.2485	50.73	7.45	14.79	27.03
C20H35N509	489.53	489.2435	49.07	7.21	14.31	29.41
C20H36N201	320.52	320.2827	74.95	11.32	8.74	4.99
C20H36N202	336.52	336.2777	71.38	10.78	8.32	9.51
C20H36N203	352.52	352.2726	68.14	10.29	7.95	13.62
C20H36N204	368.52	368.2675	65.19	9.85	7.60	17.37
C20H36N205	384.52	384.2624	62.47	9.44	7.29	20.80
C20H36N206	400.52	400.2573	59.98	9.06	6.99	23.97
C20H36N207	416.52	416.2522	57.67	8.71	6.73	26.89
C20H36N208	432.52	432.2471	55.54	8.39	6.48	29.59
C20H36N209	448.52	448.2421	53.56	8.09	6.25	32.10
C20H36N401	348.54	348.2889	68.92	10.41	16.07	4.59
C20H36N402	364.54	364.2838	65.90	9.95	15.37	8.78
C20H36N403	380.53	380.2787	63.13	9.54	14.72	12.61
C20H36N404	396.53	396.2736	60.58	9.15	14.13	16.14
C20H36N405	412.53	412.2686	58.23	8.80	13.58	19.39
C20H36N406	428.53	428.2635	56.06	8.47	13.07	22.40
C20H36N407	444.53	444.2584	54.04	8.16	12.60	25.19
C20H36N408	460.53	460.2533	52.16	7.88	12.17	27.79
C20H36N409	476.53	476.2482	50.41	7.61	11.76	30.22
C20H36N601	376.55	376.2950	63.80	9.64	22.32	4.25
C20H36N602	392.55	392.2900	61.20	9.24	21.41	8.15
C20H36N603	408.55	408.2849	58.80	8.88	20.57	11.75
C20H36N604	424.55	424.2798	56.58	8.55	19.80	15.07
C20H36N605	440.55	440.2747	54.53	8.24	19.08	18.16
C20H36N606	456.55	456.2696	52.62	7.95	18.41	21.03
C20H36N607	472.55	472.2645	50.84	7.68	17.78	23.70
C20H36N608	488.55	488.2594	49.17	7.43	17.20	26.20
C20H36N609	504.54	504.2544	47.61	7.19	16.66	28.54
C20H37N101	307.52	307.2875	78.12	12.13	4.55	5.20
C20H37N102	323.52	323.2824	74.25	11.53	4.33	9.89
C20H37N103	339.52	339.2773	70.75	10.98	4.13	14.14
C20H37N104	355.52	355.2722	67.57	10.49	3.94	18.00
C20H37N105	371.52	371.2672	64.66	10.04	3.77	21.53
C20H37N106	387.52	387.2621	61.99	9.62	3.61	24.77
C20H37N107	403.52	403.2570	59.53	9.24	3.47	27.75
C20H37N108	419.52	419.2519	57.26	8.89	3.34	30.51
C20H37N109	435.52	435.2468	55.16	8.56	3.22	33.06
C20H37N301	335.54	335.2936	71.59	11.11	12.52	4.77
C20H37N302	351.54	351.2886	68.34	10.61	11.95	9.10
C20H37N303	367.54	367.2835	65.36	10.15	11.43	13.06
C20H37N304	383.54	383.2784	62.63	9.72	10.96	16.69
C20H37N305	399.53	399.2733	60.13	9.33	10.52	20.02
C20H37N306	415.53	415.2682	57.81	8.98	10.11	23.10
C20H37N307	431.53	431.2631	55.67	8.64	9.74	25.95
C20H37N308	447.53	447.2580	53.68	8.33	9.39	28.60
C20H37N309	463.53	463.2530	51.82	8.05	9.07	31.06
C20H37N501	363.55	363.2998	66.08	10.26	19.26	4.40
C20H37N502	379.55	379.2947	63.29	9.83	18.45	8.43
C20H37N503	395.55	395.2896	60.73	9.43	17.71	12.13
C20H37N504	411.55	411.2845	58.37	9.06	17.02	15.55
C20H37N505	427.55	427.2794	56.19	8.72	16.38	18.71
C20H37N506	443.55	443.2744	54.16	8.41	15.79	21.64
C20H37N507	459.55	459.2693	52.27	8.12	15.24	24.37
C20H37N508	475.55	475.2642	50.52	7.84	14.73	26.92
C20H37N509	491.55	491.2591	48.87	7.59	14.25	29.29
C20H38N201	322.54	322.2984	74.48	11.88	8.69	4.96
C20H38N202	338.54	338.2933	70.96	11.31	8.27	9.45
C20H38N203	354.54	354.2882	67.76	10.80	7.90	13.54
C20H38N204	370.54	370.2831	64.83	10.34	7.56	17.27
C20H38N205	386.54	386.2781	62.15	9.91	7.25	20.70
C20H38N206	402.54	402.2730	59.68	9.52	6.96	23.85
C20H38N207	418.54	418.2679	57.40	9.15	6.69	26.76
C20H38N208	434.53	434.2628	55.28	8.81	6.45	29.46
C20H38N209	450.53	450.2577	53.32	8.50	6.22	31.96
C20H38N401	350.55	350.3045	68.53	10.93	15.98	4.56
C20H38N402	366.55	366.2995	65.54	10.45	15.28	8.73
C20H38N403	382.55	382.2944	62.80	10.01	14.65	12.55
C20H38N404	398.55	398.2893	60.27	9.61	14.06	16.06
C20H38N405	414.55	414.2842	57.95	9.24	13.52	19.30
C20H38N406	430.55	430.2791	55.79	8.90	13.01	22.30
C20H38N407	446.55	446.2740	53.80	8.58	12.55	25.08
C20H38N408	462.55	462.2689	51.93	8.28	12.11	27.67
C20H38N409	478.55	478.2639	50.20	8.00	11.71	30.09
C20H38N601	378.57	378.3107	63.46	10.12	22.20	4.23
C20H38N602	394.56	394.3056	60.88	9.71	21.30	8.11
C20H38N603	410.56	410.3005	58.51	9.33	20.47	11.69
C20H38N604	426.56	426.2954	56.32	8.98	19.70	15.00
C20H38N605	442.56	442.2903	54.28	8.65	18.99	18.08
C20H38N606	458.56	458.2853	52.39	8.35	18.33	20.93
C20H38N607	474.56	474.2802	50.62	8.07	17.71	23.60
C20H38N608	490.56	490.2751	48.97	7.81	17.13	26.09
C20H38N609	506.56	506.2700	47.42	7.56	16.59	28.43
C20H39N101	309.54	309.3031	77.61	12.70	4.53	5.17
C20H39N102	325.54	325.2981	73.79	12.08	4.30	9.83
C20H39N103	341.54	341.2930	70.34	11.51	4.10	14.05
C20H39N104	357.54	357.2879	67.19	10.99	3.92	17.90
C20H39N105	373.54	373.2828	64.31	10.52	3.75	21.42
C20H39N106	389.54	389.2777	61.67	10.09	3.60	24.64
C20H39N107	405.54	405.2726	59.24	9.69	3.45	27.62
C20H39N108	421.54	421.2675	56.99	9.33	3.32	30.36
C20H39N109	437.54	437.2625	54.90	8.98	3.20	32.91
C20H39N301	337.55	337.3093	71.17	11.65	12.45	4.74
C20H39N302	353.55	353.3042	67.95	11.12	11.89	9.05
C20H39N303	369.55	369.2991	65.00	10.64	11.37	12.99
C20H39N304	385.55	385.2940	62.31	10.20	10.90	16.60
C20H39N305	401.55	401.2890	59.82	9.79	10.46	19.92
C20H39N306	417.55	417.2839	57.53	9.41	10.06	22.99
C20H39N307	433.55	433.2788	55.41	9.07	9.69	25.83

C20H39N3O8	449.55	449.2737	53.44	8.74	9.35	28.47
C20H39N3O9	465.55	465.2686	51.60	8.44	9.03	30.93
C20H39N5O1	365.57	365.3154	65.71	10.75	19.16	4.38
C20H39N5O2	381.57	381.3104	62.96	10.30	18.35	8.39
C20H39N5O3	397.57	397.3053	60.42	9.89	17.62	12.07
C20H39N5O4	413.56	413.3002	58.05	9.51	16.93	15.47
C20H39N5O5	429.56	429.2951	55.92	9.15	16.30	18.62
C20H39N5O6	445.56	445.2900	53.91	8.82	15.72	21.54
C20H39N5O7	461.56	461.2849	52.05	8.52	15.17	24.26
C20H39N5O8	477.56	477.2798	50.30	8.23	14.66	26.80
C20H39N5O9	493.56	493.2748	48.67	7.96	14.19	29.17
C20H40N2O1	324.55	324.3140	74.02	12.42	8.63	4.93
C20H40N2O2	340.55	340.3090	70.54	11.84	8.23	9.40
C20H40N2O3	356.55	356.3039	67.37	11.31	7.86	13.46
C20H40N2O4	372.55	372.2988	64.48	10.82	7.52	17.18
C20H40N2O5	388.55	388.2937	61.83	10.38	7.21	20.59
C20H40N2O6	404.55	404.2886	59.38	9.97	6.92	23.73
C20H40N2O7	420.55	420.2835	57.12	9.59	6.66	26.63
C20H40N2O8	436.55	436.2784	55.03	9.24	6.42	29.32
C20H40N2O9	452.55	452.2734	53.08	8.91	6.19	31.82
C20H40N4O1	352.57	352.3202	68.14	11.44	15.89	4.54
C20H40N4O2	368.57	368.3151	65.18	10.94	15.20	8.68
C20H40N4O3	384.57	384.3100	62.47	10.48	14.57	12.48
C20H40N4O4	400.57	400.3049	59.97	10.07	13.99	15.98
C20H40N4O5	416.57	416.2998	57.67	9.68	13.45	19.20
C20H40N4O6	432.56	432.2948	55.53	9.32	12.95	22.19
C20H40N4O7	448.56	448.2897	53.55	8.99	12.49	24.97
C20H40N4O8	464.56	464.2846	51.71	8.68	12.06	27.55
C20H40N4O9	480.56	480.2795	49.99	8.39	11.66	29.96
C20H40N6O1	380.58	380.3263	63.12	10.59	22.08	4.20
C20H40N6O2	396.58	396.3213	60.57	10.17	21.19	8.07
C20H40N6O3	412.58	412.3162	58.22	9.77	20.37	11.63
C20H40N6O4	428.58	428.3111	56.05	9.41	19.61	14.93
C20H40N6O5	444.58	444.3060	54.03	9.07	18.90	17.99
C20H40N6O6	460.58	460.3009	52.16	8.75	18.25	20.84
C20H40N6O7	476.58	476.2958	50.41	8.46	17.63	23.50
C20H40N6O8	492.58	492.2907	48.77	8.19	17.06	25.98
C20H40N6O9	508.58	508.2857	47.23	7.93	16.52	28.31
C20H41N1O1	311.56	311.3188	77.10	13.26	4.50	5.14
C20H41N1O2	327.56	327.3137	73.34	12.62	4.28	9.77
C20H41N1O3	343.55	343.3086	69.92	12.03	4.08	13.97
C20H41N1O4	359.55	359.3035	66.81	11.49	3.90	17.80
C20H41N1O5	375.55	375.2985	63.97	11.00	3.73	21.30
C20H41N1O6	391.55	391.2934	61.35	10.55	3.58	24.52
C20H41N1O7	407.55	407.2883	58.94	10.14	3.44	27.48
C20H41N1O8	423.55	423.2832	56.72	9.76	3.31	30.22
C20H41N1O9	439.55	439.2781	54.65	9.40	3.19	32.76
C20H41N3O1	339.57	339.3249	70.74	12.17	12.37	4.71
C20H41N3O2	355.57	355.3199	67.56	11.62	11.82	9.00
C20H41N3O3	371.57	371.3148	64.65	11.12	11.31	12.92
C20H41N3O4	387.57	387.3097	61.98	10.66	10.84	16.51
C20H41N3O5	403.57	403.3046	59.52	10.24	10.41	19.82
C20H41N3O6	419.57	419.2995	57.26	9.85	10.02	22.88
C20H41N3O7	435.57	435.2944	55.15	9.49	9.65	25.71
C20H41N3O8	451.57	451.2893	53.20	9.15	9.31	28.34
C20H41N3O9	467.56	467.2843	51.38	8.84	8.99	30.80
C20H41N5O1	367.58	367.3311	65.35	11.24	19.05	4.35
C20H41N5O2	383.58	383.3260	62.63	10.77	18.26	8.34
C20H41N5O3	399.58	399.3209	60.12	10.34	17.53	12.01
C20H41N5O4	415.58	415.3158	57.80	9.94	16.85	15.40

C20H41N5O5	431.58	431.3107	55.66	9.58	16.23	18.54
C20H41N5O6	447.58	447.3057	53.67	9.23	15.65	21.45
C20H41N5O7	463.58	463.3006	51.82	8.91	15.11	24.16
C20H41N5O8	479.58	479.2955	50.09	8.62	14.60	26.69
C20H41N5O9	495.58	495.2904	48.47	8.34	14.13	29.06
C20H42N2O1	326.57	326.3297	73.56	12.96	8.58	4.90
C20H42N2O2	342.57	342.3246	70.12	12.36	8.18	9.34
C20H42N2O3	358.57	358.3195	66.99	11.81	7.81	13.39
C20H42N2O4	374.57	374.3144	64.13	11.30	7.48	17.09
C20H42N2O5	390.57	390.3094	61.51	10.84	7.17	20.48
C20H42N2O6	406.57	406.3043	59.09	10.41	6.89	23.61
C20H42N2O7	422.57	422.2992	56.85	10.02	6.63	26.50
C20H42N2O8	438.57	438.2941	54.77	9.65	6.39	29.18
C20H42N2O9	454.57	454.2890	52.85	9.31	6.16	31.68
C20H42N4O1	354.58	354.3358	67.75	11.94	15.80	4.51
C20H42N4O2	370.58	370.3308	64.82	11.42	15.12	8.63
C20H42N4O3	386.58	386.3257	62.14	10.95	14.49	12.42
C20H42N4O4	402.58	402.3206	59.67	10.52	13.92	15.90
C20H42N4O5	418.58	418.3155	57.39	10.11	13.38	19.11
C20H42N4O6	434.58	434.3104	55.28	9.74	12.89	22.09
C20H42N4O7	450.58	450.3053	53.31	9.40	12.43	24.86
C20H42N4O8	466.58	466.3002	51.49	9.07	12.01	27.43
C20H42N4O9	482.58	482.2952	49.78	8.77	11.61	29.84
C20H42N6O1	382.60	382.3420	62.79	11.07	21.97	4.18
C20H42N6O2	398.60	398.3369	60.27	10.62	21.08	8.03
C20H42N6O3	414.60	414.3318	57.94	10.21	20.27	11.58
C20H42N6O4	430.60	430.3267	55.79	9.83	19.52	14.86
C20H42N6O5	446.59	446.3216	53.79	9.48	18.82	17.91
C20H42N6O6	462.59	462.3166	51.93	9.15	18.17	20.75
C20H42N6O7	478.59	478.3115	50.19	8.85	17.56	23.40
C20H42N6O8	494.59	494.3064	48.57	8.56	16.99	25.88
C20H42N6O9	510.59	510.3013	47.05	8.29	16.46	28.20
C20H43N1O1	313.57	313.3344	76.61	13.82	4.47	5.10
C20H43N1O2	329.57	329.3294	72.89	13.15	4.25	9.71
C20H43N1O3	345.57	345.3243	69.51	12.54	4.05	13.89
C20H43N1O4	361.57	361.3192	66.44	11.99	3.87	17.70
C20H43N1O5	377.57	377.3141	63.62	11.48	3.71	21.19
C20H43N1O6	393.57	393.3090	61.04	11.01	3.56	24.39
C20H43N1O7	409.57	409.3039	58.65	10.58	3.42	27.34
C20H43N1O8	425.57	425.2988	56.45	10.18	3.29	30.08
C20H43N1O9	441.57	441.2938	54.40	9.82	3.17	32.61
C20H43N3O1	341.59	341.3406	70.33	12.69	12.30	4.68
C20H43N3O2	357.58	357.3355	67.18	12.12	11.75	8.95
C20H43N3O3	373.58	373.3304	64.30	11.60	11.25	12.85
C20H43N3O4	389.58	389.3253	61.66	11.13	10.79	16.43
C20H43N3O5	405.58	405.3202	59.23	10.69	10.36	19.72
C20H43N3O6	421.58	421.3152	56.98	10.28	9.97	22.77
C20H43N3O7	437.58	437.3101	54.90	9.91	9.60	25.59
C20H43N3O8	453.58	453.3050	52.96	9.56	9.26	28.22
C20H43N3O9	469.58	469.2999	51.16	9.23	8.95	30.66
C20H43N5O1	369.60	369.3467	65.00	11.73	18.95	4.33
C20H43N5O2	385.60	385.3417	62.30	11.24	18.16	8.30
C20H43N5O3	401.60	401.3366	59.82	10.79	17.44	11.95
C20H43N5O4	417.60	417.3315	57.53	10.38	16.77	15.33
C20H43N5O5	433.60	433.3264	55.40	10.00	16.15	18.45
C20H43N5O6	449.60	449.3213	53.43	9.64	15.58	21.35
C20H43N5O7	465.60	465.3162	51.59	9.31	15.04	24.05
C20H43N5O8	481.59	481.3111	49.88	9.00	14.54	26.58
C20H43N5O9	497.59	497.3061	48.28	8.71	14.07	28.94
C20H44N2O1	328.59	328.3453	73.11	13.50	8.53	4.87

C20H44N2O2	344.59	344.3403	69.71	12.87	8.13	9.29
C20H44N2O3	360.59	360.3352	66.62	12.30	7.77	13.31
C20H44N2O4	376.58	376.3301	63.79	11.78	7.44	16.99
C20H44N2O5	392.58	392.3250	61.19	11.30	7.14	20.38
C20H44N2O6	408.58	408.3199	58.79	10.85	6.86	23.49
C20H44N2O7	424.58	424.3148	56.58	10.45	6.60	26.38
C20H44N2O8	440.58	440.3097	54.52	10.07	6.36	29.05
C20H44N2O9	456.58	456.3047	52.61	9.71	6.14	31.54
C20H44N4O1	356.60	356.3515	67.36	12.44	15.71	4.49
C20H44N4O2	372.60	372.3464	64.47	11.90	15.04	8.59
C20H44N4O3	388.60	388.3413	61.82	11.41	14.42	12.35
C20H44N4O4	404.60	404.3362	59.37	10.96	13.85	15.82
C20H44N4O5	420.60	420.3311	57.11	10.54	13.32	19.02
C20H44N4O6	436.60	436.3261	55.02	10.16	12.83	21.99
C20H44N4O7	452.60	452.3210	53.08	9.80	12.38	24.75
C20H44N4O8	468.60	468.3159	51.26	9.46	11.96	27.31
C20H44N4O9	484.60	484.3108	49.57	9.15	11.56	29.71
C20H44N6O1	384.61	384.3576	62.46	11.53	21.85	4.16
C20H44N6O2	400.61	400.3526	59.96	11.07	20.98	7.99
C20H44N6O3	416.61	416.3475	57.66	10.65	20.17	11.52
C20H44N6O4	432.61	432.3424	55.53	10.25	19.43	14.79
C20H44N6O5	448.61	448.3373	53.55	9.89	18.73	17.83
C20H44N6O6	464.61	464.3322	51.70	9.55	18.09	20.66
C20H44N6O7	480.61	480.3271	49.98	9.23	17.49	23.30
C20H44N6O8	496.61	496.3220	48.37	8.93	16.92	25.77
C20H44N6O9	512.61	512.3170	46.86	8.65	16.39	28.09
C20H45N3O1	343.60	343.3562	69.91	13.20	12.23	4.66
C20H45N3O2	359.60	359.3512	66.80	12.61	11.69	8.90
C20H45N3O3	375.60	375.3461	63.96	12.08	11.19	12.78
C20H45N3O4	391.60	391.3410	61.34	11.58	10.73	16.34
C20H45N3O5	407.60	407.3359	58.94	11.13	10.31	19.63
C20H45N3O6	423.60	423.3308	56.71	10.71	9.92	22.66
C20H45N3O7	439.60	439.3257	54.65	10.32	9.56	25.48
C20H45N3O8	455.60	455.3206	52.73	9.96	9.22	28.09
C20H45N3O9	471.60	471.3156	50.94	9.62	8.91	30.53
C20H45N5O1	371.61	371.3624	64.64	12.21	18.85	4.31
C20H45N5O2	387.61	387.3573	61.97	11.70	18.07	8.26
C20H45N5O3	403.61	403.3522	59.52	11.24	17.35	11.89
C20H45N5O4	419.61	419.3471	57.25	10.81	16.69	15.25
C20H45N5O5	435.61	435.3420	55.15	10.41	16.08	18.36
C20H45N5O6	451.61	451.3370	53.19	10.04	15.51	21.26
C20H45N5O7	467.61	467.3319	51.37	9.70	14.98	23.95
C20H45N5O8	483.61	483.3268	49.67	9.38	14.48	26.47
C20H45N5O9	499.61	499.3217	48.08	9.08	14.02	28.82
C20H46N4O1	358.62	358.3671	66.99	12.93	15.62	4.46
C20H46N4O2	374.62	374.3621	64.13	12.38	14.96	8.54
C20H46N4O3	390.61	390.3570	61.50	11.87	14.34	12.29
C20H46N4O4	406.61	406.3519	59.08	11.40	13.78	15.74
C20H46N4O5	422.61	422.3468	56.84	10.97	13.26	18.93
C20H46N4O6	438.61	438.3417	54.77	10.57	12.77	21.89
C20H46N4O7	454.61	454.3366	52.84	10.20	12.32	24.64
C20H46N4O8	470.61	470.3315	51.04	9.85	11.91	27.20
C20H46N4O9	486.61	486.3265	49.37	9.53	11.51	29.59
C20H46N6O1	386.63	386.3733	62.13	11.99	21.74	4.14
C20H46N6O2	402.63	402.3682	59.66	11.52	20.87	7.95
C20H46N6O3	418.63	418.3631	57.38	11.08	20.08	11.47
C20H46N6O4	434.63	434.3580	55.27	10.67	19.34	14.72
C20H46N6O5	450.63	450.3529	53.31	10.29	18.65	17.75
C20H46N6O6	466.63	466.3479	51.48	9.94	18.01	20.57
C20H46N6O7	482.63	482.3428	49.77	9.61	17.41	23.21

C20H46N6O8	498.63	498.3377	48.18	9.30	16.85	25.67
C20H46N6O9	514.62	514.3326	46.68	9.01	16.33	27.98
C20H47N5O1	373.63	373.3780	64.29	12.68	18.74	4.28
C20H47N5O2	389.63	389.3730	61.65	12.16	17.97	8.21
C20H47N5O3	405.63	405.3679	59.22	11.68	17.27	11.83
C20H47N5O4	421.63	421.3628	56.98	11.24	16.61	15.18
C20H47N5O5	437.63	437.3577	54.89	10.83	16.00	18.28
C20H47N5O6	453.63	453.3526	52.96	10.44	15.44	21.16
C20H47N5O7	469.63	469.3475	51.15	10.09	14.91	23.85
C20H47N5O8	485.63	485.3424	49.47	9.76	14.42	26.36
C20H47N5O9	501.63	501.3374	47.89	9.44	13.96	28.71
C20H48N6O1	388.65	388.3889	61.81	12.45	21.62	4.12
C20H48N6O2	404.64	404.3839	59.37	11.96	20.77	7.91
C20H48N6O3	420.64	420.3788	57.11	11.50	19.98	11.41
C20H48N6O4	436.64	436.3737	55.02	11.08	19.25	14.66
C20H48N6O5	452.64	452.3686	53.07	10.69	18.57	17.67
C20H48N6O6	468.64	468.3635	51.26	10.32	17.93	20.48
C20H48N6O7	484.64	484.3584	49.57	9.98	17.34	23.11
C20H48N6O8	500.64	500.3533	47.98	9.66	16.79	25.57
C20H48N6O9	516.64	516.3483	46.50	9.36	16.27	27.87
C21H10N2O1	306.33	306.0793	82.34	3.29	9.14	5.22
C21H10N2O2	322.33	322.0742	78.25	3.13	8.69	9.93
C21H10N2O3	338.33	338.0691	74.55	2.98	8.28	14.19
C21H10N2O4	354.32	354.0640	71.19	2.84	7.91	18.06
C21H10N2O5	370.32	370.0590	68.11	2.72	7.56	21.60
C21H10N2O6	386.32	386.0539	65.29	2.61	7.25	24.85
C21H10N2O7	402.32	402.0488	62.69	2.51	6.96	27.84
C21H10N2O8	418.32	418.0437	60.30	2.41	6.70	30.60
C21H10N2O9	434.32	434.0386	58.08	2.32	6.45	33.15
C21H10N4O1	334.34	334.0855	75.44	3.01	16.76	4.79
C21H10N4O2	350.34	350.0804	72.00	2.88	15.99	9.13
C21H10N4O3	366.34	366.0753	68.85	2.75	15.29	13.10
C21H10N4O4	382.34	382.0702	65.97	2.64	14.65	16.74
C21H10N4O5	398.34	398.0651	63.32	2.53	14.07	20.08
C21H10N4O6	414.34	414.0600	60.88	2.43	13.52	23.17
C21H10N4O7	430.34	430.0549	58.61	2.34	13.02	26.03
C21H10N4O8	446.34	446.0499	56.51	2.26	12.55	28.68
C21H10N4O9	462.34	462.0448	54.56	2.18	12.12	31.15
C21H10N6O1	362.35	362.0916	69.61	2.78	23.19	4.42
C21H10N6O2	378.35	378.0865	66.67	2.66	22.21	8.46
C21H10N6O3	394.35	394.0814	63.96	2.56	21.31	12.17
C21H10N6O4	410.35	410.0763	61.47	2.46	20.48	15.60
C21H10N6O5	426.35	426.0713	59.16	2.36	19.71	18.76
C21H10N6O6	442.35	442.0662	57.02	2.28	19.00	21.70
C21H10N6O7	458.35	458.0611	55.03	2.20	18.34	24.43
C21H10N6O8	474.35	474.0560	53.17	2.12	17.72	26.98
C21H10N6O9	490.35	490.0509	51.44	2.06	17.14	29.37
C21H11N1O1	293.33	293.0841	85.99	3.78	4.78	5.45
C21H11N1O2	309.33	309.0790	81.54	3.58	4.53	10.34
C21H11N1O3	325.33	325.0739	77.53	3.41	4.31	14.75
C21H11N1O4	341.33	341.0688	73.90	3.25	4.10	18.75
C21H11N1O5	357.33	357.0637	70.59	3.10	3.92	22.39
C21H11N1O6	373.32	373.0586	67.56	2.97	3.75	25.71
C21H11N1O7	389.32	389.0535	64.79	2.85	3.60	28.77
C21H11N1O8	405.32	405.0485	62.23	2.74	3.46	31.58
C21H11N1O9	421.32	421.0434	59.87	2.63	3.32	34.18
C21H11N3O1	321.34	321.0902	78.49	3.45	13.08	4.98
C21H11N3O2	337.34	337.0851	74.77	3.29	12.46	9.49
C21H11N3O3	353.34	353.0800	71.39	3.14	11.89	13.58
C21H11N3O4	369.34	369.0749	68.29	3.00	11.38	17.33

120

C21H11N305	385.34	385.0699	65.46	2.88	10.90	20.76
C21H11N306	401.34	401.0648	62.85	2.76	10.47	23.92
C21H11N307	417.34	417.0597	60.44	2.66	10.07	26.84
C21H11N308	433.34	433.0546	58.21	2.56	9.70	29.54
C21H11N309	449.34	449.0495	56.13	2.47	9.35	32.05
C21H11N501	349.35	349.0964	72.20	3.17	20.05	4.58
C21H11N502	365.35	365.0913	69.04	3.03	19.17	8.76
C21H11N503	381.35	381.0862	66.14	2.91	18.36	12.59
C21H11N504	397.35	397.0811	63.48	2.79	17.63	16.11
C21H11N505	413.35	413.0760	61.02	2.68	16.94	19.35
C21H11N506	429.35	429.0709	58.75	2.58	16.31	22.36
C21H11N507	445.35	445.0658	56.64	2.49	15.73	25.15
C21H11N508	461.35	461.0608	54.67	2.40	15.18	27.74
C21H11N509	477.35	477.0557	52.84	2.32	14.67	30.17
C21H12N201	308.34	308.0950	81.80	3.92	9.09	5.19
C21H12N202	324.34	324.0899	77.77	3.73	8.64	9.87
C21H12N203	340.34	340.0848	74.11	3.55	8.23	14.10
C21H12N204	356.34	356.0797	70.78	3.39	7.86	17.96
C21H12N205	372.34	372.0746	67.74	3.25	7.52	21.48
C21H12N206	388.34	388.0695	64.95	3.11	7.21	24.72
C21H12N207	404.34	404.0644	62.38	2.99	6.93	27.70
C21H12N208	420.34	420.0594	60.01	2.88	6.66	30.45
C21H12N209	436.34	436.0543	57.81	2.77	6.42	33.00
C21H12N401	336.36	336.1011	74.99	3.60	16.66	4.76
C21H12N402	352.36	352.0960	71.59	3.43	15.90	9.08
C21H12N403	368.35	368.0909	68.48	3.28	15.21	13.03
C21H12N404	384.35	384.0858	65.63	3.15	14.58	16.65
C21H12N405	400.35	400.0808	63.00	3.02	13.99	19.98
C21H12N406	416.35	416.0757	60.58	2.91	13.46	23.06
C21H12N407	432.35	432.0706	58.34	2.80	12.96	25.90
C21H12N408	448.35	448.0655	56.26	2.70	12.50	28.55
C21H12N409	464.35	464.0604	54.32	2.60	12.07	31.01
C21H12N601	364.37	364.1073	69.22	3.32	23.06	4.39
C21H12N602	380.37	380.1022	66.31	3.18	22.09	8.41
C21H12N603	396.37	396.0971	63.64	3.05	21.20	12.11
C21H12N604	412.37	412.0920	61.17	2.93	20.38	15.52
C21H12N605	428.37	428.0869	58.88	2.82	19.62	18.67
C21H12N606	444.37	444.0818	56.76	2.72	18.91	21.60
C21H12N607	460.37	460.0767	54.79	2.63	18.26	24.33
C21H12N608	476.37	476.0717	52.95	2.54	17.64	26.87
C21H12N609	492.36	492.0666	51.23	2.46	17.07	29.25
C21H13N101	295.34	295.0997	85.40	4.44	4.74	5.42
C21H13N102	311.34	311.0946	81.01	4.21	4.50	10.28
C21H13N103	327.34	327.0895	77.06	4.00	4.28	14.66
C21H13N104	343.34	343.0844	73.46	3.82	4.08	18.64
C21H13N105	359.34	359.0794	70.19	3.65	3.90	22.26
C21H13N106	375.34	375.0743	67.20	3.49	3.73	25.58
C21H13N107	391.34	391.0692	64.45	3.35	3.58	28.62
C21H13N108	407.34	407.0641	61.92	3.22	3.44	31.42
C21H13N109	423.34	423.0590	59.58	3.10	3.31	34.01
C21H13N301	323.36	323.1059	78.00	4.05	12.99	4.95
C21H13N302	339.36	339.1008	74.33	3.86	12.38	9.43
C21H13N303	355.36	355.0957	70.98	3.69	11.82	13.51
C21H13N304	371.36	371.0906	67.92	3.53	11.32	17.23
C21H13N305	387.35	387.0855	65.12	3.38	10.85	20.65
C21H13N306	403.35	403.0804	62.53	3.25	10.42	23.80
C21H13N307	419.35	419.0753	60.15	3.12	10.02	26.71
C21H13N308	435.35	435.0703	57.94	3.01	9.65	29.40
C21H13N309	451.35	451.0652	55.88	2.90	9.31	31.90
C21H13N501	351.37	351.1120	71.79	3.73	19.93	4.55
C21H13N502	367.37	367.1069	68.66	3.57	19.06	8.71
C21H13N503	383.37	383.1018	65.79	3.42	18.27	12.52
C21H13N504	399.37	399.0967	63.16	3.28	17.54	16.02
C21H13N505	415.37	415.0917	60.73	3.15	16.86	19.26
C21H13N506	431.37	431.0866	58.47	3.04	16.24	22.25
C21H13N507	447.37	447.0815	56.38	2.93	15.65	25.03
C21H13N508	463.37	463.0764	54.44	2.83	15.11	27.62
C21H13N509	479.37	479.0713	52.62	2.73	14.61	30.04
C21H14N201	310.36	310.1106	81.27	4.55	9.03	5.16
C21H14N202	326.36	326.1055	77.29	4.32	8.58	9.80
C21H14N203	342.36	342.1004	73.68	4.12	8.18	14.02
C21H14N204	358.36	358.0953	70.39	3.94	7.82	17.86
C21H14N205	374.36	374.0903	67.38	3.77	7.48	21.37
C21H14N206	390.36	390.0852	64.62	3.62	7.18	24.59
C21H14N207	406.35	406.0801	62.07	3.47	6.89	27.56
C21H14N208	422.35	422.0750	59.72	3.34	6.63	30.31
C21H14N209	438.35	438.0699	57.54	3.22	6.39	32.85
C21H14N401	338.37	338.1168	74.54	4.17	16.56	4.73
C21H14N402	354.37	354.1117	71.18	3.98	15.81	9.03
C21H14N403	370.37	370.1066	68.10	3.81	15.13	12.96
C21H14N404	386.37	386.1015	65.28	3.65	14.50	16.56
C21H14N405	402.37	402.0964	62.69	3.51	13.92	19.88
C21H14N406	418.37	418.0913	60.29	3.37	13.39	22.95
C21H14N407	434.37	434.0862	58.07	3.25	12.90	25.78
C21H14N408	450.37	450.0812	56.01	3.13	12.44	28.42
C21H14N409	466.37	466.0761	54.08	3.03	12.01	30.88
C21H14N601	366.39	366.1229	68.84	3.85	22.94	4.37
C21H14N602	382.38	382.1178	65.96	3.69	21.98	8.37
C21H14N603	398.38	398.1127	63.31	3.54	21.10	12.05
C21H14N604	414.38	414.1076	60.87	3.41	20.28	15.44
C21H14N605	430.38	430.1026	58.61	3.28	19.53	18.59
C21H14N606	446.38	446.0975	56.51	3.16	18.83	21.51
C21H14N607	462.38	462.0924	54.55	3.05	18.18	24.22
C21H14N608	478.38	478.0873	52.73	2.95	17.57	26.76
C21H14N609	494.38	494.0822	51.02	2.85	17.00	29.13
C21H15N101	297.36	297.1154	84.82	5.08	4.71	5.38
C21H15N102	313.36	313.1103	80.49	4.82	4.47	10.21
C21H15N103	329.36	329.1052	76.58	4.59	4.25	14.57
C21H15N104	345.36	345.1001	73.04	4.38	4.06	18.53
C21H15N105	361.36	361.0950	69.80	4.18	3.88	22.14
C21H15N106	377.36	377.0899	66.84	4.01	3.71	25.44
C21H15N107	393.36	393.0848	64.12	3.84	3.56	28.47
C21H15N108	409.36	409.0798	61.62	3.69	3.42	31.27
C21H15N109	425.36	425.0747	59.30	3.55	3.29	33.85
C21H15N301	325.37	325.1215	77.52	4.65	12.91	4.92
C21H15N302	341.37	341.1164	73.89	4.43	12.31	9.37
C21H15N303	357.37	357.1113	70.58	4.23	11.76	13.43
C21H15N304	373.37	373.1062	67.56	4.05	11.25	17.14
C21H15N305	389.37	389.1012	64.78	3.88	10.79	20.55
C21H15N306	405.37	405.0961	62.22	3.73	10.37	23.68
C21H15N307	421.37	421.0910	59.86	3.59	9.97	26.58
C21H15N308	437.37	437.0859	57.67	3.46	9.61	29.26
C21H15N309	453.37	453.0808	55.64	3.33	9.27	31.76
C21H15N501	353.39	353.1277	71.38	4.28	19.82	4.53
C21H15N502	369.39	369.1226	68.28	4.09	18.96	8.66
C21H15N503	385.39	385.1175	65.45	3.92	18.17	12.45
C21H15N504	401.38	401.1124	62.84	3.77	17.45	15.94
C21H15N505	417.38	417.1073	60.43	3.62	16.78	19.17
C21H15N506	433.38	433.1022	58.20	3.49	16.16	22.15
C21H15N507	449.38	449.0971	56.13	3.36	15.58	24.92

121

C21H15N508	465.38	465.0921	54.20	3.25	15.05	27.50
C21H15N509	481.38	481.0870	52.40	3.14	14.55	29.91
C21H16N201	312.37	312.1263	80.75	5.16	8.97	5.12
C21H16N202	328.37	328.1212	76.81	4.91	8.53	9.74
C21H16N203	344.37	344.1161	73.24	4.68	8.13	13.94
C21H16N204	360.37	360.1110	69.99	4.48	7.77	17.76
C21H16N205	376.37	376.1059	67.02	4.28	7.44	21.25
C21H16N206	392.37	392.1008	64.28	4.11	7.14	24.47
C21H16N207	408.37	408.0957	61.77	3.95	6.86	27.43
C21H16N208	424.37	424.0907	59.44	3.80	6.60	30.16
C21H16N209	440.37	440.0856	57.28	3.66	6.36	32.70
C21H16N401	340.39	340.1324	74.10	4.74	16.46	4.70
C21H16N402	356.39	356.1273	70.78	4.53	15.72	8.98
C21H16N403	372.39	372.1222	67.73	4.33	15.05	12.89
C21H16N404	388.39	388.1171	64.94	4.15	14.43	16.48
C21H16N405	404.39	404.1121	62.37	3.99	13.85	19.78
C21H16N406	420.38	420.1070	60.00	3.84	13.33	22.84
C21H16N407	436.38	436.1019	57.80	3.70	12.84	25.66
C21H16N408	452.38	452.0968	55.76	3.57	12.38	28.29
C21H16N409	468.38	468.0917	53.85	3.44	11.96	30.74
C21H16N601	368.40	368.1386	68.47	4.38	22.81	4.34
C21H16N602	384.40	384.1335	65.62	4.20	21.86	8.32
C21H16N603	400.40	400.1284	63.00	4.03	20.99	11.99
C21H16N604	416.40	416.1233	60.58	3.87	20.18	15.37
C21H16N605	432.40	432.1182	58.33	3.73	19.44	18.50
C21H16N606	448.40	448.1131	56.25	3.60	18.74	21.41
C21H16N607	464.40	464.1080	54.31	3.47	18.10	24.12
C21H16N608	480.40	480.1029	52.51	3.36	17.49	26.64
C21H16N609	496.40	496.0979	50.81	3.25	16.93	29.01
C21H17N101	299.38	299.1310	84.25	5.72	4.68	5.34
C21H17N102	315.38	315.1259	79.98	5.43	4.44	10.15
C21H17N103	331.37	331.1208	76.12	5.17	4.23	14.48
C21H17N104	347.37	347.1157	72.61	4.93	4.03	18.42
C21H17N105	363.37	363.1107	69.41	4.72	3.85	22.02
C21H17N106	379.37	379.1056	66.49	4.52	3.69	25.30
C21H17N107	395.37	395.1005	63.80	4.33	3.54	28.33
C21H17N108	411.37	411.0954	61.32	4.17	3.40	31.11
C21H17N109	427.37	427.0903	59.02	4.01	3.28	33.69
C21H17N301	327.39	327.1372	77.04	5.23	12.83	4.89
C21H17N302	343.39	343.1321	73.45	4.99	12.24	9.32
C21H17N303	359.39	359.1270	70.18	4.77	11.69	13.36
C21H17N304	375.39	375.1219	67.19	4.56	11.19	17.05
C21H17N305	391.39	391.1168	64.45	4.38	10.74	20.44
C21H17N306	407.39	407.1117	61.92	4.21	10.31	23.56
C21H17N307	423.39	423.1066	59.58	4.05	9.92	26.45
C21H17N308	439.38	439.1016	57.41	3.90	9.56	29.13
C21H17N309	455.38	455.0965	55.39	3.76	9.23	31.62
C21H17N501	355.40	355.1433	70.97	4.82	19.71	4.50
C21H17N502	371.40	371.1382	67.91	4.61	18.86	8.62
C21H17N503	387.40	387.1331	65.11	4.42	18.08	12.39
C21H17N504	403.40	403.1280	62.53	4.25	17.36	15.86
C21H17N505	419.40	419.1230	60.14	4.09	16.70	19.07
C21H17N506	435.40	435.1179	57.93	3.94	16.08	22.05
C21H17N507	451.40	451.1128	55.88	3.80	15.51	24.81
C21H17N508	467.40	467.1077	53.97	3.67	14.98	27.38
C21H17N509	483.40	483.1026	52.18	3.54	14.49	29.79
C21H18N201	314.39	314.1419	80.23	5.77	8.91	5.09
C21H18N202	330.39	330.1368	76.34	5.49	8.48	9.69
C21H18N203	346.39	346.1317	72.82	5.24	8.09	13.86
C21H18N204	362.39	362.1266	69.60	5.01	7.73	17.66
C21H18N205	378.39	378.1216	66.66	4.79	7.40	21.14
C21H18N206	394.39	394.1165	63.96	4.60	7.10	24.34
C21H18N207	410.39	410.1114	61.46	4.42	6.83	27.29
C21H18N208	426.39	426.1063	59.16	4.26	6.57	30.02
C21H18N209	442.39	442.1012	57.02	4.10	6.33	32.55
C21H18N401	342.40	342.1481	73.67	5.30	16.36	4.67
C21H18N402	358.40	358.1430	70.38	5.06	15.63	8.93
C21H18N403	374.40	374.1379	67.37	4.85	14.96	12.82
C21H18N404	390.40	390.1328	64.61	4.65	14.35	16.39
C21H18N405	406.40	406.1277	62.07	4.46	13.79	19.68
C21H18N406	422.40	422.1226	59.71	4.30	13.26	22.73
C21H18N407	438.40	438.1175	57.54	4.14	12.78	25.55
C21H18N408	454.40	454.1125	55.51	3.99	12.33	28.17
C21H18N409	470.40	470.1074	53.62	3.86	11.91	30.61
C21H18N601	370.42	370.1542	68.09	4.90	22.69	4.32
C21H18N602	386.42	386.1491	65.28	4.70	21.75	8.28
C21H18N603	402.42	402.1440	62.68	4.51	20.88	11.93
C21H18N604	418.42	418.1389	60.28	4.34	20.09	15.30
C21H18N605	434.41	434.1339	58.06	4.18	19.35	18.41
C21H18N606	450.41	450.1288	56.00	4.03	18.66	21.31
C21H18N607	466.41	466.1237	54.08	3.89	18.02	24.01
C21H18N608	482.41	482.1186	52.29	3.76	17.42	26.53
C21H18N609	498.41	498.1135	50.61	3.64	16.86	28.89
C21H19N101	301.39	301.1467	83.69	6.35	4.65	5.31
C21H19N102	317.39	317.1416	79.47	6.03	4.41	10.08
C21H19N103	333.39	333.1365	75.66	5.74	4.20	14.40
C21H19N104	349.39	349.1314	72.19	5.48	4.01	18.32
C21H19N105	365.39	365.1263	69.03	5.24	3.83	21.89
C21H19N106	381.39	381.1212	66.14	5.02	3.67	25.17
C21H19N107	397.39	397.1161	63.47	4.82	3.52	28.18
C21H19N108	413.39	413.1111	61.02	4.63	3.39	30.96
C21H19N109	429.39	429.1060	58.74	4.46	3.26	33.53
C21H19N301	329.41	329.1528	76.57	5.81	12.76	4.86
C21H19N302	345.40	345.1477	73.03	5.54	12.17	9.26
C21H19N303	361.40	361.1426	69.79	5.30	11.63	13.28
C21H19N304	377.40	377.1375	66.83	5.07	11.13	16.96
C21H19N305	393.40	393.1325	64.12	4.87	10.68	20.33
C21H19N306	409.40	409.1274	61.61	4.68	10.26	23.45
C21H19N307	425.40	425.1223	59.29	4.50	9.88	26.33
C21H19N308	441.40	441.1172	57.14	4.34	9.52	29.00
C21H19N309	457.40	457.1121	55.15	4.19	9.19	31.48
C21H19N501	357.42	357.1590	70.57	5.36	19.59	4.48
C21H19N502	373.42	373.1539	67.55	5.13	18.75	8.57
C21H19N503	389.42	389.1488	64.77	4.92	17.98	12.33
C21H19N504	405.42	405.1437	62.22	4.72	17.27	15.79
C21H19N505	421.42	421.1386	59.85	4.54	16.62	18.98
C21H19N506	437.42	437.1335	57.66	4.38	16.01	21.95
C21H19N507	453.41	453.1284	55.63	4.22	15.45	24.70
C21H19N508	469.41	469.1233	53.73	4.08	14.92	27.27
C21H19N509	485.41	485.1183	51.96	3.95	14.43	29.66
C21H20N201	316.41	316.1576	79.72	6.37	8.85	5.06
C21H20N202	332.41	332.1525	75.88	6.06	8.43	9.63
C21H20N203	348.41	348.1474	72.40	5.79	8.04	13.78
C21H20N204	364.40	364.1423	69.22	5.53	7.69	17.56
C21H20N205	380.40	380.1372	66.31	5.30	7.36	21.03
C21H20N206	396.40	396.1321	63.63	5.09	7.07	24.22
C21H20N207	412.40	412.1270	61.16	4.89	6.79	27.16
C21H20N208	428.40	428.1220	58.88	4.71	6.54	29.88
C21H20N209	444.40	444.1169	56.76	4.54	6.30	32.40
C21H20N401	344.42	344.1637	73.23	5.85	16.27	4.65

C21H20N4O2	360.42	360.1586	69.98	5.59	15.54	8.88
C21H20N4O3	376.42	376.1535	67.01	5.36	14.88	12.75
C21H20N4O4	392.42	392.1484	64.28	5.14	14.28	16.31
C21H20N4O5	408.42	408.1434	61.76	4.94	13.72	19.59
C21H20N4O6	424.42	424.1383	59.43	4.75	13.20	22.62
C21H20N4O7	440.42	440.1332	57.27	4.58	12.72	25.43
C21H20N4O8	456.42	456.1281	55.26	4.42	12.28	28.04
C21H20N4O9	472.41	472.1230	53.39	4.27	11.86	30.48
C21H20N6O1	372.43	372.1698	67.73	5.41	22.57	4.30
C21H20N6O2	388.43	388.1648	64.94	5.19	21.64	8.24
C21H20N6O3	404.43	404.1597	62.37	4.98	20.78	11.87
C21H20N6O4	420.43	420.1546	59.99	4.79	19.99	15.22
C21H20N6O5	436.43	436.1495	57.79	4.62	19.26	18.33
C21H20N6O6	452.43	452.1444	55.75	4.46	18.58	21.22
C21H20N6O7	468.43	468.1393	53.85	4.30	17.94	23.91
C21H20N6O8	484.43	484.1342	52.07	4.16	17.35	26.42
C21H20N6O9	500.43	500.1292	50.40	4.03	16.79	28.77
C21H21N1O1	303.41	303.1623	83.13	6.98	4.62	5.27
C21H21N1O2	319.41	319.1572	78.97	6.63	4.39	10.02
C21H21N1O3	335.41	335.1521	75.20	6.31	4.18	14.31
C21H21N1O4	351.41	351.1470	71.78	6.02	3.99	18.21
C21H21N1O5	367.41	367.1420	68.65	5.76	3.81	21.77
C21H21N1O6	383.40	383.1369	65.79	5.52	3.65	25.04
C21H21N1O7	399.40	399.1318	63.15	5.30	3.51	28.04
C21H21N1O8	415.40	415.1267	60.72	5.10	3.37	30.81
C21H21N1O9	431.40	431.1216	58.47	4.91	3.25	33.38
C21H21N3O1	331.42	331.1685	76.11	6.39	12.68	4.83
C21H21N3O2	347.42	347.1634	72.60	6.09	12.09	9.21
C21H21N3O3	363.42	363.1583	69.41	5.82	11.56	13.21
C21H21N3O4	379.42	379.1532	66.48	5.58	11.07	16.87
C21H21N3O5	395.42	395.1481	63.79	5.35	10.63	20.23
C21H21N3O6	411.42	411.1430	61.31	5.14	10.21	23.33
C21H21N3O7	427.42	427.1379	59.01	4.95	9.83	26.20
C21H21N3O8	443.42	443.1329	56.88	4.77	9.48	28.87
C21H21N3O9	459.42	459.1278	54.90	4.61	9.15	31.34
C21H21N5O1	359.43	359.1746	70.18	5.89	19.48	4.45
C21H21N5O2	375.43	375.1695	67.18	5.64	18.65	8.52
C21H21N5O3	391.43	391.1644	64.44	5.41	17.89	12.26
C21H21N5O4	407.43	407.1593	61.91	5.20	17.19	15.71
C21H21N5O5	423.43	423.1543	59.57	5.00	16.54	18.89
C21H21N5O6	439.43	439.1492	57.40	4.82	15.94	21.85
C21H21N5O7	455.43	455.1441	55.38	4.65	15.38	24.59
C21H21N5O8	471.43	471.1390	53.50	4.49	14.86	27.15
C21H21N5O9	487.43	487.1339	51.75	4.34	14.37	29.54
C21H22N2O1	318.42	318.1732	79.21	6.96	8.80	5.02
C21H22N2O2	334.42	334.1681	75.42	6.63	8.38	9.57
C21H22N2O3	350.42	350.1630	71.98	6.33	7.99	13.70
C21H22N2O4	366.42	366.1579	68.84	6.05	7.65	17.47
C21H22N2O5	382.42	382.1529	65.96	5.80	7.33	20.92
C21H22N2O6	398.42	398.1478	63.31	5.57	7.03	24.09
C21H22N2O7	414.42	414.1427	60.86	5.35	6.76	27.02
C21H22N2O8	430.42	430.1376	58.60	5.15	6.51	29.74
C21H22N2O9	446.42	446.1325	56.50	4.97	6.28	32.26
C21H22N4O1	346.44	346.1794	72.81	6.40	16.17	4.62
C21H22N4O2	362.44	362.1743	69.59	6.12	15.46	8.83
C21H22N4O3	378.43	378.1692	66.65	5.86	14.80	12.68
C21H22N4O4	394.43	394.1641	63.95	5.62	14.20	16.23
C21H22N4O5	410.43	410.1590	61.46	5.40	13.65	19.49
C21H22N4O6	426.43	426.1539	59.15	5.20	13.14	22.51
C21H22N4O7	442.43	442.1488	57.01	5.01	12.66	25.31
C21H22N4O8	458.43	458.1437	55.02	4.84	12.22	27.92
C21H22N4O9	474.43	474.1387	53.17	4.67	11.81	30.35
C21H22N6O1	374.45	374.1855	67.36	5.92	22.44	4.27
C21H22N6O2	390.45	390.1804	64.60	5.68	21.52	8.20
C21H22N6O3	406.45	406.1753	62.06	5.46	20.68	11.81
C21H22N6O4	422.45	422.1702	59.71	5.25	19.89	15.15
C21H22N6O5	438.45	438.1652	57.53	5.06	19.17	18.25
C21H22N6O6	454.45	454.1601	55.50	4.88	18.49	21.12
C21H22N6O7	470.45	470.1550	53.62	4.71	17.86	23.81
C21H22N6O8	486.44	486.1499	51.85	4.56	17.28	26.31
C21H22N6O9	502.44	502.1448	50.20	4.41	16.73	28.66
C21H23N1O1	305.42	305.1780	82.59	7.59	4.59	5.24
C21H23N1O2	321.42	321.1729	78.47	7.21	4.36	9.96
C21H23N1O3	337.42	337.1678	74.75	6.87	4.15	14.22
C21H23N1O4	353.42	353.1627	71.37	6.56	3.96	18.11
C21H23N1O5	369.42	369.1576	68.28	6.28	3.79	21.65
C21H23N1O6	385.42	385.1525	65.44	6.02	3.63	24.91
C21H23N1O7	401.42	401.1474	62.84	5.78	3.49	27.90
C21H23N1O8	417.42	417.1424	60.43	5.55	3.36	30.66
C21H23N1O9	433.42	433.1373	58.20	5.35	3.23	33.22
C21H23N3O1	333.44	333.1841	75.65	6.95	12.60	4.80
C21H23N3O2	349.44	349.1790	72.18	6.63	12.03	9.16
C21H23N3O3	365.44	365.1739	69.02	6.34	11.50	13.13
C21H23N3O4	381.44	381.1688	66.13	6.08	11.02	16.78
C21H23N3O5	397.43	397.1638	63.47	5.83	10.57	20.13
C21H23N3O6	413.43	413.1587	61.01	5.61	10.16	23.22
C21H23N3O7	429.43	429.1536	58.74	5.40	9.79	26.08
C21H23N3O8	445.43	445.1485	56.63	5.20	9.43	28.74
C21H23N3O9	461.43	461.1434	54.66	5.02	9.11	31.21
C21H23N5O1	361.45	361.1902	69.78	6.41	19.38	4.43
C21H23N5O2	377.45	377.1852	66.83	6.14	18.55	8.48
C21H23N5O3	393.45	393.1801	64.11	5.89	17.80	12.20
C21H23N5O4	409.45	409.1750	61.60	5.66	17.10	15.63
C21H23N5O5	425.45	425.1699	59.29	5.45	16.46	18.80
C21H23N5O6	441.45	441.1648	57.14	5.25	15.86	21.75
C21H23N5O7	457.45	457.1597	55.14	5.07	15.31	24.48
C21H23N5O8	473.45	473.1546	53.28	4.90	14.79	27.03
C21H23N5O9	489.45	489.1496	51.53	4.74	14.31	29.42
C21H24N2O1	320.44	320.1889	78.72	7.55	8.74	4.99
C21H24N2O2	336.44	336.1838	74.97	7.19	8.33	9.51
C21H24N2O3	352.44	352.1787	71.57	6.86	7.95	13.62
C21H24N2O4	368.44	368.1736	68.46	6.57	7.60	17.37
C21H24N2O5	384.44	384.1685	65.61	6.29	7.29	20.81
C21H24N2O6	400.44	400.1634	62.99	6.04	7.00	23.97
C21H24N2O7	416.43	416.1583	60.57	5.81	6.73	26.89
C21H24N2O8	432.43	432.1533	58.33	5.59	6.48	29.60
C21H24N2O9	448.43	448.1482	56.25	5.39	6.25	32.11
C21H24N4O1	348.45	348.1950	72.39	6.94	16.08	4.59
C21H24N4O2	364.45	364.1899	69.21	6.64	15.37	8.78
C21H24N4O3	380.45	380.1848	66.30	6.36	14.73	12.62
C21H24N4O4	396.45	396.1797	63.62	6.10	14.13	16.14
C21H24N4O5	412.45	412.1747	61.16	5.87	13.58	19.40
C21H24N4O6	428.45	428.1696	58.87	5.65	13.08	22.41
C21H24N4O7	444.45	444.1645	56.75	5.44	12.61	25.20
C21H24N4O8	460.45	460.1594	54.78	5.25	12.17	27.80
C21H24N4O9	476.45	476.1543	52.94	5.08	11.76	30.22
C21H24N6O1	376.47	376.2011	67.00	6.43	22.32	4.25
C21H24N6O2	392.46	392.1961	64.27	6.16	21.41	8.15
C21H24N6O3	408.46	408.1910	61.75	5.92	20.57	11.75
C21H24N6O4	424.46	424.1859	59.42	5.70	19.80	15.08

C21H24N6O5	440.46	440.1808	57.27	5.49	19.08	18.16
C21H24N6O6	456.46	456.1757	55.26	5.30	18.41	21.03
C21H24N6O7	472.46	472.1706	53.39	5.12	17.79	23.70
C21H24N6O8	488.46	488.1655	51.64	4.95	17.21	26.20
C21H24N6O9	504.46	504.1605	50.00	4.80	16.66	28.54
C21H25N1O1	307.44	307.1936	82.04	8.20	4.56	5.20
C21H25N1O2	323.44	323.1885	77.99	7.79	4.33	9.89
C21H25N1O3	339.44	339.1834	74.31	7.42	4.13	14.14
C21H25N1O4	355.44	355.1783	70.96	7.09	3.94	18.01
C21H25N1O5	371.44	371.1733	67.91	6.78	3.77	21.54
C21H25N1O6	387.44	387.1682	65.10	6.50	3.62	24.78
C21H25N1O7	403.44	403.1631	62.52	6.25	3.47	27.76
C21H25N1O8	419.44	419.1580	60.14	6.01	3.34	30.52
C21H25N1O9	435.43	435.1529	57.93	5.79	3.22	33.07
C21H25N3O1	335.45	335.1998	75.19	7.51	12.53	4.77
C21H25N3O2	351.45	351.1947	71.77	7.17	11.96	9.10
C21H25N3O3	367.45	367.1896	68.64	6.86	11.44	13.06
C21H25N3O4	383.45	383.1845	65.78	6.57	10.96	16.69
C21H25N3O5	399.45	399.1794	63.15	6.31	10.52	20.03
C21H25N3O6	415.45	415.1743	60.71	6.07	10.11	23.11
C21H25N3O7	431.45	431.1692	58.46	5.84	9.74	25.96
C21H25N3O8	447.45	447.1641	56.37	5.63	9.39	28.61
C21H25N3O9	463.45	463.1591	54.43	5.44	9.07	31.07
C21H25N5O1	363.47	363.2059	69.40	6.93	19.27	4.40
C21H25N5O2	379.47	379.2008	66.47	6.64	18.46	8.43
C21H25N5O3	395.47	395.1957	63.78	6.37	17.71	12.14
C21H25N5O4	411.46	411.1906	61.30	6.12	17.02	15.55
C21H25N5O5	427.46	427.1856	59.01	5.90	16.38	18.71
C21H25N5O6	443.46	443.1805	56.88	5.68	15.79	21.65
C21H25N5O7	459.46	459.1754	54.90	5.48	15.24	24.38
C21H25N5O8	475.46	475.1703	53.05	5.30	14.73	26.92
C21H25N5O9	491.46	491.1652	51.32	5.13	14.25	29.30
C21H26N2O1	322.45	322.2045	78.22	8.13	8.69	4.96
C21H26N2O2	338.45	338.1994	74.53	7.74	8.28	9.45
C21H26N2O3	354.45	354.1943	71.16	7.39	7.90	13.54
C21H26N2O4	370.45	370.1892	68.09	7.07	7.56	17.28
C21H26N2O5	386.45	386.1842	65.27	6.78	7.25	20.70
C21H26N2O6	402.45	402.1791	62.67	6.51	6.96	23.85
C21H26N2O7	418.45	418.1740	60.28	6.26	6.69	26.76
C21H26N2O8	434.45	434.1689	58.06	6.03	6.45	29.46
C21H26N2O9	450.45	450.1638	56.00	5.82	6.22	31.97
C21H26N4O1	350.47	350.2106	71.97	7.48	15.99	4.57
C21H26N4O2	366.47	366.2056	68.83	7.15	15.29	8.73
C21H26N4O3	382.47	382.2005	65.95	6.85	14.65	12.55
C21H26N4O4	398.47	398.1954	63.30	6.58	14.06	16.06
C21H26N4O5	414.47	414.1903	60.86	6.32	13.52	19.30
C21H26N4O6	430.46	430.1852	58.60	6.09	13.02	22.30
C21H26N4O7	446.46	446.1801	56.50	5.87	12.55	25.09
C21H26N4O8	462.46	462.1750	54.54	5.67	12.11	27.68
C21H26N4O9	478.46	478.1700	52.72	5.48	11.71	30.10
C21H26N6O1	378.48	378.2168	66.64	6.92	22.20	4.23
C21H26N6O2	394.48	394.2117	63.94	6.64	21.30	8.11
C21H26N6O3	410.48	410.2066	61.45	6.38	20.47	11.69
C21H26N6O4	426.48	426.2015	59.14	6.15	19.71	15.01
C21H26N6O5	442.48	442.1965	57.00	5.92	18.99	18.08
C21H26N6O6	458.48	458.1914	55.02	5.72	18.33	20.94
C21H26N6O7	474.48	474.1863	53.16	5.52	17.71	23.60
C21H26N6O8	490.48	490.1812	51.43	5.34	17.13	26.10
C21H26N6O9	506.48	506.1761	49.80	5.17	16.59	28.43
C21H27N1O1	309.46	309.2093	81.51	8.79	4.53	5.17
C21H27N1O2	325.45	325.2042	77.50	8.36	4.30	9.83
C21H27N1O3	341.45	341.1991	73.87	7.97	4.10	14.06
C21H27N1O4	357.45	357.1940	70.56	7.61	3.92	17.90
C21H27N1O5	373.45	373.1889	67.54	7.29	3.75	21.42
C21H27N1O6	389.45	389.1838	64.77	6.99	3.60	24.65
C21H27N1O7	405.45	405.1787	62.21	6.71	3.45	27.62
C21H27N1O8	421.45	421.1737	59.85	6.46	3.32	30.37
C21H27N1O9	437.45	437.1686	57.66	6.22	3.20	32.92
C21H27N3O1	337.47	337.2154	74.74	8.06	12.45	4.74
C21H27N3O2	353.47	353.2103	71.36	7.70	11.89	9.05
C21H27N3O3	369.47	369.2052	68.27	7.37	11.37	12.99
C21H27N3O4	385.47	385.2001	65.44	7.06	10.90	16.60
C21H27N3O5	401.47	401.1951	62.83	6.78	10.47	19.93
C21H27N3O6	417.47	417.1900	60.42	6.52	10.07	23.00
C21H27N3O7	433.47	433.1849	58.19	6.28	9.69	25.84
C21H27N3O8	449.46	449.1798	56.12	6.06	9.35	28.48
C21H27N3O9	465.46	465.1747	54.19	5.85	9.03	30.94
C21H27N5O1	365.48	365.2215	69.01	7.45	19.16	4.38
C21H27N5O2	381.48	381.2165	66.12	7.13	18.36	8.39
C21H27N5O3	397.48	397.2114	63.46	6.85	17.62	12.08
C21H27N5O4	413.48	413.2063	61.00	6.58	16.94	15.48
C21H27N5O5	429.48	429.2012	58.73	6.34	16.31	18.63
C21H27N5O6	445.48	445.1961	56.62	6.11	15.72	21.55
C21H27N5O7	461.48	461.1910	54.66	5.90	15.18	24.27
C21H27N5O8	477.48	477.1859	52.83	5.70	14.67	26.81
C21H27N5O9	493.48	493.1809	51.11	5.51	14.19	29.18
C21H28N2O1	324.47	324.2202	77.74	8.70	8.63	4.93
C21H28N2O2	340.47	340.2151	74.08	8.29	8.23	9.40
C21H28N2O3	356.47	356.2100	70.76	7.92	7.86	13.46
C21H28N2O4	372.47	372.2049	67.72	7.58	7.52	17.18
C21H28N2O5	388.47	388.1998	64.93	7.27	7.21	20.59
C21H28N2O6	404.47	404.1947	62.36	6.98	6.93	23.73
C21H28N2O7	420.47	420.1896	59.99	6.71	6.66	26.64
C21H28N2O8	436.47	436.1845	57.79	6.47	6.42	29.33
C21H28N2O9	452.47	452.1795	55.75	6.24	6.19	31.82
C21H28N4O1	352.48	352.2263	71.56	8.01	15.89	4.54
C21H28N4O2	368.48	368.2212	68.45	7.66	15.20	8.68
C21H28N4O3	384.48	384.2161	65.60	7.34	14.57	12.48
C21H28N4O4	400.48	400.2110	62.98	7.05	13.99	15.98
C21H28N4O5	416.48	416.2060	60.56	6.78	13.45	19.21
C21H28N4O6	432.48	432.2009	58.32	6.53	12.95	22.20
C21H28N4O7	448.48	448.1958	56.24	6.29	12.49	24.97
C21H28N4O8	464.48	464.1907	54.30	6.08	12.06	27.56
C21H28N4O9	480.48	480.1856	52.50	5.87	11.66	29.97
C21H28N6O1	380.50	380.2324	66.29	7.42	22.09	4.20
C21H28N6O2	396.50	396.2274	63.62	7.12	21.20	8.07
C21H28N6O3	412.50	412.2223	61.15	6.84	20.37	11.64
C21H28N6O4	428.50	428.2172	58.87	6.59	19.61	14.94
C21H28N6O5	444.49	444.2121	56.75	6.35	18.91	18.00
C21H28N6O6	460.49	460.2070	54.77	6.13	18.25	20.85
C21H28N6O7	476.49	476.2019	52.94	5.92	17.64	23.50
C21H28N6O8	492.49	492.1968	51.22	5.73	17.06	25.99
C21H28N6O9	508.49	508.1918	49.60	5.55	16.53	28.32
C21H29N1O1	311.47	311.2249	80.98	9.38	4.50	5.14
C21H29N1O2	327.47	327.2198	77.02	8.93	4.28	9.77
C21H29N1O3	343.47	343.2147	73.44	8.51	4.08	13.97
C21H29N1O4	359.47	359.2096	70.17	8.13	3.90	17.80
C21H29N1O5	375.47	375.2046	67.18	7.79	3.73	21.31
C21H29N1O6	391.47	391.1995	64.43	7.47	3.58	24.52
C21H29N1O7	407.47	407.1944	61.90	7.17	3.44	27.49

C21H29N108	423.47	423.1893	59.56	6.90	3.31	30.23
C21H29N109	439.47	439.1842	57.40	6.65	3.19	32.77
C21H29N301	339.48	339.2310	74.30	8.61	12.38	4.71
C21H29N302	355.48	355.2260	70.96	8.22	11.82	9.00
C21H29N303	371.48	371.2209	67.90	7.87	11.31	12.92
C21H29N304	387.48	387.2158	65.10	7.54	10.84	16.52
C21H29N305	403.48	403.2107	62.51	7.24	10.41	19.83
C21H29N306	419.48	419.2056	60.13	6.97	10.02	22.88
C21H29N307	435.48	435.2005	57.92	6.71	9.65	25.72
C21H29N308	451.48	451.1954	55.87	6.47	9.31	28.35
C21H29N309	467.48	467.1904	53.96	6.25	8.99	30.80
C21H29N501	367.50	367.2372	68.64	7.95	19.06	4.35
C21H29N502	383.50	383.2321	65.77	7.62	18.26	8.34
C21H29N503	399.50	399.2270	63.14	7.32	17.53	12.01
C21H29N504	415.50	415.2219	60.71	7.04	16.86	15.40
C21H29N505	431.50	431.2169	58.46	6.77	16.23	18.54
C21H29N506	447.50	447.2118	56.37	6.53	15.65	21.45
C21H29N507	463.49	463.2067	54.42	6.31	15.11	24.16
C21H29N508	479.49	479.2016	52.60	6.10	14.61	26.69
C21H29N509	495.49	495.1965	50.91	5.90	14.13	29.06
C21H30N201	326.49	326.2358	77.26	9.26	8.58	4.90
C21H30N202	342.49	342.2307	73.65	8.83	8.18	9.34
C21H30N203	358.48	358.2256	70.36	8.44	7.81	13.39
C21H30N204	374.48	374.2205	67.36	8.07	7.48	17.09
C21H30N205	390.48	390.2155	64.60	7.74	7.17	20.49
C21H30N206	406.48	406.2104	62.05	7.44	6.89	23.62
C21H30N207	422.48	422.2053	59.70	7.16	6.63	26.51
C21H30N208	438.48	438.2002	57.52	6.90	6.39	29.19
C21H30N209	454.48	454.1951	55.50	6.65	6.16	31.68
C21H30N401	354.50	354.2419	71.15	8.53	15.80	4.51
C21H30N402	370.50	370.2369	68.08	8.16	15.12	8.64
C21H30N403	386.50	386.2318	65.26	7.82	14.50	12.42
C21H30N404	402.50	402.2267	62.67	7.51	13.92	15.90
C21H30N405	418.50	418.2216	60.27	7.23	13.39	19.12
C21H30N406	434.50	434.2165	58.05	6.96	12.89	22.09
C21H30N407	450.50	450.2114	55.99	6.71	12.44	24.86
C21H30N408	466.50	466.2063	54.07	6.48	12.01	27.44
C21H30N409	482.49	482.2013	52.28	6.27	11.61	29.84
C21H30N601	382.51	382.2481	65.94	7.91	21.97	4.18
C21H30N602	398.51	398.2430	63.29	7.59	21.09	8.03
C21H30N603	414.51	414.2379	60.85	7.30	20.27	11.58
C21H30N604	430.51	430.2328	58.59	7.02	19.52	14.87
C21H30N605	446.51	446.2278	56.49	6.77	18.82	17.92
C21H30N606	462.51	462.2227	54.54	6.54	18.17	20.76
C21H30N607	478.51	478.2176	52.71	6.32	17.56	23.41
C21H30N608	494.51	494.2125	51.01	6.11	16.99	25.88
C21H30N609	510.51	510.2074	49.41	5.92	16.46	28.21
C21H31N101	313.49	313.2406	80.46	9.97	4.47	5.10
C21H31N102	329.49	329.2355	76.55	9.48	4.25	9.71
C21H31N103	345.49	345.2304	73.01	9.04	4.05	13.89
C21H31N104	361.49	361.2253	69.78	8.64	3.87	17.70
C21H31N105	377.48	377.2202	66.82	8.28	3.71	21.19
C21H31N106	393.48	393.2151	64.10	7.94	3.56	24.40
C21H31N107	409.48	409.2100	61.60	7.63	3.42	27.35
C21H31N108	425.48	425.2049	59.28	7.34	3.29	30.08
C21H31N109	441.48	441.1999	57.13	7.08	3.17	32.62
C21H31N301	341.50	341.2467	73.86	9.15	12.30	4.69
C21H31N302	357.50	357.2416	70.55	8.74	11.75	8.95
C21H31N303	373.50	373.2365	67.53	8.37	11.25	12.85
C21H31N304	389.50	389.2314	64.76	8.02	10.79	16.43

C21H31N305	405.50	405.2264	62.20	7.71	10.36	19.73
C21H31N306	421.50	421.2213	59.84	7.41	9.97	22.78
C21H31N307	437.50	437.2162	57.65	7.14	9.60	25.60
C21H31N308	453.50	453.2111	55.62	6.89	9.27	28.22
C21H31N309	469.50	469.2060	53.72	6.66	8.95	30.67
C21H31N501	369.51	369.2528	68.26	8.46	18.95	4.33
C21H31N502	385.51	385.2478	65.43	8.11	18.17	8.30
C21H31N503	401.51	401.2427	62.82	7.78	17.44	11.95
C21H31N504	417.51	417.2376	60.41	7.48	16.77	15.33
C21H31N505	433.51	433.2325	58.18	7.21	16.15	18.45
C21H31N506	449.51	449.2274	56.11	6.95	15.58	21.36
C21H31N507	465.51	465.2223	54.18	6.71	15.04	24.06
C21H31N508	481.51	481.2172	52.38	6.49	14.54	26.58
C21H31N509	497.51	497.2122	50.70	6.28	14.08	28.94
C21H32N201	328.50	328.2514	76.78	9.82	8.53	4.87
C21H32N202	344.50	344.2464	73.22	9.36	8.13	9.29
C21H32N203	360.50	360.2413	69.97	8.95	7.77	13.31
C21H32N204	376.50	376.2362	66.99	8.57	7.44	17.00
C21H32N205	392.50	392.2311	64.26	8.22	7.14	20.38
C21H32N206	408.50	408.2260	61.75	7.90	6.86	23.50
C21H32N207	424.50	424.2209	59.42	7.60	6.60	26.38
C21H32N208	440.50	440.2158	57.26	7.32	6.36	29.06
C21H32N209	456.50	456.2108	55.25	7.07	6.14	31.54
C21H32N401	356.52	356.2576	70.75	9.05	15.72	4.49
C21H32N402	372.51	372.2525	67.71	8.66	15.04	8.59
C21H32N403	388.51	388.2474	64.92	8.30	14.42	12.35
C21H32N404	404.51	404.2423	62.35	7.97	13.85	15.82
C21H32N405	420.51	420.2373	59.98	7.67	13.32	19.02
C21H32N406	436.51	436.2322	57.78	7.39	12.84	21.99
C21H32N407	452.51	452.2271	55.74	7.13	12.38	24.75
C21H32N408	468.51	468.2220	53.84	6.88	11.96	27.32
C21H32N409	484.51	484.2169	52.06	6.66	11.56	29.72
C21H32N601	384.53	384.2637	65.60	8.39	21.86	4.16
C21H32N602	400.53	400.2587	62.98	8.05	20.98	7.99
C21H32N603	416.53	416.2536	60.56	7.74	20.18	11.52
C21H32N604	432.53	432.2485	58.32	7.46	19.43	14.80
C21H32N605	448.53	448.2434	56.24	7.19	18.74	17.84
C21H32N606	464.53	464.2383	54.30	6.94	18.09	20.67
C21H32N607	480.53	480.2332	52.49	6.71	17.49	23.31
C21H32N608	496.52	496.2281	50.80	6.50	16.93	25.78
C21H32N609	512.52	512.2231	49.21	6.29	16.40	28.10
C21H33N101	315.50	315.2562	79.95	10.54	4.44	5.07
C21H33N102	331.50	331.2511	76.09	10.03	4.23	9.65
C21H33N103	347.50	347.2460	72.58	9.57	4.03	13.81
C21H33N104	363.50	363.2409	69.39	9.15	3.85	17.61
C21H33N105	379.50	379.2359	66.46	8.76	3.69	21.08
C21H33N106	395.50	395.2308	63.78	8.41	3.54	24.27
C21H33N107	411.50	411.2257	61.30	8.08	3.40	27.22
C21H33N108	427.50	427.2206	59.00	7.78	3.28	29.94
C21H33N109	443.50	443.2155	56.87	7.50	3.16	32.47
C21H33N301	343.52	343.2623	73.43	9.68	12.23	4.66
C21H33N302	359.52	359.2573	70.16	9.25	11.69	8.90
C21H33N303	375.52	375.2522	67.17	8.86	11.19	12.78
C21H33N304	391.51	391.2471	64.43	8.50	10.73	16.35
C21H33N305	407.51	407.2420	61.90	8.16	10.31	19.63
C21H33N306	423.51	423.2369	59.56	7.85	9.92	22.67
C21H33N307	439.51	439.2318	57.39	7.57	9.56	25.48
C21H33N308	455.51	455.2267	55.37	7.30	9.22	28.10
C21H33N309	471.51	471.2217	53.49	7.05	8.91	30.54
C21H33N501	371.53	371.2685	67.89	8.95	18.85	4.31

C21H33N502	387.53	387.2634	65.09	8.58	18.07	8.26
C21H33N503	403.53	403.2583	62.51	8.24	17.36	11.89
C21H33N504	419.53	419.2532	60.12	7.93	16.69	15.25
C21H33N505	435.53	435.2482	57.91	7.64	16.08	18.37
C21H33N506	451.53	451.2431	55.86	7.37	15.51	21.26
C21H33N507	467.53	467.2380	53.95	7.11	14.98	23.95
C21H33N508	483.53	483.2329	52.17	6.88	14.48	26.47
C21H33N509	499.53	499.2278	50.49	6.66	14.02	28.83
C21H34N201	330.52	330.2671	76.31	10.37	8.48	4.84
C21H34N202	346.52	346.2620	72.79	9.89	8.08	9.23
C21H34N203	362.52	362.2569	69.58	9.45	7.73	13.24
C21H34N204	378.52	378.2518	66.64	9.05	7.40	16.91
C21H34N205	394.52	394.2468	63.94	8.69	7.10	20.28
C21H34N206	410.51	410.2417	61.44	8.35	6.82	23.38
C21H34N207	426.51	426.2366	59.14	8.04	6.57	26.26
C21H34N208	442.51	442.2315	57.00	7.74	6.33	28.92
C21H34N209	458.51	458.2264	55.01	7.47	6.11	31.40
C21H34N401	358.53	358.2732	70.35	9.56	15.63	4.46
C21H34N402	374.53	374.2682	67.35	9.15	14.96	8.54
C21H34N403	390.53	390.2631	64.59	8.78	14.35	12.29
C21H34N404	406.53	406.2580	62.05	8.43	13.78	15.74
C21H34N405	422.53	422.2529	59.70	8.11	13.26	18.93
C21H34N406	438.53	438.2478	57.52	7.81	12.78	21.89
C21H34N407	454.53	454.2427	55.49	7.54	12.33	24.64
C21H34N408	470.53	470.2376	53.61	7.28	11.91	27.20
C21H34N409	486.53	486.2326	51.84	7.04	11.52	29.60
C21H34N601	386.54	386.2794	65.25	8.87	21.74	4.14
C21H34N602	402.54	402.2743	62.66	8.51	20.88	7.95
C21H34N603	418.54	418.2692	60.26	8.19	20.08	11.47
C21H34N604	434.54	434.2641	58.05	7.89	19.34	14.73
C21H34N605	450.54	450.2590	55.98	7.61	18.65	17.76
C21H34N606	466.54	466.2540	54.06	7.35	18.01	20.58
C21H34N607	482.54	482.2489	52.27	7.10	17.42	23.21
C21H34N608	498.54	498.2438	50.59	6.87	16.86	25.67
C21H34N609	514.54	514.2387	49.02	6.66	16.33	27.99
C21H35N101	317.52	317.2718	79.44	11.11	4.41	5.04
C21H35N102	333.52	333.2668	75.63	10.58	4.20	9.59
C21H35N103	349.52	349.2617	72.17	10.09	4.01	13.73
C21H35N104	365.52	365.2566	69.01	9.65	3.83	17.51
C21H35N105	381.52	381.2515	66.11	9.25	3.67	20.97
C21H35N106	397.52	397.2464	63.45	8.87	3.52	24.15
C21H35N107	413.52	413.2413	61.00	8.53	3.39	27.08
C21H35N108	429.52	429.2362	58.73	8.21	3.26	29.80
C21H35N109	445.51	445.2312	56.62	7.92	3.14	32.32
C21H35N301	345.53	345.2780	73.00	10.21	12.16	4.63
C21H35N302	361.53	361.2729	69.77	9.76	11.62	8.85
C21H35N303	377.53	377.2678	66.81	9.34	11.13	12.71
C21H35N304	393.53	393.2627	64.10	8.96	10.68	16.26
C21H35N305	409.53	409.2577	61.59	8.61	10.26	19.53
C21H35N306	425.53	425.2526	59.28	8.29	9.87	22.56
C21H35N307	441.53	441.2475	57.13	7.99	9.52	25.37
C21H35N308	457.53	457.2424	55.13	7.71	9.18	27.98
C21H35N309	473.53	473.2373	53.27	7.45	8.87	30.41
C21H35N501	373.55	373.2841	67.52	9.44	18.75	4.28
C21H35N502	389.55	389.2791	64.75	9.06	17.98	8.21
C21H35N503	405.54	405.2740	62.20	8.70	17.27	11.84
C21H35N504	421.54	421.2689	59.84	8.37	16.61	15.18
C21H35N505	437.54	437.2638	57.65	8.06	16.01	18.28
C21H35N506	453.54	453.2587	55.61	7.78	15.44	21.17
C21H35N507	469.54	469.2536	53.72	7.51	14.92	23.85

C21H35N508	485.54	485.2485	51.95	7.27	14.42	26.36
C21H35N509	501.54	501.2435	50.29	7.03	13.96	28.71
C21H36N201	332.53	332.2827	75.85	10.91	8.42	4.81
C21H36N202	348.53	348.2777	72.37	10.41	8.04	9.18
C21H36N203	364.53	364.2726	69.19	9.95	7.68	13.17
C21H36N204	380.53	380.2675	66.28	9.54	7.36	16.82
C21H36N205	396.53	396.2624	63.61	9.15	7.06	20.17
C21H36N206	412.53	412.2573	61.14	8.80	6.79	23.27
C21H36N207	428.53	428.2522	58.86	8.47	6.54	26.13
C21H36N208	444.53	444.2471	56.74	8.16	6.30	28.79
C21H36N209	460.53	460.2421	54.77	7.88	6.08	31.27
C21H36N401	360.55	360.2889	69.96	10.06	15.54	4.44
C21H36N402	376.55	376.2838	66.99	9.64	14.88	8.50
C21H36N403	392.55	392.2787	64.26	9.24	14.27	12.23
C21H36N404	408.55	408.2736	61.74	8.88	13.71	15.66
C21H36N405	424.54	424.2686	59.41	8.55	13.20	18.84
C21H36N406	440.54	440.2635	57.26	8.24	12.72	21.79
C21H36N407	456.54	456.2584	55.25	7.95	12.27	24.53
C21H36N408	472.54	472.2533	53.38	7.68	11.86	27.09
C21H36N409	488.54	488.2482	51.63	7.43	11.47	29.47
C21H36N601	388.56	388.2950	64.91	9.34	21.63	4.12
C21H36N602	404.56	404.2900	62.35	8.97	20.77	7.91
C21H36N603	420.56	420.2849	59.98	8.63	19.98	11.41
C21H36N604	436.56	436.2798	57.78	8.31	19.25	14.66
C21H36N605	452.56	452.2747	55.74	8.02	18.57	17.68
C21H36N606	468.56	468.2696	53.83	7.74	17.94	20.49
C21H36N607	484.56	484.2645	52.05	7.49	17.34	23.11
C21H36N608	500.56	500.2594	50.39	7.25	16.79	25.57
C21H36N609	516.56	516.2544	48.83	7.02	16.27	27.88
C21H37N101	319.54	319.2875	78.94	11.67	4.38	5.01
C21H37N102	335.53	335.2824	75.17	11.12	4.17	9.54
C21H37N103	351.53	351.2773	71.75	10.61	3.98	13.65
C21H37N104	367.53	367.2722	68.63	10.15	3.81	17.41
C21H37N105	383.53	383.2672	65.77	9.72	3.65	20.86
C21H37N106	399.53	399.2621	63.13	9.33	3.51	24.03
C21H37N107	415.53	415.2570	60.70	8.98	3.37	26.95
C21H37N108	431.53	431.2519	58.45	8.64	3.25	29.66
C21H37N109	447.53	447.2468	56.36	8.33	3.13	32.18
C21H37N301	347.55	347.2936	72.58	10.73	12.09	4.60
C21H37N302	363.55	363.2886	69.38	10.26	11.56	8.80
C21H37N303	379.55	379.2835	66.46	9.83	11.07	12.65
C21H37N304	395.55	395.2784	63.77	9.43	10.62	16.18
C21H37N305	411.55	411.2733	61.29	9.06	10.21	19.44
C21H37N306	427.55	427.2682	59.00	8.72	9.83	22.45
C21H37N307	443.54	443.2631	56.87	8.41	9.47	25.25
C21H37N308	459.54	459.2580	54.89	8.12	9.14	27.85
C21H37N309	475.54	475.2530	53.04	7.84	8.84	30.28
C21H37N501	375.56	375.2998	67.16	9.93	18.65	4.26
C21H37N502	391.56	391.2947	64.42	9.52	17.89	8.17
C21H37N503	407.56	407.2896	61.89	9.15	17.18	11.78
C21H37N504	423.56	423.2845	59.55	8.81	16.53	15.11
C21H37N505	439.56	439.2794	57.38	8.48	15.93	18.20
C21H37N506	455.56	455.2744	55.37	8.19	15.37	21.07
C21H37N507	471.56	471.2693	53.49	7.91	14.85	23.75
C21H37N508	487.56	487.2642	51.73	7.65	14.36	26.25
C21H37N509	503.56	503.2591	50.09	7.41	13.91	28.60
C21H38N201	334.55	334.2984	75.40	11.45	8.37	4.78
C21H38N202	350.55	350.2933	71.95	10.93	7.99	9.13
C21H38N203	366.55	366.2882	68.81	10.45	7.64	13.09
C21H38N204	382.55	382.2831	65.94	10.01	7.32	16.73

C21H38N205	398.55	398.2781	63.29	9.61	7.03	20.07
C21H38N206	414.55	414.2730	60.85	9.24	6.76	23.16
C21H38N207	430.55	430.2679	58.58	8.90	6.51	26.01
C21H38N208	446.55	446.2628	56.49	8.58	6.27	28.66
C21H38N209	462.55	462.2577	54.53	8.28	6.06	31.13
C21H38N401	362.56	362.3045	69.57	10.56	15.45	4.41
C21H38N402	378.56	378.2995	66.63	10.12	14.80	8.45
C21H38N403	394.56	394.2944	63.93	9.71	14.20	12.16
C21H38N404	410.56	410.2893	61.44	9.33	13.65	15.59
C21H38N405	426.56	426.2842	59.13	8.98	13.13	18.75
C21H38N406	442.56	442.2791	56.99	8.65	12.66	21.69
C21H38N407	458.56	458.2740	55.01	8.35	12.22	24.42
C21H38N408	474.56	474.2689	53.15	8.07	11.81	26.97
C21H38N409	490.56	490.2639	51.42	7.81	11.42	29.35
C21H38N601	390.58	390.3107	64.58	9.81	21.52	4.10
C21H38N602	406.58	406.3056	62.04	9.42	20.67	7.87
C21H38N603	422.58	422.3005	59.69	9.06	19.89	11.36
C21H38N604	438.57	438.2954	57.51	8.73	19.16	14.59
C21H38N605	454.57	454.2903	55.49	8.43	18.49	17.60
C21H38N606	470.57	470.2853	53.60	8.14	17.86	20.40
C21H38N607	486.57	486.2802	51.84	7.87	17.27	23.02
C21H38N608	502.57	502.2751	50.19	7.62	16.72	25.47
C21H38N609	518.57	518.2700	48.64	7.39	16.21	27.77
C21H39N101	321.55	321.3031	78.44	12.23	4.36	4.98
C21H39N102	337.55	337.2981	74.72	11.65	4.15	9.48
C21H39N103	353.55	353.2930	71.34	11.12	3.96	13.58
C21H39N104	369.55	369.2879	68.25	10.64	3.79	17.32
C21H39N105	385.55	385.2828	65.42	10.20	3.63	20.75
C21H39N106	401.55	401.2777	62.82	9.79	3.49	23.91
C21H39N107	417.55	417.2726	60.41	9.41	3.35	26.82
C21H39N108	433.55	433.2675	58.18	9.07	3.23	29.52
C21H39N109	449.55	449.2625	56.11	8.74	3.12	32.03
C21H39N301	349.56	349.3093	72.16	11.25	12.02	4.58
C21H39N302	365.56	365.3042	69.00	10.75	11.49	8.75
C21H39N303	381.56	381.2991	66.11	10.30	11.01	12.58
C21H39N304	397.56	397.2940	63.45	9.89	10.57	16.10
C21H39N305	413.56	413.2890	60.99	9.51	10.16	19.34
C21H39N306	429.56	429.2839	58.72	9.15	9.78	22.35
C21H39N307	445.56	445.2788	56.61	8.82	9.43	25.14
C21H39N308	461.56	461.2737	54.65	8.52	9.10	27.73
C21H39N309	477.56	477.2686	52.82	8.23	8.80	30.15
C21H39N501	377.58	377.3154	66.80	10.41	18.55	4.24
C21H39N502	393.58	393.3104	64.09	9.99	17.79	8.13
C21H39N503	409.58	409.3053	61.58	9.60	17.10	11.72
C21H39N504	425.58	425.3002	59.27	9.24	16.46	15.04
C21H39N505	441.58	441.2951	57.12	8.90	15.86	18.12
C21H39N506	457.57	457.2900	55.12	8.59	15.31	20.98
C21H39N507	473.57	473.2849	53.26	8.30	14.79	23.65
C21H39N508	489.57	489.2798	51.52	8.03	14.30	26.14
C21H39N509	505.57	505.2748	49.89	7.78	13.85	28.48
C21H40N201	336.57	336.3140	74.94	11.98	8.32	4.75
C21H40N202	352.57	352.3090	71.54	11.44	7.95	9.08
C21H40N203	368.56	368.3039	68.44	10.94	7.60	13.02
C21H40N204	384.56	384.2988	65.59	10.48	7.28	16.64
C21H40N205	400.56	400.2937	62.97	10.07	6.99	19.97
C21H40N206	416.56	416.2886	60.55	9.68	6.72	23.04
C21H40N207	432.56	432.2835	58.31	9.32	6.48	25.89
C21H40N208	448.56	448.2784	56.23	8.99	6.25	28.53
C21H40N209	464.56	464.2734	54.30	8.68	6.03	31.00
C21H40N401	364.58	364.3202	69.19	11.06	15.37	4.39

C21H40N402	380.58	380.3151	66.28	10.59	14.72	8.41
C21H40N403	396.58	396.3100	63.60	10.17	14.13	12.10
C21H40N404	412.58	412.3049	61.14	9.77	13.58	15.51
C21H40N405	428.58	428.2998	58.85	9.41	13.07	18.67
C21H40N406	444.58	444.2948	56.74	9.07	12.60	21.59
C21H40N407	460.58	460.2897	54.76	8.75	12.16	24.32
C21H40N408	476.57	476.2846	52.93	8.46	11.76	26.86
C21H40N409	492.57	492.2795	51.21	8.19	11.37	29.23
C21H40N601	392.59	392.3263	64.25	10.27	21.41	4.08
C21H40N602	408.59	408.3213	61.73	9.87	20.57	7.83
C21H40N603	424.59	424.3162	59.41	9.50	19.79	11.30
C21H40N604	440.59	440.3111	57.25	9.15	19.07	14.53
C21H40N605	456.59	456.3060	55.24	8.83	18.41	17.52
C21H40N606	472.59	472.3009	53.37	8.53	17.78	20.31
C21H40N607	488.59	488.2958	51.63	8.25	17.20	22.92
C21H40N608	504.59	504.2907	49.99	7.99	16.66	25.37
C21H40N609	520.59	520.2857	48.45	7.74	16.14	27.66
C21H41N101	323.57	323.3188	77.95	12.77	4.33	4.94
C21H41N102	339.57	339.3137	74.28	12.17	4.12	9.42
C21H41N103	355.57	355.3086	70.94	11.62	3.94	13.50
C21H41N104	371.57	371.3035	67.88	11.12	3.77	17.22
C21H41N105	387.56	387.2985	65.08	10.66	3.61	20.64
C21H41N106	403.56	403.2934	62.50	10.24	3.47	23.79
C21H41N107	419.56	419.2883	60.12	9.85	3.34	26.69
C21H41N108	435.56	435.2832	57.91	9.49	3.22	29.39
C21H41N109	451.56	451.2781	55.86	9.15	3.10	31.89
C21H41N301	351.58	351.3249	71.74	11.75	11.95	4.55
C21H41N302	367.58	367.3199	68.62	11.24	11.43	8.71
C21H41N303	383.58	383.3148	65.76	10.77	10.95	12.51
C21H41N304	399.58	399.3097	63.13	10.34	10.52	16.02
C21H41N305	415.58	415.3046	60.69	9.94	10.11	19.25
C21H41N306	431.58	431.2995	58.44	9.58	9.74	22.24
C21H41N307	447.58	447.2944	56.36	9.23	9.39	25.02
C21H41N308	463.58	463.2893	54.41	8.91	9.06	27.61
C21H41N309	479.58	479.2843	52.60	8.62	8.76	30.03
C21H41N501	379.59	379.3311	66.45	10.89	18.45	4.21
C21H41N502	395.59	395.3260	63.76	10.45	17.70	8.09
C21H41N503	411.59	411.3209	61.28	10.04	17.02	11.66
C21H41N504	427.59	427.3158	58.99	9.67	16.38	14.97
C21H41N505	443.59	443.3107	56.86	9.32	15.79	18.03
C21H41N506	459.59	459.3057	54.88	8.99	15.24	20.89
C21H41N507	475.59	475.3006	53.04	8.69	14.73	23.55
C21H41N508	491.59	491.2955	51.31	8.41	14.25	26.04
C21H41N509	507.59	507.2904	49.69	8.14	13.80	28.37
C21H42N201	338.58	338.3297	74.50	12.50	8.27	4.73
C21H42N202	354.58	354.3246	71.14	11.94	7.90	9.02
C21H42N203	370.58	370.3195	68.06	11.42	7.56	12.95
C21H42N204	386.58	386.3144	65.25	10.95	7.25	16.55
C21H42N205	402.58	402.3094	62.65	10.52	6.96	19.87
C21H42N206	418.58	418.3043	60.26	10.11	6.69	22.93
C21H42N207	434.58	434.2992	58.04	9.74	6.45	25.77
C21H42N208	450.58	450.2941	55.98	9.40	6.22	28.41
C21H42N209	466.58	466.2890	54.06	9.07	6.00	30.86
C21H42N401	366.60	366.3358	68.80	11.55	15.28	4.36
C21H42N402	382.59	382.3308	65.93	11.07	14.64	8.36
C21H42N403	398.59	398.3257	63.28	10.62	14.06	12.04
C21H42N404	414.59	414.3206	60.84	10.21	13.51	15.44
C21H42N405	430.59	430.3155	58.58	9.83	13.01	18.58
C21H42N406	446.59	446.3104	56.48	9.48	12.55	21.50
C21H42N407	462.59	462.3053	54.53	9.15	12.11	24.21

C21H42N408	478.59	478.3002	52.70	8.85	11.71	26.74
C21H42N409	494.59	494.2952	51.00	8.56	11.33	29.11
C21H42N601	394.61	394.3420	63.92	10.73	21.30	4.05
C21H42N602	410.61	410.3369	61.43	10.31	20.47	7.79
C21H42N603	426.61	426.3318	59.13	9.92	19.70	11.25
C21H42N604	442.61	442.3267	56.99	9.56	18.99	14.46
C21H42N605	458.61	458.3216	55.00	9.23	18.33	17.44
C21H42N606	474.61	474.3166	53.15	8.92	17.71	20.23
C21H42N607	490.60	490.3115	51.41	8.63	17.13	22.83
C21H42N608	506.60	506.3064	49.79	8.36	16.59	25.27
C21H42N609	522.60	522.3013	48.26	8.10	16.08	27.55
C21H43N101	325.58	325.3344	77.47	13.31	4.30	4.91
C21H43N102	341.58	341.3294	73.84	12.69	4.10	9.37
C21H43N103	357.58	357.3243	70.54	12.12	3.92	13.42
C21H43N104	373.58	373.3192	67.52	11.60	3.75	17.13
C21H43N105	389.58	389.3141	64.75	11.13	3.60	20.53
C21H43N106	405.58	405.3090	62.19	10.69	3.45	23.67
C21H43N107	421.58	421.3039	59.83	10.28	3.32	26.57
C21H43N108	437.58	437.2988	57.64	9.91	3.20	29.25
C21H43N109	453.58	453.2938	55.61	9.56	3.09	31.75
C21H43N301	353.60	353.3406	71.33	12.26	11.88	4.52
C21H43N302	369.60	369.3355	68.25	11.73	11.37	8.66
C21H43N303	385.60	385.3304	65.41	11.24	10.90	12.45
C21H43N304	401.59	401.3253	62.81	10.79	10.46	15.94
C21H43N305	417.59	417.3202	60.40	10.38	10.06	19.16
C21H43N306	433.59	433.3152	58.17	10.00	9.69	22.14
C21H43N307	449.59	449.3101	56.10	9.64	9.35	24.91
C21H43N308	465.59	465.3050	54.17	9.31	9.03	27.49
C21H43N309	481.59	481.2999	52.38	9.00	8.73	29.90
C21H43N501	381.61	381.3467	66.10	11.36	18.35	4.19
C21H43N502	397.61	397.3417	63.44	10.90	17.61	8.05
C21H43N503	413.61	413.3366	60.98	10.48	16.93	11.60
C21H43N504	429.61	429.3315	58.71	10.09	16.30	14.90
C21H43N505	445.61	445.3264	56.60	9.73	15.72	17.95
C21H43N506	461.61	461.3213	54.64	9.39	15.17	20.80
C21H43N507	477.61	477.3162	52.81	9.07	14.66	23.45
C21H43N508	493.61	493.3111	51.10	8.78	14.19	25.93
C21H43N509	509.60	509.3061	49.50	8.51	13.74	28.26
C21H44N201	340.60	340.3453	74.06	13.02	8.22	4.70
C21H44N202	356.60	356.3403	70.73	12.44	7.86	8.97
C21H44N203	372.60	372.3352	67.70	11.90	7.52	12.88
C21H44N204	388.60	388.3301	64.91	11.41	7.21	16.47
C21H44N205	404.60	404.3250	62.34	10.96	6.92	19.77
C21H44N206	420.59	420.3199	59.97	10.54	6.66	22.82
C21H44N207	436.59	436.3148	57.77	10.16	6.42	25.65
C21H44N208	452.59	452.3097	55.73	9.80	6.19	28.28
C21H44N209	468.59	468.3047	53.83	9.46	5.98	30.73
C21H44N401	368.61	368.3515	68.43	12.03	15.20	4.34
C21H44N402	384.61	384.3464	65.58	11.53	14.57	8.32
C21H44N403	400.61	400.3413	62.96	11.07	13.99	11.98
C21H44N404	416.61	416.3362	60.54	10.65	13.45	15.36
C21H44N405	432.61	432.3311	58.31	10.25	12.95	18.49
C21H44N406	448.61	448.3261	56.23	9.89	12.49	21.40
C21H44N407	464.61	464.3210	54.29	9.55	12.06	24.11
C21H44N408	480.61	480.3159	52.48	9.23	11.66	26.63
C21H44N409	496.61	496.3108	50.79	8.93	11.28	29.00
C21H44N601	396.62	396.3576	63.60	11.18	21.19	4.03
C21H44N602	412.62	412.3526	61.13	10.75	20.37	7.75
C21H44N603	428.62	428.3475	58.85	10.35	19.61	11.20
C21H44N604	444.62	444.3424	56.73	9.97	18.90	14.39

C21H44N605	460.62	460.3373	54.76	9.63	18.24	17.37
C21H44N606	476.62	476.3322	52.92	9.31	17.63	20.14
C21H44N607	492.62	492.3271	51.20	9.00	17.06	22.73
C21H44N608	508.62	508.3220	49.59	8.72	16.52	25.17
C21H44N609	524.62	524.3170	48.08	8.45	16.02	27.45
C21H45N101	327.60	327.3501	76.99	13.85	4.28	4.88
C21H45N102	343.60	343.3450	73.41	13.20	4.08	9.31
C21H45N103	359.60	359.3399	70.14	12.61	3.90	13.35
C21H45N104	375.60	375.3348	67.16	12.08	3.73	17.04
C21H45N105	391.60	391.3298	64.41	11.58	3.58	20.43
C21H45N106	407.60	407.3247	61.88	11.13	3.44	23.55
C21H45N107	423.60	423.3196	59.55	10.71	3.31	26.44
C21H45N108	439.59	439.3145	57.38	10.32	3.19	29.12
C21H45N109	455.59	455.3094	55.36	9.96	3.07	31.61
C21H45N301	355.61	355.3562	70.93	12.76	11.82	4.50
C21H45N302	371.61	371.3512	67.88	12.21	11.31	8.61
C21H45N303	387.61	387.3461	65.07	11.70	10.84	12.38
C21H45N304	403.61	403.3410	62.49	11.24	10.41	15.86
C21H45N305	419.61	419.3359	60.11	10.81	10.01	19.06
C21H45N306	435.61	435.3308	57.90	10.41	9.65	22.04
C21H45N307	451.61	451.3257	55.85	10.04	9.30	24.80
C21H45N308	467.61	467.3206	53.94	9.70	8.99	27.37
C21H45N309	483.61	483.3156	52.16	9.38	8.69	29.78
C21H45N501	383.63	383.3624	65.75	11.82	18.26	4.17
C21H45N502	399.63	399.3573	63.12	11.35	17.52	8.01
C21H45N503	415.62	415.3522	60.69	10.91	16.85	11.55
C21H45N504	431.62	431.3471	58.44	10.51	16.23	14.83
C21H45N505	447.62	447.3420	56.35	10.13	15.65	17.87
C21H45N506	463.62	463.3370	54.41	9.78	15.11	20.71
C21H45N507	479.62	479.3319	52.59	9.46	14.60	23.35
C21H45N508	495.62	495.3268	50.89	9.15	14.13	25.83
C21H45N509	511.62	511.3217	49.30	8.87	13.69	28.14
C21H46N201	342.61	342.3610	73.62	13.53	8.18	4.67
C21H46N202	358.61	358.3559	70.34	12.93	7.81	8.92
C21H46N203	374.61	374.3508	67.33	12.38	7.48	12.81
C21H46N204	390.61	390.3457	64.57	11.87	7.17	16.38
C21H46N205	406.61	406.3406	62.03	11.40	6.89	19.67
C21H46N206	422.61	422.3356	59.68	10.97	6.63	22.72
C21H46N207	438.61	438.3305	57.51	10.57	6.39	25.53
C21H46N208	454.61	454.3254	55.48	10.20	6.16	28.15
C21H46N209	470.61	470.3203	53.60	9.85	5.95	30.60
C21H46N401	370.63	370.3671	68.06	12.51	15.12	4.32
C21H46N402	386.63	386.3621	65.24	11.99	14.49	8.28
C21H46N403	402.63	402.3570	62.65	11.52	13.92	11.92
C21H46N404	418.63	418.3519	60.25	11.08	13.38	15.29
C21H46N405	434.62	434.3468	58.03	10.67	12.89	18.41
C21H46N406	450.62	450.3417	55.97	10.29	12.43	21.30
C21H46N407	466.62	466.3366	54.06	9.94	12.01	24.00
C21H46N408	482.62	482.3315	52.26	9.61	11.61	26.52
C21H46N409	498.62	498.3265	50.59	9.30	11.24	28.88
C21H46N601	398.64	398.3733	63.27	11.63	21.08	4.01
C21H46N602	414.64	414.3682	60.83	11.18	20.27	7.72
C21H46N603	430.64	430.3631	58.57	10.77	19.52	11.15
C21H46N604	446.64	446.3580	56.47	10.38	18.82	14.33
C21H46N605	462.64	462.3529	54.52	10.02	18.17	17.29
C21H46N606	478.64	478.3479	52.70	9.69	17.56	20.06
C21H46N607	494.64	494.3428	50.99	9.37	16.99	22.64
C21H46N608	510.64	510.3377	49.40	9.08	16.46	25.07
C21H46N609	526.64	526.3326	47.90	8.80	15.96	27.34
C21H47N301	357.63	357.3719	70.53	13.25	11.75	4.47

C21H47N302	373.63	373.3668	67.51	12.68	11.25	8.56
C21H47N303	389.63	389.3617	64.74	12.16	10.78	12.32
C21H47N304	405.63	405.3566	62.18	11.68	10.36	15.78
C21H47N305	421.63	421.3515	59.82	11.24	9.97	18.97
C21H47N306	437.63	437.3465	57.64	10.83	9.60	21.94
C21H47N307	453.62	453.3414	55.60	10.44	9.26	24.69
C21H47N308	469.62	469.3363	53.71	10.09	8.95	27.25
C21H47N309	485.62	485.3312	51.94	9.76	8.65	29.65
C21H47N501	385.64	385.3780	65.41	12.28	18.16	4.15
C21H47N502	401.64	401.3730	62.80	11.80	17.44	7.97
C21H47N503	417.64	417.3679	60.40	11.34	16.77	11.49
C21H47N504	433.64	433.3628	58.17	10.92	16.15	14.76
C21H47N505	449.64	449.3577	56.10	10.54	15.58	17.79
C21H47N506	465.64	465.3526	54.17	10.17	15.04	20.62
C21H47N507	481.64	481.3475	52.37	9.84	14.54	23.25
C21H47N508	497.64	497.3424	50.69	9.52	14.07	25.72
C21H47N509	513.64	513.3374	49.11	9.22	13.63	28.03
C21H48N401	372.64	372.3828	67.69	12.98	15.03	4.29
C21H48N402	388.64	388.3777	64.90	12.45	14.42	8.23
C21H48N403	404.64	404.3726	62.34	11.96	13.85	11.86
C21H48N404	420.64	420.3675	59.96	11.50	13.32	15.21
C21H48N405	436.64	436.3624	57.77	11.08	12.83	18.32
C21H48N406	452.64	452.3574	55.73	10.69	12.38	21.21
C21H48N407	468.64	468.3523	53.82	10.32	11.96	23.90
C21H48N408	484.64	484.3472	52.05	9.98	11.56	26.41
C21H48N409	500.64	500.3421	50.38	9.66	11.19	28.76
C21H48N601	400.66	400.3889	62.96	12.08	20.98	3.99
C21H48N602	416.66	416.3839	60.54	11.61	20.17	7.68
C21H48N603	432.66	432.3788	58.30	11.18	19.42	11.09
C21H48N604	448.65	448.3737	56.22	10.78	18.73	14.26
C21H48N605	464.65	464.3686	54.28	10.41	18.09	17.22
C21H48N606	480.65	480.3635	52.48	10.07	17.48	19.97
C21H48N607	496.65	496.3584	50.79	9.74	16.92	22.55
C21H48N608	512.65	512.3533	49.20	9.44	16.39	24.97
C21H48N609	528.65	528.3483	47.71	9.15	15.90	27.24
C21H49N501	387.66	387.3937	65.07	12.74	18.07	4.13
C21H49N502	403.66	403.3886	62.49	12.24	17.35	7.93
C21H49N503	419.66	419.3835	60.10	11.77	16.69	11.44
C21H49N504	435.66	435.3784	57.90	11.34	16.08	14.69
C21H49N505	451.66	451.3733	55.85	10.94	15.51	17.71
C21H49N506	467.65	467.3683	53.94	10.56	14.98	20.53
C21H49N507	483.65	483.3632	52.15	10.21	14.48	23.16
C21H49N508	499.65	499.3581	50.48	9.88	14.02	25.62
C21H49N509	515.65	515.3530	48.92	9.58	13.58	27.92
C21H50N601	402.67	402.4046	62.64	12.52	20.87	3.97
C21H50N602	418.67	418.3995	60.25	12.04	20.07	7.64
C21H50N603	434.67	434.3944	58.03	11.59	19.33	11.04
C21H50N604	450.67	450.3893	55.97	11.18	18.65	14.20
C21H50N605	466.67	466.3842	54.05	10.80	18.01	17.14
C21H50N606	482.67	482.3792	52.26	10.44	17.41	19.89
C21H50N607	498.67	498.3741	50.58	10.11	16.85	22.46
C21H50N608	514.67	514.3690	49.01	9.79	16.33	24.87
C21H50N609	530.67	530.3639	47.53	9.50	15.84	27.13
C22H11N101	305.34	305.0841	86.54	3.63	4.59	5.24
C22H11N102	321.34	321.0790	82.23	3.45	4.36	9.96
C22H11N103	337.34	337.0739	78.33	3.29	4.15	14.23
C22H11N104	353.34	353.0688	74.79	3.14	3.96	18.11
C22H11N105	369.34	369.0637	71.55	3.00	3.79	21.66
C22H11N106	385.34	385.0586	68.58	2.88	3.63	24.91
C22H11N107	401.34	401.0535	65.84	2.76	3.49	27.91
C22H11N108	417.33	417.0485	63.32	2.66	3.36	30.67
C22H11N109	433.33	433.0434	60.98	2.56	3.23	33.23
C22H11N301	333.35	333.0902	79.27	3.33	12.61	4.80
C22H11N302	349.35	349.0851	75.64	3.17	12.03	9.16
C22H11N303	365.35	365.0800	72.33	3.03	11.50	13.14
C22H11N304	381.35	381.0749	69.29	2.91	11.02	16.78
C22H11N305	397.35	397.0699	66.50	2.79	10.58	20.13
C22H11N306	413.35	413.0648	63.93	2.68	10.17	23.22
C22H11N307	429.35	429.0597	61.55	2.58	9.79	26.09
C22H11N308	445.35	445.0546	59.33	2.49	9.44	28.74
C22H11N309	461.35	461.0495	57.28	2.40	9.11	31.21
C22H11N501	361.37	361.0964	73.12	3.07	19.38	4.43
C22H11N502	377.37	377.0913	70.02	2.94	18.56	8.48
C22H11N503	393.36	393.0862	67.18	2.82	17.80	12.20
C22H11N504	409.36	409.0811	64.55	2.71	17.11	15.63
C22H11N505	425.36	425.0760	62.12	2.61	16.46	18.81
C22H11N506	441.36	441.0709	59.87	2.51	15.87	21.75
C22H11N507	457.36	457.0658	57.78	2.42	15.31	24.49
C22H11N508	473.36	473.0608	55.82	2.34	14.79	27.04
C22H11N509	489.36	489.0557	54.00	2.27	14.31	29.43
C22H12N201	320.35	320.0950	82.49	3.78	8.74	4.99
C22H12N202	336.35	336.0899	78.56	3.60	8.33	9.51
C22H12N203	352.35	352.0848	74.99	3.43	7.95	13.62
C22H12N204	368.35	368.0797	71.74	3.28	7.61	17.37
C22H12N205	384.35	384.0746	68.75	3.15	7.29	20.81
C22H12N206	400.35	400.0695	66.00	3.02	7.00	23.98
C22H12N207	416.35	416.0644	63.47	2.91	6.73	26.90
C22H12N208	432.35	432.0594	61.12	2.80	6.48	29.60
C22H12N209	448.35	448.0543	58.94	2.70	6.25	32.12
C22H12N401	348.37	348.1011	75.85	3.47	16.08	4.59
C22H12N402	364.37	364.0960	72.52	3.32	15.38	8.78
C22H12N403	380.37	380.0909	69.47	3.18	14.73	12.62
C22H12N404	396.37	396.0858	66.67	3.05	14.14	16.15
C22H12N405	412.36	412.0808	64.08	2.93	13.59	19.40
C22H12N406	428.36	428.0757	61.69	2.82	13.08	22.41
C22H12N407	444.36	444.0706	59.47	2.72	12.61	25.20
C22H12N408	460.36	460.0655	57.40	2.63	12.17	27.80
C22H12N409	476.36	476.0604	55.47	2.54	11.76	30.23
C22H12N601	376.38	376.1073	70.21	3.21	22.33	4.25
C22H12N602	392.38	392.1022	67.34	3.08	21.42	8.16
C22H12N603	408.38	408.0971	64.71	2.96	20.58	11.75
C22H12N604	424.38	424.0920	62.27	2.85	19.80	15.08
C22H12N605	440.38	440.0869	60.00	2.75	19.08	18.17
C22H12N606	456.38	456.0818	57.90	2.65	18.41	21.03
C22H12N607	472.38	472.0767	55.94	2.56	17.79	23.71
C22H12N608	488.38	488.0717	54.11	2.48	17.21	26.21
C22H12N609	504.38	504.0666	52.39	2.40	16.66	28.55
C22H13N101	307.36	307.0997	85.97	4.26	4.56	5.21
C22H13N102	323.35	323.0946	81.72	4.05	4.33	9.90
C22H13N103	339.35	339.0895	77.87	3.86	4.13	14.14
C22H13N104	355.35	355.0844	74.36	3.69	3.94	18.01
C22H13N105	371.35	371.0794	71.16	3.53	3.77	21.54
C22H13N106	387.35	387.0743	68.22	3.38	3.62	24.78
C22H13N107	403.35	403.0692	65.51	3.25	3.47	27.77
C22H13N108	419.35	419.0641	63.01	3.12	3.34	30.52
C22H13N109	435.35	435.0590	60.70	3.01	3.22	33.08
C22H13N301	335.37	335.1059	78.79	3.91	12.53	4.77
C22H13N302	351.37	351.1008	75.20	3.73	11.96	9.11
C22H13N303	367.37	367.0957	71.93	3.57	11.44	13.07
C22H13N304	383.37	383.0906	68.93	3.42	10.96	16.69

C22H13N305	399.37	399.0855	66.17	3.28	10.52	20.03
C22H13N306	415.37	415.0804	63.62	3.15	10.12	23.11
C22H13N307	431.36	431.0753	61.26	3.04	9.74	25.96
C22H13N308	447.36	447.0703	59.07	2.93	9.39	28.61
C22H13N309	463.36	463.0652	57.03	2.83	9.07	31.08
C22H13N501	363.38	363.1120	72.72	3.61	19.27	4.40
C22H13N502	379.38	379.1069	69.65	3.45	18.46	8.43
C22H13N503	395.38	395.1018	66.83	3.31	17.71	12.14
C22H13N504	411.38	411.0967	64.23	3.19	17.02	15.56
C22H13N505	427.38	427.0917	61.83	3.07	16.39	18.72
C22H13N506	443.38	443.0866	59.60	2.96	15.80	21.65
C22H13N507	459.38	459.0815	57.52	2.85	15.25	24.38
C22H13N508	475.38	475.0764	55.59	2.76	14.73	26.92
C22H13N509	491.38	491.0713	53.78	2.67	14.25	29.30
C22H14N201	322.37	322.1106	81.97	4.38	8.69	4.96
C22H14N202	338.37	338.1055	78.09	4.17	8.28	9.46
C22H14N203	354.37	354.1004	74.57	3.98	7.91	13.54
C22H14N204	370.37	370.0953	71.35	3.81	7.56	17.28
C22H14N205	386.37	386.0903	68.39	3.65	7.25	20.70
C22H14N206	402.37	402.0852	65.67	3.51	6.96	23.86
C22H14N207	418.37	418.0801	63.16	3.37	6.70	26.77
C22H14N208	434.37	434.0750	60.83	3.25	6.45	29.47
C22H14N209	450.36	450.0699	58.67	3.13	6.22	31.97
C22H14N401	350.38	350.1168	75.42	4.03	15.99	4.57
C22H14N402	366.38	366.1117	72.12	3.85	15.29	8.73
C22H14N403	382.38	382.1066	69.11	3.69	14.65	12.55
C22H14N404	398.38	398.1015	66.33	3.54	14.06	16.06
C22H14N405	414.38	414.0964	63.77	3.41	13.52	19.31
C22H14N406	430.38	430.0913	61.40	3.28	13.02	22.31
C22H14N407	446.38	446.0862	59.20	3.16	12.55	25.09
C22H14N408	462.38	462.0812	57.15	3.05	12.12	27.68
C22H14N409	478.38	478.0761	55.24	2.95	11.71	30.10
C22H14N601	378.40	378.1229	69.83	3.73	22.21	4.23
C22H14N602	394.40	394.1178	67.00	3.58	21.31	8.11
C22H14N603	410.40	410.1127	64.39	3.44	20.48	11.70
C22H14N604	426.39	426.1076	61.97	3.31	19.71	15.01
C22H14N605	442.39	442.1026	59.73	3.19	19.00	18.08
C22H14N606	458.39	458.0975	57.65	3.08	18.33	20.94
C22H14N607	474.39	474.0924	55.70	2.97	17.72	23.61
C22H14N608	490.39	490.0873	53.88	2.88	17.14	26.10
C22H14N609	506.39	506.0822	52.18	2.79	16.60	28.44
C22H15N101	309.37	309.1154	85.41	4.89	4.53	5.17
C22H15N102	325.37	325.1103	81.21	4.65	4.30	9.83
C22H15N103	341.37	341.1052	77.41	4.43	4.10	14.06
C22H15N104	357.37	357.1001	73.94	4.23	3.92	17.91
C22H15N105	373.37	373.0950	70.77	4.05	3.75	21.43
C22H15N106	389.37	389.0899	67.87	3.88	3.60	24.65
C22H15N107	405.37	405.0848	65.19	3.73	3.46	27.63
C22H15N108	421.37	421.0798	62.71	3.59	3.32	30.38
C22H15N109	437.37	437.0747	60.42	3.46	3.20	32.92
C22H15N301	337.38	337.1215	78.32	4.48	12.45	4.74
C22H15N302	353.38	353.1164	74.78	4.28	11.89	9.05
C22H15N303	369.38	369.1113	71.54	4.09	11.38	12.99
C22H15N304	385.38	385.1062	68.57	3.92	10.90	16.61
C22H15N305	401.38	401.1012	65.83	3.77	10.47	19.93
C22H15N306	417.38	417.0961	63.31	3.62	10.07	23.00
C22H15N307	433.38	433.0910	60.97	3.49	9.70	25.84
C22H15N308	449.38	449.0859	58.80	3.36	9.35	28.48
C22H15N309	465.38	465.0808	56.78	3.25	9.03	30.94
C22H15N501	365.40	365.1277	72.32	4.14	19.17	4.38
C22H15N502	381.40	381.1226	69.28	3.96	18.36	8.39
C22H15N503	397.40	397.1175	66.49	3.80	17.62	12.08
C22H15N504	413.40	413.1124	63.92	3.66	16.94	15.48
C22H15N505	429.40	429.1073	61.54	3.52	16.31	18.63
C22H15N506	445.39	445.1022	59.33	3.39	15.72	21.55
C22H15N507	461.39	461.0971	57.27	3.28	15.18	24.27
C22H15N508	477.39	477.0921	55.35	3.17	14.67	26.81
C22H15N509	493.39	493.0870	53.56	3.06	14.19	29.18
C22H16N201	324.39	324.1263	81.46	4.97	8.64	4.93
C22H16N202	340.39	340.1212	77.63	4.74	8.23	9.40
C22H16N203	356.38	356.1161	74.15	4.53	7.86	13.47
C22H16N204	372.38	372.1110	70.96	4.33	7.52	17.19
C22H16N205	388.38	388.1059	68.04	4.15	7.21	20.60
C22H16N206	404.38	404.1008	65.35	3.99	6.93	23.74
C22H16N207	420.38	420.0957	62.86	3.84	6.66	26.64
C22H16N208	436.38	436.0907	60.55	3.70	6.42	29.33
C22H16N209	452.38	452.0856	58.41	3.57	6.19	31.83
C22H16N401	352.40	352.1324	74.98	4.58	15.90	4.54
C22H16N402	368.40	368.1273	71.73	4.38	15.21	8.69
C22H16N403	384.40	384.1222	68.74	4.20	14.58	12.49
C22H16N404	400.40	400.1171	66.00	4.03	13.99	15.98
C22H16N405	416.40	416.1121	63.46	3.87	13.46	19.21
C22H16N406	432.40	432.1070	61.11	3.73	12.96	22.20
C22H16N407	448.40	448.1019	58.93	3.60	12.49	24.98
C22H16N408	464.39	464.0968	56.90	3.47	12.06	27.56
C22H16N409	480.39	480.0917	55.01	3.36	11.66	29.97
C22H16N601	380.41	380.1386	69.46	4.24	22.09	4.21
C22H16N602	396.41	396.1335	66.66	4.07	21.20	8.07
C22H16N603	412.41	412.1284	64.07	3.91	20.38	11.64
C22H16N604	428.41	428.1233	61.68	3.76	19.62	14.94
C22H16N605	444.41	444.1182	59.46	3.63	18.91	18.00
C22H16N606	460.41	460.1131	57.39	3.50	18.25	20.85
C22H16N607	476.41	476.1080	55.47	3.39	17.64	23.51
C22H16N608	492.41	492.1029	53.66	3.28	17.07	25.99
C22H16N609	508.41	508.0979	51.98	3.17	16.53	28.32
C22H17N101	311.39	311.1310	84.86	5.50	4.50	5.14
C22H17N102	327.39	327.1259	80.71	5.23	4.28	9.77
C22H17N103	343.39	343.1208	76.95	4.99	4.08	13.98
C22H17N104	359.39	359.1157	73.53	4.77	3.90	17.81
C22H17N105	375.38	375.1107	70.39	4.56	3.73	21.31
C22H17N106	391.38	391.1056	67.52	4.38	3.58	24.53
C22H17N107	407.38	407.1005	64.86	4.21	3.44	27.49
C22H17N108	423.38	423.0954	62.41	4.05	3.31	30.23
C22H17N109	439.38	439.0903	60.14	3.90	3.19	32.77
C22H17N301	339.40	339.1372	77.86	5.05	12.38	4.71
C22H17N302	355.40	355.1321	74.35	4.82	11.82	9.00
C22H17N303	371.40	371.1270	71.15	4.61	11.31	12.92
C22H17N304	387.40	387.1219	68.21	4.42	10.85	16.52
C22H17N305	403.40	403.1168	65.50	4.25	10.42	19.83
C22H17N306	419.40	419.1117	63.01	4.09	10.02	22.89
C22H17N307	435.40	435.1066	60.69	3.94	9.65	25.72
C22H17N308	451.40	451.1016	58.54	3.80	9.31	28.36
C22H17N309	467.40	467.0965	56.54	3.67	8.99	30.81
C22H17N501	367.41	367.1433	71.92	4.66	19.06	4.35
C22H17N502	383.41	383.1382	68.92	4.47	18.27	8.35
C22H17N503	399.41	399.1331	66.16	4.29	17.53	12.02
C22H17N504	415.41	415.1280	63.61	4.12	16.86	15.41
C22H17N505	431.41	431.1230	61.25	3.97	16.23	18.54
C22H17N506	447.41	447.1179	59.06	3.83	15.65	21.46
C22H17N507	463.41	463.1128	57.02	3.70	15.11	24.17

130

C22H17N508	479.41	479.1077	55.12	3.57	14.61	26.70
C22H17N509	495.41	495.1026	53.34	3.46	14.14	29.07
C22H18N201	326.40	326.1419	80.96	5.56	8.58	4.90
C22H18N202	342.40	342.1368	77.17	5.30	8.18	9.35
C22H18N203	358.40	358.1317	73.73	5.06	7.82	13.39
C22H18N204	374.40	374.1266	70.58	4.85	7.48	17.09
C22H18N205	390.40	390.1216	67.69	4.65	7.18	20.49
C22H18N206	406.40	406.1165	65.02	4.46	6.89	23.62
C22H18N207	422.40	422.1114	62.56	4.30	6.63	26.51
C22H18N208	438.40	438.1063	60.28	4.14	6.39	29.20
C22H18N209	454.40	454.1012	58.15	3.99	6.16	31.69
C22H18N401	354.41	354.1481	74.56	5.12	15.81	4.51
C22H18N402	370.41	370.1430	71.34	4.90	15.13	8.64
C22H18N403	386.41	386.1379	68.38	4.70	14.50	12.42
C22H18N404	402.41	402.1328	65.67	4.51	13.92	15.90
C22H18N405	418.41	418.1277	63.15	4.34	13.39	19.12
C22H18N406	434.41	434.1226	60.83	4.18	12.90	22.10
C22H18N407	450.41	450.1175	58.67	4.03	12.44	24.87
C22H18N408	466.41	466.1125	56.66	3.89	12.01	27.44
C22H18N409	482.41	482.1074	54.78	3.76	11.61	29.85
C22H18N601	382.43	382.1542	69.10	4.74	21.98	4.18
C22H18N602	398.43	398.1491	66.32	4.55	21.09	8.03
C22H18N603	414.43	414.1440	63.76	4.38	20.28	11.58
C22H18N604	430.43	430.1389	61.39	4.22	19.52	14.87
C22H18N605	446.43	446.1339	59.15	4.06	18.83	17.92
C22H18N606	462.43	462.1288	57.14	3.92	18.17	20.76
C22H18N607	478.42	478.1237	55.23	3.79	17.57	23.41
C22H18N608	494.42	494.1186	53.45	3.67	17.00	25.89
C22H18N609	510.42	510.1135	51.77	3.55	16.46	28.21
C22H19N101	313.40	313.1467	84.31	6.11	4.47	5.11
C22H19N102	329.40	329.1416	80.22	5.81	4.25	9.71
C22H19N103	345.40	345.1365	76.50	5.54	4.06	13.90
C22H19N104	361.40	361.1314	73.12	5.30	3.88	17.71
C22H19N105	377.40	377.1263	70.02	5.07	3.71	21.20
C22H19N106	393.40	393.1212	67.17	4.87	3.56	24.40
C22H19N107	409.40	409.1161	64.54	4.68	3.42	27.36
C22H19N108	425.40	425.1111	62.12	4.50	3.29	30.09
C22H19N109	441.40	441.1060	59.87	4.34	3.17	32.62
C22H19N301	341.42	341.1528	77.40	5.61	12.31	4.69
C22H19N302	357.42	357.1477	73.93	5.36	11.76	8.95
C22H19N303	373.42	373.1426	70.76	5.13	11.25	12.85
C22H19N304	389.41	389.1375	67.86	4.92	10.79	16.43
C22H19N305	405.41	405.1325	65.18	4.72	10.36	19.73
C22H19N306	421.41	421.1274	62.70	4.54	9.97	22.78
C22H19N307	437.41	437.1223	60.41	4.38	9.61	25.60
C22H19N308	453.41	453.1172	58.28	4.22	9.27	28.23
C22H19N309	469.41	469.1121	56.29	4.08	8.95	30.68
C22H19N501	369.43	369.1590	71.53	5.18	18.96	4.33
C22H19N502	385.43	385.1539	68.56	4.97	18.17	8.30
C22H19N503	401.43	401.1488	65.83	4.77	17.45	11.96
C22H19N504	417.43	417.1437	63.30	4.59	16.78	15.33
C22H19N505	433.43	433.1386	60.97	4.42	16.16	18.46
C22H19N506	449.43	449.1335	58.80	4.26	15.58	21.36
C22H19N507	465.43	465.1284	56.77	4.11	15.05	24.06
C22H19N508	481.43	481.1233	54.89	3.98	14.55	26.59
C22H19N509	497.42	497.1183	53.12	3.85	14.08	28.95
C22H20N201	328.42	328.1576	80.46	6.14	8.53	4.87
C22H20N202	344.42	344.1525	76.72	5.85	8.13	9.29
C22H20N203	360.42	360.1474	73.32	5.59	7.77	13.32
C22H20N204	376.42	376.1423	70.20	5.36	7.44	17.00
C22H20N205	392.42	392.1372	67.34	5.14	7.14	20.39
C22H20N206	408.41	408.1321	64.70	4.94	6.86	23.50
C22H20N207	424.41	424.1270	62.26	4.75	6.60	26.39
C22H20N208	440.41	440.1220	60.00	4.58	6.36	29.06
C22H20N209	456.41	456.1169	57.90	4.42	6.14	31.55
C22H20N401	356.43	356.1637	74.14	5.66	15.72	4.49
C22H20N402	372.43	372.1586	70.95	5.41	15.04	8.59
C22H20N403	388.43	388.1535	68.03	5.19	14.42	12.36
C22H20N404	404.43	404.1484	65.34	4.98	13.85	15.82
C22H20N405	420.43	420.1434	62.85	4.79	13.33	19.03
C22H20N406	436.43	436.1383	60.55	4.62	12.84	22.00
C22H20N407	452.43	452.1332	58.41	4.46	12.38	24.75
C22H20N408	468.43	468.1281	56.41	4.30	11.96	27.32
C22H20N409	484.43	484.1230	54.55	4.16	11.57	29.72
C22H20N601	384.44	384.1698	68.73	5.24	21.86	4.16
C22H20N602	400.44	400.1648	65.99	5.03	20.99	7.99
C22H20N603	416.44	416.1597	63.45	4.84	20.18	11.53
C22H20N604	432.44	432.1546	61.11	4.66	19.43	14.80
C22H20N605	448.44	448.1495	58.93	4.50	18.74	17.84
C22H20N606	464.44	464.1444	56.90	4.34	18.09	20.67
C22H20N607	480.44	480.1393	55.00	4.20	17.49	23.31
C22H20N608	496.44	496.1342	53.23	4.06	16.93	25.78
C22H20N609	512.44	512.1292	51.57	3.93	16.40	28.10
C22H21N101	315.42	315.1623	83.78	6.71	4.44	5.07
C22H21N102	331.42	331.1572	79.73	6.39	4.23	9.66
C22H21N103	347.42	347.1521	76.06	6.09	4.03	13.82
C22H21N104	363.42	363.1470	72.71	5.82	3.85	17.61
C22H21N105	379.42	379.1420	69.65	5.58	3.69	21.08
C22H21N106	395.42	395.1369	66.83	5.35	3.54	24.28
C22H21N107	411.42	411.1318	64.23	5.15	3.40	27.22
C22H21N108	427.41	427.1267	61.82	4.95	3.28	29.95
C22H21N109	443.41	443.1216	59.59	4.77	3.16	32.47
C22H21N301	343.43	343.1685	76.94	6.16	12.24	4.66
C22H21N302	359.43	359.1634	73.52	5.89	11.69	8.90
C22H21N303	375.43	375.1583	70.38	5.64	11.19	12.78
C22H21N304	391.43	391.1532	67.51	5.41	10.74	16.35
C22H21N305	407.43	407.1481	64.86	5.20	10.31	19.63
C22H21N306	423.43	423.1430	62.41	5.00	9.92	22.67
C22H21N307	439.43	439.1379	60.13	4.82	9.56	25.49
C22H21N308	455.43	455.1329	58.02	4.65	9.23	28.10
C22H21N309	471.43	471.1278	56.05	4.49	8.91	30.54
C22H21N501	371.45	371.1746	71.14	5.70	18.85	4.31
C22H21N502	387.44	387.1695	68.20	5.46	18.08	8.26
C22H21N503	403.44	403.1644	65.50	5.25	17.36	11.90
C22H21N504	419.44	419.1593	63.00	5.05	16.70	15.26
C22H21N505	435.44	435.1543	60.68	4.86	16.08	18.37
C22H21N506	451.44	451.1492	58.53	4.69	15.51	21.26
C22H21N507	467.44	467.1441	56.53	4.53	14.98	23.96
C22H21N508	483.44	483.1390	54.66	4.38	14.49	26.48
C22H21N509	499.44	499.1339	52.91	4.24	14.02	28.83
C22H22N201	330.43	330.1732	79.97	6.71	8.48	4.84
C22H22N202	346.43	346.1681	76.28	6.40	8.09	9.24
C22H22N203	362.43	362.1630	72.91	6.12	7.73	13.24
C22H22N204	378.43	378.1579	69.83	5.86	7.40	16.91
C22H22N205	394.43	394.1529	66.99	5.62	7.10	20.28
C22H22N206	410.43	410.1478	64.38	5.40	6.83	23.39
C22H22N207	426.43	426.1427	61.97	5.20	6.57	26.26
C22H22N208	442.43	442.1376	59.73	5.01	6.33	28.93
C22H22N209	458.43	458.1325	57.64	4.84	6.11	31.41
C22H22N401	358.45	358.1794	73.72	6.19	15.63	4.46

131

C22H22N402	374.45	374.1743	70.57	5.92	14.96	8.55
C22H22N403	390.45	390.1692	67.68	5.68	14.35	12.29
C22H22N404	406.45	406.1641	65.01	5.46	13.78	15.75
C22H22N405	422.44	422.1590	62.55	5.25	13.26	18.94
C22H22N406	438.44	438.1539	60.27	5.06	12.78	21.89
C22H22N407	454.44	454.1488	58.15	4.88	12.33	24.64
C22H22N408	470.44	470.1437	56.17	4.71	11.91	27.21
C22H22N409	486.44	486.1387	54.32	4.56	11.52	29.60
C22H22N601	386.46	386.1855	68.38	5.74	21.75	4.14
C22H22N602	402.46	402.1804	65.66	5.51	20.88	7.95
C22H22N603	418.46	418.1753	63.15	5.30	20.08	11.47
C22H22N604	434.46	434.1702	60.82	5.10	19.34	14.73
C22H22N605	450.46	450.1652	58.66	4.92	18.66	17.76
C22H22N606	466.46	466.1601	56.65	4.75	18.02	20.58
C22H22N607	482.46	482.1550	54.77	4.60	17.42	23.21
C22H22N608	498.46	498.1499	53.01	4.45	16.86	25.68
C22H22N609	514.46	514.1448	51.36	4.31	16.34	27.99
C22H23N101	317.43	317.1780	83.24	7.30	4.41	5.04
C22H23N102	333.43	333.1729	79.25	6.95	4.20	9.60
C22H23N103	349.43	349.1678	75.62	6.63	4.01	13.74
C22H23N104	365.43	365.1627	72.31	6.34	3.83	17.51
C22H23N105	381.43	381.1576	69.28	6.08	3.67	20.97
C22H23N106	397.43	397.1525	66.49	5.83	3.52	24.15
C22H23N107	413.43	413.1474	63.92	5.61	3.39	27.09
C22H23N108	429.43	429.1424	61.53	5.40	3.26	29.81
C22H23N109	445.43	445.1373	59.32	5.20	3.14	32.33
C22H23N301	345.45	345.1841	76.49	6.71	12.16	4.63
C22H23N302	361.45	361.1790	73.11	6.41	11.63	8.85
C22H23N303	377.45	377.1739	70.01	6.14	11.13	12.72
C22H23N304	393.45	393.1688	67.16	5.89	10.68	16.27
C22H23N305	409.45	409.1638	64.54	5.66	10.26	19.54
C22H23N306	425.45	425.1587	62.11	5.45	9.88	22.56
C22H23N307	441.44	441.1536	59.86	5.25	9.52	25.37
C22H23N308	457.44	457.1485	57.77	5.07	9.19	27.98
C22H23N309	473.44	473.1434	55.81	4.90	8.88	30.41
C22H23N501	373.46	373.1902	70.76	6.21	18.75	4.28
C22H23N502	389.46	389.1852	67.85	5.95	17.98	8.22
C22H23N503	405.46	405.1801	65.17	5.72	17.27	11.84
C22H23N504	421.46	421.1750	62.70	5.50	16.62	15.18
C22H23N505	437.46	437.1699	60.40	5.30	16.01	18.29
C22H23N506	453.46	453.1648	58.27	5.11	15.44	21.17
C22H23N507	469.46	469.1597	56.29	4.94	14.92	23.86
C22H23N508	485.46	485.1546	54.43	4.78	14.43	26.37
C22H23N509	501.46	501.1496	52.70	4.62	13.97	28.72
C22H24N201	332.45	332.1889	79.48	7.28	8.43	4.81
C22H24N202	348.45	348.1838	75.83	6.94	8.04	9.18
C22H24N203	364.45	364.1787	72.51	6.64	7.69	13.17
C22H24N204	380.45	380.1736	69.46	6.36	7.36	16.82
C22H24N205	396.45	396.1685	66.65	6.10	7.07	20.18
C22H24N206	412.45	412.1634	64.07	5.87	6.79	23.27
C22H24N207	428.45	428.1583	61.68	5.65	6.54	26.14
C22H24N208	444.45	444.1533	59.46	5.44	6.30	28.80
C22H24N209	460.44	460.1482	57.39	5.25	6.08	31.27
C22H24N401	360.46	360.1950	73.31	6.71	15.54	4.44
C22H24N402	376.46	376.1899	70.19	6.43	14.88	8.50
C22H24N403	392.46	392.1848	67.33	6.16	14.28	12.23
C22H24N404	408.46	408.1797	64.69	5.92	13.72	15.67
C22H24N405	424.46	424.1747	62.25	5.70	13.20	18.85
C22H24N406	440.46	440.1696	59.99	5.49	12.72	21.79
C22H24N407	456.46	456.1645	57.89	5.30	12.27	24.54
C22H24N408	472.46	472.1594	55.93	5.12	11.86	27.09
C22H24N409	488.46	488.1543	54.10	4.95	11.47	29.48
C22H24N601	388.48	388.2011	68.02	6.23	21.63	4.12
C22H24N602	404.48	404.1961	65.33	5.98	20.78	7.91
C22H24N603	420.47	420.1910	62.84	5.75	19.99	11.42
C22H24N604	436.47	436.1859	60.54	5.54	19.25	14.66
C22H24N605	452.47	452.1808	58.40	5.35	18.57	17.68
C22H24N606	468.47	468.1757	56.41	5.16	17.94	20.49
C22H24N607	484.47	484.1706	54.54	4.99	17.35	23.12
C22H24N608	500.47	500.1655	52.80	4.83	16.79	25.57
C22H24N609	516.47	516.1605	51.16	4.68	16.27	27.88
C22H25N101	319.45	319.1936	82.72	7.89	4.38	5.01
C22H25N102	335.45	335.1885	78.77	7.51	4.18	9.54
C22H25N103	351.45	351.1834	75.19	7.17	3.99	13.66
C22H25N104	367.45	367.1783	71.91	6.86	3.81	17.42
C22H25N105	383.45	383.1733	68.91	6.57	3.65	20.86
C22H25N106	399.45	399.1682	66.15	6.31	3.51	24.03
C22H25N107	415.45	415.1631	63.61	6.07	3.37	26.96
C22H25N108	431.45	431.1580	61.25	5.84	3.25	29.67
C22H25N109	447.45	447.1529	59.06	5.63	3.13	32.18
C22H25N301	347.46	347.1998	76.05	7.25	12.09	4.60
C22H25N302	363.46	363.1947	72.70	6.93	11.56	8.80
C22H25N303	379.46	379.1896	69.64	6.64	11.07	12.65
C22H25N304	395.46	395.1845	66.82	6.37	10.63	16.18
C22H25N305	411.46	411.1794	64.22	6.12	10.21	19.44
C22H25N306	427.46	427.1743	61.82	5.90	9.83	22.46
C22H25N307	443.46	443.1692	59.59	5.68	9.48	25.25
C22H25N308	459.46	459.1641	57.51	5.48	9.15	27.86
C22H25N309	475.46	475.1591	55.58	5.30	8.84	30.29
C22H25N501	375.48	375.2059	70.38	6.71	18.65	4.26
C22H25N502	391.48	391.2008	67.50	6.44	17.89	8.17
C22H25N503	407.48	407.1957	64.85	6.18	17.19	11.78
C22H25N504	423.48	423.1906	62.40	5.95	16.54	15.11
C22H25N505	439.48	439.1856	60.13	5.73	15.94	18.20
C22H25N506	455.47	455.1805	58.02	5.53	15.38	21.08
C22H25N507	471.47	471.1754	56.05	5.34	14.85	23.75
C22H25N508	487.47	487.1703	54.21	5.17	14.37	26.26
C22H25N509	503.47	503.1652	52.48	5.01	13.91	28.60
C22H26N201	334.47	334.2045	79.01	7.84	8.38	4.78
C22H26N202	350.46	350.1994	75.40	7.48	7.99	9.13
C22H26N203	366.46	366.1943	72.11	7.15	7.64	13.10
C22H26N204	382.46	382.1892	69.09	6.85	7.32	16.73
C22H26N205	398.46	398.1842	66.32	6.58	7.03	20.08
C22H26N206	414.46	414.1791	63.76	6.32	6.76	23.16
C22H26N207	430.46	430.1740	61.39	6.09	6.51	26.02
C22H26N208	446.46	446.1689	59.19	5.87	6.27	28.67
C22H26N209	462.46	462.1638	57.14	5.67	6.06	31.14
C22H26N401	362.48	362.2106	72.90	7.23	15.46	4.41
C22H26N402	378.48	378.2056	69.82	6.92	14.80	8.45
C22H26N403	394.48	394.2005	66.99	6.64	14.20	12.17
C22H26N404	410.48	410.1954	64.38	6.38	13.65	15.59
C22H26N405	426.48	426.1903	61.96	6.15	13.14	18.76
C22H26N406	442.48	442.1852	59.72	5.92	12.66	21.70
C22H26N407	458.48	458.1801	57.64	5.72	12.22	24.43
C22H26N408	474.47	474.1750	55.69	5.52	11.81	26.98
C22H26N409	490.47	490.1700	53.88	5.34	11.42	29.36
C22H26N601	390.49	390.2168	67.67	6.71	21.52	4.10
C22H26N602	406.49	406.2117	65.01	6.45	20.67	7.87
C22H26N603	422.49	422.2066	62.54	6.20	19.89	11.36
C22H26N604	438.49	438.2015	60.26	5.98	19.17	14.59

C22H26N605	454.49	454.1965	58.14	5.77	18.49	17.60
C22H26N606	470.49	470.1914	56.16	5.57	17.86	20.40
C22H26N607	486.49	486.1863	54.32	5.39	17.27	23.02
C22H26N608	502.49	502.1812	52.59	5.22	16.72	25.47
C22H26N609	518.49	518.1761	50.96	5.05	16.21	27.77
C22H27N101	321.47	321.2093	82.20	8.47	4.36	4.98
C22H27N102	337.47	337.2042	78.30	8.06	4.15	9.48
C22H27N103	353.47	353.1991	74.76	7.70	3.96	13.58
C22H27N104	369.46	369.1940	71.52	7.37	3.79	17.32
C22H27N105	385.46	385.1889	68.55	7.06	3.63	20.75
C22H27N106	401.46	401.1838	65.82	6.78	3.49	23.91
C22H27N107	417.46	417.1787	63.30	6.52	3.36	26.83
C22H27N108	433.46	433.1737	60.96	6.28	3.23	29.53
C22H27N109	449.46	449.1686	58.79	6.06	3.12	32.04
C22H27N301	349.48	349.2154	75.61	7.79	12.02	4.58
C22H27N302	365.48	365.2103	72.30	7.45	11.50	8.76
C22H27N303	381.48	381.2052	69.27	7.13	11.02	12.58
C22H27N304	397.48	397.2001	66.48	6.85	10.57	16.10
C22H27N305	413.48	413.1951	63.91	6.58	10.16	19.35
C22H27N306	429.48	429.1900	61.53	6.34	9.78	22.35
C22H27N307	445.48	445.1849	59.32	6.11	9.43	25.14
C22H27N308	461.48	461.1798	57.26	5.90	9.11	27.74
C22H27N309	477.48	477.1747	55.34	5.70	8.80	30.16
C22H27N501	377.49	377.2215	70.00	7.21	18.55	4.24
C22H27N502	393.49	393.2165	67.15	6.92	17.80	8.13
C22H27N503	409.49	409.2114	64.53	6.65	17.10	11.72
C22H27N504	425.49	425.2063	62.10	6.40	16.46	15.04
C22H27N505	441.49	441.2012	59.85	6.16	15.86	18.12
C22H27N506	457.49	457.1961	57.76	5.95	15.31	20.98
C22H27N507	473.49	473.1910	55.81	5.75	14.79	23.65
C22H27N508	489.49	489.1859	53.98	5.56	14.31	26.15
C22H27N509	505.49	505.1809	52.28	5.38	13.85	28.49
C22H28N201	336.48	336.2202	78.53	8.39	8.33	4.75
C22H28N202	352.48	352.2151	74.97	8.01	7.95	9.08
C22H28N203	368.48	368.2100	71.71	7.66	7.60	13.03
C22H28N204	384.48	384.2049	68.73	7.34	7.29	16.65
C22H28N205	400.48	400.1998	65.98	7.05	6.99	19.98
C22H28N206	416.48	416.1947	63.45	6.78	6.73	23.05
C22H28N207	432.48	432.1896	61.10	6.53	6.48	25.90
C22H28N208	448.48	448.1845	58.92	6.29	6.25	28.54
C22H28N209	464.48	464.1795	56.89	6.08	6.03	31.00
C22H28N401	364.49	364.2263	72.50	7.74	15.37	4.39
C22H28N402	380.49	380.2212	69.45	7.42	14.72	8.41
C22H28N403	396.49	396.2161	66.65	7.12	14.13	12.11
C22H28N404	412.49	412.2110	64.06	6.84	13.58	15.51
C22H28N405	428.49	428.2060	61.67	6.59	13.08	18.67
C22H28N406	444.49	444.2009	59.45	6.35	12.60	21.60
C22H28N407	460.49	460.1958	57.38	6.13	12.17	24.32
C22H28N408	476.49	476.1907	55.46	5.92	11.76	26.86
C22H28N409	492.49	492.1856	53.65	5.73	11.38	29.24
C22H28N601	392.51	392.2324	67.32	7.19	21.41	4.08
C22H28N602	408.51	408.2274	64.69	6.91	20.57	7.83
C22H28N603	424.51	424.2223	62.25	6.65	19.80	11.31
C22H28N604	440.51	440.2172	59.99	6.41	19.08	14.53
C22H28N605	456.51	456.2121	57.88	6.18	18.41	17.52
C22H28N606	472.51	472.2070	55.92	5.97	17.79	20.32
C22H28N607	488.50	488.2019	54.09	5.78	17.20	22.93
C22H28N608	504.50	504.1968	52.38	5.59	16.66	25.37
C22H28N609	520.50	520.1918	50.77	5.42	16.15	27.66
C22H29N101	323.48	323.2249	81.69	9.04	4.33	4.95
C22H29N102	339.48	339.2198	77.84	8.61	4.13	9.43
C22H29N103	355.48	355.2147	74.33	8.22	3.94	13.50
C22H29N104	371.48	371.2096	71.13	7.87	3.77	17.23
C22H29N105	387.48	387.2046	68.20	7.54	3.61	20.65
C22H29N106	403.48	403.1995	65.49	7.24	3.47	23.79
C22H29N107	419.48	419.1944	62.99	6.97	3.34	26.70
C22H29N108	435.48	435.1893	60.68	6.71	3.22	29.39
C22H29N109	451.48	451.1842	58.53	6.47	3.10	31.89
C22H29N301	351.50	351.2310	75.18	8.32	11.95	4.55
C22H29N302	367.50	367.2260	71.90	7.95	11.43	8.71
C22H29N303	383.49	383.2209	68.90	7.62	10.96	12.52
C22H29N304	399.49	399.2158	66.14	7.32	10.52	16.02
C22H29N305	415.49	415.2107	63.60	7.04	10.11	19.25
C22H29N306	431.49	431.2056	61.24	6.77	9.74	22.25
C22H29N307	447.49	447.2005	59.05	6.53	9.39	25.03
C22H29N308	463.49	463.1954	57.01	6.31	9.07	27.62
C22H29N309	479.49	479.1904	55.11	6.10	8.76	30.03
C22H29N501	379.51	379.2372	69.63	7.70	18.45	4.22
C22H29N502	395.51	395.2321	66.81	7.39	17.71	8.09
C22H29N503	411.51	411.2270	64.21	7.10	17.02	11.66
C22H29N504	427.51	427.2219	61.81	6.84	16.38	14.97
C22H29N505	443.51	443.2169	59.58	6.59	15.79	18.04
C22H29N506	459.51	459.2118	57.51	6.36	15.24	20.89
C22H29N507	475.51	475.2067	55.57	6.15	14.73	23.55
C22H29N508	491.51	491.2016	53.76	5.95	14.25	26.04
C22H29N509	507.50	507.1965	52.07	5.76	13.80	28.37
C22H30N201	338.50	338.2358	78.06	8.93	8.28	4.73
C22H30N202	354.50	354.2307	74.54	8.53	7.90	9.03
C22H30N203	370.50	370.2256	71.32	8.16	7.56	12.96
C22H30N204	386.50	386.2205	68.37	7.82	7.25	16.56
C22H30N205	402.49	402.2155	65.65	7.51	6.96	19.88
C22H30N206	418.49	418.2104	63.14	7.23	6.69	22.94
C22H30N207	434.49	434.2053	60.82	6.96	6.45	25.78
C22H30N208	450.49	450.2002	58.66	6.71	6.22	28.41
C22H30N209	466.49	466.1951	56.65	6.48	6.01	30.87
C22H30N401	366.51	366.2419	72.10	8.25	15.29	4.37
C22H30N402	382.51	382.2369	69.08	7.91	14.65	8.37
C22H30N403	398.51	398.2318	66.31	7.59	14.06	12.04
C22H30N404	414.51	414.2267	63.75	7.30	13.52	15.44
C22H30N405	430.51	430.2216	61.38	7.02	13.01	18.58
C22H30N406	446.51	446.2165	59.18	6.77	12.55	21.50
C22H30N407	462.51	462.2114	57.13	6.54	12.11	24.21
C22H30N408	478.51	478.2063	55.22	6.32	11.71	26.75
C22H30N409	494.51	494.2013	53.44	6.12	11.33	29.12
C22H30N601	394.52	394.2481	66.98	7.66	21.30	4.06
C22H30N602	410.52	410.2430	64.37	7.37	20.47	7.79
C22H30N603	426.52	426.2379	61.95	7.09	19.70	11.25
C22H30N604	442.52	442.2328	59.71	6.83	18.99	14.46
C22H30N605	458.52	458.2278	57.63	6.59	18.33	17.45
C22H30N606	474.52	474.2227	55.69	6.37	17.71	20.23
C22H30N607	490.52	490.2176	53.87	6.16	17.13	22.83
C22H30N608	506.52	506.2125	52.17	5.97	16.59	25.27
C22H30N609	522.52	522.2074	50.57	5.79	16.08	27.56
C22H31N101	325.50	325.2406	81.18	9.60	4.30	4.92
C22H31N102	341.50	341.2355	77.38	9.15	4.10	9.37
C22H31N103	357.50	357.2304	73.92	8.74	3.92	13.43
C22H31N104	373.50	373.2253	70.75	8.37	3.75	17.13
C22H31N105	389.50	389.2202	67.84	8.02	3.60	20.54
C22H31N106	405.50	405.2151	65.17	7.71	3.45	23.67
C22H31N107	421.49	421.2100	62.69	7.41	3.32	26.57

C22H31N108	437.49	437.2049	60.40	7.14	3.20	29.26
C22H31N109	453.49	453.1999	58.27	6.89	3.09	31.75
C22H31N301	353.51	353.2467	74.75	8.84	11.89	4.53
C22H31N302	369.51	369.2416	71.51	8.46	11.37	8.66
C22H31N303	385.51	385.2365	68.54	8.11	10.90	12.45
C22H31N304	401.51	401.2314	65.81	7.78	10.47	15.94
C22H31N305	417.51	417.2264	63.29	7.48	10.06	19.16
C22H31N306	433.51	433.2213	60.95	7.21	9.69	22.14
C22H31N307	449.51	449.2162	58.79	6.95	9.35	24.92
C22H31N308	465.51	465.2111	56.76	6.71	9.03	27.50
C22H31N309	481.51	481.2060	54.88	6.49	8.73	29.90
C22H31N501	381.53	381.2528	69.26	8.19	18.36	4.19
C22H31N502	397.52	397.2478	66.47	7.86	17.62	8.05
C22H31N503	413.52	413.2427	63.90	7.56	16.94	11.61
C22H31N504	429.52	429.2376	61.52	7.27	16.30	14.90
C22H31N505	445.52	445.2325	59.31	7.01	15.72	17.96
C22H31N506	461.52	461.2274	57.26	6.77	15.17	20.80
C22H31N507	477.52	477.2223	55.34	6.54	14.67	23.45
C22H31N508	493.52	493.2172	53.54	6.33	14.19	25.94
C22H31N509	509.52	509.2122	51.86	6.13	13.74	28.26
C22H32N201	340.51	340.2514	77.60	9.47	8.23	4.70
C22H32N202	356.51	356.2464	74.12	9.05	7.86	8.98
C22H32N203	372.51	372.2413	70.94	8.66	7.52	12.89
C22H32N204	388.51	388.2362	68.01	8.30	7.21	16.47
C22H32N205	404.51	404.2311	65.32	7.97	6.93	19.78
C22H32N206	420.51	420.2260	62.84	7.67	6.66	22.83
C22H32N207	436.51	436.2209	60.54	7.39	6.42	25.66
C22H32N208	452.51	452.2158	58.40	7.13	6.19	28.29
C22H32N209	468.51	468.2108	56.40	6.88	5.98	30.73
C22H32N401	368.53	368.2576	71.70	8.75	15.20	4.34
C22H32N402	384.53	384.2525	68.72	8.39	14.57	8.32
C22H32N403	400.53	400.2474	65.97	8.05	13.99	11.98
C22H32N404	416.52	416.2423	63.44	7.74	13.45	15.36
C22H32N405	432.52	432.2373	61.09	7.46	12.95	18.50
C22H32N406	448.52	448.2322	58.91	7.19	12.49	21.40
C22H32N407	464.52	464.2271	56.89	6.94	12.06	24.11
C22H32N408	480.52	480.2220	54.99	6.71	11.66	26.64
C22H32N409	496.52	496.2169	53.22	6.50	11.28	29.00
C22H32N601	396.54	396.2637	66.64	8.13	21.19	4.03
C22H32N602	412.54	412.2587	64.05	7.82	20.37	7.76
C22H32N603	428.54	428.2536	61.66	7.53	19.61	11.20
C22H32N604	444.54	444.2485	59.44	7.26	18.91	14.40
C22H32N605	460.54	460.2434	57.38	7.00	18.25	17.37
C22H32N606	476.54	476.2383	55.45	6.77	17.64	20.14
C22H32N607	492.54	492.2332	53.65	6.55	17.06	22.74
C22H32N608	508.54	508.2281	51.96	6.34	16.53	25.17
C22H32N609	524.54	524.2231	50.38	6.15	16.02	27.45
C22H33N101	327.51	327.2562	80.68	10.16	4.28	4.89
C22H33N102	343.51	343.2511	76.92	9.68	4.08	9.32
C22H33N103	359.51	359.2460	73.50	9.25	3.90	13.35
C22H33N104	375.51	375.2409	70.37	8.86	3.73	17.04
C22H33N105	391.51	391.2359	67.49	8.50	3.58	20.43
C22H33N106	407.51	407.2308	64.84	8.16	3.44	23.56
C22H33N107	423.51	423.2257	62.39	7.85	3.31	26.44
C22H33N108	439.51	439.2206	60.12	7.57	3.19	29.12
C22H33N109	455.51	455.2155	58.01	7.30	3.07	31.61
C22H33N301	355.53	355.2623	74.32	9.36	11.82	4.50
C22H33N302	371.53	371.2573	71.12	8.95	11.31	8.61
C22H33N303	387.53	387.2522	68.19	8.58	10.84	12.39
C22H33N304	403.53	403.2471	65.48	8.24	10.41	15.86
C22H33N305	419.53	419.2420	62.99	7.93	10.02	19.07
C22H33N306	435.52	435.2369	60.67	7.64	9.65	22.04
C22H33N307	451.52	451.2318	58.52	7.37	9.31	24.80
C22H33N308	467.52	467.2267	56.52	7.11	8.99	27.38
C22H33N309	483.52	483.2217	54.65	6.88	8.69	29.78
C22H33N501	383.54	383.2685	68.90	8.67	18.26	4.17
C22H33N502	399.54	399.2634	66.14	8.33	17.53	8.01
C22H33N503	415.54	415.2583	63.59	8.00	16.85	11.55
C22H33N504	431.54	431.2532	61.23	7.71	16.23	14.83
C22H33N505	447.54	447.2482	59.04	7.43	15.65	17.87
C22H33N506	463.54	463.2431	57.01	7.18	15.11	20.71
C22H33N507	479.54	479.2380	55.10	6.94	14.60	23.35
C22H33N508	495.54	495.2329	53.33	6.71	14.13	25.83
C22H33N509	511.54	511.2278	51.66	6.50	13.69	28.15
C22H34N201	342.53	342.2671	77.15	10.01	8.18	4.67
C22H34N202	358.53	358.2620	73.70	9.56	7.81	8.93
C22H34N203	374.53	374.2569	70.55	9.15	7.48	12.82
C22H34N204	390.53	390.2518	67.66	8.78	7.17	16.39
C22H34N205	406.53	406.2468	65.00	8.43	6.89	19.68
C22H34N206	422.53	422.2417	62.54	8.11	6.63	22.72
C22H34N207	438.53	438.2366	60.26	7.82	6.39	25.54
C22H34N208	454.52	454.2315	58.14	7.54	6.16	28.16
C22H34N209	470.52	470.2264	56.16	7.28	5.95	30.60
C22H34N401	370.54	370.2732	71.31	9.25	15.12	4.32
C22H34N402	386.54	386.2682	68.36	8.87	14.49	8.28
C22H34N403	402.54	402.2631	65.64	8.51	13.92	11.92
C22H34N404	418.54	418.2580	63.13	8.19	13.39	15.29
C22H34N405	434.54	434.2529	60.81	7.89	12.89	18.41
C22H34N406	450.54	450.2478	58.65	7.61	12.44	21.31
C22H34N407	466.54	466.2427	56.64	7.35	12.01	24.01
C22H34N408	482.54	482.2376	54.76	7.10	11.61	26.53
C22H34N409	498.54	498.2326	53.00	6.87	11.24	28.88
C22H34N601	398.56	398.2794	66.30	8.60	21.09	4.01
C22H34N602	414.56	414.2743	63.74	8.27	20.27	7.72
C22H34N603	430.55	430.2692	61.37	7.96	19.52	11.15
C22H34N604	446.55	446.2641	59.17	7.67	18.82	14.33
C22H34N605	462.55	462.2590	57.13	7.41	18.17	17.29
C22H34N606	478.55	478.2540	55.22	7.16	17.56	20.06
C22H34N607	494.55	494.2489	53.43	6.93	16.99	22.65
C22H34N608	510.55	510.2438	51.76	6.71	16.46	25.07
C22H34N609	526.55	526.2387	50.18	6.51	15.96	27.35
C22H35N101	329.53	329.2718	80.19	10.71	4.25	4.86
C22H35N102	345.53	345.2668	76.48	10.21	4.05	9.26
C22H35N103	361.53	361.2617	73.09	9.76	3.87	13.28
C22H35N104	377.53	377.2566	69.99	9.34	3.71	16.95
C22H35N105	393.53	393.2515	67.15	8.96	3.56	20.33
C22H35N106	409.53	409.2464	64.52	8.61	3.42	23.44
C22H35N107	425.53	425.2413	62.10	8.29	3.29	26.32
C22H35N108	441.53	441.2362	59.85	7.99	3.17	28.99
C22H35N109	457.53	457.2312	57.76	7.71	3.06	31.47
C22H35N301	357.54	357.2780	73.91	9.87	11.75	4.47
C22H35N302	373.54	373.2729	70.74	9.44	11.25	8.57
C22H35N303	389.54	389.2678	67.83	9.06	10.79	12.32
C22H35N304	405.54	405.2627	65.16	8.70	10.36	15.78
C22H35N305	421.54	421.2577	62.69	8.37	9.97	18.98
C22H35N306	437.54	437.2526	60.39	8.06	9.60	21.94
C22H35N307	453.54	453.2475	58.26	7.78	9.26	24.69
C22H35N308	469.54	469.2424	56.28	7.51	8.95	27.26
C22H35N309	485.54	485.2373	54.42	7.27	8.65	29.66
C22H35N501	385.56	385.2841	68.54	9.15	18.16	4.15

134

C22H35N5O2	401.56	401.2791	65.81	8.79	17.44	7.97
C22H35N5O3	417.56	417.2740	63.28	8.45	16.77	11.50
C22H35N5O4	433.56	433.2689	60.95	8.14	16.15	14.76
C22H35N5O5	449.55	449.2638	58.78	7.85	15.58	17.79
C22H35N5O6	465.55	465.2587	56.76	7.58	15.04	20.62
C22H35N5O7	481.55	481.2536	54.87	7.33	14.54	23.26
C22H35N5O8	497.55	497.2485	53.11	7.09	14.08	25.72
C22H35N5O9	513.55	513.2435	51.45	6.87	13.64	28.04
C22H36N2O1	344.55	344.2827	76.69	10.53	8.13	4.64
C22H36N2O2	360.54	360.2777	73.29	10.06	7.77	8.88
C22H36N2O3	376.54	376.2726	70.18	9.64	7.44	12.75
C22H36N2O4	392.54	392.2675	67.32	9.24	7.14	16.30
C22H36N2O5	408.54	408.2624	64.68	8.88	6.86	19.58
C22H36N2O6	424.54	424.2573	62.24	8.55	6.60	22.61
C22H36N2O7	440.54	440.2522	59.98	8.24	6.36	25.42
C22H36N2O8	456.54	456.2471	57.88	7.95	6.14	28.04
C22H36N2O9	472.54	472.2421	55.92	7.68	5.93	30.47
C22H36N4O1	372.56	372.2889	70.93	9.74	15.04	4.29
C22H36N4O2	388.56	388.2838	68.01	9.34	14.42	8.24
C22H36N4O3	404.56	404.2787	65.32	8.97	13.85	11.86
C22H36N4O4	420.56	420.2736	62.83	8.63	13.32	15.22
C22H36N4O5	436.56	436.2686	60.53	8.31	12.83	18.32
C22H36N4O6	452.56	452.2635	58.39	8.02	12.38	21.21
C22H36N4O7	468.55	468.2584	56.40	7.74	11.96	23.90
C22H36N4O8	484.55	484.2533	54.53	7.49	11.56	26.42
C22H36N4O9	500.55	500.2482	52.79	7.25	11.19	28.77
C22H36N6O1	400.57	400.2950	65.97	9.06	20.98	3.99
C22H36N6O2	416.57	416.2900	63.43	8.71	20.17	7.68
C22H36N6O3	432.57	432.2849	61.09	8.39	19.43	11.10
C22H36N6O4	448.57	448.2798	58.91	8.09	18.74	14.27
C22H36N6O5	464.57	464.2747	56.88	7.81	18.09	17.22
C22H36N6O6	480.57	480.2696	54.99	7.55	17.49	19.98
C22H36N6O7	496.57	496.2645	53.21	7.31	16.92	22.55
C22H36N6O8	512.57	512.2594	51.55	7.08	16.40	24.97
C22H36N6O9	528.57	528.2544	49.99	6.87	15.90	27.24
C22H37N1O1	331.55	331.2875	79.70	11.25	4.22	4.83
C22H37N1O2	347.55	347.2824	76.03	10.73	4.03	9.21
C22H37N1O3	363.55	363.2773	72.69	10.26	3.85	13.20
C22H37N1O4	379.54	379.2722	69.62	9.83	3.69	16.86
C22H37N1O5	395.54	395.2672	66.81	9.43	3.54	20.22
C22H37N1O6	411.54	411.2621	64.21	9.06	3.40	23.33
C22H37N1O7	427.54	427.2570	61.81	8.72	3.28	26.20
C22H37N1O8	443.54	443.2519	59.58	8.41	3.16	28.86
C22H37N1O9	459.54	459.2468	57.50	8.12	3.05	31.33
C22H37N3O1	359.56	359.2936	73.49	10.37	11.69	4.45
C22H37N3O2	375.56	375.2886	70.36	9.93	11.19	8.52
C22H37N3O3	391.56	391.2835	67.49	9.52	10.73	12.26
C22H37N3O4	407.56	407.2784	64.84	9.15	10.31	15.70
C22H37N3O5	423.56	423.2733	62.35	8.81	9.92	18.89
C22H37N3O6	439.56	439.2682	60.12	8.48	9.56	21.84
C22H37N3O7	455.56	455.2631	58.00	8.19	9.22	24.58
C22H37N3O8	471.56	471.2580	56.04	7.91	8.91	27.14
C22H37N3O9	487.55	487.2530	54.20	7.65	8.62	29.53
C22H37N5O1	387.57	387.2998	68.18	9.62	18.07	4.13
C22H37N5O2	403.57	403.2947	65.48	9.24	17.35	7.93
C22H37N5O3	419.57	419.2896	62.98	8.89	16.69	11.44
C22H37N5O4	435.57	435.2845	60.67	8.56	16.08	14.69
C22H37N5O5	451.57	451.2794	58.52	8.26	15.51	17.72
C22H37N5O6	467.57	467.2744	56.51	7.98	14.98	20.53
C22H37N5O7	483.57	483.2693	54.64	7.71	14.48	23.16
C22H37N5O8	499.57	499.2642	52.89	7.47	14.02	25.62
C22H37N5O9	515.57	515.2591	51.25	7.23	13.58	27.93
C22H38N2O1	346.56	346.2984	76.25	11.05	8.08	4.62
C22H38N2O2	362.56	362.2933	72.88	10.56	7.73	8.83
C22H38N2O3	378.56	378.2882	69.80	10.12	7.40	12.68
C22H38N2O4	394.56	394.2831	66.97	9.71	7.10	16.22
C22H38N2O5	410.56	410.2781	64.36	9.33	6.82	19.48
C22H38N2O6	426.56	426.2730	61.95	8.98	6.57	22.50
C22H38N2O7	442.56	442.2679	59.71	8.65	6.33	25.31
C22H38N2O8	458.56	458.2628	57.63	8.35	6.11	27.91
C22H38N2O9	474.56	474.2577	55.68	8.07	5.90	30.34
C22H38N4O1	374.57	374.3045	70.55	10.23	14.96	4.27
C22H38N4O2	390.57	390.2995	67.66	9.81	14.34	8.19
C22H38N4O3	406.57	406.2944	64.99	9.42	13.78	11.81
C22H38N4O4	422.57	422.2893	62.53	9.06	13.26	15.14
C22H38N4O5	438.57	438.2842	60.25	8.73	12.77	18.24
C22H38N4O6	454.57	454.2791	58.13	8.43	12.33	21.12
C22H38N4O7	470.57	470.2740	56.15	8.14	11.91	23.80
C22H38N4O8	486.57	486.2689	54.31	7.87	11.51	26.31
C22H38N4O9	502.57	502.2639	52.58	7.62	11.15	28.65
C22H38N6O1	402.59	402.3107	65.64	9.51	20.88	3.97
C22H38N6O2	418.59	418.3056	63.13	9.15	20.08	7.64
C22H38N6O3	434.59	434.3005	60.80	8.81	19.34	11.04
C22H38N6O4	450.59	450.2954	58.64	8.50	18.65	14.20
C22H38N6O5	466.59	466.2903	56.63	8.21	18.01	17.15
C22H38N6O6	482.58	482.2853	54.76	7.94	17.41	19.89
C22H38N6O7	498.58	498.2802	53.00	7.68	16.86	22.46
C22H38N6O8	514.58	514.2751	51.35	7.44	16.33	24.87
C22H38N6O9	530.58	530.2700	49.80	7.22	15.84	27.14
C22H39N1O1	333.56	333.3031	79.22	11.79	4.20	4.80
C22H39N1O2	349.56	349.2981	75.59	11.25	4.01	9.15
C22H39N1O3	365.56	365.2930	72.28	10.75	3.83	13.13
C22H39N1O4	381.56	381.2879	69.25	10.30	3.67	16.77
C22H39N1O5	397.56	397.2828	66.47	9.89	3.52	20.12
C22H39N1O6	413.56	413.2777	63.90	9.51	3.39	23.21
C22H39N1O7	429.56	429.2726	61.52	9.15	3.26	26.07
C22H39N1O8	445.56	445.2675	59.31	8.82	3.14	28.73
C22H39N1O9	461.56	461.2625	57.25	8.52	3.03	31.20
C22H39N3O1	361.58	361.3093	73.08	10.87	11.62	4.42
C22H39N3O2	377.58	377.3042	69.98	10.41	11.13	8.47
C22H39N3O3	393.57	393.2991	67.14	9.99	10.68	12.20
C22H39N3O4	409.57	409.2940	64.52	9.60	10.26	15.63
C22H39N3O5	425.57	425.2890	62.09	9.24	9.87	18.80
C22H39N3O6	441.57	441.2839	59.84	8.90	9.52	21.74
C22H39N3O7	457.57	457.2788	57.75	8.59	9.18	24.48
C22H39N3O8	473.57	473.2737	55.80	8.30	8.87	27.03
C22H39N3O9	489.57	489.2686	53.97	8.03	8.58	29.41
C22H39N5O1	389.59	389.3154	67.83	10.09	17.98	4.11
C22H39N5O2	405.59	405.3104	65.15	9.69	17.27	7.89
C22H39N5O3	421.59	421.3053	62.68	9.32	16.61	11.39
C22H39N5O4	437.59	437.3002	60.39	8.98	16.00	14.63
C22H39N5O5	453.59	453.2951	58.26	8.67	15.44	17.64
C22H39N5O6	469.59	469.2900	56.27	8.37	14.91	20.44
C22H39N5O7	485.59	485.2849	54.42	8.10	14.42	23.06
C22H39N5O8	501.58	501.2798	52.68	7.84	13.96	25.52
C22H39N5O9	517.58	517.2748	51.05	7.60	13.53	27.82
C22H40N2O1	348.58	348.3140	75.81	11.57	8.04	4.59
C22H40N2O2	364.58	364.3090	72.48	11.06	7.68	8.78
C22H40N2O3	380.58	380.3039	69.43	10.59	7.36	12.61
C22H40N2O4	396.58	396.2988	66.63	10.17	7.06	16.14

135

C22H40N2O5	412.57	412.2937	64.05	9.77	6.79	19.39
C22H40N2O6	428.57	428.2886	61.66	9.41	6.54	22.40
C22H40N2O7	444.57	444.2835	59.44	9.07	6.30	25.19
C22H40N2O8	460.57	460.2784	57.37	8.75	6.08	27.79
C22H40N2O9	476.57	476.2734	55.45	8.46	5.88	30.21
C22H40N4O1	376.59	376.3202	70.17	10.71	14.88	4.25
C22H40N4O2	392.59	392.3151	67.31	10.27	14.27	8.15
C22H40N4O3	408.59	408.3100	64.67	9.87	13.71	11.75
C22H40N4O4	424.59	424.3049	62.24	9.50	13.20	15.07
C22H40N4O5	440.59	440.2998	59.98	9.15	12.72	18.16
C22H40N4O6	456.59	456.2948	57.87	8.83	12.27	21.02
C22H40N4O7	472.59	472.2897	55.91	8.53	11.86	23.70
C22H40N4O8	488.59	488.2846	54.08	8.25	11.47	26.20
C22H40N4O9	504.59	504.2795	52.37	7.99	11.10	28.54
C22H40N6O1	404.60	404.3263	65.31	9.97	20.77	3.95
C22H40N6O2	420.60	420.3213	62.83	9.59	19.98	7.61
C22H40N6O3	436.60	436.3162	60.52	9.23	19.25	10.99
C22H40N6O4	452.60	452.3111	58.38	8.91	18.57	14.14
C22H40N6O5	468.60	468.3060	56.39	8.60	17.93	17.07
C22H40N6O6	484.60	484.3009	54.53	8.32	17.34	19.81
C22H40N6O7	500.60	500.2958	52.79	8.05	16.79	22.37
C22H40N6O8	516.60	516.2907	51.15	7.80	16.27	24.78
C22H40N6O9	532.60	532.2857	49.61	7.57	15.78	27.04
C22H41N1O1	335.58	335.3188	78.74	12.32	4.17	4.77
C22H41N1O2	351.58	351.3137	75.16	11.75	3.98	9.10
C22H41N1O3	367.58	367.3086	71.89	11.24	3.81	13.06
C22H41N1O4	383.58	383.3035	68.89	10.77	3.65	16.68
C22H41N1O5	399.58	399.2985	66.13	10.34	3.51	20.02
C22H41N1O6	415.58	415.2934	63.59	9.94	3.37	23.10
C22H41N1O7	431.57	431.2883	61.23	9.58	3.25	25.95
C22H41N1O8	447.57	447.2832	59.04	9.23	3.13	28.60
C22H41N1O9	463.57	463.2781	57.00	8.91	3.02	31.06
C22H41N3O1	363.59	363.3249	72.68	11.37	11.56	4.40
C22H41N3O2	379.59	379.3199	69.61	10.89	11.07	8.43
C22H41N3O3	395.59	395.3148	66.80	10.45	10.62	12.13
C22H41N3O4	411.59	411.3097	64.20	10.04	10.21	15.55
C22H41N3O5	427.59	427.3046	61.80	9.67	9.83	18.71
C22H41N3O6	443.59	443.2995	59.57	9.32	9.47	21.64
C22H41N3O7	459.59	459.2944	57.50	8.99	9.14	24.37
C22H41N3O8	475.59	475.2893	55.56	8.69	8.84	26.91
C22H41N3O9	491.59	491.2843	53.75	8.41	8.55	29.29
C22H41N5O1	391.60	391.3311	67.48	10.55	17.88	4.09
C22H41N5O2	407.60	407.3260	64.83	10.14	17.18	7.85
C22H41N5O3	423.60	423.3209	62.38	9.76	16.53	11.33
C22H41N5O4	439.60	439.3158	60.11	9.40	15.93	14.56
C22H41N5O5	455.60	455.3107	58.00	9.07	15.37	17.56
C22H41N5O6	471.60	471.3057	56.03	8.76	14.85	20.36
C22H41N5O7	487.60	487.3006	54.19	8.48	14.36	22.97
C22H41N5O8	503.60	503.2955	52.47	8.21	13.91	25.42
C22H41N5O9	519.60	519.2904	50.86	7.95	13.48	27.71
C22H42N2O1	350.59	350.3297	75.37	12.08	7.99	4.56
C22H42N2O2	366.59	366.3246	72.08	11.55	7.64	8.73
C22H42N2O3	382.59	382.3195	69.07	11.07	7.32	12.55
C22H42N2O4	398.59	398.3144	66.29	10.62	7.03	16.06
C22H42N2O5	414.59	414.3094	63.74	10.21	6.76	19.30
C22H42N2O6	430.59	430.3043	61.37	9.83	6.51	22.29
C22H42N2O7	446.59	446.2992	59.17	9.48	6.27	25.08
C22H42N2O8	462.59	462.2941	57.12	9.15	6.06	27.67
C22H42N2O9	478.59	478.2890	55.21	8.85	5.85	30.09
C22H42N4O1	378.61	378.3358	69.79	11.18	14.80	4.23
C22H42N4O2	394.61	394.3308	66.96	10.73	14.20	8.11
C22H42N4O3	410.61	410.3257	64.36	10.31	13.64	11.69
C22H42N4O4	426.60	426.3206	61.94	9.92	13.13	15.00
C22H42N4O5	442.60	442.3155	59.70	9.56	12.66	18.07
C22H42N4O6	458.60	458.3104	57.62	9.23	12.22	20.93
C22H42N4O7	474.60	474.3053	55.68	8.92	11.80	23.60
C22H42N4O8	490.60	490.3002	53.86	8.63	11.42	26.09
C22H42N4O9	506.60	506.2952	52.16	8.36	11.06	28.42
C22H42N6O1	406.62	406.3420	64.99	10.41	20.67	3.93
C22H42N6O2	422.62	422.3369	62.53	10.02	19.89	7.57
C22H42N6O3	438.62	438.3318	60.24	9.65	19.16	10.94
C22H42N6O4	454.62	454.3267	58.12	9.31	18.49	14.08
C22H42N6O5	470.62	470.3216	56.15	9.00	17.86	17.00
C22H42N6O6	486.62	486.3166	54.30	8.70	17.27	19.73
C22H42N6O7	502.62	502.3115	52.57	8.42	16.72	22.28
C22H42N6O8	518.62	518.3064	50.95	8.16	16.20	24.68
C22H42N6O9	534.61	534.3013	49.43	7.92	15.72	26.93
C22H43N1O1	337.59	337.3344	78.27	12.84	4.15	4.74
C22H43N1O2	353.59	353.3294	74.73	12.26	3.96	9.05
C22H43N1O3	369.59	369.3243	71.50	11.73	3.79	12.99
C22H43N1O4	385.59	385.3192	68.53	11.24	3.63	16.60
C22H43N1O5	401.59	401.3141	65.80	10.79	3.49	19.92
C22H43N1O6	417.59	417.3090	63.28	10.38	3.35	22.99
C22H43N1O7	433.59	433.3039	60.94	10.00	3.23	25.83
C22H43N1O8	449.59	449.2988	58.77	9.64	3.12	28.47
C22H43N1O9	465.59	465.2938	56.76	9.31	3.01	30.93
C22H43N3O1	365.61	365.3406	72.28	11.85	11.49	4.38
C22H43N3O2	381.61	381.3355	69.25	11.36	11.01	8.39
C22H43N3O3	397.61	397.3304	66.46	10.90	10.57	12.07
C22H43N3O4	413.61	413.3253	63.89	10.48	10.16	15.47
C22H43N3O5	429.61	429.3202	61.51	10.09	9.78	18.62
C22H43N3O6	445.60	445.3152	59.30	9.73	9.43	21.54
C22H43N3O7	461.60	461.3101	57.25	9.39	9.10	24.26
C22H43N3O8	477.60	477.3050	55.33	9.08	8.80	26.80
C22H43N3O9	493.60	493.2999	53.53	8.78	8.51	29.17
C22H43N5O1	393.62	393.3467	67.13	11.01	17.79	4.06
C22H43N5O2	409.62	409.3417	64.51	10.58	17.10	7.81
C22H43N5O3	425.62	425.3366	62.08	10.18	16.45	11.28
C22H43N5O4	441.62	441.3315	59.84	9.81	15.86	14.49
C22H43N5O5	457.62	457.3264	57.74	9.47	15.30	17.48
C22H43N5O6	473.62	473.3213	55.79	9.15	14.79	20.27
C22H43N5O7	489.62	489.3162	53.97	8.85	14.30	22.87
C22H43N5O8	505.62	505.3111	52.26	8.57	13.85	25.31
C22H43N5O9	521.62	521.3061	50.66	8.31	13.43	27.61
C22H44N2O1	352.61	352.3453	74.94	12.58	7.94	4.54
C22H44N2O2	368.61	368.3403	71.69	12.03	7.60	8.68
C22H44N2O3	384.61	384.3352	68.71	11.53	7.28	12.48
C22H44N2O4	400.61	400.3301	65.96	11.07	6.99	15.98
C22H44N2O5	416.61	416.3250	63.43	10.65	6.72	19.20
C22H44N2O6	432.61	432.3199	61.08	10.25	6.48	22.19
C22H44N2O7	448.61	448.3148	58.90	9.89	6.24	24.97
C22H44N2O8	464.60	464.3097	56.88	9.55	6.03	27.55
C22H44N2O9	480.60	480.3047	54.98	9.23	5.83	29.96
C22H44N4O1	380.62	380.3515	69.42	11.65	14.72	4.20
C22H44N4O2	396.62	396.3464	66.62	11.18	14.13	8.07
C22H44N4O3	412.62	412.3413	64.04	10.75	13.58	11.63
C22H44N4O4	428.62	428.3362	61.65	10.35	13.07	14.93
C22H44N4O5	444.62	444.3311	59.43	9.97	12.60	17.99
C22H44N4O6	460.62	460.3261	57.37	9.63	12.16	20.84
C22H44N4O7	476.62	476.3210	55.44	9.31	11.76	23.50

C22H44N408	492.62	492.3159	53.64	9.00	11.37	25.98
C22H44N409	508.62	508.3108	51.95	8.72	11.02	28.31
C22H44N601	408.64	408.3576	64.67	10.85	20.57	3.92
C22H44N602	424.63	424.3526	62.23	10.44	19.79	7.54
C22H44N603	440.63	440.3475	59.97	10.07	19.07	10.89
C22H44N604	456.63	456.3424	57.87	9.71	18.40	14.02
C22H44N605	472.63	472.3373	55.91	9.38	17.78	16.93
C22H44N606	488.63	488.3322	54.08	9.08	17.20	19.65
C22H44N607	504.63	504.3271	52.36	8.79	16.65	22.19
C22H44N608	520.63	520.3220	50.75	8.52	16.14	24.58
C22H44N609	536.63	536.3170	49.24	8.26	15.66	26.83
C22H45N101	339.61	339.3501	77.81	13.36	4.12	4.71
C22H45N102	355.61	355.3450	74.31	12.76	3.94	9.00
C22H45N103	371.61	371.3399	71.11	12.21	3.77	12.92
C22H45N104	387.61	387.3348	68.17	11.70	3.61	16.51
C22H45N105	403.61	403.3298	65.47	11.24	3.47	19.82
C22H45N106	419.61	419.3247	62.97	10.81	3.34	22.88
C22H45N107	435.61	435.3196	60.66	10.41	3.22	25.71
C22H45N108	451.61	451.3145	58.51	10.04	3.10	28.34
C22H45N109	467.61	467.3094	56.51	9.70	3.00	30.79
C22H45N301	367.62	367.3562	71.88	12.34	11.43	4.35
C22H45N302	383.62	383.3512	68.88	11.82	10.95	8.34
C22H45N303	399.62	399.3461	66.12	11.35	10.51	12.01
C22H45N304	415.62	415.3410	63.58	10.91	10.11	15.40
C22H45N305	431.62	431.3359	61.22	10.51	9.74	18.53
C22H45N306	447.62	447.3308	59.03	10.13	9.39	21.45
C22H45N307	463.62	463.3257	57.00	9.78	9.06	24.16
C22H45N308	479.62	479.3206	55.09	9.46	8.76	26.69
C22H45N309	495.62	495.3156	53.32	9.15	8.48	29.05
C22H45N501	395.64	395.3624	66.79	11.46	17.70	4.04
C22H45N502	411.64	411.3573	64.19	11.02	17.01	7.77
C22H45N503	427.64	427.3522	61.79	10.61	16.38	11.22
C22H45N504	443.64	443.3471	59.56	10.22	15.79	14.43
C22H45N505	459.63	459.3420	57.49	9.87	15.24	17.40
C22H45N506	475.63	475.3370	55.56	9.54	14.72	20.18
C22H45N507	491.63	491.3319	53.75	9.23	14.25	22.78
C22H45N508	507.63	507.3268	52.05	8.94	13.80	25.21
C22H45N509	523.63	523.3217	50.46	8.66	13.37	27.50
C22H46N201	354.62	354.3610	74.51	13.07	7.90	4.51
C22H46N202	370.62	370.3559	71.30	12.51	7.56	8.63
C22H46N203	386.62	386.3508	68.35	11.99	7.25	12.41
C22H46N204	402.62	402.3457	65.63	11.52	6.96	15.90
C22H46N205	418.62	418.3406	63.12	11.08	6.69	19.11
C22H46N206	434.62	434.3356	60.80	10.67	6.45	22.09
C22H46N207	450.62	450.3305	58.64	10.29	6.22	24.85
C22H46N208	466.62	466.3254	56.63	9.94	6.00	27.43
C22H46N209	482.62	482.3203	54.75	9.61	5.80	29.84
C22H46N401	382.64	382.3671	69.06	12.12	14.64	4.18
C22H46N402	398.64	398.3621	66.29	11.63	14.05	8.03
C22H46N403	414.64	414.3570	63.73	11.18	13.51	11.58
C22H46N404	430.64	430.3519	61.36	10.77	13.01	14.86
C22H46N405	446.64	446.3468	59.16	10.38	12.54	17.91
C22H46N406	462.64	462.3417	57.12	10.02	12.11	20.75
C22H46N407	478.63	478.3366	55.21	9.69	11.71	23.40
C22H46N408	494.63	494.3315	53.42	9.37	11.33	25.88
C22H46N409	510.63	510.3265	51.75	9.08	10.97	28.20
C22H46N601	410.65	410.3733	64.35	11.29	20.47	3.90
C22H46N602	426.65	426.3682	61.93	10.87	19.70	7.50
C22H46N603	442.65	442.3631	59.70	10.47	18.99	10.84
C22H46N604	458.65	458.3580	57.61	10.11	18.32	13.95
C22H46N605	474.65	474.3529	55.67	9.77	17.71	16.85
C22H46N606	490.65	490.3479	53.86	9.45	17.13	19.57
C22H46N607	506.65	506.3428	52.16	9.15	16.59	22.11
C22H46N608	522.65	522.3377	50.56	8.87	16.08	24.49
C22H46N609	538.65	538.3326	49.06	8.61	15.60	26.73
C22H47N101	341.63	341.3657	77.35	13.87	4.10	4.68
C22H47N102	357.63	357.3607	73.89	13.25	3.92	8.95
C22H47N103	373.62	373.3556	70.72	12.68	3.75	12.85
C22H47N104	389.62	389.3505	67.82	12.16	3.59	16.43
C22H47N105	405.62	405.3454	65.15	11.68	3.45	19.72
C22H47N106	421.62	421.3403	62.67	11.24	3.32	22.77
C22H47N107	437.62	437.3352	60.38	10.83	3.20	25.59
C22H47N108	453.62	453.3301	58.25	10.44	3.09	28.22
C22H47N109	469.62	469.3251	56.27	10.09	2.98	30.66
C22H47N301	369.64	369.3719	71.49	12.82	11.37	4.33
C22H47N302	385.64	385.3668	68.52	12.28	10.90	8.30
C22H47N303	401.64	401.3617	65.79	11.80	10.46	11.95
C22H47N304	417.64	417.3566	63.27	11.34	10.06	15.32
C22H47N305	433.64	433.3515	60.94	10.92	9.69	18.45
C22H47N306	449.64	449.3465	58.77	10.54	9.35	21.35
C22H47N307	465.64	465.3414	56.75	10.17	9.02	24.05
C22H47N308	481.64	481.3363	54.86	9.84	8.72	26.58
C22H47N309	497.63	497.3312	53.10	9.52	8.44	28.94
C22H47N501	397.65	397.3780	66.45	11.91	17.61	4.02
C22H47N502	413.65	413.3730	63.88	11.45	16.93	7.74
C22H47N503	429.65	429.3679	61.50	11.03	16.30	11.17
C22H47N504	445.65	445.3628	59.29	10.63	15.71	14.36
C22H47N505	461.65	461.3577	57.24	10.26	15.17	17.33
C22H47N506	477.65	477.3526	55.32	9.92	14.66	20.10
C22H47N507	493.65	493.3475	53.53	9.60	14.19	22.69
C22H47N508	509.65	509.3424	51.85	9.30	13.74	25.11
C22H47N509	525.65	525.3374	50.27	9.01	13.32	27.39
C22H48N201	356.64	356.3766	74.09	13.57	7.85	4.49
C22H48N202	372.64	372.3716	70.91	12.98	7.52	8.59
C22H48N203	388.64	388.3665	67.99	12.45	7.21	12.35
C22H48N204	404.64	404.3614	65.30	11.96	6.92	15.82
C22H48N205	420.64	420.3563	62.82	11.50	6.66	19.02
C22H48N206	436.64	436.3512	60.52	11.08	6.42	21.99
C22H48N207	452.64	452.3461	58.38	10.69	6.19	24.74
C22H48N208	468.64	468.3410	56.39	10.32	5.98	27.31
C22H48N209	484.64	484.3360	54.52	9.98	5.78	29.71
C22H48N401	384.65	384.3828	68.70	12.58	14.57	4.16
C22H48N402	400.65	400.3777	65.95	12.08	13.98	7.99
C22H48N403	416.65	416.3726	63.42	11.61	13.45	11.52
C22H48N404	432.65	432.3675	61.08	11.18	12.95	14.79
C22H48N405	448.65	448.3624	58.90	10.78	12.49	17.83
C22H48N406	464.65	464.3574	56.87	10.41	12.06	20.66
C22H48N407	480.65	480.3523	54.98	10.07	11.66	23.30
C22H48N408	496.65	496.3472	53.21	9.74	11.28	25.77
C22H48N409	512.65	512.3421	51.55	9.44	10.93	28.09
C22H48N601	412.67	412.3889	64.03	11.72	20.37	3.88
C22H48N602	428.67	428.3839	61.64	11.29	19.61	7.46
C22H48N603	444.67	444.3788	59.43	10.88	18.90	10.79
C22H48N604	460.67	460.3737	57.36	10.50	18.24	13.89
C22H48N605	476.67	476.3686	55.44	10.15	17.63	16.78
C22H48N606	492.66	492.3635	53.64	9.82	17.06	19.49
C22H48N607	508.66	508.3584	51.95	9.51	16.52	22.02
C22H48N608	524.66	524.3533	50.36	9.22	16.02	24.40
C22H48N609	540.66	540.3483	48.87	8.95	15.54	26.63
C22H49N301	371.66	371.3875	71.10	13.29	11.31	4.30

C22H49N302	387.65	387.3825	68.17	12.74	10.84	8.25
C22H49N303	403.65	403.3774	65.46	12.24	10.41	11.89
C22H49N304	419.65	419.3723	62.97	11.77	10.01	15.25
C22H49N305	435.65	435.3672	60.66	11.34	9.65	18.36
C22H49N306	451.65	451.3621	58.51	10.94	9.30	21.25
C22H49N307	467.65	467.3570	56.50	10.56	8.99	23.95
C22H49N308	483.65	483.3519	54.64	10.21	8.69	26.46
C22H49N309	499.65	499.3469	52.89	9.89	8.41	28.82
C22H49N501	399.67	399.3937	66.12	12.36	17.52	4.00
C22H49N502	415.67	415.3886	63.57	11.88	16.85	7.70
C22H49N503	431.67	431.3835	61.22	11.44	16.22	11.12
C22H49N504	447.67	447.3784	59.03	11.03	15.64	14.30
C22H49N505	463.67	463.3733	56.99	10.65	15.10	17.25
C22H49N506	479.67	479.3683	55.09	10.30	14.60	20.01
C22H49N507	495.67	495.3632	53.31	9.96	14.13	22.60
C22H49N508	511.66	511.3581	51.64	9.65	13.69	25.02
C22H49N509	527.66	527.3530	50.08	9.36	13.27	27.29
C22H50N401	386.67	386.3984	68.34	13.03	14.49	4.14
C22H50N402	402.67	402.3934	65.62	12.52	13.91	7.95
C22H50N403	418.67	418.3883	63.12	12.04	13.38	11.46
C22H50N404	434.67	434.3832	60.79	11.59	12.89	14.72
C22H50N405	450.67	450.3781	58.63	11.18	12.43	17.75
C22H50N406	466.67	466.3730	56.62	10.80	12.01	20.57
C22H50N407	482.67	482.3679	54.75	10.44	11.61	23.20
C22H50N408	498.67	498.3628	52.99	10.11	11.24	25.67
C22H50N409	514.67	514.3578	51.34	9.79	10.89	27.98
C22H50N601	414.68	414.4046	63.72	12.15	20.27	3.86
C22H50N602	430.68	430.3995	61.35	11.70	19.51	7.43
C22H50N603	446.68	446.3944	59.16	11.28	18.81	10.75
C22H50N604	462.68	462.3893	57.11	10.89	18.16	13.83
C22H50N605	478.68	478.3842	55.20	10.53	17.56	16.71
C22H50N606	494.68	494.3792	53.42	10.19	16.99	19.41
C22H50N607	510.68	510.3741	51.74	9.87	16.46	21.93
C22H50N608	526.68	526.3690	50.17	9.57	15.96	24.30
C22H50N609	542.68	542.3639	48.69	9.29	15.49	26.53
C22H51N501	401.68	401.4093	65.78	12.80	17.43	3.98
C22H51N502	417.68	417.4043	63.26	12.31	16.77	7.66
C22H51N503	433.68	433.3992	60.93	11.85	16.15	11.07
C22H51N504	449.68	449.3941	58.76	11.43	15.57	14.23
C22H51N505	465.68	465.3890	56.74	11.04	15.04	17.18
C22H51N506	481.68	481.3839	54.86	10.67	14.54	19.93
C22H51N507	497.68	497.3788	53.10	10.33	14.07	22.50
C22H51N508	513.68	513.3737	51.44	10.01	13.63	24.92
C22H51N509	529.68	529.3687	49.89	9.71	13.22	27.19
C22H52N601	416.70	416.4202	63.41	12.58	20.17	3.84
C22H52N602	432.70	432.4152	61.07	12.11	19.42	7.40
C22H52N603	448.70	448.4101	58.89	11.68	18.73	10.70
C22H52N604	464.70	464.4050	56.86	11.28	18.08	13.77
C22H52N605	480.70	480.3999	54.97	10.90	17.48	16.64
C22H52N606	496.70	496.3948	53.20	10.55	16.92	19.33
C22H52N607	512.70	512.3897	51.54	10.22	16.39	21.84
C22H52N608	528.70	528.3846	49.98	9.91	15.90	24.21
C22H52N609	544.69	544.3795	48.51	9.62	15.43	26.44
C23H11N101	317.35	317.0841	87.05	3.49	4.41	5.04
C23H11N102	333.35	333.0790	82.87	3.33	4.20	9.60
C23H11N103	349.35	349.0739	79.08	3.17	4.01	13.74
C23H11N104	365.35	365.0688	75.61	3.03	3.83	17.52
C23H11N105	381.35	381.0637	72.44	2.91	3.67	20.98
C23H11N106	397.35	397.0586	69.53	2.79	3.53	24.16
C23H11N107	413.35	413.0535	66.83	2.68	3.39	27.09
C23H11N108	429.35	429.0485	64.34	2.58	3.26	29.81
C23H11N109	445.35	445.0434	62.03	2.49	3.15	32.33
C23H11N301	345.36	345.0902	79.99	3.21	12.17	4.63
C23H11N302	361.36	361.0851	76.45	3.07	11.63	8.86
C23H11N303	377.36	377.0800	73.21	2.94	11.14	12.72
C23H11N304	393.36	393.0749	70.23	2.82	10.68	16.27
C23H11N305	409.36	409.0699	67.48	2.71	10.26	19.54
C23H11N306	425.36	425.0648	64.95	2.61	9.08	22.57
C23H11N307	441.36	441.0597	62.59	2.51	9.52	25.38
C23H11N308	457.36	457.0546	60.40	2.42	9.19	27.99
C23H11N309	473.36	473.0495	58.36	2.34	8.88	30.42
C23H11N501	373.38	373.0964	73.99	2.97	18.76	4.29
C23H11N502	389.38	389.0913	70.95	2.85	17.99	8.22
C23H11N503	405.38	405.0862	68.15	2.74	17.28	11.84
C23H11N504	421.38	421.0811	65.56	2.63	16.62	15.19
C23H11N505	437.37	437.0760	63.16	2.54	16.01	18.29
C23H11N506	453.37	453.0709	60.93	2.45	15.45	21.17
C23H11N507	469.37	469.0658	58.86	2.36	14.92	23.86
C23H11N508	485.37	485.0608	56.92	2.28	14.43	26.37
C23H11N509	501.37	501.0557	55.10	2.21	13.97	28.72
C23H12N201	332.36	332.0950	83.12	3.64	8.43	4.81
C23H12N202	348.36	348.0899	79.30	3.47	8.04	9.19
C23H12N203	364.36	364.0848	75.82	3.32	7.69	13.17
C23H12N204	380.36	380.0797	72.63	3.18	7.36	16.83
C23H12N205	396.36	396.0746	69.70	3.05	7.07	20.18
C23H12N206	412.36	412.0695	66.99	2.93	6.79	23.28
C23H12N207	428.36	428.0644	64.49	2.82	6.54	26.15
C23H12N208	444.36	444.0594	62.17	2.72	6.30	28.80
C23H12N209	460.36	460.0543	60.01	2.63	6.09	31.28
C23H12N401	360.38	360.1011	76.66	3.36	15.55	4.44
C23H12N402	376.38	376.0960	73.40	3.21	14.89	8.50
C23H12N403	392.38	392.0909	70.41	3.08	14.28	12.23
C23H12N404	408.38	408.0858	67.65	2.96	13.72	15.67
C23H12N405	424.38	424.0808	65.10	2.85	13.20	18.85
C23H12N406	440.38	440.0757	62.73	2.75	12.72	21.80
C23H12N407	456.37	456.0706	60.53	2.65	12.28	24.54
C23H12N408	472.37	472.0655	58.48	2.56	11.86	27.10
C23H12N409	488.37	488.0604	56.57	2.48	11.47	29.48
C23H12N601	388.39	388.1073	71.13	3.11	21.64	4.12
C23H12N602	404.39	404.1022	68.31	2.99	20.78	7.91
C23H12N603	420.39	420.0971	65.71	2.88	19.99	11.42
C23H12N604	436.39	436.0920	63.30	2.77	19.26	14.67
C23H12N605	452.39	452.0869	61.07	2.67	18.58	17.68
C23H12N606	468.39	468.0818	58.98	2.58	17.94	20.50
C23H12N607	484.39	484.0767	57.03	2.50	17.35	23.12
C23H12N608	500.39	500.0717	55.21	2.42	16.80	25.58
C23H12N609	516.39	516.0666	53.50	2.34	16.27	27.89
C23H13N101	319.37	319.0997	86.50	4.10	4.39	5.01
C23H13N102	335.37	335.0946	82.37	3.91	4.18	9.54
C23H13N103	351.36	351.0895	78.62	3.73	3.99	13.66
C23H13N104	367.36	367.0844	75.20	3.57	3.81	17.42
C23H13N105	383.36	383.0794	72.06	3.42	3.65	20.87
C23H13N106	399.36	399.0743	69.17	3.28	3.51	24.04
C23H13N107	415.36	415.0692	66.51	3.15	3.37	26.96
C23H13N108	431.36	431.0641	64.04	3.04	3.25	29.67
C23H13N109	447.36	447.0590	61.75	2.93	3.13	32.19
C23H13N301	347.38	347.1059	79.53	3.77	12.10	4.61
C23H13N302	363.38	363.1008	76.02	3.61	11.56	8.81
C23H13N303	379.38	379.0957	72.82	3.45	11.08	12.65
C23H13N304	395.38	395.0906	69.87	3.31	10.63	16.19

138

C23H13N305	411.38	411.0855	67.15	3.19	10.21	19.45
C23H13N306	427.38	427.0804	64.64	3.07	9.83	22.46
C23H13N307	443.38	443.0753	62.31	2.96	9.48	25.26
C23H13N308	459.38	459.0703	60.14	2.85	9.15	27.86
C23H13N309	475.37	475.0652	58.11	2.76	8.84	30.29
C23H13N501	375.39	375.1120	73.59	3.49	18.66	4.26
C23H13N502	391.39	391.1069	70.58	3.35	17.89	8.18
C23H13N503	407.39	407.1018	67.81	3.22	17.19	11.78
C23H13N504	423.39	423.0967	65.25	3.09	16.54	15.12
C23H13N505	439.39	439.0917	62.87	2.98	15.94	18.21
C23H13N506	455.39	455.0866	60.66	2.88	15.38	21.08
C23H13N507	471.39	471.0815	58.60	2.78	14.86	23.76
C23H13N508	487.39	487.0764	56.68	2.69	14.37	26.26
C23H13N509	503.39	503.0713	54.88	2.60	13.91	28.61
C23H14N201	334.38	334.1106	82.62	4.22	8.38	4.78
C23H14N202	350.38	350.1055	78.84	4.03	8.00	9.13
C23H14N203	366.38	366.1004	75.40	3.85	7.65	13.10
C23H14N204	382.38	382.0953	72.25	3.69	7.33	16.74
C23H14N205	398.38	398.0903	69.35	3.54	7.03	20.08
C23H14N206	414.38	414.0852	66.67	3.41	6.76	23.17
C23H14N207	430.38	430.0801	64.19	3.28	6.51	26.02
C23H14N208	446.38	446.0750	61.89	3.16	6.28	28.67
C23H14N209	462.38	462.0699	59.75	3.05	6.06	31.14
C23H14N401	362.39	362.1168	76.23	3.89	15.46	4.41
C23H14N402	378.39	378.1117	73.01	3.73	14.81	8.46
C23H14N403	394.39	394.1066	70.05	3.58	14.21	12.17
C23H14N404	410.39	410.1015	67.32	3.44	13.65	15.59
C23H14N405	426.39	426.0964	64.79	3.31	13.14	18.76
C23H14N406	442.39	442.0913	62.45	3.19	12.66	21.70
C23H14N407	458.39	458.0862	60.27	3.08	12.22	24.43
C23H14N408	474.39	474.0812	58.23	2.97	11.81	26.98
C23H14N409	490.39	490.0761	56.33	2.88	11.42	29.36
C23H14N601	390.41	390.1229	70.76	3.61	21.53	4.10
C23H14N602	406.41	406.1178	67.98	3.47	20.68	7.87
C23H14N603	422.41	422.1127	65.40	3.34	19.90	11.36
C23H14N604	438.41	438.1076	63.01	3.22	19.17	14.60
C23H14N605	454.41	454.1026	60.80	3.11	18.49	17.60
C23H14N606	470.40	470.0975	58.73	3.00	17.87	20.41
C23H14N607	486.40	486.0924	56.80	2.90	17.28	23.03
C23H14N608	502.40	502.0873	54.99	2.81	16.73	25.48
C23H14N609	518.40	518.0822	53.29	2.72	16.21	27.78
C23H15N101	321.38	321.1154	85.96	4.70	4.36	4.98
C23H15N102	337.38	337.1103	81.88	4.48	4.15	9.48
C23H15N103	353.38	353.1052	78.18	4.28	3.96	13.58
C23H15N104	369.38	369.1001	74.79	4.09	3.79	17.33
C23H15N105	385.38	385.0950	71.68	3.92	3.63	20.76
C23H15N106	401.38	401.0899	68.83	3.77	3.49	23.92
C23H15N107	417.38	417.0848	66.19	3.62	3.36	26.83
C23H15N108	433.38	433.0798	63.74	3.49	3.23	29.53
C23H15N109	449.38	449.0747	61.48	3.36	3.12	32.04
C23H15N301	349.40	349.1215	79.07	4.33	12.03	4.58
C23H15N302	365.39	365.1164	75.60	4.14	11.50	8.76
C23H15N303	381.39	381.1113	72.43	3.96	11.02	12.58
C23H15N304	397.39	397.1062	69.52	3.80	10.57	16.10
C23H15N305	413.39	413.1012	66.83	3.66	10.16	19.35
C23H15N306	429.39	429.0961	64.34	3.52	9.79	22.36
C23H15N307	445.39	445.0910	62.03	3.39	9.43	25.15
C23H15N308	461.39	461.0859	59.87	3.28	9.11	27.74
C23H15N309	477.39	477.0808	57.87	3.17	8.80	30.16
C23H15N501	377.41	377.1277	73.20	4.01	18.56	4.24
C23H15N502	393.41	393.1226	70.22	3.84	17.80	8.13
C23H15N503	409.41	409.1175	67.48	3.69	17.11	11.72
C23H15N504	425.41	425.1124	64.94	3.55	16.46	15.04
C23H15N505	441.41	441.1073	62.59	3.43	15.87	18.12
C23H15N506	457.41	457.1022	60.40	3.31	15.31	20.99
C23H15N507	473.41	473.0971	58.36	3.19	14.79	23.66
C23H15N508	489.40	489.0921	56.45	3.09	14.31	26.15
C23H15N509	505.40	505.0870	54.66	2.99	13.86	28.49
C23H16N201	336.40	336.1263	82.12	4.79	8.33	4.76
C23H16N202	352.40	352.1212	78.39	4.58	7.95	9.08
C23H16N203	368.40	368.1161	74.99	4.38	7.60	13.03
C23H16N204	384.39	384.1110	71.87	4.20	7.29	16.65
C23H16N205	400.39	400.1059	69.00	4.03	7.00	19.98
C23H16N206	416.39	416.1008	66.35	3.87	6.73	23.05
C23H16N207	432.39	432.0957	63.89	3.73	6.48	25.90
C23H16N208	448.39	448.0907	61.61	3.60	6.25	28.55
C23H16N209	464.39	464.0856	59.49	3.47	6.03	31.01
C23H16N401	364.41	364.1324	75.81	4.43	15.37	4.39
C23H16N402	380.41	380.1273	72.62	4.24	14.73	8.41
C23H16N403	396.41	396.1222	69.69	4.07	14.13	12.11
C23H16N404	412.41	412.1171	66.99	3.91	13.59	15.52
C23H16N405	428.41	428.1121	64.48	3.76	13.08	18.67
C23H16N406	444.41	444.1070	62.16	3.63	12.61	21.60
C23H16N407	460.41	460.1019	60.00	3.50	12.17	24.33
C23H16N408	476.41	476.0968	57.99	3.39	11.76	26.87
C23H16N409	492.41	492.0917	56.10	3.28	11.38	29.24
C23H16N601	392.42	392.1386	70.40	4.11	21.42	4.08
C23H16N602	408.42	408.1335	67.64	3.95	20.58	7.83
C23H16N603	424.42	424.1284	65.09	3.80	19.80	11.31
C23H16N604	440.42	440.1233	62.73	3.66	19.08	14.53
C23H16N605	456.42	456.1182	60.53	3.53	18.41	17.53
C23H16N606	472.42	472.1131	58.48	3.41	17.79	20.32
C23H16N607	488.42	488.1080	56.56	3.30	17.21	22.93
C23H16N608	504.42	504.1029	54.77	3.20	16.66	25.37
C23H16N609	520.42	520.0979	53.08	3.10	16.15	27.67
C23H17N101	323.40	323.1310	85.42	5.30	4.33	4.95
C23H17N102	339.40	339.1259	81.40	5.05	4.13	9.43
C23H17N103	355.40	355.1208	77.73	4.82	3.94	13.51
C23H17N104	371.40	371.1157	74.38	4.61	3.77	17.23
C23H17N105	387.40	387.1107	71.31	4.42	3.62	20.65
C23H17N106	403.40	403.1056	68.48	4.25	3.47	23.80
C23H17N107	419.39	419.1005	65.87	4.09	3.34	26.70
C23H17N108	435.39	435.0954	63.45	3.94	3.22	29.40
C23H17N109	451.39	451.0903	61.20	3.80	3.10	31.90
C23H17N301	351.41	351.1372	78.61	4.88	11.96	4.55
C23H17N302	367.41	367.1321	75.19	4.66	11.44	8.71
C23H17N303	383.41	383.1270	72.05	4.47	10.96	12.52
C23H17N304	399.41	399.1219	69.17	4.29	10.52	16.02
C23H17N305	415.41	415.1168	66.50	4.12	10.12	19.26
C23H17N306	431.41	431.1117	64.04	3.97	9.74	22.25
C23H17N307	447.41	447.1066	61.75	3.83	9.39	25.03
C23H17N308	463.41	463.1016	59.61	3.70	9.07	27.62
C23H17N309	479.41	479.0965	57.62	3.57	8.77	30.04
C23H17N501	379.42	379.1433	72.81	4.52	18.46	4.22
C23H17N502	395.42	395.1382	69.86	4.33	17.71	8.09
C23H17N503	411.42	411.1331	67.15	4.16	17.02	11.67
C23H17N504	427.42	427.1280	64.63	4.01	16.39	14.97
C23H17N505	443.42	443.1230	62.30	3.86	15.79	18.04
C23H17N506	459.42	459.1179	60.13	3.73	15.24	20.90
C23H17N507	475.42	475.1128	58.11	3.60	14.73	23.56

139

C23H17N508	491.42	491.1077	56.22	3.49	14.25	26.05
C23H17N509	507.42	507.1026	54.44	3.38	13.80	28.38
C23H18N201	338.41	338.1419	81.63	5.36	8.28	4.73
C23H18N202	354.41	354.1368	77.95	5.12	7.90	9.03
C23H18N203	370.41	370.1317	74.58	4.90	7.56	12.96
C23H18N204	386.41	386.1266	71.49	4.70	7.25	16.56
C23H18N205	402.41	402.1216	68.65	4.51	6.96	19.88
C23H18N206	418.41	418.1165	66.03	4.34	6.70	22.94
C23H18N207	434.41	434.1114	63.59	4.18	6.45	25.78
C23H18N208	450.41	450.1063	61.33	4.03	6.22	28.42
C23H18N209	466.41	466.1012	59.23	3.89	6.01	30.87
C23H18N401	366.43	366.1481	75.39	4.95	15.29	4.37
C23H18N402	382.43	382.1430	72.24	4.74	14.65	8.37
C23H18N403	398.42	398.1379	69.34	4.55	14.06	12.05
C23H18N404	414.42	414.1328	66.66	4.38	13.52	15.44
C23H18N405	430.42	430.1277	64.18	4.22	13.02	18.59
C23H18N406	446.42	446.1226	61.88	4.06	12.55	21.50
C23H18N407	462.42	462.1175	59.74	3.92	12.12	24.22
C23H18N408	478.42	478.1125	57.74	3.79	11.71	26.75
C23H18N409	494.42	494.1074	55.87	3.67	11.33	29.12
C23H18N601	394.44	394.1542	70.04	4.60	21.31	4.06
C23H18N602	410.44	410.1491	67.31	4.42	20.48	7.80
C23H18N603	426.44	426.1440	64.78	4.25	19.71	11.26
C23H18N604	442.44	442.1389	62.44	4.10	18.99	14.46
C23H18N605	458.44	458.1339	60.26	3.96	18.33	17.45
C23H18N606	474.44	474.1288	58.23	3.82	17.71	20.23
C23H18N607	490.44	490.1237	56.33	3.70	17.14	22.84
C23H18N608	506.44	506.1186	54.55	3.58	16.59	25.27
C23H18N609	522.43	522.1135	52.88	3.47	16.09	27.56
C23H19N101	325.41	325.1467	84.89	5.89	4.30	4.92
C23H19N102	341.41	341.1416	80.92	5.61	4.10	9.37
C23H19N103	357.41	357.1365	77.29	5.36	3.92	13.43
C23H19N104	373.41	373.1314	73.98	5.13	3.75	17.14
C23H19N105	389.41	389.1263	70.94	4.92	3.60	20.54
C23H19N106	405.41	405.1212	68.14	4.72	3.45	23.68
C23H19N107	421.41	421.1161	65.56	4.54	3.32	26.58
C23H19N108	437.41	437.1111	63.16	4.38	3.20	29.26
C23H19N109	453.41	453.1060	60.93	4.22	3.09	31.76
C23H19N301	353.43	353.1528	78.16	5.42	11.89	4.53
C23H19N302	369.43	369.1477	74.78	5.18	11.37	8.66
C23H19N303	385.43	385.1426	71.68	4.97	10.90	12.45
C23H19N304	401.43	401.1375	68.82	4.77	10.47	15.94
C23H19N305	417.42	417.1325	66.18	4.59	10.07	19.16
C23H19N306	433.42	433.1274	63.74	4.42	9.69	22.15
C23H19N307	449.42	449.1223	61.47	4.26	9.35	24.92
C23H19N308	465.42	465.1172	59.36	4.11	9.03	27.50
C23H19N309	481.42	481.1121	57.38	3.98	8.73	29.91
C23H19N501	381.44	381.1590	72.42	5.02	18.36	4.19
C23H19N502	397.44	397.1539	69.51	4.82	17.62	8.05
C23H19N503	413.44	413.1488	66.82	4.63	16.94	11.61
C23H19N504	429.44	429.1437	64.33	4.46	16.31	14.90
C23H19N505	445.44	445.1386	62.02	4.30	15.72	17.96
C23H19N506	461.44	461.1335	59.87	4.15	15.18	20.80
C23H19N507	477.44	477.1284	57.86	4.01	14.67	23.46
C23H19N508	493.44	493.1233	55.99	3.88	14.19	25.94
C23H19N509	509.44	509.1183	54.23	3.76	13.75	28.27
C23H20N201	340.43	340.1576	81.15	5.92	8.23	4.70
C23H20N202	356.43	356.1525	77.51	5.66	7.86	8.98
C23H20N203	372.43	372.1474	74.18	5.41	7.52	12.89
C23H20N204	388.43	388.1423	71.12	5.19	7.21	16.48
C23H20N205	404.43	404.1372	68.31	4.98	6.93	19.78
C23H20N206	420.43	420.1321	65.71	4.79	6.66	22.83
C23H20N207	436.43	436.1270	63.30	4.62	6.42	25.66
C23H20N208	452.42	452.1220	61.06	4.46	6.19	28.29
C23H20N209	468.42	468.1169	58.98	4.30	5.98	30.74
C23H20N401	368.44	368.1637	74.98	5.47	15.21	4.34
C23H20N402	384.44	384.1586	71.86	5.24	14.57	8.32
C23H20N403	400.44	400.1535	68.99	5.03	13.99	11.99
C23H20N404	416.44	416.1484	66.34	4.84	13.45	15.37
C23H20N405	432.44	432.1434	63.88	4.66	12.96	18.50
C23H20N406	448.44	448.1383	61.60	4.50	12.49	21.41
C23H20N407	464.44	464.1332	59.48	4.34	12.06	24.11
C23H20N408	480.44	480.1281	57.50	4.20	11.66	26.64
C23H20N409	496.44	496.1230	55.65	4.06	11.29	29.01
C23H20N601	396.46	396.1698	69.68	5.08	21.20	4.04
C23H20N602	412.45	412.1648	66.98	4.89	20.38	7.76
C23H20N603	428.45	428.1597	64.48	4.71	19.61	11.20
C23H20N604	444.45	444.1546	62.16	4.54	18.91	14.40
C23H20N605	460.45	460.1495	60.00	4.38	18.25	17.37
C23H20N606	476.45	476.1444	57.98	4.23	17.64	20.15
C23H20N607	492.45	492.1393	56.10	4.09	17.07	22.74
C23H20N608	508.45	508.1342	54.33	3.96	16.53	25.17
C23H20N609	524.45	524.1292	52.68	3.84	16.02	27.46
C23H21N101	327.43	327.1623	84.37	6.46	4.28	4.89
C23H21N102	343.43	343.1572	80.44	6.16	4.08	9.32
C23H21N103	359.43	359.1521	76.86	5.89	3.90	13.35
C23H21N104	375.43	375.1470	73.58	5.64	3.73	17.05
C23H21N105	391.43	391.1420	70.58	5.41	3.58	20.44
C23H21N106	407.43	407.1369	67.81	5.20	3.44	23.56
C23H21N107	423.43	423.1318	65.24	5.00	3.31	26.45
C23H21N108	439.43	439.1267	62.87	4.82	3.19	29.13
C23H21N109	455.43	455.1216	60.66	4.65	3.08	31.62
C23H21N301	355.44	355.1685	77.72	5.96	11.82	4.50
C23H21N302	371.44	371.1634	74.37	5.70	11.31	8.61
C23H21N303	387.44	387.1583	71.30	5.46	10.85	12.39
C23H21N304	403.44	403.1532	68.47	5.25	10.42	15.86
C23H21N305	419.44	419.1481	65.86	5.05	10.02	19.07
C23H21N306	435.44	435.1430	63.44	4.86	9.65	22.05
C23H21N307	451.44	451.1379	61.19	4.69	9.31	24.81
C23H21N308	467.44	467.1329	59.10	4.53	8.99	27.38
C23H21N309	483.44	483.1278	57.14	4.38	8.69	29.79
C23H21N501	383.46	383.1746	72.04	5.52	18.26	4.17
C23H21N502	399.46	399.1695	69.16	5.30	17.53	8.01
C23H21N503	415.46	415.1644	66.49	5.09	16.86	11.55
C23H21N504	431.45	431.1593	64.03	4.91	16.23	14.83
C23H21N505	447.45	447.1543	61.74	4.73	15.65	17.88
C23H21N506	463.45	463.1492	59.61	4.57	15.11	20.71
C23H21N507	479.45	479.1441	57.62	4.41	14.61	23.36
C23H21N508	495.45	495.1390	55.76	4.27	14.14	25.83
C23H21N509	511.45	511.1339	54.01	4.14	13.69	28.15
C23H22N201	342.44	342.1732	80.67	6.48	8.18	4.67
C23H22N202	358.44	358.1681	77.07	6.19	7.82	8.93
C23H22N203	374.44	374.1630	73.78	5.92	7.48	12.82
C23H22N204	390.44	390.1579	70.75	5.68	7.17	16.39
C23H22N205	406.44	406.1529	67.97	5.46	6.89	19.68
C23H22N206	422.44	422.1478	65.40	5.25	6.63	22.72
C23H22N207	438.44	438.1427	63.01	5.06	6.39	25.54
C23H22N208	454.44	454.1376	60.79	4.88	6.16	28.17
C23H22N209	470.44	470.1325	58.72	4.71	5.95	30.61
C23H22N401	370.46	370.1794	74.57	5.99	15.12	4.32

C23H22N4O2	386.46	386.1743	71.48	5.74	14.50	8.28
C23H22N4O3	402.46	402.1692	68.64	5.51	13.92	11.93
C23H22N4O4	418.46	418.1641	66.02	5.30	13.39	15.29
C23H22N4O5	434.46	434.1590	63.59	5.10	12.90	18.41
C23H22N4O6	450.45	450.1539	61.33	4.92	12.44	21.31
C23H22N4O7	466.45	466.1488	59.22	4.75	12.01	24.01
C23H22N4O8	482.45	482.1437	57.26	4.60	11.61	26.53
C23H22N4O9	498.45	498.1387	55.42	4.45	11.24	28.89
C23H22N6O1	398.47	398.1855	69.33	5.57	21.09	4.02
C23H22N6O2	414.47	414.1804	66.65	5.35	20.28	7.72
C23H22N6O3	430.47	430.1753	64.18	5.15	19.52	11.15
C23H22N6O4	446.47	446.1702	61.88	4.97	18.82	14.33
C23H22N6O5	462.47	462.1652	59.74	4.79	18.17	17.30
C23H22N6O6	478.47	478.1601	57.74	4.63	17.56	20.06
C23H22N6O7	494.47	494.1550	55.87	4.48	17.00	22.65
C23H22N6O8	510.47	510.1499	54.12	4.34	16.46	25.07
C23H22N6O9	526.47	526.1448	52.47	4.21	15.96	27.35
C23H23N1O1	329.45	329.1780	83.85	7.04	4.25	4.86
C23H23N1O2	345.45	345.1729	79.97	6.71	4.05	9.26
C23H23N1O3	361.44	361.1678	76.43	6.41	3.88	13.28
C23H23N1O4	377.44	377.1627	73.19	6.14	3.71	16.96
C23H23N1O5	393.44	393.1576	70.22	5.89	3.56	20.33
C23H23N1O6	409.44	409.1525	67.47	5.66	3.42	23.45
C23H23N1O7	425.44	425.1474	64.93	5.45	3.29	26.32
C23H23N1O8	441.44	441.1424	62.58	5.25	3.17	28.99
C23H23N1O9	457.44	457.1373	60.39	5.07	3.06	31.48
C23H23N3O1	357.46	357.1841	77.28	6.49	11.76	4.48
C23H23N3O2	373.46	373.1790	73.97	6.21	11.25	8.57
C23H23N3O3	389.46	389.1739	70.93	5.95	10.79	12.32
C23H23N3O4	405.46	405.1688	68.13	5.72	10.36	15.78
C23H23N3O5	421.46	421.1638	65.55	5.50	9.97	18.98
C23H23N3O6	437.46	437.1587	63.15	5.30	9.61	21.94
C23H23N3O7	453.46	453.1536	60.92	5.11	9.27	24.70
C23H23N3O8	469.46	469.1485	58.85	4.94	8.95	27.26
C23H23N3O9	485.45	485.1434	56.91	4.78	8.66	29.66
C23H23N5O1	385.47	385.1902	71.67	6.01	18.17	4.15
C23H23N5O2	401.47	401.1852	68.81	5.77	17.44	7.97
C23H23N5O3	417.47	417.1801	66.17	5.55	16.78	11.50
C23H23N5O4	433.47	433.1750	63.73	5.35	16.16	14.76
C23H23N5O5	449.47	449.1699	61.46	5.16	15.58	17.80
C23H23N5O6	465.47	465.1648	59.35	4.98	15.05	20.62
C23H23N5O7	481.47	481.1597	57.38	4.82	14.55	23.26
C23H23N5O8	497.47	497.1546	55.53	4.66	14.08	25.73
C23H23N5O9	513.47	513.1496	53.80	4.52	13.64	28.04
C23H24N2O1	344.46	344.1889	80.20	7.02	8.13	4.64
C23H24N2O2	360.46	360.1838	76.64	6.71	7.77	8.88
C23H24N2O3	376.46	376.1787	73.38	6.43	7.44	12.75
C23H24N2O4	392.46	392.1736	70.39	6.16	7.14	16.31
C23H24N2O5	408.46	408.1685	67.63	5.92	6.86	19.59
C23H24N2O6	424.46	424.1634	65.08	5.70	6.60	22.62
C23H24N2O7	440.46	440.1583	62.72	5.49	6.36	25.43
C23H24N2O8	456.46	456.1533	60.52	5.30	6.14	28.04
C23H24N2O9	472.46	472.1482	58.47	5.12	5.93	30.48
C23H24N4O1	372.47	372.1950	74.17	6.49	15.04	4.30
C23H24N4O2	388.47	388.1899	71.11	6.23	14.42	8.24
C23H24N4O3	404.47	404.1848	68.30	5.98	13.85	11.87
C23H24N4O4	420.47	420.1797	65.70	5.75	13.32	15.22
C23H24N4O5	436.47	436.1747	63.29	5.54	12.84	18.33
C23H24N4O6	452.47	452.1696	61.06	5.35	12.38	21.22
C23H24N4O7	468.47	468.1645	58.97	5.16	11.96	23.91

C23H24N4O8	484.47	484.1594	57.02	4.99	11.56	26.42
C23H24N4O9	500.47	500.1543	55.20	4.83	11.19	28.77
C23H24N6O1	400.49	400.2011	68.98	6.04	20.98	3.99
C23H24N6O2	416.49	416.1961	66.33	5.81	20.18	7.68
C23H24N6O3	432.49	432.1910	63.88	5.59	19.43	11.10
C23H24N6O4	448.49	448.1859	61.60	5.39	18.74	14.27
C23H24N6O5	464.48	464.1808	59.48	5.21	18.09	17.22
C23H24N6O6	480.48	480.1757	57.50	5.03	17.49	19.98
C23H24N6O7	496.48	496.1706	55.64	4.87	16.93	22.56
C23H24N6O8	512.48	512.1655	53.91	4.72	16.40	24.98
C23H24N6O9	528.48	528.1605	52.27	4.58	15.90	27.25
C23H25N1O1	331.46	331.1936	83.34	7.60	4.23	4.83
C23H25N1O2	347.46	347.1885	79.51	7.25	4.03	9.21
C23H25N1O3	363.46	363.1834	76.01	6.93	3.85	13.21
C23H25N1O4	379.46	379.1783	72.80	6.64	3.69	16.87
C23H25N1O5	395.46	395.1733	69.86	6.37	3.54	20.23
C23H25N1O6	411.46	411.1682	67.14	6.12	3.40	23.33
C23H25N1O7	427.46	427.1631	64.63	5.90	3.28	26.20
C23H25N1O8	443.46	443.1580	62.30	5.68	3.16	28.86
C23H25N1O9	459.46	459.1529	60.13	5.48	3.05	31.34
C23H25N3O1	359.48	359.1998	76.85	7.01	11.69	4.45
C23H25N3O2	375.47	375.1947	73.58	6.71	11.19	8.52
C23H25N3O3	391.47	391.1896	70.57	6.44	10.73	12.26
C23H25N3O4	407.47	407.1845	67.80	6.18	10.31	15.71
C23H25N3O5	423.47	423.1794	65.24	5.95	9.92	18.89
C23H25N3O6	439.47	439.1743	62.86	5.73	9.56	21.84
C23H25N3O7	455.47	455.1692	60.65	5.53	9.23	24.59
C23H25N3O8	471.47	471.1641	58.59	5.34	8.91	27.15
C23H25N3O9	487.47	487.1591	56.67	5.17	8.62	29.54
C23H25N5O1	387.49	387.2059	71.29	6.50	18.07	4.13
C23H25N5O2	403.49	403.2008	68.47	6.25	17.36	7.93
C23H25N5O3	419.49	419.1957	65.86	6.01	16.70	11.44
C23H25N5O4	435.49	435.1906	63.44	5.79	16.08	14.70
C23H25N5O5	451.49	451.1856	61.19	5.58	15.51	17.72
C23H25N5O6	467.49	467.1805	59.09	5.39	14.98	20.53
C23H25N5O7	483.48	483.1754	57.14	5.21	14.49	23.16
C23H25N5O8	499.48	499.1703	55.31	5.05	14.02	25.63
C23H25N5O9	515.48	515.1652	53.59	4.89	13.59	27.93
C23H26N2O1	346.48	346.2045	79.73	7.56	8.09	4.62
C23H26N2O2	362.48	362.1994	76.21	7.23	7.73	8.83
C23H26N2O3	378.48	378.1943	72.99	6.92	7.40	12.68
C23H26N2O4	394.47	394.1892	70.03	6.64	7.10	16.22
C23H26N2O5	410.47	410.1842	67.30	6.38	6.82	19.49
C23H26N2O6	426.47	426.1791	64.78	6.15	6.57	22.51
C23H26N2O7	442.47	442.1740	62.43	5.92	6.33	25.31
C23H26N2O8	458.47	458.1689	60.26	5.72	6.11	27.92
C23H26N2O9	474.47	474.1638	58.22	5.52	5.90	30.35
C23H26N4O1	374.49	374.2106	73.77	7.00	14.96	4.27
C23H26N4O2	390.49	390.2056	70.75	6.71	14.35	8.19
C23H26N4O3	406.49	406.2005	67.96	6.45	13.78	11.81
C23H26N4O4	422.49	422.1954	65.39	6.20	13.26	15.15
C23H26N4O5	438.49	438.1903	63.00	5.98	12.78	18.24
C23H26N4O6	454.49	454.1852	60.78	5.77	12.33	21.12
C23H26N4O7	470.49	470.1801	58.72	5.57	11.91	23.80
C23H26N4O8	486.49	486.1750	56.79	5.39	11.52	26.31
C23H26N4O9	502.49	502.1700	54.98	5.22	11.15	28.66
C23H26N6O1	402.50	402.2168	68.63	6.51	20.88	3.97
C23H26N6O2	418.50	418.2117	66.01	6.26	20.08	7.65
C23H26N6O3	434.50	434.2066	63.58	6.03	19.34	11.05
C23H26N6O4	450.50	450.2015	61.32	5.82	18.65	14.21

C23H26N6O5	466.50	466.1965	59.22	5.62	18.02	17.15
C23H26N6O6	482.50	482.1914	57.26	5.43	17.42	19.90
C23H26N6O7	498.50	498.1863	55.42	5.26	16.86	22.47
C23H26N6O8	514.50	514.1812	53.69	5.09	16.33	24.88
C23H26N6O9	530.50	530.1761	52.07	4.94	15.84	27.14
C23H27N1O1	333.48	333.2093	82.84	8.16	4.20	4.80
C23H27N1O2	349.48	349.2042	79.05	7.79	4.01	9.16
C23H27N1O3	365.48	365.1991	75.59	7.45	3.83	13.13
C23H27N1O4	381.48	381.1940	72.42	7.13	3.67	16.78
C23H27N1O5	397.48	397.1889	69.50	6.85	3.52	20.13
C23H27N1O6	413.47	413.1838	66.81	6.58	3.39	23.22
C23H27N1O7	429.47	429.1787	64.32	6.34	3.26	26.08
C23H27N1O8	445.47	445.1737	62.01	6.11	3.14	28.73
C23H27N1O9	461.47	461.1686	59.86	5.90	3.04	31.20
C23H27N3O1	361.49	361.2154	76.42	7.53	11.62	4.43
C23H27N3O2	377.49	377.2103	73.18	7.21	11.13	8.48
C23H27N3O3	393.49	393.2052	70.21	6.92	10.68	12.20
C23H27N3O4	409.49	409.2001	67.46	6.65	10.26	15.63
C23H27N3O5	425.49	425.1951	64.93	6.40	9.88	18.80
C23H27N3O6	441.49	441.1900	62.57	6.16	9.52	21.74
C23H27N3O7	457.49	457.1849	60.39	5.95	9.18	24.48
C23H27N3O8	473.49	473.1798	58.35	5.75	8.87	27.03
C23H27N3O9	489.49	489.1747	56.44	5.56	8.58	29.42
C23H27N5O1	389.50	389.2215	70.93	6.99	17.98	4.11
C23H27N5O2	405.50	405.2165	68.13	6.71	17.27	7.89
C23H27N5O3	421.50	421.2114	65.54	6.46	16.62	11.39
C23H27N5O4	437.50	437.2063	63.14	6.22	16.01	14.63
C23H27N5O5	453.50	453.2012	60.92	6.00	15.44	17.64
C23H27N5O6	469.50	469.1961	58.84	5.80	14.92	20.45
C23H27N5O7	485.50	485.1910	56.90	5.61	14.42	23.07
C23H27N5O8	501.50	501.1859	55.09	5.43	13.96	25.52
C23H27N5O9	517.50	517.1809	53.38	5.26	13.53	27.83
C23H28N2O1	348.49	348.2202	79.27	8.10	8.04	4.59
C23H28N2O2	364.49	364.2151	75.79	7.74	7.69	8.78
C23H28N2O3	380.49	380.2100	72.61	7.42	7.36	12.61
C23H28N2O4	396.49	396.2049	69.68	7.12	7.07	16.14
C23H28N2O5	412.49	412.1998	66.97	6.84	6.79	19.39
C23H28N2O6	428.49	428.1947	64.47	6.59	6.54	22.40
C23H28N2O7	444.49	444.1896	62.15	6.35	6.30	25.20
C23H28N2O8	460.49	460.1845	59.99	6.13	6.08	27.80
C23H28N2O9	476.49	476.1795	57.98	5.92	5.88	30.22
C23H28N4O1	376.51	376.2263	73.37	7.50	14.88	4.25
C23H28N4O2	392.51	392.2212	70.38	7.19	14.27	8.15
C23H28N4O3	408.50	408.2161	67.63	6.91	13.72	11.75
C23H28N4O4	424.50	424.2110	65.08	6.65	13.20	15.08
C23H28N4O5	440.50	440.2060	62.71	6.41	12.72	18.16
C23H28N4O6	456.50	456.2009	60.52	6.18	12.27	21.03
C23H28N4O7	472.50	472.1958	58.47	5.97	11.86	23.70
C23H28N4O8	488.50	488.1907	56.55	5.78	11.47	26.20
C23H28N4O9	504.50	504.1856	54.76	5.59	11.11	28.54
C23H28N6O1	404.52	404.2324	68.29	6.98	20.78	3.96
C23H28N6O2	420.52	420.2274	65.69	6.71	19.98	7.61
C23H28N6O3	436.52	436.2223	63.29	6.47	19.25	11.00
C23H28N6O4	452.52	452.2172	61.05	6.24	18.57	14.14
C23H28N6O5	468.52	468.2121	58.96	6.02	17.94	17.07
C23H28N6O6	484.52	484.2070	57.02	5.83	17.35	19.81
C23H28N6O7	500.52	500.2019	55.19	5.64	16.79	22.38
C23H28N6O8	516.52	516.1968	53.48	5.46	16.27	24.78
C23H28N6O9	532.51	532.1918	51.88	5.30	15.78	27.04
C23H29N1O1	335.49	335.2249	82.34	8.71	4.17	4.77

C23H29N1O2	351.49	351.2198	78.60	8.32	3.98	9.10
C23H29N1O3	367.49	367.2147	75.17	7.95	3.81	13.06
C23H29N1O4	383.49	383.2096	72.04	7.62	3.65	16.69
C23H29N1O5	399.49	399.2046	69.15	7.32	3.51	20.02
C23H29N1O6	415.49	415.1995	66.49	7.04	3.37	23.10
C23H29N1O7	431.49	431.1944	64.02	6.77	3.25	25.96
C23H29N1O8	447.49	447.1893	61.73	6.53	3.13	28.60
C23H29N1O9	463.49	463.1842	59.60	6.31	3.02	31.07
C23H29N3O1	363.51	363.2310	76.00	8.04	11.56	4.40
C23H29N3O2	379.51	379.2260	72.79	7.70	11.07	8.43
C23H29N3O3	395.51	395.2209	69.85	7.39	10.62	12.14
C23H29N3O4	411.51	411.2158	67.13	7.10	10.21	15.55
C23H29N3O5	427.50	427.2107	64.62	6.84	9.83	18.71
C23H29N3O6	443.50	443.2056	62.29	6.59	9.47	21.64
C23H29N3O7	459.50	459.2005	60.12	6.36	9.14	24.37
C23H29N3O8	475.50	475.1954	58.10	6.15	8.84	26.92
C23H29N3O9	491.50	491.1904	56.21	5.95	8.55	29.30
C23H29N5O1	391.52	391.2372	70.56	7.47	17.89	4.09
C23H29N5O2	407.52	407.2321	67.79	7.17	17.19	7.85
C23H29N5O3	423.52	423.2270	65.23	6.90	16.54	11.33
C23H29N5O4	439.52	439.2219	62.85	6.65	15.93	14.56
C23H29N5O5	455.52	455.2169	60.65	6.42	15.37	17.56
C23H29N5O6	471.52	471.2118	58.59	6.20	14.85	20.36
C23H29N5O7	487.52	487.2067	56.67	6.00	14.37	22.97
C23H29N5O8	503.52	503.2016	54.87	5.81	13.91	25.42
C23H29N5O9	519.52	519.1965	53.18	5.63	13.48	27.72
C23H30N2O1	350.51	350.2358	78.82	8.63	7.99	4.56
C23H30N2O2	366.51	366.2307	75.38	8.25	7.64	8.73
C23H30N2O3	382.51	382.2256	72.22	7.91	7.32	12.55
C23H30N2O4	398.51	398.2205	69.32	7.59	7.03	16.06
C23H30N2O5	414.51	414.2155	66.65	7.30	6.76	19.30
C23H30N2O6	430.51	430.2104	64.17	7.02	6.51	22.30
C23H30N2O7	446.50	446.2053	61.87	6.77	6.27	25.08
C23H30N2O8	462.50	462.2002	59.73	6.54	6.06	27.67
C23H30N2O9	478.50	478.1951	57.73	6.32	5.85	30.09
C23H30N4O1	378.52	378.2419	72.98	7.99	14.80	4.23
C23H30N4O2	394.52	394.2369	70.02	7.66	14.20	8.11
C23H30N4O3	410.52	410.2318	67.29	7.37	13.65	11.69
C23H30N4O4	426.52	426.2267	64.77	7.09	13.14	15.00
C23H30N4O5	442.52	442.2216	62.43	6.83	12.66	18.08
C23H30N4O6	458.52	458.2165	60.25	6.59	12.22	20.94
C23H30N4O7	474.52	474.2114	58.22	6.37	11.81	23.60
C23H30N4O8	490.52	490.2063	56.32	6.16	11.42	26.09
C23H30N4O9	506.52	506.2013	54.54	5.97	11.06	28.43
C23H30N6O1	406.54	406.2481	67.95	7.44	20.67	3.94
C23H30N6O2	422.53	422.2430	65.38	7.16	19.89	7.57
C23H30N6O3	438.53	438.2379	63.00	6.90	19.16	10.95
C23H30N6O4	454.53	454.2328	60.78	6.65	18.49	14.08
C23H30N6O5	470.53	470.2278	58.71	6.43	17.86	17.00
C23H30N6O6	486.53	486.2227	56.78	6.22	17.27	19.73
C23H30N6O7	502.53	502.2176	54.97	6.02	16.72	22.29
C23H30N6O8	518.53	518.2125	53.28	5.83	16.21	24.68
C23H30N6O9	534.53	534.2074	51.68	5.66	15.72	26.94
C23H31N1O1	337.51	337.2406	81.85	9.26	4.15	4.74
C23H31N1O2	353.51	353.2355	78.15	8.84	3.96	9.05
C23H31N1O3	369.51	369.2304	74.76	8.46	3.79	12.99
C23H31N1O4	385.51	385.2253	71.66	8.11	3.63	16.60
C23H31N1O5	401.51	401.2202	68.80	7.78	3.49	19.92
C23H31N1O6	417.51	417.2151	66.17	7.48	3.35	22.99
C23H31N1O7	433.51	433.2100	63.73	7.21	3.23	25.83

C23H31N108	449.51	449.2049	61.46	6.95	3.12	28.47
C23H31N109	465.50	465.1999	59.35	6.71	3.01	30.93
C23H31N301	365.52	365.2467	75.58	8.55	11.50	4.38
C23H31N302	381.52	381.2416	72.41	8.19	11.01	8.39
C23H31N303	397.52	397.2365	69.49	7.86	10.57	12.07
C23H31N304	413.52	413.2314	66.81	7.56	10.16	15.48
C23H31N305	429.52	429.2264	64.32	7.27	9.78	18.62
C23H31N306	445.52	445.2213	62.01	7.01	9.43	21.55
C23H31N307	461.52	461.2162	59.86	6.77	9.10	24.27
C23H31N308	477.52	477.2111	57.85	6.54	8.80	26.80
C23H31N309	493.52	493.2060	55.98	6.33	8.51	29.18
C23H31N501	393.54	393.2528	70.20	7.94	17.80	4.07
C23H31N502	409.54	409.2478	67.46	7.63	17.10	7.81
C23H31N503	425.54	425.2427	64.92	7.34	16.46	11.28
C23H31N504	441.53	441.2376	62.57	7.08	15.86	14.49
C23H31N505	457.53	457.2325	60.38	6.83	15.31	17.48
C23H31N506	473.53	473.2274	58.34	6.60	14.79	20.27
C23H31N507	489.53	489.2223	56.43	6.38	14.31	22.88
C23H31N508	505.53	505.2172	54.65	6.18	13.85	25.32
C23H31N509	521.53	521.2122	52.97	5.99	13.43	27.61
C23H32N201	352.52	352.2514	78.37	9.15	7.95	4.54
C23H32N202	368.52	368.2464	74.96	8.75	7.60	8.68
C23H32N203	384.52	384.2413	71.84	8.39	7.29	12.48
C23H32N204	400.52	400.2362	68.97	8.05	6.99	15.98
C23H32N205	416.52	416.2311	66.32	7.74	6.73	19.21
C23H32N206	432.52	432.2260	63.87	7.46	6.48	22.19
C23H32N207	448.52	448.2209	61.59	7.19	6.25	24.97
C23H32N208	464.52	464.2158	59.47	6.94	6.03	27.55
C23H32N209	480.52	480.2108	57.49	6.71	5.83	29.97
C23H32N401	380.54	380.2576	72.60	8.48	14.72	4.20
C23H32N402	396.54	396.2525	69.67	8.13	14.13	8.07
C23H32N403	412.54	412.2474	66.97	7.82	13.58	11.63
C23H32N404	428.54	428.2423	64.47	7.53	13.07	14.93
C23H32N405	444.54	444.2373	62.14	7.26	12.60	18.00
C23H32N406	460.53	460.2322	59.99	7.00	12.17	20.84
C23H32N407	476.53	476.2271	57.97	6.77	11.76	23.50
C23H32N408	492.53	492.2220	56.09	6.55	11.38	25.99
C23H32N409	508.53	508.2169	54.32	6.34	11.02	28.32
C23H32N601	408.55	408.2637	67.62	7.89	20.57	3.92
C23H32N602	424.55	424.2587	65.07	7.60	19.80	7.54
C23H32N603	440.55	440.2536	62.71	7.32	19.08	10.90
C23H32N604	456.55	456.2485	60.51	7.06	18.41	14.02
C23H32N605	472.55	472.2434	58.46	6.83	17.78	16.93
C23H32N606	488.55	488.2383	56.55	6.60	17.20	19.65
C23H32N607	504.55	504.2332	54.75	6.39	16.66	22.20
C23H32N608	520.55	520.2281	53.07	6.20	16.14	24.59
C23H32N609	536.55	536.2231	51.49	6.01	15.66	26.84
C23H33N101	339.53	339.2562	81.37	9.80	4.13	4.71
C23H33N102	355.52	355.2511	77.70	9.36	3.94	9.00
C23H33N103	371.52	371.2460	74.36	8.95	3.77	12.92
C23H33N104	387.52	387.2409	71.29	8.58	3.61	16.51
C23H33N105	403.52	403.2359	68.46	8.24	3.47	19.82
C23H33N106	419.52	419.2308	65.85	7.93	3.34	22.88
C23H33N107	435.52	435.2257	63.43	7.64	3.22	25.72
C23H33N108	451.52	451.2206	61.18	7.37	3.10	28.35
C23H33N109	467.52	467.2155	59.09	7.11	3.00	30.80
C23H33N301	367.54	367.2623	75.16	9.05	11.43	4.35
C23H33N302	383.54	383.2573	72.03	8.67	10.96	8.34
C23H33N303	399.54	399.2522	69.14	8.33	10.52	12.01
C23H33N304	415.54	415.2471	66.48	8.00	10.11	15.40
C23H33N305	431.54	431.2420	64.02	7.71	9.74	18.54
C23H33N306	447.54	447.2369	61.73	7.43	9.39	21.45
C23H33N307	463.54	463.2318	59.60	7.18	9.07	24.16
C23H33N308	479.53	479.2267	57.61	6.94	8.76	26.69
C23H33N309	495.53	495.2217	55.75	6.71	8.48	29.06
C23H33N501	395.55	395.2685	69.84	8.41	17.71	4.04
C23H33N502	411.55	411.2634	67.13	8.08	17.02	7.78
C23H33N503	427.55	427.2583	64.61	7.78	16.38	11.23
C23H33N504	443.55	443.2532	62.28	7.50	15.79	14.43
C23H33N505	459.55	459.2482	60.11	7.24	15.24	17.41
C23H33N506	475.55	475.2431	58.09	6.99	14.73	20.19
C23H33N507	491.55	491.2380	56.20	6.77	14.25	22.78
C23H33N508	507.55	507.2329	54.43	6.55	13.80	25.22
C23H33N509	523.55	523.2278	52.77	6.35	13.38	27.50
C23H34N201	354.54	354.2671	77.92	9.67	7.90	4.51
C23H34N202	370.54	370.2620	74.56	9.25	7.56	8.64
C23H34N203	386.54	386.2569	71.47	8.87	7.25	12.42
C23H34N204	402.54	402.2518	68.63	8.51	6.96	15.90
C23H34N205	418.54	418.2468	66.01	8.19	6.69	19.11
C23H34N206	434.54	434.2417	63.57	7.89	6.45	22.09
C23H34N207	450.54	450.2366	61.32	7.61	6.22	24.86
C23H34N208	466.54	466.2315	59.21	7.35	6.00	27.44
C23H34N209	482.54	482.2264	57.25	7.10	5.81	29.84
C23H34N401	382.55	382.2732	72.21	8.96	14.65	4.18
C23H34N402	398.55	398.2682	69.31	8.60	14.06	8.03
C23H34N403	414.55	414.2631	66.64	8.27	13.52	11.58
C23H34N404	430.55	430.2580	64.16	7.96	13.01	14.86
C23H34N405	446.55	446.2529	61.86	7.67	12.55	17.91
C23H34N406	462.55	462.2478	59.72	7.41	12.11	20.75
C23H34N407	478.55	478.2427	57.73	7.16	11.71	23.40
C23H34N408	494.55	494.2376	55.86	6.93	11.33	25.88
C23H34N409	510.55	510.2326	54.11	6.71	10.97	28.20
C23H34N601	410.57	410.2794	67.29	8.35	20.47	3.90
C23H34N602	426.57	426.2743	64.76	8.03	19.70	7.50
C23H34N603	442.57	442.2692	62.42	7.74	18.99	10.85
C23H34N604	458.57	458.2641	60.24	7.47	18.33	13.96
C23H34N605	474.56	474.2590	58.21	7.22	17.71	16.86
C23H34N606	490.56	490.2540	56.31	6.99	17.13	19.57
C23H34N607	506.56	506.2489	54.54	6.77	16.59	22.11
C23H34N608	522.56	522.2438	52.87	6.56	16.08	24.49
C23H34N609	538.56	538.2387	51.30	6.36	15.60	26.74
C23H35N101	341.54	341.2718	80.89	10.33	4.10	4.68
C23H35N102	357.54	357.2668	77.27	9.87	3.92	8.95
C23H35N103	373.54	373.2617	73.96	9.44	3.75	12.85
C23H35N104	389.54	389.2566	70.92	9.06	3.60	16.43
C23H35N105	405.54	405.2515	68.12	8.70	3.45	19.73
C23H35N106	421.54	421.2464	65.54	8.37	3.32	22.77
C23H35N107	437.54	437.2413	63.14	8.06	3.20	25.60
C23H35N108	453.54	453.2362	60.91	7.78	3.09	28.22
C23H35N109	469.54	469.2312	58.84	7.51	2.98	30.67
C23H35N301	369.55	369.2780	74.75	9.55	11.37	4.33
C23H35N302	385.55	385.2729	71.65	9.15	10.90	8.30
C23H35N303	401.55	401.2678	68.80	8.79	10.46	11.95
C23H35N304	417.55	417.2627	66.16	8.45	10.06	15.33
C23H35N305	433.55	433.2577	63.72	8.14	9.69	18.45
C23H35N306	449.55	449.2526	61.45	7.85	9.35	21.35
C23H35N307	465.55	465.2475	59.34	7.58	9.03	24.06
C23H35N308	481.55	481.2424	57.37	7.33	8.73	26.58
C23H35N309	497.55	497.2373	55.52	7.09	8.45	28.94
C23H35N501	397.57	397.2841	69.49	8.87	17.62	4.02

C23H35N502	413.57	413.2791	66.80	8.53	16.93	7.74
C23H35N503	429.57	429.2740	64.31	8.21	16.30	11.17
C23H35N504	445.57	445.2689	62.00	7.92	15.72	14.36
C23H35N505	461.57	461.2638	59.85	7.64	15.17	17.33
C23H35N506	477.57	477.2587	57.85	7.39	14.66	20.10
C23H35N507	493.56	493.2536	55.97	7.15	14.19	22.69
C23H35N508	509.56	509.2485	54.21	6.92	13.74	25.12
C23H35N509	525.56	525.2435	52.56	6.71	13.33	27.40
C23H36N201	356.56	356.2827	77.48	10.18	7.86	4.49
C23H36N202	372.56	372.2777	74.15	9.74	7.52	8.59
C23H36N203	388.55	388.2726	71.10	9.34	7.21	12.35
C23H36N204	404.55	404.2675	68.29	8.97	6.92	15.82
C23H36N205	420.55	420.2624	65.69	8.63	6.66	19.02
C23H36N206	436.55	436.2573	63.28	8.31	6.42	21.99
C23H36N207	452.55	452.2522	61.04	8.02	6.19	24.75
C23H36N208	468.55	468.2471	58.96	7.74	5.98	27.32
C23H36N209	484.55	484.2421	57.01	7.49	5.78	29.72
C23H36N401	384.57	384.2889	71.84	9.44	14.57	4.16
C23H36N402	400.57	400.2838	68.97	9.06	13.99	7.99
C23H36N403	416.57	416.2787	66.32	8.71	13.45	11.52
C23H36N404	432.57	432.2736	63.86	8.39	12.95	14.79
C23H36N405	448.57	448.2686	61.59	8.09	12.49	17.83
C23H36N406	464.57	464.2635	59.47	7.81	12.06	20.66
C23H36N407	480.57	480.2584	57.49	7.55	11.66	23.30
C23H36N408	496.57	496.2533	55.63	7.31	11.28	25.78
C23H36N409	512.56	512.2482	53.90	7.08	10.93	28.09
C23H36N601	412.58	412.2950	66.96	8.80	20.37	3.88
C23H36N602	428.58	428.2900	64.46	8.47	19.61	7.47
C23H36N603	444.58	444.2849	62.14	8.16	18.90	10.80
C23H36N604	460.58	460.2798	59.98	7.88	18.25	13.89
C23H36N605	476.58	476.2747	57.97	7.61	17.63	16.79
C23H36N606	492.58	492.2696	56.08	7.37	17.06	19.49
C23H36N607	508.58	508.2645	54.32	7.13	16.52	22.02
C23H36N608	524.58	524.2594	52.66	6.92	16.02	24.40
C23H36N609	540.58	540.2544	51.10	6.71	15.55	26.64
C23H37N101	343.56	343.2875	80.41	10.86	4.08	4.66
C23H37N102	359.56	359.2824	76.83	10.37	3.90	8.90
C23H37N103	375.56	375.2773	73.56	9.93	3.73	12.78
C23H37N104	391.56	391.2722	70.55	9.52	3.58	16.34
C23H37N105	407.56	407.2672	67.78	9.15	3.44	19.63
C23H37N106	423.55	423.2621	65.22	8.81	3.31	22.66
C23H37N107	439.55	439.2570	62.85	8.48	3.19	25.48
C23H37N108	455.55	455.2519	60.64	8.19	3.07	28.10
C23H37N109	471.55	471.2468	58.58	7.91	2.97	30.54
C23H37N301	371.57	371.2936	74.35	10.04	11.31	4.31
C23H37N302	387.57	387.2886	71.28	9.62	10.84	8.26
C23H37N303	403.57	403.2835	68.45	9.24	10.41	11.89
C23H37N304	419.57	419.2784	65.84	8.89	10.02	15.25
C23H37N305	435.57	435.2733	63.42	8.56	9.65	18.37
C23H37N306	451.57	451.2682	61.18	8.26	9.31	21.26
C23H37N307	467.57	467.2631	59.08	7.98	8.99	23.95
C23H37N308	483.57	483.2580	57.13	7.71	8.69	26.47
C23H37N309	499.57	499.2530	55.30	7.47	8.41	28.82
C23H37N501	399.58	399.2998	69.14	9.33	17.53	4.00
C23H37N502	415.58	415.2947	66.47	8.97	16.85	7.70
C23H37N503	431.58	431.2896	64.01	8.64	16.23	11.12
C23H37N504	447.58	447.2845	61.72	8.33	15.65	14.30
C23H37N505	463.58	463.2794	59.59	8.04	15.11	17.26
C23H37N506	479.58	479.2744	57.60	7.78	14.60	20.02
C23H37N507	495.58	495.2693	55.74	7.53	14.13	22.60
C23H37N508	511.58	511.2642	54.00	7.29	13.69	25.02
C23H37N509	527.58	527.2591	52.36	7.07	13.27	27.29
C23H38N201	358.57	358.2984	77.04	10.68	7.81	4.46
C23H38N202	374.57	374.2933	73.75	10.23	7.48	8.54
C23H38N203	390.57	390.2882	70.73	9.81	7.17	12.29
C23H38N204	406.57	406.2831	67.95	9.42	6.89	15.74
C23H38N205	422.57	422.2781	65.38	9.06	6.63	18.93
C23H38N206	438.57	438.2730	62.99	8.73	6.39	21.89
C23H38N207	454.57	454.2679	60.77	8.43	6.16	24.64
C23H38N208	470.57	470.2628	58.71	8.14	5.95	27.20
C23H38N209	486.57	486.2577	56.78	7.87	5.76	29.59
C23H38N401	386.59	386.3045	71.46	9.91	14.49	4.14
C23H38N402	402.58	402.2995	68.62	9.51	13.92	7.95
C23H38N403	418.58	418.2944	66.00	9.15	13.38	11.47
C23H38N404	434.58	434.2893	63.57	8.81	12.89	14.73
C23H38N405	450.58	450.2842	61.31	8.50	12.43	17.75
C23H38N406	466.58	466.2791	59.21	8.21	12.01	20.57
C23H38N407	482.58	482.2740	57.25	7.94	11.61	23.21
C23H38N408	498.58	498.2689	55.41	7.68	11.24	25.67
C23H38N409	514.58	514.2639	53.69	7.44	10.89	27.98
C23H38N601	414.60	414.3107	66.63	9.24	20.27	3.86
C23H38N602	430.60	430.3056	64.16	8.90	19.52	7.43
C23H38N603	446.60	446.3005	61.86	8.58	18.82	10.75
C23H38N604	462.60	462.2954	59.72	8.28	18.17	13.83
C23H38N605	478.60	478.2903	57.72	8.00	17.56	16.71
C23H38N606	494.60	494.2853	55.85	7.74	16.99	19.41
C23H38N607	510.60	510.2802	54.10	7.50	16.46	21.93
C23H38N608	526.59	526.2751	52.46	7.27	15.96	24.31
C23H38N609	542.59	542.2700	50.91	7.06	15.49	26.54
C23H39N101	345.57	345.3031	79.94	11.38	4.05	4.63
C23H39N102	361.57	361.2981	76.40	10.87	3.87	8.85
C23H39N103	377.57	377.2930	73.17	10.41	3.71	12.71
C23H39N104	393.57	393.2879	70.19	9.99	3.56	16.26
C23H39N105	409.57	409.2828	67.45	9.60	3.42	19.53
C23H39N106	425.57	425.2777	64.91	9.24	3.29	22.56
C23H39N107	441.57	441.2726	62.56	8.90	3.17	25.36
C23H39N108	457.57	457.2675	60.37	8.59	3.06	27.97
C23H39N109	473.57	473.2625	58.34	8.30	2.96	30.41
C23H39N301	373.59	373.3093	73.95	10.52	11.25	4.28
C23H39N302	389.59	389.3042	70.91	10.09	10.79	8.21
C23H39N303	405.59	405.2991	68.11	9.69	10.36	11.83
C23H39N304	421.58	421.2940	65.53	9.32	9.97	15.18
C23H39N305	437.58	437.2890	63.13	8.98	9.60	18.28
C23H39N306	453.58	453.2839	60.91	8.67	9.26	21.16
C23H39N307	469.58	469.2788	58.83	8.37	8.95	23.85
C23H39N308	485.58	485.2737	56.89	8.10	8.65	26.36
C23H39N309	501.58	501.2686	55.08	7.84	8.38	28.71
C23H39N501	401.60	401.3154	68.79	9.79	17.44	3.98
C23H39N502	417.60	417.3104	66.15	9.41	16.77	7.66
C23H39N503	433.60	433.3053	63.71	9.07	16.15	11.07
C23H39N504	449.60	449.3002	61.45	8.74	15.58	14.23
C23H39N505	465.60	465.2951	59.33	8.44	15.04	17.18
C23H39N506	481.60	481.2900	57.36	8.16	14.54	19.93
C23H39N507	497.60	497.2849	55.52	7.90	14.07	22.51
C23H39N508	513.60	513.2798	53.79	7.65	13.64	24.92
C23H39N509	529.60	529.2748	52.16	7.42	13.22	27.19
C23H40N201	360.59	360.3140	76.61	11.18	7.77	4.44
C23H40N202	376.59	376.3090	73.36	10.71	7.44	8.50
C23H40N203	392.59	392.3039	70.37	10.27	7.14	12.23
C23H40N204	408.59	408.2988	67.61	9.87	6.86	15.66

C23H40N205	424.59	424.2937	65.06	9.50	6.60	18.84
C23H40N206	440.59	440.2886	62.70	9.15	6.36	21.79
C23H40N207	456.58	456.2835	60.51	8.83	6.14	24.53
C23H40N208	472.58	472.2784	58.46	8.53	5.93	27.08
C23H40N209	488.58	488.2734	56.54	8.25	5.73	29.47
C23H40N401	388.60	388.3202	71.09	10.38	14.42	4.12
C23H40N402	404.60	404.3151	68.28	9.97	13.85	7.91
C23H40N403	420.60	420.3100	65.68	9.59	13.32	11.41
C23H40N404	436.60	436.3049	63.27	9.23	12.83	14.66
C23H40N405	452.60	452.2998	61.04	8.91	12.38	17.68
C23H40N406	468.60	468.2948	58.95	8.60	11.96	20.49
C23H40N407	484.60	484.2897	57.01	8.32	11.56	23.11
C23H40N408	500.60	500.2846	55.19	8.05	11.19	25.57
C23H40N409	516.60	516.2795	53.48	7.80	10.85	27.87
C23H40N601	416.61	416.3263	66.31	9.68	20.17	3.84
C23H40N602	432.61	432.3213	63.86	9.32	19.43	7.40
C23H40N603	448.61	448.3162	61.58	8.99	18.73	10.70
C23H40N604	464.61	464.3111	59.46	8.68	18.09	13.77
C23H40N605	480.61	480.3060	57.48	8.39	17.49	16.64
C23H40N606	496.61	496.3009	55.63	8.12	16.92	19.33
C23H40N607	512.61	512.2958	53.89	7.87	16.39	21.85
C23H40N608	528.61	528.2907	52.26	7.63	15.90	24.21
C23H40N609	544.61	544.2857	50.73	7.40	15.43	26.44
C23H41N101	347.59	347.3188	79.48	11.89	4.03	4.60
C23H41N102	363.59	363.3137	75.98	11.37	3.85	8.80
C23H41N103	379.59	379.3086	72.78	10.89	3.69	12.64
C23H41N104	395.59	395.3035	69.83	10.45	3.54	16.18
C23H41N105	411.59	411.2985	67.12	10.04	3.40	19.44
C23H41N106	427.59	427.2934	64.61	9.67	3.28	22.45
C23H41N107	443.59	443.2883	62.28	9.32	3.16	25.25
C23H41N108	459.59	459.2832	60.11	8.99	3.05	27.85
C23H41N109	475.58	475.2781	58.09	8.69	2.95	30.28
C23H41N301	375.60	375.3249	73.55	11.00	11.19	4.26
C23H41N302	391.60	391.3199	70.55	10.55	10.73	8.17
C23H41N303	407.60	407.3148	67.78	10.14	10.31	11.78
C23H41N304	423.60	423.3097	65.22	9.76	9.92	15.11
C23H41N305	439.60	439.3046	62.84	9.40	9.56	18.20
C23H41N306	455.60	455.2995	60.64	9.07	9.22	21.07
C23H41N307	471.60	471.2944	58.58	8.76	8.91	23.75
C23H41N308	487.60	487.2893	56.66	8.48	8.62	26.25
C23H41N309	503.60	503.2843	54.86	8.21	8.34	28.59
C23H41N501	403.62	403.3311	68.45	10.24	17.35	3.96
C23H41N502	419.62	419.3260	65.84	9.85	16.69	7.63
C23H41N503	435.61	435.3209	63.42	9.49	16.08	11.02
C23H41N504	451.61	451.3158	61.17	9.15	15.51	14.17
C23H41N505	467.61	467.3107	59.08	8.84	14.98	17.11
C23H41N506	483.61	483.3057	57.12	8.55	14.48	19.85
C23H41N507	499.61	499.3006	55.29	8.27	14.02	22.42
C23H41N508	515.61	515.2955	53.58	8.02	13.58	24.82
C23H41N509	531.61	531.2904	51.97	7.77	13.17	27.09
C23H42N201	362.60	362.3297	76.19	11.68	7.73	4.41
C23H42N202	378.60	378.3246	72.97	11.18	7.40	8.45
C23H42N203	394.60	394.3195	70.01	10.73	7.10	12.16
C23H42N204	410.60	410.3144	67.28	10.31	6.82	15.59
C23H42N205	426.60	426.3094	64.76	9.92	6.57	18.75
C23H42N206	442.60	442.3043	62.42	9.56	6.33	21.69
C23H42N207	458.60	458.2992	60.24	9.23	6.11	24.42
C23H42N208	474.60	474.2941	58.21	8.92	5.90	26.97
C23H42N209	490.60	490.2890	56.31	8.63	5.71	29.35
C23H42N401	390.62	390.3358	70.72	10.84	14.34	4.10

C23H42N402	406.62	406.3308	67.94	10.41	13.78	7.87
C23H42N403	422.62	422.3257	65.37	10.02	13.26	11.36
C23H42N404	438.62	438.3206	62.98	9.65	12.77	14.59
C23H42N405	454.61	454.3155	60.77	9.31	12.32	17.60
C23H42N406	470.61	470.3104	58.70	9.00	11.91	20.40
C23H42N407	486.61	486.3053	56.77	8.70	11.51	23.02
C23H42N408	502.61	502.3002	54.96	8.42	11.15	25.47
C23H42N409	518.61	518.2952	53.27	8.16	10.80	27.77
C23H42N601	418.63	418.3420	65.99	10.11	20.08	3.82
C23H42N602	434.63	434.3369	63.56	9.74	19.34	7.36
C23H42N603	450.63	450.3318	61.30	9.39	18.65	10.65
C23H42N604	466.63	466.3267	59.20	9.07	18.01	13.71
C23H42N605	482.63	482.3216	57.24	8.77	17.41	16.58
C23H42N606	498.63	498.3166	55.40	8.49	16.85	19.25
C23H42N607	514.63	514.3115	53.68	8.23	16.33	21.76
C23H42N608	530.63	530.3064	52.06	7.98	15.84	24.12
C23H42N609	546.63	546.3013	50.54	7.74	15.37	26.34
C23H43N101	349.61	349.3344	79.02	12.40	4.01	4.58
C23H43N102	365.60	365.3294	75.56	11.86	3.83	8.75
C23H43N103	381.60	381.3243	72.39	11.36	3.67	12.58
C23H43N104	397.60	397.3192	69.48	10.90	3.52	16.10
C23H43N105	413.60	413.3141	66.79	10.48	3.39	19.34
C23H43N106	429.60	429.3090	64.31	10.09	3.26	22.35
C23H43N107	445.60	445.3039	62.00	9.73	3.14	25.13
C23H43N108	461.60	461.2988	59.85	9.39	3.03	27.73
C23H43N109	477.60	477.2938	57.84	9.08	2.93	30.15
C23H43N301	377.62	377.3406	73.16	11.48	11.13	4.24
C23H43N302	393.62	393.3355	70.18	11.01	10.68	8.13
C23H43N303	409.62	409.3304	67.44	10.58	10.26	11.72
C23H43N304	425.62	425.3253	64.91	10.18	9.87	15.04
C23H43N305	441.62	441.3202	62.56	9.81	9.52	18.11
C23H43N306	457.62	457.3152	60.37	9.47	9.18	20.98
C23H43N307	473.62	473.3101	58.33	9.15	8.87	23.65
C23H43N308	489.61	489.3050	56.42	8.85	8.58	26.14
C23H43N309	505.61	505.2999	54.64	8.57	8.31	28.48
C23H43N501	405.63	405.3467	68.11	10.69	17.27	3.94
C23H43N502	421.63	421.3417	65.52	10.28	16.61	7.59
C23H43N503	437.63	437.3366	63.13	9.90	16.00	10.97
C23H43N504	453.63	453.3315	60.90	9.55	15.44	14.11
C23H43N505	469.63	469.3264	58.82	9.23	14.91	17.03
C23H43N506	485.63	485.3213	56.89	8.93	14.42	19.77
C23H43N507	501.63	501.3162	55.07	8.64	13.96	22.33
C23H43N508	517.63	517.3111	53.37	8.37	13.53	24.73
C23H43N509	533.63	533.3061	51.77	8.12	13.12	26.98
C23H44N201	364.62	364.3453	75.77	12.16	7.68	4.39
C23H44N202	380.62	380.3403	72.58	11.65	7.36	8.41
C23H44N203	396.62	396.3352	69.65	11.18	7.06	12.10
C23H44N204	412.62	412.3301	66.95	10.75	6.79	15.51
C23H44N205	428.62	428.3250	64.45	10.35	6.54	18.66
C23H44N206	444.62	444.3199	62.13	9.98	6.30	21.59
C23H44N207	460.62	460.3148	59.98	9.63	6.08	24.31
C23H44N208	476.62	476.3097	57.96	9.31	5.88	26.86
C23H44N209	492.62	492.3047	56.08	9.00	5.69	29.23
C23H44N401	392.63	392.3515	70.36	11.30	14.27	4.07
C23H44N402	408.63	408.3464	67.61	10.85	13.71	7.83
C23H44N403	424.63	424.3413	65.06	10.44	13.19	11.30
C23H44N404	440.63	440.3362	62.70	10.07	12.72	14.52
C23H44N405	456.63	456.3311	60.50	9.71	12.27	17.52
C23H44N406	472.63	472.3261	58.45	9.38	11.85	20.31
C23H44N407	488.63	488.3210	56.54	9.08	11.47	22.92

C23H44N408	504.63	504.3159	54.74	8.79	11.10	25.36
C23H44N409	520.63	520.3108	53.06	8.52	10.76	27.66
C23H44N601	420.65	420.3576	65.67	10.54	19.98	3.80
C23H44N602	436.65	436.3526	63.27	10.16	19.25	7.33
C23H44N603	452.65	452.3475	61.03	9.80	18.57	10.60
C23H44N604	468.64	468.3424	58.95	9.46	17.93	13.66
C23H44N605	484.64	484.3373	57.00	9.15	17.34	16.51
C23H44N606	500.64	500.3322	55.18	8.86	16.79	19.17
C23H44N607	516.64	516.3271	53.47	8.58	16.27	21.68
C23H44N608	532.64	532.3220	51.87	8.33	15.78	24.03
C23H44N609	548.64	548.3170	50.35	8.08	15.32	26.25
C23H45N101	351.62	351.3501	78.57	12.90	3.98	4.55
C23H45N102	367.62	367.3450	75.15	12.34	3.81	8.70
C23H45N103	383.62	383.3399	72.01	11.82	3.65	12.51
C23H45N104	399.62	399.3348	69.13	11.35	3.51	16.01
C23H45N105	415.62	415.3298	66.47	10.91	3.37	19.25
C23H45N106	431.62	431.3247	64.00	10.51	3.25	22.24
C23H45N107	447.62	447.3196	61.72	10.13	3.13	25.02
C23H45N108	463.62	463.3145	59.59	9.78	3.02	27.61
C23H45N109	479.62	479.3094	57.60	9.46	2.92	30.02
C23H45N301	379.63	379.3562	72.77	11.95	11.07	4.21
C23H45N302	395.63	395.3512	69.83	11.46	10.62	8.09
C23H45N303	411.63	411.3461	67.11	11.02	10.21	11.66
C23H45N304	427.63	427.3410	64.60	10.61	9.83	14.97
C23H45N305	443.63	443.3359	62.27	10.22	9.47	18.03
C23H45N306	459.63	459.3308	60.10	9.87	9.14	20.89
C23H45N307	475.63	475.3257	58.08	9.54	8.83	23.55
C23H45N308	491.63	491.3206	56.19	9.23	8.55	26.03
C23H45N309	507.63	507.3156	54.42	8.94	8.28	28.37
C23H45N501	407.65	407.3624	67.77	11.13	17.18	3.92
C23H45N502	423.65	423.3573	65.21	10.71	16.53	7.55
C23H45N503	439.65	439.3522	62.84	10.32	15.93	10.92
C23H45N504	455.65	455.3471	60.63	9.95	15.37	14.05
C23H45N505	471.65	471.3420	58.57	9.62	14.85	16.96
C23H45N506	487.64	487.3370	56.65	9.30	14.36	19.69
C23H45N507	503.64	503.3319	54.85	9.01	13.91	22.24
C23H45N508	519.64	519.3268	53.16	8.73	13.48	24.63
C23H45N509	535.64	535.3217	51.57	8.47	13.07	26.88
C23H46N201	366.64	366.3610	75.35	12.65	7.64	4.36
C23H46N202	382.64	382.3559	72.20	12.12	7.32	8.36
C23H46N203	398.63	398.3508	69.30	11.63	7.03	12.04
C23H46N204	414.63	414.3457	66.63	11.18	6.76	15.43
C23H46N205	430.63	430.3406	64.15	10.77	6.51	18.58
C23H46N206	446.63	446.3356	61.85	10.38	6.27	21.49
C23H46N207	462.63	462.3305	59.71	10.02	6.06	24.21
C23H46N208	478.63	478.3254	57.72	9.69	5.85	26.74
C23H46N209	494.63	494.3203	55.85	9.37	5.66	29.11
C23H46N401	394.65	394.3671	70.00	11.75	14.20	4.05
C23H46N402	410.65	410.3621	67.27	11.29	13.64	7.79
C23H46N403	426.65	426.3570	64.75	10.87	13.13	11.25
C23H46N404	442.65	442.3519	62.41	10.47	12.66	14.46
C23H46N405	458.65	458.3468	60.23	10.11	12.22	17.44
C23H46N406	474.65	474.3417	58.20	9.77	11.80	20.22
C23H46N407	490.65	490.3366	56.30	9.45	11.42	22.83
C23H46N408	506.65	506.3315	54.53	9.15	11.06	25.26
C23H46N409	522.64	522.3265	52.86	8.87	10.72	27.55
C23H46N601	422.66	422.3733	65.36	10.97	19.88	3.79
C23H46N602	438.66	438.3682	62.98	10.57	19.16	7.29
C23H46N603	454.66	454.3631	60.76	10.20	18.48	10.56
C23H46N604	470.66	470.3580	58.70	9.85	17.86	13.60

C23H46N605	486.66	486.3529	56.77	9.53	17.27	16.44
C23H46N606	502.66	502.3479	54.96	9.22	16.72	19.10
C23H46N607	518.66	518.3428	53.26	8.94	16.20	21.59
C23H46N608	534.66	534.3377	51.67	8.67	15.72	23.94
C23H46N609	550.66	550.3326	50.17	8.42	15.26	26.15
C23H47N101	353.64	353.3657	78.12	13.40	3.96	4.52
C23H47N102	369.64	369.3607	74.74	12.82	3.79	8.66
C23H47N103	385.64	385.3556	71.64	12.28	3.63	12.45
C23H47N104	401.64	401.3505	68.78	11.80	3.49	15.93
C23H47N105	417.63	417.3454	66.15	11.34	3.35	19.15
C23H47N106	433.63	433.3403	63.71	10.93	3.23	22.14
C23H47N107	449.63	449.3352	61.44	10.54	3.12	24.91
C23H47N108	465.63	465.3301	59.33	10.17	3.01	27.49
C23H47N109	481.63	481.3251	57.36	9.84	2.91	29.90
C23H47N301	381.65	381.3719	72.38	12.41	11.01	4.19
C23H47N302	397.65	397.3668	69.47	11.91	10.57	8.05
C23H47N303	413.65	413.3617	66.79	11.45	10.16	11.60
C23H47N304	429.65	429.3566	64.30	11.03	9.78	14.90
C23H47N305	445.65	445.3515	61.99	10.63	9.43	17.95
C23H47N306	461.65	461.3465	59.84	10.26	9.10	20.79
C23H47N307	477.65	477.3414	57.84	9.92	8.80	23.45
C23H47N308	493.65	493.3363	55.96	9.60	8.51	25.93
C23H47N309	509.65	509.3312	54.21	9.30	8.24	28.25
C23H47N501	409.66	409.3780	67.43	11.56	17.10	3.91
C23H47N502	425.66	425.3730	64.90	11.13	16.45	7.52
C23H47N503	441.66	441.3679	62.55	10.73	15.86	10.87
C23H47N504	457.66	457.3628	60.36	10.35	15.30	13.98
C23H47N505	473.66	473.3577	58.32	10.00	14.79	16.89
C23H47N506	489.66	489.3526	56.42	9.67	14.30	19.60
C23H47N507	505.66	505.3475	54.63	9.37	13.85	22.15
C23H47N508	521.66	521.3424	52.96	9.08	13.43	24.54
C23H47N509	537.66	537.3374	51.38	8.81	13.03	26.78
C23H48N201	368.65	368.3766	74.94	13.12	7.60	4.34
C23H48N202	384.65	384.3716	71.82	12.58	7.28	8.32
C23H48N203	400.65	400.3665	68.95	12.08	6.99	11.98
C23H48N204	416.65	416.3614	66.30	11.61	6.72	15.36
C23H48N205	432.65	432.3563	63.85	11.18	6.47	18.49
C23H48N206	448.65	448.3512	61.58	10.78	6.24	21.40
C23H48N207	464.65	464.3461	59.45	10.41	6.03	24.10
C23H48N208	480.65	480.3410	57.48	10.07	5.83	26.63
C23H48N209	496.65	496.3360	55.62	9.74	5.64	28.99
C23H48N401	396.67	396.3828	69.64	12.20	14.12	4.03
C23H48N402	412.66	412.3777	66.94	11.72	13.58	7.75
C23H48N403	428.66	428.3726	64.45	11.29	13.07	11.20
C23H48N404	444.66	444.3675	62.13	10.88	12.60	14.39
C23H48N405	460.66	460.3624	59.97	10.50	12.16	17.37
C23H48N406	476.66	476.3574	57.96	10.15	11.75	20.14
C23H48N407	492.66	492.3523	56.07	9.82	11.37	22.73
C23H48N408	508.66	508.3472	54.31	9.51	11.01	25.16
C23H48N409	524.66	524.3421	52.65	9.22	10.68	27.45
C23H48N601	424.68	424.3889	65.05	11.39	19.79	3.77
C23H48N602	440.68	440.3839	62.69	10.98	19.07	7.26
C23H48N603	456.68	456.3788	60.49	10.59	18.40	10.51
C23H48N604	472.68	472.3737	58.45	10.24	17.78	13.54
C23H48N605	488.68	488.3686	56.53	9.90	17.20	16.37
C23H48N606	504.68	504.3635	54.74	9.59	16.65	19.02
C23H48N607	520.68	520.3584	53.06	9.29	16.14	21.51
C23H48N608	536.67	536.3533	51.48	9.02	15.66	23.85
C23H48N609	552.67	552.3483	49.99	8.75	15.21	26.05
C23H49N101	355.65	355.3814	77.68	13.89	3.94	4.50

C23H49N1O2	371.65	371.3763	74.33	13.29	3.77	8.61
C23H49N1O3	387.65	387.3712	71.26	12.74	3.61	12.38
C23H49N1O4	403.65	403.3661	68.44	12.24	3.47	15.85
C23H49N1O5	419.65	419.3610	65.83	11.77	3.34	19.06
C23H49N1O6	435.65	435.3560	63.41	11.34	3.22	22.04
C23H49N1O7	451.65	451.3509	61.17	10.94	3.10	24.80
C23H49N1O8	467.65	467.3458	59.07	10.56	3.00	27.37
C23H49N1O9	483.65	483.3407	57.12	10.21	2.90	29.77
C23H49N3O1	383.67	383.3875	72.00	12.87	10.95	4.17
C23H49N3O2	399.67	399.3825	69.12	12.36	10.51	8.01
C23H49N3O3	415.67	415.3774	66.46	11.88	10.11	11.55
C23H49N3O4	431.66	431.3723	64.00	11.44	9.73	14.83
C23H49N3O5	447.66	447.3672	61.71	11.03	9.39	17.87
C23H49N3O6	463.66	463.3621	59.58	10.65	9.06	20.70
C23H49N3O7	479.66	479.3570	57.59	10.30	8.76	23.35
C23H49N3O8	495.66	495.3519	55.73	9.96	8.48	25.82
C23H49N3O9	511.66	511.3469	53.99	9.65	8.21	28.14
C23H49N5O1	411.68	411.3937	67.10	12.00	17.01	3.89
C23H49N5O2	427.68	427.3886	64.59	11.55	16.38	7.48
C23H49N5O3	443.68	443.3835	62.26	11.13	15.78	10.82
C23H49N5O4	459.68	459.3784	60.10	10.74	15.24	13.92
C23H49N5O5	475.68	475.3733	58.08	10.38	14.72	16.82
C23H49N5O6	491.68	491.3683	56.19	10.05	14.24	19.52
C23H49N5O7	507.68	507.3632	54.42	9.73	13.79	22.06
C23H49N5O8	523.68	523.3581	52.75	9.43	13.37	24.44
C23H49N5O9	539.68	539.3530	51.19	9.15	12.98	26.68
C23H50N2O1	370.67	370.3923	74.53	13.60	7.56	4.32
C23H50N2O2	386.67	386.3872	71.45	13.03	7.24	8.28
C23H50N2O3	402.67	402.3821	68.61	12.52	6.96	11.92
C23H50N2O4	418.67	418.3770	65.98	12.04	6.69	15.29
C23H50N2O5	434.67	434.3719	63.56	11.59	6.44	18.40
C23H50N2O6	450.66	450.3669	61.30	11.18	6.22	21.30
C23H50N2O7	466.66	466.3618	59.20	10.80	6.00	24.00
C23H50N2O8	482.66	482.3567	57.24	10.44	5.80	26.52
C23H50N2O9	498.66	498.3516	55.40	10.11	5.62	28.88
C23H50N4O1	398.68	398.3984	69.29	12.64	14.05	4.01
C23H50N4O2	414.68	414.3934	66.62	12.15	13.51	7.72
C23H50N4O3	430.68	430.3883	64.14	11.70	13.01	11.14
C23H50N4O4	446.68	446.3832	61.85	11.28	12.54	14.33
C23H50N4O5	462.68	462.3781	59.71	10.89	12.11	17.29
C23H50N4O6	478.68	478.3730	57.71	10.53	11.70	20.05
C23H50N4O7	494.68	494.3679	55.85	10.19	11.33	22.64
C23H50N4O8	510.68	510.3628	54.10	9.87	10.97	25.06
C23H50N4O9	526.68	526.3578	52.45	9.57	10.64	27.34
C23H50N6O1	426.69	426.4046	64.74	11.81	19.70	3.75
C23H50N6O2	442.69	442.3995	62.40	11.38	18.98	7.23
C23H50N6O3	458.69	458.3944	60.23	10.99	18.32	10.46
C23H50N6O4	474.69	474.3893	58.20	10.62	17.70	13.48
C23H50N6O5	490.69	490.3842	56.30	10.27	17.13	16.30
C23H50N6O6	506.69	506.3792	54.52	9.95	16.59	18.95
C23H50N6O7	522.69	522.3741	52.85	9.64	16.08	21.43
C23H50N6O8	538.69	538.3690	51.28	9.36	15.60	23.76
C23H50N6O9	554.69	554.3639	49.80	9.09	15.15	25.96
C23H51N3O1	385.68	385.4032	71.63	13.33	10.90	4.15
C23H51N3O2	401.68	401.3981	68.77	12.80	10.46	7.97
C23H51N3O3	417.68	417.3930	66.14	12.31	10.06	11.49
C23H51N3O4	433.68	433.3879	63.70	11.85	9.69	14.76
C23H51N3O5	449.68	449.3828	61.43	11.43	9.34	17.79
C23H51N3O6	465.68	465.3778	59.32	11.04	9.02	20.61
C23H51N3O7	481.68	481.3727	57.35	10.67	8.72	23.25
C23H51N3O8	497.68	497.3676	55.51	10.33	8.44	25.72
C23H51N3O9	513.68	513.3625	53.78	10.01	8.18	28.03
C23H51N5O1	413.70	413.4093	66.78	12.43	16.93	3.87
C23H51N5O2	429.70	429.4043	64.29	11.96	16.30	7.45
C23H51N5O3	445.69	445.3992	61.98	11.53	15.71	10.77
C23H51N5O4	461.69	461.3941	59.84	11.13	15.17	13.86
C23H51N5O5	477.69	477.3890	57.83	10.76	14.66	16.75
C23H51N5O6	493.69	493.3839	55.96	10.41	14.19	19.44
C23H51N5O7	509.69	509.3788	54.20	10.09	13.74	21.97
C23H51N5O8	525.69	525.3737	52.55	9.78	13.32	24.35
C23H51N5O9	541.69	541.3687	51.00	9.49	12.93	26.58
C23H52N4O1	400.70	400.4141	68.94	13.08	13.98	3.99
C23H52N4O2	416.70	416.4090	66.30	12.58	13.45	7.68
C23H52N4O3	432.70	432.4039	63.85	12.11	12.95	11.09
C23H52N4O4	448.70	448.3988	61.57	11.68	12.49	14.26
C23H52N4O5	464.69	464.3937	59.45	11.28	12.06	17.21
C23H52N4O6	480.69	480.3887	57.47	10.90	11.66	19.97
C23H52N4O7	496.69	496.3836	55.62	10.55	11.28	22.55
C23H52N4O8	512.69	512.3785	53.88	10.22	10.93	24.97
C23H52N4O9	528.69	528.3734	52.25	9.91	10.60	27.24
C23H52N6O1	428.71	428.4202	64.44	12.23	19.60	3.73
C23H52N6O2	444.71	444.4152	62.12	11.79	18.90	7.20
C23H52N6O3	460.71	460.4101	59.96	11.38	18.24	10.42
C23H52N6O4	476.71	476.4050	57.95	11.00	17.63	13.42
C23H52N6O5	492.71	492.3999	56.07	10.64	17.06	16.24
C23H52N6O6	508.71	508.3948	54.31	10.30	16.52	18.87
C23H52N6O7	524.71	524.3897	52.65	9.99	16.02	21.34
C23H52N6O8	540.71	540.3846	51.09	9.69	15.54	23.67
C23H52N6O9	556.71	556.3795	49.62	9.42	15.10	25.87
C23H53N5O1	415.71	415.4250	66.45	12.85	16.85	3.85
C23H53N5O2	431.71	431.4199	63.99	12.37	16.22	7.41
C23H53N5O3	447.71	447.4148	61.70	11.93	15.64	10.72
C23H53N5O4	463.71	463.4097	59.58	11.52	15.10	13.80
C23H53N5O5	479.71	479.4046	57.59	11.14	14.60	16.68
C23H53N5O6	495.71	495.3996	55.73	10.78	14.13	19.37
C23H53N5O7	511.71	511.3945	53.99	10.44	13.69	21.89
C23H53N5O8	527.71	527.3894	52.35	10.12	13.27	24.25
C23H53N5O9	543.71	543.3843	50.81	9.83	12.88	26.48
C23H54N6O1	430.73	430.4359	64.14	12.64	19.51	3.71
C23H54N6O2	446.73	446.4308	61.84	12.18	18.81	7.16
C23H54N6O3	462.73	462.4257	59.70	11.76	18.16	10.37
C23H54N6O4	478.72	478.4206	57.71	11.37	17.56	13.37
C23H54N6O5	494.72	494.4155	55.84	11.00	16.99	16.17
C23H54N6O6	510.72	510.4105	54.09	10.66	16.46	18.80
C23H54N6O7	526.72	526.4054	52.45	10.33	15.96	21.26
C23H54N6O8	542.72	542.4003	50.90	10.03	15.48	23.58
C23H54N6O9	558.72	558.3952	49.44	9.74	15.04	25.77
C24H12N2O1	344.38	344.0950	83.71	3.51	8.13	4.65
C24H12N2O2	360.38	360.0899	79.99	3.36	7.77	8.88
C24H12N2O3	376.37	376.0848	76.59	3.21	7.44	12.75
C24H12N2O4	392.37	392.0797	73.47	3.08	7.14	16.31
C24H12N2O5	408.37	408.0746	70.59	2.96	6.86	19.59
C24H12N2O6	424.37	424.0695	67.93	2.85	6.60	22.62
C24H12N2O7	440.37	440.0644	65.46	2.75	6.36	25.43
C24H12N2O8	456.37	456.0594	63.17	2.65	6.14	28.05
C24H12N2O9	472.37	472.0543	61.03	2.56	5.93	30.48
C24H12N4O1	372.39	372.1011	77.41	3.25	15.05	4.30
C24H12N4O2	388.39	388.0960	74.22	3.11	14.43	8.24
C24H12N4O3	404.39	404.0909	71.28	2.99	13.85	11.87
C24H12N4O4	420.39	420.0858	68.57	2.88	13.33	15.22

C24H12N4O5	436.39	436.0808	66.06	2.77	12.84	18.33
C24H12N4O6	452.39	452.0757	63.72	2.67	12.38	21.22
C24H12N4O7	468.39	468.0706	61.54	2.58	11.96	23.91
C24H12N4O8	484.39	484.0655	59.51	2.50	11.57	26.42
C24H12N4O9	500.38	500.0604	57.61	2.42	11.20	28.78
C24H12N6O1	400.40	400.1073	71.99	3.02	20.99	4.00
C24H12N6O2	416.40	416.1022	69.23	2.90	20.18	7.68
C24H12N6O3	432.40	432.0971	66.67	2.80	19.44	11.10
C24H12N6O4	448.40	448.0920	64.29	2.70	18.74	14.27
C24H12N6O5	464.40	464.0869	62.07	2.60	18.10	17.23
C24H12N6O6	480.40	480.0818	60.01	2.52	17.49	19.98
C24H12N6O7	496.40	496.0767	58.07	2.44	16.93	22.56
C24H12N6O8	512.40	512.0717	56.26	2.36	16.40	24.98
C24H12N6O9	528.40	528.0666	54.56	2.29	15.90	27.25
C24H13N1O1	331.38	331.0597	86.99	3.95	4.23	4.83
C24H13N1O2	347.38	347.0946	82.98	3.77	4.03	9.21
C24H13N1O3	363.38	363.0895	79.33	3.61	3.85	13.21
C24H13N1O4	379.38	379.0844	75.98	3.45	3.69	16.87
C24H13N1O5	395.37	395.0794	72.91	3.31	3.54	20.23
C24H13N1O6	411.37	411.0743	70.07	3.19	3.40	23.34
C24H13N1O7	427.37	427.0692	67.45	3.07	3.28	26.21
C24H13N1O8	443.37	443.0641	65.02	2.96	3.16	28.87
C24H13N1O9	459.37	459.0590	62.75	2.85	3.05	31.35
C24H13N3O1	359.39	359.1059	80.21	3.65	11.69	4.45
C24H13N3O2	375.39	375.1008	76.79	3.49	11.19	8.52
C24H13N3O3	391.39	391.0957	73.65	3.35	10.74	12.26
C24H13N3O4	407.39	407.0906	70.76	3.22	10.31	15.71
C24H13N3O5	423.39	423.0855	68.09	3.09	9.92	18.89
C24H13N3O6	439.39	439.0804	65.61	2.98	9.56	21.85
C24H13N3O7	455.39	455.0753	63.30	2.88	9.23	24.59
C24H13N3O8	471.39	471.0703	61.15	2.78	8.91	27.15
C24H13N3O9	487.39	487.0652	59.15	2.69	8.62	29.54
C24H13N5O1	387.40	387.1120	74.41	3.38	18.08	4.13
C24H13N5O2	403.40	403.1069	71.46	3.25	17.36	7.93
C24H13N5O3	419.40	419.1018	68.73	3.12	16.70	11.44
C24H13N5O4	435.40	435.0967	66.21	3.01	16.08	14.70
C24H13N5O5	451.40	451.0917	63.86	2.90	15.51	17.72
C24H13N5O6	467.40	467.0866	61.67	2.80	14.98	20.54
C24H13N5O7	483.40	483.0815	59.63	2.71	14.49	23.17
C24H13N5O8	499.40	499.0764	57.72	2.62	14.02	25.63
C24H13N5O9	515.40	515.0713	55.93	2.54	13.59	27.94
C24H14N2O1	346.39	346.1106	83.22	4.07	8.09	4.62
C24H14N2O2	362.39	362.1055	79.55	3.89	7.73	8.83
C24H14N2O3	378.39	378.1004	76.18	3.73	7.40	12.68
C24H14N2O4	394.39	394.0953	73.09	3.58	7.10	16.23
C24H14N2O5	410.39	410.0903	70.24	3.44	6.83	19.49
C24H14N2O6	426.39	426.0852	67.61	3.31	6.57	22.51
C24H14N2O7	442.39	442.0801	65.16	3.19	6.33	25.32
C24H14N2O8	458.39	458.0750	62.89	3.08	6.11	27.92
C24H14N2O9	474.39	474.0699	60.77	2.97	5.91	30.35
C24H14N4O1	374.41	374.1168	76.99	3.77	14.96	4.27
C24H14N4O2	390.40	390.1117	73.84	3.61	14.35	8.20
C24H14N4O3	406.40	406.1066	70.93	3.47	13.79	11.81
C24H14N4O4	422.40	422.1015	68.24	3.34	13.26	15.15
C24H14N4O5	438.40	438.0964	65.75	3.22	12.78	18.25
C24H14N4O6	454.40	454.0913	63.44	3.11	12.33	21.13
C24H14N4O7	470.40	470.0862	61.28	3.00	11.91	23.81
C24H14N4O8	486.40	486.0812	59.27	2.90	11.52	26.31
C24H14N4O9	502.40	502.0761	57.38	2.81	11.15	28.66
C24H14N6O1	402.42	402.1229	71.63	3.51	20.88	3.98
C24H14N6O2	418.42	418.1178	68.89	3.37	20.09	7.65
C24H14N6O3	434.42	434.1127	66.36	3.25	19.35	11.05
C24H14N6O4	450.42	450.1076	64.00	3.13	18.66	14.21
C24H14N6O5	466.42	466.1026	61.80	3.03	18.02	17.15
C24H14N6O6	482.42	482.0975	59.76	2.93	17.42	19.90
C24H14N6O7	498.42	498.0924	57.84	2.83	16.86	22.47
C24H14N6O8	514.41	514.0873	56.04	2.74	16.34	24.88
C24H14N6O9	530.41	530.0822	54.35	2.66	15.84	27.15
C24H15N1O1	333.39	333.1154	86.46	4.54	4.20	4.80
C24H15N1O2	349.39	349.1103	82.51	4.33	4.01	9.16
C24H15N1O3	365.39	365.1052	78.89	4.14	3.83	13.14
C24H15N1O4	381.39	381.1001	75.58	3.96	3.67	16.78
C24H15N1O5	397.39	397.0950	72.54	3.80	3.52	20.13
C24H15N1O6	413.39	413.0899	69.73	3.66	3.39	23.22
C24H15N1O7	429.39	429.0848	67.13	3.52	3.26	26.08
C24H15N1O8	445.39	445.0798	64.72	3.39	3.14	28.74
C24H15N1O9	461.39	461.0747	62.48	3.28	3.04	31.21
C24H15N3O1	361.41	361.1215	79.76	4.18	11.63	4.43
C24H15N3O2	377.41	377.1164	76.38	4.01	11.13	8.48
C24H15N3O3	393.41	393.1113	73.27	3.84	10.68	12.20
C24H15N3O4	409.40	409.1062	70.41	3.69	10.26	15.63
C24H15N3O5	425.40	425.1012	67.76	3.55	9.88	18.80
C24H15N3O6	441.40	441.0961	65.31	3.43	9.52	21.75
C24H15N3O7	457.40	457.0910	63.02	3.31	9.19	24.49
C24H15N3O8	473.40	473.0859	60.89	3.19	8.88	27.04
C24H15N3O9	489.40	489.0808	58.90	3.09	8.59	29.42
C24H15N5O1	389.42	389.1277	74.02	3.88	17.98	4.11
C24H15N5O2	405.42	405.1226	71.10	3.73	17.27	7.89
C24H15N5O3	421.42	421.1175	68.40	3.59	16.62	11.39
C24H15N5O4	437.42	437.1124	65.90	3.46	16.01	14.63
C24H15N5O5	453.42	453.1073	63.58	3.33	15.45	17.64
C24H15N5O6	469.42	469.1022	61.41	3.22	14.92	20.45
C24H15N5O7	485.42	485.0971	59.39	3.11	14.43	23.07
C24H15N5O8	501.42	501.0921	57.49	3.02	13.97	25.53
C24H15N5O9	517.42	517.0870	55.71	2.92	13.54	27.83
C24H16N2O1	348.41	348.1263	82.74	4.63	8.04	4.59
C24H16N2O2	364.41	364.1212	79.11	4.43	7.69	8.78
C24H16N2O3	380.41	380.1161	75.78	4.24	7.36	12.62
C24H16N2O4	396.41	396.1110	72.72	4.07	7.07	16.14
C24H16N2O5	412.41	412.1059	69.90	3.91	6.79	19.40
C24H16N2O6	428.40	428.1008	67.29	3.76	6.54	22.41
C24H16N2O7	444.40	444.0957	64.87	3.63	6.30	25.20
C24H16N2O8	460.40	460.0907	62.61	3.50	6.08	27.80
C24H16N2O9	476.40	476.0856	60.51	3.39	5.88	30.23
C24H16N4O1	376.42	376.1324	76.58	4.28	14.88	4.25
C24H16N4O2	392.42	392.1273	73.46	4.11	14.28	8.15
C24H16N4O3	408.42	408.1222	70.58	3.95	13.72	11.75
C24H16N4O4	424.42	424.1171	67.92	3.80	13.20	15.08
C24H16N4O5	440.42	440.1121	65.45	3.66	12.72	18.16
C24H16N4O6	456.42	456.1070	63.16	3.53	12.28	21.03
C24H16N4O7	472.42	472.1019	61.02	3.41	11.86	23.71
C24H16N4O8	488.42	488.0968	59.02	3.30	11.47	26.21
C24H16N4O9	504.42	504.0917	57.15	3.20	11.11	28.55
C24H16N6O1	404.43	404.1386	71.28	3.99	20.78	3.96
C24H16N6O2	420.43	420.1335	68.56	3.84	19.99	7.61
C24H16N6O3	436.43	436.1284	66.05	3.70	19.26	11.00
C24H16N6O4	452.43	452.1233	63.71	3.56	18.58	14.15
C24H16N6O5	468.43	468.1182	61.54	3.44	17.94	17.08
C24H16N6O6	484.43	484.1131	59.51	3.33	17.35	19.82
C24H16N6O7	500.43	500.1080	57.60	3.22	16.79	22.38

C24H16N608	516.43	516.1029	55.82	3.12	16.27	24.78
C24H16N609	532.43	532.0979	54.14	3.03	15.78	27.04
C24H17N101	335.41	335.1310	85.95	5.11	4.18	4.77
C24H17N102	351.41	351.1259	82.03	4.88	3.99	9.11
C24H17N103	367.41	367.1208	78.46	4.66	3.81	13.06
C24H17N104	383.41	383.1157	75.19	4.47	3.65	16.69
C24H17N105	399.41	399.1107	72.17	4.29	3.51	20.03
C24H17N106	415.41	415.1056	69.39	4.12	3.37	23.11
C24H17N107	431.41	431.1005	66.82	3.97	3.25	25.96
C24H17N108	447.40	447.0954	64.43	3.83	3.13	28.61
C24H17N109	463.40	463.0903	62.21	3.70	3.02	31.07
C24H17N301	363.42	363.1372	79.32	4.72	11.56	4.40
C24H17N302	379.42	379.1321	75.98	4.52	11.07	8.43
C24H17N303	395.42	395.1270	72.90	4.33	10.63	12.14
C24H17N304	411.42	411.1219	70.07	4.16	10.21	15.56
C24H17N305	427.42	427.1168	67.44	4.01	9.83	18.72
C24H17N306	443.42	443.1117	65.01	3.86	9.48	21.65
C24H17N307	459.42	459.1066	62.75	3.73	9.15	24.38
C24H17N308	475.42	475.1016	60.63	3.60	8.84	26.92
C24H17N309	491.42	491.0965	58.66	3.49	8.55	29.30
C24H17N501	391.44	391.1433	73.64	4.38	17.89	4.09
C24H17N502	407.44	407.1382	70.75	4.21	17.19	7.85
C24H17N503	423.43	423.1331	68.08	4.05	16.54	11.34
C24H17N504	439.43	439.1280	65.60	3.90	15.94	14.56
C24H17N505	455.43	455.1230	63.30	3.76	15.38	17.57
C24H17N506	471.43	471.1179	61.15	3.63	14.86	20.36
C24H17N507	487.43	487.1128	59.14	3.52	14.37	22.98
C24H17N508	503.43	503.1077	57.26	3.40	13.91	25.42
C24H17N509	519.43	519.1026	55.50	3.30	13.48	27.72
C24H18N201	350.42	350.1419	82.26	5.18	7.99	4.57
C24H18N202	366.42	366.1368	78.67	4.95	7.65	8.73
C24H18N203	382.42	382.1317	75.38	4.74	7.33	12.55
C24H18N204	398.42	398.1266	72.35	4.55	7.03	16.06
C24H18N205	414.42	414.1216	69.56	4.38	6.76	19.30
C24H18N206	430.42	430.1165	66.97	4.22	6.51	22.30
C24H18N207	446.42	446.1114	64.57	4.06	6.28	25.09
C24H18N208	462.42	462.1063	62.34	3.92	6.06	27.68
C24H18N209	478.42	478.1012	60.25	3.79	5.86	30.10
C24H18N401	378.44	378.1481	76.17	4.79	14.80	4.23
C24H18N402	394.44	394.1430	73.08	4.60	14.20	8.11
C24H18N403	410.44	410.1379	70.23	4.42	13.65	11.69
C24H18N404	426.44	426.1328	67.60	4.25	13.14	15.01
C24H18N405	442.43	442.1277	65.15	4.10	12.66	18.08
C24H18N406	458.43	458.1226	62.88	3.96	12.22	20.94
C24H18N407	474.43	474.1175	60.76	3.82	11.81	23.61
C24H18N408	490.43	490.1125	58.78	3.70	11.42	26.10
C24H18N409	506.43	506.1074	56.92	3.58	11.06	28.43
C24H18N601	406.45	406.1542	70.92	4.46	20.68	3.94
C24H18N602	422.45	422.1491	68.24	4.29	19.89	7.57
C24H18N603	438.45	438.1440	65.75	4.14	19.17	10.95
C24H18N604	454.45	454.1389	63.43	3.99	18.49	14.08
C24H18N605	470.45	470.1339	61.28	3.86	17.86	17.00
C24H18N606	486.45	486.1288	59.26	3.73	17.28	19.73
C24H18N607	502.45	502.1237	57.37	3.61	16.73	22.29
C24H18N608	518.45	518.1186	55.60	3.50	16.21	24.69
C24H18N609	534.45	534.1135	53.94	3.39	15.72	26.94
C24H19N101	337.43	337.1467	85.43	5.68	4.15	4.74
C24H19N102	353.42	353.1416	81.56	5.42	3.96	9.05
C24H19N103	369.42	369.1365	78.03	5.18	3.79	12.99
C24H19N104	385.42	385.1314	74.79	4.97	3.63	16.60
C24H19N105	401.42	401.1263	71.81	4.77	3.49	19.93
C24H19N106	417.42	417.1212	69.06	4.59	3.36	23.00
C24H19N107	433.42	433.1161	66.51	4.42	3.23	25.84
C24H19N108	449.42	449.1111	64.14	4.26	3.12	28.48
C24H19N109	465.42	465.1060	61.94	4.11	3.01	30.94
C24H19N301	365.44	365.1528	78.88	5.24	11.50	4.38
C24H19N302	381.44	381.1477	75.57	5.02	11.02	8.39
C24H19N303	397.44	397.1426	72.53	4.82	10.57	12.08
C24H19N304	413.44	413.1375	69.72	4.63	10.16	15.48
C24H19N305	429.44	429.1325	67.13	4.46	9.78	18.63
C24H19N306	445.44	445.1274	64.72	4.30	9.43	21.55
C24H19N307	461.43	461.1223	62.47	4.15	9.11	24.27
C24H19N308	477.43	477.1172	60.38	4.01	8.80	26.81
C24H19N309	493.43	493.1121	58.42	3.88	8.52	29.18
C24H19N501	393.45	393.1590	73.27	4.87	17.80	4.07
C24H19N502	409.45	409.1539	70.40	4.68	17.10	7.82
C24H19N503	425.45	425.1488	67.76	4.50	16.46	11.28
C24H19N504	441.45	441.1437	65.30	4.34	15.86	14.50
C24H19N505	457.45	457.1386	63.02	4.19	15.31	17.49
C24H19N506	473.45	473.1335	60.89	4.05	14.79	20.28
C24H19N507	489.45	489.1284	58.90	3.91	14.31	22.88
C24H19N508	505.45	505.1233	57.03	3.79	13.86	25.32
C24H19N509	521.45	521.1183	55.28	3.67	13.43	27.61
C24H20N201	352.44	352.1576	81.79	5.72	7.95	4.54
C24H20N202	368.44	368.1525	78.24	5.47	7.60	8.68
C24H20N203	384.44	384.1474	74.98	5.24	7.29	12.49
C24H20N204	400.44	400.1423	71.99	5.03	7.00	15.98
C24H20N205	416.44	416.1372	69.22	4.84	6.73	19.21
C24H20N206	432.44	432.1321	66.66	4.66	6.48	22.20
C24H20N207	448.44	448.1270	64.28	4.50	6.25	24.97
C24H20N208	464.44	464.1220	62.07	4.34	6.03	27.56
C24H20N209	480.44	480.1169	60.00	4.20	5.83	29.97
C24H20N401	380.45	380.1637	75.77	5.30	14.73	4.21
C24H20N402	396.45	396.1586	72.71	5.08	14.13	8.07
C24H20N403	412.45	412.1535	69.89	4.89	13.58	11.64
C24H20N404	428.45	428.1484	67.28	4.71	13.08	14.94
C24H20N405	444.45	444.1434	64.86	4.54	12.61	18.00
C24H20N406	460.45	460.1383	62.61	4.38	12.17	20.85
C24H20N407	476.45	476.1332	60.50	4.23	11.76	23.51
C24H20N408	492.45	492.1281	58.54	4.09	11.38	25.99
C24H20N409	508.45	508.1230	56.70	3.96	11.02	28.32
C24H20N601	408.47	408.1698	70.57	4.94	20.57	3.92
C24H20N602	424.47	424.1648	67.91	4.75	19.80	7.54
C24H20N603	440.47	440.1597	65.45	4.58	19.08	10.90
C24H20N604	456.46	456.1546	63.15	4.42	18.41	14.02
C24H20N605	472.46	472.1495	61.01	4.27	17.79	16.93
C24H20N606	488.46	488.1444	59.02	4.13	17.21	19.65
C24H20N607	504.46	504.1393	57.14	4.00	16.66	22.20
C24H20N608	520.46	520.1342	55.39	3.87	16.15	24.59
C24H20N609	536.46	536.1292	53.73	3.76	15.67	26.84
C24H21N101	339.44	339.1623	84.92	6.24	4.13	4.71
C24H21N102	355.44	355.1572	81.10	5.96	3.94	9.00
C24H21N103	371.44	371.1521	77.61	5.70	3.77	12.92
C24H21N104	387.44	387.1470	74.40	5.46	3.62	16.52
C24H21N105	403.44	403.1420	71.45	5.25	3.47	19.83
C24H21N106	419.44	419.1369	68.73	5.05	3.34	22.89
C24H21N107	435.44	435.1318	66.20	4.86	3.22	25.72
C24H21N108	451.44	451.1267	63.86	4.69	3.10	28.35
C24H21N109	467.44	467.1216	61.67	4.53	3.00	30.81
C24H21N301	367.45	367.1685	78.45	5.76	11.44	4.35

149

C24H21N302	383.45	383.1634	75.18	5.52	10.96	8.34
C24H21N303	399.45	399.1583	72.17	5.30	10.52	12.02
C24H21N304	415.45	415.1532	69.39	5.10	10.11	15.40
C24H21N305	431.45	431.1481	66.81	4.91	9.74	18.54
C24H21N306	447.45	447.1430	64.42	4.73	9.39	21.45
C24H21N307	463.45	463.1379	62.20	4.57	9.07	24.17
C24H21N308	479.45	479.1329	60.12	4.41	8.76	26.70
C24H21N309	495.45	495.1278	58.18	4.27	8.48	29.06
C24H21N501	395.47	395.1746	72.89	5.35	17.71	4.05
C24H21N502	411.47	411.1695	70.06	5.14	17.02	7.78
C24H21N503	427.47	427.1644	67.44	4.95	16.38	11.23
C24H21N504	443.47	443.1593	65.00	4.77	15.79	14.43
C24H21N505	459.47	459.1543	62.74	4.61	15.24	17.41
C24H21N506	475.46	475.1492	60.63	4.45	14.73	20.19
C24H21N507	491.46	491.1441	58.65	4.31	14.25	22.79
C24H21N508	507.46	507.1390	56.81	4.17	13.80	25.22
C24H21N509	523.46	523.1339	55.07	4.04	13.38	27.51
C24H22N201	354.46	354.1732	81.33	6.26	7.90	4.51
C24H22N202	370.46	370.1681	77.81	5.99	7.56	8.64
C24H22N203	386.45	386.1630	74.59	5.74	7.25	12.42
C24H22N204	402.45	402.1579	71.63	5.51	6.96	15.90
C24H22N205	418.45	418.1529	68.89	5.30	6.69	19.12
C24H22N206	434.45	434.1478	66.35	5.10	6.45	22.10
C24H22N207	450.45	450.1427	64.00	4.92	6.22	24.86
C24H22N208	466.45	466.1376	61.80	4.75	6.01	27.44
C24H22N209	482.45	482.1325	59.75	4.60	5.81	29.85
C24H22N401	382.47	382.1794	75.37	5.80	14.65	4.18
C24H22N402	398.47	398.1743	72.34	5.57	14.06	8.03
C24H22N403	414.47	414.1692	69.55	5.35	13.52	11.58
C24H22N404	430.47	430.1641	66.97	5.15	13.02	14.87
C24H22N405	446.47	446.1590	64.57	4.97	12.55	17.92
C24H22N406	462.47	462.1539	62.33	4.80	12.11	20.76
C24H22N407	478.47	478.1488	60.25	4.63	11.71	23.41
C24H22N408	494.46	494.1437	58.30	4.48	11.33	25.89
C24H22N409	510.46	510.1387	56.47	4.34	10.98	28.21
C24H22N601	410.48	410.1855	70.23	5.40	20.47	3.90
C24H22N602	426.48	426.1804	67.59	5.20	19.71	7.50
C24H22N603	442.48	442.1753	65.15	5.01	18.99	10.85
C24H22N604	458.48	458.1702	62.87	4.84	18.33	13.9[illegible]
C24H22N605	474.48	474.1652	60.75	4.67	17.71	16.86
C24H22N606	490.48	490.1601	58.77	4.52	17.13	19.57
C24H22N607	506.48	506.1550	56.92	4.38	16.59	22.11
C24H22N608	522.48	522.1499	55.17	4.24	16.08	24.50
C24H22N609	538.48	538.1448	53.53	4.12	15.61	26.74
C24H23N101	341.46	341.1780	84.42	6.79	4.10	4.69
C24H23N102	357.46	357.1729	80.64	6.49	3.92	8.95
C24H23N103	373.46	373.1678	77.19	6.21	3.75	12.85
C24H23N104	389.46	389.1627	74.02	5.95	3.60	16.43
C24H23N105	405.45	405.1576	71.10	5.72	3.45	19.73
C24H23N106	421.45	421.1525	68.40	5.50	3.32	22.78
C24H23N107	437.45	437.1474	65.90	5.30	3.20	25.60
C24H23N108	453.45	453.1424	63.57	5.11	3.09	28.23
C24H23N109	469.45	469.1373	61.41	4.94	2.98	30.67
C24H23N301	369.47	369.1841	78.02	6.27	11.37	4.33
C24H23N302	385.47	385.1790	74.78	6.01	10.90	8.30
C24H23N303	401.47	401.1739	71.80	5.77	10.47	11.96
C24H23N304	417.47	417.1688	69.05	5.55	10.07	15.33
C24H23N305	433.47	433.1638	66.50	5.35	9.69	18.46
C24H23N306	449.47	449.1587	64.14	5.16	9.35	21.36
C24H23N307	465.47	465.1536	61.93	4.98	9.03	24.06
C24H23N308	481.47	481.1485	59.87	4.82	8.73	26.58
C24H23N309	497.47	497.1434	57.95	4.66	8.45	28.95
C24H23N501	397.48	397.1902	72.52	5.83	17.62	4.03
C24H23N502	413.48	413.1852	69.72	5.61	16.94	7.74
C24H23N503	429.48	429.1801	67.12	5.40	16.31	11.18
C24H23N504	445.48	445.1750	64.71	5.20	15.72	14.37
C24H23N505	461.48	461.1699	62.47	5.02	15.18	17.33
C24H23N506	477.48	477.1648	60.37	4.86	14.67	20.10
C24H23N507	493.48	493.1597	58.42	4.70	14.19	22.70
C24H23N508	509.48	509.1546	56.58	4.55	13.75	25.12
C24H23N509	525.48	525.1496	54.86	4.41	13.33	27.40
C24H24N201	356.47	356.1889	80.87	6.79	7.86	4.49
C24H24N202	372.47	372.1838	77.39	6.49	7.52	8.59
C24H24N203	388.47	388.1787	74.21	6.23	7.21	12.36
C24H24N204	404.47	404.1736	71.27	5.98	6.93	15.82
C24H24N205	420.47	420.1685	68.56	5.75	6.66	19.03
C24H24N206	436.47	436.1634	66.05	5.54	6.42	21.99
C24H24N207	452.47	452.1583	63.71	5.35	6.19	24.75
C24H24N208	468.47	468.1533	61.53	5.16	5.98	27.32
C24H24N209	484.47	484.1482	59.50	4.99	5.78	29.72
C24H24N401	384.49	384.1950	74.97	6.29	14.57	4.16
C24H24N402	400.48	400.1899	71.98	6.04	13.99	7.99
C24H24N403	416.48	416.1848	69.21	5.81	13.45	11.52
C24H24N404	432.48	432.1797	66.65	5.59	12.95	14.80
C24H24N405	448.48	448.1747	64.28	5.39	12.49	17.84
C24H24N406	464.48	464.1696	62.06	5.21	12.06	20.67
C24H24N407	480.48	480.1645	60.00	5.03	11.66	23.31
C24H24N408	496.48	496.1594	58.06	4.87	11.28	25.78
C24H24N409	512.48	512.1543	56.25	4.72	10.93	28.10
C24H24N601	412.50	412.2011	69.88	5.86	20.37	3.88
C24H24N602	428.50	428.1961	67.27	5.65	19.61	7.47
C24H24N603	444.50	444.1910	64.85	5.44	18.91	10.80
C24H24N604	460.50	460.1859	62.60	5.25	18.25	13.90
C24H24N605	476.50	476.1808	60.50	5.08	17.64	16.79
C24H24N606	492.50	492.1757	58.53	4.91	17.06	19.49
C24H24N607	508.49	508.1706	56.69	4.76	16.53	22.02
C24H24N608	524.49	524.1655	54.96	4.61	16.02	24.40
C24H24N609	540.49	540.1605	53.33	4.48	15.55	26.64
C24H25N101	343.47	343.1936	83.93	7.34	4.08	4.66
C24H25N102	359.47	359.1885	80.19	7.01	3.90	8.90
C24H25N103	375.47	375.1834	76.77	6.71	3.73	12.78
C24H25N104	391.47	391.1783	73.64	6.44	3.58	16.35
C24H25N105	407.47	407.1733	70.75	6.18	3.44	19.63
C24H25N106	423.47	423.1682	68.07	5.95	3.31	22.67
C24H25N107	439.47	439.1631	65.59	5.73	3.19	25.48
C24H25N108	455.47	455.1580	63.29	5.53	3.08	28.10
C24H25N109	471.47	471.1529	61.14	5.34	2.97	30.54
C24H25N301	371.49	371.1998	77.60	6.78	11.31	4.31
C24H25N302	387.49	387.1947	74.39	6.50	10.84	8.26
C24H25N303	403.49	403.1896	71.44	6.25	10.41	11.90
C24H25N304	419.48	419.1845	68.72	6.01	10.02	15.26
C24H25N305	435.48	435.1794	66.19	5.79	9.65	18.37
C24H25N306	451.48	451.1743	63.85	5.58	9.31	21.26
C24H25N307	467.48	467.1692	61.66	5.39	8.99	23.96
C24H25N308	483.48	483.1641	59.62	5.21	8.69	26.47
C24H25N309	499.48	499.1591	57.71	5.05	8.41	28.83
C24H25N501	399.50	399.2059	72.16	6.31	17.53	4.06
C24H25N502	415.50	415.2008	69.38	6.06	16.86	7.70
C24H25N503	431.50	431.1957	66.81	5.84	16.23	11.12
C24H25N504	447.50	447.1906	64.42	5.63	15.65	14.30

150

C24H25N5O5	463.50	463.1856	62.19	5.44	15.11	17.26
C24H25N5O6	479.50	479.1805	60.12	5.26	14.61	20.02
C24H25N5O7	495.50	495.1754	58.18	5.09	14.13	22.60
C24H25N5O8	511.50	511.1703	56.36	4.93	13.69	25.02
C24H25N5O9	527.49	527.1652	54.65	4.78	13.28	27.30
C24H26N2O1	358.49	358.2045	80.41	7.31	7.81	4.46
C24H26N2O2	374.49	374.1994	76.98	7.00	7.48	8.54
C24H26N2O3	390.49	390.1943	73.82	6.71	7.17	12.29
C24H26N2O4	406.49	406.1892	70.92	6.45	6.89	15.74
C24H26N2O5	422.49	422.1842	68.23	6.20	6.63	18.93
C24H26N2O6	438.48	438.1791	65.74	5.98	6.39	21.89
C24H26N2O7	454.48	454.1740	63.43	5.77	6.16	24.64
C24H26N2O8	470.48	470.1689	61.27	5.57	5.95	27.21
C24H26N2O9	486.48	486.1638	59.26	5.39	5.76	29.60
C24H26N4O1	386.50	386.2106	74.58	6.78	14.50	4.14
C24H26N4O2	402.50	402.2056	71.62	6.51	13.92	7.95
C24H26N4O3	418.50	418.2005	68.88	6.26	13.39	11.47
C24H26N4O4	434.50	434.1954	66.34	6.03	12.89	14.73
C24H26N4O5	450.50	450.1903	63.99	5.82	12.44	17.76
C24H26N4O6	466.50	466.1852	61.79	5.62	12.01	20.58
C24H26N4O7	482.50	482.1801	59.74	5.43	11.61	23.21
C24H26N4O8	498.50	498.1750	57.83	5.26	11.24	25.68
C24H26N4O9	514.50	514.1700	56.03	5.09	10.89	27.99
C24H26N6O1	414.51	414.2168	69.54	6.32	20.27	3.86
C24H26N6O2	430.51	430.2117	66.96	6.09	19.52	7.43
C24H26N6O3	446.51	446.2066	64.56	5.87	18.82	10.75
C24H26N6O4	462.51	462.2015	62.33	5.67	18.17	13.84
C24H26N6O5	478.51	478.1965	60.24	5.48	17.56	16.72
C24H26N6O6	494.51	494.1914	58.29	5.30	16.99	19.41
C24H26N6O7	510.51	510.1863	56.47	5.13	16.46	21.94
C24H26N6O8	526.51	526.1812	54.75	4.98	15.96	24.31
C24H26N6O9	542.51	542.1761	53.14	4.83	15.49	26.54
C24H27N1O1	345.49	345.2093	83.44	7.88	4.05	4.63
C24H27N1O2	361.49	361.2042	79.74	7.53	3.87	8.85
C24H27N1O3	377.49	377.1991	76.36	7.21	3.71	12.72
C24H27N1O4	393.49	393.1940	73.26	6.92	3.56	16.26
C24H27N1O5	409.49	409.1889	70.40	6.65	3.42	19.54
C24H27N1O6	425.49	425.1838	67.75	6.40	3.29	22.56
C24H27N1O7	441.49	441.1787	65.29	6.16	3.17	25.37
C24H27N1O8	457.48	457.1737	63.01	5.95	3.06	27.98
C24H27N1O9	473.48	473.1686	60.88	5.75	2.96	30.41
C24H27N3O1	373.50	373.2154	77.18	7.29	11.25	4.28
C24H27N3O2	389.50	389.2103	74.01	6.99	10.79	8.22
C24H27N3O3	405.50	405.2052	71.09	6.71	10.36	11.84
C24H27N3O4	421.50	421.2001	68.39	6.46	9.97	15.18
C24H27N3O5	437.50	437.1951	65.89	6.22	9.60	18.29
C24H27N3O6	453.50	453.1900	63.57	6.00	9.27	21.17
C24H27N3O7	469.50	469.1849	61.40	5.80	8.95	23.85
C24H27N3O8	485.50	485.1798	59.38	5.61	8.66	26.36
C24H27N3O9	501.50	501.1747	57.48	5.43	8.38	28.71
C24H27N5O1	401.52	401.2215	71.79	6.78	17.44	3.98
C24H27N5O2	417.52	417.2165	69.04	6.52	16.77	7.66
C24H27N5O3	433.51	433.2114	66.50	6.28	16.15	11.07
C24H27N5O4	449.51	449.2063	64.13	6.05	15.58	14.24
C24H27N5O5	465.51	465.2012	61.92	5.85	15.04	17.18
C24H27N5O6	481.51	481.1961	59.87	5.65	14.54	19.94
C24H27N5O7	497.51	497.1910	57.94	5.47	14.08	22.51
C24H27N5O8	513.51	513.1859	56.14	5.30	13.64	24.93
C24H27N5O9	529.51	529.1809	54.44	5.14	13.23	27.19
C24H28N2O1	360.50	360.2202	79.96	7.83	7.77	4.44
C24H28N2O2	376.50	376.2151	76.56	7.50	7.44	8.50
C24H28N2O3	392.50	392.2100	73.44	7.19	7.14	12.23
C24H28N2O4	408.50	408.2049	70.57	6.91	6.86	15.67
C24H28N2O5	424.50	424.1998	67.91	6.65	6.60	18.84
C24H28N2O6	440.50	440.1947	65.44	6.41	6.36	21.79
C24H28N2O7	456.50	456.1896	63.15	6.18	6.14	24.53
C24H28N2O8	472.50	472.1845	61.01	5.97	5.93	27.09
C24H28N2O9	488.50	488.1795	59.01	5.78	5.73	29.48
C24H28N4O1	388.52	388.2263	74.20	7.26	14.42	4.12
C24H28N4O2	404.52	404.2212	71.26	6.98	13.85	7.91
C24H28N4O3	420.52	420.2161	68.55	6.71	13.32	11.41
C24H28N4O4	436.52	436.2110	66.04	6.47	12.84	14.66
C24H28N4O5	452.51	452.2060	63.70	6.24	12.38	17.68
C24H28N4O6	468.51	468.2009	61.53	6.02	11.96	20.49
C24H28N4O7	484.51	484.1958	59.50	5.83	11.56	23.12
C24H28N4O8	500.51	500.1907	57.59	5.64	11.19	25.57
C24H28N4O9	516.51	516.1856	55.81	5.46	10.85	27.88
C24H28N6O1	416.53	416.2324	69.21	6.78	20.18	3.84
C24H28N6O2	432.53	432.2274	66.65	6.53	19.43	7.40
C24H28N6O3	448.53	448.2223	64.27	6.29	18.74	10.70
C24H28N6O4	464.53	464.2172	62.06	6.08	18.09	13.78
C24H28N6O5	480.53	480.2121	59.99	5.87	17.49	16.65
C24H28N6O6	496.53	496.2070	58.06	5.68	16.93	19.33
C24H28N6O7	512.53	512.2019	56.24	5.51	16.40	21.85
C24H28N6O8	528.53	528.1968	54.54	5.34	15.90	24.22
C24H28N6O9	544.53	544.1918	52.94	5.18	15.43	26.44
C24H29N1O1	347.50	347.2249	82.95	8.41	4.03	4.60
C24H29N1O2	363.50	363.2198	79.30	8.04	3.85	8.80
C24H29N1O3	379.50	379.2147	75.96	7.70	3.69	12.65
C24H29N1O4	395.50	395.2096	72.89	7.39	3.54	16.18
C24H29N1O5	411.50	411.2046	70.05	7.10	3.40	19.44
C24H29N1O6	427.50	427.1995	67.43	6.84	3.28	22.46
C24H29N1O7	443.50	443.1944	65.00	6.59	3.16	25.25
C24H29N1O8	459.50	459.1893	62.73	6.36	3.05	27.86
C24H29N1O9	475.50	475.1842	60.62	6.15	2.95	30.28
C24H29N3O1	375.52	375.2310	76.77	7.78	11.19	4.26
C24H29N3O2	391.52	391.2260	73.63	7.47	10.73	8.17
C24H29N3O3	407.52	407.2209	70.74	7.17	10.31	11.78
C24H29N3O4	423.52	423.2158	68.07	6.90	9.92	15.11
C24H29N3O5	439.52	439.2107	65.59	6.65	9.56	18.20
C24H29N3O6	455.52	455.2056	63.28	6.42	9.22	21.07
C24H29N3O7	471.51	471.2005	61.14	6.20	8.91	23.75
C24H29N3O8	487.51	487.1954	59.13	6.00	8.62	26.25
C24H29N3O9	503.51	503.1904	57.25	5.81	8.35	28.60
C24H29N5O1	403.53	403.2372	71.44	7.24	17.36	3.96
C24H29N5O2	419.53	419.2321	68.71	6.97	16.69	7.63
C24H29N5O3	435.53	435.2270	66.19	6.71	16.08	11.02
C24H29N5O4	451.53	451.2219	63.84	6.47	15.51	14.17
C24H29N5O5	467.53	467.2169	61.66	6.25	14.98	17.11
C24H29N5O6	483.53	483.2118	59.62	6.05	14.48	19.85
C24H29N5O7	499.53	499.2067	57.71	5.85	14.02	22.42
C24H29N5O8	515.53	515.2016	55.92	5.67	13.58	24.83
C24H29N5O9	531.53	531.1965	54.23	5.50	13.18	27.09
C24H30N2O1	362.52	362.2358	79.52	8.34	7.73	4.41
C24H30N2O2	378.52	378.2307	76.16	7.99	7.40	8.45
C24H30N2O3	394.52	394.2256	73.07	7.66	7.10	12.17
C24H30N2O4	410.52	410.2205	70.22	7.37	6.82	15.59
C24H30N2O5	426.52	426.2155	67.59	7.09	6.57	18.76
C24H30N2O6	442.52	442.2104	65.14	6.83	6.33	21.69
C24H30N2O7	458.52	458.2053	62.87	6.59	6.11	24.43

151

C24H30N2O8	474.52	474.2002	60.75	6.37	5.90	26.97
C24H30N2O9	490.51	490.1951	58.77	6.16	5.71	29.36
C24H30N4O1	390.53	390.2419	73.81	7.74	14.35	4.10
C24H30N4O2	406.53	406.2369	70.91	7.44	13.78	7.87
C24H30N4O3	422.53	422.2318	68.22	7.16	13.26	11.36
C24H30N4O4	438.53	438.2267	65.78	6.90	12.78	14.59
C24H30N4O5	454.53	454.2216	63.42	6.65	12.33	17.60
C24H30N4O6	470.53	470.2165	61.26	6.43	11.91	20.40
C24H30N4O7	486.53	486.2114	59.25	6.22	11.52	23.02
C24H30N4O8	502.53	502.2063	57.36	6.02	11.15	25.47
C24H30N4O9	518.53	518.2013	55.59	5.83	10.80	27.77
C24H30N6O1	418.55	418.2481	68.87	7.22	20.08	3.82
C24H30N6O2	434.55	434.2430	66.34	6.96	19.34	7.36
C24H30N6O3	450.55	450.2379	63.98	6.71	18.65	10.65
C24H30N6O4	466.54	466.2328	61.79	6.48	18.01	13.72
C24H30N6O5	482.54	482.2278	59.74	6.27	17.42	16.58
C24H30N6O6	498.54	498.2227	57.82	6.07	16.86	19.26
C24H30N6O7	514.54	514.2176	56.02	5.88	16.33	21.77
C24H30N6O8	530.54	530.2125	54.33	5.70	15.84	24.13
C24H30N6O9	546.54	546.2074	52.74	5.53	15.38	26.35
C24H31N1O1	349.52	349.2406	82.48	8.94	4.01	4.58
C24H31N1O2	365.52	365.2355	78.87	8.55	3.83	8.75
C24H31N1O3	381.52	381.2304	75.56	8.19	3.67	12.58
C24H31N1O4	397.52	397.2253	72.52	7.86	3.52	16.10
C24H31N1O5	413.52	413.2202	69.71	7.56	3.39	19.35
C24H31N1O6	429.52	429.2151	67.11	7.27	3.26	22.35
C24H31N1O7	445.52	445.2100	64.70	7.01	3.14	25.14
C24H31N1O8	461.52	461.2049	62.46	6.77	3.03	27.73
C24H31N1O9	477.52	477.1999	60.37	6.54	2.93	30.15
C24H31N3O1	377.53	377.2467	76.36	8.28	11.13	4.24
C24H31N3O2	393.53	393.2416	73.25	7.94	10.68	8.13
C24H31N3O3	409.53	409.2365	70.39	7.63	10.26	11.72
C24H31N3O4	425.53	425.2314	67.74	7.34	9.87	15.04
C24H31N3O5	441.53	441.2264	65.29	7.08	9.52	18.12
C24H31N3O6	457.53	457.2213	63.01	6.83	9.18	20.98
C24H31N3O7	473.53	473.2162	60.88	6.60	8.87	23.65
C24H31N3O8	489.53	489.2111	58.89	6.38	8.58	26.15
C24H31N3O9	505.53	505.2060	57.02	6.18	8.31	28.48
C24H31N5O1	405.55	405.2528	71.08	7.70	17.27	3.95
C24H31N5O2	421.55	421.2478	68.38	7.41	16.61	7.59
C24H31N5O3	437.55	437.2427	65.88	7.14	16.01	10.97
C24H31N5O4	453.55	453.2376	63.56	6.89	15.44	14.11
C24H31N5O5	469.55	469.2325	61.39	6.65	14.92	17.04
C24H31N5O6	485.54	485.2274	59.37	6.44	14.42	19.77
C24H31N5O7	501.54	501.2223	57.48	6.23	13.96	22.33
C24H31N5O8	517.54	517.2172	55.70	6.04	13.53	24.73
C24H31N5O9	533.54	533.2122	54.03	5.86	13.13	26.99
C24H32N2O1	364.54	364.2514	79.08	8.85	7.68	4.39
C24H32N2O2	380.53	380.2464	75.75	8.48	7.36	8.41
C24H32N2O3	396.53	396.2413	72.70	8.13	7.06	12.10
C24H32N2O4	412.53	412.2362	69.88	7.82	6.79	15.51
C24H32N2O5	428.53	428.2311	67.27	7.53	6.54	18.67
C24H32N2O6	444.53	444.2260	64.85	7.26	6.30	21.59
C24H32N2O7	460.53	460.2209	62.59	7.00	6.08	24.32
C24H32N2O8	476.53	476.2158	60.49	6.77	5.88	26.86
C24H32N2O9	492.53	492.2108	58.53	6.55	5.69	29.24
C24H32N4O1	392.55	392.2576	73.43	8.22	14.27	4.08
C24H32N4O2	408.55	408.2525	70.56	7.90	13.71	7.83
C24H32N4O3	424.55	424.2474	67.90	7.60	13.20	11.31
C24H32N4O4	440.55	440.2423	65.43	7.32	12.72	14.53
C24H32N4O5	456.55	456.2373	63.14	7.07	12.27	17.52
C24H32N4O6	472.55	472.2322	61.00	6.83	11.86	20.31
C24H32N4O7	488.55	488.2271	59.01	6.60	11.47	22.92
C24H32N4O8	504.54	504.2220	57.13	6.39	11.10	25.37
C24H32N4O9	520.54	520.2169	55.38	6.20	10.76	27.66
C24H32N6O1	420.56	420.2637	68.54	7.67	19.98	3.80
C24H32N6O2	436.56	436.2587	66.03	7.39	19.25	7.33
C24H32N6O3	452.56	452.2536	63.70	7.13	18.57	10.61
C24H32N6O4	468.56	468.2485	61.52	6.88	17.94	13.66
C24H32N6O5	484.56	484.2434	59.49	6.66	17.34	16.51
C24H32N6O6	500.56	500.2383	57.59	6.44	16.79	19.18
C24H32N6O7	516.56	516.2332	55.81	6.24	16.27	21.68
C24H32N6O8	532.56	532.2281	54.13	6.06	15.78	24.03
C24H32N6O9	548.56	548.2231	52.55	5.88	15.32	26.25
C24H33N1O1	351.54	351.2562	82.00	9.46	3.98	4.55
C24H33N1O2	367.54	367.2511	78.43	9.05	3.81	8.71
C24H33N1O3	383.54	383.2460	75.16	8.67	3.65	12.51
C24H33N1O4	399.53	399.2409	72.15	8.33	3.51	16.02
C24H33N1O5	415.53	415.2359	69.37	8.00	3.37	19.25
C24H33N1O6	431.53	431.2308	66.80	7.71	3.25	22.25
C24H33N1O7	447.53	447.2257	64.41	7.43	3.13	25.03
C24H33N1O8	463.53	463.2206	62.19	7.18	3.02	27.61
C24H33N1O9	479.53	479.2155	60.11	6.94	2.92	30.03
C24H33N3O1	379.55	379.2623	75.95	8.76	11.07	4.22
C24H33N3O2	395.55	395.2573	72.88	8.41	10.62	8.09
C24H33N3O3	411.55	411.2522	70.04	8.08	10.21	11.66
C24H33N3O4	427.55	427.2471	67.42	7.78	9.83	14.97
C24H33N3O5	443.55	443.2420	64.99	7.50	9.47	18.04
C24H33N3O6	459.55	459.2369	62.73	7.24	9.14	20.89
C24H33N3O7	475.55	475.2318	60.62	6.99	8.84	23.55
C24H33N3O8	491.55	491.2267	58.65	6.77	8.55	26.04
C24H33N3O9	507.55	507.2217	56.80	6.55	8.28	28.37
C24H33N5O1	407.56	407.2685	70.73	8.16	17.18	3.93
C24H33N5O2	423.56	423.2634	68.06	7.85	16.53	7.55
C24H33N5O3	439.56	439.2583	65.58	7.57	15.93	10.92
C24H33N5O4	455.56	455.2532	63.28	7.30	15.37	14.05
C24H33N5O5	471.56	471.2482	61.13	7.05	14.85	16.96
C24H33N5O6	487.56	487.2431	59.12	6.82	14.36	19.69
C24H33N5O7	503.56	503.2380	57.25	6.61	13.91	22.24
C24H33N5O8	519.56	519.2329	55.48	6.40	13.48	24.64
C24H33N5O9	535.56	535.2278	53.83	6.21	13.08	26.89
C24H34N2O1	366.55	366.2671	78.64	9.35	7.64	4.36
C24H34N2O2	382.55	382.2620	75.35	8.96	7.32	8.36
C24H34N2O3	398.55	398.2569	72.33	8.60	7.03	12.04
C24H34N2O4	414.55	414.2518	69.54	8.27	6.76	15.44
C24H34N2O5	430.55	430.2468	66.95	7.96	6.51	18.58
C24H34N2O6	446.55	446.2417	64.55	7.67	6.27	21.50
C24H34N2O7	462.55	462.2366	62.32	7.41	6.06	24.21
C24H34N2O8	478.55	478.2315	60.24	7.16	5.85	26.75
C24H34N2O9	494.55	494.2264	58.29	6.93	5.66	29.12
C24H34N4O1	394.56	394.2732	73.06	8.69	14.20	4.05
C24H34N4O2	410.56	410.2682	70.21	8.35	13.65	7.79
C24H34N4O3	426.56	426.2631	67.58	8.03	13.13	11.25
C24H34N4O4	442.56	442.2580	65.14	7.74	12.66	14.46
C24H34N4O5	458.56	458.2529	62.86	7.47	12.22	17.45
C24H34N4O6	474.56	474.2478	60.74	7.22	11.81	20.23
C24H34N4O7	490.56	490.2427	58.76	6.99	11.42	22.83
C24H34N4O8	506.56	506.2376	56.91	6.77	11.06	25.27
C24H34N4O9	522.56	522.2326	55.16	6.56	10.72	27.56
C24H34N6O1	422.58	422.2794	68.22	8.11	19.89	3.79

C24H34N602	438.58	438.2743	65.73	7.81	19.16	7.30
C24H34N603	454.58	454.2692	63.41	7.54	18.49	10.56
C24H34N604	470.58	470.2641	61.26	7.28	17.86	13.60
C24H34N605	486.58	486.2590	59.24	7.04	17.27	16.44
C24H34N606	502.58	502.2540	57.36	6.82	16.72	19.10
C24H34N607	518.57	518.2489	55.59	6.61	16.21	21.60
C24H34N608	534.57	534.2438	53.92	6.41	15.72	23.94
C24H34N609	550.57	550.2387	52.36	6.22	15.26	26.15
C24H35N101	353.55	353.2718	81.53	9.98	3.96	4.53
C24H35N102	369.55	369.2668	78.00	9.55	3.79	8.66
C24H35N103	385.55	385.2617	74.77	9.15	3.63	12.45
C24H35N104	401.55	401.2566	71.79	8.79	3.49	15.94
C24H35N105	417.55	417.2515	69.04	8.45	3.35	19.16
C24H35N106	433.55	433.2464	66.49	8.14	3.23	22.14
C24H35N107	449.55	449.2413	64.12	7.85	3.12	24.91
C24H35N108	465.55	465.2362	61.92	7.58	3.01	27.49
C24H35N109	481.55	481.2312	59.86	7.33	2.91	29.90
C24H35N301	381.57	381.2780	75.55	9.25	11.01	4.19
C24H35N302	397.57	397.2729	72.51	8.87	10.57	8.05
C24H35N303	413.56	413.2678	69.70	8.53	10.16	11.61
C24H35N304	429.56	429.2627	67.11	8.21	9.78	14.90
C24H35N305	445.56	445.2577	64.70	7.92	9.43	17.95
C24H35N306	461.56	461.2526	62.45	7.64	9.10	20.80
C24H35N307	477.56	477.2475	60.36	7.39	8.80	23.45
C24H35N308	493.56	493.2424	58.41	7.15	8.51	25.93
C24H35N309	509.56	509.2373	56.57	6.92	8.25	28.26
C24H35N501	409.58	409.2841	70.38	8.61	17.10	3.91
C24H35N502	425.58	425.2791	67.74	8.29	16.46	7.52
C24H35N503	441.58	441.2740	65.28	7.99	15.86	10.87
C24H35N504	457.58	457.2689	63.00	7.71	15.31	13.99
C24H35N505	473.58	473.2638	60.87	7.45	14.79	16.89
C24H35N506	489.58	489.2587	58.88	7.21	14.30	19.61
C24H35N507	505.58	505.2536	57.02	6.98	13.85	22.15
C24H35N508	521.58	521.2485	55.27	6.76	13.43	24.54
C24H35N509	537.57	537.2435	53.62	6.56	13.03	26.79
C24H36N201	368.57	368.2827	78.21	9.85	7.60	4.34
C24H36N202	384.57	384.2777	74.96	9.44	7.28	8.32
C24H36N203	400.57	400.2726	71.97	9.06	6.99	11.98
C24H36N204	416.57	416.2675	69.20	8.71	6.72	15.36
C24H36N205	432.56	432.2624	66.64	8.39	6.48	18.49
C24H36N206	448.56	448.2573	64.26	8.09	6.25	21.40
C24H36N207	464.56	464.2522	62.05	7.81	6.03	24.11
C24H36N208	480.56	480.2471	59.99	7.55	5.83	26.63
C24H36N209	496.56	496.2421	58.05	7.31	5.64	29.00
C24H36N401	396.58	396.2889	72.69	9.15	14.13	4.03
C24H36N402	412.58	412.2838	69.87	8.80	13.58	7.76
C24H36N403	428.58	428.2787	67.26	8.47	13.07	11.20
C24H36N404	444.58	444.2736	64.84	8.16	12.60	14.40
C24H36N405	460.58	460.2686	62.59	7.88	12.16	17.37
C24H36N406	476.58	476.2635	60.49	7.61	11.76	20.14
C24H36N407	492.58	492.2584	58.52	7.37	11.37	22.74
C24H36N408	508.58	508.2533	56.68	7.13	11.02	25.17
C24H36N409	524.58	524.2482	54.95	6.92	10.68	27.45
C24H36N601	424.59	424.2950	67.89	8.55	19.79	3.77
C24H36N602	440.59	440.2900	65.43	8.24	19.07	7.26
C24H36N603	456.59	456.2849	63.13	7.95	18.41	10.51
C24H36N604	472.59	472.2798	61.00	7.68	17.78	13.54
C24H36N605	488.59	488.2747	59.00	7.43	17.20	16.37
C24H36N606	504.59	504.2696	57.13	7.19	16.66	19.02
C24H36N607	520.59	520.2645	55.37	6.97	16.14	21.51
C24H36N608	536.59	536.2594	53.72	6.76	15.66	23.85
C24H36N609	552.59	552.2544	52.17	6.57	15.21	26.06
C24H37N101	355.57	355.2875	81.07	10.49	3.94	4.50
C24H37N102	371.57	371.2824	77.58	10.04	3.77	8.61
C24H37N103	387.57	387.2773	74.38	9.62	3.61	12.38
C24H37N104	403.57	403.2722	71.43	9.24	3.47	15.86
C24H37N105	419.57	419.2672	68.71	8.89	3.34	19.07
C24H37N106	435.57	435.2621	66.18	8.56	3.22	22.04
C24H37N107	451.56	451.2570	63.84	8.26	3.10	24.80
C24H37N108	467.56	467.2519	61.65	7.98	3.00	27.37
C24H37N109	483.56	483.2468	59.61	7.71	2.90	29.78
C24H37N301	383.58	383.2936	75.15	9.72	10.95	4.17
C24H37N302	399.58	399.2886	72.14	9.33	10.52	8.01
C24H37N303	415.58	415.2835	69.36	8.97	10.11	11.55
C24H37N304	431.58	431.2784	66.79	8.64	9.74	14.83
C24H37N305	447.58	447.2733	64.41	8.33	9.39	17.87
C24H37N306	463.58	463.2682	62.18	8.04	9.06	20.71
C24H37N307	479.58	479.2631	60.11	7.78	8.76	23.35
C24H37N308	495.58	495.2580	58.17	7.53	8.48	25.83
C24H37N309	511.58	511.2530	56.35	7.29	8.21	28.15
C24H37N501	411.60	411.2998	70.04	9.06	17.02	3.89
C24H37N502	427.59	427.2947	67.42	8.72	16.38	7.48
C24H37N503	443.59	443.2896	64.98	8.41	15.79	10.82
C24H37N504	459.59	459.2845	62.72	8.11	15.24	13.92
C24H37N505	475.59	475.2794	60.61	7.84	14.73	16.82
C24H37N506	491.59	491.2744	58.64	7.59	14.25	19.53
C24H37N507	507.59	507.2693	56.79	7.35	13.80	22.06
C24H37N508	523.59	523.2642	55.06	7.12	13.38	24.45
C24H37N509	539.59	539.2591	53.42	6.91	12.98	26.69
C24H38N201	370.58	370.2984	77.79	10.34	7.56	4.32
C24H38N202	386.58	386.2933	74.57	9.91	7.25	8.28
C24H38N203	402.58	402.2882	71.60	9.51	6.96	11.92
C24H38N204	418.58	418.2831	68.87	9.15	6.69	15.29
C24H38N205	434.58	434.2781	66.33	8.81	6.45	18.41
C24H38N206	450.58	450.2730	63.99	8.50	6.22	21.31
C24H38N207	466.58	466.2679	61.78	8.21	6.00	24.00
C24H38N208	482.58	482.2628	59.73	7.94	5.80	26.52
C24H38N209	498.58	498.2577	57.82	7.68	5.62	28.88
C24H38N401	398.60	398.3045	72.32	9.61	14.06	4.01
C24H38N402	414.60	414.2995	69.53	9.24	13.51	7.72
C24H38N403	430.60	430.2944	66.95	8.90	13.01	11.15
C24H38N404	446.59	446.2893	64.55	8.58	12.55	14.33
C24H38N405	462.59	462.2842	62.32	8.28	12.11	17.29
C24H38N406	478.59	478.2791	60.23	8.00	11.71	20.06
C24H38N407	494.59	494.2740	58.28	7.74	11.33	22.64
C24H38N408	510.59	510.2689	56.46	7.50	10.97	25.07
C24H38N409	526.59	526.2639	54.74	7.27	10.64	27.34
C24H38N601	426.61	426.3107	67.57	8.98	19.70	3.75
C24H38N602	442.61	442.3056	65.13	8.65	18.99	7.23
C24H38N603	458.61	458.3005	62.86	8.35	18.33	10.47
C24H38N604	474.61	474.2954	60.74	8.07	17.71	13.48
C24H38N605	490.61	490.2903	58.76	7.81	17.13	16.31
C24H38N606	506.61	506.2853	56.90	7.56	16.59	18.95
C24H38N607	522.61	522.2802	55.16	7.33	16.08	21.43
C24H38N608	538.61	538.2751	53.52	7.11	15.60	23.76
C24H38N609	554.61	554.2700	51.98	6.91	15.15	25.96
C24H39N101	357.58	357.3031	80.62	10.99	3.92	4.47
C24H39N102	373.58	373.2981	77.16	10.52	3.75	8.57
C24H39N103	389.58	389.2930	73.99	10.09	3.60	12.32
C24H39N104	405.58	405.2879	71.07	9.69	3.45	15.78

C24H39N105	421.58	421.2828	68.38	9.32	3.32	18.98
C24H39N106	437.58	437.2777	65.88	8.98	3.20	21.94
C24H39N107	453.58	453.2726	63.55	8.67	3.09	24.69
C24H39N108	469.58	469.2675	61.39	8.37	2.98	27.26
C24H39N109	485.58	485.2625	59.37	8.10	2.88	29.65
C24H39N301	385.60	385.3093	74.76	10.19	10.90	4.15
C24H39N302	401.60	401.3042	71.78	9.79	10.46	7.97
C24H39N303	417.60	417.2991	69.03	9.41	10.06	11.49
C24H39N304	433.60	433.2940	66.48	9.07	9.69	14.76
C24H39N305	449.60	449.2890	64.12	8.74	9.35	17.79
C24H39N306	465.59	465.2839	61.91	8.44	9.03	20.62
C24H39N307	481.59	481.2788	59.86	8.16	8.73	23.26
C24H39N308	497.59	497.2737	57.93	7.90	8.44	25.72
C24H39N309	513.59	513.2686	56.13	7.65	8.18	28.04
C24H39N501	413.61	413.3154	69.70	9.50	16.93	3.87
C24H39N502	429.61	429.3104	67.10	9.15	16.30	7.45
C24H39N503	445.61	445.3053	64.69	8.82	15.72	10.77
C24H39N504	461.61	461.3002	62.45	8.52	15.17	13.86
C24H39N505	477.61	477.2951	60.36	8.23	14.66	16.75
C24H39N506	493.61	493.2900	58.40	7.96	14.19	19.45
C24H39N507	509.61	509.2849	56.57	7.71	13.74	21.98
C24H39N508	525.61	525.2798	54.84	7.48	13.32	24.35
C24H39N509	541.61	541.2748	53.22	7.26	12.93	26.59
C24H40N201	372.60	372.3140	77.37	10.82	7.52	4.29
C24H40N202	388.60	388.3090	74.18	10.38	7.21	8.23
C24H40N203	404.60	404.3039	71.25	9.97	6.92	11.86
C24H40N204	420.60	420.2988	68.54	9.59	6.66	15.22
C24H40N205	436.60	436.2937	66.03	9.23	6.42	18.32
C24H40N206	452.60	452.2886	63.69	8.91	6.19	21.21
C24H40N207	468.60	468.2835	61.52	8.60	5.98	23.90
C24H40N208	484.60	484.2784	59.49	8.32	5.78	26.41
C24H40N209	500.59	500.2734	57.59	8.05	5.60	28.76
C24H40N401	400.61	400.3202	71.96	10.06	13.99	3.99
C24H40N402	416.61	416.3151	69.19	9.68	13.45	7.68
C24H40N403	432.61	432.3100	66.63	9.32	12.95	11.09
C24H40N404	448.61	448.3049	64.26	8.99	12.49	14.27
C24H40N405	464.61	464.2998	62.05	8.68	12.06	17.22
C24H40N406	480.61	480.2948	59.98	8.39	11.66	19.97
C24H40N407	496.61	496.2897	58.05	8.12	11.28	22.55
C24H40N408	512.61	512.2846	56.24	7.87	10.93	24.97
C24H40N409	528.61	528.2795	54.53	7.63	10.60	27.24
C24H40N601	428.63	428.3263	67.25	9.41	19.61	3.73
C24H40N602	444.63	444.3213	64.83	9.07	18.90	7.20
C24H40N603	460.62	460.3162	62.58	8.75	18.24	10.42
C24H40N604	476.62	476.3111	60.48	8.46	17.63	13.43
C24H40N605	492.62	492.3060	58.52	8.18	17.06	16.24
C24H40N606	508.62	508.3009	56.68	7.93	16.52	18.87
C24H40N607	524.62	524.2958	54.95	7.69	16.02	21.35
C24H40N608	540.62	540.2907	53.32	7.46	15.55	23.68
C24H40N609	556.62	556.2857	51.79	7.24	15.10	25.87
C24H41N101	359.60	359.3188	80.16	11.49	3.90	4.45
C24H41N102	375.60	375.3137	76.75	11.00	3.73	8.52
C24H41N103	391.60	391.3086	73.61	10.55	3.58	12.26
C24H41N104	407.60	407.3035	70.72	10.14	3.44	15.70
C24H41N105	423.60	423.2985	68.05	9.76	3.31	18.89
C24H41N106	439.60	439.2934	65.58	9.40	3.19	21.84
C24H41N107	455.60	455.2883	63.27	9.07	3.07	24.58
C24H41N108	471.60	471.2832	61.13	8.76	2.97	27.14
C24H41N109	487.60	487.2781	59.12	8.48	2.87	29.53
C24H41N301	387.61	387.3249	74.37	10.66	10.84	4.13
C24H41N302	403.61	403.3199	71.42	10.24	10.41	7.93
C24H41N303	419.61	419.3148	68.70	9.85	10.01	11.44
C24H41N304	435.61	435.3097	66.18	9.49	9.65	14.69
C24H41N305	451.61	451.3046	63.83	9.15	9.30	17.71
C24H41N306	467.61	467.2995	61.65	8.84	8.99	20.53
C24H41N307	483.61	483.2944	59.61	8.55	8.69	23.16
C24H41N308	499.61	499.2893	57.70	8.27	8.41	25.62
C24H41N309	515.61	515.2843	55.91	8.02	8.15	27.93
C24H41N501	415.63	415.3311	69.36	9.94	16.85	3.85
C24H41N502	431.63	431.3260	66.79	9.57	16.23	7.41
C24H41N503	447.63	447.3209	64.40	9.23	15.65	10.72
C24H41N504	463.63	463.3158	62.18	8.91	15.11	13.80
C24H41N505	479.62	479.3107	60.10	8.62	14.60	16.68
C24H41N506	495.62	495.3057	58.16	8.34	14.13	19.37
C24H41N507	511.62	511.3006	56.34	8.08	13.69	21.89
C24H41N508	527.62	527.2955	54.64	7.83	13.27	24.26
C24H41N509	543.62	543.2904	53.03	7.60	12.88	26.49
C24H42N201	374.62	374.3297	76.95	11.30	7.48	4.27
C24H42N202	390.61	390.3246	73.80	10.84	7.17	8.19
C24H42N203	406.61	406.3195	70.89	10.41	6.89	11.80
C24H42N204	422.61	422.3144	68.21	10.02	6.63	15.14
C24H42N205	438.61	438.3094	65.72	9.65	6.39	18.24
C24H42N206	454.61	454.3043	63.41	9.31	6.16	21.12
C24H42N207	470.61	470.2992	61.25	9.00	5.95	23.80
C24H42N208	486.61	486.2941	59.24	8.70	5.76	26.30
C24H42N209	502.61	502.2890	57.35	8.42	5.57	28.65
C24H42N401	402.63	402.3358	71.60	10.51	13.92	3.97
C24H42N402	418.63	418.3308	68.86	10.11	13.38	7.64
C24H42N403	434.63	434.3257	66.33	9.74	12.89	11.04
C24H42N404	450.63	450.3206	63.97	9.39	12.43	14.20
C24H42N405	466.63	466.3155	61.78	9.07	12.01	17.14
C24H42N406	482.63	482.3104	59.73	8.77	11.61	19.89
C24H42N407	498.62	498.3053	57.81	8.49	11.24	22.46
C24H42N408	514.62	514.3002	56.02	8.23	10.89	24.87
C24H42N409	530.62	530.2952	54.33	7.98	10.56	27.14
C24H42N601	430.64	430.3420	66.94	9.83	19.52	3.72
C24H42N602	446.64	446.3369	64.54	9.48	18.82	7.16
C24H42N603	462.64	462.3318	62.31	9.15	18.17	10.37
C24H42N604	478.64	478.3267	60.23	8.84	17.56	13.37
C24H42N605	494.64	494.3216	58.28	8.56	16.99	16.17
C24H42N606	510.64	510.3166	56.45	8.29	16.46	18.80
C24H42N607	526.64	526.3115	54.74	8.04	15.96	21.27
C24H42N608	542.64	542.3064	53.12	7.80	15.49	23.59
C24H42N609	558.64	558.3013	51.60	7.58	15.04	25.78
C24H43N101	361.62	361.3344	79.72	11.99	3.87	4.42
C24H43N102	377.62	377.3294	76.34	11.48	3.71	8.47
C24H43N103	393.62	393.3243	73.24	11.01	3.56	12.19
C24H43N104	409.61	409.3192	70.38	10.58	3.42	15.62
C24H43N105	425.61	425.3141	67.73	10.18	3.29	18.80
C24H43N106	441.61	441.3090	65.28	9.81	3.17	21.74
C24H43N107	457.61	457.3039	62.99	9.47	3.06	24.47
C24H43N108	473.61	473.2988	60.87	9.15	2.96	27.03
C24H43N109	489.61	489.2938	58.88	8.85	2.86	29.41
C24H43N301	389.63	389.3406	73.98	11.12	10.78	4.11
C24H43N302	405.63	405.3355	71.07	10.69	10.36	7.89
C24H43N303	421.63	421.3304	68.37	10.28	9.97	11.38
C24H43N304	437.63	437.3253	65.87	9.90	9.60	14.62
C24H43N305	453.63	453.3202	63.55	9.55	9.26	17.63
C24H43N306	469.63	469.3152	61.38	9.23	8.95	20.44
C24H43N307	485.63	485.3101	59.36	8.93	8.65	23.06

C24H43N308	501.63	501.3050	57.47	8.64	8.38	25.52	C24H45N505	483.66	483.3420	59.60	9.38	14.48	16.54
C24H43N309	517.63	517.2999	55.69	8.37	8.12	27.82	C24H45N506	499.66	499.3370	57.69	9.08	14.02	19.21
C24H43N501	417.64	417.3467	69.02	10.38	16.77	3.83	C24H45N507	515.66	515.3319	55.90	8.80	13.58	21.72
C24H43N502	433.64	433.3417	66.48	10.00	16.15	7.38	C24H45N508	531.65	531.3268	54.22	8.53	13.17	24.07
C24H43N503	449.64	449.3366	64.11	9.64	15.58	10.67	C24H45N509	547.65	547.3217	52.64	8.28	12.79	26.29
C24H43N504	465.64	465.3315	61.91	9.31	15.04	13.74	C24H46N201	378.65	378.3610	76.13	12.25	7.40	4.23
C24H43N505	481.64	481.3264	59.85	9.00	14.54	16.61	C24H46N202	394.65	394.3559	73.04	11.75	7.10	8.11
C24H43N506	497.64	497.3213	57.93	8.71	14.07	19.29	C24H46N203	410.65	410.3508	70.20	11.29	6.82	11.69
C24H43N507	513.64	513.3162	56.12	8.44	13.63	21.80	C24H46N204	426.65	426.3457	67.57	10.87	6.57	15.00
C24H43N508	529.64	529.3111	54.43	8.18	13.22	24.17	C24H46N205	442.64	442.3406	65.12	10.47	6.33	18.07
C24H43N509	545.64	545.3061	52.83	7.94	12.84	26.39	C24H46N206	458.64	458.3356	62.85	10.11	6.11	20.93
C24H44N201	376.63	376.3453	76.54	11.78	7.44	4.25	C24H46N207	474.64	474.3305	60.73	9.77	5.90	23.60
C24H44N202	392.63	392.3403	73.42	11.30	7.13	8.15	C24H46N208	490.64	490.3254	58.75	9.45	5.71	26.09
C24H44N203	408.63	408.3352	70.54	10.85	6.86	11.75	C24H46N209	506.64	506.3203	56.90	9.15	5.53	28.42
C24H44N204	424.63	424.3301	67.89	10.44	6.60	15.07	C24H46N401	406.66	406.3671	70.89	11.40	13.78	3.93
C24H44N205	440.63	440.3250	65.42	10.07	6.36	18.16	C24H46N402	422.66	422.3621	68.20	10.97	13.26	7.57
C24H44N206	456.63	456.3199	63.13	9.71	6.13	21.02	C24H46N403	438.66	438.3570	65.72	10.57	12.77	10.94
C24H44N207	472.63	472.3148	60.99	9.38	5.93	23.70	C24H46N404	454.66	454.3519	63.40	10.20	12.32	14.08
C24H44N208	488.63	488.3097	59.00	9.08	5.73	26.19	C24H46N405	470.66	470.3468	61.25	9.85	11.90	17.00
C24H44N209	504.63	504.3047	57.12	8.79	5.55	28.53	C24H46N406	486.66	486.3417	59.23	9.53	11.51	19.73
C24H44N401	404.64	404.3515	71.24	10.96	13.85	3.95	C24H46N407	502.66	502.3366	57.35	9.22	11.15	22.28
C24H44N402	420.64	420.3464	68.53	10.54	13.32	7.61	C24H46N408	518.66	518.3315	55.58	8.94	10.80	24.68
C24H44N403	436.64	436.3413	66.02	10.16	12.83	10.99	C24H46N409	534.66	534.3265	53.92	8.67	10.48	26.93
C24H44N404	452.64	452.3362	63.69	9.80	12.38	14.14	C24H46N601	434.67	434.3733	66.32	10.67	19.33	3.68
C24H44N405	468.64	468.3311	61.51	9.46	11.96	17.07	C24H46N602	450.67	450.3682	63.96	10.29	18.65	7.10
C24H44N406	484.64	484.3261	59.48	9.15	11.56	19.81	C24H46N603	466.67	466.3631	61.77	9.94	18.01	10.29
C24H44N407	500.64	500.3210	57.58	8.86	11.19	22.37	C24H46N604	482.67	482.3580	59.72	9.61	17.41	13.26
C24H44N408	516.64	516.3159	55.80	8.58	10.84	24.77	C24H46N605	498.67	498.3529	57.81	9.30	16.85	16.04
C24H44N409	532.64	532.3108	54.12	8.33	10.52	27.03	C24H46N606	514.67	514.3479	56.01	9.01	16.33	18.65
C24H44N601	432.66	432.3576	66.63	10.25	19.42	3.70	C24H46N607	530.67	530.3428	54.32	8.74	15.84	21.10
C24H44N602	448.66	448.3526	64.25	9.89	18.73	7.13	C24H46N608	546.67	546.3377	52.73	8.48	15.37	23.41
C24H44N603	464.66	464.3475	62.04	9.54	18.09	10.33	C24H46N609	562.67	562.3326	51.23	8.24	14.94	25.59
C24H44N604	480.66	480.3424	59.97	9.23	17.48	13.31	C24H47N101	365.65	365.3657	78.84	12.96	3.83	4.38
C24H44N605	496.66	496.3373	58.04	8.93	16.92	16.11	C24H47N102	381.65	381.3607	75.53	12.41	3.67	8.38
C24H44N606	512.65	512.3322	56.23	8.65	16.39	18.73	C24H47N103	397.65	397.3556	72.49	11.91	3.52	12.07
C24H44N607	528.65	528.3271	54.53	8.39	15.90	21.19	C24H47N104	413.65	413.3505	69.69	11.45	3.39	15.47
C24H44N608	544.65	544.3220	52.93	8.14	15.43	23.50	C24H47N105	429.65	429.3454	67.09	11.03	3.26	18.62
C24H44N609	560.65	560.3170	51.42	7.91	14.99	25.68	C24H47N106	445.65	445.3403	64.69	10.63	3.14	21.54
C24H45N101	363.63	363.3501	79.27	12.47	3.85	4.40	C24H47N107	461.64	461.3352	62.44	10.26	3.03	24.26
C24H45N102	379.63	379.3450	75.93	11.95	3.69	8.43	C24H47N108	477.64	477.3301	60.35	9.92	2.93	26.80
C24H45N103	395.63	395.3399	72.86	11.46	3.54	12.13	C24H47N109	493.64	493.3251	58.40	9.60	2.84	29.17
C24H45N104	411.63	411.3348	70.03	11.02	3.40	15.55	C24H47N301	393.66	393.3719	73.23	12.03	10.67	4.06
C24H45N105	427.63	427.3298	67.41	10.61	3.28	18.71	C24H47N302	409.66	409.3668	70.37	11.56	10.26	7.81
C24H45N106	443.63	443.3247	64.98	10.22	3.16	21.64	C24H47N303	425.66	425.3617	67.72	11.13	9.87	11.28
C24H45N107	459.63	459.3196	62.72	9.87	3.05	24.37	C24H47N304	441.66	441.3566	65.27	10.73	9.51	14.49
C24H45N108	475.63	475.3145	60.61	9.54	2.94	26.91	C24H47N305	457.66	457.3515	62.99	10.35	9.18	17.48
C24H45N109	491.63	491.3094	58.64	9.23	2.85	29.29	C24H47N306	473.66	473.3465	60.86	10.00	8.87	20.27
C24H45N301	391.65	391.3562	73.60	11.58	10.73	4.09	C24H47N307	489.66	489.3414	58.87	9.68	8.58	22.87
C24H45N302	407.65	407.3512	70.72	11.13	10.31	7.85	C24H47N308	505.66	505.3363	57.01	9.37	8.31	25.31
C24H45N303	423.64	423.3461	68.04	10.71	9.92	11.33	C24H47N309	521.66	521.3312	55.26	9.08	8.06	27.60
C24H45N304	439.64	439.3410	65.57	10.32	9.56	14.56	C24H47N501	421.68	421.3780	68.36	11.23	16.61	3.79
C24H45N305	455.64	455.3359	63.27	9.95	9.22	17.56	C24H47N502	437.67	437.3730	65.86	10.82	16.00	7.31
C24H45N306	471.64	471.3308	61.12	9.62	8.91	20.35	C24H47N503	453.67	453.3679	63.54	10.44	15.44	10.58
C24H45N307	487.64	487.3257	59.11	9.30	8.62	22.97	C24H47N504	469.67	469.3628	61.38	10.09	14.91	13.63
C24H45N308	503.64	503.3206	57.24	9.01	8.34	25.41	C24H47N505	485.67	485.3577	59.35	9.75	14.42	16.47
C24H45N309	519.64	519.3156	55.47	8.73	8.09	27.71	C24H47N506	501.67	501.3526	57.46	9.44	13.96	19.14
C24H45N501	419.66	419.3624	68.69	10.81	16.69	3.81	C24H47N507	517.67	517.3475	55.69	9.15	13.53	21.63
C24H45N502	435.66	435.3573	66.17	10.41	16.08	7.34	C24H47N508	533.67	533.3424	54.02	8.88	13.12	23.98
C24H45N503	451.66	451.3522	63.82	10.04	15.51	10.63	C24H47N509	549.67	549.3374	52.44	8.62	12.74	26.20
C24H45N504	467.66	467.3471	61.64	9.70	14.98	13.68	C24H48N201	380.66	380.3766	75.73	12.71	7.36	4.20

C24H48N202	396.66	396.3716	72.67	12.20	7.06	8.07	C24H50N208	494.67	494.3567	58.27	10.19	5.66	25.87
C24H48N203	412.66	412.3665	69.86	11.72	6.79	11.63	C24H50N209	510.67	510.3516	56.45	9.87	5.49	28.20
C24H48N204	428.66	428.3614	67.25	11.29	6.54	14.93	C24H50N401	410.69	410.3984	70.19	12.27	13.64	3.90
C24H48N205	444.66	444.3563	64.83	10.88	6.30	17.99	C24H50N402	426.69	426.3934	67.56	11.81	13.13	7.50
C24H48N206	460.66	460.3512	62.58	10.50	6.08	20.84	C24H50N403	442.69	442.3883	65.12	11.38	12.66	10.84
C24H48N207	476.66	476.3461	60.48	10.15	5.88	23.50	C24H50N404	458.69	458.3832	62.85	10.99	12.21	13.95
C24H48N208	492.66	492.3410	58.51	9.82	5.69	25.98	C24H50N405	474.69	474.3781	60.73	10.62	11.80	16.85
C24H48N209	508.66	508.3360	56.67	9.51	5.51	28.31	C24H50N406	490.69	490.3730	58.75	10.27	11.42	19.56
C24H48N401	408.68	408.3828	70.54	11.84	13.71	3.91	C24H50N407	506.69	506.3679	56.89	9.95	11.06	22.10
C24H48N402	424.68	424.3777	67.88	11.39	13.19	7.53	C24H50N408	522.69	522.3628	55.15	9.64	10.72	24.49
C24H48N403	440.68	440.3726	65.41	10.98	12.71	10.89	C24H50N409	538.69	538.3578	53.51	9.36	10.40	26.73
C24H48N404	456.67	456.3675	63.12	10.59	12.27	14.01	C24H50N601	438.71	438.4046	65.71	11.49	19.16	3.65
C24H48N405	472.67	472.3624	60.99	10.24	11.85	16.92	C24H50N602	454.71	454.3995	63.40	11.08	18.48	7.04
C24H48N406	488.67	488.3574	58.99	9.90	11.47	19.64	C24H50N603	470.70	470.3944	61.24	10.71	17.85	10.20
C24H48N407	504.67	504.3523	57.12	9.59	11.10	22.19	C24H50N604	486.70	486.3893	59.23	10.36	17.27	13.15
C24H48N408	520.67	520.3472	55.36	9.29	10.76	24.58	C24H50N605	502.70	502.3842	57.34	10.03	16.72	15.91
C24H48N409	536.67	536.3421	53.71	9.02	10.44	26.83	C24H50N606	518.70	518.3792	55.57	9.72	16.20	18.51
C24H48N601	436.69	436.3889	66.01	11.08	19.24	3.66	C24H50N607	534.70	534.3741	53.91	9.43	15.72	20.95
C24H48N602	452.69	452.3839	63.68	10.69	18.56	7.07	C24H50N608	550.70	550.3690	52.35	9.15	15.26	23.24
C24H48N603	468.69	468.3788	61.51	10.32	17.93	10.24	C24H50N609	566.70	566.3639	50.87	8.89	14.83	25.41
C24H48N604	484.69	484.3737	59.47	9.98	17.34	13.20	C24H51N101	369.68	369.3970	77.98	13.91	3.79	4.33
C24H48N605	500.69	500.3686	57.57	9.66	16.78	15.98	C24H51N102	385.68	385.3920	74.74	13.33	3.63	8.30
C24H48N606	516.69	516.3635	55.79	9.36	16.27	18.58	C24H51N103	401.68	401.3869	71.77	12.80	3.49	11.95
C24H48N607	532.69	532.3584	54.12	9.08	15.78	21.02	C24H51N104	417.68	417.3818	69.02	12.31	3.35	15.32
C24H48N608	548.69	548.3533	52.54	8.82	15.32	23.33	C24H51N105	433.68	433.3767	66.47	11.85	3.23	18.45
C24H48N609	564.68	564.3483	51.05	8.57	14.88	25.50	C24H51N106	449.68	449.3716	64.11	11.43	3.11	21.35
C24H49N101	367.66	367.3814	78.41	13.43	3.81	4.35	C24H51N107	465.68	465.3665	61.90	11.04	3.01	24.05
C24H49N102	383.66	383.3763	75.14	12.87	3.65	8.34	C24H51N108	481.68	481.3614	59.85	10.67	2.91	26.57
C24H49N103	399.66	399.3712	72.13	12.36	3.50	12.01	C24H51N109	497.68	497.3564	57.92	10.33	2.81	28.93
C24H49N104	415.66	415.3661	69.35	11.88	3.37	15.40	C24H51N301	397.69	397.4032	72.48	12.93	10.57	4.02
C24H49N105	431.66	431.3610	66.78	11.44	3.24	18.53	C24H51N302	413.69	413.3981	69.68	12.43	10.16	7.73
C24H49N106	447.66	447.3560	64.39	11.03	3.13	21.44	C24H51N303	429.69	429.3930	67.09	11.96	9.78	11.17
C24H49N107	463.66	463.3509	62.17	10.65	3.02	24.15	C24H51N304	445.69	445.3879	64.68	11.53	9.43	14.36
C24H49N108	479.66	479.3458	60.10	10.30	2.92	26.68	C24H51N305	461.69	461.3828	62.44	11.13	9.10	17.33
C24H49N109	495.66	495.3407	58.16	9.96	2.83	29.05	C24H51N306	477.69	477.3778	60.35	10.76	8.80	20.10
C24H49N301	395.68	395.3875	72.85	12.48	10.62	4.04	C24H51N307	493.69	493.3727	58.39	10.41	8.51	22.69
C24H49N302	411.68	411.3825	70.02	12.00	10.21	7.77	C24H51N308	509.69	509.3676	56.56	10.09	8.24	25.11
C24H49N303	427.68	427.3774	67.40	11.55	9.83	11.22	C24H51N309	525.69	525.3625	54.84	9.78	7.99	27.39
C24H49N304	443.68	443.3723	64.97	11.13	9.47	14.42	C24H51N501	425.71	425.4093	67.72	12.08	16.45	3.76
C24H49N305	459.68	459.3672	62.71	10.74	9.14	17.40	C24H51N502	441.71	441.4043	65.26	11.64	15.86	7.24
C24H49N306	475.67	475.3621	60.60	10.38	8.83	20.18	C24H51N503	457.71	457.3992	62.98	11.23	15.30	10.49
C24H49N307	491.67	491.3570	58.63	10.05	8.55	22.78	C24H51N504	473.71	473.3941	60.85	10.85	14.78	13.51
C24H49N308	507.67	507.3519	56.78	9.73	8.28	25.21	C24H51N505	489.70	489.3890	58.87	10.50	14.30	16.34
C24H49N309	523.67	523.3469	55.05	9.43	8.02	27.50	C24H51N506	505.70	505.3839	57.00	10.17	13.85	18.98
C24H49N501	423.69	423.3937	68.04	11.66	16.53	3.78	C24H51N507	521.70	521.3788	55.26	9.85	13.42	21.47
C24H49N502	439.69	439.3886	65.56	11.23	15.93	7.28	C24H51N508	537.70	537.3737	53.61	9.56	13.02	23.80
C24H49N503	455.69	455.3835	63.26	10.84	15.37	10.53	C24H51N509	553.70	553.3687	52.06	9.28	12.65	26.01
C24H49N504	471.69	471.3784	61.11	10.47	14.85	13.57	C24H52N201	384.69	384.4079	74.93	13.62	7.28	4.16
C24H49N505	487.69	487.3733	59.11	10.13	14.36	16.40	C24H52N202	400.69	400.4029	71.94	13.08	6.99	7.99
C24H49N506	503.69	503.3683	57.23	9.81	13.90	19.06	C24H52N203	416.69	416.3978	69.18	12.58	6.72	11.52
C24H49N507	519.69	519.3632	55.47	9.50	13.48	21.55	C24H52N204	432.69	432.3927	66.62	12.11	6.47	14.79
C24H49N508	535.69	535.3581	53.81	9.22	13.07	23.89	C24H52N205	448.69	448.3876	64.25	11.68	6.24	17.83
C24H49N509	551.69	551.3530	52.25	8.95	12.69	26.10	C24H52N206	464.69	464.3825	62.03	11.28	6.03	20.66
C24H50N201	382.68	382.3923	75.33	13.17	7.32	4.18	C24H52N207	480.69	480.3774	59.97	10.90	5.83	23.30
C24H50N202	398.68	398.3872	72.31	12.64	7.03	8.03	C24H52N208	496.69	496.3723	58.04	10.55	5.64	25.77
C24H50N203	414.68	414.3821	69.52	12.15	6.76	11.57	C24H52N209	512.69	512.3673	56.23	10.22	5.46	28.09
C24H50N204	430.68	430.3770	66.93	11.70	6.50	14.86	C24H52N401	412.71	412.4141	69.85	12.70	13.58	3.88
C24H50N205	446.68	446.3719	64.54	11.28	6.27	17.91	C24H52N402	428.71	428.4090	67.24	12.23	13.07	7.46
C24H50N206	462.68	462.3669	62.30	10.89	6.05	20.75	C24H52N403	444.71	444.4039	64.82	11.79	12.60	10.79
C24H50N207	478.68	478.3618	60.22	10.53	5.85	23.40	C24H52N404	460.71	460.3988	62.57	11.38	12.16	13.89

C24H52N405	476.71	476.3937	60.47	11.00	11.75	16.78
C24H52N406	492.71	492.3887	58.51	10.64	11.37	19.48
C24H52N407	508.70	508.3836	56.67	10.30	11.01	22.02
C24H52N408	524.70	524.3785	54.94	9.99	10.68	24.39
C24H52N409	540.70	540.3734	53.31	9.69	10.36	26.63
C24H52N601	440.72	440.4202	65.41	11.89	19.07	3.63
C24H52N602	456.72	456.4152	63.12	11.48	18.40	7.01
C24H52N603	472.72	472.4101	60.98	11.09	17.78	10.15
C24H52N604	488.72	488.4050	58.98	10.72	17.20	13.09
C24H52N605	504.72	504.3999	57.11	10.38	16.65	15.85
C24H52N606	520.72	520.3948	55.36	10.07	16.14	18.44
C24H52N607	536.72	536.3897	53.71	9.77	15.66	20.87
C24H52N608	552.72	552.3846	52.15	9.48	15.20	23.16
C24H52N609	568.72	568.3795	50.69	9.22	14.78	25.32
C24H53N301	399.71	399.4188	72.12	13.37	10.51	4.00
C24H53N302	415.71	415.4138	69.34	12.85	10.11	7.70
C24H53N303	431.71	431.4087	66.77	12.37	9.73	11.12
C24H53N304	447.71	447.4036	64.39	11.93	9.39	14.29
C24H53N305	463.71	463.3985	62.17	11.52	9.06	17.25
C24H53N306	479.71	479.3934	60.09	11.14	8.76	20.01
C24H53N307	495.71	495.3883	58.15	10.78	8.48	22.59
C24H53N308	511.71	511.3832	56.33	10.44	8.21	25.01
C24H53N309	527.70	527.3782	54.63	10.12	7.96	27.29
C24H53N501	427.72	427.4250	67.40	12.49	16.37	3.74
C24H53N502	443.72	443.4199	64.97	12.04	15.78	7.21
C24H53N503	459.72	459.4148	62.70	11.62	15.23	10.44
C24H53N504	475.72	475.4097	60.60	11.23	14.72	13.45
C24H53N505	491.72	491.4046	58.62	10.86	14.24	16.27
C24H53N506	507.72	507.3996	56.78	10.52	13.79	18.91
C24H53N507	523.72	523.3945	55.04	10.20	13.37	21.38
C24H53N508	539.72	539.3894	53.41	9.90	12.98	23.72
C24H53N509	555.72	555.3843	51.87	9.61	12.60	25.91
C24H54N401	414.72	414.4297	69.51	13.12	13.51	3.86
C24H54N402	430.72	430.4247	66.93	12.64	13.01	7.43
C24H54N403	446.72	446.4196	64.53	12.18	12.54	10.74
C24H54N404	462.72	462.4145	62.30	11.76	12.11	13.83
C24H54N405	478.72	478.4094	60.22	11.37	11.70	16.71
C24H54N406	494.72	494.4043	58.27	11.00	11.32	19.40
C24H54N407	510.72	510.3992	56.44	10.66	10.97	21.93
C24H54N408	526.72	526.3941	54.73	10.33	10.64	24.30
C24H54N409	542.72	542.3891	53.12	10.03	10.32	26.53
C24H54N601	442.74	442.4359	65.11	12.29	18.98	3.61
C24H54N602	458.74	458.4308	62.84	11.87	18.32	6.98
C24H54N603	474.74	474.4257	60.72	11.47	17.70	10.11
C24H54N604	490.74	490.4206	58.74	11.09	17.13	13.04
C24H54N605	506.74	506.4155	56.89	10.74	16.58	15.79
C24H54N606	522.73	522.4105	55.15	10.41	16.08	18.36
C24H54N607	538.73	538.4054	53.51	10.10	15.60	20.79
C24H54N608	554.73	554.4003	51.97	9.81	15.15	23.07
C24H54N609	570.73	570.3952	50.51	9.54	14.72	25.23
C24H55N501	429.74	429.4406	67.08	12.90	16.30	3.72
C24H55N502	445.74	445.4356	64.67	12.44	15.71	7.18
C24H55N503	461.74	461.4305	62.43	12.01	15.17	10.40
C24H55N504	477.74	477.4254	60.34	11.60	14.66	13.40
C24H55N505	493.74	493.4203	58.38	11.23	14.18	16.20
C24H55N506	509.74	509.4152	56.55	10.88	13.74	18.83
C24H55N507	525.74	525.4101	54.83	10.54	13.32	21.30
C24H55N508	541.73	541.4050	53.21	10.23	12.93	23.63
C24H55N509	557.73	557.3999	51.69	9.94	12.56	25.82
C24H56N601	444.75	444.4515	64.82	12.69	18.90	3.60
C24H56N602	460.75	460.4464	62.56	12.25	18.24	6.94
C24H56N603	476.75	476.4414	60.46	11.84	17.63	10.07
C24H56N604	492.75	492.4363	58.50	11.46	17.06	12.99
C24H56N605	508.75	508.4312	56.66	11.10	16.52	15.72
C24H56N606	524.75	524.4261	54.93	10.76	16.02	18.29
C24H56N607	540.75	540.4210	53.31	10.44	15.54	20.71
C24H56N608	556.75	556.4159	51.78	10.14	15.09	22.99
C24H56N609	572.75	572.4108	50.33	9.86	14.67	25.14
C25H12N201	356.39	356.0950	84.26	3.39	7.86	4.49
C25H12N202	372.39	372.0899	80.64	3.25	7.52	8.59
C25H12N203	388.39	388.0848	77.31	3.11	7.21	12.36
C25H12N204	404.39	404.0797	74.26	2.99	6.93	15.83
C25H12N205	420.38	420.0746	71.43	2.88	6.66	19.03
C25H12N206	436.38	436.0695	68.81	2.77	6.42	22.00
C25H12N207	452.38	452.0644	66.38	2.67	6.19	24.76
C25H12N208	468.38	468.0594	64.11	2.58	5.98	27.33
C25H12N209	484.38	484.0543	61.99	2.50	5.78	29.73
C25H12N401	384.40	384.1011	78.12	3.15	14.58	4.16
C25H12N402	400.40	400.0960	74.99	3.02	13.99	7.99
C25H12N403	416.40	416.0909	72.11	2.90	13.46	11.53
C25H12N404	432.40	432.0858	69.44	2.80	12.96	14.80
C25H12N405	448.40	448.0808	66.97	2.70	12.49	17.84
C25H12N406	464.40	464.0757	64.66	2.60	12.06	20.67
C25H12N407	480.40	480.0706	62.51	2.52	11.66	23.31
C25H12N408	496.40	496.0655	60.49	2.44	11.29	25.78
C25H12N409	512.40	512.0604	58.60	2.36	10.93	28.10
C25H12N601	412.41	412.1073	72.81	2.93	20.38	3.88
C25H12N602	428.41	428.1022	70.09	2.82	19.62	7.47
C25H12N603	444.41	444.0971	67.57	2.72	18.91	10.80
C25H12N604	460.41	460.0920	65.22	2.63	18.25	13.90
C25H12N605	476.41	476.0869	63.03	2.54	17.64	16.79
C25H12N606	492.41	492.0818	60.98	2.46	17.07	19.50
C25H12N607	508.41	508.0767	59.06	2.38	16.53	22.03
C25H12N608	524.41	524.0717	57.26	2.31	16.03	24.41
C25H12N609	540.41	540.0666	55.57	2.24	15.55	26.65
C25H13N101	343.39	343.0997	87.45	3.82	4.08	4.66
C25H13N102	359.39	359.0946	83.55	3.65	3.90	8.90
C25H13N103	375.39	375.0895	79.99	3.49	3.73	12.79
C25H13N104	391.39	391.0844	76.72	3.35	3.58	16.35
C25H13N105	407.39	407.0794	73.71	3.22	3.44	19.64
C25H13N106	423.39	423.0743	70.92	3.09	3.31	22.67
C25H13N107	439.38	439.0692	68.34	2.98	3.19	25.49
C25H13N108	455.38	455.0641	65.94	2.88	3.08	28.11
C25H13N109	471.38	471.0590	63.70	2.78	2.97	30.55
C25H13N301	371.40	371.1059	80.85	3.53	11.31	4.31
C25H13N302	387.40	387.1008	77.51	3.38	10.85	8.26
C25H13N303	403.40	403.0957	74.44	3.25	10.42	11.90
C25H13N304	419.40	419.0906	71.60	3.12	10.02	15.26
C25H13N305	435.40	435.0855	68.97	3.01	9.65	18.37
C25H13N306	451.40	451.0804	66.52	2.90	9.31	21.27
C25H13N307	467.40	467.0753	64.24	2.80	8.99	23.96
C25H13N308	483.40	483.0703	62.12	2.71	8.69	26.48
C25H13N309	499.40	499.0652	60.13	2.62	8.41	28.83
C25H13N501	399.42	399.1120	75.18	3.28	17.53	4.01
C25H13N502	415.41	415.1069	72.28	3.15	16.86	7.70
C25H13N503	431.41	431.1018	69.60	3.04	16.23	11.13
C25H13N504	447.41	447.0967	67.11	2.93	15.65	14.30
C25H13N505	463.41	463.0917	64.80	2.83	15.11	17.26
C25H13N506	479.41	479.0866	62.63	2.73	14.61	20.02
C25H13N507	495.41	495.0815	60.61	2.64	14.14	22.61

C25H13N508	511.41	511.0764	58.72	2.56	13.69	25.03
C25H13N509	527.41	527.0713	56.93	2.48	13.28	27.30
C25H14N201	358.40	358.1106	83.78	3.94	7.82	4.46
C25H14N202	374.40	374.1055	80.20	3.77	7.48	8.55
C25H14N203	390.40	390.1004	76.92	3.61	7.18	12.29
C25H14N204	406.40	406.0953	73.89	3.47	6.89	15.75
C25H14N205	422.40	422.0903	71.09	3.34	6.63	18.94
C25H14N206	438.40	438.0852	68.49	3.22	6.39	21.90
C25H14N207	454.40	454.0801	66.08	3.11	6.16	24.65
C25H14N208	470.40	470.0750	63.83	3.00	5.96	27.21
C25H14N209	486.40	486.0699	61.74	2.90	5.76	29.60
C25H14N401	386.42	386.1168	77.71	3.65	14.50	4.14
C25H14N402	402.42	402.1117	74.62	3.51	13.92	7.95
C25H14N403	418.42	418.1066	71.77	3.37	13.39	11.47
C25H14N404	434.41	434.1015	69.12	3.25	12.90	14.73
C25H14N405	450.41	450.0964	66.67	3.13	12.44	17.76
C25H14N406	466.41	466.0913	64.38	3.03	12.01	20.58
C25H14N407	482.41	482.0862	62.25	2.93	11.61	23.22
C25H14N408	498.41	498.0812	60.25	2.83	11.24	25.68
C25H14N409	514.41	514.0761	58.37	2.74	10.89	27.99
C25H14N601	414.43	414.1229	72.46	3.41	20.28	3.86
C25H14N602	430.43	430.1178	69.76	3.28	19.52	7.43
C25H14N603	446.43	446.1127	67.26	3.16	18.82	10.75
C25H14N604	462.43	462.1076	64.94	3.05	18.17	13.84
C25H14N605	478.43	478.1026	62.76	2.95	17.57	16.72
C25H14N606	494.43	494.0975	60.73	2.85	17.00	19.42
C25H14N607	510.43	510.0924	58.83	2.76	16.46	21.94
C25H14N608	526.43	526.0873	57.04	2.68	15.96	24.31
C25H14N609	542.43	542.0822	55.36	2.60	15.49	26.55
C25H15N101	345.40	345.1154	86.94	4.38	4.06	4.63
C25H15N102	361.40	361.1103	83.09	4.18	3.88	8.85
C25H15N103	377.40	377.1052	79.56	4.01	3.71	12.72
C25H15N104	393.40	393.1001	76.33	3.84	3.56	16.27
C25H15N105	409.40	409.0950	73.35	3.69	3.42	19.54
C25H15N106	425.40	425.0899	70.59	3.55	3.29	22.57
C25H15N107	441.40	441.0848	68.03	3.43	3.17	25.37
C25H15N108	457.40	457.0798	65.65	3.31	3.06	27.98
C25H15N109	473.40	473.0747	63.43	3.19	2.96	30.42
C25H15N301	373.42	373.1215	80.41	4.05	11.25	4.28
C25H15N302	389.42	389.1164	77.11	3.88	10.79	8.22
C25H15N303	405.42	405.1113	74.07	3.73	10.36	11.84
C25H15N304	421.42	421.1062	71.25	3.59	9.97	15.19
C25H15N305	437.42	437.1012	68.65	3.46	9.61	18.29
C25H15N306	453.41	453.0961	66.23	3.33	9.27	21.17
C25H15N307	469.41	469.0910	63.97	3.22	8.95	23.86
C25H15N308	485.41	485.0859	61.86	3.11	8.66	26.37
C25H15N309	501.41	501.0808	59.89	3.02	8.38	28.72
C25H15N501	401.43	401.1277	74.80	3.77	17.45	3.99
C25H15N502	417.43	417.1226	71.94	3.62	16.78	7.67
C25H15N503	433.43	433.1175	69.28	3.49	16.16	11.07
C25H15N504	449.43	449.1124	66.81	3.36	15.58	14.24
C25H15N505	465.43	465.1073	64.52	3.25	15.05	17.19
C25H15N506	481.43	481.1022	62.37	3.14	14.55	19.94
C25H15N507	497.43	497.0971	60.37	3.04	14.08	22.51
C25H15N508	513.43	513.0921	58.49	2.94	13.64	24.93
C25H15N509	529.43	529.0870	56.72	2.86	13.23	27.20
C25H16N201	360.42	360.1263	83.31	4.47	7.77	4.44
C25H16N202	376.42	376.1212	79.77	4.28	7.44	8.50
C25H16N203	392.42	392.1161	76.52	4.11	7.14	12.23
C25H16N204	408.42	408.1110	73.52	3.95	6.86	15.67
C25H16N205	424.42	424.1059	70.75	3.80	6.60	18.85
C25H16N206	440.42	440.1008	68.18	3.66	6.36	21.80
C25H16N207	456.42	456.0957	65.79	3.53	6.14	24.54
C25H16N208	472.41	472.0907	63.56	3.41	5.93	27.09
C25H16N209	488.41	488.0856	61.48	3.30	5.74	29.48
C25H16N401	388.43	388.1324	77.31	4.15	14.42	4.12
C25H16N402	404.43	404.1273	74.25	3.99	13.85	7.91
C25H16N403	420.43	420.1222	71.42	3.84	13.33	11.42
C25H16N404	436.43	436.1171	68.80	3.70	12.84	14.66
C25H16N405	452.43	452.1121	66.37	3.56	12.38	17.68
C25H16N406	468.43	468.1070	64.10	3.44	11.96	20.49
C25H16N407	484.43	484.1019	61.99	3.33	11.57	23.12
C25H16N408	500.43	500.0968	60.00	3.22	11.20	25.58
C25H16N409	516.43	516.0917	58.15	3.12	10.85	27.88
C25H16N601	416.45	416.1386	72.11	3.87	20.18	3.84
C25H16N602	432.45	432.1335	69.44	3.73	19.43	7.40
C25H16N603	448.44	448.1284	66.96	3.60	18.74	10.70
C25H16N604	464.44	464.1233	64.65	3.47	18.09	13.78
C25H16N605	480.44	480.1182	62.50	3.36	17.49	16.65
C25H16N606	496.44	496.1131	60.49	3.25	16.93	19.34
C25H16N607	512.44	512.1080	58.60	3.15	16.40	21.86
C25H16N608	528.44	528.1029	56.82	3.05	15.90	24.22
C25H16N609	544.44	544.0979	55.15	2.96	15.44	26.45
C25H17N101	347.42	347.1310	86.43	4.93	4.03	4.61
C25H17N102	363.42	363.1259	82.63	4.72	3.85	8.80
C25H17N103	379.42	379.1208	79.14	4.52	3.69	12.65
C25H17N104	395.42	395.1157	75.94	4.33	3.54	16.18
C25H17N105	411.42	411.1107	72.99	4.16	3.40	19.44
C25H17N106	427.42	427.1056	70.25	4.01	3.28	22.46
C25H17N107	443.42	443.1005	67.72	3.86	3.16	25.26
C25H17N108	459.42	459.0954	65.36	3.73	3.05	27.86
C25H17N109	475.42	475.0903	63.16	3.60	2.95	30.29
C25H17N301	375.43	375.1372	79.98	4.56	11.19	4.26
C25H17N302	391.43	391.1321	76.71	4.38	10.73	8.17
C25H17N303	407.43	407.1270	73.70	4.21	10.31	11.78
C25H17N304	423.43	423.1219	70.92	4.05	9.92	15.11
C25H17N305	439.43	439.1168	68.33	3.90	9.56	18.20
C25H17N306	455.43	455.1117	65.93	3.76	9.23	21.08
C25H17N307	471.43	471.1066	63.70	3.63	8.91	23.76
C25H17N308	487.43	487.1016	61.60	3.52	8.62	26.26
C25H17N309	503.43	503.0965	59.65	3.40	8.35	28.60
C25H17N501	403.45	403.1433	74.43	4.25	17.36	3.97
C25H17N502	419.45	419.1382	71.59	4.09	16.70	7.63
C25H17N503	435.45	435.1331	68.96	3.94	16.08	11.02
C25H17N504	451.45	451.1280	66.51	3.80	15.51	14.18
C25H17N505	467.44	467.1230	64.24	3.67	14.98	17.11
C25H17N506	483.44	483.1179	62.11	3.54	14.49	19.86
C25H17N507	499.44	499.1128	60.12	3.43	14.02	22.42
C25H17N508	515.44	515.1077	58.26	3.32	13.59	24.83
C25H17N509	531.44	531.1026	56.50	3.22	13.18	27.10
C25H18N201	362.44	362.1419	82.85	5.01	7.73	4.41
C25H18N202	378.43	378.1368	79.35	4.79	7.40	8.46
C25H18N203	394.43	394.1317	76.13	4.60	7.10	12.17
C25H18N204	410.43	410.1266	73.16	4.42	6.83	15.59
C25H18N205	426.43	426.1216	70.42	4.25	6.57	18.76
C25H18N206	442.43	442.1165	67.87	4.10	6.33	21.70
C25H18N207	458.43	458.1114	65.50	3.96	6.11	24.43
C25H18N208	474.43	474.1063	63.29	3.82	5.90	26.98
C25H18N209	490.43	490.1012	61.23	3.70	5.71	29.36
C25H18N401	390.45	390.1481	76.91	4.65	14.35	4.10

158

C25H18N402	406.45	406.1430	73.88	4.46	13.78	7.87
C25H18N403	422.45	422.1379	71.08	4.29	13.26	11.36
C25H18N404	438.45	438.1328	68.49	4.14	12.78	14.60
C25H18N405	454.45	454.1277	66.08	3.99	12.33	17.60
C25H18N406	470.45	470.1226	63.83	3.86	11.91	20.41
C25H18N407	486.44	486.1175	61.73	3.73	11.52	23.02
C25H18N408	502.44	502.1125	59.76	3.61	11.15	25.47
C25H18N409	518.44	518.1074	57.92	3.50	10.81	27.77
C25H18N601	418.46	418.1542	71.76	4.34	20.08	3.82
C25H18N602	434.46	434.1491	69.12	4.18	19.34	7.37
C25H18N603	450.46	450.1440	66.66	4.03	18.66	10.66
C25H18N604	466.46	466.1389	64.37	3.89	18.02	13.72
C25H18N605	482.46	482.1339	62.24	3.76	17.42	16.58
C25H18N606	498.46	498.1288	60.24	3.64	16.86	19.26
C25H18N607	514.46	514.1237	58.37	3.53	16.34	21.77
C25H18N608	530.46	530.1186	56.61	3.42	15.84	24.13
C25H18N609	546.46	546.1135	54.95	3.32	15.38	26.35
C25H19N101	349.44	349.1467	85.93	5.48	4.01	4.58
C25H19N102	365.44	365.1416	82.17	5.24	3.83	8.76
C25H19N103	381.44	381.1365	78.72	5.02	3.67	12.58
C25H19N104	397.43	397.1314	75.55	4.82	3.52	16.10
C25H19N105	413.43	413.1263	72.63	4.63	3.39	19.35
C25H19N106	429.43	429.1212	69.92	4.46	3.26	22.35
C25H19N107	445.43	445.1161	67.41	4.30	3.14	25.14
C25H19N108	461.43	461.1111	65.08	4.15	3.04	27.74
C25H19N109	477.43	477.1060	62.89	4.01	2.93	30.16
C25H19N301	377.45	377.1528	79.55	5.07	11.13	4.24
C25H19N302	393.45	393.1477	76.32	4.87	10.68	8.13
C25H19N303	409.45	409.1426	73.34	4.68	10.26	11.72
C25H19N304	425.45	425.1375	70.58	4.50	9.88	15.04
C25H19N305	441.45	441.1325	68.02	4.34	9.52	18.12
C25H19N306	457.45	457.1274	65.64	4.19	9.19	20.99
C25H19N307	473.45	473.1223	63.42	4.05	8.88	23.66
C25H19N308	489.45	489.1172	61.35	3.91	8.59	26.15
C25H19N309	505.44	505.1121	59.41	3.79	8.31	28.49
C25H19N501	405.46	405.1590	74.06	4.72	17.27	3.95
C25H19N502	421.46	421.1539	71.25	4.54	16.62	7.59
C25H19N503	437.46	437.1488	68.64	4.38	16.01	10.97
C25H19N504	453.46	453.1437	66.22	4.22	15.44	14.11
C25H19N505	469.46	469.1386	63.96	4.08	14.92	17.04
C25H19N506	485.46	485.1335	61.85	3.95	14.43	19.77
C25H19N507	501.46	501.1284	59.88	3.82	13.97	22.33
C25H19N508	517.46	517.1233	58.03	3.70	13.53	24.74
C25H19N509	533.46	533.1183	56.29	3.59	13.13	26.99
C25H20N201	364.45	364.1576	82.39	5.53	7.69	4.39
C25H20N202	380.45	380.1525	78.93	5.30	7.36	8.41
C25H20N203	396.45	396.1474	75.74	5.08	7.07	12.11
C25H20N204	412.45	412.1423	72.80	4.89	6.79	15.52
C25H20N205	428.45	428.1372	70.09	4.71	6.54	18.67
C25H20N206	444.45	444.1321	67.56	4.54	6.30	21.60
C25H20N207	460.45	460.1270	65.21	4.38	6.08	24.32
C25H20N208	476.45	476.1220	63.02	4.23	5.88	26.86
C25H20N209	492.45	492.1169	60.98	4.09	5.69	29.24
C25H20N401	392.46	392.1637	76.51	5.14	14.28	4.08
C25H20N402	408.46	408.1586	73.51	4.94	13.72	7.83
C25H20N403	424.46	424.1535	70.74	4.75	13.20	11.31
C25H20N404	440.46	440.1484	68.17	4.58	12.72	14.53
C25H20N405	456.46	456.1434	65.78	4.42	12.27	17.53
C25H20N406	472.46	472.1383	63.56	4.27	11.86	20.32
C25H20N407	488.46	488.1332	61.47	4.13	11.47	22.93

C25H20N408	504.46	504.1281	59.52	4.00	11.11	25.37
C25H20N409	520.46	520.1230	57.69	3.87	10.76	27.67
C25H20N601	420.48	420.1698	71.41	4.79	19.99	3.81
C25H20N602	436.48	436.1648	68.80	4.62	19.25	7.33
C25H20N603	452.48	452.1597	66.36	4.46	18.57	10.61
C25H20N604	468.48	468.1546	64.10	4.30	17.94	13.66
C25H20N605	484.48	484.1495	61.98	4.16	17.35	16.51
C25H20N606	500.47	500.1444	60.00	4.03	16.79	19.18
C25H20N607	516.47	516.1393	58.14	3.90	16.27	21.68
C25H20N608	532.47	532.1342	56.39	3.79	15.78	24.04
C25H20N609	548.47	548.1292	54.75	3.68	15.32	26.25
C25H21N101	351.45	351.1623	85.44	6.02	3.99	4.55
C25H21N102	367.45	367.1572	81.72	5.76	3.81	8.71
C25H21N103	383.45	383.1521	78.31	5.52	3.65	12.52
C25H21N104	399.45	399.1470	75.17	5.30	3.51	16.02
C25H21N105	415.45	415.1420	72.28	5.10	3.37	19.26
C25H21N106	431.45	431.1369	69.60	4.91	3.25	22.25
C25H21N107	447.45	447.1318	67.11	4.73	3.13	25.03
C25H21N108	463.45	463.1267	64.79	4.57	3.02	27.62
C25H21N109	479.45	479.1216	62.63	4.41	2.92	30.03
C25H21N301	379.47	379.1685	79.13	5.58	11.07	4.22
C25H21N302	395.47	395.1634	75.93	5.35	10.63	8.09
C25H21N303	411.46	411.1583	72.98	5.14	10.21	11.67
C25H21N304	427.46	427.1532	70.25	4.95	9.83	14.97
C25H21N305	443.46	443.1481	67.71	4.77	9.48	18.04
C25H21N306	459.46	459.1430	65.35	4.61	9.15	20.89
C25H21N307	475.46	475.1379	63.16	4.45	8.84	23.56
C25H21N308	491.46	491.1329	61.10	4.31	8.55	26.04
C25H21N309	507.46	507.1278	59.17	4.17	8.28	28.38
C25H21N501	407.48	407.1746	73.69	5.19	17.19	3.93
C25H21N502	423.48	423.1695	70.91	5.00	16.54	7.56
C25H21N503	439.48	439.1644	68.33	4.82	15.94	10.92
C25H21N504	455.48	455.1593	65.93	4.65	15.38	14.05
C25H21N505	471.48	471.1543	63.69	4.49	14.85	16.97
C25H21N506	487.48	487.1492	61.60	4.34	14.37	19.69
C25H21N507	503.48	503.1441	59.64	4.20	13.91	22.24
C25H21N508	519.47	519.1390	57.80	4.07	13.48	24.64
C25H21N509	535.47	535.1339	56.08	3.95	13.08	26.89
C25H22N201	366.47	366.1732	81.94	6.05	7.64	4.37
C25H22N202	382.47	382.1681	78.51	5.80	7.32	8.37
C25H22N203	398.47	398.1630	75.36	5.57	7.03	12.05
C25H22N204	414.47	414.1579	72.45	5.35	6.76	15.44
C25H22N205	430.46	430.1529	69.76	5.15	6.51	18.58
C25H22N206	446.46	446.1478	67.26	4.97	6.27	21.50
C25H22N207	462.46	462.1427	64.93	4.80	6.06	24.22
C25H22N208	478.46	478.1376	62.76	4.63	5.85	26.75
C25H22N209	494.46	494.1325	60.73	4.48	5.67	29.12
C25H22N401	394.48	394.1794	76.12	5.62	14.20	4.06
C25H22N402	410.48	410.1743	73.15	5.40	13.65	7.80
C25H22N403	426.48	426.1692	70.41	5.20	13.14	11.25
C25H22N404	442.48	442.1641	67.86	5.01	12.66	14.46
C25H22N405	458.48	458.1590	65.49	4.84	12.22	17.45
C25H22N406	474.48	474.1539	63.29	4.67	11.81	20.23
C25H22N407	490.48	490.1488	61.22	4.52	11.42	22.83
C25H22N408	506.48	506.1437	59.29	4.38	11.06	25.27
C25H22N409	522.48	522.1387	57.47	4.24	10.72	27.56
C25H22N601	422.49	422.1855	71.07	5.25	19.89	3.79
C25H22N602	438.49	438.1804	68.48	5.06	19.17	7.30
C25H22N603	454.49	454.1753	66.07	4.88	18.49	10.56
C25H22N604	470.49	470.1702	63.82	4.71	17.86	13.60

159

C25H22N605	486.49	486.1652	61.72	4.56	17.27	16.44
C25H22N606	502.49	502.1601	59.76	4.41	16.72	19.10
C25H22N607	518.49	518.1550	57.91	4.28	16.21	21.60
C25H22N608	534.49	534.1499	56.18	4.15	15.72	23.95
C25H22N609	550.49	550.1448	54.55	4.03	15.27	26.16
C25H23N101	353.47	353.1780	84.95	6.56	3.96	4.53
C25H23N102	369.47	369.1729	81.27	6.27	3.79	8.66
C25H23N103	385.47	385.1678	77.90	6.01	3.63	12.45
C25H23N104	401.47	401.1627	74.80	5.77	3.49	15.94
C25H23N105	417.47	417.1576	71.93	5.55	3.36	19.16
C25H23N106	433.47	433.1525	69.27	5.35	3.23	22.15
C25H23N107	449.46	449.1474	66.81	5.16	3.12	24.92
C25H23N108	465.46	465.1424	64.51	4.98	3.01	27.50
C25H23N109	481.46	481.1373	62.37	4.82	2.91	29.91
C25H23N301	381.48	381.1841	78.71	6.08	11.01	4.19
C25H23N302	397.48	397.1790	75.55	5.83	10.57	8.05
C25H23N303	413.48	413.1739	72.62	5.61	10.16	11.61
C25H23N304	429.48	429.1688	69.92	5.40	9.78	14.90
C25H23N305	445.48	445.1638	67.41	5.20	9.43	17.96
C25H23N306	461.48	461.1587	65.07	5.02	9.11	20.80
C25H23N307	477.48	477.1536	62.89	4.86	8.80	23.46
C25H23N308	493.48	493.1485	60.85	4.70	8.52	25.94
C25H23N309	509.48	509.1434	58.94	4.55	8.25	28.26
C25H23N501	409.49	409.1902	73.33	5.66	17.10	3.91
C25H23N502	425.49	425.1852	70.57	5.45	16.46	7.52
C25H23N503	441.49	441.1801	68.01	5.25	15.86	10.87
C25H23N504	457.49	457.1750	65.64	5.07	15.31	13.99
C25H23N505	473.49	473.1699	63.42	4.90	14.79	16.90
C25H23N506	489.49	489.1648	61.34	4.74	14.31	19.61
C25H23N507	505.49	505.1597	59.40	4.59	13.85	22.16
C25H23N508	521.49	521.1546	57.58	4.45	13.43	24.54
C25H23N509	537.49	537.1496	55.87	4.31	13.03	26.79
C25H24N201	368.48	368.1889	81.49	6.57	7.60	4.34
C25H24N202	384.48	384.1838	78.10	6.29	7.29	8.32
C25H24N203	400.48	400.1787	74.98	6.04	6.99	11.99
C25H24N204	416.48	416.1736	72.10	5.81	6.73	15.37
C25H24N205	432.48	432.1685	69.43	5.59	6.48	18.50
C25H24N206	448.48	448.1634	66.95	5.39	6.25	21.40
C25H24N207	464.48	464.1583	64.65	5.21	6.03	24.11
C25H24N208	480.48	480.1533	62.50	5.03	5.83	26.64
C25H24N209	496.48	496.1482	60.48	4.87	5.64	29.00
C25H24N401	396.50	396.1950	75.73	6.10	14.13	4.04
C25H24N402	412.50	412.1899	72.80	5.86	13.58	7.76
C25H24N403	428.50	428.1848	70.08	5.65	13.08	11.20
C25H24N404	444.49	444.1797	67.56	5.44	12.60	14.40
C25H24N405	460.49	460.1747	65.21	5.25	12.17	17.37
C25H24N406	476.49	476.1696	63.02	5.08	11.76	20.15
C25H24N407	492.49	492.1645	60.97	4.91	11.38	22.74
C25H24N408	508.49	508.1594	59.05	4.76	11.02	25.17
C25H24N409	524.49	524.1543	57.25	4.61	10.68	27.45
C25H24N601	424.51	424.2011	70.74	5.70	19.80	3.77
C25H24N602	440.51	440.1961	68.17	5.49	19.08	7.26
C25H24N603	456.51	456.1910	65.78	5.30	18.41	10.51
C25H24N604	472.51	472.1859	63.55	5.12	17.79	13.54
C25H24N605	488.51	488.1808	61.47	4.95	17.20	16.38
C25H24N606	504.51	504.1757	59.52	4.80	16.66	19.03
C25H24N607	520.51	520.1706	57.69	4.65	16.15	21.52
C25H24N608	536.51	536.1655	55.97	4.51	15.66	23.86
C25H24N609	552.50	552.1605	54.35	4.38	15.21	26.06
C25H25N101	355.48	355.1936	84.47	7.09	3.94	4.50

C25H25N102	371.48	371.1885	80.83	6.78	3.77	8.61
C25H25N103	387.48	387.1834	77.49	6.50	3.61	12.39
C25H25N104	403.48	403.1783	74.42	6.25	3.47	15.86
C25H25N105	419.48	419.1733	71.58	6.01	3.34	19.07
C25H25N106	435.48	435.1682	68.95	5.79	3.22	22.04
C25H25N107	451.48	451.1631	66.51	5.58	3.10	24.81
C25H25N108	467.48	467.1580	64.23	5.39	3.00	27.38
C25H25N109	483.48	483.1529	62.11	5.21	2.90	29.78
C25H25N301	383.50	383.1998	78.30	6.57	10.96	4.17
C25H25N302	399.50	399.1947	75.16	6.31	10.52	8.01
C25H25N303	415.50	415.1896	72.27	6.06	10.11	11.55
C25H25N304	431.50	431.1845	69.59	5.84	9.74	14.83
C25H25N305	447.50	447.1794	67.10	5.63	9.39	17.88
C25H25N306	463.49	463.1743	64.79	5.44	9.07	20.71
C25H25N307	479.49	479.1692	62.62	5.26	8.76	23.36
C25H25N308	495.49	495.1641	60.60	5.09	8.48	25.83
C25H25N309	511.49	511.1591	58.71	4.93	8.22	28.15
C25H25N501	411.51	411.2059	72.97	6.12	17.02	3.89
C25H25N502	427.51	427.2008	70.24	5.89	16.38	7.48
C25H25N503	443.51	443.1957	67.71	5.68	15.79	10.82
C25H25N504	459.51	459.1906	65.35	5.48	15.24	13.93
C25H25N505	475.51	475.1856	63.15	5.30	14.73	16.82
C25H25N506	491.51	491.1805	61.09	5.13	14.25	19.53
C25H25N507	507.51	507.1754	59.17	4.97	13.80	22.07
C25H25N508	523.51	523.1703	57.36	4.81	13.38	24.45
C25H25N509	539.51	539.1652	55.66	4.67	12.98	26.69
C25H26N201	370.50	370.2045	81.05	7.07	7.56	4.32
C25H26N202	386.50	386.1994	77.69	6.78	7.25	8.28
C25H26N203	402.50	402.1943	74.60	6.51	6.96	11.93
C25H26N204	418.50	418.1892	71.75	6.26	6.69	15.29
C25H26N205	434.50	434.1842	69.11	6.03	6.45	18.41
C25H26N206	450.50	450.1791	66.66	5.82	6.22	21.31
C25H26N207	466.50	466.1740	64.37	5.62	6.01	24.01
C25H26N208	482.49	482.1689	62.23	5.43	5.81	26.53
C25H26N209	498.49	498.1638	60.24	5.26	5.62	28.89
C25H26N401	398.51	398.2106	75.35	6.58	14.06	4.01
C25H26N402	414.51	414.2056	72.44	6.32	13.52	7.72
C25H26N403	430.51	430.2005	69.75	6.09	13.01	11.15
C25H26N404	446.51	446.1954	67.25	5.87	12.55	14.33
C25H26N405	462.51	462.1903	64.92	5.67	12.11	17.30
C25H26N406	478.51	478.1852	62.75	5.48	11.71	20.06
C25H26N407	494.51	494.1801	60.72	5.30	11.33	22.65
C25H26N408	510.51	510.1750	58.82	5.13	10.97	25.07
C25H26N409	526.51	526.1700	57.03	4.98	10.64	27.35
C25H26N601	426.53	426.2168	70.40	6.14	19.70	3.75
C25H26N602	442.52	442.2117	67.86	5.92	18.99	7.23
C25H26N603	458.52	458.2066	65.49	5.72	18.33	10.47
C25H26N604	474.52	474.2015	63.28	5.52	17.71	13.49
C25H26N605	490.52	490.1965	61.22	5.34	17.13	16.31
C25H26N606	506.52	506.1914	59.28	5.17	16.59	18.95
C25H26N607	522.52	522.1863	57.47	5.02	16.08	21.43
C25H26N608	538.52	538.1812	55.76	4.87	15.61	23.77
C25H26N609	554.52	554.1761	54.15	4.73	15.16	25.97
C25H27N101	357.50	357.2093	83.99	7.61	3.92	4.48
C25H27N102	373.50	373.2042	80.40	7.29	3.75	8.57
C25H27N103	389.50	389.1991	77.09	6.99	3.60	12.32
C25H27N104	405.50	405.1940	74.05	6.71	3.45	15.78
C25H27N105	421.50	421.1889	71.24	6.46	3.32	18.98
C25H27N106	437.50	437.1838	68.64	6.22	3.20	21.94
C25H27N107	453.50	453.1787	66.21	6.00	3.09	24.70

C25H27N108	469.50	469.1737	63.96	5.80	2.98	27.26
C25H27N109	485.50	485.1686	61.85	5.61	2.89	29.66
C25H27N301	385.51	385.2154	77.89	7.06	10.90	4.15
C25H27N302	401.51	401.2103	74.79	6.78	10.47	7.97
C25H27N303	417.51	417.2052	71.92	6.52	10.06	11.50
C25H27N304	433.51	433.2001	69.27	6.28	9.69	14.76
C25H27N305	449.51	449.1951	66.80	6.05	9.35	17.80
C25H27N306	465.51	465.1900	64.51	5.85	9.03	20.62
C25H27N307	481.51	481.1849	62.36	5.65	8.73	23.26
C25H27N308	497.51	497.1798	60.36	5.47	8.45	25.73
C25H27N309	513.51	513.1747	58.48	5.30	8.18	28.04
C25H27N501	413.53	413.2215	72.61	6.58	16.94	3.87
C25H27N502	429.53	429.2165	69.91	6.34	16.30	7.45
C25H27N503	445.53	445.2114	67.40	6.11	15.72	10.77
C25H27N504	461.53	461.2063	65.06	5.90	15.17	13.87
C25H27N505	477.52	477.2012	62.88	5.70	14.67	16.75
C25H27N506	493.52	493.1961	60.84	5.51	14.19	19.45
C25H27N507	509.52	509.1910	58.93	5.34	13.74	21.98
C25H27N508	525.52	525.1859	57.14	5.18	13.33	24.36
C25H27N509	541.52	541.1809	55.45	5.03	12.93	26.59
C25H28N201	372.51	372.2202	80.61	7.58	7.52	4.29
C25H28N202	388.51	388.2151	77.29	7.26	7.21	8.24
C25H28N203	404.51	404.2100	74.23	6.98	6.93	11.87
C25H28N204	420.51	420.2049	71.41	6.71	6.66	15.22
C25H28N205	436.51	436.1998	68.79	6.47	6.42	18.33
C25H28N206	452.51	452.1947	66.36	6.24	6.19	21.21
C25H28N207	468.51	468.1896	64.09	6.02	5.98	23.90
C25H28N208	484.51	484.1845	61.98	5.83	5.78	26.42
C25H28N209	500.51	500.1795	59.99	5.64	5.60	28.77
C25H28N401	400.53	400.2263	74.97	7.05	13.99	3.99
C25H28N402	416.53	416.2212	72.09	6.78	13.45	7.68
C25H28N403	432.53	432.2161	69.42	6.53	12.95	11.10
C25H28N404	448.53	448.2110	66.95	6.29	12.49	14.27
C25H28N405	464.53	464.2060	64.64	6.08	12.06	17.22
C25H28N406	480.53	480.2009	62.49	5.87	11.66	19.98
C25H28N407	496.52	496.1958	60.48	5.68	11.28	22.56
C25H28N408	512.52	512.1907	58.59	5.51	10.93	24.97
C25H28N409	528.52	528.1856	56.81	5.34	10.60	27.24
C25H28N601	428.54	428.2324	70.07	6.59	19.61	3.73
C25H28N602	444.54	444.2274	67.55	6.35	18.90	7.20
C25H28N603	460.54	460.2223	65.20	6.13	18.25	10.42
C25H28N604	476.54	476.2172	63.01	5.92	17.64	13.43
C25H28N605	492.54	492.2121	60.97	5.73	17.06	16.24
C25H28N606	508.54	508.2070	59.05	5.55	16.53	18.88
C25H28N607	524.54	524.2019	57.25	5.38	16.02	21.35
C25H28N608	540.54	540.1968	55.55	5.22	15.55	23.68
C25H28N609	556.54	556.1918	53.95	5.07	15.10	25.87
C25H29N101	359.52	359.2249	83.52	8.13	3.90	4.45
C25H29N102	375.52	375.2198	79.96	7.78	3.73	8.52
C25H29N103	391.51	391.2147	76.70	7.47	3.58	12.26
C25H29N104	407.51	407.2096	73.69	7.17	3.44	15.70
C25H29N105	423.51	423.2046	70.90	6.90	3.31	18.89
C25H29N106	439.51	439.1995	68.32	6.65	3.19	21.84
C25H29N107	455.51	455.1944	65.92	6.42	3.07	24.59
C25H29N108	471.51	471.1893	63.68	6.20	2.97	27.15
C25H29N109	487.51	487.1842	61.59	6.00	2.87	29.54
C25H29N301	387.53	387.2310	77.49	7.54	10.84	4.13
C25H29N302	403.53	403.2260	74.41	7.24	10.41	7.93
C25H29N303	419.53	419.2209	71.58	6.97	10.02	11.44
C25H29N304	435.53	435.2158	68.95	6.71	9.65	14.69

C25H29N305	451.53	451.2107	66.50	6.47	9.31	17.72
C25H29N306	467.53	467.2056	64.23	6.25	8.99	20.53
C25H29N307	483.53	483.2005	62.10	6.05	8.69	23.16
C25H29N308	499.53	499.1954	60.11	5.85	8.41	25.62
C25H29N309	515.52	515.1904	58.25	5.67	8.15	27.93
C25H29N501	415.54	415.2372	72.26	7.03	16.85	3.85
C25H29N502	431.54	431.2321	69.58	6.77	16.23	7.41
C25H29N503	447.54	447.2270	67.10	6.53	15.65	10.72
C25H29N504	463.54	463.2219	64.78	6.31	15.11	13.81
C25H29N505	479.54	479.2169	62.62	6.10	14.60	16.68
C25H29N506	495.54	495.2118	60.60	5.90	14.13	19.37
C25H29N507	511.54	511.2067	58.70	5.71	13.69	21.89
C25H29N508	527.54	527.2016	56.92	5.54	13.28	24.26
C25H29N509	543.54	543.1965	55.25	5.38	12.88	26.49
C25H30N201	374.53	374.2358	80.17	8.07	7.48	4.27
C25H30N202	390.53	390.2307	76.89	7.74	7.17	8.19
C25H30N203	406.53	406.2256	73.86	7.44	6.89	11.81
C25H30N204	422.53	422.2205	71.07	7.16	6.63	15.15
C25H30N205	438.53	438.2155	68.47	6.90	6.39	18.24
C25H30N206	454.53	454.2104	66.06	6.65	6.16	21.12
C25H30N207	470.53	470.2053	63.82	6.43	5.95	23.80
C25H30N208	486.53	486.2002	61.72	6.22	5.76	26.31
C25H30N209	502.53	502.1951	59.75	6.02	5.57	28.65
C25H30N401	402.54	402.2419	74.60	7.51	13.92	3.97
C25H30N402	418.54	418.2369	71.74	7.22	13.39	7.65
C25H30N403	434.54	434.2318	69.10	6.96	12.89	11.05
C25H30N404	450.54	450.2267	66.65	6.71	12.44	14.20
C25H30N405	466.54	466.2216	64.36	6.48	12.01	17.15
C25H30N406	482.54	482.2165	62.23	6.27	11.61	19.89
C25H30N407	498.54	498.2114	60.23	6.07	11.24	22.46
C25H30N408	514.54	514.2063	58.36	5.88	10.89	24.88
C25H30N409	530.54	530.2013	56.60	5.70	10.56	27.14
C25H30N601	430.56	430.2481	69.74	7.02	19.52	3.72
C25H30N602	446.56	446.2430	67.24	6.77	18.82	7.17
C25H30N603	462.56	462.2379	64.92	6.54	18.17	10.38
C25H30N604	478.56	478.2328	62.75	6.32	17.56	13.37
C25H30N605	494.56	494.2278	60.72	6.11	16.99	16.18
C25H30N606	510.55	510.2227	58.81	5.92	16.46	18.80
C25H30N607	526.55	526.2176	57.03	5.74	15.96	21.27
C25H30N608	542.55	542.2125	55.35	5.57	15.49	23.59
C25H30N609	558.55	558.2074	53.76	5.41	15.05	25.78
C25H31N101	361.53	361.2406	83.06	8.64	3.87	4.43
C25H31N102	377.53	377.2355	79.54	8.28	3.71	8.48
C25H31N103	393.53	393.2304	76.30	7.94	3.56	12.20
C25H31N104	409.53	409.2253	73.32	7.63	3.42	15.63
C25H31N105	425.53	425.2202	70.57	7.34	3.29	18.80
C25H31N106	441.53	441.2151	68.01	7.08	3.17	21.74
C25H31N107	457.53	457.2100	65.63	6.83	3.06	24.48
C25H31N108	473.53	473.2049	63.41	6.60	2.96	27.03
C25H31N109	489.53	489.1999	61.34	6.38	2.86	29.42
C25H31N301	389.55	389.2467	77.08	8.02	10.79	4.11
C25H31N302	405.54	405.2416	74.04	7.70	10.36	7.89
C25H31N303	421.54	421.2365	71.23	7.41	9.97	11.39
C25H31N304	437.54	437.2314	68.63	7.14	9.60	14.63
C25H31N305	453.54	453.2264	66.21	6.89	9.26	17.64
C25H31N306	469.54	469.2213	63.95	6.65	8.95	20.44
C25H31N307	485.54	485.2162	61.84	6.44	8.65	23.07
C25H31N308	501.54	501.2111	59.87	6.23	8.38	25.52
C25H31N309	517.54	517.2060	58.02	6.04	8.12	27.82
C25H31N501	417.56	417.2528	71.91	7.48	16.77	3.83

C25H31N502	433.56	433.2478	69.26	7.21	16.15	7.38
C25H31N503	449.56	449.2427	66.79	6.95	15.58	10.68
C25H31N504	465.56	465.2376	64.50	6.71	15.04	13.75
C25H31N505	481.56	481.2325	62.36	6.49	14.54	16.61
C25H31N506	497.56	497.2274	60.35	6.28	14.08	19.29
C25H31N507	513.56	513.2223	58.47	6.08	13.64	21.81
C25H31N508	529.55	529.2172	56.70	5.90	13.22	24.17
C25H31N509	545.55	545.2122	55.04	5.73	12.84	26.39
C25H32N201	376.55	376.2514	79.75	8.57	7.44	4.25
C25H32N202	392.55	392.2464	76.50	8.22	7.14	8.15
C25H32N203	408.55	408.2413	73.50	7.90	6.86	11.75
C25H32N204	424.54	424.2362	70.73	7.60	6.60	15.07
C25H32N205	440.54	440.2311	68.16	7.32	6.36	18.16
C25H32N206	456.54	456.2260	65.77	7.07	6.14	21.03
C25H32N207	472.54	472.2209	63.55	6.83	5.93	23.70
C25H32N208	488.54	488.2158	61.46	6.60	5.73	26.20
C25H32N209	504.54	504.2108	59.52	6.39	5.55	28.54
C25H32N401	404.56	404.2576	74.22	7.97	13.85	3.95
C25H32N402	420.56	420.2525	71.40	7.67	13.32	7.61
C25H32N403	436.56	436.2474	68.78	7.39	12.83	10.99
C25H32N404	452.56	452.2423	66.35	7.13	12.38	14.14
C25H32N405	468.56	468.2373	64.09	6.88	11.96	17.07
C25H32N406	484.56	484.2322	61.97	6.66	11.56	19.81
C25H32N407	500.56	500.2271	59.99	6.44	11.19	22.37
C25H32N408	516.56	516.2220	58.13	6.24	10.85	24.78
C25H32N409	532.56	532.2169	56.38	6.06	10.52	27.04
C25H32N601	432.57	432.2637	69.42	7.46	19.43	3.70
C25H32N602	448.57	448.2587	66.94	7.19	18.74	7.13
C25H32N603	464.57	464.2536	64.64	6.94	18.09	10.33
C25H32N604	480.57	480.2485	62.48	6.71	17.49	13.32
C25H32N605	496.57	496.2434	60.47	6.50	16.92	16.11
C25H32N606	512.57	512.2383	58.58	6.29	16.40	18.73
C25H32N607	528.57	528.2332	56.81	6.10	15.90	21.19
C25H32N608	544.57	544.2281	55.14	5.92	15.43	23.50
C25H32N609	560.57	560.2231	53.57	5.75	14.99	25.69
C25H33N101	363.55	363.2562	82.60	9.15	3.85	4.40
C25H33N102	379.55	379.2511	79.11	8.76	3.69	8.43
C25H33N103	395.55	395.2460	75.91	8.41	3.54	12.13
C25H33N104	411.55	411.2409	72.96	8.08	3.40	15.55
C25H33N105	427.55	427.2359	70.23	7.78	3.28	18.71
C25H33N106	443.54	443.2308	67.70	7.50	3.16	21.64
C25H33N107	459.54	459.2257	65.34	7.24	3.05	24.37
C25H33N108	475.54	475.2206	63.14	6.99	2.95	26.92
C25H33N109	491.54	491.2155	61.09	6.77	2.85	29.29
C25H33N301	391.56	391.2623	76.69	8.49	10.73	4.09
C25H33N302	407.56	407.2573	73.68	8.16	10.31	7.85
C25H33N303	423.56	423.2522	70.89	7.85	9.92	11.33
C25H33N304	439.56	439.2471	68.31	7.57	9.56	14.56
C25H33N305	455.56	455.2420	65.91	7.30	9.22	17.56
C25H33N306	471.56	471.2369	63.68	7.05	8.91	20.36
C25H33N307	487.56	487.2318	61.59	6.82	8.62	22.97
C25H33N308	503.56	503.2267	59.63	6.61	8.34	25.42
C25H33N309	519.56	519.2217	57.80	6.40	8.09	27.71
C25H33N501	419.57	419.2685	71.57	7.93	16.69	3.81
C25H33N502	435.57	435.2634	68.94	7.64	16.08	7.35
C25H33N503	451.57	451.2583	66.50	7.37	15.51	10.63
C25H33N504	467.57	467.2532	64.22	7.11	14.98	13.69
C25H33N505	483.57	483.2482	62.10	6.88	14.48	16.54
C25H33N506	499.57	499.2431	60.11	6.66	14.02	19.22
C25H33N507	515.57	515.2380	58.24	6.45	13.58	21.72

C25H33N508	531.57	531.2329	56.49	6.26	13.17	24.08
C25H33N509	547.57	547.2278	54.84	6.07	12.79	26.30
C25H34N201	378.56	378.2671	79.32	9.05	7.40	4.23
C25H34N202	394.56	394.2620	76.10	8.69	7.10	8.11
C25H34N203	410.56	410.2569	73.14	8.35	6.82	11.69
C25H34N204	426.56	426.2518	70.40	8.03	6.57	15.00
C25H34N205	442.56	442.2468	67.85	7.74	6.33	18.08
C25H34N206	458.56	458.2417	65.48	7.47	6.11	20.93
C25H34N207	474.56	474.2366	63.28	7.22	5.90	23.60
C25H34N208	490.56	490.2315	61.21	6.99	5.71	26.09
C25H34N209	506.56	506.2264	59.28	6.77	5.53	28.43
C25H34N401	406.58	406.2732	73.86	8.43	13.78	3.94
C25H34N402	422.58	422.2682	71.06	8.11	13.26	7.57
C25H34N403	438.57	438.2631	68.47	7.81	12.77	10.94
C25H34N404	454.57	454.2580	66.06	7.54	12.33	14.08
C25H34N405	470.57	470.2529	63.81	7.28	11.91	17.00
C25H34N406	486.57	486.2478	61.71	7.04	11.51	19.73
C25H34N407	502.57	502.2427	59.75	6.82	11.15	22.28
C25H34N408	518.57	518.2376	57.90	6.61	10.80	24.68
C25H34N409	534.57	534.2326	56.17	6.41	10.48	26.94
C25H34N601	434.59	434.2794	69.09	7.89	19.34	3.68
C25H34N602	450.59	450.2743	66.64	7.61	18.65	7.10
C25H34N603	466.59	466.2692	64.36	7.35	18.01	10.29
C25H34N604	482.59	482.2641	62.22	7.10	17.41	13.26
C25H34N605	498.59	498.2590	60.23	6.87	16.86	16.04
C25H34N606	514.59	514.2540	58.35	6.66	16.33	18.66
C25H34N607	530.59	530.2489	56.59	6.46	15.84	21.11
C25H34N608	546.59	546.2438	54.94	6.27	15.38	23.42
C25H34N609	562.58	562.2387	53.37	6.09	14.94	25.60
C25H35N101	365.56	365.2718	82.14	9.65	3.83	4.38
C25H35N102	381.56	381.2668	78.70	9.25	3.67	8.39
C25H35N103	397.56	397.2617	75.53	8.87	3.52	12.07
C25H35N104	413.56	413.2566	72.61	8.53	3.39	15.47
C25H35N105	429.56	429.2515	69.90	8.21	3.26	18.62
C25H35N106	445.56	445.2464	67.39	7.92	3.14	21.55
C25H35N107	461.56	461.2413	65.06	7.64	3.03	24.26
C25H35N108	477.56	477.2362	62.88	7.39	2.93	26.80
C25H35N109	493.56	493.2312	60.84	7.15	2.84	29.17
C25H35N301	393.58	393.2780	76.29	8.96	10.68	4.07
C25H35N302	409.58	409.2729	73.31	8.61	10.26	7.81
C25H35N303	425.58	425.2678	70.56	8.29	9.87	11.28
C25H35N304	441.58	441.2627	68.00	7.99	9.52	14.49
C25H35N305	457.57	457.2577	65.62	7.71	9.18	17.48
C25H35N306	473.57	473.2526	63.41	7.45	8.87	20.27
C25H35N307	489.57	489.2475	61.33	7.21	8.58	22.88
C25H35N308	505.57	505.2424	59.39	6.98	8.31	25.32
C25H35N309	521.57	521.2373	57.57	6.76	8.06	27.61
C25H35N501	421.59	421.2841	71.23	8.37	16.61	3.80
C25H35N502	437.59	437.2791	68.62	8.06	16.00	7.31
C25H35N503	453.59	453.2740	66.20	7.78	15.44	10.58
C25H35N504	469.59	469.2689	63.95	7.51	14.91	13.63
C25H35N505	485.59	485.2638	61.84	7.27	14.42	16.47
C25H35N506	501.59	501.2587	59.87	7.03	13.96	19.14
C25H35N507	517.59	517.2536	58.02	6.82	13.53	21.64
C25H35N508	533.59	533.2485	56.28	6.61	13.13	23.99
C25H35N509	549.59	549.2435	54.64	6.42	12.74	26.20
C25H36N201	380.58	380.2827	78.90	9.53	7.36	4.20
C25H36N202	396.58	396.2777	75.72	9.15	7.06	8.07
C25H36N203	412.58	412.2726	72.78	8.80	6.79	11.63
C25H36N204	428.58	428.2675	70.06	8.47	6.54	14.93

C25H36N2O5	444.58	444.2624	67.54	8.16	6.30	17.99
C25H36N2O6	460.58	460.2573	65.20	7.88	6.08	20.84
C25H36N2O7	476.57	476.2522	63.01	7.61	5.88	23.50
C25H36N2O8	492.57	492.2471	60.96	7.37	5.69	25.98
C25H36N2O9	508.57	508.2421	59.04	7.14	5.51	28.31
C25H36N4O1	408.59	408.2889	73.49	8.88	13.71	3.92
C25H36N4O2	424.59	424.2838	70.72	8.55	13.20	7.54
C25H36N4O3	440.59	440.2787	68.15	8.24	12.72	10.89
C25H36N4O4	456.59	456.2736	65.77	7.95	12.27	14.02
C25H36N4O5	472.59	472.2686	63.54	7.68	11.86	16.93
C25H36N4O6	488.59	488.2635	61.46	7.43	11.47	19.65
C25H36N4O7	504.59	504.2584	59.51	7.19	11.10	22.20
C25H36N4O8	520.59	520.2533	57.68	6.97	10.76	24.59
C25H36N4O9	536.59	536.2482	55.96	6.76	10.44	26.84
C25H36N6O1	436.61	436.2950	68.78	8.31	19.25	3.66
C25H36N6O2	452.60	452.2900	66.34	8.02	18.57	7.07
C25H36N6O3	468.60	468.2849	64.08	7.74	17.93	10.24
C25H36N6O4	484.60	484.2798	61.96	7.49	17.34	13.21
C25H36N6O5	500.60	500.2747	59.98	7.25	16.79	15.98
C25H36N6O6	516.60	516.2696	58.13	7.02	16.27	18.58
C25H36N6O7	532.60	532.2645	56.38	6.81	15.78	21.03
C25H36N6O8	548.60	548.2594	54.74	6.61	15.32	23.33
C25H36N6O9	564.60	564.2544	53.18	6.43	14.88	25.50
C25H37N1O1	367.58	367.2875	81.69	10.15	3.81	4.35
C25H37N1O2	383.58	383.2824	78.28	9.72	3.65	8.34
C25H37N1O3	399.58	399.2773	75.15	9.33	3.51	12.01
C25H37N1O4	415.58	415.2722	72.26	8.97	3.37	15.40
C25H37N1O5	431.58	431.2672	69.58	8.64	3.25	18.54
C25H37N1O6	447.58	447.2621	67.09	8.33	3.13	21.45
C25H37N1O7	463.58	463.2570	64.77	8.05	3.02	24.16
C25H37N1O8	479.58	479.2519	62.61	7.78	2.92	26.69
C25H37N1O9	495.57	495.2468	60.59	7.53	2.83	29.06
C25H37N3O1	395.59	395.2936	75.91	9.43	10.62	4.04
C25H37N3O2	411.59	411.2886	72.96	9.06	10.21	7.77
C25H37N3O3	427.59	427.2835	70.23	8.72	9.83	11.23
C25H37N3O4	443.59	443.2784	67.69	8.41	9.47	14.43
C25H37N3O5	459.59	459.2733	65.34	8.11	9.14	17.41
C25H37N3O6	475.59	475.2682	63.14	7.84	8.84	20.18
C25H37N3O7	491.59	491.2631	61.08	7.59	8.55	22.78
C25H37N3O8	507.59	507.2580	59.16	7.35	8.28	25.22
C25H37N3O9	523.59	523.2530	57.35	7.12	8.03	27.50
C25H37N5O1	423.61	423.2998	70.89	8.80	16.53	3.78
C25H37N5O2	439.61	439.2947	68.31	8.48	15.93	7.28
C25H37N5O3	455.61	455.2896	65.91	8.19	15.37	10.54
C25H37N5O4	471.60	471.2845	63.67	7.91	14.85	13.57
C25H37N5O5	487.60	487.2794	61.58	7.65	14.36	16.41
C25H37N5O6	503.60	503.2744	59.63	7.41	13.91	19.06
C25H37N5O7	519.60	519.2693	57.79	7.18	13.48	21.55
C25H37N5O8	535.60	535.2642	56.06	6.96	13.08	23.90
C25H37N5O9	551.60	551.2591	54.44	6.76	12.70	26.10
C25H38N2O1	382.59	382.2984	78.48	10.01	7.32	4.18
C25H38N2O2	398.59	398.2933	75.33	9.61	7.03	8.03
C25H38N2O3	414.59	414.2882	72.43	9.24	6.76	11.58
C25H38N2O4	430.59	430.2831	69.74	8.90	6.51	14.86
C25H38N2O5	446.59	446.2781	67.24	8.58	6.27	17.91
C25H38N2O6	462.59	462.2730	64.91	8.28	6.06	20.75
C25H38N2O7	478.59	478.2679	62.74	8.00	5.85	23.40
C25H38N2O8	494.59	494.2628	60.71	7.74	5.66	25.88
C25H38N2O9	510.59	510.2577	58.81	7.50	5.49	28.20
C25H38N4O1	410.61	410.3045	73.13	9.33	13.64	3.90
C25H38N4O2	426.61	426.2995	70.39	8.98	13.13	7.50
C25H38N4O3	442.61	442.2944	67.84	8.65	12.66	10.84
C25H38N4O4	458.61	458.2893	65.48	8.35	12.22	13.95
C25H38N4O5	474.61	474.2842	63.27	8.07	11.80	16.86
C25H38N4O6	490.60	490.2791	61.21	7.81	11.42	19.57
C25H38N4O7	506.60	506.2740	59.27	7.56	11.06	22.11
C25H38N4O8	522.60	522.2689	57.46	7.33	10.72	24.49
C25H38N4O9	538.60	538.2639	55.75	7.11	10.40	26.73
C25H38N6O1	438.62	438.3107	68.46	8.73	19.16	3.65
C25H38N6O2	454.62	454.3056	66.05	8.43	18.49	7.04
C25H38N6O3	470.62	470.3005	63.80	8.14	17.86	10.20
C25H38N6O4	486.62	486.2954	61.71	7.87	17.27	13.15
C25H38N6O5	502.62	502.2903	59.74	7.62	16.72	15.92
C25H38N6O6	518.62	518.2853	57.90	7.39	16.20	18.51
C25H38N6O7	534.62	534.2802	56.17	7.16	15.72	20.95
C25H38N6O8	550.62	550.2751	54.53	6.96	15.26	23.25
C25H38N6O9	566.62	566.2700	53.00	6.76	14.83	25.41
C25H39N1O1	369.60	369.3031	81.25	10.64	3.79	4.33
C25H39N1O2	385.60	385.2981	77.87	10.19	3.63	8.30
C25H39N1O3	401.59	401.2930	74.77	9.79	3.49	11.95
C25H39N1O4	417.59	417.2879	71.91	9.41	3.35	15.33
C25H39N1O5	433.59	433.2828	69.25	9.07	3.23	18.45
C25H39N1O6	449.59	449.2777	66.79	8.74	3.12	21.35
C25H39N1O7	465.59	465.2726	64.49	8.44	3.01	24.05
C25H39N1O8	481.59	481.2675	62.35	8.16	2.91	26.58
C25H39N1O9	497.59	497.2625	60.35	7.90	2.81	28.94
C25H39N3O1	397.61	397.3093	75.52	9.89	10.57	4.02
C25H39N3O2	413.61	413.3042	72.60	9.50	10.16	7.74
C25H39N3O3	429.61	429.2991	69.90	9.15	9.78	11.17
C25H39N3O4	445.61	445.2940	67.39	8.82	9.43	14.36
C25H39N3O5	461.61	461.2890	65.05	8.52	9.10	17.33
C25H39N3O6	477.61	477.2839	62.87	8.23	8.80	20.10
C25H39N3O7	493.61	493.2788	60.83	7.96	8.51	22.69
C25H39N3O8	509.60	509.2737	58.92	7.71	8.25	25.12
C25H39N3O9	525.60	525.2686	57.13	7.48	7.99	27.40
C25H39N5O1	425.62	425.3154	70.55	9.24	16.45	3.76
C25H39N5O2	441.62	441.3104	67.99	8.90	15.86	7.25
C25H39N5O3	457.62	457.3053	65.62	8.59	15.30	10.49
C25H39N5O4	473.62	473.3002	63.40	8.30	14.79	13.51
C25H39N5O5	489.62	489.2951	61.33	8.03	14.30	16.34
C25H39N5O6	505.62	505.2900	59.39	7.77	13.85	18.99
C25H39N5O7	521.62	521.2849	57.57	7.54	13.43	21.47
C25H39N5O8	537.62	537.2798	55.85	7.31	13.03	23.81
C25H39N5O9	553.62	553.2748	54.24	7.10	12.65	26.01
C25H40N2O1	384.61	384.3140	78.07	10.48	7.28	4.16
C25H40N2O2	400.61	400.3090	74.96	10.06	6.99	7.99
C25H40N2O3	416.61	416.3039	72.08	9.68	6.72	11.52
C25H40N2O4	432.61	432.2988	69.41	9.32	6.48	14.79
C25H40N2O5	448.61	448.2937	66.94	8.99	6.24	17.83
C25H40N2O6	464.61	464.2886	64.63	8.68	6.03	20.66
C25H40N2O7	480.61	480.2835	62.48	8.39	5.83	23.30
C25H40N2O8	496.61	496.2784	60.47	8.12	5.64	25.77
C25H40N2O9	512.61	512.2734	58.58	7.87	5.46	28.09
C25H40N4O1	412.62	412.3202	72.77	9.77	13.58	3.88
C25H40N4O2	428.62	428.3151	70.06	9.41	13.07	7.47
C25H40N4O3	444.62	444.3100	67.54	9.07	12.60	10.80
C25H40N4O4	460.62	460.3049	65.19	8.75	12.16	13.89
C25H40N4O5	476.62	476.2998	63.00	8.46	11.75	16.78
C25H40N4O6	492.62	492.2948	60.96	8.18	11.37	19.49
C25H40N4O7	508.62	508.2897	59.04	7.93	11.02	22.02

C25H40N4O8	524.62	524.2846	57.24	7.69	10.68	24.40
C25H40N4O9	540.62	540.2795	55.54	7.46	10.36	26.64
C25H40N6O1	440.64	440.3263	68.15	9.15	19.07	3.63
C25H40N6O2	456.64	456.3213	65.76	8.83	18.40	7.01
C25H40N6O3	472.64	472.3162	63.53	8.53	17.78	10.16
C25H40N6O4	488.64	488.3111	61.45	8.25	17.20	13.10
C25H40N6O5	504.63	504.3060	59.50	7.99	16.65	15.85
C25H40N6O6	520.63	520.3009	57.68	7.74	16.14	18.44
C25H40N6O7	536.63	536.2958	55.96	7.51	15.66	20.87
C25H40N6O8	552.63	552.2907	54.34	7.30	15.21	23.16
C25H40N6O9	568.63	568.2857	52.81	7.09	14.78	25.32
C25H41N1O1	371.61	371.3188	80.80	11.12	3.77	4.31
C25H41N1O2	387.61	387.3137	77.47	10.66	3.61	8.26
C25H41N1O3	403.61	403.3086	74.40	10.24	3.47	11.89
C25H41N1O4	419.61	419.3035	71.56	9.85	3.34	15.25
C25H41N1O5	435.61	435.2985	68.93	9.49	3.22	18.36
C25H41N1O6	451.61	451.2934	66.49	9.15	3.10	21.26
C25H41N1O7	467.61	467.2883	64.22	8.84	3.00	23.95
C25H41N1O8	483.61	483.2832	62.09	8.55	2.90	26.47
C25H41N1O9	499.61	499.2781	60.10	8.27	2.80	28.82
C25H41N3O1	399.63	399.3249	75.14	10.34	10.51	4.00
C25H41N3O2	415.62	415.3199	72.25	9.94	10.11	7.70
C25H41N3O3	431.62	431.3148	69.57	9.57	9.74	11.12
C25H41N3O4	447.62	447.3097	67.08	9.23	9.39	14.30
C25H41N3O5	463.62	463.3046	64.77	8.91	9.06	17.25
C25H41N3O6	479.62	479.2995	62.61	8.62	8.76	20.02
C25H41N3O7	495.62	495.2944	60.59	8.34	8.48	22.60
C25H41N3O8	511.62	511.2893	58.69	8.08	8.21	25.02
C25H41N3O9	527.62	527.2843	56.91	7.83	7.96	27.29
C25H41N5O1	427.64	427.3311	70.22	9.66	16.38	3.74
C25H41N5O2	443.64	443.3260	67.69	9.32	15.79	7.21
C25H41N5O3	459.64	459.3209	65.33	8.99	15.24	10.44
C25H41N5O4	475.64	475.3158	63.13	8.69	14.72	13.46
C25H41N5O5	491.64	491.3107	61.08	8.41	14.24	16.27
C25H41N5O6	507.64	507.3057	59.15	8.14	13.80	18.91
C25H41N5O7	523.63	523.3006	57.35	7.89	13.37	21.39
C25H41N5O8	539.63	539.2955	55.64	7.66	12.98	23.72
C25H41N5O9	555.63	555.2904	54.04	7.44	12.60	25.92
C25H42N2O1	386.63	386.3297	77.67	10.95	7.25	4.14
C25H42N2O2	402.63	402.3246	74.58	10.51	6.96	7.95
C25H42N2O3	418.63	418.3195	71.73	10.11	6.69	11.47
C25H42N2O4	434.62	434.3144	69.09	9.74	6.45	14.72
C25H42N2O5	450.62	450.3094	66.64	9.39	6.22	17.75
C25H42N2O6	466.62	466.3043	64.35	9.07	6.00	20.57
C25H42N2O7	482.62	482.2992	62.22	8.77	5.80	23.21
C25H42N2O8	498.62	498.2941	60.22	8.49	5.62	25.67
C25H42N2O9	514.62	514.2890	58.35	8.23	5.44	27.98
C25H42N4O1	414.64	414.3358	72.42	10.21	13.51	3.86
C25H42N4O2	430.64	430.3308	69.73	9.83	13.01	7.43
C25H42N4O3	446.64	446.3257	67.23	9.48	12.54	10.75
C25H42N4O4	462.64	462.3206	64.91	9.15	12.11	13.83
C25H42N4O5	478.64	478.3155	62.74	8.84	11.71	16.71
C25H42N4O6	494.64	494.3104	60.71	8.56	11.33	19.41
C25H42N4O7	510.64	510.3053	58.80	8.29	10.97	21.93
C25H42N4O8	526.64	526.3002	57.02	8.04	10.64	24.30
C25H42N4O9	542.63	542.2952	55.34	7.80	10.32	26.54
C25H42N6O1	442.65	442.3420	67.84	9.56	18.99	3.61
C25H42N6O2	458.65	458.3369	65.47	9.23	18.32	6.98
C25H42N6O3	474.65	474.3318	63.26	8.92	17.71	10.11
C25H42N6O4	490.65	490.3267	61.20	8.63	17.13	13.04
C25H42N6O5	506.65	506.3216	59.27	8.36	16.59	15.79
C25H42N6O6	522.65	522.3166	57.45	8.10	16.08	18.37
C25H42N6O7	538.65	538.3115	55.75	7.86	15.60	20.79
C25H42N6O8	554.65	554.3064	54.14	7.63	15.15	23.08
C25H42N6O9	570.65	570.3013	52.62	7.42	14.73	25.23
C25H43N1O1	373.63	373.3344	80.37	11.60	3.75	4.28
C25H43N1O2	389.63	389.3294	77.07	11.12	3.59	8.21
C25H43N1O3	405.63	405.3243	74.03	10.69	3.45	11.83
C25H43N1O4	421.63	421.3192	71.22	10.28	3.32	15.18
C25H43N1O5	437.63	437.3141	68.62	9.90	3.20	18.28
C25H43N1O6	453.62	453.3090	66.20	9.55	3.09	21.16
C25H43N1O7	469.62	469.3039	63.94	9.23	2.98	23.85
C25H43N1O8	485.62	485.2988	61.83	8.93	2.88	26.36
C25H43N1O9	501.62	501.2938	59.86	8.64	2.79	28.71
C25H43N3O1	401.64	401.3406	74.76	10.79	10.46	3.98
C25H43N3O2	417.64	417.3355	71.90	10.38	10.06	7.66
C25H43N3O3	433.64	433.3304	69.25	10.00	9.69	11.07
C25H43N3O4	449.64	449.3253	66.78	9.64	9.35	14.23
C25H43N3O5	465.64	465.3202	64.49	9.31	9.02	17.18
C25H43N3O6	481.64	481.3152	62.35	9.00	8.72	19.93
C25H43N3O7	497.64	497.3101	60.34	8.71	8.44	22.51
C25H43N3O8	513.64	513.3050	58.46	8.44	8.18	24.92
C25H43N3O9	529.64	529.2999	56.70	8.18	7.93	27.19
C25H43N5O1	429.65	429.3467	69.89	10.09	16.30	3.72
C25H43N5O2	445.65	445.3417	67.38	9.73	15.71	7.18
C25H43N5O3	461.65	461.3366	65.04	9.39	15.17	10.40
C25H43N5O4	477.65	477.3315	62.87	9.07	14.66	13.40
C25H43N5O5	493.65	493.3264	60.83	8.78	14.19	16.21
C25H43N5O6	509.65	509.3213	58.92	8.50	13.74	18.84
C25H43N5O7	525.65	525.3162	57.13	8.25	13.32	21.31
C25H43N5O8	541.65	541.3111	55.44	8.00	12.93	23.63
C25H43N5O9	557.65	557.3061	53.85	7.77	12.56	25.82
C25H44N2O1	388.64	388.3453	77.26	11.41	7.21	4.12
C25H44N2O2	404.64	404.3403	74.21	10.96	6.92	7.91
C25H44N2O3	420.64	420.3352	71.39	10.54	6.66	11.41
C25H44N2O4	436.64	436.3301	68.77	10.16	6.42	14.66
C25H44N2O5	452.64	452.3250	66.34	9.80	6.19	17.67
C25H44N2O6	468.64	468.3199	64.07	9.46	5.98	20.48
C25H44N2O7	484.64	484.3148	61.96	9.15	5.78	23.11
C25H44N2O8	500.64	500.3097	59.98	8.86	5.60	25.57
C25H44N2O9	516.64	516.3047	58.12	8.58	5.42	27.87
C25H44N4O1	416.66	416.3515	72.07	10.64	13.45	3.84
C25H44N4O2	432.66	432.3464	69.40	10.25	12.95	7.40
C25H44N4O3	448.65	448.3413	66.93	9.89	12.49	10.70
C25H44N4O4	464.65	464.3362	64.62	9.54	12.06	13.77
C25H44N4O5	480.65	480.3311	62.47	9.23	11.66	16.64
C25H44N4O6	496.65	496.3261	60.46	8.93	11.28	19.33
C25H44N4O7	512.65	512.3210	58.57	8.65	10.93	21.85
C25H44N4O8	528.65	528.3159	56.80	8.39	10.60	24.21
C25H44N4O9	544.65	544.3108	55.13	8.14	10.29	26.44
C25H44N6O1	444.67	444.3576	67.53	9.97	18.90	3.60
C25H44N6O2	460.67	460.3526	65.18	9.63	18.24	6.95
C25H44N6O3	476.67	476.3475	63.00	9.30	17.63	10.07
C25H44N6O4	492.67	492.3424	60.95	9.00	17.06	12.99
C25H44N6O5	508.67	508.3373	59.03	8.72	16.52	15.73
C25H44N6O6	524.67	524.3322	57.23	8.45	16.02	18.30
C25H44N6O7	540.67	540.3271	55.54	8.20	15.54	20.71
C25H44N6O8	556.66	556.3220	53.94	7.97	15.10	22.99
C25H44N6O9	572.66	572.3170	52.44	7.74	14.68	25.14
C25H45N1O1	375.64	375.3501	79.94	12.07	3.73	4.26

C25H45N1O2	391.64	391.3450	76.67	11.58	3.58	8.17
C25H45N1O3	407.64	407.3399	73.66	11.13	3.44	11.77
C25H45N1O4	423.64	423.3348	70.88	10.71	3.31	15.11
C25H45N1O5	439.64	439.3298	68.30	10.32	3.19	18.20
C25H45N1O6	455.64	455.3247	65.90	9.95	3.07	21.07
C25H45N1O7	471.64	471.3196	63.67	9.62	2.97	23.75
C25H45N1O8	487.64	487.3145	61.58	9.30	2.87	26.25
C25H45N1O9	503.64	503.3094	59.62	9.01	2.78	28.59
C25H45N3O1	403.66	403.3562	74.39	11.24	10.41	3.96
C25H45N3O2	419.66	419.3512	71.55	10.81	10.01	7.63
C25H45N3O3	435.66	435.3461	68.93	10.41	9.65	11.02
C25H45N3O4	451.66	451.3410	66.48	10.04	9.30	14.17
C25H45N3O5	467.65	467.3359	64.21	9.70	8.99	17.11
C25H45N3O6	483.65	483.3308	62.09	9.38	8.69	19.85
C25H45N3O7	499.65	499.3257	60.10	9.08	8.41	22.41
C25H45N3O8	515.65	515.3206	58.23	8.80	8.15	24.82
C25H45N3O9	531.65	531.3156	56.48	8.53	7.90	27.08
C25H45N5O1	431.67	431.3624	69.56	10.51	16.22	3.71
C25H45N5O2	447.67	447.3573	67.08	10.13	15.64	7.15
C25H45N5O3	463.67	463.3522	64.76	9.78	15.10	10.35
C25H45N5O4	479.67	479.3471	62.60	9.46	14.60	13.34
C25H45N5O5	495.67	495.3420	60.58	9.15	14.13	16.14
C25H45N5O6	511.67	511.3370	58.69	8.86	13.69	18.76
C25H45N5O7	527.67	527.3319	56.91	8.60	13.27	21.22
C25H45N5O8	543.67	543.3268	55.23	8.34	12.88	23.54
C25H45N5O9	559.67	559.3217	53.65	8.10	12.51	25.73
C25H46N2O1	390.66	390.3610	76.86	11.87	7.17	4.10
C25H46N2O2	406.66	406.3559	73.84	11.40	6.89	7.87
C25H46N2O3	422.66	422.3508	71.05	10.97	6.63	11.36
C25H46N2O4	438.66	438.3457	68.45	10.57	6.39	14.59
C25H46N2O5	454.66	454.3406	66.05	10.20	6.16	17.60
C25H46N2O6	470.66	470.3356	63.80	9.85	5.95	20.40
C25H46N2O7	486.65	486.3305	61.70	9.53	5.76	23.01
C25H46N2O8	502.65	502.3254	59.74	9.22	5.57	25.46
C25H46N2O9	518.65	518.3203	57.90	8.94	5.40	27.76
C25H46N4O1	418.67	418.3671	71.72	11.07	13.38	3.82
C25H46N4O2	434.67	434.3621	69.08	10.67	12.89	7.36
C25H46N4O3	450.67	450.3570	66.63	10.29	12.43	10.65
C25H46N4O4	466.67	466.3519	64.35	9.94	12.01	13.71
C25H46N4O5	482.67	482.3468	62.21	9.61	11.61	16.57
C25H46N4O6	498.67	498.3417	60.22	9.30	11.24	19.25
C25H46N4O7	514.67	514.3366	58.34	9.01	10.89	21.76
C25H46N4O8	530.67	530.3315	56.59	8.74	10.56	24.12
C25H46N4O9	546.67	546.3265	54.93	8.48	10.25	26.34
C25H46N6O1	446.68	446.3733	67.22	10.38	18.81	3.58
C25H46N6O2	462.68	462.3682	64.90	10.02	18.16	6.92
C25H46N6O3	478.68	478.3631	62.73	9.69	17.56	10.03
C25H46N6O4	494.68	494.3580	60.70	9.37	16.99	12.94
C25H46N6O5	510.68	510.3529	58.80	9.08	16.46	15.66
C25H46N6O6	526.68	526.3479	57.01	8.80	15.96	18.23
C25H46N6O7	542.68	542.3428	55.33	8.54	15.49	20.64
C25H46N6O8	558.68	558.3377	53.75	8.30	15.04	22.91
C25H46N6O9	574.68	574.3326	52.25	8.07	14.62	25.06
C25H47N1O1	377.66	377.3657	79.51	12.54	3.71	4.24
C25H47N1O2	393.66	393.3607	76.28	12.03	3.56	8.13
C25H47N1O3	409.66	409.3556	73.30	11.56	3.42	11.72
C25H47N1O4	425.66	425.3505	70.54	11.13	3.29	15.03
C25H47N1O5	441.66	441.3454	67.99	10.73	3.17	18.11
C25H47N1O6	457.66	457.3403	65.61	10.35	3.06	20.98
C25H47N1O7	473.66	473.3352	63.40	10.00	2.96	23.64

C25H47N1O8	489.66	489.3301	61.32	9.68	2.86	26.14
C25H47N1O9	505.65	505.3251	59.38	9.37	2.77	28.48
C25H47N3O1	405.67	405.3719	74.02	11.68	10.36	3.94
C25H47N3O2	421.67	421.3668	71.21	11.23	9.97	7.59
C25H47N3O3	437.67	437.3617	68.61	10.82	9.60	10.97
C25H47N3O4	453.67	453.3566	66.19	10.44	9.26	14.11
C25H47N3O5	469.67	469.3515	63.93	10.09	8.95	17.03
C25H47N3O6	485.67	485.3465	61.83	9.75	8.65	19.77
C25H47N3O7	501.67	501.3414	59.86	9.44	8.38	22.32
C25H47N3O8	517.67	517.3363	58.01	9.15	8.12	24.73
C25H47N3O9	533.67	533.3312	56.27	8.88	7.87	26.98
C25H47N5O1	433.69	433.3780	69.24	10.92	16.15	3.69
C25H47N5O2	449.69	449.3730	66.78	10.54	15.57	7.12
C25H47N5O3	465.69	465.3679	64.48	10.17	15.04	10.31
C25H47N5O4	481.68	481.3628	62.34	9.84	14.54	13.29
C25H47N5O5	497.68	497.3577	60.34	9.52	14.07	16.07
C25H47N5O6	513.68	513.3526	58.46	9.22	13.63	18.69
C25H47N5O7	529.68	529.3475	56.69	8.94	13.22	21.14
C25H47N5O8	545.68	545.3424	55.03	8.68	12.83	23.46
C25H47N5O9	561.68	561.3374	53.46	8.43	12.47	25.64
C25H48N2O1	392.67	392.3766	76.47	12.32	7.13	4.07
C25H48N2O2	408.67	408.3716	73.48	11.84	6.85	7.83
C25H48N2O3	424.67	424.3665	70.71	11.39	6.60	11.30
C25H48N2O4	440.67	440.3614	68.14	10.98	6.36	14.52
C25H48N2O5	456.67	456.3563	65.75	10.59	6.13	17.52
C25H48N2O6	472.67	472.3512	63.53	10.24	5.93	20.31
C25H48N2O7	488.67	488.3461	61.45	9.90	5.73	22.92
C25H48N2O8	504.67	504.3410	59.50	9.59	5.55	25.36
C25H48N2O9	520.67	520.3360	57.67	9.29	5.38	27.66
C25H48N4O1	420.69	420.3828	71.38	11.50	13.32	3.80
C25H48N4O2	436.69	436.3777	68.76	11.08	12.83	7.33
C25H48N4O3	452.69	452.3726	66.33	10.69	12.38	10.60
C25H48N4O4	468.69	468.3675	64.07	10.32	11.95	13.65
C25H48N4O5	484.69	484.3624	61.95	9.98	11.56	16.50
C25H48N4O6	500.68	500.3574	59.97	9.66	11.19	19.17
C25H48N4O7	516.68	516.3523	58.12	9.36	10.84	21.68
C25H48N4O8	532.68	532.3472	56.37	9.08	10.52	24.03
C25H48N4O9	548.68	548.3421	54.73	8.82	10.21	26.24
C25H48N6O1	448.70	448.3889	66.92	10.78	18.73	3.57
C25H48N6O2	464.70	464.3839	64.62	10.41	18.08	6.89
C25H48N6O3	480.70	480.3788	62.47	10.07	17.48	9.99
C25H48N6O4	496.70	496.3737	60.45	9.74	16.92	12.88
C25H48N6O5	512.70	512.3686	58.57	9.44	16.39	15.60
C25H48N6O6	528.70	528.3635	56.80	9.15	15.90	18.16
C25H48N6O7	544.70	544.3584	55.13	8.88	15.43	20.56
C25H48N6O8	560.70	560.3533	53.55	8.63	14.99	22.83
C25H48N6O9	576.70	576.3483	52.07	8.39	14.57	24.97
C25H49N1O1	379.68	379.3814	79.09	13.01	3.69	4.21
C25H49N1O2	395.67	395.3763	75.89	12.48	3.54	8.09
C25H49N1O3	411.67	411.3712	72.94	12.00	3.40	11.66
C25H49N1O4	427.67	427.3661	70.21	11.55	3.28	14.96
C25H49N1O5	443.67	443.3610	67.68	11.13	3.16	18.03
C25H49N1O6	459.67	459.3560	65.32	10.74	3.05	20.88
C25H49N1O7	475.67	475.3509	63.13	10.38	2.94	23.54
C25H49N1O8	491.67	491.3458	61.07	10.05	2.85	26.03
C25H49N1O9	507.67	507.3407	59.15	9.73	2.76	28.36
C25H49N3O1	407.69	407.3875	73.65	12.11	10.31	3.92
C25H49N3O2	423.69	423.3825	70.87	11.66	9.92	7.55
C25H49N3O3	439.69	439.3774	68.29	11.23	9.56	10.92
C25H49N3O4	455.69	455.3723	65.90	10.84	9.22	14.04

165

C25H49N3O5	471.69	471.3672	63.66	10.47	8.91	16.96
C25H49N3O6	487.69	487.3621	61.57	10.13	8.62	19.68
C25H49N3O7	503.69	503.3570	59.62	9.81	8.34	22.24
C25H49N3O8	519.68	519.3519	57.78	9.50	8.09	24.63
C25H49N3O9	535.68	535.3469	56.06	9.22	7.84	26.88
C25H49N5O1	435.70	435.3937	68.92	11.34	16.07	3.67
C25H49N5O2	451.70	451.3886	66.48	10.93	15.50	7.08
C25H49N5O3	467.70	467.3835	64.20	10.56	14.97	10.26
C25H49N5O4	483.70	483.3784	62.08	10.21	14.48	13.23
C25H49N5O5	499.70	499.3733	60.09	9.88	14.02	16.01
C25H49N5O6	515.70	515.3683	58.23	9.58	13.58	18.61
C25H49N5O7	531.70	531.3632	56.48	9.29	13.17	21.06
C25H49N5O8	547.70	547.3581	54.83	9.02	12.79	23.37
C25H49N5O9	563.70	563.3530	53.27	8.76	12.42	25.54
C25H50N2O1	394.69	394.3923	76.08	12.77	7.10	4.05
C25H50N2O2	410.69	410.3872	73.12	12.27	6.82	7.79
C25H50N2O3	426.69	426.3821	70.37	11.81	6.57	11.25
C25H50N2O4	442.69	442.3770	67.83	11.38	6.33	14.46
C25H50N2O5	458.69	458.3719	65.46	10.99	6.11	17.44
C25H50N2O6	474.69	474.3669	63.26	10.62	5.90	20.22
C25H50N2O7	490.69	490.3618	61.20	10.27	5.71	22.82
C25H50N2O8	506.69	506.3567	59.26	9.95	5.53	25.26
C25H50N2O9	522.69	522.3516	57.45	9.64	5.36	27.55
C25H50N4O1	422.70	422.3984	71.04	11.92	13.25	3.79
C25H50N4O2	438.70	438.3934	68.45	11.49	12.77	7.29
C25H50N4O3	454.70	454.3883	66.04	11.08	12.32	10.56
C25H50N4O4	470.70	470.3832	63.79	10.71	11.90	13.60
C25H50N4O5	486.70	486.3781	61.70	10.36	11.51	16.44
C25H50N4O6	502.70	502.3730	59.73	10.03	11.15	19.10
C25H50N4O7	518.70	518.3679	57.89	9.72	10.80	21.59
C25H50N4O8	534.70	534.3628	56.16	9.43	10.48	23.94
C25H50N4O9	550.70	550.3578	54.53	9.15	10.17	26.15
C25H50N6O1	450.72	450.4046	66.62	11.18	18.65	3.55
C25H50N6O2	466.72	466.3995	64.34	10.80	18.01	6.86
C25H50N6O3	482.72	482.3944	62.21	10.44	17.41	9.94
C25H50N6O4	498.72	498.3893	60.21	10.11	16.85	12.83
C25H50N6O5	514.71	514.3842	58.34	9.79	16.33	15.54
C25H50N6O6	530.71	530.3792	56.58	9.50	15.84	18.09
C25H50N6O7	546.71	546.3741	54.92	9.22	15.37	20.49
C25H50N6O8	562.71	562.3690	53.36	8.96	14.93	22.75
C25H50N6O9	578.71	578.3639	51.89	8.71	14.52	24.88
C25H51N1O1	381.69	381.3970	78.67	13.47	3.67	4.19
C25H51N1O2	397.69	397.3920	75.51	12.93	3.52	8.05
C25H51N1O3	413.69	413.3869	72.59	12.43	3.39	11.60
C25H51N1O4	429.69	429.3818	69.88	11.96	3.26	14.89
C25H51N1O5	445.69	445.3767	67.37	11.53	3.14	17.95
C25H51N1O6	461.69	461.3716	65.04	11.13	3.03	20.79
C25H51N1O7	477.69	477.3665	62.86	10.76	2.93	23.45
C25H51N1O8	493.69	493.3614	60.82	10.41	2.84	25.93
C25H51N1O9	509.69	509.3564	58.91	10.09	2.75	28.25
C25H51N3O1	409.70	409.4032	73.29	12.55	10.26	3.91
C25H51N3O2	425.70	425.3981	70.54	12.08	9.87	7.52
C25H51N3O3	441.70	441.3930	67.98	11.64	9.51	10.87
C25H51N3O4	457.70	457.3879	65.61	11.23	9.18	13.98
C25H51N3O5	473.70	473.3828	63.39	10.85	8.87	16.89
C25H51N3O6	489.70	489.3778	61.32	10.50	8.58	19.60
C25H51N3O7	505.70	505.3727	59.38	10.17	8.31	22.15
C25H51N3O8	521.70	521.3676	57.56	9.85	8.05	24.53
C25H51N3O9	537.70	537.3625	55.85	9.56	7.81	26.78
C25H51N5O1	437.72	437.4093	68.60	11.74	16.00	3.66

C25H51N5O2	453.72	453.4043	66.18	11.33	15.44	7.05
C25H51N5O3	469.72	469.3992	63.93	10.94	14.91	10.22
C25H51N5O4	485.72	485.3941	61.82	10.58	14.42	13.18
C25H51N5O5	501.72	501.3890	59.85	10.25	13.96	15.94
C25H51N5O6	517.72	517.3839	58.00	9.93	13.53	18.54
C25H51N5O7	533.71	533.3788	56.26	9.63	13.12	20.98
C25H51N5O8	549.71	549.3737	54.62	9.35	12.74	23.28
C25H51N5O9	565.71	565.3687	53.08	9.09	12.38	25.45
C25H52N2O1	396.71	396.4079	75.69	13.21	7.06	4.03
C25H52N2O2	412.71	412.4029	72.76	12.70	6.79	7.75
C25H52N2O3	428.70	428.3978	70.04	12.23	6.53	11.20
C25H52N2O4	444.70	444.3927	67.52	11.79	6.30	14.39
C25H52N2O5	460.70	460.3876	65.18	11.38	6.08	17.36
C25H52N2O6	476.70	476.3825	62.99	11.00	5.88	20.14
C25H52N2O7	492.70	492.3774	60.95	10.64	5.69	22.73
C25H52N2O8	508.70	508.3723	59.03	10.30	5.51	25.16
C25H52N2O9	524.70	524.3673	57.23	9.99	5.34	27.44
C25H52N4O1	424.72	424.4141	70.70	12.34	13.19	3.77
C25H52N4O2	440.72	440.4090	68.13	11.89	12.71	7.26
C25H52N4O3	456.72	456.4039	65.75	11.48	12.27	10.51
C25H52N4O4	472.72	472.3988	63.52	11.09	11.85	13.54
C25H52N4O5	488.72	488.3937	61.44	10.72	11.46	16.37
C25H52N4O6	504.72	504.3887	59.49	10.38	11.10	19.02
C25H52N4O7	520.72	520.3836	57.67	10.07	10.76	21.51
C25H52N4O8	536.72	536.3785	55.95	9.77	10.44	23.85
C25H52N4O9	552.71	552.3734	54.33	9.48	10.14	26.05
C25H52N6O1	452.73	452.4202	66.33	11.58	18.56	3.53
C25H52N6O2	468.73	468.4152	64.06	11.18	17.93	6.83
C25H52N6O3	484.73	484.4101	61.95	10.81	17.34	9.90
C25H52N6O4	500.73	500.4050	59.97	10.47	16.78	12.78
C25H52N6O5	516.73	516.3999	58.11	10.14	16.26	15.48
C25H52N6O6	532.73	532.3948	56.37	9.84	15.78	18.02
C25H52N6O7	548.73	548.3897	54.72	9.55	15.32	20.41
C25H52N6O8	564.73	564.3846	53.17	9.28	14.88	22.66
C25H52N6O9	580.73	580.3795	51.71	9.03	14.47	24.80
C25H53N1O1	383.71	383.4127	78.26	13.92	3.65	4.17
C25H53N1O2	399.71	399.4076	75.12	13.37	3.50	8.01
C25H53N1O3	415.71	415.4025	72.23	12.85	3.37	11.55
C25H53N1O4	431.71	431.3974	69.56	12.37	3.24	14.82
C25H53N1O5	447.70	447.3923	67.07	11.93	3.13	17.87
C25H53N1O6	463.70	463.3873	64.76	11.52	3.02	20.70
C25H53N1O7	479.70	479.3822	62.60	11.14	2.92	23.35
C25H53N1O8	495.70	495.3771	60.58	10.78	2.83	25.82
C25H53N1O9	511.70	511.3720	58.68	10.44	2.74	28.14
C25H53N3O1	411.72	411.4188	72.93	12.98	10.21	3.89
C25H53N3O2	427.72	427.4138	70.20	12.49	9.82	7.48
C25H53N3O3	443.72	443.4087	67.67	12.04	9.47	10.82
C25H53N3O4	459.72	459.4036	65.32	11.62	9.14	13.92
C25H53N3O5	475.72	475.3985	63.12	11.23	8.83	16.82
C25H53N3O6	491.72	491.3934	61.07	10.86	8.55	19.52
C25H53N3O7	507.72	507.3883	59.14	10.52	8.28	22.06
C25H53N3O8	523.72	523.3832	57.34	10.20	8.02	24.44
C25H53N3O9	539.72	539.3782	55.64	9.90	7.79	26.68
C25H53N5O1	439.73	439.4250	68.29	12.15	15.93	3.64
C25H53N5O2	455.73	455.4199	65.89	11.72	15.37	7.02
C25H53N5O3	471.73	471.4148	63.65	11.32	14.85	10.17
C25H53N5O4	487.73	487.4097	61.57	10.95	14.36	13.12
C25H53N5O5	503.73	503.4046	59.61	10.61	13.90	15.88
C25H53N5O6	519.73	519.3996	57.78	10.28	13.47	18.47
C25H53N5O7	535.73	535.3945	56.05	9.97	13.07	20.91

166

C25H53N508	551.73	551.3894	54.42	9.68	12.69	23.20
C25H53N509	567.73	567.3843	52.89	9.41	12.34	25.36
C25H54N201	398.72	398.4236	75.31	13.65	7.03	4.01
C25H54N202	414.72	414.4185	72.40	13.12	6.75	7.72
C25H54N203	430.72	430.4134	69.72	12.64	6.50	11.14
C25H54N204	446.72	446.4083	67.22	12.18	6.27	14.33
C25H54N205	462.72	462.4032	64.85	11.76	6.05	17.29
C25H54N206	478.72	478.3982	62.73	11.37	5.85	20.05
C25H54N207	494.72	494.3931	60.70	11.00	5.66	22.64
C25H54N208	510.72	510.3880	58.80	10.66	5.49	25.06
C25H54N209	526.72	526.3829	57.01	10.33	5.32	27.34
C25H54N401	426.74	426.4297	70.37	12.76	13.13	3.75
C25H54N402	442.73	442.4247	67.82	12.29	12.65	7.23
C25H54N403	458.73	458.4196	65.46	11.87	12.21	10.46
C25H54N404	474.73	474.4145	63.25	11.47	11.80	13.48
C25H54N405	490.73	490.4094	61.19	11.09	11.42	16.30
C25H54N406	506.73	506.4043	59.26	10.74	11.06	18.94
C25H54N407	522.73	522.3992	57.44	10.41	10.72	21.43
C25H54N408	538.73	538.3941	55.74	10.10	10.40	23.76
C25H54N409	554.73	554.3891	54.13	9.81	10.10	25.96
C25H54N601	454.75	454.4359	66.03	11.97	18.48	3.52
C25H54N602	470.75	470.4308	63.79	11.56	17.85	6.80
C25H54N603	486.75	486.4257	61.69	11.18	17.27	9.86
C25H54N604	502.75	502.4206	59.73	10.83	16.72	12.73
C25H54N605	518.75	518.4155	57.89	10.49	16.20	15.42
C25H54N606	534.75	534.4105	56.15	10.18	15.72	17.95
C25H54N607	550.75	550.4054	54.52	9.88	15.26	20.34
C25H54N608	566.74	566.4003	52.98	9.60	14.83	22.58
C25H54N609	582.74	582.3952	51.53	9.34	14.42	24.71
C25H55N301	413.74	413.4345	72.58	13.40	10.16	3.87
C25H55N302	429.74	429.4294	69.88	12.90	9.78	7.45
C25H55N303	445.74	445.4243	67.37	12.44	9.43	10.77
C25H55N304	461.73	461.4192	65.03	12.01	9.10	13.86
C25H55N305	477.73	477.4141	62.85	11.60	8.80	16.75
C25H55N306	493.73	493.4091	60.82	11.23	8.51	19.44
C25H55N307	509.73	509.4040	58.91	10.88	8.24	21.97
C25H55N308	525.73	525.3989	57.12	10.54	7.99	24.35
C25H55N309	541.73	541.3938	55.43	10.23	7.76	26.58
C25H55N501	441.75	441.4406	67.97	12.55	15.85	3.62
C25H55N502	457.75	457.4356	65.60	12.11	15.30	6.99
C25H55N503	473.75	473.4305	63.38	11.70	14.78	10.13
C25H55N504	489.75	489.4254	61.31	11.32	14.30	13.07
C25H55N505	505.75	505.4203	59.37	10.96	13.85	15.82
C25H55N506	521.75	521.4152	57.55	10.63	13.42	18.40
C25H55N507	537.75	537.4101	55.84	10.31	13.02	20.83
C25H55N508	553.75	553.4050	54.23	10.01	12.65	23.11
C25H55N509	569.75	569.3999	52.70	9.73	12.29	25.27
C25H56N401	428.75	428.4454	70.04	13.17	13.07	3.73
C25H56N402	444.75	444.4403	67.52	12.69	12.60	7.19
C25H56N403	460.75	460.4352	65.17	12.25	12.16	10.42
C25H56N404	476.75	476.4301	62.98	11.84	11.75	13.42
C25H56N405	492.75	492.4250	60.94	11.46	11.37	16.23
C25H56N406	508.75	508.4200	59.02	11.10	11.01	18.87
C25H56N407	524.75	524.4149	57.22	10.76	10.68	21.34
C25H56N408	540.75	540.4098	55.53	10.44	10.36	23.67
C25H56N409	556.75	556.4047	53.93	10.14	10.06	25.86
C25H56N601	456.76	456.4515	65.74	12.36	18.40	3.50
C25H56N602	472.76	472.4464	63.52	11.94	17.78	6.77
C25H56N603	488.76	488.4414	61.44	11.55	17.19	9.82
C25H56N604	504.76	504.4363	59.49	11.18	16.65	12.68
C25H56N605	520.76	520.4312	57.66	10.84	16.14	15.36
C25H56N606	536.76	536.4261	55.94	10.52	15.66	17.88
C25H56N607	552.76	552.4210	54.32	10.21	15.20	20.26
C25H56N608	568.76	568.4159	52.80	9.92	14.78	22.50
C25H56N609	584.76	584.4108	51.35	9.65	14.37	24.62
C25H57N501	443.77	443.4563	67.67	12.95	15.78	3.61
C25H57N502	459.77	459.4512	65.31	12.50	15.23	6.96
C25H57N503	475.76	475.4461	63.11	12.08	14.72	10.09
C25H57N504	491.76	491.4410	61.06	11.68	14.24	13.01
C25H57N505	507.76	507.4359	59.14	11.32	13.79	15.75
C25H57N506	523.76	523.4309	57.33	10.97	13.37	18.33
C25H57N507	539.76	539.4258	55.63	10.64	12.97	20.75
C25H57N508	555.76	555.4207	54.03	10.34	12.60	23.03
C25H57N509	571.76	571.4156	52.52	10.05	12.25	25.18
C25H58N601	458.78	458.4672	65.45	12.74	18.32	3.49
C25H58N602	474.78	474.4621	63.25	12.31	17.70	6.74
C25H58N603	490.78	490.4570	61.18	11.91	17.12	9.78
C25H58N604	506.78	506.4519	59.25	11.54	16.58	12.63
C25H58N605	522.78	522.4468	57.44	11.18	16.08	15.30
C25H58N606	538.78	538.4418	55.73	10.85	15.60	17.82
C25H58N607	554.78	554.4367	54.13	10.54	15.15	20.19
C25H58N608	570.78	570.4316	52.61	10.24	14.72	22.42
C25H58N609	586.78	586.4265	51.17	9.96	14.32	24.54
C26H13N101	355.40	355.0997	87.87	3.69	3.94	4.50
C26H13N102	371.40	371.0946	84.08	3.53	3.77	8.62
C26H13N103	387.40	387.0895	80.61	3.38	3.62	12.39
C26H13N104	403.40	403.0844	77.41	3.25	3.47	15.86
C26H13N105	419.40	419.0794	74.46	3.12	3.34	19.07
C26H13N106	435.40	435.0743	71.73	3.01	3.22	22.05
C26H13N107	451.40	451.0692	69.18	2.90	3.10	24.81
C26H13N108	467.40	467.0641	66.81	2.80	3.00	27.38
C26H13N109	483.39	483.0590	64.60	2.71	2.90	29.79
C26H13N301	383.41	383.1059	81.45	3.42	10.96	4.17
C26H13N302	399.41	399.1008	78.19	3.28	10.52	8.01
C26H13N303	415.41	415.0957	75.18	3.15	10.12	11.55
C26H13N304	431.41	431.0906	72.39	3.04	9.74	14.83
C26H13N305	447.41	447.0855	69.80	2.93	9.39	17.88
C26H13N306	463.41	463.0804	67.39	2.83	9.07	20.72
C26H13N307	479.41	479.0753	65.14	2.73	8.76	23.36
C26H13N308	495.41	495.0703	63.04	2.65	8.48	25.84
C26H13N309	511.41	511.0652	61.06	2.56	8.22	28.16
C26H13N501	411.43	411.1120	75.90	3.18	17.02	3.89
C26H13N502	427.43	427.1069	73.06	3.07	16.38	7.49
C26H13N503	443.43	443.1018	70.43	2.96	15.79	10.82
C26H13N504	459.42	459.0967	67.97	2.85	15.24	13.93
C26H13N505	475.42	475.0917	65.69	2.76	14.73	16.83
C26H13N506	491.42	491.0866	63.55	2.67	14.25	19.53
C26H13N507	507.42	507.0815	61.54	2.58	13.80	22.07
C26H13N508	523.42	523.0764	59.66	2.50	13.38	24.45
C26H13N509	539.42	539.0713	57.89	2.43	12.98	26.69
C26H14N201	370.41	370.1106	84.31	3.81	7.56	4.32
C26H14N202	386.41	386.1055	80.82	3.65	7.25	8.28
C26H14N203	402.41	402.1004	77.60	3.51	6.96	11.93
C26H14N204	418.41	418.0953	74.64	3.37	6.70	15.30
C26H14N205	434.41	434.0903	71.89	3.25	6.45	18.42
C26H14N206	450.41	450.0852	69.33	3.13	6.22	21.31
C26H14N207	466.41	466.0801	66.96	3.03	6.01	24.01
C26H14N208	482.41	482.0750	64.74	2.93	5.81	26.53
C26H14N209	498.41	498.0699	62.66	2.83	5.62	28.89
C26H14N401	398.43	398.1168	78.38	3.54	14.06	4.02

167

C26H14N402	414.43	414.1117	75.35	3.41	13.52	7.72
C26H14N403	430.43	430.1066	72.55	3.28	13.02	11.15
C26H14N404	446.43	446.1015	69.95	3.16	12.55	14.34
C26H14N405	462.43	462.0964	67.53	3.05	12.12	17.30
C26H14N406	478.42	478.0913	65.27	2.95	11.71	20.07
C26H14N407	494.42	494.0862	63.16	2.85	11.33	22.65
C26H14N408	510.42	510.0812	61.18	2.76	10.98	25.08
C26H14N409	526.42	526.0761	59.32	2.68	10.64	27.35
C26H14N601	426.44	426.1229	73.23	3.31	19.71	3.75
C26H14N602	442.44	442.1178	70.58	3.19	18.99	7.23
C26H14N603	458.44	458.1127	68.12	3.08	18.33	10.47
C26H14N604	474.44	474.1076	65.82	2.97	17.71	13.49
C26H14N605	490.44	490.1026	63.68	2.88	17.14	16.31
C26H14N606	506.44	506.0975	61.66	2.79	16.59	18.96
C26H14N607	522.44	522.0924	59.78	2.70	16.09	21.44
C26H14N608	538.44	538.0873	58.00	2.62	15.61	23.77
C26H14N609	554.44	554.0822	56.33	2.55	15.16	25.97
C26H15N101	357.42	357.1154	87.37	4.23	3.92	4.48
C26H15N102	373.41	373.1103	83.63	4.05	3.75	8.57
C26H15N103	389.41	389.1052	80.19	3.88	3.60	12.33
C26H15N104	405.41	405.1001	77.03	3.73	3.45	15.79
C26H15N105	421.41	421.0950	74.11	3.59	3.32	18.98
C26H15N106	437.41	437.0899	71.39	3.46	3.20	21.95
C26H15N107	453.41	453.0848	68.88	3.33	3.09	24.70
C26H15N108	469.41	469.0798	66.53	3.22	2.98	27.27
C26H15N109	485.41	485.0747	64.34	3.11	2.89	29.66
C26H15N301	385.43	385.1215	81.02	3.92	10.90	4.15
C26H15N302	401.43	401.1164	77.79	3.77	10.47	7.97
C26H15N303	417.43	417.1113	74.81	3.62	10.07	11.50
C26H15N304	433.43	433.1062	72.05	3.49	9.69	14.77
C26H15N305	449.43	449.1012	69.49	3.36	9.35	17.80
C26H15N306	465.43	465.0961	67.10	3.25	9.03	20.63
C26H15N307	481.43	481.0910	64.87	3.14	8.73	23.26
C26H15N308	497.42	497.0859	62.78	3.04	8.45	25.73
C26H15N309	513.42	513.0808	60.82	2.94	8.18	28.05
C26H15N501	413.44	413.1277	75.53	3.66	16.94	3.87
C26H15N502	429.44	429.1226	72.72	3.52	16.31	7.45
C26H15N503	445.44	445.1175	70.11	3.39	15.72	10.78
C26H15N504	461.44	461.1124	67.68	3.28	15.18	13.87
C26H15N505	477.44	477.1073	65.41	3.17	14.67	16.76
C26H15N506	493.44	493.1022	63.29	3.06	14.19	19.45
C26H15N507	509.44	509.0971	61.30	2.97	13.75	21.98
C26H15N508	525.44	525.0921	59.43	2.88	13.33	24.36
C26H15N509	541.44	541.0870	57.68	2.79	12.93	26.59
C26H16N201	372.43	372.1263	83.85	4.33	7.52	4.30
C26H16N202	388.43	388.1212	80.40	4.15	7.21	8.24
C26H16N203	404.43	404.1161	77.22	3.99	6.93	11.87
C26H16N204	420.43	420.1110	74.28	3.84	6.66	15.22
C26H16N205	436.43	436.1059	71.56	3.70	6.42	18.33
C26H16N206	452.43	452.1008	69.03	3.56	6.19	21.22
C26H16N207	468.43	468.0957	66.67	3.44	5.98	23.91
C26H16N208	484.43	484.0907	64.47	3.33	5.78	26.42
C26H16N209	500.43	500.0856	62.40	3.22	5.60	28.77
C26H16N401	400.44	400.1324	77.99	4.03	13.99	4.00
C26H16N402	416.44	416.1273	74.99	3.87	13.45	7.68
C26H16N403	432.44	432.1222	72.22	3.73	12.96	11.10
C26H16N404	448.44	448.1171	69.64	3.60	12.49	14.27
C26H16N405	464.44	464.1121	67.24	3.47	12.06	17.22
C26H16N406	480.44	480.1070	65.00	3.36	11.66	19.98
C26H16N407	496.44	496.1019	62.91	3.25	11.29	22.56
C26H16N408	512.44	512.0968	60.94	3.15	10.93	24.98
C26H16N409	528.44	528.0917	59.10	3.05	10.60	27.25
C26H16N601	428.46	428.1386	72.89	3.76	19.61	3.73
C26H16N602	444.46	444.1335	70.26	3.63	18.91	7.20
C26H16N603	460.46	460.1284	67.82	3.50	18.25	10.42
C26H16N604	476.46	476.1233	65.54	3.38	17.64	13.43
C26H16N605	492.45	492.1182	63.41	3.27	17.07	16.24
C26H16N606	508.45	508.1131	61.42	3.17	16.53	18.88
C26H16N607	524.45	524.1080	59.55	3.08	16.02	21.35
C26H16N608	540.45	540.1029	57.78	2.98	15.55	23.68
C26H16N609	556.45	556.0979	56.12	2.90	15.10	25.88
C26H17N101	359.43	359.1310	86.88	4.77	3.90	4.45
C26H17N102	375.43	375.1259	83.18	4.56	3.73	8.52
C26H17N103	391.43	391.1208	79.78	4.38	3.58	12.26
C26H17N104	407.43	407.1157	76.65	4.21	3.44	15.71
C26H17N105	423.43	423.1107	73.75	4.05	3.31	18.89
C26H17N106	439.43	439.1056	71.07	3.90	3.19	21.85
C26H17N107	455.43	455.1005	68.57	3.76	3.08	24.59
C26H17N108	471.43	471.0954	66.24	3.63	2.97	27.15
C26H17N109	487.43	487.0903	64.07	3.52	2.87	29.54
C26H17N301	387.44	387.1372	80.60	4.42	10.85	4.13
C26H17N302	403.44	403.1321	77.41	4.25	10.42	7.93
C26H17N303	419.44	419.1270	74.45	4.09	10.02	11.44
C26H17N304	435.44	435.1219	71.72	3.94	9.65	14.70
C26H17N305	451.44	451.1168	69.18	3.80	9.31	17.72
C26H17N306	467.44	467.1117	66.81	3.67	8.99	20.54
C26H17N307	483.44	483.1066	64.60	3.54	8.69	23.17
C26H17N308	499.44	499.1016	62.53	3.43	8.41	25.63
C26H17N309	515.44	515.0965	60.59	3.32	8.15	27.94
C26H17N501	415.46	415.1433	75.17	4.12	16.86	3.85
C26H17N502	431.46	431.1382	72.38	3.97	16.23	7.42
C26H17N503	447.46	447.1331	69.79	3.83	15.65	10.73
C26H17N504	463.46	463.1280	67.38	3.70	15.11	13.81
C26H17N505	479.46	479.1230	65.13	3.57	14.61	16.68
C26H17N506	495.46	495.1179	63.03	3.46	14.14	19.38
C26H17N507	511.45	511.1128	61.06	3.35	13.69	21.90
C26H17N508	527.45	527.1077	59.21	3.25	13.28	24.27
C26H17N509	543.45	543.1026	57.46	3.15	12.89	26.50
C26H18N201	374.45	374.1419	83.40	4.85	7.48	4.27
C26H18N202	390.45	390.1368	79.98	4.65	7.17	8.20
C26H18N203	406.44	406.1317	76.83	4.46	6.89	11.81
C26H18N204	422.44	422.1266	73.92	4.29	6.63	15.15
C26H18N205	438.44	438.1216	71.23	4.14	6.39	18.25
C26H18N206	454.44	454.1165	68.72	3.99	6.16	21.12
C26H18N207	470.44	470.1114	66.38	3.86	5.95	23.81
C26H18N208	486.44	486.1063	64.20	3.73	5.76	26.31
C26H18N209	502.44	502.1012	62.15	3.61	5.58	28.66
C26H18N401	402.46	402.1481	77.60	4.51	13.92	3.98
C26H18N402	418.46	418.1430	74.63	4.34	13.39	7.65
C26H18N403	434.46	434.1379	71.88	4.18	12.90	11.05
C26H18N404	450.46	450.1328	69.33	4.03	12.44	14.21
C26H18N405	466.46	466.1277	66.95	3.89	12.01	17.15
C26H18N406	482.46	482.1226	64.73	3.76	11.61	19.90
C26H18N407	498.46	498.1175	62.65	3.64	11.24	22.47
C26H18N408	514.46	514.1125	60.70	3.53	10.89	24.88
C26H18N409	530.45	530.1074	58.87	3.42	10.56	27.15
C26H18N601	430.47	430.1542	72.55	4.21	19.52	3.72
C26H18N602	446.47	446.1491	69.95	4.06	18.82	7.17
C26H18N603	462.47	462.1440	67.53	3.92	18.17	10.38
C26H18N604	478.47	478.1389	65.27	3.79	17.56	13.38

C26H18N6O5	494.47	494.1339	63.16	3.67	17.00	16.18
C26H18N6O6	510.47	510.1288	61.18	3.55	16.46	18.81
C26H18N6O7	526.47	526.1237	59.32	3.45	15.96	21.27
C26H18N6O8	542.47	542.1186	57.57	3.34	15.49	23.59
C26H18N6O9	558.47	558.1135	55.92	3.25	15.05	25.78
C26H19N1O1	361.45	361.1467	86.40	5.30	3.88	4.43
C26H19N1O2	377.45	377.1416	82.74	5.07	3.71	8.48
C26H19N1O3	393.45	393.1365	79.37	4.87	3.56	12.20
C26H19N1O4	409.45	409.1314	76.27	4.68	3.42	15.63
C26H19N1O5	425.45	425.1263	73.40	4.50	3.29	18.80
C26H19N1O6	441.44	441.1212	70.74	4.34	3.17	21.75
C26H19N1O7	457.44	457.1161	68.27	4.19	3.06	24.48
C26H19N1O8	473.44	473.1111	65.96	4.05	2.96	27.03
C26H19N1O9	489.44	489.1060	63.81	3.91	2.86	29.42
C26H19N3O1	389.46	389.1528	80.19	4.92	10.79	4.11
C26H19N3O2	405.46	405.1477	77.02	4.72	10.36	7.89
C26H19N3O3	421.46	421.1426	74.10	4.54	9.97	11.39
C26H19N3O4	437.46	437.1375	71.39	4.38	9.61	14.63
C26H19N3O5	453.46	453.1325	68.87	4.22	9.27	17.64
C26H19N3O6	469.46	469.1274	66.52	4.08	8.95	20.45
C26H19N3O7	485.46	485.1223	64.33	3.95	8.66	23.07
C26H19N3O8	501.46	501.1172	62.28	3.82	8.38	25.52
C26H19N3O9	517.46	517.1121	60.35	3.70	8.12	27.83
C26H19N5O1	417.47	417.1590	74.80	4.59	16.78	3.83
C26H19N5O2	433.47	433.1539	72.04	4.42	16.16	7.38
C26H19N5O3	449.47	449.1488	69.48	4.26	15.58	10.68
C26H19N5O4	465.47	465.1437	67.09	4.11	15.05	13.75
C26H19N5O5	481.47	481.1386	64.86	3.98	14.55	16.62
C26H19N5O6	497.47	497.1335	62.78	3.85	14.08	19.30
C26H19N5O7	513.47	513.1284	60.82	3.73	13.64	21.81
C26H19N5O8	529.47	529.1233	58.98	3.62	13.23	24.17
C26H19N5O9	545.47	545.1183	57.25	3.51	12.84	26.40
C26H20N2O1	376.46	376.1576	82.95	5.35	7.44	4.25
C26H20N2O2	392.46	392.1525	79.57	5.14	7.14	8.15
C26H20N2O3	408.46	408.1474	76.46	4.94	6.86	11.75
C26H20N2O4	424.46	424.1423	73.57	4.75	6.60	15.08
C26H20N2O5	440.46	440.1372	70.90	4.58	6.36	18.16
C26H20N2O6	456.46	456.1321	68.42	4.42	6.14	21.03
C26H20N2O7	472.46	472.1270	66.10	4.27	5.93	23.70
C26H20N2O8	488.46	488.1220	63.93	4.13	5.74	26.20
C26H20N2O9	504.46	504.1169	61.91	4.00	5.55	28.54
C26H20N4O1	404.48	404.1637	77.21	4.98	13.85	3.96
C26H20N4O2	420.47	420.1586	74.27	4.79	13.32	7.61
C26H20N4O3	436.47	436.1535	71.55	4.62	12.84	11.00
C26H20N4O4	452.47	452.1484	69.02	4.46	12.38	14.14
C26H20N4O5	468.47	468.1434	66.66	4.30	11.96	17.08
C26H20N4O6	484.47	484.1383	64.46	4.16	11.56	19.81
C26H20N4O7	500.47	500.1332	62.40	4.03	11.19	22.38
C26H20N4O8	516.47	516.1281	60.47	3.90	10.85	24.78
C26H20N4O9	532.47	532.1230	58.65	3.79	10.52	27.04
C26H20N6O1	432.49	432.1698	72.21	4.66	19.43	3.70
C26H20N6O2	448.49	448.1648	69.63	4.49	18.74	7.13
C26H20N6O3	464.49	464.1597	67.23	4.34	18.09	10.33
C26H20N6O4	480.49	480.1546	64.99	4.20	17.49	13.32
C26H20N6O5	496.49	496.1495	62.90	4.06	16.93	16.11
C26H20N6O6	512.49	512.1444	60.94	3.93	16.40	18.73
C26H20N6O7	528.49	528.1393	59.09	3.81	15.90	21.19
C26H20N6O8	544.48	544.1342	57.36	3.70	15.43	23.51
C26H20N6O9	560.48	560.1292	55.72	3.60	14.99	25.69
C26H21N1O1	363.46	363.1623	85.92	5.82	3.85	4.40

C26H21N1O2	379.46	379.1572	82.30	5.58	3.69	8.43
C26H21N1O3	395.46	395.1521	78.97	5.35	3.54	12.14
C26H21N1O4	411.46	411.1470	75.90	5.14	3.40	15.55
C26H21N1O5	427.46	427.1420	73.06	4.95	3.28	18.71
C26H21N1O6	443.46	443.1369	70.42	4.77	3.16	21.65
C26H21N1O7	459.46	459.1318	67.97	4.61	3.05	24.38
C26H21N1O8	475.46	475.1267	65.68	4.45	2.95	26.92
C26H21N1O9	491.46	491.1216	63.54	4.31	2.85	29.30
C26H21N3O1	391.48	391.1685	79.77	5.41	10.73	4.09
C26H21N3O2	407.48	407.1634	76.64	5.19	10.31	7.85
C26H21N3O3	423.48	423.1583	73.74	5.00	9.92	11.33
C26H21N3O4	439.47	439.1532	71.06	4.82	9.56	14.56
C26H21N3O5	455.47	455.1481	68.56	4.65	9.23	17.56
C26H21N3O6	471.47	471.1430	66.24	4.49	8.91	20.36
C26H21N3O7	487.47	487.1379	64.06	4.34	8.62	22.97
C26H21N3O8	503.47	503.1329	62.03	4.20	8.35	25.42
C26H21N3O9	519.47	519.1278	60.12	4.07	8.09	27.72
C26H21N5O1	419.49	419.1746	74.45	5.05	16.69	3.81
C26H21N5O2	435.49	435.1695	71.71	4.86	16.08	7.35
C26H21N5O3	451.49	451.1644	69.17	4.69	15.51	10.63
C26H21N5O4	467.49	467.1593	66.80	4.53	14.98	13.69
C26H21N5O5	483.49	483.1543	64.59	4.38	14.49	16.55
C26H21N5O6	499.49	499.1492	62.52	4.24	14.02	19.22
C26H21N5O7	515.49	515.1441	60.58	4.11	13.59	21.73
C26H21N5O8	531.49	531.1390	58.76	3.98	13.18	24.08
C26H21N5O9	547.49	547.1339	57.04	3.87	12.79	26.30
C26H22N2O1	378.48	378.1732	82.51	5.86	7.40	4.23
C26H22N2O2	394.48	394.1681	79.17	5.62	7.10	8.11
C26H22N2O3	410.48	410.1630	76.08	5.40	6.82	11.69
C26H22N2O4	426.48	426.1579	73.23	5.20	6.57	15.01
C26H22N2O5	442.48	442.1529	70.58	5.01	6.33	18.08
C26H22N2O6	458.48	458.1478	68.11	4.84	6.11	20.94
C26H22N2O7	474.47	474.1427	65.82	4.67	5.90	23.60
C26H22N2O8	490.47	490.1376	63.67	4.52	5.71	26.10
C26H22N2O9	506.47	506.1325	61.66	4.38	5.53	28.43
C26H22N4O1	406.49	406.1794	76.83	5.46	13.78	3.94
C26H22N4O2	422.49	422.1743	73.92	5.25	13.26	7.57
C26H22N4O3	438.49	438.1692	71.22	5.06	12.78	10.95
C26H22N4O4	454.49	454.1641	68.71	4.88	12.33	14.08
C26H22N4O5	470.49	470.1590	66.38	4.71	11.91	17.00
C26H22N4O6	486.49	486.1539	64.19	4.56	11.52	19.73
C26H22N4O7	502.49	502.1488	62.15	4.41	11.15	22.29
C26H22N4O8	518.49	518.1437	60.23	4.28	10.81	24.69
C26H22N4O9	534.49	534.1387	58.43	4.15	10.48	26.94
C26H22N6O1	434.50	434.1855	71.87	5.10	19.34	3.68
C26H22N6O2	450.50	450.1804	69.32	4.92	18.65	7.10
C26H22N6O3	466.50	466.1753	66.94	4.75	18.01	10.29
C26H22N6O4	482.50	482.1702	64.72	4.60	17.42	13.26
C26H22N6O5	498.50	498.1652	62.65	4.45	16.86	16.05
C26H22N6O6	514.50	514.1601	60.70	4.31	16.33	18.66
C26H22N6O7	530.50	530.1550	58.87	4.18	15.84	21.11
C26H22N6O8	546.50	546.1499	57.14	4.06	15.38	23.42
C26H22N6O9	562.50	562.1448	55.52	3.94	14.94	25.60
C26H23N1O1	365.48	365.1780	85.45	6.34	3.83	4.38
C26H23N1O2	381.48	381.1729	81.86	6.08	3.67	8.39
C26H23N1O3	397.48	397.1678	78.57	5.83	3.52	12.08
C26H23N1O4	413.48	413.1627	75.53	5.61	3.39	15.48
C26H23N1O5	429.48	429.1576	72.71	5.40	3.26	18.63
C26H23N1O6	445.48	445.1525	70.10	5.20	3.14	21.55
C26H23N1O7	461.48	461.1474	67.67	5.02	3.04	24.27

C26H23N1O8	477.48	477.1424	65.40	4.86	2.93	26.81
C26H23N1O9	493.47	493.1373	63.28	4.70	2.84	29.18
C26H23N3O1	393.49	393.1841	79.36	5.89	10.68	4.07
C26H23N3O2	409.49	409.1790	76.26	5.66	10.26	7.81
C26H23N3O3	425.49	425.1739	73.40	5.45	9.88	11.28
C26H23N3O4	441.49	441.1688	70.74	5.25	9.52	14.50
C26H23N3O5	457.49	457.1638	68.26	5.07	9.18	17.49
C26H23N3O6	473.49	473.1587	65.95	4.90	8.87	20.27
C26H23N3O7	489.49	489.1536	63.80	4.74	8.58	22.88
C26H23N3O8	505.49	505.1485	61.78	4.59	8.31	25.32
C26H23N3O9	521.49	521.1434	59.88	4.45	8.06	27.61
C26H23N5O1	421.51	421.1902	74.09	5.50	16.62	3.80
C26H23N5O2	437.51	437.1852	71.38	5.30	16.01	7.31
C26H23N5O3	453.50	453.1801	68.86	5.11	15.44	10.58
C26H23N5O4	469.50	469.1750	66.51	4.94	14.92	13.63
C26H23N5O5	485.50	485.1699	64.32	4.78	14.42	16.48
C26H23N5O6	501.50	501.1648	62.27	4.62	13.96	19.14
C26H23N5O7	517.50	517.1597	60.35	4.48	13.53	21.64
C26H23N5O8	533.50	533.1546	58.54	4.35	13.13	23.99
C26H23N5O9	549.50	549.1496	56.83	4.22	12.74	26.20
C26H24N2O1	380.49	380.1889	82.07	6.36	7.36	4.20
C26H24N2O2	396.49	396.1838	78.76	6.10	7.07	8.07
C26H24N2O3	412.49	412.1787	75.71	5.86	6.79	11.64
C26H24N2O4	428.49	428.1736	72.88	5.65	6.54	14.94
C26H24N2O5	444.49	444.1685	70.26	5.44	6.30	18.00
C26H24N2O6	460.49	460.1634	67.82	5.25	6.08	20.85
C26H24N2O7	476.49	476.1583	65.54	5.08	5.88	23.50
C26H24N2O8	492.49	492.1533	63.41	4.91	5.69	25.99
C26H24N2O9	508.49	508.1482	61.42	4.76	5.51	28.32
C26H24N4O1	408.51	408.1950	76.45	5.92	13.72	3.92
C26H24N4O2	424.51	424.1899	73.57	5.70	13.20	7.54
C26H24N4O3	440.51	440.1848	70.89	5.49	12.72	10.90
C26H24N4O4	456.51	456.1797	68.41	5.30	12.27	14.02
C26H24N4O5	472.50	472.1747	66.09	5.12	11.86	16.93
C26H24N4O6	488.50	488.1696	63.93	4.95	11.47	19.65
C26H24N4O7	504.50	504.1645	61.90	4.80	11.11	22.20
C26H24N4O8	520.50	520.1594	60.00	4.65	10.76	24.59
C26H24N4O9	536.50	536.1543	58.21	4.51	10.44	26.84
C26H24N6O1	436.52	436.2011	71.54	5.54	19.25	3.67
C26H24N6O2	452.52	452.1961	69.01	5.35	18.57	7.07
C26H24N6O3	468.52	468.1910	66.65	5.16	17.94	10.24
C26H24N6O4	484.52	484.1859	64.45	4.99	17.35	13.21
C26H24N6O5	500.52	500.1808	62.39	4.83	16.79	15.98
C26H24N6O6	516.52	516.1757	60.46	4.68	16.27	18.59
C26H24N6O7	532.52	532.1706	58.64	4.54	15.78	21.03
C26H24N6O8	548.52	548.1655	56.93	4.41	15.32	23.33
C26H24N6O9	564.52	564.1605	55.32	4.29	14.89	25.51
C26H25N1O1	367.50	367.1936	84.98	6.86	3.81	4.35
C26H25N1O2	383.49	383.1885	81.43	6.57	3.65	8.34
C26H25N1O3	399.49	399.1834	78.17	6.31	3.51	12.01
C26H25N1O4	415.49	415.1783	75.16	6.06	3.37	15.40
C26H25N1O5	431.49	431.1733	72.37	5.84	3.25	18.54
C26H25N1O6	447.49	447.1682	69.79	5.63	3.13	21.45
C26H25N1O7	463.49	463.1631	67.38	5.44	3.02	24.16
C26H25N1O8	479.49	479.1580	65.13	5.26	2.92	26.69
C26H25N1O9	495.49	495.1529	63.03	5.09	2.83	29.06
C26H25N3O1	395.51	395.1998	78.96	6.37	10.62	4.05
C26H25N3O2	411.51	411.1947	75.89	6.12	10.21	7.78
C26H25N3O3	427.51	427.1896	73.05	5.89	9.83	11.23
C26H25N3O4	443.51	443.1845	70.41	5.68	9.47	14.43

C26H25N3O5	459.51	459.1794	67.96	5.48	9.14	17.41
C26H25N3O6	475.51	475.1743	65.68	5.30	8.84	20.19
C26H25N3O7	491.51	491.1692	63.54	5.13	8.55	22.79
C26H25N3O8	507.50	507.1641	61.53	4.97	8.28	25.22
C26H25N3O9	523.50	523.1591	59.65	4.81	8.03	27.51
C26H25N5O1	423.52	423.2059	73.74	5.95	16.54	3.78
C26H25N5O2	439.52	439.2008	71.05	5.73	15.93	7.28
C26H25N5O3	455.52	455.1957	68.56	5.53	15.37	10.54
C26H25N5O4	471.52	471.1906	66.23	5.34	14.85	13.57
C26H25N5O5	487.52	487.1856	64.06	5.17	14.37	16.41
C26H25N5O6	503.52	503.1805	62.02	5.00	13.91	19.07
C26H25N5O7	519.52	519.1754	60.11	4.85	13.48	21.56
C26H25N5O8	535.52	535.1703	58.32	4.71	13.08	23.90
C26H25N5O9	551.52	551.1652	56.62	4.57	12.70	26.11
C26H26N2O1	382.51	382.2045	81.64	6.85	7.32	4.18
C26H26N2O2	398.51	398.1994	78.36	6.58	7.03	8.03
C26H26N2O3	414.51	414.1943	75.34	6.32	6.76	11.58
C26H26N2O4	430.51	430.1892	72.54	6.09	6.51	14.87
C26H26N2O5	446.51	446.1842	69.94	5.87	6.27	17.92
C26H26N2O6	462.51	462.1791	67.52	5.67	6.06	20.76
C26H26N2O7	478.51	478.1740	65.26	5.48	5.85	23.41
C26H26N2O8	494.51	494.1689	63.15	5.30	5.66	25.88
C26H26N2O9	510.51	510.1638	61.17	5.13	5.49	28.21
C26H26N4O1	410.52	410.2106	76.07	6.38	13.65	3.90
C26H26N4O2	426.52	426.2056	73.22	6.14	13.14	7.50
C26H26N4O3	442.52	442.2005	70.57	5.92	12.66	10.85
C26H26N4O4	458.52	458.1954	68.11	5.72	12.22	13.96
C26H26N4O5	474.52	474.1903	65.81	5.52	11.81	16.86
C26H26N4O6	490.52	490.1852	63.67	5.34	11.42	19.57
C26H26N4O7	506.52	506.1801	61.65	5.17	11.06	22.11
C26H26N4O8	522.52	522.1750	59.77	5.02	10.72	24.50
C26H26N4O9	538.52	538.1700	57.99	4.87	10.40	26.74
C26H26N6O1	438.54	438.2168	71.21	5.98	19.16	3.65
C26H26N6O2	454.54	454.2117	68.71	5.77	18.49	7.04
C26H26N6O3	470.54	470.2066	66.37	5.57	17.86	10.20
C26H26N6O4	486.53	486.2015	64.19	5.39	17.27	13.15
C26H26N6O5	502.53	502.1965	62.14	5.22	16.72	15.92
C26H26N6O6	518.53	518.1914	60.23	5.05	16.21	18.51
C26H26N6O7	534.53	534.1863	58.42	4.90	15.72	20.95
C26H26N6O8	550.53	550.1812	56.73	4.76	15.27	23.25
C26H26N6O9	566.53	566.1761	55.12	4.63	14.83	25.42
C26H27N1O1	369.51	369.2093	84.51	7.37	3.79	4.33
C26H27N1O2	385.51	385.2042	81.01	7.06	3.63	8.30
C26H27N1O3	401.51	401.1991	77.78	6.78	3.49	11.95
C26H27N1O4	417.51	417.1940	74.80	6.52	3.35	15.33
C26H27N1O5	433.51	433.1889	72.04	6.28	3.23	18.45
C26H27N1O6	449.51	449.1838	69.47	6.05	3.12	21.36
C26H27N1O7	465.51	465.1787	67.09	5.85	3.01	24.06
C26H27N1O8	481.51	481.1737	64.86	5.65	2.91	26.58
C26H27N1O9	497.51	497.1686	62.77	5.47	2.82	28.94
C26H27N3O1	397.52	397.2154	78.56	6.85	10.57	4.02
C26H27N3O2	413.52	413.2103	75.52	6.58	10.16	7.74
C26H27N3O3	429.52	429.2052	72.71	6.34	9.78	11.17
C26H27N3O4	445.52	445.2001	70.10	6.11	9.43	14.36
C26H27N3O5	461.52	461.1951	67.67	5.90	9.10	17.33
C26H27N3O6	477.52	477.1900	65.40	5.70	8.80	20.10
C26H27N3O7	493.52	493.1849	63.28	5.51	8.51	22.69
C26H27N3O8	509.52	509.1798	61.29	5.34	8.25	25.12
C26H27N3O9	525.52	525.1747	59.42	5.18	8.00	27.40
C26H27N5O1	425.54	425.2215	73.39	6.40	16.46	3.76

C26H27N5O2	441.54	441.2165	70.73	6.16	15.86	7.25
C26H27N5O3	457.54	457.2114	68.25	5.95	15.31	10.49
C26H27N5O4	473.54	473.2063	65.95	5.75	14.79	13.51
C26H27N5O5	489.54	489.2012	63.79	5.56	14.31	16.34
C26H27N5O6	505.53	505.1961	61.77	5.38	13.85	18.99
C26H27N5O7	521.53	521.1910	59.88	5.22	13.43	21.47
C26H27N5O8	537.53	537.1859	58.10	5.06	13.03	23.81
C26H27N5O9	553.53	553.1809	56.42	4.92	12.65	26.01
C26H28N2O1	384.53	384.2202	81.21	7.34	7.29	4.16
C26H28N2O2	400.53	400.2151	77.97	7.05	6.99	7.99
C26H28N2O3	416.52	416.2100	74.98	6.78	6.73	11.52
C26H28N2O4	432.52	432.2049	72.20	6.53	6.48	14.80
C26H28N2O5	448.52	448.1998	69.63	6.29	6.25	17.84
C26H28N2O6	464.52	464.1947	67.23	6.08	6.03	20.67
C26H28N2O7	480.52	480.1896	64.99	5.87	5.83	23.31
C26H28N2O8	496.52	496.1845	62.90	5.68	5.64	25.78
C26H28N2O9	512.52	512.1795	60.93	5.51	5.47	28.10
C26H28N4O1	412.54	412.2263	75.70	6.84	13.58	3.88
C26H28N4O2	428.54	428.2212	72.87	6.59	13.07	7.47
C26H28N4O3	444.54	444.2161	70.25	6.35	12.60	10.80
C26H28N4O4	460.54	460.2110	67.81	6.13	12.17	13.90
C26H28N4O5	476.54	476.2060	65.53	5.92	11.76	16.79
C26H28N4O6	492.54	492.2009	63.40	5.73	11.38	19.49
C26H28N4O7	508.54	508.1958	61.41	5.55	11.02	22.02
C26H28N4O8	524.54	524.1907	59.54	5.38	10.68	24.40
C26H28N4O9	540.53	540.1856	57.77	5.22	10.37	26.64
C26H28N6O1	440.55	440.2324	70.89	6.41	19.08	3.63
C26H28N6O2	456.55	456.2274	68.40	6.18	18.41	7.01
C26H28N6O3	472.55	472.2223	66.09	5.97	17.78	10.16
C26H28N6O4	488.55	488.2172	63.92	5.78	17.20	13.10
C26H28N6O5	504.55	504.2121	61.89	5.59	16.66	15.86
C26H28N6O6	520.55	520.2070	59.99	5.42	16.14	18.44
C26H28N6O7	536.55	536.2019	58.20	5.26	15.66	20.87
C26H28N6O8	552.55	552.1968	56.52	5.11	15.21	23.16
C26H28N6O9	568.55	568.1918	54.93	4.96	14.78	25.33
C26H29N1O1	371.53	371.2249	84.06	7.87	3.77	4.31
C26H29N1O2	387.53	387.2198	80.59	7.54	3.61	8.26
C26H29N1O3	403.53	403.2147	77.39	7.24	3.47	11.89
C26H29N1O4	419.53	419.2096	74.44	6.97	3.34	15.25
C26H29N1O5	435.52	435.2046	71.70	6.71	3.22	18.37
C26H29N1O6	451.52	451.1995	69.16	6.47	3.10	21.26
C26H29N1O7	467.52	467.1944	66.80	6.25	3.00	23.96
C26H29N1O8	483.52	483.1893	64.59	6.05	2.90	26.47
C26H29N1O9	499.52	499.1842	62.52	5.85	2.80	28.83
C26H29N3O1	399.54	399.2310	78.16	7.32	10.52	4.00
C26H29N3O2	415.54	415.2260	75.15	7.03	10.11	7.70
C26H29N3O3	431.54	431.2209	72.37	6.77	9.74	11.12
C26H29N3O4	447.54	447.2158	69.78	6.53	9.39	14.30
C26H29N3O5	463.54	463.2107	67.37	6.31	9.07	17.26
C26H29N3O6	479.54	479.2056	65.12	6.10	8.76	20.02
C26H29N3O7	495.54	495.2005	63.02	5.90	8.48	22.60
C26H29N3O8	511.54	511.1954	61.05	5.71	8.21	25.02
C26H29N3O9	527.54	527.1904	59.20	5.54	7.97	27.30
C26H29N5O1	427.55	427.2372	73.04	6.84	16.38	3.74
C26H29N5O2	443.55	443.2321	70.41	6.59	15.79	7.21
C26H29N5O3	459.55	459.2270	67.96	6.36	15.24	10.44
C26H29N5O4	475.55	475.2219	65.67	6.15	14.73	13.46
C26H29N5O5	491.55	491.2169	63.53	5.95	14.25	16.27
C26H29N5O6	507.55	507.2118	61.53	5.76	13.80	18.91
C26H29N5O7	523.55	523.2067	59.65	5.58	13.38	21.39
C26H29N5O8	539.55	539.2016	57.88	5.42	12.98	23.72
C26H29N5O9	555.55	555.1965	56.21	5.26	12.61	25.92
C26H30N2O1	386.54	386.2358	80.79	7.82	7.25	4.14
C26H30N2O2	402.54	402.2307	77.58	7.51	6.96	7.95
C26H30N2O3	418.54	418.2256	74.61	7.22	6.69	11.47
C26H30N2O4	434.54	434.2205	71.87	6.96	6.45	14.73
C26H30N2O5	450.54	450.2155	69.31	6.71	6.22	17.76
C26H30N2O6	466.54	466.2104	66.94	6.48	6.00	20.58
C26H30N2O7	482.54	482.2053	64.72	6.27	5.81	23.21
C26H30N2O8	498.54	498.2002	62.64	6.07	5.62	25.67
C26H30N2O9	514.54	514.1951	60.69	5.88	5.44	27.99
C26H30N4O1	414.56	414.2419	75.33	7.29	13.51	3.86
C26H30N4O2	430.55	430.2369	72.53	7.02	13.01	7.43
C26H30N4O3	446.55	446.2318	69.93	6.77	12.55	10.75
C26H30N4O4	462.55	462.2267	67.51	6.54	12.11	13.84
C26H30N4O5	478.55	478.2216	65.26	6.32	11.71	16.72
C26H30N4O6	494.55	494.2165	63.15	6.11	11.33	19.41
C26H30N4O7	510.55	510.2114	61.17	5.92	10.97	21.94
C26H30N4O8	526.55	526.2063	59.31	5.74	10.64	24.31
C26H30N4O9	542.55	542.2013	57.56	5.57	10.33	26.54
C26H30N6O1	442.57	442.2481	70.56	6.83	18.99	3.62
C26H30N6O2	458.57	458.2430	68.10	6.59	18.33	6.98
C26H30N6O3	474.57	474.2379	65.81	6.37	17.71	10.11
C26H30N6O4	490.57	490.2328	63.66	6.16	17.13	13.05
C26H30N6O5	506.57	506.2278	61.65	5.97	16.59	15.79
C26H30N6O6	522.57	522.2227	59.76	5.79	16.08	18.37
C26H30N6O7	538.56	538.2176	57.99	5.61	15.60	20.80
C26H30N6O8	554.56	554.2125	56.31	5.45	15.15	23.08
C26H30N6O9	570.56	570.2074	54.73	5.30	14.73	25.24
C26H31N1O1	373.54	373.2406	83.60	8.37	3.75	4.28
C26H31N1O2	389.54	389.2355	80.17	8.02	3.60	8.21
C26H31N1O3	405.54	405.2304	77.01	7.71	3.45	11.84
C26H31N1O4	421.54	421.2253	74.08	7.41	3.32	15.18
C26H31N1O5	437.54	437.2202	71.37	7.14	3.20	18.28
C26H31N1O6	453.54	453.2151	68.86	6.89	3.09	21.17
C26H31N1O7	469.54	469.2100	66.51	6.65	2.98	23.85
C26H31N1O8	485.54	485.2049	64.32	6.44	2.88	26.36
C26H31N1O9	501.54	501.1999	62.27	6.23	2.79	28.71
C26H31N3O1	401.56	401.2467	77.77	7.78	10.46	3.98
C26H31N3O2	417.56	417.2416	74.79	7.48	10.06	7.66
C26H31N3O3	433.56	433.2365	72.03	7.21	9.69	11.07
C26H31N3O4	449.55	449.2314	69.47	6.95	9.35	14.24
C26H31N3O5	465.55	465.2264	67.08	6.71	9.03	17.18
C26H31N3O6	481.55	481.2213	64.85	6.49	8.73	19.93
C26H31N3O7	497.55	497.2162	62.77	6.28	8.45	22.51
C26H31N3O8	513.55	513.2111	60.81	6.08	8.18	24.92
C26H31N3O9	529.55	529.2060	58.97	5.90	7.94	27.19
C26H31N5O1	429.57	429.2528	72.70	7.27	16.30	3.72
C26H31N5O2	445.57	445.2478	70.09	7.01	15.72	7.18
C26H31N5O3	461.57	461.2427	67.66	6.77	15.17	10.40
C26H31N5O4	477.57	477.2376	65.39	6.54	14.66	13.40
C26H31N5O5	493.57	493.2325	63.27	6.33	14.19	16.21
C26H31N5O6	509.57	509.2274	61.29	6.13	13.74	18.84
C26H31N5O7	525.57	525.2223	59.42	5.95	13.33	21.31
C26H31N5O8	541.57	541.2172	57.66	5.77	12.93	23.63
C26H31N5O9	557.57	557.2122	56.01	5.60	12.56	25.83
C26H32N2O1	388.56	388.2514	80.37	8.30	7.21	4.12
C26H32N2O2	404.56	404.2464	77.19	7.97	6.92	7.91
C26H32N2O3	420.56	420.2413	74.26	7.67	6.66	11.41
C26H32N2O4	436.56	436.2362	71.53	7.39	6.42	14.66

C26H32N2O5	452.56	452.2311	69.01	7.13	6.19	17.68
C26H32N2O6	468.55	468.2260	66.65	6.88	5.98	20.49
C26H32N2O7	484.55	484.2209	64.45	6.66	5.78	23.11
C26H32N2O8	500.55	500.2158	62.39	6.44	5.60	25.57
C26H32N2O9	516.55	516.2108	60.46	6.24	5.42	27.88
C26H32N4O1	416.57	416.2576	74.97	7.74	13.45	3.84
C26H32N4O2	432.57	432.2525	72.19	7.46	12.95	7.40
C26H32N4O3	448.57	448.2474	69.62	7.19	12.49	10.70
C26H32N4O4	464.57	464.2423	67.22	6.94	12.06	13.78
C26H32N4O5	480.57	480.2373	64.98	6.71	11.66	16.65
C26H32N4O6	496.57	496.2322	62.89	6.50	11.28	19.33
C26H32N4O7	512.57	512.2271	60.93	6.29	10.93	21.85
C26H32N4O8	528.57	528.2220	59.08	6.10	10.60	24.22
C26H32N4O9	544.57	544.2169	57.35	5.92	10.29	26.44
C26H32N6O1	444.58	444.2637	70.24	7.26	18.90	3.60
C26H32N6O2	460.58	460.2587	67.80	7.00	18.25	6.95
C26H32N6O3	476.58	476.2536	65.53	6.77	17.63	10.07
C26H32N6O4	492.58	492.2485	63.40	6.55	17.06	12.99
C26H32N6O5	508.58	508.2434	61.40	6.34	16.52	15.73
C26H32N6O6	524.58	524.2383	59.53	6.15	16.02	18.30
C26H32N6O7	540.58	540.2332	57.77	5.97	15.55	20.72
C26H32N6O8	556.58	556.2281	56.11	5.80	15.10	23.00
C26H32N6O9	572.58	572.2231	54.54	5.63	14.68	25.15
C26H33N1O1	375.56	375.2562	83.15	8.86	3.73	4.26
C26H33N1O2	391.56	391.2511	79.76	8.50	3.58	8.17
C26H33N1O3	407.56	407.2460	76.62	8.16	3.44	11.78
C26H33N1O4	423.56	423.2409	73.73	7.85	3.31	15.11
C26H33N1O5	439.56	439.2359	71.05	7.57	3.19	18.20
C26H33N1O6	455.56	455.2308	68.55	7.30	3.07	21.07
C26H33N1O7	471.56	471.2257	66.23	7.05	2.97	23.75
C26H33N1O8	487.55	487.2206	64.05	6.82	2.87	26.25
C26H33N1O9	503.55	503.2155	62.02	6.61	2.78	28.60
C26H33N3O1	403.57	403.2623	77.38	8.24	10.41	3.96
C26H33N3O2	419.57	419.2573	74.43	7.93	10.01	7.63
C26H33N3O3	435.57	435.2522	71.70	7.64	9.65	11.02
C26H33N3O4	451.57	451.2471	69.16	7.37	9.31	14.17
C26H33N3O5	467.57	467.2420	66.79	7.11	8.99	17.11
C26H33N3O6	483.57	483.2369	64.58	6.88	8.69	19.85
C26H33N3O7	499.57	499.2318	62.51	6.66	8.41	22.42
C26H33N3O8	515.57	515.2267	60.57	6.45	8.15	24.83
C26H33N3O9	531.57	531.2217	58.75	6.26	7.90	27.09
C26H33N5O1	431.59	431.2685	72.36	7.71	16.23	3.71
C26H33N5O2	447.59	447.2634	69.77	7.43	15.65	7.15
C26H33N5O3	463.58	463.2583	67.36	7.18	15.11	10.35
C26H33N5O4	479.58	479.2532	65.12	6.94	14.60	13.34
C26H33N5O5	495.58	495.2482	63.01	6.71	14.13	16.14
C26H33N5O6	511.58	511.2431	61.04	6.50	13.69	18.76
C26H33N5O7	527.58	527.2380	59.19	6.30	13.27	21.23
C26H33N5O8	543.58	543.2329	57.45	6.12	12.88	23.55
C26H33N5O9	559.58	559.2278	55.81	5.94	12.52	25.73
C26H34N2O1	390.57	390.2671	79.96	8.77	7.17	4.10
C26H34N2O2	406.57	406.2620	76.81	8.43	6.89	7.87
C26H34N2O3	422.57	422.2569	73.90	8.11	6.63	11.36
C26H34N2O4	438.57	438.2518	71.21	7.81	6.39	14.59
C26H34N2O5	454.57	454.2468	68.70	7.54	6.16	17.60
C26H34N2O6	470.57	470.2417	66.36	7.28	5.95	20.40
C26H34N2O7	486.57	486.2366	64.18	7.04	5.76	23.02
C26H34N2O8	502.57	502.2315	62.14	6.82	5.57	25.47
C26H34N2O9	518.57	518.2264	60.22	6.61	5.40	27.77
C26H34N4O1	418.59	418.2732	74.61	8.19	13.38	3.82
C26H34N4O2	434.59	434.2682	71.86	7.89	12.89	7.36
C26H34N4O3	450.59	450.2631	69.31	7.61	12.43	10.65
C26H34N4O4	466.59	466.2580	66.93	7.35	12.01	13.72
C26H34N4O5	482.58	482.2529	64.71	7.10	11.61	16.58
C26H34N4O6	498.58	498.2478	62.64	6.87	11.24	19.25
C26H34N4O7	514.58	514.2427	60.69	6.66	10.89	21.76
C26H34N4O8	530.58	530.2376	58.86	6.46	10.56	24.12
C26H34N4O9	546.58	546.2326	57.14	6.27	10.25	26.34
C26H34N6O1	446.60	446.2794	69.93	7.67	18.82	3.58
C26H34N6O2	462.60	462.2743	67.51	7.41	18.17	6.92
C26H34N6O3	478.60	478.2692	65.25	7.16	17.56	10.03
C26H34N6O4	494.60	494.2641	63.14	6.93	16.99	12.94
C26H34N6O5	510.60	510.2590	61.16	6.71	16.46	15.67
C26H34N6O6	526.60	526.2540	59.30	6.51	15.96	18.23
C26H34N6O7	542.60	542.2489	57.55	6.32	15.49	20.64
C26H34N6O8	558.60	558.2438	55.91	6.14	15.04	22.91
C26H34N6O9	574.60	574.2387	54.35	5.96	14.63	25.06
C26H35N1O1	377.57	377.2718	82.71	9.34	3.71	4.24
C26H35N1O2	393.57	393.2668	79.35	8.96	3.56	8.13
C26H35N1O3	409.57	409.2617	76.25	8.61	3.42	11.72
C26H35N1O4	425.57	425.2566	73.38	8.29	3.29	15.04
C26H35N1O5	441.57	441.2515	70.72	7.99	3.17	18.12
C26H35N1O6	457.57	457.2464	68.25	7.71	3.06	20.98
C26H35N1O7	473.57	473.2413	65.94	7.45	2.96	23.65
C26H35N1O8	489.57	489.2362	63.79	7.21	2.86	26.14
C26H35N1O9	505.57	505.2312	61.77	6.98	2.77	28.48
C26H35N3O1	405.59	405.2780	77.00	8.70	10.36	3.94
C26H35N3O2	421.59	421.2729	74.07	8.37	9.97	7.59
C26H35N3O3	437.59	437.2678	71.37	8.06	9.60	10.97
C26H35N3O4	453.59	453.2627	68.85	7.78	9.26	14.11
C26H35N3O5	469.59	469.2577	66.50	7.51	8.95	17.04
C26H35N3O6	485.59	485.2526	64.31	7.27	8.65	19.77
C26H35N3O7	501.58	501.2475	62.26	7.03	8.38	22.33
C26H35N3O8	517.58	517.2424	60.34	6.82	8.12	24.73
C26H35N3O9	533.58	533.2373	58.53	6.61	7.88	26.99
C26H35N5O1	433.60	433.2841	72.02	8.14	16.15	3.69
C26H35N5O2	449.60	449.2791	69.46	7.85	15.58	7.12
C26H35N5O3	465.60	465.2740	67.07	7.58	15.04	10.31
C26H35N5O4	481.60	481.2689	64.84	7.33	14.54	13.29
C26H35N5O5	497.60	497.2638	62.76	7.09	14.07	16.08
C26H35N5O6	513.60	513.2587	60.80	6.87	13.64	18.69
C26H35N5O7	529.60	529.2536	58.97	6.66	13.22	21.15
C26H35N5O8	545.60	545.2485	57.24	6.47	12.84	23.46
C26H35N5O9	561.60	561.2435	55.61	6.28	12.47	25.64
C26H36N2O1	392.59	392.2827	79.55	9.24	7.14	4.08
C26H36N2O2	408.59	408.2777	76.43	8.88	6.86	7.83
C26H36N2O3	424.59	424.2726	73.55	8.55	6.60	11.30
C26H36N2O4	440.59	440.2675	70.88	8.24	6.36	14.53
C26H36N2O5	456.59	456.2624	68.40	7.95	6.14	17.52
C26H36N2O6	472.59	472.2573	66.08	7.68	5.93	20.31
C26H36N2O7	488.59	488.2522	63.92	7.43	5.73	22.92
C26H36N2O8	504.59	504.2471	61.89	7.19	5.55	25.37
C26H36N2O9	520.58	520.2421	59.99	6.97	5.38	27.66
C26H36N4O1	420.60	420.2889	74.25	8.63	13.32	3.80
C26H36N4O2	436.60	436.2838	71.53	8.31	12.83	7.33
C26H36N4O3	452.60	452.2787	69.00	8.02	12.38	10.60
C26H36N4O4	468.60	468.2736	66.64	7.74	11.96	13.66
C26H36N4O5	484.60	484.2686	64.44	7.49	11.56	16.51
C26H36N4O6	500.60	500.2635	62.38	7.25	11.19	19.18
C26H36N4O7	516.60	516.2584	60.45	7.02	10.85	21.68

C26H36N4O8	532.60	532.2533	58.64	6.81	10.52	24.03
C26H36N4O9	548.60	548.2482	56.93	6.61	10.21	26.25
C26H36N6O1	448.62	448.2950	69.61	8.09	18.73	3.57
C26H36N6O2	464.62	464.2900	67.21	7.81	18.09	6.89
C26H36N6O3	480.62	480.2849	64.98	7.55	17.49	9.99
C26H36N6O4	496.61	496.2798	62.88	7.31	16.92	12.89
C26H36N6O5	512.61	512.2747	60.92	7.08	16.39	15.61
C26H36N6O6	528.61	528.2696	59.08	6.86	15.90	18.16
C26H36N6O7	544.61	544.2645	57.34	6.66	15.43	20.56
C26H36N6O8	560.61	560.2594	55.71	6.47	14.99	22.83
C26H36N6O9	576.61	576.2544	54.16	6.29	14.57	24.97
C26H37N1O1	379.59	379.2875	82.27	9.83	3.69	4.21
C26H37N1O2	395.59	395.2824	78.94	9.43	3.54	8.09
C26H37N1O3	411.59	411.2773	75.87	9.06	3.40	11.66
C26H37N1O4	427.59	427.2722	73.04	8.72	3.28	14.97
C26H37N1O5	443.59	443.2672	70.40	8.41	3.16	18.03
C26H37N1O6	459.59	459.2621	67.95	8.11	3.05	20.89
C26H37N1O7	475.59	475.2570	65.66	7.84	2.95	23.55
C26H37N1O8	491.59	491.2519	63.53	7.59	2.85	26.04
C26H37N1O9	507.59	507.2468	61.52	7.35	2.76	28.37
C26H37N3O1	407.60	407.2936	76.62	9.15	10.31	3.93
C26H37N3O2	423.60	423.2886	73.72	8.80	9.92	7.55
C26H37N3O3	439.60	439.2835	71.04	8.48	9.56	10.92
C26H37N3O4	455.60	455.2784	68.54	8.19	9.22	14.05
C26H37N3O5	471.60	471.2733	66.22	7.91	8.91	16.96
C26H37N3O6	487.60	487.2682	64.05	7.65	8.62	19.69
C26H37N3O7	503.60	503.2631	62.01	7.41	8.34	22.24
C26H37N3O8	519.60	519.2580	60.10	7.18	8.09	24.63
C26H37N3O9	535.60	535.2530	58.31	6.96	7.85	26.88
C26H37N5O1	435.62	435.2998	71.69	8.56	16.08	3.67
C26H37N5O2	451.62	451.2947	69.15	8.26	15.51	7.09
C26H37N5O3	467.62	467.2896	66.78	7.98	14.98	10.26
C26H37N5O4	483.62	483.2845	64.57	7.71	14.48	13.23
C26H37N5O5	499.62	499.2794	62.51	7.46	14.02	16.01
C26H37N5O6	515.61	515.2744	60.57	7.23	13.58	18.62
C26H37N5O7	531.61	531.2693	58.74	7.02	13.17	21.07
C26H37N5O8	547.61	547.2642	57.03	6.81	12.79	23.37
C26H37N5O9	563.61	563.2591	55.41	6.62	12.43	25.55
C26H38N2O1	394.61	394.2984	79.14	9.71	7.10	4.05
C26H38N2O2	410.60	410.2933	76.06	9.33	6.82	7.79
C26H38N2O3	426.60	426.2882	73.20	8.98	6.57	11.25
C26H38N2O4	442.60	442.2831	70.56	8.65	6.33	14.46
C26H38N2O5	458.60	458.2781	68.10	8.35	6.11	17.44
C26H38N2O6	474.60	474.2730	65.80	8.07	5.90	20.23
C26H38N2O7	490.60	490.2679	63.65	7.81	5.71	22.83
C26H38N2O8	506.60	506.2628	61.64	7.56	5.53	25.27
C26H38N2O9	522.60	522.2577	59.76	7.33	5.36	27.55
C26H38N4O1	422.62	422.3045	73.89	9.06	13.26	3.79
C26H38N4O2	438.62	438.2995	71.20	8.73	12.77	7.30
C26H38N4O3	454.62	454.2944	68.69	8.43	12.32	10.56
C26H38N4O4	470.62	470.2893	66.36	8.14	11.90	13.60
C26H38N4O5	486.62	486.2842	64.18	7.87	11.51	16.44
C26H38N4O6	502.62	502.2791	62.13	7.62	11.15	19.10
C26H38N4O7	518.62	518.2740	60.22	7.39	10.80	21.60
C26H38N4O8	534.61	534.2689	58.41	7.16	10.48	23.94
C26H38N4O9	550.61	550.2639	56.72	6.96	10.18	26.15
C26H38N6O1	450.63	450.3107	69.30	8.50	18.65	3.55
C26H38N6O2	466.63	466.3056	66.92	8.21	18.01	6.86
C26H38N6O3	482.63	482.3005	64.71	7.94	17.41	9.95
C26H38N6O4	498.63	498.2954	62.63	7.68	16.85	12.83

C26H38N6O5	514.63	514.2903	60.68	7.44	16.33	15.54
C26H38N6O6	530.63	530.2853	58.85	7.22	15.84	18.09
C26H38N6O7	546.63	546.2802	57.13	7.01	15.37	20.49
C26H38N6O8	562.63	562.2751	55.51	6.81	14.94	22.75
C26H38N6O9	578.63	578.2700	53.97	6.62	14.52	24.89
C26H39N1O1	381.61	381.3031	81.84	10.30	3.67	4.19
C26H39N1O2	397.61	397.2981	78.54	9.89	3.52	8.05
C26H39N1O3	413.61	413.2930	75.50	9.50	3.39	11.60
C26H39N1O4	429.61	429.2879	72.69	9.15	3.26	14.90
C26H39N1O5	445.60	445.2828	70.08	8.82	3.14	17.95
C26H39N1O6	461.60	461.2777	67.65	8.52	3.03	20.80
C26H39N1O7	477.60	477.2726	65.39	8.23	2.93	23.45
C26H39N1O8	493.60	493.2675	63.27	7.96	2.84	25.93
C26H39N1O9	509.60	509.2625	61.28	7.71	2.75	28.26
C26H39N3O1	409.62	409.3093	76.24	9.60	10.26	3.91
C26H39N3O2	425.62	425.3042	73.37	9.24	9.87	7.52
C26H39N3O3	441.62	441.2991	70.71	8.90	9.52	10.87
C26H39N3O4	457.62	457.2940	68.24	8.59	9.18	13.98
C26H39N3O5	473.62	473.2890	65.94	8.30	8.87	16.89
C26H39N3O6	489.62	489.2839	63.78	8.03	8.58	19.61
C26H39N3O7	505.62	505.2788	61.76	7.77	8.31	22.15
C26H39N3O8	521.62	521.2737	59.87	7.54	8.06	24.54
C26H39N3O9	537.62	537.2686	58.09	7.31	7.82	26.78
C26H39N5O1	437.63	437.3154	71.36	8.98	16.00	3.66
C26H39N5O2	453.63	453.3104	68.84	8.67	15.44	7.05
C26H39N5O3	469.63	469.3053	66.50	8.37	14.91	10.22
C26H39N5O4	485.63	485.3002	64.31	8.09	14.42	13.18
C26H39N5O5	501.63	501.2951	62.25	7.84	13.96	15.95
C26H39N5O6	517.63	517.2900	60.33	7.59	13.53	18.55
C26H39N5O7	533.63	533.2849	58.52	7.37	13.12	20.99
C26H39N5O8	549.63	549.2798	56.82	7.15	12.74	23.29
C26H39N5O9	565.63	565.2748	55.21	6.95	12.38	25.46
C26H40N2O1	396.62	396.3140	78.74	10.17	7.06	4.03
C26H40N2O2	412.62	412.3090	75.68	9.77	6.79	7.76
C26H40N2O3	428.62	428.3039	72.86	9.41	6.54	11.20
C26H40N2O4	444.62	444.2988	70.24	9.07	6.30	14.39
C26H40N2O5	460.62	460.2937	67.80	8.75	6.08	17.37
C26H40N2O6	476.62	476.2886	65.52	8.46	5.88	20.14
C26H40N2O7	492.62	492.2835	63.39	8.18	5.69	22.73
C26H40N2O8	508.62	508.2784	61.40	7.93	5.51	25.17
C26H40N2O9	524.62	524.2734	59.53	7.69	5.34	27.45
C26H40N4O1	424.63	424.3202	73.54	9.49	13.19	3.77
C26H40N4O2	440.63	440.3151	70.87	9.15	12.72	7.26
C26H40N4O3	456.63	456.3100	68.39	8.83	12.27	10.51
C26H40N4O4	472.63	472.3049	66.07	8.53	11.85	13.54
C26H40N4O5	488.63	488.2998	63.91	8.25	11.47	16.37
C26H40N4O6	504.63	504.2948	61.88	7.99	11.10	19.02
C26H40N4O7	520.63	520.2897	59.98	7.74	10.76	21.51
C26H40N4O8	536.63	536.2846	58.19	7.51	10.44	23.85
C26H40N4O9	552.63	552.2795	56.51	7.30	10.14	26.06
C26H40N6O1	452.65	452.3263	68.99	8.91	18.57	3.53
C26H40N6O2	468.65	468.3213	66.64	8.60	17.93	6.83
C26H40N6O3	484.65	484.3162	64.44	8.32	17.34	9.90
C26H40N6O4	500.65	500.3111	62.38	8.05	16.79	12.78
C26H40N6O5	516.65	516.3060	60.45	7.80	16.27	15.48
C26H40N6O6	532.65	532.3009	58.63	7.57	15.78	18.02
C26H40N6O7	548.64	548.2958	56.92	7.35	15.32	20.41
C26H40N6O8	564.64	564.2907	55.31	7.14	14.88	22.67
C26H40N6O9	580.64	580.2857	53.78	6.94	14.47	24.80
C26H41N1O1	383.62	383.3188	81.41	10.77	3.65	4.17

C26H41N1O2	399.62	399.3137	78.15	10.34	3.50	8.01
C26H41N1O3	415.62	415.3086	75.14	9.94	3.37	11.55
C26H41N1O4	431.62	431.3035	72.35	9.57	3.25	14.83
C26H41N1O5	447.62	447.2985	69.77	9.23	3.13	17.87
C26H41N1O6	463.62	463.2934	67.36	8.91	3.02	20.71
C26H41N1O7	479.62	479.2883	65.11	8.62	2.92	23.35
C26H41N1O8	495.62	495.2832	63.01	8.34	2.83	25.83
C26H41N1O9	511.62	511.2781	61.04	8.08	2.74	28.14
C26H41N3O1	411.64	411.3249	75.87	10.04	10.21	3.89
C26H41N3O2	427.64	427.3199	73.03	9.66	9.83	7.48
C26H41N3O3	443.63	443.3148	70.39	9.32	9.47	10.82
C26H41N3O4	459.63	459.3097	67.94	8.99	9.14	13.92
C26H41N3O5	475.63	475.3046	65.66	8.69	8.83	16.82
C26H41N3O6	491.63	491.2995	63.52	8.41	8.55	19.53
C26H41N3O7	507.63	507.2944	61.52	8.14	8.28	22.06
C26H41N3O8	523.63	523.2893	59.64	7.89	8.02	24.44
C26H41N3O9	539.63	539.2843	57.87	7.66	7.79	26.68
C26H41N5O1	439.65	439.3311	71.03	9.40	15.93	3.64
C26H41N5O2	455.65	455.3260	68.54	9.07	15.37	7.02
C26H41N5O3	471.65	471.3209	66.21	8.76	14.85	10.18
C26H41N5O4	487.65	487.3158	64.04	8.47	14.36	13.12
C26H41N5O5	503.65	503.3107	62.01	8.21	13.91	15.88
C26H41N5O6	519.65	519.3057	60.10	7.95	13.48	18.47
C26H41N5O7	535.65	535.3006	58.30	7.72	13.07	20.91
C26H41N5O8	551.65	551.2955	56.61	7.49	12.70	23.20
C26H41N5O9	567.64	567.2904	55.02	7.28	12.34	25.37
C26H42N2O1	398.64	398.3297	78.34	10.62	7.03	4.01
C26H42N2O2	414.64	414.3246	75.32	10.21	6.76	7.72
C26H42N2O3	430.64	430.3195	72.52	9.83	6.51	11.15
C26H42N2O4	446.64	446.3144	69.92	9.48	6.27	14.33
C26H42N2O5	462.64	462.3094	67.50	9.15	6.06	17.29
C26H42N2O6	478.63	478.3043	65.25	8.84	5.85	20.06
C26H42N2O7	494.63	494.2992	63.14	8.56	5.66	22.64
C26H42N2O8	510.63	510.2941	61.16	8.29	5.49	25.07
C26H42N2O9	526.63	526.2890	59.30	8.04	5.32	27.34
C26H42N4O1	426.65	426.3358	73.20	9.92	13.13	3.75
C26H42N4O2	442.65	442.3308	70.55	9.56	12.66	7.23
C26H42N4O3	458.65	458.3257	68.09	9.23	12.22	10.47
C26H42N4O4	474.65	474.3206	65.79	8.92	11.80	13.48
C26H42N4O5	490.65	490.3155	63.65	8.63	11.42	16.30
C26H42N4O6	506.65	506.3104	61.64	8.36	11.06	18.95
C26H42N4O7	522.65	522.3053	59.75	8.10	10.72	21.43
C26H42N4O8	538.65	538.3002	57.98	7.86	10.40	23.76
C26H42N4O9	554.65	554.2952	56.30	7.63	10.10	25.96
C26H42N6O1	454.66	454.3420	68.69	9.31	18.48	3.52
C26H42N6O2	470.66	470.3369	66.35	8.99	17.86	6.80
C26H42N6O3	486.66	486.3318	64.17	8.70	17.27	9.86
C26H42N6O4	502.66	502.3267	62.13	8.42	16.72	12.73
C26H42N6O5	518.66	518.3216	60.21	8.16	16.20	15.42
C26H42N6O6	534.66	534.3166	58.41	7.92	15.72	17.95
C26H42N6O7	550.66	550.3115	56.71	7.69	15.26	20.34
C26H42N6O8	566.66	566.3064	55.11	7.47	14.83	22.59
C26H42N6O9	582.66	582.3013	53.60	7.27	14.42	24.71
C26H43N1O1	385.64	385.3344	80.98	11.24	3.63	4.15
C26H43N1O2	401.64	401.3294	77.75	10.79	3.49	7.97
C26H43N1O3	417.64	417.3243	74.78	10.38	3.35	11.49
C26H43N1O4	433.64	433.3192	72.02	10.00	3.23	14.76
C26H43N1O5	449.64	449.3141	69.45	9.64	3.12	17.79
C26H43N1O6	465.64	465.3090	67.07	9.31	3.01	20.62
C26H43N1O7	481.64	481.3039	64.84	9.00	2.91	23.25

C26H43N1O8	497.63	497.2988	62.75	8.71	2.81	25.72
C26H43N1O9	513.63	513.2938	60.80	8.44	2.73	28.03
C26H43N3O1	413.65	413.3406	75.50	10.48	10.16	3.87
C26H43N3O2	429.65	429.3355	72.68	10.09	9.78	7.45
C26H43N3O3	445.65	445.3304	70.08	9.73	9.43	10.77
C26H43N3O4	461.65	461.3253	67.65	9.39	9.10	13.86
C26H43N3O5	477.65	477.3202	65.38	9.07	8.80	16.75
C26H43N3O6	493.65	493.3152	63.26	8.78	8.51	19.45
C26H43N3O7	509.65	509.3101	61.28	8.50	8.24	21.98
C26H43N3O8	525.65	525.3050	59.41	8.25	7.99	24.35
C26H43N3O9	541.65	541.2999	57.66	8.00	7.76	26.58
C26H43N5O1	441.67	441.3467	70.71	9.81	15.86	3.62
C26H43N5O2	457.66	457.3417	68.24	9.47	15.30	6.99
C26H43N5O3	473.66	473.3366	65.93	9.15	14.79	10.13
C26H43N5O4	489.66	489.3315	63.78	8.85	14.30	13.07
C26H43N5O5	505.66	505.3264	61.76	8.57	13.85	15.82
C26H43N5O6	521.66	521.3213	59.86	8.31	13.43	18.40
C26H43N5O7	537.66	537.3162	58.08	8.06	13.03	20.83
C26H43N5O8	553.66	553.3111	56.40	7.83	12.65	23.12
C26H43N5O9	569.66	569.3061	54.82	7.61	12.29	25.28
C26H44N2O1	400.65	400.3453	77.95	11.07	6.99	3.99
C26H44N2O2	416.65	416.3403	74.95	10.64	6.72	7.68
C26H44N2O3	432.65	432.3352	72.18	10.25	6.47	11.09
C26H44N2O4	448.65	448.3301	69.61	9.89	6.24	14.26
C26H44N2O5	464.65	464.3250	67.21	9.54	6.03	17.22
C26H44N2O6	480.65	480.3199	64.97	9.23	5.83	19.97
C26H44N2O7	496.65	496.3148	62.88	8.93	5.64	22.55
C26H44N2O8	512.65	512.3097	60.92	8.65	5.46	24.97
C26H44N2O9	528.65	528.3047	59.07	8.39	5.30	27.24
C26H44N4O1	428.67	428.3515	72.85	10.35	13.07	3.73
C26H44N4O2	444.67	444.3464	70.23	9.97	12.60	7.20
C26H44N4O3	460.67	460.3413	67.79	9.63	12.16	10.42
C26H44N4O4	476.66	476.3362	65.52	9.30	11.75	13.43
C26H44N4O5	492.66	492.3311	63.39	9.00	11.37	16.24
C26H44N4O6	508.66	508.3261	61.39	8.72	11.01	18.87
C26H44N4O7	524.66	524.3210	59.52	8.45	10.68	21.35
C26H44N4O8	540.66	540.3159	57.76	8.20	10.36	23.67
C26H44N4O9	556.66	556.3108	56.10	7.97	10.06	25.87
C26H44N6O1	456.68	456.3576	68.38	9.71	18.40	3.50
C26H44N6O2	472.68	472.3526	66.07	9.38	17.78	6.77
C26H44N6O3	488.68	488.3475	63.90	9.08	17.20	9.82
C26H44N6O4	504.68	504.3424	61.88	8.79	16.65	12.68
C26H44N6O5	520.68	520.3373	59.98	8.52	16.14	15.36
C26H44N6O6	536.68	536.3322	58.19	8.26	15.66	17.89
C26H44N6O7	552.68	552.3271	56.50	8.02	15.21	20.26
C26H44N6O8	568.68	568.3220	54.92	7.80	14.78	22.51
C26H44N6O9	584.68	584.3170	53.41	7.59	14.37	24.63
C26H45N1O1	387.65	387.3501	80.56	11.70	3.61	4.13
C26H45N1O2	403.65	403.3450	77.37	11.24	3.47	7.93
C26H45N1O3	419.65	419.3399	74.42	10.81	3.34	11.44
C26H45N1O4	435.65	435.3348	71.68	10.41	3.22	14.69
C26H45N1O5	451.65	451.3298	69.14	10.04	3.10	17.71
C26H45N1O6	467.65	467.3247	66.78	9.70	3.00	20.53
C26H45N1O7	483.65	483.3196	64.57	9.38	2.90	23.16
C26H45N1O8	499.65	499.3145	62.50	9.08	2.80	25.62
C26H45N1O9	515.65	515.3094	60.56	8.80	2.72	27.92
C26H45N3O1	415.67	415.3562	75.13	10.91	10.11	3.85
C26H45N3O2	431.67	431.3512	72.35	10.51	9.73	7.41
C26H45N3O3	447.67	447.3461	69.76	10.13	9.39	10.72
C26H45N3O4	463.67	463.3410	67.35	9.78	9.06	13.80

C26H45N3O5	479.67	479.3359	65.11	9.46	8.76	16.68
C26H45N3O6	495.67	495.3308	63.00	9.15	8.48	19.37
C26H45N3O7	511.66	511.3257	61.03	8.86	8.21	21.89
C26H45N3O8	527.66	527.3206	59.18	8.60	7.96	24.26
C26H45N3O9	543.66	543.3156	57.44	8.34	7.73	26.49
C26H45N5O1	443.68	443.3624	70.39	10.22	15.78	3.61
C26H45N5O2	459.68	459.3573	67.94	9.87	15.24	6.96
C26H45N5O3	475.68	475.3522	65.65	9.54	14.72	10.09
C26H45N5O4	491.68	491.3471	63.51	9.23	14.24	13.02
C26H45N5O5	507.68	507.3420	61.51	8.93	13.79	15.76
C26H45N5O6	523.68	523.3370	59.63	8.66	13.37	18.33
C26H45N5O7	539.68	539.3319	57.87	8.40	12.98	20.75
C26H45N5O8	555.68	555.3268	56.20	8.16	12.60	23.03
C26H45N5O9	571.68	571.3217	54.63	7.93	12.25	25.19
C26H46N2O1	402.67	402.3610	77.55	11.51	6.96	3.97
C26H46N2O2	418.67	418.3559	74.59	11.07	6.69	7.64
C26H46N2O3	434.67	434.3508	71.85	10.67	6.44	11.04
C26H46N2O4	450.67	450.3457	69.29	10.29	6.22	14.20
C26H46N2O5	466.67	466.3406	66.92	9.94	6.00	17.14
C26H46N2O6	482.67	482.3356	64.70	9.61	5.80	19.89
C26H46N2O7	498.67	498.3305	62.63	9.30	5.62	22.46
C26H46N2O8	514.67	514.3254	60.68	9.01	5.44	24.87
C26H46N2O9	530.66	530.3203	58.85	8.74	5.28	27.13
C26H46N4O1	430.68	430.3671	72.51	10.77	13.01	3.71
C26H46N4O2	446.68	446.3621	69.91	10.38	12.54	7.16
C26H46N4O3	462.68	462.3570	67.50	10.02	12.11	10.37
C26H46N4O4	478.68	478.3519	65.24	9.69	11.70	13.37
C26H46N4O5	494.68	494.3468	63.13	9.37	11.33	16.17
C26H46N4O6	510.68	510.3417	61.15	9.08	10.97	18.80
C26H46N4O7	526.68	526.3366	59.29	8.80	10.64	21.26
C26H46N4O8	542.68	542.3315	57.55	8.54	10.32	23.59
C26H46N4O9	558.68	558.3265	55.90	8.30	10.03	25.77
C26H46N6O1	458.70	458.3733	68.08	10.11	18.32	3.49
C26H46N6O2	474.70	474.3682	65.79	9.77	17.70	6.74
C26H46N6O3	490.69	490.3631	63.64	9.45	17.13	9.78
C26H46N6O4	506.69	506.3580	61.63	9.15	16.59	12.63
C26H46N6O5	522.69	522.3529	59.75	8.87	16.08	15.30
C26H46N6O6	538.69	538.3479	57.97	8.61	15.60	17.82
C26H46N6O7	554.69	554.3428	56.30	8.36	15.15	20.19
C26H46N6O8	570.69	570.3377	54.72	8.12	14.73	22.43
C26H46N6O9	586.69	586.3326	53.23	7.90	14.32	24.54
C26H47N1O1	389.67	389.3657	80.14	12.16	3.59	4.11
C26H47N1O2	405.67	405.3607	76.98	11.68	3.45	7.89
C26H47N1O3	421.67	421.3556	74.06	11.24	3.32	11.38
C26H47N1O4	437.67	437.3505	71.35	10.82	3.20	14.62
C26H47N1O5	453.67	453.3454	68.84	10.44	3.09	17.63
C26H47N1O6	469.67	469.3403	66.49	10.09	2.98	20.44
C26H47N1O7	485.67	485.3352	64.30	9.75	2.88	23.06
C26H47N1O8	501.67	501.3301	62.25	9.44	2.79	25.51
C26H47N1O9	517.67	517.3251	60.33	9.15	2.71	27.82
C26H47N3O1	417.68	417.3719	74.77	11.34	10.06	3.83
C26H47N3O2	433.68	433.3668	72.01	10.92	9.69	7.38
C26H47N3O3	449.68	449.3617	69.45	10.54	9.34	10.67
C26H47N3O4	465.68	465.3566	67.06	10.17	9.02	13.74
C26H47N3O5	481.68	481.3515	64.83	9.84	8.72	16.61
C26H47N3O6	497.68	497.3465	62.75	9.52	8.44	19.29
C26H47N3O7	513.68	513.3414	60.79	9.22	8.18	21.80
C26H47N3O8	529.68	529.3363	58.96	8.94	7.93	24.16
C26H47N3O9	545.68	545.3312	57.23	8.68	7.70	26.39
C26H47N5O1	445.70	445.3780	70.07	10.63	15.71	3.59

C26H47N5O2	461.70	461.3730	67.64	10.26	15.17	6.93
C26H47N5O3	477.70	477.3679	65.37	9.92	14.66	10.05
C26H47N5O4	493.70	493.3628	63.26	9.60	14.19	12.96
C26H47N5O5	509.69	509.3577	61.27	9.29	13.74	15.70
C26H47N5O6	525.69	525.3526	59.41	9.01	13.32	18.26
C26H47N5O7	541.69	541.3475	57.65	8.75	12.93	20.68
C26H47N5O8	557.69	557.3424	56.00	8.49	12.56	22.95
C26H47N5O9	573.69	573.3374	54.44	8.26	12.21	25.10
C26H48N2O1	404.69	404.3766	77.17	11.96	6.92	3.95
C26H48N2O2	420.68	420.3716	74.23	11.50	6.66	7.61
C26H48N2O3	436.68	436.3665	71.51	11.08	6.42	10.99
C26H48N2O4	452.68	452.3614	68.99	10.69	6.19	14.14
C26H48N2O5	468.68	468.3563	66.63	10.32	5.98	17.07
C26H48N2O6	484.68	484.3512	64.43	9.98	5.78	19.81
C26H48N2O7	500.68	500.3461	62.37	9.66	5.60	22.37
C26H48N2O8	516.68	516.3410	60.44	9.36	5.42	24.77
C26H48N2O9	532.68	532.3360	58.63	9.08	5.26	27.03
C26H48N4O1	432.70	432.3828	72.17	11.18	12.95	3.70
C26H48N4O2	448.70	448.3777	69.60	10.78	12.49	7.13
C26H48N4O3	464.70	464.3726	67.20	10.41	12.06	10.33
C26H48N4O4	480.70	480.3675	64.97	10.07	11.66	13.31
C26H48N4O5	496.70	496.3624	62.87	9.74	11.28	16.11
C26H48N4O6	512.70	512.3574	60.91	9.44	10.93	18.72
C26H48N4O7	528.70	528.3523	59.07	9.15	10.60	21.18
C26H48N4O8	544.69	544.3472	57.33	8.88	10.29	23.50
C26H48N4O9	560.69	560.3421	55.70	8.63	9.99	25.68
C26H48N6O1	460.71	460.3889	67.78	10.50	18.24	3.47
C26H48N6O2	476.71	476.3839	65.51	10.15	17.63	6.71
C26H48N6O3	492.71	492.3788	63.38	9.82	17.06	9.74
C26H48N6O4	508.71	508.3737	61.39	9.51	16.52	12.58
C26H48N6O5	524.71	524.3686	59.52	9.22	16.02	15.25
C26H48N6O6	540.71	540.3635	57.76	8.95	15.54	17.75
C26H48N6O7	556.71	556.3584	56.10	8.69	15.10	20.12
C26H48N6O8	572.71	572.3533	54.53	8.45	14.67	22.35
C26H48N6O9	588.71	588.3483	53.05	8.22	14.28	24.46
C26H49N1O1	391.69	391.3814	79.73	12.61	3.58	4.08
C26H49N1O2	407.69	407.3763	76.60	12.11	3.44	7.85
C26H49N1O3	423.69	423.3712	73.71	11.66	3.31	11.33
C26H49N1O4	439.68	439.3661	71.03	11.23	3.19	14.56
C26H49N1O5	455.68	455.3610	68.53	10.84	3.07	17.56
C26H49N1O6	471.68	471.3560	66.21	10.47	2.97	20.35
C26H49N1O7	487.68	487.3509	64.04	10.13	2.87	22.96
C26H49N1O8	503.68	503.3458	62.00	9.81	2.78	25.41
C26H49N1O9	519.68	519.3407	60.09	9.50	2.70	27.71
C26H49N3O1	419.70	419.3875	74.41	11.77	10.01	3.81
C26H49N3O2	435.70	435.3825	71.68	11.34	9.64	7.34
C26H49N3O3	451.70	451.3774	69.14	10.93	9.30	10.63
C26H49N3O4	467.70	467.3723	66.77	10.56	8.98	13.68
C26H49N3O5	483.70	483.3672	64.56	10.21	8.69	16.54
C26H49N3O6	499.70	499.3621	62.50	9.88	8.41	19.21
C26H49N3O7	515.70	515.3570	60.56	9.58	8.15	21.72
C26H49N3O8	531.70	531.3519	58.73	9.29	7.90	24.07
C26H49N3O9	547.70	547.3469	57.02	9.02	7.67	26.29
C26H49N5O1	447.71	447.3937	69.75	11.03	15.64	3.57
C26H49N5O2	463.71	463.3886	67.35	10.65	15.10	6.90
C26H49N5O3	479.71	479.3835	65.10	10.30	14.60	10.01
C26H49N5O4	495.71	495.3784	63.00	9.96	14.13	12.91
C26H49N5O5	511.71	511.3733	61.03	9.65	13.69	15.63
C26H49N5O6	527.71	527.3683	59.18	9.36	13.27	18.19
C26H49N5O7	543.71	543.3632	57.44	9.08	12.88	20.60

C26H49N5O8	559.71	559.3581	55.80	8.82	12.51	22.87
C26H49N5O9	575.71	575.3530	54.24	8.58	12.16	25.01
C26H50N2O1	406.70	406.3923	76.79	12.39	6.89	3.93
C26H50N2O2	422.70	422.3872	73.88	11.92	6.63	7.57
C26H50N2O3	438.70	438.3821	71.19	11.49	6.39	10.94
C26H50N2O4	454.70	454.3770	68.68	11.08	6.16	14.07
C26H50N2O5	470.70	470.3719	66.35	10.71	5.95	17.00
C26H50N2O6	486.70	486.3669	64.16	10.36	5.76	19.72
C26H50N2O7	502.70	502.3618	62.12	10.03	5.57	22.28
C26H50N2O8	518.70	518.3567	60.21	9.72	5.40	24.68
C26H50N2O9	534.70	534.3516	58.41	9.43	5.24	26.93
C26H50N4O1	434.71	434.3984	71.84	11.59	12.89	3.68
C26H50N4O2	450.71	450.3934	69.29	11.18	12.43	7.10
C26H50N4O3	466.71	466.3883	66.91	10.80	12.00	10.28
C26H50N4O4	482.71	482.3832	64.69	10.44	11.61	13.26
C26H50N4O5	498.71	498.3781	62.62	10.11	11.23	16.04
C26H50N4O6	514.71	514.3730	60.67	9.79	10.89	18.65
C26H50N4O7	530.71	530.3679	58.84	9.50	10.56	21.10
C26H50N4O8	546.71	546.3628	57.12	9.22	10.25	23.41
C26H50N4O9	562.71	562.3578	55.50	8.96	9.96	25.59
C26H50N6O1	462.73	462.4046	67.49	10.89	18.16	3.46
C26H50N6O2	478.73	478.3995	65.23	10.53	17.55	6.68
C26H50N6O3	494.73	494.3944	63.12	10.19	16.99	9.70
C26H50N6O4	510.73	510.3893	61.15	9.87	16.46	12.53
C26H50N6O5	526.73	526.3842	59.29	9.57	15.96	15.19
C26H50N6O6	542.72	542.3792	57.54	9.29	15.48	17.69
C26H50N6O7	558.72	558.3741	55.89	9.02	15.04	20.04
C26H50N6O8	574.72	574.3690	54.34	8.77	14.62	22.27
C26H50N6O9	590.72	590.3639	52.87	8.53	14.23	24.38
C26H51N1O1	393.70	393.3970	79.32	13.06	3.56	4.06
C26H51N1O2	409.70	409.3920	76.22	12.55	3.42	7.81
C26H51N1O3	425.70	425.3869	73.36	12.08	3.29	11.28
C26H51N1O4	441.70	441.3818	70.70	11.64	3.17	14.49
C26H51N1O5	457.70	457.3767	68.23	11.23	3.06	17.48
C26H51N1O6	473.70	473.3716	65.93	10.85	2.96	20.27
C26H51N1O7	489.70	489.3665	63.77	10.50	2.86	22.87
C26H51N1O8	505.70	505.3614	61.75	10.17	2.77	25.31
C26H51N1O9	521.70	521.3564	59.86	9.85	2.68	27.60
C26H51N3O1	421.72	421.4032	74.05	12.19	9.96	3.79
C26H51N3O2	437.72	437.3981	71.35	11.74	9.60	7.31
C26H51N3O3	453.71	453.3930	68.83	11.33	9.26	10.58
C26H51N3O4	469.71	469.3879	66.49	10.94	8.95	13.62
C26H51N3O5	485.71	485.3828	64.30	10.58	8.65	16.47
C26H51N3O6	501.71	501.3778	62.24	10.25	8.38	19.13
C26H51N3O7	517.71	517.3727	60.32	9.93	8.12	21.63
C26H51N3O8	533.71	533.3676	58.51	9.63	7.87	23.98
C26H51N3O9	549.71	549.3625	56.81	9.35	7.64	26.19
C26H51N5O1	449.73	449.4093	69.44	11.43	15.57	3.56
C26H51N5O2	465.73	465.4043	67.05	11.04	15.04	6.87
C26H51N5O3	481.73	481.3992	64.83	10.67	14.54	9.96
C26H51N5O4	497.73	497.3941	62.74	10.33	14.07	12.86
C26H51N5O5	513.73	513.3890	60.79	10.01	13.63	15.57
C26H51N5O6	529.73	529.3839	58.95	9.70	13.22	18.12
C26H51N5O7	545.73	545.3788	57.22	9.42	12.83	20.52
C26H51N5O8	561.73	561.3737	55.59	9.15	12.47	22.79
C26H51N5O9	577.72	577.3687	54.06	8.90	12.12	24.92
C26H52N2O1	408.72	408.4079	76.41	12.82	6.85	3.91
C26H52N2O2	424.72	424.4029	73.53	12.34	6.60	7.53
C26H52N2O3	440.72	440.3978	70.86	11.89	6.36	10.89
C26H52N2O4	456.72	456.3927	68.38	11.48	6.13	14.01

C26H52N2O5	472.71	472.3876	66.06	11.09	5.93	16.92
C26H52N2O6	488.71	488.3825	63.90	10.72	5.73	19.64
C26H52N2O7	504.71	504.3774	61.87	10.38	5.55	22.19
C26H52N2O8	520.71	520.3723	59.97	10.07	5.38	24.58
C26H52N2O9	536.71	536.3673	58.19	9.77	5.22	26.83
C26H52N4O1	436.73	436.4141	71.51	12.00	12.83	3.66
C26H52N4O2	452.73	452.4090	68.98	11.58	12.38	7.07
C26H52N4O3	468.73	468.4039	66.62	11.18	11.95	10.24
C26H52N4O4	484.73	484.3988	64.43	10.81	11.56	13.20
C26H52N4O5	500.73	500.3937	62.37	10.47	11.19	15.98
C26H52N4O6	516.73	516.3887	60.44	10.14	10.84	18.58
C26H52N4O7	532.73	532.3836	58.62	9.84	10.52	21.02
C26H52N4O8	548.73	548.3785	56.91	9.55	10.21	23.33
C26H52N4O9	564.73	564.3734	55.30	9.28	9.92	25.50
C26H52N6O1	464.74	464.4202	67.20	11.28	18.08	3.44
C26H52N6O2	480.74	480.4152	64.96	10.90	17.48	6.66
C26H52N6O3	496.74	496.4101	62.87	10.55	16.92	9.66
C26H52N6O4	512.74	512.4050	60.91	10.22	16.39	12.48
C26H52N6O5	528.74	528.3999	59.06	9.91	15.89	15.13
C26H52N6O6	544.74	544.3948	57.33	9.62	15.43	17.62
C26H52N6O7	560.74	560.3897	55.69	9.35	14.99	19.97
C26H52N6O8	576.74	576.3846	54.15	9.09	14.57	22.19
C26H52N6O9	592.74	592.3795	52.69	8.84	14.18	24.29
C26H53N1O1	395.72	395.4127	78.92	13.50	3.54	4.04
C26H53N1O2	411.72	411.4076	75.85	12.98	3.40	7.77
C26H53N1O3	427.72	427.4025	73.01	12.49	3.27	11.22
C26H53N1O4	443.72	443.3974	70.38	12.04	3.16	14.42
C26H53N1O5	459.72	459.3923	67.93	11.62	3.05	17.40
C26H53N1O6	475.72	475.3873	65.65	11.23	2.94	20.18
C26H53N1O7	491.71	491.3822	63.51	10.86	2.85	22.78
C26H53N1O8	507.71	507.3771	61.51	10.52	2.76	25.21
C26H53N1O9	523.71	523.3720	59.63	10.20	2.67	27.49
C26H53N3O1	423.73	423.4188	73.70	12.61	9.92	3.78
C26H53N3O2	439.73	439.4138	71.02	12.15	9.56	7.28
C26H53N3O3	455.73	455.4087	68.53	11.72	9.22	10.53
C26H53N3O4	471.73	471.4036	66.20	11.32	8.91	13.57
C26H53N3O5	487.73	487.3985	64.03	10.95	8.62	16.40
C26H53N3O6	503.73	503.3934	62.00	10.61	8.34	19.06
C26H53N3O7	519.73	519.3883	60.09	10.28	8.09	21.55
C26H53N3O8	535.73	535.3832	58.29	9.97	7.84	23.89
C26H53N3O9	551.73	551.3782	56.60	9.68	7.62	26.10
C26H53N5O1	451.75	451.4250	69.13	11.83	15.50	3.54
C26H53N5O2	467.74	467.4199	66.77	11.42	14.97	6.84
C26H53N5O3	483.74	483.4148	64.56	11.04	14.48	9.92
C26H53N5O4	499.74	499.4097	62.49	10.69	14.01	12.81
C26H53N5O5	515.74	515.4046	60.55	10.36	13.58	15.51
C26H53N5O6	531.74	531.3996	58.73	10.05	13.17	18.05
C26H53N5O7	547.74	547.3945	57.01	9.75	12.79	20.45
C26H53N5O8	563.74	563.3894	55.40	9.48	12.42	22.70
C26H53N5O9	579.74	579.3843	53.87	9.21	12.08	24.84
C26H54N2O1	410.73	410.4236	76.03	13.25	6.82	3.90
C26H54N2O2	426.73	426.4185	73.18	12.76	6.56	7.50
C26H54N2O3	442.73	442.4134	70.54	12.29	6.33	10.84
C26H54N2O4	458.73	458.4083	68.08	11.87	6.11	13.95
C26H54N2O5	474.73	474.4032	65.78	11.47	5.90	16.85
C26H54N2O6	490.73	490.3982	63.64	11.09	5.71	19.56
C26H54N2O7	506.73	506.3931	61.63	10.74	5.53	22.10
C26H54N2O8	522.73	522.3880	59.74	10.41	5.36	24.49
C26H54N2O9	538.73	538.3829	57.97	10.10	5.20	26.73
C26H54N4O1	438.75	438.4297	71.18	12.41	12.77	3.65

C26H54N402	454.75	454.4247	68.67	11.97	12.32	7.04	C26H56N408	552.76	552.4098	56.50	10.21	10.14	23.16
C26H54N403	470.75	470.4196	66.34	11.56	11.90	10.20	C26H56N409	568.76	568.4047	54.91	9.92	9.85	25.32
C26H54N404	486.74	486.4145	64.16	11.18	11.51	13.15	C26H56N601	468.78	468.4515	66.62	12.04	17.93	3.41
C26H54N405	502.74	502.4094	62.12	10.83	11.14	15.91	C26H56N602	484.78	484.4464	64.42	11.64	17.34	6.60
C26H54N406	518.74	518.4043	60.20	10.49	10.80	18.51	C26H56N603	500.77	500.4414	62.36	11.27	16.78	9.58
C26H54N407	534.74	534.3992	58.40	10.18	10.48	20.94	C26H56N604	516.77	516.4363	60.43	10.92	16.26	12.38
C26H54N408	550.74	550.3941	56.70	9.88	10.17	23.24	C26H56N605	532.77	532.4312	58.62	10.59	15.77	15.02
C26H54N409	566.74	566.3891	55.10	9.60	9.89	25.41	C26H56N606	548.77	548.4261	56.91	10.29	15.31	17.49
C26H54N601	466.76	466.4359	66.91	11.66	18.01	3.43	C26H56N607	564.77	564.4210	55.29	9.99	14.88	19.83
C26H54N602	482.76	482.4308	64.69	11.27	17.41	6.63	C26H56N608	580.77	580.4159	53.77	9.72	14.47	22.04
C26H54N603	498.76	498.4257	62.61	10.91	16.85	9.62	C26H56N609	596.77	596.4108	52.33	9.46	14.08	24.13
C26H54N604	514.76	514.4206	60.67	10.57	16.33	12.43	C26H57N301	427.76	427.4501	73.01	13.43	9.82	3.74
C26H54N605	530.76	530.4155	58.84	10.26	15.83	15.07	C26H57N302	443.76	443.4451	70.37	12.95	9.47	7.21
C26H54N606	546.76	546.4105	57.12	9.96	15.37	17.56	C26H57N303	459.76	459.4400	67.92	12.50	9.14	10.44
C26H54N607	562.76	562.4054	55.49	9.67	14.93	19.90	C26H57N304	475.76	475.4349	65.64	12.08	8.83	13.45
C26H54N608	578.76	578.4003	53.96	9.40	14.52	22.12	C26H57N305	491.76	491.4298	63.50	11.68	8.54	16.27
C26H54N609	594.76	594.3952	52.51	9.15	14.13	24.21	C26H57N306	507.76	507.4247	61.50	11.32	8.28	18.91
C26H55N101	397.73	397.4283	78.52	13.94	3.52	4.02	C26H57N307	523.76	523.4196	59.62	10.97	8.02	21.38
C26H55N102	413.73	413.4233	75.48	13.40	3.39	7.73	C26H57N308	539.76	539.4145	57.86	10.64	7.78	23.71
C26H55N103	429.73	429.4182	72.67	12.90	3.26	11.17	C26H57N309	555.76	555.4095	56.19	10.34	7.56	25.91
C26H55N104	445.73	445.4131	70.06	12.44	3.14	14.36	C26H57N501	455.78	455.4563	68.52	12.61	15.37	3.51
C26H55N105	461.73	461.4080	67.63	12.01	3.03	17.33	C26H57N502	471.78	471.4512	66.19	12.18	14.84	6.78
C26H55N106	477.73	477.4029	65.37	11.60	2.93	20.09	C26H57N503	487.78	487.4461	64.02	11.78	14.36	9.84
C26H55N107	493.73	493.3978	63.25	11.23	2.84	22.68	C26H57N504	503.78	503.4410	61.99	11.40	13.90	12.70
C26H55N108	509.73	509.3927	61.27	10.88	2.75	25.11	C26H57N505	519.77	519.4359	60.08	11.05	13.47	15.39
C26H55N109	525.73	525.3877	59.40	10.55	2.66	27.39	C26H57N506	535.77	535.4309	58.29	10.72	13.07	17.92
C26H55N301	425.75	425.4345	73.35	13.02	9.87	3.76	C26H57N507	551.77	551.4258	56.60	10.41	12.69	20.30
C26H55N302	441.75	441.4294	70.69	12.55	9.51	7.24	C26H57N508	567.77	567.4207	55.00	10.12	12.33	22.54
C26H55N303	457.75	457.4243	68.22	12.11	9.18	10.49	C26H57N509	583.77	583.4156	53.50	9.84	12.00	24.67
C26H55N304	473.75	473.4192	65.92	11.70	8.87	13.51	C26H58N401	442.78	442.4610	70.53	13.20	12.65	3.61
C26H55N305	489.75	489.4141	63.77	11.32	8.58	16.33	C26H58N402	458.78	458.4560	68.07	12.74	12.21	6.97
C26H55N306	505.74	505.4091	61.75	10.96	8.31	18.98	C26H58N403	474.78	474.4509	65.78	12.31	11.80	10.11
C26H55N307	521.74	521.4040	59.85	10.63	8.05	21.47	C26H58N404	490.78	490.4458	63.63	11.91	11.42	13.04
C26H55N308	537.74	537.3989	58.07	10.31	7.81	23.80	C26H58N405	506.78	506.4407	61.62	11.54	11.06	15.79
C26H55N309	553.74	553.3938	56.40	10.01	7.59	26.00	C26H58N406	522.78	522.4356	59.74	11.18	10.72	18.36
C26H55N501	453.76	453.4406	68.82	12.22	15.43	3.53	C26H58N407	538.77	538.4305	57.96	10.85	10.40	20.79
C26H55N502	469.76	469.4356	66.48	11.80	14.91	6.81	C26H58N408	554.77	554.4254	56.29	10.54	10.10	23.07
C26H55N503	485.76	485.4305	64.29	11.41	14.42	9.88	C26H58N409	570.77	570.4203	54.71	10.24	9.82	25.23
C26H55N504	501.76	501.4254	62.24	11.05	13.96	12.75	C26H58N601	470.79	470.4672	66.33	12.42	17.85	3.40
C26H55N505	517.76	517.4203	60.32	10.71	13.53	15.45	C26H58N602	486.79	486.4621	64.15	12.01	17.26	6.57
C26H55N506	533.76	533.4152	58.51	10.39	13.12	17.98	C26H58N603	502.79	502.4570	62.11	11.63	16.71	9.55
C26H55N507	549.76	549.4101	56.81	10.08	12.74	20.37	C26H58N604	518.79	518.4519	60.20	11.27	16.20	12.34
C26H55N508	565.76	565.4050	55.20	9.80	12.38	22.62	C26H58N605	534.79	534.4468	58.39	10.93	15.71	14.96
C26H55N509	581.76	581.3999	53.68	9.53	12.04	24.75	C26H58N606	550.79	550.4418	56.70	10.61	15.26	17.43
C26H56N201	412.75	412.4392	75.66	13.68	6.79	3.88	C26H58N607	566.79	566.4367	55.10	10.31	14.83	19.76
C26H56N202	428.75	428.4342	72.84	13.17	6.53	7.46	C26H58N608	582.79	582.4316	53.59	10.03	14.42	21.96
C26H56N203	444.75	444.4291	70.22	12.69	6.30	10.79	C26H58N609	598.79	598.4265	52.15	9.76	14.04	24.05
C26H56N204	460.75	460.4240	67.78	12.25	6.08	13.89	C26H59N501	457.79	457.4719	68.22	12.99	15.30	3.49
C26H56N205	476.75	476.4189	65.50	11.84	5.88	16.78	C26H59N502	473.79	473.4668	65.91	12.55	14.78	6.75
C26H56N206	492.75	492.4138	63.38	11.46	5.69	19.48	C26H59N503	489.79	489.4618	63.76	12.14	14.30	9.80
C26H56N207	508.75	508.4087	61.38	11.10	5.51	22.01	C26H59N504	505.79	505.4567	61.74	11.76	13.85	12.65
C26H56N208	524.74	524.4036	59.51	10.76	5.34	24.39	C26H59N505	521.79	521.4516	59.85	11.40	13.42	15.33
C26H56N209	540.74	540.3986	57.75	10.44	5.18	26.63	C26H59N506	537.79	537.4465	58.07	11.06	13.02	17.85
C26H56N401	440.76	440.4454	70.85	12.81	12.71	3.63	C26H59N507	553.79	553.4414	56.39	10.74	12.65	20.22
C26H56N402	456.76	456.4403	68.37	12.36	12.27	7.01	C26H59N508	569.79	569.4363	54.81	10.44	12.29	22.46
C26H56N403	472.76	472.4352	66.06	11.94	11.85	10.15	C26H59N509	585.79	585.4312	53.31	10.15	11.96	24.58
C26H56N404	488.76	488.4301	63.89	11.55	11.46	13.09	C26H60N601	472.81	472.4828	66.05	12.79	17.77	3.38
C26H56N405	504.76	504.4250	61.87	11.18	11.10	15.85	C26H60N602	488.81	488.4777	63.89	12.37	17.19	6.55
C26H56N406	520.76	520.4200	59.97	10.84	10.76	18.43	C26H60N603	504.81	504.4727	61.86	11.98	16.65	9.51
C26H56N407	536.76	536.4149	58.18	10.52	10.44	20.87	C26H60N604	520.81	520.4676	59.96	11.61	16.14	12.29

C26H60N605	536.81	536.4625	58.18	11.27	15.66	14.90	C27H15N102	385.43	385.1103	84.14	3.92	3.63	8.30
C26H60N606	552.80	552.4574	56.49	10.94	15.20	17.37	C27H15N103	401.43	401.1052	80.79	3.77	3.49	11.96
C26H60N607	568.80	568.4523	54.90	10.63	14.77	19.69	C27H15N104	417.42	417.1001	77.69	3.62	3.36	15.33
C26H60N608	584.80	584.4472	53.40	10.34	14.37	21.89	C27H15N105	433.42	433.0950	74.82	3.49	3.23	18.46
C26H60N609	600.80	600.4421	51.98	10.07	13.99	23.97	C27H15N106	449.42	449.0899	72.16	3.36	3.12	21.36
C27H13N101	367.41	367.0997	88.27	3.57	3.81	4.35	C27H15N107	465.42	465.0848	69.68	3.25	3.01	24.06
C27H13N102	383.41	383.0946	84.58	3.42	3.65	8.35	C27H15N108	481.42	481.0798	67.36	3.14	2.91	26.59
C27H13N103	399.41	399.0895	81.20	3.28	3.51	12.02	C27H15N109	497.42	497.0747	65.20	3.04	2.82	28.95
C27H13N104	415.41	415.0844	78.07	3.15	3.37	15.41	C27H15N301	397.44	397.1215	81.60	3.80	10.57	4.03
C27H13N105	431.41	431.0794	75.17	3.04	3.25	18.54	C27H15N302	413.44	413.1164	78.44	3.66	10.16	7.74
C27H13N106	447.41	447.0743	72.48	2.93	3.13	21.46	C27H15N303	429.44	429.1113	75.52	3.52	9.78	11.18
C27H13N107	463.41	463.0692	69.98	2.83	3.02	24.17	C27H15N304	445.44	445.1062	72.80	3.39	9.43	14.37
C27H13N108	479.41	479.0641	67.65	2.73	2.92	26.70	C27H15N305	461.44	461.1012	70.28	3.28	9.11	17.34
C27H13N109	495.41	495.0590	65.46	2.65	2.83	29.07	C27H15N306	477.44	477.0961	67.93	3.17	8.80	20.11
C27H13N301	395.42	395.1059	82.01	3.31	10.63	4.05	C27H15N307	493.44	493.0910	65.72	3.06	8.52	22.70
C27H13N302	411.42	411.1008	78.82	3.18	10.21	7.78	C27H15N308	509.44	509.0859	63.66	2.97	8.25	25.12
C27H13N303	427.42	427.0957	75.87	3.07	9.83	11.23	C27H15N309	525.44	525.0808	61.72	2.88	8.00	27.40
C27H13N304	443.42	443.0906	73.14	2.96	9.48	14.43	C27H15N501	425.45	425.1277	76.22	3.55	16.46	3.76
C27H13N305	459.42	459.0855	70.59	2.85	9.15	17.41	C27H15N502	441.45	441.1226	73.46	3.42	15.86	7.25
C27H13N306	475.42	475.0804	68.21	2.76	8.84	20.19	C27H15N503	457.45	457.1175	70.89	3.31	15.31	10.49
C27H13N307	491.42	491.0753	65.99	2.67	8.55	22.79	C27H15N504	473.45	473.1124	68.50	3.19	14.79	13.52
C27H13N308	507.42	507.0703	63.91	2.58	8.28	25.22	C27H15N505	489.45	489.1073	66.26	3.09	14.31	16.34
C27H13N309	523.42	523.0652	61.96	2.50	8.03	27.51	C27H15N506	505.45	505.1022	64.16	2.99	13.86	18.99
C27H13N501	423.44	423.1120	76.59	3.09	16.54	3.78	C27H15N507	521.45	521.0971	62.19	2.90	13.43	21.48
C27H13N502	439.44	439.1069	73.80	2.98	15.94	7.28	C27H15N508	537.45	537.0921	60.34	2.81	13.03	23.82
C27H13N503	455.44	455.1018	71.21	2.88	15.38	10.54	C27H15N509	553.45	553.0870	58.60	2.73	12.65	26.02
C27H13N504	471.44	471.0967	68.79	2.78	14.86	13.58	C27H16N201	384.44	384.1263	84.36	4.20	7.29	4.16
C27H13N505	487.44	487.0917	66.53	2.69	14.37	16.41	C27H16N202	400.44	400.1212	80.99	4.03	7.00	7.99
C27H13N506	503.43	503.0866	64.42	2.60	13.91	19.07	C27H16N203	416.44	416.1161	77.87	3.87	6.73	11.53
C27H13N507	519.43	519.0815	62.43	2.52	13.48	21.56	C27H16N204	432.44	432.1110	74.99	3.73	6.48	14.80
C27H13N508	535.43	535.0764	60.57	2.45	13.08	23.90	C27H16N205	448.44	448.1059	72.32	3.60	6.25	17.84
C27H13N509	551.43	551.0713	58.81	2.38	12.70	26.11	C27H16N206	464.44	464.1008	69.83	3.47	6.03	20.67
C27H14N201	382.43	382.1106	84.80	3.69	7.33	4.18	C27H16N207	480.44	480.0957	67.50	3.36	5.83	23.31
C27H14N202	398.42	398.1055	81.40	3.54	7.03	8.03	C27H16N208	496.44	496.0907	65.33	3.25	5.64	25.78
C27H14N203	414.42	414.1004	78.25	3.41	6.76	11.58	C27H16N209	512.44	512.0856	63.29	3.15	5.47	28.10
C27H14N204	430.42	430.0953	75.34	3.28	6.51	14.87	C27H16N401	412.45	412.1324	78.63	3.91	13.58	3.88
C27H14N205	446.42	446.0903	72.64	3.16	6.28	17.92	C27H16N402	428.45	428.1273	75.69	3.76	13.08	7.47
C27H14N206	462.42	462.0852	70.13	3.05	6.06	20.76	C27H16N403	444.45	444.1222	72.97	3.63	12.61	10.80
C27H14N207	478.42	478.0801	67.79	2.95	5.86	23.41	C27H16N404	460.45	460.1171	70.43	3.50	12.17	13.90
C27H14N208	494.42	494.0750	65.59	2.85	5.67	25.89	C27H16N405	476.45	476.1121	68.07	3.38	11.76	16.79
C27H14N209	510.42	510.0699	63.54	2.76	5.49	28.21	C27H16N406	492.45	492.1070	65.85	3.27	11.38	19.49
C27H14N401	410.44	410.1168	79.01	3.44	13.65	3.90	C27H16N407	508.45	508.1019	63.78	3.17	11.02	22.03
C27H14N402	426.44	426.1117	76.05	3.31	13.14	7.50	C27H16N408	524.45	524.0968	61.84	3.08	10.68	24.41
C27H14N403	442.44	442.1066	73.30	3.19	12.66	10.85	C27H16N409	540.45	540.0917	60.01	2.98	10.37	26.64
C27H14N404	458.44	458.1015	70.74	3.08	12.22	13.96	C27H16N601	440.47	440.1386	73.63	3.66	19.08	3.63
C27H14N405	474.44	474.0964	68.36	2.97	11.81	16.86	C27H16N602	456.47	456.1335	71.05	3.53	18.41	7.01
C27H14N406	490.44	490.0913	66.13	2.88	11.42	19.57	C27H16N603	472.47	472.1284	68.64	3.41	17.79	10.16
C27H14N407	506.44	506.0862	64.04	2.79	11.06	22.11	C27H16N604	488.47	488.1233	66.39	3.30	17.20	13.10
C27H14N408	522.43	522.0812	62.07	2.70	10.72	24.50	C27H16N605	504.47	504.1182	64.29	3.20	16.66	15.86
C27H14N409	538.43	538.0761	60.23	2.62	10.41	26.74	C27H16N606	520.47	520.1131	62.31	3.10	16.15	18.44
C27H14N601	438.45	438.1229	73.96	3.22	19.17	3.65	C27H16N607	536.46	536.1080	60.45	3.01	15.67	20.88
C27H14N602	454.45	454.1178	71.36	3.11	18.49	7.04	C27H16N608	552.46	552.1029	58.70	2.92	15.21	23.17
C27H14N603	470.45	470.1127	68.93	3.00	17.86	10.20	C27H16N609	568.46	568.0979	57.05	2.84	14.78	25.33
C27H14N604	486.45	486.1076	66.67	2.90	17.28	13.16	C27H17N101	371.44	371.1310	87.31	4.61	3.77	4.31
C27H14N605	502.45	502.1026	64.54	2.81	16.73	15.92	C27H17N102	387.44	387.1259	83.70	4.42	3.62	8.26
C27H14N606	518.45	518.0975	62.55	2.72	16.21	18.52	C27H17N103	403.44	403.1208	80.38	4.25	3.47	11.90
C27H14N607	534.45	534.0924	60.68	2.64	15.72	20.96	C27H17N104	419.44	419.1157	77.32	4.09	3.34	15.26
C27H14N608	550.45	550.0873	58.92	2.56	15.27	23.25	C27H17N105	435.44	435.1107	74.48	3.94	3.22	18.37
C27H14N609	566.45	566.0822	57.25	2.49	14.84	25.42	C27H17N106	451.44	451.1056	71.84	3.80	3.10	21.26
C27H15N101	369.43	369.1154	87.78	4.09	3.79	4.33	C27H17N107	467.44	467.1005	69.38	3.67	3.00	23.96

C27H17N108	483.44	483.0954	67.08	3.54	2.90	26.48
C27H17N109	499.44	499.0903	64.93	3.43	2.80	28.83
C27H17N301	399.46	399.1372	81.19	4.29	10.52	4.01
C27H17N302	415.46	415.1321	78.06	4.12	10.11	7.70
C27H17N303	431.45	431.1270	75.16	3.97	9.74	11.12
C27H17N304	447.45	447.1219	72.48	3.83	9.39	14.30
C27H17N305	463.45	463.1168	69.97	3.70	9.07	17.26
C27H17N306	479.45	479.1117	67.64	3.57	8.76	20.02
C27H17N307	495.45	495.1066	65.46	3.46	8.48	22.60
C27H17N308	511.45	511.1016	63.41	3.35	8.22	25.03
C27H17N309	527.45	527.0965	61.48	3.25	7.97	27.30
C27H17N501	427.47	427.1433	75.87	4.01	16.38	3.74
C27H17N502	443.47	443.1382	73.13	3.86	15.79	7.22
C27H17N503	459.47	459.1331	70.58	3.73	15.24	10.45
C27H17N504	475.47	475.1280	68.21	3.60	14.73	13.46
C27H17N505	491.47	491.1230	65.99	3.49	14.25	16.28
C27H17N506	507.47	507.1179	63.91	3.38	13.80	18.92
C27H17N507	523.47	523.1128	61.95	3.27	13.38	21.40
C27H17N508	539.47	539.1077	60.12	3.18	12.98	23.73
C27H17N509	555.46	555.1026	58.38	3.08	12.61	25.92
C27H18N201	386.46	386.1419	83.92	4.69	7.25	4.14
C27H18N202	402.46	402.1368	80.58	4.51	6.96	7.95
C27H18N203	418.46	418.1317	77.50	4.34	6.69	11.47
C27H18N204	434.46	434.1266	74.65	4.18	6.45	14.73
C27H18N205	450.45	450.1216	71.99	4.03	6.22	17.76
C27H18N206	466.45	466.1165	69.52	3.89	6.01	20.58
C27H18N207	482.45	482.1114	67.22	3.76	5.81	23.21
C27H18N208	498.45	498.1063	65.06	3.64	5.62	25.68
C27H18N209	514.45	514.1012	63.04	3.53	5.45	27.99
C27H18N401	414.47	414.1481	78.24	4.38	13.52	3.86
C27H18N402	430.47	430.1430	75.34	4.21	13.02	7.43
C27H18N403	446.47	446.1379	72.64	4.06	12.55	10.75
C27H18N404	462.47	462.1328	70.12	3.92	12.11	13.84
C27H18N405	478.47	478.1277	67.78	3.79	11.71	16.72
C27H18N406	494.47	494.1226	65.59	3.67	11.33	19.41
C27H18N407	510.47	510.1175	63.53	3.55	10.98	21.94
C27H18N408	526.47	526.1125	61.60	3.45	10.64	24.31
C27H18N409	542.47	542.1074	59.78	3.34	10.33	26.54
C27H18N601	442.48	442.1542	73.29	4.10	18.99	3.62
C27H18N602	458.48	458.1491	70.73	3.96	18.33	6.98
C27H18N603	474.48	474.1440	68.35	3.82	17.71	10.12
C27H18N604	490.48	490.1389	66.12	3.70	17.13	13.05
C27H18N605	506.48	506.1339	64.03	3.58	16.59	15.79
C27H18N606	522.48	522.1288	62.07	3.47	16.08	18.37
C27H18N607	538.48	538.1237	60.23	3.37	15.61	20.80
C27H18N608	554.48	554.1186	58.49	3.27	15.16	23.08
C27H18N609	570.48	570.1135	56.85	3.18	14.73	25.24
C27H19N101	373.46	373.1467	86.84	5.13	3.75	4.28
C27H19N102	389.46	389.1416	83.27	4.92	3.60	8.22
C27H19N103	405.46	405.1365	79.98	4.72	3.45	11.84
C27H19N104	421.46	421.1314	76.95	4.54	3.32	15.18
C27H19N105	437.46	437.1263	74.13	4.38	3.20	18.29
C27H19N106	453.46	453.1212	71.52	4.22	3.09	21.17
C27H19N107	469.45	469.1161	69.08	4.08	2.98	23.86
C27H19N108	485.45	485.1111	66.80	3.95	2.89	26.37
C27H19N109	501.45	501.1060	64.67	3.82	2.79	28.72
C27H19N301	401.47	401.1528	80.78	4.77	10.47	3.99
C27H19N302	417.47	417.1477	77.68	4.59	10.07	7.66
C27H19N303	433.47	433.1426	74.81	4.42	9.69	11.07
C27H19N304	449.47	449.1375	72.15	4.26	9.35	14.24
C27H19N305	465.47	465.1325	69.67	4.11	9.03	17.19
C27H19N306	481.47	481.1274	67.36	3.98	8.73	19.94
C27H19N307	497.47	497.1223	65.19	3.85	8.45	22.51
C27H19N308	513.47	513.1172	63.16	3.73	8.18	24.93
C27H19N309	529.47	529.1121	61.25	3.62	7.94	27.20
C27H19N501	429.49	429.1590	75.51	4.46	16.31	3.73
C27H19N502	445.48	445.1539	72.80	4.30	15.72	7.18
C27H19N503	461.48	461.1488	70.27	4.15	15.18	10.40
C27H19N504	477.48	477.1437	67.92	4.01	14.67	13.40
C27H19N505	493.48	493.1386	65.72	3.88	14.19	16.21
C27H19N506	509.48	509.1335	63.65	3.76	13.75	18.84
C27H19N507	525.48	525.1284	61.71	3.64	13.33	21.31
C27H19N508	541.48	541.1233	59.89	3.54	12.93	23.64
C27H19N509	557.48	557.1183	58.17	3.44	12.56	25.83
C27H20N201	388.47	388.1576	83.48	5.19	7.21	4.12
C27H20N202	404.47	404.1525	80.18	4.98	6.93	7.91
C27H20N203	420.47	420.1474	77.13	4.79	6.66	11.42
C27H20N204	436.47	436.1423	74.30	4.62	6.42	14.66
C27H20N205	452.47	452.1372	71.67	4.46	6.19	17.68
C27H20N206	468.47	468.1321	69.23	4.30	5.98	20.49
C27H20N207	484.47	484.1270	66.94	4.16	5.78	23.12
C27H20N208	500.47	500.1220	64.80	4.03	5.60	25.58
C27H20N209	516.47	516.1169	62.79	3.90	5.42	27.88
C27H20N401	416.49	416.1637	77.87	4.84	13.45	3.84
C27H20N402	432.49	432.1586	74.99	4.66	12.95	7.40
C27H20N403	448.49	448.1535	72.31	4.49	12.49	10.70
C27H20N404	464.48	464.1484	69.82	4.34	12.06	13.78
C27H20N405	480.48	480.1434	67.49	4.20	11.66	16.65
C27H20N406	496.48	496.1383	65.32	4.06	11.28	19.34
C27H20N407	512.48	512.1332	63.28	3.93	10.93	21.85
C27H20N408	528.48	528.1281	61.36	3.81	10.60	24.22
C27H20N409	544.48	544.1230	59.56	3.70	10.29	26.45
C27H20N601	444.50	444.1698	72.96	4.54	18.91	3.60
C27H20N602	460.50	460.1648	70.42	4.38	18.25	6.95
C27H20N603	476.50	476.1597	68.06	4.23	17.64	10.07
C27H20N604	492.50	492.1546	65.85	4.09	17.06	12.99
C27H20N605	508.50	508.1495	63.78	3.96	16.53	15.73
C27H20N606	524.50	524.1444	61.83	3.84	16.02	18.30
C27H20N607	540.50	540.1393	60.00	3.73	15.55	20.72
C27H20N608	556.50	556.1342	58.28	3.62	15.10	23.00
C27H20N609	572.50	572.1292	56.65	3.52	14.68	25.15
C27H21N101	375.47	375.1623	86.37	5.64	3.73	4.26
C27H21N102	391.47	391.1572	82.84	5.41	3.58	8.17
C27H21N103	407.47	407.1521	79.59	5.19	3.44	11.78
C27H21N104	423.47	423.1470	76.58	5.00	3.31	15.11
C27H21N105	439.47	439.1420	73.79	4.82	3.19	18.20
C27H21N106	455.47	455.1369	71.20	4.65	3.08	21.08
C27H21N107	471.47	471.1318	68.78	4.49	2.97	23.75
C27H21N108	487.47	487.1267	66.53	4.34	2.87	26.26
C27H21N109	503.47	503.1216	64.41	4.20	2.78	28.60
C27H21N301	403.49	403.1685	80.37	5.25	10.41	3.97
C27H21N302	419.49	419.1634	77.31	5.05	10.02	7.63
C27H21N303	435.49	435.1583	74.47	4.86	9.65	11.02
C27H21N304	451.49	451.1532	71.83	4.69	9.31	14.17
C27H21N305	467.49	467.1481	69.37	4.53	8.99	17.11
C27H21N306	483.48	483.1430	67.08	4.38	8.69	19.86
C27H21N307	499.48	499.1379	64.93	4.24	8.41	22.42
C27H21N308	515.48	515.1329	62.91	4.11	8.15	24.83
C27H21N309	531.48	531.1278	61.02	3.98	7.91	27.09
C27H21N501	431.50	431.1746	75.16	4.91	16.23	3.71

C27H21N502	447.50	447.1695	72.47	4.73	15.65	7.15
C27H21N503	463.50	463.1644	69.97	4.57	15.11	10.36
C27H21N504	479.50	479.1593	67.63	4.41	14.61	13.35
C27H21N505	495.50	495.1543	65.45	4.27	14.13	16.14
C27H21N506	511.50	511.1492	63.40	4.14	13.69	18.77
C27H21N507	527.50	527.1441	61.48	4.01	13.28	21.23
C27H21N508	543.50	543.1390	59.67	3.89	12.89	23.55
C27H21N509	559.50	559.1339	57.96	3.78	12.52	25.74
C27H22N201	390.49	390.1732	83.05	5.68	7.17	4.10
C27H22N202	406.49	406.1681	79.78	5.46	6.89	7.87
C27H22N203	422.49	422.1630	76.76	5.25	6.63	11.36
C27H22N204	438.49	438.1579	73.96	5.06	6.39	14.60
C27H22N205	454.49	454.1529	71.36	4.88	6.16	17.60
C27H22N206	470.49	470.1478	68.93	4.71	5.95	20.40
C27H22N207	486.49	486.1427	66.66	4.56	5.76	23.02
C27H22N208	502.48	502.1376	64.54	4.41	5.57	25.47
C27H22N209	518.48	518.1325	62.55	4.28	5.40	27.77
C27H22N401	418.50	418.1794	77.49	5.30	13.39	3.82
C27H22N402	434.50	434.1743	74.64	5.10	12.89	7.36
C27H22N403	450.50	450.1692	71.99	4.92	12.44	10.65
C27H22N404	466.50	466.1641	69.52	4.75	12.01	13.72
C27H22N405	482.50	482.1590	67.21	4.60	11.61	16.58
C27H22N406	498.50	498.1539	65.06	4.45	11.24	19.26
C27H22N407	514.50	514.1488	63.03	4.31	10.89	21.77
C27H22N408	530.50	530.1437	61.13	4.18	10.56	24.13
C27H22N409	546.50	546.1387	59.34	4.06	10.25	26.35
C27H22N601	446.52	446.1855	72.63	4.97	18.82	3.58
C27H22N602	462.52	462.1804	70.12	4.79	18.17	6.92
C27H22N603	478.51	478.1753	67.77	4.63	17.56	10.03
C27H22N604	494.51	494.1702	65.58	4.48	16.99	12.94
C27H22N605	510.51	510.1652	63.52	4.34	16.46	15.67
C27H22N606	526.51	526.1601	61.59	4.21	15.96	18.23
C27H22N607	542.51	542.1550	59.78	4.09	15.49	20.64
C27H22N608	558.51	558.1499	58.07	3.97	15.05	22.92
C27H22N609	574.51	574.1448	56.45	3.86	14.63	25.06
C27H23N101	377.49	377.1780	85.91	6.14	3.71	4.24
C27H23N102	393.49	393.1729	82.42	5.89	3.56	8.13
C27H23N103	409.49	409.1678	79.20	5.66	3.42	11.72
C27H23N104	425.49	425.1627	76.22	5.45	3.29	15.04
C27H23N105	441.49	441.1576	73.46	5.25	3.17	18.12
C27H23N106	457.49	457.1525	70.89	5.07	3.06	20.98
C27H23N107	473.49	473.1474	68.49	4.90	2.96	23.65
C27H23N108	489.49	489.1424	66.25	4.74	2.86	26.15
C27H23N109	505.49	505.1373	64.16	4.59	2.77	28.49
C27H23N301	405.50	405.1841	79.97	5.72	10.36	3.95
C27H23N302	421.50	421.1790	76.94	5.50	9.97	7.59
C27H23N303	437.50	437.1739	74.13	5.30	9.60	10.97
C27H23N304	453.50	453.1688	71.51	5.11	9.27	14.11
C27H23N305	469.50	469.1638	69.07	4.94	8.95	17.04
C27H23N306	485.50	485.1587	66.80	4.78	8.66	19.77
C27H23N307	501.50	501.1536	64.67	4.62	8.38	22.33
C27H23N308	517.50	517.1485	62.67	4.48	8.12	24.73
C27H23N309	533.50	533.1434	60.79	4.35	7.88	26.99
C27H23N501	433.52	433.1902	74.81	5.35	16.15	3.69
C27H23N502	449.52	449.1852	72.14	5.16	15.58	7.12
C27H23N503	465.52	465.1801	69.66	4.98	15.04	10.31
C27H23N504	481.52	481.1750	67.35	4.81	14.54	13.29
C27H23N505	497.51	497.1699	65.18	4.66	14.08	16.08
C27H23N506	513.51	513.1648	63.15	4.51	13.64	18.69
C27H23N507	529.51	529.1597	61.25	4.38	13.23	21.15
C27H23N508	545.51	545.1546	59.45	4.25	12.84	23.46
C27H23N509	561.51	561.1496	57.75	4.13	12.47	25.64
C27H24N201	392.51	392.1889	82.62	6.16	7.14	4.08
C27H24N202	408.50	408.1838	79.39	5.92	6.86	7.83
C27H24N203	424.50	424.1787	76.40	5.70	6.60	11.31
C27H24N204	440.50	440.1736	73.62	5.49	6.36	14.53
C27H24N205	456.50	456.1685	71.04	5.30	6.14	17.52
C27H24N206	472.50	472.1634	68.63	5.12	5.93	20.32
C27H24N207	488.50	488.1583	66.39	4.95	5.73	22.93
C27H24N208	504.50	504.1533	64.28	4.80	5.55	25.37
C27H24N209	520.50	520.1482	62.31	4.65	5.38	27.66
C27H24N401	420.52	420.1950	77.12	5.75	13.32	3.80
C27H24N402	436.52	436.1899	74.29	5.54	12.83	7.33
C27H24N403	452.52	452.1848	71.67	5.35	12.38	10.61
C27H24N404	468.52	468.1797	69.22	5.16	11.96	13.66
C27H24N405	484.52	484.1747	66.93	4.99	11.56	16.51
C27H24N406	500.52	500.1696	64.79	4.83	11.19	19.18
C27H24N407	516.51	516.1645	62.79	4.68	10.85	21.68
C27H24N408	532.51	532.1594	60.90	4.54	10.52	24.04
C27H24N409	548.51	548.1543	59.12	4.41	10.21	26.25
C27H24N601	448.53	448.2011	72.30	5.39	18.74	3.57
C27H24N602	464.53	464.1961	69.81	5.21	18.09	6.89
C27H24N603	480.53	480.1910	67.49	5.03	17.49	9.99
C27H24N604	496.53	496.1859	65.31	4.87	16.93	12.89
C27H24N605	512.53	512.1808	63.27	4.72	16.40	15.61
C27H24N606	528.53	528.1757	61.36	4.58	15.90	18.16
C27H24N607	544.53	544.1706	59.56	4.44	15.43	20.57
C27H24N608	560.53	560.1655	57.86	4.32	14.99	22.83
C27H24N609	576.53	576.1605	56.25	4.20	14.58	24.98
C27H25N101	379.51	379.1936	85.45	6.64	3.69	4.22
C27H25N102	395.51	395.1885	82.00	6.37	3.54	8.09
C27H25N103	411.51	411.1834	78.81	6.12	3.40	11.66
C27H25N104	427.50	427.1783	75.86	5.89	3.28	14.97
C27H25N105	443.50	443.1733	73.12	5.68	3.16	18.04
C27H25N106	459.50	459.1682	70.58	5.48	3.05	20.89
C27H25N107	475.50	475.1631	68.20	5.30	2.95	23.55
C27H25N108	491.50	491.1580	65.98	5.13	2.85	26.04
C27H25N109	507.50	507.1529	63.90	4.97	2.76	28.37
C27H25N301	407.52	407.1998	79.58	6.18	10.31	3.93
C27H25N302	423.52	423.1947	76.57	5.95	9.92	7.56
C27H25N303	439.52	439.1896	73.79	5.73	9.56	10.92
C27H25N304	455.52	455.1845	71.19	5.53	9.22	14.05
C27H25N305	471.52	471.1794	68.78	5.34	8.91	16.97
C27H25N306	487.52	487.1743	66.52	5.17	8.62	19.69
C27H25N307	503.52	503.1692	64.41	5.00	8.35	22.24
C27H25N308	519.52	519.1641	62.42	4.85	8.09	24.64
C27H25N309	535.52	535.1591	60.56	4.71	7.85	26.89
C27H25N501	435.53	435.2059	74.46	5.79	16.08	3.67
C27H25N502	451.53	451.2008	71.82	5.58	15.51	7.09
C27H25N503	467.53	467.1957	69.36	5.39	14.98	10.27
C27H25N504	483.53	483.1906	67.07	5.21	14.48	13.24
C27H25N505	499.53	499.1856	64.92	5.04	14.02	16.01
C27H25N506	515.53	515.1805	62.91	4.89	13.58	18.62
C27H25N507	531.53	531.1754	61.01	4.74	13.18	21.07
C27H25N508	547.53	547.1703	59.23	4.60	12.79	23.38
C27H25N509	563.53	563.1652	57.55	4.47	12.43	25.55
C27H26N201	394.52	394.2045	82.20	6.64	7.10	4.06
C27H26N202	410.52	410.1994	79.00	6.38	6.82	7.79
C27H26N203	426.52	426.1943	76.03	6.14	6.57	11.25
C27H26N204	442.52	442.1892	73.29	5.92	6.33	14.46

C27H26N2O5	458.52	458.1842	70.73	5.72	6.11	17.45
C27H26N2O6	474.52	474.1791	68.34	5.52	5.90	20.23
C27H26N2O7	490.52	490.1740	66.11	5.34	5.71	22.83
C27H26N2O8	506.52	506.1689	64.03	5.17	5.53	25.27
C27H26N2O9	522.52	522.1638	62.07	5.02	5.36	27.56
C27H26N4O1	422.53	422.2106	76.75	6.20	13.26	3.79
C27H26N4O2	438.53	438.2056	73.95	5.98	12.78	7.30
C27H26N4O3	454.53	454.2005	71.35	5.77	12.33	10.56
C27H26N4O4	470.53	470.1954	68.92	5.57	11.91	13.60
C27H26N4O5	486.53	486.1903	66.66	5.39	11.52	16.44
C27H26N4O6	502.53	502.1852	64.53	5.22	11.15	19.10
C27H26N4O7	518.53	518.1801	62.54	5.05	10.80	21.60
C27H26N4O8	534.53	534.1750	60.67	4.90	10.48	23.95
C27H26N4O9	550.53	550.1700	58.91	4.76	10.18	26.16
C27H26N6O1	450.55	450.2168	71.98	5.82	18.65	3.55
C27H26N6O2	466.55	466.2117	69.51	5.62	18.01	6.86
C27H26N6O3	482.55	482.2066	67.21	5.43	17.42	9.95
C27H26N6O4	498.55	498.2015	65.05	5.26	16.86	12.84
C27H26N6O5	514.55	514.1965	63.03	5.09	16.33	15.55
C27H26N6O6	530.54	530.1914	61.13	4.94	15.84	18.09
C27H26N6O7	546.54	546.1863	59.34	4.80	15.38	20.49
C27H26N6O8	562.54	562.1812	57.65	4.66	14.94	22.75
C27H26N6O9	578.54	578.1761	56.05	4.53	14.53	24.89
C27H27N1O1	381.52	381.2093	85.00	7.13	3.67	4.19
C27H27N1O2	397.52	397.2042	81.58	6.85	3.52	8.05
C27H27N1O3	413.52	413.1991	78.42	6.58	3.39	11.61
C27H27N1O4	429.52	429.1940	75.50	6.34	3.26	14.90
C27H27N1O5	445.52	445.1889	72.79	6.11	3.14	17.96
C27H27N1O6	461.52	461.1838	70.27	5.90	3.03	20.80
C27H27N1O7	477.52	477.1787	67.91	5.70	2.93	23.45
C27H27N1O8	493.52	493.1737	65.71	5.51	2.84	25.94
C27H27N1O9	509.52	509.1686	63.65	5.34	2.75	28.26
C27H27N3O1	409.54	409.2154	79.19	6.65	10.26	3.91
C27H27N3O2	425.54	425.2103	76.21	6.40	9.87	7.52
C27H27N3O3	441.53	441.2052	73.45	6.16	9.52	10.87
C27H27N3O4	457.53	457.2001	70.88	5.95	9.18	13.99
C27H27N3O5	473.53	473.1951	68.49	5.75	8.87	16.89
C27H27N3O6	489.53	489.1900	66.25	5.56	8.58	19.61
C27H27N3O7	505.53	505.1849	64.15	5.38	8.31	22.15
C27H27N3O8	521.53	521.1798	62.18	5.22	8.06	24.54
C27H27N3O9	537.53	537.1747	60.33	5.06	7.82	26.79
C27H27N5O1	437.55	437.2215	74.12	6.22	16.01	3.66
C27H27N5O2	453.55	453.2165	71.50	6.00	15.44	7.06
C27H27N5O3	469.55	469.2114	69.07	5.80	14.92	10.22
C27H27N5O4	485.55	485.2063	66.79	5.61	14.42	13.18
C27H27N5O5	501.55	501.2012	64.66	5.43	13.96	15.95
C27H27N5O6	517.55	517.1961	62.66	5.26	13.53	18.55
C27H27N5O7	533.55	533.1910	60.78	5.10	13.13	20.99
C27H27N5O8	549.54	549.1859	59.01	4.95	12.74	23.29
C27H27N5O9	565.54	565.1809	57.34	4.81	12.38	25.46
C27H28N2O1	396.54	396.2202	81.78	7.12	7.06	4.03
C27H28N2O2	412.54	412.2151	78.61	6.84	6.79	7.76
C27H28N2O3	428.54	428.2100	75.68	6.59	6.54	11.20
C27H28N2O4	444.54	444.2049	72.95	6.35	6.30	14.40
C27H28N2O5	460.53	460.1998	70.42	6.13	6.08	17.37
C27H28N2O6	476.53	476.1947	68.05	5.92	5.88	20.14
C27H28N2O7	492.53	492.1896	65.84	5.73	5.69	22.74
C27H28N2O8	508.53	508.1845	63.77	5.55	5.51	25.17
C27H28N2O9	524.53	524.1795	61.83	5.38	5.34	27.45
C27H28N4O1	424.55	424.2263	76.39	6.65	13.20	3.77
C27H28N4O2	440.55	440.2212	73.61	6.41	12.72	7.26
C27H28N4O3	456.55	456.2161	71.03	6.18	12.27	10.51
C27H28N4O4	472.55	472.2110	68.63	5.97	11.86	13.54
C27H28N4O5	488.55	488.2060	66.38	5.78	11.47	16.37
C27H28N4O6	504.55	504.2009	64.28	5.59	11.10	19.03
C27H28N4O7	520.55	520.1958	62.30	5.42	10.76	21.52
C27H28N4O8	536.55	536.1907	60.44	5.26	10.44	23.86
C27H28N4O9	552.55	552.1856	58.69	5.11	10.14	26.06
C27H28N6O1	452.56	452.2324	71.66	6.24	18.57	3.54
C27H28N6O2	468.56	468.2274	69.21	6.02	17.94	6.83
C27H28N6O3	484.56	484.2223	66.93	5.82	17.34	9.91
C27H28N6O4	500.56	500.2172	64.79	5.64	16.79	12.79
C27H28N6O5	516.56	516.2121	62.78	5.46	16.27	15.49
C27H28N6O6	532.56	532.2070	60.89	5.30	15.78	18.03
C27H28N6O7	548.56	548.2019	59.12	5.14	15.32	20.42
C27H28N6O8	564.56	564.1968	57.44	5.00	14.89	22.67
C27H28N6O9	580.56	580.1918	55.86	4.86	14.48	24.80
C27H29N1O1	383.54	383.2249	84.56	7.62	3.65	4.17
C27H29N1O2	399.54	399.2198	81.17	7.32	3.51	8.01
C27H29N1O3	415.54	415.2147	78.04	7.03	3.37	11.55
C27H29N1O4	431.54	431.2096	75.15	6.77	3.25	14.83
C27H29N1O5	447.54	447.2046	72.46	6.53	3.13	17.87
C27H29N1O6	463.54	463.1995	69.96	6.31	3.02	20.71
C27H29N1O7	479.53	479.1944	67.63	6.10	2.92	23.36
C27H29N1O8	495.53	495.1893	65.44	5.90	2.83	25.83
C27H29N1O9	511.53	511.1842	63.40	5.71	2.74	28.15
C27H29N3O1	411.55	411.2310	78.80	7.10	10.21	3.89
C27H29N3O2	427.55	427.2260	75.85	6.84	9.83	7.48
C27H29N3O3	443.55	443.2209	73.11	6.59	9.47	10.82
C27H29N3O4	459.55	459.2158	70.57	6.36	9.14	13.93
C27H29N3O5	475.55	475.2107	68.20	6.15	8.84	16.82
C27H29N3O6	491.55	491.2056	65.98	5.95	8.55	19.53
C27H29N3O7	507.55	507.2005	63.90	5.76	8.28	22.07
C27H29N3O8	523.55	523.1954	61.94	5.58	8.03	24.45
C27H29N3O9	539.55	539.1904	60.11	5.42	7.79	26.69
C27H29N5O1	439.57	439.2372	73.78	6.65	15.93	3.64
C27H29N5O2	455.56	455.2321	71.19	6.42	15.37	7.02
C27H29N5O3	471.56	471.2270	68.77	6.20	14.85	10.18
C27H29N5O4	487.56	487.2219	66.51	6.00	14.36	13.13
C27H29N5O5	503.56	503.2169	64.40	5.80	13.91	15.89
C27H29N5O6	519.56	519.2118	62.42	5.63	13.48	18.48
C27H29N5O7	535.56	535.2067	60.55	5.46	13.08	20.91
C27H29N5O8	551.56	551.2016	58.80	5.30	12.70	23.21
C27H29N5O9	567.56	567.1965	57.14	5.15	12.34	25.37
C27H30N2O1	398.55	398.2358	81.37	7.59	7.03	4.01
C27H30N2O2	414.55	414.2307	78.23	7.29	6.76	7.72
C27H30N2O3	430.55	430.2256	75.32	7.02	6.51	11.15
C27H30N2O4	446.55	446.2205	72.62	6.77	6.27	14.33
C27H30N2O5	462.55	462.2155	70.11	6.54	6.06	17.29
C27H30N2O6	478.55	478.2104	67.77	6.32	5.85	20.06
C27H30N2O7	494.55	494.2053	65.58	6.11	5.66	22.65
C27H30N2O8	510.55	510.2002	63.52	5.92	5.49	25.07
C27H30N2O9	526.55	526.1951	61.59	5.74	5.32	27.35
C27H30N4O1	426.57	426.2419	76.03	7.09	13.13	3.75
C27H30N4O2	442.57	442.2369	73.28	6.83	12.66	7.23
C27H30N4O3	458.57	458.2318	70.72	6.59	12.22	10.47
C27H30N4O4	474.56	474.2267	68.34	6.37	11.81	13.49
C27H30N4O5	490.56	490.2216	66.11	6.16	11.42	16.31
C27H30N4O6	506.56	506.2165	64.02	5.97	11.06	18.95
C27H30N4O7	522.56	522.2114	62.06	5.79	10.72	21.43

C27H30N4O8	538.56	538.2063	60.22	5.61	10.40	23.77
C27H30N4O9	554.56	554.2013	58.48	5.45	10.10	25.97
C27H30N6O1	454.58	454.2481	71.34	6.65	18.49	3.52
C27H30N6O2	470.58	470.2430	68.92	6.43	17.86	6.80
C27H30N6O3	486.58	486.2379	66.65	6.21	17.27	9.86
C27H30N6O4	502.58	502.2328	64.53	6.02	16.72	12.73
C27H30N6O5	518.58	518.2278	62.54	5.83	16.21	15.43
C27H30N6O6	534.58	534.2227	60.67	5.66	15.72	17.96
C27H30N6O7	550.58	550.2176	58.90	5.49	15.26	20.34
C27H30N6O8	566.58	566.2125	57.24	5.34	14.83	22.59
C27H30N6O9	582.57	582.2074	55.67	5.19	14.43	24.72
C27H31N1O1	385.55	385.2406	84.11	8.10	3.63	4.15
C27H31N1O2	401.55	401.2355	80.76	7.78	3.49	7.97
C27H31N1O3	417.55	417.2304	77.67	7.48	3.35	11.50
C27H31N1O4	433.55	433.2253	74.80	7.21	3.23	14.76
C27H31N1O5	449.55	449.2202	72.14	6.95	3.12	17.79
C27H31N1O6	465.55	465.2151	69.66	6.71	3.01	20.62
C27H31N1O7	481.55	481.2100	67.35	6.49	2.91	23.26
C27H31N1O8	497.55	497.2049	65.18	6.28	2.82	25.73
C27H31N1O9	513.55	513.1999	63.15	6.08	2.73	28.04
C27H31N3O1	413.57	413.2467	78.42	7.56	10.16	3.87
C27H31N3O2	429.57	429.2416	75.49	7.27	9.78	7.45
C27H31N3O3	445.57	445.2365	72.78	7.01	9.43	10.77
C27H31N3O4	461.57	461.2314	70.26	6.77	9.10	13.87
C27H31N3O5	477.57	477.2264	67.91	6.54	8.80	16.75
C27H31N3O6	493.56	493.2213	65.71	6.33	8.51	19.45
C27H31N3O7	509.56	509.2162	63.64	6.13	8.25	21.98
C27H31N3O8	525.56	525.2111	61.71	5.95	8.00	24.35
C27H31N3O9	541.56	541.2060	59.88	5.77	7.76	26.59
C27H31N5O1	441.58	441.2528	73.44	7.08	15.86	3.62
C27H31N5O2	457.58	457.2478	70.87	6.83	15.31	6.99
C27H31N5O3	473.58	473.2427	68.48	6.60	14.79	10.14
C27H31N5O4	489.58	489.2376	66.24	6.38	14.30	13.07
C27H31N5O5	505.58	505.2325	64.14	6.18	13.85	15.82
C27H31N5O6	521.58	521.2274	62.18	5.99	13.43	18.40
C27H31N5O7	537.58	537.2223	60.33	5.81	13.03	20.83
C27H31N5O8	553.58	553.2172	58.58	5.64	12.65	23.12
C27H31N5O9	569.58	569.2122	56.94	5.49	12.30	25.28
C27H32N2O1	400.57	400.2514	80.96	8.05	6.99	3.99
C27H32N2O2	416.57	416.2464	77.85	7.74	6.72	7.68
C27H32N2O3	432.57	432.2413	74.97	7.46	6.48	11.10
C27H32N2O4	448.57	448.2362	72.30	7.19	6.25	14.27
C27H32N2O5	464.57	464.2311	69.81	6.94	6.03	17.22
C27H32N2O6	480.57	480.2260	67.48	6.71	5.83	19.98
C27H32N2O7	496.57	496.2209	65.31	6.50	5.64	22.55
C27H32N2O8	512.56	512.2158	63.27	6.29	5.47	24.97
C27H32N2O9	528.56	528.2108	61.36	6.10	5.30	27.24
C27H32N4O1	428.58	428.2576	75.67	7.53	13.07	3.73
C27H32N4O2	444.58	444.2525	72.95	7.26	12.60	7.20
C27H32N4O3	460.58	460.2474	70.41	7.00	12.16	10.42
C27H32N4O4	476.58	476.2423	68.05	6.77	11.76	13.43
C27H32N4O5	492.58	492.2373	65.84	6.55	11.37	16.24
C27H32N4O6	508.58	508.2322	63.77	6.34	11.02	18.88
C27H32N4O7	524.58	524.2271	61.82	6.15	10.68	21.35
C27H32N4O8	540.58	540.2220	59.99	5.97	10.36	23.68
C27H32N4O9	556.58	556.2169	58.27	5.80	10.07	25.87
C27H32N6O1	456.60	456.2637	71.03	7.06	18.41	3.50
C27H32N6O2	472.60	472.2587	68.62	6.83	17.78	6.77
C27H32N6O3	488.59	488.2536	66.37	6.60	17.20	9.82
C27H32N6O4	504.59	504.2485	64.27	6.39	16.66	12.68
C27H32N6O5	520.59	520.2434	62.29	6.20	16.14	15.37
C27H32N6O6	536.59	536.2383	60.44	6.01	15.66	17.89
C27H32N6O7	552.59	552.2332	58.69	5.84	15.21	20.27
C27H32N6O8	568.59	568.2281	57.04	5.67	14.78	22.51
C27H32N6O9	584.59	584.2231	55.47	5.52	14.38	24.63
C27H33N1O1	387.57	387.2562	83.68	8.58	3.61	4.13
C27H33N1O2	403.57	403.2511	80.36	8.24	3.47	7.93
C27H33N1O3	419.57	419.2460	77.29	7.93	3.34	11.44
C27H33N1O4	435.57	435.2409	74.45	7.64	3.22	14.69
C27H33N1O5	451.57	451.2359	71.82	7.37	3.10	17.72
C27H33N1O6	467.57	467.2308	69.36	7.11	3.00	20.53
C27H33N1O7	483.57	483.2257	67.06	6.88	2.90	23.16
C27H33N1O8	499.57	499.2206	64.92	6.66	2.80	25.62
C27H33N1O9	515.57	515.2155	62.90	6.45	2.72	27.93
C27H33N3O1	415.58	415.2623	78.04	8.00	10.11	3.85
C27H33N3O2	431.58	431.2573	75.14	7.71	9.74	7.41
C27H33N3O3	447.58	447.2522	72.46	7.43	9.39	10.72
C27H33N3O4	463.58	463.2471	69.96	7.18	9.06	13.81
C27H33N3O5	479.58	479.2420	67.62	6.94	8.76	16.68
C27H33N3O6	495.58	495.2369	65.44	6.71	8.48	19.37
C27H33N3O7	511.58	511.2318	63.39	6.50	8.21	21.89
C27H33N3O8	527.58	527.2267	61.47	6.30	7.96	24.26
C27H33N3O9	543.58	543.2217	59.66	6.12	7.73	26.49
C27H33N5O1	443.60	443.2685	73.11	7.50	15.79	3.61
C27H33N5O2	459.60	459.2634	70.56	7.24	15.24	6.96
C27H33N5O3	475.60	475.2583	68.19	6.99	14.73	10.09
C27H33N5O4	491.60	491.2532	65.97	6.77	14.25	13.02
C27H33N5O5	507.59	507.2482	63.89	6.55	13.80	15.76
C27H33N5O6	523.59	523.2431	61.94	6.35	13.38	18.33
C27H33N5O7	539.59	539.2380	60.10	6.16	12.98	20.76
C27H33N5O8	555.59	555.2329	58.37	5.99	12.61	23.04
C27H33N5O9	571.59	571.2278	56.74	5.82	12.25	25.19
C27H34N2O1	402.58	402.2671	80.55	8.51	6.96	3.97
C27H34N2O2	418.58	418.2620	77.48	8.19	6.69	7.64
C27H34N2O3	434.58	434.2569	74.62	7.89	6.45	11.04
C27H34N2O4	450.58	450.2518	71.97	7.61	6.22	14.20
C27H34N2O5	466.58	466.2468	69.51	7.35	6.00	17.15
C27H34N2O6	482.58	482.2417	67.20	7.10	5.80	19.89
C27H34N2O7	498.58	498.2366	65.04	6.87	5.62	22.46
C27H34N2O8	514.58	514.2315	63.02	6.66	5.44	24.87
C27H34N2O9	530.58	530.2264	61.12	6.46	5.28	27.14
C27H34N4O1	430.60	430.2732	75.31	7.96	13.01	3.72
C27H34N4O2	446.60	446.2682	72.62	7.67	12.55	7.17
C27H34N4O3	462.60	462.2631	70.10	7.41	12.11	10.38
C27H34N4O4	478.60	478.2580	67.76	7.16	11.71	13.37
C27H34N4O5	494.60	494.2529	65.57	6.93	11.33	16.17
C27H34N4O6	510.60	510.2478	63.51	6.71	10.97	18.80
C27H34N4O7	526.59	526.2427	61.58	6.51	10.64	21.27
C27H34N4O8	542.59	542.2376	59.77	6.32	10.33	23.59
C27H34N4O9	558.59	558.2326	58.06	6.14	10.03	25.78
C27H34N6O1	458.61	458.2794	70.71	7.47	18.32	3.49
C27H34N6O2	474.61	474.2743	68.33	7.22	17.71	6.74
C27H34N6O3	490.61	490.2692	66.10	6.99	17.13	9.78
C27H34N6O4	506.61	506.2641	64.01	6.76	16.59	12.63
C27H34N6O5	522.61	522.2590	62.05	6.56	16.08	15.31
C27H34N6O6	538.61	538.2540	60.21	6.36	15.60	17.82
C27H34N6O7	554.61	554.2489	58.47	6.18	15.15	20.19
C27H34N6O8	570.61	570.2438	56.83	6.01	14.73	22.43
C27H34N6O9	586.61	586.2387	55.28	5.84	14.33	24.55
C27H35N1O1	389.59	389.2718	83.24	9.06	3.60	4.11

182

C27H35N102	405.59	405.2668	79.96	8.70	3.45	7.89
C27H35N103	421.58	421.2617	76.92	8.37	3.32	11.39
C27H35N104	437.58	437.2566	74.11	8.06	3.20	14.63
C27H35N105	453.58	453.2515	71.50	7.78	3.09	17.64
C27H35N106	469.58	469.2464	69.06	7.51	2.98	20.44
C27H35N107	485.58	485.2413	66.79	7.27	2.88	23.06
C27H35N108	501.58	501.2362	64.66	7.03	2.79	25.52
C27H35N109	517.58	517.2312	62.66	6.82	2.71	27.82
C27H35N301	417.60	417.2780	77.66	8.45	10.06	3.83
C27H35N302	433.60	433.2729	74.79	8.14	9.69	7.38
C27H35N303	449.60	449.2678	72.13	7.85	9.35	10.68
C27H35N304	465.60	465.2627	69.65	7.58	9.02	13.75
C27H35N305	481.60	481.2577	67.34	7.33	8.73	16.61
C27H35N306	497.60	497.2526	65.17	7.09	8.44	19.29
C27H35N307	513.60	513.2475	63.14	6.87	8.18	21.81
C27H35N308	529.60	529.2424	61.24	6.66	7.93	24.17
C27H35N309	545.59	545.2373	59.44	6.47	7.70	26.39
C27H35N501	445.61	445.2841	72.78	7.92	15.72	3.59
C27H35N502	461.61	461.2791	70.25	7.64	15.17	6.93
C27H35N503	477.61	477.2740	67.90	7.39	14.66	10.05
C27H35N504	493.61	493.2689	65.70	7.15	14.19	12.97
C27H35N505	509.61	509.2638	63.64	6.92	13.74	15.70
C27H35N506	525.61	525.2587	61.70	6.71	13.32	18.26
C27H35N507	541.61	541.2536	59.88	6.51	12.93	20.68
C27H35N508	557.61	557.2485	58.16	6.33	12.56	22.95
C27H35N509	573.61	573.2435	56.54	6.15	12.21	25.10
C27H36N201	404.60	404.2827	80.15	8.97	6.92	3.95
C27H36N202	420.60	420.2777	77.10	8.63	6.66	7.61
C27H36N203	436.60	436.2726	74.28	8.31	6.42	10.99
C27H36N204	452.60	452.2675	71.65	8.02	6.19	14.14
C27H36N205	468.60	468.2624	69.21	7.74	5.98	17.07
C27H36N206	484.60	484.2573	66.92	7.49	5.78	19.81
C27H36N207	500.60	500.2522	64.78	7.25	5.60	22.37
C27H36N208	516.60	516.2471	62.78	7.02	5.42	24.78
C27H36N209	532.60	532.2421	60.89	6.81	5.26	27.04
C27H36N401	432.61	432.2889	74.96	8.39	12.95	3.70
C27H36N402	448.61	448.2838	72.29	8.09	12.49	7.13
C27H36N403	464.61	464.2787	69.80	7.81	12.06	10.33
C27H36N404	480.61	480.2736	67.48	7.55	11.66	13.32
C27H36N405	496.61	496.2686	65.30	7.31	11.28	16.11
C27H36N406	512.61	512.2635	63.26	7.08	10.93	18.73
C27H36N407	528.61	528.2584	61.35	6.86	10.60	21.19
C27H36N408	544.61	544.2533	59.55	6.66	10.29	23.50
C27H36N409	560.61	560.2482	57.85	6.47	9.99	25.69
C27H36N601	460.63	460.2950	70.40	7.88	18.24	3.47
C27H36N602	476.63	476.2900	68.04	7.61	17.63	6.71
C27H36N603	492.63	492.2849	65.83	7.37	17.06	9.74
C27H36N604	508.63	508.2798	63.76	7.13	16.52	12.58
C27H36N605	524.63	524.2747	61.82	6.92	16.02	15.25
C27H36N606	540.62	540.2696	59.99	6.71	15.55	17.76
C27H36N607	556.62	556.2645	58.26	6.52	15.10	20.12
C27H36N608	572.62	572.2594	56.63	6.34	14.68	22.35
C27H36N609	588.62	588.2544	55.09	6.16	14.28	24.46
C27H37N101	391.60	391.2875	82.81	9.52	3.58	4.09
C27H37N102	407.60	407.2824	79.56	9.15	3.44	7.85
C27H37N103	423.60	423.2773	76.56	8.80	3.31	11.33
C27H37N104	439.60	439.2722	73.77	8.48	3.19	14.56
C27H37N105	455.60	455.2672	71.18	8.19	3.07	17.56
C27H37N106	471.60	471.2621	68.77	7.91	2.97	20.36
C27H37N107	487.60	487.2570	66.51	7.65	2.87	22.97
C27H37N108	503.60	503.2519	64.40	7.41	2.78	25.42
C27H37N109	519.60	519.2468	62.41	7.18	2.70	27.71
C27H37N301	419.62	419.2936	77.29	8.89	10.01	3.81
C27H37N302	435.61	435.2886	74.45	8.56	9.65	7.35
C27H37N303	451.61	451.2835	71.81	8.26	9.30	10.63
C27H37N304	467.61	467.2784	69.35	7.98	8.99	13.69
C27H37N305	483.61	483.2733	67.06	7.71	8.69	16.54
C27H37N306	499.61	499.2682	64.91	7.46	8.41	19.21
C27H37N307	515.61	515.2631	62.90	7.23	8.15	21.72
C27H37N308	531.61	531.2580	61.00	7.02	7.90	24.08
C27H37N309	547.61	547.2530	59.22	6.81	7.67	26.30
C27H37N501	447.63	447.2998	72.45	8.33	15.65	3.57
C27H37N502	463.63	463.2947	69.95	8.04	15.11	6.90
C27H37N503	479.63	479.2896	67.62	7.78	14.60	10.01
C27H37N504	495.63	495.2845	65.43	7.52	14.13	12.91
C27H37N505	511.63	511.2794	63.39	7.29	13.69	15.64
C27H37N506	527.63	527.2744	61.46	7.07	13.27	18.19
C27H37N507	543.63	543.2693	59.66	6.86	12.88	20.60
C27H37N508	559.62	559.2642	57.95	6.66	12.51	22.87
C27H37N509	575.62	575.2591	56.34	6.48	12.17	25.02
C27H38N201	406.62	406.2984	79.76	9.42	6.89	3.93
C27H38N202	422.62	422.2933	76.74	9.06	6.63	7.57
C27H38N203	438.62	438.2882	73.94	8.73	6.39	10.94
C27H38N204	454.61	454.2831	71.34	8.43	6.16	14.08
C27H38N205	470.61	470.2781	68.91	8.14	5.95	17.00
C27H38N206	486.61	486.2730	66.64	7.87	5.76	19.73
C27H38N207	502.61	502.2679	64.52	7.62	5.57	22.28
C27H38N208	518.61	518.2628	62.53	7.39	5.40	24.68
C27H38N209	534.61	534.2577	60.66	7.16	5.24	26.93
C27H38N401	434.63	434.3045	74.62	8.81	12.89	3.68
C27H38N402	450.63	450.2995	71.97	8.50	12.43	7.10
C27H38N403	466.63	466.2944	69.50	8.21	12.01	10.29
C27H38N404	482.63	482.2893	67.19	7.94	11.61	13.26
C27H38N405	498.63	498.2842	65.04	7.68	11.24	16.04
C27H38N406	514.63	514.2791	63.02	7.44	10.89	18.65
C27H38N407	530.63	530.2740	61.12	7.22	10.56	21.11
C27H38N408	546.63	546.2689	59.33	7.01	10.25	23.42
C27H38N409	562.63	562.2639	57.64	6.81	9.96	25.59
C27H38N601	462.64	462.3107	70.10	8.28	18.17	3.46
C27H38N602	478.64	478.3056	67.75	8.00	17.56	6.69
C27H38N603	494.64	494.3005	65.56	7.74	16.99	9.70
C27H38N604	510.64	510.2954	63.51	7.50	16.46	12.53
C27H38N605	526.64	526.2903	61.58	7.27	15.96	15.19
C27H38N606	542.64	542.2853	59.76	7.06	15.49	17.69
C27H38N607	558.64	558.2802	58.05	6.86	15.04	20.05
C27H38N608	574.64	574.2751	56.44	6.67	14.62	22.27
C27H38N609	590.64	590.2700	54.91	6.48	14.23	24.38
C27H39N101	393.62	393.3031	82.39	9.99	3.56	4.06
C27H39N102	409.62	409.2981	79.17	9.60	3.42	7.81
C27H39N103	425.62	425.2930	76.20	9.24	3.29	11.28
C27H39N104	441.62	441.2879	73.44	8.90	3.17	14.49
C27H39N105	457.62	457.2828	70.87	8.59	3.06	17.48
C27H39N106	473.61	473.2777	68.47	8.30	2.96	20.27
C27H39N107	489.61	489.2726	66.24	8.03	2.86	22.87
C27H39N108	505.61	505.2675	64.14	7.77	2.77	25.31
C27H39N109	521.61	521.2625	62.17	7.54	2.69	27.61
C27H39N301	421.63	421.3093	76.92	9.32	9.97	3.79
C27H39N302	437.63	437.3042	74.10	8.98	9.60	7.31
C27H39N303	453.63	453.2991	71.49	8.67	9.26	10.58
C27H39N304	469.63	469.2940	69.05	8.37	8.95	13.63

183

C27H39N305	485.63	485.2890	66.78	8.09	8.65	16.47
C27H39N306	501.63	501.2839	64.65	7.84	8.38	19.14
C27H39N307	517.63	517.2788	62.65	7.59	8.12	21.64
C27H39N308	533.63	533.2737	60.77	7.37	7.87	23.99
C27H39N309	549.63	549.2686	59.00	7.15	7.65	26.20
C27H39N501	449.64	449.3154	72.12	8.74	15.58	3.56
C27H39N502	465.64	465.3104	69.65	8.44	15.04	6.87
C27H39N503	481.64	481.3053	67.33	8.16	14.54	9.97
C27H39N504	497.64	497.3002	65.17	7.90	14.07	12.86
C27H39N505	513.64	513.2951	63.14	7.65	13.63	15.57
C27H39N506	529.64	529.2900	61.23	7.42	13.22	18.12
C27H39N507	545.64	545.2849	59.43	7.20	12.84	20.53
C27H39N508	561.64	561.2798	57.74	7.00	12.47	22.79
C27H39N509	577.64	577.2748	56.14	6.81	12.12	24.93
C27H40N201	408.63	408.3140	79.36	9.87	6.86	3.92
C27H40N202	424.63	424.3090	76.37	9.49	6.60	7.54
C27H40N203	440.63	440.3039	73.60	9.15	6.36	10.89
C27H40N204	456.63	456.2988	71.02	8.83	6.13	14.02
C27H40N205	472.63	472.2937	68.62	8.53	5.93	16.93
C27H40N206	488.63	488.2886	66.37	8.25	5.73	19.65
C27H40N207	504.63	504.2835	64.27	7.99	5.55	22.19
C27H40N208	520.63	520.2784	62.29	7.74	5.38	24.58
C27H40N209	536.63	536.2734	60.43	7.51	5.22	26.83
C27H40N401	436.65	436.3202	74.27	9.2.	12.83	3.66
C27H40N402	452.65	452.3151	71.65	8.91	12.38	7.07
C27H40N403	468.64	468.3100	69.20	8.60	11.96	10.24
C27H40N404	484.64	484.3049	66.92	8.32	11.56	13.21
C27H40N405	500.64	500.2998	64.78	8.05	11.19	15.98
C27H40N406	516.64	516.2948	62.77	7.80	10.84	18.58
C27H40N407	532.64	532.2897	60.89	7.57	10.52	21.03
C27H40N408	548.64	548.2846	59.11	7.35	10.21	23.33
C27H40N409	564.64	564.2795	57.43	7.14	9.92	25.50
C27H40N601	464.66	464.3263	69.79	8.68	18.09	3.44
C27H40N602	480.66	480.3213	67.47	8.39	17.48	6.66
C27H40N603	496.66	496.3162	65.30	8.12	16.92	9.66
C27H40N604	512.66	512.3111	63.26	7.86	16.39	12.48
C27H40N605	528.66	528.3060	61.34	7.63	15.90	15.13
C27H40N606	544.66	544.3009	59.54	7.40	15.43	17.63
C27H40N607	560.66	560.2958	57.84	7.19	14.99	19.98
C27H40N608	576.66	576.2907	56.24	6.99	14.57	22.20
C27H40N609	592.65	592.2857	54.72	6.80	14.18	24.30
C27H41N101	395.63	395.3188	81.97	10.45	3.54	4.04
C27H41N102	411.63	411.3137	78.78	10.04	3.40	7.77
C27H41N103	427.63	427.3086	75.84	9.66	3.28	11.22
C27H41N104	443.63	443.3035	73.10	9.32	3.16	14.43
C27H41N105	459.63	459.2985	70.56	8.99	3.05	17.40
C27H41N106	475.63	475.2934	68.18	8.69	2.94	20.18
C27H41N107	491.63	491.2883	65.96	8.41	2.85	22.78
C27H41N108	507.63	507.2832	63.89	8.14	2.76	25.21
C27H41N109	523.63	523.2781	61.93	7.89	2.67	27.50
C27H41N301	423.65	423.3249	76.55	9.75	9.92	3.78
C27H41N302	439.65	439.3199	73.76	9.40	9.56	7.28
C27H41N303	455.65	455.3148	71.17	9.07	9.22	10.53
C27H41N304	471.65	471.3097	68.76	8.76	8.91	13.57
C27H41N305	487.64	487.3046	66.50	8.47	8.62	16.40
C27H41N306	503.64	503.2995	64.39	8.21	8.34	19.06
C27H41N307	519.64	519.2944	62.41	7.95	8.09	21.55
C27H41N308	535.64	535.2893	60.54	7.72	7.84	23.90
C27H41N309	551.64	551.2843	58.79	7.49	7.62	26.10
C27H41N501	451.66	451.3311	71.80	9.15	15.51	3.54
C27H41N502	467.66	467.3260	69.35	8.84	14.98	6.84
C27H41N503	483.66	483.3209	67.05	8.54	14.48	9.92
C27H41N504	499.66	499.3158	64.90	8.27	14.02	12.81
C27H41N505	515.66	515.3107	62.89	8.01	13.58	15.51
C27H41N506	531.66	531.3057	61.00	7.77	13.17	18.06
C27H41N507	547.66	547.3006	59.22	7.55	12.79	20.45
C27H41N508	563.66	563.2955	57.54	7.33	12.42	22.71
C27H41N509	579.66	579.2904	55.95	7.13	12.08	24.84
C27H42N201	410.65	410.3297	78.97	10.31	6.82	3.90
C27H42N202	426.65	426.3246	76.01	9.92	6.57	7.50
C27H42N203	442.65	442.3195	73.26	9.56	6.33	10.84
C27H42N204	458.65	458.3144	70.71	9.23	6.11	13.95
C27H42N205	474.65	474.3094	68.32	8.92	5.90	16.85
C27H42N206	490.65	490.3043	66.10	8.63	5.71	19.57
C27H42N207	506.64	506.2992	64.01	8.36	5.53	22.11
C27H42N208	522.64	522.2941	62.05	8.10	5.36	24.49
C27H42N209	538.64	538.2890	60.21	7.86	5.20	26.73
C27H42N401	438.66	438.3358	73.93	9.65	12.77	3.65
C27H42N402	454.66	454.3308	71.33	9.31	12.32	7.04
C27H42N403	470.66	470.3257	68.90	8.99	11.90	10.20
C27H42N404	486.66	486.3206	66.64	8.70	11.51	13.15
C27H42N405	502.66	502.3155	64.52	8.42	11.15	15.91
C27H42N406	518.66	518.3104	62.53	8.16	10.80	18.51
C27H42N407	534.66	534.3053	60.66	7.92	10.48	20.95
C27H42N408	550.66	550.3002	58.89	7.69	10.17	23.24
C27H42N409	566.66	566.2952	57.23	7.47	9.89	25.41
C27H42N601	466.68	466.3420	69.49	9.07	18.01	3.43
C27H42N602	482.67	482.3369	67.19	8.77	17.41	6.63
C27H42N603	498.67	498.3318	65.03	8.49	16.85	9.63
C27H42N604	514.67	514.3267	63.01	8.23	16.33	12.43
C27H42N605	530.67	530.3216	61.11	7.98	15.84	15.07
C27H42N606	546.67	546.3166	59.32	7.74	15.37	17.56
C27H42N607	562.67	562.3115	57.64	7.52	14.94	19.90
C27H42N608	578.67	578.3064	56.04	7.32	14.52	22.12
C27H42N609	594.67	594.3013	54.53	7.12	14.13	24.21
C27H43N101	397.65	397.3344	81.55	10.90	3.52	4.02
C27H43N102	413.65	413.3294	78.40	10.48	3.39	7.74
C27H43N103	429.65	429.3243	75.48	10.09	3.26	11.17
C27H43N104	445.65	445.3192	72.77	9.73	3.14	14.36
C27H43N105	461.65	461.3141	70.25	9.39	3.03	17.33
C27H43N106	477.65	477.3090	67.90	9.07	2.93	20.10
C27H43N107	493.65	493.3039	65.70	8.78	2.84	22.69
C27H43N108	509.65	509.2988	63.63	8.50	2.75	25.11
C27H43N109	525.65	525.2938	61.70	8.25	2.66	27.39
C27H43N301	425.66	425.3406	76.19	10.18	9.87	3.76
C27H43N302	441.66	441.3355	73.43	9.81	9.51	7.25
C27H43N303	457.66	457.3304	70.86	9.47	9.18	10.49
C27H43N304	473.66	473.3253	68.47	9.15	8.87	13.51
C27H43N305	489.66	489.3202	66.23	8.85	8.58	16.34
C27H43N306	505.66	505.3152	64.13	8.57	8.31	18.98
C27H43N307	521.66	521.3101	62.17	8.31	8.06	21.47
C27H43N308	537.66	537.3050	60.32	8.06	7.82	23.81
C27H43N309	553.66	553.2999	58.57	7.83	7.59	26.01
C27H43N501	453.68	453.3467	71.48	9.55	15.44	3.53
C27H43N502	469.68	469.3417	69.05	9.23	14.91	6.81
C27H43N503	485.68	485.3366	66.77	8.92	14.42	9.88
C27H43N504	501.67	501.3315	64.64	8.64	13.96	12.76
C27H43N505	517.67	517.3264	62.65	8.37	13.53	15.45
C27H43N506	533.67	533.3213	60.77	8.12	13.12	17.99
C27H43N507	549.67	549.3162	59.00	7.89	12.74	20.37

C27H43N508	565.67	565.3111	57.33	7.66	12.38	22.63
C27H43N509	581.67	581.3061	55.75	7.45	12.04	24.76
C27H44N201	412.66	412.3453	78.59	10.75	6.79	3.88
C27H44N202	428.66	428.3403	75.65	10.35	6.54	7.46
C27H44N203	444.66	444.3352	72.93	9.97	6.30	10.79
C27H44N204	460.66	460.3301	70.40	9.63	6.08	13.89
C27H44N205	476.66	476.3250	68.04	9.30	5.88	16.78
C27H44N206	492.66	492.3199	65.83	9.00	5.69	19.49
C27H44N207	508.66	508.3148	63.76	8.72	5.51	22.02
C27H44N208	524.66	524.3097	61.81	8.45	5.34	24.40
C27H44N209	540.66	540.3047	59.98	8.20	5.18	26.63
C27H44N401	440.68	440.3515	73.59	10.06	12.71	3.63
C27H44N402	456.68	456.3464	71.01	9.71	12.27	7.01
C27H44N403	472.68	472.3413	68.61	9.38	11.85	10.15
C27H44N404	488.68	488.3362	66.36	9.08	11.47	13.10
C27H44N405	504.68	504.3311	64.26	8.79	11.10	15.85
C27H44N406	520.67	520.3261	62.28	8.52	10.76	18.44
C27H44N407	536.67	536.3210	60.43	8.26	10.44	20.87
C27H44N408	552.67	552.3159	58.68	8.02	10.14	23.16
C27H44N409	568.67	568.3108	57.03	7.80	9.85	25.32
C27H44N601	468.69	468.3576	69.19	9.46	17.93	3.41
C27H44N602	484.69	484.3526	66.91	9.15	17.34	6.60
C27H44N603	500.69	500.3475	64.77	8.86	16.78	9.59
C27H44N604	516.69	516.3424	62.77	8.58	16.27	12.39
C27H44N605	532.69	532.3373	60.88	8.33	15.78	15.02
C27H44N606	548.69	548.3322	59.10	8.08	15.32	17.50
C27H44N607	564.69	564.3271	57.43	7.85	14.88	19.83
C27H44N608	580.69	580.3220	55.85	7.64	14.47	22.04
C27H44N609	596.69	596.3170	54.35	7.43	14.08	24.13
C27H45N101	399.67	399.3501	81.14	11.35	3.50	4.00
C27H45N102	415.67	415.3450	78.02	10.91	3.37	7.70
C27H45N103	431.66	431.3399	75.13	10.51	3.24	11.12
C27H45N104	447.66	447.3348	72.44	10.13	3.13	14.30
C27H45N105	463.66	463.3298	69.94	9.78	3.02	17.25
C27H45N106	479.66	479.3247	67.61	9.46	2.92	20.01
C27H45N107	495.66	495.3196	65.43	9.15	2.83	22.60
C27H45N108	511.66	511.3145	63.38	8.86	2.74	25.02
C27H45N109	527.66	527.3094	61.46	8.60	2.65	27.29
C27H45N301	427.68	427.3562	75.83	10.61	9.83	3.74
C27H45N302	443.68	443.3512	73.09	10.22	9.47	7.21
C27H45N303	459.68	459.3461	70.55	9.87	9.14	10.44
C27H45N304	475.68	475.3410	68.18	9.54	8.83	13.45
C27H45N305	491.68	491.3359	65.96	9.23	8.55	16.27
C27H45N306	507.68	507.3308	63.88	8.93	8.28	18.91
C27H45N307	523.68	523.3257	61.93	8.66	8.02	21.39
C27H45N308	539.67	539.3206	60.09	8.40	7.79	23.72
C27H45N309	555.67	555.3156	58.36	8.16	7.56	25.91
C27H45N501	455.69	455.3624	71.17	9.95	15.37	3.51
C27H45N502	471.69	471.3573	68.75	9.62	14.85	6.78
C27H45N503	487.69	487.3522	66.50	9.30	14.36	9.84
C27H45N504	503.69	503.3471	64.38	9.01	13.90	12.71
C27H45N505	519.69	519.3420	62.40	8.73	13.48	15.39
C27H45N506	535.69	535.3370	60.54	8.47	13.07	17.92
C27H45N507	551.69	551.3319	58.78	8.22	12.69	20.30
C27H45N508	567.69	567.3268	57.13	7.99	12.34	22.55
C27H45N509	583.69	583.3217	55.56	7.77	12.00	24.67
C27H46N201	414.68	414.3610	78.21	11.18	6.76	3.86
C27H46N202	430.68	430.3559	75.30	10.77	6.50	7.43
C27H46N203	446.68	446.3508	72.60	10.38	6.27	10.75
C27H46N204	462.68	462.3457	70.09	10.02	6.05	13.83
C27H46N205	478.68	478.3406	67.75	9.69	5.85	16.71
C27H46N206	494.68	494.3356	65.56	9.37	5.66	19.41
C27H46N207	510.68	510.3305	63.50	9.08	5.[illegible]9	21.93
C27H46N208	526.68	526.3254	61.58	8.80	5.32	24.30
C27H46N209	542.68	542.3203	59.76	8.54	5.16	26.53
C27H46N401	442.69	442.3671	73.26	10.47	12.66	3.61
C27H46N402	458.69	458.3621	70.70	10.11	12.21	6.98
C27H46N403	474.69	474.3570	68.32	9.77	11.80	10.11
C27H46N404	490.69	490.3519	66.09	9.45	11.42	13.04
C27H46N405	506.69	506.3468	64.00	9.15	11.06	15.79
C27H46N406	522.69	522.3417	62.04	8.87	10.72	18.37
C27H46N407	538.69	538.3366	60.20	8.61	10.40	20.79
C27H46N408	554.69	554.3315	58.47	8.36	10.10	23.08
C27H46N409	570.69	570.3265	56.83	8.12	9.82	25.23
C27H46N601	470.71	470.3733	68.90	9.85	17.85	3.40
C27H46N602	486.71	486.3682	66.63	9.53	17.27	6.57
C27H46N603	502.71	502.3631	64.51	9.22	16.72	9.55
C27H46N604	518.71	518.3580	62.52	8.94	16.20	12.34
C27H46N605	534.70	534.3529	60.65	8.67	15.72	14.96
C27H46N606	550.70	550.3479	58.89	8.42	15.26	17.43
C27H46N607	566.70	566.3428	57.23	8.18	14.83	19.76
C27H46N608	582.70	582.3377	55.65	7.96	14.42	21.97
C27H46N609	598.70	598.3326	54.17	7.74	14.04	24.05
C27H47N101	401.68	401.3657	80.74	11.79	3.49	3.98
C27H47N102	417.68	417.3607	77.64	11.34	3.35	7.66
C27H47N103	433.68	433.3556	74.78	10.92	3.23	11.07
C27H47N104	449.68	449.3505	72.12	10.54	3.11	14.23
C27H47N105	465.68	465.3454	69.64	10.17	3.01	17.18
C27H47N106	481.68	481.3403	67.33	9.84	2.91	19.93
C27H47N107	497.68	497.3352	65.16	9.52	2.81	22.50
C27H47N108	513.68	513.3301	63.13	9.22	2.73	24.92
C27H47N109	529.68	529.3251	61.23	8.94	2.64	27.19
C27H47N301	429.70	429.3719	75.47	11.03	9.78	3.72
C27H47N302	445.69	445.3668	72.76	10.63	9.43	7.18
C27H47N303	461.69	461.3617	70.24	10.26	9.10	10.40
C27H47N304	477.69	477.3566	67.89	9.92	8.80	13.40
C27H47N305	493.69	493.3515	65.69	9.60	8.51	16.20
C27H47N306	509.69	509.3465	63.63	9.29	8.24	18.83
C27H47N307	525.69	525.3414	61.69	9.01	7.99	21.30
C27H47N308	541.69	541.3363	59.87	8.75	7.76	23.63
C27H47N309	557.69	557.3312	58.15	8.49	7.53	25.82
C27H47N501	457.71	457.3780	70.85	10.35	15.30	3.50
C27H47N502	473.71	473.3730	68.46	10.00	14.78	6.75
C27H47N503	489.71	489.3679	66.22	9.67	14.30	9.80
C27H47N504	505.71	505.3628	64.13	9.37	13.85	12.66
C27H47N505	521.71	521.3577	62.16	9.08	13.42	15.33
C27H47N506	537.71	537.3526	60.31	8.81	13.02	17.85
C27H47N507	553.70	553.3475	58.57	8.56	12.65	20.23
C27H47N508	569.70	569.3424	56.92	8.32	12.29	22.47
C27H47N509	585.70	585.3374	55.37	8.09	11.96	24.58
C27H48N201	416.70	416.3766	77.83	11.61	6.72	3.84
C27H48N202	432.70	432.3716	74.95	11.18	6.47	7.40
C27H48N203	448.70	448.3665	72.28	10.78	6.24	10.70
C27H48N204	464.69	464.3614	69.79	10.41	6.03	13.77
C27H48N205	480.69	480.3563	67.47	10.07	5.83	16.64
C27H48N206	496.69	496.3512	65.29	9.74	5.64	19.33
C27H48N207	512.69	512.3461	63.25	9.44	5.46	21.84
C27H48N208	528.69	528.3410	61.34	9.15	5.30	24.21
C27H48N209	544.69	544.3360	59.54	8.88	5.14	26.44
C27H48N401	444.71	444.3828	72.92	10.88	12.60	3.60

C27H48N402	460.71	460.3777	70.39	10.50	12.16	6.95
C27H48N403	476.71	476.3726	68.03	10.15	11.75	10.07
C27H48N404	492.71	492.3675	65.82	9.82	11.37	12.99
C27H48N405	508.71	508.3624	63.75	9.51	11.01	15.73
C27H48N406	524.71	524.3574	61.81	9.22	10.68	18.30
C27H48N407	540.71	540.3523	59.98	8.95	10.36	20.71
C27H48N408	556.71	556.3472	58.25	8.69	10.06	22.99
C27H48N409	572.71	572.3421	56.63	8.45	9.78	25.14
C27H48N601	472.72	472.3889	68.60	10.23	17.78	3.38
C27H48N602	488.72	488.3839	66.36	9.90	17.20	6.55
C27H48N603	504.72	504.3788	64.25	9.59	16.65	9.51
C27H48N604	520.72	520.3737	62.28	9.29	16.14	12.29
C27H48N605	536.72	536.3686	60.42	9.01	15.66	14.90
C27H48N606	552.72	552.3635	58.67	8.75	15.20	17.37
C27H48N607	568.72	568.3584	57.02	8.51	14.78	19.69
C27H48N608	584.72	584.3533	55.46	8.27	14.37	21.89
C27H48N609	600.72	600.3483	53.99	8.05	13.99	23.97
C27H49N101	403.70	403.3814	80.33	12.23	3.47	3.96
C27H49N102	419.70	419.3763	77.27	11.77	3.34	7.62
C27H49N103	435.70	435.3712	74.43	11.34	3.21	11.02
C27H49N104	451.70	451.3661	71.80	10.93	3.10	14.17
C27H49N105	467.70	467.3610	69.34	10.56	2.99	17.10
C27H49N106	483.69	483.3560	67.05	10.21	2.90	19.85
C27H49N107	499.69	499.3509	64.90	9.88	2.80	22.41
C27H49N108	515.69	515.3458	62.89	9.58	2.72	24.82
C27H49N109	531.69	531.3407	60.99	9.29	2.63	27.08
C27H49N301	431.71	431.3875	75.12	11.44	9.73	3.71
C27H49N302	447.71	447.3825	72.44	11.03	9.39	7.15
C27H49N303	463.71	463.3774	69.94	10.65	9.06	10.35
C27H49N304	479.71	479.3723	67.60	10.30	8.76	13.34
C27H49N305	495.71	495.3672	65.42	9.96	8.48	16.14
C27H49N306	511.71	511.3621	63.38	9.65	8.21	18.76
C27H49N307	527.71	527.3570	61.45	9.36	7.96	21.22
C27H49N308	543.71	543.3519	59.65	9.08	7.73	23.54
C27H49N309	559.71	559.3469	57.94	8.82	7.51	25.73
C27H49N501	459.72	459.3937	70.54	10.74	15.23	3.48
C27H49N502	475.72	475.3886	68.17	10.38	14.72	6.73
C27H49N503	491.72	491.3835	65.95	10.04	14.24	9.76
C27H49N504	507.72	507.3784	63.87	9.73	13.79	12.60
C27H49N505	523.72	523.3733	61.92	9.43	13.37	15.27
C27H49N506	539.72	539.3683	60.09	9.15	12.98	17.79
C27H49N507	555.72	555.3632	58.36	8.89	12.60	20.15
C27H49N508	571.72	571.3581	56.72	8.64	12.25	22.39
C27H49N509	587.72	587.3530	55.18	8.40	11.92	24.50
C27H50N201	418.71	418.3923	77.45	12.04	6.69	3.82
C27H50N202	434.71	434.3872	74.60	11.59	6.44	7.36
C27H50N203	450.71	450.3821	71.95	11.18	6.22	10.65
C27H50N204	466.71	466.3770	69.49	10.80	6.00	13.71
C27H50N205	482.71	482.3719	67.18	10.44	5.80	16.57
C27H50N206	498.71	498.3669	65.03	10.11	5.62	19.25
C27H50N207	514.71	514.3618	63.01	9.79	5.44	21.76
C27H50N208	530.71	530.3567	61.11	9.50	5.28	24.12
C27H50N209	546.71	546.3516	59.32	9.22	5.12	26.34
C27H50N401	446.73	446.3984	72.60	11.28	12.54	3.58
C27H50N402	462.73	462.3934	70.09	10.89	12.11	6.92
C27H50N403	478.72	478.3883	67.74	10.53	11.70	10.03
C27H50N404	494.72	494.3832	65.55	10.19	11.32	12.94
C27H50N405	510.72	510.3781	63.50	9.87	10.97	15.66
C27H50N406	526.72	526.3730	61.57	9.57	10.64	18.23
C27H50N407	542.72	542.3679	59.75	9.29	10.32	20.64
C27H50N408	558.72	558.3628	58.04	9.02	10.03	22.91
C27H50N409	574.72	574.3578	56.43	8.77	9.75	25.05
C27H50N601	474.74	474.4046	68.31	10.62	17.70	3.37
C27H50N602	490.74	490.3995	66.08	10.27	17.13	6.52
C27H50N603	506.74	506.3944	64.00	9.95	16.58	9.47
C27H50N604	522.74	522.3893	62.04	9.64	16.08	12.24
C27H50N605	538.74	538.3842	60.20	9.35	15.60	14.85
C27H50N606	554.74	554.3792	58.46	9.09	15.15	17.30
C27H50N607	570.74	570.3741	56.82	8.83	14.72	19.62
C27H50N608	586.73	586.3690	55.27	8.59	14.32	21.81
C27H50N609	602.73	602.3639	53.80	8.36	13.94	23.89
C27H51N101	405.71	405.3970	79.93	12.67	3.45	3.94
C27H51N102	421.71	421.3920	76.90	12.19	3.32	7.59
C27H51N103	437.71	437.3869	74.09	11.74	3.20	10.97
C27H51N104	453.71	453.3818	71.48	11.33	3.09	14.11
C27H51N105	469.71	469.3767	69.04	10.94	2.98	17.03
C27H51N106	485.71	485.3716	66.77	10.58	2.88	19.76
C27H51N107	501.71	501.3665	64.64	10.25	2.79	22.32
C27H51N108	517.71	517.3614	62.64	9.93	2.71	24.72
C27H51N109	533.71	533.3564	60.76	9.63	2.62	26.98
C27H51N301	433.73	433.4032	74.77	11.85	9.69	3.69
C27H51N302	449.73	449.3981	72.11	11.43	9.34	7.12
C27H51N303	465.73	465.3930	69.63	11.04	9.02	10.31
C27H51N304	481.73	481.3879	67.32	10.67	8.72	13.29
C27H51N305	497.72	497.3828	65.16	10.33	8.44	16.07
C27H51N306	513.72	513.3778	63.13	10.01	8.18	18.69
C27H51N307	529.72	529.3727	61.22	9.70	7.93	21.14
C27H51N308	545.72	545.3676	59.43	9.42	7.70	23.45
C27H51N309	561.72	561.3625	57.73	9.15	7.48	25.63
C27H51N501	461.74	461.4093	70.23	11.13	15.17	3.47
C27H51N502	477.74	477.4043	67.88	10.76	14.66	6.70
C27H51N503	493.74	493.3992	65.68	10.41	14.18	9.72
C27H51N504	509.74	509.3941	63.62	10.08	13.74	12.55
C27H51N505	525.74	525.3890	61.68	9.78	13.32	15.22
C27H51N506	541.74	541.3839	59.86	9.49	12.93	17.72
C27H51N507	557.74	557.3788	58.15	9.22	12.56	20.08
C27H51N508	573.74	573.3737	56.52	8.96	12.21	22.31
C27H51N509	589.74	589.3687	54.99	8.72	11.88	24.42
C27H52N201	420.73	420.4079	77.08	12.46	6.66	3.80
C27H52N202	436.73	436.4029	74.26	12.00	6.41	7.33
C27H52N203	452.73	452.3978	71.63	11.58	6.19	10.60
C27H52N204	468.73	468.3927	69.19	11.18	5.98	13.65
C27H52N205	484.73	484.3876	66.90	10.81	5.78	16.50
C27H52N206	500.73	500.3825	64.77	10.47	5.59	19.17
C27H52N207	516.72	516.3774	62.76	10.14	5.42	21.67
C27H52N208	532.72	532.3723	60.88	9.84	5.26	24.03
C27H52N209	548.72	548.3673	59.10	9.55	5.11	26.24
C27H52N401	448.74	448.4141	72.27	11.68	12.49	3.57
C27H52N402	464.74	464.4090	69.78	11.28	12.06	6.89
C27H52N403	480.74	480.4039	67.46	10.90	11.65	9.98
C27H52N404	496.74	496.3988	65.29	10.55	11.28	12.88
C27H52N405	512.74	512.3937	63.25	10.22	10.93	15.60
C27H52N406	528.74	528.3887	61.33	9.91	10.60	18.16
C27H52N407	544.74	544.3836	59.53	9.62	10.29	20.56
C27H52N408	560.74	560.3785	57.83	9.35	9.99	22.83
C27H52N409	576.74	576.3734	56.23	9.09	9.71	24.97
C27H52N601	476.76	476.4202	68.02	10.99	17.63	3.36
C27H52N602	492.75	492.4152	65.81	10.64	17.06	6.49
C27H52N603	508.75	508.4101	63.74	10.30	16.52	9.43
C27H52N604	524.75	524.4050	61.80	9.99	16.02	12.20

C27H52N605	540.75	540.3999	59.97	9.69	15.54	14.79
C27H52N606	556.75	556.3948	58.25	9.41	15.09	17.24
C27H52N607	572.75	572.3897	56.62	9.15	14.67	19.55
C27H52N608	588.75	588.3846	55.08	8.90	14.27	21.74
C27H52N609	604.75	604.3795	53.63	8.67	13.90	23.81
C27H53N101	407.73	407.4127	79.54	13.10	3.44	3.92
C27H53N102	423.73	423.4076	76.54	12.61	3.31	7.55
C27H53N103	439.73	439.4025	73.75	12.15	3.19	10.92
C27H53N104	455.73	455.3974	71.16	11.72	3.07	14.04
C27H53N105	471.73	471.3923	68.75	11.32	2.97	16.96
C27H53N106	487.73	487.3873	66.49	10.95	2.87	19.68
C27H53N107	503.73	503.3822	64.38	10.61	2.78	22.23
C27H53N108	519.73	519.3771	62.40	10.28	2.70	24.63
C27H53N109	535.72	535.3720	60.54	9.97	2.61	26.88
C27H53N301	435.74	435.4188	74.42	12.26	9.64	3.67
C27H53N302	451.74	451.4138	71.79	11.83	9.30	7.08
C27H53N303	467.74	467.4087	69.33	11.42	8.98	10.26
C27H53N304	483.74	483.4036	67.04	11.04	8.69	13.23
C27H53N305	499.74	499.3985	64.89	10.69	8.41	16.01
C27H53N306	515.74	515.3934	62.88	10.36	8.15	18.61
C27H53N307	531.74	531.3883	60.99	10.05	7.90	21.06
C27H53N308	547.74	547.3832	59.21	9.75	7.67	23.37
C27H53N309	563.74	563.3782	57.53	9.48	7.45	25.54
C27H53N501	463.76	463.4250	69.93	11.52	15.10	3.45
C27H53N502	479.76	479.4199	67.60	11.14	14.60	6.67
C27H53N503	495.76	495.4148	65.42	10.78	14.13	9.68
C27H53N504	511.75	511.4097	63.37	10.44	13.68	12.51
C27H53N505	527.75	527.4046	61.45	10.12	13.27	15.16
C27H53N506	543.75	543.3996	59.64	9.82	12.88	17.65
C27H53N507	559.75	559.3945	57.94	9.54	12.51	20.01
C27H53N508	575.75	575.3894	56.33	9.28	12.16	22.23
C27H53N509	591.75	591.3843	54.80	9.03	11.83	24.33
C27H54N201	422.74	422.4236	76.71	12.88	6.63	3.78
C27H54N202	438.74	438.4185	73.92	12.41	6.38	7.29
C27H54N203	454.74	454.4134	71.32	11.97	6.16	10.56
C27H54N204	470.74	470.4083	68.89	11.56	5.95	13.60
C27H54N205	486.74	486.4032	66.63	11.18	5.76	16.44
C27H54N206	502.74	502.3982	64.51	10.83	5.57	19.09
C27H54N207	518.74	518.3931	62.52	10.49	5.40	21.59
C27H54N208	534.74	534.3880	60.65	10.18	5.24	23.94
C27H54N209	550.74	550.3829	58.88	9.88	5.09	26.15
C27H54N401	450.76	450.4297	71.95	12.08	12.43	3.55
C27H54N402	466.76	466.4247	69.48	11.66	12.00	6.86
C27H54N403	482.76	482.4196	67.18	11.27	11.61	9.94
C27H54N404	498.76	498.4145	65.02	10.91	11.23	12.83
C27H54N405	514.76	514.4094	63.00	10.57	10.88	15.54
C27H54N406	530.75	530.4043	61.10	10.26	10.56	18.09
C27H54N407	546.75	546.3992	59.31	9.96	10.25	20.48
C27H54N408	562.75	562.3941	57.63	9.67	9.96	22.74
C27H54N409	578.75	578.3891	56.03	9.40	9.68	24.88
C27H54N601	478.77	478.4359	67.74	11.37	17.55	3.34
C27H54N602	494.77	494.4308	65.55	11.00	16.99	6.47
C27H54N603	510.77	510.4257	63.49	10.66	16.45	9.40
C27H54N604	526.77	526.4206	61.56	10.33	15.95	12.15
C27H54N605	542.77	542.4155	59.75	10.03	15.48	14.74
C27H54N606	558.77	558.4105	58.04	9.74	15.04	17.18
C27H54N607	574.77	574.4054	56.42	9.47	14.62	19.49
C27H54N608	590.77	590.4003	54.89	9.21	14.23	21.67
C27H54N609	606.77	606.3952	53.45	8.97	13.85	23.73
C27H55N101	409.75	409.4283	79.15	13.53	3.42	3.90

C27H55N102	425.74	425.4233	76.17	13.02	3.29	7.52
C27H55N103	441.74	441.4182	73.41	12.55	3.17	10.87
C27H55N104	457.74	457.4131	70.85	12.11	3.06	13.98
C27H55N105	473.74	473.4080	68.46	11.70	2.96	16.89
C27H55N106	489.74	489.4029	66.22	11.32	2.86	19.60
C27H55N107	505.74	505.3978	64.12	10.96	2.77	22.14
C27H55N108	521.74	521.3927	62.16	10.63	2.68	24.53
C27H55N109	537.74	537.3877	60.31	10.31	2.60	26.78
C27H55N301	437.76	437.4345	74.08	12.66	9.60	3.65
C27H55N302	453.76	453.4294	71.47	12.22	9.26	7.05
C27H55N303	469.76	469.4243	69.04	11.80	8.95	10.22
C27H55N304	485.76	485.4192	66.76	11.41	8.65	13.17
C27H55N305	501.76	501.4141	64.63	11.05	8.37	15.94
C27H55N306	517.76	517.4091	62.64	10.71	8.12	18.54
C27H55N307	533.76	533.4040	60.76	10.39	7.87	20.98
C27H55N308	549.75	549.3989	58.99	10.08	7.64	23.28
C27H55N309	565.75	565.3938	57.32	9.80	7.43	25.45
C27H55N501	465.77	465.4406	69.63	11.90	15.04	3.44
C27H55N502	481.77	481.4356	67.31	11.51	14.54	6.64
C27H55N503	497.77	497.4305	65.15	11.14	14.07	9.64
C27H55N504	513.77	513.4254	63.12	10.79	13.63	12.46
C27H55N505	529.77	529.4203	61.22	10.46	13.22	15.10
C27H55N506	545.77	545.4152	59.42	10.16	12.83	17.59
C27H55N507	561.77	561.4101	57.73	9.87	12.47	19.94
C27H55N508	577.77	577.4050	56.13	9.60	12.12	22.15
C27H55N509	593.77	593.3999	54.62	9.34	11.79	24.25
C27H56N201	424.76	424.4392	76.35	13.29	6.60	3.77
C27H56N202	440.76	440.4342	73.58	12.81	6.36	7.26
C27H56N203	456.76	456.4291	71.00	12.36	6.13	10.51
C27H56N204	472.76	472.4240	68.60	11.94	5.93	13.54
C27H56N205	488.76	488.4189	66.35	11.55	5.73	16.37
C27H56N206	504.76	504.4138	64.25	11.18	5.55	19.02
C27H56N207	520.76	520.4087	62.27	10.84	5.38	21.51
C27H56N208	536.76	536.4036	60.42	10.52	5.22	23.85
C27H56N209	552.76	552.3986	58.67	10.21	5.07	26.05
C27H56N401	452.77	452.4454	71.63	12.47	12.37	3.53
C27H56N402	468.77	468.4403	69.18	12.04	11.95	6.83
C27H56N403	484.77	484.4352	66.90	11.64	11.56	9.90
C27H56N404	500.77	500.4301	64.76	11.27	11.19	12.78
C27H56N405	516.77	516.4250	62.76	10.92	10.84	15.48
C27H56N406	532.77	532.4200	60.87	10.59	10.52	18.02
C27H56N407	548.77	548.4149	59.10	10.29	10.21	20.41
C27H56N408	564.77	564.4098	57.42	9.99	9.92	22.66
C27H56N409	580.77	580.4047	55.84	9.72	9.65	24.79
C27H56N601	480.79	480.4515	67.45	11.74	17.48	3.33
C27H56N602	496.79	496.4464	65.28	11.36	16.92	6.44
C27H56N603	512.79	512.4414	63.24	11.01	16.39	9.36
C27H56N604	528.79	528.4363	61.33	10.67	15.89	12.10
C27H56N605	544.78	544.4312	59.53	10.36	15.43	14.68
C27H56N606	560.78	560.4261	57.83	10.07	14.99	17.12
C27H56N607	576.78	576.4210	56.23	9.79	14.57	19.42
C27H56N608	592.78	592.4159	54.71	9.52	14.18	21.59
C27H56N609	608.78	608.4108	53.27	9.27	13.80	23.65
C27H57N101	411.76	411.4440	78.76	13.95	3.40	3.89
C27H57N102	427.76	427.4389	75.81	13.43	3.27	7.48
C27H57N103	443.76	443.4338	73.08	12.95	3.16	10.82
C27H57N104	459.76	459.4287	70.54	12.50	3.05	13.92
C27H57N105	475.76	475.4236	68.16	12.08	2.94	16.81
C27H57N106	491.76	491.4186	65.95	11.68	2.85	19.52
C27H57N107	507.76	507.4135	63.87	11.32	2.76	22.06

C27H57N108	523.76	523.4084	61.92	10.97	2.67	24.44
C27H57N109	539.76	539.4033	60.08	10.64	2.60	26.68
C27H57N301	439.77	439.4501	73.74	13.06	9.55	3.64
C27H57N302	455.77	455.4451	71.15	12.61	9.22	7.02
C27H57N303	471.77	471.4400	68.74	12.18	8.91	10.17
C27H57N304	487.77	487.4349	66.49	11.78	8.61	13.12
C27H57N305	503.77	503.4298	64.37	11.40	8.34	15.88
C27H57N306	519.77	519.4247	62.39	11.05	8.08	18.47
C27H57N307	535.77	535.4196	60.53	10.72	7.84	20.90
C27H57N308	551.77	551.4145	58.77	10.41	7.62	23.20
C27H57N309	567.77	567.4095	57.12	10.12	7.40	25.36
C27H57N501	467.79	467.4563	69.33	12.28	14.97	3.42
C27H57N502	483.79	483.4512	67.03	11.88	14.48	6.61
C27H57N503	499.79	499.4461	64.89	11.50	14.01	9.60
C27H57N504	515.79	515.4410	62.88	11.14	13.58	12.41
C27H57N505	531.79	531.4359	60.98	10.80	13.17	15.04
C27H57N506	547.79	547.4309	59.20	10.49	12.78	17.52
C27H57N507	563.78	563.4258	57.52	10.19	12.42	19.86
C27H57N508	579.78	579.4207	55.93	9.91	12.08	22.08
C27H57N509	595.78	595.4156	54.43	9.64	11.75	24.17
C27H58N201	426.78	426.4549	75.99	13.70	6.56	3.75
C27H58N202	442.78	442.4498	73.24	13.20	6.33	7.23
C27H58N203	458.77	458.4447	70.69	12.74	6.11	10.46
C27H58N204	474.77	474.4396	68.31	12.31	5.90	13.48
C27H58N205	490.77	490.4345	66.08	11.91	5.71	16.30
C27H58N206	506.77	506.4295	63.99	11.54	5.53	18.94
C27H58N207	522.77	522.4244	62.03	11.18	5.36	21.42
C27H58N208	538.77	538.4193	60.19	10.85	5.20	23.76
C27H58N209	554.77	554.4142	58.46	10.54	5.05	25.96
C27H58N401	454.79	454.4610	71.31	12.85	12.32	3.52
C27H58N402	470.79	470.4560	68.88	12.42	11.90	6.80
C27H58N403	486.79	486.4509	66.62	12.01	11.51	9.86
C27H58N404	502.79	502.4458	64.50	11.63	11.14	12.73
C27H58N405	518.79	518.4407	62.51	11.27	10.80	15.42
C27H58N406	534.79	534.4356	60.64	10.93	10.48	17.95
C27H58N407	550.79	550.4305	58.88	10.61	10.17	20.33
C27H58N408	566.79	566.4254	57.22	10.31	9.89	22.58
C27H58N409	582.78	582.4203	55.65	10.03	9.61	24.71
C27H58N601	482.80	482.4672	67.17	12.11	17.41	3.31
C27H58N602	498.80	498.4621	65.02	11.72	16.85	6.42
C27H58N603	514.80	514.4570	63.00	11.36	16.32	9.32
C27H58N604	530.80	530.4519	61.10	11.01	15.83	12.06
C27H58N605	546.80	546.4468	59.31	10.69	15.37	14.63
C27H58N606	562.80	562.4418	57.62	10.39	14.93	17.06
C27H58N607	578.80	578.4367	56.03	10.10	14.52	19.35
C27H58N608	594.80	594.4316	54.52	9.83	14.13	21.52
C27H58N609	610.80	610.4265	53.09	9.57	13.76	23.57
C27H59N301	441.79	441.4658	73.41	13.46	9.51	3.62
C27H59N302	457.79	457.4607	70.84	12.99	9.18	6.99
C27H59N303	473.79	473.4556	68.45	12.55	8.87	10.13
C27H59N304	489.79	489.4505	66.21	12.14	8.58	13.07
C27H59N305	505.79	505.4454	64.12	11.76	8.31	15.82
C27H59N306	521.79	521.4404	62.15	11.40	8.05	18.40
C27H59N307	537.79	537.4353	60.30	11.06	7.81	20.83
C27H59N308	553.79	553.4302	58.56	10.74	7.59	23.11
C27H59N309	569.79	569.4251	56.92	10.44	7.37	25.27
C27H59N501	469.80	469.4719	69.03	12.66	14.91	3.41
C27H59N502	485.80	485.4668	66.76	12.24	14.42	6.59
C27H59N503	501.80	501.4618	64.63	11.85	13.96	9.57
C27H59N504	517.80	517.4567	62.63	11.49	13.53	12.36

C27H59N505	533.80	533.4516	60.75	11.14	13.12	14.99
C27H59N506	549.80	549.4465	58.99	10.82	12.74	17.46
C27H59N507	565.80	565.4414	57.32	10.51	12.38	19.79
C27H59N508	581.80	581.4363	55.74	10.22	12.04	22.00
C27H59N509	597.80	597.4312	54.25	9.95	11.72	24.09
C27H60N401	456.81	456.4767	70.99	13.24	12.26	3.50
C27H60N402	472.80	472.4716	68.59	12.79	11.85	6.77
C27H60N403	488.80	488.4665	66.35	12.37	11.46	9.82
C27H60N404	504.80	504.4614	64.24	11.98	11.10	12.68
C27H60N405	520.80	520.4563	62.27	11.61	10.76	15.36
C27H60N406	536.80	536.4513	60.41	11.27	10.44	17.88
C27H60N407	552.80	552.4462	58.66	10.94	10.14	20.26
C27H60N408	568.80	568.4411	57.01	10.63	9.85	22.50
C27H60N409	584.80	584.4360	55.45	10.34	9.58	24.62
C27H60N601	484.82	484.4828	66.89	12.47	17.33	3.30
C27H60N602	500.82	500.4777	64.75	12.08	16.78	6.39
C27H60N603	516.82	516.4727	62.75	11.70	16.26	9.29
C27H60N604	532.82	532.4676	60.87	11.35	15.77	12.01
C27H60N605	548.82	548.4625	59.09	11.02	15.31	14.58
C27H60N606	564.82	564.4574	57.42	10.71	14.88	17.00
C27H60N607	580.82	580.4523	55.84	10.41	14.47	19.28
C27H60N608	596.81	596.4472	54.34	10.13	14.08	21.45
C27H60N609	612.81	612.4421	52.92	9.87	13.71	23.50
C27H61N501	471.82	471.4876	68.73	13.03	14.84	3.39
C27H61N502	487.82	487.4825	66.48	12.60	14.36	6.56
C27H61N503	503.82	503.4774	64.37	12.20	13.90	9.53
C27H61N504	519.82	519.4723	62.39	11.83	13.47	12.31
C27H61N505	535.82	535.4672	60.52	11.48	13.07	14.93
C27H61N506	551.82	551.4622	58.77	11.14	12.69	17.40
C27H61N507	567.82	567.4571	57.11	10.83	12.33	19.72
C27H61N508	583.82	583.4520	55.55	10.53	12.00	21.92
C27H61N509	599.82	599.4469	54.07	10.25	11.68	24.01
C27H62N601	486.83	486.4985	66.61	12.84	17.26	3.29
C27H62N602	502.83	502.4934	64.49	12.43	16.71	6.36
C27H62N603	518.83	518.4883	62.51	12.05	16.20	9.25
C27H62N604	534.83	534.4832	60.64	11.68	15.71	11.97
C27H62N605	550.83	550.4781	58.87	11.35	15.26	14.52
C27H62N606	566.83	566.4731	57.21	11.03	14.83	16.94
C27H62N607	582.83	582.4680	55.64	10.72	14.42	19.22
C27H62N608	598.83	598.4629	54.16	10.44	14.03	21.37
C27H62N609	614.83	614.4578	52.75	10.16	13.67	23.42
C28H14N201	394.44	394.1106	85.26	3.58	7.10	4.06
C28H14N202	410.44	410.1055	81.94	3.44	6.83	7.80
C28H14N203	426.44	426.1004	78.87	3.31	6.57	11.26
C28H14N204	442.43	442.0953	76.01	3.19	6.33	14.46
C28H14N205	458.43	458.0903	73.36	3.08	6.11	17.45
C28H14N206	474.43	474.0852	70.89	2.97	5.90	20.23
C28H14N207	490.43	490.0801	68.57	2.88	5.71	22.84
C28H14N208	506.43	506.0750	66.41	2.79	5.53	25.27
C28H14N209	522.43	522.0699	64.37	2.70	5.36	27.56
C28H14N401	422.45	422.1168	79.61	3.34	13.26	3.79
C28H14N402	438.45	438.1117	76.70	3.22	12.78	7.30
C28H14N403	454.45	454.1066	74.00	3.11	12.33	10.56
C28H14N404	470.45	470.1015	71.49	3.00	11.91	13.60
C28H14N405	486.45	486.0964	69.14	2.90	11.52	16.45
C28H14N406	502.45	502.0913	66.93	2.81	11.15	19.11
C28H14N407	518.45	518.0862	64.87	2.72	10.81	21.60
C28H14N408	534.45	534.0812	62.93	2.64	10.48	23.95
C28H14N409	550.45	550.0761	61.10	2.56	10.18	26.16
C28H14N601	450.46	450.1229	74.66	3.13	18.66	3.55

C28H14N6O2	466.46	466.1178	72.10	3.03	18.02	6.86
C28H14N6O3	482.46	482.1127	69.71	2.92	17.42	9.95
C28H14N6O4	498.46	498.1076	67.47	2.83	16.86	12.84
C28H14N6O5	514.46	514.1026	65.37	2.74	16.34	15.55
C28H14N6O6	530.46	530.0975	63.40	2.66	15.84	18.10
C28H14N6O7	546.46	546.0924	61.54	2.58	15.38	20.49
C28H14N6O8	562.46	562.0873	59.79	2.51	14.94	22.76
C28H14N6O9	578.46	578.0822	58.14	2.44	14.53	24.89
C28H15N1O1	381.44	381.1154	88.17	3.96	3.67	4.19
C28H15N1O2	397.44	397.1103	84.62	3.80	3.52	8.05
C28H15N1O3	413.44	413.1052	81.35	3.66	3.39	11.61
C28H15N1O4	429.44	429.1001	78.31	3.52	3.26	14.90
C28H15N1O5	445.44	445.0950	75.50	3.39	3.14	17.96
C28H15N1O6	461.43	461.0899	72.88	3.28	3.04	20.80
C28H15N1O7	477.43	477.0848	70.44	3.17	2.93	23.46
C28H15N1O8	493.43	493.0798	68.16	3.06	2.84	25.94
C28H15N1O9	509.43	509.0747	66.02	2.97	2.75	28.27
C28H15N3O1	409.45	409.1215	82.14	3.69	10.26	3.91
C28H15N3O2	425.45	425.1164	79.05	3.55	9.88	7.52
C28H15N3O3	441.45	441.1113	76.18	3.42	9.52	10.87
C28H15N3O4	457.45	457.1062	73.52	3.31	9.19	13.99
C28H15N3O5	473.45	473.1012	71.03	3.19	8.88	16.90
C28H15N3O6	489.45	489.0961	68.71	3.09	8.59	19.61
C28H15N3O7	505.45	505.0910	66.54	2.99	8.31	22.16
C28H15N3O8	521.45	521.0859	64.50	2.90	8.06	24.55
C28H15N3O9	537.45	537.0808	62.58	2.81	7.82	26.79
C28H15N5O1	437.46	437.1277	76.88	3.46	16.01	3.66
C28H15N5O2	453.46	453.1226	74.17	3.33	15.44	7.06
C28H15N5O3	469.46	469.1175	71.64	3.22	14.92	10.22
C28H15N5O4	485.46	485.1124	69.28	3.11	14.43	13.18
C28H15N5O5	501.46	501.1073	67.07	3.02	13.97	15.95
C28H15N5O6	517.46	517.1022	64.99	2.92	13.53	18.55
C28H15N5O7	533.46	533.0971	63.04	2.83	13.13	20.99
C28H15N5O8	549.46	549.0921	61.21	2.75	12.75	23.29
C28H15N5O9	565.46	565.0870	59.48	2.67	12.39	25.47
C28H16N2O1	396.45	396.1263	84.83	4.07	7.07	4.04
C28H16N2O2	412.45	412.1212	81.54	3.91	6.79	7.76
C28H16N2O3	428.45	428.1161	78.49	3.76	6.54	11.20
C28H16N2O4	444.45	444.1110	75.67	3.63	6.30	14.40
C28H16N2O5	460.45	460.1059	73.04	3.50	6.08	17.37
C28H16N2O6	476.45	476.1008	70.59	3.38	5.88	20.15
C28H16N2O7	492.45	492.0957	68.29	3.27	5.69	22.74
C28H16N2O8	508.45	508.0907	66.14	3.17	5.51	25.17
C28H16N2O9	524.45	524.0856	64.13	3.08	5.34	27.46
C28H16N4O1	424.47	424.1324	79.23	3.80	13.20	3.77
C28H16N4O2	440.47	440.1273	76.35	3.66	12.72	7.26
C28H16N4O3	456.46	456.1222	73.68	3.53	12.27	10.52
C28H16N4O4	472.46	472.1171	71.18	3.41	11.86	13.55
C28H16N4O5	488.46	488.1121	68.85	3.30	11.47	16.38
C28H16N4O6	504.46	504.1070	66.67	3.20	11.11	19.03
C28H16N4O7	520.46	520.1019	64.62	3.10	10.76	21.52
C28H16N4O8	536.46	536.0968	62.69	3.01	10.44	23.86
C28H16N4O9	552.46	552.0917	60.88	2.92	10.14	26.06
C28H16N6O1	452.48	452.1386	74.33	3.56	18.57	3.54
C28H16N6O2	468.48	468.1335	71.79	3.44	17.94	6.83
C28H16N6O3	484.48	484.1284	69.42	3.33	17.35	9.91
C28H16N6O4	500.48	500.1233	67.20	3.22	16.79	12.79
C28H16N6O5	516.48	516.1182	65.12	3.12	16.27	15.49
C28H16N6O6	532.48	532.1131	63.16	3.03	15.78	18.03
C28H16N6O7	548.48	548.1080	61.32	2.94	15.32	20.42
C28H16N6O8	564.48	564.1029	59.58	2.86	14.89	22.68
C28H16N6O9	580.47	580.0979	57.94	2.78	14.48	24.81
C28H17N1O1	383.45	383.1310	87.71	4.47	3.65	4.17
C28H17N1O2	399.45	399.1259	84.19	4.29	3.51	8.01
C28H17N1O3	415.45	415.1208	80.95	4.12	3.37	11.55
C28H17N1O4	431.45	431.1157	77.95	3.97	3.25	14.83
C28H17N1O5	447.45	447.1107	75.16	3.83	3.13	17.88
C28H17N1O6	463.45	463.1056	72.57	3.70	3.02	20.71
C28H17N1O7	479.45	479.1005	70.15	3.57	2.92	23.36
C28H17N1O8	495.45	495.0954	67.88	3.46	2.83	25.83
C28H17N1O9	511.45	511.0903	65.76	3.35	2.74	28.15
C28H17N3O1	411.47	411.1372	81.73	4.16	10.21	3.89
C28H17N3O2	427.47	427.1321	78.68	4.01	9.83	7.49
C28H17N3O3	443.47	443.1270	75.84	3.86	9.48	10.82
C28H17N3O4	459.47	459.1219	73.20	3.73	9.15	13.93
C28H17N3O5	475.46	475.1168	70.73	3.60	8.84	16.83
C28H17N3O6	491.46	491.1117	68.43	3.49	8.55	19.53
C28H17N3O7	507.46	507.1066	66.27	3.38	8.28	22.07
C28H17N3O8	523.46	523.1016	64.25	3.27	8.03	24.45
C28H17N3O9	539.46	539.0965	62.34	3.18	7.79	26.69
C28H17N5O1	439.48	439.1433	76.52	3.90	15.94	3.64
C28H17N5O2	455.48	455.1382	73.84	3.76	15.38	7.03
C28H17N5O3	471.48	471.1331	71.33	3.63	14.85	10.18
C28H17N5O4	487.48	487.1280	68.99	3.52	14.37	13.13
C28H17N5O5	503.48	503.1230	66.80	3.40	13.91	15.89
C28H17N5O6	519.48	519.1179	64.74	3.30	13.48	18.48
C28H17N5O7	535.48	535.1128	62.81	3.20	13.08	20.92
C28H17N5O8	551.48	551.1077	60.98	3.11	12.70	23.21
C28H17N5O9	567.48	567.1026	59.26	3.02	12.34	25.37
C28H18N2O1	398.47	398.1419	84.40	4.55	7.03	4.02
C28H18N2O2	414.47	414.1368	81.14	4.38	6.76	7.72
C28H18N2O3	430.47	430.1317	78.13	4.21	6.51	11.15
C28H18N2O4	446.47	446.1266	75.33	4.06	6.27	14.33
C28H18N2O5	462.47	462.1216	72.72	3.92	6.06	17.30
C28H18N2O6	478.47	478.1165	70.29	3.79	5.85	20.06
C28H18N2O7	494.46	494.1114	68.02	3.67	5.67	22.65
C28H18N2O8	510.46	510.1063	65.88	3.55	5.49	25.07
C28H18N2O9	526.46	526.1012	63.88	3.45	5.32	27.35
C28H18N4O1	426.48	426.1481	78.86	4.25	13.14	3.75
C28H18N4O2	442.48	442.1430	76.01	4.10	12.66	7.23
C28H18N4O3	458.48	458.1379	73.35	3.96	12.22	10.47
C28H18N4O4	474.48	474.1328	70.88	3.82	11.81	13.49
C28H18N4O5	490.48	490.1277	68.57	3.70	11.42	16.31
C28H18N4O6	506.48	506.1226	66.40	3.58	11.06	18.95
C28H18N4O7	522.48	522.1175	64.37	3.47	10.72	21.44
C28H18N4O8	538.48	538.1125	62.46	3.37	10.40	23.77
C28H18N4O9	554.48	554.1074	60.65	3.27	10.10	25.97
C28H18N6O1	454.50	454.1542	74.00	3.99	18.49	3.52
C28H18N6O2	470.49	470.1491	71.48	3.86	17.86	6.80
C28H18N6O3	486.49	486.1440	69.13	3.73	17.27	9.87
C28H18N6O4	502.49	502.1389	66.93	3.61	16.72	12.74
C28H18N6O5	518.49	518.1339	64.86	3.50	16.21	15.43
C28H18N6O6	534.49	534.1288	62.92	3.39	15.72	17.96
C28H18N6O7	550.49	550.1237	61.09	3.30	15.27	20.34
C28H18N6O8	566.49	566.1186	59.37	3.20	14.84	22.59
C28H18N6O9	582.49	582.1135	57.74	3.11	14.43	24.72
C28H19N1O1	385.47	385.1467	87.25	4.97	3.63	4.15
C28H19N1O2	401.47	401.1416	83.77	4.77	3.49	7.97
C28H19N1O3	417.47	417.1365	80.56	4.59	3.36	11.50
C28H19N1O4	433.47	433.1314	77.59	4.42	3.23	14.76

189

C28H19N1O5	449.47	449.1263	74.82	4.26	3.12	17.80
C28H19N1O6	465.47	465.1212	72.25	4.11	3.01	20.62
C28H19N1O7	481.47	481.1161	69.85	3.98	2.91	23.26
C28H19N1O8	497.47	497.1111	67.61	3.85	2.82	25.73
C28H19N1O9	513.46	513.1060	65.50	3.73	2.73	28.04
C28H19N3O1	413.48	413.1528	81.34	4.63	10.16	3.87
C28H19N3O2	429.48	429.1477	78.31	4.46	9.78	7.45
C28H19N3O3	445.48	445.1426	75.49	4.30	9.43	10.77
C28H19N3O4	461.48	461.1375	72.88	4.15	9.11	13.87
C28H19N3O5	477.48	477.1325	70.43	4.01	8.80	16.75
C28H19N3O6	493.48	493.1274	68.15	3.88	8.52	19.45
C28H19N3O7	509.48	509.1223	66.01	3.76	8.25	21.98
C28H19N3O8	525.48	525.1172	64.00	3.64	8.00	24.36
C28H19N3O9	541.48	541.1121	62.11	3.54	7.76	26.59
C28H19N5O1	441.50	441.1590	76.18	4.34	15.86	3.62
C28H19N5O2	457.50	457.1539	73.51	4.19	15.31	6.99
C28H19N5O3	473.50	473.1488	71.03	4.04	14.79	10.14
C28H19N5O4	489.49	489.1437	68.71	3.91	14.31	13.07
C28H19N5O5	505.49	505.1386	66.53	3.79	13.85	15.83
C28H19N5O6	521.49	521.1335	64.49	3.67	13.43	18.41
C28H19N5O7	537.49	537.1284	62.57	3.56	13.03	20.84
C28H19N5O8	553.49	553.1233	60.76	3.46	12.65	23.13
C28H19N5O9	569.49	569.1183	59.05	3.36	12.30	25.28
C28H20N2O1	400.48	400.1576	83.98	5.03	6.99	4.00
C28H20N2O2	416.48	416.1525	80.75	4.84	6.73	7.68
C28H20N2O3	432.48	432.1474	77.76	4.66	6.48	11.10
C28H20N2O4	448.48	448.1423	74.99	4.50	6.25	14.27
C28H20N2O5	464.48	464.1372	72.41	4.34	6.03	17.22
C28H20N2O6	480.48	480.1321	69.99	4.20	5.83	19.98
C28H20N2O7	496.48	496.1270	67.74	4.06	5.64	22.56
C28H20N2O8	512.48	512.1220	65.62	3.93	5.47	24.98
C28H20N2O9	528.48	528.1169	63.64	3.81	5.30	27.25
C28H20N4O1	428.50	428.1637	78.49	4.70	13.08	3.73
C28H20N4O2	444.50	444.1586	75.66	4.54	12.60	7.20
C28H20N4O3	460.50	460.1535	73.03	4.38	12.17	10.42
C28H20N4O4	476.50	476.1484	70.58	4.23	11.76	13.43
C28H20N4O5	492.50	492.1434	68.29	4.09	11.38	16.24
C28H20N4O6	508.49	508.1383	66.14	3.96	11.02	18.88
C28H20N4O7	524.49	524.1332	64.12	3.84	10.68	21.35
C28H20N4O8	540.49	540.1281	62.22	3.73	10.37	23.68
C28H20N4O9	556.49	556.1230	60.43	3.62	10.07	25.88
C28H20N6O1	456.51	456.1698	73.67	4.42	18.41	3.50
C28H20N6O2	472.51	472.1648	71.18	4.27	17.79	6.77
C28H20N6O3	488.51	488.1597	68.84	4.13	17.20	9.83
C28H20N6O4	504.51	504.1546	66.66	4.00	16.66	12.69
C28H20N6O5	520.51	520.1495	64.61	3.87	16.15	15.37
C28H20N6O6	536.51	536.1444	62.69	3.76	15.66	17.89
C28H20N6O7	552.51	552.1393	60.87	3.65	15.21	20.27
C28H20N6O8	568.51	568.1342	59.16	3.55	14.78	22.51
C28H20N6O9	584.51	584.1292	57.54	3.45	14.38	24.64
C28H21N1O1	387.49	387.1623	86.79	5.46	3.61	4.13
C28H21N1O2	403.49	403.1572	83.35	5.25	3.47	7.93
C28H21N1O3	419.48	419.1521	80.17	5.05	3.34	11.44
C28H21N1O4	435.48	435.1470	77.23	4.86	3.22	14.70
C28H21N1O5	451.48	451.1420	74.49	4.69	3.10	17.72
C28H21N1O6	467.48	467.1369	71.94	4.53	3.00	20.53
C28H21N1O7	483.48	483.1318	69.56	4.38	2.90	23.16
C28H21N1O8	499.48	499.1267	67.33	4.24	2.80	25.63
C28H21N1O9	515.48	515.1216	65.24	4.11	2.72	27.93
C28H21N3O1	415.50	415.1685	80.94	5.09	10.11	3.85
C28H21N3O2	431.50	431.1634	77.94	4.91	9.74	7.42
C28H21N3O3	447.50	447.1583	75.15	4.73	9.39	10.73
C28H21N3O4	463.50	463.1532	72.56	4.57	9.07	13.81
C28H21N3O5	479.50	479.1481	70.14	4.41	8.76	16.68
C28H21N3O6	495.50	495.1430	67.87	4.27	8.48	19.37
C28H21N3O7	511.50	511.1379	65.75	4.14	8.22	21.90
C28H21N3O8	527.49	527.1329	63.76	4.01	7.97	24.26
C28H21N3O9	543.49	543.1278	61.88	3.89	7.73	26.49
C28H21N5O1	443.51	443.1746	75.83	4.77	15.79	3.61
C28H21N5O2	459.51	459.1695	73.19	4.61	15.24	6.96
C28H21N5O3	475.51	475.1644	70.73	4.45	14.73	10.09
C28H21N5O4	491.51	491.1593	68.42	4.31	14.25	13.02
C28H21N5O5	507.51	507.1543	66.27	4.17	13.80	15.76
C28H21N5O6	523.51	523.1492	64.24	4.04	13.38	18.34
C28H21N5O7	539.51	539.1441	62.34	3.92	12.98	20.76
C28H21N5O8	555.51	555.1390	60.54	3.81	12.61	23.04
C28H21N5O9	571.51	571.1339	58.85	3.70	12.25	25.20
C28H22N2O1	402.50	402.1732	83.56	5.51	6.96	3.98
C28H22N2O2	418.50	418.1681	80.36	5.30	6.69	7.65
C28H22N2O3	434.50	434.1630	77.40	5.10	6.45	11.05
C28H22N2O4	450.50	450.1579	74.65	4.92	6.22	14.21
C28H22N2O5	466.50	466.1529	72.09	4.75	6.01	17.15
C28H22N2O6	482.50	482.1478	69.70	4.60	5.81	19.90
C28H22N2O7	498.50	498.1427	67.47	4.45	5.62	22.47
C28H22N2O8	514.50	514.1376	65.37	4.31	5.44	24.88
C28H22N2O9	530.50	530.1325	63.40	4.18	5.28	27.14
C28H22N4O1	430.51	430.1794	78.12	5.15	13.01	3.72
C28H22N4O2	446.51	446.1743	75.32	4.97	12.55	7.17
C28H22N4O3	462.51	462.1692	72.71	4.79	12.11	10.38
C28H22N4O4	478.51	478.1641	70.28	4.63	11.71	13.37
C28H22N4O5	494.51	494.1590	68.01	4.48	11.33	16.18
C28H22N4O6	510.51	510.1539	65.88	4.34	10.97	18.80
C28H22N4O7	526.51	526.1488	63.88	4.21	10.64	21.27
C28H22N4O8	542.51	542.1437	61.99	4.09	10.33	23.59
C28H22N4O9	558.51	558.1387	60.22	3.97	10.03	25.78
C28H22N6O1	458.53	458.1855	73.35	4.84	18.33	3.49
C28H22N6O2	474.53	474.1804	70.87	4.67	17.71	6.74
C28H22N6O3	490.53	490.1753	68.56	4.52	17.13	9.79
C28H22N6O4	506.53	506.1702	66.40	4.38	16.59	12.63
C28H22N6O5	522.52	522.1652	64.36	4.24	16.08	15.31
C28H22N6O6	538.52	538.1601	62.45	4.12	15.61	17.83
C28H22N6O7	554.52	554.1550	60.65	4.00	15.16	20.20
C28H22N6O8	570.52	570.1499	58.95	3.89	14.73	22.43
C28H22N6O9	586.52	586.1448	57.34	3.78	14.33	24.55
C28H23N1O1	389.50	389.1780	86.34	5.95	3.60	4.11
C28H23N1O2	405.50	405.1729	82.94	5.72	3.45	7.89
C28H23N1O3	421.50	421.1678	79.79	5.50	3.32	11.39
C28H23N1O4	437.50	437.1627	76.87	5.30	3.20	14.63
C28H23N1O5	453.50	453.1576	74.16	5.11	3.09	17.64
C28H23N1O6	469.50	469.1525	71.63	4.94	2.98	20.45
C28H23N1O7	485.50	485.1474	69.27	4.78	2.89	23.07
C28H23N1O8	501.50	501.1424	67.06	4.62	2.79	25.52
C28H23N1O9	517.50	517.1373	64.99	4.48	2.71	27.83
C28H23N3O1	417.52	417.1841	80.55	5.55	10.06	3.83
C28H23N3O2	433.51	433.1790	77.58	5.35	9.69	7.38
C28H23N3O3	449.51	449.1739	74.82	5.16	9.35	10.68
C28H23N3O4	465.51	465.1688	72.25	4.98	9.03	13.75
C28H23N3O5	481.51	481.1638	69.84	4.81	8.73	16.61
C28H23N3O6	497.51	497.1587	67.60	4.66	8.45	19.30
C28H23N3O7	513.51	513.1536	65.49	4.51	8.18	21.81

C28H23N308	529.51	529.1485	63.51	4.[illegible]8	7.94	24.17
C28H23N309	545.51	545.1434	61.65	4.25	7.70	26.40
C28H23N501	445.53	445.1902	75.49	5.20	15.72	3.59
C28H23N502	461.53	461.1852	72.87	5.02	15.17	6.93
C28H23N503	477.53	477.1801	70.43	4.85	14.67	10.05
C28H23N504	493.53	493.1750	68.14	4.70	14.19	12.97
C28H23N505	509.53	509.1699	66.00	4.55	13.74	15.70
C28H23N506	525.53	525.1648	64.00	4.41	13.33	18.27
C28H23N507	541.52	541.1597	62.10	4.28	12.93	20.68
C28H23N508	557.52	557.1546	60.32	4.16	12.56	22.96
C28H23N509	573.52	573.1496	58.64	4.04	12.21	25.11
C28H24N201	404.52	404.1889	83.14	5.98	6.93	3.96
C28H24N202	420.52	420.1838	79.98	5.75	6.66	7.61
C28H24N203	436.52	436.1787	77.04	5.54	6.42	11.00
C28H24N204	452.51	452.1736	74.32	5.35	6.19	14.14
C28H24N205	468.51	468.1685	71.78	5.16	5.98	17.07
C28H24N206	484.51	484.1634	69.41	4.99	5.78	19.81
C28H24N207	500.51	500.1583	67.19	4.83	5.60	22.38
C28H24N208	516.51	516.1533	65.11	4.68	5.42	24.78
C28H24N209	532.51	532.1482	63.16	4.54	5.26	27.04
C28H24N401	432.53	432.1950	77.75	5.59	12.95	3.70
C28H24N402	448.53	448.1899	74.98	5.39	12.49	7.13
C28H24N403	464.53	464.1848	72.40	5.21	12.06	10.33
C28H24N404	480.53	480.1797	69.99	5.03	11.66	13.32
C28H24N405	496.53	496.1747	67.73	4.87	11.28	16.11
C28H24N406	512.53	512.1696	65.62	4.72	10.93	18.73
C28H24N407	528.53	528.1645	63.63	4.58	10.60	21.19
C28H24N408	544.53	544.1594	61.76	4.44	10.29	23.51
C28H24N409	560.52	560.1543	60.00	4.32	10.00	25.69
C28H24N601	460.54	460.2011	73.03	5.25	18.25	3.47
C28H24N602	476.54	476.1961	70.57	5.08	17.64	6.71
C28H24N603	492.54	492.1910	68.28	4.91	17.06	9.74
C28H24N604	508.54	508.1859	66.13	4.76	16.53	12.58
C28H24N605	524.54	524.1808	64.12	4.61	16.02	15.25
C28H24N606	540.54	540.1757	62.22	4.48	15.55	17.76
C28H24N607	556.54	556.1706	60.43	4.35	15.10	20.12
C28H24N608	572.54	572.1655	58.74	4.23	14.68	22.36
C28H24N609	588.54	588.1605	57.14	4.11	14.28	24.47
C28H25N101	391.52	391.1936	85.90	6.44	3.58	4.09
C28H25N102	407.52	407.1885	82.53	6.18	3.44	7.85
C28H25N103	423.52	423.1834	79.41	5.95	3.31	11.33
C28H25N104	439.52	439.1783	76.52	5.73	3.19	14.56
C28H25N105	455.52	455.1733	73.83	5.53	3.07	17.56
C28H25N106	471.51	471.1682	71.33	5.34	2.97	20.36
C28H25N107	487.51	487.1631	68.99	5.17	2.87	22.97
C28H25N108	503.51	503.1580	66.79	5.00	2.78	25.42
C28H25N109	519.51	519.1529	64.74	4.85	2.70	27.72
C28H25N301	419.53	419.1998	80.16	6.01	10.02	3.81
C28H25N302	435.53	435.1947	77.22	5.79	9.65	7.35
C28H25N303	451.53	451.1896	74.48	5.58	9.31	10.63
C28H25N304	467.53	467.1845	71.93	5.39	8.99	13.69
C28H25N305	483.53	483.1794	69.55	5.21	8.69	16.54
C28H25N306	499.53	499.1743	67.33	5.04	8.41	19.22
C28H25N307	515.53	515.1692	65.24	4.89	8.15	21.72
C28H25N308	531.53	531.1641	63.27	4.74	7.91	24.08
C28H25N309	547.53	547.1591	61.42	4.60	7.67	26.30
C28H25N501	447.54	447.2059	75.15	5.63	15.65	3.57
C28H25N502	463.54	463.2008	72.55	5.44	15.11	6.90
C28H25N503	479.54	479.1957	70.13	5.25	14.60	10.01
C28H25N504	495.54	495.1906	67.87	5.09	14.13	12.91
C28H25N505	511.54	511.1856	65.74	4.93	13.69	15.64
C28H25N506	527.54	527.1805	63.75	4.78	13.28	18.20
C28H25N507	543.54	543.1754	61.87	4.64	12.88	20.60
C28H25N508	559.54	559.1703	60.11	4.50	12.52	22.88
C28H25N509	575.54	575.1652	58.43	4.38	12.17	25.02
C28H26N201	406.53	406.2045	82.73	6.45	6.89	3.94
C28H26N202	422.53	422.1994	79.59	6.20	6.63	7.57
C28H26N203	438.53	438.1943	76.69	5.98	6.39	10.95
C28H26N204	454.53	454.1892	73.99	5.77	6.16	14.08
C28H26N205	470.53	470.1842	71.48	5.57	5.95	17.00
C28H26N206	486.53	486.1791	69.12	5.39	5.76	19.73
C28H26N207	502.53	502.1740	66.92	5.22	5.57	22.29
C28H26N208	518.53	518.1689	64.86	5.05	5.40	24.68
C28H26N209	534.53	534.1638	62.92	4.90	5.24	26.94
C28H26N401	434.55	434.2106	77.39	6.03	12.89	3.68
C28H26N402	450.55	450.2056	74.65	5.82	12.44	7.10
C28H26N403	466.54	466.2005	72.09	5.62	12.01	10.29
C28H26N404	482.54	482.1954	69.70	5.43	11.61	13.26
C28H26N405	498.54	498.1903	67.46	5.26	11.24	16.05
C28H26N406	514.54	514.1852	65.36	5.09	10.89	18.66
C28H26N407	530.54	530.1801	63.39	4.94	10.56	21.11
C28H26N408	546.54	546.1750	61.53	4.80	10.25	23.42
C28H26N409	562.54	562.1700	59.78	4.66	9.96	25.60
C28H26N601	462.56	462.2168	72.71	5.67	18.17	3.46
C28H26N602	478.56	478.2117	70.28	5.48	17.56	6.69
C28H26N603	494.56	494.2066	68.00	5.30	16.99	9.71
C28H26N604	510.56	510.2015	65.87	5.13	16.46	12.53
C28H26N605	526.56	526.1965	63.87	4.98	15.96	15.19
C28H26N606	542.56	542.1914	61.99	4.83	15.49	17.69
C28H26N607	558.56	558.1863	60.21	4.69	15.05	20.05
C28H26N608	574.55	574.1812	58.53	4.56	14.63	22.28
C28H26N609	590.55	590.1761	56.95	4.44	14.23	24.38
C28H27N101	393.53	393.2093	85.46	6.92	3.56	4.07
C28H27N102	409.53	409.2042	82.12	6.65	3.42	7.81
C28H27N103	425.53	425.1991	79.03	6.40	3.29	11.28
C28H27N104	441.53	441.1940	76.17	6.16	3.17	14.49
C28H27N105	457.53	457.1889	73.51	5.95	3.06	17.48
C28H27N106	473.53	473.1838	71.02	5.75	2.96	20.27
C28H27N107	489.53	489.1787	68.70	5.56	2.86	22.88
C28H27N108	505.53	505.1737	66.53	5.38	2.77	25.32
C28H27N109	521.53	521.1686	64.49	5.22	2.69	27.61
C28H27N301	421.55	421.2154	79.78	6.46	9.97	3.80
C28H27N302	437.55	437.2103	76.86	6.22	9.60	7.31
C28H27N303	453.55	453.2052	74.15	6.00	9.26	10.58
C28H27N304	469.55	469.2001	71.63	5.80	8.95	13.63
C28H27N305	485.54	485.1951	69.26	5.61	8.65	16.48
C28H27N306	501.54	501.1900	67.06	5.43	8.38	19.14
C28H27N307	517.54	517.1849	64.98	5.26	8.12	21.64
C28H27N308	533.54	533.1798	63.03	5.10	7.88	23.99
C28H27N309	549.54	549.1747	61.20	4.95	7.65	26.20
C28H27N501	449.56	449.2215	74.81	6.05	15.58	3.56
C28H27N502	465.56	465.2165	72.24	5.85	15.04	6.87
C28H27N503	481.56	481.2114	69.84	5.65	14.54	9.97
C28H27N504	497.56	497.2063	67.59	5.47	14.08	12.86
C28H27N505	513.56	513.2012	65.49	5.30	13.64	15.58
C28H27N506	529.56	529.1961	63.51	5.14	13.22	18.13
C28H27N507	545.56	545.1910	61.65	4.99	12.84	20.53
C28H27N508	561.56	561.1859	59.89	4.85	12.47	22.79
C28H27N509	577.56	577.1809	58.23	4.71	12.13	24.93
C28H28N201	408.55	408.2202	82.32	6.91	6.86	3.92

C28H28N202	424.55	424.2151	79.22	6.65	6.60	7.54
C28H28N203	440.55	440.2100	76.34	6.41	6.36	10.90
C28H28N204	456.55	456.2049	73.66	6.18	6.14	14.02
C28H28N205	472.55	472.1998	71.17	5.97	5.93	16.93
C28H28N206	488.55	488.1947	68.84	5.78	5.73	19.65
C28H28N207	504.54	504.1896	66.66	5.59	5.55	22.20
C28H28N208	520.54	520.1845	64.61	5.42	5.38	24.59
C28H28N209	536.54	536.1795	62.68	5.26	5.22	26.84
C28H28N401	436.56	436.2263	77.04	6.46	12.83	3.66
C28H28N402	452.56	452.2212	74.31	6.24	12.38	7.07
C28H28N403	468.56	468.2161	71.78	6.02	11.96	10.24
C28H28N404	484.56	484.2110	69.41	5.82	11.56	13.21
C28H28N405	500.56	500.2060	67.19	5.64	11.19	15.98
C28H28N406	516.56	516.2009	65.11	5.46	10.85	18.58
C28H28N407	532.56	532.1958	63.15	5.30	10.52	21.03
C28H28N408	548.56	548.1907	61.31	5.14	10.21	23.33
C28H28N409	564.56	564.1856	59.57	5.00	9.92	25.51
C28H28N601	464.57	464.2324	72.39	6.08	18.09	3.44
C28H28N602	480.57	480.2274	69.98	5.87	17.49	6.66
C28H28N603	496.57	496.2223	67.73	5.68	16.92	9.67
C28H28N604	512.57	512.2172	65.61	5.51	16.40	12.49
C28H28N605	528.57	528.2121	63.63	5.34	15.90	15.13
C28H28N606	544.57	544.2070	61.76	5.18	15.43	17.63
C28H28N607	560.57	560.2019	59.99	5.03	14.99	19.98
C28H28N608	576.57	576.1968	58.33	4.90	14.58	22.20
C28H28N609	592.57	592.1918	56.75	4.76	14.18	24.30
C28H29N101	395.55	395.2249	85.02	7.39	3.54	4.04
C28H29N102	411.55	411.2198	81.72	7.10	3.40	7.78
C28H29N103	427.55	427.2147	78.66	6.84	3.28	11.23
C28H29N104	443.55	443.2096	75.82	6.59	3.16	14.43
C28H29N105	459.55	459.2046	73.18	6.36	3.05	17.41
C28H29N106	475.55	475.1995	70.72	6.15	2.95	20.19
C28H29N107	491.55	491.1944	68.42	5.95	2.85	22.78
C28H29N108	507.55	507.1893	66.26	5.76	2.76	25.22
C28H29N109	523.54	523.1842	64.24	5.58	2.68	27.50
C28H29N301	423.56	423.2310	79.40	6.90	9.92	3.78
C28H29N302	439.56	439.2260	76.51	6.65	9.56	7.28
C28H29N303	455.56	455.2209	73.82	6.42	9.22	10.54
C28H29N304	471.56	471.2158	71.32	6.20	8.91	13.57
C28H29N305	487.56	487.2107	68.98	6.00	8.62	16.41
C28H29N306	503.56	503.2056	66.79	5.80	8.34	19.06
C28H29N307	519.56	519.2005	64.73	5.63	8.09	21.56
C28H29N308	535.56	535.1954	62.80	5.46	7.85	23.90
C28H29N309	551.56	551.1904	60.97	5.30	7.62	26.11
C28H29N501	451.58	451.2372	74.48	6.47	15.51	3.54
C28H29N502	467.58	467.2321	71.93	6.25	14.98	6 34
C28H29N503	483.58	483.2270	69.55	6.04	14.48	9.93
C28H29N504	499.57	499.2219	67.32	5.85	14.02	12.81
C28H29N505	515.57	515.2169	65.23	5.67	13.58	15.52
C28H29N506	531.57	531.2118	63.27	5.50	13.17	18.06
C28H29N507	547.57	547.2067	61.42	5.34	12.79	20.45
C28H29N508	563.57	563.2016	59.68	5.19	12.43	22.71
C28H29N509	579.57	579.1965	58.03	5.04	12.08	24.85
C28H30N201	410.56	410.2358	81.91	7.37	6.82	3.90
C28H30N202	426.56	426.2307	78.84	7.09	6.57	7.50
C28H30N203	442.56	442.2256	75.99	6.83	6.33	10.85
C28H30N204	458.56	458.2205	73.34	6.59	6.11	13.96
C28H30N205	474.56	474.2155	70.87	6.37	5.90	16.86
C28H30N206	490.56	490.2104	68.56	6.16	5.71	19.57
C28H30N207	506.56	506.2053	66.39	5.97	5.53	22.11
C28H30N208	522.56	522.2002	64.36	5.79	5.36	24.49
C28H30N209	538.56	538.1951	62.45	5.61	5.20	26.74
C28H30N401	438.58	438.2419	76.68	6.89	12.77	3.65
C28H30N402	454.58	454.2369	73.98	6.65	12.33	7.04
C28H30N403	470.58	470.2318	71.47	6.43	11.91	10.20
C28H30N404	486.58	486.2267	69.12	6.21	11.51	13.15
C28H30N405	502.58	502.2216	66.92	6.02	11.15	15.92
C28H30N406	518.57	518.2165	64.85	5.83	10.80	18.51
C28H30N407	534.57	534.2114	62.91	5.66	10.48	20.95
C28H30N408	550.57	550.2063	61.08	5.49	10.18	23.25
C28H30N409	566.57	566.2013	59.36	5.34	9.89	25.42
C28H30N601	466.59	466.2481	72.08	6.48	18.01	3.43
C28H30N602	482.59	482.2430	69.69	6.27	17.41	6.63
C28H30N603	498.59	498.2379	67.45	6.06	16.86	9.63
C28H30N604	514.59	514.2328	65.36	5.88	16.33	12.44
C28H30N605	530.59	530.2278	63.38	5.70	15.84	15.08
C28H30N606	546.59	546.2227	61.53	5.53	15.38	17.56
C28H30N607	562.59	562.2176	59.78	5.38	14.94	19.91
C28H30N608	578.59	578.2125	58.13	5.23	14.53	22.12
C28H30N609	594.59	594.2074	56.56	5.09	14.13	24.22
C28H31N101	397.57	397.2406	84.59	7.86	3.52	4.02
C28H31N102	413.56	413.2355	81.32	7.56	3.39	7.74
C28H31N103	429.56	429.2304	78.29	7.27	3.26	11.17
C28H31N104	445.56	445.2253	75.48	7.01	3.14	14.36
C28H31N105	461.56	461.2202	72.86	6.77	3.03	17.33
C28H31N106	477.56	477.2151	70.42	6.54	2.93	20.10
C28H31N107	493.56	493.2100	68.14	6.33	2.84	22.69
C28H31N108	509.56	509.2049	66.00	6.13	2.75	25.12
C28H31N109	525.56	525.1999	63.99	5.95	2.67	27.40
C28H31N301	425.58	425.2467	79.02	7.34	9.87	3.76
C28H31N302	441.58	441.2416	76.16	7.08	9.52	7.25
C28H31N303	457.58	457.2365	73.50	6.83	9.18	10.49
C28H31N304	473.58	473.2314	71.02	6.60	8.87	13.51
C28H31N305	489.58	489.2264	68.69	6.38	8.58	16.34
C28H31N306	505.58	505.2213	66.52	6.18	8.31	18.99
C28H31N307	521.58	521.2162	64.48	5.99	8.06	21.47
C28H31N308	537.57	537.2111	62.56	5.81	7.82	23.81
C28H31N309	553.57	553.2060	60.75	5.64	7.59	26.01
C28H31N501	453.59	453.2528	74.14	6.89	15.44	3.53
C28H31N502	469.59	469.2478	71.62	6.65	14.91	6.81
C28H31N503	485.59	485.2427	69.26	6.43	14.42	9.88
C28H31N504	501.59	501.2376	67.05	6.23	13.96	12.76
C28H31N505	517.59	517.2325	64.98	6.04	13.53	15.46
C28H31N506	533.59	533.2274	63.03	5.86	13.12	17.99
C28H31N507	549.59	549.2223	61.19	5.69	12.74	20.38
C28H31N508	565.59	565.2172	59.46	5.52	12.38	22.63
C28H31N509	581.59	581.2122	57.83	5.37	12.04	24.76
C28H32N201	412.58	412.2514	81.51	7.82	6.79	3.88
C28H32N202	428.58	428.2464	78.47	7.53	6.54	7.47
C28H32N203	444.58	444.2413	75.65	7.26	6.30	10.80
C28H32N204	460.58	460.2362	73.02	7.00	6.08	13.90
C28H32N205	476.58	476.2311	70.57	6.77	5.88	16.79
C28H32N206	492.58	492.2260	68.28	6.55	5.69	19.49
C28H32N207	508.58	508.2209	66.13	6.34	5.51	22.02
C28H32N208	524.58	524.2158	64.11	6.15	5.34	24.40
C28H32N209	540.58	540.2108	62.21	5.97	5.18	26.64
C28H32N401	440.59	440.2576	76.33	7.32	12.72	3.63
C28H32N402	456.59	456.2525	73.66	7.06	12.27	7.01
C28H32N403	472.59	472.2474	71.16	6.83	11.86	10.16
C28H32N404	488.59	488.2423	68.83	6.60	11.47	13.10

C28H32N4O5	504.59	504.2373	66.65	6.39	11.10	15.85
C28H32N4O6	520.59	520.2322	64.60	6.20	10.76	18.44
C28H32N4O7	536.59	536.2271	62.68	6.01	10.44	20.87
C28H32N4O8	552.59	552.2220	60.86	5.84	10.14	23.16
C28H32N4O9	568.59	568.2169	59.15	5.67	9.85	25.32
C28H32N6O1	468.61	468.2637	71.77	6.88	17.93	3.41
C28H32N6O2	484.61	484.2587	69.40	6.66	17.34	6.60
C28H32N6O3	500.61	500.2536	67.18	6.44	16.79	9.59
C28H32N6O4	516.61	516.2485	65.10	6.24	16.27	12.39
C28H32N6O5	532.60	532.2434	63.14	6.06	15.78	15.02
C28H32N6O6	548.60	548.2383	61.30	5.88	15.32	17.50
C28H32N6O7	564.60	564.2332	59.57	5.71	14.88	19.84
C28H32N6O8	580.60	580.2281	57.92	5.56	14.47	22.05
C28H32N6O9	596.60	596.2231	56.37	5.41	14.09	24.14
C28H33N1O1	399.58	399.2562	84.17	8.32	3.51	4.00
C28H33N1O2	415.58	415.2511	80.93	8.00	3.37	7.70
C28H33N1O3	431.58	431.2460	77.93	7.71	3.25	11.12
C28H33N1O4	447.58	447.2409	75.14	7.43	3.13	14.30
C28H33N1O5	463.58	463.2359	72.55	7.18	3.02	17.26
C28H33N1O6	479.58	479.2308	70.13	6.94	2.92	20.02
C28H33N1O7	495.58	495.2257	67.86	6.71	2.83	22.60
C28H33N1O8	511.58	511.2206	65.74	6.50	2.74	25.02
C28H33N1O9	527.58	527.2155	63.75	6.30	2.65	27.29
C28H33N3O1	427.59	427.2623	78.65	7.78	9.83	3.74
C28H33N3O2	443.59	443.2573	75.82	7.50	9.47	7.21
C28H33N3O3	459.59	459.2522	73.18	7.24	9.14	10.44
C28H33N3O4	475.59	475.2471	70.71	6.99	8.84	13.46
C28H33N3O5	491.59	491.2420	68.41	6.77	8.55	16.27
C28H33N3O6	507.59	507.2369	66.26	6.55	8.28	18.91
C28H33N3O7	523.59	523.2318	64.23	6.35	8.03	21.39
C28H33N3O8	539.59	539.2267	62.33	6.16	7.79	23.72
C28H33N3O9	555.59	555.2217	60.53	5.99	7.56	25.92
C28H33N5O1	455.61	455.2685	73.82	7.30	15.37	3.51
C28H33N5O2	471.61	471.2634	71.31	7.05	14.85	6.79
C28H33N5O3	487.61	487.2583	68.97	6.82	14.36	9.84
C28H33N5O4	503.61	503.2532	66.78	6.60	13.91	12.71
C28H33N5O5	519.61	519.2482	64.72	6.40	13.48	15.40
C28H33N5O6	535.61	535.2431	62.79	6.21	13.08	17.92
C28H33N5O7	551.60	551.2380	60.97	6.03	12.70	20.30
C28H33N5O8	567.60	567.2329	59.25	5.86	12.34	22.55
C28H33N5O9	583.60	583.2278	57.63	5.70	12.00	24.67
C28H34N2O1	414.60	414.2671	81.12	8.27	6.76	3.86
C28H34N2O2	430.60	430.2620	78.10	7.96	6.51	7.43
C28H34N2O3	446.59	446.2569	75.31	7.67	6.27	10.75
C28H34N2O4	462.59	462.2518	72.70	7.41	6.06	13.83
C28H34N2O5	478.59	478.2468	70.27	7.16	5.85	16.72
C28H34N2O6	494.59	494.2417	68.00	6.93	5.66	19.41
C28H34N2O7	510.59	510.2366	65.87	6.71	5.49	21.93
C28H34N2O8	526.59	526.2315	63.87	6.51	5.32	24.31
C28H34N2O9	542.59	542.2264	61.98	6.32	5.16	26.54
C28H34N4O1	442.61	442.2732	75.98	7.74	12.66	3.61
C28H34N4O2	458.61	458.2682	73.33	7.47	12.22	6.98
C28H34N4O3	474.61	474.2631	70.86	7.22	11.80	10.11
C28H34N4O4	490.61	490.2580	68.55	6.99	11.42	13.04
C28H34N4O5	506.61	506.2529	66.39	6.76	11.06	15.79
C28H34N4O6	522.61	522.2478	64.35	6.56	10.72	18.37
C28H34N4O7	538.61	538.2427	62.44	6.36	10.40	20.79
C28H34N4O8	554.61	554.2376	60.64	6.18	10.10	23.08
C28H34N4O9	570.60	570.2326	58.94	6.01	9.82	25.24
C28H34N6O1	470.62	470.2794	71.46	7.28	17.86	3.40

C28H34N6O2	486.62	486.2743	69.11	7.04	17.27	6.58
C28H34N6O3	502.62	502.2692	66.91	6.82	16.72	9.55
C28H34N6O4	518.62	518.2641	64.85	6.61	16.20	12.34
C28H34N6O5	534.62	534.2590	62.91	6.41	15.72	14.96
C28H34N6O6	550.62	550.2540	61.08	6.22	15.26	17.43
C28H34N6O7	566.62	566.2489	59.35	6.05	14.83	19.77
C28H34N6O8	582.62	582.2438	57.72	5.88	14.42	21.97
C28H34N6O9	598.62	598.2387	56.18	5.73	14.04	24.05
C28H35N1O1	401.60	401.2718	83.74	8.78	3.49	3.98
C28H35N1O2	417.60	417.2668	80.54	8.45	3.35	7.66
C28H35N1O3	433.60	433.2617	77.56	8.14	3.23	11.07
C28H35N1O4	449.60	449.2566	74.80	7.85	3.12	14.23
C28H35N1O5	465.59	465.2515	72.23	7.58	3.01	17.18
C28H35N1O6	481.59	481.2464	69.83	7.33	2.91	19.93
C28H35N1O7	497.59	497.2413	67.59	7.09	2.81	22.51
C28H35N1O8	513.59	513.2362	65.48	6.87	2.73	24.92
C28H35N1O9	529.59	529.2312	63.50	6.66	2.64	27.19
C28H35N3O1	429.61	429.2780	78.28	8.21	9.78	3.72
C28H35N3O2	445.61	445.2729	75.47	7.92	9.43	7.18
C28H35N3O3	461.61	461.2678	72.86	7.64	9.10	10.40
C28H35N3O4	477.61	477.2627	70.42	7.39	8.80	13.40
C28H35N3O5	493.61	493.2577	68.13	7.15	8.51	16.21
C28H35N3O6	509.61	509.2526	65.99	6.92	8.25	18.84
C28H35N3O7	525.61	525.2475	63.99	6.71	7.99	21.31
C28H35N3O8	541.61	541.2424	62.10	6.51	7.76	23.63
C28H35N3O9	557.61	557.2373	60.31	6.33	7.54	25.82
C28H35N5O1	457.62	457.2841	73.49	7.71	15.30	3.50
C28H35N5O2	473.62	473.2791	71.01	7.45	14.79	6.76
C28H35N5O3	489.62	489.2740	68.69	7.21	14.30	9.80
C28H35N5O4	505.62	505.2689	66.51	6.98	13.85	12.66
C28H35N5O5	521.62	521.2638	64.47	6.76	13.43	15.34
C28H35N5O6	537.62	537.2587	62.56	6.56	13.03	17.86
C28H35N5O7	553.62	553.2536	60.75	6.37	12.65	20.23
C28H35N5O8	569.62	569.2485	59.04	6.19	12.29	22.47
C28H35N5O9	585.62	585.2435	57.43	6.02	11.96	24.59
C28H36N2O1	416.61	416.2827	80.73	8.71	6.72	3.84
C28H36N2O2	432.61	432.2777	77.74	8.39	6.48	7.40
C28H36N2O3	448.61	448.2726	74.97	8.09	6.24	10.70
C28H36N2O4	464.61	464.2675	72.39	7.81	6.03	13.77
C28H36N2O5	480.61	480.2624	69.98	7.55	5.83	16.64
C28H36N2O6	496.61	496.2573	67.72	7.31	5.64	19.33
C28H36N2O7	512.61	512.2522	65.61	7.08	5.46	21.85
C28H36N2O8	528.61	528.2471	63.62	6.86	5.30	24.21
C28H36N2O9	544.61	544.2421	61.75	6.66	5.14	26.44
C28H36N4O1	444.63	444.2889	75.64	8.16	12.60	3.60
C28H36N4O2	460.62	460.2838	73.01	7.88	12.16	6.95
C28H36N4O3	476.62	476.2787	70.56	7.61	11.75	10.07
C28H36N4O4	492.62	492.2736	68.27	7.37	11.37	12.99
C28H36N4O5	508.62	508.2686	66.12	7.13	11.02	15.73
C28H36N4O6	524.62	524.2635	64.11	6.92	10.68	18.30
C28H36N4O7	540.62	540.2584	62.21	6.71	10.36	20.72
C28H36N4O8	556.62	556.2533	60.42	6.52	10.07	23.00
C28H36N4O9	572.62	572.2482	58.73	6.34	9.78	25.15
C28H36N6O1	472.64	472.2950	71.16	7.68	17.78	3.39
C28H36N6O2	488.64	488.2900	68.83	7.43	17.20	6.55
C28H36N6O3	504.64	504.2849	66.64	7.19	16.65	9.51
C28H36N6O4	520.64	520.2798	64.60	6.97	16.14	12.29
C28H36N6O5	536.64	536.2747	62.67	6.76	15.66	14.91
C28H36N6O6	552.64	552.2696	60.86	6.57	15.21	17.37
C28H36N6O7	568.64	568.2645	59.14	6.38	14.78	19.70

C28H36N6O8	584.63	584.2594	57.53	6.21	14.37	21.89
C28H36N6O9	600.63	600.2544	55.99	6.04	13.99	23.97
C28H37N1O1	403.61	403.2875	83.33	9.24	3.47	3.96
C28H37N1O2	419.61	419.2824	80.15	8.89	3.34	7.63
C28H37N1O3	435.61	435.2773	77.20	8.56	3.22	11.02
C28H37N1O4	451.61	451.2722	74.47	8.26	3.10	14.17
C28H37N1O5	467.61	467.2672	71.92	7.98	3.00	17.11
C28H37N1O6	483.61	483.2621	69.54	7.71	2.90	19.85
C28H37N1O7	499.61	499.2570	67.32	7.46	2.80	22.42
C28H37N1O8	515.61	515.2519	65.23	7.23	2.72	24.82
C28H37N1O9	531.61	531.2468	63.26	7.02	2.63	27.09
C28H37N3O1	431.63	431.2936	77.92	8.64	9.74	3.71
C28H37N3O2	447.63	447.2886	75.13	8.33	9.39	7.15
C28H37N3O3	463.63	463.2835	72.54	8.04	9.06	10.35
C28H37N3O4	479.62	479.2784	70.12	7.78	8.76	13.34
C28H37N3O5	495.62	495.2733	67.86	7.52	8.48	16.14
C28H37N3O6	511.62	511.2682	65.73	7.29	8.21	18.76
C28H37N3O7	527.62	527.2631	63.74	7.07	7.96	21.23
C28H37N3O8	543.62	543.2580	61.87	6.86	7.73	23.54
C28H37N3O9	559.62	559.2530	60.10	6.66	7.51	25.73
C28H37N5O1	459.64	459.2998	73.17	8.11	15.24	3.48
C28H37N5O2	475.64	475.2947	70.71	7.84	14.72	6.73
C28H37N5O3	491.64	491.2896	68.41	7.59	14.24	9.76
C28H37N5O4	507.64	507.2845	66.25	7.35	13.80	12.61
C28H37N5O5	523.64	523.2794	64.23	7.12	13.37	15.28
C28H37N5O6	539.64	539.2744	62.32	6.91	12.98	17.79
C28H37N5O7	555.64	555.2693	60.53	6.71	12.60	20.16
C28H37N5O8	571.64	571.2642	58.83	6.52	12.25	22.39
C28H37N5O9	587.64	587.2591	57.23	6.35	11.92	24.50
C28H38N2O1	418.63	418.2984	80.34	9.15	6.69	3.82
C28H38N2O2	434.63	434.2933	77.38	8.81	6.45	7.36
C28H38N2O3	450.63	450.2882	74.63	8.50	6.22	10.65
C28H38N2O4	466.63	466.2831	72.07	8.21	6.00	13.71
C28H38N2O5	482.63	482.2781	69.68	7.94	5.80	16.58
C28H38N2O6	498.62	498.2730	67.45	7.68	5.62	19.25
C28H38N2O7	514.62	514.2679	65.35	7.44	5.44	21.76
C28H38N2O8	530.62	530.2628	63.38	7.22	5.28	24.12
C28H38N2O9	546.62	546.2577	61.53	7.01	5.12	26.34
C28H38N4O1	446.64	446.3045	75.30	8.58	12.54	3.58
C28H38N4O2	462.64	462.2995	72.69	8.28	12.11	6.92
C28H38N4O3	478.64	478.2944	70.26	8.00	11.71	10.03
C28H38N4O4	494.64	494.2893	67.99	7.74	11.33	12.94
C28H38N4O5	510.64	510.2842	65.86	7.50	10.97	15.67
C28H38N4O6	526.64	526.2791	63.86	7.27	10.64	18.23
C28H38N4O7	542.64	542.2740	61.98	7.06	10.32	20.64
C28H38N4O8	558.64	558.2689	60.20	6.86	10.03	22.91
C28H38N4O9	574.64	574.2639	58.53	6.67	9.75	25.06
C28H38N6O1	474.65	474.3107	70.85	8.07	17.71	3.37
C28H38N6O2	490.65	490.3056	68.54	7.81	17.13	6.52
C28H38N6O3	506.65	506.3005	66.38	7.56	16.59	9.47
C28H38N6O4	522.65	522.2954	64.35	7.33	16.08	12.24
C28H38N6O5	538.65	538.2903	62.44	7.11	15.60	14.85
C28H38N6O6	554.65	554.2853	60.63	6.91	15.15	17.31
C28H38N6O7	570.65	570.2802	58.93	6.71	14.73	19.63
C28H38N6O8	586.65	586.2751	57.33	6.53	14.33	21.82
C28H38N6O9	602.65	602.2700	55.81	6.36	13.95	23.89
C28H39N1O1	405.63	405.3031	82.91	9.69	3.45	3.94
C28H39N1O2	421.63	421.2981	79.77	9.32	3.32	7.59
C28H39N1O3	437.63	437.2930	76.85	8.98	3.20	10.97
C28H39N1O4	453.63	453.2879	74.14	8.67	3.09	14.11

C28H39N1O5	469.63	469.2828	71.61	8.37	2.98	17.03
C28H39N1O6	485.63	485.2777	69.25	8.09	2.88	19.77
C28H39N1O7	501.63	501.2726	67.04	7.84	2.79	22.33
C28H39N1O8	517.62	517.2675	64.97	7.59	2.71	24.73
C28H39N1O9	533.62	533.2625	63.02	7.37	2.62	26.98
C28H39N3O1	433.64	433.3093	77.56	9.07	9.69	3.69
C28H39N3O2	449.64	449.3042	74.80	8.74	9.35	7.12
C28H39N3O3	465.64	465.2991	72.23	8.44	9.02	10.31
C28H39N3O4	481.64	481.2940	69.83	8.16	8.72	13.29
C28H39N3O5	497.64	497.2890	67.58	7.90	8.44	16.08
C28H39N3O6	513.64	513.2839	65.48	7.65	8.18	18.69
C28H39N3O7	529.64	529.2788	63.50	7.42	7.93	21.15
C28H39N3O8	545.64	545.2737	61.64	7.20	7.70	23.46
C28H39N3O9	561.64	561.2686	59.88	7.00	7.48	25.64
C28H39N5O1	461.66	461.3154	72.85	8.52	15.17	3.47
C28H39N5O2	477.66	477.3104	70.41	8.23	14.66	6.70
C28H39N5O3	493.65	493.3053	68.13	7.96	14.19	9.72
C28H39N5O4	509.65	509.3002	65.99	7.71	13.74	12.56
C28H39N5O5	525.65	525.2951	63.98	7.48	13.32	15.22
C28H39N5O6	541.65	541.2900	62.09	7.26	12.93	17.72
C28H39N5O7	557.65	557.2849	60.31	7.05	12.56	20.08
C28H39N5O8	573.65	573.2798	58.63	6.85	12.21	22.31
C28H39N5O9	589.65	589.2748	57.04	6.67	11.88	24.42
C28H40N2O1	420.64	420.3140	79.95	9.59	6.66	3.80
C28H40N2O2	436.64	436.3090	77.02	9.23	6.42	7.33
C28H40N2O3	452.64	452.3039	74.30	8.91	6.19	10.60
C28H40N2O4	468.64	468.2988	71.76	8.60	5.98	13.66
C28H40N2O5	484.64	484.2937	69.39	8.32	5.78	16.51
C28H40N2O6	500.64	500.2886	67.18	8.05	5.60	19.17
C28H40N2O7	516.64	516.2835	65.10	7.80	5.42	21.68
C28H40N2O8	532.64	532.2784	63.14	7.57	5.26	24.03
C28H40N2O9	548.64	548.2734	61.30	7.35	5.11	26.25
C28H40N4O1	448.66	448.3202	74.96	8.99	12.49	3.57
C28H40N4O2	464.66	464.3151	72.38	8.68	12.06	6.89
C28H40N4O3	480.66	480.3100	69.97	8.39	11.66	9.99
C28H40N4O4	496.66	496.3049	67.72	8.12	11.28	12.89
C28H40N4O5	512.65	512.2998	65.60	7.86	10.93	15.60
C28H40N4O6	528.65	528.2948	63.62	7.63	10.60	18.16
C28H40N4O7	544.65	544.2897	61.75	7.40	10.29	20.56
C28H40N4O8	560.65	560.2846	59.99	7.19	9.99	22.83
C28H40N4O9	576.65	576.2795	58.32	6.99	9.72	24.97
C28H40N6O1	476.67	476.3263	70.55	8.46	17.63	3.36
C28H40N6O2	492.67	492.3213	68.26	8.18	17.06	6.49
C28H40N6O3	508.67	508.3162	66.12	7.93	16.52	9.44
C28H40N6O4	524.67	524.3111	64.10	7.68	16.02	12.20
C28H40N6O5	540.67	540.3060	62.20	7.46	15.54	14.80
C28H40N6O6	556.67	556.3009	60.42	7.24	15.10	17.24
C28H40N6O7	572.67	572.2958	58.73	7.04	14.68	19.56
C28H40N6O8	588.67	588.2907	57.13	6.85	14.28	21.74
C28H40N6O9	604.67	604.2857	55.62	6.67	13.90	23.81
C28H41N1O1	407.65	407.3188	82.50	10.14	3.44	3.92
C28H41N1O2	423.64	423.3137	79.39	9.76	3.31	7.55
C28H41N1O3	439.64	439.3086	76.50	9.40	3.19	10.92
C28H41N1O4	455.64	455.3035	73.81	9.07	3.07	14.05
C28H41N1O5	471.64	471.2985	71.31	8.76	2.97	16.96
C28H41N1O6	487.64	487.2934	68.97	8.47	2.87	19.69
C28H41N1O7	503.64	503.2883	66.78	8.21	2.78	22.24
C28H41N1O8	519.64	519.2832	64.72	7.95	2.70	24.63
C28H41N1O9	535.64	535.2781	62.79	7.72	2.61	26.88
C28H41N3O1	435.66	435.3249	77.20	9.49	9.65	3.67

C28H41N302	451.66	451.3199	74.46	9.15	9.30	7.08
C28H41N303	467.66	467.3148	71.91	8.84	8.99	10.26
C28H41N304	483.66	483.3097	69.54	8.54	8.69	13.23
C28H41N305	499.66	499.3046	67.31	8.27	8.41	16.01
C28H41N306	515.66	515.2995	65.22	8.01	8.15	18.62
C28H41N307	531.65	531.2944	63.26	7.77	7.90	21.07
C28H41N308	547.65	547.2893	61.41	7.55	7.67	23.37
C28H41N309	563.65	563.2843	59.67	7.33	7.45	25.55
C28H41N501	463.67	463.3311	72.53	8.91	15.10	3.45
C28H41N502	479.67	479.3260	70.11	8.62	14.60	6.67
C28H41N503	495.67	495.3209	67.85	8.34	14.13	9.68
C28H41N504	511.67	511.3158	65.73	8.08	13.69	12.51
C28H41N505	527.67	527.3107	63.74	7.83	13.27	15.16
C28H41N506	543.67	543.3057	61.86	7.60	12.88	17.66
C28H41N507	559.67	559.3006	60.09	7.38	12.51	20.01
C28H41N508	575.67	575.2955	58.42	7.18	12.17	22.23
C28H41N509	591.67	591.2904	56.84	6.98	11.84	24.34
C28H42N201	422.66	422.3297	79.57	10.02	6.63	3.79
C28H42N202	438.66	438.3246	76.67	9.65	6.39	7.29
C28H42N203	454.66	454.3195	73.97	9.31	6.16	10.56
C28H42N204	470.66	470.3144	71.46	8.99	5.95	13.60
C28H42N205	486.66	486.3094	69.11	8.70	5.76	16.44
C28H42N206	502.66	502.3043	66.91	8.42	5.57	19.10
C28H42N207	518.66	518.2992	64.84	8.16	5.40	21.59
C28H42N208	534.66	534.2941	62.90	7.92	5.24	23.94
C28H42N209	550.65	550.2890	61.07	7.69	5.09	26.15
C28H42N401	450.67	450.3358	74.62	9.39	12.43	3.55
C28H42N402	466.67	466.3308	72.07	9.07	12.01	6.86
C28H42N403	482.67	482.3257	69.68	8.77	11.61	9.94
C28H42N404	498.67	498.3206	67.44	8.49	11.24	12.83
C28H42N405	514.67	514.3155	65.35	8.23	10.89	15.54
C28H42N406	530.67	530.3104	63.37	7.98	10.56	18.09
C28H42N407	546.67	546.3053	61.52	7.74	10.25	20.49
C28H42N408	562.67	562.3002	59.77	7.52	9.96	22.75
C28H42N409	578.67	578.2952	58.12	7.32	9.68	24.88
C28H42N601	478.69	478.3420	70.26	8.84	17.56	3.34
C28H42N602	494.69	494.3369	67.98	8.56	16.99	6.47
C28H42N603	510.69	510.3318	65.86	8.29	16.46	9.40
C28H42N604	526.68	526.3267	63.85	8.04	15.96	12.15
C28H42N605	542.68	542.3216	61.97	7.80	15.49	14.74
C28H42N606	558.68	558.3166	60.20	7.58	15.04	17.18
C28H42N607	574.68	574.3115	58.52	7.37	14.62	19.49
C28H42N608	590.68	590.3064	56.94	7.17	14.23	21.67
C28H42N609	606.68	606.3013	55.43	6.98	13.85	23.73
C28H43N101	409.66	409.3344	82.10	10.58	3.42	3.91
C28H43N102	425.66	425.3294	79.01	10.18	3.29	7.52
C28H43N103	441.66	441.3243	76.15	9.81	3.17	10.87
C28H43N104	457.66	457.3192	73.49	9.47	3.06	13.98
C28H43N105	473.66	473.3141	71.00	9.15	2.96	16.89
C28H43N106	489.66	489.3090	68.68	8.85	2.86	19.60
C28H43N107	505.66	505.3039	66.51	8.57	2.77	22.15
C28H43N108	521.66	521.2988	64.47	8.31	2.69	24.54
C28H43N109	537.66	537.2938	62.55	8.06	2.61	26.78
C28H43N301	437.67	437.3406	76.84	9.90	9.60	3.66
C28H43N302	453.67	453.3355	74.13	9.55	9.26	7.05
C28H43N303	469.67	469.3304	71.61	9.23	8.95	10.22
C28H43N304	485.67	485.3253	69.25	8.92	8.65	13.18
C28H43N305	501.67	501.3202	67.04	8.64	8.38	15.95
C28H43N306	517.67	517.3152	64.97	8.37	8.12	18.54
C28H43N307	533.67	533.3101	63.02	8.12	7.87	20.99

C28H43N308	549.67	549.3050	61.18	7.89	7.64	23.29
C28H43N309	565.67	565.2999	59.45	7.66	7.43	25.46
C28H43N501	465.69	465.3467	72.22	9.31	15.04	3.44
C28H43N502	481.69	481.3417	69.82	9.00	14.54	6.64
C28H43N503	497.69	497.3366	67.58	8.71	14.07	9.64
C28H43N504	513.69	513.3315	65.47	8.44	13.63	12.46
C28H43N505	529.69	529.3264	63.49	8.18	13.22	15.10
C28H43N506	545.68	545.3213	61.63	7.94	12.83	17.59
C28H43N507	561.68	561.3162	59.88	7.72	12.47	19.94
C28H43N508	577.68	577.3111	58.22	7.50	12.12	22.16
C28H43N509	593.68	593.3061	56.65	7.30	11.80	24.25
C28H44N201	424.68	424.3453	79.19	10.44	6.60	3.77
C28H44N202	440.68	440.3403	76.32	10.06	6.36	7.26
C28H44N203	456.67	456.3352	73.64	9.71	6.13	10.51
C28H44N204	472.67	472.3301	71.15	9.38	5.93	13.54
C28H44N205	488.67	488.3250	68.82	9.08	5.73	16.37
C28H44N206	504.67	504.3199	66.64	8.79	5.55	19.02
C28H44N207	520.67	520.3148	64.59	8.52	5.38	21.51
C28H44N208	536.67	536.3097	62.67	8.26	5.22	23.85
C28H44N209	552.67	552.3047	60.85	8.02	5.07	26.05
C28H44N401	452.69	452.3515	74.29	9.80	12.38	3.53
C28H44N402	468.69	468.3464	71.76	9.46	11.95	6.83
C28H44N403	484.69	484.3413	69.39	9.15	11.56	9.90
C28H44N404	500.69	500.3362	67.17	8.86	11.19	12.78
C28H44N405	516.69	516.3311	65.09	8.58	10.84	15.48
C28H44N406	532.69	532.3261	63.14	8.33	10.52	18.02
C28H44N407	548.69	548.3210	61.29	8.08	10.21	20.41
C28H44N408	564.68	564.3159	59.56	7.85	9.92	22.67
C28H44N409	580.68	580.3108	57.92	7.64	9.65	24.80
C28H44N601	480.70	480.3576	69.96	9.23	17.48	3.33
C28H44N602	496.70	496.3526	67.71	8.93	16.92	6.44
C28H44N603	512.70	512.3475	65.60	8.65	16.39	9.36
C28H44N604	528.70	528.3424	63.61	8.39	15.90	12.10
C28H44N605	544.70	544.3373	61.74	8.14	15.43	14.69
C28H44N606	560.70	560.3322	59.98	7.91	14.99	17.12
C28H44N607	576.70	576.3271	58.32	7.69	14.57	19.42
C28H44N608	592.70	592.3220	56.74	7.48	14.18	21.60
C28H44N609	608.70	608.3170	55.25	7.29	13.81	23.66
C28H45N101	411.68	411.3501	81.69	11.02	3.40	3.89
C28H45N102	427.68	427.3450	78.64	10.61	3.28	7.48
C28H45N103	443.68	443.3399	75.80	10.22	3.16	10.82
C28H45N104	459.68	459.3348	73.16	9.87	3.05	13.92
C28H45N105	475.67	475.3298	70.70	9.54	2.94	16.82
C28H45N106	491.67	491.3247	68.40	9.23	2.85	19.52
C28H45N107	507.67	507.3196	66.25	8.93	2.76	22.06
C28H45N108	523.67	523.3145	64.22	8.66	2.67	24.44
C28H45N109	539.67	539.3094	62.32	8.40	2.60	26.68
C28H45N301	439.69	439.3562	76.49	10.32	9.56	3.64
C28H45N302	455.69	455.3512	73.80	9.95	9.22	7.02
C28H45N303	471.69	471.3461	71.30	9.62	8.91	10.18
C28H45N304	487.69	487.3410	68.96	9.30	8.62	13.12
C28H45N305	503.69	503.3359	66.77	9.01	8.34	15.88
C28H45N306	519.69	519.3308	64.71	8.73	8.09	18.47
C28H45N307	535.69	535.3257	62.78	8.47	7.84	20.91
C28H45N308	551.69	551.3206	60.96	8.22	7.62	23.20
C28H45N309	567.69	567.3156	59.24	7.99	7.40	25.37
C28H45N501	467.70	467.3624	71.91	9.70	14.97	3.42
C28H45N502	483.70	483.3573	69.53	9.38	14.48	6.62
C28H45N503	499.70	499.3522	67.30	9.08	14.02	9.61
C28H45N504	515.70	515.3471	65.21	8.80	13.58	12.41

C28H45N505	531.70	531.3420	63.25	8.53	13.17	15.05
C28H45N506	547.70	547.3370	61.40	8.28	12.79	17.53
C28H45N507	563.70	563.3319	59.66	8.05	12.42	19.87
C28H45N508	579.70	579.3268	58.01	7.82	12.08	22.08
C28H45N509	595.70	595.3217	56.46	7.61	11.76	24.17
C28H46N201	426.69	426.3610	78.82	10.87	6.57	3.75
C28H46N202	442.69	442.3559	75.97	10.47	6.33	7.23
C28H46N203	458.69	458.3508	73.32	10.11	6.11	10.46
C28H46N204	474.69	474.3457	70.85	9.77	5.90	13.48
C28H46N205	490.69	490.3406	68.54	9.45	5.71	16.30
C28H46N206	506.69	506.3356	66.37	9.15	5.53	18.95
C28H46N207	522.69	522.3305	64.34	8.87	5.36	21.43
C28H46N208	538.69	538.3254	62.43	8.61	5.20	23.76
C28H46N209	554.69	554.3203	60.63	8.36	5.05	25.96
C28H46N401	454.71	454.3671	73.96	10.20	12.32	3.52
C28H46N402	470.70	470.3621	71.45	9.85	11.90	6.80
C28H46N403	486.70	486.3570	69.10	9.53	11.51	9.86
C28H46N404	502.70	502.3519	66.90	9.22	11.15	12.73
C28H46N405	518.70	518.3468	64.84	8.94	10.80	15.42
C28H46N406	534.70	534.3417	62.90	8.67	10.48	17.95
C28H46N407	550.70	550.3366	61.07	8.42	10.17	20.34
C28H46N408	566.70	566.3315	59.35	8.18	9.89	22.59
C28H46N409	582.70	582.3265	57.72	7.96	9.62	24.71
C28H46N601	482.72	482.3733	69.67	9.61	17.41	3.31
C28H46N602	498.72	498.3682	67.44	9.30	16.85	6.42
C28H46N603	514.72	514.3631	65.34	9.01	16.33	9.33
C28H46N604	530.72	530.3580	63.37	8.74	15.84	12.06
C28H46N605	546.72	546.3529	61.51	8.48	15.37	14.63
C28H46N606	562.72	562.3479	59.77	8.24	14.93	17.06
C28H46N607	578.71	578.3428	58.11	8.01	14.52	19.35
C28H46N608	594.71	594.3377	56.55	7.80	14.13	21.52
C28H46N609	610.71	610.3326	55.07	7.59	13.76	23.58
C28H47N101	413.69	413.3657	81.30	11.45	3.39	3.87
C28H47N102	429.69	429.3607	78.27	11.03	3.26	7.45
C28H47N103	445.69	445.3556	75.46	10.63	3.14	10.77
C28H47N104	461.69	461.3505	72.84	10.26	3.03	13.86
C28H47N105	477.69	477.3454	70.40	9.92	2.93	16.75
C28H47N106	493.69	493.3403	68.12	9.60	2.84	19.44
C28H47N107	509.69	509.3352	65.98	9.29	2.75	21.97
C28H47N108	525.69	525.3301	63.98	9.01	2.66	24.35
C28H47N109	541.69	541.3251	62.09	8.75	2.59	26.58
C28H47N301	441.71	441.3719	76.14	10.73	9.51	3.62
C28H47N302	457.71	457.3668	73.48	10.35	9.18	6.99
C28H47N303	473.71	473.3617	71.00	10.00	8.87	10.13
C28H47N304	489.70	489.3566	68.68	9.67	8.58	13.07
C28H47N305	505.70	505.3515	66.50	9.37	8.31	15.82
C28H47N306	521.70	521.3465	64.46	9.08	8.05	18.40
C28H47N307	537.70	537.3414	62.55	8.81	7.81	20.83
C28H47N308	553.70	553.3363	60.74	8.56	7.59	23.12
C28H47N309	569.70	569.3312	59.03	8.32	7.38	25.28
C28H47N501	469.72	469.3780	71.60	10.09	14.91	3.41
C28H47N502	485.72	485.3730	69.24	9.75	14.42	6.59
C28H47N503	501.72	501.3679	67.03	9.44	13.96	9.57
C28H47N504	517.72	517.3628	64.96	9.15	13.53	12.36
C28H47N505	533.72	533.3577	63.01	8.88	13.12	14.99
C28H47N506	549.72	549.3526	61.18	8.62	12.74	17.46
C28H47N507	565.72	565.3475	59.45	8.37	12.38	19.80
C28H47N508	581.72	581.3424	57.81	8.14	12.04	22.00
C28H47N509	597.71	597.3374	56.27	7.93	11.72	24.09
C28H48N201	428.71	428.3766	78.45	11.29	6.53	3.73

C28H48N202	444.71	444.3716	75.63	10.88	6.30	7.20
C28H48N203	460.71	460.3665	73.00	10.50	6.08	10.42
C28H48N204	476.71	476.3614	70.55	10.15	5.88	13.42
C28H48N205	492.71	492.3563	68.26	9.82	5.69	16.24
C28H48N206	508.70	508.3512	66.11	9.51	5.51	18.87
C28H48N207	524.70	524.3461	64.10	9.22	5.34	21.34
C28H48N208	540.70	540.3410	62.20	8.95	5.18	23.67
C28H48N209	556.70	556.3360	60.41	8.69	5.03	25.87
C28H48N401	456.72	456.3828	73.64	10.59	12.27	3.50
C28H48N402	472.72	472.3777	71.14	10.23	11.85	6.77
C28H48N403	488.72	488.3726	68.81	9.90	11.46	9.82
C28H48N404	504.72	504.3675	66.63	9.59	11.10	12.68
C28H48N405	520.72	520.3624	64.59	9.29	10.76	15.36
C28H48N406	536.72	536.3574	62.66	9.01	10.44	17.89
C28H48N407	552.72	552.3523	60.85	8.75	10.14	20.26
C28H48N408	568.72	568.3472	59.14	8.51	9.85	22.51
C28H48N409	584.72	584.3421	57.52	8.27	9.58	24.63
C28H48N601	484.73	484.3889	69.38	9.98	17.34	3.30
C28H48N602	500.73	500.3839	67.16	9.66	16.78	6.39
C28H48N603	516.73	516.3788	65.08	9.36	16.26	9.29
C28H48N604	532.73	532.3737	63.13	9.08	15.78	12.01
C28H48N605	548.73	548.3686	61.29	8.82	15.32	14.58
C28H48N606	564.73	564.3635	59.55	8.57	14.88	17.00
C28H48N607	580.73	580.3584	57.91	8.33	14.47	19.29
C28H48N608	596.73	596.3533	56.36	8.11	14.08	21.45
C28H48N609	612.73	612.3483	54.89	7.90	13.72	23.50
C28H49N101	415.71	415.3814	80.90	11.88	3.37	3.85
C28H49N102	431.71	431.3763	77.90	11.44	3.24	7.41
C28H49N103	447.71	447.3712	75.12	11.03	3.13	10.72
C28H49N104	463.71	463.3661	72.53	10.65	3.02	13.80
C28H49N105	479.71	479.3610	70.11	10.30	2.92	16.68
C28H49N106	495.71	495.3560	67.85	9.96	2.83	19.37
C28H49N107	511.71	511.3509	65.72	9.65	2.74	21.89
C28H49N108	527.70	527.3458	63.73	9.36	2.65	24.26
C28H49N109	543.70	543.3407	61.86	9.08	2.58	26.48
C28H49N301	443.72	443.3875	75.79	11.13	9.47	3.61
C28H49N302	459.72	459.3825	73.16	10.74	9.14	6.96
C28H49N303	475.72	475.3774	70.70	10.38	8.83	10.09
C28H49N304	491.72	491.3723	68.40	10.04	8.55	13.02
C28H49N305	507.72	507.3672	66.24	9.73	8.28	15.76
C28H49N306	523.72	523.3621	64.22	9.43	8.02	18.33
C28H49N307	539.72	539.3570	62.31	9.15	7.79	20.75
C28H49N308	555.72	555.3519	60.52	8.89	7.56	23.03
C28H49N309	571.72	571.3469	58.82	8.64	7.35	25.19
C28H49N501	471.74	471.3937	71.29	10.47	14.85	3.39
C28H49N502	487.74	487.3886	68.95	10.13	14.36	6.56
C28H49N503	503.73	503.3835	66.76	9.80	13.90	9.53
C28H49N504	519.73	519.3784	64.71	9.50	13.47	12.31
C28H49N505	535.73	535.3733	62.78	9.22	13.07	14.93
C28H49N506	551.73	551.3683	60.96	8.95	12.69	17.40
C28H49N507	567.73	567.3632	59.24	8.70	12.34	19.73
C28H49N508	583.73	583.3581	57.61	8.46	12.00	21.93
C28H49N509	599.73	599.3530	56.08	8.24	11.68	24.01
C28H50N201	430.72	430.3923	78.08	11.70	6.50	3.71
C28H50N202	446.72	446.3872	75.28	11.28	6.27	7.16
C28H50N203	462.72	462.3821	72.68	10.89	6.05	10.37
C28H50N204	478.72	478.3770	70.25	10.53	5.85	13.37
C28H50N205	494.72	494.3719	67.98	10.19	5.66	16.17
C28H50N206	510.72	510.3669	65.85	9.87	5.49	18.80
C28H50N207	526.72	526.3618	63.85	9.57	5.32	21.26

C28H50N208	542.72	542.3567	61.97	9.29	5.16	23.58
C28H50N209	558.72	558.3516	60.19	9.02	5.01	25.77
C28H50N401	458.74	458.3984	73.31	10.99	12.21	3.49
C28H50N402	474.74	474.3934	70.84	10.62	11.80	6.74
C28H50N403	490.74	490.3883	68.53	10.27	11.42	9.78
C28H50N404	506.74	506.3832	66.37	9.95	11.06	12.63
C28H50N405	522.73	522.3781	64.34	9.64	10.72	15.30
C28H50N406	538.73	538.3730	62.43	9.35	10.40	17.82
C28H50N407	554.73	554.3679	60.63	9.09	10.10	20.19
C28H50N408	570.73	570.3628	58.93	8.83	9.82	22.43
C28H50N409	586.73	586.3578	57.32	8.59	9.55	24.54
C28H50N601	486.75	486.4046	69.09	10.35	17.27	3.29
C28H50N602	502.75	502.3995	66.89	10.02	16.72	6.36
C28H50N603	518.75	518.3944	64.83	9.72	16.20	9.25
C28H50N604	534.75	534.3893	62.89	9.42	15.72	11.97
C28H50N605	550.75	550.3842	61.06	9.15	15.26	14.53
C28H50N606	566.75	566.3792	59.34	8.89	14.83	16.94
C28H50N607	582.75	582.3741	57.71	8.65	14.42	19.22
C28H50N608	598.75	598.3690	56.17	8.42	14.04	21.38
C28H50N609	614.75	614.3639	54.71	8.20	13.67	23.42
C28H51N101	417.72	417.3970	80.51	12.31	3.35	3.83
C28H51N102	433.72	433.3920	77.54	11.85	3.23	7.38
C28H51N103	449.72	449.3869	74.78	11.43	3.11	10.67
C28H51N104	465.72	465.3818	72.21	11.04	3.01	13.74
C28H51N105	481.72	481.3767	69.81	10.67	2.91	16.61
C28H51N106	497.72	497.3716	67.57	10.33	2.81	19.29
C28H51N107	513.72	513.3665	65.47	10.01	2.73	21.80
C28H51N108	529.72	529.3614	63.49	9.70	2.64	24.16
C28H51N109	545.72	545.3564	61.63	9.42	2.57	26.39
C28H51N301	445.74	445.4032	75.45	11.53	9.43	3.59
C28H51N302	461.74	461.3981	72.84	11.13	9.10	6.93
C28H51N303	477.74	477.3930	70.40	10.76	8.80	10.05
C28H51N304	493.74	493.3879	68.12	10.41	8.51	12.96
C28H51N305	509.74	509.3828	65.98	10.08	8.24	15.69
C28H51N306	525.74	525.3778	63.97	9.78	7.99	18.26
C28H51N307	541.73	541.3727	62.08	9.49	7.76	20.67
C28H51N308	557.73	557.3676	60.30	9.22	7.53	22.95
C28H51N309	573.73	573.3625	58.62	8.96	7.32	25.10
C28H51N501	473.75	473.4093	70.99	10.85	14.78	3.38
C28H51N502	489.75	489.4043	68.67	10.50	14.30	6.53
C28H51N503	505.75	505.3992	66.50	10.16	13.85	9.49
C28H51N504	521.75	521.3941	64.46	9.85	13.42	12.27
C28H51N505	537.75	537.3890	62.54	9.56	13.02	14.88
C28H51N506	553.75	553.3839	60.73	9.28	12.65	17.34
C28H51N507	569.75	569.3788	59.03	9.02	12.29	19.66
C28H51N508	585.75	585.3737	57.42	8.78	11.96	21.85
C28H51N509	601.75	601.3687	55.89	8.54	11.64	23.93
C28H52N201	432.74	432.4079	77.72	12.11	6.47	3.70
C28H52N202	448.74	448.4029	74.95	11.68	6.24	7.13
C28H52N203	464.74	464.3978	72.37	11.28	6.03	10.33
C28H52N204	480.74	480.3927	69.96	10.90	5.83	13.31
C28H52N205	496.74	496.3876	67.70	10.55	5.64	16.10
C28H52N206	512.74	512.3825	65.59	10.22	5.46	18.72
C28H52N207	528.74	528.3774	63.61	9.91	5.30	21.18
C28H52N208	544.74	544.3723	61.74	9.62	5.14	23.50
C28H52N209	560.73	560.3673	59.98	9.35	5.00	25.68
C28H52N401	460.75	460.4141	72.99	11.38	12.16	3.47
C28H52N402	476.75	476.4090	70.54	10.99	11.75	6.71
C28H52N403	492.75	492.4039	68.25	10.64	11.37	9.74
C28H52N404	508.75	508.3988	66.11	10.30	11.01	12.58
C28H52N405	524.75	524.3937	64.09	9.99	10.68	15.24
C28H52N406	540.75	540.3887	62.19	9.69	10.36	17.75
C28H52N407	556.75	556.3836	60.41	9.41	10.06	20.12
C28H52N408	572.75	572.3785	58.72	9.15	9.78	22.35
C28H52N409	588.75	588.3734	57.12	8.90	9.52	24.46
C28H52N601	488.77	488.4202	68.81	10.72	17.19	3.27
C28H52N602	504.77	504.4152	66.63	10.38	16.65	6.34
C28H52N603	520.77	520.4101	64.58	10.06	16.14	9.22
C28H52N604	536.76	536.4050	62.66	9.76	15.66	11.92
C28H52N605	552.76	552.3999	60.84	9.48	15.20	14.47
C28H52N606	568.76	568.3948	59.13	9.22	14.78	16.88
C28H52N607	584.76	584.3897	57.51	8.96	14.37	19.15
C28H52N608	600.76	600.3846	55.98	8.72	13.99	21.31
C28H52N609	616.76	616.3795	54.53	8.50	13.63	23.35
C28H53N101	419.74	419.4127	80.12	12.73	3.34	3.81
C28H53N102	435.74	435.4076	77.18	12.26	3.21	7.34
C28H53N103	451.74	451.4025	74.45	11.83	3.10	10.63
C28H53N104	467.74	467.3974	71.90	11.42	2.99	13.68
C28H53N105	483.74	483.3923	69.52	11.04	2.90	16.54
C28H53N106	499.74	499.3873	67.30	10.69	2.80	19.21
C28H53N107	515.74	515.3822	65.21	10.36	2.72	21.72
C28H53N108	531.74	531.3771	63.25	10.05	2.63	24.07
C28H53N109	547.74	547.3720	61.40	9.75	2.56	26.29
C28H53N301	447.75	447.4188	75.11	11.93	9.38	3.57
C28H53N302	463.75	463.4138	72.52	11.52	9.06	6.90
C28H53N303	479.75	479.4087	70.10	11.14	8.76	10.00
C28H53N304	495.75	495.4036	67.84	10.78	8.48	12.91
C28H53N305	511.75	511.3985	65.72	10.44	8.21	15.63
C28H53N306	527.75	527.3934	63.73	10.12	7.96	18.19
C28H53N307	543.75	543.3883	61.85	9.82	7.73	20.60
C28H53N308	559.75	559.3832	60.08	9.54	7.51	22.87
C28H53N309	575.75	575.3782	58.41	9.28	7.30	25.01
C28H53N501	475.77	475.4250	70.69	11.23	14.72	3.36
C28H53N502	491.77	491.4199	68.39	10.86	14.24	6.51
C28H53N503	507.77	507.4148	66.23	10.52	13.79	9.45
C28H53N504	523.77	523.4097	64.21	10.20	13.37	12.22
C28H53N505	539.77	539.4046	62.31	9.90	12.97	14.82
C28H53N506	555.76	555.3996	60.51	9.61	12.60	17.27
C28H53N507	571.76	571.3945	58.82	9.34	12.25	19.59
C28H53N508	587.76	587.3894	57.22	9.09	11.92	21.78
C28H53N509	603.76	603.3843	55.70	8.85	11.60	23.85
C28H54N201	434.76	434.4236	77.36	12.52	6.44	3.68
C28H54N202	450.75	450.4185	74.61	12.08	6.21	7.10
C28H54N203	466.75	466.4134	72.05	11.66	6.00	10.28
C28H54N204	482.75	482.4083	69.67	11.27	5.80	13.26
C28H54N205	498.75	498.4032	67.43	10.91	5.62	16.04
C28H54N206	514.75	514.3982	65.33	10.57	5.44	18.65
C28H54N207	530.75	530.3931	63.37	10.26	5.28	21.10
C28H54N208	546.75	546.3880	61.51	9.96	5.12	23.41
C28H54N209	562.75	562.3829	59.76	9.67	4.98	25.59
C28H54N401	462.77	462.4297	72.67	11.76	12.11	3.46
C28H54N402	478.77	478.4247	70.25	11.37	11.70	6.68
C28H54N403	494.77	494.4196	67.97	11.00	11.32	9.70
C28H54N404	510.77	510.4145	65.84	10.66	10.97	12.53
C28H54N405	526.77	526.4094	63.84	10.33	10.64	15.19
C28H54N406	542.77	542.4043	61.96	10.03	10.32	17.69
C28H54N407	558.77	558.3992	60.19	9.74	10.03	20.04
C28H54N408	574.76	574.3941	58.51	9.47	9.75	22.27
C28H54N409	590.76	590.3891	56.93	9.21	9.48	24.37
C28H54N601	490.78	490.4359	68.53	11.09	17.12	3.26

C28H54N602	506.78	506.4308	66.36	10.74	16.58	6.31
C28H54N603	522.78	522.4257	64.33	10.41	16.08	9.18
C28H54N604	538.78	538.4206	62.42	10.10	15.60	11.88
C28H54N605	554.78	554.4155	60.62	9.81	15.15	14.42
C28H54N606	570.78	570.4105	58.92	9.54	14.72	16.82
C28H54N607	586.78	586.4054	57.32	9.28	14.32	19.09
C28H54N608	602.78	602.4003	55.79	9.03	13.94	21.23
C28H54N609	618.78	618.3952	54.35	8.80	13.58	23.27
C28H55N101	421.76	421.4283	79.74	13.14	3.32	3.79
C28H55N102	437.76	437.4233	76.83	12.66	3.20	7.31
C28H55N103	453.76	453.4182	74.12	12.22	3.09	10.58
C28H55N104	469.75	469.4131	71.59	11.80	2.98	13.62
C28H55N105	485.75	485.4080	69.24	11.41	2.88	16.47
C28H55N106	501.75	501.4029	67.03	11.05	2.79	19.13
C28H55N107	517.75	517.3978	64.96	10.71	2.71	21.63
C28H55N108	533.75	533.3927	63.01	10.39	2.62	23.98
C28H55N109	549.75	549.3877	61.18	10.08	2.55	26.19
C28H55N301	449.77	449.4345	74.77	12.33	9.34	3.56
C28H55N302	465.77	465.4294	72.21	11.90	9.02	6.87
C28H55N303	481.77	481.4243	69.81	11.51	8.72	9.96
C28H55N304	497.77	497.4192	67.56	11.14	8.44	12.86
C28H55N305	513.77	513.4141	65.46	10.79	8.18	15.57
C28H55N306	529.77	529.4091	63.48	10.46	7.93	18.12
C28H55N307	545.77	545.4040	61.62	10.16	7.70	20.52
C28H55N308	561.77	561.3989	59.87	9.87	7.48	22.78
C28H55N309	577.77	577.3938	58.21	9.60	7.27	24.92
C28H55N501	477.78	477.4406	70.39	11.60	14.66	3.35
C28H55N502	493.78	493.4356	68.11	11.23	14.18	6.48
C28H55N503	509.78	509.4305	65.97	10.87	13.74	9.42
C28H55N504	525.78	525.4254	63.96	10.54	13.32	12.17
C28H55N505	541.78	541.4203	62.08	10.23	12.93	14.77
C28H55N506	557.78	557.4152	60.29	9.94	12.56	17.21
C28H55N507	573.78	573.4101	58.61	9.66	12.21	19.52
C28H55N508	589.78	589.4050	57.02	9.40	11.87	21.70
C28H55N509	605.78	605.3999	55.52	9.15	11.56	23.77
C28H56N201	436.77	436.4392	77.00	12.92	6.41	3.66
C28H56N202	452.77	452.4342	74.28	12.47	6.19	7.07
C28H56N203	468.77	468.4291	71.74	12.04	5.98	10.24
C28H56N204	484.77	484.4240	69.38	11.64	5.78	13.20
C28H56N205	500.77	500.4189	67.16	11.27	5.59	15.97
C28H56N206	516.77	516.4138	65.08	10.92	5.42	18.58
C28H56N207	532.77	532.4087	63.13	10.59	5.26	21.02
C28H56N208	548.77	548.4036	61.29	10.29	5.10	23.32
C28H56N209	564.77	564.3986	59.55	9.99	4.96	25.50
C28H56N401	464.78	464.4454	72.36	12.14	12.05	3.44
C28H56N402	480.78	480.4403	69.95	11.74	11.65	6.66
C28H56N403	496.78	496.4352	67.70	11.36	11.28	9.66
C28H56N404	512.78	512.4301	65.59	11.01	10.93	12.48
C28H56N405	528.78	528.4250	63.60	10.67	10.60	15.13
C28H56N406	544.78	544.4200	61.73	10.36	10.28	17.62
C28H56N407	560.78	560.4149	59.97	10.07	9.99	19.97
C28H56N408	576.78	576.4098	58.31	9.79	9.71	22.19
C28H56N409	592.78	592.4047	56.73	9.52	9.45	24.29
C28H56N601	492.80	492.4515	68.25	11.45	17.05	3.25
C28H56N602	508.80	508.4464	66.10	11.09	16.52	6.29
C28H56N603	524.80	524.4414	64.08	10.76	16.01	9.15
C28H56N604	540.80	540.4363	62.19	10.44	15.54	11.83
C28H56N605	556.80	556.4312	60.40	10.14	15.09	14.37
C28H56N606	572.80	572.4261	58.71	9.85	14.67	16.76
C28H56N607	588.79	588.4210	57.12	9.59	14.27	19.02
C28H56N608	604.79	604.4159	55.61	9.33	13.90	21.16
C28H56N609	620.79	620.4108	54.17	9.09	13.54	23.20
C28H57N101	423.77	423.4440	79.36	13.56	3.31	3.78
C28H57N102	439.77	439.4389	76.47	13.06	3.18	7.28
C28H57N103	455.77	455.4338	73.79	12.61	3.07	10.53
C28H57N104	471.77	471.4287	71.29	12.18	2.97	13.57
C28H57N105	487.77	487.4236	68.95	11.78	2.87	16.40
C28H57N106	503.77	503.4186	66.76	11.40	2.78	19.06
C28H57N107	519.77	519.4135	64.70	11.05	2.69	21.55
C28H57N108	535.77	535.4084	62.77	10.72	2.61	23.89
C28H57N109	551.77	551.4033	60.95	10.41	2.54	26.10
C28H57N301	451.79	451.4501	74.44	12.72	9.30	3.54
C28H57N302	467.79	467.4451	71.89	12.28	8.98	6.84
C28H57N303	483.78	483.4400	69.52	11.88	8.69	9.92
C28H57N304	499.78	499.4349	67.29	11.50	8.41	12.81
C28H57N305	515.78	515.4298	65.20	11.14	8.15	15.51
C28H57N306	531.78	531.4247	63.24	10.80	7.90	18.05
C28H57N307	547.78	547.4196	61.40	10.49	7.67	20.45
C28H57N308	563.78	563.4145	59.65	10.19	7.45	22.70
C28H57N309	579.78	579.4095	58.01	9.91	7.25	24.84
C28H57N501	479.80	479.4563	70.09	11.97	14.60	3.33
C28H57N502	495.80	495.4512	67.83	11.59	14.13	6.45
C28H57N503	511.80	511.4461	65.71	11.23	13.68	9.38
C28H57N504	527.80	527.4410	63.72	10.89	13.27	12.13
C28H57N505	543.80	543.4359	61.85	10.57	12.88	14.71
C28H57N506	559.80	559.4309	60.08	10.26	12.51	17.15
C28H57N507	575.80	575.4258	58.41	9.98	12.16	19.45
C28H57N508	591.80	591.4207	56.83	9.71	11.83	21.63
C28H57N509	607.79	607.4156	55.33	9.45	11.52	23.69
C28H58N201	438.79	438.4549	76.65	13.32	6.38	3.65
C28H58N202	454.79	454.4498	73.95	12.85	6.16	7.04
C28H58N203	470.79	470.4447	71.44	12.42	5.95	10.20
C28H58N204	486.79	486.4396	69.09	12.01	5.75	13.15
C28H58N205	502.78	502.4345	66.89	11.63	5.57	15.91
C28H58N206	518.78	518.4295	64.83	11.27	5.40	18.50
C28H58N207	534.78	534.4244	62.89	10.93	5.24	20.94
C28H58N208	550.78	550.4193	61.06	10.61	5.09	23.24
C28H58N209	566.78	566.4142	59.34	10.31	4.94	25.41
C28H58N401	466.80	466.4610	72.05	12.52	12.00	3.43
C28H58N402	482.80	482.4560	69.66	12.11	11.60	6.63
C28H58N403	498.80	498.4509	67.42	11.72	11.23	9.62
C28H58N404	514.80	514.4458	65.33	11.36	10.88	12.43
C28H58N405	530.80	530.4407	63.36	11.01	10.56	15.07
C28H58N406	546.80	546.4356	61.51	10.69	10.25	17.56
C28H58N407	562.80	562.4305	59.76	10.39	9.96	19.90
C28H58N408	578.80	578.4254	58.11	10.10	9.68	22.11
C28H58N409	594.80	594.4203	56.54	9.83	9.42	24.21
C28H58N601	494.81	494.4672	67.97	11.81	16.98	3.23
C28H58N602	510.81	510.4621	65.84	11.44	16.45	6.26
C28H58N603	526.81	526.4570	63.84	11.10	15.95	9.11
C28H58N604	542.81	542.4519	61.96	10.77	15.48	11.79
C28H58N605	558.81	558.4468	60.18	10.46	15.04	14.32
C28H58N606	574.81	574.4418	58.51	10.17	14.62	16.70
C28H58N607	590.81	590.4367	56.92	9.90	14.22	18.96
C28H58N608	606.81	606.4316	55.42	9.63	13.85	21.09
C28H58N609	622.81	622.4265	54.00	9.39	13.49	23.12
C28H59N101	425.79	425.4596	78.99	13.97	3.29	3.76
C28H59N102	441.79	441.4546	76.13	13.46	3.17	7.24
C28H59N103	457.79	457.4495	73.46	12.99	3.06	10.48
C28H59N104	473.79	473.4444	70.98	12.55	2.96	13.51

C28H59N105	489.79	489.4393	68.67	12.14	2.86	16.33
C28H59N106	505.79	505.4342	66.45	11.76	2.77	18.98
C28H59N107	521.78	521.4291	64.45	11.40	2.68	21.46
C28H59N108	537.78	537.4240	62.54	11.06	2.60	23.80
C28H59N109	553.78	553.4190	60.73	10.74	2.53	26.00
C28H59N301	453.80	453.4658	74.11	13.10	9.26	3.53
C28H59N302	469.80	469.4607	71.59	12.66	8.94	6.81
C28H59N303	485.80	485.4556	69.23	12.24	8.65	9.88
C28H59N304	501.80	501.4505	67.02	11.85	8.37	12.75
C28H59N305	517.80	517.4454	64.95	11.49	8.12	15.45
C28H59N306	533.80	533.4404	63.00	11.14	7.87	17.98
C28H59N307	549.80	549.4353	61.17	10.82	7.64	20.37
C28H59N308	565.80	565.4302	59.44	10.51	7.43	22.62
C28H59N309	581.80	581.4251	57.81	10.22	7.22	24.75
C28H59N501	481.82	481.4719	69.80	12.34	14.54	3.32
C28H59N502	497.81	497.4668	67.56	11.95	14.07	6.43
C28H59N503	513.81	513.4618	65.45	11.57	13.63	9.34
C28H59N504	529.81	529.4567	63.48	11.22	13.22	12.08
C28H59N505	545.81	545.4516	61.62	10.90	12.83	14.66
C28H59N506	561.81	561.4465	59.86	10.59	12.47	17.09
C28H59N507	577.81	577.4414	58.20	10.29	12.12	19.38
C28H59N508	593.81	593.4363	56.64	10.02	11.79	21.55
C28H59N509	609.81	609.4312	55.15	9.75	11.48	23.61
C28H60N201	440.80	440.4705	76.30	13.72	6.36	3.63
C28H60N202	456.80	456.4655	73.62	13.24	6.13	7.00
C28H60N203	472.80	472.4604	71.13	12.79	5.92	10.15
C28H60N204	488.80	488.4553	68.80	12.37	5.73	13.09
C28H60N205	504.80	504.4502	66.62	11.98	5.55	15.85
C28H60N206	520.80	520.4451	64.58	11.61	5.38	18.43
C28H60N207	536.80	536.4400	62.65	11.27	5.22	20.86
C28H60N208	552.80	552.4349	60.84	10.94	5.07	23.15
C28H60N209	568.80	568.4299	59.13	10.63	4.93	25.32
C28H60N401	468.82	468.4767	71.74	12.90	11.95	3.41
C28H60N402	484.82	484.4716	69.37	12.47	11.56	6.60
C28H60N403	500.82	500.4665	67.15	12.08	11.19	9.58
C28H60N404	516.81	516.4614	65.07	11.70	10.84	12.38
C28H60N405	532.81	532.4563	63.12	11.35	10.52	15.01
C28H60N406	548.81	548.4513	61.28	11.02	10.21	17.49
C28H60N407	564.81	564.4462	59.54	10.71	9.92	19.83
C28H60N408	580.81	580.4411	57.90	10.41	9.65	22.04
C28H60N409	596.81	596.4360	56.35	10.13	9.39	24.13
C28H60N601	496.83	496.4828	67.69	12.17	16.92	3.22
C28H60N602	512.83	512.4777	65.58	11.79	16.39	6.24
C28H60N603	528.83	528.4727	63.60	11.44	15.89	9.08
C28H60N604	544.83	544.4676	61.73	11.10	15.43	11.75
C28H60N605	560.83	560.4625	59.97	10.78	14.99	14.26
C28H60N606	576.83	576.4574	58.30	10.48	14.57	16.64
C28H60N607	592.83	592.4523	56.73	10.20	14.18	18.89
C28H60N608	608.83	608.4472	55.24	9.93	13.80	21.02
C28H60N609	624.83	624.4421	53.83	9.68	13.45	23.05
C28H61N301	455.82	455.4814	73.78	13.49	9.22	3.51
C28H61N302	471.82	471.4764	71.28	13.03	8.91	6.78
C28H61N303	487.82	487.4713	68.94	12.60	8.61	9.84
C28H61N304	503.82	503.4662	66.75	12.20	8.34	12.70
C28H61N305	519.82	519.4611	64.70	11.83	8.08	15.39
C28H61N306	535.81	535.4560	62.77	11.48	7.84	17.92
C28H61N307	551.81	551.4509	60.95	11.14	7.61	20.30
C28H61N308	567.81	567.4458	59.23	10.83	7.40	22.54
C28H61N309	583.81	583.4407	57.61	10.53	7.20	24.66
C28H61N501	483.83	483.4876	69.51	12.71	14.47	3.31

C28H61N502	499.83	499.4825	67.29	12.30	14.01	6.40
C28H61N503	515.83	515.4774	65.20	11.92	13.58	9.31
C28H61N504	531.83	531.4723	63.24	11.56	13.17	12.03
C28H61N505	547.83	547.4672	61.39	11.22	12.78	14.60
C28H61N506	563.83	563.4622	59.65	10.91	12.42	17.03
C28H61N507	579.83	579.4571	58.00	10.60	12.08	19.32
C28H61N508	595.83	595.4520	56.44	10.32	11.75	21.48
C28H61N509	611.83	611.4469	54.97	10.05	11.45	23.54
C28H62N401	470.83	470.4923	71.43	13.27	11.90	3.40
C28H62N402	486.83	486.4872	69.08	12.84	11.51	6.57
C28H62N403	502.83	502.4822	66.88	12.43	11.14	9.55
C28H62N404	518.83	518.4771	64.82	12.05	10.80	12.33
C28H62N405	534.83	534.4720	62.88	11.68	10.48	14.96
C28H62N406	550.83	550.4669	61.06	11.35	10.17	17.43
C28H62N407	566.83	566.4618	59.33	11.03	9.88	19.76
C28H62N408	582.83	582.4567	57.70	10.72	9.61	21.96
C28H62N409	598.83	598.4516	56.16	10.44	9.36	24.05
C28H62N601	498.85	498.4985	67.42	12.53	16.85	3.21
C28H62N602	514.85	514.4934	65.32	12.14	16.32	6.22
C28H62N603	530.84	530.4883	63.35	11.77	15.83	9.04
C28H62N604	546.84	546.4832	61.50	11.43	15.37	11.70
C28H62N605	562.84	562.4781	59.75	11.10	14.93	14.21
C28H62N606	578.84	578.4731	58.10	10.80	14.52	16.58
C28H62N607	594.84	594.4680	56.54	10.51	14.13	18.83
C28H62N608	610.84	610.4629	55.06	10.23	13.76	20.95
C28H62N609	626.84	626.4578	53.65	9.97	13.41	22.97
C28H63N501	485.85	485.5032	69.22	13.07	14.41	3.29
C28H63N502	501.85	501.4982	67.01	12.65	13.96	6.38
C28H63N503	517.85	517.4931	64.94	12.26	13.52	9.27
C28H63N504	533.85	533.4880	63.00	11.90	13.12	11.99
C28H63N505	549.84	549.4829	61.16	11.55	12.74	14.55
C28H63N506	565.84	565.4778	59.44	11.22	12.38	16.97
C28H63N507	581.84	581.4727	57.80	10.91	12.04	19.25
C28H63N508	597.84	597.4676	56.25	10.62	11.71	21.41
C28H63N509	613.84	613.4625	54.79	10.35	11.41	23.46
C28H64N601	500.86	500.5141	67.15	12.88	16.78	3.19
C28H64N602	516.86	516.5090	65.07	12.48	16.26	6.19
C28H64N603	532.86	532.5040	63.11	12.11	15.77	9.01
C28H64N604	548.86	548.4989	61.27	11.75	15.31	11.66
C28H64N605	564.86	564.4938	59.54	11.42	14.88	14.16
C28H64N606	580.86	580.4887	57.90	11.11	14.47	16.53
C28H64N607	596.86	596.4836	56.35	10.81	14.08	18.76
C28H64N608	612.86	612.4785	54.88	10.53	13.71	20.88
C28H64N609	628.86	628.4734	53.48	10.26	13.36	22.90
C29H14N201	406.45	406.1106	85.70	3.47	6.89	3.94
C29H14N202	422.45	422.1055	82.45	3.34	6.63	7.57
C29H14N203	438.45	438.1004	79.44	3.22	6.39	10.95
C29H14N204	454.45	454.0953	76.65	3.11	6.16	14.08
C29H14N205	470.45	470.0903	74.04	3.00	5.95	17.00
C29H14N206	486.44	486.0852	71.61	2.90	5.76	19.73
C29H14N207	502.44	502.0801	69.33	2.81	5.58	22.29
C29H14N208	518.44	518.0750	67.19	2.72	5.40	24.69
C29H14N209	534.44	534.0699	65.18	2.64	5.24	26.94
C29H14N401	434.46	434.1168	80.17	3.25	12.90	3.68
C29H14N402	450.46	450.1117	77.33	3.13	12.44	7.10
C29H14N403	466.46	466.1066	74.67	3.03	12.01	10.29
C29H14N404	482.46	482.1015	72.20	2.92	11.61	13.26
C29H14N405	498.46	498.0964	69.88	2.83	11.24	16.05
C29H14N406	514.46	514.0913	67.71	2.74	10.89	18.66
C29H14N407	530.46	530.0862	65.66	2.66	10.56	21.11

C29H14N408	546.46	546.0812	63.74	2.58	10.25	23.42
C29H14N409	562.46	562.0761	61.93	2.51	9.96	25.60
C29H14N601	462.47	462.1229	75.32	3.05	18.17	3.46
C29H14N602	478.47	478.1178	72.80	2.95	17.56	6.69
C29H14N603	494.47	494.1127	70.44	2.85	17.00	9.71
C29H14N604	510.47	510.1076	68.24	2.76	16.46	12.54
C29H14N605	526.47	526.1026	66.16	2.68	15.96	15.19
C29H14N606	542.47	542.0975	64.21	2.60	15.49	17.70
C29H14N607	558.47	558.0924	62.37	2.53	15.05	20.05
C29H14N608	574.47	574.0873	60.63	2.46	14.63	22.28
C29H14N609	590.47	590.0822	58.99	2.39	14.23	24.39
C29H15N101	393.45	393.1154	88.53	3.84	3.56	4.07
C29H15N102	409.45	409.1103	85.07	3.69	3.42	7.82
C29H15N103	425.45	425.1052	81.87	3.55	3.29	11.28
C29H15N104	441.45	441.1001	78.90	3.42	3.17	14.50
C29H15N105	457.45	457.0950	76.15	3.31	3.06	17.49
C29H15N106	473.45	473.0899	73.57	3.19	2.96	20.28
C29H15N107	489.45	489.0848	71.17	3.09	2.86	22.88
C29H15N108	505.44	505.0798	68.91	2.99	2.77	25.32
C29H15N109	521.44	521.0747	66.80	2.90	2.69	27.61
C29H15N301	421.46	421.1215	82.65	3.59	9.97	3.80
C29H15N302	437.46	437.1164	79.62	3.46	9.61	7.31
C29H15N303	453.46	453.1113	76.81	3.33	9.27	10.58
C29H15N304	469.46	469.1062	74.20	3.22	8.95	13.63
C29H15N305	485.46	485.1012	71.75	3.11	8.66	16.48
C29H15N306	501.46	501.0961	69.46	3.02	8.38	19.14
C29H15N307	517.46	517.0910	67.31	2.92	8.12	21.64
C29H15N308	533.46	533.0859	65.30	2.83	7.88	23.99
C29H15N309	549.46	549.0808	63.39	2.75	7.65	26.21
C29H15N501	449.48	449.1277	77.50	3.36	15.58	3.56
C29H15N502	465.48	465.1226	74.83	3.25	15.05	6.87
C29H15N503	481.47	481.1175	72.35	3.14	14.55	9.97
C29H15N504	497.47	497.1124	70.02	3.04	14.08	12.86
C29H15N505	513.47	513.1073	67.84	2.94	13.64	15.58
C29H15N506	529.47	529.1022	65.79	2.86	13.23	18.13
C29H15N507	545.47	545.0971	63.86	2.77	12.84	20.53
C29H15N508	561.47	561.0921	62.04	2.69	12.47	22.80
C29H15N509	577.47	577.0870	60.32	2.62	12.13	24.94
C29H16N201	408.46	408.1263	85.28	3.95	6.86	3.92
C29H16N202	424.46	424.1212	82.06	3.80	6.60	7.54
C29H16N203	440.46	440.1161	79.08	3.66	6.36	10.90
C29H16N204	456.46	456.1110	76.31	3.53	6.14	14.02
C29H16N205	472.46	472.1059	73.73	3.41	5.93	16.93
C29H16N206	488.46	488.1008	71.31	3.30	5.74	19.65
C29H16N207	504.46	504.0957	69.05	3.20	5.55	22.20
C29H16N208	520.46	520.0907	66.93	3.10	5.38	24.59
C29H16N209	536.46	536.0856	64.93	3.01	5.22	26.84
C29H16N401	436.48	436.1324	79.80	3.69	12.84	3.67
C29H16N402	452.48	452.1273	76.98	3.56	12.38	7.07
C29H16N403	468.48	468.1222	74.35	3.44	11.96	10.25
C29H16N404	484.48	484.1171	71.90	3.33	11.56	13.21
C29H16N405	500.47	500.1121	69.60	3.22	11.19	15.98
C29H16N406	516.47	516.1070	67.44	3.12	10.85	18.59
C29H16N407	532.47	532.1019	65.42	3.03	10.52	21.03
C29H16N408	548.47	548.0968	63.51	2.94	10.22	23.34
C29H16N409	564.47	564.0917	61.71	2.86	9.93	25.51
C29H16N601	464.49	464.1386	74.99	3.47	18.09	3.44
C29H16N602	480.49	480.1335	72.49	3.36	17.49	6.66
C29H16N603	496.49	496.1284	70.16	3.25	16.93	9.67
C29H16N604	512.49	512.1233	67.97	3.15	16.40	12.49

C29H16N605	528.49	528.1182	65.91	3.05	15.90	15.14
C29H16N606	544.49	544.1131	63.97	2.96	15.43	17.63
C29H16N607	560.49	560.1080	62.15	2.88	14.99	19.98
C29H16N608	576.49	576.1029	60.42	2.80	14.58	22.20
C29H16N609	592.49	592.0979	58.79	2.72	14.18	24.30
C29H17N101	395.46	395.1310	88.08	4.33	3.54	4.05
C29H17N102	411.46	411.1259	84.65	4.16	3.40	7.78
C29H17N103	427.46	427.1208	81.49	4.01	3.28	11.23
C29H17N104	443.46	443.1157	78.55	3.86	3.16	14.43
C29H17N105	459.46	459.1107	75.81	3.73	3.05	17.41
C29H17N106	475.46	475.1056	73.26	3.60	2.95	20.19
C29H17N107	491.46	491.1005	70.88	3.49	2.85	22.79
C29H17N108	507.46	507.0954	68.64	3.38	2.76	25.22
C29H17N109	523.46	523.0903	66.54	3.27	2.68	27.51
C29H17N301	423.48	423.1372	82.25	4.05	9.92	3.78
C29H17N302	439.48	439.1321	79.26	3.90	9.56	7.28
C29H17N303	455.48	455.1270	76.47	3.76	9.23	10.54
C29H17N304	471.48	471.1219	73.88	3.63	8.91	13.57
C29H17N305	487.48	487.1168	71.45	3.52	8.62	16.41
C29H17N306	503.48	503.1117	69.18	3.40	8.35	19.07
C29H17N307	519.47	519.1066	67.05	3.30	8.09	21.56
C29H17N308	535.47	535.1016	65.05	3.20	7.85	23.90
C29H17N309	551.47	551.0965	63.16	3.11	7.62	26.11
C29H17N501	451.49	451.1433	77.15	3.80	15.51	3.54
C29H17N502	467.49	467.1382	74.51	3.67	14.98	6.84
C29H17N503	483.49	483.1331	72.04	3.54	14.48	9.93
C29H17N504	499.49	499.1280	69.74	3.43	14.02	12.81
C29H17N505	515.49	515.1230	67.57	3.32	13.59	15.52
C29H17N506	531.49	531.1179	65.54	3.22	13.18	18.06
C29H17N507	547.49	547.1128	63.62	3.13	12.79	20.46
C29H17N508	563.49	563.1077	61.82	3.04	12.43	22.71
C29H17N509	579.49	579.1026	60.11	2.96	12.09	24.85
C29H18N201	410.48	410.1419	84.86	4.42	6.82	3.90
C29H18N202	426.48	426.1368	81.67	4.25	6.57	7.50
C29H18N203	442.48	442.1317	78.72	4.10	6.33	10.85
C29H18N204	458.48	458.1266	75.97	3.96	6.11	13.96
C29H18N205	474.48	474.1216	73.41	3.82	5.90	16.86
C29H18N206	490.48	490.1165	71.02	3.70	5.71	19.57
C29H18N207	506.48	506.1114	68.77	3.58	5.53	22.11
C29H18N208	522.48	522.1063	66.67	3.47	5.36	24.50
C29H18N209	538.47	538.1012	64.69	3.37	5.20	26.74
C29H18N401	438.49	438.1481	79.44	4.14	12.78	3.65
C29H18N402	454.49	454.1430	76.64	3.99	12.33	7.04
C29H18N403	470.49	470.1379	74.03	3.86	11.91	10.20
C29H18N404	486.49	486.1328	71.60	3.73	11.52	13.15
C29H18N405	502.49	502.1277	69.32	3.61	11.15	15.92
C29H18N406	518.49	518.1226	67.18	3.50	10.81	18.51
C29H18N407	534.49	534.1175	65.17	3.39	10.48	20.95
C29H18N408	550.49	550.1125	63.28	3.30	10.18	23.25
C29H18N409	566.49	566.1074	61.49	3.20	9.89	25.42
C29H18N601	466.51	466.1542	74.67	3.89	18.01	3.43
C29H18N602	482.51	482.1491	72.19	3.76	17.42	6.63
C29H18N603	498.51	498.1440	69.87	3.64	16.86	9.63
C29H18N604	514.50	514.1389	67.70	3.53	16.33	12.44
C29H18N605	530.50	530.1339	65.66	3.42	15.84	15.08
C29H18N606	546.50	546.1288	63.74	3.32	15.38	17.57
C29H18N607	562.50	562.1237	61.92	3.23	14.94	19.91
C29H18N608	578.50	578.1186	60.21	3.14	14.53	22.13
C29H18N609	594.50	594.1135	58.59	3.05	14.14	24.22
C29H19N101	397.48	397.1467	87.63	4.82	3.52	4.03

C29H19N102	413.48	413.1416	84.24	4.63	3.39	7.74
C29H19N103	429.48	429.1365	81.10	4.46	3.26	11.18
C29H19N104	445.48	445.1314	78.19	4.30	3.14	14.37
C29H19N105	461.48	461.1263	75.48	4.15	3.04	17.33
C29H19N106	477.48	477.1212	72.95	4.01	2.93	20.10
C29H19N107	493.48	493.1161	70.59	3.88	2.84	22.70
C29H19N108	509.48	509.1111	68.37	3.76	2.75	25.12
C29H19N109	525.48	525.1060	66.29	3.64	2.67	27.40
C29H19N301	425.49	425.1528	81.86	4.50	9.88	3.76
C29H19N302	441.49	441.1477	78.90	4.34	9.52	7.25
C29H19N303	457.49	457.1426	76.14	4.19	9.18	10.49
C29H19N304	473.49	473.1375	73.56	4.04	8.87	13.52
C29H19N305	489.49	489.1325	71.16	3.91	8.58	16.34
C29H19N306	505.49	505.1274	68.91	3.79	8.31	18.99
C29H19N307	521.49	521.1223	66.79	3.67	8.06	21.48
C29H19N308	537.49	537.1172	64.81	3.56	7.82	23.81
C29H19N309	553.49	553.1121	62.93	3.46	7.59	26.02
C29H19N501	453.51	453.1590	76.81	4.22	15.44	3.53
C29H19N502	469.51	469.1539	74.19	4.08	14.92	6.82
C29H19N503	485.51	485.1488	71.74	3.94	14.42	9.89
C29H19N504	501.51	501.1437	69.46	3.82	13.96	12.76
C29H19N505	517.51	517.1386	67.31	3.70	13.53	15.46
C29H19N506	533.50	533.1335	65.29	3.59	13.13	17.99
C29H19N507	549.50	549.1284	63.39	3.49	12.74	20.38
C29H19N508	565.50	565.1233	61.60	3.39	12.38	22.63
C29H19N509	581.50	581.1183	59.90	3.29	12.04	24.76
C29H20N201	412.50	412.1576	84.44	4.89	6.79	3.88
C29H20N202	428.49	428.1525	81.29	4.70	6.54	7.47
C29H20N203	444.49	444.1474	78.36	4.54	6.30	10.80
C29H20N204	460.49	460.1423	75.64	4.38	6.08	13.90
C29H20N205	476.49	476.1372	73.10	4.23	5.88	16.79
C29H20N206	492.49	492.1321	70.73	4.09	5.69	19.49
C29H20N207	508.49	508.1270	68.50	3.96	5.51	22.03
C29H20N208	524.49	524.1220	66.41	3.84	5.34	24.40
C29H20N209	540.49	540.1169	64.45	3.73	5.18	26.64
C29H20N401	440.51	440.1637	79.07	4.58	12.72	3.63
C29H20N402	456.51	456.1586	76.30	4.42	12.27	7.01
C29H20N403	472.51	472.1535	73.72	4.27	11.86	10.16
C29H20N404	488.51	488.1484	71.30	4.13	11.47	13.10
C29H20N405	504.51	504.1434	69.04	4.00	11.11	15.86
C29H20N406	520.51	520.1383	66.92	3.87	10.76	18.44
C29H20N407	536.51	536.1332	64.92	3.76	10.44	20.88
C29H20N408	552.50	552.1281	63.04	3.65	10.14	23.17
C29H20N409	568.50	568.1230	61.27	3.55	9.86	25.33
C29H20N601	468.52	468.1698	74.35	4.30	17.94	3.41
C29H20N602	484.52	484.1648	71.89	4.16	17.34	6.60
C29H20N603	500.52	500.1597	69.59	4.03	16.79	9.59
C29H20N604	516.52	516.1546	67.44	3.90	16.27	12.39
C29H20N605	532.52	532.1495	65.41	3.79	15.78	15.02
C29H20N606	548.52	548.1444	63.50	3.68	15.32	17.50
C29H20N607	564.52	564.1393	61.70	3.57	14.89	19.84
C29H20N608	580.52	580.1342	60.00	3.47	14.48	22.05
C29H20N609	596.52	596.1292	58.39	3.38	14.09	24.14
C29H21N101	399.50	399.1623	87.19	5.30	3.51	4.00
C29H21N102	415.50	415.1572	83.83	5.09	3.37	7.70
C29H21N103	431.50	431.1521	80.72	4.91	3.25	11.12
C29H21N104	447.50	447.1470	77.84	4.73	3.13	14.30
C29H21N105	463.49	463.1420	75.15	4.57	3.02	17.26
C29H21N106	479.49	479.1369	72.64	4.41	2.92	20.02
C29H21N107	495.49	495.1318	70.30	4.27	2.83	22.60
C29H21N108	511.49	511.1267	68.10	4.14	2.74	25.02
C29H21N109	527.49	527.1216	66.03	4.01	2.66	27.30
C29H21N301	427.51	427.1685	81.48	4.95	9.83	3.74
C29H21N302	443.51	443.1634	78.54	4.77	9.47	7.21
C29H21N303	459.51	459.1583	75.80	4.61	9.14	10.45
C29H21N304	475.51	475.1532	73.25	4.45	8.84	13.46
C29H21N305	491.51	491.1481	70.87	4.31	8.55	16.28
C29H21N306	507.51	507.1430	68.63	4.17	8.28	18.92
C29H21N307	523.51	523.1379	66.54	4.04	8.03	21.39
C29H21N308	539.51	539.1329	64.56	3.92	7.79	23.72
C29H21N309	555.51	555.1278	62.70	3.81	7.56	25.92
C29H21N501	455.52	455.1746	76.47	4.65	15.37	3.51
C29H21N502	471.52	471.1695	73.87	4.49	14.85	6.79
C29H21N503	487.52	487.1644	71.45	4.34	14.37	9.85
C29H21N504	503.52	503.1593	69.18	4.20	13.91	12.71
C29H21N505	519.52	519.1543	67.05	4.07	13.48	15.40
C29H21N506	535.52	535.1492	65.04	3.95	13.08	17.93
C29H21N507	551.52	551.1441	63.16	3.84	12.70	20.31
C29H21N508	567.52	567.1390	61.38	3.73	12.34	22.55
C29H21N509	583.52	583.1339	59.69	3.63	12.00	24.68
C29H22N201	414.51	414.1732	84.03	5.35	6.76	3.86
C29H22N202	430.51	430.1681	80.91	5.15	6.51	7.43
C29H22N203	446.51	446.1630	78.01	4.97	6.27	10.75
C29H22N204	462.51	462.1579	75.31	4.79	6.06	13.84
C29H22N205	478.51	478.1529	72.79	4.63	5.85	16.72
C29H22N206	494.51	494.1478	70.44	4.48	5.66	19.41
C29H22N207	510.51	510.1427	68.23	4.34	5.49	21.94
C29H22N208	526.51	526.1376	66.16	4.21	5.32	24.31
C29H22N209	542.51	542.1325	64.21	4.09	5.16	26.54
C29H22N401	442.52	442.1794	78.71	5.01	12.66	3.62
C29H22N402	458.52	458.1743	75.97	4.84	12.22	6.98
C29H22N403	474.52	474.1692	73.40	4.67	11.81	10.12
C29H22N404	490.52	490.1641	71.01	4.52	11.42	13.05
C29H22N405	506.52	506.1590	68.77	4.38	11.06	15.79
C29H22N406	522.52	522.1539	66.66	4.24	10.72	18.37
C29H22N407	538.52	538.1488	64.68	4.12	10.40	20.80
C29H22N408	554.52	554.1437	62.82	4.00	10.10	23.08
C29H22N409	570.52	570.1387	61.05	3.89	9.82	25.24
C29H22N601	470.54	470.1855	74.03	4.71	17.86	3.40
C29H22N602	486.54	486.1804	71.59	4.56	17.27	6.58
C29H22N603	502.54	502.1753	69.31	4.41	16.72	9.55
C29H22N604	518.54	518.1702	67.17	4.28	16.21	12.34
C29H22N605	534.54	534.1652	65.16	4.15	15.72	14.97
C29H22N606	550.54	550.1601	63.27	4.03	15.27	17.44
C29H22N607	566.53	566.1550	61.48	3.91	14.83	19.77
C29H22N608	582.53	582.1499	59.79	3.81	14.43	21.97
C29H22N609	598.53	598.1448	58.20	3.70	14.04	24.06
C29H23N101	401.51	401.1780	86.75	5.77	3.49	3.98
C29H23N102	417.51	417.1729	83.43	5.55	3.35	7.66
C29H23N103	433.51	433.1678	80.35	5.35	3.23	11.07
C29H23N104	449.51	449.1627	77.49	5.16	3.12	14.24
C29H23N105	465.51	465.1576	74.83	4.98	3.01	17.18
C29H23N106	481.51	481.1525	72.34	4.81	2.91	19.94
C29H23N107	497.51	497.1474	70.01	4.66	2.82	22.51
C29H23N108	513.51	513.1424	67.83	4.51	2.73	24.93
C29H23N109	529.51	529.1373	65.78	4.38	2.65	27.19
C29H23N301	429.53	429.1841	81.09	5.40	9.78	3.72
C29H23N302	445.53	445.1790	78.18	5.20	9.43	7.18
C29H23N303	461.52	461.1739	75.47	5.02	9.10	10.40
C29H23N304	477.52	477.1688	72.94	4.85	8.80	13.40

201

C29H23N305	493.52	493.1638	70.58	4.70	8.51	16.21
C29H23N306	509.52	509.1587	68.36	4.55	8.25	18.84
C29H23N307	525.52	525.1536	66.28	4.41	8.00	21.31
C29H23N308	541.52	541.1485	64.32	4.28	7.76	23.64
C29H23N309	557.52	557.1434	62.48	4.16	7.54	25.83
C29H23N501	457.54	457.1902	76.13	5.07	15.31	3.50
C29H23N502	473.54	473.1852	73.56	4.90	14.79	6.76
C29H23N503	489.54	489.1801	71.15	4.74	14.31	9.80
C29H23N504	505.54	505.1750	68.90	4.59	13.85	12.66
C29H23N505	521.54	521.1699	66.79	4.45	13.43	15.34
C29H23N506	537.54	537.1648	64.80	4.31	13.03	17.86
C29H23N507	553.54	553.1597	62.93	4.19	12.65	20.23
C29H23N508	569.54	569.1546	61.16	4.07	12.30	22.47
C29H23N509	585.53	585.1496	59.49	3.96	11.96	24.59
C29H24N201	416.53	416.1889	83.63	5.81	6.73	3.84
C29H24N202	432.53	432.1838	80.53	5.59	6.48	7.40
C29H24N203	448.53	448.1787	77.66	5.39	6.25	10.70
C29H24N204	464.53	464.1736	74.98	5.21	6.03	13.78
C29H24N205	480.53	480.1685	72.49	5.03	5.83	16.65
C29H24N206	496.52	496.1634	70.15	4.87	5.64	19.33
C29H24N207	512.52	512.1583	67.96	4.72	5.47	21.85
C29H24N208	528.52	528.1533	65.91	4.58	5.30	24.22
C29H24N209	544.52	544.1482	63.97	4.44	5.14	26.44
C29H24N401	444.54	444.1950	78.36	5.44	12.60	3.60
C29H24N402	460.54	460.1899	75.63	5.25	12.17	6.95
C29H24N403	476.54	476.1848	73.09	5.08	11.76	10.07
C29H24N404	492.54	492.1797	70.72	4.91	11.38	12.99
C29H24N405	508.54	508.1747	68.49	4.76	11.02	15.73
C29H24N406	524.54	524.1696	66.41	4.61	10.68	18.30
C29H24N407	540.54	540.1645	64.44	4.48	10.37	20.72
C29H24N408	556.54	556.1594	62.59	4.35	10.07	23.00
C29H24N409	572.54	572.1543	60.84	4.23	9.79	25.15
C29H24N601	472.55	472.2011	73.71	5.12	17.78	3.39
C29H24N602	488.55	488.1961	71.30	4.95	17.20	6.55
C29H24N603	504.55	504.1910	69.04	4.79	16.66	9.51
C29H24N604	520.55	520.1859	66.91	4.65	16.14	12.29
C29H24N605	536.55	536.1808	64.92	4.51	15.66	14.91
C29H24N606	552.55	552.1757	63.04	4.38	15.21	17.37
C29H24N607	568.55	568.1706	61.27	4.25	14.78	19.70
C29H24N608	584.55	584.1655	59.59	4.14	14.38	21.90
C29H24N609	600.55	600.1605	58.00	4.03	13.99	23.98
C29H25N101	403.53	403.1936	86.32	6.24	3.47	3.96
C29H25N102	419.53	419.1885	83.03	6.01	3.34	7.63
C29H25N103	435.53	435.1834	79.98	5.79	3.22	11.02
C29H25N104	451.53	451.1783	77.14	5.58	3.10	14.17
C29H25N105	467.53	467.1733	74.50	5.39	3.00	17.11
C29H25N106	483.53	483.1682	72.04	5.21	2.90	19.85
C29H25N107	499.53	499.1631	69.73	5.04	2.80	22.42
C29H25N108	515.52	515.1580	67.57	4.89	2.72	24.83
C29H25N109	531.52	531.1529	65.53	4.74	2.64	27.09
C29H25N301	431.54	431.1998	80.72	5.84	9.74	3.71
C29H25N302	447.54	447.1947	77.83	5.63	9.39	7.15
C29H25N303	463.54	463.1896	75.14	5.44	9.07	10.35
C29H25N304	479.54	479.1845	72.64	5.25	8.76	13.35
C29H25N305	495.54	495.1794	70.29	5.09	8.48	16.14
C29H25N306	511.54	511.1743	68.09	4.93	8.21	18.77
C29H25N307	527.54	527.1692	66.03	4.78	7.97	21.23
C29H25N308	543.54	543.1641	64.08	4.64	[illegible]	23.55
C29H25N309	559.54	559.1591	62.25	4.50	7.51	25.73
C29H25N501	459.56	459.2059	75.80	5.48	15.24	3.48
C29H25N502	475.55	475.2008	73.25	5.30	14.73	6.73
C29H25N503	491.55	491.1957	70.86	5.13	14.25	9.76
C29H25N504	507.55	507.1906	68.63	4.96	13.80	12.61
C29H25N505	523.55	523.1856	66.53	4.81	13.38	15.28
C29H25N506	539.55	539.1805	64.56	4.67	12.98	17.79
C29H25N507	555.55	555.1754	62.70	4.54	12.61	20.16
C29H25N508	571.55	571.1703	60.94	4.41	12.25	22.39
C29H25N509	587.55	587.1652	59.28	4.29	11.92	24.51
C29H26N201	418.54	418.2045	83.22	6.26	6.69	3.82
C29H26N202	434.54	434.1994	80.16	6.03	6.45	7.36
C29H26N203	450.54	450.1943	77.31	5.82	6.22	10.65
C29H26N204	466.54	466.1892	74.66	5.62	6.00	13.72
C29H26N205	482.54	482.1842	72.19	5.43	5.81	16.58
C29H26N206	498.54	498.1791	69.87	5.26	5.62	19.26
C29H26N207	514.54	514.1740	67.70	5.09	5.44	21.77
C29H26N208	530.54	530.1689	65.65	4.94	5.28	24.13
C29H26N209	546.54	546.1638	63.73	4.80	5.13	26.35
C29H26N401	446.56	446.2106	78.00	5.87	12.55	3.58
C29H26N402	462.56	462.2056	75.30	5.67	12.11	6.92
C29H26N403	478.56	478.2005	72.79	5.48	11.71	10.03
C29H26N404	494.55	494.1954	70.43	5.30	11.33	12.94
C29H26N405	510.55	510.1903	68.22	5.13	10.97	15.67
C29H26N406	526.55	526.1852	66.15	4.98	10.64	18.23
C29H26N407	542.55	542.1801	64.20	4.83	10.33	20.64
C29H26N408	558.55	558.1750	62.36	4.69	10.03	22.92
C29H26N409	574.55	574.1700	60.63	4.56	9.75	25.06
C29H26N601	474.57	474.2168	73.40	5.52	17.71	3.37
C29H26N602	490.57	490.2117	71.00	5.34	17.13	6.52
C29H26N603	506.57	506.2066	68.76	5.17	16.59	9.48
C29H26N604	522.57	522.2015	66.66	5.02	16.08	12.25
C29H26N605	538.57	538.1965	64.68	4.87	15.60	14.85
C29H26N606	554.57	554.1914	62.81	4.73	15.15	17.31
C29H26N607	570.57	570.1863	61.05	4.59	14.73	19.63
C29H26N608	586.57	586.1812	59.38	4.47	14.33	21.82
C29H26N609	602.57	602.1761	57.81	4.35	13.95	23.90
C29H27N101	405.54	405.2093	85.89	6.71	3.45	3.95
C29H27N102	421.54	421.2042	82.63	6.46	3.32	7.59
C29H27N103	437.54	437.1991	79.61	6.22	3.20	10.97
C29H27N104	453.54	453.1940	76.80	6.00	3.09	14.11
C29H27N105	469.54	469.1889	74.18	5.80	2.98	17.04
C29H27N106	485.54	485.1838	71.74	5.61	2.88	19.77
C29H27N107	501.54	501.1787	69.45	5.43	2.79	22.33
C29H27N108	517.54	517.1737	67.30	5.26	2.71	24.73
C29H27N109	533.54	533.1686	65.29	5.10	2.63	26.99
C29H27N301	433.56	433.2154	80.34	6.28	9.69	3.69
C29H27N302	449.56	449.2103	77.48	6.05	9.35	7.12
C29H27N303	465.56	465.2052	74.82	5.85	9.03	10.31
C29H27N304	481.56	481.2001	72.33	5.65	8.73	13.29
C29H27N305	497.56	497.1951	70.01	5.47	8.45	16.08
C29H27N306	513.56	513.1900	67.83	5.30	8.18	18.69
C29H27N307	529.55	529.1849	65.78	5.14	7.93	21.15
C29H27N308	545.55	545.1798	63.85	4.99	7.70	23.46
C29H27N309	561.55	561.1747	62.03	4.85	7.48	25.64
C29H27N501	461.57	461.2215	75.46	5.90	15.17	3.47
C29H27N502	477.57	477.2165	72.94	5.70	14.66	6.70
C29H27N503	493.57	493.2114	70.57	5.51	14.19	9.72
C29H27N504	509.57	509.2063	68.36	5.34	13.74	12.56
C29H27N505	525.57	525.2012	66.28	5.18	13.33	15.22
C29H27N506	541.57	541.1961	64.32	5.03	12.93	17.73
C29H27N507	557.57	557.1910	62.47	4.88	12.56	20.09

C29H27N5O8	573.57	573.1859	60.73	4.74	12.21	22.32		C29H30N2O5	486.57	486.2155	71.59	6.21	5.76	16.44	
C29H27N5O9	589.57	589.1809	59.08	4.62	11.88	24.42		C29H30N2O6	502.57	502.2104	69.31	6.02	5.57	19.10	
C29H28N2O1	420.56	420.2202	82.82	6.71	6.66	3.80		C29H30N2O7	518.57	518.2053	67.17	5.83	5.40	21.60	
C29H28N2O2	436.56	436.2151	79.79	6.46	6.42	7.33		C29H30N2O8	534.57	534.2002	65.16	5.66	5.24	23.94	
C29H28N2O3	452.56	452.2100	76.97	6.24	6.19	10.61		C29H30N2O9	550.57	550.1951	63.27	5.49	5.09	26.15	
C29H28N2O4	468.56	468.2049	74.34	6.02	5.98	13.66		C29H30N4O1	450.59	450.2419	77.30	6.71	12.43	3.55	
C29H28N2O5	484.56	484.1998	71.88	5.82	5.78	16.51		C29H30N4O2	466.59	466.2369	74.65	6.48	12.01	6.86	
C29H28N2O6	500.56	500.1947	69.59	5.64	5.60	19.18		C29H30N4O3	482.59	482.2318	72.18	6.27	11.61	9.95	
C29H28N2O7	516.56	516.1896	67.43	5.46	5.42	21.68		C29H30N4O4	498.59	498.2267	69.86	6.06	11.24	12.84	
C29H28N2O8	532.56	532.1845	65.41	5.30	5.26	24.03		C29H30N4O5	514.59	514.2216	67.69	5.88	10.89	15.55	
C29H28N2O9	548.55	548.1795	63.50	5.15	5.11	26.25		C29H30N4O6	530.59	530.2165	65.65	5.70	10.56	18.09	
C29H28N4O1	448.57	448.2263	77.65	6.29	12.49	3.57		C29H30N4O7	546.59	546.2114	63.73	5.53	10.25	20.49	
C29H28N4O2	464.57	464.2212	74.98	6.08	12.06	6.89		C29H30N4O8	562.58	562.2063	61.91	5.38	9.96	22.75	
C29H28N4O3	480.57	480.2161	72.48	5.87	11.66	9.99		C29H30N4O9	578.58	578.2013	60.20	5.23	9.68	24.89	
C29H28N4O4	496.57	496.2110	70.15	5.68	11.28	12.89		C29H30N6O1	478.60	478.2481	72.78	6.32	17.56	3.34	
C29H28N4O5	512.57	512.2060	67.96	5.51	10.93	15.61		C29H30N6O2	494.60	494.2430	70.43	6.11	16.99	6.47	
C29H28N4O6	528.57	528.2009	65.90	5.34	10.60	18.16		C29H30N6O3	510.60	510.2379	68.22	5.92	16.46	9.40	
C29H28N4O7	544.57	544.1958	63.96	5.18	10.29	20.57		C29H30N6O4	526.60	526.2328	66.15	5.74	15.96	12.15	
C29H28N4O8	560.57	560.1907	62.14	5.03	9.99	22.83		C29H30N6O5	542.60	542.2278	64.20	5.57	15.49	14.74	
C29H28N4O9	576.57	576.1856	60.41	4.90	9.72	24.97		C29H30N6O6	558.60	558.2227	62.36	5.41	15.04	17.19	
C29H28N6O1	476.59	476.2324	73.09	5.92	17.63	3.36		C29H30N6O7	574.60	574.2176	60.62	5.26	14.63	19.49	
C29H28N6O2	492.59	492.2274	70.71	5.73	17.06	6.50		C29H30N6O8	590.60	590.2125	58.98	5.12	14.23	21.67	
C29H28N6O3	508.58	508.2223	68.49	5.55	16.52	9.44		C29H30N6O9	606.60	606.2074	57.42	4.99	13.85	23.74	
C29H28N6O4	524.58	524.2172	66.40	5.38	16.02	12.20		C29H31N1O1	409.58	409.2406	85.04	7.63	3.42	3.91	
C29H28N6O5	540.58	540.2121	64.43	5.22	15.55	14.80		C29H31N1O2	425.58	425.2355	81.85	7.34	3.29	7.52	
C29H28N6O6	556.58	556.2070	62.58	5.07	15.10	17.25		C29H31N1O3	441.58	441.2304	78.88	7.08	3.17	10.87	
C29H28N6O7	572.58	572.2019	60.83	4.93	14.68	19.56		C29H31N1O4	457.57	457.2253	76.12	6.83	3.06	13.99	
C29H28N6O8	588.58	588.1968	59.18	4.80	14.28	21.75		C29H31N1O5	473.57	473.2202	73.55	6.60	2.96	16.89	
C29H28N6O9	604.58	604.1918	57.61	4.67	13.90	23.82		C29H31N1O6	489.57	489.2151	71.15	6.38	2.86	19.61	
C29H29N1O1	407.56	407.2249	85.47	7.17	3.44	3.93		C29H31N1O7	505.57	505.2100	68.90	6.18	2.77	22.15	
C29H29N1O2	423.56	423.2198	82.24	6.90	3.31	7.55		C29H31N1O8	521.57	521.2049	66.78	5.99	2.69	24.54	
C29H29N1O3	439.56	439.2147	79.24	6.65	3.19	10.92		C29H31N1O9	537.57	537.1999	64.80	5.81	2.61	26.79	
C29H29N1O4	455.56	455.2096	76.46	6.42	3.07	14.05		C29H31N3O1	437.59	437.2467	79.60	7.14	9.60	3.66	
C29H29N1O5	471.56	471.2046	73.87	6.20	2.97	16.96		C29H31N3O2	453.59	453.2416	76.79	6.89	9.26	7.05	
C29H29N1O6	487.56	487.1995	71.44	6.00	2.87	19.69		C29H31N3O3	469.59	469.2365	74.18	6.65	8.95	10.22	
C29H29N1O7	503.56	503.1944	69.17	5.80	2.78	22.24		C29H31N3O4	485.59	485.2314	71.73	6.43	8.65	13.18	
C29H29N1O8	519.56	519.1893	67.04	5.63	2.70	24.64		C29H31N3O5	501.59	501.2264	69.44	6.23	8.38	15.95	
C29H29N1O9	535.56	535.1842	65.04	5.46	2.62	26.89		C29H31N3O6	517.59	517.2213	67.30	6.04	8.12	18.55	
C29H29N3O1	435.57	435.2310	79.97	6.71	9.65	3.67		C29H31N3O7	533.59	533.2162	65.28	5.86	7.88	20.99	
C29H29N3O2	451.57	451.2260	77.14	6.47	9.31	7.09		C29H31N3O8	549.59	549.2111	63.38	5.69	7.65	23.29	
C29H29N3O3	467.57	467.2209	74.50	6.25	8.99	10.27		C29H31N3O9	565.59	565.2060	61.59	5.52	7.43	25.46	
C29H29N3O4	483.57	483.2158	72.03	6.04	8.69	13.23		C29H31N5O1	465.60	465.2528	74.81	6.71	15.04	3.44	
C29H29N3O5	499.57	499.2107	69.72	5.85	8.41	16.01		C29H31N5O2	481.60	481.2478	72.33	6.49	14.54	6.64	
C29H29N3O6	515.57	515.2056	67.56	5.67	8.15	18.62		C29H31N5O3	497.60	497.2427	70.00	6.28	14.07	9.65	
C29H29N3O7	531.57	531.2005	65.53	5.50	7.90	21.07		C29H31N5O4	513.60	513.2376	67.82	6.08	13.64	12.46	
C29H29N3O8	547.57	547.1954	63.61	5.34	7.67	23.38		C29H31N5O5	529.60	529.2325	65.77	5.90	13.22	15.11	
C29H29N3O9	563.57	563.1904	61.81	5.19	7.46	25.55		C29H31N5O6	545.60	545.2274	63.84	5.73	12.84	17.59	
C29H29N5O1	463.59	463.2372	75.14	6.31	15.11	3.45		C29H31N5O7	561.60	561.2223	62.02	5.56	12.47	19.94	
C29H29N5O2	479.59	479.2321	72.63	6.10	14.60	6.67		C29H31N5O8	577.60	577.2172	60.31	5.41	12.12	22.16	
C29H29N5O3	495.59	495.2270	70.29	5.90	14.13	9.69		C29H31N5O9	593.60	593.2122	58.68	5.26	11.80	24.26	
C29H29N5O4	511.59	511.2219	68.09	5.71	13.69	12.51		C29H32N2O1	424.59	424.2514	82.04	7.60	6.60	3.77	
C29H29N5O5	527.58	527.2169	66.02	5.54	13.27	15.16		C29H32N2O2	440.59	440.2464	79.06	7.32	6.36	7.26	
C29H29N5O6	543.58	543.2118	64.08	5.38	12.88	17.66		C29H32N2O3	456.59	456.2413	76.29	7.06	6.14	10.51	
C29H29N5O7	559.58	559.2067	62.25	5.22	12.52	20.01		C29H32N2O4	472.59	472.2362	73.71	6.83	5.93	13.54	
C29H29N5O8	575.58	575.2016	60.52	5.08	12.17	22.24		C29H32N2O5	488.59	488.2311	71.29	6.60	5.73	16.37	
C29H29N5O9	591.58	591.1965	58.88	4.94	11.84	24.34		C29H32N2O6	504.59	504.2260	69.03	6.39	5.55	19.02	
C29H30N2O1	422.58	422.2358	82.43	7.16	6.63	3.79		C29H32N2O7	520.59	520.2209	66.91	6.20	5.38	21.51	
C29H30N2O2	438.57	438.2307	79.42	6.89	6.39	7.30		C29H32N2O8	536.59	536.2158	64.91	6.01	5.22	23.85	
C29H30N2O3	454.57	454.2256	76.63	6.65	6.16	10.56		C29H32N2O9	552.59	552.2108	63.04	5.84	5.07	26.06	
C29H30N2O4	470.57	470.2205	74.02	6.43	5.95	13.60		C29H32N4O1	452.60	452.2576	76.96	7.13	12.38	3.53	

C29H32N4O2	468.60	468.2525	74.33	6.88	11.96	6.83		C29H34N4O8	566.62	566.2376	61.47	6.05	9.89	22.59	
C29H32N4O3	484.60	484.2474	71.88	6.66	11.56	9.90		C29H34N4O9	582.62	582.2326	59.79	5.88	9.62	24.72	
C29H32N4O4	500.60	500.2423	69.58	6.44	11.19	12.78		C29H34N6O1	482.63	482.2794	72.17	7.10	17.41	3.32	
C29H32N4O5	516.60	516.2373	67.45	6.24	10.85	15.49		C29H34N6O2	498.63	498.2743	69.86	6.87	16.85	6.42	
C29H32N4O6	532.60	532.2322	65.40	6.06	10.52	18.02		C29H34N6O3	514.63	514.2692	67.68	6.66	16.33	9.33	
C29H32N4O7	548.60	548.2271	63.49	5.88	10.21	20.41		C29H34N6O4	530.63	530.2641	65.64	6.46	15.84	12.06	
C29H32N4O8	564.60	564.2220	61.69	5.71	9.92	22.67		C29H34N6O5	546.63	546.2590	63.72	6.27	15.37	14.63	
C29H32N4O9	580.60	580.2169	59.99	5.56	9.65	24.80		C29H34N6O6	562.63	562.2540	61.91	6.09	14.94	17.06	
C29H32N6O1	480.62	480.2637	72.47	6.71	17.49	3.33		C29H34N6O7	578.63	578.2489	60.20	5.92	14.52	19.36	
C29H32N6O2	496.62	496.2587	70.14	6.49	16.92	6.44		C29H34N6O8	594.63	594.2438	58.58	5.76	14.13	21.53	
C29H32N6O3	512.62	512.2536	67.95	6.29	16.39	9.36		C29H34N6O9	610.63	610.2387	57.04	5.61	13.76	23.58	
C29H32N6O4	528.62	528.2485	65.89	6.10	15.90	12.11		C29H35N1O1	413.61	413.2718	84.22	8.53	3.39	3.87	
C29H32N6O5	544.62	544.2434	63.96	5.92	15.43	14.69		C29H35N1O2	429.61	429.2668	81.08	8.21	3.26	7.45	
C29H32N6O6	560.61	560.2383	62.13	5.75	14.99	17.12		C29H35N1O3	445.61	445.2617	78.17	7.92	3.14	10.77	
C29H32N6O7	576.61	576.2332	60.41	5.59	14.57	19.42		C29H35N1O4	461.61	461.2566	75.46	7.64	3.03	13.86	
C29H32N6O8	592.61	592.2281	58.78	5.44	14.18	21.60		C29H35N1O5	477.61	477.2515	72.93	7.39	2.93	16.75	
C29H32N6O9	608.61	608.2231	57.23	5.30	13.81	23.66		C29H35N1O6	493.61	493.2464	70.57	7.15	2.84	19.45	
C29H33N1O1	411.59	411.2562	84.63	8.08	3.40	3.89		C29H35N1O7	509.60	509.2413	68.35	6.92	2.75	21.98	
C29H33N1O2	427.59	427.2511	81.46	7.78	3.28	7.48		C29H35N1O8	525.60	525.2362	66.27	6.71	2.66	24.35	
C29H33N1O3	443.59	443.2460	78.52	7.50	3.16	10.82		C29H35N1O9	541.60	541.2312	64.31	6.51	2.59	26.59	
C29H33N1O4	459.59	459.2409	75.79	7.24	3.05	13.92		C29H35N3O1	441.62	441.2780	78.87	7.99	9.51	3.62	
C29H33N1O5	475.59	475.2359	73.24	6.99	2.95	16.82		C29H35N3O2	457.62	457.2729	76.12	7.71	9.18	6.99	
C29H33N1O6	491.59	491.2308	70.86	6.77	2.85	19.53		C29H35N3O3	473.62	473.2678	73.54	7.45	8.87	10.13	
C29H33N1O7	507.59	507.2257	68.62	6.55	2.76	22.06		C29H35N3O4	489.62	489.2627	71.14	7.21	8.58	13.07	
C29H33N1O8	523.59	523.2206	66.53	6.35	2.68	24.45		C29H35N3O5	505.62	505.2577	68.89	6.98	8.31	15.82	
C29H33N1O9	539.59	539.2155	64.55	6.16	2.60	26.69		C29H35N3O6	521.62	521.2526	66.78	6.76	8.06	18.40	
C29H33N3O1	439.61	439.2623	79.24	7.57	9.56	3.64		C29H35N3O7	537.62	537.2475	64.79	6.56	7.82	20.83	
C29H33N3O2	455.61	455.2573	76.45	7.30	9.22	7.02		C29H35N3O8	553.62	553.2424	62.92	6.37	7.59	23.12	
C29H33N3O3	471.60	471.2522	73.86	7.05	8.91	10.18		C29H35N3O9	569.62	569.2373	61.15	6.19	7.38	25.28	
C29H33N3O4	487.60	487.2471	71.44	6.82	8.62	13.12		C29H35N5O1	469.64	469.2841	74.17	7.51	14.91	3.41	
C29H33N3O5	503.60	503.2420	69.17	6.61	8.34	15.88		C29H35N5O2	485.63	485.2791	71.73	7.26	14.42	6.59	
C29H33N3O6	519.60	519.2369	67.04	6.40	8.09	18.47		C29H35N5O3	501.63	501.2740	69.44	7.03	13.96	9.57	
C29H33N3O7	535.60	535.2318	65.03	6.21	7.85	20.91		C29H35N5O4	517.63	517.2689	67.29	6.82	13.53	12.36	
C29H33N3O8	551.60	551.2267	63.15	6.03	7.62	23.20		C29H35N5O5	533.63	533.2638	65.27	6.61	13.12	14.99	
C29H33N3O9	567.60	567.2217	61.37	5.86	7.40	25.37		C29H35N5O6	549.63	549.2587	63.37	6.42	12.74	17.47	
C29H33N5O1	467.62	467.2685	74.49	7.11	14.98	3.42		C29H35N5O7	565.63	565.2536	61.58	6.24	12.38	19.80	
C29H33N5O2	483.62	483.2634	72.02	6.88	14.48	6.62		C29H35N5O8	581.63	581.2485	59.89	6.07	12.04	22.01	
C29H33N5O3	499.62	499.2583	69.72	6.66	14.02	9.61		C29H35N5O9	597.63	597.2435	58.28	5.90	11.72	24.09	
C29H33N5O4	515.62	515.2532	67.55	6.45	13.58	12.41		C29H36N2O1	428.62	428.2827	81.27	8.47	6.54	3.73	
C29H33N5O5	531.62	531.2482	65.52	6.26	13.17	15.05		C29H36N2O2	444.62	444.2777	78.34	8.16	6.30	7.20	
C29H33N5O6	547.62	547.2431	63.61	6.07	12.79	17.53		C29H36N2O3	460.62	460.2726	75.62	7.88	6.08	10.42	
C29H33N5O7	563.62	563.2380	61.80	5.90	12.43	19.87		C29H36N2O4	476.62	476.2675	73.08	7.61	5.88	13.43	
C29H33N5O8	579.62	579.2329	60.10	5.74	12.08	22.08		C29H36N2O5	492.62	492.2624	70.71	7.37	5.69	16.24	
C29H33N5O9	595.61	595.2278	58.48	5.58	11.76	24.18		C29H36N2O6	508.62	508.2573	68.48	7.13	5.51	18.87	
C29H34N2O1	426.61	426.2671	81.65	8.03	6.57	3.75		C29H36N2O7	524.62	524.2522	66.40	6.92	5.34	21.35	
C29H34N2O2	442.61	442.2620	78.70	7.74	6.33	7.23		C29H36N2O8	540.62	540.2471	64.43	6.71	5.18	23.68	
C29H34N2O3	458.61	458.2569	75.95	7.47	6.11	10.47		C29H36N2O9	556.62	556.2421	62.58	6.52	5.03	25.87	
C29H34N2O4	474.61	474.2518	73.39	7.22	5.90	13.48		C29H36N4O1	456.64	456.2889	76.28	7.95	12.27	3.50	
C29H34N2O5	490.60	490.2468	71.00	6.99	5.71	16.31		C29H36N4O2	472.64	472.2838	73.70	7.68	11.85	6.77	
C29H34N2O6	506.60	506.2417	68.76	6.76	5.53	18.95		C29H36N4O3	488.64	488.2787	71.28	7.43	11.47	9.82	
C29H34N2O7	522.60	522.2366	66.65	6.56	5.36	21.43		C29H36N4O4	504.63	504.2736	69.02	7.19	11.10	12.68	
C29H34N2O8	538.60	538.2315	64.67	6.36	5.20	23.76		C29H36N4O5	520.63	520.2686	66.90	6.97	10.76	15.37	
C29H34N2O9	554.60	554.2264	62.81	6.18	5.05	25.96		C29H36N4O6	536.63	536.2635	64.91	6.76	10.44	17.89	
C29H34N4O1	454.62	454.2732	76.62	7.54	12.32	3.52		C29H36N4O7	552.63	552.2584	63.03	6.57	10.14	20.27	
C29H34N4O2	470.62	470.2682	74.01	7.28	11.90	6.80		C29H36N4O8	568.63	568.2533	61.26	6.38	9.85	22.51	
C29H34N4O3	486.62	486.2631	71.58	7.04	11.51	9.86		C29H36N4O9	584.63	584.2482	59.58	6.21	9.58	24.63	
C29H34N4O4	502.62	502.2580	69.30	6.82	11.15	12.73		C29H36N6O1	484.65	484.2950	71.87	7.49	17.34	3.30	
C29H34N4O5	518.62	518.2529	67.16	6.61	10.80	15.43		C29H36N6O2	500.65	500.2900	69.57	7.25	16.79	6.39	
C29H34N4O6	534.62	534.2478	65.15	6.41	10.48	17.96		C29H36N6O3	516.65	516.2849	67.42	7.02	16.27	9.29	
C29H34N4O7	550.62	550.2427	63.26	6.22	10.18	20.34		C29H36N6O4	532.65	532.2798	65.39	6.81	15.78	12.01	

C29H36N605	548.65	548.2747	63.49	6.61	15.32	14.58
C29H36N606	564.65	564.2696	61.69	6.43	14.88	17.00
C29H36N607	580.65	580.2645	59.99	6.25	14.47	19.29
C29H36N608	596.65	596.2594	58.38	6.08	14.09	21.45
C29H36N609	612.65	612.2544	56.86	5.92	13.72	23.50
C29H37N101	415.62	415.2875	83.81	8.97	3.37	3.85
C29H37N102	431.62	431.2824	80.70	8.64	3.25	7.41
C29H37N103	447.62	447.2773	77.82	8.33	3.13	10.72
C29H37N104	463.62	463.2722	75.13	8.04	3.02	13.80
C29H37N105	479.62	479.2672	72.62	7.78	2.92	16.68
C29H37N106	495.62	495.2621	70.28	7.52	2.83	19.37
C29H37N107	511.62	511.2570	68.08	7.29	2.74	21.89
C29H37N108	527.62	527.2519	66.02	7.07	2.65	24.26
C29H37N109	543.62	543.2468	64.07	6.86	2.58	26.49
C29H37N301	443.64	443.2936	78.52	8.41	9.47	3.61
C29H37N302	459.64	459.2886	75.78	8.11	9.14	6.96
C29H37N303	475.64	475.2835	73.23	7.84	8.83	10.09
C29H37N304	491.64	491.2784	70.85	7.59	8.55	13.02
C29H37N305	507.64	507.2733	68.62	7.35	8.28	15.76
C29H37N306	523.63	523.2682	66.52	7.12	8.02	18.33
C29H37N307	539.63	539.2631	64.55	6.91	7.79	20.75
C29H37N308	555.63	555.2580	62.69	6.71	7.56	23.04
C29H37N309	571.63	571.2530	60.93	6.52	7.35	25.19
C29H37N501	471.65	471.2998	73.85	7.91	14.85	3.39
C29H37N502	487.65	487.2947	71.43	7.65	14.36	6.56
C29H37N503	503.65	503.2896	69.16	7.40	13.91	9.53
C29H37N504	519.65	519.2845	67.03	7.18	13.48	12.32
C29H37N505	535.65	535.2794	65.03	6.96	13.07	14.93
C29H37N506	551.65	551.2744	63.14	6.76	12.70	17.40
C29H37N507	567.65	567.2693	61.36	6.57	12.34	19.73
C29H37N508	583.65	583.2642	59.68	6.39	12.00	21.93
C29H37N509	599.65	599.2591	58.09	6.22	11.68	24.01
C29H38N201	430.64	430.2984	80.89	8.89	6.51	3.72
C29H38N202	446.64	446.2933	77.99	8.58	6.27	7.16
C29H38N203	462.64	462.2882	75.29	8.28	6.06	10.37
C29H38N204	478.64	478.2831	72.77	8.00	5.85	13.37
C29H38N205	494.64	494.2781	70.42	7.74	5.66	16.17
C29H38N206	510.64	510.2730	68.21	7.50	5.49	18.80
C29H38N207	526.64	526.2679	66.14	7.27	5.32	21.27
C29H38N208	542.63	542.2628	64.19	7.06	5.16	23.59
C29H38N209	558.63	558.2577	62.35	6.86	5.01	25.78
C29H38N401	458.65	458.3045	75.94	8.35	12.22	3.49
C29H38N402	474.65	474.2995	73.39	8.07	11.80	6.74
C29H38N403	490.65	490.2944	70.99	7.81	11.42	9.78
C29H38N404	506.65	506.2893	68.75	7.56	11.06	12.63
C29H38N405	522.65	522.2842	66.65	7.33	10.72	15.31
C29H38N406	538.65	538.2791	64.67	7.11	10.40	17.82
C29H38N407	554.65	554.2740	62.80	6.91	10.10	20.19
C29H38N408	570.65	570.2689	61.04	6.71	9.82	22.43
C29H38N409	586.65	586.2639	59.38	6.53	9.55	24.55
C29H38N601	486.67	486.3107	71.57	7.87	17.27	3.29
C29H38N602	502.67	502.3056	69.30	7.62	16.72	6.37
C29H38N603	518.66	518.3005	67.16	7.38	16.20	9.25
C29H38N604	534.66	534.2954	65.15	7.16	15.72	11.97
C29H38N605	550.66	550.2903	63.26	6.96	15.26	14.53
C29H38N606	566.66	566.2853	61.47	6.76	14.83	16.94
C29H38N607	582.66	582.2802	59.78	6.57	14.42	19.22
C29H38N608	598.66	598.2751	58.18	6.40	14.04	21.38
C29H38N609	614.66	614.2700	56.67	6.23	13.67	23.43
C29H39N101	417.64	417.3031	83.40	9.41	3.35	3.83
C29H39N102	433.64	433.2981	80.33	9.07	3.23	7.38
C29H39N103	449.64	449.2930	77.47	8.74	3.12	10.67
C29H39N104	465.64	465.2879	74.81	8.44	3.01	13.74
C29H39N105	481.64	481.2828	72.32	8.16	2.91	16.61
C29H39N106	497.64	497.2777	70.00	7.90	2.81	19.29
C29H39N107	513.64	513.2726	67.82	7.65	2.73	21.80
C29H39N108	529.64	529.2675	65.77	7.42	2.64	24.17
C29H39N109	545.64	545.2625	63.84	7.20	2.57	26.39
C29H39N301	445.65	445.3093	78.16	8.82	9.43	3.59
C29H39N302	461.65	461.3042	75.45	8.52	9.10	6.93
C29H39N303	477.65	477.2991	72.92	8.23	8.80	10.05
C29H39N304	493.65	493.2940	70.56	7.96	8.51	12.96
C29H39N305	509.65	509.2890	68.35	7.71	8.24	15.70
C29H39N306	525.65	525.2839	66.27	7.48	7.99	18.26
C29H39N307	541.65	541.2788	64.31	7.26	7.76	20.68
C29H39N308	557.65	557.2737	62.46	7.05	7.54	22.95
C29H39N309	573.65	573.2686	60.72	6.85	7.33	25.10
C29H39N501	473.67	473.3154	73.54	8.30	14.79	3.38
C29H39N502	489.67	489.3104	71.13	8.03	14.30	6.53
C29H39N503	505.67	505.3053	68.88	7.77	13.85	9.49
C29H39N504	521.67	521.3002	66.77	7.54	13.42	12.27
C29H39N505	537.66	537.2951	64.78	7.31	13.03	14.88
C29H39N506	553.66	553.2900	62.91	7.10	12.65	17.34
C29H39N507	569.66	569.2849	61.15	6.90	12.29	19.66
C29H39N508	585.66	585.2798	59.48	6.71	11.96	21.85
C29H39N509	601.66	601.2748	57.89	6.53	11.64	23.93
C29H40N201	432.65	432.3140	80.51	9.32	6.47	3.70
C29H40N202	448.65	448.3090	77.64	8.99	6.24	7.13
C29H40N203	464.65	464.3039	74.96	8.68	6.03	10.33
C29H40N204	480.65	480.2988	72.47	8.39	5.83	13.31
C29H40N205	496.65	496.2937	70.13	8.12	5.64	16.11
C29H40N206	512.65	512.2886	67.95	7.86	5.46	18.73
C29H40N207	528.65	528.2835	65.89	7.63	5.30	21.19
C29H40N208	544.65	544.2784	63.95	7.40	5.14	23.50
C29H40N209	560.65	560.2734	62.13	7.19	5.00	25.68
C29H40N401	460.67	460.3202	75.61	8.75	12.16	3.47
C29H40N402	476.67	476.3151	73.07	8.46	11.75	6.71
C29H40N403	492.67	492.3100	70.70	8.18	11.37	9.74
C29H40N404	508.67	508.3049	68.48	7.93	11.01	12.58
C29H40N405	524.67	524.2998	66.39	7.68	10.68	15.25
C29H40N406	540.67	540.2948	64.42	7.46	10.36	17.76
C29H40N407	556.66	556.2897	62.57	7.24	10.06	20.12
C29H40N408	572.66	572.2846	60.83	7.04	9.78	22.35
C29H40N409	588.66	588.2795	59.17	6.85	9.52	24.46
C29H40N601	488.68	488.3263	71.28	8.25	17.20	3.27
C29H40N602	504.68	504.3213	69.02	7.99	16.65	6.34
C29H40N603	520.68	520.3162	66.90	7.74	16.14	9.22
C29H40N604	536.68	536.3111	64.90	7.51	15.66	11.92
C29H40N605	552.68	552.3060	63.02	7.30	15.21	14.47
C29H40N606	568.68	568.3009	61.25	7.09	14.78	16.88
C29H40N607	584.68	584.2958	59.58	6.90	14.37	19.16
C29H40N608	600.68	600.2907	57.99	6.71	13.99	21.31
C29H40N609	616.68	616.2857	56.48	6.54	13.63	23.35
C29H41N101	419.66	419.3188	83.00	9.85	3.34	3.81
C29H41N102	435.66	435.3137	79.95	9.49	3.22	7.34
C29H41N103	451.66	451.3086	77.12	9.15	3.10	10.63
C29H41N104	467.65	467.3035	74.48	8.84	3.00	13.68
C29H41N105	483.65	483.2985	72.02	8.54	2.90	16.54
C29H41N106	499.65	499.2934	69.71	8.27	2.80	19.21
C29H41N107	515.65	515.2883	67.55	8.01	2.72	21.72

C29H41N108	531.65	531.2832	65.52	7.77	2.63	24.07
C29H41N109	547.65	547.2781	63.60	7.55	2.56	26.29
C29H41N301	447.67	447.3249	77.81	9.23	9.39	3.57
C29H41N302	463.67	463.3199	75.12	8.91	9.06	6.90
C29H41N303	479.67	479.3148	72.62	8.62	8.76	10.01
C29H41N304	495.67	495.3097	70.27	8.34	8.48	12.91
C29H41N305	511.67	511.3046	68.08	8.08	8.21	15.63
C29H41N306	527.67	527.2995	66.01	7.83	7.96	18.19
C29H41N307	543.67	543.2944	64.07	7.60	7.73	20.60
C29H41N308	559.67	559.2893	62.24	7.38	7.51	22.87
C29H41N309	575.66	575.2843	60.51	7.18	7.30	25.01
C29H41N501	475.68	475.3311	73.23	8.69	14.72	3.36
C29H41N502	491.68	491.3260	70.84	8.41	14.24	6.51
C29H41N503	507.68	507.3209	68.61	8.14	13.79	9.45
C29H41N504	523.68	523.3158	66.51	7.89	13.37	12.22
C29H41N505	539.68	539.3107	64.54	7.66	12.98	14.82
C29H41N506	555.68	555.3057	62.68	7.44	12.60	17.28
C29H41N507	571.68	571.3006	60.93	7.23	12.25	19.59
C29H41N508	587.68	587.2955	59.27	7.03	11.92	21.78
C29H41N509	603.68	603.2904	57.70	6.85	11.60	23.85
C29H42N201	434.67	434.3297	80.13	9.74	6.44	3.68
C29H42N202	450.67	450.3246	77.29	9.39	6.22	7.10
C29H42N203	466.67	466.3195	74.64	9.07	6.00	10.29
C29H42N204	482.67	482.3144	72.17	8.77	5.80	13.26
C29H42N205	498.67	498.3094	69.85	8.49	5.62	16.04
C29H42N206	514.67	514.3043	67.68	8.23	5.44	18.65
C29H42N207	530.67	530.2992	65.64	7.98	5.28	21.10
C29H42N208	546.67	546.2941	63.72	7.74	5.12	23.41
C29H42N209	562.67	562.2890	61.91	7.52	4.98	25.59
C29H42N401	462.68	462.3358	75.28	9.15	12.11	3.46
C29H42N402	478.68	478.3308	72.77	8.84	11.70	6.68
C29H42N403	494.68	494.3257	70.41	8.56	11.33	9.70
C29H42N404	510.68	510.3206	68.21	8.29	10.97	12.53
C29H42N405	526.68	526.3155	66.14	8.04	10.64	15.19
C29H42N406	542.68	542.3104	64.19	7.80	10.32	17.69
C29H42N407	558.68	558.3053	62.35	7.58	10.03	20.05
C29H42N408	574.68	574.3002	60.61	7.37	9.75	22.27
C29H42N409	590.68	590.2952	58.97	7.17	9.49	24.38
C29H42N601	490.70	490.3420	70.99	8.63	17.13	3.26
C29H42N602	506.70	506.3369	68.74	8.36	16.59	6.32
C29H42N603	522.70	522.3318	66.64	8.10	16.08	9.18
C29H42N604	538.70	538.3267	64.66	7.86	15.60	11.88
C29H42N605	554.70	554.3216	62.80	7.63	15.15	14.42
C29H42N606	570.69	570.3166	61.03	7.42	14.73	16.82
C29H42N607	586.69	586.3115	59.37	7.22	14.32	19.09
C29H42N608	602.69	602.3064	57.79	7.02	13.94	21.24
C29H42N609	618.69	618.3013	56.30	6.84	13.58	23.27
C29H43N101	421.67	421.3344	82.61	10.28	3.32	3.79
C29H43N102	437.67	437.3294	79.59	9.90	3.20	7.31
C29H43N103	453.67	453.3243	76.78	9.55	3.09	10.58
C29H43N104	469.67	469.3192	74.16	9.23	2.98	13.63
C29H43N105	485.67	485.3141	71.72	8.92	2.88	16.47
C29H43N106	501.67	501.3090	69.43	8.64	2.79	19.14
C29H43N107	517.67	517.3039	67.29	8.37	2.71	21.63
C29H43N108	533.67	533.2988	65.27	8.12	2.62	23.98
C29H43N109	549.67	549.2938	63.37	7.89	2.55	26.20
C29H43N301	449.69	449.3406	77.46	9.64	9.34	3.56
C29H43N302	465.68	465.3355	74.80	9.31	9.02	6.87
C29H43N303	481.68	481.3304	72.31	9.00	8.72	9.96
C29H43N304	497.68	497.3253	69.99	8.71	8.44	12.86
C29H43N305	513.68	513.3202	67.81	8.44	8.18	15.57
C29H43N306	529.68	529.3152	65.76	8.18	7.93	18.12
C29H43N307	545.68	545.3101	63.83	7.94	7.70	20.52
C29H43N308	561.68	561.3050	62.01	7.72	7.48	22.79
C29H43N309	577.68	577.2999	60.30	7.50	7.27	24.93
C29H43N501	477.70	477.3467	72.92	9.07	14.66	3.35
C29H43N502	493.70	493.3417	70.55	8.78	14.19	6.48
C29H43N503	509.70	509.3366	68.34	8.50	13.74	9.42
C29H43N504	525.70	525.3315	66.26	8.24	13.32	12.17
C29H43N505	541.70	541.3264	64.30	8.00	12.93	14.77
C29H43N506	557.70	557.3213	62.46	7.77	12.56	17.21
C29H43N507	573.70	573.3162	60.72	7.56	12.21	19.52
C29H43N508	589.69	589.3111	59.07	7.35	11.88	21.71
C29H43N509	605.69	605.3061	57.51	7.16	11.56	23.77
C29H44N201	436.69	436.3453	79.77	10.16	6.41	3.66
C29H44N202	452.69	452.3403	76.95	9.80	6.19	7.07
C29H44N203	468.69	468.3352	74.32	9.46	5.98	10.24
C29H44N204	484.69	484.3301	71.87	9.15	5.78	13.20
C29H44N205	500.68	500.3250	69.57	8.86	5.60	15.98
C29H44N206	516.68	516.3199	67.42	8.58	5.42	18.58
C29H44N207	532.68	532.3148	65.39	8.33	5.26	21.02
C29H44N208	548.68	548.3097	63.48	8.08	5.11	23.33
C29H44N209	564.68	564.3047	61.68	7.85	4.96	25.50
C29H44N401	464.70	464.3515	74.96	9.54	12.06	3.44
C29H44N402	480.70	480.3464	72.46	9.23	11.66	6.66
C29H44N403	496.70	496.3413	70.13	8.93	11.28	9.66
C29H44N404	512.70	512.3362	67.94	8.65	10.93	12.48
C29H44N405	528.70	528.3311	65.88	8.39	10.60	15.13
C29H44N406	544.70	544.3261	63.95	8.14	10.29	17.62
C29H44N407	560.70	560.3210	62.12	7.91	9.99	19.97
C29H44N408	576.70	576.3159	60.40	7.69	9.72	22.19
C29H44N409	592.70	592.3108	58.77	7.48	9.45	24.29
C29H44N601	492.71	492.3576	70.69	9.00	17.06	3.25
C29H44N602	508.71	508.3526	68.47	8.72	16.52	6.29
C29H44N603	524.71	524.3475	66.38	8.45	16.02	9.15
C29H44N604	540.71	540.3424	64.42	8.20	15.54	11.84
C29H44N605	556.71	556.3373	62.57	7.97	15.10	14.37
C29H44N606	572.71	572.3322	60.82	7.74	14.67	16.76
C29H44N607	588.71	588.3271	59.17	7.53	14.28	19.02
C29H44N608	604.71	604.3220	57.60	7.33	13.90	21.17
C29H44N609	620.71	620.3170	56.12	7.15	13.54	23.20
C29H45N101	423.69	423.3501	82.21	10.71	3.31	3.78
C29H45N102	439.69	439.3450	79.22	10.32	3.19	7.28
C29H45N103	455.69	455.3399	76.44	9.95	3.07	10.53
C29H45N104	471.69	471.3348	73.85	9.62	2.97	13.57
C29H45N105	487.69	487.3298	71.42	9.30	2.87	16.40
C29H45N106	503.69	503.3247	69.15	9.01	2.78	19.06
C29H45N107	519.68	519.3196	67.03	8.73	2.70	21.55
C29H45N108	535.68	535.3145	65.02	8.47	2.61	23.89
C29H45N109	551.68	551.3094	63.14	8.22	2.54	26.10
C29H45N301	451.70	451.3562	77.11	10.04	9.30	3.54
C29H45N302	467.70	467.3512	74.48	9.70	8.98	6.84
C29H45N303	483.70	483.3461	72.01	9.38	8.69	9.92
C29H45N304	499.70	499.3410	69.71	9.08	8.41	12.81
C29H45N305	515.70	515.3359	67.54	8.80	8.15	15.51
C29H45N306	531.70	531.3308	65.51	8.53	7.90	18.05
C29H45N307	547.70	547.3257	63.60	8.28	7.67	20.45
C29H45N308	563.70	563.3206	61.79	8.05	7.45	22.71
C29H45N309	579.70	579.3156	60.09	7.82	7.25	24.84
C29H45N501	479.71	479.3624	72.61	9.46	14.60	3.34

C29H45N502	495.71	495.3573	70.27	9.15	14.13	6.46
C29H45N503	511.71	511.3522	68.07	8.86	13.69	9.38
C29H45N504	527.71	527.3471	66.01	8.60	13.27	12.13
C29H45N505	543.71	543.3420	64.06	8.34	12.88	14.71
C29H45N506	559.71	559.3370	62.23	8.10	12.51	17.15
C29H45N507	575.71	575.3319	60.50	7.88	12.16	19.45
C29H45N508	591.71	591.3268	58.87	7.67	11.84	21.63
C29H45N509	607.71	607.3217	57.32	7.46	11.52	23.69
C29H46N201	438.70	438.3610	79.40	10.57	6.39	3.65
C29H46N202	454.70	454.3559	76.60	10.20	6.16	7.04
C29H46N203	470.70	470.3508	74.00	9.85	5.95	10.20
C29H46N204	486.70	486.3457	71.57	9.53	5.76	13.15
C29H46N205	502.70	502.3406	69.29	9.22	5.57	15.91
C29H46N206	518.70	518.3356	67.15	8.94	5.40	18.51
C29H46N207	534.70	534.3305	65.14	8.67	5.24	20.95
C29H46N208	550.70	550.3254	63.25	8.42	5.09	23.24
C29H46N209	566.70	566.3203	61.47	8.18	4.94	25.41
C29H46N401	466.72	466.3671	74.63	9.93	12.00	3.43
C29H46N402	482.72	482.3621	72.16	9.61	11.61	6.63
C29H46N403	498.71	498.3570	69.84	9.30	11.23	9.62
C29H46N404	514.71	514.3519	67.67	9.01	10.89	12.43
C29H46N405	530.71	530.3468	65.63	8.74	10.56	15.07
C29H46N406	546.71	546.3417	63.71	8.48	10.25	17.56
C29H46N407	562.71	562.3366	61.90	8.24	9.96	19.90
C29H46N408	578.71	578.3315	60.19	8.01	9.68	22.12
C29H46N409	594.71	594.3265	58.57	7.80	9.42	24.21
C29H46N601	494.73	494.3733	70.41	9.37	16.99	3.23
C29H46N602	510.73	510.3682	68.20	9.08	16.45	6.27
C29H46N603	526.73	526.3631	66.13	8.80	15.96	9.11
C29H46N604	542.73	542.3580	64.18	8.54	15.48	11.79
C29H46N605	558.73	558.3529	62.34	8.30	15.04	14.32
C29H46N606	574.73	574.3479	60.61	8.07	14.62	16.70
C29H46N607	590.73	590.3428	58.97	7.85	14.23	18.96
C29H46N608	606.73	606.3377	57.41	7.64	13.85	21.10
C29H46N609	622.72	622.3326	55.94	7.45	13.50	23.12
C29H47N101	425.70	425.3657	81.82	11.13	3.29	3.76
C29H47N102	441.70	441.3607	78.86	10.73	3.17	7.24
C29H47N103	457.70	457.3556	76.10	10.35	3.06	10.49
C29H47N104	473.70	473.3505	73.53	10.00	2.96	13.51
C29H47N105	489.70	489.3454	71.13	9.67	2.86	16.34
C29H47N106	505.70	505.3403	68.88	9.37	2.77	18.98
C29H47N107	521.70	521.3352	66.77	9.08	2.68	21.47
C29H47N108	537.70	537.3301	64.78	8.81	2.60	23.80
C29H47N109	553.70	553.3251	62.91	8.56	2.53	26.01
C29H47N301	453.72	453.3719	76.77	10.44	9.26	3.53
C29H47N302	469.72	469.3668	74.16	10.09	8.95	6.81
C29H47N303	485.72	485.3617	71.71	9.75	8.65	9.88
C29H47N304	501.72	501.3566	69.43	9.44	8.38	12.76
C29H47N305	517.72	517.3515	67.28	9.15	8.12	15.45
C29H47N306	533.71	533.3465	65.26	8.88	7.87	17.99
C29H47N307	549.71	549.3414	63.36	8.62	7.64	20.37
C29H47N308	565.71	565.3363	61.57	8.37	7.43	22.63
C29H47N309	581.71	581.3312	59.88	8.14	7.22	24.75
C29H47N501	481.73	481.3780	72.31	9.83	14.54	3.32
C29H47N502	497.73	497.3730	69.98	9.52	14.07	6.43
C29H47N503	513.73	513.3679	67.80	9.22	13.63	9.34
C29H47N504	529.73	529.3628	65.76	8.94	13.22	12.08
C29H47N505	545.73	545.3577	63.83	8.68	12.83	14.66
C29H47N506	561.73	561.3526	62.01	8.43	12.47	17.09
C29H47N507	577.73	577.3475	60.29	8.20	12.12	19.39

C29H47N508	593.73	593.3424	58.67	7.98	11.80	21.56
C29H47N509	609.73	609.3374	57.13	7.77	11.49	23.62
C29H48N201	440.72	440.3766	79.04	10.98	6.36	3.63
C29H48N202	456.72	456.3716	76.27	10.59	6.13	7.01
C29H48N203	472.72	472.3665	73.69	10.23	5.93	10.15
C29H48N204	488.72	488.3614	71.27	9.90	5.73	13.10
C29H48N205	504.72	504.3563	69.01	9.59	5.55	15.85
C29H48N206	520.72	520.3512	66.89	9.29	5.38	18.44
C29H48N207	536.72	536.3461	64.90	9.01	5.22	20.87
C29H48N208	552.71	552.3410	63.02	8.75	5.07	23.16
C29H48N209	568.71	568.3360	61.25	8.51	4.93	25.32
C29H48N401	468.73	468.3828	74.31	10.32	11.95	3.41
C29H48N402	484.73	484.3777	71.86	9.98	11.56	6.60
C29H48N403	500.73	500.3726	69.56	9.66	11.19	9.59
C29H48N404	516.73	516.3675	67.41	9.36	10.84	12.39
C29H48N405	532.73	532.3624	65.38	9.08	10.52	15.02
C29H48N406	548.73	548.3574	63.48	8.82	10.21	17.49
C29H48N407	564.73	564.3523	61.68	8.57	9.92	19.83
C29H48N408	580.73	580.3472	59.98	8.33	9.65	22.04
C29H48N409	596.73	596.3421	58.37	8.11	9.39	24.13
C29H48N601	496.75	496.3889	70.12	9.74	16.92	3.22
C29H48N602	512.74	512.3839	67.93	9.44	16.39	6.24
C29H48N603	528.74	528.3788	65.88	9.15	15.89	9.08
C29H48N604	544.74	544.3737	63.94	8.88	15.43	11.75
C29H48N605	560.74	560.3686	62.12	8.63	14.99	14.27
C29H48N606	576.74	576.3635	60.39	8.39	14.57	16.64
C29H48N607	592.74	592.3584	58.76	8.16	14.18	18.89
C29H48N608	608.74	608.3533	57.22	7.95	13.81	21.03
C29H48N609	624.74	624.3483	55.75	7.74	13.45	23.05
C29H49N101	427.72	427.3814	81.44	11.55	3.27	3.74
C29H49N102	443.72	443.3763	78.50	11.13	3.16	7.21
C29H49N103	459.72	459.3712	75.77	10.74	3.05	10.44
C29H49N104	475.72	475.3661	73.22	10.38	2.94	13.45
C29H49N105	491.72	491.3610	70.84	10.04	2.85	16.27
C29H49N106	507.72	507.3560	68.61	9.73	2.76	18.91
C29H49N107	523.72	523.3509	66.51	9.43	2.67	21.38
C29H49N108	539.72	539.3458	64.54	9.15	2.60	23.72
C29H49N109	555.72	555.3407	62.68	8.89	2.52	25.91
C29H49N301	455.73	455.3875	76.43	10.84	9.22	3.51
C29H49N302	471.73	471.3825	73.84	10.47	8.91	6.78
C29H49N303	487.73	487.3774	71.42	10.13	8.62	9.84
C29H49N304	503.73	503.3723	69.15	9.80	8.34	12.70
C29H49N305	519.73	519.3672	67.02	9.50	8.08	15.39
C29H49N306	535.73	535.3621	65.02	9.22	7.84	17.92
C29H49N307	551.73	551.3570	63.13	8.95	7.62	20.30
C29H49N308	567.73	567.3519	61.35	8.70	7.40	22.55
C29H49N309	583.73	583.3469	59.67	8.46	7.20	24.67
C29H49N501	483.75	483.3937	72.01	10.21	14.48	3.31
C29H49N502	499.75	499.3886	69.70	9.88	14.01	6.40
C29H49N503	515.75	515.3835	67.54	9.58	13.58	9.31
C29H49N504	531.74	531.3784	65.51	9.29	13.17	12.04
C29H49N505	547.74	547.3733	63.59	9.02	12.79	14.60
C29H49N506	563.74	563.3683	61.79	8.76	12.42	17.03
C29H49N507	579.74	579.3632	60.08	8.52	12.08	19.32
C29H49N508	595.74	595.3581	58.47	8.29	11.76	21.48
C29H49N509	611.74	611.3530	56.94	8.07	11.45	23.54
C29H50N201	442.73	442.3923	78.68	11.38	6.33	3.61
C29H50N202	458.73	458.3872	75.93	10.99	6.11	6.98
C29H50N203	474.73	474.3821	73.37	10.62	5.90	10.11
C29H50N204	490.73	490.3770	70.98	10.27	5.71	13.04

C29H50N205	506.73	506.3719	68.74	9.95	5.53	15.79
C29H50N206	522.73	522.3669	66.64	9.64	5.36	18.36
C29H50N207	538.73	538.3618	64.66	9.36	5.20	20.79
C29H50N208	554.73	554.3567	62.79	9.09	5.05	23.07
C29H50N209	570.73	570.3516	61.03	8.83	4.91	25.23
C29H50N401	470.75	470.3984	73.99	10.71	11.90	3.40
C29H50N402	486.75	486.3934	71.56	10.35	11.51	6.57
C29H50N403	502.75	502.3883	69.28	10.02	11.14	9.55
C29H50N404	518.75	518.3832	67.15	9.72	10.80	12.34
C29H50N405	534.75	534.3781	65.14	9.42	10.48	14.96
C29H50N406	550.75	550.3730	63.25	9.15	10.17	17.43
C29H50N407	566.74	566.3679	61.46	8.89	9.89	19.76
C29H50N408	582.74	582.3628	59.77	8.65	9.61	21.96
C29H50N409	598.74	598.3578	58.18	8.42	9.36	24.05
C29H50N601	498.76	498.4046	69.84	10.10	16.85	3.21
C29H50N602	514.76	514.3995	67.67	9.79	16.33	6.22
C29H50N603	530.76	530.3944	65.63	9.50	15.83	9.04
C29H50N604	546.76	546.3893	63.71	9.22	15.37	11.70
C29H50N605	562.76	562.3842	61.90	8.96	14.93	14.22
C29H50N606	578.76	578.3792	60.18	8.71	14.52	16.59
C29H50N607	594.76	594.3741	58.57	8.47	14.13	18.83
C29H50N608	610.76	610.3690	57.03	8.25	13.76	20.96
C29H50N609	626.76	626.3639	55.58	8.04	13.41	22.97
C29H51N101	429.74	429.3970	81.06	11.96	3.26	3.72
C29H51N102	445.74	445.3920	78.15	11.53	3.14	7.18
C29H51N103	461.73	461.3869	75.44	11.13	3.03	10.40
C29H51N104	477.73	477.3818	72.91	10.76	2.93	13.40
C29H51N105	493.73	493.3767	70.55	10.41	2.84	16.20
C29H51N106	509.73	509.3716	68.33	10.08	2.75	18.83
C29H51N107	525.73	525.3665	66.25	9.78	2.66	21.30
C29H51N108	541.73	541.3614	64.30	9.49	2.59	23.63
C29H51N109	557.73	557.3564	62.45	9.22	2.51	25.82
C29H51N301	457.75	457.4032	76.09	11.23	9.18	3.50
C29H51N302	473.75	473.3981	73.52	10.85	8.87	6.75
C29H51N303	489.75	489.3930	71.12	10.50	8.58	9.80
C29H51N304	505.75	505.3879	68.87	10.16	8.31	12.65
C29H51N305	521.75	521.3828	66.76	9.85	8.05	15.33
C29H51N306	537.75	537.3778	64.77	9.56	7.81	17.85
C29H51N307	553.75	553.3727	62.90	9.28	7.59	20.23
C29H51N308	569.75	569.3676	61.14	9.02	7.38	22.47
C29H51N309	585.74	585.3625	59.47	8.78	7.17	24.58
C29H51N501	485.76	485.4093	71.71	10.58	14.42	3.29
C29H51N502	501.76	501.4043	69.42	10.25	13.96	6.38
C29H51N503	517.76	517.3992	67.27	9.93	13.53	9.27
C29H51N504	533.76	533.3941	65.26	9.63	13.12	11.99
C29H51N505	549.76	549.3890	63.36	9.35	12.74	14.55
C29H51N506	565.76	565.3839	61.57	9.09	12.38	16.97
C29H51N507	581.76	581.3788	59.87	8.84	12.04	19.25
C29H51N508	597.76	597.3737	58.27	8.60	11.72	21.41
C29H51N509	613.76	613.3687	56.75	8.38	11.41	23.46
C29H52N201	444.75	444.4079	78.32	11.79	6.30	3.60
C29H52N202	460.75	460.4029	75.60	11.38	6.08	6.94
C29H52N203	476.75	476.3978	73.06	10.99	5.88	10.07
C29H52N204	492.75	492.3927	70.69	10.64	5.69	12.99
C29H52N205	508.75	508.3876	68.47	10.30	5.51	15.72
C29H52N206	524.75	524.3825	66.38	9.99	5.34	18.29
C29H52N207	540.75	540.3774	64.42	9.69	5.18	20.71
C29H52N208	556.75	556.3723	62.56	9.41	5.03	22.99
C29H52N209	572.75	572.3673	60.82	9.15	4.89	25.14
C29H52N401	472.76	472.4141	73.68	11.09	11.85	3.38

C29H52N402	488.76	488.4090	71.27	10.72	11.46	6.55
C29H52N403	504.76	504.4039	69.01	10.38	11.10	9.51
C29H52N404	520.76	520.3988	66.89	10.06	10.76	12.29
C29H52N405	536.76	536.3937	64.89	9.76	10.44	14.90
C29H52N406	552.76	552.3887	63.02	9.48	10.14	17.37
C29H52N407	568.76	568.3836	61.24	9.22	9.85	19.69
C29H52N408	584.76	584.3785	59.57	8.96	9.58	21.89
C29H52N409	600.76	600.3734	57.98	8.72	9.33	23.97
C29H52N601	500.78	500.4202	69.56	10.47	16.78	3.19
C29H52N602	516.78	516.4152	67.40	10.14	16.26	6.19
C29H52N603	532.78	532.4101	65.38	9.84	15.77	9.01
C29H52N604	548.78	548.4050	63.47	9.55	15.31	11.66
C29H52N605	564.77	564.3999	61.67	9.28	14.88	14.16
C29H52N606	580.77	580.3948	59.98	9.02	14.47	16.53
C29H52N607	596.77	596.3897	58.37	8.78	14.08	18.77
C29H52N608	612.77	612.3846	56.84	8.55	13.71	20.89
C29H52N609	628.77	628.3795	55.40	8.34	13.37	22.90
C29H53N101	431.75	431.4127	80.68	12.37	3.24	3.71
C29H53N102	447.75	447.4076	77.79	11.93	3.13	7.15
C29H53N103	463.75	463.4025	75.11	11.52	3.02	10.35
C29H53N104	479.75	479.3974	72.61	11.14	2.92	13.34
C29H53N105	495.75	495.3923	70.26	10.78	2.83	16.14
C29H53N106	511.75	511.3873	68.07	10.44	2.74	18.76
C29H53N107	527.75	527.3822	66.00	10.12	2.65	21.22
C29H53N108	543.75	543.3771	64.06	9.82	2.58	23.54
C29H53N109	559.75	559.3720	62.23	9.54	2.50	25.72
C29H53N301	459.77	459.4188	75.76	11.62	9.14	3.48
C29H53N302	475.76	475.4138	73.21	11.23	8.83	6.73
C29H53N303	491.76	491.4087	70.83	10.86	8.54	9.76
C29H53N304	507.76	507.4036	68.60	10.52	8.28	12.60
C29H53N305	523.76	523.3985	66.50	10.20	8.02	15.27
C29H53N306	539.76	539.3934	64.53	9.90	7.78	17.78
C29H53N307	555.76	555.3883	62.67	9.61	7.56	20.15
C29H53N308	571.76	571.3832	60.92	9.34	7.35	22.39
C29H53N309	587.76	587.3782	59.26	9.09	7.15	24.50
C29H53N501	487.78	487.4250	71.41	10.95	14.36	3.28
C29H53N502	503.78	503.4199	69.14	10.60	13.90	6.35
C29H53N503	519.78	519.4148	67.01	10.28	13.47	9.23
C29H53N504	535.78	535.4097	65.01	9.97	13.07	11.94
C29H53N505	551.78	551.4046	63.13	9.68	12.69	14.50
C29H53N506	567.78	567.3996	61.35	9.41	12.33	16.91
C29H53N507	583.78	583.3945	59.67	9.15	12.00	19.18
C29H53N508	599.77	599.3894	58.08	8.91	11.68	21.34
C29H53N509	615.77	615.3843	56.57	8.68	11.37	23.38
C29H54N201	446.77	446.4236	77.97	12.18	6.27	3.58
C29H54N202	462.77	462.4185	75.27	11.76	6.05	6.91
C29H54N203	478.77	478.4134	72.75	11.37	5.85	10.03
C29H54N204	494.76	494.4083	70.40	11.00	5.66	12.93
C29H54N205	510.76	510.4032	68.20	10.66	5.48	15.66
C29H54N206	526.76	526.3982	66.13	10.33	5.32	18.22
C29H54N207	542.76	542.3931	64.18	10.03	5.16	20.63
C29H54N208	558.76	558.3880	62.34	9.74	5.01	22.91
C29H54N209	574.76	574.3829	60.60	9.47	4.87	25.05
C29H54N401	474.78	474.4297	73.37	11.46	11.80	3.37
C29H54N402	490.78	490.4247	70.97	11.09	11.42	6.52
C29H54N403	506.78	506.4196	68.73	10.74	11.06	9.47
C29H54N404	522.78	522.4145	66.63	10.41	10.72	12.24
C29H54N405	538.78	538.4094	64.65	10.10	10.40	14.85
C29H54N406	554.78	554.4043	62.79	9.81	10.10	17.30
C29H54N407	570.78	570.3992	61.03	9.54	9.82	19.62

C29H54N408	586.78	586.3941	59.36	9.28	9.55	21.81
C29H54N409	602.78	602.3891	57.79	9.03	9.29	23.89
C29H54N601	502.79	502.4359	69.28	10.83	16.71	3.18
C29H54N602	518.79	518.4308	67.14	10.49	16.20	6.17
C29H54N603	534.79	534.4257	65.13	10.18	15.71	8.98
C29H54N604	550.79	550.4206	63.24	9.88	15.26	11.62
C29H54N605	566.79	566.4155	61.46	9.60	14.83	14.11
C29H54N606	582.79	582.4105	59.77	9.34	14.42	16.47
C29H54N607	598.79	598.4054	58.17	9.09	14.04	18.70
C29H54N608	614.79	614.4003	56.66	8.85	13.67	20.82
C29H54N609	630.79	630.3952	55.22	8.63	13.32	22.83
C29H55N101	433.77	433.4283	80.30	12.78	3.23	3.69
C29H55N102	449.77	449.4233	77.45	12.33	3.11	7.11
C29H55N103	465.77	465.4182	74.78	11.90	3.01	10.31
C29H55N104	481.77	481.4131	72.30	11.51	2.91	13.28
C29H55N105	497.77	497.4080	69.98	11.14	2.81	16.07
C29H55N106	513.76	513.4029	67.80	10.79	2.73	18.68
C29H55N107	529.76	529.3978	65.75	10.46	2.64	21.14
C29H55N108	545.76	545.3927	63.82	10.16	2.57	23.45
C29H55N109	561.76	561.3877	62.01	9.87	2.49	25.63
C29H55N301	461.78	461.4345	75.43	12.01	9.10	3.46
C29H55N302	477.78	477.4294	72.90	11.60	8.79	6.70
C29H55N303	493.78	493.4243	70.54	11.23	8.51	9.72
C29H55N304	509.78	509.4192	68.33	10.87	8.24	12.55
C29H55N305	525.78	525.4141	66.25	10.54	7.99	15.21
C29H55N306	541.78	541.4091	64.29	10.23	7.76	17.72
C29H55N307	557.78	557.4040	62.45	9.94	7.53	20.08
C29H55N308	573.78	573.3989	60.71	9.66	7.32	22.31
C29H55N309	589.78	589.3938	59.06	9.40	7.12	24.42
C29H55N501	489.79	489.4406	71.12	11.32	14.30	3.27
C29H55N502	505.79	505.4356	68.87	10.96	13.85	6.33
C29H55N503	521.79	521.4305	66.76	10.62	13.42	9.20
C29H55N504	537.79	537.4254	64.77	10.31	13.02	11.90
C29H55N505	553.79	553.4203	62.90	10.01	12.65	14.45
C29H55N506	569.79	569.4152	61.13	9.73	12.29	16.85
C29H55N507	585.79	585.4101	59.46	9.46	11.96	19.12
C29H55N508	601.79	601.4050	57.88	9.21	11.64	21.27
C29H55N509	617.79	617.3999	56.38	8.97	11.34	23.31
C29H56N201	448.78	448.4392	77.62	12.58	6.24	3.57
C29H56N202	464.78	464.4342	74.94	12.14	6.03	6.88
C29H56N203	480.78	480.4291	72.45	11.74	5.83	9.98
C29H56N204	496.78	496.4240	70.12	11.36	5.64	12.88
C29H56N205	512.78	512.4189	67.93	11.01	5.46	15.60
C29H56N206	528.78	528.4138	65.87	10.67	5.30	18.15
C29H56N207	544.78	544.4087	63.94	10.36	5.14	20.56
C29H56N208	560.78	560.4036	62.11	10.07	5.00	22.82
C29H56N209	576.78	576.3986	60.39	9.79	4.86	24.97
C29H56N401	476.80	476.4454	73.06	11.84	11.75	3.36
C29H56N402	492.80	492.4403	70.68	11.45	11.37	6.49
C29H56N403	508.79	508.4352	68.46	11.09	11.01	9.43
C29H56N404	524.79	524.4301	66.37	10.76	10.68	12.19
C29H56N405	540.79	540.4250	64.41	10.44	10.36	14.79
C29H56N406	556.79	556.4200	62.56	10.14	10.06	17.24
C29H56N407	572.79	572.4149	60.81	9.85	9.78	19.55
C29H56N408	588.79	588.4098	59.16	9.59	9.52	21.74
C29H56N409	604.79	604.4047	57.59	9.33	9.26	23.81
C29H56N601	504.81	504.4515	69.00	11.18	16.65	3.17
C29H56N602	520.81	520.4464	66.88	10.84	16.14	6.14
C29H56N603	536.81	536.4414	64.89	10.52	15.66	8.94
C29H56N604	552.81	552.4363	63.01	10.21	15.20	11.58

C29H56N605	568.81	568.4312	61.24	9.92	14.77	14.06
C29H56N606	584.81	584.4261	59.56	9.65	14.37	16.42
C29H56N607	600.81	600.4210	57.98	9.40	13.99	18.64
C29H56N608	616.81	616.4159	56.47	9.15	13.63	20.75
C29H56N609	632.80	632.4108	55.04	8.92	13.28	22.75
C29H57N101	435.78	435.4440	79.93	13.18	3.21	3.67
C29H57N102	451.78	451.4389	77.10	12.72	3.10	7.08
C29H57N103	467.78	467.4338	74.46	12.28	2.99	10.26
C29H57N104	483.78	483.4287	72.00	11.88	2.90	13.23
C29H57N105	499.78	499.4236	69.70	11.50	2.80	16.01
C29H57N106	515.78	515.4186	67.53	11.14	2.72	18.61
C29H57N107	531.78	531.4135	65.50	10.80	2.63	21.06
C29H57N108	547.78	547.4084	63.59	10.49	2.56	23.37
C29H57N109	563.78	563.4033	61.78	10.19	2.48	25.54
C29H57N301	463.80	463.4501	75.10	12.39	9.06	3.45
C29H57N302	479.80	479.4451	72.60	11.97	8.76	6.67
C29H57N303	495.80	495.4400	70.26	11.59	8.48	9.68
C29H57N304	511.80	511.4349	68.06	11.23	8.21	12.50
C29H57N305	527.79	527.4298	66.00	10.89	7.96	15.16
C29H57N306	543.79	543.4247	64.05	10.57	7.73	17.65
C29H57N307	559.79	559.4196	62.22	10.26	7.51	20.01
C29H57N308	575.79	575.4145	60.49	9.98	7.30	22.23
C29H57N309	591.79	591.4095	58.86	9.71	7.10	24.33
C29H57N501	491.81	491.4563	70.82	11.68	14.24	3.25
C29H57N502	507.81	507.4512	68.59	11.31	13.79	6.30
C29H57N503	523.81	523.4461	66.50	10.97	13.37	9.16
C29H57N504	539.81	539.4410	64.53	10.64	12.97	11.86
C29H57N505	555.81	555.4359	62.67	10.34	12.60	14.39
C29H57N506	571.81	571.4309	60.92	10.05	12.25	16.79
C29H57N507	587.81	587.4258	59.26	9.77	11.91	19.05
C29H57N508	603.81	603.4207	57.69	9.52	11.60	21.20
C29H57N509	619.81	619.4156	56.20	9.27	11.30	23.23
C29H58N201	450.80	450.4549	77.27	12.97	6.21	3.55
C29H58N202	466.80	466.4498	74.62	12.52	6.00	6.85
C29H58N203	482.80	482.4447	72.15	12.11	5.80	9.94
C29H58N204	498.80	498.4396	69.83	11.72	5.62	12.83
C29H58N205	514.80	514.4345	67.66	11.36	5.44	15.54
C29H58N206	530.80	530.4295	65.62	11.01	5.28	18.09
C29H58N207	546.79	546.4244	63.70	10.69	5.12	20.48
C29H58N208	562.79	562.4193	61.89	10.39	4.98	22.74
C29H58N209	578.79	578.4142	60.18	10.10	4.84	24.88
C29H58N401	478.81	478.4610	72.75	12.21	11.70	3.34
C29H58N402	494.81	494.4560	70.40	11.82	11.32	6.47
C29H58N403	510.81	510.4509	68.19	11.44	10.97	9.40
C29H58N404	526.81	526.4458	66.12	11.10	10.64	12.15
C29H58N405	542.81	542.4407	64.17	10.77	10.32	14.74
C29H58N406	558.81	558.4356	62.33	10.46	10.03	17.18
C29H58N407	574.81	574.4305	60.60	10.17	9.75	19.48
C29H58N408	590.81	590.4254	58.96	9.90	9.48	21.66
C29H58N409	606.81	606.4203	57.40	9.63	9.23	23.73
C29H58N601	506.83	506.4672	68.73	11.53	16.58	3.16
C29H58N602	522.82	522.4621	66.62	11.18	16.07	6.12
C29H58N603	538.82	538.4570	64.65	10.85	15.60	8.91
C29H58N604	554.82	554.4519	62.78	10.54	15.15	11.53
C29H58N605	570.82	570.4468	61.02	10.24	14.72	14.01
C29H58N606	586.82	586.4418	59.36	9.96	14.32	16.36
C29H58N607	602.82	602.4367	57.78	9.70	13.94	18.58
C29H58N608	618.82	618.4316	56.29	9.45	13.58	20.68
C29H58N609	634.82	634.4265	54.87	9.21	13.24	22.68
C29H59N101	437.80	437.4596	79.56	13.58	3.20	3.65

C29H59N102	453.80	453.4546	76.76	13.10	3.09	7.05
C29H59N103	469.80	469.4495	74.14	12.66	2.98	10.22
C29H59N104	485.80	485.4444	71.70	12.24	2.88	13.17
C29H59N105	501.80	501.4393	69.42	11.85	2.79	15.94
C29H59N106	517.80	517.4342	67.27	11.49	2.71	18.54
C29H59N107	533.80	533.4291	65.25	11.14	2.62	20.98
C29H59N108	549.80	549.4240	63.36	10.82	2.55	23.28
C29H59N109	565.79	565.4190	61.56	10.51	2.48	25.45
C29H59N301	465.81	465.4658	74.78	12.77	9.02	3.43
C29H59N302	481.81	481.4607	72.29	12.34	8.72	6.64
C29H59N303	497.81	497.4556	69.97	11.95	8.44	9.64
C29H59N304	513.81	513.4505	67.79	11.57	8.18	12.46
C29H59N305	529.81	529.4454	65.74	11.22	7.93	15.10
C29H59N306	545.81	545.4404	63.82	10.90	7.70	17.59
C29H59N307	561.81	561.4353	62.00	10.59	7.48	19.93
C29H59N308	577.81	577.4302	60.28	10.29	7.27	22.15
C29H59N309	593.81	593.4251	58.66	10.02	7.08	24.25
C29H59N501	493.83	493.4719	70.54	12.04	14.18	3.24
C29H59N502	509.83	509.4668	68.32	11.66	13.74	6.28
C29H59N503	525.83	525.4618	66.24	11.31	13.32	9.13
C29H59N504	541.82	541.4567	64.29	10.98	12.93	11.81
C29H59N505	557.82	557.4516	62.44	10.66	12.55	14.34
C29H59N506	573.82	573.4465	60.70	10.36	12.20	16.73
C29H59N507	589.82	589.4414	59.06	10.08	11.87	18.99
C29H59N508	605.82	605.4363	57.50	9.82	11.56	21.13
C29H59N509	621.82	621.4312	56.02	9.56	11.26	23.16
C29H60N201	452.81	452.4705	76.92	13.36	6.19	3.53
C29H60N202	468.81	468.4655	74.30	12.90	5.98	6.83
C29H60N203	484.81	484.4604	71.85	12.47	5.78	9.90
C29H60N204	500.81	500.4553	69.55	12.08	5.59	12.78
C29H60N205	516.81	516.4502	67.40	11.70	5.42	15.48
C29H60N206	532.81	532.4451	65.37	11.35	5.26	18.02
C29H60N207	548.81	548.4400	63.47	11.02	5.10	20.41
C29H60N208	564.81	564.4349	61.67	10.71	4.96	22.66
C29H60N209	580.81	580.4299	59.97	10.41	4.82	24.79
C29H60N401	480.83	480.4767	72.44	12.58	11.65	3.33
C29H60N402	496.83	496.4716	70.11	12.17	11.28	6.44
C29H60N403	512.83	512.4665	67.92	11.79	10.93	9.36
C29H60N404	528.83	528.4614	65.87	11.44	10.59	12.10
C29H60N405	544.83	544.4563	63.93	11.10	10.28	14.68
C29H60N406	560.82	560.4513	62.11	10.78	9.99	17.12
C29H60N407	576.82	576.4462	60.39	10.48	9.71	19.42
C29H60N408	592.82	592.4411	58.76	10.20	9.45	21.59
C29H60N409	608.82	608.4360	57.21	9.93	9.20	23.65
C29H60N601	508.84	508.4828	68.45	11.89	16.52	3.14
C29H60N602	524.84	524.4777	66.37	11.52	16.01	6.10
C29H60N603	540.84	540.4727	64.40	11.18	15.54	8.87
C29H60N604	556.84	556.4676	62.55	10.86	15.09	11.49
C29H60N605	572.84	572.4625	60.81	10.56	14.67	13.97
C29H60N606	588.84	588.4574	59.15	10.27	14.27	16.30
C29H60N607	604.84	604.4523	57.59	10.00	13.89	18.52
C29H60N608	620.84	620.4472	56.11	9.74	13.54	20.62
C29H60N609	636.84	636.4421	54.70	9.50	13.20	22.61
C29H61N101	439.82	439.4753	79.20	13.98	3.18	3.64
C29H61N102	455.82	455.4702	76.42	13.49	3.07	7.02
C29H61N103	471.81	471.4651	73.83	13.03	2.97	10.17
C29H61N104	487.81	487.4600	71.40	12.60	2.87	13.12
C29H61N105	503.81	503.4549	69.14	12.20	2.78	15.88
C29H61N106	519.81	519.4499	67.01	11.83	2.69	18.47
C29H61N107	535.81	535.4448	65.01	11.48	2.61	20.90

C29H61N108	551.81	551.4397	63.12	11.14	2.54	23.20
C29H61N109	567.81	567.4346	61.34	10.83	2.47	25.36
C29H61N301	467.83	467.4814	74.46	13.14	8.98	3.42
C29H61N302	483.83	483.4764	71.99	12.71	8.68	6.61
C29H61N303	499.83	499.4713	69.69	12.30	8.41	9.60
C29H61N304	515.83	515.4662	67.53	11.92	8.15	12.41
C29H61N305	531.83	531.4611	65.50	11.56	7.90	15.04
C29H61N306	547.83	547.4560	63.58	11.22	7.67	17.52
C29H61N307	563.83	563.4509	61.78	10.91	7.45	19.86
C29H61N308	579.82	579.4458	60.07	10.60	7.25	22.07
C29H61N309	595.82	595.4407	58.46	10.32	7.05	24.17
C29H61N501	495.84	495.4876	70.25	12.40	14.12	3.23
C29H61N502	511.84	511.4825	68.05	12.01	13.68	6.25
C29H61N503	527.84	527.4774	65.99	11.65	13.27	9.09
C29H61N504	543.84	543.4723	64.05	11.31	12.88	11.77
C29H61N505	559.84	559.4672	62.22	10.98	12.51	14.29
C29H61N506	575.84	575.4622	60.49	10.68	12.16	16.67
C29H61N507	591.84	591.4571	58.85	10.39	11.83	18.92
C29H61N508	607.84	607.4520	57.31	10.12	11.52	21.06
C29H61N509	623.84	623.4469	55.84	9.86	11.23	23.08
C29H62N201	454.83	454.4862	76.58	13.74	6.16	3.52
C29H62N202	470.83	470.4811	73.98	13.27	5.95	6.80
C29H62N203	486.83	486.4760	71.55	12.84	5.75	9.86
C29H62N204	502.83	502.4709	69.27	12.43	5.57	12.73
C29H62N205	518.83	518.4658	67.14	12.05	5.40	15.42
C29H62N206	534.83	534.4608	65.13	11.68	5.24	17.95
C29H62N207	550.83	550.4557	63.24	11.35	5.09	20.33
C29H62N208	566.83	566.4506	61.45	11.03	4.94	22.58
C29H62N209	582.83	582.4455	59.76	10.72	4.81	24.71
C29H62N401	482.84	482.4923	72.14	12.94	11.60	3.31
C29H62N402	498.84	498.4872	69.83	12.53	11.23	6.41
C29H62N403	514.84	514.4822	67.66	12.14	10.88	9.32
C29H62N404	530.84	530.4771	65.62	11.77	10.55	12.06
C29H62N405	546.84	546.4720	63.70	11.43	10.25	14.63
C29H62N406	562.84	562.4669	61.89	11.10	9.95	17.06
C29H62N407	578.84	578.4618	60.18	10.80	9.68	19.35
C29H62N408	594.84	594.4567	58.56	10.51	9.42	21.52
C29H62N409	610.84	610.4516	57.02	10.23	9.17	23.57
C29H62N601	510.86	510.4985	68.18	12.23	16.45	3.13
C29H62N602	526.86	526.4934	66.11	11.86	15.95	6.07
C29H62N603	542.86	542.4883	64.16	11.51	15.48	8.84
C29H62N604	558.86	558.4832	62.33	11.18	15.04	11.45
C29H62N605	574.85	574.4781	60.59	10.87	14.62	13.92
C29H62N606	590.85	590.4731	58.95	10.58	14.22	16.25
C29H62N607	606.85	606.4680	57.40	10.30	13.85	18.46
C29H62N608	622.85	622.4629	55.92	10.03	13.49	20.55
C29H62N609	638.85	638.4578	54.52	9.78	13.15	22.54
C29H63N301	469.84	469.4971	74.14	13.52	8.94	3.41
C29H63N302	485.84	485.4920	71.69	13.07	8.65	6.59
C29H63N303	501.84	501.4869	69.41	12.65	8.37	9.56
C29H63N304	517.84	517.4818	67.26	12.26	8.11	12.36
C29H63N305	533.84	533.4767	65.25	11.90	7.87	14.99
C29H63N306	549.84	549.4717	63.35	11.55	7.64	17.46
C29H63N307	565.84	565.4666	61.56	11.22	7.43	19.79
C29H63N308	581.84	581.4615	59.87	10.91	7.22	22.00
C29H63N309	597.84	597.4564	58.26	10.62	7.03	24.09
C29H63N501	497.86	497.5032	69.96	12.76	14.07	3.21
C29H63N502	513.86	513.4981	67.79	12.36	13.63	6.23
C29H63N503	529.86	529.4931	65.74	11.98	13.22	9.06
C29H63N504	545.86	545.4880	63.81	11.63	12.83	11.72

C29H63N505	561.86	561.4829	62.00	11.30	12.46	14.24
C29H63N506	577.86	577.4778	60.28	10.99	12.12	16.61
C29H63N507	593.85	593.4727	58.65	10.69	11.79	18.86
C29H63N508	609.85	609.4676	57.12	10.41	11.48	20.99
C29H63N509	625.85	625.4625	55.66	10.15	11.19	23.01
C29H64N401	484.86	484.5080	71.84	13.30	11.56	3.30
C29H64N402	500.86	500.5029	69.55	12.88	11.19	6.39
C29H64N403	516.86	516.4978	67.39	12.48	10.84	9.29
C29H64N404	532.86	532.4927	65.37	12.11	10.51	12.01
C29H64N405	548.86	548.4876	63.46	11.75	10.21	14.58
C29H64N406	564.86	564.4826	61.67	11.42	9.92	16.99
C29H64N407	580.86	580.4775	59.97	11.11	9.65	19.28
C29H64N408	596.86	596.4724	58.36	10.81	9.39	21.44
C29H64N409	612.85	612.4673	56.84	10.53	9.14	23.50
C29H64N601	512.87	512.5141	67.92	12.58	16.39	3.12
C29H64N602	528.87	528.5090	65.86	12.20	15.89	6.05
C29H64N603	544.87	544.5040	63.93	11.84	15.42	8.81
C29H64N604	560.87	560.4989	62.10	11.50	14.98	11.41
C29H64N605	576.87	576.4938	60.38	11.18	14.57	13.87
C29H64N606	592.87	592.4887	58.75	10.88	14.18	16.19
C29H64N607	608.87	608.4836	57.21	10.60	13.80	18.39
C29H64N608	624.87	624.4785	55.74	10.32	13.45	20.48
C29H64N609	640.87	640.4734	54.35	10.07	13.11	22.47
C29H65N501	499.87	499.5189	69.68	13.11	14.01	3.20
C29H65N502	515.87	515.5138	67.52	12.70	13.58	6.20
C29H65N503	531.87	531.5087	65.49	12.32	13.17	9.02
C29H65N504	547.87	547.5036	63.58	11.96	12.78	11.68
C29H65N505	563.87	563.4985	61.77	11.62	12.42	14.19
C29H65N506	579.87	579.4935	60.07	11.30	12.08	16.55
C29H65N507	595.87	595.4884	58.46	11.00	11.75	18.80
C29H65N508	611.87	611.4833	56.93	10.71	11.45	20.92
C29H65N509	627.87	627.4782	55.48	10.43	11.15	22.93
C29H66N601	514.89	514.5298	67.65	12.92	16.32	3.11
C29H66N602	530.89	530.5247	65.61	12.53	15.83	6.03
C29H66N603	546.89	546.5196	63.69	12.16	15.37	8.78
C29H66N604	562.89	562.5145	61.88	11.82	14.93	11.37
C29H66N605	578.89	578.5094	60.17	11.49	14.52	13.82
C29H66N606	594.89	594.5044	58.55	11.18	14.13	16.14
C29H66N607	610.89	610.4993	57.02	10.89	13.76	18.33
C29H66N608	626.88	626.4942	55.56	10.61	13.41	20.42
C29H66N609	642.88	642.4891	54.18	10.35	13.07	22.40
C30H15N101	405.46	405.1154	88.87	3.73	3.45	3.95
C30H15N102	421.46	421.1103	85.50	3.59	3.32	7.59
C30H15N103	437.46	437.1052	82.37	3.46	3.20	10.97
C30H15N104	453.46	453.1001	79.46	3.33	3.09	14.11
C30H15N105	469.46	469.0950	76.76	3.22	2.98	17.04
C30H15N106	485.46	485.0899	74.23	3.11	2.89	19.77
C30H15N107	501.46	501.0848	71.86	3.02	2.79	22.33
C30H15N108	517.46	517.0798	69.64	2.92	2.71	24.74
C30H15N109	533.46	533.0747	67.55	2.83	2.63	26.99
C30H15N301	433.47	433.1215	83.13	3.49	9.69	3.69
C30H15N302	449.47	449.1164	80.17	3.36	9.35	7.12
C30H15N303	465.47	465.1113	77.41	3.25	9.03	10.31
C30H15N304	481.47	481.1062	74.84	3.14	8.73	13.29
C30H15N305	497.47	497.1012	72.43	3.04	8.45	16.08
C30H15N306	513.47	513.0961	70.18	2.94	8.18	18.70
C30H15N307	529.47	529.0910	68.06	2.86	7.94	21.15
C30H15N308	545.47	545.0859	66.06	2.77	7.70	23.47
C30H15N309	561.47	561.0808	64.18	2.69	7.48	25.65
C30H15N501	461.49	461.1277	78.08	3.28	15.18	3.47

C30H15N502	477.49	477.1226	75.46	3.17	14.67	6.70
C30H15N503	493.49	493.1175	73.02	3.06	14.19	9.73
C30H15N504	509.49	509.1124	70.73	2.97	13.75	12.56
C30H15N505	525.48	525.1073	68.57	2.88	13.33	15.22
C30H15N506	541.48	541.1022	66.55	2.79	12.93	17.73
C30H15N507	557.48	557.0971	64.64	2.71	12.56	20.09
C30H15N508	573.48	573.0921	62.83	2.64	12.21	22.32
C30H15N509	589.48	589.0870	61.13	2.56	11.88	24.43
C30H16N201	420.47	420.1263	85.70	3.84	6.66	3.81
C30H16N202	436.47	436.1212	82.56	3.69	6.42	7.33
C30H16N203	452.47	452.1161	79.64	3.56	6.19	10.61
C30H16N204	468.47	468.1110	76.92	3.44	5.98	13.66
C30H16N205	484.47	484.1059	74.38	3.33	5.78	16.51
C30H16N206	500.47	500.1008	72.00	3.22	5.60	19.18
C30H16N207	516.47	516.0957	69.77	3.12	5.42	21.68
C30H16N208	532.47	532.0907	67.67	3.03	5.26	24.04
C30H16N209	548.47	548.0856	65.70	2.94	5.11	26.25
C30H16N401	448.49	448.1324	80.34	3.60	12.49	3.57
C30H16N402	464.49	464.1273	77.58	3.47	12.06	6.89
C30H16N403	480.49	480.1222	74.99	3.36	11.66	9.99
C30H16N404	496.49	496.1171	72.58	3.25	11.28	12.89
C30H16N405	512.49	512.1121	70.31	3.15	10.93	15.61
C30H16N406	528.49	528.1070	68.18	3.05	10.60	18.16
C30H16N407	544.48	544.1019	66.18	2.96	10.29	20.57
C30H16N408	560.48	560.0968	64.29	2.88	10.00	22.84
C30H16N409	576.48	576.0917	62.51	2.80	9.72	24.98
C30H16N601	476.50	476.1386	75.62	3.38	17.64	3.36
C30H16N602	492.50	492.1335	73.16	3.27	17.06	6.50
C30H16N603	508.50	508.1284	70.86	3.17	16.53	9.44
C30H16N604	524.50	524.1233	68.70	3.07	16.02	12.20
C30H16N605	540.50	540.1182	66.67	2.98	15.55	14.80
C30H16N606	556.50	556.1131	64.75	2.90	15.10	17.25
C30H16N607	572.50	572.1080	62.94	2.82	14.68	19.56
C30H16N608	588.50	588.1029	61.23	2.74	14.28	21.75
C30H16N609	604.50	604.0979	59.61	2.67	13.90	23.82
C30H17N101	407.48	407.1310	88.43	4.21	3.44	3.93
C30H17N102	423.48	423.1259	85.09	4.05	3.31	7.56
C30H17N103	439.47	439.1208	81.99	3.90	3.19	10.92
C30H17N104	455.47	455.1157	79.11	3.76	3.08	14.05
C30H17N105	471.47	471.1107	76.43	3.63	2.97	16.97
C30H17N106	487.47	487.1056	73.92	3.52	2.87	19.69
C30H17N107	503.47	503.1005	71.57	3.40	2.78	22.24
C30H17N108	519.47	519.0954	69.37	3.30	2.70	24.64
C30H17N109	535.47	535.0903	67.29	3.20	2.62	26.89
C30H17N301	435.49	435.1372	82.74	3.93	9.65	3.67
C30H17N302	451.49	451.1321	79.81	3.80	9.31	7.09
C30H17N303	467.49	467.1270	77.08	3.67	8.99	10.27
C30H17N304	483.49	483.1219	74.53	3.54	8.69	13.24
C30H17N305	499.49	499.1168	72.14	3.43	8.41	16.02
C30H17N306	515.49	515.1117	69.90	3.32	8.15	18.62
C30H17N307	531.49	531.1066	67.80	3.22	7.91	21.07
C30H17N308	547.49	547.1016	65.82	3.13	7.68	23.38
C30H17N309	563.48	563.0965	63.95	3.04	7.46	25.55
C30H17N501	463.50	463.1433	77.74	3.70	15.11	3.45
C30H17N502	479.50	479.1382	75.15	3.57	14.61	6.67
C30H17N503	495.50	495.1331	72.72	3.46	14.13	9.69
C30H17N504	511.50	511.1280	70.45	3.35	13.69	12.51
C30H17N505	527.50	527.1230	68.31	3.25	13.28	15.17
C30H17N506	543.50	543.1179	66.30	3.15	12.89	17.66
C30H17N507	559.50	559.1128	64.40	3.06	12.52	20.02

C30H17N508	575.50	575.1077	62.61	2.98	12.17	22.24
C30H17N509	591.50	591.1026	60.92	2.90	11.84	24.34
C30H18N201	422.49	422.1419	85.29	4.29	6.63	3.79
C30H18N202	438.49	438.1368	82.18	4.14	6.39	7.30
C30H18N203	454.49	454.1317	79.28	3.99	6.16	10.56
C30H18N204	470.49	470.1266	76.59	3.86	5.95	13.60
C30H18N205	486.49	486.1216	74.07	3.73	5.76	16.44
C30H18N206	502.49	502.1165	71.71	3.61	5.57	19.10
C30H18N207	518.49	518.1114	69.50	3.50	5.40	21.60
C30H18N208	534.49	534.1063	67.42	3.39	5.24	23.95
C30H18N209	550.49	550.1012	65.46	3.30	5.09	26.16
C30H18N401	450.50	450.1481	79.98	4.03	12.44	3.55
C30H18N402	466.50	466.1430	77.24	3.89	12.01	6.86
C30H18N403	482.50	482.1379	74.68	3.76	11.61	9.95
C30H18N404	498.50	498.1328	72.28	3.64	11.24	12.84
C30H18N405	514.50	514.1277	70.04	3.53	10.89	15.55
C30H18N406	530.50	530.1226	67.92	3.42	10.56	18.10
C30H18N407	546.50	546.1175	65.93	3.32	10.25	20.49
C30H18N408	562.50	562.1125	64.06	3.23	9.96	22.75
C30H18N409	578.50	578.1074	62.29	3.14	9.68	24.89
C30H18N601	478.52	478.1542	75.30	3.79	17.56	3.34
C30H18N602	494.52	494.1491	72.87	3.67	16.99	6.47
C30H18N603	510.52	510.1440	70.58	3.55	16.46	9.40
C30H18N604	526.52	526.1389	68.44	3.45	15.96	12.15
C30H18N605	542.52	542.1339	66.42	3.34	15.49	14.75
C30H18N606	558.51	558.1288	64.52	3.25	15.05	17.19
C30H18N607	574.51	574.1237	62.72	3.16	14.63	19.49
C30H18N608	590.51	590.1186	61.02	3.07	14.23	21.68
C30H18N609	606.51	606.1135	59.41	2.99	13.86	23.74
C30H19N101	409.49	409.1467	88.00	4.68	3.42	3.91
C30H19N102	425.49	425.1416	84.69	4.50	3.29	7.52
C30H19N103	441.49	441.1365	81.62	4.34	3.17	10.87
C30H19N104	457.49	457.1314	78.76	4.19	3.06	13.99
C30H19N105	473.49	473.1263	76.10	4.04	2.96	16.90
C30H19N106	489.49	489.1212	73.61	3.91	2.86	19.61
C30H19N107	505.49	505.1161	71.28	3.79	2.77	22.16
C30H19N108	521.49	521.1111	69.10	3.67	2.69	24.54
C30H19N109	537.49	537.1060	67.04	3.56	2.61	26.79
C30H19N301	437.51	437.1528	82.36	4.38	9.60	3.66
C30H19N302	453.50	453.1477	79.46	4.22	9.27	7.06
C30H19N303	469.50	469.1426	76.75	4.08	8.95	10.22
C30H19N304	485.50	485.1375	74.22	3.94	8.65	13.18
C30H19N305	501.50	501.1325	71.85	3.82	8.38	15.95
C30H19N306	517.50	517.1274	69.63	3.70	8.12	18.55
C30H19N307	533.50	533.1223	67.54	3.59	7.88	20.99
C30H19N308	549.50	549.1172	65.57	3.49	7.65	23.29
C30H19N309	565.50	565.1121	63.72	3.39	7.43	25.46
C30H19N501	465.52	465.1590	77.40	4.11	15.04	3.44
C30H19N502	481.52	481.1539	74.83	3.98	14.54	6.65
C30H19N503	497.52	497.1488	72.43	3.85	14.08	9.65
C30H19N504	513.52	513.1437	70.17	3.73	13.64	12.46
C30H19N505	529.52	529.1386	68.05	3.62	13.23	15.11
C30H19N506	545.52	545.1335	66.05	3.51	12.84	17.60
C30H19N507	561.52	561.1284	64.17	3.41	12.47	19.95
C30H19N508	577.51	577.1233	62.39	3.32	12.13	22.16
C30H19N509	593.51	593.1183	60.71	3.23	11.80	24.26
C30H20N201	424.51	424.1576	84.88	4.75	6.60	3.77
C30H20N202	440.51	440.1525	81.80	4.58	6.36	7.26
C30H20N203	456.51	456.1474	78.93	4.42	6.14	10.51
C30H20N204	472.50	472.1423	76.26	4.27	5.93	13.54

C30H20N205	488.50	488.1372	73.76	4.13	5.73	16.38
C30H20N206	504.50	504.1321	71.42	4.00	5.55	19.03
C30H20N207	520.50	520.1270	69.23	3.87	5.38	21.52
C30H20N208	536.50	536.1220	67.16	3.76	5.22	23.86
C30H20N209	552.50	552.1169	65.22	3.65	5.07	26.06
C30H20N401	452.52	452.1637	79.63	4.45	12.38	3.54
C30H20N402	468.52	468.1586	76.91	4.30	11.96	6.83
C30H20N403	484.52	484.1535	74.37	4.16	11.56	9.91
C30H20N404	500.52	500.1484	71.99	4.03	11.19	12.79
C30H20N405	516.52	516.1434	69.76	3.90	10.85	15.49
C30H20N406	532.52	532.1383	67.67	3.79	10.52	18.03
C30H20N407	548.52	548.1332	65.69	3.68	10.21	20.42
C30H20N408	564.52	564.1281	63.83	3.57	9.92	22.67
C30H20N409	580.52	580.1230	62.07	3.47	9.65	24.80
C30H20N601	480.53	480.1698	74.99	4.20	17.49	3.33
C30H20N602	496.53	496.1648	72.57	4.06	16.93	6.44
C30H20N603	512.53	512.1597	70.30	3.93	16.40	9.36
C30H20N604	528.53	528.1546	68.18	3.81	15.90	12.11
C30H20N605	544.53	544.1495	66.17	3.70	15.43	14.69
C30H20N606	560.53	560.1444	64.28	3.60	14.99	17.13
C30H20N607	576.53	576.1393	62.50	3.50	14.58	19.43
C30H20N608	592.53	592.1342	60.81	3.40	14.18	21.60
C30H20N609	608.53	608.1292	59.21	3.31	13.81	23.66
C30H21N101	411.51	411.1623	87.56	5.14	3.40	3.89
C30H21N102	427.51	427.1572	84.29	4.95	3.28	7.48
C30H21N103	443.51	443.1521	81.25	4.77	3.16	10.82
C30H21N104	459.51	459.1470	78.42	4.61	3.05	13.93
C30H21N105	475.51	475.1420	75.78	4.45	2.95	16.82
C30H21N106	491.50	491.1369	73.31	4.31	2.85	19.53
C30H21N107	507.50	507.1318	71.00	4.17	2.76	22.07
C30H21N108	523.50	523.1267	68.83	4.04	2.68	24.45
C30H21N109	539.50	539.1216	66.79	3.92	2.60	26.69
C30H21N301	439.52	439.1685	81.98	4.82	9.56	3.64
C30H21N302	455.52	455.1634	79.10	4.65	9.22	7.02
C30H21N303	471.52	471.1583	76.42	4.49	8.91	10.18
C30H21N304	487.52	487.1532	73.91	4.34	8.62	13.13
C30H21N305	503.52	503.1481	71.56	4.20	8.35	15.89
C30H21N306	519.52	519.1430	69.36	4.07	8.09	18.48
C30H21N307	535.52	535.1379	67.29	3.95	7.85	20.91
C30H21N308	551.52	551.1329	65.34	3.84	7.62	23.21
C30H21N309	567.52	567.1278	63.49	3.73	7.40	25.37
C30H21N501	467.53	467.1746	77.07	4.53	14.98	3.42
C30H21N502	483.53	483.1695	74.52	4.38	14.48	6.62
C30H21N503	499.53	499.1644	72.13	4.24	14.02	9.61
C30H21N504	515.53	515.1593	69.90	4.11	13.58	12.41
C30H21N505	531.53	531.1543	67.79	3.98	13.18	15.05
C30H21N506	547.53	547.1492	65.81	3.87	12.79	17.53
C30H21N507	563.53	563.1441	63.94	3.76	12.43	19.87
C30H21N508	579.53	579.1390	62.18	3.65	12.08	22.09
C30H21N509	595.53	595.1339	60.51	3.55	11.76	24.18
C30H22N201	426.52	426.1732	84.48	5.20	6.57	3.75
C30H22N202	442.52	442.1681	81.43	5.01	6.33	7.23
C30H22N203	458.52	458.1630	78.59	4.84	6.11	10.47
C30H22N204	474.52	474.1579	75.94	4.67	5.90	13.49
C30H22N205	490.52	490.1529	73.46	4.52	5.71	16.31
C30H22N206	506.52	506.1478	71.14	4.38	5.53	18.95
C30H22N207	522.52	522.1427	68.96	4.24	5.36	21.43
C30H22N208	538.52	538.1376	66.91	4.12	5.20	23.77
C30H22N209	554.52	554.1325	64.98	4.00	5.05	25.97
C30H22N401	454.54	454.1794	79.28	4.88	12.33	3.52

C30H22N402	470.54	470.1743	76.58	4.71	11.91	6.80
C30H22N403	486.53	486.1692	74.06	4.56	11.52	9.87
C30H22N404	502.53	502.1641	71.70	4.41	11.15	12.73
C30H22N405	518.53	518.1590	69.49	4.28	10.80	15.43
C30H22N406	534.53	534.1539	67.41	4.15	10.48	17.96
C30H22N407	550.53	550.1488	65.45	4.03	10.18	20.34
C30H22N408	566.53	566.1437	63.60	3.91	9.89	22.59
C30H22N409	582.53	582.1387	61.86	3.81	9.62	24.72
C30H22N601	482.55	482.1855	74.67	4.60	17.42	3.32
C30H22N602	498.55	498.1804	72.28	4.45	16.86	6.42
C30H22N603	514.55	514.1753	70.03	4.31	16.33	9.33
C30H22N604	530.55	530.1702	67.92	4.18	15.84	12.06
C30H22N605	546.55	546.1652	65.93	4.06	15.38	14.64
C30H22N606	562.55	562.1601	64.05	3.94	14.94	17.06
C30H22N607	578.55	578.1550	62.28	3.83	14.53	19.36
C30H22N608	594.55	594.1499	60.61	3.73	14.14	21.53
C30H22N609	610.54	610.1448	59.02	3.63	13.76	23.58
C30H23N101	413.52	413.1780	87.14	5.61	3.39	3.87
C30H23N102	429.52	429.1729	83.89	5.40	3.26	7.45
C30H23N103	445.52	445.1678	80.88	5.20	3.14	10.77
C30H23N104	461.52	461.1627	78.08	5.02	3.03	13.87
C30H23N105	477.52	477.1576	75.46	4.85	2.93	16.75
C30H23N106	493.52	493.1525	73.01	4.70	2.84	19.45
C30H23N107	509.52	509.1474	70.72	4.55	2.75	21.98
C30H23N108	525.52	525.1424	68.57	4.41	2.67	24.36
C30H23N109	541.52	541.1373	66.54	4.28	2.59	26.59
C30H23N301	441.54	441.1841	81.61	5.25	9.52	3.62
C30H23N302	457.54	457.1790	78.76	5.07	9.18	6.99
C30H23N303	473.54	473.1739	76.09	4.90	8.87	10.14
C30H23N304	489.54	489.1688	73.61	4.74	8.58	13.07
C30H23N305	505.53	505.1638	71.28	4.59	8.31	15.82
C30H23N306	521.53	521.1587	69.09	4.45	8.06	18.41
C30H23N307	537.53	537.1536	67.03	4.31	7.82	20.84
C30H23N308	553.53	553.1485	65.10	4.19	7.59	23.12
C30H23N309	569.53	569.1434	63.27	4.07	7.38	25.28
C30H23N501	469.55	469.1902	76.74	4.94	14.92	3.41
C30H23N502	485.55	485.1852	74.21	4.77	14.42	6.59
C30H23N503	501.55	501.1801	71.84	4.62	13.96	9.57
C30H23N504	517.55	517.1750	69.62	4.48	13.53	12.37
C30H23N505	533.55	533.1699	67.54	4.35	13.13	14.99
C30H23N506	549.55	549.1648	65.57	4.22	12.74	17.47
C30H23N507	565.55	565.1597	63.71	4.10	12.38	19.80
C30H23N508	581.55	581.1546	61.96	3.99	12.04	22.01
C30H23N509	597.55	597.1496	60.30	3.88	11.72	24.10
C30H24N201	428.54	428.1889	84.08	5.65	6.54	3.73
C30H24N202	444.54	444.1838	81.06	5.44	6.30	7.20
C30H24N203	460.54	460.1787	78.24	5.25	6.08	10.42
C30H24N204	476.54	476.1736	75.62	5.08	5.88	13.43
C30H24N205	492.54	492.1685	73.16	4.91	5.69	16.24
C30H24N206	508.54	508.1634	70.86	4.76	5.51	18.88
C30H24N207	524.53	524.1583	68.70	4.61	5.34	21.35
C30H24N208	540.53	540.1533	66.66	4.48	5.18	23.68
C30H24N209	556.53	556.1482	64.75	4.35	5.03	25.87
C30H24N401	456.55	456.1950	78.93	5.30	12.27	3.50
C30H24N402	472.55	472.1899	76.25	5.12	11.86	6.77
C30H24N403	488.55	488.1848	73.76	4.95	11.47	9.82
C30H24N404	504.55	504.1797	71.42	4.79	11.10	12.68
C30H24N405	520.55	520.1747	69.22	4.65	10.76	15.37
C30H24N406	536.55	536.1696	67.16	4.51	10.44	17.89
C30H24N407	552.55	552.1645	65.21	4.38	10.14	20.27

C30H24N408	568.55	568.1594	63.38	4.25	9.85	22.51
C30H24N409	584.55	584.1543	61.64	4.14	9.58	24.63
C30H24N601	484.57	484.2011	74.36	4.99	17.34	3.30
C30H24N602	500.56	500.1961	71.99	4.83	16.79	6.39
C30H24N603	516.56	516.1910	69.76	4.68	16.27	9.29
C30H24N604	532.56	532.1859	67.66	4.54	15.78	12.02
C30H24N605	548.56	548.1808	65.69	4.41	15.32	14.58
C30H24N606	564.56	564.1757	63.83	4.28	14.89	17.00
C30H24N607	580.56	580.1706	62.07	4.17	14.48	19.29
C30H24N608	596.56	596.1655	60.40	4.06	14.09	21.46
C30H24N609	612.56	612.1605	58.82	3.95	13.72	23.51
C30H25N101	415.54	415.1936	86.71	6.06	3.37	3.85
C30H25N102	431.54	431.1885	83.50	5.84	3.25	7.42
C30H25N103	447.54	447.1834	80.51	5.63	3.13	10.72
C30H25N104	463.54	463.1783	77.74	5.44	3.02	13.81
C30H25N105	479.54	479.1733	75.14	5.25	2.92	16.68
C30H25N106	495.54	495.1682	72.72	5.09	2.83	19.37
C30H25N107	511.54	511.1631	70.44	4.93	2.74	21.89
C30H25N108	527.54	527.1580	68.31	4.78	2.66	24.26
C30H25N109	543.54	543.1529	66.29	4.64	2.58	26.49
C30H25N301	443.55	443.1998	81.24	5.68	9.47	3.61
C30H25N302	459.55	459.1947	78.41	5.48	9.14	6.96
C30H25N303	475.55	475.1896	75.77	5.30	8.84	10.09
C30H25N304	491.55	491.1845	73.31	5.13	8.55	13.02
C30H25N305	507.55	507.1794	70.99	4.96	8.28	15.76
C30H25N306	523.55	523.1743	68.83	4.81	8.03	18.34
C30H25N307	539.55	539.1692	66.78	4.67	7.79	20.76
C30H25N308	555.55	555.1641	64.86	4.54	7.56	23.04
C30H25N309	571.55	571.1591	63.05	4.41	7.35	25.19
C30H25N501	471.57	471.2059	76.41	5.34	14.85	3.39
C30H25N502	487.57	487.2008	73.90	5.17	14.36	6.56
C30H25N503	503.57	503.1957	71.56	5.00	13.91	9.53
C30H25N504	519.56	519.1906	69.35	4.85	13.48	12.32
C30H25N505	535.56	535.1856	67.28	4.71	13.08	14.94
C30H25N506	551.56	551.1805	65.33	4.57	12.70	17.40
C30H25N507	567.56	567.1754	63.49	4.44	12.34	19.73
C30H25N508	583.56	583.1703	61.75	4.32	12.00	21.93
C30H25N509	599.56	599.1652	60.10	4.20	11.68	24.02
C30H26N201	430.55	430.2045	83.69	6.09	6.51	3.72
C30H26N202	446.55	446.1994	80.69	5.87	6.27	7.17
C30H26N203	462.55	462.1943	77.90	5.67	6.06	10.38
C30H26N204	478.55	478.1892	75.30	5.48	5.85	13.37
C30H26N205	494.55	494.1842	72.86	5.30	5.66	16.18
C30H26N206	510.55	510.1791	70.58	5.13	5.49	18.80
C30H26N207	526.55	526.1740	68.43	4.98	5.32	21.27
C30H26N208	542.55	542.1689	66.41	4.83	5.16	23.59
C30H26N209	558.55	558.1638	64.51	4.69	5.02	25.78
C30H26N401	458.57	458.2106	78.58	5.72	12.22	3.49
C30H26N402	474.57	474.2056	75.93	5.52	11.81	6.74
C30H26N403	490.57	490.2005	73.45	5.34	11.42	9.78
C30H26N404	506.57	506.1954	71.13	5.17	11.06	12.63
C30H26N405	522.57	522.1903	68.95	5.02	10.72	15.31
C30H26N406	538.56	538.1852	66.91	4.87	10.40	17.82
C30H26N407	554.56	554.1801	64.98	4.73	10.10	20.20
C30H26N408	570.56	570.1750	63.15	4.59	9.82	22.43
C30H26N409	586.56	586.1700	61.43	4.47	9.55	24.55
C30H26N601	486.58	486.2168	74.05	5.39	17.27	3.29
C30H26N602	502.58	502.2117	71.70	5.21	16.72	6.37
C30H26N603	518.58	518.2066	69.48	5.05	16.21	9.26
C30H26N604	534.58	534.2015	67.41	4.90	15.72	11.97

C30H26N605	550.58	550.1965	65.45	4.76	15.26	14.53
C30H26N606	566.58	566.1914	63.60	4.63	14.83	16.94
C30H26N607	582.58	582.1863	61.85	4.50	14.43	19.22
C30H26N608	598.58	598.1812	60.20	4.38	14.04	21.38
C30H26N609	614.58	614.1761	58.63	4.26	13.67	23.43
C30H27N101	417.56	417.2093	86.30	6.52	3.35	3.83
C30H27N102	433.56	433.2042	83.11	6.28	3.23	7.38
C30H27N103	449.55	449.1991	80.15	6.05	3.12	10.68
C30H27N104	465.55	465.1940	77.40	5.85	3.01	13.75
C30H27N105	481.55	481.1889	74.83	5.65	2.91	16.61
C30H27N106	497.55	497.1838	72.42	5.47	2.82	19.29
C30H27N107	513.55	513.1787	70.17	5.30	2.73	21.81
C30H27N108	529.55	529.1737	68.05	5.14	2.65	24.17
C30H27N109	545.55	545.1686	66.05	4.99	2.57	26.39
C30H27N301	445.57	445.2154	80.87	6.11	9.43	3.59
C30H27N302	461.57	461.2103	78.07	5.90	9.10	6.93
C30H27N303	477.57	477.2052	75.45	5.70	8.80	10.05
C30H27N304	493.57	493.2001	73.01	5.51	8.51	12.97
C30H27N305	509.57	509.1951	70.71	5.34	8.25	15.70
C30H27N306	525.57	525.1900	68.56	5.18	8.00	18.27
C30H27N307	541.57	541.1849	66.54	5.03	7.76	20.68
C30H27N308	557.56	557.1798	64.63	4.88	7.54	22.96
C30H27N309	573.56	573.1747	62.82	4.74	7.33	25.11
C30H27N501	473.58	473.2215	76.09	5.75	14.79	3.38
C30H27N502	489.58	489.2165	73.60	5.56	14.30	6.54
C30H27N503	505.58	505.2114	71.27	5.38	13.85	9.49
C30H27N504	521.58	521.2063	69.09	5.22	13.43	12.27
C30H27N505	537.58	537.2012	67.03	5.06	13.03	14.88
C30H27N506	553.58	553.1961	65.09	4.92	12.65	17.34
C30H27N507	569.58	569.1910	63.26	4.78	12.30	19.66
C30H27N508	585.58	585.1859	61.53	4.65	11.96	21.86
C30H27N509	601.58	601.1809	59.90	4.52	11.64	23.94
C30H28N201	432.57	432.2202	83.30	6.52	6.48	3.70
C30H28N202	448.57	448.2151	80.33	6.29	6.25	7.13
C30H28N203	464.57	464.2100	77.56	6.08	6.03	10.33
C30H28N204	480.57	480.2049	74.98	5.87	5.83	13.32
C30H28N205	496.57	496.1998	72.56	5.68	5.64	16.11
C30H28N206	512.57	512.1947	70.30	5.51	5.47	18.73
C30H28N207	528.57	528.1896	68.17	5.34	5.30	21.19
C30H28N208	544.57	544.1845	66.17	5.18	5.14	23.50
C30H28N209	560.57	560.1795	64.28	5.03	5.00	25.69
C30H28N401	460.58	460.2263	78.23	6.13	12.16	3.47
C30H28N402	476.58	476.2212	75.61	5.92	11.76	6.71
C30H28N403	492.58	492.2161	73.15	5.73	11.37	9.74
C30H28N404	508.58	508.2110	70.85	5.55	11.02	12.58
C30H28N405	524.58	524.2060	68.69	5.38	10.68	15.25
C30H28N406	540.58	540.2009	66.66	5.22	10.36	17.76
C30H28N407	556.58	556.1958	64.74	5.07	10.07	20.12
C30H28N408	572.58	572.1907	62.93	4.93	9.78	22.35
C30H28N409	588.58	588.1856	61.22	4.80	9.52	24.46
C30H28N601	488.60	488.2324	73.75	5.78	17.20	3.27
C30H28N602	504.60	504.2274	71.41	5.59	16.65	6.34
C30H28N603	520.60	520.2223	69.22	5.42	16.14	9.22
C30H28N604	536.60	536.2172	67.15	5.26	15.66	11.93
C30H28N605	552.59	552.2121	65.21	5.11	15.21	14.48
C30H28N606	568.59	568.2070	63.37	4.96	14.78	16.88
C30H28N607	584.59	584.2019	61.64	4.83	14.38	19.16
C30H28N608	600.59	600.1968	60.00	4.70	13.99	21.31
C30H28N609	616.59	616.1918	58.44	4.58	13.63	23.35
C30H29N101	419.57	419.2249	85.88	6.97	3.34	3.81

C30H29N102	435.57	435.2198	82.73	6.71	3.22	7.35
C30H29N103	451.57	451.2147	79.80	6.47	3.10	10.63
C30H29N104	467.57	467.2096	77.07	6.25	3.00	13.69
C30H29N105	483.57	483.2046	74.52	6.04	2.90	16.54
C30H29N106	499.57	499.1995	72.13	5.85	2.80	19.22
C30H29N107	515.57	515.1944	69.89	5.67	2.72	21.72
C30H29N108	531.57	531.1893	67.79	5.50	2.63	24.08
C30H29N109	547.57	547.1842	65.81	5.34	2.56	26.30
C30H29N301	447.59	447.2310	80.51	6.53	9.39	3.57
C30H29N302	463.58	463.2260	77.73	6.31	9.06	6.90
C30H29N303	479.58	479.2209	75.13	6.10	8.76	10.01
C30H29N304	495.58	495.2158	72.71	5.90	8.48	12.91
C30H29N305	511.58	511.2107	70.44	5.71	8.21	15.64
C30H29N306	527.58	527.2056	68.30	5.54	7.96	18.20
C30H29N307	543.58	543.2005	66.29	5.38	7.73	20.60
C30H29N308	559.58	559.1954	64.39	5.22	7.51	22.87
C30H29N309	575.58	575.1904	62.60	5.08	7.30	25.02
C30H29N501	475.60	475.2372	75.76	6.15	14.73	3.36
C30H29N502	491.60	491.2321	73.30	5.95	14.25	6.51
C30H29N503	507.60	507.2270	70.99	5.76	13.80	9.46
C30H29N504	523.60	523.2219	68.82	5.58	13.38	12.22
C30H29N505	539.60	539.2169	66.78	5.42	12.98	14.83
C30H29N506	555.60	555.2118	64.86	5.26	12.61	17.28
C30H29N507	571.59	571.2067	63.04	5.11	12.25	19.59
C30H29N508	587.59	587.2016	61.32	4.97	11.92	21.78
C30H29N509	603.59	603.1965	59.70	4.84	11.60	23.86
C30H30N201	434.59	434.2358	82.91	6.96	6.45	3.68
C30H30N202	450.59	450.2307	79.97	6.71	6.22	7.10
C30H30N203	466.59	466.2256	77.23	6.48	6.00	10.29
C30H30N204	482.58	482.2205	74.67	6.27	5.80	13.26
C30H30N205	498.58	498.2155	72.27	6.06	5.62	16.04
C30H30N206	514.58	514.2104	70.02	5.88	5.44	18.66
C30H30N207	530.58	530.2053	67.91	5.70	5.28	21.11
C30H30N208	546.58	546.2002	65.93	5.53	5.13	23.42
C30H30N209	562.58	562.1951	64.05	5.38	4.98	25.60
C30H30N401	462.60	462.2419	77.89	6.54	12.11	3.46
C30H30N402	478.60	478.2369	75.29	6.32	11.71	6.69
C30H30N403	494.60	494.2318	72.85	6.11	11.33	9.70
C30H30N404	510.60	510.2267	70.57	5.92	10.97	12.53
C30H30N405	526.60	526.2216	68.43	5.74	10.64	15.19
C30H30N406	542.60	542.2165	66.41	5.57	10.33	17.69
C30H30N407	558.60	558.2114	64.51	5.41	10.03	20.05
C30H30N408	574.60	574.2063	62.71	5.26	9.75	22.28
C30H30N409	590.59	590.2013	61.01	5.12	9.49	24.38
C30H30N601	490.61	490.2481	73.45	6.16	17.13	3.26
C30H30N602	506.61	506.2430	71.13	5.97	16.59	6.32
C30H30N603	522.61	522.2379	68.95	5.79	16.08	9.18
C30H30N604	538.61	538.2328	66.90	5.61	15.60	11.88
C30H30N605	554.61	554.2278	64.97	5.45	15.15	14.42
C30H30N606	570.61	570.2227	63.15	5.30	14.73	16.82
C30H30N607	586.61	586.2176	61.43	5.15	14.33	19.09
C30H30N608	602.61	602.2125	59.80	5.02	13.95	21.24
C30H30N609	618.61	618.2074	58.25	4.89	13.59	23.28
C30H31N101	421.59	421.2406	85.47	7.41	3.32	3.80
C30H31N102	437.59	437.2355	82.35	7.14	3.20	7.31
C30H31N103	453.59	453.2304	79.44	6.89	3.09	10.58
C30H31N104	469.59	469.2253	76.73	6.65	2.98	13.63
C30H31N105	485.59	485.2202	74.21	6.43	2.88	16.47
C30H31N106	501.58	501.2151	71.84	6.23	2.79	19.14
C30H31N107	517.58	517.2100	69.62	6.04	2.71	21.64

C30H31N108	533.58	533.2049	67.53	5.86	2.63	23.99
C30H31N109	549.58	549.1999	65.57	5.69	2.55	26.20
C30H31N301	449.60	449.2467	80.15	6.95	9.35	3.56
C30H31N302	465.60	465.2416	77.39	6.71	9.02	6.87
C30H31N303	481.60	481.2365	74.82	6.49	8.73	9.97
C30H31N304	497.60	497.2314	72.41	6.28	8.44	12.86
C30H31N305	513.60	513.2264	70.16	6.08	8.18	15.58
C30H31N306	529.60	529.2213	68.04	5.90	7.93	18.13
C30H31N307	545.60	545.2162	66.04	5.73	7.70	20.53
C30H31N308	561.60	561.2111	64.16	5.56	7.48	22.79
C30H31N309	577.60	577.2060	62.39	5.41	7.27	24.93
C30H31N501	477.61	477.2528	75.44	6.54	14.66	3.35
C30H31N502	493.61	493.2478	73.00	6.33	14.19	6.48
C30H31N503	509.61	509.2427	70.71	6.13	13.74	9.42
C30H31N504	525.61	525.2376	68.56	5.94	13.32	12.18
C30H31N505	541.61	541.2325	66.53	5.77	12.93	14.77
C30H31N506	557.61	557.2274	64.62	5.60	12.56	17.22
C30H31N507	573.61	573.2223	62.82	5.45	12.21	19.52
C30H31N508	589.61	589.2172	61.11	5.30	11.88	21.71
C30H31N509	605.61	605.2122	59.50	5.16	11.56	23.78
C30H32N201	436.60	436.2514	82.53	7.39	6.42	3.66
C30H32N202	452.60	452.2464	79.61	7.13	6.19	7.07
C30H32N203	468.60	468.2413	76.90	6.88	5.98	10.24
C30H32N204	484.60	484.2362	74.36	6.66	5.78	13.21
C30H32N205	500.60	500.2311	71.98	6.44	5.60	15.98
C30H32N206	516.60	516.2260	69.75	6.24	5.42	18.58
C30H32N207	532.60	532.2209	67.66	6.06	5.26	21.03
C30H32N208	548.60	548.2158	65.68	5.88	5.11	23.33
C30H32N209	564.60	564.2108	63.82	5.71	4.96	25.50
C30H32N401	464.62	464.2576	77.56	6.94	12.06	3.44
C30H32N402	480.62	480.2525	74.97	6.71	11.66	6.66
C30H32N403	496.61	496.2474	72.56	6.49	11.28	9.67
C30H32N404	512.61	512.2423	70.29	6.29	10.93	12.48
C30H32N405	528.61	528.2373	68.17	6.10	10.60	15.13
C30H32N406	544.61	544.2322	66.16	5.92	10.29	17.63
C30H32N407	560.61	560.2271	64.28	5.75	9.99	19.98
C30H32N408	576.61	576.2220	62.49	5.59	9.72	22.20
C30H32N409	592.61	592.2169	60.80	5.44	9.45	24.30
C30H32N601	492.63	492.2637	73.15	6.55	17.06	3.25
C30H32N602	508.63	508.2587	70.84	6.34	16.52	6.29
C30H32N603	524.63	524.2536	68.68	6.15	16.02	9.15
C30H32N604	540.63	540.2485	66.65	5.97	15.54	11.84
C30H32N605	556.63	556.2434	64.74	5.79	15.10	14.37
C30H32N606	572.63	572.2383	62.93	5.63	14.68	16.76
C30H32N607	588.63	588.2332	61.22	5.48	14.28	19.03
C30H32N608	604.62	604.2281	59.60	5.33	13.90	21.17
C30H32N609	620.62	620.2231	58.06	5.20	13.54	23.20
C30H33N101	423.60	423.2562	85.06	7.85	3.31	3.78
C30H33N102	439.60	439.2511	81.97	7.57	3.19	7.28
C30H33N103	455.60	455.2460	79.09	7.30	3.07	10.54
C30H33N104	471.60	471.2409	76.41	7.05	2.97	13.57
C30H33N105	487.60	487.2359	73.90	6.82	2.87	16.41
C30H33N106	503.60	503.2308	71.55	6.61	2.78	19.06
C30H33N107	519.60	519.2257	69.35	6.40	2.70	21.55
C30H33N108	535.60	535.2206	67.28	6.21	2.62	23.90
C30H33N109	551.60	551.2155	65.33	6.03	2.54	26.10
C30H33N301	451.62	451.2623	79.79	7.37	9.30	3.54
C30H33N302	467.62	467.2573	77.06	7.11	8.99	6.84
C30H33N303	483.62	483.2522	74.51	6.88	8.69	9.92
C30H33N304	499.62	499.2471	72.12	6.66	8.41	12.81
C30H33N305	515.61	515.2420	69.88	6.45	8.15	15.51
C30H33N306	531.61	531.2369	67.78	6.26	7.90	18.06
C30H33N307	547.61	547.2318	65.80	6.07	7.67	20.45
C30H33N308	563.61	563.2267	63.93	5.90	7.46	22.71
C30H33N309	579.61	579.2217	62.17	5.74	7.25	24.84
C30H33N501	479.63	479.2685	75.13	6.94	14.60	3.34
C30H33N502	495.63	495.2634	72.70	6.71	14.13	6.46
C30H33N503	511.63	511.2583	70.43	6.50	13.69	9.38
C30H33N504	527.63	527.2532	68.29	6.30	13.27	12.13
C30H33N505	543.63	543.2482	66.28	6.12	12.88	14.72
C30H33N506	559.63	559.2431	64.39	5.94	12.51	17.15
C30H33N507	575.63	575.2380	62.60	5.78	12.17	19.46
C30H33N508	591.63	591.2329	60.91	5.62	11.84	21.63
C30H33N509	607.63	607.2278	59.30	5.47	11.53	23.70
C30H34N201	438.62	438.2671	82.15	7.81	6.39	3.65
C30H34N202	454.62	454.2620	79.26	7.54	6.16	7.04
C30H34N203	470.62	470.2569	76.57	7.28	5.95	10.20
C30H34N204	486.62	486.2518	74.05	7.04	5.76	13.15
C30H34N205	502.62	502.2468	71.69	6.82	5.57	15.92
C30H34N206	518.62	518.2417	69.48	6.61	5.40	18.51
C30H34N207	534.61	534.2366	67.40	6.41	5.24	20.95
C30H34N208	550.61	550.2315	65.44	6.22	5.09	23.25
C30H34N209	566.61	566.2264	63.59	6.05	4.94	25.41
C30H34N401	466.63	466.2732	77.22	7.34	12.01	3.43
C30H34N402	482.63	482.2682	74.66	7.10	11.61	6.63
C30H34N403	498.63	498.2631	72.26	6.87	11.24	9.63
C30H34N404	514.63	514.2580	70.02	6.66	10.89	12.44
C30H34N405	530.63	530.2529	67.91	6.46	10.56	15.08
C30H34N406	546.63	546.2478	65.92	6.27	10.25	17.56
C30H34N407	562.63	562.2427	64.04	6.09	9.96	19.91
C30H34N408	578.63	578.2376	62.27	5.92	9.68	22.12
C30H34N409	594.63	594.2326	60.60	5.76	9.42	24.22
C30H34N601	494.65	494.2794	72.85	6.93	16.99	3.23
C30H34N602	510.64	510.2743	70.56	6.71	16.46	6.27
C30H34N603	526.64	526.2692	68.42	6.51	15.96	9.11
C30H34N604	542.64	542.2641	66.40	6.32	15.49	11.79
C30H34N605	558.64	558.2590	64.50	6.13	15.04	14.32
C30H34N606	574.64	574.2540	62.71	5.96	14.62	16.71
C30H34N607	590.64	590.2489	61.01	5.80	14.23	18.96
C30H34N608	606.64	606.2438	59.40	5.65	13.85	21.10
C30H34N609	622.64	622.2387	57.87	5.50	13.50	23.13
C30H35N101	425.62	425.2718	84.66	8.29	3.29	3.76
C30H35N102	441.62	441.2668	81.59	7.99	3.17	7.25
C30H35N103	457.62	457.2617	78.74	7.71	3.06	10.49
C30H35N104	473.62	473.2566	76.08	7.45	2.96	13.51
C30H35N105	489.62	489.2515	73.60	7.21	2.86	16.34
C30H35N106	505.62	505.2464	71.27	6.98	2.77	18.99
C30H35N107	521.62	521.2413	69.08	6.76	2.69	21.47
C30H35N108	537.62	537.2362	67.02	6.56	2.61	23.81
C30H35N109	553.61	553.2312	65.09	6.37	2.53	26.01
C30H35N301	453.63	453.2780	79.43	7.78	9.26	3.53
C30H35N302	469.63	469.2729	76.73	7.51	8.95	6.81
C30H35N303	485.63	485.2678	74.20	7.26	8.65	9.88
C30H35N304	501.63	501.2627	71.83	7.03	8.38	12.76
C30H35N305	517.63	517.2577	69.61	6.82	8.12	15.45
C30H35N306	533.63	533.2526	67.53	6.61	7.87	17.99
C30H35N307	549.63	549.2475	65.56	6.42	7.65	20.38
C30H35N308	565.63	565.2424	63.71	6.24	7.43	22.63
C30H35N309	581.63	581.2373	61.95	6.07	7.22	24.76
C30H35N501	481.65	481.2841	74.81	7.32	14.54	3.32

C30H35N502	497.65	497.2791	72.41	7.09	14.07	6.43
C30H35N503	513.65	513.2740	70.15	6.87	13.63	9.34
C30H35N504	529.64	529.2689	68.03	6.66	13.22	12.08
C30H35N505	545.64	545.2638	66.04	6.47	12.84	14.66
C30H35N506	561.64	561.2587	64.16	6.28	12.47	17.09
C30H35N507	577.64	577.2536	62.38	6.11	12.12	19.39
C30H35N508	593.64	593.2485	60.70	5.94	11.80	21.56
C30H35N509	609.64	609.2435	59.11	5.79	11.49	23.62
C30H36N201	440.63	440.2827	81.78	8.24	6.36	3.63
C30H36N202	456.63	456.2777	78.91	7.95	6.13	7.01
C30H36N203	472.63	472.2726	76.24	7.68	5.93	10.16
C30H36N204	488.63	488.2675	73.74	7.43	5.73	13.10
C30H36N205	504.63	504.2624	71.41	7.19	5.55	15.85
C30H36N206	520.63	520.2573	69.21	6.97	5.38	18.44
C30H36N207	536.63	536.2522	67.15	6.76	5.22	20.87
C30H36N208	552.63	552.2471	65.20	6.57	5.07	23.16
C30H36N209	568.63	568.2421	63.37	6.38	4.93	25.32
C30H36N401	468.65	468.2889	76.89	7.74	11.95	3.41
C30H36N402	484.65	484.2838	74.35	7.49	11.56	6.60
C30H36N403	500.65	500.2787	71.97	7.25	11.19	9.59
C30H36N404	516.65	516.2736	69.74	7.02	10.84	12.39
C30H36N405	532.65	532.2686	67.65	6.81	10.52	15.02
C30H36N406	548.64	548.2635	65.68	6.61	10.21	17.50
C30H36N407	564.64	564.2584	63.82	6.43	9.92	19.83
C30H36N408	580.64	580.2533	62.06	6.25	9.65	22.04
C30H36N409	596.64	596.2482	60.39	6.08	9.39	24.13
C30H36N601	496.66	496.2950	72.55	7.31	16.92	3.22
C30H36N602	512.66	512.2900	70.29	7.08	16.39	6.24
C30H36N603	528.66	528.2849	68.16	6.86	15.90	9.08
C30H36N604	544.66	544.2798	66.16	6.66	15.43	11.75
C30H36N605	560.66	560.2747	64.27	6.47	14.99	14.27
C30H36N606	576.66	576.2696	62.49	6.29	14.57	16.65
C30H36N607	592.66	592.2645	60.80	6.12	14.18	18.90
C30H36N608	608.66	608.2594	59.20	5.96	13.81	21.03
C30H36N609	624.66	624.2544	57.69	5.81	13.45	23.05
C30H37N101	427.64	427.2875	84.26	8.72	3.28	3.74
C30H37N102	443.63	443.2824	81.22	8.41	3.16	7.21
C30H37N103	459.63	459.2773	78.40	8.11	3.05	10.44
C30H37N104	475.63	475.2722	75.76	7.84	2.94	13.46
C30H37N105	491.63	491.2672	73.29	7.59	2.85	16.27
C30H37N106	507.63	507.2621	70.98	7.35	2.76	18.91
C30H37N107	523.63	523.2570	68.81	7.12	2.67	21.39
C30H37N108	539.63	539.2519	66.77	6.91	2.60	23.72
C30H37N109	555.63	555.2468	64.85	6.71	2.52	25.92
C30H37N301	455.65	455.2936	79.08	8.19	9.22	3.51
C30H37N302	471.65	471.2886	76.40	7.91	8.91	6.78
C30H37N303	487.65	487.2835	73.89	7.65	8.62	9.84
C30H37N304	503.65	503.2784	71.55	7.40	8.34	12.71
C30H37N305	519.65	519.2733	69.34	7.18	8.09	15.39
C30H37N306	535.65	535.2682	67.27	6.96	7.84	17.92
C30H37N307	551.65	551.2631	65.32	6.76	7.62	20.30
C30H37N308	567.64	567.2580	63.48	6.57	7.40	22.55
C30H37N309	583.64	583.2530	61.74	6.39	7.20	24.67
C30H37N501	483.66	483.2998	74.50	7.71	14.48	3.31
C30H37N502	499.66	499.2947	72.12	7.46	14.02	6.40
C30H37N503	515.66	515.2896	69.88	7.23	13.58	9.31
C30H37N504	531.66	531.2845	67.78	7.01	13.17	12.04
C30H37N505	547.66	547.2794	65.80	6.81	12.79	14.61
C30H37N506	563.66	563.2744	63.93	6.62	12.42	17.03
C30H37N507	579.66	579.2693	62.16	6.43	12.08	19.32
C30H37N508	595.66	595.2642	60.49	6.26	11.76	21.49
C30H37N509	611.66	611.2591	58.91	6.10	11.45	23.54
C30H38N201	442.65	442.2984	81.40	8.65	6.33	3.61
C30H38N202	458.65	458.2933	78.56	8.35	6.11	6.98
C30H38N203	474.65	474.2882	75.92	8.07	5.90	10.11
C30H38N204	490.65	490.2831	73.44	7.81	5.71	13.04
C30H38N205	506.65	506.2781	71.12	7.56	5.53	15.79
C30H38N206	522.65	522.2730	68.94	7.33	5.36	18.37
C30H38N207	538.65	538.2679	66.90	7.11	5.20	20.79
C30H38N208	554.65	554.2628	64.97	6.91	5.05	23.08
C30H38N209	570.65	570.2577	63.15	6.71	4.91	25.23
C30H38N401	470.66	470.3045	76.56	8.14	11.90	3.40
C30H38N402	486.66	486.2995	74.04	7.87	11.51	6.58
C30H38N403	502.66	502.2944	71.69	7.62	11.15	9.55
C30H38N404	518.66	518.2893	69.47	7.38	10.80	12.34
C30H38N405	534.66	534.2842	67.39	7.16	10.48	14.96
C30H38N406	550.66	550.2791	65.44	6.96	10.17	17.43
C30H38N407	566.66	566.2740	63.59	6.76	9.89	19.76
C30H38N408	582.66	582.2689	61.84	6.57	9.62	21.97
C30H38N409	598.66	598.2639	60.19	6.40	9.36	24.05
C30H38N601	498.68	498.3107	72.26	7.68	16.85	3.21
C30H38N602	514.68	514.3056	70.01	7.44	16.33	6.22
C30H38N603	530.68	530.3005	67.90	7.22	15.84	9.04
C30H38N604	546.68	546.2954	65.91	7.01	15.37	11.71
C30H38N605	562.67	562.2903	64.04	6.81	14.94	14.22
C30H38N606	578.67	578.2853	62.27	6.62	14.52	16.59
C30H38N607	594.67	594.2802	60.59	6.44	14.13	18.83
C30H38N608	610.67	610.2751	59.01	6.27	13.76	20.96
C30H38N609	626.67	626.2700	57.50	6.11	13.41	22.98
C30H39N101	429.65	429.3031	83.87	9.15	3.26	3.72
C30H39N102	445.65	445.2981	80.86	8.82	3.14	7.18
C30H39N103	461.65	461.2930	78.05	8.52	3.03	10.40
C30H39N104	477.65	477.2879	75.44	8.23	2.93	13.40
C30H39N105	493.65	493.2828	72.99	7.96	2.84	16.21
C30H39N106	509.65	509.2777	70.70	7.71	2.75	18.84
C30H39N107	525.65	525.2726	68.55	7.48	2.66	21.31
C30H39N108	541.65	541.2675	66.53	7.26	2.59	23.63
C30H39N109	557.65	557.2625	64.62	7.05	2.51	25.82
C30H39N301	457.66	457.3093	78.73	8.59	9.18	3.50
C30H39N302	473.66	473.3042	76.07	8.30	8.87	6.76
C30H39N303	489.66	489.2991	73.59	8.03	8.58	9.80
C30H39N304	505.66	505.2940	71.26	7.77	8.31	12.66
C30H39N305	521.66	521.2890	69.07	7.54	8.06	15.34
C30H39N306	537.66	537.2839	67.02	7.31	7.82	17.85
C30H39N307	553.66	553.2788	65.08	7.10	7.59	20.23
C30H39N308	569.66	569.2737	63.25	6.90	7.38	22.47
C30H39N309	585.66	585.2686	61.53	6.71	7.17	24.59
C30H39N501	485.68	485.3154	74.19	8.09	14.42	3.29
C30H39N502	501.68	501.3104	71.83	7.84	13.96	6.38
C30H39N503	517.68	517.3053	69.61	7.59	13.53	9.27
C30H39N504	533.68	533.3002	67.52	7.37	13.12	11.99
C30H39N505	549.68	549.2951	65.55	7.15	12.74	14.55
C30H39N506	565.68	565.2900	63.70	6.95	12.38	16.97
C30H39N507	581.67	581.2849	61.95	6.76	12.04	19.25
C30H39N508	597.67	597.2798	60.29	6.58	11.72	21.42
C30H39N509	613.67	613.2748	58.72	6.41	11.41	23.46
C30H40N201	444.67	444.3140	81.03	9.07	6.30	3.60
C30H40N202	460.67	460.3090	78.22	8.75	6.08	6.95
C30H40N203	476.66	476.3039	75.59	8.46	5.88	10.07
C30H40N204	492.66	492.2988	73.14	8.18	5.69	12.99

C30H40N205	508.66	508.2937	70.84	7.93	5.51	15.73
C30H40N206	524.66	524.2886	68.68	7.68	5.34	18.30
C30H40N207	540.66	540.2835	66.65	7.46	5.18	20.71
C30H40N208	556.66	556.2784	64.73	7.24	5.03	22.99
C30H40N209	572.66	572.2734	62.92	7.04	4.89	25.14
C30H40N401	472.68	472.3202	76.23	8.53	11.85	3.38
C30H40N402	488.68	488.3151	73.74	8.25	11.46	6.55
C30H40N403	504.68	504.3100	71.40	7.99	11.10	9.51
C30H40N404	520.68	520.3049	69.20	7.74	10.76	12.29
C30H40N405	536.68	536.2998	67.14	7.51	10.44	14.91
C30H40N406	552.68	552.2948	65.20	7.30	10.14	17.37
C30H40N407	568.68	568.2897	63.36	7.09	9.85	19.69
C30H40N408	584.68	584.2846	61.63	6.90	9.58	21.89
C30H40N409	600.67	600.2795	59.99	6.71	9.33	23.97
C30H40N601	500.69	500.3263	71.97	8.05	16.78	3.20
C30H40N602	516.69	516.3213	69.74	7.80	16.27	6.19
C30H40N603	532.69	532.3162	67.64	7.57	15.78	9.01
C30H40N604	548.69	548.3111	65.67	7.35	15.32	11.66
C30H40N605	564.69	564.3060	63.81	7.14	14.88	14.17
C30H40N606	580.69	580.3009	62.05	6.94	14.47	16.53
C30H40N607	596.69	596.2958	60.39	6.76	14.08	18.77
C30H40N608	612.69	612.2907	58.81	6.58	13.72	20.89
C30H40N609	628.69	628.2857	57.32	6.41	13.37	22.90
C30H41N101	431.67	431.3188	83.48	9.57	3.24	3.71
C30H41N102	447.67	447.3137	80.49	9.23	3.13	7.15
C30H41N103	463.67	463.3086	77.71	8.91	3.02	10.35
C30H41N104	479.67	479.3035	75.12	8.62	2.92	13.34
C30H41N105	495.66	495.2985	72.70	8.34	2.83	16.14
C30H41N106	511.66	511.2934	70.42	8.08	2.74	18.76
C30H41N107	527.66	527.2883	68.29	7.83	2.65	21.22
C30H41N108	543.66	543.2832	66.28	7.60	2.58	23.54
C30H41N109	559.66	559.2781	64.38	7.38	2.50	25.73
C30H41N301	459.68	459.3249	78.39	8.99	9.14	3.48
C30H41N302	475.68	475.3199	75.75	8.69	8.83	6.73
C30H41N303	491.68	491.3148	73.29	8.41	8.55	9.76
C30H41N304	507.68	507.3097	70.98	8.14	8.28	12.61
C30H41N305	523.68	523.3046	68.81	7.89	8.02	15.28
C30H41N306	539.68	539.2995	66.77	7.66	7.79	17.79
C30H41N307	555.68	555.2944	64.85	7.44	7.56	20.15
C30H41N308	571.68	571.2893	63.03	7.23	7.35	22.39
C30H41N309	587.68	587.2843	61.32	7.03	7.15	24.50
C30H41N501	487.69	487.3311	73.89	8.47	14.36	3.28
C30H41N502	503.69	503.3260	71.54	8.20	13.90	6.35
C30H41N503	519.69	519.3209	69.34	7.95	13.48	9.24
C30H41N504	535.69	535.3158	67.27	7.71	13.07	11.95
C30H41N505	551.69	551.3107	65.31	7.49	12.69	14.50
C30H41N506	567.69	567.3057	63.47	7.28	12.34	16.91
C30H41N507	583.69	583.3006	61.73	7.08	12.00	19.19
C30H41N508	599.69	599.2955	60.09	6.89	11.68	21.34
C30H41N509	615.69	615.2904	58.53	6.71	11.37	23.39
C30H42N201	446.68	446.3297	80.67	9.48	6.27	3.58
C30H42N202	462.68	462.3246	77.88	9.15	6.05	6.92
C30H42N203	478.68	478.3195	75.28	8.84	5.85	10.03
C30H42N204	494.68	494.3144	72.84	8.56	5.66	12.94
C30H42N205	510.68	510.3094	70.56	8.29	5.49	15.66
C30H42N206	526.68	526.3043	68.42	8.04	5.32	18.23
C30H42N207	542.68	542.2992	66.40	7.80	5.16	20.64
C30H42N208	558.68	558.2941	64.50	7.58	5.01	22.91
C30H42N209	574.68	574.2890	62.70	7.37	4.87	25.06
C30H42N401	474.70	474.3358	75.91	8.92	11.80	3.37
C30H42N402	490.69	490.3308	73.43	8.63	11.42	6.52
C30H42N403	506.69	506.3257	71.11	8.36	11.06	9.47
C30H42N404	522.69	522.3206	68.94	8.10	10.72	12.24
C30H42N405	538.69	538.3155	66.89	7.86	10.40	14.85
C30H42N406	554.69	554.3104	64.96	7.63	10.10	17.31
C30H42N407	570.69	570.3053	63.14	7.42	9.82	19.62
C30H42N408	586.69	586.3002	61.42	7.22	9.55	21.82
C30H42N409	602.69	602.2952	59.79	7.02	9.30	23.89
C30H42N601	502.71	502.3420	71.68	8.42	16.72	3.18
C30H42N602	518.71	518.3369	69.47	8.16	16.20	6.17
C30H42N603	534.71	534.3318	67.39	7.92	15.72	8.98
C30H42N604	550.71	550.3267	65.43	7.69	15.26	11.62
C30H42N605	566.71	566.3216	63.58	7.47	14.83	14.12
C30H42N606	582.71	582.3166	61.84	7.27	14.42	16.47
C30H42N607	598.71	598.3115	60.19	7.07	14.04	18.71
C30H42N608	614.70	614.3064	58.62	6.89	13.67	20.82
C30H42N609	630.70	630.3013	57.13	6.71	13.32	22.83
C30H43N101	433.68	433.3344	83.09	9.99	3.23	3.69
C30H43N102	449.68	449.3294	80.13	9.64	3.11	7.12
C30H43N103	465.68	465.3243	77.38	9.31	3.01	10.31
C30H43N104	481.68	481.3192	74.81	9.00	2.91	13.29
C30H43N105	497.68	497.3141	72.40	8.71	2.81	16.07
C30H43N106	513.68	513.3090	70.15	8.44	2.73	18.69
C30H43N107	529.68	529.3039	68.03	8.18	2.64	21.14
C30H43N108	545.68	545.2988	66.03	7.94	2.57	23.46
C30H43N109	561.68	561.2938	64.15	7.72	2.49	25.64
C30H43N301	461.70	461.3406	78.05	9.39	9.10	3.47
C30H43N302	477.70	477.3355	75.43	9.07	8.80	6.70
C30H43N303	493.70	493.3304	72.99	8.78	8.51	9.72
C30H43N304	509.69	509.3253	70.70	8.50	8.24	12.56
C30H43N305	525.69	525.3202	68.54	8.24	7.99	15.22
C30H43N306	541.69	541.3152	66.52	8.00	7.76	17.72
C30H43N307	557.69	557.3101	64.61	7.77	7.53	20.08
C30H43N308	573.69	573.3050	62.81	7.56	7.32	22.31
C30H43N309	589.69	589.2999	61.11	7.35	7.13	24.42
C30H43N501	489.71	489.3467	73.58	8.85	14.30	3.27
C30H43N502	505.71	505.3417	71.25	8.57	13.85	6.33
C30H43N503	521.71	521.3366	69.07	8.31	13.42	9.20
C30H43N504	537.71	537.3315	67.01	8.06	13.02	11.90
C30H43N505	553.71	553.3264	65.08	7.83	12.65	14.45
C30H43N506	569.71	569.3213	63.25	7.61	12.29	16.85
C30H43N507	585.71	585.3162	61.52	7.40	11.96	19.12
C30H43N508	601.71	601.3111	59.89	7.20	11.64	21.27
C30H43N509	617.71	617.3061	58.33	7.02	11.34	23.31
C30H44N201	448.70	448.3453	80.31	9.88	6.24	3.57
C30H44N202	464.70	464.3403	77.54	9.54	6.03	6.89
C30H44N203	480.70	480.3352	74.96	9.23	5.83	9.99
C30H44N204	496.70	496.3301	72.55	8.93	5.64	12.88
C30H44N205	512.70	512.3250	70.28	8.65	5.46	15.60
C30H44N206	528.69	528.3199	68.16	8.39	5.30	18.16
C30H44N207	544.69	544.3148	66.15	8.14	5.14	20.56
C30H44N208	560.69	560.3097	64.27	7.91	5.00	22.83
C30H44N209	576.69	576.3047	62.48	7.69	4.86	24.97
C30H44N401	476.71	476.3515	75.59	9.30	11.75	3.36
C30H44N402	492.71	492.3464	73.13	9.00	11.37	6.49
C30H44N403	508.71	508.3413	70.83	8.72	11.01	9.44
C30H44N404	524.71	524.3362	68.67	8.45	10.68	12.20
C30H44N405	540.71	540.3311	66.64	8.20	10.36	14.79
C30H44N406	556.71	556.3261	64.73	7.97	10.06	17.24
C30H44N407	572.71	572.3210	62.92	7.74	9.78	19.56

C30H44N408	588.71	588.3159	61.21	7.53	9.52	21.74
C30H44N409	604.71	604.3108	59.59	7.33	9.27	23.81
C30H44N601	504.72	504.3576	71.39	8.79	16.65	3.17
C30H44N602	520.72	520.3526	69.20	8.52	16.14	6.15
C30H44N603	536.72	536.3475	67.14	8.26	15.66	8.94
C30H44N604	552.72	552.3424	65.19	8.02	15.20	11.58
C30H44N605	568.72	568.3373	63.36	7.80	14.78	14.07
C30H44N606	584.72	584.3322	61.62	7.58	14.37	16.42
C30H44N607	600.72	600.3271	59.98	7.38	13.99	18.64
C30H44N608	616.72	616.3220	58.43	7.19	13.63	20.75
C30H44N609	632.72	632.3170	56.95	7.01	13.28	22.76
C30H45N101	435.70	435.3501	82.70	10.41	3.21	3.67
C30H45N102	451.70	451.3450	79.77	10.04	3.10	7.08
C30H45N103	467.70	467.3399	77.04	9.70	2.99	10.26
C30H45N104	483.70	483.3348	74.50	9.38	2.90	13.23
C30H45N105	499.70	499.3298	72.11	9.08	2.80	16.01
C30H45N106	515.70	515.3247	69.87	8.80	2.72	18.61
C30H45N107	531.70	531.3196	67.77	8.53	2.63	21.06
C30H45N108	547.70	547.3145	65.79	8.28	2.56	23.37
C30H45N109	563.69	563.3094	63.92	8.05	2.48	25.54
C30H45N301	463.71	463.3562	77.71	9.78	9.06	3.45
C30H45N302	479.71	479.3512	75.11	9.46	8.76	6.67
C30H45N303	495.71	495.3461	72.69	9.15	8.48	9.68
C30H45N304	511.71	511.3410	70.42	8.86	8.21	12.51
C30H45N305	527.71	527.3359	68.28	8.60	7.96	15.16
C30H45N306	543.71	543.3308	66.27	8.34	7.73	17.66
C30H45N307	559.71	559.3257	64.38	8.10	7.51	20.01
C30H45N308	575.71	575.3206	62.59	7.88	7.30	22.23
C30H45N309	591.71	591.3156	60.90	7.67	7.10	24.34
C30H45N501	491.73	491.3624	73.28	9.22	14.24	3.25
C30H45N502	507.73	507.3573	70.97	8.93	13.79	6.30
C30H45N503	523.72	523.3522	68.80	8.66	13.37	9.16
C30H45N504	539.72	539.3471	66.76	8.40	12.98	11.86
C30H45N505	555.72	555.3420	64.84	8.16	12.60	14.40
C30H45N506	571.72	571.3370	63.03	7.93	12.25	16.79
C30H45N507	587.72	587.3319	61.31	7.72	11.92	19.06
C30H45N508	603.72	603.3268	59.69	7.51	11.60	21.20
C30H45N509	619.72	619.3217	58.14	7.32	11.30	23.24
C30H46N201	450.71	450.3610	79.95	10.29	6.22	3.55
C30H46N202	466.71	466.3559	77.21	9.93	6.00	6.86
C30H46N203	482.71	482.3508	74.65	9.61	5.80	9.94
C30H46N204	498.71	498.3457	72.25	9.30	5.62	12.83
C30H46N205	514.71	514.3406	70.01	9.01	5.44	15.54
C30H46N206	530.71	530.3356	67.90	8.74	5.28	18.09
C30H46N207	546.71	546.3305	65.91	8.48	5.12	20.49
C30H46N208	562.71	562.3254	64.04	8.24	4.98	22.75
C30H46N209	578.71	578.3203	62.27	8.01	4.84	24.88
C30H46N401	478.73	478.3671	75.27	9.69	11.70	3.34
C30H46N402	494.73	494.3621	72.84	9.37	11.32	6.47
C30H46N403	510.73	510.3570	70.55	9.08	10.97	9.40
C30H46N404	526.73	526.3519	68.41	8.80	10.64	12.15
C30H46N405	542.72	542.3468	66.39	8.54	10.32	14.74
C30H46N406	558.72	558.3417	64.49	8.30	10.03	17.18
C30H46N407	574.72	574.3366	62.70	8.07	9.75	19.49
C30H46N408	590.72	590.3315	61.00	7.85	9.48	21.67
C30H46N409	606.72	606.3265	59.39	7.64	9.23	23.73
C30H46N601	506.74	506.3733	71.11	9.15	16.58	3.16
C30H46N602	522.74	522.3682	68.93	8.87	16.08	6.12
C30H46N603	538.74	538.3631	66.88	8.61	15.60	8.91
C30H46N604	554.74	554.3580	64.96	8.36	15.15	11.54
C30H46N605	570.74	570.3529	63.13	8.12	14.72	14.02
C30H46N606	586.74	586.3479	61.41	7.90	14.32	16.36
C30H46N607	602.74	602.3428	59.78	7.69	13.94	18.58
C30H46N608	618.74	618.3377	58.24	7.49	13.58	20.69
C30H46N609	634.74	634.3326	56.77	7.30	13.24	22.69
C30H47N101	437.72	437.3657	82.32	10.82	3.20	3.66
C30H47N102	453.71	453.3607	79.42	10.44	3.09	7.05
C30H47N103	469.71	469.3556	76.71	10.09	2.98	10.22
C30H47N104	485.71	485.3505	74.19	9.75	2.88	13.18
C30H47N105	501.71	501.3454	71.82	9.44	2.79	15.94
C30H47N106	517.71	517.3403	69.60	9.15	2.71	18.54
C30H47N107	533.71	533.3352	67.51	8.88	2.62	20.98
C30H47N108	549.71	549.3301	65.55	8.62	2.55	23.28
C30H47N109	565.71	565.3251	63.70	8.37	2.48	25.45
C30H47N301	465.73	465.3719	77.37	10.17	9.02	3.44
C30H47N302	481.73	481.3668	74.80	9.83	8.72	6.64
C30H47N303	497.73	497.3617	72.40	9.52	8.44	9.64
C30H47N304	513.73	513.3566	70.14	9.22	8.18	12.46
C30H47N305	529.73	529.3515	68.02	8.94	7.93	15.10
C30H47N306	545.73	545.3465	66.03	8.68	7.70	17.59
C30H47N307	561.72	561.3414	64.15	8.43	7.48	19.94
C30H47N308	577.72	577.3363	62.37	8.20	7.27	22.16
C30H47N309	593.72	593.3312	60.69	7.98	7.08	24.25
C30H47N501	493.74	493.3780	72.98	9.60	14.18	3.24
C30H47N502	509.74	509.3730	70.69	9.29	13.74	6.28
C30H47N503	525.74	525.3679	68.54	9.01	13.32	9.13
C30H47N504	541.74	541.3628	66.51	8.74	12.93	11.81
C30H47N505	557.74	557.3577	64.61	8.49	12.56	14.34
C30H47N506	573.74	573.3526	62.80	8.26	12.21	16.73
C30H47N507	589.74	589.3475	61.10	8.03	11.88	18.99
C30H47N508	605.74	605.3424	59.49	7.82	11.56	21.13
C30H47N509	621.74	621.3374	57.96	7.62	11.26	23.16
C30H48N201	452.73	452.3766	79.59	10.69	6.19	3.53
C30H48N202	468.73	468.3716	76.87	10.32	5.98	6.83
C30H48N203	484.73	484.3665	74.34	9.98	5.78	9.90
C30H48N204	500.73	500.3614	71.96	9.66	5.59	12.78
C30H48N205	516.73	516.3563	69.73	9.36	5.42	15.48
C30H48N206	532.73	532.3512	67.64	9.08	5.26	18.02
C30H48N207	548.73	548.3461	65.67	8.82	5.11	20.41
C30H48N208	564.73	564.3410	63.81	8.57	4.96	22.67
C30H48N209	580.73	580.3360	62.05	8.33	4.82	24.80
C30H48N401	480.74	480.3828	74.95	10.06	11.65	3.33
C30H48N402	496.74	496.3777	72.54	9.74	11.28	6.44
C30H48N403	512.74	512.3726	70.28	9.44	10.93	9.36
C30H48N404	528.74	528.3675	68.15	9.15	10.60	12.10
C30H48N405	544.74	544.3624	66.15	8.88	10.29	14.69
C30H48N406	560.74	560.3574	64.26	8.63	9.99	17.12
C30H48N407	576.74	576.3523	62.48	8.39	9.71	19.42
C30H48N408	592.74	592.3472	60.79	8.16	9.45	21.59
C30H48N409	608.74	608.3421	59.19	7.95	9.20	23.65
C30H48N601	508.76	508.3889	70.83	9.51	16.52	3.14
C30H48N602	524.76	524.3839	68.67	9.22	16.02	6.10
C30H48N603	540.76	540.3788	66.64	8.95	15.54	8.88
C30H48N604	556.75	556.3737	64.72	8.69	15.09	11.49
C30H48N605	572.75	572.3686	62.91	8.45	14.67	13.97
C30H48N606	588.75	588.3635	61.20	8.22	14.27	16.31
C30H48N607	604.75	604.3584	59.58	8.00	13.90	18.52
C30H48N608	620.75	620.3533	58.05	7.79	13.54	20.62
C30H48N609	636.75	636.3483	56.59	7.60	13.20	22.61
C30H49N101	439.73	439.3814	81.94	11.23	3.19	3.64

218

C30H49N1O2	455.73	455.3763	79.07	10.84	3.07	7.02
C30H49N1O3	471.73	471.3712	76.39	10.47	2.97	10.17
C30H49N1O4	487.73	487.3661	73.88	10.13	2.87	13.12
C30H49N1O5	503.73	503.3610	71.53	9.80	2.78	15.88
C30H49N1O6	519.73	519.3560	69.33	9.50	2.70	18.47
C30H49N1O7	535.73	535.3509	67.26	9.22	2.61	20.91
C30H49N1O8	551.73	551.3458	65.31	8.95	2.54	23.20
C30H49N1O9	567.73	567.3407	63.47	8.70	2.47	25.36
C30H49N3O1	467.74	467.3875	77.04	10.56	8.98	3.42
C30H49N3O2	483.74	483.3825	74.49	10.21	8.69	6.61
C30H49N3O3	499.74	499.3774	72.10	9.88	8.41	9.60
C30H49N3O4	515.74	515.3723	69.87	9.58	8.15	12.41
C30H49N3O5	531.74	531.3672	67.76	9.29	7.90	15.04
C30H49N3O6	547.74	547.3621	65.79	9.02	7.67	17.53
C30H49N3O7	563.74	563.3570	63.92	8.76	7.45	19.87
C30H49N3O8	579.74	579.3519	62.15	8.52	7.25	22.08
C30H49N3O9	595.74	595.3469	60.49	8.29	7.05	24.17
C30H49N5O1	495.76	495.3937	72.68	9.96	14.13	3.23
C30H49N5O2	511.76	511.3886	70.41	9.65	13.68	6.25
C30H49N5O3	527.76	527.3835	68.28	9.36	13.27	9.09
C30H49N5O4	543.76	543.3784	66.27	9.08	12.88	11.77
C30H49N5O5	559.76	559.3733	64.37	8.82	12.51	14.29
C30H49N5O6	575.75	575.3683	62.58	8.58	12.16	16.67
C30H49N5O7	591.75	591.3632	60.89	8.35	11.83	18.93
C30H49N5O8	607.75	607.3581	59.29	8.13	11.52	21.06
C30H49N5O9	623.75	623.3530	57.77	7.92	11.23	23.09
C30H50N2O1	454.75	454.3923	79.24	11.08	6.16	3.52
C30H50N2O2	470.75	470.3872	76.55	10.71	5.95	6.80
C30H50N2O3	486.74	486.3821	74.03	10.35	5.76	9.86
C30H50N2O4	502.74	502.3770	71.67	10.02	5.57	12.73
C30H50N2O5	518.74	518.3719	69.46	9.72	5.40	15.42
C30H50N2O6	534.74	534.3669	67.38	9.42	5.24	17.95
C30H50N2O7	550.74	550.3618	65.43	9.15	5.09	20.34
C30H50N2O8	566.74	566.3567	63.58	8.89	4.94	22.58
C30H50N2O9	582.74	582.3516	61.83	8.65	4.81	24.71
C30H50N4O1	482.76	482.3984	74.64	10.44	11.61	3.31
C30H50N4O2	498.76	498.3934	72.25	10.10	11.23	6.42
C30H50N4O3	514.76	514.3883	70.00	9.79	10.88	9.32
C30H50N4O4	530.76	530.3832	67.89	9.50	10.56	12.06
C30H50N4O5	546.76	546.3781	65.90	9.22	10.25	14.63
C30H50N4O6	562.76	562.3730	64.03	8.96	9.96	17.06
C30H50N4O7	578.76	578.3679	62.26	8.71	9.68	19.35
C30H50N4O8	594.75	594.3628	60.55	8.47	9.42	21.52
C30H50N4O9	610.75	610.3578	59.00	8.25	9.17	23.58
C30H50N6O1	510.76	510.4046	70.55	9.87	16.45	3.13
C30H50N6O2	526.77	526.3995	68.40	9.57	15.95	6.07
C30H50N6O3	542.77	542.3944	66.39	9.29	15.48	8.84
C30H50N6O4	558.77	558.3893	64.49	9.02	15.04	11.45
C30H50N6O5	574.77	574.3842	62.69	8.77	14.62	13.92
C30H50N6O6	590.77	590.3792	60.99	8.53	14.23	16.25
C30H50N6O7	606.77	606.3741	59.39	8.31	13.85	18.46
C30H50N6O8	622.77	622.3690	57.86	8.09	13.49	20.55
C30H50N6O9	638.77	638.3639	56.41	7.89	13.16	22.54
C30H51N1O1	441.75	441.3970	81.57	11.64	3.17	3.62
C30H51N1O2	457.75	457.3920	78.72	11.23	3.06	6.99
C30H51N1O3	473.75	473.3869	76.06	10.85	2.96	10.13
C30H51N1O4	489.75	489.3818	73.58	10.50	2.86	13.07
C30H51N1O5	505.74	505.3767	71.25	10.16	2.77	15.82
C30H51N1O6	521.74	521.3716	69.06	9.85	2.68	18.40
C30H51N1O7	537.74	537.3665	67.01	9.56	2.60	20.83

C30H51N1O8	553.74	553.3614	65.07	9.28	2.53	23.11
C30H51N1O9	569.74	569.3564	63.25	9.02	2.46	25.27
C30H51N3O1	469.76	469.4032	76.71	10.94	8.95	3.41
C30H51N3O2	485.76	485.3981	74.18	10.58	8.65	6.59
C30H51N3O3	501.76	501.3930	71.81	10.25	8.37	9.57
C30H51N3O4	517.76	517.3879	69.60	9.93	8.12	12.36
C30H51N3O5	533.76	533.3828	67.51	9.63	7.87	14.99
C30H51N3O6	549.76	549.3778	65.54	9.35	7.64	17.46
C30H51N3O7	565.76	565.3727	63.69	9.09	7.43	19.80
C30H51N3O8	581.76	581.3676	61.94	8.84	7.22	22.00
C30H51N3O9	597.76	597.3625	60.28	8.60	7.03	24.09
C30H51N5O1	497.77	497.4093	72.39	10.33	14.07	3.21
C30H51N5O2	513.77	513.4043	70.13	10.01	13.63	6.23
C30H51N5O3	529.77	529.3992	68.02	9.70	13.22	9.06
C30H51N5O4	545.77	545.3941	66.02	9.42	12.83	11.73
C30H51N5O5	561.77	561.3890	64.14	9.15	12.47	14.24
C30H51N5O6	577.77	577.3839	62.37	8.90	12.12	16.61
C30H51N5O7	593.77	593.3788	60.69	8.66	11.79	18.86
C30H51N5O8	609.77	609.3737	59.09	8.43	11.49	20.99
C30H51N5O9	625.77	625.3687	57.58	8.21	11.19	23.01
C30H52N2O1	456.76	456.4079	78.89	11.48	6.13	3.50
C30H52N2O2	472.76	472.4029	76.22	11.09	5.93	6.77
C30H52N2O3	488.76	488.3978	73.72	10.72	5.73	9.82
C30H52N2O4	504.76	504.3927	71.39	10.38	5.55	12.68
C30H52N2O5	520.76	520.3876	69.19	10.07	5.38	15.36
C30H52N2O6	536.76	536.3825	67.13	9.76	5.22	17.88
C30H52N2O7	552.76	552.3774	65.19	9.48	5.07	20.26
C30H52N2O8	568.76	568.3723	63.35	9.22	4.93	22.50
C30H52N2O9	584.76	584.3673	61.62	8.96	4.79	24.62
C30H52N4O1	484.78	484.4141	74.33	10.81	11.56	3.30
C30H52N4O2	500.77	500.4090	71.96	10.47	11.19	6.39
C30H52N4O3	516.77	516.4039	69.73	10.14	10.84	9.29
C30H52N4O4	532.77	532.3988	67.63	9.84	10.52	12.01
C30H52N4O5	548.77	548.3937	65.66	9.55	10.21	14.58
C30H52N4O6	564.77	564.3887	63.80	9.28	9.92	17.00
C30H52N4O7	580.77	580.3836	62.04	9.02	9.65	19.28
C30H52N4O8	596.77	596.3785	60.38	8.78	9.39	21.45
C30H52N4O9	612.77	612.3734	58.80	8.55	9.14	23.50
C30H52N6O1	512.79	512.4202	70.27	10.22	16.39	3.12
C30H52N6O2	528.79	528.4152	68.14	9.91	15.89	6.05
C30H52N6O3	544.79	544.4101	66.14	9.62	15.43	8.81
C30H52N6O4	560.79	560.4050	64.26	9.35	14.99	11.41
C30H52N6O5	576.79	576.3999	62.47	9.09	14.57	13.87
C30H52N6O6	592.79	592.3948	60.79	8.84	14.18	16.19
C30H52N6O7	608.78	608.3897	59.19	8.61	13.80	18.40
C30H52N6O8	624.78	624.3846	57.67	8.39	13.45	20.49
C30H52N6O9	640.78	640.3795	56.23	8.18	13.12	22.47
C30H53N1O1	443.76	443.4127	81.20	12.04	3.16	3.61
C30H53N1O2	459.76	459.4076	78.37	11.62	3.05	6.96
C30H53N1O3	475.76	475.4025	75.74	11.23	2.94	10.09
C30H53N1O4	491.76	491.3974	73.27	10.86	2.85	13.01
C30H53N1O5	507.76	507.3923	70.97	10.52	2.76	15.75
C30H53N1O6	523.76	523.3873	68.80	10.20	2.67	18.33
C30H53N1O7	539.76	539.3822	66.76	9.90	2.59	20.75
C30H53N1O8	555.76	555.3771	64.84	9.61	2.52	23.03
C30H53N1O9	571.76	571.3720	63.02	9.34	2.45	25.18
C30H53N3O1	471.78	471.4188	76.38	11.32	8.91	3.39
C30H53N3O2	487.78	487.4138	73.87	10.95	8.61	6.56
C30H53N3O3	503.78	503.4087	71.53	10.60	8.34	9.53
C30H53N3O4	519.77	519.4036	69.33	10.28	8.08	12.31

219

C30H53N3O5	535.77	535.3985	67.25	9.97	7.84	14.93
C30H53N3O6	551.77	551.3934	65.30	9.68	7.62	17.40
C30H53N3O7	567.77	567.3883	63.46	9.41	7.40	19.73
C30H53N3O8	583.77	583.3832	61.73	9.15	7.20	21.93
C30H53N3O9	599.77	599.3782	60.08	8.91	7.01	24.01
C30H53N5O1	499.79	499.4250	72.10	10.69	14.01	3.20
C30H53N5O2	515.79	515.4199	69.86	10.36	13.58	6.20
C30H53N5O3	531.79	531.4148	67.76	10.05	13.17	9.03
C30H53N5O4	547.79	547.4097	65.78	9.75	12.78	11.68
C30H53N5O5	563.79	563.4046	63.91	9.48	12.42	14.19
C30H53N5O6	579.79	579.3996	62.15	9.21	12.08	16.56
C30H53N5O7	595.79	595.3945	60.48	8.97	11.75	18.80
C30H53N5O8	611.79	611.3894	58.90	8.73	11.45	20.92
C30H53N5O9	627.79	627.3843	57.40	8.51	11.16	22.94
C30H54N2O1	458.78	458.4236	78.54	11.86	6.11	3.49
C30H54N2O2	474.78	474.4185	75.90	11.46	5.90	6.74
C30H54N2O3	490.78	490.4134	73.42	11.09	5.71	9.78
C30H54N2O4	506.78	506.4083	71.10	10.74	5.53	12.63
C30H54N2O5	522.78	522.4032	68.93	10.41	5.36	15.30
C30H54N2O6	538.77	538.3982	66.88	10.10	5.20	17.82
C30H54N2O7	554.77	554.3931	64.95	9.81	5.05	20.19
C30H54N2O8	570.77	570.3880	63.13	9.54	4.91	22.42
C30H54N2O9	586.77	586.3829	61.41	9.28	4.77	24.54
C30H54N4O1	486.79	486.4297	74.02	11.18	11.51	3.29
C30H54N4O2	502.79	502.4247	71.67	10.83	11.14	6.36
C30H54N4O3	518.79	518.4196	69.46	10.49	10.80	9.25
C30H54N4O4	534.79	534.4145	67.38	10.18	10.48	11.97
C30H54N4O5	550.79	550.4094	65.42	9.88	10.17	14.52
C30H54N4O6	566.79	566.4043	63.57	9.60	9.88	16.94
C30H54N4O7	582.79	582.3992	61.83	9.34	9.61	19.22
C30H54N4O8	598.79	598.3941	60.18	9.09	9.36	21.38
C30H54N4O9	614.79	614.3891	58.61	8.85	9.11	23.42
C30H54N6O1	514.80	514.4359	69.99	10.57	16.32	3.11
C30H54N6O2	530.80	530.4308	67.88	10.25	15.83	6.03
C30H54N6O3	546.80	546.4257	65.90	9.95	15.37	8.78
C30H54N6O4	562.80	562.4206	64.03	9.67	14.93	11.37
C30H54N6O5	578.80	578.4155	62.26	9.40	14.52	13.82
C30H54N6O6	594.80	594.4105	60.58	9.15	14.13	16.14
C30H54N6O7	610.80	610.4054	58.99	8.91	13.76	18.34
C30H54N6O8	626.80	626.4003	57.49	8.68	13.41	20.42
C30H54N6O9	642.80	642.3952	56.06	8.47	13.07	22.40
C30H55N1O1	445.78	445.4283	80.83	12.44	3.14	3.59
C30H55N1O2	461.78	461.4233	78.03	12.01	3.03	6.93
C30H55N1O3	477.78	477.4182	75.42	11.60	2.93	10.05
C30H55N1O4	493.78	493.4131	72.98	11.23	2.84	12.96
C30H55N1O5	509.78	509.4080	70.68	10.88	2.75	15.69
C30H55N1O6	525.78	525.4029	68.53	10.54	2.66	18.26
C30H55N1O7	541.78	541.3978	66.51	10.23	2.59	20.67
C30H55N1O8	557.77	557.3927	64.60	9.94	2.51	22.95
C30H55N1O9	573.77	573.3877	62.80	9.66	2.44	25.10
C30H55N3O1	473.79	473.4345	76.05	11.70	8.87	3.38
C30H55N3O2	489.79	489.4294	73.57	11.32	8.58	6.53
C30H55N3O3	505.79	505.4243	71.24	10.96	8.31	9.49
C30H55N3O4	521.79	521.4192	69.06	10.62	8.05	12.26
C30H55N3O5	537.79	537.4141	67.00	10.31	7.81	14.88
C30H55N3O6	553.79	553.4091	65.07	10.01	7.59	17.33
C30H55N3O7	569.79	569.4040	63.24	9.73	7.37	19.66
C30H55N3O8	585.79	585.3989	61.51	9.46	7.17	21.85
C30H55N3O9	601.79	601.3938	59.88	9.21	6.98	23.93
C30H55N5O1	501.81	501.4406	71.81	11.05	13.96	3.19

C30H55N5O2	517.81	517.4356	69.59	10.71	13.53	6.18
C30H55N5O3	533.80	533.4305	67.50	10.39	13.12	8.99
C30H55N5O4	549.80	549.4254	65.54	10.08	12.74	11.64
C30H55N5O5	565.80	565.4203	63.69	9.80	12.38	14.14
C30H55N5O6	581.80	581.4152	61.93	9.53	12.04	16.50
C30H55N5O7	597.80	597.4101	60.28	9.27	11.72	18.73
C30H55N5O8	613.80	613.4050	58.71	9.03	11.41	20.85
C30H55N5O9	629.80	629.3999	57.21	8.80	11.12	22.86
C30H56N2O1	460.79	460.4392	78.20	12.25	6.08	3.47
C30H56N2O2	476.79	476.4342	75.57	11.84	5.88	6.71
C30H56N2O3	492.79	492.4291	73.12	11.45	5.68	9.74
C30H56N2O4	508.79	508.4240	70.82	11.09	5.51	12.58
C30H56N2O5	524.79	524.4189	68.66	10.76	5.34	15.24
C30H56N2O6	540.79	540.4138	66.63	10.44	5.18	17.75
C30H56N2O7	556.79	556.4087	64.72	10.14	5.03	20.11
C30H56N2O8	572.79	572.4036	62.91	9.85	4.89	22.35
C30H56N2O9	588.79	588.3986	61.20	9.59	4.76	24.46
C30H56N4O1	488.81	488.4454	73.72	11.55	11.46	3.27
C30H56N4O2	504.81	504.4403	71.38	11.18	11.10	6.34
C30H56N4O3	520.81	520.4352	69.19	10.84	10.76	9.22
C30H56N4O4	536.81	536.4301	67.13	10.52	10.44	11.92
C30H56N4O5	552.80	552.4250	65.18	10.21	10.14	14.47
C30H56N4O6	568.80	568.4200	63.35	9.92	9.85	16.88
C30H56N4O7	584.80	584.4149	61.62	9.65	9.58	19.15
C30H56N4O8	600.80	600.4098	59.98	9.40	9.33	21.30
C30H56N4O9	616.80	616.4047	58.42	9.15	9.08	23.35
C30H56N6O1	516.82	516.4515	69.72	10.92	16.26	3.10
C30H56N6O2	532.82	532.4464	67.63	10.59	15.77	6.01
C30H56N6O3	548.82	548.4414	65.66	10.29	15.31	8.75
C30H56N6O4	564.82	564.4363	63.80	9.99	14.88	11.33
C30H56N6O5	580.82	580.4312	62.04	9.72	14.47	13.77
C30H56N6O6	596.82	596.4261	60.38	9.46	14.08	16.08
C30H56N6O7	612.82	612.4210	58.80	9.21	13.71	18.28
C30H56N6O8	628.82	628.4159	57.30	8.98	13.36	20.35
C30H56N6O9	644.82	644.4108	55.88	8.75	13.03	22.33
C30H57N1O1	447.79	447.4440	80.47	12.83	3.13	3.57
C30H57N1O2	463.79	463.4389	77.69	12.39	3.02	6.90
C30H57N1O3	479.79	479.4338	75.10	11.97	2.92	10.00
C30H57N1O4	495.79	495.4287	72.68	11.59	2.83	12.91
C30H57N1O5	511.79	511.4236	70.41	11.23	2.74	15.63
C30H57N1O6	527.79	527.4186	68.27	10.89	2.65	18.19
C30H57N1O7	543.79	543.4135	66.26	10.57	2.58	20.60
C30H57N1O8	559.79	559.4084	64.37	10.26	2.50	22.86
C30H57N1O9	575.79	575.4033	62.58	9.98	2.43	25.01
C30H57N3O1	475.81	475.4501	75.73	12.08	8.83	3.36
C30H57N3O2	491.81	491.4451	73.27	11.68	8.54	6.51
C30H57N3O3	507.81	507.4400	70.96	11.31	8.27	9.45
C30H57N3O4	523.81	523.4349	68.79	10.97	8.02	12.22
C30H57N3O5	539.81	539.4298	66.75	10.64	7.78	14.82
C30H57N3O6	555.81	555.4247	64.83	10.34	7.56	17.27
C30H57N3O7	571.80	571.4196	63.02	10.05	7.35	19.59
C30H57N3O8	587.80	587.4145	61.30	9.77	7.15	21.78
C30H57N3O9	603.80	603.4095	59.68	9.52	6.96	23.85
C30H57N5O1	503.82	503.4563	71.52	11.40	13.90	3.18
C30H57N5O2	519.82	519.4512	69.32	11.05	13.47	6.16
C30H57N5O3	535.82	535.4461	67.25	10.72	13.07	8.96
C30H57N5O4	551.82	551.4410	65.30	10.41	12.69	11.60
C30H57N5O5	567.82	567.4359	63.46	10.12	12.33	14.09
C30H57N5O6	583.82	583.4309	61.72	9.84	12.00	16.44
C30H57N5O7	599.82	599.4258	60.07	9.58	11.68	18.67

C30H57N508	615.82	615.4207	58.51	9.33	11.37	20.78
C30H57N509	631.82	631.4156	57.03	9.09	11.08	22.79
C30H58N201	462.81	462.4549	77.86	12.63	6.05	3.46
C30H58N202	478.81	478.4498	75.26	12.21	5.85	6.68
C30H58N203	494.81	494.4447	72.82	11.82	5.66	9.70
C30H58N204	510.81	510.4396	70.54	11.45	5.48	12.53
C30H58N205	526.81	526.4345	68.40	11.10	5.32	15.19
C30H58N206	542.81	542.4295	66.38	10.77	5.16	17.69
C30H58N207	558.81	558.4244	64.48	10.46	5.01	20.04
C30H58N208	574.81	574.4193	62.69	10.17	4.87	22.27
C30H58N209	590.80	590.4142	60.99	9.90	4.74	24.37
C30H58N401	490.82	490.4610	73.41	11.91	11.41	3.26
C30H58N402	506.82	506.4560	71.10	11.54	11.05	6.31
C30H58N403	522.82	522.4509	68.92	11.18	10.72	9.18
C30H58N404	538.82	538.4458	66.87	10.85	10.40	11.88
C30H58N405	554.82	554.4407	64.95	10.54	10.10	14.42
C30H58N406	570.82	570.4356	63.13	10.24	9.82	16.82
C30H58N407	586.82	586.4305	61.40	9.96	9.55	19.09
C30H58N408	602.82	602.4254	59.77	9.70	9.29	21.23
C30H58N409	618.82	618.4203	58.23	9.45	9.05	23.27
C30H58N601	518.84	518.4672	69.45	11.27	16.20	3.08
C30H58N602	534.84	534.4621	67.37	10.93	15.71	5.98
C30H58N603	550.84	550.4570	65.42	10.61	15.26	8.71
C30H58N604	566.83	566.4519	63.57	10.31	14.83	11.29
C30H58N605	582.83	582.4468	61.82	10.03	14.42	13.73
C30H58N606	598.83	598.4418	60.17	9.76	14.03	16.03
C30H58N607	614.83	614.4367	58.61	9.51	13.67	18.22
C30H58N608	630.83	630.4316	57.12	9.27	13.32	20.29
C30H58N609	646.83	646.4265	55.71	9.04	12.99	22.26
C30H59N101	449.81	449.4596	80.11	13.22	3.11	3.56
C30H59N102	465.81	465.4546	77.36	12.77	3.01	6.87
C30H59N103	481.81	481.4495	74.79	12.34	2.91	9.96
C30H59N104	497.81	497.4444	72.38	11.95	2.81	12.86
C30H59N105	513.81	513.4393	70.13	11.57	2.73	15.57
C30H59N106	529.81	529.4342	68.01	11.22	2.64	18.12
C30H59N107	545.81	545.4291	66.02	10.90	2.57	20.52
C30H59N108	561.81	561.4240	64.14	10.59	2.49	22.78
C30H59N109	577.81	577.4190	62.36	10.29	2.42	24.92
C30H59N301	477.82	477.4658	75.41	12.45	8.79	3.35
C30H59N302	493.82	493.4607	72.97	12.04	8.51	6.48
C30H59N303	509.82	509.4556	70.68	11.66	8.24	9.41
C30H59N304	525.82	525.4505	68.53	11.31	7.99	12.17
C30H59N305	541.82	541.4454	66.50	10.98	7.76	14.76
C30H59N306	557.82	557.4404	64.60	10.66	7.53	17.21
C30H59N307	573.82	573.4353	62.80	10.36	7.32	19.52
C30H59N308	589.82	589.4302	61.09	10.08	7.12	21.70
C30H59N309	605.82	605.4251	59.48	9.82	6.94	23.77
C30H59N501	505.84	505.4719	71.24	11.76	13.85	3.16
C30H59N502	521.84	521.4668	69.05	11.40	13.42	6.13
C30H59N503	537.84	537.4618	67.00	11.06	13.02	8.92
C30H59N504	553.84	553.4567	65.06	10.74	12.65	11.56
C30H59N505	569.84	569.4516	63.23	10.44	12.29	14.04
C30H59N506	585.83	585.4465	61.51	10.15	11.95	16.39
C30H59N507	601.83	601.4414	59.87	9.88	11.64	18.61
C30H59N508	617.83	617.4363	58.32	9.63	11.34	20.72
C30H59N509	633.83	633.4312	56.85	9.38	11.05	22.72
C30H60N201	464.83	464.4705	77.52	13.01	6.03	3.44
C30H60N202	480.82	480.4655	74.94	12.58	5.83	6.65
C30H60N203	496.82	496.4604	72.53	12.17	5.64	9.66
C30H60N204	512.82	512.4553	70.26	11.79	5.46	12.48
C30H60N205	528.82	528.4502	68.14	11.44	5.30	15.13
C30H60N206	544.82	544.4451	66.14	11.10	5.14	17.62
C30H60N207	560.82	560.4400	64.25	10.78	5.00	19.97
C30H60N208	576.82	576.4349	62.47	10.48	4.86	22.19
C30H60N209	592.82	592.4299	60.78	10.20	4.73	24.29
C30H60N401	492.84	492.4767	73.11	12.27	11.37	3.25
C30H60N402	508.84	508.4716	70.82	11.89	11.01	6.29
C30H60N403	524.84	524.4665	68.66	11.52	10.68	9.15
C30H60N404	540.84	540.4614	66.63	11.18	10.36	11.83
C30H60N405	556.84	556.4563	64.71	10.86	10.06	14.37
C30H60N406	572.84	572.4513	62.90	10.56	9.78	16.76
C30H60N407	588.84	588.4462	61.19	10.27	9.51	19.02
C30H60N408	604.83	604.4411	59.58	10.00	9.26	21.16
C30H60N409	620.83	620.4360	58.04	9.74	9.02	23.19
C30H60N601	520.85	520.4828	69.18	11.61	16.14	3.07
C30H60N602	536.85	536.4777	67.12	11.27	15.65	5.96
C30H60N603	552.85	552.4727	65.18	10.94	15.20	8.68
C30H60N604	568.85	568.4676	63.34	10.63	14.77	11.25
C30H60N605	584.85	584.4625	61.61	10.34	14.37	13.68
C30H60N606	600.85	600.4574	59.97	10.07	13.99	15.98
C30H60N607	616.85	616.4523	58.42	9.80	13.62	18.16
C30H60N608	632.85	632.4472	56.94	9.56	13.28	20.23
C30H60N609	648.85	648.4421	55.53	9.32	12.95	22.19
C30H61N101	451.83	451.4753	79.75	13.61	3.10	3.54
C30H61N102	467.83	467.4702	77.02	13.14	2.99	6.84
C30H61N103	483.83	483.4651	74.48	12.71	2.89	9.92
C30H61N104	499.82	499.4600	72.09	12.30	2.80	12.80
C30H61N105	515.82	515.4549	69.86	11.92	2.72	15.51
C30H61N106	531.82	531.4499	67.75	11.56	2.63	18.05
C30H61N107	547.82	547.4448	65.78	11.22	2.56	20.44
C30H61N108	563.82	563.4397	63.91	10.91	2.48	22.70
C30H61N109	579.82	579.4346	62.15	10.60	2.42	24.83
C30H61N301	479.84	479.4814	75.09	12.81	8.76	3.33
C30H61N302	495.84	495.4764	72.67	12.40	8.47	6.45
C30H61N303	511.84	511.4713	70.40	12.01	8.21	9.38
C30H61N304	527.84	527.4662	68.27	11.65	7.96	12.12
C30H61N305	543.84	543.4611	66.26	11.31	7.73	14.71
C30H61N306	559.84	559.4560	64.36	10.98	7.51	17.15
C30H61N307	575.84	575.4509	62.58	10.68	7.30	19.45
C30H61N308	591.84	591.4458	60.88	10.39	7.10	21.63
C30H61N309	607.84	607.4407	59.28	10.12	6.91	23.69
C30H61N501	507.85	507.4876	70.95	12.11	13.79	3.15
C30H61N502	523.85	523.4825	68.79	11.74	13.37	6.11
C30H61N503	539.85	539.4774	66.75	11.39	12.97	8.89
C30H61N504	555.85	555.4723	64.83	11.06	12.60	11.51
C30H61N505	571.85	571.4672	63.01	10.75	12.25	13.99
C30H61N506	587.85	587.4622	61.30	10.46	11.91	16.33
C30H61N507	603.85	603.4571	59.67	10.18	11.60	18.55
C30H61N508	619.85	619.4520	58.13	9.92	11.30	20.65
C30H61N509	635.85	635.4469	56.67	9.67	11.01	22.65
C30H62N201	466.84	466.4862	77.19	13.39	6.00	3.43
C30H62N202	482.84	482.4811	74.63	12.94	5.80	6.63
C30H62N203	498.84	498.4760	72.23	12.53	5.62	9.62
C30H62N204	514.84	514.4709	69.99	12.14	5.44	12.43
C30H62N205	530.84	530.4658	67.88	11.77	5.28	15.07
C30H62N206	546.84	546.4608	65.89	11.43	5.12	17.55
C30H62N207	562.84	562.4557	64.02	11.10	4.98	19.90
C30H62N208	578.84	578.4506	62.25	10.80	4.84	22.11
C30H62N209	594.84	594.4455	60.58	10.51	4.71	24.21
C30H62N401	494.85	494.4923	72.82	12.63	11.32	3.23

C30H62N402	510.85	510.4872	70.54	12.23	10.97	6.26
C30H62N403	526.85	526.4822	68.39	11.86	10.63	9.11
C30H62N404	542.85	542.4771	66.38	11.51	10.32	11.79
C30H62N405	558.85	558.4720	64.48	11.18	10.03	14.31
C30H62N406	574.85	574.4669	62.68	10.87	9.75	16.70
C30H62N407	590.85	590.4618	60.99	10.58	9.48	18.95
C30H62N408	606.85	606.4567	59.38	10.30	9.23	21.09
C30H62N409	622.85	622.4516	57.85	10.03	9.00	23.12
C30H62N601	522.87	522.4985	68.91	11.95	16.07	3.06
C30H62N602	538.87	538.4934	66.87	11.60	15.60	5.94
C30H62N603	554.87	554.4883	64.94	11.26	15.15	8.65
C30H62N604	570.87	570.4832	63.12	10.95	14.72	11.21
C30H62N605	586.87	586.4781	61.40	10.65	14.32	13.63
C30H62N606	602.87	602.4731	59.77	10.37	13.94	15.92
C30H62N607	618.86	618.4680	58.23	10.10	13.58	18.10
C30H62N608	634.86	634.4629	56.76	9.84	13.24	20.16
C30H62N609	650.86	650.4578	55.36	9.60	12.91	22.12
C30H63N101	453.84	453.4909	79.40	13.99	3.09	3.53
C30H63N102	469.84	469.4859	76.69	13.52	2.98	6.81
C30H63N103	485.84	485.4808	74.17	13.07	2.88	9.88
C30H63N104	501.84	501.4757	71.80	12.65	2.79	12.75
C30H63N105	517.84	517.4706	69.58	12.26	2.70	15.45
C30H63N106	533.84	533.4655	67.50	11.90	2.62	17.98
C30H63N107	549.84	549.4604	65.53	11.55	2.55	20.37
C30H63N108	565.84	565.4553	63.68	11.22	2.48	22.62
C30H63N109	581.84	581.4503	61.93	10.91	2.41	24.75
C30H63N301	481.86	481.4971	74.78	13.18	8.72	3.32
C30H63N302	497.86	497.4920	72.38	12.76	8.44	6.43
C30H63N303	513.85	513.4869	70.12	12.36	8.18	9.34
C30H63N304	529.85	529.4818	68.01	11.98	7.93	12.08
C30H63N305	545.85	545.4767	66.01	11.63	7.70	14.66
C30H63N306	561.85	561.4717	64.13	11.30	7.48	17.09
C30H63N307	577.85	577.4666	62.36	10.99	7.27	19.38
C30H63N308	593.85	593.4615	60.68	10.69	7.08	21.55
C30H63N309	609.85	609.4564	59.09	10.41	6.89	23.61
C30H63N501	509.87	509.5032	70.67	12.45	13.74	3.14
C30H63N502	525.87	525.4981	68.52	12.08	13.32	6.08
C30H63N503	541.87	541.4931	66.50	11.72	12.92	8.86
C30H63N504	557.87	557.4880	64.59	11.38	12.55	11.47
C30H63N505	573.87	573.4829	62.79	11.07	12.20	13.94
C30H63N506	589.87	589.4778	61.09	10.77	11.87	16.27
C30H63N507	605.87	605.4727	59.47	10.48	11.56	18.49
C30H63N508	621.87	621.4676	57.94	10.21	11.26	20.58
C30H63N509	637.86	637.4625	56.49	9.96	10.98	22.57
C30H64N201	468.86	468.5018	76.85	13.76	5.97	3.41
C30H64N202	484.86	484.4968	74.32	13.30	5.78	6.60
C30H64N203	500.86	500.4917	71.94	12.88	5.59	9.58
C30H64N204	516.86	516.4866	69.72	12.48	5.42	12.38
C30H64N205	532.85	532.4815	67.62	12.11	5.26	15.01
C30H64N206	548.85	548.4764	65.65	11.75	5.10	17.49
C30H64N207	564.85	564.4713	63.79	11.42	4.96	19.83
C30H64N208	580.85	580.4662	62.04	11.11	4.82	22.04
C30H64N209	596.85	596.4611	60.37	10.81	4.69	24.13
C30H64N401	496.87	496.5080	72.52	12.98	11.28	3.22
C30H64N402	512.87	512.5029	70.26	12.58	10.92	6.24
C30H64N403	528.87	528.4978	68.13	12.20	10.59	9.08
C30H64N404	544.87	544.4927	66.13	11.84	10.28	11.75
C30H64N405	560.87	560.4876	64.25	11.50	9.99	14.26
C30H64N406	576.87	576.4826	62.46	11.18	9.71	16.64
C30H64N407	592.87	592.4775	60.78	10.88	9.45	18.89
C30H64N408	608.87	608.4724	59.18	10.60	9.20	21.02
C30H64N409	624.87	624.4673	57.67	10.32	8.97	23.04
C30H64N601	524.88	524.5141	68.65	12.29	16.01	3.05
C30H64N602	540.88	540.5090	66.62	11.93	15.54	5.92
C30H64N603	556.88	556.5040	64.71	11.58	15.09	8.62
C30H64N604	572.88	572.4989	62.90	11.26	14.67	11.17
C30H64N605	588.88	588.4938	61.19	10.95	14.27	13.58
C30H64N606	604.88	604.4887	59.57	10.66	13.89	15.87
C30H64N607	620.88	620.4836	58.04	10.39	13.54	18.04
C30H64N608	636.88	636.4785	56.58	10.13	13.20	20.10
C30H64N609	652.88	652.4734	55.19	9.88	12.87	22.06
C30H65N301	483.87	483.5127	74.47	13.54	8.68	3.31
C30H65N302	499.87	499.5076	72.09	13.11	8.41	6.40
C30H65N303	515.87	515.5026	69.85	12.70	8.15	9.30
C30H65N304	531.87	531.4975	67.75	12.32	7.90	12.03
C30H65N305	547.87	547.4924	65.77	11.96	7.67	14.60
C30H65N306	563.87	563.4873	63.90	11.62	7.45	17.02
C30H65N307	579.87	579.4822	62.14	11.30	7.25	19.31
C30H65N308	595.87	595.4771	60.47	11.00	7.05	21.48
C30H65N309	611.87	611.4720	58.89	10.71	6.87	23.53
C30H65N501	511.89	511.5189	70.39	12.80	13.68	3.13
C30H65N502	527.88	527.5138	68.26	12.41	13.27	6.06
C30H65N503	543.88	543.5087	66.25	12.05	12.88	8.83
C30H65N504	559.88	559.5036	64.36	11.70	12.51	11.43
C30H65N505	575.88	575.4985	62.57	11.38	12.16	13.89
C30H65N506	591.88	591.4935	60.88	11.07	11.83	16.22
C30H65N507	607.88	607.4884	59.28	10.78	11.52	18.42
C30H65N508	623.88	623.4833	57.76	10.50	11.23	20.52
C30H65N509	639.88	639.4782	56.31	10.24	10.94	22.50
C30H66N401	498.89	498.5236	72.23	13.33	11.23	3.21
C30H66N402	514.89	514.5185	69.98	12.92	10.88	6.21
C30H66N403	530.89	530.5135	67.87	12.53	10.55	9.04
C30H66N404	546.88	546.5084	65.89	12.16	10.24	11.70
C30H66N405	562.88	562.5033	64.02	11.82	9.95	14.21
C30H66N406	578.88	578.4982	62.25	11.49	9.68	16.58
C30H66N407	594.88	594.4931	60.57	11.18	9.42	18.83
C30H66N408	610.88	610.4880	58.99	10.89	9.17	20.95
C30H66N409	626.88	626.4829	57.48	10.61	8.94	22.97
C30H66N601	526.90	526.5298	68.39	12.63	15.95	3.04
C30H66N602	542.90	542.5247	66.37	12.25	15.48	5.89
C30H66N603	558.90	558.5196	64.47	11.90	15.04	8.59
C30H66N604	574.90	574.5145	62.68	11.57	14.62	11.13
C30H66N605	590.90	590.5094	60.98	11.26	14.22	13.54
C30H66N606	606.90	606.5044	59.37	10.96	13.85	15.82
C30H66N607	622.90	622.4993	57.85	10.68	13.49	17.98
C30H66N608	638.90	638.4942	56.40	10.41	13.15	20.03
C30H66N609	654.90	654.4891	55.02	10.16	12.83	21.99
C30H67N501	513.90	513.5345	70.12	13.14	13.63	3.11
C30H67N502	529.90	529.5294	68.00	12.74	13.22	6.04
C30H67N503	545.90	545.5244	66.01	12.37	12.83	8.79
C30H67N504	561.90	561.5193	64.13	12.02	12.46	11.39
C30H67N505	577.90	577.5142	62.35	11.69	12.12	13.84
C30H67N506	593.90	593.5091	60.67	11.37	11.79	16.16
C30H67N507	609.90	609.5040	59.08	11.07	11.48	18.36
C30H67N508	625.90	625.4989	57.57	10.79	11.19	20.45
C30H67N509	641.90	641.4938	56.14	10.52	10.91	22.43
C30H68N601	528.92	528.5454	68.13	12.96	15.89	3.02
C30H68N602	544.92	544.5403	66.13	12.58	15.42	5.87
C30H68N603	560.91	560.5353	64.24	12.22	14.98	8.56
C30H68N604	576.91	576.5302	62.46	11.88	14.57	11.09

C30H68N605	592.91	592.5251	60.77	11.56	14.17	13.49
C30H68N606	608.91	608.5200	59.18	11.26	13.80	15.77
C30H68N607	624.91	624.5149	57.66	10.97	13.45	17.92
C30H68N608	640.91	640.5098	56.22	10.69	13.11	19.97
C30H68N609	656.91	656.5047	54.85	10.43	12.79	21.92

00 HOURS, 11 MINUTES, 38 SECONDS.
END JOB 03863. 16.10.22

TOTAL NUMBER OF PAGES 222

10. CARBON, HYDROGEN, NITROGEN AND PHOSPHORUS

10. KOHLENSTOFF, WASSERSTOFF, STICKSTOFF UND PHOSPHOR

10. CARBONE, HYDROGENE, AZOTE ET PHOSPHORE

MOLECULAR WEIGHTS AND PERCENTAGE COMPOSITIONS OF COMPOUNDS
CONTAINING CARBON, HYDROGEN, NITROGEN
AND PHOSPHORUS

FORMULA	MOL. WEIGHT	ISOMOL. WEIGHT	PERC. C	PERC. H	PERC. N	PERC. P
C 1H 0N1P1	56.99	56.9768	21.08	0	24.58	54.35
C 1H 0N2P2	101.97	101.9537	11.78	0	27.47	60.75
C 1H 0N3P1	85.01	84.9830	14.13	0	49.43	36.44
C 1H 0N4P2	129.99	129.9598	9.24	0	43.10	47.66
C 1H 0N5P1	113.02	112.9891	10.63	0	61.97	27.41
C 1H 0N6P2	158.00	157.9660	7.60	0	53.19	39.21
C 1H 1N1P2	88.97	88.9584	13.50	1.13	15.74	69.62
C 1H 1N2P1	72.01	71.9877	16.68	1.40	38.90	43.02
C 1H 1N3P2	116.99	116.9646	10.27	.86	35.92	52.95
C 1H 1N4P1	100.02	99.9939	12.01	1.01	56.02	30.97
C 1H 1N5P2	145.00	144.9707	8.28	.70	48.30	42.72
C 1H 1N6P1	128.03	128.0000	9.38	.79	65.64	24.19
C 1H 2N1P1	59.01	58.9925	20.36	3.42	23.74	52.49
C 1H 2N2P2	103.99	103.9693	11.55	1.94	26.94	59.57
C 1H 2N3P1	87.02	86.9986	13.80	2.32	48.29	35.59
C 1H 2N4P2	132.00	131.9755	9.10	1.53	42.44	46.93
C 1H 2N5P1	115.03	115.0048	10.44	1.75	60.88	26.93
C 1H 2N6P2	160.01	159.9816	7.51	1.26	52.52	38.71
C 1H 3N1P2	90.99	90.9741	13.20	3.32	15.39	68.08
C 1H 3N2P1	74.02	74.0034	16.23	4.09	37.84	41.84
C 1H 3N3P2	119.00	118.9802	10.09	2.54	35.31	52.06
C 1H 3N4P1	102.04	102.0095	11.77	2.96	54.91	30.36
C 1H 3N5P2	147.02	146.9864	8.17	2.06	47.64	42.14
C 1H 3N6P1	130.05	130.0157	9.24	2.33	64.62	23.82
C 1H 4N1P1	61.02	61.0081	19.68	6.61	22.95	50.76
C 1H 4N2P2	106.00	105.9850	11.33	3.80	26.43	58.44
C 1H 4N3P1	89.04	89.0143	13.49	4.53	47.19	34.79
C 1H 4N4P2	134.02	133.9911	8.96	3.01	41.81	46.22
C 1H 4N5P1	117.05	117.0204	10.26	3.44	59.83	26.46
C 1H 4N6P2	162.03	161.9973	7.41	2.49	51.87	38.23
C 1H 5N1P2	93.01	92.9897	12.91	5.42	15.06	66.61
C 1H 5N2P1	76.04	76.0190	15.80	6.63	36.84	40.73
C 1H 5N3P2	121.02	120.9959	9.93	4.16	34.72	51.19
C 1H 5N4P1	104.05	104.0252	11.54	4.84	53.85	29.77
C 1H 5N5P2	149.03	149.0020	8.06	3.38	46.99	41.57
C 1H 5N6P1	132.07	132.0313	9.09	3.82	63.64	23.45
C 1H 6N1P1	63.04	63.0238	19.05	9.59	22.22	49.13
C 1H 6N2P2	108.02	108.0006	11.12	5.60	25.93	57.35
C 1H 6N3P1	91.05	91.0299	13.19	6.64	46.15	34.02
C 1H 6N4P2	136.03	136.0068	8.83	4.45	41.19	45.54
C 1H 6N5P1	119.07	119.0361	10.09	5.08	58.82	26.01
C 1H 6N6P2	164.05	164.0129	7.32	3.69	51.23	37.76
C 1H 7N1P2	95.02	95.0054	12.64	7.43	14.74	65.19
C 1H 7N2P1	78.05	78.0347	15.39	9.04	35.89	39.68
C 1H 7N3P2	123.03	123.0115	9.76	5.73	34.15	50.35
C 1H 7N4P1	106.07	106.0408	11.32	6.65	52.82	29.20
C 1H 7N5P2	151.05	151.0177	7.95	4.67	46.37	41.01
C 1H 7N6P1	134.08	134.0470	8.96	5.26	62.68	23.10
C 1H 8N2P2	110.04	110.0163	10.92	7.33	25.46	56.30
C 1H 8N3P1	93.07	93.0456	12.91	8.66	45.15	33.28
C 1H 8N4P2	138.05	138.0224	8.70	5.84	40.58	44.87
C 1H 8N5P1	121.08	121.0517	9.92	6.66	57.84	25.58
C 1H 8N6P2	166.06	166.0286	7.23	4.86	50.61	37.30
C 1H 9N3P2	125.05	125.0272	9.61	7.25	33.60	49.54
C 1H 9N4P1	108.08	108.0565	11.11	8.39	51.84	28.66
C 1H 9N5P2	153.05	153.0333	7.85	5.93	45.75	40.47
C 1H 9N6P1	136.10	136.0626	8.83	6.67	61.75	22.76
C 1H10N4P2	140.07	140.0381	8.58	7.20	40.00	44.23
C 1H10N5P1	123.10	123.0674	9.76	8.19	56.89	25.16
C 1H10N6P2	168.08	168.0442	7.15	6.00	50.00	36.86
C 1H11N5P2	155.08	155.0490	7.75	7.15	45.16	39.95
C 1H11N6P1	138.11	138.0783	8.70	8.03	60.85	22.43
C 1H12N6P2	170.09	170.0599	7.06	7.11	49.41	36.42
C 2H 1N1P2	100.98	100.9584	23.79	1.00	13.87	61.34
C 2H 1N2P1	84.02	83.9877	28.59	1.20	33.34	36.87
C 2H 1N3P2	129.00	128.9646	18.62	.78	32.57	48.02
C 2H 1N4P1	112.03	111.9939	21.44	.90	50.01	27.65
C 2H 1N5P2	157.01	156.9707	15.30	.64	44.60	39.45
C 2H 1N6P1	140.04	140.0000	17.15	.72	60.01	22.12
C 2H 2N1P1	71.02	70.9925	33.83	2.84	19.72	43.61
C 2H 2N2P2	116.00	115.9693	20.71	1.74	24.15	53.40
C 2H 2N3P1	99.03	98.9986	24.26	2.04	42.43	31.28
C 2H 2N4P2	144.01	143.9755	16.68	1.40	38.90	43.02
C 2H 2N5P1	127.05	127.0048	18.91	1.59	55.12	24.38
C 2H 2N6P2	172.03	171.9816	13.96	1.17	48.85	36.01
C 2H 3N1P2	103.00	102.9741	23.32	2.94	13.60	60.14
C 2H 3N2P1	86.03	86.0034	27.92	3.51	32.56	36.00
C 2H 3N3P2	131.01	130.9802	18.34	2.31	32.07	47.28
C 2H 3N4P1	114.05	114.0095	21.06	2.65	49.13	27.16
C 2H 3N5P2	159.03	158.9864	15.11	1.90	44.04	38.95
C 2H 3N6P1	142.06	142.0157	16.91	2.13	59.16	21.80
C 2H 4N1P1	73.03	73.0081	32.89	5.52	19.18	42.41
C 2H 4N2P2	118.02	117.9850	20.36	3.42	23.74	52.49
C 2H 4N3P1	101.05	101.0143	23.77	3.99	41.58	30.65
C 2H 4N4P2	146.03	145.9911	16.45	2.76	38.37	42.42
C 2H 4N5P1	129.06	129.0204	18.61	3.12	54.26	24.00
C 2H 4N6P2	174.04	173.9973	13.80	2.32	48.29	35.59
C 2H 5N1P2	105.02	104.9897	22.87	4.80	13.34	58.99
C 2H 5N2P1	88.05	88.0190	27.28	5.72	31.82	35.18
C 2H 5N3P2	133.03	132.9959	18.06	3.79	31.59	46.57
C 2H 5N4P1	116.06	116.0252	20.70	4.34	48.27	26.69
C 2H 5N5P2	161.04	161.0020	14.92	3.13	43.49	38.47
C 2H 5N6P1	144.08	144.0313	16.67	3.50	58.33	21.50
C 2H 6N1P1	75.05	75.0238	32.01	8.06	18.66	41.27
C 2H 6N2P2	120.03	120.0006	20.01	5.04	23.34	51.61
C 2H 6N3P1	103.06	103.0299	23.31	5.87	40.77	30.05
C 2H 6N4P2	148.04	148.0068	16.23	4.09	37.84	41.84
C 2H 6N5P1	131.08	131.0361	18.33	4.61	53.43	23.63
C 2H 6N6P2	176.06	176.0129	13.64	3.44	47.73	35.19
C 2H 7N1P2	107.03	107.0054	22.44	6.59	13.09	57.88

054

FORMULA	MOL. WEIGHT	ISOMOL. WEIGHT	PERC. C	PERC. H	PERC. N	PERC. P
C 2H 7N2P1	90.07	90.0347	26.67	7.83	31.10	34.39
C 2H 7N3P2	135.05	135.0115	17.79	5.22	31.12	45.87
C 2H 7N4P1	118.08	118.0408	20.34	5.98	47.45	26.23
C 2H 7N5P2	163.06	163.0177	14.73	4.33	42.95	37.99
C 2H 7N6P1	146.09	146.0470	16.44	4.83	57.53	21.20
C 2H 8N1P1	77.07	77.0394	31.17	10.46	18.17	40.19
C 2H 8N2P2	122.05	122.0163	19.68	6.61	22.95	50.76
C 2H 8N3P1	105.08	105.0456	22.86	7.67	39.99	29.48
C 2H 8N4P2	150.06	150.0224	16.01	5.37	37.34	41.28
C 2H 8N5P1	133.09	133.0517	18.05	6.06	52.62	23.27
C 2H 8N6P2	178.07	178.0286	13.49	4.53	47.19	34.79
C 2H 9N1P2	109.05	109.0210	22.03	8.32	12.84	56.81
C 2H 9N2P1	92.08	92.0503	26.09	9.85	30.42	33.64
C 2H 9N3P2	137.06	137.0272	17.53	6.62	30.66	45.20
C 2H 9N4P1	120.09	120.0565	20.00	7.55	46.65	25.79
C 2H 9N5P2	165.08	165.0333	14.55	5.50	42.43	37.53
C 2H 9N6P1	148.11	148.0626	16.22	6.13	56.74	20.91
C 2H10N2P2	124.06	124.0319	19.36	8.12	22.58	49.93
C 2H10N3P1	107.10	107.0612	22.43	9.41	39.24	28.92
C 2H10N4P2	152.08	152.0381	15.80	6.63	36.84	40.73
C 2H10N5P1	135.11	135.0674	17.78	7.46	51.83	22.92
C 2H10N6P2	180.09	180.0442	13.34	5.60	46.67	34.40
C 2H11N3P2	139.08	139.0428	17.27	7.97	30.21	44.54
C 2H11N4P1	122.11	122.0721	19.67	9.08	45.88	25.37
C 2H11N5P2	167.09	167.0490	14.38	6.64	41.91	37.07
C 2H11N6P1	150.12	150.0783	16.00	7.39	55.98	20.63
C 2H12N4P2	154.09	154.0537	15.59	7.85	36.36	40.20
C 2H12N5P1	137.13	137.0830	17.52	8.82	51.07	22.59
C 2H12N6P2	182.11	182.0599	13.19	6.64	46.15	34.02
C 2H13N5P2	169.11	169.0646	14.21	7.75	41.41	36.63
C 2H13N6P1	152.14	152.0939	15.79	8.61	55.24	20.36
C 2H14N6P2	184.12	184.0755	13.05	7.66	45.64	33.64
C 3H 1N1P2	113.00	112.9584	31.89	.89	12.40	54.82
C 3H 1N2P1	96.03	95.9877	37.52	1.05	29.17	32.25
C 3H 1N3P2	141.01	140.9646	25.55	.71	29.80	43.93
C 3H 1N4P1	124.04	123.9939	29.05	.81	45.17	24.97
C 3H 1N5P2	169.02	168.9707	21.32	.60	41.43	36.65
C 3H 1N6P1	152.06	152.0000	23.70	.66	55.27	20.37
C 3H 2N1P1	83.03	82.9925	43.40	2.43	16.87	37.30
C 3H 2N2P2	128.01	127.9693	28.15	1.57	21.88	48.39
C 3H 2N3P1	111.04	110.9986	32.45	1.82	37.84	27.89
C 3H 2N4P2	156.02	155.9755	23.09	1.29	35.91	39.70
C 3H 2N5P1	139.06	139.0048	25.91	1.45	50.36	22.27
C 3H 2N6P2	184.04	183.9816	19.58	1.10	45.66	33.66
C 3H 3N1P2	115.01	114.9741	31.33	2.63	12.18	53.86
C 3H 3N2P1	98.04	98.0034	36.75	3.08	28.57	31.59
C 3H 3N3P2	143.03	142.9802	25.19	2.11	29.38	43.31
C 3H 3N4P1	126.06	126.0095	28.58	2.40	44.45	24.57
C 3H 3N5P2	171.04	170.9864	21.07	1.77	40.95	36.22
C 3H 3N6P1	154.07	154.0157	23.39	1.96	54.55	20.10
C 3H 4N1P1	85.05	85.0081	42.37	4.74	16.47	36.42
C 3H 4N2P2	130.03	129.9850	27.71	3.10	21.54	47.64
C 3H 4N3P1	113.06	113.0143	31.87	3.57	37.17	27.40
C 3H 4N4P2	158.04	157.9911	22.80	2.55	35.45	39.20
C 3H 4N5P1	141.07	141.0204	25.54	2.86	49.64	21.96
C 3H 4N6P2	186.05	185.9973	19.37	2.17	45.17	33.30
C 3H 5N1P2	117.03	116.9897	30.79	4.31	11.97	52.93
C 3H 5N2P1	100.06	100.0190	36.01	5.04	28.00	30.96
C 3H 5N3P2	145.04	144.9959	24.84	3.47	28.97	42.71
C 3H 5N4P1	128.07	128.0252	28.13	3.94	43.75	24.18
C 3H 5N5P2	173.05	173.0020	20.82	2.91	40.47	35.80
C 3H 5N6P1	156.09	156.0313	23.09	3.23	53.84	19.84
C 3H 6N1P1	87.06	87.0238	41.39	6.95	16.09	35.58
C 3H 6N2P2	132.04	132.0006	27.29	4.58	21.22	46.91
C 3H 6N3P1	115.08	115.0299	31.31	5.26	36.52	26.92
C 3H 6N4P2	160.06	160.0068	22.51	3.78	35.00	38.70
C 3H 6N5P1	143.09	143.0361	25.18	4.23	48.94	21.65
C 3H 6N6P2	188.07	188.0129	19.16	3.22	44.69	32.94
C 3H 7N1P2	119.04	119.0054	30.27	5.93	11.77	52.04
C 3H 7N2P1	102.08	102.0347	35.30	6.91	27.44	30.34
C 3H 7N3P2	147.06	147.0115	24.50	4.80	28.57	42.12
C 3H 7N4P1	130.09	130.0408	27.70	5.42	43.07	23.81
C 3H 7N5P2	175.07	175.0177	20.58	4.03	40.00	35.38
C 3H 7N6P1	158.10	158.0470	22.79	4.46	53.16	19.59
C 3H 8N1P1	89.08	89.0394	40.45	9.05	15.72	34.77
C 3H 8N2P2	134.06	134.0163	26.88	6.02	20.90	46.21
C 3H 8N3P1	117.09	117.0456	30.77	6.89	35.89	26.45
C 3H 8N4P2	162.07	162.0224	22.23	4.98	34.57	38.22
C 3H 8N5P1	145.10	145.0517	24.83	5.56	48.26	21.35
C 3H 8N6P2	190.09	190.0286	18.96	4.24	44.21	32.59
C 3H 9N1P2	121.06	121.0210	29.77	7.49	11.57	51.17
C 3H 9N2P1	104.09	104.0503	34.62	8.72	26.91	29.76
C 3H 9N3P2	149.07	149.0272	24.17	6.09	28.19	41.56
C 3H 9N4P1	132.11	132.0565	27.28	6.87	42.41	23.45
C 3H 9N5P2	177.09	177.0333	20.35	5.12	39.55	34.98
C 3H 9N6P1	160.12	160.0626	22.50	5.67	52.49	19.34
C 3H10N1P1	91.09	91.0551	39.56	11.07	15.38	34.00
C 3H10N2P2	136.07	136.0319	26.48	7.41	20.59	45.52
C 3H10N3P1	119.11	119.0612	30.25	8.46	35.28	26.01
C 3H10N4P2	164.09	164.0381	21.96	6.14	34.14	37.75
C 3H10N5P1	147.12	147.0674	24.49	6.85	47.60	21.05
C 3H10N6P2	192.10	192.0442	18.76	5.25	43.75	32.25
C 3H11N1P2	123.08	123.0367	29.28	9.01	11.38	50.33
C 3H11N2P1	106.11	106.0660	33.96	10.45	26.40	29.19
C 3H11N3P2	151.09	151.0428	23.85	7.34	27.81	41.00
C 3H11N4P1	134.12	134.0721	26.87	8.27	41.77	23.09
C 3H11N5P2	179.10	179.0490	20.12	6.19	39.10	34.59
C 3H11N6P1	162.14	162.0783	22.22	6.84	51.83	19.10
C 3H12N2P2	138.09	138.0476	26.09	8.76	20.29	44.86
C 3H12N3P1	121.12	121.0769	29.75	9.99	34.69	25.57
C 3H12N4P2	166.10	166.0537	21.69	7.28	33.73	37.29
C 3H12N5P1	149.14	149.0830	24.16	8.11	46.96	20.77
C 3H12N6P2	194.12	194.0599	18.56	6.23	43.29	31.91
C 3H13N3P2	153.10	153.0585	23.54	8.56	27.45	40.46
C 3H13N4P1	136.14	136.0878	26.47	9.63	41.15	22.75
C 3H13N5P2	181.12	181.0646	19.89	7.23	38.67	34.20
C 3H13N6P1	164.15	164.0939	21.95	7.98	51.20	18.87
C 3H14N4P2	168.12	168.0694	21.43	8.39	33.33	36.85
C 3H14N5P1	151.15	151.0987	23.84	9.34	46.33	20.49
C 3H14N6P2	196.13	196.0755	18.37	7.19	42.85	31.58
C 3H15N5P2	183.13	183.0803	19.68	8.26	38.24	33.83
C 3H15N6P1	166.17	166.1096	21.69	9.10	50.58	18.64
C 3H16N6P2	198.15	198.0912	18.19	8.14	42.41	31.26
C 4H 2N1P1	95.04	94.9925	50.55	2.12	14.74	32.59
C 4H 2N2P2	140.02	139.9693	34.31	1.44	20.01	44.24
C 4H 2N3P1	123.05	122.9986	39.04	1.64	34.15	25.17
C 4H 2N4P2	168.03	167.9755	28.59	1.20	33.34	36.87
C 4H 2N5P1	151.07	151.0048	31.80	1.33	46.36	20.50
C 4H 2N6P2	196.05	195.9816	24.51	1.03	42.87	31.60
C 4H 3N1P2	127.02	126.9741	37.82	2.38	11.03	48.77

055

C 4H 3N2P1	110.06	110.0034	43.65	2.75	25.45	28.14
C 4H 3N3P2	155.04	154.9802	30.99	1.95	27.10	39.96
C 4H 3N4P1	138.07	138.0095	34.80	2.19	40.58	22.43
C 4H 3N5P2	183.05	182.9864	26.25	1.65	38.26	33.84
C 4H 3N6P1	166.08	166.0157	28.93	1.82	50.60	18.65
C 4H 4N1P1	97.06	97.0081	49.50	4.15	14.43	31.91
C 4H 4N2P2	142.04	141.9850	33.83	2.84	19.72	43.61
C 4H 4N3P1	125.07	125.0143	38.41	3.22	33.60	24.77
C 4H 4N4P2	170.05	169.9911	28.25	2.37	32.95	36.43
C 4H 4N5P1	153.08	153.0204	31.38	2.63	45.75	20.23
C 4H 4N6P2	198.06	197.9973	24.26	2.04	42.43	31.28
C 4H 5N1P2	129.04	128.9897	37.23	3.91	10.85	48.01
C 4H 5N2P1	112.07	112.0190	42.87	4.50	25.00	27.64
C 4H 5N3P2	157.05	156.9959	30.59	3.21	26.76	39.44
C 4H 5N4P1	140.09	140.0252	34.30	3.60	39.99	22.11
C 4H 5N5P2	185.07	185.0020	25.96	2.72	37.84	33.47
C 4H 5N6P1	168.10	168.0313	28.58	3.00	49.99	18.43
C 4H 6N1P1	99.07	99.0238	48.49	6.10	14.14	31.26
C 4H 6N2P2	144.05	144.0006	33.35	4.20	19.45	43.00
C 4H 6N3P1	127.09	127.0299	37.80	4.76	33.06	24.37
C 4H 6N4P2	172.07	172.0068	27.92	3.51	32.56	36.00
C 4H 6N5P1	155.10	155.0361	30.98	3.90	45.15	19.97
C 4H 6N6P2	200.08	200.0129	24.01	3.02	42.00	30.96
C 4H 7N1P2	131.05	131.0054	36.66	5.38	10.69	47.27
C 4H 7N2P1	114.09	114.0347	42.11	6.18	24.55	27.15
C 4H 7N3P2	159.07	159.0115	30.20	4.44	26.42	38.94
C 4H 7N4P1	142.10	142.0408	33.81	4.97	39.43	21.80
C 4H 7N5P2	187.08	187.0177	25.68	3.77	37.43	33.11
C 4H 7N6P1	170.11	170.0470	28.24	4.15	49.40	18.21
C 4H 8N1P1	101.09	101.0394	47.53	7.98	13.86	30.64
C 4H 8N2P2	146.07	146.0163	32.89	5.52	19.18	42.41
C 4H 8N3P1	129.10	129.0456	37.21	6.25	32.55	23.99
C 4H 8N4P2	174.08	174.0224	27.60	4.63	32.18	35.59
C 4H 8N5P1	157.12	157.0517	30.58	5.13	44.57	19.71
C 4H 8N6P2	202.10	202.0286	23.77	3.99	41.58	30.65
C 4H 9N1P2	133.07	133.0210	36.10	6.82	10.53	46.55
C 4H 9N2P1	116.10	116.0503	41.38	7.81	24.13	26.68
C 4H 9N3P2	161.08	161.0272	29.83	5.63	26.09	38.46
C 4H 9N4P1	144.12	144.0565	33.34	6.29	38.88	21.49
C 4H 9N5P2	189.10	189.0333	25.41	4.80	37.04	32.76
C 4H 9N6P1	172.13	172.0626	27.91	5.27	48.82	17.99
C 4H10N1P1	103.10	103.0551	46.60	9.78	13.58	30.04
C 4H10N2P2	148.09	148.0319	32.44	6.81	18.92	41.83
C 4H10N3P1	131.12	131.0612	36.64	7.69	32.05	23.62
C 4H10N4P2	176.10	176.0381	27.28	5.72	31.82	35.18
C 4H10N5P1	159.13	159.0674	30.19	6.33	44.01	19.46
C 4H10N6P2	204.11	204.0442	23.54	4.94	41.17	30.35
C 4H11N1P2	135.09	135.0367	35.57	8.21	10.37	45.86
C 4H11N2P1	118.12	118.0660	40.67	9.39	23.72	26.22
C 4H11N3P2	163.10	163.0428	29.46	6.80	25.76	37.98
C 4H11N4P1	146.13	146.0721	32.88	7.59	38.34	21.20
C 4H11N5P2	191.11	191.0490	25.14	5.80	36.65	32.41
C 4H11N6P1	174.15	174.0783	27.59	6.37	48.26	17.79
C 4H12N1P1	105.12	105.0707	45.70	11.51	13.32	29.46
C 4H12N2P2	150.10	150.0476	32.01	8.06	18.66	41.27
C 4H12N3P1	133.13	133.0769	36.09	9.09	31.56	23.27
C 4H12N4P2	178.11	178.0537	26.97	6.79	31.46	34.78
C 4H12N5P1	161.15	161.0830	29.81	7.51	43.46	19.22
C 4H12N6P2	206.13	206.0599	23.31	5.87	40.77	30.05
C 4H13N1P2	137.10	137.0523	35.04	9.56	10.22	45.18
C 4H13N2P1	120.14	120.0816	39.99	10.91	23.32	25.78
C 4H13N3P2	165.12	165.0585	29.10	7.94	25.45	37.52
C 4H13N4P1	148.15	148.0878	32.43	8.84	37.82	20.91
C 4H13N5P2	193.13	193.0646	24.88	6.78	36.26	32.08
C 4H13N6P1	176.16	176.0939	27.27	7.44	47.71	17.58
C 4H14N2P2	152.12	152.0632	31.58	9.28	18.42	40.72
C 4H14N3P1	135.15	135.0925	35.55	10.44	31.09	22.92
C 4H14N4P2	180.13	180.0694	26.67	7.83	31.10	34.39
C 4H14N5P1	163.16	163.0987	29.45	8.65	42.92	18.98
C 4H14N6P2	208.14	208.0755	23.08	6.78	40.38	29.76
C 4H15N3P2	167.13	167.0741	28.75	9.05	25.14	37.07
C 4H15N4P1	150.16	150.1034	31.99	10.07	37.31	20.63
C 4H15N5P2	195.15	195.0803	24.62	7.75	35.89	31.74
C 4H15N6P1	178.18	178.1096	26.96	8.49	47.17	17.38
C 4H16N4P2	182.15	182.0850	26.38	8.85	30.76	34.01
C 4H16N5P1	165.18	165.1143	29.09	9.76	42.40	18.75
C 4H16N6P2	210.16	210.0912	22.86	7.67	39.99	29.48
C 4H17N5P2	197.16	197.0959	24.37	8.69	35.52	31.42
C 4H17N6P1	180.19	180.1252	26.66	9.51	46.64	17.19
C 4H18N6P2	212.18	212.1068	22.64	8.55	39.61	29.20
C 5H 2N1P1	107.05	106.9925	56.10	1.88	13.08	28.93
C 5H 2N2P2	152.03	151.9693	39.50	1.33	18.43	40.75
C 5H 2N3P1	135.07	134.9986	44.46	1.49	31.11	22.93
C 5H 2N4P2	180.05	179.9755	33.36	1.12	31.12	34.41
C 5H 2N5P1	163.08	163.0048	36.83	1.24	42.94	18.99
C 5H 2N6P2	208.06	207.9816	28.86	.97	40.39	29.77
C 5H 3N1P2	139.03	138.9741	43.20	2.17	10.07	44.56
C 5H 3N2P1	122.07	122.0034	49.20	2.48	22.95	25.37
C 5H 3N3P2	167.05	166.9802	35.95	1.81	25.15	37.08
C 5H 3N4P1	150.08	150.0095	40.02	2.01	37.33	20.64
C 5H 3N5P2	195.06	194.9864	30.79	1.55	35.90	31.76
C 5H 3N6P1	178.09	178.0157	33.72	1.70	47.19	17.39
C 5H 4N1P1	109.07	109.0081	55.06	3.70	12.84	28.40
C 5H 4N2P2	154.05	153.9850	38.98	2.62	18.18	40.21
C 5H 4N3P1	137.08	137.0143	43.81	2.94	30.65	22.60
C 5H 4N4P2	182.06	181.9911	32.99	2.21	30.77	34.03
C 5H 4N5P1	165.09	165.0204	36.38	2.44	42.42	18.76
C 5H 4N6P2	210.08	209.9973	28.59	1.92	40.00	29.49
C 5H 5N1P2	141.05	140.9897	42.58	3.57	9.93	43.92
C 5H 5N2P1	124.08	124.0190	48.40	4.06	22.58	24.96
C 5H 5N3P2	169.06	168.9959	35.52	2.98	24.85	36.64
C 5H 5N4P1	152.10	152.0252	39.49	3.31	36.84	20.36
C 5H 5N5P2	197.08	197.0020	30.47	2.56	35.54	31.43
C 5H 5N6P1	180.11	180.0313	33.34	2.80	46.66	17.20
C 5H 6N1P1	111.08	111.0238	54.06	5.44	12.61	27.88
C 5H 6N2P2	156.06	156.0006	38.48	3.88	17.95	39.69
C 5H 6N3P1	139.10	139.0299	43.18	4.35	30.21	22.27
C 5H 6N4P2	184.08	184.0068	32.63	3.29	30.44	33.65
C 5H 6N5P1	167.11	167.0361	35.94	3.62	41.91	18.53
C 5H 6N6P2	212.09	212.0129	28.32	2.85	39.62	29.21
C 5H 7N1P2	143.07	143.0054	41.98	4.93	9.79	43.30
C 5H 7N2P1	126.10	126.0347	47.63	5.60	22.22	24.56
C 5H 7N3P2	171.08	171.0115	35.10	4.12	24.56	36.21
C 5H 7N4P1	154.11	154.0408	38.97	4.58	36.35	20.10
C 5H 7N5P2	199.09	199.0177	30.16	3.54	35.18	31.11
C 5H 7N6P1	182.13	182.0470	32.97	3.87	46.14	17.01
C 5H 8N1P1	113.10	113.0394	53.10	7.13	12.38	27.39
C 5H 8N2P2	158.08	158.0163	37.99	5.10	17.72	39.19
C 5H 8N3P1	141.11	141.0456	42.56	5.71	29.78	21.95
C 5H 8N4P2	186.09	186.0224	32.27	4.33	30.11	33.29

056

C 5H 8N5P1	169.13	169.0517	35.51	4.77	41.41	18.31
C 5H 8N6P2	214.11	214.0286	28.05	3.77	39.25	28.93
C 5H 9N1P2	145.08	145.0210	41.39	6.25	9.65	42.70
C 5H 9N2P1	128.11	128.0503	46.88	7.08	21.87	24.18
C 5H 9N3P2	173.10	173.0272	34.70	5.24	24.28	35.79
C 5H 9N4P1	156.13	156.0565	38.47	5.81	35.89	19.84
C 5H 9N5P2	201.11	201.0333	29.86	4.51	34.82	30.80
C 5H 9N6P1	184.14	184.0626	32.61	4.93	45.64	16.82
C 5H10N1P1	115.12	115.0551	52.17	8.76	12.17	26.91
C 5H10N2P2	160.10	160.0319	37.51	6.30	17.50	38.69
C 5H10N3P1	143.13	143.0612	41.96	7.04	29.36	21.64
C 5H10N4P2	188.11	188.0381	31.93	5.36	29.78	32.93
C 5H10N5P1	171.14	171.0674	35.09	5.89	40.92	18.10
C 5H10N6P2	216.12	216.0442	27.79	4.66	38.89	28.66
C 5H11N1P2	147.10	147.0367	40.83	7.54	9.52	42.11
C 5H11N2P1	130.13	130.0660	46.15	8.52	21.53	23.80
C 5H11N3P2	175.11	175.0428	34.30	6.33	24.00	35.38
C 5H11N4P1	158.14	158.0721	37.98	7.01	35.43	19.59
C 5H11N5P2	203.12	203.0490	29.57	5.46	34.48	30.50
C 5H11N6P1	186.16	186.0783	32.26	5.96	45.14	16.64
C 5H12N1P1	117.13	117.0707	51.27	10.33	11.96	26.44
C 5H12N2P2	162.11	162.0476	37.05	7.46	17.28	38.21
C 5H12N3P1	145.15	145.0769	41.38	8.33	28.95	21.34
C 5H12N4P2	190.13	190.0537	31.59	6.36	29.47	32.58
C 5H12N5P1	173.16	173.0830	34.68	6.99	40.44	17.89
C 5H12N6P2	218.14	218.0599	27.53	5.54	38.53	28.40
C 5H13N1P2	149.11	149.0523	40.28	8.79	9.39	41.54
C 5H13N2P1	132.15	132.0816	45.45	9.92	21.20	23.44
C 5H13N3P2	177.13	177.0585	33.91	7.40	23.72	34.97
C 5H13N4P1	160.16	160.0878	37.50	8.18	34.98	19.34
C 5H13N5P2	205.14	205.0646	29.28	6.39	34.14	30.20
C 5H13N6P1	188.17	188.0939	31.92	6.96	44.66	16.46
C 5H14N1P1	119.15	119.0864	50.40	11.84	11.76	26.00
C 5H14N2P2	164.13	164.0632	36.59	8.60	17.07	37.74
C 5H14N3P1	147.16	147.0925	40.81	9.59	28.55	21.05
C 5H14N4P2	192.14	192.0694	31.26	7.34	29.16	32.24
C 5H14N5P1	175.17	175.0987	34.28	8.06	39.98	17.68
C 5H14N6P2	220.16	220.0755	27.28	6.41	38.17	28.14
C 5H15N1P2	151.13	151.0680	39.74	10.00	9.27	40.99
C 5H15N2P1	134.16	134.0973	44.76	11.27	20.88	23.09
C 5H15N3P2	179.14	179.0741	33.52	8.44	23.46	34.58
C 5H15N4P1	162.18	162.1034	37.03	9.32	34.55	19.10
C 5H15N5P2	207.16	207.0803	28.99	7.30	33.81	29.90
C 5H15N6P1	190.19	190.1096	31.58	7.95	44.19	16.29
C 5H16N2P2	166.14	166.0789	36.15	9.71	16.86	37.29
C 5H16N3P1	149.18	149.1082	40.26	10.81	28.17	20.76
C 5H16N4P2	194.16	194.0850	30.93	8.31	28.86	31.91
C 5H16N5P1	177.19	177.1143	33.89	9.10	39.52	17.48
C 5H16N6P2	222.17	222.0912	27.03	7.26	37.83	27.88
C 5H17N3P2	181.16	181.0898	33.15	9.46	23.20	34.20
C 5H17N4P1	164.19	164.1191	36.58	10.44	34.12	18.86
C 5H17N5P2	209.17	209.0959	28.71	8.19	33.48	29.62
C 5H17N6P1	192.21	192.1252	31.25	8.92	43.72	16.11
C 5H18N4P2	196.17	196.1007	30.61	9.25	28.56	31.58
C 5H18N5P1	179.21	179.1300	33.51	10.12	39.08	17.28
C 5H18N6P2	224.19	224.1068	26.79	8.09	37.49	27.63
C 5H19N5P2	211.19	211.1116	28.44	9.07	33.16	29.33
C 5H19N6P1	194.22	194.1409	30.92	9.86	43.27	15.95
C 5H20N6P2	226.20	226.1225	26.55	8.91	37.15	27.39
C 6H 3N1P2	151.05	150.9741	47.71	2.00	9.27	41.01
C 6H 3N2P1	134.08	134.0034	53.75	2.26	20.89	23.10
C 6H 3N3P2	179.06	178.9802	40.25	1.69	23.47	34.60
C 6H 3N4P1	162.09	162.0095	44.46	1.87	34.56	19.11
C 6H 3N5P2	207.07	206.9864	34.80	1.46	33.82	29.92
C 6H 3N6P1	190.10	190.0157	37.91	1.59	44.21	16.29
C 6H 4N1P1	121.08	121.0081	59.52	3.33	11.57	25.58
C 6H 4N2P2	166.06	165.9850	43.40	2.43	16.87	37.30
C 6H 4N3P1	149.09	149.0143	48.34	2.70	28.18	20.77
C 6H 4N4P2	194.07	193.9911	37.13	2.08	28.87	31.92
C 6H 4N5P1	177.11	177.0204	40.69	2.28	39.54	17.49
C 6H 4N6P2	222.09	221.9973	32.45	1.82	37.84	27.89
C 6H 5N1P2	153.06	152.9897	47.08	3.29	9.15	40.47
C 6H 5N2P1	136.09	136.0190	52.95	3.70	20.58	22.76
C 6H 5N3P2	181.07	180.9959	39.80	2.78	23.21	34.21
C 6H 5N4P1	164.11	164.0252	43.91	3.07	34.14	18.87
C 6H 5N5P2	209.09	209.0020	34.47	2.41	33.49	29.63
C 6H 5N6P1	192.12	192.0313	37.51	2.62	43.74	16.12
C 6H 6N1P1	123.10	123.0238	58.55	4.91	11.38	25.16
C 6H 6N2P2	168.08	168.0006	42.88	3.60	16.67	36.86
C 6H 6N3P1	151.11	151.0299	47.69	4.00	27.81	20.50
C 6H 6N4P2	196.09	196.0068	36.75	3.08	28.57	31.59
C 6H 6N5P1	179.12	179.0361	40.23	3.38	39.10	17.29
C 6H 6N6P2	224.10	224.0129	32.16	2.70	37.50	27.64
C 6H 7N1P2	155.08	155.0054	46.47	4.55	9.03	39.95
C 6H 7N2P1	138.11	138.0347	52.18	5.11	20.28	22.43
C 6H 7N3P2	183.09	183.0115	39.36	3.85	22.95	33.83
C 6H 7N4P1	166.12	166.0408	43.38	4.25	33.73	18.65
C 6H 7N5P2	211.10	211.0177	34.14	3.34	33.17	29.34
C 6H 7N6P1	194.14	194.0470	37.12	3.63	43.29	15.95
C 6H 8N1P1	125.11	125.0394	57.60	6.45	11.20	24.76
C 6H 8N2P2	170.09	170.0163	42.37	4.74	16.47	36.42
C 6H 8N3P1	153.12	153.0456	47.06	5.27	27.44	20.23
C 6H 8N4P2	198.11	198.0224	36.38	4.07	28.28	31.27
C 6H 8N5P1	181.14	181.0517	39.79	4.45	38.66	17.10
C 6H 8N6P2	226.12	226.0286	31.87	3.57	37.17	27.40
C 6H 9N1P2	157.09	157.0210	45.88	5.77	8.92	39.43
C 6H 9N2P1	140.13	140.0503	51.43	6.47	19.99	22.10
C 6H 9N3P2	185.11	185.0272	38.93	4.90	22.70	33.47
C 6H 9N4P1	168.14	168.0565	42.86	5.40	33.32	18.42
C 6H 9N5P2	213.12	213.0333	33.82	4.26	32.86	29.07
C 6H 9N6P1	196.15	196.0626	36.74	4.62	42.84	15.79
C 6H10N1P1	127.13	127.0551	56.69	7.93	11.02	24.36
C 6H10N2P2	172.11	172.0319	41.87	5.86	16.28	35.99
C 6H10N3P1	155.14	155.0612	46.45	6.50	27.09	19.96
C 6H10N4P2	200.12	200.0381	36.01	5.04	28.00	30.96
C 6H10N5P1	183.15	183.0674	39.35	5.50	38.24	16.91
C 6H10N6P2	228.13	228.0442	31.59	4.42	36.84	27.15
C 6H11N1P2	159.11	159.0367	45.29	6.97	8.80	38.93
C 6H11N2P1	142.14	142.0660	50.70	7.80	19.71	21.79
C 6H11N3P2	187.12	187.0428	38.51	5.93	22.46	33.11
C 6H11N4P1	170.16	170.0721	42.35	6.52	32.93	18.20
C 6H11N5P2	215.14	215.0490	33.50	5.15	32.55	28.79
C 6H11N6P1	198.17	198.0783	36.37	5.60	42.41	15.63
C 6H12N1P1	129.14	129.0707	55.80	9.37	10.85	23.98
C 6H12N2P2	174.12	174.0476	41.39	6.95	16.09	35.58
C 6H12N3P1	157.16	157.0769	45.86	7.70	26.74	19.71
C 6H12N4P2	202.14	202.0537	35.65	5.98	27.72	30.65
C 6H12N5P1	185.17	185.0830	38.92	6.53	37.82	16.73
C 6H12N6P2	230.15	230.0599	31.31	5.26	36.52	26.92
C 6H13N1P2	161.12	161.0523	44.73	8.13	8.69	38.45

C 6H13N2P1	144.16	144.0816	49.99	9.09	19.43	21.49
C 6H13N3P2	189.14	189.0585	38.10	6.93	22.22	32.75
C 6H13N4P1	172.17	172.0878	41.86	7.61	32.54	17.99
C 6H13N5P2	217.15	217.0646	33.19	6.03	32.25	28.53
C 6H13N6P1	200.18	200.0939	36.00	6.55	41.98	15.47
C 6H14N1P1	131.16	131.0864	54.95	10.76	10.68	23.62
C 6H14N2P2	176.14	176.0632	40.91	8.01	15.90	35.17
C 6H14N3P1	159.17	159.0925	45.28	8.87	26.40	19.46
C 6H14N4P2	204.15	204.0694	35.30	6.91	27.44	30.34
C 6H14N5P1	187.19	187.0987	38.50	7.54	37.41	16.55
C 6H14N6P2	232.17	232.0755	31.04	6.08	36.20	26.68
C 6H15N1P2	163.14	163.0680	44.17	9.27	8.59	37.97
C 6H15N2P1	146.17	146.0973	49.30	10.34	19.16	21.19
C 6H15N3P2	191.15	191.0741	37.70	7.91	21.98	32.41
C 6H15N4P1	174.19	174.1034	41.37	8.68	32.16	17.78
C 6H15N5P2	219.17	219.0803	32.88	6.90	31.95	28.26
C 6H15N6P1	202.20	202.1096	35.64	7.48	41.56	15.32
C 6H16N1P1	133.17	133.1020	54.11	12.11	10.52	23.26
C 6H16N2P2	178.16	178.0789	40.45	9.05	15.72	34.77
C 6H16N3P1	161.19	161.1082	44.71	10.01	26.07	19.22
C 6H16N4P2	206.17	206.0850	34.96	7.82	27.18	30.05
C 6H16N5P1	189.20	189.1143	38.09	8.52	37.02	16.37
C 6H16N6P2	234.18	234.0912	30.77	6.89	35.89	26.45
C 6H17N1P2	165.16	165.0836	43.64	10.38	8.48	37.51
C 6H17N2P1	148.19	148.1129	48.63	11.56	18.90	20.90
C 6H17N3P2	193.17	193.0898	37.31	8.87	21.75	32.07
C 6H17N4P1	176.20	176.1191	40.90	9.72	31.80	17.58
C 6H17N5P2	221.18	221.0959	32.58	7.75	31.66	28.01
C 6H17N6P1	204.22	204.1252	35.29	8.39	41.15	15.17
C 6H18N2P2	180.17	180.0945	40.00	10.07	15.55	34.38
C 6H18N3P1	163.20	163.1238	44.16	11.12	25.75	18.98
C 6H18N4P2	208.18	208.1007	34.62	8.72	26.91	29.76
C 6H18N5P1	191.22	191.1300	37.69	9.49	36.63	16.20
C 6H18N6P2	236.20	236.1068	30.51	7.68	35.58	26.23
C 6H19N3P2	195.19	195.1054	36.92	9.81	21.53	31.74
C 6H19N4P1	178.22	178.1347	40.44	10.75	31.44	17.38
C 6H19N5P2	223.20	223.1116	32.29	8.58	31.38	27.75
C 6H19N6P1	206.23	206.1409	34.94	9.29	40.75	15.02
C 6H20N4P2	210.20	210.1163	34.28	9.59	26.65	29.47
C 6H20N5P1	193.23	193.1456	37.30	10.43	36.24	16.03
C 6H20N6P2	238.21	238.1225	30.25	8.46	35.28	26.01
C 6H21N5P2	225.22	225.1272	32.00	9.40	31.10	27.51
C 6H21N6P1	208.25	208.1565	34.61	10.16	40.36	14.87
C 6H22N6P2	240.23	240.1381	30.00	9.23	34.98	25.79
C 7H 3N1P2	163.06	162.9741	51.56	1.85	8.59	37.99
C 7H 3N2P1	146.09	146.0034	57.55	2.07	19.18	21.20
C 7H 3N3P2	191.07	190.9802	44.00	1.58	21.99	32.42
C 7H 3N4P1	174.10	174.0095	48.29	1.74	32.18	17.79
C 7H 3N5P2	219.08	218.9864	38.38	1.38	31.97	28.28
C 7H 3N6P1	202.12	202.0157	41.60	1.50	41.58	15.32
C 7H 4N1P1	133.09	133.0081	63.17	3.03	10.52	23.27
C 7H 4N2P2	178.07	177.9850	47.22	2.26	15.73	34.79
C 7H 4N3P1	161.10	161.0143	52.19	2.50	26.08	19.23
C 7H 4N4P2	206.08	205.9911	40.80	1.96	27.19	30.06
C 7H 4N5P1	189.12	189.0204	44.46	2.13	37.03	16.38
C 7H 4N6P2	234.10	233.9973	35.92	1.72	35.90	26.46
C 7H 5N1P2	165.07	164.9897	50.93	3.05	8.49	37.53
C 7H 5N2P1	148.11	148.0190	56.77	3.40	18.91	20.91
C 7H 5N3P2	193.09	192.9959	43.54	2.61	21.76	32.08
C 7H 5N4P1	176.12	176.0252	47.74	2.86	31.81	17.59
C 7H 5N5P2	221.10	221.0020	38.03	2.28	31.68	28.02
C 7H 5N6P1	204.13	204.0313	41.19	2.47	41.17	15.17
C 7H 6N1P1	135.11	135.0238	62.23	4.48	10.37	22.93
C 7H 6N2P2	180.09	180.0006	46.69	3.36	15.56	34.40
C 7H 6N3P1	163.12	163.0299	51.54	3.71	25.76	18.99
C 7H 6N4P2	208.10	208.0068	40.40	2.91	26.92	29.77
C 7H 6N5P1	191.13	191.0361	43.99	3.16	36.64	16.21
C 7H 6N6P2	236.11	236.0129	35.61	2.56	35.59	26.24
C 7H 7N1P2	167.09	167.0054	50.32	4.22	8.38	37.07
C 7H 7N2P1	150.12	150.0347	56.01	4.70	18.66	20.63
C 7H 7N3P2	195.10	195.0115	43.09	3.62	21.54	31.75
C 7H 7N4P1	178.13	178.0408	47.20	3.96	31.45	17.39
C 7H 7N5P2	223.11	223.0177	37.68	3.16	31.39	27.76
C 7H 7N6P1	206.15	206.0470	40.79	3.42	40.77	15.03
C 7H 8N1P1	137.12	137.0394	61.32	5.88	10.21	22.59
C 7H 8N2P2	182.10	182.0163	46.17	4.43	15.38	34.02
C 7H 8N3P1	165.14	165.0456	50.91	4.88	25.45	18.76
C 7H 8N4P2	210.12	210.0224	40.02	3.84	26.66	29.48
C 7H 8N5P1	193.15	193.0517	43.53	4.17	36.26	16.04
C 7H 8N6P2	238.13	238.0286	35.31	3.39	35.29	26.01
C 7H 9N1P2	169.10	169.0210	49.72	5.36	8.28	36.63
C 7H 9N2P1	152.14	152.0503	55.26	5.96	18.41	20.36
C 7H 9N3P2	197.12	197.0272	42.65	4.60	21.32	31.43
C 7H 9N4P1	180.15	180.0565	46.67	5.04	31.10	17.19
C 7H 9N5P2	225.13	225.0333	37.35	4.03	31.11	27.52
C 7H 9N6P1	208.16	208.0626	40.39	4.36	40.37	14.88
C 7H10N1P1	139.14	139.0551	60.43	7.24	10.07	22.26
C 7H10N2P2	184.12	184.0319	45.67	5.47	15.21	33.65
C 7H10N3P1	167.15	167.0612	50.30	6.03	25.14	18.53
C 7H10N4P2	212.13	212.0381	39.63	4.75	26.41	29.20
C 7H10N5P1	195.17	195.0674	43.08	5.16	35.88	15.87
C 7H10N6P2	240.15	240.0442	35.01	4.20	35.00	25.80
C 7H11N1P2	171.12	171.0367	49.13	6.48	8.19	36.20
C 7H11N2P1	154.15	154.0660	54.54	7.19	18.17	20.09
C 7H11N3P2	199.13	199.0428	42.22	5.57	21.10	31.11
C 7H11N4P1	182.17	182.0721	46.15	6.09	30.76	17.00
C 7H11N5P2	227.15	227.0490	37.01	4.88	30.83	27.27
C 7H11N6P1	210.18	210.0783	40.00	5.28	39.98	14.74
C 7H12N1P1	141.15	141.0707	59.56	8.57	9.92	21.94
C 7H12N2P2	186.13	186.0476	45.17	6.50	15.05	33.28
C 7H12N3P1	169.17	169.0769	49.70	7.15	24.84	18.31
C 7H12N4P2	214.15	214.0537	39.26	5.65	26.16	28.93
C 7H12N5P1	197.18	197.0830	42.64	6.13	35.52	15.71
C 7H12N6P2	242.16	242.0599	34.72	4.99	34.70	25.58
C 7H13N1P2	173.14	173.0523	48.56	7.57	8.09	35.78
C 7H13N2P1	156.17	156.0816	53.84	8.39	17.94	19.83
C 7H13N3P2	201.15	201.0585	41.80	6.51	20.89	30.80
C 7H13N4P1	184.18	184.0878	45.65	7.11	30.42	16.82
C 7H13N5P2	229.16	229.0646	36.69	5.72	30.56	27.03
C 7H13N6P1	212.20	212.0939	39.62	6.18	39.61	14.60
C 7H14N1P1	143.17	143.0864	58.73	9.86	9.78	21.63
C 7H14N2P2	188.15	188.0632	44.69	7.50	14.89	32.92
C 7H14N3P1	171.18	171.0925	49.12	8.24	24.55	18.09
C 7H14N4P2	216.16	216.0694	38.90	6.53	25.92	28.66
C 7H14N5P1	199.20	199.0987	42.21	7.08	35.16	15.55
C 7H14N6P2	244.18	244.0755	34.43	5.78	34.42	25.37
C 7H15N1P2	175.15	175.0680	48.00	8.63	8.00	35.37
C 7H15N2P1	158.18	158.0973	53.15	9.56	17.71	19.58
C 7H15N3P2	203.17	203.0741	41.38	7.44	20.68	30.49
C 7H15N4P1	186.20	186.1034	45.16	8.12	30.09	16.63

058

C 7H15N5P2	231.18	231.0803	36.37	6.54	30.29	26.80
C 7H15N6P1	214.21	214.1096	39.25	7.06	39.23	14.46
C 7H16N1P1	145.19	145.1020	57.91	11.11	9.65	21.33
C 7H16N2P2	190.17	190.0789	44.21	8.48	14.73	32.58
C 7H16N3P1	173.20	173.1082	48.54	9.31	24.26	17.88
C 7H16N4P2	218.18	218.0850	38.54	7.39	25.68	28.39
C 7H16N5P1	201.21	201.1143	41.79	8.02	34.81	15.39
C 7H16N6P2	246.19	246.0912	34.15	6.55	34.14	25.16
C 7H17N1P2	177.17	177.0836	47.46	9.67	7.91	34.97
C 7H17N2P1	160.20	160.1129	52.48	10.70	17.49	19.33
C 7H17N3P2	205.18	205.0898	40.98	8.35	20.48	30.19
C 7H17N4P1	188.21	188.1191	44.67	9.10	29.77	16.46
C 7H17N5P2	233.19	233.0959	36.05	7.35	30.03	26.56
C 7H17N6P1	216.23	216.1252	38.88	7.92	38.87	14.32
C 7H18N1P1	147.20	147.1177	57.12	12.33	9.52	21.04
C 7H18N2P2	192.18	192.0945	43.75	9.44	14.58	32.23
C 7H18N3P1	175.22	175.1238	47.99	10.35	23.98	17.68
C 7H18N4P2	220.20	220.1007	38.18	8.24	25.44	28.13
C 7H18N5P1	203.23	203.1300	41.37	8.93	34.46	15.24
C 7H18N6P2	248.21	248.1068	33.87	7.31	33.86	24.96
C 7H19N1P2	179.18	179.0993	46.92	10.69	7.82	34.57
C 7H19N2P1	162.22	162.1286	51.83	11.81	17.27	19.09
C 7H19N3P2	207.20	207.1054	40.58	9.24	20.28	29.90
C 7H19N4P1	190.23	190.1347	44.20	10.07	29.45	16.28
C 7H19N5P2	235.21	235.1116	35.75	8.14	29.77	26.34
C 7H19N6P1	218.24	218.1409	38.52	8.78	38.51	14.19
C 7H20N2P2	194.20	194.1102	43.29	10.38	14.43	31.90
C 7H20N3P1	177.23	177.1395	47.44	11.37	23.71	17.48
C 7H20N4P2	222.21	222.1163	37.84	9.07	25.21	27.88
C 7H20N5P1	205.24	205.1456	40.96	9.82	34.12	15.09
C 7H20N6P2	250.23	250.1225	33.60	8.06	33.59	24.76
C 7H21N3P2	209.21	209.1211	40.19	10.12	20.08	29.61
C 7H21N4P1	192.25	192.1504	43.73	11.01	29.14	16.11
C 7H21N5P2	237.23	237.1272	35.44	8.92	29.52	26.11
C 7H21N6P1	220.26	220.1565	38.17	9.61	38.16	14.06
C 7H22N4P2	224.23	224.1320	37.50	9.89	24.99	27.63
C 7H22N5P1	207.26	207.1613	40.57	10.70	33.79	14.94
C 7H22N6P2	252.24	252.1381	33.33	8.79	33.32	24.56
C 7H23N5P2	239.24	239.1429	35.14	9.69	29.27	25.89
C 7H23N6P1	222.28	222.1722	37.83	10.43	37.81	13.93
C 7H24N6P2	254.26	254.1538	33.07	9.51	33.05	24.36
C 8H 4N1P1	145.10	145.0081	66.22	2.78	9.65	21.35
C 8H 4N2P2	190.08	189.9850	50.55	2.12	14.74	32.59
C 8H 4N3P1	173.11	173.0143	55.51	2.33	24.27	17.89
C 8H 4N4P2	218.10	217.9911	44.06	1.85	25.69	28.40
C 8H 4N5P1	201.13	201.0204	47.78	2.00	34.82	15.40
C 8H 4N6P2	246.11	245.9973	39.04	1.64	34.15	25.17
C 8H 5N1P2	177.08	176.9897	54.26	2.85	7.91	34.98
C 8H 5N2P1	160.12	160.0190	60.01	3.15	17.50	19.34
C 8H 5N3P2	205.10	204.9959	46.85	2.46	20.49	30.20
C 8H 5N4P1	188.13	188.0252	51.08	2.68	29.78	16.46
C 8H 5N5P2	233.11	233.0020	41.22	2.16	30.04	26.57
C 8H 5N6P1	216.14	216.0313	44.46	2.33	38.88	14.33
C 8H 6N1P1	147.12	147.0238	65.31	4.11	9.52	21.05
C 8H 6N2P2	192.10	192.0006	50.02	3.15	14.58	32.25
C 8H 6N3P1	175.13	175.0299	54.87	3.45	23.99	17.69
C 8H 6N4P2	220.11	220.0068	43.65	2.75	25.45	28.14
C 8H 6N5P1	203.14	203.0361	47.30	2.98	34.47	15.25
C 8H 6N6P2	248.12	248.0129	38.73	2.44	33.87	24.97
C 8H 7N1P2	179.10	179.0054	53.65	3.94	7.82	34.59
C 8H 7N2P1	162.13	162.0347	59.27	4.35	17.28	19.10
C 8H 7N3P2	207.11	207.0115	46.39	3.41	20.29	29.91
C 8H 7N4P1	190.15	190.0408	50.53	3.71	29.47	16.29
C 8H 7N5P2	235.13	235.0177	40.87	3.00	29.79	26.35
C 8H 7N6P1	218.16	218.0470	44.05	3.23	38.52	14.20
C 8H 8N1P1	149.13	149.0394	64.43	5.41	9.39	20.77
C 8H 8N2P2	194.11	194.0163	49.50	4.15	14.43	31.91
C 8H 8N3P1	177.15	177.0456	54.24	4.55	23.72	17.48
C 8H 8N4P2	222.13	222.0224	43.26	3.63	25.22	27.89
C 8H 8N5P1	205.16	205.0517	46.84	3.93	34.14	15.10
C 8H 8N6P2	250.14	250.0286	38.41	3.22	33.60	24.77
C 8H 9N1P2	181.12	181.0210	53.05	5.01	7.73	34.20
C 8H 9N2P1	164.15	164.0503	58.54	5.53	17.07	18.87
C 8H 9N3P2	209.13	209.0272	45.95	4.34	20.09	29.62
C 8H 9N4P1	192.16	192.0565	50.00	4.72	29.16	16.12
C 8H 9N5P2	237.14	237.0333	40.52	3.83	29.53	26.12
C 8H 9N6P1	220.17	220.0626	43.64	4.12	38.17	14.07
C 8H10N1P1	151.15	151.0551	63.57	6.67	9.27	20.49
C 8H10N2P2	196.13	196.0319	48.99	5.14	14.28	31.58
C 8H10N3P1	179.16	179.0612	53.63	5.63	23.45	17.29
C 8H10N4P2	224.14	224.0381	42.87	4.50	25.00	27.64
C 8H10N5P1	207.18	207.0674	46.38	4.87	33.80	14.95
C 8H10N6P2	252.16	252.0442	38.11	4.00	33.33	24.57
C 8H11N1P2	183.13	183.0367	52.47	6.05	7.65	33.83
C 8H11N2P1	166.16	166.0660	57.83	6.67	16.86	18.64
C 8H11N3P2	211.14	211.0428	45.51	5.25	19.90	29.34
C 8H11N4P1	194.18	194.0721	49.49	5.71	28.85	15.95
C 8H11N5P2	239.16	239.0490	40.18	4.64	29.28	25.90
C 8H11N6P1	222.19	222.0783	43.25	4.99	37.82	13.94
C 8H12N1P1	153.17	153.0707	62.74	7.90	9.14	20.22
C 8H12N2P2	198.15	198.0476	48.49	6.10	14.14	31.26
C 8H12N3P1	181.18	181.0769	53.04	6.68	23.19	17.10
C 8H12N4P2	226.16	226.0537	42.49	5.35	24.77	27.39
C 8H12N5P1	209.19	209.0830	45.93	5.78	33.48	14.81
C 8H12N6P2	254.17	254.0599	37.80	4.76	33.06	24.37
C 8H13N1P2	185.15	185.0523	51.90	7.08	7.57	33.46
C 8H13N2P1	168.18	168.0816	57.13	7.79	16.66	18.42
C 8H13N3P2	213.16	213.0585	45.08	6.15	19.71	29.06
C 8H13N4P1	196.19	196.0878	48.98	6.68	28.56	15.79
C 8H13N5P2	241.17	241.0646	39.84	5.43	29.04	25.69
C 8H13N6P1	224.21	224.0939	42.86	5.84	37.48	13.81
C 8H14N1P1	155.18	155.0864	61.92	9.09	9.03	19.96
C 8H14N2P2	200.16	200.0632	48.01	7.05	14.00	30.95
C 8H14N3P1	183.19	183.0925	52.45	7.70	22.94	16.91
C 8H14N4P2	228.18	228.0694	42.11	6.18	24.55	27.15
C 8H14N5P1	211.21	211.0987	45.50	6.68	33.16	14.67
C 8H14N6P2	256.19	256.0755	37.51	5.51	32.80	24.18
C 8H15N1P2	187.16	187.0680	51.34	8.08	7.48	33.10
C 8H15N2P1	170.20	170.0973	56.46	8.88	16.46	18.20
C 8H15N3P2	215.18	215.0741	44.66	7.03	19.53	28.79
C 8H15N4P1	198.21	198.1034	48.48	7.63	28.27	15.63
C 8H15N5P2	243.19	243.0803	39.51	6.22	28.80	25.47
C 8H15N6P1	226.22	226.1096	42.48	6.68	37.15	13.69
C 8H16N1P1	157.20	157.1020	61.13	10.26	8.91	19.70
C 8H16N2P2	202.18	202.0789	47.53	7.98	13.86	30.64
C 8H16N3P1	185.21	185.1082	51.88	8.71	22.69	16.72
C 8H16N4P2	230.19	230.0850	41.74	7.01	24.34	26.91
C 8H16N5P1	213.22	213.1143	45.06	7.56	32.85	14.53
C 8H16N6P2	258.20	258.0912	37.21	6.25	32.55	23.99
C 8H17N1P2	189.18	189.0836	50.79	9.06	7.40	32.75

059

C 8H17N2P1	172.21	172.1129	55.80	9.95	16.27	17.99
C 8H17N3P2	217.19	217.0898	44.24	7.89	19.35	28.52
C 8H17N4P1	200.23	200.1191	47.99	8.56	27.98	15.47
C 8H17N5P2	245.21	245.0959	39.19	6.99	28.56	25.26
C 8H17N6P1	228.24	228.1252	42.10	7.51	36.82	13.57
C 8H18N1P1	159.21	159.1177	60.35	11.40	8.80	19.45
C 8H18N2P2	204.19	204.0945	47.06	8.89	13.72	30.34
C 8H18N3P1	187.23	187.1238	51.32	9.69	22.44	16.54
C 8H18N4P2	232.21	232.1007	41.38	7.81	24.13	26.68
C 8H18N5P1	215.24	215.1300	44.64	8.43	32.54	14.39
C 8H18N6P2	260.22	260.1068	36.93	6.97	32.30	23.81
C 8H19N1P2	191.19	191.0993	50.26	10.02	7.33	32.40
C 8H19N2P1	174.23	174.1286	55.15	10.99	16.08	17.78
C 8H19N3P2	219.21	219.1054	43.83	8.74	19.17	28.26
C 8H19N4P1	202.24	202.1347	47.51	9.47	27.70	15.32
C 8H19N5P2	247.22	247.1116	38.87	7.75	28.33	25.06
C 8H19N6P1	230.25	230.1409	41.73	8.32	36.50	13.45
C 8H20N1P1	161.23	161.1333	59.60	12.50	8.69	19.21
C 8H20N2P2	206.21	206.1102	46.60	9.78	13.58	30.04
C 8H20N3P1	189.24	189.1395	50.78	10.65	22.20	16.37
C 8H20N4P2	234.22	234.1163	41.02	8.61	23.92	26.45
C 8H20N5P1	217.26	217.1456	44.23	9.28	32.24	14.26
C 8H20N6P2	262.24	262.1225	36.64	7.69	32.05	23.62
C 8H21N1P2	193.21	193.1149	49.73	10.96	7.25	32.06
C 8H21N2P1	176.24	176.1442	54.52	12.01	15.89	17.57
C 8H21N3P2	221.22	221.1211	43.44	9.57	18.99	28.00
C 8H21N4P1	204.26	204.1504	47.04	10.36	27.43	15.16
C 8H21N5P2	249.24	249.1272	38.55	8.49	28.10	24.85
C 8H21N6P1	232.27	232.1565	41.37	9.11	36.18	13.34
C 8H22N2P2	208.23	208.1258	46.15	10.65	13.45	29.75
C 8H22N3P1	191.26	191.1551	50.24	11.59	21.97	16.19
C 8H22N4P2	236.24	236.1320	40.67	9.39	23.72	26.22
C 8H22N5P1	219.27	219.1613	43.82	10.11	31.94	14.13
C 8H22N6P2	264.25	264.1381	36.36	8.39	31.80	23.44
C 8H23N3P2	223.24	223.1367	43.04	10.38	18.82	27.75
C 8H23N4P1	206.27	206.1660	46.58	11.24	27.16	15.02
C 8H23N5P2	251.25	251.1429	38.24	9.23	27.87	24.66
C 8H23N6P1	234.29	234.1722	41.01	9.90	35.87	13.22
C 8H24N4P2	238.25	238.1476	40.33	10.15	23.52	26.00
C 8H24N5P1	221.29	221.1769	43.42	10.93	31.65	14.00
C 8H24N6P2	266.27	266.1538	36.09	9.09	31.56	23.27
C 8H25N5P2	253.27	253.1585	37.94	9.95	27.65	24.46
C 8H25N6P1	236.30	236.1878	40.66	10.66	35.56	13.11
C 8H26N6P2	268.28	268.1694	35.82	9.77	31.33	23.09
C 9H 4N1P1	157.11	157.0081	68.80	2.57	8.92	19.71
C 9H 4N2P2	202.09	201.9850	53.49	2.00	13.86	30.65
C 9H 4N3P1	185.13	185.0143	58.39	2.18	22.70	16.73
C 9H 4N4P2	230.11	229.9911	46.98	1.75	24.35	26.92
C 9H 4N5P1	213.14	213.0204	50.72	1.89	32.86	14.53
C 9H 4N6P2	258.12	257.9973	41.88	1.56	32.56	24.00
C 9H 5N1P2	189.09	188.9897	57.17	2.67	7.41	32.76
C 9H 5N2P1	172.13	172.0190	62.80	2.93	16.27	17.99
C 9H 5N3P2	217.11	216.9959	49.79	2.32	19.35	28.53
C 9H 5N4P1	200.14	200.0252	54.01	2.52	27.99	15.48
C 9H 5N5P2	245.12	245.0020	44.10	2.06	28.57	25.27
C 9H 5N6P1	228.15	228.0313	47.38	2.21	36.83	13.58
C 9H 6N1P1	159.13	159.0238	67.93	3.80	8.80	19.46
C 9H 6N2P2	204.11	204.0006	52.96	2.96	13.72	30.35
C 9H 6N3P1	187.14	187.0299	57.76	3.23	22.45	16.55
C 9H 6N4P2	232.12	232.0068	46.57	2.61	24.14	26.69

C 9H 6N5P1	215.16	215.0361	50.24	2.81	32.55	14.40
C 9H 6N6P2	260.14	260.0129	41.56	2.32	32.31	23.81
C 9H 7N1P2	191.11	191.0054	56.56	3.69	7.33	32.41
C 9H 7N2P1	174.14	174.0347	62.08	4.05	16.09	17.79
C 9H 7N3P2	219.12	219.0115	49.33	3.22	19.18	28.27
C 9H 7N4P1	202.16	202.0408	53.47	3.49	27.71	15.32
C 9H 7N5P2	247.14	247.0177	43.74	2.86	28.34	25.07
C 9H 7N6P1	230.17	230.0470	46.97	3.07	36.51	13.46
C 9H 8N1P1	161.14	161.0394	67.08	5.00	8.69	19.22
C 9H 8N2P2	206.13	206.0163	52.44	3.91	13.59	30.05
C 9H 8N3P1	189.16	189.0456	57.15	4.26	22.21	16.37
C 9H 8N4P2	234.14	234.0224	46.17	3.44	23.93	26.46
C 9H 8N5P1	217.17	217.0517	49.78	3.71	32.25	14.26
C 9H 8N6P2	262.15	262.0286	41.24	3.08	32.06	23.63
C 9H 9N1P2	193.13	193.0210	55.97	4.70	7.25	32.08
C 9H 9N2P1	176.16	176.0503	61.37	5.15	15.90	17.58
C 9H 9N3P2	221.14	221.0272	48.88	4.10	19.00	28.01
C 9H 9N4P1	204.17	204.0565	52.95	4.44	27.44	15.17
C 9H 9N5P2	249.15	249.0333	43.39	3.64	28.11	24.86
C 9H 9N6P1	232.19	232.0626	46.56	3.91	36.20	13.34
C 9H10N1P1	163.16	163.0551	66.25	6.18	8.58	18.98
C 9H10N2P2	208.14	208.0319	51.94	4.84	13.46	29.76
C 9H10N3P1	191.17	191.0612	56.55	5.27	21.98	16.20
C 9H10N4P2	236.15	236.0381	45.78	4.27	23.72	26.23
C 9H10N5P1	219.19	219.0674	49.32	4.60	31.95	14.13
C 9H10N6P2	264.17	264.0442	40.92	3.82	31.81	23.45
C 9H11N1P2	195.14	195.0367	55.40	5.68	7.18	31.74
C 9H11N2P1	178.18	178.0660	60.67	6.22	15.72	17.38
C 9H11N3P2	223.16	223.0428	48.44	4.97	18.83	27.76
C 9H11N4P1	206.19	206.0721	52.43	5.38	27.17	15.02
C 9H11N5P2	251.17	251.0490	43.04	4.41	27.88	24.66
C 9H11N6P1	234.20	234.0783	46.16	4.73	35.88	13.23
C 9H12N1P1	165.18	165.0707	65.45	7.32	8.48	18.75
C 9H12N2P2	210.16	210.0476	51.44	5.76	13.33	29.48
C 9H12N3P1	193.19	193.0769	55.96	6.26	21.75	16.03
C 9H12N4P2	238.17	238.0537	45.39	5.08	23.52	26.01
C 9H12N5P1	221.20	221.0830	48.87	5.47	31.66	14.00
C 9H12N6P2	266.18	266.0599	40.61	4.54	31.57	23.27
C 9H13N1P2	197.16	197.0523	54.83	6.65	7.10	31.42
C 9H13N2P1	180.19	180.0816	59.99	7.27	15.55	17.19
C 9H13N3P2	225.17	225.0585	48.01	5.82	18.66	27.51
C 9H13N4P1	208.20	208.0878	51.92	6.29	26.91	14.88
C 9H13N5P2	253.19	253.0646	42.70	5.18	27.66	24.47
C 9H13N6P1	236.22	236.0939	45.76	5.55	35.58	13.11
C 9H14N1P1	167.19	167.0864	64.66	8.44	8.38	18.53
C 9H14N2P2	212.17	212.0632	50.95	6.65	13.20	29.20
C 9H14N3P1	195.21	195.0925	55.38	7.23	21.53	15.87
C 9H14N4P2	240.19	240.0694	45.01	5.88	23.33	25.79
C 9H14N5P1	223.22	223.0987	48.43	6.32	31.37	13.88
C 9H14N6P2	268.20	268.0755	40.31	5.26	31.33	23.10
C 9H15N1P2	199.17	199.0680	54.27	7.59	7.03	31.10
C 9H15N2P1	182.21	182.0973	59.33	8.30	15.37	17.00
C 9H15N3P2	227.19	227.0741	47.58	6.66	18.50	27.27
C 9H15N4P1	210.22	210.1034	51.42	7.19	26.65	14.73
C 9H15N5P2	255.20	255.0803	42.36	5.92	27.44	24.27
C 9H15N6P1	238.23	238.1096	45.38	6.35	35.28	13.00
C 9H16N1P1	169.21	169.1020	63.89	9.53	8.28	18.31
C 9H16N2P2	214.19	214.0789	50.47	7.53	13.08	28.92
C 9H16N3P1	197.22	197.1082	54.81	8.18	21.31	15.71
C 9H16N4P2	242.20	242.0850	44.63	6.66	23.13	25.58

060

C 9H16N5P1	225.24	225.1143	47.99	7.16	31.09	13.75
C 9H16N6P2	270.22	270.0912	40.01	5.97	31.10	22.93
C 9H17N1P2	201.19	201.0836	53.73	8.52	6.96	30.79
C 9H17N2P1	184.22	184.1129	58.68	9.30	15.21	16.81
C 9H17N3P2	229.20	229.0898	47.16	7.48	18.33	27.03
C 9H17N4P1	212.24	212.1191	50.93	8.07	26.40	14.59
C 9H17N5P2	257.22	257.0959	42.03	6.66	27.23	24.08
C 9H17N6P1	240.25	240.1252	44.99	7.13	34.98	12.89
C 9H18N1P1	171.22	171.1177	63.13	10.60	8.18	18.09
C 9H18N2P2	216.20	216.0945	50.00	8.39	12.96	28.65
C 9H18N3P1	199.24	199.1238	54.26	9.11	21.09	15.55
C 9H18N4P2	244.22	244.1007	44.26	7.43	22.94	25.37
C 9H18N5P1	227.25	227.1300	47.57	7.98	30.82	13.63
C 9H18N6P2	272.23	272.1068	39.71	6.66	30.87	22.76
C 9H19N1P2	203.21	203.0993	53.20	9.42	6.89	30.49
C 9H19N2P1	186.24	186.1286	58.04	10.28	15.04	16.63
C 9H19N3P2	231.22	231.1054	46.75	8.28	18.17	26.79
C 9H19N4P1	214.25	214.1347	50.45	8.94	26.15	14.46
C 9H19N5P2	259.23	259.1116	41.70	7.39	27.02	23.90
C 9H19N6P1	242.27	242.1409	44.62	7.91	34.69	12.79
C 9H20N1P1	173.24	173.1333	62.40	11.64	8.09	17.88
C 9H20N2P2	218.22	218.1102	49.54	9.24	12.84	28.39
C 9H20N3P1	201.25	201.1395	53.71	10.02	20.88	15.39
C 9H20N4P2	246.23	246.1163	43.90	8.19	22.75	25.16
C 9H20N5P1	229.27	229.1456	47.15	8.79	30.55	13.51
C 9H20N6P2	274.25	274.1225	39.42	7.35	30.64	22.59
C 9H21N1P2	205.22	205.1149	52.67	10.31	6.83	30.19
C 9H21N2P1	188.25	188.1442	57.42	11.24	14.88	16.45
C 9H21N3P2	233.24	233.1211	46.35	9.08	18.02	26.56
C 9H21N4P1	216.27	216.1504	49.98	9.79	25.91	14.32
C 9H21N5P2	261.25	261.1272	41.38	8.10	26.81	23.71
C 9H21N6P1	244.28	244.1565	44.25	8.67	34.40	12.68
C 9H22N1P1	175.26	175.1490	61.68	12.65	7.99	17.67
C 9H22N2P2	220.24	220.1258	49.08	10.07	12.72	28.13
C 9H22N3P1	203.27	203.1551	53.18	10.91	20.67	15.24
C 9H22N4P2	248.25	248.1320	43.54	8.93	22.57	24.95
C 9H22N5P1	231.28	231.1613	46.74	9.59	30.28	13.39
C 9H22N6P2	276.26	276.1381	39.13	8.03	30.42	22.42
C 9H23N1P2	207.24	207.1306	52.16	11.19	6.76	29.89
C 9H23N2P1	190.27	190.1599	56.81	12.18	14.72	16.28
C 9H23N3P2	235.25	235.1367	45.95	9.85	17.86	26.33
C 9H23N4P1	218.28	218.1660	49.52	10.62	25.67	14.19
C 9H23N5P2	263.26	263.1429	41.06	8.81	26.60	23.53
C 9H23N6P1	246.30	246.1722	43.89	9.41	34.12	12.58
C 9H24N2P2	222.25	222.1415	48.64	10.88	12.60	27.87
C 9H24N3P1	205.29	205.1708	52.66	11.78	20.47	15.09
C 9H24N4P2	250.27	250.1476	43.19	9.67	22.39	24.75
C 9H24N5P1	233.30	233.1769	46.34	10.37	30.02	13.28
C 9H24N6P2	278.28	278.1538	38.85	8.69	30.20	22.26
C 9H25N3P2	237.27	237.1524	45.56	10.62	17.71	26.11
C 9H25N4P1	220.30	220.1817	49.07	11.44	25.43	14.06
C 9H25N5P2	265.28	265.1585	40.75	9.50	26.40	23.35
C 9H25N6P1	248.31	248.1878	43.53	10.15	33.84	12.47
C 9H26N4P2	252.28	252.1633	42.85	10.39	22.21	24.55
C 9H26N5P1	235.31	235.1926	45.94	11.14	29.76	13.16
C 9H26N6P2	280.30	280.1694	38.57	9.35	29.98	22.10
C 9H27N5P2	267.30	267.1742	40.44	10.18	26.20	23.18
C 9H27N6P1	250.33	250.2035	43.18	10.87	33.57	12.37
C 9H28N6P2	282.31	282.1851	38.29	10.00	29.77	21.94
C10H 5N1P2	201.11	200.9897	59.73	2.51	6.96	30.80

C10H 5N2P1	184.14	184.0190	65.23	2.74	15.21	16.82
C10H 5N3P2	229.12	228.9959	52.42	2.20	18.34	27.04
C10H 5N4P1	212.15	212.0252	56.62	2.38	26.41	14.60
C10H 5N5P2	257.13	257.0020	46.71	1.96	27.24	24.09
C10H 5N6P1	240.17	240.0313	50.01	2.10	34.99	12.90
C10H 6N1P1	171.14	171.0238	70.18	3.53	8.18	18.10
C10H 6N2P2	216.12	216.0006	55.58	2.80	12.96	28.66
C10H 6N3P1	199.15	199.0299	60.31	3.04	21.10	15.55
C10H 6N4P2	244.13	244.0068	49.20	2.48	22.95	25.37
C10H 6N5P1	227.17	227.0361	52.87	2.66	30.83	13.63
C10H 6N6P2	272.15	272.0129	44.13	2.22	30.88	22.76
C10H 7N1P2	203.12	203.0054	59.13	3.47	6.90	30.50
C10H 7N2P1	186.15	186.0347	64.52	3.79	15.05	16.64
C10H 7N3P2	231.13	231.0115	51.97	3.05	18.18	26.80
C10H 7N4P1	214.17	214.0408	56.08	3.29	26.16	14.46
C10H 7N5P2	259.15	259.0177	46.35	2.72	27.02	23.90
C10H 7N6P1	242.18	242.0470	49.60	2.91	34.70	12.79
C10H 8N1P1	173.16	173.0394	69.37	4.66	8.09	17.89
C10H 8N2P2	218.14	218.0163	55.06	3.70	12.84	28.40
C10H 8N3P1	201.17	201.0456	59.71	4.01	20.89	15.40
C10H 8N4P2	246.15	246.0224	48.80	3.28	22.76	25.17
C10H 8N5P1	229.18	229.0517	52.41	3.52	30.56	13.51
C10H 8N6P2	274.16	274.0286	43.81	2.94	30.65	22.60
C10H 9N1P2	205.14	205.0210	58.55	4.42	6.83	30.20
C10H 9N2P1	188.17	188.0503	63.83	4.82	14.89	16.46
C10H 9N3P2	233.15	233.0272	51.52	3.89	18.02	26.57
C10H 9N4P1	216.18	216.0565	55.56	4.20	25.92	14.33
C10H 9N5P2	261.16	261.0333	45.99	3.47	26.82	23.72
C10H 9N6P1	244.20	244.0626	49.19	3.71	34.41	12.68
C10H10N1P1	175.17	175.0551	68.57	5.75	8.00	17.68
C10H10N2P2	220.15	220.0319	54.56	4.58	12.72	28.14
C10H10N3P1	203.19	203.0612	59.11	4.96	20.68	15.24
C10H10N4P2	248.17	248.0381	48.40	4.06	22.58	24.96
C10H10N5P1	231.20	231.0674	51.95	4.36	30.29	13.40
C10H10N6P2	276.18	276.0442	43.49	3.65	30.43	22.43
C10H11N1P2	207.15	207.0367	57.98	5.35	6.76	29.90
C10H11N2P1	190.19	190.0660	63.15	5.83	14.73	16.29
C10H11N3P2	235.17	235.0428	51.08	4.71	17.87	26.34
C10H11N4P1	218.20	218.0721	55.05	5.08	25.68	14.20
C10H11N5P2	263.18	263.0490	45.64	4.21	26.61	23.54
C10H11N6P1	246.21	246.0783	48.78	4.50	34.13	12.58
C10H12N1P1	177.19	177.0707	67.79	6.83	7.91	17.48
C10H12N2P2	222.17	222.0476	54.06	5.44	12.61	27.88
C10H12N3P1	205.20	205.0769	58.53	5.89	20.48	15.09
C10H12N4P2	250.18	250.0537	48.01	4.83	22.39	24.76
C10H12N5P1	233.21	233.0830	51.50	5.19	30.03	13.28
C10H12N6P2	278.19	278.0599	43.18	4.35	30.21	22.27
C10H13N1P2	209.17	209.0523	57.42	6.26	6.70	29.62
C10H13N2P1	192.20	192.0816	62.49	6.82	14.57	16.12
C10H13N3P2	237.18	237.0585	50.64	5.52	17.72	26.12
C10H13N4P1	220.22	220.0878	54.54	5.95	25.44	14.07
C10H13N5P2	265.20	265.0646	45.29	4.94	26.41	23.36
C10H13N6P1	248.23	248.0939	48.39	5.28	33.86	12.48
C10H14N1P1	179.20	179.0864	67.03	7.87	7.82	17.28
C10H14N2P2	224.18	224.0632	53.58	6.29	12.50	27.63
C10H14N3P1	207.22	207.0925	57.96	6.81	20.28	14.95
C10H14N4P2	252.20	252.0694	47.63	5.60	22.22	24.56
C10H14N5P1	235.23	235.0987	51.06	6.00	29.77	13.17
C10H14N6P2	280.21	280.0755	42.86	5.04	29.99	22.11
C10H15N1P2	211.19	211.0680	56.87	7.16	6.63	29.33

C10H15N2P1	194.22	194.0973	61.84	7.78	14.42	15.95
C10H15N3P2	239.20	239.0741	50.21	6.32	17.57	25.90
C10H15N4P1	222.23	222.1034	54.05	6.80	25.21	13.94
C10H15N5P2	267.21	267.0803	44.95	5.66	26.21	23.18
C10H15N6P1	250.25	250.1096	48.00	6.04	33.58	12.38
C10H16N1P1	181.22	181.1020	66.28	8.90	7.73	17.09
C10H16N2P2	226.20	226.0789	53.10	7.13	12.38	27.39
C10H16N3P1	209.23	209.1082	57.41	7.71	20.08	14.80
C10H16N4P2	254.21	254.0850	47.25	6.34	22.04	24.37
C10H16N5P1	237.25	237.1143	50.63	6.80	29.52	13.06
C10H16N6P2	282.23	282.0912	42.56	5.71	29.78	21.95
C10H17N1P2	213.20	213.0836	56.34	8.04	6.57	29.06
C10H17N2P1	196.23	196.1129	61.21	8.73	14.28	15.78
C10H17N3P2	241.21	241.0898	49.79	7.10	17.42	25.68
C10H17N4P1	224.25	224.1191	53.56	7.64	24.98	13.81
C10H17N5P2	269.23	269.0959	44.61	6.36	26.01	23.01
C10H17N6P1	252.26	252.1252	47.61	6.79	33.31	12.28
C10H18N1P1	183.24	183.1177	65.55	9.90	7.64	16.90
C10H18N2P2	228.22	228.0945	52.63	7.95	12.27	27.14
C10H18N3P1	211.25	211.1238	56.86	8.59	19.89	14.66
C10H18N4P2	256.23	256.1007	46.88	7.08	21.87	24.18
C10H18N5P1	239.26	239.1300	50.20	7.58	29.27	12.95
C10H18N6P2	284.24	284.1068	42.26	6.38	29.57	21.79
C10H19N1P2	215.22	215.0993	55.81	8.90	6.51	28.78
C10H19N2P1	198.25	198.1286	60.59	9.66	14.13	15.62
C10H19N3P2	243.23	243.1054	49.38	7.87	17.28	25.47
C10H19N4P1	226.26	226.1347	53.08	8.46	24.76	13.69
C10H19N5P2	271.24	271.1116	44.28	7.06	25.82	22.84
C10H19N6P1	254.28	254.1409	47.24	7.53	33.05	12.18
C10H20N1P1	185.25	185.1333	64.84	10.88	7.56	16.72
C10H20N2P2	230.23	230.1102	52.17	8.76	12.17	26.91
C10H20N3P1	213.26	213.1395	56.32	9.45	19.70	14.52
C10H20N4P2	258.25	258.1163	46.51	7.81	21.70	23.99
C10H20N5P1	241.28	241.1456	49.78	8.36	29.03	12.84
C10H20N6P2	286.26	286.1225	41.96	7.04	29.36	21.64
C10H21N1P2	217.23	217.1149	55.29	9.74	6.45	28.52
C10H21N2P1	200.27	200.1442	59.98	10.57	13.99	15.47
C10H21N3P2	245.25	245.1211	48.98	8.63	17.13	25.26
C10H21N4P1	228.28	228.1504	52.62	9.27	24.54	13.57
C10H21N5P2	273.26	273.1272	43.96	7.75	25.63	22.67
C10H21N6P1	256.29	256.1565	46.86	8.26	32.79	12.09
C10H22N1P1	187.27	187.1490	64.14	11.84	7.48	16.54
C10H22N2P2	232.25	232.1258	51.72	9.55	12.06	26.67
C10H22N3P1	215.28	215.1551	55.79	10.30	19.52	14.39
C10H22N4P2	260.26	260.1320	46.15	8.52	21.53	23.80
C10H22N5P1	243.29	243.1613	49.37	9.11	28.79	12.73
C10H22N6P2	288.27	288.1381	41.67	7.69	29.15	21.49
C10H23N1P2	219.25	219.1306	54.78	10.57	6.39	28.25
C10H23N2P1	202.28	202.1599	59.38	11.46	13.85	15.31
C10H23N3P2	247.26	247.1367	48.58	9.38	16.99	25.05
C10H23N4P1	230.30	230.1660	52.16	10.07	24.33	13.45
C10H23N5P2	275.28	275.1429	43.63	8.42	25.44	22.50
C10H23N6P1	258.31	258.1722	46.50	8.98	32.53	11.99
C10H24N1P1	189.28	189.1646	63.46	12.78	7.40	16.36
C10H24N2P2	234.26	234.1415	51.27	10.33	11.96	26.44
C10H24N3P1	217.30	217.1708	55.28	11.13	19.34	14.25
C10H24N4P2	262.28	262.1476	45.80	9.22	21.36	23.62
C10H24N5P1	245.31	245.1769	48.96	9.86	28.55	12.63
C10H24N6P2	290.29	290.1538	41.38	8.33	28.95	21.34
C10H25N1P2	221.27	221.1462	54.28	11.39	6.33	28.00

C10H25N2P1	204.30	204.1755	58.79	12.33	13.71	15.16
C10H25N3P2	249.28	249.1524	48.18	10.11	16.86	24.85
C10H25N4P1	232.31	232.1817	51.70	10.85	24.12	13.33
C10H25N5P2	277.29	277.1585	43.32	9.09	25.26	22.34
C10H25N6P1	260.32	260.1878	46.14	9.68	32.28	11.90
C10H26N2P2	236.28	236.1571	50.83	11.09	11.86	26.22
C10H26N3P1	219.31	219.1864	54.77	11.95	19.16	14.12
C10H26N4P2	264.29	264.1633	45.45	9.92	21.20	23.44
C10H26N5P1	247.33	247.1926	48.56	10.60	28.32	12.52
C10H26N6P2	292.31	292.1694	41.09	8.97	28.75	21.19
C10H27N3P2	251.29	251.1680	47.80	10.83	16.72	24.65
C10H27N4P1	234.33	234.1973	51.26	11.61	23.91	13.22
C10H27N5P2	279.31	279.1742	43.00	9.74	25.07	22.18
C10H27N6P1	262.34	262.2035	45.78	10.37	32.03	11.81
C10H28N4P2	266.31	266.1789	45.10	10.60	21.04	23.26
C10H28N5P1	249.34	249.2082	48.17	11.32	28.09	12.42
C10H28N6P2	294.32	294.1851	40.81	9.59	28.55	21.05
C10H29N5P2	281.32	281.1898	42.70	10.39	24.89	22.02
C10H29N6P1	264.36	264.2191	45.44	11.06	31.79	11.72
C10H30N6P2	296.34	296.2007	40.53	10.20	28.36	20.90
C11H 5N1P2	213.12	212.9897	62.00	2.36	6.57	29.07
C11H 5N2P1	196.15	196.0190	67.36	2.57	14.28	15.79
C11H 5N3P2	241.13	240.9959	54.79	2.09	17.43	25.69
C11H 5N4P1	224.16	224.0252	58.94	2.25	24.99	13.82
C11H 5N5P2	269.14	269.0020	49.09	1.87	26.02	23.02
C11H 5N6P1	252.18	252.0313	52.39	2.00	33.33	12.28
C11H 6N1P1	183.15	183.0238	72.14	3.30	7.65	16.91
C11H 6N2P2	228.13	228.0006	57.92	2.65	12.28	27.15
C11H 6N3P1	211.16	211.0299	62.57	2.86	19.90	14.67
C11H 6N4P2	256.14	256.0068	51.58	2.36	21.87	24.18
C11H 6N5P1	239.18	239.0361	55.24	2.53	29.28	12.95
C11H 6N6P2	284.16	284.0129	46.50	2.13	29.58	21.80
C11H 7N1P2	215.13	215.0054	61.41	3.28	6.51	28.80
C11H 7N2P1	198.17	198.0347	66.67	3.56	14.14	15.63
C11H 7N3P2	243.15	243.0115	54.34	2.90	17.28	25.48
C11H 7N4P1	226.18	226.0408	58.42	3.12	24.77	13.69
C11H 7N5P2	271.16	271.0177	48.73	2.60	25.83	22.85
C11H 7N6P1	254.19	254.0470	51.98	2.78	33.06	12.19
C11H 8N1P1	185.17	185.0394	71.35	4.35	7.56	16.73
C11H 8N2P2	230.15	230.0163	57.41	3.50	12.17	26.92
C11H 8N3P1	213.18	213.0456	61.98	3.78	19.71	14.53
C11H 8N4P2	258.16	258.0224	51.18	3.12	21.70	24.00
C11H 8N5P1	241.19	241.0517	54.78	3.34	29.04	12.84
C11H 8N6P2	286.17	286.0286	46.17	2.82	29.37	21.65
C11H 9N1P2	217.15	217.0210	60.84	4.18	6.45	28.53
C11H 9N2P1	200.18	200.0503	66.00	4.53	13.99	15.47
C11H 9N3P2	245.16	245.0272	53.89	3.70	17.14	25.27
C11H 9N4P1	228.19	228.0565	57.90	3.98	24.55	13.57
C11H 9N5P2	273.18	273.0333	48.37	3.32	25.64	22.68
C11H 9N6P1	256.21	256.0626	51.57	3.54	32.80	12.09
C11H10N1P1	187.18	187.0551	70.58	5.38	7.48	16.55
C11H10N2P2	232.16	232.0319	56.91	4.34	12.07	26.68
C11H10N3P1	215.20	215.0612	61.40	4.68	19.53	14.39
C11H10N4P2	260.18	260.0381	50.78	3.87	21.53	23.81
C11H10N5P1	243.21	243.0674	54.32	4.14	28.80	12.74
C11H10N6P2	288.19	288.0442	45.85	3.50	29.16	21.50
C11H11N1P2	219.16	219.0367	60.28	5.06	6.39	28.27
C11H11N2P1	202.20	202.0660	65.34	5.48	13.85	15.32
C11H11N3P2	247.18	247.0428	53.45	4.49	17.00	25.06
C11H11N4P1	230.21	230.0721	57.39	4.82	24.34	13.45

062

C11H11N5P2	275.19	275.0490	48.01	4.03	25.45	22.51
C11H11N6P1	258.22	258.0783	51.17	4.29	32.55	11.99
C11H12N1P1	189.20	189.0707	69.83	6.39	7.40	16.37
C11H12N2P2	234.18	234.0476	56.42	5.17	11.96	26.45
C11H12N3P1	217.21	217.0769	60.83	5.57	19.35	14.26
C11H12N4P2	262.19	262.0537	50.39	4.61	21.37	23.63
C11H12N5P1	245.23	245.0830	53.88	4.93	28.56	12.63
C11H12N6P2	290.21	290.0599	45.53	4.17	28.96	21.35
C11H13N1P2	221.18	221.0523	59.74	5.92	6.33	28.01
C11H13N2P1	204.21	204.0816	64.70	6.42	13.72	15.17
C11H13N3P2	249.19	249.0585	53.02	5.26	16.86	24.86
C11H13N4P1	232.23	232.0878	56.89	5.64	24.13	13.34
C11H13N5P2	277.21	277.0646	47.66	4.73	25.26	22.35
C11H13N6P1	260.24	260.0939	50.77	5.04	32.29	11.90
C11H14N1P1	191.21	191.0864	69.10	7.38	7.33	16.20
C11H14N2P2	236.20	236.0632	55.94	5.97	11.86	26.23
C11H14N3P1	219.23	219.0925	60.27	6.44	19.17	14.13
C11H14N4P2	264.21	264.0694	50.01	5.34	21.21	23.45
C11H14N5P1	247.24	247.0987	53.44	5.71	28.33	12.53
C11H14N6P2	292.22	292.0755	45.21	4.83	28.76	21.20
C11H15N1P2	223.20	223.0680	59.20	6.77	6.28	27.75
C11H15N2P1	206.23	206.0973	64.07	7.33	13.58	15.02
C11H15N3P2	251.21	251.0741	52.59	6.02	16.73	24.66
C11H15N4P1	234.24	234.1034	56.40	6.45	23.92	13.22
C11H15N5P2	279.22	279.0803	47.32	5.41	25.08	22.19
C11H15N6P1	262.26	262.1096	50.38	5.77	32.05	11.81
C11H16N1P1	193.23	193.1020	68.38	8.35	7.25	16.03
C11H16N2P2	238.21	238.0789	55.46	6.77	11.76	26.01
C11H16N3P1	221.24	221.1082	59.72	7.29	18.99	14.00
C11H16N4P2	266.22	266.0850	49.63	6.06	21.04	23.27
C11H16N5P1	249.26	249.1143	53.01	6.47	28.10	12.43
C11H16N6P2	294.24	294.0912	44.90	5.48	28.56	21.05
C11H17N1P2	225.21	225.0836	58.67	7.61	6.22	27.51
C11H17N2P1	208.25	208.1129	63.45	8.23	13.45	14.87
C11H17N3P2	253.23	253.0898	52.18	6.77	16.59	24.46
C11H17N4P1	236.26	236.1191	55.92	7.25	23.71	13.11
C11H17N5P2	281.24	281.0959	46.98	6.09	24.90	22.03
C11H17N6P1	264.27	264.1252	49.95	6.48	31.80	11.72
C11H18N1P1	195.25	195.1177	67.67	9.29	7.17	15.86
C11H18N2P2	240.23	240.0945	55.00	7.55	11.66	25.79
C11H18N3P1	223.26	223.1238	59.18	8.13	18.82	13.87
C11H18N4P2	268.24	268.1007	49.26	6.76	20.89	23.09
C11H18N5P1	251.27	251.1300	52.58	7.22	27.87	12.33
C11H18N6P2	296.25	296.1068	44.60	6.12	28.37	20.91
C11H19N1P2	227.23	227.0993	58.15	8.43	6.16	27.26
C11H19N2P1	210.26	210.1286	62.84	9.11	13.32	14.73
C11H19N3P2	255.24	255.1054	51.76	7.50	16.46	24.27
C11H19N4P1	238.27	238.1347	55.45	8.04	23.51	13.00
C11H19N5P2	283.26	283.1116	46.64	6.76	24.72	21.87
C11H19N6P1	266.29	266.1409	49.62	7.19	31.56	11.63
C11H20N1P1	197.26	197.1333	66.98	10.22	7.10	15.70
C11H20N2P2	242.24	242.1102	54.54	8.32	11.56	25.57
C11H20N3P1	225.28	225.1395	58.65	8.95	18.65	13.75
C11H20N4P2	270.26	270.1163	48.89	7.46	20.73	22.92
C11H20N5P1	253.29	253.1456	52.16	7.96	27.65	12.23
C11H20N6P2	298.27	298.1225	44.30	6.76	28.18	20.77
C11H21N1P2	229.24	229.1149	57.63	9.23	6.11	27.02
C11H21N2P1	212.28	212.1442	62.24	9.97	13.20	14.59
C11H21N3P2	257.26	257.1211	51.36	8.23	16.33	24.08
C11H21N4P1	240.29	240.1504	54.98	8.81	23.32	12.89

C11H21N5P2	285.27	285.1272	46.31	7.42	24.55	21.72
C11H21N6P1	268.30	268.1565	49.24	7.89	31.32	11.54
C11H22N1P1	199.28	199.1490	66.30	11.13	7.03	15.54
C11H22N2P2	244.26	244.1258	54.09	9.08	11.47	25.36
C11H22N3P1	227.29	227.1551	58.13	9.76	18.49	13.63
C11H22N4P2	272.27	272.1320	48.53	8.14	20.58	22.75
C11H22N5P1	255.31	255.1613	51.75	8.69	27.43	12.13
C11H22N6P2	300.29	300.1381	44.00	7.38	27.99	20.63
C11H23N1P2	231.26	231.1306	57.13	10.02	6.06	26.79
C11H23N2P1	214.29	214.1599	61.66	10.82	13.07	14.45
C11H23N3P2	259.27	259.1367	50.96	8.94	16.21	23.89
C11H23N4P1	242.31	242.1660	54.53	9.57	23.12	12.78
C11H23N5P2	287.29	287.1429	45.99	8.07	24.38	21.56
C11H23N6P1	270.32	270.1722	48.88	8.58	31.09	11.46
C11H24N1P1	201.29	201.1646	65.64	12.02	6.96	15.39
C11H24N2P2	246.27	246.1415	53.65	9.82	11.37	25.15
C11H24N3P1	229.31	229.1708	57.62	10.55	18.32	13.51
C11H24N4P2	274.29	274.1476	48.17	8.82	20.43	22.58
C11H24N5P1	257.32	257.1769	51.35	9.40	27.22	12.04
C11H24N6P2	302.30	302.1538	43.71	8.00	27.80	20.49
C11H25N1P2	233.28	233.1462	56.64	10.80	6.00	26.56
C11H25N2P1	216.31	216.1755	61.08	11.65	12.95	14.32
C11H25N3P2	261.29	261.1524	50.57	9.64	16.08	23.71
C11H25N4P1	244.32	244.1817	54.08	10.31	22.93	12.68
C11H25N5P2	289.30	289.1585	45.67	8.71	24.21	21.41
C11H25N6P1	272.34	272.1878	48.51	9.25	30.86	11.37
C11H26N1P1	203.31	203.1803	64.99	12.89	6.89	15.23
C11H26N2P2	248.29	248.1571	53.21	10.56	11.28	24.95
C11H26N3P1	231.32	231.1864	57.12	11.33	18.17	13.39
C11H26N4P2	276.30	276.1633	47.82	9.48	20.28	22.42
C11H26N5P1	259.34	259.1926	50.95	10.11	27.00	11.94
C11H26N6P2	304.32	304.1694	43.42	8.61	27.62	20.36
C11H27N1P2	235.29	235.1619	56.15	11.57	5.95	26.33
C11H27N2P1	218.33	218.1912	60.52	12.47	12.83	14.19
C11H27N3P2	263.31	263.1680	50.18	10.34	15.96	23.53
C11H27N4P1	246.34	246.1973	53.63	11.05	22.74	12.57
C11H27N5P2	291.32	291.1742	45.35	9.34	24.04	21.26
C11H27N6P1	274.35	274.2035	48.16	9.92	30.63	11.29
C11H28N2P2	250.31	250.1728	52.78	11.28	11.19	24.75
C11H28N3P1	233.34	233.2021	56.62	12.10	18.01	13.27
C11H28N4P2	278.32	278.1789	47.47	10.14	20.13	22.26
C11H28N5P1	261.35	261.2082	50.55	10.80	26.80	11.85
C11H28N6P2	306.33	306.1851	43.13	9.21	27.43	20.22
C11H29N3P2	265.32	265.1837	49.80	11.02	15.84	23.35
C11H29N4P1	248.35	248.2130	53.20	11.77	22.56	12.47
C11H29N5P2	293.33	293.1898	45.04	9.97	23.87	21.12
C11H29N6P1	276.37	276.2191	47.81	10.58	30.41	11.21
C11H30N4P2	280.34	280.1946	47.13	10.79	19.99	22.10
C11H30N5P1	263.37	263.2239	50.17	11.48	26.59	11.76
C11H30N6P2	308.35	308.2007	42.85	9.81	27.25	20.09
C11H31N5P2	295.35	295.2055	44.73	10.58	23.71	20.97
C11H31N6P1	278.38	278.2348	47.46	11.22	30.19	11.13
C11H32N6P2	310.37	310.2164	42.57	10.39	27.08	19.96
C12H 6N1P1	195.16	195.0238	73.85	3.10	7.18	15.87
C12H 6N2P2	240.14	240.0006	60.02	2.52	11.67	25.80
C12H 6N3P1	223.18	223.0299	64.58	2.71	18.83	13.88
C12H 6N4P2	268.16	268.0068	53.75	2.26	20.89	23.10
C12H 6N5P1	251.19	251.0361	57.38	2.41	27.88	12.33
C12H 6N6P2	296.17	296.0129	48.67	2.04	28.38	20.92
C12H 7N1P2	227.14	227.0054	63.45	3.11	6.17	27.27

063

C12H 7N2P1	210.18	210.0347	68.58	3.36	13.33	14.74
C12H 7N3P2	255.16	255.0115	56.49	2.77	16.47	24.28
C12H 7N4P1	238.19	238.0408	60.51	2.96	23.52	13.00
C12H 7N5P2	283.17	283.0177	50.90	2.49	24.73	21.88
C12H 7N6P1	266.20	266.0470	54.14	2.65	31.57	11.64
C12H 8N1P1	197.18	197.0394	73.10	4.09	7.10	15.71
C12H 8N2P2	242.16	242.0163	59.52	3.33	11.57	25.58
C12H 8N3P1	225.19	225.0456	64.01	3.58	18.66	13.75
C12H 8N4P2	270.17	270.0224	53.35	2.98	20.74	22.93
C12H 8N5P1	253.20	253.0517	56.92	3.18	27.66	12.23
C12H 8N6P2	298.19	298.0286	48.34	2.70	28.18	20.77
C12H 9N1P2	229.16	229.0210	62.90	3.96	6.11	27.03
C12H 9N2P1	212.19	212.0503	67.93	4.28	13.20	14.60
C12H 9N3P2	257.17	257.0272	56.05	3.53	16.34	24.09
C12H 9N4P1	240.21	240.0565	60.00	3.78	23.32	12.89
C12H 9N5P2	285.19	285.0333	50.54	3.18	24.56	21.72
C12H 9N6P1	268.22	268.0626	53.74	3.38	31.33	11.55
C12H10N1P1	199.19	199.0551	72.36	5.06	7.03	15.55
C12H10N2P2	244.17	244.0319	59.03	4.13	11.47	25.37
C12H10N3P1	227.21	227.0612	63.44	4.44	18.49	13.63
C12H10N4P2	272.19	272.0381	52.95	3.70	20.58	22.76
C12H10N5P1	255.22	255.0674	56.47	3.95	27.44	12.14
C12H10N6P2	300.20	300.0442	48.01	3.36	27.99	20.64
C12H11N1P2	231.18	231.0367	62.35	4.80	6.06	26.80
C12H11N2P1	214.21	214.0660	67.29	5.18	13.08	14.46
C12H11N3P2	259.19	259.0428	55.61	4.28	16.21	23.90
C12H11N4P1	242.22	242.0721	59.50	4.58	23.13	12.79
C12H11N5P2	287.20	287.0490	50.19	3.86	24.38	21.57
C12H11N6P1	270.24	270.0783	53.34	4.10	31.10	11.46
C12H12N1P1	201.21	201.0707	71.63	6.01	6.96	15.39
C12H12N2P2	246.19	246.0476	58.55	4.91	11.38	25.16
C12H12N3P1	229.22	229.0769	62.88	5.28	18.33	13.51
C12H12N4P2	274.20	274.0537	52.56	4.41	20.43	22.59
C12H12N5P1	257.24	257.0830	56.03	4.70	27.23	12.04
C12H12N6P2	302.22	302.0599	47.69	4.00	27.81	20.50
C12H13N1P2	233.19	233.0523	61.81	5.62	6.01	26.57
C12H13N2P1	216.22	216.0816	66.66	6.06	12.96	14.32
C12H13N3P2	261.21	261.0585	55.18	5.02	16.09	23.72
C12H13N4P1	244.24	244.0878	59.01	5.37	22.94	12.68
C12H13N5P2	289.22	289.0646	49.84	4.53	24.21	21.42
C12H13N6P1	272.25	272.0939	52.94	4.81	30.87	11.38
C12H14N1P1	203.23	203.0864	70.92	6.94	6.89	15.24
C12H14N2P2	248.21	248.0632	58.07	5.69	11.29	24.96
C12H14N3P1	231.24	231.0925	62.33	6.10	18.17	13.39
C12H14N4P2	276.22	276.0694	52.18	5.11	20.28	22.43
C12H14N5P1	259.25	259.0987	55.60	5.44	27.01	11.95
C12H14N6P2	304.23	304.0755	47.38	4.64	27.62	20.36
C12H15N1P2	235.21	235.0680	61.28	6.43	5.96	26.34
C12H15N2P1	218.24	218.0973	66.04	6.93	12.84	14.19
C12H15N3P2	263.22	263.0741	54.76	5.74	15.96	23.53
C12H15N4P1	246.25	246.1034	58.53	6.14	22.75	12.58
C12H15N5P2	291.23	291.0803	49.49	5.19	24.05	21.27
C12H15N6P1	274.27	274.1096	52.55	5.51	30.64	11.29
C12H16N1P1	205.24	205.1020	70.23	7.86	6.82	15.09
C12H16N2P2	250.22	250.0789	57.60	6.45	11.20	24.76
C12H16N3P1	233.26	233.1082	61.79	6.91	18.01	13.28
C12H16N4P2	278.24	278.0850	51.80	5.80	20.14	22.26
C12H16N5P1	261.27	261.1143	55.17	6.17	26.81	11.86
C12H16N6P2	306.25	306.0912	47.06	5.27	27.44	20.23
C12H17N1P2	237.22	237.0836	60.76	7.22	5.90	26.11

C12H17N2P1	220.26	220.1129	65.44	7.78	12.72	14.06
C12H17N3P2	265.24	265.0898	54.34	6.46	15.84	23.36
C12H17N4P1	248.27	248.1191	58.06	6.90	22.57	12.48
C12H17N5P2	293.25	293.0959	49.15	5.84	23.88	21.12
C12H17N6P1	276.28	276.1252	52.17	6.20	30.42	11.21
C12H18N1P1	207.26	207.1177	69.54	8.75	6.76	14.94
C12H18N2P2	252.24	252.0945	57.14	7.19	11.11	24.56
C12H18N3P1	235.27	235.1238	61.26	7.71	17.86	13.17
C12H18N4P2	280.25	280.1007	51.43	6.47	19.99	22.10
C12H18N5P1	263.28	263.1300	54.74	6.89	26.60	11.76
C12H18N6P2	308.27	308.1068	46.76	5.89	27.26	20.10
C12H19N1P2	239.24	239.0993	60.25	8.01	5.85	25.89
C12H19N2P1	222.27	222.1286	64.85	8.62	12.60	13.94
C12H19N3P2	267.25	267.1054	53.93	7.17	15.72	23.18
C12H19N4P1	250.29	250.1347	57.59	7.65	22.39	12.38
C12H19N5P2	295.27	295.1116	48.81	6.49	23.72	20.98
C12H19N6P1	278.30	278.1409	51.79	6.88	30.20	11.13
C12H20N1P1	209.27	209.1333	68.87	9.63	6.69	14.80
C12H20N2P2	254.25	254.1102	56.69	7.93	11.02	24.36
C12H20N3P1	237.29	237.1395	60.74	8.50	17.71	13.05
C12H20N4P2	282.27	282.1163	51.06	7.14	19.85	21.95
C12H20N5P1	265.30	265.1456	54.33	7.60	26.40	11.67
C12H20N6P2	310.28	310.1225	46.45	6.50	27.09	19.96
C12H21N1P2	241.26	241.1149	59.74	8.77	5.81	25.68
C12H21N2P1	224.29	224.1442	64.26	9.44	12.49	13.81
C12H21N3P2	269.27	269.1211	53.53	7.86	15.61	23.01
C12H21N4P1	252.30	252.1504	57.13	8.39	22.21	12.28
C12H21N5P2	297.28	297.1272	48.48	7.12	23.56	20.84
C12H21N6P1	280.32	280.1565	51.42	7.55	29.98	11.05
C12H22N1P1	211.29	211.1490	68.22	10.50	6.63	14.66
C12H22N2P2	256.27	256.1258	56.24	8.65	10.93	24.17
C12H22N3P1	239.30	239.1551	60.23	9.27	17.56	12.94
C12H22N4P2	284.28	284.1320	50.70	7.80	19.71	21.79
C12H22N5P1	267.32	267.1613	53.92	8.30	26.20	11.59
C12H22N6P2	312.30	312.1381	46.15	7.10	26.91	19.84
C12H23N1P2	243.27	243.1306	59.25	9.53	5.76	25.46
C12H23N2P1	226.30	226.1599	63.69	10.24	12.38	13.69
C12H23N3P2	271.28	271.1367	53.13	8.55	15.49	22.83
C12H23N4P1	254.32	254.1660	56.67	9.12	22.03	12.18
C12H23N5P2	299.30	299.1429	48.16	7.75	23.40	20.70
C12H23N6P1	282.33	282.1722	51.05	8.21	29.77	10.97
C12H24N1P1	213.31	213.1646	67.57	11.34	6.57	14.52
C12H24N2P2	258.29	258.1415	55.80	9.37	10.85	23.98
C12H24N3P1	241.32	241.1708	59.73	10.02	17.41	12.84
C12H24N4P2	286.30	286.1476	50.34	8.45	19.57	21.64
C12H24N5P1	269.33	269.1769	53.52	8.98	26.00	11.50
C12H24N6P2	314.31	314.1538	45.86	7.70	26.74	19.71
C12H25N1P2	245.29	245.1462	58.76	10.27	5.71	25.26
C12H25N2P1	228.32	228.1755	63.13	11.04	12.27	13.57
C12H25N3P2	273.30	273.1524	52.74	9.22	15.38	22.67
C12H25N4P1	256.33	256.1817	56.23	9.83	21.86	12.08
C12H25N5P2	301.31	301.1585	47.84	8.36	23.24	20.56
C12H25N6P1	284.35	284.1878	50.69	8.86	29.56	10.89
C12H26N1P1	215.32	215.1803	66.94	12.17	6.51	14.38
C12H26N2P2	260.30	260.1571	55.37	10.07	10.76	23.80
C12H26N3P1	243.33	243.1864	59.23	10.77	17.27	12.73
C12H26N4P2	288.32	288.1633	49.99	9.09	19.43	21.49
C12H26N5P1	271.35	271.1926	53.12	9.66	25.81	11.41
C12H26N6P2	316.33	316.1694	45.56	8.28	26.57	19.58
C12H27N1P2	247.30	247.1619	58.28	11.00	5.66	25.05

064

C12H27N2P1	230.34	230.1912	62.58	11.82	12.16	13.45
C12H27N3P2	275.32	275.1680	52.35	9.89	15.26	22.50
C12H27N4P1	258.35	258.1973	55.79	10.53	21.69	11.99
C12H27N5P2	303.33	303.1742	47.52	8.97	23.09	20.42
C12H27N6P1	286.36	286.2035	50.33	9.50	29.35	10.82
C12H28N1P1	217.34	217.1959	66.32	12.99	6.44	14.25
C12H28N2P2	262.32	262.1728	54.95	10.76	10.68	23.62
C12H28N3P1	245.35	245.2021	58.75	11.50	17.13	12.62
C12H28N4P2	290.33	290.1789	49.64	9.72	19.30	21.34
C12H28N5P1	273.36	273.2082	52.73	10.32	25.62	11.33
C12H28N6P2	318.34	318.1851	45.28	8.87	26.40	19.46
C12H29N1P2	249.32	249.1775	57.81	11.72	5.62	24.85
C12H29N2P1	232.35	232.2068	62.03	12.58	12.06	13.33
C12H29N3P2	277.33	277.1837	51.97	10.54	15.15	22.34
C12H29N4P1	260.37	260.2130	55.36	11.23	21.52	11.90
C12H29N5P2	305.35	305.1898	47.20	9.57	22.94	20.29
C12H29N6P1	288.38	288.2191	49.98	10.14	29.14	10.74
C12H30N2P2	264.33	264.1884	54.53	11.44	10.60	23.44
C12H30N3P1	247.37	247.2177	58.27	12.22	16.99	12.52
C12H30N4P2	292.35	292.1946	49.30	10.34	19.16	21.19
C12H30N5P1	275.38	275.2239	52.34	10.98	25.43	11.25
C12H30N6P2	320.36	320.2007	44.99	9.44	26.23	19.34
C12H31N3P2	279.35	279.1993	51.60	11.19	15.04	22.18
C12H31N4P1	262.38	262.2286	54.93	11.91	21.35	11.80
C12H31N5P2	307.36	307.2055	46.89	10.17	22.79	20.15
C12H31N6P1	290.39	290.2348	49.63	10.76	28.94	10.67
C12H32N4P2	294.36	294.2102	48.96	10.96	19.03	21.04
C12H32N5P1	277.40	277.2395	51.96	11.63	25.25	11.17
C12H32N6P2	322.38	322.2164	44.71	10.01	26.07	19.22
C12H33N5P2	309.38	309.2211	46.59	10.75	22.64	20.02
C12H33N6P1	292.41	292.2504	49.29	11.38	28.74	10.59
C12H34N6P2	324.39	324.2320	44.43	10.56	25.91	19.10
C13H 6N1P1	207.17	207.0238	75.37	2.92	6.76	14.95
C13H 6N2P2	252.15	252.0006	61.92	2.40	11.11	24.57
C13H 6N3P1	235.19	235.0299	66.39	2.57	17.87	13.17
C13H 6N4P2	280.17	280.0068	55.73	2.16	20.00	22.11
C13H 6N5P1	263.20	263.0361	59.33	2.30	26.61	11.77
C13H 6N6P2	308.18	308.0129	50.67	1.96	27.27	20.10
C13H 7N1P2	239.16	239.0054	65.29	2.95	5.86	25.90
C13H 7N2P1	222.19	222.0347	70.28	3.18	12.61	13.94
C13H 7N3P2	267.17	267.0115	58.44	2.64	15.73	23.19
C13H 7N4P1	250.20	250.0408	62.41	2.82	22.39	12.38
C13H 7N5P2	295.18	295.0177	52.90	2.39	23.73	20.99
C13H 7N6P1	278.21	278.0470	56.12	2.54	30.21	11.13
C13H 8N1P1	209.19	209.0394	74.64	3.85	6.70	14.81
C13H 8N2P2	254.17	254.0163	61.43	3.17	11.02	24.37
C13H 8N3P1	237.20	237.0456	65.83	3.40	17.71	13.06
C13H 8N4P2	282.18	282.0224	55.33	2.86	19.85	21.95
C13H 8N5P1	265.22	265.0517	58.87	3.04	26.41	11.68
C13H 8N6P2	310.20	310.0286	50.34	2.60	27.09	19.97
C13H 9N1P2	241.17	241.0210	64.74	3.76	5.81	25.69
C13H 9N2P1	224.20	224.0503	69.64	4.05	12.49	13.82
C13H 9N3P2	269.18	269.0272	58.01	3.37	15.61	23.01
C13H 9N4P1	252.22	252.0565	61.91	3.60	22.21	12.28
C13H 9N5P2	297.20	297.0333	52.54	3.05	23.56	20.84
C13H 9N6P1	280.23	280.0626	55.72	3.24	29.99	11.05
C13H10N1P1	211.21	211.0551	73.93	4.77	6.63	14.67
C13H10N2P2	256.19	256.0319	60.95	3.93	10.93	24.18
C13H10N3P1	239.22	239.0612	65.27	4.21	17.57	12.95
C13H10N4P2	284.20	284.0381	54.94	3.55	19.71	21.80

C13H10N5P1	267.23	267.0674	58.43	3.77	26.21	11.59
C13H10N6P2	312.21	312.0442	50.01	3.23	26.92	19.84
C13H11N1P2	243.19	243.0367	64.21	4.56	5.76	25.47
C13H11N2P1	226.22	226.0660	69.02	4.90	12.38	13.69
C13H11N3P2	271.20	271.0428	57.58	4.09	15.49	22.84
C13H11N4P1	254.23	254.0721	61.42	4.36	22.04	12.18
C13H11N5P2	299.21	299.0490	52.19	3.71	23.41	20.70
C13H11N6P1	282.25	282.0783	55.32	3.93	29.78	10.97
C13H12N1P1	213.22	213.0707	73.23	5.67	6.57	14.53
C13H12N2P2	258.20	258.0476	60.47	4.68	10.85	23.99
C13H12N3P1	241.23	241.0769	64.73	5.01	17.42	12.84
C13H12N4P2	286.21	286.0537	54.56	4.23	19.58	21.64
C13H12N5P1	269.25	269.0830	57.99	4.49	26.01	11.50
C13H12N6P2	314.23	314.0599	49.69	3.85	26.74	19.71
C13H13N1P2	245.20	245.0523	63.68	5.34	5.71	25.26
C13H13N2P1	228.24	228.0816	68.41	5.74	12.27	13.57
C13H13N3P2	273.22	273.0585	57.15	4.80	15.38	22.67
C13H13N4P1	256.25	256.0878	60.93	5.11	21.86	12.09
C13H13N5P2	301.23	301.0646	51.84	4.35	23.25	20.56
C13H13N6P1	284.26	284.0939	54.93	4.61	29.56	10.90
C13H14N1P1	215.24	215.0864	72.55	6.56	6.51	14.39
C13H14N2P2	260.22	260.0632	60.01	5.42	10.77	23.81
C13H14N3P1	243.25	243.0925	64.19	5.80	17.27	12.73
C13H14N4P2	288.23	288.0694	54.17	4.90	19.44	21.49
C13H14N5P1	271.26	271.0987	57.56	5.20	25.82	11.42
C13H14N6P2	316.24	316.0755	49.37	4.46	26.57	19.59
C13H15N1P2	247.22	247.0680	63.16	6.12	5.67	25.06
C13H15N2P1	230.25	230.0973	67.81	6.57	12.17	13.45
C13H15N3P2	275.23	275.0741	56.73	5.49	15.27	22.51
C13H15N4P1	258.27	258.1034	60.46	5.85	21.69	11.99
C13H15N5P2	303.25	303.0803	51.49	4.99	23.09	20.43
C13H15N6P1	286.28	286.1096	54.54	5.28	29.36	10.82
C13H16N1P1	217.25	217.1020	71.87	7.42	6.45	14.26
C13H16N2P2	262.23	262.0789	59.54	6.15	10.68	23.62
C13H16N3P1	245.27	245.1082	63.66	6.58	17.13	12.63
C13H16N4P2	290.25	290.0850	53.80	5.56	19.30	21.34
C13H16N5P1	273.28	273.1143	57.14	5.90	25.63	11.33
C13H16N6P2	318.26	318.0912	49.06	5.07	26.41	19.46
C13H17N1P2	249.23	249.0836	62.65	6.88	5.62	24.86
C13H17N2P1	232.27	232.1129	67.23	7.38	12.06	13.34
C13H17N3P2	277.25	277.0898	56.32	6.18	15.16	22.34
C13H17N4P1	260.28	260.1191	59.99	6.58	21.53	11.90
C13H17N5P2	305.26	305.0959	51.15	5.61	22.94	20.29
C13H17N6P1	288.29	288.1252	54.16	5.94	29.15	10.74
C13H18N1P1	219.27	219.1177	71.21	8.27	6.39	14.13
C13H18N2P2	264.25	264.0945	59.09	6.87	10.60	23.44
C13H18N3P1	247.28	247.1238	63.14	7.34	16.99	12.53
C13H18N4P2	292.26	292.1007	53.43	6.21	19.17	21.20
C13H18N5P1	275.30	275.1300	56.72	6.59	25.44	11.25
C13H18N6P2	320.28	320.1068	48.75	5.66	26.24	19.34
C13H19N1P2	251.25	251.0993	62.15	7.62	5.57	24.66
C13H19N2P1	234.28	234.1286	66.65	8.17	11.96	13.22
C13H19N3P2	279.26	279.1054	55.91	6.86	15.05	22.18
C13H19N4P1	262.30	262.1347	59.53	7.30	21.36	11.81
C13H19N5P2	307.28	307.1116	50.82	6.23	22.79	20.16
C13H19N6P1	290.31	290.1409	53.79	6.60	28.95	10.67
C13H20N1P1	221.28	221.1333	70.56	9.11	6.33	14.00
C13H20N2P2	266.27	266.1102	58.64	7.57	10.52	23.27
C13H20N3P1	249.30	249.1395	62.63	8.09	16.86	12.42
C13H20N4P2	294.28	294.1163	53.06	6.85	19.04	21.05

C13H20N5P1	277.31	277.1456	56.31	7.27	25.25	11.17
C13H20N6P2	322.29	322.1225	48.45	6.26	26.08	19.22
C13H21N1P2	253.27	253.1149	61.65	8.36	5.53	24.46
C13H21N2P1	236.30	236.1442	66.08	8.96	11.86	13.11
C13H21N3P2	281.28	281.1211	55.51	7.53	14.94	22.02
C13H21N4P1	264.31	264.1504	59.08	8.01	21.20	11.72
C13H21N5P2	309.29	309.1272	50.48	6.84	22.64	20.03
C13H21N6P1	292.33	292.1565	53.41	7.24	28.75	10.60
C13H22N1P1	223.30	223.1490	69.93	9.93	6.27	13.87
C13H22N2P2	268.28	268.1258	58.20	8.27	10.44	23.09
C13H22N3P1	251.31	251.1551	62.13	8.82	16.72	12.32
C13H22N4P2	296.29	296.1320	52.70	7.48	18.91	20.91
C13H22N5P1	279.33	279.1613	55.90	7.94	25.07	11.09
C13H22N6P2	324.31	324.1381	48.15	6.84	25.91	19.10
C13H23N1P2	255.28	255.1306	61.17	9.08	5.49	24.27
C13H23N2P1	238.32	238.1599	65.52	9.73	11.75	13.00
C13H23N3P2	283.30	283.1367	55.12	8.18	14.83	21.87
C13H23N4P1	266.33	266.1660	58.63	8.70	21.04	11.63
C13H23N5P2	311.31	311.1429	50.16	7.45	22.50	19.90
C13H23N6P1	294.34	294.1722	53.05	7.88	28.55	10.52
C13H24N1P1	225.32	225.1646	69.30	10.74	6.22	13.75
C13H24N2P2	270.30	270.1415	57.77	8.95	10.36	22.92
C13H24N3P1	253.33	253.1708	61.64	9.55	16.59	12.23
C13H24N4P2	298.31	298.1476	52.34	8.11	18.78	20.77
C13H24N5P1	281.34	281.1769	55.50	8.60	24.89	11.01
C13H24N6P2	326.32	326.1538	47.85	7.41	25.75	18.98
C13H25N1P2	257.30	257.1462	60.69	9.79	5.44	24.08
C13H25N2P1	240.33	240.1755	64.97	10.49	11.66	12.89
C13H25N3P2	285.31	285.1524	54.73	8.83	14.73	21.71
C13H25N4P1	268.34	268.1817	58.19	9.39	20.88	11.54
C13H25N5P2	313.33	313.1585	49.83	8.04	22.35	19.77
C13H25N6P1	296.36	296.1878	52.69	8.50	28.36	10.45
C13H26N1P1	227.33	227.1803	68.69	11.53	6.16	13.62
C13H26N2P2	272.31	272.1571	57.34	9.62	10.29	22.75
C13H26N3P1	255.35	255.1864	61.15	10.26	16.46	12.13
C13H26N4P2	300.33	300.1633	51.99	8.73	18.66	20.63
C13H26N5P1	283.36	283.1926	55.10	9.25	24.72	10.93
C13H26N6P2	328.34	328.1694	47.56	7.98	25.60	18.87
C13H27N1P2	259.31	259.1619	60.21	10.50	5.40	23.89
C13H27N2P1	242.35	242.1912	64.43	11.23	11.56	12.78
C13H27N3P2	287.33	287.1680	54.34	9.47	14.62	21.56
C13H27N4P1	270.36	270.1973	57.75	10.07	20.72	11.46
C13H27N5P2	315.34	315.1742	49.52	8.63	22.21	19.64
C13H27N6P1	298.37	298.2035	52.33	9.12	28.17	10.38
C13H28N1P1	229.35	229.1959	68.08	12.31	6.11	13.51
C13H28N2P2	274.33	274.1728	56.92	10.29	10.21	22.58
C13H28N3P1	257.36	257.2021	60.67	10.97	16.33	12.04
C13H28N4P2	302.34	302.1789	51.65	9.33	18.53	20.49
C13H28N5P1	285.38	285.2082	54.72	9.89	24.54	10.85
C13H28N6P2	330.36	330.1851	47.27	8.54	25.44	18.75
C13H29N1P2	261.33	261.1775	59.75	11.19	5.36	23.70
C13H29N2P1	244.36	244.2068	63.90	11.96	11.46	12.68
C13H29N3P2	289.34	289.1837	53.97	10.10	14.52	21.41
C13H29N4P1	272.38	272.2130	57.33	10.73	20.57	11.37
C13H29N5P2	317.36	317.1898	49.20	9.21	22.07	19.52
C13H29N6P1	300.39	300.2191	51.98	9.73	27.98	10.31
C13H30N1P1	231.36	231.2116	67.49	13.07	6.05	13.39
C13H30N2P2	276.35	276.1884	56.50	10.94	10.14	22.42
C13H30N3P1	259.38	259.2177	60.20	11.66	16.20	11.94
C13H30N4P2	304.36	304.1946	51.30	9.94	18.41	20.35

C13H30N5P1	287.39	287.2239	54.33	10.52	24.37	10.78
C13H30N6P2	332.37	332.2007	46.98	9.10	25.28	18.64
C13H31N1P2	263.35	263.1932	59.29	11.87	5.32	23.52
C13H31N2P1	246.38	246.2225	63.38	12.68	11.37	12.57
C13H31N3P2	291.36	291.1993	53.59	10.72	14.42	21.26
C13H31N4P1	274.39	274.2286	56.91	11.39	20.42	11.29
C13H31N5P2	319.37	319.2055	48.89	9.78	21.93	19.40
C13H31N6P1	302.41	302.2348	51.63	10.33	27.79	10.24
C13H32N2P2	278.36	278.2041	56.09	11.59	10.06	22.25
C13H32N3P1	261.39	261.2334	59.74	12.34	16.08	11.85
C13H32N4P2	306.37	306.2102	50.97	10.53	18.29	20.22
C13H32N5P1	289.41	289.2395	53.95	11.15	24.20	10.70
C13H32N6P2	334.39	334.2164	46.70	9.65	25.13	18.53
C13H33N3P2	293.38	293.2150	53.22	11.34	14.32	21.12
C13H33N4P1	276.41	276.2443	56.49	12.03	20.27	11.21
C13H33N5P2	321.39	321.2211	48.58	10.35	21.79	19.27
C13H33N6P1	304.42	304.2504	51.29	10.93	27.61	10.17
C13H34N4P2	308.39	308.2259	50.63	11.11	18.17	20.09
C13H34N5P1	291.42	291.2552	53.58	11.76	24.03	10.63
C13H34N6P2	336.40	336.2320	46.42	10.19	24.98	18.41
C13H35N5P2	323.41	323.2368	48.28	10.91	21.66	19.15
C13H35N6P1	306.44	306.2661	50.95	11.51	27.42	10.11
C13H36N6P2	338.42	338.2477	46.14	10.72	24.83	18.30
C14H 7N1P2	251.17	251.0054	66.95	2.81	5.58	24.66
C14H 7N2P1	234.20	234.0347	71.80	3.01	11.96	13.23
C14H 7N3P2	279.18	279.0115	60.23	2.53	15.05	22.19
C14H 7N4P1	262.21	262.0408	64.13	2.69	21.37	11.81
C14H 7N5P2	307.19	307.0177	54.74	2.30	22.80	20.17
C14H 7N6P1	290.23	290.0470	57.94	2.43	28.96	10.67
C14H 8N1P1	221.20	221.0394	76.02	3.65	6.33	14.00
C14H 8N2P2	266.18	266.0163	63.17	3.03	10.52	23.27
C14H 8N3P1	249.21	249.0456	67.47	3.24	16.86	12.43
C14H 8N4P2	294.19	294.0224	57.16	2.74	19.04	21.06
C14H 8N5P1	277.23	277.0517	60.66	2.91	25.26	11.17
C14H 8N6P2	322.21	322.0286	52.19	2.50	26.08	19.23
C14H 9N1P2	253.18	253.0210	66.42	3.58	5.53	24.47
C14H 9N2P1	236.22	236.0503	71.19	3.84	11.86	13.11
C14H 9N3P2	281.20	281.0272	59.80	3.23	14.94	22.03
C14H 9N4P1	264.23	264.0565	63.64	3.43	21.20	11.72
C14H 9N5P2	309.21	309.0333	54.38	2.93	22.65	20.03
C14H 9N6P1	292.24	292.0626	57.54	3.10	28.76	10.60
C14H10N1P1	223.22	223.0551	75.33	4.52	6.27	13.88
C14H10N2P2	268.20	268.0319	62.70	3.76	10.45	23.10
C14H10N3P1	251.23	251.0612	66.93	4.01	16.73	12.33
C14H10N4P2	296.21	296.0381	56.77	3.40	18.91	20.91
C14H10N5P1	279.24	279.0674	60.22	3.61	25.08	11.09
C14H10N6P2	324.22	324.0442	51.86	3.11	25.92	19.11
C14H11N1P2	255.20	255.0367	65.89	4.34	5.49	24.27
C14H11N2P1	238.23	238.0660	70.59	4.65	11.76	13.00
C14H11N3P2	283.21	283.0428	59.37	3.91	14.84	21.87
C14H11N4P1	266.24	266.0721	63.16	4.16	21.04	11.63
C14H11N5P2	311.22	311.0490	54.03	3.56	22.50	19.90
C14H11N6P1	294.26	294.0783	57.15	3.77	28.56	10.53
C14H12N1P1	225.23	225.0707	74.66	5.37	6.22	13.75
C14H12N2P2	270.21	270.0476	62.23	4.48	10.37	22.93
C14H12N3P1	253.25	253.0769	66.40	4.78	16.59	12.23
C14H12N4P2	298.23	298.0537	56.39	4.06	18.79	20.77
C14H12N5P1	281.26	281.0830	59.79	4.30	24.90	11.01
C14H12N6P2	326.24	326.0599	51.54	3.71	25.76	18.99
C14H13N1P2	257.21	257.0523	65.38	5.09	5.45	24.08

066

C14H13N2P1	240.25	240.0816	69.99	5.45	11.66	12.89
C14H13N3P2	285.23	285.0585	58.96	4.59	14.73	21.72
C14H13N4P1	268.26	268.0878	62.68	4.88	20.89	11.55
C14H13N5P2	313.24	313.0646	53.68	4.18	22.36	19.78
C14H13N6P1	296.27	296.0939	56.76	4.42	28.37	10.45
C14H14N1P1	227.25	227.0864	74.00	6.21	6.16	13.63
C14H14N2P2	272.23	272.0632	61.77	5.18	10.29	22.76
C14H14N3P1	255.26	255.0925	65.88	5.53	16.46	12.13
C14H14N4P2	300.24	300.0694	56.01	4.70	18.66	20.63
C14H14N5P1	283.27	283.0987	59.36	4.98	24.72	10.93
C14H14N6P2	328.26	328.0755	51.23	4.30	25.60	18.87
C14H15N1P2	259.23	259.0680	64.87	5.83	5.40	23.90
C14H15N2P1	242.26	242.0973	69.41	6.24	11.56	12.79
C14H15N3P2	287.24	287.0741	58.54	5.26	14.63	21.57
C14H15N4P1	270.28	270.1034	62.22	5.59	20.73	11.46
C14H15N5P2	315.26	315.0803	53.34	4.80	22.21	19.65
C14H15N6P1	298.29	298.1096	56.37	5.07	28.17	10.38
C14H16N1P1	229.26	229.1020	73.35	7.03	6.11	13.51
C14H16N2P2	274.24	274.0789	61.32	5.88	10.21	22.59
C14H16N3P1	257.28	257.1082	65.36	6.27	16.33	12.04
C14H16N4P2	302.26	302.0850	55.63	5.34	18.54	20.49
C14H16N5P1	285.29	285.1143	58.94	5.65	24.55	10.86
C14H16N6P2	330.27	330.0912	50.91	4.88	25.45	18.76
C14H17N1P2	261.25	261.0836	64.37	6.56	5.36	23.71
C14H17N2P1	244.28	244.1129	68.84	7.01	11.47	12.68
C14H17N3P2	289.26	289.0898	58.13	5.92	14.53	21.42
C14H17N4P1	272.29	272.1191	61.76	6.29	20.58	11.38
C14H17N5P2	317.27	317.0959	53.00	5.40	22.07	19.53
C14H17N6P1	300.31	300.1252	55.99	5.71	27.98	10.31
C14H18N1P1	231.28	231.1177	72.71	7.84	6.06	13.39
C14H18N2P2	276.26	276.0945	60.87	6.57	10.14	22.42
C14H18N3P1	259.29	259.1238	64.85	7.00	16.21	11.95
C14H18N4P2	304.27	304.1007	55.26	5.96	18.41	20.36
C14H18N5P1	287.31	287.1300	58.53	6.32	24.38	10.78
C14H18N6P2	332.29	332.1068	50.61	5.46	25.29	18.64
C14H19N1P2	263.26	263.0993	63.87	7.27	5.32	23.53
C14H19N2P1	246.29	246.1286	68.27	7.78	11.37	12.58
C14H19N3P2	291.28	291.1054	57.73	6.58	14.43	21.27
C14H19N4P1	274.31	274.1347	61.30	6.98	20.42	11.29
C14H19N5P2	319.29	319.1116	52.67	6.00	21.93	19.40
C14H19N6P1	302.32	302.1409	55.62	6.33	27.80	10.25
C14H20N1P1	233.30	233.1333	72.08	8.64	6.00	13.28
C14H20N2P2	278.28	278.1102	60.43	7.24	10.07	22.26
C14H20N3P1	261.31	261.1395	64.35	7.71	16.08	11.85
C14H20N4P2	306.29	306.1163	54.90	6.58	18.29	20.23
C14H20N5P1	289.32	289.1456	58.12	6.97	24.21	10.71
C14H20N6P2	334.30	334.1225	50.30	6.03	25.14	18.53
C14H21N1P2	265.28	265.1149	63.39	7.98	5.28	23.35
C14H21N2P1	248.31	248.1442	67.72	8.52	11.28	12.47
C14H21N3P2	293.29	293.1211	57.33	7.22	14.33	21.12
C14H21N4P1	276.32	276.1504	60.85	7.66	20.28	11.21
C14H21N5P2	321.30	321.1272	52.34	6.59	21.80	19.28
C14H21N6P1	304.34	304.1565	55.25	6.96	27.61	10.18
C14H22N1P1	235.31	235.1490	71.46	9.42	5.95	13.16
C14H22N2P2	280.29	280.1258	59.99	7.91	9.99	22.10
C14H22N3P1	263.33	263.1551	63.86	8.42	15.96	11.76
C14H22N4P2	308.31	308.1320	54.54	7.19	18.17	20.09
C14H22N5P1	291.34	291.1613	57.72	7.61	24.04	10.63
C14H22N6P2	336.32	336.1381	50.00	6.59	24.99	18.42
C14H23N1P2	267.29	267.1306	62.91	8.67	5.24	23.18

C14H23N2P1	250.33	250.1599	67.17	9.26	11.19	12.37
C14H23N3P2	295.31	295.1367	56.94	7.85	14.23	20.98
C14H23N4P1	278.34	278.1660	60.41	8.33	20.13	11.13
C14H23N5P2	323.32	323.1429	52.01	7.17	21.66	19.16
C14H23N6P1	306.35	306.1722	54.89	7.57	27.43	10.11
C14H24N1P1	237.33	237.1646	70.85	10.19	5.90	13.05
C14H24N2P2	282.31	282.1415	59.56	8.57	9.92	21.94
C14H24N3P1	265.34	265.1708	63.37	9.12	15.84	11.67
C14H24N4P2	310.32	310.1476	54.19	7.80	18.05	19.96
C14H24N5P1	293.35	293.1769	57.32	8.25	23.87	10.56
C14H24N6P2	338.34	338.1538	49.70	7.15	24.84	18.31
C14H25N1P2	269.31	269.1462	62.44	9.36	5.20	23.00
C14H25N2P1	252.34	252.1755	66.64	9.99	11.10	12.27
C14H25N3P2	297.32	297.1524	56.56	8.48	14.13	20.84
C14H25N4P1	280.36	280.1817	59.98	8.99	19.98	11.05
C14H25N5P2	325.34	325.1585	51.69	7.75	21.53	19.04
C14H25N6P1	308.37	308.1878	54.53	8.17	27.25	10.04
C14H26N1P1	239.34	239.1803	70.26	10.95	5.85	12.94
C14H26N2P2	284.32	284.1571	59.14	9.22	9.85	21.79
C14H26N3P1	267.36	267.1864	62.90	9.80	15.72	11.59
C14H26N4P2	312.34	312.1633	53.84	8.39	17.94	19.83
C14H26N5P1	295.37	295.1926	56.93	8.87	23.71	10.49
C14H26N6P2	340.35	340.1694	49.41	7.70	24.69	18.20
C14H27N1P2	271.33	271.1619	61.98	10.03	5.16	22.83
C14H27N2P1	254.36	254.1912	66.11	10.70	11.01	12.18
C14H27N3P2	299.34	299.1680	56.18	9.09	14.04	20.69
C14H27N4P1	282.37	282.1973	59.55	9.64	19.84	10.97
C14H27N5P2	327.35	327.1742	51.37	8.31	21.39	18.92
C14H27N6P1	310.39	310.2035	54.18	8.77	27.08	9.98
C14H28N1P1	241.36	241.1959	69.67	11.69	5.80	12.83
C14H28N2P2	286.34	286.1728	58.73	9.86	9.78	21.63
C14H28N3P1	269.37	269.2021	62.42	10.48	15.60	11.50
C14H28N4P2	314.35	314.1789	53.49	8.98	17.82	19.71
C14H28N5P1	297.39	297.2082	56.54	9.49	23.55	10.42
C14H28N6P2	342.37	342.1851	49.12	8.24	24.55	18.09
C14H29N1P2	273.34	273.1775	61.52	10.69	5.12	22.66
C14H29N2P1	256.37	256.2068	65.59	11.40	10.93	12.08
C14H29N3P2	301.35	301.1837	55.80	9.70	13.94	20.56
C14H29N4P1	284.39	284.2130	59.13	10.28	19.70	10.89
C14H29N5P2	329.37	329.1898	51.05	8.87	21.26	18.81
C14H29N6P1	312.40	312.2191	53.83	9.36	26.90	9.91
C14H30N1P1	243.38	243.2116	69.09	12.42	5.76	12.73
C14H30N2P2	288.36	288.1884	58.32	10.49	9.71	21.48
C14H30N3P1	271.39	271.2177	61.96	11.14	15.48	11.41
C14H30N4P2	316.37	316.1946	53.15	9.56	17.71	19.58
C14H30N5P1	299.40	299.2239	56.16	10.10	23.39	10.35
C14H30N6P2	344.38	344.2007	48.83	8.78	24.40	17.99
C14H31N1P2	275.36	275.1932	61.07	11.35	5.09	22.50
C14H31N2P1	258.39	258.2225	65.08	12.09	10.84	11.99
C14H31N3P2	303.37	303.1993	55.43	10.30	13.85	20.42
C14H31N4P1	286.40	286.2286	58.71	10.91	19.56	10.81
C14H31N5P2	331.38	331.2055	50.74	9.43	21.13	18.69
C14H31N6P1	314.42	314.2348	53.48	9.94	26.73	9.85
C14H32N1P1	245.39	245.2272	68.53	13.14	5.71	12.62
C14H32N2P2	290.37	290.2041	57.91	11.11	9.65	21.33
C14H32N3P1	273.41	273.2334	61.50	11.80	15.37	11.33
C14H32N4P2	318.39	318.2102	52.82	10.13	17.60	19.46
C14H32N5P1	301.42	301.2395	55.79	10.70	23.23	10.28
C14H32N6P2	346.40	346.2164	48.54	9.31	24.26	17.88
C14H33N1P2	277.37	277.2088	60.62	11.99	5.05	22.33

067

C14H33N2P1	260.41	260.2381	64.57	12.77	10.76	11.89
C14H33N3P2	305.39	305.2150	55.06	10.89	13.76	20.28
C14H33N4P1	288.42	288.2443	58.30	11.53	19.43	10.74
C14H33N5P2	333.40	333.2211	50.44	9.98	21.01	18.58
C14H33N6P1	316.43	316.2504	53.14	10.51	26.56	9.79
C14H34N2P2	292.39	292.2197	57.51	11.72	9.58	21.19
C14H34N3P1	275.42	275.2490	61.05	12.44	15.26	11.25
C14H34N4P2	320.40	320.2259	52.48	10.70	17.49	19.33
C14H34N5P1	303.43	303.2552	55.42	11.29	23.08	10.21
C14H34N6P2	348.41	348.2320	48.26	9.84	24.12	17.78
C14H35N3P2	307.40	307.2306	54.70	11.48	13.67	20.15
C14H35N4P1	290.44	290.2599	57.90	12.15	19.29	10.66
C14H35N5P2	335.42	335.2368	50.13	10.52	20.88	18.47
C14H35N6P1	318.45	318.2661	52.80	11.08	26.39	9.73
C14H36N4P2	322.42	322.2415	52.15	11.25	17.38	19.21
C14H36N5P1	305.45	305.2708	55.05	11.88	22.93	10.14
C14H36N6P2	350.43	350.2477	47.99	10.35	23.98	17.68
C14H37N5P2	337.43	337.2524	49.83	11.05	20.75	18.36
C14H37N6P1	320.46	320.2817	52.47	11.64	26.22	9.67
C14H38N6P2	352.45	352.2633	47.71	10.87	23.84	17.58
C15H 7N1P2	263.18	263.0054	68.46	2.68	5.32	23.54
C15H 7N2P1	246.21	246.0347	73.18	2.87	11.38	12.58
C15H 7N3P2	291.19	291.0115	61.87	2.42	14.43	21.27
C15H 7N4P1	274.22	274.0408	65.70	2.57	20.43	11.30
C15H 7N5P2	319.20	319.0177	56.44	2.21	21.94	19.41
C15H 7N6P1	302.24	302.0470	59.61	2.33	27.81	10.25
C15H 8N1P1	233.21	233.0394	77.25	3.46	6.01	13.28
C15H 8N2P2	278.19	278.0163	64.76	2.90	10.07	22.27
C15H 8N3P1	261.22	261.0456	68.97	3.09	16.09	11.86
C15H 8N4P2	306.21	306.0224	58.84	2.63	18.30	20.23
C15H 8N5P1	289.24	289.0517	62.29	2.79	24.21	10.71
C15H 8N6P2	334.22	334.0286	53.91	2.41	25.15	18.54
C15H 9N1P2	265.19	265.0210	67.94	3.42	5.28	23.36
C15H 9N2P1	248.23	248.0503	72.58	3.65	11.29	12.48
C15H 9N3P2	293.21	293.0272	61.45	3.09	14.33	21.13
C15H 9N4P1	276.24	276.0565	65.22	3.28	20.28	11.21
C15H 9N5P2	321.22	321.0333	56.09	2.82	21.80	19.29
C15H 9N6P1	304.25	304.0626	59.22	2.98	27.62	10.18
C15H10N1P1	235.23	235.0551	76.59	4.29	5.95	13.17
C15H10N2P2	280.21	280.0319	64.30	3.60	10.00	22.11
C15H10N3P1	263.24	263.0612	68.44	3.83	15.96	11.77
C15H10N4P2	308.22	308.0381	58.45	3.27	18.18	20.10
C15H10N5P1	291.25	291.0674	61.86	3.46	24.05	10.63
C15H10N6P2	336.23	336.0442	53.58	3.00	24.99	18.42
C15H11N1P2	267.21	267.0367	67.43	4.15	5.24	23.18
C15H11N2P1	250.24	250.0660	72.00	4.43	11.19	12.38
C15H11N3P2	295.22	295.0428	61.03	3.76	14.23	20.98
C15H11N4P1	278.26	278.0721	64.75	3.98	20.14	11.13
C15H11N5P2	323.24	323.0490	55.74	3.43	21.67	19.16
C15H11N6P1	306.27	306.0783	58.83	3.62	27.44	10.11
C15H12N1P1	237.24	237.0707	75.94	5.10	5.90	13.06
C15H12N2P2	282.22	282.0476	63.84	4.29	9.93	21.95
C15H12N3P1	265.26	265.0769	67.92	4.56	15.84	11.68
C15H12N4P2	310.24	310.0537	58.07	3.90	18.06	19.97
C15H12N5P1	293.27	293.0830	61.43	4.12	23.88	10.56
C15H12N6P2	338.25	338.0599	53.26	3.58	24.85	18.31
C15H13N1P2	269.23	269.0523	66.92	4.87	5.20	23.01
C15H13N2P1	252.26	252.0816	71.42	5.19	11.11	12.28
C15H13N3P2	297.24	297.0585	60.61	4.41	14.14	20.84
C15H13N4P1	280.27	280.0878	64.28	4.68	19.99	11.05
C15H13N5P2	325.25	325.0646	55.39	4.03	21.53	19.05
C15H13N6P1	308.28	308.0939	58.44	4.25	27.26	10.05
C15H14N1P1	239.26	239.0864	75.30	5.90	5.85	12.95
C15H14N2P2	284.24	284.0632	63.39	4.96	9.86	21.79
C15H14N3P1	267.27	267.0925	67.41	5.28	15.72	11.59
C15H14N4P2	312.25	312.0694	57.70	4.52	17.94	19.84
C15H14N5P1	295.29	295.0987	61.01	4.78	23.72	10.49
C15H14N6P2	340.27	340.0755	52.95	4.15	24.70	18.21
C15H15N1P2	271.24	271.0680	66.42	5.57	5.16	22.84
C15H15N2P1	254.27	254.0973	70.86	5.95	11.02	12.18
C15H15N3P2	299.25	299.0741	60.21	5.05	14.04	20.70
C15H15N4P1	282.29	282.1034	63.82	5.36	19.85	10.97
C15H15N5P2	327.27	327.0803	55.05	4.62	21.40	18.93
C15H15N6P1	310.30	310.1096	58.06	4.87	27.08	9.98
C15H16N1P1	241.28	241.1020	74.67	6.68	5.81	12.84
C15H16N2P2	286.26	286.0789	62.94	5.63	9.79	21.64
C15H16N3P1	269.29	269.1082	66.90	5.99	15.60	11.50
C15H16N4P2	314.27	314.0850	57.33	5.13	17.83	19.71
C15H16N5P1	297.30	297.1143	60.60	5.42	23.56	10.42
C15H16N6P2	342.28	342.0912	52.64	4.71	24.55	18.10
C15H17N1P2	273.26	273.0836	65.93	6.27	5.13	22.67
C15H17N2P1	256.29	256.1129	70.30	6.69	10.93	12.09
C15H17N3P2	301.27	301.0898	59.80	5.69	13.95	20.56
C15H17N4P1	284.30	284.1191	63.37	6.03	19.71	10.89
C15H17N5P2	329.28	329.0959	54.71	5.20	21.27	18.81
C15H17N6P1	312.32	312.1252	57.69	5.49	26.91	9.92
C15H18N1P1	243.29	243.1177	74.05	7.46	5.76	12.73
C15H18N2P2	288.27	288.0945	62.50	6.29	9.72	21.49
C15H18N3P1	271.30	271.1238	66.41	6.69	15.49	11.42
C15H18N4P2	316.29	316.1007	56.96	5.74	17.71	19.59
C15H18N5P1	299.32	299.1300	60.19	6.06	23.40	10.35
C15H18N6P2	344.30	344.1068	52.33	5.27	24.41	17.99
C15H19N1P2	275.27	275.0993	65.45	6.96	5.09	22.50
C15H19N2P1	258.31	258.1286	69.75	7.41	10.85	11.99
C15H19N3P2	303.29	303.1054	59.40	6.31	13.85	20.43
C15H19N4P1	286.32	286.1347	62.93	6.69	19.57	10.82
C15H19N5P2	331.30	331.1116	54.38	5.78	21.14	18.70
C15H19N6P1	314.33	314.1409	57.32	6.09	26.74	9.85
C15H20N1P1	245.31	245.1333	73.45	8.22	5.71	12.63
C15H20N2P2	290.29	290.1102	62.07	6.94	9.65	21.34
C15H20N3P1	273.32	273.1395	65.92	7.38	15.37	11.33
C15H20N4P2	318.30	318.1163	56.60	6.33	17.60	19.46
C15H20N5P1	301.33	301.1456	59.79	6.69	23.24	10.28
C15H20N6P2	346.31	346.1225	52.02	5.82	24.27	17.89
C15H21N1P2	277.29	277.1149	64.97	7.63	5.05	22.34
C15H21N2P1	260.32	260.1442	69.21	8.13	10.76	11.90
C15H21N3P2	305.30	305.1211	59.01	6.93	13.76	20.29
C15H21N4P1	288.34	288.1504	62.49	7.34	19.43	10.74
C15H21N5P2	333.32	333.1272	54.05	6.35	21.01	18.59
C15H21N6P1	316.35	316.1565	56.95	6.69	26.57	9.79
C15H22N1P1	247.32	247.1490	72.85	8.97	5.66	12.52
C15H22N2P2	292.30	292.1258	61.64	7.59	9.58	21.19
C15H22N3P1	275.34	275.1551	65.44	8.05	15.26	11.25
C15H22N4P2	320.32	320.1320	56.25	6.92	17.49	19.34
C15H22N5P1	303.35	303.1613	59.39	7.31	23.09	10.21
C15H22N6P2	348.33	348.1381	51.72	6.37	24.13	17.78
C15H23N1P2	279.30	279.1306	64.51	8.30	5.01	22.18
C15H23N2P1	262.34	262.1599	68.68	8.84	10.68	11.81
C15H23N3P2	307.32	307.1367	58.63	7.54	13.67	20.16
C15H23N4P1	290.35	290.1660	62.05	7.98	19.30	10.67

068

C15H23N5P2	335.33	335.1429	53.73	6.91	20.88	18.47
C15H23N6P1	318.36	318.1722	56.59	7.28	26.40	9.73
C15H24N1P1	249.34	249.1646	72.26	9.70	5.62	12.42
C15H24N2P2	294.32	294.1415	61.21	8.22	9.52	21.05
C15H24N3P1	277.35	277.1708	64.96	8.72	15.15	11.17
C15H24N4P2	322.33	322.1476	55.89	7.51	17.38	19.22
C15H24N5P1	305.37	305.1769	59.00	7.92	22.93	10.14
C15H24N6P2	350.35	350.1538	51.43	6.90	23.99	17.68
C15H25N1P2	281.32	281.1462	64.04	8.96	4.98	22.02
C15H25N2P1	264.35	264.1755	68.15	9.53	10.60	11.72
C15H25N3P2	309.33	309.1524	58.24	8.15	13.58	20.03
C15H25N4P1	292.37	292.1817	61.62	8.62	19.16	10.59
C15H25N5P2	337.35	337.1585	53.41	7.47	20.76	18.36
C15H25N6P1	320.38	320.1878	56.24	7.87	26.23	9.67
C15H26N1P1	251.35	251.1803	71.68	10.43	5.57	12.32
C15H26N2P2	296.34	296.1571	60.80	8.84	9.45	20.90
C15H26N3P1	279.37	279.1864	64.49	9.38	15.04	11.09
C15H26N4P2	324.35	324.1633	55.55	8.08	17.27	19.10
C15H26N5P1	307.38	307.1926	58.61	8.53	22.78	10.08
C15H26N6P2	352.36	352.1694	51.13	7.44	23.85	17.58
C15H27N1P2	283.34	283.1619	63.59	9.61	4.94	21.86
C15H27N2P1	266.37	266.1912	67.64	10.22	10.52	11.63
C15H27N3P2	311.35	311.1680	57.87	8.74	13.50	19.90
C15H27N4P1	294.38	294.1973	61.20	9.24	19.03	10.52
C15H27N5P2	339.36	339.1742	53.09	8.02	20.64	18.25
C15H27N6P1	322.40	322.2035	55.88	8.44	26.07	9.61
C15H28N1P1	253.37	253.1959	71.11	11.14	5.53	12.22
C15H28N2P2	298.35	298.1728	60.39	9.46	9.39	20.76
C15H28N3P1	281.38	281.2021	64.03	10.03	14.93	11.01
C15H28N4P2	326.36	326.1789	55.20	8.65	17.17	18.98
C15H28N5P1	309.40	309.2082	58.23	9.12	22.64	10.01
C15H28N6P2	354.38	354.1851	50.84	7.96	23.71	17.48
C15H29N1P2	285.35	285.1775	63.14	10.24	4.91	21.71
C15H29N2P1	268.39	268.2068	67.13	10.89	10.44	11.54
C15H29N3P2	313.37	313.1837	57.49	9.33	13.41	19.77
C15H29N4P1	296.40	296.2130	60.79	9.86	18.90	10.45
C15H29N5P2	341.38	341.1898	52.78	8.56	20.51	18.15
C15H29N6P1	324.41	324.2191	55.54	9.01	25.91	9.55
C15H30N1P1	255.39	255.2116	70.55	11.84	5.48	12.13
C15H30N2P2	300.37	300.1884	59.98	10.07	9.33	20.62
C15H30N3P1	283.40	283.2177	63.57	10.67	14.83	10.93
C15H30N4P2	328.38	328.1946	54.87	9.21	17.06	18.86
C15H30N5P1	311.41	311.2239	57.85	9.71	22.49	9.95
C15H30N6P2	356.39	356.2007	50.55	8.48	23.58	17.38
C15H31N1P2	287.37	287.1932	62.70	10.87	4.87	21.56
C15H31N2P1	270.40	270.2225	66.63	11.56	10.36	11.45
C15H31N3P2	315.38	315.1993	57.13	9.91	13.32	19.64
C15H31N4P1	298.41	298.2286	60.37	10.47	18.77	10.38
C15H31N5P2	343.40	343.2055	52.47	9.10	20.39	18.04
C15H31N6P1	326.43	326.2348	55.19	9.57	25.75	9.49
C15H32N1P1	257.40	257.2272	69.99	12.53	5.44	12.03
C15H32N2P2	302.38	302.2041	59.58	10.67	9.26	20.49
C15H32N3P1	285.42	285.2334	63.12	11.30	14.72	10.85
C15H32N4P2	330.40	330.2102	54.53	9.76	16.96	18.75
C15H32N5P1	313.43	313.2395	57.48	10.29	22.34	9.88
C15H32N6P2	358.41	358.2164	50.27	9.00	23.45	17.28
C15H33N1P2	289.38	289.2088	62.26	11.49	4.84	21.41
C15H33N2P1	272.42	272.2381	66.14	12.21	10.28	11.37
C15H33N3P2	317.40	317.2150	56.76	10.48	13.24	19.52
C15H33N4P1	300.43	300.2443	59.97	11.07	18.65	10.31
C15H33N5P2	345.41	345.2211	52.16	9.63	20.28	17.93
C15H33N6P1	328.44	328.2504	54.85	10.13	25.59	9.43
C15H34N1P1	259.42	259.2429	69.45	13.21	5.40	11.94
C15H34N2P2	304.40	304.2197	59.19	11.26	9.20	20.35
C15H34N3P1	287.43	287.2490	62.68	11.92	14.62	10.78
C15H34N4P2	332.41	332.2259	54.20	10.31	16.85	18.64
C15H34N5P1	315.45	315.2552	57.12	10.86	22.20	9.82
C15H34N6P2	360.43	360.2320	49.99	9.51	23.32	17.19
C15H35N1P2	291.40	291.2245	61.83	12.11	4.81	21.26
C15H35N2P1	274.43	274.2538	65.65	12.86	10.21	11.29
C15H35N3P2	319.41	319.2306	56.41	11.04	13.16	19.39
C15H35N4P1	302.45	302.2599	59.57	11.66	18.52	10.24
C15H35N5P2	347.43	347.2368	51.86	10.15	20.16	17.83
C15H35N6P1	330.46	330.2661	54.52	10.68	25.43	9.37
C15H36N2P2	306.42	306.2354	58.80	11.84	9.14	20.22
C15H36N3P1	289.45	289.2647	62.25	12.54	14.52	10.70
C15H36N4P2	334.43	334.2415	53.87	10.85	16.75	18.52
C15H36N5P1	317.46	317.2708	56.75	11.43	22.06	9.76
C15H36N6P2	362.44	362.2477	49.71	10.01	23.19	17.09
C15H37N3P2	321.43	321.2463	56.05	11.60	13.07	19.27
C15H37N4P1	304.46	304.2756	59.18	12.25	18.40	10.17
C15H37N5P2	349.44	349.2524	51.56	10.67	20.04	17.73
C15H37N6P1	332.48	332.2817	54.19	11.22	25.28	9.32
C15H38N4P2	336.44	336.2572	53.55	11.38	16.65	18.41
C15H38N5P1	319.48	319.2865	56.39	11.99	21.92	9.70
C15H38N6P2	364.46	364.2633	49.43	10.51	23.06	17.00
C15H39N5P2	351.46	351.2681	51.26	11.19	19.93	17.63
C15H39N6P1	334.49	334.2974	53.86	11.75	25.12	9.26
C15H40N6P2	366.47	366.2790	49.16	11.00	22.93	16.90
C16H 8N1P1	245.22	245.0394	78.37	3.29	5.71	12.63
C16H 8N2P2	290.20	290.0163	66.22	2.78	9.65	21.35
C16H 8N3P1	273.24	273.0456	70.33	2.95	15.38	11.34
C16H 8N4P2	318.22	318.0224	60.39	2.53	17.61	19.47
C16H 8N5P1	301.25	301.0517	63.79	2.68	23.25	10.28
C16H 8N6P2	346.23	346.0286	55.51	2.33	24.27	17.89
C16H 9N1P2	277.20	277.0210	69.33	3.27	5.05	22.35
C16H 9N2P1	260.24	260.0503	73.85	3.49	10.76	11.90
C16H 9N3P2	305.22	305.0272	62.96	2.97	13.77	20.30
C16H 9N4P1	288.25	288.0565	66.67	3.15	19.44	10.75
C16H 9N5P2	333.23	333.0333	57.67	2.72	21.02	18.59
C16H 9N6P1	316.26	316.0626	60.77	2.87	26.57	9.79
C16H10N1P1	247.24	247.0551	77.73	4.08	5.67	12.53
C16H10N2P2	292.22	292.0319	65.77	3.45	9.59	21.20
C16H10N3P1	275.25	275.0612	69.82	3.66	15.27	11.25
C16H10N4P2	320.23	320.0381	60.01	3.15	17.50	19.34
C16H10N5P1	303.27	303.0674	63.37	3.32	23.09	10.21
C16H10N6P2	348.25	348.0442	55.18	2.89	24.13	17.79
C16H11N1P2	279.22	279.0367	68.83	3.97	5.02	22.19
C16H11N2P1	262.25	262.0660	73.28	4.23	10.68	11.81
C16H11N3P2	307.23	307.0428	62.55	3.61	13.68	20.16
C16H11N4P1	290.27	290.0721	66.21	3.82	19.30	10.67
C16H11N5P2	335.25	335.0490	57.32	3.31	20.89	18.48
C16H11N6P1	318.28	318.0783	60.38	3.48	26.40	9.73
C16H12N1P1	249.25	249.0707	77.10	4.85	5.62	12.43
C16H12N2P2	294.24	294.0476	65.31	4.11	9.52	21.05
C16H12N3P1	277.27	277.0769	69.31	4.36	15.16	11.17
C16H12N4P2	322.25	322.0537	59.64	3.75	17.39	19.22
C16H12N5P1	305.28	305.0830	62.95	3.96	22.94	10.15
C16H12N6P2	350.26	350.0599	54.87	3.45	23.99	17.69
C16H13N1P2	281.24	281.0523	68.33	4.66	4.98	22.03

C16H13N2P1	264.27	264.0816	72.72	4.96	10.60	11.72
C16H13N3P2	309.25	309.0585	62.14	4.24	13.59	20.03
C16H13N4P1	292.28	292.0878	65.75	4.48	19.17	10.60
C16H13N5P2	337.26	337.0646	56.98	3.89	20.77	18.37
C16H13N6P1	320.30	320.0939	60.00	4.09	26.24	9.67
C16H14N1P1	251.27	251.0864	76.48	5.62	5.57	12.33
C16H14N2P2	296.25	296.0632	64.87	4.76	9.46	20.91
C16H14N3P1	279.28	279.0925	68.81	5.05	15.05	11.09
C16H14N4P2	324.26	324.0694	59.27	4.35	17.28	19.10
C16H14N5P1	307.30	307.0987	62.54	4.59	22.79	10.08
C16H14N6P2	352.28	352.0755	54.55	4.01	23.86	17.58
C16H15N1P2	283.25	283.0680	67.85	5.34	4.94	21.87
C16H15N2P1	266.29	266.0973	72.17	5.68	10.52	11.63
C16H15N3P2	311.27	311.0741	61.74	4.86	13.50	19.90
C16H15N4P1	294.30	294.1034	65.30	5.14	19.04	10.52
C16H15N5P2	339.28	339.0803	56.64	4.46	20.64	18.26
C16H15N6P1	322.31	322.1096	59.62	4.69	26.07	9.61
C16H16N1P1	253.29	253.1020	75.87	6.37	5.53	12.23
C16H16N2P2	298.27	298.0789	64.43	5.41	9.39	20.77
C16H16N3P1	281.30	281.1082	68.32	5.73	14.94	11.01
C16H16N4P2	326.28	326.0850	58.90	4.94	17.17	18.99
C16H16N5P1	309.31	309.1143	62.13	5.21	22.64	10.01
C16H16N6P2	354.29	354.0912	54.24	4.55	23.72	17.48
C16H17N1P2	285.27	285.0836	67.37	6.01	4.91	21.72
C16H17N2P1	268.30	268.1129	71.63	6.39	10.44	11.54
C16H17N3P2	313.28	313.0898	61.34	5.47	13.41	19.77
C16H17N4P1	296.31	296.1191	64.86	5.78	18.91	10.45
C16H17N5P2	341.29	341.0959	56.31	5.02	20.52	18.15
C16H17N6P1	324.33	324.1252	59.25	5.28	25.91	9.55
C16H18N1P1	255.30	255.1177	75.27	7.11	5.49	12.13
C16H18N2P2	300.28	300.0945	64.00	6.04	9.33	20.63
C16H18N3P1	283.32	283.1238	67.83	6.40	14.83	10.93
C16H18N4P2	328.30	328.1007	58.54	5.53	17.07	18.87
C16H18N5P1	311.33	311.1300	61.73	5.83	22.50	9.95
C16H18N6P2	356.31	356.1068	53.94	5.09	23.59	17.39
C16H19N1P2	287.28	287.0993	66.89	6.67	4.88	21.56
C16H19N2P1	270.32	270.1286	71.09	7.08	10.36	11.46
C16H19N3P2	315.30	315.1054	60.95	6.07	13.33	19.65
C16H19N4P1	298.33	298.1347	64.42	6.42	18.78	10.38
C16H19N5P2	343.31	343.1116	55.98	5.58	20.40	18.04
C16H19N6P1	326.34	326.1409	58.89	5.87	25.75	9.49
C16H20N1P1	257.32	257.1333	74.69	7.83	5.44	12.04
C16H20N2P2	302.30	302.1102	63.57	6.67	9.27	20.49
C16H20N3P1	285.33	285.1395	67.35	7.07	14.73	10.86
C16H20N4P2	330.31	330.1163	58.18	6.10	16.96	18.75
C16H20N5P1	313.35	313.1456	61.33	6.43	22.35	9.88
C16H20N6P2	358.33	358.1225	53.63	5.63	23.45	17.29
C16H21N1P2	289.30	289.1149	66.43	7.32	4.84	21.41
C16H21N2P1	272.33	272.1442	70.57	7.77	10.29	11.37
C16H21N3P2	317.31	317.1211	60.56	6.67	13.24	19.52
C16H21N4P1	300.35	300.1504	63.99	7.05	18.65	10.31
C16H21N5P2	345.33	345.1272	55.65	6.13	20.28	17.94
C16H21N6P1	328.36	328.1565	58.53	6.45	25.59	9.43
C16H22N1P1	259.33	259.1490	74.10	8.55	5.40	11.94
C16H22N2P2	304.31	304.1258	63.15	7.29	9.21	20.36
C16H22N3P1	287.35	287.1551	66.88	7.72	14.62	10.78
C16H22N4P2	332.33	332.1320	57.83	6.67	16.86	18.64
C16H22N5P1	315.36	315.1613	60.94	7.03	22.21	9.82
C16H22N6P2	360.34	360.1381	53.33	6.15	23.32	17.19
C16H23N1P2	291.32	291.1306	65.97	7.96	4.81	21.26
C16H23N2P1	274.35	274.1599	70.05	8.45	10.21	11.29
C16H23N3P2	319.33	319.1367	60.18	7.26	13.16	19.40
C16H23N4P1	302.36	302.1660	63.56	7.67	18.53	10.24
C16H23N5P2	347.34	347.1429	55.33	6.67	20.16	17.83
C16H23N6P1	330.38	330.1722	58.17	7.02	25.44	9.38
C16H24N1P1	261.35	261.1646	73.53	9.26	5.36	11.85
C16H24N2P2	306.33	306.1415	62.74	7.90	9.14	20.22
C16H24N3P1	289.36	289.1708	66.41	8.36	14.52	10.70
C16H24N4P2	334.34	334.1476	57.48	7.24	16.76	18.53
C16H24N5P1	317.38	317.1769	60.55	7.62	22.07	9.76
C16H24N6P2	362.36	362.1538	53.04	6.68	23.19	17.10
C16H25N1P2	293.33	293.1462	65.52	8.59	4.78	21.12
C16H25N2P1	276.36	276.1755	69.54	9.12	10.14	11.21
C16H25N3P2	321.35	321.1524	59.80	7.84	13.08	19.28
C16H25N4P1	304.38	304.1817	63.14	8.28	18.41	10.18
C16H25N5P2	349.36	349.1585	55.01	7.21	20.05	17.73
C16H25N6P1	332.39	332.1878	57.82	7.58	25.28	9.32
C16H26N1P1	263.37	263.1803	72.97	9.95	5.32	11.76
C16H26N2P2	308.35	308.1571	62.33	8.50	9.09	20.09
C16H26N3P1	291.38	291.1864	65.95	8.99	14.42	10.63
C16H26N4P2	336.36	336.1633	57.13	7.79	16.66	18.42
C16H26N5P1	319.39	319.1926	60.17	8.21	21.93	9.70
C16H26N6P2	364.37	364.1694	52.74	7.19	23.06	17.00
C16H27N1P2	295.35	295.1619	65.07	9.21	4.74	20.97
C16H27N2P1	278.38	278.1912	69.03	9.78	10.06	11.13
C16H27N3P2	323.36	323.1680	59.43	8.42	12.99	19.16
C16H27N4P1	306.39	306.1973	62.72	8.88	18.29	10.11
C16H27N5P2	351.37	351.1742	54.69	7.75	19.93	17.63
C16H27N6P1	334.41	334.2035	57.47	8.14	25.13	9.26
C16H28N1P1	265.38	265.1959	72.42	10.63	5.28	11.67
C16H28N2P2	310.36	310.1728	61.92	9.09	9.03	19.96
C16H28N3P1	293.40	293.2021	65.50	9.62	14.32	10.56
C16H28N4P2	338.38	338.1789	56.79	8.34	16.56	18.31
C16H28N5P1	321.41	321.2082	59.79	8.78	21.79	9.64
C16H28N6P2	366.39	366.1851	52.45	7.70	22.94	16.91
C16H29N1P2	297.36	297.1775	64.63	9.83	4.71	20.83
C16H29N2P1	280.40	280.2068	68.54	10.42	9.99	11.05
C16H29N3P2	325.38	325.1837	59.06	8.98	12.91	19.04
C16H29N4P1	308.41	308.2130	62.31	9.48	18.17	10.04
C16H29N5P2	353.39	353.1898	54.38	8.27	19.82	17.53
C16H29N6P1	336.42	336.2191	57.12	8.69	24.98	9.21
C16H30N1P1	267.40	267.2116	71.87	11.31	5.24	11.58
C16H30N2P2	312.38	312.1884	61.52	9.68	8.97	19.83
C16H30N3P1	295.41	295.2177	65.05	10.24	14.22	10.48
C16H30N4P2	340.39	340.1946	56.46	8.88	16.46	18.20
C16H30N5P1	323.42	323.2239	59.42	9.35	21.65	9.58
C16H30N6P2	368.41	368.2007	52.16	8.21	22.81	16.82
C16H31N1P2	299.38	299.1932	64.19	10.44	4.68	20.69
C16H31N2P1	282.41	282.2225	68.05	11.06	9.92	10.97
C16H31N3P2	327.39	327.1993	58.70	9.54	12.83	18.92
C16H31N4P1	310.43	310.2286	61.91	10.07	18.05	9.98
C16H31N5P2	355.41	355.2055	54.07	8.79	19.71	17.43
C16H31N6P1	338.44	338.2348	56.78	9.23	24.83	9.15
C16H32N1P1	269.41	269.2272	71.33	11.97	5.20	11.50
C16H32N2P2	314.39	314.2041	61.13	10.26	8.91	19.70
C16H32N3P1	297.43	297.2334	64.61	10.84	14.13	10.41
C16H32N4P2	342.41	342.2102	56.13	9.42	16.36	18.09
C16H32N5P1	325.44	325.2395	59.05	9.91	21.52	9.52
C16H32N6P2	370.42	370.2164	51.88	8.71	22.69	16.72
C16H33N1P2	301.40	301.2088	63.76	11.04	4.65	20.55

C16H33N2P1	284.43	284.2381	67.57	11.69	9.85	10.89
C16H33N3P2	329.41	329.2150	58.34	10.10	12.76	18.81
C16H33N4P1	312.44	312.2443	61.51	10.65	17.93	9.91
C16H33N5P2	357.42	357.2211	53.77	9.31	19.59	17.33
C16H33N6P1	340.46	340.2504	56.45	9.77	24.68	9.10
C16H34N1P1	271.43	271.2429	70.80	12.63	5.16	11.41
C16H34N2P2	316.41	316.2197	60.74	10.83	8.85	19.58
C16H34N3P1	299.44	299.2490	64.18	11.44	14.03	10.34
C16H34N4P2	344.42	344.2259	55.80	9.95	16.27	17.99
C16H34N5P1	327.46	327.2552	58.69	10.47	21.39	9.46
C16H34N6P2	372.44	372.2320	51.60	9.20	22.56	16.63
C16H35N1P2	303.41	303.2245	63.34	11.63	4.62	20.42
C16H35N2P1	286.44	286.2538	67.09	12.32	9.78	10.81
C16H35N3P2	331.43	331.2306	57.99	10.64	12.68	18.69
C16H35N4P1	314.46	314.2599	61.11	11.22	17.82	9.85
C16H35N5P2	359.44	359.2368	53.47	9.82	19.48	17.23
C16H35N6P1	342.47	342.2661	56.12	10.30	24.54	9.04
C16H36N1P1	273.45	273.2585	70.28	13.27	5.12	11.33
C16H36N2P2	318.43	318.2354	60.35	11.40	8.80	19.45
C16H36N3P1	301.46	301.2647	63.75	12.04	13.94	10.27
C16H36N4P2	346.44	346.2415	55.47	10.47	16.17	17.88
C16H36N5P1	329.47	329.2708	58.33	11.01	21.26	9.40
C16H36N6P2	374.45	374.2477	51.32	9.69	22.44	16.54
C16H37N1P2	305.43	305.2401	62.92	12.21	4.59	20.28
C16H37N2P1	288.46	288.2694	66.62	12.93	9.71	10.74
C16H37N3P2	333.44	333.2463	57.63	11.18	12.60	18.58
C16H37N4P1	316.47	316.2756	60.72	11.78	17.70	9.79
C16H37N5P2	361.45	361.2524	53.17	10.32	19.38	17.14
C16H37N6P1	344.49	344.2817	55.79	10.83	24.40	8.99
C16H38N2P2	320.44	320.2510	59.97	11.95	8.74	19.33
C16H38N3P1	303.48	303.2803	63.33	12.62	13.85	10.21
C16H38N4P2	348.46	348.2572	55.15	10.99	16.08	17.78
C16H38N5P1	331.49	331.2865	57.97	11.55	21.13	9.34
C16H38N6P2	376.47	376.2633	51.05	10.17	22.32	16.45
C16H39N3P2	335.46	335.2619	57.29	11.72	12.53	18.47
C16H39N4P1	318.49	318.2912	60.34	12.34	17.59	9.73
C16H39N5P2	363.47	363.2681	52.87	10.82	19.27	17.04
C16H39N6P1	346.50	346.2974	55.46	11.35	24.25	8.94
C16H40N4P2	350.47	350.2728	54.83	11.50	15.99	17.68
C16H40N5P1	333.50	333.3021	57.62	12.09	21.00	9.29
C16H40N6P2	378.49	378.2790	50.78	10.65	22.20	16.37
C16H41N5P2	365.49	365.2837	52.58	11.31	19.16	16.95
C16H41N6P1	348.52	348.3130	55.14	11.86	24.11	8.89
C16H42N6P2	380.50	380.2946	50.51	11.13	22.09	16.28
C17H 8N1P1	257.23	257.0394	79.38	3.13	5.45	12.04
C17H 8N2P2	302.21	302.0163	67.56	2.67	9.27	20.50
C17H 8N3P1	285.25	285.0456	71.58	2.83	14.73	10.86
C17H 8N4P2	330.23	330.0224	61.83	2.44	16.97	18.76
C17H 8N5P1	313.26	313.0517	65.18	2.57	22.36	9.89
C17H 8N6P2	358.24	358.0286	57.00	2.25	23.46	17.29
C17H 9N1P2	289.22	289.0210	70.60	3.14	4.84	21.42
C17H 9N2P1	272.25	272.0503	75.00	3.33	10.29	11.38
C17H 9N3P2	317.23	317.0272	64.37	2.86	13.25	19.53
C17H 9N4P1	300.26	300.0565	68.00	3.02	18.66	10.32
C17H 9N5P2	345.24	345.0333	59.14	2.63	20.29	17.94
C17H 9N6P1	328.28	328.0626	62.20	2.76	25.60	9.44
C17H10N1P1	259.25	259.0551	78.76	3.89	5.40	11.95
C17H10N2P2	304.23	304.0319	67.12	3.31	9.21	20.36
C17H10N3P1	287.26	287.0612	71.08	3.51	14.63	10.78
C17H10N4P2	332.24	332.0381	61.46	3.03	16.86	18.65
C17H10N5P1	315.28	315.0674	64.77	3.20	22.21	9.82
C17H10N6P2	360.26	360.0442	56.68	2.80	23.33	17.20
C17H11N1P2	291.23	291.0367	70.11	3.81	4.81	21.27
C17H11N2P1	274.26	274.0660	74.45	4.04	10.21	11.29
C17H11N3P2	319.24	319.0428	63.96	3.47	13.16	19.40
C17H11N4P1	302.28	302.0721	67.55	3.67	18.53	10.25
C17H11N5P2	347.26	347.0490	58.80	3.19	20.17	17.84
C17H11N6P1	330.29	330.0783	61.82	3.36	25.44	9.38
C17H12N1P1	261.27	261.0707	78.15	4.63	5.36	11.86
C17H12N2P2	306.25	306.0476	66.67	3.95	9.15	20.23
C17H12N3P1	289.28	289.0769	70.59	4.18	14.53	10.71
C17H12N4P2	334.26	334.0537	61.09	3.62	16.76	18.53
C17H12N5P1	317.29	317.0830	64.35	3.81	22.07	9.76
C17H12N6P2	362.27	362.0599	56.36	3.34	23.20	17.10
C17H13N1P2	293.25	293.0523	69.63	4.47	4.78	21.12
C17H13N2P1	276.28	276.0816	73.91	4.74	10.14	11.21
C17H13N3P2	321.26	321.0585	63.56	4.08	13.08	19.28
C17H13N4P1	304.29	304.0878	67.10	4.31	18.41	10.18
C17H13N5P2	349.27	349.0646	58.46	3.75	20.05	17.74
C17H13N6P1	332.31	332.0939	61.45	3.94	25.29	9.32
C17H14N1P1	263.28	263.0864	77.56	5.36	5.32	11.76
C17H14N2P2	308.26	308.0632	66.24	4.58	9.09	20.10
C17H14N3P1	291.30	291.0925	70.10	4.84	14.43	10.63
C17H14N4P2	336.28	336.0694	60.72	4.20	16.66	18.42
C17H14N5P1	319.31	319.0987	63.95	4.42	21.93	9.70
C17H14N6P2	364.29	364.0755	56.05	3.87	23.07	17.01
C17H15N1P2	295.26	295.0680	69.16	5.12	4.74	20.98
C17H15N2P1	278.30	278.0973	73.37	5.43	10.07	11.13
C17H15N3P2	323.28	323.0741	63.16	4.68	13.00	19.16
C17H15N4P1	306.31	306.1034	66.66	4.94	18.29	10.11
C17H15N5P2	351.29	351.0803	58.13	4.30	19.94	17.63
C17H15N6P1	334.32	334.1096	61.08	4.52	25.14	9.26
C17H16N1P1	265.30	265.1020	76.97	6.08	5.28	11.68
C17H16N2P2	310.28	310.0789	65.81	5.20	9.03	19.97
C17H16N3P1	293.31	293.1082	69.62	5.50	14.33	10.56
C17H16N4P2	338.29	338.0850	60.36	4.77	16.56	18.31
C17H16N5P1	321.32	321.1143	63.55	5.02	21.80	9.64
C17H16N6P2	366.30	366.0912	55.74	4.40	22.94	16.91
C17H17N1P2	297.28	297.0836	68.69	5.76	4.71	20.84
C17H17N2P1	280.31	280.1129	72.84	6.11	9.99	11.05
C17H17N3P2	325.29	325.0898	62.77	5.27	12.92	19.04
C17H17N4P1	308.33	308.1191	66.23	5.56	18.17	10.05
C17H17N5P2	353.31	353.0959	57.79	4.85	19.82	17.53
C17H17N6P1	336.34	336.1252	60.71	5.09	24.99	9.21
C17H18N1P1	267.31	267.1177	76.39	6.79	5.24	11.59
C17H18N2P2	312.29	312.0945	65.38	5.81	8.97	19.84
C17H18N3P1	295.33	295.1238	69.14	6.14	14.23	10.49
C17H18N4P2	340.31	340.1007	60.00	5.33	16.46	18.20
C17H18N5P1	323.34	323.1300	63.15	5.61	21.66	9.58
C17H18N6P2	368.32	368.1068	55.44	4.93	22.82	16.82
C17H19N1P2	299.30	299.0993	68.22	6.40	4.68	20.70
C17H19N2P1	282.33	282.1286	72.32	6.78	9.92	10.97
C17H19N3P2	327.31	327.1054	62.38	5.85	12.84	18.93
C17H19N4P1	310.34	310.1347	65.80	6.17	18.05	9.98
C17H19N5P2	355.32	355.1116	57.47	5.39	19.71	17.43
C17H19N6P1	338.35	338.1409	60.35	5.66	24.84	9.15
C17H20N1P1	269.33	269.1333	75.81	7.49	5.20	11.50
C17H20N2P2	314.31	314.1102	64.96	6.41	8.91	19.71
C17H20N3P1	297.34	297.1395	68.67	6.78	14.13	10.42
C17H20N4P2	342.32	342.1163	59.65	5.89	16.37	18.10

071

C17H20N5P1	325.36	325.1456	62.76	6.20	21.53	9.52
C17H20N6P2	370.34	370.1225	55.14	5.44	22.69	16.73
C17H21N1P2	301.31	301.1149	67.77	7.03	4.65	20.56
C17H21N2P1	284.34	284.1442	71.81	7.44	9.85	10.89
C17H21N3P2	329.32	329.1211	62.00	6.43	12.76	18.81
C17H21N4P1	312.36	312.1504	65.37	6.78	17.94	9.92
C17H21N5P2	357.34	357.1272	57.14	5.92	19.60	17.34
C17H21N6P1	340.37	340.1565	59.99	6.22	24.69	9.10
C17H22N1P1	271.35	271.1490	75.25	8.17	5.16	11.41
C17H22N2P2	316.33	316.1258	64.55	7.01	8.86	19.58
C17H22N3P1	299.36	299.1551	68.21	7.41	14.04	10.35
C17H22N4P2	344.34	344.1320	59.30	6.44	16.27	17.99
C17H22N5P1	327.37	327.1613	62.37	6.77	21.39	9.46
C17H22N6P2	372.35	372.1381	54.84	5.96	22.57	16.64
C17H23N1P2	303.33	303.1306	67.32	7.64	4.62	20.42
C17H23N2P1	286.36	286.1599	71.31	8.10	9.78	10.82
C17H23N3P2	331.34	331.1367	61.63	7.00	12.68	18.70
C17H23N4P1	314.37	314.1660	64.95	7.37	17.82	9.85
C17H23N5P2	359.35	359.1429	56.82	6.45	19.49	17.24
C17H23N6P1	342.39	342.1722	59.64	6.77	24.55	9.05
C17H24N1P1	273.36	273.1646	74.70	8.85	5.12	11.33
C17H24N2P2	318.34	318.1415	64.14	7.60	8.80	19.46
C17H24N3P1	301.37	301.1708	67.75	8.03	13.94	10.28
C17H24N4P2	346.36	346.1476	58.95	6.98	16.18	17.89
C17H24N5P1	329.39	329.1769	61.99	7.34	21.26	9.40
C17H24N6P2	374.37	374.1538	54.54	6.46	22.45	16.55
C17H25N1P2	305.34	305.1462	66.87	8.25	4.59	20.29
C17H25N2P1	288.38	288.1755	70.81	8.74	9.71	10.74
C17H25N3P2	333.36	333.1524	61.25	7.56	12.61	18.58
C17H25N4P1	316.39	316.1817	64.54	7.96	17.71	9.79
C17H25N5P2	361.37	361.1585	56.50	6.97	19.38	17.14
C17H25N6P1	344.40	344.1878	59.29	7.32	24.40	8.99
C17H26N1P1	275.38	275.1803	74.15	9.52	5.09	11.25
C17H26N2P2	320.36	320.1571	63.74	8.18	8.74	19.34
C17H26N3P1	303.39	303.1864	67.30	8.64	13.85	10.21
C17H26N4P2	348.37	348.1633	58.61	7.52	16.08	17.78
C17H26N5P1	331.40	331.1926	61.61	7.91	21.13	9.35
C17H26N6P2	376.38	376.1694	54.25	6.96	22.33	16.46
C17H27N1P2	307.36	307.1619	66.43	8.85	4.56	20.15
C17H27N2P1	290.39	290.1912	70.32	9.37	9.65	10.67
C17H27N3P2	335.37	335.1680	60.88	8.11	12.53	18.47
C17H27N4P1	318.41	318.1973	64.13	8.55	17.60	9.73
C17H27N5P2	363.39	363.1742	56.19	7.49	19.27	17.05
C17H27N6P1	346.42	346.2035	58.94	7.86	24.26	8.94
C17H28N1P1	277.39	277.1959	73.61	10.17	5.05	11.17
C17H28N2P2	322.37	322.1728	63.34	8.75	8.69	19.22
C17H28N3P1	305.41	305.2021	66.86	9.24	13.76	10.14
C17H28N4P2	350.39	350.1789	58.28	8.05	15.99	17.68
C17H28N5P1	333.42	333.2082	61.24	8.46	21.00	9.29
C17H28N6P2	378.40	378.1851	53.96	7.46	22.21	16.37
C17H29N1P2	309.37	309.1775	66.00	9.45	4.53	20.02
C17H29N2P1	292.41	292.2068	69.83	10.00	9.58	10.59
C17H29N3P2	337.39	337.1837	60.52	8.66	12.45	18.36
C17H29N4P1	320.42	320.2130	63.73	9.12	17.49	9.67
C17H29N5P2	365.40	365.1898	55.88	8.00	19.17	16.95
C17H29N6P1	348.43	348.2191	58.60	8.39	24.12	8.89
C17H30N1P1	279.41	279.2116	73.08	10.82	5.01	11.09
C17H30N2P2	324.39	324.1884	62.95	9.32	8.64	19.10
C17H30N3P1	307.42	307.2177	66.42	9.84	13.67	10.08
C17H30N4P2	352.40	352.1946	57.94	8.58	15.90	17.58

C17H30N5P1	335.44	335.2239	60.87	9.01	20.88	9.23
C17H30N6P2	380.42	380.2007	53.68	7.95	22.09	16.28
C17H31N1P2	311.39	311.1932	65.57	10.03	4.50	19.89
C17H31N2P1	294.42	294.2225	69.35	10.61	9.51	10.52
C17H31N3P2	339.40	339.1993	60.16	9.21	12.38	18.25
C17H31N4P1	322.44	322.2286	63.33	9.69	17.38	9.61
C17H31N5P2	367.42	367.2055	55.57	8.50	19.06	16.86
C17H31N6P1	350.45	350.2348	58.26	8.92	23.98	8.84
C17H32N1P1	281.43	281.2272	72.56	11.46	4.98	11.01
C17H32N2P2	326.41	326.2041	62.56	9.88	8.58	18.98
C17H32N3P1	309.44	309.2334	65.99	10.42	13.58	10.01
C17H32N4P2	354.42	354.2102	57.61	9.10	15.81	17.48
C17H32N5P1	337.45	337.2395	60.51	9.56	20.75	9.18
C17H32N6P2	382.43	382.2164	53.39	8.43	21.98	16.20
C17H33N1P2	313.41	313.2088	65.15	10.61	4.47	19.77
C17H33N2P1	296.44	296.2381	68.88	11.22	9.45	10.45
C17H33N3P2	341.42	341.2150	59.81	9.74	12.31	18.14
C17H33N4P1	324.45	324.2443	62.93	10.25	17.27	9.55
C17H33N5P2	369.43	369.2211	55.27	9.00	18.96	16.77
C17H33N6P1	352.47	352.2504	57.93	9.44	23.84	8.79
C17H34N1P1	283.44	283.2429	72.04	12.09	4.94	10.93
C17H34N2P2	328.42	328.2197	62.17	10.44	8.53	18.86
C17H34N3P1	311.45	311.2490	65.56	11.00	13.49	9.94
C17H34N4P2	356.43	356.2259	57.29	9.61	15.72	17.38
C17H34N5P1	339.47	339.2552	60.15	10.10	20.63	9.12
C17H34N6P2	384.45	384.2320	53.11	8.91	21.86	16.11
C17H35N1P2	315.42	315.2245	64.74	11.18	4.44	19.64
C17H35N2P1	298.46	298.2538	68.42	11.82	9.39	10.38
C17H35N3P2	343.44	343.2306	59.45	10.27	12.24	18.04
C17H35N4P1	326.47	326.2599	62.54	10.81	17.16	9.49
C17H35N5P2	371.45	371.2368	54.97	9.50	18.85	16.68
C17H35N6P1	354.48	354.2661	57.60	9.95	23.71	8.74
C17H36N1P1	285.46	285.2585	71.53	12.71	4.91	10.85
C17H36N2P2	330.44	330.2354	61.79	10.98	8.48	18.75
C17H36N3P1	313.47	313.2647	65.14	11.58	13.40	9.88
C17H36N4P2	358.45	358.2415	56.96	10.12	15.63	17.28
C17H36N5P1	341.48	341.2708	59.79	10.63	20.51	9.07
C17H36N6P2	386.46	386.2477	52.84	9.39	21.75	16.03
C17H37N1P2	317.44	317.2401	64.32	11.75	4.41	19.51
C17H37N2P1	300.47	300.2694	67.96	12.41	9.32	10.31
C17H37N3P2	345.45	345.2463	59.11	10.80	12.16	17.93
C17H37N4P1	328.49	328.2756	62.16	11.35	17.06	9.43
C17H37N5P2	373.47	373.2524	54.67	9.99	18.75	16.59
C17H37N6P1	356.50	356.2817	57.28	10.46	23.57	8.69
C17H38N1P1	287.47	287.2742	71.03	13.32	4.87	10.77
C17H38N2P2	332.45	332.2510	61.42	11.52	8.43	18.63
C17H38N3P1	315.49	315.2803	64.72	12.14	13.32	9.82
C17H38N4P2	360.47	360.2572	56.65	10.63	15.54	17.19
C17H38N5P1	343.50	343.2865	59.44	11.15	20.39	9.02
C17H38N6P2	388.48	388.2633	52.56	9.86	21.63	15.95
C17H39N1P2	319.45	319.2558	63.92	12.31	4.38	19.39
C17H39N2P1	302.49	302.2851	67.50	13.00	9.26	10.24
C17H39N3P2	347.47	347.2619	58.76	11.31	12.09	17.83
C17H39N4P1	330.50	330.2912	61.78	11.89	16.95	9.37
C17H39N5P2	375.48	375.2681	54.38	10.47	18.65	16.50
C17H39N6P1	358.51	358.2974	56.95	10.96	23.44	8.64
C17H40N2P2	334.47	334.2667	61.05	12.05	8.38	18.52
C17H40N3P1	317.50	317.2960	64.31	12.70	13.23	9.76
C17H40N4P2	362.48	362 2728	56.33	11.12	15.46	17.09
C17H40N5P1	345.52	345.3021	59.10	11.67	20.27	8.96

072

C17H40N6P2	390.50	390.2790	52.29	10.33	21.52	15.86
C17H41N3P2	349.48	349.2776	58.43	11.83	12.02	17.73
C17H41N4P1	332.52	332.3069	61.41	12.43	16.85	9.31
C17H41N5P2	377.50	377.2837	54.09	10.95	18.55	16.41
C17H41N6P1	360.53	360.3130	56.64	11.46	23.31	8.59
C17H42N4P2	364.50	364.2885	56.02	11.61	15.37	17.00
C17H42N5P1	347.53	347.3178	58.75	12.18	20.15	8.91
C17H42N6P2	392.51	392.2946	52.02	10.79	21.41	15.78
C17H43N5P2	379.51	379.2994	53.80	11.42	18.45	16.32
C17H43N6P1	362.55	362.3287	56.32	11.96	23.18	8.54
C17H44N6P2	394.53	394.3103	51.76	11.24	21.30	15.70
C18H 9N1P2	301.23	301.0210	71.77	3.01	4.65	20.57
C18H 9N2P1	284.26	284.0503	76.06	3.19	9.85	10.90
C18H 9N3P2	329.24	329.0272	65.67	2.76	12.76	18.82
C18H 9N4P1	312.27	312.0565	69.23	2.91	17.94	9.92
C18H 9N5P2	357.25	357.0333	60.52	2.54	19.60	17.34
C18H 9N6P1	340.29	340.0626	63.53	2.67	24.70	9.10
C18H10N1P1	271.26	271.0551	79.70	3.72	5.16	11.42
C18H10N2P2	316.24	316.0319	68.37	3.19	8.86	19.59
C18H10N3P1	299.27	299.0612	72.24	3.37	14.04	10.35
C18H10N4P2	344.25	344.0381	62.80	2.93	16.27	17.99
C18H10N5P1	327.29	327.0674	66.06	3.08	21.40	9.46
C18H10N6P2	372.27	372.0442	58.08	2.71	22.58	16.64
C18H11N1P2	303.24	303.0367	71.30	3.66	4.62	20.43
C18H11N2P1	286.28	286.0660	75.52	3.87	9.79	10.82
C18H11N3P2	331.26	331.0428	65.27	3.35	12.69	18.70
C18H11N4P1	314.29	314.0721	68.79	3.53	17.83	9.86
C18H11N5P2	359.27	359.0490	60.18	3.09	19.49	17.24
C18H11N6P1	342.30	342.0783	63.16	3.24	24.55	9.05
C18H12N1P1	273.28	273.0707	79.11	4.43	5.13	11.33
C18H12N2P2	318.26	318.0476	67.93	3.80	8.80	19.46
C18H12N3P1	301.29	301.0769	71.76	4.01	13.95	10.28
C18H12N4P2	346.27	346.0537	62.44	3.49	16.18	17.89
C18H12N5P1	329.30	329.0830	65.65	3.67	21.27	9.41
C18H12N6P2	374.28	374.0599	57.76	3.23	22.45	16.55
C18H13N1P2	305.26	305.0523	70.83	4.29	4.59	20.29
C18H13N2P1	288.29	288.0816	74.99	4.55	9.72	10.74
C18H13N3P2	333.27	333.0585	64.87	3.93	12.61	18.59
C18H13N4P1	316.30	316.0878	68.35	4.14	17.71	9.79
C18H13N5P2	361.29	361.0646	59.84	3.63	19.38	17.15
C18H13N6P1	344.32	344.0939	62.79	3.81	24.41	9.00
C18H14N1P1	275.29	275.0864	78.53	5.13	5.09	11.25
C18H14N2P2	320.27	320.0632	67.51	4.41	8.75	19.34
C18H14N3P1	303.31	303.0925	71.28	4.65	13.85	10.21
C18H14N4P2	348.29	348.0694	62.08	4.05	16.09	17.79
C18H14N5P1	331.32	331.0987	65.25	4.26	21.14	9.35
C18H14N6P2	376.30	376.0755	57.45	3.75	22.33	16.46
C18H15N1P2	307.27	307.0680	70.36	4.92	4.56	20.16
C18H15N2P1	290.31	290.0973	74.47	5.21	9.65	10.67
C18H15N3P2	335.29	335.0741	64.48	4.51	12.53	18.48
C18H15N4P1	318.32	318.1034	67.92	4.75	17.60	9.73
C18H15N5P2	363.30	363.0803	59.51	4.16	19.28	17.05
C18H15N6P1	346.33	346.1096	62.43	4.37	24.27	8.94
C18H16N1P1	277.31	277.1020	77.96	5.82	5.05	11.17
C18H16N2P2	322.29	322.0789	67.08	5.00	8.69	19.22
C18H16N3P1	305.32	305.1082	70.81	5.28	13.76	10.14
C18H16N4P2	350.30	350.0850	61.72	4.60	15.99	17.68
C18H16N5P1	333.34	333.1143	64.86	4.84	21.01	9.29
C18H16N6P2	378.32	378.0912	57.15	4.26	22.21	16.37
C18H17N1P2	309.29	309.0836	69.90	5.54	4.53	20.03

C18H17N2P1	292.32	292.1129	73.96	5.86	9.58	10.60
C18H17N3P2	337.30	337.0898	64.10	5.08	12.46	18.37
C18H17N4P1	320.34	320.1191	67.49	5.35	17.49	9.67
C18H17N5P2	365.32	365.0959	59.18	4.69	19.17	16.96
C18H17N6P1	348.35	348.1252	62.06	4.92	24.13	8.89
C18H18N1P1	279.32	279.1177	77.40	6.50	5.01	11.09
C18H18N2P2	324.31	324.0945	66.67	5.59	8.64	19.10
C18H18N3P1	307.34	307.1238	70.35	5.90	13.67	10.08
C18H18N4P2	352.32	352.1007	61.37	5.15	15.90	17.58
C18H18N5P1	335.35	335.1300	64.47	5.41	20.88	9.24
C18H18N6P2	380.33	380.1068	56.85	4.77	22.10	16.29
C18H19N1P2	311.31	311.0993	69.45	6.15	4.50	19.90
C18H19N2P1	294.34	294.1286	73.45	6.51	9.52	10.52
C18H19N3P2	339.32	339.1054	63.72	5.64	12.38	18.26
C18H19N4P1	322.35	322.1347	67.07	5.94	17.38	9.61
C18H19N5P2	367.33	367.1116	58.86	5.21	19.07	16.86
C18H19N6P1	350.37	350.1409	61.71	5.47	23.99	8.84
C18H20N1P1	281.34	281.1333	76.85	7.17	4.98	11.01
C18H20N2P2	326.32	326.1102	66.25	6.18	8.58	18.98
C18H20N3P1	309.35	309.1395	69.89	6.52	13.58	10.01
C18H20N4P2	354.33	354.1163	61.02	5.69	15.81	17.48
C18H20N5P1	337.37	337.1456	64.08	5.98	20.76	9.18
C18H20N6P2	382.35	382.1225	56.55	5.27	21.98	16.20
C18H21N1P2	313.32	313.1149	69.00	6.76	4.47	19.77
C18H21N2P1	296.36	296.1442	72.95	7.14	9.45	10.45
C18H21N3P2	341.34	341.1211	63.34	6.20	12.31	18.15
C18H21N4P1	324.37	324.1504	66.65	6.53	17.27	9.55
C18H21N5P2	369.35	369.1272	58.54	5.73	18.96	16.77
C18H21N6P1	352.38	352.1565	61.35	6.01	23.85	8.79
C18H22N1P1	283.36	283.1490	76.30	7.83	4.94	10.93
C18H22N2P2	328.34	328.1258	65.85	6.75	8.53	18.87
C18H22N3P1	311.37	311.1551	69.44	7.12	13.50	9.95
C18H22N4P2	356.35	356.1320	60.67	6.22	15.72	17.38
C18H22N5P1	339.38	339.1613	63.70	6.53	20.64	9.13
C18H22N6P2	384.36	384.1381	56.25	5.77	21.86	16.12
C18H23N1P2	315.34	315.1306	68.56	7.35	4.44	19.64
C18H23N2P1	298.37	298.1599	72.46	7.77	9.39	10.38
C18H23N3P2	343.35	343.1367	62.97	6.75	12.24	18.04
C18H23N4P1	326.38	326.1660	66.24	7.10	17.17	9.49
C18H23N5P2	371.37	371.1429	58.22	6.24	18.86	16.68
C18H23N6P1	354.40	354.1722	61.01	6.54	23.71	8.74
C18H24N1P1	285.37	285.1646	75.76	8.48	4.91	10.85
C18H24N2P2	330.35	330.1415	65.45	7.32	8.48	18.75
C18H24N3P1	313.39	313.1708	68.99	7.72	13.41	9.88
C18H24N4P2	358.37	358.1476	60.33	6.75	15.63	17.29
C18H24N5P1	341.40	341.1769	63.33	7.09	20.51	9.07
C18H24N6P2	386.38	386.1538	55.96	6.26	21.75	16.03
C18H25N1P2	317.35	317.1462	68.13	7.94	4.41	19.52
C18H25N2P1	300.39	300.1755	71.97	8.39	9.33	10.31
C18H25N3P2	345.37	345.1524	62.60	7.30	12.17	17.94
C18H25N4P1	328.40	328.1817	65.83	7.67	17.06	9.43
C18H25N5P2	373.38	373.1585	57.90	6.75	18.76	16.59
C18H25N6P1	356.41	356.1878	60.66	7.07	23.58	8.69
C18H26N1P1	287.39	287.1803	75.23	9.12	4.87	10.78
C18H26N2P2	332.37	332.1571	65.05	7.88	8.43	18.64
C18H26N3P1	315.40	315.1864	68.55	8.31	13.32	9.82
C18H26N4P2	360.38	360.1633	59.99	7.27	15.55	17.19
C18H26N5P1	343.42	343.1926	62.96	7.63	20.39	9.02
C18H26N6P2	388.40	388.1694	55.67	6.75	21.64	15.95
C18H27N1P2	319.37	319.1619	67.70	8.52	4.39	19.40

C18H27N2P1	302.40	302.1912	71.49	9.00	9.26	10.24
C18H27N3P2	347.38	347.1680	62.24	7.83	12.10	17.83
C18H27N4P1	330.42	330.1973	65.43	8.24	16.96	9.37
C18H27N5P2	375.40	375.1742	57.59	7.25	18.66	16.50
C18H27N6P1	358.43	358.2035	60.32	7.59	23.45	8.64
C18H28N1P1	289.40	289.1959	74.71	9.75	4.84	10.70
C18H28N2P2	334.38	334.1728	64.66	8.44	8.38	18.53
C18H28N3P1	317.42	317.2021	68.11	8.89	13.24	9.76
C18H28N4P2	362.40	362.1789	59.66	7.79	15.46	17.09
C18H28N5P1	345.43	345.2082	62.59	8.17	20.27	8.97
C18H28N6P2	390.41	390.1851	55.38	7.23	21.53	15.87
C18H29N1P2	321.39	321.1775	67.27	9.10	4.36	19.28
C18H29N2P1	304.42	304.2068	71.02	9.60	9.20	10.17
C18H29N3P2	349.40	349.1837	61.88	8.37	12.03	17.73
C18H29N4P1	332.43	332.2130	65.04	8.79	16.85	9.32
C18H29N5P2	377.41	377.1898	57.28	7.75	18.56	16.41
C18H29N6P1	360.45	360.2191	59.98	8.11	23.32	8.59
C18H30N1P1	291.42	291.2116	74.19	10.38	4.81	10.63
C18H30N2P2	336.40	336.1884	64.27	8.99	8.33	18.41
C18H30N3P1	319.43	319.2177	67.68	9.47	13.15	9.70
C18H30N4P2	364.41	364.1946	59.33	8.30	15.37	17.00
C18H30N5P1	347.45	347.2239	62.23	8.70	20.16	8.91
C18H30N6P2	392.43	392.2007	55.09	7.71	21.42	15.79
C18H31N1P2	323.40	323.1932	66.85	9.66	4.33	19.15
C18H31N2P1	306.43	306.2225	70.55	10.20	9.14	10.11
C18H31N3P2	351.42	351.1993	61.52	8.89	11.96	17.63
C18H31N4P1	334.45	334.2286	64.64	9.34	16.75	9.26
C18H31N5P2	379.43	379.2055	56.98	8.24	18.46	16.33
C18H31N6P1	362.46	362.2348	59.65	8.62	23.19	8.55
C18H32N1P1	293.44	293.2272	73.68	10.99	4.77	10.56
C18H32N2P2	338.42	338.2041	63.89	9.53	8.28	18.31
C18H32N3P1	321.45	321.2334	67.26	10.03	13.07	9.64
C18H32N4P2	366.43	366.2102	59.00	8.80	15.29	16.91
C18H32N5P1	349.46	349.2395	61.87	9.23	20.04	8.86
C18H32N6P2	394.44	394.2164	54.81	8.18	21.31	15.71
C18H33N1P2	325.42	325.2088	66.44	10.22	4.30	19.04
C18H33N2P1	308.45	308.2381	70.09	10.78	9.08	10.04
C18H33N3P2	353.43	353.2150	61.17	9.41	11.89	17.53
C18H33N4P1	336.46	336.2443	64.26	9.89	16.65	9.21
C18H33N5P2	381.44	381.2211	56.68	8.72	18.36	16.24
C18H33N6P1	364.48	364.2504	59.32	9.13	23.06	8.50
C18H34N1P1	295.45	295.2429	73.18	11.60	4.74	10.48
C18H34N2P2	340.43	340.2197	63.51	10.07	8.23	18.20
C18H34N3P1	323.47	323.2490	66.84	10.59	12.99	9.58
C18H34N4P2	368.45	368.2259	58.68	9.30	15.21	16.81
C18H34N5P1	351.48	351.2552	61.51	9.75	19.93	8.81
C18H34N6P2	396.46	396.2320	54.53	8.64	21.20	15.63
C18H35N1P2	327.43	327.2245	66.03	10.77	4.28	18.92
C18H35N2P1	310.47	310.2538	69.64	11.36	9.02	9.98
C18H35N3P2	355.45	355.2306	60.82	9.93	11.82	17.43
C18H35N4P1	338.48	338.2599	63.87	10.42	16.55	9.15
C18H35N5P2	383.46	383.2368	56.38	9.20	18.26	16.15
C18H35N6P1	366.49	366.2661	58.99	9.63	22.93	8.45
C18H36N1P1	297.47	297.2585	72.68	12.20	4.71	10.41
C18H36N2P2	342.45	342.2354	63.13	10.60	8.18	18.09
C18H36N3P1	325.48	325.2647	66.42	11.15	12.91	9.52
C18H36N4P2	370.46	370.2415	58.36	9.80	15.12	16.72
C18H36N5P1	353.49	353.2708	61.16	10.27	19.81	8.76
C18H36N6P2	398.48	398.2477	54.26	9.11	21.09	15.55
C18H37N1P2	329.45	329.2401	65.62	11.32	4.25	18.80
C18H37N2P1	312.48	312.2694	69.19	11.94	8.96	9.91
C18H37N3P2	357.46	357.2463	60.48	10.43	11.76	17.33
C18H37N4P1	340.50	340.2756	63.50	10.95	16.45	9.10
C18H37N5P2	385.48	385.2524	56.09	9.68	18.17	16.07
C18H37N6P1	368.51	368.2817	58.67	10.12	22.81	8.41
C18H38N1P1	299.48	299.2742	72.19	12.79	4.68	10.34
C18H38N2P2	344.46	344.2510	62.76	11.12	8.13	17.98
C18H38N3P1	327.50	327.2803	66.02	11.70	12.83	9.46
C18H38N4P2	372.48	372.2572	58.04	10.28	15.04	16.63
C18H38N5P1	355.51	355.2865	60.81	10.77	19.70	8.71
C18H38N6P2	400.49	400.2633	53.98	9.56	20.98	15.47
C18H39N1P2	331.47	331.2558	65.23	11.86	4.23	18.69
C18H39N2P1	314.50	314.2851	68.74	12.50	8.91	9.85
C18H39N3P2	359.48	359.2619	60.14	10.94	11.69	17.23
C18H39N4P1	342.51	342.2912	63.12	11.48	16.36	9.04
C18H39N5P2	387.49	387.2681	55.79	10.14	18.07	15.99
C18H39N6P1	370.53	370.2974	58.35	10.61	22.68	8.36
C18H40N1P1	301.50	301.2898	71.71	13.37	4.65	10.27
C18H40N2P2	346.48	346.2667	62.40	11.64	8.09	17.88
C18H40N3P1	329.51	329.2960	65.61	12.24	12.75	9.40
C18H40N4P2	374.49	374.2728	57.73	10.77	14.96	16.54
C18H40N5P1	357.53	357.3021	60.47	11.28	19.59	8.66
C18H40N6P2	402.51	402.2790	53.71	10.02	20.88	15.39
C18H41N1P2	333.48	333.2714	64.83	12.39	4.20	18.58
C18H41N2P1	316.51	316.3007	68.31	13.06	8.85	9.79
C18H41N3P2	361.50	361.2776	59.81	11.43	11.62	17.14
C18H41N4P1	344.53	344.3069	62.75	12.00	16.26	8.99
C18H41N5P2	389.51	389.2837	55.51	10.61	17.98	15.90
C18H41N6P1	372.54	372.3130	58.03	11.09	22.56	8.31
C18H42N2P2	348.50	348.2823	62.04	12.15	8.04	17.78
C18H42N3P1	331.53	331.3116	65.21	12.77	12.67	9.34
C18H42N4P2	376.51	376.2885	57.42	11.24	14.88	16.45
C18H42N5P1	359.54	359.3178	60.13	11.77	19.48	8.61
C18H42N6P2	404.52	404.2946	53.45	10.47	20.78	15.31
C18H43N3P2	363.51	363.2932	59.48	11.92	11.56	17.04
C18H43N4P1	346.54	346.3225	62.39	12.51	16.17	8.94
C18H43N5P2	391.52	391.2994	55.22	11.07	17.89	15.82
C18H43N6P1	374.56	374.3287	57.72	11.57	22.44	8.27
C18H44N4P2	378.53	378.3041	57.12	11.72	14.80	16.37
C18H44N5P1	361.56	361.3334	59.80	12.27	19.37	8.57
C18H44N6P2	406.54	406.3103	53.18	10.91	20.67	15.24
C18H45N5P2	393.54	393.3150	54.94	11.53	17.80	15.74
C18H45N6P1	376.57	376.3443	57.41	12.05	22.32	8.23
C18H46N6P2	408.56	408.3259	52.92	11.35	20.57	15.16
C19H 9N1P2	313.24	313.0210	72.86	2.90	4.47	19.78
C19H 9N2P1	296.27	296.0503	77.03	3.06	9.46	10.45
C19H 9N3P2	341.25	341.0272	66.88	2.66	12.31	18.15
C19H 9N4P1	324.28	324.0565	70.37	2.80	17.28	9.55
C19H 9N5P2	369.26	369.0333	61.80	2.46	18.97	16.78
C19H 9N6P1	352.30	352.0626	64.78	2.58	23.85	8.79
C19H10N1P1	283.27	283.0551	80.56	3.56	4.94	10.93
C19H10N2P2	328.25	328.0319	69.52	3.07	8.53	18.87
C19H10N3P1	311.29	311.0612	73.31	3.24	13.50	9.95
C19H10N4P2	356.27	356.0381	64.06	2.83	15.73	17.39
C19H10N5P1	339.30	339.0674	67.26	2.97	20.64	9.13
C19H10N6P2	384.28	384.0442	59.39	2.62	21.87	16.12
C19H11N1P2	315.25	315.0367	72.39	3.52	4.44	19.65
C19H11N2P1	298.29	298.0660	76.51	3.72	9.39	10.38
C19H11N3P2	343.27	343.0428	66.48	3.23	12.24	18.05
C19H11N4P1	326.30	326.0721	69.94	3.40	17.17	9.49

C19H11N5P2	371.28	371.0490	61.47	2.99	18.86	16.68
C19H11N6P1	354.31	354.0783	64.41	3.13	23.72	8.74
C19H12N1P1	285.29	285.0707	79.99	4.24	4.91	10.86
C19H12N2P2	330.27	330.0476	69.10	3.66	8.48	18.76
C19H12N3P1	313.30	313.0769	72.84	3.86	13.41	9.89
C19H12N4P2	358.28	358.0537	63.70	3.38	15.64	17.29
C19H12N5P1	341.31	341.0830	66.86	3.54	20.52	9.07
C19H12N6P2	386.30	386.0599	59.08	3.13	21.76	16.04
C19H13N1P2	317.27	317.0523	71.93	4.13	4.41	19.53
C19H13N2P1	300.30	300.0816	75.99	4.36	9.33	10.31
C19H13N3P2	345.28	345.0585	66.09	3.80	12.17	17.94
C19H13N4P1	328.32	328.0878	69.51	3.99	17.06	9.43
C19H13N5P2	373.30	373.0646	61.13	3.51	18.76	16.59
C19H13N6P1	356.33	356.0939	64.05	3.68	23.58	8.69
C19H14N1P1	287.30	287.0864	79.43	4.91	4.88	10.78
C19H14N2P2	332.28	332.0632	68.68	4.25	8.43	18.64
C19H14N3P1	315.32	315.0925	72.38	4.48	13.33	9.82
C19H14N4P2	360.30	360.0694	63.34	3.92	15.55	17.19
C19H14N5P1	343.33	343.0987	66.47	4.11	20.40	9.02
C19H14N6P2	388.31	388.0755	58.77	3.63	21.64	15.95
C19H15N1P2	319.29	319.0680	71.48	4.74	4.39	19.40
C19H15N2P1	302.32	302.0973	75.49	5.00	9.27	10.25
C19H15N3P2	347.30	347.0741	65.71	4.35	12.10	17.84
C19H15N4P1	330.33	330.1034	69.09	4.58	16.96	9.38
C19H15N5P2	375.31	375.0803	60.81	4.03	18.66	16.51
C19H15N6P1	358.35	358.1096	63.68	4.22	23.45	8.64
C19H16N1P1	289.32	289.1020	78.88	5.57	4.84	10.71
C19H16N2P2	334.30	334.0789	68.27	4.82	8.38	18.53
C19H16N3P1	317.33	317.1082	71.92	5.08	13.24	9.76
C19H16N4P2	362.31	362.0850	62.99	4.45	15.46	17.10
C19H16N5P1	345.35	345.1143	66.08	4.67	20.28	8.97
C19H16N6P2	390.33	390.0912	58.47	4.13	21.53	15.87
C19H17N1P2	321.30	321.0836	71.03	5.33	4.36	19.28
C19H17N2P1	304.33	304.1129	74.99	5.63	9.20	10.18
C19H17N3P2	349.32	349.0898	65.33	4.91	12.03	17.73
C19H17N4P1	332.35	332.1191	68.67	5.16	16.86	9.32
C19H17N5P2	377.33	377.0959	60.48	4.54	18.56	16.42
C19H17N6P1	360.36	360.1252	63.33	4.76	23.32	8.60
C19H18N1P1	291.34	291.1177	78.33	6.23	4.81	10.63
C19H18N2P2	336.32	336.0945	67.86	5.39	8.33	18.42
C19H18N3P1	319.35	319.1238	71.46	5.68	13.16	9.70
C19H18N4P2	364.33	364.1007	62.64	4.98	15.38	17.00
C19H18N5P1	347.36	347.1300	65.70	5.22	20.16	8.92
C19H18N6P2	392.34	392.1068	58.17	4.62	21.42	15.79
C19H19N1P2	323.32	323.0993	70.58	5.92	4.33	19.16
C19H19N2P1	306.35	306.1286	74.49	6.25	9.14	10.11
C19H19N3P2	351.33	351.1054	64.96	5.45	11.96	17.63
C19H19N4P1	334.36	334.1347	68.25	5.73	16.76	9.26
C19H19N5P2	379.34	379.1116	60.16	5.05	18.46	16.33
C19H19N6P1	362.38	362.1409	62.98	5.28	23.19	8.55
C19H20N1P1	293.35	293.1333	77.79	6.87	4.77	10.56
C19H20N2P2	338.33	338.1102	67.45	5.96	8.28	18.31
C19H20N3P1	321.37	321.1395	71.01	6.27	13.08	9.64
C19H20N4P2	366.35	366.1163	62.29	5.50	15.29	16.91
C19H20N5P1	349.38	349.1456	65.32	5.77	20.05	8.87
C19H20N6P2	394.36	394.1225	57.87	5.11	21.31	15.71
C19H21N1P2	325.33	325.1149	70.15	6.51	4.31	19.04
C19H21N2P1	308.37	308.1442	74.01	6.86	9.08	10.04
C19H21N3P2	353.35	353.1211	64.59	5.99	11.89	17.53
C19H21N4P1	336.38	336.1504	67.84	6.29	16.66	9.21
C19H21N5P2	381.36	381.1272	59.84	5.55	18.36	16.24
C19H21N6P1	364.39	364.1565	62.63	5.81	23.06	8.50
C19H22N1P1	295.37	295.1490	77.26	7.51	4.74	10.49
C19H22N2P2	340.35	340.1258	67.05	6.52	8.23	18.20
C19H22N3P1	323.38	323.1551	70.57	6.86	12.99	9.58
C19H22N4P2	368.36	368.1320	61.95	6.02	15.21	16.82
C19H22N5P1	351.39	351.1613	64.94	6.31	19.93	8.81
C19H22N6P2	396.37	396.1381	57.57	5.59	21.20	15.63
C19H23N1P2	327.35	327.1306	69.72	7.08	4.28	18.92
C19H23N2P1	310.38	310.1599	73.53	7.47	9.03	9.98
C19H23N3P2	355.36	355.1367	64.22	6.52	11.82	17.43
C19H23N4P1	338.40	338.1660	67.44	6.85	16.56	9.15
C19H23N5P2	383.38	383.1429	59.53	6.05	18.27	16.16
C19H23N6P1	366.41	366.1722	62.28	6.33	22.94	8.45
C19H24N1P1	297.38	297.1646	76.74	8.13	4.71	10.42
C19H24N2P2	342.36	342.1415	66.66	7.07	8.18	18.09
C19H24N3P1	325.40	325.1708	70.13	7.43	12.91	9.52
C19H24N4P2	370.38	370.1476	61.62	6.53	15.13	16.73
C19H24N5P1	353.41	353.1769	64.57	6.85	19.82	8.76
C19H24N6P2	398.39	398.1538	57.28	6.07	21.09	15.55
C19H25N1P2	329.37	329.1462	69.29	7.65	4.25	18.81
C19H25N2P1	312.40	312.1755	73.05	8.07	8.97	9.91
C19H25N3P2	357.38	357.1524	63.86	7.05	11.76	17.33
C19H25N4P1	340.41	340.1817	67.04	7.40	16.46	9.10
C19H25N5P2	385.39	385.1585	59.22	6.54	18.17	16.07
C19H25N6P1	368.43	368.1878	61.94	6.84	22.81	8.41
C19H26N1P1	299.40	299.1803	76.22	8.75	4.68	10.35
C19H26N2P2	344.38	344.1571	66.27	7.61	8.13	17.99
C19H26N3P1	327.41	327.1864	69.70	8.00	12.83	9.46
C19H26N4P2	372.39	372.1633	61.28	7.04	15.05	16.63
C19H26N5P1	355.43	355.1926	64.21	7.37	19.70	8.71
C19H26N6P2	400.41	400.1694	56.99	6.55	20.99	15.47
C19H27N1P2	331.38	331.1619	68.87	8.21	4.23	18.69
C19H27N2P1	314.41	314.1912	72.58	8.66	8.91	9.85
C19H27N3P2	359.39	359.1680	63.50	7.57	11.69	17.24
C19H27N4P1	342.43	342.1973	66.65	7.95	16.36	9.05
C19H27N5P2	387.41	387.1742	58.91	7.02	18.08	15.99
C19H27N6P1	370.44	370.2035	61.61	7.35	22.69	8.36
C19H28N1P1	301.42	301.1959	75.71	9.36	4.65	10.28
C19H28N2P2	346.40	346.1728	65.88	8.15	8.09	17.88
C19H28N3P1	329.43	329.2021	69.27	8.57	12.76	9.40
C19H28N4P2	374.41	374.1789	60.95	7.54	14.96	16.55
C19H28N5P1	357.44	357.2082	63.85	7.90	19.59	8.67
C19H28N6P2	402.42	402.1851	56.71	7.01	20.88	15.39
C19H29N1P2	333.40	333.1775	68.45	8.77	4.20	18.58
C19H29N2P1	316.43	316.2068	72.12	9.24	8.85	9.79
C19H29N3P2	361.41	361.1837	63.14	8.09	11.63	17.14
C19H29N4P1	344.44	344.2130	66.26	8.49	16.27	8.99
C19H29N5P2	389.42	389.1898	58.60	7.51	17.98	15.91
C19H29N6P1	372.46	372.2191	61.27	7.85	22.56	8.32
C19H30N1P1	303.43	303.2116	75.21	9.97	4.62	10.21
C19H30N2P2	348.41	348.1884	65.50	8.68	8.04	17.78
C19H30N3P1	331.44	331.2177	68.85	9.12	12.68	9.35
C19H30N4P2	376.43	376.1946	60.63	8.03	14.88	16.46
C19H30N5P1	359.46	359.2239	63.49	8.41	19.48	8.62
C19H30N6P2	404.44	404.2007	56.43	7.48	20.78	15.32
C19H31N1P2	335.41	335.1932	68.04	9.32	4.18	18.47
C19H31N2P1	318.45	318.2225	71.66	9.81	8.80	9.73
C19H31N3P2	363.43	363.1993	62.79	8.60	11.56	17.05
C19H31N4P1	346.46	346.2286	65.87	9.02	16.17	8.94

075

C19H31N5P2	391.44	391.2055	58.30	7.98	17.89	15.83
C19H31N6P1	374.47	374.2348	60.94	8.34	22.44	8.27
C19H32N1P1	305.45	305.2272	74.71	10.56	4.59	10.14
C19H32N2P2	350.43	350.2041	65.12	9.20	7.99	17.68
C19H32N3P1	333.46	333.2334	68.44	9.67	12.60	9.29
C19H32N4P2	378.44	378.2102	60.30	8.52	14.80	16.37
C19H32N5P1	361.47	361.2395	63.13	8.92	19.37	8.57
C19H32N6P2	406.45	406.2164	56.15	7.94	20.68	15.24
C19H33N1P2	337.43	337.2088	67.63	9.86	4.15	18.36
C19H33N2P1	320.46	320.2381	71.21	10.38	8.74	9.67
C19H33N3P2	365.44	365.2150	62.45	9.10	11.50	16.95
C19H33N4P1	348.48	348.2443	65.49	9.55	16.08	8.89
C19H33N5P2	393.46	393.2211	58.00	8.45	17.80	15.74
C19H33N6P1	376.49	376.2504	60.62	8.84	22.32	8.23
C19H34N1P1	307.46	307.2429	74.22	11.15	4.56	10.07
C19H34N2P2	352.44	352.2197	64.75	9.72	7.95	17.58
C19H34N3P1	335.48	335.2490	68.03	10.22	12.53	9.23
C19H34N4P2	380.46	380.2259	59.98	9.01	14.73	16.28
C19H34N5P1	363.49	363.2552	62.78	9.43	19.27	8.52
C19H34N6P2	408.47	408.2320	55.87	8.39	20.57	15.17
C19H35N1P2	339.45	339.2245	67.23	10.39	4.13	18.25
C19H35N2P1	322.48	322.2538	70.77	10.94	8.69	9.60
C19H35N3P2	367.46	367.2306	62.11	9.60	11.44	16.86
C19H35N4P1	350.49	350.2599	65.11	10.07	15.99	8.84
C19H35N5P2	395.47	395.2368	57.71	8.92	17.71	15.66
C19H35N6P1	378.50	378.2661	60.29	9.32	22.20	8.18
C19H36N1P1	309.48	309.2585	73.74	11.73	4.53	10.01
C19H36N2P2	354.46	354.2354	64.38	10.24	7.90	17.48
C19H36N3P1	337.49	337.2647	67.62	10.75	12.45	9.18
C19H36N4P2	382.47	382.2415	59.67	9.49	14.65	16.20
C19H36N5P1	365.51	365.2708	62.44	9.93	19.16	8.47
C19H36N6P2	410.49	410.2477	55.60	8.84	20.47	15.09
C19H37N1P2	341.46	341.2401	66.83	10.92	4.10	18.14
C19H37N2P1	324.49	324.2694	70.33	11.49	8.63	9.55
C19H37N3P2	369.47	369.2463	61.77	10.09	11.37	16.77
C19H37N4P1	352.51	352.2756	64.74	10.58	15.89	8.79
C19H37N5P2	397.49	397.2524	57.41	9.38	17.62	15.58
C19H37N6P1	380.52	380.2817	59.97	9.80	22.09	8.14
C19H38N1P1	311.50	311.2742	73.26	12.30	4.50	9.94
C19H38N2P2	356.48	356.2510	64.02	10.74	7.86	17.38
C19H38N3P1	339.51	339.2803	67.22	11.28	12.38	9.12
C19H38N4P2	384.49	384.2572	59.35	9.96	14.57	16.11
C19H38N5P1	367.52	367.2865	62.09	10.42	19.06	8.43
C19H38N6P2	412.50	412.2633	55.32	9.29	20.37	15.02
C19H39N1P2	343.48	343.2558	66.44	11.44	4.08	18.04
C19H39N2P1	326.51	326.2851	69.89	12.04	8.58	9.49
C19H39N3P2	371.49	371.2619	61.43	10.58	11.31	16.68
C19H39N4P1	354.52	354.2912	64.37	11.09	15.80	8.74
C19H39N5P2	399.50	399.2681	57.12	9.84	17.53	15.51
C19H39N6P1	382.54	382.2974	59.66	10.28	21.97	8.10
C19H40N1P1	313.51	313.2898	72.79	12.86	4.47	9.88
C19H40N2P2	358.49	358.2667	63.66	11.25	7.81	17.28
C19H40N3P1	341.52	341.2960	66.82	11.81	12.30	9.07
C19H40N4P2	386.51	386.2728	59.04	10.43	14.50	16.03
C19H40N5P1	369.54	369.3021	61.76	10.91	18.95	8.38
C19H40N6P2	414.52	414.2790	55.05	9.73	20.27	14.94
C19H41N1P2	345.49	345.2714	66.05	11.96	4.05	17.93
C19H41N2P1	328.53	328.3007	69.47	12.58	8.53	9.43
C19H41N3P2	373.51	373.2776	61.10	11.06	11.25	16.59
C19H41N4P1	356.54	356.3069	64.01	11.59	15.71	8.69

C19H41N5P2	401.52	401.2837	56.84	10.29	17.44	15.43
C19H41N6P1	384.55	384.3130	59.34	10.75	21.85	8.05
C19H42N1P1	315.53	315.3055	72.33	13.42	4.44	9.82
C19H42N2P2	360.51	360.2823	63.30	11.74	7.77	17.18
C19H42N3P1	343.54	343.3116	66.43	12.32	12.23	9.02
C19H42N4P2	388.52	388.2885	58.74	10.90	14.42	15.94
C19H42N5P1	371.55	371.3178	61.42	11.39	18.85	8.34
C19H42N6P2	416.53	416.2946	54.79	10.16	20.18	14.87
C19H43N1P2	347.51	347.2871	65.67	12.47	4.03	17.83
C19H43N2P1	330.54	330.3164	69.04	13.11	8.47	9.37
C19H43N3P2	375.52	375.2932	60.77	11.54	11.19	16.50
C19H43N4P1	358.56	358.3225	63.65	12.09	15.63	8.64
C19H43N5P2	403.54	403.2994	56.55	10.74	17.35	15.35
C19H43N6P1	386.57	386.3287	59.04	11.21	21.74	8.01
C19H44N2P2	362.52	362.2980	62.95	12.23	7.73	17.09
C19H44N3P1	345.56	345.3273	66.04	12.83	12.16	8.96
C19H44N4P2	390.54	390.3041	58.44	11.36	14.35	15.86
C19H44N5P1	373.57	373.3334	61.09	11.87	18.75	8.29
C19H44N6P2	418.55	418.3103	54.52	10.60	20.08	14.80
C19H45N3P2	377.54	377.3089	60.45	12.01	11.13	16.41
C19H45N4P1	360.57	360.3382	63.29	12.58	15.54	8.59
C19H45N5P2	405.55	405.3150	56.27	11.18	17.27	15.27
C19H45N6P1	388.58	388.3443	58.73	11.67	21.63	7.97
C19H46N4P2	392.55	392.3198	58.14	11.81	14.27	15.78
C19H46N5P1	375.59	375.3491	60.76	12.35	18.65	8.25
C19H46N6P2	420.57	420.3259	54.26	11.02	19.98	14.73
C19H47N5P2	407.57	407.3307	55.99	11.62	17.18	15.20
C19H47N6P1	390.60	390.3600	58.43	12.13	21.52	7.93
C19H48N6P2	422.58	422.3416	54.00	11.45	19.89	14.66
C20H10N1P1	295.28	295.0551	81.35	3.41	4.74	10.49
C20H10N2P2	340.26	340.0319	70.60	2.96	8.23	18.21
C20H10N3P1	323.30	323.0612	74.30	3.12	13.00	9.58
C20H10N4P2	368.28	368.0381	65.23	2.74	15.21	16.82
C20H10N5P1	351.31	351.0674	68.38	2.87	19.93	8.82
C20H10N6P2	396.29	396.0442	60.62	2.54	21.21	15.63
C20H11N1P2	327.26	327.0367	73.40	3.39	4.28	18.93
C20H11N2P1	310.30	310.0660	77.42	3.57	9.03	9.98
C20H11N3P2	355.28	355.0428	67.62	3.12	11.83	17.44
C20H11N4P1	338.31	338.0721	71.01	3.28	16.56	9.16
C20H11N5P2	383.29	383.0490	62.67	2.89	18.27	16.16
C20H11N6P1	366.32	366.0783	65.58	3.03	22.94	8.46
C20H12N1P1	297.30	297.0707	80.80	4.07	4.71	10.42
C20H12N2P2	342.28	342.0476	70.18	3.53	8.18	18.10
C20H12N3P1	325.31	325.0769	73.84	3.72	12.92	9.52
C20H12N4P2	370.29	370.0537	64.87	3.27	15.13	16.73
C20H12N5P1	353.33	353.0830	67.99	3.42	19.82	8.77
C20H12N6P2	398.31	398.0599	60.31	3.04	21.10	15.55
C20H13N1P2	329.28	329.0523	72.95	3.98	4.25	18.81
C20H13N2P1	312.31	312.0816	76.92	4.20	8.97	9.92
C20H13N3P2	357.29	357.0585	67.23	3.67	11.76	17.34
C20H13N4P1	340.33	340.0878	70.59	3.85	16.46	9.10
C20H13N5P2	385.31	385.0646	62.35	3.40	18.18	16.08
C20H13N6P1	368.34	368.0939	65.22	3.56	22.82	8.41
C20H14N1P1	299.32	299.0864	80.26	4.71	4.68	10.35
C20H14N2P2	344.30	344.0632	69.77	4.10	8.14	17.99
C20H14N3P1	327.33	327.0925	73.39	4.31	12.84	9.46
C20H14N4P2	372.31	372.0694	64.52	3.79	15.05	16.64
C20H14N5P1	355.34	355.0987	67.60	3.97	19.71	8.87
C20H14N6P2	400.32	400.0755	60.01	3.53	20.99	15.47
C20H15N1P2	331.30	331.0680	72.51	4.56	4.23	18.70

076

C20H15N2P1	314.33	314.0973	76.42	4.81	8.91	9.85
C20H15N3P2	359.31	359.0741	66.86	4.21	11.69	17.24
C20H15N4P1	342.34	342.1034	70.17	4.42	16.37	9.05
C20H15N5P2	387.32	387.0803	62.02	3.90	18.08	15.99
C20H15N6P1	370.36	370.1096	64.86	4.08	22.69	8.36
C20H16N1P1	301.33	301.1020	79.72	5.35	4.65	10.28
C20H16N2P2	346.31	346.0789	69.37	4.66	8.09	17.89
C20H16N3P1	329.34	329.1082	72.94	4.90	12.76	9.40
C20H16N4P2	374.32	374.0850	64.17	4.31	14.97	16.55
C20H16N5P1	357.36	357.1143	67.22	4.51	19.60	8.67
C20H16N6P2	402.34	402.0912	59.71	4.01	20.89	15.40
C20H17N1P2	333.31	333.0836	72.07	5.14	4.20	18.59
C20H17N2P1	316.35	316.1129	75.94	5.42	8.86	9.79
C20H17N3P2	361.33	361.0898	66.48	4.74	11.63	17.14
C20H17N4P1	344.36	344.1191	69.76	4.98	16.27	8.99
C20H17N5P2	389.34	389.0959	61.70	4.40	17.99	15.91
C20H17N6P1	372.37	372.1252	64.51	4.60	22.57	8.32
C20H18N1P1	303.35	303.1177	79.19	5.98	4.62	10.21
C20H18N2P2	348.33	348.0945	68.96	5.21	8.04	17.78
C20H18N3P1	331.36	331.1238	72.50	5.48	12.68	9.35
C20H18N4P2	376.34	376.1007	63.83	4.82	14.89	16.46
C20H18N5P1	359.37	359.1300	66.84	5.05	19.49	8.62
C20H18N6P2	404.35	404.1068	59.41	4.49	20.78	15.32
C20H19N1P2	335.33	335.0993	71.64	5.71	4.18	18.47
C20H19N2P1	318.36	318.1286	75.46	6.02	8.80	9.73
C20H19N3P2	363.34	363.1054	66.11	5.27	11.56	17.05
C20H19N4P1	346.38	346.1347	69.35	5.53	16.18	8.94
C20H19N5P2	391.36	391.1116	61.38	4.89	17.90	15.83
C20H19N6P1	374.39	374.1409	64.16	5.12	22.45	8.27
C20H20N1P1	305.36	305.1333	78.67	6.60	4.59	10.14
C20H20N2P2	350.34	350.1102	68.57	5.75	8.00	17.68
C20H20N3P1	333.38	333.1395	72.06	6.05	12.60	9.29
C20H20N4P2	378.36	378.1163	63.49	5.33	14.81	16.37
C20H20N5P1	361.39	361.1456	66.47	5.58	19.38	8.57
C20H20N6P2	406.37	406.1225	59.11	4.96	20.68	15.24
C20H21N1P2	337.34	337.1149	71.21	6.27	4.15	18.36
C20H21N2P1	320.38	320.1442	74.98	6.61	8.74	9.67
C20H21N3P2	365.36	365.1211	65.75	5.79	11.50	16.96
C20H21N4P1	348.39	348.1504	68.95	6.08	16.08	8.89
C20H21N5P2	393.37	393.1272	61.07	5.38	17.80	15.75
C20H21N6P1	376.40	376.1565	63.82	5.62	22.33	8.23
C20H22N1P1	307.38	307.1490	78.15	7.21	4.56	10.08
C20H22N2P2	352.36	352.1258	68.18	6.29	7.95	17.58
C20H22N3P1	335.39	335.1551	71.62	6.61	12.53	9.24
C20H22N4P2	380.37	380.1320	63.15	5.83	14.73	16.29
C20H22N5P1	363.41	363.1613	66.10	6.10	19.27	8.52
C20H22N6P2	408.39	408.1381	58.82	5.43	20.58	15.17
C20H23N1P2	339.36	339.1306	70.79	6.83	4.13	18.25
C20H23N2P1	322.39	322.1599	74.51	7.19	8.69	9.61
C20H23N3P2	367.37	367.1367	65.39	6.31	11.44	16.86
C20H23N4P1	350.41	350.1660	68.56	6.62	15.99	8.84
C20H23N5P2	395.39	395.1429	60.76	5.86	17.71	15.67
C20H23N6P1	378.42	378.1722	63.48	6.13	22.21	8.19
C20H24N1P1	309.39	309.1646	77.64	7.82	4.53	10.01
C20H24N2P2	354.38	354.1415	67.79	6.83	7.91	17.48
C20H24N3P1	337.41	337.1708	71.20	7.17	12.45	9.18
C20H24N4P2	382.39	382.1476	62.82	6.33	14.65	16.20
C20H24N5P1	365.42	365.1769	65.74	6.62	19.17	8.48
C20H24N6P2	410.40	410.1538	58.53	5.89	20.48	15.09
C20H25N1P2	341.38	341.1462	70.37	7.38	4.10	18.15

C20H25N2P1	324.41	324.1755	74.05	7.77	8.64	9.55
C20H25N3P2	369.39	369.1524	65.03	6.82	11.38	16.77
C20H25N4P1	352.42	352.1817	68.16	7.15	15.90	8.79
C20H25N5P2	397.40	397.1585	60.45	6.34	17.62	15.59
C20H25N6P1	380.44	380.1878	63.14	6.62	22.09	8.14
C20H26N1P1	311.41	311.1803	77.14	8.42	4.50	9.95
C20H26N2P2	356.39	356.1571	67.40	7.35	7.86	17.38
C20H26N3P1	339.42	339.1864	70.77	7.72	12.38	9.13
C20H26N4P2	384.40	384.1633	62.49	6.82	14.57	16.12
C20H26N5P1	367.44	367.1926	65.38	7.13	19.06	8.43
C20H26N6P2	412.42	412.1694	58.25	6.35	20.38	15.02
C20H27N1P2	343.39	343.1619	69.96	7.93	4.08	18.04
C20H27N2P1	326.43	326.1912	73.59	8.34	8.58	9.49
C20H27N3P2	371.41	371.1680	64.68	7.33	11.31	16.68
C20H27N4P1	354.44	354.1973	67.78	7.68	15.81	8.74
C20H27N5P2	399.42	399.1742	60.14	6.81	17.53	15.51
C20H27N6P1	382.45	382.2035	62.81	7.12	21.97	8.10
C20H28N1P1	313.43	313.1959	76.64	9.00	4.47	9.88
C20H28N2P2	358.41	358.1728	67.03	7.87	7.82	17.28
C20H28N3P1	341.44	341.2021	70.36	8.27	12.31	9.07
C20H28N4P2	386.42	386.1789	62.17	7.30	14.50	16.03
C20H28N5P1	369.45	369.2082	65.02	7.64	18.96	8.38
C20H28N6P2	414.43	414.1851	57.96	6.81	20.28	14.95
C20H29N1P2	345.41	345.1775	69.55	8.46	4.06	17.93
C20H29N2P1	328.44	328.2068	73.14	8.90	8.53	9.43
C20H29N3P2	373.42	373.1837	64.33	7.83	11.25	16.59
C20H29N4P1	356.45	356.2130	67.39	8.20	15.72	8.69
C20H29N5P2	401.44	401.1898	59.84	7.28	17.45	15.43
C20H29N6P1	384.47	384.2191	62.48	7.60	21.86	8.06
C20H30N1P1	315.44	315.2116	76.15	9.59	4.44	9.82
C20H30N2P2	360.42	360.1884	66.65	8.39	7.77	17.19
C20H30N3P1	343.46	343.2177	69.94	8.80	12.23	9.02
C20H30N4P2	388.44	388.1946	61.84	7.78	14.42	15.95
C20H30N5P1	371.47	371.2239	64.67	8.14	18.85	8.34
C20H30N6P2	416.45	416.2007	57.68	7.26	20.18	14.88
C20H31N1P2	347.42	347.1932	69.14	8.99	4.03	17.83
C20H31N2P1	330.46	330.2225	72.69	9.46	8.48	9.37
C20H31N3P2	375.44	375.1993	63.98	8.32	11.19	16.50
C20H31N4P1	358.47	358.2286	67.01	8.72	15.63	8.64
C20H31N5P2	403.45	403.2055	59.54	7.74	17.36	15.35
C20H31N6P1	386.48	386.2348	62.16	8.08	21.74	8.01
C20H32N1P1	317.46	317.2272	75.67	10.16	4.41	9.76
C20H32N2P2	362.44	362.2041	66.28	8.90	7.73	17.09
C20H32N3P1	345.47	345.2334	69.53	9.34	12.16	8.97
C20H32N4P2	390.45	390.2102	61.52	8.26	14.35	15.87
C20H32N5P1	373.49	373.2395	64.32	8.64	18.75	8.29
C20H32N6P2	418.47	418.2164	57.41	7.71	20.08	14.80
C20H33N1P2	349.44	349.2088	68.75	9.52	4.01	17.73
C20H33N2P1	332.47	332.2381	72.25	10.00	8.43	9.32
C20H33N3P2	377.45	377.2150	63.64	8.81	11.13	16.41
C20H33N4P1	360.49	360.2443	66.64	9.23	15.54	8.59
C20H33N5P2	405.47	405.2211	59.25	8.20	17.27	15.28
C20H33N6P1	388.50	388.2504	61.83	8.56	21.63	7.97
C20H34N1P1	319.47	319.2429	75.19	10.73	4.38	9.70
C20H34N2P2	364.45	364.2197	65.91	9.40	7.69	17.00
C20H34N3P1	347.49	347.2490	69.13	9.86	12.09	8.91
C20H34N4P2	392.47	392.2259	61.21	8.73	14.28	15.78
C20H34N5P1	375.50	375.2552	63.97	9.13	18.65	8.25
C20H34N6P2	420.48	420.2320	57.13	8.15	19.99	14.73
C20H35N1P2	351.46	351.2245	68.35	10.04	3.99	17.63

C20H35N2P1	334.49	334.2538	71.82	10.55	8.37	9.26
C20H35N3P2	379.47	379.2306	63.30	9.30	11.07	16.32
C20H35N4P1	362.50	362.2599	66.27	9.73	15.46	8.54
C20H35N5P2	407.48	407.2368	58.95	8.66	17.19	15.20
C20H35N6P1	390.52	390.2661	61.51	9.03	21.52	7.93
C20H36N1P1	321.49	321.2585	74.72	11.29	4.36	9.63
C20H36N2P2	366.47	366.2354	65.55	9.90	7.64	16.90
C20H36N3P1	349.50	349.2647	68.73	10.38	12.02	8.86
C20H36N4P2	394.48	394.2415	60.90	9.20	14.20	15.70
C20H36N5P1	377.52	377.2708	63.63	9.61	18.55	8.20
C20H36N6P2	422.50	422.2477	56.86	8.59	19.89	14.66
C20H37N1P2	353.47	353.2401	67.96	10.55	3.96	17.53
C20H37N2P1	336.51	336.2694	71.39	11.08	8.32	9.20
C20H37N3P2	381.49	381.2463	62.97	9.78	11.01	16.24
C20H37N4P1	364.52	364.2756	65.90	10.23	15.37	8.50
C20H37N5P2	409.50	409.2524	58.66	9.11	17.10	15.13
C20H37N6P1	392.53	392.2817	61.20	9.50	21.41	7.89
C20H38N1P1	323.51	323.2742	74.26	11.84	4.33	9.57
C20H38N2P2	368.49	368.2510	65.19	10.39	7.60	16.81
C20H38N3P1	351.52	351.2803	68.34	10.90	11.95	8.81
C20H38N4P2	396.50	396.2572	60.59	9.66	14.13	15.62
C20H38N5P1	379.53	379.2865	63.29	10.09	18.45	8.16
C20H38N6P2	424.51	424.2633	56.59	9.02	19.80	14.59
C20H39N1P2	355.49	355.2558	67.58	11.06	3.94	17.43
C20H39N2P1	338.52	338.2851	70.96	11.61	8.28	9.15
C20H39N3P2	383.50	383.2619	62.64	10.25	10.96	16.15
C20H39N4P1	366.53	366.2912	65.54	10.73	15.29	8.45
C20H39N5P2	411.51	411.2681	58.38	9.55	17.02	15.05
C20H39N6P1	394.55	394.2974	60.89	9.96	21.30	7.85
C20H40N1P1	325.52	325.2898	73.80	12.39	4.30	9.52
C20H40N2P2	370.50	370.2667	64.84	10.88	7.56	16.72
C20H40N3P1	353.54	353.2960	67.95	11.40	11.89	8.76
C20H40N4P2	398.52	398.2728	60.28	10.12	14.06	15.54
C20H40N5P1	381.55	381.3021	62.96	10.57	18.36	8.12
C20H40N6P2	426.53	426.2790	56.32	9.45	19.70	14.52
C20H41N1P2	357.50	357.2714	67.19	11.56	3.92	17.33
C20H41N2P1	340.54	340.3007	70.54	12.14	8.23	9.10
C20H41N3P2	385.52	385.2776	62.31	10.72	10.90	16.07
C20H41N4P1	368.55	368.3069	65.18	11.21	15.20	8.40
C20H41N5P2	413.53	413.2837	58.09	9.99	16.94	14.98
C20H41N6P1	396.56	396.3130	60.58	10.42	21.19	7.81
C20H42N1P1	327.54	327.3055	73.34	12.93	4.28	9.46
C20H42N2P2	372.52	372.2823	64.49	11.36	7.52	16.63
C20H42N3P1	355.55	355.3116	67.56	11.91	11.82	8.71
C20H42N4P2	400.53	400.2885	59.98	10.57	13.99	15.47
C20H42N5P1	383.57	383.3178	62.63	11.04	18.26	8.08
C20H42N6P2	428.55	428.2946	56.06	9.88	19.61	14.46
C20H43N1P2	359.52	359.2871	66.82	12.06	3.90	17.23
C20H43N2P1	342.55	342.3164	70.13	12.65	8.18	9.04
C20H43N3P2	387.53	387.2932	61.99	11.18	10.84	15.99
C20H43N4P1	370.57	370.3225	64.83	11.70	15.12	8.36
C20H43N5P2	415.55	415.2994	57.81	10.43	16.85	14.91
C20H43N6P1	398.58	398.3287	60.27	10.87	21.08	7.77
C20H44N1P1	329.55	329.3211	72.89	13.46	4.25	9.40
C20H44N2P2	374.53	374.2980	64.14	11.84	7.48	16.54
C20H44N3P1	357.57	357.3273	67.18	12.40	11.75	8.66
C20H44N4P2	402.55	402.3041	59.68	11.02	13.92	15.39
C20H44N5P1	385.58	385.3334	62.30	11.50	18.16	8.03
C20H44N6P2	430.56	430.3103	55.79	10.30	19.52	14.39
C20H45N1P2	361.54	361.3027	66.45	12.55	3.87	17.13

C20H45N2P1	344.57	344.3320	69.72	13.16	8.13	8.99
C20H45N3P2	389.55	389.3089	61.67	11.64	10.79	15.90
C20H45N4P1	372.58	372.3382	64.48	12.17	15.04	8.31
C20H45N5P2	417.56	417.3150	57.53	10.86	16.77	14.84
C20H45N6P1	400.60	400.3443	59.97	11.32	20.98	7.73
C20H46N2P2	376.55	376.3136	63.80	12.31	7.44	16.45
C20H46N3P1	359.58	359.3429	66.81	12.89	11.69	8.61
C20H46N4P2	404.56	404.3198	59.38	11.46	13.85	15.31
C20H46N5P1	387.60	387.3491	61.98	11.96	18.07	7.99
C20H46N6P2	432.58	432.3259	55.53	10.72	19.43	14.32
C20H47N3P2	391.57	391.3245	61.35	12.10	10.73	15.82
C20H47N4P1	374.60	374.3538	64.13	12.65	14.96	8.27
C20H47N5P2	419.58	419.3307	57.25	11.29	16.69	14.76
C20H47N6P1	402.61	402.3600	59.67	11.77	20.87	7.69
C20H48N4P2	406.58	406.3354	59.08	11.90	13.78	15.24
C20H48N5P1	389.61	389.3647	61.66	12.42	17.98	7.95
C20H48N6P2	434.59	434.3416	55.28	11.13	19.34	14.25
C20H49N5P2	421.59	421.3463	56.98	11.72	16.61	14.69
C20H49N6P1	404.63	404.3756	59.37	12.21	20.77	7.65
C20H50N6P2	436.61	436.3572	55.02	11.54	19.25	14.19
C21H10N1P1	307.29	307.0551	82.08	3.28	4.56	10.08
C21H10N2P2	352.27	352.0319	71.60	2.86	7.95	17.59
C21H10N3P1	335.31	335.0612	75.22	3.01	12.53	9.24
C21H10N4P2	380.29	380.0381	66.33	2.65	14.73	16.29
C21H10N5P1	363.32	363.0674	69.42	2.77	19.28	8.53
C21H10N6P2	408.30	408.0442	61.78	2.47	20.58	15.17
C21H11N1P2	339.28	339.0367	74.34	3.27	4.13	18.26
C21H11N2P1	322.31	322.0660	78.26	3.44	8.69	9.61
C21H11N3P2	367.29	367.0428	68.67	3.02	11.44	16.87
C21H11N4P1	350.32	350.0721	72.00	3.16	15.99	8.84
C21H11N5P2	395.30	395.0490	63.81	2.80	17.72	15.67
C21H11N6P1	378.34	378.0783	66.67	2.93	22.21	8.19
C21H12N1P1	309.31	309.0707	81.55	3.91	4.53	10.01
C21H12N2P2	354.29	354.0476	71.19	3.41	7.91	17.48
C21H12N3P1	337.32	337.0769	74.78	3.59	12.46	9.18
C21H12N4P2	382.30	382.0537	65.98	3.16	14.66	16.20
C21H12N5P1	365.34	365.0830	69.04	3.31	19.17	8.48
C21H12N6P2	410.32	410.0599	61.47	2.95	20.48	15.10
C21H13N1P2	341.29	341.0523	73.91	3.84	4.10	18.15
C21H13N2P1	324.32	324.0816	77.77	4.04	8.64	9.55
C21H13N3P2	369.31	369.0585	68.30	3.55	11.38	16.77
C21H13N4P1	352.34	352.0878	71.59	3.72	15.90	8.79
C21H13N5P2	397.32	397.0646	63.48	3.30	17.63	15.59
C21H13N6P1	380.35	380.0939	66.32	3.45	22.10	8.14
C21H14N1P1	311.33	311.0864	81.02	4.53	4.50	9.95
C21H14N2P2	356.31	356.0632	70.79	3.96	7.86	17.39
C21H14N3P1	339.34	339.0925	74.33	4.16	12.38	9.13
C21H14N4P2	384.32	384.0694	65.63	3.67	14.58	16.12
C21H14N5P1	367.35	367.0987	68.66	3.84	19.06	8.43
C21H14N6P2	412.33	412.0755	61.17	3.42	20.38	15.02
C21H15N1P2	343.31	343.0680	73.47	4.40	4.08	18.04
C21H15N2P1	326.34	326.0973	77.29	4.63	8.58	9.49
C21H15N3P2	371.32	371.0741	67.93	4.07	11.32	16.68
C21H15N4P1	354.35	354.1034	71.18	4.27	15.81	8.74
C21H15N5P2	399.33	399.0803	63.16	3.79	17.54	15.51
C21H15N6P1	382.37	382.1096	65.97	3.95	21.98	8.10
C21H16N1P1	313.34	313.1020	80.50	5.15	4.47	9.88
C21H16N2P2	358.32	358.0789	70.39	4.50	7.82	17.29
C21H16N3P1	341.36	341.1082	73.89	4.72	12.31	9.07
C21H16N4P2	386.34	386.0850	65.29	4.17	14.50	16.03

078

C21H16N5P1	369.37	369.1143	68.29	4.37	18.96	8.39
C21H16N6P2	414.35	414.0912	60.87	3.89	20.28	14.95
C21H17N1P2	345.32	345.0836	73.04	4.96	4.06	17.94
C21H17N2P1	328.36	328.1129	76.82	5.22	8.53	9.43
C21H17N3P2	373.34	373.0898	67.56	4.59	11.26	16.59
C21H17N4P1	356.37	356.1191	70.78	4.81	15.72	8.69
C21H17N5P2	401.35	401.0959	62.85	4.27	17.45	15.43
C21H17N6P1	384.38	384.1252	65.62	4.46	21.86	8.06
C21H18N1P1	315.36	315.1177	79.98	5.75	4.44	9.82
C21H18N2P2	360.34	360.0945	70.00	5.04	7.77	17.19
C21H18N3P1	343.37	343.1238	73.46	5.28	12.24	9.02
C21H18N4P2	388.35	388.1007	64.95	4.67	14.43	15.95
C21H18N5P1	371.38	371.1300	67.92	4.89	18.86	8.34
C21H18N6P2	416.37	416.1068	60.58	4.36	20.18	14.88
C21H19N1P2	347.34	347.0993	72.62	5.51	4.03	17.83
C21H19N2P1	330.37	330.1286	76.35	5.80	8.48	9.38
C21H19N3P2	375.35	375.1054	67.20	5.10	11.19	16.50
C21H19N4P1	358.39	358.1347	70.38	5.34	15.63	8.64
C21H19N5P2	403.37	403.1116	62.53	4.75	17.36	15.36
C21H19N6P1	386.40	386.1409	65.28	4.96	21.75	8.02
C21H20N1P1	317.37	317.1333	79.48	6.35	4.41	9.76
C21H20N2P2	362.35	362.1102	69.61	5.56	7.73	17.10
C21H20N3P1	345.39	345.1395	73.03	5.84	12.17	8.97
C21H20N4P2	390.37	390.1163	64.61	5.16	14.35	15.87
C21H20N5P1	373.40	373.1456	67.55	5.40	18.76	8.30
C21H20N6P2	418.38	418.1225	60.29	4.82	20.09	14.81
C21H21N1P2	349.36	349.1149	72.20	6.06	4.01	17.73
C21H21N2P1	332.39	332.1442	75.89	6.37	8.43	9.32
C21H21N3P2	377.37	377.1211	66.84	5.61	11.14	16.42
C21H21N4P1	360.40	360.1504	69.99	5.87	15.55	8.59
C21H21N5P2	405.38	405.1272	62.22	5.22	17.28	15.28
C21H21N6P1	388.42	388.1565	64.94	5.45	21.64	7.97
C21H22N1P1	319.39	319.1490	78.97	6.94	4.39	9.70
C21H22N2P2	364.37	364.1258	69.22	6.09	7.69	17.00
C21H22N3P1	347.40	347.1551	72.61	6.38	12.10	8.92
C21H22N4P2	392.38	392.1320	64.28	5.65	14.28	15.79
C21H22N5P1	375.42	375.1613	67.19	5.91	18.65	8.25
C21H22N6P2	420.40	420.1381	60.00	5.27	19.99	14.74
C21H23N1P2	351.37	351.1306	71.79	6.60	3.99	17.63
C21H23N2P1	334.40	334.1599	75.43	6.93	8.38	9.26
C21H23N3P2	379.39	379.1367	66.48	6.11	11.08	16.33
C21H23N4P1	362.42	362.1660	69.60	6.40	15.46	8.55
C21H23N5P2	407.40	407.1429	61.91	5.69	17.19	15.21
C21H23N6P1	390.43	390.1722	64.60	5.94	21.52	7.93
C21H24N1P1	321.41	321.1646	78.48	7.53	4.36	9.64
C21H24N2P2	366.39	366.1415	68.84	6.60	7.65	16.91
C21H24N3P1	349.42	349.1708	72.19	6.92	12.03	8.86
C21H24N4P2	394.40	394.1476	63.95	6.13	14.21	15.71
C21H24N5P1	377.43	377.1769	66.83	6.41	18.56	8.21
C21H24N6P2	422.41	422.1538	59.71	5.73	19.90	14.67
C21H25N1P2	353.39	353.1462	71.38	7.13	3.96	17.53
C21H25N2P1	336.42	336.1755	74.98	7.49	8.33	9.21
C21H25N3P2	381.40	381.1524	66.13	6.61	11.02	16.24
C21H25N4P1	364.43	364.1817	69.21	6.91	15.37	8.50
C21H25N5P2	409.41	409.1585	61.61	6.15	17.11	15.13
C21H25N6P1	392.45	392.1878	64.27	6.42	21.41	7.89
C21H26N1P1	323.42	323.1803	77.99	8.10	4.33	9.58
C21H26N2P2	368.40	368.1571	68.47	7.11	7.60	16.82
C21H26N3P1	351.44	351.1864	71.77	7.46	11.96	8.81
C21H26N4P2	396.42	396.1633	63.63	6.61	14.13	15.63

C21H26N5P1	379.45	379.1926	66.47	6.91	18.46	8.16
C21H26N6P2	424.43	424.1694	59.43	6.17	19.80	14.60
C21H27N1P2	355.40	355.1619	70.97	7.66	3.94	17.43
C21H27N2P1	338.44	338.1912	74.53	8.04	8.28	9.15
C21H27N3P2	383.42	383.1680	65.79	7.10	10.96	16.16
C21H27N4P1	366.45	366.1973	68.83	7.43	15.29	8.45
C21H27N5P2	411.43	411.1742	61.31	6.61	17.02	15.06
C21H27N6P1	394.46	394.2035	63.94	6.90	21.30	7.85
C21H28N1P1	325.44	325.1959	77.51	8.67	4.30	9.52
C21H28N2P2	370.42	370.1728	68.09	7.62	7.56	16.72
C21H28N3P1	353.45	353.2021	71.36	7.99	11.89	8.76
C21H28N4P2	398.43	398.1789	63.31	7.08	14.06	15.55
C21H28N5P1	381.46	381.2082	66.12	7.40	18.36	8.12
C21H28N6P2	426.45	426.1851	59.15	6.62	19.71	14.53
C21H29N1P2	357.42	357.1775	70.57	8.18	3.92	17.33
C21H29N2P1	340.45	340.2068	74.09	8.59	8.23	9.10
C21H29N3P2	385.43	385.1837	65.44	7.58	10.90	16.07
C21H29N4P1	368.47	368.2130	68.46	7.93	15.21	8.41
C21H29N5P2	413.45	413.1898	61.01	7.07	16.94	14.98
C21H29N6P1	396.48	396.2191	63.62	7.37	21.20	7.81
C21H30N1P1	327.45	327.2116	77.03	9.23	4.28	9.46
C21H30N2P2	372.43	372.1884	67.73	8.12	7.52	16.63
C21H30N3P1	355.47	355.2177	70.96	8.51	11.82	8.71
C21H30N4P2	400.45	400.1946	62.99	7.55	13.99	15.47
C21H30N5P1	383.48	383.2239	65.77	7.89	18.26	8.08
C21H30N6P2	428.46	428.2007	58.87	7.06	19.61	14.46
C21H31N1P2	359.44	359.1932	70.18	8.69	3.90	17.23
C21H31N2P1	342.47	342.2225	73.65	9.12	8.18	9.04
C21H31N3P2	387.45	387.1993	65.10	8.06	10.85	15.99
C21H31N4P1	370.48	370.2286	68.08	8.43	15.12	8.36
C21H31N5P2	415.46	415.2055	60.71	7.52	16.86	14.91
C21H31N6P1	398.50	398.2348	63.30	7.84	21.09	7.77
C21H32N1P1	329.47	329.2272	76.56	9.79	4.25	9.40
C21H32N2P2	374.45	374.2041	67.36	8.61	7.48	16.54
C21H32N3P1	357.48	357.2334	70.56	9.02	11.75	8.66
C21H32N4P2	402.46	402.2102	62.67	8.01	13.92	15.39
C21H32N5P1	385.50	385.2395	65.43	8.37	18.17	8.03
C21H32N6P2	430.48	430.2164	58.59	7.49	19.52	14.39
C21H33N1P2	361.45	361.2088	69.78	9.20	3.88	17.14
C21H33N2P1	344.48	344.2381	73.22	9.66	8.13	8.99
C21H33N3P2	389.46	389.2150	64.76	8.54	10.79	15.91
C21H33N4P1	372.50	372.2443	67.71	8.93	15.04	8.32
C21H33N5P2	417.48	417.2211	60.42	7.97	16.78	14.84
C21H33N6P1	400.51	400.2504	62.98	8.31	20.98	7.73
C21H34N1P1	331.49	331.2429	76.09	10.34	4.23	9.34
C21H34N2P2	376.47	376.2197	67.00	9.10	7.44	16.46
C21H34N3P1	359.50	359.2490	70.16	9.53	11.69	8.62
C21H34N4P2	404.48	404.2259	62.36	8.47	13.85	15.32
C21H34N5P1	387.51	387.2552	65.09	8.84	18.07	7.99
C21H34N6P2	432.49	432.2320	58.32	7.92	19.43	14.32
C21H35N1P2	363.47	363.2245	69.40	9.71	3.85	17.04
C21H35N2P1	346.50	346.2538	72.79	10.18	8.08	8.94
C21H35N3P2	391.48	391.2306	64.43	9.01	10.73	15.82
C21H35N4P1	374.51	374.2599	67.35	9.42	14.96	8.27
C21H35N5P2	419.49	419.2368	60.13	8.41	16.69	14.77
C21H35N6P1	402.53	402.2661	62.66	8.76	20.88	7.69
C21H36N1P1	333.50	333.2585	75.63	10.88	4.20	9.29
C21H36N2P2	378.48	378.2354	66.64	9.59	7.40	16.37
C21H36N3P1	361.51	361.2647	69.77	10.04	11.62	8.57
C21H36N4P2	406.50	406.2415	62.05	8.93	13.78	15.24

079

C21H36N5P1	389.53	389.2708	64.75	9.32	17.98	7.95
C21H36N6P2	434.51	434.2477	58.05	8.35	19.34	14.26
C21H37N1P2	365.48	365.2401	69.01	10.20	3.83	16.95
C21H37N2P1	348.52	348.2694	72.37	10.70	8.04	8.89
C21H37N3P2	393.50	393.2463	64.10	9.48	10.68	15.74
C21H37N4P1	376.53	376.2756	66.99	9.90	14.88	8.23
C21H37N5P2	421.51	421.2524	59.84	8.85	16.61	14.70
C21H37N6P1	404.54	404.2817	62.35	9.22	20.77	7.66
C21H38N1P1	335.52	335.2742	75.18	11.42	4.17	9.23
C21H38N2P2	380.50	380.2510	66.29	10.07	7.36	16.28
C21H38N3P1	363.53	363.2803	69.38	10.54	11.56	8.52
C21H38N4P2	408.51	408.2572	61.74	9.38	13.71	15.16
C21H38N5P1	391.54	391.2865	64.42	9.78	17.89	7.91
C21H38N6P2	436.52	436.2633	57.78	8.77	19.25	14.19
C21H39N1P2	367.50	367.2558	68.64	10.70	3.81	16.86
C21H39N2P1	350.53	350.2851	71.96	11.21	7.99	8.84
C21H39N3P2	395.51	395.2619	63.77	9.94	10.62	15.66
C21H39N4P1	378.55	378.2912	66.63	10.38	14.80	8.18
C21H39N5P2	423.53	423.2681	59.56	9.28	16.54	14.63
C21H39N6P1	406.56	406.2974	62.04	9.67	20.67	7.62
C21H40N1P1	337.53	337.2898	74.73	11.95	4.15	9.18
C21H40N2P2	382.51	382.2667	65.94	10.54	7.32	16.19
C21H40N3P1	365.55	365.2960	69.00	11.03	11.50	8.47
C21H40N4P2	410.53	410.2728	61.44	9.82	13.65	15.09
C21H40N5P1	393.56	393.3021	64.09	10.24	17.79	7.87
C21H40N6P2	438.54	438.2790	57.52	9.19	19.16	14.13
C21H41N1P2	369.52	369.2714	68.26	11.18	3.79	16.76
C21H41N2P1	352.55	352.3007	71.55	11.72	7.95	8.79
C21H41N3P2	397.53	397.2776	63.45	10.40	10.57	15.58
C21H41N4P1	380.56	380.3069	66.28	10.86	14.72	8.14
C21H41N5P2	425.54	425.2837	59.27	9.71	16.46	14.56
C21H41N6P1	408.57	408.3130	61.74	10.11	20.57	7.58
C21H42N1P1	339.55	339.3055	74.28	12.47	4.13	9.12
C21H42N2P2	384.53	384.2823	65.60	11.01	7.29	16.11
C21H42N3P1	367.56	367.3116	68.62	11.52	11.43	8.43
C21H42N4P2	412.54	412.2885	61.14	10.26	13.58	15.02
C21H42N5P1	395.58	395.3178	63.76	10.70	17.70	7.83
C21H42N6P2	440.56	440.2946	57.25	9.61	19.08	14.06
C21H43N1P2	371.53	371.2871	67.89	11.67	3.77	16.67
C21H43N2P1	354.56	354.3164	71.14	12.22	7.90	8.74
C21H43N3P2	399.54	399.2932	63.13	10.85	10.52	15.50
C21H43N4P1	382.58	382.3225	65.93	11.33	14.64	8.10
C21H43N5P2	427.56	427.2994	58.99	10.14	16.38	14.49
C21H43N6P1	410.59	410.3287	61.43	10.56	20.47	7.54
C21H44N1P1	341.57	341.3211	73.85	12.98	4.10	9.07
C21H44N2P2	386.55	386.2980	65.25	11.47	7.25	16.03
C21H44N3P1	369.58	369.3273	68.25	12.00	11.37	8.38
C21H44N4P2	414.56	414.3041	60.84	10.70	13.51	14.94
C21H44N5P1	397.59	397.3334	63.44	11.15	17.61	7.79
C21H44N6P2	442.57	442.3103	56.99	10.02	18.99	14.00
C21H45N1P2	373.55	373.3027	67.52	12.14	3.75	16.58
C21H45N2P1	356.58	356.3320	70.74	12.72	7.86	8.69
C21H45N3P2	401.56	401.3089	62.81	11.30	10.46	15.43
C21H45N4P1	384.59	384.3382	65.58	11.79	14.57	8.05
C21H45N5P2	429.57	429.3150	58.72	10.56	16.30	14.42
C21H45N6P1	412.61	412.3443	61.13	10.99	20.37	7.51
C21H46N1P1	343.58	343.3368	73.41	13.50	4.08	9.01
C21H46N2P2	388.56	388.3136	64.91	11.93	7.21	15.94
C21H46N3P1	371.59	371.3429	67.88	12.48	11.31	8.34
C21H46N4P2	416.58	416.3198	60.55	11.13	13.45	14.87
C21H46N5P1	399.61	399.3491	63.12	11.60	17.53	7.75
C21H46N6P2	444.59	444.3259	56.73	10.43	18.90	13.93
C21H47N1P2	375.56	375.3184	67.16	12.61	3.73	16.49
C21H47N2P1	358.60	358.3477	70.34	13.21	7.81	8.64
C21H47N3P2	403.58	403.3245	62.50	11.74	10.41	15.35
C21H47N4P1	386.61	386.3538	65.24	12.25	14.49	8.01
C21H47N5P2	431.59	431.3307	58.44	10.98	16.23	14.35
C21H47N6P1	414.62	414.3600	60.83	11.43	20.27	7.47
C21H48N2P2	390.58	390.3293	64.58	12.39	7.17	15.86
C21H48N3P1	373.61	373.3586	67.51	12.95	11.25	8.29
C21H48N4P2	418.59	418.3354	60.26	11.56	13.38	14.80
C21H48N5P1	401.62	401.3647	62.80	12.05	17.44	7.71
C21H48N6P2	446.60	446.3416	56.48	10.83	18.82	13.87
C21H49N3P2	405.59	405.3402	62.19	12.18	10.36	15.27
C21H49N4P1	388.63	388.3695	64.90	12.71	14.42	7.97
C21H49N5P2	433.61	433.3463	58.17	11.39	16.15	14.29
C21H49N6P1	416.64	416.3756	60.54	11.85	20.17	7.43
C21H50N4P2	420.61	420.3511	59.97	11.98	13.32	14.73
C21H50N5P1	403.64	403.3804	62.49	12.49	17.35	7.67
C21H50N6P2	448.62	448.3572	56.22	11.23	18.73	13.81
C21H51N5P2	435.62	435.3620	57.90	11.80	16.08	14.22
C21H51N6P1	418.65	418.3913	60.25	12.28	20.07	7.40
C21H52N6P2	450.64	450.3729	55.97	11.63	18.65	13.75
C22H11N1P2	351.29	351.0367	75.22	3.16	3.99	17.63
C22H11N2P1	334.32	334.0660	79.04	3.32	8.38	9.26
C22H11N3P2	379.30	379.0428	69.67	2.92	11.08	16.33
C22H11N4P1	362.33	362.0721	72.93	3.06	15.46	8.55
C22H11N5P2	407.31	407.0490	64.88	2.72	17.19	15.21
C22H11N6P1	390.35	390.0783	67.69	2.84	21.53	7.93
C22H12N1P1	321.32	321.0707	82.24	3.76	4.36	9.64
C22H12N2P2	366.30	366.0476	72.14	3.30	7.65	16.91
C22H12N3P1	349.33	349.0769	75.64	3.46	12.03	8.87
C22H12N4P2	394.32	394.0537	67.01	3.07	14.21	15.71
C22H12N5P1	377.35	377.0830	70.03	3.21	18.56	8.21
C22H12N6P2	422.33	422.0599	62.57	2.86	19.90	14.67
C22H13N1P2	353.30	353.0523	74.79	3.71	3.96	17.53
C22H13N2P1	336.34	336.0816	78.57	3.90	8.33	9.21
C22H13N3P2	381.32	381.0585	69.30	3.44	11.02	16.25
C22H13N4P1	364.35	364.0878	72.53	3.60	15.38	8.50
C22H13N5P2	409.33	409.0646	64.56	3.20	17.11	15.13
C22H13N6P1	392.36	392.0939	67.35	3.34	21.42	7.89
C22H14N1P1	323.34	323.0864	81.72	4.36	4.33	9.58
C22H14N2P2	368.32	368.0632	71.74	3.83	7.61	16.82
C22H14N3P1	351.35	351.0925	75.21	4.02	11.96	8.82
C22H14N4P2	396.33	396.0694	66.67	3.56	14.14	15.63
C22H14N5P1	379.36	379.0987	69.65	3.72	18.46	8.16
C22H14N6P2	424.34	424.0755	62.27	3.33	19.80	14.60
C22H15N1P2	355.32	355.0680	74.37	4.26	3.94	17.43
C22H15N2P1	338.35	338.0973	78.10	4.47	8.28	9.15
C22H15N3P2	383.33	383.0741	68.93	3.94	10.96	16.16
C22H15N4P1	366.37	366.1034	72.13	4.13	15.29	8.45
C22H15N5P2	411.35	411.0803	64.24	3.68	17.03	15.06
C22H15N6P1	394.38	394.1096	67.00	3.83	21.31	7.85
C22H16N1P1	325.35	325.1020	81.22	4.96	4.31	9.52
C22H16N2P2	370.33	370.0789	71.35	4.35	7.56	16.73
C22H16N3P1	353.37	353.1082	74.78	4.56	11.89	8.77
C22H16N4P2	398.35	398.0850	66.34	4.05	14.06	15.55
C22H16N5P1	381.38	381.1143	69.29	4.23	18.36	8.12
C22H16N6P2	426.36	426.0912	61.98	3.78	19.71	14.53
C22H17N1P2	357.34	357.0836	73.95	4.80	3.92	17.34

080

C22H17N2P1	340.37	340.1129	77.64	5.03	8.23	9.10
C22H17N3P2	385.35	385.0898	68.57	4.45	10.90	16.08
C22H17N4P1	368.38	368.1191	71.73	4.65	15.21	8.41
C22H17N5P2	413.36	413.0959	63.93	4.15	16.94	14.99
C22H17N6P1	396.39	396.1252	66.66	4.32	21.20	7.81
C22H18N1P1	327.37	327.1177	80.72	5.54	4.28	9.46
C22H18N2P2	372.35	372.0945	70.97	4.87	7.52	16.64
C22H18N3P1	355.38	355.1238	74.36	5.11	11.82	8.72
C22H18N4P2	400.36	400.1007	66.00	4.53	13.99	15.47
C22H18N5P1	383.40	383.1300	68.92	4.73	18.27	8.08
C22H18N6P2	428.38	428.1068	61.69	4.24	19.62	14.46
C22H19N1P2	359.35	359.0993	73.53	5.33	3.90	17.24
C22H19N2P1	342.38	342.1286	77.18	5.59	8.18	9.05
C22H19N3P2	387.36	387.1054	68.22	4.94	10.85	15.99
C22H19N4P1	370.40	370.1347	71.34	5.17	15.13	8.36
C22H19N5P2	415.38	415.1116	63.62	4.61	16.86	14.91
C22H19N6P1	398.41	398.1409	66.32	4.81	21.09	7.77
C22H20N1P1	329.39	329.1333	80.22	6.12	4.25	9.40
C22H20N2P2	374.37	374.1102	70.58	5.38	7.48	16.55
C22H20N3P1	357.40	357.1395	73.94	5.64	11.76	8.67
C22H20N4P2	402.38	402.1163	65.67	5.01	13.92	15.40
C22H20N5P1	385.41	385.1456	68.56	5.23	18.17	8.04
C22H20N6P2	430.39	430.1225	61.40	4.68	19.53	14.39
C22H21N1P2	361.37	361.1149	73.12	5.86	3.88	17.14
C22H21N2P1	344.40	344.1442	76.73	6.15	8.13	8.99
C22H21N3P2	389.38	389.1211	67.86	5.44	10.79	15.91
C22H21N4P1	372.41	372.1504	70.95	5.68	15.04	8.32
C22H21N5P2	417.39	417.1272	63.31	5.07	16.78	14.84
C22H21N6P1	400.43	400.1565	65.99	5.29	20.99	7.74
C22H22N1P1	331.40	331.1490	79.74	6.69	4.23	9.35
C22H22N2P2	376.38	376.1258	70.21	5.89	7.44	16.46
C22H22N3P1	359.41	359.1551	73.52	6.17	11.69	8.62
C22H22N4P2	404.40	404.1320	65.34	5.48	13.85	15.32
C22H22N5P1	387.43	387.1613	68.21	5.72	18.08	7.99
C22H22N6P2	432.41	432.1381	61.11	5.13	19.44	14.33
C22H23N1P2	363.38	363.1306	72.72	6.38	3.85	17.05
C22H23N2P1	346.42	346.1599	76.28	6.69	8.09	8.94
C22H23N3P2	391.40	391.1367	67.51	5.92	10.74	15.83
C22H23N4P1	374.43	374.1660	70.57	6.19	14.96	8.27
C22H23N5P2	419.41	419.1429	63.00	5.53	16.70	14.77
C22H23N6P1	402.44	402.1722	65.66	5.76	20.88	7.70
C22H24N1P1	333.42	333.1646	79.25	7.26	4.20	9.29
C22H24N2P2	378.40	378.1415	69.83	6.39	7.40	16.37
C22H24N3P1	361.43	361.1708	73.11	6.69	11.63	8.57
C22H24N4P2	406.41	406.1476	65.02	5.95	13.79	15.24
C22H24N5P1	389.44	389.1769	67.85	6.21	17.98	7.95
C22H24N6P2	434.42	434.1538	60.83	5.57	19.35	14.26
C22H25N1P2	365.40	365.1462	72.32	6.90	3.83	16.95
C22H25N2P1	348.43	348.1755	75.84	7.23	8.04	8.89
C22H25N3P2	393.41	393.1524	67.17	6.41	10.68	15.75
C22H25N4P1	376.45	376.1817	70.19	6.69	14.88	8.23
C22H25N5P2	421.43	421.1585	62.70	5.98	16.62	14.70
C22H25N6P1	404.46	404.1878	65.33	6.23	20.78	7.66
C22H26N1P1	335.43	335.1803	78.78	7.81	4.18	9.23
C22H26N2P2	380.41	380.1571	69.46	6.89	7.36	16.28
C22H26N3P1	363.45	363.1864	72.71	7.21	11.56	8.52
C22H26N4P2	408.43	408.1633	64.70	6.42	13.72	15.17
C22H26N5P1	391.46	391.1926	67.50	6.69	17.89	7.91
C22H26N6P2	436.44	436.1694	60.55	6.00	19.26	14.19
C22H27N1P2	367.41	367.1619	71.92	7.41	3.81	16.86
C22H27N2P1	350.45	350.1912	75.40	7.77	7.99	8.84
C22H27N3P2	395.43	395.1680	66.83	6.88	10.63	15.67
C22H27N4P1	378.46	378.1973	69.82	7.19	14.80	8.18
C22H27N5P2	423.44	423.1742	62.40	6.43	16.54	14.63
C22H27N6P1	406.47	406.2035	65.01	6.70	20.68	7.62
C22H28N1P1	337.45	337.1959	78.31	8.36	4.15	9.18
C22H28N2P2	382.43	382.1728	69.10	7.38	7.33	16.20
C22H28N3P1	365.46	365.2021	72.30	7.72	11.50	8.48
C22H28N4P2	410.44	410.1789	64.38	6.88	13.65	15.09
C22H28N5P1	393.48	393.2082	67.16	7.17	17.80	7.87
C22H28N6P2	438.46	438.1851	60.27	6.44	19.17	14.13
C22H29N1P2	369.43	369.1775	71.53	7.91	3.79	16.77
C22H29N2P1	352.46	352.2068	74.97	8.29	7.95	8.79
C22H29N3P2	397.44	397.1837	66.49	7.35	10.57	15.59
C22H29N4P1	380.48	380.2130	69.45	7.68	14.73	8.14
C22H29N5P2	425.46	425.1898	62.11	6.87	16.46	14.56
C22H29N6P1	408.49	408.2191	64.69	7.16	20.57	7.58
C22H30N1P1	339.46	339.2116	77.84	8.91	4.13	9.12
C22H30N2P2	384.45	384.1884	68.73	7.87	7.29	16.11
C22H30N3P1	367.48	367.2177	71.91	8.23	11.43	8.43
C22H30N4P2	412.46	412.1946	64.07	7.33	13.58	15.02
C22H30N5P1	395.49	395.2239	66.81	7.65	17.71	7.83
C22H30N6P2	440.47	440.2007	59.99	6.87	19.08	14.06
C22H31N1P2	371.45	371.1932	71.14	8.41	3.77	16.68
C22H31N2P1	354.48	354.2225	74.54	8.81	7.90	8.74
C22H31N3P2	399.46	399.1993	66.15	7.82	10.52	15.51
C22H31N4P1	382.49	382.2286	69.09	8.17	14.65	8.10
C22H31N5P2	427.47	427.2055	61.82	7.31	16.38	14.49
C22H31N6P1	410.51	410.2348	64.37	7.61	20.47	7.55
C22H32N1P1	341.48	341.2272	77.38	9.45	4.10	9.07
C22H32N2P2	386.46	386.2041	68.38	8.35	7.25	16.03
C22H32N3P1	369.49	369.2334	71.52	8.73	11.37	8.38
C22H32N4P2	414.47	414.2102	63.75	7.78	13.52	14.95
C22H32N5P1	397.51	397.2395	66.48	8.11	17.62	7.79
C22H32N6P2	442.49	442.2164	59.72	7.29	18.99	14.00
C22H33N1P2	373.46	373.2088	70.76	8.91	3.75	16.59
C22H33N2P1	356.50	356.2381	74.12	9.33	7.86	8.69
C22H33N3P2	401.48	401.2150	65.82	8.29	10.47	15.43
C22H33N4P1	384.51	384.2443	68.72	8.65	14.57	8.06
C22H33N5P2	429.49	429.2211	61.53	7.74	16.31	14.42
C22H33N6P1	412.52	412.2504	64.06	8.06	20.37	7.51
C22H34N1P1	343.50	343.2429	76.93	9.98	4.08	9.02
C22H34N2P2	388.48	388.2197	68.02	8.82	7.21	15.95
C22H34N3P1	371.51	371.2490	71.13	9.22	11.31	8.34
C22H34N4P2	416.49	416.2259	63.45	8.23	13.45	14.87
C22H34N5P1	399.52	399.2552	66.14	8.58	17.53	7.75
C22H34N6P2	444.50	444.2320	59.45	7.71	18.91	13.94
C22H35N1P2	375.48	375.2245	70.38	9.40	3.73	16.50
C22H35N2P1	358.51	358.2538	73.71	9.84	7.81	8.64
C22H35N3P2	403.49	403.2306	65.49	8.74	10.41	15.35
C22H35N4P1	386.52	386.2599	68.36	9.13	14.50	8.01
C22H35N5P2	431.51	431.2368	61.24	8.18	16.23	14.36
C22H35N6P1	414.54	414.2661	63.74	8.51	20.27	7.47
C22H36N1P1	345.51	345.2585	76.48	10.50	4.05	8.96
C22H36N2P2	390.49	390.2354	67.67	9.29	7.17	15.86
C22H36N3P1	373.53	373.2647	70.74	9.71	11.25	8.29
C22H36N4P2	418.51	418.2415	63.14	8.67	13.39	14.80
C22H36N5P1	401.54	401.2708	65.81	9.04	17.44	7.71
C22H36N6P2	446.52	446.2477	59.18	8.13	18.82	13.87
C22H37N1P2	377.49	377.2401	70.00	9.88	3.71	16.41

C22H37N2P1	360.53	360.2694	73.29	10.34	7.77	8.59
C22H37N3P2	405.51	405.2463	65.16	9.20	10.36	15.28
C22H37N4P1	388.54	388.2756	68.01	9.60	14.42	7.97
C22H37N5P2	433.52	433.2524	60.95	8.60	16.15	14.29
C22H37N6P1	416.55	416.2817	63.44	8.95	20.18	7.44
C22H38N1P1	347.53	347.2742	76.04	11.02	4.03	8.91
C22H38N2P2	392.51	392.2510	67.32	9.76	7.14	15.78
C22H38N3P1	375.54	375.2803	70.36	10.20	11.19	8.25
C22H38N4P2	420.52	420.2572	62.84	9.11	13.32	14.73
C22H38N5P1	403.56	403.2865	65.48	9.49	17.35	7.68
C22H38N6P2	448.54	448.2633	58.91	8.54	18.74	13.81
C22H39N1P2	379.51	379.2558	69.63	10.36	3.69	16.32
C22H39N2P1	362.54	362.2851	72.89	10.84	7.73	8.54
C22H39N3P2	407.52	407.2619	64.84	9.65	10.31	15.20
C22H39N4P1	390.56	390.2912	67.66	10.07	14.35	7.93
C22H39N5P2	435.54	435.2681	60.67	9.03	16.08	14.22
C22H39N6P1	418.57	418.2974	63.13	9.39	20.08	7.40
C22H40N1P1	349.54	349.2898	75.60	11.53	4.01	8.86
C22H40N2P2	394.53	394.2667	66.98	10.22	7.10	15.70
C22H40N3P1	377.56	377.2960	69.99	10.68	11.13	8.20
C22H40N4P2	422.54	422.2728	62.54	9.54	13.26	14.66
C22H40N5P1	405.57	405.3021	65.15	9.94	17.27	7.64
C22H40N6P2	450.55	450.2790	58.65	8.95	18.65	13.75
C22H41N1P2	381.53	381.2714	69.26	10.83	3.67	16.24
C22H41N2P1	364.56	364.3007	72.48	11.34	7.68	8.50
C22H41N3P2	409.54	409.2776	64.52	10.09	10.26	15.13
C22H41N4P1	392.57	392.3069	67.31	10.53	14.27	7.89
C22H41N5P2	437.55	437.2837	60.39	9.44	16.01	14.16
C22H41N6P1	420.59	420.3130	62.83	9.83	19.98	7.36
C22H42N1P1	351.56	351.3055	75.16	12.04	3.98	8.81
C22H42N2P2	396.54	396.2823	66.64	10.68	7.06	15.62
C22H42N3P1	379.57	379.3116	69.62	11.15	11.07	8.16
C22H42N4P2	424.55	424.2885	62.24	9.97	13.20	14.59
C22H42N5P1	407.59	407.3178	64.83	10.39	17.18	7.60
C22H42N6P2	452.57	452.2946	58.39	9.35	18.57	13.69
C22H43N1P2	383.54	383.2871	68.90	11.30	3.65	16.15
C22H43N2P1	366.58	366.3164	72.08	11.82	7.64	8.45
C22H43N3P2	411.56	411.2932	64.21	10.53	10.21	15.05
C22H43N4P1	394.59	394.3225	66.97	10.98	14.20	7.85
C22H43N5P2	439.57	439.2994	60.11	9.86	15.93	14.09
C22H43N6P1	422.60	422.3287	62.53	10.26	19.89	7.33
C22H44N1P1	353.58	353.3211	74.73	12.54	3.96	8.76
C22H44N2P2	398.56	398.2980	66.30	11.13	7.03	15.54
C22H44N3P1	381.59	381.3273	69.25	11.62	11.01	8.12
C22H44N4P2	426.57	426.3041	61.95	10.40	13.13	14.52
C22H44N5P1	409.60	409.3334	64.51	10.83	17.10	7.56
C22H44N6P2	454.58	454.3103	58.13	9.76	18.49	13.63
C22H45N1P2	385.56	385.3027	68.54	11.76	3.63	16.07
C22H45N2P1	368.59	368.3320	71.65	12.31	7.60	8.40
C22H45N3P2	413.57	413.3089	63.89	10.97	10.16	14.98
C22H45N4P1	396.60	396.3382	66.63	11.44	14.13	7.81
C22H45N5P2	441.59	441.3150	59.84	10.27	15.86	14.03
C22H45N6P1	424.62	424.3443	62.23	10.68	19.79	7.29
C22H46N1P1	355.59	355.3368	74.31	13.04	3.94	8.71
C22H46N2P2	400.57	400.3136	65.97	11.58	6.99	15.46
C22H46N3P1	383.61	383.3429	68.88	12.09	10.95	8.07
C22H46N4P2	428.59	428.3198	61.66	10.82	13.07	14.45
C22H46N5P1	411.62	411.3491	64.20	11.26	17.01	7.52
C22H46N6P2	456.60	456.3259	57.87	10.15	18.41	13.57
C22H47N1P2	387.57	387.3184	68.18	12.22	3.61	15.98

C22H47N2P1	370.61	370.3477	71.30	12.78	7.56	8.36
C22H47N3P2	415.59	415.3245	63.58	11.40	10.11	14.91
C22H47N4P1	398.62	398.3538	66.29	11.88	14.06	7.77
C22H47N5P2	443.60	443.3307	59.57	10.68	15.79	13.96
C22H47N6P1	426.63	426.3600	61.94	11.10	19.70	7.26
C22H48N1P1	357.61	357.3524	73.89	13.53	3.92	8.66
C22H48N2P2	402.59	402.3293	65.64	12.02	6.96	15.39
C22H48N3P1	385.62	385.3586	68.52	12.55	10.90	8.03
C22H48N4P2	430.60	430.3354	61.37	11.24	13.01	14.39
C22H48N5P1	413.64	413.3647	63.88	11.70	16.93	7.49
C22H48N6P2	458.62	458.3416	57.62	10.55	18.32	13.51
C22H49N1P2	389.59	389.3340	67.83	12.68	3.60	15.90
C22H49N2P1	372.62	372.3633	70.91	13.25	7.52	8.31
C22H49N3P2	417.60	417.3402	63.28	11.83	10.06	14.83
C22H49N4P1	400.64	400.3695	65.96	12.33	13.98	7.73
C22H49N5P2	445.62	445.3463	59.30	11.08	15.72	13.90
C22H49N6P1	428.65	428.3756	61.65	11.52	19.61	7.23
C22H50N2P2	404.60	404.3449	65.31	12.46	6.92	15.31
C22H50N3P1	387.64	387.3742	68.17	13.00	10.84	7.99
C22H50N4P2	432.62	432.3511	61.08	11.65	12.95	14.32
C22H50N5P1	415.65	415.3804	63.57	12.13	16.85	7.45
C22H50N6P2	460.63	460.3572	57.37	10.94	18.24	13.45
C22H51N3P2	419.62	419.3558	62.97	12.25	10.01	14.76
C22H51N4P1	402.65	402.3851	65.63	12.77	13.91	7.69
C22H51N5P2	447.63	447.3620	59.03	11.48	15.65	13.84
C22H51N6P1	430.67	430.3913	61.36	11.94	19.51	7.19
C22H52N4P2	434.63	434.3667	60.80	12.06	12.89	14.25
C22H52N5P1	417.67	417.3960	63.27	12.55	16.77	7.42
C22H52N6P2	462.65	462.3729	57.12	11.33	18.17	13.39
C22H53N5P2	449.65	449.3776	58.77	11.88	15.58	13.78
C22H53N6P1	432.68	432.4069	61.07	12.35	19.42	7.16
C22H54N6P2	464.66	464.3885	56.87	11.71	18.09	13.33
C23H11N1P2	363.30	363.0367	76.04	3.05	3.86	17.05
C23H11N2P1	346.33	346.0660	79.77	3.20	8.09	8.94
C23H11N3P2	391.31	391.0428	70.60	2.83	10.74	15.83
C23H11N4P1	374.34	374.0721	73.80	2.96	14.97	8.27
C23H11N5P2	419.33	419.0490	65.88	2.64	16.70	14.77
C23H11N6P1	402.36	402.0783	68.66	2.76	20.89	7.70
C23H12N1P1	333.33	333.0707	82.88	3.63	4.20	9.29
C23H12N2P2	378.31	378.0476	73.02	3.20	7.40	16.37
C23H12N3P1	361.35	361.0769	76.45	3.35	11.63	8.57
C23H12N4P2	406.33	406.0537	67.99	2.98	13.79	15.25
C23H12N5P1	389.36	389.0830	70.95	3.11	17.99	7.96
C23H12N6P2	434.34	434.0599	63.60	2.78	19.35	14.26
C23H13N1P2	365.31	365.0523	75.62	3.59	3.83	16.96
C23H13N2P1	348.35	348.0816	79.30	3.76	8.04	8.89
C23H13N3P2	393.33	393.0585	70.24	3.33	10.68	15.75
C23H13N4P1	376.36	376.0878	73.40	3.48	14.89	8.23
C23H13N5P2	421.34	421.0646	65.57	3.11	16.62	14.70
C23H13N6P1	404.37	404.0939	68.32	3.24	20.78	7.66
C23H14N1P1	335.35	335.0864	82.38	4.21	4.18	9.24
C23H14N2P2	380.33	380.0632	72.64	3.71	7.37	16.29
C23H14N3P1	363.36	363.0925	76.03	3.88	11.56	8.52
C23H14N4P2	408.34	408.0694	67.65	3.46	13.72	15.17
C23H14N5P1	391.38	391.0987	70.59	3.61	17.89	7.91
C23H14N6P2	436.36	436.0755	63.31	3.23	19.26	14.20
C23H15N1P2	367.33	367.0680	75.21	4.12	3.81	16.86
C23H15N2P1	350.36	350.0973	78.85	4.32	8.00	8.84
C23H15N3P2	395.34	395.0741	69.88	3.82	10.63	15.67
C23H15N4P1	378.38	378.1034	73.01	4.00	14.81	8.19

C23H15N5P2	423.36	423.0803	65.25	3.57	16.54	14.63
C23H15N6P1	406.39	406.1096	67.98	3.72	20.68	7.62
C23H16N1P1	337.36	337.1020	81.89	4.78	4.15	9.18
C23H16N2P2	382.34	382.0789	72.25	4.22	7.33	16.20
C23H16N3P1	365.38	365.1082	75.61	4.41	11.50	8.48
C23H16N4P2	410.36	410.0850	67.32	3.93	13.65	15.10
C23H16N5P1	393.39	393.1143	70.22	4.10	17.80	7.87
C23H16N6P2	438.37	438.0912	63.02	3.68	19.17	14.13
C23H17N1P2	369.35	369.0836	74.80	4.64	3.79	16.77
C23H17N2P1	352.38	352.1129	78.40	4.86	7.95	8.79
C23H17N3P2	397.36	397.0898	69.52	4.31	10.57	15.59
C23H17N4P1	380.39	380.1191	72.62	4.50	14.73	8.14
C23H17N5P2	425.37	425.0959	64.94	4.03	16.46	14.56
C23H17N6P1	408.41	408.1252	67.64	4.20	20.58	7.58
C23H18N1P1	339.38	339.1177	81.40	5.35	4.13	9.13
C23H18N2P2	384.36	384.0945	71.87	4.72	7.29	16.12
C23H18N3P1	367.39	367.1238	75.19	4.94	11.44	8.43
C23H18N4P2	412.37	412.1007	66.99	4.40	13.59	15.02
C23H18N5P1	395.41	395.1300	69.87	4.59	17.71	7.83
C23H18N6P2	440.39	440.1068	62.73	4.12	19.08	14.07
C23H19N1P2	371.36	371.0993	74.39	5.16	3.77	16.68
C23H19N2P1	354.40	354.1286	77.95	5.40	7.90	8.74
C23H19N3P2	399.38	399.1054	69.17	4.80	10.52	15.51
C23H19N4P1	382.41	382.1347	72.24	5.01	14.65	8.10
C23H19N5P2	427.39	427.1116	64.64	4.48	16.39	14.49
C23H19N6P1	410.42	410.1409	67.31	4.67	20.48	7.55
C23H20N1P1	341.40	341.1333	80.92	5.90	4.10	9.07
C23H20N2P2	386.38	386.1102	71.50	5.22	7.25	16.03
C23H20N3P1	369.41	369.1395	74.78	5.46	11.37	8.38
C23H20N4P2	414.39	414.1163	66.67	4.86	13.52	14.95
C23H20N5P1	397.42	397.1456	69.51	5.07	17.62	7.79
C23H20N6P2	442.40	442.1225	62.44	4.56	19.00	14.00
C23H21N1P2	373.38	373.1149	73.99	5.67	3.75	16.59
C23H21N2P1	356.41	356.1442	77.51	5.94	7.86	8.69
C23H21N3P2	401.39	401.1211	68.82	5.27	10.47	15.43
C23H21N4P1	384.42	384.1504	71.86	5.51	14.57	8.06
C23H21N5P2	429.40	429.1272	64.33	4.93	16.31	14.43
C23H21N6P1	412.44	412.1565	66.98	5.13	20.38	7.51
C23H22N1P1	343.41	343.1490	80.44	6.46	4.08	9.02
C23H22N2P2	388.39	388.1258	71.13	5.71	7.21	15.95
C23H22N3P1	371.43	371.1551	74.38	5.97	11.31	8.34
C23H22N4P2	416.41	416.1320	66.34	5.33	13.45	14.88
C23H22N5P1	399.44	399.1613	69.16	5.55	17.53	7.75
C23H22N6P2	444.42	444.1381	62.16	4.99	18.91	13.94
C23H23N1P2	375.39	375.1306	73.59	6.18	3.73	16.50
C23H23N2P1	358.43	358.1599	77.07	6.47	7.82	8.64
C23H23N3P2	403.41	403.1367	68.48	5.75	10.42	15.36
C23H23N4P1	386.44	386.1660	71.49	6.00	14.50	8.02
C23H23N5P2	431.42	431.1429	64.03	5.37	16.23	14.36
C23H23N6P1	414.45	414.1722	66.66	5.59	20.28	7.47
C23H24N1P1	345.43	345.1646	79.98	7.00	4.05	8.97
C23H24N2P2	390.41	390.1415	70.76	6.20	7.18	15.87
C23H24N3P1	373.44	373.1708	73.98	6.48	11.25	8.29
C23H24N4P2	418.42	418.1476	66.02	5.78	13.39	14.81
C23H24N5P1	401.46	401.1769	68.81	6.03	17.44	7.72
C23H24N6P2	446.44	446.1538	61.88	5.42	18.82	13.88
C23H25N1P2	377.41	377.1462	73.20	6.68	3.71	16.41
C23H25N2P1	360.44	360.1755	76.64	6.99	7.77	8.59
C23H25N3P2	405.42	405.1524	68.14	6.22	10.36	15.28
C23H25N4P1	388.46	388.1817	71.12	6.49	14.42	7.97

C23H25N5P2	433.44	433.1585	63.74	5.81	16.16	14.29
C23H25N6P1	416.47	416.1878	66.33	6.05	20.18	7.44
C23H26N1P1	347.44	347.1803	79.51	7.54	4.03	8.91
C23H26N2P2	392.42	392.1571	70.40	6.68	7.14	15.79
C23H26N3P1	375.46	375.1864	73.58	6.98	11.19	8.25
C23H26N4P2	420.44	420.1633	65.71	6.23	13.33	14.73
C23H26N5P1	403.47	403.1926	68.47	6.50	17.36	7.68
C23H26N6P2	448.45	448.1694	61.60	5.84	18.74	13.81
C23H27N1P2	379.43	379.1619	72.81	7.17	3.69	16.33
C23H27N2P1	362.46	362.1912	76.22	7.51	7.73	8.55
C23H27N3P2	407.44	407.1680	67.80	6.68	10.31	15.20
C23H27N4P1	390.47	390.1973	70.75	6.97	14.35	7.93
C23H27N5P2	435.45	435.1742	63.44	6.25	16.08	14.23
C23H27N6P1	418.49	418.2035	66.01	6.50	20.08	7.40
C23H28N1P1	349.46	349.1959	79.05	8.08	4.01	8.86
C23H28N2P2	394.44	394.1728	70.04	7.16	7.10	15.71
C23H28N3P1	377.47	377.2021	73.19	7.48	11.13	8.21
C23H28N4P2	422.45	422.1789	65.39	6.68	13.26	14.66
C23H28N5P1	405.49	405.2082	68.13	6.96	17.27	7.64
C23H28N6P2	450.47	450.1851	61.33	6.27	18.66	13.75
C23H29N1P2	381.44	381.1775	72.42	7.66	3.67	16.24
C23H29N2P1	364.47	364.2068	75.80	8.02	7.69	8.50
C23H29N3P2	409.46	409.1837	67.47	7.14	10.26	15.13
C23H29N4P1	392.49	392.2130	70.39	7.45	14.27	7.89
C23H29N5P2	437.47	437.1898	63.15	6.68	16.01	14.16
C23H29N6P1	420.50	420.2191	65.70	6.95	19.99	7.37
C23H30N1P1	351.48	351.2116	78.60	8.60	3.99	8.81
C23H30N2P2	396.46	396.1884	69.68	7.63	7.07	15.63
C23H30N3P1	379.49	379.2177	72.80	7.97	11.07	8.16
C23H30N4P2	424.47	424.1946	65.08	7.12	13.20	14.59
C23H30N5P1	407.50	407.2239	67.79	7.42	17.19	7.60
C23H30N6P2	452.48	452.2007	61.05	6.68	18.57	13.69
C23H31N1P2	383.46	383.1932	72.04	8.15	3.65	16.15
C23H31N2P1	366.49	366.2225	75.38	8.53	7.64	8.45
C23H31N3P2	411.47	411.1993	67.14	7.59	10.21	15.06
C23H31N4P1	394.50	394.2286	70.03	7.92	14.20	7.85
C23H31N5P2	439.48	439.2055	62.86	7.11	15.94	14.10
C23H31N6P1	422.52	422.2348	65.38	7.40	19.89	7.33
C23H32N1P1	353.49	353.2272	78.15	9.12	3.96	8.76
C23H32N2P2	398.47	398.2041	69.33	8.09	7.03	15.55
C23H32N3P1	381.51	381.2334	72.41	8.45	11.01	8.12
C23H32N4P2	426.49	426.2102	64.78	7.56	13.14	14.53
C23H32N5P1	409.52	409.2395	67.46	7.88	17.10	7.56
C23H32N6P2	454.50	454.2164	60.78	7.10	18.49	13.63
C23H33N1P2	385.47	385.2088	71.67	8.63	3.63	16.07
C23H33N2P1	368.51	368.2381	74.97	9.03	7.60	8.41
C23H33N3P2	413.49	413.2150	66.81	8.04	10.16	14.98
C23H33N4P1	396.52	396.2443	69.67	8.39	14.13	7.81
C23H33N5P2	441.50	441.2211	62.57	7.53	15.86	14.03
C23H33N6P1	424.53	424.2504	65.07	7.84	19.80	7.30
C23H34N1P1	355.51	355.2429	77.71	9.64	3.94	8.71
C23H34N2P2	400.49	400.2197	68.98	8.56	6.99	15.47
C23H34N3P1	383.52	383.2490	72.03	8.94	10.96	8.08
C23H34N4P2	428.50	428.2259	64.47	8.00	13.08	14.46
C23H34N5P1	411.53	411.2552	67.13	8.33	17.02	7.53
C23H34N6P2	456.52	456.2320	60.51	7.51	18.41	13.57
C23H35N1P2	387.49	387.2245	71.29	9.10	3.61	15.99
C23H35N2P1	370.52	370.2538	74.56	9.52	7.56	8.36
C23H35N3P2	415.50	415.2306	66.49	8.49	10.11	14.91
C23H35N4P1	398.54	398.2599	69.32	8.85	14.06	7.77

083

C23H35N5P2	443.52	443.2368	62.29	7.95	15.79	13.97
C23H35N6P1	426.55	426.2661	64.77	8.27	19.70	7.26
C23H36N1P1	357.52	357.2585	77.27	10.15	3.92	8.66
C23H36N2P2	402.50	402.2354	68.63	9.02	6.96	15.39
C23H36N3P1	385.54	385.2647	71.65	9.41	10.90	8.03
C23H36N4P2	430.52	430.2415	64.17	8.43	13.01	14.39
C23H36N5P1	413.55	413.2708	66.80	8.77	16.93	7.49
C23H36N6P2	458.53	458.2477	60.25	7.91	18.33	13.51
C23H37N1P2	389.51	389.2401	70.92	9.57	3.60	15.90
C23H37N2P1	372.54	372.2694	74.16	10.01	7.52	8.31
C23H37N3P2	417.52	417.2463	66.17	8.93	10.06	14.84
C23H37N4P1	400.55	400.2756	68.97	9.31	13.99	7.73
C23H37N5P2	445.53	445.2524	62.01	8.37	15.72	13.90
C23H37N6P1	428.57	428.2817	64.46	8.70	19.61	7.23
C23H38N1P1	359.54	359.2742	76.84	10.65	3.90	8.61
C23H38N2P2	404.52	404.2510	68.29	9.47	6.93	15.31
C23H38N3P1	387.55	387.2803	71.28	9.88	10.84	7.99
C23H38N4P2	432.53	432.2572	63.87	8.86	12.95	14.32
C23H38N5P1	415.57	415.2865	66.48	9.22	16.85	7.45
C23H38N6P2	460.55	460.2633	59.98	8.32	18.25	13.45
C23H39N1P2	391.52	391.2558	70.56	10.04	3.58	15.82
C23H39N2P1	374.55	374.2851	73.76	10.50	7.48	8.27
C23H39N3P2	419.53	419.2619	65.85	9.37	10.02	14.77
C23H39N4P1	402.57	402.2912	68.62	9.77	13.92	7.69
C23H39N5P2	447.55	447.2681	61.73	8.78	15.65	13.84
C23H39N6P1	430.58	430.2974	64.16	9.13	19.52	7.19
C23H40N1P1	361.56	361.2898	76.41	11.15	3.87	8.57
C23H40N2P2	406.54	406.2667	67.95	9.92	6.89	15.24
C23H40N3P1	389.57	389.2960	70.91	10.35	10.79	7.95
C23H40N4P2	434.55	434.2728	63.57	9.28	12.89	14.26
C23H40N5P1	417.58	417.3021	66.16	9.66	16.77	7.42
C23H40N6P2	462.56	462.2790	59.72	8.72	18.17	13.39
C23H41N1P2	393.54	393.2714	70.20	10.50	3.56	15.74
C23H41N2P1	376.57	376.3007	73.36	10.97	7.44	8.23
C23H41N3P2	421.55	421.2776	65.53	9.80	9.97	14.70
C23H41N4P1	404.58	404.3069	68.28	10.21	13.85	7.66
C23H41N5P2	449.56	449.2837	61.45	9.19	15.58	13.78
C23H41N6P1	432.60	432.3130	63.86	9.55	19.43	7.16
C23H42N1P1	363.57	363.3055	75.98	11.64	3.85	8.52
C23H42N2P2	408.55	408.2823	67.62	10.36	6.86	15.16
C23H42N3P1	391.59	391.3116	70.55	10.81	10.73	7.91
C23H42N4P2	436.57	436.2885	63.28	9.70	12.83	14.19
C23H42N5P1	419.60	419.3178	65.84	10.09	16.69	7.38
C23H42N6P2	464.58	464.2946	59.46	9.11	18.09	13.33
C23H43N1P2	395.55	395.2871	69.84	10.96	3.54	15.66
C23H43N2P1	378.59	378.3164	72.97	11.45	7.40	8.18
C23H43N3P2	423.57	423.2932	65.22	10.23	9.92	14.63
C23H43N4P1	406.60	406.3225	67.94	10.66	13.78	7.62
C23H43N5P2	451.58	451.2994	61.18	9.60	15.51	13.72
C23H43N6P1	434.61	434.3287	63.56	9.97	19.34	7.13
C23H44N1P1	365.59	365.3211	75.57	12.13	3.83	8.47
C23H44N2P2	410.57	410.2980	67.29	10.80	6.82	15.09
C23H44N3P1	393.60	393.3273	70.19	11.27	10.68	7.87
C23H44N4P2	438.58	438.3041	62.99	10.11	12.77	14.12
C23H44N5P1	421.61	421.3334	65.52	10.52	16.61	7.35
C23H44N6P2	466.59	466.3103	59.21	9.51	18.01	13.28
C23H45N1P2	397.57	397.3027	69.49	11.41	3.52	15.58
C23H45N2P1	380.60	380.3320	72.58	11.92	7.36	8.14
C23H45N3P2	425.58	425.3089	64.91	10.66	9.87	14.56
C23H45N4P1	408.62	408.3382	67.61	11.10	13.71	7.58

C23H45N5P2	453.60	453.3150	60.90	10.00	15.44	13.66
C23H45N6P1	436.63	436.3443	63.27	10.39	19.25	7.09
C23H46N1P1	367.60	367.3368	75.15	12.61	3.81	8.43
C23H46N2P2	412.58	412.3136	66.96	11.24	6.79	15.01
C23H46N3P1	395.62	395.3429	69.83	11.72	10.62	7.83
C23H46N4P2	440.60	440.3198	62.70	10.52	12.72	14.06
C23H46N5P1	423.63	423.3491	65.21	10.95	16.53	7.31
C23H46N6P2	468.61	468.3259	58.95	9.89	17.93	13.22
C23H47N1P2	399.59	399.3184	69.14	11.86	3.51	15.50
C23H47N2P1	382.62	382.3477	72.20	12.38	7.32	8.10
C23H47N3P2	427.60	427.3245	64.61	11.08	9.83	14.49
C23H47N4P1	410.63	410.3538	67.28	11.54	13.64	7.54
C23H47N5P2	455.61	455.3307	60.63	10.40	15.37	13.60
C23H47N6P1	438.65	438.3600	62.98	10.80	19.16	7.06
C23H48N1P1	369.62	369.3524	74.74	13.09	3.79	8.38
C23H48N2P2	414.60	414.3293	66.63	11.67	6.76	14.94
C23H48N3P1	397.63	397.3586	69.48	12.17	10.57	7.79
C23H48N4P2	442.61	442.3354	62.41	10.93	12.66	14.00
C23H48N5P1	425.65	425.3647	64.90	11.37	16.45	7.28
C23H48N6P2	470.63	470.3416	58.70	10.28	17.86	13.16
C23H49N1P2	401.60	401.3340	68.79	12.30	3.49	15.43
C23H49N2P1	384.63	384.3633	71.82	12.84	7.28	8.05
C23H49N3P2	429.61	429.3402	64.30	11.50	9.78	14.42
C23H49N4P1	412.65	412.3695	66.95	11.97	13.58	7.51
C23H49N5P2	457.63	457.3463	60.37	10.79	15.30	13.54
C23H49N6P1	440.66	440.3756	62.69	11.21	19.07	7.03
C23H50N1P1	371.64	371.3681	74.34	13.56	3.77	8.33
C23H50N2P2	416.62	416.3449	66.31	12.10	6.72	14.87
C23H50N3P1	399.65	399.3742	69.12	12.61	10.51	7.75
C23H50N4P2	444.63	444.3511	62.13	11.33	12.60	13.93
C23H50N5P1	427.66	427.3804	64.60	11.78	16.38	7.24
C23H50N6P2	472.64	472.3572	58.45	10.66	17.78	13.11
C23H51N1P2	403.62	403.3497	68.45	12.74	3.47	15.35
C23H51N2P1	386.65	386.3790	71.45	13.30	7.25	8.01
C23H51N3P2	431.63	431.3558	64.00	11.91	9.74	14.35
C23H51N4P1	414.66	414.3851	66.62	12.40	13.51	7.47
C23H51N5P2	459.64	459.3620	60.10	11.18	15.24	13.48
C23H51N6P1	442.68	442.3913	62.41	11.61	18.98	7.00
C23H52N2P2	418.63	418.3606	65.99	12.52	6.69	14.80
C23H52N3P1	401.66	401.3899	68.78	13.05	10.46	7.71
C23H52N4P2	446.65	446.3667	61.85	11.74	12.54	13.87
C23H52N5P1	429.68	429.3960	64.29	12.20	16.30	7.21
C23H52N6P2	474.66	474.3729	58.20	11.04	17.71	13.05
C23H53N3P2	433.65	433.3715	63.71	12.32	9.69	14.29
C23H53N4P1	416.68	416.4008	66.30	12.82	13.45	7.43
C23H53N5P2	461.66	461.3776	59.84	11.57	15.17	13.42
C23H53N6P1	444.69	444.4069	62.12	12.01	18.90	6.97
C23H54N4P2	448.66	448.3824	61.57	12.13	12.49	13.81
C23H54N5P1	431.69	431.4117	63.99	12.61	16.22	7.17
C23H54N6P2	476.67	476.3885	57.95	11.42	17.63	13.00
C23H55N5P2	463.68	463.3933	59.58	11.96	15.10	13.36
C23H55N6P1	446.71	446.4226	61.84	12.41	18.81	6.93
C23H56N6P2	478.69	478.4041	57.71	11.79	17.56	12.94
C24H12N1P1	345.34	345.0707	83.47	3.50	4.06	8.97
C24H12N2P2	390.32	390.0476	73.85	3.10	7.18	15.87
C24H12N3P1	373.36	373.0769	77.21	3.24	11.25	8.30
C24H12N4P2	418.34	418.0537	68.91	2.89	13.39	14.81
C24H12N5P1	401.37	401.0830	71.82	3.01	17.45	7.72
C24H12N6P2	446.35	446.0599	64.58	2.71	18.83	13.88
C24H13N1P2	377.33	377.0523	76.40	3.47	3.71	16.42

084

C24H13N2P1	360.36	360.0816	79.99	3.64	7.77	8.60
C24H13N3P2	405.34	405.0585	71.12	3.23	10.37	15.28
C24H13N4P1	388.37	388.0878	74.22	3.37	14.43	7.98
C24H13N5P2	433.35	433.0646	66.52	3.02	16.16	14.29
C24H13N6P1	416.39	416.0939	69.23	3.15	20.18	7.44
C24H14N1P1	347.36	347.0864	82.99	4.06	4.03	8.92
C24H14N2P2	392.34	392.0632	73.47	3.60	7.14	15.79
C24H14N3P1	375.37	375.0925	76.79	3.76	11.19	8.25
C24H14N4P2	420.35	420.0694	68.58	3.36	13.33	14.74
C24H14N5P1	403.39	403.0987	71.46	3.50	17.36	7.68
C24H14N6P2	448.37	448.0755	64.29	3.15	18.74	13.82
C24H15N1P2	379.34	379.0680	75.99	3.99	3.69	16.33
C24H15N2P1	362.37	362.0973	79.55	4.17	7.73	8.55
C24H15N3P2	407.35	407.0741	70.77	3.71	10.32	15.21
C24H15N4P1	390.39	390.1034	73.84	3.87	14.35	7.93
C24H15N5P2	435.37	435.0803	66.21	3.47	16.09	14.23
C24H15N6P1	418.40	418.1096	68.90	3.61	20.09	7.40
C24H16N1P1	349.38	349.1020	82.51	4.62	4.01	8.87
C24H16N2P2	394.36	394.0789	73.10	4.09	7.10	15.71
C24H16N3P1	377.39	377.1082	76.38	4.27	11.13	8.21
C24H16N4P2	422.37	422.0850	68.25	3.82	13.26	14.67
C24H16N5P1	405.40	405.1143	71.11	3.98	17.28	7.64
C24H16N6P2	450.38	450.0912	64.01	3.58	18.66	13.75
C24H17N1P2	381.36	381.0836	75.59	4.49	3.67	16.24
C24H17N2P1	364.39	364.1129	79.11	4.70	7.69	8.50
C24H17N3P2	409.37	409.0898	70.42	4.19	10.26	15.13
C24H17N4P1	392.40	392.1191	73.46	4.37	14.28	7.89
C24H17N5P2	437.38	437.0959	65.91	3.92	16.01	14.16
C24H17N6P1	420.42	420.1252	68.57	4.08	19.99	7.37
C24H18N1P1	351.39	351.1177	82.04	5.16	3.99	8.81
C24H18N2P2	396.37	396.0945	72.73	4.58	7.07	15.63
C24H18N3P1	379.40	379.1238	75.98	4.78	11.08	8.16
C24H18N4P2	424.39	424.1007	67.93	4.28	13.20	14.60
C24H18N5P1	407.42	407.1300	70.75	4.45	17.19	7.60
C24H18N6P2	452.40	452.1068	63.72	4.01	18.58	13.69
C24H19N1P2	383.37	383.0993	75.19	5.00	3.65	16.16
C24H19N2P1	366.41	366.1286	78.67	5.23	7.65	8.45
C24H19N3P2	411.39	411.1054	70.07	4.66	10.21	15.06
C24H19N4P1	394.42	394.1347	73.05	4.86	14.20	7.85
C24H19N5P2	439.40	439.1116	65.60	4.36	15.94	14.10
C24H19N6P1	422.43	422.1409	68.24	4.53	19.89	7.33
C24H20N1P1	353.41	353.1333	81.57	5.70	3.96	8.76
C24H20N2P2	398.39	398.1102	72.36	5.06	7.03	15.55
C24H20N3P1	381.42	381.1395	75.58	5.29	11.02	8.12
C24H20N4P2	426.40	426.1163	67.60	4.73	13.14	14.53
C24H20N5P1	409.43	409.1456	70.41	4.92	17.10	7.57
C24H20N6P2	454.41	454.1225	63.44	4.44	18.49	13.63
C24H21N1P2	385.39	385.1149	74.80	5.49	3.63	16.07
C24H21N2P1	368.42	368.1442	78.24	5.75	7.60	8.41
C24H21N3P2	413.40	413.1211	69.73	5.12	10.16	14.98
C24H21N4P1	396.44	396.1504	72.71	5.34	14.13	7.81
C24H21N5P2	441.42	441.1272	65.31	4.80	15.87	14.03
C24H21N6P1	424.45	424.1565	67.92	4.99	19.80	7.30
C24H22N1P1	355.42	355.1490	81.11	6.24	3.94	8.71
C24H22N2P2	400.40	400.1258	71.99	5.54	7.00	15.47
C24H22N3P1	383.44	383.1551	75.18	5.78	10.96	8.08
C24H22N4P2	428.42	428.1320	67.29	5.18	13.08	14.46
C24H22N5P1	411.45	411.1613	70.06	5.39	17.02	7.53
C24H22N6P2	456.43	456.1381	63.16	4.86	18.41	13.57
C24H23N1P2	387.41	387.1306	74.41	5.98	3.62	15.99

C24H23N2P1	370.44	370.1599	77.82	6.26	7.56	8.36
C24H23N3P2	415.42	415.1367	69.39	5.58	10.12	14.91
C24H23N4P1	398.45	398.1660	72.35	5.82	14.06	7.77
C24H23N5P2	443.43	443.1429	65.01	5.23	15.79	13.97
C24H23N6P1	426.46	426.1722	67.59	5.44	19.71	7.26
C24H24N1P1	357.44	357.1646	80.65	6.77	3.92	8.67
C24H24N2P2	402.42	402.1415	71.63	6.01	6.96	15.39
C24H24N3P1	385.45	385.1708	74.79	6.28	10.90	8.04
C24H24N4P2	430.43	430.1476	66.97	5.62	13.02	14.39
C24H24N5P1	413.47	413.1769	69.72	5.85	16.94	7.49
C24H24N6P2	458.45	458.1538	62.88	5.28	18.33	13.51
C24H25N1P2	389.42	389.1462	74.02	6.47	3.60	15.91
C24H25N2P1	372.45	372.1755	77.40	6.77	7.52	8.32
C24H25N3P2	417.43	417.1524	69.06	6.04	10.07	14.84
C24H25N4P1	400.47	400.1817	71.98	6.29	13.99	7.73
C24H25N5P2	445.45	445.1585	64.71	5.66	15.72	13.91
C24H25N6P1	428.48	428.1878	67.28	5.88	19.61	7.23
C24H26N1P1	359.46	359.1803	80.20	7.29	3.90	8.62
C24H26N2P2	404.44	404.1571	71.28	6.48	6.93	15.32
C24H26N3P1	387.47	387.1864	74.40	6.76	10.84	7.99
C24H26N4P2	432.45	432.1633	66.66	6.06	12.96	14.32
C24H26N5P1	415.48	415.1926	69.38	6.31	16.86	7.45
C24H26N6P2	460.46	460.1694	62.60	5.69	18.25	13.45
C24H27N1P2	391.44	391.1619	73.64	6.95	3.58	15.83
C24H27N2P1	374.47	374.1912	76.98	7.27	7.48	8.27
C24H27N3P2	419.45	419.1680	68.73	6.49	10.02	14.77
C24H27N4P1	402.48	402.1973	71.62	6.76	13.92	7.70
C24H27N5P2	447.46	447.1742	64.42	6.08	15.65	13.84
C24H27N6P1	430.50	430.2035	66.96	6.32	19.52	7.19
C24H28N1P1	361.47	361.1959	79.75	7.81	3.87	8.57
C24H28N2P2	406.45	406.1728	70.92	6.94	6.89	15.24
C24H28N3P1	389.48	389.2021	74.01	7.25	10.79	7.95
C24H28N4P2	434.47	434.1789	66.35	6.50	12.90	14.26
C24H28N5P1	417.50	417.2082	69.05	6.76	16.77	7.42
C24H28N6P2	462.48	462.1851	62.33	6.10	18.17	13.39
C24H29N1P2	393.45	393.1775	73.27	7.43	3.56	15.74
C24H29N2P1	376.49	376.2068	76.57	7.76	7.44	8.23
C24H29N3P2	421.47	421.1837	68.40	6.94	9.97	14.70
C24H29N4P1	404.50	404.2130	71.27	7.23	13.85	7.66
C24H29N5P2	449.48	449.1898	64.13	6.50	15.58	13.78
C24H29N6P1	432.51	432.2191	66.65	6.76	19.43	7.16
C24H30N1P1	363.49	363.2116	79.31	8.32	3.85	8.52
C24H30N2P2	408.47	408.1884	70.57	7.40	6.86	15.17
C24H30N3P1	391.50	391.2177	73.63	7.72	10.73	7.91
C24H30N4P2	436.48	436.1946	66.04	6.93	12.84	14.19
C24H30N5P1	419.51	419.2239	68.71	7.21	16.69	7.38
C24H30N6P2	464.49	464.2007	62.06	6.51	18.09	13.34
C24H31N1P2	395.47	395.1932	72.89	7.90	3.54	15.66
C24H31N2P1	378.50	378.2225	76.16	8.26	7.40	8.18
C24H31N3P2	423.48	423.1993	68.07	7.38	9.92	14.63
C24H31N4P1	406.52	406.2286	70.91	7.69	13.78	7.62
C24H31N5P2	451.50	451.2055	63.85	6.92	15.51	13.72
C24H31N6P1	434.53	434.2348	66.34	7.19	19.34	7.13
C24H32N1P1	365.50	365.2272	78.87	8.82	3.83	8.47
C24H32N2P2	410.48	410.2041	70.23	7.86	6.82	15.09
C24H32N3P1	393.52	393.2334	73.25	8.20	10.68	7.87
C24H32N4P2	438.50	438.2102	65.74	7.36	12.78	14.13
C24H32N5P1	421.53	421.2395	68.39	7.65	16.61	7.35
C24H32N6P2	466.51	466.2164	61.79	6.91	18.01	13.28
C24H33N1P2	397.48	397.2088	72.52	8.37	3.52	15.58

085

C24H33N2P1	380.52	380.2381	75.76	8.74	7.36	8.14
C24H33N3P2	425.50	425.2150	67.75	7.82	9.88	14.56
C24H33N4P1	408.53	408.2443	70.56	8.14	13.71	7.58
C24H33N5P2	453.51	453.2211	63.56	7.33	15.44	13.66
C24H33N6P1	436.54	436.2504	66.03	7.62	19.25	7.10
C24H34N1P1	367.52	367.2429	78.44	9.32	3.81	8.43
C24H34N2P2	412.50	412.2197	69.88	8.31	6.79	15.02
C24H34N3P1	395.53	395.2490	72.88	8.66	10.62	7.83
C24H34N4P2	440.51	440.2259	65.44	7.78	12.72	14.06
C24H34N5P1	423.55	423.2552	68.06	8.09	16.54	7.31
C24H34N6P2	468.53	468.2320	61.53	7.31	17.94	13.22
C24H35N1P2	399.50	399.2245	72.16	8.83	3.51	15.51
C24H35N2P1	382.53	382.2538	75.36	9.22	7.32	8.10
C24H35N3P2	427.51	427.2306	67.43	8.25	9.83	14.49
C24H35N4P1	410.55	410.2599	70.22	8.59	13.65	7.54
C24H35N5P2	455.53	455.2368	63.28	7.74	15.37	13.60
C24H35N6P1	438.56	438.2661	65.73	8.04	19.16	7.06
C24H36N1P1	369.54	369.2585	78.01	9.82	3.79	8.38
C24H36N2P2	414.52	414.2354	69.54	8.75	6.76	14.94
C24H36N3P1	397.55	397.2647	72.51	9.13	10.57	7.79
C24H36N4P2	442.53	442.2415	65.14	8.20	12.66	14.00
C24H36N5P1	425.56	425.2708	67.74	8.53	16.46	7.28
C24H36N6P2	470.54	470.2477	61.26	7.71	17.86	13.17
C24H37N1P2	401.52	401.2401	71.79	9.29	3.49	15.43
C24H37N2P1	384.55	384.2694	74.96	9.70	7.28	8.05
C24H37N3P2	429.53	429.2463	67.11	8.68	9.78	14.42
C24H37N4P1	412.56	412.2756	69.87	9.04	13.58	7.51
C24H37N5P2	457.54	457.2524	63.00	8.15	15.31	13.54
C24H37N6P1	440.58	440.2817	65.43	8.47	19.08	7.03
C24H38N1P1	371.55	371.2742	77.58	10.31	3.77	8.34
C24H38N2P2	416.53	416.2510	69.21	9.20	6.73	14.87
C24H38N3P1	399.56	399.2803	72.15	9.59	10.52	7.75
C24H38N4P2	444.54	444.2572	64.85	8.62	12.60	13.94
C24H38N5P1	427.58	427.2865	67.42	8.96	16.38	7.24
C24H38N6P2	472.56	472.2633	61.00	8.11	17.78	13.11
C24H39N1P2	403.53	403.2558	71.44	9.74	3.47	15.35
C24H39N2P1	386.57	386.2851	74.57	10.17	7.25	8.01
C24H39N3P2	431.55	431.2619	66.80	9.11	9.74	14.35
C24H39N4P1	414.58	414.2912	69.53	9.48	13.51	7.47
C24H39N5P2	459.56	459.2681	62.73	8.55	15.24	13.48
C24H39N6P1	442.59	442.2974	65.13	8.88	18.99	7.00
C24H40N1P1	373.57	373.2898	77.17	10.79	3.75	8.29
C24H40N2P2	418.55	418.2667	68.87	9.63	6.69	14.80
C24H40N3P1	401.58	401.2960	71.78	10.04	10.46	7.71
C24H40N4P2	446.56	446.2728	64.55	9.03	12.55	13.87
C24H40N5P1	429.59	429.3021	67.10	9.39	16.30	7.21
C24H40N6P2	474.57	474.2790	60.74	8.50	17.71	13.05
C24H41N1P2	405.55	405.2714	71.08	10.19	3.45	15.28
C24H41N2P1	388.58	388.3007	74.18	10.64	7.21	7.97
C24H41N3P2	433.56	433.2776	66.49	9.53	9.69	14.29
C24H41N4P1	416.59	416.3069	69.20	9.92	13.45	7.43
C24H41N5P2	461.58	461.2837	62.45	8.95	15.17	13.42
C24H41N6P1	444.61	444.3130	64.84	9.30	18.90	6.97
C24H42N1P1	375.58	375.3055	76.75	11.27	3.73	8.25
C24H42N2P2	420.56	420.2823	68.54	10.07	6.66	14.73
C24H42N3P1	403.60	403.3116	71.42	10.49	10.41	7.67
C24H42N4P2	448.58	448.2885	64.26	9.44	12.49	13.81
C24H42N5P1	431.61	431.3178	66.79	9.81	16.23	7.18
C24H42N6P2	476.59	476.2946	60.49	8.88	17.63	13.00
C24H43N1P2	407.56	407.2871	70.73	10.63	3.44	15.20
C24H43N2P1	390.60	390.3164	73.80	11.10	7.17	7.93
C24H43N3P2	435.58	435.2932	66.18	9.95	9.65	14.22
C24H43N4P1	418.61	418.3225	68.86	10.35	13.38	7.40
C24H43N5P2	463.59	463.2994	62.18	9.35	15.11	13.36
C24H43N6P1	446.62	446.3287	64.54	9.70	18.82	6.94
C24H44N1P1	377.60	377.3211	76.34	11.75	3.71	8.20
C24H44N2P2	422.58	422.2980	68.22	10.50	6.63	14.66
C24H44N3P1	405.61	405.3273	71.07	10.93	10.36	7.64
C24H44N4P2	450.59	450.3041	63.98	9.84	12.43	13.75
C24H44N5P1	433.63	433.3334	66.48	10.23	16.15	7.14
C24H44N6P2	478.61	478.3103	60.23	9.27	17.56	12.94
C24H45N1P2	409.58	409.3027	70.38	11.07	3.42	15.12
C24H45N2P1	392.61	392.3320	73.42	11.55	7.14	7.89
C24H45N3P2	437.59	437.3089	65.88	10.37	9.60	14.16
C24H45N4P1	420.63	420.3382	68.53	10.78	13.32	7.36
C24H45N5P2	465.61	465.3150	61.91	9.74	15.04	13.30
C24H45N6P1	448.64	448.3443	64.25	10.11	18.73	6.90
C24H46N1P1	379.61	379.3368	75.94	12.21	3.69	8.16
C24H46N2P2	424.60	424.3136	67.89	10.92	6.60	14.59
C24H46N3P1	407.63	407.3429	70.72	11.37	10.31	7.60
C24H46N4P2	452.61	452.3198	63.69	10.24	12.38	13.69
C24H46N5P1	435.64	435.3491	66.17	10.64	16.08	7.11
C24H46N6P2	480.62	480.3259	59.98	9.65	17.49	12.89
C24H47N1P2	411.60	411.3184	70.04	11.51	3.40	15.05
C24H47N2P1	394.63	394.3477	73.05	12.00	7.10	7.85
C24H47N3P2	439.61	439.3245	65.57	10.78	9.56	14.09
C24H47N4P1	422.64	422.3538	68.21	11.21	13.26	7.33
C24H47N5P2	467.62	467.3307	61.65	10.13	14.98	13.25
C24H47N6P1	450.66	450.3600	63.97	10.51	18.65	6.87
C24H48N1P1	381.63	381.3524	75.54	12.68	3.67	8.12
C24H48N2P2	426.61	426.3293	67.57	11.34	6.57	14.52
C24H48N3P1	409.64	409.3586	70.37	11.81	10.26	7.56
C24H48N4P2	454.62	454.3354	63.41	10.64	12.32	13.63
C24H48N5P1	437.66	437.3647	65.87	11.05	16.00	7.08
C24H48N6P2	482.64	482.3416	59.73	10.02	17.41	12.84
C24H49N1P2	413.61	413.3340	69.70	11.94	3.39	14.98
C24H49N2P1	396.65	396.3633	72.68	12.45	7.06	7.81
C24H49N3P2	441.63	441.3402	65.27	11.18	9.51	14.03
C24H49N4P1	424.66	424.3695	67.88	11.63	13.19	7.29
C24H49N5P2	469.64	469.3463	61.38	10.52	14.91	13.19
C24H49N6P1	452.67	452.3756	63.68	10.91	18.57	6.84
C24H50N1P1	383.65	383.3681	75.14	13.14	3.65	8.07
C24H50N2P2	428.63	428.3449	67.25	11.76	6.54	14.45
C24H50N3P1	411.66	411.3742	70.03	12.24	10.21	7.52
C24H50N4P2	456.64	456.3511	63.13	11.04	12.27	13.57
C24H50N5P1	439.67	439.3804	65.56	11.46	15.93	7.04
C24H50N6P2	484.65	484.3572	59.48	10.40	17.34	12.78
C24H51N1P2	415.63	415.3497	69.36	12.37	3.37	14.90
C24H51N2P1	398.66	398.3790	72.31	12.89	7.03	7.77
C24H51N3P2	443.64	443.3558	64.98	11.59	9.47	13.96
C24H51N4P1	426.67	426.3851	67.56	12.05	13.13	7.26
C24H51N5P2	471.66	471.3620	61.12	10.90	14.85	13.13
C24H51N6P1	454.69	454.3913	63.40	11.31	18.48	6.81
C24H52N1P1	385.66	385.3837	74.75	13.59	3.63	8.03
C24H52N2P2	430.64	430.3606	66.94	12.17	6.51	14.38
C24H52N3P1	413.68	413.3899	69.68	12.67	10.16	7.49
C24H52N4P2	458.66	458.3667	62.85	11.43	12.22	13.51
C24H52N5P1	441.69	441.3960	65.26	11.87	15.86	7.01
C24H52N6P2	486.67	486.3729	59.23	10.77	17.27	12.73
C24H53N1P2	417.64	417.3653	69.02	12.79	3.35	14.83

086

C24H53N2P1	400.68	400.3946	71.95	13.33	6.99	7.73
C24H53N3P2	445.66	445.3715	64.68	11.99	9.43	13.90
C24H53N4P1	428.69	428.4008	67.24	12.46	13.07	7.23
C24H53N5P2	473.67	473.3776	60.86	11.28	14.79	13.08
C24H53N6P1	456.70	456.4069	63.12	11.70	18.40	6.78
C24H54N2P2	432.66	432.3762	66.63	12.58	6.47	14.32
C24H54N3P1	415.69	415.4055	69.35	13.09	10.11	7.45
C24H54N4P2	460.67	460.3824	62.58	11.82	12.16	13.45
C24H54N5P1	443.71	443.4117	64.97	12.27	15.78	6.98
C24H54N6P2	488.69	488.3885	58.99	11.14	17.20	12.68
C24H55N3P2	447.67	447.3871	64.39	12.38	9.39	13.84
C24H55N4P1	430.71	430.4164	66.93	12.87	13.01	7.19
C24H55N5P2	475.69	475.3933	60.60	11.65	14.72	13.02
C24H55N6P1	458.72	458.4226	62.84	12.09	18.32	6.75
C24H56N4P2	462.69	462.3980	62.30	12.20	12.11	13.39
C24H56N5P1	445.72	445.4273	64.67	12.66	15.71	6.95
C24H56N6P2	490.70	490.4041	58.75	11.50	17.13	12.62
C24H57N5P2	477.70	477.4089	60.34	12.03	14.66	12.97
C24H57N6P1	460.74	460.4382	62.57	12.47	18.24	6.72
C24H58N6P2	492.72	492.4198	58.51	11.87	17.06	12.57
C25H12N1P1	357.35	357.0707	84.03	3.38	3.92	8.67
C25H12N2P2	402.34	402.0476	74.63	3.01	6.96	15.40
C25H12N3P1	385.37	385.0769	77.92	3.14	10.90	8.04
C25H12N4P2	430.35	430.0537	69.78	2.81	13.02	14.39
C25H12N5P1	413.38	413.0830	72.64	2.93	16.94	7.49
C25H12N6P2	458.36	458.0599	65.51	2.64	18.33	13.51
C25H13N1P2	389.34	389.0523	77.13	3.37	3.60	15.91
C25H13N2P1	372.37	372.0816	80.64	3.52	7.52	8.32
C25H13N3P2	417.35	417.0585	71.95	3.14	10.07	14.84
C25H13N4P1	400.38	400.0878	75.00	3.27	13.99	7.74
C25H13N5P2	445.36	445.0646	67.42	2.94	15.73	13.91
C25H13N6P1	428.40	428.0939	70.09	3.06	19.62	7.23
C25H14N1P1	359.37	359.0864	83.56	3.93	3.90	8.62
C25H14N2P2	404.35	404.0632	74.26	3.49	6.93	15.32
C25H14N3P1	387.38	387.0925	77.51	3.64	10.85	8.00
C25H14N4P2	432.36	432.0694	69.45	3.26	12.96	14.33
C25H14N5P1	415.40	415.0987	72.29	3.40	16.86	7.46
C25H14N6P2	460.38	460.0755	65.22	3.07	18.25	13.46
C25H15N1P2	391.35	391.0680	76.73	3.86	3.58	15.83
C25H15N2P1	374.39	374.0973	80.21	4.04	7.48	8.27
C25H15N3P2	419.37	419.0741	71.60	3.61	10.02	14.77
C25H15N4P1	402.40	402.1034	74.62	3.76	13.92	7.70
C25H15N5P2	447.38	447.0803	67.12	3.38	15.65	13.85
C25H15N6P1	430.41	430.1096	69.77	3.51	19.53	7.20
C25H16N1P1	361.39	361.1020	83.09	4.46	3.88	8.57
C25H16N2P2	406.37	406.0789	73.89	3.97	6.89	15.24
C25H16N3P1	389.40	389.1082	77.11	4.14	10.79	7.95
C25H16N4P2	434.38	434.0850	69.13	3.71	12.90	14.26
C25H16N5P1	417.41	417.1143	71.94	3.86	16.78	7.42
C25H16N6P2	462.39	462.0912	64.94	3.49	18.18	13.40
C25H17N1P2	393.37	393.0836	76.34	4.36	3.56	15.75
C25H17N2P1	376.40	376.1129	79.78	4.55	7.44	8.23
C25H17N3P2	421.38	421.0898	71.26	4.07	9.97	14.70
C25H17N4P1	404.41	404.1191	74.25	4.24	13.85	7.66
C25H17N5P2	449.40	449.0959	66.82	3.81	15.58	13.78
C25H17N6P1	432.43	432.1252	69.44	3.96	19.43	7.16
C25H18N1P1	363.40	363.1177	82.63	4.99	3.85	8.52
C25H18N2P2	408.38	408.0945	73.53	4.44	6.86	15.17
C25H18N3P1	391.42	391.1238	76.72	4.64	10.74	7.91
C25H18N4P2	436.40	436.1007	68.81	4.16	12.84	14.20
C25H18N5P1	419.43	419.1300	71.59	4.33	16.70	7.38
C25H18N6P2	464.41	464.1068	64.66	3.91	18.10	13.34
C25H19N1P2	395.38	395.0993	75.95	4.84	3.54	15.67
C25H19N2P1	378.42	378.1286	79.35	5.06	7.40	8.19
C25H19N3P2	423.40	423.1054	70.92	4.52	9.92	14.63
C25H19N4P1	406.43	406.1347	73.88	4.71	13.79	7.62
C25H19N5P2	451.41	451.1116	66.52	4.24	15.51	13.72
C25H19N6P1	434.44	434.1409	69.12	4.41	19.34	7.13
C25H20N1P1	365.42	365.1333	82.17	5.52	3.83	8.48
C25H20N2P2	410.40	410.1102	73.17	4.91	6.83	15.09
C25H20N3P1	393.43	393.1395	76.32	5.12	10.68	7.87
C25H20N4P2	438.41	438.1163	68.49	4.60	12.78	14.13
C25H20N5P1	421.45	421.1456	71.25	4.78	16.62	7.35
C25H20N6P2	466.43	466.1225	64.38	4.32	18.02	13.28
C25H21N1P2	397.40	397.1149	75.56	5.33	3.52	15.59
C25H21N2P1	380.43	380.1442	78.93	5.56	7.36	8.14
C25H21N3P2	425.41	425.1211	70.59	4.98	9.88	14.56
C25H21N4P1	408.45	408.1504	73.52	5.18	13.72	7.58
C25H21N5P2	453.43	453.1272	66.22	4.67	15.45	13.66
C25H21N6P1	436.46	436.1565	68.80	4.85	19.25	7.10
C25H22N1P1	367.43	367.1490	81.72	6.04	3.81	8.43
C25H22N2P2	412.42	412.1258	72.81	5.38	6.79	15.02
C25H22N3P1	395.45	395.1551	75.93	5.61	10.63	7.83
C25H22N4P2	440.43	440.1320	68.18	5.03	12.72	14.07
C25H22N5P1	423.46	423.1613	70.91	5.24	16.54	7.31
C25H22N6P2	468.44	468.1381	64.10	4.73	17.94	13.22
C25H23N1P2	399.42	399.1306	75.18	5.80	3.51	15.51
C25H23N2P1	382.45	382.1599	78.51	6.06	7.32	8.10
C25H23N3P2	427.43	427.1367	70.25	5.42	9.83	14.49
C25H23N4P1	410.46	410.1660	73.16	5.65	13.65	7.55
C25H23N5P2	455.44	455.1429	65.93	5.09	15.38	13.60
C25H23N6P1	438.48	438.1722	68.48	5.29	19.17	7.06
C25H24N1P1	369.45	369.1646	81.28	6.55	3.79	8.38
C25H24N2P2	414.43	414.1415	72.46	5.84	6.76	14.95
C25H24N3P1	397.46	397.1708	75.55	6.09	10.57	7.79
C25H24N4P2	442.44	442.1476	67.87	5.47	12.66	14.00
C25H24N5P1	425.48	425.1769	70.57	5.69	16.46	7.28
C25H24N6P2	470.46	470.1538	63.83	5.14	17.86	13.17
C25H25N1P2	401.43	401.1462	74.80	6.28	3.49	15.43
C25H25N2P1	384.47	384.1755	78.10	6.55	7.29	8.06
C25H25N3P2	429.45	429.1524	69.92	5.87	9.78	14.43
C25H25N4P1	412.48	412.1817	72.80	6.11	13.58	7.51
C25H25N5P2	457.46	457.1585	65.64	5.51	15.31	13.54
C25H25N6P1	440.49	440.1878	68.17	5.72	19.08	7.03
C25H26N1P1	371.47	371.1803	80.84	7.06	3.77	8.34
C25H26N2P2	416.45	416.1571	72.10	6.29	6.73	14.88
C25H26N3P1	399.48	399.1864	75.17	6.56	10.52	7.75
C25H26N4P2	444.46	444.1633	67.56	5.90	12.61	13.94
C25H26N5P1	427.49	427.1926	70.24	6.13	16.38	7.25
C25H26N6P2	472.47	472.1694	63.55	5.55	17.79	13.11
C25H27N1P2	403.45	403.1619	74.43	6.75	3.47	15.35
C25H27N2P1	386.48	386.1912	77.70	7.04	7.25	8.01
C25H27N3P2	431.46	431.1680	69.60	6.31	9.74	14.36
C25H27N4P1	414.49	414.1973	72.44	6.57	13.52	7.47
C25H27N5P2	459.48	459.1742	65.35	5.92	15.24	13.48
C25H27N6P1	442.51	442.2035	67.86	6.15	18.99	7.00
C25H28N1P1	373.48	373.1959	80.40	7.56	3.75	8.29
C25H28N2P2	418.46	418.1728	71.76	6.74	6.69	14.80
C25H28N3P1	401.50	401.2021	74.79	7.03	10.47	7.71
C25H28N4P2	446.48	446.1789	67.26	6.32	12.55	13.87

087

C25H28N5P1	429.51	429.2082	69.91	6.57	16.31	7.21
C25H28N6P2	474.49	474.1851	63.28	5.95	17.71	13.06
C25H29N1P2	405.46	405.1775	74.06	7.21	3.45	15.28
C25H29N2P1	388.50	388.2068	77.29	7.52	7.21	7.97
C25H29N3P2	433.48	433.1837	69.27	6.74	9.69	14.29
C25H29N4P1	416.51	416.2130	72.09	7.02	13.45	7.44
C25H29N5P2	461.49	461.1898	65.07	6.33	15.18	13.42
C25H29N6P1	444.52	444.2191	67.55	6.58	18.91	6.97
C25H30N1P1	375.50	375.2116	79.97	8.05	3.73	8.25
C25H30N2P2	420.48	420.1884	71.41	7.19	6.66	14.73
C25H30N3P1	403.51	403.2177	74.42	7.49	10.41	7.68
C25H30N4P2	448.49	448.1946	66.95	6.74	12.49	13.81
C25H30N5P1	431.53	431.2239	69.59	7.01	16.23	7.18
C25H30N6P2	476.51	476.2007	63.02	6.35	17.64	13.00
C25H31N1P2	407.48	407.1932	73.69	7.67	3.44	15.20
C25H31N2P1	390.51	390.2225	76.89	8.00	7.17	7.93
C25H31N3P2	435.49	435.1993	68.95	7.18	9.65	14.22
C25H31N4P1	418.53	418.2286	71.75	7.47	13.39	7.40
C25H31N5P2	463.51	463.2055	64.78	6.74	15.11	13.36
C25H31N6P1	446.54	446.2348	67.25	7.00	18.82	6.94
C25H32N1P1	377.51	377.2272	79.54	8.54	3.71	8.20
C25H32N2P2	422.49	422.2041	71.07	7.63	6.63	14.66
C25H32N3P1	405.53	405.2334	74.05	7.95	10.36	7.64
C25H32N4P2	450.51	450.2102	66.65	7.16	12.44	13.75
C25H32N5P1	433.54	433.2395	69.26	7.44	16.15	7.14
C25H32N6P2	478.52	478.2164	62.75	6.74	17.56	12.95
C25H33N1P2	409.50	409.2088	73.33	8.12	3.42	15.13
C25H33N2P1	392.53	392.2381	76.50	8.47	7.14	7.89
C25H33N3P2	437.51	437.2150	68.63	7.60	9.60	14.16
C25H33N4P1	420.54	420.2443	71.40	7.91	13.32	7.37
C25H33N5P2	465.52	465.2211	64.50	7.15	15.04	13.31
C25H33N6P1	448.56	448.2504	66.94	7.42	18.74	6.91
C25H34N1P1	379.53	379.2429	79.12	9.03	3.69	8.16
C25H34N2P2	424.51	424.2197	70.74	8.07	6.60	14.59
C25H34N3P1	407.54	407.2490	73.68	8.41	10.31	7.60
C25H34N4P2	452.52	452.2259	66.36	7.57	12.38	13.69
C25H34N5P1	435.56	435.2552	68.94	7.87	16.08	7.11
C25H34N6P2	480.54	480.2320	62.49	7.13	17.49	12.89
C25H35N1P2	411.51	411.2245	72.97	8.57	3.40	15.05
C25H35N2P1	394.54	394.2538	76.11	8.94	7.10	7.85
C25H35N3P2	439.53	439.2306	68.32	8.03	9.56	14.09
C25H35N4P1	422.56	422.2599	71.06	8.35	13.26	7.33
C25H35N5P2	467.54	467.2368	64.23	7.55	14.98	13.25
C25H35N6P1	450.57	450.2661	66.64	7.83	18.65	6.87
C25H36N1P1	381.55	381.2585	78.70	9.51	3.67	8.12
C25H36N2P2	426.53	426.2354	70.40	8.51	6.57	14.52
C25H36N3P1	409.56	409.2647	73.32	8.86	10.26	7.56
C25H36N4P2	454.54	454.2415	66.06	7.98	12.33	13.63
C25H36N5P1	437.57	437.2708	68.62	8.29	16.00	7.08
C25H36N6P2	482.55	482.2477	62.23	7.52	17.42	12.84
C25H37N1P2	413.53	413.2401	72.61	9.02	3.39	14.98
C25H37N2P1	396.56	396.2694	75.72	9.40	7.06	7.81
C25H37N3P2	441.54	441.2463	68.01	8.45	9.52	14.03
C25H37N4P1	424.57	424.2756	70.72	8.78	13.20	7.30
C25H37N5P2	469.55	469.2524	63.95	7.94	14.91	13.19
C25H37N6P1	452.59	452.2817	66.35	8.24	18.57	6.84
C25H38N1P1	383.56	383.2742	78.29	9.99	3.65	8.08
C25H38N2P2	428.54	428.2510	70.07	8.94	6.54	14.46
C25H38N3P1	411.58	411.2803	72.96	9.31	10.21	7.53
C25H38N4P2	456.56	456.2572	65.77	8.39	12.27	13.57

C25H38N5P1	439.59	439.2865	68.31	8.71	15.93	7.05
C25H38N6P2	484.57	484.2633	61.97	7.90	17.34	12.78
C25H39N1P2	415.54	415.2558	72.26	9.46	3.37	14.91
C25H39N2P1	398.58	398.2851	75.34	9.86	7.03	7.77
C25H39N3P2	443.56	443.2619	67.70	8.86	9.47	13.97
C25H39N4P1	426.59	426.2912	70.39	9.22	13.13	7.26
C25H39N5P2	471.57	471.2681	63.68	8.34	14.85	13.14
C25H39N6P1	454.60	454.2974	66.05	8.65	18.49	6.81
C25H40N1P1	385.58	385.2898	77.88	10.46	3.63	8.03
C25H40N2P2	430.56	430.2667	69.74	9.36	6.51	14.39
C25H40N3P1	413.59	413.2960	72.60	9.75	10.16	7.49
C25H40N4P2	458.57	458.2728	65.48	8.79	12.22	13.51
C25H40N5P1	441.60	441.3021	68.00	9.13	15.86	7.01
C25H40N6P2	486.59	486.2790	61.71	8.29	17.27	12.73
C25H41N1P2	417.56	417.2714	71.91	9.90	3.35	14.84
C25H41N2P1	400.59	400.3007	74.96	10.32	6.99	7.73
C25H41N3P2	445.57	445.2776	67.39	9.27	9.43	13.90
C25H41N4P1	428.61	428.3069	70.06	9.64	13.07	7.23
C25H41N5P2	473.59	473.2837	63.41	8.73	14.79	13.08
C25H41N6P1	456.62	456.3130	65.76	9.05	18.40	6.78
C25H42N1P1	387.59	387.3055	77.47	10.92	3.61	7.99
C25H42N2P2	432.57	432.2823	69.42	9.79	6.48	14.32
C25H42N3P1	415.61	415.3116	72.25	10.19	10.11	7.45
C25H42N4P2	460.59	460.2885	65.19	9.19	12.16	13.45
C25H42N5P1	443.62	443.3178	67.69	9.54	15.79	6.98
C25H42N6P2	488.60	488.2946	61.46	8.66	17.20	12.68
C25H43N1P2	419.58	419.2871	71.57	10.33	3.34	14.76
C25H43N2P1	402.61	402.3164	74.58	10.77	6.96	7.69
C25H43N3P2	447.59	447.2932	67.09	9.68	9.39	13.84
C25H43N4P1	430.62	430.3225	69.73	10.07	13.01	7.19
C25H43N5P2	475.60	475.2994	63.14	9.11	14.73	13.03
C25H43N6P1	458.64	458.3287	65.47	9.45	18.32	6.75
C25H44N1P1	389.61	389.3211	77.07	11.38	3.60	7.95
C25H44N2P2	434.59	434.2980	69.09	10.21	6.45	14.25
C25H44N3P1	417.62	417.3273	71.90	10.62	10.06	7.42
C25H44N4P2	462.60	462.3041	64.91	9.59	12.11	13.39
C25H44N5P1	445.64	445.3334	67.38	9.95	15.72	6.95
C25H44N6P2	490.62	490.3103	61.20	9.04	17.13	12.63
C25H45N1P2	421.59	421.3027	71.23	10.76	3.32	14.69
C25H45N2P1	404.62	404.3320	74.21	11.21	6.92	7.65
C25H45N3P2	449.61	449.3089	66.79	10.09	9.35	13.78
C25H45N4P1	432.64	432.3382	69.41	10.48	12.95	7.16
C25H45N5P2	477.62	477.3150	62.87	9.50	14.66	12.97
C25H45N6P1	460.65	460.3443	65.19	9.85	18.24	6.72
C25H46N1P1	391.63	391.3368	76.67	11.84	3.58	7.91
C25H46N2P2	436.61	436.3136	68.78	10.62	6.42	14.19
C25H46N3P1	419.64	419.3429	71.56	11.05	10.01	7.38
C25H46N4P2	464.62	464.3198	64.63	9.98	12.06	13.33
C25H46N5P1	447.65	447.3491	67.08	10.36	15.64	6.92
C25H46N6P2	492.63	492.3259	60.95	9.41	17.06	12.57
C25H47N1P2	423.61	423.3184	70.89	11.18	3.31	14.62
C25H47N2P1	406.64	406.3477	73.84	11.65	6.89	7.62
C25H47N3P2	451.62	451.3245	66.49	10.49	9.30	13.72
C25H47N4P1	434.65	434.3538	69.08	10.90	12.89	7.13
C25H47N5P2	479.63	479.3307	62.61	9.88	14.60	12.92
C25H47N6P1	462.67	462.3600	64.90	10.24	18.16	6.69
C25H48N1P1	393.64	393.3524	76.28	12.29	3.56	7.87
C25H48N2P2	438.62	438.3293	68.46	11.03	6.39	14.12
C25H48N3P1	421.66	421.3586	71.21	11.47	9.97	7.35
C25H48N4P2	466.64	466.3354	64.35	10.37	12.01	13.28

088

C25H48N5P1	449.67	449.3647	66.78	10.76	15.57	6.89
C25H48N6P2	494.65	494.3416	60.71	9.78	16.99	12.52
C25H49N1P2	425.62	425.3340	70.55	11.60	3.29	14.55
C25H49N2P1	408.66	408.3633	73.48	12.09	6.85	7.58
C25H49N3P2	453.64	453.3402	66.19	10.89	9.26	13.66
C25H49N4P1	436.67	436.3695	68.77	11.31	12.83	7.09
C25H49N5P2	481.65	481.3463	62.34	10.25	14.54	12.86
C25H49N6P1	464.68	464.3756	64.62	10.63	18.09	6.67
C25H50N1P1	395.66	395.3681	75.89	12.74	3.54	7.83
C25H50N2P2	440.64	440.3449	68.15	11.44	6.36	14.06
C25H50N3P1	423.67	423.3742	70.88	11.90	9.92	7.31
C25H50N4P2	468.65	468.3511	64.07	10.75	11.95	13.22
C25H50N5P1	451.68	451.3804	66.48	11.16	15.50	6.86
C25H50N6P2	496.67	496.3572	60.46	10.15	16.92	12.47
C25H51N1P2	427.64	427.3497	70.22	12.02	3.28	14.49
C25H51N2P1	410.67	410.3790	73.12	12.52	6.82	7.54
C25H51N3P2	455.65	455.3558	65.90	11.28	9.22	13.60
C25H51N4P1	438.69	438.3851	68.45	11.72	12.77	7.06
C25H51N5P2	483.67	483.3620	62.08	10.63	14.48	12.81
C25H51N6P1	466.70	466.3913	64.34	11.01	18.01	6.64
C25H52N1P1	397.67	397.3837	75.51	13.18	3.52	7.79
C25H52N2P2	442.65	442.3606	67.84	11.84	6.33	13.99
C25H52N3P1	425.69	425.3899	70.54	12.31	9.87	7.28
C25H52N4P2	470.67	470.3667	63.80	11.14	11.90	13.16
C25H52N5P1	453.70	453.3960	66.18	11.55	15.44	6.83
C25H52N6P2	498.68	498.3729	60.21	10.51	16.85	12.42
C25H53N1P2	429.66	429.3653	69.89	12.43	3.26	14.42
C25H53N2P1	412.69	412.3946	72.76	12.94	6.79	7.51
C25H53N3P2	457.67	457.3715	65.61	11.67	9.18	13.54
C25H53N4P1	440.70	440.4008	68.14	12.12	12.71	7.03
C25H53N5P2	485.68	485.3776	61.83	11.00	14.42	12.75
C25H53N6P1	468.72	468.4069	64.06	11.40	17.93	6.61
C25H54N1P1	399.69	399.3994	75.13	13.62	3.50	7.75
C25H54N2P2	444.67	444.3762	67.53	12.24	6.30	13.93
C25H54N3P1	427.70	427.4055	70.21	12.73	9.82	7.24
C25H54N4P2	472.68	472.3824	63.53	11.52	11.85	13.11
C25H54N5P1	455.72	455.4117	65.89	11.94	15.37	6.80
C25H54N6P2	500.70	500.3885	59.97	10.87	16.78	12.37
C25H55N1P2	431.67	431.3810	69.56	12.84	3.24	14.35
C25H55N2P1	414.70	414.4103	72.41	13.37	6.76	7.47
C25H55N3P2	459.68	459.3871	65.32	12.06	9.14	13.48
C25H55N4P1	442.72	442.4164	67.83	12.52	12.66	7.00
C25H55N5P2	487.70	487.3933	61.57	11.37	14.36	12.70
C25H55N6P1	470.73	470.4226	63.79	11.78	17.85	6.58
C25H56N2P2	446.69	446.3919	67.22	12.64	6.27	13.87
C25H56N3P1	429.72	429.4212	69.88	13.14	9.78	7.21
C25H56N4P2	474.70	474.3980	63.26	11.89	11.80	13.05
C25H56N5P1	457.73	457.4273	65.60	12.33	15.30	6.77
C25H56N6P2	502.71	502.4041	59.73	11.23	16.72	12.32
C25H57N3P2	461.70	461.4028	65.04	12.44	9.10	13.42
C25H57N4P1	444.73	444.4321	67.52	12.92	12.60	6.96
C25H57N5P2	489.71	489.4089	61.32	11.73	14.30	12.65
C25H57N6P1	472.75	472.4382	63.52	12.15	17.78	6.55
C25H58N4P2	476.72	476.4137	62.99	12.26	11.75	12.99
C25H58N5P1	459.75	459.4430	65.31	12.72	15.23	6.74
C25H58N6P2	504.73	504.4198	59.49	11.58	16.65	12.27
C25H59N5P2	491.73	491.4245	61.07	12.09	14.24	12.60
C25H59N6P1	474.76	474.4539	63.25	12.53	17.70	6.52
C25H60N6P2	506.74	506.4354	59.26	11.93	16.58	12.22
C26H13N1P2	401.35	401.0523	77.81	3.26	3.49	15.43

C26H13N2P1	384.38	384.0816	81.24	3.41	7.29	8.06
C26H13N3P2	429.36	429.0585	72.73	3.05	9.79	14.43
C26H13N4P1	412.39	412.0878	75.73	3.18	13.59	7.51
C26H13N5P2	457.37	457.0646	68.28	2.86	15.31	13.54
C26H13N6P1	440.41	440.0939	70.91	2.98	19.08	7.03
C26H14N1P1	371.38	371.0864	84.09	3.80	3.77	8.34
C26H14N2P2	416.36	416.0632	75.00	3.39	6.73	14.88
C26H14N3P1	399.40	399.0925	78.19	3.53	10.52	7.76
C26H14N4P2	444.38	444.0694	70.28	3.18	12.61	13.94
C26H14N5P1	427.41	427.0987	73.07	3.30	16.39	7.25
C26H14N6P2	472.39	472.0755	66.11	2.99	17.79	13.11
C26H15N1P2	403.36	403.0680	77.42	3.75	3.47	15.36
C26H15N2P1	386.40	386.0973	80.82	3.91	7.25	8.02
C26H15N3P2	431.38	431.0741	72.39	3.50	9.74	14.36
C26H15N4P1	414.41	414.1034	75.36	3.65	13.52	7.47
C26H15N5P2	459.39	459.0803	67.98	3.29	15.24	13.48
C26H15N6P1	442.42	442.1096	70.59	3.42	19.00	7.00
C26H16N1P1	373.40	373.1020	83.63	4.32	3.75	8.30
C26H16N2P2	418.38	418.0789	74.64	3.85	6.70	14.81
C26H16N3P1	401.41	401.1082	77.80	4.02	10.47	7.72
C26H16N4P2	446.39	446.0850	69.96	3.61	12.55	13.88
C26H16N5P1	429.42	429.1143	72.72	3.76	16.31	7.21
C26H16N6P2	474.41	474.0912	65.83	3.40	17.71	13.06
C26H17N1P2	405.38	405.0836	77.04	4.23	3.46	15.28
C26H17N2P1	388.41	388.1129	80.40	4.41	7.21	7.97
C26H17N3P2	433.39	433.0898	72.06	3.95	9.70	14.29
C26H17N4P1	416.43	416.1191	74.99	4.11	13.45	7.44
C26H17N5P2	461.41	461.0959	67.68	3.71	15.18	13.43
C26H17N6P1	444.44	444.1252	70.27	3.86	18.91	6.97
C26H18N1P1	375.41	375.1177	83.19	4.83	3.73	8.25
C26H18N2P2	420.39	420.0945	74.28	4.32	6.66	14.74
C26H18N3P1	403.43	403.1238	77.41	4.50	10.42	7.68
C26H18N4P2	448.41	448.1007	69.64	4.05	12.49	13.82
C26H18N5P1	431.44	431.1300	72.38	4.21	16.23	7.18
C26H18N6P2	476.42	476.1068	65.55	3.81	17.64	13.00
C26H19N1P2	407.40	407.0993	76.66	4.70	3.44	15.21
C26H19N2P1	390.43	390.1286	79.99	4.91	7.18	7.93
C26H19N3P2	435.41	435.1054	71.72	4.40	9.65	14.23
C26H19N4P1	418.44	418.1347	74.63	4.58	13.39	7.40
C26H19N5P2	463.42	463.1116	67.39	4.13	15.11	13.37
C26H19N6P1	446.46	446.1409	69.95	4.29	18.82	6.94
C26H20N1P1	377.43	377.1333	82.74	5.34	3.71	8.21
C26H20N2P2	422.41	422.1102	73.93	4.77	6.63	14.67
C26H20N3P1	405.44	405.1395	77.02	4.97	10.36	7.64
C26H20N4P2	450.42	450.1163	69.33	4.48	12.44	13.75
C26H20N5P1	433.46	433.1456	72.05	4.65	16.16	7.15
C26H20N6P2	478.44	478.1225	65.27	4.21	17.57	12.95
C26H21N1P2	409.41	409.1149	76.28	5.17	3.42	15.13
C26H21N2P1	392.44	392.1442	79.58	5.39	7.14	7.89
C26H21N3P2	437.42	437.1211	71.39	4.84	9.61	14.16
C26H21N4P1	420.46	420.1504	74.27	5.03	13.33	7.37
C26H21N5P2	465.44	465.1272	67.10	4.55	15.05	13.31
C26H21N6P1	448.47	448.1565	69.63	4.72	18.74	6.91
C26H22N1P1	379.45	379.1490	82.30	5.84	3.69	8.16
C26H22N2P2	424.43	424.1258	73.58	5.22	6.60	14.60
C26H22N3P1	407.46	407.1551	76.64	5.44	10.31	7.60
C26H22N4P2	452.44	452.1320	69.02	4.90	12.38	13.69
C26H22N5P1	435.47	435.1613	71.71	5.09	16.08	7.11
C26H22N6P2	480.45	480.1381	65.00	4.62	17.49	12.89
C26H23N1P2	411.43	411.1306	75.90	5.63	3.40	15.06

C26H23N2P1	394.46	394.1599	79.17	5.88	7.10	7.85
C26H23N3P2	439.44	439.1367	71.07	5.28	9.56	14.10
C26H23N4P1	422.47	422.1660	73.92	5.49	13.26	7.33
C26H23N5P2	467.45	467.1429	66.81	4.96	14.98	13.25
C26H23N6P1	450.49	450.1722	69.32	5.15	18.66	6.88
C26H24N1P1	381.46	381.1646	81.87	6.34	3.67	8.12
C26H24N2P2	426.44	426.1415	73.23	5.67	6.57	14.53
C26H24N3P1	409.48	409.1708	76.27	5.91	10.26	7.56
C26H24N4P2	454.46	454.1476	68.72	5.32	12.33	13.63
C26H24N5P1	437.49	437.1769	71.38	5.53	16.01	7.08
C26H24N6P2	482.47	482.1538	64.73	5.01	17.42	12.84
C26H25N1P2	413.44	413.1462	75.53	6.09	3.39	14.98
C26H25N2P1	396.48	396.1755	78.77	6.36	7.07	7.81
C26H25N3P2	441.46	441.1524	70.74	5.71	9.52	14.03
C26H25N4P1	424.49	424.1817	73.57	5.94	13.20	7.30
C26H25N5P2	469.47	469.1585	66.52	5.37	14.92	13.20
C26H25N6P1	452.50	452.1878	69.01	5.57	18.57	6.84
C26H26N1P1	383.48	383.1803	81.44	6.83	3.65	8.08
C26H26N2P2	428.46	428.1571	72.89	6.12	6.54	14.46
C26H26N3P1	411.49	411.1864	75.89	6.37	10.21	7.53
C26H26N4P2	456.47	456.1633	68.41	5.74	12.27	13.57
C26H26N5P1	439.50	439.1926	71.06	5.96	15.93	7.05
C26H26N6P2	484.48	484.1694	64.46	5.41	17.35	12.79
C26H27N1P2	415.46	415.1619	75.17	6.55	3.37	14.91
C26H27N2P1	398.49	398.1912	78.37	6.83	7.03	7.77
C26H27N3P2	443.47	443.1680	70.42	6.14	9.48	13.97
C26H27N4P1	426.51	426.1973	73.22	6.38	13.14	7.26
C26H27N5P2	471.49	471.1742	66.24	5.77	14.85	13.14
C26H27N6P1	454.52	454.2035	68.71	5.99	18.49	6.81
C26H28N1P1	385.49	385.1959	81.01	7.32	3.63	8.03
C26H28N2P2	430.47	430.1728	72.55	6.56	6.51	14.39
C26H28N3P1	413.51	413.2021	75.52	6.83	10.16	7.49
C26H28N4P2	458.49	458.1789	68.11	6.16	12.22	13.51
C26H28N5P1	441.52	441.2082	70.73	6.39	15.86	7.02
C26H28N6P2	486.50	486.1851	64.19	5.80	17.27	12.73
C26H29N1P2	417.48	417.1775	74.80	7.00	3.36	14.84
C26H29N2P1	400.51	400.2068	77.97	7.30	6.99	7.73
C26H29N3P2	445.49	445.1837	70.10	6.56	9.43	13.91
C26H29N4P1	428.52	428.2130	72.88	6.82	13.07	7.23
C26H29N5P2	473.50	473.1898	65.95	6.17	14.79	13.08
C26H29N6P1	456.54	456.2191	68.40	6.40	18.41	6.78
C26H30N1P1	387.51	387.2116	80.59	7.80	3.61	7.99
C26H30N2P2	432.49	432.1884	72.21	6.99	6.48	14.32
C26H30N3P1	415.52	415.2177	75.16	7.28	10.11	7.45
C26H30N4P2	460.50	460.1946	67.81	6.57	12.17	13.45
C26H30N5P1	443.54	443.2239	70.41	6.82	15.79	6.98
C26H30N6P2	488.52	488.2007	63.93	6.19	17.20	12.68
C26H31N1P2	419.49	419.1932	74.44	7.45	3.34	14.77
C26H31N2P1	402.52	402.2225	77.58	7.76	6.96	7.69
C26H31N3P2	447.50	447.1993	69.78	6.98	9.39	13.84
C26H31N4P1	430.54	430.2286	72.53	7.26	13.01	7.19
C26H31N5P2	475.52	475.2055	65.67	6.57	14.73	13.03
C26H31N6P1	458.55	458.2348	68.10	6.81	18.33	6.75
C26H32N1P1	389.53	389.2272	80.17	8.28	3.60	7.95
C26H32N2P2	434.51	434.2041	71.87	7.42	6.45	14.26
C26H32N3P1	417.54	417.2334	74.79	7.73	10.06	7.42
C26H32N4P2	462.52	462.2102	67.52	6.97	12.11	13.39
C26H32N5P1	445.55	445.2395	70.09	7.24	15.72	6.95
C26H32N6P2	490.53	490.2164	63.66	6.58	17.13	12.63
C26H33N1P2	421.51	421.2088	74.09	7.89	3.32	14.70
C26H33N2P1	404.54	404.2381	77.20	8.22	6.92	7.66
C26H33N3P2	449.52	449.2150	69.47	7.40	9.35	13.78
C26H33N4P1	432.55	432.2443	72.20	7.69	12.95	7.16
C26H33N5P2	477.53	477.2211	65.40	6.97	14.67	12.97
C26H33N6P1	460.57	460.2504	67.81	7.22	18.25	6.73
C26H34N1P1	391.54	391.2429	79.76	8.75	3.58	7.91
C26H34N2P2	436.52	436.2197	71.54	7.85	6.42	14.19
C26H34N3P1	419.55	419.2490	74.43	8.17	10.02	7.38
C26H34N4P2	464.54	464.2259	67.23	7.38	12.06	13.34
C26H34N5P1	447.57	447.2552	69.77	7.66	15.65	6.92
C26H34N6P2	492.55	492.2320	63.40	6.96	17.06	12.58
C26H35N1P2	423.52	423.2245	73.74	8.33	3.31	14.63
C26H35N2P1	406.56	406.2538	76.81	8.68	6.89	7.62
C26H35N3P2	451.54	451.2306	69.16	7.81	9.31	13.72
C26H35N4P1	434.57	434.2599	71.86	8.12	12.89	7.13
C26H35N5P2	479.55	479.2368	65.12	7.36	14.60	12.92
C26H35N6P1	462.58	462.2661	67.51	7.63	18.17	6.70
C26H36N1P1	393.56	393.2585	79.35	9.22	3.56	7.87
C26H36N2P2	438.54	438.2354	71.21	8.27	6.39	14.13
C26H36N3P1	421.57	421.2647	74.08	8.61	9.97	7.35
C26H36N4P2	466.55	466.2415	66.94	7.78	12.01	13.28
C26H36N5P1	449.58	449.2708	69.46	8.07	15.58	6.89
C26H36N6P2	494.56	494.2477	63.14	7.34	16.99	12.53
C26H37N1P2	425.54	425.2401	73.39	8.76	3.29	14.56
C26H37N2P1	408.57	408.2694	76.43	9.13	6.86	7.58
C26H37N3P2	453.55	453.2463	68.85	8.22	9.26	13.66
C26H37N4P1	436.59	436.2756	71.53	8.54	12.83	7.09
C26H37N5P2	481.57	481.2524	64.85	7.74	14.54	12.86
C26H37N6P1	464.60	464.2817	67.22	8.03	18.09	6.67
C26H38N1P1	395.57	395.2742	78.95	9.68	3.54	7.83
C26H38N2P2	440.55	440.2510	70.89	8.69	6.36	14.06
C26H38N3P1	423.59	423.2803	73.73	9.04	9.92	7.31
C26H38N4P2	468.57	468.2572	66.65	8.17	11.96	13.22
C26H38N5P1	451.60	451.2865	69.15	8.48	15.51	6.86
C26H38N6P2	496.58	496.2633	62.89	7.71	16.92	12.47
C26H39N1P2	427.56	427.2558	73.04	9.19	3.28	14.49
C26H39N2P1	410.59	410.2851	76.06	9.57	6.82	7.54
C26H39N3P2	455.57	455.2619	68.55	8.63	9.22	13.60
C26H39N4P1	438.60	438.2912	71.20	8.96	12.77	7.06
C26H39N5P2	483.58	483.2681	64.58	8.13	14.48	12.81
C26H39N6P1	466.61	466.2974	66.93	8.42	18.01	6.64
C26H40N1P1	397.59	397.2898	78.55	10.14	3.52	7.79
C26H40N2P2	442.57	442.2667	70.56	9.11	6.33	14.00
C26H40N3P1	425.60	425.2960	73.38	9.47	9.87	7.28
C26H40N4P2	470.58	470.2728	66.36	8.57	11.91	13.16
C26H40N5P1	453.62	453.3021	68.84	8.89	15.44	6.83
C26H40N6P2	498.60	498.2790	62.63	8.09	16.86	12.42
C26H41N1P2	429.57	429.2714	72.70	9.62	3.26	14.42
C26H41N2P1	412.60	412.3007	75.69	10.02	6.79	7.51
C26H41N3P2	457.58	457.2776	68.25	9.03	9.18	13.54
C26H41N4P1	440.62	440.3069	70.88	9.38	12.72	7.03
C26H41N5P2	485.60	485.2837	64.31	8.51	14.42	12.76
C26H41N6P1	468.63	468.3130	66.64	8.82	17.93	6.61
C26H42N1P1	399.61	399.3055	78.15	10.59	3.51	7.75
C26H42N2P2	444.59	444.2823	70.24	9.52	6.30	13.93
C26H42N3P1	427.62	427.3116	73.03	9.90	9.83	7.24
C26H42N4P2	472.60	472.2885	66.08	8.96	11.86	13.11
C26H42N5P1	455.63	455.3178	68.54	9.29	15.37	6.80
C26H42N6P2	500.61	500.2946	62.38	8.46	16.79	12.37
C26H43N1P2	431.59	431.2871	72.36	10.04	3.25	14.35

090

C26H43N2P1	414.62	414.3164	75.32	10.45	6.76	7.47
C26H43N3P2	459.60	459.2932	67.95	9.43	9.14	13.48
C26H43N4P1	442.63	442.3225	70.55	9.79	12.66	7.00
C26H43N5P2	487.61	487.2994	64.04	8.89	14.36	12.70
C26H43N6P1	470.65	470.3287	66.35	9.21	17.86	6.58
C26H44N1P1	401.62	401.3211	77.76	11.04	3.49	7.71
C26H44N2P2	446.60	446.2980	69.93	9.93	6.27	13.87
C26H44N3P1	429.63	429.3273	72.69	10.32	9.78	7.21
C26H44N4P2	474.61	474.3041	65.80	9.34	11.80	13.05
C26H44N5P1	457.65	457.3334	68.24	9.69	15.30	6.77
C26H44N6P2	502.63	502.3103	62.13	8.82	16.72	12.32
C26H45N1P2	433.60	433.3027	72.02	10.46	3.23	14.29
C26H45N2P1	416.64	416.3320	74.96	10.89	6.72	7.43
C26H45N3P2	461.62	461.3089	67.65	9.83	9.10	13.42
C26H45N4P1	444.65	444.3382	70.23	10.20	12.60	6.97
C26H45N5P2	489.63	489.3150	63.78	9.26	14.30	12.65
C26H45N6P1	472.66	472.3443	66.07	9.60	17.78	6.55
C26H46N1P1	403.64	403.3368	77.37	11.49	3.47	7.67
C26H46N2P2	448.62	448.3136	69.61	10.34	6.24	13.81
C26H46N3P1	431.65	431.3429	72.35	10.74	9.73	7.18
C26H46N4P2	476.63	476.3198	65.52	9.73	11.75	13.00
C26H46N5P1	459.66	459.3491	67.94	10.09	15.24	6.74
C26H46N6P2	504.64	504.3259	61.88	9.19	16.65	12.28
C26H47N1P2	435.62	435.3184	71.69	10.88	3.22	14.22
C26H47N2P1	418.65	418.3477	74.59	11.32	6.69	7.40
C26H47N3P2	463.63	463.3245	67.36	10.22	9.06	13.36
C26H47N4P1	446.67	446.3538	69.92	10.61	12.54	6.93
C26H47N5P2	491.65	491.3307	63.52	9.64	14.24	12.60
C26H47N6P1	474.68	474.3600	65.79	9.98	17.70	6.53
C26H48N1P1	405.65	405.3524	76.98	11.93	3.45	7.64
C26H48N2P2	450.63	450.3293	69.30	10.74	6.22	13.75
C26H48N3P1	433.67	433.3586	72.01	11.16	9.69	7.14
C26H48N4P2	478.65	478.3354	65.24	10.11	11.71	12.94
C26H48N5P1	461.68	461.3647	67.64	10.48	15.17	6.71
C26H48N6P2	506.66	506.3416	61.64	9.55	16.59	12.23
C26H49N1P2	437.63	437.3340	71.36	11.29	3.20	14.16
C26H49N2P1	420.67	420.3633	74.24	11.74	6.66	7.36
C26H49N3P2	465.65	465.3402	67.07	10.61	9.02	13.30
C26H49N4P1	448.68	448.3695	69.60	11.01	12.49	6.90
C26H49N5P2	493.66	493.3463	63.26	10.00	14.19	12.55
C26H49N6P1	476.69	476.3756	65.51	10.36	17.63	6.50
C26H50N1P1	407.67	407.3681	76.60	12.36	3.44	7.60
C26H50N2P2	452.65	452.3449	68.99	11.13	6.19	13.69
C26H50N3P1	435.68	435.3742	71.68	11.57	9.64	7.11
C26H50N4P2	480.66	480.3511	64.97	10.49	11.66	12.89
C26H50N5P1	463.70	463.3804	67.35	10.87	15.10	6.68
C26H50N6P2	508.68	508.3572	61.39	9.91	16.52	12.18
C26H51N1P2	439.65	439.3497	71.03	11.69	3.19	14.09
C26H51N2P1	422.68	422.3790	73.88	12.16	6.63	7.33
C26H51N3P2	467.66	467.3558	66.78	10.99	8.99	13.25
C26H51N4P1	450.70	450.3851	69.29	11.41	12.43	6.87
C26H51N5P2	495.68	495.3620	63.00	10.37	14.13	12.50
C26H51N6P1	478.71	478.3913	65.24	10.74	17.56	6.47
C26H52N1P1	409.68	409.3837	76.23	12.79	3.42	7.56
C26H52N2P2	454.67	454.3606	68.69	11.53	6.16	13.62
C26H52N3P1	437.70	437.3899	71.35	11.98	9.60	7.08
C26H52N4P2	482.68	482.3667	64.70	10.86	11.61	12.83
C26H52N5P1	465.71	465.3960	67.06	11.25	15.04	6.65
C26H52N6P2	510.69	510.3729	61.15	10.26	16.46	12.13
C26H53N1P2	441.67	441.3653	70.71	12.10	3.17	14.03
C26H53N2P1	424.70	424.3946	73.53	12.58	6.60	7.29
C26H53N3P2	469.68	469.3715	66.49	11.37	8.95	13.19
C26H53N4P1	452.71	452.4008	68.98	11.80	12.38	6.84
C26H53N5P2	497.69	497.3776	62.75	10.73	14.07	12.45
C26H53N6P1	480.73	480.4069	64.96	11.11	17.48	6.44
C26H54N1P1	411.70	411.3994	75.85	13.22	3.40	7.52
C26H54N2P2	456.68	456.3762	68.38	11.92	6.13	13.56
C26H54N3P1	439.71	439.4055	71.02	12.38	9.56	7.04
C26H54N4P2	484.69	484.3824	64.43	11.23	11.56	12.78
C26H54N5P1	467.73	467.4117	66.77	11.64	14.97	6.62
C26H54N6P2	512.71	512.3885	60.91	10.62	16.39	12.08
C26H55N1P2	443.68	443.3810	70.39	12.50	3.16	13.96
C26H55N2P1	426.72	426.4103	73.18	12.99	6.56	7.26
C26H55N3P2	471.70	471.3871	66.21	11.75	8.91	13.13
C26H55N4P1	454.73	454.4164	68.68	12.19	12.32	6.81
C26H55N5P2	499.71	499.3933	62.49	11.09	14.01	12.40
C26H55N6P1	482.74	482.4226	64.69	11.48	17.41	6.42
C26H56N1P1	413.72	413.4150	75.48	13.64	3.39	7.49
C26H56N2P2	458.70	458.3919	68.08	12.31	6.11	13.51
C26H56N3P1	441.73	441.4212	70.70	12.78	9.51	7.01
C26H56N4P2	486.71	486.3980	64.16	11.60	11.51	12.73
C26H56N5P1	469.74	469.4273	66.48	12.02	14.91	6.59
C26H56N6P2	514.72	514.4041	60.67	10.97	16.33	12.04
C26H57N1P2	445.70	445.3966	70.07	12.89	3.14	13.90
C26H57N2P1	428.73	428.4259	72.84	13.40	6.53	7.22
C26H57N3P2	473.71	473.4028	65.92	12.13	8.87	13.08
C26H57N4P1	456.74	456.4321	68.37	12.58	12.27	6.78
C26H57N5P2	501.73	501.4089	62.24	11.45	13.96	12.35
C26H57N6P1	484.76	484.4382	64.42	11.85	17.34	6.39
C26H58N2P2	460.71	460.4075	67.78	12.69	6.08	13.45
C26H58N3P1	443.75	443.4368	70.38	13.17	9.47	6.98
C26H58N4P2	488.73	488.4137	63.90	11.96	11.46	12.68
C26H58N5P1	471.76	471.4430	66.20	12.39	14.85	6.57
C26H58N6P2	516.74	516.4198	60.43	11.31	16.26	11.99
C26H59N3P2	475.73	475.4184	65.64	12.50	8.83	13.02
C26H59N4P1	458.76	458.4477	68.07	12.96	12.21	6.75
C26H59N5P2	503.74	503.4245	61.99	11.81	13.90	12.30
C26H59N6P1	486.77	486.4539	64.15	12.22	17.26	6.36
C26H60N4P2	490.74	490.4293	63.64	12.32	11.42	12.62
C26H60N5P1	473.78	473.4586	65.92	12.77	14.78	6.54
C26H60N6P2	518.76	518.4354	60.20	11.66	16.20	11.94
C26H61N5P2	505.76	505.4402	61.75	12.16	13.85	12.25
C26H61N6P1	488.79	488.4695	63.89	12.58	17.19	6.34
C26H62N6P2	520.77	520.4511	59.97	12.00	16.14	11.90
C27H13N1P2	413.36	413.0523	78.46	3.17	3.39	14.99
C27H13N2P1	396.39	396.0816	81.81	3.31	7.07	7.81
C27H13N3P2	441.37	441.0585	73.48	2.97	9.52	14.04
C27H13N4P1	424.41	424.0878	76.41	3.09	13.20	7.30
C27H13N5P2	469.39	469.0646	69.09	2.79	14.92	13.20
C27H13N6P1	452.42	452.0939	71.68	2.90	18.58	6.85
C27H14N1P1	383.39	383.0864	84.59	3.68	3.65	8.08
C27H14N2P2	428.37	428.0632	75.71	3.29	6.54	14.46
C27H14N3P1	411.41	411.0925	78.83	3.43	10.21	7.53
C27H14N4P2	456.39	456.0694	71.06	3.09	12.28	13.57
C27H14N5P1	439.42	439.0987	73.80	3.21	15.94	7.05
C27H14N6P2	484.40	484.0755	66.95	2.91	17.35	12.79
C27H15N1P2	415.37	415.0680	78.07	3.64	3.37	14.91
C27H15N2P1	398.41	398.0973	81.40	3.79	7.03	7.77
C27H15N3P2	443.39	443.0741	73.14	3.41	9.48	13.97
C27H15N4P1	426.42	426.1034	76.05	3.55	13.14	7.26

091

C27H15N5P2	471.40	471.0803	68.80	3.21	14.86	13.14
C27H15N6P1	454.43	454.1096	71.36	3.33	18.49	6.82
C27H16N1P1	385.41	385.1020	84.14	4.18	3.63	8.04
C27H16N2P2	430.39	430.0789	75.35	3.75	6.51	14.39
C27H16N3P1	413.42	413.1082	78.44	3.90	10.16	7.49
C27H16N4P2	458.40	458.0850	70.75	3.52	12.22	13.51
C27H16N5P1	441.44	441.1143	73.47	3.65	15.86	7.02
C27H16N6P2	486.42	486.0912	66.67	3.32	17.28	12.74
C27H17N1P2	417.39	417.0836	77.70	4.11	3.36	14.84
C27H17N2P1	400.42	400.1129	80.99	4.28	7.00	7.74
C27H17N3P2	445.40	445.0898	72.81	3.85	9.43	13.91
C27H17N4P1	428.44	428.1191	75.69	4.00	13.08	7.23
C27H17N5P2	473.42	473.0959	68.50	3.62	14.79	13.09
C27H17N6P1	456.45	456.1252	71.05	3.75	18.41	6.79
C27H18N1P1	387.43	387.1177	83.71	4.68	3.62	7.99
C27H18N2P2	432.41	432.0945	75.00	4.20	6.48	14.33
C27H18N3P1	415.44	415.1238	78.06	4.37	10.11	7.46
C27H18N4P2	460.42	460.1007	70.44	3.94	12.17	13.45
C27H18N5P1	443.45	443.1300	73.13	4.09	15.79	6.98
C27H18N6P2	488.43	488.1068	66.40	3.71	17.21	12.68
C27H19N1P2	419.41	419.0993	77.32	4.57	3.34	14.77
C27H19N2P1	402.44	402.1286	80.58	4.76	6.96	7.70
C27H19N3P2	447.42	447.1054	72.48	4.28	9.39	13.85
C27H19N4P1	430.45	430.1347	75.34	4.45	13.02	7.20
C27H19N5P2	475.43	475.1116	68.21	4.03	14.73	13.03
C27H19N6P1	458.47	458.1409	70.74	4.18	18.33	6.76
C27H20N1P1	389.44	389.1333	83.27	5.18	3.60	7.95
C27H20N2P2	434.42	434.1102	74.65	4.64	6.45	14.26
C27H20N3P1	417.45	417.1395	77.69	4.83	10.07	7.42
C27H20N4P2	462.43	462.1163	70.13	4.36	12.12	13.40
C27H20N5P1	445.47	445.1456	72.80	4.53	15.72	6.95
C27H20N6P2	490.45	490.1225	66.12	4.11	17.14	12.63
C27H21N1P2	421.42	421.1149	76.95	5.02	3.32	14.70
C27H21N2P1	404.46	404.1442	80.18	5.23	6.93	7.66
C27H21N3P2	449.44	449.1211	72.16	4.71	9.35	13.78
C27H21N4P1	432.47	432.1504	74.99	4.89	12.96	7.16
C27H21N5P2	477.45	477.1272	67.92	4.43	14.67	12.97
C27H21N6P1	460.48	460.1565	70.43	4.60	18.25	6.73
C27H22N1P1	391.46	391.1490	82.84	5.66	3.58	7.91
C27H22N2P2	436.44	436.1258	74.31	5.08	6.42	14.19
C27H22N3P1	419.47	419.1551	77.31	5.29	10.02	7.38
C27H22N4P2	464.45	464.1320	69.82	4.77	12.06	13.34
C27H22N5P1	447.48	447.1613	72.47	4.96	15.65	6.92
C27H22N6P2	492.46	492.1381	65.85	4.50	17.07	12.58
C27H23N1P2	423.44	423.1306	76.59	5.48	3.31	14.63
C27H23N2P1	406.47	406.1599	79.78	5.70	6.89	7.62
C27H23N3P2	451.45	451.1367	71.84	5.14	9.31	13.72
C27H23N4P1	434.48	434.1660	74.64	5.34	12.89	7.13
C27H23N5P2	479.47	479.1429	67.64	4.84	14.61	12.92
C27H23N6P1	462.50	462.1722	70.12	5.01	18.17	6.70
C27H24N1P1	393.47	393.1646	82.42	6.15	3.56	7.87
C27H24N2P2	438.45	438.1415	73.96	5.52	6.39	14.13
C27H24N3P1	421.49	421.1708	76.94	5.74	9.97	7.35
C27H24N4P2	466.47	466.1476	69.52	5.19	12.01	13.28
C27H24N5P1	449.50	449.1769	72.15	5.38	15.58	6.89
C27H24N6P2	494.48	494.1538	65.58	4.89	17.00	12.53
C27H25N1P2	425.45	425.1462	76.22	5.92	3.29	14.56
C27H25N2P1	408.49	408.1755	79.39	6.17	6.86	7.58
C27H25N3P2	453.47	453.1524	71.52	5.56	9.27	13.66
C27H25N4P1	436.50	436.1817	74.30	5.77	12.84	7.10

C27H25N5P2	481.48	481.1585	67.35	5.23	14.55	12.87
C27H25N6P1	464.51	464.1878	69.82	5.42	18.09	6.67
C27H26N1P1	395.49	395.1803	82.00	6.63	3.54	7.83
C27H26N2P2	440.47	440.1571	73.63	5.95	6.36	14.06
C27H26N3P1	423.50	423.1864	76.58	6.19	9.92	7.31
C27H26N4P2	468.48	468.1633	69.22	5.59	11.96	13.22
C27H26N5P1	451.52	451.1926	71.82	5.80	15.51	6.86
C27H26N6P2	496.50	496.1694	65.32	5.28	16.93	12.48
C27H27N1P2	427.47	427.1619	75.87	6.37	3.28	14.49
C27H27N2P1	410.50	410.1912	79.00	6.63	6.82	7.55
C27H27N3P2	455.48	455.1680	71.20	5.98	9.23	13.60
C27H27N4P1	438.52	438.1973	73.95	6.21	12.78	7.06
C27H27N5P2	483.50	483.1742	67.07	5.63	14.48	12.81
C27H27N6P1	466.53	466.2035	69.51	5.83	18.01	6.64
C27H28N1P1	397.50	397.1959	81.58	7.10	3.52	7.79
C27H28N2P2	442.49	442.1728	73.29	6.38	6.33	14.00
C27H28N3P1	425.52	425.2021	76.21	6.63	9.88	7.28
C27H28N4P2	470.50	470.1789	68.93	6.00	11.91	13.17
C27H28N5P1	453.53	453.2082	71.51	6.22	15.44	6.83
C27H28N6P2	498.51	498.1851	65.05	5.66	16.86	12.43
C27H29N1P2	429.49	429.1775	75.51	6.81	3.26	14.42
C27H29N2P1	412.52	412.2068	78.61	7.09	6.79	7.51
C27H29N3P2	457.50	457.1837	70.89	6.39	9.18	13.54
C27H29N4P1	440.53	440.2130	73.62	6.64	12.72	7.03
C27H29N5P2	485.51	485.1898	66.80	6.02	14.42	12.76
C27H29N6P1	468.55	468.2191	69.21	6.24	17.94	6.61
C27H30N1P1	399.52	399.2116	81.17	7.57	3.51	7.75
C27H30N2P2	444.50	444.1884	72.96	6.80	6.30	13.94
C27H30N3P1	427.53	427.2177	75.85	7.07	9.83	7.24
C27H30N4P2	472.51	472.1946	68.63	6.40	11.86	13.11
C27H30N5P1	455.55	455.2239	71.19	6.64	15.37	6.80
C27H30N6P2	500.53	500.2007	64.79	6.04	16.79	12.38
C27H31N1P2	431.50	431.1932	75.16	7.24	3.25	14.36
C27H31N2P1	414.54	414.2225	78.23	7.54	6.76	7.47
C27H31N3P2	459.52	459.1993	70.57	6.80	9.14	13.48
C27H31N4P1	442.55	442.2286	73.28	7.06	12.66	7.00
C27H31N5P2	487.53	487.2055	66.52	6.41	14.36	12.71
C27H31N6P1	470.56	470.2348	68.92	6.64	17.86	6.58
C27H32N1P1	401.54	401.2272	80.77	8.03	3.49	7.71
C27H32N2P2	446.52	446.2041	72.63	7.22	6.27	13.87
C27H32N3P1	429.55	429.2334	75.50	7.51	9.78	7.21
C27H32N4P2	474.53	474.2102	68.34	6.80	11.81	13.05
C27H32N5P1	457.56	457.2395	70.88	7.05	15.31	6.77
C27H32N6P2	502.54	502.2164	64.53	6.42	16.72	12.33
C27H33N1P2	433.52	433.2088	74.81	7.67	3.23	14.29
C27H33N2P1	416.55	416.2381	77.85	7.99	6.73	7.44
C27H33N3P2	461.53	461.2150	70.27	7.21	9.10	13.42
C27H33N4P1	444.56	444.2443	72.95	7.48	12.60	6.97
C27H33N5P2	489.55	489.2211	66.25	6.79	14.31	12.65
C27H33N6P1	472.58	472.2504	68.62	7.04	17.78	6.55
C27H34N1P1	403.55	403.2429	80.36	8.49	3.47	7.68
C27H34N2P2	448.53	448.2197	72.30	7.64	6.25	13.81
C27H34N3P1	431.57	431.2490	75.15	7.94	9.74	7.18
C27H34N4P2	476.55	476.2259	68.05	7.19	11.76	13.00
C27H34N5P1	459.58	459.2552	70.56	7.46	15.24	6.74
C27H34N6P2	504.56	504.2320	64.27	6.79	16.66	12.28
C27H35N1P2	435.53	435.2245	74.46	8.10	3.22	14.22
C27H35N2P1	418.57	418.2538	77.48	8.43	6.69	7.40
C27H35N3P2	463.55	463.2306	69.96	7.61	9.06	13.36
C27H35N4P1	446.58	446.2599	72.62	7.90	12.55	6.94

092

C27H35N5P2	491.56	491.2368	65.97	7.18	14.25	12.60
C27H35N6P1	474.59	474.2661	68.33	7.43	17.71	6.53
C27H36N1P1	405.57	405.2585	79.96	8.95	3.45	7.64
C27H36N2P2	450.55	450.2354	71.98	8.05	6.22	13.75
C27H36N3P1	433.58	433.2647	74.80	8.37	9.69	7.14
C27H36N4P2	478.56	478.2415	67.77	7.58	11.71	12.94
C27H36N5P1	461.60	461.2708	70.26	7.86	15.17	6.71
C27H36N6P2	506.58	506.2477	64.02	7.16	16.59	12.23
C27H37N1P2	437.55	437.2401	74.12	8.52	3.20	14.16
C27H37N2P1	420.58	420.2694	77.11	8.87	6.66	7.36
C27H37N3P2	465.56	465.2463	69.66	8.01	9.03	13.31
C27H37N4P1	448.60	448.2756	72.29	8.31	12.49	6.90
C27H37N5P2	493.58	493.2524	65.70	7.56	14.19	12.55
C27H37N6P1	476.61	476.2817	68.04	7.83	17.63	6.50
C27H38N1P1	407.58	407.2742	79.57	9.40	3.44	7.60
C27H38N2P2	452.56	452.2510	71.66	8.46	6.19	13.69
C27H38N3P1	435.60	435.2803	74.45	8.79	9.65	7.11
C27H38N4P2	480.58	480.2572	67.48	7.97	11.66	12.89
C27H38N5P1	463.61	463.2865	69.95	8.26	15.11	6.68
C27H38N6P2	508.59	508.2633	63.76	7.53	16.52	12.18
C27H39N1P2	439.57	439.2558	73.78	8.94	3.19	14.09
C27H39N2P1	422.60	422.2851	76.74	9.30	6.63	7.33
C27H39N3P2	467.58	467.2619	69.36	8.41	8.99	13.25
C27H39N4P1	450.61	450.2912	71.97	8.72	12.43	6.87
C27H39N5P2	495.59	495.2681	65.44	7.93	14.13	12.50
C27H39N6P1	478.63	478.2974	67.76	8.21	17.56	6.47
C27H40N1P1	409.60	409.2898	79.17	9.84	3.42	7.56
C27H40N2P2	454.58	454.2667	71.34	8.87	6.16	13.63
C27H40N3P1	437.61	437.2960	74.11	9.21	9.60	7.08
C27H40N4P2	482.59	482.2728	67.20	8.35	11.61	12.84
C27H40N5P1	465.63	465.3021	69.65	8.66	15.04	6.65
C27H40N6P2	510.61	510.2790	63.51	7.90	16.46	12.13
C27H41N1P2	441.58	441.2714	73.44	9.36	3.17	14.03
C27H41N2P1	424.62	424.3007	76.38	9.73	6.60	7.29
C27H41N3P2	469.60	469.2776	69.06	8.80	8.95	13.19
C27H41N4P1	452.63	452.3069	71.65	9.13	12.38	6.84
C27H41N5P2	497.61	497.2837	65.17	8.31	14.07	12.45
C27H41N6P1	480.64	480.3130	67.47	8.60	17.48	6.44
C27H42N1P1	411.62	411.3055	78.79	10.29	3.40	7.52
C27H42N2P2	456.60	456.2823	71.03	9.27	6.14	13.57
C27H42N3P1	439.63	439.3116	73.77	9.63	9.56	7.05
C27H42N4P2	484.61	484.2885	66.92	8.74	11.56	12.78
C27H42N5P1	467.64	467.3178	69.35	9.05	14.98	6.62
C27H42N6P2	512.62	512.2946	63.26	8.26	16.39	12.08
C27H43N1P2	443.60	443.2871	73.11	9.77	3.16	13.96
C27H43N2P1	426.63	426.3164	76.01	10.16	6.57	7.26
C27H43N3P2	471.61	471.2932	68.76	9.19	8.91	13.14
C27H43N4P1	454.64	454.3225	71.33	9.53	12.32	6.81
C27H43N5P2	499.62	499.2994	64.91	8.68	14.02	12.40
C27H43N6P1	482.66	482.3287	67.19	8.98	17.41	6.42
C27H44N1P1	413.63	413.3211	78.40	10.72	3.39	7.49
C27H44N2P2	458.61	458.2980	70.71	9.67	6.11	13.51
C27H44N3P1	441.65	441.3273	73.43	10.04	9.51	7.01
C27H44N4P2	486.63	486.3041	66.64	9.11	11.51	12.73
C27H44N5P1	469.66	469.3334	69.05	9.44	14.91	6.59
C27H44N6P2	514.64	514.3103	63.02	8.62	16.33	12.04
C27H45N1P2	445.61	445.3027	72.78	10.18	3.14	13.90
C27H45N2P1	428.65	428.3320	75.66	10.58	6.54	7.23
C27H45N3P2	473.63	473.3089	68.47	9.58	8.87	13.08
C27H45N4P1	456.66	456.3382	71.02	9.93	12.27	6.78

C27H45N5P2	501.64	501.3150	64.65	9.04	13.96	12.35
C27H45N6P1	484.67	484.3443	66.91	9.36	17.34	6.39
C27H46N1P1	415.65	415.3368	78.02	11.16	3.37	7.45
C27H46N2P2	460.63	460.3136	70.40	10.07	6.08	13.45
C27H46N3P1	443.66	443.3429	73.10	10.45	9.47	6.98
C27H46N4P2	488.64	488.3198	66.37	9.49	11.47	12.68
C27H46N5P1	471.67	471.3491	68.76	9.83	14.85	6.57
C27H46N6P2	516.66	516.3259	62.77	8.97	16.27	11.99
C27H47N1P2	447.63	447.3184	72.45	10.58	3.13	13.84
C27H47N2P1	430.66	430.3477	75.30	11.00	6.50	7.19
C27H47N3P2	475.64	475.3245	68.18	9.96	8.83	13.02
C27H47N4P1	458.68	458.3538	70.70	10.33	12.21	6.75
C27H47N5P2	503.66	503.3307	64.39	9.41	13.91	12.30
C27H47N6P1	486.69	486.3600	66.63	9.73	17.27	6.36
C27H48N1P1	417.66	417.3524	77.65	11.58	3.35	7.42
C27H48N2P2	462.64	462.3293	70.10	10.46	6.06	13.39
C27H48N3P1	445.68	445.3586	72.77	10.86	9.43	6.95
C27H48N4P2	490.66	490.3354	66.10	9.86	11.42	12.63
C27H48N5P1	473.69	473.3647	68.46	10.21	14.78	6.54
C27H48N6P2	518.67	518.3416	62.53	9.33	16.20	11.94
C27H49N1P2	449.65	449.3340	72.12	10.98	3.12	13.78
C27H49N2P1	432.68	432.3633	74.95	11.42	6.47	7.16
C27H49N3P2	477.66	477.3402	67.89	10.34	8.80	12.97
C27H49N4P1	460.69	460.3695	70.39	10.72	12.16	6.72
C27H49N5P2	505.67	505.3463	64.13	9.77	13.85	12.25
C27H49N6P1	488.71	488.3756	66.36	10.11	17.20	6.34
C27H50N1P1	419.68	419.3681	77.27	12.01	3.34	7.38
C27H50N2P2	464.66	464.3449	69.79	10.85	6.03	13.33
C27H50N3P1	447.69	447.3742	72.44	11.26	9.39	6.92
C27H50N4P2	492.67	492.3511	65.82	10.23	11.37	12.57
C27H50N5P1	475.71	475.3804	68.17	10.59	14.72	6.51
C27H50N6P2	520.69	520.3572	62.28	9.68	16.14	11.90
C27H51N1P2	451.66	451.3497	71.80	11.38	3.10	13.72
C27H51N2P1	434.69	434.3790	74.60	11.83	6.44	7.13
C27H51N3P2	479.68	479.3558	67.61	10.72	8.76	12.91
C27H51N4P1	462.71	462.3851	70.09	11.11	12.11	6.69
C27H51N5P2	507.69	507.3620	63.88	10.13	13.79	12.20
C27H51N6P1	490.72	490.3913	66.09	10.48	17.13	6.31
C27H52N1P1	421.70	421.3837	76.90	12.43	3.32	7.35
C27H52N2P2	466.68	466.3606	69.49	11.23	6.00	13.27
C27H52N3P1	449.71	449.3899	72.11	11.66	9.34	6.89
C27H52N4P2	494.69	494.3667	65.56	10.60	11.33	12.52
C27H52N5P1	477.72	477.3960	67.88	10.97	14.66	6.48
C27H52N6P2	522.70	522.3729	62.04	10.03	16.08	11.85
C27H53N1P2	453.68	453.3653	71.48	11.78	3.09	13.65
C27H53N2P1	436.71	436.3946	74.26	12.23	6.41	7.09
C27H53N3P2	481.69	481.3715	67.33	11.09	8.72	12.86
C27H53N4P1	464.72	464.4008	69.78	11.50	12.06	6.66
C27H53N5P2	509.70	509.3776	63.63	10.48	13.74	12.15
C27H53N6P1	492.74	492.4069	65.82	10.84	17.06	6.29
C27H54N1P1	423.71	423.3994	76.54	12.85	3.31	7.31
C27H54N2P2	468.69	468.3762	69.19	11.61	5.98	13.22
C27H54N3P1	451.73	451.4055	71.79	12.05	9.30	6.86
C27H54N4P2	496.71	496.3824	65.29	10.96	11.28	12.47
C27H54N5P1	479.74	479.4117	67.60	11.35	14.60	6.46
C27H54N6P2	524.72	524.3885	61.80	10.37	16.02	11.81
C27H55N1P2	455.69	455.3810	71.17	12.17	3.07	13.59
C27H55N2P1	438.73	438.4103	73.92	12.64	6.39	7.06
C27H55N3P2	483.71	483.3871	67.04	11.46	8.69	12.81
C27H55N4P1	466.74	466.4164	69.48	11.88	12.00	6.64

C27H55N5P2	511.72	511.3933	63.37	10.83	13.69	12.11
C27H55N6P1	494.75	494.4226	65.55	11.21	16.99	6.26
C27H56N1P1	425.73	425.4150	76.18	13.26	3.29	7.28
C27H56N2P2	470.71	470.3919	68.90	11.99	5.95	13.16
C27H56N3P1	453.74	453.4212	71.47	12.44	9.26	6.83
C27H56N4P2	498.72	498.3980	65.03	11.32	11.23	12.42
C27H56N5P1	481.75	481.4273	67.32	11.72	14.54	6.43
C27H56N6P2	526.74	526.4041	61.57	10.72	15.95	11.76
C27H57N1P2	457.71	457.3966	70.85	12.55	3.06	13.53
C27H57N2P1	440.74	440.4259	73.58	13.04	6.36	7.03
C27H57N3P2	485.72	485.4028	66.77	11.83	8.65	12.75
C27H57N4P1	468.76	468.4321	69.18	12.26	11.95	6.61
C27H57N5P2	513.74	513.4089	63.13	11.18	13.63	12.06
C27H57N6P1	496.77	496.4382	65.28	11.57	16.92	6.24
C27H58N1P1	427.74	427.4307	75.82	13.67	3.27	7.24
C27H58N2P2	472.72	472.4075	68.60	12.37	5.93	13.10
C27H58N3P1	455.76	455.4368	71.16	12.83	9.22	6.80
C27H58N4P2	500.74	500.4137	64.76	11.68	11.19	12.37
C27H58N5P1	483.77	483.4430	67.04	12.08	14.48	6.40
C27H58N6P2	528.75	528.4198	61.33	11.06	15.89	11.72
C27H59N1P2	459.73	459.4123	70.54	12.94	3.05	13.47
C27H59N2P1	442.76	442.4416	73.25	13.43	6.33	7.00
C27H59N3P2	487.74	487.4184	66.49	12.19	8.62	12.70
C27H59N4P1	470.77	470.4477	68.89	12.63	11.90	6.58
C27H59N5P2	515.75	515.4245	62.88	11.53	13.58	12.01
C27H59N6P1	498.79	498.4539	65.02	11.92	16.85	6.21
C27H60N2P2	474.74	474.4232	68.31	12.74	5.90	13.05
C27H60N3P1	457.77	457.4525	70.84	13.21	9.18	6.77
C27H60N4P2	502.75	502.4293	64.50	12.03	11.14	12.32
C27H60N5P1	485.79	485.4586	66.76	12.45	14.42	6.38
C27H60N6P2	530.77	530.4354	61.10	11.39	15.83	11.67
C27H61N3P2	489.75	489.4341	66.22	12.55	8.58	12.65
C27H61N4P1	472.79	472.4634	68.59	13.01	11.85	6.55
C27H61N5P2	517.77	517.4402	62.63	11.88	13.53	11.96
C27H61N6P1	500.80	500.4695	64.76	12.28	16.78	6.18
C27H62N4P2	504.77	504.4449	64.25	12.38	11.10	12.27
C27H62N5P1	487.80	487.4743	66.48	12.81	14.36	6.35
C27H62N6P2	532.78	532.4511	60.87	11.73	15.77	11.63
C27H63N5P2	519.78	519.4558	62.39	12.22	13.47	11.92
C27H63N6P1	502.82	502.4852	64.50	12.63	16.71	6.16
C27H64N6P2	534.80	534.4667	60.64	12.06	15.71	11.58
C28H14N1P1	395.40	395.0864	85.06	3.57	3.54	7.83
C28H14N2P2	440.38	440.0632	76.37	3.20	6.36	14.07
C28H14N3P1	423.42	423.0925	79.43	3.33	9.92	7.32
C28H14N4P2	468.40	468.0694	71.80	3.01	11.96	13.23
C28H14N5P1	451.43	451.0987	74.50	3.13	15.51	6.86
C28H14N6P2	496.41	496.0755	67.75	2.84	16.93	12.48
C28H15N1P2	427.39	427.0680	78.69	3.54	3.28	14.49
C28H15N2P1	410.42	410.0973	81.94	3.68	6.83	7.55
C28H15N3P2	455.40	455.0741	73.85	3.32	9.23	13.60
C28H15N4P1	438.43	438.1034	76.71	3.45	12.78	7.06
C28H15N5P2	483.41	483.0803	69.57	3.13	14.49	12.81
C28H15N6P1	466.45	466.1096	72.10	3.24	18.02	6.64
C28H16N1P1	397.42	397.1020	84.62	4.06	3.52	7.79
C28H16N2P2	442.40	442.0789	76.02	3.65	6.33	14.00
C28H16N3P1	425.43	425.1082	79.05	3.79	9.88	7.28
C28H16N4P2	470.41	470.0850	71.49	3.43	11.91	13.17
C28H16N5P1	453.45	453.1143	74.17	3.56	15.44	6.83
C28H16N6P2	498.43	498.0912	67.47	3.24	16.86	12.43
C28H17N1P2	429.40	429.0836	78.32	3.99	3.26	14.43

C28H17N2P1	412.43	412.1129	81.54	4.15	6.79	7.51
C28H17N3P2	457.42	457.0898	73.52	3.75	9.19	13.54
C28H17N4P1	440.45	440.1191	76.36	3.89	12.72	7.03
C28H17N5P2	485.43	485.0959	69.28	3.53	14.43	12.76
C28H17N6P1	468.46	468.1252	71.79	3.66	17.94	6.61
C28H18N1P1	399.44	399.1177	84.20	4.54	3.51	7.75
C28H18N2P2	444.42	444.0945	75.67	4.08	6.30	13.94
C28H18N3P1	427.45	427.1238	78.68	4.24	9.83	7.25
C28H18N4P2	472.43	472.1007	71.19	3.84	11.86	13.11
C28H18N5P1	455.46	455.1300	73.84	3.98	15.38	6.80
C28H18N6P2	500.44	500.1068	67.20	3.63	16.79	12.38
C28H19N1P2	431.42	431.0993	77.96	4.44	3.25	14.36
C28H19N2P1	414.45	414.1286	81.15	4.62	6.76	7.47
C28H19N3P2	459.43	459.1054	73.20	4.17	9.15	13.48
C28H19N4P1	442.46	442.1347	76.01	4.33	12.66	7.00
C28H19N5P2	487.44	487.1116	68.99	3.93	14.37	12.71
C28H19N6P1	470.48	470.1409	71.48	4.07	17.86	6.58
C28H20N1P1	401.45	401.1333	83.77	5.02	3.49	7.72
C28H20N2P2	446.43	446.1102	75.33	4.52	6.27	13.88
C28H20N3P1	429.47	429.1395	78.31	4.69	9.78	7.21
C28H20N4P2	474.45	474.1163	70.89	4.25	11.81	13.06
C28H20N5P1	457.48	457.1456	73.51	4.41	15.31	6.77
C28H20N6P2	502.46	502.1225	66.93	4.01	16.73	12.33
C28H21N1P2	433.43	433.1149	77.59	4.88	3.23	14.29
C28H21N2P1	416.47	416.1442	80.75	5.08	6.73	7.44
C28H21N3P2	461.45	461.1211	72.88	4.59	9.11	13.42
C28H21N4P1	444.48	444.1504	75.66	4.76	12.61	6.97
C28H21N5P2	489.46	489.1272	68.71	4.32	14.31	12.66
C28H21N6P1	472.49	472.1565	71.18	4.48	17.79	6.56
C28H22N1P1	403.47	403.1490	83.36	5.50	3.47	7.68
C28H22N2P2	448.45	448.1258	74.99	4.94	6.25	13.81
C28H22N3P1	431.48	431.1551	77.94	5.14	9.74	7.18
C28H22N4P2	476.46	476.1320	70.59	4.65	11.76	13.00
C28H22N5P1	459.49	459.1613	73.19	4.83	15.24	6.74
C28H22N6P2	504.48	504.1381	66.67	4.40	16.66	12.28
C28H23N1P2	435.45	435.1306	77.23	5.32	3.22	14.23
C28H23N2P1	418.48	418.1599	80.36	5.54	6.69	7.40
C28H23N3P2	463.46	463.1367	72.57	5.00	9.07	13.37
C28H23N4P1	446.50	446.1660	75.32	5.19	12.55	6.94
C28H23N5P2	491.48	491.1429	68.43	4.72	14.25	12.60
C28H23N6P1	474.51	474.1722	70.88	4.89	17.71	6.53
C28H24N1P1	405.48	405.1646	82.94	5.97	3.45	7.64
C28H24N2P2	450.46	450.1415	74.66	5.37	6.22	13.75
C28H24N3P1	433.50	433.1708	77.58	5.58	9.69	7.15
C28H24N4P2	478.48	478.1476	70.29	5.06	11.71	12.95
C28H24N5P1	461.51	461.1769	72.87	5.24	15.17	6.71
C28H24N6P2	506.49	506.1538	66.40	4.78	16.59	12.23
C28H25N1P2	437.47	437.1462	76.88	5.76	3.20	14.16
C28H25N2P1	420.50	420.1755	79.98	5.99	6.66	7.37
C28H25N3P2	465.48	465.1524	72.25	5.41	9.03	13.31
C28H25N4P1	448.51	448.1817	74.98	5.62	12.49	6.91
C28H25N5P2	493.49	493.1585	68.15	5.11	14.19	12.55
C28H25N6P1	476.53	476.1878	70.58	5.29	17.64	6.50
C28H26N1P1	407.50	407.1803	82.53	6.43	3.44	7.60
C28H26N2P2	452.48	452.1571	74.33	5.79	6.19	13.69
C28H26N3P1	435.51	435.1864	77.22	6.02	9.65	7.11
C28H26N4P2	480.49	480.1633	69.99	5.45	11.66	12.89
C28H26N5P1	463.53	463.1926	72.56	5.65	15.11	6.68
C28H26N6P2	508.51	508.1694	66.14	5.15	16.53	12.18
C28H27N1P2	439.48	439.1619	76.52	6.19	3.19	14.10

C28H27N2P1	422.51	422.1912	79.60	6.44	6.63	7.33
C28H27N3P2	467.50	467.1680	71.94	5.82	8.99	13.25
C28H27N4P1	450.53	450.1973	74.65	6.04	12.44	6.88
C28H27N5P2	495.51	495.1742	67.87	5.49	14.13	12.50
C28H27N6P1	478.54	478.2035	70.28	5.69	17.56	6.47
C28H28N1P1	409.52	409.1959	82.12	6.89	3.42	7.56
C28H28N2P2	454.50	454.1728	74.00	6.21	6.16	13.63
C28H28N3P1	437.53	437.2021	76.87	6.45	9.60	7.08
C28H28N4P2	482.51	482.1789	69.70	5.85	11.61	12.84
C28H28N5P1	465.54	465.2082	72.24	6.06	15.04	6.65
C28H28N6P2	510.52	510.1851	65.88	5.53	16.46	12.13
C28H29N1P2	441.50	441.1775	76.18	6.62	3.17	14.03
C28H29N2P1	424.53	424.2068	79.22	6.89	6.60	7.30
C28H29N3P2	469.51	469.1837	71.63	6.23	8.95	13.19
C28H29N4P1	452.54	452.2130	74.32	6.46	12.38	6.84
C28H29N5P2	497.52	497.1898	67.60	5.88	14.08	12.45
C28H29N6P1	480.56	480.2191	69.98	6.08	17.49	6.45
C28H30N1P1	411.53	411.2116	81.72	7.35	3.40	7.53
C28H30N2P2	456.51	456.1884	73.67	6.62	6.14	13.57
C28H30N3P1	439.55	439.2177	76.51	6.88	9.56	7.05
C28H30N4P2	484.53	484.1946	69.41	6.24	11.56	12.79
C28H30N5P1	467.56	467.2239	71.93	6.47	14.98	6.62
C28H30N6P2	512.54	512.2007	65.62	5.90	16.40	12.09
C28H31N1P2	443.51	443.1932	75.83	7.05	3.16	13.97
C28H31N2P1	426.55	426.2225	78.85	7.33	6.57	7.26
C28H31N3P2	471.53	471.1993	71.32	6.63	8.91	13.14
C28H31N4P1	454.56	454.2286	73.99	6.87	12.33	6.81
C28H31N5P2	499.54	499.2055	67.32	6.26	14.02	12.40
C28H31N6P1	482.57	482.2348	69.69	6.48	17.42	6.42
C28H32N1P1	413.55	413.2272	81.32	7.80	3.39	7.49
C28H32N2P2	458.53	458.2041	73.35	7.03	6.11	13.51
C28H32N3P1	441.56	441.2334	76.16	7.30	9.52	7.01
C28H32N4P2	486.54	486.2102	69.12	6.63	11.52	12.73
C28H32N5P1	469.57	469.2395	71.62	6.87	14.91	6.60
C28H32N6P2	514.56	514.2164	65.36	6.27	16.33	12.04
C28H33N1P2	445.53	445.2088	75.49	7.47	3.14	13.90
C28H33N2P1	428.56	428.2381	78.47	7.76	6.54	7.23
C28H33N3P2	473.54	473.2150	71.02	7.02	8.87	13.08
C28H33N4P1	456.58	456.2443	73.66	7.29	12.27	6.78
C28H33N5P2	501.56	501.2211	67.05	6.63	13.96	12.35
C28H33N6P1	484.59	484.2504	69.40	6.86	17.34	6.39
C28H34N1P1	415.56	415.2429	80.93	8.25	3.37	7.45
C28H34N2P2	460.54	460.2197	73.02	7.44	6.08	13.45
C28H34N3P1	443.58	443.2490	75.82	7.73	9.47	6.98
C28H34N4P2	488.56	488.2259	68.84	7.01	11.47	12.68
C28H34N5P1	471.59	471.2552	71.31	7.27	14.85	6.57
C28H34N6P2	516.57	516.2320	65.10	6.63	16.27	11.99
C28H35N1P2	447.55	447.2245	75.15	7.88	3.13	13.84
C28H35N2P1	430.58	430.2538	78.11	8.19	6.51	7.19
C28H35N3P2	475.56	475.2306	70.72	7.42	8.84	13.03
C28H35N4P1	458.59	458.2599	73.34	7.69	12.22	6.75
C28H35N5P2	503.57	503.2368	66.79	7.01	13.91	12.30
C28H35N6P1	486.61	486.2661	69.11	7.25	17.27	6.37
C28H36N1P1	417.58	417.2585	80.54	8.69	3.35	7.42
C28H36N2P2	462.56	462.2354	72.71	7.84	6.06	13.39
C28H36N3P1	445.59	445.2647	75.48	8.14	9.43	6.95
C28H36N4P2	490.57	490.2415	68.55	7.40	11.42	12.63
C28H36N5P1	473.61	473.2708	71.01	7.66	14.79	6.54
C28H36N6P2	518.59	518.2477	64.85	7.00	16.21	11.95
C28H37N1P2	449.56	449.2401	74.81	8.30	3.12	13.78

C28H37N2P1	432.59	432.2694	77.74	8.62	6.48	7.16
C28H37N3P2	477.57	477.2463	70.42	7.81	8.80	12.97
C28H37N4P1	460.61	460.2756	73.01	8.10	12.16	6.72
C28H37N5P2	505.59	505.2524	66.52	7.38	13.85	12.25
C28H37N6P1	488.62	488.2817	68.83	7.63	17.20	6.34
C28H38N1P1	419.60	419.2742	80.15	9.13	3.34	7.38
C28H38N2P2	464.58	464.2510	72.39	8.24	6.03	13.33
C28H38N3P1	447.61	447.2803	75.14	8.56	9.39	6.92
C28H38N4P2	492.59	492.2572	68.27	7.78	11.37	12.58
C28H38N5P1	475.62	475.2865	70.71	8.05	14.72	6.51
C28H38N6P2	520.60	520.2633	64.60	7.36	16.14	11.90
C28H39N1P2	451.58	451.2558	74.47	8.71	3.10	13.72
C28H39N2P1	434.61	434.2851	77.38	9.05	6.45	7.13
C28H39N3P2	479.59	479.2619	70.12	8.20	8.76	12.92
C28H39N4P1	462.62	462.2912	72.70	8.50	12.11	6.70
C28H39N5P2	507.60	507.2681	66.25	7.74	13.80	12.20
C28H39N6P1	490.64	490.2974	68.55	8.01	17.13	6.31
C28H40N1P1	421.61	421.2898	79.77	9.56	3.32	7.35
C28H40N2P2	466.59	466.2667	72.08	8.64	6.00	13.28
C28H40N3P1	449.62	449.2960	74.80	8.97	9.35	6.89
C28H40N4P2	494.61	494.2728	68.00	8.15	11.33	12.52
C28H40N5P1	477.64	477.3021	70.41	8.44	14.66	6.48
C28H40N6P2	522.62	522.2790	64.35	7.71	16.08	11.85
C28H41N1P2	453.59	453.2714	74.14	9.11	3.09	13.66
C28H41N2P1	436.63	436.3007	77.03	9.47	6.42	7.09
C28H41N3P2	481.61	481.2776	69.83	8.58	8.72	12.86
C28H41N4P1	464.64	464.3069	72.38	8.89	12.06	6.67
C28H41N5P2	509.62	509.2837	65.99	8.11	13.74	12.16
C28H41N6P1	492.65	492.3130	68.27	8.39	17.06	6.29
C28H42N1P1	423.63	423.3055	79.39	9.99	3.31	7.31
C28H42N2P2	468.61	468.2823	71.77	9.03	5.98	13.22
C28H42N3P1	451.64	451.3116	74.46	9.37	9.30	6.86
C28H42N4P2	496.62	496.2885	67.72	8.52	11.28	12.47
C28H42N5P1	479.65	479.3178	70.12	8.83	14.60	6.46
C28H42N6P2	524.63	524.2946	64.10	8.07	16.02	11.81
C28H43N1P2	455.61	455.2871	73.82	9.51	3.07	13.60
C28H43N2P1	438.64	438.3164	76.67	9.88	6.39	7.06
C28H43N3P2	483.62	483.2932	69.54	8.96	8.69	12.81
C28H43N4P1	466.66	466.3225	72.07	9.29	12.01	6.64
C28H43N5P2	511.64	511.2994	65.73	8.47	13.69	12.11
C28H43N6P1	494.67	494.3287	67.99	8.76	16.99	6.26
C28H44N1P1	425.64	425.3211	79.01	10.42	3.29	7.28
C28H44N2P2	470.62	470.2980	71.46	9.42	5.95	13.16
C28H44N3P1	453.66	453.3273	74.13	9.78	9.26	6.83
C28H44N4P2	498.64	498.3041	67.45	8.89	11.24	12.42
C28H44N5P1	481.67	481.3334	69.82	9.21	14.54	6.43
C28H44N6P2	526.65	526.3103	63.86	8.42	15.96	11.76
C28H45N1P2	457.63	457.3027	73.49	9.91	3.06	13.54
C28H45N2P1	440.66	440.3320	76.32	10.29	6.36	7.03
C28H45N3P2	485.64	485.3089	69.25	9.34	8.65	12.76
C28H45N4P1	468.67	468.3382	71.76	9.68	11.95	6.61
C28H45N5P2	513.65	513.3150	65.47	8.83	13.63	12.06
C28H45N6P1	496.68	496.3443	67.71	9.13	16.92	6.24
C28H46N1P1	427.66	427.3368	78.64	10.84	3.28	7.24
C28H46N2P2	472.64	472.3136	71.16	9.81	5.93	13.11
C28H46N3P1	455.67	455.3429	73.81	10.18	9.22	6.80
C28H46N4P2	500.65	500.3198	67.17	9.26	11.19	12.37
C28H46N5P1	483.69	483.3491	69.53	9.59	14.48	6.40
C28H46N6P2	528.67	528.3259	63.62	8.77	15.90	11.72
C28H47N1P2	459.64	459.3184	73.17	10.31	3.05	13.48

095

C28H47N2P1	442.67	442.3477	75.97	10.70	6.33	7.00
C28H47N3P2	487.65	487.3245	68.97	9.71	8.62	12.70
C28H47N4P1	470.69	470.3538	71.45	10.06	11.90	6.58
C28H47N5P2	515.67	515.3307	65.22	9.19	13.58	12.01
C28H47N6P1	498.70	498.3600	67.44	9.50	16.85	6.21
C28H48N1P1	429.68	429.3524	78.27	11.26	3.26	7.21
C28H48N2P2	474.66	474.3293	70.85	10.19	5.90	13.05
C28H48N3P1	457.69	457.3586	73.48	10.57	9.18	6.77
C28H48N4P2	502.67	502.3354	66.91	9.63	11.15	12.32
C28H48N5P1	485.70	485.3647	69.24	9.96	14.42	6.38
C28H48N6P2	530.68	530.3416	63.37	9.12	15.84	11.67
C28H49N1P2	461.66	461.3340	72.85	10.70	3.03	13.42
C28H49N2P1	444.69	444.3633	75.63	11.11	6.30	6.97
C28H49N3P2	489.67	489.3402	68.68	10.09	8.58	12.65
C28H49N4P1	472.70	472.3695	71.15	10.45	11.85	6.55
C28H49N5P2	517.68	517.3463	64.96	9.54	13.53	11.97
C28H49N6P1	500.72	500.3756	67.17	9.86	16.78	6.19
C28H50N1P1	431.69	431.3681	77.91	11.67	3.24	7.17
C28H50N2P2	476.67	476.3449	70.55	10.57	5.88	13.00
C28H50N3P1	459.70	459.3742	73.16	10.96	9.14	6.74
C28H50N4P2	504.69	504.3511	66.64	9.99	11.10	12.27
C28H50N5P1	487.72	487.3804	68.96	10.33	14.36	6.35
C28H50N6P2	532.70	532.3572	63.13	9.46	15.78	11.63
C28H51N1P2	463.67	463.3497	72.53	11.09	3.02	13.36
C28H51N2P1	446.71	446.3790	75.29	11.51	6.27	6.93
C28H51N3P2	491.69	491.3558	68.40	10.46	8.55	12.60
C28H51N4P1	474.72	474.3851	70.84	10.83	11.80	6.52
C28H51N5P2	519.70	519.3620	64.71	9.89	13.48	11.92
C28H51N6P1	502.73	502.3913	66.90	10.23	16.72	6.16
C28H52N1P1	433.71	433.3837	77.54	12.09	3.23	7.14
C28H52N2P2	478.69	478.3606	70.26	10.95	5.85	12.94
C28H52N3P1	461.72	461.3899	72.84	11.35	9.10	6.71
C28H52N4P2	506.70	506.3667	66.37	10.34	11.06	12.23
C28H52N5P1	489.73	489.3960	68.67	10.70	14.30	6.32
C28H52N6P2	534.71	534.3729	62.90	9.80	15.72	11.59
C28H53N1P2	465.69	465.3653	72.22	11.47	3.01	13.30
C28H53N2P1	448.72	448.3946	74.95	11.91	6.24	6.90
C28H53N3P2	493.70	493.3715	68.12	10.82	8.51	12.55
C28H53N4P1	476.74	476.4008	70.54	11.21	11.75	6.50
C28H53N5P2	521.72	521.3776	64.46	10.24	13.42	11.87
C28H53N6P1	504.75	504.4069	66.63	10.58	16.65	6.14
C28H54N1P1	435.72	435.3994	77.18	12.49	3.21	7.11
C28H54N2P2	480.70	480.3762	69.96	11.32	5.83	12.89
C28H54N3P1	463.74	463.4055	72.52	11.74	9.06	6.68
C28H54N4P2	508.72	508.3824	66.11	10.70	11.01	12.18
C28H54N5P1	491.75	491.4117	68.39	11.07	14.24	6.30
C28H54N6P2	536.73	536.3885	62.66	10.14	15.66	11.54
C28H55N1P2	467.70	467.3810	71.91	11.85	2.99	13.25
C28H55N2P1	450.74	450.4103	74.61	12.30	6.22	6.87
C28H55N3P2	495.72	495.3871	67.84	11.18	8.48	12.50
C28H55N4P1	478.75	478.4164	70.25	11.58	11.70	6.47
C28H55N5P2	523.73	523.3933	64.21	10.59	13.37	11.83
C28H55N6P1	506.76	506.4226	66.36	10.94	16.58	6.11
C28H56N1P1	437.74	437.4150	76.83	12.89	3.20	7.08
C28H56N2P2	482.72	482.3919	69.67	11.69	5.80	12.83
C28H56N3P1	465.75	465.4212	72.21	12.12	9.02	6.65
C28H56N4P2	510.73	510.3980	65.85	11.05	10.97	12.13
C28H56N5P1	493.77	493.4273	68.11	11.43	14.18	6.27
C28H56N6P2	538.75	538.4041	62.42	10.48	15.60	11.50
C28H57N1P2	469.72	469.3966	71.60	12.23	2.98	13.19
C28H57N2P1	452.75	452.4259	74.28	12.69	6.19	6.84
C28H57N3P2	497.73	497.4028	67.57	11.54	8.44	12.45
C28H57N4P1	480.77	480.4321	69.95	11.95	11.65	6.44
C28H57N5P2	525.75	525.4089	63.97	10.93	13.32	11.78
C28H57N6P1	508.78	508.4382	66.10	11.29	16.52	6.09
C28H58N1P1	439.75	439.4307	76.48	13.29	3.19	7.04
C28H58N2P2	484.74	484.4075	69.38	12.06	5.78	12.78
C28H58N3P1	467.77	467.4368	71.90	12.50	8.98	6.62
C28H58N4P2	512.75	512.4137	65.59	11.40	10.93	12.08
C28H58N5P1	495.78	495.4430	67.83	11.79	14.13	6.25
C28H58N6P2	540.76	540.4198	62.19	10.81	15.54	11.46
C28H59N1P2	471.74	471.4123	71.29	12.61	2.97	13.13
C28H59N2P1	454.77	454.4416	73.95	13.08	6.16	6.81
C28H59N3P2	499.75	499.4184	67.30	11.90	8.41	12.40
C28H59N4P1	482.78	482.4477	69.66	12.32	11.60	6.42
C28H59N5P2	527.76	527.4245	63.72	11.27	13.27	11.74
C28H59N6P1	510.80	510.4539	65.84	11.64	16.45	6.06
C28H60N1P1	441.77	441.4463	76.13	13.69	3.17	7.01
C28H60N2P2	486.75	486.4232	69.09	12.42	5.76	12.73
C28H60N3P1	469.78	469.4525	71.59	12.87	8.94	6.59
C28H60N4P2	514.76	514.4293	65.33	11.75	10.88	12.03
C28H60N5P1	497.80	497.4586	67.56	12.15	14.07	6.22
C28H60N6P2	542.78	542.4354	61.96	11.14	15.48	11.41
C28H61N1P2	473.75	473.4279	70.99	12.98	2.96	13.08
C28H61N2P1	456.79	456.4572	73.63	13.46	6.13	6.78
C28H61N3P2	501.77	501.4341	67.03	12.25	8.37	12.35
C28H61N4P1	484.80	484.4634	69.37	12.68	11.56	6.39
C28H61N5P2	529.78	529.4402	63.48	11.61	13.22	11.69
C28H61N6P1	512.81	512.4695	65.58	11.99	16.39	6.04
C28H62N2P2	488.77	488.4388	68.81	12.79	5.73	12.67
C28H62N3P1	471.80	471.4681	71.28	13.25	8.91	6.57
C28H62N4P2	516.78	516.4449	65.08	12.09	10.84	11.99
C28H62N5P1	499.81	499.4743	67.29	12.50	14.01	6.20
C28H62N6P2	544.79	544.4511	61.73	11.47	15.43	11.37
C28H63N3P2	503.78	503.4497	66.76	12.61	8.34	12.30
C28H63N4P1	486.81	486.4790	69.08	13.04	11.51	6.36
C28H63N5P2	531.80	531.4558	63.24	11.94	13.17	11.65
C28H63N6P1	514.83	514.4852	65.33	12.33	16.32	6.02
C28H64N4P2	518.80	518.4606	64.83	12.43	10.80	11.94
C28H64N5P1	501.83	501.4899	67.02	12.85	13.96	6.17
C28H64N6P2	546.81	546.4667	61.50	11.80	15.37	11.33
C28H65N5P2	533.81	533.4715	63.00	12.27	13.12	11.60
C28H65N6P1	516.84	516.5008	65.07	12.68	16.26	5.99
C28H66N6P2	548.83	548.4824	61.28	12.12	15.31	11.29
C29H14N1P1	407.42	407.0864	85.50	3.46	3.44	7.60
C29H14N2P2	452.40	452.0632	77.00	3.12	6.19	13.69
C29H14N3P1	435.43	435.0925	80.00	3.24	9.65	7.11
C29H14N4P2	480.41	480.0694	72.51	2.94	11.66	12.89
C29H14N5P1	463.44	463.0987	75.16	3.04	15.11	6.68
C29H14N6P2	508.42	508.0755	68.51	2.78	16.53	12.18
C29H15N1P2	439.40	439.0680	79.27	3.44	3.19	14.10
C29H15N2P1	422.43	422.0973	82.46	3.58	6.63	7.33
C29H15N3P2	467.41	467.0741	74.52	3.23	8.99	13.25
C29H15N4P1	450.44	450.1034	77.33	3.36	12.44	6.88
C29H15N5P2	495.42	495.0803	70.31	3.05	14.14	12.50
C29H15N6P1	478.46	478.1096	72.80	3.16	17.56	6.47
C29H16N1P1	409.43	409.1020	85.07	3.94	3.42	7.57
C29H16N2P2	454.41	454.0789	76.65	3.55	6.16	13.63
C29H16N3P1	437.44	437.1082	79.63	3.69	9.61	7.08
C29H16N4P2	482.43	482.0850	72.20	3.34	11.61	12.84

096

C29H16N5P1	465.46	465.1143	74.83	3.46	15.05	6.65
C29H16N6P2	510.44	510.0912	68.24	3.16	16.46	12.14
C29H17N1P2	441.41	441.0836	78.91	3.88	3.17	14.03
C29H17N2P1	424.45	424.1129	82.07	4.04	6.60	7.30
C29H17N3P2	469.43	469.0898	74.20	3.65	8.95	13.20
C29H17N4P1	452.46	452.1191	76.98	3.79	12.38	6.85
C29H17N5P2	497.44	497.0959	70.02	3.44	14.08	12.45
C29H17N6P1	480.47	480.1252	72.50	3.57	17.49	6.45
C29H18N1P1	411.45	411.1177	84.66	4.41	3.40	7.53
C29H18N2P2	456.43	456.0945	76.32	3.98	6.14	13.57
C29H18N3P1	439.46	439.1238	79.26	4.13	9.56	7.05
C29H18N4P2	484.44	484.1007	71.90	3.75	11.57	12.79
C29H18N5P1	467.47	467.1300	74.51	3.88	14.98	6.63
C29H18N6P2	512.45	512.1068	67.97	3.54	16.40	12.09
C29H19N1P2	443.43	443.0993	78.55	4.32	3.16	13.97
C29H19N2P1	426.46	426.1286	81.68	4.49	6.57	7.26
C29H19N3P2	471.44	471.1054	73.88	4.06	8.91	13.14
C29H19N4P1	454.48	454.1347	76.64	4.21	12.33	6.82
C29H19N5P2	499.46	499.1116	69.74	3.83	14.02	12.40
C29H19N6P1	482.49	482.1409	72.19	3.97	17.42	6.42
C29H20N1P1	413.46	413.1333	84.25	4.88	3.39	7.49
C29H20N2P2	458.44	458.1102	75.98	4.40	6.11	13.51
C29H20N3P1	441.48	441.1395	78.90	4.57	9.52	7.02
C29H20N4P2	486.46	486.1163	71.60	4.14	11.52	12.73
C29H20N5P1	469.49	469.1456	74.19	4.29	14.92	6.60
C29H20N6P2	514.47	514.1225	67.71	3.92	16.34	12.04
C29H21N1P2	445.45	445.1149	78.20	4.75	3.14	13.91
C29H21N2P1	428.48	428.1442	81.29	4.94	6.54	7.23
C29H21N3P2	473.46	473.1211	73.57	4.47	8.88	13.08
C29H21N4P1	456.49	456.1504	76.30	4.64	12.27	6.79
C29H21N5P2	501.47	501.1272	69.46	4.22	13.97	12.35
C29H21N6P1	484.50	484.1565	71.89	4.37	17.35	6.39
C29H22N1P1	415.48	415.1490	83.84	5.34	3.37	7.45
C29H22N2P2	460.46	460.1258	75.65	4.82	6.08	13.45
C29H22N3P1	443.49	443.1551	78.54	5.00	9.47	6.98
C29H22N4P2	488.47	488.1320	71.31	4.54	11.47	12.68
C29H22N5P1	471.51	471.1613	73.87	4.70	14.85	6.57
C29H22N6P2	516.49	516.1381	67.44	4.29	16.27	11.99
C29H23N1P2	447.46	447.1306	77.84	5.18	3.13	13.84
C29H23N2P1	430.49	430.1599	80.91	5.39	6.51	7.19
C29H23N3P2	475.47	475.1367	73.26	4.88	8.84	13.03
C29H23N4P1	458.51	458.1660	75.97	5.06	12.22	6.76
C29H23N5P2	503.49	503.1429	69.18	4.60	13.91	12.30
C29H23N6P1	486.52	486.1722	71.59	4.77	17.27	6.37
C29H24N1P1	417.50	417.1646	83.43	5.79	3.35	7.42
C29H24N2P2	462.48	462.1415	75.32	5.23	6.06	13.39
C29H24N3P1	445.51	445.1708	78.19	5.43	9.43	6.95
C29H24N4P2	490.49	490.1476	71.02	4.93	11.42	12.63
C29H24N5P1	473.52	473.1769	73.56	5.11	14.79	6.54
C29H24N6P2	518.50	518.1538	67.18	4.67	16.21	11.95
C29H25N1P2	449.48	449.1462	77.50	5.61	3.12	13.78
C29H25N2P1	432.51	432.1755	80.54	5.83	6.48	7.16
C29H25N3P2	477.49	477.1524	72.95	5.28	8.80	12.97
C29H25N4P1	460.52	460.1817	75.64	5.47	12.17	6.73
C29H25N5P2	505.50	505.1585	68.91	4.98	13.85	12.25
C29H25N6P1	488.54	488.1878	71.30	5.16	17.20	6.34
C29H26N1P1	419.51	419.1803	83.03	6.25	3.34	7.38
C29H26N2P2	464.49	464.1571	74.99	5.64	6.03	13.34
C29H26N3P1	447.52	447.1864	77.83	5.86	9.39	6.92
C29H26N4P2	492.50	492.1633	70.72	5.32	11.38	12.58
C29H26N5P1	475.54	475.1926	73.25	5.51	14.73	6.51
C29H26N6P2	520.52	520.1694	66.92	5.03	16.15	11.90
C29H27N1P2	451.49	451.1619	77.15	6.03	3.10	13.72
C29H27N2P1	434.53	434.1912	80.16	6.26	6.45	7.13
C29H27N3P2	479.51	479.1680	72.64	5.68	8.76	12.92
C29H27N4P1	462.54	462.1973	75.31	5.88	12.11	6.70
C29H27N5P2	507.52	507.1742	68.63	5.36	13.80	12.21
C29H27N6P1	490.55	490.2035	71.01	5.55	17.13	6.31
C29H28N1P1	421.53	421.1959	82.63	6.70	3.32	7.35
C29H28N2P2	466.51	466.1728	74.67	6.05	6.00	13.28
C29H28N3P1	449.54	449.2021	77.48	6.28	9.35	6.89
C29H28N4P2	494.52	494.1789	70.44	5.71	11.33	12.53
C29H28N5P1	477.55	477.2082	72.94	5.91	14.67	6.49
C29H28N6P2	522.53	522.1851	66.66	5.40	16.08	11.86
C29H29N1P2	453.51	453.1775	76.81	6.45	3.09	13.66
C29H29N2P1	436.54	436.2068	79.79	6.70	6.42	7.10
C29H29N3P2	481.52	481.1837	72.34	6.07	8.73	12.86
C29H29N4P1	464.56	464.2130	74.98	6.29	12.06	6.67
C29H29N5P2	509.54	509.1898	68.36	5.74	13.74	12.16
C29H29N6P1	492.57	492.2191	70.72	5.93	17.06	6.29
C29H30N1P1	423.54	423.2116	82.24	7.14	3.31	7.31
C29H30N2P2	468.52	468.1884	74.34	6.45	5.98	13.22
C29H30N3P1	451.56	451.2177	77.14	6.70	9.31	6.86
C29H30N4P2	496.54	496.1946	70.15	6.09	11.28	12.48
C29H30N5P1	479.57	479.2239	72.63	6.31	14.60	6.46
C29H30N6P2	524.55	524.2007	66.40	5.76	16.02	11.81
C29H31N1P2	455.52	455.1932	76.47	6.86	3.07	13.60
C29H31N2P1	438.56	438.2225	79.42	7.12	6.39	7.06
C29H31N3P2	483.54	483.1993	72.04	6.46	8.69	12.81
C29H31N4P1	466.57	466.2286	74.66	6.70	12.01	6.64
C29H31N5P2	511.55	511.2055	68.09	6.11	13.69	12.11
C29H31N6P1	494.58	494.2348	70.43	6.32	16.99	6.26
C29H32N1P1	425.56	425.2272	81.85	7.58	3.29	7.28
C29H32N2P2	470.54	470.2041	74.03	6.85	5.95	13.17
C29H32N3P1	453.57	453.2334	76.80	7.11	9.26	6.83
C29H32N4P2	498.55	498.2102	69.87	6.47	11.24	12.43
C29H32N5P1	481.59	481.2395	72.33	6.70	14.54	6.43
C29H32N6P2	526.57	526.2164	66.15	6.13	15.96	11.76
C29H33N1P2	457.54	457.2088	76.13	7.27	3.06	13.54
C29H33N2P1	440.57	440.2381	79.06	7.55	6.36	7.03
C29H33N3P2	485.55	485.2150	71.74	6.85	8.65	12.76
C29H33N4P1	468.59	468.2443	74.33	7.10	11.96	6.61
C29H33N5P2	513.57	513.2211	67.82	6.48	13.64	12.06
C29H33N6P1	496.60	496.2504	70.14	6.70	16.92	6.24
C29H34N1P1	427.57	427.2429	81.46	8.02	3.28	7.24
C29H34N2P2	472.56	472.2197	73.71	7.25	5.93	13.11
C29H34N3P1	455.59	455.2490	76.46	7.52	9.22	6.80
C29H34N4P2	500.57	500.2259	69.59	6.85	11.19	12.38
C29H34N5P1	483.60	483.2552	72.03	7.09	14.48	6.40
C29H34N6P2	528.58	528.2320	65.90	6.48	15.90	11.72
C29H35N1P2	459.56	459.2245	75.80	7.68	3.05	13.48
C29H35N2P1	442.59	442.2538	78.70	7.97	6.33	7.00
C29H35N3P2	487.57	487.2306	71.44	7.24	8.62	12.71
C29H35N4P1	470.60	470.2599	74.02	7.50	11.91	6.58
C29H35N5P2	515.58	515.2368	67.56	6.84	13.58	12.02
C29H35N6P1	498.62	498.2661	69.86	7.08	16.85	6.21
C29H36N1P1	429.59	429.2585	81.08	8.45	3.26	7.21
C29H36N2P2	474.57	474.2354	73.40	7.65	5.90	13.05
C29H36N3P1	457.60	457.2647	76.12	7.93	9.18	6.77
C29H36N4P2	502.58	502.2415	69.31	7.22	11.15	12.33

C29H36N5P1	485.62	485.2708	71.73	7.47	14.42	6.38
C29H36N6P2	530.60	530.2477	65.65	6.84	15.84	11.68
C29H37N1P2	461.57	461.2401	75.46	8.08	3.03	13.42
C29H37N2P1	444.61	444.2694	78.34	8.39	6.30	6.97
C29H37N3P2	489.59	489.2463	71.15	7.62	8.58	12.65
C29H37N4P1	472.62	472.2756	73.70	7.89	11.85	6.55
C29H37N5P2	517.60	517.2524	67.30	7.21	13.53	11.97
C29H37N6P1	500.63	500.2817	69.58	7.45	16.79	6.19
C29H38N1P1	431.61	431.2742	80.70	8.87	3.25	7.18
C29H38N2P2	476.59	476.2510	73.09	8.04	5.88	13.00
C29H38N3P1	459.62	459.2803	75.79	8.33	9.14	6.74
C29H38N4P2	504.60	504.2572	69.03	7.59	11.10	12.28
C29H38N5P1	487.63	487.2865	71.43	7.85	14.36	6.35
C29H38N6P2	532.61	532.2633	65.40	7.19	15.78	11.63
C29H39N1P2	463.59	463.2558	75.14	8.48	3.02	13.36
C29H39N2P1	446.62	446.2851	77.99	8.80	6.27	6.94
C29H39N3P2	491.60	491.2619	70.85	8.00	8.55	12.60
C29H39N4P1	474.63	474.2912	73.39	8.28	11.80	6.53
C29H39N5P2	519.62	519.2681	67.03	7.57	13.48	11.92
C29H39N6P1	502.65	502.2974	69.30	7.82	16.72	6.16
C29H40N1P1	433.62	433.2898	80.33	9.30	3.23	7.14
C29H40N2P2	478.60	478.2667	72.78	8.42	5.85	12.94
C29H40N3P1	461.64	461.2960	75.45	8.73	9.10	6.71
C29H40N4P2	506.62	506.2728	68.75	7.96	11.06	12.23
C29H40N5P1	489.65	489.3021	71.14	8.23	14.30	6.33
C29H40N6P2	534.63	534.2790	65.15	7.54	15.72	11.59
C29H41N1P2	465.60	465.2714	74.81	8.88	3.01	13.30
C29H41N2P1	448.64	448.3007	77.64	9.21	6.24	6.90
C29H41N3P2	493.62	493.2776	70.57	8.37	8.51	12.55
C29H41N4P1	476.65	476.3069	73.08	8.67	11.75	6.50
C29H41N5P2	521.63	521.2837	66.78	7.92	13.43	11.88
C29H41N6P1	504.66	504.3130	69.02	8.19	16.65	6.14
C29H42N1P1	435.64	435.3055	79.96	9.72	3.22	7.11
C29H42N2P2	480.62	480.2823	72.47	8.81	5.83	12.89
C29H42N3P1	463.65	463.3116	75.13	9.13	9.06	6.68
C29H42N4P2	508.63	508.2885	68.48	8.32	11.02	12.18
C29H42N5P1	491.67	491.3178	70.85	8.61	14.24	6.30
C29H42N6P2	536.65	536.2946	64.91	7.89	15.66	11.54
C29H43N1P2	467.62	467.2871	74.49	9.27	3.00	13.25
C29H43N2P1	450.65	450.3164	77.29	9.62	6.22	6.87
C29H43N3P2	495.63	495.2932	70.28	8.74	8.48	12.50
C29H43N4P1	478.67	478.3225	72.77	9.05	11.70	6.47
C29H43N5P2	523.65	523.2994	66.52	8.28	13.37	11.83
C29H43N6P1	506.68	506.3287	68.75	8.55	16.59	6.11
C29H44N1P1	437.65	437.3211	79.59	10.13	3.20	7.08
C29H44N2P2	482.64	482.2980	72.17	9.19	5.80	12.84
C29H44N3P1	465.67	465.3273	74.80	9.52	9.02	6.65
C29H44N4P2	510.65	510.3041	68.21	8.69	10.97	12.13
C29H44N5P1	493.68	493.3334	70.56	8.98	14.19	6.27
C29H44N6P2	538.66	538.3103	64.66	8.23	15.60	11.50
C29H45N1P2	469.64	469.3027	74.17	9.66	2.98	13.19
C29H45N2P1	452.67	452.3320	76.95	10.02	6.19	6.84
C29H45N3P2	497.65	497.3089	69.99	9.11	8.44	12.45
C29H45N4P1	480.68	480.3382	72.46	9.44	11.66	6.44
C29H45N5P2	525.66	525.3150	66.26	8.63	13.32	11.78
C29H45N6P1	508.70	508.3443	68.47	8.92	16.52	6.09
C29H46N1P1	439.67	439.3368	79.22	10.55	3.19	7.04
C29H46N2P2	484.65	484.3136	71.87	9.57	5.78	12.78
C29H46N3P1	467.68	467.3429	74.48	9.91	8.98	6.62
C29H46N4P2	512.66	512.3198	67.94	9.04	10.93	12.08

C29H46N5P1	495.70	495.3491	70.27	9.35	14.13	6.25
C29H46N6P2	540.68	540.3259	64.42	8.58	15.54	11.46
C29H47N1P2	471.65	471.3184	73.85	10.04	2.97	13.13
C29H47N2P1	454.69	454.3477	76.61	10.42	6.16	6.81
C29H47N3P2	499.67	499.3245	69.71	9.48	8.41	12.40
C29H47N4P1	482.70	482.3538	72.16	9.81	11.61	6.42
C29H47N5P2	527.68	527.3307	66.01	8.98	13.27	11.74
C29H47N6P1	510.71	510.3600	68.20	9.28	16.46	6.06
C29H48N1P1	441.69	441.3524	78.86	10.95	3.17	7.01
C29H48N2P2	486.67	486.3293	71.57	9.94	5.76	12.73
C29H48N3P1	469.70	469.3586	74.16	10.30	8.95	6.59
C29H48N4P2	514.68	514.3354	67.68	9.40	10.89	12.04
C29H48N5P1	497.71	497.3647	69.98	9.72	14.07	6.22
C29H48N6P2	542.69	542.3416	64.18	8.92	15.49	11.41
C29H49N1P2	473.67	473.3340	73.54	10.43	2.96	13.08
C29H49N2P1	456.70	456.3633	76.27	10.81	6.13	6.78
C29H49N3P2	501.68	501.3402	69.43	9.84	8.38	12.35
C29H49N4P1	484.71	484.3695	71.86	10.19	11.56	6.39
C29H49N5P2	529.69	529.3463	65.76	9.32	13.22	11.69
C29H49N6P1	512.73	512.3756	67.94	9.63	16.39	6.04
C29H50N1P1	443.70	443.3681	78.50	11.36	3.16	6.98
C29H50N2P2	488.68	488.3449	71.28	10.31	5.73	12.68
C29H50N3P1	471.72	471.3742	73.84	10.68	8.91	6.57
C29H50N4P2	516.70	516.3511	67.41	9.75	10.84	11.99
C29H50N5P1	499.73	499.3804	69.70	10.09	14.01	6.20
C29H50N6P2	544.71	544.3572	63.95	9.25	15.43	11.37
C29H51N1P2	475.68	475.3497	73.23	10.81	2.94	13.02
C29H51N2P1	458.72	458.3790	75.93	11.21	6.11	6.75
C29H51N3P2	503.70	503.3558	69.15	10.21	8.34	12.30
C29H51N4P1	486.73	486.3851	71.56	10.56	11.51	6.36
C29H51N5P2	531.71	531.3620	65.51	9.67	13.17	11.65
C29H51N6P1	514.74	514.3913	67.67	9.99	16.33	6.02
C29H52N1P1	445.72	445.3837	78.15	11.76	3.14	6.95
C29H52N2P2	490.70	490.3606	70.99	10.68	5.71	12.62
C29H52N3P1	473.73	473.3899	73.53	11.06	8.87	6.54
C29H52N4P2	518.71	518.3667	67.15	10.10	10.80	11.94
C29H52N5P1	501.75	501.3960	69.42	10.45	13.96	6.17
C29H52N6P2	546.73	546.3729	63.71	9.59	15.37	11.33
C29H53N1P2	477.70	477.3653	72.92	11.18	2.93	12.97
C29H53N2P1	460.73	460.3946	75.60	11.60	6.08	6.72
C29H53N3P2	505.71	505.3715	68.88	10.56	8.31	12.25
C29H53N4P1	488.75	488.4008	71.27	10.93	11.46	6.34
C29H53N5P2	533.73	533.3776	65.26	10.01	13.12	11.61
C29H53N6P1	516.76	516.4069	67.41	10.34	16.26	5.99
C29H54N1P1	447.73	447.3994	77.80	12.16	3.13	6.92
C29H54N2P2	492.71	492.3762	70.69	11.05	5.69	12.57
C29H54N3P1	475.75	475.4055	73.22	11.44	8.83	6.51
C29H54N4P2	520.73	520.3824	66.89	10.45	10.76	11.90
C29H54N5P1	503.76	503.4117	69.14	10.80	13.90	6.15
C29H54N6P2	548.74	548.3885	63.48	9.92	15.32	11.29
C29H55N1P2	479.72	479.3810	72.61	11.56	2.92	12.91
C29H55N2P1	462.75	462.4103	75.27	11.98	6.05	6.69
C29H55N3P2	507.73	507.3871	68.60	10.92	8.28	12.20
C29H55N4P1	490.76	490.4164	70.98	11.30	11.42	6.31
C29H55N5P2	535.74	535.3933	65.02	10.35	13.07	11.56
C29H55N6P1	518.78	518.4226	67.14	10.69	16.20	5.97
C29H56N1P1	449.75	449.4150	77.45	12.55	3.11	6.89
C29H56N2P2	494.73	494.3919	70.41	11.41	5.66	12.52
C29H56N3P1	477.76	477.4212	72.91	11.81	8.80	6.48
C29H56N4P2	522.74	522.3980	66.63	10.80	10.72	11.85

098

C29H56N5P1	505.78	505.4273	68.87	11.16	13.85	6.12
C29H56N6P2	550.76	550.4041	63.24	10.25	15.26	11.25
C29H57N1P2	481.73	481.3966	72.31	11.93	2.91	12.86
C29H57N2P1	464.76	464.4259	74.95	12.36	6.03	6.66
C29H57N3P2	509.75	509.4028	68.33	11.27	8.24	12.15
C29H57N4P1	492.78	492.4321	70.69	11.66	11.37	6.29
C29H57N5P2	537.76	537.4089	64.77	10.68	13.02	11.52
C29H57N6P1	520.79	520.4382	66.88	11.03	16.14	5.95
C29H58N1P1	451.77	451.4307	77.10	12.94	3.10	6.86
C29H58N2P2	496.75	496.4075	70.12	11.77	5.64	12.47
C29H58N3P1	479.78	479.4368	72.60	12.19	8.76	6.46
C29H58N4P2	524.76	524.4137	66.38	11.14	10.68	11.80
C29H58N5P1	507.79	507.4430	68.60	11.51	13.79	6.10
C29H58N6P2	552.77	552.4198	63.01	10.58	15.20	11.21
C29H59N1P2	483.75	483.4123	72.01	12.29	2.90	12.81
C29H59N2P1	466.78	466.4416	74.62	12.74	6.00	6.64
C29H59N3P2	511.76	511.4184	68.06	11.62	8.21	12.10
C29H59N4P1	494.79	494.4477	70.40	12.02	11.32	6.26
C29H59N5P2	539.77	539.4245	64.53	11.02	12.97	11.48
C29H59N6P1	522.81	522.4539	66.63	11.38	16.07	5.92
C29H60N1P1	453.78	453.4463	76.76	13.33	3.09	6.83
C29H60N2P2	498.76	498.4232	69.84	12.13	5.62	12.42
C29H60N3P1	481.80	481.4525	72.30	12.55	8.72	6.43
C29H60N4P2	526.78	526.4293	66.12	11.48	10.64	11.76
C29H60N5P1	509.81	509.4586	68.32	11.86	13.74	6.08
C29H60N6P2	554.79	554.4354	62.78	10.90	15.15	11.17
C29H61N1P2	485.76	485.4279	71.71	12.66	2.88	12.75
C29H61N2P1	468.80	468.4572	74.30	13.12	5.98	6.61
C29H61N3P2	513.78	513.4340	67.80	11.97	8.18	12.06
C29H61N4P1	496.81	496.4634	70.11	12.38	11.28	6.23
C29H61N5P2	541.79	541.4402	64.29	11.35	12.93	11.43
C29H61N6P1	524.82	524.4695	66.37	11.72	16.01	5.90
C29H62N1P1	455.80	455.4620	76.42	13.71	3.07	6.80
C29H62N2P2	500.78	500.4388	69.56	12.48	5.59	12.37
C29H62N3P1	483.81	483.4681	72.00	12.92	8.69	6.40
C29H62N4P2	528.79	528.4449	65.87	11.82	10.60	11.71
C29H62N5P1	511.82	511.4743	68.06	12.21	13.68	6.05
C29H62N6P2	556.81	556.4511	62.56	11.22	15.09	11.13
C29H63N1P2	487.78	487.4436	71.41	13.02	2.87	12.70
C29H63N2P1	470.81	470.4729	73.98	13.49	5.95	6.58
C29H63N3P2	515.79	515.4497	67.53	12.31	8.15	12.01
C29H63N4P1	498.83	498.4790	69.83	12.73	11.23	6.21
C29H63N5P2	543.81	543.4558	64.05	11.68	12.88	11.39
C29H63N6P1	526.84	526.4852	66.12	12.05	15.95	5.88
C29H64N2P2	502.79	502.4545	69.28	12.83	5.57	12.32
C29H64N3P1	485.83	485.4838	71.70	13.28	8.65	6.38
C29H64N4P2	530.81	530.4606	65.62	12.15	10.56	11.67
C29H64N5P1	513.84	513.4899	67.79	12.55	13.63	6.03
C29H64N6P2	558.82	558.4667	62.33	11.54	15.04	11.09
C29H65N3P2	517.81	517.4653	67.27	12.65	8.11	11.96
C29H65N4P1	500.84	500.4947	69.55	13.08	11.19	6.18
C29H65N5P2	545.82	545.4715	63.82	12.00	12.83	11.35
C29H65N6P1	528.86	528.5008	65.86	12.39	15.89	5.86
C29H66N4P2	532.82	532.4762	65.37	12.49	10.52	11.63
C29H66N5P1	515.86	515.5056	67.52	12.90	13.58	6.00
C29H66N6P2	560.84	560.4824	62.11	11.86	14.98	11.05
C29H67N5P2	547.84	547.4871	63.58	12.33	12.78	11.31
C29H67N6P1	530.87	530.5165	65.61	12.72	15.83	5.83
C29H68N6P2	562.85	562.4980	61.89	12.18	14.93	11.01
C30H15N1P2	451.41	451.0680	79.82	3.35	3.10	13.72

C30H15N2P1	434.44	434.0973	82.94	3.48	6.45	7.13
C30H15N3P2	479.42	479.0741	75.16	3.15	8.76	12.92
C30H15N4P1	462.45	462.1034	77.92	3.27	12.12	6.70
C30H15N5P2	507.44	507.0803	71.01	2.98	13.80	12.21
C30H15N6P1	490.47	490.1096	73.47	3.08	17.13	6.32
C30H16N1P1	421.44	421.1020	85.50	3.83	3.32	7.35
C30H16N2P2	466.42	466.0789	77.25	3.46	6.01	13.28
C30H16N3P1	449.46	449.1082	80.17	3.59	9.35	6.89
C30H16N4P2	494.44	494.0850	72.88	3.26	11.33	12.53
C30H16N5P1	477.47	477.1143	75.47	3.38	14.67	6.49
C30H16N6P2	522.45	522.0912	68.97	3.09	16.09	11.86
C30H17N1P2	453.42	453.0836	79.47	3.78	3.09	13.66
C30H17N2P1	436.46	436.1129	82.56	3.93	6.42	7.10
C30H17N3P2	481.44	481.0898	74.85	3.56	8.73	12.87
C30H17N4P1	464.47	464.1191	77.58	3.69	12.06	6.67
C30H17N5P2	509.45	509.0959	70.73	3.36	13.75	12.16
C30H17N6P1	492.48	492.1252	73.17	3.48	17.06	6.29
C30H18N1P1	423.46	423.1177	85.09	4.28	3.31	7.31
C30H18N2P2	468.44	468.0945	76.92	3.87	5.98	13.22
C30H18N3P1	451.47	451.1238	79.81	4.02	9.31	6.86
C30H18N4P2	496.45	496.1007	72.58	3.65	11.29	12.48
C30H18N5P1	479.49	479.1300	75.15	3.78	14.61	6.46
C30H18N6P2	524.47	524.1068	68.71	3.46	16.02	11.81
C30H19N1P2	455.44	455.0993	79.12	4.21	3.08	13.60
C30H19N2P1	438.47	438.1286	82.18	4.37	6.39	7.06
C30H19N3P2	483.45	483.1054	74.53	3.96	8.69	12.81
C30H19N4P1	466.49	466.1347	77.24	4.11	12.01	6.64
C30H19N5P2	511.47	511.1116	70.45	3.74	13.69	12.11
C30H19N6P1	494.50	494.1409	72.87	3.87	16.99	6.26
C30H20N1P1	425.47	425.1333	84.69	4.74	3.29	7.28
C30H20N2P2	470.45	470.1102	76.59	4.29	5.95	13.17
C30H20N3P1	453.49	453.1395	79.46	4.45	9.27	6.83
C30H20N4P2	498.47	498.1163	72.29	4.04	11.24	12.43
C30H20N5P1	481.50	481.1456	74.84	4.19	14.54	6.43
C30H20N6P2	526.48	526.1225	68.44	3.83	15.96	11.77
C30H21N1P2	457.46	457.1149	78.77	4.63	3.06	13.54
C30H21N2P1	440.49	440.1442	81.80	4.81	6.36	7.03
C30H21N3P2	485.47	485.1211	74.22	4.36	8.66	12.76
C30H21N4P1	468.50	468.1504	76.91	4.52	11.96	6.61
C30H21N5P2	513.48	513.1272	70.17	4.12	13.64	12.06
C30H21N6P1	496.52	496.1565	72.57	4.26	16.93	6.24
C30H22N1P1	427.49	427.1490	84.29	5.19	3.28	7.25
C30H22N2P2	472.47	472.1258	76.27	4.69	5.93	13.11
C30H22N3P1	455.50	455.1551	79.11	4.87	9.22	6.80
C30H22N4P2	500.48	500.1320	72.00	4.43	11.19	12.38
C30H22N5P1	483.52	483.1613	74.52	4.59	14.48	6.41
C30H22N6P2	528.50	528.1381	68.18	4.20	15.90	11.72
C30H23N1P2	459.47	459.1306	78.42	5.05	3.05	13.48
C30H23N2P1	442.51	442.1599	81.43	5.24	6.33	7.00
C30H23N3P2	487.49	487.1367	73.92	4.76	8.62	12.71
C30H23N4P1	470.52	470.1660	76.58	4.93	11.91	6.58
C30H23N5P2	515.50	515.1429	69.90	4.50	13.59	12.02
C30H23N6P1	498.53	498.1722	72.28	4.65	16.86	6.21
C30H24N1P1	429.51	429.1646	83.90	5.63	3.26	7.21
C30H24N2P2	474.49	474.1415	75.94	5.10	5.90	13.06
C30H24N3P1	457.52	457.1708	78.76	5.29	9.18	6.77
C30H24N4P2	502.50	502.1476	71.71	4.81	11.15	12.33
C30H24N5P1	485.53	485.1769	74.21	4.98	14.42	6.38
C30H24N6P2	530.51	530.1538	67.92	4.56	15.84	11.68
C30H25N1P2	461.49	461.1462	78.08	5.46	3.04	13.42

099

C30H25N2P1	444.52	444.1755	81.06	5.67	6.30	6.97
C30H25N3P2	489.50	489.1524	73.61	5.15	8.58	12.66
C30H25N4P1	472.53	472.1817	76.26	5.33	11.86	6.55
C30H25N5P2	517.51	517.1585	69.63	4.87	13.53	11.97
C30H25N6P1	500.55	500.1878	71.99	5.03	16.79	6.19
C30H26N1P1	431.52	431.1803	83.50	6.07	3.25	7.18
C30H26N2P2	476.50	476.1571	75.62	5.50	5.88	13.00
C30H26N3P1	459.54	459.1864	78.41	5.70	9.14	6.74
C30H26N4P2	504.52	504.1633	71.42	5.19	11.11	12.28
C30H26N5P1	487.55	487.1926	73.91	5.38	14.36	6.35
C30H26N6P2	532.53	532.1694	67.66	4.92	15.78	11.63
C30H27N1P2	463.50	463.1619	77.74	5.87	3.02	13.37
C30H27N2P1	446.54	446.1912	80.70	6.09	6.27	6.94
C30H27N3P2	491.52	491.1680	73.31	5.54	8.55	12.60
C30H27N4P1	474.55	474.1973	75.93	5.73	11.81	6.53
C30H27N5P2	519.53	519.1742	69.36	5.24	13.48	11.92
C30H27N6P1	502.56	502.2035	71.70	5.42	16.72	6.16
C30H28N1P1	433.54	433.1959	83.11	6.51	3.23	7.14
C30H28N2P2	478.52	478.1728	75.30	5.90	5.85	12.95
C30H28N3P1	461.55	461.2021	78.07	6.11	9.10	6.71
C30H28N4P2	506.53	506.1789	71.14	5.57	11.06	12.23
C30H28N5P1	489.56	489.2082	73.60	5.76	14.31	6.33
C30H28N6P2	534.55	534.1851	67.41	5.28	15.72	11.59
C30H29N1P2	465.52	465.1775	77.40	6.28	3.01	13.31
C30H29N2P1	448.55	448.2068	80.33	6.52	6.25	6.91
C30H29N3P2	493.53	493.1837	73.01	5.92	8.51	12.55
C30H29N4P1	476.57	476.2130	75.61	6.13	11.76	6.50
C30H29N5P2	521.55	521.1898	69.09	5.60	13.43	11.88
C30H29N6P1	504.58	504.2191	71.41	5.79	16.66	6.14
C30H30N1P1	435.55	435.2116	82.73	6.94	3.22	7.11
C30H30N2P2	480.53	480.1884	74.99	6.29	5.83	12.89
C30H30N3P1	463.57	463.2177	77.73	6.52	9.06	6.68
C30H30N4P2	508.55	508.1946	70.86	5.95	11.02	12.18
C30H30N5P1	491.58	491.2239	73.30	6.15	14.25	6.30
C30H30N6P2	536.56	536.2007	67.16	5.64	15.66	11.55
C30H31N1P2	467.54	467.1932	77.07	6.68	3.00	13.25
C30H31N2P1	450.57	450.2225	79.97	6.94	6.22	6.87
C30H31N3P2	495.55	495.1993	72.71	6.31	8.48	12.50
C30H31N4P1	478.58	478.2286	75.29	6.53	11.71	6.47
C30H31N5P2	523.56	523.2055	68.82	5.97	13.38	11.83
C30H31N6P1	506.60	506.2348	71.13	6.17	16.59	6.11
C30H32N1P1	437.57	437.2272	82.35	7.37	3.20	7.08
C30H32N2P2	482.55	482.2041	74.67	6.68	5.81	12.84
C30H32N3P1	465.58	465.2334	77.39	6.93	9.03	6.65
C30H32N4P2	510.56	510.2102	70.58	6.32	10.97	12.13
C30H32N5P1	493.60	493.2395	73.00	6.53	14.19	6.28
C30H32N6P2	538.58	538.2164	66.90	5.99	15.60	11.50
C30H33N1P2	469.55	469.2088	76.74	7.08	2.98	13.19
C30H33N2P1	452.58	452.2381	79.62	7.35	6.19	6.84
C30H33N3P2	497.57	497.2150	72.42	6.69	8.45	12.45
C30H33N4P1	480.60	480.2443	74.98	6.92	11.66	6.44
C30H33N5P2	525.58	525.2211	68.56	6.33	13.33	11.79
C30H33N6P1	508.61	508.2504	70.85	6.54	16.52	6.09
C30H34N1P1	439.59	439.2429	81.97	7.80	3.19	7.05
C30H34N2P2	484.57	484.2197	74.37	7.07	5.78	12.78
C30H34N3P1	467.60	467.2490	77.06	7.33	8.99	6.62
C30H34N4P2	512.58	512.2259	70.30	6.69	10.93	12.09
C30H34N5P1	495.61	495.2552	72.70	6.91	14.13	6.25
C30H34N6P2	540.59	540.2320	66.66	6.34	15.55	11.46
C30H35N1P2	471.57	471.2245	76.41	7.48	2.97	13.14
C30H35N2P1	454.60	454.2538	79.26	7.76	6.16	6.81
C30H35N3P2	499.58	499.2306	72.13	7.06	8.41	12.40
C30H35N4P1	482.61	482.2599	74.66	7.31	11.61	6.42
C30H35N5P2	527.59	527.2368	68.30	6.69	13.27	11.74
C30H35N6P1	510.63	510.2661	70.57	6.91	16.46	6.07
C30H36N1P1	441.60	441.2585	81.60	8.22	3.17	7.01
C30H36N2P2	486.58	486.2354	74.05	7.46	5.76	12.73
C30H36N3P1	469.62	469.2647	76.73	7.73	8.95	6.60
C30H36N4P2	514.60	514.2415	70.02	7.05	10.89	12.04
C30H36N5P1	497.63	497.2708	72.41	7.29	14.07	6.22
C30H36N6P2	542.61	542.2477	66.41	6.69	15.49	11.42
C30H37N1P2	473.58	473.2401	76.09	7.88	2.96	13.08
C30H37N2P1	456.62	456.2694	78.91	8.17	6.13	6.78
C30H37N3P2	501.60	501.2463	71.84	7.44	8.38	12.35
C30H37N4P1	484.63	484.2756	74.35	7.70	11.56	6.39
C30H37N5P2	529.61	529.2524	68.04	7.04	13.22	11.70
C30H37N6P1	512.64	512.2817	70.29	7.28	16.39	6.04
C30H38N1P1	443.62	443.2742	81.23	8.63	3.16	6.98
C30H38N2P2	488.60	488.2510	73.75	7.84	5.73	12.68
C30H38N3P1	471.63	471.2803	76.40	8.12	8.91	6.57
C30H38N4P2	516.61	516.2572	69.75	7.41	10.85	11.99
C30H38N5P1	499.64	499.2865	72.12	7.67	14.02	6.20
C30H38N6P2	544.63	544.2633	66.16	7.03	15.43	11.37
C30H39N1P2	475.60	475.2558	75.76	8.27	2.95	13.03
C30H39N2P1	458.63	458.2851	78.57	8.57	6.11	6.75
C30H39N3P2	503.61	503.2619	71.55	7.81	8.34	12.30
C30H39N4P1	486.65	486.2912	74.04	8.08	11.51	6.36
C30H39N5P2	531.63	531.2681	67.78	7.39	13.17	11.65
C30H39N6P1	514.66	514.2974	70.01	7.64	16.33	6.02
C30H40N1P1	445.63	445.2898	80.86	9.05	3.14	6.95
C30H40N2P2	490.61	490.2667	73.45	8.22	5.71	12.63
C30H40N3P1	473.65	473.2960	76.08	8.51	8.87	6.54
C30H40N4P2	518.63	518.2728	69.48	7.77	10.80	11.94
C30H40N5P1	501.66	501.3021	71.83	8.04	13.96	6.17
C30H40N6P2	546.64	546.2790	65.92	7.38	15.37	11.33
C30H41N1P2	477.62	477.2714	75.44	8.65	2.93	12.97
C30H41N2P1	460.65	460.3007	78.22	8.97	6.08	6.72
C30H41N3P2	505.63	505.2776	71.26	8.17	8.31	12.25
C30H41N4P1	488.66	488.3069	73.74	8.46	11.47	6.34
C30H41N5P2	533.64	533.2837	67.52	7.74	13.12	11.61
C30H41N6P1	516.68	516.3130	69.74	8.00	16.27	5.99
C30H42N1P1	447.65	447.3055	80.49	9.46	3.13	6.92
C30H42N2P2	492.63	492.2823	73.15	8.59	5.69	12.57
C30H42N3P1	475.66	475.3116	75.75	8.90	8.83	6.51
C30H42N4P2	520.64	520.2885	69.21	8.13	10.76	11.90
C30H42N5P1	503.68	503.3178	71.54	8.41	13.90	6.15
C30H42N6P2	548.66	548.2946	65.68	7.72	15.32	11.29
C30H43N1P2	479.63	479.2871	75.13	9.04	2.92	12.92
C30H43N2P1	462.66	462.3164	77.88	9.37	6.05	6.69
C30H43N3P2	507.64	507.2932	70.98	8.54	8.28	12.20
C30H43N4P1	490.68	490.3225	73.44	8.83	11.42	6.31
C30H43N5P2	535.66	535.2994	67.27	8.09	13.07	11.56
C30H43N6P1	518.69	518.3287	69.47	8.36	16.20	5.97
C30H44N1P1	449.67	449.3211	80.13	9.86	3.11	6.89
C30H44N2P2	494.65	494.2980	72.85	8.97	5.66	12.52
C30H44N3P1	477.68	477.3273	75.43	9.28	8.80	6.48
C30H44N4P2	522.66	522.3041	68.94	8.49	10.72	11.85
C30H44N5P1	505.69	505.3334	71.26	8.77	13.85	6.13
C30H44N6P2	550.67	550.3103	65.44	8.05	15.26	11.25
C30H45N1P2	481.65	481.3027	74.81	9.42	2.91	12.86

100

C30H45N2P1	464.68	464.3320	77.54	9.76	6.03	6.67
C30H45N3P2	509.66	509.3089	70.70	8.90	8.24	12.15
C30H45N4P1	492.69	492.3382	73.14	9.21	11.37	6.29
C30H45N5P2	537.67	537.3150	67.02	8.44	13.03	11.52
C30H45N6P1	520.71	520.3443	69.20	8.71	16.14	5.95
C30H46N1P1	451.68	451.3368	79.78	10.27	3.10	6.86
C30H46N2P2	496.66	496.3136	72.55	9.34	5.64	12.47
C30H46N3P1	479.70	479.3429	75.12	9.67	8.76	6.46
C30H46N4P2	524.68	524.3198	68.68	8.84	10.68	11.81
C30H46N5P1	507.71	507.3491	70.97	9.13	13.79	6.10
C30H46N6P2	552.69	552.3259	65.20	8.39	15.21	11.21
C30H47N1P2	483.66	483.3184	74.50	9.79	2.90	12.81
C30H47N2P1	466.70	466.3477	77.21	10.15	6.00	6.64
C30H47N3P2	511.68	511.3245	70.42	9.26	8.21	12.11
C30H47N4P1	494.71	494.3538	72.84	9.58	11.33	6.26
C30H47N5P2	539.69	539.3307	66.77	8.78	12.98	11.48
C30H47N6P1	522.72	522.3600	68.93	9.06	16.08	5.93
C30H48N1P1	453.70	453.3524	79.42	10.66	3.09	6.83
C30H48N2P2	498.68	498.3293	72.26	9.70	5.62	12.42
C30H48N3P1	481.71	481.3586	74.80	10.04	8.72	6.43
C30H48N4P2	526.69	526.3354	68.41	9.19	10.64	11.76
C30H48N5P1	509.72	509.3647	70.69	9.49	13.74	6.08
C30H48N6P2	554.70	554.3416	64.96	8.72	15.15	11.17
C30H49N1P2	485.68	485.3340	74.19	10.17	2.88	12.75
C30H49N2P1	468.71	468.3633	76.88	10.54	5.98	6.61
C30H49N3P2	513.69	513.3402	70.15	9.61	8.18	12.06
C30H49N4P1	496.73	496.3695	72.54	9.94	11.28	6.24
C30H49N5P2	541.71	541.3463	66.52	9.12	12.93	11.44
C30H49N6P1	524.74	524.3756	68.67	9.41	16.02	5.90
C30H50N1P1	455.71	455.3681	79.07	11.06	3.07	6.80
C30H50N2P2	500.69	500.3449	71.97	10.07	5.59	12.37
C30H50N3P1	483.73	483.3742	74.49	10.42	8.69	6.40
C30H50N4P2	528.71	528.3511	68.15	9.53	10.60	11.72
C30H50N5P1	511.74	511.3804	70.41	9.85	13.69	6.05
C30H50N6P2	556.72	556.3572	64.72	9.05	15.10	11.13
C30H51N1P2	487.70	487.3497	73.89	10.54	2.87	12.70
C30H51N2P1	470.73	470.3790	76.55	10.92	5.95	6.58
C30H51N3P2	515.71	515.3558	69.87	9.97	8.15	12.01
C30H51N4P1	498.74	498.3851	72.25	10.31	11.23	6.21
C30H51N5P2	543.72	543.3620	66.27	9.45	12.88	11.39
C30H51N6P1	526.75	526.3913	68.41	9.76	15.95	5.88
C30H52N1P1	457.73	457.3837	78.72	11.45	3.06	6.77
C30H52N2P2	502.71	502.3606	71.68	10.43	5.57	12.32
C30H52N3P1	485.74	485.3899	74.18	10.79	8.65	6.38
C30H52N4P2	530.72	530.3667	67.89	9.88	10.56	11.67
C30H52N5P1	513.76	513.3960	70.14	10.20	13.63	6.03
C30H52N6P2	558.74	558.3729	64.49	9.38	15.04	11.09
C30H53N1P2	489.71	489.3653	73.58	10.91	2.86	12.65
C30H53N2P1	472.74	472.3946	76.22	11.30	5.93	6.55
C30H53N3P2	517.72	517.3715	69.60	10.32	8.12	11.97
C30H53N4P1	500.76	500.4008	71.96	10.67	11.19	6.19
C30H53N5P2	545.74	545.3776	66.03	9.79	12.83	11.35
C30H53N6P1	528.77	528.4069	68.15	10.10	15.89	5.86
C30H54N1P1	459.75	459.3994	78.38	11.84	3.05	6.74
C30H54N2P2	504.73	504.3762	71.39	10.78	5.55	12.27
C30H54N3P1	487.76	487.4055	73.88	11.16	8.61	6.35
C30H54N4P2	532.74	532.3824	67.64	10.22	10.52	11.63
C30H54N5P1	515.77	515.4117	69.86	10.55	13.58	6.01
C30H54N6P2	560.75	560.3885	64.26	9.71	14.99	11.05
C30H55N1P2	491.73	491.3810	73.28	11.27	2.85	12.60
C30H55N2P1	474.76	474.4103	75.90	11.68	5.90	6.52
C30H55N3P2	519.74	519.3871	69.33	10.67	8.08	11.92
C30H55N4P1	502.77	502.4164	71.67	11.03	11.14	6.16
C30H55N5P2	547.75	547.3933	65.78	10.12	12.79	11.31
C30H55N6P1	530.79	530.4226	67.89	10.44	15.83	5.84
C30H56N1P1	461.76	461.4150	78.03	12.22	3.03	6.71
C30H56N2P2	506.74	506.3919	71.11	11.14	5.53	12.22
C30H56N3P1	489.77	489.4212	73.57	11.52	8.58	6.32
C30H56N4P2	534.76	534.3980	67.38	10.56	10.48	11.58
C30H56N5P1	517.79	517.4273	69.59	10.90	13.53	5.98
C30H56N6P2	562.77	562.4041	64.03	10.03	14.93	11.01
C30H57N1P2	493.74	493.3966	72.98	11.64	2.84	12.55
C30H57N2P1	476.78	476.4259	75.58	12.05	5.88	6.50
C30H57N3P2	521.76	521.4028	69.06	11.01	8.05	11.87
C30H57N4P1	504.79	504.4321	71.38	11.38	11.10	6.14
C30H57N5P2	549.77	549.4089	65.54	10.45	12.74	11.27
C30H57N6P1	532.80	532.4382	67.63	10.78	15.77	5.81
C30H58N1P1	463.78	463.4307	77.70	12.61	3.02	6.68
C30H58N2P2	508.76	508.4075	70.83	11.49	5.51	12.18
C30H58N3P1	491.79	491.4368	73.27	11.89	8.54	6.30
C30H58N4P2	536.77	536.4137	67.13	10.89	10.44	11.54
C30H58N5P1	519.80	519.4430	69.32	11.25	13.47	5.96
C30H58N6P2	564.78	564.4198	63.80	10.35	14.88	10.97
C30H59N1P2	495.76	495.4123	72.68	12.00	2.83	12.50
C30H59N2P1	478.79	478.4416	75.26	12.42	5.85	6.47
C30H59N3P2	523.77	523.4184	68.80	11.35	8.02	11.83
C30H59N4P1	506.81	506.4477	71.10	11.73	11.05	6.11
C30H59N5P2	551.79	551.4245	65.30	10.78	12.69	11.23
C30H59N6P1	534.82	534.4539	67.38	11.12	15.71	5.79
C30H60N1P1	465.79	465.4463	77.36	12.98	3.01	6.65
C30H60N2P2	510.77	510.4232	70.55	11.84	5.48	12.13
C30H60N3P1	493.81	493.4525	72.97	12.25	8.51	6.27
C30H60N4P2	538.79	538.4293	66.88	11.22	10.40	11.50
C30H60N5P1	521.82	521.4586	69.05	11.59	13.42	5.94
C30H60N6P2	566.80	566.4354	63.57	10.67	14.83	10.93
C30H61N1P2	497.77	497.4279	72.39	12.35	2.81	12.44
C30H61N2P1	480.81	480.4572	74.94	12.79	5.83	6.44
C30H61N3P2	525.79	525.4340	68.53	11.69	7.99	11.78
C30H61N4P1	508.82	508.4634	70.82	12.08	11.01	6.09
C30H61N5P2	553.80	553.4402	65.07	11.10	12.65	11.19
C30H61N6P1	536.83	536.4695	67.12	11.45	15.65	5.77
C30H62N1P1	467.81	467.4620	77.03	13.36	2.99	6.62
C30H62N2P2	512.79	512.4388	70.27	12.19	5.46	12.08
C30H62N3P1	495.82	495.4681	72.67	12.60	8.47	6.25
C30H62N4P2	540.80	540.4449	66.63	11.56	10.36	11.45
C30H62N5P1	523.84	523.4743	68.79	11.93	13.37	5.91
C30H62N6P2	568.82	568.4511	63.35	10.99	14.77	10.89
C30H63N1P2	499.79	499.4436	72.10	12.71	2.80	12.39
C30H63N2P1	482.82	482.4729	74.63	13.15	5.80	6.42
C30H63N3P2	527.80	527.4497	68.27	12.03	7.96	11.74
C30H63N4P1	510.84	510.4790	70.54	12.43	10.97	6.06
C30H63N5P2	555.82	555.4558	64.83	11.42	12.60	11.15
C30H63N6P1	538.85	538.4852	66.87	11.78	15.60	5.75
C30H64N1P1	469.83	469.4776	76.70	13.73	2.98	6.59
C30H64N2P2	514.81	514.4545	69.99	12.53	5.44	12.03
C30H64N3P1	497.84	497.4838	72.38	12.96	8.44	6.22
C30H64N4P2	542.82	542.4606	66.38	11.88	10.32	11.41
C30H64N5P1	525.85	525.4899	68.52	12.27	13.32	5.89
C30H64N6P2	570.83	570.4667	63.12	11.30	14.72	10.85
C30H65N1P2	501.81	501.4592	71.81	13.06	2.79	12.34

C30H65N2P1	484.84	484.4885	74.32	13.51	5.78	6.39
C30H65N3P2	529.82	529.4653	68.01	12.37	7.93	11.69
C30H65N4P1	512.85	512.4947	70.26	12.78	10.92	6.04
C30H65N5P2	557.83	557.4715	64.60	11.75	12.55	11.11
C30H65N6P1	540.87	540.5008	66.62	12.11	15.54	5.73
C30H66N2P2	516.82	516.4701	69.72	12.87	5.42	11.99
C30H66N3P1	499.85	499.4994	72.09	13.31	8.41	6.20
C30H66N4P2	544.83	544.4762	66.14	12.21	10.28	11.37
C30H66N5P1	527.87	527.5056	68.26	12.60	13.27	5.87
C30H66N6P2	572.85	572.4824	62.90	11.61	14.67	10.81
C30H67N3P2	531.84	531.4810	67.75	12.70	7.90	11.65
C30H67N4P1	514.87	514.5103	69.99	13.12	10.88	6.02
C30H67N5P2	559.85	559.4871	64.36	12.06	12.51	11.07
C30H67N6P1	542.88	542.5165	66.37	12.44	15.48	5.71
C30H68N4P2	546.85	546.4919	65.89	12.53	10.25	11.33
C30H68N5P1	529.88	529.5212	68.00	12.94	13.22	5.85
C30H68N6P2	574.86	574.4980	62.68	11.92	14.62	10.78
C30H69N5P2	561.87	561.5028	64.13	12.38	12.46	11.03
C30H69N6P1	544.90	544.5321	66.13	12.76	15.42	5.68
C30H70N6P2	576.88	576.5137	62.46	12.23	14.57	10.74

11. CARBON, HYDROGEN, OXYGEN AND PHOSPHORUS

11. KOHLENSTOFF, WASSERSTOFF, SAUERSTOFF UND PHOSPHOR

11. CARBONE, HYDROGENE, OXYGENE ET PHOSPHORE

MOLECULAR WEIGHTS AND PERCENTAGE COMPOSITIONS OF COMPOUNDS CONTAINING CARBON, HYDROGEN, OXYGEN AND PHOSPHORUS

FORMULA	MOL. WEIGHT	ISOMOL. WEIGHT	PERC. C	PERC. H	PERC. O	PERC. P	FORMULA	MOL. WEIGHT	ISOMOL. WEIGHT	PERC. C	PERC. H	PERC. O	PERC. P
C 1H 001P2	89.96	89.9424	13.35	0	17.79	68.86	C 1H 5O6P1	144.02	143.9824	8.34	3.50	66.65	21.51
C 1H 002P2	105.96	105.9374	11.34	0	30.20	58.46	C 1H 5O7P1	160.02	159.9773	7.51	3.15	69.99	19.36
C 1H 003P2	121.96	121.9323	9.85	0	39.36	50.79	C 1H 5O8P1	176.02	175.9722	6.82	2.86	72.72	17.60
C 1H 004P2	137.96	137.9272	8.71	0	46.39	44.90	C 1H 5O9P1	192.02	191.9671	6.26	2.62	74.99	16.13
C 1H 005P2	153.96	153.9221	7.80	0	51.96	40.24	C 1H 6O1P2	96.01	95.9894	12.51	6.30	16.67	64.52
C 1H 006P2	169.96	169.9170	7.07	0	56.48	36.45	C 1H 6O2P2	112.01	111.9843	10.72	5.40	28.57	55.31
C 1H 007P2	185.95	185.9119	6.46	0	60.23	33.31	C 1H 6O3P2	128.00	127.9792	9.38	4.72	37.50	48.39
C 1H 008P2	201.95	201.9068	5.95	0	63.38	30.67	C 1H 6O4P2	144.00	143.9741	8.34	4.20	44.44	43.02
C 1H 009P2	217.95	217.9018	5.51	0	66.07	28.42	C 1H 6O5P2	160.00	159.9690	7.51	3.78	50.00	38.72
C 1H 101P1	59.99	59.9765	20.02	1.68	26.67	51.63	C 1H 6O6P2	176.00	175.9640	6.82	3.44	54.54	35.20
C 1H 102P1	75.99	75.9714	15.81	1.33	42.11	40.76	C 1H 6O7P2	192.00	191.9589	6.26	3.15	58.33	32.26
C 1H 103P1	91.99	91.9663	13.06	1.10	52.18	33.67	C 1H 6O8P2	208.00	207.9538	5.77	2.91	61.54	29.78
C 1H 104P1	107.99	107.9612	11.12	.93	59.26	28.68	C 1H 6O9P2	224.00	223.9487	5.36	2.70	64.28	27.66
C 1H 105P1	123.99	123.9562	9.69	.81	64.52	24.98	C 2H 101P1	72.00	71.9765	33.36	1.40	22.22	43.02
C 1H 106P1	139.99	139.9511	8.58	.72	68.57	22.13	C 2H 102P1	88.00	87.9714	27.30	1.15	36.36	35.20
C 1H 107P1	155.99	155.9460	7.70	.65	71.80	19.86	C 2H 103P1	104.00	103.9663	23.10	.97	46.15	29.78
C 1H 108P1	171.99	171.9409	6.98	.59	74.42	18.01	C 2H 104P1	120.00	119.9612	20.02	.84	53.33	25.81
C 1H 109P1	187.99	187.9358	6.39	.54	76.60	16.48	C 2H 105P1	136.00	135.9562	17.66	.74	58.82	22.77
C 1H 201P2	91.97	91.9581	13.06	2.19	17.40	67.35	C 2H 106P1	152.00	151.9511	15.80	.66	63.16	20.38
C 1H 202P2	107.97	107.9530	11.12	1.87	29.64	57.37	C 2H 107P1	168.00	167.9460	14.30	.60	66.66	18.44
C 1H 203P2	123.97	123.9479	9.69	1.63	38.72	49.97	C 2H 108P1	184.00	183.9409	13.06	.55	69.56	16.83
C 1H 204P2	139.97	139.9428	8.58	1.44	45.72	44.26	C 2H 109P1	200.00	199.9358	12.01	.50	72.00	15.49
C 1H 205P2	155.97	155.9377	7.70	1.29	51.29	39.72	C 2H 201P2	103.99	103.9581	23.10	1.94	15.39	59.57
C 1H 206P2	171.97	171.9327	6.98	1.17	55.82	36.02	C 2H 202P2	119.98	119.9530	20.02	1.68	26.67	51.63
C 1H 207P2	187.97	187.9276	6.39	1.07	59.58	32.96	C 2H 203P2	135.98	135.9479	17.67	1.48	35.30	45.56
C 1H 208P2	203.97	203.9225	5.89	.99	62.75	30.37	C 2H 204P2	151.98	151.9428	15.81	1.33	42.11	40.76
C 1H 209P2	219.97	219.9174	5.46	.92	65.46	28.16	C 2H 205P2	167.98	167.9377	14.30	1.20	47.62	36.88
C 1H 301P1	62.01	61.9922	19.37	4.88	25.80	49.95	C 2H 206P2	183.98	183.9327	13.06	1.10	52.18	33.67
C 1H 302P1	78.01	77.9871	15.40	3.88	41.02	39.71	C 2H 207P2	199.98	199.9276	12.01	1.01	56.00	30.98
C 1H 303P1	94.01	93.9820	12.78	3.22	51.06	32.95	C 2H 208P2	215.98	215.9225	11.12	.93	59.26	28.68
C 1H 304P1	110.01	109.9769	10.92	2.75	58.18	28.16	C 2H 209P2	231.98	231.9174	10.36	.87	62.07	26.70
C 1H 305P1	126.01	125.9718	9.53	2.40	63.49	24.58	C 2H 301P1	74.02	73.9922	32.45	4.09	21.62	41.85
C 1H 306P1	142.01	141.9667	8.46	2.13	67.60	21.81	C 2H 302P1	90.02	89.9871	26.69	3.36	35.55	34.41
C 1H 307P1	158.00	157.9616	7.60	1.91	70.88	19.60	C 2H 303P1	106.02	105.9820	22.66	2.85	45.27	29.22
C 1H 308P1	174.00	173.9566	6.90	1.74	73.56	17.80	C 2H 304P1	122.02	121.9769	19.69	2.48	52.45	25.38
C 1H 309P1	190.00	189.9515	6.32	1.59	75.79	16.30	C 2H 305P1	138.02	137.9718	17.41	2.19	57.96	22.44
C 1H 401P2	93.99	93.9737	12.78	4.29	17.02	65.91	C 2H 306P1	154.02	153.9667	15.60	1.96	62.33	20.11
C 1H 402P2	109.99	109.9687	10.92	3.67	29.09	56.32	C 2H 307P1	170.02	169.9616	14.13	1.78	65.87	18.22
C 1H 403P2	125.99	125.9636	9.53	3.20	38.10	49.17	C 2H 308P1	186.02	185.9566	12.91	1.63	68.81	16.65
C 1H 404P2	141.99	141.9585	8.46	2.84	45.07	43.63	C 2H 309P1	202.01	201.9515	11.89	1.50	71.28	15.33
C 1H 405P2	157.99	157.9534	7.60	2.55	50.63	39.21	C 2H 401P2	106.00	105.9737	22.66	3.80	15.09	58.44
C 1H 406P2	173.99	173.9483	6.90	2.32	55.17	35.60	C 2H 402P2	122.00	121.9687	19.69	3.30	26.23	50.78
C 1H 407P2	189.99	189.9432	6.32	2.12	58.95	32.61	C 2H 403P2	138.00	137.9636	17.41	2.92	34.78	44.89
C 1H 408P2	205.99	205.9381	5.83	1.96	62.14	30.07	C 2H 404P2	154.00	153.9585	15.60	2.62	41.56	40.23
C 1H 409P2	221.99	221.9331	5.41	1.82	64.87	27.91	C 2H 405P2	170.00	169.9534	14.13	2.37	47.06	36.44
C 1H 501P1	64.02	64.0078	18.76	7.87	24.99	48.38	C 2H 406P2	186.00	185.9483	12.92	2.17	51.61	33.31
C 1H 502P1	80.02	80.0027	15.01	6.30	39.99	38.71	C 2H 407P2	202.00	201.9432	11.89	2.00	55.44	30.67
C 1H 503P1	96.02	95.9976	12.51	5.25	49.99	32.26	C 2H 408P2	218.00	217.9381	11.02	1.85	58.71	28.42
C 1H 504P1	112.02	111.9925	10.72	4.50	57.13	27.65	C 2H 409P2	234.00	233.9331	10.27	1.72	61.54	26.47
C 1H 505P1	128.02	127.9875	9.38	3.94	62.49	24.19	C 2H 501P1	76.04	76.0078	31.59	6.63	21.04	40.74

103

C 2H 502P1	92.03	92.0027	26.10	5.48	34.77	33.65	C 3H 308P1	198.03	197.9566	18.20	1.53	64.64	15.64
C 2H 503P1	108.03	107.9976	22.24	4.67	44.43	28.67	C 3H 309P1	214.03	213.9515	16.84	1.41	67.28	14.47
C 2H 504P1	124.03	123.9925	19.37	4.06	51.60	24.97	C 3H 401P2	118.01	117.9737	30.53	3.42	13.56	52.49
C 2H 505P1	140.03	139.9875	17.15	3.60	57.13	22.12	C 3H 402P2	134.01	133.9687	26.89	3.01	23.88	46.23
C 2H 506P1	156.03	155.9824	15.40	3.23	61.52	19.85	C 3H 403P2	150.01	149.9636	24.02	2.69	32.00	41.30
C 2H 507P1	172.03	171.9773	13.96	2.93	65.10	18.00	C 3H 404P2	166.01	165.9585	21.71	2.43	38.55	37.32
C 2H 508P1	188.03	187.9722	12.78	2.68	68.07	16.47	C 3H 405P2	182.01	181.9534	19.80	2.22	43.95	34.04
C 2H 509P1	204.03	203.9671	11.77	2.47	70.58	15.18	C 3H 406P2	198.01	197.9483	18.20	2.04	48.48	31.29
C 2H 601P2	108.02	107.9894	22.24	5.60	14.81	57.35	C 3H 407P2	214.01	213.9432	16.84	1.88	52.33	28.95
C 2H 602P2	124.02	123.9843	19.37	4.88	25.80	49.95	C 3H 408P2	230.01	229.9381	15.67	1.75	55.65	26.93
C 2H 603P2	140.02	139.9792	17.16	4.32	34.28	44.24	C 3H 409P2	246.01	245.9331	14.65	1.64	58.53	25.18
C 2H 604P2	156.02	155.9741	15.40	3.88	41.02	39.71	C 3H 501P1	88.05	88.0078	40.93	5.72	18.17	35.18
C 2H 605P2	172.01	171.9690	13.97	3.52	46.51	36.01	C 3H 502P1	104.05	104.0027	34.63	4.84	30.75	29.77
C 2H 606P2	188.01	187.9640	12.78	3.22	51.06	32.95	C 3H 503P1	120.05	119.9976	30.02	4.20	39.98	25.80
C 2H 607P2	204.01	203.9589	11.77	2.96	54.90	30.36	C 3H 504P1	136.04	135.9925	26.49	3.70	47.04	22.77
C 2H 608P2	220.01	219.9538	10.92	2.75	58.18	28.16	C 3H 505P1	152.04	151.9875	23.70	3.31	52.61	20.37
C 2H 609P2	236.01	235.9487	10.18	2.56	61.01	26.25	C 3H 506P1	168.04	167.9824	21.44	3.00	57.13	18.43
C 2H 701P1	78.05	78.0235	30.78	9.04	20.50	39.68	C 3H 507P1	184.04	183.9773	19.58	2.74	60.85	16.83
C 2H 702P1	94.05	94.0184	25.54	7.50	34.02	32.93	C 3H 508P1	200.04	199.9722	18.01	2.52	63.98	15.48
C 2H 703P1	110.05	110.0133	21.83	6.41	43.61	28.15	C 3H 509P1	216.04	215.9671	16.68	2.33	66.65	14.34
C 2H 704P1	126.05	126.0082	19.06	5.60	50.77	24.57	C 3H 601P2	120.03	119.9894	30.02	5.04	13.33	51.61
C 2H 705P1	142.05	142.0031	16.91	4.97	56.32	21.81	C 3H 602P2	136.03	135.9843	26.49	4.45	23.52	45.54
C 2H 706P1	158.05	157.9980	15.20	4.46	60.74	19.60	C 3H 603P2	152.03	151.9792	23.70	3.98	31.57	40.75
C 2H 707P1	174.05	173.9929	13.80	4.05	64.35	17.80	C 3H 604P2	168.03	167.9741	21.45	3.60	38.09	36.87
C 2H 708P1	190.05	189.9878	12.64	3.71	67.35	16.30	C 3H 605P2	184.03	183.9690	19.58	3.29	43.47	33.66
C 2H 709P1	206.05	205.9828	11.66	3.42	69.88	15.03	C 3H 606P2	200.03	199.9640	18.01	3.02	47.99	30.97
C 2H 801P2	110.03	110.0050	21.83	7.33	14.54	56.30	C 3H 607P2	216.02	215.9589	16.68	2.80	51.84	28.68
C 2H 802P2	126.03	126.0000	19.06	6.40	25.39	49.15	C 3H 608P2	232.02	231.9538	15.53	2.61	55.16	26.70
C 2H 803P2	142.03	141.9949	16.91	5.68	33.79	43.62	C 3H 609P2	248.02	247.9487	14.53	2.44	58.06	24.98
C 2H 804P2	158.03	157.9898	15.20	5.10	40.50	39.20	C 3H 701P1	90.06	90.0235	40.01	7.83	17.76	34.39
C 2H 805P2	174.03	173.9847	13.80	4.63	45.97	35.60	C 3H 702P1	106.06	106.0184	33.97	6.65	30.17	29.20
C 2H 806P2	190.03	189.9796	12.64	4.24	50.52	32.60	C 3H 703P1	122.06	122.0133	29.52	5.78	39.32	25.38
C 2H 807P2	206.03	205.9745	11.66	3.91	54.36	30.07	C 3H 704P1	138.06	138.0082	26.10	5.11	46.35	22.43
C 2H 808P2	222.03	221.9694	10.82	3.63	57.65	27.90	C 3H 705P1	154.06	154.0031	23.39	4.58	51.93	20.11
C 2H 809P2	238.03	237.9644	10.09	3.39	60.49	26.03	C 3H 706P1	170.06	169.9980	21.19	4.15	56.45	18.21
C 3H 101P1	84.01	83.9765	42.89	1.20	19.04	36.87	C 3H 707P1	186.06	185.9929	19.37	3.79	60.19	16.65
C 3H 102P1	100.01	99.9714	36.03	1.01	31.99	30.97	C 3H 708P1	202.06	201.9878	17.83	3.49	63.35	15.33
C 3H 103P1	116.01	115.9663	31.06	.87	41.37	26.70	C 3H 709P1	218.06	217.9828	16.52	3.24	66.04	14.20
C 3H 104P1	132.01	131.9612	27.30	.76	48.48	23.46	C 3H 801P2	122.04	122.0050	29.52	6.61	13.11	50.76
C 3H 105P1	148.01	147.9562	24.34	.68	54.05	20.93	C 3H 802P2	138.04	138.0000	26.10	5.84	23.18	44.88
C 3H 106P1	164.01	163.9511	21.97	.61	58.53	18.89	C 3H 803P2	154.04	153.9949	23.39	5.23	31.16	40.21
C 3H 107P1	180.01	179.9460	20.02	.56	62.22	17.21	C 3H 804P2	170.04	169.9898	21.19	4.74	37.64	36.43
C 3H 108P1	196.01	195.9409	18.38	.51	65.30	15.80	C 3H 805P2	186.04	185.9847	19.37	4.33	43.00	33.30
C 3H 109P1	212.01	211.9358	17.00	.48	67.92	14.61	C 3H 806P2	202.04	201.9796	17.83	3.99	47.51	30.66
C 3H 201P2	116.00	115.9581	31.06	1.74	13.79	53.40	C 3H 807P2	218.04	217.9745	16.53	3.70	51.36	28.41
C 3H 202P2	132.00	131.9530	27.30	1.53	24.24	46.93	C 3H 808P2	234.04	233.9694	15.40	3.45	54.69	26.47
C 3H 203P2	148.00	147.9479	24.35	1.36	32.43	41.86	C 3H 809P2	250.04	249.9644	14.41	3.22	57.59	24.78
C 3H 204P2	163.99	163.9428	21.97	1.23	39.02	37.77	C 3H 901P1	92.08	92.0391	39.13	9.85	17.38	33.64
C 3H 205P2	179.99	179.9377	20.02	1.12	44.44	34.42	C 3H 902P1	108.08	108.0340	33.34	8.39	29.61	28.66
C 3H 206P2	195.99	195.9327	18.39	1.03	48.98	31.61	C 3H 903P1	124.08	124.0289	29.04	7.31	38.68	24.96
C 3H 207P2	211.99	211.9276	17.00	.95	52.83	29.22	C 3H 904P1	140.08	140.0238	25.72	6.48	45.69	22.11
C 3H 208P2	227.99	227.9225	15.80	.88	56.14	27.17	C 3H 905P1	156.08	156.0188	23.09	5.81	51.26	19.85
C 3H 209P2	243.99	243.9174	14.77	.83	59.02	25.39	C 3H 906P1	172.08	172.0137	20.94	5.27	55.79	18.00
C 3H 301P1	86.03	85.9922	41.88	3.51	18.60	36.00	C 3H 907P1	188.07	188.0086	19.16	4.82	59.55	16.47
C 3H 302P1	102.03	101.9871	35.32	2.96	31.36	30.36	C 3H 908P1	204.07	204.0035	17.66	4.45	62.72	15.18
C 3H 303P1	118.03	117.9820	30.53	2.56	40.67	26.24	C 3H 909P1	220.07	219.9984	16.37	4.12	65.43	14.07
C 3H 304P1	134.03	133.9769	26.88	2.26	47.75	23.11	C 3H10O1P2	124.06	124.0207	29.05	8.12	12.90	49.93
C 3H 305P1	150.03	149.9718	24.02	2.02	53.32	20.65	C 3H10O2P2	140.06	140.0156	25.73	7.20	22.85	44.23
C 3H 306P1	166.03	165.9667	21.70	1.82	57.82	18.66	C 3H10O3P2	156.06	156.0105	23.09	6.46	30.76	39.69
C 3H 307P1	182.03	181.9616	19.80	1.66	61.53	17.02	C 3H10O4P2	172.06	172.0054	20.94	5.86	37.20	36.00

104

C 3H1005P2	188.06	188.0003	19.16	5.36	42.54	32.94
C 3H1006P2	204.06	203.9953	17.66	4.94	47.04	30.36
C 3H1007P2	220.06	219.9902	16.37	4.58	50.89	28.15
C 3H1008P2	236.06	235.9851	15.26	4.27	54.22	26.24
C 3H1009P2	252.06	251.9800	14.30	4.00	57.13	24.58
C 4H 201P2	128.01	127.9581	37.53	1.57	12.50	48.39
C 4H 202P2	144.01	143.9530	33.36	1.40	22.22	43.02
C 4H 203P2	160.01	159.9479	30.03	1.26	30.00	38.72
C 4H 204P2	176.01	175.9428	27.30	1.15	36.36	35.20
C 4H 205P2	192.01	191.9377	25.02	1.05	41.66	32.26
C 4H 206P2	208.00	207.9327	23.10	.97	46.15	29.78
C 4H 207P2	224.00	223.9276	21.45	.90	50.00	27.65
C 4H 208P2	240.00	239.9225	20.02	.84	53.33	25.81
C 4H 209P2	256.00	255.9174	18.77	.79	56.25	24.20
C 4H 301P1	98.04	97.9922	49.00	3.08	16.32	31.59
C 4H 302P1	114.04	113.9871	42.13	2.65	28.06	27.16
C 4H 303P1	130.04	129.9820	36.95	2.33	36.91	23.82
C 4H 304P1	146.04	145.9769	32.90	2.07	43.82	21.21
C 4H 305P1	162.04	161.9718	29.65	1.87	49.37	19.11
C 4H 306P1	178.04	177.9667	26.99	1.70	53.92	17.40
C 4H 307P1	194.04	193.9616	24.76	1.56	57.72	15.96
C 4H 308P1	210.04	209.9566	22.87	1.44	60.94	14.75
C 4H 309P1	226.04	225.9515	21.26	1.34	63.70	13.70
C 4H 401P2	130.02	129.9737	36.95	3.10	12.31	47.64
C 4H 402P2	146.02	145.9687	32.90	2.76	21.91	42.42
C 4H 403P2	162.02	161.9636	29.65	2.49	29.62	38.23
C 4H 404P2	178.02	177.9585	26.99	2.26	35.95	34.80
C 4H 405P2	194.02	193.9534	24.76	2.08	41.23	31.93
C 4H 406P2	210.02	209.9483	22.88	1.92	45.71	29.50
C 4H 407P2	226.02	225.9432	21.26	1.78	49.55	27.41
C 4H 408P2	242.02	241.9381	19.85	1.67	52.89	25.60
C 4H 409P2	258.02	257.9331	18.62	1.56	55.81	24.01
C 4H 501P1	100.06	100.0078	48.02	5.04	15.99	30.96
C 4H 502P1	116.06	116.0027	41.40	4.34	27.57	26.69
C 4H 503P1	132.06	131.9976	36.38	3.82	36.35	23.45
C 4H 504P1	148.06	147.9925	32.45	3.40	43.23	20.92
C 4H 505P1	164.06	163.9875	29.29	3.07	48.76	18.88
C 4H 506P1	180.05	179.9824	26.68	2.80	53.32	17.20
C 4H 507P1	196.05	195.9773	24.51	2.57	57.12	15.80
C 4H 508P1	212.05	211.9722	22.66	2.38	60.36	14.61
C 4H 509P1	228.05	227.9671	21.07	2.21	63.14	13.58
C 4H 601P2	132.04	131.9894	36.39	4.58	12.12	46.92
C 4H 602P2	148.04	147.9843	32.45	4.09	21.62	41.85
C 4H 603P2	164.04	163.9792	29.29	3.69	29.26	37.76
C 4H 604P2	180.04	179.9741	26.69	3.36	35.55	34.41
C 4H 605P2	196.04	195.9690	24.51	3.09	40.81	31.60
C 4H 606P2	212.04	211.9640	22.66	2.85	45.27	29.22
C 4H 607P2	228.04	227.9589	21.07	2.65	49.11	27.17
C 4H 608P2	244.04	243.9538	19.69	2.48	52.45	25.38
C 4H 609P2	260.03	259.9487	18.48	2.33	55.38	23.82
C 4H 701P1	102.07	102.0235	47.07	6.91	15.67	30.34
C 4H 702P1	118.07	118.0184	40.69	5.98	27.10	26.23
C 4H 703P1	134.07	134.0133	35.83	5.26	35.80	23.10
C 4H 704P1	150.07	150.0082	32.01	4.70	42.64	20.64
C 4H 705P1	166.07	166.0031	28.93	4.25	48.17	18.65
C 4H 706P1	182.07	181.9980	26.39	3.88	52.72	17.01
C 4H 707P1	198.07	197.9929	24.26	3.56	56.54	15.64
C 4H 708P1	214.07	213.9878	22.44	3.30	59.79	14.47
C 4H 709P1	230.07	229.9828	20.88	3.07	62.59	13.46
C 4H 801P2	134.06	134.0050	35.84	6.02	11.93	46.21

C 4H 802P2	150.05	150.0000	32.02	5.37	21.32	41.28
C 4H 803P2	166.05	165.9949	28.93	4.86	28.91	37.31
C 4H 804P2	182.05	181.9898	26.39	4.43	35.15	34.03
C 4H 805P2	198.05	197.9847	24.26	4.07	40.39	31.28
C 4H 806P2	214.05	213.9796	22.45	3.77	44.85	28.94
C 4H 807P2	230.05	229.9745	20.88	3.51	48.68	26.93
C 4H 808P2	246.05	245.9694	19.53	3.28	52.02	25.18
C 4H 809P2	262.05	261.9644	18.33	3.08	54.95	23.64
C 4H 901P1	104.09	104.0391	46.16	8.72	15.37	29.76
C 4H 902P1	120.09	120.0340	40.01	7.55	26.65	25.79
C 4H 903P1	136.09	136.0289	35.30	6.67	35.27	22.76
C 4H 904P1	152.09	152.0238	31.59	5.96	42.08	20.37
C 4H 905P1	168.09	168.0188	28.58	5.40	47.59	18.43
C 4H 906P1	184.09	184.0137	26.10	4.93	52.15	16.83
C 4H 907P1	200.09	200.0086	24.01	4.53	55.97	15.48
C 4H 908P1	216.09	216.0035	22.23	4.20	59.23	14.33
C 4H 909P1	232.08	231.9984	20.70	3.91	62.04	13.35
C 4H1001P2	136.07	136.0207	35.31	7.41	11.76	45.53
C 4H1002P2	152.07	152.0156	31.59	6.63	21.04	40.74
C 4H1003P2	168.07	168.0105	28.59	6.00	28.56	36.86
C 4H1004P2	184.07	184.0054	26.10	5.48	34.77	33.65
C 4H1005P2	200.07	200.0003	24.01	5.04	39.98	30.96
C 4H1006P2	216.07	215.9953	22.24	4.67	44.43	28.67
C 4H1007P2	232.07	231.9902	20.70	4.34	48.26	26.69
C 4H1008P2	248.07	247.9851	19.37	4.06	51.60	24.97
C 4H1009P2	264.07	263.9800	18.19	3.82	54.53	23.46
C 4H1101P1	106.11	106.0547	45.28	10.45	15.08	29.19
C 4H1102P1	122.10	122.0497	39.35	9.08	26.21	25.37
C 4H1103P1	138.10	138.0446	34.79	8.03	34.76	22.43
C 4H1104P1	154.10	154.0395	31.18	7.19	41.53	20.10
C 4H1105P1	170.10	170.0344	28.24	6.52	47.03	18.21
C 4H1106P1	186.10	186.0293	25.82	5.96	51.58	16.64
C 4H1107P1	202.10	202.0242	23.77	5.49	55.42	15.33
C 4H1108P1	218.10	218.0191	22.03	5.08	58.69	14.20
C 4H1109P1	234.10	234.0141	20.52	4.74	61.51	13.23
C 4H1201P2	138.09	138.0363	34.79	8.76	11.59	44.86
C 4H1202P2	154.09	154.0313	31.18	7.85	20.77	40.20
C 4H1203P2	170.09	170.0262	28.25	7.11	28.22	36.42
C 4H1204P2	186.09	186.0211	25.82	6.50	34.39	33.29
C 4H1205P2	202.08	202.0160	23.77	5.99	39.59	30.65
C 4H1206P2	218.08	218.0109	22.03	5.55	44.02	28.41
C 4H1207P2	234.08	234.0058	20.52	5.17	47.84	26.46
C 4H1208P2	250.08	250.0007	19.21	4.84	51.18	24.77
C 4H1209P2	266.08	265.9957	18.06	4.55	54.12	23.28
C 5H 201P2	140.02	139.9581	42.89	1.44	11.43	44.24
C 5H 202P2	156.02	155.9530	38.49	1.29	20.51	39.71
C 5H 203P2	172.02	171.9479	34.91	1.17	27.90	36.01
C 5H 204P2	188.02	187.9428	31.94	1.07	34.04	32.95
C 5H 205P2	204.02	203.9377	29.44	.99	39.21	30.36
C 5H 206P2	220.02	219.9327	27.30	.92	43.63	28.16
C 5H 207P2	236.02	235.9276	25.45	.85	47.45	26.25
C 5H 208P2	252.01	251.9225	23.83	.80	50.79	24.58
C 5H 209P2	268.01	267.9174	22.41	.75	53.73	23.11
C 5H 301P1	110.05	109.9922	54.57	2.75	14.54	28.14
C 5H 302P1	126.05	125.9871	47.64	2.40	25.39	24.57
C 5H 303P1	142.05	141.9820	42.28	2.13	33.79	21.80
C 5H 304P1	158.05	157.9769	38.00	1.91	40.49	19.60
C 5H 305P1	174.05	173.9718	34.50	1.74	45.96	17.80
C 5H 306P1	190.05	189.9667	31.60	1.59	50.51	16.30
C 5H 307P1	206.05	205.9616	29.15	1.47	54.35	15.03

105

C 5H 308P1	222.05	221.9566	27.05	1.36	57.64	13.95
C 5H 309P1	238.05	237.9515	25.23	1.27	60.49	13.01
C 5H 401P2	142.03	141.9737	42.28	2.84	11.26	43.61
C 5H 402P2	158.03	157.9687	38.00	2.55	20.25	39.20
C 5H 403P2	174.03	173.9636	34.51	2.32	27.58	35.60
C 5H 404P2	190.03	189.9585	31.60	2.12	33.68	32.60
C 5H 405P2	206.03	205.9534	29.15	1.96	38.83	30.07
C 5H 406P2	222.03	221.9483	27.05	1.82	43.24	27.90
C 5H 407P2	238.03	237.9432	25.23	1.69	47.05	26.03
C 5H 408P2	254.03	253.9381	23.64	1.59	50.39	24.39
C 5H 409P2	270.03	269.9331	22.24	1.49	53.33	22.94
C 5H 501P1	112.07	112.0078	53.59	4.50	14.28	27.64
C 5H 502P1	128.07	128.0027	46.89	3.94	24.99	24.19
C 5H 503P1	144.07	143.9976	41.69	3.50	33.32	21.50
C 5H 504P1	160.07	159.9925	37.52	3.15	39.98	19.35
C 5H 505P1	176.07	175.9875	34.11	2.86	45.44	17.59
C 5H 506P1	192.07	191.9824	31.27	2.62	49.98	16.13
C 5H 507P1	208.07	207.9773	28.86	2.42	53.83	14.89
C 5H 508P1	224.06	223.9722	26.80	2.25	57.12	13.82
C 5H 509P1	240.06	239.9671	25.02	2.10	59.98	12.90
C 5H 601P2	144.05	143.9894	41.69	4.20	11.11	43.00
C 5H 602P2	160.05	159.9843	37.52	3.78	19.99	38.71
C 5H 603P2	176.05	175.9792	34.11	3.44	27.26	35.19
C 5H 604P2	192.05	191.9741	31.27	3.15	33.32	32.26
C 5H 605P2	208.05	207.9690	28.87	2.91	38.45	29.78
C 5H 606P2	224.05	223.9640	26.80	2.70	42.85	27.65
C 5H 607P2	240.05	239.9589	25.02	2.52	46.66	25.81
C 5H 608P2	256.05	255.9538	23.46	2.36	49.99	24.19
C 5H 609P2	272.05	271.9487	22.00	2.22	52.93	22.77
C 5H 701P1	114.08	114.0235	52.64	6.18	14.02	27.15
C 5H 702P1	130.08	130.0184	46.17	5.42	24.60	23.81
C 5H 703P1	146.08	146.0133	41.11	4.83	32.86	21.20
C 5H 704P1	162.08	162.0082	37.05	4.35	39.48	19.11
C 5H 705P1	178.08	178.0031	33.72	3.96	44.92	17.39
C 5H 706P1	194.08	193.9980	30.94	3.64	49.46	15.96
C 5H 707P1	210.08	209.9929	28.59	3.36	53.31	14.74
C 5H 708P1	226.08	225.9878	26.56	3.12	56.61	13.70
C 5H 709P1	242.08	241.9828	24.81	2.91	59.48	12.79
C 5H 801P2	146.07	146.0050	41.12	5.52	10.95	42.41
C 5H 802P2	162.07	162.0000	37.06	4.98	19.74	38.22
C 5H 803P2	178.07	177.9949	33.73	4.53	26.96	34.79
C 5H 804P2	194.06	193.9898	30.95	4.16	32.98	31.92
C 5H 805P2	210.06	209.9847	28.59	3.84	38.08	29.49
C 5H 806P2	226.06	225.9796	26.57	3.57	42.46	27.40
C 5H 807P2	242.06	241.9745	24.81	3.33	46.27	25.59
C 5H 808P2	258.06	257.9694	23.27	3.12	49.60	24.00
C 5H 809P2	274.06	273.9644	21.91	2.94	52.54	22.60
C 5H 901P1	116.10	116.0391	51.73	7.81	13.78	26.68
C 5H 902P1	132.10	132.0340	45.46	6.87	24.22	23.45
C 5H 903P1	148.10	148.0289	40.55	6.13	32.41	20.91
C 5H 904P1	164.10	164.0238	36.60	5.53	39.00	18.88
C 5H 905P1	180.10	180.0188	33.35	5.04	44.42	17.20
C 5H 906P1	196.10	196.0137	30.63	4.63	48.95	15.80
C 5H 907P1	212.10	212.0086	28.32	4.28	52.80	14.60
C 5H 908P1	228.10	228.0035	26.33	3.98	56.11	13.58
C 5H 909P1	244.10	243.9984	24.60	3.72	58.99	12.69
C 5H1001P2	148.08	148.0207	40.56	6.81	10.80	41.83
C 5H1002P2	164.08	164.0156	36.60	6.14	19.50	37.75
C 5H1003P2	180.08	180.0105	33.35	5.60	26.65	34.40
C 5H1004P2	196.08	196.0054	30.63	5.14	32.64	31.59

C 5H1005P2	212.08	212.0003	28.32	4.75	37.72	29.21
C 5H1006P2	228.08	227.9953	26.33	4.42	42.09	27.16
C 5H1007P2	244.08	243.9902	24.61	4.13	45.89	25.38
C 5H1008P2	260.08	259.9851	23.09	3.88	49.21	23.82
C 5H1009P2	276.08	275.9800	21.75	3.65	52.16	22.44
C 5H1101P1	118.12	118.0547	50.84	9.39	13.55	26.22
C 5H1102P1	134.12	134.0497	44.78	8.27	23.86	23.09
C 5H1103P1	150.12	150.0446	40.01	7.39	31.97	20.63
C 5H1104P1	166.11	166.0395	36.15	6.67	38.53	18.65
C 5H1105P1	182.11	182.0344	32.98	6.09	43.93	17.01
C 5H1106P1	198.11	198.0293	30.31	5.60	48.46	15.63
C 5H1107P1	214.11	214.0242	28.05	5.18	52.31	14.47
C 5H1108P1	230.11	230.0191	26.10	4.82	55.62	13.46
C 5H1109P1	246.11	246.0141	24.40	4.51	58.51	12.59
C 5H1201P2	150.10	150.0363	40.01	8.06	10.66	41.27
C 5H1202P2	166.10	166.0313	36.16	7.28	19.27	37.30
C 5H1203P2	182.10	182.0262	32.98	6.64	26.36	34.02
C 5H1204P2	198.10	198.0211	30.32	6.11	32.31	31.27
C 5H1205P2	214.10	214.0160	28.05	5.65	37.37	28.93
C 5H1206P2	230.10	230.0109	26.10	5.26	41.72	26.92
C 5H1207P2	246.09	246.0058	24.40	4.92	45.51	25.17
C 5H1208P2	262.09	262.0007	22.91	4.61	48.84	23.64
C 5H1209P2	278.09	277.9957	21.60	4.35	51.78	22.28
C 5H1301P1	120.13	120.0704	49.99	10.91	13.32	25.78
C 5H1302P1	136.13	136.0653	44.12	9.63	23.51	22.75
C 5H1303P1	152.13	152.0602	39.48	8.61	31.55	20.36
C 5H1304P1	168.13	168.0551	35.72	7.79	38.06	18.42
C 5H1305P1	184.13	184.0501	32.62	7.12	43.45	16.82
C 5H1306P1	200.13	200.0450	30.01	6.55	47.97	15.48
C 5H1307P1	216.13	216.0399	27.79	6.06	51.82	14.33
C 5H1308P1	232.13	232.0348	25.87	5.64	55.14	13.34
C 5H1309P1	248.13	248.0297	24.20	5.28	58.03	12.48
C 5H1401P2	152.11	152.0520	39.48	9.28	10.52	40.72
C 5H1402P2	168.11	168.0469	35.72	8.39	19.03	36.85
C 5H1403P2	184.11	184.0418	32.62	7.66	26.07	33.65
C 5H1404P2	200.11	200.0367	30.01	7.05	31.98	30.96
C 5H1405P2	216.11	216.0316	27.79	6.53	37.02	28.66
C 5H1406P2	232.11	232.0266	25.87	6.08	41.36	26.69
C 5H1407P2	248.11	248.0215	24.21	5.69	45.14	24.97
C 5H1408P2	264.11	264.0164	22.74	5.34	48.46	23.46
C 5H1409P2	280.11	280.0113	21.44	5.04	51.41	22.12
C 6H 301P1	122.06	121.9922	59.04	2.48	13.11	25.38
C 6H 302P1	138.06	137.9871	52.20	2.19	23.18	22.43
C 6H 303P1	154.06	153.9820	46.78	1.96	31.15	20.10
C 6H 304P1	170.06	169.9769	42.38	1.78	37.63	18.21
C 6H 305P1	186.06	185.9718	38.73	1.63	42.99	16.65
C 6H 306P1	202.06	201.9667	35.67	1.50	47.51	15.33
C 6H 307P1	218.06	217.9616	33.05	1.39	51.36	14.20
C 6H 308P1	234.06	233.9566	30.79	1.29	54.68	13.23
C 6H 309P1	250.06	249.9515	28.82	1.21	57.58	12.39
C 6H 401P2	154.05	153.9737	46.78	2.62	10.39	40.21
C 6H 402P2	170.05	169.9687	42.38	2.37	18.82	36.43
C 6H 403P2	186.04	185.9636	38.74	2.17	25.80	33.30
C 6H 404P2	202.04	201.9585	35.67	2.00	31.68	30.66
C 6H 405P2	218.04	217.9534	33.05	1.85	36.69	28.41
C 6H 406P2	234.04	233.9483	30.79	1.72	41.02	26.47
C 6H 407P2	250.04	249.9432	28.82	1.61	44.79	24.77
C 6H 408P2	266.04	265.9381	27.09	1.52	48.11	23.28
C 6H 409P2	282.04	281.9331	25.55	1.43	51.05	21.96
C 6H 501P1	124.08	124.0078	58.08	4.06	12.89	24.96

106

C 6H 502P1	140.08	140.0027	51.45	3.60	22.84	22.11
C 6H 503P1	156.08	155.9976	46.17	3.23	30.75	19.84
C 6H 504P1	172.08	171.9925	41.88	2.93	37.19	18.00
C 6H 505P1	188.08	187.9875	38.32	2.68	42.53	16.47
C 6H 506P1	204.08	203.9824	35.31	2.47	47.04	15.18
C 6H 507P1	220.08	219.9773	32.75	2.29	50.89	14.07
C 6H 508P1	236.08	235.9722	30.53	2.13	54.22	13.12
C 6H 509P1	252.08	251.9671	28.59	2.00	57.12	12.29
C 6H 601P2	156.06	155.9894	46.18	3.88	10.25	39.69
C 6H 602P2	172.06	171.9843	41.88	3.51	18.60	36.00
C 6H 603P2	188.06	187.9792	38.32	3.22	25.52	32.94
C 6H 604P2	204.06	203.9741	35.32	2.96	31.36	30.36
C 6H 605P2	220.06	219.9690	32.75	2.75	36.35	28.15
C 6H 606P2	236.06	235.9640	30.53	2.56	40.67	26.24
C 6H 607P2	252.06	251.9589	28.59	2.40	44.43	24.58
C 6H 608P2	268.06	267.9538	26.88	2.26	47.75	23.11
C 6H 609P2	284.06	283.9487	25.37	2.13	50.69	21.81
C 6H 701P1	126.10	126.0235	57.15	5.60	12.69	24.56
C 6H 702P1	142.10	142.0184	50.72	4.97	22.52	21.80
C 6H 703P1	158.09	158.0133	45.58	4.46	30.36	19.59
C 6H 704P1	174.09	174.0082	41.40	4.05	36.76	17.79
C 6H 705P1	190.09	190.0031	37.91	3.71	42.08	16.29
C 6H 706P1	206.09	205.9980	34.97	3.42	46.58	15.03
C 6H 707P1	222.09	221.9929	32.45	3.18	50.43	13.95
C 6H 708P1	238.09	237.9878	30.27	2.96	53.76	13.01
C 6H 709P1	254.09	253.9828	28.36	2.78	56.67	12.19
C 6H 801P2	158.08	158.0050	45.59	5.10	10.12	39.19
C 6H 802P2	174.08	174.0000	41.40	4.63	18.38	35.59
C 6H 803P2	190.08	189.9949	37.91	4.24	25.25	32.59
C 6H 804P2	206.08	205.9898	34.97	3.91	31.06	30.06
C 6H 805P2	222.08	221.9847	32.45	3.63	36.02	27.89
C 6H 806P2	238.07	237.9796	30.27	3.39	40.32	26.02
C 6H 807P2	254.07	253.9745	28.36	3.17	44.08	24.38
C 6H 808P2	270.07	269.9694	26.68	2.99	47.39	22.94
C 6H 809P2	286.07	285.9644	25.19	2.82	50.33	21.65
C 6H 901P1	128.11	128.0391	56.25	7.08	12.49	24.18
C 6H 902P1	144.11	144.0340	50.01	6.29	22.20	21.49
C 6H 903P1	160.11	160.0289	45.01	5.67	29.98	19.35
C 6H 904P1	176.11	176.0238	40.92	5.15	36.34	17.59
C 6H 905P1	192.11	192.0188	37.51	4.72	41.64	16.12
C 6H 906P1	208.11	208.0137	34.63	4.36	46.13	14.88
C 6H 907P1	224.11	224.0086	32.16	4.05	49.97	13.82
C 6H 908P1	240.11	240.0035	30.01	3.78	53.31	12.90
C 6H 909P1	256.11	255.9984	28.14	3.54	56.22	12.09
C 6H1001P2	160.09	160.0207	45.02	6.30	9.99	38.69
C 6H1002P2	176.09	176.0156	40.93	5.72	18.17	35.18
C 6H1003P2	192.09	192.0105	37.52	5.25	24.99	32.25
C 6H1004P2	208.09	208.0054	34.63	4.84	30.75	29.77
C 6H1005P2	224.09	224.0003	32.16	4.50	35.70	27.64
C 6H1006P2	240.09	239.9953	30.02	4.20	39.98	25.80
C 6H1007P2	256.09	255.9902	28.14	3.94	43.73	24.19
C 6H1008P2	272.09	271.9851	26.49	3.70	47.04	22.77
C 6H1009P2	288.09	287.9800	25.02	3.50	49.98	21.50
C 6H1101P1	130.13	130.0547	55.38	8.52	12.30	23.80
C 6H1102P1	146.13	146.0497	49.32	7.59	21.90	21.20
C 6H1103P1	162.13	162.0446	44.45	6.84	29.61	19.10
C 6H1104P1	178.13	178.0395	40.46	6.22	35.93	17.39
C 6H1105P1	194.13	194.0344	37.12	5.71	41.21	15.96
C 6H1106P1	210.12	210.0293	34.30	5.28	45.69	14.74
C 6H1107P1	226.12	226.0242	31.87	4.90	49.53	13.70

C 6H1108P1	242.12	242.0191	29.76	4.58	52.86	12.79
C 6H1109P1	258.12	258.0141	27.92	4.30	55.79	12.00
C 6H1201P2	162.11	162.0363	44.46	7.46	9.87	38.21
C 6H1202P2	178.11	178.0313	40.46	6.79	17.97	34.78
C 6H1203P2	194.11	194.0262	37.13	6.23	24.73	31.91
C 6H1204P2	210.11	210.0211	34.30	5.76	30.46	29.48
C 6H1205P2	226.11	226.0160	31.87	5.35	35.38	27.40
C 6H1206P2	242.11	242.0109	29.77	5.00	39.65	25.59
C 6H1207P2	258.11	258.0058	27.92	4.69	43.39	24.00
C 6H1208P2	274.11	274.0007	26.29	4.41	46.70	22.60
C 6H1209P2	290.10	289.9957	24.84	4.17	49.64	21.35
C 6H1301P1	132.14	132.0704	54.54	9.92	12.11	23.44
C 6H1302P1	148.14	148.0653	48.65	8.85	21.60	20.91
C 6H1303P1	164.14	164.0602	43.91	7.98	29.24	18.87
C 6H1304P1	180.14	180.0551	40.01	7.27	35.53	17.19
C 6H1305P1	196.14	196.0501	36.74	6.68	40.79	15.79
C 6H1306P1	212.14	212.0450	33.97	6.18	45.25	14.60
C 6H1307P1	228.14	228.0399	31.59	5.74	49.09	13.58
C 6H1308P1	244.14	244.0348	29.52	5.37	52.43	12.69
C 6H1309P1	260.14	260.0297	27.70	5.04	55.35	11.91
C 6H1401P2	164.13	164.0520	43.91	8.60	9.75	37.74
C 6H1402P2	180.12	180.0469	40.01	7.83	17.76	34.39
C 6H1403P2	196.12	196.0418	36.75	7.20	24.47	31.59
C 6H1404P2	212.12	212.0367	33.97	6.65	30.17	29.20
C 6H1405P2	228.12	228.0316	31.59	6.19	35.07	27.16
C 6H1406P2	244.12	244.0266	29.52	5.78	39.32	25.38
C 6H1407P2	260.12	260.0215	27.71	5.42	43.06	23.81
C 6H1408P2	276.12	276.0164	26.10	5.11	46.35	22.43
C 6H1409P2	292.12	292.0113	24.67	4.83	49.29	21.21
C 6H1501P1	134.16	134.0860	53.72	11.27	11.93	23.09
C 6H1502P1	150.16	150.0810	47.99	10.07	21.31	20.63
C 6H1503P1	166.16	166.0759	43.37	9.10	28.89	18.64
C 6H1504P1	182.16	182.0708	39.56	8.30	35.13	17.00
C 6H1505P1	198.16	198.0657	36.37	7.63	40.37	15.63
C 6H1506P1	214.16	214.0606	33.65	7.06	44.83	14.46
C 6H1507P1	230.16	230.0555	31.31	6.57	48.66	13.46
C 6H1508P1	246.16	246.0504	29.28	6.14	52.00	12.58
C 6H1509P1	262.15	262.0454	27.49	5.77	54.93	11.82
C 6H1601P2	166.14	166.0676	43.38	9.71	9.63	37.29
C 6H1602P2	182.14	182.0625	39.57	8.85	17.57	34.01
C 6H1603P2	198.14	198.0575	36.37	8.14	24.22	31.26
C 6H1604P2	214.14	214.0524	33.65	7.53	29.89	28.93
C 6H1605P2	230.14	230.0473	31.31	7.01	34.76	26.92
C 6H1606P2	246.14	246.0422	29.28	6.55	39.00	25.17
C 6H1607P2	262.14	262.0371	27.49	6.15	42.72	23.63
C 6H1608P2	278.14	278.0320	25.91	5.80	46.02	22.27
C 6H1609P2	294.14	294.0269	24.50	5.48	48.96	21.06
C 7H 301P1	134.08	133.9922	62.71	2.26	11.93	23.10
C 7H 302P1	150.07	149.9871	56.02	2.01	21.32	20.64
C 7H 303P1	166.07	165.9820	50.63	1.82	28.90	18.65
C 7H 304P1	182.07	181.9769	46.18	1.66	35.15	17.01
C 7H 305P1	198.07	197.9718	42.45	1.53	40.39	15.64
C 7H 306P1	214.07	213.9667	39.28	1.41	44.84	14.47
C 7H 307P1	230.07	229.9616	36.54	1.31	48.68	13.46
C 7H 308P1	246.07	245.9566	34.17	1.23	52.02	12.59
C 7H 309P1	262.07	261.9515	32.08	1.15	54.95	11.82
C 7H 401P2	166.06	165.9737	50.63	2.43	9.63	37.31
C 7H 402P2	182.06	181.9687	46.18	2.21	17.58	34.03
C 7H 403P2	198.06	197.9636	42.45	2.04	24.23	31.28
C 7H 404P2	214.06	213.9585	39.28	1.88	29.90	28.94

107

C 7H 405P2	230.05	229.9534	36.55	1.75	34.77	26.93
C 7H 406P2	246.05	245.9483	34.17	1.64	39.01	25.18
C 7H 407P2	262.05	261.9432	32.08	1.54	42.74	23.64
C 7H 408P2	278.05	277.9381	30.24	1.45	46.03	22.28
C 7H 409P2	294.05	293.9331	28.59	1.37	48.97	21.07
C 7H 501P1	136.09	136.0078	61.78	3.70	11.76	22.76
C 7H 502P1	152.09	152.0027	55.28	3.31	21.04	20.37
C 7H 503P1	168.09	167.9976	50.02	3.00	28.56	18.43
C 7H 504P1	184.09	183.9925	45.67	2.74	34.76	16.83
C 7H 505P1	200.09	199.9875	42.02	2.52	39.98	15.48
C 7H 506P1	216.09	215.9824	38.91	2.33	44.42	14.33
C 7H 507P1	232.09	231.9773	36.23	2.17	48.26	13.35
C 7H 508P1	248.09	247.9722	33.89	2.03	51.59	12.49
C 7H 509P1	264.09	263.9671	31.84	1.91	54.53	11.73
C 7H 601P2	168.07	167.9894	50.02	3.60	9.52	36.86
C 7H 602P2	184.07	183.9843	45.68	3.29	17.38	33.65
C 7H 603P2	200.07	199.9792	42.02	3.02	23.99	30.96
C 7H 604P2	216.07	215.9741	38.91	2.80	29.62	28.67
C 7H 605P2	232.07	231.9690	36.23	2.61	34.47	26.69
C 7H 606P2	248.07	247.9640	33.89	2.44	38.70	24.97
C 7H 607P2	264.07	263.9589	31.84	2.29	42.41	23.46
C 7H 608P2	280.07	279.9538	30.02	2.16	45.70	22.12
C 7H 609P2	296.07	295.9487	28.40	2.04	48.64	20.92
C 7H 701P1	138.11	138.0235	60.88	5.11	11.58	22.43
C 7H 702P1	154.11	154.0184	54.56	4.58	20.76	20.10
C 7H 703P1	170.11	170.0133	49.43	4.15	28.22	18.21
C 7H 704P1	186.11	186.0082	45.18	3.79	34.39	16.64
C 7H 705P1	202.10	202.0031	41.60	3.49	39.58	15.33
C 7H 706P1	218.10	217.9980	38.55	3.24	44.01	14.20
C 7H 707P1	234.10	233.9929	35.91	3.01	47.84	13.23
C 7H 708P1	250.10	249.9878	33.62	2.82	51.18	12.38
C 7H 709P1	266.10	265.9828	31.60	2.65	54.11	11.64
C 7H 801P2	170.09	170.0050	49.43	4.74	9.41	36.42
C 7H 802P2	186.09	186.0000	45.18	4.33	17.20	33.29
C 7H 803P2	202.09	201.9949	41.60	3.99	23.75	30.65
C 7H 804P2	218.09	217.9898	38.55	3.70	29.34	28.40
C 7H 805P2	234.09	233.9847	35.92	3.44	34.17	26.46
C 7H 806P2	250.09	249.9796	33.62	3.22	38.39	24.77
C 7H 807P2	266.09	265.9745	31.60	3.03	42.09	23.28
C 7H 808P2	282.08	281.9694	29.81	2.86	45.37	21.96
C 7H 809P2	298.08	297.9644	28.21	2.71	48.31	20.78
C 7H 901P1	140.12	140.0391	60.00	6.47	11.42	22.10
C 7H 902P1	156.12	156.0340	53.85	5.81	20.50	19.84
C 7H 903P1	172.12	172.0289	48.85	5.27	27.89	18.00
C 7H 904P1	188.12	188.0238	44.69	4.92	34.02	16.46
C 7H 905P1	204.12	204.0188	41.19	4.44	39.19	15.17
C 7H 906P1	220.12	220.0137	38.20	4.12	43.61	14.07
C 7H 907P1	236.12	236.0086	35.61	3.84	47.43	13.12
C 7H 908P1	252.12	252.0035	33.35	3.60	50.77	12.29
C 7H 909P1	268.12	267.9984	31.36	3.38	53.71	11.55
C 7H1001P2	172.10	172.0207	48.85	5.86	9.30	35.99
C 7H1002P2	188.10	188.0156	44.70	5.36	17.01	32.93
C 7H1003P2	204.10	204.0105	41.19	4.94	23.52	30.35
C 7H1004P2	220.10	220.0054	38.20	4.58	29.08	28.14
C 7H1005P2	236.10	236.0003	35.61	4.27	33.88	26.24
C 7H1006P2	252.10	251.9953	33.35	4.00	38.08	24.57
C 7H1007P2	268.10	267.9902	31.36	3.76	41.77	23.11
C 7H1008P2	284.10	283.9851	29.59	3.55	45.05	21.80
C 7H1009P2	300.10	299.9800	28.02	3.36	47.98	20.64
C 7H1101P1	142.14	142.0547	59.15	7.80	11.26	21.79

C 7H1102P1	158.14	158.0497	53.17	7.01	20.23	19.59
C 7H1103P1	174.14	174.0446	48.28	6.37	27.56	17.79
C 7H1104P1	190.14	190.0395	44.22	5.83	33.66	16.29
C 7H1105P1	206.14	206.0344	40.79	5.38	38.81	15.03
C 7H1106P1	222.14	222.0293	37.85	4.99	43.22	13.94
C 7H1107P1	238.14	238.0242	35.31	4.66	47.03	13.01
C 7H1108P1	254.13	254.0191	33.08	4.36	50.37	12.19
C 7H1109P1	270.13	270.0141	31.12	4.10	53.30	11.47
C 7H1201P2	174.12	174.0363	48.29	6.95	9.19	35.58
C 7H1202P2	190.12	190.0313	44.22	6.36	16.83	32.58
C 7H1203P2	206.12	206.0262	40.79	5.87	23.29	30.05
C 7H1204P2	222.12	222.0211	37.85	5.45	28.81	27.89
C 7H1205P2	238.12	238.0160	35.31	5.08	33.60	26.02
C 7H1206P2	254.12	254.0109	33.09	4.76	37.78	24.38
C 7H1207P2	270.12	270.0058	31.13	4.48	41.46	22.93
C 7H1208P2	286.12	286.0007	29.39	4.23	44.74	21.65
C 7H1209P2	302.12	301.9957	27.83	4.00	47.66	20.50
C 7H1301P1	144.15	144.0704	58.32	9.09	11.10	21.49
C 7H1302P1	160.15	160.0653	52.50	8.18	19.98	19.34
C 7H1303P1	176.15	176.0602	47.73	7.44	27.25	17.58
C 7H1304P1	192.15	192.0551	43.76	6.82	33.31	16.12
C 7H1305P1	208.15	208.0501	40.39	6.30	38.43	14.88
C 7H1306P1	224.15	224.0450	37.51	5.85	42.83	13.82
C 7H1307P1	240.15	240.0399	35.01	5.46	46.64	12.90
C 7H1308P1	256.15	256.0348	32.82	5.12	49.97	12.09
C 7H1309P1	272.15	272.0297	30.89	4.81	52.91	11.38
C 7H1401P2	176.14	176.0520	47.73	8.01	9.08	35.17
C 7H1402P2	192.14	192.0469	43.76	7.34	16.65	32.24
C 7H1403P2	208.14	208.0418	40.40	6.78	23.06	29.76
C 7H1404P2	224.13	224.0367	37.51	6.30	28.55	27.64
C 7H1405P2	240.13	240.0316	35.01	5.88	33.31	25.80
C 7H1406P2	256.13	256.0266	32.83	5.51	37.48	24.19
C 7H1407P2	272.13	272.0215	30.90	5.19	41.15	22.76
C 7H1408P2	288.13	288.0164	29.18	4.90	44.42	21.50
C 7H1409P2	304.13	304.0113	27.65	4.64	47.35	20.37
C 7H1501P1	146.17	146.0860	57.52	10.34	10.95	21.19
C 7H1502P1	162.17	162.0810	51.85	9.32	19.73	19.10
C 7H1503P1	178.17	178.0759	47.19	8.49	26.94	17.38
C 7H1504P1	194.17	194.0708	43.30	7.79	32.96	15.95
C 7H1505P1	210.17	210.0657	40.01	7.19	38.06	14.74
C 7H1506P1	226.17	226.0606	37.18	6.69	42.44	13.70
C 7H1507P1	242.17	242.0555	34.72	6.24	46.25	12.79
C 7H1508P1	258.17	258.0504	32.57	5.86	49.58	12.00
C 7H1509P1	274.17	274.0454	30.67	5.51	52.52	11.30
C 7H1601P2	178.15	178.0676	47.19	9.05	8.98	34.77
C 7H1602P2	194.15	194.0625	43.31	8.31	16.48	31.91
C 7H1603P2	210.15	210.0575	40.01	7.67	22.84	29.48
C 7H1604P2	226.15	226.0524	37.18	7.13	28.30	27.39
C 7H1605P2	242.15	242.0473	34.72	6.66	33.04	25.58
C 7H1606P2	258.15	258.0422	32.57	6.25	37.19	24.00
C 7H1607P2	274.15	274.0371	30.67	5.88	40.85	22.60
C 7H1608P2	290.15	290.0320	28.98	5.56	44.11	21.35
C 7H1609P2	306.15	306.0269	27.46	5.27	47.03	20.23
C 7H1701P1	148.19	148.1017	56.74	11.56	10.80	20.90
C 7H1702P1	164.19	164.0966	51.21	10.44	19.49	18.87
C 7H1703P1	180.19	180.0915	46.66	9.51	26.64	17.19
C 7H1704P1	196.18	196.0864	42.86	8.73	32.62	15.79
C 7H1705P1	212.18	212.0814	39.63	8.08	37.70	14.60
C 7H1706P1	228.18	228.0763	36.85	7.51	42.07	13.57
C 7H1707P1	244.18	244.0712	34.43	7.02	45.87	12.68

108

C 7H1708P1	260.18	260.0661	32.32	6.59	49.19	11.90
C 7H1709P1	276.18	276.0610	30.44	6.20	52.14	11.21
C 7H1801P2	180.17	180.0833	46.67	10.07	8.88	34.38
C 7H1802P2	196.17	196.0782	42.86	9.25	16.31	31.58
C 7H1803P2	212.17	212.0731	39.63	8.55	22.62	29.20
C 7H1804P2	228.17	228.0680	36.85	7.95	28.05	27.15
C 7H1805P2	244.17	244.0629	34.43	7.43	32.76	25.37
C 7H1806P2	260.17	260.0579	32.32	6.97	36.90	23.81
C 7H1807P2	276.16	276.0528	30.44	6.57	40.55	22.43
C 7H1808P2	292.16	292.0477	28.78	6.21	43.81	21.20
C 7H1809P2	308.16	308.0426	27.28	5.89	46.73	20.10
C 8H 401P2	178.07	177.9737	53.96	2.26	8.98	34.79
C 8H 402P2	194.07	193.9687	49.51	2.08	16.49	31.92
C 8H 403P2	210.07	209.9636	45.74	1.92	22.85	29.49
C 8H 404P2	226.07	225.9585	42.50	1.78	28.31	27.40
C 8H 405P2	242.07	241.9534	39.70	1.67	33.05	25.59
C 8H 406P2	258.07	257.9483	37.23	1.56	37.20	24.00
C 8H 407P2	274.06	273.9432	35.06	1.47	40.86	22.60
C 8H 408P2	290.06	289.9381	33.13	1.39	44.13	21.36
C 8H 409P2	306.06	305.9331	31.40	1.32	47.05	20.24
C 8H 501P1	148.10	148.0078	64.88	3.40	10.80	20.91
C 8H 502P1	164.10	164.0027	58.55	3.07	19.50	18.87
C 8H 503P1	180.10	179.9976	53.35	2.80	26.65	17.20
C 8H 504P1	196.10	195.9925	49.00	2.57	32.64	15.79
C 8H 505P1	212.10	211.9875	45.30	2.38	37.72	14.60
C 8H 506P1	228.10	227.9824	42.13	2.21	42.09	13.58
C 8H 507P1	244.10	243.9773	39.36	2.06	45.88	12.69
C 8H 508P1	260.10	259.9722	36.94	1.94	49.21	11.91
C 8H 509P1	276.10	275.9671	34.80	1.83	52.15	11.22
C 8H 601P2	180.08	179.9894	53.36	3.36	8.88	34.40
C 8H 602P2	196.08	195.9843	49.00	3.08	16.32	31.59
C 8H 603P2	212.08	211.9792	45.31	2.85	22.63	29.21
C 8H 604P2	228.08	227.9741	42.13	2.65	28.06	27.16
C 8H 605P2	244.08	243.9690	39.37	2.48	32.77	25.38
C 8H 606P2	260.08	259.9640	36.95	2.33	36.91	23.82
C 8H 607P2	276.08	275.9589	34.80	2.19	40.57	22.44
C 8H 608P2	292.08	291.9538	32.90	2.07	43.82	21.21
C 8H 609P2	308.08	307.9487	31.19	1.96	46.74	20.11
C 8H 701P1	150.12	150.0235	64.01	4.70	10.66	20.63
C 8H 702P1	166.12	166.0184	57.84	4.25	19.26	18.65
C 8H 703P1	182.12	182.0133	52.76	3.87	26.36	17.01
C 8H 704P1	198.12	198.0082	48.50	3.56	32.30	15.63
C 8H 705P1	214.12	214.0031	44.88	3.30	37.36	14.47
C 8H 706P1	230.12	229.9980	41.76	3.07	41.72	13.46
C 8H 707P1	246.11	245.9929	39.04	2.87	45.51	12.59
C 8H 708P1	262.11	261.9878	36.66	2.69	48.83	11.82
C 8H 709P1	278.11	277.9828	34.55	2.54	51.78	11.14
C 8H 801P2	182.10	182.0050	52.77	4.43	8.79	34.02
C 8H 802P2	198.10	198.0000	48.51	4.07	16.15	31.27
C 8H 803P2	214.10	213.9949	44.88	3.77	22.42	28.93
C 8H 804P2	230.10	229.9898	41.76	3.50	27.81	26.92
C 8H 805P2	246.10	245.9847	39.05	3.28	32.51	25.17
C 8H 806P2	262.10	261.9796	36.66	3.08	36.63	23.64
C 8H 807P2	278.10	277.9745	34.55	2.90	40.27	22.28
C 8H 808P2	294.10	293.9694	32.67	2.74	43.52	21.06
C 8H 809P2	310.10	309.9644	30.99	2.60	46.44	19.98
C 8H 901P1	152.13	152.0391	63.16	5.96	10.52	20.36
C 8H 902P1	168.13	168.0340	57.15	5.40	19.03	18.42
C 8H 903P1	184.13	184.0289	52.18	4.93	26.07	16.82
C 8H 904P1	200.13	200.0238	48.01	4.53	31.98	15.48

C 8H 905P1	216.13	216.0188	44.46	4.20	37.01	14.33
C 8H 906P1	232.13	232.0137	41.39	3.91	41.35	13.34
C 8H 907P1	248.13	248.0086	38.73	3.66	45.14	12.48
C 8H 908P1	264.13	264.0035	36.38	3.43	48.46	11.73
C 8H 909P1	280.13	279.9984	34.30	3.24	51.40	11.06
C 8H1001P2	184.12	184.0207	52.19	5.47	8.69	33.65
C 8H1002P2	200.12	200.0156	48.02	5.04	15.99	30.96
C 8H1003P2	216.11	216.0105	44.46	4.66	22.21	28.66
C 8H1004P2	232.11	232.0054	41.40	4.34	27.57	26.69
C 8H1005P2	248.11	248.0003	38.73	4.06	32.24	24.97
C 8H1006P2	264.11	263.9953	36.38	3.82	36.35	23.45
C 8H1007P2	280.11	279.9902	34.30	3.60	39.98	22.12
C 8H1008P2	296.11	295.9851	32.45	3.40	43.23	20.92
C 8H1009P2	312.11	311.9800	30.79	3.23	46.14	19.85
C 8H1101P1	154.15	154.0547	62.33	7.19	10.38	20.09
C 8H1102P1	170.15	170.0497	56.47	6.52	18.81	18.20
C 8H1103P1	186.15	186.0446	51.62	5.96	25.78	16.64
C 8H1104P1	202.15	202.0395	47.53	5.48	31.66	15.32
C 8H1105P1	218.15	218.0344	44.05	5.08	36.67	14.20
C 8H1106P1	234.15	234.0293	41.04	4.74	41.00	13.23
C 8H1107P1	250.15	250.0242	38.41	4.43	44.77	12.38
C 8H1108P1	266.15	266.0191	36.10	4.17	48.09	11.64
C 8H1109P1	282.15	282.0141	34.06	3.93	51.04	10.98
C 8H1201P2	186.13	186.0363	51.62	6.50	8.60	33.28
C 8H1202P2	202.13	202.0313	47.54	5.98	15.83	30.65
C 8H1203P2	218.13	218.0262	44.05	5.55	22.00	28.40
C 8H1204P2	234.13	234.0211	41.04	5.17	27.33	26.46
C 8H1205P2	250.13	250.0160	38.42	4.84	31.98	24.77
C 8H1206P2	266.13	266.0109	36.11	4.55	36.07	23.28
C 8H1207P2	282.13	282.0058	34.06	4.29	39.70	21.96
C 8H1208P2	298.13	298.0007	32.23	4.06	42.93	20.78
C 8H1209P2	314.13	313.9957	30.59	3.85	45.84	19.72
C 8H1301P1	156.17	156.0704	61.53	8.39	10.25	19.83
C 8H1302P1	172.17	172.0653	55.81	7.61	18.59	17.99
C 8H1303P1	188.16	188.0602	51.07	6.96	25.51	16.46
C 8H1304P1	204.16	204.0551	47.06	6.42	31.35	15.17
C 8H1305P1	220.16	220.0501	43.64	5.95	36.34	14.07
C 8H1306P1	236.16	236.0450	40.69	5.55	40.65	13.12
C 8H1307P1	252.16	252.0399	38.11	5.20	44.41	12.28
C 8H1308P1	268.16	268.0348	35.83	4.89	47.73	11.55
C 8H1309P1	284.16	284.0297	33.82	4.61	50.67	10.90
C 8H1401P2	188.15	188.0520	51.07	7.50	8.50	32.92
C 8H1402P2	204.15	204.0469	47.07	6.91	15.67	30.34
C 8H1403P2	220.15	220.0418	43.65	6.41	21.80	28.14
C 8H1404P2	236.15	236.0367	40.69	5.98	27.10	26.23
C 8H1405P2	252.15	252.0316	38.11	5.60	31.73	24.57
C 8H1406P2	268.14	268.0266	35.83	5.26	35.80	23.10
C 8H1407P2	284.14	284.0215	33.82	4.97	39.42	21.80
C 8H1408P2	300.14	300.0164	32.01	4.70	42.64	20.64
C 8H1409P2	316.14	316.0113	30.39	4.46	45.55	19.59
C 8H1501P1	158.18	158.0860	60.75	9.56	10.11	19.58
C 8H1502P1	174.18	174.0810	55.17	8.68	18.37	17.78
C 8H1503P1	190.18	190.0759	50.53	7.95	25.24	16.29
C 8H1504P1	206.18	206.0708	46.60	7.33	31.04	15.02
C 8H1505P1	222.18	222.0657	43.25	6.81	36.01	13.94
C 8H1506P1	238.18	238.0606	40.34	6.35	40.30	13.00
C 8H1507P1	254.18	254.0555	37.80	5.95	44.06	12.19
C 8H1508P1	270.18	270.0504	35.57	5.60	47.37	11.46
C 8H1509P1	286.18	286.0454	33.58	5.28	50.32	10.82
C 8H1601P2	190.16	190.0676	50.53	8.48	8.41	32.58

109

C 8H1602P2	206.16	206.0625	46.61	7.82	15.52	30.05
C 8H1603P2	222.16	222.0575	43.25	7.26	21.60	27.88
C 8H1604P2	238.16	238.0524	40.35	6.77	26.87	26.01
C 8H1605P2	254.16	254.0473	37.81	6.35	31.47	24.37
C 8H1606P2	270.16	270.0422	35.57	5.97	35.53	22.93
C 8H1607P2	286.16	286.0371	33.58	5.64	39.14	21.65
C 8H1608P2	302.16	302.0320	31.80	5.34	42.36	20.50
C 8H1609P2	318.16	318.0269	30.20	5.07	45.26	19.47
C 8H1701P1	160.20	160.1017	59.98	10.70	9.99	19.33
C 8H1702P1	176.20	176.0966	54.54	9.73	18.16	17.58
C 8H1703P1	192.20	192.0915	50.00	8.92	24.97	16.12
C 8H1704P1	208.20	208.0864	46.15	8.23	30.74	14.88
C 8H1705P1	224.20	224.0814	42.86	7.64	35.68	13.82
C 8H1706P1	240.19	240.0763	40.00	7.13	39.97	12.90
C 8H1707P1	256.19	256.0712	37.51	6.69	43.72	12.09
C 8H1708P1	272.19	272.0661	35.30	6.30	47.02	11.38
C 8H1709P1	288.19	288.0610	33.34	5.95	49.96	10.75
C 8H1801P2	192.18	192.0833	50.00	9.44	8.33	32.23
C 8H1802P2	208.18	208.0782	46.16	8.72	15.37	29.76
C 8H1803P2	224.18	224.0731	42.86	8.09	21.41	27.63
C 8H1804P2	240.18	240.0680	40.01	7.55	26.65	25.79
C 8H1805P2	256.18	256.0629	37.51	7.08	31.23	24.18
C 8H1806P2	272.18	272.0579	35.30	6.67	35.27	22.76
C 8H1807P2	288.18	288.0528	33.34	6.30	38.86	21.50
C 8H1808P2	304.18	304.0477	31.59	5.96	42.08	20.37
C 8H1809P2	320.17	320.0426	30.01	5.67	44.97	19.35
C 8H1901P1	162.21	162.1173	59.24	11.81	9.86	19.09
C 8H1902P1	178.21	178.1123	53.92	10.75	17.96	17.38
C 8H1903P1	194.21	194.1072	49.48	9.86	24.71	15.95
C 8H1904P1	210.21	210.1021	45.71	9.11	30.44	14.73
C 8H1905P1	226.21	226.0970	42.48	8.47	35.36	13.69
C 8H1906P1	242.21	242.0919	39.67	7.91	39.63	12.79
C 8H1907P1	258.21	258.0868	37.21	7.42	43.37	12.00
C 8H1908P1	274.21	274.0817	35.04	6.98	46.68	11.30
C 8H1909P1	290.21	290.0767	33.11	6.60	49.62	10.67
C 8H2001P2	194.20	194.0989	49.48	10.38	8.24	31.90
C 8H2002P2	210.20	210.0938	45.71	9.59	15.22	29.47
C 8H2003P2	226.19	226.0888	42.48	8.91	21.22	27.39
C 8H2004P2	242.19	242.0837	39.67	8.32	26.42	25.58
C 8H2005P2	258.19	258.0786	37.22	7.81	30.98	23.99
C 8H2006P2	274.19	274.0735	35.04	7.35	35.01	22.59
C 8H2007P2	290.19	290.0684	33.11	6.95	38.59	21.35
C 8H2008P2	306.19	306.0633	31.38	6.58	41.80	20.23
C 8H2009P2	322.19	322.0582	29.82	6.26	44.69	19.23
C 9H 401P2	190.08	189.9737	56.87	2.12	8.42	32.59
C 9H 402P2	206.08	205.9687	52.46	1.96	15.53	30.06
C 9H 403P2	222.08	221.9636	48.68	1.82	21.61	27.89
C 9H 404P2	238.08	237.9585	45.41	1.69	26.88	26.02
C 9H 405P2	254.08	253.9534	42.55	1.59	31.49	24.38
C 9H 406P2	270.08	269.9483	40.03	1.49	35.54	22.94
C 9H 407P2	286.08	285.9432	37.79	1.41	39.15	21.65
C 9H 408P2	302.08	301.9381	35.79	1.33	42.37	20.51
C 9H 409P2	318.07	317.9331	33.99	1.27	45.27	19.48
C 9H 501P1	160.11	160.0078	67.51	3.15	9.99	19.34
C 9H 502P1	176.11	176.0027	61.38	2.86	18.17	17.59
C 9H 503P1	192.11	191.9976	56.27	2.62	24.98	16.12
C 9H 504P1	208.11	207.9925	51.94	2.42	30.75	14.88
C 9H 505P1	224.11	223.9875	48.24	2.25	35.70	13.82
C 9H 506P1	240.11	239.9824	45.02	2.10	39.98	12.90
C 9H 507P1	256.11	255.9773	42.21	1.97	43.73	12.09

C 9H 508P1	272.11	271.9722	39.73	1.85	47.04	11.38
C 9H 509P1	288.11	287.9671	37.52	1.75	49.98	10.75
C 9H 601P2	192.10	191.9894	56.27	3.15	8.33	32.25
C 9H 602P2	208.09	207.9843	51.95	2.91	15.38	29.77
C 9H 603P2	224.09	223.9792	48.24	2.70	21.42	27.64
C 9H 604P2	240.09	239.9741	45.02	2.52	26.66	25.80
C 9H 605P2	256.09	255.9690	42.21	2.36	31.24	24.19
C 9H 606P2	272.09	271.9640	39.73	2.22	35.28	22.77
C 9H 607P2	288.09	287.9589	37.52	2.10	38.88	21.50
C 9H 608P2	304.09	303.9538	35.55	1.99	42.09	20.37
C 9H 609P2	320.09	319.9487	33.77	1.89	44.99	19.35
C 9H 701P1	162.13	162.0235	66.68	4.35	9.87	19.10
C 9H 702P1	178.13	178.0184	60.69	3.96	17.96	17.39
C 9H 703P1	194.13	194.0133	55.69	3.63	24.73	15.96
C 9H 704P1	210.13	210.0082	51.45	3.36	30.46	14.74
C 9H 705P1	226.13	226.0031	47.81	3.12	35.38	13.70
C 9H 706P1	242.13	241.9980	44.65	2.91	39.65	12.79
C 9H 707P1	258.13	257.9929	41.88	2.73	43.39	12.00
C 9H 708P1	274.13	273.9878	39.43	2.57	46.69	11.30
C 9H 709P1	290.12	289.9828	37.26	2.43	49.63	10.68
C 9H 801P2	194.11	194.0050	55.69	4.15	8.24	31.91
C 9H 802P2	210.11	210.0000	51.45	3.84	15.23	29.48
C 9H 803P2	226.11	225.9949	47.81	3.57	21.23	27.40
C 9H 804P2	242.11	241.9898	44.65	3.33	26.43	25.59
C 9H 805P2	258.11	257.9847	41.88	3.12	30.99	24.00
C 9H 806P2	274.11	273.9796	39.44	2.94	35.02	22.60
C 9H 807P2	290.11	289.9745	37.26	2.78	38.60	21.35
C 9H 808P2	306.11	305.9694	35.31	2.63	41.81	20.24
C 9H 809P2	322.11	321.9644	33.56	2.50	44.70	19.23
C 9H 901P1	164.15	164.0391	65.86	5.53	9.75	18.87
C 9H 902P1	180.14	180.0340	60.01	5.04	17.76	17.19
C 9H 903P1	196.14	196.0289	55.11	4.63	24.47	15.79
C 9H 904P1	212.14	212.0238	50.96	4.28	30.17	14.60
C 9H 905P1	228.14	228.0188	47.38	3.98	35.06	13.58
C 9H 906P1	244.14	244.0137	44.28	3.72	39.32	12.69
C 9H 907P1	260.14	260.0086	41.55	3.49	43.05	11.91
C 9H 908P1	276.14	276.0035	39.15	3.29	46.35	11.22
C 9H 909P1	292.14	291.9984	37.00	3.11	49.29	10.60
C 9H1001P2	196.13	196.0207	55.12	5.14	8.16	31.59
C 9H1002P2	212.13	212.0156	50.96	4.75	15.08	29.20
C 9H1003P2	228.13	228.0105	47.39	4.42	21.04	27.16
C 9H1004P2	244.13	244.0054	44.28	4.13	26.22	25.38
C 9H1005P2	260.12	260.0003	41.56	3.87	30.75	23.81
C 9H1006P2	276.12	275.9953	39.15	3.65	34.77	22.43
C 9H1007P2	292.12	291.9902	37.01	3.45	38.34	21.21
C 9H1008P2	308.12	307.9851	35.08	3.27	41.54	20.10
C 9H1009P2	324.12	323.9800	33.35	3.11	44.43	19.11
C 9H1101P1	166.16	166.0547	65.06	6.67	9.63	18.64
C 9H1102P1	182.16	182.0497	59.34	6.09	17.57	17.00
C 9H1103P1	198.16	198.0446	54.55	5.60	24.22	15.63
C 9H1104P1	214.16	214.0395	50.48	5.18	29.88	14.46
C 9H1105P1	230.16	230.0344	46.97	4.82	34.76	13.46
C 9H1106P1	246.16	246.0293	43.91	4.50	39.00	12.58
C 9H1107P1	262.16	262.0242	41.23	4.23	42.72	11.81
C 9H1108P1	278.16	278.0191	38.86	3.99	46.02	11.14
C 9H1109P1	294.16	294.0141	36.75	3.77	48.95	10.53
C 9H1201P2	198.14	198.0363	54.56	6.10	8.07	31.26
C 9H1202P2	214.14	214.0313	50.48	5.65	14.94	28.93
C 9H1203P2	230.14	230.0262	46.97	5.26	20.86	26.92
C 9H1204P2	246.14	246.0211	43.92	4.91	26.00	25.17

C 9H12O5P2	262.14	262.0160	41.24	4.61	30.52	23.63
C 9H12O6P2	278.14	278.0109	38.87	4.35	34.51	22.27
C 9H12O7P2	294.14	294.0058	36.75	4.11	38.08	21.06
C 9H12O8P2	310.14	310.0007	34.86	3.90	41.27	19.97
C 9H12O9P2	326.14	325.9957	33.15	3.71	44.15	18.99
C 9H13O1P1	168.18	168.0704	64.28	7.79	9.51	18.42
C 9H13O2P1	184.18	184.0653	58.69	7.11	17.37	16.82
C 9H13O3P1	200.18	200.0602	54.00	6.55	23.98	15.47
C 9H13O4P1	216.18	216.0551	50.01	6.06	29.60	14.33
C 9H13O5P1	232.17	232.0501	46.56	5.64	34.46	13.34
C 9H13O6P1	248.17	248.0450	43.56	5.28	38.68	12.48
C 9H13O7P1	264.17	264.0399	40.92	4.96	42.39	11.72
C 9H13O8P1	280.17	280.0348	38.58	4.68	45.68	11.06
C 9H13O9P1	296.17	296.0297	36.50	4.42	48.62	10.46
C 9H14O1P2	200.16	200.0520	54.01	7.05	7.99	30.95
C 9H14O2P2	216.16	216.0469	50.01	6.53	14.80	28.66
C 9H14O3P2	232.16	232.0418	46.56	6.08	20.67	26.68
C 9H14O4P2	248.16	248.0367	43.56	5.69	25.79	24.96
C 9H14O5P2	264.16	264.0316	40.92	5.34	30.28	23.45
C 9H14O6P2	280.16	280.0266	38.59	5.04	34.27	22.11
C 9H14O7P2	296.16	296.0215	36.50	4.76	37.82	20.92
C 9H14O8P2	312.15	312.0164	34.63	4.52	41.00	19.85
C 9H14O9P2	328.15	328.0113	32.94	4.30	43.88	18.88
C 9H15O1P1	170.19	170.0860	63.52	8.88	9.40	18.20
C 9H15O2P1	186.19	186.0810	58.06	8.12	17.19	16.64
C 9H15O3P1	202.19	202.0759	53.46	7.48	23.74	15.32
C 9H15O4P1	218.19	218.0708	49.54	6.93	29.33	14.20
C 9H15O5P1	234.19	234.0657	46.16	6.46	34.16	13.23
C 9H15O6P1	250.19	250.0606	43.21	6.04	38.37	12.38
C 9H15O7P1	266.19	266.0555	40.61	5.68	42.07	11.64
C 9H15O8P1	282.19	282.0504	38.31	5.36	45.36	10.98
C 9H15O9P1	298.19	298.0454	36.25	5.07	48.29	10.39
C 9H16O1P2	202.17	202.0676	53.47	7.98	7.91	30.64
C 9H16O2P2	218.17	218.0625	49.55	7.39	14.67	28.39
C 9H16O3P2	234.17	234.0575	46.16	6.89	20.50	26.45
C 9H16O4P2	250.17	250.0524	43.21	6.45	25.58	24.76
C 9H16O5P2	266.17	266.0473	40.61	6.06	30.05	23.27
C 9H16O6P2	282.17	282.0422	38.31	5.72	34.02	21.95
C 9H16O7P2	298.17	298.0371	36.25	5.41	37.56	20.78
C 9H16O8P2	314.17	314.0320	34.41	5.13	40.74	19.72
C 9H16O9P2	330.17	330.0269	32.74	4.88	43.61	18.76
C 9H17O1P1	172.21	172.1017	62.77	9.95	9.29	17.99
C 9H17O2P1	188.21	188.0966	57.44	9.10	17.00	16.46
C 9H17O3P1	204.21	204.0915	52.94	8.39	23.50	15.17
C 9H17O4P1	220.21	220.0864	49.09	7.78	29.06	14.07
C 9H17O5P1	236.21	236.0814	45.77	7.25	33.87	13.11
C 9H17O6P1	252.21	252.0763	42.86	6.79	38.06	12.28
C 9H17O7P1	268.21	268.0712	40.31	6.39	41.76	11.55
C 9H17O8P1	284.20	284.0661	38.04	6.03	45.04	10.90
C 9H17O9P1	300.20	300.0610	36.01	5.71	47.97	10.32
C 9H18O1P2	204.19	204.0833	52.94	8.89	7.84	30.34
C 9H18O2P2	220.19	220.0782	49.09	8.24	14.53	28.13
C 9H18O3P2	236.19	236.0731	45.77	7.68	20.32	26.23
C 9H18O4P2	252.19	252.0680	42.86	7.19	25.38	24.56
C 9H18O5P2	268.19	268.0629	40.31	6.77	29.83	23.10
C 9H18O6P2	284.19	284.0579	38.04	6.38	33.78	21.80
C 9H18O7P2	300.19	300.0528	36.01	6.04	37.31	20.64
C 9H18O8P2	316.19	316.0477	34.19	5.74	40.48	19.59
C 9H18O9P2	332.19	332.0426	32.54	5.46	43.35	18.65
C 9H19O1P1	174.22	174.1173	62.05	10.99	9.18	17.78

C 9H19O2P1	190.22	190.1123	56.83	10.07	16.82	16.28
C 9H19O3P1	206.22	206.1072	52.42	9.29	23.27	15.02
C 9H19O4P1	222.22	222.1021	48.64	8.62	28.80	13.94
C 9H19O5P1	238.22	238.0970	45.38	8.04	33.58	13.00
C 9H19O6P1	254.22	254.0919	42.52	7.53	37.76	12.18
C 9H19O7P1	270.22	270.0868	40.00	7.09	41.45	11.46
C 9H19O8P1	286.22	286.0817	37.77	6.69	44.72	10.82
C 9H19O9P1	302.22	302.0767	35.77	6.34	47.65	10.25
C 9H20O1P2	206.21	206.0989	52.42	9.78	7.76	30.04
C 9H20O2P2	222.21	222.0938	48.65	9.07	14.40	27.88
C 9H20O3P2	238.21	238.0888	45.38	8.46	20.15	26.01
C 9H20O4P2	254.20	254.0837	42.52	7.93	25.18	24.37
C 9H20O5P2	270.20	270.0786	40.01	7.46	29.61	22.93
C 9H20O6P2	286.20	286.0735	37.77	7.04	33.54	21.64
C 9H20O7P2	302.20	302.0684	35.77	6.67	37.06	20.50
C 9H20O8P2	318.20	318.0633	33.97	6.34	40.22	19.47
C 9H20O9P2	334.20	334.0582	32.35	6.03	43.09	18.54
C 9H21O1P1	176.24	176.1330	61.34	12.01	9.08	17.57
C 9H21O2P1	192.24	192.1279	56.23	11.01	16.65	16.11
C 9H21O3P1	208.24	208.1228	51.91	10.16	23.05	14.87
C 9H21O4P1	224.24	224.1177	48.21	9.44	28.54	13.81
C 9H21O5P1	240.24	240.1127	45.00	8.81	33.30	12.89
C 9H21O6P1	256.24	256.1076	42.19	8.26	37.46	12.09
C 9H21O7P1	272.24	272.1025	39.71	7.78	41.14	11.38
C 9H21O8P1	288.24	288.0974	37.50	7.34	44.41	10.75
C 9H21O9P1	304.24	304.0923	35.53	6.96	47.33	10.18
C 9H22O1P2	208.22	208.1146	51.92	10.65	7.68	29.75
C 9H22O2P2	224.22	224.1095	48.21	9.89	14.27	27.63
C 9H22O3P2	240.22	240.1044	45.00	9.23	19.98	25.79
C 9H22O4P2	256.22	256.0993	42.19	8.65	24.98	24.18
C 9H22O5P2	272.22	272.0942	39.71	8.15	29.39	22.76
C 9H22O6P2	288.22	288.0892	37.51	7.69	33.31	21.49
C 9H22O7P2	304.22	304.0841	35.53	7.29	36.81	20.36
C 9H22O8P2	320.22	320.0790	33.76	6.93	39.97	19.35
C 9H22O9P2	336.22	336.0739	32.15	6.60	42.83	18.42
C10H 5O1P1	172.12	172.0078	69.78	2.93	9.30	17.99
C10H 5O2P1	188.12	188.0027	63.85	2.68	17.01	16.46
C10H 5O3P1	204.12	203.9976	58.84	2.47	23.51	15.17
C10H 5O4P1	220.12	219.9925	54.57	2.29	29.07	14.07
C10H 5O5P1	236.12	235.9875	50.87	2.13	33.88	13.12
C10H 5O6P1	252.12	251.9824	47.64	2.00	38.08	12.29
C10H 5O7P1	268.12	267.9773	44.80	1.88	41.77	11.55
C10H 5O8P1	284.12	283.9722	42.27	1.77	45.05	10.90
C10H 5O9P1	300.12	299.9671	40.02	1.68	47.98	10.32
C10H 6O1P2	204.11	203.9894	58.85	2.96	7.84	30.35
C10H 6O2P2	220.11	219.9843	54.57	2.75	14.54	28.14
C10H 6O3P2	236.11	235.9792	50.87	2.56	20.33	26.24
C10H 6O4P2	252.10	251.9741	47.64	2.40	25.39	24.57
C10H 6O5P2	268.10	267.9690	44.80	2.26	29.84	23.11
C10H 6O6P2	284.10	283.9640	42.28	2.13	33.79	21.80
C10H 6O7P2	300.10	299.9589	40.02	2.02	37.32	20.64
C10H 6O8P2	316.10	315.9538	38.00	1.91	40.49	19.60
C10H 6O9P2	332.10	331.9487	36.17	1.82	43.36	18.65
C10H 7O1P1	174.14	174.0235	68.97	4.05	9.19	17.79
C10H 7O2P1	190.14	190.0184	63.17	3.71	16.83	16.29
C10H 7O3P1	206.14	206.0133	58.27	3.42	23.28	15.03
C10H 7O4P1	222.14	222.0082	54.07	3.18	28.81	13.94
C10H 7O5P1	238.14	238.0031	50.44	2.96	33.59	13.01
C10H 7O6P1	254.14	253.9980	47.26	2.78	37.77	12.19
C10H 7O7P1	270.14	269.9929	44.46	2.61	41.46	11.47

C10H 7O8P1	286.14	285.9878	41.98	2.47	44.73	10.82
C10H 7O9P1	302.14	301.9828	39.75	2.34	47.66	10.25
C10H 8O1P2	206.12	206.0050	58.27	3.91	7.76	30.05
C10H 8O2P2	222.12	222.0000	54.07	3.63	14.41	27.89
C10H 8O3P2	238.12	237.9949	50.44	3.39	20.16	26.02
C10H 8O4P2	254.12	253.9898	47.27	3.17	25.18	24.38
C10H 8O5P2	270.12	269.9847	44.47	2.99	29.62	22.93
C10H 8O6P2	286.12	285.9796	41.98	2.82	33.55	21.65
C10H 8O7P2	302.12	301.9745	39.76	2.67	37.07	20.50
C10H 8O8P2	318.12	317.9694	37.76	2.53	40.24	19.47
C10H 8O9P2	334.12	333.9644	35.95	2.41	43.10	18.54
C10H 9O1P1	176.16	176.0391	68.18	5.15	9.08	17.58
C10H 9O2P1	192.16	192.0340	62.51	4.72	16.65	16.12
C10H 9O3P1	208.16	208.0289	57.70	4.36	23.06	14.88
C10H 9O4P1	224.15	224.0238	53.58	4.05	28.55	13.82
C10H 9O5P1	240.15	240.0188	50.01	3.78	33.31	12.90
C10H 9O6P1	256.15	256.0137	46.89	3.54	37.48	12.09
C10H 9O7P1	272.15	272.0086	44.13	3.33	41.15	11.38
C10H 9O8P1	288.15	288.0035	41.68	3.15	44.42	10.75
C10H 9O9P1	304.15	303.9984	39.49	2.98	47.34	10.18
C10H10O1P2	208.14	208.0207	57.71	4.84	7.69	29.76
C10H10O2P2	224.14	224.0156	53.59	4.50	14.28	27.64
C10H10O3P2	240.14	240.0105	50.02	4.20	19.99	25.80
C10H10O4P2	256.14	256.0054	46.89	3.94	24.99	24.19
C10H10O5P2	272.14	272.0003	44.14	3.70	29.40	22.76
C10H10O6P2	288.14	287.9953	41.69	3.50	33.32	21.50
C10H10O7P2	304.13	303.9902	39.49	3.31	36.82	20.37
C10H10O8P2	320.13	319.9851	37.52	3.15	39.98	19.35
C10H10O9P2	336.13	335.9800	35.73	3.00	42.84	18.43
C10H11O1P1	178.17	178.0547	67.41	6.22	8.98	17.38
C10H11O2P1	194.17	194.0497	61.86	5.71	16.48	15.95
C10H11O3P1	210.17	210.0446	57.15	5.28	22.84	14.74
C10H11O4P1	226.17	226.0395	53.11	4.90	28.30	13.69
C10H11O5P1	242.17	242.0344	49.60	4.58	33.03	12.79
C10H11O6P1	258.17	258.0293	46.52	4.29	37.18	12.00
C10H11O7P1	274.17	274.0242	43.81	4.04	40.85	11.30
C10H11O8P1	290.17	290.0191	41.39	3.82	44.11	10.67
C10H11O9P1	306.17	306.0141	39.23	3.62	47.03	10.12
C10H12O1P2	210.15	210.0363	57.15	5.76	7.61	29.48
C10H12O2P2	226.15	226.0313	53.11	5.35	14.15	27.39
C10H12O3P2	242.15	242.0262	49.60	5.00	19.82	25.58
C10H12O4P2	258.15	258.0211	46.53	4.69	24.79	24.00
C10H12O5P2	274.15	274.0160	43.81	4.41	29.18	22.60
C10H12O6P2	290.15	290.0109	41.40	4.17	33.08	21.35
C10H12O7P2	306.15	306.0058	39.23	3.95	36.58	20.23
C10H12O8P2	322.15	322.0007	37.28	3.75	39.73	19.23
C10H12O9P2	338.15	337.9957	35.52	3.58	42.58	18.32
C10H13O1P1	180.19	180.0704	66.66	7.27	8.88	17.19
C10H13O2P1	196.19	196.0653	61.22	6.68	16.31	15.79
C10H13O3P1	212.19	212.0602	56.61	6.18	22.62	14.60
C10H13O4P1	228.19	228.0551	52.64	5.74	28.05	13.57
C10H13O5P1	244.19	244.0501	49.19	5.37	32.76	12.68
C10H13O6P1	260.19	260.0450	46.16	5.04	36.90	11.90
C10H13O7P1	276.18	276.0399	43.49	4.74	40.55	11.21
C10H13O8P1	292.18	292.0348	41.11	4.48	43.81	10.60
C10H13O9P1	308.18	308.0297	38.97	4.25	46.72	10.05
C10H14O1P2	212.17	212.0520	56.61	6.65	7.54	29.20
C10H14O2P2	228.17	228.0469	52.64	6.18	14.02	27.15
C10H14O3P2	244.17	244.0418	49.19	5.78	19.66	25.37
C10H14O4P2	260.17	260.0367	46.17	5.42	24.60	23.81

C10H14O5P2	276.17	276.0316	43.49	5.11	28.97	22.43
C10H14O6P2	292.17	292.0266	41.11	4.83	32.86	21.20
C10H14O7P2	308.17	308.0215	38.98	4.58	36.34	20.10
C10H14O8P2	324.17	324.0164	37.05	4.35	39.48	19.11
C10H14O9P2	340.17	340.0113	35.31	4.15	42.33	18.21
C10H15O1P1	182.20	182.0860	65.92	8.30	8.78	17.00
C10H15O2P1	198.20	198.0810	60.60	7.63	16.14	15.63
C10H15O3P1	214.20	214.0759	56.07	7.06	22.41	14.46
C10H15O4P1	230.20	230.0708	52.18	6.57	27.80	13.46
C10H15O5P1	246.20	246.0657	48.79	6.14	32.49	12.58
C10H15O6P1	262.20	262.0606	45.81	5.77	36.61	11.81
C10H15O7P1	278.20	278.0555	43.17	5.43	40.26	11.13
C10H15O8P1	294.20	294.0504	40.83	5.14	43.51	10.53
C10H15O9P1	310.20	310.0454	38.72	4.87	46.42	9.99
C10H16O1P2	214.19	214.0676	56.08	7.53	7.47	28.92
C10H16O2P2	230.19	230.0625	52.18	7.01	13.90	26.91
C10H16O3P2	246.18	246.0575	48.79	6.55	19.50	25.16
C10H16O4P2	262.18	262.0524	45.81	6.15	24.41	23.63
C10H16O5P2	278.18	278.0473	43.18	5.80	28.76	22.27
C10H16O6P2	294.18	294.0422	40.83	5.48	32.63	21.06
C10H16O7P2	310.18	310.0371	38.72	5.20	36.11	19.97
C10H16O8P2	326.18	326.0320	36.82	4.94	39.24	18.99
C10H16O9P2	342.18	342.0269	35.10	4.71	42.08	18.10
C10H17O1P1	184.22	184.1017	65.20	9.30	8.68	16.81
C10H17O2P1	200.22	200.0966	59.99	8.56	15.98	15.47
C10H17O3P1	216.22	216.0915	55.55	7.93	22.20	14.33
C10H17O4P1	232.22	232.0864	51.72	7.38	27.56	13.34
C10H17O5P1	248.22	248.0814	48.39	6.90	32.23	12.48
C10H17O6P1	264.22	264.0763	45.46	6.49	36.33	11.72
C10H17O7P1	280.22	280.0712	42.86	6.12	39.97	11.05
C10H17O8P1	296.22	296.0661	40.55	5.78	43.21	10.46
C10H17O9P1	312.22	312.0610	38.47	5.49	46.12	9.92
C10H18O1P2	216.20	216.0833	55.56	8.39	7.40	28.65
C10H18O2P2	232.20	232.0782	51.73	7.81	13.78	26.68
C10H18O3P2	248.20	248.0731	48.39	7.31	19.34	24.96
C10H18O4P2	264.20	264.0680	45.46	6.87	24.22	23.45
C10H18O5P2	280.20	280.0629	42.87	6.48	28.55	22.11
C10H18O6P2	296.20	296.0579	40.55	6.13	32.41	20.91
C10H18O7P2	312.20	312.0528	38.47	5.81	35.87	19.84
C10H18O8P2	328.20	328.0477	36.60	5.53	39.00	18.88
C10H18O9P2	344.20	344.0426	34.90	5.27	41.83	18.00
C10H19O1P1	186.24	186.1173	64.49	10.28	8.59	16.63
C10H19O2P1	202.24	202.1123	59.39	9.47	15.82	15.32
C10H19O3P1	218.23	218.1072	55.04	8.78	21.99	14.19
C10H19O4P1	234.23	234.1021	51.28	8.18	27.32	13.22
C10H19O5P1	250.23	250.0970	48.00	7.65	31.97	12.38
C10H19O6P1	266.23	266.0919	45.12	7.19	36.06	11.63
C10H19O7P1	282.23	282.0868	42.56	6.79	39.68	10.97
C10H19O8P1	298.23	298.0817	40.27	6.42	42.92	10.39
C10H19O9P1	314.23	314.0767	38.22	6.09	45.82	9.86
C10H20O1P2	218.22	218.0989	55.04	9.24	7.33	28.39
C10H20O2P2	234.22	234.0938	51.28	8.61	13.66	26.45
C10H20O3P2	250.22	250.0888	48.00	8.06	19.18	24.76
C10H20O4P2	266.22	266.0837	45.12	7.57	24.04	23.27
C10H20O5P2	282.22	282.0786	42.56	7.14	28.35	21.95
C10H20O6P2	298.21	298.0735	40.28	6.76	32.19	20.77
C10H20O7P2	314.21	314.0684	38.23	6.42	35.64	19.72
C10H20O8P2	330.21	330.0633	36.37	6.10	38.76	18.76
C10H20O9P2	346.21	346.0582	34.69	5.82	41.59	17.89
C10H21O1P1	188.25	188.1330	63.80	11.24	8.50	16.45

112

C10H2102P1	204.25	204.1279	58.81	10.36	15.67	15.16	C11H 708P1	298.15	297.9878	44.31	2.37	42.93	10.39
C10H2103P1	220.25	220.1228	54.53	9.61	21.79	14.06	C11H 709P1	314.15	313.9828	42.06	2.25	45.84	9.86
C10H2104P1	236.25	236.1177	50.84	8.96	27.09	13.11	C11H 801P2	218.13	218.0050	60.57	3.70	7.33	28.40
C10H2105P1	252.25	252.1127	47.62	8.39	31.71	12.28	C11H 802P2	234.13	234.0000	56.43	3.44	13.67	26.46
C10H2106P1	268.25	268.1076	44.78	7.89	35.79	11.55	C11H 803P2	250.13	249.9949	52.82	3.22	19.19	24.77
C10H2107P1	284.25	284.1025	42.26	7.45	39.40	10.90	C11H 804P2	266.13	265.9898	49.65	3.03	24.05	23.28
C10H2108P1	300.25	300.0974	40.00	7.05	42.63	10.32	C11H 805P2	282.13	281.9847	46.83	2.86	28.35	21.96
C10H2109P1	316.25	316.0923	37.98	6.69	45.53	9.79	C11H 806P2	298.13	297.9796	44.32	2.70	32.20	20.78
C10H2201P2	220.23	220.1146	54.54	10.07	7.26	28.13	C11H 807P2	314.13	313.9745	42.06	2.57	35.65	19.72
C10H2202P2	236.23	236.1095	50.84	9.39	13.55	26.22	C11H 808P2	330.13	329.9694	40.02	2.44	38.77	18.76
C10H2203P2	252.23	252.1044	47.62	8.79	19.03	24.56	C11H 809P2	346.13	345.9644	38.17	2.33	41.60	17.90
C10H2204P2	268.23	268.0993	44.78	8.27	23.86	23.09	C11H 901P1	188.17	188.0391	70.22	4.82	8.50	16.46
C10H2205P2	284.23	284.0942	42.26	7.80	28.15	21.79	C11H 902P1	204.17	204.0340	64.71	4.44	15.67	15.17
C10H2206P2	300.23	300.0892	40.01	7.39	31.97	20.63	C11H 903P1	220.17	220.0289	60.01	4.12	21.80	14.07
C10H2207P2	316.23	316.0841	37.98	7.01	35.42	19.59	C11H 904P1	236.17	236.0238	55.94	3.84	27.10	13.12
C10H2208P2	332.23	332.0790	36.15	6.67	38.53	18.65	C11H 905P1	252.17	252.0188	52.40	3.60	31.72	12.28
C10H2209P2	348.23	348.0739	34.49	6.37	41.35	17.79	C11H 906P1	268.16	268.0137	49.27	3.38	35.80	11.55
C10H2301P1	190.27	190.1486	63.13	12.18	8.41	16.28	C11H 907P1	284.16	284.0086	46.50	3.19	39.41	10.90
C10H2302P1	206.27	206.1436	58.23	11.24	15.51	15.02	C11H 908P1	300.16	300.0035	44.02	3.02	42.64	10.32
C10H2303P1	222.27	222.1385	54.04	10.43	21.59	13.94	C11H 909P1	316.16	315.9984	41.79	2.87	45.54	9.80
C10H2304P1	238.27	238.1334	50.41	9.73	26.86	13.00	C11H1001P2	220.15	220.0207	60.02	4.58	7.27	28.14
C10H2305P1	254.27	254.1283	47.24	9.12	31.46	12.18	C11H1002P2	236.15	236.0156	55.95	4.27	13.55	26.23
C10H2306P1	270.27	270.1232	44.44	8.58	35.52	11.46	C11H1003P2	252.15	252.0105	52.40	4.00	19.04	24.57
C10H2307P1	286.26	286.1181	41.96	8.10	39.12	10.82	C11H1004P2	268.15	268.0054	49.27	3.76	23.87	23.10
C10H2308P1	302.26	302.1130	39.74	7.67	42.35	10.25	C11H1005P2	284.15	284.0003	46.50	3.55	28.15	21.80
C10H2309P1	318.26	318.1080	37.74	7.28	45.24	9.73	C11H1006P2	300.15	299.9953	44.02	3.36	31.98	20.64
C10H2401P2	222.25	222.1302	54.04	10.88	7.20	27.87	C11H1007P2	316.15	315.9902	41.79	3.19	35.43	19.59
C10H2402P2	238.25	238.1251	50.41	10.15	13.43	26.00	C11H1008P2	332.15	331.9851	39.78	3.03	38.54	18.65
C10H2403P2	254.25	254.1201	47.24	9.51	18.88	24.36	C11H1009P2	348.14	347.9800	37.95	2.90	41.36	17.79
C10H2404P2	270.25	270.1150	44.44	8.95	23.68	22.92	C11H1101P1	190.18	190.0547	69.47	5.83	8.41	16.29
C10H2405P2	286.25	286.1099	41.96	8.45	27.95	21.64	C11H1102P1	206.18	206.0497	64.08	5.38	15.52	15.02
C10H2406P2	302.25	302.1048	39.74	8.00	31.76	20.50	C11H1103P1	222.18	222.0446	59.47	4.99	21.60	13.94
C10H2407P2	318.25	318.0997	37.74	7.60	35.19	19.47	C11H1104P1	238.18	238.0395	55.47	4.66	26.87	13.00
C10H2408P2	334.25	334.0946	35.94	7.24	38.29	18.53	C11H1105P1	254.18	254.0344	51.98	4.36	31.47	12.19
C10H2409P2	350.24	350.0895	34.29	6.91	41.11	17.69	C11H1106P1	270.18	270.0293	48.90	4.10	35.53	11.46
C11H 501P1	184.14	184.0078	71.75	2.74	8.69	16.82	C11H1107P1	286.18	286.0242	46.17	3.87	39.13	10.82
C11H 502P1	200.14	200.0027	66.02	2.52	15.99	15.48	C11H1108P1	302.18	302.0191	43.72	3.67	42.36	10.25
C11H 503P1	216.13	215.9976	61.13	2.33	22.21	14.33	C11H1109P1	318.18	318.0141	41.52	3.48	45.26	9.73
C11H 504P1	232.13	231.9925	56.92	2.17	27.57	13.34	C11H1201P2	222.17	222.0363	59.47	5.44	7.20	27.88
C11H 505P1	248.13	247.9875	53.25	2.03	32.24	12.48	C11H1202P2	238.16	238.0313	55.48	5.08	13.44	26.01
C11H 506P1	264.13	263.9824	50.02	1.91	36.34	11.73	C11H1203P2	254.16	254.0262	51.98	4.76	18.88	24.37
C11H 507P1	280.13	279.9773	47.16	1.80	39.98	11.06	C11H1204P2	270.16	270.0211	48.90	4.48	23.69	22.93
C11H 508P1	296.13	295.9722	44.62	1.70	43.22	10.46	C11H1205P2	286.16	286.0160	46.17	4.23	27.96	21.65
C11H 509P1	312.13	311.9671	42.33	1.61	46.13	9.92	C11H1206P2	302.16	302.0109	43.73	4.00	31.77	20.50
C11H 601P2	216.12	215.9894	61.13	2.80	7.40	28.66	C11H1207P2	318.16	318.0058	41.53	3.80	35.20	19.47
C11H 602P2	232.12	231.9843	56.92	2.61	13.79	26.69	C11H1208P2	334.16	334.0007	39.54	3.62	38.30	18.54
C11H 603P2	248.12	247.9792	53.25	2.44	19.35	24.97	C11H1209P2	350.16	349.9957	37.73	3.45	41.12	17.69
C11H 604P2	264.12	263.9741	50.02	2.29	24.23	23.45	C11H1301P1	192.20	192.0704	68.74	6.82	8.32	16.12
C11H 605P2	280.12	279.9690	47.17	2.16	28.56	22.12	C11H1302P1	208.20	208.0653	63.46	6.29	15.37	14.88
C11H 606P2	296.11	295.9640	44.62	2.04	32.42	20.92	C11H1303P1	224.20	224.0602	58.93	5.84	21.41	13.82
C11H 607P2	312.11	311.9589	42.33	1.94	35.88	19.85	C11H1304P1	240.20	240.0551	55.01	5.46	26.64	12.90
C11H 608P2	328.11	327.9538	40.27	1.84	39.01	18.88	C11H1305P1	256.20	256.0501	51.57	5.11	31.22	12.09
C11H 609P2	344.11	343.9487	38.40	1.76	41.85	18.00	C11H1306P1	272.20	272.0450	48.54	4.81	35.27	11.38
C11H 701P1	186.15	186.0235	70.98	3.79	8.59	16.64	C11H1307P1	288.20	288.0399	45.84	4.55	38.86	10.75
C11H 702P1	202.15	202.0184	65.36	3.49	15.83	15.32	C11H1308P1	304.20	304.0348	43.43	4.31	42.08	10.18
C11H 703P1	218.15	218.0133	60.56	3.23	22.00	14.20	C11H1309P1	320.19	320.0297	41.26	4.09	44.97	9.67
C11H 704P1	234.15	234.0082	56.43	3.01	27.33	13.23	C11H1401P2	224.18	224.0520	58.94	6.29	7.14	27.63
C11H 705P1	250.15	250.0031	52.82	2.82	31.98	12.38	C11H1402P2	240.18	240.0469	55.01	5.88	13.32	25.79
C11H 706P1	266.15	265.9980	49.64	2.65	36.07	11.64	C11H1403P2	256.18	256.0418	51.57	5.51	18.74	24.18
C11H 707P1	282.15	281.9929	46.83	2.50	39.69	10.98	C11H1404P2	272.18	272.0367	48.54	5.18	23.51	22.76

113

C11H1405P2	288.18	288.0316	45.85	4.90	27.76	21.50	C11H2102P1	216.26	216.1279	61.09	9.79	14.80	14.32
C11H1406P2	304.18	304.0266	43.44	4.64	31.56	20.37	C11H2103P1	232.26	232.1228	56.89	9.11	20.67	13.34
C11H1407P2	320.18	320.0215	41.27	4.41	34.98	19.35	C11H2104P1	248.26	248.1177	53.22	8.53	25.78	12.48
C11H1408P2	336.18	336.0164	39.30	4.20	38.07	18.43	C11H2105P1	264.26	264.1127	50.00	8.01	30.27	11.72
C11H1409P2	352.18	352.0113	37.52	4.01	40.89	17.59	C11H2106P1	280.26	280.1076	47.14	7.55	34.25	11.05
C11H1501P1	194.22	194.0860	68.03	7.78	8.24	15.95	C11H2107P1	296.26	296.1025	44.60	7.14	37.80	10.45
C11H1502P1	210.21	210.0810	62.85	7.19	15.22	14.73	C11H2108P1	312.26	312.0974	42.31	6.78	40.99	9.92
C11H1503P1	226.21	226.0759	58.41	6.68	21.22	13.69	C11H2109P1	328.26	328.0923	40.25	6.45	43.87	9.44
C11H1504P1	242.21	242.0708	54.55	6.24	26.42	12.79	C11H2201P2	232.24	232.1146	56.89	9.55	6.89	26.67
C11H1505P1	258.21	258.0657	51.17	5.86	30.98	12.00	C11H2202P2	248.24	248.1095	53.22	8.93	12.89	24.95
C11H1506P1	274.21	274.0606	48.18	5.51	35.01	11.30	C11H2203P2	264.24	264.1044	50.00	8.39	18.16	23.44
C11H1507P1	290.21	290.0555	45.53	5.21	38.59	10.67	C11H2204P2	280.24	280.0993	47.15	7.91	22.84	22.10
C11H1508P1	306.21	306.0504	43.15	4.94	41.80	10.12	C11H2205P2	296.24	296.0942	44.60	7.49	27.00	20.91
C11H1509P1	322.21	322.0454	41.01	4.69	44.69	9.61	C11H2206P2	312.24	312.0892	42.31	7.10	30.74	19.84
C11H1601P2	226.20	226.0676	58.41	7.13	7.07	27.39	C11H2207P2	328.24	328.0841	40.25	6.76	34.12	18.87
C11H1602P2	242.20	242.0625	54.55	6.66	13.21	25.58	C11H2208P2	344.24	344.0790	38.38	6.44	37.18	18.00
C11H1603P2	258.20	258.0575	51.17	6.25	18.59	23.99	C11H2209P2	360.24	360.0739	36.68	6.16	39.97	17.20
C11H1604P2	274.20	274.0524	48.19	5.88	23.34	22.59	C11H2301P1	202.28	202.1486	65.32	11.46	7.91	15.31
C11H1605P2	290.19	290.0473	45.53	5.56	27.57	21.35	C11H2302P1	218.28	218.1436	60.53	10.62	14.66	14.19
C11H1606P2	306.19	306.0422	43.15	5.27	31.35	20.23	C11H2303P1	234.28	234.1385	56.40	9.90	20.49	13.22
C11H1607P2	322.19	322.0371	41.01	5.01	34.76	19.23	C11H2304P1	250.28	250.1334	52.79	9.26	25.57	12.38
C11H1608P2	338.19	338.0320	39.07	4.77	37.85	18.32	C11H2305P1	266.28	266.1283	49.62	8.71	30.04	11.63
C11H1609P2	354.19	354.0269	37.30	4.55	40.65	17.49	C11H2306P1	282.28	282.1232	46.81	8.21	34.01	10.97
C11H1701P1	196.23	196.1017	67.33	8.73	8.15	15.78	C11H2307P1	298.28	298.1181	44.30	7.77	37.55	10.38
C11H1702P1	212.23	212.0966	62.25	8.07	15.08	14.59	C11H2308P1	314.27	314.1130	42.04	7.38	40.73	9.86
C11H1703P1	228.23	228.0915	57.89	7.51	21.03	13.57	C11H2309P1	330.27	330.1080	40.00	7.02	43.60	9.38
C11H1704P1	244.23	244.0864	54.10	7.02	26.20	12.68	C11H2401P2	234.26	234.1302	56.40	10.33	6.83	26.44
C11H1705P1	260.23	260.0814	50.77	6.58	30.74	11.90	C11H2402P2	250.26	250.1251	52.79	9.67	12.79	24.75
C11H1706P1	276.23	276.0763	47.83	6.20	34.75	11.21	C11H2403P2	266.26	266.1201	49.62	9.09	18.03	23.27
C11H1707P1	292.23	292.0712	45.21	5.86	38.32	10.60	C11H2404P2	282.26	282.1150	46.81	8.57	22.67	21.95
C11H1708P1	308.23	308.0661	42.87	5.56	41.53	10.05	C11H2405P2	298.26	298.1099	44.30	8.11	26.82	20.77
C11H1709P1	324.23	324.0610	40.75	5.29	44.41	9.55	C11H2406P2	314.26	314.1048	42.04	7.70	30.55	19.71
C11H1801P2	228.21	228.0833	57.89	7.95	7.01	27.14	C11H2407P2	330.26	330.0997	40.01	7.32	33.91	18.76
C11H1802P2	244.21	244.0782	54.10	7.43	13.10	25.37	C11H2408P2	346.26	346.0946	38.16	6.99	36.97	17.89
C11H1803P2	260.21	260.0731	50.78	6.97	18.45	23.81	C11H2409P2	362.26	362.0895	36.47	6.68	39.75	17.10
C11H1804P2	276.21	276.0680	47.83	6.57	23.17	22.43	C11H2501P1	204.30	204.1643	64.67	12.33	7.83	15.16
C11H1805P2	292.21	292.0629	45.21	6.21	27.38	21.20	C11H2502P1	220.29	220.1592	59.98	11.44	14.53	14.06
C11H1806P2	308.21	308.0579	42.87	5.89	31.15	20.10	C11H2503P1	236.29	236.1541	55.91	10.66	20.31	13.11
C11H1807P2	324.21	324.0528	40.75	5.60	34.54	19.11	C11H2504P1	252.29	252.1490	52.37	9.99	25.37	12.28
C11H1808P2	340.21	340.0477	38.84	5.33	37.62	18.21	C11H2505P1	268.29	268.1439	49.25	9.39	29.82	11.54
C11H1809P2	356.21	356.0426	37.09	5.09	40.42	17.39	C11H2506P1	284.29	284.1389	46.47	8.86	33.77	10.90
C11H1901P1	198.25	198.1173	66.65	9.66	8.07	15.62	C11H2507P1	300.29	300.1338	44.00	8.39	37.30	10.31
C11H1902P1	214.25	214.1123	61.67	8.94	14.94	14.46	C11H2508P1	316.29	316.1287	41.77	7.97	40.47	9.79
C11H1903P1	230.25	230.1072	57.38	8.32	20.85	13.45	C11H2509P1	332.29	332.1236	39.76	7.58	43.33	9.32
C11H1904P1	246.25	246.1021	53.65	7.78	25.99	12.58	C11H2601P2	236.28	236.1459	55.92	11.09	6.77	26.22
C11H1905P1	262.24	262.0970	50.38	7.30	30.50	11.81	C11H2602P2	252.28	252.1408	52.37	10.39	12.68	24.56
C11H1906P1	278.24	278.0919	47.48	6.88	34.50	11.13	C11H2603P2	268.28	268.1357	49.25	9.77	17.89	23.09
C11H1907P1	294.24	294.0868	44.90	6.51	38.06	10.53	C11H2604P2	284.28	284.1306	46.48	9.22	22.51	21.79
C11H1908P1	310.24	310.0817	42.59	6.17	41.26	9.98	C11H2605P2	300.27	300.1255	44.00	8.73	26.64	20.63
C11H1909P1	326.24	326.0767	40.50	5.87	44.14	9.49	C11H2606P2	316.27	316.1205	41.77	8.29	30.35	19.59
C11H2001P2	230.23	230.0989	57.39	8.76	6.95	26.91	C11H2607P2	332.27	332.1154	39.76	7.89	33.71	18.64
C11H2002P2	246.23	246.0938	53.66	8.19	13.00	25.16	C11H2608P2	348.27	348.1103	37.94	7.52	36.75	17.79
C11H2003P2	262.23	262.0888	50.38	7.69	18.30	23.62	C11H2609P2	364.27	364.1052	36.27	7.19	39.53	17.01
C11H2004P2	278.23	278.0837	47.49	7.25	23.00	22.27	C12H 601P2	228.13	227.9894	63.18	2.65	7.01	27.15
C11H2005P2	294.23	294.0786	44.91	6.85	27.19	21.05	C12H 602P2	244.13	243.9843	59.04	2.48	13.11	25.38
C11H2006P2	310.23	310.0735	42.59	6.50	30.94	19.97	C12H 603P2	260.13	259.9792	55.41	2.32	18.45	23.81
C11H2007P2	326.23	326.0684	40.50	6.18	34.33	18.99	C12H 604P2	276.13	275.9741	52.20	2.19	23.18	22.43
C11H2008P2	342.22	342.0633	38.61	5.89	37.40	18.10	C12H 605P2	292.13	291.9690	49.34	2.07	27.38	21.21
C11H2009P2	358.22	358.0582	36.88	5.63	40.20	17.29	C12H 606P2	308.13	307.9640	46.78	1.96	31.15	20.10
C11H2101P1	200.26	200.1330	65.97	10.57	7.99	15.47	C12H 607P2	324.13	323.9589	44.47	1.87	34.55	19.11

C12H 608P2	340.12	339.9538	42.38	1.78	37.63	18.21
C12H 609P2	356.12	355.9487	40.47	1.70	40.43	17.39
C12H 701P1	198.16	198.0235	72.74	3.56	8.07	15.63
C12H 702P1	214.16	214.0184	67.30	3.29	14.94	14.46
C12H 703P1	230.16	230.0133	62.62	3.07	20.85	13.46
C12H 704P1	246.16	246.0082	58.55	2.87	26.00	12.58
C12H 705P1	262.16	262.0031	54.98	2.69	30.51	11.81
C12H 706P1	278.16	277.9980	51.82	2.54	34.51	11.14
C12H 707P1	294.16	293.9929	49.00	2.40	38.07	10.53
C12H 708P1	310.16	309.9878	46.47	2.27	41.27	9.99
C12H 709P1	326.16	325.9828	44.19	2.16	44.15	9.50
C12H 801P2	230.14	230.0050	62.63	3.50	6.95	26.92
C12H 802P2	246.14	246.0000	58.56	3.28	13.00	25.17
C12H 803P2	262.14	261.9949	54.98	3.08	18.31	23.63
C12H 804P2	278.14	277.9898	51.82	2.90	23.01	22.27
C12H 805P2	294.14	293.9847	49.00	2.74	27.20	21.06
C12H 806P2	310.14	309.9796	46.47	2.60	30.95	19.97
C12H 807P2	326.14	325.9745	44.19	2.47	34.34	18.99
C12H 808P2	342.14	341.9694	42.13	2.36	37.41	18.11
C12H 809P2	358.14	357.9644	40.25	2.25	40.21	17.30
C12H 901P1	200.18	200.0391	72.00	4.53	7.99	15.47
C12H 902P1	216.18	216.0340	66.67	4.20	14.80	14.33
C12H 903P1	232.18	232.0289	62.08	3.91	20.67	13.34
C12H 904P1	248.18	248.0238	58.08	3.66	25.79	12.48
C12H 905P1	264.18	264.0188	54.56	3.43	30.28	11.72
C12H 906P1	280.18	280.0137	51.44	3.24	34.26	11.06
C12H 907P1	296.18	296.0086	48.67	3.06	37.81	10.46
C12H 908P1	312.17	312.0035	46.17	2.91	41.00	9.92
C12H 909P1	328.17	327.9984	43.92	2.76	43.88	9.44
C12H1001P2	232.16	232.0207	62.08	4.34	6.89	26.68
C12H1002P2	248.16	248.0156	58.08	4.06	12.89	24.96
C12H1003P2	264.16	264.0105	54.56	3.82	18.17	23.45
C12H1004P2	280.16	280.0054	51.45	3.60	22.84	22.11
C12H1005P2	296.16	296.0003	48.67	3.40	27.01	20.92
C12H1006P2	312.16	311.9953	46.17	3.23	30.75	19.84
C12H1007P2	328.16	327.9902	43.92	3.07	34.13	18.88
C12H1008P2	344.16	343.9851	41.88	2.93	37.19	18.00
C12H1009P2	360.16	359.9800	40.02	2.80	39.98	17.20
C12H1101P1	202.19	202.0547	71.28	5.48	7.91	15.32
C12H1102P1	218.19	218.0497	66.06	5.08	14.67	14.20
C12H1103P1	234.19	234.0446	61.54	4.73	20.50	13.23
C12H1104P1	250.19	250.0395	57.61	4.43	25.58	12.38
C12H1105P1	266.19	266.0344	54.15	4.17	30.05	11.64
C12H1106P1	282.19	282.0293	51.08	3.93	34.02	10.98
C12H1107P1	298.19	298.0242	48.34	3.72	37.56	10.39
C12H1108P1	314.19	314.0191	45.87	3.53	40.74	9.86
C12H1109P1	330.19	330.0141	43.65	3.36	43.61	9.38
C12H1201P2	234.18	234.0363	61.55	5.17	6.83	26.45
C12H1202P2	250.18	250.0313	57.61	4.83	12.79	24.76
C12H1203P2	266.18	266.0262	54.15	4.54	18.03	23.27
C12H1204P2	282.17	282.0211	51.08	4.29	22.68	21.95
C12H1205P2	298.17	298.0160	48.34	4.06	26.83	20.78
C12H1206P2	314.17	314.0109	45.88	3.85	30.56	19.72
C12H1207P2	330.17	330.0058	43.65	3.66	33.92	18.76
C12H1208P2	346.17	346.0007	41.64	3.49	36.97	17.90
C12H1209P2	362.17	361.9957	39.80	3.34	39.76	17.10
C12H1301P1	204.21	204.0704	70.58	6.42	7.83	15.17
C12H1302P1	220.21	220.0653	65.45	5.95	14.53	14.07
C12H1303P1	236.21	236.0602	61.02	5.55	20.32	13.11
C12H1304P1	252.21	252.0551	57.15	5.20	25.37	12.28

C12H1305P1	268.21	268.0501	53.74	4.89	29.83	11.55
C12H1306P1	284.21	284.0450	50.71	4.61	33.78	10.90
C12H1307P1	300.21	300.0399	48.01	4.36	37.31	10.32
C12H1308P1	316.21	316.0348	45.58	4.14	40.48	9.80
C12H1309P1	332.21	332.0297	43.39	3.94	43.34	9.32
C12H1401P2	236.19	236.0520	61.02	5.97	6.77	26.23
C12H1402P2	252.19	252.0469	57.15	5.60	12.69	24.56
C12H1403P2	268.19	268.0418	53.74	5.26	17.90	23.10
C12H1404P2	284.19	284.0367	50.72	4.97	22.52	21.80
C12H1405P2	300.19	300.0316	48.01	4.70	26.65	20.64
C12H1406P2	316.19	316.0266	45.58	4.46	30.36	19.59
C12H1407P2	332.19	332.0215	43.39	4.25	33.71	18.65
C12H1408P2	348.19	348.0164	41.40	4.05	36.76	17.79
C12H1409P2	364.19	364.0113	39.58	3.87	39.54	17.01
C12H1501P1	206.23	206.0860	69.89	7.33	7.76	15.02
C12H1502P1	222.23	222.0810	64.86	6.80	14.40	13.94
C12H1503P1	238.23	238.0759	60.50	6.35	20.15	13.00
C12H1504P1	254.22	254.0708	56.70	5.95	25.17	12.18
C12H1505P1	270.22	270.0657	53.34	5.60	29.60	11.46
C12H1506P1	286.22	286.0606	50.36	5.28	33.54	10.82
C12H1507P1	302.22	302.0555	47.69	5.00	37.06	10.25
C12H1508P1	318.22	318.0504	45.29	4.75	40.22	9.73
C12H1509P1	334.22	334.0454	43.13	4.52	43.08	9.27
C12H1601P2	238.21	238.0676	60.51	6.77	6.72	26.01
C12H1602P2	254.21	254.0625	56.70	6.34	12.59	24.37
C12H1603P2	270.21	270.0575	53.34	5.97	17.76	22.93
C12H1604P2	286.21	286.0524	50.36	5.63	22.36	21.64
C12H1605P2	302.21	302.0473	47.69	5.34	26.47	20.50
C12H1606P2	318.21	318.0422	45.30	5.07	30.17	19.47
C12H1607P2	334.20	334.0371	43.13	4.83	33.51	18.54
C12H1608P2	350.20	350.0320	41.16	4.61	36.55	17.69
C12H1609P2	366.20	366.0269	39.36	4.40	39.32	16.92
C12H1701P1	208.24	208.1017	69.21	8.23	7.68	14.87
C12H1702P1	224.24	224.0966	64.28	7.64	14.27	13.81
C12H1703P1	240.24	240.0915	60.00	7.13	19.98	12.89
C12H1704P1	256.24	256.0864	56.25	6.69	24.98	12.09
C12H1705P1	272.24	272.0814	52.94	6.29	29.38	11.38
C12H1706P1	288.24	288.0763	50.00	5.94	33.30	10.75
C12H1707P1	304.24	304.0712	47.38	5.63	36.81	10.18
C12H1708P1	320.24	320.0661	45.01	5.35	39.97	9.67
C12H1709P1	336.24	336.0610	42.87	5.10	42.83	9.21
C12H1801P2	240.22	240.0833	60.00	7.55	6.66	25.79
C12H1802P2	256.22	256.0782	56.25	7.08	12.49	24.18
C12H1803P2	272.22	272.0731	52.95	6.66	17.63	22.76
C12H1804P2	288.22	288.0680	50.01	6.29	22.20	21.49
C12H1805P2	304.22	304.0629	47.38	5.96	26.30	20.36
C12H1806P2	320.22	320.0579	45.01	5.67	29.98	19.35
C12H1807P2	336.22	336.0528	42.87	5.40	33.31	18.42
C12H1808P2	352.22	352.0477	40.92	5.15	36.34	17.59
C12H1809P2	368.22	368.0426	39.14	4.93	39.11	16.82
C12H1901P1	210.26	210.1173	68.55	9.11	7.61	14.73
C12H1902P1	226.26	226.1123	63.70	8.46	14.14	13.69
C12H1903P1	242.26	242.1072	59.50	7.91	19.81	12.79
C12H1904P1	258.26	258.1021	55.81	7.42	24.78	11.99
C12H1905P1	274.26	274.0970	52.55	6.98	29.17	11.29
C12H1906P1	290.26	290.0919	49.66	6.60	33.07	10.67
C12H1907P1	306.25	306.0868	47.06	6.25	36.57	10.11
C12H1908P1	322.25	322.0817	44.73	5.94	39.72	9.61
C12H1909P1	338.25	338.0767	42.61	5.66	42.57	9.16
C12H2001P2	242.24	242.0989	59.50	8.32	6.60	25.57

C12H2002P2	258.24	258.0938	55.81	7.81	12.39	23.99
C12H2003P2	274.24	274.0888	52.56	7.35	17.50	22.59
C12H2004P2	290.24	290.0837	49.66	6.95	22.05	21.34
C12H2005P2	306.24	306.0786	47.07	6.58	26.12	20.23
C12H2006P2	322.24	322.0735	44.73	6.26	29.79	19.22
C12H2007P2	338.24	338.0684	42.61	5.96	33.11	18.31
C12H2008P2	354.24	354.0633	40.69	5.69	36.13	17.49
C12H2009P2	370.24	370.0582	38.93	5.45	38.89	16.73
C12H2101P1	212.27	212.1330	67.90	9.97	7.54	14.59
C12H2102P1	228.27	228.1279	63.14	9.27	14.02	13.57
C12H2103P1	244.27	244.1228	59.01	8.67	19.65	12.68
C12H2104P1	260.27	260.1177	55.38	8.13	24.59	11.90
C12H2105P1	276.27	276.1127	52.17	7.66	28.96	11.21
C12H2106P1	292.27	292.1076	49.32	7.24	32.84	10.60
C12H2107P1	308.27	308.1025	46.76	6.87	36.33	10.05
C12H2108P1	324.27	324.0974	44.45	6.53	39.47	9.55
C12H2109P1	340.27	340.0923	42.36	6.22	42.32	9.10
C12H2201P2	244.26	244.1146	59.01	9.08	6.55	25.36
C12H2202P2	260.26	260.1095	55.38	8.52	12.30	23.80
C12H2203P2	276.25	276.1044	52.17	8.03	17.37	22.42
C12H2204P2	292.25	292.0993	49.32	7.59	21.90	21.20
C12H2205P2	308.25	308.0942	46.76	7.19	25.95	20.10
C12H2206P2	324.25	324.0892	44.45	6.84	29.61	19.10
C12H2207P2	340.25	340.0841	42.36	6.52	32.92	18.21
C12H2208P2	356.25	356.0790	40.46	6.22	35.93	17.39
C12H2209P2	372.25	372.0739	38.72	5.96	38.68	16.64
C12H2301P1	214.29	214.1486	67.26	10.82	7.47	14.45
C12H2302P1	230.29	230.1436	62.59	10.07	13.90	13.45
C12H2303P1	246.29	246.1385	58.52	9.41	19.49	12.58
C12H2304P1	262.29	262.1334	54.95	8.84	24.40	11.81
C12H2305P1	278.29	278.1283	51.79	8.33	28.75	11.13
C12H2306P1	294.29	294.1232	48.98	7.88	32.62	10.53
C12H2307P1	310.29	310.1181	46.45	7.47	36.09	9.98
C12H2308P1	326.29	326.1130	44.17	7.11	39.23	9.49
C12H2309P1	342.29	342.1080	42.11	6.77	42.07	9.05
C12H2401P2	246.27	246.1302	58.53	9.82	6.50	25.15
C12H2402P2	262.27	262.1251	54.96	9.22	12.20	23.62
C12H2403P2	278.27	278.1201	51.80	8.69	17.25	22.26
C12H2404P2	294.27	294.1150	48.98	8.22	21.75	21.05
C12H2405P2	310.27	310.1099	46.45	7.80	25.78	19.97
C12H2406P2	326.27	326.1048	44.18	7.41	29.42	18.99
C12H2407P2	342.27	342.0997	42.11	7.07	32.72	18.10
C12H2408P2	358.27	358.0946	40.23	6.75	35.73	17.29
C12H2409P2	374.27	374.0895	38.51	6.46	38.47	16.55
C12H2501P1	216.31	216.1643	66.63	11.65	7.40	14.32
C12H2502P1	232.31	232.1592	62.04	10.85	13.77	13.33
C12H2503P1	248.31	248.1541	58.05	10.15	19.33	12.47
C12H2504P1	264.30	264.1490	54.53	9.53	24.21	11.72
C12H2505P1	280.30	280.1439	51.42	8.99	28.54	11.05
C12H2506P1	296.30	296.1389	48.64	8.50	32.40	10.45
C12H2507P1	312.30	312.1338	46.15	8.07	35.86	9.92
C12H2508P1	328.30	328.1287	43.90	7.68	38.99	9.43
C12H2509P1	344.30	344.1236	41.86	7.32	41.82	9.00
C12H2601P2	248.29	248.1459	58.05	10.56	6.44	24.95
C12H2602P2	264.29	264.1408	54.54	9.92	12.11	23.44
C12H2603P2	280.29	280.1357	51.42	9.35	17.12	22.10
C12H2604P2	296.29	296.1306	48.65	8.85	21.60	20.91
C12H2605P2	312.29	312.1255	46.15	8.39	25.62	19.84
C12H2606P2	328.29	328.1205	43.91	7.98	29.24	18.87
C12H2607P2	344.28	344.1154	41.86	7.61	32.53	17.99

C12H2608P2	360.28	360.1103	40.01	7.27	35.53	17.19
C12H2609P2	376.28	376.1052	38.30	6.96	38.27	16.46
C12H2701P1	218.32	218.1799	66.02	12.47	7.33	14.19
C12H2702P1	234.32	234.1749	61.51	11.61	13.66	13.22
C12H2703P1	250.32	250.1698	57.58	10.87	19.17	12.37
C12H2704P1	266.32	266.1647	54.12	10.22	24.03	11.63
C12H2705P1	282.32	282.1596	51.05	9.64	28.34	10.97
C12H2706P1	298.32	298.1545	48.32	9.12	32.18	10.38
C12H2707P1	314.32	314.1494	45.86	8.66	35.63	9.85
C12H2708P1	330.32	330.1443	43.63	8.24	38.75	9.38
C12H2709P1	346.32	346.1393	41.62	7.86	41.58	8.94
C12H2801P2	250.30	250.1615	57.58	11.28	6.39	24.75
C12H2802P2	266.30	266.1564	54.12	10.60	12.02	23.26
C12H2803P2	282.30	282.1514	51.06	10.00	17.00	21.94
C12H2804P2	298.30	298.1463	48.32	9.46	21.45	20.77
C12H2805P2	314.30	314.1412	45.86	8.98	25.45	19.71
C12H2806P2	330.30	330.1361	43.64	8.54	29.06	18.75
C12H2807P2	346.30	346.1310	41.62	8.15	32.34	17.89
C12H2808P2	362.30	362.1259	39.78	7.79	35.33	17.10
C12H2809P2	378.30	378.1208	38.10	7.46	38.06	16.38
C13H 601P2	240.14	239.9894	65.02	2.52	6.66	25.80
C13H 602P2	256.14	255.9843	60.96	2.36	12.49	24.19
C13H 603P2	272.14	271.9792	57.38	2.22	17.64	22.76
C13H 604P2	288.14	287.9741	54.19	2.10	22.21	21.50
C13H 605P2	304.14	303.9690	51.34	1.99	26.30	20.37
C13H 606P2	320.14	319.9640	48.77	1.89	29.99	19.35
C13H 607P2	336.14	335.9589	46.45	1.80	33.32	18.43
C13H 608P2	352.14	351.9538	44.34	1.72	36.35	17.59
C13H 609P2	368.13	367.9487	42.42	1.64	39.11	16.83
C13H 701P1	210.17	210.0235	74.29	3.36	7.61	14.74
C13H 702P1	226.17	226.0184	69.04	3.12	14.15	13.69
C13H 703P1	242.17	242.0133	64.48	2.91	19.82	12.79
C13H 704P1	258.17	258.0082	60.48	2.73	24.79	12.00
C13H 705P1	274.17	274.0031	56.95	2.57	29.18	11.30
C13H 706P1	290.17	289.9980	53.81	2.43	33.08	10.67
C13H 707P1	306.17	305.9929	51.00	2.30	36.58	10.12
C13H 708P1	322.17	321.9878	48.47	2.19	39.73	9.61
C13H 709P1	338.17	337.9828	46.17	2.09	42.58	9.16
C13H 801P2	242.16	242.0050	64.48	3.33	6.61	25.58
C13H 802P2	258.16	258.0000	60.48	3.12	12.40	24.00
C13H 803P2	274.15	273.9949	56.96	2.94	17.51	22.60
C13H 804P2	290.15	289.9898	53.81	2.78	22.06	21.35
C13H 805P2	306.15	305.9847	51.00	2.63	26.13	20.23
C13H 806P2	322.15	321.9796	48.47	2.50	29.80	19.23
C13H 807P2	338.15	337.9745	46.18	2.38	33.12	18.32
C13H 808P2	354.15	353.9694	44.09	2.28	36.14	17.49
C13H 809P2	370.15	369.9644	42.18	2.18	38.90	16.74
C13H 901P1	212.19	212.0391	73.59	4.28	7.54	14.60
C13H 902P1	228.19	228.0340	68.43	3.98	14.02	13.57
C13H 903P1	244.19	244.0289	63.94	3.72	19.66	12.68
C13H 904P1	260.19	260.0238	60.01	3.49	24.60	11.90
C13H 905P1	276.19	276.0188	56.54	3.28	28.96	11.21
C13H 906P1	292.19	292.0137	53.44	3.10	32.85	10.60
C13H 907P1	308.19	308.0086	50.67	2.94	36.34	10.05
C13H 908P1	324.19	324.0035	48.17	2.80	39.48	9.55
C13H 909P1	340.19	339.9984	45.90	2.67	42.33	9.10
C13H1001P2	244.17	244.0207	63.95	4.13	6.55	25.37
C13H1002P2	260.17	260.0156	60.02	3.87	12.30	23.81
C13H1003P2	276.17	276.0105	56.54	3.65	17.38	22.43
C13H1004P2	292.17	292.0054	53.44	3.45	21.90	21.20

C13H10O5P2	308.17	308.0003	50.67	3.27	25.96	20.10
C13H10O6P2	324.17	323.9953	48.17	3.11	29.61	19.11
C13H10O7P2	340.17	339.9902	45.90	2.96	32.92	18.21
C13H10O8P2	356.17	355.9851	43.84	2.83	35.94	17.39
C13H10O9P2	372.17	371.9800	41.96	2.71	38.69	16.65
C13H11O1P1	214.21	214.0547	72.89	5.18	7.47	14.46
C13H11O2P1	230.21	230.0497	67.83	4.82	13.90	13.45
C13H11O3P1	246.20	246.0446	63.42	4.50	19.50	12.58
C13H11O4P1	262.20	262.0395	59.55	4.23	24.41	11.81
C13H11O5P1	278.20	278.0344	56.13	3.99	28.75	11.13
C13H11O6P1	294.20	294.0293	53.07	3.77	32.63	10.53
C13H11O7P1	310.20	310.0242	50.34	3.57	36.10	9.99
C13H11O8P1	326.20	326.0191	47.87	3.40	39.24	9.50
C13H11O9P1	342.20	342.0141	45.63	3.24	42.08	9.05
C13H12O1P2	246.19	246.0363	63.43	4.91	6.50	25.16
C13H12O2P2	262.19	262.0313	59.55	4.61	12.20	23.63
C13H12O3P2	278.19	278.0262	56.13	4.35	17.25	22.27
C13H12O4P2	294.19	294.0211	53.08	4.11	21.75	21.06
C13H12O5P2	310.19	310.0160	50.34	3.90	25.79	19.97
C13H12O6P2	326.18	326.0109	47.87	3.71	29.43	18.99
C13H12O7P2	342.18	342.0058	45.63	3.53	32.73	18.10
C13H12O8P2	358.18	358.0007	43.59	3.38	35.73	17.29
C13H12O9P2	374.18	373.9957	41.73	3.23	38.48	16.56
C13H13O1P1	216.22	216.0704	72.22	6.06	7.40	14.33
C13H13O2P1	232.22	232.0653	67.24	5.64	13.78	13.34
C13H13O3P1	248.22	248.0602	62.91	5.28	19.34	12.48
C13H13O4P1	264.22	264.0551	59.10	4.96	24.22	11.72
C13H13O5P1	280.22	280.0501	55.72	4.68	28.55	11.05
C13H13O6P1	296.22	296.0450	52.71	4.42	32.41	10.46
C13H13O7P1	312.22	312.0399	50.01	4.20	35.87	9.92
C13H13O8P1	328.22	328.0348	47.57	3.99	39.00	9.44
C13H13O9P1	344.22	344.0297	45.36	3.81	41.83	9.00
C13H14O1P2	248.20	248.0520	62.91	5.69	6.45	24.96
C13H14O2P2	264.20	264.0469	59.10	5.34	12.11	23.45
C13H14O3P2	280.20	280.0418	55.73	5.04	17.13	22.11
C13H14O4P2	296.20	296.0367	52.72	4.76	21.61	20.91
C13H14O5P2	312.20	312.0316	50.01	4.52	25.62	19.84
C13H14O6P2	328.20	328.0266	47.58	4.30	29.25	18.87
C13H14O7P2	344.20	344.0215	45.36	4.10	32.54	18.00
C13H14O8P2	360.20	360.0164	43.35	3.92	35.53	17.20
C13H14O9P2	376.20	376.0113	41.51	3.75	38.28	16.47
C13H15O1P1	218.24	218.0860	71.55	6.93	7.33	14.19
C13H15O2P1	234.24	234.0810	66.66	6.45	13.66	13.22
C13H15O3P1	250.24	250.0759	62.40	6.04	19.18	12.38
C13H15O4P1	266.24	266.0708	58.65	5.68	24.04	11.63
C13H15O5P1	282.24	282.0657	55.32	5.36	28.34	10.97
C13H15O6P1	298.23	298.0606	52.36	5.07	32.19	10.39
C13H15O7P1	314.23	314.0555	49.69	4.81	35.64	9.86
C13H15O8P1	330.23	330.0504	47.28	4.58	38.76	9.38
C13H15O9P1	346.23	346.0454	45.10	4.37	41.59	8.95
C13H16O1P2	250.22	250.0676	62.40	6.45	6.39	24.76
C13H16O2P2	266.22	266.0625	58.65	6.06	12.02	23.27
C13H16O3P2	282.22	282.0575	55.33	5.71	17.01	21.95
C13H16O4P2	298.22	298.0524	52.36	5.41	21.46	20.77
C13H16O5P2	314.22	314.0473	49.69	5.13	25.46	19.71
C13H16O6P2	330.22	330.0422	47.29	4.88	29.07	18.76
C13H16O7P2	346.22	346.0371	45.10	4.66	32.35	17.89
C13H16O8P2	362.22	362.0320	43.11	4.45	35.34	17.10
C13H16O9P2	378.21	378.0269	41.28	4.26	38.07	16.38
C13H17O1P1	220.25	220.1017	70.89	7.78	7.26	14.06

C13H17O2P1	236.25	236.0966	66.09	7.25	13.54	13.11
C13H17O3P1	252.25	252.0915	61.90	6.79	19.03	12.28
C13H17O4P1	268.25	268.0864	58.21	6.39	23.86	11.55
C13H17O5P1	284.25	284.0814	54.93	6.03	28.14	10.90
C13H17O6P1	300.25	300.0763	52.00	5.71	31.97	10.32
C13H17O7P1	316.25	316.0712	49.37	5.42	35.41	9.79
C13H17O8P1	332.25	332.0661	47.00	5.16	38.52	9.32
C13H17O9P1	348.25	348.0610	44.84	4.92	41.35	8.89
C13H18O1P2	252.24	252.0833	61.90	7.19	6.34	24.56
C13H18O2P2	268.23	268.0782	58.21	6.76	11.93	23.09
C13H18O3P2	284.23	284.0731	54.94	6.38	16.89	21.79
C13H18O4P2	300.23	300.0680	52.01	6.04	21.32	20.63
C13H18O5P2	316.23	316.0629	49.38	5.74	25.30	19.59
C13H18O6P2	332.23	332.0579	47.00	5.46	28.89	18.65
C13H18O7P2	348.23	348.0528	44.84	5.21	32.16	17.79
C13H18O8P2	364.23	364.0477	42.87	4.98	35.14	17.01
C13H18O9P2	380.23	380.0426	41.07	4.77	37.87	16.29
C13H19O1P1	222.27	222.1173	70.25	8.62	7.20	13.94
C13H19O2P1	238.27	238.1123	65.53	8.04	13.43	13.00
C13H19O3P1	254.27	254.1072	61.41	7.53	18.88	12.18
C13H19O4P1	270.27	270.1021	57.77	7.09	23.68	11.46
C13H19O5P1	286.27	286.0970	54.55	6.69	27.94	10.82
C13H19O6P1	302.27	302.0919	51.66	6.34	31.76	10.25
C13H19O7P1	318.27	318.0868	49.06	6.02	35.19	9.73
C13H19O8P1	334.27	334.0817	46.71	5.73	38.29	9.27
C13H19O9P1	350.26	350.0767	44.58	5.47	41.11	8.84
C13H20O1P2	254.25	254.0989	61.41	7.93	6.29	24.36
C13H20O2P2	270.25	270.0938	57.78	7.46	11.84	22.92
C13H20O3P2	286.25	286.0888	54.55	7.04	16.77	21.64
C13H20O4P2	302.25	302.0837	51.66	6.67	21.17	20.50
C13H20O5P2	318.25	318.0786	49.06	6.33	25.14	19.47
C13H20O6P2	334.25	334.0735	46.72	6.03	28.72	18.53
C13H20O7P2	350.25	350.0684	44.58	5.76	31.98	17.69
C13H20O8P2	366.25	366.0633	42.63	5.50	34.95	16.91
C13H20O9P2	382.25	382.0582	40.85	5.27	37.67	16.21
C13H21O1P1	224.29	224.1330	69.62	9.44	7.13	13.81
C13H21O2P1	240.28	240.1279	64.98	8.81	13.32	12.89
C13H21O3P1	256.28	256.1228	60.93	8.26	18.73	12.09
C13H21O4P1	272.28	272.1177	57.35	7.77	23.50	11.38
C13H21O5P1	288.28	288.1127	54.16	7.34	27.75	10.74
C13H21O6P1	304.28	304.1076	51.32	6.96	31.55	10.18
C13H21O7P1	320.28	320.1025	48.75	6.61	34.97	9.67
C13H21O8P1	336.28	336.0974	46.43	6.29	38.06	9.21
C13H21O9P1	352.28	352.0923	44.32	6.01	40.87	8.79
C13H22O1P2	256.27	256.1146	60.93	8.65	6.24	24.17
C13H22O2P2	272.27	272.1095	57.35	8.14	11.75	22.75
C13H22O3P2	288.27	288.1044	54.17	7.69	16.65	21.49
C13H22O4P2	304.27	304.0993	51.32	7.29	21.03	20.36
C13H22O5P2	320.26	320.0942	48.75	6.92	24.98	19.34
C13H22O6P2	336.26	336.0892	46.44	6.59	28.55	18.42
C13H22O7P2	352.26	352.0841	44.33	6.30	31.79	17.59
C13H22O8P2	368.26	368.0790	42.40	6.02	34.76	16.82
C13H22O9P2	384.26	384.0739	40.63	5.77	37.47	16.12
C13H23O1P1	226.30	226.1486	69.00	10.24	7.07	13.69
C13H23O2P1	242.30	242.1436	64.44	9.57	13.21	12.78
C13H23O3P1	258.30	258.1385	60.45	8.98	18.58	11.99
C13H23O4P1	274.30	274.1334	56.92	8.45	23.33	11.29
C13H23O5P1	290.30	290.1283	53.79	7.99	27.56	10.67
C13H23O6P1	306.30	306.1232	50.98	7.57	31.34	10.11
C13H23O7P1	322.30	322.1181	48.45	7.19	34.75	9.61

C13H23O8P1	338.30	338.1130	46.16	6.85	37.84	9.16
C13H23O9P1	354.30	354.1080	44.07	6.54	40.64	8.74
C13H24O1P2	258.28	258.1302	60.45	9.37	6.19	23.98
C13H24O2P2	274.28	274.1251	56.93	8.82	11.67	22.59
C13H24O3P2	290.28	290.1201	53.79	8.33	16.54	21.34
C13H24O4P2	306.28	306.1150	50.98	7.90	20.90	20.23
C13H24O5P2	322.28	322.1099	48.45	7.51	24.82	19.22
C13H24O6P2	338.28	338.1048	46.16	7.15	28.38	18.31
C13H24O7P2	354.28	354.0997	44.07	6.83	31.61	17.49
C13H24O8P2	370.28	370.0946	42.17	6.53	34.57	16.73
C13H24O9P2	386.28	386.0895	40.42	6.26	37.28	16.04
C13H25O1P1	228.32	228.1643	68.39	11.04	7.01	13.57
C13H25O2P1	244.32	244.1592	63.91	10.31	13.10	12.68
C13H25O3P1	260.32	260.1541	59.98	9.68	18.44	11.90
C13H25O4P1	276.32	276.1490	56.51	9.12	23.16	11.21
C13H25O5P1	292.32	292.1439	53.42	8.62	27.37	10.60
C13H25O6P1	308.31	308.1389	50.64	8.17	31.14	10.05
C13H25O7P1	324.31	324.1338	48.15	7.77	34.53	9.55
C13H25O8P1	340.31	340.1287	45.88	7.40	37.61	9.10
C13H25O9P1	356.31	356.1236	43.82	7.07	40.41	8.69
C13H26O1P2	260.30	260.1459	59.99	10.07	6.15	23.80
C13H26O2P2	276.30	276.1408	56.51	9.49	11.58	22.42
C13H26O3P2	292.30	292.1357	53.42	8.97	16.42	21.19
C13H26O4P2	308.30	308.1306	50.65	8.50	20.76	20.09
C13H26O5P2	324.30	324.1255	48.15	8.08	24.67	19.10
C13H26O6P2	340.30	340.1205	45.89	7.70	28.21	18.20
C13H26O7P2	356.30	356.1154	43.82	7.36	31.43	17.39
C13H26O8P2	372.29	372.1103	41.94	7.04	34.38	16.64
C13H26O9P2	388.29	388.1052	40.21	6.75	37.08	15.95
C13H27O1P1	230.33	230.1799	67.79	11.82	6.95	13.45
C13H27O2P1	246.33	246.1749	63.39	11.05	12.99	12.57
C13H27O3P1	262.33	262.1698	59.52	10.37	18.30	11.81
C13H27O4P1	278.33	278.1647	56.10	9.78	22.99	11.13
C13H27O5P1	294.33	294.1596	53.05	9.25	27.18	10.52
C13H27O6P1	310.33	310.1545	50.32	8.77	30.93	9.98
C13H27O7P1	326.33	326.1494	47.85	8.34	34.32	9.49
C13H27O8P1	342.33	342.1443	45.61	7.95	37.39	9.05
C13H27O9P1	358.33	358.1393	43.58	7.60	40.19	8.64
C13H28O1P2	262.32	262.1615	59.53	10.76	6.10	23.62
C13H28O2P2	278.31	278.1564	56.10	10.14	11.50	22.26
C13H28O3P2	294.31	294.1514	53.05	9.59	16.31	21.05
C13H28O4P2	310.31	310.1463	50.32	9.10	20.62	19.96
C13H28O5P2	326.31	326.1412	47.85	8.65	24.52	18.98
C13H28O6P2	342.31	342.1361	45.61	8.24	28.04	18.10
C13H28O7P2	358.31	358.1310	43.58	7.88	31.26	17.29
C13H28O8P2	374.31	374.1259	41.72	7.54	34.19	16.55
C13H28O9P2	390.31	390.1208	40.01	7.23	36.89	15.87
C13H29O1P1	232.35	232.1956	67.20	12.58	6.89	13.33
C13H29O2P1	248.35	248.1905	62.87	11.77	12.88	12.47
C13H29O3P1	264.35	264.1854	59.07	11.06	18.16	11.72
C13H29O4P1	280.35	280.1803	55.70	10.43	22.83	11.05
C13H29O5P1	296.35	296.1752	52.69	9.86	26.99	10.45
C13H29O6P1	312.35	312.1702	49.99	9.36	30.73	9.92
C13H29O7P1	328.35	328.1651	47.56	8.90	34.11	9.43
C13H29O8P1	344.35	344.1600	45.35	8.49	37.17	8.99
C13H29O9P1	360.34	360.1549	43.33	8.11	39.96	8.60
C13H30O1P2	264.33	264.1772	59.07	11.44	6.05	23.44
C13H30O2P2	280.33	280.1721	55.70	10.79	11.41	22.10
C13H30O3P2	296.33	296.1670	52.69	10.20	16.20	20.90
C13H30O4P2	312.33	312.1619	49.99	9.68	20.49	19.83

C13H30O5P2	328.33	328.1568	47.56	9.21	24.36	18.87
C13H30O6P2	344.33	344.1518	45.35	8.78	27.88	17.99
C13H30O7P2	360.33	360.1467	43.33	8.39	31.08	17.19
C13H30O8P2	376.33	376.1416	41.49	8.04	34.01	16.46
C13H30O9P2	392.33	392.1365	39.80	7.71	36.70	15.79
C14H 7O1P1	222.19	222.0235	75.68	3.18	7.20	13.94
C14H 7O2P1	238.18	238.0184	70.60	2.96	13.43	13.00
C14H 7O3P1	254.18	254.0133	66.16	2.78	18.88	12.19
C14H 7O4P1	270.18	270.0082	62.24	2.61	23.69	11.46
C14H 7O5P1	286.18	286.0031	58.76	2.47	27.95	10.82
C14H 7O6P1	302.18	301.9980	55.65	2.33	31.77	10.25
C14H 7O7P1	318.18	317.9929	52.85	2.22	35.20	9.73
C14H 7O8P1	334.18	333.9878	50.32	2.11	38.30	9.27
C14H 7O9P1	350.18	349.9828	48.02	2.01	41.12	8.85
C14H 8O1P2	254.17	254.0050	66.16	3.17	6.29	24.37
C14H 8O2P2	270.17	270.0000	62.24	2.98	11.84	22.93
C14H 8O3P2	286.17	285.9949	58.76	2.82	16.77	21.65
C14H 8O4P2	302.17	301.9898	55.65	2.67	21.18	20.50
C14H 8O5P2	318.16	317.9847	52.85	2.53	25.14	19.47
C14H 8O6P2	334.16	333.9796	50.32	2.41	28.73	18.54
C14H 8O7P2	350.16	349.9745	48.02	2.30	31.98	17.69
C14H 8O8P2	366.16	365.9694	45.92	2.20	34.96	16.92
C14H 8O9P2	382.16	381.9644	44.00	2.11	37.68	16.21
C14H 9O1P1	224.20	224.0391	75.00	4.05	7.14	13.82
C14H 9O2P1	240.20	240.0340	70.01	3.78	13.32	12.89
C14H 9O3P1	256.20	256.0289	65.63	3.54	18.73	12.09
C14H 9O4P1	272.20	272.0238	61.78	3.33	23.51	11.38
C14H 9O5P1	288.20	288.0188	58.35	3.15	27.76	10.75
C14H 9O6P1	304.20	304.0137	55.28	2.98	31.56	10.18
C14H 9O7P1	320.20	320.0086	52.52	2.83	34.98	9.67
C14H 9O8P1	336.20	336.0035	50.02	2.70	38.07	9.21
C14H 9O9P1	352.20	351.9984	47.75	2.58	40.88	8.79
C14H10O1P2	256.18	256.0207	65.64	3.93	6.25	24.18
C14H10O2P2	272.18	272.0156	61.78	3.70	11.76	22.76
C14H10O3P2	288.18	288.0105	58.35	3.50	16.66	21.50
C14H10O4P2	304.18	304.0054	55.28	3.31	21.04	20.37
C14H10O5P2	320.18	320.0003	52.52	3.15	24.98	19.35
C14H10O6P2	336.18	335.9953	50.02	3.00	28.56	18.43
C14H10O7P2	352.18	351.9902	47.75	2.86	31.80	17.59
C14H10O8P2	368.18	367.9851	45.67	2.74	34.76	16.83
C14H10O9P2	384.18	383.9800	43.77	2.62	37.48	16.12
C14H11O1P1	226.22	226.0547	74.33	4.90	7.07	13.69
C14H11O2P1	242.22	242.0497	69.42	4.58	13.21	12.79
C14H11O3P1	258.22	258.0446	65.12	4.29	18.59	12.00
C14H11O4P1	274.22	274.0395	61.32	4.04	23.34	11.30
C14H11O5P1	290.21	290.0344	57.94	3.82	27.56	10.67
C14H11O6P1	306.21	306.0293	54.91	3.62	31.35	10.12
C14H11O7P1	322.21	322.0242	52.19	3.44	34.76	9.61
C14H11O8P1	338.21	338.0191	49.72	3.28	37.84	9.16
C14H11O9P1	354.21	354.0141	47.47	3.13	40.65	8.74
C14H12O1P2	258.20	258.0363	65.13	4.68	6.20	23.99
C14H12O2P2	274.20	274.0313	61.33	4.41	11.67	22.59
C14H12O3P2	290.20	290.0262	57.95	4.17	16.54	21.35
C14H12O4P2	306.20	306.0211	54.92	3.95	20.90	20.23
C14H12O5P2	322.20	322.0160	52.19	3.75	24.83	19.23
C14H12O6P2	338.20	338.0109	49.72	3.58	28.38	18.32
C14H12O7P2	354.20	354.0058	47.48	3.41	31.62	17.49
C14H12O8P2	370.19	370.0007	45.42	3.27	34.58	16.73
C14H12O9P2	386.19	385.9957	43.54	3.13	37.29	16.04
C14H13O1P1	228.23	228.0704	73.68	5.74	7.01	13.57

C14H1302P1	244.23	244.0653	68.85	5.37	13.10	12.68
C14H1303P1	260.23	260.0602	64.62	5.04	18.44	11.90
C14H1304P1	276.23	276.0551	60.88	4.74	23.17	11.21
C14H1305P1	292.23	292.0501	57.54	4.48	27.37	10.60
C14H1306P1	308.23	308.0450	54.56	4.25	31.14	10.05
C14H1307P1	324.23	324.0399	51.86	4.04	34.54	9.55
C14H1308P1	340.23	340.0348	49.42	3.85	37.62	9.10
C14H1309P1	356.23	356.0297	47.20	3.68	40.42	8.69
C14H1401P2	260.21	260.0520	64.62	5.42	6.15	23.81
C14H1402P2	276.21	276.0469	60.88	5.11	11.58	22.43
C14H1403P2	292.21	292.0418	57.55	4.83	16.43	21.20
C14H1404P2	308.21	308.0367	54.56	4.58	20.76	20.10
C14H1405P2	324.21	324.0316	51.87	4.35	24.67	19.11
C14H1406P2	340.21	340.0266	49.43	4.15	28.22	18.21
C14H1407P2	356.21	356.0215	47.21	3.96	31.44	17.39
C14H1408P2	372.21	372.0164	45.18	3.79	34.39	16.64
C14H1409P2	388.21	388.0113	43.32	3.64	37.09	15.96
C14H1501P1	230.25	230.0860	73.03	6.57	6.95	13.45
C14H1502P1	246.25	246.0810	68.29	6.14	12.99	12.58
C14H1503P1	262.25	262.0759	64.12	5.77	18.30	11.81
C14H1504P1	278.25	278.0708	60.43	5.43	23.00	11.13
C14H1505P1	294.25	294.0657	57.15	5.14	27.19	10.53
C14H1506P1	310.25	310.0606	54.20	4.87	30.94	9.98
C14H1507P1	326.25	326.0555	51.54	4.63	34.33	9.49
C14H1508P1	342.24	342.0504	49.13	4.42	37.40	9.05
C14H1509P1	358.24	358.0454	46.94	4.22	40.19	8.65
C14H1601P2	262.23	262.0676	64.13	6.15	6.10	23.62
C14H1602P2	278.23	278.0625	60.44	5.80	11.50	22.26
C14H1603P2	294.23	294.0575	57.15	5.48	16.31	21.05
C14H1604P2	310.23	310.0524	54.20	5.20	20.63	19.97
C14H1605P2	326.23	326.0473	51.55	4.94	24.52	18.99
C14H1606P2	342.23	342.0422	49.14	4.71	28.05	18.10
C14H1607P2	358.23	358.0371	46.94	4.50	31.26	17.29
C14H1608P2	374.23	374.0320	44.93	4.31	34.20	16.55
C14H1609P2	390.23	390.0269	43.09	4.13	36.90	15.87
C14H1701P1	232.26	232.1017	72.40	7.38	6.89	13.34
C14H1702P1	248.26	248.0966	67.73	6.90	12.89	12.48
C14H1703P1	264.26	264.0915	63.63	6.48	18.16	11.72
C14H1704P1	280.26	280.0864	60.00	6.11	22.83	11.05
C14H1705P1	296.26	296.0814	56.76	5.78	27.00	10.45
C14H1706P1	312.26	312.0763	53.85	5.49	30.74	9.92
C14H1707P1	328.26	328.0712	51.23	5.22	34.12	9.44
C14H1708P1	344.26	344.0661	48.85	4.98	37.18	9.00
C14H1709P1	360.26	360.0610	46.68	4.76	39.97	8.60
C14H1801P2	264.25	264.0833	63.64	6.87	6.05	23.44
C14H1802P2	280.25	280.0782	60.00	6.47	11.42	22.10
C14H1803P2	296.25	296.0731	56.76	6.12	16.20	20.91
C14H1804P2	312.24	312.0680	53.85	5.81	20.50	19.84
C14H1805P2	328.24	328.0629	51.23	5.53	24.37	18.87
C14H1806P2	344.24	344.0579	48.85	5.27	27.89	18.00
C14H1807P2	360.24	360.0528	46.68	5.04	31.09	17.20
C14H1808P2	376.24	376.0477	44.69	4.82	34.02	16.46
C14H1809P2	392.24	392.0426	42.87	4.63	36.71	15.79
C14H1901P1	234.28	234.1173	71.78	8.17	6.83	13.22
C14H1902P1	250.28	250.1123	67.19	7.65	12.79	12.38
C14H1903P1	266.28	266.1072	63.15	7.19	18.03	11.63
C14H1904P1	282.28	282.1021	59.57	6.78	22.67	10.97
C14H1905P1	298.28	298.0970	56.38	6.42	26.82	10.38
C14H1906P1	314.28	314.0919	53.51	6.09	30.55	9.86
C14H1907P1	330.28	330.0868	50.91	5.80	33.91	9.38

C14H1908P1	346.28	346.0817	48.56	5.53	36.96	8.94
C14H1909P1	362.28	362.0767	46.42	5.29	39.75	8.55
C14H2001P2	266.26	266.0989	63.15	7.57	6.01	23.27
C14H2002P2	282.26	282.0938	59.57	7.14	11.34	21.95
C14H2003P2	298.26	298.0888	56.38	6.76	16.09	20.77
C14H2004P2	314.26	314.0837	53.51	6.41	20.36	19.71
C14H2005P2	330.26	330.0786	50.92	6.10	24.22	18.76
C14H2006P2	346.26	346.0735	48.56	5.82	27.72	17.89
C14H2007P2	362.26	362.0684	46.42	5.56	30.92	17.10
C14H2008P2	378.26	378.0633	44.46	5.33	33.84	16.38
C14H2009P2	394.26	394.0582	42.65	5.11	36.52	15.71
C14H2101P1	236.30	236.1330	71.16	8.96	6.77	13.11
C14H2102P1	252.30	252.1279	66.65	8.39	12.68	12.28
C14H2103P1	268.30	268.1228	62.68	7.89	17.89	11.54
C14H2104P1	284.29	284.1177	59.15	7.45	22.51	10.89
C14H2105P1	300.29	300.1127	56.00	7.05	26.64	10.31
C14H2106P1	316.29	316.1076	53.16	6.69	30.35	9.79
C14H2107P1	332.29	332.1025	50.60	6.37	33.70	9.32
C14H2108P1	348.29	348.0974	48.28	6.08	36.75	8.89
C14H2109P1	364.29	364.0923	46.16	5.81	39.53	8.50
C14H2201P2	268.28	268.1146	62.68	8.27	5.96	23.09
C14H2202P2	284.28	284.1095	59.15	7.80	11.26	21.79
C14H2203P2	300.28	300.1044	56.00	7.38	15.98	20.63
C14H2204P2	316.28	316.0993	53.17	7.01	20.23	19.59
C14H2205P2	332.28	332.0942	50.61	6.67	24.08	18.64
C14H2206P2	348.28	348.0892	48.28	6.37	27.56	17.79
C14H2207P2	364.27	364.0841	46.16	6.09	30.74	17.01
C14H2208P2	380.27	380.0790	44.22	5.83	33.66	16.29
C14H2209P2	396.27	396.0739	42.43	5.60	36.34	15.63
C14H2301P1	238.31	238.1486	70.56	9.73	6.71	13.00
C14H2302P1	254.31	254.1436	66.12	9.12	12.58	12.18
C14H2303P1	270.31	270.1385	62.21	8.58	17.76	11.46
C14H2304P1	286.31	286.1334	58.73	8.10	22.35	10.82
C14H2305P1	302.31	302.1283	55.62	7.67	26.46	10.25
C14H2306P1	318.31	318.1232	52.83	7.28	30.16	9.73
C14H2307P1	334.31	334.1181	50.30	6.93	33.50	9.27
C14H2308P1	350.31	350.1130	48.00	6.62	36.54	8.84
C14H2309P1	366.31	366.1080	45.91	6.33	39.31	8.46
C14H2401P2	270.29	270.1302	62.21	8.95	5.92	22.92
C14H2402P2	286.29	286.1251	58.74	8.45	11.18	21.64
C14H2403P2	302.29	302.1201	55.63	8.00	15.88	20.49
C14H2404P2	318.29	318.1150	52.83	7.60	20.11	19.46
C14H2405P2	334.29	334.1099	50.30	7.24	23.93	18.53
C14H2406P2	350.29	350.1048	48.00	6.91	27.40	17.68
C14H2407P2	366.29	366.0997	45.91	6.60	30.58	16.91
C14H2408P2	382.29	382.0946	43.99	6.33	33.48	16.20
C14H2409P2	398.29	398.0895	42.22	6.07	36.15	15.55
C14H2501P1	240.33	240.1643	69.97	10.49	6.66	12.89
C14H2502P1	256.33	256.1592	65.60	9.83	12.48	12.08
C14H2503P1	272.33	272.1541	61.75	9.25	17.63	11.37
C14H2504P1	288.33	288.1490	58.32	8.74	22.20	10.74
C14H2505P1	304.33	304.1439	55.26	8.28	26.29	10.18
C14H2506P1	320.33	320.1389	52.50	7.87	29.97	9.67
C14H2507P1	336.32	336.1338	50.00	7.49	33.30	9.21
C14H2508P1	352.32	352.1287	47.73	7.15	36.33	8.79
C14H2509P1	368.32	368.1236	45.65	6.84	39.09	8.41
C14H2601P2	272.31	272.1459	61.75	9.62	5.88	22.75
C14H2602P2	288.31	288.1408	58.32	9.09	11.10	21.49
C14H2603P2	304.31	304.1357	55.26	8.61	15.77	20.36
C14H2604P2	320.31	320.1306	52.50	8.18	19.98	19.34

C14H2605P2	336.31	336.1255	50.00	7.79	23.79	18.42
C14H2606P2	352.31	352.1205	47.73	7.44	27.25	17.58
C14H2607P2	368.31	368.1154	45.66	7.12	30.41	16.82
C14H2608P2	384.31	384.1103	43.76	6.82	33.31	16.12
C14H2609P2	400.31	400.1052	42.01	6.55	35.97	15.48
C14H2701P1	242.34	242.1799	69.39	11.23	6.60	12.78
C14H2702P1	258.34	258.1749	65.09	10.53	12.39	11.99
C14H2703P1	274.34	274.1698	61.29	9.92	17.50	11.29
C14H2704P1	290.34	290.1647	57.92	9.37	22.04	10.67
C14H2705P1	306.34	306.1596	54.89	8.88	26.11	10.11
C14H2706P1	322.34	322.1545	52.17	8.44	29.78	9.61
C14H2707P1	338.34	338.1494	49.70	8.04	33.10	9.15
C14H2708P1	354.34	354.1443	47.46	7.68	36.12	8.74
C14H2709P1	370.34	370.1393	45.41	7.35	38.88	8.36
C14H2801P2	274.33	274.1615	61.30	10.29	5.83	22.58
C14H2802P2	290.33	290.1564	57.92	9.72	11.02	21.34
C14H2803P2	306.33	306.1514	54.89	9.21	15.67	20.22
C14H2804P2	322.32	322.1463	52.17	8.76	19.86	19.22
C14H2805P2	338.32	338.1412	49.70	8.34	23.65	18.31
C14H2806P2	354.32	354.1361	47.46	7.97	27.09	17.48
C14H2807P2	370.32	370.1310	45.41	7.62	30.24	16.73
C14H2808P2	386.32	386.1259	43.53	7.31	33.13	16.04
C14H2809P2	402.32	402.1208	41.80	7.02	35.79	15.40
C14H2901P1	244.36	244.1956	68.81	11.96	6.55	12.68
C14H2902P1	260.36	260.1905	64.59	11.23	12.29	11.90
C14H2903P1	276.36	276.1854	60.85	10.58	17.37	11.21
C14H2904P1	292.36	292.1803	57.52	10.00	21.89	10.59
C14H2905P1	308.36	308.1752	54.53	9.48	25.94	10.04
C14H2906P1	324.36	324.1702	51.84	9.01	29.60	9.55
C14H2907P1	340.36	340.1651	49.41	8.59	32.91	9.10
C14H2908P1	356.36	356.1600	47.19	8.20	35.92	8.69
C14H2909P1	372.36	372.1549	45.16	7.85	38.67	8.32
C14H3001P2	276.34	276.1772	60.85	10.94	5.79	22.42
C14H3002P2	292.34	292.1721	57.52	10.34	10.95	21.19
C14H3003P2	308.34	308.1670	54.54	9.81	15.57	20.09
C14H3004P2	324.34	324.1619	51.85	9.32	19.73	19.10
C14H3005P2	340.34	340.1568	49.41	8.88	23.51	18.20
C14H3006P2	356.34	356.1518	47.19	8.49	26.94	17.38
C14H3007P2	372.34	372.1467	45.16	8.12	30.08	16.64
C14H3008P2	388.34	388.1416	43.30	7.79	32.96	15.95
C14H3009P2	404.34	404.1365	41.59	7.48	35.61	15.32
C14H3101P1	246.38	246.2112	68.25	12.68	6.49	12.57
C14H3102P1	262.38	262.2062	64.09	11.91	12.20	11.81
C14H3103P1	278.38	278.2011	60.41	11.22	17.24	11.13
C14H3104P1	294.37	294.1960	57.12	10.61	21.74	10.52
C14H3105P1	310.37	310.1909	54.18	10.07	25.77	9.98
C14H3106P1	326.37	326.1858	51.52	9.57	29.41	9.49
C14H3107P1	342.37	342.1807	49.11	9.13	32.71	9.05
C14H3108P1	358.37	358.1756	46.92	8.72	35.72	8.64
C14H3109P1	374.37	374.1706	44.92	8.35	38.46	8.27
C14H3201P2	278.36	278.1928	60.41	11.59	5.75	22.25
C14H3202P2	294.36	294.1877	57.13	10.96	10.87	21.05
C14H3203P2	310.36	310.1827	54.18	10.39	15.47	19.96
C14H3204P2	326.36	326.1776	51.53	9.88	19.61	18.98
C14H3205P2	342.36	342.1725	49.12	9.42	23.37	18.09
C14H3206P2	358.36	358.1674	46.92	9.00	26.79	17.29
C14H3207P2	374.35	374.1623	44.92	8.62	29.92	16.55
C14H3208P2	390.35	390.1572	43.08	8.26	32.79	15.87
C14H3209P2	406.35	406.1521	41.38	7.94	35.44	15.24
C15H 701P1	234.20	234.0235	76.93	3.01	6.83	13.23

C15H 702P1	250.20	250.0184	72.01	2.82	12.79	12.38
C15H 703P1	266.20	266.0133	67.68	2.65	18.03	11.64
C15H 704P1	282.19	282.0082	63.85	2.50	22.68	10.98
C15H 705P1	298.19	298.0031	60.42	2.37	26.83	10.39
C15H 706P1	314.19	313.9980	57.34	2.25	30.55	9.86
C15H 707P1	330.19	329.9929	54.56	2.14	33.92	9.38
C15H 708P1	346.19	345.9878	52.04	2.04	36.97	8.95
C15H 709P1	362.19	361.9828	49.74	1.95	39.76	8.55
C15H 801P2	266.18	266.0050	67.69	3.03	6.01	23.27
C15H 802P2	282.18	282.0000	63.85	2.86	11.34	21.95
C15H 803P2	298.18	297.9949	60.42	2.70	16.10	20.78
C15H 804P2	314.18	313.9898	57.35	2.57	20.37	19.72
C15H 805P2	330.18	329.9847	54.57	2.44	24.23	18.76
C15H 806P2	346.18	345.9796	52.05	2.33	27.73	17.89
C15H 807P2	362.17	361.9745	49.75	2.23	30.92	17.10
C15H 808P2	378.17	377.9694	47.64	2.13	33.85	16.38
C15H 809P2	394.17	393.9644	45.71	2.05	36.53	15.72
C15H 901P1	236.21	236.0391	76.27	3.84	6.77	13.11
C15H 902P1	252.21	252.0340	71.43	3.60	12.69	12.28
C15H 903P1	268.21	268.0289	67.17	3.38	17.90	11.55
C15H 904P1	284.21	284.0238	63.39	3.19	22.52	10.90
C15H 905P1	300.21	300.0188	60.01	3.02	26.65	10.32
C15H 906P1	316.21	316.0137	56.98	2.87	30.36	9.80
C15H 907P1	332.21	332.0086	54.23	2.73	33.71	9.32
C15H 908P1	348.21	348.0035	51.74	2.61	36.76	8.90
C15H 909P1	364.21	363.9984	49.47	2.49	39.54	8.50
C15H1001P2	268.19	268.0207	67.18	3.76	5.97	23.10
C15H1002P2	284.19	284.0156	63.40	3.55	11.26	21.80
C15H1003P2	300.19	300.0105	60.02	3.36	15.99	20.64
C15H1004P2	316.19	316.0054	56.98	3.19	20.24	19.59
C15H1005P2	332.19	332.0003	54.24	3.03	24.08	18.65
C15H1006P2	348.19	347.9953	51.74	2.89	27.57	17.79
C15H1007P2	364.19	363.9902	49.47	2.77	30.75	17.01
C15H1008P2	380.19	379.9851	47.39	2.65	33.67	16.29
C15H1009P2	396.19	395.9800	45.48	2.54	36.34	15.64
C15H1101P1	238.23	238.0547	75.63	4.65	6.72	13.00
C15H1102P1	254.23	254.0497	70.87	4.36	12.59	12.18
C15H1103P1	270.23	270.0446	66.67	4.10	17.76	11.46
C15H1104P1	286.23	286.0395	62.95	3.87	22.36	10.82
C15H1105P1	302.23	302.0344	59.61	3.67	26.47	10.25
C15H1106P1	318.23	318.0293	56.62	3.48	30.17	9.73
C15H1107P1	334.22	334.0242	53.91	3.32	33.51	9.27
C15H1108P1	350.22	350.0191	51.44	3.17	36.55	8.84
C15H1109P1	366.22	366.0141	49.20	3.03	39.32	8.46
C15H1201P2	270.21	270.0363	66.68	4.48	5.92	22.93
C15H1202P2	286.21	286.0313	62.95	4.23	11.18	21.64
C15H1203P2	302.21	302.0262	59.62	4.00	15.88	20.50
C15H1204P2	318.21	318.0211	56.62	3.80	20.11	19.47
C15H1205P2	334.21	334.0160	53.91	3.62	23.94	18.54
C15H1206P2	350.21	350.0109	51.45	3.45	27.41	17.69
C15H1207P2	366.21	366.0058	49.20	3.30	30.58	16.92
C15H1208P2	382.21	382.0007	47.14	3.16	33.49	16.21
C15H1209P2	398.21	397.9957	45.24	3.04	36.16	15.56
C15H1301P1	240.24	240.0704	74.99	5.45	6.66	12.89
C15H1302P1	256.24	256.0653	70.31	5.11	12.49	12.09
C15H1303P1	272.24	272.0602	66.18	4.81	17.63	11.38
C15H1304P1	288.24	288.0551	62.51	4.55	22.20	10.75
C15H1305P1	304.24	304.0501	59.22	4.31	26.29	10.18
C15H1306P1	320.24	320.0450	56.26	4.09	29.98	9.67
C15H1307P1	336.24	336.0399	53.58	3.90	33.31	9.21

120

C15H1308P1	352.24	352.0348	51.15	3.72	36.34	8.79
C15H1309P1	368.24	368.0297	48.93	3.56	39.10	8.41
C15H1401P2	272.23	272.0520	66.18	5.18	5.88	22.76
C15H1402P2	288.23	288.0469	62.51	4.90	11.10	21.49
C15H1403P2	304.22	304.0418	59.22	4.64	15.78	20.36
C15H1404P2	320.22	320.0367	56.26	4.41	19.99	19.35
C15H1405P2	336.22	336.0316	53.59	4.20	23.79	18.42
C15H1406P2	352.22	352.0266	51.15	4.01	27.25	17.59
C15H1407P2	368.22	368.0215	48.93	3.83	30.42	16.82
C15H1408P2	384.22	384.0164	46.89	3.67	33.31	16.12
C15H1409P2	400.22	400.0113	45.02	3.53	35.98	15.48
C15H1501P1	242.26	242.0860	74.37	6.24	6.60	12.79
C15H1502P1	258.26	258.0810	69.76	5.85	12.39	11.99
C15H1503P1	274.26	274.0759	65.69	5.51	17.50	11.29
C15H1504P1	290.26	290.0708	62.07	5.21	22.05	10.67
C15H1505P1	306.26	306.0657	58.83	4.94	26.12	10.11
C15H1506P1	322.26	322.0606	55.91	4.69	29.79	9.61
C15H1507P1	338.26	338.0555	53.26	4.47	33.11	9.16
C15H1508P1	354.26	354.0504	50.86	4.27	36.13	8.74
C15H1509P1	370.26	370.0454	48.66	4.08	38.89	8.37
C15H1601P2	274.24	274.0676	65.70	5.88	5.83	22.59
C15H1602P2	290.24	290.0625	62.08	5.56	11.02	21.34
C15H1603P2	306.24	306.0575	58.83	5.27	15.67	20.23
C15H1604P2	322.24	322.0524	55.91	5.00	19.86	19.22
C15H1605P2	338.24	338.0473	53.27	4.77	23.65	18.31
C15H1606P2	354.24	354.0422	50.86	4.55	27.10	17.49
C15H1607P2	370.24	370.0371	48.66	4.36	30.25	16.73
C15H1608P2	386.24	386.0320	46.65	4.18	33.14	16.04
C15H1609P2	402.24	402.0269	44.79	4.01	35.80	15.40
C15H1701P1	244.28	244.1017	73.76	7.01	6.55	12.68
C15H1702P1	260.28	260.0966	69.22	6.58	12.29	11.90
C15H1703P1	276.27	276.0915	65.21	6.20	17.37	11.21
C15H1704P1	292.27	292.0864	61.64	5.86	21.90	10.60
C15H1705P1	308.27	308.0814	58.44	5.56	25.95	10.05
C15H1706P1	324.27	324.0763	55.56	5.28	29.60	9.55
C15H1707P1	340.27	340.0712	52.95	5.04	32.91	9.10
C15H1708P1	356.27	356.0661	50.57	4.81	35.93	8.69
C15H1709P1	372.27	372.0610	48.40	4.60	38.68	8.32
C15H1801P2	276.26	276.0833	65.22	6.57	5.79	22.42
C15H1802P2	292.26	292.0782	61.65	6.21	10.95	21.20
C15H1803P2	308.26	308.0731	58.45	5.89	15.57	20.10
C15H1804P2	324.26	324.0680	55.56	5.60	19.74	19.10
C15H1805P2	340.26	340.0629	52.95	5.33	23.51	18.21
C15H1806P2	356.25	356.0579	50.57	5.09	26.95	17.39
C15H1807P2	372.25	372.0528	48.40	4.87	30.09	16.64
C15H1808P2	388.25	388.0477	46.40	4.67	32.97	15.96
C15H1809P2	404.25	404.0426	44.57	4.49	35.62	15.32
C15H1901P1	246.29	246.1173	73.15	7.78	6.50	12.58
C15H1902P1	262.29	262.1123	68.69	7.30	12.20	11.81
C15H1903P1	278.29	278.1072	64.74	6.88	17.25	11.13
C15H1904P1	294.29	294.1021	61.22	6.51	21.75	10.52
C15H1905P1	310.29	310.0970	58.06	6.17	25.78	9.98
C15H1906P1	326.29	326.0919	55.22	5.87	29.42	9.49
C15H1907P1	342.29	342.0868	52.64	5.60	32.72	9.05
C15H1908P1	358.29	358.0817	50.29	5.35	35.72	8.64
C15H1909P1	374.29	374.0767	48.14	5.12	38.47	8.28
C15H2001P2	278.27	278.0989	64.74	7.24	5.75	22.26
C15H2002P2	294.27	294.0938	61.22	6.85	10.87	21.05
C15H2003P2	310.27	310.0888	58.07	6.50	15.47	19.97
C15H2004P2	326.27	326.0837	55.22	6.18	19.61	18.99
C15H2005P2	342.27	342.0786	52.64	5.89	23.37	18.10
C15H2006P2	358.27	358.0735	50.29	5.63	26.79	17.29
C15H2007P2	374.27	374.0684	48.14	5.39	29.92	16.55
C15H2008P2	390.27	390.0633	46.16	5.17	32.80	15.87
C15H2009P2	406.27	406.0582	44.35	4.96	35.44	15.25
C15H2101P1	248.31	248.1330	72.56	8.52	6.44	12.47
C15H2102P1	264.31	264.1279	68.17	8.01	12.11	11.72
C15H2103P1	280.31	280.1228	64.28	7.55	17.12	11.05
C15H2104P1	296.31	296.1177	60.80	7.14	21.60	10.45
C15H2105P1	312.31	312.1127	57.69	6.78	25.61	9.92
C15H2106P1	328.30	328.1076	54.88	6.45	29.24	9.43
C15H2107P1	344.30	344.1025	52.33	6.15	32.53	9.00
C15H2108P1	360.30	360.0974	50.00	5.87	35.52	8.60
C15H2109P1	376.30	376.0923	47.88	5.63	38.27	8.23
C15H2201P2	280.29	280.1146	64.28	7.91	5.71	22.10
C15H2202P2	296.29	296.1095	60.81	7.48	10.80	20.91
C15H2203P2	312.29	312.1044	57.69	7.10	15.37	19.84
C15H2204P2	328.29	328.0993	54.88	6.75	19.49	18.87
C15H2205P2	344.29	344.0942	52.33	6.44	23.24	17.99
C15H2206P2	360.29	360.0892	50.01	6.15	26.64	17.19
C15H2207P2	376.29	376.0841	47.88	5.89	29.76	16.46
C15H2208P2	392.29	392.0790	45.93	5.65	32.63	15.79
C15H2209P2	408.28	408.0739	44.13	5.43	35.27	15.17
C15H2301P1	250.32	250.1486	71.97	9.26	6.39	12.37
C15H2302P1	266.32	266.1436	67.65	8.70	12.02	11.63
C15H2303P1	282.32	282.1385	63.82	8.21	17.00	10.97
C15H2304P1	298.32	298.1334	60.39	7.77	21.45	10.38
C15H2305P1	314.32	314.1283	57.32	7.38	25.45	9.85
C15H2306P1	330.32	330.1232	54.54	7.02	29.06	9.38
C15H2307P1	346.32	346.1181	52.02	6.69	32.34	8.94
C15H2308P1	362.32	362.1130	49.73	6.40	35.33	8.55
C15H2309P1	378.32	378.1080	47.62	6.13	38.06	8.19
C15H2401P2	282.31	282.1302	63.82	8.57	5.67	21.94
C15H2402P2	298.30	298.1251	60.40	8.11	10.73	20.77
C15H2403P2	314.30	314.1201	57.32	7.70	15.27	19.71
C15H2404P2	330.30	330.1150	54.55	7.32	19.38	18.75
C15H2405P2	346.30	346.1099	52.03	6.99	23.10	17.89
C15H2406P2	362.30	362.1048	49.73	6.68	26.50	17.10
C15H2407P2	378.30	378.0997	47.63	6.39	29.60	16.38
C15H2408P2	394.30	394.0946	45.69	6.14	32.46	15.71
C15H2409P2	410.30	410.0895	43.91	5.90	35.09	15.10
C15H2501P1	252.34	252.1643	71.40	9.99	6.34	12.27
C15H2502P1	268.34	268.1592	67.14	9.39	11.92	11.54
C15H2503P1	284.34	284.1541	63.36	8.86	16.88	10.89
C15H2504P1	300.34	300.1490	59.99	8.39	21.31	10.31
C15H2505P1	316.34	316.1439	56.95	7.97	25.29	9.79
C15H2506P1	332.34	332.1389	54.21	7.58	28.89	9.32
C15H2507P1	348.34	348.1338	51.72	7.23	32.15	8.89
C15H2508P1	364.34	364.1287	49.45	6.92	35.13	8.50
C15H2509P1	380.33	380.1236	47.37	6.63	37.86	8.14
C15H2601P2	284.32	284.1459	63.37	9.22	5.63	21.79
C15H2602P2	300.32	300.1408	59.99	8.73	10.65	20.63
C15H2603P2	316.32	316.1357	56.96	8.29	15.17	19.58
C15H2604P2	332.32	332.1306	54.22	7.89	19.26	18.64
C15H2605P2	348.32	348.1255	51.72	7.52	22.97	17.78
C15H2606P2	364.32	364.1205	49.45	7.19	26.35	17.00
C15H2607P2	380.32	380.1154	47.37	6.89	29.45	16.29
C15H2608P2	396.32	396.1103	45.46	6.61	32.30	15.63
C15H2609P2	412.32	412.1052	43.70	6.36	34.92	15.02
C15H2701P1	254.36	254.1799	70.83	10.70	6.29	12.18

121

C15H2702P1	270.36	270.1749	66.64	10.07	11.84	11.46
C15H2703P1	286.35	286.1698	62.92	9.50	16.76	10.82
C15H2704P1	302.35	302.1647	59.59	9.00	21.17	10.24
C15H2705P1	318.35	318.1596	56.59	8.55	25.13	9.73
C15H2706P1	334.35	334.1545	53.89	8.14	28.71	9.26
C15H2707P1	350.35	350.1494	51.42	7.77	31.97	8.84
C15H2708P1	366.35	366.1443	49.18	7.43	34.94	8.45
C15H2709P1	382.35	382.1393	47.12	7.12	37.66	8.10
C15H2801P2	286.34	286.1615	62.92	9.86	5.59	21.63
C15H2802P2	302.34	302.1564	59.59	9.34	10.58	20.49
C15H2803P2	318.34	318.1514	56.60	8.87	15.08	19.46
C15H2804P2	334.34	334.1463	53.89	8.44	19.14	18.53
C15H2805P2	350.34	350.1412	51.43	8.06	22.83	17.68
C15H2806P2	366.33	366.1361	49.18	7.70	26.20	16.91
C15H2807P2	382.33	382.1310	47.12	7.38	29.29	16.20
C15H2808P2	398.33	398.1259	45.23	7.09	32.13	15.55
C15H2809P2	414.33	414.1208	43.48	6.81	34.75	14.95
C15H2901P1	256.37	256.1956	70.28	11.40	6.24	12.08
C15H2902P1	272.37	272.1905	66.15	10.73	11.75	11.37
C15H2903P1	288.37	288.1854	62.48	10.14	16.64	10.74
C15H2904P1	304.37	304.1803	59.19	9.60	21.03	10.18
C15H2905P1	320.37	320.1752	56.24	9.12	24.97	9.67
C15H2906P1	336.37	336.1702	53.56	8.69	28.54	9.21
C15H2907P1	352.37	352.1651	51.13	8.30	31.78	8.79
C15H2908P1	368.37	368.1600	48.91	7.94	34.75	8.41
C15H2909P1	384.37	384.1549	46.87	7.61	37.46	8.06
C15H3001P2	288.35	288.1772	62.48	10.49	5.55	21.48
C15H3002P2	304.35	304.1721	59.20	9.94	10.51	20.35
C15H3003P2	320.35	320.1670	56.24	9.44	14.98	19.34
C15H3004P2	336.35	336.1619	53.57	8.99	19.03	18.42
C15H3005P2	352.35	352.1568	51.13	8.58	22.70	17.58
C15H3006P2	368.35	368.1518	48.91	8.21	26.06	16.82
C15H3007P2	384.35	384.1467	46.88	7.87	29.14	16.12
C15H3008P2	400.35	400.1416	45.00	7.55	31.97	15.47
C15H3009P2	416.35	416.1365	43.27	7.26	34.59	14.88
C15H3101P1	258.39	258.2112	69.73	12.09	6.19	11.99
C15H3102P1	274.39	274.2062	65.66	11.39	11.66	11.29
C15H3103P1	290.39	290.2011	62.04	10.76	16.53	10.67
C15H3104P1	306.39	306.1960	58.80	10.20	20.89	10.11
C15H3105P1	322.39	322.1909	55.89	9.69	24.81	9.61
C15H3106P1	338.38	338.1858	53.24	9.23	28.37	9.15
C15H3107P1	354.38	354.1807	50.84	8.82	31.60	8.74
C15H3108P1	370.38	370.1756	48.64	8.44	34.56	8.36
C15H3109P1	386.38	386.1706	46.63	8.09	37.27	8.02
C15H3201P2	290.37	290.1928	62.05	11.11	5.51	21.33
C15H3202P2	306.37	306.1877	58.81	10.53	10.44	20.22
C15H3203P2	322.37	322.1827	55.89	10.01	14.89	19.22
C15H3204P2	338.37	338.1776	53.25	9.53	18.91	18.31
C15H3205P2	354.37	354.1725	50.84	9.10	22.57	17.48
C15H3206P2	370.37	370.1674	48.65	8.71	25.92	16.73
C15H3207P2	386.37	386.1623	46.63	8.35	28.99	16.03
C15H3208P2	402.37	402.1572	44.78	8.02	31.81	15.40
C15H3209P2	418.36	418.1521	43.06	7.71	34.42	14.81
C15H3301P1	260.40	260.2269	69.19	12.77	6.14	11.89
C15H3302P1	276.40	276.2218	65.18	12.03	11.58	11.21
C15H3303P1	292.40	292.2167	61.62	11.38	16.42	10.59
C15H3304P1	308.40	308.2116	58.42	10.79	20.75	10.04
C15H3305P1	324.40	324.2065	55.54	10.25	24.66	9.55
C15H3306P1	340.40	340.2015	52.93	9.77	28.20	9.10
C15H3307P1	356.40	356.1964	50.55	9.33	31.42	8.69
C15H3308P1	372.40	372.1913	48.38	8.93	34.37	8.32
C15H3309P1	388.40	388.1862	46.39	8.56	37.07	7.97
C15H3401P2	292.39	292.2085	61.62	11.72	5.47	21.19
C15H3402P2	308.38	308.2034	58.42	11.11	10.38	20.09
C15H3403P2	324.38	324.1983	55.54	10.56	14.80	19.10
C15H3404P2	340.38	340.1932	52.93	10.07	18.80	18.20
C15H3405P2	356.38	356.1881	50.55	9.62	22.45	17.38
C15H3406P2	372.38	372.1830	48.38	9.20	25.78	16.64
C15H3407P2	388.38	388.1780	46.39	8.82	28.84	15.95
C15H3408P2	404.38	404.1729	44.55	8.47	31.65	15.32
C15H3409P2	420.38	420.1678	42.86	8.15	34.25	14.74
C16H 801P2	278.19	278.0050	69.08	2.90	5.75	22.27
C16H 802P2	294.19	294.0000	65.32	2.74	10.88	21.06
C16H 803P2	310.19	309.9949	61.96	2.60	15.47	19.97
C16H 804P2	326.19	325.9898	58.92	2.47	19.62	18.99
C16H 805P2	342.19	341.9847	56.16	2.36	23.38	18.10
C16H 806P2	358.19	357.9796	53.65	2.25	26.80	17.29
C16H 807P2	374.19	373.9745	51.36	2.16	29.93	16.56
C16H 808P2	390.18	389.9694	49.25	2.07	32.80	15.88
C16H 809P2	406.18	405.9644	47.31	1.99	35.45	15.25
C16H 901P1	248.22	248.0391	77.42	3.65	6.45	12.48
C16H 902P1	264.22	264.0340	72.73	3.43	12.11	11.72
C16H 903P1	280.22	280.0289	68.58	3.24	17.13	11.05
C16H 904P1	296.22	296.0238	64.88	3.06	21.60	10.46
C16H 905P1	312.22	312.0188	61.55	2.91	25.62	9.92
C16H 906P1	328.22	328.0137	58.55	2.76	29.25	9.44
C16H 907P1	344.22	344.0086	55.83	2.64	32.54	9.00
C16H 908P1	360.22	360.0035	53.35	2.52	35.53	8.60
C16H 909P1	376.22	375.9984	51.08	2.41	38.27	8.23
C16H1001P2	280.21	280.0207	68.58	3.60	5.71	22.11
C16H1002P2	296.20	296.0156	64.88	3.40	10.80	20.91
C16H1003P2	312.20	312.0105	61.56	3.23	15.37	19.84
C16H1004P2	328.20	328.0054	58.55	3.07	19.50	18.87
C16H1005P2	344.20	344.0003	55.83	2.93	23.24	18.00
C16H1006P2	360.20	359.9953	53.35	2.80	26.65	17.20
C16H1007P2	376.20	375.9902	51.08	2.68	29.77	16.47
C16H1008P2	392.20	391.9851	49.00	2.57	32.64	15.79
C16H1009P2	408.20	407.9800	47.08	2.47	35.28	15.18
C16H1101P1	250.24	250.0547	76.80	4.43	6.39	12.38
C16H1102P1	266.24	266.0497	72.18	4.16	12.02	11.63
C16H1103P1	282.24	282.0446	68.09	3.93	17.01	10.97
C16H1104P1	298.24	298.0395	64.44	3.72	21.46	10.39
C16H1105P1	314.24	314.0344	61.16	3.53	25.46	9.86
C16H1106P1	330.24	330.0293	58.19	3.36	29.07	9.38
C16H1107P1	346.24	346.0242	55.51	3.20	32.35	8.95
C16H1108P1	362.24	362.0191	53.05	3.06	35.33	8.55
C16H1109P1	378.23	378.0141	50.81	2.93	38.07	8.19
C16H1201P2	282.22	282.0363	68.09	4.29	5.67	21.95
C16H1202P2	298.22	298.0313	64.44	4.06	10.73	20.77
C16H1203P2	314.22	314.0262	61.16	3.85	15.28	19.71
C16H1204P2	330.22	330.0211	58.20	3.66	19.38	18.76
C16H1205P2	346.22	346.0160	55.51	3.49	23.11	17.89
C16H1206P2	362.22	362.0109	53.06	3.34	26.50	17.10
C16H1207P2	378.22	378.0058	50.81	3.20	29.61	16.38
C16H1208P2	394.22	394.0007	48.75	3.07	32.47	15.71
C16H1209P2	410.22	409.9957	46.85	2.95	35.10	15.10
C16H1301P1	252.26	252.0704	76.18	5.19	6.34	12.28
C16H1302P1	268.25	268.0653	71.64	4.88	11.93	11.55
C16H1303P1	284.25	284.0602	67.61	4.61	16.89	10.90
C16H1304P1	300.25	300.0551	64.01	4.36	21.31	10.32

C16H1305P1	316.25	316.0501	60.77	4.14	25.30	9.79
C16H1306P1	332.25	332.0450	57.84	3.94	28.89	9.32
C16H1307P1	348.25	348.0399	55.18	3.76	32.16	8.89
C16H1308P1	364.25	364.0348	52.76	3.60	35.14	8.50
C16H1309P1	380.25	380.0297	50.54	3.45	37.87	8.15
C16H1401P2	284.24	284.0520	67.61	4.96	5.63	21.79
C16H1402P2	300.24	300.0469	64.01	4.70	10.66	20.63
C16H1403P2	316.24	316.0418	60.77	4.46	15.18	19.59
C16H1404P2	332.24	332.0367	57.84	4.25	19.26	18.65
C16H1405P2	348.23	348.0316	55.19	4.05	22.97	17.79
C16H1406P2	364.23	364.0266	52.76	3.87	26.36	17.01
C16H1407P2	380.23	380.0215	50.54	3.71	29.45	16.29
C16H1408P2	396.23	396.0164	48.50	3.56	32.30	15.63
C16H1409P2	412.23	412.0113	46.62	3.42	34.93	15.03
C16H1501P1	254.27	254.0860	75.58	5.95	6.29	12.18
C16H1502P1	270.27	270.0810	71.11	5.59	11.84	11.46
C16H1503P1	286.27	286.0759	67.13	5.28	16.77	10.82
C16H1504P1	302.27	302.0708	63.58	5.00	21.17	10.25
C16H1505P1	318.27	318.0657	60.38	4.75	25.14	9.73
C16H1506P1	334.27	334.0606	57.49	4.52	28.72	9.27
C16H1507P1	350.27	350.0555	54.87	4.32	31.97	8.84
C16H1508P1	366.27	366.0504	52.47	4.13	34.95	8.46
C16H1509P1	382.27	382.0454	50.27	3.96	37.67	8.10
C16H1601P2	286.25	286.0676	67.14	5.63	5.59	21.64
C16H1602P2	302.25	302.0625	63.58	5.34	10.59	20.50
C16H1603P2	318.25	318.0575	60.39	5.07	15.08	19.46
C16H1604P2	334.25	334.0524	57.50	4.82	19.15	18.53
C16H1605P2	350.25	350.0473	54.87	4.60	22.84	17.69
C16H1606P2	366.25	366.0422	52.47	4.40	26.21	16.91
C16H1607P2	382.25	382.0371	50.28	4.22	29.30	16.21
C16H1608P2	398.25	398.0320	48.26	4.05	32.14	15.56
C16H1609P2	414.25	414.0269	46.39	3.89	34.76	14.95
C16H1701P1	256.29	256.1017	74.99	6.69	6.24	12.09
C16H1702P1	272.29	272.0966	70.58	6.29	11.75	11.38
C16H1703P1	288.29	288.0915	66.66	5.94	16.65	10.74
C16H1704P1	304.29	304.0864	63.16	5.63	21.03	10.18
C16H1705P1	320.28	320.0814	60.00	5.35	24.98	9.67
C16H1706P1	336.28	336.0763	57.15	5.10	28.55	9.21
C16H1707P1	352.28	352.0712	54.55	4.86	31.79	8.79
C16H1708P1	368.28	368.0661	52.18	4.65	34.75	8.41
C16H1709P1	384.28	384.0610	50.01	4.46	37.47	8.06
C16H1801P2	288.27	288.0833	66.67	6.29	5.55	21.49
C16H1802P2	304.27	304.0782	63.16	5.96	10.52	20.36
C16H1803P2	320.27	320.0731	60.01	5.67	14.99	19.34
C16H1804P2	336.27	336.0680	57.15	5.40	19.03	18.42
C16H1805P2	352.27	352.0629	54.55	5.15	22.71	17.59
C16H1806P2	368.27	368.0579	52.18	4.93	26.07	16.82
C16H1807P2	384.27	384.0528	50.01	4.72	29.15	16.12
C16H1808P2	400.26	400.0477	48.01	4.53	31.98	15.48
C16H1809P2	416.26	416.0426	46.17	4.36	34.59	14.88
C16H1901P1	258.30	258.1173	74.40	7.41	6.19	11.99
C16H1902P1	274.30	274.1123	70.06	6.98	11.67	11.29
C16H1903P1	290.30	290.1072	66.20	6.60	16.53	10.67
C16H1904P1	306.30	306.1021	62.74	6.25	20.89	10.11
C16H1905P1	322.30	322.0970	59.63	5.94	24.82	9.61
C16H1906P1	338.30	338.0919	56.81	5.66	28.38	9.16
C16H1907P1	354.30	354.0868	54.24	5.41	31.61	8.74
C16H1908P1	370.30	370.0817	51.90	5.17	34.57	8.36
C16H1909P1	386.30	386.0767	49.75	4.96	37.28	8.02
C16H2001P2	290.28	290.0989	66.20	6.94	5.51	21.34

C16H2002P2	306.28	306.0938	62.75	6.58	10.45	20.23
C16H2003P2	322.28	322.0888	59.63	6.26	14.89	19.22
C16H2004P2	338.28	338.0837	56.81	5.96	18.92	18.31
C16H2005P2	354.28	354.0786	54.24	5.69	22.58	17.49
C16H2006P2	370.28	370.0735	51.90	5.44	25.93	16.73
C16H2007P2	386.28	386.0684	49.75	5.22	28.99	16.04
C16H2008P2	402.28	402.0633	47.77	5.01	31.82	15.40
C16H2009P2	418.28	418.0582	45.94	4.82	34.43	14.81
C16H2101P1	260.32	260.1330	73.82	8.13	6.15	11.90
C16H2102P1	276.32	276.1279	69.55	7.66	11.58	11.21
C16H2103P1	292.32	292.1228	65.74	7.24	16.42	10.60
C16H2104P1	308.32	308.1177	62.33	6.87	20.76	10.05
C16H2105P1	324.32	324.1127	59.26	6.53	24.67	9.55
C16H2106P1	340.32	340.1076	56.47	6.22	28.21	9.10
C16H2107P1	356.32	356.1025	53.93	5.94	31.43	8.69
C16H2108P1	372.31	372.0974	51.62	5.69	34.38	8.32
C16H2109P1	388.31	388.0923	49.49	5.45	37.08	7.98
C16H2201P2	292.30	292.1146	65.75	7.59	5.47	21.19
C16H2202P2	308.30	308.1095	62.33	7.19	10.38	20.09
C16H2203P2	324.30	324.1044	59.26	6.84	14.80	19.10
C16H2204P2	340.30	340.0993	56.47	6.52	18.81	18.20
C16H2205P2	356.30	356.0942	53.94	6.22	22.45	17.39
C16H2206P2	372.30	372.0892	51.62	5.96	25.78	16.64
C16H2207P2	388.30	388.0841	49.49	5.71	28.84	15.95
C16H2208P2	404.30	404.0790	47.53	5.48	31.66	15.32
C16H2209P2	420.30	420.0739	45.72	5.28	34.26	14.74
C16H2301P1	262.33	262.1486	73.26	8.84	6.10	11.81
C16H2302P1	278.33	278.1436	69.05	8.33	11.50	11.13
C16H2303P1	294.33	294.1385	65.29	7.88	16.31	10.52
C16H2304P1	310.33	310.1334	61.93	7.47	20.62	9.98
C16H2305P1	326.33	326.1283	58.89	7.10	24.51	9.49
C16H2306P1	342.33	342.1232	56.14	6.77	28.04	9.05
C16H2307P1	358.33	358.1181	53.63	6.47	31.25	8.64
C16H2308P1	374.33	374.1130	51.34	6.19	34.19	8.27
C16H2309P1	390.33	390.1080	49.23	5.94	36.89	7.94
C16H2401P2	294.32	294.1302	65.30	8.22	5.44	21.05
C16H2402P2	310.32	310.1251	61.93	7.80	10.31	19.96
C16H2403P2	326.32	326.1201	58.89	7.41	14.71	18.98
C16H2404P2	342.31	342.1150	56.14	7.07	18.70	18.10
C16H2405P2	358.31	358.1099	53.63	6.75	22.33	17.29
C16H2406P2	374.31	374.1048	51.34	6.46	25.65	16.55
C16H2407P2	390.31	390.0997	49.24	6.20	28.69	15.87
C16H2408P2	406.31	406.0946	47.30	5.95	31.50	15.25
C16H2409P2	422.31	422.0895	45.51	5.73	34.10	14.67
C16H2501P1	264.35	264.1643	72.70	9.53	6.05	11.72
C16H2502P1	280.35	280.1592	68.55	8.99	11.41	11.05
C16H2503P1	296.35	296.1541	64.85	8.50	16.20	10.45
C16H2504P1	312.35	312.1490	61.53	8.07	20.49	9.92
C16H2505P1	328.35	328.1439	58.53	7.67	24.36	9.43
C16H2506P1	344.35	344.1389	55.81	7.32	27.88	8.99
C16H2507P1	360.35	360.1338	53.33	6.99	31.08	8.60
C16H2508P1	376.35	376.1287	51.06	6.70	34.01	8.23
C16H2509P1	392.35	392.1236	48.98	6.42	36.70	7.89
C16H2601P2	296.33	296.1459	64.85	8.84	5.40	20.90
C16H2602P2	312.33	312.1408	61.53	8.39	10.25	19.83
C16H2603P2	328.33	328.1357	58.53	7.98	14.62	18.87
C16H2604P2	344.33	344.1306	55.81	7.61	18.59	17.99
C16H2605P2	360.33	360.1255	53.33	7.27	22.20	17.19
C16H2606P2	376.33	376.1205	51.07	6.96	25.51	16.46
C16H2607P2	392.33	392.1154	48.98	6.68	28.55	15.79

123

C16H2608P2	408.33	408.1103	47.06	6.42	31.35	15.17
C16H2609P2	424.33	424.1052	45.29	6.18	33.93	14.60
C16H2701P1	266.37	266.1799	72.15	10.22	6.01	11.63
C16H2702P1	282.37	282.1749	68.06	9.64	11.33	10.97
C16H2703P1	298.37	298.1698	64.41	9.12	16.09	10.38
C16H2704P1	314.36	314.1647	61.13	8.66	20.36	9.85
C16H2705P1	330.36	330.1596	58.17	8.24	24.21	9.38
C16H2706P1	346.36	346.1545	55.48	7.86	27.72	8.94
C16H2707P1	362.36	362.1494	53.03	7.51	30.91	8.55
C16H2708P1	378.36	378.1443	50.79	7.19	33.83	8.19
C16H2709P1	394.36	394.1393	48.73	6.90	36.51	7.85
C16H2801P2	298.35	298.1615	64.41	9.46	5.36	20.76
C16H2802P2	314.35	314.1564	61.14	8.98	10.18	19.71
C16H2803P2	330.35	330.1514	58.17	8.54	14.53	18.75
C16H2804P2	346.35	346.1463	55.49	8.15	18.48	17.89
C16H2805P2	362.35	362.1412	53.04	7.79	22.08	17.10
C16H2806P2	378.35	378.1361	50.79	7.46	25.37	16.37
C16H2807P2	394.34	394.1310	48.73	7.16	28.40	15.71
C16H2808P2	410.34	410.1259	46.83	6.88	31.19	15.10
C16H2809P2	426.34	426.1208	45.08	6.62	33.77	14.53
C16H2901P1	268.38	268.1956	71.61	10.89	5.96	11.54
C16H2902P1	284.38	284.1905	67.58	10.28	11.25	10.89
C16H2903P1	300.38	300.1854	63.98	9.73	15.98	10.31
C16H2904P1	316.38	316.1803	60.74	9.24	20.23	9.79
C16H2905P1	332.38	332.1752	57.82	8.79	24.07	9.32
C16H2906P1	348.38	348.1702	55.16	8.39	27.56	8.89
C16H2907P1	364.38	364.1651	52.74	8.02	30.74	8.50
C16H2908P1	380.38	380.1600	50.52	7.68	33.65	8.14
C16H2909P1	396.38	396.1549	48.48	7.37	36.33	7.81
C16H3001P2	300.36	300.1772	63.98	10.07	5.33	20.62
C16H3002P2	316.36	316.1721	60.75	9.56	10.11	19.58
C16H3003P2	332.36	332.1670	57.82	9.10	14.44	18.64
C16H3004P2	348.36	348.1619	55.17	8.68	18.37	17.78
C16H3005P2	364.36	364.1568	52.74	8.30	21.96	17.00
C16H3006P2	380.36	380.1518	50.53	7.95	25.24	16.29
C16H3007P2	396.36	396.1467	48.49	7.63	28.26	15.63
C16H3008P2	412.36	412.1416	46.60	7.33	31.04	15.02
C16H3009P2	428.36	428.1365	44.86	7.06	33.62	14.46
C16H3101P1	270.40	270.2112	71.07	11.56	5.92	11.45
C16H3102P1	286.40	286.2062	67.10	10.91	11.17	10.81
C16H3103P1	302.40	302.2011	63.55	10.33	15.87	10.24
C16H3104P1	318.40	318.1960	60.36	9.81	20.10	9.73
C16H3105P1	334.40	334.1909	57.47	9.34	23.92	9.26
C16H3106P1	350.40	350.1858	54.85	8.92	27.40	8.84
C16H3107P1	366.40	366.1807	52.45	8.53	30.57	8.45
C16H3108P1	382.39	382.1756	50.26	8.17	33.47	8.10
C16H3109P1	398.39	398.1706	48.24	7.84	36.14	7.77
C16H3201P2	302.38	302.1928	63.56	10.67	5.29	20.49
C16H3202P2	318.38	318.1877	60.36	10.13	10.05	19.46
C16H3203P2	334.38	334.1827	57.47	9.65	14.35	18.53
C16H3204P2	350.38	350.1776	54.85	9.21	18.27	17.68
C16H3205P2	366.38	366.1725	52.45	8.80	21.83	16.91
C16H3206P2	382.38	382.1674	50.26	8.44	25.11	16.20
C16H3207P2	398.38	398.1623	48.24	8.10	28.11	15.55
C16H3208P2	414.38	414.1572	46.38	7.78	30.89	14.95
C16H3209P2	430.38	430.1521	44.65	7.49	33.46	14.39
C16H3301P1	272.41	272.2269	70.55	12.21	5.87	11.37
C16H3302P1	288.41	288.2218	66.63	11.53	11.09	10.74
C16H3303P1	304.41	304.2167	63.13	10.93	15.77	10.17
C16H3304P1	320.41	320.2116	59.98	10.38	19.97	9.67

C16H3305P1	336.41	336.2065	57.13	9.89	23.78	9.21
C16H3306P1	352.41	352.2015	54.53	9.44	27.24	8.79
C16H3307P1	368.41	368.1964	52.16	9.03	30.40	8.41
C16H3308P1	384.41	384.1913	49.99	8.65	33.30	8.06
C16H3309P1	400.41	400.1862	48.00	8.31	35.96	7.74
C16H3401P2	304.40	304.2085	63.13	11.26	5.26	20.35
C16H3402P2	320.40	320.2034	59.98	10.70	9.99	19.33
C16H3403P2	336.40	336.1983	57.13	10.19	14.27	18.42
C16H3404P2	352.39	352.1932	54.54	9.73	18.16	17.58
C16H3405P2	368.39	368.1881	52.17	9.30	21.72	16.82
C16H3406P2	384.39	384.1830	50.00	8.92	24.97	16.12
C16H3407P2	400.39	400.1780	48.00	8.56	27.97	15.47
C16H3408P2	416.39	416.1729	46.15	8.23	30.74	14.88
C16H3409P2	432.39	432.1678	44.45	7.93	33.30	14.33
C16H3501P1	274.43	274.2425	70.03	12.86	5.83	11.29
C16H3502P1	290.43	290.2375	66.17	12.15	11.02	10.66
C16H3503P1	306.43	306.2324	62.72	11.51	15.66	10.11
C16H3504P1	322.43	322.2273	59.60	10.94	19.85	9.61
C16H3505P1	338.43	338.2222	56.79	10.42	23.64	9.15
C16H3506P1	354.43	354.2171	54.22	9.95	27.08	8.74
C16H3507P1	370.43	370.2120	51.88	9.52	30.23	8.36
C16H3508P1	386.43	386.2069	49.73	9.13	33.12	8.02
C16H3509P1	402.43	402.2019	47.75	8.77	35.78	7.70
C16H3601P2	306.41	306.2241	62.72	11.84	5.22	20.22
C16H3602P2	322.41	322.2190	59.61	11.25	9.92	19.21
C16H3603P2	338.41	338.2140	56.79	10.72	14.18	18.31
C16H3604P2	354.41	354.2089	54.22	10.24	18.06	17.48
C16H3605P2	370.41	370.2038	51.88	9.80	21.60	16.72
C16H3606P2	386.41	386.1987	49.73	9.39	24.84	16.03
C16H3607P2	402.41	402.1936	47.76	9.02	27.83	15.39
C16H3608P2	418.41	418.1885	45.93	8.67	30.59	14.81
C16H3609P2	434.41	434.1834	44.24	8.35	33.15	14.26
C17H 801P2	290.20	290.0050	70.36	2.78	5.51	21.35
C17H 802P2	306.20	306.0000	66.69	2.63	10.45	20.23
C17H 803P2	322.20	321.9949	63.37	2.50	14.90	19.23
C17H 804P2	338.20	337.9898	60.38	2.38	18.92	18.32
C17H 805P2	354.20	353.9847	57.65	2.28	22.59	17.49
C17H 806P2	370.20	369.9796	55.16	2.18	25.93	16.73
C17H 807P2	386.20	385.9745	52.87	2.09	29.00	16.04
C17H 808P2	402.20	401.9694	50.77	2.00	31.82	15.40
C17H 809P2	418.20	417.9644	48.83	1.93	34.43	14.81
C17H 901P1	260.23	260.0391	78.46	3.49	6.15	11.90
C17H 902P1	276.23	276.0340	73.92	3.28	11.58	11.21
C17H 903P1	292.23	292.0289	69.87	3.10	16.42	10.60
C17H 904P1	308.23	308.0238	66.25	2.94	20.76	10.05
C17H 905P1	324.23	324.0188	62.98	2.80	24.67	9.55
C17H 906P1	340.23	340.0137	60.01	2.67	28.22	9.10
C17H 907P1	356.23	356.0086	57.32	2.55	31.44	8.69
C17H 908P1	372.23	372.0035	54.86	2.44	34.39	8.32
C17H 909P1	388.23	387.9984	52.60	2.34	37.09	7.98
C17H1001P2	292.22	292.0207	69.88	3.45	5.48	21.20
C17H1002P2	308.22	308.0156	66.25	3.27	10.38	20.10
C17H1003P2	324.22	324.0105	62.98	3.11	14.80	19.11
C17H1004P2	340.21	340.0054	60.02	2.96	18.81	18.21
C17H1005P2	356.21	356.0003	57.32	2.83	22.46	17.39
C17H1006P2	372.21	371.9953	54.86	2.71	25.79	16.64
C17H1007P2	388.21	387.9902	52.60	2.60	28.85	15.96
C17H1008P2	404.21	403.9851	50.52	2.49	31.67	15.33
C17H1009P2	420.21	419.9800	48.59	2.40	34.27	14.74
C17H1101P1	262.25	262.0547	77.86	4.23	6.10	11.81

124

C17H1102P1	278.25	278.0497	73.38	3.98	11.50	11.13
C17H1103P1	294.25	294.0446	69.39	3.77	16.31	10.53
C17H1104P1	310.25	310.0395	65.81	3.57	20.63	9.98
C17H1105P1	326.25	326.0344	62.59	3.40	24.52	9.49
C17H1106P1	342.25	342.0293	59.66	3.24	28.05	9.05
C17H1107P1	358.25	358.0242	57.00	3.09	31.26	8.65
C17H1108P1	374.25	374.0191	54.56	2.96	34.20	8.28
C17H1109P1	390.25	390.0141	52.32	2.84	36.90	7.94
C17H1201P2	294.23	294.0363	69.40	4.11	5.44	21.05
C17H1202P2	310.23	310.0313	65.82	3.90	10.31	19.97
C17H1203P2	326.23	326.0262	62.59	3.71	14.71	18.99
C17H1204P2	342.23	342.0211	59.66	3.53	18.70	18.10
C17H1205P2	358.23	358.0160	57.00	3.38	22.33	17.29
C17H1206P2	374.23	374.0109	54.56	3.23	25.65	16.55
C17H1207P2	390.23	390.0058	52.33	3.10	28.70	15.87
C17H1208P2	406.23	406.0007	50.26	2.98	31.51	15.25
C17H1209P2	422.23	421.9957	48.36	2.86	34.10	14.67
C17H1301P1	264.27	264.0704	77.27	4.96	6.05	11.72
C17H1302P1	280.27	280.0653	72.86	4.68	11.42	11.05
C17H1303P1	296.27	296.0602	68.92	4.42	16.20	10.45
C17H1304P1	312.26	312.0551	65.39	4.20	20.49	9.92
C17H1305P1	328.26	328.0501	62.20	3.99	24.37	9.44
C17H1306P1	344.26	344.0450	59.31	3.81	27.88	9.00
C17H1307P1	360.26	360.0399	56.68	3.64	31.09	8.60
C17H1308P1	376.26	376.0348	54.27	3.48	34.02	8.23
C17H1309P1	392.26	392.0297	52.05	3.34	36.71	7.90
C17H1401P2	296.25	296.0520	68.93	4.76	5.40	20.91
C17H1402P2	312.25	312.0469	65.39	4.52	10.25	19.84
C17H1403P2	328.25	328.0418	62.21	4.30	14.62	18.87
C17H1404P2	344.25	344.0367	59.31	4.10	18.59	18.00
C17H1405P2	360.25	360.0316	56.68	3.92	22.21	17.20
C17H1406P2	376.25	376.0266	54.27	3.75	25.51	16.46
C17H1407P2	392.24	392.0215	52.06	3.60	28.55	15.79
C17H1408P2	408.24	408.0164	50.02	3.46	31.35	15.17
C17H1409P2	424.24	424.0113	48.13	3.33	33.94	14.60
C17H1501P1	266.28	266.0860	76.68	5.68	6.01	11.63
C17H1502P1	282.28	282.0810	72.34	5.36	11.34	10.97
C17H1503P1	298.28	298.0759	68.46	5.07	16.09	10.38
C17H1504P1	314.28	314.0708	64.97	4.81	20.36	9.86
C17H1505P1	330.28	330.0657	61.82	4.58	24.22	9.38
C17H1506P1	346.28	346.0606	58.97	4.37	27.72	8.94
C17H1507P1	362.28	362.0555	56.36	4.17	30.91	8.55
C17H1508P1	378.28	378.0504	53.98	4.00	33.84	8.19
C17H1509P1	394.28	394.0454	51.79	3.83	36.52	7.86
C17H1601P2	298.26	298.0676	68.46	5.41	5.36	20.77
C17H1602P2	314.26	314.0625	64.97	5.13	10.18	19.71
C17H1603P2	330.26	330.0575	61.83	4.88	14.53	18.76
C17H1604P2	346.26	346.0524	58.97	4.66	18.48	17.89
C17H1605P2	362.26	362.0473	56.37	4.45	22.08	17.10
C17H1606P2	378.26	378.0422	53.98	4.26	25.38	16.38
C17H1607P2	394.26	394.0371	51.79	4.09	28.41	15.71
C17H1608P2	410.26	410.0320	49.77	3.93	31.20	15.10
C17H1609P2	426.26	426.0269	47.90	3.78	33.78	14.53
C17H1701P1	268.30	268.1017	76.11	6.39	5.96	11.54
C17H1702P1	284.30	284.0966	71.82	6.03	11.26	10.89
C17H1703P1	300.30	300.0915	68.00	5.71	15.98	10.31
C17H1704P1	316.30	316.0864	64.56	5.42	20.23	9.79
C17H1705P1	332.30	332.0814	61.45	5.16	24.07	9.32
C17H1706P1	348.30	348.0763	58.63	4.92	27.56	8.89
C17H1707P1	364.29	364.0712	56.05	4.70	30.74	8.50

C17H1708P1	380.29	380.0661	53.69	4.51	33.66	8.14
C17H1709P1	396.29	396.0610	51.52	4.32	36.34	7.82
C17H1801P2	300.28	300.0833	68.00	6.04	5.33	20.63
C17H1802P2	316.28	316.0782	64.56	5.74	10.12	19.59
C17H1803P2	332.28	332.0731	61.45	5.46	14.45	18.64
C17H1804P2	348.28	348.0680	58.63	5.21	18.38	17.79
C17H1805P2	364.28	364.0629	56.05	4.98	21.96	17.01
C17H1806P2	380.28	380.0579	53.69	4.77	25.24	16.29
C17H1807P2	396.28	396.0528	51.53	4.58	28.26	15.63
C17H1808P2	412.28	412.0477	49.53	4.40	31.05	15.03
C17H1809P2	428.28	428.0426	47.68	4.24	33.62	14.46
C17H1901P1	270.31	270.1173	75.54	7.08	5.92	11.46
C17H1902P1	286.31	286.1123	71.32	6.69	11.18	10.82
C17H1903P1	302.31	302.1072	67.54	6.33	15.88	10.25
C17H1904P1	318.31	318.1021	64.15	6.02	20.11	9.73
C17H1905P1	334.31	334.0970	61.08	5.73	23.93	9.26
C17H1906P1	350.31	350.0919	58.29	5.47	27.40	8.84
C17H1907P1	366.31	366.0868	55.74	5.23	30.57	8.46
C17H1908P1	382.31	382.0817	53.41	5.01	33.48	8.10
C17H1909P1	398.31	398.0767	51.26	4.81	36.15	7.78
C17H2001P2	302.30	302.0989	67.55	6.67	5.29	20.49
C17H2002P2	318.30	318.0938	64.15	6.33	10.05	19.46
C17H2003P2	334.29	334.0888	61.08	6.03	14.36	18.53
C17H2004P2	350.29	350.0837	58.29	5.75	18.27	17.68
C17H2005P2	366.29	366.0786	55.74	5.50	21.84	16.91
C17H2006P2	382.29	382.0735	53.41	5.27	25.11	16.20
C17H2007P2	398.29	398.0684	51.27	5.06	28.12	15.55
C17H2008P2	414.29	414.0633	49.29	4.87	30.89	14.95
C17H2009P2	430.29	430.0582	47.45	4.69	33.46	14.40
C17H2101P1	272.33	272.1330	74.98	7.77	5.88	11.37
C17H2102P1	288.33	288.1279	70.82	7.34	11.10	10.74
C17H2103P1	304.33	304.1228	67.10	6.96	15.77	10.18
C17H2104P1	320.33	320.1177	63.74	6.61	19.98	9.67
C17H2105P1	336.33	336.1127	60.71	6.29	23.79	9.21
C17H2106P1	352.33	352.1076	57.95	6.01	27.25	8.79
C17H2107P1	368.33	368.1025	55.44	5.75	30.41	8.41
C17H2108P1	384.33	384.0974	53.13	5.51	33.30	8.06
C17H2109P1	400.33	400.0923	51.01	5.29	35.97	7.74
C17H2201P2	304.31	304.1146	67.10	7.29	5.26	20.36
C17H2202P2	320.31	320.1095	63.75	6.92	9.99	19.34
C17H2203P2	336.31	336.1044	60.71	6.59	14.27	18.42
C17H2204P2	352.31	352.0993	57.96	6.29	18.17	17.58
C17H2205P2	368.31	368.0942	55.44	6.02	21.72	16.82
C17H2206P2	384.31	384.0892	53.13	5.77	24.98	16.12
C17H2207P2	400.31	400.0841	51.01	5.54	27.98	15.47
C17H2208P2	416.31	416.0790	49.05	5.33	30.75	14.88
C17H2209P2	432.31	432.0739	47.23	5.13	33.31	14.33
C17H2301P1	274.35	274.1486	74.43	8.45	5.83	11.29
C17H2302P1	290.35	290.1436	70.33	7.98	11.02	10.67
C17H2303P1	306.34	306.1385	66.65	7.57	15.67	10.11
C17H2304P1	322.34	322.1334	63.35	7.19	19.85	9.61
C17H2305P1	338.34	338.1283	60.35	6.85	23.64	9.15
C17H2306P1	354.34	354.1232	57.62	6.54	27.09	8.74
C17H2307P1	370.34	370.1181	55.14	6.26	30.24	8.36
C17H2308P1	386.34	386.1130	52.85	6.00	33.13	8.02
C17H2309P1	402.34	402.1080	50.75	5.76	35.79	7.70
C17H2401P2	306.33	306.1302	66.66	7.90	5.22	20.22
C17H2402P2	322.33	322.1251	63.35	7.51	9.93	19.22
C17H2403P2	338.33	338.1201	60.35	7.15	14.19	18.31
C17H2404P2	354.33	354.1150	57.63	6.83	18.06	17.48

125

C17H2405P2	370.33	370.1099	55.14	6.53	21.60	16.73
C17H2406P2	386.32	386.1048	52.85	6.26	24.85	16.04
C17H2407P2	402.32	402.0997	50.75	6.01	27.84	15.40
C17H2408P2	418.32	418.0946	48.81	5.78	30.60	14.81
C17H2409P2	434.32	434.0895	47.01	5.57	33.15	14.26
C17H2501P1	276.36	276.1643	73.88	9.12	5.79	11.21
C17H2502P1	292.36	292.1592	69.84	8.62	10.94	10.59
C17H2503P1	308.36	308.1541	66.22	8.17	15.57	10.04
C17H2504P1	324.36	324.1490	62.95	7.77	19.73	9.55
C17H2505P1	340.36	340.1439	59.99	7.40	23.50	9.10
C17H2506P1	356.36	356.1389	57.30	7.07	26.94	8.69
C17H2507P1	372.36	372.1338	54.84	6.77	30.08	8.32
C17H2508P1	388.36	388.1287	52.58	6.49	32.96	7.98
C17H2509P1	404.36	404.1236	50.50	6.23	35.61	7.66
C17H2601P2	308.34	308.1459	66.22	8.50	5.19	20.09
C17H2602P2	324.34	324.1408	62.95	8.08	9.87	19.10
C17H2603P2	340.34	340.1357	60.00	7.70	14.10	18.20
C17H2604P2	356.34	356.1306	57.30	7.35	17.96	17.38
C17H2605P2	372.34	372.1255	54.84	7.04	21.48	16.64
C17H2606P2	388.34	388.1205	52.58	6.75	24.72	15.95
C17H2607P2	404.34	404.1154	50.50	6.48	27.70	15.32
C17H2608P2	420.34	420.1103	48.58	6.23	30.45	14.74
C17H2609P2	436.34	436.1052	46.80	6.01	33.00	14.20
C17H2701P1	278.38	278.1799	73.35	9.78	5.75	11.13
C17H2702P1	294.38	294.1749	69.36	9.25	10.87	10.52
C17H2703P1	310.38	310.1698	65.79	8.77	15.46	9.98
C17H2704P1	326.38	326.1647	62.56	8.34	19.61	9.49
C17H2705P1	342.38	342.1596	59.64	7.95	23.37	9.05
C17H2706P1	358.37	358.1545	56.98	7.59	26.79	8.64
C17H2707P1	374.37	374.1494	54.54	7.27	29.92	8.27
C17H2708P1	390.37	390.1443	52.31	6.97	32.79	7.93
C17H2709P1	406.37	406.1393	50.25	6.70	35.43	7.62
C17H2801P2	310.36	310.1615	65.79	9.09	5.16	19.96
C17H2802P2	326.36	326.1564	62.57	8.65	9.80	18.98
C17H2803P2	342.36	342.1514	59.64	8.24	14.02	18.09
C17H2804P2	358.36	358.1463	56.98	7.88	17.86	17.29
C17H2805P2	374.36	374.1412	54.54	7.54	21.37	16.55
C17H2806P2	390.36	390.1361	52.31	7.23	24.59	15.87
C17H2807P2	406.36	406.1310	50.25	6.95	27.56	15.24
C17H2808P2	422.36	422.1259	48.35	6.68	30.31	14.67
C17H2809P2	438.35	438.1208	46.58	6.44	32.85	14.13
C17H2901P1	280.39	280.1956	72.82	10.43	5.71	11.05
C17H2902P1	296.39	296.1905	68.89	9.86	10.80	10.45
C17H2903P1	312.39	312.1854	65.36	9.36	15.36	9.92
C17H2904P1	328.39	328.1803	62.18	8.90	19.49	9.43
C17H2905P1	344.39	344.1752	59.29	8.49	23.23	8.99
C17H2906P1	360.39	360.1702	56.66	8.11	26.64	8.59
C17H2907P1	376.39	376.1651	54.25	7.77	29.76	8.23
C17H2908P1	392.39	392.1600	52.04	7.45	32.62	7.89
C17H2909P1	408.39	408.1549	50.00	7.16	35.26	7.58
C17H3001P2	312.38	312.1772	65.37	9.68	5.12	19.83
C17H3002P2	328.38	328.1721	62.18	9.21	9.74	18.86
C17H3003P2	344.37	344.1670	59.29	8.78	13.94	17.99
C17H3004P2	360.37	360.1619	56.66	8.39	17.76	17.19
C17H3005P2	376.37	376.1568	54.25	8.03	21.25	16.46
C17H3006P2	392.37	392.1518	52.04	7.71	24.47	15.79
C17H3007P2	408.37	408.1467	50.00	7.40	27.42	15.17
C17H3008P2	424.37	424.1416	48.12	7.13	30.16	14.60
C17H3009P2	440.37	440.1365	46.37	6.87	32.70	14.07
C17H3101P1	282.41	282.2112	72.30	11.06	5.67	10.97

C17H3102P1	298.41	298.2062	68.43	10.47	10.72	10.38
C17H3103P1	314.41	314.2011	64.94	9.94	15.27	9.85
C17H3104P1	330.41	330.1960	61.80	9.46	19.37	9.37
C17H3105P1	346.41	346.1909	58.94	9.02	23.09	8.94
C17H3106P1	362.41	362.1858	56.34	8.62	26.49	8.55
C17H3107P1	378.41	378.1807	53.96	8.26	29.60	8.19
C17H3108P1	394.41	394.1756	51.77	7.92	32.45	7.85
C17H3109P1	410.41	410.1706	49.75	7.61	35.09	7.55
C17H3201P2	314.39	314.1928	64.95	10.26	5.09	19.70
C17H3202P2	330.39	330.1877	61.80	9.76	9.69	18.75
C17H3203P2	346.39	346.1827	58.95	9.31	13.86	17.88
C17H3204P2	362.39	362.1776	56.35	8.90	17.66	17.09
C17H3205P2	378.39	378.1725	53.96	8.52	21.14	16.37
C17H3206P2	394.39	394.1674	51.77	8.18	24.34	15.71
C17H3207P2	410.39	410.1623	49.76	7.86	27.29	15.09
C17H3208P2	426.39	426.1572	47.89	7.56	30.02	14.53
C17H3209P2	442.39	442.1521	46.16	7.29	32.55	14.00
C17H3301P1	284.43	284.2269	71.79	11.69	5.63	10.89
C17H3302P1	300.43	300.2218	67.97	11.07	10.65	10.31
C17H3303P1	316.42	316.2167	64.53	10.51	15.17	9.79
C17H3304P1	332.42	332.2116	61.42	10.01	19.25	9.32
C17H3305P1	348.42	348.2065	58.60	9.55	22.96	8.89
C17H3306P1	364.42	364.2015	56.03	9.13	26.34	8.50
C17H3307P1	380.42	380.1964	53.67	8.74	29.44	8.14
C17H3308P1	396.42	396.1913	51.51	8.39	32.29	7.81
C17H3309P1	412.42	412.1862	49.51	8.07	34.91	7.51
C17H3401P2	316.41	316.2085	64.53	10.83	5.06	19.58
C17H3402P2	332.41	332.2034	61.43	10.31	9.63	18.64
C17H3403P2	348.41	348.1983	58.61	9.84	13.78	17.78
C17H3404P2	364.41	364.1932	56.03	9.40	17.56	17.00
C17H3405P2	380.41	380.1881	53.68	9.01	21.03	16.28
C17H3406P2	396.40	396.1830	51.51	8.65	24.22	15.63
C17H3407P2	412.40	412.1780	49.51	8.31	27.16	15.02
C17H3408P2	428.40	428.1729	47.66	8.00	29.88	14.46
C17H3409P2	444.40	444.1678	45.95	7.71	32.40	13.94
C17H3501P1	286.44	286.2425	71.28	12.32	5.59	10.81
C17H3502P1	302.44	302.2375	67.51	11.66	10.58	10.24
C17H3503P1	318.44	318.2324	64.12	11.08	15.07	9.73
C17H3504P1	334.44	334.2273	61.05	10.55	19.14	9.26
C17H3505P1	350.44	350.2222	58.27	10.07	22.83	8.84
C17H3506P1	366.44	366.2171	55.72	9.63	26.20	8.45
C17H3507P1	382.44	382.2120	53.39	9.22	29.28	8.10
C17H3508P1	398.44	398.2069	51.25	8.85	32.12	7.77
C17H3509P1	414.44	414.2019	49.27	8.51	34.74	7.47
C17H3601P2	318.42	318.2241	64.13	11.40	5.02	19.45
C17H3602P2	334.42	334.2190	61.06	10.85	9.57	18.52
C17H3603P2	350.42	350.2140	58.27	10.36	13.70	17.68
C17H3604P2	366.42	366.2089	55.73	9.90	17.47	16.91
C17H3605P2	382.42	382.2038	53.39	9.49	20.92	16.20
C17H3606P2	398.42	398.1987	51.25	9.11	24.09	15.55
C17H3607P2	414.42	414.1936	49.27	8.76	27.02	14.95
C17H3608P2	430.42	430.1885	47.44	8.43	29.74	14.39
C17H3609P2	446.42	446.1834	45.74	8.13	32.26	13.88
C17H3701P1	288.46	288.2582	70.79	12.93	5.55	10.74
C17H3702P1	304.46	304.2531	67.07	12.25	10.51	10.17
C17H3703P1	320.46	320.2480	63.72	11.64	14.98	9.67
C17H3704P1	336.46	336.2429	60.69	11.08	19.02	9.21
C17H3705P1	352.46	352.2378	57.93	10.58	22.70	8.79
C17H3706P1	368.45	368.2328	55.42	10.12	26.05	8.41
C17H3707P1	384.45	384.2277	53.11	9.70	29.13	8.06

C17H3708P1	400.45	400.2226	50.95	9.31	31.96	7.73	C18H1405P2	372.26	372.0316	58.08	3.79	21.49	16.64
C17H3709P1	416.45	416.2175	49.03	8.96	34.58	7.44	C18H1406P2	388.26	388.0266	55.69	3.63	24.73	15.96
C17H3801P2	320.44	320.2398	63.72	11.95	4.99	19.33	C18H1407P2	404.26	404.0215	53.48	3.49	27.70	15.32
C17H3802P2	336.44	336.2347	60.65	11.38	9.51	18.41	C18H1408P2	420.26	420.0164	51.45	3.36	30.46	14.74
C17H3803P2	352.44	352.2296	57.94	10.87	13.62	17.58	C18H1409P2	436.25	436.0113	49.56	3.23	33.01	14.20
C17H3804P2	368.44	368.2245	55.42	10.40	17.37	16.81	C18H1501P1	278.29	278.0860	77.69	5.43	5.75	11.13
C17H3805P2	384.44	384.2194	53.11	9.96	20.81	16.11	C18H1502P1	294.29	294.0810	73.46	5.14	10.87	10.52
C17H3806P2	400.44	400.2143	50.99	9.57	23.97	15.47	C18H1503P1	310.29	310.0759	69.68	4.87	15.47	9.98
C17H3807P2	416.44	416.2093	49.03	9.20	26.89	14.88	C18H1504P1	326.29	326.0708	66.26	4.63	19.61	9.49
C17H3808P2	432.44	432.2042	47.22	8.86	29.60	14.33	C18H1505P1	342.29	342.0657	63.16	4.42	23.37	9.05
C17H3809P2	448.43	448.1991	45.53	8.54	32.11	13.81	C18H1506P1	358.29	358.0606	60.34	4.22	26.79	8.64
C18H 901P1	272.25	272.0391	79.41	3.33	5.88	11.38	C18H1507P1	374.29	374.0555	57.76	4.04	29.92	8.28
C18H 902P1	288.25	288.0340	75.01	3.15	11.10	10.75	C18H1508P1	390.29	390.0504	55.39	3.87	32.79	7.94
C18H 903P1	304.24	304.0289	71.06	2.98	15.78	10.18	C18H1509P1	406.29	406.0454	53.21	3.72	35.44	7.62
C18H 904P1	320.24	320.0238	67.51	2.83	19.98	9.67	C18H1601P2	310.28	310.0676	69.68	5.20	5.16	19.97
C18H 905P1	336.24	336.0188	64.30	2.70	23.79	9.21	C18H1602P2	326.27	326.0625	66.26	4.94	9.81	18.99
C18H 906P1	352.24	352.0137	61.38	2.58	27.25	8.79	C18H1603P2	342.27	342.0575	63.17	4.71	14.02	18.10
C18H 907P1	368.24	368.0086	58.71	2.46	30.41	8.41	C18H1604P2	358.27	358.0524	60.35	4.50	17.86	17.29
C18H 908P1	384.24	384.0035	56.27	2.36	33.31	8.06	C18H1605P2	374.27	374.0473	57.77	4.31	21.37	16.55
C18H 909P1	400.24	399.9984	54.02	2.27	35.98	7.74	C18H1606P2	390.27	390.0422	55.40	4.13	24.60	15.87
C18H1001P2	304.23	304.0207	71.07	3.31	5.26	20.36	C18H1607P2	406.27	406.0371	53.22	3.97	27.57	15.25
C18H1002P2	320.23	320.0156	67.51	3.15	9.99	19.34	C18H1608P2	422.27	422.0320	51.20	3.82	30.31	14.67
C18H1003P2	336.23	336.0105	64.30	3.00	14.28	18.42	C18H1609P2	438.27	438.0269	49.33	3.68	32.86	14.13
C18H1004P2	352.23	352.0054	61.38	2.86	18.17	17.59	C18H1701P1	280.31	280.1017	77.13	6.11	5.71	11.05
C18H1005P2	368.23	368.0003	58.71	2.74	21.73	16.82	C18H1702P1	296.31	296.0966	72.96	5.78	10.80	10.45
C18H1006P2	384.22	383.9953	56.27	2.62	24.98	16.12	C18H1703P1	312.31	312.0915	69.23	5.49	15.37	9.92
C18H1007P2	400.22	399.9902	54.02	2.52	27.98	15.48	C18H1704P1	328.31	328.0864	65.85	5.22	19.49	9.43
C18H1008P2	416.22	415.9851	51.94	2.42	30.75	14.88	C18H1705P1	344.31	344.0814	62.79	4.98	23.23	9.00
C18H1009P2	432.22	431.9800	50.02	2.33	33.31	14.33	C18H1706P1	360.31	360.0763	60.00	4.76	26.64	8.60
C18H1101P1	274.26	274.0547	78.83	4.04	5.83	11.29	C18H1707P1	376.31	376.0712	57.45	4.55	29.76	8.23
C18H1102P1	290.26	290.0497	74.48	3.82	11.02	10.67	C18H1708P1	392.31	392.0661	55.11	4.37	32.63	7.90
C18H1103P1	306.26	306.0446	70.59	3.62	15.67	10.11	C18H1709P1	408.30	408.0610	52.95	4.20	35.27	7.59
C18H1104P1	322.26	322.0395	67.09	3.44	19.86	9.61	C18H1801P2	312.29	312.0833	69.23	5.81	5.12	19.84
C18H1105P1	338.26	338.0344	63.92	3.28	23.65	9.16	C18H1802P2	328.29	328.0782	65.86	5.53	9.75	18.87
C18H1106P1	354.26	354.0293	61.03	3.13	27.10	8.74	C18H1803P2	344.29	344.0731	62.80	5.27	13.94	17.99
C18H1107P1	370.26	370.0242	58.39	2.99	30.25	8.37	C18H1804P2	360.29	360.0680	60.01	5.04	17.76	17.19
C18H1108P1	386.26	386.0191	55.97	2.87	33.14	8.02	C18H1805P2	376.29	376.0629	57.46	4.82	21.26	16.46
C18H1109P1	402.26	402.0141	53.75	2.76	35.80	7.70	C18H1806P2	392.29	392.0579	55.11	4.63	24.47	15.79
C18H1201P2	306.24	306.0363	70.60	3.95	5.22	20.23	C18H1807P2	408.29	408.0528	52.95	4.44	27.43	15.17
C18H1202P2	322.24	322.0313	67.09	3.75	9.93	19.22	C18H1808P2	424.29	424.0477	50.96	4.28	30.17	14.60
C18H1203P2	338.24	338.0262	63.92	3.58	14.19	18.31	C18H1809P2	440.29	440.0426	49.10	4.12	32.70	14.07
C18H1204P2	354.24	354.0211	61.03	3.41	18.07	17.49	C18H1901P1	282.33	282.1173	76.58	6.78	5.67	10.97
C18H1205P2	370.24	370.0160	58.39	3.27	21.61	16.73	C18H1902P1	298.32	298.1123	72.47	6.42	10.73	10.38
C18H1206P2	386.24	386.0109	55.98	3.13	24.85	16.04	C18H1903P1	314.32	314.1072	68.78	6.09	15.27	9.85
C18H1207P2	402.24	402.0058	53.75	3.01	27.84	15.40	C18H1904P1	330.32	330.1021	65.45	5.80	19.37	9.38
C18H1208P2	418.24	418.0007	51.69	2.89	30.60	14.81	C18H1905P1	346.32	346.0970	62.43	5.53	23.10	8.94
C18H1209P2	434.24	433.9957	49.79	2.79	33.16	14.27	C18H1906P1	362.32	362.0919	59.67	5.29	26.49	8.55
C18H1301P1	276.28	276.0704	78.25	4.74	5.79	11.21	C18H1907P1	378.32	378.0868	57.15	5.06	29.60	8.19
C18H1302P1	292.28	292.0653	73.97	4.48	10.95	10.60	C18H1908P1	394.32	394.0817	54.83	4.86	32.46	7.85
C18H1303P1	308.28	308.0602	70.13	4.25	15.57	10.05	C18H1909P1	410.32	410.0767	52.69	4.67	35.09	7.55
C18H1304P1	324.28	324.0551	66.67	4.04	19.74	9.55	C18H2001P2	314.31	314.0989	68.79	6.41	5.09	19.71
C18H1305P1	340.28	340.0501	63.54	3.85	23.51	9.10	C18H2002P2	330.31	330.0938	65.45	6.10	9.69	18.75
C18H1306P1	356.27	356.0450	60.68	3.68	26.94	8.69	C18H2003P2	346.31	346.0888	62.43	5.82	13.86	17.89
C18H1307P1	372.27	372.0399	58.08	3.52	30.08	8.32	C18H2004P2	362.31	362.0837	59.67	5.56	17.66	17.10
C18H1308P1	388.27	388.0348	55.68	3.37	32.97	7.98	C18H2005P2	378.30	378.0786	57.15	5.33	21.15	16.38
C18H1309P1	404.27	404.0297	53.48	3.24	35.62	7.66	C18H2006P2	394.30	394.0735	54.83	5.11	24.35	15.71
C18H1401P2	308.26	308.0520	70.14	4.58	5.19	20.10	C18H2007P2	410.30	410.0684	52.69	4.91	27.30	15.10
C18H1402P2	324.26	324.0469	66.68	4.35	9.87	19.10	C18H2008P2	426.30	426.0633	50.72	4.73	30.02	14.53
C18H1403P2	340.26	340.0418	63.54	4.15	14.11	18.21	C18H2009P2	442.30	442.0582	48.88	4.56	32.56	14.01
C18H1404P2	356.26	356.0367	60.69	3.96	17.96	17.39	C18H2101P1	284.34	284.1330	76.04	7.44	5.63	10.89

C18H2102P1	300.34	300.1279	71.99	7.05	10.65	10.31	C18H2708P1	402.38	402.1443	53.73	6.76	31.81	7.70
C18H2103P1	316.34	316.1228	68.34	6.69	15.17	9.79	C18H2709P1	418.38	418.1393	51.68	6.50	34.42	7.40
C18H2104P1	332.34	332.1177	65.05	6.37	19.26	9.32	C18H2801P2	322.37	322.1615	67.07	8.75	4.96	19.22
C18H2105P1	348.34	348.1127	62.07	6.08	22.97	8.89	C18H2802P2	338.37	338.1564	63.89	8.34	9.46	18.31
C18H2106P1	364.34	364.1076	59.34	5.81	26.35	8.50	C18H2803P2	354.37	354.1514	61.01	7.96	13.54	17.48
C18H2107P1	380.34	380.1025	56.84	5.57	29.45	8.14	C18H2804P2	370.37	370.1463	58.37	7.62	17.28	16.73
C18H2108P1	396.34	396.0974	54.55	5.34	32.29	7.82	C18H2805P2	386.37	386.1412	55.96	7.30	20.70	16.03
C18H2109P1	412.34	412.0923	52.43	5.13	34.92	7.51	C18H2806P2	402.37	402.1361	53.73	7.01	23.86	15.40
C18H2201P2	316.32	316.1146	68.35	7.01	5.06	19.58	C18H2807P2	418.37	418.1310	51.68	6.75	26.77	14.81
C18H2202P2	332.32	332.1095	65.06	6.67	9.63	18.64	C18H2808P2	434.37	434.1259	49.77	6.50	29.47	14.26
C18H2203P2	348.32	348.1044	62.07	6.37	13.78	17.78	C18H2809P2	450.37	450.1208	48.01	6.27	31.97	13.75
C18H2204P2	364.32	364.0993	59.34	6.09	17.57	17.00	C18H2901P1	292.41	292.1956	73.94	10.00	5.47	10.59
C18H2205P2	380.32	380.0942	56.85	5.83	21.03	16.29	C18H2902P1	308.40	308.1905	70.10	9.48	10.38	10.04
C18H2206P2	396.32	396.0892	54.55	5.60	24.22	15.63	C18H2903P1	324.40	324.1854	66.65	9.01	14.80	9.55
C18H2207P2	412.32	412.0841	52.44	5.38	27.16	15.02	C18H2904P1	340.40	340.1803	63.51	8.59	18.80	9.10
C18H2208P2	428.32	428.0790	50.48	5.18	29.88	14.46	C18H2905P1	356.40	356.1752	60.66	8.20	22.45	8.69
C18H2209P2	444.32	444.0739	48.66	4.99	32.41	13.94	C18H2906P1	372.40	372.1702	58.06	7.85	25.78	8.32
C18H2301P1	286.36	286.1486	75.50	8.10	5.59	10.82	C18H2907P1	388.40	388.1651	55.66	7.53	28.84	7.97
C18H2302P1	302.36	302.1436	71.51	7.67	10.58	10.24	C18H2908P1	404.40	404.1600	53.46	7.23	31.65	7.66
C18H2303P1	318.36	318.1385	67.91	7.28	15.08	9.73	C18H2909P1	420.40	420.1549	51.43	6.95	34.25	7.37
C18H2304P1	334.36	334.1334	64.66	6.93	19.14	9.26	C18H3001P2	324.39	324.1772	66.65	9.32	4.93	19.10
C18H2305P1	350.35	350.1283	61.71	6.62	22.83	8.84	C18H3002P2	340.39	340.1721	63.52	8.88	9.40	18.20
C18H2306P1	366.35	366.1232	59.01	6.33	26.20	8.45	C18H3003P2	356.39	356.1670	60.66	8.48	13.47	17.38
C18H2307P1	382.35	382.1181	56.54	6.06	29.29	8.10	C18H3004P2	372.38	372.1619	58.06	8.12	17.19	16.64
C18H2308P1	398.35	398.1130	54.27	5.82	32.13	7.78	C18H3005P2	388.38	388.1568	55.67	7.79	20.60	15.95
C18H2309P1	414.35	414.1080	52.18	5.60	34.75	7.48	C18H3006P2	404.38	404.1518	53.46	7.48	23.74	15.32
C18H2401P2	318.34	318.1302	67.92	7.60	5.03	19.46	C18H3007P2	420.38	420.1467	51.43	7.19	26.64	14.74
C18H2402P2	334.34	334.1251	64.67	7.24	9.57	18.53	C18H3008P2	436.38	436.1416	49.54	6.93	29.33	14.20
C18H2403P2	350.34	350.1201	61.71	6.91	13.70	17.68	C18H3009P2	452.38	452.1365	47.79	6.68	31.83	13.69
C18H2404P2	366.34	366.1150	59.02	6.60	17.47	16.91	C18H3101P1	294.42	294.2112	73.43	10.61	5.43	10.52
C18H2405P2	382.34	382.1099	56.55	6.33	20.92	16.20	C18H3102P1	310.42	310.2062	69.65	10.07	10.31	9.98
C18H2406P2	398.34	398.1048	54.28	6.07	24.10	15.55	C18H3103P1	326.42	326.2011	66.23	9.57	14.70	9.49
C18H2407P2	414.34	414.0997	52.18	5.84	27.03	14.95	C18H3104P1	342.42	342.1960	63.14	9.13	18.69	9.05
C18H2408P2	430.33	430.0946	50.24	5.62	29.74	14.40	C18H3105P1	358.42	358.1909	60.32	8.72	22.32	8.64
C18H2409P2	446.33	446.0895	48.44	5.42	32.26	13.88	C18H3106P1	374.42	374.1858	57.74	8.35	25.64	8.27
C18H2501P1	288.37	288.1643	74.97	8.74	5.55	10.74	C18H3107P1	390.42	390.1807	55.38	8.00	28.69	7.93
C18H2502P1	304.37	304.1592	71.03	8.28	10.51	10.18	C18H3108P1	406.42	406.1756	53.20	7.69	31.49	7.62
C18H2503P1	320.37	320.1541	67.48	7.87	14.98	9.67	C18H3109P1	422.42	422.1706	51.18	7.40	34.09	7.33
C18H2504P1	336.37	336.1490	64.27	7.49	19.03	9.21	C18H3201P2	326.40	326.1928	66.24	9.88	4.90	18.98
C18H2505P1	352.37	352.1439	61.36	7.15	22.70	8.79	C18H3202P2	342.40	342.1877	63.14	9.42	9.35	18.09
C18H2506P1	368.37	368.1389	58.65	6.84	26.06	8.41	C18H3203P2	358.40	358.1827	60.32	9.00	13.39	17.28
C18H2507P1	384.37	384.1338	56.25	6.56	29.14	8.06	C18H3204P2	374.40	374.1776	57.75	8.62	17.09	16.55
C18H2508P1	400.37	400.1287	54.00	6.29	31.97	7.74	C18H3205P2	390.40	390.1725	55.38	8.26	20.49	15.87
C18H2509P1	416.37	416.1236	51.93	6.05	34.58	7.44	C18H3206P2	406.40	406.1674	53.20	7.94	23.62	15.24
C18H2601P2	320.35	320.1459	67.49	8.18	4.99	19.34	C18H3207P2	422.40	422.1623	51.18	7.64	26.51	14.67
C18H2602P2	336.35	336.1408	64.28	7.79	9.51	18.42	C18H3208P2	438.40	438.1572	49.32	7.36	29.20	14.13
C18H2603P2	352.35	352.1357	61.36	7.44	13.62	17.58	C18H3209P2	454.40	454.1521	47.58	7.10	31.69	13.63
C18H2604P2	368.35	368.1306	58.65	7.11	17.37	16.82	C18H3301P1	296.44	296.2269	72.93	11.22	5.40	10.45
C18H2605P2	384.35	384.1255	56.25	6.82	20.81	16.12	C18H3302P1	312.44	312.2218	69.20	10.65	10.24	9.91
C18H2606P2	400.35	400.1205	54.00	6.55	23.98	15.47	C18H3303P1	328.44	328.2167	65.83	10.13	14.61	9.43
C18H2607P2	416.35	416.1154	51.93	6.29	26.90	14.88	C18H3304P1	344.44	344.2116	62.77	9.66	18.58	8.99
C18H2608P2	432.35	432.1103	50.01	6.06	29.60	14.33	C18H3305P1	360.43	360.2065	59.98	9.23	22.19	8.59
C18H2609P2	448.35	448.1052	48.22	5.85	32.12	13.82	C18H3306P1	376.43	376.2015	57.43	8.84	25.50	8.23
C18H2701P1	290.39	290.1799	74.45	9.37	5.51	10.67	C18H3307P1	392.43	392.1964	55.09	8.48	28.54	7.89
C18H2702P1	306.39	306.1749	70.56	8.88	10.44	10.11	C18H3308P1	408.43	408.1913	52.93	8.14	31.34	7.58
C18H2703P1	322.39	322.1698	67.06	8.44	14.89	9.61	C18H3309P1	424.43	424.1862	50.94	7.84	33.93	7.30
C18H2704P1	338.39	338.1647	63.89	8.04	18.91	9.15	C18H3401P2	328.42	328.2085	65.83	10.44	4.87	18.86
C18H2705P1	354.39	354.1596	61.01	7.68	22.57	8.74	C18H3402P2	344.42	344.2034	62.77	9.95	9.29	17.99
C18H2706P1	370.39	370.1545	58.37	7.35	25.92	8.36	C18H3403P2	360.42	360.1983	59.99	9.51	13.32	17.19
C18H2707P1	386.39	386.1494	55.95	7.04	28.99	8.02	C18H3404P2	376.42	376.1932	57.44	9.10	17.00	16.46

C18H34O5P2	392.42	392.1881	55.09	8.73	20.39	15.79
C18H34O6P2	408.42	408.1830	52.94	8.39	23.50	15.17
C18H34O7P2	424.42	424.1780	50.94	8.07	26.39	14.60
C18H34O8P2	440.41	440.1729	49.09	7.78	29.06	14.07
C18H34O9P2	456.41	456.1678	47.37	7.51	31.55	13.57
C18H35O1P1	298.45	298.2425	72.44	11.82	5.36	10.38
C18H35O2P1	314.45	314.2375	68.75	11.22	10.18	9.85
C18H35O3P1	330.45	330.2324	65.43	10.68	14.53	9.37
C18H35O4P1	346.45	346.2273	62.40	10.18	18.47	8.94
C18H35O5P1	362.45	362.2222	59.65	9.73	22.07	8.55
C18H35O6P1	378.45	378.2171	57.13	9.32	25.37	8.18
C18H35O7P1	394.45	394.2120	54.81	8.94	28.39	7.85
C18H35O8P1	410.45	410.2069	52.67	8.60	31.18	7.55
C18H35O9P1	426.45	426.2019	50.70	8.27	33.77	7.26
C18H36O1P2	330.43	330.2241	65.43	10.98	4.84	18.75
C18H36O2P2	346.43	346.2190	62.41	10.47	9.24	17.88
C18H36O3P2	362.43	362.2140	59.65	10.01	13.24	17.09
C18H36O4P2	378.43	378.2089	57.13	9.59	16.91	16.37
C18H36O5P2	394.43	394.2038	54.81	9.20	20.28	15.71
C18H36O6P2	410.43	410.1987	52.68	8.84	23.39	15.09
C18H36O7P2	426.43	426.1936	50.70	8.51	26.26	14.53
C18H36O8P2	442.43	442.1885	48.87	8.20	28.93	14.00
C18H36O9P2	458.43	458.1834	47.16	7.92	31.41	13.51
C18H37O1P1	300.47	300.2582	71.95	12.41	5.32	10.31
C18H37O2P1	316.47	316.2531	68.32	11.78	10.11	9.79
C18H37O3P1	332.47	332.2480	65.03	11.22	14.44	9.32
C18H37O4P1	348.47	348.2429	62.04	10.70	18.37	8.89
C18H37O5P1	364.47	364.2378	59.32	10.23	21.95	8.50
C18H37O6P1	380.47	380.2328	56.83	9.80	25.23	8.14
C18H37O7P1	396.47	396.2277	54.53	9.41	28.25	7.81
C18H37O8P1	412.46	412.2226	52.42	9.04	31.03	7.51
C18H37O9P1	428.46	428.2175	50.46	8.70	33.61	7.23
C18H38O1P2	332.45	332.2398	65.03	11.52	4.81	18.63
C18H38O2P2	348.45	348.2347	62.05	10.99	9.18	17.78
C18H38O3P2	364.45	364.2296	59.32	10.51	13.17	17.00
C18H38O4P2	380.45	380.2245	56.83	10.07	16.82	16.28
C18H38O5P2	396.45	396.2194	54.53	9.66	20.18	15.63
C18H38O6P2	412.45	412.2143	52.42	9.29	23.27	15.02
C18H38O7P2	428.45	428.2093	50.46	8.94	26.14	14.46
C18H38O8P2	444.45	444.2042	48.64	8.62	28.80	13.94
C18H38O9P2	460.45	460.1991	46.95	8.32	31.27	13.45
C18H39O1P1	302.48	302.2738	71.47	13.00	5.29	10.24
C18H39O2P1	318.48	318.2688	67.88	12.34	10.05	9.73
C18H39O3P1	334.48	334.2637	64.64	11.75	14.35	9.26
C18H39O4P1	350.48	350.2586	61.69	11.22	18.26	8.84
C18H39O5P1	366.48	366.2535	58.99	10.73	21.83	8.45
C18H39O6P1	382.48	382.2484	56.53	10.28	25.10	8.10
C18H39O7P1	398.48	398.2433	54.26	9.87	28.11	7.77
C18H39O8P1	414.48	414.2382	52.16	9.48	30.88	7.47
C18H39O9P1	430.48	430.2332	50.22	9.13	33.45	7.20
C18H40O1P2	334.47	334.2554	64.64	12.05	4.78	18.52
C18H40O2P2	350.47	350.2503	61.69	11.50	9.13	17.68
C18H40O3P2	366.47	366.2453	59.00	11.00	13.10	16.90
C18H40O4P2	382.46	382.2402	56.53	10.54	16.73	16.20
C18H40O5P2	398.46	398.2351	54.26	10.12	20.08	15.55
C18H40O6P2	414.46	414.2300	52.16	9.73	23.16	14.95
C18H40O7P2	430.46	430.2249	50.23	9.37	26.02	14.39
C18H40O8P2	446.46	446.2198	48.43	9.03	28.67	13.88
C18H40O9P2	462.46	462.2147	46.75	8.72	31.14	13.40
C19H 9O1P1	284.26	284.0391	80.28	3.19	5.63	10.90

C19H 9O2P1	300.26	300.0340	76.01	3.02	10.66	10.32
C19H 9O3P1	316.26	316.0289	72.16	2.87	15.18	9.79
C19H 9O4P1	332.25	332.0238	68.69	2.73	19.26	9.32
C19H 9O5P1	348.25	348.0188	65.53	2.60	22.97	8.89
C19H 9O6P1	364.25	364.0137	62.65	2.49	26.35	8.50
C19H 9O7P1	380.25	380.0086	60.02	2.39	29.45	8.15
C19H 9O8P1	396.25	396.0035	57.59	2.29	32.30	7.82
C19H 9O9P1	412.25	411.9984	55.36	2.20	34.93	7.51
C19H10O1P2	316.24	316.0207	72.16	3.19	5.06	19.59
C19H10O2P2	332.24	332.0156	68.69	3.03	9.63	18.65
C19H10O3P2	348.24	348.0105	65.53	2.89	13.78	17.79
C19H10O4P2	364.24	364.0054	62.65	2.77	17.57	17.01
C19H10O5P2	380.24	380.0003	60.02	2.65	21.04	16.29
C19H10O6P2	396.24	395.9953	57.59	2.54	24.23	15.63
C19H10O7P2	412.23	411.9902	55.36	2.45	27.17	15.03
C19H10O8P2	428.23	427.9851	53.29	2.35	29.89	14.47
C19H10O9P2	444.23	443.9800	51.37	2.27	32.41	13.94
C19H11O1P1	286.27	286.0547	79.72	3.87	5.59	10.82
C19H11O2P1	302.27	302.0497	75.50	3.67	10.59	10.25
C19H11O3P1	318.27	318.0446	71.70	3.48	15.08	9.73
C19H11O4P1	334.27	334.0395	68.27	3.32	19.15	9.27
C19H11O5P1	350.27	350.0344	65.15	3.17	22.84	8.84
C19H11O6P1	366.27	366.0293	62.31	3.03	26.21	8.46
C19H11O7P1	382.27	382.0242	59.70	2.90	29.30	8.10
C19H11O8P1	398.27	398.0191	57.30	2.78	32.14	7.78
C19H11O9P1	414.27	414.0141	55.09	2.68	34.76	7.48
C19H12O1P2	318.25	318.0363	71.71	3.80	5.03	19.46
C19H12O2P2	334.25	334.0313	68.28	3.62	9.57	18.53
C19H12O3P2	350.25	350.0262	65.16	3.45	13.70	17.69
C19H12O4P2	366.25	366.0211	62.31	3.30	17.47	16.91
C19H12O5P2	382.25	382.0160	59.70	3.16	20.93	16.21
C19H12O6P2	398.25	398.0109	57.30	3.04	24.10	15.55
C19H12O7P2	414.25	414.0058	55.09	2.92	27.04	14.95
C19H12O8P2	430.25	430.0007	53.04	2.81	29.75	14.40
C19H12O9P2	446.25	445.9957	51.14	2.71	32.27	13.88
C19H13O1P1	288.29	288.0704	79.16	4.55	5.55	10.74
C19H13O2P1	304.29	304.0653	75.00	4.31	10.52	10.18
C19H13O3P1	320.29	320.0602	71.25	4.09	14.99	9.67
C19H13O4P1	336.29	336.0551	67.86	3.90	19.03	9.21
C19H13O5P1	352.29	352.0501	64.78	3.72	22.71	8.79
C19H13O6P1	368.29	368.0450	61.97	3.56	26.07	8.41
C19H13O7P1	384.29	384.0399	59.39	3.41	29.14	8.06
C19H13O8P1	400.28	400.0348	57.01	3.27	31.98	7.74
C19H13O9P1	416.28	416.0297	54.82	3.15	34.59	7.44
C19H14O1P2	320.27	320.0520	71.26	4.41	5.00	19.34
C19H14O2P2	336.27	336.0469	67.87	4.20	9.52	18.42
C19H14O3P2	352.27	352.0418	64.78	4.01	13.63	17.59
C19H14O4P2	368.27	368.0367	61.97	3.83	17.38	16.82
C19H14O5P2	384.27	384.0316	59.39	3.67	20.82	16.12
C19H14O6P2	400.27	400.0266	57.01	3.53	23.98	15.48
C19H14O7P2	416.27	416.0215	54.82	3.39	26.90	14.88
C19H14O8P2	432.27	432.0164	52.79	3.26	29.61	14.33
C19H14O9P2	448.27	448.0113	50.91	3.15	32.12	13.82
C19H15O1P1	290.30	290.0860	78.61	5.21	5.51	10.67
C19H15O2P1	306.30	306.0810	74.51	4.94	10.45	10.11
C19H15O3P1	322.30	322.0759	70.81	4.69	14.89	9.61
C19H15O4P1	338.30	338.0708	67.46	4.47	18.92	9.16
C19H15O5P1	354.30	354.0657	64.41	4.27	22.58	8.74
C19H15O6P1	370.30	370.0606	61.63	4.08	25.92	8.36
C19H15O7P1	386.30	386.0555	59.08	3.91	28.99	8.02

C19H15O8P1	402.30	402.0504	56.73	3.76	31.82	7.70
C19H15O9P1	418.30	418.0454	54.56	3.61	34.42	7.40
C19H16O1P2	322.29	322.0676	70.81	5.00	4.96	19.22
C19H16O2P2	338.29	338.0625	67.46	4.77	9.46	18.31
C19H16O3P2	354.29	354.0575	64.41	4.55	13.55	17.49
C19H16O4P2	370.28	370.0524	61.63	4.36	17.28	16.73
C19H16O5P2	386.28	386.0473	59.08	4.18	20.71	16.04
C19H16O6P2	402.28	402.0422	56.73	4.01	23.86	15.40
C19H16O7P2	418.28	418.0371	54.56	3.86	26.78	14.81
C19H16O8P2	434.28	434.0320	52.55	3.71	29.47	14.26
C19H16O9P2	450.28	450.0269	50.68	3.58	31.98	13.76
C19H17O1P1	292.32	292.1017	78.07	5.86	5.47	10.60
C19H17O2P1	308.32	308.0966	74.02	5.56	10.38	10.05
C19H17O3P1	324.32	324.0915	70.37	5.28	14.80	9.55
C19H17O4P1	340.32	340.0864	67.06	5.04	18.81	9.10
C19H17O5P1	356.32	356.0814	64.05	4.81	22.45	8.69
C19H17O6P1	372.32	372.0763	61.29	4.60	25.78	8.32
C19H17O7P1	388.32	388.0712	58.77	4.41	28.84	7.98
C19H17O8P1	404.32	404.0661	56.44	4.24	31.66	7.66
C19H17O9P1	420.32	420.0610	54.30	4.08	34.26	7.37
C19H18O1P2	324.30	324.0833	70.37	5.59	4.93	19.10
C19H18O2P2	340.30	340.0782	67.06	5.33	9.40	18.20
C19H18O3P2	356.30	356.0731	64.05	5.09	13.47	17.39
C19H18O4P2	372.30	372.0680	61.30	4.87	17.19	16.64
C19H18O5P2	388.30	388.0629	58.77	4.67	20.60	15.95
C19H18O6P2	404.30	404.0579	56.45	4.49	23.74	15.32
C19H18O7P2	420.30	420.0528	54.30	4.32	26.65	14.74
C19H18O8P2	436.30	436.0477	52.31	4.16	29.34	14.20
C19H18O9P2	452.30	452.0426	50.46	4.01	31.84	13.70
C19H19O1P1	294.34	294.1173	77.53	6.51	5.44	10.52
C19H19O2P1	310.34	310.1123	73.54	6.17	10.31	9.98
C19H19O3P1	326.34	326.1072	69.93	5.87	14.71	9.49
C19H19O4P1	342.33	342.1021	66.66	5.59	18.69	9.05
C19H19O5P1	358.33	358.0970	63.69	5.34	22.32	8.64
C19H19O6P1	374.33	374.0919	60.96	5.12	25.64	8.27
C19H19O7P1	390.33	390.0868	58.47	4.91	28.69	7.94
C19H19O8P1	406.33	406.0817	56.16	4.71	31.50	7.62
C19H19O9P1	422.33	422.0767	54.04	4.53	34.10	7.33
C19H20O1P2	326.32	326.0989	69.94	6.18	4.90	18.98
C19H20O2P2	342.32	342.0938	66.67	5.89	9.35	18.10
C19H20O3P2	358.32	358.0888	63.65	5.63	13.40	17.29
C19H20O4P2	374.32	374.0837	60.97	5.39	17.10	16.55
C19H20O5P2	390.32	390.0786	58.47	5.16	20.50	15.87
C19H20O6P2	406.32	406.0735	56.17	4.96	23.63	15.25
C19H20O7P2	422.31	422.0684	54.04	4.77	26.52	14.67
C19H20O8P2	438.31	438.0633	52.07	4.60	29.20	14.13
C19H20O9P2	454.31	454.0582	50.23	4.44	31.69	13.64
C19H21O1P1	296.35	296.1330	77.01	7.14	5.40	10.45
C19H21O2P1	312.35	312.1279	73.06	6.78	10.24	9.92
C19H21O3P1	328.35	328.1228	69.50	6.45	14.62	9.43
C19H21O4P1	344.35	344.1177	66.27	6.15	18.59	8.99
C19H21O5P1	360.35	360.1127	63.33	5.87	22.20	8.60
C19H21O6P1	376.35	376.1076	60.64	5.62	25.51	8.23
C19H21O7P1	392.35	392.1025	58.17	5.40	28.54	7.89
C19H21O8P1	408.35	408.0974	55.89	5.18	31.34	7.59
C19H21O9P1	424.35	424.0923	53.78	4.99	33.93	7.30
C19H22O1P2	328.33	328.1146	69.51	6.75	4.87	18.87
C19H22O2P2	344.33	344.1095	66.28	6.44	9.29	17.99
C19H22O3P2	360.33	360.1044	63.33	6.15	13.32	17.19
C19H22O4P2	376.33	376.0993	60.64	5.89	17.01	16.46

C19H22O5P2	392.33	392.0942	58.17	5.65	20.39	15.79
C19H22O6P2	408.33	408.0892	55.89	5.43	23.51	15.17
C19H22O7P2	424.33	424.0841	53.78	5.23	26.39	14.60
C19H22O8P2	440.33	440.0790	51.83	5.04	29.07	14.07
C19H22O9P2	456.33	456.0739	50.01	4.86	31.55	13.58
C19H23O1P1	298.37	298.1486	76.49	7.77	5.36	10.38
C19H23O2P1	314.37	314.1436	72.59	7.37	10.18	9.85
C19H23O3P1	330.37	330.1385	69.08	7.02	14.53	9.38
C19H23O4P1	346.37	346.1334	65.89	6.69	18.48	8.94
C19H23O5P1	362.37	362.1283	62.98	6.40	22.08	8.55
C19H23O6P1	378.37	378.1232	60.32	6.13	25.37	8.19
C19H23O7P1	394.36	394.1181	57.87	5.88	28.40	7.85
C19H23O8P1	410.36	410.1130	55.61	5.65	31.19	7.55
C19H23O9P1	426.36	426.1080	53.53	5.44	33.77	7.26
C19H24O1P2	330.35	330.1302	69.08	7.32	4.84	18.75
C19H24O2P2	346.35	346.1251	65.89	6.98	9.24	17.89
C19H24O3P2	362.35	362.1201	62.98	6.68	13.25	17.10
C19H24O4P2	378.35	378.1150	60.32	6.39	16.91	16.37
C19H24O5P2	394.35	394.1099	57.87	6.13	20.29	15.71
C19H24O6P2	410.35	410.1048	55.61	5.90	23.39	15.10
C19H24O7P2	426.35	426.0997	53.53	5.67	26.27	14.53
C19H24O8P2	442.35	442.0946	51.59	5.47	28.94	14.00
C19H24O9P2	458.35	458.0895	49.79	5.28	31.42	13.52
C19H25O1P1	300.38	300.1643	75.97	8.39	5.33	10.31
C19H25O2P1	316.38	316.1592	72.13	7.96	10.11	9.79
C19H25O3P1	332.38	332.1541	68.66	7.58	14.44	9.32
C19H25O4P1	348.38	348.1490	65.51	7.23	18.37	8.89
C19H25O5P1	364.38	364.1439	62.63	6.92	21.95	8.50
C19H25O6P1	380.38	380.1389	60.00	6.62	25.24	8.14
C19H25O7P1	396.38	396.1338	57.57	6.36	28.25	7.81
C19H25O8P1	412.38	412.1287	55.34	6.11	31.04	7.51
C19H25O9P1	428.38	428.1236	53.27	5.88	33.61	7.23
C19H26O1P2	332.37	332.1459	68.66	7.89	4.81	18.64
C19H26O2P2	348.37	348.1408	65.51	7.52	9.19	17.78
C19H26O3P2	364.36	364.1357	62.63	7.19	13.17	17.00
C19H26O4P2	380.36	380.1306	60.00	6.89	16.83	16.29
C19H26O5P2	396.36	396.1255	57.58	6.61	20.18	15.63
C19H26O6P2	412.36	412.1205	55.34	6.36	23.28	15.02
C19H26O7P2	428.36	428.1154	53.28	6.12	26.15	14.46
C19H26O8P2	444.36	444.1103	51.36	5.90	28.80	13.94
C19H26O9P2	460.36	460.1052	49.57	5.69	31.28	13.46
C19H27O1P1	302.40	302.1799	75.47	9.00	5.29	10.24
C19H27O2P1	318.40	318.1749	71.67	8.55	10.05	9.73
C19H27O3P1	334.40	334.1698	68.25	8.14	14.35	9.26
C19H27O4P1	350.40	350.1647	65.13	7.77	18.26	8.84
C19H27O5P1	366.40	366.1596	62.29	7.43	21.83	8.45
C19H27O6P1	382.40	382.1545	59.68	7.12	25.10	8.10
C19H27O7P1	398.40	398.1494	57.28	6.83	28.11	7.77
C19H27O8P1	414.40	414.1443	55.07	6.57	30.89	7.47
C19H27O9P1	430.40	430.1393	53.02	6.32	33.46	7.20
C19H28O1P2	334.38	334.1615	68.25	8.44	4.78	18.53
C19H28O2P2	350.38	350.1564	65.13	8.05	9.13	17.68
C19H28O3P2	366.38	366.1514	62.29	7.70	13.10	16.91
C19H28O4P2	382.38	382.1463	59.68	7.38	16.74	16.20
C19H28O5P2	398.38	398.1412	57.29	7.08	20.08	15.55
C19H28O6P2	414.38	414.1361	55.07	6.81	23.17	14.95
C19H28O7P2	430.38	430.1310	53.03	6.56	26.02	14.39
C19H28O8P2	446.38	446.1259	51.13	6.32	28.67	13.88
C19H28O9P2	462.38	462.1208	49.36	6.10	31.14	13.40
C19H29O1P1	304.42	304.1956	74.97	9.60	5.26	10.17

C19H2902P1	320.42	320.1905	71.22	9.12	9.99	9.67	C19H3508P1	422.46	422.2069	54.02	8.35	30.30	7.33
C19H2903P1	336.41	336.1854	67.84	8.69	14.27	9.21	C19H3509P1	438.46	438.2019	52.05	8.05	32.84	7.06
C19H2904P1	352.41	352.1803	64.76	8.29	18.16	8.79	C19H3601P2	342.45	342.2241	66.64	10.60	4.67	18.09
C19H2905P1	368.41	368.1752	61.94	7.93	21.71	8.41	C19H3602P2	358.45	358.2190	63.67	10.12	8.93	17.28
C19H2906P1	384.41	384.1702	59.37	7.60	24.97	8.06	C19H3603P2	374.44	374.2140	60.95	9.69	12.82	16.54
C19H2907P1	400.41	400.1651	56.99	7.30	27.97	7.74	C19H3604P2	390.44	390.2089	58.45	9.29	16.39	15.87
C19H2908P1	416.41	416.1600	54.80	7.02	30.74	7.44	C19H3605P2	406.44	406.2038	56.15	8.93	19.68	15.24
C19H2909P1	432.41	432.1549	52.78	6.76	33.30	7.16	C19H3606P2	422.44	422.1987	54.02	8.59	22.72	14.66
C19H3001P2	336.40	336.1772	67.84	8.99	4.76	18.41	C19H3607P2	438.44	438.1936	52.05	8.28	25.54	14.13
C19H3002P2	352.40	352.1721	64.76	8.58	9.08	17.58	C19H3608P2	454.44	454.1885	50.22	7.98	28.17	13.63
C19H3003P2	368.40	368.1670	61.95	8.21	13.03	16.82	C19H3609P2	470.44	470.1834	48.51	7.71	30.61	13.17
C19H3004P2	384.40	384.1619	59.37	7.87	16.65	16.12	C19H3701P1	312.48	312.2582	73.03	11.94	5.12	9.91
C19H3005P2	400.40	400.1568	57.00	7.55	19.98	15.47	C19H3702P1	328.48	328.2531	69.48	11.35	9.74	9.43
C19H3006P2	416.39	416.1518	54.81	7.26	23.05	14.88	C19H3703P1	344.48	344.2480	66.25	10.83	13.93	8.99
C19H3007P2	432.39	432.1467	52.78	6.99	25.90	14.33	C19H3704P1	360.48	360.2429	63.31	10.35	17.75	8.59
C19H3008P2	448.39	448.1416	50.90	6.74	28.55	13.82	C19H3705P1	376.48	376.2378	60.62	9.91	21.25	8.23
C19H3009P2	464.39	464.1365	49.14	6.51	31.01	13.34	C19H3706P1	392.48	392.2328	58.15	9.50	24.46	7.89
C19H3101P1	306.43	306.2112	74.47	10.20	5.22	10.11	C19H3707P1	408.48	408.2277	55.87	9.13	27.42	7.58
C19H3102P1	322.43	322.2062	70.78	9.69	9.92	9.61	C19H3708P1	424.48	424.2226	53.76	8.79	30.15	7.30
C19H3103P1	338.43	338.2011	67.43	9.23	14.18	9.15	C19H3709P1	440.48	440.2175	51.81	8.47	32.69	7.03
C19H3104P1	354.43	354.1960	64.39	8.82	18.06	8.74	C19H3801P2	344.46	344.2398	66.25	11.12	4.64	17.98
C19H3105P1	370.43	370.1909	61.61	8.44	21.60	8.36	C19H3802P2	360.46	360.2347	63.31	10.63	8.88	17.19
C19H3106P1	386.43	386.1858	59.06	8.09	24.84	8.02	C19H3803P2	376.46	376.2296	60.62	10.17	12.75	16.46
C19H3107P1	402.43	402.1807	56.71	7.76	27.83	7.70	C19H3804P2	392.46	392.2245	58.15	9.76	16.31	15.78
C19H3108P1	418.43	418.1756	54.54	7.47	30.59	7.40	C19H3805P2	408.46	408.2194	55.87	9.38	19.59	15.17
C19H3109P1	434.43	434.1706	52.53	7.19	33.15	7.13	C19H3806P2	424.46	424.2143	53.77	9.02	22.62	14.59
C19H3201P2	338.41	338.1928	67.44	9.53	4.73	18.31	C19H3807P2	440.46	440.2093	51.81	8.70	25.43	14.06
C19H3202P2	354.41	354.1877	64.39	9.10	9.03	17.48	C19H3808P2	456.46	456.2042	50.00	8.39	28.04	13.57
C19H3203P2	370.41	370.1827	61.61	8.71	12.96	16.72	C19H3809P2	472.46	472.1991	48.30	8.11	30.48	13.11
C19H3204P2	386.41	386.1776	59.06	8.35	16.56	16.03	C19H3901P1	314.50	314.2738	72.56	12.50	5.09	9.85
C19H3205P2	402.41	402.1725	56.71	8.02	19.88	15.39	C19H3902P1	330.50	330.2688	69.05	11.89	9.68	9.37
C19H3206P2	418.41	418.1674	54.54	7.71	22.94	14.81	C19H3903P1	346.49	346.2637	65.86	11.35	13.85	8.94
C19H3207P2	434.41	434.1623	52.53	7.43	25.78	14.26	C19H3904P1	362.49	362.2586	62.96	10.84	17.65	8.54
C19H3208P2	450.41	450.1572	50.67	7.16	28.42	13.75	C19H3905P1	378.49	378.2535	60.29	10.39	21.14	8.18
C19H3209P2	466.41	466.1521	48.93	6.92	30.87	13.28	C19H3906P1	394.49	394.2484	57.85	9.96	24.33	7.85
C19H3301P1	308.45	308.2269	73.99	10.78	5.19	10.04	C19H3907P1	410.49	410.2433	55.59	9.58	27.28	7.55
C19H3302P1	324.45	324.2218	70.34	10.25	9.86	9.55	C19H3908P1	426.49	426.2382	53.51	9.22	30.01	7.26
C19H3303P1	340.45	340.2167	67.03	9.77	14.10	9.10	C19H3909P1	442.49	442.2332	51.57	8.88	32.54	7.00
C19H3304P1	356.45	356.2116	64.02	9.33	17.95	8.69	C19H4001P2	346.48	346.2554	65.87	11.64	4.62	17.88
C19H3305P1	372.45	372.2065	61.27	8.93	21.48	8.32	C19H4002P2	362.48	362.2503	62.96	11.12	8.83	17.09
C19H3306P1	388.45	388.2015	58.75	8.56	24.71	7.97	C19H4003P2	378.48	378.2453	60.30	10.65	12.68	16.37
C19H3307P1	404.44	404.1964	56.43	8.22	27.69	7.66	C19H4004P2	394.48	394.2402	57.85	10.22	16.22	15.70
C19H3308P1	420.44	420.1913	54.28	7.91	30.44	7.37	C19H4005P2	410.48	410.2351	55.60	9.82	19.49	15.09
C19H3309P1	436.44	436.1862	52.29	7.62	32.99	7.10	C19H4006P2	426.47	426.2300	53.51	9.45	22.51	14.53
C19H3401P2	340.43	340.2085	67.04	10.07	4.70	18.20	C19H4007P2	442.47	442.2249	51.58	9.11	25.31	14.00
C19H3402P2	356.43	356.2034	64.03	9.62	8.98	17.38	C19H4008P2	458.47	458.2198	49.78	8.79	27.92	13.51
C19H3403P2	372.43	372.1983	61.28	9.20	12.89	16.63	C19H4009P2	474.47	474.2147	48.10	8.50	30.35	13.06
C19H3404P2	388.43	388.1932	58.75	8.82	16.48	15.95	C19H4101P1	316.51	316.2895	72.10	13.06	5.05	9.79
C19H3405P2	404.43	404.1881	56.43	8.47	19.78	15.32	C19H4102P1	332.51	332.2844	68.63	12.43	9.62	9.32
C19H3406P2	420.43	420.1830	54.28	8.15	22.83	14.73	C19H4103P1	348.51	348.2793	65.48	11.86	13.77	8.89
C19H3407P2	436.43	436.1780	52.29	7.85	25.66	14.19	C19H4104P1	364.51	364.2742	62.61	11.34	17.56	8.50
C19H3408P2	452.43	452.1729	50.44	7.57	28.29	13.69	C19H4105P1	380.51	380.2691	59.98	10.86	21.02	8.14
C19H3409P2	468.43	468.1678	48.72	7.32	30.74	13.22	C19H4106P1	396.51	396.2641	57.56	10.42	24.21	7.81
C19H3501P1	310.46	310.2425	73.51	11.36	5.15	9.98	C19H4107P1	412.51	412.2590	55.32	10.02	27.15	7.51
C19H3502P1	326.46	326.2375	69.90	10.81	9.80	9.49	C19H4108P1	428.51	428.2539	53.26	9.64	29.87	7.23
C19H3503P1	342.46	342.2324	66.64	10.30	14.02	9.04	C19H4109P1	444.51	444.2488	51.34	9.30	32.39	6.97
C19H3504P1	358.46	358.2273	63.66	9.84	17.85	8.64	C19H4201P2	348.49	348.2711	65.49	12.15	4.59	17.78
C19H3505P1	374.46	374.2222	60.94	9.42	21.36	8.27	C19H4202P2	364.49	364.2660	62.61	11.61	8.78	17.00
C19H3506P1	390.46	390.2171	58.45	9.04	24.59	7.93	C19H4203P2	380.49	380.2609	59.98	11.13	12.61	16.28
C19H3507P1	406.46	406.2120	56.15	8.68	27.55	7.62	C19H4204P2	396.49	396.2558	57.56	10.68	16.14	15.62

C19H4205P2	412.49	412.2507	55.33	10.26	19.39	15.02	C20H1602P2	350.30	350.0625	68.58	4.60	9.13	17.68
C19H4206P2	428.49	428.2456	53.26	9.88	22.40	14.46	C20H1603P2	366.30	366.0575	65.58	4.40	13.10	16.91
C19H4207P2	444.49	444.2406	51.34	9.52	25.20	13.94	C20H1604P2	382.30	382.0524	62.84	4.22	16.74	16.20
C19H4208P2	460.49	460.2355	49.56	9.19	27.80	13.45	C20H1605P2	398.30	398.0473	60.31	4.05	20.08	15.55
C19H4209P2	476.49	476.2304	47.89	8.88	30.22	13.00	C20H1606P2	414.29	414.0422	57.98	3.89	23.17	14.95
C20H1001P2	328.25	328.0207	73.18	3.07	4.87	18.87	C20H1607P2	430.29	430.0371	55.83	3.75	26.03	14.40
C20H1002P2	344.25	344.0156	69.78	2.93	9.30	17.99	C20H1608P2	446.29	446.0320	53.83	3.61	28.68	13.88
C20H1003P2	360.25	360.0105	66.68	2.80	13.32	17.20	C20H1609P2	462.29	462.0269	51.96	3.49	31.15	13.40
C20H1004P2	376.25	376.0054	63.85	2.68	17.01	16.46	C20H1701P1	304.33	304.1017	78.93	5.63	5.26	10.18
C20H1005P2	392.25	392.0003	61.24	2.57	20.39	15.79	C20H1702P1	320.33	320.0966	74.99	5.35	9.99	9.67
C20H1006P2	408.25	407.9953	58.84	2.47	23.51	15.17	C20H1703P1	336.33	336.0915	71.42	5.09	14.27	9.21
C20H1007P2	424.25	423.9902	56.62	2.38	26.40	14.60	C20H1704P1	352.33	352.0864	68.18	4.86	18.16	8.79
C20H1008P2	440.25	439.9851	54.57	2.29	29.07	14.07	C20H1705P1	368.33	368.0814	65.22	4.65	21.72	8.41
C20H1009P2	456.24	455.9800	52.65	2.21	31.56	13.58	C20H1706P1	384.33	384.0763	62.50	4.46	24.98	8.06
C20H1101P1	298.28	298.0547	80.54	3.72	5.36	10.38	C20H1707P1	400.33	400.0712	60.01	4.28	27.98	7.74
C20H1102P1	314.28	314.0497	76.44	3.53	10.18	9.86	C20H1708P1	416.33	416.0661	57.70	4.12	30.74	7.44
C20H1103P1	330.28	330.0446	72.73	3.36	14.53	9.38	C20H1709P1	432.33	432.0610	55.57	3.96	33.31	7.16
C20H1104P1	346.28	346.0395	69.37	3.20	18.48	8.94	C20H1801P2	336.31	336.0833	71.43	5.39	4.76	18.42
C20H1105P1	362.28	362.0344	66.31	3.06	22.08	8.55	C20H1802P2	352.31	352.0782	68.18	5.15	9.08	17.58
C20H1106P1	378.28	378.0293	63.50	2.93	25.38	8.19	C20H1803P2	368.31	368.0731	65.22	4.93	13.03	16.82
C20H1107P1	394.28	394.0242	60.93	2.81	28.41	7.86	C20H1804P2	384.31	384.0680	62.51	4.72	16.65	16.12
C20H1108P1	410.28	410.0191	58.55	2.70	31.20	7.55	C20H1805P2	400.31	400.0629	60.01	4.53	19.98	15.47
C20H1109P1	426.28	426.0141	56.35	2.60	33.78	7.27	C20H1806P2	416.31	416.0579	57.70	4.36	23.06	14.88
C20H1201P2	330.27	330.0363	72.74	3.66	4.84	18.76	C20H1807P2	432.31	432.0528	55.57	4.20	25.91	14.33
C20H1202P2	346.27	346.0313	69.38	3.49	9.24	17.89	C20H1808P2	448.31	448.0477	53.58	4.05	28.55	13.82
C20H1203P2	362.26	362.0262	66.31	3.34	13.25	17.10	C20H1809P2	464.31	464.0426	51.74	3.91	31.01	13.34
C20H1204P2	378.26	378.0211	63.51	3.20	16.92	16.38	C20H1901P1	306.35	306.1173	78.42	6.25	5.22	10.11
C20H1205P2	394.26	394.0160	60.93	3.07	20.29	15.71	C20H1902P1	322.35	322.1123	74.52	5.94	9.93	9.61
C20H1206P2	410.26	410.0109	58.55	2.95	23.40	15.10	C20H1903P1	338.35	338.1072	71.00	5.66	14.19	9.15
C20H1207P2	426.26	426.0058	56.36	2.84	26.27	14.53	C20H1904P1	354.35	354.1021	67.79	5.40	18.06	8.74
C20H1208P2	442.26	442.0007	54.32	2.73	28.94	14.01	C20H1905P1	370.35	370.0970	64.86	5.17	21.60	8.36
C20H1209P2	458.26	457.9957	52.42	2.64	31.42	13.52	C20H1906P1	386.34	386.0919	62.18	4.96	24.85	8.02
C20H1301P1	300.30	300.0704	79.99	4.36	5.33	10.31	C20H1907P1	402.34	402.0868	59.71	4.76	27.84	7.70
C20H1302P1	316.30	316.0653	75.95	4.14	10.12	9.79	C20H1908P1	418.34	418.0817	57.42	4.58	30.60	7.40
C20H1303P1	332.30	332.0602	72.29	3.94	14.44	9.32	C20H1909P1	434.34	434.0767	55.31	4.41	33.15	7.13
C20H1304P1	348.30	348.0551	68.97	3.76	18.37	8.89	C20H2001P2	338.33	338.0989	71.00	5.96	4.73	18.31
C20H1305P1	364.30	364.0501	65.94	3.60	21.96	8.50	C20H2002P2	354.33	354.0938	67.80	5.69	9.03	17.48
C20H1306P1	380.30	380.0450	63.17	3.45	25.24	8.14	C20H2003P2	370.33	370.0888	64.87	5.44	12.96	16.73
C20H1307P1	396.30	396.0399	60.62	3.31	28.26	7.82	C20H2004P2	386.33	386.0837	62.18	5.22	16.57	16.03
C20H1308P1	412.30	412.0348	58.26	3.18	31.04	7.51	C20H2005P2	402.33	402.0786	59.71	5.01	19.88	15.40
C20H1309P1	428.30	428.0297	56.09	3.06	33.62	7.23	C20H2006P2	418.33	418.0735	57.42	4.82	22.95	14.81
C20H1401P2	332.28	332.0520	72.30	4.25	4.82	18.64	C20H2007P2	434.33	434.0684	55.31	4.64	25.79	14.26
C20H1402P2	348.28	348.0469	68.97	4.05	9.19	17.79	C20H2008P2	450.33	450.0633	53.34	4.48	28.42	13.76
C20H1403P2	364.28	364.0418	65.94	3.87	13.18	17.01	C20H2009P2	466.32	466.0582	51.51	4.32	30.88	13.28
C20H1404P2	380.28	380.0367	63.17	3.71	16.83	16.29	C20H2101P1	308.36	308.1330	77.90	6.86	5.19	10.04
C20H1405P2	396.28	396.0316	60.62	3.56	20.19	15.63	C20H2102P1	324.36	324.1279	74.06	6.53	9.87	9.55
C20H1406P2	412.28	412.0266	58.27	3.42	23.28	15.03	C20H2103P1	340.36	340.1228	70.58	6.22	14.10	9.10
C20H1407P2	428.28	428.0215	56.09	3.29	26.15	14.46	C20H2104P1	356.36	356.1177	67.41	5.94	17.96	8.69
C20H1408P2	444.28	444.0164	54.07	3.18	28.81	13.94	C20H2105P1	372.36	372.1127	64.51	5.68	21.48	8.32
C20H1409P2	460.28	460.0113	52.19	3.07	31.28	13.46	C20H2106P1	388.36	388.1076	61.86	5.45	24.72	7.98
C20H1501P1	302.32	302.0860	79.46	5.00	5.29	10.25	C20H2107P1	404.36	404.1025	59.41	5.23	27.70	7.66
C20H1502P1	318.32	318.0810	75.47	4.75	10.05	9.73	C20H2108P1	420.36	420.0974	57.15	5.04	30.45	7.37
C20H1503P1	334.31	334.0759	71.86	4.52	14.36	9.26	C20H2109P1	436.36	436.0923	55.05	4.85	33.00	7.10
C20H1504P1	350.31	350.0708	68.57	4.32	18.27	8.84	C20H2201P2	340.35	340.1146	70.58	6.52	4.70	18.20
C20H1505P1	366.31	366.0657	65.58	4.13	21.84	8.46	C20H2202P2	356.34	356.1095	67.41	6.22	8.98	17.38
C20H1506P1	382.31	382.0606	62.83	3.95	25.11	8.10	C20H2203P2	372.34	372.1044	64.52	5.96	12.89	16.64
C20H1507P1	398.31	398.0555	60.31	3.80	28.12	7.78	C20H2204P2	388.34	388.0993	61.86	5.71	16.48	15.95
C20H1508P1	414.31	414.0504	57.98	3.65	30.89	7.48	C20H2205P2	404.34	404.0942	59.41	5.48	19.78	15.32
C20H1509P1	430.31	430.0454	55.83	3.51	33.46	7.20	C20H2206P2	420.34	420.0892	57.15	5.28	22.84	14.74
C20H1601P2	334.30	334.0676	71.86	4.82	4.79	18.53	C20H2207P2	436.34	436.0841	55.05	5.08	25.67	14.20

C20H2208P2	452.34	452.0790	53.11	4.90	28.30	13.69
C20H2209P2	468.34	468.0739	51.29	4.73	30.75	13.23
C20H2301P1	310.38	310.1486	77.40	7.47	5.15	9.98
C20H2302P1	326.38	326.1436	73.60	7.10	9.80	9.49
C20H2303P1	342.38	342.1385	70.16	6.77	14.02	9.05
C20H2304P1	358.38	358.1334	67.03	6.47	17.86	8.64
C20H2305P1	374.38	374.1283	64.17	6.19	21.37	8.27
C20H2306P1	390.38	390.1232	61.54	5.94	24.59	7.93
C20H2307P1	406.38	406.1181	59.11	5.70	27.56	7.62
C20H2308P1	422.38	422.1130	56.87	5.49	30.30	7.33
C20H2309P1	438.37	438.1080	54.80	5.29	32.85	7.07
C20H2401P2	342.36	342.1302	70.17	7.07	4.67	18.09
C20H2402P2	358.36	358.1251	67.03	6.75	8.93	17.29
C20H2403P2	374.36	374.1201	64.17	6.46	12.82	16.55
C20H2404P2	390.36	390.1150	61.54	6.20	16.39	15.87
C20H2405P2	406.36	406.1099	59.12	5.95	19.69	15.24
C20H2406P2	422.36	422.1048	56.88	5.73	22.73	14.67
C20H2407P2	438.36	438.0997	54.80	5.52	25.55	14.13
C20H2408P2	454.36	454.0946	52.87	5.32	28.17	13.63
C20H2409P2	470.36	470.0895	51.07	5.14	30.61	13.17
C20H2501P1	312.40	312.1643	76.90	8.07	5.12	9.91
C20H2502P1	328.39	328.1592	73.15	7.67	9.74	9.43
C20H2503P1	344.39	344.1541	69.75	7.32	13.94	8.99
C20H2504P1	360.39	360.1490	66.66	6.99	17.76	8.59
C20H2505P1	376.39	376.1439	63.82	6.69	21.25	8.23
C20H2506P1	392.39	392.1389	61.22	6.42	24.46	7.89
C20H2507P1	408.39	408.1338	58.82	6.17	27.42	7.58
C20H2508P1	424.39	424.1287	56.60	5.94	30.16	7.30
C20H2509P1	440.39	440.1236	54.55	5.72	32.70	7.03
C20H2601P2	344.38	344.1459	69.76	7.61	4.65	17.99
C20H2602P2	360.38	360.1408	66.66	7.27	8.88	17.19
C20H2603P2	376.38	376.1357	63.83	6.96	12.75	16.46
C20H2604P2	392.38	392.1306	61.22	6.68	16.31	15.79
C20H2605P2	408.37	408.1255	58.82	6.42	19.59	15.17
C20H2606P2	424.37	424.1205	56.61	6.18	22.62	14.60
C20H2607P2	440.37	440.1154	54.55	5.95	25.43	14.07
C20H2608P2	456.37	456.1103	52.64	5.74	28.05	13.57
C20H2609P2	472.37	472.1052	50.85	5.55	30.48	13.11
C20H2701P1	314.41	314.1799	76.40	8.66	5.09	9.85
C20H2702P1	330.41	330.1749	72.70	8.24	9.68	9.37
C20H2703P1	346.41	346.1698	69.35	7.86	13.86	8.94
C20H2704P1	362.41	362.1647	66.28	7.51	17.66	8.55
C20H2705P1	378.41	378.1596	63.48	7.19	21.14	8.19
C20H2706P1	394.41	394.1545	60.91	6.90	24.34	7.85
C20H2707P1	410.41	410.1494	58.53	6.63	27.29	7.55
C20H2708P1	426.41	426.1443	56.34	6.38	30.02	7.26
C20H2709P1	442.41	442.1393	54.30	6.15	32.55	7.00
C20H2801P2	346.39	346.1615	69.35	8.15	4.62	17.88
C20H2802P2	362.39	362.1564	66.29	7.79	8.83	17.09
C20H2803P2	378.39	378.1514	63.49	7.46	12.68	16.37
C20H2804P2	394.39	394.1463	60.91	7.16	16.23	15.71
C20H2805P2	410.39	410.1412	58.54	6.88	19.49	15.09
C20H2806P2	426.39	426.1361	56.34	6.62	22.51	14.53
C20H2807P2	442.39	442.1310	54.30	6.38	25.32	14.00
C20H2808P2	458.39	458.1259	52.41	6.16	27.92	13.51
C20H2809P2	474.39	474.1208	50.64	5.95	30.35	13.06
C20H2901P1	316.43	316.1956	75.92	9.24	5.06	9.79
C20H2902P1	332.43	332.1905	72.26	8.79	9.63	9.32
C20H2903P1	348.43	348.1854	68.95	8.39	13.78	8.89
C20H2904P1	364.43	364.1803	65.92	8.02	17.56	8.50
C20H2905P1	380.42	380.1752	63.15	7.68	21.03	8.14
C20H2906P1	396.42	396.1702	60.60	7.37	24.22	7.81
C20H2907P1	412.42	412.1651	58.25	7.09	27.16	7.51
C20H2908P1	428.42	428.1600	56.07	6.82	29.88	7.23
C20H2909P1	444.42	444.1549	54.05	6.58	32.40	6.97
C20H3001P2	348.41	348.1772	68.95	8.68	4.59	17.78
C20H3002P2	364.41	364.1721	65.92	8.30	8.78	17.00
C20H3003P2	380.41	380.1670	63.15	7.95	12.62	16.28
C20H3004P2	396.41	396.1619	60.60	7.63	16.14	15.63
C20H3005P2	412.41	412.1568	58.25	7.33	19.40	15.02
C20H3006P2	428.41	428.1518	56.07	7.06	22.41	14.46
C20H3007P2	444.41	444.1467	54.05	6.80	25.20	13.94
C20H3008P2	460.40	460.1416	52.18	6.57	27.80	13.46
C20H3009P2	476.40	476.1365	50.42	6.35	30.23	13.00
C20H3101P1	318.44	318.2112	75.44	9.81	5.02	9.73
C20H3102P1	334.44	334.2062	71.83	9.34	9.57	9.26
C20H3103P1	350.44	350.2011	68.55	8.92	13.70	8.84
C20H3104P1	366.44	366.1960	65.56	8.53	17.46	8.45
C20H3105P1	382.44	382.1909	62.81	8.17	20.92	8.10
C20H3106P1	398.44	398.1858	60.29	7.84	24.09	7.77
C20H3107P1	414.44	414.1807	57.96	7.54	27.02	7.47
C20H3108P1	430.44	430.1756	55.81	7.26	29.74	7.20
C20H3109P1	446.44	446.1706	53.81	7.00	32.25	6.94
C20H3201P2	350.43	350.1928	68.55	9.20	4.57	17.68
C20H3202P2	366.42	366.1877	65.56	8.80	8.73	16.91
C20H3203P2	382.42	382.1827	62.82	8.43	12.55	16.20
C20H3204P2	398.42	398.1776	60.29	8.10	16.06	15.55
C20H3205P2	414.42	414.1725	57.97	7.78	19.30	14.95
C20H3206P2	430.42	430.1674	55.81	7.49	22.30	14.39
C20H3207P2	446.42	446.1623	53.81	7.23	25.09	13.88
C20H3208P2	462.42	462.1572	51.95	6.98	27.68	13.40
C20H3209P2	478.42	478.1521	50.21	6.74	30.10	12.95
C20H3301P1	320.46	320.2269	74.96	10.38	4.99	9.67
C20H3302P1	336.46	336.2218	71.40	9.89	9.51	9.21
C20H3303P1	352.46	352.2167	68.16	9.44	13.62	8.79
C20H3304P1	368.46	368.2116	65.20	9.03	17.37	8.41
C20H3305P1	384.46	384.2065	62.48	8.65	20.81	8.06
C20H3306P1	400.46	400.2015	59.99	8.31	23.97	7.73
C20H3307P1	416.46	416.1964	57.68	7.99	26.89	7.44
C20H3308P1	432.46	432.1913	55.55	7.69	29.60	7.16
C20H3309P1	448.45	448.1862	53.57	7.42	32.11	6.91
C20H3401P2	352.44	352.2085	68.16	9.72	4.54	17.58
C20H3402P2	368.44	368.2034	65.20	9.30	8.68	16.81
C20H3403P2	384.44	384.1983	62.49	8.91	12.49	16.11
C20H3404P2	400.44	400.1932	59.99	8.56	15.98	15.47
C20H3405P2	416.44	416.1881	57.69	8.23	19.21	14.88
C20H3406P2	432.44	432.1830	55.55	7.93	22.20	14.33
C20H3407P2	448.44	448.1780	53.57	7.64	24.97	13.81
C20H3408P2	464.44	464.1729	51.72	7.38	27.56	13.34
C20H3409P2	480.44	480.1678	50.00	7.13	29.97	12.89
C20H3501P1	322.48	322.2425	74.49	10.94	4.96	9.61
C20H3502P1	338.47	338.2375	70.97	10.42	9.45	9.15
C20H3503P1	354.47	354.2324	67.77	9.95	13.54	8.74
C20H3504P1	370.47	370.2273	64.84	9.52	17.27	8.36
C20H3505P1	386.47	386.2222	62.16	9.13	20.70	8.01
C20H3506P1	402.47	402.2171	59.69	8.77	23.85	7.70
C20H3507P1	418.47	418.2120	57.40	8.43	26.76	7.40
C20H3508P1	434.47	434.2069	55.29	8.12	29.46	7.13
C20H3509P1	450.47	450.2019	53.33	7.83	31.97	6.88
C20H3601P2	354.46	354.2241	67.77	10.24	4.51	17.48

C20H3602P2	370.46	370.2190	64.85	9.80	8.64	16.72
C20H3603P2	386.46	386.2140	62.16	9.39	12.42	16.03
C20H3604P2	402.46	402.2089	59.69	9.02	15.90	15.39
C20H3605P2	418.45	418.2038	57.41	8.67	19.12	14.80
C20H3606P2	434.45	434.1987	55.29	8.35	22.10	14.26
C20H3607P2	450.45	450.1936	53.33	8.06	24.86	13.75
C20H3608P2	466.45	466.1885	51.50	7.78	27.44	13.28
C20H3609P2	482.45	482.1834	49.79	7.52	29.85	12.84
C20H3701P1	324.49	324.2582	74.03	11.49	4.93	9.55
C20H3702P1	340.49	340.2531	70.55	10.95	9.40	9.10
C20H3703P1	356.49	356.2480	67.39	10.46	13.46	8.69
C20H3704P1	372.49	372.2429	64.49	10.01	17.18	8.32
C20H3705P1	388.49	388.2378	61.84	9.60	20.59	7.97
C20H3706P1	404.49	404.2328	59.39	9.22	23.73	7.66
C20H3707P1	420.49	420.2277	57.13	8.87	26.63	7.37
C20H3708P1	436.49	436.2226	55.04	8.54	29.32	7.10
C20H3709P1	452.49	452.2175	53.09	8.24	31.82	6.85
C20H3801P2	356.47	356.2398	67.39	10.74	4.49	17.38
C20H3802P2	372.47	372.2347	64.49	10.28	8.59	16.63
C20H3803P2	388.47	388.2296	61.84	9.86	12.36	15.95
C20H3804P2	404.47	404.2245	59.39	9.47	15.82	15.32
C20H3805P2	420.47	420.2194	57.13	9.11	19.03	14.73
C20H3806P2	436.47	436.2143	55.04	8.78	21.99	14.19
C20H3807P2	452.47	452.2093	53.09	8.47	24.75	13.69
C20H3808P2	468.47	468.2042	51.28	8.18	27.32	13.22
C20H3809P2	484.47	484.1991	49.58	7.91	29.72	12.79
C20H3901P1	326.51	326.2738	73.57	12.04	4.90	9.49
C20H3902P1	342.51	342.2688	70.14	11.48	9.34	9.04
C20H3903P1	358.51	358.2637	67.01	10.97	13.39	8.64
C20H3904P1	374.51	374.2586	64.14	10.50	17.09	8.27
C20H3905P1	390.50	390.2535	61.52	10.07	20.49	7.93
C20H3906P1	406.50	406.2484	59.09	9.67	23.62	7.62
C20H3907P1	422.50	422.2433	56.86	9.30	26.51	7.33
C20H3908P1	438.50	438.2382	54.78	8.96	29.19	7.06
C20H3909P1	454.50	454.2332	52.85	8.65	31.68	6.81
C20H4001P2	358.49	358.2554	67.01	11.25	4.46	17.28
C20H4002P2	374.49	374.2503	64.15	10.77	8.54	16.54
C20H4003P2	390.49	390.2453	61.52	10.33	12.29	15.86
C20H4004P2	406.49	406.2402	59.10	9.92	15.74	15.24
C20H4005P2	422.49	422.2351	56.86	9.54	18.93	14.66
C20H4006P2	438.49	438.2300	54.78	9.20	21.89	14.13
C20H4007P2	454.49	454.2249	52.86	8.87	24.64	13.63
C20H4008P2	470.48	470.2198	51.06	8.57	27.20	13.17
C20H4009P2	486.48	486.2147	49.38	8.29	29.60	12.73
C20H4101P1	328.52	328.2895	73.12	12.58	4.87	9.43
C20H4102P1	344.52	344.2844	69.73	12.00	9.29	8.99
C20H4103P1	360.52	360.2793	66.63	11.46	13.31	8.59
C20H4104P1	376.52	376.2742	63.80	10.98	17.00	8.23
C20H4105P1	392.52	392.2691	61.20	10.53	20.38	7.89
C20H4106P1	408.52	408.2641	58.80	10.12	23.50	7.58
C20H4107P1	424.52	424.2590	56.59	9.73	26.38	7.30
C20H4108P1	440.52	440.2539	54.53	9.38	29.06	7.03
C20H4109P1	456.52	456.2488	52.62	9.05	31.54	6.78
C20H4201P2	360.50	360.2711	66.64	11.74	4.44	17.18
C20H4202P2	376.50	376.2660	63.80	11.24	8.50	16.45
C20H4203P2	392.50	392.2609	61.20	10.79	12.23	15.78
C20H4204P2	408.50	408.2558	58.81	10.36	15.67	15.16
C20H4205P2	424.50	424.2507	56.59	9.97	18.84	14.59
C20H4206P2	440.50	440.2456	54.53	9.61	21.79	14.06
C20H4207P2	456.50	456.2406	52.62	9.27	24.53	13.57
C20H4208P2	472.50	472.2355	50.84	8.96	27.09	13.11
C20H4209P2	488.50	488.2304	49.18	8.67	29.48	12.68
C20H4301P1	330.54	330.3051	72.68	13.11	4.84	9.37
C20H4302P1	346.54	346.3000	69.32	12.51	9.23	8.94
C20H4303P1	362.54	362.2950	66.26	11.96	13.24	8.54
C20H4304P1	378.54	378.2899	63.46	11.45	16.91	8.18
C20H4305P1	394.54	394.2848	60.89	10.99	20.28	7.85
C20H4306P1	410.54	410.2797	58.51	10.56	23.38	7.54
C20H4307P1	426.54	426.2746	56.32	10.16	26.26	7.26
C20H4308P1	442.53	442.2695	54.28	9.79	28.92	7.00
C20H4309P1	458.53	458.2644	52.39	9.45	31.40	6.75
C20H4401P2	362.52	362.2867	66.26	12.23	4.41	17.09
C20H4402P2	378.52	378.2816	63.46	11.72	8.45	16.37
C20H4403P2	394.52	394.2766	60.89	11.24	12.17	15.70
C20H4404P2	410.52	410.2715	58.52	10.80	15.59	15.09
C20H4405P2	426.52	426.2664	56.32	10.40	18.76	14.52
C20H4406P2	442.52	442.2613	54.29	10.02	21.69	14.00
C20H4407P2	458.52	458.2562	52.39	9.67	24.43	13.51
C20H4408P2	474.52	474.2511	50.62	9.35	26.97	13.05
C20H4409P2	490.52	490.2460	48.97	9.04	29.36	12.63
C21H1001P2	340.26	340.0207	74.13	2.96	4.70	18.21
C21H1002P2	356.26	356.0156	70.80	2.83	8.98	17.39
C21H1003P2	372.26	372.0105	67.76	2.71	12.89	16.64
C21H1004P2	388.26	388.0054	64.97	2.60	16.48	15.96
C21H1005P2	404.26	404.0003	62.39	2.49	19.79	15.32
C21H1006P2	420.26	419.9953	60.02	2.40	22.84	14.74
C21H1007P2	436.26	435.9902	57.82	2.31	25.67	14.20
C21H1008P2	452.26	451.9851	55.77	2.23	28.30	13.70
C21H1009P2	468.26	467.9800	53.87	2.15	30.75	13.23
C21H1101P1	310.30	310.0547	81.29	3.57	5.16	9.98
C21H1102P1	326.29	326.0497	77.30	3.40	9.81	9.49
C21H1103P1	342.29	342.0446	73.69	3.24	14.02	9.05
C21H1104P1	358.29	358.0395	70.40	3.09	17.86	8.64
C21H1105P1	374.29	374.0344	67.39	2.96	21.37	8.28
C21H1106P1	390.29	390.0293	64.63	2.84	24.60	7.94
C21H1107P1	406.29	406.0242	62.08	2.73	27.57	7.62
C21H1108P1	422.29	422.0191	59.73	2.63	30.31	7.33
C21H1109P1	438.29	438.0141	57.55	2.53	32.85	7.07
C21H1201P2	342.28	342.0363	73.69	3.53	4.67	18.10
C21H1202P2	358.28	358.0313	70.40	3.38	8.93	17.29
C21H1203P2	374.28	374.0262	67.39	3.23	12.82	16.55
C21H1204P2	390.27	390.0211	64.63	3.10	16.40	15.87
C21H1205P2	406.27	406.0160	62.08	2.98	19.69	15.25
C21H1206P2	422.27	422.0109	59.73	2.86	22.73	14.67
C21H1207P2	438.27	438.0058	57.55	2.76	25.55	14.13
C21H1208P2	454.27	454.0007	55.52	2.66	28.18	13.64
C21H1209P2	470.27	469.9957	53.64	2.57	30.62	13.17
C21H1301P1	312.31	312.0704	80.76	4.20	5.12	9.92
C21H1302P1	328.31	328.0653	76.83	3.99	9.75	9.43
C21H1303P1	344.31	344.0602	73.26	3.81	13.94	9.00
C21H1304P1	360.31	360.0551	70.00	3.64	17.76	8.60
C21H1305P1	376.31	376.0501	67.03	3.48	21.26	8.23
C21H1306P1	392.31	392.0450	64.29	3.34	24.47	7.90
C21H1307P1	408.31	408.0399	61.78	3.21	27.43	7.59
C21H1308P1	424.31	424.0348	59.45	3.09	30.17	7.30
C21H1309P1	440.31	440.0297	57.29	2.98	32.70	7.03
C21H1401P2	344.29	344.0520	73.26	4.10	4.65	17.99
C21H1402P2	360.29	360.0469	70.01	3.92	8.88	17.19
C21H1403P2	376.29	376.0418	67.03	3.75	12.76	16.46
C21H1404P2	392.29	392.0367	64.30	3.60	16.31	15.79

C21H1405P2	408.29	408.0316	61.78	3.46	19.59	15.17
C21H1406P2	424.29	424.0266	59.45	3.33	22.63	14.60
C21H1407P2	440.29	440.0215	57.29	3.21	25.44	14.07
C21H1408P2	456.29	456.0164	55.28	3.09	28.05	13.58
C21H1409P2	472.29	472.0113	53.41	2.99	30.49	13.12
C21H1501P1	314.33	314.0860	80.25	4.81	5.09	9.85
C21H1502P1	330.33	330.0810	76.36	4.58	9.69	9.38
C21H1503P1	346.33	346.0759	72.83	4.37	13.86	8.94
C21H1504P1	362.33	362.0708	69.62	4.17	17.66	8.55
C21H1505P1	378.32	378.0657	66.67	4.00	21.15	8.19
C21H1506P1	394.32	394.0606	63.97	3.83	24.34	7.85
C21H1507P1	410.32	410.0555	61.47	3.68	27.29	7.55
C21H1508P1	426.32	426.0504	59.17	3.55	30.02	7.27
C21H1509P1	442.32	442.0454	57.02	3.42	32.55	7.00
C21H1601P2	346.31	346.0676	72.84	4.66	4.62	17.89
C21H1602P2	362.31	362.0625	69.62	4.45	8.83	17.10
C21H1603P2	378.31	378.0575	66.67	4.26	12.69	16.37
C21H1604P2	394.31	394.0524	63.97	4.09	16.23	15.71
C21H1605P2	410.31	410.0473	61.47	3.93	19.50	15.10
C21H1606P2	426.31	426.0422	59.17	3.78	22.52	14.53
C21H1607P2	442.31	442.0371	57.03	3.65	25.32	14.01
C21H1608P2	458.30	458.0320	55.04	3.52	27.93	13.52
C21H1609P2	474.30	474.0269	53.18	3.40	30.36	13.06
C21H1701P1	316.34	316.1017	79.73	5.42	5.06	9.79
C21H1702P1	332.34	332.0966	75.90	5.16	9.63	9.32
C21H1703P1	348.34	348.0915	72.41	4.92	13.78	8.89
C21H1704P1	364.34	364.0864	69.23	4.70	17.57	8.50
C21H1705P1	380.34	380.0814	66.32	4.51	21.03	8.14
C21H1706P1	396.34	396.0763	63.64	4.32	24.22	7.81
C21H1707P1	412.34	412.0712	61.17	4.16	27.16	7.51
C21H1708P1	428.34	428.0661	58.89	4.00	29.88	7.23
C21H1709P1	444.34	444.0610	56.77	3.86	32.41	6.97
C21H1801P2	348.32	348.0833	72.41	5.21	4.59	17.78
C21H1802P2	364.32	364.0782	69.23	4.98	8.78	17.00
C21H1803P2	380.32	380.0731	66.32	4.77	12.62	16.29
C21H1804P2	396.32	396.0680	63.64	4.58	16.15	15.63
C21H1805P2	412.32	412.0629	61.17	4.40	19.40	15.02
C21H1806P2	428.32	428.0579	58.89	4.24	22.41	14.46
C21H1807P2	444.32	444.0528	56.77	4.08	25.21	13.94
C21H1808P2	460.32	460.0477	54.80	3.94	27.81	13.46
C21H1809P2	476.32	476.0426	52.95	3.81	30.23	13.01
C21H1901P1	318.36	318.1173	79.23	6.02	5.03	9.73
C21H1902P1	334.36	334.1123	75.44	5.73	9.57	9.26
C21H1903P1	350.36	350.1072	71.99	5.47	13.70	8.84
C21H1904P1	366.36	366.1021	68.85	5.23	17.47	8.45
C21H1905P1	382.36	382.0970	65.97	5.01	20.92	8.10
C21H1906P1	398.36	398.0919	63.32	4.81	24.10	7.78
C21H1907P1	414.36	414.0868	60.87	4.62	27.03	7.48
C21H1908P1	430.35	430.0817	58.61	4.45	29.74	7.20
C21H1909P1	446.35	446.0767	56.51	4.29	32.26	6.94
C21H2001P2	350.34	350.0989	72.00	5.75	4.57	17.68
C21H2002P2	366.34	366.0938	68.85	5.50	8.73	16.91
C21H2003P2	382.34	382.0888	65.97	5.27	12.55	16.20
C21H2004P2	398.34	398.0837	63.32	5.06	16.07	15.55
C21H2005P2	414.34	414.0786	60.88	4.87	19.31	14.95
C21H2006P2	430.34	430.0735	58.61	4.68	22.31	14.40
C21H2007P2	446.34	446.0684	56.51	4.52	25.09	13.88
C21H2008P2	462.34	462.0633	54.56	4.36	27.68	13.40
C21H2009P2	478.34	478.0582	52.73	4.21	30.10	12.95
C21H2101P1	320.37	320.1330	78.73	6.61	4.99	9.67

C21H2102P1	336.37	336.1279	74.99	6.29	9.51	9.21
C21H2103P1	352.37	352.1228	71.58	6.01	13.62	8.79
C21H2104P1	368.37	368.1177	68.47	5.75	17.37	8.41
C21H2105P1	384.37	384.1127	65.62	5.51	20.81	8.06
C21H2106P1	400.37	400.1076	63.00	5.29	23.98	7.74
C21H2107P1	416.37	416.1025	60.58	5.08	26.90	7.44
C21H2108P1	432.37	432.0974	58.34	4.90	29.60	7.16
C21H2109P1	448.37	448.0923	56.26	4.72	32.12	6.91
C21H2201P2	352.36	352.1146	71.58	6.29	4.54	17.58
C21H2202P2	368.36	368.1095	68.48	6.02	8.69	16.82
C21H2203P2	384.36	384.1044	65.63	5.77	12.49	16.12
C21H2204P2	400.35	400.0993	63.00	5.54	15.99	15.47
C21H2205P2	416.35	416.0942	60.58	5.33	19.21	14.88
C21H2206P2	432.35	432.0892	58.34	5.13	22.20	14.33
C21H2207P2	448.35	448.0841	56.26	4.95	24.98	13.82
C21H2208P2	464.35	464.0790	54.32	4.78	27.56	13.34
C21H2209P2	480.35	480.0739	52.51	4.62	29.98	12.90
C21H2301P1	322.39	322.1486	78.24	7.19	4.96	9.61
C21H2302P1	338.39	338.1436	74.54	6.85	9.46	9.15
C21H2303P1	354.39	354.1385	71.17	6.54	13.54	8.74
C21H2304P1	370.39	370.1334	68.10	6.26	17.28	8.36
C21H2305P1	386.39	386.1283	65.28	6.00	20.70	8.02
C21H2306P1	402.39	402.1232	62.68	5.76	23.86	7.70
C21H2307P1	418.39	418.1181	60.29	5.54	26.77	7.40
C21H2308P1	434.39	434.1130	58.07	5.34	29.47	7.13
C21H2309P1	450.39	450.1080	56.00	5.15	31.97	6.88
C21H2401P2	354.37	354.1302	71.18	6.83	4.51	17.48
C21H2402P2	370.37	370.1251	68.10	6.53	8.64	16.73
C21H2403P2	386.37	386.1201	65.28	6.26	12.42	16.03
C21H2404P2	402.37	402.1150	62.69	6.01	15.91	15.40
C21H2405P2	418.37	418.1099	60.29	5.78	19.12	14.81
C21H2406P2	434.37	434.1048	58.07	5.57	22.10	14.26
C21H2407P2	450.37	450.0997	56.01	5.37	24.87	13.75
C21H2408P2	466.37	466.0946	54.08	5.19	27.45	13.28
C21H2409P2	482.37	482.0895	52.29	5.02	29.85	12.84
C21H2501P1	324.41	324.1643	77.75	7.77	4.93	9.55
C21H2502P1	340.41	340.1592	74.10	7.40	9.40	9.10
C21H2503P1	356.41	356.1541	70.77	7.07	13.47	8.69
C21H2504P1	372.40	372.1490	67.73	6.77	17.18	8.32
C21H2505P1	388.40	388.1439	64.94	6.49	20.60	7.97
C21H2506P1	404.40	404.1389	62.37	6.23	23.74	7.66
C21H2507P1	420.40	420.1338	60.00	5.99	26.64	7.37
C21H2508P1	436.40	436.1287	57.80	5.77	29.33	7.10
C21H2509P1	452.40	452.1236	55.75	5.57	31.83	6.85
C21H2601P2	356.39	356.1459	70.78	7.35	4.49	17.38
C21H2602P2	372.39	372.1408	67.73	7.04	8.59	16.64
C21H2603P2	388.39	388.1357	64.94	6.75	12.36	15.95
C21H2604P2	404.39	404.1306	62.37	6.48	15.83	15.32
C21H2605P2	420.39	420.1255	60.00	6.23	19.03	14.74
C21H2606P2	436.39	436.1205	57.80	6.01	22.00	14.20
C21H2607P2	452.38	452.1154	55.76	5.79	24.76	13.69
C21H2608P2	468.38	468.1103	53.85	5.60	27.33	13.23
C21H2609P2	484.38	484.1052	52.07	5.41	29.73	12.79
C21H2701P1	326.42	326.1799	77.27	8.34	4.90	9.49
C21H2702P1	342.42	342.1749	73.66	7.95	9.34	9.05
C21H2703P1	358.42	358.1698	70.37	7.59	13.39	8.64
C21H2704P1	374.42	374.1647	67.37	7.27	17.09	8.27
C21H2705P1	390.42	390.1596	64.61	6.97	20.49	7.93
C21H2706P1	406.42	406.1545	62.06	6.70	23.62	7.62
C21H2707P1	422.42	422.1494	59.71	6.44	26.51	7.33

135

C21H2708P1	438.42	438.1443	57.53	6.21	29.19	7.06
C21H2709P1	454.42	454.1393	55.51	5.99	31.69	6.82
C21H2801P2	358.40	358.1615	70.38	7.87	4.46	17.28
C21H2802P2	374.40	374.1564	67.37	7.54	8.55	16.55
C21H2803P2	390.40	390.1514	64.61	7.23	12.29	15.87
C21H2804P2	406.40	406.1463	62.07	6.94	15.75	15.24
C21H2805P2	422.40	422.1412	59.71	6.68	18.94	14.67
C21H2806P2	438.40	438.1361	57.53	6.44	21.90	14.13
C21H2807P2	454.40	454.1310	55.51	6.21	24.65	13.63
C21H2808P2	470.40	470.1259	53.62	6.00	27.21	13.17
C21H2809P2	486.40	486.1208	51.86	5.80	29.60	12.74
C21H2901P1	328.44	328.1956	76.80	8.90	4.87	9.43
C21H2902P1	344.44	344.1905	73.23	8.49	9.29	8.99
C21H2903P1	360.44	360.1854	69.98	8.11	13.32	8.59
C21H2904P1	376.44	376.1803	67.01	7.77	17.00	8.23
C21H2905P1	392.44	392.1752	64.27	7.45	20.38	7.89
C21H2906P1	408.44	408.1702	61.76	7.16	23.50	7.58
C21H2907P1	424.43	424.1651	59.43	6.89	26.39	7.30
C21H2908P1	440.43	440.1600	57.27	6.64	29.06	7.03
C21H2909P1	456.43	456.1549	55.26	6.40	31.55	6.79
C21H3001P2	360.42	360.1772	69.98	8.39	4.44	17.19
C21H3002P2	376.42	376.1721	67.01	8.03	8.50	16.46
C21H3003P2	392.42	392.1670	64.28	7.71	12.23	15.79
C21H3004P2	408.42	408.1619	61.76	7.40	15.67	15.17
C21H3005P2	424.42	424.1568	59.43	7.12	18.85	14.60
C21H3006P2	440.42	440.1518	57.27	6.87	21.80	14.07
C21H3007P2	456.42	456.1467	55.26	6.63	24.54	13.57
C21H3008P2	472.42	472.1416	53.39	6.40	27.09	13.11
C21H3009P2	488.42	488.1365	51.64	6.19	29.48	12.68
C21H3101P1	330.45	330.2112	76.33	9.46	4.84	9.37
C21H3102P1	346.45	346.2062	72.80	9.02	9.24	8.94
C21H3103P1	362.45	362.2011	69.59	8.62	13.24	8.55
C21H3104P1	378.45	378.1960	66.65	8.26	16.91	8.18
C21H3105P1	394.45	394.1909	63.95	7.92	20.28	7.85
C21H3106P1	410.45	410.1858	61.45	7.61	23.39	7.55
C21H3107P1	426.45	426.1807	59.15	7.33	26.26	7.26
C21H3108P1	442.45	442.1756	57.01	7.06	28.93	7.00
C21H3109P1	458.45	458.1706	55.02	6.82	31.41	6.76
C21H3201P2	362.44	362.1928	69.59	8.90	4.41	17.09
C21H3202P2	378.44	378.1877	66.65	8.52	8.46	16.37
C21H3203P2	394.43	394.1827	63.95	8.18	12.17	15.71
C21H3204P2	410.43	410.1776	61.46	7.86	15.59	15.09
C21H3205P2	426.43	426.1725	59.15	7.56	18.76	14.53
C21H3206P2	442.43	442.1674	57.01	7.29	21.70	14.00
C21H3207P2	458.43	458.1623	55.02	7.04	24.43	13.51
C21H3208P2	474.43	474.1572	53.17	6.80	26.98	13.06
C21H3209P2	490.43	490.1521	51.43	6.58	29.36	12.63
C21H3301P1	332.47	332.2269	75.87	10.00	4.81	9.32
C21H3302P1	348.47	348.2218	72.38	9.55	9.18	8.89
C21H3303P1	364.47	364.2167	69.21	9.13	13.17	8.50
C21H3304P1	380.47	380.2116	66.30	8.74	16.82	8.14
C21H3305P1	396.47	396.2065	63.62	8.39	20.18	7.81
C21H3306P1	412.47	412.2015	61.15	8.06	23.27	7.51
C21H3307P1	428.47	428.1964	58.87	7.76	26.14	7.23
C21H3308P1	444.47	444.1913	56.75	7.48	28.80	6.97
C21H3309P1	460.47	460.1862	54.78	7.22	31.27	6.73
C21H3401P2	364.45	364.2085	69.21	9.40	4.39	17.00
C21H3402P2	380.45	380.2034	66.30	9.01	8.41	16.28
C21H3403P2	396.45	396.1983	63.62	8.64	12.11	15.63
C21H3404P2	412.45	412.1932	61.16	8.31	15.52	15.02

C21H3405P2	428.45	428.1881	58.87	8.00	18.67	14.46
C21H3406P2	444.45	444.1830	56.75	7.71	21.60	13.94
C21H3407P2	460.45	460.1780	54.78	7.44	24.32	13.45
C21H3408P2	476.45	476.1729	52.94	7.19	26.86	13.00
C21H3409P2	492.45	492.1678	51.22	6.96	29.24	12.58
C21H3501P1	334.49	334.2425	75.41	10.55	4.78	9.26
C21H3502P1	350.49	350.2375	71.97	10.07	9.13	8.84
C21H3503P1	366.49	366.2324	68.83	9.63	13.10	8.45
C21H3504P1	382.48	382.2273	65.95	9.22	16.73	8.10
C21H3505P1	398.48	398.2222	63.30	8.85	20.08	7.77
C21H3506P1	414.48	414.2171	60.86	8.51	23.16	7.47
C21H3507P1	430.48	430.2120	58.59	8.20	26.02	7.20
C21H3508P1	446.48	446.2069	56.49	7.90	28.67	6.94
C21H3509P1	462.48	462.2019	54.54	7.63	31.14	6.70
C21H3601P2	366.47	366.2241	68.83	9.90	4.37	16.90
C21H3602P2	382.47	382.2190	65.95	9.49	8.37	16.20
C21H3603P2	398.47	398.2140	63.30	9.11	12.05	15.55
C21H3604P2	414.47	414.2089	60.86	8.76	15.44	14.95
C21H3605P2	430.47	430.2038	58.60	8.43	18.58	14.39
C21H3606P2	446.47	446.1987	56.50	8.13	21.50	13.88
C21H3607P2	462.46	462.1936	54.54	7.85	24.22	13.40
C21H3608P2	478.46	478.1885	52.72	7.58	26.75	12.95
C21H3609P2	494.46	494.1834	51.01	7.34	29.12	12.53
C21H3701P1	336.50	336.2582	74.96	11.08	4.75	9.20
C21H3702P1	352.50	352.2531	71.56	10.58	9.08	8.79
C21H3703P1	368.50	368.2480	68.45	10.12	13.03	8.41
C21H3704P1	384.50	384.2429	65.60	9.70	16.64	8.06
C21H3705P1	400.50	400.2378	62.98	9.31	19.97	7.73
C21H3706P1	416.50	416.2328	60.56	8.95	23.05	7.44
C21H3707P1	432.50	432.2277	58.32	8.62	25.90	7.16
C21H3708P1	448.50	448.2226	56.24	8.32	28.54	6.91
C21H3709P1	464.50	464.2175	54.30	8.03	31.00	6.67
C21H3801P2	368.48	368.2398	68.45	10.39	4.34	16.81
C21H3802P2	384.48	384.2347	65.60	9.96	8.32	16.11
C21H3803P2	400.48	400.2296	62.98	9.56	11.99	15.47
C21H3804P2	416.48	416.2245	60.56	9.20	15.37	14.87
C21H3805P2	432.48	432.2194	58.32	8.86	18.50	14.32
C21H3806P2	448.48	448.2143	56.24	8.54	21.40	13.81
C21H3807P2	464.48	464.2093	54.30	8.25	24.11	13.34
C21H3808P2	480.48	480.2042	52.50	7.97	26.64	12.89
C21H3809P2	496.48	496.1991	50.80	7.71	29.00	12.48
C21H3901P1	338.52	338.2738	74.51	11.61	4.73	9.15
C21H3902P1	354.52	354.2688	71.15	11.09	9.03	8.74
C21H3903P1	370.52	370.2637	68.08	10.61	12.95	8.36
C21H3904P1	386.52	386.2586	65.26	10.17	16.56	8.01
C21H3905P1	402.52	402.2535	62.66	9.77	19.87	7.70
C21H3906P1	418.52	418.2484	60.27	9.39	22.94	7.40
C21H3907P1	434.51	434.2433	58.05	9.05	25.77	7.13
C21H3908P1	450.51	450.2382	55.99	8.73	28.41	6.88
C21H3909P1	466.51	466.2332	54.07	8.43	30.87	6.64
C21H4001P2	370.50	370.2554	68.08	10.88	4.32	16.72
C21H4002P2	386.50	386.2503	65.26	10.43	8.28	16.03
C21H4003P2	402.50	402.2453	62.67	10.02	11.93	15.39
C21H4004P2	418.50	418.2402	60.27	9.63	15.29	14.80
C21H4005P2	434.50	434.2351	58.05	9.28	18.41	14.26
C21H4006P2	450.50	450.2300	55.99	8.95	21.31	13.75
C21H4007P2	466.50	466.2249	54.07	8.64	24.01	13.28
C21H4008P2	482.50	482.2198	52.28	8.36	26.53	12.84
C21H4009P2	498.50	498.2147	50.60	8.09	28.89	12.43
C21H4101P1	340.53	340.2895	74.07	12.14	4.70	9.10

136

C21H4102P1	356.53	356.2844	70.75	11.59	8.97	8.69
C21H4103P1	372.53	372.2793	67.71	11.09	12.88	8.31
C21H4104P1	388.53	388.2742	64.92	10.64	16.47	7.97
C21H4105P1	404.53	404.2691	62.35	10.22	19.78	7.66
C21H4106P1	420.53	420.2641	59.98	9.83	22.83	7.37
C21H4107P1	436.53	436.2590	57.78	9.47	25.66	7.10
C21H4108P1	452.53	452.2539	55.74	9.13	28.28	6.84
C21H4109P1	468.53	468.2488	53.84	8.82	30.73	6.61
C21H4201P2	372.52	372.2711	67.71	11.36	4.29	16.63
C21H4202P2	388.52	388.2660	64.92	10.90	8.24	15.94
C21H4203P2	404.51	404.2609	62.35	10.47	11.87	15.31
C21H4204P2	420.51	420.2558	59.98	10.07	15.22	14.73
C21H4205P2	436.51	436.2507	57.78	9.70	18.33	14.19
C21H4206P2	452.51	452.2456	55.74	9.36	21.21	13.69
C21H4207P2	468.51	468.2406	53.84	9.04	23.90	13.22
C21H4208P2	484.51	484.2355	52.06	8.74	26.42	12.79
C21H4209P2	500.51	500.2304	50.40	8.46	28.77	12.38
C21H4301P1	342.55	342.3051	73.63	12.65	4.67	9.04
C21H4302P1	358.55	358.3000	70.35	12.09	8.92	8.64
C21H4303P1	374.55	374.2950	67.34	11.57	12.81	8.27
C21H4304P1	390.55	390.2899	64.58	11.10	16.39	7.93
C21H4305P1	406.55	406.2848	62.04	10.66	19.68	7.62
C21H4306P1	422.55	422.2797	59.69	10.26	22.72	7.33
C21H4307P1	438.55	438.2746	57.52	9.88	25.54	7.06
C21H4308P1	454.55	454.2695	55.49	9.54	28.16	6.81
C21H4309P1	470.55	470.2644	53.60	9.21	30.60	6.58
C21H4401P2	374.53	374.2867	67.35	11.84	4.27	16.54
C21H4402P2	390.53	390.2816	64.59	11.36	8.19	15.86
C21H4403P2	406.53	406.2766	62.05	10.91	11.81	15.24
C21H4404P2	422.53	422.2715	59.70	10.50	15.15	14.66
C21H4405P2	438.53	438.2664	57.52	10.11	18.24	14.13
C21H4406P2	454.53	454.2613	55.49	9.76	21.12	13.63
C21H4407P2	470.53	470.2562	53.61	9.43	23.80	13.17
C21H4408P2	486.53	486.2511	51.84	9.12	26.31	12.73
C21H4409P2	502.53	502.2460	50.19	8.83	28.65	12.33
C21H4501P1	344.57	344.3208	73.20	13.16	4.64	8.99
C21H4502P1	360.57	360.3157	69.96	12.58	8.87	8.59
C21H4503P1	376.56	376.3106	66.98	12.05	12.75	8.23
C21H4504P1	392.56	392.3055	64.25	11.55	16.30	7.89
C21H4505P1	408.56	408.3004	61.74	11.10	19.58	7.58
C21H4506P1	424.56	424.2954	59.41	10.68	22.61	7.30
C21H4507P1	440.56	440.2903	57.25	10.30	25.42	7.03
C21H4508P1	456.56	456.2852	55.25	9.93	28.03	6.78
C21H4509P1	472.56	472.2801	53.38	9.60	30.47	6.55
C21H4601P2	376.55	376.3024	66.99	12.31	4.25	16.45
C21H4602P2	392.55	392.2973	64.26	11.81	8.15	15.78
C21H4603P2	408.55	408.2922	61.74	11.35	11.75	15.16
C21H4604P2	424.55	424.2871	59.41	10.92	15.07	14.59
C21H4605P2	440.55	440.2820	57.25	10.52	18.16	14.06
C21H4606P2	456.54	456.2769	55.25	10.16	21.03	13.57
C21H4607P2	472.54	472.2719	53.38	9.81	23.70	13.11
C21H4608P2	488.54	488.2668	51.63	9.49	26.20	12.68
C21H4609P2	504.54	504.2617	49.99	9.19	28.54	12.28
C22H1101P1	322.31	322.0547	81.99	3.44	4.96	9.61
C22H1102P1	338.31	338.0497	78.11	3.28	9.46	9.16
C22H1103P1	354.30	354.0446	74.58	3.13	13.55	8.74
C22H1104P1	370.30	370.0395	71.36	2.99	17.28	8.36
C22H1105P1	386.30	386.0344	68.40	2.87	20.71	8.02
C22H1106P1	402.30	402.0293	65.68	2.76	23.86	7.70
C22H1107P1	418.30	418.0242	63.17	2.65	26.77	7.40
C22H11O8P1	434.30	434.0191	60.84	2.55	29.47	7.13
C22H11O9P1	450.30	450.0141	58.68	2.46	31.98	6.88
C22H12O1P2	354.29	354.0363	74.58	3.41	4.52	17.49
C22H12O2P2	370.29	370.0313	71.36	3.27	8.64	16.73
C22H12O3P2	386.29	386.0262	68.41	3.13	12.43	16.04
C22H12O4P2	402.29	402.0211	65.69	3.01	15.91	15.40
C22H12O5P2	418.29	418.0160	63.17	2.89	19.12	14.81
C22H12O6P2	434.28	434.0109	60.85	2.79	22.10	14.26
C22H12O7P2	450.28	450.0058	58.68	2.69	24.87	13.76
C22H12O8P2	466.28	466.0007	56.67	2.59	27.45	13.29
C22H12O9P2	482.28	481.9957	54.79	2.51	29.86	12.84
C22H13O1P1	324.32	324.0704	81.48	4.04	4.93	9.55
C22H13O2P1	340.32	340.0653	77.65	3.85	9.40	9.10
C22H13O3P1	356.32	356.0602	74.16	3.68	13.47	8.69
C22H1304P1	372.32	372.0551	70.97	3.52	17.19	8.32
C22H13O5P1	388.32	388.0501	68.05	3.37	20.60	7.98
C22H13O6P1	404.32	404.0450	65.36	3.24	23.74	7.66
C22H13O7P1	420.32	420.0399	62.87	3.12	26.65	7.37
C22H1308P1	436.32	436.0348	60.56	3.00	29.34	7.10
C22H13O9P1	452.32	452.0297	58.42	2.90	31.83	6.85
C22H14O1P2	356.30	356.0520	74.16	3.96	4.49	17.39
C22H14O2P2	372.30	372.0469	70.98	3.79	8.59	16.64
C22H14O3P2	388.30	388.0418	68.05	3.63	12.36	15.95
C22H14O4P2	404.30	404.0367	65.36	3.49	15.83	15.32
C22H14O5P2	420.30	420.0316	62.87	3.36	19.03	14.74
C22H14O6P2	436.30	436.0266	60.56	3.23	22.00	14.20
C22H14O7P2	452.30	452.0215	58.42	3.12	24.76	13.70
C22H1408P2	468.30	468.0164	56.43	3.01	27.33	13.23
C22H14O9P2	484.30	484.0113	54.56	2.91	29.73	12.79
C22H15O1P1	326.34	326.0860	80.97	4.63	4.90	9.49
C22H15O2P1	342.34	342.0810	77.19	4.42	9.35	9.05
C22H15O3P1	358.34	358.0759	73.74	4.22	13.39	8.64
C22H15O4P1	374.34	374.0708	70.59	4.04	17.10	8.27
C22H15O5P1	390.34	390.0657	67.70	3.87	20.49	7.94
C22H15O6P1	406.34	406.0606	65.03	3.72	23.62	7.62
C22H15O7P1	422.33	422.0555	62.57	3.58	26.52	7.33
C22H15O8P1	438.33	438.0504	60.28	3.45	29.20	7.07
C22H15O9P1	454.33	454.0454	58.16	3.33	31.69	6.82
C22H16O1P2	358.32	358.0676	73.75	4.50	4.47	17.29
C22H16O2P2	374.32	374.0625	70.59	4.31	8.55	16.55
C22H16O3P2	390.32	390.0575	67.70	4.13	12.30	15.87
C22H16O4P2	406.32	406.0524	65.03	3.97	15.75	15.25
C22H16O5P2	422.32	422.0473	62.57	3.82	18.94	14.67
C22H16O6P2	438.32	438.0422	60.29	3.68	21.90	14.13
C22H16O7P2	454.32	454.0371	58.16	3.55	24.65	13.64
C22H16O8P2	470.32	470.0320	56.18	3.43	27.21	13.17
C22H16O9P2	486.32	486.0269	54.34	3.32	29.61	12.74
C22H17O1P1	328.35	328.1017	80.48	5.22	4.87	9.43
C22H17O2P1	344.35	344.0966	76.74	4.98	9.29	8.99
C22H17O3P1	360.35	360.0915	73.33	4.76	13.32	8.60
C22H17O4P1	376.35	376.0864	70.21	4.55	17.00	8.23
C22H17O5P1	392.35	392.0814	67.35	4.37	20.39	7.89
C22H1706P1	408.35	408.0763	64.71	4.20	23.51	7.59
C22H17O7P1	424.35	424.0712	62.27	4.04	26.39	7.30
C22H1708P1	440.35	440.0661	60.01	3.89	29.07	7.03
C22H1709P1	456.35	456.0610	57.90	3.75	31.55	6.79
C22H1801P2	360.34	360.0833	73.33	5.04	4.44	17.19
C22H18O2P2	376.34	376.0782	70.22	4.82	8.50	16.46
C22H18O3P2	392.33	392.0731	67.35	4.62	12.23	15.79
C22H18O4P2	408.33	408.0680	64.71	4.44	15.67	15.17

137

C22H1805P2	424.33	424.0629	62.27	4.28	18.85	14.60
C22H1806P2	440.33	440.0579	60.01	4.12	21.80	14.07
C22H1807P2	456.33	456.0528	57.91	3.98	24.54	13.58
C22H1808P2	472.33	472.0477	55.94	3.84	27.10	13.12
C22H1809P2	488.33	488.0426	54.11	3.72	29.49	12.69
C22H1901P1	330.37	330.1173	79.98	5.80	4.84	9.38
C22H1902P1	346.37	346.1123	76.29	5.53	9.24	8.94
C22H1903P1	362.37	362.1072	72.92	5.29	13.25	8.55
C22H1904P1	378.37	378.1021	69.84	5.06	16.91	8.19
C22H1905P1	394.37	394.0970	67.00	4.86	20.28	7.85
C22H1906P1	410.37	410.0919	64.39	4.67	23.39	7.55
C22H1907P1	426.37	426.0868	61.98	4.49	26.27	7.26
C22H1908P1	442.37	442.0817	59.73	4.33	28.93	7.00
C22H1909P1	458.37	458.0767	57.65	4.18	31.41	6.76
C22H2001P2	362.35	362.0989	72.93	5.56	4.42	17.10
C22H2002P2	378.35	378.0938	69.84	5.33	8.46	16.37
C22H2003P2	394.35	394.0888	67.01	5.11	12.17	15.71
C22H2004P2	410.35	410.0837	64.40	4.91	15.60	15.10
C22H2005P2	426.35	426.0786	61.98	4.73	18.76	14.53
C22H2006P2	442.35	442.0735	59.74	4.56	21.70	14.00
C22H2007P2	458.35	458.0684	57.65	4.40	24.43	13.52
C22H2008P2	474.35	474.0633	55.71	4.25	26.98	13.06
C22H2009P2	490.35	490.0582	53.89	4.11	29.37	12.63
C22H2101P1	332.39	332.1330	79.50	6.37	4.81	9.32
C22H2102P1	348.39	348.1279	75.85	6.08	9.18	8.89
C22H2103P1	364.38	364.1228	72.52	5.81	13.17	8.50
C22H2104P1	380.38	380.1177	69.47	5.56	16.82	8.14
C22H2105P1	396.38	396.1127	66.66	5.34	20.18	7.81
C22H2106P1	412.38	412.1076	64.08	5.13	23.28	7.51
C22H2107P1	428.38	428.1025	61.68	4.94	26.14	7.23
C22H2108P1	444.38	444.0974	59.46	4.76	28.80	6.97
C22H2109P1	460.38	460.0923	57.40	4.60	31.28	6.73
C22H2201P2	364.37	364.1146	72.52	6.09	4.39	17.00
C22H2202P2	380.37	380.1095	69.47	5.83	8.41	16.29
C22H2203P2	396.37	396.1044	66.67	5.59	12.11	15.63
C22H2204P2	412.37	412.0993	64.08	5.38	15.52	15.02
C22H2205P2	428.37	428.0942	61.69	5.18	18.67	14.46
C22H2206P2	444.36	444.0892	59.47	4.99	21.60	13.94
C22H2207P2	460.36	460.0841	57.40	4.82	24.33	13.46
C22H2208P2	476.36	476.0790	55.47	4.66	26.87	13.00
C22H2209P2	492.36	492.0739	53.67	4.50	29.25	12.58
C22H2301P1	334.40	334.1486	79.02	6.93	4.78	9.26
C22H2302P1	350.40	350.1436	75.41	6.62	9.13	8.84
C22H2303P1	366.40	366.1385	72.12	6.33	13.10	8.45
C22H2304P1	382.40	382.1334	69.10	6.06	16.74	8.10
C22H2305P1	398.40	398.1283	66.33	5.82	20.08	7.77
C22H2306P1	414.40	414.1232	63.77	5.59	23.17	7.47
C22H2307P1	430.40	430.1181	61.40	5.39	26.02	7.20
C22H2308P1	446.40	446.1130	59.20	5.19	28.67	6.94
C22H2309P1	462.40	462.1080	57.15	5.01	31.14	6.70
C22H2401P2	366.38	366.1302	72.12	6.60	4.37	16.91
C22H2402P2	382.38	382.1251	69.10	6.33	8.37	16.20
C22H2403P2	398.38	398.1201	66.33	6.07	12.05	15.55
C22H2404P2	414.38	414.1150	63.77	5.84	15.44	14.95
C22H2405P2	430.38	430.1099	61.40	5.62	18.59	14.39
C22H2406P2	446.38	446.1048	59.20	5.42	21.51	13.88
C22H2407P2	462.38	462.0997	57.15	5.23	24.22	13.40
C22H2408P2	478.38	478.0946	55.24	5.06	26.76	12.95
C22H2409P2	494.38	494.0895	53.45	4.89	29.13	12.53
C22H2501P1	336.42	336.1643	78.55	7.49	4.76	9.21
C22H25O2P1	352.42	352.1592	74.98	7.15	9.08	8.79
C22H25O3P1	368.42	368.1541	71.72	6.84	13.03	8.41
C22H25O4P1	384.42	384.1490	68.74	6.56	16.65	8.06
C22H25O5P1	400.42	400.1439	65.99	6.29	19.98	7.74
C22H25O6P1	416.41	416.1389	63.46	6.05	23.05	7.44
C22H25O7P1	432.41	432.1338	61.11	5.83	25.90	7.16
C22H25O8P1	448.41	448.1287	58.93	5.62	28.54	6.91
C22H2509P1	464.41	464.1236	56.90	5.43	31.01	6.67
C22H26O1P2	368.40	368.1459	71.73	7.11	4.34	16.82
C22H26O2P2	384.40	384.1408	68.74	6.82	8.32	16.12
C22H26O3P2	400.40	400.1357	66.00	6.55	11.99	15.47
C22H26O4P2	416.40	416.1306	63.46	6.29	15.37	14.88
C22H26O5P2	432.40	432.1255	61.11	6.06	18.50	14.33
C22H26O6P2	448.40	448.1205	58.93	5.84	21.41	13.82
C22H26O7P2	464.40	464.1154	56.90	5.64	24.12	13.34
C22H26O8P2	480.40	480.1103	55.01	5.46	26.64	12.90
C22H26O9P2	496.39	496.1052	53.23	5.28	29.01	12.48
C22H2701P1	338.43	338.1799	78.08	8.04	4.73	9.15
C22H27O2P1	354.43	354.1749	74.55	7.68	9.03	8.74
C22H27O3P1	370.43	370.1698	71.33	7.35	12.96	8.36
C22H2704P1	386.43	386.1647	68.38	7.04	16.56	8.02
C22H27O5P1	402.43	402.1596	65.66	6.76	19.88	7.70
C22H27O6P1	418.43	418.1545	63.15	6.50	22.94	7.40
C22H27O7P1	434.43	434.1494	60.83	6.26	25.78	7.13
C22H27O8P1	450.43	450.1443	58.67	6.04	28.42	6.88
C22H2709P1	466.43	466.1393	56.65	5.83	30.87	6.64
C22H28O1P2	370.42	370.1615	71.34	7.62	4.32	16.72
C22H28O2P2	386.41	386.1564	68.38	7.30	8.28	16.03
C22H28O3P2	402.41	402.1514	65.66	7.01	11.93	15.39
C22H28O4P2	418.41	418.1463	63.15	6.75	15.30	14.81
C22H2805P2	434.41	434.1412	60.83	6.50	18.41	14.26
C22H28O6P2	450.41	450.1361	58.67	6.27	21.31	13.75
C22H28O7P2	466.41	466.1310	56.65	6.05	24.01	13.28
C22H28O8P2	482.41	482.1259	54.78	5.85	26.53	12.84
C22H28O9P2	498.41	498.1208	53.02	5.66	28.89	12.43
C22H29O1P1	340.45	340.1956	77.62	8.59	4.70	9.10
C22H29O2P1	356.45	356.1905	74.13	8.20	8.98	8.69
C22H29O3P1	372.45	372.1854	70.95	7.85	12.89	8.32
C22H29O4P1	388.45	388.1803	68.03	7.53	16.48	7.97
C22H29O5P1	404.45	404.1752	65.33	7.23	19.78	7.66
C22H29O6P1	420.45	420.1702	62.85	6.95	22.83	7.37
C22H29O7P1	436.45	436.1651	60.54	6.70	25.66	7.10
C22H29O8P1	452.45	452.1600	58.40	6.46	28.29	6.85
C22H29O9P1	468.44	468.1549	56.41	6.24	30.74	6.61
C22H30O1P2	372.43	372.1772	70.95	8.12	4.30	16.63
C22H30O2P2	388.43	388.1721	68.03	7.78	8.24	15.95
C22H30O3P2	404.43	404.1670	65.34	7.48	11.87	15.32
C22H30O4P2	420.43	420.1619	62.85	7.19	15.22	14.73
C22H30O5P2	436.43	436.1568	60.55	6.93	18.33	14.19
C22H30O6P2	452.43	452.1518	58.41	6.68	21.22	13.69
C22H30O7P2	468.43	468.1467	56.41	6.46	23.91	13.22
C22H30O8P2	484.43	484.1416	54.55	6.24	26.42	12.79
C22H30O9P2	500.43	500.1365	52.80	6.04	28.77	12.38
C22H3101P1	342.47	342.2112	77.16	9.12	4.67	9.04
C22H3102P1	358.46	358.2062	73.72	8.72	8.93	8.64
C22H31O3P1	374.46	374.2011	70.57	8.34	12.82	8.27
C22H31O4P1	390.46	390.1960	67.67	8.00	16.39	7.93
C22H31O5P1	406.46	406.1909	65.01	7.69	19.68	7.62
C22H31O6P1	422.46	422.1858	62.55	7.40	22.72	7.33
C22H31O7P1	438.46	438.1807	60.27	7.13	25.54	7.06

C22H3108P1	454.46	454.1756	58.14	6.88	28.16	6.82
C22H3109P1	470.46	470.1706	56.17	6.64	30.61	6.58
C22H3201P2	374.45	374.1928	70.57	8.61	4.27	16.54
C22H3202P2	390.45	390.1877	67.68	8.26	8.20	15.87
C22H3203P2	406.45	406.1827	65.01	7.94	11.81	15.24
C22H3204P2	422.45	422.1776	62.55	7.64	15.15	14.66
C22H3205P2	438.44	438.1725	60.27	7.36	18.25	14.13
C22H3206P2	454.44	454.1674	58.15	7.10	21.12	13.63
C22H3207P2	470.44	470.1623	56.17	6.86	23.81	13.17
C22H3208P2	486.44	486.1572	54.32	6.63	26.31	12.73
C22H3209P2	502.44	502.1521	52.59	6.42	28.66	12.33
C22H3301P1	344.48	344.2269	76.71	9.66	4.64	8.99
C22H3302P1	360.48	360.2218	73.30	9.23	8.88	8.59
C22H3303P1	376.48	376.2167	70.19	8.84	12.75	8.23
C22H3304P1	392.48	392.2116	67.33	8.48	16.31	7.89
C22H3305P1	408.48	408.2065	64.69	8.14	19.58	7.58
C22H3306P1	424.48	424.2015	62.25	7.84	22.62	7.30
C22H3307P1	440.48	440.1964	59.99	7.55	25.43	7.03
C22H3308P1	456.48	456.1913	57.89	7.29	28.04	6.79
C22H3309P1	472.48	472.1862	55.93	7.04	30.48	6.56
C22H3401P2	376.46	376.2085	70.19	9.10	4.25	16.46
C22H3402P2	392.46	392.2034	67.33	8.73	8.15	15.78
C22H3403P2	408.46	408.1983	64.69	8.39	11.75	15.17
C22H3404P2	424.46	424.1932	62.25	8.07	15.08	14.59
C22H3405P2	440.46	440.1881	59.99	7.78	18.16	14.06
C22H3406P2	456.46	456.1830	57.89	7.51	21.03	13.57
C22H3407P2	472.46	472.1780	55.93	7.25	23.70	13.11
C22H3408P2	488.46	488.1729	54.10	7.02	26.20	12.68
C22H3409P2	504.46	504.1678	52.38	6.79	28.54	12.28
C22H3501P1	346.50	346.2425	76.26	10.18	4.62	8.94
C22H3502P1	362.50	362.2375	72.90	9.73	8.83	8.54
C22H3503P1	378.50	378.2324	69.81	9.32	12.68	8.18
C22H3504P1	394.50	394.2273	66.98	8.94	16.22	7.85
C22H3505P1	410.50	410.2222	64.37	8.59	19.49	7.55
C22H3506P1	426.49	426.2171	61.96	8.27	22.51	7.26
C22H3507P1	442.49	442.2120	59.72	7.97	25.31	7.00
C22H3508P1	458.49	458.2069	57.63	7.69	27.92	6.76
C22H3509P1	474.49	474.2019	55.69	7.44	30.35	6.53
C22H3601P2	378.48	378.2241	69.82	9.59	4.23	16.37
C22H3602P2	394.48	394.2190	66.99	9.20	8.11	15.70
C22H3603P2	410.48	410.2140	64.38	8.84	11.69	15.09
C22H3604P2	426.48	426.2089	61.96	8.51	15.01	14.53
C22H3605P2	442.48	442.2038	59.72	8.20	18.08	14.00
C22H3606P2	458.48	458.1987	57.64	7.91	20.94	13.51
C22H3607P2	474.48	474.1936	55.69	7.65	23.60	13.06
C22H3608P2	490.48	490.1885	53.88	7.40	26.10	12.63
C22H3609P2	506.47	506.1834	52.17	7.16	28.43	12.23
C22H3701P1	348.51	348.2582	75.82	10.70	4.59	8.89
C22H3702P1	364.51	364.2531	72.49	10.23	8.78	8.50
C22H3703P1	380.51	380.2480	69.44	9.80	12.61	8.14
C22H3704P1	396.51	396.2429	66.64	9.41	16.14	7.81
C22H3705P1	412.51	412.2378	64.06	9.04	19.39	7.51
C22H3706P1	428.51	428.2328	61.67	8.70	22.40	7.23
C22H3707P1	444.51	444.2277	59.45	8.39	25.20	6.97
C22H3708P1	460.51	460.2226	57.38	8.10	27.79	6.73
C22H3709P1	476.51	476.2175	55.45	7.83	30.22	6.50
C22H3801P2	380.50	380.2398	69.45	10.07	4.20	16.28
C22H3802P2	396.49	396.2347	66.65	9.66	8.07	15.62
C22H3803P2	412.49	412.2296	64.06	9.29	11.64	15.02
C22H3804P2	428.49	428.2245	61.67	8.94	14.94	14.46

C22H3805P2	444.49	444.2194	59.45	8.62	18.00	13.94
C22H3806P2	460.49	460.2143	57.38	8.32	20.85	13.45
C22H3807P2	476.49	476.2093	55.46	8.04	23.50	13.00
C22H3808P2	492.49	492.2042	53.65	7.78	25.99	12.58
C22H3809P2	508.49	508.1991	51.97	7.53	28.32	12.18
C22H3901P1	350.53	350.2738	75.38	11.21	4.56	8.84
C22H3902P1	366.53	366.2688	72.09	10.73	8.73	8.45
C22H3903P1	382.53	382.2637	69.08	10.28	12.55	8.10
C22H3904P1	398.53	398.2586	66.31	9.86	16.06	7.77
C22H3905P1	414.53	414.2535	63.75	9.48	19.30	7.47
C22H3906P1	430.53	430.2484	61.38	9.13	22.30	7.19
C22H3907P1	446.53	446.2433	59.18	8.80	25.08	6.94
C22H3908P1	462.53	462.2382	57.13	8.50	27.67	6.70
C22H3909P1	478.52	478.2332	55.22	8.22	30.09	6.47
C22H4001P2	382.51	382.2554	69.08	10.54	4.18	16.19
C22H4002P2	398.51	398.2503	66.31	10.12	8.03	15.54
C22H4003P2	414.51	414.2453	63.75	9.73	11.58	14.94
C22H4004P2	430.51	430.2402	61.38	9.37	14.87	14.39
C22H4005P2	446.51	446.2351	59.18	9.03	17.92	13.87
C22H4006P2	462.51	462.2300	57.13	8.72	20.76	13.39
C22H4007P2	478.51	478.2249	55.22	8.43	23.41	12.95
C22H4008P2	494.51	494.2198	53.44	8.15	25.88	12.53
C22H4009P2	510.51	510.2147	51.76	7.90	28.21	12.13
C22H4101P1	352.55	352.2895	74.95	11.72	4.54	8.79
C22H4102P1	368.54	368.2844	71.70	11.21	8.68	8.40
C22H4103P1	384.54	384.2793	68.72	10.75	12.48	8.05
C22H4104P1	400.54	400.2742	65.97	10.32	15.98	7.73
C22H4105P1	416.54	416.2691	63.44	9.92	19.20	7.44
C22H4106P1	432.54	432.2641	61.09	9.55	22.19	7.16
C22H4107P1	448.54	448.2590	58.91	9.21	24.97	6.91
C22H4108P1	464.54	464.2539	56.88	8.90	27.55	6.67
C22H4109P1	480.54	480.2488	54.99	8.60	29.97	6.45
C22H4201P2	384.53	384.2711	68.72	11.01	4.16	16.11
C22H4202P2	400.53	400.2660	65.97	10.57	7.99	15.47
C22H4203P2	416.53	416.2609	63.44	10.16	11.52	14.87
C22H4204P2	432.53	432.2558	61.09	9.79	14.80	14.32
C22H4205P2	448.52	448.2507	58.91	9.44	17.84	13.81
C22H4206P2	464.52	464.2456	56.89	9.11	20.67	13.34
C22H4207P2	480.52	480.2406	54.99	8.81	23.31	12.89
C22H4208P2	496.52	496.2355	53.22	8.53	25.78	12.48
C22H4209P2	512.52	512.2304	51.56	8.26	28.10	12.09
C22H4301P1	354.56	354.3051	74.53	12.22	4.51	8.74
C22H4302P1	370.56	370.3001	71.31	11.70	8.64	8.36
C22H4303P1	386.56	386.2950	68.36	11.21	12.42	8.01
C22H4304P1	402.56	402.2899	65.64	10.77	15.90	7.69
C22H4305P1	418.56	418.2848	63.13	10.36	19.11	7.40
C22H4306P1	434.56	434.2797	60.81	9.97	22.09	7.13
C22H4307P1	450.56	450.2746	58.65	9.62	24.86	6.87
C22H4308P1	466.56	466.2695	56.64	9.29	27.43	6.64
C22H4309P1	482.56	482.2644	54.76	8.98	29.84	6.42
C22H4401P2	386.54	386.2867	68.36	11.47	4.14	16.03
C22H4402P2	402.54	402.2816	65.64	11.02	7.95	15.39
C22H4403P2	418.54	418.2766	63.13	10.60	11.47	14.80
C22H4404P2	434.54	434.2715	60.81	10.21	14.73	14.26
C22H4405P2	450.54	450.2664	58.65	9.84	17.76	13.75
C22H4406P2	466.54	466.2613	56.64	9.51	20.58	13.28
C22H4407P2	482.54	482.2562	54.76	9.19	23.21	12.84
C22H4408P2	498.54	498.2511	53.00	8.90	25.67	12.43
C22H4409P2	514.54	514.2460	51.36	8.62	27.99	12.04
C22H4501P1	356.58	356.3208	74.11	12.72	4.49	8.69

C22H4502P1	372.58	372.3157	70.92	12.17	8.59	8.31
C22H4503P1	388.58	388.3106	68.00	11.67	12.35	7.97
C22H4504P1	404.58	404.3055	65.31	11.21	15.82	7.66
C22H4505P1	420.57	420.3004	62.83	10.78	19.02	7.36
C22H4506P1	436.57	436.2954	60.53	10.39	21.99	7.09
C22H4507P1	452.57	452.2903	58.39	10.02	24.75	6.84
C22H4508P1	468.57	468.2852	56.39	9.68	27.32	6.61
C22H4509P1	484.57	484.2801	54.53	9.36	29.72	6.39
C22H4601P2	388.56	388.3024	68.01	11.93	4.12	15.94
C22H4602P2	404.56	404.2973	65.32	11.46	7.91	15.31
C22H4603P2	420.56	420.2922	62.83	11.03	11.41	14.73
C22H4604P2	436.56	436.2871	60.53	10.62	14.66	14.19
C22H4605P2	452.56	452.2820	58.39	10.25	17.68	13.69
C22H4606P2	468.56	468.2769	56.40	9.90	20.49	13.22
C22H4607P2	484.56	484.2719	54.53	9.57	23.11	12.78
C22H4608P2	500.55	500.2668	52.79	9.26	25.57	12.38
C22H4609P2	516.55	516.2617	51.16	8.98	27.88	11.99
C22H4701P1	358.59	358.3364	73.69	13.21	4.46	8.64
C22H4702P1	374.59	374.3313	70.54	12.65	8.54	8.27
C22H4703P1	390.59	390.3263	67.65	12.13	12.29	7.93
C22H4704P1	406.59	406.3212	64.99	11.65	15.74	7.62
C22H4705P1	422.59	422.3161	62.53	11.21	18.93	7.33
C22H4706P1	438.59	438.3110	60.25	10.80	21.89	7.06
C22H4707P1	454.59	454.3059	58.13	10.42	24.64	6.81
C22H4708P1	470.59	470.3008	56.15	10.07	27.20	6.58
C22H4709P1	486.59	486.2957	54.31	9.74	29.59	6.37
C22H4801P2	390.57	390.3180	67.66	12.39	4.10	15.86
C22H4802P2	406.57	406.3129	64.99	11.90	7.87	15.24
C22H4803P2	422.57	422.3079	62.53	11.45	11.36	14.66
C22H4804P2	438.57	438.3028	60.25	11.03	14.59	14.12
C22H4805P2	454.57	454.2977	58.13	10.64	17.60	13.63
C22H4806P2	470.57	470.2926	56.15	10.28	20.40	13.16
C22H4807P2	486.57	486.2875	54.31	9.94	23.02	12.73
C22H4808P2	502.57	502.2824	52.58	9.63	25.47	12.33
C22H4809P2	518.57	518.2773	50.96	9.33	27.77	11.95
C23H1101P1	334.32	334.0547	82.63	3.32	4.79	9.26
C23H1102P1	350.32	350.0497	78.86	3.17	9.13	8.84
C23H1103P1	366.32	366.0446	75.41	3.03	13.10	8.46
C23H1104P1	382.32	382.0395	72.26	2.90	16.74	8.10
C23H1105P1	398.31	398.0344	69.36	2.78	20.08	7.78
C23H1106P1	414.31	414.0293	66.68	2.68	23.17	7.48
C23H1107P1	430.31	430.0242	64.20	2.58	26.03	7.20
C23H1108P1	446.31	446.0191	61.90	2.48	28.68	6.94
C23H1109P1	462.31	462.0141	59.76	2.40	31.15	6.70
C23H1201P2	366.30	366.0363	75.42	3.30	4.37	16.91
C23H1202P2	382.30	382.0313	72.26	3.16	8.37	16.20
C23H1203P2	398.30	398.0262	69.36	3.04	12.05	15.55
C23H1204P2	414.30	414.0211	66.68	2.92	15.45	14.95
C23H1205P2	430.30	430.0160	64.20	2.81	18.59	14.40
C23H1206P2	446.30	446.0109	61.90	2.71	21.51	13.88
C23H1207P2	462.30	462.0058	59.76	2.62	24.23	13.40
C23H1208P2	478.29	478.0007	57.76	2.53	26.76	12.95
C23H1209P2	494.29	493.9957	55.89	2.45	29.13	12.53
C23H1301P1	336.33	336.0704	82.14	3.90	4.76	9.21
C23H1302P1	352.33	352.0653	78.41	3.72	9.08	8.79
C23H1303P1	368.33	368.0602	75.00	3.56	13.03	8.41
C23H1304P1	384.33	384.0551	71.88	3.41	16.65	8.06
C23H1305P1	400.33	400.0501	69.01	3.27	19.98	7.74
C23H1306P1	416.33	416.0450	66.36	3.15	23.06	7.44
C23H1307P1	432.33	432.0399	63.90	3.03	25.91	7.16

C23H1308P1	448.33	448.0348	61.62	2.92	28.55	6.91
C23H1309P1	464.33	464.0297	59.50	2.82	31.01	6.67
C23H1401P2	368.32	368.0520	75.01	3.83	4.34	16.82
C23H1402P2	384.31	384.0469	71.88	3.67	8.33	16.12
C23H1403P2	400.31	400.0418	69.01	3.53	11.99	15.47
C23H1404P2	416.31	416.0367	66.36	3.39	15.37	14.88
C23H1405P2	432.31	432.0316	63.90	3.26	18.50	14.33
C23H1406P2	448.31	448.0266	61.62	3.15	21.41	13.82
C23H1407P2	464.31	464.0215	59.50	3.04	24.12	13.34
C23H1408P2	480.31	480.0164	57.52	2.94	26.65	12.90
C23H1409P2	496.31	496.0113	55.66	2.84	29.01	12.48
C23H1501P1	338.35	338.0860	81.65	4.47	4.73	9.15
C23H1502P1	354.35	354.0810	77.96	4.27	9.03	8.74
C23H1503P1	370.35	370.0759	74.59	4.08	12.96	8.36
C23H1504P1	386.35	386.0708	71.50	3.91	16.56	8.02
C23H1505P1	402.35	402.0657	68.66	3.76	19.88	7.70
C23H1506P1	418.35	418.0606	66.04	3.61	22.95	7.40
C23H1507P1	434.35	434.0555	63.60	3.48	25.78	7.13
C23H1508P1	450.34	450.0504	61.34	3.36	28.42	6.88
C23H1509P1	466.34	466.0454	59.24	3.24	30.88	6.64
C23H1601P2	370.33	370.0676	74.60	4.35	4.32	16.73
C23H1602P2	386.33	386.0625	71.51	4.17	8.28	16.03
C23H1603P2	402.33	402.0575	68.66	4.01	11.93	15.40
C23H1604P2	418.33	418.0524	66.04	3.86	15.30	14.81
C23H1605P2	434.33	434.0473	63.61	3.71	18.42	14.26
C23H1606P2	450.33	450.0422	61.35	3.58	21.32	13.76
C23H1607P2	466.33	466.0371	59.24	3.46	24.02	13.28
C23H1608P2	482.33	482.0320	57.28	3.34	26.54	12.84
C23H1609P2	498.33	498.0269	55.44	3.24	28.90	12.43
C23H1701P1	340.37	340.1017	81.16	5.03	4.70	9.10
C23H1702P1	356.36	356.0966	77.52	4.81	8.98	8.69
C23H1703P1	372.36	372.0915	74.19	4.60	12.89	8.32
C23H1704P1	388.36	388.0864	71.13	4.41	16.48	7.98
C23H1705P1	404.36	404.0814	68.32	4.24	19.78	7.66
C23H1706P1	420.36	420.0763	65.72	4.08	22.84	7.37
C23H1707P1	436.36	436.0712	63.31	3.93	25.67	7.10
C23H1708P1	452.36	452.0661	61.07	3.79	28.29	6.85
C23H1709P1	468.36	468.0610	58.98	3.66	30.74	6.61
C23H1801P2	372.35	372.0833	74.19	4.87	4.30	16.64
C23H1802P2	388.35	388.0782	71.14	4.67	8.24	15.95
C23H1803P2	404.35	404.0731	68.32	4.49	11.87	15.32
C23H1804P2	420.35	420.0680	65.72	4.32	15.23	14.74
C23H1805P2	436.34	436.0629	63.31	4.16	18.33	14.20
C23H1806P2	452.34	452.0579	61.07	4.01	21.22	13.69
C23H1807P2	468.34	468.0528	58.99	3.87	23.91	13.23
C23H1808P2	484.34	484.0477	57.04	3.75	26.43	12.79
C23H1809P2	500.34	500.0426	55.21	3.63	28.78	12.38
C23H1901P1	342.38	342.1173	80.69	5.59	4.67	9.05
C23H1902P1	358.38	358.1123	77.08	5.34	8.93	8.64
C23H1903P1	374.38	374.1072	73.79	5.12	12.82	8.27
C23H1904P1	390.38	390.1021	70.77	4.91	16.39	7.93
C23H1905P1	406.38	406.0970	67.98	4.71	19.69	7.62
C23H1906P1	422.38	422.0919	65.41	4.53	22.73	7.33
C23H1907P1	438.38	438.0868	63.02	4.37	25.55	7.07
C23H1908P1	454.38	454.0817	60.80	4.21	28.17	6.82
C23H1909P1	470.38	470.0767	58.73	4.07	30.61	6.58
C23H2001P2	374.36	374.0989	73.79	5.38	4.27	16.55
C23H2002P2	390.36	390.0938	70.77	5.16	8.20	15.87
C23H2003P2	406.36	406.0888	67.98	4.96	11.81	15.24
C23H2004P2	422.36	422.0837	65.41	4.77	15.15	14.67

140

C23H2005P2	438.36	438.0786	63.02	4.60	18.25	14.13
C23H2006P2	454.36	454.0735	60.80	4.44	21.13	13.63
C23H2007P2	470.36	470.0684	58.73	4.29	23.81	13.17
C23H2008P2	486.36	486.0633	56.80	4.14	26.32	12.74
C23H2009P2	502.36	502.0582	54.99	4.01	28.66	12.33
C23H2101P1	344.40	344.1330	80.21	6.15	4.65	8.99
C23H2102P1	360.40	360.1279	76.65	5.87	8.88	8.59
C23H2103P1	376.40	376.1228	73.40	5.62	12.75	8.23
C23H2104P1	392.40	392.1177	70.40	5.39	16.31	7.89
C23H2105P1	408.39	408.1127	67.64	5.18	19.59	7.58
C23H2106P1	424.39	424.1076	65.09	4.99	22.62	7.30
C23H2107P1	440.39	440.1025	62.73	4.81	25.43	7.03
C23H2108P1	456.39	456.0974	60.53	4.64	28.04	6.79
C23H2109P1	472.39	472.0923	58.48	4.48	30.48	6.56
C23H2201P2	376.38	376.1146	73.40	5.89	4.25	16.46
C23H2202P2	392.38	392.1095	70.41	5.65	8.16	15.79
C23H2203P2	408.38	408.1044	67.65	5.43	11.75	15.17
C23H2204P2	424.38	424.0993	65.10	5.23	15.08	14.60
C23H2205P2	440.38	440.0942	62.73	5.04	18.17	14.07
C23H2206P2	456.38	456.0892	60.53	4.86	21.03	13.57
C23H2207P2	472.38	472.0841	58.48	4.69	23.71	13.11
C23H2208P2	488.37	488.0790	56.57	4.54	26.21	12.68
C23H2209P2	504.37	504.0739	54.77	4.40	28.55	12.28
C23H2301P1	346.41	346.1486	79.75	6.69	4.62	8.94
C23H2302P1	362.41	362.1436	76.23	6.40	8.83	8.55
C23H2303P1	378.41	378.1385	73.00	6.13	12.68	8.19
C23H2304P1	394.41	394.1334	70.04	5.88	16.23	7.85
C23H2305P1	410.41	410.1283	67.31	5.65	19.49	7.55
C23H2306P1	426.41	426.1232	64.79	5.44	22.51	7.26
C23H2307P1	442.41	442.1181	62.44	5.24	25.31	7.00
C23H2308P1	458.41	458.1130	60.26	5.06	27.92	6.76
C23H2309P1	474.41	474.1080	58.23	4.89	30.35	6.53
C23H2401P2	378.39	378.1302	73.01	6.39	4.23	16.37
C23H2402P2	394.39	394.1251	70.05	6.13	8.11	15.71
C23H2403P2	410.39	410.1201	67.32	5.89	11.70	15.09
C23H2404P2	426.39	426.1150	64.79	5.67	15.01	14.53
C23H2405P2	442.39	442.1099	62.45	5.47	18.08	14.00
C23H2406P2	458.39	458.1048	60.27	5.28	20.94	13.51
C23H2407P2	474.39	474.0997	58.23	5.10	23.61	13.06
C23H2408P2	490.39	490.0946	56.33	4.93	26.10	12.63
C23H2409P2	506.39	506.0895	54.55	4.78	28.44	12.23
C23H2501P1	348.43	348.1643	79.29	7.23	4.59	8.89
C23H2502P1	364.43	364.1592	75.81	6.91	8.78	8.50
C23H2503P1	380.43	380.1541	72.62	6.62	12.62	8.14
C23H2504P1	396.43	396.1490	69.69	6.36	16.14	7.81
C23H2505P1	412.43	412.1439	66.98	6.11	19.40	7.51
C23H2506P1	428.43	428.1389	64.48	5.88	22.41	7.23
C23H2507P1	444.43	444.1338	62.16	5.67	25.20	6.97
C23H2508P1	460.42	460.1287	60.00	5.47	27.80	6.73
C23H2509P1	476.42	476.1236	57.99	5.29	30.22	6.50
C23H2601P2	380.41	380.1459	72.62	6.89	4.21	16.28
C23H2602P2	396.41	396.1408	69.69	6.61	8.07	15.63
C23H2603P2	412.41	412.1357	66.99	6.35	11.64	15.02
C23H2604P2	428.41	428.1306	64.48	6.12	14.94	14.46
C23H2605P2	444.41	444.1255	62.16	5.90	18.00	13.94
C23H2606P2	460.41	460.1205	60.00	5.69	20.85	13.45
C23H2607P2	476.41	476.1154	57.99	5.50	23.51	13.00
C23H2608P2	492.41	492.1103	56.10	5.32	25.99	12.58
C23H2609P2	508.41	508.1052	54.34	5.15	28.32	12.18
C23H2701P1	350.44	350.1799	78.83	7.77	4.57	8.84

C23H2702P1	366.44	366.1749	75.39	7.43	8.73	8.45
C23H2703P1	382.44	382.1698	72.23	7.12	12.55	8.10
C23H2704P1	398.44	398.1647	69.33	6.83	16.06	7.77
C23H2705P1	414.44	414.1596	66.66	6.57	19.30	7.47
C23H2706P1	430.44	430.1545	64.18	6.32	22.30	7.20
C23H2707P1	446.44	446.1494	61.88	6.10	25.09	6.94
C23H2708P1	462.44	462.1443	59.74	5.89	27.68	6.70
C23H2709P1	478.44	478.1393	57.74	5.69	30.10	6.47
C23H2801P2	382.43	382.1615	72.24	7.38	4.18	16.20
C23H2802P2	398.43	398.1564	69.34	7.08	8.03	15.55
C23H2803P2	414.43	414.1514	66.66	6.81	11.58	14.95
C23H2804P2	430.42	430.1463	64.18	6.56	14.87	14.39
C23H2805P2	446.42	446.1412	61.88	6.32	17.92	13.88
C23H2806P2	462.42	462.1361	59.74	6.10	20.76	13.40
C23H2807P2	478.42	478.1310	57.74	5.90	23.41	12.95
C23H2808P2	494.42	494.1259	55.87	5.71	25.89	12.53
C23H2809P2	510.42	510.1208	54.12	5.53	28.21	12.14
C23H2901P1	352.46	352.1956	78.38	8.29	4.54	8.79
C23H2902P1	368.46	368.1905	74.98	7.93	8.68	8.41
C23H2903P1	384.46	384.1854	71.86	7.60	12.48	8.06
C23H2904P1	400.46	400.1803	68.98	7.30	15.98	7.73
C23H2905P1	416.46	416.1752	66.33	7.02	19.21	7.44
C23H2906P1	432.46	432.1702	63.88	6.76	22.20	7.16
C23H2907P1	448.46	448.1651	61.60	6.52	24.97	6.91
C23H2908P1	464.46	464.1600	59.48	6.29	27.56	6.67
C23H2909P1	480.46	480.1549	57.50	6.08	29.97	6.45
C23H3001P2	384.44	384.1772	71.86	7.87	4.16	16.11
C23H3002P2	400.44	400.1721	68.99	7.55	7.99	15.47
C23H3003P2	416.44	416.1670	66.34	7.26	11.53	14.88
C23H3004P2	432.44	432.1619	63.88	6.99	14.80	14.33
C23H3005P2	448.44	448.1568	61.60	6.74	17.84	13.81
C23H3006P2	464.44	464.1518	59.48	6.51	20.67	13.34
C23H3007P2	480.44	480.1467	57.50	6.29	23.31	12.89
C23H3008P2	496.44	496.1416	55.65	6.09	25.78	12.48
C23H3009P2	512.44	512.1365	53.91	5.90	28.10	12.09
C23H3101P1	354.48	354.2112	77.93	8.81	4.51	8.74
C23H3102P1	370.48	370.2062	74.57	8.43	8.64	8.36
C23H3103P1	386.48	386.2011	71.48	8.09	12.42	8.01
C23H3104P1	402.47	402.1960	68.64	7.76	15.90	7.70
C23H3105P1	418.47	418.1909	66.02	7.47	19.12	7.40
C23H3106P1	434.47	434.1858	63.58	7.19	22.09	7.13
C23H3107P1	450.47	450.1807	61.33	6.94	24.86	6.88
C23H3108P1	466.47	466.1756	59.22	6.70	27.44	6.64
C23H3109P1	482.47	482.1706	57.26	6.48	29.85	6.42
C23H3201P2	386.46	386.1928	71.48	8.35	4.14	16.03
C23H3202P2	402.46	402.1877	68.64	8.01	7.95	15.39
C23H3203P2	418.46	418.1827	66.02	7.71	11.47	14.80
C23H3204P2	434.46	434.1776	63.59	7.42	14.73	14.26
C23H3205P2	450.46	450.1725	61.33	7.16	17.76	13.75
C23H3206P2	466.46	466.1674	59.22	6.91	20.58	13.28
C23H3207P2	482.45	482.1623	57.26	6.69	23.21	12.84
C23H3208P2	498.45	498.1572	55.42	6.47	25.68	12.43
C23H3209P2	514.45	514.1521	53.70	6.27	27.99	12.04
C23H3301P1	356.49	356.2269	77.49	9.33	4.49	8.69
C23H3302P1	372.49	372.2218	74.16	8.93	8.59	8.32
C23H3303P1	388.49	388.2167	71.11	8.56	12.36	7.97
C23H3304P1	404.49	404.2116	68.30	8.22	15.82	7.66
C23H3305P1	420.49	420.2065	65.70	7.91	19.02	7.37
C23H3306P1	436.49	436.2015	63.29	7.62	21.99	7.10
C23H3307P1	452.49	452.1964	61.05	7.35	24.75	6.85

141

C23H3308P1	468.49	468.1913	58.97	7.10	27.32	6.61
C23H3309P1	484.49	484.1862	57.02	6.87	29.72	6.39
C23H3401P2	388.47	388.2085	71.11	8.82	4.12	15.95
C23H3402P2	404.47	404.2034	68.30	8.47	7.91	15.32
C23H3403P2	420.47	420.1983	65.70	8.15	11.42	14.73
C23H3404P2	436.47	436.1932	63.29	7.85	14.66	14.19
C23H3405P2	452.47	452.1881	61.05	7.57	17.68	13.69
C23H3406P2	468.47	468.1830	58.97	7.32	20.49	13.22
C23H3407P2	484.47	484.1780	57.02	7.07	23.12	12.79
C23H3408P2	500.47	500.1729	55.20	6.85	25.57	12.38
C23H3409P2	516.47	516.1678	53.49	6.64	27.88	11.99
C23H3501P1	358.51	358.2425	77.06	9.84	4.46	8.64
C23H3502P1	374.51	374.2375	73.77	9.42	8.54	8.27
C23H3503P1	390.51	390.2324	70.74	9.03	12.29	7.93
C23H3504P1	406.51	406.2273	67.96	8.68	15.74	7.62
C23H3505P1	422.51	422.2222	65.39	8.35	18.93	7.33
C23H3506P1	438.51	438.2171	63.00	8.05	21.89	7.06
C23H3507P1	454.51	454.2120	60.78	7.76	24.64	6.81
C23H3508P1	470.50	470.2069	58.71	7.50	27.20	6.58
C23H3509P1	486.50	486.2019	56.78	7.25	29.60	6.37
C23H3601P2	390.49	390.2241	70.75	9.29	4.10	15.86
C23H3602P2	406.49	406.2190	67.96	8.93	7.87	15.24
C23H3603P2	422.49	422.2140	65.39	8.59	11.36	14.66
C23H3604P2	438.49	438.2089	63.00	8.28	14.60	14.13
C23H3605P2	454.49	454.2038	60.78	7.98	17.60	13.63
C23H3606P2	470.49	470.1987	58.72	7.71	20.40	13.17
C23H3607P2	486.49	486.1936	56.79	7.46	23.02	12.73
C23H3608P2	502.49	502.1885	54.98	7.22	25.47	12.33
C23H3609P2	518.49	518.1834	53.28	7.00	27.77	11.95
C23H3701P1	360.52	360.2582	76.63	10.34	4.44	8.59
C23H3702P1	376.52	376.2531	73.37	9.91	8.50	8.23
C23H3703P1	392.52	392.2480	70.38	9.50	12.23	7.89
C23H3704P1	408.52	408.2429	67.62	9.13	15.67	7.58
C23H3705P1	424.52	424.2378	65.07	8.79	18.84	7.30
C23H3706P1	440.52	440.2328	62.71	8.47	21.79	7.03
C23H3707P1	456.52	456.2277	60.51	8.17	24.53	6.78
C23H3708P1	472.52	472.2226	58.46	7.89	27.09	6.56
C23H3709P1	488.52	488.2175	56.55	7.63	29.48	6.34
C23H3801P2	392.51	392.2398	70.38	9.76	4.08	15.78
C23H3802P2	408.51	408.2347	67.63	9.38	7.83	15.16
C23H3803P2	424.51	424.2296	65.08	9.02	11.31	14.59
C23H3804P2	440.50	440.2245	62.71	8.70	14.53	14.06
C23H3805P2	456.50	456.2194	60.52	8.39	17.52	13.57
C23H3806P2	472.50	472.2143	58.47	8.11	20.32	13.11
C23H3807P2	488.50	488.2093	56.55	7.84	22.93	12.68
C23H3808P2	504.50	504.2042	54.76	7.59	25.37	12.28
C23H3809P2	520.50	520.1991	53.08	7.36	27.66	11.90
C23H3901P1	362.54	362.2738	76.20	10.84	4.41	8.54
C23H3902P1	378.54	378.2688	72.98	10.38	8.45	8.18
C23H3903P1	394.54	394.2637	70.02	9.96	12.17	7.85
C23H3904P1	410.54	410.2586	67.29	9.58	15.59	7.54
C23H3905P1	426.54	426.2535	64.77	9.22	18.75	7.26
C23H3906P1	442.54	442.2484	62.43	8.88	21.69	7.00
C23H3907P1	458.54	458.2433	60.25	8.57	24.42	6.75
C23H3908P1	474.54	474.2382	58.22	8.28	26.97	6.53
C23H3909P1	490.54	490.2332	56.32	8.01	29.35	6.31
C23H4001P2	394.52	394.2554	70.02	10.22	4.06	15.70
C23H4002P2	410.52	410.2503	67.29	9.82	7.79	15.09
C23H4003P2	426.52	426.2453	64.77	9.45	11.25	14.52
C23H4004P2	442.52	442.2402	62.43	9.11	14.46	14.00

C23H4005P2	458.52	458.2351	60.25	8.79	17.45	13.51
C23H4006P2	474.52	474.2300	58.22	8.50	20.23	13.05
C23H4007P2	490.52	490.2249	56.32	8.22	22.83	12.63
C23H4008P2	506.52	506.2198	54.54	7.96	25.27	12.23
C23H4009P2	522.52	522.2147	52.87	7.72	27.56	11.86
C23H4101P1	364.56	364.2895	75.78	11.34	4.39	8.50
C23H4102P1	380.56	380.2844	72.59	10.86	8.41	8.14
C23H4103P1	396.56	396.2793	69.66	10.42	12.10	7.81
C23H4104P1	412.55	412.2742	66.96	10.02	15.51	7.51
C23H4105P1	428.55	428.2691	64.46	9.64	18.67	7.23
C23H4106P1	444.55	444.2641	62.14	9.30	21.59	6.97
C23H4107P1	460.55	460.2590	59.98	8.97	24.32	6.73
C23H4108P1	476.55	476.2539	57.97	8.67	26.86	6.50
C23H4109P1	492.55	492.2488	56.09	8.39	29.23	6.29
C23H4201P2	396.54	396.2711	69.67	10.68	4.03	15.62
C23H4202P2	412.54	412.2660	66.97	10.26	7.76	15.02
C23H4203P2	428.54	428.2609	64.47	9.88	11.20	14.46
C23H4204P2	444.54	444.2558	62.14	9.52	14.40	13.94
C23H4205P2	460.54	460.2507	59.99	9.19	17.37	13.45
C23H4206P2	476.54	476.2456	57.97	8.88	20.14	13.00
C23H4207P2	492.53	492.2406	56.09	8.60	22.74	12.58
C23H4208P2	508.53	508.2355	54.32	8.32	25.17	12.18
C23H4209P2	524.53	524.2304	52.67	8.07	27.45	11.81
C23H4301P1	366.57	366.3051	75.36	11.82	4.36	8.45
C23H4302P1	382.57	382.3001	72.21	11.33	8.36	8.10
C23H4303P1	398.57	398.2950	69.31	10.87	12.04	7.77
C23H4304P1	414.57	414.2899	66.64	10.45	15.44	7.47
C23H4305P1	430.57	430.2848	64.16	10.07	18.58	7.19
C23H4306P1	446.57	446.2797	61.86	9.71	21.50	6.94
C23H4307P1	462.57	462.2746	59.72	9.37	24.21	6.70
C23H4308P1	478.57	478.2695	57.73	9.06	26.75	6.47
C23H4309P1	494.57	494.2644	55.86	8.76	29.12	6.26
C23H4401P2	398.55	398.2867	69.31	11.13	4.01	15.54
C23H4402P2	414.55	414.2816	66.64	10.70	7.72	14.94
C23H4403P2	430.55	430.2766	64.16	10.30	11.15	14.39
C23H4404P2	446.55	446.2715	61.86	9.93	14.33	13.87
C23H4405P2	462.55	462.2664	59.72	9.59	17.29	13.39
C23H4406P2	478.55	478.2613	57.73	9.27	20.06	12.94
C23H4407P2	494.55	494.2562	55.86	8.97	22.65	12.53
C23H4408P2	510.55	510.2511	54.11	8.69	25.07	12.13
C23H4409P2	526.55	526.2460	52.47	8.42	27.35	11.76
C23H4501P1	368.59	368.3208	74.95	12.31	4.34	8.40
C23H4502P1	384.59	384.3157	71.83	11.79	8.32	8.05
C23H4503P1	400.59	400.3106	68.96	11.32	11.98	7.73
C23H4504P1	416.59	416.3055	66.31	10.89	15.36	7.44
C23H4505P1	432.59	432.3004	63.86	10.49	18.49	7.16
C23H4506P1	448.59	448.2954	61.58	10.11	21.40	6.90
C23H4507P1	464.58	464.2903	59.46	9.76	24.11	6.67
C23H4508P1	480.58	480.2852	57.48	9.44	26.63	6.45
C23H4509P1	496.58	496.2801	55.63	9.13	29.00	6.24
C23H4601P2	400.57	400.3024	68.97	11.58	3.99	15.46
C23H4602P2	416.57	416.2973	66.32	11.13	7.68	14.87
C23H4603P2	432.57	432.2922	63.86	10.72	11.10	14.32
C23H4604P2	448.57	448.2871	61.59	10.34	14.27	13.81
C23H4605P2	464.57	464.2820	59.47	9.98	17.22	13.33
C23H4606P2	480.57	480.2769	57.49	9.65	19.98	12.89
C23H4607P2	496.57	496.2719	55.63	9.34	22.55	12.48
C23H4608P2	512.57	512.2668	53.90	9.05	24.97	12.09
C23H4609P2	528.57	528.2617	52.27	8.77	27.24	11.72
C23H4701P1	370.60	370.3364	74.54	12.78	4.32	8.36

C23H4702P1	386.60	386.3313	71.46	12.25	8.28	8.01
C23H4703P1	402.60	402.3263	68.62	11.77	11.92	7.69
C23H4704P1	418.60	418.3212	65.99	11.32	15.29	7.40
C23H4705P1	434.60	434.3161	63.57	10.90	18.41	7.13
C23H4706P1	450.60	450.3110	61.31	10.51	21.30	6.87
C23H4707P1	466.60	466.3059	59.21	10.15	24.00	6.64
C23H4708P1	482.60	482.3008	57.24	9.82	26.52	6.42
C23H4709P1	498.60	498.2957	55.41	9.50	28.88	6.21
C23H4801P2	402.59	402.3180	68.62	12.02	3.97	15.39
C23H4802P2	418.59	418.3129	66.00	11.56	7.64	14.80
C23H4803P2	434.58	434.3079	63.57	11.13	11.04	14.25
C23H4804P2	450.58	450.3028	61.31	10.74	14.20	13.75
C23H4805P2	466.58	466.2977	59.21	10.37	17.15	13.28
C23H4806P2	482.58	482.2926	57.25	10.03	19.89	12.84
C23H4807P2	498.58	498.2875	55.41	9.70	22.46	12.42
C23H4808P2	514.58	514.2824	53.69	9.40	24.87	12.04
C23H4809P2	530.58	530.2773	52.07	9.12	27.14	11.68
C23H4901P1	372.62	372.3521	74.14	13.25	4.29	8.31
C23H4902P1	388.62	388.3470	71.09	12.71	8.23	7.97
C23H4903P1	404.62	404.3419	68.28	12.21	11.86	7.66
C23H4904P1	420.62	420.3368	65.68	11.74	15.22	7.36
C23H4905P1	436.62	436.3317	63.27	11.31	18.32	7.09
C23H4906P1	452.62	452.3267	61.04	10.91	21.21	6.84
C23H4907P1	468.62	468.3216	58.95	10.54	23.90	6.61
C23H4908P1	484.62	484.3165	57.01	10.19	26.41	6.39
C23H4909P1	500.62	500.3114	55.18	9.87	28.76	6.19
C23H5001P2	404.60	404.3337	68.28	12.46	3.95	15.31
C23H5002P2	420.60	420.3286	65.68	11.98	7.61	14.73
C23H5003P2	436.60	436.3235	63.27	11.54	10.99	14.19
C23H5004P2	452.60	452.3184	61.04	11.14	14.14	13.69
C23H5005P2	468.60	468.3133	58.95	10.76	17.07	13.22
C23H5006P2	484.60	484.3082	57.01	10.40	19.81	12.78
C23H5007P2	500.60	500.3032	55.19	10.07	22.37	12.37
C23H5008P2	516.60	516.2981	53.48	9.76	24.78	11.99
C23H5009P2	532.60	532.2930	51.87	9.46	27.04	11.63
C24H1201P2	378.31	378.0363	76.20	3.20	4.23	16.37
C24H1202P2	394.31	394.0313	73.11	3.07	8.12	15.71
C24H1203P2	410.31	410.0262	70.26	2.95	11.70	15.10
C24H1204P2	426.31	426.0211	67.62	2.84	15.01	14.53
C24H1205P2	442.31	442.0160	65.17	2.73	18.09	14.01
C24H1206P2	458.31	458.0109	62.90	2.64	20.95	13.52
C24H1207P2	474.31	474.0058	60.78	2.55	23.61	13.06
C24H1208P2	490.31	490.0007	58.79	2.47	26.11	12.63
C24H1209P2	506.31	505.9957	56.94	2.39	28.44	12.24
C24H1301P1	348.34	348.0704	82.75	3.76	4.59	8.89
C24H1302P1	364.34	364.0653	79.12	3.60	8.78	8.50
C24H1303P1	380.34	380.0602	75.79	3.45	12.62	8.14
C24H1304P1	396.34	396.0551	72.73	3.31	16.15	7.81
C24H1305P1	412.34	412.0501	69.91	3.18	19.40	7.51
C24H1306P1	428.34	428.0450	67.30	3.06	22.41	7.23
C24H1307P1	444.34	444.0399	64.88	2.95	25.20	6.97
C24H1308P1	460.34	460.0348	62.62	2.85	27.80	6.73
C24H1309P1	476.34	476.0297	60.52	2.75	30.23	6.50
C24H1401P2	380.33	380.0520	75.79	3.71	4.21	16.29
C24H1402P2	396.33	396.0469	72.74	3.56	8.07	15.63
C24H1403P2	412.32	412.0418	69.91	3.42	11.64	15.02
C24H1404P2	428.32	428.0367	67.30	3.29	14.94	14.46
C24H1405P2	444.32	444.0316	64.88	3.18	18.00	13.94
C24H1406P2	460.32	460.0266	62.62	3.07	20.85	13.46
C24H1407P2	476.32	476.0215	60.52	2.96	23.51	13.01

C24H1408P2	492.32	492.0164	58.55	2.87	26.00	12.58
C24H1409P2	508.32	508.0113	56.71	2.78	28.33	12.19
C24H1501P1	350.36	350.0860	82.28	4.32	4.57	8.84
C24H1502P1	366.36	366.0810	78.68	4.13	8.73	8.45
C24H1503P1	382.36	382.0759	75.39	3.95	12.55	8.10
C24H1504P1	398.36	398.0708	72.36	3.80	16.07	7.78
C24H1505P1	414.36	414.0657	69.57	3.65	19.31	7.48
C24H1506P1	430.36	430.0606	66.98	3.51	22.31	7.20
C24H1507P1	446.36	446.0555	64.58	3.39	25.09	6.94
C24H1508P1	462.36	462.0504	62.35	3.27	27.68	6.70
C24H1509P1	478.36	478.0454	60.26	3.16	30.10	6.48
C24H1601P2	382.34	382.0676	75.40	4.22	4.18	16.20
C24H1602P2	398.34	398.0625	72.37	4.05	8.03	15.55
C24H1603P2	414.34	414.0575	69.57	3.89	11.58	14.95
C24H1604P2	430.34	430.0524	66.99	3.75	14.87	14.40
C24H1605P2	446.34	446.0473	64.58	3.61	17.92	13.88
C24H1606P2	462.34	462.0422	62.35	3.49	20.76	13.40
C24H1607P2	478.34	478.0371	60.26	3.37	23.41	12.95
C24H1608P2	494.34	494.0320	58.31	3.26	25.89	12.53
C24H1609P2	510.34	510.0269	56.49	3.16	28.22	12.14
C24H1701P1	352.38	352.1017	81.81	4.86	4.54	8.79
C24H1702P1	368.38	368.0966	78.25	4.65	8.69	8.41
C24H1703P1	384.38	384.0915	75.00	4.46	12.49	8.06
C24H1704P1	400.37	400.0864	72.00	4.28	15.98	7.74
C24H1705P1	416.37	416.0814	69.23	4.12	19.21	7.44
C24H1706P1	432.37	432.0763	66.67	3.96	22.20	7.16
C24H1707P1	448.37	448.0712	64.29	3.82	24.98	6.91
C24H1708P1	464.37	464.0661	62.08	3.69	27.56	6.67
C24H1709P1	480.37	480.0610	60.01	3.57	29.98	6.45
C24H1801P2	384.36	384.0833	75.00	4.72	4.16	16.12
C24H1802P2	400.36	400.0782	72.00	4.53	7.99	15.47
C24H1803P2	416.36	416.0731	69.24	4.36	11.53	14.88
C24H1804P2	432.36	432.0680	66.67	4.20	14.80	14.33
C24H1805P2	448.36	448.0629	64.29	4.05	17.84	13.82
C24H1806P2	464.36	464.0579	62.08	3.91	20.67	13.34
C24H1807P2	480.35	480.0528	60.01	3.78	23.32	12.90
C24H1808P2	496.35	496.0477	58.08	3.66	25.79	12.48
C24H1809P2	512.35	512.0426	56.26	3.54	28.10	12.09
C24H1901P1	354.39	354.1173	81.34	5.40	4.51	8.74
C24H1902P1	370.39	370.1123	77.83	5.17	8.64	8.36
C24H1903P1	386.39	386.1072	74.61	4.96	12.42	8.02
C24H1904P1	402.39	402.1021	71.64	4.76	15.90	7.70
C24H1905P1	418.39	418.0970	68.90	4.58	19.12	7.40
C24H1906P1	434.39	434.0919	66.36	4.41	22.10	7.13
C24H1907P1	450.39	450.0868	64.00	4.25	24.87	6.88
C24H1908P1	466.39	466.0817	61.81	4.11	27.44	6.64
C24H1909P1	482.39	482.0767	59.76	3.97	29.85	6.42
C24H2001P2	386.37	386.0989	74.61	5.22	4.14	16.03
C24H2002P2	402.37	402.0938	71.64	5.01	7.95	15.40
C24H2003P2	418.37	418.0888	68.90	4.82	11.47	14.81
C24H2004P2	434.37	434.0837	66.36	4.64	14.73	14.26
C24H2005P2	450.37	450.0786	64.01	4.48	17.76	13.75
C24H2006P2	466.37	466.0735	61.81	4.32	20.58	13.28
C24H2007P2	482.37	482.0684	59.76	4.18	23.22	12.84
C24H2008P2	498.37	498.0633	57.84	4.05	25.68	12.43
C24H2009P2	514.37	514.0582	56.04	3.92	27.99	12.04
C24H2101P1	356.41	356.1330	80.88	5.94	4.49	8.69
C24H2102P1	372.41	372.1279	77.41	5.68	8.59	8.32
C24H2103P1	388.41	388.1228	74.22	5.45	12.36	7.97
C24H2104P1	404.41	404.1177	71.28	5.23	15.83	7.66

C24H2105P1	420.41	420.1127	68.57	5.03	19.03	7.37
C24H2106P1	436.41	436.1076	66.06	4.85	22.00	7.10
C24H2107P1	452.40	452.1025	63.72	4.68	24.76	6.85
C24H2108P1	468.40	468.0974	61.54	4.52	27.33	6.61
C24H2109P1	484.40	484.0923	59.51	4.37	29.73	6.39
C24H2201P2	388.39	388.1146	74.22	5.71	4.12	15.95
C24H2202P2	404.39	404.1095	71.28	5.48	7.91	15.32
C24H2203P2	420.39	420.1044	68.57	5.27	11.42	14.74
C24H2204P2	436.39	436.0993	66.06	5.08	14.67	14.20
C24H2205P2	452.39	452.0942	63.72	4.90	17.68	13.69
C24H2206P2	468.39	468.0892	61.54	4.73	20.50	13.23
C24H2207P2	484.39	484.0841	59.51	4.58	23.12	12.79
C24H2208P2	500.39	500.0790	57.61	4.43	25.58	12.38
C24H2209P2	516.39	516.0739	55.82	4.29	27.89	12.00
C24H2301P1	358.42	358.1486	80.43	6.47	4.46	8.64
C24H2302P1	374.42	374.1436	76.99	6.19	8.55	8.27
C24H2303P1	390.42	390.1385	73.83	5.94	12.29	7.93
C24H2304P1	406.42	406.1334	70.93	5.70	15.75	7.62
C24H2305P1	422.42	422.1283	68.24	5.49	18.94	7.33
C24H2306P1	438.42	438.1232	65.75	5.29	21.90	7.06
C24H2307P1	454.42	454.1181	63.44	5.10	24.65	6.82
C24H2308P1	470.42	470.1130	61.28	4.93	27.21	6.58
C24H2309P1	486.42	486.1080	59.26	4.77	29.60	6.37
C24H2401P2	390.41	390.1302	73.84	6.20	4.10	15.87
C24H2402P2	406.41	406.1251	70.93	5.95	7.87	15.24
C24H2403P2	422.40	422.1201	68.24	5.73	11.36	14.67
C24H2404P2	438.40	438.1150	65.75	5.52	14.60	14.13
C24H2405P2	454.40	454.1099	63.44	5.32	17.60	13.63
C24H2406P2	470.40	470.1048	61.28	5.14	20.41	13.17
C24H2407P2	486.40	486.0997	59.27	4.97	23.03	12.74
C24H2408P2	502.40	502.0946	57.38	4.82	25.48	12.33
C24H2409P2	518.40	518.0895	55.61	4.67	27.78	11.95
C24H2501P1	360.44	360.1643	79.98	6.99	4.44	8.59
C24H2502P1	376.44	376.1592	76.58	6.69	8.50	8.23
C24H2503P1	392.44	392.1541	73.46	6.42	12.23	7.89
C24H2504P1	408.44	408.1490	70.58	6.17	15.67	7.58
C24H2505P1	424.44	424.1439	67.92	5.94	18.85	7.30
C24H2506P1	440.44	440.1389	65.45	5.72	21.80	7.03
C24H2507P1	456.44	456.1338	63.16	5.52	24.54	6.79
C24H2508P1	472.44	472.1287	61.02	5.33	27.09	6.56
C24H2509P1	488.44	488.1236	59.02	5.16	29.48	6.34
C24H2601P2	392.42	392.1459	73.46	6.68	4.08	15.79
C24H2602P2	408.42	408.1408	70.58	6.42	7.83	15.17
C24H2603P2	424.42	424.1357	67.92	6.17	11.31	14.60
C24H2604P2	440.42	440.1306	65.45	5.95	14.53	14.07
C24H2605P2	456.42	456.1255	63.16	5.74	17.53	13.57
C24H2606P2	472.42	472.1205	61.02	5.55	20.32	13.11
C24H2607P2	488.42	488.1154	59.02	5.37	22.93	12.68
C24H2608P2	504.42	504.1103	57.15	5.20	25.37	12.28
C24H2609P2	520.42	520.1052	55.39	5.04	27.67	11.90
C24H2701P1	362.46	362.1799	79.53	7.51	4.41	8.55
C24H2702P1	378.46	378.1749	76.17	7.19	8.46	8.18
C24H2703P1	394.45	394.1698	73.08	6.90	12.17	7.85
C24H2704P1	410.45	410.1647	70.23	6.63	15.59	7.55
C24H2705P1	426.45	426.1596	67.60	6.38	18.76	7.26
C24H2706P1	442.45	442.1545	65.15	6.15	21.70	7.00
C24H2707P1	458.45	458.1494	62.88	5.94	24.43	6.76
C24H2708P1	474.45	474.1443	60.76	5.74	26.98	6.53
C24H2709P1	490.45	490.1393	58.78	5.55	29.36	6.32
C24H2801P2	394.44	394.1615	73.08	7.16	4.06	15.71

C24H2802P2	410.44	410.1564	70.23	6.88	7.80	15.09
C24H2803P2	426.44	426.1514	67.60	6.62	11.26	14.53
C24H2804P2	442.44	442.1463	65.15	6.38	14.46	14.00
C24H2805P2	458.44	458.1412	62.88	6.16	17.45	13.51
C24H2806P2	474.43	474.1361	60.76	5.95	20.23	13.06
C24H2807P2	490.43	490.1310	58.78	5.75	22.84	12.63
C24H2808P2	506.43	506.1259	56.92	5.57	25.27	12.23
C24H2809P2	522.43	522.1208	55.18	5.40	27.56	11.86
C24H2901P1	364.47	364.1956	79.09	8.02	4.39	8.50
C24H2902P1	380.47	380.1905	75.77	7.68	8.41	8.14
C24H2903P1	396.47	396.1854	72.71	7.37	12.11	7.81
C24H2904P1	412.47	412.1803	69.89	7.09	15.52	7.51
C24H2905P1	428.47	428.1752	67.28	6.82	18.67	7.23
C24H2906P1	444.47	444.1702	64.86	6.58	21.60	6.97
C24H2907P1	460.47	460.1651	62.60	6.35	24.32	6.73
C24H2908P1	476.47	476.1600	60.50	6.13	26.86	6.50
C24H2909P1	492.47	492.1549	58.54	5.94	29.24	6.29
C24H3001P2	396.45	396.1772	72.71	7.63	4.04	15.63
C24H3002P2	412.45	412.1721	69.89	7.33	7.76	15.02
C24H3003P2	428.45	428.1670	67.28	7.06	11.20	14.46
C24H3004P2	444.45	444.1619	64.86	6.80	14.40	13.94
C24H3005P2	460.45	460.1568	62.61	6.57	17.37	13.45
C24H3006P2	476.45	476.1518	60.50	6.35	20.15	13.00
C24H3007P2	492.45	492.1467	58.54	6.14	22.74	12.58
C24H3008P2	508.45	508.1416	56.70	5.95	25.17	12.18
C24H3009P2	524.45	524.1365	54.97	5.77	27.46	11.81
C24H3101P1	366.49	366.2112	78.66	8.53	4.37	8.45
C24H3102P1	382.49	382.2062	75.37	8.17	8.37	8.10
C24H3103P1	398.49	398.2011	72.34	7.84	12.05	7.77
C24H3104P1	414.49	414.1960	69.55	7.54	15.44	7.47
C24H3105P1	430.49	430.1909	66.96	7.26	18.58	7.20
C24H3106P1	446.48	446.1858	64.56	7.00	21.50	6.94
C24H3107P1	462.48	462.1807	62.33	6.76	24.22	6.70
C24H3108P1	478.48	478.1756	60.25	6.53	26.75	6.47
C24H3109P1	494.48	494.1706	58.30	6.32	29.12	6.26
C24H3201P2	398.47	398.1928	72.34	8.09	4.02	15.55
C24H3202P2	414.47	414.1877	69.55	7.78	7.72	14.95
C24H3203P2	430.47	430.1827	66.97	7.49	11.15	14.39
C24H3204P2	446.47	446.1776	64.57	7.22	14.33	13.88
C24H3205P2	462.47	462.1725	62.33	6.97	17.30	13.40
C24H3206P2	478.47	478.1674	60.25	6.74	20.06	12.95
C24H3207P2	494.47	494.1623	58.30	6.52	22.65	12.53
C24H3208P2	510.47	510.1572	56.47	6.32	25.07	12.14
C24H3209P2	526.46	526.1521	54.76	6.13	27.35	11.77
C24H3301P1	368.50	368.2269	78.23	9.03	4.34	8.41
C24H3302P1	384.50	384.2218	74.97	8.65	8.32	8.06
C24H3303P1	400.50	400.2167	71.98	8.31	11.98	7.73
C24H3304P1	416.50	416.2116	69.21	7.99	15.37	7.44
C24H3305P1	432.50	432.2065	66.65	7.69	18.50	7.16
C24H3306P1	448.50	448.2015	64.27	7.42	21.40	6.91
C24H3307P1	464.50	464.1964	62.06	7.16	24.11	6.67
C24H3308P1	480.50	480.1913	59.99	6.92	26.64	6.45
C24H3309P1	496.50	496.1862	58.06	6.70	29.00	6.24
C24H3401P2	400.49	400.2085	71.98	8.56	4.00	15.47
C24H3402P2	416.48	416.2034	69.21	8.23	7.68	14.87
C24H3403P2	432.48	432.1983	66.65	7.92	11.10	14.32
C24H3404P2	448.48	448.1932	64.28	7.64	14.27	13.81
C24H3405P2	464.48	464.1881	62.06	7.38	17.22	13.34
C24H3406P2	480.48	480.1830	60.00	7.13	19.98	12.89
C24H3407P2	496.48	496.1780	58.06	6.90	22.56	12.48

144

C24H3408P2	512.48	512.1729	56.25	6.69	24.98	12.09
C24H3409P2	528.48	528.1678	54.55	6.48	27.25	11.72
C24H3501P1	370.52	370.2425	77.80	9.52	4.32	8.36
C24H3502P1	386.52	386.2375	74.58	9.13	8.28	8.01
C24H3503P1	402.52	402.2324	71.62	8.76	11.92	7.69
C24H3504P1	418.52	418.2273	68.88	8.43	15.29	7.40
C24H3505P1	434.52	434.2222	66.34	8.12	18.41	7.13
C24H3506P1	450.52	450.2171	63.99	7.83	21.31	6.88
C24H3507P1	466.52	466.2120	61.79	7.56	24.01	6.64
C24H3508P1	482.52	482.2069	59.74	7.31	26.53	6.42
C24H3509P1	498.51	498.2019	57.83	7.08	28.88	6.21
C24H3601P2	402.50	402.2241	71.62	9.02	3.97	15.39
C24H3602P2	418.50	418.2190	68.88	8.67	7.65	14.80
C24H3603P2	434.50	434.2140	66.34	8.35	11.05	14.26
C24H3604P2	450.50	450.2089	63.99	8.05	14.21	13.75
C24H3605P2	466.50	466.2038	61.79	7.78	17.15	13.28
C24H3606P2	482.50	482.1987	59.74	7.52	19.90	12.84
C24H3607P2	498.50	498.1936	57.83	7.28	22.47	12.43
C24H3608P2	514.50	514.1885	56.03	7.05	24.88	12.04
C24H3609P2	530.50	530.1834	54.34	6.84	27.14	11.68
C24H3701P1	372.54	372.2582	77.38	10.01	4.29	8.31
C24H3702P1	388.54	388.2531	74.19	9.60	8.24	7.97
C24H3703P1	404.53	404.2480	71.26	9.22	11.87	7.66
C24H3704P1	420.53	420.2429	68.55	8.87	15.22	7.37
C24H3705P1	436.53	436.2378	66.04	8.54	18.33	7.10
C24H3706P1	452.53	452.2328	63.70	8.24	21.21	6.84
C24H3707P1	468.53	468.2277	61.53	7.96	23.90	6.61
C24H3708P1	484.53	484.2226	59.49	7.70	26.42	6.39
C24H3709P1	500.53	500.2175	57.59	7.45	28.77	6.19
C24H3801P2	404.52	404.2398	71.26	9.47	3.96	15.31
C24H3802P2	420.52	420.2347	68.55	9.11	7.61	14.73
C24H3803P2	436.52	436.2296	66.04	8.77	11.00	14.19
C24H3804P2	452.52	452.2245	63.70	8.46	14.14	13.69
C24H3805P2	468.52	468.2194	61.53	8.18	17.07	13.22
C24H3806P2	484.51	484.2143	59.50	7.91	19.81	12.79
C24H3807P2	500.51	500.2093	57.59	7.65	22.38	12.38
C24H3808P2	516.51	516.2042	55.81	7.42	24.78	11.99
C24H3809P2	532.51	532.1991	54.13	7.19	27.04	11.63
C24H3901P1	374.55	374.2738	76.96	10.50	4.27	8.27
C24H3902P1	390.55	390.2688	73.81	10.07	8.19	7.93
C24H3903P1	406.55	406.2637	70.91	9.67	11.81	7.62
C24H3904P1	422.55	422.2586	68.22	9.30	15.15	7.33
C24H3905P1	438.55	438.2535	65.73	8.96	18.24	7.06
C24H3906P1	454.55	454.2484	63.42	8.65	21.12	6.81
C24H3907P1	470.55	470.2433	61.26	8.35	23.80	6.58
C24H3908P1	486.55	486.2382	59.25	8.08	26.31	6.37
C24H3909P1	502.55	502.2332	57.36	7.82	28.65	6.16
C24H4001P2	406.53	406.2554	70.91	9.92	3.94	15.24
C24H4002P2	422.53	422.2503	68.22	9.54	7.57	14.66
C24H4003P2	438.53	438.2453	65.73	9.19	10.95	14.13
C24H4004P2	454.53	454.2402	63.42	8.87	14.08	13.63
C24H4005P2	470.53	470.2351	61.26	8.57	17.00	13.17
C24H4006P2	486.53	486.2300	59.25	8.29	19.73	12.73
C24H4007P2	502.53	502.2249	57.36	8.02	22.29	12.33
C24H4008P2	518.53	518.2198	55.59	7.78	24.68	11.95
C24H4009P2	534.53	534.2147	53.93	7.54	26.94	11.59
C24H4101P1	376.57	376.2895	76.55	10.97	4.25	8.23
C24H4102P1	392.57	392.2844	73.43	10.53	8.15	7.89
C24H4103P1	408.57	408.2793	70.56	10.12	11.75	7.58
C24H4104P1	424.57	424.2742	67.90	9.73	15.07	7.30
C24H4105P1	440.57	440.2691	65.43	9.38	18.16	7.03
C24H4106P1	456.56	456.2641	63.14	9.05	21.03	6.78
C24H4107P1	472.56	472.2590	61.00	8.75	23.70	6.55
C24H4108P1	488.56	488.2539	59.00	8.46	26.20	6.34
C24H4109P1	504.56	504.2488	57.13	8.19	28.54	6.14
C24H4201P2	408.55	408.2711	70.56	10.36	3.92	15.16
C24H4202P2	424.55	424.2660	67.90	9.97	7.54	14.59
C24H4203P2	440.55	440.2609	65.43	9.61	10.90	14.06
C24H4204P2	456.55	456.2558	63.14	9.27	14.02	13.57
C24H4205P2	472.55	472.2507	61.00	8.96	16.93	13.11
C24H4206P2	488.55	488.2456	59.01	8.67	19.65	12.68
C24H4207P2	504.55	504.2406	57.13	8.39	22.20	12.28
C24H4208P2	520.55	520.2355	55.38	8.13	24.59	11.90
C24H4209P2	536.54	536.2304	53.73	7.89	26.84	11.55
C24H4301P1	378.58	378.3051	76.14	11.45	4.23	8.18
C24H4302P1	394.58	394.3001	73.06	10.98	8.11	7.85
C24H4303P1	410.58	410.2950	70.21	10.56	11.69	7.54
C24H4304P1	426.58	426.2899	67.58	10.16	15.00	7.26
C24H4305P1	442.58	442.2848	65.13	9.79	18.08	7.00
C24H4306P1	458.58	458.2797	62.86	9.45	20.93	6.75
C24H4307P1	474.58	474.2746	60.74	9.13	23.60	6.53
C24H4308P1	490.58	490.2695	58.76	8.84	26.09	6.31
C24H4309P1	506.58	506.2644	56.90	8.56	28.42	6.11
C24H4401P2	410.57	410.2867	70.21	10.80	3.90	15.09
C24H4402P2	426.56	426.2816	67.58	10.40	7.50	14.52
C24H4403P2	442.56	442.2766	65.14	10.02	10.85	14.00
C24H4404P2	458.56	458.2715	62.86	9.67	13.96	13.51
C24H4405P2	474.56	474.2664	60.74	9.35	16.86	13.05
C24H4406P2	490.56	490.2613	58.76	9.04	19.57	12.63
C24H4407P2	506.56	506.2562	56.91	8.76	22.11	12.23
C24H4408P2	522.56	522.2511	55.16	8.49	24.49	11.85
C24H4409P2	538.56	538.2460	53.53	8.24	26.74	11.50
C24H4501P1	380.60	380.3208	75.74	11.92	4.20	8.14
C24H4502P1	396.60	396.3157	72.68	11.44	8.07	7.81
C24H4503P1	412.60	412.3106	69.87	10.99	11.63	7.51
C24H4504P1	428.60	428.3055	67.26	10.58	14.93	7.23
C24H4505P1	444.60	444.3004	64.84	10.20	17.99	6.97
C24H4506P1	460.60	460.2954	62.59	9.85	20.84	6.72
C24H4507P1	476.60	476.2903	60.48	9.52	23.50	6.50
C24H4508P1	492.60	492.2852	58.52	9.21	25.98	6.29
C24H4509P1	508.59	508.2801	56.68	8.92	28.31	6.09
C24H4601P2	412.58	412.3024	69.87	11.24	3.88	15.01
C24H4602P2	428.58	428.2973	67.26	10.82	7.47	14.45
C24H4603P2	444.58	444.2922	64.84	10.43	10.80	13.93
C24H4604P2	460.58	460.2871	62.59	10.07	13.90	13.45
C24H4605P2	476.58	476.2820	60.49	9.73	16.79	13.00
C24H4606P2	492.58	492.2769	58.52	9.41	19.49	12.58
C24H4607P2	508.58	508.2719	56.68	9.12	22.02	12.18
C24H4608P2	524.58	524.2668	54.95	8.84	24.40	11.81
C24H4609P2	540.58	540.2617	53.33	8.58	26.64	11.46
C24H4701P1	382.62	382.3364	75.34	12.38	4.18	8.10
C24H4702P1	398.61	398.3313	72.32	11.88	8.03	7.77
C24H4703P1	414.61	414.3263	69.53	11.43	11.58	7.47
C24H4704P1	430.61	430.3212	66.94	11.00	14.86	7.19
C24H4705P1	446.61	446.3161	64.55	10.61	17.91	6.94
C24H4706P1	462.61	462.3110	62.31	10.24	20.75	6.70
C24H4707P1	478.61	478.3059	60.23	9.90	23.40	6.47
C24H4708P1	494.61	494.3008	58.28	9.58	25.88	6.26
C24H4709P1	510.61	510.2957	56.46	9.28	28.20	6.07
C24H4801P2	414.60	414.3180	69.53	11.67	3.86	14.94

145

C24H4802P2	430.60	430.3129	66.95	11.24	7.43	14.39
C24H4803P2	446.60	446.3079	64.55	10.83	10.75	13.87
C24H4804P2	462.60	462.3028	62.32	10.46	13.83	13.39
C24H4805P2	478.59	478.2977	60.23	10.11	16.71	12.94
C24H4806P2	494.59	494.2926	58.28	9.78	19.41	12.52
C24H4807P2	510.59	510.2875	56.46	9.48	21.93	12.13
C24H4808P2	526.59	526.2824	54.74	9.19	24.31	11.76
C24H4809P2	542.59	542.2773	53.13	8.92	26.54	11.42
C24H4901P1	384.63	384.3521	74.95	12.84	4.16	8.05
C24H4902P1	400.63	400.3470	71.95	12.33	7.99	7.73
C24H4903P1	416.63	416.3419	69.19	11.85	11.52	7.43
C24H4904P1	432.63	432.3368	66.63	11.42	14.79	7.16
C24H4905P1	448.63	448.3317	64.26	11.01	17.83	6.90
C24H4906P1	464.63	464.3267	62.04	10.63	20.66	6.67
C24H4907P1	480.63	480.3216	59.98	10.28	23.30	6.44
C24H4908P1	496.63	496.3165	58.05	9.95	25.77	6.24
C24H4909P1	512.63	512.3114	56.23	9.63	28.09	6.04
C24H5001P2	416.61	416.3337	69.19	12.10	3.84	14.87
C24H5002P2	432.61	432.3286	66.63	11.65	7.40	14.32
C24H5003P2	448.61	448.3235	64.26	11.23	10.70	13.81
C24H5004P2	464.61	464.3184	62.04	10.85	13.77	13.33
C24H5005P2	480.61	480.3133	59.98	10.49	16.64	12.89
C24H5006P2	496.61	496.3082	58.05	10.15	19.33	12.47
C24H5007P2	512.61	512.3032	56.24	9.83	21.85	12.08
C24H5008P2	528.61	528.2981	54.53	9.53	24.21	11.72
C24H5009P2	544.61	544.2930	52.93	9.25	26.44	11.37
C24H5101P1	386.65	386.3677	74.56	13.30	4.14	8.01
C24H5102P1	402.65	402.3626	71.59	12.77	7.95	7.69
C24H5103P1	418.65	418.3576	68.86	12.28	11.47	7.40
C24H5104P1	434.65	434.3525	66.32	11.83	14.72	7.13
C24H5105P1	450.64	450.3474	63.97	11.41	17.75	6.87
C24H5106P1	466.64	466.3423	61.77	11.02	20.57	6.64
C24H5107P1	482.64	482.3372	59.73	10.65	23.20	6.42
C24H5108P1	498.64	498.3321	57.81	10.31	25.67	6.21
C24H5109P1	514.64	514.3270	56.01	9.99	27.98	6.02
C24H5201P2	418.63	418.3493	68.86	12.52	3.82	14.80
C24H5202P2	434.63	434.3442	66.33	12.06	7.36	14.25
C24H5203P2	450.63	450.3391	63.97	11.63	10.65	13.75
C24H5204P2	466.63	466.3341	61.78	11.23	13.71	13.28
C24H5205P2	482.63	482.3290	59.73	10.86	16.58	12.84
C24H5206P2	498.63	498.3239	57.81	10.51	19.25	12.42
C24H5207P2	514.63	514.3188	56.02	10.18	21.76	12.04
C24H5208P2	530.62	530.3137	54.33	9.88	24.12	11.67
C24H5209P2	546.62	546.3086	52.74	9.59	26.34	11.33
C25H1201P2	390.32	390.0363	76.93	3.10	4.10	15.87
C25H1202P2	406.32	406.0313	73.90	2.98	7.88	15.25
C25H1203P2	422.32	422.0262	71.10	2.86	11.37	14.67
C25H1204P2	438.32	438.0211	68.51	2.76	14.60	14.13
C25H1205P2	454.32	454.0160	66.09	2.66	17.61	13.64
C25H1206P2	470.32	470.0109	63.85	2.57	20.41	13.17
C25H1207P2	486.32	486.0058	61.75	2.49	23.03	12.74
C25H1208P2	502.32	502.0007	59.78	2.41	25.48	12.33
C25H1209P2	518.32	517.9957	57.93	2.33	27.78	11.95
C25H1301P1	360.35	360.0704	83.33	3.64	4.44	8.60
C25H1302P1	376.35	376.0653	79.79	3.48	8.50	8.23
C25H1303P1	392.35	392.0602	76.53	3.34	12.23	7.89
C25H1304P1	408.35	408.0551	73.53	3.21	15.67	7.59
C25H1305P1	424.35	424.0501	70.76	3.09	18.85	7.30
C25H1306P1	440.35	440.0450	68.19	2.98	21.80	7.03
C25H1307P1	456.35	456.0399	65.80	2.87	24.54	6.79
C25H1308P1	472.35	472.0348	63.57	2.77	27.10	6.56
C25H1309P1	488.35	488.0297	61.49	2.68	29.49	6.34
C25H1401P2	392.34	392.0520	76.54	3.60	4.08	15.79
C25H1402P2	408.34	408.0469	73.54	3.46	7.84	15.17
C25H1403P2	424.34	424.0418	70.76	3.33	11.31	14.60
C25H1404P2	440.34	440.0367	68.19	3.20	14.53	14.07
C25H1405P2	456.33	456.0316	65.80	3.09	17.53	13.58
C25H1406P2	472.33	472.0266	63.57	2.99	20.32	13.12
C25H1407P2	488.33	488.0215	61.49	2.89	22.93	12.69
C25H1408P2	504.33	504.0164	59.54	2.80	25.38	12.28
C25H1409P2	520.33	520.0113	57.71	2.71	27.67	11.91
C25H1501P1	362.37	362.0860	82.86	4.17	4.42	8.55
C25H1502P1	378.37	378.0810	79.36	4.00	8.46	8.19
C25H1503P1	394.37	394.0759	76.14	3.83	12.17	7.85
C25H1504P1	410.37	410.0708	73.17	3.68	15.60	7.55
C25H1505P1	426.37	426.0657	70.43	3.55	18.76	7.26
C25H1506P1	442.37	442.0606	67.88	3.42	21.70	7.00
C25H1507P1	458.37	458.0555	65.51	3.30	24.43	6.76
C25H1508P1	474.37	474.0504	63.30	3.19	26.98	6.53
C25H1509P1	490.37	490.0454	61.24	3.08	29.36	6.32
C25H1601P2	394.35	394.0676	76.14	4.09	4.06	15.71
C25H1602P2	410.35	410.0625	73.18	3.93	7.80	15.10
C25H1603P2	426.35	426.0575	70.43	3.78	11.26	14.53
C25H1604P2	442.35	442.0524	67.88	3.65	14.47	14.00
C25H1605P2	458.35	458.0473	65.51	3.52	17.45	13.52
C25H1606P2	474.35	474.0422	63.30	3.40	20.24	13.06
C25H1607P2	490.35	490.0371	61.24	3.29	22.84	12.63
C25H1608P2	506.35	506.0320	59.30	3.19	25.28	12.23
C25H1609P2	522.35	522.0269	57.49	3.09	27.57	11.86
C25H1701P1	364.39	364.1017	82.41	4.70	4.39	8.50
C25H1702P1	380.39	380.0966	78.94	4.50	8.41	8.14
C25H1703P1	396.39	396.0915	75.75	4.32	12.11	7.81
C25H1704P1	412.39	412.0864	72.82	4.16	15.52	7.51
C25H1705P1	428.39	428.0814	70.10	4.00	18.67	7.23
C25H1706P1	444.38	444.0763	67.57	3.86	21.60	6.97
C25H1707P1	460.38	460.0712	65.22	3.72	24.33	6.73
C25H1708P1	476.38	476.0661	63.03	3.60	26.87	6.50
C25H1709P1	492.38	492.0610	60.98	3.48	29.24	6.29
C25H1801P2	396.37	396.0833	75.76	4.58	4.04	15.63
C25H1802P2	412.37	412.0782	72.82	4.40	7.76	15.02
C25H1803P2	428.37	428.0731	70.10	4.24	11.20	14.46
C25H1804P2	444.37	444.0680	67.57	4.08	14.40	13.94
C25H1805P2	460.37	460.0629	65.23	3.94	17.38	13.46
C25H1806P2	476.37	476.0579	63.04	3.81	20.15	13.00
C25H1807P2	492.37	492.0528	60.99	3.68	22.75	12.58
C25H1808P2	508.37	508.0477	59.07	3.57	25.18	12.19
C25H1809P2	524.36	524.0426	57.27	3.46	27.46	11.81
C25H1901P1	366.40	366.1173	81.95	5.23	4.37	8.45
C25H1902P1	382.40	382.1123	78.52	5.01	8.37	8.10
C25H1903P1	398.40	398.1072	75.37	4.81	12.05	7.77
C25H1904P1	414.40	414.1021	72.46	4.62	15.44	7.47
C25H1905P1	430.40	430.0970	69.77	4.45	18.59	7.20
C25H1906P1	446.40	446.0919	67.27	4.29	21.50	6.94
C25H1907P1	462.40	462.0868	64.94	4.14	24.22	6.70
C25H1908P1	478.40	478.0817	62.77	4.00	26.75	6.47
C25H1909P1	494.40	494.0767	60.74	3.87	29.13	6.26
C25H2001P2	398.39	398.0989	75.37	5.06	4.02	15.55
C25H2002P2	414.38	414.0938	72.46	4.86	7.72	14.95
C25H2003P2	430.38	430.0888	69.77	4.68	11.15	14.39
C25H2004P2	446.38	446.0837	67.27	4.52	14.34	13.88

C25H2005P2	462.38	462.0786	64.94	4.36	17.30	13.40
C25H2006P2	478.38	478.0735	62.77	4.21	20.07	12.95
C25H2007P2	494.38	494.0684	60.74	4.08	22.65	12.53
C25H2008P2	510.38	510.0633	58.83	3.95	25.08	12.14
C25H2009P2	526.38	526.0582	57.05	3.83	27.36	11.77
C25H2101P1	368.42	368.1330	81.50	5.75	4.34	8.41
C25H2102P1	384.42	384.1279	78.11	5.51	8.32	8.06
C25H2103P1	400.42	400.1228	74.99	5.29	11.99	7.74
C25H2104P1	416.42	416.1177	72.11	5.08	15.37	7.44
C25H2105P1	432.42	432.1127	69.44	4.90	18.50	7.16
C25H2106P1	448.42	448.1076	66.96	4.72	21.41	6.91
C25H2107P1	464.42	464.1025	64.66	4.56	24.12	6.67
C25H2108P1	480.42	480.0974	62.50	4.41	26.64	6.45
C25H2109P1	496.41	496.0923	60.49	4.26	29.01	6.24
C25H2201P2	400.40	400.1146	74.99	5.54	4.00	15.47
C25H2202P2	416.40	416.1095	72.11	5.33	7.68	14.88
C25H2203P2	432.40	432.1044	69.44	5.13	11.10	14.33
C25H2204P2	448.40	448.0993	66.97	4.95	14.27	13.82
C25H2205P2	464.40	464.0942	64.66	4.78	17.23	13.34
C25H2206P2	480.40	480.0892	62.51	4.62	19.98	12.90
C25H2207P2	496.40	496.0841	60.49	4.47	22.56	12.48
C25H2208P2	512.40	512.0790	58.60	4.33	24.98	12.09
C25H2209P2	528.40	528.0739	56.83	4.20	27.25	11.72
C25H2301P1	370.44	370.1486	81.06	6.26	4.32	8.36
C25H2302P1	386.43	386.1436	77.70	6.00	8.28	8.02
C25H2303P1	402.43	402.1385	74.62	5.76	11.93	7.70
C25H2304P1	418.43	418.1334	71.76	5.54	15.29	7.40
C25H2305P1	434.43	434.1283	69.12	5.34	18.41	7.13
C25H2306P1	450.43	450.1232	66.66	5.15	21.31	6.88
C25H2307P1	466.43	466.1181	64.38	4.97	24.01	6.64
C25H2308P1	482.43	482.1130	62.24	4.81	26.53	6.42
C25H2309P1	498.43	498.1080	60.24	4.65	28.89	6.21
C25H2401P2	402.42	402.1302	74.62	6.01	3.98	15.39
C25H2402P2	418.42	418.1251	71.77	5.78	7.65	14.81
C25H2403P2	434.42	434.1201	69.12	5.57	11.05	14.26
C25H2404P2	450.42	450.1150	66.67	5.37	14.21	13.75
C25H2405P2	466.41	466.1099	64.38	5.19	17.15	13.28
C25H2406P2	482.41	482.1048	62.25	5.01	19.90	12.84
C25H2407P2	498.41	498.0997	60.25	4.85	22.47	12.43
C25H2408P2	514.41	514.0946	58.37	4.70	24.88	12.04
C25H2409P2	530.41	530.0895	56.61	4.56	27.15	11.68
C25H2501P1	372.45	372.1643	80.62	6.77	4.30	8.32
C25H2502P1	388.45	388.1592	77.30	6.49	8.24	7.97
C25H2503P1	404.45	404.1541	74.24	6.23	11.87	7.66
C25H2504P1	420.45	420.1490	71.42	5.99	15.22	7.37
C25H2505P1	436.45	436.1439	68.80	5.77	18.33	7.10
C25H2506P1	452.45	452.1389	66.37	5.57	21.22	6.85
C25H2507P1	468.45	468.1338	64.10	5.38	23.91	6.61
C25H2508P1	484.45	484.1287	61.98	5.20	26.42	6.39
C25H2509P1	500.45	500.1236	60.00	5.04	28.77	6.19
C25H2601P2	404.43	404.1459	74.25	6.48	3.96	15.32
C25H2602P2	420.43	420.1408	71.42	6.23	7.61	14.73
C25H2603P2	436.43	436.1357	68.80	6.00	11.00	14.19
C25H2604P2	452.43	452.1306	66.37	5.79	14.15	13.69
C25H2605P2	468.43	468.1255	64.10	5.59	17.08	13.22
C25H2606P2	484.43	484.1205	61.99	5.41	19.82	12.79
C25H2607P2	500.43	500.1154	60.00	5.24	22.38	12.38
C25H2608P2	516.43	516.1103	58.15	5.07	24.78	12.00
C25H2609P2	532.43	532.1052	56.40	4.92	27.04	11.63
C25H2701P1	374.47	374.1799	80.19	7.27	4.27	8.27

C25H2702P1	390.47	390.1749	76.90	6.97	8.20	7.93
C25H2703P1	406.47	406.1698	73.88	6.70	11.81	7.62
C25H2704P1	422.47	422.1647	71.08	6.44	15.15	7.33
C25H2705P1	438.46	438.1596	68.48	6.21	18.24	7.06
C25H2706P1	454.46	454.1545	66.07	5.99	21.12	6.82
C25H2707P1	470.46	470.1494	63.83	5.78	23.81	6.58
C25H2708P1	486.46	486.1443	61.73	5.59	26.31	6.37
C25H2709P1	502.46	502.1393	59.76	5.42	28.66	6.16
C25H2801P2	406.45	406.1615	73.88	6.94	3.94	15.24
C25H2802P2	422.45	422.1564	71.08	6.68	7.57	14.66
C25H2803P2	438.45	438.1514	68.49	6.44	10.95	14.13
C25H2804P2	454.45	454.1463	66.08	6.21	14.08	13.63
C25H2805P2	470.45	470.1412	63.83	6.00	17.00	13.17
C25H2806P2	486.45	486.1361	61.73	5.80	19.73	12.73
C25H2807P2	502.45	502.1310	59.76	5.62	22.29	12.33
C25H2808P2	518.44	518.1259	57.92	5.44	24.69	11.95
C25H2809P2	534.44	534.1208	56.19	5.28	26.94	11.59
C25H2901P1	376.48	376.1956	79.76	7.76	4.25	8.23
C25H2902P1	392.48	392.1905	76.51	7.45	8.15	7.89
C25H2903P1	408.48	408.1854	73.51	7.16	11.75	7.58
C25H2904P1	424.48	424.1803	70.74	6.89	15.08	7.30
C25H2905P1	440.48	440.1752	68.17	6.64	18.16	7.03
C25H2906P1	456.48	456.1702	65.78	6.40	21.03	6.79
C25H2907P1	472.48	472.1651	63.55	6.19	23.70	6.56
C25H2908P1	488.48	488.1600	61.47	5.98	26.20	6.34
C25H2909P1	504.48	504.1549	59.52	5.79	28.54	6.14
C25H3001P2	408.46	408.1772	73.51	7.40	3.92	15.17
C25H3002P2	424.46	424.1721	70.74	7.12	7.54	14.59
C25H3003P2	440.46	440.1670	68.17	6.87	10.90	14.06
C25H3004P2	456.46	456.1619	65.78	6.62	14.02	13.57
C25H3005P2	472.46	472.1568	63.56	6.40	16.93	13.11
C25H3006P2	488.46	488.1518	61.47	6.19	19.65	12.68
C25H3007P2	504.46	504.1467	59.52	5.99	22.20	12.28
C25H3008P2	520.46	520.1416	57.69	5.81	24.59	11.90
C25H3009P2	536.46	536.1365	55.97	5.64	26.84	11.55
C25H3101P1	378.50	378.2112	79.33	8.26	4.23	8.18
C25H3102P1	394.50	394.2062	76.12	7.92	8.11	7.85
C25H3103P1	410.50	410.2011	73.15	7.61	11.69	7.55
C25H3104P1	426.50	426.1960	70.41	7.33	15.01	7.26
C25H3105P1	442.50	442.1909	67.86	7.06	18.08	7.00
C25H3106P1	458.50	458.1858	65.49	6.82	20.94	6.76
C25H3107P1	474.50	474.1807	63.28	6.59	23.60	6.53
C25H3108P1	490.49	490.1756	61.22	6.37	26.10	6.31
C25H3109P1	506.49	506.1706	59.29	6.17	28.43	6.12
C25H3201P2	410.48	410.1928	73.15	7.86	3.90	15.09
C25H3202P2	426.48	426.1877	70.41	7.56	7.50	14.53
C25H3203P2	442.48	442.1827	67.86	7.29	10.85	14.00
C25H3204P2	458.48	458.1776	65.49	7.04	13.96	13.51
C25H3205P2	474.48	474.1725	63.29	6.80	16.86	13.06
C25H3206P2	490.48	490.1674	61.22	6.58	19.57	12.63
C25H3207P2	506.48	506.1623	59.29	6.37	22.11	12.23
C25H3208P2	522.48	522.1572	57.47	6.17	24.50	11.86
C25H3209P2	538.48	538.1521	55.76	5.99	26.74	11.50
C25H3301P1	380.51	380.2269	78.91	8.74	4.20	8.14
C25H3302P1	396.51	396.2218	75.73	8.39	8.07	7.81
C25H3303P1	412.51	412.2167	72.79	8.06	11.64	7.51
C25H3304P1	428.51	428.2116	70.07	7.76	14.93	7.23
C25H3305P1	444.51	444.2065	67.55	7.48	18.00	6.97
C25H3306P1	460.51	460.2015	65.21	7.22	20.85	6.73
C25H3307P1	476.51	476.1964	63.02	6.98	23.50	6.50

C25H3308P1	492.51	492.1913	60.97	6.75	25.99	6.29
C25H3309P1	508.51	508.1862	59.05	6.54	28.32	6.09
C25H3401P2	412.50	412.2085	72.80	8.31	3.88	15.02
C25H3402P2	428.50	428.2034	70.08	8.00	7.47	14.46
C25H3403P2	444.50	444.1983	67.55	7.71	10.80	13.94
C25H3404P2	460.49	460.1932	65.21	7.44	13.90	13.45
C25H3405P2	476.49	476.1881	63.02	7.19	16.79	13.00
C25H3406P2	492.49	492.1830	60.97	6.96	19.49	12.58
C25H3407P2	508.49	508.1780	59.05	6.74	22.03	12.18
C25H3408P2	524.49	524.1729	57.25	6.53	24.40	11.81
C25H3409P2	540.49	540.1678	55.56	6.34	26.64	11.46
C25H3501P1	382.53	382.2425	78.50	9.22	4.18	8.10
C25H3502P1	398.53	398.2375	75.35	8.85	8.03	7.77
C25H3503P1	414.53	414.2324	72.44	8.51	11.58	7.47
C25H3504P1	430.53	430.2273	69.75	8.19	14.86	7.19
C25H3505P1	446.53	446.2222	67.25	7.90	17.92	6.94
C25H3506P1	462.53	462.2171	64.92	7.63	20.75	6.70
C25H3507P1	478.53	478.2120	62.75	7.37	23.40	6.47
C25H3508P1	494.53	494.2069	60.72	7.13	25.88	6.26
C25H3509P1	510.53	510.2019	58.82	6.91	28.21	6.07
C25H3601P2	414.51	414.2241	72.44	8.75	3.86	14.94
C25H3602P2	430.51	430.2190	69.75	8.43	7.43	14.39
C25H3603P2	446.51	446.2140	67.25	8.13	10.75	13.87
C25H3604P2	462.51	462.2089	64.92	7.85	13.84	13.39
C25H3605P2	478.51	478.2038	62.75	7.58	16.72	12.95
C25H3606P2	494.51	494.1987	60.72	7.34	19.41	12.53
C25H3607P2	510.51	510.1936	58.82	7.11	21.94	12.13
C25H3608P2	526.51	526.1885	57.03	6.89	24.31	11.77
C25H3609P2	542.51	542.1834	55.35	6.69	26.54	11.42
C25H3701P1	384.55	384.2582	78.09	9.70	4.16	8.05
C25H3702P1	400.55	400.2531	74.97	9.31	7.99	7.73
C25H3703P1	416.55	416.2480	72.09	8.95	11.52	7.44
C25H3704P1	432.55	432.2429	69.42	8.62	14.80	7.16
C25H3705P1	448.54	448.2378	66.95	8.31	17.83	6.91
C25H3706P1	464.54	464.2328	64.64	8.03	20.66	6.67
C25H3707P1	480.54	480.2277	62.49	7.76	23.31	6.45
C25H3708P1	496.54	496.2226	60.47	7.51	25.78	6.24
C25H3709P1	512.54	512.2175	58.59	7.28	28.09	6.04
C25H3801P2	416.53	416.2398	72.09	9.20	3.84	14.87
C25H3802P2	432.53	432.2347	69.42	8.86	7.40	14.32
C25H3803P2	448.53	448.2296	66.95	8.54	10.70	13.81
C25H3804P2	464.53	464.2245	64.64	8.25	13.78	13.34
C25H3805P2	480.53	480.2194	62.49	7.97	16.65	12.89
C25H3806P2	496.53	496.2143	60.48	7.71	19.33	12.48
C25H3807P2	512.53	512.2093	58.59	7.47	21.85	12.09
C25H3808P2	528.52	528.2042	56.81	7.25	24.22	11.72
C25H3809P2	544.52	544.1991	55.15	7.03	26.44	11.38
C25H3901P1	386.56	386.2738	77.68	10.17	4.14	8.01
C25H3902P1	402.56	402.2688	74.59	9.77	7.95	7.69
C25H3903P1	418.56	418.2637	71.74	9.39	11.47	7.40
C25H3904P1	434.56	434.2586	69.10	9.05	14.73	7.13
C25H3905P1	450.56	450.2535	66.65	8.72	17.76	6.87
C25H3906P1	466.56	466.2484	64.36	8.43	20.58	6.64
C25H3907P1	482.56	482.2433	62.23	8.15	23.21	6.42
C25H3908P1	498.56	498.2382	60.23	7.88	25.67	6.21
C25H3909P1	514.56	514.2332	58.36	7.64	27.98	6.02
C25H4001P2	418.54	418.2554	71.74	9.63	3.82	14.80
C25H4002P2	434.54	434.2503	69.10	9.28	7.36	14.26
C25H4003P2	450.54	450.2453	66.65	8.95	10.65	13.75
C25H4004P2	466.54	466.2402	64.36	8.64	13.72	13.28

C25H4005P2	482.54	482.2351	62.23	8.36	16.58	12.84
C25H4006P2	498.54	498.2300	60.23	8.09	19.26	12.43
C25H4007P2	514.54	514.2249	58.36	7.84	21.77	12.04
C25H4008P2	530.54	530.2198	56.60	7.60	24.13	11.68
C25H4009P2	546.54	546.2147	54.94	7.38	26.35	11.33
C25H4101P1	388.58	388.2895	77.28	10.64	4.12	7.97
C25H4102P1	404.58	404.2844	74.22	10.21	7.91	7.66
C25H4103P1	420.58	420.2793	71.40	9.83	11.41	7.36
C25H4104P1	436.58	436.2742	68.78	9.47	14.66	7.09
C25H4105P1	452.58	452.2691	66.35	9.13	17.68	6.84
C25H4106P1	468.58	468.2641	64.08	8.82	20.49	6.61
C25H4107P1	484.58	484.2590	61.97	8.53	23.11	6.39
C25H4108P1	500.57	500.2539	59.99	8.26	25.57	6.19
C25H4109P1	516.57	516.2488	58.13	8.00	27.87	6.00
C25H4201P2	420.56	420.2711	71.40	10.07	3.80	14.73
C25H4202P2	436.56	436.2660	68.78	9.70	7.33	14.19
C25H4203P2	452.56	452.2609	66.35	9.35	10.61	13.69
C25H4204P2	468.56	468.2558	64.09	9.04	13.66	13.22
C25H4205P2	484.56	484.2507	61.97	8.74	16.51	12.78
C25H4206P2	500.56	500.2456	59.99	8.46	19.18	12.38
C25H4207P2	516.56	516.2406	58.13	8.20	21.68	11.99
C25H4208P2	532.56	532.2355	56.38	7.95	24.03	11.63
C25H4209P2	548.56	548.2304	54.74	7.72	26.25	11.29
C25H4301P1	390.59	390.3051	76.88	11.10	4.10	7.93
C25H4302P1	406.59	406.3001	73.85	10.66	7.87	7.62
C25H4303P1	422.59	422.2950	71.06	10.26	11.36	7.33
C25H4304P1	438.59	438.2899	68.46	9.88	14.59	7.06
C25H4305P1	454.59	454.2848	66.05	9.53	17.60	6.81
C25H4306P1	470.59	470.2797	63.81	9.21	20.40	6.58
C25H4307P1	486.59	486.2746	61.71	8.91	23.02	6.37
C25H4308P1	502.59	502.2695	59.75	8.62	25.47	6.16
C25H4309P1	518.59	518.2644	57.90	8.36	27.77	5.97
C25H4401P2	422.58	422.2867	71.06	10.50	3.79	14.66
C25H4402P2	438.58	438.2816	68.47	10.11	7.30	14.12
C25H4403P2	454.58	454.2766	66.06	9.76	10.56	13.63
C25H4404P2	470.57	470.2715	63.81	9.42	13.60	13.16
C25H4405P2	486.57	486.2664	61.71	9.11	16.44	12.73
C25H4406P2	502.57	502.2613	59.75	8.82	19.10	12.33
C25H4407P2	518.57	518.2562	57.90	8.55	21.60	11.95
C25H4408P2	534.57	534.2511	56.17	8.30	23.94	11.59
C25H4409P2	550.57	550.2460	54.54	8.06	26.15	11.25
C25H4501P1	392.61	392.3208	76.48	11.55	4.08	7.89
C25H4502P1	408.61	408.3157	73.49	11.10	7.83	7.58
C25H4503P1	424.61	424.3106	70.72	10.68	11.30	7.29
C25H4504P1	440.61	440.3055	68.15	10.29	14.52	7.03
C25H4505P1	456.61	456.3004	65.76	9.93	17.52	6.78
C25H4506P1	472.61	472.2954	63.54	9.60	20.31	6.55
C25H4507P1	488.61	488.2903	61.46	9.28	22.92	6.34
C25H4508P1	504.61	504.2852	59.51	8.99	25.37	6.14
C25H4509P1	520.61	520.2801	57.68	8.71	27.66	5.95
C25H4601P2	424.59	424.3024	70.72	10.92	3.77	14.59
C25H4602P2	440.59	440.2973	68.15	10.52	7.26	14.06
C25H4603P2	456.59	456.2922	65.77	10.15	10.51	13.57
C25H4604P2	472.59	472.2871	63.54	9.81	13.54	13.11
C25H4605P2	488.59	488.2820	61.46	9.49	16.37	12.68
C25H4606P2	504.59	504.2769	59.51	9.19	19.02	12.28
C25H4607P2	520.59	520.2719	57.68	8.91	21.51	11.90
C25H4608P2	536.59	536.2668	55.96	8.64	23.85	11.54
C25H4609P2	552.59	552.2617	54.34	8.39	26.06	11.21
C25H4701P1	394.63	394.3364	76.09	12.00	4.05	7.85

148

C25H4702P1	410.63	410.3313	73.13	11.54	7.79	7.54
C25H4703P1	426.63	426.3263	70.38	11.10	11.25	7.26
C25H4704P1	442.62	442.3212	67.84	10.70	14.46	7.00
C25H4705P1	458.62	458.3161	65.47	10.33	17.44	6.75
C25H4706P1	474.62	474.3110	63.27	9.98	20.23	6.53
C25H4707P1	490.62	490.3059	61.20	9.66	22.83	6.31
C25H4708P1	506.62	506.3008	59.27	9.35	25.26	6.11
C25H4709P1	522.62	522.2957	57.46	9.06	27.55	5.93
C25H4801P2	426.61	426.3180	70.39	11.34	3.75	14.52
C25H4802P2	442.61	442.3129	67.84	10.93	7.23	14.00
C25H4803P2	458.61	458.3079	65.48	10.55	10.47	13.51
C25H4804P2	474.61	474.3028	63.27	10.19	13.48	13.05
C25H4805P2	490.61	490.2977	61.21	9.86	16.31	12.63
C25H4806P2	506.61	506.2926	59.27	9.55	18.95	12.23
C25H4807P2	522.60	522.2875	57.46	9.26	21.43	11.85
C25H4808P2	538.60	538.2824	55.75	8.98	23.76	11.50
C25H4809P2	554.60	554.2773	54.14	8.72	25.96	11.17
C25H4901P1	396.64	396.3521	75.71	12.45	4.03	7.81
C25H4902P1	412.64	412.3470	72.77	11.97	7.75	7.51
C25H4903P1	428.64	428.3419	70.05	11.52	11.20	7.23
C25H4904P1	444.64	444.3368	67.53	11.11	14.39	6.97
C25H4905P1	460.64	460.3317	65.19	10.72	17.37	6.72
C25H4906P1	476.64	476.3267	63.00	10.36	20.14	6.50
C25H4907P1	492.64	492.3216	60.95	10.03	22.73	6.29
C25H4908P1	508.64	508.3165	59.04	9.71	25.16	6.09
C25H4909P1	524.64	524.3114	57.24	9.41	27.45	5.90
C25H5001P2	428.62	428.3337	70.06	11.76	3.73	14.45
C25H5002P2	444.62	444.3286	67.54	11.34	7.20	13.93
C25H5003P2	460.62	460.3235	65.19	10.94	10.42	13.45
C25H5004P2	476.62	476.3184	63.00	10.57	13.43	13.00
C25H5005P2	492.62	492.3133	60.96	10.23	16.24	12.58
C25H5006P2	508.62	508.3082	59.04	9.91	18.87	12.18
C25H5007P2	524.62	524.3032	57.24	9.61	21.35	11.81
C25H5008P2	540.62	540.2981	55.54	9.32	23.68	11.46
C25H5009P2	556.62	556.2930	53.95	9.05	25.87	11.13
C25H5101P1	398.66	398.3677	75.32	12.89	4.01	7.77
C25H5102P1	414.66	414.3626	72.42	12.40	7.72	7.47
C25H5103P1	430.66	430.3576	69.73	11.94	11.15	7.19
C25H5104P1	446.66	446.3525	67.23	11.51	14.33	6.93
C25H5105P1	462.66	462.3474	64.90	11.11	17.29	6.69
C25H5106P1	478.66	478.3423	62.73	10.74	20.06	6.47
C25H5107P1	494.65	494.3372	60.70	10.39	22.64	6.26
C25H5108P1	510.65	510.3321	58.80	10.07	25.06	6.07
C25H5109P1	526.65	526.3270	57.02	9.76	27.34	5.88
C25H5201P2	430.64	430.3493	69.73	12.17	3.72	14.39
C25H5202P2	446.64	446.3442	67.23	11.74	7.16	13.87
C25H5203P2	462.64	462.3391	64.91	11.33	10.37	13.39
C25H5204P2	478.64	478.3341	62.74	10.95	13.37	12.94
C25H5205P2	494.64	494.3290	60.71	10.60	16.17	12.52
C25H5206P2	510.64	510.3239	58.80	10.26	18.80	12.13
C25H5207P2	526.64	526.3188	57.02	9.95	21.27	11.76
C25H5208P2	542.64	542.3137	55.34	9.66	23.59	11.42
C25H5209P2	558.64	558.3086	53.75	9.38	25.78	11.09
C25H5301P1	400.67	400.3834	74.94	13.33	3.99	7.73
C25H5302P1	416.67	416.3783	72.07	12.82	7.68	7.43
C25H5303P1	432.67	432.3732	69.40	12.35	11.09	7.16
C25H5304P1	448.67	448.3681	66.93	11.91	14.26	6.90
C25H5305P1	464.67	464.3630	64.62	11.50	17.22	6.67
C25H5306P1	480.67	480.3580	62.47	11.11	19.97	6.44
C25H5307P1	496.67	496.3529	60.46	10.76	22.55	6.24
C25H5308P1	512.67	512.3478	58.57	10.42	24.97	6.04
C25H5309P1	528.67	528.3427	56.80	10.11	27.24	5.86
C25H5401P2	432.66	432.3650	69.40	12.58	3.70	14.32
C25H5402P2	448.66	448.3599	66.93	12.13	7.13	13.81
C25H5403P2	464.65	464.3548	64.62	11.71	10.33	13.33
C25H5404P2	480.65	480.3497	62.47	11.32	13.31	12.89
C25H5405P2	496.65	496.3446	60.46	10.96	16.11	12.47
C25H5406P2	512.65	512.3395	58.57	10.62	18.73	12.08
C25H5407P2	528.65	528.3345	56.80	10.30	21.19	11.72
C25H5408P2	544.65	544.3294	55.13	9.99	23.50	11.37
C25H5409P2	560.65	560.3243	53.56	9.71	25.68	11.05
C26H1301P1	372.37	372.0704	83.87	3.52	4.30	8.32
C26H1302P1	388.37	388.0653	80.41	3.37	8.24	7.98
C26H1303P1	404.37	404.0602	77.23	3.24	11.87	7.66
C26H1304P1	420.36	420.0551	74.29	3.12	15.22	7.37
C26H1305P1	436.36	436.0501	71.57	3.00	18.33	7.10
C26H1306P1	452.36	452.0450	69.04	2.90	21.22	6.85
C26H1307P1	468.36	468.0399	66.68	2.80	23.91	6.61
C26H1308P1	484.36	484.0348	64.47	2.71	26.43	6.39
C26H1309P1	500.36	500.0297	62.41	2.62	28.78	6.19
C26H1401P2	404.35	404.0520	77.23	3.49	3.96	15.32
C26H1402P2	420.35	420.0469	74.29	3.36	7.61	14.74
C26H1403P2	436.35	436.0418	71.57	3.23	11.00	14.20
C26H1404P2	452.35	452.0367	69.04	3.12	14.15	13.69
C26H1405P2	468.35	468.0316	66.68	3.01	17.08	13.23
C26H1406P2	484.35	484.0266	64.48	2.91	19.82	12.79
C26H1407P2	500.34	500.0215	62.41	2.82	22.38	12.38
C26H1408P2	516.34	516.0164	60.48	2.73	24.79	12.00
C26H1409P2	532.34	532.0113	58.66	2.65	27.05	11.64
C26H1501P1	374.38	374.0860	83.41	4.04	4.27	8.27
C26H1502P1	390.38	390.0810	80.00	3.87	8.20	7.93
C26H1503P1	406.38	406.0759	76.85	3.72	11.81	7.62
C26H1504P1	422.38	422.0708	73.94	3.58	15.15	7.33
C26H1505P1	438.38	438.0657	71.24	3.45	18.25	7.07
C26H1506P1	454.38	454.0606	68.73	3.33	21.13	6.82
C26H1507P1	470.38	470.0555	66.39	3.21	23.81	6.58
C26H1508P1	486.38	486.0504	64.21	3.11	26.32	6.37
C26H1509P1	502.38	502.0454	62.16	3.01	28.66	6.17
C26H1601P2	406.36	406.0676	76.85	3.97	3.94	15.24
C26H1602P2	422.36	422.0625	73.94	3.82	7.58	14.67
C26H1603P2	438.36	438.0575	71.24	3.68	10.95	14.13
C26H1604P2	454.36	454.0524	68.73	3.55	14.09	13.63
C26H1605P2	470.36	470.0473	66.39	3.43	17.01	13.17
C26H1606P2	486.36	486.0422	64.21	3.32	19.74	12.74
C26H1607P2	502.36	502.0371	62.16	3.21	22.29	12.33
C26H1608P2	518.36	518.0320	60.25	3.11	24.69	11.95
C26H1609P2	534.36	534.0269	58.44	3.02	26.95	11.59
C26H1701P1	376.40	376.1017	82.97	4.55	4.25	8.23
C26H1702P1	392.40	392.0966	79.58	4.37	8.15	7.89
C26H1703P1	408.40	408.0915	76.47	4.20	11.75	7.58
C26H1704P1	424.40	424.0864	73.58	4.04	15.08	7.30
C26H1705P1	440.40	440.0814	70.91	3.89	18.16	7.03
C26H1706P1	456.40	456.0763	68.43	3.75	21.03	6.79
C26H1707P1	472.39	472.0712	66.11	3.63	23.71	6.56
C26H1708P1	488.39	488.0661	63.94	3.51	26.21	6.34
C26H1709P1	504.39	504.0610	61.91	3.40	28.55	6.14
C26H1801P2	408.38	408.0833	76.47	4.44	3.92	15.17
C26H1802P2	424.38	424.0782	73.59	4.28	7.54	14.60
C26H1803P2	440.38	440.0731	70.91	4.12	10.90	14.07
C26H1804P2	456.38	456.0680	68.43	3.98	14.02	13.57

149

C26H1805P2	472.38	472.0629	66.11	3.84	16.93	13.11
C26H1806P2	488.38	488.0579	63.94	3.72	19.66	12.68
C26H1807P2	504.38	504.0528	61.92	3.60	22.20	12.28
C26H1808P2	520.38	520.0477	60.01	3.49	24.60	11.90
C26H1809P2	536.38	536.0426	58.22	3.38	26.85	11.55
C26H1901P1	378.41	378.1173	82.53	5.06	4.23	8.19
C26H1902P1	394.41	394.1123	79.18	4.86	8.11	7.85
C26H1903P1	410.41	410.1072	76.09	4.67	11.70	7.55
C26H1904P1	426.41	426.1021	73.24	4.49	15.01	7.26
C26H1905P1	442.41	442.0970	70.59	4.33	18.08	7.00
C26H1906P1	458.41	458.0919	68.12	4.18	20.94	6.76
C26H1907P1	474.41	474.0868	65.83	4.04	23.61	6.53
C26H1908P1	490.41	490.0817	63.68	3.91	26.10	6.32
C26H1909P1	506.41	506.0767	61.67	3.78	28.43	6.12
C26H2001P2	410.40	410.0989	76.09	4.91	3.90	15.09
C26H2002P2	426.40	426.0938	73.24	4.73	7.50	14.53
C26H2003P2	442.40	442.0888	70.59	4.56	10.85	14.00
C26H2004P2	458.39	458.0837	68.13	4.40	13.96	13.51
C26H2005P2	474.39	474.0786	65.83	4.25	16.86	13.06
C26H2006P2	490.39	490.0735	63.68	4.11	19.58	12.63
C26H2007P2	506.39	506.0684	61.67	3.98	22.12	12.23
C26H2008P2	522.39	522.0633	59.78	3.86	24.50	11.86
C26H2009P2	538.39	538.0582	58.00	3.74	26.75	11.51
C26H2101P1	380.43	380.1330	82.09	5.56	4.21	8.14
C26H2102P1	396.43	396.1279	78.78	5.34	8.07	7.81
C26H2103P1	412.43	412.1228	75.72	5.13	11.64	7.51
C26H2104P1	428.43	428.1177	72.89	4.94	14.94	7.23
C26H2105P1	444.43	444.1127	70.27	4.76	18.00	6.97
C26H2106P1	460.43	460.1076	67.83	4.60	20.85	6.73
C26H2107P1	476.43	476.1025	65.55	4.44	23.51	6.50
C26H2108P1	492.43	492.0974	63.42	4.30	25.99	6.29
C26H2109P1	508.43	508.0923	61.42	4.16	28.32	6.09
C26H2201P2	412.41	412.1146	75.72	5.38	3.88	15.02
C26H2202P2	428.41	428.1095	72.89	5.18	7.47	14.46
C26H2203P2	444.41	444.1044	70.27	4.99	10.80	13.94
C26H2204P2	460.41	460.0993	67.83	4.82	13.90	13.45
C26H2205P2	476.41	476.0942	65.55	4.65	16.79	13.00
C26H2206P2	492.41	492.0892	63.42	4.50	19.50	12.58
C26H2207P2	508.41	508.0841	61.42	4.36	22.03	12.18
C26H2208P2	524.41	524.0790	59.55	4.23	24.41	11.81
C26H2209P2	540.41	540.0739	57.79	4.10	26.65	11.46
C26H2301P1	382.45	382.1486	81.66	6.06	4.18	8.10
C26H2302P1	398.45	398.1436	78.38	5.82	8.03	7.77
C26H2303P1	414.45	414.1385	75.35	5.59	11.58	7.47
C26H2304P1	430.44	430.1334	72.55	5.39	14.87	7.20
C26H2305P1	446.44	446.1283	69.95	5.19	17.92	6.94
C26H2306P1	462.44	462.1232	67.53	5.01	20.76	6.70
C26H2307P1	478.44	478.1181	65.27	4.85	23.41	6.47
C26H2308P1	494.44	494.1130	63.16	4.69	25.89	6.26
C26H2309P1	510.44	510.1080	61.18	4.54	28.21	6.07
C26H2401P2	414.43	414.1302	75.35	5.84	3.86	14.95
C26H2402P2	430.43	430.1251	72.55	5.62	7.43	14.39
C26H2403P2	446.43	446.1201	69.95	5.42	10.75	13.88
C26H2404P2	462.43	462.1150	67.53	5.23	13.84	13.40
C26H2405P2	478.43	478.1099	65.27	5.06	16.72	12.95
C26H2406P2	494.43	494.1048	63.16	4.89	19.42	12.53
C26H2407P2	510.42	510.0997	61.18	4.74	21.94	12.14
C26H2408P2	526.42	526.0946	59.32	4.60	24.31	11.77
C26H2409P2	542.42	542.0895	57.57	4.46	26.55	11.42
C26H2501P1	384.46	384.1643	81.23	6.55	4.16	8.06
C26H2502P1	400.46	400.1592	77.98	6.29	7.99	7.73
C26H2503P1	416.46	416.1541	74.99	6.05	11.53	7.44
C26H2504P1	432.46	432.1490	72.21	5.83	14.80	7.16
C26H2505P1	448.46	448.1439	69.64	5.62	17.84	6.91
C26H2506P1	464.46	464.1389	67.24	5.43	20.67	6.67
C26H2507P1	480.46	480.1338	65.00	5.24	23.31	6.45
C26H2508P1	496.46	496.1287	62.90	5.08	25.78	6.24
C26H2509P1	512.46	512.1236	60.94	4.92	28.10	6.04
C26H2601P2	416.44	416.1459	74.99	6.29	3.84	14.88
C26H2602P2	432.44	432.1408	72.22	6.06	7.40	14.33
C26H2603P2	448.44	448.1357	69.64	5.84	10.70	13.81
C26H2604P2	464.44	464.1306	67.24	5.64	13.78	13.34
C26H2605P2	480.44	480.1255	65.00	5.45	16.65	12.89
C26H2606P2	496.44	496.1205	62.91	5.28	19.34	12.48
C26H2607P2	512.44	512.1154	60.94	5.11	21.86	12.09
C26H2608P2	528.44	528.1103	59.10	4.96	24.22	11.72
C26H2609P2	544.44	544.1052	57.36	4.81	26.45	11.38
C26H2701P1	386.48	386.1799	80.80	7.04	4.14	8.01
C26H2702P1	402.48	402.1749	77.59	6.76	7.95	7.70
C26H2703P1	418.48	418.1698	74.63	6.50	11.47	7.40
C26H2704P1	434.48	434.1647	71.88	6.26	14.73	7.13
C26H2705P1	450.48	450.1596	69.32	6.04	17.76	6.88
C26H2706P1	466.48	466.1545	66.95	5.83	20.58	6.64
C26H2707P1	482.47	482.1494	64.73	5.64	23.21	6.42
C26H2708P1	498.47	498.1443	62.65	5.46	25.68	6.21
C26H2709P1	514.47	514.1393	60.70	5.29	27.99	6.02
C26H2801P2	418.46	418.1615	74.63	6.74	3.82	14.80
C26H2802P2	434.46	434.1564	71.88	6.50	7.37	14.26
C26H2803P2	450.46	450.1514	69.33	6.27	10.66	13.75
C26H2804P2	466.46	466.1463	66.95	6.05	13.72	13.28
C26H2805P2	482.46	482.1412	64.73	5.85	16.58	12.84
C26H2806P2	498.46	498.1361	62.65	5.66	19.26	12.43
C26H2807P2	514.46	514.1310	60.70	5.49	21.77	12.04
C26H2808P2	530.46	530.1259	58.87	5.32	24.13	11.68
C26H2809P2	546.46	546.1208	57.15	5.16	26.35	11.34
C26H2901P1	388.49	388.1956	80.38	7.52	4.12	7.97
C26H2902P1	404.49	404.1905	77.21	7.23	7.91	7.66
C26H2903P1	420.49	420.1854	74.27	6.95	11.41	7.37
C26H2904P1	436.49	436.1803	71.55	6.70	14.66	7.10
C26H2905P1	452.49	452.1752	69.02	6.46	17.68	6.85
C26H2906P1	468.49	468.1702	66.66	6.24	20.49	6.61
C26H2907P1	484.49	484.1651	64.46	6.03	23.12	6.39
C26H2908P1	500.49	500.1600	62.40	5.84	25.57	6.19
C26H2909P1	516.49	516.1549	60.46	5.66	27.88	6.00
C26H3001P2	420.48	420.1772	74.27	7.19	3.81	14.73
C26H3002P2	436.48	436.1721	71.55	6.93	7.33	14.19
C26H3003P2	452.47	452.1670	69.02	6.68	10.61	13.69
C26H3004P2	468.47	468.1619	66.66	6.45	13.66	13.22
C26H3005P2	484.47	484.1568	64.46	6.24	16.51	12.79
C26H3006P2	500.47	500.1518	62.40	6.04	19.18	12.38
C26H3007P2	516.47	516.1467	60.47	5.85	21.68	11.99
C26H3008P2	532.47	532.1416	58.65	5.68	24.04	11.63
C26H3009P2	548.47	548.1365	56.94	5.51	26.25	11.29
C26H3101P1	390.51	390.2112	79.97	8.00	4.10	7.93
C26H3102P1	406.51	406.2062	76.82	7.69	7.87	7.62
C26H3103P1	422.51	422.2011	73.91	7.40	11.36	7.33
C26H3104P1	438.51	438.1960	71.22	7.13	14.59	7.06
C26H3105P1	454.51	454.1909	68.71	6.87	17.60	6.81
C26H3106P1	470.51	470.1858	66.37	6.64	20.40	6.58
C26H3107P1	486.51	486.1807	64.19	6.42	23.02	6.37

C26H3108P1	502.51	502.1756	62.15	6.22	25.47	6.16
C26H3109P1	518.51	518.1706	60.23	6.03	27.77	5.97
C26H3201P2	422.49	422.1928	73.92	7.63	3.79	14.66
C26H3202P2	438.49	438.1877	71.22	7.36	7.30	14.13
C26H3203P2	454.49	454.1827	68.71	7.10	10.56	13.63
C26H3204P2	470.49	470.1776	66.38	6.86	13.60	13.17
C26H3205P2	486.49	486.1725	64.19	6.63	16.44	12.73
C26H3206P2	502.49	502.1674	62.15	6.42	19.10	12.33
C26H3207P2	518.49	518.1623	60.23	6.22	21.60	11.95
C26H3208P2	534.49	534.1572	58.43	6.03	23.95	11.59
C26H3209P2	550.49	550.1521	56.73	5.86	26.16	11.25
C26H3301P1	392.53	392.2269	79.56	8.47	4.08	7.89
C26H3302P1	408.53	408.2218	76.44	8.14	7.83	7.58
C26H3303P1	424.52	424.2167	73.56	7.84	11.31	7.30
C26H3304P1	440.52	440.2116	70.89	7.55	14.53	7.03
C26H3305P1	456.52	456.2065	68.41	7.29	17.52	6.78
C26H3306P1	472.52	472.2015	66.09	7.04	20.32	6.55
C26H3307P1	488.52	488.1964	63.93	6.81	22.93	6.34
C26H3308P1	504.52	504.1913	61.90	6.59	25.37	6.14
C26H3309P1	520.52	520.1862	60.00	6.39	27.66	5.95
C26H3401P2	424.51	424.2085	73.57	8.07	3.77	14.59
C26H3402P2	440.51	440.2034	70.89	7.78	7.26	14.06
C26H3403P2	456.51	456.1983	68.41	7.51	10.51	13.57
C26H3404P2	472.51	472.1932	66.09	7.25	13.54	13.11
C26H3405P2	488.51	488.1881	63.93	7.02	16.38	12.68
C26H3406P2	504.50	504.1830	61.90	6.79	19.03	12.28
C26H3407P2	520.50	520.1780	60.00	6.58	21.52	11.90
C26H3408P2	536.50	536.1729	58.21	6.39	23.86	11.55
C26H3409P2	552.50	552.1678	56.52	6.20	26.06	11.21
C26H3501P1	394.54	394.2425	79.15	8.94	4.06	7.85
C26H3502P1	410.54	410.2375	76.07	8.59	7.79	7.54
C26H3503P1	426.54	426.2324	73.21	8.27	11.25	7.26
C26H3504P1	442.54	442.2273	70.57	7.97	14.46	7.00
C26H3505P1	458.54	458.2222	68.11	7.69	17.45	6.75
C26H3506P1	474.54	474.2171	65.81	7.43	20.23	6.53
C26H3507P1	490.54	490.2120	63.66	7.19	22.83	6.31
C26H3508P1	506.54	506.2069	61.65	6.96	25.27	6.11
C26H3509P1	522.54	522.2019	59.76	6.75	27.56	5.93
C26H3601P2	426.52	426.2241	73.22	8.51	3.75	14.52
C26H3602P2	442.52	442.2190	70.57	8.20	7.23	14.00
C26H3603P2	458.52	458.2140	68.11	7.91	10.47	13.51
C26H3604P2	474.52	474.2089	65.81	7.65	13.49	13.05
C26H3605P2	490.52	490.2038	63.66	7.40	16.31	12.63
C26H3606P2	506.52	506.1987	61.65	7.16	18.95	12.23
C26H3607P2	522.52	522.1936	59.77	6.94	21.43	11.86
C26H3608P2	538.52	538.1885	57.99	6.74	23.77	11.50
C26H3609P2	554.52	554.1834	56.32	6.54	25.97	11.17
C26H3701P1	396.56	396.2582	78.75	9.40	4.03	7.81
C26H3702P1	412.56	412.2531	75.70	9.04	7.76	7.51
C26H3703P1	428.56	428.2480	72.87	8.70	11.20	7.23
C26H3704P1	444.56	444.2429	70.25	8.39	14.40	6.97
C26H3705P1	460.56	460.2378	67.81	8.10	17.37	6.73
C26H3706P1	476.55	476.2328	65.53	7.83	20.14	6.50
C26H3707P1	492.55	492.2277	63.40	7.57	22.74	6.29
C26H3708P1	508.55	508.2226	61.41	7.33	25.17	6.09
C26H3709P1	524.55	524.2175	59.53	7.11	27.45	5.90
C26H3801P2	428.54	428.2398	72.87	8.94	3.73	14.46
C26H3802P2	444.54	444.2347	70.25	8.62	7.20	13.94
C26H3803P2	460.54	460.2296	67.81	8.32	10.42	13.45
C26H3804P2	476.54	476.2245	65.53	8.04	13.43	13.00
C26H3805P2	492.54	492.2194	63.40	7.78	16.24	12.58
C26H3806P2	508.54	508.2143	61.41	7.53	18.88	12.18
C26H3807P2	524.54	524.2093	59.54	7.30	21.35	11.81
C26H3808P2	540.54	540.2042	57.77	7.09	23.68	11.46
C26H3809P2	556.53	556.1991	56.11	6.88	25.87	11.13
C26H3901P1	398.57	398.2738	78.35	9.86	4.01	7.77
C26H3902P1	414.57	414.2688	75.33	9.48	7.72	7.47
C26H3903P1	430.57	430.2637	72.53	9.13	11.15	7.19
C26H3904P1	446.57	446.2586	69.93	8.80	14.33	6.94
C26H3905P1	462.57	462.2535	67.51	8.50	17.29	6.70
C26H3906P1	478.57	478.2484	65.25	8.21	20.06	6.47
C26H3907P1	494.57	494.2433	63.14	7.95	22.65	6.26
C26H3908P1	510.57	510.2382	61.16	7.70	25.07	6.07
C26H3909P1	526.57	526.2332	59.31	7.47	27.35	5.88
C26H4001P2	430.56	430.2554	72.53	9.36	3.72	14.39
C26H4002P2	446.56	446.2503	69.93	9.03	7.17	13.87
C26H4003P2	462.55	462.2453	67.51	8.72	10.38	13.39
C26H4004P2	478.55	478.2402	65.26	8.43	13.37	12.94
C26H4005P2	494.55	494.2351	63.15	8.15	16.18	12.53
C26H4006P2	510.55	510.2300	61.17	7.90	18.80	12.13
C26H4007P2	526.55	526.2249	59.31	7.66	21.27	11.76
C26H4008P2	542.55	542.2198	57.56	7.43	23.59	11.42
C26H4009P2	558.55	558.2147	55.91	7.22	25.78	11.09
C26H4101P1	400.59	400.2895	77.96	10.32	3.99	7.73
C26H4102P1	416.59	416.2844	74.96	9.92	7.68	7.44
C26H4103P1	432.59	432.2793	72.19	9.55	11.10	7.16
C26H4104P1	448.59	448.2742	69.62	9.21	14.27	6.90
C26H4105P1	464.59	464.2691	67.22	8.90	17.22	6.67
C26H4106P1	480.59	480.2641	64.98	8.60	19.97	6.44
C26H4107P1	496.59	496.2590	62.89	8.32	22.55	6.24
C26H4108P1	512.59	512.2539	60.92	8.06	24.97	6.04
C26H4109P1	528.59	528.2488	59.08	7.82	27.24	5.86
C26H4201P2	432.57	432.2711	72.19	9.79	3.70	14.32
C26H4202P2	448.57	448.2660	69.62	9.44	7.13	13.81
C26H4203P2	464.57	464.2609	67.22	9.11	10.33	13.33
C26H4204P2	480.57	480.2558	64.98	8.81	13.32	12.89
C26H4205P2	496.57	496.2507	62.89	8.53	16.11	12.48
C26H4206P2	512.57	512.2456	60.93	8.26	18.73	12.09
C26H4207P2	528.57	528.2406	59.08	8.01	21.19	11.72
C26H4208P2	544.57	544.2355	57.35	7.77	23.50	11.38
C26H4209P2	560.57	560.2304	55.71	7.55	25.69	11.05
C26H4301P1	402.61	402.3051	77.57	10.77	3.97	7.69
C26H4302P1	418.61	418.3001	74.60	10.35	7.64	7.40
C26H4303P1	434.60	434.2950	71.86	9.97	11.04	7.13
C26H4304P1	450.60	450.2899	69.30	9.62	14.20	6.87
C26H4305P1	466.60	466.2848	66.93	9.29	17.14	6.64
C26H4306P1	482.60	482.2797	64.71	8.98	19.89	6.42
C26H4307P1	498.60	498.2746	62.63	8.69	22.46	6.21
C26H4308P1	514.60	514.2695	60.69	8.42	24.87	6.02
C26H4309P1	530.60	530.2644	58.86	8.17	27.14	5.84
C26H4401P2	434.59	434.2867	71.86	10.21	3.68	14.25
C26H4402P2	450.59	450.2816	69.31	9.84	7.10	13.75
C26H4403P2	466.59	466.2766	66.93	9.51	10.29	13.28
C26H4404P2	482.59	482.2715	64.71	9.19	13.26	12.84
C26H4405P2	498.59	498.2664	62.64	8.90	16.04	12.42
C26H4406P2	514.58	514.2613	60.69	8.62	18.66	12.04
C26H4407P2	530.58	530.2562	58.86	8.36	21.11	11.68
C26H4408P2	546.58	546.2511	57.13	8.11	23.42	11.33
C26H4409P2	562.58	562.2460	55.51	7.88	25.60	11.01
C26H4501P1	404.62	404.3208	77.18	11.21	3.95	7.66

C26H4502P1	420.62	420.3157	74.24	10.78	7.61	7.36
C26H4503P1	436.62	436.3106	71.52	10.39	10.99	7.09
C26H4504P1	452.62	452.3055	69.00	10.02	14.14	6.84
C26H4505P1	468.62	468.3004	66.64	9.68	17.07	6.61
C26H4506P1	484.62	484.2954	64.44	9.36	19.81	6.39
C26H4507P1	500.62	500.2903	62.38	9.06	22.37	6.19
C26H4508P1	516.62	516.2852	60.45	8.78	24.78	6.00
C26H4509P1	532.62	532.2801	58.63	8.52	27.04	5.82
C26H4601P2	436.60	436.3024	71.53	10.62	3.66	14.19
C26H4602P2	452.60	452.2973	69.00	10.24	7.07	13.69
C26H4603P2	468.60	468.2922	66.64	9.89	10.24	13.22
C26H4604P2	484.60	484.2871	64.44	9.57	13.21	12.78
C26H4605P2	500.60	500.2820	62.38	9.26	15.98	12.37
C26H4606P2	516.60	516.2769	60.45	8.98	18.58	11.99
C26H4607P2	532.60	532.2719	58.63	8.71	21.03	11.63
C26H4608P2	548.60	548.2668	56.92	8.45	23.33	11.29
C26H4609P2	564.60	564.2617	55.31	8.21	25.50	10.97
C26H4701P1	406.64	406.3364	76.80	11.65	3.93	7.62
C26H4702P1	422.64	422.3313	73.89	11.21	7.57	7.33
C26H4703P1	438.64	438.3263	71.20	10.80	10.94	7.06
C26H4704P1	454.64	454.3212	68.69	10.42	14.08	6.81
C26H4705P1	470.64	470.3161	66.35	10.07	17.00	6.58
C26H4706P1	486.63	486.3110	64.17	9.74	19.73	6.36
C26H4707P1	502.63	502.3059	62.13	9.43	22.28	6.16
C26H4708P1	518.63	518.3008	60.21	9.13	24.68	5.97
C26H4709P1	534.63	534.2957	58.41	8.86	26.93	5.79
C26H4801P2	438.62	438.3180	71.20	11.03	3.65	14.12
C26H4802P2	454.62	454.3129	68.69	10.64	7.04	13.63
C26H4803P2	470.62	470.3079	66.36	10.28	10.20	13.16
C26H4804P2	486.62	486.3028	64.18	9.94	13.15	12.73
C26H4805P2	502.62	502.2977	62.13	9.63	15.92	12.33
C26H4806P2	518.62	518.2926	60.22	9.33	18.51	11.94
C26H4807P2	534.62	534.2875	58.41	9.05	20.95	11.59
C26H4808P2	550.62	550.2824	56.72	8.79	23.25	11.25
C26H4809P2	566.61	566.2773	55.12	8.54	25.41	10.93
C26H4901P1	408.65	408.3521	76.42	12.09	3.92	7.58
C26H4902P1	424.65	424.3470	73.54	11.63	7.54	7.29
C26H4903P1	440.65	440.3419	70.87	11.21	10.89	7.03
C26H4904P1	456.65	456.3368	68.39	10.82	14.01	6.78
C26H4905P1	472.65	472.3317	66.07	10.45	16.93	6.55
C26H4906P1	488.65	488.3267	63.91	10.11	19.65	6.34
C26H4907P1	504.65	504.3216	61.88	9.79	22.19	6.14
C26H4908P1	520.65	520.3165	59.98	9.49	24.58	5.95
C26H4909P1	536.65	536.3114	58.19	9.20	26.83	5.77
C26H5001P2	440.64	440.3337	70.87	11.44	3.63	14.06
C26H5002P2	456.63	456.3286	68.39	11.04	7.01	13.57
C26H5003P2	472.63	472.3235	66.07	10.66	10.16	13.11
C26H5004P2	488.63	488.3184	63.91	10.31	13.10	12.68
C26H5005P2	504.63	504.3133	61.88	9.99	15.85	12.28
C26H5006P2	520.63	520.3082	59.98	9.68	18.44	11.90
C26H5007P2	536.63	536.3032	58.19	9.39	20.87	11.54
C26H5008P2	552.63	552.2981	56.51	9.12	23.16	11.21
C26H5009P2	568.63	568.2930	54.92	8.86	25.32	10.89
C26H5101P1	410.67	410.3677	76.04	12.52	3.90	7.54
C26H5102P1	426.67	426.3626	73.19	12.05	7.50	7.26
C26H5103P1	442.67	442.3576	70.55	11.61	10.84	7.00
C26H5104P1	458.67	458.3525	68.09	11.21	13.95	6.75
C26H5105P1	474.67	474.3474	65.79	10.83	16.85	6.53
C26H5106P1	490.67	490.3423	63.65	10.48	19.56	6.31
C26H5107P1	506.67	506.3372	61.64	10.15	22.10	6.11
C26H5108P1	522.67	522.3321	59.75	9.84	24.49	5.93
C26H5109P1	538.66	538.3270	57.97	9.54	26.73	5.75
C26H5201P2	442.65	442.3493	70.55	11.84	3.61	13.99
C26H5202P2	458.65	458.3442	68.09	11.43	6.98	13.51
C26H5203P2	474.65	474.3391	65.79	11.04	10.11	13.05
C26H5204P2	490.65	490.3341	63.65	10.68	13.04	12.63
C26H5205P2	506.65	506.3290	61.64	10.35	15.79	12.23
C26H5206P2	522.65	522.3239	59.75	10.03	18.37	11.85
C26H5207P2	538.65	538.3188	57.98	9.73	20.79	11.50
C26H5208P2	554.65	554.3137	56.30	9.45	23.08	11.17
C26H5209P2	570.65	570.3086	54.73	9.19	25.23	10.86
C26H5301P1	412.69	412.3834	75.67	12.95	3.88	7.51
C26H5302P1	428.68	428.3783	72.85	12.46	7.46	7.23
C26H5303P1	444.68	444.3732	70.23	12.01	10.79	6.97
C26H5304P1	460.68	460.3681	67.79	11.60	13.89	6.72
C26H5305P1	476.68	476.3630	65.51	11.21	16.78	6.50
C26H5306P1	492.68	492.3580	63.39	10.84	19.48	6.29
C26H5307P1	508.68	508.3529	61.39	10.50	22.02	6.09
C26H5308P1	524.68	524.3478	59.52	10.18	24.39	5.90
C26H5309P1	540.68	540.3427	57.76	9.88	26.63	5.73
C26H5401P2	444.67	444.3650	70.23	12.24	3.60	13.93
C26H5402P2	460.67	460.3599	67.79	11.82	6.95	13.45
C26H5403P2	476.67	476.3548	65.52	11.42	10.07	13.00
C26H5404P2	492.67	492.3497	63.39	11.05	12.99	12.57
C26H5405P2	508.66	508.3446	61.39	10.70	15.73	12.18
C26H5406P2	524.66	524.3395	59.52	10.37	18.30	11.81
C26H5407P2	540.66	540.3345	57.76	10.07	20.71	11.46
C26H5408P2	556.66	556.3294	56.10	9.78	22.99	11.13
C26H5409P2	572.66	572.3243	54.53	9.50	25.14	10.82
C26H5501P1	414.70	414.3990	75.30	13.37	3.86	7.47
C26H5502P1	430.70	430.3939	72.51	12.87	7.43	7.19
C26H5503P1	446.70	446.3889	69.91	12.41	10.75	6.93
C26H5504P1	462.70	462.3838	67.49	11.98	13.83	6.69
C26H5505P1	478.70	478.3787	65.24	11.58	16.71	6.47
C26H5506P1	494.70	494.3736	63.13	11.21	19.41	6.26
C26H5507P1	510.70	510.3685	61.15	10.86	21.93	6.06
C26H5508P1	526.70	526.3634	59.29	10.53	24.30	5.88
C26H5509P1	542.70	542.3583	57.54	10.22	26.53	5.71
C26H5601P2	446.68	446.3806	69.91	12.64	3.58	13.87
C26H5602P2	462.68	462.3755	67.50	12.20	6.92	13.39
C26H5603P2	478.68	478.3704	65.24	11.79	10.03	12.94
C26H5604P2	494.68	494.3654	63.13	11.41	12.94	12.52
C26H5605P2	510.68	510.3603	61.15	11.05	15.66	12.13
C26H5606P2	526.68	526.3552	59.29	10.72	18.23	11.76
C26H5607P2	542.68	542.3501	57.55	10.40	20.64	11.42
C26H5608P2	558.68	558.3450	55.90	10.10	22.91	11.09
C26H5609P2	574.68	574.3399	54.34	9.82	25.06	10.78
C27H1301P1	384.38	384.0704	84.37	3.41	4.16	8.06
C27H1302P1	400.38	400.0653	81.00	3.27	7.99	7.74
C27H1303P1	416.38	416.0602	77.89	3.15	11.53	7.44
C27H1304P1	432.38	432.0551	75.00	3.03	14.80	7.16
C27H1305P1	448.38	448.0501	72.33	2.92	17.84	6.91
C27H1306P1	464.37	464.0450	69.84	2.82	20.67	6.67
C27H1307P1	480.37	480.0399	67.51	2.73	23.31	6.45
C27H1308P1	496.37	496.0348	65.33	2.64	25.79	6.24
C27H1309P1	512.37	512.0297	63.29	2.56	28.10	6.05
C27H1401P2	416.36	416.0520	77.89	3.39	3.84	14.88
C27H1402P2	432.36	432.0469	75.01	3.26	7.40	14.33
C27H1403P2	448.36	448.0418	72.33	3.15	10.71	13.82
C27H1404P2	464.36	464.0367	69.84	3.04	13.78	13.34

C27H1405P2	480.36	480.0316	67.51	2.94	16.65	12.90
C27H1406P2	496.36	496.0266	65.34	2.84	19.34	12.48
C27H1407P2	512.36	512.0215	63.30	2.75	21.86	12.09
C27H1408P2	528.36	528.0164	61.38	2.67	24.23	11.72
C27H1409P2	544.35	544.0113	59.58	2.59	26.45	11.38
C27H1501P1	386.39	386.0860	83.93	3.91	4.14	8.02
C27H1502P1	402.39	402.0810	80.59	3.76	7.95	7.70
C27H1503P1	418.39	418.0759	77.51	3.61	11.47	7.40
C27H1504P1	434.39	434.0708	74.66	3.48	14.73	7.13
C27H1505P1	450.39	450.0657	72.00	3.36	17.76	6.88
C27H1506P1	466.39	466.0606	69.53	3.24	20.58	6.64
C27H1507P1	482.39	482.0555	67.23	3.13	23.22	6.42
C27H1508P1	498.39	498.0504	65.07	3.03	25.68	6.21
C27H1509P1	514.39	514.0454	63.05	2.94	27.99	6.02
C27H1601P2	418.38	418.0676	77.51	3.85	3.82	14.81
C27H1602P2	434.37	434.0625	74.66	3.71	7.37	14.26
C27H1603P2	450.37	450.0575	72.01	3.58	10.66	13.75
C27H1604P2	466.37	466.0524	69.54	3.46	13.72	13.28
C27H1605P2	482.37	482.0473	67.23	3.34	16.58	12.84
C27H1606P2	498.37	498.0422	65.07	3.24	19.26	12.43
C27H1607P2	514.37	514.0371	63.05	3.14	21.77	12.04
C27H1608P2	530.37	530.0320	61.15	3.04	24.13	11.68
C27H1609P2	546.37	546.0269	59.36	2.95	26.35	11.34
C27H1701P1	388.41	388.1017	83.49	4.41	4.12	7.97
C27H1702P1	404.41	404.0966	80.19	4.24	7.91	7.66
C27H1703P1	420.41	420.0915	77.14	4.08	11.42	7.37
C27H1704P1	436.41	436.0864	74.31	3.93	14.66	7.10
C27H1705P1	452.41	452.0814	71.68	3.79	17.68	6.85
C27H1706P1	468.41	468.0763	69.23	3.66	20.49	6.61
C27H1707P1	484.41	484.0712	66.95	3.54	23.12	6.39
C27H1708P1	500.41	500.0661	64.81	3.42	25.58	6.19
C27H1709P1	516.40	516.0610	62.80	3.32	27.88	6.00
C27H1801P2	420.39	420.0833	77.14	4.32	3.81	14.74
C27H1802P2	436.39	436.0782	74.31	4.16	7.33	14.20
C27H1803P2	452.39	452.0731	71.69	4.01	10.61	13.69
C27H1804P2	468.39	468.0680	69.24	3.87	13.66	13.23
C27H1805P2	484.39	484.0629	66.95	3.75	16.52	12.79
C27H1806P2	500.39	500.0579	64.81	3.63	19.18	12.38
C27H1807P2	516.39	516.0528	62.80	3.51	21.69	12.00
C27H1808P2	532.39	532.0477	60.91	3.41	24.04	11.64
C27H1809P2	548.39	548.0426	59.14	3.31	26.26	11.30
C27H1901P1	390.43	390.1173	83.06	4.91	4.10	7.93
C27H1902P1	406.43	406.1123	79.79	4.71	7.87	7.62
C27H1903P1	422.42	422.1072	76.77	4.53	11.36	7.33
C27H1904P1	438.42	438.1021	73.97	4.37	14.60	7.06
C27H1905P1	454.42	454.0970	71.37	4.21	17.60	6.82
C27H1906P1	470.42	470.0919	68.94	4.07	20.41	6.58
C27H1907P1	486.42	486.0868	66.67	3.94	23.02	6.37
C27H1908P1	502.42	502.0817	64.55	3.81	25.48	6.16
C27H1909P1	518.42	518.0767	62.56	3.69	27.78	5.97
C27H2001P2	422.41	422.0989	76.77	4.77	3.79	14.67
C27H2002P2	438.41	438.0938	73.97	4.60	7.30	14.13
C27H2003P2	454.41	454.0888	71.37	4.44	10.56	13.63
C27H2004P2	470.41	470.0837	68.94	4.29	13.60	13.17
C27H2005P2	486.41	486.0786	66.67	4.14	16.45	12.74
C27H2006P2	502.40	502.0735	64.55	4.01	19.11	12.33
C27H2007P2	518.40	518.0684	62.56	3.89	21.60	11.95
C27H2008P2	534.40	534.0633	60.68	3.77	23.95	11.59
C27H2009P2	550.40	550.0582	58.92	3.66	26.16	11.25
C27H2101P1	392.44	392.1330	82.64	5.39	4.08	7.89

C27H2102P1	408.44	408.1279	79.40	5.18	7.83	7.58
C27H2103P1	424.44	424.1228	76.41	4.99	11.31	7.30
C27H2104P1	440.44	440.1177	73.63	4.81	14.53	7.03
C27H2105P1	456.44	456.1127	71.05	4.64	17.53	6.79
C27H2106P1	472.44	472.1076	68.64	4.48	20.32	6.56
C27H2107P1	488.44	488.1025	66.40	4.33	22.93	6.34
C27H2108P1	504.44	504.0974	64.29	4.20	25.37	6.14
C27H2109P1	520.44	520.0923	62.31	4.07	27.67	5.95
C27H2201P2	424.42	424.1146	76.41	5.22	3.77	14.60
C27H2202P2	440.42	440.1095	73.63	5.04	7.27	14.07
C27H2203P2	456.42	456.1044	71.05	4.86	10.52	13.57
C27H2204P2	472.42	472.0993	68.65	4.69	13.55	13.11
C27H2205P2	488.42	488.0942	66.40	4.54	16.38	12.68
C27H2206P2	504.42	504.0892	64.29	4.40	19.03	12.28
C27H2207P2	520.42	520.0841	62.32	4.26	21.52	11.90
C27H2208P2	536.42	536.0790	60.46	4.13	23.86	11.55
C27H2209P2	552.42	552.0739	58.71	4.01	26.07	11.21
C27H2301P1	394.46	394.1486	82.21	5.88	4.06	7.85
C27H2302P1	410.46	410.1436	79.01	5.65	7.80	7.55
C27H2303P1	426.46	426.1385	76.05	5.44	11.26	7.26
C27H2304P1	442.46	442.1334	73.30	5.24	14.46	7.00
C27H2305P1	458.46	458.1283	70.74	5.06	17.45	6.76
C27H2306P1	474.45	474.1232	68.35	4.89	20.23	6.53
C27H2307P1	490.45	490.1181	66.12	4.73	22.84	6.32
C27H2308P1	506.45	506.1130	64.03	4.58	25.27	6.12
C27H2309P1	522.45	522.1080	62.07	4.44	27.56	5.93
C27H2401P2	426.44	426.1302	76.05	5.67	3.75	14.53
C27H2402P2	442.44	442.1251	73.30	5.47	7.23	14.00
C27H2403P2	458.44	458.1201	70.74	5.28	10.47	13.51
C27H2404P2	474.44	474.1150	68.35	5.10	13.49	13.06
C27H2405P2	490.44	490.1099	66.12	4.93	16.31	12.63
C27H2406P2	506.44	506.1048	64.04	4.78	18.96	12.23
C27H2407P2	522.44	522.0997	62.07	4.63	21.44	11.86
C27H2408P2	538.44	538.0946	60.23	4.49	23.77	11.51
C27H2409P2	554.43	554.0895	58.49	4.36	25.97	11.17
C27H2501P1	396.47	396.1643	81.80	6.36	4.04	7.81
C27H2502P1	412.47	412.1592	78.62	6.11	7.76	7.51
C27H2503P1	428.47	428.1541	75.69	5.88	11.20	7.23
C27H2504P1	444.47	444.1490	72.96	5.67	14.40	6.97
C27H2505P1	460.47	460.1439	70.43	5.47	17.37	6.73
C27H2506P1	476.47	476.1389	68.06	5.29	20.15	6.50
C27H2507P1	492.47	492.1338	65.85	5.12	22.74	6.29
C27H2508P1	508.47	508.1287	63.78	4.96	25.17	6.09
C27H2509P1	524.47	524.1236	61.83	4.80	27.46	5.91
C27H2601P2	428.46	428.1459	75.69	6.12	3.73	14.46
C27H2602P2	444.45	444.1408	72.97	5.90	7.20	13.94
C27H2603P2	460.45	460.1357	70.43	5.69	10.42	13.45
C27H2604P2	476.45	476.1306	68.07	5.50	13.43	13.00
C27H2605P2	492.45	492.1255	65.85	5.32	16.24	12.58
C27H2606P2	508.45	508.1205	63.78	5.15	18.88	12.18
C27H2607P2	524.45	524.1154	61.84	5.00	21.35	11.81
C27H2608P2	540.45	540.1103	60.01	4.85	23.68	11.46
C27H2609P2	556.45	556.1052	58.28	4.71	25.88	11.13
C27H2701P1	398.49	398.1799	81.38	6.83	4.02	7.77
C27H2702P1	414.49	414.1749	78.24	6.57	7.72	7.47
C27H2703P1	430.49	430.1698	75.33	6.32	11.15	7.20
C27H2704P1	446.49	446.1647	72.63	6.10	14.33	6.94
C27H2705P1	462.49	462.1596	70.12	5.88	17.30	6.70
C27H2706P1	478.49	478.1545	67.78	5.69	20.06	6.47
C27H2707P1	494.49	494.1494	65.58	5.50	22.65	6.26

153

C27H2708P1	510.49	510.1443	63.53	5.33	25.07	6.07
C27H2709P1	526.48	526.1393	61.60	5.17	27.35	5.88
C27H2801P2	430.47	430.1615	75.34	6.56	3.72	14.39
C27H2802P2	446.47	446.1564	72.64	6.32	7.17	13.87
C27H2803P2	462.47	462.1514	70.12	6.10	10.38	13.39
C27H2804P2	478.47	478.1463	67.78	5.90	13.38	12.95
C27H2805P2	494.47	494.1412	65.59	5.71	16.18	12.53
C27H2806P2	510.47	510.1361	63.53	5.53	18.81	12.14
C27H2807P2	526.47	526.1310	61.60	5.36	21.27	11.77
C27H2808P2	542.47	542.1259	59.78	5.20	23.60	11.42
C27H2809P2	558.47	558.1208	58.07	5.05	25.78	11.09
C27H2901P1	400.51	400.1956	80.97	7.30	3.99	7.73
C27H2902P1	416.50	416.1905	77.86	7.02	7.68	7.44
C27H2903P1	432.50	432.1854	74.98	6.76	11.10	7.16
C27H2904P1	448.50	448.1803	72.31	6.52	14.27	6.91
C27H2905P1	464.50	464.1752	69.82	6.29	17.22	6.67
C27H2906P1	480.50	480.1702	67.49	6.08	19.98	6.45
C27H2907P1	496.50	496.1651	65.32	5.89	22.56	6.24
C27H2908P1	512.50	512.1600	63.28	5.70	24.97	6.04
C27H2909P1	528.50	528.1549	61.36	5.53	27.25	5.86
C27H3001P2	432.49	432.1772	74.99	6.99	3.70	14.32
C27H3002P2	448.49	448.1721	72.31	6.74	7.13	13.81
C27H3003P2	464.49	464.1670	69.82	6.51	10.33	13.34
C27H3004P2	480.49	480.1619	67.49	6.29	13.32	12.89
C27H3005P2	496.48	496.1568	65.32	6.09	16.11	12.48
C27H3006P2	512.48	512.1518	63.28	5.90	18.73	12.09
C27H3007P2	528.48	528.1467	61.36	5.72	21.19	11.72
C27H3008P2	544.48	544.1416	59.56	5.55	23.51	11.38
C27H3009P2	560.48	560.1365	57.86	5.40	25.69	11.05
C27H3101P1	402.52	402.2112	80.57	7.76	3.97	7.69
C27H3102P1	418.52	418.2062	77.49	7.47	7.65	7.40
C27H3103P1	434.52	434.2011	74.63	7.19	11.05	7.13
C27H3104P1	450.52	450.1960	71.98	6.94	14.21	6.88
C27H3105P1	466.52	466.1909	69.52	6.70	17.15	6.64
C27H3106P1	482.52	482.1858	67.21	6.48	19.89	6.42
C27H3107P1	498.52	498.1807	65.05	6.27	22.47	6.21
C27H3108P1	514.52	514.1756	63.03	6.07	24.88	6.02
C27H3109P1	530.52	530.1706	61.13	5.89	27.14	5.84
C27H3201P2	434.50	434.1928	74.64	7.42	3.68	14.26
C27H3202P2	450.50	450.1877	71.99	7.16	7.10	13.75
C27H3203P2	466.50	466.1827	69.52	6.91	10.29	13.28
C27H3204P2	482.50	482.1776	67.21	6.68	13.26	12.84
C27H3205P2	498.50	498.1725	65.06	6.47	16.05	12.43
C27H3206P2	514.50	514.1674	63.03	6.27	18.66	12.04
C27H3207P2	530.50	530.1623	61.13	6.08	21.11	11.68
C27H3208P2	546.50	546.1572	59.34	5.90	23.42	11.34
C27H3209P2	562.50	562.1521	57.65	5.73	25.60	11.01
C27H3301P1	404.54	404.2269	80.17	8.22	3.95	7.66
C27H3302P1	420.54	420.2218	77.12	7.91	7.61	7.37
C27H3303P1	436.54	436.2167	74.29	7.62	11.00	7.10
C27H3304P1	452.54	452.2116	71.66	7.35	14.14	6.84
C27H3305P1	468.53	468.2065	69.22	7.10	17.07	6.61
C27H3306P1	484.53	484.2015	66.93	6.86	19.81	6.39
C27H3307P1	500.53	500.1964	64.79	6.65	22.38	6.19
C27H3308P1	516.53	516.1913	62.78	6.44	24.78	6.00
C27H3309P1	532.53	532.1862	60.90	6.25	27.04	5.82
C27H3401P2	436.52	436.2085	74.29	7.85	3.67	14.19
C27H3402P2	452.52	452.2034	71.67	7.57	7.07	13.69
C27H3403P2	468.52	468.1983	69.22	7.31	10.24	13.22
C27H3404P2	484.52	484.1932	66.93	7.07	13.21	12.79

C27H3405P2	500.52	500.1881	64.79	6.85	15.98	12.38
C27H3406P2	516.52	516.1830	62.79	6.64	18.59	11.99
C27H3407P2	532.52	532.1780	60.90	6.44	21.03	11.63
C27H3408P2	548.51	548.1729	59.12	6.25	23.33	11.29
C27H3409P2	564.51	564.1678	57.45	6.07	25.51	10.97
C27H3501P1	406.55	406.2425	79.77	8.68	3.94	7.62
C27H3502P1	422.55	422.2375	76.75	8.35	7.57	7.33
C27H3503P1	438.55	438.2324	73.95	8.04	10.94	7.06
C27H3504P1	454.55	454.2273	71.35	7.76	14.08	6.81
C27H3505P1	470.55	470.2222	68.92	7.50	17.00	6.58
C27H3506P1	486.55	486.2171	66.65	7.25	19.73	6.37
C27H3507P1	502.55	502.2120	64.53	7.02	22.29	6.16
C27H3508P1	518.55	518.2069	62.54	6.80	24.68	5.97
C27H3509P1	534.55	534.2019	60.67	6.60	26.94	5.79
C27H3601P2	438.53	438.2241	73.95	8.27	3.65	14.13
C27H3602P2	454.53	454.2190	71.35	7.98	7.04	13.63
C27H3603P2	470.53	470.2140	68.92	7.71	10.20	13.17
C27H3604P2	486.53	486.2089	66.66	7.46	13.15	12.73
C27H3605P2	502.53	502.2038	64.53	7.22	15.92	12.33
C27H3606P2	518.53	518.1987	62.54	7.00	18.51	11.95
C27H3607P2	534.53	534.1936	60.67	6.79	20.95	11.59
C27H3608P2	550.53	550.1885	58.91	6.59	23.25	11.25
C27H3609P2	566.53	566.1834	57.24	6.41	25.42	10.93
C27H3701P1	408.57	408.2582	79.37	9.13	3.92	7.58
C27H3702P1	424.57	424.2531	76.38	8.78	7.54	7.30
C27H3703P1	440.57	440.2480	73.61	8.47	10.89	7.03
C27H3704P1	456.57	456.2429	71.03	8.17	14.02	6.78
C27H3705P1	472.57	472.2378	68.63	7.89	16.93	6.55
C27H3706P1	488.57	488.2328	66.38	7.63	19.65	6.34
C27H3707P1	504.57	504.2277	64.27	7.39	22.20	6.14
C27H3708P1	520.56	520.2226	62.30	7.16	24.59	5.95
C27H3709P1	536.56	536.2175	60.44	6.95	26.84	5.77
C27H3801P2	440.55	440.2398	73.61	8.69	3.63	14.06
C27H3802P2	456.55	456.2347	71.03	8.39	7.01	13.57
C27H3803P2	472.55	472.2296	68.63	8.11	10.16	13.11
C27H3804P2	488.55	488.2245	66.38	7.84	13.10	12.68
C27H3805P2	504.55	504.2194	64.28	7.59	15.86	12.28
C27H3806P2	520.55	520.2143	62.30	7.36	18.44	11.90
C27H3807P2	536.55	536.2093	60.44	7.14	20.87	11.55
C27H3808P2	552.55	552.2042	58.69	6.93	23.16	11.21
C27H3809P2	568.55	568.1991	57.04	6.74	25.33	10.90
C27H3901P1	410.59	410.2738	78.99	9.57	3.90	7.54
C27H3902P1	426.58	426.2688	76.02	9.22	7.50	7.26
C27H3903P1	442.58	442.2637	73.27	8.88	10.84	7.00
C27H3904P1	458.58	458.2586	70.72	8.57	13.96	6.75
C27H3905P1	474.58	474.2535	68.33	8.28	16.86	6.53
C27H3906P1	490.58	490.2484	66.11	8.01	19.57	6.31
C27H3907P1	506.58	506.2433	64.02	7.76	22.11	6.11
C27H3908P1	522.58	522.2382	62.06	7.52	24.49	5.93
C27H3909P1	538.58	538.2332	60.21	7.30	26.74	5.75
C27H4001P2	442.57	442.2554	73.28	9.11	3.62	14.00
C27H4002P2	458.57	458.2503	70.72	8.79	6.98	13.51
C27H4003P2	474.57	474.2453	68.34	8.50	10.11	13.05
C27H4004P2	490.57	490.2402	66.11	8.22	13.05	12.63
C27H4005P2	506.56	506.2351	64.02	7.96	15.79	12.23
C27H4006P2	522.56	522.2300	62.06	7.72	18.37	11.85
C27H4007P2	538.56	538.2249	60.22	7.49	20.80	11.50
C27H4008P2	554.56	554.2198	58.48	7.27	23.08	11.17
C27H4009P2	570.56	570.2147	56.84	7.07	25.24	10.86
C27H4101P1	412.60	412.2895	78.60	10.02	3.88	7.51

C27H4102P1	428.60	428.2844	75.67	9.64	7.47	7.23
C27H4103P1	444.60	444.2793	72.94	9.30	10.80	6.97
C27H4104P1	460.60	460.2742	70.41	8.97	13.89	6.72
C27H4105P1	476.60	476.2691	68.04	8.67	16.78	6.50
C27H4106P1	492.60	492.2641	65.83	8.39	19.49	6.29
C27H4107P1	508.60	508.2590	63.76	8.13	22.02	6.09
C27H4108P1	524.60	524.2539	61.82	7.88	24.40	5.90
C27H4109P1	540.60	540.2488	59.99	7.64	26.64	5.73
C27H4201P2	444.58	444.2711	72.95	9.52	3.60	13.93
C27H4202P2	460.58	460.2660	70.41	9.19	6.95	13.45
C27H4203P2	476.58	476.2609	68.05	8.88	10.07	13.00
C27H4204P2	492.58	492.2558	65.84	8.59	12.99	12.58
C27H4205P2	508.58	508.2507	63.77	8.32	15.73	12.18
C27H4206P2	524.58	524.2456	61.82	8.07	18.30	11.81
C27H4207P2	540.58	540.2406	59.99	7.83	20.72	11.46
C27H4208P2	556.58	556.2355	58.27	7.61	23.00	11.13
C27H4209P2	572.58	572.2304	56.64	7.39	25.15	10.82
C27H4301P1	414.62	414.3051	78.22	10.45	3.86	7.47
C27H4302P1	430.62	430.3001	75.31	10.07	7.43	7.19
C27H4303P1	446.62	446.2950	72.61	9.70	10.75	6.94
C27H4304P1	462.62	462.2899	70.10	9.37	13.83	6.70
C27H4305P1	478.61	478.2848	67.76	9.06	16.71	6.47
C27H4306P1	494.61	494.2797	65.57	8.76	19.41	6.26
C27H4307P1	510.61	510.2746	63.51	8.49	21.93	6.07
C27H4308P1	526.61	526.2695	61.58	8.23	24.31	5.88
C27H4309P1	542.61	542.2644	59.77	7.99	26.54	5.71
C27H4401P2	446.60	446.2867	72.62	9.93	3.58	13.87
C27H4402P2	462.60	462.2816	70.10	9.59	6.92	13.39
C27H4403P2	478.60	478.2766	67.76	9.27	10.03	12.94
C27H4404P2	494.60	494.2715	65.57	8.97	12.94	12.52
C27H4405P2	510.60	510.2664	63.51	8.69	15.67	12.13
C27H4406P2	526.60	526.2613	61.58	8.42	18.23	11.76
C27H4407P2	542.60	542.2562	59.77	8.17	20.64	11.42
C27H4408P2	558.59	558.2511	58.06	7.94	22.91	11.09
C27H4409P2	574.59	574.2460	56.44	7.72	25.06	10.78
C27H4501P1	416.63	416.3208	77.84	10.89	3.84	7.43
C27H4502P1	432.63	432.3157	74.96	10.48	7.40	7.16
C27H4503P1	448.63	448.3106	72.29	10.11	10.70	6.90
C27H4504P1	464.63	464.3055	69.80	9.76	13.77	6.67
C27H4505P1	480.63	480.3004	67.47	9.44	16.64	6.44
C27H4506P1	496.63	496.2954	65.30	9.13	19.33	6.24
C27H4507P1	512.63	512.2903	63.26	8.85	21.85	6.04
C27H4508P1	528.63	528.2852	61.35	8.58	24.21	5.86
C27H4509P1	544.63	544.2801	59.55	8.33	26.44	5.69
C27H4601P2	448.61	448.3024	72.29	10.34	3.57	13.81
C27H4602P2	464.61	464.2973	69.80	9.98	6.89	13.33
C27H4603P2	480.61	480.2922	67.48	9.65	9.99	12.89
C27H4604P2	496.61	496.2871	65.30	9.34	12.89	12.47
C27H4605P2	512.61	512.2820	63.26	9.05	15.61	12.08
C27H4606P2	528.61	528.2769	61.35	8.77	18.16	11.72
C27H4607P2	544.61	544.2719	59.55	8.51	20.56	11.37
C27H4608P2	560.61	560.2668	57.85	8.27	22.83	11.05
C27H4609P2	576.61	576.2617	56.24	8.04	24.97	10.74
C27H4701P1	418.65	418.3364	77.46	11.32	3.82	7.40
C27H4702P1	434.65	434.3313	74.61	10.90	7.36	7.13
C27H4703P1	450.65	450.3263	71.96	10.51	10.65	6.87
C27H4704P1	466.65	466.3212	69.50	10.15	13.71	6.64
C27H4705P1	482.65	482.3161	67.19	9.82	16.57	6.42
C27H4706P1	498.65	498.3110	65.04	9.50	19.25	6.21
C27H4707P1	514.65	514.3059	63.01	9.21	21.76	6.02

C27H4708P1	530.64	530.3008	61.11	8.93	24.12	5.84
C27H4709P1	546.64	546.2957	59.33	8.67	26.34	5.67
C27H4801P2	450.63	450.3180	71.97	10.74	3.55	13.75
C27H4802P2	466.63	466.3129	69.50	10.37	6.86	13.28
C27H4803P2	482.63	482.3079	67.19	10.02	9.95	12.84
C27H4804P2	498.63	498.3028	65.04	9.70	12.83	12.42
C27H4805P2	514.63	514.2977	63.02	9.40	15.54	12.04
C27H4806P2	530.63	530.2926	61.12	9.12	18.09	11.67
C27H4807P2	546.63	546.2875	59.33	8.85	20.49	11.33
C27H4808P2	562.63	562.2824	57.64	8.60	22.75	11.01
C27H4809P2	578.63	578.2773	56.05	8.36	24.89	10.71
C27H4901P1	420.66	420.3521	77.09	11.74	3.80	7.36
C27H4902P1	436.66	436.3470	74.27	11.31	7.33	7.09
C27H4903P1	452.66	452.3419	71.64	10.91	10.60	6.84
C27H4904P1	468.66	468.3368	69.20	10.54	13.66	6.61
C27H4905P1	484.66	484.3317	66.91	10.19	16.51	6.39
C27H4906P1	500.66	500.3267	64.77	9.87	19.17	6.19
C27H4907P1	516.66	516.3216	62.77	9.56	21.68	5.99
C27H4908P1	532.66	532.3165	60.88	9.27	24.03	5.81
C27H4909P1	548.66	548.3114	59.11	9.00	26.24	5.65
C27H5001P2	452.65	452.3337	71.65	11.13	3.53	13.69
C27H5002P2	468.65	468.3286	69.20	10.75	6.83	13.22
C27H5003P2	484.65	484.3235	66.92	10.40	9.90	12.78
C27H5004P2	500.64	500.3184	64.78	10.07	12.78	12.37
C27H5005P2	516.64	516.3133	62.77	9.75	15.48	11.99
C27H5006P2	532.64	532.3082	60.89	9.46	18.02	11.63
C27H5007P2	548.64	548.3032	59.11	9.19	20.41	11.29
C27H5008P2	564.64	564.2981	57.43	8.93	22.67	10.97
C27H5009P2	580.64	580.2930	55.85	8.68	24.80	10.67
C27H5101P1	422.68	422.3677	76.72	12.16	3.79	7.33
C27H5102P1	438.68	438.3626	73.93	11.72	7.29	7.06
C27H5103P1	454.68	454.3576	71.33	11.31	10.56	6.81
C27H5104P1	470.68	470.3525	68.90	10.92	13.60	6.58
C27H5105P1	486.68	486.3474	66.64	10.56	16.44	6.36
C27H5106P1	502.68	502.3423	64.51	10.23	19.10	6.16
C27H5107P1	518.68	518.3372	62.52	9.91	21.59	5.97
C27H5108P1	534.68	534.3321	60.65	9.61	23.94	5.79
C27H5109P1	550.68	550.3270	58.89	9.34	26.15	5.62
C27H5201P2	454.66	454.3493	71.33	11.53	3.52	13.62
C27H5202P2	470.66	470.3442	68.90	11.14	6.80	13.16
C27H5203P2	486.66	486.3391	66.64	10.77	9.86	12.73
C27H5204P2	502.66	502.3341	64.52	10.43	12.73	12.32
C27H5205P2	518.66	518.3290	62.53	10.11	15.42	11.94
C27H5206P2	534.66	534.3239	60.66	9.80	17.95	11.59
C27H5207P2	550.66	550.3188	58.89	9.52	20.34	11.25
C27H5208P2	566.66	566.3137	57.23	9.25	22.59	10.93
C27H5209P2	582.66	582.3086	55.66	9.00	24.71	10.63
C27H5301P1	424.70	424.3834	76.36	12.58	3.77	7.29
C27H5302P1	440.70	440.3783	73.59	12.12	7.26	7.03
C27H5303P1	456.70	456.3732	71.01	11.70	10.51	6.78
C27H5304P1	472.69	472.3681	68.61	11.30	13.54	6.55
C27H5305P1	488.69	488.3630	66.36	10.93	16.37	6.34
C27H5306P1	504.69	504.3580	64.26	10.59	19.02	6.14
C27H5307P1	520.69	520.3529	62.28	10.26	21.51	5.95
C27H5308P1	536.69	536.3478	60.43	9.95	23.85	5.77
C27H5309P1	552.69	552.3427	58.68	9.67	26.05	5.60
C27H5401P2	456.68	456.3650	71.01	11.92	3.50	13.56
C27H5402P2	472.68	472.3599	68.61	11.52	6.77	13.11
C27H5403P2	488.68	488.3548	66.36	11.14	9.82	12.68
C27H5404P2	504.68	504.3497	64.26	10.79	12.68	12.27

C27H5405P2	520.68	520.3446	62.28	10.45	15.36	11.90
C27H5406P2	536.68	536.3395	60.43	10.14	17.89	11.54
C27H5407P2	552.67	552.3345	58.68	9.85	20.26	11.21
C27H5408P2	568.67	568.3294	57.03	9.57	22.51	10.89
C27H5409P2	584.67	584.3243	55.47	9.31	24.63	10.60
C27H5501P1	426.71	426.3990	76.00	12.99	3.75	7.26
C27H5502P1	442.71	442.3939	73.25	12.52	7.23	7.00
C27H5503P1	458.71	458.3889	70.70	12.09	10.46	6.75
C27H5504P1	474.71	474.3838	68.32	11.68	13.48	6.52
C27H5505P1	490.71	490.3787	66.09	11.30	16.30	6.31
C27H5506P1	506.71	506.3736	64.00	10.94	18.95	6.11
C27H5507P1	522.71	522.3685	62.04	10.61	21.43	5.93
C27H5508P1	538.71	538.3634	60.20	10.29	23.76	5.75
C27H5509P1	554.71	554.3583	58.46	9.99	25.96	5.58
C27H5601P2	458.69	458.3806	70.70	12.31	3.49	13.51
C27H5602P2	474.69	474.3755	68.32	11.89	6.74	13.05
C27H5603P2	490.69	490.3704	66.09	11.50	9.78	12.62
C27H5604P2	506.69	506.3654	64.00	11.14	12.63	12.23
C27H5605P2	522.69	522.3603	62.04	10.80	15.30	11.85
C27H5606P2	538.69	538.3552	60.20	10.48	17.82	11.50
C27H5607P2	554.69	554.3501	58.47	10.18	20.19	11.17
C27H5608P2	570.69	570.3450	56.83	9.89	22.43	10.85
C27H5609P2	586.69	586.3399	55.28	9.62	24.54	10.56
C27H5701P1	428.73	428.4147	75.64	13.40	3.73	7.22
C27H5702P1	444.73	444.4096	72.92	12.92	7.20	6.96
C27H5703P1	460.73	460.4045	70.39	12.47	10.42	6.72
C27H5704P1	476.73	476.3994	68.03	12.05	13.42	6.50
C27H5705P1	492.73	492.3943	65.82	11.66	16.24	6.29
C27H5706P1	508.73	508.3893	63.75	11.29	18.87	6.09
C27H5707P1	524.72	524.3842	61.80	10.95	21.34	5.90
C27H5708P1	540.72	540.3791	59.98	10.63	23.67	5.73
C27H5709P1	556.72	556.3740	58.25	10.32	25.86	5.56
C27H5801P2	460.71	460.3963	70.39	12.69	3.47	13.45
C27H5802P2	476.71	476.3912	68.03	12.26	6.71	12.99
C27H5803P2	492.71	492.3861	65.82	11.87	9.74	12.57
C27H5804P2	508.71	508.3810	63.75	11.49	12.58	12.18
C27H5805P2	524.71	524.3759	61.81	11.14	15.25	11.81
C27H5806P2	540.71	540.3708	59.98	10.81	17.75	11.46
C27H5807P2	556.71	556.3658	58.25	10.50	20.12	11.13
C27H5808P2	572.71	572.3607	56.63	10.21	22.35	10.82
C27H5809P2	588.71	588.3556	55.09	9.93	24.46	10.52
C28H1401P2	428.37	428.0520	78.51	3.29	3.73	14.46
C28H1402P2	444.37	444.0469	75.68	3.18	7.20	13.94
C28H1403P2	460.37	460.0418	73.05	3.07	10.43	13.46
C28H1404P2	476.37	476.0367	70.60	2.96	13.43	13.00
C28H1405P2	492.37	492.0316	68.30	2.87	16.25	12.58
C28H1406P2	508.37	508.0266	66.16	2.78	18.88	12.19
C28H1407P2	524.37	524.0215	64.14	2.69	21.36	11.81
C28H1408P2	540.37	540.0164	62.24	2.61	23.69	11.46
C28H1409P2	556.37	556.0113	60.45	2.54	25.88	11.13
C28H1501P1	398.40	398.0860	84.41	3.80	4.02	7.77
C28H1502P1	414.40	414.0810	81.16	3.65	7.72	7.47
C28H1503P1	430.40	430.0759	78.14	3.51	11.15	7.20
C28H1504P1	446.40	446.0708	75.34	3.39	14.34	6.94
C28H1505P1	462.40	462.0657	72.73	3.27	17.30	6.70
C28H1506P1	478.40	478.0606	70.30	3.16	20.07	6.47
C28H1507P1	494.40	494.0555	68.02	3.06	22.65	6.26
C28H1508P1	510.40	510.0504	65.89	2.96	25.08	6.07
C28H1509P1	526.40	526.0454	63.89	2.87	27.35	5.88
C28H1601P2	430.39	430.0676	78.14	3.75	3.72	14.39

C28H1602P2	446.39	446.0625	75.34	3.61	7.17	13.88
C28H1603P2	462.39	462.0575	72.73	3.49	10.38	13.40
C28H1604P2	478.38	478.0524	70.30	3.37	13.38	12.95
C28H1605P2	494.38	494.0473	68.03	3.26	16.18	12.53
C28H1606P2	510.38	510.0422	65.89	3.16	18.81	12.14
C28H1607P2	526.38	526.0371	63.89	3.06	21.28	11.77
C28H1608P2	542.38	542.0320	62.01	2.97	23.60	11.42
C28H1609P2	558.38	558.0269	60.23	2.89	25.79	11.09
C28H1701P1	400.42	400.1017	83.99	4.28	4.00	7.74
C28H1702P1	416.42	416.0966	80.76	4.11	7.68	7.44
C28H1703P1	432.42	432.0915	77.77	3.96	11.10	7.16
C28H1704P1	448.42	448.0864	75.00	3.82	14.27	6.91
C28H1705P1	464.42	464.0814	72.42	3.69	17.23	6.67
C28H1706P1	480.42	480.0763	70.00	3.57	19.98	6.45
C28H1707P1	496.42	496.0712	67.75	3.45	22.56	6.24
C28H1708P1	512.42	512.0661	65.63	3.34	24.98	6.04
C28H1709P1	528.42	528.0610	63.65	3.24	27.25	5.86
C28H1801P2	432.40	432.0833	77.78	4.20	3.70	14.33
C28H1802P2	448.40	448.0782	75.00	4.05	7.14	13.82
C28H1803P2	464.40	464.0731	72.42	3.91	10.34	13.34
C28H1804P2	480.40	480.0680	70.01	3.78	13.32	12.89
C28H1805P2	496.40	496.0629	67.75	3.66	16.12	12.48
C28H1806P2	512.40	512.0579	65.63	3.54	18.73	12.09
C28H1807P2	528.40	528.0528	63.65	3.43	21.20	11.72
C28H1808P2	544.40	544.0477	61.78	3.33	23.51	11.38
C28H1809P2	560.40	560.0426	60.01	3.24	25.70	11.05
C28H1901P1	402.44	402.1173	83.57	4.76	3.98	7.70
C28H1902P1	418.44	418.1123	80.37	4.58	7.65	7.40
C28H1903P1	434.44	434.1072	77.41	4.41	11.05	7.13
C28H1904P1	450.44	450.1021	74.66	4.25	14.21	6.88
C28H1905P1	466.43	466.0970	72.10	4.11	17.15	6.64
C28H1906P1	482.43	482.0919	69.71	3.97	19.90	6.42
C28H1907P1	498.43	498.0868	67.47	3.84	22.47	6.21
C28H1908P1	514.43	514.0817	65.38	3.72	24.88	6.02
C28H1909P1	530.43	530.0767	63.40	3.61	27.15	5.84
C28H2001P2	434.42	434.0989	77.42	4.64	3.68	14.26
C28H2002P2	450.42	450.0938	74.67	4.48	7.10	13.75
C28H2003P2	466.42	466.0888	72.11	4.32	10.29	13.28
C28H2004P2	482.42	482.0837	69.71	4.18	13.27	12.84
C28H2005P2	498.42	498.0786	67.48	4.04	16.05	12.43
C28H2006P2	514.42	514.0735	65.38	3.92	18.66	12.04
C28H2007P2	530.41	530.0684	63.41	3.80	21.11	11.68
C28H2008P2	546.41	546.0633	61.55	3.69	23.42	11.34
C28H2009P2	562.41	562.0582	59.80	3.58	25.60	11.01
C28H2101P1	404.45	404.1330	83.15	5.23	3.96	7.66
C28H2102P1	420.45	420.1279	79.99	5.03	7.61	7.37
C28H2103P1	436.45	436.1228	77.06	4.85	11.00	7.10
C28H2104P1	452.45	452.1177	74.33	4.68	14.14	6.85
C28H2105P1	468.45	468.1127	71.79	4.52	17.08	6.61
C28H2106P1	484.45	484.1076	69.42	4.37	19.82	6.39
C28H2107P1	500.45	500.1025	67.20	4.23	22.38	6.19
C28H2108P1	516.45	516.0974	65.12	4.10	24.78	6.00
C28H2109P1	532.45	532.0923	63.16	3.98	27.04	5.82
C28H2201P2	436.43	436.1146	77.06	5.08	3.67	14.19
C28H2202P2	452.43	452.1095	74.33	4.90	7.07	13.69
C28H2203P2	468.43	468.1044	71.80	4.73	10.25	13.22
C28H2204P2	484.43	484.0993	69.42	4.58	13.21	12.79
C28H2205P2	500.43	500.0942	67.20	4.43	15.99	12.38
C28H2206P2	516.43	516.0892	65.12	4.29	18.59	12.00
C28H2207P2	532.43	532.0841	63.17	4.16	21.03	11.63

156

C28H2208P2	548.43	548.0790	61.32	4.04	23.34	11.30
C28H2209P2	564.43	564.0739	59.58	3.93	25.51	10.98
C28H2301P1	406.47	406.1486	82.74	5.70	3.94	7.62
C28H2302P1	422.47	422.1436	79.61	5.49	7.57	7.33
C28H2303P1	438.47	438.1385	76.70	5.29	10.95	7.06
C28H2304P1	454.47	454.1334	74.00	5.10	14.08	6.82
C28H2305P1	470.47	470.1283	71.48	4.93	17.00	6.58
C28H2306P1	486.47	486.1232	69.13	4.77	19.73	6.37
C28H2307P1	502.47	502.1181	66.93	4.61	22.29	6.16
C28H2308P1	518.46	518.1130	64.87	4.47	24.69	5.97
C28H2309P1	534.46	534.1080	62.93	4.34	26.94	5.80
C28H2401P2	438.45	438.1302	76.70	5.52	3.65	14.13
C28H2402P2	454.45	454.1251	74.00	5.32	7.04	13.63
C28H2403P2	470.45	470.1201	71.49	5.14	10.20	13.17
C28H2404P2	486.45	486.1150	69.14	4.97	13.16	12.73
C28H2405P2	502.45	502.1099	66.93	4.81	15.92	12.33
C28H2406P2	518.45	518.1048	64.87	4.67	18.52	11.95
C28H2407P2	534.45	534.0997	62.93	4.53	20.96	11.59
C28H2408P2	550.45	550.0946	61.10	4.39	23.25	11.25
C28H2409P2	566.45	566.0895	59.37	4.27	25.42	10.94
C28H2501P1	408.48	408.1643	82.33	6.17	3.92	7.58
C28H2502P1	424.48	424.1592	79.23	5.94	7.54	7.30
C28H2503P1	440.48	440.1541	76.35	5.72	10.90	7.03
C28H2504P1	456.48	456.1490	73.67	5.52	14.02	6.79
C28H2505P1	472.48	472.1439	71.18	5.33	16.93	6.56
C28H2506P1	488.48	488.1389	68.85	5.16	19.65	6.34
C28H2507P1	504.48	504.1338	66.66	5.00	22.20	6.14
C28H2508P1	520.48	520.1287	64.62	4.84	24.59	5.95
C28H2509P1	536.48	536.1236	62.69	4.70	26.84	5.77
C28H2601P2	440.47	440.1459	76.35	5.95	3.63	14.06
C28H2602P2	456.47	456.1408	73.68	5.74	7.01	13.57
C28H2603P2	472.47	472.1357	71.18	5.55	10.16	13.11
C28H2604P2	488.46	488.1306	68.85	5.37	13.10	12.68
C28H2605P2	504.46	504.1255	66.67	5.20	15.86	12.28
C28H2606P2	520.46	520.1205	64.62	5.04	18.44	11.90
C28H2607P2	536.46	536.1154	62.69	4.89	20.88	11.55
C28H2608P2	552.46	552.1103	60.88	4.74	23.17	11.21
C28H2609P2	568.46	568.1052	59.16	4.61	25.33	10.90
C28H2701P1	410.50	410.1799	81.93	6.63	3.90	7.55
C28H2702P1	426.50	426.1749	78.85	6.38	7.50	7.26
C28H2703P1	442.50	442.1698	76.00	6.15	10.85	7.00
C28H2704P1	458.50	458.1647	73.35	5.94	13.96	6.76
C28H2705P1	474.50	474.1596	70.88	5.74	16.86	6.53
C28H2706P1	490.50	490.1545	68.57	5.55	19.57	6.31
C28H2707P1	506.50	506.1494	66.40	5.37	22.11	6.12
C28H2708P1	522.50	522.1443	64.37	5.21	24.50	5.93
C28H2709P1	538.50	538.1393	62.45	5.05	26.74	5.75
C28H2801P2	442.48	442.1615	76.01	6.38	3.62	14.00
C28H2802P2	458.48	458.1564	73.35	6.16	6.98	13.51
C28H2803P2	474.48	474.1514	70.88	5.95	10.12	13.06
C28H2804P2	490.48	490.1463	68.57	5.75	13.05	12.63
C28H2805P2	506.48	506.1412	66.40	5.57	15.79	12.23
C28H2806P2	522.48	522.1361	64.37	5.40	18.37	11.86
C28H2807P2	538.48	538.1310	62.46	5.24	20.80	11.50
C28H2808P2	554.48	554.1259	60.65	5.09	23.08	11.17
C28H2809P2	570.48	570.1208	58.95	4.95	25.24	10.86
C28H2901P1	412.52	412.1956	81.53	7.09	3.88	7.51
C28H2902P1	428.52	428.1905	78.48	6.82	7.47	7.23
C28H2903P1	444.52	444.1854	75.66	6.58	10.80	6.97
C28H2904P1	460.51	460.1803	73.03	6.35	13.90	6.73
C28H2905P1	476.51	476.1752	70.58	6.13	16.79	6.50
C28H2906P1	492.51	492.1702	68.28	5.94	19.49	6.29
C28H2907P1	508.51	508.1651	66.14	5.75	22.02	6.09
C28H2908P1	524.51	524.1600	64.12	5.57	24.40	5.91
C28H2909P1	540.51	540.1549	62.22	5.41	26.64	5.73
C28H3001P2	444.50	444.1772	75.66	6.80	3.60	13.94
C28H3002P2	460.50	460.1721	73.03	6.57	6.95	13.45
C28H3003P2	476.50	476.1670	70.58	6.35	10.07	13.00
C28H3004P2	492.50	492.1619	68.29	6.14	12.99	12.58
C28H3005P2	508.50	508.1568	66.14	5.95	15.73	12.18
C28H3006P2	524.50	524.1518	64.12	5.77	18.30	11.81
C28H3007P2	540.49	540.1467	62.22	5.59	20.72	11.46
C28H3008P2	556.49	556.1416	60.43	5.43	23.00	11.13
C28H3009P2	572.49	572.1365	58.75	5.28	25.15	10.82
C28H3101P1	414.53	414.2112	81.13	7.54	3.86	7.47
C28H3102P1	430.53	430.2062	78.12	7.26	7.43	7.19
C28H3103P1	446.53	446.2011	75.32	7.00	10.75	6.94
C28H3104P1	462.53	462.1960	72.71	6.76	13.84	6.70
C28H3105P1	478.53	478.1909	70.28	6.53	16.72	6.47
C28H3106P1	494.53	494.1858	68.01	6.32	19.41	6.26
C28H3107P1	510.53	510.1807	65.88	6.12	21.94	6.07
C28H3108P1	526.53	526.1756	63.87	5.93	24.31	5.88
C28H3109P1	542.53	542.1706	61.99	5.76	26.54	5.71
C28H3201P2	446.51	446.1928	75.32	7.22	3.58	13.87
C28H3202P2	462.51	462.1877	72.71	6.97	6.92	13.39
C28H3203P2	478.51	478.1827	70.28	6.74	10.03	12.95
C28H3204P2	494.51	494.1776	68.01	6.52	12.94	12.53
C28H3205P2	510.51	510.1725	65.88	6.32	15.67	12.13
C28H3206P2	526.51	526.1674	63.88	6.13	18.23	11.77
C28H3207P2	542.51	542.1623	61.99	5.95	20.64	11.42
C28H3208P2	558.51	558.1572	60.22	5.78	22.92	11.09
C28H3209P2	574.51	574.1521	58.54	5.61	25.06	10.78
C28H3301P1	416.55	416.2269	80.74	7.99	3.84	7.44
C28H3302P1	432.55	432.2218	77.75	7.69	7.40	7.16
C28H3303P1	448.55	448.2167	74.98	7.42	10.70	6.91
C28H3304P1	464.55	464.2116	72.40	7.16	13.78	6.67
C28H3305P1	480.55	480.2065	69.99	6.92	16.65	6.45
C28H3306P1	496.55	496.2015	67.73	6.70	19.33	6.24
C28H3307P1	512.54	512.1964	65.62	6.49	21.85	6.04
C28H3308P1	528.54	528.1913	63.63	6.29	24.22	5.86
C28H3309P1	544.54	544.1862	61.76	6.11	26.44	5.69
C28H3401P2	448.53	448.2085	74.98	7.64	3.57	13.81
C28H3402P2	464.53	464.2034	72.40	7.38	6.89	13.34
C28H3403P2	480.53	480.1983	69.99	7.13	9.99	12.89
C28H3404P2	496.53	496.1932	67.73	6.90	12.89	12.48
C28H3405P2	512.53	512.1881	65.62	6.69	15.61	12.09
C28H3406P2	528.53	528.1830	63.63	6.48	18.16	11.72
C28H3407P2	544.53	544.1780	61.76	6.29	20.57	11.38
C28H3408P2	560.53	560.1729	60.00	6.11	22.83	11.05
C28H3409P2	576.53	576.1678	58.33	5.94	24.98	10.74
C28H3501P1	418.56	418.2425	80.35	8.43	3.82	7.40
C28H3502P1	434.56	434.2375	77.39	8.12	7.36	7.13
C28H3503P1	450.56	450.2324	74.64	7.83	10.65	6.87
C28H3504P1	466.56	466.2273	72.08	7.56	13.72	6.64
C28H3505P1	482.56	482.2222	69.69	7.31	16.58	6.42
C28H3506P1	498.56	498.2171	67.46	7.08	19.25	6.21
C28H3507P1	514.56	514.2120	65.36	6.86	21.77	6.02
C28H3508P1	530.56	530.2069	63.39	6.65	24.12	5.84
C28H3509P1	546.56	546.2019	61.53	6.45	26.35	5.67
C28H3601P2	450.55	450.2241	74.65	8.05	3.55	13.75

157

C28H3602P2	466.55	466.2190	72.09	7.78	6.86	13.28
C28H3603P2	482.54	482.2140	69.70	7.52	9.95	12.84
C28H3604P2	498.54	498.2089	67.46	7.28	12.84	12.43
C28H3605P2	514.54	514.2038	65.36	7.05	15.55	12.04
C28H3606P2	530.54	530.1987	63.39	6.84	18.09	11.68
C28H3607P2	546.54	546.1936	61.53	6.64	20.49	11.33
C28H3608P2	562.54	562.1885	59.78	6.45	22.75	11.01
C28H3609P2	578.54	578.1834	58.13	6.27	24.89	10.71
C28H3701P1	420.58	420.2582	79.96	8.87	3.80	7.36
C28H3702P1	436.58	436.2531	77.03	8.54	7.33	7.09
C28H3703P1	452.58	452.2480	74.31	8.24	10.61	6.84
C28H3704P1	468.58	468.2429	71.77	7.96	13.66	6.61
C28H3705P1	484.58	484.2378	69.40	7.70	16.51	6.39
C28H3706P1	500.58	500.2328	67.18	7.45	19.18	6.19
C28H3707P1	516.58	516.2277	65.10	7.22	21.68	6.00
C28H3708P1	532.58	532.2226	63.15	7.00	24.03	5.82
C28H3709P1	548.58	548.2175	61.31	6.80	26.25	5.65
C28H3801P2	452.56	452.2398	74.31	8.46	3.54	13.69
C28H3802P2	468.56	468.2347	71.78	8.17	6.83	13.22
C28H3803P2	484.56	484.2296	69.41	7.90	9.91	12.78
C28H3804P2	500.56	500.2245	67.19	7.65	12.79	12.38
C28H3805P2	516.56	516.2194	65.11	7.41	15.49	11.99
C28H3806P2	532.56	532.2143	63.15	7.19	18.03	11.63
C28H3807P2	548.56	548.2093	61.31	6.98	20.42	11.29
C28H3808P2	564.56	564.2042	59.57	6.78	22.67	10.97
C28H3809P2	580.56	580.1991	57.93	6.60	24.80	10.67
C28H3901P1	422.60	422.2738	79.58	9.30	3.79	7.33
C28H3902P1	438.60	438.2688	76.68	8.96	7.30	7.06
C28H3903P1	454.60	454.2637	73.98	8.65	10.56	6.81
C28H3904P1	470.59	470.2586	71.47	8.35	13.60	6.58
C28H3905P1	486.59	486.2535	69.12	8.08	16.44	6.37
C28H3906P1	502.59	502.2484	66.92	7.82	19.10	6.16
C28H3907P1	518.59	518.2433	64.85	7.58	21.60	5.97
C28H3908P1	534.59	534.2382	62.91	7.35	23.94	5.79
C28H3909P1	550.59	550.2332	61.08	7.14	26.15	5.63
C28H4001P2	454.58	454.2554	73.98	8.87	3.52	13.63
C28H4002P2	470.58	470.2503	71.47	8.57	6.80	13.16
C28H4003P2	486.58	486.2453	69.12	8.29	9.86	12.73
C28H4004P2	502.58	502.2402	66.92	8.02	12.73	12.33
C28H4005P2	518.58	518.2351	64.85	7.77	15.43	11.95
C28H4006P2	534.57	534.2300	62.91	7.54	17.96	11.59
C28H4007P2	550.57	550.2249	61.08	7.32	20.34	11.25
C28H4008P2	566.57	566.2198	59.36	7.12	22.59	10.93
C28H4009P2	582.57	582.2147	57.73	6.92	24.72	10.63
C28H4101P1	424.61	424.2895	79.20	9.73	3.77	7.29
C28H4102P1	440.61	440.2844	76.33	9.38	7.26	7.03
C28H4103P1	456.61	456.2793	73.65	9.05	10.51	6.78
C28H4104P1	472.61	472.2742	71.16	8.74	13.54	6.55
C28H4105P1	488.61	488.2691	68.83	8.46	16.37	6.34
C28H4106P1	504.61	504.2641	66.65	8.19	19.02	6.14
C28H4107P1	520.61	520.2590	64.60	7.94	21.51	5.95
C28H4108P1	536.61	536.2539	62.67	7.70	23.85	5.77
C28H4109P1	552.61	552.2488	60.86	7.48	26.06	5.61
C28H4201P2	456.59	456.2711	73.66	9.27	3.50	13.57
C28H4202P2	472.59	472.2660	71.16	8.96	6.77	13.11
C28H4203P2	488.59	488.2609	68.83	8.66	9.82	12.68
C28H4204P2	504.59	504.2558	66.65	8.39	12.68	12.28
C28H4205P2	520.59	520.2507	64.60	8.13	15.37	11.90
C28H4206P2	536.59	536.2456	62.68	7.89	17.89	11.54
C28H4207P2	552.59	552.2406	60.86	7.66	20.27	11.21
C28H4208P2	568.59	568.2355	59.15	7.45	22.51	10.89
C28H4209P2	584.59	584.2304	57.53	7.24	24.63	10.60
C28H4301P1	426.63	426.3051	78.83	10.16	3.75	7.26
C28H4302P1	442.63	442.3001	75.98	9.79	7.23	7.00
C28H4303P1	458.63	458.2950	73.33	9.45	10.47	6.75
C28H4304P1	474.63	474.2899	70.86	9.13	13.48	6.53
C28H4305P1	490.63	490.2848	68.55	8.83	16.31	6.31
C28H4306P1	506.63	506.2797	66.38	8.56	18.95	6.11
C28H4307P1	522.62	522.2746	64.35	8.29	21.43	5.93
C28H4308P1	538.62	538.2695	62.44	8.05	23.76	5.75
C28H4309P1	554.62	554.2644	60.64	7.81	25.96	5.58
C28H4401P2	458.61	458.2867	73.33	9.67	3.49	13.51
C28H4402P2	474.61	474.2816	70.86	9.34	6.74	13.05
C28H4403P2	490.61	490.2766	68.55	9.04	9.78	12.63
C28H4404P2	506.61	506.2715	66.39	8.75	12.63	12.23
C28H4405P2	522.61	522.2664	64.35	8.49	15.31	11.85
C28H4406P2	538.61	538.2613	62.44	8.23	17.82	11.50
C28H4407P2	554.61	554.2562	60.64	8.00	20.19	11.17
C28H4408P2	570.61	570.2511	58.94	7.77	22.43	10.86
C28H4409P2	586.61	586.2460	57.33	7.56	24.55	10.56
C28H4501P1	428.64	428.3208	78.46	10.58	3.73	7.23
C28H4502P1	444.64	444.3157	75.64	10.20	7.20	6.97
C28H4503P1	460.64	460.3106	73.01	9.85	10.42	6.72
C28H4504P1	476.64	476.3055	70.56	9.52	13.43	6.50
C28H4505P1	492.64	492.3004	68.27	9.21	16.24	6.29
C28H4506P1	508.64	508.2954	66.12	8.92	18.87	6.09
C28H4507P1	524.64	524.2903	64.10	8.65	21.35	5.90
C28H4508P1	540.64	540.2852	62.21	8.39	23.67	5.73
C28H4509P1	556.64	556.2801	60.42	8.15	25.87	5.56
C28H4601P2	460.63	460.3024	73.01	10.07	3.47	13.45
C28H4602P2	476.63	476.2973	70.56	9.73	6.71	13.00
C28H4603P2	492.62	492.2922	68.27	9.41	9.74	12.58
C28H4604P2	508.62	508.2871	66.12	9.12	12.58	12.18
C28H4605P2	524.62	524.2820	64.11	8.84	15.25	11.81
C28H4606P2	540.62	540.2769	62.21	8.58	17.76	11.46
C28H4607P2	556.62	556.2719	60.42	8.33	20.12	11.13
C28H4608P2	572.62	572.2668	58.73	8.10	22.35	10.82
C28H4609P2	588.62	588.2617	57.14	7.88	24.46	10.52
C28H4701P1	430.66	430.3364	78.09	11.00	3.72	7.19
C28H4702P1	446.66	446.3313	75.30	10.61	7.16	6.93
C28H4703P1	462.66	462.3263	72.69	10.24	10.37	6.69
C28H4704P1	478.66	478.3212	70.26	9.90	13.37	6.47
C28H4705P1	494.66	494.3161	67.99	9.58	16.17	6.26
C28H4706P1	510.66	510.3110	65.86	9.28	18.80	6.07
C28H4707P1	526.66	526.3059	63.86	9.00	21.27	5.88
C28H4708P1	542.66	542.3008	61.98	8.73	23.59	5.71
C28H4709P1	558.66	558.2957	60.20	8.48	25.78	5.54
C28H4801P2	462.64	462.3180	72.69	10.46	3.46	13.39
C28H4802P2	478.64	478.3129	70.26	10.11	6.69	12.94
C28H4803P2	494.64	494.3079	67.99	9.78	9.70	12.52
C28H4804P2	510.64	510.3028	65.86	9.47	12.53	12.13
C28H4805P2	526.64	526.2977	63.86	9.19	15.19	11.76
C28H4806P2	542.64	542.2926	61.98	8.92	17.69	11.42
C28H4807P2	558.64	558.2875	60.20	8.66	20.05	11.09
C28H4808P2	574.64	574.2824	58.53	8.42	22.27	10.78
C28H4809P2	590.64	590.2773	56.94	8.19	24.38	10.49
C28H4901P1	432.68	432.3521	77.73	11.42	3.70	7.16
C28H4902P1	448.68	448.3470	74.96	11.01	7.13	6.90
C28H4903P1	464.67	464.3419	72.38	10.63	10.33	6.67
C28H4904P1	480.67	480.3368	69.97	10.28	13.31	6.44

C28H4905P1	496.67	496.3317	67.71	9.94	16.11	6.24
C28H4906P1	512.67	512.3267	65.60	9.63	18.72	6.04
C28H4907P1	528.67	528.3216	63.61	9.34	21.18	5.86
C28H4908P1	544.67	544.3165	61.75	9.07	23.50	5.69
C28H4909P1	560.67	560.3114	59.98	8.81	25.68	5.52
C28H5001P2	464.66	464.3337	72.38	10.85	3.44	13.33
C28H5002P2	480.66	480.3286	69.97	10.49	6.66	12.89
C28H5003P2	496.66	496.3235	67.72	10.15	9.66	12.47
C28H5004P2	512.66	512.3184	65.60	9.83	12.48	12.08
C28H5005P2	528.66	528.3133	63.62	9.53	15.13	11.72
C28H5006P2	544.65	544.3082	61.75	9.25	17.63	11.37
C28H5007P2	560.65	560.3032	59.99	8.99	19.98	11.05
C28H5008P2	576.65	576.2981	58.32	8.74	22.20	10.74
C28H5009P2	592.65	592.2930	56.75	8.50	24.30	10.45
C28H5101P1	434.69	434.3677	77.37	11.83	3.68	7.13
C28H5102P1	450.69	450.3626	74.62	11.41	7.10	6.87
C28H5103P1	466.69	466.3576	72.06	11.02	10.28	6.64
C28H5104P1	482.69	482.3525	69.67	10.65	13.26	6.42
C28H5105P1	498.69	498.3474	67.44	10.31	16.04	6.21
C28H5106P1	514.69	514.3423	65.34	9.99	18.65	6.02
C28H5107P1	530.69	530.3372	63.37	9.69	21.10	5.84
C28H5108P1	546.69	546.3321	61.52	9.40	23.41	5.67
C28H5109P1	562.69	562.3270	59.77	9.14	25.59	5.50
C28H5201P2	466.67	466.3493	72.07	11.23	3.43	13.27
C28H5202P2	482.67	482.3442	69.68	10.86	6.63	12.83
C28H5203P2	498.67	498.3391	67.44	10.51	9.63	12.42
C28H5204P2	514.67	514.3341	65.34	10.18	12.43	12.04
C28H5205P2	530.67	530.3290	63.37	9.88	15.07	11.67
C28H5206P2	546.67	546.3239	61.52	9.59	17.56	11.33
C28H5207P2	562.67	562.3188	59.77	9.32	19.90	11.01
C28H5208P2	578.67	578.3137	58.12	9.06	22.12	10.71
C28H5209P2	594.67	594.3086	56.55	8.81	24.21	10.42
C28H5301P1	436.71	436.3834	77.01	12.23	3.66	7.09
C28H5302P1	452.71	452.3783	74.29	11.80	7.07	6.84
C28H5303P1	468.71	468.3732	71.75	11.40	10.24	6.61
C28H5304P1	484.71	484.3681	69.38	11.02	13.20	6.39
C28H5305P1	500.71	500.3630	67.17	10.67	15.98	6.19
C28H5306P1	516.70	516.3580	65.09	10.34	18.58	5.99
C28H5307P1	532.70	532.3529	63.13	10.03	21.02	5.81
C28H5308P1	548.70	548.3478	61.29	9.74	23.33	5.64
C28H5309P1	564.70	564.3427	59.56	9.46	25.50	5.48
C28H5401P2	468.69	468.3650	71.76	11.61	3.41	13.22
C28H5402P2	484.69	484.3599	69.39	11.23	6.60	12.78
C28H5403P2	500.69	500.3548	67.17	10.87	9.59	12.37
C28H5404P2	516.69	516.3497	65.09	10.53	12.39	11.99
C28H5405P2	532.69	532.3446	63.14	10.22	15.02	11.63
C28H5406P2	548.69	548.3395	61.29	9.92	17.50	11.29
C28H5407P2	564.69	564.3345	59.56	9.64	19.83	10.97
C28H5408P2	580.69	580.3294	57.92	9.37	22.04	10.67
C28H5409P2	596.68	596.3243	56.36	9.12	24.13	10.38
C28H5501P1	438.72	438.3990	76.66	12.64	3.65	7.06
C28H5502P1	454.72	454.3939	73.96	12.19	7.04	6.81
C28H5503P1	470.72	470.3889	71.45	11.78	10.20	6.58
C28H5504P1	486.72	486.3838	69.10	11.39	13.15	6.36
C28H5505P1	502.72	502.3787	66.90	11.03	15.91	6.16
C28H5506P1	518.72	518.3736	64.83	10.69	18.51	5.97
C28H5507P1	534.72	534.3685	62.89	10.37	20.94	5.79
C28H5508P1	550.72	550.3634	61.07	10.07	23.24	5.62
C28H5509P1	566.72	566.3583	59.34	9.78	25.41	5.47
C28H5601P2	470.71	470.3806	71.45	11.99	3.40	13.16
C28H5602P2	486.70	486.3755	69.10	11.60	6.57	12.73
C28H5603P2	502.70	502.3704	66.90	11.23	9.55	12.32
C28H5604P2	518.70	518.3654	64.84	10.88	12.34	11.94
C28H5605P2	534.70	534.3603	62.90	10.56	14.96	11.59
C28H5606P2	550.70	550.3552	61.07	10.25	17.43	11.25
C28H5607P2	566.70	566.3501	59.35	9.96	19.76	10.93
C28H5608P2	582.70	582.3450	57.72	9.69	21.97	10.63
C28H5609P2	598.70	598.3399	56.17	9.43	24.05	10.35
C28H5701P1	440.74	440.4147	76.31	13.04	3.63	7.03
C28H5702P1	456.74	456.4096	73.63	12.58	7.01	6.78
C28H5703P1	472.74	472.4045	71.14	12.15	10.15	6.55
C28H5704P1	488.74	488.3994	68.81	11.76	13.09	6.34
C28H5705P1	504.74	504.3943	66.63	11.38	15.85	6.14
C28H5706P1	520.74	520.3893	64.58	11.03	18.43	5.95
C28H5707P1	536.74	536.3842	62.66	10.70	20.87	5.77
C28H5708P1	552.74	552.3791	60.85	10.39	23.16	5.60
C28H5709P1	568.73	568.3740	59.13	10.10	25.32	5.45
C28H5801P2	472.72	472.3963	71.14	12.37	3.38	13.10
C28H5802P2	488.72	488.3912	68.81	11.96	6.55	12.68
C28H5803P2	504.72	504.3861	66.63	11.58	9.51	12.27
C28H5804P2	520.72	520.3810	64.59	11.23	12.29	11.90
C28H5805P2	536.72	536.3759	62.66	10.89	14.90	11.54
C28H5806P2	552.72	552.3708	60.85	10.58	17.37	11.21
C28H5807P2	568.72	568.3658	59.14	10.28	19.69	10.89
C28H5808P2	584.72	584.3607	57.52	10.00	21.89	10.59
C28H5809P2	600.72	600.3556	55.99	9.73	23.97	10.31
C28H5901P1	442.76	442.4303	75.96	13.43	3.61	7.00
C28H5902P1	458.76	458.4252	73.31	12.96	6.98	6.75
C28H5903P1	474.75	474.4202	70.84	12.53	10.11	6.52
C28H5904P1	490.75	490.4151	68.53	12.12	13.04	6.31
C28H5905P1	506.75	506.4100	66.37	11.74	15.79	6.11
C28H5906P1	522.75	522.4049	64.33	11.38	18.36	5.93
C28H5907P1	538.75	538.3998	62.42	11.04	20.79	5.75
C28H5908P1	554.75	554.3947	60.62	10.72	23.07	5.58
C28H5909P1	570.75	570.3896	58.92	10.42	25.23	5.43
C28H6001P2	474.74	474.4119	70.84	12.74	3.37	13.05
C28H6002P2	490.74	490.4068	68.53	12.32	6.52	12.62
C28H6003P2	506.74	506.4017	66.37	11.93	9.47	12.22
C28H6004P2	522.74	522.3967	64.34	11.57	12.24	11.85
C28H6005P2	538.73	538.3916	62.43	11.23	14.85	11.50
C28H6006P2	554.73	554.3865	60.63	10.90	17.30	11.17
C28H6007P2	570.73	570.3814	58.93	10.60	19.62	10.85
C28H6008P2	586.73	586.3763	57.32	10.31	21.81	10.56
C28H6009P2	602.73	602.3712	55.80	10.03	23.89	10.28
C29H1401P2	440.38	440.0520	79.10	3.20	3.63	14.07
C29H1402P2	456.38	456.0469	76.32	3.09	7.01	13.57
C29H1403P2	472.38	472.0418	73.74	2.99	10.16	13.11
C29H1404P2	488.38	488.0367	71.32	2.89	13.10	12.68
C29H1405P2	504.38	504.0316	69.06	2.80	15.86	12.28
C29H1406P2	520.38	520.0266	66.94	2.71	18.45	11.90
C29H1407P2	536.38	536.0215	64.94	2.63	20.88	11.55
C29H1408P2	552.38	552.0164	63.06	2.55	23.17	11.21
C29H1409P2	568.38	568.0113	61.28	2.48	25.33	10.90
C29H1501P1	410.42	410.0860	84.87	3.68	3.90	7.55
C29H1502P1	426.42	426.0810	81.69	3.55	7.50	7.26
C29H1503P1	442.41	442.0759	78.73	3.42	10.85	7.00
C29H1504P1	458.41	458.0708	75.98	3.30	13.96	6.76
C29H1505P1	474.41	474.0657	73.42	3.19	16.86	6.53
C29H1506P1	490.41	490.0606	71.03	3.08	19.57	6.32
C29H1507P1	506.41	506.0555	68.78	2.99	22.12	6.12

159

C29H1508P1	522.41	522.0504	66.68	2.89	24.50	5.93
C29H1509P1	538.41	538.0454	64.69	2.81	26.74	5.75
C29H1601P2	442.40	442.0676	78.74	3.65	3.62	14.00
C29H1602P2	458.40	458.0625	75.99	3.52	6.98	13.51
C29H1603P2	474.40	474.0575	73.42	3.40	10.12	13.06
C29H1604P2	490.40	490.0524	71.03	3.29	13.05	12.63
C29H1605P2	506.40	506.0473	68.78	3.18	15.80	12.23
C29H1606P2	522.39	522.0422	66.68	3.09	18.38	11.86
C29H1607P2	538.39	538.0371	64.70	3.00	20.80	11.51
C29H1608P2	554.39	554.0320	62.83	2.91	23.09	11.17
C29H1609P2	570.39	570.0269	61.07	2.83	25.24	10.86
C29H1701P1	412.43	412.1017	84.46	4.15	3.88	7.51
C29H1702P1	428.43	428.0966	81.30	4.00	7.47	7.23
C29H1703P1	444.43	444.0915	78.38	3.86	10.80	6.97
C29H1704P1	460.43	460.0864	75.65	3.72	13.90	6.73
C29H1705P1	476.43	476.0814	73.11	3.60	16.79	6.50
C29H1706P1	492.43	492.0763	70.74	3.48	19.49	6.29
C29H1707P1	508.43	508.0712	68.51	3.37	22.03	6.09
C29H1708P1	524.43	524.0661	66.42	3.27	24.41	5.91
C29H1709P1	540.43	540.0610	64.45	3.17	26.64	5.73
C29H1801P2	444.41	444.0833	78.38	4.08	3.60	13.94
C29H1802P2	460.41	460.0782	75.65	3.94	6.95	13.45
C29H1803P2	476.41	476.0731	73.11	3.81	10.07	13.00
C29H1804P2	492.41	492.0680	70.74	3.68	13.00	12.58
C29H1805P2	508.41	508.0629	68.51	3.57	15.73	12.18
C29H1806P2	524.41	524.0579	66.42	3.46	18.31	11.81
C29H1807P2	540.41	540.0528	64.46	3.36	20.72	11.46
C29H1808P2	556.41	556.0477	62.60	3.26	23.00	11.13
C29H1809P2	572.41	572.0426	60.85	3.17	25.16	10.82
C29H1901P1	414.45	414.1173	84.05	4.62	3.86	7.47
C29H1902P1	430.45	430.1123	80.92	4.45	7.43	7.20
C29H1903P1	446.45	446.1072	78.02	4.29	10.75	6.94
C29H1904P1	462.45	462.1021	75.32	4.14	13.84	6.70
C29H1905P1	478.45	478.0970	72.80	4.00	16.72	6.47
C29H1906P1	494.44	494.0919	70.45	3.87	19.41	6.26
C29H1907P1	510.44	510.0868	68.24	3.75	21.94	6.07
C29H1908P1	526.44	526.0817	66.17	3.64	24.31	5.88
C29H1909P1	542.44	542.0767	64.21	3.53	26.55	5.71
C29H2001P2	446.43	446.0989	78.02	4.52	3.58	13.88
C29H2002P2	462.43	462.0938	75.32	4.36	6.92	13.40
C29H2003P2	478.43	478.0888	72.81	4.21	10.03	12.95
C29H2004P2	494.43	494.0837	70.45	4.08	12.94	12.53
C29H2005P2	510.43	510.0786	68.24	3.95	15.67	12.14
C29H2006P2	526.43	526.0735	66.17	3.83	18.24	11.77
C29H2007P2	542.43	542.0684	64.22	3.72	20.65	11.42
C29H2008P2	558.43	558.0633	62.38	3.61	22.92	11.09
C29H2009P2	574.42	574.0582	60.64	3.51	25.07	10.78
C29H2101P1	416.46	416.1330	83.64	5.08	3.84	7.44
C29H2102P1	432.46	432.1279	80.54	4.89	7.40	7.16
C29H2103P1	448.46	448.1228	77.67	4.72	10.70	6.91
C29H2104P1	464.46	464.1177	74.99	4.56	13.78	6.67
C29H2105P1	480.46	480.1127	72.50	4.41	16.65	6.45
C29H2106P1	496.46	496.1076	70.16	4.26	19.34	6.24
C29H2107P1	512.46	512.1025	67.97	4.13	21.85	6.04
C29H2108P1	528.46	528.0974	65.91	4.01	24.22	5.86
C29H2109P1	544.46	544.0923	63.98	3.89	26.45	5.69
C29H2201P2	448.45	448.1146	77.67	4.94	3.57	13.81
C29H2202P2	464.45	464.1095	75.00	4.77	6.89	13.34
C29H2203P2	480.44	480.1044	72.50	4.62	9.99	12.89
C29H2204P2	496.44	496.0993	70.16	4.47	12.89	12.48
C29H2205P2	512.44	512.0942	67.97	4.33	15.61	12.09
C29H2206P2	528.44	528.0892	65.92	4.20	18.17	11.72
C29H2207P2	544.44	544.0841	63.98	4.07	20.57	11.38
C29H2208P2	560.44	560.0790	62.15	3.96	22.84	11.05
C29H2209P2	576.44	576.0739	60.43	3.85	24.98	10.75
C29H2301P1	418.48	418.1486	83.24	5.54	3.82	7.40
C29H2302P1	434.48	434.1436	80.17	5.34	7.36	7.13
C29H2303P1	450.48	450.1385	77.32	5.15	10.65	6.88
C29H2304P1	466.48	466.1334	74.67	4.97	13.72	6.64
C29H2305P1	482.48	482.1283	72.19	4.81	16.58	6.42
C29H2306P1	498.48	498.1232	69.88	4.65	19.26	6.21
C29H2307P1	514.48	514.1181	67.70	4.51	21.77	6.02
C29H2308P1	530.48	530.1130	65.66	4.37	24.13	5.84
C29H2309P1	546.48	546.1080	63.74	4.24	26.35	5.67
C29H2401P2	450.46	450.1302	77.33	5.37	3.55	13.75
C29H2402P2	466.46	466.1251	74.67	5.19	6.86	13.28
C29H2403P2	482.46	482.1201	72.20	5.01	9.95	12.84
C29H2404P2	498.46	498.1150	69.88	4.85	12.84	12.43
C29H2405P2	514.46	514.1099	67.71	4.70	15.55	12.04
C29H2406P2	530.46	530.1048	65.66	4.56	18.10	11.68
C29H2407P2	546.46	546.0997	63.74	4.43	20.49	11.34
C29H2408P2	562.46	562.0946	61.93	4.30	22.76	11.01
C29H2409P2	578.46	578.0895	60.22	4.18	24.89	10.71
C29H2501P1	420.50	420.1643	82.84	5.99	3.80	7.37
C29H2502P1	436.50	436.1592	79.80	5.77	7.33	7.10
C29H2503P1	452.49	452.1541	76.98	5.57	10.61	6.85
C29H2504P1	468.49	468.1490	74.35	5.38	13.66	6.61
C29H2505P1	484.49	484.1439	71.89	5.20	16.51	6.39
C29H2506P1	500.49	500.1389	69.60	5.03	19.18	6.19
C29H2507P1	516.49	516.1338	67.44	4.88	21.68	6.00
C29H2508P1	532.49	532.1287	65.41	4.73	24.04	5.82
C29H2509P1	548.49	548.1236	63.51	4.59	26.25	5.65
C29H2601P2	452.48	452.1459	76.98	5.79	3.54	13.69
C29H2602P2	468.48	468.1408	74.35	5.59	6.83	13.22
C29H2603P2	484.48	484.1357	71.90	5.41	9.91	12.79
C29H2604P2	500.48	500.1306	69.60	5.24	12.79	12.38
C29H2605P2	516.48	516.1255	67.44	5.07	15.49	11.99
C29H2606P2	532.47	532.1205	65.42	4.92	18.03	11.63
C29H2607P2	548.47	548.1154	63.51	4.78	20.42	11.29
C29H2608P2	564.47	564.1103	61.71	4.64	22.68	10.97
C29H2609P2	580.47	580.1052	60.01	4.51	24.81	10.67
C29H2701P1	422.51	422.1799	82.44	6.44	3.79	7.33
C29H2702P1	438.51	438.1749	79.43	6.21	7.30	7.06
C29H2703P1	454.51	454.1698	76.64	5.99	10.56	6.81
C29H2704P1	470.51	470.1647	74.03	5.78	13.60	6.58
C29H2705P1	486.51	486.1596	71.60	5.59	16.44	6.37
C29H2706P1	502.51	502.1545	69.32	5.42	19.10	6.16
C29H2707P1	518.51	518.1494	67.18	5.25	21.60	5.97
C29H2708P1	534.51	534.1443	65.17	5.09	23.95	5.79
C29H2709P1	550.51	550.1393	63.27	4.94	26.16	5.63
C29H2801P2	454.49	454.1615	76.64	6.21	3.52	13.63
C29H2802P2	470.49	470.1564	74.03	6.00	6.80	13.17
C29H2803P2	486.49	486.1514	71.60	5.80	9.87	12.73
C29H2804P2	502.49	502.1463	69.32	5.62	12.74	12.33
C29H2805P2	518.49	518.1412	67.18	5.44	15.43	11.95
C29H2806P2	534.49	534.1361	65.17	5.28	17.96	11.59
C29H2807P2	550.49	550.1310	63.28	5.13	20.34	11.25
C29H2808P2	566.49	566.1259	61.49	4.98	22.59	10.94
C29H2809P2	582.49	582.1208	59.80	4.85	24.72	10.63
C29H2901P1	424.53	424.1956	82.05	6.89	3.77	7.30

C29H2902P1	440.53	440.1905	79.07	6.64	7.26	7.03
C29H2903P1	456.53	456.1854	76.30	6.40	10.51	6.78
C29H2904P1	472.53	472.1803	73.72	6.19	13.54	6.55
C29H2905P1	488.53	488.1752	71.30	5.98	16.38	6.34
C29H2906P1	504.52	504.1702	69.04	5.79	19.03	6.14
C29H2907P1	520.52	520.1651	66.92	5.62	21.52	5.95
C29H2908P1	536.52	536.1600	64.92	5.45	23.86	5.77
C29H2909P1	552.52	552.1549	63.04	5.29	26.06	5.61
C29H3001P2	456.51	456.1772	76.30	6.62	3.50	13.57
C29H3002P2	472.51	472.1721	73.72	6.40	6.77	13.11
C29H3003P2	488.51	488.1670	71.30	6.19	9.83	12.68
C29H3004P2	504.51	504.1619	69.04	5.99	12.69	12.28
C29H3005P2	520.51	520.1568	66.92	5.81	15.37	11.90
C29H3006P2	536.51	536.1518	64.92	5.64	17.89	11.55
C29H3007P2	552.51	552.1467	63.04	5.47	20.27	11.21
C29H3008P2	568.51	568.1416	61.27	5.32	22.51	10.90
C29H3009P2	584.50	584.1365	59.59	5.17	24.64	10.60
C29H3101P1	426.54	426.2112	81.66	7.33	3.75	7.26
C29H3102P1	442.54	442.2062	78.71	7.06	7.23	7.00
C29H3103P1	458.54	458.2011	75.96	6.81	10.47	6.75
C29H3104P1	474.54	474.1960	73.40	6.58	13.49	6.53
C29H3105P1	490.54	490.1909	71.01	6.37	16.31	6.31
C29H3106P1	506.54	506.1858	68.77	6.17	18.95	6.11
C29H3107P1	522.54	522.1807	66.66	5.98	21.43	5.93
C29H3108P1	538.54	538.1756	64.68	5.80	23.77	5.75
C29H3109P1	554.54	554.1706	62.81	5.63	25.97	5.59
C29H3201P2	458.53	458.1928	75.97	7.03	3.49	13.51
C29H3202P2	474.52	474.1877	73.40	6.80	6.74	13.05
C29H3203P2	490.52	490.1827	71.01	6.58	9.79	12.63
C29H3204P2	506.52	506.1776	68.77	6.37	12.63	12.23
C29H3205P2	522.52	522.1725	66.66	6.17	15.31	11.86
C29H3206P2	538.52	538.1674	64.68	5.99	17.83	11.50
C29H3207P2	554.52	554.1623	62.82	5.82	20.20	11.17
C29H3208P2	570.52	570.1572	61.05	5.65	22.43	10.86
C29H3209P2	586.52	586.1521	59.39	5.50	24.55	10.56
C29H3301P1	428.56	428.2269	81.28	7.76	3.73	7.23
C29H3302P1	444.56	444.2218	78.35	7.48	7.20	6.97
C29H3303P1	460.56	460.2167	75.63	7.22	10.42	6.73
C29H3304P1	476.56	476.2116	73.09	6.98	13.43	6.50
C29H3305P1	492.56	492.2065	70.72	6.75	16.24	6.29
C29H3306P1	508.56	508.2015	68.49	6.54	18.88	6.09
C29H3307P1	524.56	524.1964	66.40	6.34	21.35	5.90
C29H3308P1	540.56	540.1913	64.44	6.15	23.68	5.73
C29H3309P1	556.55	556.1862	62.59	5.98	25.87	5.57
C29H3401P2	460.54	460.2085	75.63	7.44	3.47	13.45
C29H3402P2	476.54	476.2034	73.09	7.19	6.71	13.00
C29H3403P2	492.54	492.1983	70.72	6.96	9.75	12.58
C29H3404P2	508.54	508.1932	68.49	6.74	12.58	12.18
C29H3405P2	524.54	524.1881	66.41	6.53	15.25	11.81
C29H3406P2	540.54	540.1830	64.44	6.34	17.76	11.46
C29H3407P2	556.54	556.1780	62.59	6.16	20.12	11.13
C29H3408P2	572.54	572.1729	60.84	5.99	22.36	10.82
C29H3409P2	588.54	588.1678	59.18	5.82	24.47	10.53
C29H3501P1	430.58	430.2425	80.90	8.19	3.72	7.19
C29H3502P1	446.57	446.2375	78.00	7.90	7.17	6.94
C29H3503P1	462.57	462.2324	75.30	7.63	10.38	6.70
C29H3504P1	478.57	478.2273	72.78	7.37	13.37	6.47
C29H3505P1	494.57	494.2222	70.43	7.13	16.17	6.26
C29H3506P1	510.57	510.2171	68.22	6.91	18.80	6.07
C29H3507P1	526.57	526.2120	66.15	6.70	21.27	5.88

C29H3508P1	542.57	542.2069	64.20	6.50	23.59	5.71
C29H3509P1	558.57	558.2019	62.36	6.32	25.78	5.55
C29H3601P2	462.56	462.2241	75.30	7.84	3.46	13.39
C29H3602P2	478.56	478.2190	72.79	7.58	6.69	12.94
C29H3603P2	494.56	494.2140	70.43	7.34	9.71	12.53
C29H3604P2	510.56	510.2089	68.22	7.11	12.53	12.13
C29H3605P2	526.55	526.2038	66.15	6.89	15.19	11.76
C29H3606P2	542.55	542.1987	64.20	6.69	17.69	11.42
C29H3607P2	558.55	558.1936	62.36	6.50	20.05	11.09
C29H3608P2	574.55	574.1885	60.63	6.32	22.28	10.78
C29H3609P2	590.55	590.1834	58.98	6.14	24.38	10.49
C29H3701P1	432.59	432.2582	80.52	8.62	3.70	7.16
C29H3702P1	448.59	448.2531	77.65	8.31	7.13	6.90
C29H3703P1	464.59	464.2480	74.97	8.03	10.33	6.67
C29H3704P1	480.59	480.2429	72.48	7.76	13.32	6.44
C29H3705P1	496.59	496.2378	70.14	7.51	16.11	6.24
C29H3706P1	512.59	512.2328	67.95	7.28	18.73	6.04
C29H3707P1	528.59	528.2277	65.90	7.06	21.19	5.86
C29H3708P1	544.59	544.2226	63.96	6.85	23.50	5.69
C29H3709P1	560.59	560.2175	62.14	6.65	25.69	5.53
C29H3801P2	464.57	464.2398	74.98	8.24	3.44	13.33
C29H3802P2	480.57	480.2347	72.48	7.97	6.66	12.89
C29H3803P2	496.57	496.2296	70.15	7.71	9.67	12.48
C29H3804P2	512.57	512.2245	67.96	7.47	12.49	12.09
C29H3805P2	528.57	528.2194	65.90	7.25	15.13	11.72
C29H3806P2	544.57	544.2143	63.96	7.03	17.63	11.38
C29H3807P2	560.57	560.2093	62.14	6.83	19.98	11.05
C29H3808P2	576.57	576.2042	60.41	6.64	22.20	10.74
C29H3809P2	592.57	592.1991	58.78	6.46	24.30	10.45
C29H3901P1	434.61	434.2738	80.15	9.05	3.68	7.13
C29H3902P1	450.61	450.2688	77.30	8.72	7.10	6.87
C29H3903P1	466.61	466.2637	74.65	8.42	10.29	6.64
C29H3904P1	482.61	482.2586	72.18	8.15	13.26	6.42
C29H3905P1	498.60	498.2535	69.86	7.88	16.04	6.21
C29H3906P1	514.60	514.2484	67.69	7.64	18.65	6.02
C29H3907P1	530.60	530.2433	65.65	7.41	21.11	5.84
C29H3908P1	546.60	546.2382	63.73	7.19	23.42	5.67
C29H3909P1	562.60	562.2332	61.91	6.99	25.59	5.51
C29H4001P2	466.59	466.2554	74.65	8.64	3.43	13.28
C29H4002P2	482.59	482.2503	72.18	8.35	6.63	12.84
C29H4003P2	498.59	498.2453	69.86	8.09	9.63	12.42
C29H4004P2	514.59	514.2402	67.69	7.84	12.44	12.04
C29H4005P2	530.59	530.2351	65.65	7.60	15.08	11.68
C29H4006P2	546.59	546.2300	63.73	7.38	17.56	11.33
C29H4007P2	562.59	562.2249	61.91	7.17	19.91	11.01
C29H4008P2	578.58	578.2198	60.20	6.97	22.12	10.71
C29H4009P2	594.58	594.2147	58.58	6.78	24.22	10.42
C29H4101P1	436.62	436.2895	79.78	9.47	3.66	7.09
C29H4102P1	452.62	452.2844	76.96	9.13	7.07	6.84
C29H4103P1	468.62	468.2793	74.33	8.82	10.24	6.61
C29H4104P1	484.62	484.2742	71.88	8.53	13.21	6.39
C29H4105P1	500.62	500.2691	69.58	8.26	15.98	6.19
C29H4106P1	516.62	516.2641	67.42	8.00	18.58	6.00
C29H4107P1	532.62	532.2590	65.40	7.76	21.03	5.82
C29H4108P1	548.62	548.2539	63.49	7.53	23.33	5.65
C29H4109P1	564.62	564.2488	61.69	7.32	25.50	5.49
C29H4201P2	468.61	468.2711	74.33	9.03	3.41	13.22
C29H4202P2	484.60	484.2660	71.88	8.74	6.60	12.78
C29H4203P2	500.60	500.2609	69.58	8.46	9.59	12.37
C29H4204P2	516.60	516.2558	67.43	8.19	12.39	11.99

161

C29H4205P2	532.60	532.2507	65.40	7.95	15.02	11.63
C29H4206P2	548.60	548.2456	63.49	7.72	17.50	11.29
C29H4207P2	564.60	564.2406	61.69	7.50	19.84	10.97
C29H4208P2	580.60	580.2355	59.99	7.29	22.05	10.67
C29H4209P2	596.60	596.2304	58.38	7.10	24.14	10.38
C29H4301P1	438.64	438.3051	79.41	9.88	3.65	7.06
C29H4302P1	454.64	454.3001	76.62	9.53	7.04	6.81
C29H4303P1	470.64	470.2950	74.01	9.21	10.20	6.58
C29H4304P1	486.64	486.2899	71.58	8.91	13.15	6.36
C29H4305P1	502.64	502.2848	69.30	8.62	15.92	6.16
C29H4306P1	518.64	518.2797	67.16	8.36	18.51	5.97
C29H4307P1	534.64	534.2746	65.15	8.11	20.95	5.79
C29H4308P1	550.64	550.2695	63.26	7.87	23.25	5.63
C29H4309P1	566.63	566.2644	61.47	7.65	25.41	5.47
C29H4401P2	470.62	470.2867	74.01	9.42	3.40	13.16
C29H4402P2	486.62	486.2816	71.58	9.11	6.58	12.73
C29H4403P2	502.62	502.2766	69.30	8.82	9.55	12.32
C29H4404P2	518.62	518.2715	67.16	8.55	12.34	11.94
C29H4405P2	534.62	534.2664	65.15	8.30	14.96	11.59
C29H4406P2	550.62	550.2613	63.26	8.05	17.43	11.25
C29H4407P2	566.62	566.2562	61.47	7.83	19.77	10.93
C29H4408P2	582.62	582.2511	59.79	7.61	21.97	10.63
C29H4409P2	598.62	598.2460	58.19	7.41	24.05	10.35
C29H4501P1	440.66	440.3208	79.05	10.29	3.63	7.03
C29H4502P1	456.65	456.3157	76.28	9.93	7.01	6.78
C29H4503P1	472.65	472.3106	73.70	9.60	10.16	6.55
C29H4504P1	488.65	488.3055	71.28	9.28	13.10	6.34
C29H4505P1	504.65	504.3004	69.02	8.99	15.85	6.14
C29H4506P1	520.65	520.2954	66.90	8.71	18.44	5.95
C29H4507P1	536.65	536.2903	64.91	8.45	20.87	5.77
C29H4508P1	552.65	552.2852	63.03	8.21	23.16	5.60
C29H4509P1	568.65	568.2801	61.25	7.98	25.32	5.45
C29H4601P2	472.64	472.3024	73.70	9.81	3.39	13.11
C29H4602P2	488.64	488.2973	71.28	9.49	6.55	12.68
C29H4603P2	504.64	504.2922	69.02	9.19	9.51	12.28
C29H4604P2	520.64	520.2871	66.90	8.91	12.29	11.90
C29H4605P2	536.63	536.2820	64.91	8.64	14.91	11.54
C29H4606P2	552.63	552.2769	63.03	8.39	17.37	11.21
C29H4607P2	568.63	568.2719	61.26	8.15	19.70	10.89
C29H4608P2	584.63	584.2668	59.58	7.93	21.89	10.60
C29H4609P2	600.63	600.2617	57.99	7.72	23.97	10.31
C29H4701P1	442.67	442.3364	78.69	10.70	3.61	7.00
C29H4702P1	458.67	458.3313	75.94	10.33	6.98	6.75
C29H4703P1	474.67	474.3263	73.38	9.98	10.11	6.53
C29H4704P1	490.67	490.3212	70.99	9.66	13.04	6.31
C29H4705P1	506.67	506.3161	68.75	9.35	15.79	6.11
C29H4706P1	522.67	522.3110	66.64	9.06	18.37	5.93
C29H4707P1	538.67	538.3059	64.66	8.79	20.79	5.75
C29H4708P1	554.67	554.3008	62.80	8.54	23.08	5.58
C29H4709P1	570.67	570.2957	61.04	8.30	25.23	5.43
C29H4801P2	474.65	474.3180	73.38	10.19	3.37	13.05
C29H4802P2	490.65	490.3129	70.99	9.86	6.52	12.63
C29H4803P2	506.65	506.3079	68.75	9.55	9.47	12.23
C29H4804P2	522.65	522.3028	66.65	9.26	12.24	11.85
C29H4805P2	538.65	538.2977	64.67	8.98	14.85	11.50
C29H4806P2	554.65	554.2926	62.80	8.72	17.31	11.17
C29H4807P2	570.65	570.2875	61.04	8.48	19.63	10.86
C29H4808P2	586.65	586.2824	59.38	8.25	21.82	10.56
C29H4809P2	602.65	602.2773	57.80	8.03	23.89	10.28
C29H4901P1	444.69	444.3521	78.33	11.11	3.60	6.97

C29H4902P1	460.69	460.3470	75.61	10.72	6.95	6.72
C29H4903P1	476.69	476.3419	73.07	10.36	10.07	6.50
C29H4904P1	492.69	492.3368	70.70	10.02	12.99	6.29
C29H4905P1	508.68	508.3317	68.48	9.71	15.73	6.09
C29H4906P1	524.68	524.3267	66.39	9.41	18.30	5.90
C29H4907P1	540.68	540.3216	64.42	9.13	20.71	5.73
C29H4908P1	556.68	556.3165	62.57	8.87	22.99	5.56
C29H4909P1	572.68	572.3114	60.82	8.62	25.14	5.41
C29H5001P2	476.67	476.3337	73.07	10.57	3.36	13.00
C29H5002P2	492.67	492.3286	70.70	10.23	6.49	12.57
C29H5003P2	508.67	508.3235	68.48	9.91	9.44	12.18
C29H5004P2	524.67	524.3184	66.39	9.61	12.20	11.81
C29H5005P2	540.67	540.3133	64.42	9.32	14.80	11.46
C29H5006P2	556.67	556.3082	62.57	9.05	17.24	11.13
C29H5007P2	572.67	572.3032	60.82	8.80	19.56	10.82
C29H5008P2	588.66	588.2981	59.17	8.56	21.74	10.52
C29H5009P2	604.66	604.2930	57.61	8.33	23.81	10.24
C29H5101P1	446.70	446.3677	77.98	11.51	3.58	6.93
C29H5102P1	462.70	462.3626	75.28	11.11	6.92	6.69
C29H5103P1	478.70	478.3576	72.76	10.74	10.03	6.47
C29H5104P1	494.70	494.3525	70.41	10.39	12.94	6.26
C29H5105P1	510.70	510.3474	68.20	10.07	15.66	6.06
C29H5106P1	526.70	526.3423	66.13	9.76	18.23	5.88
C29H5107P1	542.70	542.3372	64.18	9.47	20.64	5.71
C29H5108P1	558.70	558.3321	62.35	9.20	22.91	5.54
C29H5109P1	574.70	574.3270	60.61	8.94	25.06	5.39
C29H5201P2	478.68	478.3493	72.77	10.95	3.34	12.94
C29H5202P2	494.68	494.3442	70.41	10.60	6.47	12.52
C29H5203P2	510.68	510.3391	68.21	10.26	9.40	12.13
C29H5204P2	526.68	526.3341	66.14	9.95	12.15	11.76
C29H5205P2	542.68	542.3290	64.19	9.66	14.74	11.42
C29H5206P2	558.68	558.3239	62.35	9.38	17.18	11.09
C29H5207P2	574.68	574.3188	60.61	9.12	19.49	10.78
C29H5208P2	590.68	590.3137	58.97	8.87	21.67	10.49
C29H5209P2	606.68	606.3086	57.41	8.64	23.73	10.21
C29H5301P1	448.72	448.3834	77.63	11.91	3.57	6.90
C29H5302P1	464.72	464.3783	74.95	11.50	6.89	6.67
C29H5303P1	480.72	480.3732	72.46	11.11	9.98	6.44
C29H5304P1	496.72	496.3681	70.13	10.76	12.88	6.24
C29H5305P1	512.72	512.3630	67.94	10.42	15.60	6.04
C29H5306P1	528.72	528.3580	65.88	10.10	18.16	5.86
C29H5307P1	544.72	544.3529	63.95	9.81	20.56	5.69
C29H5308P1	560.71	560.3478	62.12	9.53	22.83	5.52
C29H5309P1	576.71	576.3427	60.40	9.26	24.97	5.37
C29H5401P2	480.70	480.3650	72.46	11.32	3.33	12.89
C29H5402P2	496.70	496.3599	70.13	10.96	6.44	12.47
C29H5403P2	512.70	512.3548	67.94	10.62	9.36	12.08
C29H5404P2	528.70	528.3497	65.88	10.30	12.10	11.72
C29H5405P2	544.70	544.3446	63.95	9.99	14.69	11.37
C29H5406P2	560.70	560.3395	62.12	9.71	17.12	11.05
C29H5407P2	576.70	576.3345	60.40	9.44	19.42	10.74
C29H5408P2	592.70	592.3294	58.77	9.18	21.60	10.45
C29H5409P2	608.70	608.3243	57.22	8.94	23.66	10.18
C29H5501P1	450.73	450.3990	77.28	12.30	3.55	6.87
C29H5502P1	466.73	466.3939	74.63	11.88	6.86	6.64
C29H5503P1	482.73	482.3889	72.16	11.48	9.94	6.42
C29H5504P1	498.73	498.3838	69.84	11.12	12.83	6.21
C29H5505P1	514.73	514.3787	67.67	10.77	15.54	6.02
C29H5506P1	530.73	530.3736	65.63	10.45	18.09	5.84
C29H5507P1	546.73	546.3685	63.71	10.14	20.48	5.67

C29H5508P1	562.73	562.3634	61.90	9.85	22.75	5.50
C29H5509P1	578.73	578.3583	60.19	9.58	24.88	5.35
C29H5601P2	482.72	482.3806	72.16	11.69	3.31	12.83
C29H5602P2	498.72	498.3755	69.84	11.32	6.42	12.42
C29H5603P2	514.72	514.3704	67.67	10.97	9.33	12.04
C29H5604P2	530.71	530.3654	65.63	10.64	12.06	11.67
C29H5605P2	546.71	546.3603	63.71	10.32	14.63	11.33
C29H5606P2	562.71	562.3552	61.90	10.03	17.06	11.01
C29H5607P2	578.71	578.3501	60.19	9.75	19.35	10.70
C29H5608P2	594.71	594.3450	58.57	9.49	21.52	10.42
C29H5609P2	610.71	610.3399	57.04	9.24	23.58	10.14
C29H5701P1	452.75	452.4147	76.93	12.69	3.53	6.84
C29H5702P1	468.75	468.4096	74.31	12.26	6.83	6.61
C29H5703P1	484.75	484.4045	71.86	11.85	9.90	6.39
C29H5704P1	500.75	500.3994	69.56	11.47	12.78	6.19
C29H5705P1	516.75	516.3943	67.41	11.12	15.48	5.99
C29H5706P1	532.75	532.3893	65.38	10.78	18.02	5.81
C29H5707P1	548.75	548.3842	63.48	10.47	20.41	5.64
C29H5708P1	564.75	564.3791	61.68	10.17	22.66	5.48
C29H5709P1	580.75	580.3740	59.98	9.89	24.79	5.33
C29H5801P2	484.73	484.3963	71.86	12.06	3.30	12.78
C29H5802P2	500.73	500.3912	69.56	11.68	6.39	12.37
C29H5803P2	516.73	516.3861	67.41	11.31	9.29	11.99
C29H5804P2	532.73	532.3810	65.38	10.97	12.01	11.63
C29H5805P2	548.73	548.3759	63.48	10.65	14.58	11.29
C29H5806P2	564.73	564.3708	61.68	10.35	17.00	10.97
C29H5807P2	580.73	580.3658	59.98	10.07	19.29	10.67
C29H5808P2	596.73	596.3607	58.37	9.80	21.45	10.38
C29H5809P2	612.73	612.3556	56.85	9.54	23.50	10.11
C29H5901P1	454.77	454.4303	76.59	13.08	3.52	6.81
C29H5902P1	470.77	470.4252	73.99	12.63	6.80	6.58
C29H5903P1	486.77	486.4202	71.56	12.22	9.86	6.36
C29H5904P1	502.76	502.4151	69.28	11.83	12.73	6.16
C29H5905P1	518.76	518.4100	67.14	11.46	15.42	5.97
C29H5906P1	534.76	534.4049	65.14	11.12	17.95	5.79
C29H5907P1	550.76	550.3998	63.24	10.80	20.33	5.62
C29H5908P1	566.76	566.3947	61.46	10.49	22.58	5.47
C29H5909P1	582.76	582.3896	59.77	10.20	24.71	5.32
C29H6001P2	486.75	486.4119	71.56	12.42	3.29	12.73
C29H6002P2	502.75	502.4068	69.28	12.03	6.36	12.32
C29H6003P2	518.75	518.4017	67.15	11.66	9.25	11.94
C29H6004P2	534.75	534.3967	65.14	11.31	11.97	11.58
C29H6005P2	550.75	550.3916	63.25	10.98	14.53	11.25
C29H6006P2	566.75	566.3865	61.46	10.67	16.94	10.93
C29H6007P2	582.74	582.3814	59.77	10.38	19.22	10.63
C29H6008P2	598.74	598.3763	58.18	10.10	21.38	10.35
C29H6009P2	614.74	614.3712	56.66	9.84	23.42	10.08
C29H6101P1	456.78	456.4460	76.26	13.46	3.50	6.78
C29H6102P1	472.78	472.4409	73.68	13.01	6.77	6.55
C29H6103P1	488.78	488.4358	71.26	12.58	9.82	6.34
C29H6104P1	504.78	504.4307	69.00	12.18	12.68	6.14
C29H6105P1	520.78	520.4256	66.88	11.81	15.36	5.95
C29H6106P1	536.78	536.4205	64.89	11.45	17.88	5.77
C29H6107P1	552.78	552.4155	63.01	11.12	20.26	5.60
C29H6108P1	568.78	568.4104	61.24	10.81	22.50	5.45
C29H6109P1	584.78	584.4053	59.57	10.51	24.62	5.30
C29H6201P2	488.76	488.4276	71.27	12.79	3.27	12.67
C29H6202P2	504.76	504.4225	69.01	12.38	6.34	12.27
C29H6203P2	520.76	520.4174	66.89	12.00	9.22	11.90
C29H6204P2	536.76	536.4123	64.89	11.64	11.92	11.54
C29H6205P2	552.76	552.4072	63.02	11.31	14.47	11.21
C29H6206P2	568.76	568.4021	61.24	10.99	16.88	10.89
C29H6207P2	584.76	584.3971	59.57	10.69	19.15	10.59
C29H6208P2	600.76	600.3920	57.98	10.40	21.31	10.31
C29H6209P2	616.76	616.3869	56.48	10.13	23.35	10.04
C30H1501P1	422.43	422.0860	85.30	3.58	3.79	7.33
C30H1502P1	438.43	438.0810	82.19	3.45	7.30	7.06
C30H1503P1	454.43	454.0759	79.29	3.33	10.56	6.82
C30H1504P1	470.43	470.0708	76.60	3.21	13.60	6.58
C30H1505P1	486.42	486.0657	74.08	3.11	16.45	6.37
C30H1506P1	502.42	502.0606	71.72	3.01	19.11	6.16
C30H1507P1	518.42	518.0555	69.51	2.92	21.60	5.97
C30H1508P1	534.42	534.0504	67.42	2.83	23.95	5.80
C30H1509P1	550.42	550.0454	65.47	2.75	26.16	5.63
C30H1601P2	454.41	454.0676	79.30	3.55	3.52	13.63
C30H1602P2	470.41	470.0625	76.60	3.43	6.80	13.17
C30H1603P2	486.41	486.0575	74.08	3.32	9.87	12.74
C30H1604P2	502.41	502.0524	71.72	3.21	12.74	12.33
C30H1605P2	518.41	518.0473	69.51	3.11	15.43	11.95
C30H1606P2	534.41	534.0422	67.43	3.02	17.96	11.59
C30H1607P2	550.41	550.0371	65.47	2.93	20.35	11.25
C30H1608P2	566.40	566.0320	63.62	2.85	22.60	10.94
C30H1609P2	582.40	582.0269	61.87	2.77	24.72	10.64
C30H1701P1	424.44	424.1017	84.90	4.04	3.77	7.30
C30H1702P1	440.44	440.0966	81.81	3.89	7.27	7.03
C30H1703P1	456.44	456.0915	78.94	3.75	10.52	6.79
C30H1704P1	472.44	472.0864	76.27	3.63	13.55	6.56
C30H1705P1	488.44	488.0814	73.77	3.51	16.38	6.34
C30H1706P1	504.44	504.0763	71.43	3.40	19.03	6.14
C30H1707P1	520.44	520.0712	69.24	3.29	21.52	5.95
C30H1708P1	536.44	536.0661	67.17	3.19	23.86	5.77
C30H1709P1	552.44	552.0610	65.23	3.10	26.07	5.61
C30H1801P2	456.42	456.0833	78.95	3.98	3.51	13.57
C30H1802P2	472.42	472.0782	76.27	3.84	6.77	13.11
C30H1803P2	488.42	488.0731	73.77	3.71	9.83	12.68
C30H1804P2	504.42	504.0680	71.43	3.60	12.69	12.28
C30H1805P2	520.42	520.0629	69.24	3.49	15.37	11.90
C30H1806P2	536.42	536.0579	67.17	3.38	17.90	11.55
C30H1807P2	552.42	552.0528	65.23	3.28	20.27	11.21
C30H1808P2	568.42	568.0477	63.39	3.19	22.52	10.90
C30H1809P2	584.42	584.0426	61.66	3.10	24.64	10.60
C30H1901P1	426.46	426.1173	84.49	4.49	3.75	7.26
C30H1902P1	442.46	442.1123	81.44	4.33	7.23	7.00
C30H1903P1	458.46	458.1072	78.60	4.18	10.47	6.76
C30H1904P1	474.46	474.1021	75.95	4.04	13.49	6.53
C30H1905P1	490.46	490.0970	73.47	3.90	16.31	6.32
C30H1906P1	506.46	506.0919	71.15	3.78	18.95	6.12
C30H1907P1	522.46	522.0868	68.97	3.67	21.44	5.93
C30H1908P1	538.45	538.0817	66.92	3.56	23.77	5.75
C30H1909P1	554.45	554.0767	64.99	3.45	25.97	5.59
C30H2001P1	458.44	458.0989	78.60	4.40	3.49	13.51
C30H2002P2	474.44	474.0938	75.95	4.25	6.74	13.06
C30H2003P2	490.44	490.0888	73.47	4.11	9.79	12.63
C30H2004P2	506.44	506.0837	71.15	3.98	12.64	12.23
C30H2005P2	522.44	522.0786	68.97	3.86	15.31	11.86
C30H2006P2	538.44	538.0735	66.92	3.74	17.83	11.51
C30H2007P2	554.44	554.0684	64.99	3.64	20.20	11.17
C30H2008P2	570.44	570.0633	63.17	3.53	22.44	10.86
C30H2009P2	586.44	586.0582	61.44	3.44	24.55	10.56
C30H2101P1	428.48	428.1330	84.10	4.94	3.73	7.23

C30H2102P1	444.47	444.1279	81.07	4.76	7.20	6.97
C30H2103P1	460.47	460.1228	78.25	4.60	10.42	6.73
C30H2104P1	476.47	476.1177	75.63	4.44	13.43	6.50
C30H2105P1	492.47	492.1127	73.17	4.30	16.24	6.29
C30H2106P1	508.47	508.1076	70.87	4.16	18.88	6.09
C30H2107P1	524.47	524.1025	68.70	4.04	21.35	5.91
C30H2108P1	540.47	540.0974	66.67	3.92	23.68	5.73
C30H2109P1	556.47	556.0923	64.75	3.80	25.88	5.57
C30H2201P2	460.46	460.1146	78.26	4.82	3.47	13.45
C30H2202P2	476.46	476.1095	75.63	4.65	6.72	13.00
C30H2203P2	492.46	492.1044	73.17	4.50	9.75	12.58
C30H2204P2	508.46	508.0993	70.87	4.36	12.59	12.18
C30H2205P2	524.45	524.0942	68.71	4.23	15.25	11.81
C30H2206P2	540.45	540.0892	66.67	4.10	17.76	11.46
C30H2207P2	556.45	556.0841	64.76	3.99	20.13	11.13
C30H2208P2	572.45	572.0790	62.95	3.87	22.36	10.82
C30H2209P2	588.45	588.0739	61.23	3.77	24.47	10.53
C30H2301P1	430.49	430.1486	83.70	5.39	3.72	7.19
C30H2302P1	446.49	446.1436	80.70	5.19	7.17	6.94
C30H2303P1	462.49	462.1385	77.91	5.01	10.38	6.70
C30H2304P1	478.49	478.1334	75.31	4.85	13.37	6.47
C30H2305P1	494.49	494.1283	72.87	4.69	16.18	6.26
C30H2306P1	510.49	510.1232	70.59	4.54	18.80	6.07
C30H2307P1	526.49	526.1181	68.44	4.40	21.27	5.88
C30H2308P1	542.49	542.1130	66.42	4.27	23.59	5.71
C30H2309P1	558.49	558.1080	64.52	4.15	25.78	5.55
C30H2401P2	462.47	462.1302	77.91	5.23	3.46	13.39
C30H2402P2	478.47	478.1251	75.31	5.06	6.69	12.95
C30H2403P2	494.47	494.1201	72.87	4.89	9.71	12.53
C30H2404P2	510.47	510.1150	70.59	4.74	12.54	12.14
C30H2405P2	526.47	526.1099	68.44	4.59	15.19	11.77
C30H2406P2	542.47	542.1048	66.42	4.46	17.70	11.42
C30H2407P2	558.47	558.0997	64.52	4.33	20.05	11.09
C30H2408P2	574.47	574.0946	62.72	4.21	22.28	10.78
C30H2409P2	590.47	590.0895	61.03	4.10	24.39	10.49
C30H2501P1	432.51	432.1643	83.31	5.83	3.70	7.16
C30H2502P1	448.51	448.1592	80.34	5.62	7.13	6.91
C30H2503P1	464.51	464.1541	77.57	5.42	10.33	6.67
C30H2504P1	480.51	480.1490	74.99	5.24	13.32	6.45
C30H2505P1	496.50	496.1439	72.57	5.08	16.11	6.24
C30H2506P1	512.50	512.1389	70.31	4.92	18.73	6.04
C30H2507P1	528.50	528.1338	68.18	4.77	21.19	5.86
C30H2508P1	544.50	544.1287	66.18	4.63	23.51	5.69
C30H2509P1	560.50	560.1236	64.29	4.50	25.69	5.53
C30H2601P2	464.49	464.1459	77.58	5.64	3.44	13.34
C30H2602P2	480.49	480.1408	74.99	5.45	6.66	12.89
C30H2603P2	496.49	496.1357	72.58	5.28	9.67	12.48
C30H2604P2	512.49	512.1306	70.31	5.11	12.49	12.09
C30H2605P2	528.49	528.1255	68.18	4.96	15.14	11.72
C30H2606P2	544.49	544.1205	66.18	4.81	17.63	11.38
C30H2607P2	560.49	560.1154	64.29	4.68	19.98	11.05
C30H2608P2	576.48	576.1103	62.51	4.55	22.20	10.75
C30H2609P2	592.48	592.1052	60.82	4.42	24.30	10.46
C30H2701P1	434.52	434.1799	82.93	6.26	3.68	7.13
C30H2702P1	450.52	450.1749	79.98	6.04	7.10	6.88
C30H2703P1	466.52	466.1698	77.24	5.83	10.29	6.64
C30H2704P1	482.52	482.1647	74.68	5.64	13.26	6.42
C30H2705P1	498.52	498.1596	72.28	5.46	16.05	6.21
C30H2706P1	514.52	514.1545	70.03	5.29	18.66	6.02
C30H2707P1	530.52	530.1494	67.92	5.13	21.11	5.84
C30H2708P1	546.52	546.1443	65.93	4.98	23.42	5.67
C30H2709P1	562.52	562.1393	64.06	4.84	25.60	5.51
C30H2801P2	466.50	466.1615	77.24	6.05	3.43	13.28
C30H2802P2	482.50	482.1564	74.68	5.85	6.63	12.84
C30H2803P2	498.50	498.1514	72.28	5.66	9.63	12.43
C30H2804P2	514.50	514.1463	70.04	5.49	12.44	12.04
C30H2805P2	530.50	530.1412	67.92	5.32	15.08	11.68
C30H2806P2	546.50	546.1361	65.93	5.16	17.57	11.34
C30H2807P2	562.50	562.1310	64.06	5.02	19.91	11.01
C30H2808P2	578.50	578.1259	62.29	4.88	22.13	10.71
C30H2809P2	594.50	594.1208	60.61	4.75	24.22	10.42
C30H2901P1	436.54	436.1956	82.54	6.70	3.67	7.10
C30H2902P1	452.54	452.1905	79.63	6.46	7.07	6.84
C30H2903P1	468.54	468.1854	76.91	6.24	10.24	6.61
C30H2904P1	484.54	484.1803	74.37	6.03	13.21	6.39
C30H2905P1	500.54	500.1752	71.99	5.84	15.98	6.19
C30H2906P1	516.54	516.1702	69.76	5.66	18.58	6.00
C30H2907P1	532.54	532.1651	67.66	5.49	21.03	5.82
C30H2908P1	548.53	548.1600	65.69	5.33	23.33	5.65
C30H2909P1	564.53	564.1549	63.83	5.18	25.51	5.49
C30H3001P2	468.52	468.1772	76.91	6.45	3.41	13.22
C30H3002P2	484.52	484.1721	74.37	6.24	6.60	12.79
C30H3003P2	500.52	500.1670	71.99	6.04	9.59	12.38
C30H3004P2	516.52	516.1619	69.76	5.85	12.39	11.99
C30H3005P2	532.52	532.1568	67.67	5.68	15.02	11.63
C30H3006P2	548.52	548.1518	65.69	5.51	17.50	11.29
C30H3007P2	564.52	564.1467	63.83	5.36	19.84	10.97
C30H3008P2	580.52	580.1416	62.07	5.21	22.05	10.67
C30H3009P2	596.52	596.1365	60.41	5.07	24.14	10.38
C30H3101P1	438.55	438.2112	82.16	7.13	3.65	7.06
C30H3102P1	454.55	454.2062	79.27	6.87	7.04	6.81
C30H3103P1	470.55	470.2011	76.58	6.64	10.20	6.58
C30H3104P1	486.55	486.1960	74.06	6.42	13.15	6.37
C30H3105P1	502.55	502.1909	71.70	6.22	15.92	6.16
C30H3106P1	518.55	518.1858	69.49	6.03	18.51	5.97
C30H3107P1	534.55	534.1807	67.41	5.85	20.95	5.79
C30H3108P1	550.55	550.1756	65.45	5.68	23.25	5.63
C30H3109P1	566.55	566.1706	63.60	5.52	25.42	5.47
C30H3201P2	470.54	470.1928	76.58	6.85	3.40	13.17
C30H3202P2	486.54	486.1877	74.06	6.63	6.58	12.73
C30H3203P2	502.54	502.1827	71.70	6.42	9.55	12.33
C30H3204P2	518.53	518.1776	69.49	6.22	12.34	11.95
C30H3205P2	534.53	534.1725	67.41	6.03	14.97	11.59
C30H3206P2	550.53	550.1674	65.45	5.86	17.44	11.25
C30H3207P2	566.53	566.1623	63.60	5.69	19.77	10.93
C30H3208P2	582.53	582.1572	61.86	5.54	21.97	10.63
C30H3209P2	598.53	598.1521	60.20	5.39	24.06	10.35
C30H3301P1	440.57	440.2269	81.79	7.55	3.63	7.03
C30H3302P1	456.57	456.2218	78.92	7.29	7.01	6.78
C30H3303P1	472.57	472.2167	76.25	7.04	10.16	6.55
C30H3304P1	488.57	488.2116	73.75	6.81	13.10	6.34
C30H3305P1	504.57	504.2065	71.41	6.59	15.85	6.14
C30H3306P1	520.57	520.2015	69.22	6.39	18.44	5.95
C30H3307P1	536.57	536.1964	67.16	6.20	20.87	5.77
C30H3308P1	552.57	552.1913	65.21	6.02	23.16	5.61
C30H3309P1	568.57	568.1862	63.38	5.85	25.33	5.45
C30H3401P2	472.55	472.2085	76.25	7.25	3.39	13.11
C30H3402P2	488.55	488.2034	73.76	7.01	6.55	12.68
C30H3403P2	504.55	504.1983	71.42	6.79	9.51	12.28
C30H3404P2	520.55	520.1932	69.22	6.58	12.29	11.90

C30H3405P2	536.55	536.1881	67.16	6.39	14.91	11.55
C30H3406P2	552.55	552.1830	65.21	6.20	17.37	11.21
C30H3407P2	568.55	568.1780	63.38	6.03	19.70	10.90
C30H3408P2	584.55	584.1729	61.64	5.86	21.90	10.60
C30H3409P2	600.55	600.1678	60.00	5.71	23.98	10.32
C30H3501P1	442.59	442.2425	81.42	7.97	3.61	7.00
C30H3502P1	458.59	458.2375	78.58	7.69	6.98	6.75
C30H3503P1	474.59	474.2324	75.93	7.43	10.11	6.53
C30H3504P1	490.58	490.2273	73.45	7.19	13.05	6.31
C30H3505P1	506.58	506.2222	71.13	6.96	15.79	6.11
C30H3506P1	522.58	522.2171	68.95	6.75	18.37	5.93
C30H3507P1	538.58	538.2120	66.90	6.55	20.79	5.75
C30H3508P1	554.58	554.2069	64.97	6.36	23.08	5.59
C30H3509P1	570.58	570.2019	63.15	6.18	25.24	5.43
C30H3601P2	474.57	474.2241	75.93	7.65	3.37	13.05
C30H3602P2	490.57	490.2190	73.45	7.40	6.52	12.63
C30H3603P2	506.57	506.2140	71.13	7.16	9.48	12.23
C30H3604P2	522.57	522.2089	68.95	6.94	12.25	11.85
C30H3605P2	538.57	538.2038	66.91	6.74	14.85	11.50
C30H3606P2	554.57	554.1987	64.98	6.54	17.31	11.17
C30H3607P2	570.56	570.1936	63.15	6.36	19.63	10.86
C30H3608P2	586.56	586.1885	61.43	6.19	21.82	10.56
C30H3609P2	602.56	602.1834	59.80	6.02	23.90	10.28
C30H3701P1	444.60	444.2582	81.05	8.39	3.60	6.97
C30H3702P1	460.60	460.2531	78.23	8.10	6.95	6.72
C30H3703P1	476.60	476.2480	75.61	7.83	10.07	6.50
C30H3704P1	492.60	492.2429	73.15	7.57	12.99	6.29
C30H3705P1	508.60	508.2378	70.85	7.33	15.73	6.09
C30H3706P1	524.60	524.2328	68.69	7.11	18.30	5.90
C30H3707P1	540.60	540.2277	66.65	6.90	20.72	5.73
C30H3708P1	556.60	556.2226	64.74	6.70	23.00	5.56
C30H3709P1	572.60	572.2175	62.93	6.51	25.15	5.41
C30H3801P2	476.58	476.2398	75.61	8.04	3.36	13.00
C30H3802P2	492.58	492.2347	73.15	7.78	6.50	12.58
C30H3803P2	508.58	508.2296	70.85	7.53	9.44	12.18
C30H3804P2	524.58	524.2245	68.69	7.30	12.20	11.81
C30H3805P2	540.58	540.2194	66.66	7.09	14.80	11.46
C30H3806P2	556.58	556.2143	64.74	6.88	17.25	11.13
C30H3807P2	572.58	572.2093	62.93	6.69	19.56	10.82
C30H3808P2	588.58	588.2042	61.22	6.51	21.75	10.52
C30H3809P2	604.58	604.1991	59.60	6.34	23.82	10.25
C30H3901P1	446.62	446.2738	80.68	8.80	3.58	6.94
C30H3902P1	462.62	462.2688	77.89	8.50	6.92	6.70
C30H3903P1	478.62	478.2637	75.29	8.21	10.03	6.47
C30H3904P1	494.62	494.2586	72.85	7.95	12.94	6.26
C30H3905P1	510.62	510.2535	70.57	7.70	15.67	6.07
C30H3906P1	526.62	526.2484	68.42	7.46	18.23	5.88
C30H3907P1	542.61	542.2433	66.41	7.24	20.64	5.71
C30H3908P1	558.61	558.2382	64.51	7.04	22.91	5.54
C30H3909P1	574.61	574.2332	62.71	6.84	25.06	5.39
C30H4001P2	478.60	478.2554	75.29	8.42	3.34	12.94
C30H4002P2	494.60	494.2503	72.85	8.15	6.47	12.52
C30H4003P2	510.60	510.2453	70.57	7.90	9.40	12.13
C30H4004P2	526.60	526.2402	68.43	7.66	12.15	11.76
C30H4005P2	542.60	542.2351	66.41	7.43	14.74	11.42
C30H4006P2	558.60	558.2300	64.51	7.22	17.19	11.09
C30H4007P2	574.60	574.2249	62.71	7.02	19.49	10.78
C30H4008P2	590.60	590.2198	61.01	6.83	21.67	10.49
C30H4009P2	606.60	606.2147	59.40	6.65	23.74	10.21
C30H4101P1	448.63	448.2895	80.32	9.21	3.57	6.90
C30H4102P1	464.63	464.2844	77.55	8.89	6.89	6.67
C30H4103P1	480.63	480.2793	74.97	8.60	9.99	6.44
C30H4104P1	496.63	496.2742	72.56	8.32	12.89	6.24
C30H4105P1	512.63	512.2691	70.29	8.06	15.61	6.04
C30H4106P1	528.63	528.2641	68.16	7.82	18.16	5.86
C30H4107P1	544.63	544.2590	66.16	7.59	20.56	5.69
C30H4108P1	560.63	560.2539	64.27	7.37	22.83	5.52
C30H4109P1	576.63	576.2488	62.49	7.17	24.97	5.37
C30H4201P2	480.62	480.2711	74.97	8.81	3.33	12.89
C30H4202P2	496.62	496.2660	72.56	8.52	6.44	12.47
C30H4203P2	512.62	512.2609	70.29	8.26	9.36	12.08
C30H4204P2	528.61	528.2558	68.17	8.01	12.11	11.72
C30H4205P2	544.61	544.2507	66.16	7.77	14.69	11.37
C30H4206P2	560.61	560.2456	64.28	7.55	17.12	11.05
C30H4207P2	576.61	576.2406	62.49	7.34	19.42	10.74
C30H4208P2	592.61	592.2355	60.80	7.14	21.60	10.45
C30H4209P2	608.61	608.2304	59.21	6.96	23.66	10.18
C30H4301P1	450.65	450.3051	79.96	9.62	3.55	6.87
C30H4302P1	466.65	466.3001	77.22	9.29	6.86	6.64
C30H4303P1	482.65	482.2950	74.66	8.98	9.94	6.42
C30H4304P1	498.65	498.2899	72.26	8.69	12.83	6.21
C30H4305P1	514.65	514.2848	70.02	8.42	15.54	6.02
C30H4306P1	530.65	530.2797	67.90	8.17	18.09	5.84
C30H4307P1	546.65	546.2746	65.92	7.93	20.49	5.67
C30H4308P1	562.65	562.2695	64.04	7.70	22.75	5.51
C30H4309P1	578.65	578.2644	62.27	7.49	24.88	5.35
C30H4401P2	482.63	482.2867	74.66	9.19	3.32	12.84
C30H4402P2	498.63	498.2816	72.26	8.89	6.42	12.42
C30H4403P2	514.63	514.2766	70.02	8.62	9.33	12.04
C30H4404P2	530.63	530.2715	67.91	8.36	12.06	11.67
C30H4405P2	546.63	546.2664	65.92	8.11	14.63	11.33
C30H4406P2	562.63	562.2613	64.04	7.88	17.06	11.01
C30H4407P2	578.63	578.2562	62.27	7.66	19.36	10.71
C30H4408P2	594.63	594.2511	60.60	7.46	21.53	10.42
C30H4409P2	610.63	610.2460	59.01	7.26	23.58	10.14
C30H4501P1	452.67	452.3208	79.60	10.02	3.53	6.84
C30H4502P1	468.67	468.3157	76.89	9.68	6.83	6.61
C30H4503P1	484.67	484.3106	74.35	9.36	9.90	6.39
C30H4504P1	500.66	500.3055	71.97	9.06	12.78	6.19
C30H4505P1	516.66	516.3004	69.74	8.78	15.48	5.99
C30H4506P1	532.66	532.2954	67.65	8.52	18.02	5.81
C30H4507P1	548.66	548.2903	65.68	8.27	20.41	5.65
C30H4508P1	564.66	564.2852	63.81	8.03	22.67	5.49
C30H4509P1	580.66	580.2801	62.06	7.81	24.80	5.33
C30H4601P2	484.65	484.3024	74.35	9.57	3.30	12.78
C30H4602P2	500.65	500.2973	71.97	9.26	6.39	12.37
C30H4603P2	516.65	516.2922	69.74	8.97	9.29	11.99
C30H4604P2	532.65	532.2871	67.65	8.70	12.02	11.63
C30H4605P2	548.65	548.2820	65.68	8.45	14.58	11.29
C30H4606P2	564.65	564.2769	63.82	8.21	17.00	10.97
C30H4607P2	580.64	580.2719	62.06	7.99	19.29	10.67
C30H4608P2	596.64	596.2668	60.39	7.77	21.45	10.38
C30H4609P2	612.64	612.2617	58.82	7.57	23.50	10.11
C30H4701P1	454.68	454.3364	79.25	10.42	3.52	6.81
C30H4702P1	470.68	470.3313	76.56	10.07	6.80	6.58
C30H4703P1	486.68	486.3263	74.04	9.73	9.86	6.36
C30H4704P1	502.68	502.3212	71.68	9.42	12.73	6.16
C30H4705P1	518.68	518.3161	69.47	9.13	15.42	5.97
C30H4706P1	534.68	534.3110	67.39	8.86	17.95	5.79
C30H4707P1	550.68	550.3059	65.43	8.60	20.34	5.62

165

C30H4708P1	566.68	566.3008	63.59	8.36	22.59	5.47
C30H4709P1	582.68	582.2957	61.84	8.13	24.71	5.32
C30H4801P2	486.66	486.3180	74.04	9.94	3.29	12.73
C30H4802P2	502.66	502.3129	71.69	9.63	6.37	12.32
C30H4803P2	518.66	518.3079	69.47	9.33	9.25	11.94
C30H4804P2	534.66	534.3028	67.39	9.05	11.97	11.59
C30H4805P2	550.66	550.2977	65.44	8.79	14.53	11.25
C30H4806P2	566.66	566.2926	63.59	8.54	16.94	10.93
C30H4807P2	582.66	582.2875	61.84	8.30	19.22	10.63
C30H4808P2	598.66	598.2824	60.19	8.08	21.38	10.35
C30H4809P2	614.66	614.2773	58.62	7.87	23.43	10.08
C30H4901P1	456.70	456.3521	78.90	10.81	3.50	6.78
C30H4902P1	472.70	472.3470	76.23	10.45	6.77	6.55
C30H4903P1	488.70	488.3419	73.73	10.11	9.82	6.34
C30H4904P1	504.70	504.3368	71.40	9.79	12.68	6.14
C30H4905P1	520.70	520.3317	69.20	9.49	15.36	5.95
C30H4906P1	536.70	536.3267	67.14	9.20	17.89	5.77
C30H4907P1	552.69	552.3216	65.20	8.94	20.26	5.60
C30H4908P1	568.69	568.3165	63.36	8.68	22.51	5.45
C30H4909P1	584.69	584.3114	61.63	8.45	24.63	5.30
C30H5001P2	488.68	488.3337	73.74	10.31	3.27	12.68
C30H5002P2	504.68	504.3286	71.40	9.99	6.34	12.27
C30H5003P2	520.68	520.3235	69.20	9.68	9.22	11.90
C30H5004P2	536.68	536.3184	67.14	9.39	11.92	11.54
C30H5005P2	552.68	552.3133	65.20	9.12	14.47	11.21
C30H5006P2	568.68	568.3082	63.36	8.86	16.88	10.89
C30H5007P2	584.68	584.3032	61.63	8.62	19.16	10.60
C30H5008P2	600.68	600.2981	59.99	8.39	21.31	10.31
C30H5009P2	616.68	616.2930	58.43	8.17	23.35	10.05
C30H5101P1	458.71	458.3677	78.55	11.21	3.49	6.75
C30H5102P1	474.71	474.3626	75.91	10.83	6.74	6.52
C30H5103P1	490.71	490.3576	73.43	10.48	9.78	6.31
C30H5104P1	506.71	506.3525	71.11	10.15	12.63	6.11
C30H5105P1	522.71	522.3474	68.94	9.83	15.30	5.93
C30H5106P1	538.71	538.3423	66.89	9.54	17.82	5.75
C30H5107P1	554.71	554.3372	64.96	9.27	20.19	5.58
C30H5108P1	570.71	570.3321	63.14	9.01	22.43	5.43
C30H5109P1	586.71	586.3270	61.42	8.76	24.54	5.28
C30H5201P2	490.70	490.3493	73.43	10.68	3.26	12.62
C30H5202P2	506.70	506.3442	71.11	10.34	6.32	12.23
C30H5203P2	522.69	522.3391	68.94	10.03	9.18	11.85
C30H5204P2	538.69	538.3341	66.89	9.73	11.88	11.50
C30H5205P2	554.69	554.3290	64.96	9.45	14.42	11.17
C30H5206P2	570.69	570.3239	63.14	9.18	16.82	10.85
C30H5207P2	586.69	586.3188	61.42	8.93	19.09	10.56
C30H5208P2	602.69	602.3137	59.79	8.70	21.24	10.28
C30H5209P2	618.69	618.3086	58.24	8.47	23.27	10.01
C30H5301P1	460.73	460.3834	78.21	11.60	3.47	6.72
C30H5302P1	476.73	476.3783	75.58	11.21	6.71	6.50
C30H5303P1	492.73	492.3732	73.13	10.84	9.74	6.29
C30H5304P1	508.73	508.3681	70.83	10.50	12.58	6.09
C30H5305P1	524.73	524.3630	68.67	10.18	15.25	5.90
C30H5306P1	540.73	540.3580	66.64	9.88	17.75	5.73
C30H5307P1	556.73	556.3529	64.72	9.60	20.12	5.56
C30H5308P1	572.73	572.3478	62.92	9.33	22.35	5.41
C30H5309P1	588.73	588.3427	61.21	9.07	24.46	5.26
C30H5401P2	492.71	492.3650	73.13	11.05	3.25	12.57
C30H5402P2	508.71	508.3599	70.83	10.70	6.29	12.18
C30H5403P2	524.71	524.3548	68.67	10.37	9.15	11.81
C30H5404P2	540.71	540.3497	66.64	10.07	11.84	11.46
C30H5405P2	556.71	556.3446	64.73	9.78	14.37	11.13
C30H5406P2	572.71	572.3395	62.92	9.50	16.76	10.82
C30H5407P2	588.71	588.3345	61.21	9.25	19.02	10.52
C30H5408P2	604.71	604.3294	59.59	9.00	21.17	10.24
C30H5409P2	620.71	620.3243	58.05	8.77	23.20	9.98
C30H5501P1	462.75	462.3990	77.87	11.98	3.46	6.69
C30H5502P1	478.75	478.3939	75.27	11.58	6.68	6.47
C30H5503P1	494.74	494.3889	72.83	11.21	9.70	6.26
C30H5504P1	510.74	510.3838	70.55	10.85	12.53	6.06
C30H5505P1	526.74	526.3787	68.41	10.52	15.19	5.88
C30H5506P1	542.74	542.3736	66.39	10.21	17.69	5.71
C30H5507P1	558.74	558.3685	64.49	9.92	20.04	5.54
C30H5508P1	574.74	574.3634	62.70	9.65	22.27	5.39
C30H5509P1	590.74	590.3583	61.00	9.38	24.38	5.24
C30H5601P2	494.73	494.3806	72.83	11.41	3.23	12.52
C30H5602P2	510.73	510.3755	70.55	11.05	6.27	12.13
C30H5603P2	526.73	526.3704	68.41	10.72	9.11	11.76
C30H5604P2	542.73	542.3654	66.39	10.40	11.79	11.41
C30H5605P2	558.73	558.3603	64.49	10.10	14.32	11.09
C30H5606P2	574.72	574.3552	62.70	9.82	16.70	10.78
C30H5607P2	590.72	590.3501	61.00	9.56	18.96	10.49
C30H5608P2	606.72	606.3450	59.39	9.30	21.10	10.21
C30H5609P2	622.72	622.3399	57.86	9.06	23.12	9.95
C30H5701P1	464.76	464.4147	77.53	12.36	3.44	6.66
C30H5702P1	480.76	480.4096	74.95	11.95	6.66	6.44
C30H5703P1	496.76	496.4045	72.54	11.57	9.66	6.24
C30H5704P1	512.76	512.3994	70.27	11.20	12.48	6.04
C30H5705P1	528.76	528.3943	68.15	10.87	15.13	5.86
C30H5706P1	544.76	544.3893	66.15	10.55	17.62	5.69
C30H5707P1	560.76	560.3842	64.26	10.25	19.97	5.52
C30H5708P1	576.76	576.3791	62.48	9.96	22.19	5.37
C30H5709P1	592.76	592.3740	60.79	9.69	24.29	5.23
C30H5801P2	496.74	496.3963	72.54	11.77	3.22	12.47
C30H5802P2	512.74	512.3912	70.28	11.40	6.24	12.08
C30H5803P2	528.74	528.3861	68.15	11.06	9.08	11.72
C30H5804P2	544.74	544.3810	66.15	10.73	11.75	11.37
C30H5805P2	560.74	560.3759	64.26	10.43	14.27	11.05
C30H5806P2	576.74	576.3708	62.48	10.14	16.64	10.74
C30H5807P2	592.74	592.3658	60.79	9.86	18.89	10.45
C30H5808P2	608.74	608.3607	59.19	9.60	21.03	10.18
C30H5809P2	624.74	624.3556	57.68	9.36	23.05	9.92
C30H5901P1	466.78	466.4303	77.20	12.74	3.43	6.64
C30H5902P1	482.78	482.4252	74.64	12.32	6.63	6.42
C30H5903P1	498.78	498.4202	72.24	11.92	9.62	6.21
C30H5904P1	514.78	514.4151	70.00	11.55	12.43	6.02
C30H5905P1	530.78	530.4100	67.89	11.20	15.07	5.84
C30H5906P1	546.77	546.4049	65.90	10.88	17.56	5.66
C30H5907P1	562.77	562.3998	64.03	10.57	19.90	5.50
C30H5908P1	578.77	578.3947	62.26	10.28	22.11	5.35
C30H5909P1	594.77	594.3896	60.58	10.00	24.21	5.21
C30H6001P2	498.76	498.4119	72.25	12.13	3.21	12.42
C30H6002P2	514.76	514.4068	70.00	11.75	6.22	12.03
C30H6003P2	530.76	530.4017	67.89	11.39	9.04	11.67
C30H6004P2	546.76	546.3967	65.90	11.06	11.70	11.33
C30H6005P2	562.76	562.3916	64.03	10.75	14.22	11.01
C30H6006P2	578.76	578.3865	62.26	10.45	16.59	10.70
C30H6007P2	594.76	594.3814	60.59	10.17	18.83	10.42
C30H6008P2	610.76	610.3763	59.00	9.90	20.96	10.14
C30H6009P2	626.75	626.3712	57.49	9.65	22.97	9.88
C30H6101P1	468.79	468.4460	76.86	13.12	3.41	6.61

C30H6102P1	484.79	484.4409	74.33	12.68	6.60	6.39
C30H6103P1	500.79	500.4358	71.95	12.28	9.58	6.18
C30H6104P1	516.79	516.4307	69.73	11.90	12.38	5.99
C30H6105P1	532.79	532.4256	67.63	11.54	15.01	5.81
C30H6106P1	548.79	548.4205	65.66	11.20	17.49	5.64
C30H6107P1	564.79	564.4155	63.80	10.89	19.83	5.48
C30H6108P1	580.79	580.4104	62.04	10.59	22.04	5.33
C30H6109P1	596.79	596.4053	60.38	10.30	24.13	5.19
C30H6201P2	500.78	500.4276	71.96	12.48	3.19	12.37
C30H6202P2	516.78	516.4225	69.73	12.09	6.19	11.99
C30H6203P2	532.77	532.4174	67.63	11.73	9.01	11.63
C30H6204P2	548.77	548.4123	65.66	11.39	11.66	11.29
C30H6205P2	564.77	564.4072	63.80	11.07	14.16	10.97
C30H6206P2	580.77	580.4021	62.04	10.76	16.53	10.67
C30H6207P2	596.77	596.3971	60.38	10.47	18.77	10.38
C30H6208P2	612.77	612.3920	58.80	10.20	20.89	10.11
C30H6209P2	628.77	628.3869	57.31	9.94	22.90	9.85
C30H6301P1	470.81	470.4616	76.54	13.49	3.40	6.58
C30H6302P1	486.81	486.4565	74.02	13.04	6.57	6.36
C30H6303P1	502.81	502.4515	71.66	12.63	9.55	6.16
C30H6304P1	518.81	518.4464	69.45	12.24	12.34	5.97
C30H6305P1	534.81	534.4413	67.38	11.87	14.96	5.79
C30H6306P1	550.81	550.4362	65.42	11.53	17.43	5.62
C30H6307P1	566.81	566.4311	63.57	11.20	19.76	5.46
C30H6308P1	582.81	582.4260	61.83	10.90	21.96	5.31
C30H6309P1	598.81	598.4209	60.18	10.60	24.05	5.17
C30H6401P2	502.79	502.4432	71.67	12.83	3.18	12.32
C30H6402P2	518.79	518.4381	69.46	12.43	6.17	11.94
C30H6403P2	534.79	534.4330	67.38	12.06	8.98	11.58
C30H6404P2	550.79	550.4280	65.42	11.71	11.62	11.25
C30H6405P2	566.79	566.4229	63.57	11.38	14.11	10.93
C30H6406P2	582.79	582.4178	61.83	11.07	16.47	10.63
C30H6407P2	598.79	598.4127	60.18	10.77	18.70	10.35
C30H6408P2	614.79	614.4076	58.61	10.49	20.82	10.08
C30H6409P2	630.79	630.4025	57.12	10.23	22.83	9.82

00 HOURS, 09 MINUTES, 32 SECONDS.
END JOB 04990. 22.03.34

TOTAL NUMBER OF PAGES 166

APPENDIX

ANHANG

APPENDICE

TABLE OF RELATIVE ATOMIC WEIGHTS 1965*

BASED ON THE ATOMIC MASS OF $^{12}C = 12$

Name	Symbol	Atomic number	Atomic weight	Name	Symbol	Atomic number	Atomic weight
Actinium	Ac	89	—	Mercury	Hg	80	200.59
Aluminium	Al	13	26.9815	Molybdenum	Mo	42	95.94
Americium	Am	95	—	Neodymium	Nd	60	144.24
Antimony	Sb	51	121.75	Neon	Ne	10	20.183
Argon	Ar	18	39.948	Neptunium	Np	93	—
Arsenic	As	33	74.9216	Nickel	Ni	28	58.71
Astatine	At	85	—	Niobium	Nb	41	92.906
Barium	Ba	56	137.34	Nitrogen	N	7	14.0067
Berkelium	Bk	97	—	Nobelium	No	102	—
Beryllium	Be	4	9.0122	Osmium	Os	76	190.2
Bismuth	Bi	83	208.980	Oxygen	O	8	15.9994[a]
Boron	B	5	10.811[a]	Palladium	Pd	46	106.4
Bromine	Br	35	79.904[b]	Phosphorus	P	15	30.9738
Cadmium	Cd	48	112.40	Platinum	Pt	78	195.09
Caesium	Cs	55	132.905	Plutonium	Pu	94	—
Calcium	Ca	20	40.08	Polonium	Po	84	—
Californium	Cf	98	—	Potassium	K	19	39.102
Carbon	C	6	12.01115[a]	Praseodymium	Pr	59	140.907
Cerium	Ce	58	140.12	Promethium	Pm	61	—
Chlorine	Cl	17	35.453[b]	Protactinium	Pa	91	—
Chromium	Cr	24	51.996[b]	Radium	Ra	88	—
Cobalt	Co	27	58.9332	Radon	Rn	86	—
Copper	Cu	29	63.546[b]	Rhenium	Re	75	186.2
Curium	Cm	96	—	Rhodium	Rh	45	102.905
Dysprosium	Dy	66	162.50	Rubidium	Rb	37	85.47
Einsteinium	Es	99	—	Ruthenium	Ru	44	101.07
Erbium	Er	68	167.26	Samarium	Sm	62	150.35
Europium	Eu	63	151.96	Scandium	Sc	21	44.956
Fermium	Fm	100	—	Selenium	Se	34	78.96
Fluorine	F	9	18.9984	Silicon	Si	14	28.086[a]
Francium	Fr	87	—	Silver	Ag	47	107.868[b]
Gadolinium	Gd	64	157.25	Sodium	Na	11	22.9898
Gallium	Ga	31	69.72	Strontium	Sr	38	87.62
Germanium	Ge	32	72.59	Sulfur	S	16	32.064[a]
Gold	Au	79	196.967	Tantalum	Ta	73	180.948
Hafnium	Hf	72	178.49	Technetium	Tc	43	—
Helium	He	2	4.0026	Tellurium	Te	52	127.60
Holmium	Ho	67	164.930	Terbium	Tb	65	158.924
Hydrogen	H	1	1.00797[a]	Thallium	Tl	81	204.37
Indium	In	49	114.82	Thorium	Th	90	232.038
Iodine	I	53	126.9044	Thulium	Tm	69	168.934
Iridium	Ir	77	192.2	Tin	Sn	50	118.69
Iron	Fe	26	55.847[b]	Titanium	Ti	22	47.90
Krypton	Kr	36	83.80	Tungsten	W	74	183.85
Lanthanum	La	57	138.91	Uranium	U	92	238.03
Lawrencium	Lr	103	—	Vanadium	V	23	50.942
Lead	Pb	82	207.19	Xenon	Xe	54	131.30
Lithium	Li	3	6.939	Ytterbium	Yb	70	173.04
Lutetium	Lu	71	174.97	Yttrium	Y	39	88.905
Magnesium	Mg	12	24.312	Zinc	Zn	30	65.37
Manganese	Mn	25	54.9380	Zirconium	Zr	40	91.22
Mendelevium	Md	101	—				

The values for atomic weights given in the Table apply to elements as they exist in nature, without artificial alteration of their isotopic composition, and further to natural mixtures that do not include isotopes of radiogenic origin.

The figures are the same as those of the 1961 Table except for bromine, copper, and silver. Element 103, Lawrencium, (symbol Lr) has been included.

[a] Atomic weights so designated are known to be variable because of natural variations in isotopic composition. The observed ranges are:

Boron	± 0.003	Oxygen	± 0.0001
Carbon	± 0.00005	Silicon	± 0.001
Hydrogen	± 0.00001	Sulfur	± 0.003

[b] Atomic weights so designated are believed to have the following experimental uncertainties:

Bromine	± 0.001	Copper	± 0.001
Chlorine	± 0.001	Iron	± 0.003
Chromium	± 0.001	Silver	± 0.001

MULTIPLES OF ATOMIC WEIGHTS

BASED ON CARBON-12, 1965 TABLE OF ATOMIC WEIGHTS

C	
C_1	12.011
C_2	24.022
C_3	36.033
C_4	48.045
C_5	60.056
C_6	72.067
C_7	84.078
C_8	96.089
C_9	108.100
C_{10}	120.111_5
C_{11}	132.123
C_{12}	144.134
C_{13}	156.145
C_{14}	168.156
C_{15}	180.167
C_{16}	192.178
C_{17}	204.190
C_{18}	216.201
C_{19}	228.212
C_{20}	240.223
C_{21}	252.234
C_{22}	264.245
C_{23}	276.256
C_{24}	288.268
C_{25}	300.279
C_{26}	312.290
C_{27}	324.301
C_{28}	336.312
C_{29}	348.323
C_{30}	360.334_5
C_{31}	372.346
C_{32}	384.357
C_{33}	396.368
C_{34}	408.379
C_{35}	420.390
C_{36}	432.401
C_{37}	444.413
C_{38}	456.424
C_{39}	468.435
C_{40}	480.446
C_{41}	492.457
C_{42}	504.468
C_{43}	516.479
C_{44}	528.491
C_{45}	540.502
C_{46}	552.513
C_{47}	564.524
C_{48}	576.535
C_{49}	588.546

H	
H_1	1.008
H_2	2.016
H_3	3.024
H_4	4.032
H_5	5.040
H_6	6.048
H_7	7.056
H_8	8.064
H_9	9.072
H_{10}	10.080
H_{11}	11.088
H_{12}	12.096
H_{13}	13.104
H_{14}	14.112
H_{15}	15.120
H_{16}	16.128
H_{17}	17.135
H_{18}	18.143
H_{19}	19.151
H_{20}	20.159
H_{21}	21.167
H_{22}	22.175
H_{23}	23.183
H_{24}	24.191
H_{25}	25.199
H_{26}	26.207
H_{27}	27.215
H_{28}	28.223
H_{29}	29.231
H_{30}	30.239
H_{31}	31.247
H_{32}	32.255
H_{33}	33.263
H_{34}	34.271
H_{35}	35.279
H_{36}	36.287
H_{37}	37.295
H_{38}	38.303
H_{39}	39.311
H_{40}	40.319
H_{41}	41.327
H_{42}	42.335
H_{43}	43.343
H_{44}	44.351
H_{45}	45.359
H_{46}	46.367
H_{47}	47.375
H_{48}	48.383
H_{49}	49.391
H_{50}	50.398_5
H_{51}	51.406
H_{52}	52.414
H_{53}	53.422
H_{54}	54.430
H_{55}	55.438
H_{56}	56.446
H_{57}	57.454
H_{58}	58.462
H_{59}	59.470
H_{60}	60.478
H_{61}	61.486
H_{62}	62.494
H_{63}	63.502
H_{64}	64.510
H_{65}	65.518

Br	
Br_1	79.904
Br_2	159.808
Br_3	239.712
Br_4	319.616
Br_5	399.520
Br_6	479.424
Br_7	559.328
Br_8	639.232

Ca	
Ca_1	40.08
Ca_2	80.16
Ca_3	120.24
Ca_4	160.32

Cl	
Cl_1	35.453
Cl_2	70.906
Cl_3	106.359
Cl_4	141.812
Cl_5	177.265
Cl_6	212.718
Cl_7	248.171

F	
F_1	18.998
F_2	37.997
F_3	56.995
F_4	75.994
F_5	94.992
F_6	113.990
F_7	132.989
F_8	151.987
F_9	170.986
F_{10}	189.984

I	
I_1	126.904
I_2	253.809
I_3	380.713
I_4	507.618
I_5	634.522
I_6	761.426

K	
K_1	39.102
K_2	78.204
K_3	117.306
K_4	156.408
K_5	195.510

Mg	
Mg_1	24.312
Mg_2	48.624
Mg_3	72.936
Mg_4	97.248

N	
N_1	14.007
N_2	28.013
N_3	42.020
N_4	56.027
N_5	70.033_5
N_6	84.040
N_7	98.047
N_8	112.054
N_9	126.060
N_{10}	140.067
N_{11}	154.074
N_{12}	168.080
N_{13}	182.087
N_{14}	196.094
N_{15}	212.100_5

Na	
Na_1	22.990
Na_2	45,980
Na_3	68.969
Na_4	91.959
Na_5	114.949

O	
O_1	15.999
O_2	31.999
O_3	47.998
O_4	63.998
O_5	79.997
O_6	95.996
O_7	111.996
O_8	127.995
O_9	143.995
O_{10}	159.994
O_{11}	175.993
O_{12}	191.993
O_{13}	207.992
O_{14}	223.992
O_{15}	239.991

P	
P_1	30.974
P_2	61.948
P_3	92.921
P_4	123.895
P_5	154.869

S	
S_1	32.064
S_2	64.128
S_3	96.192
S_4	128.256
S_5	160.320
S_6	192.384
S_7	224.448
S_8	256.512
S_9	288.576

Si	
Si_1	28.086
Si_2	56.172
Si_3	84.258
Si_4	112.344
Si_5	140.430
Si_6	168.516

$(COOH)_n$	
$(COOH)_1$	45.018
$(COOH)_2$	90.036
$(COOH)_3$	135.054
$(COOH)_4$	180.072
$(COOH)_5$	225.090
$(COOH)_6$	270.108

$(H_2O)_n$	
$(H_2O)_{0\cdot5}$	9.008
$(H_2O)_1$	18.015
$(H_2O)_{1\cdot5}$	27.022
$(H_2O)_2$	36.031
$(H_2O)_3$	54.046
$(H_2O)_4$	72.061
$(H_2O)_5$	90.077

$(NH_2)_n$	
$(NH_2)_1$	16.023
$(NH_2)_2$	32.045
$(NH_2)_3$	48.068
$(NH_2)_4$	64.091
$(NH_2)_5$	80.113

$(OH)_n$	
$(OH)_1$	17.007
$(OH)_2$	34.015
$(OH)_3$	51.022
$(OH)_4$	68.029
$(OH)_5$	85.037
$(OH)_6$	102.044
$(OH)_7$	119.052
$(OH)_8$	136.059
$(OH)_9$	153.066
$(OH)_{10}$	170.074

$(SO_3H)_n$	
$(SO_3H)_1$	81.070
$(SO_3H)_2$	162.140
$(SO_3H)_3$	243.211
$(SO_3H)_4$	324.281
$(SO_3H)_5$	405.351
$(SO_3H)_6$	486.421

$(CONH_2)_n$	
$(CONH_2)_1$	44.033
$(CONH_2)_2$	88.066
$(CONH_2)_3$	132.100
$(CONH_2)_4$	176.133
$(CONH_2)_5$	220.166

TABELLE RELATIVER ATOMGEWICHTE 1965*

BASIERENDE AUF DER ATOM-MASSE $^{12}C = 12$

Name	Symbol	Atomzahl	Atomgewicht	Name	Symbol	Atomzahl	Atomgewicht
Actinium	Ac	89	—	Natrium	Na	11	22,9898
Aluminium	Al	13	26,9815	Neodym	Nd	60	144,24
Americium	Am	95	—	Neon	Ne	10	20,183
Antimon	Sb	51	121,75	Neptunium	Np	93	—
Argon	Ar	18	39,948	Nickel	Ni	28	58,71
Arsen	As	33	74,9216	Niob	Nb	41	92,906
Astatin	At	85	—	Nobelium	No	102	—
Barium	Ba	56	137,34	Osmium	Os	76	190,2
Berkelium	Bk	97	—	Palladium	Pd	46	106,4
Beryllium	Be	4	9,0122	Phosphor	P	15	30,9738
Blei	Pb	82	207,19	Platin	Pt	78	195,09
Bor	B	5	10,811[a]	Plutonium	Pu	94	—
Brom	Br	35	79,904[b]	Polonium	Po	84	—
Cadmium	Cd	48	112,40	Praseodymium	Pr	59	140,907
Caesium	Cs	55	132,905	Promethium	Pm	61	—
Calcium	Ca	20	40,08	Protactinium	Pa	91	—
Californium	Cf	98	—	Quecksilber	Hg	80	200,59
Cer	Ce	58	140,12	Radium	Ra	88	—
Chlor	Cl	17	35,453[b]	Radon	Rn	86	—
Chrom	Cr	24	51,996[b]	Rhenium	Re	75	186,2
Curium	Cm	96	—	Rhodium	Rh	45	102,905
Dysprosium	Dy	66	162,50	Rubidium	Rb	37	85,47
Einsteinium	Es	99	—	Ruthenium	Ru	44	101,07
Eisen	Fe	26	55,847[b]	Samarium	Sm	62	150,35
Erbium	Er	68	167,26	Sauerstoff	O	8	15,9994[a]
Europium	Eu	63	151,96	Scandium	Sc	21	44,956
Fermium	Fm	100	—	Schwefel	S	16	32,064[a]
Fluor	F	9	18,9984	Selen	Se	34	78,96
Francium	Fr	87	—	Silber	Ag	47	107,868[b]
Gadolinium	Gd	64	157,25	Silicium	Si	14	28,086[a]
Gallium	Ga	31	69,72	Stickstoff	N	7	14,0067
Germanium	Ge	32	72,59	Strontium	Sr	38	87,62
Gold	Au	79	196,967	Tantal	Ta	73	180,948
Hafnium	Hf	72	178,49	Technetium	Tc	43	—
Helium	He	2	4,0026	Tellur	Te	52	127,60
Holmium	Ho	67	164,930	Terbium	Tb	65	158,924
Indium	In	49	114,82	Thallium	Tl	81	204,37
Iridium	Ir	77	192,2	Thorium	Th	90	232,038
Jod	J	53	126,9044	Thulium	Tm	69	168,934
Kobalt	Co	27	58,9332	Titan	Ti	22	47,90
Kohlenstoff	C	6	12,01115[a]	Uran	U	92	238,03
Kalium	K	19	39,102	Vanadium	V	23	50,942
Krypton	Kr	36	83,80	Wasserstoff	H	1	1,00797[a]
Kupfer	Cu	29	63,546[b]	Wismut	Bi	83	208,980
Lanthan	La	57	138,91	Wolfram	W	74	183,85
Lawrencium	Lr	103	—	Xenon	Xe	54	131,30
Lithium	Li	3	6,939	Ytterbium	Yb	70	173,04
Lutetium	Lu	71	174,97	Yttrium	Y	39	88,905
Magnesium	Mg	12	24,312	Zink	Zn	30	65,37
Mangan	Mn	25	54,9380	Zinn	Sn	50	118,69
Mendelevium	Md	101	—	Zirkon	Zr	40	91,22
Molybdän	Mo	42	95,94				

Die in der Tabelle angegebenen Werte für die Atomgewichte beziehen sich auf die Elemente, wie sie in der Natur vorkommen, ohne künstliche Anderung der Isotopen-Zusammensetzung, sowie auf natürliche Gemische, die keine Isotope radiogenen Ursprungs enthalten.

Die Zahlen sind dieselben wie in der Tabelle 1961, ausser für Brom, Kupfer und Silber. Das Element Lawrencium, 103 (Symbol Lr) wurde neu aufgenommen.

[a] Die so gekennzeichneten Atomgewichte sind bekanntlich variabel wegen natürlicher Anderungen in ihrer isotopischen Zusammensetzung. Es wurden die folgenden Bereiche beobachtet:

Bor	±0,003	Schwefel	±0,003
Kohlenstoff	±0,00005	Silicium	±0,001
Sauerstoff	±0,0001	Wasserstoff	±0,00001

[b] Die so gekennzeichneten Atomgewichte haben wahrscheinlich die folgenden experimentellen Unsicherheiten:

Brom	±0,001	Eisen	±0,003
Chlor	±0,001	Kupfer	±0,001
Chrom	±0,001	Silber	±0,001

*Wiedergegeben mit freundlicher Benehmigung von Butterworths Scientific Publications.

MULTIPLE ATOMGEWICHTE

BASIEREND AUF KOHLENSTOFF-12 TABELLE 1965 DER ATOMGEWICHTE

C	
C_{1}	12,011
C_{2}	24,022
C_{3}	36,033
C_{4}	48,045
C_{5}	60,056
C_{6}	72,067
C_{7}	84,078
C_{8}	96,089
C_{9}	108,100
C_{10}	$120,111_{5}$
C_{11}	132,123
C_{12}	144,134
C_{13}	156,145
C_{14}	168,156
C_{15}	180,167
C_{16}	192,178
C_{17}	204,190
C_{18}	216,201
C_{19}	228,212
C_{20}	240,223
C_{21}	252,234
C_{22}	264,245
C_{23}	276,256
C_{24}	288,268
C_{25}	300,279
C_{26}	312,290
C_{27}	324,301
C_{28}	336,312
C_{29}	348,323
C_{30}	$360,334_{5}$
C_{31}	372,346
C_{32}	384,357
C_{33}	396,368
C_{34}	408,379
C_{35}	420,390
C_{36}	432,401
C_{37}	444,413
C_{38}	456,424
C_{39}	468,435
C_{40}	480,446
C_{41}	492,457
C_{42}	504,468
C_{43}	516,479
C_{44}	528,491
C_{45}	540,502
C_{46}	552,513
C_{47}	564,524
C_{48}	576,535
C_{49}	588,546

H	
H_{1}	1,008
H_{2}	2,016
H_{3}	3,024
H_{4}	4,032
H_{5}	5,040
H_{6}	6,048
H_{7}	7,056
H_{8}	8,064
H_{9}	9,072
H_{10}	10,080
H_{11}	11,088
H_{12}	12,096
H_{13}	13,104
H_{14}	14,112
H_{15}	15,120
H_{16}	16,128
H_{17}	17,135
H_{18}	18,143
H_{19}	19,151
H_{20}	20,159
H_{21}	21,167
H_{22}	22,175
H_{23}	23,183
H_{24}	24,191
H_{25}	25,199
H_{26}	26,207
H_{27}	27,215
H_{28}	28,223
H_{29}	29,231
H_{30}	30,239
H_{31}	31,247
H_{32}	32,255
H_{33}	33,263
H_{34}	34,271
H_{35}	35,279
H_{36}	36,287
H_{37}	37,295
H_{38}	38,303
H_{39}	39,311
H_{40}	40,319
H_{41}	41,327
H_{42}	42,335
H_{43}	43,343
H_{44}	44,351
H_{45}	45,359
H_{46}	46,367
H_{47}	47,375
H_{48}	48,383
H_{49}	49,391
H_{50}	$50,398_{5}$
H_{51}	51,406
H_{52}	52,414
H_{53}	53,422
H_{54}	54,430
H_{55}	55,438
H_{56}	56,446
H_{57}	57,454
H_{58}	58,462
H_{59}	59,470
H_{60}	60,478
H_{61}	61,486
H_{62}	62,494
H_{63}	63,502
H_{64}	64,510
H_{65}	65,518

Br	
Br_{1}	79,904
Br_{2}	159,808
Br_{3}	239,712
Br_{4}	319,616
Br_{5}	399,520
Br_{6}	479,424
Br_{7}	559,328
Br_{8}	639,232

Ca	
Ca_{1}	40,08
Ca_{2}	80,16
Ca_{3}	120,24
Ca_{4}	160,32

Cl	
Cl_{1}	35,453
Cl_{2}	70,906
Cl_{3}	106,359
Cl_{4}	141,812
Cl_{5}	177,265
Cl_{6}	212,718
Cl_{7}	248,171

F	
F_{1}	18,998
F_{2}	37,997
F_{3}	56,995
F_{4}	75,994
F_{5}	94,992
F_{6}	113,990
F_{7}	132,989
F_{8}	151,987
F_{9}	170,986
F_{10}	189,984

I	
I_{1}	126,904
I_{2}	253,809
I_{3}	380,713
I_{4}	507,618
I_{5}	634,522
I_{6}	761,426

K	
K_{1}	39,102
K_{2}	78,204
K_{3}	117,306
K_{4}	156,408
K_{5}	195,510

Mg	
Mg_{1}	24,312
Mg_{2}	48,624
Mg_{3}	72,936
Mg_{4}	97,248

N	
N_{1}	14,007
N_{2}	28,013
N_{3}	42,020
N_{4}	56,027
N_{5}	$70,033_{5}$
N_{6}	84,040
N_{7}	98,047
N_{8}	112,054
N_{9}	126,060
N_{10}	140,067
N_{11}	154,074
N_{12}	168,080
N_{13}	182,087
N_{14}	196,094
N_{15}	$212,100_{5}$

Na	
Na_{1}	22,990
Na_{2}	45,980
Na_{3}	68,969
Na_{4}	91,959
Na_{5}	114,949

O	
O_{1}	15,999
O_{2}	31,999
O_{3}	47,998
O_{4}	63,998
O_{5}	79,997
O_{6}	95,996
O_{7}	111,996
O_{8}	127,995
O_{9}	143,995
O_{10}	159,994
O_{11}	175,993
O_{12}	191,993
O_{13}	207,992
O_{14}	223,992
O_{15}	239,991

P	
P_{1}	30,974
P_{2}	61,948
P_{3}	92,921
P_{4}	123,895
P_{5}	154,869

S	
S_{1}	32,064
S_{2}	64,128
S_{3}	96,192
S_{4}	128,256
S_{5}	160,320
S_{6}	192,384
S_{7}	224,448
S_{8}	256,512
S_{9}	288,576

Si	
Si_{1}	28,086
Si_{2}	56,172
Si_{3}	84,258
Si_{4}	112,344
Si_{5}	140,430
Si_{6}	168,516

$(COOH)n$	
$(COOH)_{1}$	45,018
$(COOH)_{2}$	90,036
$(COOH)_{3}$	135,054
$(COOH)_{4}$	180,072
$(COOH)_{5}$	225,090
$(COOH)_{6}$	270,108

$(H_2O)n$	
$(H_2O)_{0,5}$	9,008
$(H_2O)_{1}$	18,015
$(H_2O)_{1,5}$	27,022
$(H_2O)_{2}$	36,031
$(H_2O)_{3}$	54,046
$(H_2O)_{4}$	72,061
$(H_2O)_{5}$	90,077

$(NH_2)n$	
$(NH_2)_{1}$	16,023
$(NH_2)_{2}$	32,045
$(NH_2)_{3}$	48,068
$(NH_2)_{4}$	64,091
$(NH_2)_{5}$	80,113

$(OH)n$	
$(OH)_{1}$	17,007
$(OH)_{2}$	34,015
$(OH)_{3}$	51,022
$(OH)_{4}$	68,029
$(OH)_{5}$	85,037
$(OH)_{6}$	102,044
$(OH)_{7}$	119,052
$(OH)_{8}$	136,059
$(OH)_{9}$	153,066
$(OH)_{10}$	170,074

$(SO_3H)n$	
$(SO_3H)_{1}$	81,070
$(SO_3H)_{2}$	162,140
$(SO_3H)_{3}$	243,211
$(SO_3H)_{4}$	324,281
$(SO_3H)_{5}$	405,351
$(SO_3H)_{6}$	486,421

$(CONH_2)n$	
$(CONH_2)_{1}$	44,033
$(CONH_2)_{2}$	88,066
$(CONH_2)_{3}$	132,100
$(CONH_2)_{4}$	176,133
$(CONH_2)_{5}$	220,166

TABLE DES POIDS ATOMIQUES (MASSES ATOMIQUES RELATIVES) 1965*

LE CARBONE-12 EST PRIS COMME BASE AVEC $^{12}C=12$

Nom	Symb.	No. atom.	Masse
Actinium	Ac	89	—
Aluminium	Al	13	26,9815
Américium	Am	95	—
Antimoine	Sb	51	121,75
Argent	Ag	47	107,868[b]
Argon	Ar	18	39,948
Arsenic	As	33	74,9216
Astate	At	85	—
Azote	N	7	14,0067
Baryum	Ba	56	137,34
Berkélium	Bk	97	—
Béryllium	Be	4	9,0122
Bismuth	Bi	83	208,980
Bore	B	5	10,811[a]
Brome	Br	35	79,904[b]
Cadmium	Cd	48	112,40
Caesium	Cs	55	132,905
Calcium	Ca	20	40,08
Californium	Cf	98	—
Carbone	C	6	12,01115[a]
Cérium	Ce	58	140,12
Chlore	Cl	17	35,453[b]
Chrome	Cr	24	51,996[b]
Cobalt	Co	27	58,9332
Cuivre	Cu	29	63,546[b]
Curium	Cm	96	—
Dysprosium	Dy	66	162,50
Einsteinium	Es	99	—
Erbium	Er	68	167,26
Etain	Sn	50	118,69
Europium	Eu	63	151,96
Fer	Fe	26	55,847[b]
Fermium	Fm	100	—
Fluor	F	9	18,9984
Francium	Fr	87	—
Gadolinium	Gd	64	157,25
Gallium	Ga	31	69,72
Germanium	Ge	32	72,59
Hafnium	Hf	72	178,49
Hélium	He	2	4,0026
Holmium	Ho	67	164,930
Hydrogène	H	1	1,00797[a]
Indium	In	49	114,82
Iode	I	53	126,9044
Iridium	Ir	77	192,2
Krypton	Kr	36	83,80
Lanthane	La	57	138,91
Lawrencium	Lr	103	—
Lithium	Li	3	6,939
Lutétium	Lu	71	174,97
Magnésium	Mg	12	24,312
Manganèse	Mn	25	54,9380
Mendélévium	Md	101	—
Mercure	Hg	80	200,59
Molybdène	Mo	42	95,94
Néodyme	Nd	60	144,24
Néon	Ne	10	20,183
Neptunium	Np	93	—
Nickel	Ni	28	58,71
Niobium	Nb	41	92,906
Nobélium	No	102	—
Or	Au	79	196,967
Osmium	Os	76	190,2
Oxygène	O	8	15,9994[a]
Palladium	Pd	46	106,4
Phosphore	P	15	30,9738
Platine	Pt	78	195,09
Plomb	Pb	82	207,19
Plutonium	Pu	94	—
Polonium	Po	84	—
Potassium	K	19	39,102
Praséodyme	Pr	59	140,907
Prométhium	Pm	61	—
Protactinium	Pa	91	—
Radium	Ra	88	—
Radon	Rn	86	—
Rhénium	Re	75	186,2
Rhodium	Rh	45	102,905
Rubidium	Rb	37	85,47
Ruthénium	Ru	44	101,07
Samarium	Sm	62	150,35
Scandium	Sc	21	44,956
Sélénium	Se	34	78,96
Silicium	Si	14	28,086[a]
Sodium	Na	11	22,9898
Soufre	S	16	32,064[a]
Strontium	Sr	38	87,62
Tantale	Ta	73	180,948
Technétium	Tc	43	—
Tellure	Te	52	127,60
Terbium	Tb	65	158,924
Thallium	Tl	81	204,37
Thorium	Th	90	232,038
Thulium	Tm	69	168,934
Titane	Ti	22	47,90
Tungstène	W	74	183,85
Uranium	U	92	238,03
Vanadium	V	23	50,942
Xénon	Xe	54	131,30
Ytterbium	Yb	70	173,04
Yttrium	Y	39	88,905
Zinc	Zn	30	65,37
Zirconium	Zr	40	91,22

Les valeurs des poids atomiques données dans cette table s'appliquent aux éléments tels qu'ils existent dans la nature, sans modification artificielle de leur composition isotopique, et aussi aux mélanges naturels qui ne contiennent pas d'isotopes d'origine radiogène.

Les chiffres sont les mêmes que ceux de la table de 1961 sauf pour le brome, le cuivre et l'argent. L'élément 103, Lawrencium (symbole Lr), a été inclus.

[a] Les masses atomiques ainsi indiquées sont variables en raison des variations naturelles des compositions isotopiques correspondantes. Les écarts observés sont:

Bore	± 0,003	Oxygène	± 0,0001
Carbone	± 0,00005	Silicium	± 0,001
Hydrogène	± 0,00001	Soufre	± 0,003

[b] Les masses atomiques ainsi désignées sont exactes semble-t-il à:

Argent	± 0,001	Chrome	± 0,001
Brome	± 0,001	Cuivre	± 0,001
Chlore	± 0,001	Fer	± 0,003

MULTIPLES DES POIDS ATOMIQUES

BASES SUR LE CARBONE-12, TABLE DES POIDS ATOMIQUES 1965

C

C_1	12,011
C_2	24,022
C_3	36,033
C_4	48,045
C_5	60,056
C_6	72,067
C_7	84,078
C_8	96,089
C_9	108,100
C_{10}	120,111$_5$
C_{11}	132,123
C_{12}	144,134
C_{13}	156,145
C_{14}	168,156
C_{15}	180,167
C_{16}	192,178
C_{17}	204,190
C_{18}	216,201
C_{19}	228,212
C_{20}	240,223
C_{21}	252,234
C_{22}	264,245
C_{23}	276,256
C_{24}	288,268
C_{25}	300,279
C_{26}	312,290
C_{27}	324,301
C_{28}	336,312
C_{29}	348,323
C_{30}	360,334$_5$
C_{31}	372,346
C_{32}	384,357
C_{33}	396,368
C_{34}	408,379
C_{35}	420,390
C_{36}	432,401
C_{37}	444,413
C_{38}	456,424
C_{39}	468,435
C_{40}	480,446
C_{41}	492,457
C_{42}	504,468
C_{43}	516,479
C_{44}	528,491
C_{45}	540,502
C_{46}	552,513
C_{47}	564,524
C_{48}	576,535
C_{49}	588,546

H

H_1	1,008
H_2	2,016
H_3	3,024
H_4	4,032
H_5	5,040
H_6	6,048
H_7	7,056
H_8	8,064
H_9	9,072
H_{10}	10,080
H_{11}	11,088
H_{12}	12,096
H_{13}	13,104
H_{14}	14,112
H_{15}	15,120
H_{16}	16,128
H_{17}	17,135
H_{18}	18,143
H_{19}	19,151
H_{20}	20,159
H_{21}	21,167
H_{22}	22,175
H_{23}	23,183
H_{24}	24,191
H_{25}	25,199
H_{26}	26,207
H_{27}	27,215
H_{28}	28,223
H_{29}	29,231
H_{30}	30,239
H_{31}	31,247
H_{32}	32,255
H_{33}	33,263
H_{34}	34,271
H_{35}	35,279
H_{36}	36,287
H_{37}	37,295
H_{38}	38,303
H_{39}	39,311
H_{40}	40,319
H_{41}	41,327
H_{42}	42,335
H_{43}	43,343
H_{44}	44,351
H_{45}	45,359
H_{46}	46,367
H_{47}	47,375
H_{48}	48,383
H_{49}	49,391
H_{50}	50,398$_5$
H_{51}	51,406
H_{52}	52,414
H_{53}	53,422
H_{54}	54,430
H_{55}	55,438
H_{56}	56,446
H_{57}	57,454
H_{58}	58,462
H_{59}	59,470
H_{60}	60,478
H_{61}	61,486
H_{62}	62,494
H_{63}	63,502
H_{64}	64,510
H_{65}	65,518

Br

Br_1	79,904
Br_2	159,808
Br_3	239,712
Br_4	319,616
Br_5	399,520
Br_6	479,424
Br_7	559,328
Br_8	639,232

Ca

Ca_1	40,08
Ca_2	80,16
Ca_3	120,24
Ca_4	160,32

Cl

Cl_1	35,453
Cl_2	70,906
Cl_3	106,359
Cl_4	141,812
Cl_5	177,265
Cl_6	212,718
Cl_7	248,171

F

F_1	18,998
F_2	37,997
F_3	56,995
F_4	75,994
F_5	94,992
F_6	113,990
F_7	132,989
F_8	151,987
F_9	170,986
F_{10}	189,984

I

I_1	126,904
I_2	253,809
I_3	380,713
I_4	507,618
I_5	634,522
I_6	761,426

K

K_1	39,102
K_2	78,204
K_3	117,306
K_4	156,408
K_5	195,510

Mg

Mg_1	24,312
Mg_2	48,624
Mg_3	72,936
Mg_4	97,248

N

N_1	14,007
N_2	28,013
N_3	42,020
N_4	56,027
N_5	70,033$_5$
N_6	84,040
N_7	98,047
N_8	112,054
N_9	126,060
N_{10}	140,067
N_{11}	154,074
N_{12}	168,080
N_{13}	182,087
N_{14}	196,094
N_{15}	212,100$_5$

Na

Na_1	22,990
Na_2	45,980
Na_3	68,969
Na_4	91,959
Na_5	114,949

O

O_1	15,999
O_2	31,999
O_3	47,998
O_4	63,998
O_5	79,997
O_6	95,996
O_7	111,996
O_8	127,995
O_9	143,995
O_{10}	159,994
O_{11}	175,993
O_{12}	191,993
O_{13}	207,992
O_{14}	223,992
O_{15}	239,991

P

P_1	30,974
P_2	61,948
P_3	92,921
P_4	123,895
P_5	154,869

S

S_1	32,064
S_2	64,128
S_3	96,192
S_4	128,256
S_5	160,320
S_6	192,384
S_7	224,448
S_8	256,512
S_9	288,576

Si

Si_1	28,086
Si_2	56,172
Si_3	84,258
Si_4	112,344
Si_5	140,430
Si_6	168,516

$(COOH)n$

$(COOH)_1$	45,018
$(COOH)_2$	90,036
$(COOH)_3$	135,054
$(COOH)_4$	180,072
$(COOH)_5$	225,090
$(COOH)_6$	270,108

$(H_2O)n$

$(H_2O)_{0,5}$	9,008
$(H_2O)_1$	18,015
$(H_2O)_{1,5}$	27,022
$(H_2O)_2$	36,031
$(H_2O)_3$	54,046
$(H_2O)_4$	72,061
$(H_2O)_5$	90,077

$(NH_2)n$

$(NH_2)_1$	16,023
$(NH_2)_2$	32,045
$(NH_2)_3$	48,068
$(NH_2)_4$	64,091
$(NH_2)_5$	80,113

$(OH)n$

$(OH)_1$	17,007
$(OH)_2$	34,015
$(OH)_3$	51,022
$(OH)_4$	68,029
$(OH)_5$	85,037
$(OH)_6$	102,044
$(OH)_7$	119,052
$(OH)_8$	136,059
$(OH)_9$	153,066
$(OH)_{10}$	170,074

$(SO_3H)n$

$(SO_3H)_1$	81,070
$(SO_3H)_2$	162,140
$(SO_3H)_3$	243,211
$(SO_3H)_4$	324,281
$(SO_3H)_5$	405,351
$(SO_3H)_6$	486,421

$(CONH_2)n$

$(CONH_2)_1$	44,033
$(CONH_2)_2$	88,066
$(CONH_2)_3$	132,100
$(CONH_2)_4$	176,133
$(CONH_2)_5$	220,166